ANNUAL BOOK OF ASTM STANDARDS

2001

SECTION 11

Water and Environmental Technology

VOLUME 11.04

Environmental Assessment; Hazardous Substances and Oil Spill Responses; Waste Management

Includes standards of the following committees:

D34 on Waste Management
E50 on Environmental Assessment
E51 on Environmental Risk Management (Became E50.03 10/98)
F20 on Hazardous Substances and Oil Spill Response

ASTM Stock Number: S110401

ASTM ■ **100 BARR HARBOR DRIVE, PO BOX C700, WEST CONSHOHOCKEN, PA 19428-2959**
TEL: **610-832-9500** ■ FAX: **610-832-9555** ■ EMAIL: **service@astm.org** ■ WEBSITE: **www.astm.org**

Library of Congress Catalog Card Number: 83-641658

ISBN 0-8031-3039-2 (set)
ISBN 0-8031-3013-9 (section)
ISBN 0-8031-3017-1 (volume)

ORDER INFORMATION: Additional copies of this book in print, on CD-ROM, or reprints (single or multiple copies) of individual standards may be obtained by contacting ASTM at the above address or at 610-832-9585 (phone), 610-832-9555 (fax), service@astm.org (email), or through www.astm.org (website).

Printed in Baltimore, MD, U.S.A.

Foreword

Organized in 1898, ASTM has grown into one of the largest voluntary standards development systems in the world. ASTM is a not-for-profit organization which provides a forum for producers, users, ultimate consumers, and those having a general interest (representatives of government and academia) to meet on common ground and write standards for materials, products, systems, and services.

From the work of 128 standards-writing committees, ASTM publishes more than 11,000 standards each year. These standards and other related technical information are sold throughout the world.

ASTM Headquarters has no technical research or testing facilities; such work is done voluntarily by 30,520 technically qualified ASTM members located throughout the world. Membership in the Society is open to all concerned with the fields in which ASTM is active. A membership application may be obtained from Member and Committee Services, ASTM, 100 Barr Harbor Drive, PO Box C700, West Conshohocken, PA 19428–2959; tel. 610-832-9694 or from the ASTM website, www.astm.org under Technical Committees/Membership Information.

2001 Annual Book of ASTM Standards

The 2001 *Annual Book of ASTM Standards* consists of 75 volumes, divided among 16 sections, of which this volume is one. It contains approved ASTM standards, provisional standards, and related material. These terms are defined as follows in the *Regulations Governing ASTM Technical Committees:*

Categories:

standard, n—as used in ASTM, a document that has been developed and established within the consensus principles of the Society and that meets the approval requirements of ASTM procedures and regulations.

standard, adj—as used in ASTM, a descriptive used in titles of test methods, specifications, and other documents to indicate consensus approval in accordance with ASTM procedures and regulations.

provisional standard—a document published for a limited period of time by the Society to meet a demand for more rapid issuance of specific documents, such as an emergency situation, regulatory requirements, or other special circumstances.
Discussion—Provisional standards are not full consensus documents because they require subcommittee consensus only. (See *Regulations Governing ASTM Technical Committees*, Section 14.)

Types:

The various types of ASTM documents are to provide a flexibility of form, communication, and usage for both the technical committees and the myriad users of ASTM documents. The type of ASTM document that is developed and titled is based on the technical content and intended use, not on the degree of consensus achieved. The two categories of ASTM documents (standard and provisional standard) can be of the following forms and types:

classification—a systematic arrangement or division of materials, products, systems, or services into groups based on similar characteristics such as origin, composition, properties, or use.

guide—a compendium of information or series of options that does not recommend a specific course of action.
Discussion—A guide increases the awareness of information and approaches in a given subject area.

practice—a definitive set of instructions for performing one or more specific operations or functions that does not produce a test result.
Discussion—Examples of practices include, but are not limited to: application, assessment, cleaning, collection,

decontamination, inspection, installation, preparation, sampling, screening, and training.

specification—an explicit set of requirements to be satisfied by a material, product, system, or service.
Discussion—Examples of specifications include, but are not limited to, requirements for: physical, mechanical, or chemical properties, and safety, quality, or performance criteria. A specification identifies the test methods for determining whether each of the requirements is satisfied.

terminology—a document comprising definitions of terms; explanations of symbols, abbreviations, or acronyms.

test method—a definitive procedure that produces a test result.
Discussion—Examples of test methods include, but are not limited to: identification, measurement, and evaluation of one or more qualities, characteristics, or properties. A precision and bias statement shall be reported at the end of a test method. (See *Form and Style for ASTM Standards*, Section A21, Precision and Bias.)

A new edition of the Book of Standards is published annually because of additions of new standards and significant revisions to existing standards. Approximately 30 % of each volume is new or revised. Each volume contains all actions approved by the Society at least six months before the publication date. New and revised standards approved by the Society between the annual editions of any given volume are made available as separate copies. Users are cautioned to follow the most current issue of a standard except when a specific edition of a standard is cited, for example, as in a contract.

Development and Use of ASTM Standards

ASTM believes that technically competent standards result when a full consensus of all concerned parties is achieved and rigorous due process procedures are followed. This philosophy and standards development system ensure technically competent standards having the highest credibility when critically examined and used as the basis for commercial, legal, or regulatory actions.

ASTM standards are developed voluntarily and used voluntarily. Standards become legally binding only when a government body references them in regulations, or when they are cited in a contract. Any item that is produced and marked as conforming to an ASTM standard must meet all applicable requirements of that standard.

ASTM standards are used by thousands of individuals, companies, and agencies. Purchasers and sellers incorporate standards into contracts; scientists and engineers use them in laboratories; architects and designers use them in plans; government agencies reference them in codes, regulations, and laws; and many others refer to standards for guidance.

Consideration of Comments on ASTM Standards

An ASTM standard is subject to revision at any time by the responsible technical committee and must be reviewed every five years and if not revised, either reapproved or withdrawn. Your comments are invited either for revision of any standard or for the development of new standards and should be addressed to ASTM Headquarters. Your comments will receive careful consideration at a meeting of the responsible technical committee, which you may attend. If you feel your comments have not received a fair hearing you should make your views known to the ASTM Committee on Standards, 100 Barr Harbor Drive, PO Box C700, West Conshohocken, PA 19428.

Using the Annual Book of ASTM Standards

The standards are assembled in each volume in alphanumeric sequence of their ASTM designation numbers. Volumes 03.06, 04.09, 05.05, and 06.03 are assembled first by committee, then in alphanumeric sequence. Each volume has a table of contents, listing the standards in alphanumeric sequence by ASTM designation; and a list by subjects, categorizing the standards according to subject. A subject index of the standards in each volume appears at the back of each volume.

Availability of Individual Standards

Each ASTM standard is available as a separate copy from ASTM. Standards can be ordered from the ASTM website at www.astm.org, in the store section. Standards can also be ordered from Customer Services at 610–832–9585, Monday through Friday, 8:30 AM to 4:30 PM Eastern Standard Time.

FOREWORD

Caveat Statements and Policies in Standards

ASTM caveat statements on Safety Hazards and Fire Hazards are required to appear in standards where appropriate. They are located in the scope section of applicable standards. The caveats on General Statement of ASTM Policy and Patents are contained in all standards and located at the end of each standard. For more information on the caveats see Section F2 of the *Form and Style for ASTM Standards.*

ANNUAL BOOK OF ASTM STANDARDS

Listed by Section and Volume

Section 1— Iron and Steel Products

Volume 01.01 Steel—Piping, Tubing, Fittings
Volume 01.02 Ferrous Castings; Ferroalloys
Volume 01.03 Steel—Plate, Sheet, Strip, Wire; Stainless Steel Bar
Volume 01.04 Steel—Structural, Reinforcing, Pressure Vessel, Railway
Volume 01.05 Steel—Bars, Forgings, Bearing, Chain, Springs
Volume 01.06 Coated Steel Products
Volume 01.07 Ships and Marine Technology
Volume 01.08 Fasteners

Section 2— Nonferrous Metal Products

Volume 02.01 Copper and Copper Alloys
Volume 02.02 Aluminum and Magnesium Alloys
Volume 02.03 Electrical Conductors
Volume 02.04 Nonferrous Metals—Nickel, Cobalt, Lead, Tin, Zinc, Cadmium, Precious, Reactive, Refractory Metals and Alloys; Materials for Thermostats, Electrical Heating and Resistance, Contacts, and Connectors
Volume 02.05 Metallic and Inorganic Coatings; Metal Powders, Sintered P/M Structural Parts

Section 3—Metals Test Methods and Analytical Procedures

Volume 03.01 Metals—Mechanical Testing; Elevated and Low-Temperature Tests; Metallography
Volume 03.02 Wear and Erosion; Metal Corrosion
Volume 03.03 Nondestructive Testing
Volume 03.04 Magnetic Properties
Volume 03.05 Analytical Chemistry for Metals, Ores, and Related Materials (I): E 32 - E 1724
Volume 03.06 Analytical Chemistry for Metals, Ores, and Related Materials (II): E 1763 - latest; Molecular Spectroscopy; Surface Analysis

Section 4—Construction

Volume 04.01 Cement; Lime; Gypsum
Volume 04.02 Concrete and Aggregates
Volume 04.03 Road and Paving Materials; Vehicle-Pavement Systems
Volume 04.04 Roofing, Waterproofing, and Bituminous Materials
Volume 04.05 Chemical-Resistant Nonmetallic Materials; Vitrified Clay Pipe, Concrete Pipe, Fiber-Reinforced Cement Products; Mortars and Grouts; Masonry
Volume 04.06 Thermal Insulation; Environmental Acoustics
Volume 04.07 Building Seals and Sealants; Fire Standards; Dimension Stone
Volume 04.08 Soil and Rock (I): D 420 - D 5779
Volume 04.09 Soil and Rock (II): D 5780 - latest; Geosynthetics
Volume 04.10 Wood
Volume 04.11 Building Constructions

Section 5—Petroleum Products, Lubricants, and Fossil Fuels

Volume 05.01 Petroleum Products and Lubricants (I): D 56 - D 2596
Volume 05.02 Petroleum Products and Lubricants (II): D 2597 - D 4927
Volume 05.03 Petroleum Products and Lubricants (III): D 4928 - D 5950
Volume 05.04 Petroleum Products and Lubricants (IV): D 5966 - latest
Volume 05.05 Test Methods for Rating Motor, Diesel, and Aviation Fuels; Manufactured Carbon and Graphite Products; Catalysts
Volume 05.06 Gaseous Fuels; Coal and Coke

Section 6—Paints, Related Coatings, and Aromatics

Volume 06.01 Paint—Tests for Chemical, Physical, and Optical Properties; Appearance

LISTED BY SECTION AND VOLUME

Volume 06.02 Paint—Products and Applications; Protective Coatings; Pipeline Coatings
Volume 06.03 Paint—Pigments, Drying Oils, Polymers, Resins, Naval Stores, Cellulosic Esters, and Ink Vehicles
Volume 06.04 Paint—Solvents; Aromatic Hydrocarbons

Section 7—Textiles
Volume 07.01 Textiles (I): D 76 - D 3218
Volume 07.02 Textiles (II): D 3333 - latest

Section 8—Plastics
Volume 08.01 Plastics (I): D 256 - D 2343
Volume 08.02 Plastics (II): D 2383 - D 4322
Volume 08.03 Plastics (III): D 4329 - latest
Volume 08.04 Plastic Pipe and Building Products

Section 9—Rubber
Volume 09.01 Rubber, Natural and Synthetic—General Test Methods; Carbon Black
Volume 09.02 Rubber Products, Industrial—Specifications and Related Test Methods; Gaskets; Tires

Section 10—Electrical Insulation and Electronics
Volume 10.01 Electrical Insulation (I): D 69 - D 2484
Volume 10.02 Electrical Insulation (II): D 2518 - latest
Volume 10.03 Electrical Insulating Liquids and Gases; Electrical Protective Equipment
Volume 10.04 Electronics (I)
Volume 10.05 Electronics (II)

Section 11—Water and Environmental Technology
Volume 11.01 Water (I)
Volume 11.02 Water (II)
Volume 11.03 Atmospheric Analysis; Occupational Health and Safety; Protective Clothing
Volume 11.04 Environmental Assessment; Hazardous Substances and Oil Spill Responses; Waste Management
Volume 11.05 Biological Effects and Environmental Fate; Biotechnology; Pesticides

Section 12—Nuclear, Solar, and Geothermal Energy
Volume 12.01 Nuclear Energy (I)
Volume 12.02 Nuclear (II), Solar, and Geothermal Energy

Section 13— Medical Devices and Services
Volume 13.01 Medical Devices; Emergency Medical Services

Section 14—General Methods and Instrumentation
Volume 14.01 Healthcare Informatics
Volume 14.02 General Test Methods; Forensic Sciences; Terminology; Conformity Assessment; Statistical Methods
Volume 14.03 Temperature Measurement
Volume 14.04 Laboratory Apparatus; Degradation of Materials; Filtration; SI; Oxygen Fire Safety

Section 15—General Products, Chemical Specialties, and End Use Products
Volume 15.01 Refractories; Activated Carbon; Advanced Ceramics
Volume 15.02 Glass; Ceramic Whitewares
Volume 15.03 Space Simulation; Aerospace and Aircraft; High Modulus Fibers and Composites
Volume 15.04 Soaps and Other Detergents; Polishes; Leather; Resilient Floor Coverings
Volume 15.05 Engine Coolants; Halogenated Organic Solvents and Fire Extinguishing Agents; Industrial and Specialty Chemicals
Volume 15.06 Adhesives
Volume 15.07 Sports Equipment; Safety and Traction for Footwear; Amusement Rides; Consumer Products
Volume 15.08 Sensory Evaluation; Vacuum Cleaners; Security Systems; Detention Facilities; Food Service Equipment
Volume 15.09 Paper; Packaging; Flexible Barrier Materials; Business Imaging Products

Section 00—Index
Volume 00.01 Subject Index; Alphanumeric List

ANNUAL BOOK OF ASTM STANDARDS

Listed by Subjects

LISTED BY SUBJECTS

LISTED BY SUBJECTS

Contents

2001 ANNUAL BOOK OF ASTM STANDARDS, VOLUME 11.04

STANDARDS RELATING TO ENVIRONMENTAL ASSESSMENT; HAZARDOUS SUBSTANCES AND OIL SPILL RESPONSES; WASTE MANAGEMENT

A complete Subject Index appears at the end of this volume.

Listed below are those standards included in this volume(s) and those standards that have appeared previously which have been discontinued within the past three years.

In the serial designations prefixed to the following titles, the number following the dash indicates the year of original adoption as standard or, in the case of revision, the year of last revision. Thus, standards adopted or revised during the year 2001 have as their final number, 01. A letter following this number indicates more than one revision during that year, that is, 01a indicates the second revision in 2001, 01b, the third revision, etc. Standards that have been reapproved without change are indicated by the year of last reapproval in parentheses as part of the designation number, for example, (2001). A superscript epsilon indicates an editorial change since the last revision or reapproval—$^{\epsilon 1}$ for the first change, $^{\epsilon 2}$ for the second change, etc.

§Approved for use by agencies of the Department of Defense and, if indicated on the standard, replaces corresponding Federal or Military document. Consult the DoD Index of Specifications and Standards for the specific year of issue which has been adopted by the Department of Defense.

†Although this standard has been officially withdrawn from Society approval, a brief description is included for information only.

CONTENTS

CONTENTS

CONTENTS

CONTENTS

RELATED MATERIAL

List by Subjects

2001 ANNUAL BOOK OF ASTM STANDARDS, VOLUME 11.04

ENVIRONMENTAL ASSESSMENT; HAZARDOUS SUBSTANCES AND OIL SPILL RESPONSES; WASTE MANAGEMENT

A complete Subject Index appears at the end of this volume.

Since the standards in this volume are arranged in alphanumeric sequence, no page numbers are given in this list by subjects.

ENVIRONMENTAL ASSESSMENT

Commercial Real Estate Transactions

Practices for:

E 1929 – 98 Assessment of Certification Programs for Environmental Professionals
E 1527 – 00 Environmental Site Assessments
E 1528 – 00 Environmental Site Assessments

Guides for:

E 1903 – 97 Environmental Site Assessments
E 2018 – 99 Property Condition Assessments

Environmental Risk Management/Sustainable Development/Pollution Prevention

Guides for:

E 1609 – 94 Development and Implementation of a Pollution Prevention Program
E 1991 – 98 Environmental Life Cycle Assessment of Building Materials/Products
E 2137 – 01 Estimating Monetary Costs and Liabilities for Environmental Matters
E 1984 – 98 Process of Sustainable Brownfields Redevelopment
E 1971 – 98 Stewardship for the Cleaning of Commercial and Institutional Buildings
E 1861 – 97 Use of Coal Combustion By-Products in Structural Fills
E 2060 – 00 Use of Coal Combustion Products for Solidification/Stabilization of Inorganic Wastes

Performance Standards Related to Environmental Regulatory Programs

Practices for:

E 2107 – 00 Environmental Regulatory Compliance Audits

Guides for:

E 1752 – 95(2000)$^{\epsilon 1}$ Collection of Multi-Media Field Emission and Discharge Data Associated with Glycol Dehydration Units
E 2081 – 00 Risk-Based Corrective Action
E 2091 – 00 Use of Activity and Use Limitations, Including Institutional and Engineering Controls

Storage Tanks

Practices for:

E 1526 – 93 Evaluating the Performance of Release Detection Systems for Underground Storage Tank Systems

Guides for:

E 1912 – 98 Accelerated Site Characterization for Confirmed or Suspected Petroleum Releases
E 1599 – 94 Corrective Action for Petroleum Releases
E 1990 – 98 Performing Evaluations of Underground Storage Tank Systems for Operational Conformance with 40 CFR, Part 280 Regulations
E 1943 – 98 Remediation of Ground Water by Natural Attenuation at Petroleum Release Sites
E 1739 – 95$^{\epsilon 1}$ Risk-Based Corrective Action Applied at Petroleum Release Sites
E 1430 Using Release Detection Devices with Underground Storage Tanks (Discontinued 2000†)

§Approved for use by agencies of the Department of Defense and, if indicated on the standard, replaces corresponding Federal or Military document. Consult the DoD Index of Specifications and Standards for the specific year of issue which has been adopted by the Department of Defense.

†Although this standard has been officially withdrawn from Society approval, a brief description is included for information only.

LIST BY SUBJECTS

Wetland Ecosystems

Guides for:	
E 1983 – 98	Assessment of Wetland Functions

HAZARDOUS SUBSTANCES AND OIL SPILL RESPONSE

Bioremediation

Guides for:	
F 1834 – 98	Consideration of Anaerobic Bioremediation as a Chemical Pollutant Mitigation Method on Land
F 1693 – 96	Consideration of Bioremediation as an Oil Spill Response Method on Land
F 1481 – 94(2001)	Ecological Considerations for the Use of Bioremediation in Oil Spill Response—Sand and Gravel Beaches

Control

Specifications for:	
F 962 – 99	Oil Spill Response Boom Connection
Test Methods for:	
F 715 – 95	Coated Fabrics Used for Oil Spill Control and Storage
F 1093 – 99	Tensile Strength Characteristics of Oil Spill Response Boom
Practices for:	
F 625 – 94(2000)	Classifying Water Bodies for Spill Control Systems
F 1657 – 96	Emergency Joining of Booms with Incompatible Connectors
Guides for:	
F 2084 – 01	Collecting Containment Boom Performance Data in Controlled Environments
F 1523 – 94(2001)	Selection of Booms in Accordance With Water Body Classifications
Terminology for:	
F 818 – 93(1998)$^{\epsilon 1}$	Spill Response Barriers

Disposal

Guides for:	
F 1599 – 95	Collecting Performance Data on Temporary Storage Devices
F 873 – 84(1996)	Incinerating Oil Spill Wastes at Temporary Field Locations

In-Situ Burning

Guides for:	
F 1788 – 97	In-Situ Burning of Oil Spills on Water
F 1990 – 99	In-Situ Burning of Spilled Oil

Initial Response Actions

Guides for:	
F 1011 – 86(2001)	Developing a Hazardous Materials Training Curriculum for Initial Response Personnel
F 1656 – 95	Health and Safety Training of Oil Spill Responders in the United States
F 1644 – 95	Health and Safety Training of Oil Spill Responders

Mitigation Actions

Test Methods for:	
§F 726 – 99	Sorbent Performance of Adsorbents
Guides for:	
F 1127 – 88(1996)	Containment by Emergency Response Personnel of Hazardous Material Spills
F 1524 – 95	Use of Advanced Oxidation Process for the Mitigation of Chemical Spills
F 1525 – 96	Use of Membrane Technology in Mitigating Hazardous Chemical Spills
F 1522 – 95	Use of the Steam Stripping Process in Mitigating Chemical Spills
F 1129 – 88(1995)$^{\epsilon 1}$	Using Aqueous Foams to Control the Vapor Hazard from Immiscible Volatile Liquids

Removal

Test Methods for:	
F 716 – 82(1993)$^{\epsilon 1}$	Sorbent Performance of Absorbents
Guides for:	
F 631 – 99	Collecting Skimmer Performance Data in Controlled Environments

F 1688	Estimating Oil Spill Recovery System Effectiveness (Discontinued 1999†)
F 1780 – 97	Estimating Oil Spill Recovery System Effectiveness
F 933 – 85(1994)$^{\epsilon 1}$	Evaluation of Oil Water Separation Systems for Spilled Oil Recovery Applications
F 2008 – 00	Qualitative Observations of Skimmer Performance
F 1607 – 95	Reporting of Test Performance Data for Oil Spill Response Pumps
F 1084 – 90(1995)$^{\epsilon 1}$	Sampling Oil-Water Mixtures for Oil Spill Recovery Equipment
F 1778 – 97	Selection of Skimmers for Oil-Spill Response

Shoreline Countermeasures

Guides for:

F 1686 – 97	Surveys to Document and Assess Oiling Conditions on Shorelines
F 1687 – 97	Terminology and Indices to Describe Oiling Conditions on Shorelines

Surveillance and Tracking

Practices for:

F 1779 – 97	Reporting Visual Observations of Oil on Water

Treatment

Test Methods for:

F 1738 – 96(1999)	Determination of Deposition of Aerially Applied Oil Spill Dispersants
F 2059 – 00	Laboratory Oil Spill Dispersant Effectiveness Using The Swirling Flask

Practices for:

F 1460 – 93(1999)	Calibrating Oil Spill Dispersant Application Equipment Boom and Nozzle Systems
F 2067 – 00	Development and Use of Oil-Spill Trajectory Models

Guides for:

F 990 – 86(1993)$^{\epsilon 1}$	Ecological Considerations for Use of Chemical Dispersants in Oil Spill Response-Sandy Beaches
F 931 – 85(1999)	Ecological Considerations for the Use of Chemical Dispersants in Oil Spill Response-Seagrasses
F 1010 – 86(1993)$^{\epsilon 1}$	Ecological Considerations for the Use of Chemical Dispersants in Oil Spill Response-Bird Habitats
F 932 – 85(1999)	Ecological Considerations for the Use of Chemical Dispersants in Oil Spill Response-Coral Reefs
F 999 – 86(1993)$^{\epsilon 1}$	Ecological Considerations for the Use of Chemical Dispersants in Oil Spill Response-Gravel or Cobble
F 929 – 85(1999)	Ecological Considerations for the Use of Chemical Dispersants in Oil Spill Response-Marine Mammals
F 972 – 86(1993)$^{\epsilon 1}$	Ecological Considerations for the Use of Chemical Dispersants in Oil Spill Response-Nearshore Subtid
F 1009 – 86(1999)	Ecological Considerations for the Use of Chemical Dispersants in Oil Spill Response-Offshore
F 930 – 85(1993)$^{\epsilon 1}$	Ecological Considerations for the Use of Chemical Dispersants in Oil Spill Response-Rocky Shores
F 1008 – 86(1993)$^{\epsilon 1}$	Ecological Considerations for the Use of Chemical Dispersants in Oil Spill Response-Salt Marshes
F 1012 – 86(1999)	Ecological Considerations for the Use of Chemical Dispersants in Oil Spill Response-the Arctic
F 971 – 86(1999)	Ecological Considerations for the Use of Chemical Dispersants in Oil Spill Response
F 973 – 86(1993)$^{\epsilon 1}$	Ecological Considerations for the Use of Chemical Dispersants in Oil Spill Response
F 1280 – 90(1999)	Ecological Considerations for the Use of Oilspill Dispersants in Freshwater and Other Inland Environments, Impermeable Surfaces
F 1210 – 89(1999)	Ecological Considerations for the Use of Oilspill Dispersants in Freshwater and Other Inland Environments, Lakes and Large Water Bodies
F 1279 – 90(1999)	Ecological Considerations for the Use of Oilspill Dispersants in Freshwater and Other Inland Environments, Permeable Surfaces
F 1209 – 89(1999)	Ecological Considerations for the Use of Oilspill Dispersants in Freshwater and Other Inland Environments, Ponds and Sloughs
F 1231 – 89(1999)	Ecological Considerations for the Use of Oilspill Dispersants in Freshwater and Other Inland Environments, Rivers and Creeks
F 1413 – 92(1999)	Oil Spill Dispersant Application Equipment
F 1872 – 98	Use of Chemical Shoreline Cleaning Agents
F 1737 – 96(1999)	Use of Oil Spill Dispersant Application Equipment During Spill Response

WASTE MANAGEMENT

LIST BY SUBJECTS

Analytical Methods

Test Methods for:

D 5285 – 92(1997)	24-Hour Batch-Type Measurement of Volatile Organic Sorption by Soils and Sediments
D 4646 – 87(2001)	24-h Batch-Type Measurement of Contaminant Sorption by Soils and Sediments
D 5659 – 95(2001)	Chlorophenoxy Acid Herbicides in Waste Using HPLC
D 6050 – 97	Determination of Insoluble Solids in Organic Liquid Hazardous Waste
D 5231 – 92(1998)	Determination of the Composition of Unprocessed Municipal Solid Waste
D 4842 – 90(2001)	Determining the Resistance of Solid Wastes to Freezing and Thawing
D 5232 – 92(1997)	Determining the Stability and Miscibility of a Solid, Semi-Solid, or Liquid Waste Material
D 5368 – 93(2001)	Gravimetric Determination of Total Solvent Extractable Content (TSEC) of Solid Waste Samples
D 6052 – 97	Preparation and Elemental Analysis of Liquid Hazardous Waste by Energy-Dispersive X-Ray Fluorescence
D 5233 – 92(1999)	Single Batch Extraction Method for Wastes
D 5830 – 95(2001)	Solvents Analysis in Hazardous Waste Using Gas Chromatography
D 5530 – 94(1998)	Total Moisture of Hazardous Waste Fuel by Karl Fischer Titrimetry
D 5839 – 96(2001)	Trace Element Analysis of Hazardous Waste Fuel by Energy-Dispersive X-Ray Fluorescence Spectrometry
D 4843 – 88(1999)	Wetting and Drying Test of Solid Wastes

Practices for:

D 6010 – 96(2001)	Closed Vessel Microwave Solvent Extraction of Organic Compounds from Solid Matrices
D 4503 – 86(1998)	Dissolution of Solid Waste by Lithium Metaborate Fusion
D 5369 – 93(1998)	Extraction of Solid Waste Samples for Chemical Analysis Using Soxhlet Extraction
D 5513 – 99	Microwave Digestion of Industrial Furnace Feedstreams and Waste for Trace Element Analysis
D 5198 – 92(1997)	Nitric Acid Digestion of Solid Waste

Guides for:

D 5829 – 96(2001)	Preparing a Training Program for Environmental Analytical Laboratories

Medical Waste

Test Methods for:

E 885 – 88(1996)	Analyses of Metals in Refuse-Derived Fuel by Atomic Absorption Spectroscopy
E 830 – 87(1996)	Ash in the Analysis Sample of Refuse-Derived Fuel
E 791 – 90(1996)	Calculating Refuse-Derived Fuel Analysis Data from As-Determined to Different Bases
E 777 – 87(1996)	Carbon and Hydrogen in the Analysis Sample of Refuse-Derived Fuel
E 889 – 82(1996)	Composition or Purity of a Solid Waste Materials Stream
E 828 – 81(1997)	Designating the Size of RDF-3 From Its Sieve Analysis
E 776 – 87(1996)	Forms of Chlorine in Refuse-Derived Fuel
E 711 – 87(1996)	Gross Calorific Value of Refuse-Derived Fuel by the Bomb Calorimeter
E 778 – 87(1996)	Nitrogen in the Analysis Sample of Refuse-Derived Fuel
E 926 – 94	Preparing Refuse-Derived Fuel (RDF) Samples for Analyses of Metals
E 790 – 87(1996)	Residual Moisture in a Refuse-Derived Fuel Analysis Sample
E 887 – 88(1996)	Silica in Refuse-Derived Fuel (RDF) and RDF Ash
E 955 – 88(1996)	Thermal Characteristics of Refuse-Derived Fuel Macrosamples
E 949 – 88(1996)	Total Moisture in a Refuse-Derived Fuel Laboratory Sample

Municipal Soil Waste

Guides for:

D 6346 – 98	Accepting, Segregating and Packaging Materials Collected Through Household Hazardous Waste Programs
D 6498 – 99	Household Hazardous Waste Training Outline for Household Hazardous Waste Collection Operations

Physical and Chemical Characterization (Disbanded 11/00)

Test Methods for:

D 5049	the Screening of Cyanides in Waste (Discontinued 1999†)

Planning for Sampling

Practices for:

D 6597 – 00	Assessment of Attaining Clean Up Level for Site Closure
D 6250 – 98	Derivation of Decision Point and Confidence Limit for Statistical Testing of Mean Concentration in Waste Management Decisions
D 5283 – 92(1997)	Generation of Environmental Data Related to Waste Management Activities
D 5792 – 95	Generation of Environmental Data Related to Waste Management Activities

LIST BY SUBJECTS

Guides for:

D 4844 – 88(1998)	Air Monitoring at Waste Management Facilities for Worker Protection
D 6051 – 96(2001)	Composite Sampling and Field Subsampling for Environmental Waste Management Activities
D 6233 – 98	Data Assessment for Environmental Waste Management Activities
D 4447 – 84(1997)	Disposal of Laboratory Chemicals and Samples
D 4687 – 95(2001)	General Planning of Waste Sampling
D 6311 – 98	Generation of Environmental Data Related to Waste Management Activities
D 6323 – 98	Laboratory Subsampling of Media Related to Waste Management Activities
D 6582 – 00	Ranked Set Sampling
D 6044 – 96	Representative Sampling for Management of Waste and Contaminated Media
D 5956 – 96(2001)	Sampling Strategies for Heterogeneous Wastes
D 6009 – 96(2001)	Sampling Waste Piles
D 6538 – 00	Sampling Wastewater With Automatic Samplers
D 6232 – 00	Selection of Sampling Equipment for Waste and Contaminated Media Data Collection Activities

Recovery and Reuse

Specifications for:

D 5359 – 98	Glass Cullet Recovered from Waste for Use in Manufacture of Glass Fiber
E 753 – 80(1999)	Municipal Aluminum Scrap (MAS)
E 702 – 85(1999)	Municipal Ferrous Scrap
D 5370 – 96	Pozzolanic Blended Materials in Construction Applications
E 1134 – 86(1991)	Source-Separated Steel Cans
E 708 – 79(1999)	Waste Glass as a Raw Material for the Manufacture of Glass Containers

Classifications:

E 956 – 83(1999)	Municipal-Mixed Nonferrous Metals (MNM)

Test Methods for:

E 1183 – 87(1996)	Air Drying RDF-5 for Further Analysis
E 959 – 83(1999)	Characterizing the Performance of Refuse Size-Reduction Equipment
E 868 – 82(1999)	Conducting Performance Tests on Mechanical Conveying Equipment Used in Resource Recovery Systems
E 1108 – 86(1996)	Determination of the Recovery of a Product in a Materials Separation Device
E 1109 – 86(1996)	Determining the Bulk Density of Solid Waste Fractions
E 929 – 83(1999)	Measuring Electrical Energy Requirements of Processing Equipment
E 1037 – 84(1996)	Measuring Particle Size Distribution of RDF-5
E 1107 – 86(1996)	Measuring the Throughput of Resource-Recovery Unit Operations
E 701 – 80(1999)	Municipal Ferrous Scrap
E 688 – 94(1999)	Waste Glass as a Raw Material for Glass Manufacturing

Practices for:

E 1076 – 85(1999)	Maintaining Health and Safety Records at Solid Waste Processing Facilities
E 1266 – 88(1999)	Processing Mixtures of Lime, Fly Ash, and Heavy Metal Wastes in Structural Fills and Other Construction Applications
E 1248 – 90(1997)	Shredder Explosion Protection
E 850 – 95	Use of Inorganic Process Wastes as Structural Fill
D 6270 – 98	Use of Scrap Tires in Civil Engineering Applications

Guides for:

D 5050 – 96	Commercial Use of Lime Kiln Dusts and Portland Cement Kiln Dusts

Sampling Equipment

Practices for:

D 5451 – 93(2001)	Sampling Using a Trier Sampler
D 5495 – 94(2001)	Sampling With a Composite Liquid Waste Sampler (COLIWASA)
D 5358 – 93(1998)	Sampling with a Dipper or Pond Sampler
D 5633 – 94(2001)	Sampling with a Scoop
D 6418 – 99$^{\epsilon 1}$	Using the Disposable En Core Sampler for Sampling and Storing Soil for Volatile Organic Analysis

Sampling Techniques

Practices for:

E 884 – 82(2001)	Sampling Airborne Microorganisms at Municipal Solid-Waste Processing Facilities
D 5679 – 95a(2001)	Sampling Consolidated Solids in Drums or Similar Containers
D 5743 – 97	Sampling Single or Multilayered Liquids, With or Without Solids, in Drums or Similar Containers

D 5680 – 95a(2001)	Sampling Unconsolidated Solids in Drums or Similar Containers
D 5658 – 95(2001)	Sampling Unconsolidated Waste From Trucks
D 5013 – 89(1998)	Sampling Wastes from Pipes and Other Point Discharges

Guides for:

D 4448 – 85a(1992)	Sampling Groundwater Monitoring Wells
D 4547 – 98	Sampling Waste and Soils for Volatile Organic Compounds
D 6063 – 96(2001)	Sampling of Drums and Similar Containers by Field Personnel

Sampling and Monitoring

Test Methods for:

D 5831 – 96(2001)	Screening Fuels in Soils

Practices for:

D 6640 – 01	Collection and Handling of Soils Obtained in Core Barrel Samplers for Environmental Investigations
D 6661 – 01	Field Collection of Organic Compounds from Surfaces Using Wipe Sampling

Screening Methods

Test Methods for:

D 5058 – 90(2001)	Compatibility of Screening Analysis of Waste
D 4982 – 95(2001)	Flammability Potential Screening Analysis of Waste
D 4979 – 95	Physical Description Screening Analysis in Waste
D 5057 – 90(2001)	Screening Apparent Specific Gravity and Bulk Density of Waste
D 4981 – 95	Screening of Oxidizers in Waste
D 4978 – 95(2001)	Screening of Reactive Sulfides in Waste
D 5928 – 96	Screening of Waste for Radioactivity
D 4980 – 89	Screening of pH in Waste

Site Remediation

Classifications:

D 5746 – 98	Environmental Condition of Property Area Types for Defense Base Closure and Realignment Facilities

Practices for:

D 6008 – 96	Conducting Environmental Baseline Surveys

Guides for:

D 5745 – 95(1999)	Developing and Implementing Short Term Measures or Early Actions for Site Remediation

Terminology

Terminology for:

D 5681 – 98a	Waste and Waste Management

Treatment

Test Methods for:

D 5660 – 96	Assessing the Microbial Detoxification of Chemically Contaminated Water and Soil Using a Toxicity Test with a Luminescent Marine Bacterium
D 5115 – 90(1996)	Collecting Gross Samples and Determining the Fuel Quality of RDF
D 5929 – 96	Determining Biodegradability of Materials Exposed to Municipal Solid Waste Composting Conditions by Compost Respirometry
D 5975 – 96	Determining the Stability of Compost by Measuring Oxygen Consumption
E 953 – 88(1998)	Fusibility of Refuse-Derived Fuel (RFD) Ash
D 5468 – 95	Gross Calorific and Ash Value of Waste Materials
D 5120 – 90(1995)	Inhibition of Respiration in Microbial Cultures in the Activated Sludge Process
E 775 – 87(1996)	Total Sulfur in the Analysis Sample of Refuse-Derived Fuel
E 897 – 88(1998)	Volatile Matter in the Analysis Sample of Refuse-Derived Fuel

Practices for:

E 886 – 94	Dissolution of Refuse-Derived Fuel (RDF) Ash Samples for Analyses of Metals
E 954 – 94	Packaging and Shipping of Laboratory Samples of Refuse-Derived Fuel
E 829 – 94	Preparing Refuse-Derived Fuel (RDF) Laboratory Samples for Analysis

Guides for:

D 6249 – 98	Alkaline Stabilization of Wastewater Treatment Plant Residuals
D 5759 – 95	Characterization of Coal Fly Ash and Clean Coal Combustion Fly Ash for Potential Uses
D 5177 – 91(1996)	Waste Acceptance at Hazardous Waste Incinerators

Terminology for:

E 856 – 83(1998)	Terms and Abbreviations Relating to Physical and Chemical Characteristics of Refuse-Derived Fuel

Waste Leaching Techniques

Test Methods for:

D 5744 – 96(2001)	Accelerated Weathering of Solid Materials Using a Modified Humidity Cell
D 4874 – 95(2001)	Leaching Solid Material in a Column Apparatus
D 5284 – 93(1999)	Sequential Batch Extraction of Waste with Acidic Extraction Fluid
D 4793 – 93(1999)	Sequential Batch Extraction of Waste with Water
D 6234 – 98	Shake Extraction of Mining Waste by the Synthetic Precipitation Leaching Procedure
D 3987 – 85(1999)	Shake Extraction of Solid Waste with Water

Designation: D 3987 – 85 (Reapproved 1999)

Standard Test Method for Shake Extraction of Solid Waste with Water[1]

This standard is issued under the fixed designation D 3987; the number immediately following the designation indicates the year of original adoption or, in the case of revision, the year of last revision. A number in parentheses indicates the year of last reapproval. A superscript epsilon (ε) indicates an editorial change since the last revision or reapproval.

1. Scope

1.1 This test method covers a procedure for leaching of solid waste to obtain an aqueous solution to be used to determine the materials leached under the specified testing conditions.

1.2 This test method provides for the shaking of a known weight of waste with water of specified composition and the separation of the aqueous phase for analysis.

1.3 *This standard does not purport to address all of the safety problems, if any, associated with its use. It is the responsibility of the user of this standard to establish appropriate safety and health practices and determine the applicability of regulatory limitations prior to use.*

2. Referenced Documents

2.1 *ASTM Standards:*

C 471 Test Methods for Chemical Analysis of Gypsum and Gypsum Products[2]

D 75 Practice for Sampling Aggregates[3]

D 420 Practice for Investigating and Sampling Soil and Rock for Engineering Purposes[4]

D 1129 Terminology Relating to Water[5]

D 1193 Specification for Reagent Water[5]

D 2216 Test Method for Laboratory Determination of Water (Moisture) Content of Soil and Rock[4]

D 2234 Test Methods for Collection of a Gross Sample of Coal[6]

D 3370 Practices for Sampling Water[5]

E 122 Practice for Choice of Sample Size to Estimate a Measure of Quality for a Lot or Process[7]

3. Terminology Definitions

3.1 For definitions of terms used in this test method, see Terminology D 1129.

4. Significance and Use

4.1 This test method is intended as a rapid means for obtaining an extract of solid waste. The extract may be used to estimate the release of certain constituents of the solid waste under the laboratory conditions described in this procedure.

4.2 This test method is not intended to provide an extract that is representative of the actual leachate produced from a solid waste in the field or to produce extracts to be used as the sole basis of engineering design.

4.3 This test method is not intended to simulate site-specific leaching conditions. It has not been demonstrated to simulate actual disposal site leaching conditions.

4.4 The intent of this test method is that the final pH of the extract reflect the interaction of the extractant with the buffering capacity of the solid waste.

4.5 The intent of this test method is that the water extraction simulate conditions where the solid waste is the dominant factor in determining the pH of the extract.

4.6 The test method produces an extract that is amenable to the determination of both major and minor constituents. When minor constituents are being determined, it is especially important that precautions are taken in sample storage and handling to avoid possible contamination of the samples.

4.7 This test method has been tested to determine its applicability to certain inorganic components in the solid waste. The test method has not been tested for applicability to organic substances and volatile matter (see 5.3).

4.8 The agitation technique, rate, and liquid-to-solid ratio specified in the procedure may not be suitable for extracting all types of solid wastes. (See Sections 7, 8, and the discussion in Appendix X1.)

5. Apparatus

5.1 *Agitation Equipment*, of any type that rotates about a central axis at a rate of 29 r/min, Fig. 1. (See discussion of agitation in Appendix X1.)

5.2 *Membrane Filter Assembly*—A borosilicate glass or stainless steel funnel with a flat, fritted base of the same material and membrane filters.

5.3 *Containers*, round, wide-mouth, of a composition suitable to the nature of the solid waste and the analyses to be performed, and constructed of materials that will not allow sorption of constituents of interest. One-gallon (or 4-L) containers should be used with 140-g samples and ½-gallon (or

[1] This test method is under the jurisdiction of ASTM Committee D34 on Waste Management and is the direct responsibility of Subcommittee D34.01.04 on Waste Leaching Techniques.

Current edition approved Oct. 21, 1985. Published March 1986. Originally published as D 3987 – 81. Last previous edition D 3987 – 81.

[2] *Annual Book of ASTM Standards*, Vol 04.01.

[3] *Annual Book of ASTM Standards*, Vol 04.03.

[4] *Annual Book of ASTM Standards*, Vol 04.08.

[5] *Annual Book of ASTM Standards*, Vol 11.01.

[6] *Annual Book of ASTM Standards*, Vol 05.05.

[7] *Annual Book of ASTM Standards*, Vol 14.02.

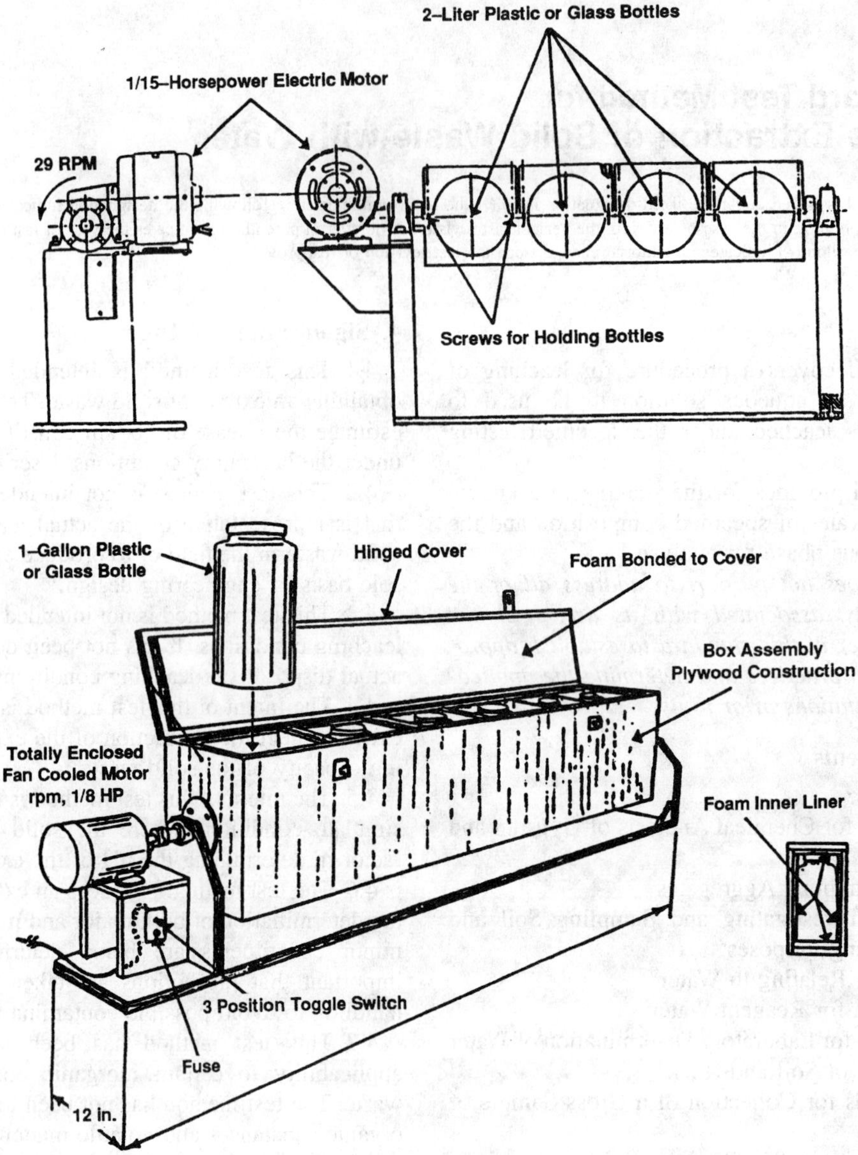

FIG. 1 Extractor

2-L) containers with 70-g samples. Multiples of these sizes may be used for larger samples. The containers should be of the same approximate geometry as the 2-L and 4-L bottles. These sizes were selected to establish suitable geometry and provide that the sample plus liquid would occupy approximately 80 to 90 % of the container. Containers must have a watertight closure. Containers for samples where gases may be released should be provided with a venting mechanism. (Note that the venting of the container has the potential to affect the concentration of volatile extracts in the extract.) Containers should be cleaned in a manner consistent with the analyses to be performed.

6. Reagents

6.1 *Purity of Reagents*—Reagent grade chemicals shall be used in all tests. Unless otherwise indicated, it is intended that all reagents shall conform to the specifications of the American Chemical Society, where such specifications are available,[8] Other grades may be used, provided it is first ascertained that the reagent is of sufficiently high purity to permit its use without lessening the accuracy of the determination.

6.2 *Purity of Water*—Unless otherwise indicated, references to water shall be understood to mean Type IV reagent water at 18 to 27°C (Specification D 1193). The method by which the Type IV water is prepared, that is, distillation, ion exchange, reverse osmosis, electrodialysis, should remain constant throughout testing.

7. Sampling

7.1 Obtain a representative sample of the solid waste to be

[8] "Reagent Chemicals, American Chemical Society Specifications," Am. Chemical Soc., Washington, DC. For suggestions on the testing of reagents not listed by the American Chemical Society, see "Reagent Chemicals and Standards," by Joseph Rosin. D. Van Nostrand Co., Inc., New York, NY, and the "United States Pharmacopeia."

tested using ASTM sample methods developed for the specific industry where available. (See Practices D 75 and D 420 and Test Methods D 2234.)

7.2 Where no specific methods are available, sampling methodology for materials of similar physical form shall be used.

7.3 A minimum sample of 5000 g shall be sent to the laboratory (see Practice E 122).

7.4 It is important that the sample of the solid waste be representative with respect to surface area, as variations in surface area would directly affect the leaching characteristics of the sample. Solid waste samples should contain a representative distribution of particles sizes.

7.5 Keep samples in closed containers appropriate to the sample type prior to the extraction in order to prevent sample contamination or constituent loss. Where it is desired to extract biologically or chemically active samples in their existing state, store the samples at 4°C (Practices D 3370) and start the extraction within 8 h. Where it is desired to extract such samples in a state representative of the results of biological or chemical activities, the samples may be specifically handled to simulate such activities. Record the storage conditions and handling procedures in the report.

8. Sample Preparation

8.1 For free-flowing particulate solid wastes, obtain a sample of the approximate size required in the test by quartering the sample (Section 7) received for testing on an impermeable sheet of glazed paper, oil cloth, or other flexible material as follows:

8.1.1 Empty the sample container into the center of the sheet.

8.1.2 Flatten out the sample gently with a suitable straightedge until it is spread uniformly to a depth at least twice the maximum particle diameter particle size.

8.1.3 Remix the sample by lifting a corner of the sheet and drawing it across, low down, to the opposite corner in a manner that the material is made to roll over and over and does not merely slide along. Continue operation with each corner, proceeding in a clockwise direction. Repeat this operation ten times.

8.1.4 Lift all four corners of the sheet towards the center and holding all four corners together, raise the entire sheet into the air to form a pocket for the sample.

8.1.5 Repeat 8.1.2.

8.1.6 With a straightedge at least as long as the flattened mound of sample (such as a thin-edged yard stick), gently divide the sample into quarters. An effort should be made to avoid using pressure on the straightedge sufficient to cause damage to the particles.

8.1.7 Discard alternate quarters.

8.1.8 If further reduction of sample size is necessary, repeat 8.1.3 through 8.1.7. A minimum sample size of 70 g is recommended for each extraction. Additional samples should be provided for determination of solids content. If smaller samples are used in the test, report this fact.

8.2 For field-cored solid wastes or castings produced in the laboratory, cut a representative section weighing approximately 70 or 140 g for testing, plus samples for determination of solids content. Shape the sample so that the leaching solution will cover the material to be leached.

8.3 For fluid solid wastes, mix thoroughly in a manner that does not incorporate air to assure uniformity before withdrawing a 70 or 140 g sample for test. Take samples for determination of solids content at the same time as the test samples.

9. Procedure

9.1 Record the physical description of the sample to be tested including particle size so far as it is known.

9.2 *Solids Content*—Determine the solids content of separate portions of the sample as follows:

9.2.1 Dry to constant weight two dishes or pans of size suitable to the solid waste being tested at 104 ± 2°C. Cool in a desiccator and weigh. Record the value to ± 0.1 g.

9.2.2 Put an appropriately sized portion of sample of the solid waste to be tested into each pan. Scale the weight used to the physical form of the solid waste tested. Use a minimum of 50 g but use larger samples where particles larger than 10-mm in average diameter are being tested. Weigh. Record the weight to ± 0.1 g.

9.2.3 Dry 16 to 20 h at 104 ± 2°C. Certain solid wastes, such as scrubber sludges, may contain compounds that are subject to calcination at the specified drying temperature. Dry these compounds at lower temperatures. For example, gypsum may be successfully dried at 45°C (Test Methods C 471) and $CaSO_3 \cdot 1/2H_2O$ wastes at 85°C. Record the actual temperature and time of the drying period.

9.2.4 Cool to room temperature in a desiccator and reweigh. Record the weight to ± 0.1 g.

9.3 *Shake Procedure*—Weigh or tare the container to be used in the shake test to the nearest or within 1 g.

9.4 Add to the container approximately 70 g of solid waste (Section 8) and determine and record the weight of sample used to ½ g. If weights other than 70 g are used, note in the report.

9.5 Add to the container a volume of test water (6.2) equal in millilitres to 20 times the weight in grams of the sample used in 9.4. See discussion of dilution ratio in Appendix (Example: 70 g sample = 1400 mL water).

9.6 Agitate continuously for 18 ± 0.25 h at 18 to 27°C.

9.7 Open the container. Observe and record any physical changes in the sample and leaching solution.

9.8 Let the sample settle for 5 min; then separate the bulk of the aqueous phase from the solid phase by decantation, centrifugation, or filtration through a coarse filter paper as appropriate. Then vacuum or pressure filter the liquid through a 0.45-μm filter. If the separation method results in prolonged filtering time, an 8-μm filter or other device may be used. Record any deviations in the report.

9.9 The filtrate obtained in 9.8 is the extract mentioned elsewhere in this test method. Measure the pH of the extract immediately, then preserve the extract in a manner consistent with the chemical analysis or biological testing procedures to be performed (Practices D 3370). If sufficient liquid phase is not available for the analyses, so indicate in the report and do not continue the procedure; or alternatively, perform the extraction procedure on additional samples of the solid waste to obtain sufficient liquid phase. Where phase separation occurs during the storage of the extract, appropriate mixing should be used to ensure the homogeneity of the extract prior to its use in such analysis or testing.

9.10 Analyze the extract for specific constituents or properties or use the extract for biological testing procedures as desired using appropriate ASTM standard methods. Where no appropriate ASTM methods exist, other methods may be used and recorded in the report.

10. Calculation

10.1 Calculate the solids content of the individual samples from the data obtained in 9.2 as follows:

$$S = A/B \quad (1)$$

where:

A = weight in grams of sample after drying,
B = original weight in grams of sample, and
S = solid content, g/g.

Average the two values obtained. Record as the solids content.

11. Report

11.1 The report shall include the following:

11.1.1 Source of the solid waste, date of sampling, and sample preservation used,

11.1.2 Description of the solid waste including physical characteristics and particle size, if known (9.1),

11.1.3 Solids content (9.2) (see Test Method D 2216),

11.1.4 Sample weight if other than 70 g,

11.1.5 Drying time and temperature if other than 16 to 20 h at 104 ± 2°C,

11.1.6 pH and results of specific analyses calculated in appropriate units. State analytical procedures used, and filter used if other than 0.45 µm,

11.1.7 Observation of changes in test material or leaching solution recorded in 9.8.

11.1.8 Date leach testing started, preservation used for extract, and date of analysis.

12. Precision and Bias

12.1 No information is presently available as to the precision or bias of the analysis of specific constituents in the extract. It is recommended that users of this test method validate the applicability of their chosen methods of detection by spiking portions of the extract, before using these test methods for the analysis of the extract.

12.2 Based on a collaborative series of tests on five solid wastes including fly ash, cutting waste, aluminum industry waste, foundry waste, and non-ferrous foundry waste, the overall precision of this test method was improved. The above data was presented at the second ASTM D-34 Solid Waste Symposium.[9]

12.3 The precision of this test method may vary depending on the solid waste being tested and on the element being extracted.

12.4 Determination of the bias of this test method is not possible, as no standard reference material exists.

[9] Contact Dr Robert Paule, National Institute of Standards and Technology, Gaithersburg, MD 20899, on availability of a copy of this paper "Statistical Analysis and Description of Factors Affecting the ASTM Leaching Test."

APPENDIX

(Nonmandatory Information)

X1. AGITATION TECHNIQUES AND RATE, AND LIQUID/SOLID RATIOS

X1.1 While a major effort relative to development of the test method has been undertaken to determine the optimum agitation rate, equipment, and liquid/solid ratios specified in the test method, it is recognized that these variables may significantly influence the results on certain solid wastes, and that they may not be adequate for certain solid wastes, such as monolithic, solidified, or organic wastes.

X1.2 The possible effects of varying the agitation technique and rate include degree of mixing, rate of release of constituents, and particle abrasion effects. The precision of the test method may also be influenced.

X1.3 The possible effects of varying the dilution ratio include degree of mixing, rate of release of constituents (and possible concentration effects, depending on availability), and particle abrasion effects.

X1.4 Efforts are underway to develop test methods that are more suitable for use with organic, monolithic, and solidified wastes.

Standard Guide for
Disposal of Laboratory Chemicals and Samples[1]

This standard is issued under the fixed designation D 4447; the number immediately following the designation indicates the year of original adoption or, in the case of revision, the year of last revision. A number in parentheses indicates the year of last reapproval. A superscript epsilon (ε) indicates an editorial change since the last revision or reapproval.

1. Scope

1.1 This guide is intended to provide the chemical laboratory manager with guidelines for the disposal of small quantities of laboratory wastes safely and in an environmentally sound manner. This guide is applicable to laboratories that generate small quantities of chemical or toxic wastes. Generally, such tasks include, but are not limited to, analytical chemistry, process control, and research or life science laboratories. It would be impossible to address the disposal of all waste from all types of laboratories. This guide is intended to address the more common laboratory waste streams.

1.2 *This standard does not purport to address all of the safety concerns, if any, associated with its use. It is the responsibility of the user of this standard to establish appropriate safety and health practices and determine the applicability of regulatory limitations prior to use.*

2. Referenced Documents

2.1 *Department of Transportation Regulations:*

40 CFR 173 Shippers—General Requirements for Shipments and Packagings[2]

40 CFR 178 Shipping Container Specifications[2]

40 CFR 179 Specifications for Tank Cars[2]

40 CFR 261.3 Definition of Hazardous Waste[2]

40 CFR 261.33 Discarded Commercial Chemical Products, Off-Specifications Species, Container Residues, and Residues Thereof[2]

40 CFR 261.5 Special Requirements for Hazardous Waste Generated by Small Quantity Generators[2]

40 CFR 761 Polychlorinated Biphenyls (PCBs) Manufacturing, Processing, Distribution in Commerce, and Use Prohibitions[2]

49 CFR 172 Hazardous Materials Tables and Hazardous Materials Communications Regulations[2]

3. Summary of Guide

3.1 The necessary classification of the waste for shipping and manifesting is addressed both by their common or generic chemical name and by Department of Transportation (DOT) guidelines.

3.2 Various types of wastes are listed and defined in a manner necessary to segregate them for recovery, pretreatment, or disposal, or both.

3.3 Procedures are not for recovery of the materials, or to render them nonhazardous and amenable to municipal landfill or in-house disposal, or to prepare them for disposal in an authorized chemical waste disposal site.

3.4 Various methods of disposal are discussed.

3.5 Each type of waste is designated a specific recovery or pretreatment and disposal method. In most cases, disposal alternatives are offered.

4. Significance and Use

4.1 Laboratories rarely generate or handle large volumes of hazardous substances. However, the safe handling and disposal of these substances are impaired by diversity, toxicity, high hazard risks, and contemptuous familiarity. With the promulgation of the Resource Conservation and Recovery Act (RCRA) of 1976, more attention is being given to the proper handling and disposal of such materials. Laboratory management should designate an individual who will be responsible for waste disposal and must review the RCRA guidelines, in particular, the definition of a hazardous waste, the specific substances listed as hazardous, generator requirements and exclusions, and proper shipping and manifesting procedures. Because many laboratory employees could be involved in the proper (and improper) treatment and disposal of laboratory chemicals and samples, it is suggested that a safety and training program be designed and presented to all regarding procedures to follow in the treatment and disposal of designated laboratory wastes.

4.2 If practical and economically feasible, it is, of course, recommended that all laboratory waste be either recovered, re-used, or disposed of in-house. The disposal of laboratory samples, especially those received in large numbers or quantities from a specific source, can often be accommodated by returning the material to the originator, so he can combine them with larger quantities for recycling or disposal. However, should this not be the case, other alternatives are presented. This guide is intended only as a suggested organized method for classification, segregation, and disposal of chemical laboratory waste.

---

[1] This guide is under the jurisdiction of ASTM Committee D34 on Waste Management and is the direct responsibility of Subcommittee D34.01.01 on Planning for Sampling.

Current edition approved Dec. 28, 1984. Published June 1985.

[2] *Draft Manual for Infectious Waste Management*, SW-957, USEPA, Washington, DC., September 1982.

4.3 Even though the small quantity generator exclusion (40 CRF 261.5) may apply to laboratories, the professional laboratory supervisor and his or her employers must balance the importance of protecting human health and the environment from the adverse impact of potential mismanagement of small quantities of hazardous waste, with the need to hold the administrative and economic burden of management of these wastes under RCRA within reasonable and practical limits. Additionally, all lab supervisors should be aware of all current local, state, and federal regulations, and specific hazardous waste management facility criteria.

5. Classification of Waste Types

5.1 The individual responsible for classification and segregation must be familiar with the waste's chemical, physical, and hazardous properties. If the waste is ultimately to be disposed of off-site, it must be segregated, packaged, and classified according to defined DOT hazard classification, as specified in the DOT hazardous materials regulations 49 CFR 172.

5.2 The chemical waste may be segregated into the following waste types,

5.2.1 Trash, inert chemicals, non-toxic, non-reactive, non-ignitable, non-corrosive solids as per RCRA or DOT guidelines,

5.2.2 Weak aqueous acid solutions (<10 % weight) and related compounds,

5.2.3 Weak aqueous alkaline solutions (<10 % weight) and related compounds,

5.2.4 Concentrated aqueous acid solutions and related compounds,

5.2.5 Concentrated aqueous alkaline solutions and related compounds,

5.2.6 Flammable (flash point, closed cup, °F < 140°), non-halogenated organic solvents and related compounds,

5.2.7 Flammable halogenated organic solvents and related compounds,

5.2.8 Non-flammable non-halogenated organic solvents and related compounds,

5.2.9 Non-flammable halogenated organic solvents and related compounds,

5.2.10 Organic acids,

5.2.11 Organic bases,

5.2.12 Inorganic oxidizers, peroxides,

5.2.13 Organic oxidizers, peroxides,

5.2.14 Toxic heavy metals,

5.2.15 Toxic poisons, herbicides, pesticides, and carcinogens,

5.2.16 Aqueous solutions of reducing agents and related compounds,

5.2.17 Pyrophoric substances,

5.2.18 Water reactive substances,

5.2.19 Cyanide, sulfide, and ammonia bearing waste,

5.2.20 Explosive materials,

5.2.21 Radioactive materials,

5.2.22 Infectious waste in life science laboratories,

5.2.23 Infectious waste in hospitals,

5.2.24 Water soluble waste of unknown origin or properties,

5.2.25 Water insoluble waste of unknown origin or properties,

5.2.26 Empty containers,

5.2.27 Asbestos or asbestos containing waste,

5.2.28 Contaminated labware and trash,

5.2.29 Polychlorinated biphenyls (PCBs).

6. Pretreatment and Recovery Methods

6.1 The following methods may be employed for the recovery or pretreatment of waste in the laboratory. All persons using chemicals in the laboratory must be aware of the toxic or hazardous properties of the substance(s) used, including consideration of the toxic properties of possible reaction products. In incorporating the following procedures, examine the possible hazards associated with each.

6.1.1 *Recovery, re-use*— Consideration should be given to distillation for the recovery of larger volumes of solvents. Many laboratories have systems for the recovery and re-use of mercury. Other recovery methods such as precipitation or crystallization may be practical.

6.1.2 *Dilution*—Many laboratory chemical wastes can be diluted to an extent to allow disposal to the sewer system. However, this procedure is not recommended for toxic substances exhibiting characteristics of eroaccumulation, persistence, or degradation to more toxic substances. Strong acid and bases should be diluted to pH 3-11 for this purpose. Some solutions of water soluble flammable solvents can be diluted enough to render them non-flammable. Small amounts of various heavy metal compounds may be diluted to an extent that does not pose a hazard to a sewer system. Consult the local waste-water treatment facility for acceptable guidelines.

6.1.3 *Neutralization*— Strong acids and bases can carefully be neutralized to pH 3-11 to render them less hazardous for disposal.

6.1.4 *Oxidation*—Compounds such as sulfides, cyanides, aldehydes, mercaptans, and phenolics can be oxidized to less toxic and less odoriferous compounds.

6.1.5 *Reduction*—In addition to oxidizers and peroxides, various organic chemicals and heavy metal solutions can be reduced to less toxic substances. Aqueous waste containing hexavalent chromium may be reduced to tri-valent using reducing agents such as bisulfite and ferrous sulfate. Mercury, lead, and silver may be removed from aqueous streams by the process of reduction/precipitation. Organo-lead compounds can be removed by the same type processes. The resulting concentrated heavy metal waste can be containerized and disposed of at an authorized hazardous waste management facility, or subjected to recovery at a treatment facility.

6.1.6 *Controlled Reactions/Processes*—Other methods for reducing the hazardous properties of waste will involve processes specific to particular waste generated by the laboratory. To be practical, the waste would have to be of sufficient volume and, for safety purposes, the process would need to be carefully studied and the resulting products identified. Examples may include evaporation, filtration, ion exchange, carbon adsorption, solvent extraction, hydrolysis, ozonolysis, and electrolysis.

7. Disposal Methods

7.1 *Containerization (Dumpster)*—This method should be used only in the disposal of inert laboratory solid waste. Each institution should have a procedure for handling solid waste to include classification, segregation, and collection. Materials disposed of in this manner must be suitable for sanitary landfill disposal and must be of no threat to the personnel handling the waste. Many materials disposed of in this manner by laboratories may be regulated by local authorities.

7.2 *Disposal to the Sewer System*—Many laboratory chemicals, with or without pretreatment by one or more of the above prescribed methods, are amenable to sewer disposal. RCRA regulations (40 CFR 261.3) grant special exemptions for laboratory effluents from hazardous waste regulations if the annualized average flow of laboratory wastewater is less than 1% of the total wastewater going to the headworks of the water treatment facility and the concentration of hazardous material is less than 1 ppm in the headwaters. Also, local regulations govern the concentrations and types of chemicals that may be let to a sewer. Laboratory supervisors must familiarize themselves and their co-workers with these regulations. In addition to the statements made earlier regarding dilution (6.1.2) and neutralization (6.1.3), it is important to emphasize that highly toxic, malodorous, or lachrymatory chemicals should not be disposed of down the drain. Laboratory drains are usually interconnected, and a substance that goes down one sink may arise as a vapor from another. Additionally, the comingling of waste from different sources in the sewer system may present definite hazards for example, the sulfide poured down one drain may contact an acid poured into another. Some simple reactions, such as ammonia plus iodine or silver nitrate plus ethanol, may produce explosions. Laboratory supervisors must be aware of the types of chemicals disposed in this manner so that the risk of potential laboratory accidents is reduced.

7.3 *Incineration, Solvent Recovery*—Waste solvents free of solids and corrosive or reactive substances should be collected, segregated, and containerized. Due to the fact that some incineration or solvent recovery sites will not handle chlorinated solvents, it is often necessary to segregate into two or three types of waste solvents. In addition to solvent recovery techniques that may be employed in the laboratory, some of these materials may be disposed of in-house by mixing with fuel oils for combustion in process boilers, power generators, etc. Additionally, outside disposal firms may be contacted for disposal, but they generally are less interested in handling small volume waste streams, particularly if inconsistent in composition. The laboratory supervisor should be aware of the chemicals collected, and ensure that incompatible materials are not comingled.

7.4 *Lab Pack*—Recent Environmental Protection Agency (EPA) regulations allow the disposal of small containers of hazardous waste (liquids and solids) in overpacked drums in secure or specially permitted landfills.

7.4.1 Each chemical is to be identified by its generic or common name, the quantity, and the DOT hazard classification. All chemicals are to be segregated and packaged according to the following classification: poisons, oxidizers, flammables, corrosives-acids, and corrosives-alkalies.

7.4.2 Many chemicals that are similarly classified will react for example, concentrated solutions of nitric acid mixed with acetic acid can cause spontaneous ignition. Therefore, an employee of the waste generator, familiar with the chemicals and their respective hazards, is to be responsible for not only segregation, but also for the documentation and packaging operations. Compatible materials, of the same classification, are to be packaged in tightly and securely sealed inside containers of the size and type specified in the DOT hazardous materials regulations 40 CFR 173, 178, and 179, if those regulations specify a particular inside container, and placed in DOT-approved open-top metal drums.

7.4.3 Vermiculite or an other inert and compatible material is to be placed around the original waste containers to avoid breakage and to act as an absorbent should any breakage or leakage occur. The chemicals are to be equally distributed within the drum with not less than an equal volume of vermiculite. The drums must be completely filled and properly sealed.

7.4.4 A list detailing the contents of each drum, including the chemicals' common or generic names, the DOT hazard classes, quantities of each, and any pertinent comments, must be available for completion of manifesting purposes and for the disposal firm.

7.5 *Solidification*— In addition to the lab pack, an alternate drum disposal method involves the solidification of compatible liquid chemical waste with vermiculite or a suitable solidification agent such as diatomaceous earth or clay.

7.5.1 A suggested procedure is as follows: A DOT-approved open-top metal drum (17 H) containing a free-standing liner is filled to approximately one-third with the adsorbent. The liquid waste is then carefully poured into the adsorbent, mixed, and allowed to stand. The liquid waste may need pretreatment (for example, neutralization, reduction, etc.) to render it compatible with the solidification agent. The remainder of the drum is then filled in the same manner. An extra layer of dry adsorbent (about 2 in.) is added to top-off the drum to contain any free liquid that may surface in transporting or handling. The drum must be completely full, contain only solids, properly sealed, and labeled.

7.5.2 A list detailing the contents of each drum, including the chemicals' common or generic names, the DOT hazard classes, quantities of each, and any pertinent information, must be available for manifesting purposes and for the disposal firm.

7.6 *Waste Exchange*— Evaluate the possibility of using a waste exchange for specific waste generated by a laboratory, especially if large quantities are produced. The laboratory's unused chemicals or waste may be another's reagent, with possibly only minor pretreatment methods.

8. Special Waste

8.1 Various types of waste cannot be readily disposed of within the laboratory, or through normal outside waste disposal contractors. Disposal alternatives for special waste are discussed.

8.1.1 *Local Fire and Police Department*—Local emergency response offices are often called upon by institutions for disposal of especially dangerous wastes, in most cases explosives. Examples of substances that may be handled in this manner are metallic sodium, picric acid, metal azides, organic azides, and organic peroxides. Often, this type of service is offered only in the larger metropolitan areas. If laboratory personnel are unable to perform controlled degradation processes, contact the local fire or police department for guidance.

8.1.2 *Radioactive Waste Disposal*—The method of disposal will depend upon the activity level, the type of waste, and the radiochemical properties of the isotopes. The two major methods for waste management are dilution for environmental disposal and containment. Dilution is the process of mixing the waste with sufficient inert material to reduce the concentration of activity below the permissible levels. An example would be the flushing of microcurie amounts of an isotope into the sewer system. Large concentrations may require precipitation or flocculation followed by filtration to remove the concentrated activity and the carrier. Nonetheless, any laboratory involved in radioactive waste disposal should be familiar with local regulations and the federal regulations set forth in the Code of Federal Regulations (CFR), Title 10, Chapter 1.

8.1.3 *Infectious Waste*— The USEPA has published a guide to infectious waste management.[2] The manual discusses environmentally acceptable techniques for infectious waste management. Topics covered include a definition of infectious waste, and packaging, transportation, treatment, storage, and disposal practices. Recommendations are presented for methods of treating different types of infectious waste.

8.1.4 *Polychlorinated Biphenyls (PCB's)(4.2)*—The disposal of PCB's is well defined in Toxic Substance and Control Act (TSCA) regulations. PCB's in concentrations less than 50 ppm are currently non-regulated, and disposal in a municipal landfill is possible with local regulatory agency approval. Also, quantities of less than 50 ppm could be comingled with compatible organic wastes destined for incineration. PCB's in concentrations of 50 to 500 ppm are designated by TSCA as "PCB contaminated wastes," and those above 500 ppm "PCB wastes." Disposal of these wastes will require detoxification, or management at a facility approved by the USEPA for the disposal (incineration or landfilling) of PCB's. It is suggested that laboratory managers concerned with PCB disposal consult the TSCA regulations (40 CFR 761) and their regional EPA office.

9. Recommended Pretreatment and Disposal Methods for Laboratory Chemical Wastes

9.1 A summary of pretreatment and disposal methods for lab chemicals is given in Table 1 by waste type, however, applicable recovery methods are not given. Recovery methods are not indicated in this section, but rather, methods by which to prepare waste chemicals for disposal and the disposal methods themselves. As mentioned above, if practical and economical, recovery and re-use is the preferred method for the management of chemical waste. Additionally, controlled reactions/processes are not referred to since they are waste specific. Small quantities of pyrophoric and hydrophoric substances, under controlled conditions in the laboratory, could possibly be reduced to less hazardous compounds, forming products more easily managed. The disposal methods cited are those generally applicable to the types of wastes mentioned, and are recommended contingent upon a proper and thorough pretreatment, if necessary, and where prescribed.

9.2 Pretreatment and disposal methods are addressed separately for the following wastes:

9.2.1 *Water-Soluble Waste of Unknown Origin or Properties (4.2)*—In order to determine a proper disposal method for these wastes, the following minimal data should be determined: radioactivity, water solubility, pH, cyanide content, ignitability, sulfide content, and reactivity. Based on this information, the waste type and corresponding pretreatment and disposal method can be determined.

9.2.2 *Water-Insoluble Wastes of Unknown Origin or Properties (4.2)*—In order to determine a proper disposal method for these wastes, the following minimal data should be determined: ignitability (flash point), organic halide content, PCB content, total solids, and ash.

TABLE 1 Pretreatment and Disposal Methods

Waste Types (4.2)	Pretreatment Methods Section5	Disposal Methods Section6
Trash		containerization
Weak acids	dilution, neutralization	sewer solidification
Weak bases	dilution, neutralization	sewer solidification
Concentrated acids	dilution, neutralization	sewer lab pack solidification
Concentrated bases	dilution, neutralization	sewer lab pak solidification
Flammable non-halogenated solvents	...	incineration lab pack solidification
Flammable halogenated solvents	...	incineration lab pack solidification
Non-flammable, non-halogenated solvents	...	incineration lab pack solidification
Non-flammable, halogenated solvents	...	incineration lab pack solidification
Organic acids	neutralization	sewer incineration lab pack
Organic bases	neutralization	sewer incineration lab pack
Inorganic oxidizers	dilution, reduction	sewer lab pack
Organic oxidizers	dilution, reduction	sewer lab pack
Toxic metals	dilution, reduction	sewer lab pack solidification
Toxic organics	dilution, reduction	sewer lab pack solidification
Aqueous solutions of reducing agents	dilution, oxidation	sewer lab pack solidification
Pyrophorics	...	fire or police department
Hydrophorics	...	fire or police department
Cyanide, sulfide, or ammonia containing waste	dilution, oxidation	sewer lab pack
Explosives	...	fire or police department
Radioactive	...	special waste
Infectious wastes	sterilization, disinfection (see section 7.1.3	incineration lab pack
PCB's	...	see 7.1.4

NOTE 1—If disposal of these materials is desired by methods other than through a hazardous waste management facility, a more complete analysis, sufficient for safe handling and disposal will be necessary.

9.2.3 *Empty Containers* (*4.2*)—Containers with less than three percent of the weight of the original contents of a hazardous substance would be classified as trash and proper disposal would be by means of containerization (6.1), unless they contained substances as identified in 40 CFR 261.33(d). If so identified, the containers must be managed as specified in 40 CFR 261.33(c).

9.2.4 *Asbestos or Asbestos Containing Waste* (*4.2*)—Asbestos or asbestos containing waste must be wetted, and sealed into a leak-tight container while wet. The containers must bear a warning label stating: "Caution, contains asbestos—avoid opening or breaking container—breathing asbestos is hazardous to your health." Disposal is permitted only in authorized landfills.

9.2.5 *Contaminated Lab Ware, Trash* (*4.2*)—Disposal for this type of waste is difficult to determine. Obviously, if the material cannot be recovered, cleaned, or used for another purpose, it must be disposed. The RCRA regulations should be reviewed to determine proper disposal (see section 40 CFR 261.3). In essence, if the waste exhibits any of the characteristics lised in 40 CFR 261 Subpart C (ignitability, reactivity, corrosivity, or toxicity), or is a listed waste or a mixture of a solid waste and one or more wastes listed in 40 CRF 261 Subpart D, it is defined as a hazardous waste.

RECOMMENDED READINGS

(1) "Hazardous Waste Management System" (RCRA), U.S. Environmental Protection Agency, Part 261, *Federal Register,* May 19, 1980, pp 33083–33133.

(2) *Prudent Practices for Handling Hazardous Chemicals in Laboratories,* National Research Council, National Academy Press, Washington, DC, 1981.

(3) *Prudent Practices for Disposing of Chemicals from Laboratories,* National Academy Press, Washington, DC, 1983.

(4) Sax, N. Irving, *Dangerous Properties of Industrial Materials,* Fifth Edition, Van Nostrand Reinhold Company, New York, NY, 1979.

Designation: D 4448 – 85a (Reapproved 1992)

Standard Guide for Sampling Groundwater Monitoring Wells[1]

This standard is issued under the fixed designation D 4448; the number immediately following the designation indicates the year of original adoption or, in the case of revision, the year of last revision. A number in parentheses indicates the year of last reapproval. A superscript epsilon (ε) indicates an editorial change since the last revision or reapproval.

1. Scope

1.1 This guide covers procedures for obtaining valid, representative samples from groundwater monitoring wells. The scope is limited to sampling and "in the field" preservation and does not include well location, depth, well development, design and construction, screening, or analytical procedures.

1.2 This guide is only intended to provide a review of many of the most commonly used methods for sampling groundwater quality monitoring wells and is not intended to serve as a groundwater monitoring plan for any specific application. Because of the large and ever increasing number of options available, no single guide can be viewed as comprehensive. The practitioner must make every effort to ensure that the methods used, whether or not they are addressed in this guide, are adequate to satisfy the monitoring objectives at each site.

1.3 *This standard does not purport to address all of the safety problems, if any, associated with its use. It is the responsibility of the user of this standard to establish appropriate safety and health practices and determine the applicability of regulatory limitations prior to use.*

2. Summary of Guide

2.1 The equipment and procedures used for sampling a monitoring well depend on many factors. These include, but are not limited to, the design and construction of the well, rate of groundwater flow, and the chemical species of interest. Sampling procedures will be different if analyzing for trace organics, volatiles, oxidizable species, or trace metals is needed. This guide considers all of these factors by discussing equipment and procedure options at each stage of the sampling sequence. For ease of organization, the sampling process can be divided into three steps: well flushing, sample withdrawal, and field preparation of samples.

2.2 Monitoring wells must be flushed prior to sampling so that the groundwater is sampled, not the stagnant water in the well casing. If the well casing can be emptied, this may be done although it may be necessary to avoid oxygen contact with the groundwater. If the well cannot be emptied, procedures must be established to demonstrate that the sample represents groundwater. Monitoring an indicative parameter such as pH during flushing is desirable if such a parameter can be identified.

2.3 The types of species that are to be monitored as well as the concentration levels are prime factors for selecting sampling devices **(1,2)**.[2] The sampling device and all materials and devices the water contacts must be constructed of materials that will not introduce contaminants or alter the analyte chemically in any way.

2.4 The method of sample withdrawal can vary with the parameters of interest. The ideal sampling scheme would employ a completely inert material, would not subject the sample to negative pressure and only moderate positive pressure, would not expose the sample to the atmosphere, or preferably, any other gaseous atmosphere before conveying it to the sample container or flow cell for on-site analysis.

2.5 The degree and type of effort and care that goes into a sampling program is always dependent on the chemical species of interest and the concentration levels of interest. As the concentration level of the chemical species of analytical interest decreases, the work and precautions necessary for sampling are increased. Therefore, the sampling objective must clearly be defined ahead of time. For example, to prepare equipment for sampling for mg/L (ppm) levels of Total Organic Carbon (TOC) in water is about an order of magnitude easier than preparing to sample for µg/L (ppb) levels of a trace organic like benzene. The specific precautions to be taken in preparing to sample for trace organics are different from those to be taken in sampling for trace metals. No final Environmental Protection Agency (EPA) protocol is available for sampling of trace organics. A short guidance manual **(3)** and an EPA document **(4)** concerning monitoring well sampling, including considerations for trace organics, are available.

2.6 Care must be taken not to cross contaminate samples or monitoring wells with sampling or pumping devices or materials. All samples, sampling devices, and containers must be protected from the environment when not in use. Water level measurements should be made before the well is flushed. Oxidation-reduction potential, pH, dissolved oxygen, and temperature measurements and filtration should all be performed on the sample in the field, if possible. All but temperature measurement must be done prior to any significant atmospheric exposure, if possible.

2.7 The sampling procedures must be well planned and all sample containers must be prepared and labeled prior to going to the field.

---

[1] This guide is under the jurisdiction of ASTM Committee D34 on Waste Management and is the direct responsibility of Subcommittee D34.01.02 on Sampling Techniques.

Current edition approved Aug. 23 and Oct. 25, 1985. Published May 1986.

[2] The boldface numbers in parentheses refer to a list of references at the end of this guide.

3. Significance and Use

3.1 The quality of groundwater has become an issue of national concern. Groundwater monitoring wells are one of the more important tools for evaluating the quality of groundwater, delineating contamination plumes, and establishing the integrity of hazardous material management facilities.

3.2 The goal in sampling groundwater monitoring wells is to obtain samples that are truly representative of the aquifer or groundwater in question. This guide discusses the advantages and disadvantages of various well flushing, sample withdrawal, and sample preservation techniques. It reviews the parameters that need to be considered in developing a valid sampling plan.

4. Well Flushing (Purging)

4.1 Water that stands within a monitoring well for a long period of time may become unrepresentative of formation water because chemical or biochemical change may cause water quality alterations and even if it is unchanged from the time it entered the well, the stored water may not be representative of formation water at the time of sampling, or both. Because the representativeness of stored water is questionable, it should be excluded from samples collected from a monitoring well.

4.2 The surest way of accomplishing this objective is to remove all stored water from the casing prior to sampling. Research with a tracer in a full scale model 2 in. PVC well **(5)** indicates that pumping 5 to 10 times the volume of the well via an inlet near the free water surface is sufficient to remove all the stored water in the casing. The volume of the well may be calculated to include the well screen and any gravel pack if natural flow through these is deemed insufficient to keep them flushed out.

4.3 In deep or large diameter wells having a volume of water so large as to make removal of all the water impractical, it may be feasible to lower a pump or pump inlet to some point well below the water surface, purge only the volume below that point then withdraw the sample from a deeper level. Research indicates this approach should avoid most contamination associated with stored water **(5, 6,7)**. Sealing the casing above the purge point with a packer may make this approach more dependable by preventing migration of stored water from above. But the packer must be above the top of the screened zone, or stagnant water from above the packer will flow into the purged zone through the well's gravel/sand pack.

4.4 In low yielding wells, the only practical way to remove all standing water may be to empty the casing. Since it is not always possible to remove all water, it may be advisable to let the well recover (refill) and empty it again at least once. If introduction of oxygen into the aquifer may be of concern, it would be best not to uncover the screen when performing the above procedures. The main disadvantage of methods designed to remove all the stored water is that large volumes may need to be pumped in certain instances. The main advantage is that the potential for contamination of samples with stored water is minimized.

4.5 Another approach to well flushing is to monitor one or more indicator parameters such as pH, temperature, or conductivity and consider the well to be flushed when the indicator(s) no longer change. The advantage of this method is that pumping can be done from any location within the casing and the volume of stored water present has no direct bearing on the volume of water that must be pumped. Obviously, in a low yielding well, the well may be emptied before the parameters stabilize. A disadvantage of this approach is that there is no assurance in all situations that the stabilized parameters represent formation water. If significant drawdown has occurred, water from some distance away may be pulled into the screen causing a steady parameter reading but not a representative reading. Also, a suitable indicator parameter and means of continuously measuring it in the field must be available.

4.6 Gibb **(4,8)** has described a time-drawdown approach using a knowledge of the well hydraulics to predict the percentage of stored water entering a pump inlet near the top of the screen at any time after flushing begins. Samples are taken when the percentage is acceptably low. As before, the advantage is that well volume has no direct effect in the duration of pumping. A current knowledge of the well's hydraulic characteristics is necessary to employ this approach. Downward migration of stored water due to effects other than drawdown (for example density differences) is not accounted for in this approach.

4.7 In any flushing approach, a withdrawal rate that minimizes drawdown while satisfying time constraints should be used. Excessive drawdown distorts the natural flow patterns around a well and can cause contaminants that were not present originally to be drawn into the well.

5. Materials and Manufacture

5.1 The choice of materials used in the construction of sampling devices should be based upon a knowledge of what compounds may be present in the sampling environment and how the sample materials may interact via leaching, adsorption, or catalysis. In some situations, PVC or some other plastic may be sufficient. In others, an all glass apparatus may be necessary.

5.2 Most analytical protocols suggest that the devices used in sampling and storing samples for trace organics analysis (µg/L levels) must be constructed of glass or TFE–fluorocarbon resin, or both. One suggestion advanced by the EPA is that the monitoring well be constructed so that only TFE–fluorocarbon tubing be used in that portion of the sampling well that extends from a few feet above the water table to the bottom of the borehole. **(3,5)** Although this type of well casing is now commercially available, PVC well casings are currently the most popular. If adhesives are avoided, PVC well casings are acceptable in many cases although their use may still lead to some problems if trace organics are of concern. At present, the type of background presented by PVC and interactions occurring between PVC and groundwater are not well understood. Tin, in the form of an organotin stabilizer added to PVC, may enter samples taken from PVC casing. **(9)**

5.3 Since the most significant problem encountered in trace organics sampling, results from the use of PVC adhesives in monitoring well construction, threaded joints might avoid the problem **(3,5)**. Milligram per litre (parts per million) levels of compounds such as tetrahydrofuran, methyl-ethyl-ketone, and toluene are found to leach into groundwater samples from monitoring well casings sealed with PVC solvent cement. Pollutant phthalate esters **(8,10)** are often found in water samples at ppb levels; the EPA has found them on occasion at ppm levels in their samples. The ubiquitous presence of these phthalate esters is unexplained, except to say that they may be leached from plastic pipes, sampling devices, and containers.

5.4 TFE–fluorocarbon resins are highly inert and have sufficient mechanical strength to permit fabrication of sampling devices and well casings. Molded parts are exposed to high temperature during fabrication which destroys any organic contaminants. The evolution of fluorinated compounds can occur during fabrication, will cease rapidly, and does not occur afterwards unless the resin is heated to its melting point.

5.5 Extruded tubing of TFE-fluorocarbon for sampling may contain surface traces of an organic solvent extrusion aid. This can be removed easily by the fabricator and, once removed by flushing, should not affect the sample. TFE-fluorocarbon FEP and TFE-fluorocarbon PFA resins do not require this extrusion aid and may be suitable for sample tubing as well. Unsintered thread-sealant tape of TFE-fluorocarbon is available in an "oxygen service" grade and contains no extrusion aid and lubricant.

5.6 Louneman, et al. **(11)** allude to problems caused by a lubricating oil used during TFE-fluorocarbon tubing extrusion. This reference also presents evidence that a fluorinated ethylene-propylene copolymer adsorbed acetone to a degree that later caused contamination of a gas sample.

5.7 Glass and stainless steel are two other materials generally considered inert in aqueous environments. Glass is probably among the best choices though it is not inconceivable it could adsorb some constituents as well as release other contaminants (for example, Na, silicate, and Fe). Of course, glass sampling equipment must be handled carefully in the field. Stainless steel is strongly and easily machined to fabricate equipment. Unfortunately, it is not totally immune to corrosion that could release metallic contaminants. Stainless steel contains various alloying metals, some of these (for example Ni) are commonly used as catalysts for various reactions. The alloyed constituents of some stainless steels can be solubilized by the pitting action of nonoxidizing anions such as chloride, fluoride, and in some instances sulfate, over a range of pH conditions. Aluminum, titanium, polyethylene, and other corrosion resistant materials have been proposed by some as acceptable materials, depending on groundwater quality and the constituents of interest.

5.8 Where temporarily installed sampling equipment is used, the sampling device that is chosen should be non-plastic (unless TFE-fluorocarbon), cleanable of trace organics, and must be cleaned between each monitoring well use in order to avoid cross-contamination of wells and samples. The only way to ensure that the device is indeed "clean" and acceptable is to analyze laboratory water blanks and field water blanks that have been soaked in and passed through the sampling device to check for the background levels that may result from the sampling materials or from field conditions. Thus, all samplings for trace materials should be accompanied by samples which represent the field background (if possible), the sampling equipment background, and the laboratory background.

5.9 Additional samples are often taken in the field and spiked (spiked-field samples) in order to verify that the sample handling procedures are valid. The American Chemical Society's committee on environmental improvement has published guidelines for data acquisition and data evaluation which should be useful in such environmental evaluations **(10,12)**.

6. Sampling Equipment

6.1 There is a fairly large choice of equipment presently available for groundwater sampling from single screened wells and well clusters. The sampling devices can be categorized into the following eight basic types.

6.1.1 *Down-Hole Collection Devices*:

6.1.1.1 Bailers, messenger bailers, or thief samplers **(13,14)** are examples of down-hole devices that probably provide valid samples once the well has been flushed. They are not practical for removal of large volumes of water. These devices can be constructed in various shapes and sizes from a variety of materials. They do not subject the sample to pressure extremes.

6.1.1.2 Bailers do expose part of the sample to the atmosphere during withdrawal. Bailers used for sampling of volatile organic compounds should have a sample cock or draft valve in or near the bottom of the sampler allowing withdrawal of a sample from the well below the exposed surface of the water or the first few inches of the sample should be discarded. Suspension lines for bailers and other samplers should be kept off the ground and free of other contaminating materials that could be carried into the well. Down-hole devices are not very practical for use in deep wells. However, potential sample oxidation during transfer of the sample into a collection vessel and time constraints for lowering and retrieval for deep sampling are the primary disadvantages.

6.1.1.3 Three down-hole devices are the single and double check valve bailers and thief samplers. A schematic of a single check valve unit is illustrated in Fig. 1. The bailer may be threaded in the middle so that additional lengths of blank casing may be added to increase the sampling volume. TFE-fluorocarbon or PVC are the most common materials used for construction **(15)**.

6.1.1.4 In operation, the single check valve bailer is lowered into the well, water enters the chamber through the bottom, and the weight of the water column closes the check valve upon bailer retrieval. The specific gravity of the ball should be about 1.4 to 2.0 so that the ball almost sits on the check valve seat during chamber filling. Upon bailer withdrawal, the ball will immediately seat without any samples loss through the check valve. A similar technique involves lowering a sealed sample container within a weighted bottle into the well. The stopper is then pulled from the bottle via a line and the entire assembly is retrieved upon filling of the container **(14,16)**.

6.1.1.5 A double check valve bailer allows point source sampling at a specific depth **(15,17)**. An example is shown in Fig. 2. In this double check valve design, water flows through the sample chamber as the unit is lowered. A venturi tapered inlet and outlet ensures that water passes freely through the unit. When a depth where the sample is to be collected is reached, the unit is retrieved. Because the difference between each ball and check valve seat is maintained by a pin that blocks vertical movement of the check ball, both check valves

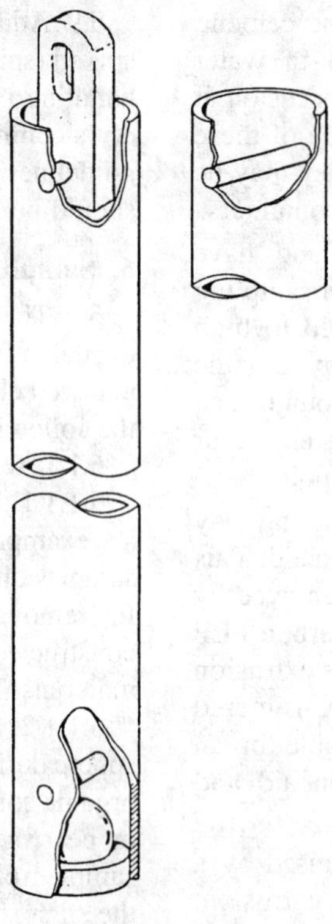

NOTE 1—Taken from Ref **(15)**.

FIG. 1 Single Check Valve Bailer

close simultaneously upon retrieval. A drainage pin is placed into the bottom of the bailer to drain the sample directly into a collection vessel to reduce the possibility of air oxidation. The acrylic model in Fig. 2 is threaded at the midsection allowing the addition of threaded casing to increase the sampling volume.

6.1.1.6 Another approach for obtaining point source samples employs a weighted messenger or pneumatic change to "trip" plugs at either end of an open tube (for example, tube water sampler or thief sampler) to close the chamber **(18)**. Foerst, Kemmerer, and Bacon samplers are of this variety **(14,17,19)**. A simple and inexpensive pneumatic sampler was recently described by Gillham **(20)**. The device (Fig. 3) consists of a disposable 50 mL plastic syringe modified by sawing off the plunger and the finger grips. The syringe is then attached to a gas-line by means of a rubber stopper assembly. The gas-line extends to the surface, and is used to drive the stem-less plunger, and to raise and lower the syringe into the hole. When the gas-line is pressurized, the rubber plunger is held at the tip of the syringe. The sampler is then lowered into the installation, and when the desired depth is reached, the pressure in the gas-line is reduced to atmospheric (or slightly less) and water enters the syringe. The sampler is then retrieved from the installation and the syringe detached from the gas-line. After the tip is sealed, the syringe is used as a short-term storage container. A number of thief or messenger devices are available in various materials and shapes.

6.1.2 *Suction Lift Pumps*:

6.1.2.1 Three types of suction lift pumps are the direct line, centrifugal, and peristaltic. A major disadvantage of any suction pump is that it is limited in its ability to raise water by the head available from atmospheric pressure. Thus, if the surface of the water is more than about 25 ft below the pump, water may not be withdrawn. The theoretical suction limit is about 34 ft, but most suction pumps are capable of maintaining a water lift of only 25 ft or less.

6.1.2.2 Many suction pumps draw the water through some sort of volute in which impellers, pistons, or other devices operate to induce a vacuum. Such pumps are probably unacceptable for most sampling purposes because they are usually constructed of common materials such as brass or mild steel and may expose samples to lubricants. They often induce very low pressures around rotating vanes or other such parts such that degassing or even cavitation may occur. They can mix air with the sample via small leaks in the casing, and they are difficult to adequately clean between uses. Such pumps are acceptable for purging of wells, but should not generally be used for sampling.

6.1.2.3 One exception to the above statements is a peristaltic pump. A peristaltic pump is a self-priming, low volume suction pump which consists of a rotor with ball bearing rollers **(21)**. Flexible tubing is inserted around the pump rotor and squeezed by heads as they revolve in a circular pattern around the rotor. One end of the tubing is placed into the well while the other end can be connected directly to a receiving vessel. As the rotor moves, a reduced pressure is created in the well tubing and an increased pressure (<40 psi) on the tube leaving the rotor head. A drive shaft connected to the rotor head can be extended so that multiple rotor heads can be attached to a single drive shaft.

6.1.2.4 The peristaltic pump moves the liquid totally within the sample tube. No part of the pump contacts the liquid. The sample may still be degassed (cavitation is unlikely) but the

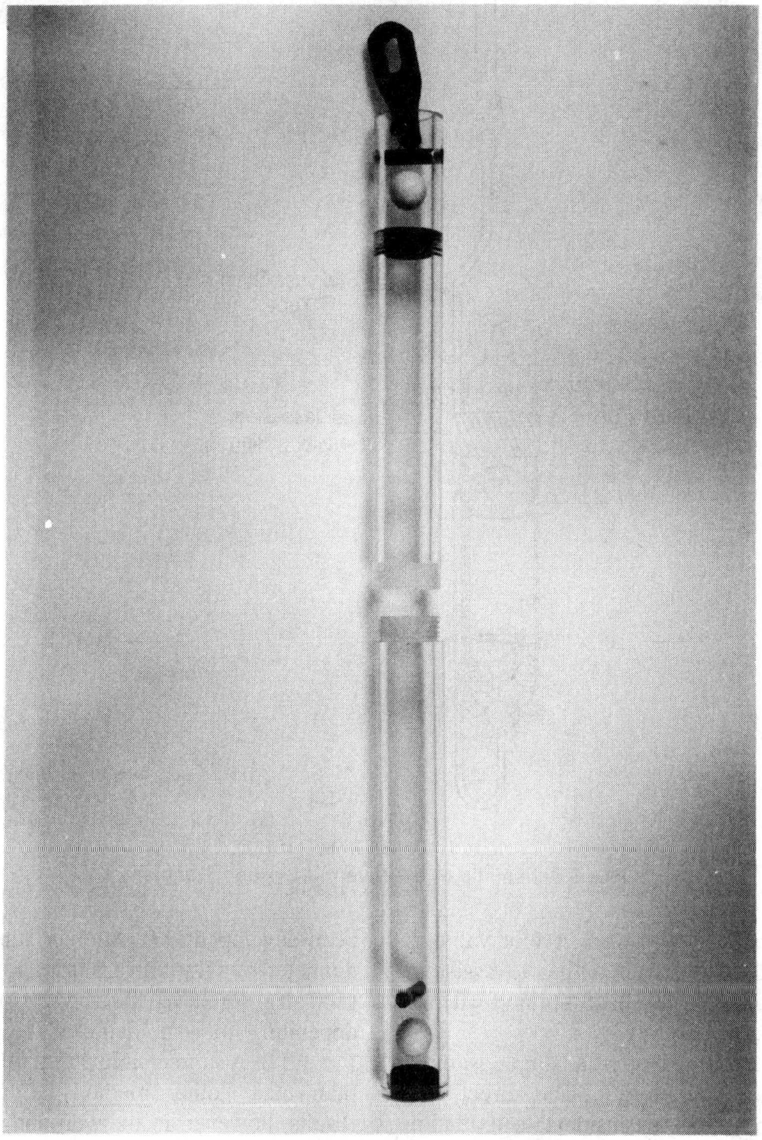

NOTE 1—Taken from Ref **(17)**.
FIG. 2 Acrylic Point Source Bailer

problems due to contact with the pump mechanism are eliminated. Peristaltic pumps do require a fairly flexible section of tubing within the pumphead itself. A section of silicone tubing is commonly used within the peristaltic pumphead, but other types of tubing can be used particularly for the sections extending into the well or from the pump to the receiving container. The National Council of the Paper Industry for Air and Stream Improvement **(22)** recommends using medical grade silicone tubing for organic sampling purposes as the standard grade uses an organic vulcanizing agent which has been shown to leach into samples. Medical grade silicone tube is, however, limited to use over a restricted range of ambient temperatures. Various manufacturers offer tubing lined with TFE-fluorocarbon or Viton[3] for use with their pumps. Gibb **(1,8)** found little difference between samples withdrawn by a peristaltic pump and those taken by a bailer.

6.1.2.5 A direct method of collecting a sample by suction consists of lowering one end of a length of plastic tubing into the well or piezometer. The opposite end of the tubing is connected to a two way stopper bottle and a hand held or mechanical vacuum pump is attached to a second tubing leaving the bottle. A check valve is attached between the two lines to maintain a constant vacuum control. A sample can then be drawn directly into the collection vessel without contacting the pump mechanism **(5,23,24)**.

6.1.2.6 A centrifugal pump can be attached to a length of

[3] Viton is a trademark of E. I. du Pont de Nemours & Co., Wilmington, DE 19898 and has been found suitable for this purpose.

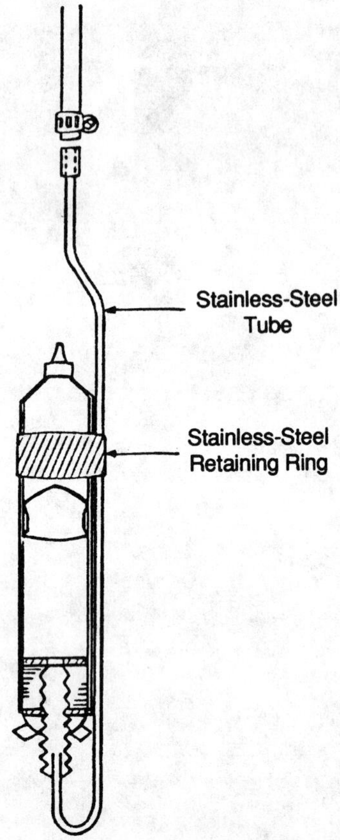

NOTE 1—Taken from Ref **(21)**.

FIG. 3 Schematic of the Inverted Syringe Sampler

plastic tubing that is lowered into the well. A foot valve is usually attached to the end of the well tubing to assist in priming the tube. The maximum lift is about 4.6 m (15 ft) for such an arrangement **(23,25,26)**.

6.1.2.7 Suction pump approaches offer a simple sample retrieval method for shallow monitoring. The direct line method is extremely portable though considerable oxidation and mixing may occur during collection. A centrifugal pump will agitate the sample to an even greater degree although pumping rates of 19 to 151 Lpm (5 to 40 gpm) can be attained. A peristaltic pump provides a lower sampling rate with less agitation than the other two pumps. The withdrawal rate of peristaltic pumps can be carefully regulated by adjustment of the rotor head revolution.

6.1.2.8 All three systems can be specially designed so that the water sample contacts only the TFE flourocarbon or silicone tubing prior to sample bottle entry. Separate tubing is recommended for each well or piezometer sampled.

6.1.3 *Electric Submersible Pumps*:

6.1.3.1 A submersible pump consists of a sealed electric motor that powers a piston or helical single thread worm at a high rpm. Water is brought to the surface through an access tube. Such pumps have been used in the water well industry for years and many designs exist **(5,26)**.

6.1.3.2 Submersible pumps provide relatively high discharge rates for water withdrawal at depths beyond suction lift capabilities. A battery operated unit 3.6 cm (1.4 in.) in diameter and with a 4.5 Lpm (1.2 gpm) flow rate at 33.5 m (110 ft) has been developed **(27)**. Another submersible pump has an outer diameter of 11.4 cm (4.5 in.) and can pump water from 91 m (300 ft). Pumping rates vary up to 53.0 Lpm (14 gpm) depending upon the depth of the pump **(28)**.

6.1.3.3 A submersible pump provides higher extraction rates than many other methods. Considerable sample agitation results, however, in the well and in the collection tube during transport. The possibility of introducing trace metals into the sample from pump materials also exists. Steam cleaning of the unit followed by rinsing with unchlorinated, deionized water is suggested between sampling when analysis for organics in the parts per million (ppm) or parts per billion (ppb) range is required **(29)**.

6.1.4 *Gas-Lift Pumps*:

6.1.4.1 Gas-lift pumps use compressed air to bring a water sample to the surface. Water is forced up an eductor pipe that may be the outer casing or a smaller diameter pipe inserted into the well annulus below the water level **(30,31)**.

6.1.4.2 A similar principle is used for a unit that consists of a small diameter plastic tube perforated in the lower end. This tube is placed within another tube of slightly larger diameter. Compressed air is injected into the inner tube; the air bubbles through the perforations, thereby lifting the water sample via the annulus between the outer and inner tubing **(32)**. In practice, the eductor line should be submerged to a depth equal to 60 % of the total submerged eductor length during pumping **(26)**. A60 % ratio is considered optimal although a 30 % submergence ratio is adequate.

6.1.4.3 The source of compressed gas may be a hand pump for depths generally less than 7.6 m (25 ft). For greater depths, air compressors, pressurized air bottles, and air compressed from an automobile engine have been used.

6.1.4.4 As already mentioned, gas-lift methods result in considerable sample agitation and mixing within the well, and cannot be used for samples which will be tested for volatile organics. The eductor pipe or weighted plastic tubing is a potential source of sample contamination. In addition, Gibb **(8)** uncovered difficulties in sampling for inorganics. These difficulties were attributed to changes in redox, pH, and species transformation due to solubility constant changes resulting from stripping, oxidation, and pressure changes.

6.1.5 *Gas Displacement Pumps*:

6.1.5.1 Gas displacement or gas drive pumps are distinguished from gas-lift pumps by the method of sample transport. Gas displacement pumps force a discrete column of water to the surface via mechanical lift without extensive mixing of the pressurized gas and water as occurs with air-lift equipment. The principle is shown schematically in Fig. 4. Water fills the chamber. A positive pressure is applied to the gas line closing the sampler check valve and forcing water up the sample line. By removing the pressure the cycle can be repeated. Vacuum can also be used in conjunction with the gas **(30)**. The device can be permanently installed in the well **(33,34,35)** or lowered into the well **(36,37)**.

6.1.5.2 A more complicated two stage design constructed of glass with check valves made of TFE-fluorocarbon has been constructed **(38,39)**. The unit was designed specifically for sample testing for trace level organics. Continuous flow rates up to 2.3 Lpm (0.6 gpm) are possible with a 5.1 cm (2 in.) diameter unit.

6.1.5.3 Gas displacement pumps have also been developed with multiple functions. The water sample in Fig. 5 provides piezometric data measurements with an internally mounted transducer **(40)**. A sample with its transducer exposed externally for piezometric measurements is illustrated in Fig. 6 **(41)**. The sensor can activate the gas source at the surface to cause sample chamber pressurization at the predetermined depth. Another design can be used as a water sampler or as a tool for injecting brine or other tracers into a well **(42)**.

6.1.5.4 Gas displacement pumps offer reasonable potential for preserving sample integrity because little of the driving gas comes in contact with the sample as the sample is conveyed to the surface by a positive pressure. There is, however, a potential loss of dissolved gasses or contamination from the driving gas and the housing materials.

6.1.6 *Bladder Pumps*:

6.1.6.1 Bladder pumps, also referred to as gas-operated squeeze pumps, consist of a flexible membrane enclosed by a rigid housing. Water enters the membrane through a check valve in the vessel bottom; compressed gas injected into the cavity between the housing and bladder forces the sample through a check valve at the top of the membrane and into a discharge line (Fig. 7). Water is prevented from re-entering the bladder by the top check valve. The process is repeated to cycle

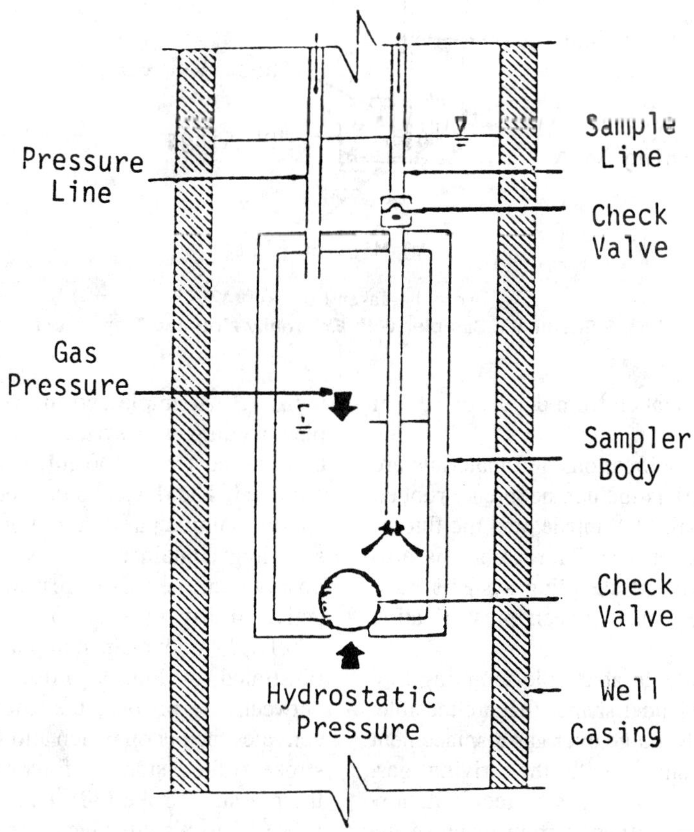

NOTE 1—Taken from Ref **(5)**.

FIG. 4 The Principal of Gas Displacement Pumping

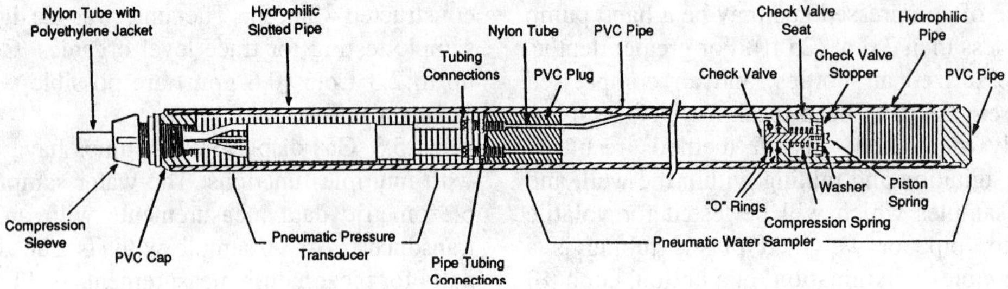

NOTE 1—Taken from Ref (**41**).

FIG. 5 Pneumatic Water Sampler With Internal Transducer

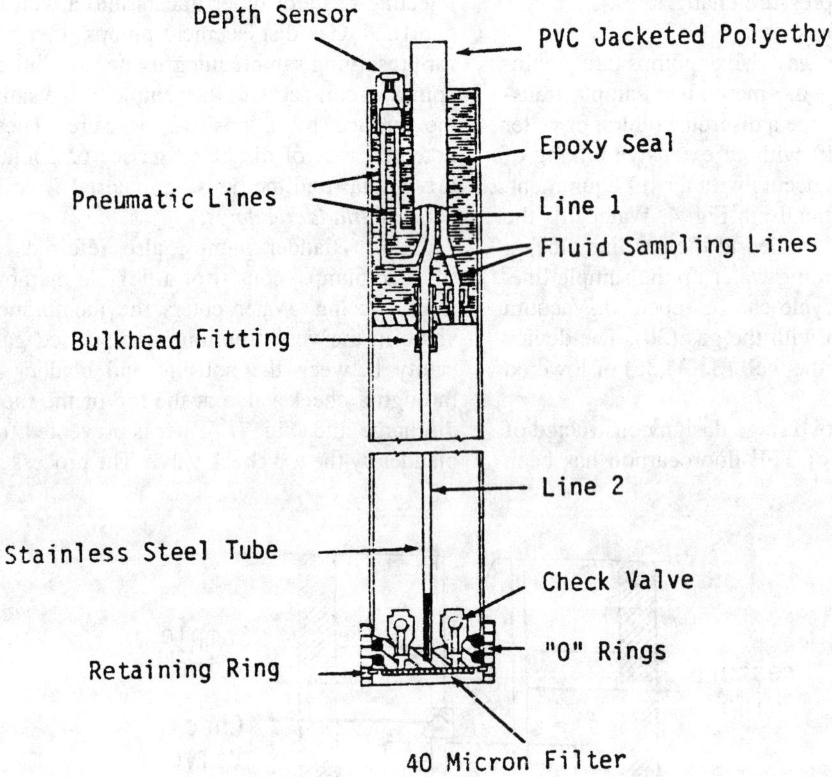

NOTE 1—Taken from Ref (**42**).

FIG. 6 Pneumatic Sampler With Externally Mounted Transducer

the water to the surface. Samples taken from depths of 30.5 m (100 ft) have been reported.

6.1.6.2 A variety of design modifications and materials are available **(43,44)**. Bladder materials include neoprene, rubber, ethylene propylene terpolymer (E.P.T.), nitrile, and the fluorocarbon Viton.[3] A bladder made of TFE-fluorocarbon is also under development **(45)**. Automated sampling systems have been developed to control the time between pressurization cycles **(46)**.

6.1.6.3 Bladder pumps provide an adaptable sampling tool due primarily to the number of bladder shapes that are feasible. These devices have a distinct advantage over gas displacement pumps in that there is no contact with the driving gas. Disadvantages include the large gas volumes required, low pumping rates, and potential contamination from many of the bladder materials, the rigid housing, or both.

6.1.7 *Gas Driven Piston Pumps*:

6.1.7.1 A simple and inexpensive example of a gas driven piston pump is a syringe pump **(47)**. The pump (Fig. 8) is constructed from a 50 mL plastic syringe with plunger stem removed. The device is connected to a gas line to the surface and the sample passes through a check valve arrangement to a sampling container at the surface. By successively applying positive and negative pressure to the gas-line, the plunger is activated driving water to the surface.

6.1.7.2 A double piston pump powered by compressed air is illustrated in Fig. 9. Pressurized gas enters the chamber between the pistons; the alternating chamber pressurization activates the piston which allows water entry during the suction stroke of the piston and forces the sample to the surface during the pressure stroke **(48)**. Pumping rates between 9.5 and 30.3 L/hr (2.5 to 8 gal/hr) have been reported from 30.5 m (100 ft). Depths in excess of 457 m (1500 ft) are possible.

6.1.7.3 The gas piston pump provides continuous sample

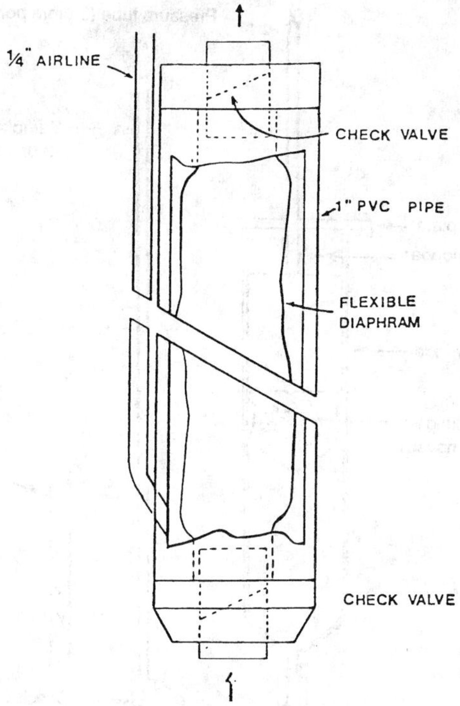

NOTE 1—Taken from Ref (**4**).

FIG. 7 Bladder Pump

withdrawal at depths greater than is possible with most other approaches. Nevertheless, contribution of trace elements from the stainless steel and brass is a potential problem and the quantity of gas used is significant.

6.1.8 *Packer Pump Arrangement*:

6.1.8.1 A packer pump arrangement provides a means by which two expandable "packers" isolate a sampling unit between two packers within a well. Since the hydraulic or pneumatic activated packers are wedged against the casing wall or screen, the sampling unit will obtain water samples only from the isolated well portion. The packers are deflated for vertical movement within the well and inflated when the desired depth is attained. Submersible, gas lift, and suction pumps can be used for sampling. The packers are usually constructed from some type of rubber or rubber compound (**48-51**). A packer pump unit consisting of a vacuum sampler positioned between two packers is illustrated in Fig. 10 (**52**).

6.1.8.2 A packer assembly allows the isolation of discrete sampling points within a well. A number of different samplers can be situated between the packers depending upon the analytical specifications for sample testing. Vertical movement of water outside the well casing during sampling is possible with packer pumps but depends upon the pumping rate and subsequent disturbance. Deterioration of the expandable materials will occur with time with the increased possibility of undesirable organic contaminants contributing to the water sample.

7. Sample Containers and Preservation

7.1 Complete and unequivocal preservation of samples, whether domestic wastewater, industrial wastes, or natural waters, is practically impossible. At best, preservation techniques only retard the chemical and biological changes that inevitably continue after the sample is removed from the source. Therefore, insuring the timely analysis of a sample should be one of the foremost considerations in the sampling plan schedule. Methods of preservation are somewhat limited and are intended to retard biological action, retard hydrolysis of chemical compounds and complexes, and reduce the volatility of constituents. Preservation methods are generally limited to pH control, chemical addition, refrigeration and freezing. For water samples, immediate refrigeration just above freezing (4°C in wet ice) is often the best preservation technique available, but it is not the only measure nor is it applicable in all cases. There may be special cases where it might be prudent to include a recording thermometer in the sample shipment to verify the maximum and minimum temperature to which the samples were exposed. Inexpensive devices for this purpose are available.

7.2 All bottles and containers must be specially pre-cleaned, pre-labelled, and organized in ice-chests (isolating samples and sampling equipment from the environment) before one goes into the field. Otherwise, in any comprehensive program utter chaos usually develops in the field or laboratory. The time in the field is very valuable and should be spent on taking field notes, measurements, and in documenting samples, not on labelling and organizing samples. Therefore, the sampling plan should include clear instructions to the sampling personnel concerning the information required in the field data record logbook (notebook), the information needed on container labels for identification, the chain-of-custody protocols, and the methods for preparing field blanks and spiked samples. Example of detailed plans and documentation procedures have

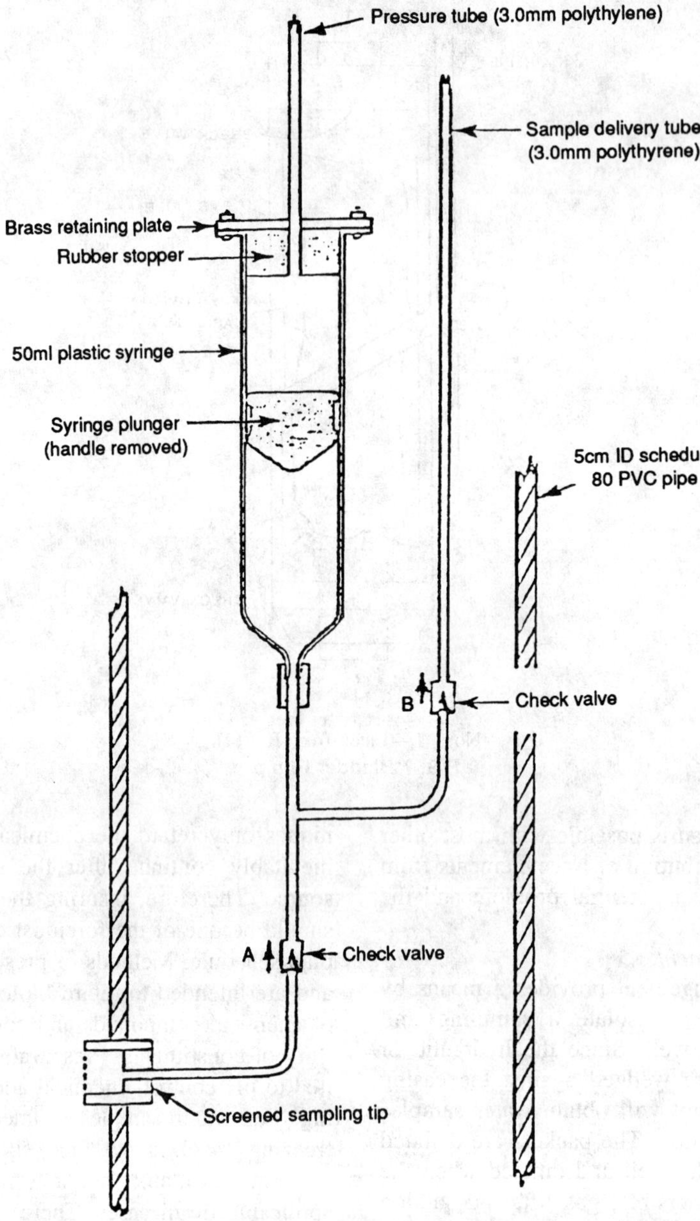

NOTE 1—Taken from Ref (**48**).

FIG. 8 Positive Displacement Syringe Pump

been published (**14,53**).

7.3 The exact requirements for the volumes of sample needed and the number of containers to use may vary from laboratory to laboratory. This will depend on the specific analyses to be performed, the concentration levels of interest, and the individual laboratory protocols. The manager of the sampling program should make no assumptions about the laboratory analyses. He should discuss the analytical requirements of the sampling program in detail with the laboratory coordinator beforehand. This is especially the case since some analyses and preservation measures must be performed at the laboratory as soon as possible after the samples arrive. Thus, appropriate arrangements must be made.

7.4 There are a number of excellent references available which list the containers and preservation techniques appropriate for water and soils (**13,14,50,54-56**. The "Handbook for Sampling and Sample Preservation of Water and Wastewater" is an excellent reference and perhaps the most comprehensive one (**14**). Some of this information is summarized in Table 1.

7.5 Sample containers for trace organic samples require special cleaning and handling considerations (**57**). The sample container for purgeable organics consist of a screw-cap vial (25 to 125 mL) fitted with a TFE-flourocarbon faced silicone septum. The vial is sealed in the laboratory immediately after cleaning and is only opened in the field just prior to pouring sample into it. The water sample then must be sealed into the vial headspace free (no air bubbles) and immediately cooled (4°C) for shipment. Multiple samples (usually about four taken from one large sample container) are taken because leakage of containers may cause losses, may allow air to enter the

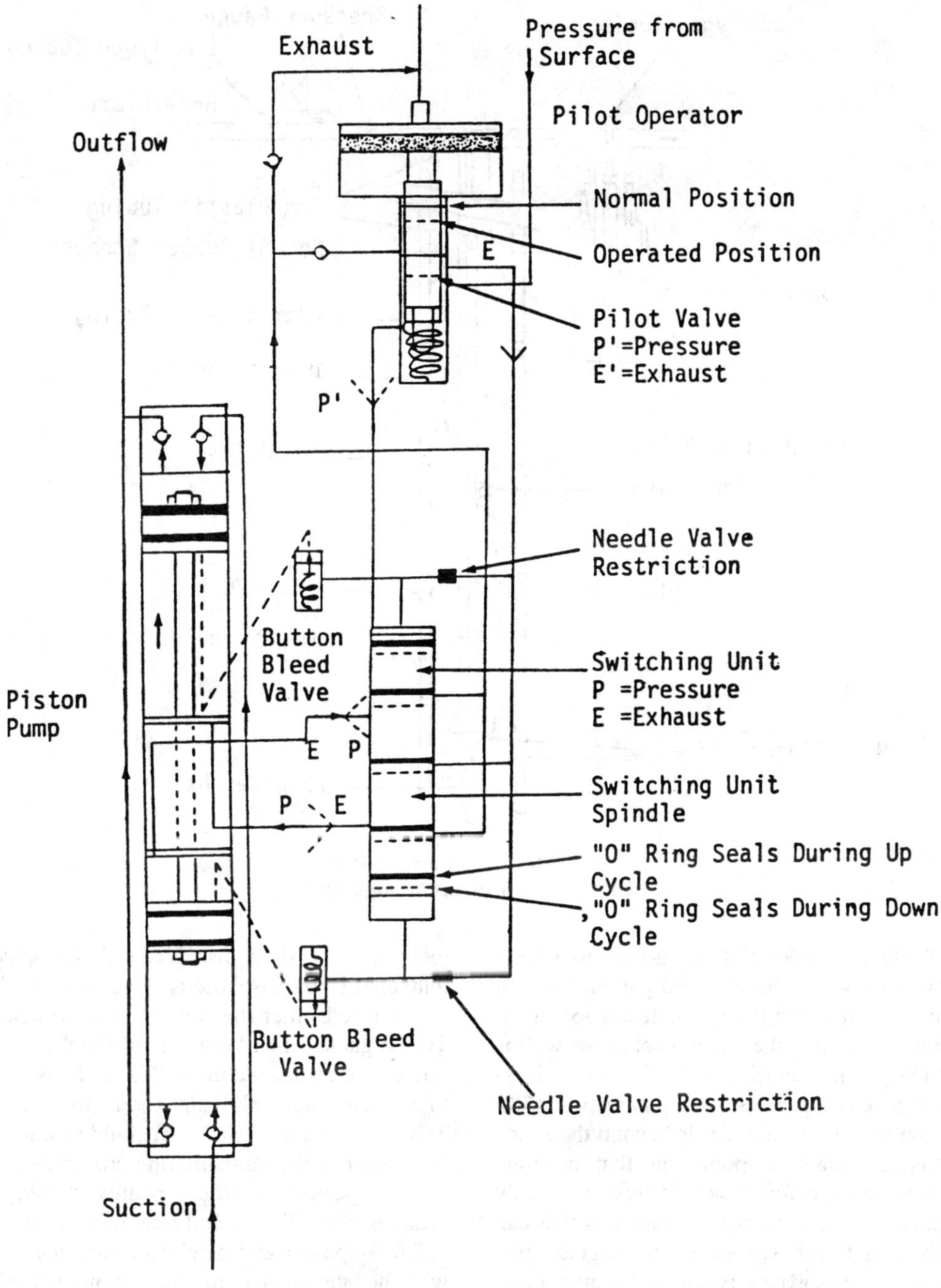

NOTE 1—Taken from Ref **(49)**.

FIG. 9 Gas Driven Piston Pump

containers, and may cause erroneous analysis of some constituents. Also, some analyses are best conducted on independent protected samples.

7.6 The purgeable samples must be analyzed by the laboratory within 14 days after collection, unless they are to be analyzed for acrolein or acrylonitrile (in which case they are to be analyzed within 3 days). For samples for solvent extractions (extractable organics-base neutrals, acids and pesticides), the sample bottles are narrow mouth, screw cap quart bottles or half-gallon bottles that have been precleaned, rinsed with the extracting organic solvent and oven dried at 105°C for at least 1 h. These bottles must be sealed with TFE-fluorocarbon lined caps (Note 1). Samples for organic extraction must be extracted within 7 days and analyzed within 30 days after extraction. Special pre-cleaned, solvent rinsed and oven-dried stainless steel beakers (one for each monitoring well) may be used for transferring samples from the sampling device to the sample containers.

NOTE 1—When collecting samples, the bottles should not be overfilled or prerinsed with sample before filling because oil and other materials may remain in the bottle. This can cause erroneously high results.

7.7 For a number of groundwater parameters, the most meaningful measurements are those made in the field at the

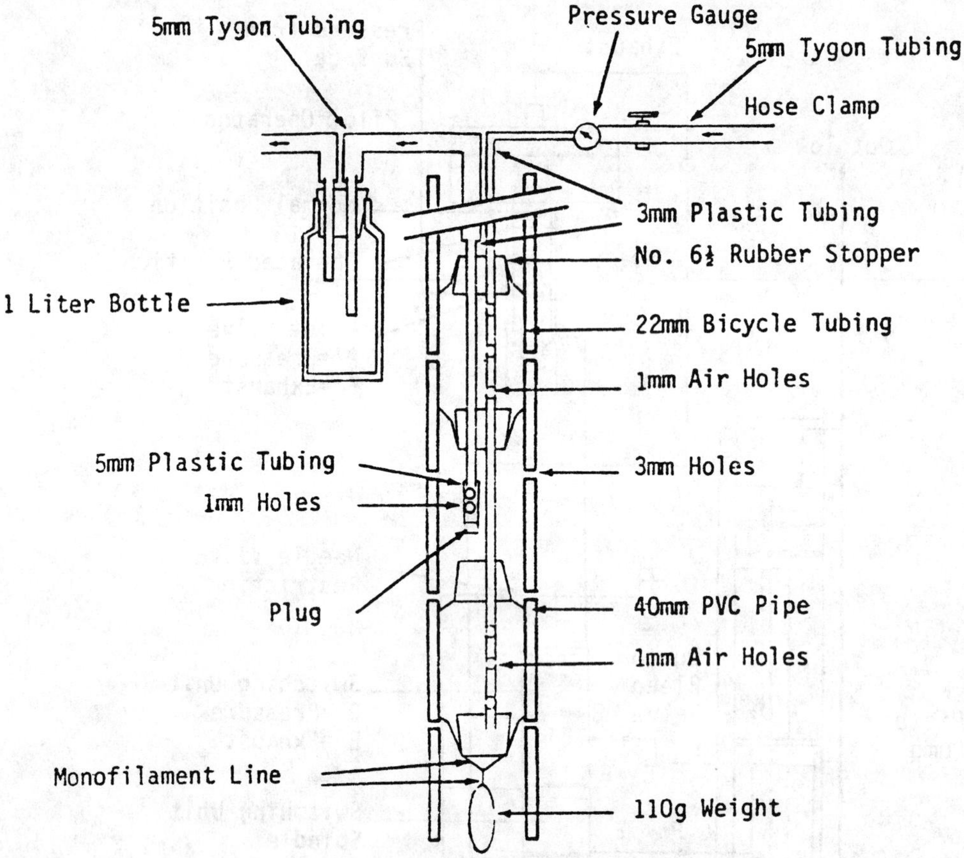

NOTE 1—Taken from Ref **(53)**.

FIG. 10 Packer Pump Arrangement

time of sample collection or at least at an on-site laboratory. These include the water level in the well and parameters that sometimes can change rapidly with storage. A discussion of the various techniques for measuring the water level in the well is contained in a NCASI publication **(5)** and detailed procedures are outlined in a U.S. Geological Survey publication **(58)**. Although a discussion of these techniques is beyond the scope of this guide, it is important to point out that accurate measurements must be made before a well is flushed or only after it has had sufficient time to recover. Parameters that can change rapidly with storage include specific conductance, pH, turbidity, redox potential, dissolved oxygen, and temperature. For some of the other parameters, the emphasis in groundwater monitoring is on the concentration of each specific dissolved component, not the total concentration of each. Samples for these types of measurements should be filtered through 0.45 µm membrane filters ideally in the field or possibly at an on-site laboratory as soon as possible. Analyses often requiring filtered samples include all metals, radioactivity parameters, total organic carbon, dissolved orthophosphate (if needed), and total dissolved phosphorous (if needed) **(13,14)**. If metals are to be analyzed, filter the sample prior to acid preservation. For TOC organics, the filter material should be tested to assure that it does not contribute to the TOC. The type or size of the filter to be used is not well understood. However, if results of metal, TOC or other parameters that could be effected by solids are to be compared, the same filtering procedure must be used in each case. Repeated analytical results should state whether the samples were filtered and how they were filtered.

7.8 Shipment and receipt of samples must be coordinated with the laboratory to minimize time in transit. All samples for organic analysis (and many other parameters), should arrive at the laboratory within one day after it is shipped and be maintained at about 4°C with wet ice. The best way to get them to the laboratory in good condition is to send them in sturdy insulated ice chests (coolers) equipped with bottle dividers. 24-h courier service is recommended, if personal delivery service is not practical.

TABLE 1 Typical Container and Preservation Requirements for a Ground-Water Monitoring Program

Sample and Measurement	Volume Required (mL)	Container P—Polyethylene G—Glass	Preservative	Maximum Holding Time
Metals As/Ba/Cd/Cr/Fe Pb/Se/ Ag/Mn/Na	1000–2000	P/G (special acid cleaning)	high purity nitric acid to pH <2	6 months
Mercury	200–300	P/G (special acid cleaning)	high purity nitric acid to pH <2 +0.05 % $K_2Cr_2O_7$	28 days
Radioactivity alpha/beta/radium	4000	P/G (special acid cleaning)	high purity nitric acid to pH< 2	6 months
Phenolics	500–1000	G	cool, 4°C H_2SO_4 to pH <2	28 days
Miscellaneous	1000–2000	P	cool, 4°C	
Fluoride	300–500	P		28 days
Chloride	50–200	P/G		28 days
Sulfate	100–500	P/G		28 days
Nitrate	100–250	P/G		48 hours
Coliform	100	P/G		6 h
Conductivity	100	P/G		on site/24 h
pH	100	P/G		on site/6 h
Turbidity	100	P/G		48 h
Total organic carbon (TOC)	25–100	P/G	cool, 4°C or cool, 4°C HCl or H_2SO_4 to pH <2	24 h 28 days
Pesticides, herbicides and total organic halogen (TOX)	1000–4000	G/TFE-fluoro-carbon lined cap solvent rinsed	cool, 4°C	7 days/ extraction +30 days/analysis
Extractable organics	1000–2000	G/TFE-fluoro-carbon-lined cap solvent rinsed	cool, 4°C	7 days/ extraction +30 days/analysis
Organic purgeables acroleln/acrylonltrlle	25–120	G/vial TFE-fluorocar-bon-lined sep-tum	cool, 4°C	14 days 3 days

REFERENCES

(1) Gibb, J. P., Schuller, R. M., Griffin, R. A., *Monitoring Well Sampling and Preservation Techniques*, EPA-600/9-80-101, 1980.

(2) Pettyjohn, W. A., Dunlap, W. J., Cosby, R. L., Keeley, J. W., "Sampling Ground Water for Organic Contaminants," *Ground Water*, Vol 19, (2), March/April 1981, pp. 180–189.

(3) Dunlap, W. J., McNabb, J. F., Scalf, M. R., Cosby, R. L., *Sampling for Organic Chemicals and Microorganisms in the Subsurface*, EPA-600/2-77-176, NTIS PB 276 679, August 1977, 35 pp.

(4) Scalf, M. R., McNabb, J. F., Dunlap, W. J., and Cosby, R. L., *Manual of Ground Water Quality Sampling Procedures*, National Water Well Association, NTIS PB-82 103 045, 1981.

(5) "A Guide to Groundwater Sampling," *NCASI Technical Bulletin*, No. 362, January 1982.

(6) Humenick, M. J., Turk, L. J., Coldrin, M.," Methodology for Monitoring Ground Water at Uranium Solution Mines," *Ground Water*, Vol 18 (3), May–June 1980, p. 262.

(7) Marsh, J. M., and Lloyd, J. W., "Details of Hydrochemical Variations in Flowing Wells," *Ground Water*, Volume 18 (4), July–August 1980, p. 366.

(8) Gibb, J. P., Schuller, R. M., Griffin, R. A., "Collection of Representative Water Quality Data from Monitoring Wells," *Proceeding of the Municipal Solid Waste Resource Recovery Symposium*, EPA-600/9-81-002A, March 1981.

(9) Boettner, E. A., Gwendolyn, L. B., Zand, H., Aquino, R., *Organic and Organotin Compounds Leached from PVC and CPVC Pipe*, NTIS P8 82-108 333, 1982.

(10) Junk, G. A., Svec, H. J., Vick, R. D., Avery, M. J., "Contamination of Water by Synthetic Polymer Tubes," *Environmental Science and Technology*, Vol 8 (13) 1100, December 1974.

(11) Louneman, W. A., Bufalini, J. J., Kuntz, R. L., and Meeks, S. A.," Contamination from Fluorocarbon Films," *Environmental Science and Technology*, Vol 15 (1), January 1981.

(12) ASC Committee on Environmental Improvement, "Guidelines for Data Acquisition and Data Quality Evaluation in Environmental Chemistry." *Analytical Chemistry*, Vol 52, 1980, pp. 2242–2249.

(13) *Procedures Manual for Ground Water Monitoring at Solid Waste Disposal Facilities*, EPA/530/SW-611, August 1977.

(14) *Handbook for Sampling and Sample Preservation of Water and Wastewater*, U.S. Dept. of Commerce NTIS PB-259 946, September 1976.

(15) Timco Manufacturing Co., Inc., "Variable Capacity Bailer," *Timco Geotechnical Catalogue*, Prairie du Sac, WI, 1982.

(16) deVera, E., Simmons, B., Stephens, R., Storm, D., *Samplers and Sampling Procedures for Hazardous Waste Streams*, Environmental Protection Agency, EPA-600/2-80-018, 1980, p. 51.

(17) Morrison, R., *Ground Water Monitoring Technology*, Timco Manufacturing Co., 1982, p. 276.

(18) Eijelkamp, "Equipment for Soil Research," *General Catalogue*, Geisbeek, The Netherlands, 1979, pp. 82–83.

(19) Wood, W., "Guidelines for Collection and Field Analysis of Ground-Water Samples for Selected Unstable Constituents," *Techniques of Water-Resources Investigations of the United States Geological*

Survey, Chapter D2, 1976, p. 24.

(20) Gilham, R. W., "Syringe Devices for Groundwater Monitoring," *Ground Water Monitoring Review*, Vol 2 (2), Spring 1982, p. 36.

(21) Masterflex, *Masterflex Pump Catalogue*, Barnant Corp., Barrington, IL, 1981.

(22) "Guidelines for Contracting Sampling and Analyses for Priority Pollutants in Pulp and Paper Industry Effluents," *NCASI Stream Improvement Technical Bulletin*, No. 335, August 1980.

(23) Allison, L., "A Simple Device for Sampling Ground Water in Auger Holes," *Soil Science Society of America Proceedings* 35: 844–45, 1971.

(24) Willardson, L., Meek, B., Huber, M., "A Flow Path Ground Water Sampler," *Soil Science Society of America Proceedings* 36: 965–66, 1972.

(25) Wilson, L., *Monitoring in the Vadose Zone: A Review of Technical Elements and Methods*, U.S. Environmental Protection Agency, EPA-60017-80-134, 1980, p. 180.

(26) *Ground Water and Wells*, Johnson, E. E., Inc., St. Paul, MN, 1980, p. 440.

(27) Keck, W. G. and Associates, *New "Keck" Submersible Water Sampling Pump for Groundwater Monitoring*, Keck, W. G. and Associates, East Lansing, MI, 1981.

(28) McMillion, L., and Keeley, J. W., "Sampling Equipment for Ground-Water Investigation," *Ground Water*, Vol 6, 1968, pp. 9–11.

(29) Industrial and Environmental Analysts, Inc., *Procedures and Equipment for Groundwater Monitoring*, Industrial and Environmental Analysts, Inc., Essex Junction, VT, 1981.

(30) Trescott, P., and Pinder, G., "Air Pump for Small-Diameter Piezometers," *Ground Water*, Vol 8, 1970, pp. 10–15.

(31) Sommerfeldt, T., and Campbell, D., "A Pneumatic System to Pump Water From Piezometers," *Ground Water*, Vol 13, p. 293.

(32) Smith, A., "Water Sampling Made Easier with New Device," *The Johnson Drillers Journal*, July–August 1976, pp. 1–2.

(33) Morrison, R., and Ross, D., "Monitoring for Groundwater Contamination at Hazardous Waste Disposal Sites," *Proceedings of 1978 National Conference on Control of Hazardous Material Spills*, April 13, Miami Beach, FL, 1968, pp. 281–286.

(34) Morrison, R., and Brewer, P. "Air-Lift Samplers for Zone-of-Saturation Monitoring," *Ground Water Monitoring Review*, Spring 1981, pp. 52–54.

(35) Morrison, R., and Timmons, R., "Groundwater Monitoring II," *Groundwater Digest*, Vol 4, 1981, pp. 21–24.

(36) Bianchi, W. C., Johnson, C., Haskell, E., "A Positive Action Pump for Sampling Small Bore Holes," *Soil Science Society of America Proceedings*, Vol 26, 1961, pp. 86–87.

(37) Timmons, R., Discussion of "An All-Teflon Bailer and An Air-Driven Pump for Evacuating Small-Diameter Ground-Water Wells" by D. Buss and K. Bandt, *Ground Water*, Vol 19, 1981, pp. 666–667.

(38) Timco Manufacturing Co., Inc., "Gas Lift Teflon Pump," *Timco Geotechnical Catalogue*, Prairie du Sac, WI, 1982.

(39) Tomson, M., King, K., Ward, C., "A Nitrogen Powered Continuous Delivery, All Glass Teflon Pumping System for Groundwater Sampling from Below 10 Meters," *Ground Water*, Vol 18, 1980, pp. 444–446.

(40) *Pneumatic Water Sampler*, Slope Indicator Co., Seattle, WA, 1982.

(41) Petur Instrument Co., Inc., *Petur Liquid Sampler*, Petur Instrument Co., Inc., Seattle, WA, 1982.

(42) Idler, G., "Modification of an Electronic Downhole Water Sampler," *Ground Water*, Vol 18, 1980, pp. 532–535.

(43) *Remote Sampler Model 200*, Markland Specialty Engineering, Ltd., Etobicoke, Ontario, Bulletin 200/78, 1978.

(44) Middleburg, R., "Methods for Sampling Small Diameter Wells for Chemical Quality Analysis," *Presented at the National Conference on Quality Assurance of Environmental Measurements*, Nov. 27–29, Denver, CO, 1978.

(45) *Air Squeeze Pump*, Leonard Mold and Die Works, Denver, CO, 1982.

(46) *Automatic Sampler Controller: Markland Model 105 and 2105*, Markland Specialty Engineering, Ltd., Etobicoke, Ontario, Bulletin 105/78, 1981.

(47) Gillham, R. W., and Johnson, P. E., "A Positive Displacement Ground-Water Sampling Device," *Ground Water Monitoring Review*, Vol 1 (2), Summer 1981, p. 33.

(48) Signor, D., "Gas-Driven Pump for Ground-Water Samples," *U.S. Geological Survey, Water Resources Investigation 78–72*, Open File Report, 1978.

(49) *Tigre Tierra HX Pneumatic Packer*, Tigre Tierra, Inc., Puyallup, WA, 1981.

(50) Cherry, R., "A Portable Sampler for Collecting Water Samples from Specific Zones in Uncased or Screened Wells," *U.S. Geological Survey*, Prof. Paper 25-C, 1965, pp. 214–216.

(51) Grisak, G., Merritt, W., Williams, D., "Fluoride Borehole Dilution Apparatus for Groundwater Velocity Measurements," *Canadian Geotechnical Journal*, Vol 14, 1977, pp. 554–561.

(52) Galgowski, C., Wright, W., "A Variable-Depth Ground-Water Sampler," *Soil Science Society of America Proceedings*, Vol 44, 1980, pp. 1120–1121.

(53) *Samplers and Sampling Procedures for Hazardous Waste Streams*, USEPA MERL Laboratory, Cincinnati, OH, EPA-600/2-80-018, January 1980.

(54) *Methods for Chemical Analysis of Water and Wastes*, EPA-600/4-79-020, USEPA EMSL Laboratory, Cincinnati, OH, March 1979.

(55) *Federal Register*, Vol 44, No. 244, Dec. 18, 1979, pp. 75050–75052.

(56) *Standard Methods for the Examination of Water and Wastewater*, APAA, 14th ed., Washington, DC, 1976, pp. 38–45.

(57) *Handbook for Analytical Quality Control in Water and Wastewater Laboratories*, EPA-600/4-79-019, USEPA EMSL Laboratory, Cincinnati, OH, March 1979.

(58) U.S. Department of Interior, "Groundwater," Chapter II, *National Handbook of Recommended Methods for Water Data Acquisition*, 1980.

Standard Practice for Dissolution of Solid Waste by Lithium Metaborate Fusion[1]

This standard is issued under the fixed designation D 4503; the number immediately following the designation indicates the year of original adoption or, in the case of revision, the year of last revision. A number in parentheses indicates the year of last reapproval. A superscript epsilon (ϵ) indicates an editorial change since the last revision or reapproval.

1. Scope

1.1 This practice covers the drying, ashing, and solubilization of solid waste using a lithium metaborate ($LiBO_2$) fusion for the subsequent determination of inorganic constituents by argon plasma emission spectroscopy or atomic absorption spectroscopy.

1.2 The following elements may be solubilized by this practice:

aluminum	chromium	silicon
barium	iron	titanium
cadmium	magnesium	vanadium
calcium	manganese	zinc
copper	nickel	

1.3 This practice has been used successfully with a bauxite ore and a neutralized metal treatment sludge. The practice may be applicable to other elements not listed above. Some metals, such as cadmium and zinc, may volatilize from some samples during the drying, ashing, or fusion steps. The analyst is responsible for determining whether the practice is applicable to the solid waste being tested.

1.4 This practice is intended for the solubilization of nonvolatile inorganic constituents in solid waste. The $LiBO_2$ fusion is appropriate for a silicate matrix or acid resistant samples.

1.5 *This standard does not purport to address all of the safety problems associated with its use. It is the responsibility of the user of this standard to establish appropriate safety and health practices and determine the applicability of regulatory limitations prior to use.* For specific hazard statements see Section 7.

2. Referenced Documents

2.1 *ASTM Standards:*

D 1193 Specification for Reagent Water[2]

D 2777 Practice for Determination of Precision and Bias of Applicable Methods of Committee D-19 on Water[2]

D 3682 Test Method for Major and Minor Elements in Coal and Coke Ash by Atomic Absorption[3]

E 50 Practices for Apparatus, Reagents, and Safety Precautions for Chemical Analysis of Metals[4]

3. Summary of Practice

3.1 The solid waste is weighed, dried, and ashed at 550°C to remove water and organic constituents, and reweighed. A known portion of the ground ash is mixed with $LiBO_2$ in a graphite crucible and fused at 1000°C. Immediately after fusion, the molten mass is poured directly into stirred dilute HNO_3 solution, dissolved, filtered, and made to appropriate volume for subsequent analysis.

4. Significance and Use

4.1 A knowledge of the inorganic constituent composition in a waste is often required for the selection of appropriate waste disposal practices. Solid waste may exist in a variety of forms and contain a range of organic and inorganic constituents. This practice describes a drying and ashing step that may be applied to remove moisture and volatile and nonvolatile organic constituents prior to determining nonvolatile metals. Generation of a dry ash concentrates the inorganic constituents of interest and makes the $LiBO_2$ fusion feasible for a greater variety of waste samples. Acidification of the $LiBO_2$ fusion mix results in a solution amenable to inductively coupled plasma (ICP) or atomic absorption spectrometry (AAS) analysis.

5. Apparatus

5.1 *Analytical Balance*, sensitive to 0.1 mg.

5.2 *Fusion Muffle Furnace*, electrically heated, capable of maintaining a temperature of 1000°C.

5.3 *Ashing Muffle Furnace*, electrically heated, capable of maintaining a temperature of 550°C± 30°C and with an adequate air circulation. This may be accomplished by connecting rubber tubing to a controlled source of clean dry air. Then, via a ceramic tube inserted into a convenient muffle opening, flow approximately 4 L/min of air into the furnace.

5.4 *Drying Oven*, capable of operating at a temperature up to 150°C.

5.5 *Evaporating/Ashing Dish*, 50 to 100-mL capacity, made of platinum, silica, or porcelain.

5.6 *Fusion Crucibles*, graphite, 28 to 30-mL capacity.

5.7 *Stirring Hot Plate*, capable of operating at a surface temperature up to 300°C with TFE-fluorocarbon-coated stir magnet.

[1] This practice is under the jurisdiction of ASTM Committee D34 on Waste Management and is the direct responsibility of Subcommittee D34.01.06 on Analytical Methods.

Current edition approved Nov. 28, 1986. Published February 1987.

[2] *Annual Book of ASTM Standards*, Vol 11.01.

[3] *Annual Book of ASTM Standards*, Vol 05.05.

[4] *Annual Book of ASTM Standards*, Vol 03.05.

5.8 *Mortar and Pestle*, agate or mullite type.

5.9 *Sieve and Pan*, ASTM U.S. Standard Testing Sieve, 200 m (75 μm opening).

5.10 *Desiccator*.

6. Reagents and Materials

6.1 *Purity of Reagents*—Reagent grade chemicals shall be used in all tests. Unless otherwise indicated, it is intendedthat all reagents shall conform to the specifications of the Committee on Analytical Reagents of the American Chemical Society, where such specifications are available.[5] Other grades may be used provided it is first ascertained that the reagent is of sufficiently high purity to permit its use without lessening the accuracy of the determination.

6.2 *Purity of Water*—Unless otherwise indicated, references to water shall be understood to mean Type II reagent water as defined in Specification D 1193.

6.3 *Lithium Metaborate*—$LiBo_2$, anhydrous powder.

6.4 *Nitric Acid, Solution* (5 ± 95)—Add while stirring 50-mL of nitric acid (HNO_3, sp. gr. 1.42) to 900 mL of water. Make to 1 L volume and store in a polyethylene bottle.

6.5 *Dessicant*—Anhydrous calcium sulfate ($CaSO_4$) or equivalent.

7. Hazards

7.1 Samples known or suspected of containing toxic or hazardous materials must be handled in a fume hood. Safety information relative to the handling of any known toxic materials must be obtained and implemented prior to any sample handling.

7.2 Flammable materials must be kept from heat, sparks, or flames.

7.3 Drying should be conducted on an explosion proof electrical heater in a fume hood if use of a conventional drying apparatus may result in toxic, flammable, or irritating vapors.

7.4 Ashing at 550°C must be conducted in a hood with adequate ventilation and shielding. Refer to Practices E 50 for additional information.

8. Procedure

8.1 Prepare the analytical sample from a thoroughly mixed representative laboratory sample.

8.2 Weigh sufficient sample to the nearest 1 mg into a tared ashing dish so that after drying and ashing at least 2 g remains for use in 8.6. Typically 5 to 10 g of waste is sufficient unless moisture and organic content are a major portion of the sample.

8.3 Dry the sample at 110 to 150°C. If appreciable free moisture or volatiles are present, continue drying until the sample is suitable for ashing.

8.4 Place the sample into an ashing furnace set at about 300°C and increase heat gradually so the furnace reaches 550°C in 1 h. Ash at 550°C until no carbonaceous matter is apparent. Stirring the sample once an hour may increase the oxidation of carbonaceous matter. The ashing time required will depend on the nature of the sample. Several hours, or even overnight, may be required by difficult-to-ash samples.

8.5 Remove the ashing dish and sample from the muffle, cool in a desiccator, and weigh to determine the combined loss on ashing and drying.

8.6 Quantitatively transfer the ash to a mortar and grind to pass a No. 200 sieve, if necessary. Transfer back to the ashing dish and reheat the ground ash at 550°C for 1 h, remove from the ashing furnace and cool in a desiccator. Transfer quantitatively to a weighing bottle. Weigh approximately 0.3 g of sample to the nearest 0.0001 g by difference into a graphite crucible containing 1.5 g of $LiBO_2$. Mix the ash and $LiBO_2$ well, then add an additional 0.5 g of $LiBO_2$ on top of the mix.

NOTE 1—Ashing at 550°C typically gives a free flowing or friable ash, so quantitative transfer is possible with careful brushing. Should a portion of the ash melt or stick to the dish so quantitative transfer is impossible, the analyst should use a lower ashing temperature or consider an alternative dissolution practice.

8.7 Place the crucible in a muffle furnace preheated to 1000°C and fuse for 20 min. Remove the crucible from the muffle, swirl to consolidate the molten bead, and pour into a 250-mL beaker containing 150 mL of HNO_3 solution (5 ± 95). The acid solution should be warm (50 to 70°C) and stirred with a stirring hot plate. Complete dissolution of the melt, other than traces of graphite particles from the crucible, should occur in 10 to 15 min.

8.8 Gravity filter the solution through a medium filter, such as Whatman No. 41, into a 250-mL volumetric flask. Quantitatively wash the beaker and filter with water. Add the wash water to the filtrate, cool, dilute to volume with water, and mix. This solution is ready for ICP or AAS analysis. Refer to Test Method D 3682 for an AAS analysis method.

8.9 Carry a $LiBO_2$ blank as in 8.6-8.8 for use as a method blank in the analytical step.

9. Precision and Bias

9.1 Six laboratories participated in a collaborative test program. Two wastes were tested with a single operator at each laboratory performing the practice once on each of 3 days. A central laboratory analyzed all the prepared solutions using an inductively coupled argon plasma spectrometer.

9.1.1 The mean, single-operator precision (S_o), and overall precision (S_T) for the elements solubilized by this practice are shown in Table 1. Practice D 2777 was used in developing these precision estimates.

9.2 Determination of the bias of this practice is not possible, as no suitable standard reference material exists.

9.2.1 Comparison of the analytical values obtained using this practice with those obtained by a single laboratory using classical dissolution and analytical methods is shown in Table 2. The data indicate a suitable degree of agreement between independent methods.

[5] Reagent Chemicals, American Chemical Society Specifications, American Chemical Society, Washington, DC. For suggestions on the testing of reagents not listed by the American Chemical Society, see *Reagent Chemicals and Standards*, Joseph Rosin, D. Van Nostrand and Co., Inc., New York, NY, and the *United States Pharmacopeia*.

TABLE 1 Mean, Single Operator Precision, and Overall Precision

Constituents	Sample 1 Bauxite Ore			Sample 2 Metal Treatment Sludge		
	$\bar{X}$, %	S_o	S_T	$\bar{X}$, %	S_o	S_T
Aluminum	31.17	1.14	2.11	3.24	0.13	0.22
Barium	0.0087	0.0003	0.0009	0.0087	0.0002	0.0008
Cadmium	0.0051	0.0003	0.0007	[A]	[A]	[A]
Calcium	0.093	0.019	0.025	24.79	0.79	1.02
Copper	0.0063	0.0011	0.0027	0.0063	0.0014	0.0030
Chromium	0.24	0.012	0.019	3.13	0.11	0.18
Iron	17.20	0.62	1.17	1.21	0.42	0.07
Magnesium	0.07	0.003	0.006	0.92	0.018	0.033
Manganese	0.25	0.009	0.013	0.02	0.0007	0.0014
Nickel	0.020	0.0009	0.0012	0.017	0.0009	0.0032
Silicon	0.73	0.020	0.071	5.58	0.26	0.52
Titanium	2.09	0.07	0.14	0.05	0.002	0.004
Vanadium	0.09	0.004	0.005	0.016	0.0004	0.001
Zinc	0.03	0.0019	0.003	0.18	0.002	0.008

[A] At or less than detection limit.

TABLE 2 Comparison of Analytical Values

Constituents	Sample 1		Sample 2	
	This Work (ICP), %	Classical,[A] %	This Work, %	Classical, %
Aluminum	31.17	33.2	3.24	3.38
Iron	17.20	18.6	1.21	1.17
Titanium	2.09	2.16	0.05	0.05
Silicon	0.73	0.86	5.58	[B]
Manganese	0.25	0.31	0.02	0.02
Chromium	0.24	0.19	3.13	3.24
Calcium	0.093	[B]	24.79	26.4
Magnesium	0.07	0.08	0.92	0.92
Nickel	0.02	0.03	0.017	[B]
Vanadium	0.09	0.07	0.016	0.015
Others[C]				

[A] Single laboratory values using gravimetric, titrimetric, and colorimetric methods are reported.
[B] ND = not determined.
[C] One collaborative laboratory determined other constituents in Sample 2 and found the following:

Sulfate (SO_4)	7.6
Organic carbon (C)	9.6
Carbonate (CO_3)	4.4
Phosphate (P_2O_5)	16.0
Fluoride (F)	1.9

Designation: D 4547 – 98

Standard Guide for Sampling Waste and Soils for Volatile Organic Compounds[1]

This standard is issued under the fixed designation D 4547; the number immediately following the designation indicates the year of original adoption or, in the case of revision, the year of last revision. A number in parentheses indicates the year of last reapproval. A superscript epsilon (ε) indicates an editorial change since the last revision or reapproval.

1. Scope

1.1 This guide describes recommended procedures for the collection, handling, and preparation of solid waste, soil, and sediment samples for subsequent determination of volatile organic compounds (VOCs). This class of compounds includes low molecular weight aromatics, hydrocarbons, halogenated hydrocarbons, ketones, acetates, nitriles, acrylates, ethers, and sulfides with boiling points below 200° Celsius (C) that are insoluble or slightly soluble in water.

1.2 Methods of sample collection, handling, and preparation for analysis are described.

1.3 This guide does not cover the details of sampling design, laboratory preparation of containers, and the analysis of the samples.

1.4 It is recommended that this guide be used in conjunction with Guide D 4687.

1.5 *This standard does not purport to address all of the safety concerns, if any, associated with its use. It is the responsibility of the user of this standard to establish appropriate safety and health practices and determine the applicability of regulatory limitations prior to use.*

2. Referenced Documents

2.1 *ASTM Standards:*

D 3550 Practice for Ring-Lined Barrel Sampling of Soils[2]

D 4687 Guide for General Planning of Waste Sampling[3]

D 4700 Guide for Soil Sampling from the Vadose Zone[2]

D 5058 Test Methods for Compatibility of Screening Analysis of Waste[3]

D 5792 Practice for Generation of Environmental Data Related to Waste Management Activities: Development of Data Quality Objectives[3]

D 6051 Guide for Composite Sampling and Field Subsampling for Environmental Waste Management Activities[3]

E 856 Definitions of Terms and Abbreviations Relating to Physical and Chemical Characteristics of Refuse-Derived Fuel[3]

2.2 *Federal Standard:*

Title 49 Transportation, Code of Federal Regulations (CFR), Part 172, List of Hazardous Substances and Reportable Quantities[4]

3. Terminology

3.1 *sample, n*—a portion of material taken from a larger quantity for the purpose of estimating properties or composition of the larger quantity. **(E 856)**

3.2 *subsample, n*—a portion of a sample taken for the purpose of estimating properties or composition of the whole sample. **(D 6051)**

3.2.1 *Discussion*—A subsample, by definition, is also a sample.

4. Summary of Guide

4.1 This guide addresses the use of tools for sample collection and transfer, conditions for sample storage, sample preservation, and two common means of sample preparation for analysis. Special attention is given to each step from sample collection to analysis to limit the loss of VOC s by volatilization of biodegradation. The sample collected and analyzed should be representative of the parent material sampled. The two methods cited for the preparation of samples for VOC analysis are methanol extraction and vapor partitioning (that is, purge-and-trap and headspace). The method of sample preparation for VOC analysis is dependent on the data quality objectives (see Practice D 5792). Although the equilibrium–driven proportions of VOC's in the different phases change during sample preparation and analysis, the overall intent of the methods in this guide are to minimize VOC losses.

5. Significance and Use

5.1 This guide describes sample collection and handling procedures designed to minimize losses of VOCs. The principal mechanisms for the loss of VOCs from materials during collection and handling are volatilization and biodegradation. Susceptibility of various VOCs to these two loss mechanisms is both compound and matrix specific. In general, compounds with higher vapor pressures are more susceptible to volatilization than compounds with lower vapor pressures. Also, aerobically degradable compounds are generally more susceptible to biodegradation than anaerobically degradable compounds. In some cases, the formation of other compounds not originally

[1] This guide is under the jurisdiction of ASTM Committee D34 on Waste Management and is the direct responsibility of Subcommittee D34.01.02 on Monitoring.

Current edition approved Sept. 10, 1998. Published November 1998. Originally published as D 4547–91. Last previous edition D 4547–91.

[2] *Annual Book of ASTM Standards*, Vol 04.08.

[3] *Annual Book of ASTM Standards*, Vol 11.04.

[4] Available from Superintendent of Documents, U.S. Government Printing Office, Washington, DC 20402.

present in the material can occur. Loss or gain of VOCs leads to analytical results that are both unrepresentative of the field conditions and ambiguous.

5.2 Ancillary information concerning sample collection and handling for VOC analysis is provided in Appendix X1 and Appendix X2. These appendixes and cited references are recommended reading for those unfamiliar with many challenges presented during the collection and handling of samples for VOC analysis.

6. Selection of Sample Preparation Method for VOC Analysis

6.1 *Introduction*:

6.1.1 Sample collection, handling, and preservation methods should be compatible with the method used to prepare the sample for VOC analysis, and meet the project's data quality objectives (see Practice D 5792). Preparation of a sample for instrumental analysis can be initiated either in the field or laboratory. In either case, the sample should be placed into a tared volatile organic analysis (VOA) vial or bottle meeting the specifications given in 7.2.2. When working with an uncharacterized solid waste, it is advisable to perform compatibility tests (see Test Methods D 5058) between the sample material and the solution (see 7.2.2.1 and 7.2.2.2) into which it will be transferred in preparation for analysis. For instance, when collecting highly contaminated soils or waste of unknown composition, or both, it is strongly recommended that preliminary testing be performed to adequately characterize the waste materials so that when the user applies the procedures cited in this guide, there will be no chemical reaction which may jeopardize the user's safety.

6.1.2 Figs. 1 and 2 are flow diagrams showing some different options for combining sample collection, handling and preparation methods for instrumental analysis.

6.2 *Methanol Extraction*:

6.2.1 This method involves the extraction of VOCs from a sample with methanol and the subsequent transfer of an aliquot of the extract to water for either purge-and-trap or headspace analysis. Advantages of methanol extraction are (*1*) large samples or composite samples, or both, can be collected to enhance representatives (see Guide D 6051), (*2*) biodegradation is inhibited, (*3*) an efficient extraction of VOCs from the matrix materials can be achieved with methanol due to its strong affinity for these compounds and favorable wetting properties, (*4*) a subsample can be analyzed several times, and (*5*) sample extracts can be archived (held up to two months, and perhaps longer, if verified that VOC losses have not occurred (see 8.2.1.2)).

6.2.2 The primary disadvantages of methanol extraction are (*1*) samples may have to be shipped as a flammable liquid depending on the amount of methanol present (U.S. DOT reg. 49CFR§172.101), (*2*) hazards to personnel due to methanol's toxicity and flammability, (*3*) detection limits are elevated due to analyte dilution, (*4*) possible interference of the methanol

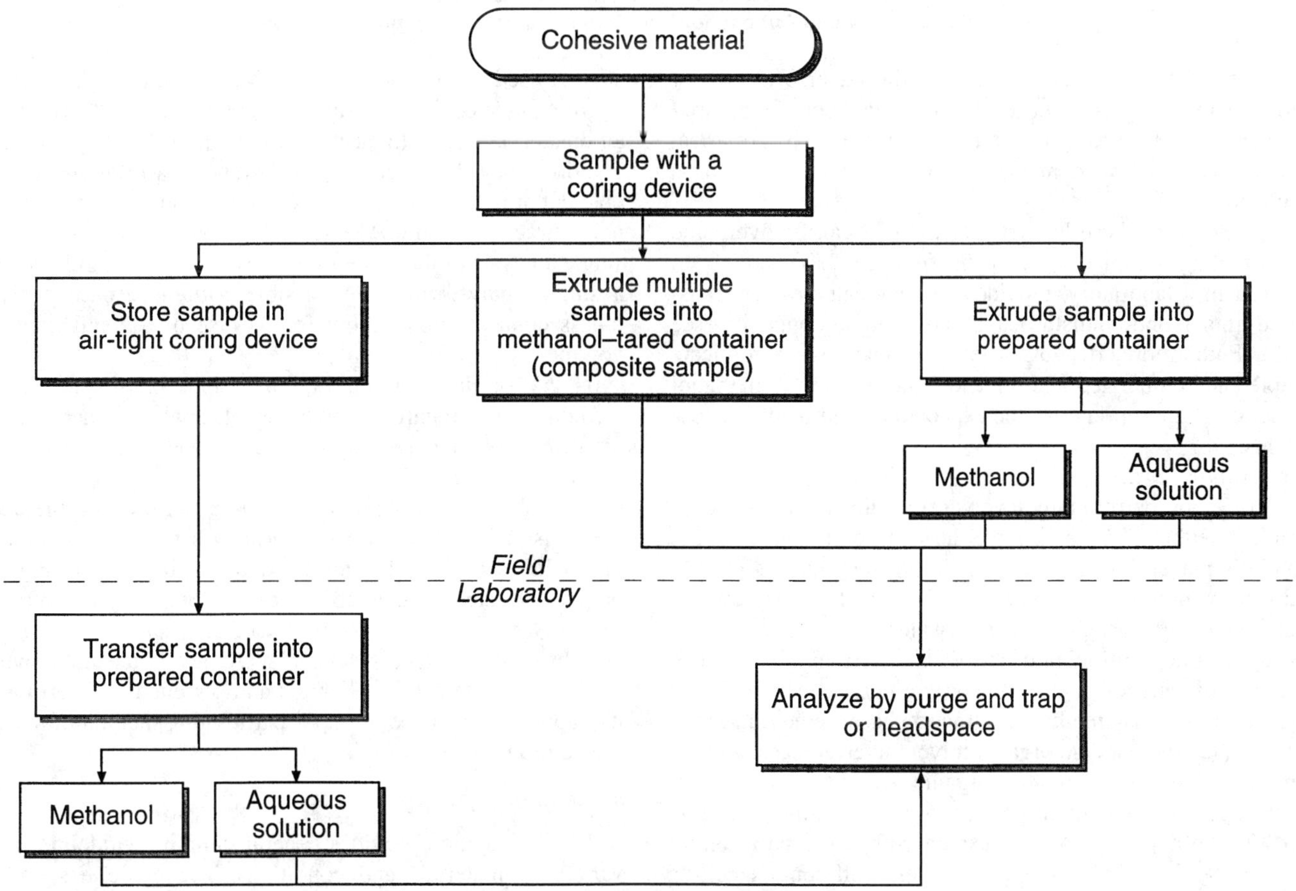

FIG. 1 Sample Handling Options for Cohesive Materials

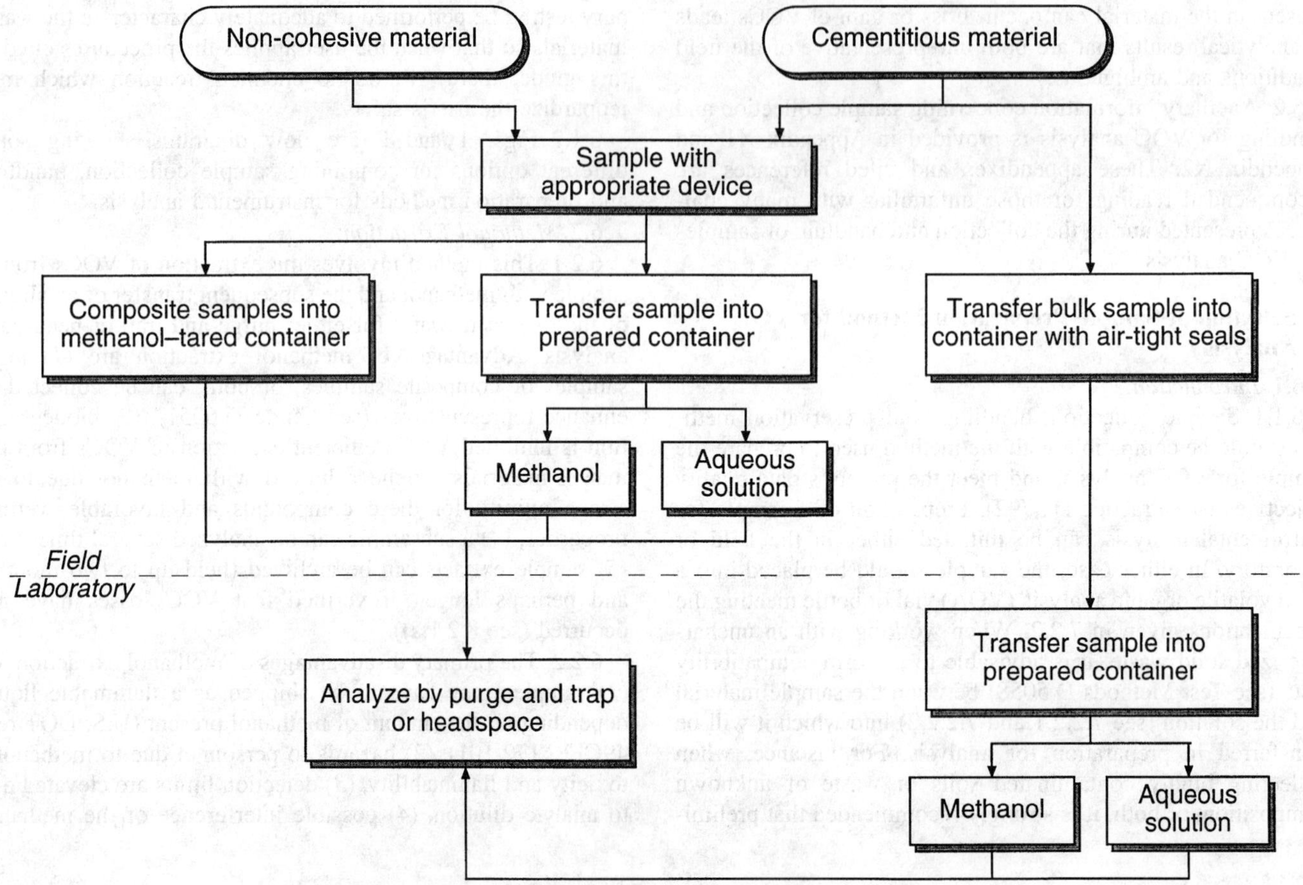

FIG. 2 Sample Handling Options for Non-Cohesive and Cementitious Materials

peak with VOCs of interest, (*5*) potential adverse impact of methanol on the performance of certain gas chromatograph/detector systems, and (*6*) in the United States of America, samples extracted with methanol must be disposed of as a regulated waste.

6.2.3 Some of these logistical challenges can be overcome by extracting samples with methanol once they have been received in a laboratory, provided that the samples are transported intact and undisturbed in an airtight container (see 8.1.1). Futhermore, if VOC levels are unknown, a replicate sample can be obtained and screened to determine if methanol extraction is appropriate for the expected contaminant concentrations.

6.3 *Vapor Partitioning*:

6.3.1 Vapor partitioning involves the direct analysis of a sample by either purge-and-trap or headspace. In both case, the sample is placed into a tared volatile organic analysis (VOA) vial from which the vapor is removed for analysis without the container being opened. Heat and water are usually used to assist with the partitioning of the VOCs from the sample. The principal advantages of this method are (*1*) it can offer lower detection limits than methanol extraction because no dilution is involved, (*2*) there are no organic solvent interferences, and (*3*) there is no use of regulated organic solvents, which may require special shipment, disposal, and field handling practices.

6.3.2 The disadvantages associated with vapor partitioning are (*1*) the VOA vial (VOA vials are different sizes for automated purge-and-trap and headspace instrumentation) or adapter used in conjunction with a VOA vial, or both, often are instrument specific, (*2*) sample size is limited (<10 g) by automated systems, (*3*) a matrix-appropriate method of preservation may be necessary, (*4*) vapor partitioning is less efficient at recovering VOCs from some materials than methanol extraction, and (*5*) when using purge-and-trap, only a single analysis of the same sample can be made; similarly only a single analysis may be possible with headspace analysis unless concentrations allow for the use of a small injection volume.

6.3.2.1 Limitations imposed by vapor phase partitioning methods with regard to number of analyses that can be performed on a single sample can be addressed by taking replicate samples.

6.3.3 When employing vapor phase partitioning methods, the logistical challenges of performing sample preparation in the field (see 7.2.2.2) can be avoided by performing the preparation step in the laboratory, so long as the sample is transported to the laboratory intact and undisturbed, in an airtight container (see 8.1.1). If VOC levels are unknown, a replicate sample can be obtained and screened to determine if it is appropriate to use of vapor partitioning method of sample preparation.

7. Sample Collection

7.1 This section provides general sampling guidelines for a variety of materials and conditions. The guidelines are intended to allow flexibility in the following:

7.1.1 The procedure for obtaining a sample from, for example, walls of test pits, surface grid locations, waste piles, building, and from bulk samples obtained with subsurface retrieval systems.

7.1.2 The design and dimensions of the equipment used to obtain and transfer samples of the appropriate size for analysis.

7.1.3 The selection and preparation of sample containers.

7.2 *Preparation for Sampling*:

7.2.1 All sampling and handling devices and vessels used to collect and store samples for analysis should be constructed of nonreactive materials that will not sorb, leach or diffuse constituents of interest. Examples of materials that meet these criteria are glass, stainless steel, and brass. Materials, such as polytetrafluorethylene and many rigid plastics also can be used, however, it should be recognized that they may have some limited adsorptive properties or allow slow diffusive passage of some VOCs. Materials which show limited reactivity can be use when they have a very short period of contact with the sample or when they are necessary for making airtight (hermetic) seals. Collection tools and storage containers made of materials other than those cited in this section should only be used after they have demonstrated equivalency. All collection tools and storage containers should be cleaned in a manner consistent with their intended use.

7.2.1.1 There are often several steps to sampling, particularly if it involves obtaining bulk materials from subsurface regions. Most of the equipment used to obtain samples from the subsurface was originally developed for the geotechnical industry. These subsurface bulk sample retrieval systems are designed to obtain intact cylindrical cores of material, ranging anywhere from 1 to 4 in. diameter, and 1 to several ft in length. Two commonly used collection devices for subsurface sample collection are the split-spoon corer that opens to expose the entire length of the material obtained for subsampling and core barrel liners (that is, ring-lined barrel sampling, see Practice D 3550) that typically are subsampled through the open ends. Core barrel liners fit snugly within a corer and come in a variety of lengths and materials (stainless steel, brass, polytetrafluorethylene, rigid plastics, etc.). Several liners, or a single long liner, can be placed within the core barrel, depending on the sampling plan. Additional information on the application and design of different types of subsurface collection systems available can be found in Guide D 4700. Subsurface materials retrieved for VOC characterization should be obtained (sampling tubes filled and brought to the surface) as quickly as possible and remain intact and undisturbed until they are subsampled. Subsampling of a bulk sample should occur within a couple of minutes of its being brought to the surface (see 7.3 through 7.6 for additional sampling guidance).

7.2.1.2 In addition to the coring devices which retrieve bulk quantities of material, there are smaller hand operated coring tools for obtaining samples of the appropriate size (for example, 5 and 25 g) for analysis. If one of these smaller coring devices is used to store the sample (see 8.1.1) for more than a few seconds, the main body should be constructed of materials that are nonreactive and have airtight seals that show limited sorption and penetration of VOCs. When the coring device is only used to rapidly transfer the sample to a VOA vial, corer material requirements are less stringent.

7.2.2 VOA vials and bottles used for storage and preparation of samples for analysis should be made of glass and have airtight seals. To achieve and airtight seal, these containers should have a thick septum cushion between the sealing material (polytetrafluoroethylene) and cap (rigid plastic screw cap or aluminum crimp top). Polytetrafluoroethylene-lined caps that do not have flexible septum backing often fail to achieve a liquid or airtight seal. Futhermore, the thickness of the polytetrafluoroethylene used for a lined septum, should be at least 10 min. These VOA vials and bottles should be prepared as described in 7.2.2.1 or 7.2.2.2 prior to adding the sample to the container.

7.2.2.1 *Preparation of Containers for Methanol Extraction*—The appropriate volume of analytical-grade methanol (high-performance liquid chromatography or spectrographic) is added to the organic-free container by the laboratory that supplies the container, by the sample collector, or by a third party. The party that adds the methanol to the container should also be responsible for providing trip blanks and introducing surrogate compounds (see Guide D 4687). Once methanol has been placed in a container, it should be opened only to add the sample(s) and to remove aliquots for analysis. The tared weight of the container with methanol should be recorded prior to adding the sample to the container. A predetermined volume of sample is added to the methanol. This sample volume corresponds to a weight in grams (g) that is equivalent to or less than the volume (mL) of methanol. The ratio (typically 1.1 to 10.1, methanol to material) between the two constituents should allow for formation of a clear layer of methanol over the sample after thorough mixing. The difference in weight of the container, measured before and after the sample is introduced, is used to establish the sample's wet weight. Because of the water (moisture) present in most samples, calculation of the recovery of surrogate concentrations in the sample should account for this source of dilution.

NOTE 1—Initially the concentration of the surrogates is influenced only by the volume of methanol. After adding a sample, however, the liquid volume is increased by the volume of water present in the soil sample. The increased liquid volume results in a reduction in the surrogate concentration due to its dilution effect. After the moisture content of a sample is known, the expected surrogate concentration can be determined by multiplying the original surrogate concentration by: volume of methanol/ total liquid volume.

7.2.2.2 *Preparation of Containers for Vapor Partitioning*—This method allows for the direct analysis of a sample by either purge-and-trap or headspace techniques. In both cases, the sample is placed into a VOA vial from which the vapor (only a portion for headspace analysis) is removed for analysis without the container being opened. Moreover, water that contains no detectable levels of VOCs is usually present in the VOA vial prior to introducing the sample. Current automated equipment for purge-and-trap and headspace systems use 44- (or 40-) and 22-mL VOA vials, respectively. The volume of water used for these two different systems is typically 10 mL or less. The addition of surrogate compounds and matrix spikes to samples prepared for vapor partitioning analysis can be problematic. Adding surrogates or matrix spikes (see Guide D 4687) to an open sample container prior to or immediately after sample collection can potentially result in the loss of VOCs from the VOA vial. A procedure for introducing surrogates or spikes that is less susceptible to VOC losses, can be performed by puncturing the septum with a needle (small gauge if done manually) after the sample has been sealed inside the container. If aromatic compounds are of concern and the sample will be held for more than 48 h before analysis, some form of chemical preservation in addition to refrigerated (4°C) storage should be used (see 8.2.2.2). When carbonates are not present, chemical preservation for aromatic compounds can be achieved by acidification to a pH of 2 or less with either sodium bisulfate or hydrochloric acid (for general guidance on testing for carbonates and the use of these acids to preserve samples, see Appendix X3). Precautions must be taken when preserving by acidification since certain compounds within the following classes: olefins, ketones, esters, ethers, and sulfides, will react under a low pH condition. Water, and if necessary, a chemical preservative, should be added to the sample container prior to adding the sample. Futhermore, the tared weight of the container should be recorded prior to adding the sample. The difference in weight of the container, measured before and after the sample is added, is used to determine the sample's wet weight.

NOTE 2—Promising alternative methods of preservation that may be considered when a low pH is undesirable or carbonates are present and longer holding times can not be avoided, are cited in Appendix X1 and Appendix X2.

7.2.3 A separate collocated sample should be collected within a couple centimeters and from the same stratum to determine percent moisture content so that analyte concentration can be expressed on a dry weight basis. The location adjacent to where the sample for VOC analysis was removed should be inspected visually and its characteristics logged. This adjacent material can also be retained for determining other relevant properties, such as general appearance, color, presence of oils, other visible signs of contamination, grain-size distribution, organic carbon content, etc. Collection of these ancillary samples should be performed after the collection of samples for VOC analysis.

7.3 *General Sampling Guidance*—After a fresh surface is exposed to the atmosphere, the sample collection process should be completed in a minimal amount of time. Removing a sample from a material should be done with at least amount of disruption (disaggregation) as possible. Additionally, rough trimming of the sampling location's surface layers should be considered if the material may have already lost VOCs (been exposed for more than a couple of minutes) or if it may be contaminated by other waste, different soil strata, or vegetation. Removal of surface layers can be accomplished by scraping the surface using a clean spatula, scoop, knife, or shovel.

7.4 *Sampling of Cohesive but Uncemented Materials Using Devices Designed to Obtain a Sample Appropriate for Analysis* :

7.4.1 Samples of the appropriate size for analysis should be collected using a metal or rigid plastic coring tool (Fig. 3). For example, coring tools for the purpose of transferring a sample can be made from disposable plastic syringes by cutting off the tapered front end and removing the rubber cap from the plunger (see Note 3). These smaller coring devices help maintain the sample structure during collection and transfer to the VOA vial or a larger bottle, as do their larger counterparts used to retrieve subsurface materials. When inserting a clean coring tool into a fresh surface for sample collection, air should not be trapped behind the sample. If air is trapped, it could either pass through the sampled material causing VOCs to be lost or cause the sample to be pushed prematurely from the coring tool. For greater ease in pushing coring tools into the solid matrix, their front edge can be sharpened. The optimum diameter of the coring tool depends on the following: size of the opening on the collection vial or bottle (tool should fit inside mouth), particle size of the solid materials (for example, gravel-size particles would require larger samplers), and volume of sample required for analysis. For example when a 5-g sample of soil is specified, only a single 3-cm^3 volume has to be collected (assuming the soil has density of 1.7 g/cm^3). Larger sample masses or composite samples may be preferred as the heterogeneity of the material increases.

NOTE 3—**Caution:** This coring device is not appropriate for sample storage.

7.4.2 An undisturbed sample is obtained by pushing the barrel of the coring tool into a freshly exposed surface and removing the corer once filled. The exterior of the barrel should be quickly wiped with a clean disposable towel. If the coring tool is used as a storage container, cap the open end after ensuring that the sealing surfaces are cleaned (see 8.1.1). If the device is solely used for collection and not storage, immediately extrude the sample into a tared VOA vial or bottle by gently pushing the plunger. During transfer of the sample into the container, care should be taken to prevent the sample from contacting the sealing surfaces which can comprise these surfaces and prevent an airtight seal. The volume of material collected should not cause excessive stress on the coring tool during intrusion into the material, or be so large that the sample easily falls apart during extrusion. Obtaining and transferring a sample should be done rapidly (<10 s) to reduce volatilization losses. If the vial or bottle contains methanol or another liquid, it should be held at an angle when extruding the sample into the container to minimize splashing. Just before capping, a visual inspection of the lip and threads of the sample vessel should be made, and any foreign debris should be removed with a clean towel, allowing an airtight seal to form.

7.5 *Devices that Can be Used for Sampling a Cemented Material*:

7.5.1 Samples of hard or cementitious material may be obtained by fragmenting a larger portion of the material using a clean chisel to generate aggregate(s) of a size that can be placed into a VOA vial or bottle. When transferring the aggregate(s), precautions must be taken to prevent compromising the sealing surfaces and threads of the container. Losses of VOCs by using this procedure are dependent on the location of the contaminant relative to the surface of the material being sampled. Therefore, caution should be taken in the interpretation of the data obtained from materials that fit this description. As a last resort when this task can not be performed on-site a large sample can be collected in a vapor-tight container and

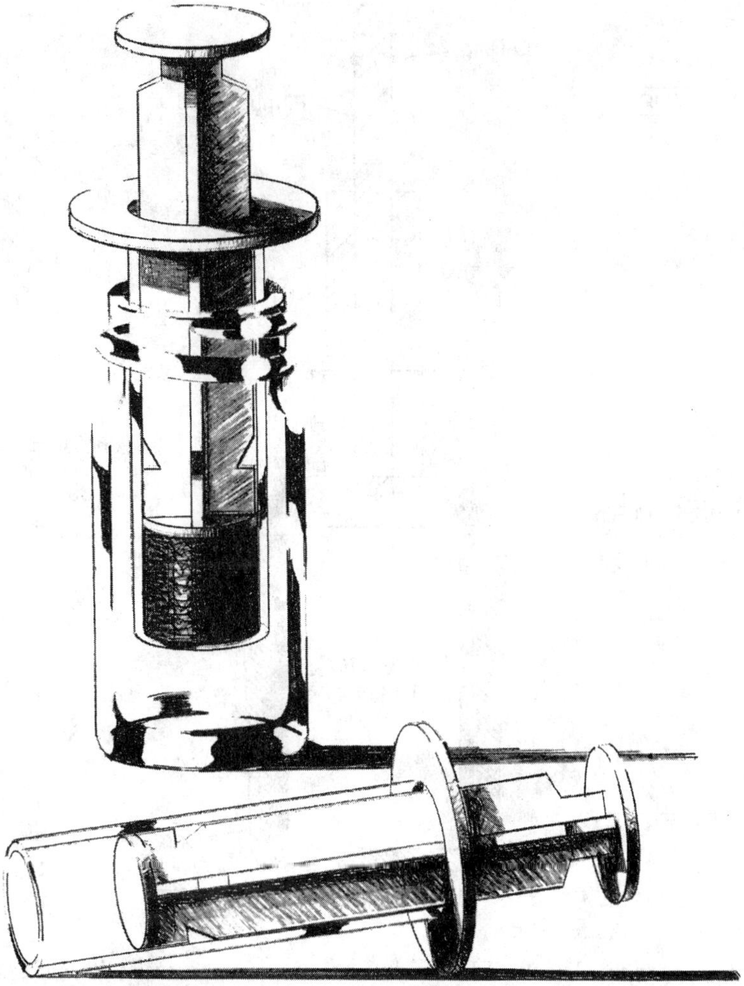

FIG. 3 A Coring Tool Made by Cutting the Tip Off a Plastic Syringe

transported to the laboratory for further processing (see 8.1.2). Collecting, fragmenting, and adding the sample to a container should be accomplished as quickly as possible.

7.6 *Devices that Can be Used for Sampling a Noncohesive Material*:

7.6.1 When sampling gravel, or a mixture of gravel and fines, that can not be easily obtained or transferred using coring tools, as a last resort, a sample can be quickly transferred using a spatula or scoop (see discussion concerning the use of a spatula or scoop to collect samples for VOC analysis, in Appendix X1). Typically the collection vial or bottle will contain methanol or an aqueous solution, therefore, samples should be dislodged with minimal splashing and without the spatula or scoop contacting the liquid contents. For some solids, a wide-bottom funnel or similar channeling device may be necessary to facilitate transfer to the container and prevent compromising of the sealing surfaces of the container. Caution should be taken in the interpretation of the data obtained from materials that fit this description. Losses of VOCs are likely because of the nature of the sampling method and the noncohesive nature of the material exposes more surface area to the atmosphere than for other types of samples. Another potential source of error during the sampling process, is the separation of coarser materials from fines, which can bias the concentration data if the different particle sizes, which have different surface areas, are not properly represented in the sample.

8. Sample Handling

8.1 *Storage*:

8.1.1 *Material Stored in Coring Device*—If the coring device also serves as a shipping container, it should have air tight seals. Presently the En Core[5] sampler, which is available for the collection, storage and transfer of approximately 5 and 25 g samples, is the only commercially available coring device that has been specifically designed to meet this requirement (see Fig. 4). Currently, it is recommended that when samples are stored in the En Core sampler, they should be held for no longer than 48 h at 4 ± 2°C prior to preparing the sample for analysis.[6] Refer to Appendix X2 for additional information concerning coring devices used as storage and shipping containers.

[5] En Core sampler, En Novative Technologies, Inc., 1241 Bellevue St, Green Bay, WI 54302. The En Core sampler is covered by a patent. Interested parties are invited to submit information regarding the identification of an alternative(s) to this patented item to the ASTM Headquarters. Your comments will receive careful consideration at a meeting of the responsible technical committee, which you may attend.

[6] SW-846, Method 5035, 3rd Update.

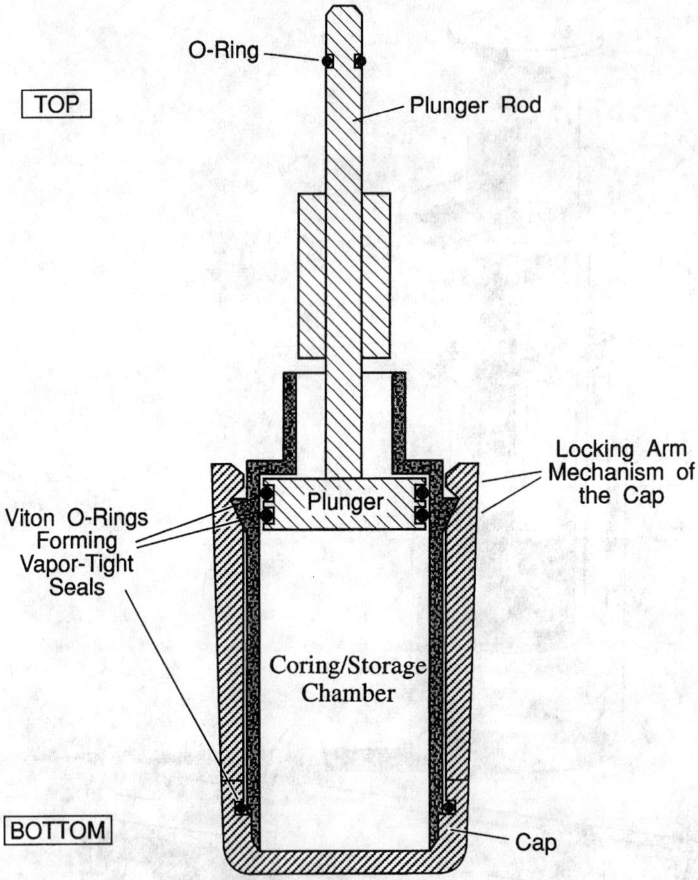

FIG. 4 Components of the En Core Sampler

8.1.2 *Material Not Stored in Coring Device*—To retard volatilization and biodegradation, samples transferred to a prepared VOC vial or bottle (see 7.2.2), whether preserved or not, should be cooled to 4 ± 2°C as soon as possible.

8.1.2.1 If a chemical preservative is not present, then samples should be held under this condition for no longer than 48 h prior to analysis.

NOTE 4—If the sample is stored in the En Core sampler for 48 h, it should be analyzed promptly or preserved in methanol or chemically preserved (for example, acidified) during the sample preparation step.

8.1.2.2 Chemically preserved (for example, acidified) subsamples awaiting vapor partitioning methods of analysis should be analyzed as soon as possible and should not be stored for more than 14 days before analysis.

8.1.2.3 When samples are preserved by immersion in methanol, the holding period can be extended beyond the typical 14–day period, to up to two months, and perhaps longer, if verified (for example, demonstrated that no VOC losses occurred (see 8.2.1.2)).

8.2 *Sample Preparation for Analysis*:

8.2.1 *Methanol Extraction*:

8.2.1.1 Samples should be completely dispersed in methanol, if possible (vortex mixing or sonication can be used), then allowed to settle so that an aliquot of clear supernatant can be removed for analysis. Some materials will show a slow release of VOCs to methanol; therefore, the dissolved VOC concentration may increase with increasing storage time (see Appendix X1). This increased extraction of VOCs over time may be particularly pronounced with cementitious materials or materials that strongly adsorb VOCs. Vials and bottles containing methanol should always be stored upright and be allowed to reach room temperature before removing aliquots appropriate for analysis. Transfer of an aliquot should be performed using a clean liquid-tight syringe (composed of glass and metal) or a clean glass pipette. For purge-and-trap analysis, methanol aliquot volumes of less than 0.2 mL are typically transferred to a vessel containing 5 mL of organic-free water. For headspace analysis, methanol aliquot volumes as large as 1.0 mL, depending on the detector and analytes of concern, can be transferred to VOA vials containing 10 mL of organic-free water.

8.2.1.2 Periodic weighing of the sample container can be used to determine if a hermetic seal is being maintained, that is, no weight loss of methanol. Analysis of the surrogate compounds in the trip blanks (see Guide D 4687) can be used to establish whether containers chosen have an airtight seal.

8.2.2 *Vapor Partitioning*:

8.2.2.1 Before being placed on the autosampler carousel, the sampled materials should be completely dispersed in water, if possible (vortex mixing or sonication can be used). This mixing of the solid material with the aqueous solution not only helps prevent the plugging of the sparging needle used by purge-and-trap systems but also assists in attaining an equilibrium state by completely exposing the sample to the partitioning solution. To enhance partitioning into the vapor phase, automated purge-and-trap and headspace analysis systems warm the sample to a selected temperature (typically 40°C for purge-and-trap, this temperature or higher for headspace) prior to removing vapors for analysis.

8.2.2.2 Chemically preserved samples awaiting vapor phase partitioning analysis can show small losses of VOCs through the polytetrafluoroethylene-lines septum caps. For this reason, chemically preserved samples prepared for vapor partitioning methods of analysis should be analyzed as soon as possible and should not be stored for more than 14 days.

9. Keywords

9.1 field sampling; methanol extraction; sample collection; sample handling; soil; solid wastes; vapor phase partitioning; volatile organic compounds

APPENDIXES

(Nonmandatory Information)

X1. ADDITIONAL INFORMATION CONCERNING SAMPLE COLLECTION AND HANDLING FOR VOC ANALYSIS.

X1.1 Volatile organic compounds (VOCs) are among the most frequently identified soil contaminants at Superfund and other hazardous waste sites. Because some VOCs are potentially mutagenic, carcinogenic, and teratogenic, they often are key factors in the risk assessment process at these sites. These risk assessments and other remedial decisions depend upon an accurate understanding of the levels and extent of VOC contamination in soils and other materials.

X1.2 In most materials, VOCs coexist as gaseous, liquid, and solid (sorbed) phases. The VOC equilibrium that exists among these phases is controlled by physiochemical properties, material properties, and environmental variables (**1, 2**). Unaccounted loss of analytes from any phase may result in rendering the sample unrepresentative of the material from which it was taken. For this reason, sample collection, handling, and analysis must be performed under conditions that maintain the accountability of all phases present (**3**).

X1.3 In general, uncontrolled losses of VOCs from materials occur through two mechanisms: volatilization and biodegradation. Volatilization losses occur whenever gaseous molecules, which have diffusion coefficients up to four orders of magnitude greater than liquid diffusion coefficients, are allowed to move freely. Therefore, whenever a new surface is exposed, VOC losses are incurred. The extent to which VOCs are lost depends on the vapor phase concentration (analyte vapor pressure), surface area exposed, duration of exposure, porosity of matrix, and perhaps meteorological conditions (**4**).

X1.4 Biological degradation of VOCs in samples is usually dominated by aerobic processes because many conventional intrusive collection methods expose the sample to the atmosphere. The rate of this biological degradation is dependent on several factors, including the indigenous microbiological population, chemical properties of the VOC, and temperature. Provided that sufficient quantities of electron acceptors, nutrients, and moisture are present, indigenous microbes continue to aerobically degrade compounds even when stored at 4°C (**5**). Aromatic compounds are quite susceptible to this loss mechanism. To inhibit biodegradation of these compounds, the sample can be immersed in methanol, acidified to a pH of 2 or less, added to 20 mM (or greater) solution of cupric sulfate, or frozen to -12 ± 3°C (applies to both samples dispersed in water within a partially filled container and samples held in a sealed En Core sampler). While the effectiveness of cupric sulfate and freezing on prohibiting microbiological activity is well known, the use of these two preservation methods with regard to samples held for VOC characterization, has not been documented. For this reason, these methods (preserving with cupric sulfate or freezing) should only be used when accompanied by a validation study. For cases where holding times of longer than 48 h are necessary and chemical preservation is not practical, the storage of samples in the En Core sampler at either 4 ± 2°C or -12 ± 3°C is an option, provided it can be demonstrated that the longer holding periods used do not influence the VOC concentrations or that the data generated from the analysis of the samples meet the data quality objectives. Additional information concerning storage of samples in this type of coring device is further discussed in Appendix X2.

X1.5 An example of a sampling procedure that fails to achieve accountability of all VOC phases is the collection of a bulk sample by using a spatula-type device to completely fill a bottle for sample storage and transportation (**4, 6-10**). Samples collected and transferred with spatual-type devices fail to control surface area exposure, for both cohesive and noncementitious materials. Moreover, in the process of filling a bulk sample bottle to capacity, the sealing surfaces often become compromised (dirty), preventing a vapor-tight seal during storage. For these reasons this procedure routinely results in concentrations that are less than 10 % of the in-situ contamination (**6, 7-9**).

X1.6 The procedures discussed in this guide are designed to limit VOC losses by volatilization and biodegradation. This is accomplished by stressing that (*1*) samples be collected only from freshly exposes surfaces, (*2*) collection and transfer of a sample be performed quickly and with minimal disruption to its physical state, (*3*) if possible, samples be held at 4 ± 2°C for not more than 48 h in an airtight vessel without taking (or initially including) additional preservation measures (chemical or perhaps freezing, particularly when aromatic compounds are of concern), (*4*) and in the case of samples collected for vapor partitioning methods of analysis (purge-and-trap or headspace), the VOA vial's airtight seal never be broken prior to analysis.

X1.7 An additional source of inaccuracy in VOC analyses is their incomplete extraction from the sampled material (**11-16**). This source of error can either be due to differences between sample preparation methods (that is, methanol extraction versus vapor partitioning) or the kinetics of the desorption processes. In general, as the indigenous total organic carbon content increases in a matrix, recoveries using vapor partitioning methods would be expected to decrease compared to methanol extraction, particularly for the more hydrophobic analytes (**16**). Heat or sonication, or both, have been shown to speed up desorption processes (that is, mass transfer of analyte from the sampled matrix to the vapor phase or methanol); therefore, these more aggressive extraction steps should be considered when a total VOC concentration is needed. In some instances, incomplete extraction due to either the method used or the kinetics associated with the release of VOCs from a given matrix can result in an under estimation (perhaps as much as an order of magnitude) of the total VOC concentration present in a sample (**11, 13, 15, 16**).

X2. INFORMATION ON THE STORAGE OF SAMPLES IN DISCRETE SAMPLERS AND COVERED CORE BARREL LINERS

X2.1 The En Core sampler is a coring tool designed to collect, store, and transfer a discrete sample for VOC analysis. This sampler has three components: (*1*) a coring/storage chamber, which is volumetrically designed to collect either a 5-g or 25-g soil sample; (*2*) a vapor-tight plunger for the nondisruptive extrusion of the sample into an appropriate container in preparation for analysis; and (*3*) a press on cap that has a vapor-tight seal and locking arm mechanism (see Fig. 4). The coring/storage chamber, plunger, and cap of the En Core sampler are constructed of an inert composite polymer. This device is certified as not having detectable levels of VOCs by the manufacturer and should not be cleaned or reused. In addition, there are two stainless steel reusable handles that can be used with the En Core sampler. One is designed for use during sample collection and the other when extruding a sample in preparation for analysis. The manufacturer instructions should be followed when using the En Core sampler.

X2.2 Preliminary tests performed with the En Core sampler have demonstrated that this tool can prevent volatilization losses for up to six days when held at 4 ± 2°C, and when stored at -12 ± 3°C can prevent both losses due to biodegradation and volatilization for up to 14 days, and perhaps longer. Although very promising, these initial results should not be used to justify the use of holding times longer than 48 h, unless it can be demonstrated that a longer storage time or different storage conditions, or both, do not influence the concentration of the analyte(s) of interest in the matrix of concern or that the data generated from analysis of the samples meet the data quality objectives.

X2.3 Core barrel liners covered with either polytetrafluoroethylene or aluminum foil, are not recommended for the storage of samples intended for VOC analysis because they do not have air-tight seals. Core barrel liners are inserts placed inside a split spoon or hallow tube sampler (see Practice D 3550). When used for storage, the sampled material exposed at the ends of these liners is often covered with thin sheets of polytetrafluoroethylene or aluminum foil which are held in place with a polyethylene cap that is some times further wrapped (edges of cap and ends of the sheet) with a sealing tape. Core barrel liners, which are effective at maintaining the sampled materials' structure often fail to retain VOCs when these barriers are used to cover the open ends (**3**). Sheets of polytetrafluoroethylene that are only 2 or 3 mil (mil = 0.001 in.) thick are easily penetrated by VOC vapors. Elastic formulations of polytetrafluoroethylene (for example, white plumber's tape) show much greater rates of VOC penetration than non-elastic semi-translucent formulations. Aluminum foil fails because an airtight seal is often not achieved when this material is folded over the end of a core liner and it sometimes corrodes allowing small holes to form. The addition of plastic caps and sealing tape does not improve the performance of either polytetrafluoroethylene or aluminum foil, since pliable plastics sorb VOCs and sealing tapes often contain VOCs in their adhesive, which serves as a potential artifact source (**3**).

X2.4 Preliminary tests with both field and laboratory treated samples have confirmed the inadequacy of these coverings to prevent significant VOC losses after only a few (two or four) days of storage at 4 ± 2°C. In general, poorer performances were observed for aluminum foil and an elastic formulation of polytetrafluoroethylene as compared to a non-elastic formulation of polytetrafluoroethylene. Overall, it is expected that the performance of core barrel liners to serve as a storage chamber is subject to several variables, for example, the materials used to cover the ends, vapor pressure of the contaminant VOC(s), length of storage, the texture of the bulk material collected, and the method used to obtain and transfer subsamples.

X3. TESTING FOR CARBONATES

X3.1 Testing for the presence of carbonates in waste and soil samples using an effervescence test:

X3.1.1 Moisten approximately 1 g of sample material that has been placed on a watch glass or similar surface with water, than stir to remove any trapped air.

X3.1.2 Add drop-wise a cold solution of 4N HC1 to the moistened material while observing for effervescence using a hand lens. If effervescence (rapid formation of bubbles) is observed, then preservation by acidification is not appropriate. If effervescence is not observed samples can be preserved by acidification. To determine the amount of hydrochloric acid or sodium bisulfate required to give a pH of 2 or less add the prescribed (usually 5 g) amount of sample material to a vessel containing 5 mL of water, mix thoroughly, then slowly add one of these acids while monitoring that pH. The volume or weight of acid determined to be necessary to achieve a pH of 2 or less should then be added to vials containing 5 mL of water, prior to adding samples of similar materials.

REFERENCES

(1) Minnich, M., Behavior and Determination of Volatile Organic Compounds in Soil: A Literature Review. EPA 600/R-93/140, National Exposure Research Laboratory, Las Vegas, NV, 1993.

(2) Lewis, T.E., Crockett, A.B., Siegrist, R.L., and Zarrabi, K., Soil Sampling and Analysis for Volatile Organic Compounds, *Environmental Monitoring and Assessment*, 1994, 30: 213-246.

(3) Hewitt, A.D., Lukash, N.J.E., Sampling for In-Vial Analysis of Volatile Organic Compounds in Soil, *Am. Environ. Lab*, 1996, 7: 15-19.

(4) Siegrist, R.L., Jenssen, P.D., Evaluation of Sampling Method Effects on Volatile Organic Compounds Measurements in Contaminated Soils, *Environ. Sci. Technol.*, 1990. 24: 1387-1392.

(5) Hewitt, A.D., Chemical Preservation of Volatile Organic Compounds in Soil, 1995, *Environ. Sci. Technol.*, 1995, 31, 67-70.

(6) Urban, M.J., Smith, J.S., Schultz, E.K., Dickinson, R.K., Volatile Organic Analysis of a Soil Sediment or Waste Sample, Fifth Annual Waste Testing and Quality Assurance Symposium, U.S. Environmental Protection Agency, Washington, DC, 1989, II-87-II-101.

(7) Illias, A.M., Jeager, C., Evaluation of Sampling Techniques for the Analysis of Volatile and Total Recoverable Petroleum Hydrocarbons (TRPH) by IR, GC, and GC/MS Methods. In: Kostecki PT, Calabrese EJ, Bonazountas M, eds, *Hydrocarbon Contaminated Soils*, Vol 3, Chelsea, MI, Lewis Publishers, 1993, 3: 147-165.

(8) Hewitt, A.D., Jenkins, T.F., Grant, C.L., Collection, Handling, and Storage: Keys to Improved Data Quality for Volatile Organic Compounds in Soil. *Am. Environ. Lab*, 1995, 7: 25-28.

(9) Liikala, T.L., Olsen, K.B., Teel, S.S., Lanigan, D.C., Volatile Organic Compounds: Comparison of Two Sample Collection and Preservation Methods. *Environ. Sci. Technol.*, 1996, 30: 3441-7.

(10) Smith, J.S., Enj, L., Comeau, J., Rose, C., Schulte, R.M., Barcelona, M.J., Kloop, K., Piligram, M.J., Minnich, M., Feenstra, S., Urban, M.J., Moore, M.B., Maskarinec, M.P., Siegrist, R., Parr, J., and Claff, R.E., Volatile Organic Compounds in Soil: Accurate and Representative Analysis, *3152-4/96/0693* American Chemical Society, 1996, pp. 693-704.

(11) Sawhney, B.L., Pignatello, J.J., and Steinberg, S.M., Determination of 1,2 dibromoethane (EDB) in Field Soils: Implications for Volatile Organic Compounds, *J. Environ. Qual.*, 1998, 17: 149-52.

(12) Hewitt, A.D., Miyares, P.H., Leggett, D.C., Jenkins, T.F., Comparison of Analytical Methods for Determination of Volatile Organic Compounds, *Environ. Sci. Technol.*, 1992, 26: 1932-8.

(13) Askari, M.D.F., Maskarinec, M.P., Smith, S.M., Beam, P.M., and Travis, C.C., Effectiveness of Purge-and-trap for Measurement of Volatile Organic Compounds in Aged Soils, *Analytical Chemistry*, 1996, 68: 3431-3433.

(14) Minnich, M.M., Zimmerman, J.H., and Schumacher, B.A., Extraction Methods for Recovery of Volatile Organic Compounds from Fortified, Dry Soils, *J. Assoc. Off. Anal. Chem.*, 1996, 79: 1198-1204.

(15) Ball, W.P., Xia, G., Durfee, D.P., Wilson, R.D., Brown, M.J., and Mackay, D.M., Hot Methanol Extraction for the Analysis of Volatile Organic Chemicals in Subsurface Core Samples from Dover Air Force Base, Delaware, *Ground Water Monitoring and Remediation*, 1997, 17: 104-21.

(16) Hewitt, A.D., Comparison of Sample Preparation Methods for the Analysis of Volatile Organic Compounds in Soil Samples: Solvent Extraction versus Vapor Partitioning, 1998, *Environ. Sci. Technol.*, 1998, 32: 143-9.

Designation: D 4646 – 87 (Reapproved 2001)

Standard Test Method for 24-h Batch-Type Measurement of Contaminant Sorption by Soils and Sediments[1]

This standard is issued under the fixed designation D 4646; the number immediately following the designation indicates the year of original adoption or, in the case of revision, the year of last revision. A number in parentheses indicates the year of last reapproval. A superscript epsilon (ε) indicates an editorial change since the last revision or reapproval.

1. Scope

1.1 This test method describes a procedure for determining the sorption affinity of waste solutes by unconsolidated geologic material in aqueous suspension. The waste solute may be derived from a variety of sources such as wells, underdrain systems, or laboratory solutions such as those produced by waste extraction tests like the Method D 3987 shake extraction method.

1.2 This test method is applicable in screening and providing relative rankings of a large number of geomedia samples for their sorption affinity in aqueous leachate/geomedia suspensions. This test method may not exactly simulate sorption characteristics that would occur in unperturbed geologic settings.

1.3 While this procedure may be applicable to both organic and inorganic constituents, care must be taken with respect to the stability of the particular constituents and their possible losses from solution by such processes as degradation by microbes, light, or hydrolysis. This test method should not be used for volatile chemical constituents (see 6.1).

1.4 *This standard does not purport to address all of the safety concerns, if any, associated with its use. It is the responsibility of the user of this standard to establish appropriate safety and health practices and determine the applicability of regulatory limitations prior to use.*

2. Referenced Documents

2.1 *ASTM Standards:*

D 1129 Terminology Relating to Water[2]

D 1193 Specification for Reagent Water[2]

D 2216 Test Method for Laboratory Determination of Water (Moisture) Content of Soil and Rock[3]

D 3987 Test Method for Shake Extraction of Solid Waste with Water[4]

D 4319 Test Method for Distribution Ratios by the Short-Term Batch Method[3]

3. Terminology

3.1 *Definitions*—For definition of terms used in this test method refer to Terminology D 1129.

3.1.1 *solute*—chemical species (for example, ion, molecule, etc.) in solution.

3.1.2 *sorbate*—chemical species sorbed by a sorbent.

3.1.3 *sorbent*—a substance that sorbs the solute from solution (for example, soil, sediment, till, etc.).

3.1.4 *sorption*—depletion of an amount of solute initially present in solution by a sorbent.

3.1.5 *sorption affinity*—the relative degree of sorption that occurs by a geomedia.

3.1.6 *unconsolidated geologic material (geomedia)*—a loosely aggregated solid natural material of geologic origin (for example, soil, sediment, till, etc.).

3.2 *Definitions of Terms Specific to This Standard:*

3.2.1 *distribution coefficient, K_d*—is defined identically to R_d, except it is considered to be an equilibrium value and independent of the concentration of solute (that is, linear sorption curve).

3.2.2 *distribution ratio (R_d)*—the ratio of the concentration of solute sorbed on the soil or other geomedia divided by its concentration in solution. A 24-h R_d is the analogous ratio evaluated after 24 h of contact of the solute with the geomedia. The R_d value is calculated as follows:

$$R_d = \frac{\text{(mass of solute sorbed per unit mass of geomedia)}}{\text{(mass of solute in solution per unit volume of solution)}} = \frac{\mu g/g}{\mu g/mL} = \frac{mL}{g} \quad (1)$$

The dimensions of R_d reduce to units of volume per mass. It is convenient to express R_d in units of millilitres (or cubic centimetres) of solution per gram of geomedia. Dissimilar R_d values may be obtained if different initial solute concentrations are used, depending on the sorption behavior of the solute and the properties of the geomedia (that is, nonlinear sorption curve). This concentration dependency may be absent where the solute concentrations are sufficiently low or the characteristics of the particular solute-sorbent combination yield R_d values that are independent of the concentration of solute (that

---

[1] This test method is under the jurisdiction of ASTM Committee D34 on Waste Management and is the direct responsibility of Subcommittee D34.01.06 on Analytical Methods.

Current edition approved March 3, 1987. Published April 1987. Originally published as ES 10 – 85. Last previous edition ES 10 – 85.

[2] *Annual Book of ASTM Standards*, Vol 11.01.

[3] *Annual Book of ASTM Standards*, Vol 04.08.

[4] *Annual Book of ASTM Standards*, Vol 11.04.

is, linear sorption curve).

4. Summary of Test Method

4.1 Distilled water, natural water, waste leachate, or other aqueous solution containing a known concentration of a solute is mixed with a known amount of unconsolidated geologic material (geomedia) for 24 h. Changes in solute concentrations are used to calculate a distribution ratio (R_d).

5. Significance and Use

5.1 This test method is meant to allow for a rapid (24 h) index of a geomedia's sorption affinity for given chemicals or leachate constituents. A large number of samples may be run using this test method to determine a comparative ranking of those samples, based upon the amount of solute sorbed by the geomedia, or by various geomedia or leachate constituents. The 24-h time is used to make the test convenient and also to minimize microbial degradation which may be a problem in longer-timed procedures. Due to this time constraint, the final (24-h) concentration should not be confused with that of an equilibrium or steady-state concentration. While R_d values are directly applicable for screening and comparative ranking purposes, their use in predictive field applications generally requires the assumption that $R_d = K_d$; the validity of this assumption must be carefully evaluated by qualified personnel.

5.2 While this test method may be useful in determining 24-h R_d values for nonvolatile organic constituents, interlaboratory testing has been carried out only for the nonvolatile inorganic species, arsenic and cadmium. However, the procedure has been tested for single laboratory precision with polychlorinated biphenyls (PCBs) and is believed to be useful for all stable and nonvolatile inorganic, and organic constituents. This test method is not considered appropriate for volatile constituents.

5.3 The 24-h time limit may be sufficient to reach a steady-state R_d. However, to report this determination as a steady-state R_d (that is, $R_d = K_d$), the relevant time studies must be carried out to document the development of steady-state conditions within the 24-h time period. Tests exceeding the 24-h time period are beyond the scope of this test method. Refer to Test Method D 4319, for an alternate procedure of longer duration.

6. Interferences

6.1 When dealing with solutes of unknown stability either in contact with the geomedia or when used as blanks, care must be taken to determine if volatilization, hydrolysis, photodegradation, microbial degradation, oxidation-reduction (that is, Cr^{3+} to Cr^{6+}) or other physicochemical processes are operating at a significant rate within the time frame of the procedure. The stability and hence loss from solution may affect the outcome of this procedure if the aforementioned reactions are significant. The compatibility of the method and the solute of interest may be assessed by determining the differences between the initial solute concentration (see 9.8) and the final blank concentration of the solute (see 9.15). If this difference is greater than the expected precision of the method (10 %), then the R_d value generated may be unreliable and must be carefully evaluated.

7. Apparatus

7.1 *Agitation Equipment*—The agitation equipment to be used is the rotary solid waste extractor[5] specified in Test Method D 3987.

7.2 *Phase Separation Equipment*—A filtration apparatus made of materials compatible with the solutions being filtered and equipped with a 0.45-μm pore size membrane filter, or a constant temperature centrifuge capable of separating particles with diameters greater than 0.1 μm (see Section 9). If organic compounds are being measured, the filtration apparatus, centrifuge tubes etc., should be compatible with the compounds being measured (that is, glass or stainless steel). Sorption of solute onto the filtration membrane may be significant for some solutes, and must be evaluated by the use of blanks through all steps of the procedure.

7.3 *Containers*—Round, wide-mouth bottles compatible with the rotary extractor (Test Method D 3987) and of composition suitable to the nature of the solute(s) under investigation and the analysis to be performed will be used. For nonvolatile inorganic constituents, high-density, linear polyethylene bottles should be used with the size of the bottle dictated by sample size, and the need for the solution to occupy 70 to 80 % of the container volume (that is, 125 mL, 250 mL, or 2-L bottles for sample sizes of 5, 10, or 70 g respectively). For nonvolatile organic constituents, TFE-fluorocarbon, glass bottles, or stainless steel containers with water-tight closures made of chemically inert materials should be used with size requirements being the same as for nonvolatile inorganics. Containers should be cleaned in a manner consistent with the analyses to be performed. Samples of the solutions to be analyzed should be stored in similar chemically compatible bottles.

7.4 *Balance*, having a minimum capacity of 70 g and a sensitivity of ±0.005 g shall be used.

8. Reagents

8.1 *Purity of Reagents*—Reagent grade chemicals shall be used in all tests. Unless otherwise indicated, it is intended that all reagents shall conform to the specifications of the American Chemical Society, where such specifications are available.[6] Other grades may be used, provided it is first ascertained that the reagent is of sufficiently high purity to permit its use without lessening the accuracy of the determination.

8.2 *Purity of Water*— Unless otherwise indicated, references to water shall be understood to mean Type IV reagent water of Specification D 1193.

9. Procedure

9.1 Geomedia samples are spread out on a flat surface, no more than 2 to 3 cm deep, and allowed to air dry for 7 days or

[5] Diamondstone, B. T., Burke, R. W., and Garner, E. L., "Improved Leach Measurements on Solid Wastes," *ASTM Standardization News*, June 1982, pp. 28–33.

[6] *Reagent Chemicals, American Chemical Society Specifications*, American Chemical Society, Washington, DC. For suggestions on the testing of reagents not listed by the American Chemical Society, see *Analar Standards for Laboratory Chemicals*, BDH Ltd., Poole, Dorset, U.K., and the *United States Pharmacopeia and National Formulary*, U.S. Pharmaceutical Convention, Inc. (USPC), Rockville, MD.

until constant weight (a change that is less than 5 %/24-h period) is achieved (do *not* oven dry).

9.2 After the sample has air dried, it is passed through a 2-mm screen sieve. Large aggregates are to be crushed, without grinding, using a clean mortar and a rubber-tipped pestle.

9.3 Mix the sieved material until the sample is homogeneous. Use a riffle splitter, or other unbiased splitting procedure, to obtain subsamples of appropriate size.

9.4 Remove subsamples and determine the moisture content of the air-dried sample (refer to Test Method D 2216).

9.5 Determine the mass of geomedia sample, corrected for moisture content:

Determination of air-dried soil mass equivalent to the desired mass of oven-dried soil:

$$A = M_s[1 + (M/100)] \quad (2)$$

where:

A = air dry soil mass
M_s = mass of oven-dried soil desired, and
M = moisture, %.

9.6 Place between 5 and 70 g (oven-dried basis) of the weighed air-dried sample into the appropriate container. The samples should be weighed to a minimum of three significant figures.

9.7 Add to the container an amount of solute solution necessary to yield a 1:20 soil-to-solution ratio. This is determined on the oven-dried basis:

Determination of solution volume needed per sample for a soil-to-solution ratio of 1:20:

$$V = (M_s \times 20)/\rho \quad (3)$$

where:

ρ = density of solution, g/cm^3,
V = volume of solution per sample, cm^3, and
M_s = mass of soil to be used, g, (oven-dried basis).

9.8 Retain a separate, appropriately preserved aliquot of the initial solute solution for analysis.

9.9 Close the container and place it on the rotary extractor (Test Method D 3987).

9.10 Agitate continuously for 24 ± 0.5 h at 29± 2 r/min at room temperature (22 ± 5°C).

9.11 Open the container. Note the temperature of the solution and any changes in the sample or solution (that is, color, odor, etc.).

9.12 Separate the solution phase from the majority of the solid phase by decantation.

9.13 Filter the solution phase through a 0.45-µm pore size membrane filter (see 7.2), or centrifuge a subsample at the predetermined rate of rotation and time for the centrifugation equipment employed at constant temperature (the temperature recorded after 24 h):

$$t = 9/2\left(\frac{\eta}{\omega^2 r_p^{\,2}(\rho_p - \rho)}\right)\ln(R_b/R_t) \quad (4)$$

where:

ω^2 = $\frac{4_\pi{}^2(r/min)^2}{60}$
= angular velocity,
r_p = particle radius, cm,
η = viscosity of water, 8.95 × 10^{-3} g/s-cm at 25°C,
ρ_p = particle density,
ρ = density of solution,
r/min = revolutions per minute,
R_t = distance from center of centrifuge rotor to top of solution in centrifuge tube, cm,
R_b = distance from center of centrifuge rotor to bottom of centrifuge tube, cm, and
t = time, min.

To remove particles >0.1-µm radius and 2.65-g/cm^3 density from solution:

$$t = \left(\frac{3.71 \times 10^8}{(r/min)^2}\right)\ln(R_b/R_t) \quad (5)$$

Note that if filtration is used, the affinity of the filtration membrane for the solute must be evaluated. Failure to do so, may lead to erroneous results.

9.14 After a clear solution has been obtained, place an aliquot in an appropriate container (see 7.3), and analyze or store in a refrigerator at 4 ± 2°C until analyzed.

9.15 Each geomedia sample is to be subjected to the procedure in three or more replicates. The number of blanks (that is, solute solution without geomedia) carried through all steps of the procedure should be a minimum of 5 % of the total number of geomedia samples, but not less than three.

10. Calculation

10.1 Calculate the distribution ratio as follows:

$$R_d = \frac{(A - B)V}{(M_s)B} \quad (6)$$

where:

A = initial concentration of the solute defined as the mean concentration of the blanks, µg/mL,
B = final concentration of the solute after 24 h in contact with the geomedia, µg/mL,
V = volume of solution used, mL,
M_s = mass of soil expressed on an oven-dried basis, g, and
R_d = distribution ratio, mL/g.

11. Report

11.1 The R_d value must be clearly marked as nonequilibrium 24-h distribution ratio.

11.2 Both the initial solute concentration (A in 10.1) and the final solute concentration (B in 10.1) must be reported.

11.3 The initial and final solute concentration for each blank (solution without geomedia) must be reported.

11.4 The mass of the sorbent (M_s in 10.1), volume of solution (V in 10.1), and the room temperature at which the rotary extractor was operated must be reported.

11.5 Report the temperature of the solution and any changes noted in 9.11.

11.6 Note and report negative R_d values when and if they occur. Negative R_d values may occur if the geomedia contains the test solute prior to the application of the method.

11.7 It is suggested that the pH of the sorbent-solute mixture be determined prior to separating the sorbent from the liquid and reported where feasible.

12. Precision and Bias

12.1 An interlaboratory round-robin test was conducted at a soil-to-solution ratio of 1:20 using 70.0 g of soil. Intralaboratory testing using a 1:20 soil-to-solution ratio was carried out with no significant loss of precision for soil masses of 5.00, 10.0, and 70.0 g. Therefore, it is specified that the soil-to-solution ratio be 1:20 with the working mass of soil (on an oven-dried basis) between 5 and 70 g.

12.2 *Precision*:

12.2.1 The precision of this test method is limited by the ability to obtain a homogeneous sample of geomedia, and the precision of the various methods used to carry out the procedure (that is, mass determinations, initial and 24-h concentration of constituents).

12.2.2 A comprehensive overall precision statement, covering all species, is not feasible. Interlaboratory testing of this procedure, using cadmium and arsenic as sorbates, with five independent laboratories, and single laboratory testing with PCBs indicated that a coefficient of variation of less than 10 % is obtainable.

12.2.3 For a summary of interlaboratory testing for the 24-h batch-type R_d determination see Table 1.

12.2.4 For a summary of single laboratory testing for an organic solute see Table 2.

TABLE 1 Summary of Interlaboratory Testing for the 24-h Batch-Type R_d Determination

Initial Concentration, μg/mL	Soil 1 10	Soil 1 200	Soil 2 100	Soil 3 100	Soil 4 100
Cadmium Carbonate (as $CdCO_3$):					
$\bar{X}$ R_d, mL/g	1568	96.3	69.50	69.74	28.94
Standard deviation, mL/g	±156	±6.32	±7.73	±7.87	±2.64
Coefficient of variation, %	9.95	6.56	11.13	11.29	9.14
Number of replicates	12	12	10	12	9
Arsenic (as KH_2AsO_4):					
$\bar{X}$ R_d, mL/g	15.42	2.75	4.95	3.25	2.99
Standard deviation, mL/g	±0.92	±0.85	±0.27	±0.26	±0.20
Coefficient of variation, %	5.99	30.9	5.53	7.88	6.52
Number of replicates	12	12	12	12	12

12.3 *Bias*—A determination of the bias for this procedure is not possible since no standard soil or alternate technique exists.

TABLE 2 Summary of Single Laboratory for an Organic Solute

	Soil 1[A]	Soil 1[B]	Soil 2[C]	Soil 2[D]
Initial Concentration, µg/mL	0.216 (0 % Acetone)	0.187 (20 % Acetone)	0.216 (0 % Acetone)	0.187 (20 % Acetone)
PCBs (as Aroclor 1242):				
$\bar{X}$ R_d, mL/g	798	78.62	23.83	2.28
Standard deviation, mL/g	±18.41	±3.45	±1.48	±0.11
Coefficient of variation, %	2.31	4.39	6.21	4.82
Number of replicates	4	4	4	4

[A] Soil 1—Catlin silt loam.
[B] Soil 2—Sangamon paleosol.
[C] Soil 3—Kaolinite clay.
[D] Soil 4—Vandalia till, unaltered phase.

Designation: D 4687 – 95 (Reapproved 2001)

Standard Guide for General Planning of Waste Sampling[1]

This standard is issued under the fixed designation D 4687; the number immediately following the designation indicates the year of original adoption or, in the case of revision, the year of last revision. A number in parentheses indicates the year of last reapproval. A superscript epsilon (ε) indicates an editorial change since the last revision or reapproval.

INTRODUCTION

The analysis and testing of solid waste requires collection of adequately sized, representative samples. Wastes are found in various locations and physical states. Therefore, each sampling routine must be tailored to fit the waste and situation. Wastes often occur as nonhomogeneous mixtures in stratified layers or as poorly mixed conglomerations. For example, wastes are commonly stored or disposed of in surface impoundments with stratified or layered sludges covered by ponded wastewater. In these situations, the collector may be faced with sampling the wastewater, the sludge, and some depth of soil beneath the sludges. Collecting representative samples in these situations requires a carefully assessed, well-planned, and well-executed sampling routine.

Currently, Subcommittee D34.01 is working on practices for sampling wastes from a variety of different sampling locations and situations. Also in progress is a practice for containerization, preservation, and holding times for waste samples. As these documents are approved by ASTM, reference to these standards will be made in this general guide on waste sampling. Further, Subcommittee D34.01 recommends this guide be used in conjunction with the new waste sampling practices when available in print by ASTM.

1. Scope

1.1 This guide provides information for formulating and planning the many aspects of waste sampling (see 1.2) which are common to most waste sampling situations.

1.2 The aspects of sampling which this guide addresses are as follows:

	Section
Safety plans	4
Sampling plans	5
Quality assurance considerations	6
General sampling considerations	7
Preservation and containerization	8
Cleaning equipment	9
Labeling and shipping procedures	10
Chain-of-custody procedure	11

1.3 This guide does not provide comprehensive sampling procedures for these aspects, nor does it serve as a guide to any specific application. It is the responsibility of the user to assure that the procedures used are proper and adequate.

1.4 *This standard does not purport to address all of the safety concerns, if any, associated with its use. It is the responsibility of the user of this standard to establish appropriate safety and health practices and determine the applicability of regulatory limitations prior to use.* For more specific precautionary statements see 3.2, 3.3, and Section 4.

2. Referenced Documents

2.1 *ASTM Standards:*

E 122 Practice for Choice of Sample Size to Estimate a Measure of Quality for a Lot or Process[2]

2.2 *Other Document:*

EPA-SW-846 Test Methods for Evaluating Solid Waste, Physical/Chemical Methods[3]

3. Significance and Use

3.1 The procedures covered in this guide are general and provide the user with information helpful for writing sampling plans, safety plans, labeling and shipping procedures, chain-of-custody procedures, general sampling procedures, general cleaning procedures, and general preservation procedures.

3.2 For purposes of this guide, it is assumed that the user has knowledge of the waste being sampled and the possible safety hazards.

3.3 This guide is not to be used when sampling sites or wastes where safety hazards are unknown. In such cases, the user must use other more appropriate procedures.

4. Hazards

4.1 Proper safety precautions must always be observed when sampling wastes. Persons collecting samples must be

[1] This guide is under the jurisdiction of ASTM Committee D34 on Waste Management and is the direct responsibility of Subcommittee D34.01.01 on Planning for Sampling.

Current edition approved March 15, 1995. Published May 1995. Originally published as D 4687 – 87. Last previous edition D 4687 – 87.

[2] *Annual Book of ASTM Standards*, Vol 14.02.

[3] Available from Superintendent of Documents, U.S. Printing Office, Washington, DC 20402.

aware that the waste can be a strong sensitizer and can be corrosive, flammable, explosive, toxic, and capable of releasing extremely poisonous gases. The background information obtained about the waste should be helpful in deciding the extent of safety precautions to be observed and in choosing protective equipment to be used. The information obtained should be checked for hazardous properties against such references as "Dangerous Properties of Industrial Materials" the "March Index," the "Condensed Chemical Dictionary," and the "Toxic and Hazardous Industrial Chemicals Safety Manual for Handling and Disposal with Toxicity and Hazardous Data."

NOTE 1—The following safety precautions are not comprehensive. Rather, they provide additional guidance on health and safety to complement professional judgment and experience.

4.2 Personnel should wear protective equipment when response activities involve known or suspected atmospheric contamination; when vapors, gases, or airborne particulates may be generated; or when direct contact with skin-affecting substances may occur. Respirators can protect lungs, gastrointestinal tract, and eyes against air toxicants. Chemical-resistant clothing can protect the skin from contact with skin-destructive and -absorbable chemicals. Good personal hygiene limits or prevents ingestion of material.

4.2.1 Equipment to protect the body against contact with known or anticipated chemical hazards has been divided into four categories according to the degree of protection afforded:

4.2.1.1 *Level A*—Should be worn when the highest level of respiratory, skin, and eye protection is needed.

4.2.1.2 *Level B*—Should be selected when the highest level of respiratory protection is needed, but a lesser level of skin protection. Level B protection is the minimum level recommended on initial site entries until the hazards have been further defined by on-site studies and appropriate personnel protection utilized.

4.2.1.3 *Level C*—Should be selected when the type(s) of airborne substance(s) is (are) known, the concentrations(s) is measured, and the criteria for using air-purifying respirators are met.

4.2.1.4 *Level D*—Should not be worn on any site with respiratory or skin hazards. It is primarily a work uniform providing minimal protection.

4.2.2 The level of Protection selected should be based primarily on the following:

4.2.2.1 Type(s) and measured concentration(s) of the chemical substance(s) in the ambient atmosphere and its toxicity and

4.2.2.2 Potential or measured exposure to substances in air, splashes of liquids, or other direct contact with material due to work being performed.

4.2.2.3 In situations where the type(s) of chemical(s), concentration(s), and possibilities of contact are not known, the appropriate Level of Protection must be selected based on professional experience and judgment until the hazards can be better characterized.

4.2.3 *Level A Protection—Personnel Protective Equipment:*

(a) (*a*) Pressure-demand, self-contained breathing apparatus, approved by the Mine Safety and Health Administration (MSHA) and National Institute of Occupational Safety and Health (NIOSH),

(b) (*b*) Fully encapsulating chemical-resistant suit,

(c) (*c*) Coveralls,[4]

(d) (*d*) Long cotton underwear,[4]

(e) (*e*) Gloves (outer), chemical-resistant,

(f) (*f*) Gloves (inner), chemical-resistant,

(g) (*g*) Boots, chemical-resistant, steel toe and shank. (Depending on suit construction, worn over or under suit boot),

(h) (*h*) Hard hat[4] (under suit),

(i) (*i*) Disposable protective suit, gloves, and boots[4] (worn over fully encapsulating suit), and

(j) (*j*) Two-way radio communications (intrinsically safe).

4.2.3.1 The fully encapsulating suit provides the highest degree of protection to skin, eyes, and respiratory system if the suit material is resistant to the chemical(s) of concern during the time the suit is worn or at the measured or anticipated concentrations, or both. While Level A provides maximum protection, the suit material may be rapidly permeated and penetrated by certain chemicals from extremely high air concentrations, splashes, or immersion of boots or gloves in concentrated liquids or sludges. These limitations should be recognized when specifying the type of chemical-resistant garment. Whenever possible, the suit material should be matched with the substance it is used to protect against.

4.2.3.2 Many toxic substances are difficult to detect or measure in the field. When such substances (especially those readily absorbed by or destructive to the skin) are known or suspected to be present and personnel contact is unavoidable, Level A protection should be worn until more accurate information can be obtained.

4.2.4 *Level B Protection—Personnel Protective Equipment*:

(a) (*a*) Pressure-demand, self-contained breathing apparatus (MSHA/NIOSH approved),

(b) (*b*) Chemical-resistant clothing (overalls and long-sleeved jacket; coveralls; hooded, one- or two-piece chemical-splash suit; disposable chemical-resistant coveralls),

(c) (*c*) Coveralls,[4]

(d) (*d*) Gloves (outer), chemical-resistant,

(e) (*e*) Gloves (inner), chemical-resistant,

(f) (*f*) Boots, chemical-resistant, steel toe and shank,

(g) (*g*) Boots (outer), chemical-resistant (disposable, worn over permanent boots),[4]

(h) (*h*) Hard hat (face shield),[4] and

(i) (*i*) Two-way radio communications (intrinsically safe).

4.2.4.1 Level B equipment provides a high level of protection to the respiratory tract, but a somewhat lower level of protection to skin. The chemical-resistant clothing required in Level B is available in a wide variety of styles, materials, construction detail, permeability, etc. These factors all affect the degree of protection afforded. Therefore, a specialist should select the most effective chemical-resistant clothing (and fully encapsulating suit) based on the known or anticipated hazards or job function, or both.

4.2.4.2 For initial site entry and reconnaissance at an open

[4] Equipment is optional.

site, approaching whenever possible from the upwind direction, Level B protection (with good quality, hooded, chemical-resistant clothing) should protect response personnel, providing the conditions described in selecting Level A are known or judged to be absent.

4.2.5 *Level C Protection—Personnel Protective Equipment*:

(a) (*a*) Full-face, air purifying, canister-equipped respirator (MSHA/NIOSH approved),

(b) (*b*) Chemical-resistant clothing (coveralls; hooded, two-piece chemical splash suit; chemical-resistant hood and apron; disposable chemical-resistant coveralls),

(c) (*c*) Coveralls,[4]

(d) (*d*) Gloves (outer), chemical-resistant,

(e) (*e*) Gloves (inner), chemical-resistant,[4]

(f) (*f*) Boots, chemical resistant, steel toe and shank,

(g) (*g*) Boots (outer), chemical-resistant (disposable, worn over permanent boots),[4]

(h) (*h*) Hard hat (face shield),[4]

(i) (*i*) Escape mask[4], and

(j) (*g*) Two-way radio communications (intrinsically safe).

4.2.5.1 Level C protection is distinguished from Level B by the equipment used to protect the respiratory system, assuming the same type of chemical-resistant clothing is used. The main selection criterion for Level C is that conditions permit wearing air-purifying devices.

4.2.5.2 Total unidentified vapor/gas concentrations of 5 ppm above background require Level B protection. Only a qualified individual should select Level C (air-purifying respirators) protection for continual use in an unidentified vapor/gas concentration of background to 5 ppm above background.

4.2.6 *Level D Protection—Personnel Protective Equipment*:

(a) (*a*) Coveralls,

(b) (*b*) Gloves,[4]

(c) (*c*) Boots/shoes, leather or chemical-resistant, steel toe and shank,

(d) (*d*) Boots, chemical-resistant (disposable worn over permanent boots),[4]

(e) (*e*) Safety glasses or chemical splash goggles,[4]

(f) (*f*) Hard hat (face shield),[4] and

(g) (*g*) Escape mask.[4]

4.2.6.1 Level D protection is primarily a work uniform. It should be worn in areas where: (1) only boots can be contaminated, or (2) there are no inhalable toxic substances.

4.3 Personnel should not eat, drink, or smoke during or after sampling until after decontamination steps are taken. Sampling personnel should be trained in safety aspects of hazardous waste sampling.

4.4 Testing air emission for determining the vapor/gas concentrations can be accomplished through the use of a portable organic vapor analyzer. The probe should be held 1 to 2 in. above the sampling point. Follow manufacturers operating instructions for proper calibration, use, and care.

5. Sampling Plans

5.1 A sampling plan is a scheme or design to locate sampling points so that suitable representative samples descriptive of the waste body can be obtained. Developmentof sampling plans requires the following:

5.1.1 Review of background information about the waste and site.

5.1.2 Knowledge of the waste location and situation.

5.1.3 Decisions as to the types of samples needed.

5.1.4 Decisions as to the sampling design required.

5.2 Background data on the waste is extremely helpful in preassessment of the waste's composition, hazards, and extent. (See Notes 2 and 3.)

NOTE 2—If after researching the available background information the user cannot obtain from the material enough information about the waste to determine the probable composition and probable hazards, then the user should use other procedures. Such situations are beyond the scope of this guide.

NOTE 3—The background information is needed to determine necessary safety equipment, safety procedures, sampling equipment and sampling design, and procedures to be used.

5.2.1 Possible sources of information on the site and waste include the following:

5.2.1.1 File searches of state and local records including waste manifests, waste approvals, land permit applications.

5.2.1.2 File searches of generator records (if the generator can be identified) including chemical analyses, safety data sheets, design drawings, and manufacturing process information.

5.2.1.3 File searches of treatment, storage, disposal, and transport facilities. Records involved with handling the waste.

5.2.1.4 Researching published data concerning the site such as scientific journal articles, EPA publications, and newspaper stories. Newspapers are the most likely source but the information is seldom very technical.

5.2.1.5 Interviews of key people such as past and present employees of the site or generator, state and local officials, residents of the area, etc.

5.2.1.6 Aerial photographs provide a historical record of the site development. Many federal agencies conduct aerial surveys that are available to the public. Some of these agencies include the following:

(a) (*a*) U. S. Department of Agriculture

(b) (*b*) Soil Conservation Service (USDA-SCS).

(c) (*c*) U. S. Geological Survey.

(d) (*d*) U. S. Forest Service.

(e) (*e*) National Air and Space Administration (NASA).

(f) (*f*) National Oceanic and Atmospheric Administration (NOAA).

(g) (*g*) National Weather Service.

(h) (*h*) Corps of Engineers.

(i) (*i*) Agricultural Stabilization and Conservation Service.

5.2.1.7 Published maps can also provide a historical record of the site development such as topographic, soil, and county maps.

5.3 Waste location and site conditions greatly influence a sampling plan. The most common waste locations may include lagoons, landfills, pipes, point discharges, piles, drums, bins, tanks, and trucks. The site conditions include the physical condition of the waste; that is, whether it is a solid (granular, consolidated, or cohesive), liquid (slurry or flowable sludge), or gas, and it describes under what conditions it was disposed; that is, does it exist as a multiphased waste in a lagoon, tank or drum; is it stratified solids in a lagoon; is it a poorly mixed concoction of municipal garbage and hazardous sludges; or a landfill containing barrels of unknown waste.

5.3.1 Based on these considerations, the collector will have to decide what must be sampled. Each situation is different and requires the best judgement of the user in writing such a plan.

5.4 The types of samples that may be collected are most commonly either composite or single samples. The sample collector must decide considering the complexity of the waste location, the situation, and the financial resources, and what types of samples will best provide representative samples for reliable measurements.

5.4.1 A composite sample, sometimes referred to as a batch sample, is a well-mixed collection of subsamples of the same waste taken from different points. A composite sample is used most commonly in determining an average measure of a parameter. Generally, composite samples are taken when differences in the waste exist because of stratification, or because of the simultaneous deposition of different wastes such as in a landfill.

5.4.2 A single sample is a well-mixed sample taken from a single point. It is used to measure a particular parameter or parameter set at a given point or within a unique homogeneous layer or throughout the strata at one or several locations.

5.5 Sampling plans or schemes should be carefully thought out, well in advance of sampling. The most common sampling schemes involve the selection of sampling points using a judgement, a coordinate system, or a grid system.

5.5.1 *Judgement Samples*—This system is commonly used when, because of resource restraints, multiple samples cannot be collected. They are collected by deciding through visual observation or knowledge of the site where a representative sample may be collected. This type of design can be very effective if the collector is familiar and knowledgeable about the site, and if the goal of sampling is merely to establish whether a waste meets some set criteria.

5.5.2 *Coordinate Sampling System*—This system uses a one or two coordinate system and involves collecting samples at random points from the origin of the coordinates. Random numbers can be generated using random number tables available in most statistic texts. The origin of the coordinate system is normally placed at some corner of the site and marked off in steps, feet, yards, etc. for sampling landfills, waste piles, and lagoons. For storage areas containing barrels, the numbers of barrels from the origin are often used as intervals along the coordinate. For sampling from a flowing stream the origin may be taken as time-zero (start), and samples are collected at random time intervals over the period of interest.

5.5.3 *Grid System*—This system also involves taking samples at regular intervals, grid points, along an imaginary grid system laid out over the site. The number of sampling points will vary with the size of the grid. Such sampling schemes are used when a statistically sound sampling program is required. They should be used only when the waste body is known to be homogenous, or when the strata have been defined. If the waste is stratified, a separate grid system may be required for each stratum.

5.6 The proper number of samples required in a statistically sound sampling program can be estimated. This can be done using Eq 1 and by estimating the sample composition and variance either from a pilot sampling effort or knowledgeable judgement. The number of samples required, n, to achieve the desired precision in waste composition is estimated using fundamental statistical concepts, as follows (financial constraints not considered):[3]

$$n = (t^2_{0.80} S^2)/d^2 \quad (1)$$

where:

n = appropriate number of samples to be collected;
$t^2_{0.80}$ = square of the tabulated value of student's t for a two-sided confidence interval and a coverage probability of 0.80 for the unknown mean, with the degrees of freedom defined for the S^2 used to estimate the population variance, σ^2;
S^2 = preliminary estimate of σ^2 obtained from previous samplings, a pilot sampling effort or other information such as the likely range of the population values;
d = deviation to be exceeded only in two cases out of ten in repeated sampling for the quantity $| \bar{X} - T |$, the difference in absolute value between the sample average and a threshold value such as a regulatory limit;
$\bar{X}$ = preliminary estimate of sample average; and
T = threshold value, often the regulatory limit.

5.6.1 The variables in Eq 1 are appropriate only for a given waste type. Therefore, the appropriate number of samples n, required to achieve the desired precision is also applicable only to that same waste type. If two or more waste types are present in the impoundment, either as strata or other segregated wastes, then a value for n should be calculated for each waste.

5.6.2 Although the use of Student's t distribution is based on an underlying normal distribution for the measurements, the robustness of the t statistic for many applications may be relied upon here. If ancillary information seems to indicate that normality may not be a good assumption, then a goodness of fit test should be performed to determine if the assumption of a normal distribution is reasonable. The Lilliefors goodness of fit test, as it applies to the pilot sampling presented here, is described in the Appendix. This test involves examining the data from a sampling and analysis program in order to test the hypothesis that the data are distributed normally. If the Lilliefors test shows the contention of normality is acceptable, it does *not* mean that the parent population is normal. But it does mean that the Student's t distribution does not appear to be an unreasonable approximation to the true unknown distribution. If the Lilliefors test shows that a normal distribution does not adequately fit the data, then further pilot sampling will be required to adequately determine the spatial distributions in the impoundment.

5.6.3 The following hypothetical example illustrates the use of Eq 1:

5.6.3.1 A preliminary study of barium levels in sludge collected from a lagoon generated values of 86, 90, 98, and 104 ppm for barium in four sludge samples. Based on these values and a knowledge of the processes producing the waste, the sludge is judged to be homogeneous (not stratified) within the lagoon. Therefore, preliminary estimates of $\bar{X}$ and s^2 are calculated as follows:

$$\bar{X} = \frac{\sum_{i=1}^{n} X_i}{n} = \frac{86 + 90 + 98 + 104}{4} = 94.50, \text{ and}$$

$$s^2 = \frac{\sum_{i=1}^{n} X_i^2 - (\sum_{i=1}^{n} X_i)^2/n}{n - 1}$$

$$= \frac{35\ 916.00 - 35\ 721.00}{3} = 65.00$$

5.6.3.2 The deviation not to be exceeded for measured barium in the sludge samples, d, is chosen as 5.50 ppm, that is, the difference between the sample average, $\bar{X}$ or 94.5, and the threshold limit, T or 100.0 for barium (assuming 100.0 is the regulatory threshold for barium) is 5.50 ppm.

5.6.3.3 The value of $t_{0.80}$ is obtained from tabulated values of Student's t, as shown in Table 1. Although an assumption of a t distribution would seem to be restrictive, it can be shown that even non-normal populations possessing bell-shaped distributions can be closely approximated by a t distribution. From the preliminary study $n = 4$, and the degrees of freedom, $n - 1$, is 3. Therefore,

$$t_{0.80} = 1.638$$

5.6.3.4 The appropriate number of sludge samples to be collected from the lagoon is,

$$n = t^2_{0.80}s^2/d^2 = \frac{(1.638^2)(65.00)}{5.50^2} = 5.77,$$

or six. That number of samples (plus extra for protection against poor preliminary estimates of $\bar{X}$ and s^2) is collected from the lagoon.

TABLE 1 Tabulated Values of Student's *t* for Evaluating Solid Wastes

Degrees of Freedom, $(n - 1)^A$	Tabulated t Value^B
1	3.078
2	1.886
3	1.638
4	1.533
5	1.476
6	1.440
7	1.415
8	1.397
9	1.383
10	1.372
11	1.363
12	1.356
13	1.350
14	1.345
15	1.341
16	1.337
17	1.333
18	1.330
19	1.328
20	1.325
21	1.323
22	1.321
23	1.319
24	1.318
25	1.316
26	1.315
27	1.314
28	1.313
29	1.311
30	1.310
40	1.303
60	1.296
120	1.289
	1.282

[A] Degrees of freedom, *df*, are equal to the number of samples, *n*, collected from a solid waste less one.

[B] Tabulated *t* values are for a two-tailed confidence interval and a probability of 0.80 (the same values are applicable to a one-tailed confidence interval and a probability of 0.90).

6. Quality Assurance Considerations

6.1 Quality assurance for solid waste sampling should include adherence to the sampling plan and safety plan and in some cases, the use of quality control samples.

6.2 The sampling and safety plans should be well formulated before any actual sampling is attempted. The plans must be consistent with the objectives of the sampling. The sampling plan must include the selected points of sampling and the intended number, volumes, and types of samples to be taken. The safety plan should address the proper clothing and protective equipment, all known hazards associated with the sampling activities, and the measures to be taken to avoid these hazards.

6.3 Four types of quality control samples relate to the quality assurance of field sampling: (1) field blanks, (2) split samples, (3) field rinsates, and (4) field spikes. The selection of the types of quality control samples to be used should be made prior to the sampling event and included in the sampling plan. The nature of the sampling, the intended uses of the data, and the material being sampled all impact upon the selection of quality control samples to be used in an event.

6.3.1 *Field Blanks*:

6.3.1.1 Field blanks are samples prepared in the laboratory using reagent water or other blank matrix and sent with the sampling team. These samples are exposed to the sampling environment and returned with the samples to the laboratory for analysis. The purpose of the field blank is to verify that none of the analytes of interest measured in the field samples resulted from contamination of the samples during sampling.

6.3.1.2 The sampling plan should normally include a minimum of one field blank for each procedure for each sampling event. These samples can be submitted blind to the laboratory to challenge their analytical system or can be shipped with the instruction to hold them unless there is a reason to suspect sample contamination. The submission of blind field blanks would normally be reserved for those situations where the competency of the analytical laboratory was unproven (that is, where a new laboratory was being utilized).

6.3.2 *Split Samples*:

6.3.2.1 Split samples are used to challenge the analytical laboratory performance. Split samples are also used when two different parties are sampling the same site and verification of analytical results is necessary. A split sample is prepared by subsampling a homogenous sample into two or more portions and submitting each portion separately to the analytical laboratory.

6.3.2.2 For liquid matrixes, the material should be placed in a large, clean container and stirred or swirled to ensure thorough mixing of the medium prior to subsampling. For solid media, a sufficient quantity must be removed, mixed with clean utensils, and subsampled. Sufficient mixing of the sample should be accomplished to ensure that stratification of analytes is avoided.

NOTE 4—**Caution:** If volatile organics are a concern, homogenizing in open containers will likely result in losses of volatiles.

6.3.2.3 Split samples are treated as separate study samples, carried through the entire sample handling procedures, and submitted to the analytical laboratory without distinguishing identification. Split samples are an indication of the precision of the analytical procedures. For comments on sampling precision, see 5.6; the definition of acceptable levels of precision is the responsibility of the user.

6.3.2.4 Where feasible, each sampling event should include a minimum of one split sample for each type of media or location sampled. Where the data are intended for demonstration of data quality to an outside agency, replicates should be included at a greater frequency, up to 10 % of the total number of samples collected.

6.3.3 *Field Rinsates*—Field rinsates are samples collected in the field by filling a sample collection vessel, such as a well bailer with reagent water or other blank matrix, and then transferring this water to the proper sample bottles. It may be necessary in some instance to fill the collection vessel a number of times to ensure enough water is collected for analysis. The purpose of a field rinsate is to ensure that sampling equipment cleaned in the field is not cross contaminating samples through improper cleaning techniques. These types of samples should be taken at least once for each procedure for each sampling event when field cleaning is performed. If only one such sample is taken it should be collected just prior to the last sample.

6.3.4 *Field Spikes*:

6.3.4.1 Field spikes are samples collected in the field and spiked with compounds of interest or related compounds. These samples are used to check on the potential for loss of analyte on shipping and for recovery of analytes from a particular medium. The field spike is prepared by adding a known amount of the spiking material to a known amount of the matrix and mixing thoroughly. Where a liquid medium is to be collected, the spiking material may be added to the collection container at the laboratory and the sample medium added to the container. For a solid matrix, the material should be added in the field and thoroughly mixed through the matrix prior to closing and sealing the container.

6.3.4.2 Field spikes are normally not required. Instances when a field spike may be desired include where preservation techniques are in question and the integrity of the analytes at the laboratory is not known, when there is a question concerning matrix effects, and when the results from the analytical laboratory for a particular analyte or class of analytes is in question.

6.3.4.3 Field spikes should be submitted blind to the laboratory in the same manner as outlined for the split samples. These samples should be carried through all stages of the sampling and sample handling process as the actual study samples to ensure that they truly indicate the integrity of the samples collected.

7. General Sampling Consideration

7.1 Sampling equipment must be selected that is chemically compatible with the type of waste and type of analyses. Generally, plastic sampling equipment is not suitable for waste containing or to be analyzed for organic parameters. Stainless steel, glass, and plastic are generally acceptable for most samples to be analyzed for inorganics. It is up to the user to ensure that the equipment will not contaminate or bias the analyses.

7.2 The sampling equipment must be capable of extracting a sample from the desired location, depth, or point and at the same time provide protection from cross-contamination during sampling. For instance, one very common problem is extracting a sludge sample from beneath a top layer of wastewater or sludge without contaminating the sample with the overlying wastewater or sludge. This situation, as well as many others, requires special equipment. The collector is therefore faced in many instances with having to fabricate the needed equipment.

7.3 Recommended sampling procedures are for collection of samples from the edge of ponds or lagoons or from piers or catwalks. Sampling from boats is not recommended and should be attempted only if the collector knows the waste poses no real health problem and every possible safety measure has been taken.

7.4 Tanks and drums containing unknown substances also pose potential health risks for the collector due to the possibility of fire, explosion, or the release of deadly gases upon opening. Therefore, it is recommended that in these situations only spark-proof, remote opening devices be used and only fully trained and experienced personnel attempt to do this.

7.5 Representative samples are intended to reflect the true makeup of the population. Composite sampling is one way to help achieve representativeness in a cost- and resource-efficient way. Frequently overlooked but important problems with composite sampling of waste materials include the following:

7.5.1 Loss of volatile components during the mixing process,

7.5.2 Reactivity of dissimilar materials combined into a single composite,

7.5.3 Collecting the correct number and size of aliquots to form the composite, and

7.5.4 Properly homogenizing and subsampling to reduce the amount of material sent to the laboratory.

7.6 The laboratory should provide guidance to the field sampling team to avoid losses due to volatilization or reactivity. Guidelines based on geostatistical principles for forming the composite from individual aliquots should be provided to sampling personnel. Sample homogenization and subsampling in the field should only be attempted by qualified personnel using appropriate equipment. If these are not available, the individual aliquots should be sent to the laboratory for processing. All composite samples or aliquots, or both, intended for forming a composite should be clearly marked so that the laboratory can mix and subsample appropriately.

7.6.1 If low chemical concentrations of a parameter to be tested are expected, take a large volume of sample.

7.6.2 If high chemical concentrations of a parameter to be tested are expected, smaller volumes will suffice.

7.7 When possible, it is recommended that sampling proceed from the least contaminated to the most contaminated areas to reduce problems of cross contamination.

8. Preservation and Containerization of Samples

8.1 Water sample preservation techniques cannot be used for waste samples but are sometimes confused with them.

NOTE 5—Subcommittee D34.01 on Sampling and Monitoring is currently balloting a draft practice on preservation of waste and when approved by the ASTM membership, it will be referenced here. Container specifications for waste samples are found in the same draft practice mentioned above and will be similarly referenced.

9. Cleaning Equipment

9.1 All sampling equipment must be cleaned before use and if possible, it is preferable to have them lab cleaned. Improper cleaning causes cross-contamination of samples. Use of disposable samplers is a very simple way of eliminating the cleaning problem. When samplers are lab cleaned, it is equally important to protect the samples from contamination by wrapping, packaging or containerizing in clean, non-contaminating material.

9.2 Wash equipment cleaned in the lab with a warm detergent solution, rinse several times with tap water, rinse with deionized water and air dry. If organic analyses are to be run, wash with a warm detergent solution (it may first be necessary to wipe with an absorbent cloth to eliminate residues), rinse several times with tap water, rinse with deionized water, rinse with an appropriate organic solvent (the solvent should be, if possible, the extracting solvent) and oven dry at 105°C for at least an hour.

9.3 Cleaning procedures in the field are the same except that tap water rinses, solvent rinses, and drying may not be practical. The user must be careful in using a solvent rinse without oven drying as it may permeate other samples if shipped in the same container or it may react with constituents in the waste. It should be kept in mind that field cleaning requires the capability to carry a large amount of water.

10. Packaging and Package Marking

10.1 An indelible label should be secured to the container identifying the sample. The label should contain or reference the following information:

10.1.1 Name and location of site,

10.1.2 Date and time of sampling,

10.1.3 Location of sampling,

10.1.4 Sample number,

10.1.5 Description and disposition of sample,

10.1.6 Name of sampling personnel,

10.1.7 If possible, full weight of sample and container upon shipping,

10.1.8 Type of preservative, and

10.1.9 Analytical requirements.

10.2 Pack the sample container securely in a shipping container. Generally, if the sample is to be analyzed for volatile organics, it should be packed in ice and cooled to 4°C (See Section 8). A minimum-maximum thermometer packed with the sample container is valuable where knowledge of temperature extremes is critical.

10.3 For shipping purposes, samples must be classified as environmental or hazardous material (waste) samples. In general, environmental samples are collected offsite (for example, from streams, ponds, or wells) and are not expected to be grossly contaminated with high levels of hazardous materials. Waste samples (for example, material from drums or bulk storage tanks; obviously contaminated lagoons, ponds, soils; and leachates from hazardous waste sites) are considered hazardous. Hazardous materials are subject to shipping regulation (for example, International Air Transportation Association (IATA) or U.S. Department of Transportation (USDOT)) and should be packaged and shipped by a person trained in those requirements. If the substance in the sample is known or can be identified, package, mark, and ship according to the specific instructions for that material. Further detail on the shipment of hazardous materials is beyond the scope of this guide, and professional assistance should be sought.

10.4 Make arrangements for handling, logging in, adequate storage, and analysis of the sample at its destination.

10.5 All information pertinent to a field survey or sampling activity must be recorded in a logbook. This should be bound, preferably with consecutively numbered waterproof pages. Where the same information is routinely collected, preprinted logbooks are suggested. Information should be recorded in indelible ink, all entry errors should be crossed out with a single line and initialed, and all entries should be dated and signed at least daily. Entries in the logbook should include information such as the folowing:

10.5.1 Identification of sampling plan (by reference),

10.5.2 Location of sampling,

10.5.3 Name and address of field contact,

10.5.4 Producer of waste and address,

10.5.5 Type of process producing waste (if known),

10.5.6 Type of waste (for example, sludge, wastewater),

10.5.7 Suspected waste composition, including concentrations,

10.5.8 Number and volume of samples taken,

10.5.9 Purpose of sampling (for example, surveillance, contract number),

10.5.10 Description of sampling point and sampling methods,

10.5.11 Date and time of collection,

10.5.12 Preservation used, if any (include ice),

10.5.13 Analytical parameter to be measured,

10.5.14 Unique sample identification number(s),

10.5.15 Sample destination, how transported, and name of transporter,

10.5.16 References, such as maps or photographs of the sampling site,

10.5.17 Field observations (for example, ambient temperature, wind conditions, or other site specific conditions),

10.5.18 Field measurements (for example, pH, explosivity) and results, and

10.5.19 Names of personnel on sampling team.

10.6 Record sufficient information so that anyone could reconstruct the sampling without reliance on the collector's memory. The contents of the logbook should be specified in the sampling plan, a standard practice or similar document. The logbook must be protected and kept in a safe place. Additional detail may be needed than listed above depending on the technical objectives of the sampling project.

11. Chain-of-Custody Procedure

11.1 A chain-of-custody procedure should be developed for samples that may be used in legal proceedings. It is recommended that legal counsel be consulted to assist in developing an individual chain-of-custody record. Chain-of-custody procedures are used to ensure sample integrity and to ensure the sample will provide legally and technically defensible data.

11.2 Samples should be collected in accordance with sampling procedures designated in the sampling plan if the sample is to maintain its legal integrity.

11.3 A legal seal should be attached to the immediate sample container in a manner that the sample cannot be opened without breaking the seal. The seal should have a unique number written across it or the signature of the sampler. Be sure to record the unique legal seal number in the log book.

11.4 The sample should be kept in view or within limited access or locked storage until custody is relinquished and formal documentation of the transfer is completed. The collector should initiate documents at the source of the sample and start the chain-of-custody procedure. Documentation should include the following:

11.4.1 Sample number and legal seal numbers,

11.4.2 Site name and location,

11.4.3 Date and time sampled,

11.4.4 Date sent to laboratory,

11.4.5 Name of sampler,

11.4.6 Information describing source of sample and sample,

11.4.7 Sampling method,

11.4.8 Preservation techniques,

11.4.9 Condition of legal seal upon delivery and name of receiving person,

11.4.10 Method of shipment,

11.4.11 Dates of all activities, and

11.4.12 Signature of persons delivering and receiving samples.

APPENDIX

(Nonmandatory Information)

X1. THE LILLIEFORS TEST FOR GOODNESS OF FIT

X1.1 The data consist of a random sample $X_1, X_2, \ldots, X_n$ of size, n, associated with some unknown distribution function, denoted by $F(x)$. Compute the sample mean:

$$\bar{n} \sum_{i=1}^{n} X_i \quad {}^{-}X = {}^{1} \tag{X1.1}$$

for use as an estimate of μ, and compute:

$$s = \sqrt{\frac{1}{n-1} \sum_{i=1}^{n} (X_i - \bar{X})^2} \tag{X1.2}$$

as an estimate of σ. Then compute the "normalized" sample values Z_i, defined as follows:

$$Z_i = \frac{X_i - \bar{X}}{s} \quad i = 1, 2 \ldots, n \tag{X1.3}$$

The test statistic is computed from the Z_i's instead of from the original random sample.

X1.2 *Assumptions*—The sample is a random sample.

X1.3 *Hypotheses:*

X1.3.1 H_0—The random sample has the normal distribution, with unspecified mean and variance.

X1.3.2 H_1—The distribution function of the X_i's is nonnormal.

X1.4 *Test Statistic*—Ordinarily the test statistic is the usual two-sided Kolmogorov test statistic, defined as the maximum vertical distance between the empirical distribution function of the X_i's and the normal distribution function with mean $\bar{X}$ and standard deviation s, as given by Eq X1.1 and X1.2. However, the following method of computing the test statistic is slightly easier, and is equivalent to the method indicated in X1.1.

X1.4.1 Draw a graph of the standard normal distribution function, and call it $F'(x)$. Table X1.1 may be of assistance. Also draw a graph of the empirical distribution function of the normalized sample, the Z_i's defined by Eq X1.3, using the same set of coordinates as was used above for $F'(x)$. Find the maximum vertical distance between the two graphs, $F'(x)$ and the empirical distribution function which we shall call $S(x)$. This distance is the test statistic. That is, the Lilliefors test statistic T_2 is defined by the following equation:

$$T_2 = \sup | F'(x) - S(x) | \tag{X1.4}$$

X1.5 *Decision Rule*—Reject H_o at the approximate level of significance of α if T_2 exceeds the $l - \alpha$ quantile as given in Table X1.2.

X1.5.1 The same data used to calculate n, the number of samples required, in Section 5 will be used to illustrate the Lilliefors test.

X1.5.2 Barium levels in four sludge samples were 86, 90, 98, and 104 ppm. The sample values, X_i, are arranged from smallest to largest, and converted to Z_i by subtracting $\bar{X}$ = 94.50, and dividing by $S = \sqrt{65.00} = 8.06$.

X_i	Z_i
86	−1.05
90	−0.56
98	0.43
104	1.18

X1.5.2.1 The null hypothesis of normality is tested with the Lilliefors test statistic

TABLE X1.1 Normal Distribution[A]

NOTE 1—This table was abridged from Tables 3 and 4 by Pearson, E. S., and Hartley, H. E., "Biometric Tables for Statisticians," Cambridge University Press, Cambridge, England, pp. 111–112, 1962.

w_p	P	w_p	P	w_p	P
−3.7190	0.0001	−0.4677	0.32	0.5244	0.70
−3.2905	0.0005	−0.4399	0.33	0.5534	0.71
−3.0902	0.001	−0.4125	0.34	0.5828	0.72
−2.5758	0.005	−0.3853	0.35	0.6128	0.73
−2.3263	0.01	−0.3585	0.36	0.6433	0.74
−2.1701	0.015	−0.3319	0.37	0.6745	0.75
−2.0537	0.02	−0.3055	0.38	0.7063	0.76
−1.9600	0.025	−0.2793	0.39	0.7388	0.77
−1.8808	0.03	−0.2533	0.40	0.7722	0.78
−1.7507	0.04	−0.2275	0.41	0.8064	0.79
−1.6449	0.05	−0.2019	0.42	0.8416	0.80
−1.5548	0.06	−0.1764	0.43	0.8779	0.81
−1.4758	0.07	−0.1510	0.44	0.9154	0.82
−1.4395	0.075	−0.1257	0.45	0.9542	0.83
−1.4051	0.08	−0.1004	0.46	0.9945	0.84
−1.3408	0.09	−0.0753	0.47	1.0364	0.85
−1.2816	0.10	−0.0502	0.48	1.0803	0.86
−1.2265	0.11	−0.0251	0.49	1.1264	0.87
−1.1750	0.12	0.0000	0.50	1.1750	0.88
−1.1264	0.13	0.0251	0.51	1.2265	0.89
−1.0803	0.14	0.0502	0.52	1.2816	0.90
−1.0364	0.15	0.0753	0.53	1.3408	0.91
−0.9945	0.16	0.1004	0.54	1.4051	0.92
−0.9542	0.17	0.1257	0.55	1.4395	0.925
−0.9154	0.18	0.1510	0.56	1.4758	0.93
−0.8779	0.19	0.1764	0.57	1.5548	0.94
−0.8416	0.20	0.2019	0.58	1.6449	0.95
−0.8064	0.21	0.2275	0.59	1.7507	0.96
−0.7722	0.22	0.2533	0.60	1.8808	0.97
−0.7388	0.23	0.2793	0.61	1.9600	0.975
−0.7063	0.24	0.3055	0.62	2.0537	0.98
−0.6745	0.25	0.3319	0.63	2.1701	0.985
−0.6433	0.26	0.3585	0.64	2.3263	0.99
−0.6128	0.27	0.3853	0.65	2.5758	0.995
−0.5828	0.28	0.4125	0.66	3.0902	0.999
−0.5534	0.29	0.4399	0.67	3.2905	0.9995
−0.5244	0.30	0.4677	0.68	3.7190	0.9999
−0.4959	0.31	0.4959	0.69		

[A] The entries in this table are quantiles w_p of the standard normal random variable *W*, selected so $P(W<w_p) = p$ and $P(W>w_p) = 1 - p$.

$$T_2 = \sup | F'(x) - S(x) | \quad \text{(X1.5)}$$

where $F'(x)$ is the standard normal distribution function and $S(x)$ is the empirical distribution function of the Z_i's. Fig. X1.1 shows the curves representing $F'(x)$ (generated from Table X1.1) and $S(x)$. The maximum vertical distance between $F'(x)$ and $S(x)$ is seen from Fig. X1.1 to occur at $x = -1.05$, where $S(x)$ equals 0 and $F'(x)$ equals 0.15, and so T_2 equals 0.15.

X1.5.2.2 The Lilliefors test calls for rejection of H_o at $\alpha = 0.05$ if T_2 exceeds its 0.95 quantile, which is given in Table X1.2 as follows:

$$W_{0.95} = \frac{0.381}{\sqrt{n}} = \frac{0.381}{\sqrt{4}} = 0.19$$

TABLE X1.2 Quantiles of the Lilliefors Test Statistic[A]

NOTE 1—This table was adapted from Table 1 of Lilliefors, H. W., "On the Kolmogorov-Smirnov Test for Normality with Mean and Variance Unknown," *Journal of American Statistical Association,* Vol 62, pp. 399–402, 1967.

	$p = 0.80$	0.85	0.90	0.95	0.99
Sample size $n = 4$	0.300	0.319	0.352	0.381	0.417
5	0.285	0.299	0.315	0.337	0.405
6	0.265	0.277	0.294	0.319	0.364
7	0.247	0.258	0.276	0.300	0.348
8	0.233	0.244	0.261	0.285	0.331
9	0.223	0.233	0.249	0.271	0.311
10	0.215	0.224	0.239	0.258	0.294
11	0.206	0.217	0.230	0.249	0.284
12	0.199	0.212	0.223	0.242	0.275
13	0.190	0.202	0.214	0.234	0.268
14	0.183	0.194	0.207	0.227	0.261
15	0.177	0.187	0.201	0.220	0.257
16	0.173	0.182	0.195	0.213	0.250
17	0.169	0.177	0.189	0.206	0.245
18	0.166	0.173	0.184	0.200	0.239
19	0.163	0.169	0.179	0.195	0.235
20	0.160	0.166	0.174	0.190	0.231
25	0.142	0.147	0.158	0.173	0.200
30	0.131	0.136	0.144	0.161	0.187
Over 30	$\frac{0.736}{\sqrt{n}}$	$\frac{0.768}{\sqrt{n}}$	$\frac{0.805}{\sqrt{n}}$	$\frac{0.886}{\sqrt{n}}$	$\frac{1.031}{\sqrt{n}}$

[A] The entries in this table are the approximate quantiles w_p of the Lilliefors test statistic T_2. Reject H_O at the level α if T_2 exceeds $w_{1-\alpha}$ for the particular sample size *n*.

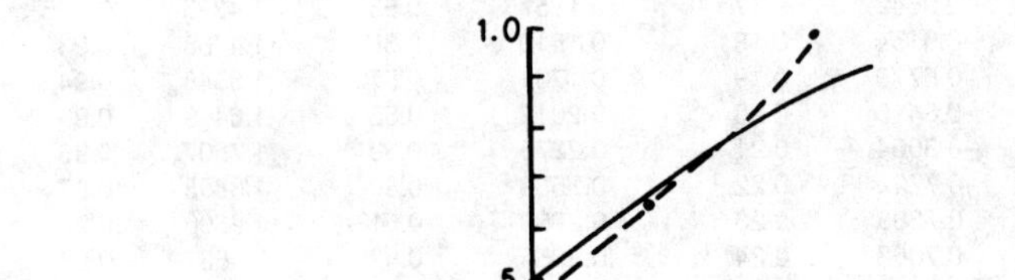

NOTE 1—Because T_2 equals 0.15, and is less than 0.19, the null hypothesis is accepted. This means that the normal distribution is a reasonable approximation of the true unknown distribution.

FIG. X1.1 Graphs of *F'*(*x*) and *S*(*x*) from Example

Designation: D 4793 – 93 (Reapproved 1999)

Standard Test Method for Sequential Batch Extraction of Waste with Water[1]

This standard is issued under the fixed designation D 4793; the number immediately following the designation indicates the year of original adoption or, in the case of revision, the year of last revision. A number in parentheses indicates the year of last reapproval. A superscript epsilon (ε) indicates an editorial change since the last revision or reapproval.

1. Scope

1.1 This test method is a procedure for the sequential leaching of a waste containing at least five % solids to generate solutions to be used to determine the constituents leached under the specified testing conditions.

1.2 This test method calls for the shaking of a known weight of waste with water of a specified purity and the separation of the aqueous phase for analysis. The procedure is conducted ten times in sequence on the same sample of waste and generates ten aqueous solutions.

1.3 This test method is intended to describe the procedure for performing sequential batch extractions only. It does not describe all types of sampling and analytical requirements that may be associated with its application.

1.4 *This standard does not purport to address all of the safety problems, if any, associated with its use. It is the responsibility of the user of this standard to establish appropriate safety and health practices and determine the applicability of regulatory limitations prior to use.*

2. Referenced Documents

2.1 *ASTM Standards:*

D 75 Practices for Sampling Aggregates[2]

D 420 Practice for Investigating and Sampling Soil and Rock for Engineering Purposes[3]

D 653 Terminology Relating to Soil, Rock, and Contained Fluids[3]

D 1129 Terminology Relating to Water[4]

D 1193 Specification for Reagent Water[4]

D 2216 Test Method for Laboratory Determination of Water (Moisture) Content of Soil, Rock, and Soil-Aggregate Mixtures[3]

D 2234 Test Method for Collection of a Gross Sample of Coal[5]

D 2777 Practice for Determination of Precision and Bias of Methods of Committee D-19 on Water[4]

D 3370 Practices for Sampling Water[4]

3. Terminology

3.1 *Definitions:*

3.1.1 For definitions of terms used in this test method, see Terminology D 1129.

3.2 *Symbols:Symbols:*

3.2.1 Variables listed in this test method are defined in the individual sections where they are discussed. A list of defined variables is also given in Section 11.

3.2.2 *Explanation of Variables*:

$\bar{X}_t$ = total mean value
$\bar{X}_a$ = analytical mean value (calculated using data from analysis of standards)
S_{tt} = total standard deviation
S_{ta} = analytical standard deviation
S_{te} = estimated standard deviation due to the extraction procedure
S_{ot} = total single operator standard deviation
S_{oa} = analytical single operator standard deviation
S_{oe} = estimated single operator standard deviation due to the extraction procedure

4. Significance and Use

4.1 This test method is intended as a means for obtaining sequential extracts of a waste. The extracts may be used to estimate the release of certain constituents of the waste under thc laboratory conditions described in this test method.

4.2 This test method is not intended to provide extracts that are representative of the actual leachate produced from a waste in the field or to produce extracts to be used as the sole basis of engineering design.

4.3 This test method is not intended to simulate site-specific leaching conditions. It has not been demonstrated to simulate actual disposal site leaching conditions.

4.4 An intent of this test method is that the final pH of each of the extracts reflects the interaction of the extractant with the buffering capacity of the waste.

4.5 An intent of this test method is that the water extractions reflect conditions where the waste is the dominant factor in determining the pH of the extracts.

4.6 This test method produces extracts that are amenable to the determination of both major and minor constituents. When minor constituents are being determined, it is especially important that precautions are taken in sample storage and handling to avoid possible contamination of the samples.

[1] This test method is under the jurisdiction of ASTM Committee D34 on Waste Management and is the direct responsibility of Subcommittee D34.01.04 on Ancillary Topics.

Current edition approved Jan. 15, 1993. Published March 1993. Originally published as D 4793 – 88. Last previous edition D 4793 – 88.

[2] *Annual Book of ASTM Standards*, Vol 04.03.

[3] *Annual Book of ASTM Standards*, Vol 04.08.

[4] *Annual Book of ASTM Standards*, Vol 11.01.

[5] *Annual Book of ASTM Standards*, Vol 05.05.

4.7 This test method has been tested to determine its applicability to certain inorganic components in the waste. This test method has not been tested for applicability to organic substances, volatile matter (see Note 3 in 5.15), or biologically active samples.

4.8 The agitation technique, rate, liquid-to-solid ratio, and filtration conditions specified in the procedure may not be suitable for extracting all types of wastes (see Sections 7, 8, and the discussion in Appendix X1).

5. Apparatus

5.1 *Straightedge* (such as a thin-edged yard stick).

5.2 *Impermeable Sheet*, of glazed paper, oil cloth, or other flexible material of a composition suitable to the analytes of interest.

5.3 *Drying Pans or Dishes*—Two per waste (for example, aluminum tins, porcelain dishes, or glass weighing pans), suitable to the waste being tested and the instructions given in 9.2.

5.4 *Drying Oven*—Any thermostatically controlled drying oven capable of maintaining a steady temperature of ±2°C in a range from 100 to 110°C.

5.5 *Desiccator*, having the capacity to hold the drying pans described in 5.3 and the crucibles described in 5.8.

5.6 *Laboratory Balance*, capable of weighing to 0.1 g.

5.7 *Pipet*, 10-mL capacity.

5.8 *Crucibles*—Two per waste, porcelain, 20-mL capacity each.

5.9 *Analytical Balance*, capable of weighing to 0.1 mg.

5.10 *Large Glass Funnel.*

5.11 *Wash Bottle*, 500-mL capacity.

5.12 *pH Meter*—Any pH meter with a readability of 0.01 units and an accuracy of ±0.05 units at 25°C is acceptable.

5.13 *Agitation Equipment*, of any type that rotates the extraction vessel in an end-over-end fashion at a rate of 30 ± 2 r/min such that the axis of rotation is horizontal and it goes through the center of the bottle, (see Fig. 1 and the discussion of agitation in Appendix X1).

NOTE 1—Similar devices having a different axial arrangement may be used if equivalency can be demonstrated.

5.14 *Pressure Filtration Assembly*—A pressure filtration device of a composition suitable to the nature of the analyses to be performed and equipped with a 0.45 or 0.8-µm pore-size filter (see Note 7, pertaining to 9.4).

5.15 *Extraction Vessels*, cylindrical, wide-mouth, of a composition suitable to the nature of the waste and analyses to be performed, constructed of materials that will not allow sorption of constituents of interest, and sturdy enough to withstand the impact of the falling sample fragments. Container size should be selected so that the sample plus extraction fluid occupy approximately 95 % of the container. Containers must have water-tight closure. Containers for samples where gases may be released should be provided with a venting mechanism.

NOTE 2—Suitable container sizes range from 4.0 to 4.5 in. (102 to 114 mm) in diameter and 8.5 to 13.0 in. (216 to 330 mm) in height.

NOTE 3—The venting of the container has the potential to affect the concentration of volatile compounds in the extracts.

5.15.1 Extraction vessels should be cleaned in a manner consistent with the analyses to be performed. See Practices D 3370, Section 13.

6. Reagents

6.1 *Purity of Reagents*—Reagent grade chemicals shall be used in all tests. Unless otherwise indicated, it is intended that all reagents shall conform to the specifications of the American Chemical Society, where such specifications are available.[6] Other grades may be used, provided it is first ascertained that the reagent is of sufficiently high purity to permit its use without lessening the accuracy of the determination.

6.2 *Purity of Water*—Unless otherwise indicated, references to water shall be understood to mean Type IV reagent water at 18 to 27°C (Specification D 1193). The method by which the water is prepared, that is, distillation, ion exchange, reverse osmosis, electrodialysis, or a combination thereof, should remain constant throughout testing.

7. Sampling

7.1 Obtain a representative sample of the waste to be tested using ASTM sampling methods developed for the specific industry where available (see Practices D 75 and D 420, Terminology D 653, and Test Method D 2234).

7.2 Where no specific methods are available, sampling methodology for material of similar physical form shall be used.

7.3 The amount of sample to be sent to the laboratory should be sufficient to perform the solids content determination as specified in 9.2 and to provide 100 g of sample on a dry weight basis for each extraction.

7.4 It is important that the sample of the waste be representative with respect to surface area, as variations in surface area would directly affect the leaching characteristics of the sample. Waste samples should contain a representative distribution of particle sizes.

NOTE 4—Information on obtaining representative samples can also be found in *Pierre Gy's Sampling Theory and Sampling Practice*, Volumes I and II, by F. Picard, CRC Press, 1989.

7.5 In order to prevent sample contamination or constituent loss prior to extraction, keep samples in closed containers appropriate to the sample type and desired analysis. See Practices D 3370 for guidance. Record the storage conditions and handling procedures in the report.

7.6 The time between collection and extraction of the sample should be determined by the nature of the sample and the information desired. See Practices D 3370 for guidance. Report the length of time between sample collection and extraction.

8. Sample Preparation

8.1 For free-flowing particulate solid wastes, obtain a

[6] "Reagent Chemicals, American Chemical Society Specifications," Am. Chemical Soc., Washington, DC. For suggestions on the testing of reagents not listed by the American Chemical Society, see "Reagent Chemicals and Standards," by Joseph Rosin, D. Van Nostrand Co., Inc., New York, NY, and the "United States Pharmacopeia."

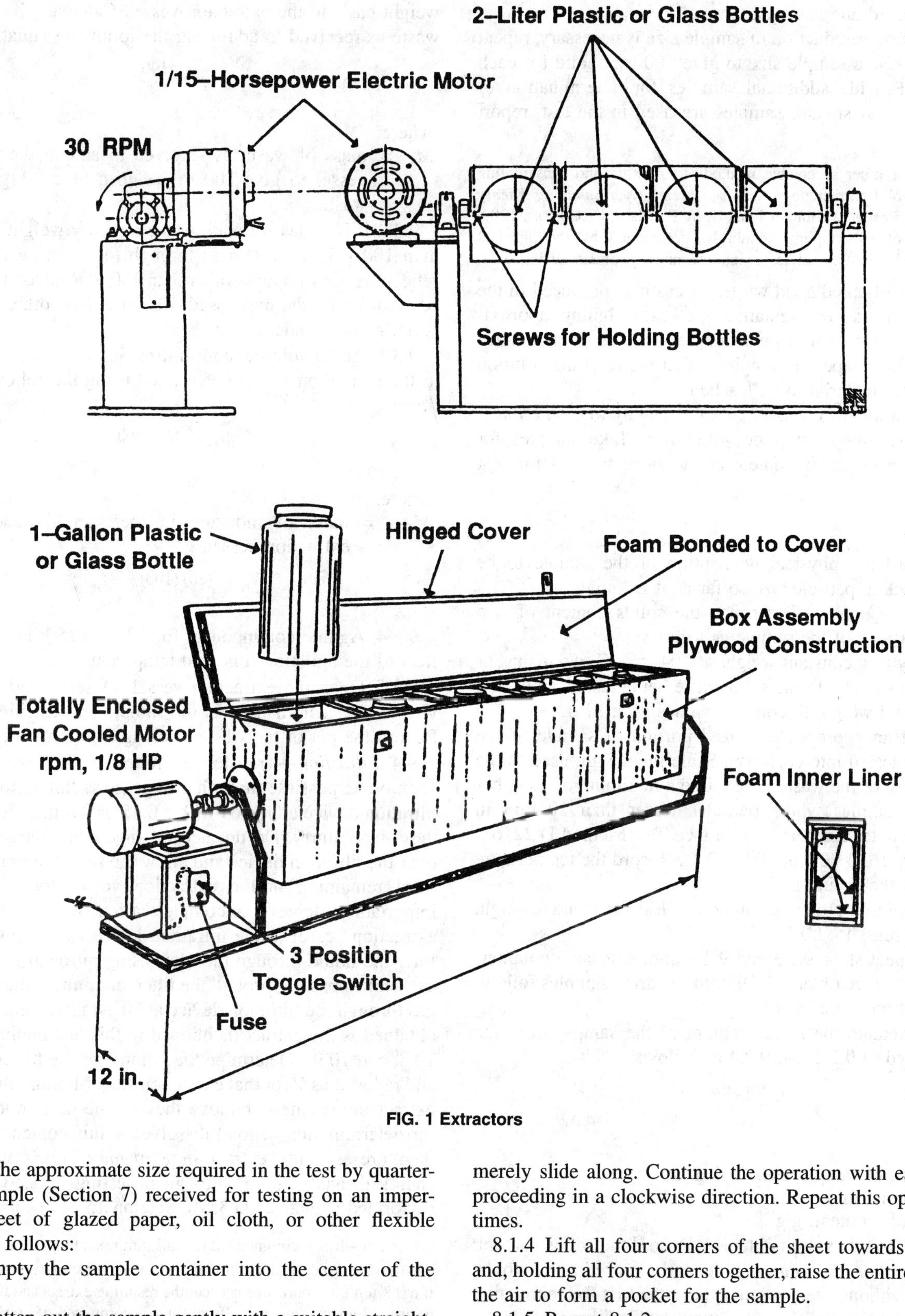

FIG. 1 Extractors

sample of the approximate size required in the test by quartering the sample (Section 7) received for testing on an impermeable sheet of glazed paper, oil cloth, or other flexible material as follows:

8.1.1 Empty the sample container into the center of the sheet.

8.1.2 Flatten out the sample gently with a suitable straightedge until it is spread uniformly to a depth at least twice the maximum particle diameter.

8.1.3 Remix the sample by lifting a corner of the sheet and drawing it across, low down, to the opposite corner in a manner that the material is made to roll over and over and does not merely slide along. Continue the operation with each corner, proceeding in a clockwise direction. Repeat this operation ten times.

8.1.4 Lift all four corners of the sheet towards the center and, holding all four corners together, raise the entire sheet into the air to form a pocket for the sample.

8.1.5 Repeat 8.1.2.

8.1.6 With a straightedge (such as a thin-edged yard stick), one at least as long as the flattened mound of sample, gently divide the sample into quarters. Make an effort to avoid using pressure on the straightedge sufficient to cause damage to the particles.

8.1.7 Discard alternate quarters.

8.1.8 If further reduction of sample size is necessary, repeat 8.1.3-8.1.7. Use a sample size to give 100 g of solid for each extraction. Provide additional samples for determination of solids content. If smaller samples are used in the test, report this fact.

NOTE 5—For other acceptable methods for mixing and subsampling free-flowing solid particulate wastes, see *Pierre Gy's Sampling Theory and Sampling Practice*, Volumes I and II, by F. Picard, CRC Press, 1989. The method of subsampling should be determined by the physical properties of the waste, analytes of interest, and equipment available.

8.2 For field-cored solid wastes or castings produced in the laboratory, cut a representative section weighing approximately 100 g for testing plus samples for determination for solids content. Shape the sample so that the leaching solution will cover the material to be leached.

8.3 For multiphasic wastes, mix thoroughly to ensure that a representative sample will be withdrawn. Take samples for determination of solids content at the same time as the test samples.

9. Procedure

9.1 Record the physical description of the sample to be tested, including particle size so far as it is known.

9.2 *Solids Content*—Determine the solids content of two separate portions of the sample as follows:

9.2.1 Dry to a constant weight at 104 ± 2°C two dishes or pans of size suitable to the solid waste being tested. Cool in a desiccator and weigh. Record the values to ±0.1 g.

9.2.2 Put an appropriately sized portion of sample of the waste to be tested into each pan. Scale the weight used to the physical form of the waste tested. Use a minimum of 50 g, but use larger samples where particles larger than 10 mm in average diameter are being tested (see Test Method D 2216).

9.2.3 Dry 16 to 20 h at 104 ± 2°C. Record the temperature and time of the drying period.

9.2.4 Cool to room temperature in a desiccator and reweigh. Record the mass to ±0.1 g.

9.2.5 Repeat steps 9.2.3 and 9.2.4 until constant container-sample masses are obtained. Discard the dried samples following completion of this step.

9.2.6 Calculate the solids content of the sample from the data obtained in 9.2.2 and 9.2.4 as follows:

$$S = A/B \tag{1}$$

where:

A = mass of sample after drying, g,
B = original mass of sample, g, and
S = solids content, g/g.

Average the two values obtained. Record the solids content.

9.3 *Extraction Procedure*—If the entire procedure cannot be conducted without interruption, at least the first four extraction sequences must be conducted without interruption.

9.3.1 Determine the mass of the extraction vessel to be used in the extraction procedure to the nearest 0.1 g. Record the mass of the extraction vessel, M_{v1}. Use one extraction vessel per waste throughout the sequence of extractions.

9.3.2 Add 100 g (weighed to ±0.1 g) of solid waste on a dry weight basis to the extraction vessel. Calculate the amount of waste as received to add using the following equation:

$$M = \frac{100}{S} \tag{2}$$

where:

M = mass of waste as received to add to the extraction vessel to give 100 g (weighed to ±0.1 g) of solid waste.

9.3.2.1 If a mass of solid waste on a dry weight basis other than 100 g is used, (Eq 2) through (4) must be modified to reflect the use of a mass other than 100 g. Replace 100 in these equations with the mass used. Use of a mass other than 100 g is not recommended.

9.3.3 Add a volume in millilitres, V_{vl}, of test water (see 6.2) to the extraction vessel determined using the following equations:

$$M_{sw} = M - 100 \tag{3}$$

where:

M_{sw} = mass of moisture in the sample added to the extraction vessel, g.

$$V_{v_1} = (20)(100) - M_{sw}. \tag{4}$$

9.3.4 Agitate continuously for 18 ± 0.25 h at 18 to 27°C. Record the agitation time and temperature.

9.3.5 Open the extraction vessel. Observe and record any visible physical changes in the sample and leaching solution. Record the pH of the waste/leaching solution slurry.

9.4 *Filtration*—Transfer as much of the waste/leaching solution as possible through a large glass funnel to a pressure filtration device equipped with a 0.45 or 0.8-μm filter. Transfer the mixed slurry. Do not decant. Invert the extraction vessel over the filtration device and allow the liquid to drain from the solid remaining in the extraction vessel for 1 min. It is important to achieve as complete a transfer of fluid from the extraction vessel to the filtration device as possible. Pressure filter the liquid through the filter using nitrogen gas. After the extract has passed through the filter, continue running nitrogen gas through the filtration device at 30 psi for 3 min. The filtrate obtained is the extract mentioned in this test method (see 9.5, 10.8, and 10.9). Determine the volume of the filtrate collected and report it as V for that extraction step. Measure the pH of the extract immediately, remove the volume of filtrate necessary for determination of total dissolved solids content in 9.5, and then preserve the extract in a manner consistent with the chemical analyses or biological testing procedures to be performed (Practices D 3370, Section 15).

NOTE 6—It is recommended that all filtrations be performed in a hood.

NOTE 7—Analytical results may be affected by the type of filter used. If a 0.8-μm filter pore size is used, the resulting extract should be digested prior to elemental analysis. If the filter is composed of material that may contaminate the extract during filtration, the filter should be washed in the filtration device in a manner consistent with the chemical analyses or biological testing procedures to be performed on the extract. For example, for elemental analysis of the extract, if a filter composed of borosilicate glass fiber is used, it should be washed in the filtration device with a dilute acid solution and rinsed with approximately 2 L of water prior to filtration to prevent contamination.

NOTE 8—Prefilters can be used only if it is absolutely necessary (if the filtrate for analysis or testing cannot be obtained unless a prefilter is used) due to loss of sample trapped in the pores of the prefilter and the possibility of the prefilter disintegrating during rinsing.

9.5 *Total Dissolved Solids Content (TDS)*—Transfer a 10.0-mL sample of the extract to each of two preweighed crucibles (weighed to ±0.1 mg), previously dried at 110± 2°C. Place the samples in a drying oven at 110 ± 2°C for 3 h. Record the drying oven temperature and drying time. Remove the crucibles and let cool in a desiccator. Reweigh the crucibles and record their weights to ±0.1 mg.

NOTE 9—Only one drying is performed to limit the contact time between the solid and the rinse water in the extraction vessel prior to the next extraction step (see 9.6 and Section 10).

9.6 Quantitatively transfer the damp solid from the filter back to the original extraction vessel, including the filter. Use water (see 6.2) from a pre-weighed wash bottle to assist in this transfer and to rinse the filtration device. No more than 500 mL of water should be used for rinsing. Use the smallest volume of wash water possible to achieve a thorough transfer. Using tweezers or a similar device, recover the filter and rinse the adhering solid into the extraction vessel with water from the pre-weighed wash bottle. Do not leave the filter in the extraction vessel. Reweigh the wash bottle to determine the amount of water used in the transfer. Record this value as M_w. Weigh the extraction vessel following the transfer described above and record this value as M_v. The extraction vessel may be sealed until a feasible time for addition of new extraction fluid. This is to enable filtration during the next sequence at a reasonable time during the day. If the slurry is stored for longer than 6 h in the extraction vessel prior to the addition of new extraction fluid, the data generated by the analysis of the extracts should be plotted to check for perturbation of the data curve.

10. Calculation

10.1 Calculate the total dissolved solids contents, TDS, in milligrams per litre of the filtrate using the following equation:

$$TDS = (M_{sc} - M_c)(100) \quad (5)$$

where:

M_{sc} = mass of the crucible and dried solids, mg, and
M_c = mass of the crucible, mg.

10.2 Calculate the mass of the solid in grams lost through dissolution, M_d, using the following equation:

$$M_d = (TDS)(V)(0.001) \quad (6)$$

where:

V = volume of filtrate collected in that extraction, L, and
M_d = mass loss through dissolution.

10.3 Calculate the mass of the solid corrected for TDS remaining for the next extraction step, M_s, using the following equation:

$$M_s = M_s^{e-1} - M_d \quad (7)$$

where:

M_s^{e-1} = mass of the solid extracted in the current extraction step, g.

NOTE 10—For example, in beginning the first extraction, M_s^{e-1} will equal 100 g, and to calculate the mass of solid remaining for the second extraction step, M_s will equal 100 g – M_d.

10.4 Calculate the combined mass of the solid and the residual liquid in the extraction vessel, M_{sl}, using the following equation:

$$M_{sl} = M_v - M_{v1} - M_w \quad (8)$$

10.5 Calculate the mass of liquid adhering to the solids in the extraction vessel, M_l, using the following equation:

$$M_l = M_{sl} - M_s \quad (9)$$

10.6 Calculate the volume in millilitres of new test water to be added to the extraction vessel, Test Water Volume, TWV, using the following equation:

$$TWV = [(M_s)(20)] - M_l - M_w \quad (10)$$

10.7 Add to the extraction vessel the amount of new test water, TWV, determined in 10.6, and repeat 9.3.4 through 10.7 so that ten extractions are done in sequence.

NOTE 11—This procedure assumes that the amount of waste that is trapped in the filters after rinsing is negligible.

10.8 Analyze the extracts for specific constituents or properties or use the extracts for biological testing procedures as desired using appropriate ASTM test methods. Where no appropriate ASTM test methods exist, other methods may be used and recorded in the report. Where phase separation occurs during the storage of the extracts, appropriate mixing should be used to ensure the homogeneity of the extracts prior to their use in such analyses or testing.

10.9 *Compensation for Carry-Over*—For each constituent in each of the extracts generated in the extraction sequence, the contribution to concentration from the residual liquid from the previous extraction step, C_j, can be calculated using the following equation:

$$C_j = [M_{li}/20(M_s^{e-1})][C_i] \quad (11)$$

where:

C_i = concentration of the constituent in the filtrate from the previous extraction step,
M_{li} = M_l from the previous extraction step, and
M_s^{e-1} = mass of solid extracted in the current extraction step (see Note 10).

11. Definition of Variables

11.1 The following variables must be determined when performing the sequential batch extraction procedure:

11.1.1 *Solids Content Determination*:

A = mass of the sample after drying in the determination of the solids content of the waste to be extracted, g,

B = original mass of the sample prior to drying in the determination of the solids content of the waste to be extracted, g, and
S = solids content of the waste to be extracted, g/g.

11.1.2 *First Extraction Step*:

M = mass of waste as received added to the extraction vessel to give 100 g (weighted to ±0.1 g) of solid on a dry weight basis for the first extraction step, g,
V_{vl} = volume of test water to be added for the first step in the extraction procedure, mL, and
M_{sw} = mass of the moisture in the sample to be extracted in the first extraction step, g.

11.1.3 *TDS Determination*:

TDS = total dissolved solids content of the filtrate, mg/L,
M_c = mass of the crucible to be used in the TDS determination, mg, and
M_{sc} = mass of the crucible and dried solids in the TDS determination, mg.

11.1.4 *Extraction Sequence*:

M_d = mass of the solid lost through dissolution during extraction, g,
V = volume of filtrate collected in that extraction, L,
M_s = mass of the solid remaining for the next extraction step, g,
M_s^{e-1} = mass of the solid extracted in the current extraction step, g,
M_{v1} = mass of the empty extraction vessel, g,
M_w = mass of the rinse water, g,
M_v = combined mass of the extraction vessel, rinse water, solid and moisture in the solid, and solid and liquid left in the extraction vessel after transfer to the filtering device, g,
M_{sl} = combined mass of the solid and the residual liquid in the extraction vessel following transfer of the moist sample cake back to the extraction vessel, g,
M_l = mass of the liquid adhering to the solids in the extraction vessel following transfer of the moist sample cake back to the extraction vessel, g, and
TWV = volume of test water to be added for the next extraction step, mL.

11.1.5 *Compensation for Carry-Over*:

C_j = contribution to a constituent's concentration in the current step from the residual liquid of the previous extraction step, mg/L,
M_{li} = M_l from the previous extraction step, g,
M_s^{e-1} = mass of solid extracted in the current step, g, and
C_i = concentration of the constituent in the filtrate from the previous extraction step, mg/L.

12. Report

12.1 Report the following information:

12.1.1 Source of the waste, date of sampling, method of sampling, method of sample preservation, storage conditions, handling procedures, and length of time between sample collection and extraction,

12.1.2 Description of the waste, including physical characteristics and particle size, if known (9.1),

12.1.3 Solids content (9.2) (see Test Method D 2216),

12.1.4 Mass of solid waste extracted if other than 100 g (8.1.8),

12.1.5 Time and temperature used in the determination of solids content and TDS,

12.1.6 Agitation temperature and time,

12.1.7 Filter pore size used and filter composition; use of a prefilter and prefilter pore size and composition,

12.1.8 Observations of changes in test material or leaching solutions (9.3.5),

12.1.9 Storage of the solid with rinse water in the extraction vessel for any period longer than 6 h,

12.1.10 pH before and after filtration and results of specific analyses calculated in appropriate units and corrected for carry-over if necessary, and

12.1.11 Dates sequential batch extraction started and completed, preservation used for extracts, and date of analyses.

NOTE 12—Fig. 2 presents a report format for recording some of the experimental data.

13. Precision and Bias [7]

13.1 *Precision*:

13.1.1 A collaborative study of this test method involving eight laboratories was conducted. Each laboratory extracted a single sample in duplicate. The extracts generated in the first, third, fifth, seventh, and tenth extraction steps were analyzed by each participant and by a reference laboratory. In addition, three standards containing high, medium, and low concentrations of the elements of interest, aluminum, calcium, copper, iron, magnesium, nickel, and zinc were analyzed by each participant in triplicate in order to determine the analytical precision. From the data generated, precision calculations were performed using Practice D 2777 as a guideline.

13.1.2 Three types of precision can be determined from the data generated. These are the total standard deviation, S_{tt}, the analytical standard deviation, S_{ta}, and the estimated standard deviation of the extraction procedure, S_{te}. The standard deviations calculated using the data generated by the individual laboratories from their analyses of the extracts are due to a combination of both the extraction procedure and the analytical errors (S_{tt}). The precision data determined from the analyses of the high, medium, and low standards represent those values due to analytical error only (S_{ta}), and the standard deviation of the extraction procedure represents the estimated error due only to the extraction method (S_{te}). The estimated standard deviation of the extraction procedure for each element of interest in Extracts 1, 3, 5, 7, and 10 was calculated using the following equation:

$$S_{te} = (S_{tt}^2 - S_{ta}^2)^{1/2} \quad (12)$$

[7] Supporting data for the precision statement and concerning analytical bias have been filed at ASTM Headquarters and may be obtained by requesting RR: D34-1005.

Sample Number: ______

In Generating Extract Number	TDS (mg/L)	V (L)	M_d (g)	M_s (g)	M_v (g)	M_{v1} (g)	M_w (g)	M_{s1} (g)	M_1 (g)	TWV (mL)
1										
2										
3										
4										
5										
6										
7										
8										
9										
10										

FIG. 2 Sequential Batch Procedure Data Sheet

These values, along with the total and analytical mean values ($\bar{X}_t$ and $\bar{X}_a$) and standard deviations, are listed in Table 1.

13.1.3 The three types of precision values discussed in 13.1.2, total, analytical, and extraction procedure, can also be calculated based on a single operator. Calculations were performed to determine the total single operator precision, S_{ot}, the single operator analytical standard deviation, S_{oa}, and the single operator estimated standard deviation of the extraction procedure, S_{oe}. The single operator estimated standard deviation of the extraction procedure was calcu-lated using the following equation:

$$S_{oe} = (S_{ot}^2 - S_{oa}^2)^{1/2} \tag{13}$$

The single operator precision values are listed in Table 2.

13.1.4 Calculation of the standard deviation of the extraction procedure can provide only an approximation due to the limited high-, medium-, and low-concentration values of the analytical standards. To calculate the precision of the extraction procedure for a particular element, the analytical standard deviation for analysis of the analytical standard containing the concentration of the element closest to its concentration in the extract was used. For some of the extracts, the elemental concentration varies significantly from the element's closest concentration among the analytical standards. Also in some cases, the analytical standard deviation values, S_{ta} and S_{oa} in Table 1 and Table 2, are larger than the total standard deviation value, S_{tt} and S_{ot}. In those particular cases, the standard deviation of the extraction procedure cannot be determined.

13.1.5 The estimated precision of this sequential batch extraction procedure varies with the concentration of each of the seven constituents of interest in the collaborative study

TABLE 1 Sequential Batch Extraction Round-Robin Study Statistical Data Summarized—Estimated Total Precision of the Extraction Procedure (µg/g)

Extract 1	Aluminum	Calcium	Copper	Iron	Magnesium	Nickel	Zinc
$\bar{X}_t$	75.4	982.0	12.3	68.2	189.0	63.3	237.0
S_{tt}	10.1	286.0	2.69	9.49	19.6	6.52	25.3
$\bar{X}_a$	54.2	1072.0	2.25	55.2	190.0	55.3	186.0
S_{ta}	5.03	326.0	0.260	4.24	16.4	4.56	20.1
S_{te}	8.75	...[A]	2.65	8.49	10.7	4.66	15.4
Extract 3							
$\bar{X}_t$	10.5	72.1	0.990	1.87	7.78	2.74	11.3
S_{tt}	9.20	30.6	1.06	0.404	2.32	1.05	5.24
$\bar{X}_a$	2.50	19.9	2.25	2.70	2.33	2.22	2.28
S_{ta}	0.765	1.67	0.260	0.450	0.327	0.331	0.183
S_{te}	9.17	30.6	0.950	...[A]	2.30	0.996	5.24
Extract 5							
$\bar{X}_t$	6.23	52.8	0.322	1.85	4.73	2.01	8.46
S_{tt}	2.17	12.4	0.100	0.516	1.09	0.470	2.38
$\bar{X}_a$	2.50	19.9	2.25	2.70	2.33	2.22	2.28
S_{ta}	0.765	1.67	0.260	0.450	0.327	0.331	0.183
S_{te}	2.03	12.3	...[A]	0.252	1.04	0.333	2.37
Extract 7							
$\bar{X}_t$	5.13	52.7	0.416	1.53	3.95	1.59	6.65
S_{tt}	1.90	4.86	0.089	0.730	1.47	0.270	0.652
$\bar{X}_a$	2.50	19.9	2.25	2.70	2.33	2.22	2.28
S_{ta}	0.765	1.67	0.260	0.450	0.327	0.331	0.183
S_{te}	1.74	4.56	...[A]	0.455	1.43	...[A]	0.626
Extract 10							
$\bar{X}_t$	1.46	62.3	0.444	1.56	2.72	1.36	6.71
S_{tt}	0.866	21.7	0.067	0.467	0.679	0.239	1.89
$\bar{X}_a$	2.50	19.9	2.25	2.70	2.33	2.22	2.28
S_{ta}	0.765	1.67	0.260	0.450	0.327	0.331	0.183
S_{te}	0.406	21.6	...[A]	0.125	0.556	...[A]	1.88

[A] See 13.1.4.

TABLE 2 Sequential Batch Extraction Round-Robin Study Statistical Data Summarized—Estimated Single Operator Precision of the Extraction Procedure (μg/g)

Extract 1	Aluminum	Calcium	Copper	Iron	Magnesium	Nickel	Zinc
$\bar{X}_t$	75.4	982.0	12.3	68.2	189.0	63.3	237.0
S_{ot}	2.31	28.0	0.687	0.877	7.90	2.03	7.19
$\bar{X}_a$	54.2	1072.0	2.25	55.2	190.0	55.3	186.0
S_{oa}	1.82	26.2	0.260	1.83	5.05	2.30	3.84
S_{oe}	1.42	9.88	0.636	...[A]	6.07	...[A]	6.08
Extract 3							
$\bar{X}_t$	10.5	72.1	0.990	1.87	7.78	2.74	11.3
S_{ot}	0.678	5.26	0.080	0.437	1.07	0.094	0.622
$\bar{X}_a$	2.50	19.9	2.25	2.70	2.33	2.22	2.28
S_{oa}	0.098	0.666	0.260	0.450	0.288	0.110	0.097
S_{oe}	0.671	5.22	...[A]	...[A]	1.03	...[A]	0.614
Extract 5							
$\bar{X}_t$	6.23	52.8	0.322	1.85	4.73	2.01	8.46
S_{ot}	1.46	12.7	0.000	0.266	0.987	0.160	0.810
$\bar{X}_a$	2.50	19.9	2.25	2.70	2.33	2.22	2.28
S_{oa}	0.098	0.666	0.260	0.450	0.288	0.110	0.097
S_{oe}	1.46	12.7	...[A]	...[A]	0.944	0.116	0.804
Extract 7							
$\bar{X}_t$	5.13	52.7	0.416	1.53	3.95	1.59	6.65
S_{ot}	0.865	2.17	0.100	0.255	0.385	0.319	0.535
$\bar{X}_a$	2.50	19.9	2.25	2.70	2.33	2.22	2.28
S_{oa}	0.098	0.666	0.260	0.450	0.288	0.110	0.097
S_{oe}	0.859	2.06	...[A]	...[A]	0.255	0.299	0.526
Extract 10							
$\bar{X}_t$	1.46	62.3	0.444	1.56	2.72	1.36	6.71
S_{ot}	0.955	8.40	0.032	0.308	0.188	0.235	0.968
$\bar{X}_a$	2.50	19.9	2.25	2.70	2.33	2.22	2.28
S_{oa}	0.098	0.666	0.260	0.450	0.288	0.110	0.097
S_{oe}	0.950	8.37	...[A]	...[A]	...[A]	0.208	0.963

[A]See 13.1.4.

according to Figs. 3-9. These are plots of the calculated percent relative standard deviation of the extraction procedure versus the total mean concentration of the constituent (data listed in Table 1).

13.1.6 For the concentration values determined in the third, fifth, seventh, and tenth extracts, there does not appear to be a relationship between elemental concentration and estimated precision of the extraction procedure. Because of the very limited data at higher concentrations, it cannot be determined if such a trend exists at higher concentration levels; however, the estimated precision of the extraction procedure is generally best for the elemental concentration values determined in the first extract.

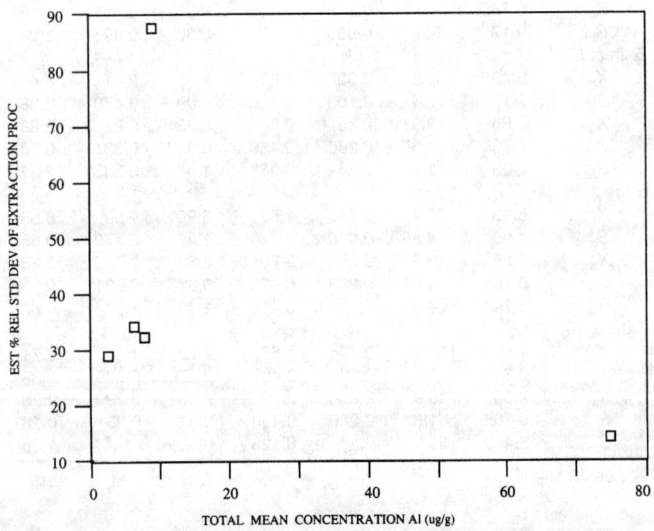

FIG. 3 Estimated Precision of Extraction Procedure—Aluminum

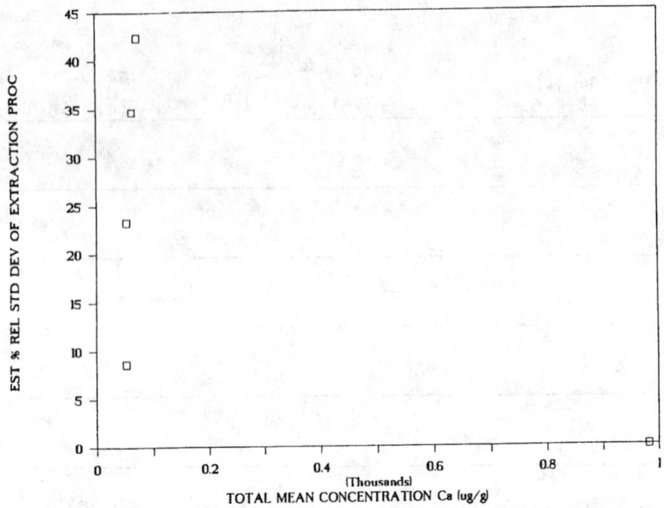

FIG. 4 Estimated Precision of Extraction Procedure—Calcium

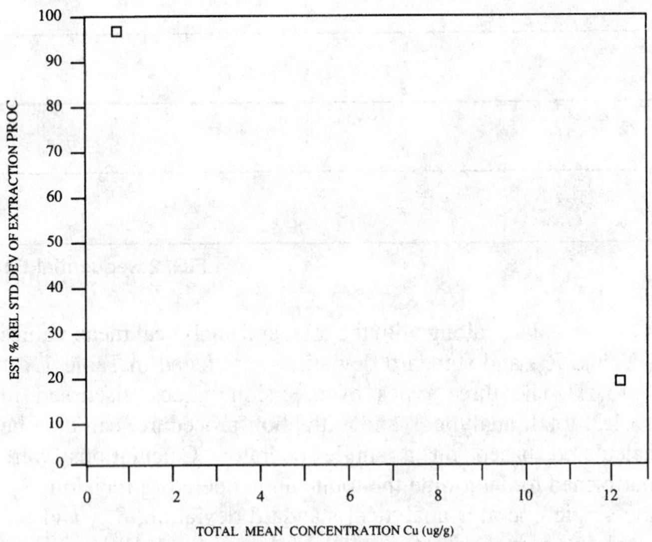

FIG. 5 Estimated Precision of Extraction Procedure—Copper

13.1.7 These collaborative test data were obtained through the extraction of a raw oil shale sample. Nitrocellulose filters having a pore size of 0.45 μm were used by each of the collaborative study laboratories for the filtering specified in 9.4. For other test materials and filter types, these data may not apply.

13.1.8 The estimated precision of the extraction proce-dure includes the increase in variability that may be attributable to field collection, laboratory crushing and sample splitting, and distribution of split samples to the various laboratories for testing. The analytical precision was calculated using data determined for the standard solutions, and as a result, it does not include variability due to various liquid matrices.

13.2 *Bias*—Determination of the bias of this test method is not possible, as no standard reference material exists. Information concerning the analytical bias determined from the collaborative study of this procedure is available in RR: D34-1005.

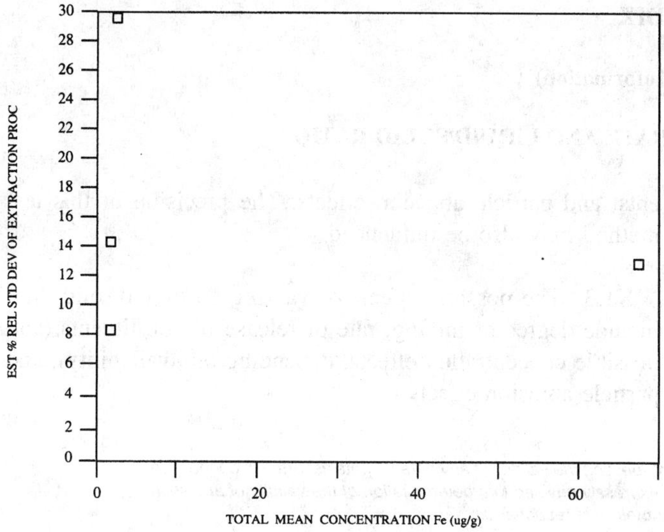

FIG. 6 Estimated Precision of Extraction Procedure—Iron

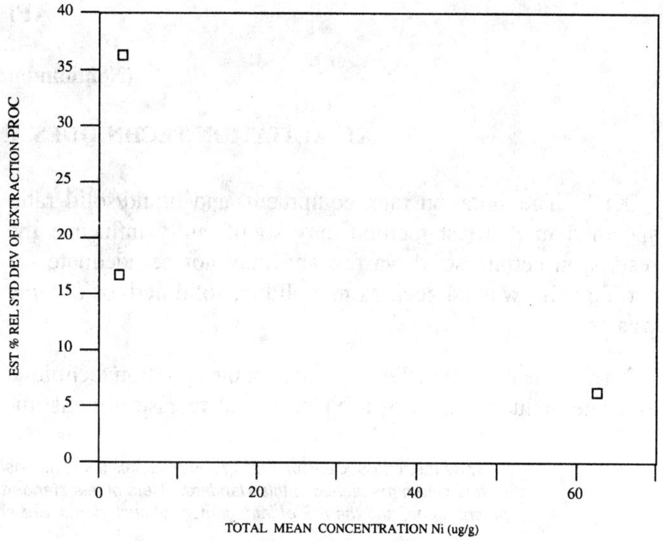

FIG. 8 Estimated Precision of Extraction Procedure—Nickel

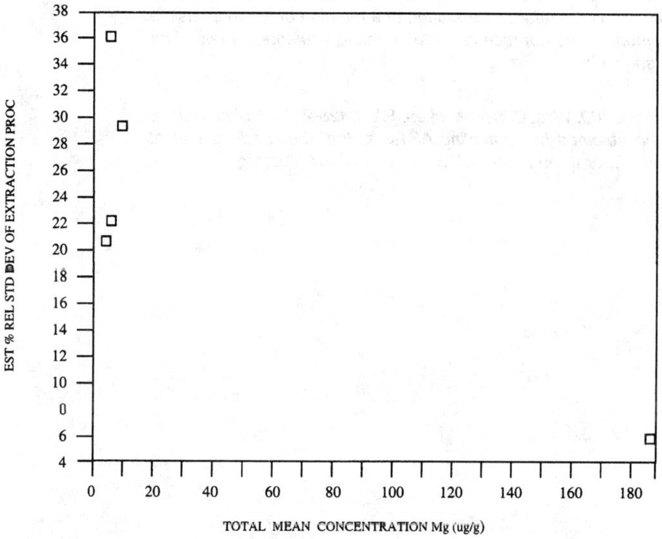

FIG. 7 Estimated Precision of Extraction Procedure—Magnesium

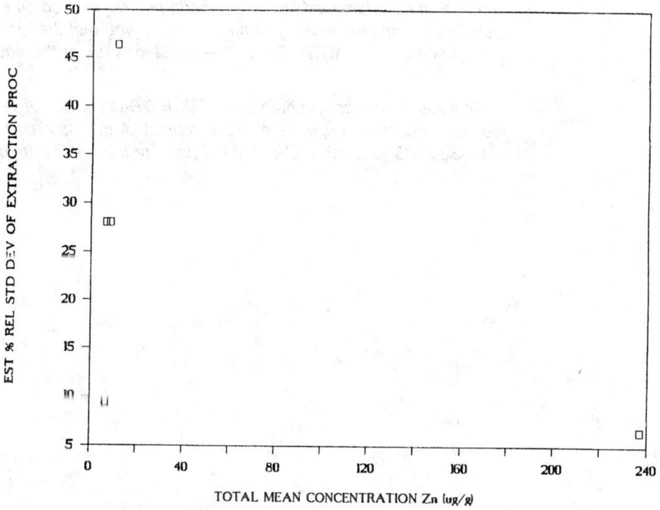

FIG. 9 Estimated Precision of Extraction Procedure—Zinc

14. Keywords

14.1 extract; extraction fluid; leaching; sequential batch extraction; waste leaching technique

APPENDIX

(Nonmandatory Information)

X1. AGITATION TECHNIQUES AND RATE AND LIQUID/SOLID RATIO

X1.1 The agitation rate, equipment, and liquid/solid ratio specified in this test method may significantly influence the results on certain solid wastes, and may not be adequate for certain solid wastes, such as monolithic, solidified, or organic wastes.

X1.2 The possible effects of varying the agitation technique and rate include degree of mixing, rate of release of constituents, and particle abrasion effects. The precision of this test method may also be influenced.

X1.3 The possible effects of varying the liquid/solid ratio include degree of mixing, rate of release of constituents (and possible concentration effects, depending on availability), and particle abrasion effects.

Designation: D 4842 – 90 (Reapproved 2001)

Standard Test Method for Determining the Resistance of Solid Wastes to Freezing and Thawing[1]

This standard is issued under the fixed designation D 4842; the number immediately following the designation indicates the year of original adoption or, in the case of revision, the year of last revision. A number in parentheses indicates the year of last reapproval. A superscript epsilon (ε) indicates an editorial change since the last revision or reapproval.

1. Scope

1.1 This test method covers procedures for determining material losses produced by repeated freezing and thawing of solid waste specimens. It also covers the visual observation of the disintegration of solid specimens.

1.2 This test method intends that the material used in the procedure be physically, chemically, and biologically representative, hence it does not address problems as a result of the inhomogeneity of specimens.

1.3 *This standard does not purport to address all of the safety concerns, if any, associated with its use. It is the responsibility of the user of this standard to establish appropriate safety and health practices and determine the applicability of regulatory limitations prior to use.*

2. Referenced Documents

2.1 *ASTM Standards:*

C 305 Practice for Mechanical Mixing of Hydraulic Cement Pastes and Mortars of Plastic Consistency[2]

D 2216 Test Method for Laboratory Determination of Water (Moisture) Content of Soil and Rock[3]

3. Significance and Use

3.1 This test method is intended for the evaluation of the freezing and thawing resistance of monolithic, solid, solidified/stabilized wastes under the testing conditions of this test method.

3.2 This test method may be used for the comparison of freezing and thawing resistance of wastes.

3.3 Data tabulated in the charts shown in Figs. 1-3 may be used to observe irregularities caused by inhomogeneity of specimens or comparison of mass loss-cycle relations of different wastes, or both, as well as to measure method-related weight losses such as matrix dissolution.

4. Apparatus

4.1 *Disposable Molds*, 44-mm inside diameter by 74-mm in length.

4.2 *Balance or Scale*, with a capacity at least 50 % greater than the weight of the specimen and beaker, and a sensitivity of 0.01 g.

4.3 *Drying Oven*, a thermostatically controlled drying oven capable of maintaining a temperature of 60 ± 2°C; to be used for drying moisture specimen and for the solids content determination.

4.4 *Freezing Cabinet*, capable of maintaining – 20 ± 3°C.

4.5 *Refrigerator*, capable of maintaining + 4 ± 3°C.

4.6 *Moisture Chamber*, a suitably covered container capable of maintaining a temperature of 20 ± 3°C and maintain 95 % relative humidity, for preconditioning and thawing specimens.

4.7 *Beakers*, 400-mL size (narrow type), to store sample and to collect particulates.

4.8 *Tongs*, to handle samples.

5. Sample Preparation

5.1 *Specimen Size*— 44-mm diameter by 74-mm in length.

5.1.1 Specimens may be cut to size from larger samples.

5.1.2 Specimens can also be molded in disposable plastic molds. When molding specimens refer to Practice C 305 (see 2.1).

NOTE 1—Practice C 305 refers to pastes and mortars. Molding materials with different consistency may require modifications and may result in different precision.

5.2 Condition samples that are not molded for this test in the moisture chamber for a period of seven days.

5.2.1 Samples molded for this test have to be cured in the moisture chamber for a period of 28 days.

6. Procedure

6.1 Select one specimen for moisture content determination.

6.2 Determine moisture content of specimen in accordance with Test Method D 2216 but revised to use a temperature of 60° ± 3°C (see 2.1).

6.3 Select three specimens for testing and three for control and mark them respectively.

6.4 Weigh specimens (to the nearest 0.01 g).

6.5 Place each specimen into a tared beaker, dried in accordance with Test Method D 2216, and weighed to the nearest 0.01 g. Cover the beaker with a watch glass or plastic wrap.

6.6 Place the three beakers with testing specimens in a freezing cabinet. Maintain temperature at – 20 ± 3°C for 24 h.

---

[1] This test method is under the jurisdiction of ASTM Committee D34 on Waste Management and is the direct responsibility of Subcommittee D34.01.06 on Analytical Methods.

Current edition approved May 25, 1990. Published July 1990.

[2] *Annual Book of ASTM Standards*, Vol 04.01.

[3] *Annual Book of ASTM Standards*, Vol 04.08.

TABLE 1 Control Group

Laboratory:
Technician:

Sample name:
Sample Id No.:
Test Start Date:

Sample 1 / Cycle No.	1					2					3				
Date	$T_{l,c,1}$	$B_{l,c,1}$	$W_{l,c,1}$	$M_{c,1}$	$R_{l,c,1}$	$T_{l,c,2}$	$B_{l,c,2}$	$W_{l,c,2}$	$M_{c,2}$	$R_{l,c,2}$	$T_{l,c,3}$	$B_{l,c,3}$	$W_{l,c,3}$	$M_{c,3}$	$R_{l,c,3}$
1.															
2.															
3.															
4.															
5.															
6.															
7.															
8.															
9.															
10.															
11.															
12.															

TABLE 2 Sample Group

Laboratory:
Technician:

Sample name:
Sample Id No.
Test Start Date:

Sample 1 / Cycle No.	1					2					3				
Date	$T_{l,s,1}$	$B_{l,s,1}$	$W_{l,s,1}$	$M_{s,1}$	$R_{l,s,1}$	$T_{l,s,2}$	$B_{l,s,2}$	$W_{l,s,2}$	$M_{s,2}$	$R_{l,s,2}$	$T_{l,s,3}$	$B_{l,s,3}$	$W_{l,s,3}$	$M_{s,3}$	$R_{l,s,3}$
1.															
2.															
3.															
4.															
5.															
6.															
7.															
8.															
9.															
10.															
11.															
12.															

6.7 Store the three beakers with the control specimens in the moisture chamber at 20°C for 24 h.

6.8 Remove the specimens from the freezing cabinet and the moisture chamber.

6.8.1 To the frozen specimens add 240 mL of distilled chilled water. This water shall be at a temperature of 4 ± 3°C.

6.8.2 To the control specimens, add 240 mL of room temperature water. This water shall be at a temperature of 20 ± 3°C.

6.8.3 Place a watch glass or plastic wrap on the beakers and store the water covered specimens at 20 ± 3°C for 23 h.

6.9 Using tongs, transfer each specimen to another dry beaker. This second set of beakers shall be prepared in accordance with 6.5.

NOTE 2—Excessive tong pressure may result in premature failure or damage to specimen.

6.10 Remove any loosely attached particulates by spraying distilled water from a wash bottle to the surface of specimen

TABLE 3 Relative Weight Loss

Laboratory:
Technician:

Sample name:
Sample Id No.:
Test Start Date:

Cycle No.	$\bar{R}_{l,s}$	$\bar{R}_{l,c}$	$\bar{C}_l$	$\bar{S}_l$	Observations
1.					
2.					
3.					
4.					
5.					
6.					
7.					
8.					
9.					
10.					
11.					
12.					
$\bar{S}$					

(10 to 20 mL distilled water). Let water drain into the beaker of origin.

6.11 Conduct visual observation on the specimens' physical deterioration including: cracking, fracturing, integrity, and surface roughness.

6.12 Determine the specimens' weight loss: the mass of the solid residue in beakers by evaporating water at 60 ± 3°C in drying oven.

6.13 Correct the average relative mass loss of samples using the average relative mass loss of control specimens.

6.14 Repeat the procedures in 6.5-6.10 eleven additional times, for a total of 12 cycles.

6.15 Terminate experiment of all specimens if the corrected cumulative mass loss of any of the specimens exceeds 30 % (failure), and note the number of cycles survived.

7. Calculation

7.1 Calculate the dry mass of specimens as follows:

$$M_s = 1 - \frac{w}{100} M_{sw} \text{ g}$$

where:

M_s = oven dry mass of specimen in g,
M_{sw} = initial mass of specimen in g, and
w = moisture content, %.

It is assumed that the moisture contents of specimens are identical. Oven dry masses of sample and control specimens are calculated on that basis.

7.2 Calculate corrected mass loss of specimens after each cycle. Express mass loss in percent of initial calculated oven-dry mass. Calculate average cumulated, corrected mass loss of specimens after each cycle as follows:

$$W_{i,s,j} = T_{i,s,j} - B_{i,s,j} \text{ g} \tag{1}$$

where:

$W_{i,s,j}$ = mass loss of sample j during cycle i, in g,
$T_{i,s,j}$ = oven-dry mass of beaker and residue of sample j after cycle i, in g, and
$B_{i,s,j}$ = oven-dry mass of beaker for sample j before cycle i, in g.

$$W_{i,c,j} = T_{i,c,j} - B_{i,c,j} \text{ g} \tag{2}$$

where:

$W_{i,c,j}$ = mass loss of control j during cycle i, in g,
$T_{i,c,j}$ = oven-dry mass of beaker and residue of control j after cycle i, in g, and
$B_{i,c,j}$ = oven dry mass of beaker for control j before cycle i, in g.

$$R_{i,s,j} = \frac{W_{i,s,j}}{M_{s,j}} \text{ \%} \tag{3}$$

where:

$R_{i,s,j}$ = relative mass loss of sample j during cycle i, %,
$W_{i,s,j}$ = mass loss of sample j during cycle i, in g, and
$M_{s,j}$ = oven-dry mass of specimen j, in g.

$$R_{i,c,j} = \frac{W_{i,c,j}}{M_{c,j}} \text{ \%} \tag{4}$$

where:

$R_{i,c,j}$ = relative mass loss of j control during cycle i, %,
$W_{i,c,j}$ = mass loss of control j during cycle i, in g, and
$M_{c,j}$ = oven-dry mass of control j, in g.

$$\bar{R}_{i,s} = \frac{\sum_{j=1-3} R_{i,s,j}}{3} \text{ \%} \tag{5}$$

where:

$\bar{R}_{i,s}$ = average relative mass loss of samples ($j = 1 - 3$) during cycle i, %, and
$R_{i,s,j}$ = relative mass loss of sample j during cycle i, %.

$$\bar{R}_{i,c} = \frac{\sum_{j=1-3} R_{i,c,j}}{3} \text{ \%} \tag{6}$$

where:

$\bar{R}_{i,c}$ = average relative mass loss of control ($j = 1 - 3$) during cycle i, %, and
$R_{i,c,j}$ = relative mass loss of control j during cycle i, %.

$$\bar{C}_i = \bar{R}_{i,s} - \bar{R}_{i,c} \text{ \%} \tag{7}$$

where:

$\bar{C}_i$ = average corrected relative mass loss of samples ($j = 1 - 3$) during cycle i, %,
$\bar{R}_{i,s}$ = average relative mass loss of samples ($j = 1 - 3$) during cycle i, %, and
$\bar{R}_{i,c}$ = average relative mass loss of control ($j = 1 - 3$) during cycle i, %.

$$\bar{S}_i = \sum_{i=1-i} \bar{C}_i \text{ \%} \tag{8}$$

where:

$\bar{S}_i$ = average cumulated, corrected relative mass loss of samples after i cycles %, and
$\bar{C}_i$ = average corrected relative mass loss of samples ($j = 1 - 3$) during cycle i, %.

$$\bar{S} = \sum_{i=1-12} \bar{C}_i \text{\%} \tag{9}$$

where:

$\bar{S}$ = average cumulated, corrected relative mass loss of samples after 12 cycles, %, and
$\bar{C}_i$ = average corrected relative mass loss of samples ($j = 1 - 3$) during cycle i, %.

8. Report

8.1 Report the following information:

8.1.1 Moisture content of specimens.

8.1.2 Average cumulative, corrected relative mass loss after 12 cycles, ($\bar{S}_i$).

8.1.3 Number of cycles survived if the specimens did not survive 12 cycles of testing.

8.1.4 Results of visual observation after each cycle (physical deterioration).

9. Precision and Bias [4]

9.1 *Precision*:

9.1.1 The precision of this test method, in terms of standard deviation, was determined in an interlaboratory experiment

[4] Supporting data are available from ASTM Headquarters. Request R.R. 1003.

involving four laboratories, two types of specimens and respective controls. Duplicates of specimens and controls were measured in each laboratory.

9.1.2 The precision of this test method can be expressed as follows:

Sample Code	Mean ($\bar{X}$)	Standard Deviation (S)
LFP	0.09	0.07
CFP	1.99	1.20

9.1.3 The precision of this test method may be dependent on the level of the properties measured.

9.2 *Bias*:

Since there is no accepted reference material suitable determining the bias for the procedure in this test method, no statement on bias is being made.

Designation: D 4843 – 88 (Reapproved 1999)

Standard Test Method for Wetting and Drying Test of Solid Wastes[1]

This standard is issued under the fixed designation D 4843; the number immediately following the designation indicates the year of original adoption or, in the case of revision, the year of last revision. A number in parentheses indicates the year of last reapproval. A superscript epsilon (ε) indicates an editorial change since the last revision or reapproval.

1. Scope

1.1 This test method covers procedures for determining material losses produced by repeated wetting and drying of solid waste specimens. It also covers the visual observation of the disintegration of solid specimens.

1.2 This test method intends that the material used in the procedure be physically, chemically, and biologically representative; hence it does not address problems as a result of the inhomogeneity of specimens.

1.3 *This standard does not purport to address all of the safety concerns, if any, associated with its use. It is the responsibility of the user of this standard to establish appropriate safety and health practices and determine the applicability of regulatory limitations prior to use.*

2. Referenced Documents

2.1 *ASTM Standards:*

C 305 Practice for Mechanical Mixing of Hydraulic Cement Pastes and Mortars of Plastic Consistency[2]

D 2216 Test Method for Laboratory Determination of Water (Moisture) Content of Soil and Rock[3]

3. Significance and Use

3.1 This test method is intended for the evaluation of the wetting and drying resistance of monolithic, solid, solidified/stabilized wastes under the testing conditions of this test method.

3.2 This test method may be used for the comparison of wetting and drying resistance of wastes.

3.3 Data tabulated in Table 1, Table 2, and Table 3 may be used to observe irregularities caused by inhomogeneity of specimens and/or comparison of mass loss-cycle relations of different wastes, as well as to measure method-related mass losses such as matrix dissolution.

4. Apparatus

4.1 *Disposable Molds*, 44 mm inside diameter by 74 mm in length.

4.2 *Balance or Scale*, with a capacity at least 50 % greater than the mass of the specimen and beaker, and a sensitivity of 0.01 g.

4.3 *Drying Oven*, a thermostatically controlled drying oven capable of maintaining a temperature of 60 ± 2°C; to be used for drying moisture specimen and for the solids content determination.

4.4 *Oven*, capable of maintaining a temperature of 60 ± 3°C; at a nitrogen purge rate specified in 4.5.

4.5 *Flow Controller*, to set nitrogen purge flow at a rate that will give 30 ± 5 min residence time.

4.6 *Moisture Chamber*, a suitably covered container capable of maintaining a temperature of 20 ± 3°C and minimum 95 % relative humidity, for preconditioning specimens.

4.7 *Beakers*, 400-mL size (narrow type), to store sample and to collect particulates.

4.8 *Tongs*, to handle samples.

5. Sample Preparation

5.1 *Specimen Size*—44 mm diameter by 74 mm in length.

5.1.1 Specimens may be cut to size from larger samples.

5.1.2 Specimens can also be molded in disposable plastic molds. When molding specimens refer to Practice C 305 (see 2.1).

NOTE 1—Practice C 305 refers to pastes and mortars. Molding materials with different consistency may require modifications and may result in different precision.

5.2 Condition samples that are not molded for this test in the moisture chamber for a period of seven days.

5.2.1 Samples molded for this test have to be cured in the moisture chamber for a period of 28 days.

6. Procedure

6.1 Select one specimen for moisture content determination.

6.2 Determine moisture content of sample with Test Method D 2216 but revised to use a temperature of 60 ± 3°C (see section 2.2).

6.3 Select three specimens for testing and three for control and mark them respectively.

6.4 Weigh specimens (accuracy to 0.01 g).

6.5 Place each specimen into a beaker of known tare mass (accuracy to 0.01 g) and cover it.

6.5.1 Use watch glass or plastic wrap.

6.5.2 The tare mass of beaker shall be determined after drying in accordance with Test Method D 2216.

6.6 Place the three beakers containing the testing specimens

---

[1] This test method is under the jurisdiction of ASTM Committee D34 on Waste Management and is the direct responsibility of Subcommittee D34.01.06 on Analytical Methods.

Current edition approved July 29, 1988. Published September 1988.

[2] *Annual Book of ASTM Standards*, Vol 04.01.

[3] *Annual Book of ASTM Standards*, Vol 04.08.

TABLE 1 Control Group

Laboratory:
Technician:

Sample name:
Sample Id No.:
Test Start Date:

Sample 1 / Cycle No.	1					2					3				
Date	$T_{i,c,1}$	$B_{i,c,1}$	$W_{i,c,1}$	$M_{c,1}$	$R_{i,c,1}$	$T_{i,c,2}$	$B_{i,c,2}$	$W_{i,c,2}$	$M_{c,2}$	$R_{i,c,2}$	$T_{i,c,3}$	$B_{i,c,3}$	$W_{i,c,3}$	$M_{c,3}$	$R_{i,c,3}$
1.															
2.															
3.															
4.															
5.															
6.															
7.															
8.															
9.															
10.															
11.															
12.															

TABLE 2 Sample Group

Laboratory:
Technician:

Sample name:
Sample Id No.:
Test Start Date:

Sample 1 / Cycle No.	1					2					3				
Date	$T_{i,s,1}$	$B_{i,s,1}$	$W_{i,s,1}$	$M_{s,1}$	$R_{i,s,1}$	$T_{i,s,2}$	$B_{i,s,2}$	$W_{i,s,2}$	$M_{s,2}$	$R_{i,s,2}$	$T_{i,s,3}$	$B_{i,s,3}$	$W_{i,s,3}$	$M_{s,3}$	$R_{i,s,3}$
1.															
2.															
3.															
4.															
5.															
6.															
7.															
8.															
9.															
10.															
11.															
12.															

in an oven. Maintain the temperature at 60 ± 3°C for 24 h while purging the oven with nitrogen gas at the controlled flow rate corresponding to 30 ± 5 min residence time.

6.6.1 In order to remove moisture from the nitrogen stream, a water-cooled condenser and condensate collection flask may be used downstream from the oven.

6.7 Store the three beakers with the control specimens in the moisture chamber at 20°C for 24 h.

6.8 Remove the specimens from the vacuum oven and the moisture chamber. Allow 1 h for the sample to cool to room temperature. Add 230 mL distilled water to the beaker to fully cover the specimens.

6.8.1 Add laboratory temperature water 20 ± 3°C.

6.9 Place a watch glass or plastic wrap on the beakers and store the water covered specimens at 20 ± 3°C for 23 h; then transfer them to new beakers prepared according to 6.5.

6.9.1 Use tongs to transfer specimens. Excessive tong pressure may result in premature failure or damage specimen.

6.10 Remove any loosely attached particulates by spraying distilled water from a wash bottle to the surface of specimen (10 to 20 mL distilled water). Let the water drain into the beaker of origin.

6.11 Conduct visual observation on the specimens' physical deterioration including: cracking, fracturing, integrity, and surface roughness.

6.12 Determine the specimens' mass loss; solid content in beakers by evaporating water at 60 ± 3°C in drying oven.

6.13 Correct the average relative mass loss of samples using the average relative mass loss of control specimens.

6.14 Repeat the procedures in 6.5 through 6.10 eleven additional times, for a total of 12 cycles.

6.15 Terminate experiment of all specimens if the corrected

TABLE 3 Relative Weight Loss

Laboratory:
Technician:
Sample name:
Sample Id No.:
Test Start Date:

Cycle No.	$\bar{R}_{i,s}$	$\bar{R}_{i,c}$	$\bar{C}_i$	$\bar{S}_i$	Observations
1.					
2.					
3.					
4.					
5.					
6.					
7.					
8.					
9.					
10.					
11.					
12.					
$\bar{S}$					

cumulative mass loss of any of the specimens exceeds 30 % (failure), and note the number of cycles survived.

7. Calculation

7.1 Calculate the dry mass of the specimens as follows:

$$M_s = 1 - \frac{w}{100} M_{sw} \quad (1)$$

where:

M_s = oven dry mass of specimen,
M_{gw} = initial mass of specimen, and
w = moisture content, %.

It is assumed that the moisture contents of specimens are identical. Oven dry masses of sample and control specimens are calculated on that basis.

7.2 Calculate corrected mass loss of specimens after each cycle. Express mass loss in percent of initial calculated oven-dry mass. Calculate average cumulated corrected mass loss of specimens after each cycle.

$$W_{i,s,j} = T_{i,s,j} - B_{i,s,j} \text{ g} \quad (2)$$

where:

$W_{i,s,j}$ = mass loss of sample j during cycle i, g,
$T_{i,s,j}$ = oven-dry mass of beaker and residue of sample j after cycle i, in g, and
$B_{i,s,j}$ = oven-dry mass of beaker for sample j before cycle i, g.

$$W_{i,c,j} = T_{i,c,j} - B_{i,c,j} \text{ g} \quad (3)$$

where:

$W_{i,c,j}$ = mass loss of control j during cycle i, in g,
$T_{i,c,j}$ = oven-dry mass of beaker and residue of control j after cycle i, in g, and
$B_{i,c,j}$ = oven-dry mass of beaker for control j before cycle i, in g.

$$R_{i,s,j} = \frac{W_{i,s,j}}{M_{s,j}} \% \quad (4)$$

where:

$R_{i,s,j}$ = relative mass loss of sample j during cycle i, %,
$W_{i,s,j}$ = mass loss of sample j during cycle i, in g, and
$M_{s,j}$ = oven-dry mass of specimen j, in g.

$$R_{i,c,j} = \frac{W_{i,c,j}}{M_{c,j}} \% \quad (5)$$

where:

$R_{i,c,j}$ = relative mass loss of control j during cycle i %,
$W_{i,c,j}$ = mass loss of control j during cycle i, in g, and
$M_{c,j}$ = oven-dry mass of control j, in g.

$$\bar{R}_{i,s} = \frac{\sum_{j=1-3} R_{i,s,j}}{3} \% \quad (6)$$

where:

$\bar{R}_{i,s}$ = average relative mass loss of samples ($j = 1 - 3$) during cycle i %, and
$R_{i,s,j}$ = relative mass loss of sample j during cycle i %.

$$\bar{R}_{i,c} = \frac{\sum_{j=1-3} R_{i,c,j}}{3} \% \quad (7)$$

where:

$\bar{R}_{i,c}$ = average relative mass loss of control ($j = 1 - 3$) during cycle i %, and
$R_{i,c,j}$ = relative mass loss of control j during cycle i %.

$$\bar{C}_i = \bar{R}_{i,s} - \bar{R}_{i,c} \% \quad (8)$$

where:

$\bar{C}_i$ = average corrected relative mass loss of samples ($j = 1 - 3$) during cycle i %,
$\bar{R}_{i,s}$ = average relative mass loss of samples ($j = 1 - 3$) during cycle i %, and
$\bar{R}_{i,c}$ = average relative mass loss of control ($j = 1 - 3$) during cycle i %.

$$\bar{S}_i = \sum_{i=1-i} \bar{C}_i \% \quad (9)$$

where:

$\bar{S}_i$ = average cumulated, corrected relative mass loss of samples after i cycles %, and
$\bar{C}_i$ = average corrected relative mass loss of samples ($j = 1 - 3$) during cycle i %.

$$\bar{S} = \sum_{I=1-12} \bar{C}_i \% \quad (10)$$

where:

$\bar{S}$ = average cumulated, corrected relative mass loss of samples after 12 cycles, %, and
$\bar{C}_i$ = average corrected relative mass loss of samples ($j = 1 - 3$) during cycle i, %.

8. Report

8.1 Report the following information:

8.1.1 Moisture content of specimens.

8.1.2 Average cumulative, corrected relative mass loss after 12 cycles. ($\bar{S}$)

8.1.3 Number of cycles survived if the specimens did not survive 12 cycles of testing.

8.1.4 Results of visual observation after each cycle (physical deterioration).

9. Precision and Bias [4]

9.1 The precision of this test method, in terms of standard deviation, was determined in an interlaboratory experiment involving five laboratories, two types of samples and respective controls. Duplicates of samples and controls were measured in each laboratory.

9.2 The precision of this test method can be expressed as follows:

Sample Code	Mean ($\bar{X}$)	Standard Deviation(s)
LFP	0.024	0.038
CFP	0.112	0.138

9.3 The precision of this test method may be dependent on the level of the properties measured.

[4] Supporting data are available from ASTM Headquarters. Request RR: D34-1004.

Standard Guide for Air Monitoring at Waste Management Facilities for Worker Protection[1]

This standard is issued under the fixed designation D 4844; the number immediately following the designation indicates the year of original adoption or, in the case of revision, the year of last revision. A number in parentheses indicates the year of last reapproval. A superscript epsilon (ε) indicates an editorial change since the last revision or reapproval.

1. Scope

1.1 This guide is intended to provide a standardized approach for establishing and carrying out an air monitoring program to protect workers at waste management facilities. This guide may apply to routine operations at an active treatment, storage, or disposal site or the extraordinary conditions that can be encountered in opening and cleaning up a remedial action site.

1.2 Any user of this guide must understand that it is impossible to predict all the difficulties that could develop at a waste management facility due to hazardous airborne emissions. Although air contaminant measurements obtained in accordance with this guide may indicate acceptable or tolerable levels of toxic agents are present, care and judgment must still be exercised before concluding that all atmospheric contaminants at the site are under control.

2. Referenced Documents

2.1 *ASTM Standards:*

D 1356 Terminology Relating to Atmospheric Sampling and Analysis[2]

D 1357 Practice for Planning the Sampling of the Ambient Atmosphere[2]

D 1605 Practices for Sampling Atmospheres for Analysis of Gases and Vapors[2]

D 3614 Guide for Evaluating Laboratories Engaged in Sampling and Analysis of Atmospheres and Emissions[2]

D 4687 Guide for General Planning of Waste Sampling[3]

E 548 Guide for General Criteria Used for Evaluating Laboratory Competence[4]

2.2 *Federal Standards:*

OSHA Analytical Methods Manual[5]

NIOSH Manual for Analytical Methods[6]

OSHA, 29 CFR Part 1910 Hazardous Waste Operations and Emergency Response; Interim Final Rule, December 1986[7]

3. Terminology

3.1 *Definitions:*

3.1.1 *General*—Terminology commonly used in air monitoring can be found in Terminology D 1356.

3.2 *Descriptions of Terms Specific to This Standard:*

3.2.1 *operating site*—an operating site is a location or facility where waste is treated, stored, or disposed as part of an on-going operation.

3.2.2 *remedial action site*—a remedial action site is a location or facility that may pose a threat to human health and the environment.

4. Summary of Guide

4.1 The procedures described in this guide address safety considerations, acute health hazards, and chronic health hazards due to airborne hazardous materials.

4.2 Monitoring concepts are described for cleanup operations at remedial action sites as well as routine activities at operational waste management sites.

5. Significance and Use

5.1 The techniques of air monitoring are many and varied. This guide is intended to describe the standard approaches that are used in designing an air monitoring program to protect waste management site workers.

5.2 When entering a remedial action site to initiate an investigation or a cleanup operation, operating personnel may be faced with the extreme hazards of fire, explosion, and acute or chronic health hazards. A thorough safety and health program, including a site-specific safety and health plan, must be in place to direct worker activity. Details for such plans can be found in the OSHA Interim Final Rule for Hazardous Waste Operations and Emergency Response and Refs **(1, 2)**.[8] Air monitoring is an integral part of such a program. This guide describes equipment and sampling procedures which can be used to evaluate the airborne hazard potential so as to gain and maintain control over the situation at the site.

---

[1] This guide is under the jurisdiction of ASTM Committee D34 on Waste Management and is the direct responsibility of Subcommittee D34.01.01 on Planning for Sampling.

Current edition approved Oct. 6, 1988. Published February 1989.

[2] *Annual Book of ASTM Standards*, Vol 11.03.

[3] *Annual Book of ASTM Standards*, Vol 11.04.

[4] *Annual Book of ASTM Standards*, Vol 14.02.

[5] 1985 manual available from Occupational Safety and Health Administration, OSHA Analytical Laboratory, Salt Lake City, UT.

[6] Third edition manual, February 1984, available from the National Institute of Occupational Safety and Health, (NIOSH), Cincinnati, OH.

[7] Available from the Superintendent of Documents, Government Printing Office, Washington, DC, 20401.

[8] The boldface numbers in parentheses refer to the list of references at the end of this guide.

5.3 Upon obtaining readings at the site, a decision must be made as to whether conditions are under control or not. That decision will depend on the nature of the contaminants (toxicity, reactivity, volatility, etc.), the extent (area affected, number of workers, etc.) of the problem and the level of worker protection available. Since all such parameters will be site specific, the necessary decision-making is beyond the range of this guide.

5.4 This guide does not include monitoring sites containing radioactive materials, nor does it cover general safety aspects, such as access to emergency equipment or medical support of emergency needs. These items should be covered in a safety and health plan.

5.5 It is recommended that this guide be used in conjunction with Guide D 4687.

6. General Considerations

6.1 That aspect of science which routinely deals with the assessment of airborne hazards to workers is known as industrial hygiene. Professional industrial hygienists, besides measuring the concentration of contaminants in air, recommend means for controlling such airborne hazards, protecting workers, and demonstrating compliance with applicable laws and regulations. A certified industrial hygienist generally offers the optimum combination of background and credentials for recognizing, evaluating, and controlling workplace health hazards. If industrial hygiene staff support is not available on site, coverage can be obtained through the use of consultants and possibly through loss prevention insurance carriers. The remainder of this guide reflects the general thought process that an industrial hygiene professional would most likely go through in establishing an air monitoring program to protect workers at a waste management site.

6.2 *Establishing a Test Protocol*:

6.2.1 Various combinations of equipment and sampling techniques are used in work place air monitoring. The best monitoring program is one that combines accuracy with timely response in a cost effective manner.

6.2.2 The particular test protocol which is selected for an industrial hygiene study depends on the nature of the contaminants and the end purpose of the monitoring effort (that is, routine monitoring, searching for worst case exposure, looking for contaminant leaks in a process).[9]

6.3 *Selecting Specific Methods*:

6.3.1 The choice of sampling method is most often tied in with the analytical method. There may be no difference in the analytical work whether it is for a 15-min ceiling sample or a 7-h full day sample. If the analytical method has poor sensitivity, however, it may be necessary to increase the pump flow rate for the short duration sample to make certain that sufficient sample is collected. Such fine adjustments must be worked out between the sampling personnel and the laboratory personnel. Extensive guidance on the latest developments in air sampling technology is available in Refs **(3, 4)**.

6.3.2 A number of sources of information are available to describe general methodology. Practice D 1605 lists some of the classic methods that have been used when sampling for gases or vapors. The American Conference of Governmental Industrial Hygienists offers a publication, Ref **(5)**, that provides a review of newer equipment and methodology. The final combination of equipment and procedures is predicted on the precision, accuracy, and sensitivity needed to support the test protocol.

6.3.3 Once the goals and protocol for the sampling program have been set, specific sampling/analytical methods must be selected. Within the *Annual Book of ASTM Standards*, Volume 11.03 is dedicated to atmospheric analysis and to occupational health and safety issues. Some applicable methods from that reference are listed in Annex A1. Other sources of health and safety support include the NIOSH Manual of Analytical Methods and the OSHA Analytical Methods Manual. The specific equipment and sampling media for a particular set of airborne contaminants are selected from sources such as these.

7. Procedures

7.1 *Operating Site*:

7.1.1 The procedures described in this section apply to air monitoring activities at an operational waste treatment, storage, or disposal site. At an operating site, controls (work practices, engineering controls, and personal protective equipment) would be in place to minimize the exposure of workers to hazardous conditions. These are defined in the health and safety plan.

7.1.2 *Knowledge of Materials*—Knowledge of the materials arriving at or present at an operating site is critical to the design of a sampling plan. If hazardous wastes are arriving, be sure that they are listed on the manifest. The results of waste sample analyses will also help to identify contaminants of greatest concern in an incoming shipment. It is also likely that specific users of the disposal site will tend to be consistent in the types of wastes they send to the site based on the generating process and history of shipment. For example, paint manufacturers will most likely send mixtures of solvents, resins, and pigments, whereas plating firms will generally send alkaline sludge of heavy metal waste; and so on. Deviation from established patterns, however, is possible and should not be discounted in sampling plan design.

7.1.3 *Worker Sampling*:

7.1.3.1 Of all the different techniques for workplace air monitoring, personal sampling of the worker's breathing zone is paramount. While some workers may be quite sedentary in an operations trailer at a control panel, others may be out covering all areas of the work site. For this reason, the assessment must be capable of following the activity of the worker.

7.1.3.2 The first order of personal monitoring is long duration time-weighted-average (TWA) sampling. For an 8-h work shift, be sure that TWA samples are at a minimum of 7-h duration either as a single sample or a series of two or more samples. For any other work hour situation, the procedure is to sample for the duration of the shift less 1 h. For workers handling organic wastes (for example, vapor degreaser solvent waste) the program would call for charcoal tube sampling with

[9] Subcommittee E34.18 of ASTM Committee E34 on Occupational Health and Safety is developing a guide for industrial hygiene air monitoring programs titled "Standard Guide to Air Sampling Strategies for Worker and Workplace Protection."

analysis for one or two of the chlorinated solvents most likely to be present in the waste. Such TWA monitoring, as well as the following information, would be repeated periodically to ensure that worker exposure is not increasing.

7.1.3.3 Another form of personal monitoring that would be carried out is for peak exposures. For example, 15-min ceiling samples might be taken while a set of containers was being opened to inspect or remove the contents. The same type of sampling might be done while pumping the contents of a truck into a holding tank. At these times, personal protective equipment (for example, respiratory protection) is often used to minimize worker exposure to vapors. Ceiling samples will help ensure that workers are using respirators having a high enough protection factor.

7.1.3.4 Ceiling samples might be the only form of monitoring for certain toxic agents. If waste acid pickling solution were to come in from a steel mill for neutralization, it might be appropriate to sample for hydrogen chloride. In that instance, only 15-min samples would be of interest, because that is how exposure to HCl is controlled by health/regulatory agencies.

7.1.3.5 New equipment has come into use to cover both TWA and peak sampling. Some personal dosimeters, worn by the employees, give an overall average exposure and also record the instantaneous exposures of the worker during the day. These units, which are read out on a portable computer, are generally good for only one particular contaminant, though all the different types are read using the same computer. These might be very useful in monitoring a heavy equipment operator for carbon monoxide or a waste treatment plant attendant for sulfur dioxide.

7.1.3.6 Another concept to be considered in both the monitoring and safety and health plans is the additive effect of certain substances. Paragraph 7.1.3.2 presented the concept of screening for only one or two solvents. When this is done, the eventual comparison with permissible exposure limits must be done using a safety factor. This safety factor is intended to take account of the possible effects of other similar compounds which are likely to be present, but are not measured routinely.

7.1.4 *Area Monitoring*:

7.1.4.1 A good complement to personal monitoring is fixed location area monitoring. This can be done with either sample collecting-type equipment, direct reading instruments, or specialized fixed-parameter monitors such as those described in 7.1.3.5. Area monitoring offers the advantage of potentially providing an early warning.

7.1.4.2 A combustible vapor meter in a solvent storage area can give warning before an employee must walk in to find a leak.

7.1.4.3 A carbon monoxide monitoring system around a pyrolyzer or incinerator can warn both the operator in the control room and workers in the loading area of a system upset.

7.1.4.4 An oxygen meter permanently mounted in a below ground pit can warn an employee of an oxygen deficient atmosphere before he enters the confined space.

7.1.4.5 Direct reading colorimetric tubes Ref **(6)**, offer a convenient means for obtaining a quick reading. Besides their suitability for qualitative checks (see Annex A2), they also provide reasonable quantitative estimates.

7.1.5 *Complex Exposure Potential*:

7.1.5.1 Although much of the sampling effort may involve monitoring for one or two particular contaminants on specific operations, there will be other times when the exposure potential is more complex. Examples of more complex monitoring might include: *1*) where a sludge is handled on site, and there is a chance of spillage and eventual spreading of the debris around the site by vehicular traffic and wind, dust samples will need to be analyzed periodically for heavy metals; *2*) where waste from a polymer plant (in particular one processing nitrile rubber or acrylonitrile butadiene styrene, (ABS) plastic) is handled on site, it may be necessary to devise a sampling protocol which looks for trace quantities of acrylonitrile in an atmosphere dominated by one or two less harmful organic vapors; *3*) where polychlorinated biphenyl, (PCB) vapor can be carried into the atmosphere by methane gas evolving from a closed site, Ref **(7)**, monitoring must cover these and perhaps other compounds; and *4*) where a wide range of similar compounds arise, such as in some organic wastes and landfill gas, the cumulative effect must be estimated rather than the potential effect of individual contaminants.

7.1.6 *Data Storage and Analysis*:

7.1.6.1 The various forms of air monitoring described in 7.1.3, 7.1.4, and 7.1.5 will result in the accumulation of a substantial amount of data by the site operator. The data need to be recorded and catalogued in a manner that provides for ready retrieval and comparison.

7.1.6.2 Store and retrieve data so that the level of airborne contamination can be reviewed over time. In this way seasonal or diurnal trends may be identified.

7.1.6.3 The site operator may want to determine if certain shipments or customers are sending waste material that is difficult to handle.

7.1.7 *Quantitative Considerations*:

7.1.7.1 The premise of most of the discussion of Section 7 is that the site operator has at least some working knowledge of the materials being handled. As stated previously, however, surprises can be expected.

7.1.7.2 An unexpected odor or phase separation may indicate an unknown is present in the waste material.

7.1.7.3 An abnormal reaction in a neutralization process may be a sign that an unexpected volatile is being emitted. Given this reality, even the best run waste site may have to analyze for unknowns in the work atmosphere. For the following reasons it is important to be aware of the qualitative aspects of air monitoring as they are described in 7.2.

7.2 *Remedial Action Site*:

7.2.1 The level of hazard that is found by air monitoring at a remedial action waste site can vary from relatively innocuous to very dangerous. The uncertainty associated with the types and composition of wastes present at these sites complicates virtually every aspect of site cleanup and monitoring. For these reasons a thorough site characterization must be made before work is started to provide data for a site-specific health and safety plan, including subsequent air monitoring requirements.

7.2.2 *Qualitative Assessment*:

7.2.2.1 From an operational viewpoint, a prime difference between a remedial action site and an operating site is the need for qualitative assessment. The unknowns at an abandoned site require a qualitative analysis of the work environment prior to any quantitative measurements.

7.2.2.2 Before starting cleanup activity, check available records to gain knowledge of the materials that might have been treated, stored, or buried on the site. Be sure that labels on drums also are checked to see if there is some information on the hazardous nature of the material contained inside.

7.2.2.3 Qualitative assessment must be conducted in a systematic fashion. Be sure that the operating areas are mapped and perhaps gridded so that the assessment can be done with minimal chance of overlooking important zones.

7.2.3 *Safety Monitoring*:

7.2.3.1 Identify life threatening situations due to fire, explosion, and toxic hazards as the first air monitoring concerns for personnel starting a cleanup action.

7.2.3.2 Enclosed spaces bring the danger of suffocation by excessive chemical vapors causing an oxygen deficiency.

7.2.3.3 For these reasons make sure that the first pieces of equipment on site, Ref **(8)** are the organic and combustible vapor meters and oxygen meters. These would be used to scan the area and map out any high hazard areas.

7.2.3.4 Personal monitoring devices are also available to warn workers of unsafe conditions immediately. This type of unit, which has a preset alarm, is worn by the worker on his belt as he goes about the site performing his work.

7.2.3.5 When a new area of the site is entered or another level is unearthed, the mapping routine for imminent danger situation would be repeated. The routine monitoring of the site-specific health and safety plan would be run from then on.

7.2.4 *Direct Reading Meters and Dosimeters*:

7.2.4.1 Once the initial safety survey has been completed, a more sensitive screening is needed to determine the presence of acute and chronic health hazards. It may be necessary to use the air data to define potentially contaminated zones where specific personal protective equipment and decontamination procedures are required.

7.2.4.2 A wide array of direct reading instruments and personal dosimeters can be used to search for classes of, or specific, chemical agents.

7.2.4.3 Most direct reading instruments either lack specificity or are susceptible to interferences. The technology used for the basic design, the relative price of the instrument within a given design, and the skill of the operator will determine the suitability of an instrument in terms of freedom from such bias.

7.2.4.4 The complexity of the chemical matrix at the site will determine the level of sophistication of instrument array that will be needed to obtain a reliable identification of the hazardous agents that are present.

7.2.4.5 A listing of such equipment along with a discussion of capabilities and limitations is presented in Annex A1. Additional information on the useful application of such equipment is available in Ref **(4)**.

7.2.5 *Qualitative Sample Collection*:

7.2.5.1 Just as with quantitative determination, the most reliable technique in qualitative sampling, for the most part, is concentration of airborne contaminants using a collection device followed by appropriate laboratory analysis.

7.2.5.2 For metal dusts (or dust containing heavy metal contaminants) a series of 37-mm diameter membrane samples or high volume filter samples would be taken at various locations around the site.

7.2.5.3 These would then be submitted for analysis via some general analytical technique such as inductively-coupled plasma spectrometry, which can determine a large number of metals per sample. Certain volatile materials, such as arsenic compounds or mercury, would have to be sampled separately.

7.2.5.4 Organic vapors can be sampled by drawing site air through an array of absorbent packed tubes. By using different tube packing materials in a manifold system, it is possible to sample for a wide range of organics concurrently.

7.2.5.5 A popular combination of packings is charcoal, magnesium silicate,[10] phenylene oxide porous polymer,[11] and silica gel. These sample tubes would be desorbed either thermally or using an appropriate mix of solvents for eventual analysis via a combination gas chromatograph/mass spectrometer or other appropriate analytical method. Decomposition can be a problem using thermal desorption.

7.2.5.6 Additional techniques, more specific in nature, can be used for other agents. If acid gases are suspect, air can be bubbled through a buffer solution which is then checked by pH meter, specific ion electrode, or ion chromatography. For pesticides and PCBs even more specific sampling/analysis combinations are available.

7.2.5.7 No direct reading meter is available for asbestos, so care is required if suspicious deposits are encountered or anticipated. Bulk samples will have to be taken for qualitative identification using a technique such as polarizing light microscopy. Any attempt to use an air sample for qualitative identification of asbestos fibers will most likely end up in the use of an extremely demanding analytical procedure, such as a scanning electron microscope with X-ray diffraction analysis. Once asbestos has been positively identified, conventional fiber counting techniques are used for quantitation in air samples.

7.2.6 *Select Toxic Agents for Monitoring*:

7.2.6.1 Previous procedures described in 7.2.1 through 7.2.5 should provide site operating or surveying personnel with an appreciable amount of qualitative information from direct reading instruments and laboratory analysis of collected air samples. In addition, data should be available from the preliminary paper studies and from laboratory results on the analysis of bulk samples taken from waste drums, lagoons, tanks, other equipment, and perhaps contaminated soil.

7.2.6.2 Select key parameters to be monitored quantitatively based on:

a) Those which must be monitored due to applicable air regulatory controls (for example, lead, benzene).

b) Those offering the most severe risk to health, even in trace quantities (for example, methyl isocyanate).

c) Those that are dominant in the site environment due to quantity and volatility or dispersion (for example, chlorinated solvents).

[10] Magnesium silicate available from various sources under the tradename Florisil, has been found suitable for this purpose.

[11] Phenylene oxide porous polymer available from various sources under the tradename Tenax, has been found suitable for this purpose.

7.2.7 *Quantitative Assessment*:

7.2.7.1 Once the hazardous agents have been identified, air sampling routines must be established to determine the concentration of those substances in the site atmosphere. This information is needed to ensure that site workers are adequately protected and for regulatory requirements.

7.2.7.2 *Operating Site Similarity*—Once the unknowns at the remedial action site are for the most part accounted for, establishment of a testing program is similar to that for an operating disposal site. Refer then to 7.1 for guidance except for the modifications that follow.

7.2.7.3 *Increased Sampling Frequency*—Conditions at a cleanup site are much more unstable than at an operating disposal site. The level of control over airborne emissions is greatly reduced in a cleanup operation. As a result, much wider inter-day and intra-day fluctuations in the concentrations of airborne contaminants can be expected. To account for such variations, increase frequency. In a cleanup operation, daily quantitative monitoring may be required.

7.2.7.4 *Timely Analytical Support*—Considering the wide range of conditions that are possible at remedial action sites, it is most important that timely analytical support be available. If an intense sampling program has been designed in attempting to correct an out-of-control condition, a 2 to 4-week wait for analytical results is out of the question. Since some samples in the program may very well be of lesser significance, it would not matter if site personnel had to wait for those results. For that reason, it is recommended that some type of priority system be established to identify those samples which are critical and those which can be put on a slower schedule.

7.2.7.5 *Periodic Qualitative*—To detect residual unknown emissions at the site, it is recommended that qualitative determinations be carried out periodically. The results of these tests may require additional quantitative monitoring. As a minimum, the results for the analysis of bulk samples should be reviewed periodically to identify any new compounds which may become airborne hazards.

8. Quality Assurance

8.1 Regardless of the end purpose of the particular air sampling effort, the work must be conducted with a firm awareness of quality control. Procedures must be in place to control the calibration of direct reading meters and sampling equipment, sample handling, analytical precision and bias, and data handling.[12]

8.2 A full-range quality assurance program can be developed using Practice D 3614.

8.3 If a certain segment of the surveillance effort, such as analytical support, is to be handled by an outside organization, then the quality program must be extended over the outside organization. One approach to this particular situation is to restrict analytical support to a professionally accredited industrial hygiene laboratory. Though the laboratory would have its own quality control program, it would still be necessary to evaluate the outside organization in the manner of Guide E 548.

---

[12] Subcommittee D34.01 is in the process of developing a standard for quality assurance in waste and environmental sampling that has an excellent discussion on the use of control samples.

ANNEXES

(Mandatory Information)

A1. INDUSTRIAL HYGIENE TEST METHODS AND PRACTICES

TABLE A1.1 ASTM Standards[2]

Test Requirements	
Asbestos	D 4240 Test Method for Airborne Asbestos Concentration in Workplace Atmosphere
bis (Chloromethyl) Ether	D 3476 Test Method for *bis* (Chloromethyl) Ether (*bis* CME) in Workplace Atmospheres (Gas Chromatography—Mass Spectrometry)
Metals	D 4185 Practice for Measurement of Metals in Workplace Atmosphere by Atomic Absorption Spectrophotometry
Hydrocarbon Gases	D 2820 Test Method for C_1 through C_5 Hydrocarbons in the Atmosphere by Gas Chromatography
Carbon Monoxide	D 3162 Test Method for Carbon Monoxide in the Atmosphere (Continuous Measurement by Nondispersive Infrared Spectrometry)
Oxides of Nitrogen	D 3824 Test Methods for Continuous Measurement of Oxides of Nitrogen in the Ambient or Workplace Atmosphere by the Chemiluminescent Method
Fluoride	D 3269 Test Methods for Fluoride Content of the Atmosphere and Plant Tissues (Manual Procedures), Analysis for
Hydrogen Sulfide	D 4323 Test Method for Hydrogen Sulfide in the Atmosphere by Rate of Change of Reflectance
Inorganic Lead	D 3413 Test Method for Lead (Inorganic) in Workplace Atmospheres by Atomic Absorption Spectrometry
Mercaptans	D 2913 Test Method for Mercaptan Content of the Atmosphere
Particulate (Coke Ovens)	D 4600 Test Method for Benzene-Soluble Particulate Matter in Workplace Atmospheres, Determination of
Respirable Dust	D 4532 Test Method for Respirable Dust in Workplace Atmospheres
Sulfur Dioxide	D 3449 Test Method for Sulfur Dioxide in Workplace Atmospheres (Barium Perchorate Method)
Organic Vapors	D 3687 Practice for Analysis of Organic Compound Vapors Collected by the Activated Charcoal Tube Adsorption Method
General Air Analyzer	D 3249 Practice for General Ambient Air Analyzer Procedures
Particulate (Hi-Vol)	D 4096 Practice for Particulate (Airborne Matter), Collection and Mass Determination of, Application of the Hi-Vol (High Volume) Sampler Method
Organic Vapors	D 3686 Practice for Sampling Atmospheres to Collect Organic Compound Vapors (Activated Charcoal Tube Adsorption Method)
Detector Tubes	D 4490 Practice for Toxic Gases or Vapors Using Detector Tubes, Measuring the Concentration of
Length-of-Stain Dosimeters	D 4599 Practice for Toxic Gases or Vapors Using Length-of-Stain Dosimeter, Measuring the Concentration of

A2. DIRECT READING AIR MONITORING EQUIPMENT AND DOSIMETERS

A2.1 *Photo-Ionization Meter-Battery Operated Device*:

A2.1.1 *Basic Design*—Draws in air and ionizes organic compounds using ultra-violet radiation; charged plates measure proportional current flow due to ionized field Ref **(2)**.

A2.1.2 *Limits*—Sensitivity down to the ppm range, but non-specific except for very broad categories (for example, aromatic versus non-aromatic) by selection of lamp with appropriate UV output.

A2.2 *Colorimetric Indicator Tubes*:

A2.2.1 *Basic Design*—Single use glass tubes containing chemical reactants which change color in presence of the pertinent analyte; length of stain is proportional to concentration Ref **(3)**.

A2.2.2 *Limits*—Convenient apparatus with poor precision but good sensitivity. Response is generally for an entire class of compounds (for example, chlorinated hydrocarbon) and sometimes susceptible to interferences (oxidizer may bleach out color).

A2.3 *Electrolytic Cell Detectors*:

A2.3.1 *Basic Design*—Battery operated monitors using solid state electrolytic cell in which conductivity of cell material changes in proportion to concentration of H_2S, CO, CH_4, SO_2 or O_2.

A2.3.2 *Limits*—Good sensitivity and freedom from interferences, but each cell is for a specific gas resulting in five different instruments to cover all five gases.

A2.4 *Portable Gas Chromatograph*:

A2.4.1 *Basic Design*—Truly portable GC equipment with choice of Flame Ionization Detector (FID), Electron Capture Detector, (ECD), or Photo Ionization Detector, (PID) and with choices of standard packed columns; though not a direct reading instrument, it is used in the field.

A2.4.2 *Limits*—Relatively good sensitivity and can separate mixtures in a gas sample, but as with GC, requires calibration with analytes of interest at time of use.

A2.5 *Portable Combustible Gas Meter*:

A2.5.1 *Basic Design*—Battery operated catalytic combustion changes electrical conductivity of detector; for H_2, CO, and any combustible organic.

A2.5.2 *Limits*—Used for higher concentrations, particularly in low % range to check for explosive atmospheres. Non-specific and can be a source of ignition.

A2.6 *Portable Infrared Analyzer*:

A2.6.1 *Basic Design*—Sometimes battery operated, senses organic or inorganic gases capable of IR absorption.

A2.6.2 *Limits*—If a non-dispersive instrument, it is not specific. Dispersive instruments not battery powered, but are selective. Requires experienced analyst when dealing with a mixture of compounds.

REFERENCES

(1) *Occupational Safety and Health Guidance Manual for Hazardous Waste Site Activities*, NIOSH/OSHA/USCG/EPA, DHHS, NIOSH Publication No. 85-115, NIOSH, 1014 Broadway, Cincinnati, OH 45202, October 1985.

(2) *Standard Operating Safety Guides*, U.S. Environmental Protection Agency, Environmental Response Branch, November 1984.

(3) *Instrumentation for Monitoring Air Quality, ASTM STP 555*, ASTM, 1973.

(4) *Sampling and Analysis of Toxic Organics in the Atmosphere, ASTM STP 721*, ASTM, 1979.

(5) *Air Sampling Instruments for Evaluation of Atmospheric Contaminants*, American Conference of Governmental Industrial Hygienists, latest edition.

(6) *Direct Reading Colorimetric Indicator Tubes Manual*, American Industrial Hygiene Association, AIHA, latest edition.

(7) Lewis, R. G., Martin, B. E., Sgontz, D. L., and Howes, Jr., J. E., "Measurement of Fugitive Atmospheric Emissions of Polychlorinated Biphenyls from Hazardous Waste Landfills," *Environmental Science & Technology*, October 1985.

(8) *Manual of Recommended Practice for Combustible Gas Indicators and Portable, Direct Reading Hydrocarbon Detectors*, AIHA, latest edition.

Standard Test Method for Leaching Solid Material in a Column Apparatus[1]

This standard is issued under the fixed designation D 4874; the number immediately following the designation indicates the year of original adoption or, in the case of revision, the year of last revision. A number in parentheses indicates the year of last reapproval. A superscript epsilon (ε) indicates an editorial change since the last revision or reapproval.

1. Scope

1.1 This test method is a standard laboratory procedure for generating aqueous leachate from materials using a column apparatus. It provides a leachate suitable for organic analysis of semivolatile and nonvolatile compounds as well as inorganic analyses.

1.2 The column apparatus is designed and constructed of materials chosen to enhance the leaching of low concentrations of semivolatile and nonvolatile organic constituents as well as to maximize the leaching of metallic species from the solid. Analysis of column effluent provides information on the leaching characteristics of material under the conditions used in the test.

1.3 This test method provides for the passage of an aqueous fluid through materials of known mass in a saturated up-flow mode.

1.4 It is intended that the sample used in the procedure be physically, chemically, and biologically representative of the material.

1.5 This test method does not produce results that can be used as the sole basis for (*1*) engineering design of a disposal site, or (*2*) the characterization of wastes based on their leaching characteristics.

1.6 *This test method has the following limitations*:

1.6.1 Maximum particle size is 10 mm (0.4 in.). Particle size reduction is not recommended. Large-diameter material (cinders, rocks, and so forth) should be removed prior to packing the column to ensure adequate compaction.

1.6.2 Test materials containing densely immiscible organic material may result in phase separation and lead to column plugging.

1.6.3 This test method does not differentiate between dissolved constituents and sub-70-μm particulates that pass through the pores of the end plates.

1.6.4 This test method is not applicable to the leachability characterization of materials with regard to volatile compounds.

1.6.5 This test method is not applicable to the characterization of materials that dissolve in water to a degree that significantly impacts the void volume in the column or the determination of the specific gravity of the material.

1.7 Application of this test method to materials with initial low permeability, or to those that lose permeability over the course of the test, may result in long testing periods.

1.8 The values stated in SI units are to be regarded as the standard. The values given in parentheses are in approximate inch-pound equivalents.

1.9 *This standard does not purport to address all of the safety concerns, if any, associated with its use. It is the responsibility of the user of this standard to establish appropriate safety and health practices and determine the applicability of regulatory limitations prior to use.*

2. Referenced Documents

2.1 *ASTM Standards:*

C 819 Test Method for Specific Surface Area of Carbon or Graphite[2]

D 422 Test Method for Particle Size Analysis of Soils[3]

D 698 Test Method for Laboratory Compaction Characteristics of Soil Using Standard Effort (12 400 ft-lbf/ft^3) (600 kN-m/m^3)[3]

D 854 Test Method for Specific Gravity of Soils[3]

D 1125 Test Methods for Electrical Conductivity and Resistivity of Water[4]

D 1129 Terminology Relating to Water[4]

D 1293 Test Methods for pH of Water[4]

D 1498 Practice for Oxidation-Reduction Potential of Water[4]

D 1888 Test Methods for Particulate and Dissolved Matter in Water[5]

D 2216 Test Method for Laboratory Determination of Water (Moisture) Content of Soil and Rock[3]

D 2434 Test Method for Permeability of Granular Soils (Constant Head)[3]

D 3370 Practices for Sampling Water[4]

D 3694 Practice for Preparation of Sample Containers and for Preservation of Organic Constituents[6]

D 4253 Test Methods for Maximum Index Density and Unit Weight of Soils Using a Vibratory Table[3]

E 691 Practice for Conducting an Interlaboratory Study to

---

[1] This test method is under the jurisdiction of ASTM Committee D34 on Waste Management and is the direct responsibility of Subcommittee D34.01.04 on Ancillary Topics.

Current edition approved Oct. 10, 1995. Published December 1995. Originally published as D 4874 – 89. Last previous edition D 4874 – 89.

[2] *Discontinued*, see *1988 Annual Book of ASTM Standards*, Vol 15.01.

[3] *Annual Book of ASTM Standards*, Vol 04.08.

[4] *Annual Book of ASTM Standards*, Vol 11.01.

[5] *Discontinued*, see *1990 Annual Book of ASTM Standards*, Vol 11.01.

[6] *Annual Book of ASTM Standards*, Vol 11.02.

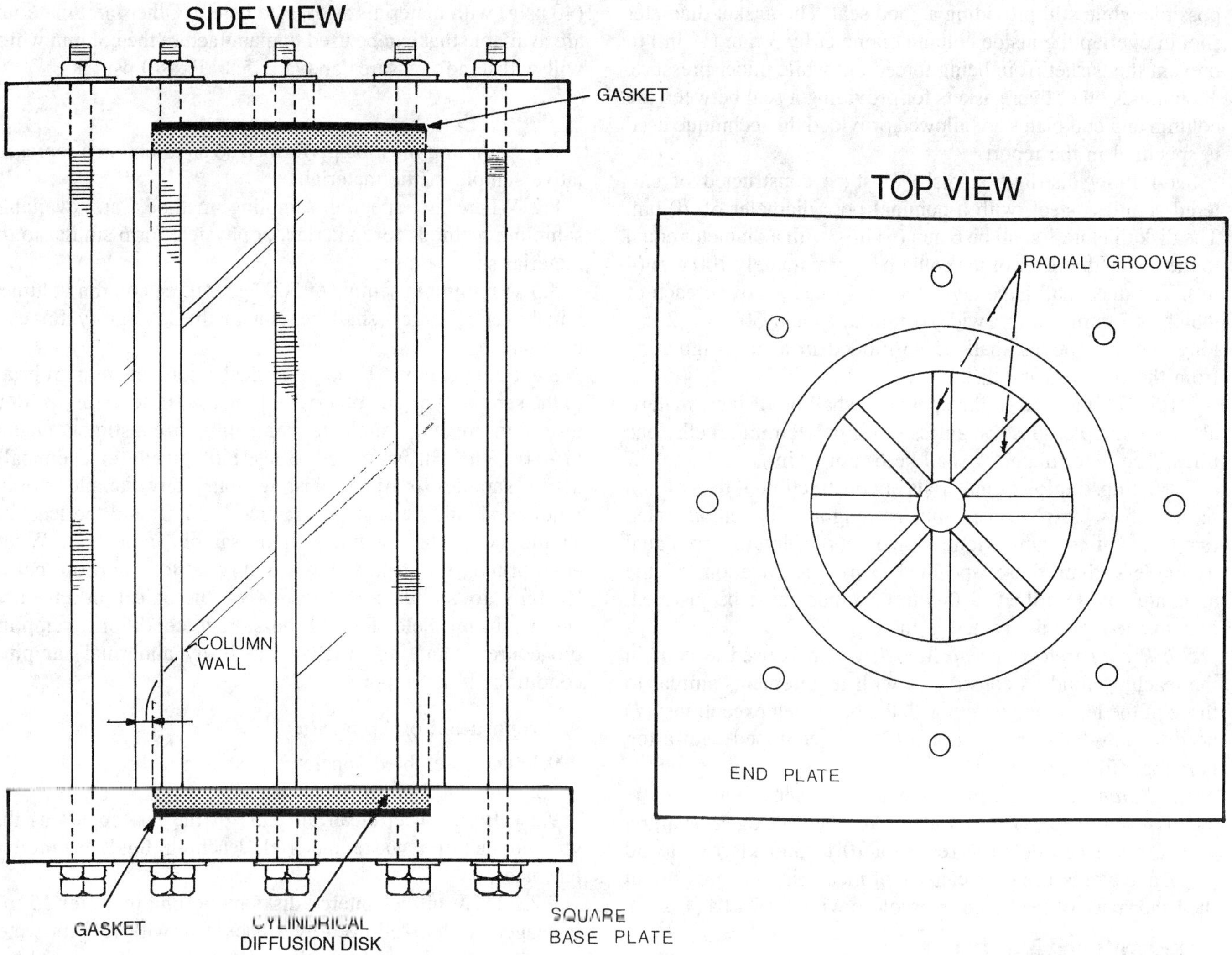

FIG. 1 Column Apparatus

Determine the Precision of a Test Method[7]

3. Terminology

3.1 *Definition:*

3.1.1 *reagent water*—as defined in SW-846, Method 1311, 5.2.[8]

3.1.2 *void volume*—the volume between the solid particles in a bed of granular material. Also called the interstitial volume.

3.2 Several terms used in this test method are defined in Terminology D 1129.

4. Significance and Use

4.1 This test method is intended to provide an aqueous leaching of a material in a dynamic partitioning manner.

4.2 Specific operating conditions for the column can be selected to satisfy the objectives of individual studies. An understanding of the fundamental principles governing column leaching is necessary in the selection of appropriate conditions. References **(1-3)**[9] provide useful information on these principles.

4.3 The column apparatus is constructed of materials that permit the generation of a leachate which is suitable for low concentration organic analysis for semivolatile and nonvolatile compounds as well as inorganic species.

5. Apparatus (See Fig. 1)

5.1 *Columns*:

5.1.1 The column body is constructed of glass pipe, 300 mm (12 in.) in length, with an inside diameter of 100 mm (4 in.). The cylinder wall must be of sufficient thickness, approximately 6 mm (¼ in.), to withstand the operating pressure.

5.1.2 End plates are constructed of stainless steel. They are attached by means of eight, 6-mm (¼-in.) threaded rods or any other means which ensures a leakproof seal.

5.1.3 Gaskets, one at each end of the column, are constructed of chemically inert materials, and are as thin as

[7] *Annual Book of ASTM Standards*, Vol 14.02.

[8] SW-846, 3rd ed., Method 1311, Available from USEPA, Office of Solid Waste and Emergency Response, Washington, DC 20460.

[9] The boldface numbers in parentheses refer to the list of references at the end of this standard.

possible while still providing a good seal. The gasket diameter should overlap the inside column diameter by 3 mm (⅛ in.) to prevent the gasket from being forced out while under pressure. Techniques other than gaskets for providing a seal between the column and end plates are allowed provided the technique used is specified in the report.

5.1.4 Flow distribution disks must be constructed of sintered stainless steel, with a nominal pore diameter of 70 µm. The disk thickness shall be 6 mm (¼ in.), with a diameter equal to the inside diameter of the column, approximately 100 mm (4 in.). The disk shall have eight evenly spaced grooves; each of which is 3 mm (⅛ in.) wide, 3 mm deep and 50 mm (2 in.) long. These grooves shall be positioned in a ray originating from the center of the disk.

5.1.5 Tubing used in the apparatus shall be of inert materials, for example, glass, stainless steel, polytetrafluoroethylene lined. The outer diameter shall be 6 mm (¼in.).

5.1.6 Substitution of materials of construction of the column or any of its parts is acceptable, as long as it is demonstrated that levels of contamination for analyte(s) of interest are equal to or less than those specified. Any modification of the apparatus as described in this test method must be justified, documented, and delineated in the report.

5.2 *Pressurized Reservoir Vessel*, which is used to contain the leaching fluid, is constructed with requirements similar to those of the leaching column with the following exceptions: (*1*) no diffusion disks are used, and (*2*) it is equipped with a top port for refilling the fluid.

5.3 *Balance*, 10-kg capacity, with a 1-g sensitivity.

5.4 *Compressed Gas Source*, prepurified nitrogen or argon with a two-stage delivery regulator (0 to 350 kPa) (0 to 50 psig), and a pressure gage capable of measuring the pressure in the head space of the liquid reservoir to within ±7 kPa (1 psig).

6. Reagents and Materials

6.1 American Chemical Society (ACS) Reagent grade chemicals or equivalent are preferred.[10]

6.2 Other grades of chemicals may be used, provided that the reagent is of sufficiently high purity to permit its use without compromising the objectives of the testing.

6.3 Demonstration of acceptability through reagent blank data at or below the quantitation limits for all analytes of interest is required.

6.4 Unless otherwise indicated, references to reagent water mean water as defined in SW-846, Method 1311, 5.2. See reagent water under the terminology section of this test method (3.1.1).

NOTE 1—Reagent water is defined in SW-846 as water in which an interferant is not observed at or above the method's detection limit of the analyte(s) of interest.

7. Safety Precaution

7.1 General operating pressure should not exceed 275.8 kPa (40 psig) with materials as defined herein. Other inert materials are available that can be used to manufacture the column which will withstand pressures above 275.8 kPa (40 psig).

8. Sampling

8.1 Sampling must be performed so as to obtain a representative sample of the material.

8.2 Where no specific sampling methods are available, sampling methods for materials of physical form similar to the material shall be used.

8.3 A minimum sample of 5000 g, or three column volumes, whichever is larger, shall be sent to the laboratory for each column.

8.4 Samples must be kept in closed containers appropriate to the sample type and otherwise protected if necessary prior to testing to prevent sample contamination or constituent change or loss. Where it is desired to test biologically or chemically active samples in their existing state, any sample storage required should be at 4°C (Practice D 3370) and the leaching should be started within 8 h of sample collection. Where appropriate, the tester may modify a test portion before leaching to simulate the results of biological or chemical activity in the field. Record the storage conditions, sampling procedures, handling practices, and any abnormal sampling conditions in the report.

9. Preparation of Apparatus

9.1 The assembled apparatus is shown in Fig. 1.

9.2 *Column Preparation*:

9.2.1 Before use, clean all parts of the test apparatus that will contact the waste material, leaching fluid, or product leachate.

9.2.1.1 Clean the sintered disks by boiling them for 15 min in reagent water, followed by a backflush with reagent water. Then saturate the disks with concentrated sulfuric acid and soak until all residues are removed. Then vacuum or pressure remove the excess acid with reagent water. Next, pump or draw either acetone or methanol through the disks, followed by either hexane or methylene chloride. Permit disks to air dry.

9.2.1.2 Clean the column apparatus by washing it with a nonionic surfactant soap and water. Rinse with tap water and follow with a reagent water rinse. Then rinse it with either acetone or methanol, followed by a rinse with either hexane or methylene chloride. Permit the apparatus to air dry.

9.3 Assemble the apparatus as depicted in Figs. 1 and 2. Weigh the dried, clean empty column, including end caps and other fittings necessary to contain the waste, and record the mass. This is the tare of the apparatus. Record the inside diameter and height of that part of the column to be filled with waste (that is, the column cylinder). If, following column saturation, the tester wishes to check the degree of saturation, it will first be necessary to determine the mass of water that can be contained in the porous flow distribution disks, end plates, and fittings.

10. Procedure

10.1 *Preconditioning*:

10.1.1 Prepare a test portion of waste in a manner that simulates the state the waste is in or will be in as it undergoes

[10] *Reagent Chemicals, American Chemical Society Specifications*, American Chemical Society, Washington, DC. For suggestions on the testing of reagents not listed by the American Chemical Society, see *Analar Standards for Laboratory Chemicals*, BDH Ltd., Poole, Dorset, U.K., and the *United States Pharmacopeia and National Formulary*, U.S. Pharmaceutical Convention, Inc. (USPC), Rockville, MD.

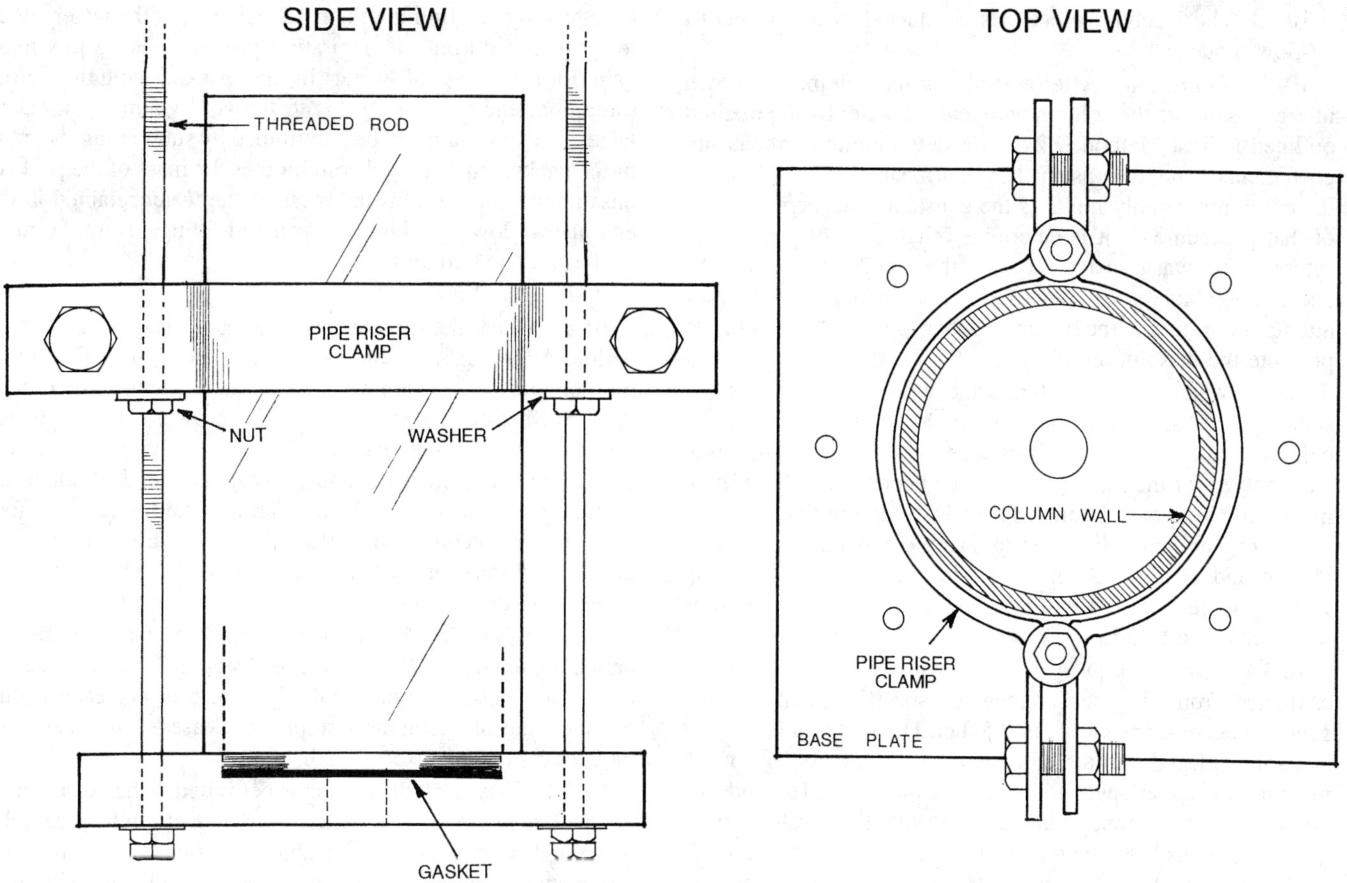

FIG. 2 Packing Configuration

leaching in the field. Preparation of the test portion may include such factors as curing, and adjustment of moisture content and density. For such adjustments, the following procedures can be used where appropriate.

10.1.2 *Moisture Content*— Adjust the moisture content to that defined in the disposal scenario by dewatering or adding reagent water to the material. Increase moisture content to that defined in the disposal scenario by addition of reagent water. Note and record the volume of reagent water added. If the concentration of any trace analyte of concern found in the reagent water exceeds the reporting (quantitation) limit for the volume added, note this on the report form. Decrease moisture content by determining scenario temperature (normally less than 60°C) and drying the material at the temperature for a specified time interval. Determine actual moisture content of the waste as it is placed in the column using Test Method D 2216 and an appropriate drying temperature. Note and record the drying temperature used. Record the moisture content.

10.1.3 *Density*—Density is to be adjusted by vibration (see Test Method D 4253) or compaction (see Test Method D 698) to the anticipated field density. Pack the waste material in the column so that uniform density is achieved. This can be checked by visual observation of the waste in the transparent column.

10.1.4 *Curing*—Aging may be necessary for those materials that undergo physical or chemical changes with time in order to obtain a specimen (test portion) representative of the desired conditions. Curing may be accomplished in the field before testing in the column to produce a physical and chemical state representative of the waste as it undergoes leaching in the field. Note and record field conditions of temperature, humidity, and atmospheric pressure. Record the duration of the curing process and report with the test data.

10.1.5 *Particle Size*— Ensure that the particle size distribution of the waste as placed in the column is representative of that expected in field placement. Interpretation of results is based in part on a knowledge of the particle size distribution and surface area of the material. Maximum particle diameter must not exceed 1⁄10 of the inside diameter of the column. Particle size reduction is not recommended. An evaluation of the particle characteristics of size distribution and surface area may be useful. Test Methods D 422 and C 819 should be used to evaluate these properties of the waste.

10.2 *Column Filling*:

10.2.1 Fill the column by compacting or vibrating the waste in five approximately equal layers. Approximate the expected field density. Scarify the top surface of each layer before adding the next increment. Do not scarify the top layer after final compaction. Fill the column to its final height, ensuring that the final height of the compacted material is equal to the height of the leaching column.

10.2.2 After filling the column, determine the mass, including the same fittings as in the determination of the tare. Record the total mass.

10.3 *Leaching*:

10.3.1 The leaching process is conducted in a continuous up-flow mode.

10.3.2 *Saturation*—After assembling the column as shown in Fig. 1, saturate the column with reagent water by the method outlined in Test Method D 2434 for determining permeability coefficients using a constant head permeameter. In this case, the column assembly replaces the constant head permeameter of that procedure. If it is determined that it is not possible to saturate the waste using Test Method D 2434, it may be necessary to increase the vacuum used in the procedure or attempt to saturate the column under pressure, or both, to promote the dissolution of gasses.

10.3.3 Water used to saturate the waste should be of the same quality as that to be used to leach the waste in the column. Collect a sample of this water at the time of saturation and analyze for the same properties and constituents as will be measured in the column effluent. An appropriate sample container, similar to that used for sample collection should be chosen and cleaned. Sample container guidance is given in 10.6.1. The test portion will be collected using the procedures for leachate collection.

10.3.4 *Determination of the Void Volume*—Void volume is estimated from the measurement of specific gravity of the solids in accordance with Test Method D 854, the mass of the wet solids placed into the column, the moisture content of the material in accordance with Test Method D 2216, and the volume of the column. Although the drying temperatures used in Test Methods D 854 and D 2216 can be selected on the basis of the specific characteristics of the material being dried, the drying temperatures used in the two procedures must be the same for the data to be used for porosity and void volume calculations.

10.3.4.1 Measure the mass of the sample material which fills the column apparatus. Note and record the moisture content of the material, as well as the specific gravity and any other specific information relevant to the material.

10.3.4.2 Calculate the void volume using the following equation:

$$V_p = V_c\{M/[(1 + w) \times S \times D]\} \quad (1)$$

where:

V_p = void volume in the column, cm^3,
V_c = volume of the column, cm^3,
M = as-packed mass of the material, including moisture, contained in the column,
w = moisture content of the material contained in the column as a mass ratio from Test Method D 2216, g water/g solids
S = specific gravity of the material as determined in Test Methods D 854, unitless, and
D = density of water, g/cm^3.

10.3.5 Estimate the degree of saturation by calculating the porosity n of the material using the following equation:

$$n = V_p/V_c \quad (2)$$

Calculate the mass gain in the saturated material. Add the value of the mass gain to the mass of the water in the packed column. Compare the value to the calculated void volume. The ratio is the degree of saturation. The degree of saturation can be estimated by determining the mass gain due to the water added to the waste during the saturation process, adding this mass gain to the mass of water in the packed column before saturation, and comparing this to the void volume just calculated. The mass gain can be determined by subtracting the mass of the packed, unsaturated column plus the mass of the packed, unsaturated column plus the mass of the water retained in the end plates, flow distribution disks, and fittings, from the mass of the saturated column.

10.4 *Effluent Flow*:

10.4.1 Once the column is saturated, begin the testing period. Adjust the pressure or hydraulic head, or both, so that one complete void volume exchange rate is accomplished in 24 ± 3 h. Maximum operating column pressure, for apparatus specified, should not exceed 275 kPa (40 psig). Read and record operating pressure, and effluent start time. If some test period other than 24 ± 3 h (an alternate void volume period) is warranted, record justification, time interval required, operating conditions, and effluent start time. Collect all column effluents as discrete void volume increments.

10.4.2 *Operating Requirements*—The column must be operated continuously. Note and record stop and restart times for all cessations. Determine total stoppage time for each occurrence per day and total daily stoppage. Cease test if, on any one day, total stoppage exceeds 1 h.

10.4.3 No type of channeling is permitted. When channeling is noted, note the time and cease the test. Determine if samples generated previously are usable. Document rationale for acceptance/rejection of previous sample effluent. Columns shall be protected from extraneous light except during column checks and sample collection intervals. Note and record all exposure periods.

10.4.4 Columns are to be run at ambient temperature, under normal test conditions. Sufficient pressure is required to maintain flow at the specified rate which will yield a void volume during the specified time interval. Temperature and pressures must be read and recorded periodically. For safety reasons, do not operate this equipment above 275 kPa (40 psig).

10.5 *Leachate Collection*:

10.5.1 The column must be inspected periodically and adjustments made if necessary to maintain the desired operating conditions.

10.5.2 Sufficient leachate from void volume periods shall be collected to establish a basis for judgment on completeness of the test. Minimally, it is recommended that Void Volumes 1, 2, 4, and 8 be collected and analyzed. Additional void volumes may provide additional useful data. The minimum number and sequence numbers of void volumes that will be collected and analyzed must be identified in advance of any collection activity.

10.5.3 Void volumes are sampled by collecting all column effluent generated over a 24 ± 3 h or other period corresponding to the void volume of interest.

10.5.4 Minimize contact of the leachate with the atmosphere. Fill effluent collection vessel head space with prepurified nitrogen or argon. Protect vessel head space with a water trap.

10.5.5 Collect effluent in appropriate containers. Containers shall be cleaned in an appropriate manner. Containers must be free from contamination for those parameters of interest in the leachate. Follow the procedures outlined in Practice D 3694 or Ref **(4)**, or both.

10.6 *Leachate Sample Preparation*:

10.6.1 Sample containers must be constructed of materials that do not alter the sample quality. Follow the procedures outlined in Practice D 3694 or Ref **(4)**, or both.

10.6.2 Sample containers must be cleaned according to Practice D 3694, methods contained in Ref **(4)**, or as appropriate for the constituents of interest.

10.6.3 If collected leachates from void volume periods are stirred or agitated before filtration or analysis, this must be noted in the report. Minimize contact between the liquid and air during this mixing. If there are solids visible in the collected leachate before or after mixing, this must be noted in the report.

10.6.4 If filtration of the leachate is desired prior to chemical analyses, then such filtration steps must be noted. Analyze the filtered material separately to ensure that the suspended solids did not contain constituents that are of environmental significance. If suspended solids are found to contain analytes of significance, then redo the column leach test, and analyze the leachate without filtration.

10.6.5 If samples are filtered before analysis, the method of filtration must be noted in the report. Filtration must be completed within 1 h of sample collection. Contact with air must be minimized.

10.6.6 Samples must be stored under refrigeration at 4°C. Sample container head space must be minimized. A tight seal between the cap and the container is expected. Guidance on preservation techniques can be found in Practices D 3694 and D 3370 as well as in Ref **(4)**.

10.7 *Sample Labeling*— At a minimum, the information regarding column source, collection date, collection time, void volume number, sampler's name, sampler's signature, and preservatives used must be recorded.

10.8 *Characterization of the Leachate*—The minimum test portion for characterization is equal to the volume of leachate produced during the void volume period. Suggested analyses for characterization of leachate include: pH, Test Methods D 1293; Conductivity, Test Methods D 1125; Redox Potential, Practice D 1498; Total Solids, Test Method D 1888; and any other analyses that will further characterize the material and provide interpretive guidance.

11. Report

11.1 The report must include at least the following information:

11.1.1 Project identification and description,

11.1.2 Description of waste as sampled,

11.1.3 Sampling procedures, waste sample handling, and storage procedures including the date and time of sampling,

11.1.4 All procedures used to condition the waste for placement in the column including curing and moisture adjustment procedures,

11.1.5 Permeability, particle size, and surface area information where available, including disposal scenario specifications; and a complete description of material as sampled,

11.1.6 Temperature used for drying in moisture determination and specific gravity procedures if different than 110 ± 5°C,

11.1.7 Materials used to construct the column apparatus if different than those specified in the test method,

11.1.8 Column washing procedures,

11.1.9 Waste moisture content as sampled and as packed in the column,

11.1.10 Characteristics of the waste as placed in the column including specific gravity of the solids, porosity, and void volume,

11.1.11 Mass of the waste placed in the column,

11.1.12 Saturation procedures used if different than Test Method D 2434, and date and time that column saturation began and ended,

11.1.13 Date and time that leaching began,

11.1.14 Temperature and pressure of column operation at the beginning of the test, and at each time samples are collected or flow rates are adjusted,

11.1.15 Means used to prevent exposure of column contents to light,

11.1.16 Total time for collection of each void volume, the measured volume of the leachate collected during the period, the date and time when measured, any corrections made to flow rates, and quantities of effluent collected between flow rate determinations,

11.1.17 Date, time, and duration of all flow stoppages,

11.1.18 Observations of unusual conditions in the column, such as density variations, cracks, gas-filled spaces, or evidence of fluid channeling along column walls,

11.1.19 Date and time that the tester began and finished collecting each void volume and the number of void volumes collected,

11.1.20 Type and material of construction of sampling containers and methods used to clean sample containers,

11.1.21 Steps taken to avoid sample contact with air where such steps are called for in the test method,

11.1.22 Results of measurements of index properties and any other analyses performed on unfiltered samples within 1 h of collecting the void volume from the column. The results of the analyses of that water used to saturate and leach the waste must also be included, and

11.1.23 All analytical data requested in disposal scenario with complete documentation and traceability through laboratory records.

12. Precision and Bias

12.1 *Precision*—The precision of this test method has been studied for inorganic materials in a three-laboratory study involving triplicate columns at each laboratory. Fly ash was used as a waste material. Because the fly ash was of very low permeability, end caps without radial grooves were used. The column effluents were analyzed by a single laboratory for metals by emission spectroscopy using an inductively coupled plasma source. Those metals that were measured at concentrations above the method detection limits were used to determine the precision of the column extraction method according to Practice E 691. The precision identified in the study was a measure of the ability of the column method to generate effluents of similar quality from a single waste material in tests conducted at different laboratories, or in tests using multiple columns at a single laboratory. A summary of the precision data is contained in Table 1. The complete interlaboratory study is described in Ref **(5)**. The pooled within-interlaboratory relative standard deviation for metals data averaged 33 %. The overall interlaboratory single test reproducibility in terms of standard deviation was 54 %. This excludes interlaboratory analytical effects because all analyses were performed by a central laboratory. The analytical precision reported by the central laboratory, at concentrations most frequently measured, was seldom greater than 5 % relative standard deviation.

12.2 Precision data for organic constituents are not available.

12.3 *Bias*—It is not possible to determine the bias of this test method because there are no accepted reference effluents generated.

13. Keywords

13.1 column leach; leachate; waste leaching technique

TABLE 1 Test Method Precision (Percent Relative Standard Deviations)

NOTE 1—The top number in each set is the pooled within-laboratory relative standard deviation while the bottom number is the overall interlaboratory variability for a single test result in terms of relative standard deviation.

NOTE 2—Because all chemical analyses were performed by a single laboratory, the total variance does not contain an interlaboratory analytical component.

Element	Number of Void Volumes of Liquid Having Passed Through the Column at the Time of Sample Collection						Average[A]
	1	2	4	8	16	32	
Aluminum	61/130	22/62[B]	44/54	30/40	29/85	10/50	35/72
Calcium	9/20	5/16	6/28	7/14	14/75	13/42	9/33
Cobalt	25/64[B]	...	...	...	...	...	...
Chromium	13/51	12/15	16/33	8/8	17/33[B]	...	12/17
Copper	25/33	8/10	12/22	18/26	33/43[B]	...	16/23
Iron	88/140	16/18	20/26	16/18	27/40[B]	...	35/51
Lithium	34/50	75/130	63/120	...	...	...	57/100
Magnesium	130/170[B]	...	...	...	...	...	...
Molybdenum	5/10	13/60	59/79	17/69	51/77[B]	...	24/55
Nickel	...	...	...	50/86[B]	36/41	21/34	61/38
Phosphorus	...	...	64/71	85/85[B]	67/73[B]	...	64/71
Potassium	24/82	28/110	63/120	93/100[B]	48/80	...	38/104
Silicon	...	...	...	...	34/48	22/29	28/39
Sodium	24/87	56/100	50/86	28/39	46/59	...	41/74
Strontium	19/42	16/18	28/52	4/22	36/75[B]	...	17/34
Vanadium	49/53[B]	10/27	30/60	16/18	24/33[B]	14/24	18/32
Zinc	99/100[B]	...	...	...	...	...	...
						Average	33/54

[A] Average values are based only on data without asterisks.

[B] Based on data sets that contain one to three values below detection limits (out of nine values). Those values below detection limits are assigned the detection limit in the calculations. If more than three values are below the detection limit, the data were not analyzed.

REFERENCES

(1) Grove, D. B., and Stollenwerk, K. G.," Modeling the Rate-Controlled Sorption of Hexavalent Chromium," *Water Resources Research*, Vol 21, No. 11, 1985, pp. 1703–1709.

(2) Parker, J. C., and Van Genuchten, M. T., "Determining Transport Parameters from Laboratory and Field Tracer Experiments," *Bulletin No. 84-3*, Virginia Agricultural Experiment Station, Virginia Polytechnical Institute and State University, Blacksburg, VA, 1984.

(3) Rao, P. S. C., Davidson, J. M., Jessup, R. E., and Selim, H. M., "Evaluation of Conceptual Models for Describing Non-Equilibrium Adsorption-Desorption of Pesticides During Steady-Flow in Soils," *Soil Science Society of America Journal*, Vol 43, 1979, pp. 22–28.

(4) *Handbook for Sampling and Sample Preservation of Water and Wastewater*, EPA 600/4-82-029, September 1982.

(5) Miner, R. A., Van Maltby, C., Dell, L. R., "The Results of an Interlaboratory Study of a Column Method for Leaching Solid Waste," *Symposium on Hazardous and Industrial Solid Waste Testing and Disposal*, 6th ed., *ASTM STP 933*, ASTM, 1986.

Designation: D 4978 – 95 (Reapproved 2001)

Standard Test Methods for Screening of Reactive Sulfides in Waste[1]

This standard is issued under the fixed designation D 4978; the number immediately following the designation indicates the year of original adoption or, in the case of revision, the year of last revision. A number in parentheses indicates the year of last reapproval. A superscript epsilon (ε) indicates an editorial change since the last revision or reapproval.

1. Scope

1.1 These test methods are applicable to the screening of reactive sulfides in wastes, liquids, sludges, semisolids, and solids by using the following methods:

	Sections
Test Method A—Lead Acetate Paper	9-14
Test Method B—Gas Detector Tube	15-19

1.2 These test methods are not applicable in determining the type and concentration of reactive sulfides.

1.3 These test methods are designed and intended as a preliminary test to complement the more sophisticated quantitative analytical techniques that may be used to determine sulfide concentration. These test methods offer, to the user, the option and the ability to screen waste for potentially hazardous levels of reactive sulfide when the sophisticated techniques are not available and the total waste composition is unknown.

1.4 *This standard does not purport to address all of the safety concerns, if any, associated with its use. It is the responsibility of the user of this standard to establish appropriate safety and health practices and determine the applicability of regulatory limitations prior to use.* Specific hazard information is given in Section 7 and 11.3.

2. Referenced Documents

2.1 *ASTM Standards:*

D 1193 Specification for Reagent Water[2]

3. Terminology

3.1 *Descriptions of Terms Specific to This Standard:*

3.1.1 *reactive sulfide*—a compound containing sulfide that readily forms hydrogen sulfide gas upon reaction with acid.

3.1.2 *screening analysis*—a *preliminary* qualitative or semiquantitative test developed from classical qualitative and quantitative techniques that is designed to efficiently give the user parameters about a waste that will aid in determining waste identification, process compatibility, and safety in handling.

4. Summary of Test Methods

4.1 *Method A: Lead Acetate Paper*—A sample of waste material is acidified. If sulfides are present in the waste, hydrogen sulfide (H_2S) is evolved. In the presence of hydrogen sulfide, lead acetate paper changes color to silvery brown or black.

4.2 *Method B: Gas Detector Tube*—A portion of the sample is acidified in a beaker to release sulfide as hydrogen sulfide gas. The gas is funneled through a detector tube creating a color stain in the tube proportionate to the concentration of sulfide gas in the vapor. A definite color change in the detector tube indicates a positive presence of sulfide.

5. Significance and Use

5.1 These test methods are intended for use by waste treatment, storage, disposal, and remedial facilities, in order to show the presence of potentially hazardous sulfide.

5.2 *Method B: Gas Detector Tube*—This proposed test method was designed to measure gases migrating into an air space at a height of 10 units (for example, 10 ft) from a receiving or mixing pit of 10 units square (for example, 100 ft^2). In essence, the total volume of the atmosphere is approximately 10 times the surface area of the sample and test solution.

6. Reagents

6.1 *Purity of Reagents*—Reagent grade chemicals shall be used in all tests. Unless otherwise indicated, it is intended that all reagents conform to the specifications of the Committee on Analytical Reagents of the American Chemical Society where such specifications are available.[3] Other grades may be used provided it is first ascertained that the reagent is of sufficiently high purity to permit its use without lessening the accuracy of the determination.

6.2 *Purity of Water*—Unless otherwise indicated, references to water shall be understood to mean reagent water as defined by Type III of Specification D 1193.

---

[1] These test methods are under the jurisdiction of ASTM Committee D34 on Waste Management and is the direct responsibility of Subcommittee D34.01.05 on Screening Methods.

Current edition approved Sept. 10, 1995. Published November 1995. Originally published as D 4978 – 89. Last previous edition D 4978 – 89.

[2] *Annual Book of ASTM Standards*, Vol 11.01.

[3] *Reagent Chemicals, American Chemical Society Specifications*, American Chemical Society, Washington, DC. For suggestions on the testing of reagents not listed by the American Chemical Society, see *Analar Standards for Laboratory Chemicals*, BDH Ltd., Poole, Dorset, U.K., and the *United States Pharmacopeia and National Formulary*, U.S. Pharmaceutical Convention, Inc. (USPC), Rockville, MD.

7. Hazards

7.1 These tests should be conducted under a fume hood.

7.2 Avoid inhalation and skin or eye contact, or both, with all hazardous material.

8. Sampling

8.1 Collect a representative sample of the waste in a sealed container.

8.2 Store the sample in an operating fume hood.

8.3 The sample should be analyzed as soon as possible.

TEST METHOD A—LEAD ACETATE PAPER

9. Interferences

9.1 Compounds evolving gases or vapors upon acidification that react with lead may interfere causing false positives.

NOTE 1—All positives should be followed up by a more detailed method, that is, distillation for total sulfides.

10. Reagents and Materials

10.1 *Beakers or Test Tubes.*

10.2 *Stirring Rod.*

10.3 *Hydrochloric Acid* (1 + 1)—Add 1 volume of hydrochloric acid (HCl, sp gr 1.19) to 1 volume water.

10.4 *Lead Acetate Paper.*

10.5 *Sodium Sulfide, Standard Solution* (1 mL = 1 mg Na_2S)—Dissolve 1.00 g sodium sulfide (Na_2S) in water and dilute to 1 L. Make this solution fresh daily.

10.6 *Quality Control Check Standard*—Dilute an appropriate volume of sodium sufide standard solution with water to produce a solution with the desired action level. Make this solution fresh daily.

11. Procedure

11.1 Place 5 to 10 g of sample into the beaker (slurry with approximately 15 mL water if the material is solid).

11.2 Wet a strip of lead acetate paper and let it adhere to the side of beaker above the sample.

11.3 Slowly and carefully adjust the pH to less than 2.0 with HCl solution 1 + 1.

NOTE 2—**Warning:** The addition of HCl to waste materials may cause violent reaction or highly toxic vapors, or both.

11.4 Stir the solution.

11.5 A silvery brown to black color on the lead acetate paper indicates the presence of sulfides and shall be recorded as positive. Compare the color observed to a blank and standard solution.

11.6 No color change to the lead acetate paper will be reported as negative.

12. Report

12.1 Report the following information:

12.1.1 Sample identification,

12.1.2 Date of test,

12.1.3 Test procedure performed, and

12.1.4 Sample classification: positive or negative.

13. Quality Control

13.1 A quality control check sample shall be tested with each batch of samples. The concentration of this sample should reflect the action level required by the laboratory. Method or reagent blanks, duplications, and fortification (spikes), should be performed at an action level specified by the laboratory at an appropriate frequency.

13.2 Method detection limits should be determined by each laboratory using the standard and at the appropriate action level.

14. Precision and Bias

14.1 No statement is made about either the precision or bias of Test Method A for measuring sulfides in wastes since the result merely states whether there is conformance to the criteria for success specified in the procedure.

TEST METHOD B—GAS DETECTOR TUBE

15. Reagents and Materials

15.1 *Buffer Solution*—Dissolve 740 g trisodium phosphate (Na_3PO_4) in 3.5 L of water. Add carefully 500 mL of phosphoric acid (H_3PO_4, 85 %) and mix. Adjust the pH to 2 with H_3PO_4 if necessary.

15.2 *pH Indicator Paper*, range: 0 to 14.

15.3 *Disposable Beaker*, 400-mL.

15.4 *Plastic Funnel*, 8-cm.

15.5 *Jack Stand.*

15.6 *Ring Stand and Clamps.*

15.7 *Gas Detector Pump* with a volume range of 0 to 100 cc.

15.8 *Gas Detector Tube* for hydrogen sulfide.[4]

15.9 *Magnetic Stirrer.*

15.10 *Magnetic Stirring Bar.*

15.11 See Fig. 1 for suggested equipment setup.

16. Procedure

16.1 Place approximately 20 g of the material to be tested into a beaker with a magnetic stirring bar. Solid samples shall be crushed into small particles <9.5 mm in size (3⁄8-in. mesh screen).

16.2 Add approximately 50 mL of the buffer solution to the sample.

16.3 With a lab jack, raise the beaker and magnetic stirrer to a point at which the funnel fits down inside the beaker opening.

16.4 The gas detector tube with the pump attached should be directly positioned through the small opening of the funnel, with the tip of the tube about midway in the funnel area.

16.5 Begin stirring the sample.

16.6 Activate the pump by pulling out the handle to draw in 100 cm^3 of atmosphere from the apparatus.

16.7 A positive sulfide is determined by a color stain in the detector tube.

16.8 Allow at least a 1-min reaction time before reading results.

17. Report

17.1 Report the following information:

[4] Precision gas detector tubes available from Matheson Safety Products, East Rutherford, NJ 07073, have been found to be satisfactory for this purpose.

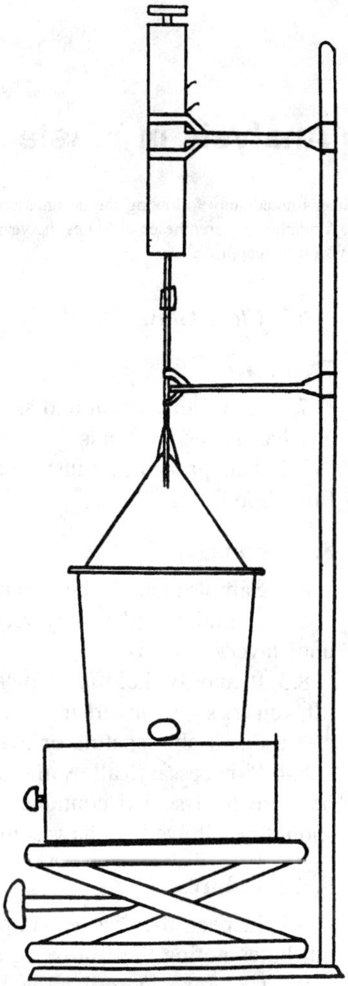

FIG. 1 Method D, Sulfide Test Apparatus

17.1.1 Sample identification,
17.1.2 Date of test,
17.1.3 Test procedure performed, and
17.1.4 Sample classification: positive or negative.

18. Quality Control

18.1 A quality control check sample shall be tested with each batch of samples. The concentration of this sample should reflect the action level required by the laboratory. Method or reagent blanks, duplications, and fortification (spikes), should be performed at an action level specified by the laboratory at an appropriate frequency.

18.2 Method detection limits should be determined by each laboratory using the standard and at the appropriate action level.

19. Precision and Bias

19.1 No statement is made about either the precision orbias of Method B for measuring sulfides in wastes since the result merely states whether there is conformance to the criteria for success specified in the procedure.

Designation: D 4979 – 95

Standard Test Method for Physical Description Screening Analysis in Waste[1]

This standard is issued under the fixed designation D 4979; the number immediately following the designation indicates the year of original adoption or, in the case of revision, the year of last revision. A number in parentheses indicates the year of last reapproval. A superscript epsilon (ε) indicates an editorial change since the last revision or reapproval.

1. Scope

1.1 This test method is used to identify wastes by describing certain physical properties. It has been developed as a rapid but effective means for visually screening wastes.

1.2 *This standard does not purport to address all of the safety concerns, if any, associated with its use. It is the responsibility of the user of this standard to establish appropriate safety and health practices and determine the applicability of regulatory limitations prior to use.* Specific hazard information is given in Section 7 and 9.1.6.

2. Terminology

2.1 *Description of Term Specific to This Standard:*

2.1.1 *screening analysis*—a preliminary qualitative or semiquantitative test that is designed to efficiently give the user specific information about a waste that will aid in determining waste identification, process compatibility, and safety in handling.

3. Summary of Test Method

3.1 Samples are inspected and the physical appearance is recorded, including color, turbidity, viscosity, physical state, layering, and any other observable attribute (for example, texture).

4. Significance and Use

4.1 This test method is intended for use by those in the waste management industries to aid in describing the physical characteristics of waste.

4.2 This test method can be used in identifying discrepancies between the waste, manifest, and historical descriptions.

5. Interferences

5.1 Opaque sample containers require removal of a representative sample sufficient for complete observation. It is recommended that a representative portion be transferred to a clean, dry, clear glass container.

6. Apparatus

6.1 *Disposable Spatula or Eye Dropper.*

6.2 *Clear Glass Sample Containers.*

7. Hazards

7.1 Avoid inhalation and skin or eye contact, or both, with any hazardous materials.

7.2 This procedure must be carried out within a laboratory fume hood.

8. Sampling

8.1 Samples should be obtained in clear glass containers.

8.2 Liquid samples may require time to stabilize (that is, until layers reform).

8.3 It may be helpful to pick up the sample container and tilt, rotate, swirl, invert it or to manipulate the sample with a clean, disposable spatula or eye dropper.

8.4 If necessary, allow the sample to come to room temperature in a sealed container. For example, frozen material should be allowed to thaw completely.

9. Procedures

9.1 Inspect the waste sample and describe the physical attributes noting the following areas.

9.1.1 *Color*—Describe the visual color of the sample. If more than one color is present, list colors in decreasing order of prominence. Additional descriptive terms may be useful (for example, purple with swirls of blue and flecks of yellow).

9.1.2 *Turbidity*—Describe liquid samples or liquid portions of samples in terms of clear (transparent), cloudy (translucent), or opaque.

9.1.3 *Viscosity*—Describe the viscosity of liquids and sludges. Describe viscosity in reference to viscosity like water, medium viscosity like syrup or motor oil, or high viscosity like molasses or warm tar. Tip the container sideways or invert to note the viscosity. If a sample is tipped 90° for 5 s and has no visible flow, report it as such.

9.1.4 *Physical State*—Describe as liquid, solid, sludge, powder, granular, etc. Note any unusual physical attributes. Note the general range of particle size, (fine to coarse) and the presence of larger chunks. The presence of any free liquids must be noted, as must the presence of (or for stabilized wastes, the absence of) sorbants (absorbants and adsorbents) as a sole treatment. Such sorbants might include sawdust, ground or whole corn cobs, or vermiculite.

9.1.5 *Layering*—Describe any layering in terms of rough percentages of the total sample. Record the color, turbidity, and

---

[1] This test method is under the jurisdiction of ASTM Committee D34 on Waste Management and is the direct responsibility of Subcommittee D34.01.05 on Screening Methods.

Current edition approved June 15, 1995. Published August 1995. Originally published as D 4979 – 89. Last previous edition D 4979 – 89.

physical state of each layer. Describe multilayered samples by listing the layers and their percentages from the top downward (that is, 5 % golden transparent medium-viscosity liquid over 90 % black opaque low-viscosity liquid over 5 % blue translucent low-viscosity liquid).

NOTE 1—The presence of oil, water, or heavier fractions such as halogenated solvents can be confirmed by noting the solubility or insolubility of each phase and, if insoluble, noting if it is heavier or lighter than water.

9.1.6 *Odor*—Note only obvious incidental odors noticed during sample handling.

NOTE 2—**Warning:** Intentional smelling of samples must *not* be performed under any circumstances!

9.2 All descriptions must be recorded in a complete and concise manner.

10. Precision and Bias

10.1 No statement is made about either the precision or bias of this test method since the results of the test are based on visual observations only.

11. Keywords

11.1 color; physical description; screening analysis; turbidity; viscosity

The American Society for Testing and Materials takes no position respecting the validity of any patent rights asserted in connection with any item mentioned in this standard. Users of this standard are expressly advised that determination of the validity of any such patent rights, and the risk of infringement of such rights, are entirely their own responsibility.

This standard is subject to revision at any time by the responsible technical committee and must be reviewed every five years and if not revised, either reapproved or withdrawn. Your comments are invited either for revision of this standard or for additional standards and should be addressed to ASTM Headquarters. Your comments will receive careful consideration at a meeting of the responsible technical committee, which you may attend. If you feel that your comments have not received a fair hearing you should make your views known to the ASTM Committee on Standards, at the address shown below.

Designation: D 4980 – 89

Standard Test Methods for Screening of pH in Waste[1]

This standard is issued under the fixed designation D 4980; the number immediately following the designation indicates the year of original adoption or, in the case of revision, the year of last revision. A number in parentheses indicates the year of last reapproval. A superscript epsilon (ϵ) indicates an editorial change since the last revision or reapproval.

1. Scope

1.1 These test methods are used to determine the pH of a hazardous waste liquid, sludge, semisolid and solid.

	Sections
Test Method A—pH Screening by pH Paper	9-14
Test Method B—pH Screening by Electrometric Measurement	15-22

1.2 Test Method A uses a wide-range pH paper for a rapid indication of pH to within about 1 pH unit.

1.3 Test Method B uses a pH meter to measure within about 0.1 pH unit.

1.4 These test methods are designed and intended as a preliminary test to complement the more sophisticated quantitative analytical techniques that may be used to determine pH. These test methods offer, to the user, the option and the ability to screen waste for potentially hazardous levels of acidity and alkalinity when the more sophisticated techniques are not available and the total waste composition is unknown.

1.5 *This standard may involve hazardous materials, operations, and equipment. This standard does not purport to address all of the safety problems associated with its use. It is the responsibility of the user of this standard to establish appropriate safety and health practices and determine the applicability of regulatory limitations prior to use.* Specific hazard information is given in Section 6.

2. Referenced Documents

2.1 *ASTM Standards:*

D 1193 Specification for Reagent Water[2]

D 1293 Test Methods for pH of Water[2]

3. Terminology

3.1 *Definitions of Terms Specific to This Standard:*

3.1.1 *screening analysis*—a preliminary qualitative or semi quantitative test developed from classical qualitative and quantitative techniques that is designed to efficiently give the user specific information about a waste that will aid in determining waste identification, process compatibility and safety in handling.

4. Significance and Use

4.1 These test methods are intended for use by those in the waste management industries to characterize waste streams by pH. These methods will identify those waste materials that may dictate a specific waste management procedure due to high acidity or alkalinity.

5. Reagents

5.1 *Purity of Reagents*—Reagent grade chemicals shall be used in all tests. Unless otherwise indicated, it is intended that all reagents conform to the specifications of the Committee on Analytical Reagents of the American Chemical Society where such specifications are available.[3] Other grades may be used, provided it is first ascertained that the reagent is of sufficiently high purity to permit its use without lessening the accuracy of the determination.

5.2 *Purity of Water*—Unless otherwise indicated, references to water shall be understood to mean reagent water as defined by Type III of ASTM Specification D 1193.

6. Hazards

6.1 Avoid inhalation and skin and eye contact of all hazardous materials.

6.2 All measurements shall be done in a laboratory fume hood.

7. Sampling

7.1 Collect a representative sample of the waste.

7.2 Samples should be analyzed as soon as possible after collection.

8. Report

8.1 The report shall include at a minimum:

8.1.1 Sample identification,

8.1.2 Date of test,

8.1.3 Reference to the procedure applied, that is, test method and if applicable, dilution ratio,

8.1.4 Analytical results, and

8.1.5 Identification of the analyst.

[1] These test methods are under the jurisdiction of ASTM Committee D34 on Waste Management and is the direct responsibility of Subcommittee D34.01.05 on Screening Methods.

Current edition approved Nov. 10, 1989. Published December 1989.

[2] *Annual Book of ASTM Standards*, Vol 11.01.

[3] "Reagent Chemicals, American Chemical Society Specifications," Am. Chemical Soc., Washington, DC. For suggestions on the testing of reagents not listed by the American Chemical Society, see "Reagent Chemicals and Standards," by Joseph Rosin, D. Van Nostrand Co., Inc., New York, NY, and the "United States Pharmacopeia."

TEST METHOD A—pH SCREENING BY pH PAPER

9. Summary of Test Method A

9.1 *pH Paper*—A small portion of the sample is introduced onto full range pH paper. The results are visually compared to the appropriate color chart and reported to ±1 pH unit.

10. Interferences

10.1 Materials that mask the pH paper, for example, oils, syrups, paint, etc. cause the visual detection of color on the paper to be difficult.

10.2 Strong dyes and solutions of deep colors can give false results.

10.3 Oxidizers can bleach the pH paper.

11. Apparatus

11.1 *Stirring Rod, Spatula, and Disposable Pipet or Eye Dropper*, for transferring sample to test paper.

11.2 *Disposable Beaker, Test Tube*, etc.

11.3 *Vortex Mixer* (optional).

11.4 *Separatory Funnel.*

12. Reagents and Materials

12.1 Full range pH paper with a stated precision of 1 pH unit and with a corresponding color chart.

12.2 Standard Buffer Solutions.

13. Procedure

13.1 *Aqueous Liquids*—Introduce a representative portion of the sample onto a strip of pH paper.

13.2 *Solids, Sludges, and Non-Aqueous Liquids*:

13.2.1 Place approximately 10 mL of water in a disposable beaker or test tube.

13.2.2 Add approximately 1 mL of waste sample and agitate for 10 s or until well mixed.

13.2.3 Let suspension settle and measure the pH of the supernatant as described in 13.1.

13.3 Visually compare the pH paper to the color chart and record the pH to the nearest whole unit. See 8.1.4.

14. Quality Control

14.1 Standard buffer solutions are used to check each new lot or shipment of pH paper.

15. Precision and Bias

15.1 Precision and Bias statements shall be supplied at a later date.

METHOD B—pH SCREENING BY ELECTROMETRIC MEASUREMENT

16. Summary of Test Method B

16.1 *pH Meter*—The pH of a sample is determined electrometrically using a pH meter with a combination pH electrode. Prior to the analysis of a sample, the instrument must be calibrated by using at least two standard buffer solutions. The results should be reported to the nearest 0.1 pH unit.

17. Interferences

17.1 The glass electrode, in general, is not subject to solution interferences from color, turbidity, colloidal matter, oxidants, reductants, or high salinity.

17.2 Sodium error at pH levels >10 can be reduced or eliminated by using a low sodium error electrode.

17.3 Coatings of oily material or particulate matter can impair electrode response. These coatings can usually be removed by gentle wiping or detergent washing, followed by dilute water rinsing. An additional treatment with hydrochloric acid 1 + 9, (1 volume HCl added to 9 volumes of H_2O) may be necessary to remove any remaining film.

17.4 Temperature effects on the electrometric determination of pH arise from two sources. The first is caused by the change in electrode output at various temperatures. This interference can be controlled with instruments having temperature compensation or by calibrating the electrode instrument system at the temperature of the samples. The second source of temperature effects is the change of pH due to changes in the sample as the temperature changes. This error is sample dependent and cannot be controlled. It should, therefore, be noted by reporting both the pH and temperature at the time of analysis.

18. Apparatus

18.1 *pH Meter, Laboratory of Field Model*—A wide variety of instruments are commercially available with various specifications and optional equipment.

18.2 *Glass Electrode.*

18.3 *Reference Electrode*—A silver-silver chloride or other reference electrode of constant potential may be used.

NOTE 1—Combination electrodes incorporating both measuring and reference functions are convenient to use and are available with solid, gel type filling materials that require minimal maintenance.

18.4 *Magnetic Stirrer and TFE-Fluorocarbon-Coated Stirring Bar.*

18.5 *Thermometer or Temperature Sensor*, for automatic compensation.

19. Reagents and Materials

19.1 *Primary Standard Buffer Salts*, are available from NIST and should be used in situations where extreme accuracy is necessary. Preparation of reference solutions from these salts requires some special precautions and handling such as low conductivity dilution water, drying ovens, and carbon-dioxide-free purge gas. These solutions should be replaced at least once each month.

19.2 *Secondary Standard Buffers*, may be prepared from NIST salts or purchased as solutions from commercial vendors. These commercially available solutions have been validated by comparison to NIST standards and are recommended for routine use.

20. Calibration and Standardization

20.1 Each instrument/electrode system must be calibrated at two points that bracket the expected pH of the samples and are approximately three pH units or more apart. Various instrument designs may involve use of a balance or standardize dial or slope adjustment as outlined in the manufacturer's instructions.

Repeat adjustments on successive portions of the two buffer solutions until readings are within 0.05 pH units of the buffer solution values.

NOTE 2—Wide variation in the stirring speed may cause fluctuations in pH measurements, but the variations will not affect test accuracy.

20.2 Place the liquid sample or buffer solution in a glass beaker using a sufficient volume to cover the sensing elements or the electrodes and to give adequate clearance for the magnetic stirring bar. If field measurements are being made, the electrodes may be immersed directly in the sample stream to an adequate depth and moved in a manner to ensure sufficient sample movement across the electrode-sensing element as indicated by drift free (less than 0.1 pH) readings during the analysis.

20.3 If the sample temperature differs by more than 2°C from the buffer solution, the measured pH values must be corrected. Instruments are equipped with automatic or manual compensators that electronically adjust for temperature differences. Refer to manufacturer's instructions.

21. Procedure

21.1 Because of the wide variety of pH meters and accessories, detailed operation procedures cannot be incorporated into this test method. Each analyst must be acquainted with the operation of each system and familiar with all instrument functions. Special attention to care of the electrodes is recommended.

21.2 Thoroughly rinse and gently wipe the electrodes prior to measuring pH of each sample. Immerse the electrodes into the sample beaker or sample stream and gently stir at a constant rate to provide homogeneity and suspension of solids (see Note 3). Note and record sample pH and temperature. Repeat measurement on successive volumes of sample until values differ by less than 0.1 pH unit. Two or three volume changes are usually sufficient.

NOTE 3—Wide variation in stirring speed may cause fluctuations in pH measurements but the variations will not adversely affect test applicability.

21.3 *Aqueous Liquid*—Insert the electrode into the aqueous portion of the waste and report the instrument reading.

21.4 *Solids, sludges, and non-aqueous liquids*:

21.4.1 Mix approximately a 10 % slurry of waste in water and take measurement on the aqueous portion. (See 8.1.4.)

21.4.2 For pH determination, refer to 21.2.

22. Quality Control

22.1 Instrument performance standards (where applicable), quality control check samples of appropriate matricies, and duplications, should be performed at an action level specified by the laboratory at an appropriate frequency.

23. Precision and Bias

23.1 The data are being accumulated and will be added to the test method when completed.

Designation: D 4981 – 95

Standard Test Method for Screening of Oxidizers in Waste[1]

This standard is issued under the fixed designation D 4981; the number immediately following the designation indicates the year of original adoption or, in the case of revision, the year of last revision. A number in parentheses indicates the year of last reapproval. A superscript epsilon (ε) indicates an editorial change since the last revision or reapproval.

1. Scope

1.1 This test method is intended for use prior to preparation of waste samples for organic analysis. Waste samples that have oxidizing compounds may react with certain reagents in the laboratory (for example, organic solvents).

1.2 This test method is applicable to the analysis of waste liquids, sludges, and solids.

1.3 This test method can neither identify specific oxidizing compounds nor measure concentrations. Since no acid or base is added in this test method, potential oxidizers that require the presence of acid or base will not be detected by this test method.

1.4 It is recommended that, prior to this test, waste samples be screened for water compatibility; see Test Methods D 5058.

1.5 This test method is designed and intended as a preliminary test to complement quantitative analytical techniques that may be used to determine the presence of oxidizers in wastes. This test method offers the ability to screen waste for potentially hazardous reactions due to oxidizer content when the more sophisticated techniques are not available or the total waste composition is unknown.

1.6 *This standard does not purport to address all of the safety concerns, if any, associated with its use. It is the responsibility of the user of this standard to establish appropriate safety and health practices and determine the applicability of regulatory limitations prior to use.* Specific hazard statement is given in Section 8 and 11.2.

2. Referenced Documents

2.1 *ASTM Standards:*

D 1193 Specification for Reagent Water[2]

D 4980 Test Method for Screening of pH in Waste[3]

D 5058 Test Methods for Compatability of Screening Analysis of Waste[3]

3. Terminology

3.1 *Description of Term Specific to This Standard:*

3.1.1 *screening analysis*—A preliminary qualitative or semiquantitative test that is designed to efficiently give the user specific information about a waste that will aid in determining waste identification, process compatibility, and safety in handling.

4. Summary of Test Method

4.1 A small portion of the sample is placed onto a strip of potassium iodide (KI) starch paper. The blue color which is the result of oxidizing the potassium iodide to iodine (I_2) in the presence of starch, indicates a positive test for oxidizers.

5. Significance and Use

5.1 This test method is intended for use by those in waste management industries to avoid potentially harmful reactions due to oxidizing compounds in wastes.

6. Interferences

6.1 Materials that mask the KI starch paper, (for example, oils, syrups, etc.) prevent reaction with the test paper or visual detection of a color change.

6.2 Samples or slurries of samples that are already dark colored prior to applying a test portion to the KI starch paper can give false results.

6.3 Oxidizers such as ferric salts may not oxidize organics; however, they may show positive reactions with KI.

6.4 Oxidizers that require the presence of acid or base will give a false negative result. To overcome this problem, run the oxidizer test at acid, base, and neutral pH conditions.

7. Reagents and Materials

7.1 *Purity of Reagents*—Reagent-grade chemicals shall be used in all tests. Unless otherwise indicated, it is intended that all reagents conform to specifications of the Committee on Analytical Reagents of the American Chemical Society where such specifications are available.[4] Other grades may be used, provided it is first ascertained that the reagent is of sufficiently high purity to permit its use without lessening the accuracy of the determination.

7.2 *Purity of Water*—Unless otherwise indicated, references to water shall be understood to mean reagent water as defined by Type III of Specification D 1193.

---

[1] This test method is under the jurisdiction of ASTM Committee D34 on Waste Management and is the direct responsibility of Subcommittee D34.01.05 on Screening Methods.

Current edition approved Sept. 10, 1995. Published November 1995. Originally published as D 4981 – 89. Last previous edition D 4981 – 89.

[2] *Annual Book of ASTM Standards*, Vol 11.01.

[3] *Annual Book of ASTM Standards*, Vol 11.04.

[4] *Reagent Chemicals, American Chemical Society Specifications*, American Chemical Society, Washington, DC. For suggestions on the testing of reagents not listed by the American Chemical Society, see *Analar Standards for Laboratory Chemicals*, BDH Ltd., Poole, Dorset, U.K., and the *United States Pharmacopeia and National Formulary*, U.S. Pharmaceutical Convention, Inc. (USPC), Rockville, MD.

7.3 *Potassium Iodide (KI) Starch Paper Strips.*

7.4 *Beakers.*

7.5 *Pipet or Droppers.*

7.6 *Hydrogen Peroxide Solution (3 % H_2O_2)*, commercial grade.

7.7 *Nitric Acid* (1 + 9)—Add 1 volume of nitric acid (HNO_3, sp. gr 1.42) to 9 volumes of water.

8. Hazards

8.1 Avoid inhalation, skin or eye contact, or both, with all hazardous material.

8.2 This analysis must always be performed in a fume hood.

9. Sampling

9.1 Collect a representative sample of the waste in a container with a sealed lid.

9.2 The sample should be analyzed as soon as possible.

9.3 Allow the samples to stabilize to room temperature.

10. Procedure

10.1 *Aqueous Samples*—Using a clean pipet or dropper, place a drop sample on a strip of KI starch paper and note color change.

10.2 *Non-Aqueous Samples (for example, solid, oil, or solvents)*:

10.2.1 In a beaker, prepare a slurry by adding 1 to 5 g of sample to an equal amount of reagent water.

10.2.2 Using a clean pipet or dropper, place a drop of slurry on a strip of KI starch paper.

10.2.3 Note any color change.

10.3 All sample results shall be checked against the quality control and blank test strips to verify positive/negative readings (see Section 11).

10.4 All positive results (a blue color) shall be reported immediately to prevent potential hazardous reactions that may occur.

NOTE 1—Analyst should be aware that the blue color on the reference strip and any positive sample test strip will slowly fade with time.

11. Quality Control

11.1 Quality control check samples, reference blanks and duplicates should be performed at an action level specified by the laboratory and at an appropriate frequency.

11.1.1 Place a drop of reagent water on a KI starch test strip to serve as a reference blank.

11.1.2 Place a drop of 3 % H_2O_2 solution (or 10 % HNO_3) on a KI starch test strip to obtain a dark blue reference color.

11.2 **Warning—**Fortifications (spikes) are not to be prepared due to possible adverse reactions.

11.3 Method detection limits should be determined by each laboratory by using a test oxidizer (7.6 or 7.7).

12. Precision and Bias

12.1 *Precision and Bias*—No information is presented about either the precision or bias of Test Method D 4981 for measuring oxidizers since the test result is nonquantitative.

13. Keywords

13.1 oxidizers; screening analysis; starch-iodide paper

Designation: D 4982 – 95 (Reapproved 2001)

Standard Test Methods for Flammability Potential Screening Analysis of Waste[1]

This standard is issued under the fixed designation D 4982; the number immediately following the designation indicates the year of original adoption or, in the case of revision, the year of last revision. A number in parentheses indicates the year of last reapproval. A superscript epsilon (ε) indicates an editorial change since the last revision or reapproval.

1. Scope

1.1 These test methods are used to indicate the fire-producing or fire-sustaining potential of wastes. The following test methods can be applied to waste liquids, sludges, or solids:

	Sections
Test Method A—Test Sample Exposed to Heat and Flame	7-9
Test Method B—Test Sample Exposed to Spark Source	10 and 11

1.2 These test methods should be used to measure and describe the properties of materials, in response to heat and flame under controlled laboratory conditions and should not be used to describe or appraise the fire hazard or fire risk of materials under actual fire conditions. However, results of these tests may be used in addition to all other factors that are pertinent to a fire hazard assessment of a particular end use.

1.3 These test methods are designed and intended as preliminary tests to complement quantitative analytical techniques that may be used to determine flammability. These test methods offer the option and the ability to screen waste for hazardous flammability potential when the analytical techniques are not available or the total waste composition is unknown.

1.4 *This standard does not purport to address all of the safety concerns, if any, associated with its use. It is the responsibility of the user of this standard to establish appropriate safety and health practices and determine the applicability of regulatory limitations prior to use.* Specific hazard information is given in Section 5 and Note 2.

2. Terminology

2.1 *Definitions of Terms Specific to This Standard:*

2.1.1 *screening analysis*—a *preliminary* qualitative or semi-quantitative test that is designed to efficiently give the user specific information about a waste that will aid in determining waste identification, process compatibility, and safety in handling.

3. Summary of Test Methods

3.1 *Method A*—A test sample is exposed to heat and flame. The sample is reported as having a positive or negative flammability potential as described in the test procedure.

3.2 *Method B*—Sparks from a flint lighter are introduced to the vapor space immediately above a representative sample of a waste, and observation is made for a flash in the vapor space or ignition of the sample. A flash in the vapor space or ignition and burning of the waste indicates a positive flammability potential at ambient temperature.

4. Significance and Use

4.1 These test methods are intended for use by those in the waste management industries to aid in identifying the flammability potential or waste materials.

5. Hazards

5.1 Avoid inhalation and skin or eye contact, or both, of any hazardous materials.

5.2 Standard laboratory hygiene practices should be followed when conducting these tests.

5.3 All tests must be performed in a laboratory hood.

5.4 Waste containing or suspected of containing highly volatile organics or peroxides should be tested using a much smaller sample than that used in 9.2.

5.5 The aluminum weighing boat should be placed on an inert, nonflammable surface.

6. Sampling

6.1 Sample containers must be kept tightly sealed until tested.

6.2 Samples should be analyzed as soon as possible after collection.

6.3 If necessary, allow the sample to come to room temperature in a tightly sealed container. For example, frozen material should be allowed to thaw completely.

TEST METHOD A—EXPOSURE TO HEAT AND FLAME

7. Interferences

7.1 Drafts in the laboratory fume hood where the test is performed could cause excessive cooling and false negative results. A properly operating fume hood with a face velocity of 100 ft/min should provide consistent, usable results.

7.2 Ignition sources that provide excessive heating rates alone or combined with a very small test portion may obscure results. That is, the sample may be decomposed, sintered, fused, evaporated, or otherwise consumed before positive or negative evidence of flammability is observed. The use of sufficiently large test portions and of heating rates typical of a

[1] These test methods are under the jurisdiction of ASTM Committee D34 on Waste Management and is the direct responsibility of Subcommittee D34.01.05 on Screening Methods.

Current edition approved Oct. 10, 1995. Published December 1995. Originally published as D 4982 – 89. Last previous edition D 4982 – 89.

bunsen burner should resolve this problem.

7.3 An inadequate source of heating could limit the volatilization of flammable components and provide false negative results.

7.4 Difficult-to-observe flames resulting from the burning of certain compounds (for example, methanol) could provide false negative results. If the presence of these compounds is suspected, the presence of flammability might be confirmed by the insertion of a watchglass just above the test material. The watchglass is then examined for products of combustion (for example, moisture and soot).

NOTE 1—The watchglass method cannot distinguish between vaporized water or water produced by combustion.

7.5 An improperly adjusted gas burner (for example, with an insufficiently aerated flame) could introduce unburnt gas into or immediately above the sample. This unburnt gas could briefly support a flame after the source of ignition is removed, providing the appearance of a flammable sample and a false positive result.

8. Apparatus

8.1 *Gas Burner*, (for example, a bunsen burner) with an adjustable air shutter and an adjustable gas orifice is needed. The gas burner and fuel supply line must be appropriate to the gas supplied: natural gas, artificial gas (including propane and butane), or liquified petroleum gas (LP gas or LPG). Where a gas supply line cannot be provided, a propane torch may be substituted.

8.2 *Lighter*, (for example, piezo lighter) for burner is required.

8.3 *Aluminum Weighing Boats* or other non-flammable containers are needed.

8.3.1 **Caution—**Weighing boats of material other than aluminum should be used if the testing materials react with aluminum, for example, caustics.

8.4 *Watchglass.*

8.5 *Large Beaker, Tongs, Asbestos-Freė High-Temperature Gloves or Mittens*, or other apparatus as needed to extinguish burning materials.

9. Procedure

9.1 Light a gas burner and adjust to a typically blue flame that is not readily blown out. A yellow flame easily affected by drafts indicates insufficient air (the air:fuel ratio is too low). A sharp, blue flame is good. (If the flame rises above the burner head, is very difficult to light, or tends to extinguish itself, indicates that too much air or too much air and fuel are being supplied to the burner.)

9.2 Place a sufficient amount (approximately 5 g) of a test sample in an aluminum weighing boat or other nonflammable container.

9.3 Using a gas burner, hold the flame immediately above and perpendicular to the test sample for 2 to 3 s without touching the visible flame to the sample.

9.3.1 If ignition (a flash or burning) is observed before or after the source of ignition (the flame of the burner) is removed, the sample is said to have a positive flammability potential. A positive result may require further investigation (see 9.4.1).

9.3.2 The confirmation of flammability may require the use of a watchglass (see 7.4).

9.3.3 If there is no ignition, proceed to 9.4.

9.4 Using a gas burner, briefly (for at least 10 s) apply the flame to the sample in an attempt to ignite the sample.

9.4.1 If the sample ignites, the sample is said to have a positive flammability potential.

9.4.1.1 When more accurate waste characterization is necessary, liquid samples may be quantified using a closed-cup flash point tester.

9.4.1.2 Solids with a positive flammability potential should be further investigated.

9.4.2 If the sample decomposes, boils (if a liquid), or otherwise fails to ignite after at least 15 s of continuous sample heating by the burner flame, the flammability potential is reported as negative.

9.4.3 Halogenated solvents typically give off visible vapors that may result in a false positive flammability potential.

NOTE 2—**Warning:** Phosgene, an extremely toxic gas, is a combustion product of halogenated compounds burned in air.

9.5 Shut off the gas burner when not in use. Extinguish a burning sample by setting an aluminum weighing boat or watchglass atop the one containing the burning sample, (or invert a spoutless beaker over the sample container and all). Use of tongs or high temperature gloves or mittens may be necessary to handle the equipment.

TEST METHOD B—EXPOSURE TO SPARK SOURCE

10. Apparatus

10.1 *Oven Gloves.*

10.2 *Flint Lighter*, the type typically used to light an air/acetylene torch is required.

10.3 *Disposable 250-mL Beaker*, of plastic is required.

10.4 *Watchglass*, 100 mm.

10.5 *Metal Vessel* (with lid), of adequate depth and diameter to contain beaker and watchglass is needed.

10.6 *Thermometer.*

11. Procedure

11.1 Place approximately 100 mL of the test sample of the material to be tested into the plastic beaker (see 6.3).

11.2 Place the plastic beaker in the steel vessel, cover the beaker with the watchglass, and allow to stand at ambient conditions for 5 min.

11.3 Remove the watchglass, place the igniter immediately above the waste and strike it several times to produce sparks.

11.4 If the material does catch fire and burn, extinguish the flames by immediately placing the lid on the steel vessel, thus smothering the fire and report as positive flammability potential.

12. Report for Methods A and B

12.1 Report the following information:

12.1.1 Sample identification,

12.1.2 Date of test,

12.1.3 Sample classification: positive or negative, and

12.1.4 Reference to the procedure applied.

13. Quality Control for Methods A and B

13.1 Quality control check samples and duplications should be performed at an action level specified by the laboratory and at an appropriate frequency.

13.2 Flammability standards should be maintained for analyst training and as reference guides. Examples are given in the following table.

Flammability Potential Screening Analysis

Methyl alcohol (methanol)	Difficult-to-see flame; flammable at or below room temperature.
Glacial acetic acid	Flammable at temperatures above normal room temperature.
Hendecane (n-undecane)	Flammable at temperatures higher than glacial acetic acid.
Kerosene (Fuel Oil No. 1)	Flammable over a range of temperatures higher than normal room temperature.
p-Xylene	Flammable at or above room temperature; freezes at 13°C. A spiked soil sample frozen at 0°C demonstrates interference.

14. Precision and Bias for Methods A & B

14.1 *Precision*—No statement is made about the precision because this screening method produces only a" Positive" or "Negative" test result.

14.2 *Bias*—No statement is made about the bias because this screening method produces only a "Positive" or "Negative" test result.

15. Keywords

15.1 flammability potential; ignition; screening analysis; wastes

Standard Practices for
Sampling Wastes from Pipes and Other Point Discharges[1]

This standard is issued under the fixed designation D 5013; the number immediately following the designation indicates the year of original adoption or, in the case of revision, the year of last revision. A number in parentheses indicates the year of last reapproval. A superscript epsilon (ϵ) indicates an editorial change since the last revision or reapproval.

1. Scope

1.1 Those practices provide guidance for obtaining samples of waste at discharge points from pipes, sluiceways, conduits, and conveyor belts. The following are included:

	Sections
Practice A—Liquid or Slurry Discharges	7 through 9
Practice B—Solid or Semisolid Discharges	10 through 12

1.2 These practices are intended for situations in which there are no other applicable ASTM sampling methods (see Practices D 140 and D 75) for the specific industry.

1.3 These practices do not address flow and time-proportional samplers and other automatic sampling devices.

1.4 Samples are taken from a flowing waste stream or moving waste mass and, therefore, are descriptive only within a certain period. The length of the period for which a sample is descriptive will depend on the sampling frequency and compositing scheme.

1.5 It is recommended that these practices be used in conjunction with Guide D 4687.

1.6 *This standard does not purport to address all of the safety concerns, if any, associated with its use. It is the responsibility of the user of this standard to establish appropriate safety and health practices and determine the applicability of regulatory limitations prior to use.* See Section 5 for more information.

2. Referenced Documents

2.1 *ASTM Standards:*

D 75 Practice for Sampling Aggregates[2]

D 140 Practice for Sampling Bituminous Materials[2]

D 4687 Guide for General Planning of Waste Sampling[3]

E 882 Guide for Accountability and Quality Control in the Chemical Analysis Laboratory[4]

2.2 *Other Document:*

EPA-SW-846 Test Methods for Evaluating Solid Waste, Physical/Chemical Methods[5]

3. Summary of Practices

3.1 The variability of the waste stream is first determined based on (*1*) knowledge of the processes producing the stream, or (*2*) the results of a preliminary investigation of the waste stream's variability. A sampling design is then developed that considers the waste stream's variability, the time frame the sample is to represent, and the precision and accuracy required for waste analysis or testing. The actual sampling procedure consists of obtaining several grab samples from the moving stream or mass for analysis or testing.

4. Significance and Use

4.1 The procedure outlined in these practices are guides for obtaining descriptive samples of solid, semisolid and liquid waste from flowing streams, and incorporate many of the same procedures and equipment covered in the Referenced Documents. These practices by themselves will not necessarily result in the collection of samples representative of the total waste mass. The degree to which samples describe a waste mass must be estimated by application of appropriate statistical methods and measures of quality assurance. It is recommended that those practices be used in conjunction with Guide D 4687.

5. Hazards

5.1 In all sampling practices, safety should be the first consideration. Personnel involved in the sampling should be fully aware of, and take precautions against, the presence of toxic or corrosive gases, the potential for contact with toxic or corrosive liquids or solids, and the dangers of moving belts, conveyors, or other mechanical equipment. Guidance on waste sampling safety can be found in Guide D 4687.

6. Sampling Design

6.1 The frequency of sampling and the number of composites required to obtain a sample of the waste will depend on the following:

6.1.1 Time variability of the waste composition,

6.1.2 Time span which the sample is to represent, and

6.1.3 Precision of waste analysis that is required, for example, if a hazardous constituent is present in the waste at levels near the regulatory limit or another limit of concern, then better precision will be required than if the levels are well below or well above the limits of concern.

6.2 The processes that produce the waste will largely dictate

---

[1] These practices are under the jurisdiction of ASTM Committee D34 on Waste Management and are the direct responsibility of Subcommittee D34.01.02 on Monitoring.

Current edition approved Nov. 24, 1989. Published January 1990.

[2] *Annual Book of ASTM Standards*, Vol 04.03.

[3] *Annual Book of ASTM Standards*, Vol 11.04.

[4] *Annual Book of ASTM Standards*, Vol 03.05.

[5] Available from Superintendent of Documents, U.S. Government Printing Office, Washington, DC 20402.

the variability in the composition of the waste. If the processes are known to be constant and reliable, then fewer samples should be required than from a highly variable process.

6.3 To obtain a descriptive sample of the waste, the concentration levels and approximate variation in the waste composition should first be estimated. In some cases, a rough estimate can be made based on knowledge of the processes that produce the waste. In other cases, results from previous sampling efforts can be used to estimate waste composition and variability. A preliminary pilot sampling effort may be necessary to establish the waste composition prior to designing the primary sampling program. Procedures for estimating sample variability and for establishing a sampling design are provided in Guide D 4687.

6.4 The sampling design should include quality assurance procedures. At the least, this should include the following:

6.4.1 Sample handling quality control by carrying a blank sample through all of the sampling and analytical steps, and

6.4.2 User should be aware of the laboratories' internal quality control procedures. More rigorous quality control/quality assurance procedures may be required depending on the particular goals of the sampling program. For further information on quality control/quality assurance, see Guide E 882 and EPA SW-846.

6.5 A sampling plan should be prepared prior to sampling. The plan should describe such things as (a) safety procedures; (*b*) sampling design, including number and location of samples; (*c*) quality assurance procedures; (*d*) apparatus; (*e*) sampling procedures; and (*f*) sampling labeling. The details of the sampling procedure should consider all aspects of the specific discharge, including pipe diameter, velocity, rate of discharge, solids content of the discharge, requirements for grab or composite samples, and ultimate use of the analytical data.

PRACTICE A—LIQUID OR SLURRY DISCHARGES

7. Apparatus

7.1 *Dipper Sampler*— For slurry and liquid discharges, a dipper type sampler should be employed. One example of this type of sampler is depicted in Fig. 1. The dipper can be varied in size depending on the flow rate from the pipe or sluiceway. This procedure should not be used for high stream flow velocities or rates (>100 gal/min) because problems will arise in physically holding the dipper in the stream. Stream dimensions, size and shape of pipe, should also be considered in addition to flow rate. The sample should be taken across the full opening of the stream in as short a time as possible. This will minimize the effect of changes in composition of the waste stream due to density differentiation, laminar flow, and the like.

8. Procedure

8.1 Clean the beaker and container for compositing sample by methods appropriate for the analysis to be performed. Cleaning the equipment is especially important to prevent cross contamination between different waste types. In some cases, it may be necessary or simpler to dedicate equipment to a specific waste type.

8.2 Assemble the sampler by clamping the beaker to the pole.

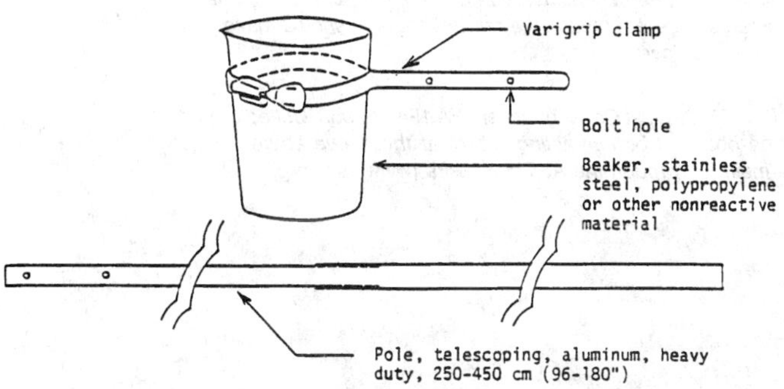

FIG. 1 Dipper Sampler (Source: Test Methods for Evaluating Solid Wastes, U.S. EPA, S2–846, July 1982)

8.3 Make sure that the sampler matches the dimensions of the discharge stream if at all possible. If not, pass the dipper in one sweeping motion through the discharge stream at a rate such that the dipper is filled in one pass. Make enough passes to cover the entire cross sectional area of the discharge stream.

8.4 If the entire discharge width cannot be covered in one pass additional passes will be needed. Begin each additional pass at the ending point of the previous pass. It may be necessary to make a few trial passes or practice runs before actually sampling the discharge. If compositing of samples is required by the sampling plan it may be preferable and sometimes necessary to have compositing performed in the laboratory. Include specific procedures for compositing in the sampling plan. If samples have not been appropriately composited in the field, clearly indicate this information on the sample label, in the field log book, and on the analytical request form.

8.5 In cases of high flow (>100 gal/min), a probe may be inserted upstream of the discharge to obtain a sample. Design the probe to collect a representative cross section of the flowing stream. See Fig. 2 for an example sampling probe.

8.6 Use preservation techniques appropriate for the analyses or testing to be conducted.

9. Packaging and Package Marking

9.1 An indelible label should be secured to the container identifying the sample. The label should contain or reference the following information:

9.1.1 Name and location of site,

9.1.2 Date and time of sampling,

9.1.3 Location of sampling,

9.1.4 Sample number,

9.1.5 Description and disposition of sample,

9.1.6 Name of sampling personnel,

9.1.7 Type of preservative, and

9.1.8 Sampling conditions, and analytical requirements.

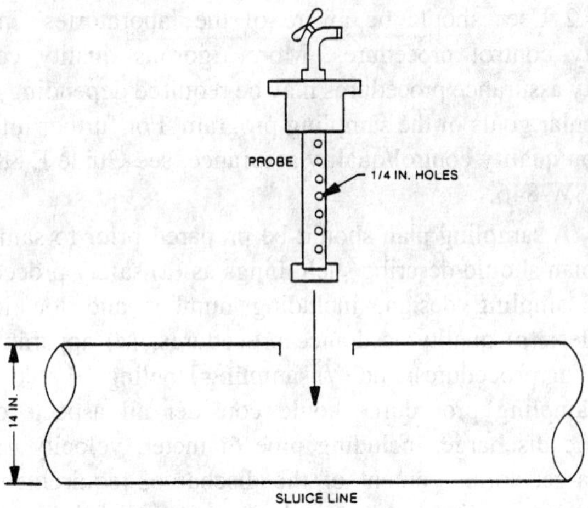

FIG. 2 In–Pipe Sampling Probe

9.2 Pack the sample container securely in a shipping container. If required for preservation of analytes, the sample container should be packed in ice and cooled to 4°C. A minimum-maximum thermometer should be packed with the samples.

9.3 Follow DOT (Department of Transportation) shipping regulations.

9.4 Make arrangements for handling, logging in, adequate storage and analysis of the sample at its destination. If warranted, follow chain-of-custody protocol.

PRACTICE B—SOLID OR SEMISOLID DISCHARGES

10. Apparatus

10.1 *Scoop or Shovel Sampler*—For solid or semisolid discharges from belts, a scoop or shovel should be used, made of material compatible with the waste stream and inert to the analytes. Where routine sampling is to be performed, a scoop may be designed to match the width and contour of the belt. In this way, a single time increment sample may be taken in one scoop.

11. Procedure

11.1 Clean the scoop or shovel as appropriate for the desired chemical analysis to prevent cross contamination. In some cases, it may be necessary to dedicate equipment to specific waste types or waste strata.

11.2 Sample at a convenient point along the belt.

11.3 Sample the waste with the scoop or shovel making sure to sample across the entire width of the belt. Be sure all fines and any liquid are also included representatively in the scooped sample.

11.4 Repeat sampling depending on the uniformity of the waste. Composite the grab samples and transfer a well-mixed portion to a chemically compatible container and seal. It may be preferable or necessary to have compositing performed in the laboratory. Include specific procedures for compositing in the sampling plan.

11.5 If compositing has not been appropriately accomplished prior to shipment, clearly indicate on the sample label, in the field log book, and on the analytical request form.

12. Sampling Labeling and Shipping

12.1 Refer to Section 9.

Last ASTM Designation: D 5049 – 90

Standard Test Method for the Screening of Cyanides in Waste

This test method is applicable to the screening of cyanides in waste liquids, sludges, semisolids, and solids.
Formerly under the jurisdiction of ASTM Committee D-34 on Waste Management, this test method was discontinued in 1999.

Designation: D 5050 – 96

Standard Guide for Commercial Use of Lime Kiln Dusts and Portland Cement Kiln Dusts[1]

This standard is issued under the fixed designation D 5050; the number immediately following the designation indicates the year of original adoption or, in the case of revision, the year of last revision. A number in parentheses indicates the year of last reapproval. A superscript epsilon (ϵ) indicates an editorial change since the last revision or reapproval.

1. Scope

1.1 This guide is intended to evaluate and describe lime kiln dusts and Portland cement kiln dusts for uses in commercial applications. This guide is not intended to cover uses for lime or Portland cement.

1.2 *This standard does not purport to address all of the safety concerns, if any, associated with its use. It is the responsibility of the user of this standard to establish appropriate safety and health practices and determine the applicability of regulatory limitations prior to use.*

2. Referenced Documents

2.1 *ASTM Standards:*

C 25 Test Methods for Chemical Analysis of Limestone, Quicklime, and Hydrated Lime[2]

C 109 Test Method for Comprehensive Strength of Hydraulic Cement Mortars (Using 2-in. or 50-mm Cube Specimens)[2]

C 110 Test Methods for Physical Testing of Quicklime, Hydrated Lime, and Limestone[2]

C 151 Test Method for Autoclave Expansion of Portland Cement[2]

C 184 Test Method for Fineness of Hydraulic Cement by the 150-μm (No. 100) and 75-μm (No. 200) Sieves[2]

C 187 Test Method for Normal Consistency of Hydraulic Cement[2]

C 266 Test Method for Time of Setting of Hydraulic Cement Paste by Gillmore Needles[2]

C 305 Practice for Mechanical Mixing of Hydraulic Cement Pastes and Mortars of Plastic Consistency[2]

C 400 Test Methods for Testing Quicklime and Hydrated Lime for Neutralization of Waste Acid[2]

C 593 Specification for Fly Ash and Other Pozzolans for Use With Lime[2]

C 602 Specification for Agricultural Liming Materials[2]

C 911 Specification for Quicklime, Hydrated Lime, and Limestone for Chemical Uses[2]

D 558 Test Method for Moisture-Density Relations of Soil-Content Mixtures[3]

D 1632 Practice for Making and Curing Soil-Cement Compression and Flexure Test Specimens in the Laboratory[3]

D 1633 Test Method for Compressive Strength of Molded Soil-Cement Cylinders[3]

D 3155 Test Method for Lime Content of Uncured Soil-Lime Mixtures[3]

D 3551 Practice for Laboratory Preparation of Soil-Lime Mixtures Using a Mechanical Mixer[3]

D 3668 Test Method for Bearing Ratio for Laboratory Compacted Soil-Lime Mixtures[3]

D 4318 Test Method for Liquid Limit, Plastic Limit, and Plasticity Index of Soils[3]

2.2 *Code of Federal Regulations*

40 CFR Part 268 Appendix 1 Toxicity Characteristics Leaching Procedure[4]

40 CFR Part 268.43(a) Treatment Standards Expressed as Waste Concentrations (reserved)[4]

3. Terminology

3.1 *Definitions of Terms Specific to This Standard:*

3.1.1 *lime kiln dust*:

3.1.1.1 Lime kiln dust is the finely-divided particulate matter carried from a lime kiln by the exhaust gases.

3.1.1.2 The composition of the dust is dependent upon the stone being processed, the processing equipment in use, and the fuel being used. Concentrations of combustion by-products, including sulfur compounds usually exceed that of the primary quicklime product. Chemical and physical properties are also dependent upon these same factors and should be evaluated for the use intended.

3.1.2 *portland cement kiln dust*:

3.1.2.1 Cement kiln dust is the finely-divided particulate matter carried from a cement kiln by the exhaust gases.

3.1.2.2 The dust is composed of variable mixtures of calcined and uncalcined feed materials, fine cement-clinker formed during the high temperature processing, fuel combustion by-products, and condensed alkali compounds. Alkalis may be concentrated in the dust through volatilization in the

---

[1] This guide is under the jurisdiction of ASTM Committee D-34 on Waste Management and is the direct responsibility of Subcommittee D34.06 on Recovery and Reuse.

Current edition approved Oct. 10, 1996. Published December 1996. Originally published as D 5050 – 90. Last previous edition D 5050 – 90.

[2] *Annual Book of ASTM Standards*, Vol 04.01.

[3] *Annual Book of ASTM Standards*, Vol 04.08.

[4] Code of Federal Regulations available from the Superintendent of Documents, United States Government Printing Office, Washington, DC 20402.

high temperature zones within the kiln then condensed in the exhaust gases as they pass through the dust collection system. The relative quantity of the aforementioned components of Portland cement kiln dust may change from one source to another because of raw material, fuel, and process variations. Some cement kiln dusts may contain large quantities of calcined food materials, alkalies and sulfur compounds, or both, while others may be primarily composed of uncalcined raw food. Many Portland cement kiln dusts are easily compacted and reactive with small quantities of water to form a low strength cementitious mass; some may show only mild reactivity and be non-consolidating.

4. Significance and Use

4.1 This guide provides a list of applications and appropriate test procedures to establish selected uses for the lime kiln dusts and Portland cement kiln dusts.

4.2 The lime kiln dusts and Portland cement kiln dusts covered by this guide may vary in composition from one source to another. It is therefore advised that the use of such products be undertaken only after their specific compositions, physical properties, performance characteristics, and the anticipated variabilities have been evaluated for the service intended. Until the degree of variability in the dust source or sources has been established, frequent performance testing is recommended.

4.3 Specifications should be prepared to facilitate uses, environmental protection, and engineering designs by the responsible persons.

5. Categories for Kiln Dust Uses

5.1 The following applications present a sampling of applications to be considered. Other uses may be considered where similar chemical or physical properties are required.

5.1.1 *Structural Purposes*—Such as additions in the manufacture of glass, brick, block, and other building materials.

5.1.2 *Stabilization Purposes*—With soils to modify the plastic limit or moisture content and in conjunction with or without fly ash to provide stabilized properties to solid and other waste construction products.

5.1.3 *Fixation Purposes*—For neutralization and fixation alone and in conjunction with other materials, both hazardous and nonhazardous.

5.1.4 *Solidification Purposes*—For dewatering waste materials to provide structural and handling properties, for economic and environmental use, and disposal.

5.1.5 *pH Control Purposes*—For use with domestic wastewater sludges to significantly reduce pathogens and to enable domestic wastewater sludges to be safely and economically applied to the land.

5.1.6 *Flocculation Purposes*—For use to separate solids from liquids and in multi-purpose processes to treat, dispose, and utilize industrial, municipal, and utility wastes.

5.1.7 *Agricultural Purposes*—A soil conditioner and nutrient source to meet the needs of the agricultural industry.

6. Determination of Materials Characteristics

6.1 Lime kiln dusts or Portland cement kiln dusts may or may not require slaking for effective use depending on the end use.

6.2 Comparative testing with applications in use or products should be undertaken before applying the dust to specific applications. Lime kiln dusts and Portland cement kiln dusts are variable and may contain inert materials. The degree of variability and the inclusion of materials detrimental to the intended use should be evaluated.

6.3 Testing lime kiln dusts or Portland cement kiln dusts that have been used as a cementitious binder alone or in conjunction with fly ash should be performed on specimens that have been cured and aged to duplicate field conditions or by appropriate specification methods.

6.4 It should be noted that some lime kiln dusts and Portland cement kiln dusts react more quickly than others depending upon the amount of calcined calcium oxide or calcined magnesium oxide clinker, or other cementitious materials and particle fineness. In addition, potassium and sodium compounds may contribute to the reactivity of cement kiln dusts.

6.5 The selection of laboratory tests to evaluate the suitability of materials for a specific application will depend upon the performance requirements of the job and the materials available. A list of test methods and material specifications is given in Table X1.1 of Appendix X1 and may be used in selecting those procedures and requirements pertinent to the application in question.

7. Keywords

7.1 construction application; lime kiln dust; Portland cement kiln dust; waste disposal

APPENDIX

(Nonmandatory Information)

X1. REPRESENTATIVE RECOMMENDED METHODS

TABLE X1.1 Representative Recommended Methods for Determining and Evaluating Characteristics of Lime Kiln Dust and Cement Kiln Dust Suitable for the Intended Purpose

Characteristics	Test Method
General Laboratory Procedures:	
Chemical Analysis of Limestone Quicklime and Hydrated Lime	C 25
Test for Compressive Strength of Hydraulic Cement Mortars	C 109
Physical Testing of Quicklime Hydrated Lime, and Limestone	C 110
Test for Autoclave Expansion of Portland Cement	C 151
Tests for Fineness of Cement	C 184
Test for Normal Consistency of Cement	C 187
Test for Setting of Hydraulic Cement by Gillmore Needle	C 266
Mechanical Mixing of Hydraulic Pastes and Mortars of Plastic Consistency	C 305
Neutralization of Acid Waste	C 400
Fly Ash and Other Pozzolans for Use with Lime	C 593
Agricultural Liming Materials	C 602
Quicklime, Hydrated Lime and Limestone for Chemical Use	C 911[A]
Moisture-Density Relations of Soil-Cement Mixtures	D 558
Soil-Cement Compression and Flexure Test Specimens, Making and Curing in the Laboratory	D 1632
Compressive Strength of Molded Soil-Cement Cylinders	D 1633
Lime Content of Uncured Soil-Lime Mixtures	D 3155
Laboratory Preparation of Soil-Lime Mixtures Using a Mechanical Mixer	D 3551
Bearing Ratio of Laboratory-Compacted Soil-Lime Mixtures	D 3668
Liquid Limit, Plastic Limit, and Plasticity Index of Soils	D 4318
Toxicity Characteristic Leaching Procedure	40 CFR Part 268 Appendix 1
Treatment Standards Expressed as Waste Concentrations (reserved)	40 CFR Part 268.43

[A] Suggest an alternative test procedure to determine basicity factor(s) when used in dry application as in Test Methods C 400.

Designation: D 5057 – 90 (Reapproved 2001)

Standard Test Method for Screening Apparent Specific Gravity and Bulk Density of Waste[1]

This standard is issued under the fixed designation D 5057; the number immediately following the designation indicates the year of original adoption or, in the case of revision, the year of last revision. A number in parentheses indicates the year of last reapproval. A superscript epsilon (ε) indicates an editorial change since the last revision or reapproval.

1. Scope

1.1 This test method covers the determination of apparent specific gravity and bulk density in waste. For the purpose of this test method, materials to be measured will be classified into three groups:

1.1.1 *Group A*—Free-flowing liquids; apparent specific gravity (ASG),

1.1.2 *Group B*—Granules, powders and water reactive liquids, solids or sludges; bulk density (BD), and

1.1.3 *Group C*—Bulk solids (such as gravel, paper or wood, etc.); apparent specific gravity (ASG).

1.2 This test method is designed and intended as a preliminary test to complement the more sophisticated quantitative analytical techniques that may be used to determine specific gravity. This test method offers to the user the option and the ability to screen waste for apparent specific gravity or bulk density when the more sophisticated techniques are not available and the total waste composition is unknown.

1.3 *This standard does not purport to address all of the safety concerns, if any, associated with its use. It is the responsibility of the user of this standard to establish appropriate safety and health practices and determine the applicability of regulatory limitations prior to use.* For a specific hazard statement, see Section 9.

2. Referenced Documents

2.1 *ASTM Standards:*

D 1192 Specification for Equipment for Sampling Water and Steam[2]

D 1193 Specification for Reagent Water[2]

D 3370 Practices for Sampling Water[2]

D 4057 Practice for Manual Sampling of Petroleum and Petroleum Products[3]

3. Terminology

3.1 *Definition of Term Specific to This Standard:*

3.1.1 *screening*—a *preliminary* qualitative or semi-quantitative test, developed from classical qualitative and quantitative techniques, that is designed to efficiently give the user specific information about a waste that will aid in determining waste identification, process compatibility, and safety in handling.

4. Summary of Test Method

4.1 The specific gravity of a material is the ratio of the masses of equal volumes of a waste and reagent water. The apparent specific gravity of materials in Groups A and C is determined by comparing the mass of a sample to the mass of the same volume of reagent water. The bulk density of wastes in Group B is determined as a direct mass/volume ratio of the sample alone and should be used for determinations on water reactive materials. The weights are used in determining mass.

5. Significance and Use

5.1 This test method is intended for use by those in the waste management industries for the determination of apparent specific gravity and bulk density of waste.

5.2 The apparent specific gravity and bulk density determined by this test method can be used for the conversion of measured volumes to weights.

5.3 The apparent specific gravity and bulk density, when correlated with other properties, can be used to indicate the character of the waste.

6. Interferences

6.1 Excessive temperatures causing loss of sample components due to vaporization could result in erroneous readings.

6.2 Large, obvious void spaces interfere in this test method and will give inaccurate results because of the false volume measured.

7. Apparatus

7.1 *Weighing Bottle*— Specific gravity bottle or equivalent container is needed.

7.2 *Spatulas.*

7.3 *Top Loader Balance*, with a sensitivity of 0.01 g is required.

8. Reagents and Materials

8.1 *Purity of Water*— Unless otherwise indicated, references to water shall be understood to mean reagent water as defined by Type IV of Specification D 1193.

[1] This test method is under the jurisdiction of ASTM Committee D34 on Waste Management and is the direct responsibility of Subcommittee D34.01.05 on Screening Methods.

Current edition approved May 25, 1990. Published July 1990.

[2] *Annual Book of ASTM Standards*, Vol 11.01.

[3] *Annual Book of ASTM Standards*, Vol 05.03.

9. Hazards

9.1 **Precaution**—Avoid inhalation of and skin and eye contact with all hazardous materials.

10. Sampling

10.1 Collect the sample in accordance with Specification D 1192 and Practices D 3370 and D 4057.

NOTE 1—Extreme temperature variations between the sample and reagent water should be avoided.

11. Procedure

11.1 Make all weight measurements to nearest 0.01 g.

11.2 Weigh the empty container (weighing bottle or specific gravity bottle) with lid on, and record weight as *W*.

11.3 Fill the container with water and place lid on container. Wipe off excess water and weigh. Record weight of water-filled container as *R*.

11.4 For free-flowing liquids (Group A):

11.4.1 Fill the empty container (see 7.1) with sample.

11.4.2 Place the lid on the container, pushing out excess sample through the hole.

11.4.3 Wipe off excess sample.

11.4.4 Weigh the sample-filled container with lid on, and record weight as *S*.

11.5 For granules, powders, and water reactive materials (Group B):

11.5.1 Add as much of the sample to the weighed container (see 7.1) as possible without exerting pressure, filling the container with sample but not allowing large void spaces (see 6.2). The container may be tapped or lightly tamped.

11.5.2 Place the lid on the container and weigh the sample and bottle and record weight as *S*.

11.6 For bulk solids such as gravel, paper or wood (Group C):

11.6.1 Add as much of the sample to the weighed container (see 7.1) as possible without exerting pressure. Place the lid on the container and weigh and record weight as *S*.

11.6.2 Fill remaining space in the container with water and place the lid on the container, taking care that air bubbles are not trapped in the material or the container.

11.6.3 Weigh and record weight of sample and water, in container with the lid on. Record weight as *Q*.

NOTE 2—The quantity *Q-S* may be erroneously high due to the water absorbed by the solid sample.

12. Calculation

12.1 Calculate the apparent specific gravity (ASG) or bulk density from the following equations, matching the appropriate equation with the procedure of choice for each group.

12.1.1 For Group A materials:

$$\text{ASG} = \frac{(S-W)}{(R-W)} \tag{1}$$

where:

W = weight of empty container (with lid),
R = weight of water-filled container (with lid), and
S = weight of sample-filled container (with lid).

12.1.2 For Group B materials:

$$\text{Bulk Density (g/mL)} = (Y)\frac{(S-W)}{(R-W)} \tag{2}$$

where

Y = 1 g/mL, the conversion of mass/volume at 4°C.

12.1.3 For Group C materials:

$$\text{ASG} = \frac{(S-W)}{(R-W)-(Q-S)} \tag{3}$$

where

Q = weight of sample and water-filled container.

13. Report

13.1 Report at a minimum the following information:

13.1.1 Sample identification and group (A, B, or C),

13.1.2 Date of test,

13.1.3 Procedure applied, and

13.1.4 Test results.

14. Quality Assurance

14.1 Instrument performance standards, quality control check samples of appropriate matrices, and duplicates should be performed at an action level specified by the laboratory and at an appropriate frequency.

15. Precision and Bias

15.1 The precision and bias of the procedure in this test method is being determined.

16. Keywords

16.1 density; specific gravity; waste screening

Designation: D 5058 – 90 (Reapproved 2001)

Standard Test Methods for Compatibility of Screening Analysis of Waste[1]

This standard is issued under the fixed designation D 5058; the number immediately following the designation indicates the year of original adoption or, in the case of revision, the year of last revision. A number in parentheses indicates the year of last reapproval. A superscript epsilon (ε) indicates an editorial change since the last revision or reapproval.

1. Scope

1.1 These test methods cover assessment of the compatibility/reactivity of waste. The individual test methods are as follows:

	Sections
Test Method A—Commingled Waste Compatibility	8-12
Test Method B—Polymerization Potential (Reaction with Triethylamine)	13-18
Test Method C—Water Compatibility	19-25

1.2 These test methods are applicable to waste liquids, sludges, semi-solids, and solids.

1.3 These test methods are designed and intended as a preliminary or supplementary test to complement the more sophisticated quantitative analytical techniques that should be used to determine waste composition and compatibilities. This standard offers the user the option and the ability to screen wastes for potentially hazardous reactions when the more sophisticated techniques are not available and the total waste composition is unknown and to screen compatibility when the composition is known.

NOTE 1—**Warning:** Delayed or slow reactions of wastes may go unnoticed.

1.4 The values stated in SI units are to be regarded as the standard. The values given in parentheses are for information only.

1.5 *This standard does not purport to address all of the safety concerns, if any, associated with its use. It is the responsibility of the user of this standard to establish appropriate safety and health practices and determine the applicability of regulatory limitations prior to use.* For specific hazard statements, see Sections 10, 15, and 23 and Notes 1 and 2.

2. Referenced Documents

2.1 *ASTM Standards:*

D 1193 Specification for Reagent Water[2]

E 1 Specification for ASTM Thermometers[3]

E 200 Practice for Preparation, Standardization, and Storage of Standard Solutions for Chemical Analysis[4]

3. Terminology

3.1 *Definition of Term Specific to This Standard:*

3.1.1 *screening*—a *preliminary* qualitative or semi-quantitative test, developed from classical qualitative and quantitative techniques, that is designed to efficiently give the user specific information about a waste that will aid in determining waste identification, process compatibility, and safety in handling.

4. Summary of Test Methods

4.1 *Test Method A*— Representative samples of waste are added to each other. The generation of heat or violent reaction is noted. In addition, the production of mists, fumes, dusts, gases, layering, polymerization, precipitation, emulsification or increase in viscosity and other chemical or physical changes are noted.

4.2 *Test Method B*— Reactivity of wastes is determined by adding an aliquot of a sample to an equal volume of reagent and observing any characteristic reaction, such as temperature increase, gas evolution, gelling, or polymerization.

4.3 *Test Method C*— Water and the waste are mixed in an approximate 10 + 1 ratio to test for compatibility. A thermometer is used to measure heat generation when applicable. Qualitative solubility and relative apparent density are observed concurrently.

5. Purity of Reagents

5.1 *Purity of Reagents*—Reagent grade chemicals shall be used in all tests. Unless otherwise indicated, it is intended that all reagents conform to the specifications of the Committee on Analytical Reagents of the American Chemical Society, where such specifications are available.[5] Other grades may be used provided it is first ascertained that the reagent is of sufficiently high purity to permit its use without lessening the accuracy of the determination (see Practice E 200).

---

[1] These test methods are under the jurisdiction of ASTM Committee D34 on Waste Management and are the direct responsibility of Subcommittee D34.01.05 on Screening Methods.

Current edition approved May 25, 1990. Published July 1990.

[2] *Annual Book of ASTM Standards*, Vol 11.01.

[3] *Annual Book of ASTM Standards*, Vol 14.03.

[4] *Annual Book of ASTM Standards*, Vol 15.05.

[5] "Reagent Chemicals, American Chemical Society Specifications," Am. Chemical Soc., Washington, DC. For suggestions on the testing of reagents not listed by the American Chemical Society, see "Reagent Chemicals and Standards," by Joseph Rosin, D. Van Nostrand Co., Inc., New York, NY, and the "United States Pharmacopeia."

5.2 *Purity of Water*— Unless otherwise indicated, references to water shall be understood to mean reagent water as defined by Type III of Specification D 1193.

6. Sampling

NOTE 2—**Precaution:** Avoid inhalation of or skin contact with any hazardous waste.

6.1 Obtain representative samples of waste. If composite samples are taken, report any generation of heat, gases or solids during compositing. If reactions are observed during compositing, then individual samples should be taken. If the waste is suspected of containing varying proportions of reactive compounds, take individual samples and conduct tests on each sample.

6.2 Allow all samples to stabilize to room temperature and analyze as soon as possible.

6.3 Always perform this procedure in a hood with the sash down as far as possible.

7. Quality Assurance

7.1 Thermometers are evaluated and verified at a frequency specified by the laboratory (see Specification E 1).

7.2 Care is taken to ensure that samples are representative of the total wastes involved.

TEST METHOD A—COMMINGLED WASTE COMPATIBILITY

8. Significance and Use

8.1 This test method is intended for use by those in the waste management industries to aid in determining the compatibility of hazardous wastes before they are commingled.

9. Apparatus

9.1 *Graduated Cylinders*, 100 mL.

9.2 *Thermometer*, 20 to 110°C or equivalent with 0.5°C divisions.

9.3 *Disposable Pipet.*

9.4 *Spatula.*

9.5 *Beakers*, 500 mL.

9.6 *Funnels.*

9.7 *Vortex Mixer* (optional).

10. Hazards

10.1 **Precaution**—Avoid inhalation of and skin and eye contact with any hazardous material.

10.2 **Precaution**—This procedure must be performed within a laboratory fume hood with the sash down as far as possible.

10.3 **Caution**—By keeping the sample size small and by first screening for very reactive wastes, the overall hazard is small. The small hazard is justified due to the much larger hazard of mishandling reactive waste at plant scale.

11. Procedure

11.1 Determine the total quantity *A* of the incoming waste to be added to the storage or treatment unit.

11.2 Determine the total quantity *B* of the waste in the storage tank or treatment unit.

11.2.1 Both quantities *A* and *B* must be stated in the same units of measure; pounds or gallons are typically used.

11.2.2 The waste in a tank can be estimated from the design volume of the tank. The volume of a tank truck can be determined from the contents' net weight and an estimate of the density. A value of 3.75 kg/L (8.34 lb/gal) can be used as an approximate density for a wide range of aqueous wastes.

11.2.3 The total volume of *A* and *B*, upon mixing, should not exceed 300 mL. The initial volume *A* (150 mL) may be adjusted proportionally to accommodate total volume specification.

NOTE 3—**Warning:** Perform a pre-test using 1 or 2 mL of each sample to reduce the risk when mixing potentially highly reactive wastes.

11.3 Place in a 500-mL beaker 150 mL of a representative sample from the storage tank or treatment unit.

11.3.1 Measure the temperature, when applicable, of the test sample and remove the thermometer.

NOTE 4—High precision thermometers may be employed to provide higher sensitivity in temperature readings.

11.4 Use the ratio A + B of wastes to determine the aliquot, *V*, in milliliters, of incoming waste to now be added. Use the following equation:

$$V = V\,(A/B) \quad (1)$$

where *V* is the volume in milliliters used in step 11.3 (150 mL), and *A* and *B* are as defined in 11.1 and 11.2 respectively.

11.5 Slowly and very carefully add the aliquot *V* of incoming waste to the test sample volume *V* already in the beaker.

11.5.1 The recommended rate of addition is approximately 1 mL/s.

11.5.2 While the addition is in progress, watch for adverse reactions.

NOTE 5—**Warning:** If a reaction is observed, stop the addition immediately and report the observation.

11.6 If after adding the aliquot *V* of incoming waste no adverse reaction is observed, mix well and immediately measure the temperature.

11.6.1 Compare the temperature here with the temperature measured in step 11.3.1. Record the difference, using (+) to indicate an increase and (–) to indicate a decrease in temperature (see Note 4).

NOTE 6—Mixing the representative waste samples at equal proportions can increase the sensitivity of reactivity and may be used as a substitute or in addition to the test based on actual proportions.

11.7 Record any generation of heat or violent reaction. Record the production of any mists, fumes, dust, or gases. Any layering, polymerization, precipitation, emulsification, increase in viscosity, bubbling, foaming, solidification, spattering, or other interaction of the commingled wastes must be observed and recorded.

11.8 If no reaction is observed, the waste passes the compatibility test. If any reaction or temperature rise is observed, the incoming waste has failed the compatibility test and is reported.

12. Precision and Bias

12.1 No statement is made about either the precision or bias of this test method since the result merely states whether there is conformance to the criteria for success specified in the procedure.

TEST METHOD B—POLYMERIZATION POTENTIAL (REACTION WITH TRIETHYLAMINE)

13. Significance and Use

13.1 This test method is significant to those in the waste management industries.

13.2 It is designed to screen wastes that have the potential of undergoing hazardous polymerization when mixed with incompatible waste streams.

13.3 This test method can be used to detect potential hazardous polymerization of waste containing or suspected of containing isocyanates such as methylene bis-phenyl isocyanate, methylene diisocyanate (MDI), or toluene diisocyanates (TDI).

14. Apparatus

14.1 *White Ceramic Spotplate.*

14.2 *Disposable Transfer Pipets.*

14.3 *Spatula.*

14.4 *10-mL Graduated Cylinder*, with stopper.

14.5 *Thermometer*, 20 to 110°C or equivalent with 0.5°C divisions.

15. Reagents and Materials

15.1 *Triethylamine (CH) N.*

16. Hazards

16.1 Use triethylamine in the hood and avoid exposure.

16.2 With samples that do not contain any reactive compounds, this test procedure does not present any other special hazards. However, samples that are reactive will fail this test and some reaction will result. The reaction could be severe.

16.3 **Caution**—By keeping the sample size small and by first screening for very reactive wastes, the overall hazard is small. The small hazard is justified due to the much larger hazard of mishandling reactive waste at plant scale.

17. Procedure

17.1 Conduct the following procedure in a fume hood:

17.1.1 Place approximately 1 mL of triethylamine reagent in the cavity of a ceramic spotplate.

17.1.2 Place approximately 1 mL of sample in the spotplate cavity with reagent. Immediately lower hood sash as protection against violent reactions.

17.1.3 Observe mixture for about 1 min and record any reaction characteristics, such as gas evolution, fuming, charring, precipitation, gelling, polymerization, or burning.

17.1.4 If any reaction characteristics are observed, then material is reactive and fails this test. Material which fails this test should not be tested using 17.2 or 17.3.

17.2 Conduct the following procedure with special care in a fume hood:

17.2.1 Add about 5 mL of reagent to a 10-mL graduated cylinder or disposable test tube.

17.2.2 Carefully add 5 mL of sample to the cylinder, stopper, and invert several times or vortex to mix well. Immediately remove stopper, insert the thermometer, and record temperature of mixture (see Note 4).

17.2.3 Continue to monitor temperature of mixture for several minutes. Observe and record any reaction characteristics, such as temperature increase, gas evolution or gelling. Note that gas evolution may be observed as tiny bubbles that consistently rise to surface (see 17.3).

17.2.4 If temperature increases significantly or any reaction characteristics are observed, then material is reactive and fails this test. Material which fails this test should not be tested using 17.3.

17.3 If gas evolution is difficult to observe during 17.2, conduct the following procedure with special care in a fume hood:

17.3.1 Add about 5 mL of reagent to 10-mL graduated cylinder or disposable test tube.

17.3.2 Carefully add 5 mL of sample to the cylinder, stopper, and invert several times or vortex to mix well. Immediately remove stopper and restopper. Lower hood sash as protection against violent reaction.

17.3.3 After several minutes, carefully remove stopper and observe mixture for gas evolution. Gas evolution will be observed as immediate venting or bubbles at surface, similar to opening a carbonated drink.

17.3.4 If gas evolution is observed, then material is reactive and fails this test. If no gas evolution or other signs of reaction are observed, the material has passed the test.

17.3.5 Record observations.

18. Precision and Bias

18.1 No statement is made about either the precision or bias of this test method since the result merely states whether there is conformance to the criteria for success specified in the procedure.

TEST METHOD C—WATER COMPATIBILITY

19. Significance and Use

19.1 This test method is intended for use by those in the waste management industries.

19.2 This test method is used to determine whether a waste has the potential to generate extreme heat or violent reactions, and produce fumes, dusts, gases, or other products when mixed with water.

19.3 This test method is designed to determine water compatibility of a waste.

19.4 This test method can be used to qualitatively judge the solubility and apparent density of waste (if immiscible) relative to water.

20. Interferences

20.1 The generation of colorless fumes or gases, pressure buildup without visible bubbling, mild effervescence, or heat may go undetected.

21. Apparatus

21.1 *Disposable Beakers*, test tubes, or similar equipment.

21.2 *Disposable Pipet* (5-mL capacity).

21.3 *Spatula*.

21.4 *Thermometer*, 20 to 110°C or equivalent with 0.5°C divisions.

21.5 *Vortex Mixer* (optional).

22. Reagents and Materials

22.1 *Reagent Water*.

23. Hazards

23.1 Avoid inhalation of and skin and eye contact with any hazardous waste.

23.2 This procedure must be performed within a laboratory fume hood with the sash down as far as possible.

23.3 **Caution**—By keeping the sample size small and by first screening for very reactive wastes, the overall hazard is small. The small hazard is justified due to the much larger hazard of mishandling reactive waste at plant scale.

24. Procedure

24.1 Keep thermometer in water at room temperature until ready for use. Note temperature of water in degrees Celsius or Fahrenheit.

24.2 Bring sample to room temperature, if necessary.

24.3 Place a small amount (approximately 10 mL) of water into a disposable beaker or test tube.

24.4 Introduce approximately 1 mL of waste into the beaker or test tube and mix well. Note any violent reactions, fumes, dusts or gases, and any precipitates or emulsions, and record observations.

24.5 If any such reactions are noted, the waste fails the water compatibility test.

24.6 Once it has been determined that no violent reaction is occurring, and as soon as possible after 24.4, place the thermometer into the beaker or test tube and note any temperature change (increase (+) or decrease (−)), recording the change in temperature in degrees Celsius or Fahrenheit (see Note 4).

24.7 Some reactions may have a latent period or accelerate as they proceed. Retain the mixed sample for 5 or 10 min, then observe it and record its temperature again.

24.8 If no reactions were observed and no significant temperature change is noted, the waste has passed the water compatibility test.

24.9 Report the miscibility and apparent density of the sample as immiscible or miscible and lighter or heavier than water.

25. Precision and Bias

25.1 No statement is made about either the precision or bias of this test method since the result merely states whether there is conformance to the criteria for success specified in the procedure.

26. Keywords

26.1 compatibility; waste screening

The American Society for Testing and Materials takes no position respecting the validity of any patent rights asserted in connection with any item mentioned in this standard. Users of this standard are expressly advised that determination of the validity of any such patent rights, and the risk of infringement of such rights, are entirely their own responsibility.

This standard is subject to revision at any time by the responsible technical committee and must be reviewed every five years and if not revised, either reapproved or withdrawn. Your comments are invited either for revision of this standard or for additional standards and should be addressed to ASTM Headquarters. Your comments will receive careful consideration at a meeting of the responsible technical committee, which you may attend. If you feel that your comments have not received a fair hearing you should make your views known to the ASTM Committee on Standards, at the address shown below.

Standard Test Method for Collecting Gross Samples and Determining the Fuel Quality of RDF[1]

This standard is issued under the fixed designation D 5115; the number immediately following the designation indicates the year of original adoption or, in the case of revision, the year of last revision. A number in parentheses indicates the year of last reapproval. A superscript epsilon (ε) indicates an editorial change since the last revision or reapproval.

INTRODUCTION

It is not possible to gather a sample of a multi-ton lot of refuse derived fuel (RDF) with assurance that the sample is representative of the lot. Therefore, this test method has been developed whereby a series of twenty samples are analyzed over the course of a month and the average of the results are an empirical estimate that is adequate to be used to represent the RDF production over that month. This test method may be used to ascertain RDF quality for purposes of contractual adjustment between producer and user.

1. Scope

1.1 This test method covers the procedure for collection of a series of 20 gross samples for analyses. The results of analysis, when averaged, are deemed representative of the lot or a production stream for a 1-month period.

1.2 This test method may be used to determine fuel quality on other than a monthly basis if 20 gross samples are gathered, analyzed and averaged.

1.3 The values stated in inch-pound units are to be regarded as the standard. The values given in parentheses are for information only.

1.4 *This standard does not purport to address all of the safety concerns, if any, associated with its use. It is the responsibility of the user of this standard to establish appropriate safety and health practices and determine the applicability of regulatory limitations prior to use.* Specific hazard statements are given in Section 7.

2. Referenced Documents

2.1 *ASTM Standards:*

E 711 Test Method for Gross Calorific Value of Refuse-Derived Fuel by a Bomb Calorimeter[2]

E 790 Test Method for Residual Moisture in a Refuse-Derived Fuel Analysis Sample[2]

E 791 Test Method for Calculating Refuse-Derived Fuel Analysis Data from As-Determined to Different Bases[2]

E 829 Test Method for Preparing RDF Laboratory Samples for Analysis[2]

E 830 Test Method for Ash in the Analysis Sample of Refuse-Derived Fuel[2]

E 856 Definitions of Terms and Abbreviations Relating to Physical and Chemical Characteristics of Refuse-Derived Fuel[2]

E 954 Test Method for Packaging and Shipping of Laboratory Samples of Refuse-Derived Fuel[2]

E 955 Test Method for Thermal Characteristics of Refuse-Derived Fuel Macrosamples[2]

3. Terminology

3.1 *Definitions*—The definitions covered in Definitions E 856 are applicable to this test method.

4. Summary of Test Method

4.1 Twenty gross samples on RDF are taken at intervals over a 1-month period and sent to the laboratory for analysis.

4.2 The analysis results are averaged to establish the fuel quality of RDF for the month.

5. Significance and Use

5.1 This test method is available to producers and users of RDF as a means of determining thermal characteristics of RDF produced over a 1-month production period for purposes of contract adjustment.

6. Apparatus

6.1 *Conveyor Belt Sample Divider*—made of two vertical wall members have a profile matching that of the top of the belt, and having a space between the two walls not less than three times the top nominal particle size of the RDF is required.

6.2 *Conveyor Stream Sampling Device*—a container with a handle or mechanism to control insertion and withdrawal, suitable for collecting a representative sample from a stream of material falling from the discharge end of a conveyor is needed. It shall have dimensions not less than three times the size of the top nominal particle size (95 % passing) of the RDF,

[1] This test method is under the jurisdiction of ASTM Committee D 34 on Waste Management and is the direct responsibility of Subcommittee D34.03 on Treatment.
Current edition approved May 21, 1990. Published July 1990.

[2] *Annual Book of ASTM Standards*, Vol 11.04.

to avoid bridging and rejection of large particles.

6.3 *Sample Container*— 3-mil polyethylene bags, double bagged, properly sealed, or noncorroding buckets or barrels with airtight covers, of sufficient size to hold the samples is required.

7. Hazards

7.1 Due to the origins of municipal solid waste, workers shall use proper safety precautions in handling samples and conducting tests. Workers shall wear gloves and safety glasses, and when appropriate, dust masks. Workers shall be cautioned to wash their hands before eating, smoking or handling genitalia.

7.2 Appropriate safety precautions shall be taken when working on or near equipment in collecting samples, due to the possibility of getting objects or clothing caught in moving equipment.

8. Sampling

8.1 Samples may be collected from a stopped belt, from a free-falling stream, or from a static pile.

8.1.1 Stopped-belt samples shall be taken over the full width of the belt between the boards, and all material, including grit, collected and immediately placed in a container, and the container sealed.

8.1.2 The free-falling stream sampling device shall make a full sweep across the stream and accept material from the full cross section of the stream.

NOTE 1—Make sure that the sweep is parallel with flow and not perpendicular to flow of the RDF stream.

8.1.3 Samples may be taken from static piles, containers or bins, provided that the samples are taken from 1 ft beneath the surface of the pile (hence not subjected to loss of moisture due to exposure to the atmosphere).

NOTE 2—Static pile sampling may lead to biased results.

8.2 *Number and Weight of Samples*:

8.2.1 Take over the monthly period, gross samples, one per day, at random times during the day, for 20 days spread over the month.

8.2.1.1 If for reasons beyond the control of the sample gathered, such as plant shutdown, as few as 15 samples may be considered acceptable for determining the monthly average.

8.2.2 Make sure that the weight of each sample is between 5 and 10 lb (2 and 4 kg), or as agreed upon.

8.3 Place the entire gross sample in a sealed container and send to the laboratory in accordance with Test Method E 954.

9. Procedure

9.1 Prepare the laboratory samples and analyzed in accord with either Test Method E 829, followed by Test Methods E 790, E 830, E 711, or E 955.

10. Calculation

10.1 Calculate percent moisture, percent ash, and higher heating value, for each laboratory sample by either Test Method E 791, or Test Method E 955.

10.2 Average the monthly laboratory sample data for moisture, ash, and heating value.

11. Report

11.1 The report shall include the following RDF average data for the month, rounded to the nearest percent or 50 Btu/lb (0.1 MJ/kg)

Moisture	A %
Ash	B %
Heating Value	C Btu/lb (MJ/kg)

where:

A = mass % moisture, "as received,"
B = mass % ash, "as received," and
C = heating value, "as received."

NOTE 3—Heating value is normally reported as higher heating value. The producer and user may agree to report lower heating value.

NOTE 4—The producer and user may agree to include range and standard deviation in the report.

12. Precision and Bias

12.1 Precision, bias, variance, and accuracy for RDF sampling is supported by EDS-18, "Determination of Relationship Between Precision and Procedures used for Collection and Division of RDF Samples.''

Designation: D 5120 – 90 (Reapproved 1995)$^{\epsilon 1}$

Standard Test Method for Inhibition of Respiration in Microbial Cultures in the Activated Sludge Process[1]

This standard is issued under the fixed designation D 5120; the number immediately following the designation indicates the year of original adoption or, in the case of revision, the year of last revision. A number in parentheses indicates the year of last reapproval. A superscript epsilon (ε) indicates an editorial change since the last revision or reapproval.

ϵ^1 NOTE—Section 12 was added editorially in September 1995.

1. Scope

1.1 This test method covers a batch procedure that evaluates the impact of selected wastewaters, materials, or specific compounds on the respiration rate of an aqueous microbial culture, such as activated sludge.

1.2 Alternative procedures for measurement of microbial activity, such as adenosine 5′ triphosphate (ATP), specific substrate utilization, etc. are not within the scope of this test method.

1.3 The results obtained are based on comparisons in a specific test series that examines a range of concentrations of the potentially inhibitory test candidate using batch methods in a laboratory. Results are completed in a short time frame (a few hours).

1.4 The test results are specific to the microbial culture used. Microbial culture from different wastewater treatment plants will differ in kinds and numbers of organisms, and performance capability. Thus, there is no basis for comparing results for microbial cultures from different treatment facilities.

1.5 *This standard does not purport to address all of the safety concerns, if any, associated with its use. It is the responsibility of the user of this standard to establish appropriate safety and health practices and determine the applicability of regulatory limitations prior to use.*

2. Referenced Documents

2.1 *ASTM Standards:*

D 4478 Test Methods for Oxygen Uptake[2]

3. Terminology

3.1 *Definitions:*

3.1.1 *respiration rate*—the quantitative consumption of oxygen by an aqueous microbial system. The consumption is generally expressed as mg O_2/L/h.

3.1.2 EC_{50}—the concentration of the test candidate in this procedure (volume percent or mg/L) that results in a reduction of respiration rate to 50 % of that observed for the control.

4. Summary of Test Method

4.1 This test method utilizes respiration rate as the indicator of microbial activity.

4.2 A batch system that contains a microbial culture (returned activated sludge from the process or a culture maintained in the laboratory), selected nutrient dose, and a dilution of a compound, substance, wastewater, etc. (test candidate) is prepared in a container in the laboratory. The batch system is called a "cell suspension."

4.3 The nutrient dose introduces a large excess of biodegradable substrate thereby putting the culture at a high metabolic rate. Inhibition of respiration by the test candidate is observed under these conditions.

4.4 The prepared cell suspension is aerated for a 2-h period. At the end of the period, the respiration rate is determined using a respirometric or an oxygen uptake technique.

4.5 A lower respiration rate for a cell suspension that has received the test candidate compared to the respiration rate of a control cell suspension indicates inhibition of respiration.

5. Significance and Use

5.1 The objectives of the respiration inhibition tests may be defined by the interests of the user, but the test method is designed primarily for examination of the inhibition response with operating microbial systems such as an activated sludge process treating domestic or industrial wastes.

5.2 Different apparatus exist that facilitate continuous or continual measurement of respiration in microbial systems and each may be used as the tool to observe respiration in this test method.

5.3 Respirometry may utilize any apparatus and technique that will achieve the determination of respiration rate. A number of devices are presented in Appendix X1. Equivalency in the experimental capability of each device is not implied. The analyst should select the respirometric approach that best suits his needs.

5.4 The inhibitory effect of a test candidate is identified more completely by examining inhibition over a range of concentrations, such as determining the EC_{50}. The use of aerated containers permits concurrent management of a series of cell suspensions. A respirometer for each cell suspension might also be used.

---

[1] This Test Method is under the jurisdiction of ASTM Committee D-34 on Waste Management and is the direct responsibility of Subcommittee D34.03 on Treatment.

Current edition approved Sept. 28, 1990. Published November 1990.

[2] *Annual Book of ASTM Standards*, Vol 11.02.

6. Interferences

6.1 This test method is most readily applied to substances which, due to water solubility and low volatility, are likely to remain in the aqueous system.

6.2 Results have been observed where cell suspensions containing the test candidate had a respiration rate greater than the blank, particularly at shorter aeration periods of the cell suspensions (less than 1 h). Thus, a minimum aeration period for the cell suspensions before determinations of respiration rate is 2 h.

6.2.1 One reason for increased oxygen uptake rate in an experimental cell suspension may be that severe physical or chemical reactions with the test candidate cause a fraction of the microbial culture to be lysed. The release of very readily biodegradable soluble organic material from the lysed cells may support a higher oxygen uptake rate by the cell suspension.

6.2.2 An alternate reason for increased oxygen uptake rate is that certain test candidates (2,4-dichlorophenol for example) may uncouple the transfer of electrons involved in the process called oxidative phosphorylation in which adenosine 5′ triphosphate (ATP) is formed by the phosphorylation of adenosine 5′ diphosphate (ADP). The result of the uncoupling is an increase in the rate of oxygen consumption that is not related to substrate stabilization.

6.2.3 A respiration rate by an experimental cell suspension that is greater than the respiration rate of the control represents microbial system damage. The degree of damage is not quantified by comparison of respiration rates for the test candidate and the control. Whether the cause is due to uncoupled electron transfer or lysis of cells can be determined by comparing the filtered Dissolved Organic Carbon (DOC) of the experimental cell suspension with the sum of the DOC of the control plus that added by the test candidate. A higher DOC represents cell lysis.

6.3 Where industrial wastewaters in the sewer system are continually introducing inhibitory components to the collective wastewaters, it may not be feasible to utilize the returned sludge from the process directly as the microbial culture. The maintenance of a protected culture of organisms in the laboratory may be necessary.

7. Apparatus

7.1 *Respirometer or an Oxygen probe*— An apparatus capable of measuring the respiration rate or oxygen uptake rate of the cell suspension.

7.1.1 *Respirometer*—A device that receives the cell suspension, or an aliquot and provides a technique for measurement of oxygen utilization to be interpreted as respiration rate (see Appendix X1).

7.1.2 *Dissolved Oxygen Probe and Instrumentation*—An alternate device for the measurement of respiration rate as oxygen uptake rate.

7.2 *Culture Tank*—If it is deemed necessary to maintain a microbial culture in the laboratory, the apparatus required is a container with adequate mixing and oxygen transfer. The container should hold at least four times the volume of culture that might be used in one day.

7.2.1 The culture tank should be adequately mixed to insure that the culture remains in suspension and that sufficient mechanical or bubble aeration occurs to maintain the desired dissolved oxygen (DO) concentration.

NOTE 1—Energy input should not be such that the biological floc is sheared to sizes smaller than that which exists in the large-scale process. Mixing and aeration provided through diffused aeration in a laboratory-sized container may result in an excessive power input. Consider controlling the power input per unit volume to approximately that which exists in the large-scale process. For example, pure oxygen for aeration in combination with mechanical mixing may be utilized to achieve a balance between oxygen transfer and mixing. Determine the mixer power input by measuring the electrical power consumed at different operating speed, and adjust the mixer speed to achieve a power input that is equivalent to that which exists in the large-scale system.

NOTE 2—Cultures grown at low (0.5 to 2 mg/L) and high (>5 mg/L) DO concentrations possess different kinetic capabilities. Thus, to maintain a laboratory culture with performance capabilities similar to those of the full-scale culture, the DO concentration should be maintained at the level appropriate for the full-scale process. The probable explanation for the difference in culture performance is that higher concentrations of oxygen penetrate more completely through the floc particles.

7.3 *A pH Probe and Instrumentation.*

7.4 *Dissolved Oxygen Probe*—If utilized, the following apparatus is needed:

7.4.1 *Biochemical Oxygen Demand (BOD) bottles.*

7.4.2 *Agitation Device*, may be used with the dissolved oxygen probe in the BOD bottle. The device must provide complete mixing of the microbial culture in the BOD bottle.

7.4.3 *Magnetic Stirrer and Magnetic Stirring Bar*, alternatively, may be used to mix the BOD bottle.

7.5 *Beakers, 2-L size*, (or other containers of suitable size).

7.6 *Clean, Oil-Free Air Supply*, to provide cell suspension mixing and aeration.

7.7 *Fritted Glass Diffusers or Pasteur-Pipets*, as air diffusers.

8. Reagents

8.1 *Microbial Culture*—The microbial culture to be used is the returned sludge from the full-scale facility. For those activated sludge system where industrial contributions regularly cause microbial inhibition, direct use of the returned sludge may be impractical. For those systems where microbial inhibition is not a continuous problem, the returned sludge may be used directly if, by observed system performance, it appears to be healthy.

8.1.1 A microbial culture may be maintained in the laboratory. The culture should be maintained at the temperature of the full-scale mixed liquor and approximately at the concentration of the full-scale process returned sludge.

8.1.2 If maintenance of a microbial culture is to be practiced in the laboratory, and if the inhibition tests are to be related to a specific activated sludge wastewater treatment process, the initial microbial culture should be taken from the process returned sludge.

8.1.2.1 Care should be taken to obtain the microbial culture when, by appearance and performance, the culture is considered to be healthy.

8.1.2.2 The microbial culture should be fed daily with the actual process wastewater if the wastewater is of suitable quality and not inhibitory. Determine the wastewater quality by following the procedure in Section 9. Prepare a control cell suspension and a wastewater cell suspension. If the wastewater cell suspension does not show inhibition, it is suitable for use as feed material.

8.1.2.3 If wastewater of good quality is not available, a synthetic feed, such as Marlene's Mix (see 8.2) or other feed similar in character to the wastewater should be used (sucrose has been used successfully with domestic wastewater activated sludge). The feed application should not be excessive. For example, preferably it should be equal to about one-half of the food-to-microorganism ratio that exists in the full-scale process.

8.1.2.4 When the full-scale system is considered to be in good condition, replenish one third to one half of the volume of the microbial culture daily with the returned sludge from the full-scale system. The replenishment will aid in maintaining a microbial culture with approximately the same population dynamics as the full-scale process.

8.1.2.5 Replenishment and feeding should be done at the end of a work day so that the culture will have an overnight period to complete the synthesis of substrate. Replace any water that has evaporated over night by adding distilled or deionized water.

8.2 *Nutrient Dose Preparation (Marlene's Mix)*—A solution of the following substances is prepared for use when conducting inhibition studies. Store the prepared solution in a refrigerator at 4°C. Warm the portion of the feed to be used in tests to the operating temperature of the test before use. Replace the solution after 14 days of storage or earlier if it becomes odorous.[3]

Ingredient	Quantity (g)
Bacto peptone[3]	32
Beef extract[3]	22
Ammonium chloride	11
Sodium chloride	1.4
Calcium chloride ($CaCl_2 \cdot 2H_2O$)	0.8
Potassium dihydrogen phosphate	3.5
Potassium monohydrogen phosphate	4.5
Distilled water	make up to 1 L

8.3 *Stock Inhibitor Solution*—Dissolve 0.5 g of 3,5-dichlorophenol in 10 mL of 1N NaOH, dilute to 30 mL with distilled water, add 1N H_2SO_4 to the point of incipient precipitation (approximately 8 mL of 1N H_2SO_4 will be required), bring the volume to 950 mL with distilled water, adjust the pH to the range of 7 to 8, and bring the volume to 1 L. The EC_{50} of 3,5-dichlorophenol for relatively non-acclimatized microbial cultures from domestic wastewater plants is about 10 to 30 mg/L but may be outside this range. For acclimatized microbial cultures, the EC_{50} will be higher, and values <200 mg/L has been reported. The stock solution may be used to check experimental technique and possibly the susceptivity of a microbial culture.

9. Procedure

9.1 Prepare experimental and control cell suspensions so that each is identical in its concentration of microbial culture and Marlene's mix.

9.1.1 When the test is related to an operating activated sludge process, the microbial culture concentration, and the initial pH, make sure that the temperature of the cell suspension (throughout the experimental period) is the same as that in the process mixed liquor. As an example of the control cell suspension, the following table applies if a microbial concentration of 2000 mg/L is desired and the microbial culture concentration is 10 000 mg/L.

Component	Percent of Preparation Volume
Microbial Culture (at 10 000 mg/L)	20[A]
Marlene's Mix	3.8
Tap Water	76.2 or as required to equal 100

[A] If the culture is not at 10 000 mg/L, adjust the volume percent to obtain the desired microbial concentration in the cell suspension.

9.1.2 Make sure that the concentration of readily biodegradable organic material in the nutrient dose (such as Marlene's Mix) is high enough that an additional increment of biodegradable organic material will not result in a significant increase in the rate of respiration. That is, during these tests the microbial culture is essentially saturated with substrate. Marlene's Mix has a soluble Chemical Oxygen Demand (COD) concentration of approximately 60 000 mg/L. The resulting COD introduced to a cell suspension by the nutrient dose is about 2300 mg/L.

9.1.2.1 If the analyst elects to use an alternative nutrient feed, it should contain a biodegradable COD concentration that will establish the desired substrate saturation. Test saturation of the microbial culture with biodegradable organic material by making respiration rate observations at different concentrations of nutrient dose. Make sure that the selected minimum nutrient dose is such that an increase in dose will not result in an increase in respiration rate.

9.1.3 In the preparation of the test cell suspensions, the volume occupied by water in the control is replaced by dilutions of the test candidate. Available quantities of nitrogen and phosphorus must meet nutrients required in biochemical reactions. If nitrogen and phosphorus concentrations in the test candidate are unknown, determine the COD and add 10 mg/L of ammonium chloride (NH_4Cl) and 3 mg/L of potassium monohydrogen phosphate/100 mg/L of COD. After nutrient addition, adjust the pH of the test candidate dilutions to the normal pH for the full-scale system mixed liquor.

9.2 The number of test cell suspensions and concentrations of test candidate are selected to meet the objectives of the study. For aqueous test candidates, the concentration is represented as the volume percent of the total cell suspension volume. For dry or pure liquid test candidates, the concentration is mg/L in the total cell suspension volume.

9.2.1 For the objective of identifying possible inhibitory effects of a test candidate on a wastewater treatment process, the minimum testing effort might be a single inhibition test in which the experimental cell suspension is at a selected concentration.

9.2.2 If a "Threshold of inhibition" or an" EC_{50}" is desired, a range of dilutions of the test candidate is prepared. A suggested minimum number of test cell suspensions is four spaced by a multiplier of not more than 3.2. For example, four systems for an EC_{50} beginning at a 4-vol % concentration and with a multiplier of 2.0 would be 4, 8, 16, and 32 %.

[3] Dehydrated Bacto Nutrient Broth, available from Difco Laboratories, P.O. Box 1058A, Detroit,MI 48732 or equivalent has been found suitable for this purpose.

9.3 Provide one 2-L beaker (or other suitably sized container) for each test cell suspension and the control cell suspension.

9.3.1 Prepare the test cell suspensions. For example, to prepare a 1-L cell suspension when the microbial culture concentration is 10 000 mg/L and the objective cell microbial cell suspension concentration is 2 000 mg/L, transfer 38 mL of Marlene's mix and the desired quantity of test candidate to the 2-L beaker, bring the volume to 800 mL with tap water, transfer 200 mL of the microbial culture and initiate aeration. The rate of air flow should result in a DO concentration in the cell suspension of at least 2 to 3 mg/L.

9.3.2 The start time for each cell suspension and control in the series is staggered by a time interval required to measure respiration rate on each cell suspension at the end of the 2-h period. The staggering in start time permits the series of cell suspensions to be managed concurrently.

9.3.3 Aerate the cell suspensions for 2 h and then determine the respiration rates of the control and test candidate cell suspensions by transferring all or a portion of the cell suspension to the respiration measuring device.

9.3.4 If a dissolved oxygen probe is used, before determining the oxygen uptake rate, increase the DO of the cell suspension aliquot to the 4 to 6 mg/L range by increased aeration rate or pouring it back and forth between containers. Determine the DO of the cell suspension aliquot at 30-s intervals until the rate is established (normally a 2.5-min period). Calculate the oxygen uptake rate by plotting the DO values versus time and calculating the slope of the line.

NOTE 3—When fitting the line through the plotted DO probe data, favor the last three readings. The first two readings may not be linear and may represent the response of the DO probe as it chases the decreasing DO concentration. Calculate the slope (mg/L/min) and multiply by 60 to obtain mg/L/h.

10. Interpretation of Results

10.1 *Percent Inhibition of Respiration*— Calculate the inhibition of respiration rate as follows:

$$\text{Percent Inhibition} = (B - A)/B \times 100 \quad (1)$$

where:

A = respiration rate for the experimental cell suspension, (mg O_2/L/hr), and

B = respiration rate for the control (mg O_2/L/h).

10.2 EC_{50}—Determine the EC_{50} by plotting the "Percent Inhibition" employing a linear Y axis versus the concentration of the test candidate on a logarithmic X axis. Fit a straight line through the apparent slope of the data (Fig. 1). Since only two or three data points are involved in the line fitting, statistically methods are not indicated. However, if the fit is made statistically, data near 0 or 100 % inhibition should be ignored. The

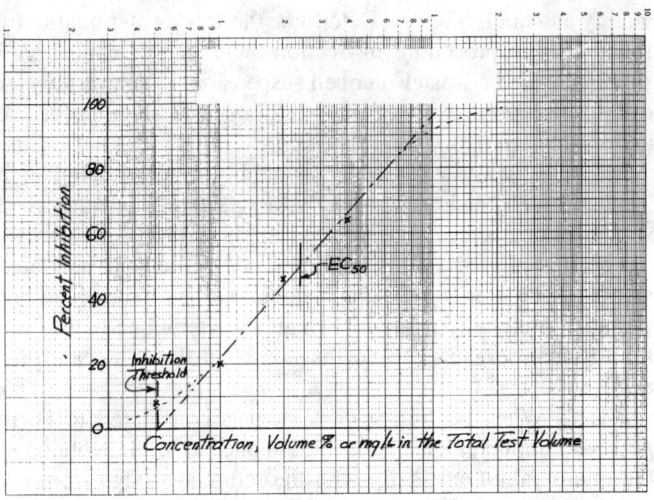

FIG. 1 Illustration of Inhibition Results

EC_{50} is the concentration of the test candidate that results in a 50 % inhibition of respiration.

10.3 *Inhibition Threshold*—The intercept of the straight line in Fig. 1 with 0 % inhibition is an indication of the inhibition threshold.

11. Precision and Bias

11.1 *Precision*—It is not practical to specify the precision of this test method for measurement of inhibition of respiration of microbial cultures because the desired results are the demonstrated impact of a test candidate on the *specific* microbial culture that exists in a *specific* activated sludge facility. Microbial cultures at different activated sludge facilities will differ in kind, numbers, and overall activity of the mixed population of microorganisms, and the resultant sensitivity to a specific test candidate. Thus, microbial cultures from different activated sludge systems are not suitable as the reference culture in measurements of interlaboratory procedural precision.

11.1.1 With respect to the microbial culture at a specific activated sludge facility, the response to a particular test candidate can vary with time due to shifts in the microbial culture's performance through acclimatization or other factors. Thus, the microbial culture at a specific activated sludge facility is not a suitable reference for the measurement of single operator precision.

11.2 *Bias*—Since there is no reference microbial culture for determining the bias of this test method for measuring inhibition of respiration of microbial cultures, no statement on bias is being made.

12. Keywords

12.1 inhibition of respiration; microbial cultures; microbial cultures in the activated sludge process; respiration

APPENDIX

(Nonmandatory Information)

X1. RESPIROMETERS FOR EXAMINATION OF MICROBIAL INHIBITION

X1.1 *Respirometers:*

X1.1.1 Respirometers are available in a variety of sizes. In some of the respirometric operations, the sample volume in the apparatus may contribute to measurement sensitivity because of the increased quantity of oxygen used per unit of time. However, smaller sized systems, such as an oxygraph[4] with 1.5 mL of sample in the measuring cell, are suitable for measurements of oxygen uptake rate in this test method.

X1.1.2 The basic principle of respirometry is the detection of oxygen removed from a gas volume that is in contact with the cell suspension. Microbial respiration consumes oxygen and releases carbon dioxide as a by-product of the biochemical reactions. If the carbon dioxide is removed from the gas volume, oxygen removal to support respiration will cause a net reduction in the gas volume. Measurement of the amount of oxygen utilized with time defines the respiration rate.

X1.1.3 The available respirometric systems reduce labor as compared to manual procedures and provide convenient electronic and computer supported procedures for interpretation and recording of results.

X1.2 *Apparatus:*

X1.2.1 The principle of all respirometers is similar. Carbon dioxide released to the gas space is removed by reaction with potassium hydroxide. The net change in pressure as oxygen is removed from the gas volume into the cell suspension is interpreted through either the pressure change (constant volume device) or the amount of oxygen required to maintain a constant pressure (constant pressure device).

X1.2.1.1 An example of a constant volume respirometer is presented by the COMPU-TOX respirometer in Fig. X1.1. A change in pressure in the gas space of the reaction bottle is referenced to a fixed pressure and oxygen is introduced to reestablish the gas space pressure. The quantity of oxygen introduced is measured and interpreted as oxygen consumption.

X1.2.1.2 A second example of a constant volume respirometer is the electrolytic biochemical oxygen demand (EBOD) respirometer in Fig. X1.2. The removal of oxygen from the gas space is sensed by the pressure change at the electrolysis cell. The cell generates oxygen to reestablish the gas space pressure. The flow of current to generate the oxygen is interpreted as weight of oxygen consumed.

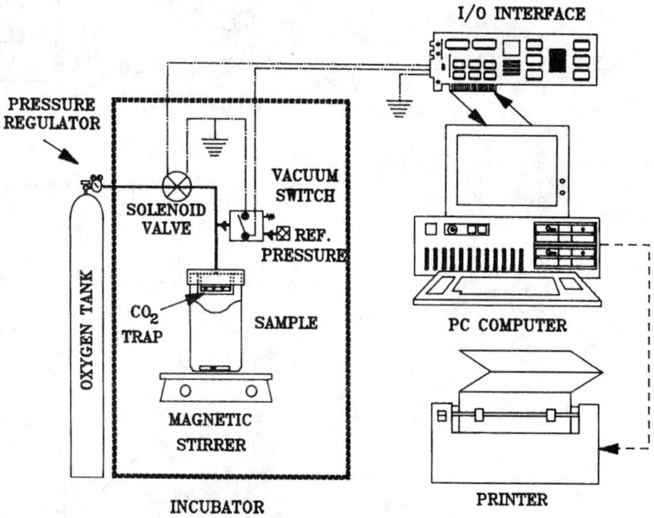

FIG. X1.1 Volumetric Respirometer

X1.3 *Interferences:*

X1.3.1 Physical conditions associated with respirometric measurement of oxygen utilization can cause errors in the interpreted respiration rate. Refer to specific operating instructions or information from manufacturers for corrective procedures to minimize errors.

X1.3.1.1 Changes in barometric pressure during the period of respiration measurement will cause an error in the apparent utilization of oxygen for those respirometers that reference internal pressure to barometric pressure.

X1.3.1.2 Temperature imbalance due to a cell suspension temperature different from the temperature of gas in the respirometer will cause an error in apparent gas space volume or pressure (equivalent oxygen utilization) until the temperatures have had a chance to equilibrate.

X1.3.1.3 When the cell suspension, at the time of introduction to the respirometer, has a low or zero oxygen concentration, a falsely high oxygen consumption rate will exist until the operating DO concentration for the cell suspension is reached.

X1.3.1.4 Dissolved oxygen concentration in the cell suspension will vary if the partial pressure in the respirometer varies during the period of uptake rate measurement. The change in dissolved oxygen concentration will represent an error in uptake rate measurement.

[4] Gilson Oxygraph, available from Gilson Medical Electronics Inc., Box 27, 3000 W. Beltline, Middletown, WI 53562 or its equivalent, has been found suitable for this purpose.

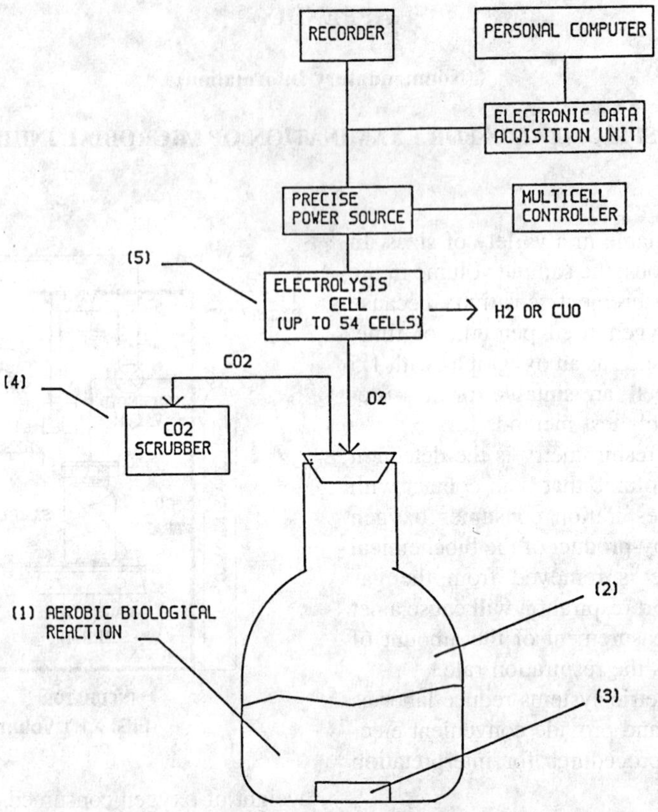

NOTE 1—The respiration of a cell suspension (1) is measured enclosed in a sealed reactor vessel (2) under continuous mixing (3). The reactor shares gas with a carbon dioxide scrubber (4) and an electrolytic cell (5) that generates oxygen in response to oxygen consumption to support respiration by the cell suspension.

FIG. X1.2 Electrolytic Respirometer

The American Society for Testing and Materials takes no position respecting the validity of any patent rights asserted in connection with any item mentioned in this standard. Users of this standard are expressly advised that determination of the validity of any such patent rights, and the risk of infringement of such rights, are entirely their own responsibility.

This standard is subject to revision at any time by the responsible technical committee and must be reviewed every five years and if not revised, either reapproved or withdrawn. Your comments are invited either for revision of this standard or for additional standards and should be addressed to ASTM Headquarters. Your comments will receive careful consideration at a meeting of the responsible technical committee, which you may attend. If you feel that your comments have not received a fair hearing you should make your views known to the ASTM Committee on Standards, at the address shown below.

Designation: D 5177 – 91 (Reapproved 1996)

Standard Guide for Waste Acceptance at Hazardous Waste Incinerators[1]

This standard is issued under the fixed designation D 5177; the number immediately following the designation indicates the year of original adoption or, in the case of revision, the year of last revision. A number in parentheses indicates the year of last reapproval. A superscript epsilon (ϵ) indicates an editorial change since the last revision or reapproval.

1. Scope

1.1 This guide addresses the acceptance of wastes at hazardous waste incinerators.

1.2 The purpose of this guide is to assist the user in determining the information and chemical and physical characteristics to be used when assessing the suitability of a waste for incineration.

1.3 This guide is intended to assist in meeting the requirements of applicable permits.

1.4 *This standard does not purport to address all of the safety concerns, if any, associated with its use. It is the responsibility of the user of this standard to establish appropriate safety and health practices and determine the applicability of regulatory limitations prior to use.*

2. Referenced Documents

2.1 *ASTM Standards:*

D 56 Test Method for Flash Point by Tag Closed Tester[2]
D 88 Test Method for Saybolt Viscosity[3]
D 92 Test Method for Flash and Fire Points by Cleveland Open Cup[2]
D 93 Test Methods for Flash Point by Pensky-Martens Closed Tester[2]
D 129 Test Method for Sulfur in Petroleum Products (General Bomb Method)[2]
D 240 Test Method for Heat of Combustion of Liquid Hydrocarbon Fuels by Bomb Calorimeter[2]
D 482 Test Method for Ash from Petroleum Products[2]
D 808 Test Method for Chlorine in New and Used Petroleum Products (Bomb Method)[2]
D 891 Test Methods for Specific Gravity of Liquid Industrial Chemicals[4]
D 1292 Test Method for Odor in Water[5]
D 1796 Test Method for Water and Sediment in Fuel Oils by the Centrifuge Method (Laboratory Procedure)[2]
D 2015 Test Method for Gross Calorific Value of Coal and Coke by the Adiabatic Bomb Calorimeter[6]
D 2196 Test Method for Viscosity Measurements and Rheological Properties of Non-Newtonian Materials by Rotational (Brookfield) Viscometer[7]
D 2361 Test Method for Chlorine in Coal[6]
D 2983 Test Method for Low-Temperature Viscosity of Automotive Fluid Lubricants Measured by Brookfield Viscometer[8]
D 3177 Test Method for Total Sulfur in the Analysis Sample of Coal and Coke[6]
D 3286 Test Method for Gross Calorific Value of Coal and Coke by the Isoperibol Bomb Calorimeter[6]
D 3761 Test Method for Total Fluorine in Coal by the Oxygen Bomb Combustion/Ion Selective Electrode Method[6]
D 3828 Test Method for Flash Point by Setaflash Closed Tester[9]
D 4007 Test Method for Water and Sediment in Crude Oil by the Centrifuge Method (Laboratory Procedure)[9]
D 4208 Test Method for Total Chlorine in Coal by the Oxygen Bomb Combustion/Ion Selective Electrode Method[6]
E 203 Test Method for Water Using Karl Fischer Reagent[4]
E 260 Practice for Packed Column Gas Chromatography[10]
E 663 Practice for Flame Atomic Absorption Analysis[11]
E 775 Test Methods for Total Sulfur in the Analysis Sample of Refuse-Derived Fuel[12]
E 776 Test Method for Forms of Chlorine in Refuse-Derived Fuel[12]
E 819 Test Method for Trace Amounts of Arsenic in Organic Industrial Chemicals[4]
E 885 Test Method for Analysis of Metals in Refuse-Derived Fuel (RDF) by Atomic Absorption Spectrophotometry[12]
E 926 Test Method for Preparing Refuse-Derived Fuel (RDF) Samples for Analyses of Metals[12]

---

[1] This guide is under the jurisdiction of ASTM Committee D-34 on Waste Managementand is the direct responsibility of Subcommittee D34.03 on Treatment.
Current edition approved Sept. 15, 1991. Published November 1991.
[2] *Annual Book of ASTM Standards*, Vol 05.01.
[3] *Annual Book of ASTM Standards*, Vol 04.04.
[4] *Annual Book of ASTM Standards*, Vol 15.05.
[5] *Annual Book of ASTM Standards*, Vol 11.01.
[6] *Annual Book of ASTM Standards*, Vol 05.05.
[7] *Annual Book of ASTM Standards*, Vol 06.01.
[8] *Annual Book of ASTM Standards*, Vol 05.02.
[9] *Annual Book of ASTM Standards*, Vol 05.03.
[10] *Annual Book of ASTM Standards*, Vol 14.01.
[11] *Annual Book of ASTM Standards*, Vol 03.06.
[12] *Annual Book of ASTM Standards*, Vol 11.04.

F 873 Guide for Incinerating Oil Spill Wastes at Temporary Field Locations[12]

2.2 *SW-846 Test Methods for Evaluating Solid Waste, Physical/Chemical Methods*[13]

1010 Test Method for Pensky-Martens Closed-Cup Method for Determining Ignitability

1020 Test Method for Setaflash Closed-Cup Method for Determining Ignitability

1110 Test Method for Corrosivity Toward Steel

3020 Test Method for Acid Digestion of Aqueous Samples and Extracts for Total Metals for Analysis by Furnace Atomic Absorption Spectroscopy

3040 Test Method for Dissolution Procedures for Oils, Greases, or Waxes

3050 Test Method for Acid Digestion of Sediments, Sludges, and Soils

6010 Inductively Coupled Plasma Method

7060 Test Method for Arsenic (AA, Furnace Technique)

7080 Test Method for Barium (AA, Direct Aspiration)

7090 Test Method for Beryllium (AA, Direct Aspiration)

7130 Test Method for Cadmium (AA, Direct Aspiration)

7190 Test Method for Chromium (AA, Direct Aspiration)

7470 Test Method for Mercury in Liquid Waste (Manual Cold-Vapor Technique)

7471 Test Method for Mercury in Solid or Semi-Solid Waste (Manual Cold-Vapor Technique)

7520 Test Method for Nickel (AA, Direct Aspiration)

7610 Test Method for Potassium (AA, Direct Aspiration)

7740 Test Method for Selenium (AA, Furnace Method)

7770 Test Method for Sodium (AA, Direct Aspiration)

7840 Test Method for Thallium (AA, Direct Aspiration)

7910 Test Method for Vanadium (AA, Direct Aspiration)

7950 Test Method for Zinc (AA, Direct Aspiration)

8010 Test Method for Halogenated Volatile Organics

8015 Test Method for Non-Halogenated Volatile Organics

8020 Test Method for Aromatic Volatile Organics

8080 Test Method for Organochlorine Pesticides and PCBs

8120 Test Method for Chlorinated Hydro-Carbons

8240 GC/MS Method for Volatile Organics

8250 GC/MS Method for Semivolatile Organics: Packed Column Technique

8270 GC/MS Method for Semivolatile Organics: Capillary Column Technique

9041 Test Method for pH Paper Method

9045 Test Method for Soil pH

2.3 *Code of Federal Regulations:*[13]

40 CFR, Protection of Environment, Parts 260, 261, 262, 263, 264, 265, 266, 267, 268, and 270.

49 CFR, Transportation, Parts 171 and 172

29 CFR, Labor, Parts 1910.100 through 1910.1046

3. Terminology

3.1 *Definitions:*

3.1.1 *incineration*—controlled burning of waste products or other combustible material (see Guide F 873).

3.1.2 *incinerator*—a device constructed for the purpose of containing a material for thermal oxidation (see Guide F 873).

4. Significance and Use

4.1 This guide has been developed to assist in establishing the information and chemical and physical characteristics upon which hazardous waste incineration acceptance criteria can be based.

4.2 This guide is not intended to serve as a set of rigid criteria since acceptable conditions will vary with the type and size of equipment. The appropriate regulatory and permit requirements must be followed for the wastes that may be incinerated.

4.3 This guide does not obviate the need to follow all local, state, and federal regulations and permits.

5. Preacceptance

5.1 The decision to accept a waste for incineration requires an evaluation of the information and chemical and physical characteristics discussed in Sections 6 and 7. This type of information should be known before actual shipments of waste are accepted at the facility. This information is necessary in order to make a determination that the facility can properly manage the waste prior to its receipt. Failure to do this can result in long delays in handling waste shipments or ultimate rejection of these wastes.

6. Required Information

6.1 The following information must be supplied prior to the receipt of a waste at the incineration site, either by the generator or developed through analysis of a sample of the waste (see section 40 CFR, Parts 264 and 265). This information is the criterion used for determining the acceptance or rejection of the waste for incineration.

6.2 *General Information provided by the Generator*:

6.2.1 Generator name, address, Federal Environmental Protection Agency (EPA) facility identification number (and state EPA facility identification number, if applicable), technical contact, and phone number. This information provides the identity and location of the waste generated. It also provides a technical person who can be contacted regarding additional information prior to or during handling of the waste. Examples are waste nonconformity, safety information, special handling, and waste composition clarification.

6.2.2 Waste name/generic description. The description identifies the specific waste from a generating location. It provides a name that describes the waste adequately.

6.2.3 Process generating the waste. This provides information used in the determination of proper regulatory classifications and may indicate properties of the waste.

6.2.4 Department of Transportation (DOT) shipping information (see section 49 CFR, Parts 171 and 172). This is used as a basis for on-site storage and compatibility. Waste arriving on-site must also be classified properly on the hazardous waste manifest. In addition, proposed shipping frequency and shipment packaging (for example, 20 or 55 gallon drums per week, etc.) should be included.

6.2.5 Resource Conservation and Recovery Act (RCRA) classifications (see section 40 CFR, Part 261). The proper EPA

[13] Available from the Superintendent of Documents, U.S. Government Printing Office, Washington, DC 20402.

and state hazardous waste numbers must be designated for each waste in order to meet regulatory and permitting requirements.

6.2.6 Waste stream profile. A general list of organic and inorganic components and their percentage ranges in the waste. For example,

oil	10–20 %
water	40–50 %
dirt	1–2 %
methylene chloride	15–28 %

This should be a breakdown of the waste constituents provided to the best of the generator's knowledge (or through analyses) and should include an estimate of the variability of the waste. It is used to determine proper RCRA classifications, toxicity, compatibility, safety and health considerations (see section 29 CFR, Part 1910), and conformity of waste shipments. In addition, it indicates waste variability, which is used in determining storage and handling.

6.2.7 Special handling. This may be information on air or water reactivity, explosivity, or shock sensitivity of the waste. This information is very useful in determining additional safety and health considerations (see section 29 CFR, Part 1910), handling, storage, packaging, method of introduction into the incinerator, and feed rates.

6.3 *Specific Information*—This information may be provided through the generator's knowledge of the waste or through analysis of a representative sample. An estimated variability of these values should be provided.

6.3.1 *Physical Characteristics*:

6.3.1.1 *Physical State*— For example, solid, liquid, semisolid, powder, multi-phased, layers. This information is used to determine the proper storage and feed type or rate.

6.3.1.2 *Flashpoint/Flammability Potential*—These values or ranges can be expressed by liquid flashpoint or the tendency of solids to burn vigorously when ignited (a subjective evaluation). They provide the basis for storage and handling (see Test Methods D 56, D 92, D 93, D 3828, 1010, and 1020).

6.3.1.3 *Specific Gravity/Apparent Bulk Density*—These data are used for volume to weight conversions, pump selections, storage, and feed rates (see Test Methods D 891).

6.3.1.4 *Viscosity*—The viscosity is useful in the determination of pump selection, storage, tank off-loading capabilities, and feed rates (see Test Methods D 88, D 2196, and D 2983).

6.3.1.5 *pH/Corrosivity*— These values or ranges provide necessary information in determining the compatibility of wastes with tank and processing equipment materials of construction. The information provides insight into the nature of the waste and its suitability for incineration. For example, mineral acids and caustics may not be suited for a particular incinerator, whereas organic acids and amines are (see Test Methods D 1292, 1110, 9045, and 9041).

6.3.1.6 *Radioactivity*— Information regarding radioactivity is critical for safe handling and processing of wastes. Many incineration facilities are not designed to store or handle wastes containing radioactivity. It is recommended that each facility perform some type of screening of waste for radioactivity. This screen must be based on the situations specific to each facility. A survey meter with an appropriate detector is useful in providing this type of screening. A facility may place this type of screening device in a location appropriate to the facility's operations (for example, in the laboratory or at the vehicle receiving gate, staging area, load area, etc.).

6.3.2 *Chemical Characteristics*:

6.3.2.1 *Heating Value*— The heating value may be necessary to meet some permit requirements. It can be used to establish waste stream feed rates and auxiliary fuel requirements (see Test Methods D 240, D2015, and D3286).

6.3.2.2 *Ash Content*— The ash content may be necessary to meet some permit requirements. It may be used to predict refractory or slag buildup and volumes of ash for disposal. It may also be used to estimate particulate generation (see Test Method D 482).

6.3.2.3 *Halogen and Sulfur Content*—The halogen and sulfur contents are necessary to meet permit requirements. They are useful in tracking refractory degradation and estimating acid gas generation, for example, HCl, SOx, etc. Individual halogens may need to be determined, depending on the type of incinerator and air pollution control systems (see Test Methods D 129, D808, D2361, D3177, D3761, D4208, E775, and E776).

6.3.2.4 *Metals*—The analysis of metals may be necessary to meet permit requirements. Specific metals and concern are dependent on the incinerator type, air pollution control system, and ash disposal requirements. Table 1 lists typical metals that are of primary concern (see Practice E 663 and Test Methods E 819, E 885, E 926, 3020, 3040, 3050, 6010, 7060, 7080, 7090, 7130, 7190, 7470, 7471, 7520, 7610, 7740, 7770, 7840, 7910, and 7950).

7. Optional Information

7.1 The following information may be provided by the generator's knowledge of the waste or through analysis of a representative sample. Estimated variability of these values should be provided.

7.1.1 *Solids Content*— This information is useful in determining pump requirements, feed rates, and storage designations. It may also be useful in determining particulate generation (see Test Methods D 1796 and D 4007).

7.1.2 *Volatile and Semi-Volatile Organics*—This information may be necessary in determining organic constituents, which is a permitting issue. It also provides additional information on the waste (see Practice E 260 and Test Methods 8010, 8015, 8020, 8080, 8120, 8240, 8250, and 8270).

TABLE 1 Typical Metals of Concern

Metals	Concerns
Arsenic	Air emissions
Barium	Ash disposal
Beryllium	Refractory degradation
Cadmium	Scrubber effluent
Chromium (hexavalent)	
Copper	
Mercury	
Potassium	
Sodium	
Nickel	
Lead	
Selenium	
Thallium	
Vanadium	
Zinc	

7.1.3 *Water Content*— These data are useful in determining storage, phase composition, compatibilities, and feed rates (see Test Method E 203).

7.1.4 *Material Safety Data Sheets*—These documents provide vital information for handling and storing out-of-date or off-specification products. Other sources may be product information and technical data sheets.

8. Incoming Shipment Evaluation

8.1 Once the decision has been made to accept a waste for incineration and the waste arrives at the facility, a plan must be in place to identify that the waste approved is indeed the waste received. The following evaluations must be made prior to commingling the waste shipment with other wastes: (*1*) sampling, (*2*) screening tests, (*3*) site procedural requirements, and (*4*) acceptance.

8.1.1 *Sampling*—At a minimum, a site sampling protocol must yield a representative sample or samples of each incoming shipment. The sample or samples should be collected in accordance with "Test Methods for the Evaluating Solid Waste Physical/Chemical Methods," SW-846, EPA, or 40 CFR, Part 261, Appendix XI.

8.1.2 *Analysis*—All or some of the information and chemical and physical characteristics in Sections 6 and 7 may be required to evaluate the acceptance of a shipment of waste. The information and chemical and physical characteristics to be considered may depend on the type and size of the operation.

8.2 If a waste must be rejected due to nonconformity, and returned to the generator, the waste must be properly DOT classified prior to shipment.

9. Keywords

9.1 hazardous waste incinerators; waste acceptance

APPENDIX

(Nonmandatory Information)

X1. EPA DOCUMENTS AND REGULATIONS

X1.1 The following EPA documents and regulations are useful in establishing the information that is necessary for hazardous waste incinerator material acceptance.

X1.2 *EPA Guidance Manuals:*

X1.2.1 *Permit Applicant's Guidance Manual for the General Facility Standards of 40 CFR, Part 264* (EPA SW-968).

X1.2.2 *Guidance Manual for Hazardous Waste Incineration Permits* (EPA SW-966).

X1.2.3 *Engineering Handbook for Hazardous Waste Incineration* (EPA SW-889).

X1.2.4 *Practical Guide—Trial Burns for Hazardous Waste Incinerators* (EPA/600/2-86/050).

X1.2.5 *Test Methods for Evaluating Solid Waste Physical/ Chemical Methods* (EPA SW-846).

X1.2.6 *Waste Analysis Plans A Guidance Manual* (September 1984).

X1.2.7 *Sampling and Analysis Methods for Hazardous Waste Combustion* (EPA Contract No. 68-01-3111).

X1.3 *Code of Federal Regulations:*

X1.3.1 40 CFR, Protection of Environment, Part 260.

X1.3.2 40 CFR, Protection of Environment, Part 262.

X1.3.3 40 CFR, Protection of Environment, Part 263.

X1.3.4 40 CFR, Protection of Environment, Part 266.

X1.3.5 40 CFR, Protection of Environment, Part 267.

X1.3.6 40 CFR, Protection of Environment, Part 268.

X1.3.7 40 CFR, Protection of Environment, Part 270.

Designation: D 5198 – 92 (Reapproved 1997)

Standard Practice for Nitric Acid Digestion of Solid Waste[1]

This standard is issued under the fixed designation D 5198; the number immediately following the designation indicates the year of original adoption or, in the case of revision, the year of last revision. A number in parentheses indicates the year of last reapproval. A superscript epsilon (ϵ) indicates an editorial change since the last revision or reapproval.

1. Scope

1.1 This practice describes the digestion of solid waste using nitric acid for the subsequent determination of inorganic constituents by argon plasma emission spectroscopy or atomic absorption spectroscopy.

1.2 The following elements may be solubilized by this practice:

aluminum	manganese
beryllium	mercury
cadmium	nickel
chromium	phosphorus
copper	vanadium
iron	zinc
lead	

1.3 This practice is to be used when the concentrations of total recoverable elements are to be determined from a waste sample. Total recoverable elements may or may not be equivalent to total elements, depending on the element sought and the sample matrix. Recovery from refractory sample matrices, such as soils, is usually significantly less than total concentrations of the elements present.

NOTE 1—This practice has been used successfully for oily sludges and a municipal digested sludge standard [Environmental Protection Agency (EPA) Sample No. 397]. The practice may be applicable to some elements not listed above, such as arsenic, barium, selenium, cobalt, magnesium, and calcium. Refractory elements such as silicon, silver, and titanium are not solubilized by this practice.

1.4 *This standard does not purport to address all of the safety concerns, if any, associated with its use. It is the responsibility of the user of this standard to establish appropriate safety and health practices and determine the applicability of regulatory limitations prior to use.* For specific hazard statements, see Section7.

2. Referenced Documents

2.1 *ASTM Standards:*

D 1193 Specification for Reagent Water[2]

D 2777 Practice for Determination of Precision and Bias of Applicable Methods of Committee D-19 on Water[2]

D 3223 Test Method for Total Mercury in Water[2]

D 3682 Test Method for Major and Minor Elements in Coal and Coke Ash by the Atomic Absorption Method[3]

D 3919 Practice for Measuring Trace Elements in Water by Graphite Furnace Atomic Absorption Spectrophotometry[2]

D 4503 Practice for the Dissolution of Solid Waste by Lithium Metaborate Fusion[4]

D 4698 Practice for Total Digestion of Sediment Samples for Chemical Analysis of Various Metals[5]

E 50 Practices for Apparatus, Reagents, and Safety Precautions for Chemical Analysis of Metals[6]

3. Summary of Practice

3.1 A weighed portion of the waste sample is mixed with 1 + 1 nitric acid (HNO_3) in an Erlenmeyer flask. The flask is heated for 2 h at 90 to 95°C to dissolve the elements of interest. After cooling, the contents of the flask are diluted with reagent water and filtered, and the filtrate is made up to appropriate volume for subsequent analysis.

4. Significance and Use

4.1 A knowledge of the inorganic composition of a waste is often required for the selection of appropriate waste disposal practices. Solid waste may exist in a variety of forms and contain a range of organic and inorganic constituents. This practice describes a digestion procedure which dissolves many of the toxic inorganic constituents and produces a solution suitable for determination by such techniques as atomic absorption spectroscopy, atomic emission spectroscopy, and so forth. The relatively large sample size aids representative sampling of heterogenous wastes. The relatively small dilution factor allows lower detection limits than most other sample digestion methods. Volatile metals, such as lead and mercury, are not lost during this digestion procedure. Hydride-forming elements, such as arsenic and selenium, may be partially lost. Samples with total metal contents greater than 5 % may give low results. The analyst is responsible for determining whether this practice is applicable to the solid waste being tested.

5. Apparatus

5.1 *Analytical Balance*, capable of weighing to 0.01 g.

5.2 *Erlenmeyer Flasks*, 125 mL.

5.3 *Graduated Cylinder*, 50 mL.

---

[1] This practice is under the jurisdiction of ASTM Committee D34 on Waste Management and is the direct responsibility of Subcommittee D34.01.06 on Analytical Methods.

Current edition approved Jan. 9, 1992. Published March 1992.

[2] *Annual Book of ASTM Standards*, Vol 11.01.

[3] *Annual Book of ASTM Standards*, Vol 05.05.

[4] *Annual Book of ASTM Standards*, Vol 11.04.

[5] *Annual Book of ASTM Standards*, Vol 11.02.

[6] *Annual Book of ASTM Standards*, Vol 03.05.

5.4 *Electric Hot Plate*, adjustable, capable of maintaining a temperature of 90 to 95°C.

5.5 *Watch Glasses*.

5.6 *Thermometer*.

5.7 *Funnels*, glass or plastic.

5.8 *Volumetric Flasks*, glass-stoppered, 200 mL.

5.9 *Filter Paper*, quantitative, medium flow rate, Whatman No. 40 or equivalent.

6. Reagents

6.1 *Purity of Reagents*—Reagent grade chemicals shall be used in all tests. Unless otherwise indicated, it is intended that all reagents shall conform to the specifications of the Committee on Analytical Reagents of the American Chemical Society, where such specifications are available.[7] Other grades may be used, provided it is first ascertained that the reagent is of sufficiently high purity to permit its use without lessening the accuracy of the determination.

6.2 *Purity of Water*— Unless otherwise indicated, reference to water shall be understood to mean reagent water as defined by Type II of Specification D 1193.

6.3 *Nitric Acid*, concentrated, reagent grade.

6.4 *Nitric Acid* (1+1)—Add slowly, with stirring, 200 mL of concentrated nitric acid (HNO_3, sp gr 1.42) to 200 mL water. Cool the mixture and store in a clean pint glass bottle.

7. Hazards

7.1 Add the nitric acid mixture slowly, with swirling, to the sample. Samples containing carbonates may foam excessively during acid addition and result in loss of sample. Nitric acid may react violently with some samples containing organic material.

7.2 Addition of acid and sample digestion must be conducted in a hood with adequate ventilation and shielding to avoid contact with nitrogen oxides, acid fumes, or toxic gases.

8. Procedure

8.1 Weigh 5 g of a thoroughly mixed waste sample to the nearest 0.01 g into a tared Erlenmeyer flask.

8.2 With a graduated cylinder, slowly add 25 mL of 1+1 nitric acid to the flask. Swirl the flask to wet the sample completely.

8.3 Carry a blank of 25 mL of 1+1 nitric acid through the procedure.

8.4 Place the flask on a cold hot plate, cover with a watch glass, and set the hot plate to maintain a temperature of 90 to 95°C.

8.5 Heat the flask and contents for 2 h, occasionally swirling the flask to wash down any sample adhering to the walls. Check the solution temperature with the thermometer and adjust the heat if necessary.

8.6 After 2 h, remove the flask from the hot plate and cool to room temperature. Add 50 mL of reagent water to the flask, washing down the flask walls during addition. Swirl the flask to mix the contents.

8.7 Filter the contents of the flask into a 200 mL volumetric flask. Rinse the flask and filter paper with several small portions of reagent water and add the rinsings to the volumetric flask.

8.8 Dilute the solution in the volumetric flask to the mark with reagent water and mix thoroughly. The solution is now ready for analysis.

9. Precision and Bias

9.1 No statement is made about either the precision or bias since this practice does not produce a test result. Appendix X1 contains representative results obtained with this practice and subsequent analysis.

10. Keywords

10.1 digestion; nitric acid; waste

[7] "Reagent Chemicals, American Chemical Society Specifications," Am. Chemical Soc., Washington, DC. For suggestions on the testing of reagents not listed by the American Chemical Society, see "Reagent Chemicals and Standards," by Joseph Rosin, D. Van Nostrand Co., Inc., New York, NY, and the "United States Pharmacopeia."

APPENDIX

(Nonmandatory Information)

X1. REPRESENTATIVE ANALYSES OF SAMPLES AFTER NITRIC ACID DIGESTION

TABLE X1.1 Nitric Acid Digestion—Analysis of Standard Sludge Sample (EPA Sample No. 397)

Element	Concentration, mg/kg		Percent Recovery
	Actual	Found	
Zn	1323	1300	98
Mn	205	235	115
Pb	519	500	96
Cd	20.8	24.1	116
Cr	204	218	107
Fe	16 155	16 400	102
V	13.0	12.7	98
Cu	1095	1130	103
Ni	198	186	94
Al	4558	4500	99
Ti	2121	95	4.5
P	11 573	11 800	102
Be	0.28	0.5	179
As	17	9	53
Ag	81	7.4	9.1
Hg	16.3	13.6	83

TABLE X1.2 Nitric Acid Digestion—Replicate Analysis of Used Motor Oil, mg/kg

Element	Run 1	Run 2	Run 3
Zn	1170	1150	1170
Mn	2.5	2.1	2.1
Pb	43	40	35
Cd	0.6	0.5	0.4
Cr	2.8	2.1	2.3
Fe	74	66	66
V	1.32	1.04	0.92
Cu	3.8	3.6	3.6
Ni	0.7 ND[A]	0.7 ND	0.7 ND
Ba	0.0	0.5	0.5
Be	0.04 ND	0.04 ND	0.04 ND
Hg	0.08 ND	0.08 ND	0.08 ND

[A] ND = not detected at the detection limit shown.

TABLE X1.3 Nitric Acid Digestion—Recovery of Used Motor Oil Matrix Spikes

Element	mg/kg		Percent Recovery
	Added	Found	
Mn	3.2	3.3	103
Pb	3.4	3.0	88
Cd	2.0	1.6	80
Cr	2.0	2.4	120
V	2.6	1.0	38
Cu	2.2	4.2	191
Ni	2.8	2.75	98
Ba	2.3	2.1	91
Be	1.6	1.3	81
Hg	0.08	0.09	113

Standard Test Method for Determination of the Composition of Unprocessed Municipal Solid Waste[1]

This standard is issued under the fixed designation D 5231; the number immediately following the designation indicates the year of original adoption or, in the case of revision, the year of last revision. A number in parentheses indicates the year of last reapproval. A superscript epsilon (ε) indicates an editorial change since the last revision or reapproval.

1. Scope

1.1 This test method describes procedures for measuring the composition of unprocessed municipal solid waste (MSW) by employing manual sorting. This test method applies to determination of the mean composition of MSW based on the collection and manual sorting of a number of samples of waste over a selected time period covering a minimum of one week.

1.2 This test method includes procedures for the collection of a representative sorting sample of unprocessed waste, manual sorting of the waste into individual waste components, data reduction, and reporting of the results.

1.3 This test method may be applied at landfill sites, waste processing and conversion facilities, and transfer stations.

1.4 The values stated in inch-pound units are to be regarded as the standard. The values given in parentheses are for information only.

1.5 *This standard does not purport to address all of the safety problems, if any, associated with its use. It is the responsibility of the user of this standard to establish appropriate safety and health practices and determine the applicability of regulatory limitations prior to use.* For specific hazard statements, see Section 6.

2. Terminology

2.1 *Definitions:*

2.1.1 *composite item*—an object in the waste composed of multiple waste components or dissimilar materials, such as disposable diapers, bi-metal beverage containers, electrical conductors composed of metallic wire encased in plastic insulation, etc.

2.1.2 *solid waste composition or waste composition*—the characterization of solid waste as represented by a breakdown of the mixture into specified waste components on the basis of mass fraction or of weight percent.

2.1.3 *sorting sample*—a 200 to 300-lb (91 to 136-kg) portion deemed to represent the characteristics of a vehicle load of MSW.

2.1.4 *unprocessed municipal solid waste*—solid waste in its discarded form, that is, waste that has not been size reduced or otherwise processed.

2.1.5 *waste component*—a category of solid waste, composed of materials of similar physical properties and chemical composition, which is used to define the composition of solid waste, for example, ferrous, glass, newsprint, yard waste, aluminum, etc.

3. Summary of Test Method

3.1 The number of samples to be sorted is calculated based on statistical criteria selected by the investigators.

3.2 Vehicle loads of waste are designated for sampling, and a sorting sample is collected from the discharged vehicle load.

3.3 The sample is sorted manually into waste components. The weight fraction of each component in the sorting sample is calculated from the weights of the components.

3.4 The mean waste composition is calculated using the results of the composition of each of the sorting samples.

4. Significance and Use

4.1 Waste composition information has widespread applications and can be used for activities such as solid waste planning, designing waste management facilities, and establishing a reference waste composition for use as a baseline standard in both facility contracts and acceptance test plans.

4.2 The method can be used to define and report the composition of MSW through the selection and manual sorting of waste samples. Where applicable, care should be taken to consider the source and seasonal variation of waste.

4.3 After performing a waste composition analysis, laboratory analyses may be performed on representative samples of waste components, or mixtures of waste components, for purposes related to the planning, management, design, testing, and operation of resource recovery facilities.

5. Apparatus

5.1 *Metal, Plastic, or Fiber Containers*, sufficient for storing and weighing each waste component, labeled accordingly. For components that will have a substantial moisture content (for example, food waste), metal or plastic containers are recommended in order to avoid absorption of moisture by the container and thus the need for a substantial number of weighings to maintain an accurate tare weight for the container.

5.2 *Mechanical or Electronic Weigh Scale*, with a capacity of at least 200 lb (91 kg) and precision of at least 0.1 lb (0.045 kg).

[1] This test method is under the jurisdiction of ASTM Committee D34 on Waste Disposal and is the direct responsibility of Subcommittee D34.01.06 on Analytical Methods.

Current edition approved July 31, 1992. Published September 1992.

5.3 *Heavy-Duty Tarps, Shovels, Rakes, Push Brooms, Dust Pans, Hand Brooms, Magnets, Sorting Table, First Aid Kit, Miscellaneous Small Tools, Traffic Cones, Traffic Vests, Leather Gloves, Hardhats, Safety Glasses, and Leather Boots.*

6. Hazards

6.1 Review the hazards and procedures with the operating and sorting personnel prior to conducting the field activities.

6.2 Sharp objects, such as nails, razor blades, hypodermic needles, and pieces of glass, are present in solid waste. Personnel should be instructed of this danger, and they should brush waste particles aside while sorting rather than projecting their hands with force into the mixture. Personnel handling and sorting solid waste should wear appropriate protection, such as heavy leather gloves, dust masks, hardhats, safety glasses, and safety boots.

6.3 During the processes of unloading waste from collection vehicles and handling waste with heavy equipment, projectiles may issue from the mass of waste. The projectiles can include flying glass particles from breaking glass containers and metal lids from plastic and metal containers that burst under pressure when run over by heavy equipment. The problem is particularly severe when the waste handling surface is of high compressive strength, for example, concrete. Personnel should be informed of this danger and wear eye and head protection if in the vicinity of either the collection vehicle unloading point or heavy equipment, or both.

6.4 Select a location for the discharge of designated loads, manual sorting activities, and weighing operations that is flat, level, and away from the normal waste handling and processing areas.

6.5 Weigh storage containers each day, or more frequently, if necessary, in order to maintain an accounting of the tare weight.

6.6 Loss of mass from the sorting sample can occur through the evaporation of water. Samples should thus be sorted as soon as possible after collection.

6.7 Containers of liquids or other potentially dangerous wastes shall be put aside and handled by the crew chief.

7. Calibration

7.1 All weigh scale equipment shall be calibrated according to the manufacturer's instructions. Take appropriate corrective action if the readings are different from those of the calibration weights.

8. Procedure

8.1 Secure a flat and level area for discharge of the vehicle load. The surface should be swept clean or covered with a clean, durable tarp prior to discharge of the load.

8.2 Position the scale on a clean, flat, level surface and adjust the level of the scale if necessary. Determine the accuracy and operation of the scale with a known (that is, reference) weight.

8.3 Weigh all empty storage containers and record the tare weights.

8.4 Determine the number of samples to be sorted. The determination is a function of the waste components to be sorted and the desired precision as applied to each component. Weights of 200 to 300 lb (91 to 136 kg) for sorting samples of unprocessed solid waste are recommended. The number of samples is determined using the calculational method described in 9.1.

8.5 A comprehensive list of waste components for sorting is given in Table 1. A description of some of the waste component categories is given in Table 2. Other waste components can be defined and sorted, depending on the purpose of the waste composition determination. The list in Table 1 is comprised of those components most commonly used to define and report the composition of solid waste. It is recommended that, at a minimum, the complement of left-justified categories in Table 1 be sorted. Similar breakdowns of solid waste composition are therefore available for purposes of comparison, if desired. Label the storage containers accordingly.

8.6 Vehicles for sampling shall be selected at random during each day of the one-week sampling period, or so as to be representative of the waste stream as agreed upon by the affected parties. With respect to the random selection of vehicles, any method is acceptable that does not introduce a bias into the selection. An acceptable method is the use of a random number generator. For a weekly sampling period of k days, the number of vehicles sampled each day shall be approximately n/k, where n is the total number of vehicle loads to be selected for the determination of waste composition. A weekly period is defined as 5 to 7 days.

8.7 Direct the designated vehicle containing the load of waste to the area secured for discharge of the load and collection of the sorting sample.

8.8 Collect any required information from the vehicle operator before the vehicle leaves the discharge area. Direct the vehicle operator to discharge the load onto the clean surface in one contiguous pile, that is, to avoid gaps in the discharged load in order to facilitate collection of the samples.

8.9 Using a front-end loader with at least a 1-yd^3 (0.765-m^3) bucket, remove the material longitudinally along one entire side of the discharged load in order to obtain a representative cross-section of the material. The mass of material shall be sufficient to form a mass of material which, on a visual basis, is at least four times the desired weight of the sorting sample (that is, approximately 1000 lb (454 kg)). Mix, cone, and quarter the material, and select one quarter to be the sorting sample, using a random method of selection or a sequence agreed by all affected parties, for the purpose of eliminating or minimizing biasing of the sample. If an oversize item (for

TABLE 1 List of Waste Component Categories

Mixed paper	Other organics
High-grade paper	Ferrous
Computer printout	Cans
Other office paper	Other ferrous
Newsprint	Aluminum
Corrugated	Cans
Plastic	Foil
PET bottles	Other aluminum
HDPE bottles	Glass
Film	Clear
Other plastic	Brown
Yard waste	Green
Food waste	Other inorganics
Wood	

TABLE 2 Descriptions of Some Waste Component Categories

Category	Description
Mixed paper	Office paper, computer paper, magazines, glossy paper, waxed paper, and other paper not fitting the categories of newsprint and corrugated
Newsprint	Newspaper
Corrugated	Corrugated medium, corrugated boxes or cartons, and brown (kraft) paper (that is, corrugated) bags
Plastic	All plastics
Yard waste	Branches, twigs, leaves, grass, and other plant material
Food waste	All food waste except bones
Wood	Lumber, wood products, pallets, and furniture
Other organics/ combustibles	Textiles, rubber, leather, and other primarily burnable materials not included in the above component categories
Ferrous	Iron, steel, tin cans, and bi-metal cans
Aluminum	Aluminum, aluminum cans, and aluminum foil
Glass	All glass
Other inorganics/ non-combustibles	Rock, sand, dirt, ceramics, plaster, non-ferrous non-aluminum metals (copper, brass, etc.), and bones

example, water heater) composes a large weight percent of the sorting sample, add a notation on the data sheet and weigh it, if possible. Unprocessed solid waste is a heterogeneous mixture of materials. Care must thus be taken during application of the procedures for sample collection in order to obtain a representative sample.

8.10 One sorting sample is selected from each collection vehicle load designated for sampling. All handling and manipulation of the discharged load and longitudinal and sorting samples shall be conducted on previously cleaned surfaces. If necessary, remove the sorting sample to a secured manual sorting area. The sorting sample may be placed on a clean table for sorting for the convenience of the sorting personnel. The sorting area shall be a previously cleaned, flat, level surface.

8.11 Position the storage containers around the sorting sample. Empty all containers from the sorting sample, such as capped jars, paper bags, and plastic bags of their contents. Segregate each waste item and place it in the appropriate storage container.

8.12 In the case of composite items found in the waste, separate the individual materials where practical, and place the individual materials into the appropriate storage containers. Where impractical, segregate the composite items for classification by the crew chief according to the following order:

8.12.1 If there are many identical composite items (for example, plastic-sheathed aluminum electrical conductor), place them into the waste component containers corresponding to the materials present in the item, and in the approximate proportions according to the estimated mass fraction of each material in the item.

8.12.2 If there are only a few of the identical composite item, place them in the storage container corresponding to the material that comprises, on a weight basis, the majority of the item (for example, place bi-metal beverage cans in the ferrous container).

8.12.3 If composite items represent substantial weight percents of the sorting sample, a separate category should be established, for example, composite roofing shingles.

8.12.4 If none of the above procedures is appropriate, place the item(s) (or proportion it (them)) in the storage container labeled" other non-combustible" or "other combustible," as appropriate.

8.13 Sorting continues until the maximum particle size of the remaining waste particles is approximately 0.5 in. (12.7 mm). At this point, apportion the remaining particles into the storage containers corresponding to the waste components represented in the remaining mixture. The apportionment shall be accomplished by making a visual estimate of the mass fraction of waste components represented in the remaining mixture.

8.14 Record the gross weights of the storage containers and of any waste items sorted but not stored in containers. The data sheet shown in Fig. 1 can be used to record both gross and tare weights.

8.15 After recording the gross weights, empty the storage containers and weigh them again, if appropriate. Re-weighing

Waste Composition Data Sheet

Day/Date: ________ Collection Company: ________
Site: ________ Vehicle Type: ________
Weather: ________ Route No: ________
Recorded by: ________

Component	Weight in Pounds		Percent of Total
	Gross	Tare	
Mixed Paper			
High Grade Paper			
Computer Printout			
Other Office Paper			
Newsprint			
Corrugated			
Plastic			
PET bottles			
HDPE bottles			
Film			
Other Plastic			
Food Waste			
Wood			
Other Organics			
Ferrous			
Cans			
Other Ferrous			
Aluminum			
Cans			
Foil			
Other Aluminum			
Glass			
Clear			
Brown			
Green			
Other Inorganics			

TOTALS ________

NOTES: ________

Lab sample taken? Yes ____ No ____

FIG. 1 Waste Composition Data Sheet

is important and necessary if the containers become moisture-laden, for example, from wet waste.

8.16 Clean the sorting site, as well as the load discharge area, of all waste materials.

9. Calculation

9.1 *Number of 200 to 300-lb (91 to 136-kg) Samples*:

9.1.1 The number of sorting samples (that is, vehicle loads) (n) required to achieve a desired level of measurement precision is a function of the component(s) under consideration and the confidence level. The governing equation for n is as follows:

$$n = (t^* \, s/e \cdot \bar{x})^2 \tag{1}$$

where:

t^* = student t statistic corresponding to the desired level of confidence,
s = estimated standard deviation,
e = desired level of precision, and
$\bar{x}$ = estimated mean.

9.1.1.1 All numerical values for the symbols are in decimal notation. For example, a precision value (e) of 20 % is represented as 0.2.

9.1.1.2 One sorting sample is chosen per vehicle load.

9.1.1.3 Suggested values of s and of $\bar{x}$ for waste components are listed in Table 3. Values of t^* are given in Table 4 for 90 and 95 % levels of confidence, respectively.

9.1.2 Estimate the number of samples (n') for the selected conditions (that is, precision and level of confidence) and components using (Eq 1). For the purposes of estimation, select from Table 4 the t^* value for $n = \infty$ for the selected level of confidence. Since the required number of samples will vary among the components for a given set of conditions, a compromise will be required in terms of selecting a sample size, that is, the number of samples that will be sorted. The component that is chosen to govern the precision of the composition measurement (and therefore the number of samples required for sorting) is termed the "governing component" for the purposes of this method.

TABLE 3 Values of Mean ($\bar{x}$) and Standard Deviation(s) for Within-Week Sampling to Determine MSW Component Composition[A]

Component	Standard Deviation(s)	Mean ($\bar{x}$)
Newsprint	0.07	0.10
Corrugated	0.06	0.14
Plastic	0.03	0.09
Yard waste	0.14	0.04
Food waste	0.03	0.10
Wood	0.06	0.06
Other organics	0.06	0.05
Ferrous	0.03	0.05
Aluminum	0.004	0.01
Glass	0.05	0.08
Other inorganics	0.03	0.06
		1.00

[A]The tabulated mean values and standard deviations are estimates based on field test data reported for MSW sampled during weekly sampling periods at several locations around the United States.

TABLE 4 Values of t Statistics (t^*) as a Function of Number of Samples and Confidence Interval

Number of Samples, n	90 %	95 %
2	6.314	12.706
3	2.920	4.303
4	2.353	3.182
5	2.132	2.776
6	2.015	2.571
7	1.943	2.447
8	1.895	2.365
9	1.860	2.306
10	1.833	2.262
11	1.812	2.228
12	1.796	2.201
13	1.782	2.179
14	1.771	2.160
15	1.761	2.145
16	1.753	2.131
17	1.746	2.120
18	1.740	2.110
19	1.734	2.101
20	1.729	2.093
21	1.725	2.086
22	1.721	2.080
23	1.717	2.074
24	1.714	2.069
25	1.711	2.064
26	1.708	2.060
27	1.706	2.056
28	1.703	2.052
29	1.701	2.048
30	1.699	2.045
31	1.697	2.042
36	1.690	2.030
41	1.684	2.021
46	1.679	2.014
51	1.676	2.009
61	1.671	2.000
71	1.667	1.994
81	1.664	1.990
91	1.662	1.987
101	1.660	1.984
121	1.658	1.980
141	1.656	1.977
161	1.654	1.975
189	1.653	1.973
201	1.653	1.972
∞	1.645	1.960

9.1.3 After determining the governing component and its corresponding number of samples (n_o), return to Table 4 and select the student t statistic (t^*_o) corresponding to n_o. Recalculate the number of samples, that is, n', using t^*_o.

9.1.4 Compare n_o to the new estimate of n, that is, n', which was calculated for the governing component. If the values differ by more than 10 %, repeat the calculations given in 9.1.2 and 9.1.3.

9.1.5 If the values are within 10 %, select the larger value as the number of samples to be sorted. Refer to Appendix X1 for a sample calculation of n.

9.2 *Component Composition*:

9.2.1 The component composition of solid waste is reported on the basis of the mass fraction (expressed as a decimal) or percent of waste component i in the solid waste mixture. The reporting is on the basis of wet weight, that is, the weight of materials immediately after sorting.

9.2.2 The mass fraction of component i, mf_i, is defined and computed as follows:

$$mf_i = \frac{w_i}{\sum_{i=1}^{j} w_i} \quad (2)$$

where:

w_i = weight of component i and

j = number of waste components.

In those cases in which a container is used to store and weigh the materials,

$$w_i = \text{gross weight} - \text{tare weight of container} \quad (3)$$

9.2.3 The percent of component i, P_i, is defined and computed as follows:

$$P_i = mf_i \times 100 \quad (4)$$

9.2.4 For the data analysis to be correct, the denominator of (Eq 2) must be unity, and

$$\sum_{i=1}^{j} P_i = 100 \quad (5)$$

9.3 The mean component composition for the one-week period is calculated using the component composition results from each of the analysis samples. The mean mass fraction of component i, $\overline{mf}_i$, is calculated as follows:

$$\overline{mf}_i = \frac{1}{n} \sum_{k=1}^{n} (mf_i)_k \quad (6)$$

and the mean percent of component i, P_i, is calculated as follows:

$$\bar{P}_i = \frac{1}{n} \sum_{k=1}^{n} (P_i)_k \quad (7)$$

where:

n = number of samples.

10. Precision and Bias

10.1 A precision and bias statement cannot be made for this test method at this time. However, the committee is interested in conducting an interlaboratory test program and encourages interested parties to contact ASTM Headquarters.[2]

11. Keywords

11.1 composition; municipal solid waste; waste characterization

[2] ASTM Headquarters, 1916 Race Street, Philadelphia, PA 19103.

APPENDIX

(Nonmandatory Information)

X1. EXAMPLE CALCULATION OF THE NUMBER OF SAMPLES FOR ANALYSIS

X1.1 *Example Assumptions:*

X1.1.1 Corrugated is selected as the governing component.

X1.1.2 A 90 % confidence level is selected.

X1.1.3 A precision of 10 % is desired.

X1.1.4 Therefore:

s = 0.06 (from Table 3),
$\bar{x}$ = 0.14 (from Table 3),
e = 0.10, and
$t^*(n = \infty)$ = 1.645 (from Table 4).

Using (Eq 1):

$$n = [t^* \, s/(e \cdot \bar{x})]^2 = \left[\frac{1.645\ (0.06)}{0.1\ (0.14)}\right]^2 = 50 = n_o \quad (X1.1)$$

Referring again to Table 4, for $n = 50$,

$$t^*_{90}\ (n = 50) = 1.677 \quad (X1.2)$$

and,

$$n = \left[\frac{1.677\ (0.06)}{0.1\ (0.14)}\right]^2 = 52 = n' \quad (X1.3)$$

Since 52 (that is, n') is within 10 % of 50 (that is, n_o), 52 samples should be selected for analysis.

Designation: D 5232 – 92 (Reapproved 1997)

Standard Test Method for Determining the Stability and Miscibility of a Solid, Semi-Solid, or Liquid Waste Material[1]

This standard is issued under the fixed designation D 5232; the number immediately following the designation indicates the year of original adoption or, in the case of revision, the year of last revision. A number in parentheses indicates the year of last reapproval. A superscript epsilon (ϵ) indicates an editorial change since the last revision or reapproval.

1. Scope

1.1 This test method is designed to determine whether a waste material reacts when it is mixed with air, water, strong acid, strong base, an oil/solvent mixture, other waste mixtures, or solid media such as a geological formation or solidification agents.

1.2 The miscibility of the waste material with the above media can also be defined.

NOTE 1—The following ASTM standards provide supplemental information: Test Methods D 4978, D 4979, D 4980, D 4981, D 4982, D 5049, D 5057, and D 5058.

1.3 *This standard does not purport to address all of the safety concerns, if any, associated with its use. It is the responsibility of the user of this standard to establish appropriate safety and health practices and determine the applicability of regulatory limitations prior to use.* For specific hazard statements, see Section 7.

2. Referenced Documents

2.1 *ASTM Standards:*

D 1193 Specification for Reagent Water[2]

D 4978 Test Methods for the Screening of Sulfides in Waste[3]

D 4979 Test Method for Physical Description Screening Analysis in Waste[3]

D 4980 Test Methods for the Screening of pH in Waste[3]

D 4981 Test Method for the Screening of Oxidizers in Waste[3]

D 4982 Test Methods for Flammability Potential Screening Analysis of Waste[3]

D 5049 Test Method for the Screening of Cyanides in Waste[3]

D 5057 Test Method for the Screening of Apparent Specific Gravity and Bulk Density of Waste[3]

D 5058 Test Methods for Compatibility Screening Analysis of Waste[3]

3. Summary of Test Method

3.1 The stability and miscibility of a waste material are observed when the waste is mixed with various media.

4. Significance and Use

4.1 This test method will identify waste materials that are potentially unstable when they come in contact with other materials at a waste treatment or disposal site.

4.2 This test method will serve to determine the miscibility of waste materials with various media, including other wastes.

4.3 This test method may not be applicable to all wastes. The appropriateness of these tests depends upon the proposed management of the waste.

4.4 Since the initiation of some chemical reactions are slow to take place, the user may wish to establish reagent-to-waste contact times prior to observing the mixes for any reactions.

5. Apparatus and Materials

5.1 *Disposable Cups*, minimum 40-mL total volume. Select plastics or other materials compatible with the reagents involved.

5.2 *Stirring Rods, Spatulas, Disposal Droppers, Watch Glasses*, and so forth.

5.3 Other waste materials to be evaluated with the test sample.

5.4 *Solid Media*, to be evaluated with the test sample.

6. Reagents

6.1 *Purity of Water*— The deionized water used to prepare the aqueous test solutions should conform to the conductivity specifications of ASTM Type II water as found in Specification D 1193.

6.2 *Hydrochloric Acid (HCl) Test Solution*, carefully mix 100 g of reagent grade hydrochloric acid (~36 %) with 900 g of deionized water.

6.3 *Sodium Hydroxide (NaOH) Test Solution*, carefully mix 200 g of reagent grade sodium hydroxide solution (~50 %) with 800 g of deionized water.

6.4 *Oil/Solvent Test Mixture*, prepare a fifty percent (50 %)

---

[1] This test method is under the jurisdiction of ASTM Committee D34 on Waste Management and is the direct responsibility of Subcommittee D34.01.06 on Analytical Methods.

Current edition approved June 2, 1992. Published August 1992.

[2] *Annual Book of ASTM Standards*, Vol 11.01.

[3] *Annual Book of ASTM Standards*, Vol 11.04.

by weight mixture of reagent grade perchloroethylene in USP grade mineral oil.

7. Hazards

7.1 Standard laboratory hygiene practices should be followed when conducting these tests. Potentially violent reactions may occur in performance of this test, so appropriate precautions may be advised.

7.2 All tests should be performed in a laboratory fume hood. Waste samples may produce toxic or noxious fumes when mixed with other substances.

7.3 The analyst should wear gloves that are resistant to acids, bases, and organic liquids.

8. Procedure

8.1 Refer to the sample's label and respective chemical composition data to determine if there is any waste component listed that could react with the agents or materials selected for testing.

8.2 *Air Stability*:

8.2.1 Open the cap of the sample container very slowly, and remove the cap slowly by sliding it across the top of the sample container.

8.2.2 If any evidence of fuming is observed when the container is first opened, close the container immediately and refer to the sample's respective chemical composition data to determine if there is any waste component listed that could be reactive in air. If no air reactive components are identified, then carefully open the container.

8.2.3 Using a glass, plastic spatula, or disposable dropper, remove a small portion of the sample and place it on a watch glass to observe any reaction of the waste in air. Note any reaction. If the waste sample is air reactive, do not proceed with any of the following steps.

8.3 *Water Stability*:

8.3.1 Mix liquid waste sample thoroughly, and place a 10-mL aliquot of the sample in a 40-mL plastic cup. For dry waste or sludges, add enough sample to fill the cup to the 10-mL mark.

8.3.2 Pour 10-mL of water slowly and carefully over the sample and observe any reaction. If no reaction is observed immediately, then mix the sample and water and note any reaction. Also estimate the volume percent solubility of the sample in water. If the waste appears to be water reactive, it is suggested that the appropriateness of the following steps be reviewed.

8.4 *Stability with Acid*:

8.4.1 Take a 10-mL aliquot of the well mixed liquid waste sample and place it in a 40-mL plastic cup. For dry waste or sludges, fill to the 10-mL mark.

8.4.2 Slowly and carefully add 10-mL of the HCl test solution or up to 10-mL until a reaction is observed. If no reaction is observed after the addition of 10-mL, then carefully mix the HCl solution with the sample until it is evenly mixed. Note any reaction that occurs. Also estimate the volume percent solubility of the sample in the acid.

8.5 *Stability with Base*:

8.5.1 Take a 10-mL aliquot of the well mixed liquid waste sample and place it in a 40-mL plastic cup. For dry waste or sludges, fill to the 10-mL mark.

8.5.2 Slowly and carefully add 10-mL of the NaOH test solution or up to 10-mL until a reaction is observed. If no reaction is observed after the addition of 10-mL, then carefully mix the NaOH solution with the sample until it is evenly mixed. Note any reaction that occurs. Also estimate the volume percent solubility of the sample in base.

8.6 *Stability with Organics*:

8.6.1 For waste material that may be mixed with other wastes for treatment and disposal, take a 10-mL aliquot of the well mixed liquid waste sample and place it in a 40-mL plastic cup. For dry waste or sludges fill to the 10-mL mark.

8.6.2 Slowly and carefully add 10-mL of oil/solvent mixture or up to 10-mL until a reaction is observed. If no reaction is observed after the addition of 10-mL, then carefully mix the oil/solvent with the sample until it is evenly mixed. Note any reaction that occurs. Also estimate the volume percent solubility of the sample with organics.

8.7 *Stability with Other Waste Materials or Solid Media*:

8.7.1 Place a 10-mL aliquot of the well mixed liquid sample in a 40-mL plastic cup. For dry waste or sludge, fill to the 10-mL mark.

8.7.2 Carefully and slowly add 10-mL of the other waste material or solid media or up to 10-mL until a reaction is observed. If no reaction is observed after the addition of the 10-mL, then carefully mix the contents of the plastic cup until evenly mixed. Note any reaction that occurs. Also note the volume percent solubility of the waste sample with the other waste material, if applicable.

9. Interpretation of Results

9.1 If the waste material reacts with any of the test media evaluated, resulting in the evolution of heat or the vigorous evolution of gases, then the waste will be termed unstable with that respective media.

9.2 The solubility of the waste material observed when mixed with the various liquid media can be an indication of the miscibility of the waste in those media.

10. Precision and Bias

10.1 *Precision*—It is not practicable to specify the precision of the procedure in Test Method D 5232 for measuring the stability and miscibility of a waste material, because no numerical data exist due to the wide variations in waste material compositions.

10.2 *Bias*—Since there is no accepted reference material suitable for measuring the bias for the procedure in Test Method D 5232 for measuring the stability and miscibility of a waste material, bias has not been determined.

11. Quality Control

11.1 Quality control samples and replicate evaluations should be performed at a level specified by the laboratory and at a predetermined frequency.

12. Keywords

12.1 miscibility; stability; waste

Designation: D 5233 – 92 (Reapproved 1999)

Standard Test Method for Single Batch Extraction Method for Wastes[1]

This standard is issued under the fixed designation D 5233; the number immediately following the designation indicates the year of original adoption or, in the case of revision, the year of last revision. A number in parentheses indicates the year of last reapproval. A superscript epsilon (ε) indicates an editorial change since the last revision or reapproval.

1. Scope

1.1 This test method is applicable to the extraction of samples of treated or untreated solid wastes or sludges, or solidified waste samples, to provide an indication of the leaching potential.

1.2 This test method is intended to provide an extract for measurement of the concentration of the analytes of concern. The measured values may be compared against set or chosen acceptance levels in some applications.

1.3 If the sole application of the test method is such a pass/fail comparison and a total analysis of the waste demonstrates that individual analytes are not present in the waste, or that the chosen acceptance concentration levels could not possibly be exceeded, the test method need not be run.

1.4 If the sole application of the test method is such a pass/fail comparison and an analysis of any one of the liquid fractions of the extract indicates that the concentration of the target analyte is so high that, even after accounting for dilution from the other fractions of the extract, it would be equal to or above an acceptance concentration level, then the waste fails the test. In such a case it may not be necessary to analyze the remaining fractions of the extract.

1.5 This test method is intended to provide an extract suitable for the measurement of the concentration of analytes that will not volatilize under the conditions of the test method.

1.6 Presence of volatile analytes may be established if an analysis of the extract obtained using this test method detects the target volatile analyte. If its concentration is equal to or exceeds an acceptance level for that analyte, the waste fails the test. However, extract from this test method shall not be used to determine the concentration of volatile organic analytes.

1.7 This test method is intended to describe only the procedure for performing a batch extraction. It does not describe all of the sampling and analytical requirements that may be associated with the application of this test method.

1.8 The values stated in either SI or inch-pound units are to be regarded as the standard. The values given in parentheses are for information only.

1.9 *This standard does not purport to address all of the safety problems, if any, associated with its use. It is the responsibility of the user of this standard to establish appropriate safety and health practices and determine the applicability of regulatory limitations prior to use.* For a specific precautionary statement, see Note 8.

2. Referenced Documents

2.1 *ASTM Standards:*

D 75 Practices for Sampling Aggregates[2]

D 420 Practice for Investigating and Sampling Soil and Rock for Engineering Purposes[3]

D 653 Terminology Relating to Soil, Rock, and Contained Fluids[3]

D 1129 Terminology Relating to Water[4]

D 1193 Specification for Reagent Water[4]

D 2234 Test Method for Collection of a Gross Sample of Coal[5]

D 3370 Practices for Sampling Water[4]

E 122 Practice for Choice of Sample Size to Estimate A Measure of Quality for a Lot or Process[6]

ES 16 Practice for the Generation of Environmental Data Related to Waste Management Activities[7]

3. Terminology

3.1 Definitions—For definitions of terms used but not defined in this test method, see Terminology D 1129.

4. Summary of Test Method (See Fig. 1)

4.1 For wastes containing less than 0.5 % dry solid material, the filtrate of the waste, after filtration through a 0.6 to 0.8-μm glass fiber filter, is defined as the method extract. Extraction of the solid is not required for such wastes.[8,9]

4.2 For wastes containing greater than or equal to 0.5 % dry solid material, the liquid, if any, is separated from the solid phase and stored for later analysis. The solid phase is extracted with an amount of extraction fluid equal to 20 times the weight of the solid phase. The extraction fluid used is a function of the alkalinity of the solid phase of the waste. Following extraction,

---

[1] This test method is under the jurisdiction of ASTM Committee D34 on Waste Disposal and is the direct responsibility of Subcommittee D34.01.06 on Analytical Methods.

Current edition approved March 15, 1992. Published May 1992.

[2] *Annual Book of ASTM Standards*, Vol 04.03.

[3] *Annual Book of ASTM Standards*, Vol 04.08.

[4] *Annual Book of ASTM Standards*, Vol 11.01.

[5] *Annual Book of ASTM Standards*, Vol 05.05.

[6] *Annual Book of ASTM Standards*, Vol 14.02.

[7] *Annual Book of ASTM Standards*, Vol 11.04 (see 1991 edition).

[8] *Federal Register*, Vol 55, No. 61, March 29, 1990. Toxicity Characteristics Revisions, Final Rule.

[9] *Federal Register*, Vol 55, No. 126, June 29, 1990. Toxicity Characteristic Revisions, Final Rule, Corrections.

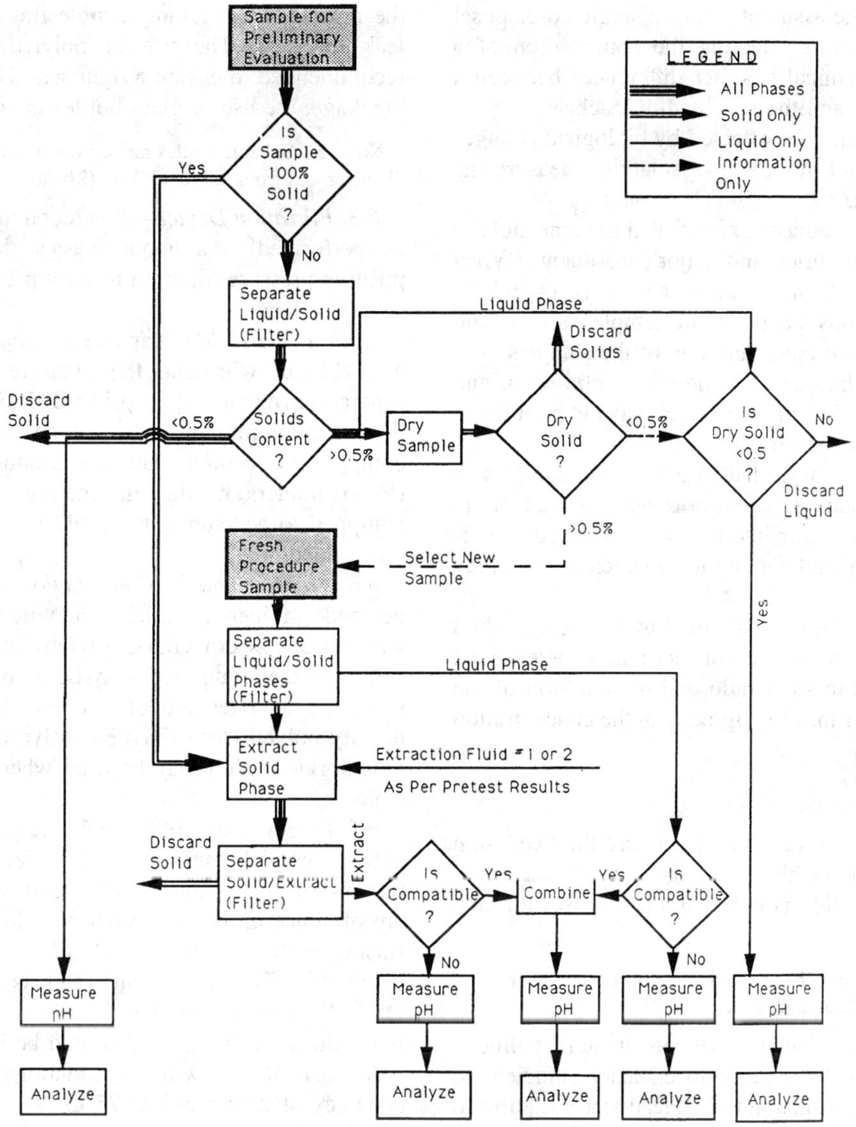

FIG. 1 Method Flow Chart

the liquid extract is separated from the solid phase by filtration through a 0.6 to 0.8-μm glass fiber filter.

4.3 If compatible (that is, multiple phases will not form upon combination), the initial liquid phase of the waste is added to the liquid extract, and these are analyzed together. If incompatible, the liquids are analyzed separately and the results are combined mathematically to yield a volume-weighted average concentration.

5. Significance and Use

5.1 This test method is intended to generate an extract with a concentration of the target analyte(s) representative of the expected release under the scenario simulated, and which can be compared with concentration levels acceptable in waste disposal, treatment, or production activities.

5.2 The extraction conditions of the test method were chosen to simulate a potential disposal scenario to which the wastes may be exposed.

5.3 One intent of this test method is that the amount of acid in the extraction fluids reflect the acid available from the leachate of a specific landfill where municipal and industrial wastes were co-disposed.[10]

5.4 One intent of this test method is to not allow the pH of the extraction fluid to be lower than that of the leachate of a specific landfill where municipal and industrial wastes were co-disposed. Therefore, the pH of the extraction fluid was chosen with the following considerations:

(1) Not to be less than 4.93 ± 0.05 for the extraction of wastes with an acid neutralization capacity of less than the acid available in the total volume of extraction fluid used in the method (Extraction Fluid No. 1).

(2) At 2.88 ± 0.05, as defined by the pH of the acid, for the extraction of wastes with an acid neutralization capacity of more than the acid available in the extraction fluid used in the method (Extraction Fluid No. 2).

5.5 The interpretation and use of the results of this test

[10] Kimmel, T. A., and Friedman, D. A., "Model Assumptions and Rationale Behind the Development of EP III," *ASTM STP 886*, J. K. Petros, et al, Eds., ASTM, Philadelphia, PA, 1986, pp. 36–53.

method are limited by the assumptions of a single co-disposal scenario and by the factors affecting the composition of a landfill leachate and chemical or other differences between a selected extraction fluid and the real landfill leachate.

5.6 This test method may be affected by biological changes in the waste, and it is not designed to isolate or measure the effect of such processes.

5.7 This test method produces extracts that are amenable to the determination of both minor and major constituents. When minor constituents are being determined, it is especially important that precautions be taken in sample storage and handling to avoid possible contamination of the samples.

5.8 The agitation technique, rate, liquid-to-solid ratio, and filtration conditions specified in the method may not be suitable for extracting all types of wastes.

5.9 This test method is intended to extract the samples in their original physical state as is, without any size reduction. However, the sample/extractor interaction is expected to correlate with the environmental conditions to which a waste may be exposed.[11]

5.10 The extraction conditions defined by this test method are expected to yield steady-state concentrations, determined by the extraction liquid-to-solid ratio and the duration of the extraction, which may or may not agree with the concentration of an equilibrium.

6. Apparatus and Materials

6.1 *Agitation Apparatus*, capable of rotating the extraction vessel in an end-over-end fashion (see Fig. 2), at 30 ± 2 r/min, such that the axis of rotation is horizontal and passes through the center of the bottle.

NOTE 1—Similar devices may be used having a different axle arrangement if equivalency can be demonstrated.

6.2 *Extraction Vessel*—Suitable vessels include cylindrically shaped, minimum 2-L size, with capacity sufficient to hold the sample and the extraction fluid. Head-space is allowed in this vessel. The extraction bottles may be constructed from various plastic materials, depending on the analytes of interest and the nature of the waste. Plastic bottles, other than polytetrafluoroethylene, shall not be used if organics are to be investigated. The bottles should be sturdy, in order to withstand the impact of the falling sample fragments, and shall have a leak-free seal. The use of polytetrafluoroethylene tape is recommended to ensure a tight seal. Due to their potential for breakage, the use of glass bottles is not recommended.

NOTE 2—Suitable bottles range from 4.0 to 4.5 in. (102 to 114 mm) in diameter and from 8.5 to 13.0 in. (216 to 330 mm) in height.

6.3 *Filtration Device*—It is recommended that all filtrations be performed in a hood. Wastes should be filtered using positive-pressure filtration using a pre-purified grade inert gas such as nitrogen.

6.3.1 *Filter Holder*, capable of supporting a glass fiber filter and able to withstand the pressure needed to accomplish separation (maximum 50 psi or 345 kPa). These devices shall have a minimum internal volume of 300 mL and shall be equipped to accommodate a minimum filter size of 47 mm. (Filter holders having an internal capacity of 2.2 L and equipped to accommodate a 142-mm diameter filter are recommended.)

6.3.1.1 *Materials of Construction*—Filtration devices shall be made of inert materials that will not leach or adsorb the analytes of concern. Glass, polytetrafluoroethylene, or type 316 stainless steel equipment may be used when both organic and inorganic analytes are of concern. Devices made of high-density polyethylene (HDPE), polypropylene (PP), or polyvinylchloride (PVC) may be used when only inorganics are of concern.

6.4 *Filters*, made of borosilicate glass fiber, containing no binder materials, and having an effective pore size of 0.6 to 0.8μ m. Pre-filters must not be used. When inorganic analytes are of concern, the filter shall be acid washed prior to use by rinsing with 1 N nitric acid followed by three consecutive rinses with Type II reagent water as defined in Specification D 1193. (A minimum of 1 L per rinse is recommended.) Glass fiber filters are fragile and should be handled with care.

6.5 *pH Meter*, with a readability of 0.01 unit and an accuracy of ±0.05 unit at 25°C.

6.6 *Laboratory Balance*, accurate to within ±0.01 g. (All weight measurements are to be within± 0.1 g.)

6.7 *Beakers or Erlenmeyer Flasks*, glass 500-mL, and 2-L.

6.8 *Watch-Glass*, with an appropriate diameter to cover the beaker or Erlenmeyer flask.

6.9 *Magnetic Stirrer.*

6.10 *Mold*, cylindrical, made of inert, non-adsorbing and non-contaminating material for casting of laboratory samples.

6.11 *Straightedge*, made of stainless steel.

6.12 *Impermeable Sheet or Glazed Paper.*

6.13 *Volumetric Flask*, 1-L size.

6.14 *Drying Oven*—Any thermostatically controlled drying oven capable of maintaining a temperature between 85 and 115°C within ±5°C.

6.15 *Graduated Pipet*, readable to 0.1 mL.

6.16 *Hot Plate*, equipped for agitation and temperature control capable of maintaining a 50 ± 3°C temperature.

6.17 *Graduated Measuring Cylinder*, with a precision of ±3 %.

7. Reagents

7.1 *Purity of Reagents*—Reagent grade chemicals shall be

[11] *Federal Register*, Vol 53, No. 100, May 24, 1988. Proposed Cage Modification of TCLP.

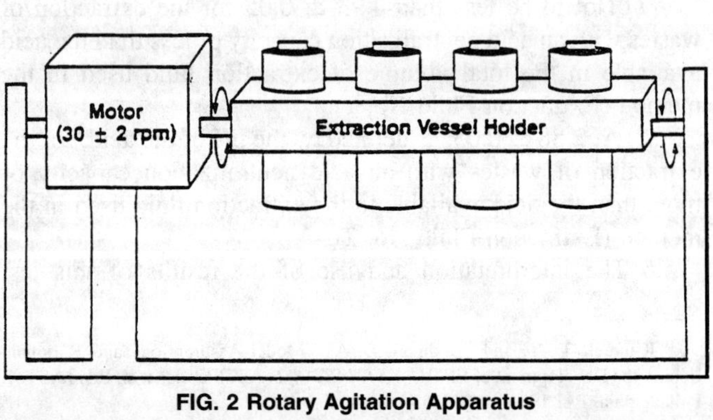

FIG. 2 Rotary Agitation Apparatus

used in all tests. Unless otherwise indicated, it is intended that all reagents shall conform to the specifications of the Committee on Analytical Reagents of the American Chemical Society, where such specifications are available.[12] Other grades may be used, provided it is first ascertained that the reagent is of sufficiently high purity to permit its use without lessening the accuracy of the determination.

7.2 *Purity of Water*—Reagent water is defined as water in which an interfering analyte is not observed at or above the method detection limit of the analyte(s) of interest. Type II of Specification D 1193 or equivalent meets the definition of reagent water.

7.3 *Hydrochloric Acid* (HCl), 1 N, made from ACS reagent grade.

7.4 *Nitric Acid* (HNO_3), 1 N, made from ACS reagent grade.

7.5 *Sodium Hydroxide* (NaOH), 1 N, made from ACS reagent grade.

7.6 *Glacial Acetic Acid* (CH_3COOH), ACS reagent grade.

7.7 *Extraction Fluids*—Several batches or multiple volumes of extraction fluids should be prepared in accordance with the number of extractions. The volume needed for an individual extraction is approximately 2 L. The extraction fluids should be monitored frequently for impurities. The pH should be examined prior to extraction to ensure that these fluids were made up accurately. If impurities are found or the pH is not within the specifications, the fluid shall be discarded and fresh extraction fluid prepared.

7.7.1 *Extraction Fluid* No. 1—Add 5.7 mL glacial acetic acid to 500 mL of reagent water, add 64.3 mL of 1 N NaOH, and dilute to a volume of 1 L. When correctly prepared, the pH of this fluid will be 4.93 ± 0.05.

7.7.2 *Extraction Fluid* No. 2—Dilute 5.7 mL glacial acetic acid with reagent water to a volume of 1 L. When correctly prepared, the pH of this fluid will be 2.88 ± 0.05.

8. Sampling

8.1 If representative samples of the waste must be tested, use ASTM Sampling methods developed for the specific industry where available (see Practices D 75, D 420, D 3370, Terminology D 653, and Method D 2234).

8.2 All samples shall be collected using an appropriate sampling plan to ensure sample integrity and representativeness (see Practice E 122).

8.3 Where no specific methods are available, sampling methodology for materials of similar physical form shall be used.

8.4 It is important that the sample of the waste be representative with respect to surface area, as variations in surface area would directly affect the extraction characteristics of the sample. Waste samples should contain a representative distribution of particle sizes.

NOTE 3—Information on obtaining representative samples can also be found in *Pierre Gy's Sampling Theory and Practice*.[13]

8.5 Approximately 100 g of solid phase samples are required for each extraction. Preliminary evaluation also requires 100 g of solid phase sample. A larger sample size may be required, depending on the solids content of the waste sample (percent solids; see 10.2.9).

8.6 Enough extract should be generated such that the volume will be sufficient to support all of the analyses required. If the volume of extract generated by the performance of a single extraction will not be sufficient to perform all of the analyses to be conducted, it is recommended that more than one extraction be performed and that the extracts from each extraction be combined and then aliquoted for analysis.

8.7 For the evaluation of solidified or stabilized wastes, or both, samples may be cast in the form of a cylinder that will fit into the extraction apparatus. Such cylinders may be used for the evaluation. The casting may be allowed to cure for 30 days before the extraction procedure is performed. For other monolithic materials, a coring may be produced that will fit into the extraction apparatus. Waste materials to which these casting and coring procedures apply include concrete materials, rock, wood, slag, and so forth.

8.8 Quality control measures may require additional samples.

9. Sample Handling and Preparation

9.1 For free-flowing particulate solid wastes, obtain a sample in accordance with the requirements of Section 8 by quartering the sample received for testing on an impermeable sheet of glazed paper, or other flexible non-contaminating material, as follows:

9.1.1 Empty the sample container into the center of the sheet.

9.1.2 Flatten out the sample gently with a suitable straightedge until it is spread uniformly to a depth at least twice the maximum particle diameter.

9.1.3 Remix the sample by lifting a corner of the sheet and drawing low across to the opposite corner in the manner that the material is made to roll over and over and does not merely slide along. Return that corner to its original position. Continue the operation with each corner, proceeding in a clockwise direction. Repeat this cycle ten times.

9.1.4 Lift all four corners of the sheet toward the center, and holding all four corners together, raise the entire sheet into the air to form a pocket for the sample.

9.1.5 Repeat 9.1.2.

9.1.6 Gently divide the sample into quarters with a straightedge, one at least as long as the flattened mound of sample. Make an effort to avoid using pressure on the straightedge sufficient to cause damage to the particles.

9.1.7 Discard alternate quarters.

9.1.8 If further reduction of the sample volume is necessary, repeat 9.1.3 through 9.1.7. Use a sample volume to give 100 g of solid for each extraction. Provide additional samples for the preliminary evaluation.

[12] "Reagent Chemicals, American Chemical Society Specifications," Am. Chemical Soc., Washington, DC. For suggestions on the testing of reagents not listed by the American Chemical Society, see "Reagent Chemicals and Standards," by Joseph Rosin, D. Van Nostrand Co., Inc., New York, NY, and the "United States Pharmacopeia."

[13] Pitard, F., *Pierre Gy's Sampling Theory and Practice*, Volumes I and II, CRC Press, 1989.

9.2 For field-cored solid wastes or castings produced in the laboratory, cut a representative section weighing approximately 100 g for each extraction plus the preliminary evaluation. Take samples for the preliminary evaluation at the same time as the test samples.

9.2.1 If necessary, shape the sample so that its largest dimension would not exceed the radius of the extraction bottle. (The material shall move freely while fully covered with the extraction fluid.)

9.3 Preservatives shall not be added to the samples prior to extraction.

9.4 For multi-phasic wastes, mix thoroughly to ensure that a representative sample will be withdrawn.

9.5 Samples shall be stored at 4°C to minimize changes due to biological processes. If precipitation occurs, the entire sample (including precipitate) of the precipitate existing at room temperature (see 4.1 and 4.2) should be used.

9.6 The method filtrates and extracts should be prepared for analysis and analyzed as soon as possible. Filtrates and extracts or their portions for metallic analyte determinations should be acidified with nitric acid to pH <2 unless precipitation occurs. To minimize losses, filtrates or extracts or their portions for organic contaminant determinations shall not be allowed to make contact with the atmosphere (that is, head-space).

9.7 Sample maximum holding times (days) are given in Table 1.

10. Preliminary Evaluations and Pre-Extraction Procedures

10.1 Perform preliminary method evaluations on a minimum 100-g aliquot of waste. This aliquot may not undergo the actual extraction. These preliminary evaluations include the determination of the following: (*1*) percent solids, (*2*) whether the waste contains dry solids in excess of 0.5 %, and (*3*) which of the two extraction fluids are to be used for extraction of the waste.

10.2 *Preliminary Determination of Percent Solids*—Percent solids is defined as that fraction of the waste sample (as a percent of the total sample w/w) from which no liquid may be forced out by an applied pressure as described below.

10.2.1 If the waste will obviously yield no free liquid when subjected to pressure filtration of this method (that is, 100 % solids), proceed to 10.4.

NOTE 4—Some materials may look like dry solids but may release liquids under pressure, for example, adsorbents, filter cakes, paint, and other sludges. If uncertain, proceed to the filtration step (10.2.2).

10.2.2 If the sample is liquid or multi-phasic, liquid/solid separation to make a preliminary evaluation of percent solids is required. This involves the filtration device described in 6.3, and the procedure is outlined in 10.2.3 through 10.2.9.

10.2.3 Pre-weigh the filter and the container that will receive the filtrate.

10.2.4 Assemble the filter holder and filter following the manufacturer's instructions. Place the filter on the support screen and secure.

10.2.5 Weigh out a sub-sample of the waste (100-g minimum) and record the weight.

10.2.6 Allow slurries to stand to permit the solid phase to settle. Wastes that settle slowly may be centrifuged prior to filtration. Centrifugation is to be used only as an aid to filtration. If used, the liquid should be decanted and filtered followed by filtration of the solid portion of the waste through the same filtration system.

10.2.7 Quantitatively transfer the waste sample to the filter holder (liquid and solid phases). Spread the waste sample evenly over the surface of the filter. Allow the sample to warm to room temperature in the device before filtering.

NOTE 5—If some waste material (>1 % of the original sample weight) has obviously adhered to the container used to transfer the sample to the filtration apparatus, determine the weight of this residue and subtract it from the sample weight determined in 10.2.5 to determine the weight of the waste sample that will be filtered.

10.2.7.1 Gradually apply gentle pressure of 1 to 10 psi (7 to 70 kPa), until the pressurizing gas moves through the filter. If this point is not reached below 10 psi (69 kPa), and if no additional liquid has passed through the filter in any 2-min interval, slowly increase the pressure in 10-psi (69-kPa) increments to a maximum of 50 psi (345 kPa). After each incremental increase of 10 psi (69 kPa), if the pressurizing gas has not moved through the filter, and if no additional liquid has passed the filter in any 2-min interval, proceed to the next 10-psi (69-kPa) increment. When the pressurizing gas begins to move through the filter, or when the liquid flow has ceased at 50 psi (345 kPa) (that is, filtration does not result in any additional filtrate within any 2-min period), stop the filtration.

NOTE 6—Instantaneous application of high pressure can degrade the glass fiber filter and may cause premature plugging.

10.2.8 The material in the filter holder is defined as the solid phase of the waste, and the filtrate is defined as the liquid phase.

NOTE 7—Some wastes, such as oily and some paint wastes, will obviously contain some material that appears to be liquid. Even after applying the pressure filtration as outlined in 10.2.7, this material may not filter. If this is the case, the material within the filtration device is defined as solid. Do not replace the original filter under any circumstances. Use only one filter.

10.2.9 Determine the weight of the liquid phase by subtracting the weight of the filtrate container (see 10.2.3) from the total weight of the container plus filtrate. Determine the weight of the solid phase of the waste sample by subtracting the

TABLE 1 Sample Maximum Holding Times, Days

Period Compound	From Field Collection to Method Extraction	From the End of Extraction to the Start of Filtration, h	From Method Extraction to Analytical Extraction	From the Analytical Extraction to the Chemical Analysis	Total Time, Days
Organics	14	2	7	40	61
Mercury	28	2	NA	28	56
Inorganics except Mercury	180	2	NA	180	360

weight of the liquid phase from the weight of the total waste sample, as determined in 10.2.5 or 10.2.7. Record the weight of the liquid and solid phases. Calculate the percent solids as follows:

$$\text{percent solids} = \frac{\text{weight of solid (10.2.9)}}{\text{total weight of waste (10.2.5 or 10.2.7)}} \times 100 \quad (1)$$

10.3 If the percent solids determined in 10.2.9 is equal to or greater than 0.5 %, proceed with the filtered solid phase to 10.3.1. If the percent solids determined in 10.2.9 is less than 0.5 %, the initial liquid phase becomes the method extract. Measure the pH and proceed with this method extract to 11.16.

10.3.1 Remove the solid phase and filter from the filtration apparatus.

10.3.2 Dry the filter and solid phase at 100 ± 20°C until two successive weighings yield the same value within 1 % of the last measurement. Record the final weight.

NOTE 8—**Caution:** Take care to ensure that the subject solid will not flash upon heating. It is recommended that the drying oven be vented to a hood or another appropriate device.

10.3.3 Calculate the percent dry solids as follows:

$$\text{percent dry solids} = \frac{(\text{weight of dry waste + filter}) - \text{tared weight of filter}}{\text{initial weight of waste (10.2.5 or 10.2.7)}} \quad (2)$$

10.3.4 If the percent dry solids is less than 0.5 %, the initial liquid phase becomes the method extract. Proceed with the extract to 11.16. Measure the pH of the method extract. If the percent dry solids is greater than or equal to 0.5 %, return to the beginning of this section and filter a small, fresh portion of the waste to determine the appropriate extraction fluid (10.4). Since the original sample was destroyed by drying and the preliminary evaluation has been completed, proceed to 11.3 to perform the extraction procedure on a fresh sample.

10.4 *Determination of Appropriate Extraction Fluid* (See Fig. 3)—If the dry solids content of the waste is greater than or equal to 0.5 %, perform the determination of the appropriate fluid (7.7) to be used for the extraction as follows:

10.4.1 Weigh out a small sub-sample of the solid phase of the waste; reduce the solid, if necessary, to a fragment size of approximately 1 mm in diameter or less; and transfer 5.0 g of the solid phase of the waste to a 500-mL beaker or an Erlenmeyer flask.

10.4.2 Add 96.5 mL of reagent water to the beaker, cover with a watch-glass, and stir vigorously for 5 min using a magnetic stirrer. Measure and record the pH. If the pH is <5.0, use Extraction Fluid No. 1. Proceed to Section 11.

10.4.3 If the pH from 10.4.2 is >5.0, add 3.5 mL 1 N HCl, slurry briefly, cover with a watch-glass, heat to 50°C, and hold at 50°C for 10 min.

10.4.4 Let the solution cool to room temperature and record the pH. If the pH is <5.0, use Extraction Fluid No. 1. If the pH is >5.0, use Extraction Fluid No. 2. Proceed to Section 11.

10.5 If the aliquot of the waste used for the preliminary evaluation (10.2 through 10.4) was determined to be 100 % solids at 10.2.1, it can be used for Section 11 extraction (assuming that at least 100 g remains). The aliquot subjected to the procedure in 10.2.7 might be appropriate for use in Section 11 if an adequate amount of solids (as determined by 10.2.9) was obtained. The amount of solids necessary is also dependent on whether a sufficient amount of extract will be produced to support the analyses for the target analytes. If an adequate amount of solids remains, proceed to 11.11.

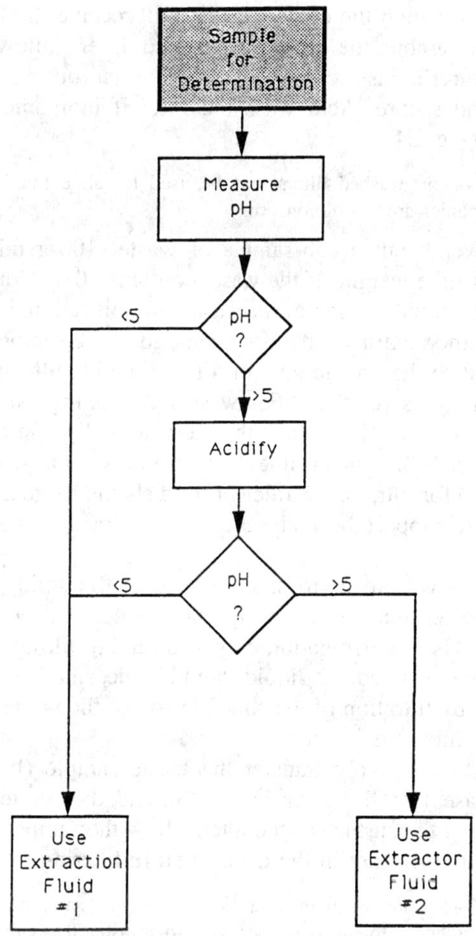

FIG. 3 Determination of the Extraction Fluid

11. Extraction

11.1 A sample size of minimally 100 g (solid and liquid phases) is required. A larger sample size may be appropriate in some cases, depending on the solid contents of the waste sample (percent solids; see 10.2); whether the initial liquid phase (filtrate) will be miscible with the aqueous extract of the solid; and whether inorganics, semivolatile organics, pesticides and herbicides are all analytes of concern. If the amount of extract generated by a single extraction will not be sufficient to perform all of the analyses, more than one extraction may be performed and the extracts from each combined and aliquoted for analysis.

11.2 If the waste will obviously yield no liquid when subjected to pressure filtration (that is, is 100 % solid; 10.2), weigh out a sub-sample of the waste (100 g minimum) and proceed to 11.10.

11.3 If the sample is liquid-like or multi-phasic, liquid-solid separation is required. This involves the filtration device described in 6.3 and is outlined in 11.4 through 11.9.

11.4 Pre-weigh the container that will receive the filtrate.

11.5 Assemble the filter holder and filter following the manufacturer's instructions. Place the filter on the support screen and secure. Acid wash the filter if inorganics are of concern (see 6.4).

NOTE 9—Acid washed filters may be used for all extractions, even when inorganics are not of concern.

11.6 Weigh out a sub-sample of waste (100-g minimum) and record the weight. If the waste contains <0.5 % dry solids (10.3), the liquid portion of the waste, after filtration, is defined as the method extract. Therefore, enough of the sample should be filtered so that the amount of filtered liquid will support all of the analyses required. For wastes containing >0.5 % dry solids (10.2 or 10.3), use the percent solids information obtained in 10.2 to determine the optimum sample size (100-g minimum) for filtration. Sufficient solid should be generated by filtration to support the analyses to be performed on the method extract.

11.7 Allow slurries to stand to permit the solid phase to settle. Wastes that settle slowly may be centrifuged prior to filtration. Use centrifugation only as an aid to filtration. If the waste is centrifuged, the liquid should be decanted and filtered followed by filtration of the solid portion of the waste through the same filtration system.

11.8 Quantitatively transfer the waste sample (liquid and solid phases) to the filter holder. Spread the waste sample evenly over the surface of the filter. Allow the sample to warm to room temperature in the device before filtering.

NOTE 10—If some waste material (>1 % of original sample weight) has obviously adhered to the container used to transfer the sample to the filtration apparatus, determine the weight of this residue and subtract it from the sample weight determined in 10.2.5 to determine the weight of the waste sample that will be filtered.

11.8.1 Gradually apply gentle pressure of 1 to 10 psi (7 to 70 kPa), until the pressurizing gas moves through the filter. If this point is not reached below 10 psi (69 kPa), and if no additional liquid has passed through the filter in any 2-min interval, slowly increase the pressure in 10-psi (69-kPa) increments to a maximum of 50 psi (345 kPa). After each incremental increase of 10 psi (69 kPa), if the pressurizing gas has not moved through the filter, and if no additional liquid has passed the filter in any 2-min interval, proceed to the next 10-psi (69-kPa) increment. When the pressurizing gas begins to move through the filter, or when the liquid flow has ceased at 50 psi (345 kPa) (that is, filtration does not result in any additional filtrate within any 2-min period), stop the filtration.

NOTE 11—Instantaneous application of high pressure can degrade the glass fiber filter and may cause premature plugging.

11.9 The material in the filter holder is defined as the solid phase of the waste, and the filtrate is defined as the liquid phase. Measure the pH of the filtrate. Measure the volume of the filtrate (V_1) if the data are to be combined mathematically. Use a graduated measuring cylinder for the volume measurement. The liquid phase may now be analyzed (see 11.13) or stored at 4°C until the time of analysis.

NOTE 12—Some wastes, such as oily and some paint wastes, will obviously contain some material that appears to be liquid. Even after applying the pressure filtration as outlined in 10.2.7, this material may not filter. If this is the case, the material within the filtration device is defined as solid. Do not replace the original filter under any circumstances. Use only one filter.

11.10 If the waste contains <0.5 % dry solids (see 10.3), proceed to 11.14. If the waste contains >0.5 % solids (see 10.2 or 10.3), proceed to 11.11.

11.11 Quantitatively transfer the solid residue retained by the filter, or the solid sample if it did not require filtration, into an extractor bottle. Include the filter if it was used to separate the initial liquid from the solid phase.

11.12 Determine the amount of extraction fluid to add to the extractor as follows:

$$\text{weight of extraction fluid} = \frac{20 \times \text{percent solids (10.2.9)} \times \text{weight of waste filter (11.6 or10.2.9)}}{100} \quad (3)$$

Slowly add this amount of appropriate extraction fluid (see 10.4) to the extractor vessel. Close the extractor bottle tightly (it is recommended that polytetrafluoroethylene tape be used to ensure a tight seal), secure in a rotary agitation device, and rotate at 30 ± 2 r/min for 18± 2 h. Ambient temperature (that is, temperature of the room in which the extraction takes place) shall be maintained at 23 ± 2°C during the extraction period.

NOTE 13—As agitation continues, pressure may build up within the extractor bottle for some types of wastes (for example, limed or calcium carbonate-containing waste may evolve gases such as carbon dioxide). To relieve excess pressure, the extractor bottle may be taken under a hood and opened carefully from time to time (for example, after 15 min, 30 min, and 1 h).

11.13 Within 2 h, following the 18 ± 2 h extraction, initiate the separation of the material in the extraction vessel into its component liquid and solid phases by filtering through a new glass fiber filter, as outlined in 11.8. For final filtration of the method extract, the glass fiber filter may be changed, if necessary, to facilitate filtration. Filter(s) shall be acid-washed (see 6.4) if inorganics are of concern.

11.14 Prepare the method extract as follows:

11.14.1 If the waste contained no initial liquid phase, the filtered liquid material obtained from 11.13 is defined as the method extract. Proceed to 11.15.

11.14.2 If compatible (for example, multi-phase waste will not result upon combination), combine the filtered liquid resulting from 11.13 with the initial liquid phase of the waste obtained in 11.8. This combined liquid is defined as the method extract. Proceed to 11.15.

11.14.3 If the initial liquid phase of the waste, as obtained from 11.8, is not or may not be compatible with the filtered liquid resulting from 11.13, do not combine these liquids. Analyze these liquids, collectively defined as the method extract, and combine the results mathematically, as described in 11.15.

11.15 Following collection of the method extract, the pH of the extract should be recorded. The volume of the extract (V_2) shall be measured if the data are to be combined mathematically. Use a graduated measuring cylinder for volume measurement. Immediately aliquot and preserve the extract for analysis. Metals aliquots must be acidified with 1 N nitric acid to pH

<2. If precipitation is observed upon the addition of nitric acid to a small aliquot of the extract, the remaining portion of the extract for metals analyses shall not be acidified and the extract shall be analyzed as soon as possible. All other aliquots must be stored under refrigeration (4°C) until analyzed. The method extract shall be prepared and analyzed according to appropriate analytical methods. The method extracts to be analyzed for metals shall be acid-digested except in those instances in which digestion causes a loss of metallic contaminants. If an analysis of the undigested extract reveals that the concentration of any regulated metallic contaminant exceeds the acceptance level, the waste fails the test and digestion of the extract is not necessary. However, data on undigested extracts alone cannot be used to demonstrate that the waste met the set acceptance level. If the individual phases are to be analyzed separately, determine the volumes of the individual phases (to ±3 %), using a separatory funnel and graduated cylinder. Conduct the appropriate analyses, and combine the results mathematically by using a simple volume-weighted average:

$$\text{final analyte concentration} = \frac{(V_1)(C_1) + (V_2)(C_2)}{V_1 + V_2} \quad (4)$$

where:
V_1 = volume of the initial phase (filtrate, L),
C_1 = concentration of the contaminant of concern in the initial phase (mg/L),
V_2 = volume of the extract (L), and
C_2 = concentration of the analyte of concern in the extract (mg/L).

NOTE 14—If either V_1 or V_2 were not detected, use the appropriate detection limit in the calculation.

11.16 Compare the analyte concentration of the method extract with the applicable comparative values. Refer to Section 13 for quality assurance requirements.

12. Report

12.1 Report the following information:

12.1.1 All pH values for the method extract(s);

12.1.2 The concentration of all analytes of concern as measured or calculated in 11.6 and 11.15, respectively;

12.1.3 The type of the extraction fluid used;

12.1.4 The solids content of the sample; and

12.1.5 The dry solids content of the sample if the solids content is > 0.5 %.

13. Quality Assurance Requirements

13.1 Maintain all data, including quality assurance data, and keep them available for reference or inspection.

13.2 All quality control measures described in the appropriate analytical methods shall be followed. If the analytical quality control requirements are not specified in the appropriate method, follow the requirements in 13.3 and 13.4 and refer to Practice ES 16.

13.3 A minimum of one blank (using the same extraction fluid and equipment as used for the samples) for every ten extractions that have been conducted shall be used as a check to determine whether any memory effects from the extraction equipment are occurring.

13.4 A matrix spike shall be performed for each waste type where the compositions of the waste matrices are significantly different.

NOTE 15—It is recommended that it be assumed that each sample has a significantly different composition unless previous data would indicate otherwise (for example, regular sampling of steady-state process streams). If more than 20 samples of the same waste are being tested, a matrix spike must be performed for every 20 samples.

13.4.1 The matrix spikes are to be added after filtration and combination, if necessary, of the method extract and before preservation.

13.4.2 Matrix spike levels should be established at the appropriate acceptance levels. If the analyte concentration in the method extract is less than one half of the acceptance level, the spike level may be as low as one-half the analyte concentration. However, it shall not be less than the quantitation limit or one-fifth of the acceptance level. In order to avoid differences in the matrix effects, the matrix spikes must be added to the same nominal volume of the method extract as that which was analyzed for the unspiked sample.

13.4.3 The purpose of the matrix spike is to monitor the performance of the analytical method used and to determine whether matrix interference exists. The use of other internal calibration methods, modifications of the analytical methods, or the use of alternate analytical methods may be needed to measure accurately the analyte concentration of the method extract when recovery of matrix spike is below the expected analytical method performance.

13.4.4 Acceptable sample holding times are outlined in 9.7. Exceeding the holding time is not acceptable to establish any compliance with the acceptance levels, but it may be used if the samples are not in compliance.

14. Precision and Bias

14.1 *Precision*—The precision of the procedure in Test Method D 5233 for measuring the sample disintegration, surface area increase, has been evaluated. The fractions retained on 9.5 mm sieve were measured in three laboratories. In each laboratory, duplicate, monolithic samples not passing a 9.5 mm sieve of four high strength and four low strength materials were extracted (see Table 2). The weight of the residual 9.5-mm size fractions was measured and expressed as a percent of the initial sample weight.

14.2 *Bias*—The procedure in Test Method D 5233 for measuring extract generation has no bias because the value of extract is defined only in terms of this method.

15. Keywords

15.1 batch; extraction; laboratory; leaching; single; sludge; solid; solidified; testing; waste

TABLE 2 Precision Data

	High Strength	Low Strength
Mean, %	21.0	1.9
Standard deviation,%	3.0	0.63

Designation: D 5283 – 92 (Reapproved 1997)

Standard Practice for Generation of Environmental Data Related to Waste Management Activities: Quality Assurance and Quality Control Planning and Implementation[1]

This standard is issued under the fixed designation D 5283; the number immediately following the designation indicates the year of original adoption or, in the case of revision, the year of last revision. A number in parentheses indicates the year of last reapproval. A superscript epsilon (ϵ) indicates an editorial change since the last revision or reapproval.

1. Scope

1.1 Environmental data generation efforts are composed of four parts: (*1*) establishment of data quality objectives (DQOs); (*2*) design of field measurement and sampling strategies and specification of laboratory analyses and data acceptance criteria; (*3*) implementation of sampling and analysis strategies; and (*4*) data quality assessment. This practice addresses the planning and implementation of the sampling and analysis aspects of environmental data generation activities (Parts (*1*) and (*2*) above).

1.2 This practice defines the criteria that must be considered to assure the quality of the field and analytical aspects of environmental data generation activities. Environmental data include, but are not limited to, the results from analyses of samples of air, soil, water, biota, waste, or any combinations thereof.

1.3 DQOs should be adopted prior to application of this practice. Data generated in accordance with this practice are subject to a final assessment to determine whether the DQOs were met. For example, many screening activities do not require all of the mandatory quality assurance (QA) and quality control (QC) steps found in this practice to generate data adequate to meet the project DQOs. The extent to which all of the requirements must be met remains a matter of technical judgement as it relates to the established DQOs.

1.4 This practice presents extensive management requirements designed to ensure high-quality environmental data. The words "must,"" shall," "may," and "should" have been selected carefully to reflect the importance placed on many of the statements made in this practice.

1.5 *This standard does not purport to address all of the safety concerns, if any, associated with its use. It is the responsibility of the user of this standard to establish appropriate safety and health practices and determine the applicability of regulatory limitations prior to use.*

NOTE 1—A complete table of contents of this practice is given in Appendix X1.

2. Referenced Documents

2.1 *ASTM Standards:*

D 1129 Terminology Relating to Water[2]

E 1187 Terminology Relating to Laboratory Accreditation[3]

2.2 *U.S. Environmental Protection Agency Documents:*[4]

SW-846, *Test Methods for Evaluating Solid Waste*, Vol 1, Third Edition (NTIS No. PB88239223/LL), November 1986

QAMS-005/80 (NTIS No. PB83170514/LL), *Interim Guidelines and Specifications for Preparing Quality Assurance Project Plans*, Office of Monitoring Systems and Quality Assurance, December 29, 1980

EPA/QAMS, *Development of Data Quality Objectives, Description of Stages I and II,* July 16, 1986

QAMS 004/80 (NTIS No. PB83219667/LL), *Guidelines and Specifications for Preparing Quality Assurance Program Plans*, Office of Monitoring Systems and Quality Assurance, September 20, 1980

2.3 Other documents related to the subject matter of this practice are cited in Appendix X2. This list is not intended to be comprehensive.

3. Terminology

3.1 *Definitions*—The terms most applicable to this practice have been defined in Terminologies D 1129 and E 1187.

3.2 *Definitions of Terms Specific to This Standard:*

3.2.1 *background sample*—a sample taken from a location on or proximate to the site of interest and used to document baseline or historical information.

3.2.2 *collocated samples*—independent samples collected as close as possible to the same point in space and time and intended to be identical.

3.2.3 *data quality objectives* (DQOs)—statements on the level of uncertainty that a decision maker is willing to accept in the results derived from environmental data (see EPA/QAMS, July 16, 1986).

---

[1] This practice is under the jurisdiction of ASTM Committee D34 on Waste Management and is the direct responsibility of Subcommittee D34.01.01 on Sampling.

Current edition approved Aug. 15, 1992. Published February 1993.

[2] *Annual Book of ASTM Standards*, Vol 11.01.

[3] *Annual Book of ASTM Standards*, Vol 14.02.

[4] Available from Superintendent of Documents, Government Printing Office, Washington, DC 20402.

3.2.4 *environmental data generation activity*— tasks associated with the production of environmental data, including planning, sampling, and analysis.

3.2.5 *equipment rinsate (equipment blank)*—a sample of analyte-free media that has been used to rinse the sampling equipment. This blank is collected after the completion of decontamination and prior to sampling and is useful for documenting the adequate decontamination of sampling equipment.

3.2.6 *field blank*—a sample of analyte-free media similar to the sample matrix that is transferred from one vessel to another or exposed to the sampling environment at the sampling site. This blank is preserved and processed in the same manner as the associated samples and is used to document contamination in the sampling and analysis process.

3.2.7 *field duplicates*—collocated samples that are analyzed independently and are useful in documenting the precision of the sampling and analytical process.

3.2.8 *laboratory control sample*—a known matrix spiked with compound(s) representative of the target analytes and used to document laboratory performance.

3.2.9 *material blank*—a sample composed of construction materials such as those used in well installation, well development, pump and flow testing, and slurry wall construction. Examples of these materials are bentonite, sand, drilling fluids, and source and purge water. This blank documents the contamination resulting from use of the construction materials.

3.2.10 *matrix duplicate*—an intralaboratory split sample used to document the precision of a procedure in a given sample matrix.

3.2.11 *matrix spike*—an aliquot of sample spiked with a known concentration of target analyte(s) and used to document the bias of an analytical process in a given sample matrix. The spiking occurs prior to sample preparation and analysis.

3.2.12 *matrix spike duplicates*—intralaboratory split samples spiked with identical concentrations of target analyte(s) and used to document the precision and bias of a procedure in a given sample matrix. The spiking occurs prior to sample preparation and analysis.

3.2.13 *method blank*—an analyte-free media, to which all reagents are added in the same volumes or proportions used in sample processing. The method blank must be carried through the complete sample preparation and analytical procedure and is used to document contamination resulting from the analytical process.

3.2.14 *project*—single or multiple data collection activities that are related through the same planning sequence.

3.2.15 *project planning documents*—all documents related to the definition of the environmental data collection activities associated with a project.

3.2.16 *quality assurance program plan (QAPP)*— an orderly assemblage of management policies, objectives, principles, and general procedures by which an organization involved in environmental data generation activities outlines how it intends to produce data of known quality.

3.2.17 *quality assurance project plan (QAPjP)*— an orderly assemblage of detailed procedures designed to produce data of sufficient quality to meet the DQOs for a specific data collection activity.

3.2.18 *reference material*—a material containing known quantities of target analytes in either solution or a homogeneous matrix and used to document the bias of the analytical process.

3.2.19 *split samples*—aliquots of sample taken from the same container and analyzed independently. These are usually taken after mixing or compositing and are used to document intra- or interlaboratory precision.

3.2.20 *standard addition*—the practice of adding a known amount of an analyte to a sample immediately prior to analysis, typically used to evaluate matrix effects.

3.2.21 *standard operating procedures (SOPs)*—the established written procedures of a given organization. Special project plans may require procedures different from the established SOPs.

3.2.22 *surrogate*—an organic compound that is similar to the target analyte(s) in chemical composition and behavior in the analytical process, but is not normally found in environmental samples.

3.2.23 *trip blank*—a sample of analyte-free media taken from the laboratory (or appropriate point of origin) to the sampling site and returned to the laboratory unopened. A trip blank is used to document the contamination attributable to shipping and field handling procedures and is also useful in documenting the contamination of volatile organics samples.

4. Summary of Practice

4.1 This practice describes the criteria and activities for field and laboratory organizations involved in generating environmental data in terms of human and physical resources, QA and QC procedures, and documentation requirements depending on the DQOs.

5. Significance and Use

5.1 Environmental data are often required for making regulatory and programmatic decisions. These data must be of known quality commensurate with their intended use.

5.2 Data generation efforts involve the following: establishment of the DQOs; design of the project plan to meet the DQOs; implementation of the project plan; and assessment of the data to determine whether the DQOs have been met.

5.3 Certain minimal criteria must be met by the field and laboratory organizations generating environmental data. Additional activities may be required based on the DQOs of the data collection effort.

5.4 This practice defines the criteria for field and laboratory organizations generating environmental data and identifies some other activities that may be required based on the DQOs.

5.5 This practice emphasizes the importance of communication among those involved in establishing DQOs, planning and implementing the sampling and analysis aspects of environmental data generation activities, and assessing data quality.

5.6 Environmental field operations are discussed in Section 7, and environmental laboratory operations are discussed in Section 8.

6. Project Specification

6.1 Project activities should be defined prior to the start of any field or laboratory activities. At a minimum, project specifications should address the following topics:

6.2 *Data Quality Objectives*—DQOs for the data generation activity should be defined prior to the initiation of field and laboratory work. It is desirable that the field and laboratory organizations be aware of the DQOs so that the personnel conducting the work are able to make informed decisions during the course of the project.

6.3 *Project Plan*— The project should be designed to meet the DQOs, and the project plan should define the following:

6.3.1 *Project Objectives*—Project objectives provide background information, state reasons for the data collection effort, identify any regulatory programs governing data collection, define specific objectives for each sampling location, and describe the intended uses for the data.

6.3.2 *Project Management*—A person(s) shall be designated as having responsibility and authority for the following: (*1*) developing project documents that implement the DQOs; (*2*) selecting field and laboratory organizations to conduct the work; (*3*) coordinating communication among the field and laboratory organizations and government agencies, as required; and (*4*) reviewing and assessing the final data.

6.3.3 *Sampling Requirements*—Sampling locations, equipment, and procedures and sample preservation and handling requirements shall be specified.

6.3.4 *Analytical Requirements*—The analytical procedures, analyte list, required detection limits, and required precision and bias values shall be specified. Regulatory requirements and DQOs shall be considered when developing the specifications.

NOTE 2—This does not imply that the specified analytical requirements can be met.

6.3.5 *Quality Assurance and Quality Control Requirements*—The QA and QC requirements shall address both field and laboratory activities. The means for controlling false positives and false negatives shall be specified. Standard practices for field and laboratory operations as described in Sections 7 and 8 of this practice shall be required.

6.3.5.1 *Field Quality Control*—The types and frequency of field QC samples to be collected, including field blanks, trip blanks, equipment rinsates, field duplicates, background samples, reference materials, material blanks, and split samples, shall be specified. Control parameters for field activities shall also be described (see 7.6.4).

6.3.5.2 *Laboratory Quality Control*—The types and frequency of use of laboratory QC samples, such as laboratory control samples, laboratory blanks, matrix spikes, matrix duplicates, and matrix spike duplicates, shall be specified. Any specific performance criteria shall be specified. Data validation criteria shall be defined.

6.4 *Project Documentation*—All documents required for planning, implementing, and evaluating the data collection effort shall be specified. These may include, although not limited to, a statement of work, technical and cost proposals, work plan, sampling and analysis plan, quality assurance project plan (QAPjP), health and safety plan, community relations plan, documents required by regulatory agencies, requirements for raw field and analytical records, technical reports assessing the environmental data, and records retention policy. Planning documents shall specify the required level of document control and identify the personnel having access. Document formats that may be required to ensure that all data needs are satisfied shall be specified. In addition, a project schedule that identifies critical milestones and completion dates should be available.

7. Standard Practices for Environmental Field Operations

7.1 *Purpose*—The field organization must conduct its operations in such a manner as to provide reliable information that meets the DQOs. To achieve this goal, certain minimum policies and procedures must be implemented in order to meet the DQOs.

7.2 *Organization*— The field organization shall be structured such that each member of the organization has a clear understanding of his or her duties and responsibilities and the relationship of those responsibilities to the total effort. The organizational structure, functional responsibilities, levels of authority, job descriptions, and lines of communication for activities shall be established and documented. One person may cover more than one organizational function.

7.2.1 *Management*—The management of the field organization is responsible for establishing organizational, operational, health and safety, and QA policies. Management shall ensure that the following requirements are met: (*1*) the appropriate methodologies are followed, as documented in the standard operating procedures (SOPs); (*2*) personnel clearly understand their duties and responsibilities; (*3*) each staff member has access to appropriate project documents; (*4*) any deviations from the project plan are communicated to project management; and (*5*) communication occurs between the field, laboratory, and project management, as specified in the project plan. Management shall foster an attitude within the organization that emphasizes the importance of quality and supports implementation of the quality assurance program plan (QAPP).

7.2.2 *Quality Assurance Function*—The organization shall appoint a person or persons to be responsible for monitoring field operations in order to ensure that the site facilities, equipment, personnel, procedures, practices, and documentation are in conformance with the organization's QAPP and any applicable QAPjP. The QA monitoring function should be entirely separate from, and independent of, personnel engaged in the work being monitored. The QA function shall be responsible for the QA review, as per 7.7.

7.2.3 *Personnel*—It is the responsibility of the organization to establish personnel qualifications and training requirements for all positions. Each member of the organization shall possess the education, training, technical knowledge, and experience, or a combination thereof, to enable that individual to perform his or her assigned functions. Personnel qualifications shall be documented in terms of education, experience, and training. Training shall be provided for all staff members, as necessary, so that they can perform their functions properly.

7.2.4 *Subcontractors*— The use of subcontractors shall not jeopardize data quality. Therefore, subcontractors shall comply with the requirements of Sections 7 and 8, as appropriate to the specific task(s) they are performing.

7.3 *Field Logistics*:

7.3.1 *General*—Sampling site facilities shall be examined prior to the start of work in order to ensure that all required items are available. The actual sampling area shall be examined to ensure that trucks, drilling equipment, and personnel have access to the site. Security, health and safety, and protection of the environment shall be controlled at the site support areas and sampling site.

7.3.2 *Field Measurements*—Project planning documents shall both address the type of field measurements to be performed and plan for the appropriate area to perform the work. Planning documents shall address ventilation, protection from extreme weather and temperatures, access to stable power, and provisions for water and gases of required purity. Plans shall be made to identify and supply applicable safety equipment, as specified in the project health and safety plan.

7.3.3 *Sample Handling, Shipping, and Storage Area*—The determination of whether sample shipping is necessary shall be made during project planning. This need is established by evaluating the analyses required, holding times, and location of the site and laboratory. Shipping or transporting of the samples to a laboratory shall be completed in a timely manner, ensuring that the laboratory is allowed sufficient time to perform its analysis within any required holding times.

7.3.3.1 Samples shall be packaged, labeled, and documented in an area that minimizes sample contamination and provides for safe storage. The level of custody and whether sample storage is required shall be outlined in the planning documents.

7.3.4 *Chemical Storage*— Safe storage areas for solvents, reagents, standards, and reference materials shall be adequate to preserve their identity, concentration, purity, and stability prior to use.

7.3.5 *Decontamination*— Decontamination of sampling equipment may be performed at the location at which sampling occurs, prior to transfer to the sampling site, or in designated areas near the sampling site. Project documentation shall specify where this work will be performed and how it will be accomplished. If decontamination is to be conducted at the site, water and solvents of appropriate purity shall be available. The method of accomplishing decontamination and the materials, solvents, and water purity shall be specified in planning documents or standard operating procedures (SOPs).

7.3.6 *Waste Storage Area*—Waste materials may be generated during both the sampling process and on-site or *in situ* analysis. Planning documents and SOPs shall outline the method for storage and disposal of these waste materials. Adequate facilities shall be provided for the collection and storage of all wastes. These facilities shall be operated so as to minimize environmental contamination. Waste storage and disposal facilities shall comply with applicable federal, state, and local regulations.

7.3.7 *Data Storage Area*—Planning documents shall specify the location of long- and short-term storage for field records. The storage environment shall be maintained to ensure the integrity of the data. Access shall be limited to authorized personnel only.

7.4 *Equipment and Instrumentation*:

7.4.1 *Equipment and Instrumentation*—The equipment, instrumentation, and supplies required at the sampling site shall be appropriate to accomplish the activities planned. The equipment and instrumentation shall meet the requirements of pertinent specifications, methods, and SOPs. Before the field staff arrives at the site, a list of required items shall be prepared and checked to ensure availability at the site.

7.4.2 *Maintenance and Calibration of Equipment and Instrumentation*—An SOP or operation and maintenance manual shall set forth the methods, materials, and schedules to be used in the routine inspection, cleaning, maintenance, testing, and calibration of the equipment and instrumentation used in performing geophysical, analytical, or *in situ* measurements. For common malfunctions, procedures or manuals may outline typical problems, methods of trouble-shooting, and possible corrective actions to be taken. Procedures shall designate a person(s) or organizations responsible for maintenance and calibration. Records of all inspections, maintenance, repairs, testing, and calibration shall be maintained.

7.5 *Standard Operating Procedures*—The organization shall have written SOPs for all procedures performed routinely that affect data quality. SOPs shall be available for the following areas and shall contain the information described:

7.5.1 *Sample Management*—These SOPs describe the numbering and labeling system, chain-of-custody procedures, and tracking of samples from collection to shipment or relinquishment to the laboratory. Sample management also includes the specification of holding times, volume of sample required by the laboratory, preservatives, and shipping requirements.

7.5.2 *Reagent and Standard Preparation*—These SOPs describe the procedures used to prepare standards and reagents. Information should be included concerning the specific grades of materials used in reagent and standard preparation, appropriate glassware and containers for preparation and storage, labeling and record keeping for stocks and dilutions, and safety precautions to be taken.

7.5.3 *Decontamination*— These SOPs describe the procedures used to clean field equipment before and during the sample collection process. The SOPs should include the cleaning materials used, the order of washing and rinsing with the cleaning materials, requirements for protecting or covering cleaned equipment, procedures for disposing of cleaning materials, and safety considerations.

7.5.4 *Sample Collection Procedures*—SOPs for sample collection procedures shall describe how the procedures are actually performed in the field and shall not be a simple reference to standard test methods, unless a procedure is performed exactly as described in the published test method. If possible, industry-recognized test methods from source documents published by the U.S. Environmental Protection Agency, ASTM, U.S. Department of the Interior, National Water Well Association, American Petroleum Institute, or other recognized organizations should be used. The SOP for sample collection procedures should include the following information:

7.5.4.1 Applicability of the procedure,

7.5.4.2 Equipment and reagents required,

7.5.4.3 Detailed description of the procedures to be followed in collecting the samples,

7.5.4.4 Common problems encountered,

7.5.4.5 Precautions to be taken, and

7.5.4.6 Health and safety considerations.

7.5.5 *Equipment Calibration and Maintenance*—These SOPs describe the procedures used to ensure that field equipment and instrumentation are in working order. The SOPs describe calibration and maintenance procedures and schedules, maintenance logs, service contracts or service arrangements for equipment, and spare parts available in-house. The calibration and maintenance of field equipment and instrumentation should generally be in accordance with manufacturers' specifications and shall be documented.

7.5.6 *Field Measurements*—These SOPs describe all methods used in the field to determine a chemical or physical parameter. The SOPs shall address criteria from Section 8, as appropriate.

7.5.7 *Corrective Action*—These SOPs describe procedures used to identify and correct deficiencies in the sample collection process. These should include specific steps to take in correcting deficiencies, such as performing additional decontamination of equipment, resampling, or additional training of field personnel in methods procedures. The SOP shall specify that each corrective action must be documented with a description of the deficiency, the corrective action taken, and the person(s) responsible for implementing the corrective action.

7.5.8 *Data Reduction and Validation*—These SOPs describe procedures used to compute the results from field measurements and to review and validate these data. They should include all formulas used to calculate the results and procedures used to verify independently that the field measurement results are correct.

7.5.9 *Reporting*—These SOPs describe the process for reporting the results of field activities.

7.5.10 *Records Management*—These SOPs describe the procedures for generating, controlling, and archiving field records. The SOPs should describe the responsibilities for record generation and control and the policies for record retention, including type, time, security, and retrieval and disposal authorities. Records should include project-specific and field operations records.

7.5.10.1 Project-specific records relate to field work performed for a group of samples. Project records may include correspondence, chain-of-custody, field notes, all reports issued as a result of the work, project planning documents, and procedural SOPs used.

7.5.10.2 Field operations records document overall field operations. These records may include equipment performance and maintenance logs, personnel files, general field SOPs, and corrective action reports.

7.5.11 *Waste Disposal*— These SOPs describe policies and procedures for the disposal of waste materials resulting from field operations. The disposal of all wastes must conform to federal, state, and local regulations, including those associated with the Resource Conservation and Recovery Act, Superfund Act Reauthorization and Amendments, Department of Transportation, and Occupational Safety and Health Administration.

7.5.12 *Health and Safety*—These SOPs describe policies and procedures designed both to provide a safe and healthy working environment for field personnel and to comply with federal and state regulations.

7.6 *Field Quality Assurance and Quality Control Requirements*:

7.6.1 *Quality Assurance Program Plan*—The field organization shall have a written QAPP that describes the organization's QA policy. The plan shall specify the responsibilities of the field management and field staff and the QA function in the areas of QA and QC, and it shall also describe the QC procedures followed by the organization (see EPA QAMS-004/80 for an example).

7.6.2 *Quality Assurance Project Plan*—Some projects, particularly those that are large or complex, require a QAPjP. The QAPjP details the QA and QC goals and protocol for a specific data collection activity to ensure that the data generated by sampling and analysis activities are of quality commensurate with their intended use. QAPjP elements should include a discussion of the quality objectives of the project, identification of those involved in the data collection and their responsibilities and authorities, enumeration of the QC procedures to be followed, and reference to the specific SOPs that will be followed for all aspects of the project. Elements may be added or removed, as required by the project or the end-user of the data (see EPA QAMS-005/80 for an example).

7.6.3 *Control Samples*— Control samples are QC samples that are introduced into a process to monitor the performance of the system. Control samples, which may include blanks, duplicates, spikes, analytical standards, and reference materials, can be used in different phases of the overall process, beginning with sampling and continuing through transportation, storage, and analysis. The types of control samples used, and the frequency of usage, are dependent on the DQOs of the data collection effort and must be specified for each project.

7.6.4 *Procedures for Establishing Acceptance Criteria*—Procedures shall be in place for establishing acceptance criteria for field activities, as required in the project planning documents. Acceptance criteria may be qualitative or quantitative. Field events or data that fall outside of the established acceptance criteria may indicate a problem with the sampling process that must be investigated.

7.6.5 *Deviations*—Any activity not performed in accordance with the SOPs or project planning documents is considered a deviation from the plan. Deviations from the plan may or may not affect data quality. All deviations from the plan shall be documented as to the extent of, and reason for, the deviation.

7.6.6 *Corrective Action*—Errors, deficiencies, deviations, or field events or data that fall outside the established acceptance criteria require investigation. Corrective action may be necessary to resolve the problem and restore proper functioning to the system in some instances. Investigation of the problem and any subsequent corrective action taken shall be documented.

7.6.7 *Data Handling Procedures*:

7.6.7.1 *Data Reduction*— All field measurement data are reduced according to protocol described in the appropriate SOP. Computer programs used for data reduction shall be validated before use and verified on a regular basis. All information used in the calculations shall be recorded to enable reconstruction of the final result at a later date.

7.6.7.2 *Data Review*— All data are reviewed according to SOPs to ensure that the calculations are correct and to detect transcription errors. Spot checks are performed on computer calculations to verify program validity.

7.6.7.3 *Data Reporting*— Data are reported in accordance with the requirements of the end-user.

7.7 *Quality Assurance Review*:

7.7.1 *General*—The QA review consists of internal and external assessments to ensure that both QA and QC procedures are in use and field staff conform to these procedures. Planning documents shall specify the requirements for internal, external, and on-site assessment. These documents shall specify the frequency and documentation of these assessments.

7.7.1.1 *Internal Assessment*—Personnel responsible for performing field activities are responsible for continually monitoring individual compliance with the QA and QC programs and planning documents. A QA officer or an appropriate management designee shall review the field results and findings for compliance with the QA and QC programs and planning documents. The results of this internal assessment should be reported to management with requirements for a plan to correct the observed deficiencies.

7.7.1.2 *External Assessment*—The field staff may be reviewed by personnel external to the organization. The results of the external assessment should be submitted to management with requirements for a plan to correct the observed deficiencies.

7.7.1.3 *On-Site Evaluation*—On-site evaluations may be conducted as part of both internal and external assessments. On-site evaluations may include, but are not limited to, a complete review of the facilities, staff, training, instrumentation, SOPs, methods, field analysis, sample collection, QA and QC policies, and procedures related to the generation of environmental data. Records of each evaluation shall be maintained until superseded or according to policy. These records should include the date of the evaluation, area or site, areas reviewed, person performing the evaluation, findings and problems, actions recommended and taken to resolve the problems, and scheduled date for re-inspection. Any problems identified that are likely to affect data integrity shall be brought to the attention of management immediately.

7.7.2 *Evaluation of Field Records*—The review of field records shall be conducted by one or more persons knowledgeable in the field activities, evaluating the following subjects at a minimum:

7.7.2.1 *Completeness of Field Reports*—This review ensures that all requirements for field activities in the planning documents have been fulfilled, that complete records exist for each field activity, and that the procedures specified in the planning documents have been implemented. The emphasis on field documentation will help assure sample integrity and sufficient technical information to recreate each field event. The results of this completeness check shall be documented, and environmental data affected by incomplete records shall be identified.

7.7.2.2 *Identification of Invalid Samples*—This review involves interpretation and evaluation of the field records to detect problems affecting the representativeness of environmental samples. Examples of items that could indicate invalid samples include improper well development, improperly screened wells, instability of pH or conductivity, and collection of volatiles near combustion engines. The field records shall be evaluated against planning documents and SOPs. The reviewer shall document the sample validity and identify the environmental data associated with poor or incorrect field work.

7.7.2.3 *Correlation of Field Test Data*—The results of field measurements obtained by more than one method shall be compared. For example, surface geophysics may be surveyed using both ground penetrating radar and a resistivity survey.

7.7.2.4 *Identification of Anomalous Field Test Data*—Anomalous field test data should be identified. For example, a water temperature for one well that is five degrees higher than any other well temperature in the same aquifer should be noted. The impact of anomalous field measurement results on the associated environmental data shall be evaluated.

7.7.2.5 *Validation of Field Analysis*—All data from field analysis that are generated *in situ* or from a mobile laboratory shall be validated per 8.7.2. The results of the validation shall be reported. The report shall discuss whether the QC checks meet the acceptance criteria and whether corrective actions were taken for any analysis performed when acceptance criteria were not met.

7.7.3 *Quality Assurance Reports to Management*—The QA program shall provide for the periodic reporting of pertinent QA and QC information to management to allow assessment of the overall effectiveness of the QA program. There are three major types of QA reports to management:

7.7.3.1 *Report on Measurement Quality Indicators*—This report shall include the assessment of QC data (such as that generated per 7.6.3) gathered over the period, the frequency of repeating work due to unacceptable performance, and corrective action taken.

7.7.3.2 *Report on Quality Assurance Assessments*—This report shall be submitted immediately following any internal or external on-site evaluations or upon receipt of the results of any performance evaluation studies. The report shall include the results of the assessment and the plan for correcting identified deficiencies.

7.7.3.3 *Report on Key Quality Assurance Activities During the Period*—A report shall be delivered to management summarizing key QA activities during the period. The report shall stress measures that are being taken to improve data quality and shall include a summary of the significant quality problems observed and corrective actions taken. The report shall also include a summary of involvements in resolution of quality issues with clients or agencies, QA organizational changes, and notice of the distribution of any revised documents controlled by the QA function.

7.8 *Field Records*— Records provide direct evidence and support for the necessary technical interpretations, judgments, and discussions concerning project activities. These records, particularly those that are anticipated for use as evidentiary data, must directly support current or ongoing technical studies and activities and must provide the historical evidence necessary for later reviews and analyses. Records shall be legible, identifiable, and retrievable and protected from damage, deterioration, or loss. Field records generally consist of bound field notebooks with prenumbered pages, sample collection forms, personnel qualification and training forms, sample location maps, equipment maintenance and calibration forms, chain-of-custody forms, sample analysis request forms, and field change request forms. All records shall be completed with black, waterproof ink. Procedures for reviewing, approving, and revising field records must be defined clearly, with the lines of authority included. At a minimum, all documentation errors shall be corrected by drawing a single line through the error and initialing by the responsible individual, along with the date of change. The correction is written adjacent to the error. Deviations from field SOPs shall be documented.

7.8.1 *Personnel Training and Qualification Records*—It is the responsibility of the organization to establish personnel qualifications and training requirements. Each staff member shall possess the education, training, technical knowledge, and experience, or a combination thereof, to enable that individual to perform his or her assigned functions. Personnel qualifications shall be documented in terms of education, experience, and training. Training shall be provided for all staff members so that they can perform their functions properly.

7.8.2 *Standard Operating Procedures*—SOPs shall be available to those performing the task outlined. Any revisions to field SOPs shall be written and distributed to all affected individuals to ensure the implementation of changes. The areas covered by SOPs are given in 7.5

7.8.3 *Quality Assurance Plans*—The QAPP and all applicable QAPjPs shall be on file.

7.8.4 *Equipment Maintenance*—Maintenance procedures shall be defined clearly and written for each measurement system and required support equipment. When maintenance is necessary, it shall be documented in either standard forms or in logbooks. A history of the maintenance record of each system serves as an indication of the adequacy of maintenance schedules and parts inventory.

7.8.5 *Calibration and Traceability of Standards and Reagents*—Calibration is a reproducible reference base to which all sample measurements can be correlated. A sound calibration program shall include provisions for documentation of the frequency, conditions, standards, and records reflecting the calibration history of a measurement system. The accuracy of calibration standards is an important point to consider because all data will be in reference to the standards used. A program for verifying and documenting the accuracy of all working standards against primary grade standards shall be followed routinely.

7.8.6 *Sample Collection and Tracking Records*—To ensure maximum utility of the sampling effort and resulting data, documentation of the sampling protocol, as performed in the field, is essential. At a minimum, sample collection records shall contain the persons conducting the activity, sample number, sample location, equipment used, climatic conditions, documentation of adherence to protocol, and unusual observations. The actual sample collection record is usually one of the following: a bound field notebook with prenumbered pages, a pre-printed form, or digitized information on a computer tape or disc.

7.8.6.1 Sample tracking records involving the possession of samples from the time at which they are obtained until they are relinquished shall be documented with the following minimum information: (*1*) project name; (*2*) signatures of the samplers; (*3*) sample number, date and time of collection, and grab or composite sample designation; (*4*) signatures of the individuals involved in sample transfer; and (*5*) the air bill or other shipping number, if applicable.

7.8.7 *Maps and Drawings*—Project planning documents and reports often contain maps. The maps are used to document the location of sample collection points and monitoring wells, and as a means of presenting environmental data. Information used to prepare maps and drawings is normally obtained through field surveys, property surveys, surveys of monitoring wells, aerial photography, or photogrammetric mapping. The final, approved maps shall have a revision number and date and shall be subject to the same controls as other project records.

7.8.8 *Results from Control Samples*—Documentation for the collection of QC samples, such as field, trip, and equipment rinsate blanks, duplicate samples, spikes, and reference materials, shall be maintained.

7.8.9 *Correspondence*— Project correspondence can provide evidence supporting technical interpretations. Correspondence pertinent to the project shall be kept and placed in the project files.

7.8.10 *Deviations*—Field changes and deviations from the planning documents shall be reviewed and approved by either the authorized personnel who performed the original technical review or their designees. All deviations from the procedural and planning documents shall be recorded in the site log.

7.8.11 *Final Report*— The final report shall summarize the field activities, data, results of deviations from the planning documents, and interpretation of the data. The planning documents shall outline the items to be included in the report, which may include any special formats required, QC reporting requirements, conclusions, and recommendations.

7.9 *Documentation Storage*:

7.9.1 *Documentation Archive*—Procedures shall be established to ensure that documents required to recreate the sampling, analysis, and reporting of information are stored. These documents may include, but are not limited to, planning documents, SOPs, logbooks, field data records, sample tags and labels, chain-of-custody records, photographs, and any other information noted in 7.8.

7.9.2 *Storage Time*— The length of storage time for field records shall comply with regulatory requirements, organizational policy, or project requirements, whichever is/are more stringent.

7.9.3 *Filing System*— The control of records is essential in providing evidence of technical adequacy and quality for all project activities. These records shall be identified, retrievable, and organized to prevent loss.

7.9.4 *Personnel Authorized to Enter Archive*—Access to project files shall be controlled to restrict unauthorized personnel from having free and open access. An authorized access list shall be prepared for the project files and shall name the personnel who have unrestricted access to the files.

8. Standard Practices for Environmental Laboratory Operations

8.1 *Purpose*—Each laboratory must conduct its operations in such a way as to provide reliable information. To achieve this goal, certain minimum policies and procedures must be implemented.

8.2 *Organization*— The laboratory shall be structured such that each member of the organization has a clear understanding of his or her duties and responsibilities and the relationship of those responsibilities to the total effort. The organizational structure, functional responsibilities, levels of authority, job descriptions, and lines of communication for activities shall be established and documented. The laboratory shall also maintain a current list of accreditations from government agencies or private associations.

8.2.1 *Management*—The management of the laboratory is responsible for establishing organizational, operational, health and safety, and QA policies. These responsibilities include the following: oversight of personnel selection, development, and training; review, selection, and approval of analysis methods; and development, implementation, and maintenance of a QA program. Management shall foster an attitude within the organization that emphasizes the importance of quality and supports implementation of the QAPP.

8.2.2 *Quality Assurance Function*—The laboratory shall appoint a person or persons to be responsible for monitoring laboratory operations to ensure that the facilities, equipment, personnel, methods, practices, and documentation are in conformance with the laboratory QAPP and any applicable QAPjP(s). The QA monitoring function shall be entirely separate from, and independent of, personnel engaged in direct supervision or performance of the work being monitored. The QA function shall inspect records to ensure that analyses have been performed correctly and within the proper time frame; maintain copies of the QAPP and QAPjPs pertaining to all analyses; perform assessments of the laboratory to ensure adherence to the QAPP; periodically submit written status reports to management, noting any problems and the corrective actions taken; ensure that any deviations from the approved QAPP, QAPjP, or SOPs have been authorized and documented properly; and ensure that the results reported reflect the raw data accurately. The responsibilities of the QA function and procedures used in conducting those responsibilities shall be in writing and shall be maintained.

8.2.3 *Personnel*—It is the responsibility of the organization to establish personnel qualifications and training requirements for all positions. Each member of the organization shall possess the education, training, technical knowledge, and experience, or a combination thereof, which enables that individual to perform his or her assigned functions. Personnel qualifications shall be documented in terms of education, experience, and training. Training shall be provided for all staff members, as necessary, so that they can perform their functions properly.

8.2.4 *Subcontractors*— The use of subcontractors shall not jeopardize data quality. Subcontractors shall therefore comply with the requirements of Sections 7 and 8, as appropriate to the specific task(s) they are performing.

8.3 *Facilities*:

8.3.1 *General*—Each laboratory shall be of a size and construction suitable to facilitate proper conduct of the analyses. Adequate bench space or working area per analyst shall be provided. The space requirement per analyst depends on the equipment or apparatus being used, the number of samples the analyst is expected to handle at any one time, and the number of operations to be performed concurrently by a single analyst. The laboratory shall be well-ventilated, adequately lit, free of dust and drafts, protected from extreme temperatures, and offer access to a stable source of power. Laboratories shall be designed so that there is adequate separation of functions in order to ensure that no laboratory activity has an adverse effect on the analyses. The laboratory may require specialized facilities such as a perchloric acid hood or glovebox.

8.3.2 *Sample Handling, Receiving, and Storage Area*—As necessary to ensure safe and secure storage and prevent contamination or misidentification, there shall be adequate facilities for the receipt and storage of samples. The level of custody required and any special requirements for storage, such as refrigeration and lighting, shall be described in the planning documents.

8.3.3 *Chemical Storage*— Storage areas for reagents, solvents, standards, and reference materials shall be adequately safe to preserve their identity, concentration, purity, and stability.

8.3.4 *Laboratory Operations Area*—Separate spaces for laboratory operations and appropriate ancillary support shall be provided, as necessary, for the performance of routine and specialized procedures.

8.3.5 *Waste Storage Area*—Adequate facilities shall be provided for the collection and storage of all wastes, and these facilities shall be operated so as to minimize environmental contamination. Waste storage and disposal facilities shall comply with applicable federal, state, and local regulations.

8.3.6 *Data Storage Area*—Space shall be provided for the storage and retrieval of all documents, as specified in 8.8. The storage environment shall be maintained to assure the integrity of the materials stored. Access shall be limited to authorized personnel only.

8.4 *Equipment and Instrumentation*:

8.4.1 *Equipment and Instrumentation*—Equipment and instrumentation shall meet the requirements and specifications of the specific test methods and other SOPs. The laboratory shall maintain an equipment and instrument description list that includes the manufacturer, model number, year of purchase, accessories, and any modifications, updates, or upgrades that have been made.

8.4.2 *Maintenance and Calibration of Equipment and Instrumentation*—Equipment and instrumentation shall be adequately inspected, cleaned, maintained, tested, and calibrated, as required in the SOP or operations manual. SOPs or manuals shall specify the identification and repair of common maintenance problems. Procedures shall designate a person(s) or organizations responsible for maintenance. Records of all inspections, maintenance, repairs, testing, and calibration shall be maintained.

8.5 *Standard Operating Procedures*—The laboratory should have written SOPs for all laboratory functions that affect data quality. Procedures and methods shall be performed in the laboratory as described in the SOPs. Any modification of an SOP made during a data collection activity must be documented. SOPs shall be available for the following areas and shall, at a minimum, contain the information described:

8.5.1 *Sample Management*—These SOPs describe the receipt, handling, and storage of samples.

8.5.1.1 *Sample Receipt and Handling*—These SOPs describe precautions to be used in opening sample shipment containers, as well as procedures used to perform the following tasks: verify that chain-of-custody has been maintained, examine samples for damage, check for proper preservatives and temperature, assign the testing program, and log samples into the laboratory sample streams.

8.5.1.2 *Sample Scheduling*—These SOPs describe procedures and criteria used for scheduling work in the laboratory, including procedures used to ensure that holding time requirements are met.

8.5.1.3 *Sample Storage*— These SOPs describe the storage conditions for all samples and procedures used both to verify and document daily storage temperature, and to ensure that custody of the samples is maintained while in the laboratory.

8.5.2 *Reagent and Standard Preparation*—These SOPs detail procedures used to prepare standards and reagent mixtures. In addition, these SOPs shall specify requirements concerning the following: purity of materials used, including water; appropriate glassware and containers; record keeping and labeling, including dating; procedures used to verify concentration, purity, and stability; and safety precautions necessary to meet the requirements of the DQOs or test methods.

8.5.3 *General Laboratory Techniques*—These SOPs detail all of the essentials of laboratory operations not addressed in other SOPs. These techniques include, but are not limited to, glassware cleaning procedures, operation of analytical balances, pipetting techniques, and use of volumetric glassware.

8.5.4 *Analytical Methods*—SOPs for analytical methods for sample analysis shall be a description of how the analysis is actually performed in the laboratory and not a simple reference to standard test methods, unless the analysis is performed exactly as described in the published test method. If possible, industry-recognized test methods from source documents published by the U.S. Environmental Protection Agency, American Public Health Association, ASTM, the National Institute for Occupational Safety and Health, or other recognized organizations should be used. The SOP for analytical methods should include the following:

8.5.4.1 Sample preparation and analysis procedures, including the applicable holding time, extraction, digestion, or preparation steps, as appropriate to the method; procedures for determining the appropriate dilution to analyze; and any other information required to perform the analysis accurately and consistently.

8.5.4.2 Instrument standardization, including the concentration and frequency of analysis of calibration standards, linear range of the method, and calibration acceptance criteria.

8.5.4.3 Raw data recording requirements and documentation, including the sample identification number, analyst, data verification analyst, date of analysis and verification, and computational method(s).

8.5.4.4 Detection and reporting limits for all analytes in the method.

8.5.4.5 Reference to the applicable QC SOPs and any specific exceptions or additions.

8.5.5 *Equipment Calibration and Maintenance*—These SOPs describe procedures used to assure or verify that the laboratory equipment and instrumentation are in working order. The SOPs describe calibration and maintenance procedures and schedules, maintenance logs, service contracts or service arrangements for all equipment, and spare parts available in-house. Calibration and maintenance of the laboratory equipment and instrumentation shall be in accordance with manufacturers' specifications and shall be documented.

8.5.6 *Quality Control Data*—These SOPs detail the type, purpose, and frequency of QC samples analyzed in the laboratory and establish acceptance criteria. They should include information on applicability of the QC sample to the analytical process, statistical treatment of the data, and responsibility of laboratory staff and management in generating and using the data.

8.5.7 *Corrective Action*—These SOPs describe procedures used to identify and correct deficiencies in the analytical process. These include specific steps to take in correcting deficiencies, such as the preparation of new standards and reagents, recalibration and restandardization of equipment, reanalysis of samples, or additional training of laboratory personnel in methods and procedures. The SOP shall specify that each corrective action must be documented with a description of the deficiency, corrective action taken, and person(s) responsible for implementing the corrective action.

8.5.8 *Data Reduction and Validation*—These SOPs describe the procedures used to review and validate the data. They should include procedures for computing and interpreting the results from QC samples and procedures used to verify independently that the analytical results are correct. In addition, routine procedures used to monitor precision and bias, including evaluations of reagent, field, and trip blanks, calibration standards, control samples, duplicate and matrix spike samples, and surrogate recovery should be detailed in an SOP.

8.5.9 *Reporting*—These SOPs describe the process for reporting the analytical results.

8.5.10 *Records Management*—These SOPs describe the procedures for generating, controlling, and archiving laboratory records. The SOPs should detail the responsibilities for record generation and control and the policies for record retention, including type, time, security, and retrieval and disposal authorities. Records shall include project-specific and laboratory operations records.

8.5.10.1 Project-specific records related to analyses performed for a group of samples shall be maintained. These records may include an index of documents, correspondence, chain-of-custody records, request for analysis, calibration records, raw and finished analytical and QC data, data reports, and project planning documents.

8.5.10.2 Laboratory operations records, which document

the overall laboratory operation, shall be maintained. These records may include the following: laboratory notebooks, instrument performance and maintenance logs in bound notebooks with prenumbered pages; laboratory benchsheets; software documentation; control charts; reference material certification; personnel files; laboratory SOPs; and corrective action reports.

8.5.11 *Waste Disposal*— These SOPs describe policies and procedures for the disposal of chemicals, including standard and reagent solutions, process waste, and samples. The disposal of these materials shall conform to federal, state, and local regulations, including those associated with the Resource Conservation and Recovery Act, Superfund Act Reauthorization and Amendments, Department of Transportation, and Occupational Safety and Health Administration.

8.5.12 *Health and Safety*—These SOPs describe policies and procedures designed both to provide a safe and healthy working environment for laboratory staff and to comply with federal and state regulations.

8.6 *Laboratory Quality Assurance and Quality Control Procedures*:

8.6.1 *Quality Assurance Program Plan*—The laboratory shall have a written QAPP that describes the organization's QA policy. The plan shall specify the responsibilities of the laboratory staff and management and the QA function in the areas of QA and QC and describe the QC procedures followed by the laboratory, including the following: (*1*) use of control samples; (*2*) statistical and mathematical basis for assigning warning and rejection limits; (*3*) detection of shifts, trends, or biases; (*4*) how an out-of-control condition is detected; (*5*) how control is re-established; (*6*) how out-of-control events and corrective actions are documented; (*7*) how reporting limits are established, including their dependence on serial dilutions and sample size; (*8*) how instrument calibration and maintenance logs, corrective action reports, and routine QC data summaries are maintained; and (*9*) QA assessment procedures used (see EPA QAMS-004/80 for an example).

8.6.2 *Quality Assurance Project Plan*—Some projects, particularly those that are large or complex, require a QAPjP. The QAPjP details the QA and QC goals and protocol for a specific data collection activity in order to ensure that the data generated by sampling and analysis activities are of quality commensurate with their intended use. QAPjP elements should include a discussion of the quality objectives of the project, identification of those involved in data collection and their responsibilities and authorities, enumeration of QC procedures to be followed, and reference to the specific SOPs that will be followed for all aspects of the project. Elements may be added or removed, as required by the project or the end-user of the data (see EPA QAMS-005/80 for an example).

8.6.3 *Method Proficiency*—The laboratory shall have procedures for demonstrating proficiency with each analytical method used in the laboratory. These shall include procedures for demonstrating the precision and bias of the method as performed by the laboratory and procedures for determining the method detection limit (MDL). All terminology, procedures, and frequency of determinations associated with the laboratory's establishment of one MDL and the reporting limit shall be well-defined and well-documented. Documented precision, bias, and MDL information shall be maintained for all methods performed in the laboratory.

8.6.4 *Laboratory Control Procedures*—The laboratory shall have procedures for demonstrating that it is in control within laboratory established limits or project specified limits during each data collection activity.

8.6.4.1 *Laboratory Control Samples*—The laboratory shall analyze control samples for each analytical method. A laboratory control sample consists of a control matrix spiked with analytes representative of the target analytes. Laboratory control sample(s) shall be analyzed with each batch of samples processed in order to verify the precision and bias of the analytical process. The results of the laboratory control sample(s) are compared to the control limits established for both precision and bias to determine the usability of the data (see 8.6.6). Analytical data generated with laboratory control samples that fall within the prescribed limits are judged to be generated while the laboratory was in control. Data generated with laboratory control samples that fall outside of the established control limits are judged to be generated during an "out of control" situation. These data are considered suspect and must be repeated or reported with qualifiers.

8.6.4.2 *Method Blank*— To assess contamination levels in the laboratory, a method blank shall be run with each batch of samples processed, unless it is not appropriate for the method. The laboratory shall have guidelines in place for accepting or rejecting data based on the level of blank contamination.

8.6.5 *Determination of Matrix Effects*—The laboratory shall have procedures for documenting the effect of the matrix on method performance. The type of matrix-specific information required for a specific data collection activity is dependent on the DQOs of the activity.

8.6.5.1 *Matrix Spikes*— Procedures shall be in place for determining the bias of the method in the matrix unless not appropriate for the method. These procedures should include the analysis of matrix spikes, use of surrogates for organic methods, and method of standard additions for metal and inorganic methods. The frequency of use of these techniques shall be based on the DQOs of the data collection activity.

8.6.5.2 *Matrix Duplicates and Matrix Spike Duplicates*—Procedures shall be in place for determining the precision of the method in the matrix. These procedures should include the analysis of matrix duplicates or matrix spike duplicates, or both. The frequency of use of these techniques shall be based on the DQOs of the data collection activity.

8.6.5.3 *Sample-Specific Detection Limit*—Procedures shall be in place for determining the MDL in a specific sample or group of samples. The frequency of use of these procedures shall be based on the DQOs of the data collection activity.

8.6.6 *Procedures for Establishing Control Limits*:

8.6.6.1 Procedures shall be in place for establishing and updating the control limits for analysis control. Control limits shall be established in order to evaluate laboratory precision and bias based on the analysis of control samples (see 8.6.4.1). Control limits for bias are typically based on the historical mean recovery plus or minus three standard deviation units, and control limits for precision range from zero (no difference between duplicate control samples) to the historical mean relative percent difference plus three standard deviation units.

8.6.6.2 Procedures shall be in place for monitoring historical performance, and they should include graphical (control charts) or tabular presentations of the data.

8.6.7 *Deviations*—Any activity not performed in accordance with laboratory SOPs or project planning documents is considered a deviation from the plan. Deviations from the plan may or may not affect data quality. All deviations from the plan shall be documented as to the extent of, and reason for, the deviation.

8.6.8 *Corrective Action*—Errors, deficiencies, deviations, or laboratory events or data that fall outside of the established acceptance criteria require investigation. Corrective action may be necessary in some instances to resolve the problem and to restore proper functioning to the analytical system. Investigation of the problem and any subsequent corrective action taken shall be documented.

8.6.9 *Data Handling Procedures*:

8.6.9.1 *Data Reduction*— Data resulting from the analyses of samples are reduced according to protocol described in the laboratory SOPs. Computer programs used for data reduction shall be validated before use and verified on a regular basis. All information used in the calculations shall be recorded in order to enable reconstruction of the final result at a later date. This information may include the weight or volume of sample used, percent dry weight for solids, extract volume, dilution factor used, and blank- or background-correction protocol followed.

8.6.9.2 *Data Review*— All data are reviewed by a second analyst or supervisor according to laboratory SOPs in order to ensure that calculations are correct and to detect transcription errors. Spot checks are performed on computer calculations to verify program validity. Errors detected in the review process are referred back to the analyst(s) for corrective action.

8.6.9.3 *Data Reporting*— Data are reported in accordance with the requirements of the end-user. The data report may include the following:

(1) Laboratory name and address;

(2) Sample information (including unique sample identification, sample collection date and time, date of sample receipt, and date(s) of sample preparation and analysis);

(3) Analytical results reported with appropriate significant figures;

(4) Detection limits that reflect dilutions, interferences, or correction for equivalent dry weight;

(5) Method reference;

(6) Appropriate QC results as described in 8.6.4 and 8.6.5; and

(7) Data qualifiers with appropriate references and narrative on the quality of the results.

8.7 *Quality Assurance Review*:

8.7.1 *General*—The QA program shall provide for routine evaluations of the effectiveness of laboratory QA and QC procedures and conformance of the laboratory to these procedures. These evaluations can be performed by persons internal or external to the organization and can involve on-site evaluations or performance evaluation studies, or both.

8.7.1.1 *Internal Assessment*—The analyst is responsible for continually monitoring individual compliance with the QA and QC programs. In addition, on some frequent basis, the QA officer, or an appropriate management designee, must review the laboratory data and operations for compliance with the QA and QC programs. The results of the internal assessment should be reported to management with requirements for a plan to correct observed deficiencies.

8.7.1.2 *External Assessment*—The laboratory should periodically be subject to evaluation by an assessor who is independent of the laboratory. This may be a representative of a government agency or other independent organization. The results of the external assessment should be submitted to management with requirements for a plan to correct the observed deficiencies.

8.7.1.3 *On-Site Evaluation*—On-site evaluations should be conducted as part of both internal and external assessments. An on-site evaluation includes a complete evaluation of the facilities, staff, instrumentation, SOPs, analytical methods, sample management procedures, and QA and QC policies and procedures as they relate to the generation of data for specific analytical and regulatory applications. Records of each evaluation shall be maintained by the laboratory. These records should include the date of the evaluation, area(s) or analyses inspected, person performing the evaluation, findings and problems, actions recommended and taken to resolve existing problems, and scheduled date for re-inspection. Any problems identified during the course of an on-site evaluation that are likely to affect data integrity shall be brought to the attention of management immediately.

8.7.1.4 *Performance Evaluation Studies*—Performance evaluation (PE) studies are used to measure the performance of the laboratory on unknown samples. PE samples are typically submitted to the laboratory as blind samples by an independent, outside source. The results are compared to predetermined acceptance limits set by the submitting agency. PE samples can also be submitted to the laboratory by the laboratory QA officer or an appropriate management designee as part of an internal assessment of laboratory performance. Records of all PE studies shall be maintained by the laboratory. Problems identified as a result of participation in PE studies shall immediately be investigated and corrected.

8.7.2 *Evaluation of Laboratory Data*—The evaluation of laboratory data shall be conducted by one or more persons knowledgeable in laboratory activities. Such evaluations often occur in the QA section of the laboratory. Data evaluation should include the following subjects:

8.7.2.1 *Completeness of Laboratory Data*—This ensures that *(1)* all samples and analyses required by the project planning documents have been processed, *(2)* complete records exist for each analysis and the associated QC samples, and *(3)* the procedures specified in project planning documents and SOPs have been implemented. The results of the completeness check should be documented.

8.7.2.2 *Evaluation of Data with Respect to Detection Limits*—Analytical results should be compared to required detection limits. Any detection limits that exceed regulatory limits or action levels, as specified in the project planning documents, should be identified.

8.7.2.3 *Evaluation of Data with Respect to Control Limits*—The results of QC and calibration check samples should be compared to control criteria. Data not within the control limits require corrective action, and reviewers should check that both corrective action reports and the results of reanalysis are available. Samples associated with out-of-control QC data should be identified in a written record or the data review, and an assessment of the utility of such analytical results should be recorded.

8.7.2.4 *Review of Holding Time Data*—Sample holding times should be compared to those required by the project planning documents, and all deviations should be noted.

8.7.2.5 *Evaluation of Performance Evaluation Results*—PE study results can be helpful in evaluating the impact of out-of-control conditions. Recurring trends or problems evident in PE studies should be documented in the review, and their effect on environmental data should be evaluated.

8.7.2.6 *Correlation of Laboratory Data*—The results of data obtained from related laboratory tests, such as purgeable organic hallides (POX) and volatile organics, shall be documented and the significance of differences discussed in the review reports.

8.7.3 *Quality Assurance Reports to Management*—The QA program shall provide for the periodic reporting of pertinent QA and MQC information to management in order to allow management to assess the overall effectiveness of the QA program. Three examples of QA reports to management are as follows:

8.7.3.1 *Report on Measurement Quality Indicators*—This report should include the assessment of QC data gathered over the period, the frequency of analyses repeated due to unacceptable QC performance, and, if possible, the reason for the unacceptable performance and corrective action taken.

8.7.3.2 *Report on Quality Assurance Assessments*—This report should be submitted immediately following any internal or external on-site evaluation or upon receipt of the results of any PE studies. The report should include the results of the assessments and the plan for correcting the identified deficiencies.

8.7.3.3 *Report on Key Quality Assurance Activities During the Period*—A report summarizing key QA activities during the period should be delivered to management. The report should stress measures that are being taken to improve data quality and include a summary of the significant quality problems observed and corrective actions taken. The report should also include a summary of any changes in certification and accreditation status; involvements in the resolution of quality issues with clients or agencies; QA organizational changes; and notice of the distribution of revised documents controlled by the QA organization (that is, SOPs, QAPP).

8.8 *Laboratory Records*:

8.8.1 All information relevant to environmental QA and QC activities relating to laboratory facilities, equipment, personnel, methods, and practices shall be documented. Copies of the SOPs, and the equipment manuals provided by the manufacturer, shall be accessible in the workplace.

8.8.2 Handwritten test data shall be recorded legibly in ink in laboratory notebooks or on designated benchsheets. Each page shall be signed and dated by the person who performed the analysis and entered the data. Corrections shall be made by drawing a single line through the information to be changed and initialling and dating the change. The reason for the change shall be indicated.

8.8.3 Strip-chart recorder printouts shall be signed by the person who performed the instrumental analysis. If corrections need to be made in computerized data, a system parallel to the corrections for handwritten data shall be in place.

8.8.4 Records of sample management shall be available to permit the re-creation of an analytical event for review in the case of an audit or investigation of a dubious result.

8.8.5 *Personnel Training and Qualification Records*—It is the responsibility of the organization to establish personnel qualifications and training requirements. Each staff member shall possess the education, training, technical knowledge, and experience, or a combination thereof, to enable that individual to perform his or her assigned functions. Personnel qualifications shall be documented in terms of education, experience, and training. Training shall be provided for all staff members so that they can perform their functions properly.

8.8.6 *Standard Operating Procedures*—SOPs shall be written and available for all methods and practices.

8.8.7 *Quality Assurance Plans*—The QAPP and all applicable QAPjPs shall be on file.

8.8.8 *Equipment Maintenance*—Maintenance and performance logs shall be kept on all equipment and instrumentation and must include the following: (*1*) name of the equipment and manufacturer; (*2*) model and serial number of the equipment; (*3*) date the equipment was placed in service; (*4*) instructions for proper maintenance procedures; (*5*) dates of the maintenance and performance checks and the names of personnel performing them; and (*6*) if nonroutine repairs were performed as a result of a malfunction, nature and cause of the malfunction, and name of the person performing the repair.

8.8.9 *Proficiency*— Proficiency information on all parameters reported shall be maintained and shall include the following: (*1*) precision; (*2*) bias; (*3*) method detection limits; (*4*) spike recovery, where applicable; (*5*) surrogate recovery, where applicable; (*6*) checks on reagent purity, where applicable; and (*7*) checks on glassware cleanliness, where applicable.

8.8.10 *Traceability*— Traceability of standards, reagents, and weights shall be documented, with the traceability to national or international sources indicated.

8.8.11 *Calibration*— Calibration and standardization data records for each instrument and method shall include the following: (*1*) data of the last calibration; (*2*) calibration history; (*3*) frequency of calibration; (*4*) outside sources of calibration, if used, for example, the manufacturer; (*5*) date of preparation, expiration date, and name of the person performing the preparation of standards; and (*6*) written procedures for instrument calibration.

8.8.12 *Sample Management*—All required records pertaining to sample management shall be maintained and updated regularly. These include chain-of-custody forms, sample receipt forms, and sample disposition records.

8.8.13 *Original Data*— The raw data and calculated results

for all samples shall be maintained in laboratory notebooks and logs, benchsheets, files, and other sample tracking or data entry forms. Instrumental output shall be stored in a computer file or a hardcopy report.

8.8.14 *Quality Control Data*—The raw data and calculated results for all QC and field samples and standards shall be maintained in the manner described in 8.8.13. QC samples include, but are not limited to, control samples, method blanks, matrix spikes, and matrix spike duplicates.

8.8.15 *Correspondence*— Project correspondence can provide evidence supporting technical interpretations. Correspondence pertinent to the project shall be kept and placed in the project files.

8.8.16 *Final Report*— A copy of the final report submitted to the sponsor of the work, or other designated individual, shall be retained.

8.9 *Document Storage*:

8.9.1 A written policy concerning the minimum length of time that documentation is to be retained shall be in place and understood by the appropriate personnel, including the sponsor of the work. Deviations from this policy shall be documented and readily available to those individuals responsible for the deletion or destruction of expired documentation.

8.9.2 Documentation shall be stored securely in a facility that adequately addresses or minimizes its deterioration for the length of time it is to be retained. A system allowing for the expedient retrieval of information shall exist.

8.9.3 Access to archived information shall be controlled to restrict unauthorized personnel from having free and open access. An authorized access list shall be prepared and shall name the personnel who have unrestricted access to the archived information.

8.9.4 All accesses to archived information shall be documented. This documentation shall include the name of the individual, date, reason for accessing the data, and all changes, deletions, or withdrawals that may have occurred.

8.9.5 If a facility conducting testing or an archive contract facility goes out of business before the time period specified in the policy (8.9.1), all documentation shall be transferred in whole to the archives of the sponsor of the work.

9. Data Quality Assessment

9.1 The assessment of environmental data occurs in two phases. Field records and analytical data are reviewed during the first phase to identify whether the data are accurate and defensible. The data are interpreted in the second phase with respect to meeting DQOs.

9.2 Technical reports of environmental data collection efforts should summarize the information contained in the field records and the results of the laboratory data review, as described in 7.7.2 and 8.7.2. This information should be used to identify clearly the data that are not representative of environmental conditions or that have been generated using poor field or laboratory practices.

9.3 The combined field and laboratory data are then subject to a final assessment to determine whether the DQOs have been met. Although the data quality assessment process is beyond the scope of this practice, a general description of the items that should be considered in the assessment process is given below.

9.4 Data quality assessment typically includes, but is not limited to the following:

9.4.1 Evaluation of field duplicate results;

9.4.2 Comparison of the results of all field blanks, trip blanks, and equipment rinsates, with the full data set to provide information concerning contaminants that may have been introduced during sampling or shipping;

9.4.3 Evaluation of matrix effects to assess performance of the analytical method with respect to the sample matrix and determine whether the data have been biased high or low due to matrix effects;

9.4.4 Integration of the field and laboratory data with geological, hydrogeological and meteorological data to provide information on the extent of contamination; and

9.4.5 Comparison of precision, bias, completeness, representativeness, and defensibility of the data generated with that required to meet the DQOs.

APPENDIXES

(Nonmandatory Information)

X1. TABLE OF CONTENTS

X2. REFERENCE LIST

X2.1 *ASTM Standards*:

D 1192 Specification for Equipment for Sampling Water and Steam[2]

D 1357 Practice for Planning the Sampling of the Ambient Atmosphere[5]

D 2777 Practice for Determination of Precision and Bias of Applicable Methods of Committee D-19 on Water[2]

D 3370 Practices for Sampling Water[2]

D 3614 Guide for Laboratories Engaged in Sampling and Analysis of Atmospheres and Emissions[5]

D 3670 Guide for Determination of Precision and Accuracy of Methods of Committee D-22[5]

D 3694 Practices for Preparation of Sample Containers and for Preservation of Organic Constituents[6]

D 3856 Guide for Evaluating Laboratories Engaged in Sampling and Analysis of Water and Waste Water[2]

D 4210 Practice for Interlaboratory Quality Control Procedures and a Discussion on Reporting Low-Level Data[2]

D 4411 Guide for Sampling Fluvial Sediment in Motion[6]

D 4447 Guide for the Disposal of Laboratory Chemicals and Samples[7]

D 4448 Guide for Sampling Groundwater Monitoring Wells[7]

D 4489 Practice for Sampling of Waterborne Oils[6]

D 4687 Guide for General Planning of Waste Sampling[7]

E 29 Practice for Using Significant Digits in Test Data to Determine Conformance with Specifications[3]

E 122 Practice for Choice of Sample Size to Estimate a Measure of Quality for a Lot or Process[3]

E 178 Practice for Dealing with Outlying Observations[3]

X2.2 *U.S. Environmental Protection Agency Documents:*[4] *A Compendium of Superfund Field Operations Methods*, **EPA/540/P-87/001 (NTIS No. PB88181557/LL), December 1987.**

Blackman, Benson, Hardy, and Fisher, "Enforcement and Safety Procedures for Evaluation of Hazardous Waste Disposal Sites," *Management of Uncontrolled Hazardous Waste Sites*, National Enforcement Investigations Center, Denver, CO, 1980.

Characterization of Hazardous Waste Sites: A Methods Manual, Vol II, *Available Sampling Methods*, EPA-600/4-83-040 (NTIS No. PB84-12692/LL), Environmental Monitoring and Support Laboratory, Las Vegas, NV, 1983.

Characterization of Hazardous Waste Sites: A Methods Manual, Vol II, *Available Sampling Methods*, 2nd Edition, EPA 600/4-84-076 (NTIS No. PB85-168771/LL), December 1984.

Data Quality Objectives for Remedial Response Activities: 1987a-OSWER Directive 9355, 0-14 and 1987-OSWER Directive 9355, 0-7A; EPA 540/G-87/004 and 003; and NTIS-PB88-131388, PB88131370/LL.

DeVera, Emil R., Simmons, Bart P., Stephens, Robert D., and Storm, David L., *Samplers and Sampling Procedures for Hazardous Waste Streams*, EPA-600/2-80-018 (NTIS No. PB80-135353/LL), Municipal Environmental Research Laboratory, Cincinnati, OH, 1980.

Field Screening Methods Catalog: User's Guide, USEPA/540-2-88/005 (NTIS PB89-134159/LL), September 1988.

Guidance for Conducting Remedial Investigations and Feasibility Studies Under CERCLA, OSWER Directive 9335.3-01, Draft (NTIS No. PB89184626/LL), 1988.

Guide for Decontamination of Buildings, Structures and Equipment at Superfund Sites, EPA-600/2-85-028 (NTIS No. PB85-201234/LL), Hazardous Waste Engineering Research Laboratory, Cincinnati, OH, 1985.

Handbook for Analytical Quality Control in Water and Wastewater Laboratories , EPA-600/4-79-019 (NTIS No. PB297451/LL), Environmental Monitoring and Support Laboratory, Washington, DC, 1979.

Handbook for Sampling and Sample Preservation of Water and Wastewater, EPA-600/4-82-029 (NTIS No. PB83124503/LL), Environmental Monitoring and Support Laboratory, Cincinnati, OH, 1982.

Kerr, R. S., *Practical Guide for Groundwater Sampling*, EPA-600/2-85/104 (NTIS No. PB86-137304/LL), Environmental Research Laboratory, Ada, OK, 1985.

Manual of Ground-Water Quality Sampling Procedures, EPA-600/2-81-160 (NTIS No. PB82103045/LL), 1981.

Manual of Water Well Construction Practices, EPA-570/9-75-001 (NTIS No. PB267371/LL), National Technical Information Service, 1977.

Methods for Chemical Analysis of Water and Wastes, EPA-600/4-79-020 (NTIS No. PB84-128677/LL), Environmental Monitoring and Support Laboratory, Cincinnati, OH, 1983.

Methods for Chemical Analysis of Water and Wastes, EPA-600/4-79-020 (NTIS PB84128677/LL), March 1979.

NPDES Compliance Inspection Manual, 4211-545/11828, 1984.

NPDES Compliance Inspection Manual (NTIS No. PB88-221098/LL), Office of Water Enforcement and Permits (EN-338), Washington, DC, May 1988.

Occupational Safety and Health Guidance Manual for

[5] *Annual Book of ASTM Standards*, Vol 11.03.

[6] *Annual Book of ASTM Standards*, Vol 11.02.

[7] *Annual Book of ASTM Standards*, Vol 11.04.

Hazardous Waste Site Activities, DHHS (NIOSH) Publication No. 85-115 (NTIS No. PB87-162855/LL), NIOSH, OSHA, USCG, EPA, October 1985.

Preparation of State Lead Remedial Investigation Quality Assurance Project Plans for Region V, Office of Monitoring Systems and Quality Assurance, Washington, DC, April 4, 1984.

RCRA Groundwater Monitoring Technical Enforcement Document, 0-172-437, 1986.

Scalf, Marion R., et al., *Manual of Ground Water Sampling Procedures*, U.S. EPA and National Well Water Association, Worthington, OH, 1981, pp. 92 and 93.

Soil Sampling Quality Assurance User's Guide, EPA-600/4-84-043 (NTIS No. PB84198621/LL), 1984.

Standard Operating Safety Guides, Office of Emergency and Remedial Response, Hazardous Response Support Division, November 1984.

X2.3 *Other Government Documents:*[4]

Barcelona, Gibb, and Miller, *A Guide to the Selection of Materials for Monitoring, Well Construction and Ground Water Sampling* (NTIS No. PB84-141779/LL), Champaign, IL, August 1983.

Chemistry Quality Assurance Handbook, Vol I, U.S. Department of Agriculture Food Safety and Inspection Service, July 2, 1987.

Environmental Survey Manual, DOE/EH-0053, Vols 1 to 5, U.S. Department of Energy, Washington, DC, January 1989.

National Handbook of Recommended Methods for Water-Data Acquisition, U.S. Department of Interior, U.S. Geological Survey, Reston, VA, 2 Vols, 1977.

Soil Survey Handbook, 0-336-718:QL3, U.S. Department of Agriculture, 1981.

Soil Taxonomy, 0-470-728, U.S. Department of Agriculture, 1975.

X2.4 *Books*:

Barth, D. S., and Mason, B. J., *Soil Sampling Quality Assurance and the Importance of an Exploratory Study, ACS Symposium Series 267: Environmental* Sampling for Hazardous Waste, ACS Publications American Chemical Society, Washington, DC, 1984.

Compton, R. R., *Manual of Field Geology*, John Wiley and Sons, Inc., New York, NY, 1962.

Driscoll, Fletcher, G., *Groundwater and Wells*, 2nd Edition, Johnson Division, St. Paul, MN, 1987, pp. 885–891.

Keith, L. D., *Principles of Environmental Sampling*, American Chemical Society, Washington, DC, 1988.

Kratochvil, B., and Taylor, J. K., "Sampling for Chemical Analysis," *Analytical Chemistry*, American Chemical Society, Washington, DC, July 1981, pp. 924A–938A.

Quality Control in Remedial Site Investigation: Hazardous and Industrial Solid Waste Testing, 5th Vol, *ASTM STP 925*, C. L. Perket, Ed., ASTM, Philadelphia, PA, 1986.

RCRA Ground Water Monitoring Technical Enforcement Guidance Document, NWWA/EPA Series, National Water Well Association, Dublin, OH, September 1986.

Taylor, J. K., *Quality Assurance of Chemical Measurements*, Lewis Publishers, Inc., Chelsea, MI, 1987.

Designation: D 5284 – 93 (Reapproved 1999)

Standard Test Method for Sequential Batch Extraction of Waste with Acidic Extraction Fluid[1]

This standard is issued under the fixed designation D 5284; the number immediately following the designation indicates the year of original adoption or, in the case of revision, the year of last revision. A number in parentheses indicates the year of last reapproval. A superscript epsilon (ε) indicates an editorial change since the last revision or reapproval.

1. Scope

1.1 This test method provides a procedure for the sequential leaching of a waste containing at least 5 % dry solids in order to generate solutions to be used to determine the constituents leached under the specified testing conditions.

1.2 This test method calls for the shaking of a known weight of waste with acidic extraction fluid of a specified composition as well as the separation of the liquid phase for analysis. The pH of the extraction fluid is to reflect the pH of acidic precipitation in the geographic region in which the waste being tested is to be disposed. The procedure is conducted ten times in sequence on the same sample of waste, and it generates ten solutions.

1.3 This test method is intended to describe the procedure for performing sequential batch extractions only. It does not describe all types of sampling and analytical requirements that may be associated with its application.

1.4 The values stated in SI units are to be regarded as the standard.

1.5 *This standard does not purport to address all of the safety problems, if any, associated with its use. It is the responsibility of the user of this standard to establish appropriate safety and health practices and determine the applicability of regulatory limitations prior to use.*

2. Referenced Documents

2.1 *ASTM Standards:*

D 75 Practice for Sampling Aggregates[2]

D 420 Guide to Site Characterization for Engineering, Design and Construction Purposes[3]

D 653 Terminology Relating to Soil, Rock, and Contained Fluids[3]

D 1129 Terminology Relating to Water[4]

D 1193 Specification for Reagent Water[4]

D 2234 Test Methods for Collection of a Gross Sample of Coal[5]

D 2777 Practice for Determination of Precision and Bias of Methods of Committee D-19 on Water[4]

D 3370 Practices for Sampling Water[4]

D 4793 Test Method for Sequential Batch Extraction of Waste with Water[6]

3. Terminology

3.1 *Definitions*—For definitions of terms used in this test method, see Terminology D 1129.

3.2 *Symbols:Symbols*—Variables listed in this test method are defined in the individual sections in which they are discussed. A list of the defined variables is also provided in Section 11.

4. Significance and Use

4.1 This test method is intended as a means for obtaining sequential extracts of a waste. The extracts may be used to estimate the release of certain constituents of the waste under the laboratory conditions described in this test method.

4.2 The pH of the extraction fluid used in this test method is to reflect the pH of acidic precipitation in the geographic region in which the waste being tested is to be disposed.

NOTE 1—Possible sources of information concerning the pH of precipitation in the geographic region of interest include state and federal environmental agencies, state universities, libraries, etc.

NOTE 2—For sequential batch extraction of waste using a nonacidic extraction fluid, see Test Method D 4793.

4.3 An intent of this test method is for the final pH of each of the extracts to reflect the interaction of the extractant with the buffering capacity of the waste.

4.4 This test method is not intended to provide extracts that are representative of the actual leachate produced from a waste in the field or to produce extracts to be used as the sole basis of engineering design.

4.5 This test method has not been demonstrated to simulate actual disposal site leaching conditions.

4.6 This test method produces extracts that are amenable to the determination of both major and minor (trace) constituents. When minor constituents are being determined, it is especially

---

[1] This test method is under the jurisdiction of ASTM Committee D34 on Waste Management and is the direct responsibility of Subcommittee D34.01.04 on Ancillary Topics.

Current edition approved Sept. 15, 1993. Published November 1993. Originally published as D 5284–92. Last previous edition D 5284–92.

[2] *Annual Book of ASTM Standards*, Vol 04.03.

[3] *Annual Book of ASTM Standards*, Vol 04.08.

[4] *Annual Book of ASTM Standards*, Vol 11.01.

[5] *Annual Book of ASTM Standards*, Vol 05.05.

[6] *Annual Book of ASTM Standards*, Vol 11.04.

important that precautions be taken in sample storage and handling to avoid possible contamination of the samples.

4.7 This test method has been tested to determine its applicability to certain inorganic components in the waste. This test method has not been tested for applicability to organic substances, volatile matter (see Note 5), or biologically active samples.

4.8 The agitation technique, rate, liquid-to-solid ratio, and filtration conditions specified in the procedure may not be suitable for extracting all types of wastes (see Sections 7 and 8 and Appendix X1).

5. Apparatus

5.1 *Straight Edge*, such as a thin-edged yardstick.

5.2 *Impermeable Sheet*, of glazed paper, oil cloth, or other flexible material of a composition suitable to the analytes of interest.

5.3 *Drying Pans or Dishes* (for example, aluminum tins, porcelain dishes, glass weighing pans), two per waste, suitable to the waste being tested and the instructions given in 9.2.

5.4 *Drying Oven*—Any thermostatically controlled drying oven capable of maintaining a steady temperature of ±2°C in a range of 100 to 110°C.

5.5 *Desiccator*, having a capacity to hold the drying pans described in 5.3 and the crucibles described in 5.16.

5.6 *Laboratory Balance*, capable of weighing to 0.1 g.

5.7 *Erlenmeyer Flask*, 2-L capacity, equipped with a magnetic stir bar.

5.8 *Magnetic Stir Plate*.

5.9 *Graduated Cylinder*, 1 or 2-L capacity.

5.10 *Pipet*, 1-mL capacity.

5.11 *Volumetric Flask*, 1-L capacity.

5.12 *Pipet*, 10-mL capacity. (Various other sized pipets, including micropipets, may be necessary for 9.3.2.)

5.13 *pH Meter*—Any pH meter with a readability of 0.01 units and an accuracy of ±0.05 units at 25°C.

5.14 *Carboy-type Container*, with spigot, 20 to 50-L capacity, of a composition suitable to the nature of the analyses to be performed (see Practices D 3370).

5.15 *Large Glass Funnel*.

5.16 *Crucibles*, porcelain, 20-mL capacity each, two per waste.

5.17 *Analytical Balance*, capable of weighing to 0.1 mg.

5.18 *Wash Bottle*, 500-mL capacity.

5.19 *Agitation Equipment*, of any type that rotates the extraction vessel in an end-over-end fashion at a rate of 30 ± 2 r/min such that the axis of rotation is horizontal and it passes through the center of the bottle (see Fig. 1 and Appendix X1).

NOTE 3—Similar devices having a different axial arrangement may be used if equivalency can be demonstrated.

5.20 *Pressure Filtration Assembly*—A pressure filtration device of a composition suitable to the nature of the analyses to be performed and equipped with a 0.45 or 0.8-μm pore size filter (see Note 8).

5.21 *Extraction Vessels*, cylindrical, wide-mouth, of a composition suitable to the nature of the waste and analyses to be performed, constructed of materials that will not allow sorption of the constituents of interest, and sturdy enough to withstand the impact of the falling sample fragments. The size of the container should be selected so that the sample plus extraction fluid occupy approximately 95 % of the container. The containers must have water-tight closures. Containers for samples in which gases may be released should be provided with venting mechanisms.

NOTE 4—Suitable container sizes range from 4.0 to 4.5 in. in diameter and 8.5 to 13.0 in. in height.

NOTE 5—The venting of the container has the potential to affect the concentration of volatile compounds in the extracts.

5.21.1 Extraction vessels should be cleaned in a manner consistent with the analyses to be performed (see Section 13 of Practice D 3370).

6. Reagents

6.1 *Purity of Reagents*—Reagent grade chemicals shall be used in all tests. Unless otherwise indicated, it is intended that all reagents shall conform to the specifications of the Committee on Analytical Reagents of the American Chemical Society, where such specifications are available.[7] Other grades may be used, provided it is first ascertained that the reagent is of sufficiently high purity to permit its use without lessening the accuracy of the determination.

6.2 *Purity of Water*—Unless otherwise indicated, references to water shall be understood to mean Type IV reagent water at 18 to 27°C conforming to Specification D 1193. The method by which the water is prepared, that is, distillation, ion exchange, reverse osmosis, electrodialysis, or a combination thereof, should remain constant throughout testing.

6.3 *Sulfuric Acid/Nitric Acid Solution*—A 60/40 weight percent (wt %) mixture prepared using 95 to 98 wt % sulfuric acid and 69 to 71 wt % nitric acid. (See 9.3 for instructions on the preparation of this solution.)

7. Sampling

7.1 Obtain a representative sample of the waste to be tested by using, where available, ASTM sampling methods developed for the specific industry (see Practice D 75, Guide D 420, Terminology D 653, and Test Method D 2234).

7.2 Sampling methodology for materials of similar physical form shall be used where no specific methods are available.

7.3 The amount of sample to be sent to the laboratory should be sufficient to perform the solids content determination as specified in 9.2, and to provide 100 g of sample on a dry weight basis for each extraction.

7.4 It is important that the sample of the waste be representative with respect to surface area, as variations in surface area would directly affect the leaching characteristics of the sample. Waste samples should contain a representative distribution of particle sizes.

NOTE 6—Information on obtaining representative samples can also be

[7] *Reagent Chemicals, American Chemical Society Specifications*, American Chemical Society, Washington, DC. For suggestions on the testing of reagents not listed by the American Chemical Society, see *Analar Standards for Laboratory Chemicals*, BDH Ltd., Poole, Dorset, U.K., and the *United States Pharmacopeia and National Formulary*, U.S. Pharmaceutical Convention, Inc. (USPC), Rockville, MD.

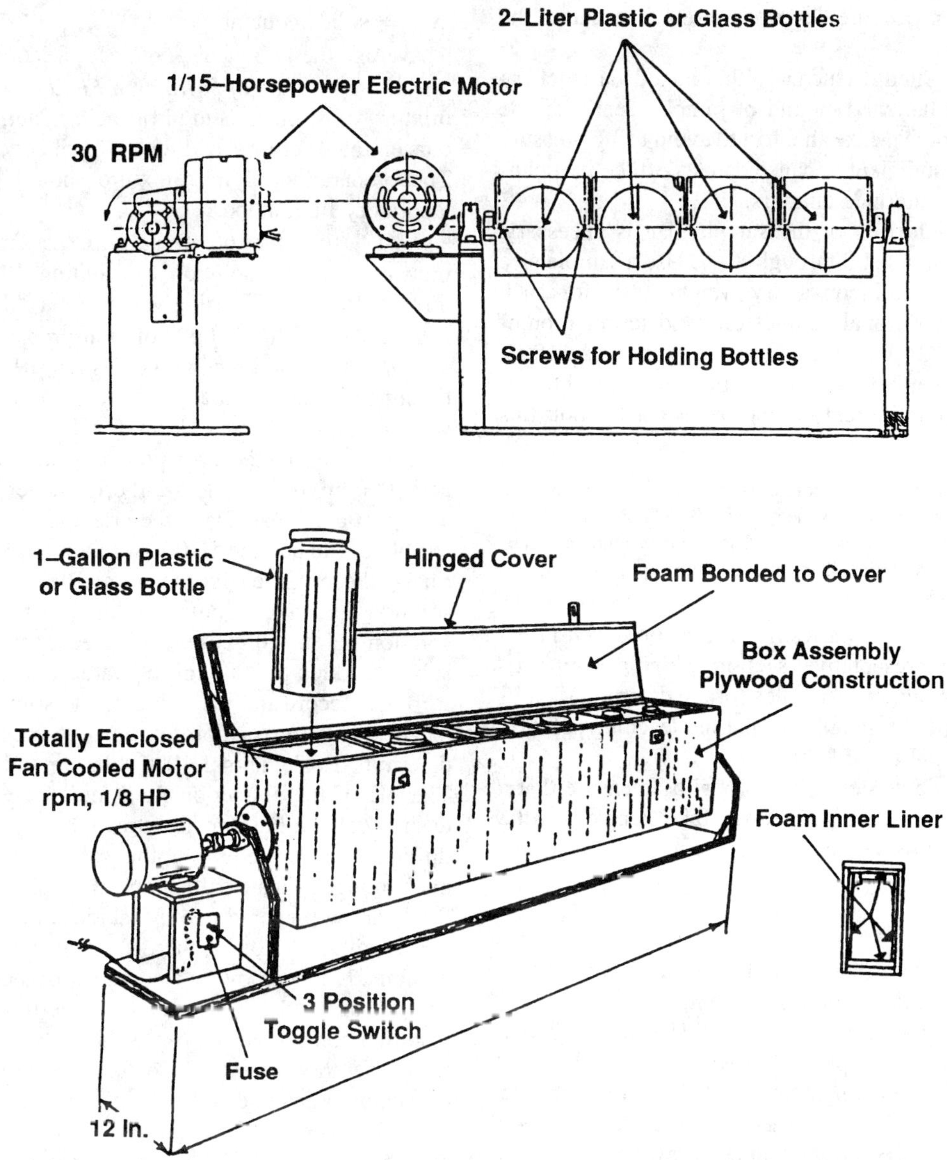

FIG. 1 Extractors

found in *Pierre Gy's Sampling Theory and Sampling Practice*.[8]

7.5 In order to prevent sample contamination or constituent loss prior to extraction, keep the samples in closed containers appropriate to sample type and desired analysis. See Practices D 3370 for guidance. Record the storage conditions and handling procedures in the report.

7.6 The time between collection and extraction of the sample should be determined by the nature of the sample and the information desired. See Practices D 3370 for guidance. Report the length of time between sample collection and extraction.

8. Sample Preparation

8.1 For free-flowing particulate solid wastes, obtain a sample of the approximate size required in the test by quartering the sample (Section 7) received for testing on an impermeable sheet of glazed paper, oil cloth, or other flexible material having a composition suitable to the analytes of interest, as follows:

8.1.1 Empty the sample container into the center of the sheet.

8.1.2 Gently flatten the sample out with a suitable straightedge until it is spread uniformly to a depth at least twice the maximum particle diameter.

8.1.3 Remix the sample by lifting a corner of the sheet and drawing it low across to the opposite corner in such a manner that the material is made to roll over and over and does not merely slide along. Continue the operation with each corner, proceeding in a clockwise direction. Repeat this operation ten times.

8.1.4 Lift all four corners of the sheet toward the center and, holding all four corners together, raise the entire sheet into the air to form a pocket for the sample.

[8] Pitard, F., *Pierre Gy's Sampling Theory and Sampling Practice*, Vols I and II, CRC Press, 1989.

8.1.5 Repeat the procedure described in 8.1.2 to flatten the sample out.

8.1.6 With a straightedge (such as a thin-edged yardstick) at least as long as the flattened mound of sample, gently divide the sample into quarters. Make an effort to avoid using pressure on the straightedge sufficient to cause damage to the particles.

8.1.7 Discard the alternate quarters.

8.1.8 If further reduction of the sample size is necessary, repeat the steps given in 8.1.3 through 8.1.7. Use a sample size to provide 100 g of solid on a dry weight basis for each extraction. Provide additional samples for the determination of solids content (see 9.2). Use of a sample size other than 100 g of solid on a dry weight basis for extraction is not recommended; however, if a different sample size is used, report this fact.

NOTE 7—For other acceptable methods of mixing and subsampling free-flowing solid particulate wastes, see *Pierre Gy's Sampling Theory and Sampling Practice*.[9] The method of subsampling should be determined by the physical properties of the waste, analytes of interest, and equipment available.

8.2 For field-cored solid wastes or castings produced in the laboratory, cut a representative section weighing approximately 100 g for testing, plus samples for the determination of solids content. Shape the sample so that the leaching solution will cover the material to be leached.

8.3 For multiphasic wastes, mix thoroughly to ensure that a representative sample will be withdrawn. Take samples for the determination of solids content at the same time that test samples are taken.

9. Procedure

9.1 Record a physical description of the sample to be tested, including particle size so far as it is known.

9.2 *Solids Content*—Determine the solids content of two separate portions of the sample as follows:

9.2.1 Dry to a constant weight, at 104 ± 2°C, two dishes or pans of size suitable to the solid waste being tested. Cool in a desiccator and weigh. Record the values to ±0.1 g.

9.2.2 Place 50 g of the waste to be tested into each pan. Record the mass of sample in each pan to ±0.1 g.

9.2.3 Dry 16 to 20 h at 104 ± 2°C. Record the temperature and time of the drying period.

9.2.4 Cool to room temperature in a desiccator and reweigh. Record the mass to ±0.1 g.

9.2.5 Repeat the steps given in 9.2.3 and 9.2.4 until constant container-sample masses are obtained. Discard the dried samples following completion of this step.

9.2.6 Calculate the solids content of the sample from the data obtained in 9.2.1, 9.2.2, and 9.2.4 as follows:

$$S = A/B \tag{1}$$

where:
A = mass of sample after drying, g,
B = original mass of sample, g, and
S = solids content, g/g.

Average the two values obtained. Record the solids content.

9.3 *Preparation of Extraction Fluid*—Prepare a 60/40 wt % mixture of sulfuric acid/nitric acid. Cautiously mix 60 g of concentrated sulfuric acid with 40 g of concentrated nitric acid. The preparation of this mixture should be performed in a laboratory fume hood.

9.3.1 Using the 60/40 sulfuric acid/nitric acid mixture, prepare a second solution by diluting 1.0 mL of the 60/40 mixture to 1000 mL using water and a 1-L volumetric flask.

9.3.2 Using the 1/1000 solution prepared in 9.3.1, prepare the extraction fluid having the desired pH ± 0.05 (see 4.2) by pipeting a volume of the 1/1000 solution into 2000 mL of water with mixing until the desired pH ± 0.05 is achieved. A recommended method for preparing the extraction fluid is to add 2000 mL of water to a 2-L erlenmeyer flask equipped with a magnetic stir bar. Place the erlenmeyer flask on a magnetic stir plate, and add the 1/1000 solution to the flask with stirring. Shake the mixture vigorously, and measure its pH once the solution is static. Continue this process until the desired solution pH ± 0.05 is reached. Record the amount of 1/1000 solution added to 2000 mL of water to achieve the desired pH ± 0.05. Record the pH value of the solution. Additional 2-L batches of the extraction fluid can be prepared by mixing the determined volume of 1/1000 solution with 2000 mL of water. The pH of the extraction fluid must be within ±0.05 of the desired value for use in the extraction procedure. For extracting different wastes requiring the same extraction fluid pH or performing replicate extractions, multiple batches of extraction fluid can be prepared and measured for correct pH, and if the pH is within ±0.05 of the desired value, the batches can be combined in a carboy-type container of a composition suitable to the nature of the analyses to be performed. The pH of the resulting solution in the carboy must be measured once again to verify the correct pH before using the solution in the extraction procedure and as a rinse solution (see the procedures given in 9.4 and 9.7). If the pH value is not within the above specification, the solution shall not be used, and fresh extraction fluid shall be prepared. Record the pH value of each batch, and of the solution in the carboy, prior to its use.

9.4 *Extraction Procedure*—If the entire procedure cannot be conducted without interruption, at least the first four extraction sequences must be conducted without interruption.

9.4.1 Determine the mass of the extraction vessel to be used in the extraction procedure to the nearest 0.1 g. Record the mass of the extraction vessel, M_{v1}. Use one extraction vessel per waste throughout the sequence of extractions.

9.4.2 Add 100 g (weighed to ±0.1 g) of solid waste on a dry weight basis to the extraction vessel. Calculate the amount of as-received waste to add using the following equation:

$$M = \frac{100}{S} \tag{2}$$

where:
S = solids content (g/g) determined in 9.2.6, and
M = mass of as-received waste (weighed to ±0.1 g) to add to the extraction vessel to yield 100 g of solid waste.

[9] Supporting data for the precision and bias information are available from ASTM Customer Service and may be obtained by requesting Report PCN:33-000004-56.

9.4.2.1 If a mass of solid waste on a dry weight basis other than 100 g is used, (Eq 2) through (4) must be modified to reflect the use of a mass other than 100 g. Replace 100 in these equations with the mass used. The use of a mass other than 100 g is not recommended.

9.4.3 Add a mass in g, M_{ef}, of extraction fluid (see 9.3) to the extraction vessel determined using the following equations:

$$M_{sw} = M - 100 \quad (3)$$

where:

M_{sw} = mass of moisture (g) in the sample added to the extraction vessel, and

$$M_{ef} = (20)\,(100) - M_{sw} \quad (4)$$

This will provide a solid-to-liquid ratio of 1:20 in the extraction vessel.

9.4.4 Agitate continuously for 18 ± 0.25 h at 18 to 27°C. Record the agitation time and temperature.

9.4.5 Open the extraction vessel. Observe and record any visible physical changes in the sample and leaching solution. Record the pH of the waste/leaching solution slurry.

9.5 *Filtration*—Transfer as much of the waste/leaching solution as possible through a large glass funnel to a pressure filtration device equipped with a 0.45 or 0.8-μm filter. Transfer the mixed slurry. Do not decant. Invert the extraction vessel over the filtration device and allow the liquid to drain for 1 min from the solid remaining in the extraction vessel. It is important to achieve as complete a transfer of fluid from the extraction vessel to the filtration device as possible. Pressure filter the liquid through the filter using nitrogen gas or an inert gas that will not contaminate or change the integrity of the sample. After the extract has passed through the filter, continue running gas through the filtration device at 30 psi for 3 min. The filtrate obtained is the extract mentioned in this test method (see 9.6 and 10.8). Determine the mass of the filtrate collected and report it as M_f for the extraction step. Measure the pH of the extract immediately, remove the amount of filtrate necessary for the determination of total dissolved solids content in 9.6, and then preserve the extract in a manner consistent with the chemical analyses or biological testing procedures to be performed (Practices D 3370, Section 15).

NOTE 8—Analytical results may be affected by the type of filter used. If a 0.8-μm filter pore size is used, the resulting extract should be digested prior to elemental analysis. The composition of the filter should also be considered. If the filter is composed of material that may contaminate the extract during filtration, the filter should be washed in the filtration device in a manner consistent with the chemical analyses or biological testing procedures to be performed on the extract. For example, for elemental analysis of the extract, if a filter composed of borosilicate glass fiber is used, in order to prevent contamination, it should be washed in the filtration device with a dilute acid solution and rinsed with approximately 2 L of water prior to filtration.

NOTE 9—Prefilters can be used only if it is absolutely necessary (filtrate for analysis or testing cannot be obtained unless a prefilter is used) due to the loss of sample trapped in the pores of the prefilter and the possibility of the prefilter disintegrating during rinsing.

NOTE 10—It is recommended that all filtrations be performed in a hood.

9.6 *Total Dissolved Solids Content (TDS)*—Add a 10.0-g aliquot of the extract to each of two 110°C± 2°C dried, preweighed crucibles. Place the samples in a drying oven at 110°C ± 2°C for 3 h. Record the drying oven temperature and drying time. Remove the crucibles and let them cool in a desiccator. Reweigh the crucibles and record their weights to ±0.1 mg.

NOTE 11—Only one drying is performed to limit the contact time between the solid and the rinse solution in the extraction vessel prior to the next extraction step (see 9.7, Section 10, and 10.7).

9.6.1 If the mass of solid lost through dissolution, M_d (see 10.2), in the first extract is less than 1 % of the mass of solid used in the first extraction step, and if the percent of solid lost through dissolution in the second extraction step is less than or equal to the percent of solid lost through dissolution in the first extraction step, the determination of TDS in the following extracts is not required, and the user can assume that TDS and M_d are equal to 0 for (Eq 6) and (7) for Extracts 3 through 10.

9.7 Quantitatively transfer the damp solid from the filter back to the original extraction vessel, including the filter. Use extraction fluid, prepared as described in 9.3, from a pre-weighed wash bottle to assist in this transfer and to rinse the filtration device. No more than 500 g of rinse solution should be used. Use the smallest amount of rinse solution possible to achieve a thorough transfer. Using tweezers or a similar device, recover the filter and rinse the adhering solid into the extraction vessel with rinse extraction fluid from the pre-weighed wash bottle. Do not leave the filter in the extraction vessel. Reweigh the wash bottle to determine the amount of rinse solution used in the transfer. Record this value as M_R. Weigh the extraction vessel following the transfer described above, and record this value as M_v. The extraction vessel may be sealed until a feasible time for the addition of new extraction fluid. This is to enable filtration during the next sequence at a reasonable time during the day. If the slurry is stored for longer than 6 h in the extraction vessel prior to the addition of new extraction fluid, the data generated by analysis of the extracts should be plotted to check for perturbation of the data curve.

10. Calculation

10.1 Calculate the TDS, in milligrams per gram of the filtrate, using the following equation:

$$TDS = (M_{sc} - M_c)/(10\ g) \quad (5)$$

where:

M_{sc} = mass of the crucible and dried solids, mg, and
M_c = mass of the crucible, mg.

10.2 Calculate the mass of the solid, in grams, lost through dissolution, M_d, using the following equation:

$$M_d = (TDS)\,(M_f)(0.001) \quad (6)$$

where:

M_f = mass of filtrate collected in that extraction, g, and
M_d = mass lost through dissolution.

10.3 Calculate the mass of the solid, in grams, corrected for TDS remaining for the next extraction step, M_s, using the following equation:

$$M_s = M_s^{e-1} - M_d \quad (7)$$

where:

M_s^{e-1} = mass of the solid extracted in the current extraction step, g.

NOTE 12—For example, in beginning the first extraction, M_s^{e-1} will equal 100 g, and to calculate the mass of solid remaining for the second extraction step, M_s will equal 100 g – M_d.

10.4 Calculate the combined mass of the solid and the residual liquid in the extraction vessel, M_{sl}, using the following equation:

$$M_{sl} = M_v - M_{vl} - M_R \quad (8)$$

10.5 Calculate the mass of liquid adhering to the solids in the extraction vessel, M_l, using the following equation:

$$M_l = M_{sl} - M_s \quad (9)$$

10.6 Calculate the mass, in grams, of extraction fluid to be added to the extraction vessel, extraction fluid mass (EFM), using the following equation:

$$EFM = [(M_s)(20)] - M_l - M_R \quad (10)$$

10.7 Add the amount of extraction fluid, EFM, determined in 10.6 to the extraction vessel and repeat the procedures described in 9.4.3 through 10.7 so that ten extractions are conducted in sequence.

NOTE 13—This procedure assumes that the amount of waste trapped in the filters after rinsing is negligible.

10.8 Analyze the extracts for specific constituents or properties, or use the extracts for biological testing procedures as desired, using appropriate ASTM test methods. Where no appropriate ASTM test methods exist, other test methods may be used and recorded in the report. Whether visible phase separation during storage of the extracts occurs or not, appropriate mixing should be used to ensure the homogeneity of the extracts prior to their use in such analyses or testing.

10.9 *Compensation for Carry-Over*—For each constituent in each of the extracts generated in the extraction sequence, the contribution to concentration from the residual liquid from the previous extraction step, C_j, can be calculated using the following equation:

$$C_j = [M_{li}/20(M_s^{e-1})][C_i] \quad (11)$$

where:

C_i = concentration of the constituent in the filtrate from the previous extraction step,
M_{li} = M_l from the previous extraction step, and
M_s^{e-1} = mass of solid extracted in the current extraction step (see Note 12).

11. Definitions of Variables

11.1 The following variables must be determined when performing the sequential batch extraction procedure:

11.1.1 *Solids Content Determination*:

A = mass of sample after drying in the determination of solids content of the waste to be extracted, g,
B = original mass of the sample prior to drying in the determination of solids content of the waste to be extracted, g, and
S = solids content of the waste to be extracted, g/g.

11.1.2 *First Extraction Step*:

M = mass of as-received waste added to the extraction vessel to yield 100 g (weighed to ± 0.1 g) of solid on a dry weight basis for the first extraction step, g,
M_{ef} = mass of extraction fluid to be added for the first step in the extraction procedure, g, and
M_{sw} = mass of moisture in the sample to be extracted in the first extraction step, g.

11.1.3 *TDS Determination*:

TDS = total dissolved solids content of the filtrate, mg/g,
M_c = mass of the crucible to be used in the TDS determination, mg, and
M_{sc} = mass of the crucible and dried solids in the TDS determination, mg.

11.1.4 *Extraction Sequence*:

M_d = mass of solid lost through dissolution during extraction, g,
M_f = mass of filtrate collected in that extraction, g,
M_s = mass of solid remaining for the next extraction step, g,
M_s^{e-1} = mass of solid extracted in the current extraction step, g,
M_{vl} = mass of the empty extraction vessel, g,
M_R = mass of the rinse solution, g,
M_v = combined mass of the extraction vessel, rinse solution, solid and moisture in the solid, and solid and liquid left in the extraction vessel after transfer to the filtering device, g,
M_{sl} = combined mass of solid and residual liquid in the extraction vessel following transfer of the moist sample cake back to the extraction vessel, g,
M_l = mass of liquid adhering to the solids in the extraction vessel following transfer of the moist sample cake back to the extraction vessel, g, and
EFM = mass of extraction fluid to be added for next extraction step, g.

11.1.5 *Compensation for Carry-Over*:

C_j = contribution to a constituent's concentration in the current step from the residual liquid of the previous extraction step, mg/L,
M_{li} = M_l from the previous extraction step, g,
M_s^{e-1} = mass of solid extracted in the current step, g, and
C_i = concentration of constituent in the filtrate from the previous extraction step, mg/L.

12. Report

12.1 Report the following information:

12.1.1 Source of information concerning the pH value of the precipitation in the geographic region of interest (see Note 1);

12.1.2 Source of the waste, date of sampling, methods of sampling and sample preservation, storage conditions, handling procedures, and length of time between sample collection and extraction;

12.1.3 Description of the waste, including its physical characteristics and particle size, if known (9.1);

12.1.4 Solids content (9.2);

12.1.5 Mass of solid waste on a dry weight basis extracted, if other than 100 g (8.1.8);

12.1.6 pH of the extraction fluid used for each extraction sequence, and as wash solution for each transfer;

12.1.7 Time and temperature used in the determination of solids content and TDS;

12.1.8 Agitation temperature and time;

12.1.9 Filter pore size used and filter composition;

12.1.10 Use of a prefilter, prefilter pore size, and composition;

12.1.11 Observations of changes in the test material or leaching solution (9.4.4);

12.1.12 Storage of the solid with rinse solution in the extraction vessel for any period longer than 6 h;

12.1.13 pH before and after filtration, and the results of specific analyses calculated in appropriate units and corrected for carry-over, if necessary; and

12.1.14 Dates on which sequential batch extraction was started and completed, preservation used for extracts, and dates of analyses.

NOTE 14—Fig. 2 presents a summary report format for recording some of the experimental data, and Figs. 3 and is a detailed laboratory worksheet that may be helpful in performing the test method.

13. Precision and Bias [9]

13.1 *Precision:*

13.1.1 A collaborative study of this test method involving ten laboratories was conducted. A spray dryer waste from an innovative clean coal technology process and a composite mining waste were extracted ten times in duplicate using this test procedure. The spray dryer waste was extracted using an extraction fluid having a pH of 4.3± 0.05, and the composite mining waste was extracted using an extraction fluid having a pH of 5.0 ± 0.05. Nitrocellulose filters having a pore size of 0.45-µm were used by the collaborative study participants for the filtering specified in 9.5. The resulting extracts were then analyzed for specific inorganic analytes. The collaborative study participants analyzed six analytical standards in triplicate to generate data for calculating analytical precision. These standards contained high, medium, and low concentrations of the elements of interest in the extracts. In addition, eight special samples were analyzed by the laboratories to generate data for evaluating the effects of filter pore size, and digestion versus nondigestion, on analytical values determined in the extracts. To prepare the special samples, extractions of four samples each of the spray dryer waste and the composite mining waste were performed using the first extraction step. Two extracts from each waste were generated by filtering the extraction slurries through 0.45-µm pore-size filters, and two extracts from each waste were generated by filtering the extraction slurries through 0.8-µm pore-size filters. One of the 0.45-µm extracts and one of the 0.8-µm extracts from each waste were digested using U.S. Environmental Protection Agency (EPA) Method 3010.[10] Portions of the 0.8-µm and the

[10] U.S. EPA, Method 3010: "Acid Digestion of Aqueous Samples and Extracts for Total Metals for Analysis by Flame Atomic Absorption Spectroscopy or Inductively Coupled Plasma Spectroscopy," *Test Methods for Evaluating Solid Waste: Physical/Chemical Methods (SW846)*, Vol 1A, 3rd Ed., 1990.

Sample Number: ____________

In Generating Extract Number	TDS (mg/g)	M_f (g)	M_d (g)	M_s (g)	M_v (g)	M_{vl} (g)	M_R (g)	M_{sl} (g)	M_l (g)	EFM (g)
1										
2										
3										
4										
5										
6										
7										
8										
9										
10										

FIG. 2 Sequential Batch Procedure Data Sheet

Sequential Extraction Laboratory Worksheet

Sample: ____________

Solids Content Determination:

Dish #1 mass after drying: ____________
Dish #2 mass after drying: ____________
Sample #1 mass before drying (B#1): ____________
Sample #2 mass before drying (B#2): ____________
Drying temperature: ____________
Drying time: ____________
Sample #1 mass after drying (A#1): ____________
Sample #2 mass after drying (A#2): ____________
Solids Content #1 (S#1): S = A#1/B#1 = ____________
Solids Content #2 (S#2): S = A#2/B#2 = ____________
Average Solids Content (S) = ____________

Extraction #: 1

Mass of empty extraction vessel (M_{vl}): ____________
Mass of sample to be extracted (M): M = 100/S = ____________
Mass of moisture in the sample to be extracted (M_{sw}: $M_{sw} = M - 100$ = ____________
Mass of extraction fluid to be added (M_{ef}): $M_{ef} = (20)(100) - M_{sw}$ = ____________
pH of extraction fluid: ____________
Extraction time and temperature: ____________
Filter type and pore-size: ____________
Mass of filtrate collected (M_f): ____________
pH of extract: ____________
Mass of wash bottle before transfer: ____________
Mass of extraction vessel after solid transfer back (M_v): ____________
Mass of wash bottle after transfer: ____________
Mass of rinse solution used in transfer (M_R): ____________

Total Dissolved Solids Content of Extract #1

Crucible #1 mass (M_c#1): ____________
Crucible #2 mass (M_c#2): ____________
Subsample #1 mass: ____________
Subsample #2 mass: ____________
Drying temperature: ____________
Drying time: ____________
Crucible #1 plus residue mass (M_{sc}#1): ____________
Crucible #2 plus residue mass (M_{sc}#2): ____________
Total Dissolved Solids Content (TDS):

TDS#1 = (M_{sc}#1 − M_c#1)/(10 g) = ____________
TDS#2 = (M_{sc}#2 − M_c#2)/(10 g) = ____________
Average TDS value: ____________

Mass of solid lost through dissolution (M_d): $M_d = (TDS)(M_f)(0.001)$

M_d = ____________

Mass of solid remaining for second extraction step (M_s): $M_s = 100 - M_d$

M_s = ____________

Mass of solid and residual liquid in the extraction vessel (M_{sl}):

$M_{sl} = M_v - M_{vl} - M_R$
M_{sl} = ____________

Mass of liquid adhering to solids in the extraction vessel (M_l):

$M_l = M_{sl} - M_s$
M_l = ____________

Mass of extraction fluid to add for the second extraction step (EFM):

$EFM = (M_s)(20) - M_l - M_R$
EFM = ____________

FIG. 3 Sequential Batch Extraction Laboratory Worksheet

0.45-μm filtrates and digestates were sent to the collaborative study participants for analysis. All of the participants in the study analyzed the regular extracts, analytical standards, and special analytical samples using inductively coupled plasma

Sequential Extraction Laboratory Worksheet—*Continued*

Sample: ____________

Extraction #: ____________ (2–10)

pH of extraction fluid: ____________

Extraction time and temperature: ____________

Filter type and pore-size: ____________

Mass of filtrate collected (M_f): ____________

pH of extract: ____________

Mass of wash bottle before transfer: ____________

Mass of extraction vessel after solid transfer back (M_v): ____________

Mass of wash bottle after transfer: ____________

Mass of rinse solution used in transfer (M_R): ____________

Total Dissolved Solids Content of Extract (if necessary)

Crucible #1 mass (M_c#1): ____________

Crucible #2 mass (M_c#2): ____________

Subsample #1 mass: ____________

Subsample #2 mass: ____________

Drying temperature: ____________

Drying time: ____________

Crucible #1 plus residue mass (M_{sc}#1): ____________

Crucible #2 plus residue mass (M_{sc}#2): ____________

Total Dissolved Solids Content (TDS):

TDS#1 = (M_{sc}#1 − M_c#1)/(10 g) = ____________

TDS#2 = (M_{sc}#2 − M_c#2)/(10 g) = ____________

Average TDS value ____________

Mass of solid lost through dissolution (M_d): M_d = (TDS) (M_f) (0.001)

M_d = ____________

Mass of solid remaining for next extraction step (M_s): $M_s = M_s^{e-1} - M_d$

M_s = ____________

Mass of solid and residual liquid in the extraction vessel (M_{sl}):

$M_{sl} = M_v - M_{vl} - M_R$

M_{sl} = ____________

Mass of liquid adhering to solids in the extraction vessel (M_l):

$M_l = M_{sl} - M_s$

M_l = ____________

Mass of extraction fluid to add for the next extraction step (EFM):

EFM = (M_s) (20) − M_l − M_R

EFM = ____________

FIG. 3 *(continued)*

spectrometry. Two of the laboratories digested their extracts and analytical standards prior to analysis. The other participants did not digest the extracts and standards. Information on the specific methods used by the participants is available.[9]

13.1.2 In the collaborative study, some of the laboratories teamed together so that eight data sets were generated by the participants. Two of the laboratories did not follow the study specifications, and as a result, six data sets were used to evaluate the precision of the extraction procedure. Practice D 2777 was used as a guideline for the statistical evaluation of the data.

13.1.3 The data generated in this collaborative study are specific to the test materials used in the study, the elements of interest, the pH values of the extraction fluids used, and 0.45-μm filter pore-size filtration. For other materials, elements, pH values, and filter types, these data may not apply.

13.1.4 The data generated in the collaborative study can be divided into three categories: (*1*) data from extraction of the test materials and analysis of the resulting extracts, which were used to determine the mean concentration for each element in each extract, $\bar{x}$, and the total standard deviation of the extraction procedure-plus-analysis of the extracts, s; (*2*) data from analysis of the analytical standards, which were used to determine the mean concentration for each level of analytical standard, $\bar{x}_a$, and the standard deviation for analysis of the analytical standards, s_a; and (*3*) data from analysis of the special analytical samples, which were used to calculate the mean concentration for each type of special sample and the standard deviation to evaluate the effects of filter pore size, and digestion versus nondigestion, on the analytical concentrations in the extracts.

13.1.5 Three types of precision can be determined from the data generated in the collaborative study. These are the total standard deviation, s, which is described in 13.1.4, the analytical standard deviation, s_a, which is also described in 13.1.4, and the estimated standard deviation of the extraction procedure, s_e, which represents the estimated error due to only the extraction method. The estimated multiple-laboratory standard deviation of the extraction procedure, s_e, for each element of interest in the two test materials was calculated using the following equation:

$$s_e = [s^2 - s_a^2]^{1/2} \qquad (12)$$

where:

s = standard deviation of the extraction procedure-plus-analysis of the extract, and

s_a = standard deviation for analysis of the analytical standard containing the concentration of the specific element closest to its concentration in the extract.

These estimated multiple-laboratory values for the elements of interest in extracts 1, 3, 5, 7, and 10 of the spray dryer waste and composite mining waste, along with the mean concentration values, $\bar{x}$ and $\bar{x}_a$, are listed in Table 1 and Table 2. These data, for all ten of the extracts of the spray dryer waste and composite mining waste, are available.[9]

13.1.6 The three types of precision values discussed in 13.1.4 and 13.1.5, total, analytical, and extraction procedure, can also be calculated based on a single operator. Calculations were performed to determine the total single-operator precision, s_o, the single-operator analytical standard deviation, s_{oa}, and the estimated single-operator precision of the extraction procedure, s_{oe}. The estimated single-operator precision of the

TABLE 1 Estimated Multiple-Laboratory Precision of the Sequential Batch Extraction Method Using Acidic Extraction Fluid for the Spray Dryer Waste, µg/g

	Aluminum	Barium	Boron	Calcium	Chromium	Silicon	Sodium	Strontium
Extract 1								
$\bar{x}$	25.5	14	1.8	3755	2.8	181	3515	214
$\bar{x}_a$	29	7	4	3927	2.8	171	3669	188
s	3.2	7	0.4	261	0.3	15	342	18
s_a	3	4	3	182	0.2	9	397	10
s_e	1	6	[A]	187	0.2	12	[A]	15
Extract 3								
$\bar{x}$	108	45	2.3	2795	[B]	106	100	61.9
$\bar{x}_a$	118	50	4	2611	[B]	107	116	46.7
s	10	19	0.9	111	[B]	8	11	5.3
s_a	9	4	3	212	[B]	8	28	2.7
s_e	4	19	[A]	[A]	[B]	0	[A]	4.6
Extract 5								
$\bar{x}$	145	41.3	3.2	2301	[B]	97	98	42.7
$\bar{x}_a$	118	50	4	2611	[B]	107	116	46.7
s	10	5.4	1.0	79	[B]	8	24	3.4
s_a	9	4	3	212	[B]	8	28	2.7
s_e	4	4	[A]	[A]	[B]	0	[A]	2.1
Extract 7								
$\bar{x}$	164	26.8	4.5	2119	[B]	80.1	50	26.8
$\bar{x}_a$	172	31	4	1983	[B]	75	48	19
s	14	2.4	1.0	79	[B]	6.8	11	1.7
s_a	11	5	3	149	[B]	6	28	1
s_e	9	[A]	[A]	[A]	[B]	3	[A]	1
Extract 10								
$\bar{x}$	183	19	5.3	1831	[B]	64.8	51	19
$\bar{x}_a$	172	31	5	1983	[B]	75	48	19
s	12	3	1.3	395	[B]	5.9	25	1
s_a	11	5	3	149	[B]	6	28	1
s_e	5	[A]	[A]	366	[B]	[A]	[A]	0

[A] The multiple-laboratory standard deviation of the extraction procedure cannot be determined. See 13.1.7.
[B] Values for extracts 3 to 10 cannot be calculated because the data sets contain "less-than" values.

TABLE 2 Estimated Multiple-Laboratory Precision of the Sequential Batch Extraction Method Using Acidic Extraction Fluid for the Composite Mining Waste, μg/g

	Barium	Calcium	Lead	Magnesium	Manganese	Silicon	Zinc
Extract 1							
$\bar{x}$	1.9	1815	63	107	36.7	55	297
$\bar{x}_a$	3	1614	91	102	39.9	44	312
s	0.3	238	8	12	5.5	21	35
s_a	4	110	4	5	1.7	3	11
s_e	A	211	7	11	5.2	21	33
Extract 3							
$\bar{x}$	2.3	448	B	12	3.6	35	13
$\bar{x}_a$	3	622	B	14	6.1	34	3.6
s	0.3	64	B	2	1.3	6	10
s_a	4	42	B	1	0.3	3	4.8
s_e	A	48	B	2	1.3	5	9
Extract 5							
$\bar{x}$	6	268	B	10	0.65	20	C
$\bar{x}_a$	7	248	B	14	1.0	13	C
s	2	16	B	2	0.15	3	C
s_a	5	11	B	1	0.1	2	C
s_e	A	12	B	2	0.1	2	C
Extract 7							
$\bar{x}$	15	228	B	5	D	14	C
$\bar{x}_a$	20	248	B	6	D	13	C
s	1	26	B	0.5	D	2	C
s_a	5	11	B	1	D	2	C
s_e	A	24	B	A	D	0	C
Extract 10							
$\bar{x}$	15	214	B	5	D	14	C
$\bar{x}_a$	20	248	B	6	D	13	C
s	1	26	B	2	D	3	C
s_a	5	11	B	1	D	2	C
s_e	A	24	B	2	D	2	C

[A] The multiple-laboratory standard deviation of the extraction procedure cannot be determined. See 13.1.7.
[B] Values for extracts 3–10 cannot be calculated because the data sets contain "less-than" values.
[C] Values for extracts 4–10 cannot be calculated because the data sets contain "less-than" values.
[D] Values for extracts 6–10 cannot be calculated because the data sets contain "less-than" values.

extraction procedure was calculated using the following equation:

$$s_{oe} = [s_o{}^2 - s_{oa}{}^2]^{1/2} \tag{13}$$

where:

s_o = single-operator standard deviation of the extraction procedure-plus-analysis of the extract, and

s_{oa} = single-operator standard deviation for analysis of the analytical standard containing the concentration of the specific element closest to its concentration in the extract.

The estimated single-operator precisions of the extraction procedure for the elements of interest in extracts 1, 3, 5, 7, and 10 of the spray dryer waste and composite mining waste are listed in Table 3 and Table 4. These values, for all ten of the extracts of the spray dryer waste and composite mining waste, are available.[9]

13.1.7 It was not economically practical to determine analytical precision using actual extracts of the wastes because of the extensive number of analyses that would have been required. The analytical standards were, therefore, used to determine analytical precision. Calculation of the standard deviation of the extraction procedure can provide only an approximation because the analytical standards do not contain the specific matrix resulting from the interaction of the extraction fluid and solid waste. In addition, the analytical standards are limited to three concentration levels for each element. To calculate the precision of the extraction procedure for a particular element, the analytical standard deviation for analysis of the analytical standard containing the concentration of the element closest to its concentration in the extract was used. For some of the extracts, the elemental concentration in the extract varies significantly from the element's closest concentration in the analytical standards. Because of the way in which the total precision and analytical precision were determined, in some cases, the analytical standard deviation values, s_a and s_{oa}, are larger than the total standard deviation values, s and s_o. In these cases, the precision of the extraction procedure cannot be determined.

13.1.8 The estimated precision of this sequential batch extraction procedure varies somewhat with the concentration of aluminum, barium, silicon, sodium, and strontium inthe spray dryer waste extracts according to Figs. 4-8, respectively, and the estimated precision of the sequential batch extraction procedure varies somewhat with the concentration of calcium, magnesium, manganese, and silicon in the composite mining waste extracts according to Figs. 9-12, respectively. Figs. 4-12 are plots of the calculated percent relative standard deviation of the extraction method versus the mean concentration of the constituent in the extract. The three-dimensional bar graphs are helpful in examining trends and interpreting the data. As expected, the relative precisions of the single-operator values are better than those of the multiple-operator values, and the single-operator relative precisions are encouragingly good. Plots for the other elements of interest in the spray dryer waste and composite mining waste can be found in PCN report.[9]

13.1.9 Collaborative study data from analysis of the special

TABLE 3 Estimated Single-Operator Precision of the Sequential Batch Extraction Method Using Acidic Extraction Fluid for the Spray Dryer Waste, μg/g

	Aluminum	Barium	Boron	Calcium	Chromium	Silicon	Sodium	Strontium
Extract 1								
$\bar{x}$	25.5	14	1.8	3755	2.8	181	3515	214
$\bar{x}_a$	29	7	4	3927	2.8	171	3669	188
s_o	1.8	1	0.3	154	0.2	6	50	6
s_{oa}	1	0.1	0.2	120	0.1	8	161	3
s_{oe}	1	1	0.2	97	0.2	[A]	[A]	5
Extract 3								
$\bar{x}$	108	45	2.3	2795	[B]	106	100	61.9
$\bar{x}_a$	118	50	4	2611	[B]	107	116	46.7
s_o	3	2	0.6	55	[B]	2	5	4.1
s_{oa}	2	1	0.2	62	[B]	3	6	1
s_{oe}	2	2	0.6	[A]	[B]	[A]	[A]	4
Extract 5								
$\bar{x}$	145	41.3	3.2	2301	[B]	97	98	42.7
$\bar{x}_a$	118	50	4	2611	[B]	107	116	46.7
s_o	2	1.2	0.5	37	[B]	3	5	1.5
s_{oa}	2	1	0.2	62	[B]	3	6	1
s_{oe}	0	1	0.5	[A]	[B]	0	[A]	1
Extract 7								
$\bar{x}$	164	26.8	4.5	2119	[B]	80.1	50	26.8
$\bar{x}_a$	172	31	4	1983	[B]	75	48	19
s_o	2	0.5	0.4	32	[B]	2.6	2	1.2
s_{oa}	9	0.5	0.2	43	[B]	3	3	0.2
s_{oe}	[A]	0	0.3	[A]	[B]	[A]	[A]	1.2
Extract 10								
$\bar{x}$	183	19	5.3	1831	[B]	64.8	51	19
$\bar{x}_a$	172	31	5	1983	[B]	75	48	19
s_o	2	1	0.5	185	[B]	3.5	4	1
s_{oa}	9	0.5	0.3	43	[B]	3	3	0.2
s_{oe}	[A]	1	0.4	180	[B]	2	3	1

[A] The single-operator standard deviation of the extraction procedure cannot be determined. See 13.1.7.

[B] Values for extracts 3–10 cannot be calculated because the data sets contain "less-than" values.

analytical samples show that filter pore size, 0.45-μm versus 0.8-μm, and digestion versus nondigestion, affect certain elemental concentrations determined in the extracts of the spray dryer waste and composite mining waste. The data also show that these effects are waste and element specific. Data from analysis of the special analytical samples are available.[9]

13.2 *Bias*—Determination of the extraction bias for thistest method is not possible, as no standard reference material exists. However, calculations using the collaborative study data were performed to evaluate analytical bias. Calculations were performed according to Practice D 2777.

13.2.1 In the collaborative study, the level of analytical accuracy for determining certain concentrations of silicon, sodium, strontium, chromium, and barium in the dilute acid solution standards (prepared using 60/40 weight percent sulfuric/nitric acid solution) having a pH of 4.3 ± 0.05 was not as good as desired based on the criterion given in Practice D 2777. For the levels of aluminum, boron, and calcium in these standards, the analytical accuracy was acceptable based on Practice D 2777 criterion. The level of analytical accuracy for determining certain concentrations of calcium, silicon, zinc, barium, lead, and magnesium in the dilute acid solution standards (prepared using 60/40 weight percent sulfuric/nitric acid solution) having a pH of 5.0 ± 0.05 was not as good as desired based on the criterion given in Practice D 2777. For the levels of manganese in these standards, the analytical accuracy was acceptable based on the same criterion. Data and additional information concerning these evaluations are available.[9]

13.3 *Comparative Data Evaluation*—Calculations using the collaborative study data were performed to evaluate the effect of extracting the test materials using deionized, distilled water versus the acidic extraction fluids. Data and information concerning these evaluations are available.[9] Elemental solubility in sequential batch water extracts versus acid solution extracts depends on the solution chemistry of the extraction slurries. As a result, the effect of using an acidic extraction fluid versus water depends on the material being tested, the pH of the extraction fluid, the element or parameter(s) of interest, and the extraction number in the sequence of the sequential batch extraction.

14. Keywords

14.1 acidic precipitation; extract; extraction fluid; leaching; sequential batch extraction; waste leaching technique

TABLE 4 Estimated Single-Operator Precision of the Sequential Batch Extraction Method Using Acidic Extraction Fluid for the Composite Mining Waste, µg/g

	Barium	Calcium	Lead	Magnesium	Manganese	Silicon	Zinc
Extract 1							
$\bar{x}$	1.9	1815	63	107	36.7	55	297
$\bar{x}_a$	3	1614	91	102	39.9	44	312
s_o	0.2	139	3	7	1.1	2	31
s_{oa}	0.2	33	3	1	0.5	1	6
s_{oe}	0	135	0	7	1.0	2	30
Extract 3							
$\bar{x}$	2.3	448	A	12	3.6	35	13
$\bar{x}_a$	3	622	A	14	6.1	34	3.6
s_o	0.2	22	A	1	0.3	3	1
s_{oa}	0.2	9	A	0.4	0.2	1	1.0
s_{oe}	0	20	A	1	0.2	3	0
Extract 5							
$\bar{x}$	6	268	A	10	0.65	20	B
$\bar{x}_a$	7	248	A	14	1.0	13	B
s_o	1	14	A	1	0.12	2	B
s_{oa}	0.05	5	A	0.4	0.03	1	B
s_{oe}	1	13	A	1	0.1	2	B
Extract 7							
$\bar{x}$	15	228	A	5	C	14	B
$\bar{x}_a$	20	248	A	6	C	13	B
s_o	0.2	9	A	0.3	C	1	B
s_{oa}	0.4	5	A	0.2	C	1	B
s_{oe}	D	7	A	0.2	C	0	B
Extract 10							
$\bar{x}$	15	214	A	5	C	14	B
$\bar{x}_a$	20	248	A	6	C	13	B
s_o	1	12	A	0.3	C	1	B
s_{oa}	0.4	5	A	0.2	C	1	B
s_{oe}	1	11	A	0.2	C	0	B

[A] Values for extracts 3–10 cannot be calculated because the data sets contain "less-than" values.
[B] Values for extracts 4–10 cannot be calculated because the data sets contain "less-than" values.
[C] Values for extracts 6–10 cannot be calculated because the data sets contain "less-than" values.
[D] The single operator standard deviation of the extraction procedure cannot be determined. See 13.1.7.

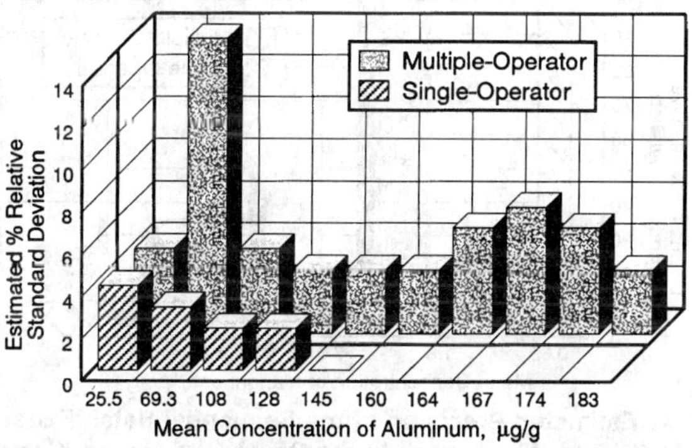

FIG. 4 Estimated Precision of the Sequential Batch Extraction Method for Aluminum in the Spray Dryer Waste

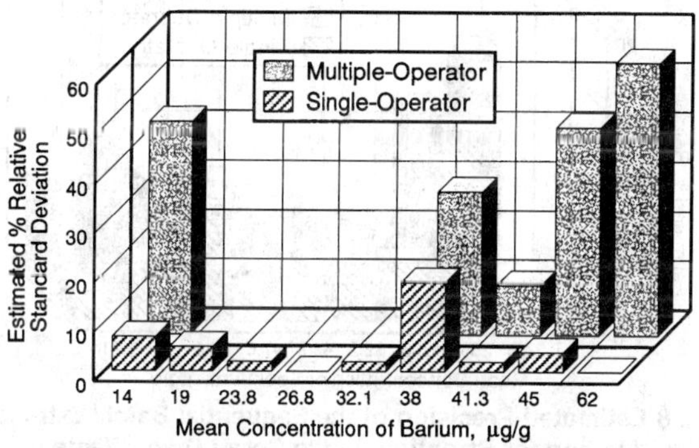

FIG. 5 Estimated Precision of the Sequential Batch Extraction Method for Barium in the Spray Dryer Waste

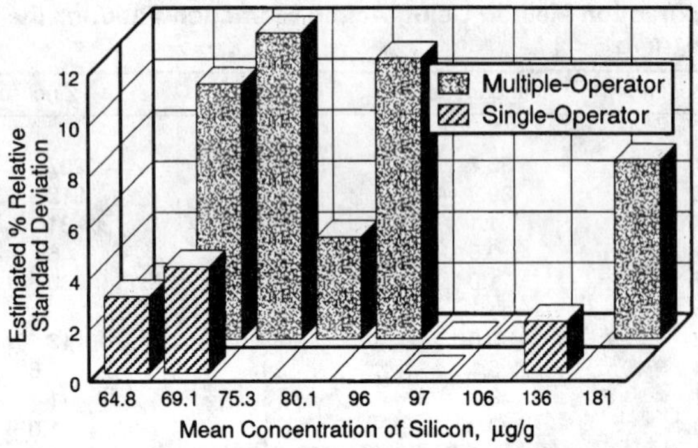

FIG. 6 Estimated Precision of the Sequential Batch Extraction Method for Silicon in the Spray Dryer Waste

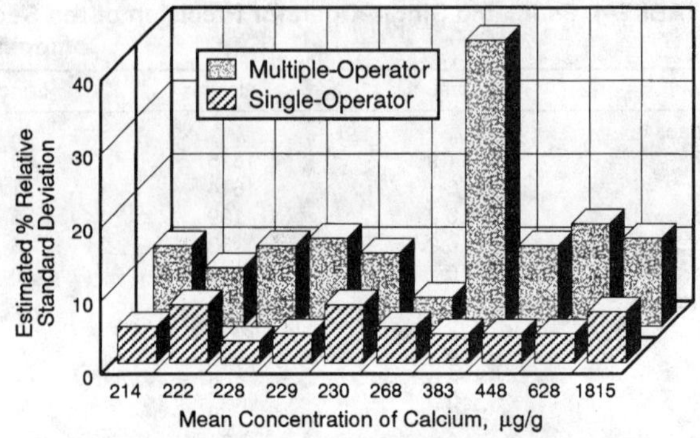

FIG. 9 Estimated Precision of the Sequential Batch Extraction Method for Calcium in the Composite Mining Waste

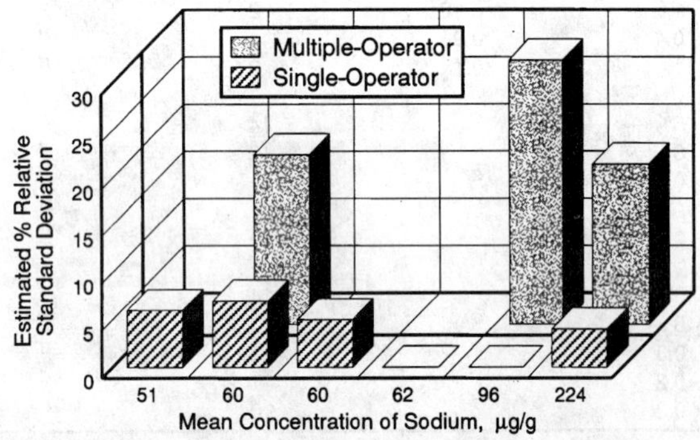

FIG. 7 Estimated Precision of the Sequential Batch Extraction Method for Sodium in the Spray Dryer Waste

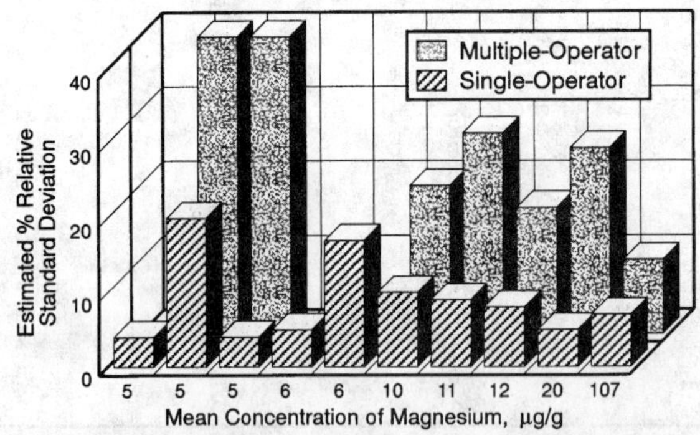

FIG. 10 Estimated Precision of the Sequential Batch Extraction Method for Magnesium in the Composite Mining Waste

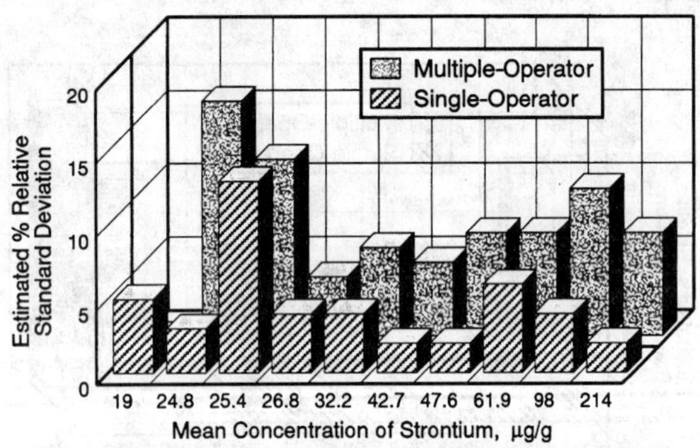

FIG. 8 Estimated Precision of the Sequential Batch Extraction Method for Strontium in the Spray Dryer Waste

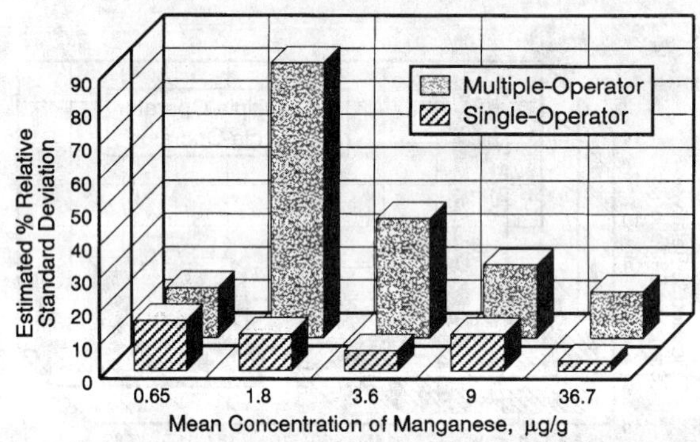

FIG. 11 Estimated Precision of the Sequential Batch Extraction Method for Manganese in the Composite Mining Waste

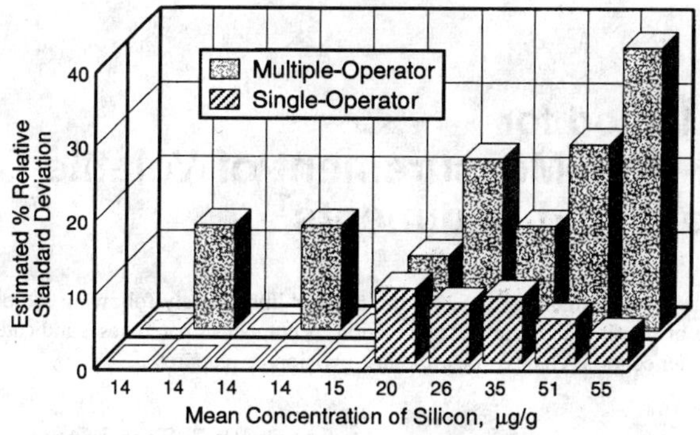

FIG. 12 Estimated Precision of the Sequential Batch Extraction Method for Silicon in the Composite Mining Waste

APPENDIX

(Nonmandatory Information)

X1. AGITATION TECHNIQUES AND RATE AND LIQUID/SOLID RATIO

X1.1 The agitation rate, equipment, and liquid/solid ratio specified in this test method may influence the results on certain solid wastes significantly, and it may not be adequate for certain solid wastes, such as monolithic, solidified, or organic wastes.

X1.2 The possible effects of varying the agitation techniques and rate include degree of mixing, rate of release of constituents, and particle abrasion effects. The precision of this test method may also be influenced.

X1.3 The possible effects of varying the liquid/solid ratio include degree of mixing, rate of release of constituents (and possible concentration effects, depending on availability), and particle abrasion effects.

Designation: D 5285 – 92 (Reapproved 1997)

Standard Test Method for 24-Hour Batch-Type Measurement of Volatile Organic Sorption by Soils and Sediments[1]

This standard is issued under the fixed designation D 5285; the number immediately following the designation indicates the year of original adoption or, in the case of revision, the year of last revision. A number in parentheses indicates the year of last reapproval. A superscript epsilon (ε) indicates an editorial change since the last revision or reapproval.

1. Scope

1.1 This test method describes a procedure for determining the sorption affinity of waste solutes by unconsolidated geologic material in aqueous suspension, for example, soils, fluvial sediments, sedimentary deposits, or any other accumulations of unconsolidated solid particles (for a companion method, for metal solute, see Test Method D 4319). The waste solute may be derived from a variety of sources such as wells, underdrain systems, or laboratory solutions like those produced by waste extraction tests (for example, Test Method D 3987).

1.2 This test method is applicable for screening and providing the relative rankings of a large number of samples for their sorption affinity in aqueous leachate/geomedia suspensions. This test method may not simulate closely the sorption characteristics that would occur in unperturbed geologic settings and under flow conditions.

1.3 While this test method is intended to be applicable for all soluble organic constituents, care must be taken with respect to the stability of the particular constituents and their possible losses from solution by such processes as volatilization or degradation by microbes, light, or hydrolysis.

1.4 The values stated in SI units are to be regarded as the standard.

1.5 *This standard does not purport to address all of the safety concerns, if any, associated with its use. It is the responsibility of the user of this standard to establish appropriate safety and health practices and determine the applicability of regulatory limitations prior to use.*

2. Referenced Documents

2.1 *ASTM Standards:*

D 653 Terminology Relating to Soil, Rock, and Contained Fluids[2]

D 1129 Terminology Relating to Water[3]

D 1193 Specification for Reagent Water[3]

D 2216 Test Method for Laboratory Determination of Water (Moisture) Content of Soil and Rock[2]

D 3987 Test Method for Shake Extraction of Solid Waste with Water[4]

D 4319 Test Method for Distribution Ratios by the Short-Term Batch Method[2]

D 4410 Terminology of Fluvial Sediment[3]

3. Terminology

3.1 Definitions—For definitions of terms used in this test method, refer to Terminology D 1129. For additional definitions of terms pertaining to soils and fluvial sediments, refer to Terminologies D 653 and D 4410.

3.1.1 *solute*—chemical species (for example, ion, molecules, etc.) in solution.

3.1.2 *sorbate*—chemical species sorbed by a sorbent.

3.1.3 *sorbent*—a solid substance that sorbs the solute from solution (for example, soil, sediment, till, etc.).

3.1.4 *sorption*—depletion of an amount of solute initially present in solution by a sorbent.

3.1.5 *unconsolidated geologic material (geomedia)*—a loosely aggregated solid natural material of geologic origin (for example, soil, sediment, till, etc.).

3.2 *Definitions of Terms Specific to This Standard:*

3.2.1 *distribution coefficient, K_d*— identical to the distribution ratio, R_d (see 3.2.2), except that it is considered to be an equilibrium value and independent of the concentration of solute (that is, linear sorption curve).

3.2.2 *distribution ratio, R_d*—the ratio of the concentration of solute sorbed on the soil or other geomedia divided by its concentration in solution. A 24-h R_d is the analogous ratio evaluated after 24 h of contact of the solute with the geomedia. The R_d value is calculated as follows:

$$R_d = \frac{\text{(mass of solute sorbed per unit mass of geomedia)}}{\text{(mass of solute in solution after 24 h per unit mass of solution)}}$$

3.2.3 *Discussion*—Dissimilar R_d values may be obtained for the same solute if different initial solute concentrations are used, depending on the sorption behavior of the solute and the properties of the geomedia (that is, nonlinear sorption curve). This concentration dependency may be absent where the solute concentrations are sufficiently low. It is absent when the characteristics of the particular solute-sorbent combination yield R_d values that are independent of the concentration of

[1] This test method is under the jurisdiction of ASTM Committee D34 on Waste Management and is the direct responsibility of Subcommittee D34.01.06 on Analytical Methods.

Current edition approved Aug. 15, 1992. Published October 1992.

[2] *Annual Book of ASTM Standards*, Vol 04.08.

[3] *Annual Book of ASTM Standards*, Vol 11.01.

[4] *Annual Book of ASTM Standards*, Vol 11.04.

solute (that is, linear sorption curve).

3.2.4 *octanol water partition coefficient, K_{ow}*—the distribution coefficient of an organic compound between *n*-octanol and water. It has been found to be useful in predicting other constants that describe the distribution of organics between organic carbon-containing media and water and is usually easy to measure experimentally.

3.2.5 *sorption affinity*—the relative degree of sorption that occurs by a geomedia.

3.2.6 *test compound constant, K_{oc}*— the distribution coefficient between water and a hypothetical geomedia that is 100 % organic carbon. As such, it is used to convert distribution coefficients (or distribution ratios) measured for one geomedia to another geomedia with a different organic carbon content.

4. Summary of Test Method

4.1 Distilled water, natural water, waste leachate, or any aqueous solution containing a known concentration of solute is mixed with a known amount of unconsolidated geologic material (geomedia) for 24 h. Changes in solute concentrations are used to calculate a distribution ratio (R_d).

5. Significance and Use

5.1 This test method is intended to allow for a rapid (24-h) index R_d of a geomedia's sorption affinity for given chemicals or leachate constituents. A large number of samples may be analyzed using this test method to determine a comparative ranking of those samples, based on the amount of solute sorbed by the geomedia, or by various geomedia or leachate constituents. The 24-h time period is used to make the test convenient as well as to minimize microbial degradation, which may be a problem in longer procedures. Due to this time constraint, the final (24-h) concentration should not be confused with that of an equilibrium or steady-state concentration. While R_d values are directly applicable for screening and comparative ranking purposes, their use in predictive field applications generally requires the assumption that $R_d = K_d$; the validity of this assumption must be evaluated carefully by qualified personnel.

5.2 The 24-h time limit may be sufficient to reach a steady-state R_d. However, to report this determination as a steady-state R_d (that is, $R_d = K_d$), the relevant time studies must be conducted to document the development of steady-state conditions within the 24-h time period. Tests exceeding 24-h are beyond the scope of this test method.

6. Interferences

6.1 When solutes of unknown stability are dealt with, when they are either in contact with the geomedia or used as blanks, care must be taken to ensure that volatilization, hydrolysis, photodegradation, microbial degradation, oxidation-reduction, or other physicochemical processes are not operating at a significant rate within the time frame of the procedure. The stability, and hence loss from solution, may affect the outcome of this procedure if the aforementioned reactions are significant. The compatibility of the test method and the solute of interest may be assessed by determining the differences between the initial solute concentration (see 9.3.3) and the silica sand blank concentration of the solute (see 9.3.7). If this difference is significant compared to the expected precision of the test method, the R_d value generated may be unreliable and thus must be evaluated carefully.

6.2 It is essential that the geomedia used for measuring distribution ratios be free of any chemical species for which the distribution ratio is to be measured. If it is suspected that the geomedia is contaminated, a procedure identical to that described in 9.3.1-9.3.7 should be followed, with Type IV water substituted for the test solution. If the concentration of a chemical in the water after equilibrating for 24 h, compared to the concentration of that chemical in the solution to be tested, is significant compared to the expected precision of the test method (± 8 %), a different geomedia should be used. Correcting the measured R_d for contaminated soil is not recommended since the preexisting contamination may not be adsorbed such that it can equilibrate readily with water.

7. Apparatus

7.1 *Agitation Equipment*—The Rotary Solid Waste Extractor[5] specified in Test Method D 3987.

7.2 *Containers*—Round, wide-mouth glass bottles that can be fitted with standard volatile organics analysis (VOA) caps and are compatible with the rotary extractor.

7.3 *Balance*, having a minimum capacity of 500 g and a sensitivity of ± 0.05 g, to be used for weighing the geomedia and solute solution. A more sensitive balance may be required for preparing analytical standards.

8. Reagents

8.1 *Purity of Reagents*—Reagent grade chemicals shall be used in all tests. Unless otherwise indicated, it is intended that all reagents shall conform to the specifications of the Committee on Analytical Reagents of the American Chemical Society, where such specifications are available.[6] Other grades may be used, provided it is first ascertained that the reagent is of sufficiently high purity to permit its use without lessening the accuracy of the determination.

8.2 *Purity of Water*— Unless otherwise indicated, references to water shall be understood to mean Type IV reagent water conforming to Specification D 1193.

8.3 *Silica Sand*— 20/40 grit sand blasting pure silica sand, to be used for the silica sand blank.

8.4 *Preparation of Analytical Standard Solutions*:

8.4.1 *Standard Stock Solutions*—Place approximately 9 to 9.8 mL of methanol into a 10-mL ground-glass, stoppered volumetric flask, and then allow the flask to stand unstoppered for a few minutes or until all methanol-wetted surfaces have dried. Weight the flask to the nearest 0.1 mg and immediately add a few drops of the test organic to the flask by using a 100-μL syringe, and then reweigh. Be sure that the drops fall directly into the methanol without contacting the neck of the

[5] Diamondstone, B. T., Burke, R. W., and Gardner, E. L., "Improved Leach Measurements on Solid Wastes," *Standardization News*, Vol 10, No. 6, June 1982, pp. 28–33.

[6] "Reagent Chemicals, American Chemical Society Specifications," Am. Chemical Soc., Washington, DC. For suggestions on the testing of reagents not listed by the American Chemical Society, see "Analar Standards for Laboratory U.K. Chemicals," BDH Ltd., Poole, Dorset, and the "United States Pharmacopeia."

flask. Finally, dilute to volume, stopper, and mix by inverting the flask several times.

8.4.2 Transfer the stock solution into a polytetrafluoroethylene-sealed screw-cap vial. Store, with minimal headspace, at approximately 4°C.

8.4.3 All standard stock solutions must be replaced after one month, or sooner, if a comparison with check standards indicates a change of concentration greater than 5 %.

8.4.4 When ready to prepare secondary standard solutions, stabilize the temperature of the vial containing the standard stock solutions. When the temperatures of the stock solutions and containers have been stabilized, transfer into a known amount of pure water using a 25-μL syringe. Keep the headspace as small as possible.

8.4.5 Aqueous standards and solutions stored with headspace are not stable and should be discarded after 1 h.

8.5 *Preparation of Test Solutions for Sorption Studies*:

8.5.1 Solutions used as the test fluid for sorption studies may be actual environmental samples, laboratory or field leachates, or laboratory-prepared solutions. The following procedure is recommended if laboratory-prepared solutions are used as the test fluid.

8.5.2 Place 990 mL of Type IV water into a 1000-mL clean amber glass bottle, and then seal with an open-top screw-cap with polytetrafluoroethylene-lined septum. Inject known amounts of the pure test components into each bottle using a 100-μL syringe. (Prepare at least two different concentration levels.) Then mix by inverting the bottle several times. The actual test concentration should be determined by gas chromatographic analysis.

8.5.3 When the sorption coefficients for weakly adsorbing organics are to be measured, the best results are obtained using test fluids that are as concentrated as possible. However, experience has shown that it is very difficult to dissolve many nonpolar organics at concentrations approaching their published solubilities. Concentrations that are approximately one-half of the published solubilities seem to be a reasonable compromise between the need for concentrated solutions and the practicality of preparing them.

9. Procedure

9.1 *Preparation of Materials to be Used as Sorbents*:

9.1.1 Samples of sorbents such as soils, clays, or sediments are spread out on a flat surface, no more than 2 to 3-cm deep, and allowed to air dry for 7 days or until constant weight (a change of less than 5 % per 24-h period) is achieved. Do *not* oven dry the samples.

9.1.2 After the sample has air dried, it is passed through a 2-mm screen sieve. Large aggregates are to be crushed without grinding the sample by using a clean mortar and a rubber-tipped pestle.

9.1.3 Mix the sieved material until the sample is homogeneous. Use a riffle splitter, or other unbiased splitting procedure, to obtain subsamples of appropriate size.

9.1.4 Remove the subsamples and determine the moisture content of the air-dried sample (refer to Test Method D 2216).

9.1.5 Determine the mass of the geomedia sample, corrected for moisture content. Determination of the air-dry soil mass equivalent to the desired mass of oven-dried soil is made as follows:

$$A = Mm \frac{100}{100 - W} \quad (1)$$

where:
A = air-dry geomedia mass,
Mm = mass of oven-dried geomedia desired, and
W = moisture percent in air-dried geomedia.

9.2 *Selection of a Geomedia: Solution Ratio*:

9.2.1 The geomedia to solution ratios used for sorption studies will depend on the distribution ratio (R_d) and the relative degree of sorption desired. The first step in selection of a geomedia to solution ratio for a test solute is to estimate the value of the test compound constant (K_{oc}) from the water solubility (S, ppm) of the test compound, or from the value of its *n*-octanol/water partition coefficient (K_{ow}). The relationship between the K_{oc} and S or K_{oc} and K_{ow} for various hydrophobic compounds has been reported by Hassett, et al.[7]

$$\log K_{oc} = 3.95 - 0.62 \log S \text{ (ppm)} \quad (2)$$

$$\log K_{oc} = 0.088 + 0.909 \log K_{ow} \quad (3)$$

9.2.2 Sorption constant values (K_{oc}) can be adjusted for the organic carbon content of the geomedia by using the percent organic carbon to predict the geomedia distribution ratio (R_d). Therefore, the individual values of the distribution ratio (R_d) can be calculated if the respective percent organic carbon content of the geomedia is known:

$$R_d = \frac{K_{oc} \times \% \, OC}{100} \quad (4)$$

9.2.3 The selection of an appropriate ratio can be based on a plot of $1/R$ (where R = g soil/g water) versus K_d (estimated) for fixed percents of solute desired to be sorbed. The relationship obtained by rearranging the definition of R_d given in 4.2 is as follows:

$$\frac{1}{R} = \left[\left(\frac{W_s^0}{W_s} \right) - 1 \right] \times R_d \quad (5)$$

where:
W_s^0 = weight of total solute initially in the aqueous solution, and
W_s = weight of solute adsorbed.

Examples of ranges of R_d (up to 25) and water/geomedia ratios (up to 30) are shown in Fig. 1. For detailed procedures, refer to "Batch-Type Adsorption Studies: Conceptual Guidelines and Experimental Procedures" (Roy, et al.).[8]

9.2.4 In theory, sorption studies can be performed at any geomedia/water ratio, but in practice it is recommended that

---

[7] Hassett, J. J., Banwart, W. L., and Griffin, R. A. ,"Correlation of Compound Properties with Sorption Characteristics of Non-Polar Organic Compounds by Soils and Sediments," *Concepts and Limitations, Environment and Solid Waste; Characterization Treatment and Disposal*, Chapter 15, Eds. C. W. Francis and S. I. Auerbach, Butterworth Publishers, Woburn, MA, 1983, pp. 161–178.

[8] Roy, W. R., Krapac, I. G., Chou, S. F. J., and Griffin, R. A., "Batch Type Adsorption Procedures for Estimating Soil Attenuation of Chemicals," *Technical Resource Document*, EPA/530-SW-87-006 (NTIS No. PB87146-155), Washington, DC, 1986.

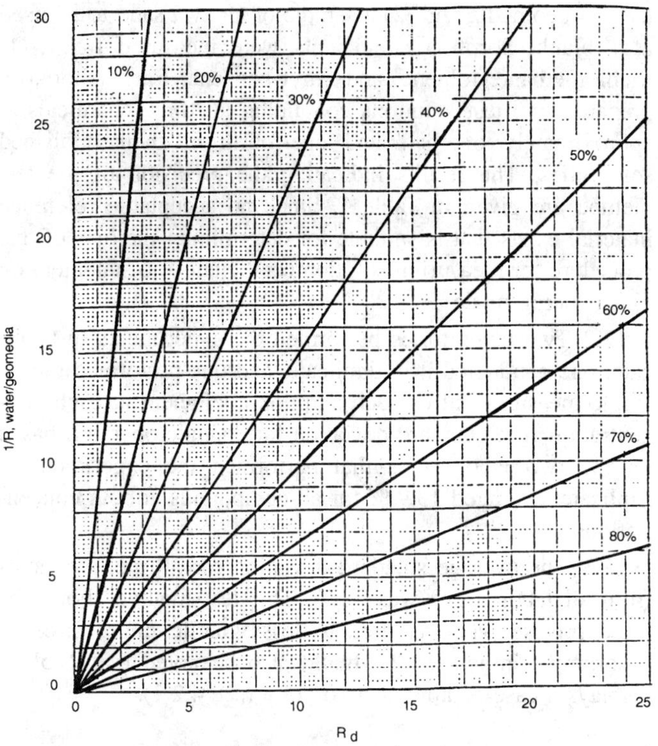

FIG. 1 Relationship Between the Water/Geomedia Ratios and R_d at Various Percents of Solute Adsorbed

the geomedia/water ratio selected fall within the range of 20 to 80 % sorption of the compound.

9.2.5 Many of the volatile organic-soil combinations of interest result in an R_d value so low that Fig. 1 is not particularly useful. In this case, it is recommended that a water to geomedia ratio (1/ R) of 2 be used as a practical lower limit.

9.3 *Reaction Container Filling*

9.3.1 Pre-weigh a reaction container (such as a 125-mL amber glass serum bottle), a polytetrafluoroethylene-faced septum and aluminum seal, and then place an appropriate amount of air-dried geomedia into the reaction container. Weigh the geomedia, bottle, polytetrafluoroethylene-faced septum, and aluminum seal combination, and calculate the weight of geomedia (oven-dried basis) to the nearest 0.1 g. The amount of geomedia used depends on the geomedia/solution ratio estimated previously.

9.3.2 Fill the container with the solute solution. The filling procedure may require shaking the geomedia/solution sample gently, to drive trapped air from the geomedia, and refilling with solution to minimize headspace. Place the polytetrafluoroethylene-faced septum and aluminum seal or the VOA cap on the container and invert to ensure that no air bubbles are present. Weigh the filled bottle and calculate the weight of solution to the nearest 0.1 g.

9.3.3 Retain a separate reaction container of the initial solute solution after filling each set of containers to determine the initial solute concentration. If the work is completed quickly and carefully, solute losses from the filling are negligible.

9.3.4 Place the filled reaction container from 9.3.2 on a rotating tumbler and rotate continuously at 29 ± 2 r/min for 24 h. Maintain the temperature at 22 ± 5°C.

9.3.5 After 24 h of tumbling, allow the solutions in the reaction containers to settle for a minimum of 1 h. If the water in the neck of the bottle is too turbid to sample with a syringe, centrifuge for a minimum of 1 h at 3000 r/min. Analyze directly from the container by removing the samples through the septum with a syringe.

9.3.6 Analyze the aqueous supernatant for the volatile organic constituents of interest. Consult U.S. EPA Test Methods 8010, 8020, and 8240 for detailed analysis procedures for volatile solutes.[9]

9.3.7 Each geomedia sample is to be subjected to the procedure in three or more replicates of at least two different concentrations. Blanks handled exactly as described above (steps given in 9.3.1-9.3.6), except that silica sand is used in place of the geomedia, are required for each solute solution, or dilution thereof, each time a series of geomedia is analyzed.

10. Calculation

10.1 Calculate the distribution ratio as follows:

$$R_d = \frac{(A - B)Ms}{(Mm)B} \quad (6)$$

where:

A = initial concentration of the solute defined as the final concentration in the retained sample (9.3.3) in g/mL,
B = final concentration of the solute after 24 h in contact with the geomedia in g/mL,
Ms = mass of solute solution in g,
Mm = mass of geomedia expressed on an oven-dried basis in g, and
R_d = distribution ratio.

11. Report

11.1 Mark the R_d value clearly as a nonequilibrium 24-h distribution ratio.

11.2 Report both the initial solute concentration (A in 10.1) and the final solute concentration (B in 10.1).

11.3 Report the final solute concentration for each blank. If a blank differs from the concentration of the retained solution by more than 10 %, repeat the procedure.

11.4 Report the mass of geomedia (Mm in 10.1), mass of solution (Ms in 10.1), and room temperature at which the extraction was conducted.

11.5 Note and report negative R_d values when and if they occur. Substantial negative R_d values may occur if the geomedia contains the test solute prior to application of the method. If this is suspected, test the geomedia with pure water extractions (see 6.2).

11.6 Where feasible, determine and report the pH of the solute solution and sorbent-solute mixture.

12. Precision and Bias [10]

12.1 Four laboratories participated in a collaborative study of this test method. Each laboratory was provided with two soil

[9] "Test Methods for Evaluating Solid Waste," EPA/SW-846, Washington, DC, November 1986.

[10] Supporting data have been filed at ASTM Headquarters/Customer Service and may be obtained by requesting: RR:PCN:33-000004-56.

TABLE 1 Volatile Organics Distribution Ratios (R_d); Four Laboratory Tests[A]

Laboratory	Compound	High Concentration	Low Concentration
	1,1,1-Trichloroethane		
1		0.21	0.21
1		0.59	0.98
2		0.50	0.48
3		0.77	...
3		0.75	...
3		0.73	...
4		0.52	1.28
4		1.00	1.02
4		1.00	1.07
		0.67 ± 0.25	0.84 ± 0.41
	Toluene		
1		0.55	1.11
1		0.65	1.09
2		0.75	0.93
3		1.57	1.73
3		1.41	1.88
3		1.50	2.09
4		1.09	1.17
4		1.19	1.07
4		1.36	1.12
		1.12 ± 0.38	1.35 ± 0.43
	Ethyl Benzene		
1		1.29	1.57
1		0.67	1.53
2		1.69	1.56
3		3.15	3.80
3		4.00	3.95
3		4.10	3.88
4		1.69	2.65
4		2.60	2.34
4		2.29	2.36
		2.39 ± 1.19	2.62 ± 1.02

[A] The uncertainty shown is one standard deviation

samples: (*1*) the A horizon of a Catlin silt loam to be used as the test geomedia and (*2*) a sample of Ottawa sand to be used as a blank. Sorption coefficients were measured for 1,1,1-trichloroethane, toluene, and ethyl benzene. Each laboratory prepared two test solutions containing all three chemicals at approximately one-half and one-fourth of their published solubilities. The results indicating the precision of the test method are given in Table 1. Since no acceptable reference material exists that is suitable for determining the bias of this procedure for measuring sorption coefficients, no statement on bias is being made.

12.2 Two laboratories performed the experiments in triplicate, one conducted it in duplicate, and one performed only a single measurement of each sorption coefficient at each concentration. The standard deviation for a single operator, based on the two laboratories that performed the experiment in triplicate, averaged 13.9 % for all compounds at both concentrations.

12.3 The average standard deviation for all measurements from all laboratories was 38 %. This test method should be more precise when applied to less volatile, more strongly adsorbing compounds. It should also be noted that Table 1 displays a case of nonlinear sorption (see 4.2.1).

Designation: D 5358 – 93 (Reapproved 1998)

Standard Practice for Sampling with a Dipper or Pond Sampler[1]

This standard is issued under the fixed designation D 5358; the number immediately following the designation indicates the year of original adoption or, in the case of revision, the year of last revision. A number in parentheses indicates the year of last reapproval. A superscript epsilon (ε) indicates an editorial change since the last revision or reapproval.

1. Scope

1.1 This practice describes the procedure and equipment for taking surface samples of water or other liquids using a dipper. A pond sampler or dipper with extension handle allows the operator to sample streams, ponds, waste pits, and lagoons as far as 15 ft from the bank or other secure footing. The dipper is useful in filling a sample bottle without contaminating the outside of the bottle.

1.2 *This standard does not purport to address all of the safety concerns, if any, associated with its use. It is the responsibility of the user of this standard to establish appropriate safety and health practices and determine the applicability of regulatory limitations prior to use.*

2. Referenced Documents

2.1 *ASTM Standards:*

D 4687 Guide for General Planning of Waste Sampling[2]

D 5088 Practice for Decontamination of Field Equipment Used at Nonradioactive Waste Sites[3]

2.2 *Other Documents:*

EPA-600/2-80-018 Samplers and Sampling Procedures for Hazardous Waste Streams[4]

EPA-600/4-84-076 Characterization of Hazardous Waste Sites-A Methods Manual: Volume II. Available Sampling Methods, Second Edition[4]

3. Summary of Practice

3.1 The dipper is lowered into the liquid and lifted out while avoiding splashing or otherwise disturbing the surface layer. The sample is then poured into a sample container.

4. Significance and Use

4.1 This practice is intended for use in the sampling of surface waters and other liquids.

4.2 Dipper equipment is uncomplicated in construction, simple to use, and relatively easy to decontaminate; however, this practice will not provide accurate results with multiphase liquids.

4.3 This practice is to be used by personnel acquiring samples.

4.4 The dipper is best used to take a *surface* sample of liquid. No attempt should be made to take subsurface samples with a dipper because mixing and dilution will occur as the dipper is brought to the surface. Subsurface layers must be sampled using a device that can be closed prior to bringing it to the surface (for example, a COLIWASA, tube sampler, or bottle sampler).

4.5 The dipper is not appropriate for sampling multiphase liquids if quantitative characterization is needed.

4.6 When volatile organic analysis (VOA) will be performed, samples obtained with a dipper should be poured into an appropriate container (VOA vial) with minimal air contact and agitation.

4.7 This practice should be used in conjunction with Guide D 4687, which covers sampling plans, safety, Quality Assurance (QA), preservation, decontamination, labeling, and chain-of-custody procedures, also Practice D 5088, which covers decontamination of field equipment used at waste sites. Other documents pertinent to this practice are EPA-600/2-80-018 and EPA-600/4-84-076.

5. Sampling Equipment

5.1 Dippers may be fabricated as shown in Fig. 1 and are also available commercially (see Fig. 2). Disposable dippers are convenient for use with hazardous materials. Dippers selected must be constructed of materials compatible with the liquid being sampled and with the tests or analyses to be performed. Light weight and rigidity are important characteristics of the extension handle.

6. Sample Containers

6.1 Plastic, glass, or other nonreactive containers should be used. Refer to Guide D 4687 for further information on containers.

7. Procedure

7.1 Have sample containers ready for use.

7.2 Assemble the dipper and, if desired, the extension handle for the pond sampler.

7.3 Lower the dipper slowly into the liquid. The dipper should be on its side so that liquid runs into the container without bubbling or swirling. Rotate the dipper so that the lip of the container is up, preventing the sample from running out. Lift the dipper from the liquid.

[1] This practice is under the jurisdiction of ASTM Committee D34 on Waste Management and is the direct responsibility of Subcommittee D34.01.03 on Sampling Preservation.

Current edition approved Jan. 15, 1993. Published March 1993.

[2] *Annual Book of ASTM Standards*, Vol 11.04.

[3] *Annual Book of ASTM Standards*, Vol 04.08.

[4] Available from Standardization Documents Order Desk, Bldg. 4 Section D, 700 Robbins Ave., Philadelphia, PA 19111-5094, Attn: NPODS.

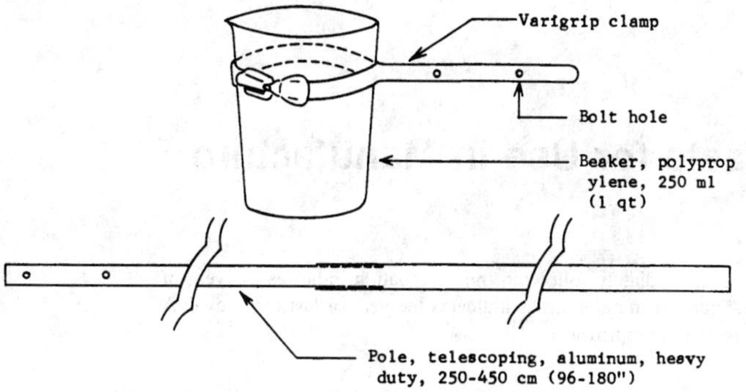

FIG. 1 Dipper with Extension Handle (also known as a Pond Sampler)

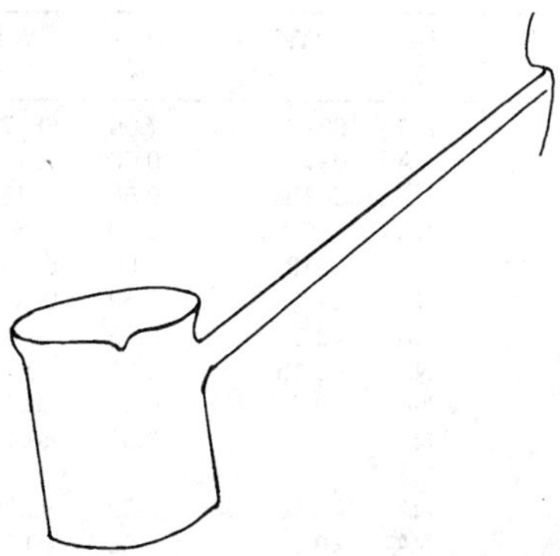

FIG. 2 Typical Commercially Available Dipper

7.4 Remove the cap from the sample container, and tilt the container slightly.

7.5 Pour the sample from the dipper down the inside of the sample container. Avoid splashing of the sample.

7.6 Leave adequate air space in the container to allow for expansion, except for VOA vials. Fill VOA vials to the lip, and seal with no headspace to prevent loss of volatiles.

7.7 Complete the field log book and chain-of-custody forms.

7.8 Decontaminate the used equipment in accordance with Practice D 5088.

8. Keywords

8.1 dipper; liquid sampling; pond sampler; sampling; waste

Designation: D 5359 – 98

Standard Specification for Glass Cullet Recovered from Waste for Use in Manufacture of Glass Fiber[1]

This standard is issued under the fixed designation D 5359; the number immediately following the designation indicates the year of original adoption or, in the case of revision, the year of last revision. A number in parentheses indicates the year of last reapproval. A superscript epsilon (ϵ) indicates an editorial change since the last revision or reapproval.

1. Scope

1.1 This specification describes glass cullet recovered from municipal waste destined for disposal. The recovered cullet is intended for use in the manufacture of glass fiber used for insulation-type products.

2. Referenced Documents

2.1 *ASTM Standards:*

C 162 Terminology of Glass and Glass Products[2]

D 4129 Test Method for Total and Organic Carbon in Water by High-Temperature Oxidation and Coulometric Detection[3]

E 688 Test Methods for Waste Glass as a Raw Material for Glass Manufacturing[4]

3. Terminology

3.1 For definitions of terms used in this specification, refer to Terminology C 162.

4. General Requirements

4.1 Glass cullet from municipal waste is primarily soda-lime bottle glass and shall be one of three grades depending upon the total usage rate requirement of the user. The three grades shall satisfy the following chemical composition, color mix, contamination, and particle size requirements as listed in Section 4:

4.2 *Chemical Composition*—See Table 1.

4.3 *Color Mix*—Color is an indicator of the oxidation state of container cullet. SO_3 gas solubility in the glass melt is a function of the glass oxidation state. Changes in the oxidation state of cullet added to the fiberglass batch can shift the glass oxidation state causing the release of dissolved SO_3 gas, which can upset the furnace. A change in the glass oxidation state also means a change in the glass FeO content. This affects the heat transfer in the melt and can affect furnace efficiency and glass quality. See Table 2.

TABLE 1 Chemical Composition

Grade	1		2		3	
Use Range	0 to 5 % in batch		5 to 15 % in batch		>15 % in batch	
Oxide	Weight %	±Range %	Weight %	±Range %	Weight %	±Range %
SiO_2	68–77	NA	68–77	1.00	68–77	1.00
Al_2O_3	0–7	NA	0–7	0.50	0–7	0.50
CaO	5–15	NA	5–15	0.50	5–15	0.50
MgO	0–5	NA	0–5	0.50	0–5	0.50
Na_2O	8–18	NA	8–18	0.50	8–18	0.50
K_2O	0–4	NA	0–4	0.50	0–4	0.50
Fe_2O_3	<0.5	NA	<0.5	0.05	<0.5	0.05
Cr_2O_3	<0.2	NA	<0.15	0.03	<0.1	0.02
SO_3	<0.4	NA	<0.3	0.03	<0.2	0.02
All other oxides	<0.5	NA	<0.3	0.05	<0.1	0.02
C[A]	<0.15	NA	<0.10	0.02	<0.05	0.01
H_2O	<0.5	NA	<0.5	0.05	<0.5	0.05
LOI	<0.45	NA	<0.30	0.05	<0.15	0.03

[A] Carbon is determined directly by instrumental method such as Coulometrics, Inc. Model 5010 Coulometer. Test Method D 4129 uses this instrumentation for total and organic carbon in water. The instrument can be readily adapted to solid materials such as cullet.

4.4 *Contaminants*—Free metals, magnetic or nonmagnetic, are not oxidized in the glass melting process and, therefore, are insoluble. Metals will pool on the furnace floor and leak through joints causing premature wear of refractories and electrical shorts, which can lead to glass leaks. Some metals will attack and destroy precious metal skimmers and thermocouples and molybdenum electrodes. Examples are silver, tin, lead, and aluminum.

4.4.1 Other inorganic materials and refractories will not melt in the glass melting process. Other inorganics can be porcelains, ceramics, or high-temperature glasses. Refractories can be remnants of furnace construction materials or minerals contained as unmelted inclusions in the cullet. See Table 3.

4.4.1.1 The container cullet mixture must not contain glass types or other materials whose composition contains anything that is either harmful to the fiberglass production process or

TABLE 2 Color Mix Ratio

Grade	1		2		3	
Use Range	0 to 5 % in batch		5 to 15 % in batch		>15 % in batch	
Color	Weight %	±Range %	Weight %	±Range %	Weight %	±Range %
Flint	0–100	NA	0–100	5	0–100	3
Green	0–100	NA	0–100	5	0–100	3
Amber	<50	NA	<30	5	<25	3

[1] This specification is under the jurisdiction of ASTM Committee D-34 on Waste Management and is the direct responsibility of Subcommittee D34.06 on Recovery and Reuse.

Current edition approved Sept. 10, 1998. Published December 1998. Originally published as D 5359–93. Last previous edition D 5359–93.

[2] *Annual Book of ASTM Standards*, Vol 15.02.

[3] *Annual Book of ASTM Standards*, Vol 11.02.

[4] *Annual Book of ASTM Standards*, Vol 11.04.

TABLE 3 Contaminants

Grade	1	2	3
Use Range	0 to 5 % in batch	5 to 15 % in batch	>15 % in batch
Contaminant Type	Weight %	Weight %	Weight %
Magnetic Material	<0.3	<0.2	<0.1
Nonmagnetic Metals	<0.01	<0.005	<0.005
Other Inorganic Material			
+ 12 Mesh	NA	0.0	0.0
+ 20 Mesh	NA	<0.2	<0.1
−20 Mesh	NA	<0.3	<0.2
Refractories			
+ 12 Mesh	NA	0.0	0.0
+ 20 Mesh	NA	<0.2	<0.1
−20 Mesh	NA	<0.3	<0.2

which affects the user's ability to meet Federal or state environmental, safety, or health laws. Examples of such materials are anything which contains elements or oxides of phosphorus, arsenic, antimony, and chlorides.

4.5 *Particle Size*—For all grades of cullet, the particle size shall be 100 % < ¼ in. and <15 % < 200 mesh. The specification for the particle size distribution between these two end points shall be agreed to on an individual basis between the cullet supplier and the cullet user.

5. Sampling and Testing

5.1 Sampling and testing shall be in accordance with Test Methods E 688.

6. Keywords

6.1 fiberglass; glass cullet; insulation; municipal solid waste; recycled glass

Designation: D 5368 – 93 (Reapproved 2001)

Standard Test Methods for Gravimetric Determination of Total Solvent Extractable Content (TSEC) of Solid Waste Samples[1]

1. Scope

1.1 These test methods describe standard procedures for gravimetrically determining the total nonvolatile and semi-volatile organic content of solvent extracts from soils or solid wastes. The following methods are included:

	Section
Method A—Micro-Determination of TSEC	11-13
Method B—Evaporating Dish Procedure	14-16
Method C—Boiling Flask Procedure	17-19

1.2 These methods are used after a solvent extract is obtained from a soil or solid waste. For these methods to be applicable, the extraction solvent must have a boiling point less than that of water at ambient pressure.

1.3 The total solvent extractable content (TSEC) of a soil, sediment, sludge, or solid waste depends upon the solvent and method used for the extraction procedure.

2. Referenced Documents

2.1 *ASTM Standards:*

D 2109 Test Methods for Nonvolatile Matter in Halogenated Organic Solvents and Their Admixtures[2]

D 2910 Practice for Concentration and Recovery of Organic Matter from Water by Activated Carbon[3]

D 3086 Test Method for Organochlorine Pesticides in Water[3]

D 3445 Test Method for Nonvolatile Matter in Trichlorotrifluoroethane[4]

D 3694 Practices for Preparation of Sample Containers and for Preservation of Organic Constituents[3]

D 4281 Test Method for Oil and Grease (Fluorocarbon Extractable Substances) by Gravimetric Determination[3]

D 5369 Practice for the Extraction of Solid Waste Samples for Chemical Analysis Using Soxhlet Extraction[5]

3. Terminology

3.1 *Definition:*

3.1.1 *total solvent extractable content (TSEC)*— the total concentration by weight (w/w) of organic materials that is extractable from a soil or solid waste by the selected solvent.

4. Summary of Methods

4.1 The sample is extracted with an organic solvent using a procedure such as described in Practices D 5369 (Soxhlet Extraction) or Test Method D 4281. The quantity of material extracted into the solvent is determined as the residue weight after solvent evaporation. Similar ASTM methods are Test Methods D 2109 and D 3445, and Practice D 2910.

4.1.1 *Method A, Micro-Determination Method*—The residue weight after solvent evaporation is determined for 200 μL of solvent extract evaporated on an aluminum weighing dish. A heat lamp is used to effect evaporation.

4.1.2 *Method B, Evaporating Dish Procedure*—The residue weight after solvent evaporation is determined for 1.0 to 75.0 mL of solvent extract. Solvent is evaporated in an evaporating dish by heating at 5°C below the solvent boiling point. A stream of nitrogen over the surface of the solvent is used to accelerate evaporation.

4.1.3 *Method C, Boiling Flask Method*—The residue weight after solvent evaporation is determined for 100 to 300 mL of solvent evaporated in a boiling flask. Solvent is evaporated in a water bath at the solvent boiling point.

5. Significance and Use

5.1 The TSEC provides a quantitative measure of the total solvent extractable organic content of the solid waste in question. Based upon the TSEC, the extract may be analyzed or further processed (that is, further cleanup or solvent concentration) in preparation for analysis. Thus, the TSEC provides a quantitative measure for optimizing the extractable organic concentration prior to chemical analysis.

5.2 The TSEC of soil or waste material may be used as a quantitative measure for the screening and selection of samples for chemical analysis.

5.3 The TSEC may be useful as a simple and relatively inexpensive quantitative indicator of changes occurring in the total extractable organic content of soil or waste materials.

5.4 The detection limit of the TSEC depends upon the specific procedures employed. Typical detection limits in solvent extracts (that is, without solvent pre-concentration) for Method A are more than 1000 μg/g. For Method B typical

[1] These test methods are under the jurisdiction of ASTM Committee D34 on Waste Management and are the direct responsibility of Subcommittee D34.01.06 on Analytical Methods.

Current edition approved Feb. 15, 1993. Published April 1993.

[2] *Annual Book of ASTM Standards*, Vol 15.05.

[3] *Annual Book of ASTM Standards*, Vol 11.02.

[4] *Discontinued 1991*; Replaced by Test Methods D 2109, *Annual Book of ASTM Standards*, Vol 15.05.

[5] *Annual Book of ASTM Standards*, Vol 11.04.

detection limits are 20 to 100 μg/g in the solvent extract. For Method C the typical detection limits are 3 to 20 μg/g for solutes in the solvent extract.

5.5 The sensitivity of the methods can be adjusted by varying the volume of solvent extract prior to gravimetric residue determination.

6. Interferences

6.1 Solvents, reagents, glassware, and other sample processing hardware may yield an unacceptable background relative to the limit of detection required for the TSEC measurement. A method blank must be performed in order to demonstrate the viability of the solvent and equipment used. Specific selection of reagents and the purification of solvents by distillation in all-glass systems may be required when low levels (that is, <100 μg/g) of TSEC are of interest.

6.1.1 Glassware should be cleaned by washing with detergent or non-chromate cleaning solution, rinsing first with tap water, then reagent water, then redistilled acetone, and finally with pesticide quality solvent (that is, the solvent used for extraction). For additional information, see Practices D 3694. If the type and size of glassware permits, it may be cleaned by heating in a muffle furnace at 400°C for 15 to 30 min. Alternatively, glassware may be oven dried at 103°C for at least 1 h, after solvent rinsing and draining. Volumetric glassware should not be heated in a muffle furnace.

6.1.2 Plastics, except PTFE-fluorocarbon, can be a source of serious interference, especially when specific organic constituents are of analytical interest. Their use must be avoided. Extracted samples should be stored in glass bottles with PTFE-fluorocarbon-lined caps. Extraction solvent volumes should be recorded and containers sealed to prevent solvent evaporation or cross contamination. Extracts should be refrigerated above the freezing point of the solvent.

6.1.3 If required, rinse glassware and utensils with extraction solvent prior to use in order to remove interferences. The method blank should be less than 20 % of the minimum reportable concentration.

6.1.4 Impurities in the extracting solvent can be a source of interferences or TSEC background. Solvent blanks should be analyzed with each new bottle of solvent. Whenever a high TSEC background, or interfering compounds are traced to the solvent, a new source of solvent should be obtained. Alternatively, impurities can often be removed by distillation in glass.

6.2 A relatively high TSEC background can also result from inorganic drying agents (that is, Na_2SO_4, $MgSO_4$, $CuSO_4$, $CaCl_2$, $CaSO_4$, K_2CO_3, KOH, BaO, CaO, H_2SO_4, P_2O_5) in the extracting solvent. A relatively high background of any drying agent used (>20 % of the minimum TSEC concentration of interest) in the solvent will preclude the application of the method for TSEC determinations in those cases. Residue ignition and inorganic analysis are suggested to evaluate this potential background.

6.3 Organic interferences can arise from the extraction of impurities from inorganic drying agents. If this is suspected, the interference may be removed by rinsing the drying agent with pure extraction solvent followed by oven drying of the inorganic drying agent.

6.4 Inorganic constituents extractable from the waste by the solvent selected will result in a positive interference in the determination of TSEC, unless removed prior to TSEC determination. This potential for interference must be determined by the analyst on a case by case basis using existing data and best judgment.

7. Selection of the Extraction Solvent

7.1 The selection of solvent for extraction and TSEC determination will depend upon many factors, including the following (see Table 1 for selected applications):

7.1.1 Boiling point of the solvent,

7.1.2 Boiling point of the compounds or class of compounds of interest,

7.1.3 Tendency of the solvent and matrix to form emulsions,

7.1.4 Solvent strength (that is, polarity, solubility of compounds of interest),

7.1.5 Safety of solvent use (that is, toxicity, flammability),

7.1.6 Purity of solvent, and

7.1.7 Solvent compatibility with analysis instrumentation.

7.2 The analyst should demonstrate a recovery using a spiking procedure in the matrix of interest before using this procedure.

7.3 The extract is exposed to temperatures approaching the boiling point of the solvent during the evaporation procedure. Consequently, one must ensure that heat-labile and more volatile solutes that may be of interest are stable and recoverable by the method and the solvent of choice. Because low-boiling fractions are lost in solvent removal, reproducible results can only be obtained by strict adherence to method details.

8. Reagents and Materials

8.1 *Purity of Reagents*—Reagent grade chemicals shall be used in all tests. Unless otherwise indicated, all reagents must conform to the specifications of the Committee on Analytical Reagents of the American Chemical Society where such specifications are available.[6] Other grades may be used, provided it is first determined that the reagent is of sufficiently high purity to permit its use without lessening the accuracy of the determination.

8.2 When low concentrations of TSEC are to be determined (that is, <100 μg/g in the waste) pesticide-grade solvents (that is, distilled in glass) are recommended. The solvent blank for TSEC and the specific constituents of interest must be below 20 % of the reportable lower limit for the analysis.

8.3 *Acetone* ($(CH_3)_2CO$), technical grade.

8.4 *Explosion-Proof Oven or Furnace*.

8.5 *Heating Mantles*.

9. Precautions

9.1 Some solvents (for example, benzene, chloroform, and carbon tetrachloride) are suspected human carcinogens and must be handled by approved methods.

[6] "Reagent Chemicals, American Chemical Society Specifications," Am. Chem. Soc., Washington, DC. For suggestions on the testing of reagents not listed by the American Chemical Society, see "Analar Standards for Laboratory U.K. Chemical," BDH Ltd., Poole, Dorset, and the "United States Pharmacopeia."

TABLE 1 Selected Applications of Soxhlet Extraction for Extraction of Organic Constituents from Solid Matrices

Sample Matrix	Solvent	Compounds or Constituents	Extraction Time, h (cycles)	Reference
1) Sediment	1,1,1-trichloro-1,2,2-trifluoroethane (freon)	oil and grease	4 (80)	**(1)**[A] Plumb (1983)
2) Sludges and similar materials	1,1,1-trichloro-1,2,2-trifluoroethane (freon)	oil and grease	4 (80)	**(2)** Standard Methods
3) Sludges from sewage	hexane then methanol	total organic C oil, grease, fats	24	**(3)** Strachan (1983)
4) Municipal wastewater suspended solids and activated carbon	hexane/dichloromethane	semi-volatile priority pollutants	24 (480)	**(4)** Harrold (1982)
5) Soil and housedust	acetone/hexane (1:1)	organochlorine insecticides	5 (60)	**(5)** EPA (1980)
6) Sediment	dichloromethane	phenols	8	**(6)** Goldberg (1980)
7) Soil	a) acetone/n-hexane (1:1)	aldrin, dieldrin	12 (554)	**(7)** Chiba (1968)
	b) acetonitrile	aldrin, dieldrin	14 (47)	
	c) 2-propanol/n-hexane (1:1)	aldrin, dieldrin	18 (108)	
8) Soil	chloroform/methanol (1:1) (other solvents also studied)	dieldrin	8 (160)	**(8)** Saha (1969)
9) Airborne particulates	methanol (cyclohexane also studied)	gross organics	2	**(9)** Hill (1977)
10) Airborne particulates	benzene	selected PAHs	4–6	**(10)** Pierce (1975)
11) Airborne particulates	numerous solvents studied	selected PAHs	6	**(11)** Stanley (1967)
12) Coke oven aerosol particulates	benzene	selected PAHs	2 (18–20)	**(12)** Broddin (1977)
13) Artificial aerosol particulates	methanol/benzene	selected PAHs	8 (80)	**(13)** Cautreels (1976)
	methanol/benzene	selected phthalates	16 (160)	
	methanol/benzene	selected aliphatics	2 (20)	
	methanol	selected nitrogen aromatics	4 (40)	
	benzene	selected nitrogen aromatics	2 (20)	
14) Activated carbon	chloroform	phenols	44 (440)	**(14)** Pahl (1973)
	chloroform/ethanol	gross organics		**(15)** Buelow (1973)
15) Glass fiber filters	26 solvents and 24 binary mixtures	total organic carbon	6	**(16)** Grosjean (1975)
16) Surface sediments	methanol then dichloromethane	total oil hydrocarbon	48 (160)	**(17)** Sporstol (1985)
17) Bottom sediment	hexane/acetone/isooctane	chlorinated benzenes	18	**(18)** Onuska (1985)
18) Environmental particulates	benzene	chlorinated dioxins	16	**(19)** Lamparski (1980)
19) Soils	hexane/acetone/methanol	DDT	12	**(20)** Nash (1972)

[A] The boldface numbers in parentheses refer to the list of references at the end of this standard.

9.2 Explosive peroxides tend to form in ether solvent. A convenient means of testing for their presence is with E. M. Quant test paper.[7] This test should be performed before evaporation of any ether-bearing extract.

9.3 The use of fume hoods with volatile and toxic solvents is mandatory.

9.4 Flammable solvents must be protected from heat, sparks, or flames. Avoid buildup of vapors and eliminate all sources of ignition, especially nonexplosion-proof electrical apparatus and heaters. Keep containers closed. Use with adequate ventilation. Store bulk solvents in safety cabinets. Remove only a one-day supply and keep it in a hood.

9.5 Avoid prolonged breathing of vapor or spray mist and avoid prolonged or repeated skin contact for all organic solvents. Consult Material Safety Data Sheets for recommended handling procedures and precautions.

9.6 Samples containing odorous, known or suspected volatile materials, must be processed in a fume hood.

9.7 Samples known or suspected of containing toxic or hazardous materials must be handled in a fume hood. Safety information relative to the handling of any known toxic materials must be obtained and implemented prior to any sample handling.

10. Sample and Sample Preparation

10.1 Remove refrigerated solvent extract sample from storage and allow it to reach room temperature. Keep sample sealed to prevent evaporation.

10.2 When sample has reached room temperature examine to ensure that the sample is homogeneous and does not contain insoluble matter. If necessary, remix and homogenize.

10.3 Record sample identification information for the solvent extract. Record or measure, if necessary, the solvent volume of the extract to be analyzed. Note any changes in the extract volume from the volumetric mark, if available, on the storage vial or container.

Method A—Micro-Determination of TSEC

11. Apparatus

11.1 *Microsyringe*, 250-μL, or alternatively, a calibrated microdispenser capable of 200-μL delivery. The delivery precision should be to 1 %.

11.2 *Aluminum Foil Weighing Pans, Micro*, approximately 1 to 6-cm diameter. Purchase or fabricate from oil-free aluminum foil.

11.3 *Analytical Semimicro Balance*, capable of reading to 0.00001 g.

11.4 *Balance Table*— An electronic load-cell type balance is recommended.

11.5 *Heating Lamp*, 250-W reflector-type bulb in a heat-resistant fixture whose height above the sample may be conveniently adjusted.

11.6 *Forceps*, 3 to 5-in., steel or nickel-plated steel.

11.7 *Desiccator with Desiccant.*

12. Procedure

12.1 Dry an aluminum weighing dish for 1 h at 105 ± 5°C. Cool in a desiccator for 30 min. Remove the weighing dish and

[7] E. M. Quant test paper, trademark of, and available from, MCB Manufacturing Chemists, Inc., 2909 Highland Avenue, Cincinnati, OH 45212 has been found satisfactory for this purpose.

allow it to equilibrate in ambient air for 30 min. Continue air equilibration until a constant weight is obtained.

12.2 With a 250-μL syringe or microdispenser accurately transfer 200 μL of solvent extract to a tared (or weighed) aluminum weighing dish.

12.3 Place the weighing dish under a heat lamp at a distance of 8 cm from the lamp for 1 min, or until dry, to allow the solvent to evaporate.

12.4 Place the weighing dish in a desiccator for 30 min, remove, and weigh to constant weight.

12.5 If the residue weight of the 200-μL solvent extract aliquot is less than 0.00010 g, then a larger volume of solvent extract must be used for the TSEC determination by the application of Methods B or C.

13. Calculation

13.1 For TSEC of the sample, where no solvent preconcentration was performed, calculate the TSEC as micrograms of residue per gram of sample (μg/g) using the following equation:

$$\text{TSEC} = \frac{(Rw)\,(Ve)\,(10^6)}{(Sw)\,(Vr)} \quad (1)$$

where:

TSEC = total solvent extractable content, μg/g,
Rw = residue weight, g (that is, difference in weight of the tared aluminum weighing dish and the dish weight after solvent evaporation),
Sw = weight of sample that was solvent extracted, g,
Ve = total volume of solvent extract, mL, and
Vr = volume of solvent evaporated in the determination of the residue weight, mL (that is, 200μ L).

Method B—Evaporating Dish Procedure

14. Apparatus

14.1 *Evaporating Dish*, 125-mL capacity, platinum, porcelain, or high-silica glass, or glass beaker of similar capacity.

14.2 *Oven*, thermostatically controlled at 105 ± 5°C.

14.3 *Analytical Balance*, capable of reading to 0.0001 g (± 0.0002 g). An electronic load-cell type balance is recommended.

14.4 *Temperature-Controlled Water Bath (± 2°C) or Temperature-Controlled Hot Plate*— Use water bath or hot plate in a laboratory fume hood.

14.5 *Metal Tongs, Evaporating Dish*, safety, 9½ in.

14.6 *Glass-Transfer Pipets*, 1, 10, 25, 50, and 75-mL volume.

14.7 *Desiccator with Desiccant.*

15. Procedure

15.1 Dry a 125-mL evaporating dish, or glass beaker, for 1 h at 105°C. Wipe the outside of the dish or beaker with a lint-free cloth. Cool in a desiccator for 1 h, remove the dish or beaker, and weigh to constant weight.

15.2 Transfer an accurately measured volume of solvent extract (1.0 to 75.0 mL) to the evaporating dish, or beaker, using a pipet.

15.3 Place the weighing dish, or beaker, on the water bath or on a warm plate with the temperature set at 5 ± 2°C below the boiling point of the extracting solvent. Accelerate solvent evaporation by passing a gentle stream of nitrogen over the sample by means of a disposable Pasteur pipet.[8] Glass, metal, or PTFE tubing and glass pipets should be used for the nitrogen, and a new pipet should be used for each sample.

15.4 When the solvent has evaporated remove the dish, or beaker, and contents from the water bath (or from the warm plate) with metal tongs. Wipe the outside of the dish, or beaker, with lint-free cloth and a small amount of acetone to remove any water adhering to the dish. If a warm plate is used to heat the sample, the dish wipe step is unnecessary.

15.5 Place the weighing dish in a desiccator for 1 h, remove, and weigh immediately to constant weight.

16. Calculation

16.1 Calculate the TSEC as micrograms of residue per gram of sample (μg/g) using the following equation:

$$\text{TSEC} = \frac{(Rw)\,(Ve)\,(10^6)}{(Sw)\,(Vr)} \quad (2)$$

where:

TSEC = total solvent extractable content, μg/g,
Rw = residue weight, g (that is, difference in weight of the tared evaporating dish and the dish weight after solvent evaporation),
Sw = weight of sample that was solvent extracted, g,
Ve = total volume of solvent extract, mL, and
Vr = volume of solvent evaporated in the determination of the residue weight, mL.

Method C—Boiling Flask Procedure

17. Apparatus

17.1 *Oven*, thermostatically controlled at 105 ± 5°C.

17.2 *Analytical Balance*, capable of weighing to 0.0001 g (± 0.0002 g). An electronic load-cell type balance is recommended.

17.3 *Temperature-Controlled Water Bath (± 2°C) or Temperature-Controlled Hot Plate*—Use water bath or hot plate in a laboratory fume hood.

17.4 *Metal Tongs, Evaporating Dish*, safety, 9½ in.

17.5 *Glass Pipets or Graduated Cylinders*, 100, 250, or 500-mL.

17.6 *Boiling Flask*, 250 or 500-mL.

17.7 *Desiccator with Desiccant.*

18. Procedure

18.1 Dry a 250 or 500-mL boiling flask for 1 h at 105 ± 5°C. Carefully wipe the outside of the flask with a lint-free cloth. Cool in a desiccator for 1 h, remove, and weigh to constant weight.

18.2 Transfer an accurately measured volume (100 to 300 mL) of the solvent extract to the boiling flask by means of a pipet.

18.3 Evaporate the solvent from the boiling flask on a water bath or warm plate set at the boiling point temperature of the

[8] Refer to Test Method D 3086 under Alternative Method for Final Concentration.

extraction solvent. When only traces of solvent remain, leave the flask on the water bath and draw air through the flask using vacuum for 5 min to remove any remaining solvent vapor or residual water. Carefully wipe the exterior of the flask with a lint-free cloth and a small amount of acetone to remove any water adhering to the flask. If a warm plate is used to heat the sample, the dish wipe step is unnecessary.

18.4 Place in a desiccator for 1 h, remove, and weigh immediately to a constant weight.

19. Calculation

19.1 Calculate the TSEC as micrograms of residue per gram of sample (μg/g) using the following equation:

$$\text{TSEC} = \frac{(Rw)\ (Ve)\ (10^6)}{(Sw)\ (Vr)} \quad (3)$$

where:

TSEC = total solvent extractable content, μg/g,
Rw = residue weight, g (that is, difference between the tare weight of flask and the weight of the flask after removal of extraction solvent),
Sw = weight of sample that was solvent extracted, g,
Ve = total volume of solvent extract, mL, and
Vr = volume of solvent evaporated in the determination of the residue weight, mL (that is, 100 to 300 mL).

REFERENCES

(**1**) Plumb, Jr., R. H., *"Procedures for Handling and Chemical Analysis of Sediment and Water Samples,"* NTIS AD/A-103 788, May 1981.
(**2**) "Standard Methods for the Examination of Water and Wastewater," by APHA, AWWA, and WPCF, 15th Ed., 1981.
(**3**) Strachan, S. D., Nelson, D. W., and Sommers, L. E., "Sewage Sludge Components Extractable with Nonaqueous Solvents," *Journal of Environmental Quality*, Vol 12, No. 1, 1983, p. 69–74.
(**4**) Harrold, D. E., and Young, J. C., "Extraction of Priority Pollutants from Solids," *Proceedings of the American Society of Civil Engineers*, Vol 108, No. EE6, December 1982, pp. 1211–1227.
(**5**) Watts, R. R., editor, "Manual of Analytical Methods for the Analysis of Pesticide Residues in Human and Environmental Samples," U.S. EPA, Environmental Toxicology Division, Research Triangle Park, North Carolina, 1980.
(**6**) Goldberg, M. C., and Weiner, E. R., "Extraction and Concentration of Phenolic Compounds from Water and Sediments," *Analytical Chimica Acta*, 115:373–378, 1980.
(**7**) Chiba, M., and Morley, H. V., "Factors Influencing Extraction of Aldrin and Dieldrin Residues from Different Soil Types," *Journal of Agricultural and Food Chemistry*, 16:916–922, 1968.
(**8**) Saha, J. G., Bhavaraju, B., Lee, Y. W., and Randell, R. L.," Factors Affecting Extraction of Dieldrin-14C from Soil," *Journal of Agricultural and Food Chemistry*, 17:877–882, 1969.
(**9**) Hill, H. H., Chan, K. W., and Karasek, F. W., "Extraction of Organic Compounds from Airborne Particulate Matter for Gas Chromatographic Analysis," *Journal of Chromatography*, 131:245–252, 1977.
(**10**) Pierce, R. C., and Katz, M., "Determination of Atmospheric Isomeric Polycyclic Arenes by Thin Layer Chromatography and Fluorescence Spectrophotometry," *Analytical Chemistry*, 47:1743–1748, 1975.
(**11**) Stanley, T. W., Meeker, J. E., and Morgan, M. J., "Extraction of Organics from Airborne Particles—Effects of Various Solvents and Conditions on the Recovery of Benzo(a)pyrene, Benz(c)acridine and 7-h-Benz(de)anthracen-7-one," *Environmental Science and Technology*, 1:927–931, 1967.
(**12**) Broddin, G., Van Vaeck, L., and Van Cauwenberghe, K., "On the Size Distribution of Polycyclic Aromatic Hydrocarbon Containing Particles from a Coke Oven Emission Source," *Atmospheric Environment*, 11:1061–1064, 1977.
(**13**) Cautreels, W., and Van Cauwenberghe, K.," Extraction of Organic Compounds from Airborne Particulate Matter," *Water, Air, Soil Pollution*, Vol 6, 1976, pp. 103–110.
(**14**) Pahl, R. H., Mayan, K. G., and Bertrand, G. L., "Organic Desorption from Carbon—II. The Effect of Solvent in the Desorption of Phenol from Wet Carbon," *Water Research*, Vol 7, 1973, pp. 1309–1322.
(**15**) Buelow, R. W., Carswell, J. K., and Symons, J. M., "Improved Method for Determining Organics by Activated Carbon Adsorption and Solvent Extraction—Part I," *Journal of the American Water Works Association*, Vol 63, 1973, pp. 57–72.
(**16**) Grosjean, D., "Solvent Extraction and Organic Carbon Determination in Atmospheric Particulate Matter: The Organic Extraction-Organic Carbon Analyzer (OE-OCA) Technique," *Analytical Chemistry*, Vol 47, No. 6, May 1975, pp. 797–805.
(**17**) Sporstol, S., Lichtenthaler, R. G., and Oreld, F., "Extraction Efficiences for Hydrocarbons from Sediments Polluted by Oil Drill Cuttings," *Analytical Chimica Acta*, Vol 169, 1985, pp. 343–347.
(**18**) Onuska, F. I., and Terry, K. A., "Determination of Chlorinated Benzenes in Bottom Sediment Samples by WCOT Column Gas Chromatography," *Analytical Chemistry*, Vol 57, 1985, pp. 801–805.
(**19**) Lamparski, L. L., and Nestrick, T. J., "Determination of Tetra-, Hepta-, and Octachlorodibenzo-p-dioxin Isomers in Particulate Samples at Parts per Trillion Levels," *Analytical Chemistry*, Vol 52, 1980, pp. 2045–2054.
(**20**) Nash, R. G., and Harris, W. G., "Soil Moisture Influence on Treatment and Extraction and Extraction of DDT from Soils," *Journal of the Association of Official Analytical Chemists*, Vol 55, No. 3, 1972, pp. 532–536.

Designation: D 5369 – 93 (Reapproved 1998)

Standard Practice for Extraction of Solid Waste Samples for Chemical Analysis Using Soxhlet Extraction[1]

This standard is issued under the fixed designation D 5369; the number immediately following the designation indicates the year of original adoption or, in the case of revision, the year of last revision. A number in parentheses indicates the year of last reapproval. A superscript epsilon (ε) indicates an editorial change since the last revision or reapproval.

1. Scope

1.1 This practice describes standard procedures for extracting nonvolatile and semivolatile organic compounds from solids such as soils, sediments, sludges, and granular wastes using Soxhlet extraction.

1.1.1 The sample must be suitable for being mixed with the sample drying agent, sodium sulfate or magnesium sulfate, to provide drying of all sample surfaces.

1.2 This practice, when used in conjunction with Test Method D 5368 is applicable to the determination of the total solvent extractable content (TSEC) of a soil, sediment, sludge, or granular solid waste and depends upon the solvent chosen for extraction.

1.3 This practice is limited to solvents having boiling points below the boiling point of water at ambient pressure.

1.4 The solvent extract obtained by this practice may be analyzed for total or specific nonvolatile and semivolatile organic compounds but may require sample clean-up procedures prior to specific compound analysis.

1.4.1 This practice provides sample extracts suitable for analysis by various techniques such as gas chromatography with flame ionization detection (GC/FID) or gas chromatography with mass spectrometric detection (GC/MS).

1.5 This practice is recommended only for solid samples that can pass through a 10-mesh sieve (approximately 2-mm openings), or are less than 2 mm in thickness.

1.6 *This standard does not purport to address all of the safety problems, if any, associated with its use. It is the responsibility of the user of this standard to establish appropriate safety and health practices and determine the applicability of regulatory limitations prior to use.* For specific precautions see Section 9.

2. Referenced Documents

2.1 *ASTM Standards:*

D 75 Practices for Sampling Aggregates[2]

D 420 Practice for Investigating and Sampling Soil and Rock for Engineering Purposes[3]

D 2234 Test Method for Collection of a Gross Sample of Coal[4]

D 2910 Practice for Concentration and Recovery of Organic Matter from Water by Activated Carbon[5]

D 3086 Test Method for Organochlorine Pesticides in Water[5]

D 3694 Practices for Preparation of Sample Containers and for Preservation of Organic Constituents[5]

D 3975 Practice for Preparation of Samples for Collaborative Testing of Methods for Analysis of Sediments[5]

D 3976 Practice for Preparation of Sediment Samples for Chemical Analysis[5]

D 4281 Test Method for Oil and Grease (Fluorocarbon Extractable Substances) by Gravimetric Determination[5]

D 5368 Test Method for the Gravimetric Determination of Total Solvent Extractable Content (TSEC) of Solid Waste Samples[6]

E 122 Practice for Choice of Sample Size to Estimate a Measure of Quality for a Lot or Process[7]

2.2 *EPA Document:*

SW 846 Method 3540 Soxhlet Extraction, Test Methods for Evaluating Solid Waste, Physical/Chemical Methods SW 846, Third Edition[8]

3. Terminology

3.1 *Definitions:*

3.1.1 *GC*—gas chromatography.

3.1.2 *GC/MS*—gas chromatography with mass spectrometric detection.

3.1.3 *TSEC*—total solvent extractable content. The total concentration by weight (w/w) of organic materials that are extractable from a soil or solid waste by the selected solvent.

4. Summary of Practice

4.1 The sample is mixed with sodium sulfate or magnesium sulfate, placed in an extraction thimble, and extracted using an appropriate solvent in a Soxhlet extractor for a 15 to 20-h period. For additional information, see Test Method D 4281.

---

[1] This practice is under the jurisdiction of ASTM Committee D34 on Waste Management and is the direct responsibility of Subcommittee D34.01.06 on Analytical Methods.

Current edition approved March 15, 1993. Published May 1993.

[2] *Annual Book of ASTM Standards*, Vol 04.03.

[3] *Annual Book of ASTM Standards*, Vol 04.08.

[4] *Annual Book of ASTM Standards*, Vol 05.05.

[5] *Annual Book of ASTM Standards*, Vol 11.02.

[6] *Annual Book of ASTM Standards*, Vol 14.02.

[7] *Annual Book of ASTM Standards*, Vol 11.04.

[8] Available from the Superintendent of Documents, U.S. Government Printing Office, Washington, D.C. 20402.

The time required may be longer or shorter than the stated period, provided the extraction time selected has been demonstrated appropriate for the compounds and matrix of interest.

4.2 The solvent extract may be further processed using sample clean-up techniques and may be analyzed using instrumental methods for specific nonvolatile or semivolatile organic compounds. This practice does not include sample extract clean-up methods.

5. Significance and Use

5.1 This practice provides a general procedure for the solvent extraction of organics from soils, sediments, sludges, and fine-grained solid wastes. It may be used as the initial step in the solvent extraction of organic constituents from waste materials for the purpose of quantifying extractable organic compounds. When the appropriate extraction solvent is used, the procedure can be applied to the determination of the total solvent extractable content (TSEC) of the sample. The extraction of nonvolatile or semivolatile organic compounds from the above solid waste should use solvents listed in Table 1 or SW846 Method 3540.

5.2 The detection limit, linear concentration range, and sensitivity of the method for specific organic compound analysis will depend upon the method used for instrumental analysis and will also depend upon the sample clean-up and solvent concentration methods used. Typical detection limits that can be achieved for GC or GC/MS are in the parts per million and sub-parts per million range.

5.2.1 The method detection limit can be adjusted by varying the volume of extract used and by additional sample clean-up prior to analysis.

5.3 Soxhlet extraction has an advantage when analyzing solid waste and soil/waste mixtures which form emulsions with more rigorous solvent mixing extraction techniques.

6. Interferences

6.1 Solvents, reagents, glassware, and other sample processing hardware may yield discrete artifacts or elevated baselines which cause misinterpretation of follow-up analyses. All of these materials must be demonstrated to be free of interference under the conditions of the follow-up analysis. Specific selection of reagents or the purification of solvents by distillation in all glass systems, or both, are required when organic component analysis follows extraction.

6.1.1 Glassware should be cleaned by washing with detergent or non-chromate cleaning solution, rinsing first with tap water, then reagent water, then redistilled acetone, and finally with pesticide quality solvent (that is, the solvent used for extraction). If the type and size of glassware permits, it may be cleaned by heating in a muffle furnace at 400°C for 15 to 30 min. Alternatively, glassware may be oven dried at 103°C for at least 1 h, after solvent rinsing and draining. Volumetric glassware should not be heated in a muffle furnace.

6.1.2 Plastics, except PTFE-fluorocarbon, can be a source of serious interference, especially when specific organic constituents are of analytical interest. Their use must be avoided. Samples should be collected in glass bottles with PTFE-fluorocarbon-lined caps. Alternatively, caps may be lined with precleaned aluminum foil when the pH of the sample is near neutral and the salt content of the sample is low. To minimize the possibility of contamination of containers after cleaning, these items should be cleaned immediately prior to use and

TABLE 1 Selected Applications of Soxhlet Extraction for Extraction of Organic Constituents from Solid Matrices

Sample Matrix	Solvent	Compounds or Constituents	Extraction Time, h (cycles)	Reference
(1) Sediment	1,1,1-trichloro-1,2,2-trifluoroethane (Freon)	oil and grease	4 (80)	**(1)**[A] Plumb (1983)
(2) Sludges and similar materials	1,1,1-trichloro-1,2,2-trifluoroethane (Freon)	oil and grease	4 (80)	**(2)** Standard Methods
(3) Sludges from sewage	hexane then methanol	total organic C oil, grease, fats	24	**(3)** Strachan (1983)
(4) Municipal wastewater suspended solids and activated carbon	hexane/dichloromethane	semivolatile priority pollutants	24 (480)	**(4)** Harrold (1982)
(5) Soil and housedust	acetone/hexane(1:1)	organochlorine insecticides	5 (60)	**(5)** EPA (1980)
(6) Sediment	dichloromethane	phenols	8	**(6)** Goldberg (1980)
(7) Soil	a) acetone/n-hexane(1:1)	aldrin, dieldrin	12 (554)	**(7)** Chiba (1968)
	b) acetonitrile	aldrin, dieldrin	14 (47)	
	c) 2-propanol/n-hexane(1:1)	aldrin, dieldrin	18 (108)	
(8) Soil	chloroform/methanol(1:1) (other solvents also studied)	dieldrin	8 (160)	**(8)** Saha (1969)
(9) Airborne particulates	methanol (cyclohexane also studied)	gross organics	2	**(9)** Hill (1977)
(10) Airborne particulates	benzene	selected PAHs	4–6	**(10)** Pierce (1975)
(11) Airborne particulates	numerous solvents studied	selected PAHs	6	**(11)** Stanley (1967)
(12) Coke oven aerosol particulates	benzene	selected PAHs	2 (18–20)	**(12)** Broddin (1977)
(13) Artificial aerosol particulates	methanol/benzene	selected PAHs	8 (80)	**(13)** Cautreels (1976)
	methanol/benzene	selected phthalates	16 (160)	
	methanol/benzene	selected aliphatics	2 (20)	
	methanol	selected nitrogen aromatics	4 (40)	
	benzene	selected nitrogen aromatics	2 (20)	
(14) Activated carbon	chloroform	phenols	44 (440)	**(14)** Pahl (1973)
	chloroform/ethanol	gross organics		**(15)** Buelow (1973)
(15) Glass fiber filters	26 solvents and 24 binary mixtures	total organic carbon	6	**(16)** Grosjean (1975)
(16) Surface sediments	methanol then dichloromethane	total oil hydrocarbon	48 (160)	**(17)** Sporstol (1985)
(17) Bottom sediment	hexane/acetone/isooctane	chlorinated benzenes	18	**(18)** Onuska (1985)
(18) Environmental particulates	benzene	chlorinated dioxins	16	**(19)** Lamparski (1980)
(19) Soils	hexane/acetone/methanol	DDT	12	**(20)** Nash (1972)

[A] The boldface numbers in parentheses refer to the list of references at the end of this practice.

capped. A glassware rinse should be performed with the extraction solvent immediately prior to use.

6.2 A blank Soxhlet extraction and analysis should be performed in order to determine the solvent, drying agent, and apparatus background. If necessary, the glass (or paper) thimble and glass wool should be leached with solvent prior to use in order to minimize contamination. Method blanks should be less than 20 % of the concentration of the minimum needed reportable concentration. A method blank(s) must be reported with method data.

6.3 A relatively high TSEC background (>20 % of the minimum TSEC of interest) can result from solubility of sodium sulfate or other drying agent in the extraction solvent. This will preclude the application of the method for TSEC determinations.

6.4 Impurities in the extracting solvent can be a source of interference or TSEC background. Solvent blanks should be analyzed with each new bottle of solvent. Whenever a high TSEC background, or interfering compounds, are traced to the solvent, a new source of solvent should be obtained. Alternatively, impurities can often be removed by distillation in glass.

6.5 If organic compound interferences are found in the anhydrous sodium sulfate or magnesium sulfate, the drying agent may be rinsed with pure extraction solvent, or alternatively, Soxhlet extracted, followed by oven drying.

6.6 Inorganic constituents extractable from the waste by the extraction solvent will result in a positive interference in the determination of TSEC. This potential for interference must be determined by the analyst on a case-by-case basis.

6.7 When specific organic compound analysis is of interest, matrix interferences may be caused by compounds that are coextracted from the sample (but are not of interest). The extent of such matrix interferences will vary considerably depending on the sample and the specific instrumental analysis method used. Matrix interferences may be reduced by a judicious choice of extracting solvent, or alternatively, by using a clean-up technique on the extract following Soxhlet extraction.

7. Selection of the Extraction Solvent

7.1 The selection of solvent for extraction will depend upon many factors, including the following (see Table 1 for selected applications):

7.1.1 Boiling point of the solvent.

7.1.2 Boiling point of the compounds or class of compounds of interest.

7.1.3 Tendency of the solvent and matrix to form emulsions.

7.1.4 Solvent strength (that is, polarity, solubility of compounds of interest).

7.1.5 Safety of solvent use (that is, toxicity, flammability).

7.1.6 Purity of solvent.

7.1.7 Solvent compatibility with analytical instrumentation.

7.2 The analyst should demonstrate a recovery using a spiking procedure in the matrix of interest before using this procedure.

7.3 Because the extraction is carried out at temperatures approaching the boiling point of the solvent for the entire extraction period, the analyst must ensure that heat-labile and more volatile solutes that may be of interest are stable and recoverable by the method. Low-boiling fractions can also be lost in the solvent evaporation steps where Kuderna-Danish solvent concentration is performed or where TSEC is determined gravimetrically after solvent evaporation.

7.4 The rate and extraction efficiency of the Soxhlet extraction process are not only a function of the solubility of the constituents of interest in solvent versus the matrix, but are also related to the dissolution process. This depends upon the ease of penetration of the solvent into the solid sample matrix. Fine particles are extracted more easily than large particles because of the increased surface area they provide. Therefore the particle nature of the sample must be evaluated and documented.

7.5 In many cases where the extraction efficiency of the constituent of interest during the extraction period is less than 100 %, the efficiency of the solvent extraction process is highly dependent upon the control of conditions during the Soxhlet extraction process. Extraction efficiency will depend upon the combined effect of the specific solvent used, the temperature at which extraction occurs, the cycle time for the Soxhlet apparatus, and the total extraction time. Therefore the rate and duration of extraction must be closely controlled.

8. Apparatus

8.1 Soxhlet extraction apparatus consisting of Soxhlet extractor, Allihn condenser, and 500-mL round-bottom boiling flask.

8.1.1 *Allihn Condenser*, bottom standard taper joint 45/50.

8.1.2 *Boiling Flask*, flat bottom, standard taper joint 24/40.

8.1.3 *Soxhlet Extractor*, 85-mL, top standard taper joint 45/50, bottom standard taper joint 24/40.

8.2 *Glass or Paper Thimble or Glass Wool* to retain sample in Soxhlet extraction device. It should drain freely and may require cleaning before use. To clean the thimbles, pre-extract them with the solvent to be used for sample extraction.

8.3 *Boiling Chips, Glass Beads, or PTFE-fluorocarbon Boiling Stones*, approximately 10/40 mesh. Boiling chips or glass beads can be cleaned by heating to 400°C for 30 min. Alternatively, Soxhlet extract with the same solvent as will be used for sample extraction.

8.4 *Heating Source*, capable of heating Soxhlet apparatus to achieve 10 solvent cycles per hour. Most commercially available heating mantles are adequate.

8.5 *Kuderna-Danish Concentrator*, fitted with graduated evaporative concentrator tube.

8.5.1 *Kuderna-Danish Concentrator Receiving Vessel*, graduated 10-mL. A ground glass stopper is used to prevent evaporation of extracts.

8.5.2 *Kuderna-Danish Evaporative Flask*, 500-mL, attached to the concentrator tube with springs.

8.5.3 *Kuderna-Danish Evaporative Flask*, 40-mL, attached to the concentrator tube with springs.

8.5.4 *Kuderna-Danish Snyder Column*, three-ball macro.

8.5.5 *Kuderna-Danish Snyder Column*, two-ball micro.

8.6 *Water Bath for Kuderna-Danish*, heated with concentric ring cover, capable of temperature control (±2°C). The bath must be used in a hood.

8.7 *Chromatographic Column*:

borosilicate, 20-mm inside diameter, approximately 400 mm

long, with coarse fritted plate on bottom and an appropriate solid phase.

8.8 *Impermeable Sheet of PTFE-fluorocarbon-Coated Flexible Material*, approximately 2 ft^2, for sample mixing, quartering, and preparation.

8.9 *Stainless Steel Utensils Coated With PTFE-fluorocarbon*, or other appropriate material for sample handling and mixing (that is, spatula, trowel, scoop, etc.).

8.10 *Stainless Steel or Brass Sieve*, 10 mesh.

9. Reagents and Materials

9.1 *Purity of Reagents*—Reagent grade chemicals shall be used in all tests. Unless otherwise indicated, all reagents shall conform to the specifications of the Committee on Analytical Reagents of the American Chemical Society where such specifications are available.[9] Other grades may be used, provided it is first ascertained that the reagent is of sufficiently high purity to permit its use without lessening the accuracy of the determination.

9.2 The extraction solvent of choice should be appropriate for the matrix and compounds of interest. This choice is dependent upon the chemical properties of the organic constituents of interest and the matrix being extracted. Literature references for solvents that have been used for Soxhlet extractions are provided in Table 1.

9.3 When specific organic compounds are to be analyzed, pesticide grade solvents (distilled in glass) are recommended. In general, the solvent blank for TSEC or for the specific compound of interest should be less than 20 % of the lowest reportable concentration required for the analysis.

9.4 *Sodium Sulfate (Na_2SO_4), or Magnesium Sulfate ($MgSO_4$)*, reagent grade, granular, anhydrous, prepared by heating at 400°C for a minimum of 4 h in a shallow tray to eliminate interfering organics.

10. Precautions

10.1 Some solvents (for example, benzene, chloroform, and carbon tetrachloride) are suspected human carcinogens and must be handled accordingly.

10.2 Explosive peroxides tend to form in ether solvents. A convenient means of testing for their presence is with E.M. Quant test paper.[10]

10.3 The use of fume hoods with volatile and toxic solvents is mandatory.

10.4 Flammable solvents must be kept from heat, sparks, or flames. Avoid buildup of vapors and eliminate all sources of ignition, especially nonexplosion-proof electrical apparatus and heaters. Keep containers closed. Use with adequate ventilation. Store bulk solvents in safety cabinets. Remove only a one-day supply and keep it in a hood.

10.5 Avoid prolonged breathing of vapor or spray mist and avoid prolonged or repeated skin contact with any organic solvent. Consult Materials Safety Data Sheets for handling instructions and precautions.

10.6 Samples containing odorous, known or suspected volatile materials must be processed in a fume hood.

10.7 Samples known or suspected of containing toxic or hazardous materials must be handled in a fume hood. Safety information relative to the handling of any known toxic materials must be taken into account prior to any sample handling.

11. Sample and Sample Preparation

11.1 Obtain a representative sample of the waste or soil to be tested using ASTM sample methods developed for the specific industry. (See Practices D 75, D 420, D 2910, D 3694, and D 3975, and Test Method D 2234.)

11.2 Where no specific methods are available, use sample methodology for materials of similar physical form.

11.3 Determine the minimum sample size for collection and delivery to the laboratory in accordance with Practice E 122.

11.4 Keep samples in closed containers appropriate to the sample type prior to testing. Store biologically active samples at 4°C and their extraction started within 8 h of receipt.

11.5 Document the physical description of the sample to be tested, including particle characteristics, as far as is known or can be determined by visual observation.

11.6 Do not use organic materials such as rubber or plastics other than clean PTFE-fluorocarbon in the storage or handling of samples.

11.7 Prepare sediment and soil samples in accordance with the instructions in Practice D 3976. This practice includes procedures for the determination of the moisture content of the sample, allowing final results to be reported on a wet or dry basis, as appropriate.

11.8 Samples may be screened to remove foreign objects prior to preparation or homogenization for chemical analysis. Large objects are removed mechanically.

11.9 For field-cored wastes or castings produced in the laboratory, cut a representative vertical section weighing approximately 100 g for extraction, plus any additional samples required for the determination of solids content or other physical properties. If the material cannot be ground into a fine granular consistency, then it is not appropriate for Soxhlet extraction.

11.10 For free-flowing particulate wastes, prepare a sample of the required size by quartering on a sheet of PTFE-fluorocarbon material or PTFE-fluorocarbon-coated material that has been demonstrated to be noncontaminating, as follows:

11.10.1 Empty the sample container onto the center of the sheet.

11.10.2 Flatten out the sample gently with a suitable straightedge until it is spread uniformly to a depth appropriate for its particle size.

11.10.3 Remix the sample by lifting a corner of the sheet and drawing it across to the opposite corner in a manner that the material is made to roll over and over and does not merely slide along. Continue operation with each corner, proceeding in a clockwise direction. Repeat this operation ten times.

11.10.4 Lift all four corners of the sheet towards the center and holding all four corners together, raise the entire sheet into

[9] "Reagent Chemicals, American Chemical Society Specifications," Am. Chemical Soc., Washington, DC. For suggestions on the testing of reagents not listed by the American Chemical Society, see "Analar Standards for Laboratory U.K. Chemicals," BDH Ltd., Poole, Dorset, and the "United States Pharmacopeia."

[10] Test paper, such as E.M. Quant, available from MCB Manufacturing Chemists, Inc., 2909 Highland Avenue, Cincinnati, OH 45212.

the air to form a pocket for the sample.

11.10.5 Repeat 11.10.2.

11.10.6 With a straightedge at least as long as the flattened mound of sample (such as a thin-edged yard stick) gently divide the sample into quarters.

11.10.7 Discard alternate quarters.

11.10.8 If further reduction of the sample size is necessary, repeat 11.10.3 through 11.10.7. A minimum sample size equivalent to 5 g, dry weight, is recommended for each Soxhlet extraction. The exact sample size used for Soxhlet extraction must be recorded.

11.11 *Sample Sieving*—The analytical sample for Soxhlet extraction is defined as that material which passes a 10-mesh sieve (approximately 2-mm openings). Stainless steel or brass sieves may be used. Materials that cannot be sieved but are of a thickness of less than 2 mm (that is, after drying in accordance with 12.3 and grinding in accordance with 12.4) or where the material is penetrable by the extracting solvent, can still be Soxhlet extracted. The material must be suitable for being mixed with the sample drying agent, sodium sulfate or magnesium sulfate, to provide drying of all sample surfaces.

11.11.1 If the purpose of extracting is to measure specific compounds, recovery experiments will have to be performed to demonstrate the efficacy of sample handling and sieving steps (or whether those steps should be omitted).

11.11.2 When appropriate, sieve dry samples without further pretreatment.

11.11.3 Vigorously stir wet samples to incorporate as much liquid as possible, thereby facilitating subsequent wet sieving.

11.11.4 Pass the wet sample, preferably as a slurry, through the sieve. The bottom of an appropriately sized Erlenmeyer flask may be used to gently press the waste sample through the sieve. Manually remove large objects that will not pass through the sieve and will interfere with the sieving operation. These objects are not considered appropriate for Soxhlet extraction.

11.11.5 Mix the sieved material by stirring, then allow it to settle for subsequent removal of the supernatant liquid. A centrifuge may be used to facilitate the separation of the liquid and solid.

12. Procedure

12.1 Decant any supernatant liquid from the sieved and settled sample. Save the supernatant liquid if separate analysis is required.

12.2 Blend 10 to 15 g of homogenized solid sample, "as received" from 12.1, with an equal weight of anhydrous sodium sulfate or magnesium sulfate in a 150-mL glass beaker. Add the drying agent slowly with continuous stirring and continue stirring until a dry powder or dry granules are formed.

12.3 Allow this mixture to stand 15 to 30 min until solidified. More drying agent may be added if needed to give the sample a dry appearance.

12.4 Remove the solidified material and grind it in a porcelain or glass mortar.

12.5 Place the ground sample in a thimble of appropriate size. The thimble should extend just above the siphon arm of the Soxhlet apparatus. Add sufficient sample blend so that at least 5 g of dry weight sample is extracted. If disintegration or floatation of sample occurs, cover with a plug of glass wool which has been previously extracted with the solvent. The glass wool will also aid in distributing the condensed solvent over the sample. Place the thimble in the extractor.

12.6 Add 275 mL of the extraction solvent into a 500-mL round-bottom flask containing a boiling chip, glass beads, or PTFE-fluorocarbon balls and attach the flask to the extractor. Initiate water flow through the cooling condenser.

12.7 Gradually heat the solvent to its boiling point. The solvent vapors will condense and drip into the thimble. The rate of extraction is varied by adjusting the heat supplied to the heating mantle. When the extraction thimble has filled to the top of the side arm, siphon the solvent automatically and drain back into the boiling flask. Adjust the heat until there is one dischargeable cycle every six minutes (that is, ten cycles per hour).

12.8 Extract the sample for 15 to 20 h. The extraction time may be shorter or longer, provided that the time selected has been demonstrated to be appropriate for the extraction (at least 80 % of the concentration which is normally extracted during the 15 to 20-h extraction procedure in the matrix of interest).

12.9 Turn off the heating mantle and allow the extract to cool. Rinse the condenser with three portions of 5 mL each of the extraction solvent and drain the Soxhlet into a graduated 500-mL glass cylinder. Record the volume of solvent recovered.

12.10 Filter and dry the extract by passing it through a 20-mm inside diameter column containing 10 cm of sodium sulfate which has been prewashed with the extracting solvent. Collect the eluent in a graduated 500-mL glass cylinder. Wash the sodium sulfate column with 25 mL of extracting solvent and add it to the graduated cylinder. Dilute the extract to exactly 300 mL with solvent.

12.11 Using a portion of the solutions from 12.10, determine the TSEC of the sample using TSEC Method A, B, or C in accordance with Test Method D 5368.

12.12 If the residue weight of the extract is less than 500 µg/g (with respect to the original solid), concentrate the solvent extract using the Kuderna-Danish apparatus prior to any other analysis (see 12.13).

12.13 Add the solvent extract to a 500-mL Kuderna-Danish concentrator flask fitted with a 10-mL graduated receiving vessel. Add one or two clean boiling chips or PTFE-fluorocarbon balls to the flask and attach a three-ball Snyder column. Prewet the Snyder column by adding about 1 mL solvent to the top. Place the Kuderna-Danish apparatus on a steam bath or hot water bath in a fume hood so that the concentrator tube and the entire rounded surface of the flask are bathed in hot water or vapor. Adjust the vertical position of the apparatus and the water temperature as required to complete the concentration in 15 to 20 min. At the proper rate of distillation the balls of the column will actively chatter, but the chamber will not flood with condensed solvent. When the apparent volume of liquid reaches 1 to 3 mL, remove the Kuderna-Danish apparatus and allow it to drain for at least 10 min while cooling.

12.14 Rinse the Kuderna-Danish apparatus with 2 to 4 mL of solvent. Remove the Snyder column and flask and adjust the sample volume to 10.0 mL with solvent.

12.15 If desired, measure the TSEC of the sample using TSEC Method A or B in accordance with Test Method D 5368. Proceed with instrumental analysis or further sample cleanup if required.

12.16 If necessary to meet the sensitivity requirements for the instrumental method used, the extract may be further concentrated 10-fold in the Kuderna-Danish receiver vessel.

12.16.1 Suspend the graduated receiver vessel in a beaker of warm water, a heating bath, or a heating block at a temperature 10 to 20°C below the boiling point of the solvent. Aid solvent evaporation by passing a gentle stream of nitrogen gas (prepurified quality or better) over the extract by means of a disposable Pasteur pipet (that is, nitrogen blowdown).[11] Use glass, metal, or PTFE-fluorocarbon tubing and glass pipets for the nitrogen, and use a new pipet for each sample.

12.16.2 When the extract has evaporated to a little less than 0.5 mL, remove the concentration vessel from the heating device and allow it to cool to room temperature. Rinse the vessel with about 0.5 mL of solvent and dilute to the 1.0-mL mark.

12.17 If required, determine the TSEC of the concentrated extract using Method A or B in accordance with Test Method D 5368.

12.18 Proceed with further analysis of the concentrate or with further sample cleanup procedures as appropriate.

[11] Refer to Test Method D 3086 under Alternative Method for Final Concentration.

REFERENCES

(**1**) Plumb, Jr., R. H., "Procedures for Handling and Chemical Analysis of Sediment and Water Samples," NT1S AD/A-103 788, May 1981.

(**2**) "Standard Methods for the Examination of Water and Wastewater," APHA, AWWA, and WPCF, 15th Edition, 1981.

(**3**) Strachan, S. D., Nelson, D. W., and Sommers, L. E., "Sewage Sludge Components Extractable with Nonaqueous Solvents," *Journal of Environmental Quality*, Vol 12, No. 1, 1983, pp. 69–74.

(**4**) Harrold, D. E., and Young, J. C.," Extraction of Priority Pollutants from Solids,'' *Proceedings of the American Society of Civil Engineers*, Vol 108, No. EE6, December 1982, pp. 1211–1227.

(**5**) Watts, R. R.," Manual of Analytical Methods for the Analysis of Pesticide Residues in Human and Environmental Samples," U.S. EPA, Environmental Toxicology Division, Research Triangle Park, North Carolina, 1980.

(**6**) Goldberg, M. C., and Weiner, E. R., "Extraction and Concentration of Phenolic Compounds from Water and Sediments," *Analytical Chimica Acta*, Vol 115, 1980, pp. 373–378.

(**7**) Chiba, M., and Morley, H. V., "Factors Influencing Extraction of Aldrin and Dieldrin Residues from Different Soil Types," *Journal of Agricultural and Food Chemistry*, Vol 16, 1968, pp. 916–922.

(**8**) Saha, J. G., Bhavaraju, B., Lee, Y. W., and Randell, R. L., "Factors Affecting Extraction of Dieldrin-14C from Soil," *Journal of Agricultural and Food Chemistry*, Vol 17, 1969, pp. 877–882.

(**9**) Hill, H. H., Chan, K. W., and Karasek, F. W., "Extraction of Organic Compounds from Airborne Particulate Matter for Gas Chromatographic Analysis," *Journal of Chromatography*, Vol 131, 1977, pp. 245–252.

(**10**) Pierce, R. C., and Katz, M., "Determination of Atmospheric Isomeric Polycyclic Arenes by Thin Layer Chromatography and Fluorescence Spectrophotometry," *Analytical Chemistry*, Vol 47, 1975, pp. 1743–1748.

(**11**) Stanley, T. W., Meeker, J. E., and Morgan, M. J., "Extraction of Organics from Airborne Particles—Effects of Various Solvents and Conditions on the Recovery of Benzo(a)pyrene, Benz(c)acridine and 7-h-Benz(de)anthracen-7-one," *Environmental Science and Technology*, Vol 1, 1967, pp. 927–931.

(**12**) Broddin, G., Van Vaeck, L., and Van Cauwenberghe, K., "On the Size Distribution of Polycyclic Aromatic Hydrocarbon Containing Particles from a Coke Oven Emission Source," *Atmospheric Environment*, Vol 11, 1977, pp. 1061–1064.

(**13**) Cautreels, W., and Van Cauwenberghe, K., "Extraction of Organic Compounds from Airborne Particulate Matter," *Water, Air, Soil Pollution*, Vol 6, 1976, pp. 103–110.

(**14**) Pahl, R. H., Mayan, K. G., and Bertrand, G. L., "Organic Desorption from Carbon—II. The Effect of Solvent in the Desorption of Phenol from Wet Carbon," *Water Research*, Vol 7, 1973, pp. 1309–1322.

(**15**) Buelow, R. W., Carswell, J. K., and Symons, J. M., "Improved Method for Determining Organics by Activated Carbon Adsorption and Solvent Extraction—Part I," *Journal of the American Water Works Association*, Vol 63, 1973, pp. 57–72.

(**16**) Grosjean, D., "Solvent Extraction and Organic Carbon Determination in Atmospheric Particulate Matter: The Organic Extraction-Organic Carbon Analyzer (OE-OCA) Technique," *Analytical Chemistry*, Vol 47, No. 6, May 1975, pp. 797–805.

(**17**) Sporstol, S., Lichtenthaler, R. G., and Oreld, F., "Extraction Efficiencies for Hydrocarbons from Sediments Polluted by Oil Drill Cuttings," *Analytical Chimica Acta*, Vol 169, 1985, pp. 343–347.

(**18**) Onuska, F. I., and Terry, K. A., "Determination of Chlorinated Benzenes in Bottom Sediment Samples by WCOT Column Gas Chromatography," *Analytical Chemistry*, Vol 57, 1985, pp. 801–805.

(**19**) Lamparski, L. L., and Nestrick, T. J., "Determination of Tetra-, Hepta-, and Octachlorodibenzo-p-dioxin Isomers in Particulate Samples at Parts per Trillion Levels," *Analytical Chemistry*, Vol 52, 1980, pp. 2045–2054.

(**20**) Nash, R. G., and Harris, W. G., "Soil Moisture Influence on Treatment and Extraction and Extraction of DDT from Soils," *Journal of the Association of Official Analytical Chemists*, Vol 55, No. 3, 1972, pp. 532–536.

Designation: D 5370 – 96

Standard Specification for Pozzolanic Blended Materials in Construction Applications[1]

This standard is issued under the fixed designation D 5370; the number immediately following the designation indicates the year of original adoption or, in the case of revision, the year of last revision. A number in parentheses indicates the year of last reapproval. A superscript epsilon (ε) indicates an editorial change since the last revision or reapproval.

1. Scope

1.1 This specification covers pozzolanic blended material for use in construction applications where the properties normally attributed to finely divided mineral admixtures may be desired.

1.2 The values stated in SI units are to be regarded as the standard. The values given in parentheses are for information only.

1.3 *This standard does not purport to address all of the safety concerns, if any, associated with its use. It is the responsibility of the user of this standard to establish appropriate safety and health practices and determine the applicability of regulatory limitations prior to use.*

2. Referenced Documents

2.1 *ASTM Standards:*

C 25 Test Methods for Chemical Analysis of Limestone, Quicklime, and Hydrated Lime[2]

C 114 Test Methods for Chemical Analysis of Hydraulic Cement[2]

C 125 Terminology Relating to Concrete and Concrete Aggregates[3]

C 219 Terminology Relating to Hydraulic Cement[2]

C 311 Test Methods for Sampling and Testing Fly Ash or Natural Pozzolans for Use as a Mineral Admixture in Portland-Cement Concrete[3]

C 593 Specification for Fly Ash and Other Pozzolans for Use with Lime[2]

C 618 Specification for Coal Fly Ash and Raw or Calcined Natural Pozzolan for Use as a Mineral Admixture in Concrete[3]

3. Terminology

3.1 *Definitions of Terms Specific to This Standard:*

3.1.1 *cement kiln dust*—the finely divided particulate matter carried from a cement kiln by the exhaust gases.

3.1.2 *pozzolanic blended material*—a pozzolan resulting from the intimate and uniform blend of two or more finely divided materials.

3.2 *Definitions*—Other terms used in this specification are defined in Terminologies C 125 or C 219 or Specification C 618.

4. Classification

4.1 The types of pozzolanic blended materials covered by this specification are as follows:

4.1.1 *Type FC*—Blended pozzolan consisting of an intimate and uniform blend of Class F fly ash and Class C fly ash, and

4.1.2 *Type FCKD*—Blended pozzolan consisting of an intimate and uniform blend of Class F fly ash and cement kiln dust.

5. Ordering Information

5.1 Include the following information in orders for blended pozzolan meeting the requirements of this specification:

5.1.1 The specification designation ASTM Specification D 5370,

5.1.2 The specification date, if other than the latest published version,

5.1.3 The quantity of blended pozzolan desired, and the classification desired,

5.1.4 A statement as to specific requirements or intended use of the blended pozzolan, and

5.1.5 Request for the manufacturers certification, if desired.

6. Materials and Manufacturer

6.1 The pozzolanic blended material is the result of an intimate and uniform blend.

NOTE 1—The attainment of an intimate and uniform blend of two types of fine material is difficult. Consequently, adequate equipment and controls must be provided by the manufacturer. The purchaser should ensure himself of the adequacy of the blending operation. See Appendix X1 for the procedure.

NOTE 2—Certain raw materials like fly ash and cement kiln dust, which blends are covered by this specification, vary from source to source, as well as within sources.

[1] This specification is under the jurisdiction of ASTM Committee D-34 on Waste Management and is the direct responsibility of Subcommittee D34.06 on Recovery and Reuse.

Current edition approved June 10, 1996. Published August 1996.

[2] *Annual Book of ASTM Standards*, Vol 04.01.

[3] *Annual Book of ASTM Standards*, Vol 04.02.

7. Blended Pozzolan for Use with Lime[4]

7.1 Ensure that blended pozzolan conforms to the requirements prescribed in Specification C 593.

8. Blended Pozzolan for Use in Concrete[5]

8.1 Ensure that blended pozzolan conforms to the chemical and physical requirements as prescribed in the tables of Specification C 618. The purchaser will specify the mineral admixture class (in Specification C 618). Carbon may substitute for loss on ignition requirement. The percentage of carbon or loss on ignition is determined by the procedure in the section on test methods.

9. Sampling

9.1 Sample the blended pozzolan in accordance with the requirements of Test Methods C 311.

10. Test Methods

10.1 *Blended Pozzolans for Use With Lime*—Use test methods as required in Specification C 593.

10.2 *Blended Pozzolans for Use in Concrete*—Use test methods as required in Test Methods C 311 with the following exception:

10.2.1 *Determination of Loss on Ignition (Corrected for CO_2) of Type FCKD Blended Pozzolan*—This test method corrects for the presence of carbon dioxide. Ignite the oven-dried blended pozzolan in a muffle furnace at a controlled temperature. Determine total loss on ignition according to Test Methods C 311 except use platinum crucibles and ignite at a temperature of 950 ± 50°C. Calculate total loss on ignition according to Test Methods C 311. The total loss in weight is assumed to be CO_2, C, and any chemically combined water.

10.2.1.1 Determine carbon dioxide on an oven-dried sample of blended pozzolan according to Test Methods C 25. Some blended pozzolans may have negligible amounts of carbon dioxide.

10.2.1.2 Report loss on ignition (corrected for CO_2) for blended pozzolan as the difference between total loss on ignition and carbon dioxide. Thus, the percentage loss on ignition (corrected for CO_2) = the percentage total loss on ignition minus percentage carbon dioxide.

10.2.2 Determination of carbon (corrected for CO_2) in blended pozzolan is an acceptable alternative to loss on ignition (corrected for CO_2).

10.2.2.1 Determine total carbon on an oven-dried sample of blended pozzolan according to Test Methods C 25.

10.2.2.2 Determine carbon dioxide on an oven-dried sample of blended pozzolan according to Test Methods C 25. Report carbon corrected for CO_2 for blended pozzolan as the difference between total carbon and carbon dioxide. Thus, the percentage carbon (corrected for CO_2) = the percentage total carbon – ($CO_2 \times 0.273$).

[4] *Kiln Dust Fly Ash Systems for Highway Bases and Subbases*, U.S. Department of Transportation-Report #FHWA/RD/167, 1982.

[5] *Evaluation of Kiln Dust as in Concrete*, Proceedings of Second International Conference in Madrid on Fly Ash, Silica Fume, Slag, Natural Pozzolans in Concrete, V. Ramakrishnan, ed. Vol 1, 1986.

11. Storage and Inspection

11.1 Store the blended pozzolan in such a manner as to permit easy access for proper inspection and identification of each shipment. The purchaser shall have access to every facility for careful sampling and inspection, either at the source or at the site of the work as may be specified by the purchaser.

12. Rejection

12.1 Material that fails to conform to the requirements of this specification may be rejected. Report rejection to the producer or supplier promptly and in writing. In case of dissatisfaction with the results of the test, the producer or supplier may make claim for a rehearing.

12.2 Packages more than 2 % below the weight marked thereon, may be rejected; and if the average weight of the packages in any shipment, as shown by weighing 50 packages taken at random, is less than that marked on the packages, the entire shipment may be rejected.

12.3 Blended pozzolan remaining in storage for a period greater than 6 months after completion of the tests may be retested and may be rejected if it fails to conform to any of the requirements of this specification.

13. Certification

13.1 When specified in the purchase order or contract, furnish the purchaser certification that samples representing each lot have been tested or inspected as directed in this specification and the requirements of this specification have been met.

13.1.1 When specified in the purchase order or contract, furnish a report of the test results that contain the results of chemical analysis of the blended pozzolan to provide the purchaser the composition of the material. Include in that analysis: loss on ignition or carbon (corrected for CO_2), moisture, SiO_2, Al_2O_3, Fe_2O_3, CaO, MgO, Total Na_2O, Total K_2O, and SO_3.

14. Manufacturer's Statement

14.1 The manufacturer will state, in writing, the nature of physical and chemical changes in the pozzolanic blended material since last purchased. Any changes in sources of raw material or in proportions of those raw materials will be reported to the purchaser or user. If requested, test data will be supplied indicating specification compliance.

15. Packaging and Package Marking

15.1 Plainly mark the class of the blended pozzolan, name and brand of the producer, and the mass of the material contained therein on each package when the blended pozzolan is delivered in packages. Provide similar information in the shipping documents accompanying the shipment of packaged or bulk blended pozzolan.

16. Keywords

16.1 cement kiln dust; fly ash; lime kiln dust; pozzolan; pozzolanic blended material

APPENDIX

(Nonmandatory Information)

X1. QUALITY ASSURANCE PROGRAM: ADEQUACY OF THE BLENDING PROCESS AT THE PRODUCTION FACILITY

X1.1 A blend of two ingredients of different absolute densities should result in a known reproducible density for each proportion. For example, a uniform blend of 80 % fly ash (density = 2.27) and 20 % kiln dust (density = 2.9) should result in a density within 5 % of the theoretical density. Take three random samples after the blending process has been completed and determine the densities of each.

X1.2 To complete the testing for blending uniformity, one can determine the distribution of a very small quantity of admixture within the products. Add fluorescein to the batch (for example, add 30 g fluorescein to a 5-lb bag of material, mix, and add the bag to the 12-ton batch) and after the blending process is completed. The concentration of fluorescein in the batch, as measured by spectrophotometric technique, should be 95 to 105 % of theoretical.

X1.3 The exact blending process must be in the QA report including batch weights, blending times, blending procedure, and other pertinent data.

X1.4 The check for the uniformity of the blend should be performed on a monthly basis or each time the blending process varies (equipment, blending time, weights, and so forth).

Designation: D 5451 – 93 (Reapproved 2001)

Standard Practice for Sampling Using a Trier Sampler[1]

This standard is issued under the fixed designation D 5451; the number immediately following the designation indicates the year of original adoption or, in the case of revision, the year of last revision. A number in parentheses indicates the year of last reapproval. A superscript epsilon (ε) indicates an editorial change since the last revision or reapproval.

1. Scope

1.1 This practice covers sampling using a trier. A trier resembles an elongated scoop as shown in Fig. 1. The trier is used to collect samples of granular or powdered materials that are moist or sticky and have a particle diameter less than one-half the diameter of the trier.

1.2 The trier can be used as a vertical coring device only when it is certain that a relatively complete and cylindrical sample can be extracted.

1.3 *This standard does not purport to address all of the safety problems, if any, associated with its use. It is the responsibility of the user of this standard to establish appropriate safety and health practices and determine the applicability of regulatory limitations prior to use.*

2. Referenced Documents

2.1 *ASTM Standards:*

D 4687 Guide for General Planning of Waste Sampling[2]

D 5088 Practice for the Decontamination of Field Equipment Used at Non-Radioactive Waste Sites[3]

D 5283 Practice for Generation of Environmental Data Related to Waste Management Activities: Quality Assurance and Quality Control Planning and Implementation[2]

3. Summary of Practice

3.1 As a coring device, the trier is pushed into the material to be sampled and is turned to cut the core. The core is then removed from the hole.

4. Significance and Use

4.1 This practice is applicable to sampling soils and similar fine-grained cohesive materials. This practice is to be used by personnel who are to acquire the samples.

4.2 This practice should be used in conjunction with Guide D 4687, which covers sampling plans, safety, quality assurance, preservation, decontamination, labeling, and chain-of-custody procedures; Practice D 5088, which covers the decontamination of field equipment used at waste sites; and Practice D 5283, which covers project specifications and practices for environmental field operations.

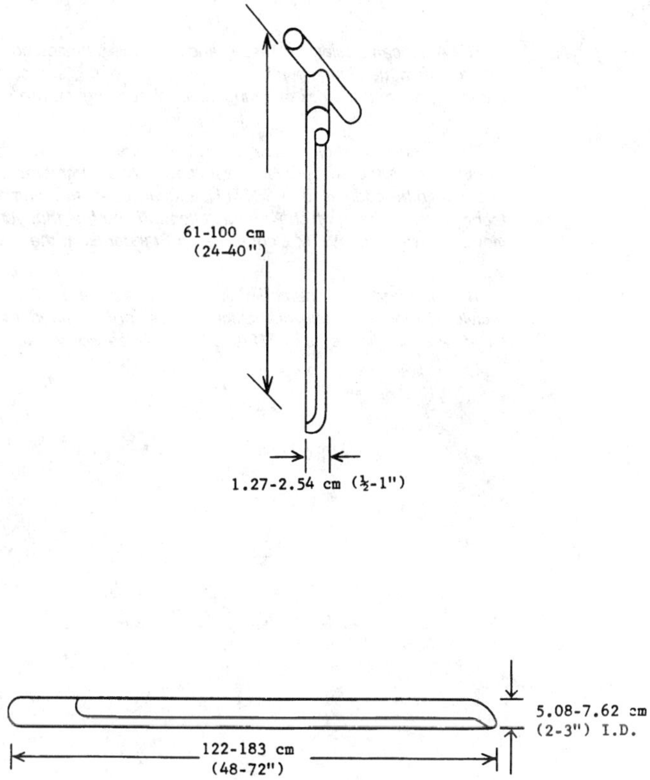

FIG. 1 Triers (Typical)

5. Sampling Equipment

5.1 The trier should be made from materials that are compatible with the substances being sampled and with the tests or analyses to be performed. Either stainless steel or polytetrafluoroethylene-coated metal will be suitable for most situations (see Fig. 1).

6. Sample Containers

6.1 Plastic, glass, or other nonreactive containers should be used. Refer to Guide D 4687 for further information on containers.

7. Procedure

7.1 Record appropriate information and observations on the sample location.

7.2 If sampling soils, remove surface vegetation or debris, or both, from the area of sample extraction.

---

[1] This practice is under the jurisdiction of ASTM Committee D34 on Waste Management and is the direct responsibility of Subcommittee D34.01.03 on Sampling Preservation.

Current edition approved Aug. 15, 1993. Published October 1993.

[2] *Annual Book of ASTM Standards*, Vol 11.04.

[3] *Annual Book of ASTM Standards*, Vol 04.08.

7.3 For core sampling, proceed as follows:

7.3.1 Insert the trier approximately perpendicular to the surface of the material and rotate (one or two times) to cut a core.

7.3.2 Pull the core out of the hole slowly. Do not allow additional overburden material to become part of the sample. Inspect the core surface, and note the appearance of any irregularities (for example, scratches, pock marks, and pebbles). Also inspect the core for breakage. If breakage has occurred and the core does not satisfy minimum length requirements, discard it and extract another from an immediately adjacent location.

7.4 Transfer the sample into a suitable container using a clean spatula. Seal the container and affix a label.

7.5 Complete the field logbook and chain-of-custody forms.

7.6 Decontaminate the used equipment in accordance with Practice D 5088.

8. Keywords

8.1 sampling; soil sampling; trier; waste

Designation: D 5468 – 95

Standard Test Method for Gross Calorific and Ash Value of Waste Materials[1]

This standard is issued under the fixed designation D 5468; the number immediately following the designation indicates the year of original adoption or, in the case of revision, the year of last revision. A number in parentheses indicates the year of last reapproval. A superscript epsilon (ϵ) indicates an editorial change since the last revision or reapproval.

1. Scope

1.1 This test method covers the determination of the gross calorific value of waste materials by either an isoperibol, aneroid, air-jacketed isoperibol, or adiabatic bomb calorimeter using electronic temperature sensors, automatic calorimeter controllers, and subsequent analysis of the residue for ash. After calorific analysis, the bomb washings may be collected for subsequent elemental analysis.

1.2 The values stated in SI units are regarded as the standard.

1.3 *This standard does not purport to address all of the safety concerns, if any, associated with its use. It is the responsibility of the user of this standard to establish appropriate safety and health practices and determine the applicability of regulatory limitations prior to use.* Specific hazard statements are given in Section 8.

2. Referenced Documents

2.1 *ASTM Standards:*

D 121 Terminology of Coal and Coke[2]

D 240 Test Method for Heat of Combustion of Liquid Hydrocarbon Fuels by Bomb Calorimeter[3]

D 1018 Test Method for Hydrogen in Petroleum Fractions[3]

D 1193 Specification for Reagent Water[4]

D 3177 Test Method for Total Sulfur in the Analysis Sample of Coal and Coke[2]

D 4239 Test Method for Sulfur in the Analysis Sample of Coal and Coke Using High Temperature Tube Furnace Combustion Methods[2]

E 144 Practice for Safe Use of Oxygen Combustion Bombs[5]

3. Terminology

3.1 *Definitions:*

3.1.1 *calorific value*—the heat produced by combustion of a unit quantity of a specimen under specified conditions. It is expressed in this test method in joules per kilogram (J/kg). Calorific value may also be expressed in calories per gram (cal/g) or in the British thermal units per pound (Btu/lb), when required. The unit equivalents are given in Table 1.

3.1.2 *calorimeter*—as used in this test method, is not only the bomb and its contents, but it also includes the bucket, electronic temperature sensing devices, ignition leads, water, and stirrer when using water.

3.1.2.1 *adiabatic calorimeter*—a calorimeter that has a jacket temperature adjusted to follow the calorimeter temperature as closely as possibly so as to maintain zero thermal head.

3.1.2.2 *calorimeter jacket*—the insulating medium surrounding the calorimeter.

3.1.2.3 *isoperibol calorimeter*—the calorimeter has a jacket of uniform and constant temperature.

3.1.3 *gross calorific value (gross heat of combustion), Qv (gross)*—the heat produced by combustion of unit quantity of a solid or liquid fuel when burned at constant volume in an oxygen bomb calorimeter under specified conditions with the resulting water condensed to a liquid.

3.1.4 *heat capacity*—the quantity of heat required to raise a system one degree in temperature either at constant volume or constant pressure.

3.1.5 *heat of formation*—the increase in heat content resulting from the formation of 1 mole of a substance from its elements at constant pressure.

3.1.6 *net calorific value (net heat of combustion at constant pressure), Qp*—the heat produced by combustion of unit quantity of a solid or liquid fuel when burned at a constant pressure of 0.1 MPa (1 atm), under conditions such that all the water in the products remain in the form of vapor.

3.1.7 *static calorimeter*—a calorimeter without a thermostated jacket.

3.2 *Definitions of Terms Specific to This Standard:*

3.2.1 *corrected temperature rise*—the temperature of the calorimeter caused by the process that occurs inside the bomb, which is, the observed temperature change corrected for various effects.

3.2.2 *heat capacity, energy equivalent, or water equivalent*—the energy required to raise the temperature of the calorimeter one arbitrary unit. This is the quantity that, when multiplied by the corrected temperature, rises, then when adjusted for extraneous heat effects and divided by the mass of the sample, gives the gross calorific value.

---

[1] This test method is under the jurisdiction of ASTM Committee D-34 on Waste Management and is the direct responsibility of Subcommittee D34.03 on Treatment.

Current edition approved March 15, 1995. Published May 1995. Originally published as D 5468 – 93. Last previous edition D 5468 – 93.

[2] *Annual Book of ASTM Standards*, Vol 05.05.

[3] *Annual Book of ASTM Standards*, Vol 05.01.

[4] *Annual Book of ASTM Standards*, Vol 11.01.

[5] *Annual Book of ASTM Standards*, Vol 14.02.

TABLE 1 Calorific Value

1 Btu = 1055.06 J	1 J/g = 0.430 Btu/lb
1 cal[A] = 4.1868 J	1 J/g = 0.239 cal/g

[A] International tables calorie.

4. Summary of Test Method

4.1 Calorific value is determined in this test method by burning a weighed sample under controlled conditions, in oxygen, in a calibrated calorimeter. The calorimeter is standardized by burning a specified amount of benzoic acid. The calorific value of the test specimen is computed from temperature observations made before, during, and after combustion and making proper allowances for heat contribution by other processes.

4.2 After disassembly of the bomb, estimate the ash content by one of two of the following procedures:

4.2.1 Dry the fuel capsule at 200°C for 5 to 10 min and weigh to provide an estimation of the ash content of the waste material.

4.2.2 Wipe any moisture from the exterior surface of the capsule and weigh to provide an estimation of the ash content of the waste material.

5. Significance and Use

5.1 This test method is used to compute the calorific value and estimated ash content of the waste material represented by the sample for incineration purposes or as fuel for cement kilns.

5.2 The gross calorific and ash value may be used for evaluating the effectiveness of any beneficiation process.

6. Apparatus and Facilities

6.1 *Test Room*—A room or area free from drafts and that can be kept at a constant temperature equal to ± 3°C, for all calorimetric tests. The apparatus shall be shielded from direct sunlight and radiation from other sources. Thermostatic control of room temperature and controlled relative humidity are desirable.

6.2 *Combustion Bomb*— Constructed of materials that are not affected by the combustion process or products sufficiently to introduce measurable heat input or alteration of end products. The bomb shall be designed so that all liquid combustion products can be completely recovered by washing the inner surfaces. There shall be no gas leakage. The bomb shall be capable of withstanding a hydrostatic pressure test of 20 MPa (3000 psig) at room temperature without stressing any part beyond its elastic limit.

6.3 *Balance*—A laboratory balance having the capability to weight the sample to the nearest 0.0001 g. The balance shall be checked for accuracy at the beginning of each week that the calorimeter tests will be made.

6.4 *Calorimeter Vessel (Bucket)*—Made of metal with a tarnish-resistant coating, with all outer surfaces highly polished. For calorimeters having a bucket it may be a separate component or integral component of the bomb. It shall be of such construction that the environment of the calorimeter's entire outer boundaries is maintained at a uniform temperature.

6.5 *Jacket*—An insulated container that keeps the inner perimeter of the jacket at constant temperature (isoperibol), or at the same temperature as the calorimeter (adiabatic), or is monitored continuously for temperature changes (aneroid).

6.6 *Temperature Sensing Device*—Thermometers such as platinum resistance, thermistors, or mercurial thermometers are satisfactory and may be used if properly calibrated.

NOTE 1—Temperature is measured in either degrees Celsius, degrees Fahrenheit, ohms, or other arbitrary units instead of degrees. Consistent units must be used in standardization and the actual calorific value determination.

6.7 *Sample Holder*— An open crucible of platinum, quartz, or acceptable base-metal alloy. The base metal crucibles should be heat-treated for 4 h at 500°C to ensure the capsules are completely oxidized.

6.8 *Ignition Fuse*— The ignition fuse shall be 100 mm of 0.16-mm diameter or smaller (No. 34 B & S gage), nickel-chromium alloy (Chromel C) alloy, platinum, cotton thread, or iron wire. The same length or mass of ignition fuse shall be used for all calibrations and calorific value determinations.

6.9 *Firing Circuit*— A 6 to 24-V alternating or direct current is required for ignition purposes. A step-down transformer connected to an alternating current lighting circuit, capacitors, or batteries may be used.

6.10 *Buret*—Used for the acid titration and shall have 0.1 mL divisions.

6.11 *Gelatin Capsules or Cellulose Tape*—Used to minimize vaporization of volatile samples. These items should be halogen free.

6.12 *pH Titrators or mV Meters*—May be used for pH titration of wash solution.

7. Reagents

7.1 *Reagent Water*— Conforming to conductivity requirements for Type II of Specification D 1193, it shall be used for preparation of reagents and washing of the bomb interior.

7.2 *Purity or Reagents*—Reagent-grade chemicals conforming to the specifications of the Committee on Analytical Reagents of the American Chemical Society shall be used in all tests.[6]

7.3 *Benzoic Acid, Standard* (C_6H_5—COOH)—Use pellets made from benzoic acid available from the National Institute of Standards and Technology (NIST) or benzoic acid calibrated against NIST standard material. The value of heat of combustion of benzoic acid, for use in the calibrations calculations, shall be in accordance with the certified value.

7.4 *Methyl Orange, Methyl Red, or Methyl Purple Indicator*—May be used to titrate the acid formed during combustion. The indicator used shall be the same for both calibration and calorific value determinations.

7.5 *Oxygen*—Shall be free of combustible matter. Only oxygen manufactured from liquid air, guaranteed to be greater than 99.5 % pure, shall be used. Oxygen made by the electrolytic process may contain a small amount of hydrogen,

[6] *Reagent Chemicals, American Chemical Society Specifications*, American Chemical Society, Washington, DC. For suggestions on the testing of reagents not listed by the American Chemical Society, see *Analar Standards for Laboratory Chemicals*, BDH Ltd., Poole, Dorset, U.K., and the *United States Pharmacopeia and National Formulary*, U.S. Pharmaceutical Convention, Inc. (USPC), Rockville, MD.

rendering it unfit without purification by passage over copper oxide at 500°C.

7.6 *Standard Solution*—Sodium carbonate (Na_2CO_3) or other suitable standard solution. Dissolve 3.76 g of sodium carbonate that has been dried for 24 h at 105°C in water and dilute to 1 L. One millilitre of this solution is equivalent to 4.2 J (1.0 cal) in the acid titration when the entire rinse solution is used as the analytical sample. Alternatively, the titrant volume is to be corrected by the volume fraction of the rinse that is used as the analytical sample. (See 10.4.2.)

8. Hazards

8.1 The following precautions are recommended for safe calorimeter operation. Additional precautions are given in Practice E 144. Also consult the calorimeter equipment manufacturer's installation and operating instructions before using the calorimeter.

8.2 The mass of waste sample and the pressure of the oxygen admitted to the bomb must not exceed the bomb manufacturer's recommendations.

8.3 Inspect the bomb parts carefully after each use. When halogenated organic compounds are being tested, replace all O-rings and valve seats after 100 firings. If no halogenated organic compounds are being tested, replace all O-rings and valve seats after 500 firings. At least every 3000 firings, check the bomb for thread wear on any closures or return to the manufacturer for these tests. These are recommendations for worst case situations. For more details, contact the manufacturer.

8.4 Equip oxygen supply cylinder with an approved type of safety device, such as a reducing valve, in addition to the needle valve and pressure gage used in regulating the oxygen feed to the bomb. Valves, gages, and gaskets must meet industry safety codes. Suitable reducing valves and adapters for 3 to 4-MPa (300 to 500-psi) discharge pressure are obtainable from commercial sources of compressed gas equipment. Check the pressure gage annually for accuracy or after any accidental over pressures that reach maximum gage pressure.

8.5 During ignition of a sample, the operator must not permit any portion of the operator's body to extend over the calorimeter.

8.6 Exercise extreme caution not to exceed the bomb manufacturer's recommendations and to avoid damage to the bomb when combustion aids are employed. Do not fire loose fluffy material, such as unpelleted benzoic acid, unless thoroughly mixed with the sample.

8.7 Do not fire the bomb if the bomb has been dropped or turned over after loading or if there is evidence of gas leakage when the bomb is submerged in the calorimeter water.

8.8 When volatile materials are present, explosive firings can result and the procedure in 12.2 should be followed.

9. Sampling

9.1 The sample shall be representative of the material of interest (see Appendix X3).

10. Standardization

10.1 The calorimeter is standardized by combustion of benzoic acid.

10.2 Determine heat capacity as the average of a minimum of 6 individual runs made over a period of not less than 3 days nor more than 5 days. To be acceptable for highprecision work, the relative standard deviation of the series shall be 0.15 % or less. See Table 2. For users of static-type calorimeters, an acceptable relative standard deviation would be 0.40 % or less. For this purpose, any individual test may be discarded only if there is evidence of incomplete combustion. If this limit is not met, the entire series shall be repeated until a series is obtained with a relative standard deviation below the acceptable limit.

10.3 *Procedure*:

10.3.1 Control the mass of the pellets of benzoic acid in each calibration series to obtain the same temperature rise as obtained with the waste specimen tested in the same laboratory. Weigh the pellet during the same test day and to the nearest 0.0001 g in the sample holder in which it is to be burned, and record the weight.

10.3.2 Rinse the bomb with water to lubricate internal seals and dry surfaces. Add approximately 1 mL of water or desired amount of other appropriate solution (see Note 2) to the bomb prior to assembly for a test.

NOTE 2—If the washing will be used for elemental analysis, then add 5 mL of a 10 % sodium carbonate solution or other solution of desired amount rather than 1 mL of water to the bomb prior to assembly for both standardization and analysis procedures.

10.3.3 Connect a measured fuse to the ignition terminals, in accordance with the manufacturer's guidelines. (See 6.8.)

10.3.4 Assemble the bomb and charge it with oxygen to a consistent pressure of 3 MPa (30 atm). Lower charging pressures may be used if they will not result in incomplete combustion as evident by visual inspection for carbon residues or the formation of carbon monoxide rather than carbon dioxide. This pressure must remain the same for each calibration and each calorific value determination. Admit the oxygen slowly into the bomb so as not to blow powdered material from the sample holder. If the pressure exceeds the specified pressure, detach the filling connection and exhaust the bomb in the usual manner, then discard the sample and bucket of water.

TABLE 2 Standard Deviations for Calorimeter Standardization

NOTE 1—Average = X = 10253.5.

NOTE 2—Variance =

$$s^2 = \frac{\text{Sum Column } C - [(\text{Sum Column } B)^2/n] = 26.34}{n-1}.$$

NOTE 3—Standard Deviation = s = $\sqrt{variance}$ = $\sqrt{237.1}$ = 5.14.

Standardization Number	Column A Energy Equivalent, J/°C	Column B Difference from Average	Column C (Column B)^2
1	10257.7	4.2	17.6
2	10249.3	4.2	17.6
3	10270.2	16.7	278.9
4	10253.5	0	0
5	10245.1	8.4	70.6
6	10249.3	4.2	17.6
7	10240.9	12.6	158.8
8	10266.0	12.5	156.3
9	10257.7	4.2	17.6
10	10245.1	8.4	70.6
SUM		75.4	805.6

Transfer the bucket, bomb, and calorimeter water to the jacket. Finish assembly of the apparatus and start the calorimeter. The calorimeter initial water temperature shall be at the same temperature ±0.5°C for every determination. The calorimeter monitors and determines whether the calorimeter has been constant over a period of 30 s. For adiabatic calorimeters with microprocessor control this stability requirement is 10^{-3}°C, and it is 10^{-3}°C/s or better for the isoperibol calorimeter with microprocessor control.

NOTE 3—The initial temperature adjustment will ensure a final temperature slightly below that of the jacket for isoperibol calorimeters.

10.3.5 *Temperature Observations*:

10.3.5.1 *Observations (Extrapolation Method)*—The controller will fire the charge, record the temperature rise, and correct the temperature rise for any appropriate heat leak corrections as recommended or defined, or both, by the instrument manufacturer. The controller may terminate the test when the observed thermal curve matches the manufacturer's thermal curve, which allows extrapolation to a final temperature. The extrapolated temperature rise should have a maximum uncertainty of ± 0.002°C.

10.3.5.2 *Observations (Full Development Method)*—The controller will fire the bomb and record the temperature rise until the temperature has stabilized for a period of 30 s.

10.3.5.3 *Manual Full Development Method*—Fire the charge at time *a* and record this temperature as t_a. For adiabatic calorimeters record subsequent readings at intervals no greater than 1 min until the difference between 3 or more pairs of successive readings are the same. For isoperibol calorimeters record the subsequent readings until the difference between 3 or more pairs of successive readings are the same within 1 millidegree per min. Record the time of stabilization as *c* and the final temperature as t_c, the first reading after the rate of change has become uniform.

10.3.6 Open the cover and remove the bomb. Release the pressure at a uniform rate, such that the operation will require not less than 1 min. Open the bomb and examine the bomb interior. Unburned samples or sooty deposits are indications of incomplete combustion possibly due to highash content or high-moisture content, or both. The results are suspect and generally require combustion aid such as ethylene glycol for water-based compounds or mineral oil for high-ash content compounds. When combustion aids are used, be sure heat liberation for any test meets the calorimeter manufacturer's guidelines. Discard the test if unburned sample or sooty deposits are found. Use the manufacturer's recommended constant value for acid correction or wash the interior of the bomb with distilled water containing the titration indicator until the washings are free of acid. Rinse only the exterior of the fuel capsule, taking care not to disturb the ash residue in the fuel capsule. Titrate the washings with standard solution as defined under reagents, or titrate with pH meter.

10.3.7 Use the manufacturer' recommended constant value or remove and measure, or weigh, the combined pieces of unburned ignition (firing) wire and subtract from the original length or weigh to determine the wire consumed in firing. If the wire is weighed, remove the ball of oxidized metal from the end of each piece of wire before weighing. When cotton thread fuse is used, use the manufacturer's recommended calorie correction.

10.4 *Calculations*:

10.4.1 *Temperature Rise*— Where applicable use data obtained as prescribed in 10.3.5; compute the corrected temperature rise, *t*, as follows:

$$t = t_c - t_a + C_r \tag{1}$$

where:

t = corrected temperature rise, °C (°F),
t_a = initial temperature reading at time of firing,
t_c = final temperature reading, and
C_r = radiation correction (see A1.1.3).

10.4.2 *Thermochemical Corrections* (see Appendix X1)—Compute the following for each test:

e_1 = correction for the heat of formation of HNO_3, in joules. Each millilitre of standard Na_2CO_3 is equivalent to 4.2 J (1.0 cal) and may be considered to be a constant value as recommended by the calorimeter manufacturer.
e_2 = correction for heat of combustion of firing fuse, in joules, or may be considered to be a constant value as recommended by the calorimeter manufacturer.
e_2 = 0.96 J/mm (0.23 cal/mm) or 5.9 J/mg (1.4 cal/mg) for No. 34 B&S gage Chromel C.
e_2 = 1.13 J/mm (0.27 cal/mm) or 7.5 J/mg (1.8 cal/mg) for No. 34 B&S gage iron wire.

10.4.3 Compute the calorimeter heat capacity by substituting in the following:

$$E = [(H_c m) + e_1 + e_2]/t \tag{2}$$

where:

E = calorimeter heat capacity,
H_c = heat of combustion of benzoic acid, as stated in the NIST Certificate, J/kg in air,
m = mass (weight in air) of benzoic acid, g,
e_1 = titration correction (10.4.2),
e_2 = fuse correction (10.4.2), and
t = corrected temperature rise (10.4.1).

For sample calculations, see Appendix X4.

10.5 Repeat the procedure for a total of ten standardizations or a minimum of six standardizations. Compute the standard deviation as illustrated in Table 2. (The percent relative standard deviation is the standard deviation times 100 and divided by the average value.)

11. Restandardization

11.1 Make checks on the heat capacity value, after changing any part of the bomb or calorimeter, at a minimum of at least once a month.

11.1.1 If a single new standardization differs from the old energy equivalent value by 0.33 % of the heat capacity, the old standard value is suspect, thereby requiring a second test.

11.1.2 The difference between the two new standardizations must not exceed 0.23 % of the heat capacity, and the average of the two new standardizations shall not differ from the old standard by more than ± 0.25 % of the heat capacity. If these requirements are met, do not change the calorimeter standard.

11.1.3 If the requirements given in 11.1.2 are not met, two more standardizations must be run. The range of the four values shall not exceed 0.35 % of the heat capacity, and the average of the four new standardizations shall not differ from the old standard value by more than ± 0.17 % of the heat capacity. If these requirements are met, do not change the calorimeter standard.

11.1.4 If the requirements given in 11.1.3 are not met, a fifth and sixth standardization shall be run. The range of the six new standardizations shall not exceed 0.44 % of the heat capacity, and the average of the six values shall not differ from the old standard by more than ±0.17 % of the heat capacity. If these requirements are met, do not change the calorimeter standard. If the requirements are not met, use the new average of the last six calibration tests.

11.2 The summary of the numerical requirements at each stage of restandardization is given in Table 3. A similar table may be configured for the static calorimeter by using a factor of 2.7.

12. Procedure for Waste Samples (Note 4)

12.1 *Weigh the Test Specimen*—Weigh a representative portion of waste and combustion aid as recommended by calorimeter manufacture (minimum 0.4 g) to the nearest 0.1 mg.

NOTE 4—For waste samples which have high-ash or moisture content, or both, or are ash samples that do not easily burn completely, one of the following procedures are recommended. (*1*) The mass of the sample may be varied to obtain good ignition. If the mass is varied, it will be necessary to recalibrate the calorimeter so that the heat capacity will be based on the same temperature rise as that obtained with the new sample weight. (2) A known amount of benzoic acid may be mixed with the sample as a spiking agent. Correction must be made for the calorific value of benzoic acid when calculating the calorific value of the sample. (3) A known amount of mineral oil, ethylene glycol, or other combustion aid may be mixed with the sample. The amount of combustion aid will be dependent on the completeness of combustion. Calorific value for the combustion aid must be determined so as to make the necessary correction for the use of the aid as a spiking agent. When using spiking agents or combustion aids, the total amount of combustibles should meet the manufacturer's guidelines.

12.2 Volatile compounds should be weighed in gelatin capsules or in the sample holders, which are sealed with halogen-free cellulose tape. Samples with metal particle content should be combusted in silica capsules.

12.3 Follow the bomb procedures and temperature observation procedures as described in 10.3.2-10.4.2 for standardization.

TABLE 3 Summary of Numerical Requirements

NOTE 1—Test values exceeding table limits require additional runs.

NOTE 2—X_1 = average of original standard; X_2 = average of check runs.

Number of Runs	Maximum Range of Results Relative to System's Heat Capacity, %	Maximum Difference Between X1 and X2 Relative to System's Heat Capacity, %
1	...	±0.33
2	0.23	±0.25
4	0.33	±0.17
6	0.40	±0.17

12.4 Remove the fuel capsule and dry on a hot plate. Weigh the fuel capsule with ash residue on an analytical balance to 0.0001 g.

13. Calculation

13.1 Compute the corrected temperature rise, *t*, as shown in 10.4.1, where applicable. (See Note 1.)

13.2 *Thermochemical Correction* (Appendix X1)—Compute the following for each test:

e_1 = correction for the heat of formation of HNO_3 (X1.2). Each millimetre of standard sodium carbonate is equivalent to 4.2 J (1 cal). This correction may be considered to be a constant value.
e_2 = correction for heat of combustion of ignition fuse (X1.3).
e_2 = 0.96 J/mm (0.23 cal/mm) or 5.9 J/mg (1.4 cal/mg) for No. 34 B & S Chromel C Wire. This correction may be considered to be a constant value.
e_2 = 1.13 J/mm (0.27 cal/mm) or 7.5 J/mg (1.8 cal/mg) for No. 34 B & S gage iron wire.
e_3 = correction for difference between heat of formation of H_2SO_4 from the heat of formation of HNO_3, J (X1.2).
e_3 = 55.2 J/g (13.17 cal/g) times percent of sulfur in sample times sample mass (for additional information, refer to Test Methods D 3177 and D 4239)
e_4 = correction for use of spiking material, ethylene glycol, mineral oil, gelatin capsule, or tape = m (g) × H_c(J/g).

14. Calorific Value (Note 4)

14.1 *Gross Calorific Value*—Calculations may be performed by a microprocessor, or one may calculate the gross calorific value (gross heat of combustion at constant volume) Q_g(gross) as follows:

$$Q_g\,(\text{gross}) = [(tE) - e_1 - e_2 - e_3 - e_4]m \tag{3}$$

where:
Q_g (gross) = gross calorific value, J/kg,
t = corrected temperature rise as calculated in 10.4.1,
E = heat capacity calculated in 10.4.3,
e_1, e_2, e_3, e_4 = corrections as described in 13.2, and
m = mass of sample, g.

For sample of calculations, reference Appendix X4.

14.2 *Net Calorific Value*—Calculate the net calorific value (net heat of combustion at a constant pressure), Q_p (net) as follows:

$$Q_n\,(\text{net}) = Q_g\,(\text{gross}) - 0.2122H \tag{4}$$

where:
Q_n (net) = net calorific value, MJ/kg,
Q_g (gross) = gross calorific value, MJ/kg, and
H = total hydrogen, mass % (for additional information, refer to Test Method D 1018).

Net calorific value formulas are taken from Test Method D 240.

NOTE 5—This calculation gives calorific value in joules per kilogram. To obtain calorific value in other units, see Table 1.

15. Ash

15.1 *Estimated Ash Content*—Calculations for estimated ash content is calculated as follows:

$$\% \text{ Ash} = (r - c)100/g \quad (5)$$

where:
r = weight of capsule and residue, g,
c = tare weight of capsule, g, and
g = mass of sample, g.

16. Precision and Bias

16.1 *Gross Calorific Value*—Repeatability, reproducibility, and bias data has not been determined at this time.

16.2 *Ash*—Repeatability, reproducibility, and bias data has not been determined at this time.

17. Keywords

17.1 ash; bomb calorimeters; calorific value; calorimeter; isoperibol bomb calorimeter

ANNEX

(Mandatory Information)

A1. THERMOMETRIC CORRECTIONS

A1.1 *Thermometer Corrections*—It is necessary to make the following individual corrections if not making the corrections would result in a heat capacity change of 11.3 J (2.7 cal) or more:

A1.1.1 *Calibration Correction*—Shall be made in accordance with the calibration certificate furnished by the calibration authority.

A1.1.2 *Radiation Corrections*—These are used to calculate heat loss to the water jacket. They are based on the Dickinson formula,[7] the Regnault-Pfaundler formula,[8] or the U.S. Bureau of Mines method.[9] The same method of determining the radiation correction must be used consistently in calibration and test measurements.

A1.1.2.1 *Dickinson Formula*:

$$C_r = -r_1(b - a) - r_2(c - b) \quad \text{(A1.1)}$$

where:
C_r = radiation correction,
r_1 = rate of rise in temperature per minute in the preliminary period,
r_2 = rate of rise of temperature per minute in the final period (if temperature is falling, r_2 is negative),
t_a = firing temperature,
t_c = final temperature, being the first temperature after which the rate of change is constant,
a = time at temperature, t_a, min,
b = time at temperature, $t_a + 0.60\ (t_c - t_a)$, min, and
c = time at temperature, t_c, min.

A1.1.2.2 *Regnault-Pfaundler Formula*:

$$C_r = nr_1 + kS \quad \text{(A1.2)}$$

where:
C_r = radiation correction,
n = number of minutes in the combustion period,
k = $(r_1 - r_2)/(t'' - t')$,
S = $t_n - 1 + (1/2)(t_i + t_f)n'$,
t' = average temperature during the preliminary period,
t'' = average temperature during the final period,
r_1, r_2 = see A1.1.2.1,
$t_1, t_2, t_3 ... t_n$ = successive temperature recorded during the combustion period, at 1-min intervals, and
$t_n - 1$ = sum of $t_1, t_2, t_3 ... t_n - 1$.

A1.1.2.3 *Burea of Mines Method*—A table of radiation correction can be established so that only the initial and final readings are required to determine the heat value of any fuel. This may be done by carrying out a series of tests utilizing the procedure described in Section 10, using the following conditions. Regulate the amount of sample burned so that a series of determinations is made in which different temperature rises are obtained. For all determinations, keep the water jacket temperature constant, fire the bomb at the same initial temperature, and have the same time, $c - a$, elapse ($\pm$ 2 s) between the initial and final readings. Determine the radiation corrections for each of the series of temperature rises using the Dickinson method (see A1.1.2.1), or the Regnault-Pfaundler method (see A1.1.2.2). These corrections are constant for a given temperature rise. From the series of readings a table or graph is plotted to show radiation correction versus temperature rise. Once the table or graph is established, the radiation corrections can be obtained from it until there is a major change in the equipment.

[7] Dickinson, H. C., *Bulletin*, U.S. Bureau of Standards, Vol 11, 1951, p. 189.
[8] Pfaundler, L., *Annalen der Physik (Leipzig)*, ANPYA, Vol 129, 1966, p. 102.
[9] "Methods of Analyzing and Testing Coal and Coke," U.S. Bureau of Mines, *Bulletin 638*, XMBUA, 1967, pp. 16–17.

APPENDIXES

(Nonmandatory Information)

X1. THERMOCHEMICAL CORRECTIONS

X1.1 *Energy of Formation of Nitric Acid*—A correction, e_1 (10.4.2 and 13.2) is applied for the acid titration. This correction is based on the assumptions (*1*) that all the acid titrated is HNO_3 formed by the following reaction: ½ N_2 (g) + ¾ O_2 (g) + ½ H_2O (I) = HNO_3 (in 500 mol H_2O), and (*2*) that the energy of formation of HNO_3 in approximately 500 mol of water under bomb conditions is minus 59.0 kJ/mol.[10] Normal convention assigns a negative value for a heat of formation that is exothermic. By definition heat released from combustion processes is expressed as a positive value. For conformity, the negative factors developed for nitric and sulfuric acid corrections are expressed as positive values in the calculations.

X1.1.1 A convenient concentration of Na_2CO_3 is 37.6 g Na_2CO_3/1000 mL, which gives $e_1 = 10 \times V$, where V is the volume of Na_2CO_3 in millilitres. When H_2SO_4 is also present, a part of the correction for H_2SO_4 is also present in the e_1 correction and the remainder in the e_3 correction.

X1.2 *Energy of Formation of Sulfuric Acid*—By definition (see Terminology D 121) the gross calorific value is obtained when the product of the combustion of sulfur in the sample is SO_2(g). However, in actual bomb combustion processes, all the sulfur is found as H_2SO_4 in the bomb washings. A correction e_3 (see 13.2) is applied for the sulfur that is converted to H_2SO_4. This correction is based upon the energy of formation of H_2SO_4 in solutions, such as will be present in the bomb at the end of a combustion. This energy is taken as − 295.0 kJ/mol.[11] When nitric and sulfuric acid are present, a correction of 2 × −59.0 kJ/mol of sulfur was applied in the e_1 correction so that the additional correction that is necessary is the difference in the heats of formation for nitric and sulfuric acid, and this correction is − 295.0 − (−2 × 59.0) = −177 kJ/mol, or − 5.52 kJ/g of sulfur in the sample (55.2 J/g × weight of sample in grams × percent sulfur in sample). This causes e_3 to be 13.17 cal/g × weight of samples in grams × percent sulfur in sample. The factor 23.7 Btu/g (= 55.2/2.326 × 1.8) for e_3 (see 13.2) is to be used for calculating calorific value in Btu/lb. The values stated are based on a sample containing about 5 % sulfur and about 5 % hydrogen. The assumption is also made that the H_2SO_4 is dissolved entirely in the water condensed during combustion of the sample.[12]

X1.2.1 If a 1-g sample is burned, the resulting H_2SO_4 condensed with water formed on the walls of the bomb will have a ratio of about 15 mol of water to 1 mol of H_2SO_4. For this concentration the energy of the reaction SO_2 (g) + ½ O_2(g) + H_2O (1) = H_2SO_4 (in 15 mol of H_2O) under the conditions of the bomb process is − 295 kJ/mol.[12] For different sample weights or sulfur content, or both, the resultant normality of acid formed may be significantly different, and therefore the normality of titrant must be adjusted accordingly. Basing the calculation upon a sample of comparatively large sulfur content reduces the possible overall errors, because, for small percentages of sulfur, the correction is smaller.

X1.3 *Fuse (Ignition)*—Calculate the energy contributed by burning the fuse in accordance with the directions furnished by the supplier of the fuse. The energy required to melt a platinum wire is constant for each experiment if the same amount of platinum wire is used. As the energy is small, its effect is essentially canceled out in the relationship between the standardization experiments and the calorific value determinations, and it can be neglected.

[10] Calculated from data in *NBS Technical Note 270-3.*

[11] Calculated from data in *NBS Circular 500.*

[12] Mott, R. A., and Parker, C., "Studies in Bomb Calorimetry IX—Formation of Sulfuric Acid," Fuel, FUELB, Vol 37, 1958, p. 371.

X2. REPORTING RESULTS IN OTHER UNITS

X2.1 The gross calorific value can be expressed in joules per kilogram, calorie per gram, or British thermal unit per pound. The relationships between these units are given in Table 1.

X3. SAMPLING INFORMATION

X3.1 SW-846 Test Methods for the Evaluation of Solid Waste, Physical/Chemical Methods, or their equivalent.[13]

X4. SAMPLE CALCULATIONS

X4.1 *Heat Capacity*—Calculation from 10.4.3:

$$E = [(H_c m) + e_1 + e_2]/t \quad \text{(X4.1)}$$

where:
H_c = 6318 cal/g,
m = 1.0047 g,
e_1 = 10.4 cal – acid correction,
e_2 = 18.2 cal – fuse correction,
t = 2.542°C,
E = [(6318 cal/g × 1.0047 g) – 10.4 cal – 18.2 cal]/ 2.6350°C, and
E = 2398.1 cal/°C.

X4.2 *Heat of Combustion*—Calculations from 14.1:

$$Qg\ \text{(gross)} = [(tE) - e_1 - e_2 - e_3 - e_4]/m \quad \text{(X4.2)}$$

where:
E = 2398.1 cal/°C
t = 2.417°C,
e_1 = 27.1 cal – acid correction,
e_2 = 17.9 cal – fuse correction,
e_3 = 13.17 × 1.24 % × 0.7423 g – sulfur correction,
e_4 = 11 000 cal/g × 0.2043 g – spike correction,
m = 0.7423 g – mass of sample,
Qg = [(2398.1 cal/°C × 2.417°C) – 27.1 cal – 17.9 cal – 12.1 cal – 2247 cal (3492.11 cal]/0.7423 g), and
Qg = 4704.01 cal/g.

The American Society for Testing and Materials takes no position respecting the validity of any patent rights asserted in connection with any item mentioned in this standard. Users of this standard are expressly advised that determination of the validity of any such patent rights, and the risk of infringement of such rights, are entirely their own responsibility.

This standard is subject to revision at any time by the responsible technical committee and must be reviewed every five years and if not revised, either reapproved or withdrawn. Your comments are invited either for revision of this standard or for additional standards and should be addressed to ASTM Headquarters. Your comments will receive careful consideration at a meeting of the responsible technical committee, which you may attend. If you feel that your comments have not received a fair hearing you should make your views known to the ASTM Committee on Standards, at the address shown below.

Designation: D 5495 – 94 (Reapproved 2001)

Standard Practice for Sampling With a Composite Liquid Waste Sampler (COLIWASA)[1]

This standard is issued under the fixed designation D 5495; the number immediately following the designation indicates the year of original adoption or, in the case of revision, the year of last revision. A number in parentheses indicates the year of last reapproval. A superscript epsilon (ε) indicates an editorial change since the last revision or reapproval.

1. Scope

1.1 This practice describes the procedure for sampling liquids with the composite liquid waste sampler, or "COLIWASA." The COLIWASA is an appropriate device for obtaining a representative sample from stratified or unstratified liquids. Its most common use is for sampling containerized liquids, such as tanks, barrels, and drums. It may also be used for pools and other open bodies of stagnant liquid.

NOTE 1—A limitation of the COLIWASA is that the stopper mechanism may not allow collection of approximately the bottom inch of material, depending on construction of the stopper.

1.2 The COLIWASA should not be used to sample flowing or moving liquids.

1.3 *This standard does not purport to address all of the safety problems, if any, associated with its use. It is the responsibility of the user of this standard to establish appropriate safety and health practices and determine the applicability of regulatory limitations prior to use.*

2. Referenced Documents

2.1 *ASTM Standards:*

D 4687 Guide for General Planning of Waste Sampling[2]

D 5088 Practice for Decontamination of Field Equipment Used at Nonradioactive Waste Sites[3]

D 5283 Practice for Generation of Environmental Data Related to Waste Management Activities: Quality Assurance and Quality Control Planning and Implementation[2]

3. Summary of Practice

3.1 A clean device is slowly lowered into the liquid to be sampled. After it has filled, the bottom of the sampling tube is closed and the device is retrieved. The contents are subsequently discharged into a sample container.

4. Significance and Use

4.1 This practice is applicable to sampling liquid wastes and other stratified liquids. The COLIWASA is used to obtain a vertical column of liquid representing an accurate cross-section of the sampled material. To obtain a representative sample of stratified liquids, the COLIWASA should be open at both ends so that material flows through it as it is slowly lowered to the desired sampling depth. The COLIWASA must not be lowered with the stopper in place. Opening the stopper after the tube is submerged will cause material to flow in from the bottom layer only, resulting in gross over-representation of that layer.

4.2 This practice is to be used by personnel acquiring samples.

4.3 This practice should be used in conjunction with Guide D 4687 which covers sampling plans, safety, QA, preservation, decontamination, labeling and chain-ofcustody procedures; Practice D 5088 which covers decontamination of field equipment used at waste sites; and Practice E 5283 which covers project specifications and practices for environmental field operations.

5. Sampling Equipment

5.1 COLIWASAs are available commercially with different types of stoppers and locking mechanisms, but they all operate using the same principle. They can also be constructed from materials such as polyvinylchloride (PVC), glass, metal, or polytetrafluoroethylene (PTFE). A traditional model of the COLIWASA is shown in Fig. 1 [de Vera et al.][4,5]; however, the design can be modified or adapted, or both, to meet the needs of the sampler. COLIWASAs must be selected that are constructed of materials compatible with the waste being sampled and with the analyses or tests to be performed. Due to the unknown nature of most containerized liquid wastes, COLIWASAs made of glass or polytetrafluoroethylene are best for general use.

6. Sample Containers

6.1 Plastic, glass or other nonreactive containers should be

[1] This practice is under the jurisdiction of ASTM Committee D34 on Waste Management and is the direct responsibility of Subcommittee D34.01.03 on Sample Preservation.

Current edition approved Jan. 15, 1994. Published March 1994.

[2] *Annual Book of ASTM Standards*, Vol 11.04.

[3] *Annual Book of ASTM Standards*, Vol 04.08.

[4] de Vera, E. R., Simmons, B. P., Stephens, R. C., and Storm, D. L., "Samplers and Sampling Procedures for Hazardous Waste Streams," EPA-600/2-80-018, January 1980.

[5] Ford, P. J., Turina, P. J., and Seeley, D. E., "Characterization of Hazardous Waste Sites—A Methods Manual: Volume II," Available Sampling Methods, Second Edition, EPA-600/4-84-076, December 1984.

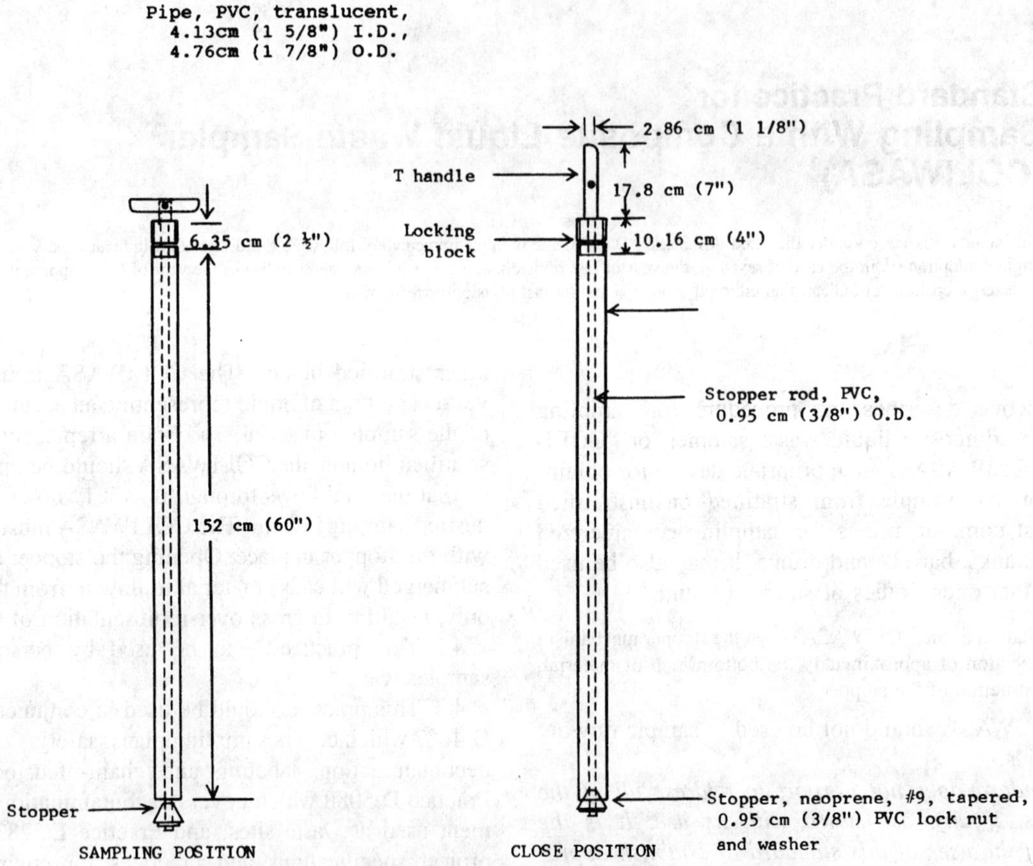

FIG. 1 Composite Liquid Waste Sampler (COLIWASA) (Typical)

used. Refer to Guide D 4687 for further information on containers.

7. Procedure

7.1 Make certain the COLIWASA is clean and functioning properly. It is essential that the stopper at the bottom of the sampling tube closes securely.

7.2 Open the COLIWASA by placing the stopper mechanism in the open position.

7.3 Lower the COLIWASA into the liquid slowly so that the levels of the liquid inside and outside the sampler tube remain about the same. If the level of the liquid in the sample tube is lower than that outside the sampler, the sampling rate is too fast and a nonrepresentative sample will result.

7.4 Use the stopper mechanism to close the COLIWASA when it reaches the desired depth in the liquid.

7.5 Withdraw the sampler from the liquid with one hand while wiping the sampler tube with a disposable cloth or rag with the other hand.

7.6 Carefully discharge the sample into a suitable container by slowly opening the stopper mechanism while the lower end of the COLIWASA is positioned in the sample container.

7.7 Seal the sample container; attach the label and seal; record in the field logbook; and complete the chain-of-custody record.

7.8 Decontaminate the used equipment in accordance with Practice D 5088.

8. Keywords

8.1 COLIWASA; drum sampling; liquid sampling; sampling; waste

Designation: D 5513 – 99

Standard Practice for Microwave Digestion of Industrial Furnace Feedstreams and Waste for Trace Element Analysis[1]

This standard is issued under the fixed designation D 5513; the number immediately following the designation indicates the year of original adoption or, in the case of revision, the year of last revision. A number in parentheses indicates the year of last reapproval. A superscript epsilon (ϵ) indicates an editorial change since the last revision or reapproval.

1. Scope

1.1 This practice describes the multi-stage microwave digestion of typical industrial furnace feedstream materials using nitric, hydrofluoric, hydrochloric, and boric acids for the subsequent determination of trace metals.

1.2 This practice has been used successfully on samples of coal, coke, cement raw feed materials, and waste-derived fuels composed primarily of waste paint-related material in preparation for measuring the following trace elements: Ag, As, Ba, Be, Cd, Cr, Hg, Pb, Sb, and Tl. This practice may be applicable to elements not previously listed.

1.3 This practice is also effective for other waste materials (for example, flyash, foundry sand, alum process residue, cement kiln dust, etc.).

1.4 *This standard does not purport to address all of the safety concerns, if any, associated with its use. It is the responsibility of the user of this standard to establish appropriate safety and health practices and determine the applicability of regulatory limitations prior to use.* Specific hazard statements are given in Section 7.

2. Referenced Documents

2.1 *ASTM Standards:*

D 1193 Specifications for Reagent Water[2]

2.2 *Other Document:*

40 CFR 266, Subpart H, Hazardous Waste Burned in Boilers and Industrial Furnaces, Latest Revision[3]

3. Summary of Practice

3.1 A weighed portion of the feedstream material is combined with concentrated nitric acid in a polytetrafluoroethylene-lined digestion vessel, and heated in a microwave digestion unit. Following a programmed heating cycle, the vessel is vented and specified quantities of hydrofluoric and hydrochloric acids are added, and the mixture undergoes further microwave heating. Following this heating cycle, the vessel is vented and a specified quantity of boric acid solution is added, and the mixture undergoes a final microwave heating. Following this final heating cycle, the vessel is vented, the contents are quantitatively transferred to a volumetric flask and brought to volume. Typically, the only undissolved material is particulate carbon. If particulate matter is observed, filtration or centrifugation may be needed. The digested sample is ready for analysis.

4. Significance and Use

4.1 The U.S. Environmental Protection Agency Regulations, 40 CFR, require that boilers, cement kilns, and other industrial furnaces utilizing waste-derived fuel adhere to specific guidelines in assessing potential metals emissions. A common approach for estimating potential emissions is performing total metals analysis on all feedstream materials. This practice describes a multi-stage microwave-assisted digestion procedure that solubilizes trace elements for spectroscopic analyses.

5. Apparatus

5.1 *Microwave Digestion Unit*—Equipped with an automatic turntable, pressure and/or temperature controller, and closed perfluoroalkoxy (PFA)-lined digestion vessels equipped with pressure relief/rupture membrane fittings or equivalent pressure relief device. The unit should comply with applicable federal or state standards, or both, for microwave leakage. The user must follow specific manufacturer's instructions for system installation.

NOTE 1—The digestion unit used in developing this practice was equipped with a pressure controller, automatic turntable, exhaust fan, and programming capacity. The unit delivers 1000 W of power at 100 % output. The lined digestion vessels consist of a high-strength polymeric vessel body and cap, inner PFA liner and rupture membrane housing, and PFA vent stem. These vessels have a maximum operating pressure of 200 psig. There are a number of suitable lab grade microwave systems available to the user that meet these minimum specifications. The user must follow specific manufacturer's instructions for using digestion vessels.

5.2 *Analytical Balance*—Capable of weighing to 0.001 g.

5.3 *Labware*—High-density volumetric polyethylene or polypropylene flasks/sample containers are recommended for this practice. The user should be mindful of the quality

[1] This practice is under the jurisdiction of ASTM Committee D34 on Waste Management and is the direct responsibility of Subcommittee D34.01.06 on Analytical Methods.

Current edition approved March 10, 1999. Published May 1999. Originally published as D 5513 – 94. Last previous edition D 5513 – 94 (1998).

[2] *Annual Book of ASTM Standards*, Vol 11.01.

[3] Available from Standardization Documents Order Desk, Bldg. 4 Section D, 700 Robbins Ave., Philadelphia, PA 19111-5094, Attn: NPODS.

limitations associated with volumetric non-glass labware.

6. Reagents and Materials

6.1 *Purity of Reagents*—Reagent grade chemicals shall be used in all tests. Unless otherwise indicated, it is intended that all reagents conform to the specifications of the Committee on Analytical Reagents of the American Chemical Society where such specifications are available.[4] Other grades may be used, provided it is first ascertained that the reagent is of sufficiently high purity to permit its use without lessening the accuracy of the determination.

6.2 *Purity of Water*—Unless otherwise indicated, references to water shall be understood to mean meeting the numerical requirements of Type II water as defined by Specifications D 1193.

6.3 *Boric Acid Solution* (20 g/L)—Dissolve 20 g of boric acid (H_3BO_3) in water and dilute to 1 L. It may be necessary to place solution on a combination hot plate/magnetic stirrer and with the aid of a stir bar, allow the solution to mix under gentle heat until boric acid is fully in solution.

6.4 *Hydrochloric Acid*, 37 %, (sp. gr. 1.200), HCl.

6.5 *Hydrofluoric Acid*, 48 %, (sp. gr. 1.150), HF.

6.6 *Nitric Acid*, 70 %, (sp. gr. 1.400), HNO_3.

7. Hazards [5]

7.1 It is recommended that all operations involving concentrated acids be performed in a laboratory fume hood.

7.2 Hydrochloric acid is a highly corrosive chemical that is reactive with metals and most alkaline chemicals. Impervious gloves and chemical goggles are required for handling. See material safety data sheet (MSDS) for additional information.

7.3 Hydrofluoric acid is a highly corrosive chemical that is reactive with metals and water or steam. Additionally, HF specifically attacks silicate glass making certain fume hood enclosures susceptible to damage. Impervious gloves and chemical goggles are required for handling. See MSDS for additional information.

7.4 Nitric acid is a highly corrosive chemical that is reactive with metals and most organic materials. Impervious gloves and chemical goggles are required for handling. See MSDS for additional information.

7.5 Some samples undergoing microwave digestion can exhibit a rapid pressure rise within the digestion vessel. The potential exists for this type of sample to rupture the rupture membrane and liberate corrosive gases. Because of this, the microwave unit must be vented to a fume hood for proper evacuation of vapors.

8. Sample

8.1 Although feedstream materials are generally pulverized powders or liquids, the homogeneity of some feedstream materials can be uncertain. The laboratory sample should be thoroughly mixed or homogenized prior to withdrawing a portion for analysis. This practice assumes that non-liquid feedstreams are pulverized powders at the time of sample preparation.

NOTE 2—If a non-liquid feedstream material is not in the form of a pulverized powder, it may need to be reduced in particle size to pass through a No. 100 sieve.

9. Calibration and Standardization

9.1 Although equipment manufacturers specify general power output ratings for microwave digestion units, it is important to verify the actual power output of a specific unit. It is recommended that this microwave power check procedure be performed monthly.

9.1.1 *Power Check Procedure at 100 % Instrument Power*:

9.1.1.1 Remove from the instrument cavity the turntable, drive lug, and all vessels.

9.1.1.2 Adjust the instrument cavity exhaust to minimum air flow (refer to the manufacturer's instructions).

9.1.1.3 Program the instrument for 4-min time and 100 % power.

9.1.1.4 Transfer 2000 ± 2 mL of room temperature (19 to 25°C) water into a 2-L polypropylene beaker.

9.1.1.5 Measure and record the initial water temperature (T_i) to the nearest 0.1°C.

9.1.1.6 Place the beaker in the right front corner of the instrument cavity (as you face the front of the instrument). This position closely approximates the position of a digestion vessel during processing.

9.1.1.7 Heat the water for the programmed time.

9.1.1.8 When the heating cycle is complete, immediately remove the beaker from the cavity, thoroughly stir the water to ensure even heat distribution, and measure the final temperature (T_f) to the nearest 0.1°C.

9.1.1.9 Calculate the delivered power in accordance with the following equations:

$$\text{Power (watts)} = \Delta T \times (35\ W/°C) \tag{1}$$

where:

ΔT = $T_f - T_i$.

$W/°C = \dfrac{K \times Cp \times M}{t}$

where:

W = watts,
K = 4.2, the factor for converting thermo-chemical calories/s to joules to watts.
Cp = 1.0, the heat capacity for water, cal g^{-1} $degree^{-1}$,
M = mass of water, g (1 mL H_2O = 1 g), and
t = time, s.

9.1.1.10 If the calculated power is not within the specifications of the unit, do a second test beginning at 9.1.1.1 for confirmation before contacting the manufacturer.

10. Procedure

NOTE 3—Many microwave digestion units are capable of handling up to 12 digestion vessels at a time. The number of vessels processed affects the required program settings. The microwave program examples given in

[4] *Reagent Chemicals, American Chemical Society Specifications*, American Chemical Society, Washington, DC. For suggestions on the testing of reagents not listed by the American Chemical Society, see *Analar Standards for Laboratory Chemicals*, BDH Ltd., Poole, Dorset, U.K., and the *United States Pharmacopeia and National Formulary*, U.S. Pharmaceutical Convention, Inc. (USPC), Rockville, MD.

[5] This information taken from the *NIOSH Guide to Chemical Hazards*, U.S. Depart. of Health and Human Services, June 1990.

Annexes A1 and A2 were developed to accommodate 10 pressure vessels have been proven effective in field testing. A maximum pressure of 140 psi for a minimum time at pressure of 10 min is generally sufficient for sample digestion. A complete digestion can be verified by the user by processing reference materials and calculating analyte recoveries.

10.1 Process coal, coke, cement raw feed, and other geologic/inorganic materials as follows:

10.1.1 Weigh 0.5 g of the feedstream material directly in a PFA digestion vessel liner and record the mass to the nearest 0.01 g.

10.1.2 Place the liner with sample in the digestion vessel body and with a pipet unit, or equivalent device, slowly add 8 mL HNO_3 to the PFA liner. If a chemical reaction is observed, allow the reaction to subside prior to performing 10.1.3.

10.1.3 Place pressure controller cap assembly on the vessel body, and hand tighten to achieve a firm fit. Do not overtighten because this can damage the sealing surface and threads.

10.1.4 Be sure the pressure relief/rupture membrane screw-type fitting houses a new rupture membrane, and is tightened to finger-tightness. Do not overtighten because this can potentially deform the rupture membrane and damage the threads. If a rupture membrane bursts during the digestion, stop the digestion and remove the vessel with the ruptured membrane from the turntable and resume digestion. Discard the sample in the removed vessel.

10.1.5 Place the vessel in the microwave turntable. Insert the pressure sensing tubing from the pressure controller into the pressure controller cap assembly and tighten to finger-tightness. Be sure that the pressure sensing tubing is completely filled with water (from the pressure controller unit to the pressure control vessel) prior to inserting this line.

10.1.6 Repeat 10.1.1 and 10.1.2 for three or five additional vessels. Cap these vessels with standard vessel cap assemblies. Be sure each vessel cap assembly is tightened to firmness. Evenly distribute the digestion vessels in the turntable to ensure uniform heating of all vessels.

10.1.7 It is recommended that a blank and a certified reference standard or a matrix spike be processed with each batch of samples of similar composition. Generally, a batch is considered to be samples that are processed together by the same method, reagents, and the same manipulation. The user will need to determine the number of samples comprising a batch based on the user's quality assurance requirements.

10.1.8 The samples are ready for microwave processing. Process the samples by following the Stage 1 microwave program described in Annex A1. Be sure the instrument cavity exhaust is set to maximum air flow.

10.1.9 Following the cooling step (0 power) in Stage 1, the vessels should be cool to the touch. Additional cooling time may be necessary if vessels remain hot. Vent the vessels in a laboratory fume hood or the microwave oven cavity by slowly unscrewing the pressure relief/rupture membrane fitting. Vent the pressure control vessel in the microwave in order to remove the pressure sensing tubing. Remove the vessel cap assemblies.

10.1.9.1 Flush the pressure sensing tubing with water to remove any trace of acid. This is an important step, because damage will occur to the pressure controller if an acidic solution from the pressure sensing tubing is allowed to migrate to the pressure controller. The flush water should be disposed of as a waste.

10.1.9.2 Using a polyethylene syringe, or other equivalent compatible device, add 4 mL of HF, and 2 mL of HCl to each vessel. If a chemical reaction is observed, allow the reaction to subside prior to performing 10.1.9.3.

NOTE 4—The user may choose to vary the amounts of HF and HCl based on prior knowledge of the sample. Samples known to be low in siliceous material may require a lesser amount of HF. In no case should the user reduce the amount of HF or HCl below a 2-mL volume for each acid.

10.1.9.3 Recap the digestion vessels, making sure the pressure relief/rupture membrane fitting is tightened to finger-tighteness. Replace the digestion vessels in the turntable, and reconnect the pressure sensing tubing to the pressure control vessel. Be sure that the pressure sensing tubing is completely filled with water. Place the remaining digestion vessels in the turntable.

10.1.9.4 The samples are ready for microwave processing. Process the samples by following the Stage 2 microwave program described in Annex A1.

10.1.9.5 Following the vessel cooling step, vent the vessels as described in 10.1.9. Remove the vessel cap assemblies, flush and fill the pressure sensing tubing with water.

10.1.9.6 Add 35.5 mL of H_3BO_3 solution to each vessel, and recap as described in 10.1.9.1.

10.1.9.7 The samples are ready for final microwave processing. Process the samples by following the Stage 3 microwave program described in Annex A1.

10.1.9.8 Following the vessel cooling step, vent the vessels as described in 10.1.9. Remove the vessel cap assemblies, flush and fill the pressure sensing tubing with water.

10.1.9.9 Quantitatively transfer the digested sample solution to a clean polyethylene 50-mL volumetric flask and bring to volume with water. If any particulate matter is observed, it may be removed by filtration or centrifugation.

NOTE 5—Typically, the only undissolved material is particulate carbon, which is readily removed by centrifugation. If filtration is used, the appropriate filter must be cleaned and pre-rinsed with a 1-*N* nitric acid solution.

10.1.9.10 Transfer the sample solution to a clean polyethylene sample container.

10.1.9.11 The sample is now ready for analysis.

10.2 *Waste-Derived Fuels, Oils, and Other High Organic Content Samples*—The procedure for the microwave-assisted digestion of waste-derived fuel samples, or other high organic content samples is the same as outlined in 10.1, except that *(1)* the user may wish to reduce the sample mass to lessen the possibility of vessel overpressurization and *(2)* the Annex A2 microwave program should be followed. It should be noted that the programs in Annex A2 include a cooling step in Stage 2, and a more gradual initial temperature ramp in Stage 1. These steps are important because some waste-derived fuels can be somewhat reactive in the presence of concentrated acids and can exhibit a rapid pressure increase resulting in the venting of the digestion vessel(s). These steps will lessen the likelihood of venting the vessels.

11. Precision and Bias

11.1 *Precision*—No statement is made about the precision since this practice does not produce a specific test result.

11.2 *Bias*—No statement is made about the bias since this practice does not produce a specific test result.

11.3 Appendix X1 contains analytical results obtained following this procedure on National Institute of Standards and Technology (NIST) standard reference materials and paraffin oil-based metals standards.

12. Keywords

12.1 cement; coal; coke; microwave digestion; trace elements; trace metals; waste; waste-derived fuel

ANNEXES

(Mandatory Information)

A1. DIGESTION PROGRAM FOR COAL, COKE, RAW FEED, AND FLY ASH

STAGE 1 – Add: sample, HNO_3	(1)	(2)	(3)
Power	65 %	60 %	0 %
PSI	140	140	0
Time	30.0	30.0	10.0
TAP^A	7.00	10.00	10.00

STAGE 2 – Add: HC1, HF	(1)	(2)	(3)
Power	60 %	45 %	0 %
PSI	140	140	0
Time	30.0	30.0	10.0
TAP^A	6.00	10.00	10.00

STAGE 3 – Add: H_3BO_3 solution	(1)	(2)	(3)
Power	65 %	58 %	0 %
PSI	50	50	0
Time	30.0	30.0	10.0
TAP^A	4.00	5.00	10.00

^{A} TAP = time at pressure.

A2. DIGESTION PROGRAM FOR HIGH ORGANIC CONTENT SAMPLES

STAGE 1 – Add: sample, HNO_3	(1)	(2)	(3)	
Power	65 %	60 %	0 %	
PSI	140	140	0	
Time	30.0	30.0	10.0	
TAP[A]	7.00	10.00	10.00	
STAGE 2 – Add: HC1, HF	**(1)**	**(2)**	**(3)**	**(3)**
Power	60 %	0	45 %	0 %
PSI	140	140	140	0
Time	30.0	10.0	30.0	10.0
TAP[A]	6.00	3.00	10.00	10.00
STAGE 3 – Add: H_3BO_3 solution	**(1)**	**(2)**	**(3)**	
Power	65 %	58 %	0 %	
PSI	50	50	0	
Time	30.0	30.0	10.0	
TAP[A]	4.00	5.00	10.00	

[A] TAP = time at pressure.

APPENDIX

(Nonmandatory Information)

X1. ANALYTICAL RESULTS OF NIST PARAFFINIC OIL-BASED STANDARDS

Element[A]	SRM 1632b Bituminous Coal Value, mg/kg	Digestion Result, mg/kg	Recovery, %	SRM 2704 Buffalo River Sediment Value, mg/kg	Digestion Result, mg/kg	Recovery, %	SRM 1633a Coal Fly Ash Value, mg/kg	Digestion Result, mg/kg	Recovery, %	Oil-based Metals Standards Value, mg/kg	Digestion Result, mg/kg	Recovery, %
Ag	...	1.6	...	...	1.9	...	...	1.12	...	500	484.6	96.9
As	3.72	4.1	110.2	23.4	21.0	89.7	145	140.9	97.2	100	97.5	97.5
Ba	67.5	58	85.9	414	319	77.1	1500	1253	83.5	500	465	93
Be	...	1.2	...	...	1.9	...	(12)	24.8	...	5000	4830	96.6
Cd	0.0573	<2.0	...	3.45	<2.0	...	1	1.1	110.0	500	489	97.8
Cr	(11)	13	118.2	135	113	83.7	196	182	92.9	500	502	100.4
Hg	...	...	...	1.44	1.55	107.6	0.16	0.12	75.0	100	91.8	91.8
Pb	3.67	<5.0	...	161	155	96.3	72.4	73	100.8	500	492	98.4
Sb	(0.24)	<0.5	...	3.79	3.7	97.6	6.8	6.2	91.2	5000	4900	98
Tl	...	...	...	1.2	1.5	125.0	5.7	5.9	103.5	...	...	...

[A]Elemental analysis by flame and graphic furnace atomic absorption. Mercury analysis by cold vapor atomic absorption.

Designation: D 5530 – 94 (Reapproved 1998)

Standard Test Method for Total Moisture of Hazardous Waste Fuel by Karl Fischer Titrimetry[1]

This standard is issued under the fixed designation D 5530; the number immediately following the designation indicates the year of original adoption or, in the case of revision, the year of last revision. A number in parentheses indicates the year of last reapproval. A superscript epsilon (ε) indicates an editorial change since the last revision or reapproval.

1. Scope

1.1 This test method covers the determination by Karl Fischer (KF) titrimetry of total moisture in solid or liquid hazardous waste fuels used by industrial furnaces.

1.2 This test method has been used successfully on numerous samples of hazardous waste fuel composed of solvents, spent oils, paints, and pigments. The expected range of applicability for this test method is between 1.0 and 100 %; however, this evaluation was limited to samples containing approximately 5 to 50 % water.

1.3 The values stated in SI units are to be regarded as the standard.

1.4 *This standard does not purport to address all of the safety concerns, if any, associated with its use. It is the responsibility of the user of this standard to establish appropriate safety and health practices and determine the applicability of regulatory limitations prior to use.*

2. Referenced Documents

2.1 *ASTM Standards:*

D 1193 Specifications for Reagent Water[2]

D 4017 Test Method for Water in Paints and Paint Materials by Karl Fischer Method[3]

E 691 Practice for Conducting an Interlaboratory Study to Determine the Precision of a Test Method[4]

3. Summary of Test Method

3.1 An amount of solvent (see 7.3) sufficient to immerse the electrode in the titration flask fully (see Note 1) is titrated to dryness as explained in 9.1, and the Karl Fischer reagent factor is determined by titration of measured amounts of water. A weighed portion of the sample is dissolved in KF solvent and titrated with reagent to dryness. If solid material interferes (see 5.3) with the electrode or does not dissolve sufficiently, an extraction using KF solvent is performed prior to introduction into the titration flask. The total moisture in the sample is then determined. The final total moisture percent is an average of two trials for each sample.

3.2 The contents of the tiration flask may be retained and used for additional analyses. The contents of the titration flask will need to be emptied and replaced with new solvent when the capacity of the flask is nearly exhausted or when solid material affects the sensing by the electrode.

4. Significance and Use

4.1 The determination of total moisture is important for assessing the quality of fuels. Water content will affect the heating value of fuels directly and can contribute to instability in the operation of an industrial furnace. Additionally, high water contents can present material handling and storage problems during winter months or in cold environments.

5. Interferences

5.1 A small number of oxidants such as ferric and chromate salts can oxidize iodide and may produce artificially low results.

5.2 Certain reductants oxidized by iodine such as mercaptans, thioacetate, thiosulfate, stannous chloride, sulfides, hydroquinone, and phenylenediamines can consume iodine and may cause artificially high results. Basic materials such as hydroxides, oxides, and inorganic carbonates may cause artificially high results by water-forming reactions.

5.3 Some types of solid material found in waste-derived fuel may interfere with the electrode by blocking its contact with the solvent. Depending on the nature of the solid material, artificially high or low results can occur.

6. Apparatus

6.1 *Karl Fischer Potentiometric Titration Unit*, automated or semi-automated, equipped with a magnetic vessel stirrer. The user must follow the manufacturer's instructions for installation and use.

NOTE 1—The Karl Fischer unit used for developing this test method was equipped with a twin platinum electrode, 25 to 80-mL capacity titration flask, magnetic stirrer, electronic piston burette, adjustable delay interval, LED display, visual and audible endpoint notification.

6.2 *Syringe*, 100-μL capacity, with needle.

6.3 *Syringe*, 1 to 5-mL capacity, without needle.

6.4 *Analytical Balance*, with minimum capacity of 160 g

[1] This test method is under the jurisdiction of ASTM Committee D34 on Waste Management and is the direct responsibility of Subcommittee D34.01.06 on Analytical Methods.

Current edition approved April 15, 1994. Published June 1994.

[2] *Annual Book of ASTM Standards*, Vol 11.01.

[3] *Annual Book of ASTM Standards*, Vol 06.01.

[4] *Annual Book of ASTM Standards*, Vol 14.02.

and capable of weighing to 0.0001 g.

7. Reagents and Materials

7.1 *Purity of Reagents*—Reagent grade chemicals shall be used in all tests. Unless otherwise indicated, it is intended that all reagents shall conform to the specifications of the Committee on Analytical Reagents of the American Chemical Society, where such specifications are available.[5] Other grades may be used, provided it is first ascertained that the reagent is of sufficiently high purity to permit its use without lessening the accuracy of the determination.

7.2 *Purity of Water*— Unless otherwise indicated, references to water shall be understood to mean reagent water as defined by Type II of Specification D 1193.

7.3 *Karl Fischer Solvent*—The solvent system must be a non-methanol formulation for the analysis of aldehydes and ketones and for general usage. The use of a strictly general purpose solvent is not acceptable because of the potential of obtaining artificially high results from water-forming reactions by chemicals with active carbonyl groups.

7.4 *Karl Fischer Reagent*—The reagent should have a standard titer value of 5 mg H_2O/mL reagent and be listed as being compatible with the solvent system.

8. Sample

8.1 Because stratification or layering of liquid samples is possible, the laboratory sample should be mixed thoroughly by shaking prior to withdrawing a portion for testing. Strongly multi-phasic samples should have each layer analyzed separately and the total moisture percent of the sample calculated as a weighted average. Solid samples should be mixed thoroughly by shaking or stirring, depending on the physical characteristics of the sample.

9. Calibration and Standardization

9.1 Determination of Karl Fischer reagent factor (*F*):

9.1.1 Determine the mg H_2O/mL reagent (*F*) for each new procurement of KF solvent and at a minimum of daily on a solvent retained for further titrations (see Section 3).

9.1.2 Add a sufficient quantity of solvent to cover the electrode tip, set the delay interval to 30 s, and engage the magnetic stirrer.

9.1.3 Bring the KF solvent to dryness by titrating with reagent to the endpoint. The endpoint is typically indicated by a visual or audible alarm, or both, depending on the equipment manufacturer.

9.1.4 Refill the burette with reagent.

9.1.5 Fill the 100-μL syringe with approximately 10 μL of water and weigh to the nearest 0.0001 g.

9.1.6 Dispense the syringe contents into the titration vessel and replace the sample port stopper immediately. Reweigh the syringe and record the water mass.

9.1.7 Titrate with reagent until the endpoint is reached. Record the titrant volume used.

9.1.8 The final value for *F* should be based on the mean of a minimum of three replicates.

9.1.9 *Calculation*:

9.1.9.1 Calculate *F* by the following formula:

$$F\ (\text{mg/mL}) = \text{water weight (mg)/reagent volume (mL)} \qquad (1)$$

9.1.9.2 The replicate values for *F* should have a relative standard deviation of not greater than 5 %. The standardization should be repeated if a greater variance is determined.

10. Procedure

10.1 Bring the KF solvent to dryness by titrating with reagent to the endpoint. The endpoint is typically indicated visually or by an audible alarm, or both, depending on the equipment manufacturer.

10.2 Refill the burette with reagent.

10.3 Draw a portion of the sample into the 1 to 5-mL syringe, and clean any residual sample material from the syringe tip. If the sample contains a large amount of solids or its viscosity is such that it cannot be drawn into the syringe easily, it may be necessary to weigh the sample and introduce it using a laboratory spatula or suitable device.

10.4 Weigh the syringe with sample to the nearest 0.0001 g.

10.5 Add the sample to the titration vessel and allow sufficient time for the material to dissolve adequately. The suggested amount of sample based on the amount of expected water content is as given in Table 1.

NOTE 2—The information given in Table 1 is described in Test Method D 4017.

10.6 Reweigh the syringe and record the sample mass added to the vessel to the nearest 0.0001 g.

10.7 If solid material is observed interfering with the sensing probe, or it is observed that the sample is not dissolved in the KF solvent sufficiently, extract the sample with solvent as follows:

10.7.1 Beginning again with a new sample, combine the sample and solvent gravimetrically at an approximate 1:1 ratio in a large test tube. Record the mass of both sample and solvent.

10.7.2 Cap the sample mixture and shake or mix on a vortex-type or orbital mixer for approximately 1 min.

10.7.3 Centrifuge the sample mixture to separate the undissolved/leached solids.

10.7.4 Analyze the supernatant beginning with 10.1.

10.7.5 In addition to the extracted sample, analyze a blank solution of virgin KF solvent beginning with 10.1. The KF solvent blank (mL of titrant used) is subtracted in 10.9.3.

TABLE 1 Specimen Guidelines

Expected water, %	Approximate Specimen Mass, g	Approximate Titrant Volume at 5 mg/mL titre, mL
0.5–1.0	5	5–10
1–3	2–5	10–20
3–10	1–2	10–20
10–30	0.4–1.0	15–25
>70	0.1	20

[5] *Reagent Chemicals, American Chemical Society Specifications*, American Chemical Society, Washington, DC. For suggestions on the testing of reagents not listed by the American Chemical Society, see *Analar Standards for Laboratory Chemicals*, BDH Ltd., Poole, Dorset, U.K., and the *United States Pharmacopeia and National Formulary*, U.S. Pharmaceutical Convention, Inc. (USPC), Rockville, MD.

10.8 Titrate with reagent until the endpoint is reached. Record the volume.

10.9 The final percentage of water should be based on the mean of a minimum of two determinations.

10.9.1 Calculation:

10.9.1.1 For unextracted samples, calculate the percentage water as follows:

$$(V)(F) = W$$

$$(W/S)\,100 = \text{water, weight\%} \quad (2)$$

where:
V = reagent volume (mL),
F = Karl Fischer reagent factor (mg/mL),
W = mass of water contained in the sample (mg), and
S = sample mass (mg).

10.9.1.2 For extracted samples, calculate the percentage water as follows:

$$(V - \text{solvent blank})(F) = W$$

$$(W/S)(\text{dilution factor}) = \%\ \text{water} \quad (3)$$

where:
V = reagent volume (mL),
F = Karl Fischer reagent factor (mg/mL),
W = mass of water contained in the sample (mg),
S = sample mass (mg), and

$$\text{dilution factor} = \frac{\text{KF solvent (mg)} + \text{sample (mg)}}{\text{sample (mg)}}$$

11. Precision and Bias [6]

11.1 *Precision*—The precision estimates are based on an interlaboratory study in which operators in six different laboratories analyzed in duplicate, on separate days, five samples of hazardous waste fuel containing between 13 and 32 % water. The results were analyzed statistically in accordance with Practice E 691.

11.1.1 *Repeatability (Within Laboratory)*—The coefficient of variation of results (each the average of duplicate determinations), obtained by the same analyst on different days, has been estimated to be 1.3 % relative at 30 degrees of freedom. Therefore, the results of two properly conducted tests, by the same operator on the same sample, should not differ by more than 3.6 % of their average.

11.1.2 *Reproducibility (Between Laboratories)*—The coefficient of variation of results (each the average of duplicate determinations), obtained by analysts in different laboratories, has been estimated to be 4.3 % relative at 5 degrees of freedom. Therefore, the results of properly conducted tests, on identical samples of the same material, by two different laboratories, should not differ from each other by more than 11.9 % of their average.

11.2 *Bias*—The bias of this test method has not been determined because there are no recognized reference standards.

12. Keywords

12.1 hazardous waste fuel; Karl Fischer titrimetry; percent water; total moisture

[6] A copy of the research report used to develop the precision statement is available from ASTM headquarters. Request RR:D 34-1009.

Designation: D 5633 – 94 (Reapproved 2001)

Standard Practice for Sampling with a Scoop[1]

This standard is issued under the fixed designation D 5633; the number immediately following the designation indicates the year of original adoption or, in the case of revision, the year of last revision. A number in parentheses indicates the year of last reapproval. A superscript epsilon (ε) indicates an editorial change since the last revision or reapproval.

1. Scope

1.1 This procedure covers the method and equipment used to collect surface and near-surface samples of soils and physically similar materials using a scoop.

1.2 *This standard does not purport to address all of the safety concerns, if any, associated with its use. It is the responsibility of the user of this standard to establish appropriate safety and health practices and determine the applicability of regulatory limitations prior to use.*

2. Referenced Documents

2.1 *ASTM Standards:*

D 4687 Guide for General Planning of Waste Sampling[2]

D 5088 Practice for Decontamination of Field Equipment Used at Nonradioactive Waste Sites[3]

D 5283 Practice for Generation of Environmental Data Related to Waste Management Activities: Quality Assurance and Quality Control Planning and Implementation[2]

2.2 *Other Documents:*

Pierre Gy's Sampling Theory and Sampling Practice, Francis F. Pitard[4]

3. Summary of Practice

3.1 The top layers of material are removed down to the required sample depth using a shovel or other suitable equipment. A clean scoop is then used to collect the actual sample, which is placed in a sample container.

4. Significance and Use

4.1 This practice is intended for use in collecting samples of contaminated soils and similar materials.

4.2 Scoops are used primarily for collecting samples near the surface. Subsurface samples can be obtained by first removing higher layers using a shovel or other suitable equipment and collecting the sample with the scoop.

4.3 Because of their simplicity, scoops are useful in taking samples of waste materials where decontamination or disposal is a problem with other types of sampling equipment. Scoops are also suitable for use in rapid screening programs, pilot studies, and other semi-quantitative investigations.

4.4 Samples should be collected in accordance with an appropriate work plan (see Practice D 5283 and Guide D 4687).

5. Sampling Equipment

5.1 A shovel or other suitable equipment can be used for the initial removal of overburden material. This equipment should be manufactured from material that is compatible with the soil or waste to be sampled. The scoop must be manufactured from material that is compatible with the soil or waste to be sampled and the required test or analysis to be performed. For most hazardous waste sampling, either a disposable plastic scoop or a reusable stainless steel or polytetrafluoroethylene-coated scoop is suitable.

5.2 The design of the scoop is important to minimize sampling error, that is, all the material intended as the sample can be collected and placed in the sample container and is not lost as the scoop is systematically lifted from the source to the sample container (see Pierre Gy's Sampling Theory and Sampling Practice and Fig. 1).

5.3 For measurement of sample depth, a ruler or tape measure can be used.

6. Sample Containers

6.1 Plastic, glass, or other nonreactive containers should be used. Refer to Guide D 4687 for information on sample containers.

7. Procedure

7.1 Record all relevant information and observations about the sample location.

7.2 Use a shovel or other suitable equipment to remove any overburden material down to the level specified in the work plan.

7.3 Measure to the depth at which the sample will be collected with a ruler or tape measure. Record this information in the field log book.

7.4 Remove the thin layer of material that was in contact with the overburden removal equipment and discard it using a clean scoop. The work plan will define if the scoop may or may not be reused to collect the actual sample.

7.5 Collect a suitable volume of sample with the scoop (the

---

[1] This practice is under the jurisdiction of ASTM Committee D34 on Waste Management and is the direct responsibility of Subcommittee D34.01.03 on Sample Preservation.

Current edition approved Nov. 15, 1994. Published January 1995.

[2] *Annual Book of ASTM Standards*, Vol 11.04.

[3] *Annual Book of ASTM Standards*, Vol 04.09.

[4] Available from CRC Press, Inc., 2000 Corporate Blvd., NW, Boca Raton, FL 33431.

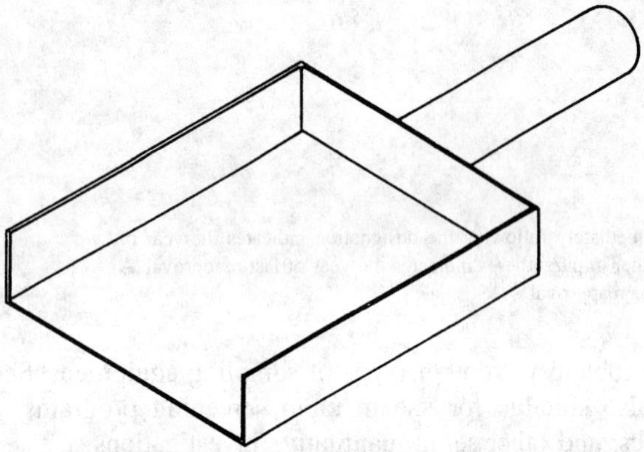

FIG. 1 Example of a Scoop

same scoop can be used to collect multiple scoopfuls to obtain sufficient volume to fill the container). Use a new (or decontaminated) scoop for each sample. Transfer the sample into the suitable container. Close the sample container and complete and attach the sample label.

7.6 Complete the field log book and chain-of-custody form.

7.7 Decontaminate the reusable equipment in accordance with the protocol specified in the work plan (see Practice D 5088).

8. Keywords

8.1 sampling; scoop; soil sampling; waste

Standard Practice for Sampling Unconsolidated Waste From Trucks[1]

This standard is issued under the fixed designation D 5658; the number immediately following the designation indicates the year of original adoption or, in the case of revision, the year of last revision. A number in parentheses indicates the year of last reapproval. A superscript epsilon (ε) indicates an editorial change since the last revision or reapproval.

1. Scope

1.1 This practice covers several methods for collecting waste samples from trucks. These methods are adapted specifically for sampling unconsolidated solid wastes in bulk loads using several types of sampling equipment.

1.2 *This standard does not purport to address all of the safety concerns, if any, associated with its use. It is the responsibility of the user of this standard to establish appropriate safety and health practices and determine the applicability of regulatory limitations prior to use.* See Section 6 for specific precautionary statements.

2. Referenced Documents

2.1 *ASTM Standards:*

D 4687 Guide for General Planning of Waste Sampling[2]

D 4700 Guide for Soil Sampling from the Vadose Zone[3]

D 5088 Practice for Decontamination of Field Equipment Used at Non-radioactive Waste Sites[4]

D 5283 Practice for Generation of Environmental Data Related to Waste Management Activities: Quality Assurance and Quality Control Planning and Implementation[2]

D 5633 Practice for Sampling with a Scoop[2]

3. Terminology

3.1 Descriptions of Terms Specific to This Standard:

3.1.1 *authoritative sampling*—a sample selected without regard to randomization.

3.1.2 *paperwork*—all required documentation, which may include manifests, waste profiles, sample labels, site forms, etc.

3.1.3 *screening analysis*—a preliminary qualitative or semiquantitative test that is designed to give the user rapid and specific information about a waste that will aid in determining waste identification, process compatibility, and safety in handling.

3.1.4 *unconsolidated*—for solid material, the characteristic of being uncemented or uncompacted, or both, and easily separated into smaller particles.

3.1.5 *waste profile*—specific information about the waste including its properties and composition, chemical constituents, waste codes, transportation information, etc.

3.1.6 *work plan*—a plan specific to a particular site, for conducting activities specified in the plan.

4. Summary of Practice

4.1 The truck and its contents are inspected and appropriate sampling equipment is selected. A clean sampling device is then used to scoop, core, or auger into the waste material. The sample or samples are collected and transferred to a sample container. The sampling device is then cleaned and decontaminated or disposed of.

5. Significance and Use

5.1 This practice is intended for use in the waste management industries to collect samples of unconsolidated waste from trucks. The sampling procedures described are general and should be used in conjunction with a site-specific work plan.

5.2 The purpose of collecting waste samples directly from a truck (rather than the waste source) is to verify (usually with screening analyses) that the waste contained in the truck is the same or similar material from a waste source that has been previously characterized and approved for treatment or disposal, or both.

6. Safety Precautions

6.1 Safety precautions must always be observed when sampling waste. The work plan must include a Worker Health and Safety section, because there are potential hazards associated with working around trucks as well as their potentially hazardous contents.

6.2 Truck sampling should be conducted from a properly designed platform to allow the sampler to safely access the truck bed with a minimum of difficulty.

7. Sampling Design

7.1 Truck sampling can be conducted for many different purposes. It is important that the purpose be integrated into the sample design. If the purpose of sampling is to characterize the waste, the sample should be collected from the waste source during the loading or unloading of the truck. This allows access to all portions of the material in the truck. If the purpose is to determine if the material in the truck conforms to a waste profile (that is, waste material that has previously been characterized), then a less rigorous sampling approach can be used. Because of the difficulties of sampling the material in the truck

---

[1] This practice is under the jurisdiction of ASTM Committee D34 on Waste Management and is the direct responsibility of Subcommittee D34.01.02 on Monitoring.

Current edition approved Jan. 15, 1995. Published March 1995.

[2] *Annual Book of ASTM Standards*, Vol 11.04.

[3] *Annual Book of ASTM Standards*, Vol 04.08.

[4] *Annual Book of ASTM Standards*, Vol 04.09.

in situ, (authoritative) grab samples are usually collected from the top portion of the material and subjected to screening type analysis. This method will quickly demonstrate that the sampled material (top portion) does or does not match the waste profile.

7.2 A work plan should be prepared describing the sampling locations, number of samples, depth of sampling and type of sampling equipment (see Practice D 5283 and Guide D 4687).

NOTE 1—Because of limited access to the truck bed for sampling, the samples collected are usually near-surface samples. There is a possibility that the material in the middle or on the bottom of the bed is different.

8. Pre-Sampling

8.1 *Basic Pre-Sampling Practices*:

8.1.1 Review all paperwork.

8.1.2 Access the truck by way of the sampling platform so that the waste can be visually inspected to confirm agreement with the paperwork and identify any obvious discrepancies (such as free liquids, etc.).

8.2 *Sampling Equipment*:

8.2.1 *Selection*:

8.2.1.1 Select the sampling equipment and sample containers appropriate for the waste in the truck, in accordance with the work plan or site-specific procedure. See Guide D 4687 for information on sample container selection.

8.2.1.2 The sampling equipment, sample preparation equipment, sample containers, etc., must be clean, dry, and inert to the material being sampled. Before use, all equipment including sample containers shall be inspected to ensure they are clear of obvious dirt and contamination and are in good working condition. Visible contamination shall be removed, and the equipment shall be decontaminated with the appropriate rinse materials. Prior to use, all cleaned equipment should be protected from contamination.

8.2.2 *Materials of Construction*:

8.2.2.1 Sampling devices are usually made of stainless steel, brass, or aluminum.

8.2.2.2 Sample containers should be made of plastic, glass, or other nonreactive materials (see Guide D 4687).

8.3 *Generic Equipment List*—The following is a general identification of equipment required for sampling unconsolidated waste from trucks.

8.3.1 *Scoop*, with extension handle.

8.3.2 *Trier.*

8.3.3 *Auger.*

8.3.4 *Concentric tube thief*, single slot, split tube, Missouri trier.

8.3.5 *Thin-walled tube.*

8.3.6 *Barrel auger.*

8.3.7 *Sample collection sheet.*

8.3.8 *Sample containers*, with lids and liners.

8.3.9 *Chain of custody forms.*

8.3.10 *Paperwork and site forms.*

8.3.11 *Sample labels.*

8.3.12 *Cloths or wipes.*

9. Sampling

9.1 *Basic Sampling Practices*:

9.1.1 Access the truck by way of the sampling platform and collect the required number of samples using techniques in accordance with 9.2.

9.1.2 Place the collected material in a sample container.

9.1.3 Close the sample container.

9.1.4 Wipe the outside of the sample container. Dispose of the wipe cloth properly.

9.1.5 Note on site forms all relevant conditions and physical characteristics associated with the collection of the sample.

9.1.6 Fill out all required paperwork for each sample, as required by the work plan.

9.1.7 Complete and attach the label to the side of the sample container after the sample has been collected.

9.2 *Sampling with a Concentric Tube Thief*:

9.2.1 *General Description*—This device consists of two tubes, one fitting snugly inside the other (see Fig. 1). The bottom end of the outer tube is fitted with a point. Oblong holes are cut through both tubes. The holes are opened or closed by rotating the inner tube. Concentric tube samplers are commercially available up to 6 ft (1.8 m) long and several inches (centimeters) in diameter.

9.2.2 Concentric tube samples have a limited application for sampling trucks. Materials that are not free-flowing such as those that are hard packed, moist, or finely powdered will not enter this type of sampler under normal field conditions. Sampling of materials containing granules or particles exceeding one third of the slot width should not be attempted because bridging may occur.

9.2.3 Insert the tube into the material and push with uniform force to the bottom of the truck or until refusal. Rotate the concentric tubes to the open position, thereby allowing the sample to flow into the inner tube. Wiggle the sampler several times and rotate the tubes to the closed position. Withdraw the sampler. Place the sampling device immediately over a sample collection sheet and release the sample by rotating the concentric tubes to the open position. A sample can normally be removed from the thief with a spatula or similar instrument (reamer) and placed in the sample container.

9.3 *Sampling with a Thin-Walled Tube Sampler*:

9.3.1 *General Description*—Tube samplers may vary in length, diameter, and material of construction (see Fig. 2). The material to be sampled must be of a physical consistency (cohesive solid material) to be cored and retrieved with the tube. Materials with particles larger than one third of the inner diameter of the tube should not be sampled with that particular device. The length of the tube will depend on the desired sampling depth (see Guide D 4700). The tube is attached to a length of solid or tubular rod. The upper end of this rod is threaded to accept a handle or extension rods. This sampler can

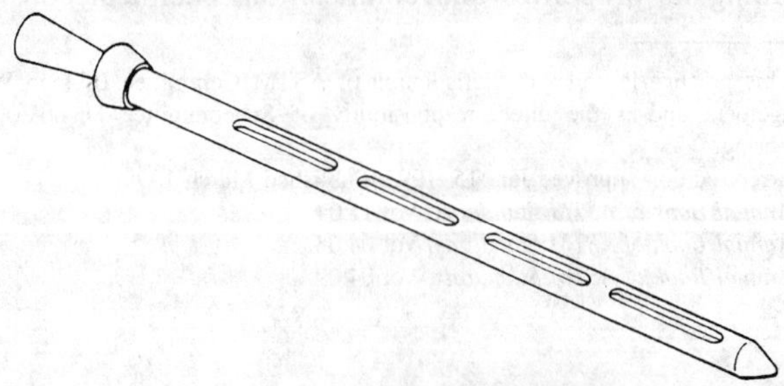

FIG. 1 Concentric Tube Thief

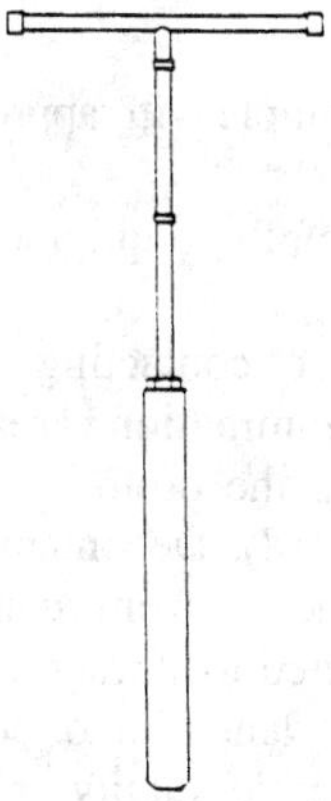

FIG. 2 Thin-walled Tube

be used to collect samples of unconsolidated clay-like materials.

9.3.2 The tube sampler is pushed into the material to be sampled by applying downward force on the unit's handle. Once the sampler has reached the bottom of the sampling interval, it is twisted to break the continuity at the tip. The sampler is pulled from the material and the sample is extruded into the sample container. Samples are extruded by forcing a rod through the tube.

9.4 *Sampling with a Trier Sampler*:

9.4.1 *General Description*—The trier is a metal or plastic tube from which one third to one half of the wall of the tube has been removed to form a slot along its entire length (see Fig. 3). This device can be up to 4 ft long (1.2 m) and should have a sharp, angled point at its lower end. The material to be sampled must have a physical consistency like a soil or similar fine-grained cohesive material. Sampling procedures can be found in Practice D 5451.

9.4.2 The trier is pushed vertically into the material and rotated one or two times to cut a core. The core is pulled out of the hole and removed from the trier with a spatula or similar instrument and placed in the sample container.

9.5 *Sampling with an Auger*:

9.5.1 *General Description*—The screw or ship auger is essentially a small diameter (for example, 1.5 in.; 3.8 cm) wood auger from which the cutting side flanges and tip have been removed. The auger is welded onto a length of solid or tubular rod. The upper end of this rod is threaded to accept a handle or extension rods (see Fig. 4 and Guide D 4700).

9.5.2 An auger can be used for collecting a disturbed sample of unconsolidated material from the truck. The auger is rotated manually or with a power source into the waste material. The operator may have to apply downward force to embed the auger; afterwards, the auger screws itself into the material. The auger is advanced to its full length, then pulled and removed. Material from the deepest interval is retained on the auger flights. The sample is collected from this extracted portion. Augers can be used to sample hard or compacted solid wastes or soil. Augers, like triers, are equipped with crossbars, facilitating the penetration of the waste.

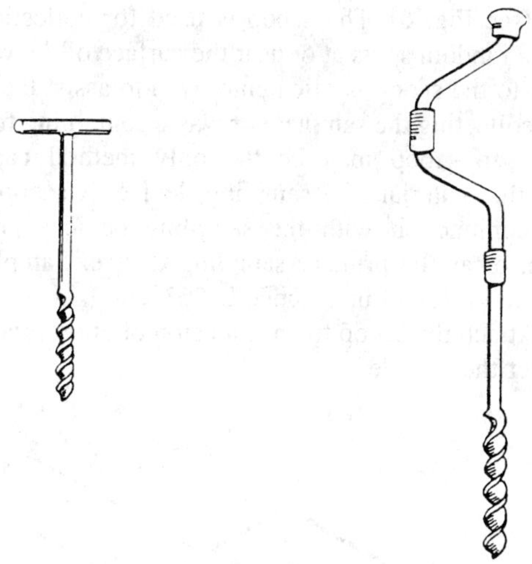

FIG. 4 Auger

9.6 *Sampling with a Barrel Auger*:

9.6.1 *General Discussion*—Dimensions and construction of a barrel auger will vary. A barrel auger typically consists of a stainless steel or carbide steel auger tip (orchard bit), a stainless steel cylinder, a bailed cap, an extension, and a cross handle (see Fig. 5 and Guide D 4700). A thin-walled internal sleeve may be used to contain the sample.

9.6.2 Barrel augers can be used for collecting discreet samples of disturbed material from various depths. The auger is rotated to advance the barrel into the truck load. The operator may have to apply downward pressure to keep the auger advancing. When the barrel is filled, the unit is withdrawn from the waste material and the sample is collected from the barrel.

9.7 *Sampling with a Scoop*:

9.7.1 *General Description*—Scoops must be of a size and shape suitable for the quantity and size of the particles to be

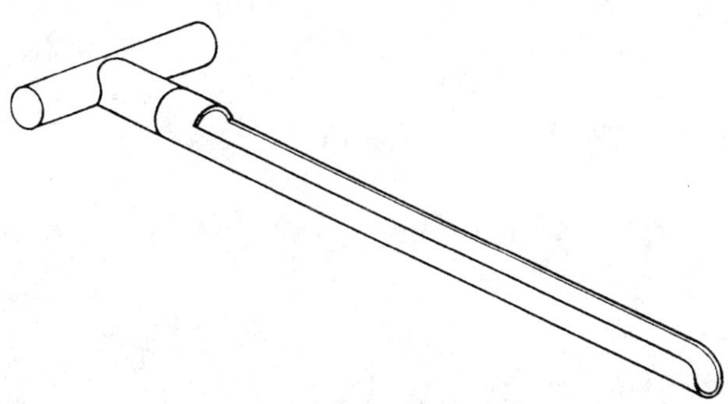

FIG. 3 Trier

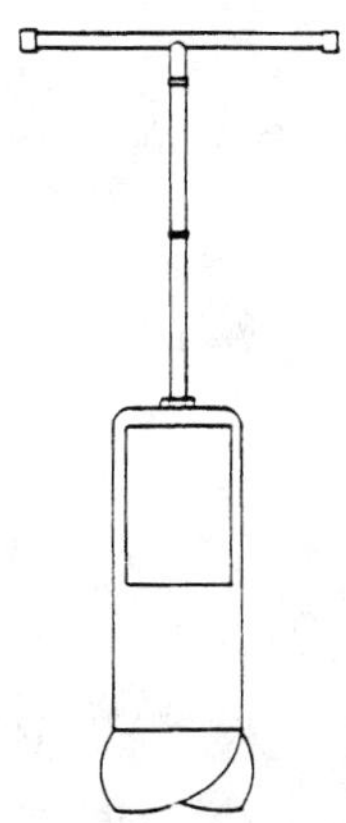

FIG. 5 Barrel Auger

sampled (see Fig. 6). The scoop is used for collecting equal portions at random spots at or near the surface of the waste. An extension to the scoop is often employed to assist the sampler in safely collecting the sample. For waste containing fragments or chunks, a scoop may be the only method capable of retaining the material in a sampling device. A scoop may be used in conjunction with the sampling devices previously described, or as the primary sampling device. Sampling procedures can be found in Practice D 5633.

9.7.2 Attach the scoop to an extension of appropriate length and collect the sample.

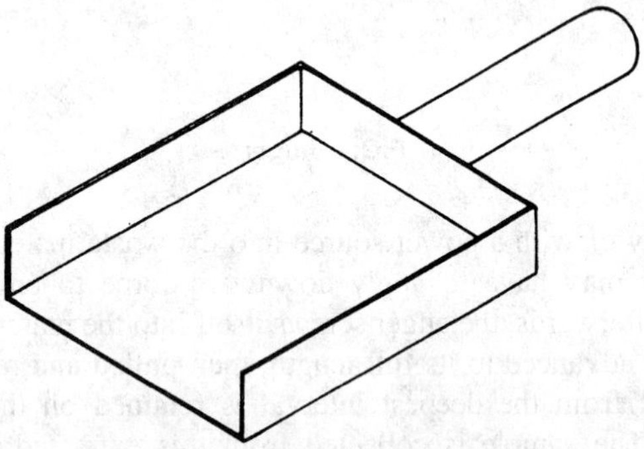

FIG. 6 Example of a Scoop

10. Post Sampling

10.1 Transport the sample with appropriate chain of custody to the laboratory.

10.2 Remove all sampling equipment from the truck sampling area.

10.3 Transfer all direct contacting reusable equipment to a pre-designated decontamination area. Decontaminate the equipment according to the protocol established in the work plan (see Practice D 5088). Decontaminated sampling equipment should be protected from contamination. This may include, but not be limited to storage in aluminum foil, plastic bags, polytetrafluoroethylene film, or other means of protection that will not impact sample quality or the intended analyses.

10.4 Properly dispose of all used (disposable) contacting equipment.

11. Data Quality Objectives

11.1 The objectives for sampling and testing unconsolidated waste from trucks should be specified in the work plan (see Practice D 5283 and Guide D 4687).

12. Keywords

12.1 auger; barrel auger; concentric tube thief; sampling; scoop; thin-walled tube sampler; trier; truck sampling; waste

Standard Test Method for Chlorophenoxy Acid Herbicides in Waste Using HPLC[1]

This standard is issued under the fixed designation D 5659; the number immediately following the designation indicates the year of original adoption or, in the case of revision, the year of last revision. A number in parentheses indicates the year of last reapproval. A superscript epsilon (ε) indicates an editorial change since the last revision or reapproval.

1. Scope

1.1 This test method covers the analysis of 2,4-dichlorophenoxyacetic acid (2,4-D), 2,4,5-trichlorophenoxyacetic acid (2,4,5-T), and 2,4,5-trichlorophenoxypropionic acid (silvex) in liquids and solids, using high performance liquid chromatography with an ultraviolet detector (HPLC/UV). This test method is applicable for a concentration range from approximately 50 to 1000 ppm. This range takes into consideration the sample preparation and dilutions outlined in Section 10. Lower detection levels can be obtained by using larger sample sizes, smaller total final volumes, or with the use of in-line or solid phase extraction, concentration, and/or cleanup.

1.2 The chlorophenoxy herbicides may be present as a variety of salts or esters, which are converted to, analyzed, and reported as their respective acids.

1.3 This test method is applicable to liquid and solid waste and waste extract matrices including aqueous, oil, spent solvent, soil, ash, leachates, etc.

1.4 This test method may be applicable to other phenoxy acid herbicides.

1.5 *This standard does not purport to address all of the safety concerns, if any, associated with its use. It is the responsibility of the user of this standard to establish appropriate safety and health practices and determine the applicability of regulatory limitations prior to use.* See Section 7 and 10.3.1 for specific precautionary statements.

2. Referenced Documents

2.1 *ASTM Standards:*

D 1193 Specification for Reagent Water[2]

D 3478 Test Method for Chlorinated Phenoxy Acid Herbicides in Water[3]

2.2 *EPA Documents:*

Method 8150 Chlorinated Herbicides, Test Methods for Evaluating Solid Waste Physical/Chemical Methods SW-846 Third Edition[4]

Method 8000A Gas Chromatography, Test Methods for Evaluating Solid Waste Physical/Chemical Methods, SW-846, 3rd Edition, Final Update 1[4]

Method 555 Determination of Chlorinated Acids in Water by High Performance Liquid Chromatography with a Photodiode Array Ultraviolet Detector EPA/600/R-92/129, Methods for the Determination of Organic Compounds in Drinking Water, Supplement No. 2[4]

3. Summary of Test Method

3.1 The chlorophenoxy acids and esters are hydrolyzed to their respective salts by heating and stirring the sample with aqueous alkali. The salts are then converted to their respective acids by the addition of HCl. The aqueous solutions of the free acids are then analyzed using High Performance Liquid Chromatography (HPLC) using ultraviolet detection.

4. Significance of Use

4.1 Phenoxy acid herbicides are used extensively for weed control. Esters and salts of 2,4-D, 2,4,5-T, and Silvex have been used for agricultural crop and lawn care.

5. Interferences

5.1 Organic liquids that have high percent levels of chlorinated organics and are denser than water may interfere with the extraction of Silvex. It is necessary to mix these samples with hexadecane, typically in a 1:1 ratio, before hydrolysis.

5.2 Phenols, especially chlorophenols interfere with the procedure, by coeluting with the analytes of interest.

5.3 Interferences may be encountered from other organic compounds that absorb UV at the specified wavelengths. Also, closely eluting compounds may complicate identification based solely on retention time. When these types of interferences are encountered, the analyst must rely on other sources of information for positive identification, such as the following:

5.3.1 Secondary confirmation wavelengths such as 227 nm or 235 nm.

5.3.2 Use of a confirmation column.

5.3.3 Use of a confirmatory chromatography program such as changing the mobile phase composition or gradient.

---

[1] This test method is under the jurisdiction of ASTM Committee D34 on Waste Management and is the direct responsibility of Subcommittee D34.01.06 on Analytical Methods.

Current edition approved Jan. 15, 1995. Published March 1995.

[2] *Annual Book of ASTM Standards*, Vol 11.01.

[3] Discontinued, 1992. See *Annual Book of ASTM Standards*, Vol 11.02.

[4] Available from the Superintendent of Documents, U.S. Government Printing Office, Washington, DC 20402.

6. Apparatus

6.1 *Analytical High-Pressure Liquid Chromatograph*, capable of achieving pressures of 4000 psi and flow rates of 3 mL/min.

6.2 *Variable Wavelength Ultraviolet Detector*, capable of monitoring at 207, 227, and 235 nm, either simultaneously or individually.

6.3 *Chromatographic Column*, C18 radial compression 8 by 100 mm, 4-μm particle size. Equivalent stainless steel or radial compression columns may be used.

6.4 *Guard Column*, C18, 4-μm particle size.

6.5 *Injector*, manual injection valve, instrument autosampler, equipped with a 500-μL sample loop, or equivalent.

6.6 *Data Systems*, data systems capable of controlling the HPLC system and for acquiring data may be used.

6.7 *Glass Vials*, 16-mL capacity with TFE-fluorocarbon-lined screw caps.

6.8 *Microsyringes*, 10, 100, and 500-μL capacity.

6.9 *Balance*, analytical, capable of accurately weighing to the nearest 0.0001 g.

6.10 *Pipets*, Pasteur, disposable glass.

6.11 *Pipets*, disposable glass, 1-mL and 10-mL, calibrated.

6.12 *pH paper*, wide range from 1 to 11.

6.13 *Hot Plate*, with multiple stirring positions.

6.14 *Water Filtration Apparatus*, used for the purification of water for HPLC use in 7.4.

6.15 *Water Filtration Filters*, 0.22-μm used in 7.4.

6.16 *Flasks*, 100-mL volumetric glass.

6.17 *Centrifuge*.

6.18 *Stir Bars*.

6.19 *Funnels*, glass.

6.20 *Filter Papers*, 15-cm hardened/ashless, fast.

7. Reagents and Materials

7.1 *Purity of Reagents*—Reagent grade chemicals shall be used in all tests. Unless otherwise indicated, it is intended that all reagents conform to the specifications of the Committee on Analytical Reagents of the American Chemical Society where such specifications are available.[5] Other grades may be used, provided it is first ascertained that the reagent is of sufficiently high purity to permit its use without lessening the accuracy of the determination.

7.2 *Purity of Water*—Unless otherwise indicated, references to water shall be understood to mean reagent water as defined by Type II of Specification D 1193.

7.3 *Acetone*, pesticide quality or equivalent, used for preparing standards. Be advised that if methanol is used for preparing the acid standards, over time the acid form of the herbicides will convert to their methyl ester form.

7.4 *Filtered Water with 0.5 % Phosphoric Acid*, (FWPA). Add 5 mL high-purity phosphoric acid to 995-mL reagent water in a volumetric flask. Filter through a 0.22-μm filter.

7.5 *Herbicide Reference Standards*—The following reference calibration compounds are required. Reference solutions can be prepared from the pure standard materials or purchased as certified solutions.

7.5.1 2,4-D, 2,4,5-T, Silvex, 2,4-D methyl ester.

7.5.2 2,4,5-T methyl ester, Silvex methyl ester.

7.6 *Hexadecane*, 99 %.

7.7 *Hydrochloric Acid* (density 1.195 g/mL), concentrated hydrochloric acid (HCl).

7.8 *Methanol*, pesticide quality or equivalent.

7.9 *Acetonitrile*, pesticide quality or equivalent. Acetonitrile may be used as the mobile phase instead of methanol.

7.10 *Phosphoric Acid, H_3PO_4*, 85 % o-phosphoric acid, HPLC Grade.

7.11 *Potassium Hydroxide 37 % Solution*—carefully add 37 g of ACS grade potassium hydroxide (KOH) pellets to 50 mL of reagent water in a 100-mL volumetric flask. Mix by swirling the flask. Bring to volume with additional reagent water.

NOTE 1—**Caution:** Observe caution when adding KOH to water. This will cause an exothermic reaction. Cooling the flask while mixing is suggested.

8. Standard Preparation

8.1 Prepare stock standard solutions as follows. For herbicide acids and methyl esters, accurately weigh 0.100 g into a 10-mL glass volumetric flask. Bring to volume in pesticide-grade acetone. Larger volumes can be used at the convenience of the analyst. If compound purity is 96 % or greater, the weight can be used without correction to calculate the concentration of the stock standard.

8.2 Transfer the stock standard solutions into TFE-fluorocarbon-sealed screw-cap vials. Store at 4°C and protect from light. Stock standard solutions should be checked frequently for signs of degradation or evaporation especially prior to being used to prepare calibration standards. Marking the meniscus level on the standard vial is recommended when monitoring for evaporation.

8.3 Stock standard solutions must be replaced after one year or sooner if comparison with check standard indicates a problem.

8.4 Prepare calibration standard solutions from stock standard solutions at a minimum of 5 concentration levels for each parameter. Transfer the appropriate volume of stock solution to a volumetric flask and dilute with FWPA. The concentration levels should correspond to the expected range of concentrations found in real samples or should define the working range of the HPLC/UV. A working range from 0.5 to 5 ppm has been found to be useful. Calibration solutions must be replaced after six months, or sooner if comparison with check standards indicates a problem. A 20 % deviation in response between the daily check and the initial calibration would indicate a potential problem.

9. Sample Collection, Preservation, and Handling

9.1 Sample collection should be in accordance with appropriate sampling protocols.

9.2 Sample extracts should be stored in glass containers. Long-term storage should be at 4°C. For an aqueous matrix, 100 mL of sample are needed and 5 g of sample are needed for

---

[5] *Reagent Chemicals, American Chemical Society Specifications*, American Chemical Society, Washington, DC. For suggestions on the testing of reagents not listed by the American Chemical Society, see *Analar Standards for Laboratory Chemicals*, BDH Ltd., Poole, Dorset, U.K. and the *United States Pharmacopeia and National Formulary*, U.S. Pharmaceutical Convention, Inc. (USPC), Rockville, MD.

solids or organic liquids.

9.3 Unused sample material, laboratory dilutions, and waste from the samples may be regulated. Consult your specialist and the regulations, or both, for guidance in the proper handling and disposal of laboratory waste.

10. Procedure

10.1 *Sample Extraction and Hydrolysis (for Aqueous Matrices)*:

10.1.1 Add 50 mL or 50 g of sample to a 100-mL volumetric or Erlenmeyer flask.

10.1.2 Add 10 mL of 37 % KOH solution to the flask. Check the pH. If the pH is not 10 or greater, add additional KOH solution. Record the KOH solution used, in millilitres, and the final pH.

10.1.3 Add a stir bar to the flask. Place the flask on a stirring hot plate and heat and mix at 70 to 90°C for 2 h.

10.1.4 Continue to 10.3.

10.2 *Sample Extraction and Hydrolysis (All Other Matrices)*:

10.2.1 Weigh 5 g of sample into a 100-mL volumetric or Erlenmeyer flask.

NOTE 2—Be cognizant of sample matrix type, especially for solid materials such as soils and ash. It may be necessary to crush or powder the sample to ensure complete extraction.

10.2.1.1 If the sample is a dense organic liquid for example, high PCB oils, chlorinated solvents, and so forth, weigh 2.5 g of sample into a 100-mL volumetric flask and add 2.5 g hexadecane. Mix by swirling the flask.

NOTE 3—It was determined through recovery studies that the chlorophenoxy herbicides remained in highly chlorinated organic liquids on the bottom of the volumetric flask after hydrolysis. Mixing the sample with hexadecane brought the sample to the top of the volumetric flask and the herbicides were extracted into and remained in the aqueous phase after hydrolysis.

10.2.1.2 Add 50 mL reagent water to the flask. Mix by swirling the flask.

10.2.1.3 Add 10 mL of 37 % KOH solution to the flask. Check the pH. If the pH is not 10 or greater add additional KOH solution. Record millilitres of KOH solution used and the final pH.

10.2.1.4 Add a stir bar to the flask. Place the flask on a stirring hot plate and heat and mix at 70 to 90°C for 2 h.

10.2.1.5 For oil samples, after stirring is complete, remove stir bar and bring volumetric flask to volume with reagent water. The water meniscus should be at the mark. Let stand until the layers separate.

10.2.1.6 Draw off and discard any oil layer that may be present on top of the water.

10.2.1.7 For solid samples, filter extract through filter paper (see 6.20) into a separate 100-mL volumetric flask. Rinse the filter paper being careful not to exceed volume. Bring to volume with reagent water.

10.2.1.8 Continue to 10.4.

NOTE 4—This is a good stopping point in the method if needed. Store sample extracts overnight in the refrigerator.

10.3 *Preparation for Analysis (Aqueous Matrices)*:

10.3.1 Add 5 mL of concentrated HCl to the volumetric flask containing the solution prepared in 10.1. Mix well. **Caution: always wear personal protective gear when adding concentrated acid to water because of the potential for heat generation.**

10.3.2 Check the pH using wide-range pH paper. The pH must be less than two before HPLC analysis.

10.3.3 If the pH is not less than two, add additional HCl to the flask. Check and record the final pH.

10.3.4 Bring the volumetric flask to volume with reagent water.

10.4 *Preparation for Analysis (All Other Matrices)*:

10.4.1 Add 8 mL of FWPA prepared in 7.4, 1 mL concentrated HCl, and 1 mL of the sample extract prepared in 10.2 to a 16-mL glass vial. Shake.

10.4.2 Check the pH using wide-range pH paper. The pH must be less than two before HPLC analysis.

10.4.3 If the pH is not less than two, re-prepare the sample as in 10.4.1 using 7 mL FWPA and 2 mL concentrated HCl. Check and record the final pH.

10.4.4 If visible suspended solids are present in the sample, centrifuge the sample before injecting into the HPLC. It is very important to limit the amount of suspended solids injected onto the column and into the HPLC system because of the nature of HPLC and the potential for problems.

10.5 *HPLC Calibration*:

10.5.1 Using the calibration standards prepared in 8.4, generate a calibration curve that defines the working range or range of interest of the HPLC system. A correlation coefficient of 0.995 with a percent RSD of less than 20 is recommended for the acceptance of a valid calibration curve.

10.5.2 Check the initial calibration curve daily by injecting a mid-level standard. If the response varies from the initial calibration by more than 20 %, recalibration is recommended. If the daily standard and initial calibration is acceptable, use the response of the daily standard for H_{std} value in Section 11.

10.6 *HPLC Analysis*—Inject the solution prepared in 10.3 or 10.4 into the HPLC. A typical chromatogram and UV spectra are shown in Figs. 1 and 2.

10.7 *HPLC Conditions*:

10.7.1 *Mobile Phase*—75 % methanol/25 % FWPA.

NOTE 5—Optimum separation may require different percentages of organic and aqueous constituents if acetonitrile is used as the mobile phase in place of methanol.

10.7.2 *Flow Rate*—2.5 mL/min.

10.7.3 *Detector Wavelength*—primary quantitation at 207 nm.

10.8 *Quality Control*:

10.8.1 It is recommended that each laboratory using this test method operate a formal quality control program. The minimum requirements of this program should consist of an initial demonstration of laboratory capability and the on-going analysis of method blanks, duplicate and spiked samples to evaluate and document quality data.

10.8.2 Before processing any samples, the analyst must demonstrate, through the analysis of a method blank, that the interferences from the analytical system, glassware, and reagents are under control. Each time a set of samples is prepared or there is a change in reagents, a method blank must be

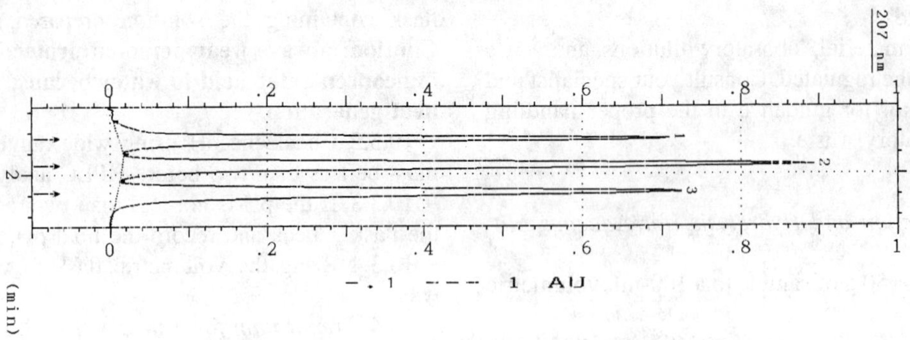

FIG. 1 Typical Chromatogram Using Conditions in 10.7

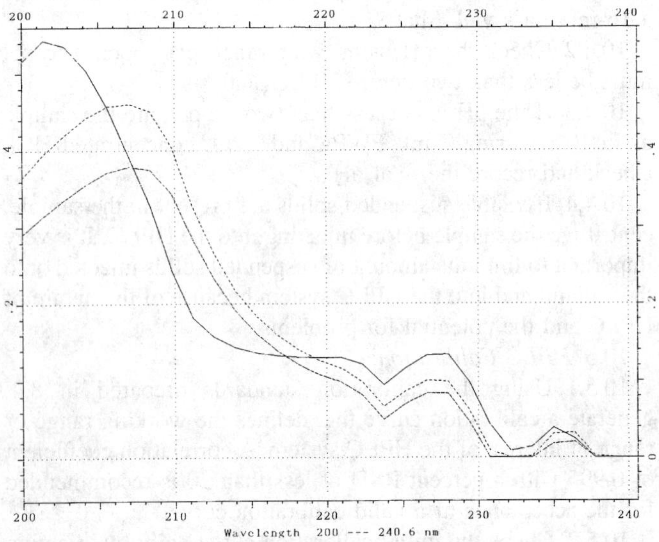

FIG. 2 UV Spectra of 2,4-D, 2,4,5-T, and Silvex Wavelength from 200 to 240 nm

processed as a safeguard against chronic laboratory contamination. The method blank must be carried through all stages of sample preparation and measurement.

10.8.3 The laboratory must, on an ongoing basis, analyze a replicate sample and a matrix spike or a matrix spike and a replicate matrix spike. This will allow the laboratory to assess method precision and accuracy.

11. Calculation

$$C_{samp} = \frac{H_{samp}}{H_{std}} \times C_{std} \times \frac{FV}{W} \quad (1)$$

where:

C_{samp} = concentration of compound in sample, µg/g,
C_{std} = concentration of compound in standard, µg/mL,
H_{samp} = peak height of compound in sample, AU,
H_{std} = peak height of compound in standard, AU, and
FV = final volume (mL).

NOTE 6—This includes the initial volume plus any subsequent dilutions. In this test method, as written for solids and organic liquids, the final volume (FV) would be 1000 mL, (100 mL for the initial volume multiplied by a 10 times dilution). This test method as written for aqueous samples would have a final volume of 100 mL.

W = weight of sample used, grams.

NOTE 7—For aqueous samples, convert millilitres to equal gram weight.

AU = absorbance units.

NOTE 8—Peak area may be substituted for peak height.

12. Precision and Bias

12.1 The data presented in Table 1 represents single-operator precision.

NOTE 9—All precision data were obtained using eight replicates.

13. Keywords

13.1 chlorophenoxy acid herbicides; 2,4-D; 2,4,5-T; chlorophenoxy acid herbicides in waste; HPLC; Silvex, 2,4,5-TP

TABLE 1 Precision and Bias

Compound	Concentration, ppm	Average Recovery, %	Single Operator Precision, % RSD
Matrix: Water/Oil Mix			
2,4-D	186	106	3.9
2,4,5-T	140	101	2.4
Silvex	195	115	1.7
Matrix: Soil			
2,4-D	199	102	3.3
2,4,5-T	185	113	2.4
Silvex	198	98	3.2
Matrix: PCB Oil			
2,4-D	199	100	5.0
2,4,5-T	185	118	2.8
Silvex	194	88	5.7
Matrix: Water Solvents			
2,4-D	953	105	1.8
2,4,5-T	943	103	2.9
Silvex	969	91	9.4

Designation: D 5660 – 96

Standard Test Method for Assessing the Microbial Detoxification of Chemically Contaminated Water and Soil Using a Toxicity Test with a Luminescent Marine Bacterium[1]

This standard is issued under the fixed designation D 5660; the number immediately following the designation indicates the year of original adoption or, in the case of revision, the year of last revision. A number in parentheses indicates the year of last reapproval. A superscript epsilon (ε) indicates an editorial change since the last revision or reapproval.

1. Scope

1.1 This test method **(1)**[2] covers a procedure for the rapid evaluation of the toxicity[3] of wastewaters and aqueous extracts from contaminated soils and sediments, to the luminescent marine bacterium *Photobacterium phosphoreum*,[4] prior to and following biological treatment. This test method is meant for use as a means to assess samples resulting from biotreatability studies. Sensitivity data for *P. phosphoreum* to over 1300 chemicals have been reported in the literature **(2)**. Some of the publications are very relevant to this test method **(3)**. The data obtained from this test method, when combined with respirometry, total organic carbon (TOC), biochemical oxygen demand (BOD), chemical oxygen demand (COD), or spectrophotometric data, can assist in the determination of the degree of biodegradability of a contaminant in water, soil, or sediment **(3)**. The percentage difference between the IC20 of treated and untreated sample is used to assess the progress of detoxification.

1.2 This test method is applicable to the evaluation of the toxicity (to a specific microbe) and its implication on the biodegradation of aqueous samples from laboratory research bio-reactors (liquid or soil), pilot-plant biological treatment systems, full-scale biological treatment systems, and land application processes (see Notes 1 and 2).

NOTE 1—If the biologically treated material is to be discharged in such a manner as to potentially impact surface waters and ground water, or both, then the user must consult appropriate regulatory guidance documents to determine the proper test species for evaluating potential environmental impact **(4)**. Correlations between data concerning reduction in toxicity produced by this test method and by procedures for acute or short-term chronic toxicity tests, or both, utilizing invertebrates and fish (see Guides E 729 and E 1192), should be established, wherever possible.

NOTE 2—Color (especially red and brown), turbidity, and suspended solids interfere with this test method by absorbing or reflecting light. In these situations data are corrected for these effects by use of an absorbance correction procedure included in this test method (see 5.3, 6.1, and 6.2).[5]

1.3 The results of this test method are reported in terms of an inhibitory concentration (IC), which is the calculated concentration of sample required to produce a specific quantitative and qualitative inhibition. The inhibition measured is the quantitative reduction in light output of luminescent marine bacteria (that is, IC20 represents the calculated concentration of sample that would produce a 20 % reduction in the light output of exposed bacteria over a specified time).

1.4 The values stated in SI units are to be regarded as the standard. The values given in parentheses are for information only.

1.5 *This standard does not purport to address all of the safety concerns, if any, associated with its use. It is the responsibility of the user of this standard to establish appropriate safety and health practices and determine the applicability of regulatory limitations prior to use.* Specific hazard statements are given in Section 9.

2. Referenced Documents

2.1 *ASTM Standards:*

D 888 Test Methods for Dissolved Oxygen in Water[6]

D 1125 Test Methods for Electrical Conductivity and Resistivity of Water[6]

D 1129 Terminology Relating to Water[6]

D 1193 Specification for Reagent Water[6]

D 1293 Test Method for pH of Water[6]

D 3370 Practices for Sampling Water[6]

E 729 Guide for Conducting Acute Toxicity Tests with

---

[1] This test method is under the jurisdiction of ASTM Committee D-34 on Waste Management and is the direct responsibility of Subcommittee D34.03 on Treatment.

Current edition approved March 10, 1996. Published May 1996. Originally published as D 5660 – 95. Last previous edition D 5660 – 95.

[2] The boldface numbers in parentheses refer to the list of references at the end of this standard.

[3] Toxicity measured as toxic inhibition of bacterial light output.

[4] Microbics Corp. is currently the only known supplier of the reagents (test organism *Photobacterium phosphoreum* strain NRRL B-11177) specific to this test method. There are two known manufacturers of analyzers that can be used to measure bioluminescence under temperature control: Microbics Corp., 2232 Rutherford Road, Carlsbad, CA 92008 (Microtox Model 500 and Model 2055 Analyzers), and Pharmacia LKB, 9319 Gaither Road, Gaithersburg, MD 20877 (LKB Wallac Model 1250 and Model 1251 Luminometers). Other instruments would be considered when they become available. Please notify ASTM Subcommittee D34.09 if you are aware of any additional systems or instruments capable of performing this testing.

[5] At present (1993) use of the color correction scheme described in this procedure is known to be effective only with the Microbics Corporation's toxicity analyzers, due to the fact that the correction mathematics involve the detailed geometry of both the ACC and the light meter. Please notify ASTM Subcommittee D34.09 if you are aware of any other source of equipment capable of providing color or turbidity correction, or both, for the *P. phosphoreum test*. Data validating the absorbance correction procedure are available from Microbics Corp.

[6] *Annual Book of ASTM Standards*, Vol 11.01.

Fishes, Macroinvertebrates, and Amphibians[7]

E 943 Terminology Relating to Biological Effects and Environmental Fate[7]

E 1192 Guide for Conducting Acute Toxicity Tests on Aqueous Effluents with Fishes, Macroinvertebrates, and Amphibians[7]

3. Terminology

3.1 *Definitions*—The IC20 is defined in terms of a modification of the definition of IC50 as it appears in Terminology E 943. The terms turbidity and volatile matter are defined in accordance with Terminology D 1129. These terms are as follows:

3.1.1 *color*—that is, the presence of dissolved matter that absorbs the light emitted by *P. phosphoreum* (that is, wavelength of 490 ± 100 nm).

3.1.2 *IC20*—a statistically or graphically estimated concentration of test material that, under specified conditions, is expected to cause a 20 % inhibition of a biological process (such as growth, reproduction, or bioluminescence) for which the data are not dichotomous.

3.1.3 *turbidity*—reduction of transparency of a sample due to the presence of particulate matter.

3.1.4 *volatile matter*—that matter that is changed under conditions of the test to the gaseous state.

4. Summary of Test Method

4.1 This test method covers the determination of acute toxicity of aqueous samples to luminescent marine bacteria, *P. phosphoreum*.

4.2 Wastewater samples are osmotically adjusted to the appropriate salinity for the test species *P. phosphoreum*. A sodium chloride (NaCl) concentration of 2 % has been found optimal for this test organism for freshwater tests, or about 3.4 % NaCl for seawater samples. This provides the necessary osmotic protection for the bacteria, which are of marine origin.

4.3 Samples should not be pH adjusted unless the user is not concerned about toxic effects related directly to pH. Altering the sample pH will usually alter the solubility of both organic and inorganic constituents of the sample. Altering the pH can also cause chemical reactions that will change the integrity of the sample, and greatly alter the exhibited toxicity of the sample. If sample pH is considered secondary to organism response, then the optimal pH for the bacterium *Photobacterium phosphoreum* is 6.7.

4.4 Comparison of inhibitory concentrations (IC20s) for untreated wastewater (or extracts of untreated soils) versus those for biologically treated wastewater (or extracts of treated soils), calculated from measured decreases in light output of exposed bacteria, allows for an assessment of the reduction in toxicity to the marine bacterium *P. phosphoreum* (see 1.1, 1.2, and Note 1).

4.5 Samples that are highly colored, or contain solids that cannot be removed without seriously compromising sample integrity, can be analyzed using an absorbance correction procedure. This procedure determines the amount of light absorbed by the wastewater at a concentration near the nominal IC20 versus the baseline light output established by measuring the light absorbed by the clear diluent.

5. Significance and Use

5.1 This test method provides a rapid means of determining the acute toxicity of an aqueous waste, or waste extract, prior to and following biological treatment, and contributes to assessing the potential biodegradability of the waste (see 1.1, 1.2, and Note 1). The change in toxicity to the marine bacterium *P. phosphoreum* with respect to time may serve as an indication of the biodegradation potential. Sample analyses are usually obtained in 45 to 60 min, with as little as 5 mL of sample required **(5)**.

5.2 Samples with high suspended solids concentrations may test nontoxic to the bacteria, while still exhibiting significant toxicity to freshwater organisms, due to those suspended solids.

5.3 The absorbance correction procedure included in this test method allows for the analysis of highly colored lightabsorbing samples, by providing a means for mathematically adjusting the light output readings to account for light lost due to absorption.[5]

6. Interferences

6.1 Some test samples that are highly colored (especially red and brown) interfere with this test method, but the absorbance correction procedure can be used to correct for this interference.[5]

6.2 Turbidity due to suspended solids interferes with this test method. The absorbance correction procedure can be used to correct for this interference and is preferable to other alternatives. Pressure filtration, or centrifuging and decanting, will also remove this interference. Some toxics may be lost through adsorption and volatilization during filtration or centrifugation, thus impacting the exhibited toxicity.[5]

7. Apparatus

7.1 *Fixed or Adjustable Volume Pipetter*, 10 µL, with disposable tips.

7.2 *Variable Volume Pipetter*, 10 to 1000 µL, with disposable tips.

7.3 *Variable Volume Pipetter*, 1 to 5 mL, with disposable tips.

7.4 *Timer or Stopwatch*.

7.5 *Glass Cuvettes*, 11.75 mm OD, 10.5 mm ID by 50 mm height, 4-mL volume.

7.6 *Absorbance Correction Cuvettes (ACC)*—Optional item, but required to analyze highly colored samples or those containing suspended particulates.[5]

7.7 *Variable Voltage Chart Recorder (optional)*—Useful when using some types of light meters.

7.8 *Computer (optional)*—Useful with some light meters, for which software is also available, to facilitate data capture and reduction.

7.9 *Light Meter*, for cuvettes listed in 7.5.[4,5]

7.10 *Temperature Control Devices* (temperature-controlled room, water bath, refrigerators, or other device)—One capable of maintaining 5.5 ± 1°C and one capable of maintaining 15 ± 0.5°C.

[7] *Annual Book of ASTM Standards*, Vol 11.05.

8. Reagents and Materials

8.1 *Test Reagents*:

8.1.1 For purposes of this test method, test reagents are defined as the reagents actually used in performance of the test method. The necessary requirement with regard to qualification of test reagents is that this test method provide acceptable results when reference toxicants are tested using the test reagents. They are then considered to be non-toxic for purposes of this test method.

8.1.2 *Microbial Reagent*—Freeze-dried *Photobacterium phosphoreum*. This is the only test reagent that is currently (1993) available from only one source.[4] While other acceptable means of preservation may become available in the future, freeze-dried *P. phosphoreum* is specified in this test method because a large number of users concur in the opinion that the strain is well standardized by this method of preservation, and that the same strain does not provide comparable response to reference toxicants when preserved by other methods, or when freshly cultured and harvested at the user's laboratory, as described by Anthony A. Bulich, et al **(1)**. Another consideration is that a large body of published results, for which freeze-dried *P. phosphoreum* was used, has accumulated since about 1980 **(1,2,3,5,6)**.

8.1.3 *Reconstitution Solution*—Nontoxic water.

8.1.4 *Diluent*—Nontoxic 2 % sodium chloride (NaCl), or 3.4 % NaCl, reconstituted seawater or sea water (depending upon the type of sample and purpose of the test). The *P. phosphoreum* test has been performed at osmotic pressures equivalent to 1 to 6 % NaCl, but has long been standardized at 2 % for freshwater samples. The major requirement is that the osmotic pressure be held constant within each test, to minimize transient variations in luminescence due to variations in osmotic pressure. The higher salinity (and osmotic pressure) of marine samples dictate the use of a diluent other than 2 % NaCl. Both reconstituted seawater and clean seawater have been used as diluent. A procedure for preparing reconstituted salt water, and formula, are given in Table 3 of Guide E 729. Actual seawater has also been collected at remote sites and used as diluent for testing aqueous samples of marine origin. The most important requirement is that the diluent must be qualified for use with this test method (see 8.1.1).

8.2 *Reagent Chemicals*—Reagent grade chemicals are recommended for use in preparation of test reagents and reference toxicants. Unless otherwise indicated, it is intended that all reagents shall conform to the specifications of the Committee on Analytical Reagents of the American Chemical Society.[8] Other grades may be used, but there will be more risk that the resulting test reagents will fail to qualify (see 8.1.1).

8.2.1 *Sodium Chloride (NaCl)*—Used in preparation of diluent, and for adjusting the osmotic pressure of samples to that of the chosen diluent.

8.2.2 *Phenol, or Other Common Organic Toxicant*—Used as a reference toxicant.

8.2.3 *Zinc Sulfate Heptahydrate, or Other Common Inorganic Toxicant*—Used as reference toxicant.

8.3 *Purity of Water*— Unless otherwise indicated, references to water shall be understood to mean reagent water conforming to Specification D 1193, Reagent Water, Type I or II, Subtype A. Test reagents prepared from reagent water are to be qualified for use with this test method (see 8.1.1).

8.4 When this test method is used in conjunction with other tests employing higher organisms, appropriate dilution water for bulk samples should meet the acceptability criteria established in Section 8 of Guide E 729. In addition, all such dilution water used for comparative testing with this test method and invertebrates and fish is to be assayed on *P. phosphoreum* (minimally once per month).

9. Hazards

9.1 The handling of wastewaters entails potential hazards due to exposure to chemical and biological contaminants. Appropriate safety measures, such as the wearing of protective clothing (gloves, apron, face shield, respirator, etc.) and maintaining proper hygiene, are utilized to minimize the chance of exposure. This test method is to be performed in a well-ventilated area.

9.2 Appropriate, environmentally safe procedures prescribed by regulatory agencies are utilized in the disposal of used waste samples.

9.3 Due to the presence of aqueous samples and electrical instrumentation in close proximity, care must be taken to prevent electrical shock.

10. Technical Precautions

10.1 Osmotic adjustment of freshwater test samples, to 2 % sodium chloride concentration, is required due to the use of a marine bacterium as a test organism. Osmotic adjustment may make some components of a wastewater less soluble, reducing concentrations in solution and altering exhibited toxic inhibition.

10.2 Samples containing highly volatile components are to be handled as little as possible to reduce losses due to stripping. Mixing procedures (see 13.8.4) are modified by performing only one pipet mixing per sample dilution versus the usual five pipet mixings. Volatile samples, which can be analyzed by UV spectrophotometry, allow the investigator to measure the average sample concentration of volatiles over the actual test period.

10.3 The addition of any preservative or other chemical agent, including acid or base to alter pH, will in all likelihood impact the exhibited toxicity of the sample. These practices should be avoided in most cases, unless the user is specifically testing to determine the effects of these sample modifications.

10.4 The use of a reference toxicant, such as phenol or zinc sulphate, is recommended for validation of data produced with different lots of test reagents (that is, bacteria, reconstitution solution, and diluent) or for individual lots used over an extended period of time. A good practice is to perform a reference toxicant analysis with each new lot of bacterial reagent received and new lots of test reagents prepared (or purchased). Under normal conditions, with reagent in good

---

[8] *Reagent Chemicals, American Chemical Society Specifications*, American Chemical Society, Washington, DC. For suggestions on the testing of reagents not listed by the American Chemical Society, see *Analar Standards for Laboratory Chemicals*, BDH Ltd., Poole, Dorset, U.K., and the *United States Pharmacopeia and National Formulary*, U.S. Pharmaceutical Convention, Inc. (USPC), Rockville, MD.

condition, tests on phenol produce an IC50 (5 min) between 13 and 26 mg phenol/L, and tests of zinc sulfate heptahydrate produce an IC50 (15 min) between 5 and 12 mg $ZnSO_4 \cdot 7H_2O$/L (or, 1.1 to 2.7 mg Zn/L). The corresponding nominal ranges are IC20 (5 min) = 3 to 6 mg phenol/L and IC20 (15 min) about 1.5 to 4.5 mg $ZnSO_4 \cdot 7H_2O$/L (or, 0.34 to 1.02 mg Zn/L).

10.5 In order to verify that changes in observed toxicity are due to treatment, it is essential to have control samples for biodegradation test systems. Typical controls would be sterilized (autoclaved) waste samples. These samples undergo toxicity assessment for comparison with the treated samples; that is, they undergo the same physical manipulations and testing as the inoculated or nutrient-enhanced treatment systems, but all microbial action has been terminated by sterilization at the outset of the test series. It is necessary to compare the toxicity (IC20s) of autoclaved and untreated samples immediately after autoclaving in order to determine changes due to autoclaving (**3**). Autoclaving of samples for use as control samples requires special consideration and sample handling techniques. The following procedure is recommended:

10.5.1 Completely fill new borosilicate jars with sample, and seal them with autoclavable caps having TFE-fluorocarbon liners, to minimize loss of volatile toxicants during autoclaving.

10.5.2 Soil and sediment samples are to be moist, for optimal effectiveness of autoclaving.

10.5.3 Bring the autoclave to 121°C and hold the sample jars there for one to two hours, then turn off the heat and allow the autoclave to cool very slowly, to avoid large transient positive pressure inside the jars, which might cause them to fracture.

10.5.4 It is recommended that the autoclaving be repeated 24 h later as a precaution against survival of spores. In addition, or alternatively, commercially available spore strips or preparations may be added to a jar of soil and included in the autoclave load as a direct means of validating the effectiveness of the autoclave cycle.

11. Sampling

11.1 Collect aqueous samples in accordance with Practices D 3370. Soil and other solid material samples, for aqueous extraction, should be sampled in such a way as to reduce the risk of loss of volatile components.

11.2 All sample containers (vials or bottles) should be made of borosilicate glass that has been thoroughly cleaned using a nontoxic soap wash, HCl wash, and water rinse (twice). All sample containers should be sealed with TFE-fluorocarbon-lined caps.

11.3 Prepare all dilutions required for a single toxicity evaluation from the same treated or untreated wastewater sample. Portions of the sample shall be stored, until needed, at a temperature of 2 to 4°C in completely filled, tightly stoppered borosilicate-type glass containers. TFE-fluorocarbon-lined caps are used to seal collection bottles to minimize adsorption or sample contamination.

11.4 Uniformly disperse (by mild agitation), any undissolved material present in a wastewater sample, before withdrawing a measured portion for osmotic adjustment and subsequent analysis. Undissolved material, which will interfere with light transmission during analysis, should be adjusted for or removed from the osmotic pressure-adjusted sample as described in Section 6. Avoid violent agitation and unnecessary exposure of the sample to the atmosphere.

12. Calibration and Standardization[9]

12.1 Use the procedure specified by the manufacturer of whatever light-measuring instrument is being utilized. The procedure should include a mechanism for zeroing the instrument for no light production and a procedure for setting the output range.

12.2 If a chart recorder is being used, it should be calibrated against either the digital reading of the photometer or the voltage output of the photometer to the recorder.

13. Procedure [10]

13.1 Samples taken from a treatment process are collected using an ASTM acceptable sampling procedure (see Section 11).

13.2 For aqueous samples, visually evaluate the sample for suspended particulates and color. Both of these factors can interfere with measured light output readings. If either of these conditions is present use one of the methods described in 6.2 to remove or account for the interference.

13.3 For solid phase samples prepare the test sample as follows:

13.3.1 Wet sediment should be centrifuged to separate the pore water. Centrifuge 50 to 100 g of sediment at 2000 to 4000 g, for 10 to 20 min at 4°C. Decant the pore water and use the resulting pellet of solids as if it were a soil sample.

13.3.2 Homogenize 10 to 50 g of representative soil sample by hand mixing with a spatula for 10 min.

13.3.3 Weigh a representative 3 to 5-g portion of the homogenized sample to the nearest 0.01 g, then air dry at 20 to 25°C for 16 h. After drying, reweigh the dried sample.

13.3.4 Take a 2-g sample from the homogenized soil or sediment and add 20 mL of the appropriate diluent.

13.3.5 Mix the soil/diluent mixture for 16 h using an orbital shaker set at 200 r/min.

13.3.6 Centrifuge the sample at 2000 to 4000 g, for 10 to 20 min at 4°C.

13.3.7 Decant 10 to 15 mL of the aqueous phase for use in the analysis of toxic inhibition.

13.4 Positive pressure filtration (through a prerinsed, glass-fiber filter) can be used to remove suspended solids, while minimizing loss of volatile organics. Rinsing the filter with nontoxic water, prior to sample filtration, reduces organic leaching from the filter. Note the potential sample alterations mentioned in 6.2.

13.5 Take 5 mL of the aqueous sample from 13.2 to 13.3 and measure the pH (Test Methods D 1293), dissolved oxygen (DO) (Test Methods D 888), conductivity (Test Methods D 1125), and salinity.

---

[9] Calibration and standardization procedures will vary depending on the instrument being used to measure the bacterial light output.

[10] This is a generic procedure that will require modification depending on the particular instrument being used to measure microbial light output.

13.6 Adjust the sample salinity to either 2 % NaCl or 3.4 % NaCl (for samples of marine origin) by adding sodium chloride to 10 mL of sample. Adjust the pH and DO only if those factors are not concerns in the process under investigation. Be aware of the potential changes in overall sample chemistry as noted in 6.2.

13.7 If the user is adjusting the sample pH to determine the effect thereof, the acid or base, or both, used for the adjustment should be noted, and the quantity required in the adjustment should be recorded. Sample dilution and chemical species changes must be taken into account if pH adjustment is necessary.

13.8 Samples of unknown toxicity are screened, prior to definitive testing, using the following range finding procedure:

13.8.1 Prepare a cuvette of bacterial reagent (*Photobacterium phosphoreum*) by adding 1 mL of nontoxic water at 5.5 ± 0.5°C to a bottle of lyophilized luminescent bacteria and transferring the reconstituted bacteria to a cuvette maintained at 5.5 ± 0.5°C.

13.8.2 Prepare 10 test cuvettes, by adding 0.5 mL of diluent and 10µ L of reconstituted bacteria. Maintain the test cuvettes at 15 ± 0.5°C.

13.8.3 Without waiting the normal 15-min temperature acclimation period, place one of the test cuvettes of bacteria into the photometer, and measure the light output for 10 to 20 s. If the instrument used allows the output value to be adjusted, adjust the output to read 90 units. Otherwise record the output value as it is.

13.8.4 Add l0 µL of the unknown sample to the cuvette being measured. Mix the contents with a 250-µL pipet by aspirating and dispensing its full volume five times, or as an alternative, mix the contents by briskly flicking the cuvette with a finger (cuvette flicking method).

13.8.5 Measure the light output of the exposed bacteria for 10 to 20 s.

13.8.6 If the loss of light output is greater than 20 % within several minutes, dilute the sample ten-fold, and repeat 13.8.3-13.8.5 with one of the unused cuvettes prepared in 13.8.2 using the diluted sample. Repeat this procedure until a sample dilution produces a loss of light of less than 20 % during the first few minutes after sample addition. Observe the bacterial response for 5 min, and then estimate graphically the 5-min toxic response. This information gives the tester a good indication of the sample concentration range which will produce a statistically sound IC20, if the sample is toxic to that extent.

13.9 The procedure for running a toxicity test using *Photobacterium phosphoreum* is as follows:

13.9.1 Place 20 clean cuvettes in a temperature-controlled area at 15 ± 0.5°C, and one additional clean cuvette at 5.5 ± 1°C. Set the cuvettes in two rows of ten, and use a labeled test tube rack or other device to identify the cuvettes as A1–A10 and B1–B10.

13.9.2 Add 1 mL of nontoxic water to the cuvette being held at 5.5°C.

13.9.3 Add the appropriate amount of diluent to Cuvettes A1–A9 (being maintained at 15°C) to obtain the desired concentrations after serial dilution (for example, for a 2:1 serial dilution, 1.5 mL of diluent is added to A1–A9). Cuvette A10 is left empty for the primary sample concentration.

13.9.4 Add 0.5 mL of diluent to Cuvettes B1–B10 (which serve as the test cuvettes).

13.9.5 Add 1.5 mL of the osmotically adjusted primary sample concentration (diluted or not) to Cuvette A10, and an appropriate amount to A9. Mix the diluted contents of A9 by aspirating and dispensing, by pipette, 500 µL of sample five times; or by briskly flicking the cuvette with a finger. Complete the serial dilution of the test sample by transferring an appropriate volume of A9 to A8 and A8 to A7 . . . A3 to A2, using one of the mixing methods previously described. In the example of a 2:1 serial dilution scheme, the dilution would be performed as follows: 1.5 mL of 100 % sample (note that the actual concentration is 91 to 100 % depending on the need for and method of salinity adjustment) added to Cuvettes A10 and A9 and mix A9, 1.5 mL of A9 to A8 and mix, 1.5 mL of A8 to A7 and mix, 1.5 mL of A7 to A6 and mix, 1.5 mL of A6 to A5 and mix, 1.5 mL of A5 to A4 and mix, 1.5 mL of A4 to A3 and mix, 1.5 mL of A3 to A2 and mix, and remove and discard 1.5 mL of A2.

13.9.6 Allow 5 to 10 min for samples to reach thermal equilibrium, then check to verify that the temperature of the reconstitution solution is 5.5 ± 1°C and that the test cuvettes have reached 15 ± 0.5°C.

13.9.7 While the prepared test cuvettes are temperature equilibrating, remove a vial of lyophilized bacteria from refrigeration and rapidly add the precooled 1-mL volume of reconstitution solution into the vial, swirl the vial to mix, and return the reconstituted bacteria to the cuvette which is replaced at a temperature of 5.5 ± 1°C. Mix the reconstituted bacteria by aspirating and dispensing 0.5 mL of solution, by pipet, 20 times. The reagent dilution is started within 5 min of bacterial reconstitution, in order to maintain maximum sensitivity.

13.9.8 Transfer 10 µL of reconstituted bacterial reagent to each Cuvette B1 through B10. Wipe the pipet tip of excess reagent before each transfer. Mix the contents of each cuvette using a 250-µL pipet to aspirate and dispense the solution five times, or by the cuvette flicking method.

13.9.9 Allow the bacteria in the test cuvettes to achieve a stable light output level by remaining undisturbed at 15°C for 15 min. This allows the bacteria to recover from the shocks of reconstitution, shift in temperature, and dilution of nutrients.

13.9.10 Cycle the cuvettes through the photometer, and adjust the light output levels to read between 80 and 100 units if possible (some units will automatically perform this task with the initial I_0 light readings). Cuvette output reading is performed in the order of B1, B2, B3 ... B10.

13.9.11 Take the initial (I_0) readings by cycling the cuvettes, one cuvette every 25 s, and recording the light output of each cuvette (B1 through B10) for 5 s. Record the time with each reading so that the 5, 15, and 30-min exposure periods are accurately timed.

13.9.12 Start the addition of the test samples (Cuvettes A1–A10) to the test cuvettes (Cuvettes B1–B10) immediately following the reading of the light output of Cuvette B10, the last cuvette in the cycle. The addition starts with 0.5 mL of A1 (the nontoxic blank) added to Cuvette B1, mixing the sample by the pipet technique or flicking technique. The sample additions proceed from low concentration to high concentration, adding 0.5 mL of A2 to B2 and continuing up to A10 to B10, allowing 25 s between each sample addition. The time of each addition is recorded so that the light output of each challenged test cuvette can be measured 5, 15, and 30 min after the sample addition.

13.9.13 The test cuvettes (B1 through B10) are cycled through the photometer 5 min after the sample additions and the light output of the bacteria is recorded for each cuvette. This procedure is repeated at 15 and 30 min to observe any time-dependent increases in toxic inhibition (that is, toxicity due to metals).

13.9.14 The recorded light outputs are used to calculate IC values by plotting or mathematical determination.

13.10 The procedure used to correct for absorbance in highly colored aqueous samples, as described in 6.1, is as follows:

13.10.1 Pipet 1.5 mL of diluent into the outer chamber of a clean absorbance correction cuvette (ACC) and place it in the photometer.

13.10.2 Pipet 1.0 mL of diluent into a standard cuvette (A1) and place it at 15°C.

13.10.3 Pipet 2.0 mL of sample of chosen concentration C_c (the concentration closest to the nominal ICxx) into each of two standard cuvettes (A2 and A3), and place them at 15°C.

13.10.4 Allow 10 min for the solutions to reach thermal equilibrium.

13.10.5 Pipet 50 μL of reconstituted bacterial reagent into Cuvette A1. Mix five times with a 500-μL pipet or flick the cuvette briskly.

13.10.6 Remove the ACC from the photometer long enough to transfer a sufficient amount of bacterial solution from cuvette A1 into the inner chamber of the ACC to get a volume level equal to that of the diluent level in the outer chamber.

13.10.7 Return the ACC to the photometer. Adjust the light output reading of the ACC to 90 units (if possible), then record the light output for 10 to 20 min until a stable baseline or steady drift baseline is established.

13.10.8 Using a clean aspirator, remove the diluent from the outer chamber of the ACC while the ACC remains in the photometer.

13.10.9 Remove as much of the diluent as possible with an aspirator. Transfer 1.5 mL of test sample from Cuvette A3 into the outer chamber of the ACC.

13.10.10 Record the light output for 10 min or more. The light levels recorded for the sixth through tenth minute will be used in data reduction.

14. Calculation

14.1 The following equations are used to determine 20 % inhibitory concentrations (IC20s) from light output readings produced using the methods described in Section 13:

14.1.1 Calculate the blank ratios (which will be used to normalize the Γ responses calculated in 14.1.2) for 5, 15, and 30 min, using the following equations:

$$R(t) = I(t)b/I(0)b \tag{1}$$

where:

$R\ (t)$ = blank ratio for time t,
$I\ (0)b$ = initial light reading for the blank cuvette (zero time, just before transferring toxicants), and
$I\ (t)b$ = final light reading for the blank cuvette (t min after transferring toxicants).

14.1.2 Calculate the 5, 15, and 30-min gamma responses, Γ (t), for each of the eight test cuvettes, normalized for reagent pipetting errors and normal drift of luminescence with time, using the following equation:

$$\Gamma(t) = \text{Light Lost/Light Remaining}$$
$$= [R(t)I\ (0) - I(t)]/I(t)$$
$$= [R(t)I\ (0)/I(t)] - 1 \tag{2}$$

where:

$I\ (0)$ = initial light reading for any given test cuvette at zero time, just before challenging the organisms,
$I(t)$ = light reading for the corresponding test cuvette at time (t),
$R(t)$ = blank ratio for time (t) as defined in 14.1.1, and
$\Gamma(t)$ = Γ effect calculated for each exposure time (t); that is, at 5, 15, and 30 min.

It should be noted that $1n\ \Gamma(t) = 1n\ (D/(1 - D))$ (see 14.1.4) is identical to Berkson's logit P/Q = logit $P/(1 - P)$ **(7)**. The method described in this test method is, therefore, a logit analysis.

14.1.3 Use linear regression[11] of $1n\ \Gamma(t)$ on $1\ n\ C$, with $1n\ \Gamma(t)$ as the dependent variable, to obtain the $1\ n$-$1n$ regresssion equation,

$$1n\ \Gamma(t) = b(1n\ C) + 1n\ a \tag{3}$$

then solve this equation for $1n\ C$ to obtain the estimating equation,

$$1n\ C = (1/b)[1n\ \Gamma(t)] - [1n\ a] \tag{4}$$

where:

C = concentration of sample,
$1n\ a$ = intercept of the $1\ n$-$1n$ regression line with the ordinate $1n\ C = 0$, which will be a constant number, but different for each exposure time (5, 15, and 30 min),
b = slope of the $1n$-$1\ n$ regression line, which will also be a constant number, but different for each exposure time (5, 15, and 30 min), and
$\Gamma(t)$ = toxic responses for corresponding concentrations, for each exposure time (5, 15, and 30 min).

14.1.4 In order to find IC20s, solve the above estimating equation for C when $\Gamma(t) = 0.25$, corresponding to 20 % reduction of light output (see 1.3), for 5, 15, and 30-min data. These concentrations (Cs) are the IC20s for 5, 15, and 30 min, respectively. The relationship between Γ and percent reduction of light output (% D) is:

[11] Standard regression analysis should be used, with care given to make certain that the quality of the data warrants the conclusions drawn. The estimating equation reserves the variables compared to the conventional dose response curve to facilitate solution of the equation for C for a specified Γ. This estimating equation is simply the regression equation rearranged to make $1n\ C$ a function of $\Gamma(t)$.

$$\Gamma = \% D/(100\,\% - \% D) \quad \text{or} \quad \% D = 100\,\% \times \Gamma/(1 + \Gamma) \tag{5}$$

It may be easily seen that IC20 (that is, % D = 20 %) corresponds to $\Gamma = 20\,\%/(100\,\% - 20\,\%) = 20\,\%/80\,\% = 0.25$. The estimating equation must be satisfied by these corresponding values of C and Γ. Substituting these specific values into the estimating equation results in the following:

$$\ln(IC20) = 1/b\ln(0.25) + 1/b\ln a = 1/b(-1.3863) + 1/b\ln a \tag{6}$$

Once the right side of the equation is reduced to a single number, say N, IC20 is the antiln of N. The antiln (N) is simply e^N, where e = 2.7182818...; that is, the base of the natural logarithms.

14.2 The following equations use data obtained in 13.9 and 13.10 to determine corrected light loss when a sample is highly colored and light absorbing or highly turbid, or both.[12]

14.2.1 Considerable labor can be saved when it is possible to calculate the values of A_c for all sample concentrations (C) based upon measurement of only one concentration (C_c) in the ACC, using the equation given in 14.2.2. When the sample is such that this approach is not applicable,[13] determine A_c for each concentration that yielded a significant Γ (that is Γ between 0.02 and 100) by direct measurement with each such concentration in the ACC. The equation in 14.2.2 must then be solved for each set of ACC data, I_0/I_F, with $C/C_C = 1$ in each case. It should be noted that A_C is considered to be zero for concentrations having Γ responses of 0.02 or less.

14.2.2 When applicable (see 14.2.1),[12] calculate absorbance due to color (A_C) for the ACC for all concentrations (C) of sample tested in the toxicity cuvettes which gave significant Γ responses, using the following equation:

$$A_C = (C/C_C)\,[3.1\ \ln(I_0/I_F)] \tag{7}$$

where:

A_C = calculated absorbance expected if concentration C were to be measured in the ACC, for each concentration tested in the toxicity test which gave a significant Γ.[12] (Alternatively, each A_C is calculated using I_0 and I_F results from direction measurements in the ACC.)
I_0 = initial light level, measured in the ACC (for diluent),
I_F = final light level, measured in the ACC (for C_C),
C_C = chosen concentration measured in the ACC (in 13.10),
C = each sample concentration tested in the toxicity cuvette, which gave a significant Γ (that is, 0.02 or larger), and
3.1 = composite factor for the ACC which corrects for geometrical differences between it and the standard test cuvette.[5]

14.2.3 Calculate the transmittance (T_C) of the toxicity cuvette for each sample concentration tested that gave a significant Γ, using the following formula:

$$T_C = (1 - e^{-A_C})/A_C \tag{8}$$

where:

T_C = unity (that is, 1.00) for concentrations having insignificant Γ responses, corresponding to A_C = zero.

14.2.4 Calculate the corrected gamma responses ($\Gamma_C(t)$) for 5, 15, and 30-min data for each concentration tested, using the following equation:

$$\Gamma_C(t) = T_C(1 + \Gamma(t)) - 1 \tag{9}$$

where:

$\Gamma(t)$ = Γ response observed for each concentration (C) in the test, at each test time (5, 15, and 30 min), and
$\Gamma_C(t)$ = color-corrected toxic response for each test time (5, 15, and 30 min).

14.2.5 Determine the color-corrected IC20 ($IC20_C$) for 5, 15, and 30-min data as described in 14.1.3, using the $\Gamma_C(t)$ values determined in 14.2.3 for each exposure time.

14.3 The following equation is used to correct the IC20s determined for soil and sediment samples in either 14.1.3 or 14.2.4 (if color/turbidity corrected) to dry-weight basis. The wet and dry weights of a representative soil/sediment sample were determined in 13.3.3.

$$IC20(t)_{DRY} = IC20(t)_{WET} \times \text{(dry weight)/(wet weight)} \tag{10}$$

15. Data Interpretation

15.1 *Choice of Exposure Time*—the exposure time of choice is, in general, that which provides the greatest sensitivity. However, the IC20 having the smallest 95 % confidence interval may be preferred in cases in which the confidence interval varies appreciably with time of exposure. Consistency of choice between control samples and treated samples is of major importance for comparative studies. Finally, it should be noted that organics generally cause fast (5 to 10 min) response, while some metals continue to affect the luminescence of *P. phosphoreum* beyond 30 min. The changes in relative IC20 for the various exposure times as treatment progresses may, therefore, provide some additional information with regard to progress of treatment or further treatability, or both.

15.2 Compare the IC20 values (calculated concentration at $\Gamma = 0.25$) for the treated and untreated sample. Any toxicity reduction of 20 % or more, compared to the untreated system control sample or the raw starting material, is considered to be significant and a potential indication of biodegradability (see 1.1, 1.2, and Note 1).

15.3 Care must be taken to account for toxicity reduction that is not due to biodegradation (that is, adsorption, volatilization, and sample preparation errors). Control samples not exposed to biodegradation are essential as part of the data validation process (see 10.5).

16. Report

16.1 The record of the test and published reports of the

[12] In samples where absorbance due to concentration does not behave in accordance with Beer's Law or the samples causing significant Γ responses (0.02 or larger) are turbid, or both, it is necessary to directly measure the absorbance in the ACC for each sample concentration toxicity tested that gave a significant Γ response, by this test method. If desired, verify conformance to Beer's Law by performing the test in 13.10 with a second concentration in the ACC, for example, the highest concentration of interest. Using the equation in 14.2.1, calculate two values of A_c for the lower concentration using both ACC results. If the ratio of the two A_c values is between 0.98 and 1.02, the deviation from Beer's Law is within acceptable limits.

results of the test should contain the following information:

16.1.1 Name of test, investigator, and laboratory; and the date the test was conducted;

16.1.2 Detailed description of the test sample including its source (detail biodegradation system used), composition (identity and concentration of major ingredients and major impurities), known physical and chemical properties, and identity and concentration of any solvents or other additives used;

16.1.3 The source of the dilution water, its chemical characteristics, and a description of any pretreatment;

16.1.4 Detailed information about the reagents used, including lot number, date received, reference toxicant data for the reagent lot, and any noted abnormalities;

16.1.5 A detailed description of the toxic inhibition analysis performed on the sample, including the test date, exposure times, test temperature, pH of sample before and after testing, all parametric data about sample, observations during test, and data reduction results (see 1.1, 1.2, and Note 1).

17. Precision and Bias

17.1 Quality data are produced when test procedures are followed as stated. The greatest source of error will be due to operator error. Errors are most likely to occur during sample preparation, salinity adjustment, filtration (if required), sample dilution, reagent dilutions, sample transfer and mixing steps, and data interpretation and resulting calculations. Use of the proper equipment and development of the appropriate skills required for using the test equipment are necessities in producing quality data.

17.2 Precision of the data may be improved by running a split sample duplicate analysis, repeating the procedures listed in 13.9 with the duplicate sample. Duplicate analyses can be performed simultaneously, or the duplicate sample can be analyzed separately. The duplicate sample must be protected from incurring further biodegradation or other physical/chemical changes. The results of the duplicate analyses are compared for any irregularities (obvious differences) in response versus exposure concentration. If such irregularities are noted, the sample should be retested if at all possible.

17.3 The raw data generated by the test procedures will determine whether an IC20 can be calculated with reasonable accuracy.

17.4 The determination of 95 % confidence intervals, using an acceptable procedure, will assist the investigator in determining the quality of generated IC20s (computer programs are available to perform these calculations).

17.5 An interlaboratory comparison study **(5)** was conducted on the toxic inhibition procedure described in this test method. The study involved 18 laboratories in four round robins, during which a total of six blind samples (five toxic and one nontoxic) were analyzed. The coefficient of variation (CV) ranged from 14.29 to 18.57 for the pooled data set, while the overall CV (regardless of sample) was calculated to be approximately 17.8 %.

17.6 The lack of an internal standard for this test method makes it impossible to determine the bias.

18. Keywords

18.1 bioluminescence; bioremediation; contaminated soil; contaminated water; detoxification; marine bacterium; toxicity

REFERENCES

(1) Bulich, A. A., Greene, M. W., and Isenberg, D. L., "Reliability of the Bacterial Luminescence Assay for Determination of the Toxicity of Pure Compounds and Complex Effluents," *Aquatic Toxicology and Hazard Assessment: Fourth Conference*, ASTM, STP 737, D. R. Branson and K. L. Dickson, Eds., ASTM, 1981, pp. 338-347; Quareshi, A. A., Flood, K. W., Thompson, S. R., Janhurst, S. M., Inniss, C. S., and Rokosh, D. A., "Comparison of a Luminescent Bacterial Test with Other Bioassays for Determining Toxicity of Pure Compounds and Complex Effluents," *Aquatic Toxicology and Hazard Assessment: Fifth Conference*, ASTM STP 766, J. G. Pearson, R. B. Foster, and W. E. Bishop, Eds., ASTM, 1982, pp. 179–195.

(2) Kaiser, K. L. E., and Palabrica, V. S.," Photobacterium Phosphoreum Toxicity Data Index," *Water Pollution Research Journal Canada*, Vol 26, No. 3, pp. 361–431, 1991.

(3) Matthews, J. E., and Bulich, A. A., "A Toxicity Reduction Test System to Assist in Predicting Land Treatability of Hazardous Organic Wastes," Presented at ASTM D-34, Washington, DC, 1984; Chang, J. C., Taylor, P. B., and Leach, F. R., "Use of the Microtox® Assay System for Environmental Samples," *Bulletin of Environmental Contamination and Toxicology*, Vol 26, 1981, pp. 150–156; Matthews, J. E., and Hastings, L., "Evaluation of Toxicity Test Procedure For Screening Treatability Potential of Waste in Soil," *Toxicity Assessment: An International Quarterly*, Vol 2, 1987, pp. 265–281, copyright John Wiley and Sons, Inc.

(4) For example, see: Pelitier, W. H., and Weber, C. I., "Methods for Measuring the Acute Toxicity of Effluents to Freshwater and Marine Organisms," EPA/600/4-85/013, EMSL—ORD, Cincinnati, OH, March 1985; Weber, C. I., et al, "Short-Term Methods for Estimating the Chronic Toxicity of Effluents and Receiving Waters to Freshwater Organisms," EPA/600/4-89/001, EMSL— ORD, Cincinnati, OH, March 1989; Weber, C. I., et al, "Short-Term Methods for Estimating the Chronic Toxicity of Effluents and Receiving Waters to Marine and Estuarine Organisms," EPA/600/4-87/028, EMSL—ORD, Cincinnati, OH, March 1988.

(5) Casseri, N. A., Ying, W., and Sojka, S. A., "Use of a Rapid Bioassay for Assessment of Industrial Wastewater Treatment Effectiveness," *Proceedings of the 38th Purdue Industrial Waste Conference*, Butterworth Publishers, Stoneham, MA, 1983, pp. 867–878.

(6) Qureshi, A. A., et al, "Microtox Interlaboratory Comparison Study (MICS)," Presented at the Third International Symposium on Toxicity Testing Using Microbial Systems, Valencia, Spain, May 1987.

(7) Berkson, Joseph, "A Statistically Precise and Relatively Simple Method of Estimating the Bioassay with Quantal Response, Based on the Logistic Function," *American Statistical Association Journal*, September 1953, p. 565.

Designation: D 5679 – 95a (Reapproved 2001)

Standard Practice for Sampling Consolidated Solids in Drums or Similar Containers[1]

This standard is issued under the fixed designation D 5679; the number immediately following the designation indicates the year of original adoption or, in the case of revision, the year of last revision. A number in parentheses indicates the year of last reapproval. A superscript epsilon (ϵ) indicates an editorial change since the last revision or reapproval.

1. Scope

1.1 This practice covers typical equipment and methods for collecting samples of consolidated solids in drums or similar containers. These methods are adapted specifically for sampling drums having a volume of 110 U.S. gal (416 L) or less. These methods are applicable to hazardous material, product, or waste. Specific sample collection and handling requirements should be described in the site-specific work plan.

1.2 The values stated in inch-pound units are to be regarded as the standard. The values given in parentheses are for information only.

1.3 *This standard does not purport to address all of the safety concerns, if any, associated with its use. It is the responsibility of the user of this standard to establish appropriate safety and health practices and determine the applicability of regulatory limitations prior to use.*

2. Referenced Documents

2.1 *ASTM Standards:*

C 702 Practice for Reducing Samples of Aggregate to Testing Size[2]

C 783 Practice for Core Sampling of Graphite Electrodes[3]

D 4547 Practice for Sampling Waste and Soils for Volatile Organics[4]

D 4687 Guide for General Planning of Waste Sampling[4]

D 4700 Guide for Soil Sampling from the Vadose Zone[5]

D 5088 Practice for the Decontamination of Field Equipment Used at Non-Radioactive Waste Sites[5]

D 5283 Practice for Generation of Environmental Data Related to Waste Management Activities: Quality Assurance and Quality Control Planning[4]

2.2 *NSC Document:*

Accident Prevention Manual for Industrial Operations, 1985[6]

2.3 *Government Documents:*[7]

Drum Handling Practices at Hazardous Waste Sites, EPA/600/2-86/013, January 1986

Occupational Safety and Health Guidance Manual for Hazardous Waste Site Activities, National Institute for Occupational Safety and Health (NIOSH), Occupational Safety and Health Administration (OSHA), U.S. Coast Guard (USCG), and U.S. Environmental Protection Agency (EPA), October 1985

3. Terminology

3.1 *Definitions:*

3.1.1 *bonding*—touching the sampling equipment to the drum to form an electrically conductive path to minimize potential electrical differences between the sampling equipment and the drum, reducing the buildup of static electricity.

3.1.2 *bung*—usually a 2-in. (5.1-cm) or ¾-in. (1.3-cm) diameter threaded plug designed specifically to close a bung hole.

3.1.3 *bung hole*—an opening in a barrel or drum through which it can be filled, emptied, or vented.

3.1.4 *consolidated*—the characteristic of being cemented or compacted, or both, and not separated easily into smaller particles.

3.1.5 *deheading*—removal of the lid of a closed-head drum; usually accomplished with a drum deheader.

3.1.6 *drum*—implies any drum, barrel, or non-bulk container of 5 to 110 U.S. gal (19 to 416 L) capacity.

3.1.7 *pail*—a small container, usually with a capacity of 5 U.S. gal (19 L). Pails typically have bungs or spouts, or the entire lid can be removed.

3.1.8 *paperwork*—all required site documentation, which may include the manifests, waste profiles, material safety data sheets (MSDS), site forms, sample labels, custody seals, and chain of custody forms.

3.1.9 *work plan*—a plan, specific to a particular site, for conducting activities specified in the plan.

4. Summary of Practice

4.1 The drum and its contents are inspected, and appropriate sampling equipment is selected. A clean device is then used to

---

[1] This practice is under the jurisdiction of ASTM Committee D34 on Waste Management and is the direct responsibility of Subcommittee D34.01.02 on Sampling Techniques.

Current edition approved June 15, 1995. Published August 1995. Originally published as D 5679 – 95. Last previous edition D 5679 – 95.

[2] *Annual Book of ASTM Standards*, Vol 04.02.

[3] *Annual Book of ASTM Standards*, Vol 15.01.

[4] *Annual Book of ASTM Standards*, Vol 11.04.

[5] *Annual Book of ASTM Standards*, Vol 04.08.

[6] Available from National Safety Council, P.O. Box 558, Itasca, IL 60143-0558.

[7] Available from Superintendent of Documents, U.S. Government Printing Office, Washington, DC 20402.

auger, chisel, chip, or core into the consolidated solid material to be sampled. The sample is collected and placed in a sample container. The sampling device is then cleaned and decontaminated or disposed of.

5. Significance and Use

5.1 This practice is intended for use in collecting samples of consolidated or compacted materials from drums or similar containers, including those that are unstable, ruptured, or compromised otherwise. Special handling procedures (for example, remote drum opening, overpressurized drum opening, drum deheading, etc.) are described in *Drum Handling Practices at Hazardous Waste Sites*.

6. Interferences

6.1 The condition of the materials to be sampled and the condition and accessibility of the drums will have a significant impact on the selection of sampling equipment.

7. Pre-Sampling

7.1 *General Principles and Precautions*:

7.1.1 Samples should be collected in accordance with an appropriate work plan (Practice D 5283 and Guide D 4687). This plan must include a worker health and safety section because there are potential hazards associated with opening drums as well as potentially hazardous contents. See *Occupational Safety and Health Guidance Manual for Hazardous Waste Site Activities* for information on health and safety at hazardous waste sites.

7.1.2 Correct sampling procedures must be applied to the conditions as they are encountered. It is impossible to specify rigid rules describing the exact manner of sample collection because of unknowns associated with each solid sampling situation. It is essential that the samples be collected by a trained and experienced sampler because the various conditions under which drummed solids must be sampled.

7.1.3 To be able to make probability or confidence statements concerning the properties of a sampled lot, the sampling procedure must allow for some element of randomness in selection because of the possible variations in the material. The sampler should always be on the alert for possible biases arising from the use of a particular sampling device or from unexpected segregation within the material.

7.1.4 All augering, chipping, or flaking sampling methods may fail a prime sampling requirement: that of random selection of sample fractions. Particles on the bottom or along the sides of the drum may consequently never have an opportunity to be included in a sample. Sample particles should be selected by techniques that will minimize variation in measured characteristics between the available fractions and the resulting sample (Practice C 702).

7.1.5 The sampling equipment, sample preparation equipment, sample containers, etc. must be clean, dry, and inert to the material being sampled. All equipment, including sample containers, must be inspected before use to ensure that they are clear of obvious dirt and contamination and in good working condition. Visible contamination must be removed, and the equipment must be decontaminated with the appropriate rinse materials. Decontaminated sampling equipment should be protected from contamination. This may include, but not be limited to, storage in aluminum foil, plastic bags, polytetrafluoroethylene (PTFE) film, or other means of protection that will not impact the sample quality or intended analysis.

7.2 *Basic Pre-Sampling Practices*:

7.2.1 Review all paperwork.

7.2.2 Select the sampling equipment and sample containers appropriate for the material in the drum, as detailed in the work plan.

7.2.3 Enter the work zone.

7.2.4 Inspect all drums to be sampled visually. Note any abnormal conditions, including rust marks, stains, bulges, or other signs of pressurization or leaks that may require special handling. The work plan should define clearly the limiting conditions under which special handling procedures shall be initiated. See *Drum Handling Practices at Hazardous Waste Sites* for information on opening overpressurized drums and the use of remotely operated drum opening equipment.

7.2.5 Stage the drums to be sampled in a designated work area if they cannot be sampled in their current location. See *Drum Handling Practices at Hazardous Waste Sites* for further information on staging drums.

7.2.5.1 Move the drums to upright stable positions if necessary. Sufficient space shall be left between drums to prevent movement hazards.

7.2.5.2 Number or identify uniquely all drums to be sampled.

7.2.6 Perform a detailed inspection of individual drums.

7.2.6.1 Record all relevant information from the drum labels, markings, data sheets, etc. in the field log book or on forms specified in the work plan.

7.2.6.2 Make sure there are no discrepancies with existing paperwork.

7.2.7 Slowly loosen the ring that secures the lid or loosen the bung, allowing any pressure or vacuum to equalize.

7.2.7.1 *Precautionary Notes:*

(1) If the drum or pail appears to be under positive or negative pressure (that is, a slight lid bulge or dimple), control the release of pressure until it has equalized. For example, if the drum or pail is equipped with bungs, loosen the smaller bung first since doing so will make it easier to control the release of pressure.

(2) If the tope of the drum is dished inward (dimpled), it may "pop" when equalizing pressure, spraying the sampler with any material that is sitting on top of the drum.

(3) If there is evidence of a chemical reaction or sudden pressure buildup, the sampler should leave the area immediately and evaluate whether remote drum opening equipment should be used.

(4) For flammable or explosive materials, the drum and sampling equipment should be grounded if the generation of static electricity while opening or sampling the drum is a possibility. The drum and sampling equipment should be grounded to a ground stake or to an existing ground (building ground, grounded water pipes, etc.). New sampling equipment may have some residual static electrical charge due to the materials in which they are packed and shipped. The work plan should specify whether grounding is necessary. See *Accident*

Prevention Manual for Industrial Operations for information on grounding and bonding.

7.2.7.2 Drums should be opened, sampled, and closed individually to minimize the risk of exposure.

7.2.7.3 *Drums (or Pails) with Bungs*—Loosen the large bung slowly. Use non-sparking tools.

7.2.7.4 *Drums with Removable Lids*—Loosen the ring slowly with a manual wrench or air impact wrench. Use non-sparking tools.

7.2.7.5 *Pails with Removable Lids (Side-Lever Lock Ring)*—Release the lever slowly.

7.2.7.6 *Pails with Removable Lids (Snap-On)*—Pry the lid loose slowly with a pail lid opener.

7.2.8 Manual or remote puncturing or deheading will be required if the drum (or pail) has a stuck bung or the lid cannot be removed. See *Drum Handling Practices at Hazardous Waste Sites* for further information on manual or remote drum opening.

7.2.9 Any discrepancy discovered (such as evidence of free liquid) upon opening the drum should be recorded in the field log book.

7.3 *Sampling Equipment—Selection*:

7.3.1 Table 1 summarizes selection criteria for equipment by the material to be sampled.

7.3.2 *Sampling Equipment, Materials of Construction*—Sampling devices will usually be made of stainless steel, brass, aluminum, or plastic. Devices using permanent coatings or liners (such as PTFE) may be subject to abrasion, leading to contamination of the sample.

7.3.3 *Generic Equipment List*—A general list of equipment used for sampling consolidated solids follows:

7.3.3.1 Scoop.

7.3.3.2 Rotating corer.

7.3.3.3 Thin-wall tube sampler.

7.3.3.4 Chipper.

7.3.3.5 Hammer and chisel.

7.3.3.6 Auger.

7.3.3.7 Pry bars.

7.3.3.8 Wipes or cloths, or both.

7.3.3.9 Spatula.

7.3.3.10 Sample containers, lids, and liners.

7.3.3.11 Sample labels.

7.3.3.12 Chain of custody forms.

7.3.3.13 Field log books.

7.3.3.14 Sample cooler.

7.3.3.15 Ice or gel ice.

7.3.3.16 Grounding cables with alligator clips and emery cloth.

TABLE 1 Selection Criteria for Equipment

Equipment	ASTM Standard	Visually Homogeneous	Heterogeneous
Auger	D 4700	X[A]	X
Chipper, hammer, and chisel	...	X	X
Rotating corer	C 783	X	X
Thin-walled tube	D 4700	N[B]	...[C]

[A] X = equipment usually may be used with this type of waste.
[B] N = not equipment of choice but may be used (for example, a clay-like material).
[C] Equipment is probably not suitable.

7.3.3.17 Portable monitoring equipment (combustible gas indicator, organic vapor detector, radiation survey meter, etc.).

7.3.4 Equipment needed to open drums should be non-sparking (brass or beryllium copper) and include, but not be limited to, the following:

7.3.4.1 Bung wrenches (one straight and one bent),

7.3.4.2 Flathead screwdriver,

7.3.4.3 Breaker bar (½ in. (1.3 cm)),

7.3.4.4 Ratchet (½in. (1.3 cm)),

7.3.4.5 Speed handle (½ in. (1.3 cm)),

7.3.4.6 Adjustable wrenches (10 and 12 in. (25 and 30 cm)).

7.3.4.7 Air impact wrench and sockets, and

7.3.4.8 Pail lid opener.

8. Sample Collection

8.1 *Basic Sampling Practices*:

8.1.1 Bond the sampling equipment to the drum, if specified in the work plan.

8.1.2 Note the physical characteristics, including any discrepancies (such as free liquid).

8.1.3 Collect the required number of samples from the drum.

8.1.3.1 See Practice D 4547 for the collection of samples for volatile analysis.

8.1.4 Place the collected material in a sample container.

8.1.5 Close the sample container.

8.1.6 Wipe the outside of the sample container. Dispose of the wipe cloth properly.

8.1.7 Record in the field log book all relevant conditions associated with the collection of each sample.

8.1.8 Fill out all required paperwork for each sample, as required by the work plan.

8.1.9 Complete and attach the label to the side of the sample container before or after sampling, as directed by the work plan. The sample label should include the following:

(1) Sample ID number,

(2) Name of sampler,

(3) Sampler's initials or signature,

(4) Date and time of sampling, and

(5) Sampling location.

8.1.9.1 The sample label can also include the following:

(1) Sampling information (for example, grab, composite, etc.),

(2) Preservative and preservation required,

(3) Special instructions, and

(4) Analysis request.

8.2 *Sampling Using a Rotating Corer*—The rotating corer can be as simple as a cylinder attached to an electric drill with the crown modified for cutting (see Fig. 1 and Practice C 783) or as complex as a double metal tube fitted onto a diamond-impregnated coring bit, mounted on a portable stand. The double metal tube corer mounted on a portable stand has the capability of collecting a full-depth core of the drum contents. This procedure describes the single metal tube corer attached to an electric drill.

8.2.1 *General Description*—The rotating corer is usually 1 to 1½ ft (0.3 to 0.5 m) long, with available diameters of approximately 2 to 6 in. (5.1 to 15.2 cm). The corer is driven by suitable equipment, such as a portable electric drill.

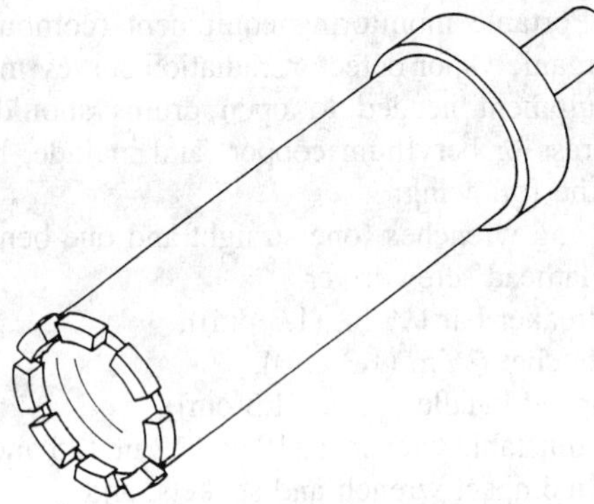

FIG. 1 Rotating Corer

8.2.2 *Operation and Use*—Place the coring apparatus over the area to be cored. Turn the coring mechanism on, and place the coring bit in continuous contact with the solid material by supplying uniform and continuous pressure. Continue the operation until the solid material is bored to the specified depth and the resulting core is forced into the tube. Withdraw the corer. Record the actual length of the core. The tube can be capped and transferred to the analytical laboratory. This is the desirable method for samples requiring volatile organics analysis. The sample core can also be extruded into a form-fitting sample container.

8.3 *Sampling with a Thin-Wall Tube Sampler*:

8.3.1 *General Description*—Thin-wall tube samplers may vary in diameter, length, and material of construction (see Fig. 2). The material to be sampled must be of a physical consistency (cohesive solid material) to be cored and retrieved with the tube. Materials with particles larger than one-third of the inner diameter of the tube should not be sampled with that particular device. The length of the tube will depend on the desired sampling depth (Guide D 4700). The tube is attached onto a length of solid or tubular rod. The upper end of this rod is threaded to accept a handle or extension rods. This sampler can be used to collect samples of consolidated clay-like materials.

8.3.2 *Operation and Use*—The sampler is pushed into the material to be sampled by applying downward force on the unit's handle. Once the sampler has reached the bottom of the sampling interval, it is twisted to break the continuity at the tip. The sampler is pulled from the material, and the sample material is extruded into the sample container. Samples are extruded by forcing a rod through the tube.

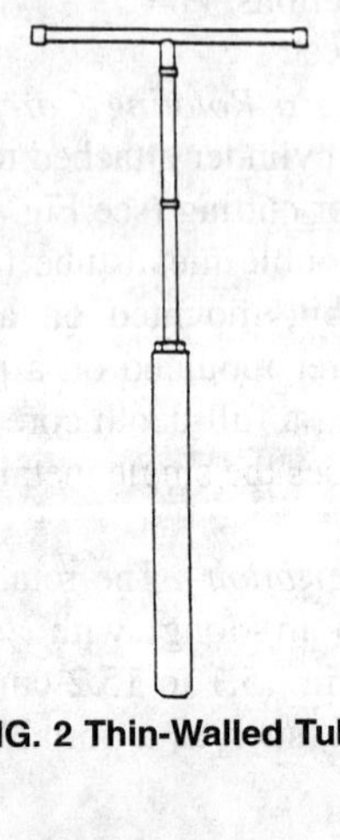

FIG. 2 Thin-Walled Tube

8.4 *Sampling with a Chipper*:

8.4.1 *General Description*—A hardened steel bit or knife is fitted to a pneumatic hammer (see Fig. 3). The hammer forces the bit or knife to break the consolidated material into chips, flakes, and chunks suitable for collection with a scoop. This method is not recommended for samples requiring volatile organics analysis.

8.4.2 *Operation and Use*—Insert the chipper into the drum. Place the chipping knife directly on the material to be sampled. Activate the pneumatic hammer. Break the consolidated material into manageably sized chips, flakes, and chunks. Remove the chipper. Collect a sample from the drum with a scoop, and transfer to the sample container.

NOTE 1—See 7.1.4 and the site work plan for information on selecting chipped particles for the sample.

8.5 *Sampling with a Hammer and Chisel*:

8.5.1 *General Description*—A hammer is used to impact a hardened steel chisel to break the consolidated material into chips, flakes, and chunks suitable for collection with a scoop.

8.5.2 *Operation and Use*—These tools are used as necessary to collect sample material from the drum. A hammer and chisel have been found useful in sampling drums for which surface sampling is required or particle size reduction is necessary. This method is not recommended for samples requiring volatile organics analysis.

NOTE 2—See 7.1.4 and the site work plan for information on selecting chipped particles for the sample.

8.6 *Sampling with an Auger*:

8.6.1 *General Description*—The screw or ship auger is essentially a small-diameter (for example, 1½-in. (3.8-cm)) wood auger from which the cutting side flanges and tip have been removed. The auger is welded onto a length of solid or tubular rod. The upper end of this rod is threaded to accept a handle or extension rods. An auger can be used for collecting a disturbed sample of consolidated material in drums (See Fig. 4 and Guide D 4700).

8.6.2 *Operation and Use*—The auger is rotated manually or with a power source into the material to be sampled. The operator may have to apply downward force to embed the auger; the auger screws itself into the material afterwards. The auger is advanced to its full length and then pulled up and removed. Material from the deepest interval is retained on the auger flights. Sample material can be collected from the flights using a spatula.

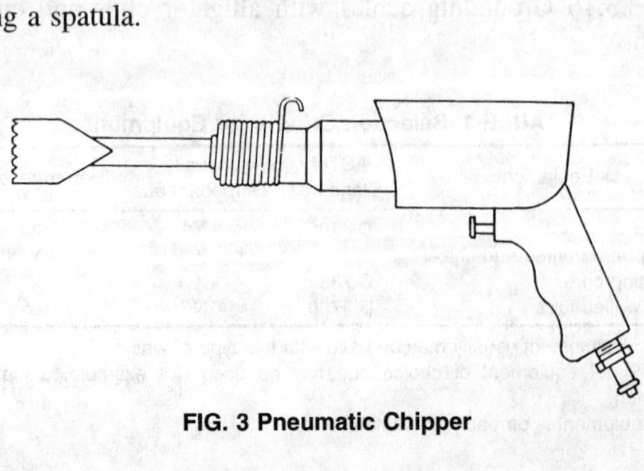

FIG. 3 Pneumatic Chipper

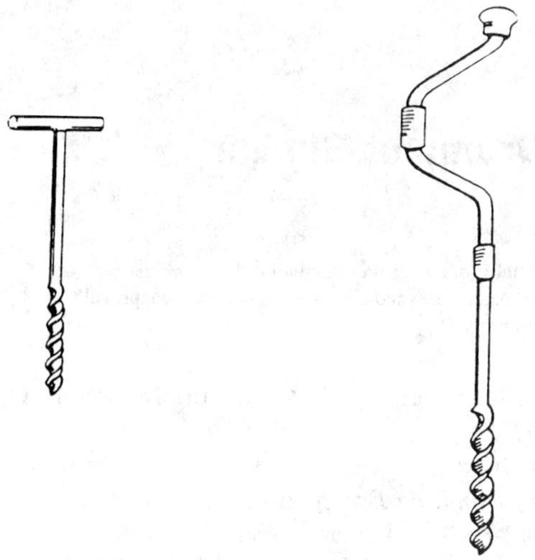

FIG. 4 Augers (Typical)

9. Post-Sampling

9.1 Remove all sampling equipment from the work zone.

9.2 Transfer all reusable equipment that was in contact with the waste to a pre-designated decontamination area. Decontaminate the equipment according to the protocol established in the work plan (Practice D 5088). Decontaminated sampling equipment should be protected from contamination. This may include, but not be limited to, storage in aluminum foil, plastic bags, PTFE film, or other means of protection that will not impact the sample quality or intended analysis.

9.3 Dispose of all used (disposable) contacting equipment.

10. Data Quality Objectives

10.1 The objectives for sampling and testing of consolidated solid material should be specified in the work plan.

11. Quality Control

11.1 Quality Control (QC) samples (for example, equipment blanks, trip blanks, and duplicates) must be collected as required by the work plan. These QC samples must be evaluated to provide a determination of the sampling quality and reliability of the resulting analytical data.

12. Keywords

12.1 auger; chipper; consolidated solids; hazardous waste; rotating corer sampling; thin-wall tube

Designation: D 5680 – 95a (Reapproved 2001)

Standard Practice for Sampling Unconsolidated Solids in Drums or Similar Containers[1]

This standard is issued under the fixed designation D 5680; the number immediately following the designation indicates the year of original adoption or, in the case of revision, the year of last revision. A number in parentheses indicates the year of last reapproval. A superscript epsilon (ε) indicates an editorial change since the last revision or reapproval.

1. Scope

1.1 This practice covers typical equipment and methods for collecting samples of unconsolidated solids in drums or similar containers. These methods are adapted specifically for sampling drums having a volume of 110 U.S. gal (416 L) or less. These methods are applicable to hazardous material, product, or waste. Specific sample collection and handling requirements should be described in the site-specific work plan.

1.2 The values stated in inch-pound units are to be regarded as the standard. The values given in parentheses are for information only.

1.3 *This standard does not purport to address all of the safety concerns, if any, associated with its use. It is the responsibility of the user of this standard to establish appropriate safety and health practices and determine the applicability of regulatory limitations prior to use.*

2. Referenced Documents

2.1 *ASTM Standards:*

C 702 Practice for Reducing Samples of Aggregate to Testing Size[2]

D 4547 Practice for Sampling Waste and Soils for Volatile Organics[3]

D 4687 Guide for General Planning of Waste Sampling[3]

D 4700 Guide for Soil Sampling from the Vadose Zone[4]

D 5088 Practice for the Decontamination of Field Equipment Used at Non-Radioactive Waste Sites[4]

D 5283 Practice for Generation of Environmental Data Related to Waste Management Activities: Quality Assurance and Quality Control Planning[3]

D 5451 Practice for Sampling with a Trier Sampler[3]

E 300 Practice for Sampling Industrial Chemicals[5]

D 5633 Practice for Sampling With a Scoop[3]

2.2 *NSC Document:*

Accident Prevention Manual for Industrial Operations, 1992[6]

2.3 *Government Documents:*[7]

Drum Handling Practices at Hazardous Waste Sites, EPA/600/2-86/013, January 1986

Soil Sampling and Analysis for Volatile Compounds, EPA/540/4-91/001, February 1991

Occupational Safety and Health Guidance Manual for Hazardous Waste Site Activities, National Institute for Occupational Safety and Health (NIOSH), Occupational Safety and Health Administration (OSHA), U.S. Coast Guard (USCG), and U.S Environmental Protection Agency (EPA), October 1985

3. Terminology

3.1 *Definitions:*

3.1.1 *bonding*—touching the sample equipment to the drum to form an electrically conductive path to minimize potential electrical differences between the sampling equipment and the drum, reducing the buildup of static electricity.

3.1.2 *bung*—usually a 2-in. (5.1-cm) or ¾-in. (1.3-cm) diameter threaded plug designed specifically to close a bung hole.

3.1.3 *bung hole*—an opening in a barrel or drum through which it can be filled, emptied, or vented.

3.1.4 *deheading*—removal of the lid of a closed-head drum; usually accomplished with a drum deheader.

3.1.5 *drum*—implies any drum, barrel, or non-bulk container of 5 to 110 U.S gal (19 to 416 L) capacity.

3.1.6 *pail*—a small container, usually with a capacity of 5 U.S gal. Pails typically have bungs or spouts, or the entire lid can be removed.

3.1.7 *paperwork*—all required site documentation, which may include the manifests, waste profiles, material safety data sheets (MSDS), site forms, sample labels, custody seals, and chain of custody forms.

3.1.8 *unconsolidated*—for solid material, the characteristic of being uncemented or uncompacted, or both, and separated easily into smaller particles.

3.1.9 *work plan*—a plan specific to a particular site; for conducting activities specified in the plan.

---

[1] This practice is under the jurisdiction of ASTM Committee D34 on Waste Management and is the direct responsibility of Subcommittee D34.01.02 on Sampling Techniques.

Current edition approved June 15, 1995. Published August 1995. Originally published as D 5680 – 95. Last previous edition D 5680 – 95.

[2] *Annual Book of ASTM Standards,* Vol 04.02.

[3] *Annual Book of ASTM Standards,* Vol 11.04.

[4] *Annual Book of ASTM Standards,* Vol 04.08.

[5] *Annual Book of ASTM Standards,* Vol 15.05.

[6] Available from National Safety Council, P.O. Box 558, Itasca, IL 60143-0558.

[7] Available from the Superintendent of Documents, U. S. Government Printing Office, Washington, DC 20402.

4. Summary of Practice

4.1 The drum and its contents are inspected, and appropriate sampling equipment is selected. A clean sampling device is then used to auger, scoop, or core into the unconsolidated solid material to be sampled. The sample is collected and placed in a sample container. The sampling device is then either disposed of or cleaned and decontaminated.

5. Significance and Use

5.1 This practice is intended for use in collecting samples of unconsolidated solid materials from drums or similar containers, including those that are unstable, ruptured, or compromised otherwise. Special handling procedures (for example, remote drum opening, overpressurized drum opening, drum deheading, etc.) are described in *Drum Handling Practices at Hazardous Waste Sites.*

6. Interferences

6.1 The condition of the materials to be sampled and the condition and accessibility of the drums will have a significant impact on the selection of sampling equipment.

7. Pre-Sampling

7.1 *General Principles and Precautions*:

7.1.1 Samples should be collected in accordance with an appropriate work plan (Practice D 5283 and Guide D 4687). This plan must include a worker health and safety section because there are potential hazards associated with opening drums as well as potentially hazardous contents. See *Occupational Safety and Health Guidance Manual for Hazardous Waste Site Activities* for information on health and safety at hazardous waste sites.

7.1.2 Correct sampling procedures must be applied to the conditions as they are encountered. It is impossible to specify rigid rules describing the exact manner of sample collection because of unknowns associated with each solid sampling situation. It is essential that the samples be collected by a trained and experienced sampler because of the various conditions under which drummed solids must be sampled.

7.1.3 To be able to make probability or confidence statements concerning the properties of a sampled lot, the sampling procedure must allow for some element of randomness in selection because of the possible variations in the material. The sampler should always be on the alert for possible biases arising from the use of a particular sampling device or from unexpected segregation within the material.

7.1.4 All auger, trier, thief, and scoop methods may fail a prime sampling requirement: that of random selection of sample fractions. Scoops are limited to use at or near the top surface. Augers, triers, and thiefs are normally inserted in a present pattern. Particles on the bottom or along the sides of the drum may consequently never have an opportunity to be included in a sample. Sample particles should be selected by techniques that will minimize variation in measured characteristics between the available fractions and the resulting sample (Practice C 702).

7.1.5 The sampling equipment, sample preparation equipment, sample containers, etc. must be clean, dry, and inert to the material being sampled. All equipment, including sample containers, shall be inspected before use to ensure that they are clear of obvious dirt and contamination and are in good working condition. Visible contamination shall be removed, and the equipment shall be decontaminated with the appropriate rinse materials. Decontaminated sampling equipment should be protected from contamination. This may include, but not be limited to, storage in aluminum foil, plastic bags, polytetrafluoroethylene (PTFE) film, or other means of protection that will not impact the sample quality of intended analysis.

7.2 *Basic Pre-Sampling Practices*:

7.2.1 Review all paperwork.

7.2.2 Select the sampling equipment and sample containers appropriate for the material in the drum, as detailed in the work plan.

7.2.3 Enter the work zone.

7.2.4 Inspect all drums to be sampled visually. Note any abnormal conditions, including rust marks, stains, bulges, or other signs of pressurization or leaks that may require special handling. The work plan should define clearly the limiting condition under which special handling procedures shall be initiated. See *Drum Handling Practices at Hazardous Waste Sites* for information on opening overpressurized drums and the use of remotely operated drum opening equipment.

7.2.5 Stage the drums to be sampled in a designated work area if they cannot be sampled in their current location. See *Drum Handling Practices at Hazardous Waste Sites* for further information on staging turns.

7.2.5.1 Move the drums to upright, stable positions if necessary. Sufficient space shall be left between drums to prevent movement hazards.

7.2.5.2 Number or identify uniquely all drums to be sampled.

7.2.6 Perform a detailed inspection of individual drums.

7.2.6.1 Record all relevant information from drum labels, markings, data sheets, etc. in the field log book or on forms specified in the work plan.

7.2.6.2 Make sure there are no discrepancies with existing paperwork.

7.2.7 Slowly loosen the ring that secures the lid, or loosen the bung allowing any pressure or vacuum to equalize.

7.2.7.1 *Precautionary Notes*:

(1) If the drum or pail appears to be under positive or negative pressure (that is, a slight bulge or dimple in the lid), control the release of pressure until it has equalized. For example, if the drum or pail is equipped with bungs, loosen the smaller bung first since doing so will make it easier to control the release of pressure.

(2) If the top of the drum is dished inward (dimpled), it may "pop" when equalizing pressure, spraying the sampler with any material that is sitting on top of the drum.

(3) If there is evidence of a chemical reaction or sudden pressure buildup, the sampler should leave the area immediately and evaluate whether remote drum opening equipment should be used.

(4) For flammable or explosive materials, the drum and sampling equipment should be grounded if the generation of static electricity while opening or sampling the drum is a

possibility. The drum and sampling equipment should be grounded to a ground stake or to an existing ground (building ground, grounded water pipes, etc.). New sampling equipment may have some residual static electrical charge due to the materials in which they are packed and shipped. The work plan should specify whether grounding is necessary. See *Accident Prevention Manual for Industrial Operations* for information on grounding and bonding.

7.2.7.2 Drums should be opened, sampled, and closed individually to minimize the risk of exposure.

7.2.7.3 *Drums (or Pails) with Bungs*—Loosen the large bung slowly. Use non-sparking tools.

7.2.7.4 *Drums with Removable Lids*—Loosen the ring slowly with a manual wrench or air impact wrench. Use non-sparking tools.

7.2.7.5 *Pails with Removable Lids (Side-Lever Lock Ring)*—Release the lever slowly.

7.2.7.6 *Pails with Removable Lids (Snap-On)*—Pry the lid loose slowly with a pail lid opener.

7.2.8 Manual or remote puncturing or deheading will be required if the drum (or pail) has a stuck bung or the lid cannot be removed. See *Drum Handling Practices at Hazardous Waste Sites* for further information on manual or remote drum opening.

7.2.9 Any discrepancy discovered (such as evidence of free liquid) upon opening the drum should be recorded in the field log book.

7.3 *Sampling Equipment—Selection*:

7.3.1 Table 1 summarizes selection criteria for equipment by the material to be sampled.

7.3.2 *Sampling Equipment, Materials of Construction*—Sampling devices will usually be made of stainless steel, brass, or aluminum. Devices using permanent coatings or liners (such as PTFE) may be subject to abrasion, leading to contamination of the sample.

7.3.3 *Generic Equipment List*—A general list of equipment used for sampling unconsolidated solids follows:

7.3.3.1 Scoop.

7.3.3.2 Trier.

7.3.3.3 Auger.

7.3.3.4 Concentric tube thief (single slot, multi-slot, grain probe, and missouri trier).

7.3.3.5 Thin-walled tube.

7.3.3.6 Scissors.

7.3.3.7 Tongs.

TABLE 1 Selection Criteria for Equipment

Equipment	ASTM Standard	Cohesive Solid	Sheet, Cloth, or Chunk Material	Dry Flowable Solids	Moist Flowable Solids
Scoop	D 5633	X[A]	X	X	X
Auger	D 4700	X	...[B]	...	...
Trier	D 5451	X	...	...	X
Thin-walled tube	D 4700	...	X[C]	X[C]	
Hammer and chisel		N[D]	N	...	...
Scissors and tongs		...	X	...	...
Concentric tube thief	E 300	...	...	X	...

[A] X = equipment may be used with this type of waste.
[B] Equipment is probably unsuitable.
[C] Sampling equipment with retaining device.
[D] N = not equipment of choice, but may be used.

7.3.3.8 Hammer and chisel.

7.3.3.9 Cloths or wipes, or both.

7.3.3.10 Spatula.

7.3.3.11 Sample containers, lids, and liners.

7.3.3.12 Chain of custody forms.

7.3.3.13 Field log books.

7.3.3.14 Sample labels.

7.3.3.15 Sample cooler.

7.3.3.16 Ice or gel ice.

7.3.3.17 Grounding cables with alligator clips and emery cloth.

7.3.3.18 Portable monitoring equipment (combustible gas indicator, organic vapor detectors, radiation survey meter, etc.).

7.3.4 Equipment needed to open drums should be non-sparking (brass or beryllium copper) and include, but not be limited to, the following:

7.3.4.1 Bung wrenches (one straight and one bent),

7.3.4.2 Flatblade screwdriver,

7.3.4.3 Breaker bar (½ in. (1.3 cm)),

7.3.4.4 Ratchet (½in. (1.3 cm)),

7.3.4.5 Speed handle (½ in. (1.3 cm)),

7.3.4.6 Adjustable wrenches (10 and 12 in. (25 and 30 cm)),

7.3.4.7 Air impact wrench and sockets, and

7.3.4.8 Pail lid opener.

8. Sample Collection

8.1 *Basic Sampling Practices*:

8.1.1 Bond the sampling equipment to the drum, if specified in the work plan.

8.1.2 Note the physical characteristics, including any discrepancies (such as free liquid).

8.1.3 Collect the required number of samples from the drum.

8.1.3.1 See Practice D 4547 and *Soil Sampling and Analysis for Volatile Compounds* for the collection of samples for volatile analysis.

8.1.4 Place the collected material in a sample container.

8.1.5 Close the sample container.

8.1.6 Wipe the outside of the sample container. Dispose of the wipe cloth properly.

8.1.7 Record in the field log book all relevant conditions and physical characteristics associated with the collection of each sample.

8.1.8 Fill out all required paperwork for each sample, as required by the work plan.

8.1.9 Complete and attach the label to the side of the sample container before or after sampling, as directed by the work plan. The sample label should include the following:

(1) Sample ID number,
(2) Name of sampler,
(3) Sampler's initials or signature,
(4) Date and time of sampling, and
(5) Sample location.

8.1.9.1 The sample label can also include the following:

(1) Sampling information (for example, grab, composite, etc.),
(2) Preservative and preservation required,
(3) Special instructions, and
(4) Analysis request.

8.2 *Sampling Using a Scoop*:

8.2.1 *General Description*—A plastic or metal scoop is used for collecting approximately equal fractions at random intervals at or near the surface of the material (see Fig. 1). This sampling tool should be of a size and shape suitable for the quantity and size of the particles to be sampled. Scoop sampling provides best results if the material is uniform.

8.2.2 *Operation and Use*—A thin layer of material is removed with the scoop and discarded. A suitable volume of material is collected with the scoop and transferred to a sample container.

8.3 *Sampling with a Thin-Walled Tube Sampler*:

8.3.1 This sampler can be used to collect samples of unconsolidated material that is usually moist or cohesive and may be powdery or granular.

8.3.2 *General Description*—Tube samplers may vary in diameter, length, and material of construction (see Fig. 2). The material to be sampled must be of a physical consistency (cohesive solid material) to be cored and retrieved with the tube. Materials with particles larger than one-third of the inner diameter of the tube should not be sampled with this particular device. The length of the tube will depend on the desired sampling depth (Guide D 4700). The tube is attached to a length of a solid or tubular rod. The upper end of this rod is threaded to accept a handle or extension rods. This sampler can be used to collect samples of unconsolidated clay-like materials.

8.3.3 *Operation and Use*—The sampler is pushed into the material to be sampled by applying downward force on the unit's handle. Once the sampler has reached the bottom of the sampling interval, it is twisted to break the continuity at the tip. The sampler is pulled from the material, and the sample material is extruded into the sample container. Samples are extruded by forcing a rod through the tube.

8.4 *Sampling with a Trier Sampler*:

8.4.1 *General Description*—The trier is a metal or plastic tube from which one third to one-half of the wall of the tube has been removed to form a slot along its entire length (see Fig. 3). This device can be up to 4 ft (1.2 m) long and should have a sharp, angled point at its lower end. The trier can be used as a coring device in drums of solids such as soils and similar fine-grained cohesive materials.

8.4.2 *Operation and Use*—The trier is pushed vertically into the material and rotated one or two times to cut a core. The core is pulled out of the hole and transferred to a sample container (Practice D 5451).

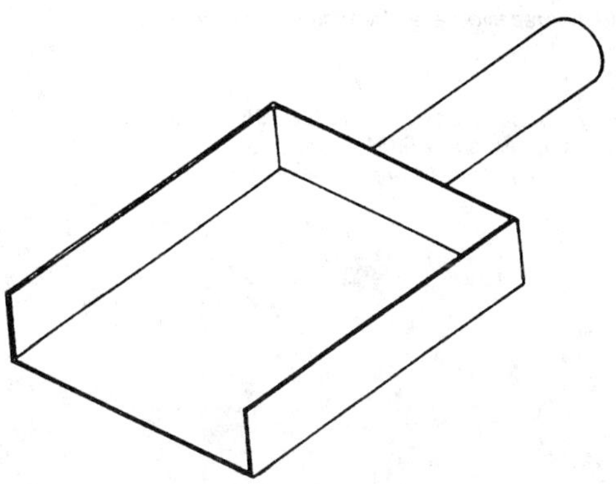

FIG. 1 Scoop

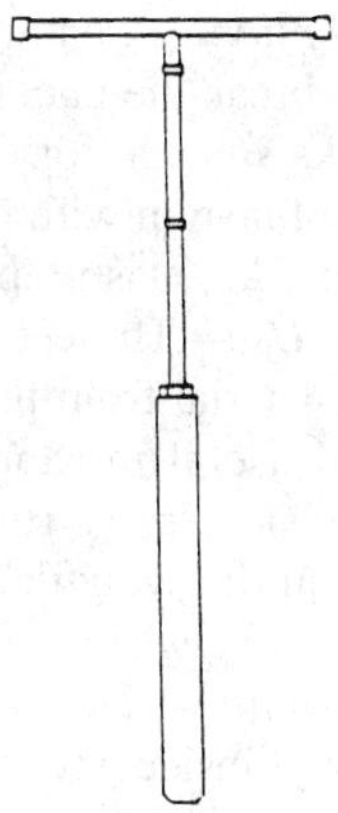

FIG. 2 Thin-Walled Tube Sampler

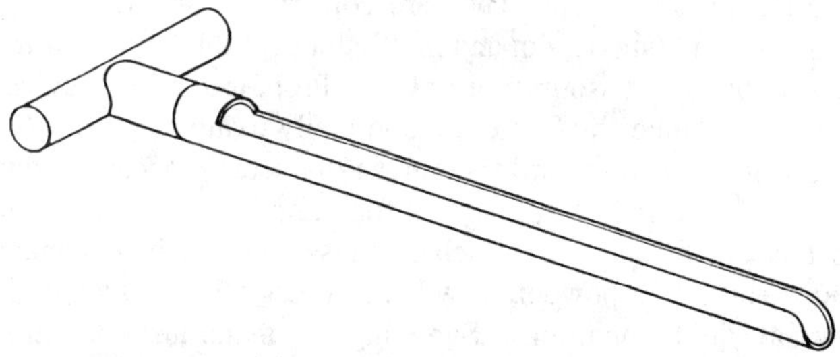

FIG. 3 Trier

8.5 *Sampling with an Auger*:

8.5.1 *General Description*—The screw or ship auger is essentially a small-diameter (for example, 1.5-in. (3.8-cm)) wood auger from which the cutting side flanges and tip have been removed. The auger is welded onto a length of solid or tubular rod. The upper end of this rod is threaded to accept a handle or extension rods. An auger can be used for collecting a disturbed sample of unconsolidated material in drums (see Fig. 4 and Guide D 4700).

8.5.2 *Operation and Use*—The auger is rotated manually or with a power source into the material to be sampled. The operator may have to apply downward force to embed the auger; the auger screws itself into the material afterwards. The auger is advanced to its full length and then pulled and removed. Material from the deepest interval is retained on the auger flights. Sample material can be collected from the flights using a spatula.

8.6 *Sampling with a Hammer and Chisel, Scissors, or Tongs*:

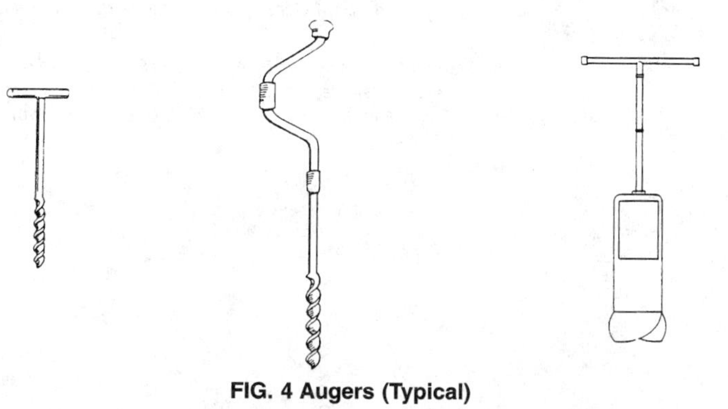

FIG. 4 Augers (Typical)

8.6.1 *General Description*—A hammer is used to impact a hardened steel chisel to break the unconsolidated material into chips, flakes, and chunks suitable for collection with a scoop. Scissors are used in combination with tongs to collect samples of material that are clothlike, elastic, paperlike, etc.

8.6.2 *Operation and Use*—These tools are used as necessary to collect sample material from the drum. A hammer and chisel have been found useful in sampling drums for which particle size reduction is necessary. This method is not recommended for samples requiring volatile organics analysis.

8.7 *Sampling with a Concentric Tube Thief*:

8.7.1 *General Description*—This device consists of two tubes, one fitting snugly inside the other (see Fig. 5). The bottom end of the outer tube is fitted with a point. Oblong holes are cut through both tubes. The holes are opened or closed by rotating the inner tube. They are constructed as either single compartment (single-slot and multi-slot tube thiefs) or multiple compartment (Missouri trier or Grain Probe) sampling devices. Concentric tube thiefs are commercially available up to 6 ft (1.8 m) long and several in. (cm) in diameter. Concentric tube thiefs have a limited application for sampling drums. Material that is not free-flowing, such as those that are hard packed, moist, or finely powdered, will not enter this sampler under normal field conditions. Sampling of materials containing granules or particles exceeding one-third of the slot width should not be attempted because bridging may occur. These devices cannot sample the bottom of a drum because of their pointed ends.

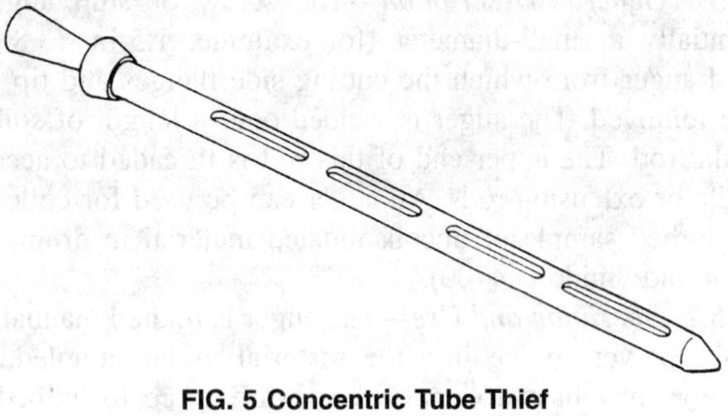

FIG. 5 Concentric Tube Thief

8.7.2 *Operation and Use*—Insert the tube into the material, and push with uniform force to the bottom of the drum or until refusal. Rotate the concentric tubes to the open position, thereby allowing the sample to flow into the inner tube. Wiggle the sampler several times, and rotate the tubes to the closed position. Withdraw the sampler. Place the sampling device immediately over a sample collection sheet, and release the sample by rotating the concentric tubes to the open position. A sample can normally be removed from the thief with a spatula or similar instrument (reamer) and placed in the sample container.

9. Post-Sampling

9.1 Remove all sampling equipment from the work zone.

9.2 Transfer all reusable equipment that was in contact with the waste to a pre-designated decontamination area. Decontaminate the equipment according to the protocol established in the work plan (Practice D 5088). Decontaminated sampling equipment should be protected from contamination. This may include, but not be limited to, storage in aluminum foil, plastic bags, PTFE film, or other means of protection that will not impact the sample quality or intended analysis.

9.3 Dispose of all used (disposable) contacting equipment.

10. Data Quality Objectives

10.1 The objectives for sampling and testing of unconsolidated solid material should be specified in the work plan.

11. Quality Control

11.1 Quality control (QC) (for example, equipment blanks, trip blanks, and duplicates) must be collected as required by the work plan. These QC samples shall be evaluated to provide a determination of the sampling quality and reliability of the resulting analytical data.

12. Keywords

12.1 Auger; concentric tube thief; scoop; thin-wall tube; trier; unconsolidated solids; waste

Designation: D 5681 – 98a

Standard Terminology for Waste and Waste Management[1]

This standard is issued under the fixed designation D 5681; the number immediately following the designation indicates the year of original adoption or, in the case of revision, the year of last revision. A number in parentheses indicates the year of last reapproval. A superscript epsilon (ϵ) indicates an editorial change since the last revision or reapproval.

1. Scope

1.1 This terminology contains standard definitions of terms used in the general area of waste and waste management. It is intended to promote understanding by providing precise technical definitions of terms used in the standards developed by Committee D-34 and its subcommittees.

1.2 Terms used only within an individual standard, and having a meaning unique to that standard, may be defined or explained in the terminology section of that individual standard.

2. Referenced Documents

2.1 *ASTM Standards:*

E 177 Practice for Use of the Terms Precision and Bias in ASTM Test Methods[2]

E 456 Terminology Relating to Quality and Statistics[2]

3. Terminology

accepts, *n*—the output stream from a materials separation device that contains the highest concentration (purity) of the components that the device is designed to separate.

adiabatic calorimeter, *n*—a calorimeter that has a jacket temperature adjusted to follow the calorimeter temperature as closely as possible so as to maintain zero thermal head.

analysis sample, *n*—the final subsample prepared from the air-dried laboratory sample but reduced in particle size by passing through a mill with a 0.5 mm (0.02–in.) size or smaller final screen.

as-determined basis, *n*—analytical data obtained from an analysis sample after conditioning and preparation which represent the numerical values obtained at the particular moisture and ash level in the sample at the time of analysis.

as-received basis, *n*—test data calculated to the condition of the sample as it arrived in the laboratory and before any laboratory processing or conditioning.

ash *n*—the residue remaining after ignition of a substance as determined by definite prescribed methods.

DISCUSSION—Ash may not be identical in composition or quantity with the inorganic substances present in the analysis sample before ignition.

calorimeter jacket, *n*—the insulating medium surrounding a calorimeter.

calorific value, *n*—the heat produced by combustion of a unit quantity of a specimen under specified conditions.

characteristic product size, *n*—the screen size corresponding to 63.2 % cumulative passing by mass.

combustibles, *n*—the portion of a sample which is consumed by oxidation upon ignition and exclusive of the moisture present in the sample.

corrected temperature rise, *n*—the increase in temperature of the calorimeter caused by the process that occurs inside the bomb; the observed temperature change corrected for various effects.

dispose, *v*—to discard, abandon, or manage as waste.

dry ash-free basis, *n*—test data calculated to a theoretical base of no moisture or ash associated with the sample.

dry basis, *n*—test data calculated to a theoretical base of no moisture associated with the sample.

duplicate analysis,*n*—paired determinations on the same sample performed by one analyst at essentially the same time.

energy equivalent, *n*—the energy required to raise the temperature of a calorimeter system 1°C (or 1°F) per gram of sample.

fill material, *n*—material used in the construction of a structural fill.

fixed carbon, *n*—the ash-free carbonous material that remains after volatile matter is driven off during the proximate analysis of a dry sample.

flint glass cullet, *n*—a particulate glass material that contains no more than 0.1 mass percent Fe_2O_3, or 0.0015 mass percent Cr_2O_3, as determined by chemical analysis.

fluid temperature, FT, *n*—in ash fusion determinations, the temperature at which a fused mass has spread out in a nearly flat layer with maximum height of 1.6 mm (1/16 in.).

gross calorific value, (gross heat of combustion), Q_v (gross), *n*—the heat produced by combustion of unit quantity of a solid or liquid specimen when burned at constant volume in an oxygen bomb calorimeter under specified conditions with the resulting water condensed to a liquid.

gross sample *n*—a sample representing one lot, normally composed of a number of increments, on which neither

[1] This terminology is under the jurisdiction of ASTM Committee D-34 on Waste Management and is the direct responsibility of Subcommittee D34.94 on Terminology.

Current edition approved June 10, 1998. Published December 1998. Originally published as D 5681–95. Last previous edition D 5681–98.

[2] *Annual Book of ASTM Standards*, Vol 14.02.

reduction nor division has been preformed.

heat capacity, ***n***—the quantity of heat required to raise a system one degree in temperature either at constant volume or constant pressure.

heat capacity (energy equivalent, or water equivalent), ***n***—the energy required to raise the temperature of a calorimeter one arbitrary unit; the quantity that when multiplied by the corrected temperature rise, then adjusted for extraneous heat effects and divided by the mass of the sample, gives the gross calorific value.

hemispherical temperature, HT, ***n***—the temperature at which a pyrometric cone has fused down to a hemispherical lump where the height is one half the width of the base.

higher heating value, HHV, ***n***—a synonym for gross calorific value.

incineration, ***n***—controlled burning of waste products or other combustible material.

incinerator, ***n***—a device constructed for the purpose of containing a material for thermal oxidation.

increment, ***n***—a portion of a lot as collected by one individual manual or mechanical sampling operation and normally combined with other increments from the lot to make a gross sample.

initial deformation temperature, IT, ***n***—the temperature at which the first rounding of the apex of a pyrometric cone occurs; shrinking or warping of the cone is ignored if the tip remains sharp.

isoperibol calorimeter, ***n***—a calorimeter that has a jacket of uniform and constant temperature.

laboratory sample, ***n***—a representative portion of a gross sample received by a laboratory for analysis.

lot, ***n***—a large designated quantity of a material which can be represented by a properly selected gross sample.

lower heating value, LHV, ***n***—a synonym for net calorific value.

metallic yield, ***n***—the mass percentage of a ferrous waste stream that is generally recoverable as metal or alloy.

microbiological aerosol, ***n***—an airborne particle partially or exclusively composed of microorganisms including bacteria and fungi.

milling, ***n***—in waste derived fuels, reduction in particle size by shearing, cutting, grinding to a suitable particle size for analysis and characterization.

municipal ferrous scrap, ***n***—ferrous waste that is collected from industrial, commercial, or household sources and destined for disposal facilities.

net calorific value (net heat of combustion at constant pressure), Q_p, ***n***—the heat produced by combustion of unit quantity of a solid or liquid specimen when burned at a constant pressure of 0.1 MPa (1 atm), under conditions such that all the water in the products remains in the form of vapor.

operating site, ***n***—in waste management, a location or facility where waste is treated, stored, of disposed as part of an on-going operation.

optimum concentration range, ***n***—in analysis of trace metals, a range, defined by limits expressed in concentration, below which scale expansion must be used and above which curve correction should be considered.

polynary separator, ***n***—a device that separates a single input feed stream into three or more output product streams.

proximate analysis, ***n***—the determination, by prescribed methods, of moisture, volatile matter, fixed carbon (by difference), and ash.

DISCUSSION—Unless otherwise specified, the term *proximate analysis* does not include determinations of chemical elements or any determinations other than those named.

recovery, percent, ***n***—the amount of a material actually recovered by an assay using a prescribed procedure, or obtained from a process, as a percentage of the as-received material.

respiration rate, ***n***—in an microbial aqueous system, the quantitative consumption of oxygen, generally expressed as mg O_2/L/h.

retainer basket, ***n***—in sampling, a one-way gate on a sampling device that minimizes loss of sample when retrieving a sampler; also called a core catcher.

significant loss, ***n***—any loss that introduces a bias in final results and that is of appreciable importance to concerned parties.

softening temperature, ST, ***n***—the temperature at which a pyrometric cone has fused down to a spherical lump in which the height is equal to the width at the base.

sorting sample, ***n***—in waste management, a 100 to 150 kg (200 to 300 lb) portion of a vehicle load of municipal solid waste that is deemed to represent the characteristics of that load.

source-separated steel cans, ***n***—post-consumer products that are generated as separated can fractions by commercial or household sources.

(a) all other steel cans, n—steel containers for food products or liquids, with a maximum capacity of 5 gal, that are not included in one of the other definitions.

(b) bi-metal beverage cans, n—steel cans with nonferrous metal convenience ends (normally made of aluminum), originally containing beer or carbonated beverages, but not including other contaminants.

(c) bi-metal food cans, n—steel cans with nonferrous metal (usually aluminum) convenience ends, originally containing snack foods, but not including other contaminants.

static calorimeter, ***n***—a calorimeter without a thermostated jacket.

subsample, ***n***—a portion of a material that is collected by subdividing or trimming of a sample.

total combustibles, ***n***—combustible materials that include paints, lacquers, coatings, plastics, and so forth, associated with an original metal product, as well as combustible materials which become associated with the product after it is manufactured.

ultimate analysis, ***n***—in analysis of combustible materials, the determination of the percentages of carbon, hydrogen, sulfur, nitrogen, chlorine, ash, and oxygen in a moisture-free sample; the percentage of oxygen is usually determined by difference.

unprocessed municipal solid waste, ***n***—municipal solid waste in its as-discarded form and that has not been size-reduced, separated, or otherwise processed.

volatile matter, ***n***—those products, exclusive of moisture, given off by a material as gas or vapor as determined by definite prescribed methods.

waste, ***n***—a material that is unwanted at its present location; that is no longer useful for its original purpose; that has been disposed, or any combination thereof.

waste composition, ***n***—of a solid waste, characterization of multi-constituent waste by a breakdown into specified waste components on the basis of mass or volume fraction or percentage. (*Syn.***solid waste composition**.)

4. Significance and Use

4.1 This terminology defines terms and specialized meanings of terms in the subject areas of waste and management of waste.

4.2 This terminology is not intended for subjects other than waste and waste management. For terms applicable to other subject areas, the appropriate terminology standard(s) should be consulted. See the current edition of the Compilation of ASTM Standard Definitions[3] and the list of terminology standards cited therein.

4.3 Standards relating to subcategories of waste or waste management may use terms defined more narrowly than those included here. The more specialized terminology standards relating to the applicable specific subcategory, or terms defined within individual standards, or both, should be consulted for the exact meaning intended within a given standard.

4.4 The Thesaurus on Resource Recovery Terminology (Special Technical Publication (STP) 832)[4] contains many terms and may be useful for those not listed in terminology standards. However, a definition in a standard terminology shall be considered governing when the term is used in the sense or meaning defined therein.

4.5 Statistical terms are not defined in this terminology to the extent that the terms, when used regarding waste and management of waste, have the same meanings as in Practice E 177 or Terminology E 456.

4.6 Regulatory terms are often developed by regulatory agencies for special regulatory purposes and may have technical content or meaning different from terms defined herein. When a regulatory term exists that differs in meaning from a term given here, the regulatory term should be considered to take precedence for regulatory matters.

[3] *Compilation of ASTM Standard Definitions*, ASTM, 8th edition, 1994.

[4] *Thesaurus on Resource Recovery Terminology*, ASTM STP 832, ASTM, 1983.

Designation: D 5743 – 97

Standard Practice for Sampling Single or Multilayered Liquids, With or Without Solids, in Drums or Similar Containers[1]

This standard is issued under the fixed designation D 5743; the number immediately following the designation indicates the year of original adoption or, in the case of revision, the year of last revision. A number in parentheses indicates the year of last reapproval. A superscript epsilon (ϵ) indicates an editorial change since the last revision or reapproval.

1. Scope

1.1 This practice covers typical equipment and methods for collecting samples of single or multilayered liquids, with or without solids, in drums or similar containers. These methods are adapted specifically for sampling drums having a volume of 110 gal (416 L) or less. These methods are applicable to hazardous material, product, or waste. Specific sample collection and handling requirements should be described in the site-specific work plan.

1.2 The values stated in inch-pound units are to be regarded as the standard. The values given in parentheses are for information only.

1.3 *This standard does not purport to address all of the safety concerns, if any, associated with its use. It is the responsibility of the user of this standard to establish appropriate safety and health practices and determine the applicability of regulatory limitations prior to use.* Specific precautionary statements are given in 7.2.7.1 and Notes 1 and 2.

2. Referenced Documents

2.1 *ASTM Standards:*

D 4687 Guide for General Planning of Waste Sampling[2]

D 5088 Practice for the Decontamination of Field Equipment Used at Non-Radioactive Waste Sites[3]

D 5283 Practice for Generation of Environmental Data Related to Waste Management Activities: Quality Assurance and Quality Control Planning and Implementation[2]

D 5495 Practice for Sampling with a Composite Liquid Waste Sampler (COLIWASA)[2]

2.2 *Other Documents:*

Drum Handling Practices at Hazardous Waste Sites, EPA/600/2-86/013, PB 165362, January 1986[4]

Accident Prevention Manual for Industrial Operations, 1992[5]

Occupational Safety and Health Guidance Manual for Hazardous Waste Site Activities, No. 85-115, October 1985[4]

3. Terminology

3.1 *Definitions:*

3.1.1 *bonding*—touching the sampling equipment to the drum to form an electrically conductive path to minimize potential electrical differences between the sampling equipment and drum, reducing the buildup of static electricity.

3.1.2 *bung*—usually a 2-in. (5.1-cm) or ¾-in. (1.3-cm) diameter threaded plug designed specifically to close a bung hole.

3.1.3 *bung hole*—an opening in a barrel or drum through which it can be filled, emptied, or vented.

3.1.4 *deheading*—removal of the lid of a closed-head drum; it is usually accomplished with a drum deheader.

3.1.5 *drum*—implicitly any drum, barrel, or non-bulk container of 5 to 110-gal (19 to 416-L) capacity.

3.1.6 *pail*—a small container, usually of 5-gal (19-L) capacity. Pails typically have bungs or spouts, or the entire lid can be removed.

3.1.7 *paperwork*—all required site documentation, which may include the manifests, waste profiles, material safety data sheets (MSDS), site forms, sample labels, seals, and chain of custody forms.

3.1.8 *sludge*—any mixture of solids that settles out of solution. Sludges contain liquids that are not apparent as free liquids.

3.1.9 *work plan*—a plan specific to a particular site; it is for conducting activities specified in the plan.

4. Summary of Practice

4.1 The drum and its contents are inspected, and appropriate sampling equipment is selected. A clean sampling device is lowered slowly into the liquid to be sampled. After the material has entered the device, it is removed from the drum. The contents of the device are discharged into a sample container. The sampling device is then either disposed of or cleaned and decontaminated.

5. Significance and Use

5.1 This practice is intended for use in collecting samples of single and multilayered liquids, with or without solids, from drums or similar containers, including those that are unstable,

---

[1] This practice is under the jurisdiction of ASTM Committee D34 on Waste Management and is the direct responsibility of Subcommittee D34.01.02 on Sampling Techniques.

Current edition approved Oct. 10, 1997. Published March 1998. Originally published as D 5743 – 95. Last previous edition D 5743 – 95.

[2] *Annual Book of ASTM Standards*, Vol 11.04.

[3] *Annual Book of ASTM Standards*, Vol 04.09.

[4] Available from National Technical Information Service, 5285 Port Royal Road, Springfield, VA 22161.

[5] Available from National Safety Council, P.O. Box 558, Hasca, IL 60143-0558.

ruptured, or otherwise compromised. Special handling procedures (for example, remote drum opening, overpressurized drum opening, drum deheading, etc.) are described in *Drum Handling Practices at Hazardous Waste Sites*.

6. Interferences

6.1 The condition of the materials to be sampled, and the condition and accessibility of the drums, will have a significant impact on the selection of sampling equipment.

7. Pre-Sampling

7.1 *General Principles and Precautions*:

7.1.1 Samples should be collected in accordance with an appropriate work plan (Practice D 5283 and Guide D 4687). This plan must include a worker health and safety section because there are potential hazards associated with opening drums as well as potentially hazardous contents. See the *Occupational Safety and Health Guidance Manual for Hazardous Waste Site Activities* for information on health and safety at hazardous waste sites.

7.1.2 Correct sampling procedures must be applied to conditions as they are encountered. It is impossible to specify rigid rules describing the precise manner of sample collection because of unknowns associated with each liquid sampling situation. It is essential that the samples be collected by a trained and experienced sampler because of the various conditions under which drummed liquids must be sampled.

7.1.3 To be able to make probability or confidence statements concerning the properties of a sampled lot, the sampling procedure must allow for some element of randomness in selection because of possible variations in the material. The sampler should always be on the alert for possible biases arising from the use of a particular sampling device or from unexpected segregation within the material.

7.1.4 The sampling equipment, sample preparation equipment, sample containers, etc. must be clean, dry, and inert to the material being sampled. All equipment, including sample containers, must be inspected before use to ensure that they are clear of obvious dirt and contamination and are in good working condition. Visible contamination must be removed, and the equipment must be decontaminated with the appropriate rinse materials. Decontaminated sampling equipment should be protected from contamination. This may include, but not be limited to, storage in aluminum foil, plastic bags, polytetrafluoroethylene (PTFE) film, or other means of protection that will not impact the sample quality or intended analysis.

7.2 *Basic Pre-Sampling Practices*:

7.2.1 Review all paperwork.

7.2.2 Select the sampling equipment and sample containers appropriate for the material in the drum, as detailed in the work plan.

7.2.3 Enter the work zone.

7.2.4 Inspect all drums to be sampled visually. Note any abnormal conditions (for example, rust marks, stains, bulges, or other signs of pressurization or leaks) that may require special handling. The work plan should clearly define the limiting conditions under which special handling procedures shall be initiated. See *Drum Handling Practices at Hazardous Waste Sites* for information on opening overpressurized drums and the use of remotely operated drum opening equipment.

7.2.5 Stage the drums to be sampled in a designated work area if they cannot be sampled in their current location. See *Drum Handling Practices at Hazardous Waste Sites* for further information on staging drums.

7.2.5.1 Move the drums to upright stable positions if necessary. Sufficient space shall be left between drums to prevent movement hazards.

7.2.5.2 Allow adequate time for the drum contents to stabilize if movement of a drum is required. The settling time is dependent on the type of material expected.

7.2.5.3 Number or identify uniquely all drums to be sampled.

7.2.6 Perform a detailed inspection of individual drums.

7.2.6.1 Record all relevant information from drum labels, markings, data sheets, and so forth, in the field log book or on forms specified in the work plan.

7.2.6.2 Verify that there are no discrepancies with existing paperwork.

7.2.6.3 Any discovered inconsistency from the paperwork (such as evidence of crystals on the drum exterior) should be noted in the field log book.

7.2.7 Slowly remove the bung or loosen the ring that secures the lid, allowing any pressure or vacuum to equalize.

7.2.7.1 *Precautionary Notes*:

(1) If the drum or pail appears to be under positive or negative pressure (that is, a slight bulge or dimple in the lid), control the release of pressure until it has equalized. For example, if the drum or pail is equipped with bungs, loosen the smaller bung first since doing so will make it easier to control the release of pressure.

(2) Pails equipped with snap-on lids may be difficult to open. Care must be exercised when opening to minimize the potential of splashing of the contents.

(3) If the top of the drum is dished inward (dimpled), it may "pop" when equalizing pressure, spraying the sampler with any material that is sitting on top of the drum.

(4) If there is evidence of a chemical reaction or sudden pressure buildup, the sampler should leave the area immediately and evaluate whether remote drum opening equipment should be used.

(5) For flammable or explosive materials, the drum and sampling equipment should be grounded if the generation of static electricity while opening or sampling the drum is a possibility. The drum and sampling equipment should be grounded to a ground stake or to an existing ground (building ground, grounded water pipes, etc.). New glass, plastic thiefs, or composite liquid waste samplers (COLIWASAs) may have some residual static electrical charge due to the materials in which they are packed and shipped. The work plan should specify whether grounding is required. See the *Accident Prevention Manual for Industrial Operations* for information on grounding and bonding.

7.2.7.2 Drums should be opened, sampled, and closed individually to minimize the risk of volatilization and exposure.

7.2.7.3 *Drums (or Pails) with Bungs*—When using a

manual bung wrench, cover it with a wipe or cloth to control potential liquid spray. Use non-sparking tools.

7.2.7.4 *Drums with Removable Lids*—Loosen the ring slowly with a manual wrench or air impact wrench. Use non-sparking tools.

7.2.7.5 *Pails with Removable Lids (Side-Lever Lock Ring)*—Release the lever slowly.

7.2.7.6 *Pails with Removable Lids (Snap-On)*—Pry the lid loose slowly with a pail lid opener.

7.2.8 Manual or remote puncturing or deheading will be required if the drum has a stuck bung or the lid cannot be removed. See *Drum Handling Practices at Hazardous Waste Sites* for further information on manual or remote drum opening.

7.2.9 If required, insert a measuring rod (graduated in litres or gallons) into the drum to measure the liquid volume and determine the presence of solids at the bottom and estimate their percentage. (If minimal disturbance of the contents is required, the measuring rod can be inserted in the vent bung hole when working with a bung-top-drum.) The rod can be graduated in litres or gallons for a specific size drum, or it can be graduated in linear units (inches, centimetres, and so forth), with the liquid depth converted to volume using an appropriate volume conversion. The measuring rod should be nonreactive to the waste being contacted.

NOTE 1—**Caution:** Before inserting the measuring rod into the drum, touch the rim gently with the rod (bonding) opposite from the bung to equalize any static charge that the drum may exhibit. The work plan should specify whether bonding is required.

7.2.9.1 For many liquids, the sampling equipment can serve as a substitute measuring device. This can be accomplished by measuring the length of the liquid column as it is being held over the drum and applying an appropriate volume conversion (for example, 1 in. (2.54 cm) equals 1.7 gal (6.43 L) in a 55-gal (208-L) drum).

NOTE 2—**Caution:** The sampling equipment or measuring rod should be at or near the temperature of the drummed liquid to minimize any reaction caused by temperature differences.

7.3 *Sampling Equipment, Selection*— Table 1 summarizes selection criteria for equipment by the material to be sampled.

7.4 *Sampling Equipment, Materials of Construction*—Each of the sampling devices listed should be constructed from materials that are inert to any materials that may be encountered at a specific site. These devices are usually made of glass, stainless steel, aluminum, brass, or plastic. Devices with permanent coatings or liners of an inert nonreactive material, such as PTFE, may be substituted, if approved by the work plan.

TABLE 1 Selection Criteria for Equipment

Equipment	ASTM Standard	One Liquid Layer	Two or More Liquid Layers	Liquid and Solid (Sludge) Layers
Drum thief		X[A]	X	X
COLIWASA	D 5495	X	X	N[B]
Syringe-type sampler		N	N	X
Coring-type sludge sampler		–[C]	–	X
Piston-Type Sampler		X	X	N

[A] X = equipment may usually be used with this type of waste.
[B] N = not the equipment of choice, but it may be used.
[C] – = equipment is probably unsuitable.

7.5 *Generic Equipment List*:

7.5.1 A list of equipment generally required for sampling liquids follows:

7.5.1.1 Sample containers, lids, and liners;

7.5.1.2 Sample labels;

7.5.1.3 COLIWASAs, drum thiefs, sludge samplers, or equivalent devices;

7.5.1.4 Measuring rods;

7.5.1.5 Chain of custody forms;

7.5.1.6 Field log books;

7.5.1.7 Sample cooler;

7.5.1.8 Wipes or cloths, or both;

7.5.1.9 Ice or gel ice;

7.5.1.10 Grounding cables with alligator clips and emery cloth; and

7.5.1.11 Portable monitoring equipment (combustible gas indicator, organic vapor detectors, radiation survey meter, etc.).

7.5.2 Equipment needed to open drums should be non-sparking (brass or beryllium copper) and include, but not be limited to, the following:

7.5.2.1 Bung wrenches (one straight and one bent),

7.5.2.2 Flathead screwdriver,

7.5.2.3 Breaker bar (½ in. (13 mm)),

7.5.2.4 Ratchet (½in. (13 mm)),

7.5.2.5 Speed handle (½ in. (13 mm)),

7.5.2.6 Adjustable wrenches (10 and 12 in. (25 and 30 cm)),

7.5.2.7 Air impact wrench and sockets, and

7.5.2.8 Pail lid opener.

8. Sample Collection

8.1 *Basic Sampling Practice*:

8.1.1 Bond the sampling equipment to the drum, if specified in the work plan.

8.1.2 Collect a sample from the drum. Whenever possible, do not sample where the measuring rod has been inserted; however, bung-type drums might not permit avoidance of the disturbed region.

8.1.3 Note the physical characteristics, including any discrepancies (such as solidified contents or crystalline material).

8.1.4 Place the collected material in a sample container.

8.1.5 Close the sample container.

8.1.6 Wipe the outside of the sample container. Dispose of the wipe cloth properly.

8.1.7 Record in the field log book all of the relevant conditions and physical characteristics associated with the sample.

8.1.8 Fill out all of the required paperwork for each sample, as required by the work plan.

8.1.9 Complete and attach a label to the side of the sample container before or after sampling, as directed by the work plan. The sample label should include the following:

(1) Sample ID number,

(2) Name of sampler,

(3) Sampler's initials or signature,

(4) Date and time of sampling, and

(5) Sample location.

The sample label can also include the following:

(1) Sampling information (for example, grab or composite),

(2) Preservative or preservation required,

(3) Special instructions, and

(4) Analysis request.

8.2 *Sampling with a Drum Thief*:

8.2.1 *General Description*—A tube of small diameter, which yields a vertical representation of the contents of a drum when lowered and sealed (see Fig. 1).

NOTE 3—When sampling liquids of high specific gravity, it may be difficult to retain the entire sample in the drum thief. A smaller-diameter drum thief may overcome this problem. The use of a COLIWASA or similar device may be necessary if the problem persists.

8.2.2 *Operation and Use*—Slowly insert the tube vertically until it reaches either the bottom of the drum or the liquid layer to be sampled. The sampling device should be lowered at a rate that permits the liquid level inside and outside the tube to be approximately the same.

NOTE 4—Multiple sample increments are usually necessary to provide enough sample volume for analysis and quality control (QC). Drum contents will become increasingly disturbed with each successive insertion of the drum thief.

8.2.2.1 Cover the top of the tube with the thumb or a rubber stopper to form a seal. Use gloves or a stopper, as described in the work plan.

8.2.2.2 Withdraw the tube carefully.

8.2.2.3 Use a clean cloth or paper towel to wipe the tube as it is being extracted from the liquid, to prevent unnecessary dripping.

8.2.2.4 Note the proportions of any layers or solids.

8.2.2.5 Place the bottom end of the tube into the sample container, and release the contents slowly.

8.3 *Sampling with a COLIWASA*:

8.3.1 *General Description*—A glass, plastic, or metal tube with an end closure that can be opened while the tube is immersed in the waste to be sampled (see Practice D 5495). The COLIWASA will yield a vertical representation of a drum's contents when immersed in the open position into a drum (see Fig. 2).

NOTE 5—Multiple sample increments are usually necessary to provide enough sample volume for analysis and QC. Drum contents will become increasingly disturbed with each successive insertion of the COLIWASA.

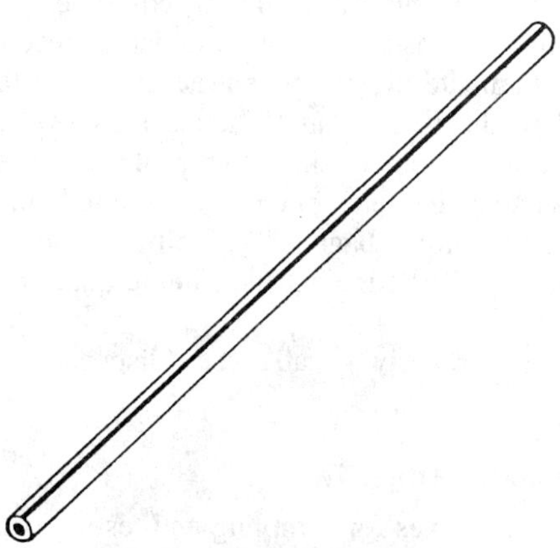

FIG. 1 Drum Thief

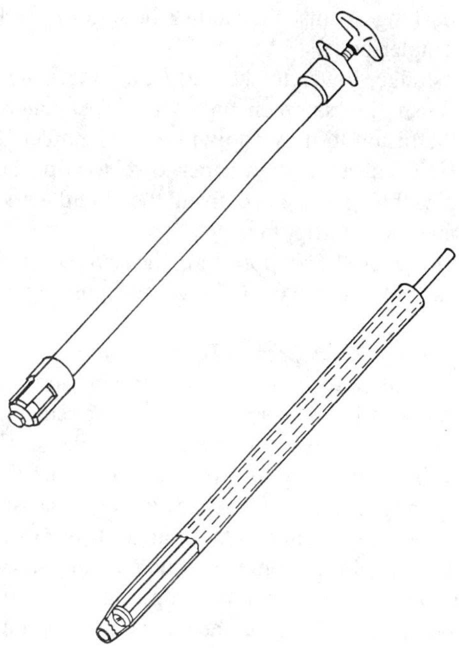

FIG. 2 COLIWASAs (Typical)

8.4 *Sampling with a Syringe-Type Sampler*:

8.4.1 *General Description*—A tube with a manually operated piston that can be used as a syringe for high-viscosity liquids or as a coring device for sludge (see Fig. 3).

8.4.2 *Operation and Use*—*(1)* For high-viscosity liquids, the tube is lowered to the sampling point and the piston is pulled out to collect the sample. *(2)* For sludge, the tube is lowered to the surface of the sludge. The sampler body is pushed into the sludge while allowing the piston to move up within the sampler body.

8.4.2.1 Assemble with the piston at the lower end of the sampler body. Attach the bottom valve (for high-viscosity liquids) or the coring tip (for sludge).

(1) For high-viscosity liquids, lower to the sampling point and withdraw the piston to collect the sample. Close the bottom

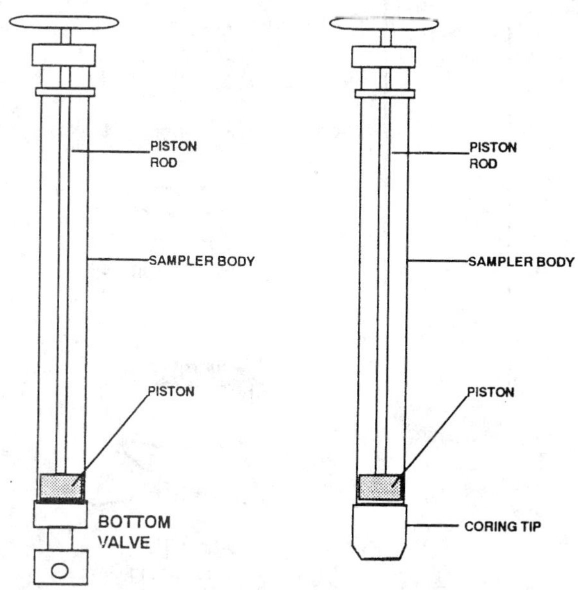

FIG. 3 Syringe-Type Sampler (Typical)

valve by pushing against the side or bottom of the container with the sampler body.

(2) For sludge, lower to the surface of the material to be sampled. Push the sampler body into the material while allowing the piston to move up within the sampler body.

8.4.2.2 Use a clean cloth or paper towel to wipe the sampler body as it is being extracted from the liquid or sludge, to prevent unnecessary dripping.

8.4.2.3 Transfer the sample into the sample container by opening the bottom valve, if fitted, and pushing the piston down.

8.5 *Sampling with a Coring-Type Sampler*:

8.5.1 *General Description*—A coring-type sampler consists of a cylinder, a coring tip (or auger tip) with a retaining device, a top cap, and an extension with a cross handle (see Fig. 4). A thin-walled internal sleeve may be used to contain the sample.

8.5.2 *Operation and Use*—The coring-type sampler is pushed (pushed and rotated with an auger tip) into the sludge to collect the sample and removed. The retaining device allows the sludge to enter the cylinder when pushing the sampler. The retaining device closes to hold the sludge in the cylinder while removing the sampler.

8.5.3 Remove the top cap and transfer the sample from the cylinder into a sample container. If equipped with an internal sleeve, remove the top cap and place an end cap on the internal sleeve. Invert, remove the internal sleeve from the cylinder, and place an end cap on the open end of the sleeve.

8.6 *Sampling with a Plunger-Type Sampler*:

8.6.1 *General Description*—A liquid sampling device that consists of a sample tube, sample line or rod, headsection, and plunger (see Fig. 5). The sample jar is connected to the headsection and the sample tube is lowered into the material to be sampled. After insertion the plunger is engaged into the sample tube lifting the sampled materials directly into a sample jar.

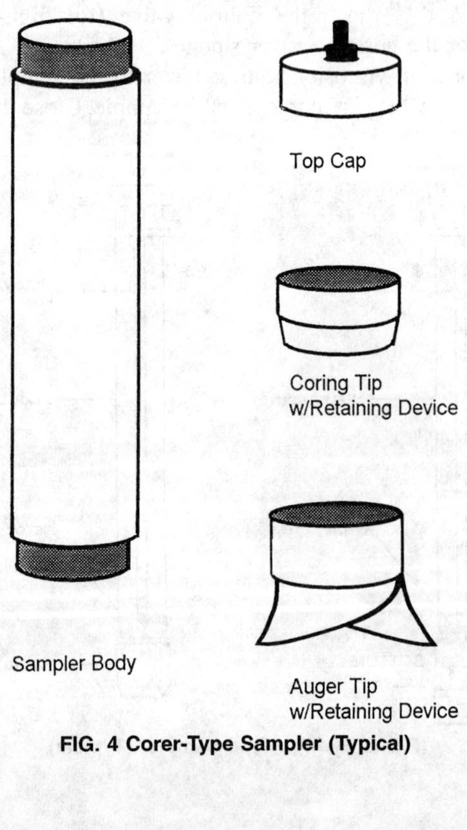

FIG. 4 Corer-Type Sampler (Typical)

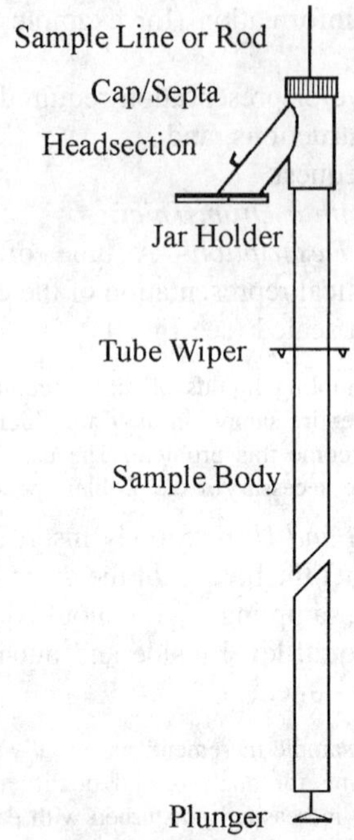

FIG. 5 Plunger-Type Sampler (Typical)

8.6.2 *Operation and Use*—Slowly insert the sampler vertically into the material to be sampled. Lower the sampler at a controlled rate to allow for representative sample collection. When the sampler reaches the bottom, or desired depth, the plunger mechanism is engaged. This secures the sample within the sample tube.

8.6.3 Without removing the sampler from the container and with the plunger lodged inside the sample tube, apply a constant and continual upward force on the sample line or rod. Retract the plunger at a rate that allows for a controlled delivery of the material into the sample jar. Continue to retract the plunger until the unit is seated into the headsection. Prior to removing the sample jar from the headsection, allow sufficient time for residual materials to drain.

9. Post-Sampling

9.1 Remove all sampling equipment from the work zone.

9.2 Transfer all reusable sampling equipment that was in contact with the waste to a predesignated decontamination area. Decontaminate the equipment according to the protocol established in the work plan (Practice D 5088). Decontaminated sampling equipment should be protected from contamination. This may include, but not be limited to, storage in aluminum foil, plastic bags, PTFE film, or other means of protection that will not impact the sample quality or intended analysis.

9.3 Dispose properly of all used (disposable) contacting equipment.

10. Data Quality Objectives

10.1 The objectives for sampling and testing of liquids and sludges should be specified in the work plan.

11. Quality Control

11.1 Quality Control (QC) samples (for example, equipment blanks, trip blanks, and duplicates) must be collected as required by the work plan. These QC samples must be evaluated to provide a determination of the quality of the sampling and reliability of the resulting analytical data.

12. Keywords

12.1 COLIWASA; drum; drum thief; liquid; pail; sampling; sludge sampler; waste

The American Society for Testing and Materials takes no position respecting the validity of any patent rights asserted in connection with any item mentioned in this standard. Users of this standard are expressly advised that determination of the validity of any such patent rights, and the risk of infringement of such rights, are entirely their own responsibility.

This standard is subject to revision at any time by the responsible technical committee and must be reviewed every five years and if not revised, either reapproved or withdrawn. Your comments are invited either for revision of this standard or for additional standards and should be addressed to ASTM Headquarters. Your comments will receive careful consideration at a meeting of the responsible technical committee, which you may attend. If you feel that your comments have not received a fair hearing you should make your views known to the ASTM Committee on Standards, at the address shown below.

Designation: D 5744 – 96 (Reapproved 2001)

Standard Test Method for Accelerated Weathering of Solid Materials Using a Modified Humidity Cell[1]

This standard is issued under the fixed designation D 5744; the number immediately following the designation indicates the year of original adoption or, in the case of revision, the year of last revision. A number in parentheses indicates the year of last reapproval. A superscript epsilon (ε) indicates an editorial change since the last revision or reapproval.

1. Scope

1.1 This test method covers a procedure that accelerates the natural weathering rate of a solid material sample so that diagnostic weathering products can be produced, collected, and quantified. Soluble weathering products are mobilized by a fixed-volume aqueous leach that is performed, collected, and analyzed weekly. When conducted in accordance with the following protocol, this laboratory test method has accelerated metal-mine waste-rock weathering rates by at least an order of magnitude greater than observed field rates (**1**).[2]

1.1.1 This test method is intended for use to meet kinetic testing regulatory requirements for mining waste and ores.

1.2 This test method is a modification of an accelerated weathering test method developed originally for mining wastes (**2-4**). However, it may have useful application wherever gaseous oxidation coupled with aqueous leaching are important mechanisms for contaminant mobility.

1.3 This test method calls for the weekly leaching of a 1000-g solid material sample, with water of a specified purity, and the collection and chemical characterization of the resulting leachate over a minimum period of 20 weeks.

1.4 As described, this test method may not be suitable for some materials containing plastics, polymers, or refined metals. These materials may be resistant to traditional particle size reduction methods.

1.5 Additionally, this test method has not been tested for applicability to organic substances and volatile matter.

1.6 This test method is not intended to provide leachates that are identical to the actual leachate produced from a solid material in the field or to produce leachates to be used as the sole basis of engineering design.

1.7 This test method is not intended to simulate site-specific leaching conditions. It has not been demonstrated to simulate actual disposal site leaching conditions.

1.8 This test method is intended to describe the procedure for performing the accelerated weathering of solid materials to generate leachates. It does not describe all types of sampling and analytical requirements that may be associated with its application.

1.9 The values stated in SI units are to be regarded as the standard. The values given in parentheses are for information only.

1.10 *This standard does not purport to address all of the safety concerns, if any, associated with its use. It is the responsibility of the user of this standard to establish appropriate safety and health practices and determine the applicability of regulatory limitations prior to use.*

2. Referenced Documents

2.1 *ASTM Standards:*

D 75 Practices for Sampling Aggregates[3]
D 276 Test Methods for Identification of Fibers in Textiles[4]
D 420 Guide to Site Characterization for Engineering, Design and Construction Purposes[5]
D 653 Terminology Relating to Soil, Rock, and Contained Fluids[5]
D 737 Test Method for Air Permeability of Textile Fabrics[6]
D 1067 Test Methods for Acidity or Alkalinity of Water[6]
D 1125 Test Methods for Electrical Conductivity and Resistivity of Water[7]
D 1193 Specification for Reagent Water[7]
D 1293 Test Methods for pH of Water[7]
D 1498 Practice for Oxidation-Reduction Potential of Water[7]
D 2234 Test Methods for Collection of a Gross Sample of Coal[8]
D 3370 Practices for Sampling Water[7]
E 276 Test Method for Particle Size or Screen Analysis at No. 4 (4.75-mm) Sieve and Finer for Metal-Bearing Ores and Related Materials[9]
E 877 Practice for Sampling and Sample Preparation of Iron Ores and Related Materials[10]

[1] This test method is under the jurisdiction of ASTM Committee D34 on Waste Management and is the direct responsibility of Subcommittee D34.01.04 on Waste Leaching Techniques.

Current edition approved March 10, 1996. Published May 1996.

[2] The boldface numbers in parentheses refer to the list of references at the end of this standard.

[3] *Annual Book of ASTM Standards*, Vol 04.03.
[4] *Annual Book of ASTM Standards*, Vol 07.01.
[5] *Annual Book of ASTM Standards*, Vol 04.08.
[6] *Discontinued*; see *1994 Annual Book of ASTM Standards*, Vol 07.01.
[7] *Annual Book of ASTM Standards*, Vol 11.01.
[8] *Annual Book of ASTM Standards* , Vol 05.05.
[9] *Annual Book of ASTM Standards*, Vol 03.05.
[10] *Annual Book of ASTM Standards*, Vol 03.06.

3. Terminology

3.1 *Definitions:*

3.1.1 *acid producing potential (AP), n*—the potential for a solid material sample to produce acidic effluent, based on the percent of sulfide contained in that sample as iron-sulfide mineral (for example, pyrite or pyrrhotite) **(3)**. The AP is commonly converted to the amount of calcium carbonate required to neutralize the resulting amount of acidic effluent produced by the oxidation of contained iron sulfide minerals; it is expressed as the equivalent tons of calcium carbonate per 1000 tons of solid material **(4)**. The AP is therefore calculated by multiplying the percent of sulfide contained in the material by a stoichiometric factor of 31.25 **(5)**.

3.1.2 *interstitial water, n*—the residual water remaining in the sample pore spaces at the completion of the fixed-volume weekly leach.

3.1.3 *leach, n*—a weekly addition of water to solid material that is performed either dropwise or by flooding for a specified time period.

3.1.4 *loading, n*—the product of the weekly concentration for a constituent of interest and the weight of solution collected that may be interpreted for water quality impacts.

3.1.5 *mill tailings, n*—finely ground mine waste (commonly passing a 150-μm (100 mesh screen) resulting from the mill processing of ore.

3.1.6 *neutralizing potential (NP), n*—the potential for a solid material sample to neutralize acidic effluent produced from the oxidation of iron-sulfide minerals, based on the amount of carbonate present in the sample. The NP is also presented in terms of tons of calcium carbonate equivalent per 1000 tons of solid material **(4)**. It is calculated by digesting the solid material with an excess of standardized acid and back-titrating with a standardized base to measure and convert the acid consumption to calcium carbonate equivalents **(3, 6)**.

3.1.6.1 *Discussion*—The AP and NP are specifically applicable to the determination of AP from mining wastes comprised of iron-sulfide and carbonate minerals. These terms may be applicable to any solid material containing iron-sulfide and carbonate minerals.

3.1.7 *run-of-mine, adj*—usage in this test method refers to ore and waste rock produced by excavation (with attendant variable particle sizes) from open pit or underground mining operations.

3.1.8 *waste rock, n*—rock produced by excavation from open pit or underground mining operations whose economic mineral content is less than a specified economic cutoff value.

4. Summary of Test Method

4.1 This accelerated weathering test method is designed to increase the geological-chemical-weathering rate for selected 1000-g solid material samples and produce a weekly effluent that can be characterized for solubilized weathering products. This test method is performed on each sample in a cylindrical cell. Multiple cells can be arranged in parallel; this configuration permits the simultaneous testing of different solid material samples. The test procedure calls for weekly cycles comprised of three days of dry air (less than 10 % relative humidity) and three days of water-saturated air (approximately 95 % relative humidity) pumped up through the sample, followed by a leach with water on Day 7. A test duration of 20 weeks is recommended **(3, 4)**.

5. Significance and Use

5.1 The purpose of this accelerated weathering procedure is to determine the following: (*1*) whether a solid material will produce an acidic, alkaline, or neutral effluent, (*2*) whether that effluent will contain diagnostic cations (including trace metals) and anions that represent solubilized weathering products formed during a specified period of time, and (*3*) the rate at which these diagnostic cations and anions will be released (from the solids in the effluent) under the closely controlled conditions of the test.

NOTE 1—Examples of products that can be produced from the test include the following: (*1*) weekly effluent acidity and alkalinity determined by titration and (*2*) weekly aqueous concentrations of cations and anions converted to their respective release rates (for example, the average release of μg sulfate ion/g of solid material sample/week, over a 20-week period). In acid drainage studies, for example, the average weekly rates of acid production (measured as μg/g/wk of sulfate released) determined from accelerated weathering tests of mine waste samples are compared with the AP present in each sample. The number of years of acidic effluent expected to be produced under laboratory accelerated weathering conditions can then be estimated from this comparison. The years of accelerated weathering required to deplete a mine waste sample's NP are calculated similarly by determining the average weekly calcium and magnesium release rates and dividing the sample's NP by the sum of those rates **(7)**.

5.2 The principle of the accelerated weathering test method is to promote more rapid oxidation of solid material constituents than can be accomplished in nature and maximize the loadings of weathering reaction products contained in the resulting weekly effluent. This is accomplished by controlling the exposure of the solid material sample to such environmental parameters as temperature, volume, and application rate of water and oxygen. Specifically, an excess amount of air pumped up through the sample during the dry- and wet-air portions of the weekly cycle ensures that oxidation reactions are not limited by low oxygen concentrations. Weekly leaches with low ionic strength water ensure the removal of leachable oxidation products produced from the previous week's weathering cycle. The purpose of the three-day dry-air portion of the weekly cycle is to evaporate water that remains in the pores of the sample after the weekly leach. Evaporation increases pore water cation/anion concentrations and may also cause increased acidity (for example, by increasing the concentration of hydrogen ion generated from previously oxidized iron sulfide). Increased acid generation will accelerate the dissolution of additional sample constituents. Precipitation occurs as evaporation continues, and the remaining water becomes over-saturated. Some of these precipitated salts are potential sources of acidity when re-solubilized (for example, melanterite, $FeSO_4{\cdot}7H_2O$; and jarosite, $K_2Fe_6(OH)_{12}(SO_4)_4$). During the dry-air portion of the cycle, the oxygen diffusion rate through the sample may increase several orders of magnitude as compared to its diffusion rate under more saturated conditions of the leach. This increase in the diffusion rate under near-dryness conditions helps to accelerate the abiotic oxidation of such constituents as iron sulfide. The wet (saturated)-air portion of the weekly cycle enhances the bacteria-catalyzed oxidation of solid material sample constituents (for example, iron sulfide) by providing a moist micro-environment throughout the available surface area of the 1000-g sample. This micro-environment promotes the diffusion of weathering products (for example, resolubilized precipitation products) and metabolic byproducts (for example, ferric iron) between the microbes and the substrate without saturating the sample and affecting oxygen diffusion adversely.

NOTE 2—Under idealized conditions (that is, infinite dilution in air and water), published oxygen diffusion rates in air are five orders of magnitude greater than in water (0.178 $cm^2 \cdot s^{-1}$ versus 2.5×10^{-5} $cm^2 \cdot s^{-1}$ at 0 and 25°C, respectively) (**8**). However, in the humidity cell setting, corresponding oxygen diffusion rates in porous media are also functions of solid phase porosity and attendant tortuosity. Actual diffusion rates will therefore be somewhat slower than five orders of magnitude.

5.3 This test method has been tested on both coal and metal mine wastes to classify their respective tendencies to produce acidic, alkaline, or neutral effluent, and to subsequently measure the concentrations of selected inorganic components leached from the waste (**2-4, 7**). The following are examples of parameters for which the weekly effluent may be analyzed:

5.3.1 pH, Eh (oxidation/reduction potential), and conductivity (see Test Methods D 1293, Practice D 1498, and Test Methods D 1125, respectively, for guidance);

5.3.2 Dissolved gaseous oxygen and carbon dioxide;

5.3.3 Alkalinity/acidity values (see Test Methods D 1067 for guidance);

5.3.4 Cation and anion concentrations; and

5.3.5 Metals and trace metals concentrations.

NOTE 3—Sulfate and iron concentrations in the weekly leachates from solid material containing iron-sulfide minerals should be monitored because their release rates are critical measurements of iron-sulfide mineral oxidation rates. Acidic effluent or acid drainage is a consequence of iron-sulfide mineral oxidation and the subsequent aqueous transport of resulting hydrogen ion and oxidation/dissolution products to the receiving environment (for example, surface and ground waters).

5.4 An assumption used in this test method is that the pH of each of the leachates reflects the progressive interaction of the interstitial water with the buffering capacity of the solid material under specified laboratory conditions.

5.5 This test method produces leachates that are amenable to the determination of both major and minor constituents. It is important that precautions be taken in sample preservation, storage, and handling to prevent possible contamination of the samples or alteration of the concentrations of constituents through sorption or precipitation.

5.6 The leaching technique, rate, liquid-to-solid ratio, and apparatus size may not be suitable for all types of solid material.

6. Apparatus

6.1 *Humidity Cell*—A modified column constructed of materials suitable to the nature of the analyses to be performed (see Practices D 3370 for guidance). Multiple humidity cells can be arranged in an array to accommodate the simultaneous accelerated weathering of different solid material types (Fig. 1). Two different sets of humidity cell dimensions are used to accommodate particle size differences present in the solid material:

6.1.1 Cells having suggested dimensions of 10.2-cm (4.0-in.) inside diameter (ID) by 20.3-cm (8.0-in.) height can be used to accommodate coarse solid material samples that have been either screened or crushed to 100 % passing 6.3 mm (¼in.).

6.1.2 Cells with suggested dimensions of 20.3-cm (8.0-in.) ID by 10.2-cm (4.0-in.) height can be used to accommodate solid material samples that pass a 150-µm (100-mesh) screen (examples would be processed mill tailings or fly ash).

6.1.3 A perforated disk (comprised of materials suitable to the nature of analyses to be performed), approximately 3.15-mm (⅛-in.) thick, with an outside diameter (OD) suitable to the suggested vessel ID (6.1.1 and 6.1.2) is elevated approximately 12.5 mm (½in.) above the cell bottom to support the solid material sample (see Fig. 1).

NOTE 4—The cell and particle size dimensions described above are those used commonly for assessing the potential of waste-rock and mill-tailings samples associated with coal and metal mining operations to produce acidic effluent. A "shoe box"-shaped cell design with similar dimensions is preferred by some researchers (**6**).

6.2 *Cylindrical Humidifier*, with suggested dimensions of 12.1-cm (4.75-in.) ID by 134.6-cm (53.0-in.) length. The following associated equipment are needed to provide saturated air for the three-day wet-air portion of the weekly cycle:

6.2.1 A thermostatically controlled heating element to maintain the water temperature at 30°C during the wet-air cycle.

6.2.2 An aeration stone (similar to aquarium-aeration equipment) or commercially available gas dispersion fritted cylinders or disks to bubble air into the humidifier water.

6.3 *Flowmeter*, capable of delivering air to each humidity cell at a rate of approximately 1 to 10 L/min/cell.

6.4 *Oil/Water Trap*, 0.01-µm, for inclusion in the feed-air line.

6.5 *BK Bacteria Filter Tube*, for inclusion in the feed-air line, which must be capable of retaining 99.99 % of 0.1-µm particles.

6.6 *Air-Exit Port Bubbler*—A 50-mL Erlenmeyer flask with a rubber stopper containing a vent and air-inlet tube (Fig. 1). The bubbler is connected to the air exit port in the humidity cell lid with flexible tubing. This helps maintain similar positive air pressure throughout all of the humidity cells.

6.7 *Flexible-Tubing Quick Disconnect*—A fitted, two-piece connection placed in the middle of the air-exit port flexible tubing so that the bubbler can be disconnected from the humidity cell to facilitate the measurement of air flow and relative humidity.

6.8 *Separatory-Funnel Rack*, capable of holding 500-mL or 1-L separatory funnels above the humidity cells.

6.9 *Desiccant Column*, 5.1-cm (2-in.) ID by 50.8-cm (20-in.) length, plastic or glass cylinder capped on both ends (one cap should be removable for desiccant replacement), with an air inlet port on the bottom and an air exit port on the top.

6.10 *Dry Air Manifold*—A line of plastic tubing exiting the desiccant column and containing multiple regularly spaced "tee" connectors to supply air to each humidity cell.

6.11 *Filter Media*, such as a 12-oz/yd^2 polypropylene felt characterized by 22-µm (0.009-in.) diameter filaments. The media should be able to transmit dry air at a rate of 20 to 30

FIG. 1 Side View of 16-Cell Array

cfm (see Test Methods D 276 and D 737 for guidance).

NOTE 5—Caution must be used in the selection of filter media materials since they may affect the effluent pH and chemistry adversely. Both pyrex wool and quartz wool retain as much as 10 to 15 g of water per g of wool (retained water tends to re-humidify the dry-air cycle to as much as 85 % relative humidity). Additionally, pyrex wool causes the neutral effluent pH to be raised by as much as 2 pH units due to leaching of the wool.[7] In addition, pyrex (borosilicate) can contribute boron if this is a constituent of interest.

6.12 *Two Riffle Splitters*, with 0.63-cm (0.25-in.) and 2.5-cm (1.0-in.) wide riffles, respectively; the riffle splitter is a commonly used device for obtaining representative splits of dry, free-flowing granular materials.

6.13 *Laboratory Balance*, capable of weighing to 0.1 g.

6.14 *Analytical Balance*, capable of weighing to 1.0 mg.

6.15 *Screen*, 6.3 mm (¼ in.).

6.16 *Screen*, 150 µm (100 mesh).

6.17 *Drying Oven*—Any thermostatically controlled drying oven capable of maintaining a steady temperature of 50 ± 2°C.

6.18 *pH Meter*—Any pH meter with a readability of 0.01 units and an accuracy of ±0.05 units at 25°C; two-channel operation (that is, pH and Eh) is desirable.

6.19 *Conductivity Meter*, capable of reading in micromohs (microseimens); calibrate at 25°C.

6.20 *Separatory Funnel*, 500 mL or 1 L, one per each humidity cell.

6.21 *Erlenmeyer Flask*, 500 mL or 1 L, one per each humidity cell.

6.22 *Volumetric Flask*, 500 mL or 1 L.

6.23 *Digital Hygrometer/Thermometer*, with a relative humidity range of 5 to 95 %, and temperature range of − 40 to

104°C (–40 to 220°F).

7. Reagents

7.1 *Purity of Reagents*—Reagent grade chemicals shall be used in all tests. Unless otherwise indicated, it is intended that all reagents conform to the specifications of the Committee on Analytical Reagents of the American Chemical Society, where such specifications are available.[11]

7.2 *Purity of Water*—Unless otherwise indicated, references to water shall be understood to mean reagent water as defined by Type III at 18 to 27°C conforming to Specification D 1193. The method by which the water is prepared, that is, distillation, ion exchange, reverse osmosis, electrodialysis, or a combination thereof, should remain constant throughout testing.

7.3 *Purity of Air*—The feed air line shall contain a 0.01-µm oil/water trap and a grade BK bacteria filter tube in advance of the flowmeter.

8. Sampling

8.1 Collect the samples using available sample methods developed for the specific industry (see Practices D 75 and E 877, Guide D 420, Terminology D 653, and Test Methods D 2234).

8.2 The sampling methodology for materials of similar physical form shall be used where no specific methods are available.

8.3 The amount of material to be sent to the laboratory should be sufficient to provide 8 to 10 kg of bulk sample for splitting and testing (see 9.3).

NOTE 6—Additional information on theory and methods for obtaining representative samples is contained in Pitard (**9**).

8.4 To prevent sample contamination or constituent loss prior to testing, store the samples in closed containers that are appropriate to the sample type and desired analyses (see Guide D 420 for guidance).

8.5 The time elapsed between sample collection and subsequent humidity cell testing should be minimized to reduce the amount of sample pre-oxidation (see Practices D 3370 for guidance). Report the length of time between sample collection and testing.

9. Sample Preparation

9.1 Air dry as-received bulk samples of solid material to prevent the additional oxidation of reactive minerals or compounds. If air drying is not practicable, oven dry the solid material at a maximum temperature of 50 ± 2°C for 24 h, or until a constant weight is reached.

9.1.1 If exploration-generated or run-of-mine solid material samples are not readily available, archived dried and crushed samples from geological exploratory or development drilling programs may be used for preliminary evaluations of ore and waste rock from new operations; this is provided that the available solid material samples are not significantly finer than 95 % passing a No. 12 (1.7-mm) sieve. Document the sample drying and preparation procedures used during the drill sampling program in order to interpret the results properly. Evaluate the effects of drying temperature on metals volatilization (for example, mercury in cinnabar vaporizes at temperatures exceeding 80 to 90°C) and mineral morphology and chemistry modifications (for example, on heating at temperatures exceeding 100°C, chalcocite changes crystal form and is oxidized subsequently from Cu_2S to CuO, $CuSO_4$, and SO_2). Especially ensure that the effects of particle size distribution changes resulting from the more finely crushed sample are considered in the interpretation (that is, the potential for increased liberation of acid-producing and acid-consuming minerals with an attendant increase in mineral surface area).

9.1.2 In mining waste evaluations, the particle size for mill tailings will be significantly finer (commonly less than 150 µm/100 mesh) than the particle size distributions from ore and waste rock. Pilot plant tailings should be used if mill tailings are not available.

9.2 Screen the air-dried bulk samples through a 6.3-mm (¼-in.) screen in accordance with Test Method E 276. Crush any oversize material so that 100 % passes the screen.

NOTE 7—**Caution:** Recent accelerated weathering studies of run-of mine waste rock from metal mines demonstrate that crushing a bulk sample so it passes a 6.3-mm (¼-in.) screen may change the character of the sample by artificially increasing liberation and consequent surface areas of acid-producing and acid-consuming minerals contained in the + 6.3-mm (¼-in.) material. A suggestion for avoiding this problem is to segregate the – 6.3-mm (¼-in.) fraction by screening rather than crushing, and to test that fraction according to the protocol and equipment described in this test method. The + 6.3-mm (¼-in.) material can be tested separately (for example, Brodie, et al (**10**) describe a large-scale humidity cell test that would accommodate – 75-mm material). Samples from the drill core and cuttings also present material sizing problems, which must be considered when interpreting drill core and cuttings accelerated weathering data. The drill core must be crushed to – 6.3-mm (¼-in.) to fit the cell described in this test method. The resulting size distribution from crushing will differ from that of run-of-mine due to differences in fracture patterns inherent to blasting practices that produce run-of-mine material. By contrast, drill cuttings size fractions are commonly less than 6.3-mm (¼-in.) due to the rotary-percussive nature of obtaining the sample.

9.3 Mix and divide the bulk sample to obtain a representative test unit with a weight in the range of 8 to 10 kg, using a riffle splitter with 1-in. (2.54-cm) chutes. Divide the test unit into eight nominal 1-kg test specimens. Seal each test specimen in a moisture-barrier bag.

NOTE 8—The dried sample should be mixed through the riffle splitter at least once before making any splits; recombine the splits resulting from the sample mixing exercise by pouring individual splits either over each other or through the splitter again. Once the actual split is made, it is wise to re-mix it (according to the above procedure) prior to making the next split.

9.4 Select one test specimen at random, and determine the moisture content by weighing and drying to constant weight at 80 ± 5°C.

9.4.1 Crush the dried test specimen so that at least 95 % passes a 1.7-mm (10-mesh) screen, in accordance with Test Method E 276.

[11] *Reagent Chemicals, American Chemical Society Specifications*, American Chemical Society, Washington, DC. For suggestions on the testing of reagents not listed by the American Chemical Society, see *Analar Standards for Laboratory Chemicals*, BDH Ltd., Poole, Dorset, U.K., and the *United States Pharmacopeia and National Formulary*, U.S. Pharmaceutical Convention, Inc. (USPC), Rockville, MD.

9.4.2 Divide the crushed test specimen in half twice, using a riffle splitter with 6.35-mm (¼-in.) chutes, and select a ¼ subsample at random.

9.4.3 Transfer the selected subsample to a ring and puck grinding mill and grind to a nominal 95 % passing a 150-µm (100-mesh) screen, in accordance with Test Method E 276. Use the subsample for chemical and mineralogical characterization of the test unit.

9.5 Select one test specimen at random, and determine the particle size distribution in accordance with Test Method E 276.

9.6 Select one test specimen at random for use in the accelerated weathering test method. Divide the test specimen into four nominal 250-g subsamples using the riffle splitter with 25.4-mm (1-in.) chutes, and label and store in vapor-barrier bags until it is time to load the humidity cells.

9.7 Reserve the remaining test specimens for replicated testing or to resolve disputed results.

10. Apparatus Assembly

10.1 The humidity cells are table-mounted at a height sufficient to accommodate the placement of both the humidifier and one Erlenmeyer flask for effluent collection from the bottom of each cell (Fig. 1). During the water-saturated and dry-air portions of each weekly cycle, feed air is metered to the bottom of each cell at the selected rate (1 to 10 L/min). Feed air for the three-day dry-air portion is routed first through a desiccant column and then to each of the cells through a dry-air manifold (Fig. 2 and Fig. 3). Feed air for the water-saturated air portion is routed through a water-filled humidifier by means of aeration stones or gas dispersion fritted cylinders/disks, and then to each humidity cell (Fig. 2). Attach a water-bubbling vessel to each humidity cell lid air exit port to prevent the short circuiting of air through cells containing more permeable solid material samples (Fig. 1). A separatory funnel rack is mounted on the table that holds the cells if the weekly water leach is applied dropwise (drip trickle). Multiple separatory funnels (one for each cell) are held in the rack during the drip trickle leach that is performed on the seventh day of each weekly cycle (Fig. 2). The separatory funnel can be used to meter the required water volume slowly down the sides of the cell wall until the sample is flooded if the weekly leach is to be a flooded leach.

11. Procedure

11.1 *Cell Loading*:

11.1.1 If more than one humidity cell is used at one time, label each with a sequential number, and use the same number for the matching collection vessel (Erlenmeyer flask).

11.1.2 Weigh each humidity cell (without its lid) and each collection vessel; record the tare weights of each to the nearest 0.1 g.

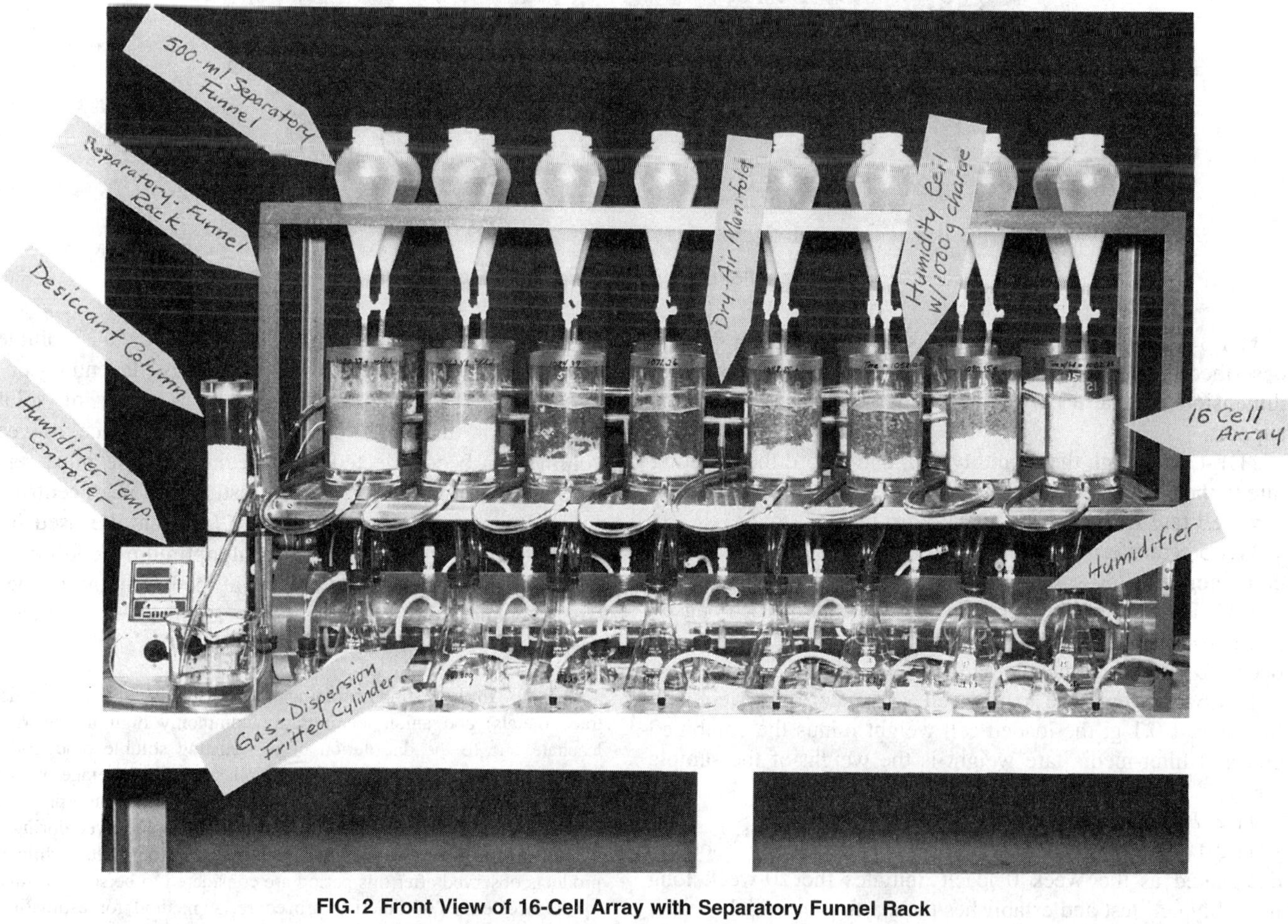

FIG. 2 Front View of 16-Cell Array with Separatory Funnel Rack

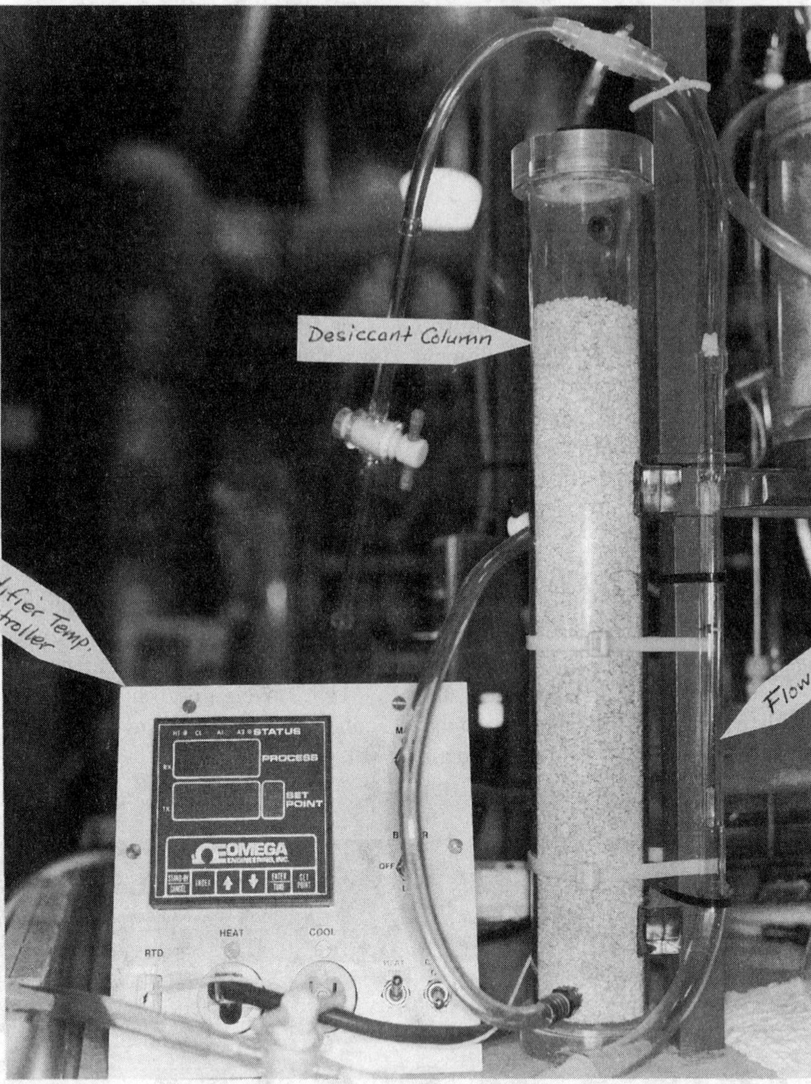

FIG. 3 Detail of Desiccant Column and Flowmeter

11.1.3 Cut the filter media (such as 12-oz/yd^2 polypropylene described in 6.11) to the humidity cell's inside diameter dimensions so that it fits snugly yet lies flat on the perforated support.

11.1.4 Re-weigh the humidity cell, and record the resulting tare to the nearest 0.1 g; the original cell tare (11.1.2) minus the new cell tare is the weight of the filter media.

11.1.5 Transfer the contents from each of the four bags containing the 250-g samples (9.6) into the humidity cell (see Fig. 4). Prior to the transfer, mix the contents of each bag by gentle rolling to eliminate possible stratification that may have occurred during sample storage.

11.1.6 Re-weigh the loaded cell, and record the weight to the nearest 0.1 g; the loaded cell weight minus the combined cell and filter-media tare weight is the weight of the sample charge.

11.2 *First Leach*:

11.2.1 The first leach (whether drip trickle or flooded), designated as the Week 0 leach, initiates the 20-week long humidity cell test and establishes the starting or initial characteristics of the leachate. Either a 500-mL or 1-L volume of water may be used for the weekly leaches, depending on the weekly pore volume desired or the quantity of solution required for analytical purposes; however, once a weekly volume has been selected, that weekly volume must remain constant throughout the 20-week testing period. A centrifuged cell culture of *Thiobacillus ferrooxidans* may be used in the first leach in order to ensure that optimum conditions for accelerated weathering are present at the beginning of the test (see Appendix X1 for the preparation of a washed cell suspension of *Thiobacillus ferrooxidans*).

NOTE 9—In the testing of mining wastes, cation (including metals and trace metals) and anion loadings are commonly high in the Week 0 leachate due to the dissolution of pre-existing soluble oxidation salts present in the sample prior to sample collection. The average number of weekly accelerated weathering cycles required to flush these pre-existing salts ranges from 3 to 5 weeks. Oxidation products observed during these 3 to 5 weeks are principally from the pre-existing salts, while those products observed after this period are considered to be solely a function of the accelerated weathering procedure. A method for estimating the

FIG. 4 Loading Humidity Cell with Filter Media and 1000-g Sample Charge

amount of pre-existing oxidation salts present in a solid material sample is described by Sobek, et al (**6**). A comparison of estimated salt storage data obtained using this method with the first three weeks of humidity cell effluent loadings from three different samples is described by White and Jeffers (**7**).

11.2.2 Fill a separatory funnel for each cell with de-ionized water using a volumetric flask. If the leach is to be performed using the drip trickle method, set each separatory funnel above its corresponding cell, and adjust the drip rate (approximately 3 to 4 mL/min) so that the solid material sample is wetted thoroughly but not flooded.

11.2.3 A minimum of 2 to 3 h is commonly required to complete the drip trickle leach.

11.2.4 If the leach is to be performed by flooding, the separatory funnel can be used to meter the selected water volume slowly down the sides of the cell wall until the sample is flooded. This application method reduces hydraulic agitation of the sample surface commonly caused by pouring liquid from an open-mouthed vessel. Alternatively, flooding may be accomplished by any application apparatus (for example, a peristaltic pump) that supplies the selected volume of leachant at a reasonable rate without causing agitation and suspension of the finer fractions contained in the sample charge.

11.2.4.1 Allow the flooded cell to sit for a period of 1 h before draining the leachate into the Erlenmeyer collection flask. The 1-h leach time commences after all of the leachant has been placed in the cell. The solid material sample should be saturated and covered with leachant to a depth sufficient to maintain sample saturation. In testing mining wastes, the observed depth of leachant cover from a 500-mL flooded leach performed in 10.2-cm (4.0-in.) ID cells is approximately 2.5 cm (1.0 in.).

11.2.5 The following is performed once the leaching process has been completed: to reduce the effects of evaporation, and to prevent the contamination of each cell by airborne contaminants, place the lids on their corresponding cells and let the cells complete the leachate draining process for the remainder of the leaching day and overnight.

11.2.6 Disconnect the cells on the day following the leach, and weigh and record the weight of each cell and Erlenmeyer collection flask. Set each filled collection flask aside for leachate analyses. (Measurements of pH and Eh and sample preservation procedures must be performed as soon as possible after leachate collection.) Return each cell, replace the filled collection flasks with clean, tared Erlenmeyer flasks, hook up all connections, and begin the dry-air cycle.

11.3 *Dry-Air Cycle*:

11.3.1 The commencement of the three-day dry-air period marks the beginning of each new weekly cycle of the accelerated weathering humidity cell test; the first full-week cycle after the first leaching is designated Week 1; subsequent weeks (commencing with the second dry-air period) are designated as Week 2, Week 3 .... Week *n*, etc.

11.3.2 To perform the dry-air cycle, feed air is metered to the humidity cell array with a flowmeter (see 6.3) set at a target rate in the range of 1 to 10 L/min per cell, depending on the objectives of the testing. The air flow rate must be checked daily and adjusted to the target value ±0.5 L/min.

11.3.3 Feed air from the flowmeter is routed first through a desiccant column and then to each of the cells through a dry-air manifold (Fig. 2). Air exiting the desiccant column should have a relative humidity of less than 10 % as measured with a hygrometer (see 6.23).

11.3.4 To maintain similar positive air pressure through the cells, attach a water-bubbling vessel to each humidity cell air exit port coming out of the humidity cell lid; a 50-mL Erlenmeyer flask with a rubber stopper containing a vent and air inlet tube serves as a simple and efficient bubbler (Fig. 1).

11.3.5 The dry air is passed through each humidity cell for three days. Air flow rates from each of the cells should be checked each day, recorded, and adjusted, if necessary. See also Note 10.

11.4 *Wet-Air Cycle*:

11.4.1 The three-day wet-air period commences on the fourth day of each weekly cycle.

11.4.2 To perform the wet-air cycle of the method, feed air is routed through a water-filled humidifier via aeration stones or gas dispersion fritted cylinders/disks and then to each humidity cell (Fig. 2).

11.4.3 The water temperature in the humidifier is maintained at 30 ± 2°C to ensure that the sparged air maintains a relative humidity of approximately 95 % as measured with a hygrometer (see 6.23) from one of the humidifier exit lines (see Fig. 1). Air flow rates to each of the cells should be checked each day, recorded, and adjusted, if necessary.

NOTE 10—It is good practice to measure the air flow rates and relative humidity of the air exiting each humidity cell during each day of the three-day dry- and wet-air periods; the measurements should be taken at the same time each day from the humidity cell air exit port; these measurements can be accomplished by installing a quick-disconnect fitting in the tubing that connects the air exit port to the bubbler (Fig. 1).

NOTE 11—Coals spoils in eastern states are commonly saturated; Caruccio **(11)** has suggested the following geographic control alternative to the dry-air versus saturated-air scheduling:

(*1*) *Eastern States Samples*—Six days of saturated air (versus three days dry/three days wet); and

(*2*) *Western States Samples*—Three days dry/three days wet.

11.5 *Subsequent Weekly Leaches*:

11.5.1 A second leach with water is performed on the day following the end of the three-day wet-air period (that is, day seven of the first weekly cycle). This leach marks the end of the first weekly cycle and is designated as the Week 1 leach.

11.5.2 Subsequent leaches are designated as Week 2, Week 3 ... Week *n*, and they mark the end of the weekly cycle for that numbered week. Perform each weekly leach as described in 11.2.2-11.2.5. Weekly weighing of the test cells is optional.

11.6 It is recommended that the weekly accelerated weathering cycles described in 11.2 11.311.4 11.5 be performed for a minimum of 20 weeks.

NOTE 12—Additional weeks of accelerated weathering may be required to demonstrate the nature of the material, depending on the chemical composition of the solid material. For some metal mining wastes, researchers have shown that as much as 60 to 120 weeks of accelerated weathering data may be required to demonstrate the complete weathering characteristics of a particular sample **(7, 12)**. The criteria for ending the testing may be site specific and should be agreed upon before initiating the testing.

11.7 *Leachate Analyses*:

11.7.1 Analyze the leachates for specific constituents or properties, or use them for biological testing procedures as desired, using (*1*) appropriate ASTM test methods or (*2*) methods accepted for the site where disposal will occur. Where no appropriate ASTM test methods exist, other test methods may be used and recorded in the report, provided that they are sufficiently sensitive to assess potential water quality impacts at the proposed disposal site. Suggested minimum weekly analyses should include pH, Eh, conductivity, and sulfate-ion concentration; acidity, alkalinity, and selected metals could be analyzed less frequently (for example, at Weeks 0, 1, 2, 4, 8, 12, 16, and 20), especially if changes in leachate chemistry are slow. Whether visible phase separation during storage of the leachates occurs or not, appropriate mixing should be used to ensure the homogeneity of the leachates prior to their use in such analyses or testing.

11.7.2 Table 1 is an example of a spreadsheet format used for recording 20 weeks of leachate analytical data.

11.7.3 Fig. 5 is an example of a method used to plot the temporal variation (by week) of leachate pH, sulfate load, and cumulative sulfate load from 21 weeks of accelerated weathering (see 12.9 for the calculation of cumulative load and release rates).

11.8 *Weathered Solid Material Analyses*:

11.8.1 Weigh the humidity cell after collection of the final effluent and completion of a three-day dry-air period.

11.8.2 Transfer the weathered residue and filter media to a clean drying pan, and dry to constant weight at 50 ± 5°C. Record the final weight.

NOTE 13—Perform any gross sample examination (for example, sample texture and weathering product mineralogic characterization) desired for the weathered residues prior to pulverization. To facilitate such an examination, empty the humidity cell contents into a clean drying pan carefully by pushing gently on the bottom of the perforated plate with a wooden dowel until the sample exits the cell mouth. The perforated plate is accessed through the humidity cell drain port (see Fig. 1).

11.8.3 Identify and mark the top versus bottom portions of

TABLE 1 Example Format for Recording 20 Weeks of Humidity Cell Leach Data

Cell 6,8C											
Week>>>	0	1	2	3	4	5	6	7	8	9	10
Concentration (μg/g)											
*Cu	9.190	0.103	0.051	0.078	0.064	0.062	0.058	0.074	0.003	0.130	0.060
*Zn	8.30	0.42	0.22	0.62	0.20	0.12	0.13	0.11	0.10	0.19	0.34
SO_4	4361	2568	1737	1763	1616	1635	1843	1424	1790	1540	1200
Liquid weight (g)	387.7	454.8	424.7	399.4	413.0	391.1	423.0	398.5	434.3	403.9	394.5
Loads (μg × 10^A – 3)											
*Cu	3.56	0.05	0.02	0.03	0.03	0.02	0.02	0.03	0.00	0.05	0.02
*Zn	3.22	0.19	0.09	0.25	0.08	0.05	0.05	0.04	0.04	0.08	0.13
SO_4	1691	1168	738	704	667	639	780	567	777	622	473
Cum (μg × 10^A – 3)											
SO_4	1690.8	2858.7	3596.4	4300.5	4967.9	5607.4	6387.1	6954.5	7731.9	8353.9	8827.3
Condition	4960	3900	2500	2290	2420	2530	2380	2300	2810	1987	1822
pH	3.160	5.020	4.730	4.5	5.030	4.560	4.76	3.97	4.37	4.28	4.32
Eh	586.9	613	582.2	597.3	521	553.3	504.7	555.8	610.7	590.3	549.3

Week>>>[A]	11	12	13	14	15	16	17	18	19	20
Concentration (μg/g)										
*Cu	0.146	0.108	0.174	0.240	0.250	0.260	0.450	0.640	0.830	1.020
*Zn	0.31	0.13	0.20	0.28	0.32	0.37	0.48	0.59	0.70	0.81
SO_4	1220	1239	1117.5	996	1040	1084	1175	1266	1357	1448
Liquid weight (g)	392.4	407.6	419.0	412.9	382.3	426.7	392.5	406.8	421.4	399.3
Loads (μg × 10^A – 3)										
*Cu	0.06	0.0	0.1	0.10	0.1	0.11	0.2	0.3	0.3	0.41
*Zn	0.12	0.1	0.1	0.11	0.1	0.16	0.2	0.2	0.3	0.32
SO_4	478.5	505.0	468.2	411.2	397.6	462.5	461.2	515.0	571.8	578.2
Cum (μg × 10^A – 3)										
SO_4	9306	9811	10 279	10 690	11 088	11 550	12 012	12 527	13 099	13 677
Condition	1872	1980	1823	1570	1905	2010	1636	2050	1751	2040
pH	3.84	4.05	4.23	3.81	3.47	4	3.4	3.37	3.21	3.02
Eh	552.2	551.3	570.1	561.6	556.1	579.8	567.5	565.2	583.1	578.2

[A] Some of these blocks did not have data and were filled by linear interpolation, or these values were affected by the linear interpolation.

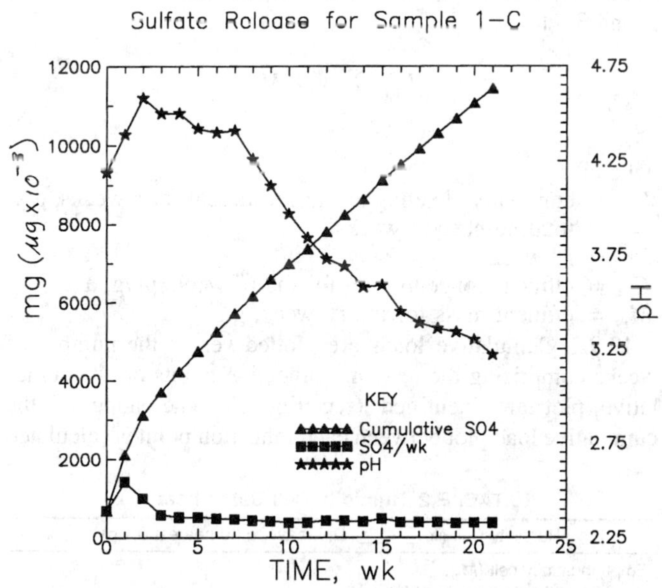

FIG. 5 Plot of Temporal Variation of pH, Sulfate Load, and Cumulative Sulfate Load from 21 Weeks

the sample for gross sampling purposes. Formations of cemented lumps of sample termed "ferricrete" that result from the accelerated weathering process are common in iron-sulfide-mineral rich samples. Depending on the sample mineralogy, the degree of "ferricrete" cementation may vary vertically within the sample, and the investigator may wish to segregate the sample into upper, middle, and lower thirds to document and characterize such changes.

11.8.4 After drying to constant weight and prior to splitting, use an instrument such as a rolling pin to break up cemented lumps in the sample (if the cemented lumps cannot be sufficiently reduced to pass through the chutes of a riffle splitter, remove, record, and weigh separately):

11.8.4.1 Split the sample into halves using a riffle splitter with 2.54 cm (1 in.) chutes, and reserve one half to determine the particle size distribution in accordance with Test Method E 276.

11.8.4.2 Split the remaining half sample into two quarters using a riffle splitter with 2.54-cm (1-in.) chutes, and submit one quarter for mineral characterization; pulverize the other quarter in either a ring-and-puck or disk-pulverizing machine to 95 % passing a 150-μm (100-mesh) screen in accordance with Test Method E 276.

11.8.5 Mix the pulverized residue in a blender or on a rolling cloth. Use the prepared residue for chemical characterization and for comparison with the pre-weathered solid material sample.

12. Calculation

12.1 Calculate the mass, in g, of the dry filter media:

$$M_f = M_{hf} - M_h \tag{1}$$

where:

M_f = mass of the filter media, g,
M_{hf} = mass of the humidity cell and filter media, g, and
M_h = mass of the humidity cell, g.

12.2 Calculate the mass, in g, of the dry solid material contained in the humidity cell:

$$M_{sd} = M_{hfsd} - M_{hf} \tag{2}$$

where:

M_{sd} = mass of the dry solid material, g,

M_{hfsd} = mass of the humidity cell, filter, and solid material, g, and

M_{hf} = mass of the humidity cell and filter media, g.

12.3 Calculate the mass, in g, of residual interstitial leachant contained in the solid material at the completion of the leach:

$$M_i = M_{hfsw} - M_{hfsd} \tag{3}$$

where:

M_i = mass of the residual interstitial leachant contained in the material, g,

M_{hfsw} = mass of the humidity cell, filter, and solid material after leach, g, and

M_{hfsd} = mass of the humidity cell, filter, and dry solid material, g.

12.4 Calculate the mass, in g, of the weekly collected effluent:

$$M_e = M_{ef} - M_{et} \tag{4}$$

where:

M_e = mass of the collected effluent, g,

M_{ef} = mass of the collection flask and collected effluent, g, and

M_{et} = mass of the tared collection flask, g.

12.5 Calculate the weekly loading, in µg, of the constituents of interest:

$$L_e = C_e \times M_e \tag{5}$$

where:

L_e = loading of the constituent of interest in the effluent, µg,

C_e = concentration of the constituent in the effluent, µg/g, and

M_e = mass of the weekly collected effluent, g.

12.5.1 If an analyte is not measured during a particular week, it may be estimated by linear interpolation between data points. Values below detection limits for the analytical method have zero loading for the affected week.

12.6 Calculate the final residue loading, in µg, of the constituents of interest:

$$L_r = C_r \times M_r \tag{6}$$

where:

L_r = loading of constituent in the residue, µg,

C_r = concentration of the constituent in the residue, µg/g, and

M_r = mass of the dried weathered residue and filter media, g.

12.7 Calculate the calculated head concentration of the constituents of interest:

$$C_h = (L_{e0} + L_{e1} + L_{e2} ... + L_{ef} + L_r)/M_{sd} \tag{7}$$

where:

L_{e0} = loading of the constituent for Week 0, µg,

L_{e1} = loading of the constituent for Week 1, µg,

L_{e2} = loading of the constituent for Week 2, µg,

L_{ef} = loading of the constituent for the final week, µg,

L_r = loading of the constituent in the residue, µg, and

M_{sd} = mass of the dry solid material at the start of the test, g.

12.8 Calculate the difference between the initial chemical analyses and the calculated head for the constituents of interest. It is recommended that if they differ by more than 10 %, the quality assurance procedures be checked, any deficiencies be subsequently corrected, and the analyses be repeated.

NOTE 14—Although agreement within 10 % is desirable, several assays must be summed to determine the total load that increases the potential for error. Moreover, determining the head concentration may be difficult since small volumes are typically analyzed. Normalization of the weekly loadings may be required based on the beginning and ending residue analyses.

NOTE 15—Table 2 and Table 3 are examples of recording formats used to record weekly humidity cell and collection flask data.

12.9 Release rates for constituents of interest (diagnostic cations and anions) are calculated in two steps:

12.9.1 Weekly loads are determined by multiplying the constituent concentrations (determined from weekly leachate analyses) by the mass of recovered leachate; cumulative constituent loads are then determined by summing the respective weekly loads (for example, the cumulative load for Week 1 is the sum of loads for Week 0 and Week 1, and the cumulative load for Week 3 is the sum of loads for Weeks 0, 1, 2, and 3, etc.):

$$L_n = \sum_{i-0}^{n} (C_i \times M_i) \tag{8}$$

where:

L_n = cumulative loading of the constituent for n weeks, µg,

n = total number of weeks,

i = i^{th} week,

C_i = effluent concentration for the ith week, µg/g, and

M_i = effluent mass for the ith week, g.

12.9.2 Cumulative loads are plotted versus the number of weeks comprising the test, and inflection points on the cumulative plot are identified (see Fig. 5). The slope of the cumulative load plot between each inflection point is calculated

TABLE 2 Humidity Cell Data Sheet

Humidity Cell No.______: Dry mass, g (to nearest 0.1 g)
Empty humidity cell (M_h):
Humidity cell + filter media (M_{hf}):
Filter media (M_f)
Humidity cell + filter + sample (M_{hfsd}):
Sample charge (M_{sd}):
Erlenmeyer collection flask (M_{et}):

Humidity Cell No. ______: Weekly Mass, g (to nearest 0.1 g)			
Week No.	Humidity Cell + Filter + Sample at:		
	End, 3-Day Dry	End, 3-Day Wet	End, Leach (M_{hfsw})
Week 0	N/A	N/A	X
Week 1	X	X	X
Week 2	X	X	X
Week ...	X	X	X
Week 20	X	X	X

TABLE 3 Collection Flask Data Sheet

Collection Flask No.^{A}______: Weekly Mass, g (to nearest 0.1 g)			
Week No.	Flask + Effluent (M_{ef})	Flask Tare (M_{et})	Effluent (M_e)
Week 0	X	X	X
Week 1	X	X	X
Week 2	X	X	X
Week ...	X	X	X
Week 20	X	X	X

Collection Flask No.^{A}______: Weekly Effluent Parameters					
Week No.	Conductivity, mohs	Eh, mV	pH	$CaCO_3$ equivalent, parts per thousand	
				Acidity	Alkalinity
Week 0					
Week 1					
Week 2					
Week ...					
Week 20					

^{A} The flask number corresponds with the humidity cell number.

and represents the release rate asμ g constituent/g sample/week for the weeks between and including the inflection points. Fig. 5 shows that the first inflection point on the cumulative sulfate plot occurs at Week 2. Note that the release rates for Weeks 0 to 2 and Weeks 2 to 21 can be calculated using (Eq 9); the results are summarized in Table 4:

$$R_n = \frac{(L_{n2} - L_{n1})}{(n_2 - n_1)} \quad (9)$$

TABLE 4 Calculated Release Rate for Weeks 0 to 2 and 2 to 21 from Cumulative Sulfate Plot, Fig. 5

n Weeks	L_{n2}, μg × 10^{-3}	L_{n1}, μg × 10^{-3}	n_2	n_1	R_n ,μg/g/week
0 to 2	3122.4	688.7	2	0	1216.9
2 to 21	11 432.6	3122.4	21	2	437.4

where:

R_n = release rate of the constituent for *n* weeks between and including the inflection points,μ g/g/week,

L_{n2} = constituent cumulative load, the final week of *n* weeks between and including the inflection points, μg,

L_{n1} = constituent cumulative load, the initial week of *n* weeks between and including the inflection points, μg,

n_2 = final week of *n* weeks between and including the inflection points, and

n_1 = initial week of *n* weeks between and including the inflection points.

13. Precision and Bias

13.1 *Precision*—The precision of the procedure for measuring the rate of accelerated weathering is currently being determined using actual waste-rock samples from western United States metal mines. No accepted reference material is currently available. Within-laboratory precision performed by a single laboratory in duplicate cells is summarized in Table 5.

13.2 *Bias*—Bias has not been determined at this time because no accepted reference material is currently available for determining the bias present in the accelerated weathering procedure.

14. Keywords

14.1 accelerated weathering; chemical weathering; humidity cell; mill tailings; ore; oxidation; solid material; waste rock

TABLE 5 Comparison of Sulfate Release Rates (in 20-Week Increments) Between Separate Humidity Cell Tests Conducted on Two Splits from Two Different Samples (Samples 1-B and 1-C)

Test Duration (20-Week Intervals)	Average Sulfate Release Rates,μ g/g/week			
	Sample 1-B		Sample 1-C	
	Cell 1A	Cell 3A	Cell 5A	Cell 7A
0–20	182.5	184.1	523.3	558.7
20–40	321.8	254.6	887.1	868.1
40–60	354.8	197.5	1985.6	1481.1
60–80	306.5	192.1	1834.5	1292.1
80–100	294.0	235.0	1783.4	1247.6
100–120	227.6	141.3	3165.0	1660.7

APPENDIX

(Nonmandatory Information)

X1. PREPARATION OF A WASHED CELL SUSPENSION OF *THIOBACILLUS FERROOXIDANS* FOR HUMIDITY CELL TESTING

X1.1 *Cell Suspension*—Prepare an iron-grown culture of *T. ferrooxidans* with a cell density of 10^6 to 10^7/mL.

X1.2 *Culture Wash Procedure:*

X1.2.1 Divide 500 mL of iron-grown culture into two 250-mL samples, and place each into a separate centrifuge bottle. Balance the two filled centrifuge bottles and centrifuge at 12 000 r/min (relative centrifugal force (RCF) 19 140) for 15 min. The iron-grown culture should be separated into a concentrated cell pellet and a nutrient- and waste-laden aqueous phase (supernatant) after centrifuging. Carefully decant and discard the resulting supernatant. This step can be repeated to increase cell density.

X1.2.2 Suspend each cell pellet in 5 to 10 mL of deionized water that has been adjusted to pH 3.0 using H_2SO_4. Combine both suspended cell pellets in a single 50-mL centrifuge tube. Place the 50-mL centrifuge tube containing both suspended cell pellets in the centrifuge, and balance it with a matching

50-mL, water-filled blank tube. Centrifuge at 20 000 r/min (RCF 36 590) for 15 min. Carefully decant and discard the supernatant.

X1.2.3 Suspend the resulting cell pellet with 20 mL of deionized water that has been adjusted to pH 3.0, and agitate until a cell suspension results. Adjust the resulting 20-mL cell suspension to a final 100-mL volume using deionized water adjusted to pH 3.0.

X1.3 *Humidity Cell Inoculation*—Obtain a 10-mL aliquot from the 100-mL volume of washed cell suspension described in X1.2.3, and bring it up to either a 500 or 1000-mL volume (depending on the leach volume selected in 11.2.1) with either 490 or 990 mL, respectively, of deionized water. Use this volume immediately to inoculate the solid material sample during the first leach cycle.

NOTE X1.1—The preparation of a replicate inoculum, as described in X1.3, is recommended; designate it as a "blank inoculum," and analyze it for all parameters along with the first leach cycle leachate.

REFERENCES

(1) Lapakko, K., *Field Dissolution of Test Piles of Duluth Complex Rock*, Report to the U.S. Bureau of Mines, Cooperative Agreement No. C0219003, Minnesota Department of Natural Resources, St. Paul, MN, 1993.

(2) Caruccio, F. T., *Proceedings of the Second Symposium on Coal Mine Drainage Research*, Bituminous Coal Research, Monroeville, PA, 1968, pp. 107–151.

(3) Lawrence, R. W., *Mining and Mineral Processing Wastes*, F. M. Doyle, ed., *Proceedings of the Western Regional Symposium on Mining and Mineral Processing Wastes*, Berkeley, CA, AIME, Littleton, CO, 1990, pp. 115–121.

(4) Ferguson, K. D., and Morin, K. A., *Proceedings, of the Second International Conference on the Abatement of Acidic Drainage*, MEND, Ottawa, Ont., 1991, pp. 83–86.

(5) Erickson, P. M., and Hedin, R. S., *Proceedings, of the Mine Drainage and Surface Mine Reclamation*, U.S. Bureau of Mines, Info. Circ. 9183, 1988, pp. 11–19.

(6) Sobek, A. A., Schuller, W. A., Freeman, J. R., and Smith, R. M., *U.S. EPA Report EPA 600/2-78-054*, 1978.

(7) White, W. W., III, and Jeffers, T. H.," The Environmental Geochemistry of Sulfide Oxidation," C. N. Alpers and D. Blowes, eds., *Proceedings of the American Chemical Society Symposium Series 550*, American Chemical Society, Washington, DC, 1994, pp. 608–630.

(8) Perry, R. H., and C. H. Chilton, 1973, "Chemical Engineer's Handbook," Fifth Edition, McGraw-Hill Book Company, pp. 3–222, 3–223, Table 3-249.

(9) Pitard, F., *Pierre Gy's Sampling Theory and Sampling Practice*, Vols I and II, CRC Press, 1989.

(10) Brodie, M. J., Broughton, L. M., and Robertson, A. M., A Conceptual Rock Classification System for Waste Management and a Laboratory Method for ARD Prediction from Rock Piles, *In Proceedings of the Second International Conference on the Abatement of Acidic Drainage*, Montreal, Canada, September 16–18, 1991, pp. 119–135.

(11) Caruccio, personal communications, May 1995.

(12) Lapakko, K., *Mining and Mineral Processing Wastes*, F. M. Doyle, ed., *Proceedings of the Western Regional Symposium on Mining and Mineral Processing Wastes*, Berkeley, CA, AIME, Littleton, CO, 1990, pp. 81–86.

Designation: D 5745 – 95 (Reapproved 1999)

Standard Guide for Developing and Implementing Short-Term Measures or Early Actions for Site Remediation[1]

This standard is issued under the fixed designation D 5745; the number immediately following the designation indicates the year of original adoption or, in the case of revision, the year of last revision. A number in parentheses indicates the year of last reapproval. A superscript epsilon (ε) indicates an editorial change since the last revision or reapproval.

1. Scope

1.1 The purpose of this guide is to provide guidance for assisting in the development, selection, design, and implementation of partial, short-term, or early action remedies undertaken at sites of waste contamination for the purpose of managing, controlling, or reducing risk posed by environmental site contamination. Early action remedies and strategies are applicable to the management of other regulatory processes (for example, state UST programs are equally applicable) in addition to the Comprehensive Environmental Response, Compensation and Liability Act (CERCLA)/NCP process. this guide identifies and describes a standard process, technical requirements, information needs, benefits, and strategy for early actions.

1.2 This guide is applicable to both nonhazardous and hazardous sites of contamination as defined by CERCLA as amended by the Superfund Amendments and Reauthorization Act of 1986 (SARA) and the Resource Conservation and Recovery Act (RCRA) as amended by the Hazardous and Solid Waste Amendments (HSWA) of 1986.

1.3 To the extent that this guide may be used for hazardous materials actions operations, it does not address the applicability of regulatory limitations and local requirements.

1.4 *This standard does not purport to address all of the safety concerns, if any, associated with its use. It is the responsibility of the user of this standard to establish appropriate safety and health practices and determine the applicability of regulatory limitations prior to use.*

2. Referenced Documents

2.1 *ASTM Standards:*

E 1528 Practice for Environmental Site Assessment: Transaction Screen Process[2]

2.2 *Code of Federal Regulations:*[3]

Corrective Action or Solid Waste Management Units at Hazardous Waste Management Facilities, Proposed Rule, 27 July 1990, 40 CFR Parts 264, 265, 270, and 271 (55 FR 30797)

Corrective Action Management Units and Temporary Units; Corrective Action Provisions; Final Rule, 16 February 1993, 58 FR 8658

National Oil and Hazardous Substances Pollution Contingency Plan, Final Rule, 8 March 1990, 40 CFR Part 300

2.3 *EPA Documents*[3]

CERCLA, Compliance with Other Laws Manual, Part I (Interim Final), August 1988, EPA/9234.1-01

CERCLA, Compliance with Other Laws Manual, Part II: Clean Air Act and Other Environmental Statutes and State Requirements, August 1989, EPA/9234.1-02

Guidance for Performing Preliminary Assessments under CERCLA, September 1991, EPA/9345.0-01A

Guidance for Performing Site Inspections under CERCLA, September 1992, EPA/9345.1-05

Data Quality Objectives for Remedial Response Activities: Development Process, EPA/540/G-87/003

Guidance for Conducting Remedial Investigations and Feasibility Studies (RI/FS) under CERCLA, October 1988, EPA/9355.3-01

RCRA Corrective Action Interim Measures Guidance, Interim Final, June 1988, EPA/9902.4

3. Terminology

3.1 *Definitions:*

3.1.1 *applicable or relevant and appropriate requirements (ARAR)*—those requirements, cleanup standards, standards of control, and other substantive environmental protection requirements, criteria, or limitations promulgated under federal or state law that show either a direct correspondence or address problems or situations sufficiently similar at a site to show that they are well suited for application.

3.1.2 *conceptual site model, n*—a mental or physical representation of the physical system and the iterative characterization of the physical and chemical processes and conditions that affect the transport of contaminants from sources through environmental media to receptors or potential receptors.

3.1.3 *contaminant, n*—any substance potentially hazardous to human health or the environment and present in the

---

[1] This guide is under the jurisdiction of ASTM Committee D-34 on Waste Management and is the direct responsibility of Subcommittee D34.05 on Site Remediation.

Current edition approved June 15, 1995. Published August 1995.

[2] *Annual Book of ASTM Standards*, Vol 11.04.

[3] Available from Superintendent of Documents, U.S. Government Printing Office, Washington, DC 20402.

environment above background concentration.

3.1.4 *early action, n*—any remedial plan initiated in advance of a complete or final characterization of a contaminated site.

3.1.5 *final remedy, n*—site restoration.

3.1.6 *interim remedial measure, n*—a remedial action that implements a partial solution prior to the selection of a final complete remedy. Interim remedial measures may be early actions, but they are often not.

3.1.7 *migration, n*—the movement of contaminant(s) away from a source through permeable subsurface media (such as the movement of a groundwater plume of contamination) or the movement of contaminant(s) by a combination of surficial and subsurface processes.

3.1.8 *partial remedy, n*—an interim or incomplete solution intended to be consistent with the expected permanent remedy for treatment, control, elimination, or management of risk associated with the release of a contaminant to the environment.

3.1.9 *potential migration pathway, n*—the route that may be taken by contaminants in the environment as they move or are transported from the source(s), usually in a downgradient direction.

3.1.10 *preliminary assessment (PA), n*—a review of existing information and an off-site reconnaissance, if appropriate, to determine whether a release may require additional investigation or action. A preliminary assessment may include an on-site reconnaissance, if appropriate. See ASTM Guidance for Transaction Screen Questionnaire (Practice E 1528).

3.1.11 *receptor, n*—humans or other species potentially at risk from exposure to contaminant(s) at the point(s) of exposure.

3.1.12 *release, n*—any spilling, leaking, pumping, emitting, emptying, discharging, injecting, escaping, leaching, dumping, and disposing into the environment (including the abandonment or discarding of barrels, containers, and other closed receptacles) of any hazardous chemical, extremely hazardous substance, or CERCLA hazardous substance.

3.1.13 *removal, n*—the cleanup or removal of released hazardous substances from the environment; such actions as may be necessary to take in the event of the threat of release of hazardous substances into the environment; such actions as may be necessary to monitor, assess, and evaluate the release or threat of release of hazardous substances; the disposal of removed material; or the taking of such other actions as may be necessary to prevent, minimize, or mitigate damage to the public health or welfare or to the environment, which may otherwise result from a release or threat of release.

3.1.14 *short-term measure, n*—an early action designed to have an authorized duration of less than one year for the effective control or management of a contaminant released to the environment.

3.1.15 *size characterization, n*—the process by which information relating to the nature, extent, potential migration pathways, and receptors of environmental contaminants is gathered, interpreted, and documented. Site characterization efforts to provide a basis for the following: (*1*) the development of a conceptual site model (CSM), (*2*) the selection and design of a site remediation plan, or (*3*) the measuring point against which the effectiveness of a remedy can be evaluated, or some combination thereof.

3.1.16 *site inspection (SI), n*—an on-site investigation to determine whether a release or potential release exists and the nature of the associated threats. The purpose is to augment the data collected in the preliminary assessment and to generate, if necessary, sampling and other field data to determine whether further action or investigation is appropriate.

3.1.17 *site remediation, n*—those actions taken in the event of a release or threatened release of a hazardous substance in to the environment, to prevent or minimize the impact of the release, or to mitigate a substantial hazard to present or future environmental conditions. This early action may or may not lead to ultimate restoration of the site.

3.1.18 *source, n*—the location at which contamination has entered the natural environment.

3.2 *Description of Terms Specific to This Standard:*

3.2.1 *significantly above background, adj*—the mean concentration of a site contaminant can be shown (by statistical analysis of other methods) to be greater than nearby background samples from the same pathway.

4. Summary of Guide

4.1 The basic activities associated with implementing an early action are as follows: (*1*) construction of a CSM and estimation of risk(s); (*2*) identification of exposure control pathways amenable to engineered control; (*3*) development of interim or partial solutions, estimation of engineered risk, and identification and negotiation of required action levels; (*4*) selection of the desired solution(s); (*5*) attainment of legal authority for implementation of the planned solution(s); (*6*) design and execution of the selected solution(s); and (*7*) post-implementation monitoring of the conceptual site model.

4.2 Five common objectives for an early action are to achieve the following: (*1*) minimize the human or environmental risk exposure, or both; (*2*) minimize the time required to implement a final remedy; (*3*) protect resources (for example, financial, mineral, and ecological); (*4*) minimize the complexity of a final remedy; or (*5*) provide a solution-oriented project focus, or combination thereof.

4.3 There are three basic types of partial, short-term or early action remedies: (*1*) source control remedies, (*2*) pathway control remedies, and (*3*) receptor control remedies. It is more common for early actions to be of the source or receptor control type since pathway controls usually require a sophisticated understanding of the dynamics of a conceptual site model.

4.4 The development of a final remedy is often an iterative process that evolves frequently with the compilation of new data in the CSM. The development and implementation of early actions that support the final remediation objectives of a project provides maximum benefit when performed as promptly as practical.

4.5 Early actions, short-term remedies, or interim remedial measures are effective risk management tools when designed and executed properly. Some common benefits derived from early actions are as follows: (*1*) human, ecological, and financial resources are protected; (*2*) the actual time required to remediate an unacceptable environmental condition is minimized or reduced; (*3*) the geometric magnitude or scale of an unacceptable environmental condition is reduced; (*4*) the complexity and scope of a final remedial solution is reduced; and (*5*) environmental projects become "solution" oriented.

4.6 A successful strategy for the application of early actions has been developed. The strategy consists of phases or steps that are as follows:

4.6.1 Development of a list of potential proactive early action remedies.

4.6.2 Identification of candidate sites for early action.

4.6.3 Identification of site-specific and easily definable CSM components(s).

4.6.3.1 Establishing and prioritizing early action objectives for each CSM component.

4.6.3.2 Identifying early action alternatives to address each objective and identifying anticipated or expected results and their impact on final regulations and remedy.

4.6.3.3 Seeking regulatory and public comment, as appropriate.

4.6.4 Establishment of funding availability for early actions.

4.6.5 Prioritization of early action solutions consistent with the objectives, public response, expected results, and funding availability.

4.6.6 Selection and integration of early action solutions.

4.6.6.1 Selecting criteria for management and measurement of the results and progress of early action.

4.6.6.2 Establishing documentation and recording procedures and requirements for early action implementation and effective final remedy implementation.

4.6.6.3 Analyzing the validation approach prior to the implementation of early action.

4.6.7 Preparation and finalization of the early action remedial plan.

4.6.8 Implementation and documentation of early action activities.

4.6.8.1 Validating early action results in comparison to the early action plan and the final remedial action frequently and periodically.

4.6.8.2 Reviewing the documentation of all early action activities frequently and periodically.

5. Significance and Use

5.1 This guide is intended to provide a systematic approach for the application and execution of early actions for purposes of remediating sites of both hazardous and non-hazardous contamination. Fundamental to the use of this guide is the iterative development of a CSM.

5.2 Anticipated users of this guide are owners or operators at sites of environmental contamination; technical professionals involved in the field of environmental site characterization and remediation; environmental regulators, property owners, employees, and residents adjacent to sites of environmental contamination; and lenders, sureties, and persons of general interest within an affected community.

5.3 This guide is not intended to replace legal requirements for remediating sites of environmental contamination. This guide should be used to supplement existing regulatory guidance and to focus remedial efforts toward final remedy solutions.

6. Procedure

6.1 *Assembling Required Information*—Assemble all available information, including the following: historical records, interviews, previous studies, environmental analytical data, permits, regulatory guidance and requirements, maps, geologic cross sections, engineering infrastructure as-built plans, and drawings. At least one site visit by technical personnel tasked with the responsibility of designing and implementing an early action is required prior to the development of a remedial plan.

6.2 *Development of the Conceptual Site Model*—An initial concept of the site(s) conceptual site model should be developed using all assembled information. The quality and accuracy of all information should be assessed both quantitatively and qualitatively, and the use of the information should be focused on the following:

6.2.1 *Identification of Contaminants*—Identify the environmental contaminants for all pathways of a conceptual site model. Particular emphasis should be placed on identifying the contaminants for any suspected exposure pathways of concern.

6.2.2 *Characterization of Background Conditions*—The natural and secondary (modified) background concentration of contaminants in all conceptual site model pathways must be characterized or estimated in order to design a useful early action. This information is necessary in order to develop appropriate action levels, identify possible synergism, estimate environmental risk, and identify and design remedial solutions.

6.2.3 *Contaminant Source Characterization*—An understanding of contaminant source characteristics is essential in developing a successful early action remedy. At a minimum, the following source characteristics should be measured or estimated for a site:

6.2.3.1 Source location, boundaries, volume, and mass;

6.2.3.2 Hazardous constituents and their concentration at a source;

6.2.3.3 Time, duration, rate of volume, and mass contaminant release from a source; and

6.2.3.4 Suspected areas (three dimensional) of contaminant migration within a pathway from a point or source release.

6.2.4 *Migration Pathway Characterization*—Potential contaminant migration pathways through the soil, surface water, air, and ground water must be identified and characterized primarily for each source of contamination at a site. The minimum information or characterization requirements for developing an early action for each migration pathway type is as follows: (*1*) an evaluation and estimate of the contaminant mass released and its release mechanism to a pathway, (*2*) identification of the transport mechanism and an estimate of contaminant transport rate or dispersion within a pathway, or both; and (*3*) identification of the human and ecological receptors at potential points of exposure above levels of acceptable risk on a contaminant migration pathway.

6.2.5 *Contaminant Mass Estimate*—An estimate of contaminant mass and contaminant distribution is required for developing successfully focused early action remedies.

6.2.6 *Receptor Exposure Characterization*—Estimates of the concentration and duration of both human and ecological contaminant exposure should be developed for each exposure point within a migration pathway.

6.2.7 *Estimation of Human and Ecological and Other Risk*—Early actions are engineered risk management solutions. An estimate or perception of unacceptable risk should exist before an early action is considered and developed. There are many categories of environmental risks; some examples are human and ecological risk, financial risk, community relations, etc.

6.3 *Identification of Early Action Strategy*—Most successful early actions or interim remedial measures incorporate a strategy that emphasize a technical approach that expediously balances and expedites the technical requirements and needs of a project risk and available resources. The elements of a proven strategy for developing and implementing early actions, as summarized in 4.6, are discussed as follows.

6.3.1 *Proactive Development of Early Action Remedies*—It is important for all affected parties to provide input within the framework of a "positive" forum to identify their concerns, risks, resources, and objectives for an early action. The development and implementation of an optimum early action will be delayed unless a proactive and technically focused environment of cooperation is developed among the parties affected by environmental contamination concerns. It is especially important for time and resource critical projects to foster proactive interaction on technical issues. ASTM advocates the early solicitation and consideration of community concerns. Some examples of early action remedies are listed in Appendix X1.

6.3.2 *Identification of Early Action Candidate Sites*—Not all sites of environmental contamination are appropriate candidates for early action. Sites that are dynamic and contain complex migration pathways commonly require sophisticated and detailed site characterization before sufficient technical information is available to design an appropriate partial remedy. Usually, the more simplistic an environmental problem, the more likely the site is to be a candidate for an early action remedy.

6.3.3 *Identification of Manageable CSM Components and Early Action Solution Alternatives*:

6.3.3.1 Each site of environmental contamination has a CSM component appropriate to manage for the control of human or ecological risk. Three examples are as follows: (*1*) as a pathway control, surface water diversion and runoff control from a contaminant release area may be a useful CSM component to manage for risk control; (*2*) source control or removal of a contaminant release to the environment may prevent migration of contaminant mass through a pathway to a receptor; and (*3*) fencing or warning signs of hazardous contaminants. Identification of the CSM components appropriate for engineered risk management is often the most critical element for developing a successful early action. Regulatory agency involvement is recommended to communicate the evaluation of the CSM components. Early agreement to the strategy by the regulatory agencies is essential.

6.3.3.2 Each CSM component identified should have well-defined risk management and mitigation objectives, each with associated desired and anticipated results from the potential early action solutions. These CSM components and objectives should be prioritized as the primary basis for evaluating alternatives and desired results. To the extent practical at this stage in the strategy, the possible impact on projected final remedies should be considered while the CSM components, objectives, and expected results are being identified and prioritized.

6.3.3.3 Public participation should be solicited and evaluated whether or not legally required. Early public/citizen participation may reveal objectives and concerns in addition to technical and site issues that could jeopardize the future success of the early action unless considered in all phases of the strategy.

6.3.3.4 At many sites where early actions have been implemented, often only one potential technical remedy was considered. The identification of several potential multiple technical solutions *targeted at the most appropriate CSM components* is essential if the most flexible, timely, and technically responsive remedy(ies) is to be developed for that site.

6.3.4 *Funding of Early Actions*—Few sites have been remediated successfully using early actions alone and seldom are all contaminant migration pathways and risks understood at the early stages of a remedial project, the time when many early actions are performed. For these reasons, it is advisable to identify and allocate (budget) only a reasonable portion of the available funding for early action, which is balanced between cost and risk management benefits. The available funding levels should be used to guide and focus the following steps toward a realistic early action solution. If the human or ecological risks identified in the CSM component(s) cannot be addressed adequately by available funding, other or additional funding alternatives should be considered.

6.3.5 *Prioritization of Early Action Solutions*—The alternative elements, including desired results and technical components, of a proposed early action should be prioritized by the affected parties. It is important that the prioritization be performed in a proactive fashion to ensure that most critical and beneficial elements of an early action are implemented. The resulting priority should be consistent with the technical and risk management objectives, public response, expected results, and available funding.

6.3.6 *Selection and Integration of Early Actions*—Based on the priority of alternative solutions, selection of the most beneficial solution should be conducted before formulating a remedial implementation plan. Performance criteria should be selected to document and measure progress toward the expected results in order to integrate the selected early action with follow-on remedial activities and a final remedy. These criteria should be incorporated into the remedial plan and include, as a minimum, recording and reporting procedures by the responsible party, interim technical objectives and schedule, budgetary objectives and constraints, reporting format for public participation, and documentation of early action activities useful for final remedy preparation and implementation. The criteria resulting from the selection process should also include an analysis to validate that the selected early action approach does, in fact, satisfy the risk management objectives and the CSM components.

6.3.7 *Preparation and Finalization of Remedial Plan*—Regulatory agencies often have format and content requirements for remedial plans; however, the regulatory agency requirements may be minimal for many of the example early actions listed in Appendix X1. The preparation of early action remedial plans must meet these regulatory requirements to receive approval. The remedial plan should be sufficiently detailed to provide guidance for implementation but simple enough to allow flexibility to respond to changing technical and site conditions. Specifically, it should be noted that site characterizations activities may be ongoing during early action activities on complicated contaminated sites with complex CSMs. This ongoing site characterization will contribute developments and refinements to the CSM that may require changes to the early action remedial plan.

6.3.8 *Implementation and Documentation of Early Action Activities*—During implementation of the plan, the results must be documented faithfully and compared to the original objectives frequently. Actual results and progress during the early action must be validated as achieving the targeted objectives. Consistency with a projected final remedy must be validated frequently during implementation of the early action.

6.3.9 At some sites where early actions have been implemented, valuable technical information has been lost or not properly documented, recorded, and reported by the responsible party. For example, early and undocumented removal and disposal of contaminated soil resulted in lack of contaminant characterization chemical data and knowledge concerning the volume of the removed soil. This lack of information made it difficult and more costly to plan and implement a final remedy. Extreme care and extra expense may be needed to ensure proper documentation. There must be proper documentation and record-keeping in order for the early action strategy to benefit the final remedy.

6.4 *Identification of Requirements for Early Actions*—Some requirements for developing as early action are site specific. The following sections discuss those requirements that must be considered for any interim remedial measure or early action. Often, although not always required, written documents describing the following topics are developed and submitted to legal entities for approval.

6.4.1 *Legal Authority*—Early actions must meet those cleanup standards, standards of control, and other substantive requirements, criteria, or limitations promulgated under federal environment or state environmental or facility citing laws that specifically address a hazardous substance, pollutant, contaminant, remedial action, location, or other circumstances found at a hazardous waste, RCRA, or CERCLA site. Only those state standards that are identified by a state in a timely manner and that are more stringent than federal requirements may be applicable. Many different legal requirements may impact the design and implementation of an early action. Early actions are commonly authorized by the following: (*1*) letters of agreement, (*2*) interim records of decision, (*3*) engineering estimates and cost analysis (EECAs), and (*4*) permit amendments. There are many other types of legal mechanisms that may also be used to authorize or approve early actions.

6.4.2 *Health and Safety Plan*—The operational health and safety aspects of implementing an early action must be considered. Typically, emergency response plans, site evacuation plans, worker safety, and alternate pathway contaminant transport (for example, soils contamination transported in the air pathway during waste excavation) control are topics that are considered and discussed by a health and safety plan.

6.4.3 *Sampling and Analysis Plan*—Most early actions incorporate some sampling and analytical testing; however, it is not always required. Samples are frequently collected to monitor treatment efficiency, characterize wastes for disposal, and characterize a site further as components of an early action. All sampling and analysis plans (SAPs) should identify the following: sampling procedures; sampling frequency; preservation, transport, and handing techniques; decontamination procedures; analytical methods; and quality assurance/quality control (QA/QC) systems that are associated with the environmental data generation process. Additional guidance for these efforts is available by reviewing the following:

6.4.4 *Early Action Plan*—All early actions must have a remedial plan. The remedial plan describes how an early action will be implemented. Operational items that should be addressed by the remedial plan are discussed as follows:

(*1*) Security,

(*2*) Mobilization/demobilization,

(*3*) Unit/system operation,

(*4*) Unit/system test/performance monitoring,

(*5*) Community relations,

(*6*) Site analysis,

(*7*) Contaminant mass balance,

(*8*) Waste characterization/management plan, and

(*9*) Permits.

Early action plans address complex technical issues affecting operation and execution of the remedy, but they are often relatively short and simple documents.

6.4.5 *Execution and Implementation of Early Action Plan*—It is usually not possible to deviate from an approved early action plant. Operations should conform to the plan unless circumstances require change and written authorization for plan modification is obtained. Documentation of all operations and activities must be maintained to verify that the early action plan was implemented correctly and fully and to demonstrate what was accomplished at a future date. Documentation such as the following is necessary so that the final remedy can be selected and implemented without delay and question:

6.4.5.1 *Records of Public Participation*—Notes of all public meetings; records of responses to public comments and meetings.

6.4.5.2 *Field Logs*—Daily records of activities, site manager; health and safety meeting notes; transportation, disposal, treatment records; field sampling records and sample identification; site entry logs; all other records to document field activities.

6.4.5.3 *Analytical Data Records*—Chain of custody records to document field sampling records: purchase orders for traceability of laboratory data and expenses to project; analytical data results and QC data/records.

6.4.5.4 *Early Action Results*—Chronological comparison of

remedial plan with actual activities; soil excavation history and verification sampling results to verify that post excavation remedial goals were met; volume and disposition (location) of contaminated material removed from site; analytical data associated with disposal and waste management activities; all waste management and disposal manifests and bill of latent (solid, sludge, liquids, etc.); documentation of all post remedial site restoration activities; copies/records of all written and verbal correspondence with property owners, public media, and regulatory agencies; recommendations/lessons learned for final remedy.

6.4.6 *Documentation Retention*—Responsibility for record keeping rests with the property owner. Documentation for early actions taken under CERCLA should be maintained in the administrative record. This documentation should be maintained for other sites as part of the legal records for the site.

6.4.7 *Post Remedy Monitoring Plan*—Many early actions will require that the success of the remedy be monitored during its operation. It is advisable to develop an operation and monitoring plan prior to, or in conjunction with, development of the remedial plan.

6.4.8 *Early Action Performance Assessment*—The success of an early action should be assessed by comparing its actual result to the predicted goal or desired objective.

6.5 *Other Considerations*—In addition to the previously discussed requirements, other factors must often be considered when developing a remedial plan for an early action. Some of the more common factors of this type are as follows: (*1*) funding limitations, (*2*) time constraints, (*3*) community acceptance, and (*4*) technology availability.

7. Keywords

7.1 conceptual site model; early action; environmental risk management; hazardous waste; interim-remedial measure; nonhazardous waste; short-term remedy; site characterization; site remediation; waste management

APPENDIX

(Nonmandatory Information)

X1. EXAMPLES OF EARLY ACTION REMEDIES

X1.1 Some examples of early action remedies are as follows: fences; site access controls; warning signs; physical security; covers; barriers; underground barrier walls; drainage controls; runoff diversion barriers; berms; dikes; impoundment areas; capping; neutralizing chemicals; removal of debris; removal of drums, tanks, containers; removal of soil or solid materials; removal of liquids; in-situ treatments; bioremediation; alternate water treatment process; provision of alternate potable water sources or supplies; and provision of alternate habitat.

Standard Classification of Environmental Condition of Property Area Types for Defense Base Closure and Realignment Facilities[1]

This standard is issued under the fixed designation D 5746; the number immediately following the designation indicates the year of original adoption or, in the case of revision, the year of last revision. A number in parentheses indicates the year of last reapproval. A superscript epsilon (ε) indicates an editorial change since the last revision or reapproval.

1. Scope

1.1 *Purpose*—The purpose of this classification is to define seven standard environmental condition of property area types for Department of Defense (DoD) real property at a closing military installation with respect to the requirements of the Comprehensive Environmental Response, Compensation and Liability Act (CERCLA) of 1980 Section 120(h), as amended by the Community Environmental Response Facilitation Act (CERFA) of 1992, and Section 331 of the National Defense Authorization Act for Fiscal Year 1997. As such, this classification is intended to permit a DoD component to classify property into seven area types, in order to facilitate and support findings of suitability to transfer (FOSTs), findings of suitability to lease (FOSLs), and uncontaminated parcel determinations pursuant to the requirements of CERFA. *Users of this classification should note that it does not address (except where noted explicitly) requirements for appropriate and timely regulatory consultation or concurrence, or both, during the identification and use of these environmental condition of property area types.*

1.1.1 *Seven Recognized Standard Environmental Condition of Property Area Types*—The goal of this classification is to permit DoD components to classify properties on closing DoD installations in order to support determinations of which properties are suitable and unsuitable for transfer by lease or by deed. The term "standard environmental condition of property area type" refers to one of the seven area types defined in this classification. An identification of an area type on an environmental condition of property map means that a DoD component has conducted sufficient studies to make a determination of the recognized environmental conditions of installation real property or has complied with the identification requirements of uncontaminated property under CERFA, or both, and has categorized the property into one of the following seven area types:

1.1.1.1 *Standard Environmental Condition of Property Area Type 1*—An area or parcel of real property where no release, or disposal of hazardous substances or petroleum products or their derivatives has occurred (including no migration of these substances from adjacent properties).

1.1.1.2 *Standard Environmental Condition of Property Area Type 2*—An area or parcel of real property where only the release or disposal of petroleum products or their derivatives has occurred.

1.1.1.3 *Standard Environmental Condition of Property Area Type 3*—An area or parcel of real property where release, disposal, or migration, or some combination thereof, of hazardous substances has occurred, but at concentrations that do not require a removal or remedial action.

1.1.1.4 *Standard Environmental Condition of Property Area Type 4*—An area or parcel of real property where release, disposal, or migration, or some combination thereof, of hazardous substances has occurred, and all remedial actions necessary to protect human health and the environment have been taken.

1.1.1.5 *Standard Environmental Condition of Property Area Type 5*—An area or parcel of real property where release, disposal, or migration, or some combination thereof, of hazardous substances has occurred and removal or remedial actions, or both, are under way, but all required actions have not yet been taken.

1.1.1.6 *Standard Environmental Condition of Property Area Type 6*—An area or parcel of real property where release, disposal, or migration, or some combination thereof, of hazardous substances has occurred, but required response actions have not yet been initiated.

1.1.1.7 *Standard Environmental Condition of Property Area Type 7*—An area or parcel of real property that is unevaluated or requires additional evaluation.

1.1.2 *CERCLA Section 120(h) Requirements*—This classification of environmental condition of property area types is consistent with CERCLA § 120(h) requirements relating to the transfer of contaminated federal real property (42 USC 9601 and following). Areas classified as Area Types 1 through 4, as defined in this classification, are suitable, with respect to CERCLA § 120(h) requirements, for deed transfer to a non-federal recipient.

1.1.3 *CERFA Requirements*—This classification of environmental condition of property area types can be used in conjunction with the reporting requirements of CERFA, which amended CERCLA (Public Law 102-426, 106 Statute 2174).

---

[1] This classification is under the jurisdiction of ASTM Committee D-34 on Waste Management and is the direct responsibility of Subcommittee D34.05 on Site Remediation.

Current edition approved Sept. 10, 1998. Published December 1998. Originally published as D 5746 - 95. Last previous edition D 5746 - 95.

As defined in this classification, areas classified as Type 1 areas are eligible for reporting as "uncontaminated property" under the provisions of CERFA. At installations listed on the national priorities list, Environmental Protection Agency (EPA) concurrence must be obtained for a parcel to be considered uncontaminated and therefore transferable under CERCLA § 120(h)(4). EPA has stated as a matter of policy that there may be instances in which it would be appropriate to concur with the military service that certain parcels can be identified as uncontaminated under CERCLA § 120(h)(4), although some limited quantity of hazardous substances or petroleum products have been stored, released, or disposed of on the parcel. If the information available indicates that the storage, release, or disposal was associated with activities that would not be expected to pose a threat to human health or the environment (for example, housing areas, petroleum-stained pavement areas, and areas having undergone routine application of pesticides), such parcels should be eligible for expeditious reuse.

1.1.4 *Petroleum Products*—Petroleum products and their derivatives are included within the scope of this classification. Under DoD policy, areas on which petroleum products and their derivatives have been released or disposed of may not be suitable for deed transfer until a response action has been completed.

1.2 *Objectives*—The objectives guiding the development of this classification are as follows: (*1*) to synthesize and put in writing a standard classification of environmental condition of property area types; (*2*) to facilitate the development of high-quality, standardized environmental condition of property maps that can be used to support FOSTs and FOSLs; (*3*) to facilitate the development of a standard practice for conducting environmental baseline surveys; and (*4*) to facilitate the development of a standard guide for preparing environmental baseline survey reports.

2. Referenced Documents

2.1 *ASTM Standards:*

E 1527 Practice for Environmental Site Assessments: Phase I Environmental Site Assessment Process[2]

E 1528 Practice for Environmental Site Assessments: Transaction Screen Process[2]

2.2 *Department of Defense Policies:*[3]

DoD Policy on the Environmental Review Process to Reach a Finding of Suitability to Lease (FOSL), September 1993

DoD Policy on the Environmental Review Process to Reach a Finding of Suitability to Transfer (FOST) for Property Where No Release or Disposal Has Occurred, June 1994

DoD Policy on the Environmental Review Process to Reach a Finding of Suitability to Transfer (FOST) for Property Where Release or Disposal Has Occurred, June 1994

DoD Policy on the Implementation of the Community Environmental Response Facilitation Act (CERFA), September 1993

DoD Clarification of "Uncontaminated" Environmental Condition of Property at Base Realignment and Closure (BRAC) Installations, October 1996

3. Terminology

3.1 This section provides definitions, descriptions of terms, and a list of acronyms for many of the words used in this classification. The terms are an integral part of this classification and are critical to an understanding of this classification and its use.

3.2 *Definitions:*

3.2.1 *environmental baseline survey (EBS)*—a survey of DoD real property based on all existing environmental information related to the storage, release, treatment, or disposal of hazardous substances or petroleum products or derivatives on the property to determine or discover the obviousness of the presence or likely presence of a release or threatened release of any hazardous substance or petroleum product. In certain cases, additional data, including sampling and analysis, may be needed in the EBS to support classification of the property into one of the standard environmental condition of property area types. Additionally, an EBS may also satisfy the uncontaminated property identification requirements of CERFA. An EBS will consider all sources of available information concerning environmentally significant current and past uses of the real property and shall, at a minimum, consist of the following: (*1*) a detailed search and review of available information and records in the possession of the DoD components or records made available by the regulatory agencies or other involved Federal agencies. DoD components are responsible for requesting and making reasonable inquiry into the existence and availability of relevant information and records to include any additional study information (for example, surveys for radioactive materials, asbestos, radon, lead-based paint, transformers containing PCB, Resource Conservation and Recovery Act Facility Assessments and Investigations (RFA and RFI), and underground storage tank cleanup program) to determine the environmental condition of the property; (*2*) a review of all reasonably obtainable Federal, state, and local government records for each adjacent facility where there has been a release or likely release of any hazardous substance or any petroleum product, and that is likely to cause or contribute to a release or threatened release of any hazardous substance or any petroleum product on the DoD real property; (*3*) an analysis of aerial photographs that may reflect prior uses of the property, which are in the possession of the Federal government or are reasonably obtainable through state or local government agencies; (*4*) interviews with current or former employees, or both, involved in operations on the real property; (*5*) visual inspections of the real property; any buildings, structures, equipment, pipe, pipeline, or other improvements on the real property; and of properties immediately adjacent to the real property, noting sewer lines, runoff patterns, evidence of environmental impacts (for example, stained soil, stressed vegetation, and dead or ill wildlife), and other observations that indicate the actual or potential release of hazardous substances or petroleum products; (*6*) the identification of sources of contamination on the installation and on adjacent properties that could migrate to the parcel during Federal government ownership; (*7*) ongoing response actions or actions that have been taken at or adjacent

[2] *Annual Book of ASTM Standards*, Vol 11.04.

[3] Available from National Technical Information Sevices, 5285 Port Royal Road, Springfield, VA 22161.

to the parcel; and (*8*) physical inspection of the property adjacent to the real property, to the extent permitted by owners or operators of such property.

3.2.2 *environmental baseline survey (EBS) report*—the written record of an EBS that includes the following: (*1*) an executive summary briefly stating the areas of real property (or parcels) evaluated and the conclusions of the EBS; (*2*) the property identification (for example, the address, assessor parcel number, or legal description); (*3*) any relevant information obtained from a detailed search of Federal government records pertaining to the property, including available maps; (*4*) any relevant information obtained from a review of the recorded chain of title documents regarding the real property. The review should address those prior ownerships and uses that could reasonably have contributed to an environmental concern, and, at a minimum, cover the preceding 60 years; (*5*) a description of past and current activities, including all past DoD uses to the extent such information is reasonably available, on the property and on adjacent properties; (*6*) a description of hazardous substances or petroleum products management practices (to include storage, release, treatment, or disposal) at the property and adjacent properties; (*7*) any relevant information obtained from records reviews and visual and physical inspections of adjacent properties; (*8*) a description of ongoing response actions or actions that have been taken at or adjacent to the property; (*9*) an evaluation of the environmental suitability of the property for an intended lease or deed transaction, if known, including the basis for determination of such suitability; and (*10*) references to key documents examined (for example, aerial photographs, spill incident reports, and investigation results).

3.2.3 *environmental condition of property map*—a map, prepared on the basis of all environmental investigation information conducted to date, that shows the environmental condition of a DoD installation's real property in terms of the seven standard environmental condition of property area types defined in this classification.

3.2.4 *hazardous substance*—a substance defined as a hazardous substance pursuant to CERCLA 42 USC § 9601(14), as interpreted by EPA regulations and the courts: "(*A*) any substance designated pursuant to section 1321(b)(2)(A) of Title 33, (*B*) any element, compound, mixture, solution, or substance designated pursuant to section 9602 of this title, (*C*) any hazardous waste having the characteristics identified under or listed pursuant to section 3001 of the Solid Waste Disposal Act (42 USC § 6921) (but not including any waste the regulation of which under the Solid Waste Disposal Act (42 USC § 6921 *et seq.*) has been suspended by Act of Congress), (*D*) any toxic pollutant listed under section 1317(a) of Title 33, (*E*) any hazardous air pollutant listed under section 112 of the Clean Air Act (42 USC § 7412), and (*F*) any imminently hazardous chemical substance or mixture with respect to which the Administrator (of EPA) has taken action pursuant to section 2606 of Title 15. The term does not include petroleum, including crude oil or any fraction thereof which is not otherwise specifically listed or designated as a hazardous substance under subparagraphs (*A*) through (*F*) of this paragraph, and the term does not include natural gas, natural gas liquids, liquefied natural gas, or synthetic gas usable for fuel (or mixtures of natural gas and such synthetic gas)."

3.2.5 *petroleum products*—those substances included within the meaning of the petroleum exclusion to CERCLA 42 USC § 9601(14) as interpreted by the courts and EPA, that is: ″petroleum, including crude oil or any fraction thereof which is not otherwise specifically listed or designated as a hazardous substance under subparagraphs (*A*) through (*F*) of this paragraph, and the term does not include natural gas, natural gas liquids, liquefied natural gas, or synthetic gas usable for fuel (or mixtures of natural gas and such synthetic gas)."

3.2.6 *property*—the real DoD property subject to classification under the classification of environmental condition of property area types.

3.2.7 *recorded land title records*—records to be searched during a chain of title search, including records of fee ownership, leases, land contracts, easements, liens, and other encumbrances on or of the property recorded in the place where land title records are recorded, by law or custom, for the local jurisdiction in which the property is located. (Such records are commonly kept by a municipal or county recorder or clerk.) Such records may be obtained from title companies or from the local government agency directly.

3.2.8 *release*—any spilling, leaking, pumping, pouring, emitting, emptying, discharging, injecting, escaping, leaching, dumping, or disposing into the environment (including the abandonment or discarding of barrels, containers, and other closed receptacles) of any hazardous chemical, extremely hazardous substance, or CERCLA hazardous substance.

3.2.9 *relevant and appropriate requirements*—those cleanup standards, standards of control, and other substantive requirements, criteria, or limitations promulgated under federal environmental or state environmental or facility siting laws that, while not "applicable" to a hazardous substance, pollutant, contaminant, remedial action, location, or other circumstance at a CERCLA site, address problems or situations sufficiently similar to those encountered at the CERCLA site that their use is well suited to the particular site. Only those state standards that are identified in a timely manner and are more stringent than federal requirements may be relevant and appropriate.

3.2.10 *remedial actions*—those actions consistent with a permanent remedy taken instead of, or in addition to, removal action in the event of a release or threatened release of a hazardous substance into the environment, to prevent or minimize the release of hazardous substances so that they do not migrate to cause substantial danger to the present or future public health or welfare or the environment.

3.2.11 *removal*—the cleanup or removal of released hazardous substances from the environment; such actions as may be necessary to take in the event of the threat of release of hazardous substances into the environment; such actions as may be necessary to monitor, assess, and evaluate the release or threat of release of hazardous substances; the disposal of removed material; or the taking of such other actions as may be necessary to prevent, minimize, or mitigate damage to the public health or welfare or to the environment, which may otherwise result from a release or threat of release.

3.2.12 *required remedial actions*—remedial actions determined necessary to comply with the requirements of CERCLA § 120(h)(3)(B)(i).

3.2.13 *required response actions*—removal or remedial actions, or both, determined necessary to comply with the requirements of CERCLA § 120(h)(3)(B)(i).

3.2.14 *risk-based criteria*—cleanup levels intended to meet a predetermined level of acceptable risk to human health or the environment.

3.2.15 *site inspection (SI)*—an on-site investigation to determine whether a release or potential release exists and the nature of the associated threats. The purpose is to augment the data collected in the preliminary assessment and to generate, if necessary, sampling and other field data to determine whether further action or investigation is appropriate.

3.3 *Descriptions of Terms Specific to This Standard:*

3.3.1 *adjacent properties*—those properties contiguous or partially contiguous to the boundaries of the property being surveyed during an EBS or other activity intended to classify the property into a standard environmental condition of property area type, or other properties relatively near the installation that could pose significant environmental concern or have a significant impact on the results of an EBS or on the classification of installation property into standard environmental condition of property area types, or both.

3.3.2 *aerial photographs*—photographs taken from an airplane or helicopter (from a low enough altitude to allow the identification of development and activities) of areas encompassing the property. Aerial photographs are commonly available from government agencies or private collections unique to a local area.

3.3.3 *all remedial action taken*—for the purposes of this classification, all remedial action, as described in CERCLA § 120(h)(3)(B)(i), has been taken if "the construction and installation of an approved remedial design has been completed, and the remedy has been demonstrated to the Administrator [of EPA] to be operating properly and successfully. The carrying out of long-term pumping and treating, or operation and maintenance, after the remedy has been demonstrated to the Administrator to be operating properly and successfully does not preclude the transfer of the property" (42 USC§ 9620(h)(3)).

3.3.4 *applicable requirements*—those cleanup standards, standards of control, and other substantive requirements, criteria, or limitations promulgated under federal environmental or state environmental or facility siting laws that specifically address a hazardous substance, pollutant, contaminant, remedial action, location, or other circumstances found at a CERCLA site. Only those state standards that are identified by a state in a timely manner and that are more stringent than federal requirements may be applicable.

3.3.5 *biased field sampling*—by any technique, field sampling of environmental media, which aids in the delineation of standard environmental condition of property area types.

3.3.6 *BRAC statutes*—Title II of the Defense Authorization Amendments and Base Closure and Realignment Act of 1988 (Public Law 100-526, 10 USC 2687) and the Defense Base Closure and Realignment Act of 1990 (Part A of Title XXIX of Public Law 101-510, 10 USC 2687), collectively.

3.3.7 *carcinogenic*—a cancer-causing substance.

3.3.8 *chain of title review*—a review of recorded land title records, conducted as part of an EBS.

3.3.9 *chemical-specific*—associated with the definition of applicable, or relevant and appropriate, requirements (ARARs), chemical-specific ARARs are those that may define acceptable exposure levels and can therefore be used in establishing primary remediation goals.

3.3.10 *closing military installation*—installations identified for closure pursuant to BRAC statutes.

3.3.11 *disposal*—the discharge, deposit, injection, dumping, spilling, leaking, or placing of any hazardous substances, or petroleum products or their derivatives, into or on any land or water so that such hazardous substances, or petroleum products or their derivatives or any constituent thereof, may enter the environment or be emitted into the air or discharged into any waters including groundwater.

3.3.12 *environmental investigation*—any investigation intended to determine the nature and extent of environmental contamination or to determine the environmental condition of property at a BRAC installation. Environmental investigations may include, but are not limited to, environmental site assessments, preliminary assessments, site inspections, remedial investigations, EBSs, Resource Conservation and Recovery Act (RCRA) facility assessments, and RCRA facility investigations.

3.3.13 *environmental site assessment*—the process by which a person or entity seeks to determine whether a particular parcel of real property (including improvements) is subject to recognized environmental conditions. This is the same meaning as provided in Practice E 1527.

3.3.14 *exposure pathway*—the route from a contaminant source to a human or any other environmental receptor.

3.3.15 *interviews*—sessions with current or former employees involved in operations on the real property, conducted to ascertain if release, treatment, or disposal of hazardous substances, petroleum products, or their derivatives has occurred or is occurring on the real property.

3.3.16 *migration*—the movement of contaminant(s) away from a source through permeable subsurface media (such as the movement of a groundwater plume of contamination), or the movement of contaminant(s) by a combination of surficial and subsurface processes.

3.3.17 *obviousness*—the condition of being plain or evident. A condition or fact that could not be ignored or overlooked by a reasonable observer while conducting a records search or while physically or visually observing the property in conjunction with an EBS.

3.3.18 *practically reviewable*—information that is practically reviewable, that is, provided by the source in a manner and form that, upon examination, yields information relevant to the property without the need for extraordinary analysis of irrelevant data. The form of the information shall be such that the user can review the records for a limited geographic area. Records that cannot be retrieved feasibly by reference to the location of the property or a geographic area in which the property is located are not generally practically reviewable.

Most data bases of public records are practically reviewable if they can be obtained from the source agency by the county, city, zip code, or other geographic area of the facilities listed in the record system. Records that are sorted, filed, organized, or maintained by the source agency only chronologically are not generally practically reviewable. This term has the same meaning as provided in Practice E 1527.

3.3.19 *preliminary assessment (PA)*—the review of existing information and an off-site reconnaissance, if appropriate, to determine whether a release or potential release may require additional investigation or action. A preliminary assessment may include an on-site reconnaissance, if appropriate.

3.3.20 *reasonably obtainable*—information that is (*1*) publicly available, (*2*) obtainable from its source within reasonable time and cost constraints, and (*3*) practically reviewable. This term has the same meaning as the term ''reasonably ascertainable," as provided in Practice E 1527.

3.3.21 *recognized environmental conditions*—the presence or likely presence of any hazardous substances or petroleum products on any DoD real property under conditions that indicate an existing release, a past release, or the material threat of a release of any hazardous substances or petroleum products into structures on the property or into the ground, groundwater, or surface water of the property. The term includes hazardous substances or petroleum products even under conditions in compliance with laws. The term is not intended to include de minimus conditions that generally do not present a material risk of harm to the public health or environment and that generally would not be the subject of an enforcement action if brought to the attention of appropriate governmental agencies.

3.3.22 *records search*—a detailed search and review of available information and records in the possession of the DoD components and records made available by the regulatory agencies or other involved Federal agencies, including, but not limited to, installation restoration program studies and analyses, surveys for radioactive materials, asbestos, radon, lead-based paint, electrical devices (that is, transformers) containing polychlorinated biphenyl (PCB), RCRA facility assessments and investigations to determine which, if any, hazardous substances or petroleum products may be present on the property. For the purposes of adjacent facilities, a records search includes the review of all reasonably obtainable Federal, state, and local government records for each adjacent facility where there has been a release or likely release of any hazardous substance or any petroleum product, and which is likely to cause or contribute to a release or threatened release of any hazardous substance or any petroleum product on the DoD real property.

3.3.23 *site property*—property within the boundaries of the DoD installation.

3.3.24 *standard environmental condition of property area type*—one of the seven environmental condition of property area types defined in this classification.

3.3.25 *standards-based criteria*—cleanup criteria intended to meet the performance standards for the selected remedial technology.

3.3.26 *storage*—the containment of hazardous substances, petroleum products, or their derivatives, either on a temporary basis or for a period of years, in such a manner as not to constitute the disposal of such hazardous substances, petroleum products, or their derivatives.

3.3.27 *unevaluated*—not previously evaluated during any type of environmental investigation. This may also be used to designate areas that are unevaluated regarding CERFA reporting requirements.

3.3.28 *visual or physical inspection, or both*—actions taken during an EBS to include observations made by vision while walking through or otherwise traversing a property and structures located on it and observations made by the sense of smell, particularly observations of noxious or foul odors.

3.4 Acronyms:

3.4.1 *ARARs*—applicable, or relevant and appropriate, requirements.

3.4.2 *ASTM*—American Society for Testing and Materials.

3.4.3 *CERCLA*—Comprehensive Environmental Response, Compensation and Liability Act of 1980, as amended (42 USC 9620 and following).

3.4.4 *CERFA*—Community Environmental Response Facilitation Act of 1992 (102 Public Law 426, 106 Statute 2174).

3.4.5 *DoD*—Department of Defense.

3.4.6 *EPA*—United States Environmental Protection Agency.

3.4.7 *FOSL*—finding of suitability to lease, as described in the applicable DoD policy.

3.4.8 *FOST*—finding of suitability to transfer, as described in the applicable DoD policy.

3.4.9 *PCBs*—polychlorinated biphenyls.

3.4.10 *RCRA*—Resource Conservation and Recovery Act, as amended, 42 USC 6901 and following.

3.4.11 *USC*—United States Code.

4. Significance and Use

4.1 *Uses*—This classification is intended for use by DoD components in order to direct EBS efforts. It is also intended for use by preparers and reviewers of environmental condition of property maps and EBS reports used to support CERFA uncontaminated parcel identifications and parcels suitable for transfer by lease or by deed. This classification should be used to facilitate standardized determinations of the environmental condition of a DoD installation's real property. Such environmental condition of property determinations are necessary to assess the progress of ongoing environmental restoration, identify areas where further response may be required, identify areas where further evaluation is necessary, and to support FOSTs and FOSLs. An environmental condition of property map, which should be prepared using this classification, provides a consolidated view of a DoD installation's environmental investigation data, including sampling information.

5. Basis of Classification

5.1 *Classification*—Classification of the seven standard environmental condition of property area types is according to statutory requirements for (*1*) the identification of uncontaminated property within the provisions of CERFA and (*2*) for designating parcels of DoD installations as being suitable or unsuitable for transfer by deed within the provisions of CERCLA § 120(h)(3)(B)(i). Standard environmental condition of property area types are ranked in order of their suitability for transfer, with Area Types 1 through 4 being suitable for transfer by deed and Area Types 5 through 6 being unsuitable for transfer by deed until all remedial actions have been taken. Areas classified as Area Type 7 are either unevaluated or require further evaluation in order to classify them as one of the other environmental condition of property area types.

6. Standard Classification of Environmental Condition of Property Area Types

6.1 *Standard Environmental Condition of Property Area Type 1*—Areas where no release, or disposal of hazardous substances or petroleum products or their derivatives has occurred, including no migration of these substances from adjacent areas. This is a geographically contiguous area or parcel of real property where the results of investigations reveal that no hazardous substances or petroleum products or their derivatives were released, or disposed of on site property. A determination of this area type cannot be made, however, unless a minimum level of information gathering and assessment has been completed. In accordance with CERCLA § 120(h)(4) , "the identification [of Type 1 areas] shall consist, at a minimum, of a review of each of the following sources of information concerning the current and previous uses of the real property: (i) A detailed search of Federal Government records pertaining to the property. (ii) Recorded chain of title documents regarding the real property. (iii) Aerial photographs that may reflect prior uses of the real property and that are reasonably obtainable through State or local government agencies. (iv) A visual inspection of the real property and any buildings, structures, equipment, pipe, pipeline, or other improvements on the real property, and a visual inspection of properties immediately adjacent to the real property. (v) A physical inspection of property adjacent to the real property, to the extent permitted by owners or operators of such property. (vi) Reasonably obtainable Federal, state, and local government records of each adjacent facility where there has been a release of any hazardous substance or any petroleum product or its derivatives, including aviation fuel and motor oil, and which is likely to cause or contribute to a release or threatened release of any hazardous substance or any petroleum product or its derivatives, including aviation fuel and motor oil, on the real property. (vii) Interviews with current or former employees involved in operations on the real property. Such identification shall also be based on sampling, if appropriate under the circumstances." These efforts (i through vii) can be functionally accomplished during the conduct of an EBS or a properly scoped preliminary assessment of the property being classified. However, if information gathered from these efforts indicates that hazardous substances, petroleum products, or their derivatives have been released, or disposed of, the property should be classified as an Area Type 2, 3, 4, 5, 6, or 7.

6.2 *Standard Environmental Condition of Property Area Type 2*—Areas where only release or disposal of petroleum products has occurred. This is a geographically contiguous area or parcel of real property where the results of investigations reveal that only the release or disposal of petroleum products, or their derivatives has occurred. In accordance with CERCLA § 120(h)(4), "the identification [of Type 2 areas] shall consist, at a minimum, of a review of each of the following sources of information concerning the current and previous uses of the real property: (i) A detailed search of Federal Government records pertaining to the property. (ii) Recorded chain of title documents regarding the real property. (iii) Aerial photographs that may reflect prior uses of the real property and that are reasonably obtainable through State or local government agencies. (iv) A visual inspection of the real property and any buildings, structures, equipment, pipe, pipeline, or other improvements on the real property, and a visual inspection of properties immediately adjacent to the real property. (v) A physical inspection of property adjacent to the real property, to the extent permitted by owners or operators of such property. (vi) Reasonably obtainable Federal, State, and local government records of each adjacent facility where there has been a release of any hazardous substance or any petroleum product or its derivatives, including aviation fuel and motor oil, and which is likely to cause or contribute to a release or threatened release of any hazardous substance or any petroleum product or its derivatives, including aviation fuel and motor oil, on the real property. (vii) Interviews with current or former employees involved in operations on the real property. Such identification shall also be based on sampling, if appropriate under the circumstances." These efforts (i through vii) can be functionally accomplished during the conduct of an EBS or a properly scoped preliminary assessment of the property being classified. However, if information gathered from these efforts indicates that hazardous substances have been released to the property, the property should be classified as an Area Type 3, 4, 5, or 6.

6.3 *Standard Environmental Condition of Property Area Type 3*—Areas where the release, disposal, or migration, or some combination thereof, of hazardous substances has occurred, but at concentrations that do not require a removal or remedial action. This is a geographically contiguous area or parcel of real property where environmental evidence demonstrates that hazardous substances have been released, or disposed of, but are present in quantities that require no response action to protect human health and the environment. Such quantities of hazardous substances or petroleum products can be below defensible detection limits or can be above detection limits but below action levels. In the absence of installation-specific risk-based or standards-based criteria that have received regulatory concurrence, below action levels means that the concentration of any hazardous substance or petroleum constituent in any medium does not exceed chemical-specific ARARs. A designation of this area type also means that risk estimates completed for contamination do not do the following: (*1*) exceed 10^{-6} for any carcinogenic hazardous substance or petroleum constituent detected in any medium; (*2*) result in a hazard index above 1 for any noncarcinogenic hazardous substance or petroleum constituent detected in any medium; (*3*) exceed 10^{-6} for all carcinogenic hazardous substances and petroleum constituents, taken together, in any exposure pathway; (*4*) result in a hazard index above 1 for all noncarcinogenic hazardous substances and petroleum constituents, taken together, in any exposure pathway; (*5*) exceed 10^{-4} for all carcinogenic hazardous substances and petroleum constituents accumulated across all pathways; or (*6*) result in a hazard index above 1 for all noncarcinogenic hazardous substances and petroleum constituents cumulated across all pathways. An Area Type 3 classification cannot be made with confidence unless a minimum level of information gathering and assessment has been completed. As such, all such determinations should be made on the basis of a site inspection or equivalent level of effort, which includes biased field sampling and laboratory analysis to support a conceptual understanding of the area. However, if information gathered from these efforts indicates that hazardous substances are on the property above action levels, the property should be classified as an Area Type 5 or 6.

6.4 *Standard Environmental Condition of Property Area Type 4*—Areas where release, disposal, or migration, or some combination thereof, of hazardous substances has occurred, and all remedial actions necessary to protect human health and the environment have been taken. This is a geographically contiguous area or parcel of real property where all remedial actions necessary to protect human health and the environment have been taken. Type 4 areas include those areas in which an EBS report or other environmental investigation documents evidence that hazardous substances are known to have been released or disposed of on the property, but all remedial actions necessary to protect human health and the environment regarding any hazardous substances remaining on the property have already been taken to meet the covenant requirements of CERCLA § 120(h)(3)(A)(ii).

6.5 *Standard Environmental Condition of Property Area Type 5*—Areas where release, disposal, or migration, or some combination thereof, of hazardous substances has occurred and removal or remedial actions, or both, are under way, but all required remedial actions have not yet been taken. This is a geographically contiguous area or parcel of real property where the presence of sources or releases of hazardous substances has been confirmed by the results of sampling and analysis efforts. The results of such sampling and analysis efforts may be contained in electronic data bases or other environmental investigation or environmental compliance reports, or both. This area type contains contaminant concentrations above action levels. Such concentrations do not meet the criteria of a Type 3 area classification. Removal actions are under way but are not yet demonstrated to have met the criteria of an Area Type 4. Remedial systems for Type 5 areas may be partially or entirely in place, but they have not been demonstrated to EPA to be "operating properly and successfully" within the meaning of CERCLA § 120(h)(3)(A)(ii).

6.6 *Standard Environmental Condition of Property Area Type 6*—Areas where release, disposal, or migration, or some combination thereof, of hazardous substances has occurred, but required response actions have not yet been initiated. This is a geographically contiguous area or parcel of real property where the presence of sources or releases of hazardous substances has been confirmed by the results of sampling and analysis efforts. The results of such sampling and analysis efforts may be contained in electronic data bases or environmental investigation or environmental compliance reports, or both. This area type contains concentrations of contaminants above action levels. Such concentrations do not meet the criteria of a Type 3 area classification. Required remedial systems or other response actions have not been initiated.

6.7 *Standard Environmental Condition of Property Area Type 7*—Areas that are unevaluated or that require additional evaluation. This is a geographically contiguous area or parcel of real property that is unevaluated, or a geographically contiguous area or parcel of real property where the presence of sources or releases of hazardous substances or petroleum products or their derivatives is suspected, but not well characterized, based on the results of a properly scoped records search, chain of title review, aerial photography review, visual inspection, set of employee interviews, and possibly sampling and analysis. They do not fit any of the previous area types with certainty because evaluation efforts have not occurred, are ongoing, or are inconclusive.

7. Keywords

7.1 Community Environmental Response Facilitation Act (CERFA); Comprehensive Environmental Response, Compensation and Liability Act; environmental baseline survey (EBS); environmental condition of property; FOSL; FOST; hazardous substance; property area type; real estate; recognized environmental conditions; remediation

Designation: D 5759 – 95

Standard Guide for Characterization of Coal Fly Ash and Clean Coal Combustion Fly Ash for Potential Uses[1]

This standard is issued under the fixed designation D 5759; the number immediately following the designation indicates the year of original adoption or, in the case of revision, the year of last revision. A number in parentheses indicates the year of last reapproval. A superscript epsilon (ε) indicates an editorial change since the last revision or reapproval.

1. Scope

1.1 This guide recommends standards for the characterization of fly ash from the combustion of coal, fly ash from coal combusted in the presence of alkaline materials, and fly ash from combusted coal in which the flue gases have been treated with alkaline materials in the presence of the fly ash.

1.2 This guide provides recommended and optional test methods for fly ash evaluation. Acceptance criteria can be negotiated between the producer and the user according to the potential end use.

1.3 The coal fly ash and clean coal combustion fly ash of this guide do not include the following:

1.3.1 Dusts from kilns producing products such as lime, portland cement, activated clays, etc.;

1.3.2 By-products of flue gas desulfurization that are not collected with the primary fly ash removal equipment such as the baghouse or electrostatic precipitator; and

1.3.3 Fly ash or other combustion products derived from the burning of waste; municipal, industrial, or commercial garbage; sewage sludge or other refuse, or both; derived fuels; wood; wood waste products; rice hulls; agriculture waste; or other non-coal fuels or other such fuels blended with coal, or some combination thereof.

1.4 Fly ash may contain some trace elements that may affect performance or potential end use.

1.5 The values stated in inch-pound units are to be regarded as the standard. The values given in parentheses are for information only.

1.6 *This standard does not purport to address all of the safety concerns, if any, associated with its use. It is the responsibility of the user of this standard to establish appropriate safety and health practices and determine the applicability of regulatory limitations prior to use.*

2. Referenced Documents

2.1 *ASTM Standards:*

C 22 Specification for Gypsum[2]

C 25 Test Methods for Chemical Analysis of Limestone, Quicklime, and Hydrated Lime[2]

C 51 Terminology Relating to Lime and Limestone (As Used by the Industry)[2]

C 109 Test Method for Compressive Strength of Hydraulic Cement Mortars (Using 2-in. or 50-mm Cube Specimens)[2]

C 110 Test Methods for Physical Testing of Quicklime, Hydrated Lime, and Limestone[2]

C 114 Test Methods for Chemical Analysis of Hydraulic Cement[2]

C 150 Specification for Portland Cement[2]

C 191 Test Method for Time of Setting of Hydraulic Cement by Vicat Needle[2]

C 311 Test Methods for Sampling and Testing Fly Ash or Natural Pozzolans for Use as a Mineral Admixture in Portland-Cement Concrete[3]

C 400 Test Method for Testing Quicklime and Hydrated Lime for Neutralization of Waste Acid[2]

C 593 Specification for Fly Ash and Other Pozzolans for Use with Lime[2]

C 595 Specification for Blended Hydraulic Cements[2]

C 602 Specification for Agricultural Liming Materials[2]

C 618 Specification for Coal Fly Ash and Raw or Calcined Natural Pozzolan for Use as a Mineral Admixture in Portland Cement Concrete[3]

D 546 Test Method for Sieve Analysis of Mineral Filler for Road and Paving Materials[4]

D 1973 Guide for Design of a Linear System for Containment of Wastes[5]

D 2795 Method for Analysis of Coal and Coke Ash[6]

D 3178 Test Method for Carbon and Hydrogen in the Analysis Sample of Coal and Coke[6]

D 3682 Test Method for Major and Minor Elements in Coal and Coke Ash by the Atomic Absorption Method[6]

D 3683 Test Method for Trace Elements in Coal and Coke Ash by the Atomic Absorption Method[6]

D 4326 Test Method for Major and Minor Elements in Coal and Coke by X-Ray Fluorescence[6]

D 5239 Practice for Characterizing Fly Ash for Use in Soil Stabilization[7]

E 1266 Practice for Processing Mixtures of Lime, Fly Ash,

---

[1] This guide is under the jurisdiction of ASTM Committee D-34 on Waste Management and is the direct responsibility of Subcommittee D34.03 on Treatment.
Current edition approved July 15, 1995. Published September 1995.

[2] *Annual Book of ASTM Standards*, Vol 04.01.

[3] *Annual Book of ASTM Standards*, Vol 04.02.

[4] *Annual Book of ASTM Standards*, Vol 04.03.

[5] *Annual Book of ASTM Standards*, Vol 11.04.

[6] *Annual Book of ASTM Standards*, Vol 05.05.

[7] *Annual Book of ASTM Standards*, Vol 04.09.

and Heavy Metal Wastes in Structural Fills and Other Construction Applications[5]

2.2 *Other Document:*

USEPA Method 9100-SW846 Falling Head or Constand Head[8]

3. Terminology

3.1 *Definitions:*

3.1.1 *clean coal combustion*—the burning of coal, coal culm, or coal fines in a furnace designed to operate to minimize emissions (that is, a fluidized bed or aerated fluidized bed, etc.) or coal burned in the presence of alkaline materials, which combine to reduce these emissions.

3.1.2 *fly ash, n*—residual material that exits a combustion chamber in the flue gas.

4. Significance and Use

4.1 This guide provides guidance for the characterization of coal fly ash or clean coal combustion fly ash for potential uses in which absorption, cementitious activity, pozzolanic activity, pH adjustment, heat rise, or stabilization and solidification properties may be desired.

5. Chemical Composition

5.1 Fly ash from coal and clean coal combustion can be characterized by the recommended chemical tests of Table 1 and may be characterized further by the optional chemical tests of Table 2. Limits may be specified by the purchaser if required for a specific application. The most recent limits established by the appropriate regulatory agency shall govern if no specific parameters are required. See Appendix X1 for possible non-mandatory information for various end uses of fly ash.

6. Physical Tests

6.1 Fly ash from coal and clean coal combustion can be tested further in accordance with the optional physical tests of Table 3, if required by the purchaser.

7. Shipments for Delivery to Purchaser

7.1 Fly ash shipped for delivery to the purchaser should be from a single combustion unit or a blend from multiple combustion units, as agreed upon between the purchaser and the supplier, such that the delivered fly ash complies with the provisions of Sections 8 and 9 herein.

8. Sampling and Testing

8.1 Take individual, representative samples of at least 0.5 lb (227 g) from each 100 tons (100 Mg) of fly ash delivered to the purchaser.

8.2 Samples are composites of 0.5-lb (227-g) individual samples of fly ash taken from each 100 tons (100 Mg) shipped. The minimum composite sample size should not be less than 8 lb (3.6 kg). Composites are to consist of nearly equal amounts from individual samples and be mixed thoroughly.

TABLE 1 Recommended Chemical Tests

Test Method	Component(s)	Limit[A]
C 114	sulfur trioxide (SO_3),%[B]	
C 311	moisture content, %	
C 311	loss on ignition, %	
D 2795, D 3682, or D 4326	calcium oxide (CaO),%	
D 2795, D 3682, or D 4326	magnesium oxide (MgO), %	
D 2795, D 3682, or D 4326	silicon dioxide (SiO_2) plus aluminum oxide (Al_2O_3) plus iron oxide (Fe_2O_3), %	

[A] On specific projects, a minimum or maximum may be applicable.
[B] Fly ash replaces hydraulic cement in method.

TABLE 2 Optional Chemical Tests[A] (Limits to be Specified Only if Applicable, by the Purchaser)

Test Method	Component(s)	Limit[B]
C 25	available lime index (ALI), %[C]	
C 311	available alkalies as Na_2O, %	
C 400	pH[D]	
C 602	calcium carbonate equivalent ($CaCO_3$), %	
D 3178	carbon (C),%	
D 3683	trace elements (totals) (for example, sulfide, sulfite, and sulfate)	

[A] Individual requirements may be specified by the purchaser if applicable to the project for which fly ash is to be used.
[B] On specific projects, a minimum or maximum may be applicable.
[C] Fly ash replaces limestone in analysis.
[D] Fly ash replaces quicklime in method.

TABLE 3 Optional Physical Tests[A] (To be Specified Only as Required by the Purchaser)

Test Method	Component(s)
C 109	compressive strength of hydraulic cement mortars, psi[B]
C 110	heat rise (slaking rate), °C[C]
C 191	time of set, min[B]
C 311	amount retained on No. 325 sieve, %
C 311	strength activity index with portland cement
	7 days, % of control
	28 days, % of control
C 311	water requirement, % of control[D]
C 311	specific gravity
C 311	increase in drying shrinkage, %
C 311	reactivity with cement alkalies, mortar expansion, % of control
C 311	soundness
C 593	amount retained on No. 200 sieve,%
C 593	amount retained on No. 30 sieve, %
C 593	lime pozzolan strength 7 days, psi
	lime pozzolan strength 28 days, psi
USEPA Method 9100-SW846	

[A] Individual requirements may be specified by the purchaser if applicable to the project for which fly ash is to be used.
[B] Modification of Test Method C 109 to approximate proportion(s) of fly ash instead of cement; or fly ash in combination with other materials to be used on the project (that is, cement, lime, etc.) should be used.
[C] Modify Test Methods C 110 to a proportion of fly ash instead of lime. The fly ash to water ratio may need to be modified further to obtain measurable results.
[D] Comparisons of water requirements to a control material, used at an equal flow, may be useful to determine the relative water requirement.

8.2.1 If insufficient fly ash has been shipped to comply with the composite sample requirements as listed in this section, an individual sample, or samples, may be used if agreed upon between the purchaser and the supplier. These samples should be at least 1 lb (454 g) and may represent less than 100 tons (100 Mg) shipped.

8.3 The recommended tests of Table 1 are performed on a sample of at least 1 lb (454 g) composited of individual shipment samples, at a frequency no less than semiannual

[8] Available from Standardization Documents Order Desk, Bldg. 4 Section D, 700 Robbins Ave., Philadelphia, PA 19111-5094, Attn: NPODS.

when at least 100 tons (100 Mg) of fly ash have been shipped to a purchaser.

8.4 For limits as required by the purchaser from Table 2 and Table 3, optional tests may be performed on the following frequency unless otherwise agreed upon between the supplier and the purchaser prior to shipment:

8.4.1 Every 2000 tons (2000 Mg) shipped,

8.4.2 A calendar month during which at least 100 tons (100 Mg) of fly ash have been shipped to a purchaser.

9. Test Method

9.1 Testing is conducted in accordance with the appropriate sections of the ASTM standards, or noted modifications, indicated in Table 1, the optional Table 2 and Table 3, and the nonmandatory information of Appendix X1, as specified by the purchaser. Individual tests may be selected from Table 2 and Table 3 and Appendix X1 according to the need of the user.

10. Inspection and Rejection

10.1 Upon agreement with the supplier, the purchaser should establish the criteria for inspection and testing. The criteria for rejection should be based on the ability of the material to comply with the appropriate specifications as referenced by the purchaser.

11. Compliance and Certification

11.1 Fly ash sampling and testing pursuant to Section 8 may be performed as the fly ash is produced or shipped, or both. Certification of compliance should be provided by the supplier with shipments of fly ash shipped from the production site, at the purchaser's request. If requested by the purchaser, test results on samples of the fly ash shipped should be sent to the purchaser as soon as they are available.

12. Retention of Test Samples

12.1 Samples shall be retained in airtight containers. Sample retention times are permissible if agreed upon between the purchaser and the supplier prior to shipment.

12.2 It is suggested that test data records should be maintained for a period of three years.

13. Keywords

13.1 characterization of clean coal combustion fly ash; characterization of clean coal fly ash; characterization of coal fly ash; fly ash for potential use

APPENDIX

(Nonmandatory Information)

X1. POTENTIAL END USE

TABLE X1.1 Potential End Use[A]

Suggested Referenced Documents[B]	Waste Treatment	Hazard Waste Treatment	Soil Amendment	Soil Stabilization	Mineral Filler	Pozzolanic Liners	Slurry Wall	Cement Product	Concrete Product	Gypsum Product
USEPA Method 9100-SW846						X	X			
Specification C 22			X			X		X		X
Test Methods C 25	X	X								
Test Method C 109	X	X				X	X		X	
Test Methods C 110	X	X			X	X				
Test Methods C 114	X	X	X	X				X		
Specification C 150								X		
Test Method C 191							X		X	
Test Methods C 311					X	X			X	
Test Method C 400	X	X	X							
Specification C 593						X				
Specification C 595								X		
Specification C 602			X							
Specification C 618									X	
Test Method D 546					X					
Guide D 1973						X				
Method D 2795	X	X								
Test Method D 3178								X		
Test Method D 3683	X	X						X		
Practice D 5239				X						
Practice E 1266	X	X				X				

[A] The use of this table is meant to assist the purchaser in selecting applicable best procedures for a specified end use. It is suggestive only.

[B] Suggested documents are to assist in characterizing fly ash as a substitute for the subject materials of these documents.

Standard Practice for Generation of Environmental Data Related to Waste Management Activities: Development of Data Quality Objectives[1]

This standard is issued under the fixed designation D 5792; the number immediately following the designation indicates the year of original adoption or, in the case of revision, the year of last revision. A number in parentheses indicates the year of last reapproval. A superscript epsilon (ε) indicates an editorial change since the last revision or reapproval.

1. Scope

1.1 This practice covers the process of development of data quality objectives (DQOs) for the acquisition of environmental data. Optimization of sampling and analysis design is a part of the DQO process. This practice describes the DQO process in detail. The various strategies for design optimization are too numerous to include in this practice. Many other documents outline alternatives for optimizing sampling and analysis design. Therefore, only an overview of design optimization is included. Some design aspects are included in the practice's examples for illustration purposes.

1.2 DQO development is the first of three parts of data generation activities. The other two aspects are (*1*) implementation of the sampling and analysis strategies and (*2*) data quality assessment. This guide should be used in concert with Practice D 5283, which outlines the quality assurance (QA) processes specified during planning and used during implementation.

1.3 Environmental data related to waste management activities include, but are not limited to, the results from the sampling and analyses of air, soil, water, biota, or waste samples, or any combinations thereof.

1.4 The DQO process should be developed and initiated prior to the application of planning, implementation, and assessment of sampling and analysis activities.

1.5 This practice presents extensive requirements of management, designed to ensure high-quality environmental data. The words "must" and "shall" (requirements), "should" (recommendation), and "may" (optional), have been selected carefully to reflect the importance placed on many of the statements in this practice. The extent to which all requirements will be met remains a matter of technical judgement.

1.6 The values stated in SI units are to be regarded as the standard. The values given in parentheses are for information only.

1.7 *This standard does not purport to address all of the safety concerns, if any, associated with its use. It is the responsibility of the user of this standard to establish appropriate safety and health practices and determine the applicability of regulatory limitations prior to use.*

2. Referenced Documents

2.1 *ASTM Standards:*

C 970 Guide for Sampling Special Nuclear Materials in Multi-Container Lots[2]

C 1215 Guide for Preparing and Interpreting Precision and Bias Statements in Test Method Standards Used in the Nuclear Industry[2]

D 5283 Practice for Generation of Environmental Data Related to Waste Management Activities: Quality Assurance and Quality Control Planning and Implementation[3]

3. Terminology

3.1 *Definitions*—Where applicable, the originating reference is associated with the definition and follows the definition in boldface type.

3.1.1 *accuracy (see bias)*—(*1*) bias; (*2*) the closeness of a measured value to the true value; (*3*) the closeness of a measured value to an accepted reference or standard value.

3.1.1.1 *Discussion*—For many investigators, *accuracy* is attained only if a procedure is both precise and unbiased (see *bias*). Because this blending of *precision* and *accuracy* can lead to confusion, ASTM requires a statement on *bias* instead of *accuracy*. **D 5283**

3.1.2 *action level*—the numerical value that causes the decision maker to choose one of the alternative actions (for example, compliance or noncompliance). It may be a regulatory threshold standard, such as maximum contaminant level for drinking water, a risk-based concentration level, a technological limitation, or reference-based standard. **EPA QA/G-4 (1)**[4]

3.1.3 *bias (see accuracy)*—the difference between the population mean of the test results and an accepted reference value.

3.1.3.1 *Discussion*—*Bias* represents a constant error as opposed to a *random error*. A method *bias* can be estimated by the difference (or relative difference) between a measured

---

[1] This practice is under the jurisdiction of ASTM Committee D34 on Waste Management and is the direct responsibility of Subcommittee D34.01.01 on Planning for Sampling.

Current edition approved Nov. 10, 1995. Published January 1996.

[2] *Annual Book of ASTM Standards*, Vol 12.01.

[3] *Annual Book of ASTM Standards*, Vol 11.04.

[4] The boldface numbers in parentheses refer to the list of references at the end of this practice.

average and an accepted standard or reference value. The data from which the estimate is obtained should be statistically analyzed to establish *bias* in the presence of *random error.* A thorough *bias* investigation of a measurement procedure requires a statistically designed experiment to repeatedly measure, under essentially the same conditions, a set of standards or reference materials of known value that cover the range of application. *Bias* often varies with the range of application and should be reported accordingly. **C 1215, D 5283**

3.1.4 *confidence interval*—an interval used to bound the value of a population parameter with a specified degree of confidence (this is an interval that has different values for different samples).

3.1.4.1 *Discussion*—When providing a *confidence interval,* analysts should give the number of observations on which the interval is based. The specified degree of confidence is usually 90, 95, or 99 %. The form of a *confidence interval* depends on underlying assumptions and intentions. *Confidence intervals* are usually taken to be symmetric, but that is not necessarily so, as in the case of *confidence intervals* for *variances.* **C 1215**

3.1.5 *confidence level*—the probability, usually expressed as a percent, that a *confidence interval* will contain the parameter of interest (see discussion of *confidence interval*).

3.1.6 *data quality objectives (DQOs)*—qualitative and quantitative statements derived from the DQO process describing the decision rules and the uncertainties of the decision(s) within the context of the problem(s).

3.1.6.1 *Discussion*—DQOs clarify the study objectives, define the most appropriate type of data to collect, determine the most appropriate conditions from which to collect the data, and establish acceptable levels of decision errors that will be used as the basis for establishing the quantity and quality of data needed to support the decision. The DQOs are used to develop a sampling and analysis design.

3.1.7 *data quality objectives process*—a quality management tool based on the scientific method and developed by the U.S. Environmental Protection Agency (EPA) to facilitate the planning of environmental data collection activities. The DQO process enables planners to focus their planning efforts by specifying the use of the data (the decision), decision criteria (action level), and decision maker's acceptable decision error rates. The products of the DQO process are the DQOs.

3.1.7.1 *Discussion*—DQOs result from an iterative process between the decision makers and the technical team to develop qualitative and quantitative statements that describe the problem and the certainty and uncertainty that decision makers are willing to accept in the results derived from the environmental data. This acceptable level of uncertainty should then be used as the basis for the design specifications for project data collection and data assessment. All of the information from the first six steps of the DQO process are used in designing the study and assessing the data adequacy. **EPA QA/G-4**

3.1.8 *decision error*

3.1.8.1 *false negative error*—this occurs when environmental data mislead decision maker(s) into not taking action specified by a decision rule when action should be taken.

3.1.8.2 *false positive error*—this occurs when environmental data mislead decision maker(s) into taking action specified by a decision rule when action should not be taken.

3.1.9 *decision rule*—a set of directions in the form of a conditional statement that specify the following: (*1*) how the sample data will be compared to the action level, (*2*) which decision will be made as a result of that comparison, and (*3*) what subsequent action will be taken based on the decisions.

3.1.10 *precision*—a generic concept used to describe the dispersion of a set of measured values.

3.1.10.1 *Discussion*—It is important that some quantitative measure be used to specify *precision.* A statement such as "the *precision* is 1.54 g" is useless. Measures frequently used to express *precision* are standard deviation, relative standard deviation, variance, repeatability, reproducibility, confidence interval, and range. In addition to specifying the measure and the *precision,* it is important that the number of repeated measurements upon which the estimated *precision* is based also be given. **D 5283**

3.1.11 *quality assurance (QA)*—an integrated system of management activities involving planning, quality control, quality assessment, reporting, and quality improvement to ensure that a process or service (for example, environmental data) meets defined standards of quality with a stated level of confidence. **EPA QA/G-4**

3.1.12 *quality control (QC)*—the overall system of technical activities whose purpose is to measure and control the quality of a product or service so that it meets the needs of users. The aim is to provide quality that is satisfactory, adequate, dependable, and economical. **EPA QA/G-4**

3.1.13 *random error*—(*1*) the chance variation encountered in all measurement work, characterized by the random occurrence of deviations from the mean value; (*2*) an error that affects each member of a set of data (measurements) in a different manner. **D 5283**

3.1.14 *risk*—the probability or expectation that an adverse effect will occur.

3.1.14.1 *Discussion*—*Risk* is frequently used to describe the adverse effect on health or on economics. Health-based *risk* is the probability of induced diseases in persons exposed to physical, chemical, biological, or radiological insults over time. This *risk* probability depends on the concentration or level of the insult, which is expressed by a mathematical model describing the dose and *risk* relationship. *Risk* is also associated with economics when decision makers have to select one action from a set of available actions. Each action has a corresponding cost. The *risk* or expected loss is the cost multiplied by the probability of the outcome of a particular action. Decision makers should adopt a strategy to select actions that minimize the expected loss.

3.1.15 *standard deviation*—the square root of the sum of the squares of the individual deviations from the sample average divided by one less than the number of results involved.

$$S = \sqrt{\frac{\sum_{j=1}^{n} (X_j - \bar{X})^2}{n - 1}}$$

where:

S = sample standard deviation,
n = number of results obtained,

X_j = jth individual result, and
$\bar{X}$ = sample average.

3.1.15.1 *Discussion*—The use of the *standard deviation* to describe *precision* implies that the uncertainty is independent of the measurement value. The practice of associating the ± symbol with *standard deviation* (or RSD) is not recommended. The ± symbol denotes an interval. The *standard deviation* is not an interval, and it should not be treated as such. **D 5283**

4. Summary of Practice

4.1 This practice describes the process of developing and documenting the DQO process and the resulting DQOs. This practice also outlines the overall environmental study process as shown in Fig. 1. It must be emphasized that any specific study scheme must be conducted in conformity with applicable agency and company guidance and procedures.

4.2 For example, the investigation of a Superfund site would include feasibility studies and community relations plans, which are not a part of this practice.

5. Significance and Use

5.1 Environmental data are often required for making regulatory and programmatic decisions. Decision makers must determine whether the levels of assurance associated with the data are sufficient in quality for their intended use.

5.2 Data generation efforts involve three parts: development of DQOs and subsequent project plan(s) to meet the DQOs, implementation and oversight of the project plan(s), and assessment of the data quality to determine whether the DQOs were met.

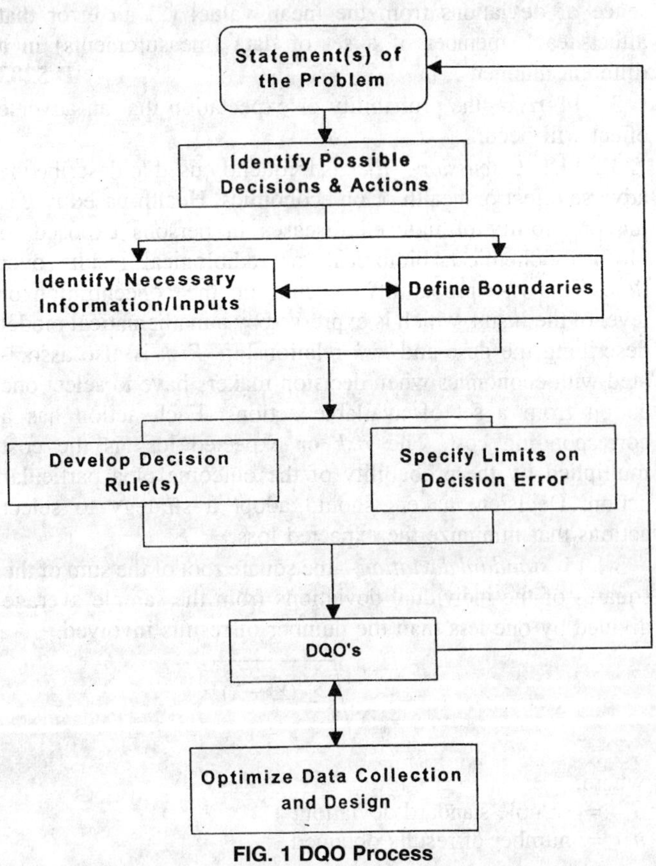

FIG. 1 DQO Process

5.3 To determine the level of assurance necessary to support the decision, an iterative process must be used by decision makers, data collectors, and users. This practice emphasizes the iterative nature of the process of DQO development. Objectives may need to be reevaluated and modified as information related to the level of data quality is gained. This means that DQOs are the product of the DQO process and are subject to change as data are gathered and assessed.

5.4 This practice defines the process of developing DQOs. Each step of the planning process is described.

5.5 This practice emphasizes the importance of communication among those involved in developing DQOs, those planning and implementing the sampling and analysis aspects of environmental data generation activities, and those assessing data quality.

5.6 The impacts of a successful DQO process on the project are as follows: (*1*) a consensus on the nature of the problem and the desired decision shared by all the decision makers, (*2*) data quality consistent with its intended use, (*3*) a more resource-efficient sampling and analysis design, (*4*) a planned approach to data collection and evaluation, (*5*) quantitative criteria for knowing when to stop sampling, and (*6*) known measure of risk for making an incorrect decision.

6. Data Quality Objective Process

6.1 The DQO process is a logical sequence of seven steps that leads to decisions with a known level of uncertainty (Fig. 1). It is a planning tool used to determine the type, quantity, and adequacy of data needed to support a decision. It allows the users to collect proper, sufficient, and appropriate information for the intended decision. The output from each step of the process is stated in clear and simple terms and agreed upon by all affected parties. The seven steps are as follows:

(*1*) Stating the problem,
(*2*) Identifying possible decisions,
(*3*) Identifying inputs to decisions,
(*4*) Defining boundaries,
(*5*) Developing decision rules,
(*6*) Specifying limits on decision errors, and
(*7*) Optimizing data collection design.

All outputs from steps one through six are assembled into an integrated package that describes the project objectives (the problem and desired decision rules). These objectives summarize the outputs from the first five steps and end with a statement of a decision rule with specified levels of the decision errors (from the sixth step). In the last step of the process, various approaches to a sampling and analysis plan for the project are developed that allow the decision makers to select a plan that balances resource allocation considerations (personnel, time, and capital) with the project's technical objectives. Taken together, the outputs from these seven steps comprise the DQO process. The relationship of the DQO process to the overall project process is shown in Fig. 2. At any stage of the project or during the field implementation phase, it may be appropriate to reiterate the DQO process, beginning with the first step based on new information. See Refs **(2, 3)** for examples of the DQO process.

6.2 *Step 1—Stating the Problem*:

6.2.1 *Purpose*—The purpose of this step is to state the

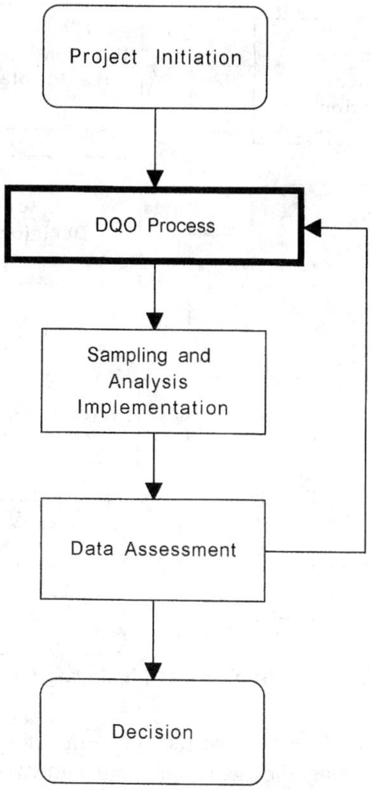

FIG. 2 DQOs Process and Overall Decision Process

problem clearly and concisely. The first indication that a problem (or issue) exists is often articulated poorly from a technical perspective. A single event or observation is usually cited to substantiate that a problem exists. The identity and roles of key decision makers and technical qualifications of the problem solving team may not be provided with the first notice. Only after the appropriate information and problem-solving team are assembled can a clear statement of the problem be made.

6.2.2 *Activities*:

6.2.2.1 *Assembling of all Pertinent Information*—The necessary first action to describe a problem is to verify the conditions that indicate a problem exists. The pertinent information should be assembled during this phase of problem definition. A key source is any historical record of events at the site where the problem is believed to exist. This enables the decision makers to understand the context of the problem. A series of questions need to be developed concerning the problem.

(1) What happened (or could happen) that suggests a problem?

(2) When did it (could it) happen?

(3) How did it (could it) happen?

(4) Where did it (could it) happen?

(5) Why did it (could it) happen?

(6) How bad is (might be) the result or situation?

(7) How fast is (might be) the situation changing?

(8) What is (could be) the impact on human health and the environment?

(9) Who was (could be) involved?

(10) Who knows (should know) about the situation?

(11) Has anything been (might anything be) done to mitigate the problem?

(12) What contaminants are (could be) involved?

(13) How reliable is the information?

(14) What regulations could or should apply?

(15) Is there any information that suggests there is not a problem?

This list of potential information is not exhaustive, and there may be other data applicable to the definition of the problem.

6.2.2.2 *Identification of the DQO Team*—Even as information is being gathered, it is necessary to begin assembling a team of decision makers and technical support personnel to organize and evaluate the information. These individuals become the core of the DQO team and may be augmented by others as information and events dictate. The identities and roles of the DQO team members are usually determined by the decision makers who have either jurisdiction over the site and personnel or financial resources that will be used in resolving the problem. The DQO team is usually made up of the following key individuals:

(1) Site Owners or Potentially Responsible Parties—These individuals have authority to commit personnel and financial resources to resolve the problem and have a vital interest in the definition of the problem and possible decisions.

(2) Representatives of Regulatory Agencies—These individuals are usually responsible for enforcing the standards that have been exceeded, leading to classifying the observations or events as a problem. Additionally, they have an active role in characterizing the extent of the problem, approving any proposed remedial action, and concurring that the action mitigated the problem.

(3) Project Manager—This individual generally has the responsibility for overseeing resolution of the problem. This person may represent either the regulatory agency or the potentially responsible parties.

(4) Technical Specialists—These individuals have the expertise to assess the information and data to determine the nature and extent of the potential problem and may become key players in the design and implementation of proposed decisions.

It is important that these individuals be assembled early in the process and remain actively involved to foster good communications and to achieve consensus among the DQO team on important decision-related issues.

6.2.3 *Outputs*:

6.2.3.1 *Statement of Problem and Context*—Once the initial information and data have been collected, organized, and evaluated, the conclusions of the DQO team should be documented. If it is determined that no problem exists, the conclusion must be supported by a summary of the existing conditions and the standards or regulatory conditions that apply to the problem.

(1) If a problem is found to exist, the reasons must be stated clearly and concisely. Any standards or regulatory conditions that apply to the situation must be cited. If the initial investigation concludes that the existing conditions are the result of a series of problems, the DQO team should attempt to define as many discrete problems (or issues) as possible.

(2) The following are examples of problem statements:

(a) A former pesticide formulation facility is for sale, but it is unknown whether it meets local environmental standards for property transfer.

(b) An industrial site is known to be contaminated with low levels of lead, but it is unknown whether levels are below risk-based standards.

(c) Most of a vacant lot is believed to be uncontaminated with PCBs (<2 ppm), but it is unknown whether abandoned, leaky transformers in the vacant lot make it necessary to remove any of the top layer of soil.

(d) The former industrial site has contaminated soil areas that may be contaminating ground water, and it is necessary to decide which type of monitoring program will satisfy local health requirements.

(e) The city would like to use local ground water on an athletic field near a Superfund site, but must know how this water will impact the health of the athletes and spectators.

(3) Complex problems should be broken down into manageable smaller problems that are linked together to form the final decision. As an example, the sale of a piece of property may involve solving the following problems:

(a) Is the site contaminated? If yes, then,

(b) Is off-site disposal required? If no, then

(c) Which of two allowable on-site treatment options should be used?

6.2.3.2 *Identification of Resources*—As the nature and magnitude of the problem is being documented, the decision makers should be conferring to determine the type and amount of resources that can be committed. Preliminary budget, personnel assignments, and schedule should be established. Preliminary milestones, timelines, and approvals should be documented and concurred upon by affected decision makers. The DQO team leader and technical specialists should be included in these discussions where possible. At a minimum, they should be kept informed of these issues so their impact can be anticipated in the definition of the problem.

(1) Fig. 3 shows the primary components of the problem statement step. After this step is completed, the DQO team moves on to the next step, where the process to resolve the problem continues.

(2) It is important to remember that the DQO process is an iterative one. New information is collected as projects proceed. The DQO team members associated with the problem-statement step should remain involved with the DQO process. If new data, unavailable to the DQO team during the development of the problem statement, demonstrates that the statement is incomplete or otherwise inadequate, the problem statement should be reconsidered.

6.3 *Step 2—Identifying Possible Decisions*:

6.3.1 *Purpose*—The purpose of this step is to identify the possible decision(s) that will address the problem once it has been stated clearly. Multiple decisions are required when the problem is complex. Information required to make decisions and to define the domain or boundaries of the decision will be determined in later steps (6.4 and 6.5, respectively). Each potential decision is tested to ensure that it is worth pursuing further in the process. A series of one or more decisions will result in actions that resolve the problem. The activities that lead to identifying the decision(s) are shown in Fig. 3 and discussed in 6.3.2.

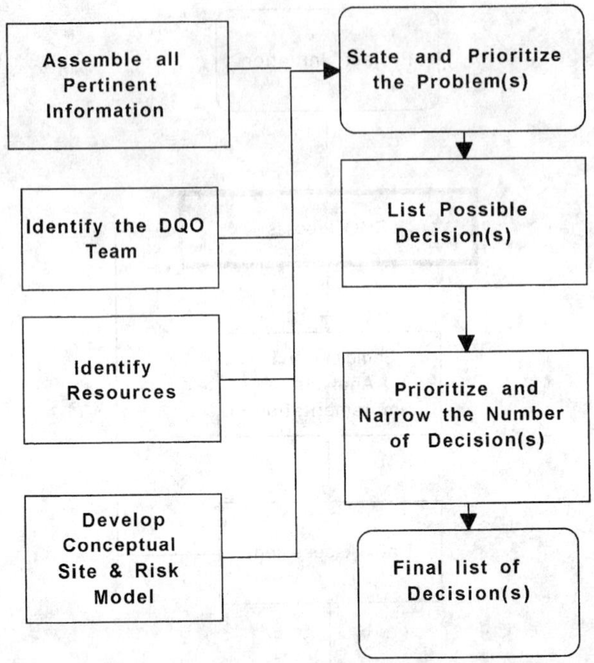

FIG. 3 Stating the Problem and Identifying the Decisions

6.3.2 *Activities*:

6.3.2.1 *Listing of Possible Decisions*—All possible decisions concerning the problem should be listed. Choices should not be eliminated at this time. Possible decision statements are presented in the form of a series of questions that, when answered, result in actions that will resolve the problem. Examples of questions related to problems given in 6.2.3 (Step 1) are as follows:

(1) Are possible contaminants on the site below regulatory thresholds?

(2) Must all of the surface soil be remediated to less than 5 ppm lead?

(3) Can only locations with PCB levels above 2 ppm be remediated?

(4) Will a ground water monitoring program at the site capable of detecting contaminants at the 5-ppm level satisfy regulatory requirements?

(5) Will a single monitoring point on or near the athletic field be sufficient?

6.3.3 *Output*—After all possible decisions that might be made have been documented, those determined to be most appropriate to resolve the problem should be prioritized by the DQO team in decreasing order of level of effort (available resources and technical challenge). Justification for the rankings should be provided. The recommended sequence in which the decisions are made should also be listed. In cases in which a complex decision statement has been broken down into a series of simpler decisions, the DQO team should identify whether the individual decisions should be addressed sequentially or in parallel. After the possible decisions have been identified, the DQO team focuses on gathering the information necessary to formulate the decision statements in Step 3 (6.4).

6.4 *Step 3—Identifying Inputs to Decisions*:

6.4.1 *Purpose*—The answers to each of the questions identified by the previous step in the DQO process must be resolved with data. Fig. 4 shows the key activities that lead to development of the data requirements. This sequence of activities must be performed for each question. Note that the limits of the study (or boundary conditions) are determined in a parallel step identified as "define boundaries" in Fig. 1. This is another type of data requirement and is discussed in 6.4.

6.4.2 *Activities*:

6.4.2.1 *Determination of Data Requirements*—At this stage of the process, it is important to carefully examine the complete set of data requirements needed to support each of the decisions. Each possible decision to be made should be considered independently of others to ensure that no omissions have occurred. After all possible questions concerning the decisions have been considered, group the data requirements together to determine overall data needs for the project. It may be possible to plan efficiencies in collecting and processing data to meet multiple needs and thereby lower overall project costs or reduce the time necessary to meet important milestones, or both.

(*1*) When considering whether specific information is needed for making a decision, test the data to ensure that it is appropriate for the decision statement. If no use of the data can be identified, it may be extraneous to the needs.

(*2*) The following list is indicative of some of the information needs that may be considered for each decision. It is not inclusive of all important data, but it provides examples common to many environmental problems.

(a) (*a*) What regulatory limits may be associated with the problem or regulatory issue?

(b) (*b*) Does contamination exceed regulatory limits?

(c) (*c*) What tests must be performed for the type of waste in question?

(d) (*d*) What are the hydrogeological considerations?

(e) (*e*) What populations are at risk?

(f) (*f*) What are the ecological considerations?

(g) (*g*) What process knowledge is available?

(h) (*h*) What historical/background data (past uses or spills) are available?

(i) (*i*) What are the budget constraints?

(j) (*j*) What is the time schedule?

(k) (*k*) What potential health, political, and social factors must be considered?

(l) (*l*) What is the potential for legal action?

(m) (*m*) Who is the end-user of the data?

(n) (*n*) What data validation criteria will be used?

(o) (*o*) What, if any, limitations exist on the data collection process (detection limits, matrix interferences, or no known measurement technology)?

6.4.3 *Outputs*:

6.4.3.1 The DQO team must specify data needs for each problem/decision that has been identified in the first two steps.

6.4.3.2 List the types of data required. Some example data types include, but are not limited to, the following:

(*1*) Chemical,

(*2*) Physical (including site hydrogeology and meteorology),

(*3*) Biological,

(*4*) Toxicological,

(*5*) Historical,

(*6*) Economic (time, budget, and manpower),

(*7*) Demographic,

(*8*) Toxicity characteristics, and

(*9*) Fate and transport model output.

6.4.3.3 *Listing of Data Generation Activities*—Determine which data can be acquired from historical records and which new data must be obtained in the field or laboratory, or both. If the DQO team determines that no new data are necessary to make a decision, they should document their reasoning. If new information is necessary, activities that will be required to generate inputs (data) affecting the decision should be listed. Examples of these include, but are not limited to, the following:

(*1*) Assembly of historical data,

(*2*) Sampling and chemical analysis,

(*3*) Physical testing, and

(*4*) Modeling.

6.4.3.4 *Definition of Data Use(s)*—Each set of data will be used for some purpose. This purpose must be defined. For example, will action levels for contaminants be determined by a risk-based calculation, by reference dose, or by pre-defined threshold values established by regulators? If so, ensure that data requirements are consistent with the criteria against which

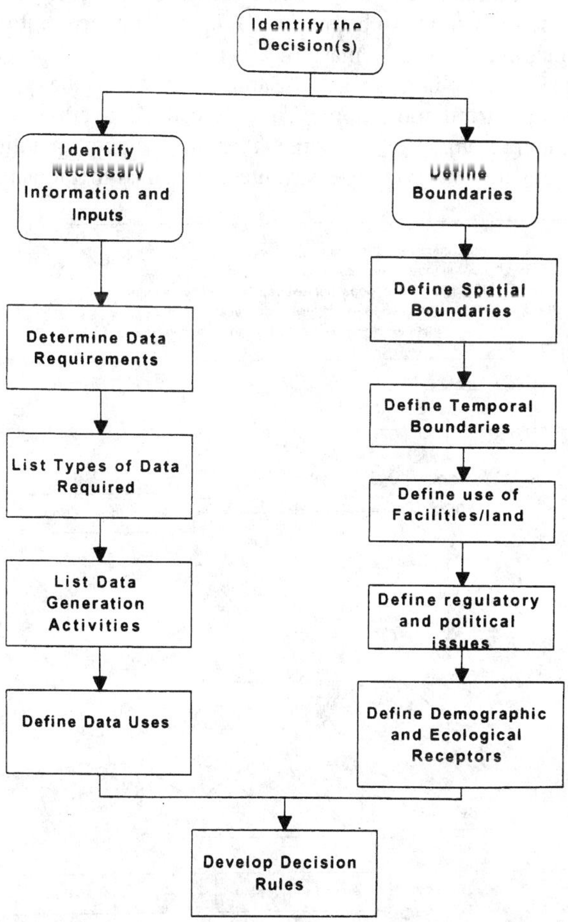

FIG. 4 Determination of Information Inputs and Study Boundaries

they will be compared. Data collected at the parts per million level may not be useful if they are to be compared to criteria at the parts per billion level.

6.5 *Step 4—Defining Boundaries*:

6.5.1 *Purpose*—This step of the DQO process determines the boundaries to which the decisions will apply. Boundaries establish limits on the data collection activities identified in Step 3 (6.4). These boundaries include, but are not limited to, spatial boundaries (physical and geographical), temporal boundaries (time periods), demographic, regulatory, political, and budget. The activities for this step of the DQO process are shown in Fig. 4.

6.5.2 *Activities*:

6.5.2.1 *Definition of Spatial Boundaries*—Define the boundaries of the total area and smallest increment of concern. Examples of items affecting the boundary definition are as follows:

(1) Horizontal or lateral areas,

(2) Vertical boundaries (depth/height),

(3) Discrete locations (hot spots),

(4) Media/matrix (air, soil, water, biota, and waste),

(5) Number of containers of waste, and

(6) Volume.

6.5.2.2 *Definition of Temporal Boundaries (Time Period)*—This activity determines the time interval over which environmental data will be collected for use in the decision-making process. If current or future real-time data are used to represent or model previous conditions, the basis of these assumptions or models must be documented and agreed upon between the decision makers and the technical team. The same constraint is also placed on the extrapolation of historical or real-time data, or both, to future time periods.

(1) The duration of new data collection activities must be established. In addition, the following factors should be considered:

(a) Availability and reliability of existing historical data,

(b) Access to the site or impacted area,

(c) Exposure potential, and

(d) Budgetary constraints.

6.5.2.3 *Definition of the Demographic Receptors*—The DQO team must frequently define the receptor population that may be effected. All affected populations and the mode of their anticipated exposure should be identified. These populations include the following:

(1) Known/Anticipated Population(s)—Human (children, adults, age, gender, and so forth), plant/animal (wetlands, endangered species, and so forth), and global;

(2) Population activity patterns; and

(3) Exposure pathway for each population.

6.5.2.4 *Definition of Nontechnical Boundaries*—Decision makers also have to consider nontechnical boundaries that can impact the resolution of the problem seriously. These nontechnical boundaries include the following:

(1) Regulatory considerations, and

(2) Political or legal action(s).

6.5.3 *Outputs*—The results from each of the activities in this step must be documented. Care must be taken to identify which boundary conditions apply to each decision being made. It may be that similar information is needed for several decisions but different boundary conditions may apply. It is important that decision makers understand and concur on the boundaries; otherwise, the ability to make decisions may be compromised.

6.6 *Step 5—Developing Decision Rules*:

6.6.1 *Purpose*:

6.6.1.1 The purpose of this step is to integrate outputs from previous steps into a set of statements that describe the logical basis for choosing among alternative outcomes/results/actions. These statements are decision rules that define the following:

(1) How the sample data will be compared to the action level,

(2) Which decision(s) will be made as a result of that comparison, and

(3) What subsequent action(s) will be taken based on the decisions.

6.6.1.2 The formats for these rules are either "if (criterion) ..., then (action)" statements or a decision tree, as shown in Fig. 5. The decision criteria should be stated as clearly and concisely as possible. The rule(s) must contain both a decision point (or action level) and an action. The decision rule is generated through a cooperative effort among the DQO team. If an acceptable decision rule cannot be formulated, the process returns to the appropriate previous step of the DQO process.

6.6.1.3 Decision rules usually contain the following elements: measurement of interest, sample statistic, action level, and a resultant action. "Measurement of interest" is the variable or attribute to be measured. It can be concentration of a contaminant, volume/mass of a waste, or physical property, such as flash point of a waste. "Sample statistic" is the quantity computed from the sample data. It can be average value, median, present/absent, or some other expression of quantity. If that data are not normally distributed, statistical methods based

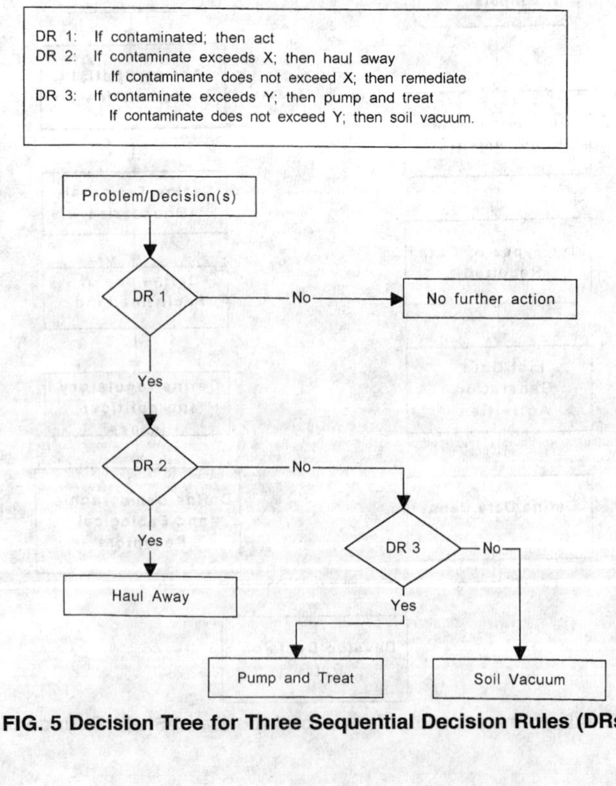

FIG. 5 Decision Tree for Three Sequential Decision Rules (DRs)

on other distributions or non-parametric methods can be used.

6.6.1.4 The "action level" is the limit against which the sample statistic will be compared. Depending on whether the action level is exceeded or not, the specified action will result. If the action level equals the regulatory threshold, the probability of a false positive error equals the probability of a false negative error. For unequal probabilities of the decision errors, the action level can be either less or greater than the regulatory threshold. The degree to which the action level is different from the regulatory threshold depends on the acceptable level of uncertainty for the decision errors that the decision makers are willing to accept. The action level is determined by the levels of false positive error, false negative error, measurement variability, and number of samples. Derivation of an action level for given level of false positive and false negative error is included as part of Appendix X1.

6.6.1.5 The decision rule is completed by stating the "resultant action" to be taken based on comparison of the sample statistic with the action level.

6.6.1.6 Two illustrations of general decision rule formats are as follows:

(1) "If the average concentration of a contaminant in waste is greater than the action level for that contaminant, then the waste will be classified as a 'hazardous' waste and will be disposed of according to the governing regulations."

(2) "If the average concentration of a contaminant in a waste is lower than the action level for that contaminant, then the waste is classified as 'nonhazardous' and there are no special limitations placed on the disposal options.

6.6.1.7 In this illustration, the measurement of interest is "concentration of a contaminant." The sample statistic is the "average concentration." The action level is some value to be specified. The resultant action is "disposal according to governing regulations." There may be separate decision rules for each medium, each domain (site), or other designated collections of data.

6.6.1.8 The action level may be an observation or occurrence in some cases. An example of this type of decision rule is as follows:

(1) If soil exhibits a visible dark spot as compared to the surrounding soil, use the portable organic monitor to screen for organics in the dark spot.

6.6.2 *Activities*—The activities that must be completed to establish a decision rule are shown in Fig. 6.

6.6.2.1 *Determination of Measurement of Interest*—A clear expression of the measurement (parameter) upon which the decision is based must be provided.

6.6.2.2 *Specification of Action Level*—The sample statistic of the measurement or observation of interest that initiates the agreed-upon action must be specified. The determination of the action level for any decision is a combination of the total variability in the data acquisition process and the level of decision errors that decision makers will accept in the final decision. The role of decision makers and decision errors is discussed in 6.7 (Step 6), and the derivation of an action level is illustrated in Appendix X1.

6.6.2.3 *Specification of Sample Statistic (if Applicable)*—Prior to the statement of a decision rule, it is necessary to determine how the sample statistic will be calculated and expressed (units of measure). The statistical approach chosen can be the average, mean, median, high, low, range, present/absent, and so forth. The unit of measurement must correspond to those of the decision criteria, and the limit of detection (measurement) must be lower than the action level. A statistic may or may not be applicable in stating observations.

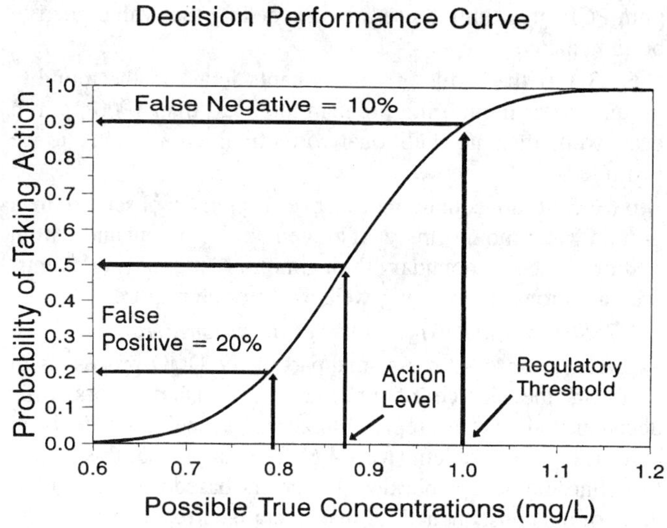

FIG. 6 Decision Rule Development

6.6.2.4 *Specification of Mode of Comparison*—After the sample statistic is derived from historical or real-time data and an action level has been identified, they must be compared. This comparison is usually stated as greater than ..., less than ..., equal to ..., or present/absent. Depending on the results of the comparison, a specific action is indicated by the decision rule.

6.6.2.5 *Specification of Action*—When the result of the comparison of the sample statistic with the action level is known, an action must be specified. It should be sufficient to resolve the problem. In complex situations, the action may direct decision makers to another problem (addressed by an additional set of DQOs) that must also be resolved. This type of logical pathway is described frequently as a decision tree. These situations should have been identified in Step 2 (6.3). Fig. 5 shows the decision tree derived from the application of a set of three sequential decision rules.

6.6.3 *Outputs*—An example showing the application of a decision rule is presented in Appendix X1. Some additional examples of decision rules that might apply to waste problems and possible actions discussed in 6.2 and 6.3, respectively, are given as follows:

6.6.3.1 If the historical record of site monitoring activities shows the absence of any regulated constituent above 1 ppm, then the site can be left as is.

NOTE 1—A value of 1 ppm selected for this example only.

6.6.3.2 If site characterization indicates that 20 % of the soil (top 30 cm) is contaminated above 5 ppm lead, then the entire soil layer (1 m) must be remediated.

6.6.3.3 If site characterization data show that 95 % of the total surface area (10 cm deep) of the site contains less than 2 ppm PCB, then only those areas exceeding that value need to be remediated.

6.6.3.4 If the levels of contaminants found in the monthly ground water monitoring program total less than 1000 ppm in each well, then no additional corrective action needs to be instituted.

6.6.3.5 If no contaminate above 1 ppm is observed in a ground water monitoring well located downgradient and within 100 m of the site boundary during monthly monitoring events, then additional monitoring wells will not be required.

6.7 *Step 6—Specifying Limits on Decision Errors*:

6.7.1 *Purpose*—An essential part of the DQO process is to establish the degree of uncertainty (decision errors) that decision makers are prepared to accept in making a decision concerning the problem **(Refs 4-6)**. The purpose of this step is to define the acceptable decision errors based on a consideration of the consequences of making an incorrect decision. The perspective of the decision makers or baseline assumption must be stated clearly, that is, the site is considered contaminated or the site is not contaminated. There are two kinds of decision errors: false positive error and false negative error.

6.7.2 *Activities*:

6.7.2.1 *Specifications of Decision Errors*—It should be understood that, when a decision is made based on empirical data, there is no way to reduce either type of decision error to zero. Furthermore, there is usually a tradeoff between the two decision errors, meaning that a lower false negative error would lead to a higher false positive error, and vice-versa (for a given amount of data or number of samples). Decision makers should understand the consequences of decision errors and the tradeoffs between a false positive error and a false negative error. Error rates (false positive and false negative errors) must be specified relative to an agreed-upon concentration regulatory threshold or health-risk level.

6.7.2.2 *Consequences of an Incorrect Decision*—The random variability for empirical data is composed of sample variability and measurement variability. Sampling variability is composed of both environmental variability (for example, spatial, temporal, matrix, and so forth) and sample collection variability. Measurement variability is a function of extraction efficiencies, matrices effects, and analyte interferences. Taken together, sample variance and measurement variance components comprise the total variability in the data that contributes to errors in the decision under consideration. Decision makers must make an a priori judgement regarding how often they are willing to be wrong because of data variability. This uncertainty is the "acceptable error" in the decision. In the context of a decision designed to be protective of human health, they can be wrong by taking a prescribed action when none was necessary (false positive error), or they can fail to take action when it was necessary (false negative error).

6.7.2.3 *False Positive Error*—If the true concentration is lower than the regulatory threshold, but the decision makers conclude that the waste is hazardous because the sample average concentration is equal to or higher than the action level, then a false positive error has been made. The consequence of this error is that the nonhazardous waste will be remediated or disposed of according to stricter requirements than required by regulations. A false positive error is undesirable because it will incur unnecessary costs and result in inefficiency.

6.7.2.4 *False Negative Error*—If the true concentration is equal to or greater than the regulatory threshold, but the decision makers conclude that the waste is nonhazardous because the sample average concentration is below the action level, then a false negative error has been made. The consequence of this error is that the waste will be disposed of by a less stringent method. This error is undesirable because the allowed waste management method may allow consequences harmful to health or the environment.

6.7.2.5 The relationship between the probability of taking action on a decision rule and the possible true value of the measurement of interest is illustrated graphically by a decision performance curve in Fig. 7 based on the example described in Appendix X1. The decision performance curve depends on the decision makers' willingness to accept false positive and false negative errors, the total variability of the measurement process, the number of samples, and a regulatory threshold. The interval between the action level and the regulatory threshold represents the range of possible true measurement values over which decision makers are willing to take more than a 50 % chance of sending a nonhazardous waste to a regulated landfill to ensure a specified false negative error. The curve is derived from the following:

(1) Acceptable errors (either a false positive error or a false negative error) agreed upon between the decision makers,

(2) Total variability of the system,

(3) Number of samples analyzed, and

(4) Statistical distribution of sample data (normal, lognormal and so forth).

6.7.2.6 In some cases, the action level may equal a regulatory level, or risk level. In these cases, all of the decision makers must understand that the value of a false positive error and false negative error associated with making a decision are equal.

6.7.2.7 Specification of false positive error and false negative error is typically made on the basis of the relative importance of the consequence of an incorrect decision of

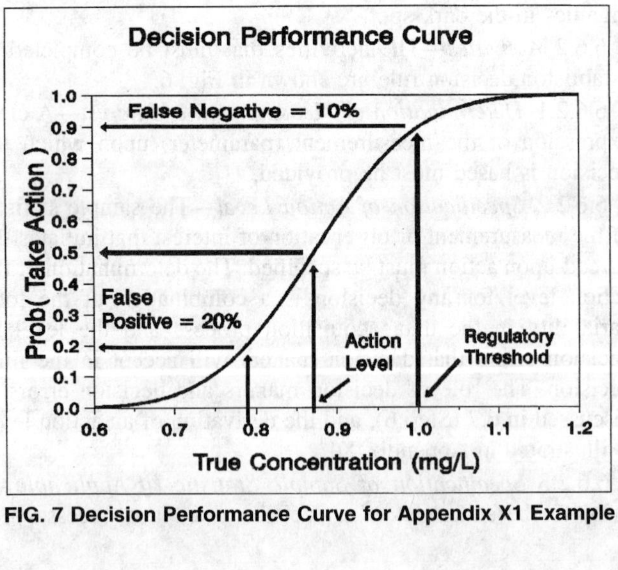

FIG. 7 Decision Performance Curve for Appendix X1 Example

either type. If the costs of environmental disposal or remediation are substantial and the potential environmental impact is relatively minor, then the emphasis may be on the control or reduction of false positive error (cost control). If the reverse is the case, then the emphasis may be on the control or reduction of the false negative error (control of environmental risk and liability). This important issue must be negotiated and resolved on a case-by-case basis for each problem identified in Step 1 by all decision makers.

6.7.2.8 This curve and several others that illustrate the relationships between these factors are discussed in the example in Appendix X1.

6.7.2.9 *Control of Decision Errors*—While decision errors cannot be eliminated, their errors can be reduced by (*1*) reducing measurement errors (sampling or analytical variabilities, or both) or (*2*) increasing the number of samples taken. These issues relate to optimization of the study design and are covered in Step 7 (see 6.8).

6.7.3 *Output*—The rational and acceptable errors for both the false positive and false negative errors for each decision from Step 1 must be documented.

6.7.4 *DQO Summary*:

6.7.4.1 *Purpose*:

(*1*) The purpose of this step is to present the results of the DQO process clearly and concisely, in a form usable for optimizing data collection design (6.8; Step 7). This presentation of the DQOs and the complete documentation of the outputs and logic from which they were derived is essential for the initiation of data collection design.

(*2*) The DQOs are derived from the outputs of all of the preceding steps in the DQO process. Each output is important. However, the uncertainty on the decision and the decision rules incorporate the decision, boundaries, and inputs required to generate a sampling design. Indeed, the uncertainties on the decisions, together with the respective decision rules, are the primary results of the DQO process for a particular problem.

6.7.4.2 *Activities*:

(*1*) Activities include the establishment of a framework in which the decision rule(s) and associated limits on decision error are expressed as the DQO(s) supported by the documented logic and outputs of the previous steps of DQO process development. Within this decision framework, the DQOs can be improved and refined through an iterative process that includes use of and further evaluation of the following:

(*a*) Problem statement,

(*b*) Possible decisions,

(*c*) Inputs,

(*d*) Definition of spatial and temporal boundaries,

(*e*) Development of decision rule(s), and

(*f*) Acceptance of limits on decision error.

(*2*) Establishment of the DQOs by integration of concise decision rule(s) with their associated limits on decision error and the documentation of the DQO process is critical in facilitating understanding of the risk of making the wrong decision by the decision makers.

6.7.4.3 *Outputs*:

(*1*) Primary outputs consist of clear and concise presentation of the DQO process and complete documentation of the logic involved in development of the decision rules and associated limits on decision errors.

(*2*) As a useful tool, the DQO process can be integrated graphically into a typical decision tree or logic flow diagram that clearly indicates actions to be taken as the result of implementation of the decision rule(s) (see Fig. X1.1). These diagrams and associated descriptive text are effective formats for use during the optimization of data collection design and are important elements in project work plans.

(*3*) For example, the following are DQO summaries from Appendix X1: To make the following decision for the "cadmium incineration waste problem" with a false positive error not to exceed 20 % and a false negative error not to exceed 10 %. If the mean cadmium concentration in the toxicity characteristic leaching procedure (TCLP) extract is equal to or >1 mg/L, then dispose of the fly ash load in a suitable landfill. If the mean cadmium concentration in the TCLP extract is <1 mg/L, then dispose of the fly ash load in a sanitary landfill.

6.7.4.4 *Application of Data Quality Objectives*:

(*1*) The DQOs are applied on a day-to-day basis by incorporating the decision errors into the action level. This makes the decision rule easier to use. To apply DQOs, statisticians apply statistical methods such as those used in the example in Appendix X1 to calculate an action level that takes into account the acceptable decision uncertainty.

(*2*) The applied DQOs from Appendix X1 are as follows:

(*a*) If the average concentration of cadmium is ≥0.87 mg/L, then dispose of the waste fly ash in a hazardous waste landfill; and

(*b*) If the average concentration of cadmium is <0.87 mg/L, then dispose of the waste fly ash in a sanitary landfill.

6.7.4.5 *Decision Tree Format*—In decision tree format, the DQOs are presented along with the actions and tasks that are required in the data collection design step (see Fig. 5).

6.8 *Step 7—Optimizing Data Collection Design*:

6.8.1 Prior to beginning this step of the process, the output from the first six steps must be assembled and provided to DQO team members who will undertake to optimize the actual sampling design for data collection. Care should be taken to separate the factual material from the DQO team's assumptions or estimates, or both, of factors important to development of the output from each step. The data collection effort must gather sufficient data to confirm (if possible/feasible) the accuracy of these assumptions.

6.8.2 *Purpose*:

6.8.2.1 The objective of this step is to generate the most resource-effective sampling design that will provide adequate data for decisions to be made. In this step, sampling designs are developed based on the outputs of the first six steps of the process, assumptions made during those steps, and applicable statistical techniques.

6.8.2.2 An understanding of the sources of variability and levels of uncertainty is essential in developing the sampling design alternatives. The focus of the DQO process is the balancing of the limits of decision errors against the resources available to complete the project. Many of the sampling design alternatives will address different strategies for balancing the different types of decision errors with the resources available (time, money, and personnel) to resolve the problem.

6.8.2.3 Once sampling designs are developed, the sampling design alternatives and required resources for each should be presented to the decision makers. These alternatives allow for an understanding of the benefits and resource commitments to each sampling design. If a resource-effective sampling design to provide adequate data for the decision rule cannot be found among the sampling design alternatives, it may be necessary to alter the decision or revise the inputs into the DQO process. This decision is the responsibility of the decision makers and requires that all DQO team members be involved. New members may be added if, in the opinion of the decision makers, their expertise is needed to develop acceptable DQOs.

6.8.3 *Activities*—The activities involved in the development of an optimal sampling design and chemical analyses are shown in Fig. 8.

6.8.3.1 *Summary of Information*—The data collectors should summarize any previous data and the outputs from the previous six steps of the DQO process. This allows data collectors to remain focused on the decision makers' needs in design optimization.

6.8.3.2 *Development of Sampling Design Alternatives*—Alternative sampling designs must be based on DQOs, which were developed with an understanding of measurement variability and the resources available for resolving the problem. Design alternatives must address the degree of representation of any one sample within the problem boundaries. This is accomplished by selecting from among the sampling designs

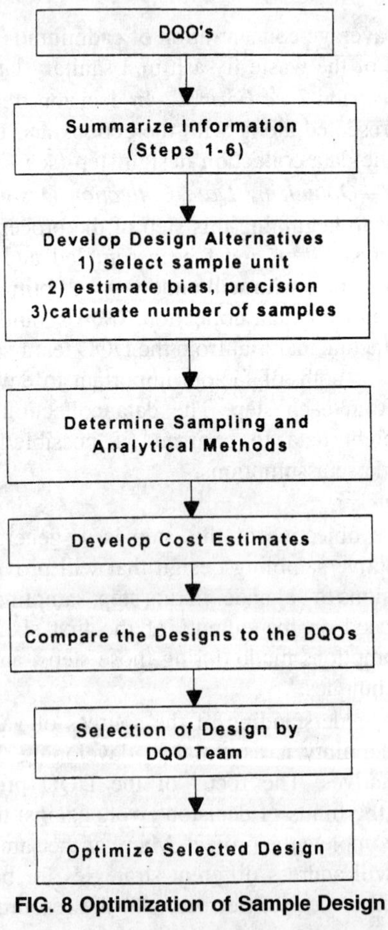

FIG. 8 Optimization of Sample Design

those that best describe the system. These include, but are not limited to, random, sequential random, systematic, and stratified sampling designs.

(1) Probabilities of selecting an appropriate sample are related to the type of sampling design. An equal probability of selecting a sample implies a random sample design. Selecting unequal probabilities for sample selection implies a stratified sample design. The more heterogeneous the sampling units, the more likely unequal probabilities will be assigned to the sample. Furthermore, the more heterogeneous the waste site, the more useful historical or process information is in assessing the sampling design alternatives. The participation of a qualified statistician is critical in this process.

(2) Variability may also be introduced during sample handling and preparation procedures that may be necessary between field sampling and analytical methods. Consideration of the important factors impacting sample variability should occur during the design process.

6.8.3.3 *Determination of Analytical Chemistry Methods*—The alternative analytical chemistry methods as documented during the DQO process must be considered. Factors that affect selecting alternative methods include, but are not limited to, the following:

(1) Detection limits versus action levels;

(2) Matrix effects on detection limits, bias, and variability; and

(3) Sample amount available (volume or weight).

6.8.3.4 *For Each Sampling Design Alternative, Selection of the Sample Unit that Satisfies the DQOs*—Sampling units include drums, tanks, an area within a grid, a boring location on a grid, a depth interval in a boring, or any other appropriate defined physical unit from which material can be obtained. Different sampling units may and often will be appropriate for different materials or locations. The sampling unit may depend on logistical and resource issues, such as whether the material will be disposed by drum or truck or the amount of material that can be excavated.

6.8.3.5 *For Each Sampling Design Alternative, Calculation of the Optimal Number of Samples that Satisfies the DQOs*—Typically, samples are collected from each sample unit for chemical analyses. Using the mathematical expressions for sampling design optimization, solve for the optimal number of samples that meet the uncertainty limits on the decision errors specified in the DQOs. Selection of the number of samples is an iterative process. Initial selection of the number of samples may be based on different project criteria (for example, budget, precision limits, and so forth). These initial calculations should be examined to determine whether they are adequate for the specified decision errors. In addition, preliminary sample designs may be required for better estimates of mean concentrations and measurement variability for optimal planning of larger sample designs.

6.8.3.6 *For Each Sampling Design Alternative, Development of Cost Estimates*—The estimates should relate the total cost of sampling and chemical analyses for alternative sampling designs. These cost functions may take into account such items as the cost of remediation or waste disposal by sample unit. This enables the decision makers to assess whether sampling and chemical analyses are more cost effective than proceeding with cleanup or disposal with minimal data collection.

6.8.4 *Outputs*—The list of sampling design alternatives is submitted to the decision makers for selection. After selection of the final sampling design, document the operational details and theoretical assumptions of the selected sampling design in a final sampling and chemical analyses plan. The documentation should include the sampling plan, sampling and analytical chemistry procedures, data assessment procedures, quality control requirements, and overall project quality assurance requirements.

7. Documentation of the Data Quality Objective Process

7.1 The following statements and information document the outputs of the specific DQO process used to develop the DQOs. The DQOs are meaningless if they are not connected with the specific problem and other qualifying information used to develop them.

7.2 DQO process documentation summaries can vary from problem to problem, but most will include information such as the following:

7.2.1 Facility name, location, and process;

7.2.2 List of decision makers, affiliations, and responsibilities for this project.

7.2.3 Statement of the problem.

7.2.4 Summary of logic for the decisions chosen for consideration. For each problem there must be at least one decision.

7.2.5 Information and inputs such as those given in 6.4.2. There should be appropriate inputs to allow generation of the data to make a decision. It may be useful to establish separate decisions for each matrix (that is, soil, sediment, and water).

7.2.6 Defined boundaries, which should be addressed for each decision. It may be useful to segregate the boundaries by matrix.

7.2.7 Decision rules, which should incorporate appropriate boundaries. The rules may be stated by matrix.

7.2.8 Limits on decision error. The rationale or assumptions upon which decision error estimates are based should be documented.

8. Keywords

8.1 data quality objectives; DQOs; project planning; waste analysis; waste testing

APPENDIX

(Nonmandatory Information)

X1. DQO CASE STUDY—CADMIUM-CONTAMINATED FLY ASH WASTE

X1.1 *Background:*

X1.1.1 A municipal waste incineration facility located in the Midwest routinely removes "fly ash" from its flue gas scrubber system and disposes it in a sanitary landfill. It was determined previously that the ash was nonhazardous under hazardous waste regulations. However, the incinerator has recently begun treating a new waste stream. As a result, a local environmental public interest group asked that the ash be retested and evaluated for hazardous waste compliance before it is disposed. The group is primarily concerned that the ash may contain hazardous levels of cadmium due to the new waste sources. The facility manager has agreed to test the ash and decided to use the DQOs process to help guide decision making throughout the project. Although not constrained by cost, the facility is interested in minimizing expenditures.

X1.1.2 The 40 CFR Part 261 RCRA toxicity characteristic criteria **(7)** for determining whether a solid waste is hazardous requires collection of a "representative portion" of the waste and performance of TCLP. During this process, the solid fly ash will be "extracted" or mixed in an acid solution for 18 h. The extraction liquid will then be subjected to tests for specific metals.

X1.1.3 Since the impact of this new waste stream is not known, a preliminary study was conducted to determine the variability of the concentration of the contaminants. Random samples were collected from the first 20 truckloads. Since process knowledge of the waste stream indicated that cadmium was the only toxicity characteristic (TC) constituent in the waste, these samples were analyzed individually for cadmium using TCLP. The results were expressed as the average concentration along with the standard deviation.

X1.2 *Data Quality Objective Development*—The following is an example of the outputs from each step in the DQO process.

X1.2.1 *Statement of the Problem*:

X1.2.1.1 *Identification of the DQO Team*—The plant manager assembled a DQO team consisting of himself and a representative of the current disposal facility staff. The two of them subsequently assembled the additional DQO team members.

(*1*) The decision makers on the DQO team included the incinerator owner and incineration plant manager, and a representative of the environmental public interest group, in which a representative of the community in which the ash is currently being disposed. The technical staff included a statistician, toxicologist, and chemist with sampling experience.

X1.2.1.2 *Statement of the Problem*—The problem is to determine whether any loads of fly ash are hazardous with cadmium under RCRA regulations using TCLP testing. If a load is hazardous, it must be disposed of in a RCRA landfill.

X1.2.2 *Identification of Possible Decisions*:

X1.2.2.1 *Decision*—Determine whether the concentration of cadmium in TCLP leachate from waste fly ash exceeds the regulatory RCRA standards.

X1.2.2.2 *Statement of the Actions that Could Result from the Decision*:

(*1*) If the average concentration of cadmium is greater than or equal to the action level, dispose of the waste fly ash in a RCRA landfill.

(*2*) If the average concentration of cadmium is less than the action level, dispose of the waste fly ash in a sanitary landfill.

X1.2.3 *Identification of Inputs to Decisions*—The DQO team identified the following inputs or information needed for the decision rules:

X1.2.3.1 *Preliminary Study Information*—Since the concern is with a new waste stream, the DQO team ordered a pilot study of the fly ash to determine the variability in the concentration of cadmium between loads of fly ash leaving the facility. They have determined that each load is fairly homogeneous. However, there is a high variability between loads due to the nature of the waste-stream. Most of the fly ash produced is not a RCRA hazardous waste and may be disposed of in a sanitary landfill. Because of this, the company has decided that testing each individual waste load before it leaves the facility would be the most economical. In that way, they could send loads of ash that exceeded the regulated cadmium concentrations to the higher-cost RCRA landfills and continue to send the others to the sanitary landfill.

(*1*) The study showed that the standard deviation of the cadmium concentration within a load was $S_w = 0.4$ mg/L, and the standard deviation of the cadmium concentration between loads was $S_b = 1.4$ mg/L. Sample and quality control data indicate that a normal distribution can be assumed.

X1.2.3.2 *Identification of Contaminants of Concern, Matrix, and Regulatory Limits*—The DQO team identified the following factors critical to the problem:

(*1*) *Contaminants of Concern*—Cadmium soluble in the TCLP extract.

(*2*) *Sample Matrix*—Fly ash.

(*3*) *Regulatory Threshold*—1 mg/L.

X1.2.3.3 *Specific Project Budget and Time Constraints*—The incinerator plant manager has requested that all stages of the operation be performed in a manner that minimizes the cost of sampling, chemical analysis, and waste disposal. However, no formal cost constraints have been implemented.

(*1*) The environmental public interest group has threatened to file a lawsuit for violation of environmental regulations if testing does not proceed within a "reasonable time-frame."

(*2*) The waste does not pose a threat to humans or the environment while contained in the trucks. Additionally, since the fly ash is not subject to change, disintegration, or alteration, the chemical properties of the waste do not warrant any temporal constraints. However, in order to expedite decision making, the DQO team has placed deadlines on sampling and reporting. The fly ash waste will be tested within 48 h of being loaded onto waste hauling trailers. The analytical results from each sampling round should be completed and reported within five working days of sampling.

X1.2.3.4 *Identification of the Testing Methods*—In this case, 40 CFR Part 261, Appendix II specified the TCLP Method SW 846, Method 1311 (**8**). The leachate must be analyzed by an appropriate method. Potential methods of characterizing the leachate for cadmium include, but are not limited to, SW 846, Methods 6010, 6020, 7130, or 7131.

X1.2.4 *Inputs to Be Determined*:

X1.2.4.1 *Method Validation and Quality Control (QC)*—The analytical method accuracy and precision and method detection limits in leachate from the fly ash matrix must be determined. The QC samples must be specified.

X1.2.4.2 *Identification of Sampling Procedure or Devices*—The following must be determined:

(*1*) Number of samples,

(*2*) Sampling methods for composite or grab samples of ash, and

(*3*) The QC requirements for sampling.

X1.2.5 *Definition of the Boundaries*—Define a detailed description of the spatial and temporal boundaries of the decision, characteristics that define the environmental media and objects or people of interest, and any practical considerations for the study.

X1.2.5.1 *Specification of the Characteristics that Define the Sample Matrix*—The fly ash should not be mixed with any other constituents except the water used for dust control.

X1.2.5.2 *Identification of Spatial Boundaries*—The variability between loads was greater than within a load; therefore, a decision will be made on each load. The waste fly ash will be tested after it has been deposited in the trailer used by the waste hauler. Separate decisions regarding the toxicity of the fly ash will be made for each load of ash leaving the incinerator facility. Each load of ash should fill the waste trailer at least 70 %. In cases in which the trailer is filled less than 70 %, the trailer must wait on-site until more ash is produced and can fill the trailer to the appropriate capacity.

X1.2.5.3 *Identification of Temporal Boundaries (Including the Time Frame Over Which the Study Should Be Conducted)*—The waste does not pose a threat to humans or the environment while contained in the trucks. However, in order to expedite decision making, the DQO team has placed deadlines for reaching a decision. The fly ash waste will be tested and a disposal decision made within 48 h of being loaded onto waste hauling trucks.

X1.2.6 *Development of Decision Rules*—The arithmetic mean of sample results will be compared to the action level.

X1.2.6.1 *Decision Rule*:

(*1*) If the average concentration of cadmium in a truck load is equal to or greater than the action level, then dispose of the waste fly ash in a RCRA landfill; or

(*2*) If the average concentration of cadmium in a truck load is less than the action level, then dispose of the waste fly ash in a sanitary landfill.

Note that the DQO team will decide that the action level is less than the regulatory level in order to meet a 10 % false negative error for concentrations at the regulatory level of 1 mg/L.

X1.2.7 *Specification of Limits on Decision Errors*:

X1.2.7.1 The decision makers specify acceptable decision errors based on the consequences of making an incorrect decision. Both types of decision errors have negative consequences.

(*1*) *False Positive Error* (declaring the load hazardous

when it is not)—If the true cadmium concentration is below 1 mg/L, but the average measured cadmium concentration is above the action level, the nonhazardous fly ash waste will be sent to a RCRA landfill. The consequence of a false positive error is that the company will have to pay additional cost to dispose of the waste with a cadmium concentration between the action level and regulatory threshold at a RCRA facility as opposed to a less expensive method of disposal in a sanitary landfill.

(2) False Negative Error (declaring the load nonhazardous when it is hazardous)—If the true cadmium concentration is equal to or greater than 1 mg/L, but the average measured cadmium concentration is below the action level, the hazardous fly ash waste will be sent to a sanitary landfill. The consequence of a false negative error is that the fly ash waste may be disposed of in a manner that will be harmful to human health or the environment. Legal consequences and subsequent remedial costs are also possible consequences.

X1.2.7.2 The purpose of this stage of the process is to specify the probabilities of making incorrect decisions that are acceptable to decision makers. The DQO team must agree on which type of decision error is of greater concern, either a false positive error or false negative error.

X1.2.7.3 For this example, the DQO team is more concerned about a false negative error because of the increased liability due to sending potentially hazardous waste to a sanitary landfill. The DQO team set a value for the false negative error of 10 % when the true concentration is 1 mg/L. The false negative error is a greater concern because of the perceived increased liability due to sending potentially hazardous waste to a sanitary landfill. This level is determined based on the comfort of the decision makers accepting the risk associated with calling a hazardous waste nonhazardous.

X1.2.7.4 *Data Quality Objective Summary*—Application of the DQOs on a day-to-day basis depends on *(1)* selecting the number of samples and (2) quantifying the action level for the decision rule. The decision performance curves are used to visually compare the desired decision errors versus the possible true cadmium concentrations for different numbers of samples.

(1) The uncertainty for the DQOs can be quantified by calculating the action level based on a false negative error of 10 % when the true cadmium concentration of a TCLP extract for a fly ash load has a value of the regulatory threshold (1 mg/L).

(2) To begin the early phases of design optimization, the DQO team determined how the environmental data should be summarized and used in the decision. The DQO team identified that the mean concentration of cadmium from each load would be compared to the action level. The background data indicated that a normal distribution can be used to calculate the action level. A normal distribution is an appropriate probability model for the preliminary data. A false negative error less than 50 % implies that an action level will be lower than the regulatory threshold.

(3) How the statisticians on the DQO team calculated the action level for the project is shown as follows. The action level is dependent on variables such as regulatory threshold, standard deviation, false negative error, and number of samples. Changing one variable will affect the value of the action level. Another iteration through the last DQO process steps must be made if any of these changes are made.

X1.2.7.5 *Concentration Range and Action Level*—The DQO team examined the concentration data from the first 20 analyses and determined that a reasonable concentration range to examine was between 0.6 and 1.3 µg/L. The DQO team agreed that the action level should be based on a 10 % false negative error at the regulatory threshold. This implies that the action level will be less than the regulatory threshold. Paragraph X1.2.8 describes the calculations for several action levels corresponding to different numbers of samples in the decision performance curve, using the standard deviation, the limits of error, and the desired false negative error. The decision performance curve will be calculated to determine the action level and review the performance of the decision rule. To calculate the decision performance curve, decision makers use the following steps:

(1) Step 1—Number of Samples:

(a) Selecting the number of samples is always difficult because imperfect knowledge is available concerning the variability of the measurement process for the selected sample matrix. All calculations for the number of samples are approximations. Different methods can be used to determine the number of samples. For the cadmium example, an initial selection of the number of samples is determined by an estimation method that specifies the precision limits on determining the concentration in the TCLP extract. Another sample size method would be based on the decision performance curve that examines the effect of a different number of samples on the decision errors. This decision method for number of samples is investigated in X1.2.8. Another method would be to calculate the number of samples for specified values of the measurement standard deviation, action level, and false positive error and false negative error. This procedure is illustrated in Guides C 970 and C 1215.

(b) For the initial fly ash waste loads, chemists on the DQO team would like to verify that their instrument is calibrated for the proper concentration range. They want to estimate the true cadmium concentration in the TCLP extract with an uncertainty of ±0.2 mg/L. In addition, the decision makers are willing to allocate resources to learn that the true cadmium concentration is in this interval with a confidence of 95 %. The number of samples for these precision limits can be approximated by a normal probability distribution. Another approximation to the number of samples could use an iterative method for a Student's *t*-distribution rather than the normal distribution. This more general assumption usually adds only two or three samples beyond the normal distribution used herein.

(c) The number of samples (*n*) is calculated by the following equations (**9, 10**), with $L = 0.2$ mg/L, $\sigma = S_w = 0.4$ mg/L, and $\alpha = 0.05$ (or $Z_{\alpha/2} = 1.960$ for a 95 % confidence level):

$$n = \left(\frac{Z_{\alpha/2}\,\sigma}{L}\right)^2$$

$$n = \left(\frac{1.960 \times 0.4}{0.2}\right)^2 \approx 16 \quad \text{(X1.1)}$$

where:

n = number of samples,
L = limit of error on the average (for example, 0.2 mg/L),
$1 - \alpha$ = probability level for the confidence interval for $\alpha = 0.05$, and then $1 - \alpha = 0.95$ confidence interval,
σ = standard deviation of the measurement process (for example, 0.4 mg/L), and
$Z_{\alpha/2}$ = $\alpha/2$ percentile point of normal probability distribution (for example, $Z_{\alpha/2} = Z_{0.025}$). Common normal percentile values are given in Table X1.1.

(2) *Step 2—Action Level*—The action level value for the decision rule is determined by controlling the false negative error established in the DQO process. The quantification of the action level used a value of 0.10 (or 10 %) for the probability of the false negative error and 16 samples to determine the average cadmium concentration from the TCLP extracts. The probability calculations are based on an approximating normal probability distribution for the cadmium concentration measurements. This approximating normal probability assumes a mean = RT = 1.0 mg/L and a standard deviation = S_w = 0.4 mg/L. The 10 % percentile point for the standardized normal probability distribution is $Z_{0.10} = 1.282$ (see Table X1.1). The probability (Pr) for the false negative error evaluated at RT is as follows:

$$\text{Pr (false negative error)} = \text{Pr (average} < AL \text{ when the true concentration} = RT) = 0.10$$

or

$$\Pr(\text{FN}) = \Pr\left[\frac{\text{average} - \text{RT}}{S_w/\sqrt{n}} < \frac{\text{AL} - \text{RT}}{S_w/\sqrt{n}}\right] = 0.10,$$

$$\frac{AL - RT}{S_w/\sqrt{n}} = -Z_{0.10},$$

$$\text{AL} = \text{RT} - Z_{0.10}\frac{S_w}{\sqrt{n}}.$$

$$\text{AL} = 1.0 \text{ mg/L} - (1.282)(0.4 \text{ mg/L})/4 = 1.0 \text{ mg/L} - 0.13 \text{ mg/L},$$
$$\text{AL} = 0.87 \text{ mg/L}. \quad \text{(X1.2)}$$

where:
AL = action level,
RT = regulatory threshold,
S_w = standard deviation of the measurement process estimated from a sufficient number of samples, and
$Z_{0.10}$ = tabulated 10 % percentile point from a standard normal distribution (see Table X1.1).

Therefore, the decision rule is as follows:

(*a*) If (average concentration of cadmium) ≥ 0.87 mg/L, the fly ash load is considered to be a RCRA waste and will be disposed of in a RCRA landfill; or

TABLE X1.1 Common Normal Percentile Points

$Z_{0.20}$	$Z_{0.10}$	$Z_{0.05}$	$Z_{0.025}$	$Z_{0.01}$	$Z_{0.005}$
0.842	1.282	1.645	1.960	2.326	2.576

(*b*) If (average concentration of cadmium) < 0.87 mg/L, the fly ash load is not considered to be a RCRA waste and will be disposed of in a sanitary landfill.

X1.2.7.6 *Decision Tree Format*—Fig. X1.1 shows the decision tree format for the DQOs, along with the action level and tasks that are required in the data-collection design step.

(*1*) *Step 3—True Concentration Corresponding to the False Positive Error:*

(*a*) Calculate the true concentration (θ mg/L < RT) that corresponds to a probability for the false positive error of 20 % using an action level of AL = 0.87 mg/L. This calculation again uses the approximating normal probability distribution for the cadmium concentration measurements. For the specified false positive error, the approximating normal probability assumes a mean = θ mg/L (to be determined), a standard deviation = S_w = 0.4 mg/L, and the number of samples = 16. The 20 % percentile point for the standardized normal probability distribution is $Z_{0.20} = 0.842$ (see Table X1.1).

$$\text{Pr (false positive error)} = \text{Pr \{average} \geq \text{AL when the true concentration} = \theta < RT\} = 0.20$$

or

$$\Pr(\text{FP}) = \Pr\left[\frac{\text{average} - \theta}{S_w/\sqrt{n}} \geq \frac{\text{AL} - \theta}{S_w/\sqrt{n}}\right] = 0.20,$$

$$\frac{AL - \theta}{S_w/\sqrt{n}} = +Z_{0.20},$$

$$\theta = AL - Z_{0.20}\frac{S_w}{\sqrt{n}}.$$

$$\theta = 0.87 \text{ mg/L} - (0.842)(0.4 \text{ mg/L})/4 = 0.87 \text{ mg/L} - 0.08 \text{ mg/L},$$
$$\theta = 0.79 \text{ mg/L}. \quad \text{(X1.3)}$$

where:
AL = action level,
RT = regulatory threshold, and
$Z_{0.20}$ = tabulated 20 % percentile point from a standard normal distribution (see Table X1.1).

(*b*) The decision performance curve would have a probability of taking an action (that is, sending fly ash waste to a RCRA landfill) of 0.20 at a true cadmium concentration of θ = 0.79 mg/L. The possible true cadmium concentration values in the

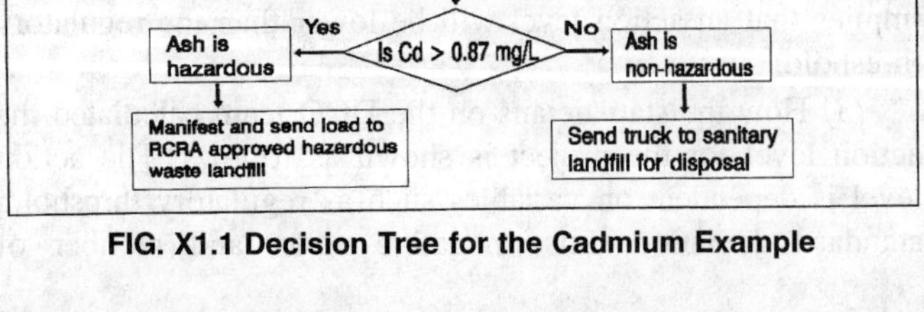

FIG. X1.1 Decision Tree for the Cadmium Example

interval (0.79 and 1.0 mg/L) represent values that cause the decision rule to send fly ash waste to a RCRA landfill even though the true concentration is below the regulatory threshold. This interval can be reduced by increasing the number of samples, changing the false negative error, or changing the false positive error.

(2) *Step 4—Drawing the Decision Performance Curve:*

(*a*) Draw the decision performance curve by using the standardized normal probability distribution. The standardized normal probability distribution is defined as a normal probability distribution with mean = 0 and standard deviation = 1.0. There are many tables and computer programs that can be used to calculate probabilities for a standardized normal random variable, Z. A normal random variable, X, with mean = μ and standard deviation = σ can be transformed to a standardized normal random variable by $Z = (X - \mu)/\sigma$.

$$\text{Prob (action)} = \text{Pr (average} \geq \text{AL when the true concentration} = \theta)$$

$$\text{Prob (action)} = 1.0 - \text{Prob}\left(Z \leq \frac{\text{AL} - \theta}{S_w/\sqrt{n}}\right),$$

$$\text{Prob (action)} = 1.0 - \text{Prob}\left(Z \leq \frac{0.87 - \theta}{0.1}\right). \qquad \text{(X1.4)}$$

(*b*) Fig. X1.2 is a plot of the decision performance curve generated by calculating a Prob (action) value using the standard normal probability distribution for each possible true concentration value θ. The decision performance curve can frequently be drawn freehand if three pairs of (concentration and probability) values are determined: ((RT, 1 − Pr (false negative error)), (AL, 0.50), and (θ, Pr (false positive error)).

X1.2.8 *Optimizing Data Collection and Design*—The decision makers will select the lowest-cost sampling design that is expected to achieve the DQOs. The series of designs for sampling the fly ash waste will be generated by the statisticians on the DQO team. The choice of sampling plan will be decided by consensus.

X1.2.8.1 *Decision Performance Curve*—The decision performance curve in Fig. X1.2 plots the probability of taking action (disposing of the waste in a RCRA landfill) versus different possible values for the true concentration in the TCLP extract. The DQO process specified a probability of 0.10 for the false negative error when the true concentration is at the RT. This specified false negative error implies that the decision performance curve will have a probability of taking action equal to 0.90 when the true concentration is equal to RT. If the true concentration value is equal to the value of the action level (0.87 mg/L), there is a probability of taking action of 0.50. The DQO team can also determine the true concentration for a specified false positive error from the decision performance curve.

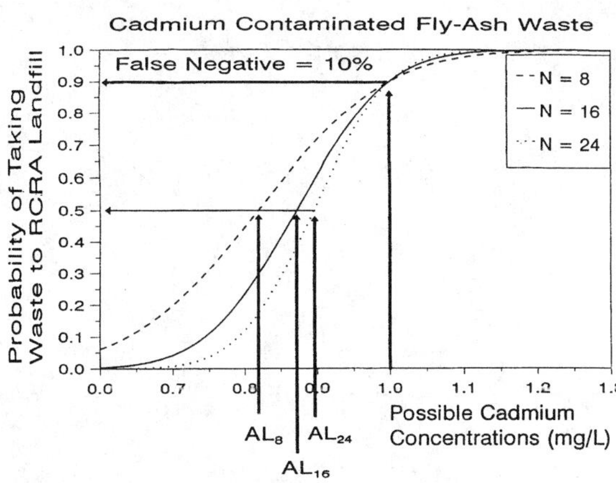

FIG. X1.2 Decision Performance Curves for Cadmium Example

(*1*) Fig. X1.2 shows three decision performance curves for three different numbers of samples (8, 16, and 24). All three decision performance curves meet the specified probability for the false negative error of 0.10 at a true concentration equal to RT. The purpose of these curves is to assess the effects of taking more or fewer samples on the action level and the false positive error. This analysis can be used to update applying the decision rule. For example, the decision makers concluded that eight additional samples (that is, 24) does not improve the AL value and false positive error sufficiently to justify the increase in cost.

X1.2.8.2 *Implementation*—Cadmium concentration values from the TCLP extracts will be collected over a long time period because this waste stream is a continuous process. The decision makers will establish a QC program to monitor the cadmium concentration values for process changes. After every 30 fly ash loads, the process variability will be reestimated and new values for the number of samples and action level will be considered. This strategy becomes part of the decision process.

X1.2.8.3 *Documentation of the Data Quality Objective Process*—The following statements and information document the outputs of the specific DQO process used to develop the above-stated DQOs. These objectives are meaningless if they are not connected with the specific problem and other qualifying information used in the DQO development.

(*1*) The DQO team required that the documentation be a concise summary of the following information:

(*a*) Facility name, location, and process;

(*b*) List of DQO team members, affiliations, and responsibilities for this project;

(*c*) Statement of the problem;

(*d*) Logic for the solutions chosen for consideration;

(*e*) Information and inputs required by the DQO team to make the decision, including sample matrix, preliminary study results, sampling methods required, and use of each input in reaching a decision.

(*f*) Defined boundaries;

(*g*) Decision logic in rule or decision tree format; and

(*h*) Assumptions made regarding the decision error and any information used to generate preliminary action levels and the number of samples.

(*2*) All meetings held by the DQO team should be documented. The meeting minutes should include the attendees, information used to generate each step of the process, and rationale used to make final agreements on the decision logic, boundaries, inputs, and decision errors.

REFERENCES

(1) *Guidance for Planning for Data Collection in Support of Environmental Decision Making Using the Data Quality Objectives Process*, EPA QA/G-4, U.S. Environmental Protection Agency, Washington, DC, 1993.

(2) Blacker, S., and Maney, J., "The DQO Process: System DQO Planning Process and the Analytical Laboratory," *Environmental Testing and Analysis*, July/August 1993.

(3) Neptune, D., Blacker, S., Fairless, B., and Ryti, R., "Application of Total Quality Principles to Environmental Data Operations," *National Energy Division Conference*, American Society of Quality Control, 1990.

(4) *A Rationale for the Assessment of Errors in the Sampling of Soils*, EPA-600/4-90-013, Environmental Monitoring Systems Laboratory, U.S. Environmental Protection Agency, Washington, DC, May 1990.

(5) *Methods for Evaluating the Attainment of Cleanup Standards Volume 1: Soils and Solid Media*, U.S. Environmental Protection Agency, Washington, DC, February 1989.

(6) *Characterizing Heterogeneous Wastes: Methods and Recommendations*, EPA 600/R-92/033, U.S. Environmental Protection Agency and Department of Energy Office of Technology, Washington, DC, February 1992.

(7) 40 Code of Federal Regulations (CFR), Part 261, 1995.

(8) *Test Methods for Evaluating Solid Waste*, 3rd ed., with Updates I, II, IIA, IIB, SW-846, Office of Solid Waste and Emergency Response, U.S. Environmental Protection Agency, Washington, DC, January 1992.

(9) Cochran, W. G., *Sampling Techniques*, 3rd ed., John Wiley and Sons, Inc., New York, NY, 1977.

(10) Desu, M. M., and Raghavarao, D., *Sample Size Methodology*, Academic Press, San Diego, CA, 1990.

Designation: D 5829 – 96 (Reapproved 2001)

Standard Guide for Preparing a Training Program for Environmental Analytical Laboratories[1]

This standard is issued under the fixed designation D 5829; the number immediately following the designation indicates the year of original adoption or, in the case of revision, the year of last revision. A number in parentheses indicates the year of last reapproval. A superscript epsilon (ε) indicates an editorial change since the last revision or reapproval.

1. Scope

1.1 This guide is intended to assist the laboratories that analyze environmental samples with the development of a documented training program. The training program should develop and increase the competence of analysts and provide a means of recording the results of all proficiency testing.

1.2 Some of the functions within a laboratory that can be addressed using this guide are as follows:

1.2.1 Analysts,

1.2.2 Technicians,

1.2.3 Quality assurance (QA),

1.2.4 Sample receiving and control, and

1.2.5 Sample procurement (sampling).

2. Referenced Documents

2.1 *EPA Standards:*

EPA Method 150.1[2]

SW 846 USEPA Test Methods for Evaluating Solid Waste—Physical/Chemical Methods, 9040 and 9045[2]

3. Summary of Guide

3.1 This guide is summarized in the following steps:

3.1.1 Define the training needs;

3.1.2 Prepare training materials;

3.1.2.1 Develop training objectives;

3.1.2.2 Develop performance evaluation materials;

3.1.2.3 Develop a detailed training outline; and,

3.1.2.4 Develop a documentation form;

3.1.3 Identify trainers competent in the areas defined in 3.1.1;

3.1.4 Determine need and frequency for retraining;

3.1.5 Assemble and store training materials; and,

3.1.6 Assign responsibility for training program.

4. Significance and Use

4.1 Training is a key component in the development of a competent staff in the environmental laboratory.

4.2 This guide will assist in providing both the organizational structure and the direction for a laboratory training program.

4.3 This guide will result in a documentation effort that will satisfy the requirements of environmental auditing groups.

5. Defining the Need for Training

5.1 Each individual who handles or analyzes environmental samples must be knowledgeable in the proper procedures for performing one's job function. Any deficiencies must be corrected by training. Training also may be extended to include those individuals who accept and record information prior to receipt of a sample or who are responsible for generating a report detailing the results of the analyses. Those involved with quality assurance functions need specialized training as well.

5.2 The first step in developing a training program is to identify the procedures or methods within the laboratory for which an individual is responsible. These may range from basic activities to very complex manipulations or interpretations.

5.2.1 Basic activities might include: pH measurement, pipetting, titrating, unloading sample shippers, or data entry.

5.2.2 Complex activities might include: operating an inductively coupled plasma (ICP), cleaning the source of a mass spectrometer, or data validation.

5.2.3 Within each activity certain tasks will have to be learned. A detailed analysis of these tasks must be performed before specific training courses can be developed.

5.2.4 Before beginning actual training, the level of training should be determined using the task analysis.

5.3 Before starting a training program, an approved Standard Operating Procedure (SOP) must be available for each method or activity.

5.3.1 A detailed manual or training video may serve as an SOP.

5.3.2 Published or externally prepared materials must be followed explicitly or an in-house document should be prepared. It is often difficult to follow commercially prepared materials due to differences in the equipment, work areas, sample type, or even level of training of the analyst. For this reason, it is advisable to prepare SOPs that detail the actual

---

[1] This guide is under the jurisdiction of ASTM Committee D34 on Waste Management and is the direct responsibility of Subcommittee D34.01.06 on Analytical Methods.

Current edition approved Jan. 10, 1996. Published March 1996.

[2] Available from the Superintendent of Documents, U.S. Government Printing Office, Washington, DC 20402.

situation that will be encountered by the trainee.

6. Preparation of Training Materials

6.1 An effective training program for environmental laboratories should use training courses that include, as a minimum, the following components. See Appendix X1 for an example of a training course employing these elements.

6.2 *Training Objectives*:

6.2.1 Each method or procedure for which training will be conducted must have a set of objectives;

6.2.2 Each objective should deal with specific aspects of the training process that require a demonstrated response; and,

6.2.3 An objective should be written to include the following components:

6.2.3.1 A statement of the desired result of the training. This statement will often take the following form: "After completing this training course on (specific topic), the trainee will be able to (specify result)." The specified result may be as simple as a new appreciation for the topic or the more difficult development of a specialized skill or the complex ability to perform problem solving activities.

6.2.3.2 A statement of the way in which the expected result will be demonstrated needs to be made. This might include discussion with a trainer, completing a written exercise, demonstrating a skill, or performing an operation without assistance. This statement should be very specific both for the trainee and the trainer. Ambiguity may lead to significantly different expectations and make the training process less effective.

6.2.3.3 A statement of the expected level of performance of the result detailed in 6.2.3.2. For discussions this may be more subjective and left up to the discretion of the trainer. For written exercises this can often be stated as the number of correct responses (7 out of 10) or as percentage of an expected score (80 %). For demonstration of skills it may be the performance on a sample of known composition within specified limits (80 to 120 % of true value).

6.3 *Performance Evaluation Tools*:

6.3.1 After a set of objectives has been formulated that clearly defines the goals of the training exercises, the tools necessary to measure the success of the training must be prepared. Often the development of these performance evaluation tools will aid in refining the objectives formulated in 6.2.

6.3.2 These tools should be carefully designed to measure exactly what has been defined in the objectives. For example, if the training objectives require the trainee to have a general knowledge of a process, a written exercise should not include a detailed discussion of that process. If the objective requires a complicated skill to be mastered, anything less than having the trainee perform that skill successfully will demonstrate inadequately the trainee's competence.

6.3.3 Performance evaluation tools must be prepared so that the trainee can use them for demonstrating competence without ambiguity or confusion.

6.3.3.1 Written exercises must be clear in their direction. Any questions must be worded in such a way that the desired response will be easily recognized by a properly trained individual.

6.3.3.2 Exercises requiring the demonstration of skills must be explicit in their directions. Any supplies or equipment called for must be readily available. Any hazards associated with the procedure must be clearly stated.

6.4 *Detailed Outline*:

6.4.1 Once objectives and performance evaluation tools have been selected, a detailed outline of each aspect of the process being trained must be developed. The development of this outline will help refine the performance evaluation tools indicated in 6.3. Much of this outline will be based on the task analysis (see 5.3). In all cases this outline must focus on accomplishing the stated objectives and providing a result that can be measured.

6.4.2 This outline should contain the general topics of:

6.4.2.1 Overview of task to be accomplished,

6.4.2.2 Definitions and terminology,

6.4.2.3 Theoretical considerations,

6.4.2.4 Safety issues,

6.4.2.5 Operational details,

6.4.2.6 Quality assurance,

6.4.2.7 Reference materials,

6.4.2.8 Documentation requirements,

6.4.2.9 Maintenance procedures, and,

6.4.2.10 Troubleshooting.

6.4.3 The amount of detail included with each of these topics will depend on the complexity of the procedure identified in the task analysis. Some topics may have only one or two items requiring training. Others may have much longer lists with several subheadings. Procedures involving modern computer driven equipment may require more detailed outlines than those processes involving only manual operations. Explicit SOPs in these areas can also reduce the amount of detail necessary in the outline.

6.5 *Documentation Form or Checklist*:

6.5.1 Following the completion of a detailed outline, develop a form to document the completion of items listed on the outline and to record the results of the performance evaluation.

6.5.2 A straightforward way of preparing this form is to use the major headings from the outline. Since this form also could be used as a training checklist, one may want to go into more detail. Space for recording the scores of oral or written examinations, or both, and performance on skills-based exercises should be customized to each task.

6.5.3 This form should have spaces to be initialed by both the trainer and the trainee and dated to show that both parties involved are in agreement as to the status of the training process. Following the completion of each item on the form or checklist and the performance evaluation, a formal statement describing the level of competency should be signed and dated by both the trainer and trainee.

6.5.4 Failure to reach agreement on the successful completion of a training element may require the intervention of a supervisor. Not everyone being trained for a particular process may be able to meet all the criteria successfully and should be considered for alternate job assignments. The standard of success should not be set at a level exceeding that necessary for the job in question.

6.5.5 All forms and scores should be placed in the individual's training or personnel file.

6.6 *Supplementary Material*:

6.6.1 Whenever a method or procedure requires information not readily available in SOPs, method manuals, or instrument operational materials, supplementary materials should be supplied.

6.6.2 These materials may consist of published textbooks, journals, etc., or information prepared specially for this training exercise. Enough information should be provided to allow the trainee to understand the method or procedure being taught. One should resist the temptation, however, to make the training course an end in itself, losing sight of the ultimate goal of an analyst performing competently.

6.6.3 These materials also may include regulatory information that gives a background for the use of an analytical method. Knowing how the results are going to be used can often emphasize the importance of performing an analysis properly.

7. Trainers

7.1 *Identification of Trainers*:

7.1.1 Trainers should meet the minimum qualifications as follows:

7.1.1.1 Demonstrated competence in method or process to be trained;

7.1.1.2 Education necessary to understand and explain the concepts involved in the method or procedure;

7.1.1.3 Ability to communicate effectively. Depending on the training needs this may require writing skills, speaking skills, or the use of other creative means to communicate concepts and activities to the trainee;

7.1.1.4 Ability to be objective;

7.1.1.5 Understanding of the training program may require training sessions for the trainers in the philosophy, goals, and practices of the training program. Trainers who are able to perform the activities outlined in this guide become more useful in assisting the ongoing development of a successful program;

7.1.1.6 Willingness to put forth the effort necessary to see that the trainee follows through to the completion of the training; and,

7.1.1.7 Sensitivity to the needs of the trainee and a willingness to adapt to meet those needs. Many of those needs will depend on a trainee's learning style.

7.1.2 In addition to those items listed in 7.1.1, effective trainers should also exhibit the following general characteristics:

7.1.2.1 Interest in training,

7.1.2.2 Be a positive role model,

7.1.2.3 Be considerate of others,

7.1.2.4 Be comfortable with people,

7.1.2.5 Be willing to listen, and,

7.1.2.6 Be a problem solver.

7.2 *Training of Trainers*:

7.2.1 If available, effective trainers should be aware of and use the following:

7.2.1.1 Leadership techniques,

7.2.1.2 Proper training techniques,

7.2.1.3 Learning styles and evaluation techniques, and,

7.2.1.4 Concepts of interpersonal relationships.

7.2.2 Effective trainers should be familiar with resources that can aid in the training process. These may include:

7.2.2.1 Standard operating procedures (SOPs),

7.2.2.2 Published methods,

7.2.2.3 Instrument manuals,

7.2.2.4 Reference materials, including books and audio or video tapes, and,

7.2.2.5 Training courses and seminars.

7.2.3 Trainers should be committed to the concepts of training and regularly participate in classes, workshops, seminars, or trade journals to improve their own effectiveness.

7.2.3.1 Many of the concepts listed in 7.2.1 may be unfamiliar to technically trained individuals. Each trainer should become familiar with these concepts through organized sessions designed to develop these skills relative to the specific training needs of the laboratory.

7.2.3.2 Technical improvement also should be encouraged among trainers. As they become more comfortable with new concepts, it will give them more confidence with the trainee.

8. Retraining

8.1 Once an individual has completed a training course and has demonstrated competency, one should be certified to perform that job for a specified period of time. Periodic review of the individual's performance should indicate the need for any additional or remedial training. If a deficiency exists, immediate action should be taken to retrain in the areas of the deficiency.

8.2 If no deficiencies are noticed over an extended period of time, for example, one year, a formal process should be in place that requires an updated set of performance data be placed in the individual's training personnel file.

8.2.1 These data may be from routine performance evaluation samples run by the laboratory.

8.2.2 If a specific set of performance evaluation tools is used to determine continued competency, the individual must be notified of the need to complete it and the required proficiency and time frame.

8.2.3 If a deficiency is noted based on any of the tools specified in 6.3, retraining should be instituted until competency can again be demonstrated.

9. Storage of Training Materials

9.1 It is recommended that the master copy of all training materials be stored in a central location. It should be available for copying and distribution at all times, but it should not be used for routine training purposes. This master file may be electronically maintained.

9.2 As new materials are prepared, they should be distributed to the appropriate trainers and the master copy placed in the central storage location.

10. Responsibilities

10.1 *Preparing Training Materials*:

10.1.1 Anyone can prepare training materials in accordance with Section 6.

10.1.2 All materials should be reviewed and approved by someone competent in the method or process and by someone familiar with the training program and its requirements. Only materials meeting both sets of requirements should be used.

10.1.3 In some laboratories, a full-time position may be created with responsibility for developing training materials, including performing the task analyses, writing the objectives, developing the evaluation tools, the training outline, the documentation form, and assembling supplementary material.

10.2 *Using the Training Materials*:

10.2.1 It is the responsibility of the laboratory management to assure that each individual within the laboratory receives the training necessary to competently perform one's job function.

10.2.2 It may be desirable to designate the QA department or a designated training director to monitor the training progress and notify individuals of their need for retraining.

10.2.3 Laboratory management should be provided with regular reports on the progress of training within the laboratory. Many organizations may want to establish training goals to keep everyone's focus on the need to improve.

11. Keywords

11.1 environmental laboratories; guide; training

APPENDIX

(Nonmandatory Information)

X1. EXAMPLE TRAINING COURSE—MEASUREMENT OF pH

X1.1 *pH Electrometric Methods Objectives*—After the completion of training for pH electrometric measurement-Tables X1.1-X1.4, the trainee should be able to:

X1.1.1 Explain the use of pH in sample testing and demonstrate proficiency by achieving a minimum score of 80 % on a written or oral evaluation to be given by the trainer;

X1.1.2 Perform a pH test measurement on a given sample of known pH with a ±0.05 unit error;

X1.1.3 Use the pH meter according to the manufacturer's manual to analyze a batch of samples, meeting the stated criteria of the SOP including quality control (QC) and any written documentation; and,

X1.1.4 Troubleshoot problems commonly encountered with a pH meter and various sample conditions.

X1.2 *Supplemental Reading for pH*—pH is a convenient measure of hydrogen ion concentration. Specifically, it is defined as the logarithm of the reciprocal of the hydrogen ion concentration, expressed in molarity. Thus a 10^{-3} M solution of hydrogen ions has a theoretical pH of 3. A 10^{-9} M solution has a theoretical pH of 9. Since the pH scale is logarithmic, a change of 1 pH unit means a ten-fold change in hydrogen ion concentration. A pH change of about 0.3 means a two-fold change in hydrogen ion concentration.

X1.2.1 The definition of pH is based on the dissociation of water, which produces hydrogen ions and hydroxide ions.

$$H_2O = H^+ + OH^-$$

Pure water has equal concentrations of hydrogen and hydroxide ions and a pH of 7. Solutions with a pH of 7 are called *neutral*. Solutions with excess hydrogen ions have a low pH and are *acidic*. Solutions with excess hydroxide ions have a high pH and are *basic* or *alkaline*.

X1.2.2 *Occurrence—Natural, Anthropogenic*—Carbon dioxide is the principal corrosive agent in natural water. It dissolves in the water to produce carbonic acid, which lowers the pH. Equilibrium between rainwater and atmospheric carbon dioxide results in a pH of about 5.7. Deionized water in the laboratory also often has a pH near this value. As rainwater infiltrates the soil, the low pH causes minerals such as limestone to dissolve. This tends to raise the pH to higher values. Unpolluted surface waters normally lie in the pH range 6.5 to 8.5. There are many factors, however, which may change the pH. Oxidation reactions (those which consume oxygen) generally lead to a decrease in pH. This can be observed in the bottom parts of deep lakes where reduction reactions such as photosynthesis, denitrification, and sulfate reduction tend to raise the pH. Photosynthesis can result in diurnal (day/night) pH changes, reaching to a pH of 9 or even 12 in poorly buffered waters. Algae and other microorganisms break down carbonate and bicarbonate when there is no other source of carbon dioxide. This also results in an increase in pH. Evaporation can affect the pH. Basins without outlets may accumulate alkaline substances resulting in pH greater than 12.

X1.2.2.1 Most groundwaters in the United States have a pH of 6.0 to 8.5, but water with lower pH is not unusual in thermal springs and in waters affected by sulfur oxidation. Groundwater unsaturated with respect to calcium carbonate (limestone) causes caves to form.

X1.2.2.2 Rain will absorb gases other than carbon dioxide. In particular, sulfur dioxide and nitrogen oxides are common air pollutants produced as by-products of the combustion of fossil fuels. These gases will dissolve in rain to form sulfuric and nitric acids. This causes very low pH and the phenomenon is known as *acid rain*. Acid rain with a pH of less than 2 has been measured.

X1.2.2.3 In areas with little or no limestone, the water remains acidic unless there are other minerals present that are capable of neutralizing the acid. In areas susceptible to acid rain, such as the Northeast United States and Scandinavia, lakes and streams can easily have pH below 4.5.

X1.2.2.4 All acids will release hydrogen ions in water and all bases will consume hydrogen ions. The pH of natural waters, therefore, can be affected by leaks and spills of industrial chemicals, as well as by the discharge of improperly treated wastes. Examples include discharges from the timber and wood industry, pulp and paper mills, acid mine drainage (for example, from coal mines), and pickling liquors from steel mills. Many natural systems will contain a certain buffering capacity, which means that, unless excess of certain chemicals occur, the system tends to resist changes in pH. *Buffering* will

I. Reading
   A. Objectives for training
   B. SOPs
   C. Methods
      1. SW846 Method 9040 and 9045
   2. EPA Method 150.1
   D. Instrument Manual
II. Theory of Operation
   A. Basic understanding of pH. Use the supplemental material that has been provided.
   B. Instrument
   1. How it detects
   2. Why it is used rather than titration, pH paper
   C. Buffers
      1. Definition
   2. Use in calibration
   D. Use of pH in the laboratory
   1. As a parameter
   2. Use for other parameters
      a. Leaching/extracting procedures
      b. Acidity procedures
      c. Alkalinity
      d. Where pH will interfere with analyses
III. Safety
   A. Safety wear
   1. Laboratory coat
   2. Safety glasses
   3. Disposable gloves
   B. Sample handling
   C. Instrument
IV. Procedures
   A. Instrument operation
   1. Set-up
   2. Function keys
   3. Calibration
   B. Buffer usage
   1. Initial calibration
   2. Continuing calibration
   C. Sample preparation
   1. Waters
   2. Soils
      a. Calcareous
   b. Non-calcareous
   D. Sample analysis
   1. Calibration
   2. Run order
   3. QC and control limits
   E. Documentation
   1. Logbook
   2. Standards log
V. Quality Control (QC)
   A. Theory
   B. QC elements
   1. Initial calibration
   2. Continuing calibration
   3. Duplicates
   C. Reporting
   1. Standards logbook
   2. Calibration
VI. Detection limits
   A. Instrument
   B. Reporting
VII. Troubleshooting
   A. Meter
   1. Programming
   2. Connections
   B. Electrodes
   1. Temperature probe
   2. pH probe
   3. KCl solution
   C. Matrix problems
   1. Oily materials
   2. Particulate matter
   3. Alternate method(s)
VIII. Maintenance
   A. Cleaning
   1. Probes
   2. Instrument

FIG. X1.1 pH Electrometric Method Training Outline

Name: ____________________

Description of Method or Process: pH—Electrometric Method

Start of Training: ____________________

Completion of Training: ____________________

Trainer(s): ____________________

	Completion Date	Trainer's Initials	Trainee's Initials
I. Reading/References			
A. Objectives for Training			
B. SOP(s)			
C. Method(s)			
D. Instrument Manual			
II. Theory			
A Understanding of pH			
B. Instrument			
C. Buffers			
D. Use of pH in the Laboratory			
III. Safety (Method Specific)			
A. Safety Wear			
B. Sample Handling			
C. Instrument			
IV. Procedures			
A. Instrument Operation			
B. Buffer			
C. Sample Preparation			
D. Sample Analysis			
E. Documentation			

IX. Competency Demonstration
   A. Written/Oral Evaluation explaining the use of pH (80 %) (Objective #1)
   B. Three Performance Evaluation Scores with ±0.05 pH unit error (Objective #2)
   C. Complete Batch Analysis (Objective #3) (Includes all quality control samples, data entry, scheduling)
   D. Troubleshooting (Objective #4)

I certify that ______ (name of trainee) is competent to perform pH—Electrometric Methods

____________________ Trainer   ____________________ Date

I acknowledge that I have been trained in pH—Electrometric Methods and I feel adequately prepared to perform the above procedure(s)

____________________ Trainer   ____________________ Date

	Completion Date	Trainer's Initials	Trainee's Initials
V. Quality Control			
A. Theory			
B. Quality Control Elements			
C. Reporting			
VI. Detection Limits			
A. Instrument			
B. Reporting			
VII. Troubleshooting			
A. Meter			
B. Electrodes			
C. Matrix Problems			
VIII. Maintenance			
A. Cleaning			

FIG. X1.2 Sample Competency Training Checklist/Documentation Form

1. How is pH defined? (2 pts.)
2. What significance does pH have in the laboratory? (3 pts.)
3. What is the usual numerical range for pH? (2 pts.)
4. How is the pH meter calibrated? (4 pts.)
5. What safety precautions are needed to perform a pH test? List three. (3 pts.)
6. What may cause an incorrect pH reading? List two. (2 pts.)
7. What are three common interferences when testing pH? (3 pts.)
8. What is a buffer and what is it used for? (2 pts.)
9. Define calcareous (2 pts.)
10. When would the pH meter need to be recalibrated? List at least two. (2 pts.)
11. At what pH should most electrodes be stored? (1 pt.)
12. After how many samples should a continuing calibration verification (CCV) be run? (1 pt.)

FIG. X1.3 Sample pH—General Evaluation (27 pts.)

1. pH is the measurement of hydrogen ion concentration.
2. It measures how much acidity or basicity a substance has in a water solution.
3. 0–14
4. By using a 2 point calibration made of two buffers and checked against a third buffer.
5. Electricity if using the electrometric methods, proper clothing, care in sample-handling.
6. Poor calibration, buffer contamination, bad electrode, temperature, KCl solution low, or a clogged reference electrode junction.
7. Oily matrices, high salinity, particulate matter.
8. A solution that controls pH at a specific level.
9. Solids that contain limestone.
10. If a buffer doesn't read correctly, or when it has been turned off.
11. 7
12. After every twenty samples.

Trainer's Use Only

PERFORMANCE SAMPLE EVALUATION SUGGESTIONS FOR pH

Water sample A (Electrometric)	4.88
Soil sample A (Electrometric)	11.44
Water sample B (Paper)	7.0

FIG. X1.4 Sample Answers to pH Evaluation (For Trainer's Use Only)

be discussed further in a later section.

X1.2.3 *Health and Environmental Significance*—Extremes of pH are toxic to plants, animals, and microorganisms. Damage to surface structures results and low pH inhibits enzyme activity. A pH between 6.5 and 9.0 is generally considered to be safe. A pH less than about 3.5 or greater than about 11 is lethal to most fish species. Some species, such as trout and salmon, are very sensitive and can be harmed by pH below 6 or above 9. Another consequence is that biological treatment of wastes is difficult or impossible without the proper pH.

X1.2.3.1 Sudden changes in pH can also have harmful effects, even if the pH always remains within safe ranges. Small changes in pH can result in a competitive advantage for some species and their eventual domination over others that were originally present in the system. Large changes in pH can cause total destruction of fish and other populations.

X1.2.3.2 Chemical speciation is highly dependent on pH. For example, unionized ammonia is present at alkaline pH and ammonium ion is present at lower pH. Ammonia is much more toxic than ammonium ion, so that even a moderate increase in pH can result in greatly increased toxicity. Complexation of metals by agents such as hydroxide, chloride, sulfate, bicarbonate, and fluoride is also affected by pH. At low pH, the free metal ions are usually present. As the pH increases, various complexes begin to form. The free metal ions are usually much more toxic than the complexes. The toxicity of metals, therefore, depends on pH. Low pH can also displace trace metals from adsorption sites on soil and sediment particles, leading to increased toxicity.

X1.2.3.3 Extremes of pH are irritating to the skin (such as bathing, swimming) and produce undesirable effects on food and beverages. Water with a low pH maybe corrosive to pipes and other fixtures.

X1.2.4 *Other Effects*—Proper pH is important in many waste treatment processes. Biological treatment depends on the health of various microorganisms, which are sensitive to pH. Metals are typically removed from waste streams by precipitating them as hydroxides. Maximum precipitation of metals is usually obtained if the pH is about 8.0 to 8.3. If the pH is too low or too high, then removal is less efficient; the metals redissolve at either extreme.

X1.2.4.1 The pH affects many chemical reactions. For example, reduction of hexavalent chromium to trivalent chromium occurs rapidly at low pH, but not at high pH. Many chemical processes used in industry depend on tight control of pH, since the chemical reactions they utilize proceed slowly or not at all if the pH is incorrect.

X1.2.5 *Regulatory Limits*—The pH of wastewater discharges is regulated under the United States Environmental Protection Agency (USEPA) Clean Water Act; the limits are usually 6.0 to 9.0. A secondary drinking water standard of 6.5 to 8.5 has been established for drinking water supplies. As a result of concern over acid rain, controls on sources of sulfur dioxide and nitrogen oxides have been proposed. pH is used under United States Resource Conservation and Recovery Act (USRCRA) regulations as one definition of corrosivity and is also one of the "indicator parameters" used in groundwater monitoring.

X1.2.6 *Analytical Methods*—The most common way of measuring pH is with an electrode. The electrode suffers from sodium effect at high pH. Samples with low ionic strength will cause sluggish response. The electrode is also temperature sensitive. It may also become coated with oil and greases destroying its sensitivity to pH.

X1.2.6.1 There are also various indicators that can be used to measure pH. These are organic dyes that change color in response to changes in pH; they can be used in liquid form or impregnated into absorbent strips. Although they are convenient to use, indicators are less accurate and are subject to numerous interferences. The electrode method is most accurate and is the method preferred by most laboratories.

X1.2.6.2 It should be noted that pH measurements are usually meaningful only for aqueous solutions; nonaqueous samples are first extracted with water using defined phase ratios and the pH of the resulting extract is measured. Dilution effects must be taken into account when interpreting the data.

X1.2.6.3 Also, the pH will change in a sample after collection due to chemical changes that occur when the water is removed from its environment. These include oxidation/reduction reactions, biological processes, and equilibration with atmospheric carbon dioxide. A change in pH of 1 unit or more between the field and laboratory is not unusual. The pH measured in the laboratory will almost certainly be different than the pH at the time of sampling.

X1.2.7 *Notes on pH*—From the definition of pH and the equilibrium constant for dissociation of water, it can be shown that:

$pH + pOH^- = 14$

or

$[H^+] \times [OH^-] = 1 \times 10^{-14}$ assuming that ion activity equals concentration.

X1.2.7.1 It is defined for aqueous system only. Although it is entirely possible to speak of hydrogen ion concentration in nonaqueous solvents, many of the conventional wisdoms do not apply. For example, neutral pH is defined as the point where $[H^+]$ and $[OH^-]$ are equal. In aqueous systems, this is pH 7 (based on the ion product of water). For nonaqueous systems, this may not be true. It is also difficult to measure pH in nonaqueous systems.

X1.2.7.2 The pH scale is logarithmic. A change of 1 pH unit means a ten-fold change in hydrogen ion concentration. A pH change of about 0.3 means a two-fold change in hydrogen ion concentration.

X1.2.7.3 *pH Electrode*—A pH electrode is constructed from specially made glass (for example, lithium silicates doped with lanthanum and barium). These types of glass have active sites that are selective almost exclusively for hydrogen ions. In a typical electrode, a thin-walled bulb of pH-sensitive glass is sealed to a thick-walled stem of non-responsive glass. This is to prevent complications arising due to changes in the depth of immersion of the electrode.

X1.2.7.4 The bulb of the pH electrode is filled with a constant pH buffer solution (usually pH 7), which is in contact with an internal solid electrode. This is connected in turn to a lead that can be connected to the pH meter.

X1.2.7.5 Hydrogen ions in the internal buffer solution diffuse into the inner surface of the glass membrane and eventually reach an equilibrium. Hydrogen ions in the sample diffuse into the outer surface of the membrane and reach a separate equilibrium. If the sample has a lower pH than the filling solution, the outer surface will have more hydrogen ions than the inner surface. If the pH is higher than the internal solution, there will be fewer hydrogen ions on the outer surface. This results in a voltage across the membrane that depends on the concentration of hydrogen ions in the sample.

X1.3 *Importance of pH:*

X1.3.1 *Natural Waters*—The most important buffering system in natural waters is that formed with carbonic acid (carbon dioxide), bicarbonate, and carbonate. In this system, carbonic acid is the predominant form present below about pH 4.5. Between 4.5 and 8.3, predominant form is bicarbonate. Above pH 8.3, it is carbonate.

X1.3.2 Water exposed to the atmosphere will absorb carbon dioxide and the following reactions occur:

$$H_2O + CO_2 \leftrightarrow H_2CO_3$$
$$H_2CO_3 \leftrightarrow H^+ + HCO_3^-$$
$$HCO_3^- \leftrightarrow H^+ + CO_3^{-2}$$

These reactions lower the pH of the water. Water in equilibrium with atmospheric carbon dioxide will have a pH of about 5.6, depending on temperature, atmospheric pressure, and other factors. Thus, rain is naturally slightly acidic.

X1.3.3 When rain falls and comes into contact with minerals in the soil, these minerals begin to dissolve. Limestone, which is a form of calcium carbonate, is a common mineral and neutralizes the carbonic acid in rain water.

$$CaCO_3 + H_2CO_3 \rightarrow Ca^{+2} + 2HCO_3^-$$

This is the mechanism by which caves are formed.

X1.3.4 *Acid Rain*—Rain will also absorb other gases; in particular, sulfur dioxide and nitrogen oxides are common pollutants and dissolve in rain to produce acid rain. Examples of these reactions are:

$$2H_2O + 2SO_2 + O_2 \rightarrow 2H_2SO_4$$
$$NO_2 + NO + H_2O \rightarrow 2HNO_2$$
$$2NO_2 + H_2O \rightarrow HNO_2 + HNO_3$$
$$N_2O_5 + H_2O \rightarrow 2HNO_3$$

Acid rain with a pH of less than two has been measured. In areas with little or no limestone, the water remains acidic unless there are other minerals present that are capable of neutralizing the acid present. With little limestone, lakes and streams can easily have pHs below 4.5. Areas particularly vulnerable to acid rain include the Northeast United States and Scandinavia.

Designation: D 5830 – 95 (Reapproved 2001)

Standard Test Method for Solvents Analysis in Hazardous Waste Using Gas Chromatography[1]

This standard is issued under the fixed designation D 5830; the number immediately following the designation indicates the year of original adoption or, in the case of revision, the year of last revision. A number in parentheses indicates the year of last reapproval. A superscript epsilon (ε) indicates an editorial change since the last revision or reapproval.

1. Scope

1.1 This test method is used to determine qualitatively and quantitatively the presence of the following compounds in waste samples using gas chromatography. This test method is designed for use as a screening method with a typical reporting level of 0.1 %.

Dichodifluoromethane	Tetrahydrofuran
Trichlorofluoromethane	Acetone
1,1,2-Trichloro-1,2,2-trifluoroethane	Methyl Ethyl Ketone
Methanol	MIBK
Ethanol	Cyclohexanone
Isopropanol	Ethyl Acetate
n-Propanol	Propyl Acetate
Isobutanol	Butyl Acetate
n-Butanol	Benzene
tert-Butanol	Toluene
Methylene Chloride	Ethylbenzene
Chloroform	Xylenes
Carbon Tetrachloride	Styrene
1,1-Dichloroethane	Chlorobenzene
1,2-Dichloroethane	Dichlorobenzenes
1,2-Dichloropropane	Nitrobenzene
1,1-Dichloroethylene	Fluorobenzene
1,2-Dichloroethene	*n*-Propyl Benzene
1,1,1-Trichloroethane	Isopropyl Benzene
Tetrachloroethylene	Isobutyl Benzene
Trichloroethylene	*n*-Butyl Benzene
Tetrachloroethane	2-Ethoxyethanol
Cyclopentane	2-Butoxyethanol
Pentane	2-Ethoxyethanol Acetate
Hexane	2-Methoxyethanol
Heptane	Bromoform
Cyclohexane	Carbitol
*Iso*octane	Ethyl Ether
Nitropropane	1,4-Dioxane
Ethanolamine	Diacetone Alcohol
Nitromethane	Acetonitrile
Ethylene Chloride	Pyridine
Benzyl Chloride	Toluidine
	Ethylene Glycol
	Propylene Glycol

1.1.1 This compound list is a compilation of hazardous solvents and other constituents that are routinely seen in hazardous waste samples.

1.2 The scope of this test method may be expanded to include other volatile and semivolatile organic constituents.

1.2.1 Hydrocarbon mixtures such as kerosene and mineral spirits.

1.2.2 High-boiling organics, defined here as compounds which boil above *n*-Hexadecane.

1.2.3 Other organics that the analyst is able to identify, either through retention time data or gas chromatography/mass spectrometric (GC/MS) analysis.

1.3 Gas chromatographic methods are recommended for use only by, or under close supervision of, an experienced analyst.

1.4 *This standard does not purport to address all of the safety concerns, if any, associated with its use. It is the responsibility of the user of this standard to establish appropriate safety and health practices and determine the applicability of regulatory limitations prior to use.*

2. Referenced Documents

2.1 *ASTM Standards:*

D 1193 Specification for Reagent Water[2]

2.2 *EPA Document:*

Gas Chromatography/Mass Spectrometry Method 8260, Test Methods for Evaluating Solid Waste Physical/Chemical Methods, SW-846, Third Edition, Final Update 1, July 1992[3]

3. Summary of Test Method

3.1 Waste samples are analyzed by direct injection, or by carbon disulfide, *M*-Pyrol, or other suitable solvent extraction and injection of the extract into a gas chromatograph. Detection is achieved using a detector which is specific for the needed application, for example, flame ionization detector (FID), electron capture detector (ECD), thermal conductivity detector (TCD), photoionization detector (PID), or mass selective detector (MSD). This test method may be expanded to utilize other detector types not previously mentioned.

4. Significance and Use

4.1 This test method is useful in identifying the major solvent constituents in hazardous waste samples. This test method is designed to support field or site assessments,

[1] This test method is under the jurisdiction of ASTM Committee D-34 on Waste Management and is the direct responsibility of Subcommittee D34.01.06 on Analytical Methods.

Current edition approved Sept. 10, 1995. Published November 1995.

[2] *Annual Book of ASTM Standards*, Vol 11.01.

[3] Available from the Superintendent of Documents, U.S. Government Printing Office, Washington, DC 20402.

recycling operations, plant operations, or pollution control programs.

5. Interferences

5.1 Interferences may be encountered from any number of organic compounds that respond in the detector. Also, closely eluting components may complicate identification based solely on retention time. When these types of interferences are encountered, the analyst must rely on other sources of information for positive identification, such as:

5.1.1 Gas chromatography/mass spectrometric (GC/MS) confirmation, see EPA Method 8260, direct injection technique;

5.1.2 Use of confirmation column, or confirmatory detector;

5.1.3 Use of varying temperature programs or standard comparison, or both;

5.1.4 Sample history, for example, any information available from the waste generator; and,

5.1.5 Physical characteristics, for example, flammability, specific gravity, or miscibility with water.

5.2 Interferences may also be encountered from syringe carryover. Immediately following each injection, the syringe should be thoroughly rinsed with carbon disulfide, or *M*-Pyrol. Other solvents such as methanol may be used as rinse solvents if sample types necessitate their use, but be aware that carryover and possible interferences may occur if the rinse solvent is not completely cleaned from the syringe before reuse. Before each injection the syringe must be thoroughly rinsed with the sample to be injected, where the first two pumps are flushed into a separate waste receptacle.

5.3 When carbon disulfide (CS_2) is used to extract solids or sludges that contain significant amounts of water, low recovery of the water miscible solvents may result.

5.4 Some grades of CS_2 may contain trace amounts of benzene.

5.5 *M*-Pyrol seems to degrade slowly with time. The low-level degradation products interfere with some late eluting compounds on some columns (approximately five small peaks).

5.6 Interference from the CS_2 solvent peak may occur if using a TCD.

5.7 When using a TCD, be aware that water, as well as oxygenated compounds, for example, MEK, MIBK, may

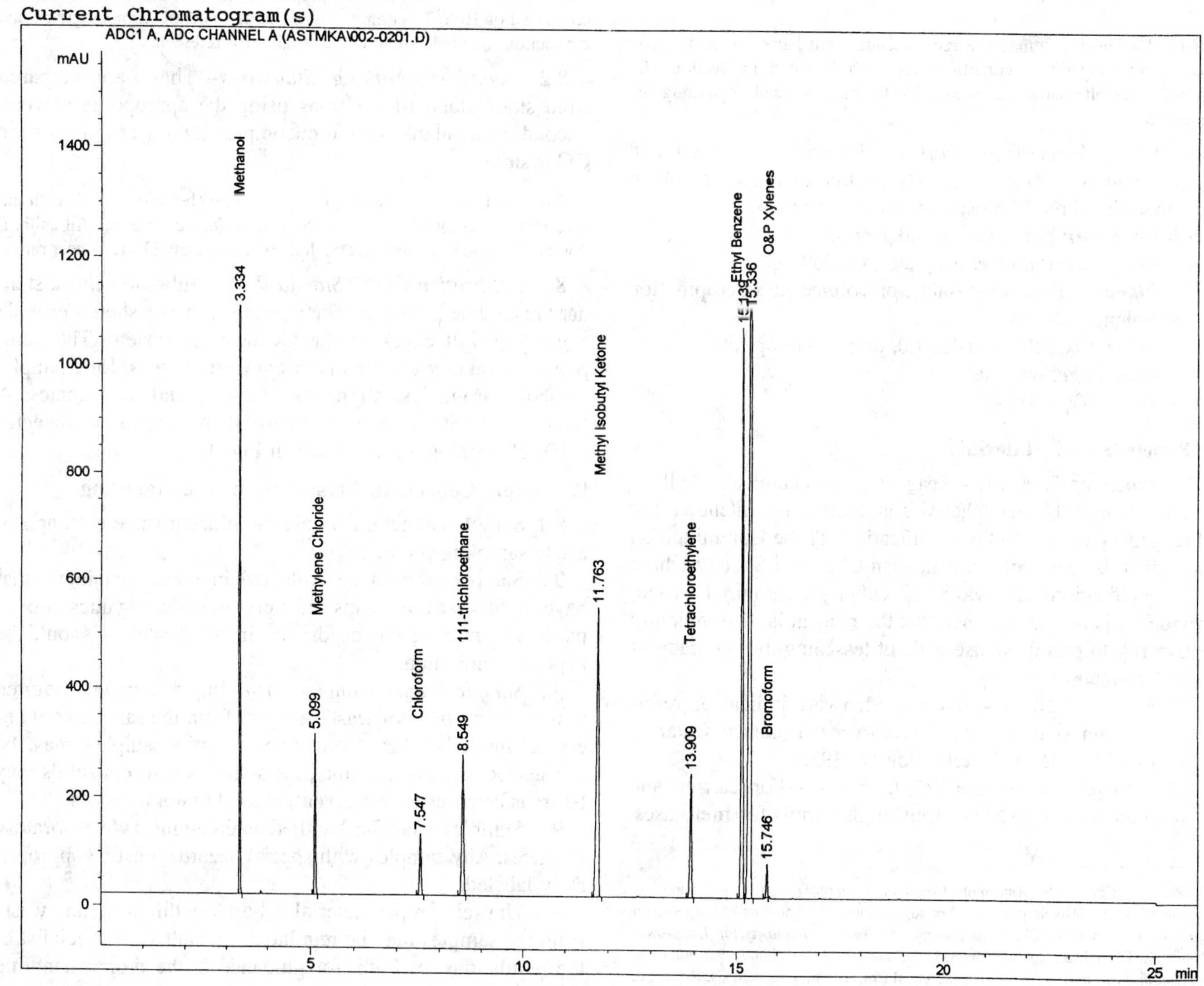

FIG. 1 Daily QC Standard FID/DB-1701

suppress detector response.

5.8 If an electrolytic conductivity detector (ECD) must be used, be aware that CS_2, *M*-Pyrol, and high concentrations of halogenated compounds may overload and possibly damage the detector. It is recommended that the ECD be used only when very low detection levels of halogenated compounds are expected and direct injection of the sample is possible.

6. Apparatus

6.1 *Gas Chromatograph System*—Equipped with capillary or packed column injection ports, or both, detector, and data system.

6.2 *Recommended Chromatographic Columns*:

6.2.1 *Capillary; Microbore or Megabore.*

6.2.1.1 DB-1701, 30M × 0.25-mm inside diameter, 0.25-μm film thickness.

6.2.1.2 DB-624, 30M × 0.3-mm inside diameter, 1.8-μm film thickness.

6.2.2 *Packed: Stainless Steel or Glass.*

6.2.2.1 1 % SP-1000, 60/80 Carbopak B, 8-ft by 1/8-in. inside diameter.

6.2.2.2 10 % SP-2100, 100/120 Chromosorb WHP, 2M × 2 mm ID.

NOTE 1—These columns are recommended and have shown to give good results. Operating conditions for each is listed in Section 10. Equivalent or alternative columns, or both, may be used depending on application.

6.3 *Glass Screw-Cap Vials or Equivalent*—To collect samples and store standards. Polytetrafluoroethylene or other inert material should be used for the cap liner.

6.4 *Microsyringes*, 1.0, 10, and 100 μL.

6.5 *Analytical Balance*, accurate to 0.0001 g.

6.6 *Pipettes*, glass, disposable, or volumetric micropipettor or equivalent.

6.7 *Microdisk Filters*, 0.45, 1.0, or 5.0 μm, optional.

6.8 *Centrifuge*, optional.

6.9 *Vortex-Type Mixer.*

7. Reagents and Materials

7.1 *Purity of Reagents*—Reagent grade chemicals shall be used in all tests. Unless otherwise indicated, it is intended that all reagents conform to the specifications of the Committee on Analytical Reagents of the American Chemical Society where such specifications are available.[4] Other grades may be used, provided it is first ascertained that the reagent is of sufficiently high purity to permit its use without lessening the accuracy of the determination.

7.2 *Purity of Water*— Unless otherwise indicated, references to water shall be understood to mean reagent water as defined by Type II of Specification D 1193.

7.3 *Nitrogen or Helium (High Purity)*—For carrier and makeup gases. Air and hydrogen (high purity) for fuel gases. Gases may be obtained from a gas generator if available, through purification of a lower grade, or from a high-purity tank supply.

7.4 *Carbon Disulfide, CS_2*—Chromatography grade.

7.5 *M-Pyrol, C_oH_oNO*—Available through several chemical suppliers and sources as 1-methyl-2-pyrrolidone.

7.6 *Individual Standards for Each Component of Interest*—99 % purity available from many vendors.

8. Standard Preparation

8.1 *Stock Standard Solutions*—Stock standards are prepared from pure standard materials. It is recommended that the standards be prepared so that each component is 5 to 10 % by weight. The stock standards must be prepared by directly weighing each component. For extremely volatile components, such as ether and freons, it is recommended that a new stock standard be prepared daily or as needed. If a dilution solvent is needed when preparing the stock standards, use the same solvent used for sample extraction or dilution in Section 7.

NOTE 2—Due to the incompatibility of some standard compounds, that is, some compounds are not miscible with each other, and also because of the number of compounds typically looked for in a single chromatographic run, it is advisable to prepare 3 or 4 standard solutions each composed of 10 to 15 compounds. A set of standard chromatograms and a retention timetable should be available for reference.

8.2 *Secondary Working Standards*—These are prepared from stock standard solutions using the appropriate solvent. Secondary standards should encompass the linear range of the GC system.

NOTE 3—Linear response and range must be established with all detectors and chromatography systems used for quantitation. All calibration and sample analysis must be done within the established linear range.

8.3 *Calibration Check Standard*—A calibration check standard should be prepared. The standard mixture should provide a good overall check of the GC/detector system. The compounds should cover the major compound types, for example, alcohols, aromatics, aliphatics, ketones, and halogenates. A typical calibration check standard flame ionization detector (FID) chromatogram is shown in Fig. 1.

9. Sample Collection, Preservation, and Handling

9.1 Sample collection should be in accordance with appropriate sampling protocols.

9.2 Samples should be collected in glass containers, that have tightly sealing caps. If very volatile organics are of particular interest, the headspace in the container should be kept to a minimum.

9.3 *Sample Transfer Implements*—Implements are required to transfer portions of waste samples from the sample containers to the laboratory containers. Liquid samples may be transferred using disposable pipets. Solids and semisolids may be transferred using a conventional laboratory spatula.

9.4 Samples shall be handled maintaining safe laboratory practices. Any samples with special hazards must be appropriately labeled.

9.5 Unused sample material, laboratory dilutions, and waste from the samples may be regulated. Consult your specialist or the regulations, or both, for guidance in the proper handling and disposal of laboratory wastes.

[4] *Reagent Chemicals, American Chemical Society Specifications*, American Chemical Society, Washington, DC. For suggestions on the testing of reagents not listed by the American Chemical Society, see *Analar Standards for Laboratory Chemicals*, BDH Ltd., Poole, Dorset, U.K., and the *United States Pharmacopeia and National Formulary*, U.S. Pharmaceutical Convention, Inc. (USPC), Rockville, MD.

10. Procedure

10.1 *Sample Preparation*:

10.1.1 Analyze liquid matrices with relatively low viscosity using direct injection into the GC, either as received or after dilution with CS_2, *M*-Pyrol, or other suitable solvent.

10.1.2 Analyze solid or semisolid samples as follows:

10.1.2.1 For carbon disulfide or *M*-Pyrol preparation, weigh 3 g of the waste sample in a 15-mL glass vial. Add 3 g of carbon disulfide or *M*-Pyrol to the vial and the mixture is vortexed vigorously. After allowing the solids to settle, inject the CS_2 or *M*-Pyrol extract into the GC.

10.1.2.2 Use alternate sample sizes and extraction solvent weights if neccssary. Actual sample size and solvent weight must be recorded in the appropriate sample preparation log book. It is essential for accurate waste sample analysis that sample size be sufficient to ensure a representative sample. If alternate sample size or extraction solvent volumes, or both, are used, this must be reflected in the calculations under the dilution factor in Section 11.

10.1.3 Multiple phases or layers are typically present in hazardous waste samples. Depending on treatment or process requirements, it may be necessary to analyze each phase or layer individually.

10.2 *Recommended GC Operating Conditions*:

10.2.1 For Capillary DB-1701 with FID

Column flow rate	1–1.5 mL/min
Make-up gas flow rate	30–60 mL/min
Split flow	60 cm^3/min
Injector temperature	250°C
Detector temperature	250°C
Airflow (FID)	Approximately 300 mL/min
Hydrogen flow (FID)	Approximately 30 mL/min
Initial oven temperature	35°C
Initial time	6 min
Level 1 rate	6°C/min
Level 1 final value	180°C
Level 2 rate	10°C/min
Level 2 final value	230°C
Run time	40 min
Threshold	4 units
Peak width	0.04 min

NOTE 4—Typical chromatograms are shown in Figs. 2-5.

10.2.2 For Capillary DB-624 with FID

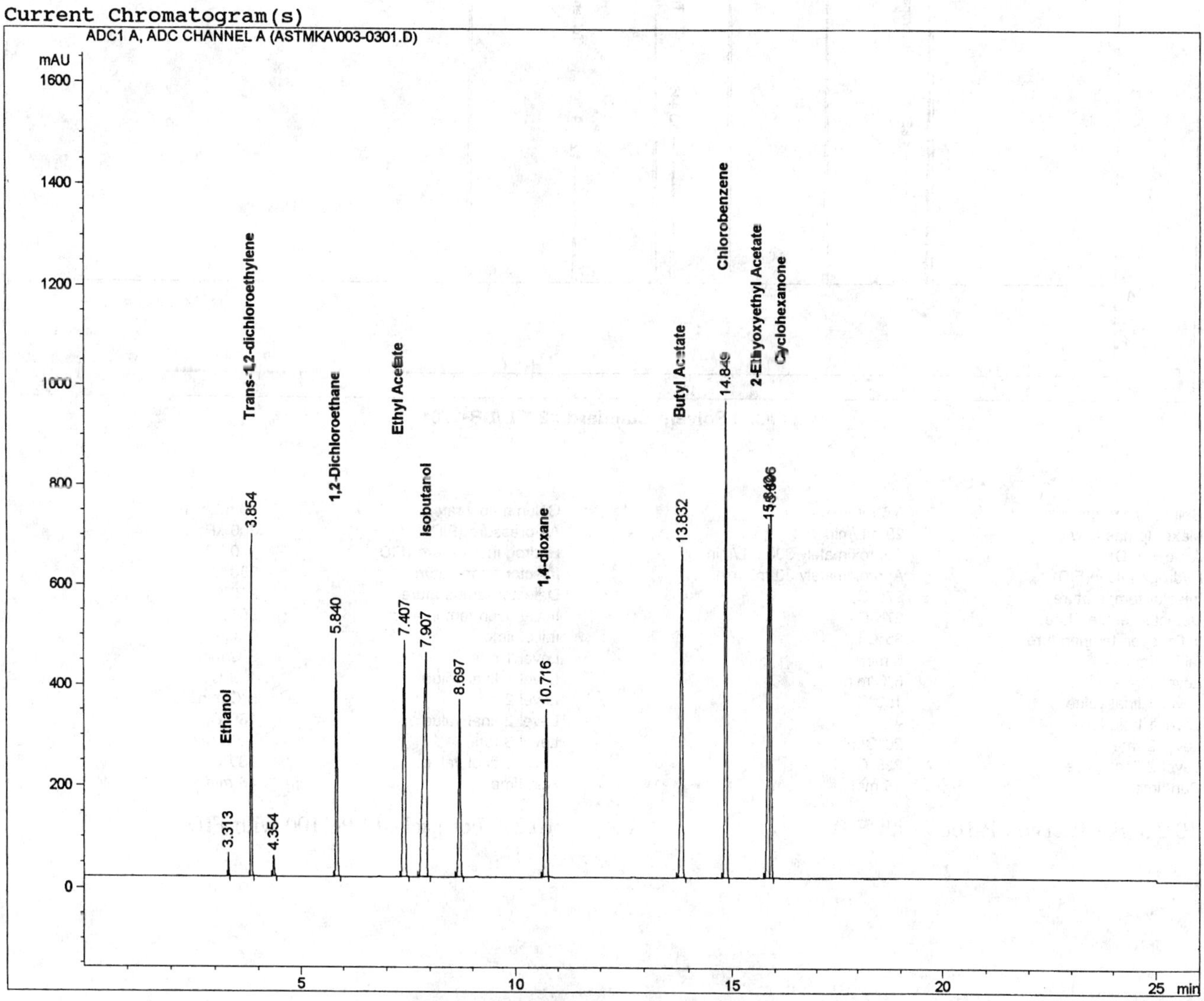

FIG. 2 Solvent Standard #1 FID/DB-1701

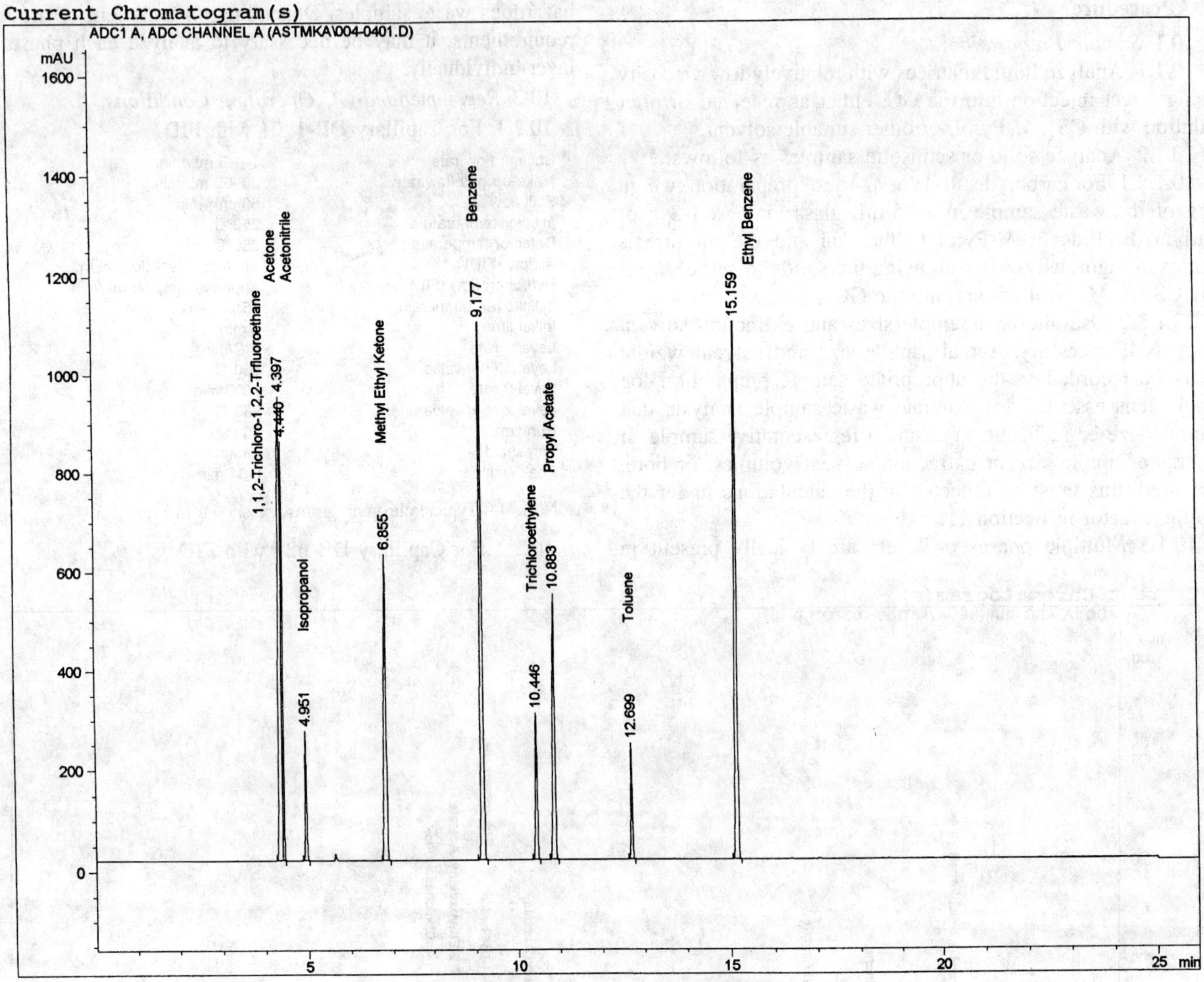

FIG. 3 Solvent Standard #2 FID/DB-1701

Column flow rate	3.5 mL/min
Make-up gas flow	29 mL/min
Airflow (FID)	Approximately 300 mL/min
Hydrogen flow (FID)	Approximately 30 mL/min
Injector temperature	275°C
Detector temperature	275°C
Initial oven temperature	35°C
Initial time	5 min
Level 1 rate	5°C/min
Level 1 final value	150°C
Level 1 hold time	4 min
Level 2 rate	20°C/min
Level 2 final value	225°C
Run time	45 min

10.2.3 For Packed SP-1000 with FID

Column flow rate	40 mL/min
Air pressure (FID)	300 kPa
Hydrogen pressure (FID)	130 kPa
Injector temperature	250°C
Detector temperature	250°C
Initial oven temperature	90°C
Initial time	6 min
Level 1 rate	3°C/min
Level 1 final value	120°C
Level 2 rate	5°C/min
Level 2 final value	180°C
Level 3 rate	10°C/min
Level 3 final value	230°C
Run time	46 min

10.2.4 For packed SP-2100 with FID

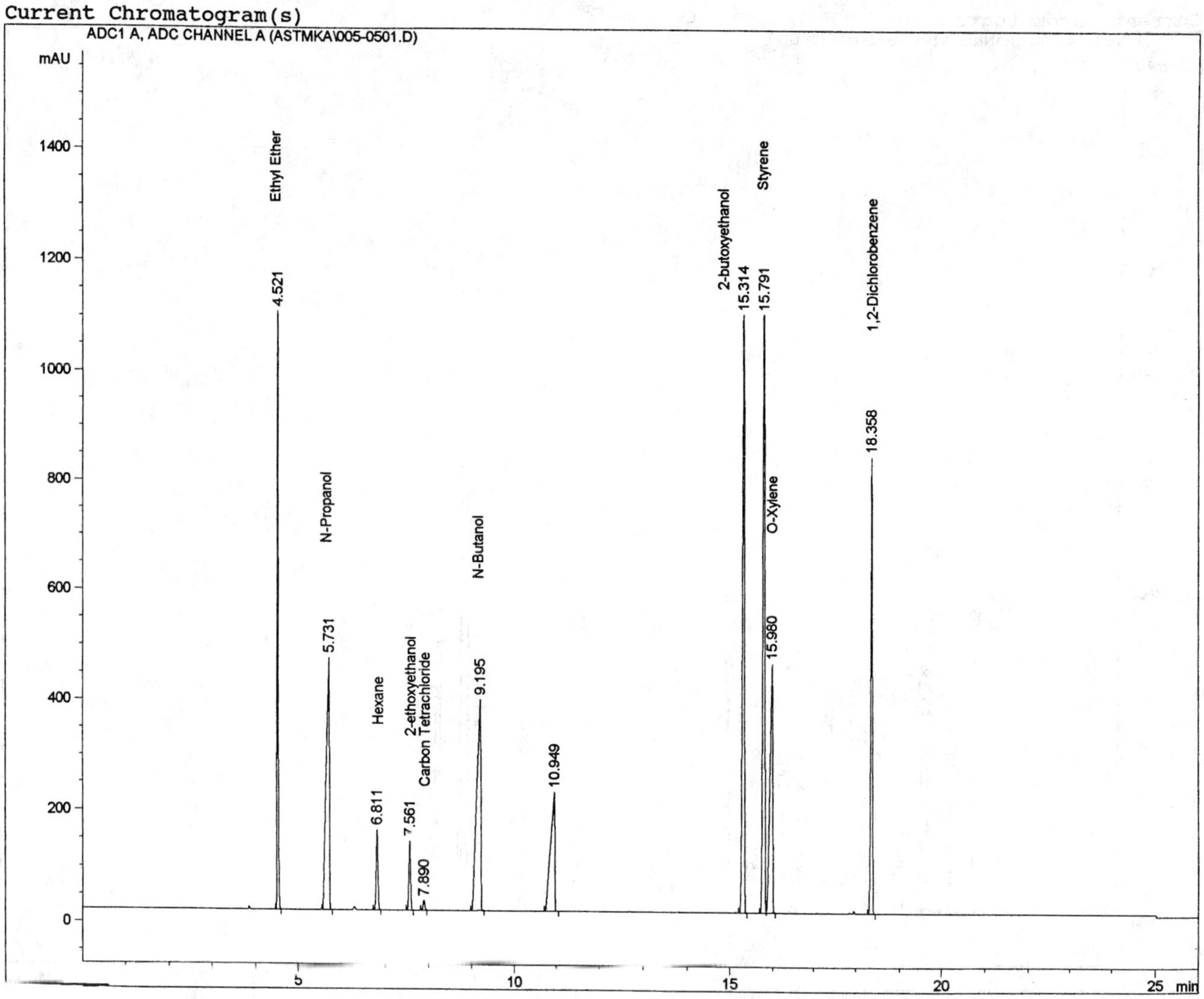

FIG. 4 Solvent Standard #3 FID/DB-1701

Carrier gas flow	30 mL/min
Injector temperatuare	250°C
Detector temperature	300°C
Airflow (FID)	Approximately 300 mL/min
Hydrogen flow (FID)	Approximately 30 mL/min
Initial oven temperature	45°C
Initial hold time	3 min
Level 1 rate	15°C/min
Level 1 final value	90°C
Level 2 rate	10°C/min
Level 2 final value	195°C
Run time	16.5 min

10.3 *Linear Range Determination*—The linearity and linear range for each compound must be established on any GC system used for quantitation. This must be done on an annual basis or after any major maintenance or alteration of the system configuration, for example, detector replacement. Final quantitation for each compound must be done within the linear range of that compound.

10.4 *Calibration (External Standard Procedure)*:

10.4.1 A single-point initial calibration of all compounds must be performed monthly. Inject 0.5 to 2.0 μL of the working standards that were prepared in 8.2. Tabulate peak area against concentration and express response factors (RF) for each component. This calculation is shown in 11.1. It is recommended for ease of calculation that the response factors be expressed as area counts per 1 % by weight.

10.4.2 The response factors must be verified daily or after every 20 samples, whichever is more frequent, by injecting 0.5 to 2.0 μL of the calibration check standard. This must be done for every column used for quantitation. If the predicted response varies by ±20 % from the initial calibration, corrective action must be taken, for example, septum, liner, or column maintenance, or a combination thereof. If the variance cannot be solved using the corrective action techniques, a new initial calibration must be performed.

10.5 *Gas Chromatographic Analysis*:

10.5.1 Liquid samples are introduced into the gas chromatograph by direct injection of 0.5 to 2.0 μL of sample. This may be done as received or after dilution or extraction with the appropriate solvent.

10.5.2 Solid and semisolids should be prepared in accordance with 10.1.2, followed by injection of 0.5 to 2.0 μL of the extract.

NOTE 5—Because of the number of different types of compounds

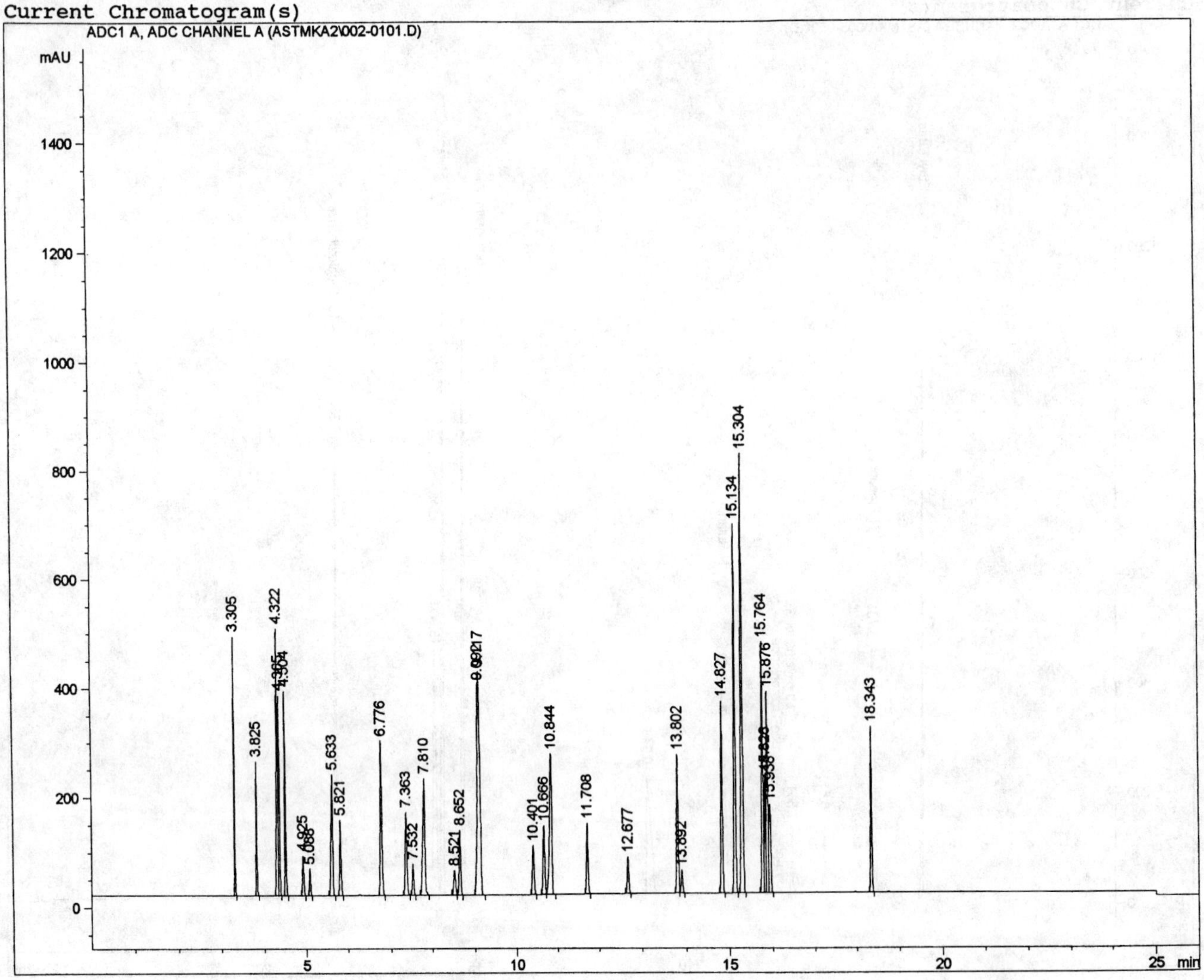

FIG. 5 Combined Solvent Standards FID/DB-1701

typically present in hazardous waste samples or looked for in a single chromatographic run, interferences from coelution may complicate identification based solely on retention time. The analyst should use the tools listed in 5.1 and any other available knowledge to correctly identify the compound.

11. Calculation

11.1 The compound concentration is determined by the following calculations:

$$Csamp = (Asamp/RF) \times D \quad (1)$$

where:
Csamp = concentration of compound in sample (wt %),
Asamp = peak area of compound in sample,
RF = response factor, and
D = dilution factor.

$$RF = Astd/Cstd \quad (2)$$

where:
Astd = peak area of compound in standard, and
Cstd = concentration of compound in standard (wt %).

This calculation assumes a reproducible injection volume. A dilution factor must be applied if additional sample or solvent amounts are used for extraction. For example, when using CS_2, if 6 g of CS_2 instead of the normal 3 g is used for extraction, a dilution factor of two must be applied to Csamp since the compounds in the 3 g of the original sample are now diluted in 6-g CS_2. This is also true if any dilutions are made prior to direct injection of the sample.

NOTE 6—If solids or semisolids are soluble in CS_2 or *M*-Pyrol, use an additional dilution factor for calculation. This is mainly true for oils and sludges.

11.2 Calculated results should be reported to one significant figure for values of 1 % or less, and to two significant figures for values greater than 1 %.

12. Quality Control

12.1 Each laboratory using this test method will operate a formal quality control program. The minimum requirements of this program must consist of an initial demonstration of laboratory capability and the ongoing analysis of method blanks, duplicate and spiked samples to evaluate, and document quality data.

12.2 Before processing any samples, the analyst must demonstrate through the analysis of a method blank, that the interferences from the analytical system, glassware, and reagents are under control. Each time a set of samples is prepared or there is a change in reagents, a method blank must be processed as a safeguard against chronic laboratory contamination. The method blank must be carried through all stages of sample preparation and measurement.

12.3 On an ongoing basis, the laboratory must analyze a replicate sample and a matrix spike or a matrix spike and a replicate matrix spike. This will allow the laboratory to assess precision and bias.

12.4 On a monthly basis, each laboratory must demonstrate the ability to identify each compound at the reporting levels. For a signal to be considered a valid peak it must have a signal-to-noise ratio greater than ten.

13. Precision and Bias

13.1 With the exception of commercial paint, the following mixtures were synthesized by weight. Carbon disulfide was the predominant solvent used for extraction and dilution in obtaining Tables 1-4.

NOTE 7—*Theo %* represents the theoretical value of the compound in the mixture as prepared by the round-robin technician. *Avg.* represents the average value reported from all the laboratories. *% RSD* represents the percent relative standard deviation between laboratories based on reported values. *Avg. Rec.* represents the average recovery based on the reported values and the theoretical values. Average recovery data is not reported for solvents in commercial paints because theoretical values are given only as ranges.

NOTE 8—All precision and bias data are based on multilaboratory round-robin analyses.

TABLE 1 Solvents in Motor Oil, Percent Detected

	Compound				
Laboratory No.	XYL	$MeCl_2$	PERC	TCA	TCE
1	1.1	3.7	3.2	62.5	2.7
2	1.7	3.9	...	69.5	...
3	0.5	2.0	1.4	39.4	...
4	2.1	4.2	2.7	76.5	2.4
5	1.2	3.0	2.7	59.7	...
6	1.3	4.0	3.0	71.1	2.5
7	1.4	3.9	3.2	67.7	2.4
8	1.7	5.1	4.1	59	3.2
9	...	4	3.3	65.9	2.9
10	1.1	4.2	5.2	37.1	2.7
11	1.8	3.5	...	50.6	...
12	1.4	4.1	3.3	59	2.3
13	1.2	3.6	2.7	...	...
Theo%	1.5	3.7	3.0	61.9	2.2
Avg.	1.4	3.8	3.2	59.8	2.6
%RSD	29.7	8.9	29.3	20.3	11.6
Avg. Rec.	92	103	105	97	117

TABLE 2 Solvents in Commercial Paint, Percent Detected

	Compound			
Laboratory No.	EB	MEK	TOL	XYL
1	...	3.2	13.6	25.4
2	5.8	...	11.2	15.8
3	...	4.4	9.3	27
4	5.3	4.1	15.1	...
5	...	2.2	7.5	15.5
6	3.6	...	8.7	17.8
7	2.7	2.7	6.5	14
8	5.5	4.1	12	27.4
9	6.2	4.9	16	...
10	...	4.5	18.2	...
11	3.7	3.7	10.4	22.8
12	4.7	3.2	11.4	22.8
13	2.9	...	7.3	23.1
Theo%	...	3–4	10–11	23–30
Avg.	4.5	3.7	11.3	21
%RSD	28.8	23.2	31.8	25

TABLE 3 VOCs in Methanol, Percent Detected

	Compound				
Laboratory No.	ChlBz	EB	TCA	TCE	TOL
1	...	0.36	0.35	0.39	0.4
2	0.13	0.39	0.35	0.42	0.44
3	...	...	...	0.3	0.3
4	0.08	0.26	0.24	0.27	0.3
5	...	...	0.2	0.23	0.23
6	0.13	0.4	...	...	...
7	...	0.23	...	...	...
8	0.1	0.3	0.28	0.32	0.35
9	0.09	0.33	0.26	0.31	0.37
Theo%	0.1	0.32	0.29	0.35	0.37
Avg.	0.11	0.32	0.28	0.32	0.34
%RSD	19.8	18.2	21.4	18.7	20.5
Avg. Rec.	102	100	97	92	92

TABLE 4 Solvents in Oil, Percent Detected

	Compound				
Laboratory No.	CCl_4	PERC	TCA	TCE	TOL
1	5.5	5.3	7.1	4.7	2.1
2	3.4	6.0	4.7	4.0	3.4
3	...	5.0	...	4.6	2.1
4	4.2	3.8	...	3.2	1.5
5	4.1	4.1	5.8	3.5	1.6
6	4.7	4.5	6.2	4.1	1.7
7	5.8	5.3	8.5	6.0	2.3
8	6.1	7.0	4.6	6.0	2.7
9	...	3.9	5.0	3.2	1.6
10	4.4	5.0	6.2	3.1	1.7
11	4.5	5.3	7.3	4.6	2.0
Theo%	5.1	5.1	7.2	4.6	1.9
Avg.	4.7	5.0	6.2	4.3	2.1
%RSD	18.7	18.9	21.1	24.1	27.2
Avg. Rec.	93	98	86	93	106

Compound Key:
ChlBz = Chlorobenzene
CCl4 = Carbon Tetrachloride
EB = Ethylbenzene
MeCl2 = Methylene Chloride
MEK = Methylethyl Ketone
TCA = 1,1,1-Trichloroethane
TCE = Trichloroethylene
TOL = Toluene
PERC = Perchloroethylene
XYL = Xylene

14. Keywords

14.1 flame ionization; gas chromatography; hazardous waste; solvents; thermal conductivity detector

Designation: D 5831 – 96 (Reapproved 2001)

Standard Test Method for Screening Fuels in Soils[1]

This standard is issued under the fixed designation D 5831; the number immediately following the designation indicates the year of original adoption or, in the case of revision, the year of last revision. A number in parentheses indicates the year of last reapproval. A superscript epsilon (ε) indicates an editorial change since the last revision or reapproval.

1. Scope

1.1 This test method is a screening procedure for determining the presence of fuels containing aromatic compounds in soils. If the contaminant fuel is available for calibration, the approximate concentration of the fuel in the soil can be calculated. If the contaminant fuel type is known, but the contaminant fuel is not available for calibration, an estimate of the concentration of the fuel in the soil can be determined using average response factors. If the nature of the contaminant fuel is unknown, this screening test method can be used to identify the possible presence of contamination.

1.2 *This standard does not purport to address all of the safety concerns, if any, associated with its use. It is the responsibility of the user of this standard to establish appropriate safety and health practices and determine the applicability of regulatory limitations prior to use.*

2. Referenced Documents

2.1 *ASTM Standards:*

D 2777 Practice for the Determination of Precision and Bias of Applicable Test Methods of Committee D-19 on Water[2]

E 131 Terminology Relating to Molecular Spectroscopy[3]

E 169 Practices for General Techniques of Ultraviolet-Visible Quantitative Analysis[3]

E 177 Practice for Use of the Terms Precision and Bias in ASTM Test Methods[4]

E 275 Practice for Describing and Measuring Performance of Ultraviolet, Visible, and Near Infrared Spectrophotometers[3]

E 691 Practice for Conducting an Interlaboratory Study to Determine the Precision of a Test Method[4]

E 925 Practice for the Periodic Calibration of Narrow Band-Pass Spectrophotometers[3]

3. Terminology

3.1 *Definitions*—For definitions of terms used in this screening test method, refer to Terminology E 131.

4. Summary of Test Method

4.1 A sample of soil is extracted with isopropyl alcohol, and the extract is filtered. The ultraviolet absorbance of the extract is measured at 254 nm. If the contaminant fuel is available for calibration, the approximate concentration of contamination is calculated. If the contaminant fuel type is known, but the contaminant fuel is not available for calibration, an estimate of the contaminant concentration is determined using average response factors. If the nature of the contaminant fuel is not known, the absorbance value is used to indicate the presence or absence of fuel contamination. Calcium oxide is added to the soil as a conditioning agent to minimize interferences from humic materials and moisture present in the soil. Particulate interferences are removed by passing the extract through a filter.

5. Significance and Use

5.1 This screening test method is intended primarily for field use to define the boundaries of soil contamination. If the contaminant fuel is available for calibration, the approximate concentration of the fuel in the soil can be calculated. If the fuel type is known, but the contaminant fuel is not available for calibration, an estimate of the contaminant fuel concentration can be calculated using average response factors. If the nature of the contaminant fuel is unknown, a contaminant concentration cannot be calculated, and the test method can only be used only to indicate the presence or absence of fuel contamination.

5.2 Fuels containing aromatic compounds, such as diesel fuel and gasoline, as well as other aromatic-containing hydrocarbon materials, such as crude oil, coal oil, and motor oil, can be determined by this test method. The quantitation limit for diesel fuel is about 75 mg/kg. Approximate quantitation limits for other aromatic-containing hydrocarbon materials that can be determined by this screening test method are given in Table 1. Quantitation limits for highly aliphatic materials, such as aviation gasoline and synthetic motor oil, are much higher than those for more aromatic materials, such as coal oil and diesel fuel.

NOTE 1—The quantitation limits listed in Table 1 are approximate values because in this test method, the quantitation limit can be influenced by the particular fuel type and soil background levels. For information on how the values given in Table 1 were determined, see Appendix X1. Data

[1] This test method is under the jurisdiction of ASTM Committee D34 on Waste Management and is the direct responsibility of Subcommittee D34.02 on Physical and Chemical Characterization.

Current edition approved Oct. 10, 1996. Published December 1996. Originally published as D 5831 – 95. Last previous edition D 5831 – 95.

[2] *Annual Book of ASTM Standards*, Vol 11.01.

[3] *Annual Book of ASTM Standards*, Vol 03.06.

[4] *Annual Book of ASTM Standards*, Vol 14.02.

TABLE 1 Approximate Quantitation Limits for Various Fuel Types in Soils Based on 0.036 AU

Material	Limit of Quantitation (LOQ), mg/kg
Coal Oil	21
Crude Oil	61
Diesel Fuel	75
Used Motor Oil	162
Weathered Gasoline	170
Unleaded Gasoline	316
Jet Fuel JP-2	378
Motor Oil	533
Aviation Gasoline	1066
Synthetic Motor Oil	1382

generated during the development of this screening test method and other information pertaining to this test method can be found in the research report.[5]

5.3 Extractable material, which scatters or absorbs light at 254 nm, is a potential interference for this screening test method.

6. Apparatus

6.1 *Glass Bottles*, wide-mouth, 125-mL (4-oz) with polytetrafluoroethylene-lined lids.

6.2 *Portable Scale*, (for field testing) or laboratory balance, capable of weighing to 0.1 g.

6.3 *Portable Stirring Device*, (for field testing) or magnetic stir bar and stirrer, which result in motion of all solids during stirring.

6.4 *Syringe*, disposable, plastic, 10-mL capacity.

6.5 *Syringe Filters*, disposable, 0.45-μm, in 25-mm polytetrafluoroethylene cartridges.

6.6 *Spectrometer Set*, at 254 nm with a 1-cm path length, quartz cell (cuvette).

6.7 *Volumetric Flasks and Pipets*, for preparing standard solutions.

6.8 *Laboratory Balance*, capable of weighing to 0.0001 g.

7. Reagents and Materials

7.1 *Purity of Reagents*—Reagent grade chemicals shall be used in all screening tests. Unless otherwise indicated, it is intended that all reagents shall conform to the specifications of the Committee on Analytical Reagents of the American Chemical Society where such specifications are available.[6] Other grades may be used provided it is first ascertained that the reagent is of sufficiently high purity to permit its use without lessening the accuracy of the determination.

7.2 *Calcium Oxide Powder, Reagent Grade*—Use calcium oxide powder, reagent grade dried at 900°C for 12 h and stored in a desiccator or tightly sealed glass container prior to use. This is a conditioning agent for removal of interferences caused by the presence of humic material or moisture, or both, in the sample.

7.3 *Isopropyl Alcohol, Reagent Grade*— Transportation of isopropyl alcohol for field testing must comply with current Department of Transportation (DOT) regulations.

8. Procedure

8.1 *Preparation of Standard Solutions*:

8.1.1 Weigh out 200 mg (weighed to ±0.1 mg) of the fuel type of interest into a 100-mL volumetric flask and dilute to volume using isopropyl alcohol. This gives a 2000-mg/L standard stock solution. Other standard solutions can be prepared as needed by appropriate dilution of this stock solution. For example, to prepare a 200-mg/L solution of the fuel type of interest, pipet 5 mL of the stock solution into a 50-mL volumetric flask and dilute to volume using isopropyl alcohol. For work in the field, a standard stock solution can be prepared by diluting 25 μL of a fuel standard (density can vary from~ 0.75–0.90 g/mL) to 100 mL with isopropyl alcohol.

8.2 *Sample Preparation*:

8.2.1 Preweigh a 125-mL (4-oz), wide-mouth, glass sample collection bottle having a polytetrafluoroethylene-lined lid. Record the mass of the empty sample collection bottle to ±0.1 g.

8.2.2 Add 5 g (weighed to ±0.1 g) of soil directly to the preweighed sample collection bottle. Weigh the sample bottle-plus-sample, and record the mass of the soil sample added to the bottle to ±0.1 g.

8.2.3 Add approximately 5 g of calcium oxide, prepared as specified in 7.3, to the soil, and stir with a spatula until a uniform dry mixture is obtained. If the soil contains an excessive amount of water (>50 %), add approximately 5 g more of calcium oxide.

8.3 *Sample Extraction and Analysis*:

8.3.1 Pour 50 mL of isopropyl alcohol into the sample bottle.

8.3.2 Stir the slurry for 3 min using a portable stirring device or magnetic stir bar and stirrer so that all solids are in motion during stirring. A shorter stirring time or hand shaking may decrease the extraction efficiency.

8.3.3 Allow the soil slurry to settle briefly, then remove the lid and draw the supernatant solution into a 10-mL disposable syringe. Attach a filter cartridge to the end of the syringe. Rinse the sample cuvette with filtered extract. Then fill the cuvette with filtered extract for analysis.

8.3.4 Calibration procedures specific to the spectrometer being used to perform the absorbance measurements must be followed. Instrument instructions for spanning from 0 to 1 absorbance unit must be followed. Calibration is to be performed using isopropyl alcohol to zero the instrument, and if a calibration line is to be established, calibration standards prepared from the standard stock solution should be used (see 8.1.1). Calibration using three standards is recommended. Calibration curves are nonlinear above 1 AU (>90 % of the light absorbed). As a result, readings must be made below this level. In addition, the extract absorbance reading must fall between the absorbance readings of two calibration standards.

[5] Schabron, J. F., Niss, N. D., Hart, B. K., and Sorini, S. S., "Remote Chemical Sensor Development: A New Field Screening Method for Soil Fuel Contamination," Laramie, WY, WRI-95-R016, 1995.

[6] *Reagent Chemicals, American Chemical Society Specifications*, American Chemical Society, Washington, DC. For suggestions on the testing of reagents not listed by the American Chemical Society, see *Analar Standards for Laboratory Chemicals*, BDH Ltd., Poole, Dorset, U.K., and the *United States Pharmacopeia and National Formulary*, U.S. Pharmacopeial Convention, Inc. (USPC), Rockville, MD.

NOTE 2—For general information on the techniques most often used in ultraviolet analysis, see Practice E 169. For additional information on the performance of ultraviolet spectrophotometers, see Practice E 275. For information on evaluating the performance of an ultraviolet spectrophotometer to verify its suitability for continued routine use, see Practice E 925.

8.3.5 Read and record the absorbance of the extract at 254 nm.

8.3.6 Determine an approximate or estimated concentration of a known fuel type in the filtered extract.

8.3.6.1 If the contaminant fuel was used for calibration, an approximate concentration of the fuel in the extract can be calculated using a calibration line. Record this approximate concentration of the fuel in the extract in milligrams/litre.

8.3.6.2 If the contaminant fuel type is known, but the contaminant fuel was not used for calibration, an estimated concentration of the fuel type in the extract can be calculated by multiplying the absorbance of the extract by the reciprocal absorptivity for that fuel type (see Table 2 and Eq 1). Record this estimated concentration of the fuel in the extract in milligrams/litre.

$$(\text{Absorbance}) \times (1/\text{Absorptivity}) = \text{Estimated concentration of the fuel in the filtered extract (mg/L)} \quad (1)$$

NOTE 3—Information pertaining to the reciprocal absorptivity values are listed in Table X1.1 of the appendix.

8.3.7 Convert the approximate or estimated concentration of fuel in the extract (see 8.3.6.1 or 8.3.6.2) to an approximate or estimated concentration of the fuel in the original soil sample in milligrams/kilograms by multiplying the concentration of the fuel in the extract in milligrams/litre by a factor representing the solvent volume in millilitres-to-sample mass in grams ratio used in the extraction, that is, a factor of ten is used for a solvent volume to soil mass ratio of 50 mL of isopropyl alcohol: 5 g of soil. If the extract is diluted, the appropriate correction must be made. Record the approximate/estimated concentration of the fuel in the soil sample in milligrams/kilograms.

8.3.8 If the nature of the fuel-type contaminant is unknown, the concentration of the contaminant can not be calculated. In this case, the absorbance of the extract at 254 nm (see 8.3.5) can be used to indicate the presence of fuel contamination in the soil.

8.3.9 To ensure that the batch of conditioning agent, plastic syringe, filter cartridge, and so forth, are not contributing to the absorbance reading, it is recommended that the procedure also should be performed as specified, except with no soil. If significant absorbance is noted, the various components should be tested individually by contacting them with isopropyl alcohol, and the problem component(s) should be replaced.

8.3.10 In this procedure, the conditioning agent inhibits the extraction of most humic materials, and there is very little, if any, background from inorganic materials. It is recommended, however, that a blank soil sample should be tested by extracting contaminant-free soil of the same type and from the same general area as the site being studied. If any absorbance is measured for the blank, the sample values should be corrected for this procedure.

8.3.11 Also, it is recommended that one spike should be run for every batch of samples or for every 20 samples, whichever is most frequent. A soil sample is spiked by adding 5 μL of diesel fuel or 25 μL of gasoline and shaking the bottle for 3 min. The extraction and analysis then are performed as outlined in 8.2.3-8.3.5. Recovery is calculated by comparing the absorbance of the extract from the spiked soil at 254 nm with the absorbance of a solution of 5 μL of diesel fuel or 25 μL of gasoline in 50 mL of isopropyl alcohol. After correction for any material appearing in the unspiked soil, the recovery should be within 20 % of the true value.

TABLE 2 Reciprocal Absorptivities at 254 nm for a 1-cm Path Length Cell

Material	1/Absorptivity, mg/L/AU
Coal Oil	58.7
Crude Oil	169
Diesel Fuel	209
Used Motor Oil	450
Weathered Gasoline	473
Unleaded Gasoline	877
Jet Fuel JP-2	1050
Motor Oil	1480
Aviation Gasoline	2960
Synthetic Motor Oil	3840

9. Record

9.1 Record the following information:

9.1.1 Type of the fuel contaminant,

9.1.2 Mass of the empty sample collection bottle, g,

9.1.3 Mass of the sample bottle plus soil sample, g,

9.1.4 Mass of the soil sample, g,

9.1.5 Volume of isopropyl alcohol (solvent) used in the extraction, mL,

9.1.6 Solvent for zeroing spectrometer,

9.1.7 Calibration standard solutions and absorbance values at 254 nm,

9.1.8 One/absorptivity for the fuel type of interest, if the contaminant fuel is not used for calibration,

9.1.9 Absorbance of the soil sample extract at 254 nm,

9.1.10 Approximate/estimated concentration of the fuel in the filtered extract, mg/L,

9.1.11 Approximate/estimated concentration of the fuel in the soil sample, mg/kg, and,

9.1.12 Suggested data recording form for performing this screening procedure as given in Fig. 1.

10. Report

10.1 Report the indicated presence or absence of fuel contamination or approximate or estimated concentration of contaminant fuel in the sample.

11. Precision and Bias [7]

11.1 *Precision*:

11.1.1 A collaborative study of this screening test method involving eight participants was conducted. Each participant tested seven materials in triplicate. The test materials were a

[7] A copy of the research report used to develop the precision statement is available from ASTM Headquarters. Request RR: D34-1011.

Site:

Date:

Operator:

Contaminant:

Calibration Solvent :

Calibration Standards/Absorbance:

1/Absorptivity for Fuel Type:

Sample ID	Sample Mass, g	Solvent Volume, mL	Solvent Volume-to-Sample Mass Ratio	Absorbance at 254 nm	Approximate/Estimate Concentration of Fuel in Extract, mg/L	Approximate/Estimate Concentration of Fuel in Soil, mg/kg

FIG. 1 Fuels in Soils Data Form

sand spiked with three different concentrations of diesel fuel (Test Materials *A, B,* and *C*), an unspiked sand (Test Material *D*), an organic soil spiked with two different concentrations of diesel fuel (Test Materials *E* and *F*), and an unspiked organic soil (Test Material *G*). The absorbance values of three calibration standards, which were prepared by the participants, were also determined for generation of a calibration line by each participant. The collaborative study materials were tested for and met a specified criterion for homogeneity prior to being sent to the collaborative study participants.

11.1.2 The collaborative study participants used the absorbance values they recorded to calculate the approximate and estimated concentrations of diesel fuel in the test materials. The approximate concentrations were determined using a calibrated line, which was generated by each participant from analysis of their calibration standards. The estimated concentrations of diesel fuel in the test materials were calculated using a response factor of 209 mg/L/AU (see Table 2).

11.1.3 In the collaborative study, to keep the identity of the samples unknown, the participants were not given any information on sample type or if any of the samples were unspiked. As a result, the participants did not know they had blank data, which could be used to correct sample values for background (see 8.3.10). Calculations to correct the approximate and estimated spiked sample concentrations for concentrations reported in the blank materials were performed by the collaborative study coordinator using the data provided by the participants. The blank-corrected approximate and estimated concentration values calculated for the test materials are listed in Table 3 and Table 4.

11.1.4 Practices D 2777, E 177, and E 691 were used as guidance in performing statistical evaluation of the data listed in Table 3 and Table 4. The index used for expressing reproducibility and repeatability of this test method is the 95 % limit on the difference between two test results. The 95 % limit means that approximately 95 % of all pairs of test results from users similar to the participants in the collaborative study can be expected to differ in absolute value by less than 1.960

TABLE 3 Tabulation of Collaborative Study Data for the Fuels in Soils Screening Method: Blank-Corrected Approximate Concentrations of Diesel Fuel in the Test Materials, mg/Kg

Participant	Material				
	A	B	C	E	F
1	153	364	761	220	714
	167	407	881	200	673
	178	371	847	220	819
2	172	340	763	101	577
	156	366	770	85	598
	158	386	762	86	574
3	157	403	830	132	587
	159	403	841	120	641
	159	405	848	122	634
4	180	405	851	101	690
	185	414	874	113	687
	167	404	793	117	685
5	168	389	751	87	593
	152	358	768	87	609
	156	375	792	101	576
6	137	341	662	95	471
	170	378	763	116	597
	153	369	768	97	555
7	101	314	801	84	472
	107	322	721	112	500
	104	301	781	76	505
8	132	380	793	97	561
	107	395	957	98	540
	114	375	764	93	607

TABLE 4 Tabulation of Collaborative Study Data for the Fuels in Soils Screening Method: Blank-Corrected Estimated Concentrations of Diesel Fuel in the Test Materials, mg/Kg

Participant	Material				
	A	B	C	E	F
1	182	435	906	262	853
	199	460	1048	234	828
	212	442	1008	264	1085
2	216	424	949	127	727
	195	457	957	106	754
	197	481	948	108	724
3	177	455	937	150	663
	180	455	949	136	724
	179	457	958	138	716
4	212	498	1029	122	833
	218	508	1056	136	829
	197	495	960	142	827
5	204	473	914	107	723
	185	436	935	107	742
	189	457	965	123	702
6	165	409	794	114	566
	204	453	915	139	716
	183	443	921	116	666
7	128	394	1006	108	595
	135	401	914	141	637
	131	385	981	98	642
8	151	528	1115	137	783
	117	552	1335	138	750
	127	521	1065	131	851

$(2)^{1/2}s$, which corresponds to 2.8 s or 2.8 CV % (percent coefficient of variation) (Practice E 177). The steps involved in the data analysis were (*1*) eliminating "outlier" participants (participants that are so consistently high or low that their results are unreasonable), (*2*) eliminating individual "outlier" data points, (*3*) calculating reproducibility (between participants) standard deviation (s_R), (*4*) calculating repeatability (within participant) standard deviation (S_r), (*5*) determining the 95 % reproducibility limit (2.8 S_R or 2.8 CV $\%_R$, and (*6*) determining the 95 % repeatability limit (2.8 S_r or 2.8 CV $\%_r$).

11.1.5 The reproducibility and repeatability precision statistics calculated for this test method using the collaborative study data are listed in Table 5 and Table 6. Based on these data, the 95 % reproducibility and repeatability limits for testing the diesel-spiked sand and diesel-spiked organic soil using this test method were determined. These limits are listed in Table 7.

11.1.6 The data listed in Table 7 give information on the reproducibility and repeatability of this screening test method when it is applied to a sand and organic soil contaminated with various concentrations of diesel fuel. The data are specific to the test materials used in the study. For other soil types and fuel contaminants, these data may not apply.

11.1.7 The information given in Table 7 shows that the reproducibility precision (95 % reproducibility limit) of this test method varies between the two test materials. The reproducibility of the test method when applied to the diesel-spiked sand varies with diesel concentration and also between approximate and estimated concentration determinations. As expected, the reproducibility precision of this test method at a lower diesel concentration in the sand (~160 mg/Kg) is less than at higher concentrations (~400 to 970 mg/Kg). The reproducibility precision of the method for testing the diesel-spiked sand is approximately two times higher at the higher

TABLE 5 Reproducibility Precision Statistics for the Screening Method for Fuels in Soils[A]

Approximate Concentration Statistics for Testing the Sand			
$\bar{x}$	s_{Ra}[B]	2.8 s_{Ra}	2.8 CV $\%_{Ra}$
156	20	56	36 %
382	22	62	16 %
802	63	176	22 %
Estimated Concentration Statistics for Testing the Sand			
$\bar{x}$	s_{Re}[C]	2.8 s_{Re}	2.8 CV $\%_{Re}$
179	32	90	50 %
459	44	123	27 %
972	73	204	21 %
Approximate Concentration Statistics for Testing the Organic Soil			
$\bar{x}$	s_{Ra}	2.8 s_{Ra}	2.8 CV $\%_{Ra}$
103	14	39	38 %
618	77	216	35 %
Estimated Concentration Statistics for Testing the Organic Soil			
$\bar{x}$	s_{Re}	2.8 s_{Re}	2.8 CV $\%_{Re}$
125	15	42	34 %
737	85	238	32 %

[A] Units are mg/Kg unless otherwise specified.

[B] Reproducibility (between paticipants) standard deviation for determining approximate concentration.

[C] Reproducibility (between participants) standard deviation for determining estimated concentration.

TABLE 6 Repeatability Precision Statistics for the Screening Method for Fuels in Soils[A]

$\bar{x}$	s_{ra}[B]	2.8 s_{ra}	2.8 CV $\%_{ra}$
Approximate Concentration Statistics for Testing the Sand			
156	11	31	20 %
382	16	45	12 %
802	54	151	19 %
Estimated Concentration Statistics for Testing the Sand			
$\bar{x}$	s_{re}[C]	2.8 s_{re}	2.8 CV $\%_{re}$
179	13	36	20 %
459	17	48	10 %
972	46	129	13 %
Approximate Concentration Statistics for Testing the Organic Soil			
$\bar{x}$	s_{ra}	2.8 s_{ra}	2.8 CV $\%_{ra}$
103	8	22	21 %
618	42	118	19 %
Estimated Concentration Statistics for Testing the Organic Soil			
$\bar{x}$	s_{re}	2.8 s_{re}	2.8 CV $\%_{re}$
125	12	34	27 %
737	38	106	14 %

[A] Units are mg/Kg unless otherwise specified.
[B] Repeatability (within participant) standard deviation for determining approximate concentration.
[C] Repeatability (within participant) standard deviation for determining estimated concentration.

TABLE 7 95 % Reproducibility[A] and Repeatability[B] Limits for Testing Diesel-Spiked Sand and Organic Soil Using the Screening Method for Fuels in Soils

Test Range, mg/Kg	95 % Reproducibility Limit (% of the test result)
Material: Diesel-Spiked Sand	
156 (approximate)	36 %
382 to 802 (approximate)	19 % (16 %, 22 %)
179 (estimated)	50 %
459 to 972 (estimated)	24 % (24 %, 27 %)
Material: Diesel-Spiked Organic Soil	
Test Range, mg/Kg	95 % Reproducibility Limit (% of the test result)
103 to 737 (approximate or estimated)	35 % (32 to 38 %)
Materials: Diesel-Spiked Sand and Organic Soil	
Test Range, mg/Kg	95 % Repeatability Limit (% of the test result)
103 to 179 (approximate or estimated)	22 % (20 to 27 %)
382 to 459 (approximate or estimated)	11 % (10 %, 12 %)
618 to 802 (approximate)	19 % (19 %, 19 %)
737 to 972 (estimated)	14 % (13 %, 14 %)

[A] Between participants.
[B] Within participant.

diesel concentrations. The data in Table 7 shows that the reproducibility precision of the screening method for testing the diesel-spiked organic soil is constant over the diesel concentration range tested (from 103 to 737 mg/Kg), and does not vary between approximate and estimated concentration determinations. It appears that the characteristics of the diesel-spiked organic soil mask any variations in the reproducibility precision of the test method related to concentration or approximate/estimated concentration determinations. If the 95 % reproducibility limits listed for the diesel-spiked sand in Table 7 are averaged, the result is an overall 95 % reproducibility limit of 32 %, which corresponds very closely to 35 %, the 95 % reproducibility limit for the diesel-spiked organic soil.

11.1.8 The repeatability precision (95 % repeatability limit) of the screening test method does not vary between the two materials (Table 7). It is slightly less at the lower approximate and estimated diesel concentrations of 103 and 179 mg/Kg, repectively. The repeatability precision of the test method does not vary between approximate and estimated concentration determinations at test levels between approximately 100 to 450 mg/Kg. However, at the higher concentrations, approximately 600 to 970 mg/Kg diesel fuel, there is a slight difference between the repeatability precision of the test method for approximate and estimated concentration determinations. Overall, the repeatability precision of the screening test method does not vary significantly with concentration nor between approximate and estimated concentration determinations for the diesel-spiked sand and organic soil.

11.2 *Bias*—Since there is no accepted reference material suitable for determining the bias for the procedure in this test method for screening fuels in soils, no statement on bias is being made. Available reference materials vary in their aromaticity, and, as a result, may not give a value when tested by this test method that corresponds to a value determined by a test method using a different measurement principle.

12. Keywords

12.1 absorbance; contamination; extraction; field screening; fuels; soils

APPENDIXES

(Nonmandatory Information)

X1. APPROXIMATE QUANTITATION LIMITS

X1.1 In the research to develop this screening test method, a series of approaches to determine the LOQ for various fuel types were taken. For example, one approach to estimate the LOQ for diesel fuel was to multiply the standard deviation of several absorbance measurements of a diesel standard in isopropyl alcohol ($s = 0.00315$) by a factor of 10[8,9] and convert the resulting absorbance value (0.0315) to concentration of diesel in soil using the equation:

$$(absorbance) \times (1/absorptivity) \times (solvent\ volume\ to\ sample\ mass\ ratio) = LOQ \quad (X1.1)$$

$$(0.0315) \times (209\ \text{mg/L}) \times (10\ \text{mL/g}) = 66\ \text{mg/kg diesel in soil as the LOQ}$$

X1.2 The 1/absorptivity value is given in Table X1.1. It is the average response factor for diesel determined using a portable photometer. A second approach was to use the 99 % confidence level one-sided t statistic.[10] The standard deviation of 34 blank absorbance values determined using a variety of soil types was calculated to be 0.016 AU. This value multiplied by a t value of 2.442 gives an absorbance of 0.039 AU, which converts to an estimated LOQ of 67 mg/kg for the particular diesel used (red can diesel listed in Table X1.1, 1/Absorptivity = 171 mg/L) or 81 mg/kg based on the average diesel response factor (Table X1.1, 1/Absorptivity = 209 mg/L). This approach depends on the soils used and the blank values measured. Next, to estimate the LOQ for diesel using the conditioning agent, calcium oxide, a series of blank extracts were generated using a variety of wet and dry soils. The standard deviation of the absorbance values of these extracts was 0.0135. Using the 99 % one-sided t statistic, $ts = 0.035$ AU (2.602×0.0135). Using this value, the LOQ for diesel based on the average response factor given in Table X1.1 (209 mg/L) is 73-mg/kg diesel. This approach, which is based on typical soil blank background levels using calcium oxide, is considered to be a more realistic estimate of the LOQ. Based on additional testing in the laboratory, a typical soil blank background absorbance reading using calcium oxide was determined to be 0.036 AU. As a result, this value and the average reciprocal absorptivity values given in Table 2 of this test method were used to calculate the approximate LOQ values for the fuel types studied. These are the values given in Table 1 of this test method. For example, for weathered gasoline, an approximate LOQ of 170 mg/kg is listed in Table 1. This value was calculated using the following equation:

$$(0.036) \times (473\ \text{mg/L}) \times (10\ \text{mL/g}) = 170-\text{mg/kg weathered gasoline in soil} \quad (X1.2)$$

Because of variability in absorptivity between fuels of the same type (see Appendix X2) and variations in soil background levels, the LOQ values given in Table 1 can only be approximate values.

---

[8] Stanko, G. H., Krochta, W. G., Stanley, A., Dawson, T. L., Hillig, K. J. D., Javik, R. A., Obrycki, R., Hughes, B. M., and Saska, F. I., "Defining the Limits," *Environmental Laboratory*, 5(5), October/November: 16, 1993.

[9] *The Hazardous Waste Consultant*, Soil Sampling and Analysis, Practices and Pitfalls, November/December, 4.1–4.26, 1992.

[10] US EPA, *Test Methods for Evaluating Solid Waste*, SW-846, Third Edition, Final Update I, US Environmental Protection Agency, Washington, DC, One-25-26, 1990.

TABLE X1.1 Reciprocal Absorptivities at 254 nm for a 1-cm Path Length Cell

Material	1/Absorptivity, mg/L per AU-Portable Photometer
	Diesel Fuel
Texaco #1	120
Texaco #2	194
Pilot	123
CG	308
4-C	371
Phillips	176
Red Can	171
Total	206
Average:	209
	Coal Oil
WY SRCII-A	55.2
WY SRCII-B	62.2
Average:	58.7
	Crude Oil
Recluse	210
S. Swan Hill	169
Wasson	129
Gach Saran	187
Gulf X-27, 683A	196
Gulf X-26, 483A	172
Wilmington	149
Prudhoe Bay	143
Swan Hill	165
Average:	169
	Used Motor Oil
Sample A	512
Sample B	389
Average:	450
	Motor Oil
Mobil 10W-40	972
Exxon 10W-40	1590
Amoco 10W-40	2140
Havoline 10W-30	445
Amoco 10W-30	1950
Castrol 5W-30	1440
Castrol SAE-30	1840
Average:	1480
	Synthetic Motor Oil
Castrol 5W-50	7040
Slick 50	1650
Penzoil 5W-50	2840
Average:	3840
	Other Fuels
Aviation Gas	2960
Jet Fuel JP-2	1050
Unleaded Gasoline	877
Weathered Gasoline	473
2-Cycle Oil	1890

X2. RELATIVE RESPONSE FACTORS

X2.1 Different aromatic-containing materials will have different absorptivities at 254 nm. In addition, because there are so many different sources of these materials, aromaticity can vary between materials of the same type, such as diesel fuels, and aromaticity can also vary between batches of the same material. Listed in Table X1.1 are reciprocal absorptivities for various aromatic-containing materials measured using a portable photometer. Similar values for these materials were also measured using a laboratory scanning spectrophotometer. Those values are given in the research report describing the development of this test method.[3] Because the fuels in the soils screening test method were developed for use in the field, the data generated using the portable photometer are given in Table 2 of this test method. The variability in absorptivity between materials of the same type is the reason the absorptivity calculation to determine the concentration of contamination (see 8.3.6.2) gives only an estimated value.

Designation: D 5839 – 96 (Reapproved 2001)

Standard Test Method for Trace Element Analysis of Hazardous Waste Fuel by Energy-Dispersive X-Ray Fluorescence Spectrometry[1]

This standard is issued under the fixed designation D 5839; the number immediately following the designation indicates the year of original adoption or, in the case of revision, the year of last revision. A number in parentheses indicates the year of last reapproval. A superscript epsilon (ε) indicates an editorial change since the last revision or reapproval.

1. Scope

1.1 This test method applies to the determination of trace element concentrations by energy-dispersive X-ray fluorescence (EDXRF) spectrometry in typical liquid hazardous waste fuels (LHWF) used by industrial furnaces.

1.2 This test method has been used successfully on numerous samples of LHWF that are mixtures of solvents, oils, paints, and pigments for the determination of the following elements: Ag, As, Ba, Cd, Cr, Hg, Ni, Pb, Sb, Se, and Tl.

1.3 This test method also may be applicable to elements not listed above and to the analysis of trace metals in organic liquids other than those used as LHWF.

1.4 *This standard does not purport to address all of the safety concerns, if any, associated with its use. It is the responsibility of the user of this standard to establish appropriate safety and health practices and determine the applicability of regulatory limitations prior to use.*

2. Referenced Documents

2.1 *ASTM Standards:*

C 982 Guide for Selecting Components for Energy-Dispersive X-Ray Fluorescence (XRF) Systems[2]

3. Summary of Test Method

3.1 A weighed portion of the sample is transferred to a porcelain evaporating dish and placed on a hot plate. The sample is heated for 15 to 30 min at a temperature not exceeding 105°C to evaporate highly volatile components. The dish is removed from the hot plate and allowed to cool. Graphite powder is blended with the evaporated sample until a homogeneous paste is produced and the sample weight is recorded. The blended sample is inserted in a disposable sample cup and placed in the X-ray spectrometer for analysis.

3.2 The K Spectral X-ray emission lines are used for Ag, As, Ba, Cd, Cr, Ni, Sb, and Se.

3.3 The L spectral X-ray emission lines are used for Hg, Pb, and Tl.

4. Significance and Use

4.1 The analysis of trace elements is often a regulatory and process specific requirement for facilities utilizing LHWF. With proper instrument standardization, set-up, and quality control, this test method provides the user an accurate, rapid, nondestructive method for trace element determinations.

5. Interferences

5.1 *Spectral Overlaps*:

5.1.1 Samples containing a mixture of elements often exhibit X-ray emission line overlap. Modern Si (Li) detectors generally provide adequate resolution to minimize the effects of spectrum overlaps on the analytical results of the LHWF. In those cases where direct emission line overlap exists, spectral deconvolution methods extract corrected analyte emission line intensities. Table 1 lists the significant line overlaps observed for the elements analyzed in LHWF. Follow the EDXRF manufacturer's recommendations concerning spectral deconvolution of the emission lines.

NOTE 1—Not all possible interferences are listed in Table 1. The LHWF samples to be analyzed may have other emission line interferences not mentioned. Consult the manufacturer's recommendations for optimum deconvolution methods.

5.2 *Matrix Interferences*:

5.2.1 Large concentration variations of metal or matrix components, or both, in LHWF samples can result in nonlinear metal X-ray intensity response at increasing metal concentrations. Untreated matrix interactions may have a deleterious effect on metal determination accuracy. Matrix interactions, if exhibited by the LHWF samples, must be accounted for by method calibration.

6. Apparatus

6.1 *Energy Dispersive X-Ray Spectrometer*, capable of measuring the wavelengths listed in Table 1. Refer to C 982 for system specifications.

6.2 *Hot Plate*, with temperature control to 105°C.

6.3 *Analytical Balance*, capable of weighing to 0.001 g.

6.4 *Porcelain Evaporating Dishes*, 70 to 150 mL capacity.

7. Reagents and Materials

7.1 *Purity of Reagents*—Use reagent grade chemicals in all tests. Unless otherwise indicated, all reagents will conform to

---

[1] This guide is under the jurisdiction of ASTM Committee D34 on Waste Management and is the direct responsibility of Subcommittee D34.01.06 on Analytical Methods.

Current edition approved Jan. 10, 1996. Published March 1996.

[2] *Annual Book of ASTM Standards*, Vol 12.01.

TABLE 1 Common Emission Line Spectral Interferences for LHWF Analysis

Element	Lines Determined	Spectral Interferences
As	Kα, Kβ	Pb Lα, Hg Lβ, Tl Lα Lβ
Cr	Kα, Kβ	V Kβ, Mn Kα
Hg	Lα, Lβ	As Kβ, Se Kα, Pb Lβ, Tl Lα, Br Kα
Ni	Kα, Kβ	Cu Kβ, Cu Kα
Pb	Lα, Lβ	As Kα, Se Kβ Hg Lβ Tl Lα, Tl Lβ, Br Kα, Br Kβ
Se	Kα, Kβ	Pb Lβ, Hg Lβ, Tl Lβ
Tl	Lα, Lβ	As Kα, As Kβ, Se Kβ, Pb Lα, Br Kα, Hg Lα, Hg Lβ

the specifications of the Committee on Analytical Reagents of the American Chemical Society where such specifications are available.[3] Other grades may be used, provided it is first ascertained that the reagent is of sufficient high purity to permit its use without lessening the accuracy of the determination.

7.2 Use graphite powder, mixing grade, 44 μm, 325 mesh.[4]

7.3 Use oil-based Ag, As, Ba, Cd, Cr, Hg, Ni, Pb, Sb and Se standards, 10 to 10 000 mg/kg depending on the user's analytical requirements. If the results of this test method are to be used for compliance purposes, standards or a commercial source must be traceable to NIST or other certifying body. Quality control samples for analyses done for compliance purposes may need to be prepared with standards from a different vendor or lot number.

7.4 Use oil or solvent-based Tl standard, 10 to 1000 mg/kg depending on the user's analytical requirements. If the results of this test method are to be used for compliance purposes, standards or a commercial source must be traceable to NIST or other certifying body.[5]

7.5 Use paraffinic base oil.

7.6 Use sample cups, vented or unvented.

7.7 Use thin-film sample support.

NOTE 2—The user should select a thin-film support that provides for maximum transmittance and is resistant to typical components in LHWF. The thin-film supports used in the development of this test method were 4 μm Prolene.

8. Sampling

8.1 Because stratification or layering of liquid samples is possible, the laboratory sample should be thoroughly mixed by shaking prior to withdrawing a portion for testing, or use a sampling method that assures all portions of the sample are represented.

9. Preparation of Apparatus

9.1 Follow the manufacturer's instructions for set-up, conditioning, preparation, and maintenance of the XRF spectrometer.

9.2 Obtain reference spectra from pure element standards for all analytes and interelement correction metals.

9.3 Address spectral interferences, as listed in 5.1.1, in accordance with manufacturer's recommendations.

10. Calibration and Standardization

10.1 Calibrate the spectrometer to an appropriate reference element at the minimum frequency specified by the manufacturer.

10.2 Analytical standards should be gravimetrically prepared by blending the elemental standards and graphite listed in Section 7. These preparations can contain single or multiple elements and should be prepared at combinations and ratios to meet the user's individual needs. For example, a Pb and Cd standard at 125 ppm each could be prepared by gravimetrically combining 2 g oil-based Pb standard (500 ppm), 2 g oil-based Cd standard (500 ppm) and 4 g graphite and blending to a homogeneous mixture. For this example, the Pb concentration is calculated as follows:

$$\text{Pb Standard Concentration} = 500\text{ ppm} \times \frac{\text{Pb Std. (g)}}{\text{graphite (g)} + \text{Cd Std. (g)} + \text{Pb Std. (g)}} \quad (1)$$

10.3 The metals standard/graphite paste is placed in an XRF sample cup and affixed with a thin-film support. The sample cup is inverted and lightly tapped on a level surface until the blended paste makes full contact (no air spaces) with the thin-film support.

10.3.1 The standard blends in the sample cups are placed into the spectrometer's designated sample holder. Avoid touching the thin-film, as this can further reduce transmittance. Initiate data acquisition for the desired elements according to the manufacturer's instructions.

10.4 Two alternative standards calibration methods may be used:

10.4.1 *Test Method A (Empirical Calibration Method)*—Organic-based metals standards containing the metals of interest are prepared as described in 10.2. The standard metal concentrations must bracket the unknown LHWF concentration levels. If the standard concentrations are less than 0.1 % by weight, a linear calibration method may be used to plot metal emission intensity response against metal concentration.

10.4.1.1 If metal concentration exceeds 0.1 % by weight, then a matrix correction technique is employed. Standard mixtures of varying metal composition must be prepared that bracket the LWHF unknown concentrations. Eq A is solved for the slope, intercept and all alpha terms:

$$C_i = B_i + K_i I_i (1 + \Sigma \alpha_{ij} C_j) \quad (2)$$

where:

C_i = concentration of analyte element i,
B_i = intercept term,
K_i = slope term,
I_i = X-ray intensity of element, i,
α_{ij} = alpha coefficient describing the matrix interaction of element j on analyte i, and
C_j = concentration of matrix element j.

10.4.2 *Test Method B (Fundamental Parameters Method)*—

[3] *Reagent Chemicals, American Chemical Society Specifications*, American Chemical Society, Washington, DC. For suggestions on the testing of reagents not listed by the American Chemical Society, see *Analar Standards for Laboratory Chemicals*, BDH Ltd., Poole, Dorset, U.K., and the *United States Pharmacopeia and National Formulary*, U.S. Pharmaceutical Convention, Inc. (USPC), Rockville, MD.

[4] This reagent, or its equivalent, is available from Chemplex Industries, Inc., 160 Marbledale Road, Tuckahoe, NY 10707.

[5] This reagent, or its equivalent, is available from Alfa Aesar, Johnson Matthey Inc., 30 Bond Street, Ward Hill, MA 01835.

Some instrument vendors provide software capable of estimating matrix interferences with a minimum number of standards. A high and low standard concentration mixture or a mid-range concentration mixture gives the fundamental parameters method an initial emission response determination. Using this information and calculations using the principles of X-ray absorption and emission, a data matrix is created sufficient to solve equations of the form in Eq 2 for all sample matrix interactions.

10.4.2.1 Follow the software manufacturer's fundamental parameters set-up recommendations on the number of standard mixtures and the nature of the matrix balance estimate (usually a stoichiometric combination of carbon, hydrogen, and oxygen) for LWHF.

10.4.3 Calibration should be repeated at least every two weeks or whenever quality control results as specified in Section 13 are outside data quality objectives as determined by the user.

11. Procedure

11.1 Mix the LHWF sample thoroughly, see 8.1.

11.1.1 To a clean, dry, and tared evaporating dish, add 10 g of the LHWF sample and weigh to the nearest 0.001 g. Record the mass.

11.1.2 Place the dish on a hot plate at a temperature such that the sample material does not exceed 105°C. Samples containing significant levels of very volatile components may need to be heated initially at a lower temperature. If spattering of the sample is observed, the sample should be removed from the hot plate and allowed to cool slightly. Adjust the hot plate to a lower temperature and replace evaporating dish and sample on the hot plate and continue heating. The heating of the sample is a concentration step, so the heating time is not a critical factor. Generally, the time required to evaporate the volatile fraction from most LHWF is approximately 15 to 30 min.

11.1.3 Remove the dish from the hot plate and allow to cool to the touch.

11.1.4 If the evaporated sample appears to contain large or variable sized solid material, it may be necessary to grind the sample in a mortar and pestle or other suitable device.

11.1.5 Mix graphite into the evaporated sample with a lab spatula or equivalent device until a homogeneous paste results that has no clumping, and when allowed to stand briefly, exhibits no visible liquid phase separation. An approximate amount of 4 to 6 g of graphite is normally sufficient. Weigh the graphite/sample mixture and record.

11.1.6 Mount the graphite/sample preparation in an XRF sample cup as described in 10.3.

11.1.7 Place the mounted sample preparation into the spectrometer's designated sample holder. Avoid touching the thin-film, as this can reduce transmittance. Initiate data acquisition for the desired elements according to the manufacturer's instructions.

11.1.8 After instrument analysis is completed, process analytical results in accordance with Section 12.

12. Calculation

12.1 The analytical results for each element obtained by either empirical calibration or fundamental parameters models in 11.1.8 should be corrected for the sample preparation dilution as follows:

$$\text{Corrected concentration (mg/kg)} = \text{Analytical Result (mg/kg)} \times \frac{\text{Mass of graphite (g) and sample mix (g)}}{\text{Initial mass of LHWF (g)}} \quad (3)$$

13. Quality Control

13.1 Before using this test method, fully investigate any specific regulatory quality control requirements.

13.2 Process a minimum of one quality control check standard, matrix spike/matrix spike duplicate and one analytical blank consisting of graphite powder/analyte-free paraffinic oil blend with each batch of LHWF samples.

13.3 Evaluate quality control sample with each batch of analyzed samples. These results will verify that user defined data quality objectives have been met.

14. Precision and Bias

14.1 *Precision*—No statement is made about precision at this time. An interlaboratory test study will be conducted in the future.

14.2 *Bias*—No statement is made about bias at this time. An interlaboratory test study will be conducted in the future.

14.3 Appendix X1 contains analytical results following this test method (10.4.2) on typical LHWF, MS/MSD, check standard, and blank. Lower Limits of Detections also are listed.

15. Keywords

15.1 EDXRF; hazardous waste fuel; spectrometry; spectroscopy; trace elements; XRF

APPENDIX

(Nonmandatory Information)

X1. ANALYTICAL AND QUALITY CONTROL RESULTS AND LOWER LIMITS OF DETECTION FOR LHWF

LHWF Sample No. 1	Element															
	V	Cr	Ti	Ni	Cu	Zn	Se	As	Tl	Pb	Hg	Ag	Cd	Sb	Ba	Sn
Blank (ppm)	<13.0	<3.0	<8.0	<6.0	<8.0	<7.0	<3.0	<16.0	<6.0	<4.0	<3.0	<3.0	<2.0	<8.0	<34.0	<4.0
Check standard recovery (%)	102.1	106.1	110.0	94.5	102.8	102.8	115.0	124.8	...	104.6	110.0	100.3	105.8	108.0	130.0	104.8
MS/MSD-RPD[1](%)	2.1	1.2	8.2	9.9	0.2	1.5	25.1	16.4	...	7.2	1.1	4.6	1.3	18.7	15.1	1.2
MS/MSD Recovery[2] (%)	75.7	103.9	102.7	93.0	100.5	102.3	145.7	91.2	...	97.3	123.4	101.9	86.5	70.1	153.2	80.2

[1] Relative Percent Difference $= 1C_1 - C_2 1 / ((C_1 + C_2) / 2) \times 100.$ (X1.1)

where:

C_1 = concentration result of the matrix spike.
C_2 = concentration result of the matrix spike duplicate.
1 1 = denotes the absolute value of the difference.

[2] $$\frac{((\text{MS} - \text{sample result}) + (\text{MSD} - \text{sample result}) / 2}{\text{Matrix Spike (Added)}} (100) = \%\ \text{Recovery} \quad (X1.2)$$

Lower Limits of Detection (LLD)[1]			
Element	LLD (ppm)	Element	LLD (ppm)
V	5.0	Tl	...
Cr	2.0	Pb	4.0
Ti	3.0	Hg	3.0
Ni	10.0	Ag	3.0
Cu	7.0	Cd	3.0
Zn	6.0	Sb	8.0
Se	4.0	Ba	26.0
As	4.0	Sn	3.0

[1] $$\text{LLD} = \frac{3\sqrt{Bi}}{Pi} \times \frac{1}{\sqrt{T}} \times \text{conc.} \quad (X1.3)$$

where:

Bi = background intensity (counts/sec) under fitted analyte peak.
Pi = fitted peak intensity (counts/sec) of analyte peak.
T = counting livetime (sec).
conc. = concentration of analyte from metallic oil standard.

Designation: D 5928 – 96

Standard Test Method for Screening of Waste for Radioactivity[1]

This standard is issued under the fixed designation D 5928; the number immediately following the designation indicates the year of original adoption or, in the case of revision, the year of last revision. A number in parentheses indicates the year of last reapproval. A superscript epsilon (ε) indicates an editorial change since the last revision or reapproval.

1. Scope

1.1 This test method covers the screening for α, β, and γ radiation above ambient background levels or user-defined criteria, or both, in liquid, sludge, or solid waste materials.

1.2 This test method is intended to be a gross screening method for determining the presence or absence of radioactive materials in liquid, sludge, or solid waste materials. It is not intended to replace more sophisticated quantitative analytical techniques, but to provide a method for rapidly screening samples for radioactivity above ambient background levels or user-defined criteria, or both, for facilities prohibited from handling radioactive waste.

1.3 This test method may not be suitable for applications such as site assessments and remediation activities.

1.4 The values stated in SI units are to be regarded as the standard.

1.5 *This standard does not purport to address all of the safety concerns, if any, associated with its use. It is the responsibility of the user of this standard to establish appropriate safety and health practices and determine the applicability of regulatory limitations prior to use.*

2. Referenced Documents

2.1 *ASTM Standards:*

C 859 Terminology Relating to Nuclear Materials[2]

3. Terminology

3.1 For terminology related to radioactive materials, see Terminology C 859.

4. Summary of Test Method

4.1 A sample is held within 6 mm of the detector window of a radiation survey meter, and the visible or audible reaction of the meter, or both, is noted. The user defines an application/project specific "negative" and "positive" result criteria. A "negative" test result indicates radiation levels are below the user-established criteria; a "positive" test result indicates the radiation levels are above the user-established criteria.

5. Significance and Use

5.1 Most facilities disposing or utilizing waste materials are prohibited from handling wastes that contain radioactive materials. This test method provides the user a rapid method for screening waste material samples in the field or laboratory for the presence or absence of radioactivity at userestablished criteria. It is important to these facilities to be able to verify generator-supplied information that radioactive or mixed wastes have not been included in shipments of waste materials.

6. Interferences

6.1 Needle deflections or audible clicks of the survey meter, or both, occur due to naturally occurring omni-directional background radiation. This level of ambient background radiation should be periodically assessed. See Section 10.

6.2 Possible sources of interference include pacemakers, X-ray generating equipment, radium-based luminescent dials, polonium-based static eliminators, and smoke detectors containing a radioactive isotope sensing mechanism. Such interferences can usually be traced to their source using the portable instrument specified in this test method.

6.3 A large amount of potassium in the waste sample may produce a positive result due to the natural presence of the radioactive isotope, Potassium-40.

6.4 The sensitivity of this test method to beta and gamma radiation may be dependent on sample volume. A small sample volume with readings near background levels may give a false negative result.

6.5 Some radioactive isotopes, such as ^{3}H and ^{14}C, may not emit radiation of sufficient energy to be detected. If suspected to be present in the waste, another procedure should be used that is appropriate to their determination.

6.6 Liquid samples, as well as moisture in solid samples, are good attenuators of radiation and will hinder detection of many radionuclides unless they emit high-energy gamma radiation. The possible inability to detect alpha particles and low-level beta emissions that may be attenuated, in many cases, should not be a serious shortcoming in this test method because these emissions are often accompanied by higher energy gamma emissions.

6.6.1 Moisture-laden Americium-241 bearing waste would be a case where there is a high probability of non-detection due to attenuation.

---

[1] This test method is under the jurisdiction of ASTM Committee D34 on Waste Management and is the direct responsibility of Subcommittee D34.01.05 on Screening Methods.

Current edition approved March 10, 1996. Published May 1996.

[2] *Annual Book of ASTM Standards*, Vol 12.01.

6.7 Survey meter contamination can cause artificially low or high results when reading standard sources or ambient background. If this occurs, the survey meter performance shall be considered unreliable and should be appropriately decontaminated by qualified personnel, or disposed of in accordance with applicable regulations.

6.8 The radiation monitor should be operated in accordance with the manufacturer's instructions.

7. Apparatus

7.1 *Radiation Survey Meter*, with a halogen-quenched uncompensated Geiger-Mueller tube with thin mica end-window. It is advisable to choose an instrument model that has an internal counter (cpm) or is capable of operating in a scaler mode.

NOTE 1—The meter used in the development of this test method was the Monitor 4.[3] A number of other survey meters[4] are suitable for this test method. Through the end-window, this unit is capable of detecting and indicating alpha radiation (down to 2.5 MeV with a typical detection efficiency of 80 % at 3.6 MeV), beta radiation (down to 150 KeV with a typical 75 % detection efficiency), gamma radiation and X-rays (down to 10 KeV), with the survey meter set on its lowest energy range of 0.1 mr/h.

8. Reagents and Materials

8.1 The check sources used in the development of this test method were manufactured by Oxford Instruments, Inc.[5] Other suitable check sources are available from a number of suppliers.[6] The check sources used in the development of this test method are federally exempt and do not require licensing. Sources other than the following (that is, Cesium 137) are appropriate for this test method. The sources, levels of activity, and half-lives used in the development of this test method are as follows:

Source	Type	Activity	Half-Life ($t_{1/2}$)
Polonium 210	alpha	0.1 µCi	138.4 days
Strontium 90	beta	0.1 µCi	28.6years
Cobalt 60	gamma	1.0 µCi	5.3years

8.2 Check sources should be replaced at least every three half lives.

9. Sample

9.1 Because stratification of liquid samples is possible, thoroughly mix the sample by shaking or stirring prior to testing. Strongly multiphasic samples should have each layer tested separately. Solid samples should be thoroughly mixed by a method appropriate to their size and physical characteristics.

10. Calibration and Standardization

10.1 Check the battery power source of the survey meter prior to use. If the battery check indicates that the battery charge is not in the optimal power range, it must be replaced prior to use.

10.2 Check the survey meter's performance at least daily using the α, β, and γ check sources. This is done by observing the survey meter response when held within 6 mm of each check standard. At a minimum, check sources should be >2 times the ambient background. The check sources listed in 9.1 far exceed this requirement.

10.3 The survey meter should be factory-calibrated at least yearly.

10.4 Assess the level of ambient background radiation periodically. This may be prior to each test, daily, weekly, or monthly depending on user requirements.

10.4.1 Set the survey meter to its X1 or most sensitive setting and enable the audible alarm, if unit is so equipped.

10.4.2 Record the number of audible meter clicks over a period of 60 s (counts per minute). Repeat this procedure two more times, and take the average of the three repetitions using the following formula:

$$\textit{Average ambient background radiation} = \frac{\textit{Measurement No. }1 + \textit{Measurement No. }2 + \textit{Measurement No. }3}{3}$$

10.5 Unexpectedly high readings for ambient background or standard sources infers that the survey meter is contaminated. See the section on Interferences.

10.6 The user must clearly define the criteria to be used for reporting "negative" or "positive" test results. These criteria may be project- or application-specific. A common defining point used is "two to three times the ambient background level."

11. Procedure

11.1 Perform the necessary steps presented in Section 10.

11.2 Allow the survey meter to stabilize for 10 to 15 s at its X1 or lowest appropriate setting.

11.3 Place the end-window of the survey meter within 6 mm of the sample. Depending on the physical characteristics of the sample container (for example, the container is such that the sample is >6 mm from the opening), it may be necessary to withdraw a representative portion of the sample to an alternate sample container, watch glass, or other appropriate vessel to allow the end-window to be placed within 6 mm of the sample. Do not allow the survey meter to come in direct contact with the sample. This may contaminate the meter and may damage the end-window.

11.4 Record the counts per minute for a period of 60 s (counts per minute).

11.5 If the counts per minute are above the user-defined criteria, then the sample may contain radioactive material.

NOTE 2—The user should recognize the potential interferences mentioned in Section 6.

11.6 If radioactivity levels above the user-defined criteria are observed, the background levels should be redetermined as described in Section 10. The sample should then be retested beginning with 11.2.

11.7 If radioactivity levels are confirmed to be above the user-defined criteria, sample testing should be discontinued, and the appropriate health and safety professional or site regulatory officer should be contacted for further instructions.

[3] Manufactured by SE International, Inc., Summertown, TN.

[4] For example, Five Level Portable Meter, manufactured by Harshaw-Bicron, Inc., Solon, OH; Ludlum Measurements, Inc., Sweetwater, TX; and Eberline Instrument Corp., Santa Fe, NM.

[5] Oxford Instruments, Inc., Oak Ridge, TN.

[6] For example, The Source, Inc., Santa Fe, NM; Ludlum Measurements, Inc., Sweetwater, TX; and Eberline Instrument Corp., Santa Fe, NM.

11.8 If radioactivity levels are below the user-defined criteria, the test result may be recorded as not detected.

12. Precision and Bias

12.1 *Precision*—No statement is made about the precision because this screening test method produces only a positive or negative test result.

12.2 *Bias*—No statement is made about the bias because this screening test method produces only a positive or negative test result.

13. Keywords

13.1 radioactivity; screening method; waste materials

Designation: D 5929 – 96

Standard Test Method for Determining Biodegradability of Materials Exposed to Municipal Solid Waste Composting Conditions by Compost Respirometry[1]

This standard is issued under the fixed designation D 5929; the number immediately following the designation indicates the year of original adoption or, in the case of revision, the year of last revision. A number in parentheses indicates the year of last reapproval. A superscript epsilon (ϵ) indicates an editorial change since the last revision or reapproval.

1. Scope

1.1 This test method covers the biodegradation properties of a material by reproducibly exposing materials to conditions typical of municipal solid waste (MSW) composting. A material is composted under controlled conditions using a synthetic compost matrix and determining the acclimation time, cumulative oxygen uptake, cumulative carbon dioxide production, and percent of theoretical biodegradation over the period of the test. This test method does not establish the suitability of the composted product for any use.

1.2 The values stated in both inch-pound and SI units are to be regarded separately as the standard. The values given in parentheses are for information only.

1.3 *This standard does not purport to address all of the safety concerns, if any, associated with its use. It is the responsibility of the user of this standard to establish appropriate safety and health practices and determine the applicability of regulatory limitations prior to use.*

2. Referenced Documents

2.1 *ASTM Standards:*

D 513 Test Methods for Total and Dissolved Carbon Dioxide in Water[2]

D 1129 Terminology Relating to Water[2]

D 1293 Test Methods for pH of Water[2]

D 2908 Practice for Measuring Volatile Organic Matter in Water by Aqueous-Injection Chromatography[3]

2.2 *APHA-AWWA-WEF Standard Methods:*

2540G Total, Fixed, and Volatile Solids in Solid and Semisolid Samples[4]

3. Terminology

3.1 *Definitions*—Definitions of terms applying to this test method appear in Terminology D 1129.

3.2 *Definitions of Terms Specific to This Standard:*

3.2.1 *acclimation time, n*—the time required for the oxygen uptake to reach 10 % of the total measured cumulative oxygen uptake.[5]

3.2.2 *oxygen uptake, n*—the cumulative oxygen consumed by the organisms during the test.

3.2.3 *theoretical carbon dioxide production (ThCDP), n*—the maximum carbon dioxide that can be produced by a material as calculated by the carbon content of the material.

3.2.4 *theoretical oxygen uptake (ThOU), n*— the maximum oxygen consumption required to fully oxidize a material based on the elemental content of the material.

3.2.5 *virgin newsprint*—nonprinted newspaper roll stock.

4. Summary of Test Method

4.1 This test method consists of the following:

4.1.1 The samples are prepared by cutting or forming the material into the form it would most likely be seen in the waste stream. A theoretical maximum carbon dioxide production and oxygen uptake are determined from an elemental analysis.

4.1.2 An inoculum is obtained from a municipal MSW or yard waste compost facility. It is procured from a static pile that has been composting for at least two months.

4.1.3 The synthetic MSW is prepared from virgin newsprint, pine bark or wood chips, corn starch, corn oil, bovine casein, and urea. A buffer/dilution water is prepared from magnesium, calcium, iron and a phosphate buffer.

4.1.4 The test material, synthetic compost, inoculum, and dilution water are combined and placed in a highly insulated reactor which monitors oxygen consumption and temperature and captures all evolved carbon dioxide.

4.1.5 The system is monitored, and oxygen uptake rates, temperature profiles, and total carbon dioxide produced are recorded.

4.1.6 The total oxygen uptake and carbon dioxide produced are compared with the theoretical values obtained from the elemental analysis, and a percent of biodegradation is generated. Possible negative effects of the material are evaluated by observing the acclimation time of the synthetic MSW and evaluating the oxygen uptake rate.

[1] This test method is under the jurisdiction of ASTM Committee D-34 on Waste Management and is the direct responsibility of Subcommittee D34.03 on Treatment.

Current edition approved March 10, 1996. Published May 1996.

[2] *Annual Book of ASTM Standards*, Vol 11.01.

[3] *Annual Book of ASTM Standards*, Vol 11.02.

[4] Available from American Public Health Assoc., 1015 15th Street, NW, Washington, DC 20005, Standard Methods for the Examination of Water and Waste Water, 18th ed., 1992.

[5] Tabak, Henry H. and Lewis, Ronald F., *CEC/OECD Ring Test of Respiration Method for Determination of Biodegradability*, U. S. Environmental Protection Agency, pp. 1–3.

5. Significance and Use

5.1 As the crisis in solid waste continues to grow, MSW composting is increasingly being considered as one component in the overall solid waste management strategy. The volume reduction achieved by composting, combined with the production of a usable end product, is resulting in increasing numbers of municipalities analyzing and selecting MSW composting as an alternative to incineration or to reduce reliance on landfill disposal. This test method will help determine the effect of materials on the compost process and establish if the material can be properly disposed through solid waste composting facilities.[6]

5.2 This test method attempts to provide a simulation of the overall compost process while maintaining reproducibility. Exposing the test material with several other types of materials that are typically in MSW provides an environment which provides the key characteristics of composting: material not in a sole carbon source environment which allows co-metabolism, compost system is self heating, and provides a direct measurement of organism respiration.

6. Apparatus

6.1 *Compost Respirometry Apparatus (see* Fig. 1*)*:

6.1.1 A minimum of six reactors, 2 to 6-L volume, with the test material in triplicate and the controls in triplicate. The reactors should be surrounded with efficient insulation to minimize heat loss and be gastight. Insulation should be 8 cm of urethane foam or equivalent.

6.1.2 *Tubing*, with high resistance to gas permeation.

6.1.3 *Peristaltic Pump*, to control and maintain gas flow through each reactor.

6.1.4 *4-L Scrubber Vessel*, for each reactor fitted with a scrubber solution sampling port.

6.1.5 *Differential Pressure Switch*, for each reactor that actuates between 2 and 5 in. (51 and 127 mm) of water.

6.1.6 *Solenoid and Mass Flowmeter*, to control and measure the addition of pure (99.997 +) oxygen to system.

6.1.7 *Temperature Probe*, situated in the middle of the compost.

6.1.8 *Data Acquisition and Control System*, for the measurement of temperature and the control and measurement of the oxygen addition.

6.2 *Miscellaneous*:

6.2.1 *Temperature Control Room*, or hood to maintain the external temperature of the apparatus at 40°C.

6.2.2 *Flow Meter*, to measure recirculation flow in each reactor (optional).

6.2.3 *Computer Control of Peristaltic Pump*, for automatic recirculation flow control (optional).

6.3 Suitable devices for the measurement of pH, dry solids (105°C), elemental analysis of material, carbon dioxide content of scrubbers, weight, and volume of the final compost material.

[6] Biocycle: Journal of Waste Recycling Staff, eds., *The Biocycle Guide to Composting Municipal Wastes*, JG Press, Inc., 1989.

7. Test Materials

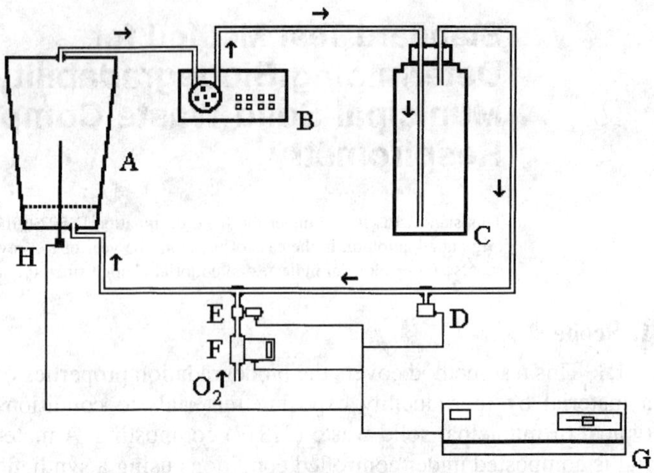

NOTE 1—The compost respirometer features a 4-L reactor vessel (A) insulated with 8 cm of urethane foam. The atmosphere is drawn through the reactor by a peristaltic pump (B) to maintain aeration. The effluent gases are passed through a 4-L scrubber vessel (C) containing 1.5 L of 5 *M* NaOH to remove any carbon dioxide from the effluent gas stream. Samples are drawn from this scrubber solution during the evaluation to determine the carbon dioxide released by the compost. As the microorganisms consume the oxygen in the system, a pressure drop occurs and is detected by a highly sensitive pressure switch (D). This signals the data acquisition and control system (G) and the oxygen is replaced with pure bottled oxygen by a solenoid (E) and the amount added is measured by a mass flowmeter (F). The gasses are then returned to the reactor. A thermocouple (H) is centered in the test reactor to monitor the temperature of the compost. The system is sealed to prevent interference from barometric fluctuations.

FIG. 1 Compost Respirometer Functional Diagram

7.1 The test materials can be in any form as long as it's dimensions do not exceed 3 by 3 by 12 cm. The test materials should be in the form that they would be seen in the waste stream. A representative sample must be obtained by using appropriate ASTM methods or other documented method.

7.2 Analyze the test materials for carbon, hydrogen, nitrogen, oxygen, phosphorus, sulfur, and any other elements that are suspected to be present at a level to effect oxygen uptake. The ThOU must be calculated for each material.

7.3 Calculate the ThCDP from the carbon content of the test material.

7.4 The nitrogen content of the synthetic MSW should be adjusted if the C/N ratio is greater than 40:1. This is accomplished by adjusting the urea content of the synthetic MSW. The synthetic MSW has adequate nitrogen to support the addition of up to 35 g of carbon before the ratio exceeds 40:1. If the urea content is adjusted, all reactors including controls must contain the same concentration of urea.

8. Reagents and Materials

8.1 *Scrubber Solution*, containing 3.25 *N* NaOH in distilled water. Store in a gas-tight plastic container. Add 30 mg of phenolphthalein to the solution to indicate scrubber exhaustion.

8.2 *Dilution/Buffer Solution*, containing the following:

Chemical	Purpose	Concentration, g/L
KH_2PO_4	phosphate buffer	6.8
$Na_2HPO_4 \cdot 7H_2O$	phosphate buffer	55.6
$MgSO_4 \cdot 7H_2O$	nutrient	0.0225
$CaCl_2$	nutrient	0.0275
$FeCl_3 \cdot 6H_2O$	nutrient	0.00025

8.3 *Synthetic Municipal Solid Waste*, containing the following:[7,8]

Constituent	Chemical Used	Dry Weight, %
Cellulosics	shredded, virgin newsprint	41
Inerts	pine bark or wood chips	39
Carbohydrates	corn starch	5.2
Lipids	corn oil	5.4
Proteins	bovine casein	2.0
Organic nitrogen	urea	1.4
Buffer/Nutrient	as listed	5.8

8.4 *Polyethylene*, or another nonbiodegradable material is the negative control material. It should be in the same form as the test materials to provide the same physical conditions in all reactors. The synthetic MSW acts as a positive control to verify the viability of the inoculum, see 13.4 for requirements.

9. Hazards

9.1 This test method requires the use of hazardous chemicals. Avoid contact with the chemicals and follow the manufacturer's instructions and Material Safety Data Sheets.

9.2 This test method does not address all of the health and safety issues related to it's use. It is the responsibility of the user to establish appropriate safety measures.

9.3 High-purity high-pressure gases can be dangerous if not handled correctly. Follow all safety precautions and monitor the system often to ensure proper operation.

10. Inoculum

10.1 The inoculum should be obtained from MSW or yard waste that has properly composted for two to four months. The compost should be screened with a <3-mm screen.

10.2 The compost can be stored at room temperature for up to 48 h before use. It should not be allowed to dry.

11. Procedure

11.1 This procedure is for twelve 4-L reactors with 4-L scrubber vessels. Other configurations will need to adjust weights and volumes to maintain proportional liquid:solid ratios of components.

11.2 *Synthetic Municipal Solid Waste*:

11.2.1 *Dilution Water*— Weigh out the ingredients for 3600 mL of dilution water. This will make enough dilution water for 13 reactors:

Compound	Quantity per Reactor	Per 3600 mL
KH_2PO_4	1.87 g	24.5 g
$Na_2HPO_4 \cdot 7H_2O$	15.29 g	200 g
$MgSO_4$	0.003 g	0.039 g
$CaCl_22H_2O$	0.0076 g	0.099 g
$FeCl_3$	0.0002 g	0.003 g
Urea	4.0 g	52 g

11.2.2 *Dry Ingredient Preparation*:

11.2.2.1 Weigh out 120 g of shredded virgin newsprint and place in a sealed plastic bag. Mark each bag with the actual weight of newsprint.

11.2.2.2 Weigh out 115 g of 2.5 by 2.5 by 0.6 cm (approximately) fresh wood chips or pine bark into a 1-L plastic beaker with 15.1 g of corn starch and 5.95 g of bovine casein. Repeat until twelve sets of dry ingredients are prepared.

11.2.3 *Inoculum Preparation*:

11.2.3.1 Obtain approximately 1 kg of mature compost from a municipal MSW or yard waste compost facility.

11.2.3.2 Screen compost with 3-mm wire mesh screen and retain the <3-mm portion that is used as the inoculum.

11.2.3.3 Weigh out 12 g of inoculum into each of 12 weighing trays.

11.3 *Sample Preparation*:

11.3.1 Determine the dry solids of the test materials and obtain the elemental analysis. Calculate the amount of test material required to provide 50 g of ThOU.

11.3.2 Prepare the control samples by using polyethylene as the material and form or cut it into the same physical size and shape as the test material.

11.4 *Reactor Loading*:

11.4.1 Mix the shredded newsprint and the dry ingredients and add the dilution water. Thoroughly mix until there are no clumps of paper or chemical.

11.4.2 Add 15.8 g of corn oil by dispensing directly to the mixture. Mix the ingredients until the oil is evenly distributed.

11.4.3 Add control or test product and mix until products are evenly distributed.

11.4.4 Add inoculum and thoroughly mix into compost. Load into reactor, taking care not to compact the compost mixture.

11.5 *Scrubber Preparation*:

11.5.1 Fill the scrubber vessels with 1.5 L of 3.25 *N* NaOH solution.

11.5.2 Add 30 mg of phenolphthalein indicator to the scrubber solution.

11.5.3 Seal the scrubber vessel to minimize atmospheric carbon dioxide absorption.

11.6 *Run Startup*:

11.6.1 Assemble test reactor system and allow system to reach ambient temperature and stabilize.

11.6.2 Sample the scrubber vessels and analyze for carbon dioxide content by using Test Method D 513 or other suitable method.

11.6.3 Set the recirculation pump to 100 mL/min or at a level that will allow adequate aeration without excessive heat loss. If the test reactor temperatures exceed 60°C increase the aeration to keep the reactor temperature below 65°C.

11.7 *Run Operation*:

11.7.1 The reactors are operated for a period of 45 days. If the environment control option is implemented, maintain ambient temperature at 40°C.

11.7.2 Set the computer data acquisition system to collect

[7] Clark, C. S., et al., "Laboratory Scale Composting: Techniques," *Journal of the Environmental Engineering Division-ASCE*, October, 1977.

[8] U. S. Environmental Protection Agency, Office of Solid Waste and Emergency Response, *Characterization of Municipal Solid Waste in the United States, 1960 to 2000 (update 1988)*, EPA/530-SW-88-033, March 1988.

and store the temperature and oxygen uptake data a minimum of once every 2 h.

11.7.3 To prevent channeling, shake the reactors vigorously once a week.

11.7.4 There must be NO leaks in the system. The tubing and fittings must be checked thoroughly at the beginning of the test and after each shaking with a soap solution. Any leaks must be corrected immediately. The test apparatus should be monitored daily to ensure proper operation.

11.8 *End of Run*:

11.8.1 Sample the scrubber vessels in duplicate and analyze for carbon dioxide content.

11.8.2 Determine the total weight of the compost material and determine the dry solids concentration.

11.8.3 Measure the pH by diluting 10 g of sample with 50 mL of distilled water (refer to Test Methods D 1293). Shake and measure pH immediately. If the pH of the compost is below 7, determine the volatile fatty acids (VFA) content of the compost (see Practice D 2908). If the VFA is >2g/kg, the reactors have soured and the results are invalid.

12. Calculation

12.1 Calculate the theoretical oxygen uptake from the percent by weight of each of the elements:

$$ThOU\ (\text{mg } O_2/\text{mg}) = C/37.5 + H/12.5 + N/29.2 + P/48.4 + S/66.7 - O/100 \quad (1)$$

12.2 The theoretical carbon dioxide production is also determined from the elemental analysis and is calculated as follows:

$$ThCDP\ (\text{mg } CO_2/\text{mg}) = (\%\ \text{by}\ weight\ carbon)\ (3.667) \quad (2)$$

12.3 Calculate the oxygen uptake due to the test material by subtracting the mean of the control reactors from each of the test reactors at each data point.

12.4 Calculate the total carbon dioxide produced by subtracting the mean of the control analytical results from each of the test reactors.

12.5 Calculate the percent of theoretical of the total oxygen uptake and carbon dioxide by dividing the measured totals by the calculated ThOU or ThCDP and multiplying by 100.

12.6 The acclimation time for the test material and compost matrix can be determined at the point which the oxygen uptake exceeded 10 % of the total oxygen uptake for that reactor.

12.7 If the test materials oxygen uptake curve shows a negative slope at the beginning of the curve, the test material acclimation time is calculated at the point were the slope of the oxygen uptake curve becomes positive for a minimum of 24 h.

13. Interpretation of Results

13.1 The acclimation time can provide valuable information on the toxicity or inhibition effects of the test material.

13.2 The percent of theoretical oxygen uptake is an excellent measure of the biodegradability of the test material since it represents the mineralization of carbon in addition to all other oxidative reactions.

13.3 The percent of theoretical carbon dioxide production can be used to measure biodegradability (with limitations) and can be used to correlate results with other carbon dioxide based test methods.

13.4 To ensure an active and viable inoculum, the total oxygen uptake for the control reactors should exceed 80 g. If this is not observed over the 45 days then the test must be regarded as invalid and should be repeated with new inoculum.

14. Report

14.1 Report the following information and data with standard deviations and 95 % confidence intervals when appropriate:

14.1.1 Information on source, dry solids, storage conditions, and age of the inoculum,

14.1.2 The ThOU and ThCDP of test materials with elemental analysis results,

14.1.3 Test material physical characterization including size, shape, weight, dry weight, initial and final reactor observations, and sampling technique used,

14.1.4 Percent of ThOU and ThCDP for each test material,

14.1.5 Total oxygen uptake and oxygen uptake versus time plots for entire reactors and test materials only. Rate plots if applicable,

14.1.6 Acclimation times for test materials,

14.1.7 Total carbon dioxide produced and carbon dioxide versus time plots if multiple samples were analyzed during the test,

14.1.8 Temperature plots of each reactor,

14.1.9 Total weight loss of reactor contents and dry weight of end reactor contents,

14.1.10 Final pH of compost, investigative results if pH is below seven, and

14.1.11 Full description of test apparatus and documentation of any variations from this test method.

15. Precision and Bias

15.1 The precision and bias of this test method is being determined.

15.2 Typical oxygen uptake and temperature plots are presented in Fig. 2 and Fig. 3. Within-test reproducibility from three different test runs are presented in Table 1. Run 1 is the control reactors with pine bark, Run 2 is with fresh wood chips, and Run 3 is with previously used wood chips.

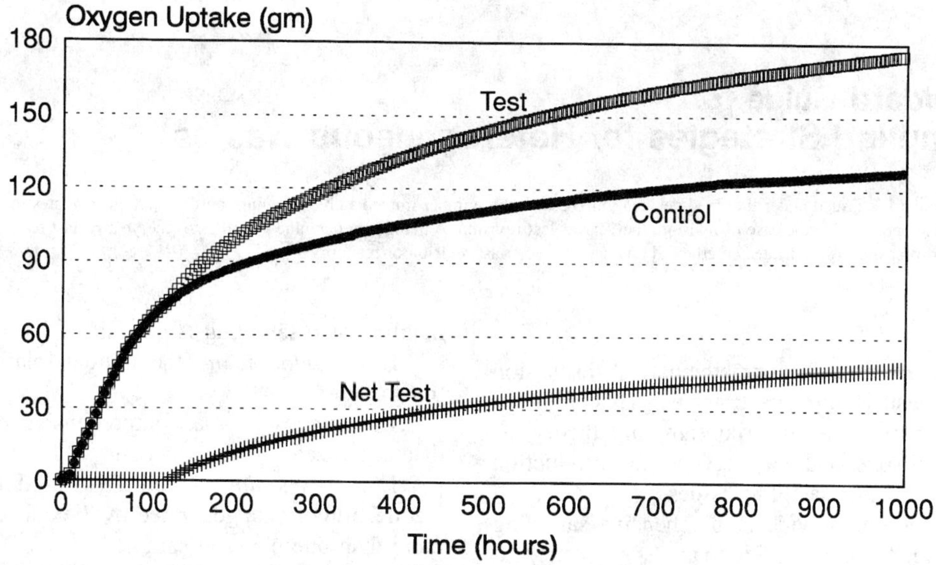

FIG. 2 Typical Control Reactor, Test Reactor, and Net Test Reactor

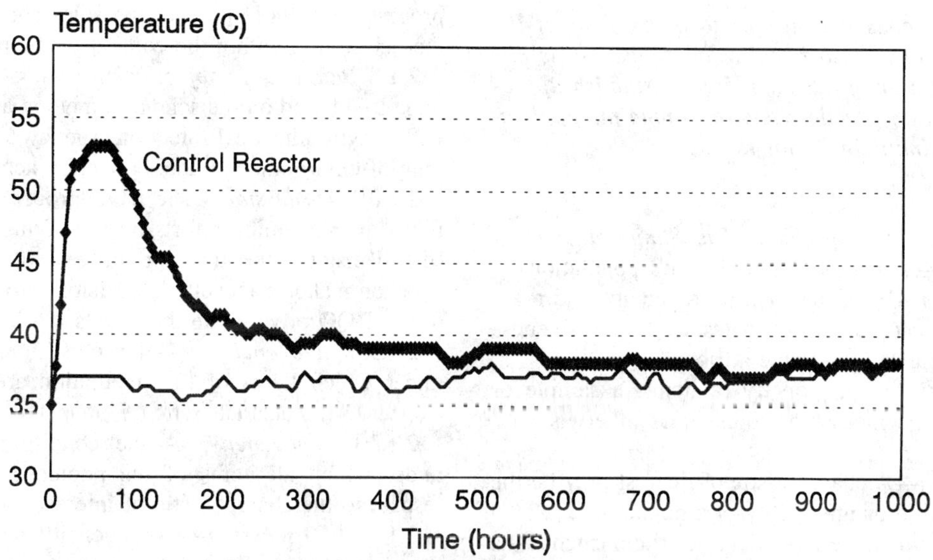

FIG. 3 Typical Temperature Plots, Ambient Set at 37°C

TABLE 1 Within-Test Reproducibility for Three Synthetic MSW Compositions

	O_2 Uptake, g	Standard Deviation, %	CO_2 Produced, g	Standard Deviation,%
Run 1, pine bark	131	2.4	111	6.2
Run 2, new chips	134	0.82	115	4.0
Run 3, used chips	115	0.87	91.4	3.7

Designation: D 5956 – 96 (Reapproved 2001)

Standard Guide for Sampling Strategies for Heterogeneous Wastes[1]

This standard is issued under the fixed designation D 5956; the number immediately following the designation indicates the year of original adoption or, in the case of revision, the year of last revision. A number in parentheses indicates the year of last reapproval. A superscript epsilon (ε) indicates an editorial change since the last revision or reapproval.

1. Scope

1.1 This guide is a practical, nonmathematical discussion for heterogeneous waste sampling strategies. This guide is consistent with the particulate material sampling theory, as well as inferential statistics, and may serve as an introduction to the statistical treatment of sampling issues.

1.2 This guide does not provide comprehensive sampling procedures, nor does it serve as a guide to any specification. It is the responsibility of the user to ensure appropriate procedures are used.

1.3 *This standard does not purport to address all of the safety concerns, if any, associated with its use. It is the responsibility of the user of this standard to establish appropriate safety and health practices and determine the applicability of regulatory limitations prior to use.*

2. Terminology

2.1 *Definitions of Terms Specific to This Standard:*

2.1.1 *attribute, n*—a quality of samples or a population.

2.1.1.1 *Discussion*—Homogeneity, heterogeneity, and practical homogeneity are population attributes. Representativeness and intersample variance are sample attributes.

2.1.2 *characteristic, n*—a property of items, a sample or population that can be measured, counted, or otherwise observed.

2.1.2.1 *Discussion*—A characteristic of interest may be the cadmium concentration or ignitability of a population.

2.1.3 *component, n*—an easily identified item such as a large crystal, an agglomerate, rod, container, block, glove, piece of wood, or concrete.

2.1.4 *composite sample, n*—a combination of two or more samples.

2.1.4.1 *Discussion*—When compositing samples to detect hot spots or whenever there may be a reason to determine which of the component samples that constitute the composite are the source of the detected contaminant, it can be helpful to composite only portions of the component samples. The remainders of the component samples then can be archived for future reference and analysis. This approach is particularly helpful when sampling is expensive, hazardous, or difficult.

2.1.5 *correlation, n*—the mutual relation of two or more things.

2.1.6 *database, n*—a comprehensive collection of related data organized for quick access.

2.1.6.1 *Discussion*—Database as used in this guide refers to a collection of data generated by the collection and analysis of more than one physical sample.

2.1.7 *data quality objectives (DQO), n*— DQOs are qualitative and quantitative statements derived from the DQO process describing the decision rules and the uncertainties of the decision(s) within the context of the problem(s).

2.1.8 *data quality objective process, n*— a quality management tool based on the scientific method and developed by the U.S. Environmental Protection Agency to facilitate the planning of environmental data collection activities.

2.1.8.1 *Discussion*—The DQO process enables planners to focus their planning efforts by specifying the use of the data (the decision), the decision criteria (action level) and the decision maker's acceptable decision error rates. The products of the DQO process are the DQOs.

2.1.9 *heterogeneity, n*—the condition of the population under which items of the population are not identical with respect to the characteristic of interest.

2.1.10 *homogeneity, n*—the condition of the population under which all items of the population are identical with respect to the characteristic of interest.

2.1.10.1 *Discussion*—Homogeneity is a word that has more than one meaning. In statistics, a population may be considered homogeneous when it has one distribution (for example, if the concentration of lead varies between the different items that constitute a population and the varying concentrations can be described by a single distribution and mean value, then the population would be considered homogeneous). A population containing different strata would not have a single distribution throughout, and in statistics, may be considered to be heterogeneous. The terms *homogeneity* and *heterogeneity* as used in this guide, however, reflect the understanding more common to chemists, geologists, and engineers. The terms are used as described in the previous definitions and refer to the similarity or dissimilarity of items that constitute the population. According to this guide, a population that has dissimilar items would be considered heterogeneous regardless of the type of distribution.

[1] This guide is under the jurisdiction of ASTM Committee D34 on Waste Management and is the direct responsibility of Subcommittee D34.01.01 on Planning for Sampling.

Current edition approved Oct. 10, 1996. Published December 1996.

2.1.11 *item, n*—a distinct part of a population (for example, microscopic particles, macroscopic particles, and 20-ft long steel beams).

2.1.11.1 *Discussion*—The term *component* defines a subset of items. Components are those items that are easily identified as being different from the remainder of items that constitute the population. The identification of components may facilitate the stratification and sampling of a highly stratified population when the presence of the characteristic of interest is correlated with a specific component.

2.1.12 *population, n*—the totality of items or units under consideration.

2.1.13 *practical homogeneity, n*—the condition of the population under which all items of the population are not identical. For the characteristic of interest, however, the differences between individual physical samples are not measurable or significant relative to project objectives.

2.1.13.1 *Discussion*—For practical purposes, the population is homogeneous.

2.1.14 *random, n*—lack of order or patterns in a population whose items have an equal probability of occurring.

2.1.14.1 *Discussion*—The word *random* is used in two different contexts in this guide. In relation to sampling, random means that all items of a population have an equal probability of being sampled. In relation to the distribution of a population characteristic, random means that the characteristic has an equal probability of occurring in any and all items of the population.

2.1.15 *representative sample, n*—a sample collected in such a manner that it reflects one or more characteristics of interest (as defined by the project objectives) of a population from which it was collected.

2.1.15.1 *Discussion*—A representative sample can be (*1*) a single sample, (*2*) a set of samples, or (*3*) one or more composite samples.

2.1.16 *sample, n*—a portion of material that is taken for testing or for record purposes.

2.1.16.1 *Discussion*—Sample is a term with numerous meanings. The scientist collecting physical samples (for example, from a landfill, drum, or waste pipe) or analyzing samples, considers a sample to be that unit of the population collected and placed in a container. In statistics, a sample is considered to be a subset of the population, and this subset may consist of one or more physical samples. To minimize confusion the term *physical sample* is a reference to the sample held in a sample container or that portion of the population that is subjected to in situ measurements. One or more physical samples, *discrete samples*, or aliquots are combined to form a *composite sample*. The term *sample size* has more than one meaning and may mean different things to the scientist and the statistician. To avoid confusion, terms such as sample mass or sample volume and number of samples are used instead of sample size.

2.1.17 *sample variance, n*—a measure of the dispersion of a set of results. Variance is the sum of the squares of the individual deviations from the sample mean divided by one less than the number of results involved. It may be expressed as $s^2 = \Sigma (x_i - \bar{x})^2/(n - 1)$.

2.1.18 *sampling, n*—obtaining a portion of the material concerned.

2.1.19 *stratum, n*—a subgroup of a population separated in space or time, or both, from the remainder of the population, being internally consistent with respect to a target constituent or property of interest, and different from adjacent portions of the population.

2.1.19.1 *Discussion*—A landfill may display spatially separated strata since old cells may contain different wastes than new cells. A waste pipe may discharge temporally separated strata if night-shift production varies from the day shift. Also, a waste may have a contaminant of interest associated with a particular component in the population, such as lead exclusively associated with a certain particle size.

2.1.19.2 *Discussion*—Highly stratified populations consist of such a large number of strata that it is not practical or effective to employ conventional sampling approaches, nor would the mean concentration of a highly stratified population be a useful predictor (that is, the level of uncertainty is too great) for an individual subset that may be subjected to evaluation, handling, storage, treatment, or disposal. *Highly stratified* is a relative term used to identify certain types of nonrandom heterogeneous populations. Classifying a population according to its level of stratification is relative to the persons planning and performing the sampling, their experience, available equipment, budgets, and sampling objectives. Under one set of circumstances a population could be considered highly stratified, while under a different context the same population may be considered stratified.

2.1.19.3 *Discussion*—The terms *stratum* and *strata* are used in two different contexts in this guide. In relation to the population of interest, *stratum* refers to the actual subgroup of the population (for example, a single truck load of lead-acid batteries dumped in the northeast corner of a landfill cell). In relation to sampling, *stratum* or *strata* refers to the subgroups or divisions of the population as assigned by the sampling team. When assigning sampling strata, the sampling team should maximize the correlation between the boundaries of the assigned sampling strata and the actual strata that exist within the population. To minimize confusion in this guide, those strata assigned by the sampling team will be referred to as *sampling strata*.

3. Significance and Use

3.1 This guide is suitable for sampling heterogeneous wastes.

3.2 The focus of this guidance is on wastes; however, the approach described in this guide may be applicable to non-waste populations, as well.

3.3 Sections 4-9 describe a guide for the sampling of heterogeneous waste according to project objectives. Appendix X1 describes an application of the guide to heterogeneous wastes. The user is strongly advised to read Annex A1 prior to reading and employing Sections 4-9 of this guide.

3.4 Annex A1 contains an introductory discussion of heterogeneity, stratification, and the relationship of samples and populations.

3.5 This guide is intended for those who manage, design, or

implement sampling and analytical plans for the characterization of heterogeneous wastes.

4. Sampling Difficulties

4.1 There are numerous difficulties that can complicate efforts to sample a population. These difficulties can be classified into four general categories:

4.1.1 Population access problems making it difficult to sample all or portions of the population;

4.1.2 Sample collection difficulties due to physical properties of the population (for example, unwieldy large items or high viscosity);

4.1.3 Planning difficulties caused by insufficient knowledge regarding population size, heterogeneity of the contaminant of interest, or item size, or a combination thereof; and,

4.1.4 Budget problems that prevent implementation of a workable, but too costly, sampling design.

4.2 The difficulties included in the first three categories are a function of the physical properties of the population being sampled. The last sampling difficulty category is a function of budget restraints that dictate a less-costly sampling approach that often results in a reduced number of samples and a reduced certainty in the estimates of population characteristics. Budget restraints can make it difficult to balance costs with the levels of confidence needed in decision making. These difficulties may be resolved by changing the objectives or sampling/analytical plans since population attributes or physical properties of the population can seldom be altered. Documents on DQOs discuss a process for balancing budgets with needed levels of confidence.

4.3 Population access and sample collection difficulties often are obvious, and therefore, more likely either to be addressed or the resulting limitations well-documented. A field notebook is likely to describe difficulties in collecting large items or the fact that the center of a waste pile could not be accessed.

4.4 Population size, heterogeneity, and item size have a substantial impact on sampling. The cost and difficulty of accurately sampling a population usually is correlated with the knowledge of these population attributes and characteristics. The least understood population attribute is heterogeneity of the characteristic of interest. If heterogeneity is not known through process knowledge, then some level of preliminary sampling or field analysis is often required prior to sampling design.

4.5 Sampling of any population may be difficult. However, with all other variables being the same, nonrandom heterogeneous populations are usually more difficult to sample. The increased difficulty in sampling nonrandom heterogeneous populations is due to the existence of unidentified or numerous strata, or both. If the existence of strata are not considered when sampling a nonrandom heterogeneous population, the resulting data will average the measured characteristics of the individual strata over the entire population. If the different strata are relatively similar in composition, then the mean characteristic of the population may be a good predictor for portions of the population and will often allow the project-specific objectives to be achieved. As the difference in composition between different strata increases, average population characteristics become less useful in predicting composition or properties of individual portions of the population. In this latter case, when possible, it is advantageous to sample the individual strata separately, and if an overall average of a population characteristic is needed, it can be calculated mathematically using the weighted averages of the sampling stratum means **(7)**.

5. Stratification

5.1 Strata can be thought of as different portions of a population, which may be separated in time or space with each portion having internally similar concentrations or properties, which are different from adjacent portions of the population (that is, concentrations/properties are correlated with space, time, component, or source). Fig. 1 is a graphical depiction of different types of strata.

5.1.1 A landfill may display spatially separated strata since old cells may contain different wastes than new cells (stratification over space);

5.1.2 A waste pipe may discharge temporally separated strata if night-shift production varies from the day shift (stratification over time);

5.1.3 Lead-acid batteries will constitute a strata separate from commingled soil if lead is the characteristic of interest (stratification by component); and,

5.1.4 Drums from an inorganic process may constitute a different strata from those co-disposed drums generated by an organic process (a subtype of stratification by component referred to as stratification by source).

5.2 Different strata often are generated by different processes or a significant variant of the same process. The different origins of the strata usually result in a different concentration distribution and mean concentration.

5.3 Highly stratified populations, a type of nonrandom heterogeneous populations, have so many strata that they become difficult to sample and characterize. Classifying a population according to its level of stratification is a relative issue pertaining to the persons planning and performing the sampling, their experience, available equipment, and budgets. Highly stratified populations are such that it is not practical or effective to employ conventional sampling approaches to generate a representative database, nor would the mean concentration of a highly stratified population be a useful predictor (that is, the level of uncertainty is too great) for an individual subset that may be subjected to evaluation, handling, storage, treatment, or disposal.

NOTE 1—An example of a highly stratified population is a landfill, a candidate for remediation, that is contaminated with the pure and very viscous Aroclor 1260 and with solutions containing varying concentrations of Aroclor 1260. (Aroclor 1260 is viscous and can exist as globules of the pure Aroclor.) The detected concentration of Aroclors in analytical subsamples would reflect a highly stratified population if some samples contained globules of pure 1260, while other samples contained soils that came in contact with solvents containing varying concentrations of 1260. Highly nonrandom heterogeneous populations have numerous strata, each of which contain different distributions of contaminants or item sizes, or both, such that an average value for the population would not be useful in predicting the composition or properties of individual portions of the waste (that is, statistically speaking, the variance and standard error of the mean will be large).

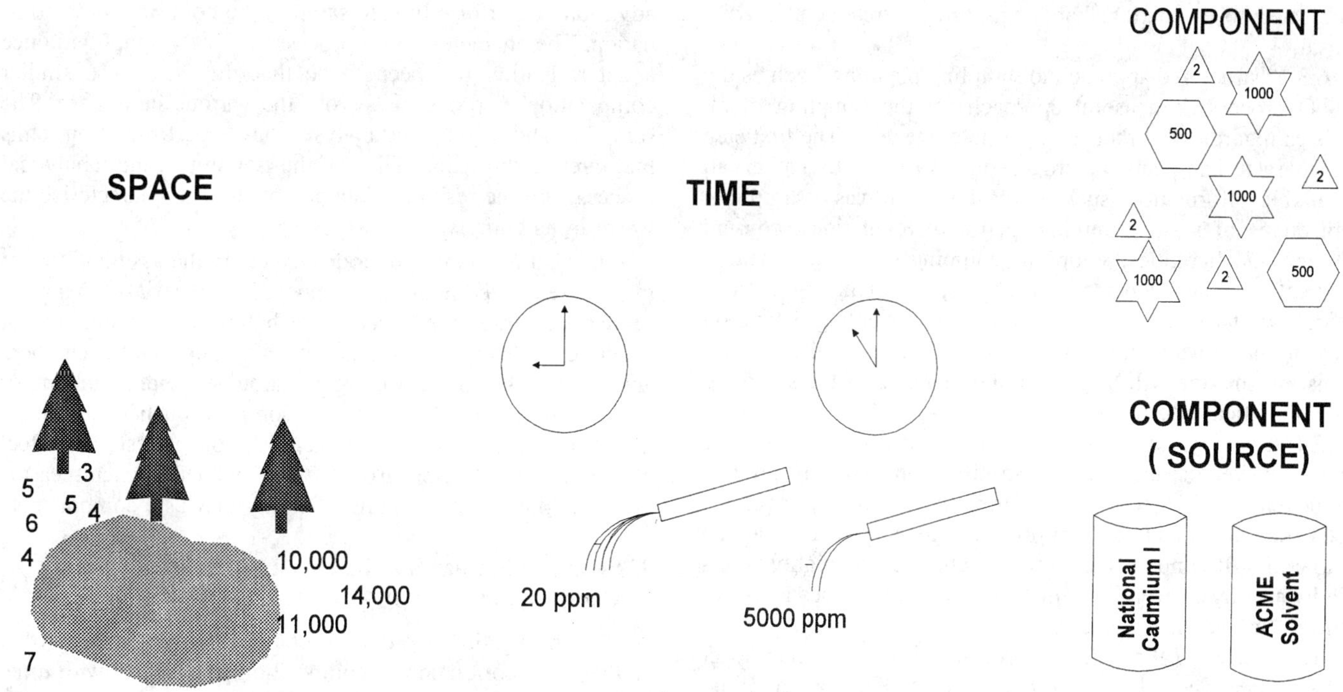

FIG. 1 Types of Stratified Heterogeneous Wastes

A second and more visually obvious example of a highly stratified population would be a landfill that is filled with unconfined sludge, building debris, laboratory packs, automobile parts, and contained liquids with the constituent of interest having different concentrations in each strata.

5.4 Certain populations do not display any obvious temporal or spatial stratification, yet the distribution of the target characteristic is excessively erratic. For these populations it may be helpful to consider stratification of the population by component. Stratification by component is applied to populations that contain easily identifiable items, such as large crystals or agglomerates, rods, blocks, gloves, pieces of wood, or concrete. Separating a population into sampling strata according to components is useful when a specific kind of component is distributed within the population and when a characteristic of interest is correlated with the component. Stratification by source (for example, organic process waste drums versus inorganic process waste drums) is a type of component stratification. Stratification by component is an important mechanism for understanding the properties of component-heterogeneous populations and for designing appropriate sampling and analytical efforts.

5.4.1 Component strata are not necessarily separated in time or space but are usually intermixed and the properties or composition of the individual components are the basis of stratification. For example, automobile batteries that are mixed in an unrelated waste would be a component that could constitute an individual strata if lead was a target characteristic. If one were to sequester the batteries, they would have a consistent distribution that was different from the rest of the waste.

5.4.2 There is usually no purpose in stratifying by component if different components have similar concentrations of the target characteristic or if the components are small enough such that the different components are represented in the chosen sample size. Even when components have similar composition, however, stratification and use of separate sampling strategies by component may be useful when the different components are so physically different that they cannot all be sampled with the same technique.

5.4.3 A primary objective for employing a stratified sampling strategy is to improve the precision of population parameters such as population means by dividing the population into homogeneous strata. The precision of the population parameters will increase as the sampling strata boundaries, chosen by the sampling team, more closely overlay the actual physical strata that exist within the population.

6. Sampling of Highly Stratified Heterogeneous Wastes

6.1 Sections 6-9 focus on the sampling of highly stratified wastes, a type of heterogeneous waste. It is strongly advised that Annex A1 be read and studied prior to the use of this guide. Annex A1 discusses heterogeneity and the relationship between samples and populations.

6.2 Nonrandom heterogeneous wastes contain two or more strata. Stratification of a waste does not always complicate the sampling process; at times, could simplify sampling. Highly stratified populations, however, contain such a large number of strata that they become difficult to sample and characterize. Use of the word *highly* and the classification of wastes according to their level of stratification is a relative issue pertaining to the persons planning and performing the sampling, their experience, available equipment, budgets, and objectives. Highly stratified wastes are such that it is not practical or effective to employ conventional sampling approaches, nor would the mean concentration of a highly stratified waste be a useful predictor (that is, the level of uncertainty is too great) for an individual subset that may be subjected to evaluation, handling, storage, treatment, or disposal.

6.3 A structured approach to sampling planning, such as the DQO process, is a useful approach for the sampling of all wastes regardless of their level of heterogeneity. The first step in characterizing any heterogeneous waste is to gather all available information, such as the need for waste sampling; objectives of waste sampling; pertinent regulations, consent orders, and liabilities; sampling, shipping, laboratory, health, and safety issues; generation, handling, treatment, and storage of the waste; existing analytical data and exacting details on how it was generated; and treatment and disposal alternatives. This information will be used in the planning of the sampling and analytical effort.

6.4 If enough information is available, the planning process may uncover the existence of stratification that may prevent achievement of objectives. If information is lacking, a preliminary sampling/analytical effort may identify and evaluate variability. It is not cost-effective to characterize highly stratified waste by conventional methods, which becomes apparent during the planning process.

6.5 Sections 7-9 consider approaches that lessen the impact of stratification and allow for more cost-effective sampling. Some of these approaches require changes in objectives, waste handling or disposal methods, and some require compromises, but all approaches require the above types of information.

6.6 Heterogeneity is a necessary condition for the existence of strata. Wastes can be heterogeneous in particle size or in composition, or both, allowing for the existence of the following:

6.6.1 Strata of different-sized items of similar composition,

6.6.2 Strata of similar-sized items of different composition, and,

6.6.3 Strata of different-sized items and different composition.

7. Strata of Different-Sized Items With Similar Composition

7.1 Wastes having stratification due only to different-sized items will by definition have the same composition or property (that is, for compositional characteristics there is no significant intersample variance and no correlation with space, time, or component) throughout its different strata. The different-sized items may be separated in space or in time. Unless one is attempting to measure particle size for which there is significant intersample variance, this type of population is the simplest of the highly stratified waste types to characterize. All items in these types of wastes usually are generated by the same process (for example, the discussion of silver nitrate powder and crystals in Annex A1), which is the reason for similar composition across all item sizes. These types of wastes, which are compositionally homogeneous and only heterogeneous in item size, are not commonly encountered.

7.2 The complexity of dealing with these types of wastes is in proving that the waste has similar composition across the varying item sizes. This determination can be made by using process knowledge or by sampling the different-sized items to determine if there are significant compositional differences. If the determination is made using knowledge of the waste, it is advisable to perform limited sampling to confirm the determination. The characterization process is greatly simplified once a determination has been made that the waste has similar composition or properties across the various item sizes. The sampling and subsequent analysis can be performed on items that are readily amenable to the sampling and analytical process, and the resulting data can be used to characterize the waste in its entirety.

7.3 It is important to periodically verify the assumption that the different-sized items are composed of materials having the same concentration levels and distributions of the contaminant of interest. This verification is especially important when there are any changes to the waste generation, storage, treatment, or disposal processes. Similarity of composition between items has to be verified for each characteristic of interest. The effect of different-sized items also must be considered when measuring properties, such as the leachability of waste components.

8. Strata of Similar-Sized Items and Different Composition

8.1 Stratification due only to composition or property (that is, there is a correlation of composition or property with time, space, or component) by definition necessitates that item sizes will be consistent across different strata. The strata may be separable in space, time, or by component or source. Identifying and sampling the individual strata may simplify the characterization process. An example of this waste type is a long-term accumulation of wastewater sludge produced by the processing of materials having different composition, through the same waste-generation process (that is, batch-processing that results in waste having uniform item size but different composition from batch to batch).

8.2 Wastes having uniform item size and different composition or properties can be sampled using the same strategy as described for waste containing strata having different composition and different item size (see Section 9).

9. Strata of Different-Sized Items and Different Composition

9.1 Wastes having excessive stratification due to both composition/property and item size (that is, particle size and composition or property, or both, are correlated with time, space, or component) are usually the most difficult wastes to characterize. The difficulty in sampling highly stratified waste can result from:

9.1.1 Various item sizes and waste consistency that makes sampling difficult and conventional sampling approaches cost prohibitive;

9.1.2 Extraordinary concentration gradients between different components or innumerable strata that lead to such excessive variance in the data, that project objectives cannot be achieved; and,

9.1.3 Wastes that exhibit the properties in 9.1.1 and 9.1.2.

9.2 Fig. 2 summarizes an approach to characterizing these types of highly stratified wastes. If a waste is highly stratified, conventional methods of sampling will not allow objectives to be achieved cost-effectively. To sample cost-effectively a

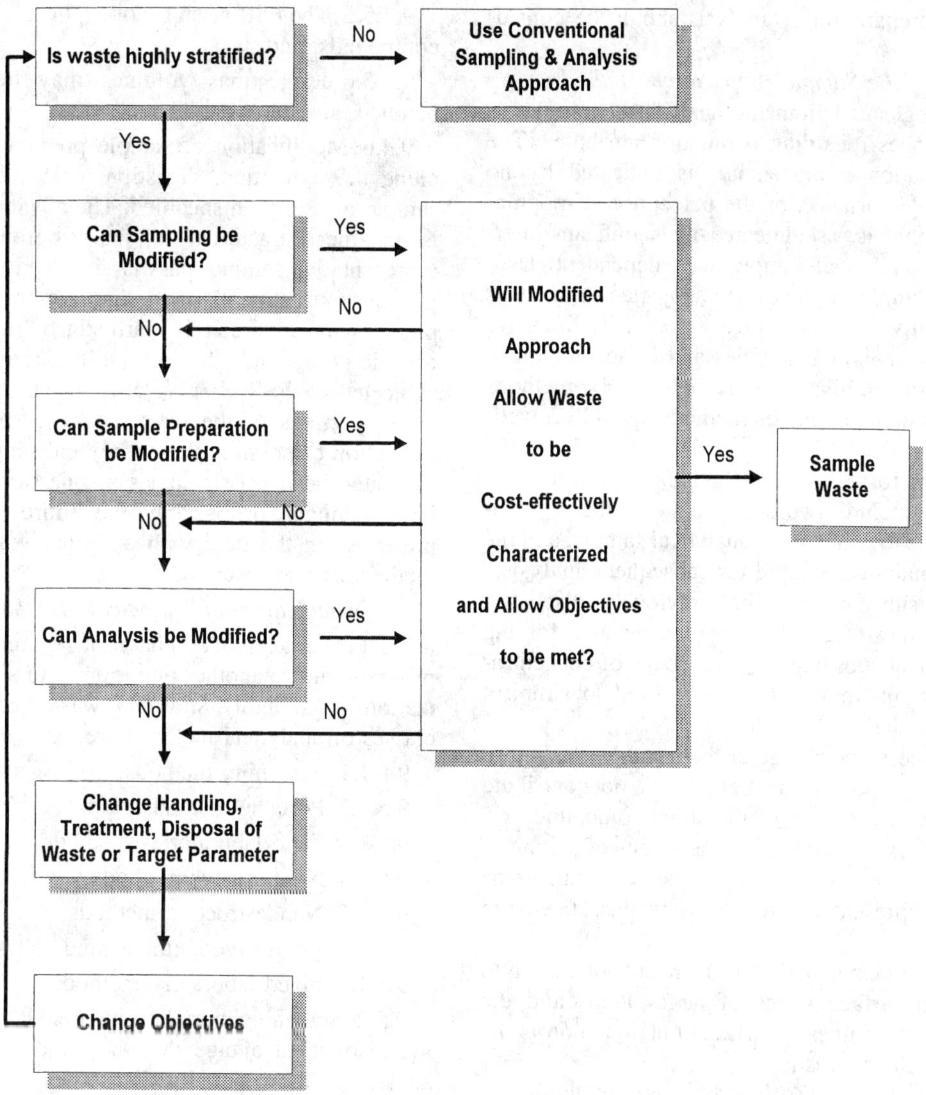

FIG. 2 Approach for the Characterization of Heterogeneous Wastes

highly stratified waste, one must use a nonconventional approach, such as modification of the sampling, sample preparation, or analytical phase of the process. If after modifying the sampling and analysis, the objectives still cannot be achieved in a cost-effective manner, then the original plan of waste handling, treatment, or disposal has to be examined and changed so the waste can be characterized according to new and achievable objectives.

9.3 *Design of the Sampling Approach*:

9.3.1 The first efforts to resolve the difficulty in characterizing a highly stratified waste are focused usually on sampling. A strategy for designing a sampling plan for such highly stratified waste may include the following five steps:

9.3.1.1 Use a planning process such as the DQO process to identify the target characteristics, the population boundaries, the statistic of interest, confidence levels, and other critical issues.

9.3.1.2 Determine whether characteristics of interest are correlated with item size, space, time, components, or sources.

9.3.1.3 Determine if any waste components or strata can be eliminated from consideration during sampling because they do not contribute significantly to the target characteristic.

9.3.1.4 Determine if small items in a stratum represent the stratum, as well as large more difficult to sample items. If yes, sample the smaller items, and only track the volume/mass contribution of the larger items.

9.3.1.5 Determine if the target characteristic is innate or surface adsorbed. Is the target characteristic surface adsorbed, which would allow the material to be sampled representatively by wipe sampling? Can large items be wiped and smaller items extracted, leached, or digested? Can waste be stratified according to impervious and nonimpervious waste and sampled and analyzed accordingly?

9.3.1.6 It is essential that all assumptions (that is, any correlations) be verified at least by knowledge of the waste, and preferably confirmed by sampling and analysis.

9.3.2 All steps taken to optimize sampling should be well-documented.

9.3.3 Appendix X1 contains a case study that applies the above process for optimizing sampling to highly stratified waste. If optimization of sampling design is not sufficient by itself to allow the project objectives to be met cost-effectively, changes to sample preparation or analysis should be considered.

9.4 *Modification of the Sample Preparation Method*:

9.4.1 Information gleaned from the analysis of samples is used to make inferences regarding population attributes. The perception of population homogeneity, as indicated by no significant intersample variance, or the perception of population heterogeneity (that is, as indicated by significant intersample variance) is analytical sample-mass dependent. Usually, the larger the sample mass/volume subjected to analysis the more representative the analytical sample. To improve representativeness of analytical samples and to accommodate large-sized items, conventional sample preparatory methods can be altered. All modifications of methods should be well-documented.

9.4.2 In the laboratory, the term *sample preparation* is commonly meant to include two separate steps: the subsampling of a field sample to generate an analytical sample, and the preparation of the analytical sample for subsequent analysis.

9.4.3 Regarding subsampling, the previously discussed logic for field sampling (see 9.3) is applicable also for the generation of analytical subsamples. Knowledge of concentration distributions within the waste can be used to simplify subsampling by considering the following:

9.4.3.1 Using process knowledge or the results of testing to eliminate any waste components or strata that do not contribute significantly to the concentration of the target compound;

9.4.3.2 Using process knowledge or the results of testing to discriminate against large items, and only select small items when small items represent the waste, as well as the large items; and,

9.4.3.3 Using process knowledge or the results of testing to restrict sampling to surface wipes of larger items and the extraction or digestion of fines if surface contamination is the source of the target characteristic.

9.4.4 If the approaches in 9.4.3.1-9.4.3.3 are not applicable to a field sample, the field sample will have to be subjected to particle size reduction (PSR) prior to subsampling or the sample preparation method will have to be modified to accommodate the entire field sample.

NOTE 2—Prior to modifying a sample preparatory method, it is advisable to consult the end user of the data to see if modifications could have any adverse affects. For example, PSR could dramatically alter leaching data.

9.4.5 The PSR is useful for handling field samples, which have items too large for proper representation in an analytical subsample. The intent of PSR is to decrease the maximum item size of the field sample so that the field sample then can be split or subsampled, or both, to generate a representative subsample. The difficulties in applying PSR to waste samples are the following:

9.4.5.1 Not all materials are easily amenable to PSR (for example, stainless steel artifacts);

9.4.5.2 Adequate PSR capabilities and capacities do not exist in all laboratories;

9.4.5.3 The PSR can change the properties of material (for example, leachability);

9.4.5.4 The PSR can be a source of cross-contamination;

9.4.5.5 The PSR often is not applicable to volatile and labile compounds; and,

9.4.5.6 Large mass/volumes may have to be shipped, handled, and disposed.

9.4.6 Modification of sample preparative methods can include the extraction, digestion, or leaching of much larger sample masses than specified. The advantage of this approach is that the characteristic of interest from a larger and more representative sample mass is dissolved into a relative homogeneous extract or digestate that is more suitable for subsampling. This approach is particularly important for volatile organic compounds that may suffer from substantial losses if subjected to PSR. For volatile organic compound analysis, larger portions of the wastes can be subjected to methanol extraction or possibly the entire field sample can be subjected to heated headspace analysis as one sample or as a series of large aliquots, or possibly the entire field sample can be preserved in the field with an equal volume of methanol or methanol/water solution.

9.5 *Modification of Analytical Method*:

9.5.1 The analytical phase of a sampling and analytical program allows another opportunity to simplify the characterization of a highly stratified waste. Examples of different classes of analytical methods are:

9.5.1.1 Screening methods,

9.5.1.2 Portable methods,

9.5.1.3 Field laboratories methods,

9.5.1.4 Nonintrusive methods,

9.5.1.5 Nondestructive methods,

9.5.1.6 Innovative methods, and

9.5.1.7 Fixed laboratory methods.

9.5.2 Screening, portable, and field laboratory methods have the distinct advantage that they allow for the cost-effective analysis of more samples. These methods generate more data, making it easier to detect correlations between concentration levels and waste strata or components. Also, some screening methods may analyze a larger sample volume than what is traditionally analyzed in a fixed laboratory.

9.5.3 Nonintrusive methods (for example, geophysical methods) can be useful when there are health and safety issues regarding exposure to the waste. These methods also may be used to evaluate large-volume wastes qualitatively or semi-quantitatively.

9.5.4 Nondestructive methods are useful in that the integrity of the samples is maintained for additional analyses or evidence, or both.

9.5.5 Innovative methods may provide more cost-effective or timely results or improve sensitivity or accommodate larger and more representative sample sizes.

9.5.6 Fixed laboratory methods usually have the advantage of regulatory approval, established quality assurance/quality control requirements and often greater sensitivity than that achievable by screening, portable, or field laboratory methods.

9.6 *Modification of the Waste Handling, Treatment, Disposal Plan*:

9.6.1 If modifications to sampling, sample preparation, and

analysis are not appropriate for a given waste, or are appropriate but still do not allow the objectives to be met cost-effectively, then the reasoning behind the original program must be reconsidered. It may be possible to achieve the program objectives by means of an alternative approach. For example, a change in waste treatment, handling, or disposal technologies may require analysis for different characteristics or may allow for simplified sampling. Alternatively, the waste population could be defined differently by employing smaller remediation or exposure units that would be sampled separately as opposed to characterizing the entire population. The need behind the waste characterization objectives has to be examined and an approach for simplifying the characterization process devised. This process is addressed in the optimization step of the planning process.

9.6.2 For example, consider a hypothetical waste that must be evaluated prior to waste disposal to determine if it is hazardous. An initial attempt to characterize the waste failed to meet the objective, indicated that the waste was highly stratified, and proved that portions of the waste are hazardous. After reviewing this preliminary information and the costs to attempt a defensible characterization of the waste, it could be decided that it is resource and cost-effective to consider all the waste hazardous and treat it as a hazardous waste by incineration. Under this scenario, the sampling and analytical requirements change, requiring simplified testing for general characteristics prior to incineration, and more comprehensive analysis of the less heterogeneous and more easily sampled incinerator ash to determine if it is within compliance.

9.7 *Changing Objectives*—If the project objectives are not met and none of the strategies can be changed or modified, the objectives need to be reconsidered. After changing the objectives, the sampling and analysis plans also should be adjusted. These iterations will continue until the project objectives can be met.

10. Keywords

10.1 analysis; heterogeneity; homogeneity; nonrandom; populations; random; sample preparation; samples; sampling; strata; stratified; stratum

ANNEX

(Mandatory Information)

A1. DISCUSSION OF HETEROGENEITY AND STRATIFICATION OF WASTES AND RELATIONSHIP OF SAMPLES AND POPULATIONS

A1.1 *Introduction*—This annex contains a practical non-mathematical discussion of issues pertinent to heterogeneous waste sampling. The discussion deals with heterogeneity, stratification, and the relationship of samples and populations in sampling design. It is consistent with sampling theory and statistics and may serve as an introduction to the statistical treatment of sampling issues (see Refs **1-10**).[2] The content of this annex is applicable to the sampling of wastes regardless of their degree of heterogeneity.

A1.2 *Population Attributes:*

A1.2.1 A population is the total collection of items to be studied. Theoretically, the classification of a population as being homogeneous or heterogeneous is straightforward. If all of the items in the population are identical, then the population is homogeneous. If one or more of the items are dissimilar, the population is heterogeneous. Theoretical homogeneity, the equivalent to nonheterogeneity, is a unique state of absolute uniformity for all items in the population while heterogeneity is a variable attribute that can range from a population, which is almost homogeneous (that is, homogeneous for applied purposes) to a population that displays dissimilarity between all items of the population.

A1.2.2 According to the theoretical definition for homogeneity, virtually all real-world populations would be heterogeneous. From a practical perspective, however, as the level of heterogeneity approaches the state of homogeneity, populations can be considered homogeneous for applied purposes. References to the homogeneity of a population are usually made in light of this applied meaning, that is, for practical purposes, the population is homogeneous (practical homogeneity).

A1.2.3 The attributes of homogeneity and heterogeneity are relative. Heterogeneity and homogeneity are a function of the specified chemical constituent, property, particle size, visual appearance, sampling objectives, and the sample mass/volume. The same population can be homogeneous with regards to one constituent or property, and at the same time be heterogeneous with regards to another constituent or property.

A1.2.3.1 Consider a nonrandom mixture of silver nitrate, some of which is a powder and the remainder is in the form of large crystals (see Fig. A1.1). The population is heterogeneous when considering particle size or homogeneous when silver content is of interest.

A1.2.3.2 Following comprehensive emission spectroscopic and titrimetric analyses of uranium metal, a chemist may find the population to be homogeneous while the nuclear chemist analyzing for U^{235} and U^{238} would find the same population to be isotopically heterogeneous (see Fig. A1.2).

A1.2.3.3 Decisions regarding heterogeneity also can be a function of the analytical method used to process samples. If one method (AAS-graphite furnace atomic absorption spectroscopy) is more sensitive and has method detection limits (MDL) that are lower than the other (X-ray fluorescence field screening), what was originally thought to be a homogeneous

[2] The boldface numbers in parentheses refer to the list of references at the end of this standard.

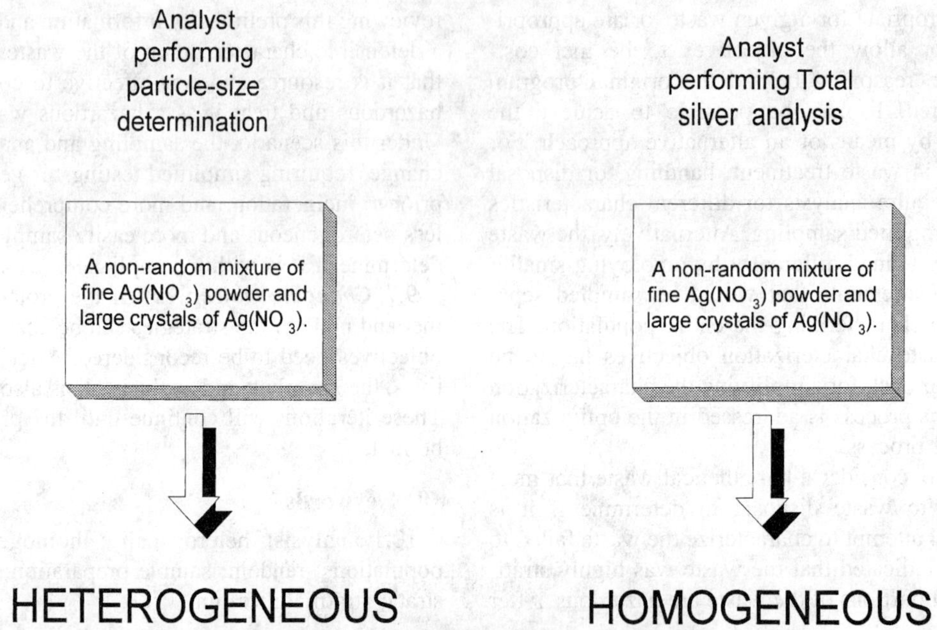

FIG. A1.1 Heterogeneity Relative to Objectives

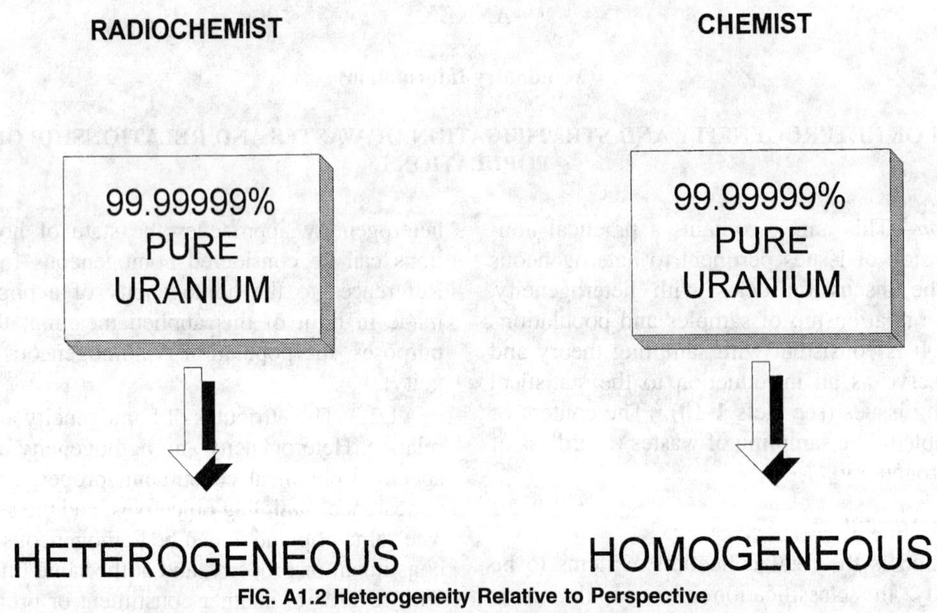

FIG. A1.2 Heterogeneity Relative to Perspective

waste may be found to be a heterogeneous waste (see Fig. A1.3).

A1.2.4 Two population attributes are the causative factors for heterogeneity. The primary attribute is referred to as compositional heterogeneity, and the secondary attribute is distributional heterogeneity.

A1.2.4.1 Compositional heterogeneity occurs when the concentration of the targeted constituent or targeted property varies from item to item. This compositional or property difference between items is a requisite for a heterogeneous population, that is, dissimilar items must be present for heterogeneity to exist.

A1.2.4.2 Distributional heterogeneity results from differences in the spatial distribution of dissimilar items resulting in microscopic or macroscopic concentration gradients or property gradients, or both. Compositional heterogeneity is a necessary condition for the existence of distributional heterogeneity. Distributional heterogeneity is a population attribute, and if a population is defined differently (that is, change the population boundaries), the distributional heterogeneity for the expanded or smaller population may differ.

A1.2.5 Compositional and distributional heterogeneity are the underlying causes for the more commonly understood types of random heterogeneity and nonrandom heterogeneity. Random and nonrandom are the terms that will be used in the remainder of this guide to describe the different types of heterogeneity. The introduction of compositional and distributional heterogeneity is to assist those who may want to

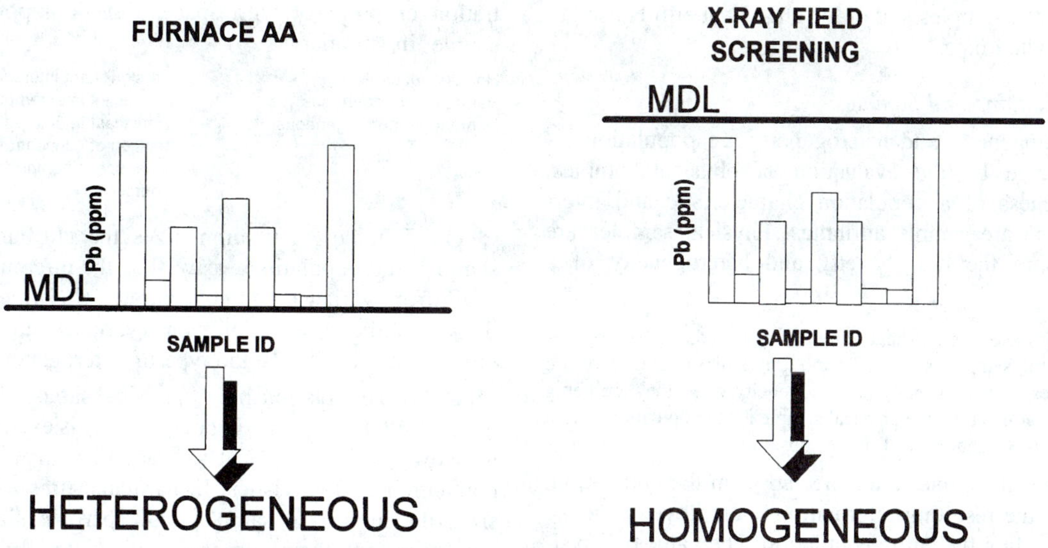

FIG. A1.3 Heterogeneity Relative to Method Detection Limits (MDLs)

investigate further the particulate material sampling theory.

A1.2.5.1 Random heterogeneity is that type of heterogeneity that occurs when dissimilar items are randomly distributed throughout the population.

A1.2.5.2 Nonrandom heterogeneity is that type of heterogeneity that occurs when dissimilar items in the population are nonrandomly distributed. In a nonrandom heterogeneous population, similar items or similar concentrations are grouped into strata. This type of population, also is referred to as a *stratified population*. The terms *stratified population* and *nonrandom heterogeneous populations* are interchangeable. Strata are separated from other strata by time or space or correlated with different components or waste sources. This guidance focuses on sampling strategies for a particular type of stratified waste referred to as *highly stratified.*

A1.3 *Physical Sample Attributes:*

A1.3.1 To characterize a population, it must be subjected to evaluation. The population can be characterized with great certainty if all population elements are evaluated for the characteristic of interest. Populations, however, are usually so large that the entire population cannot be subjected to evaluation. Practically and economically it makes more sense to collect a number of samples and compile the analytical results in a database that is used to make inferences regarding the population.

A1.3.2 Due to the different meanings assigned to the term *sample* and to minimize confusion the term *physical sample* is used throughout this discussion. Physical sample is a reference to the sample held in a sample container or that portion of the population that is subjected to in situ measurements. The term *sample size* also can have different meanings. Although use of multi-word terms can appear wordy, to avoid confusion, specific terms such as sample mass, sample volume, and number of samples are used.

A1.3.3 The accuracy of inferences made to populations are dependent on how well the physical samples represent the population characteristic of interest. The term *representativeness* usually is associated with mean concentrations. Physical samples, however, also are used to measure other statistical parameters of the population, such as variance, trends, and proportions.

A1.3.4 Sampling of a theoretically homogeneous population always results in physical samples that represent the characteristics of the population, assuming that the sampling process itself does not introduce contamination or allows for selective loss of waste constituents. The lack of variance in a homogeneous population ensures all physical samples collected from the population are identical and representative of the population.

A1.3.5 The meaning of the term *representative sample* is susceptible to misinterpretation since it connotes a single sample. The difficulty in collecting a single physical sample that represents a population increases with increasing heterogeneity. When trying to represent a heterogeneous population, it is more appropriate to collect a number of physical samples. If the physical samples are collected according to a properly designed plan, the population is better represented by the characteristics associated with the entire set of physical samples. Such a set of physical samples would be referred to as a representative set of physical samples.

A1.3.6 To properly represent a characteristic of a heterogeneous population, more than one physical sample usually is required. Samples collected from a heterogeneous population will display intersample variance. Intersample variance measured between different physical samples results from the following:

A1.3.6.1 Differences in the composition of items between sampling locations;

A1.3.6.2 Differences in how these items are distributed throughout the population; and,

A1.3.6.3 Sampling and analytical errors that ideally will be minimal so that the true intersample variance can be measured.

A1.3.7 The intersample variance may be used to make inferences about the homogeneity or heterogeneity of the population. The accuracy of these inferences will be a function of the sampling design and the quality of the sampling efforts used to collect the samples and of the analytical efforts used to generate associated data.

A1.4 *Populations and Samples:*

A1.4.1 Homogeneity and heterogeneity are population attributes estimated by the evaluation of physical samples. Representativeness of a population characteristic and intersample variance are sample attributes. Physical samples are used to measure the homogeneity and heterogeneity of a population.

NOTE A1.1—If the entire physical sample is analyzed, the heterogeneity of the physical sample is not relevant. Physical samples only are assigned attributes of heterogeneity or homogeneity when they are being subsampled since at this time the physical sample is the population whose characteristics must be represented in the subsample.

A1.4.2 Physical samples are collected from the population, evaluated, and the resulting information is employed to make inferences regarding the entire population. The value of physical samples is related directly to how accurately they represent the population characteristics of interest. The value of the inferences about a population are only as good as the associated samples. To properly represent a population characteristic, sampling location, sample mass, sample collection methods, the number of physical samples and compositing of physical samples are controlled.

A1.4.3 In a nonrandom heterogeneous population, the concentration of a target constituent (for example, arsenic) or the degree to which a property (for example, ignitability) is expressed is correlated with time, space, component, or waste source. Conversely, the constituent or property displays no correlation with time, space, component, or waste source in a random heterogeneous population.

A1.4.4 Samples collected from nonrandom heterogeneous populations, therefore, display a correlation of constituent concentrations or properties with time, space, components, or waste source and less intersample variance when samples are collected from the same stratum. Samples collected from random heterogeneous populations display a significant amount of intersample variance but no correlation of concentration or property with space, time, component, or waste source. In summary:

Homogeneous	no significant intersample variance
Random heterogeneous	significant intersample variance
Nonrandom heterogeneous	significant intersample variance and correlation of concentration/property with time, space component, or waste source

A1.4.5 Table A1.1 summarizes the attributes of physical samples and populations, as well as the inferences that can be made from variance and concentration information. Fig. A1.4 illustrates the process of using variance and concentration information to classify the type of heterogeneity.

A1.4.6 The relationship of physical samples to a population is explored in the following example. This example is designed to show the role physical samples play in the evaluation of population characteristics. In particular, this example emphasizes the impact of sample mass, particle size, and sample collection on sample representativeness and the resulting inferences for population characteristics.

A1.4.7 The population consists of a 2-L waste container that has 1-g nuggets of cadmium randomly distributed throughout an otherwise homogeneous and cadmium-free matrix. The cadmium-free matrix has a substantially smaller particle size than that of the cadmium nuggets. The cadmium nuggets constitute 37 % of the waste on a weight basis. The waste is composed of dissimilar particles resulting, in compositional differences and allowing distributional differences within the population. It is assumed that after collection, the entire physical sample is analyzed for cadmium.

Characteristics of interest: cadmium concentrations
Statistical parameters of interest: mean and standard deviation

A1.4.7.1 The following information pertains to Sampling Design No. 1 (Fig. A1.5).

Physical samples mass: 0.1 g
Sampling locations chosen randomly
Number of samples: 10
Sample collection device: a small spatula
Cadmium data: all 10 samples had concentrations less than 0.2 mg/kg
Average <0.2 mg/kg ± 0 mg/kg (<0.2 mg/kg = Method Detection Limit, MDL)

TABLE A1.1 Population and Sample Attributes

Population Attribute	Sample Description	Sample Attribute	Inference
Homogeneous (theoretical homogeneity)	All samples contain only identical items.	No significant intersample variance. No correlation of concentration or properties with time, space, component, or waste source.	Samples are representative of a homogeneous population.
Practical homogeneity	All samples contain dissimilar items, but each sample contains similar proportions.	No significant intersample variance. No correlation of concentration or properties with space, time, component, or waste source.	Samples are representative of a homogeneous population.
Random heterogeneous	All samples contain dissimilar items, each sample has different proportions, but these proportions are not correlated with time, space, or components.	Significant intersample variance. No correlation of concentration or properties with space, time, component, or waste source.	Samples are representative of a random heterogeneous population.
Nonrandom heterogeneous (stratified)	All samples contain dissimilar items, each sample has different proportions and these proportions are correlated with time, space, components, or source.	Significant intersample variance. Correlation of concentration or properties with space, time, component, or waste source.	Samples are representative of a nonrandom heterogeneous population.

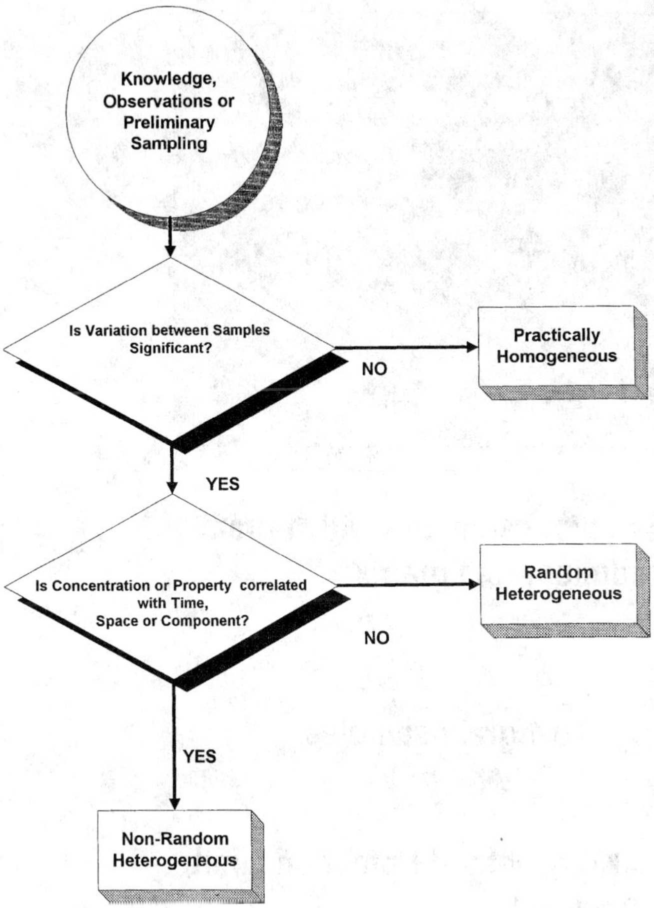

FIG. A1.4 Process for Classifying Type of Heterogeneity By Measurement or Process Knowledge

The lack of variance between physical samples (that is, no significant intersample variance) indicates falsely that the population is homogeneous with regards to cadmium. This is an incorrect evaluation since the physical samples are not representative of the population as a result of the sample collection method, which discriminated against the larger cadmium particles.

A1.4.7.2 The following information pertains to Sampling Design No. 2 (Fig. A1.6).

Physical samples mass: 1 g
Sampling locations chosen randomly
Number of samples: 10
Sample collection device: a spatula approximately 10 × larger than the one used in the previous example
Cadmium data: 100 %, <0.2 mg/kg, <0.2 mg/kg, <0.2 mg/kg, <0.2 mg/kg, 100 %, <0.2 mg/kg, <0.2 mg/kg, <0.2 mg/kg, <0.2 mg/kg (<0.2 mg/kg = Method Detection Limit, MDL)
Average: 20 ± 42 %

The variance between physical samples (that is, existence of significant intersample variance) indicates that the population is heterogeneous with regards to cadmium. Although more representative of the population characteristic than Sampling Design No. 1, this design also suffers from an sample collection error since the large cadmium particles were not collected unless they were aligned perfectly with the sampling device. Since only two samples had detected cadmium concentrations, there is a 25 % chance that these two samples could have occurred in the top half or the bottom half of the waste container (that is, there is a significant correlation of concentration to space). If this had occurred, the incorrect assumption may have been made that the population was nonrandomly heterogeneous (stratified) with a stratum of pure cadmium in half the container with the other half of the container consisting of cadmium-free material. These samples do not properly represent the population characteristic.

A1.4.7.3 The following information pertains to Sampling Design No. 3 (Fig. A1.7).

Physical samples mass: 30 g
Sampling locations chosen randomly
Number of samples: 10
Sample collection device: tube with a diameter that can easily accommodate a number of cadmium particles and can take a core from top to bottom
Cadmium data: 35 %, 50 %, 39 %, 24 %, 32 %, 47 %, 43 %, 27 %, 29 %, 44 %
Average: 37 ± 8.9 %

The sample mass required by this sampling design allowed for proper extraction and evaluation of the resulting physical samples yielding a database that was representative of the population (that is, there is significant intersample variance and that concentration is not correlated with space). The waste would be considered randomly heterogeneous with regards to cadmium.

A1.4.8 The previous three designs for sampling the same population showed how the perception of population heterogeneity can be affected by sampling design. The usefulness of using physical samples to make inferences regarding population heterogeneity varies according to the ability of the sampling device and the resulting sample mass to accommodate all the different-sized items of a population and the ability of all collected physical samples, as a set, to accommodate representative amounts of all constituents of the population.

A1.4.9 These previous sampling designs assumed that the entire physical sample was subjected to analysis. Practical experience indicates that most physical samples will be subjected to subsampling prior to analysis. Fig. A1.8 graphically depicts the common relationship between populations, physical samples, subsamples, and data. If subsampling is employed, then subsamples are the windows through which the population is viewed, and the subsamples will be used to make inferences including those regarding the homogeneity, random heterogeneity, or nonrandom heterogeneity of the population. Subsampling, when required, becomes an additional critical step that must be implemented properly to ensure the accuracy of inferences.

A1.5 *Population Attributes and Sampling Design:*

A1.5.1 The relationship between physical samples and populations clearly implies that knowledge of population attributes and use of this knowledge should decrease the bias and increase the precision of sampling. Table A1.2 and Fig. A1.9 respectively tabularize and depict the relationship between critical sampling design decisions and population attributes, planning information and specifications gleaned from planning processes such as the DQO process, and, analytical requirements.

A1.5.2 In addition to budget constraints, the following information, to the extent that it is known, should be considered during sampling design:

A1.5.2.1 *Population Attributes*:

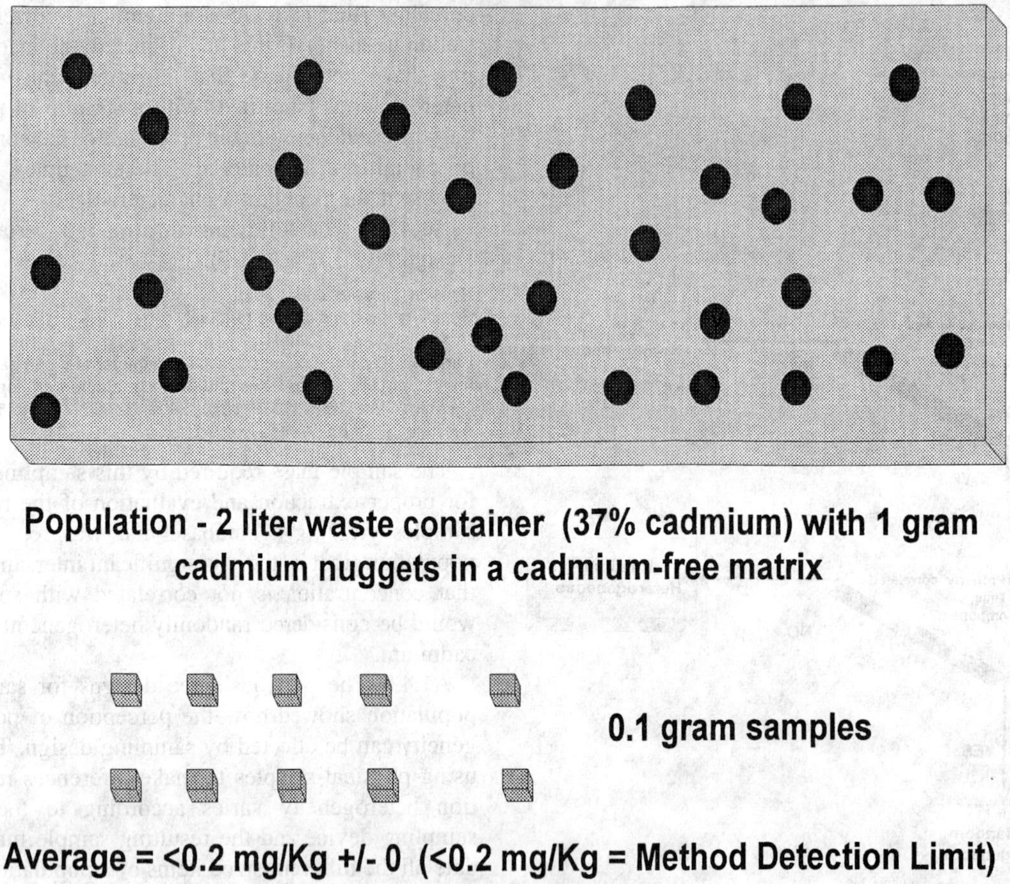

FIG. A1.5 Sampling Design No. 1

(1) Heterogeneity—Heterogeneity of the population in terms of the characteristic of interest; homogeneous, randomly heterogeneous, or nonrandomly heterogeneous (stratified).

(2) Item Size—The size of items present in the population including items that may or may not contain the characteristic of interest.

(3) Population Accessibility—The ability or inability to access all portions of the population for purposes of sampling.

A1.5.2.2 *DQOs*:

(1) Statistic—The mean, mode, variance, proportion, or other measure of a population which is of interest.

(2) Level of Confidence—The specified level of confidence that decisions will be correct. In other words, the maximum decision error rate that is acceptable to the decision maker.

(3) Boundaries—The temporal and spatial boundaries of the population that is to be studied.

A1.5.2.3 *Analytical Requirements*:

(1) Analytical Sample Volume/Mass—The sample volume/ mass needed to prepare and analyze physical samples.

(2) Analyte/Media Integrity—The handling, containerization, preservation, and shipping procedures required to maintain the physical, compositional, and legal integrity of the physical samples.

A1.5.3 *Sample Mass or Volume*—The appropriate sample mass or volume will be determined by considering the size of the largest items contained within the population, the heterogeneity of the population, and the optimum sample mass/ volume for preparation and analysis. Knowledge of item sizes contained within a population and their content of the characteristic of interest are needed to choose the correct sample mass/volume. Bias can result if certain item sizes are discriminated against during sampling. The correct sample mass/ volume will accommodate all item sizes or be chosen such that the impact of any discrimination is accounted for and understood. The variance of data caused by local heterogeneity of the population may be controlled by using a properly sized sampling device and by taking greater sample volumes or masses.

A1.5.4 *Sampling Locations*—Sampling locations are a function of the population boundaries, the accessibility of all portions of the population, and the type of heterogeneity. Other than background and other reference samples, sampling usually is restricted to those accessible areas within the population boundaries.

NOTE A1.2—Since sampling of inaccessible portions of the population is not possible, any extrapolation of sampling/analytical data to these areas must be well-documented. Extrapolation to unsampled areas is a judgment call and not a statistically valid inference. The type of heterogeneity may impact the sample locations since the existence or potential existence of strata may alter the sampling strategy for choosing sampling locations, for example, simple random versus stratified random.

A1.5.5 *Number of Samples*—The number of samples collected is determined after considering the population heterogeneity and information and specifications generated during the initial stages of the planning process, that is, the statistic of interest, the levels of uncertainty in decisionmaking and the

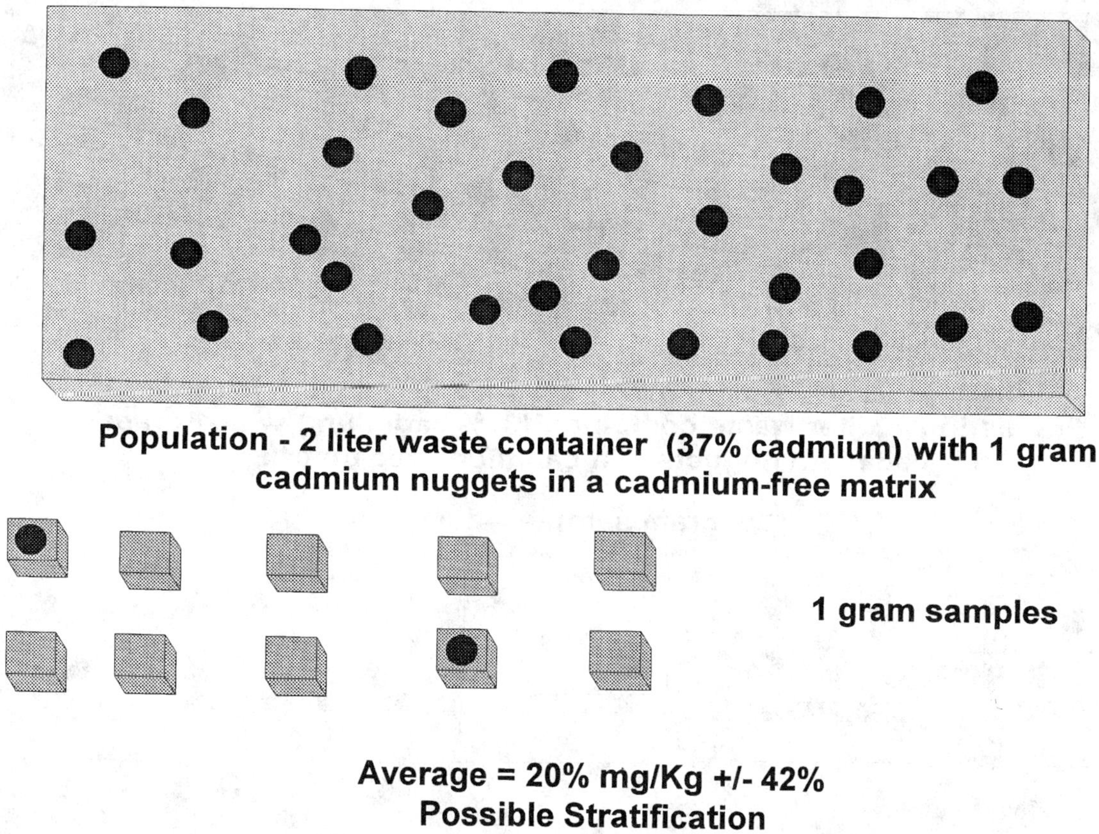

FIG. A1.6 Sampling Design No. 2

population boundaries. If a population is not substantially larger than the physical sample and the distribution of the characteristic of interest is randomly heterogeneous, it may be appropriate to collect a fewer number of samples by a random or systematic sampling procedure. If a population is relatively large as compared to the physical sample and the characteristic of interest is nonrandomly distributed (stratified), a greater number of samples and a stratified sampling approach may be needed to achieve similar levels of precision and bias and the specified confidence level.

A1.5.6 *Composite Versus Discrete*—A composite sample is made by combining one or more discrete samples (physical samples) into one sample. Compositing has the potential advantage of yielding a more accurate estimate of average concentrations or average properties. Compositing, however, has the potential disadvantage of losing pertinent variance information and the possibility of diluting hot spots such that the average results fall below thresholds or limits of detection. The chosen statistical approach for data evaluation, the acceptable level of uncertainty, the type of heterogeneity, and budgets are considered when deciding between the use of composite and discrete samples.

A1.5.7 *Sampling Devices*—The choice of sampling devices is made after determining the analytical-sample requirements, the size of the largest items that must be accommodated by the sampling device, the accessibility of sampling locations, sample integrity, and reactivity of sampling-device materials.

A1.5.8 Comprehensive knowledge of population attributes is infrequent and the degree of knowledge varies from population to population. However, the more thorough the planning process and the better the understanding of a population's attributes, the more likely that samples and associated data will be representative of the characteristic of interest.

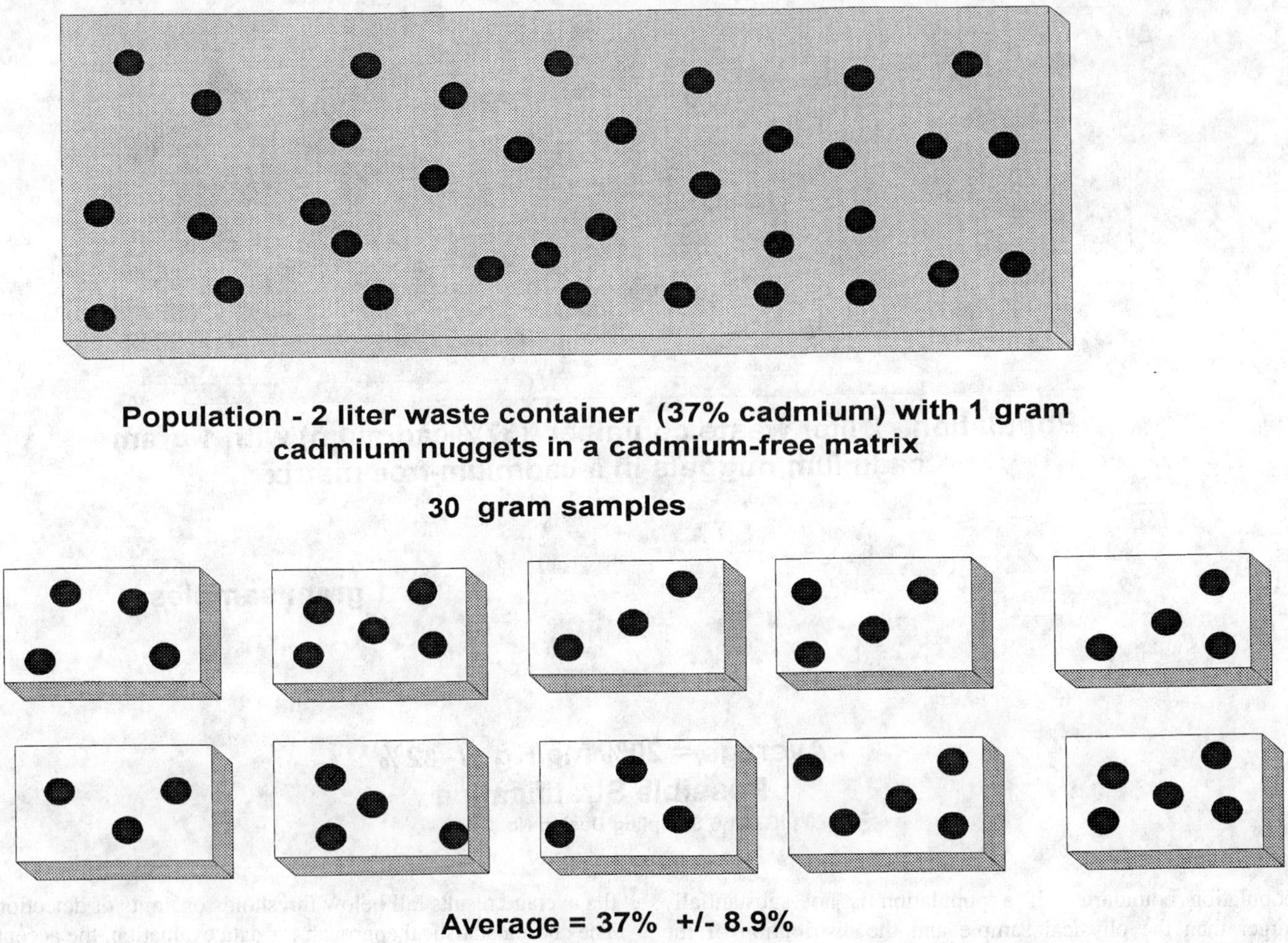

FIG. A1.7 Sampling Design No. 3

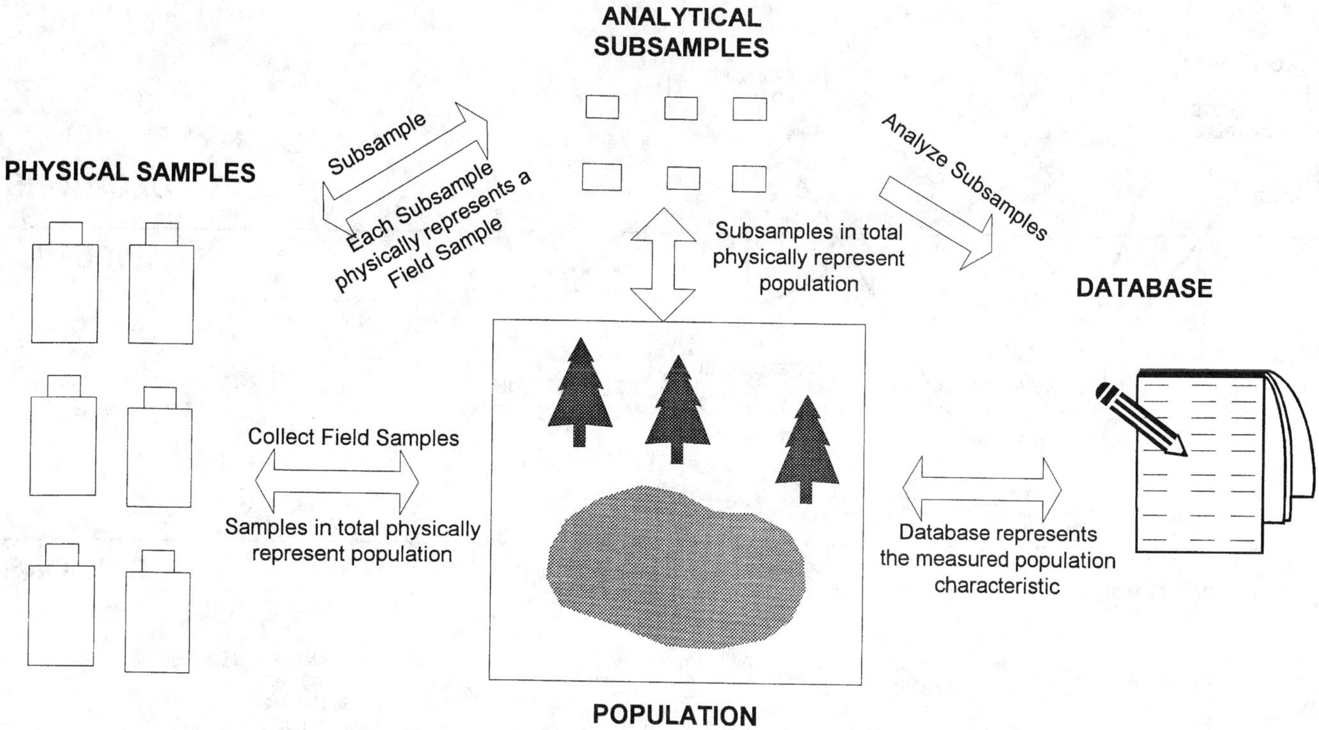

FIG. A1.8 Process of Using Physical Samples to Measure a Characteristic of the Population

TABLE A1.2 Role of Population Attributes, DQOs, and Analytical Requirements in Optimizing Sampling Designs[A]

	Inputs into Decision-Making Process						
Sampling Design Decisions	DQO Confidence Level	DQO Statistic	DQO Boundaries	Heterogeneity Type	Population Accessibility	Item Size	Analytical Requirements
Number of samples	X[B]	X	X	X			
Sampling location			X	X	X		
Sample mass/volume				X		X	X
Sampling device					X	X	X
Composite versus discrete	X	X		X			

[A] Regulatory requirements addressed during the DQO process may impact sampling design.
[B] X indicates that attribute or requirement will impact the design decision.

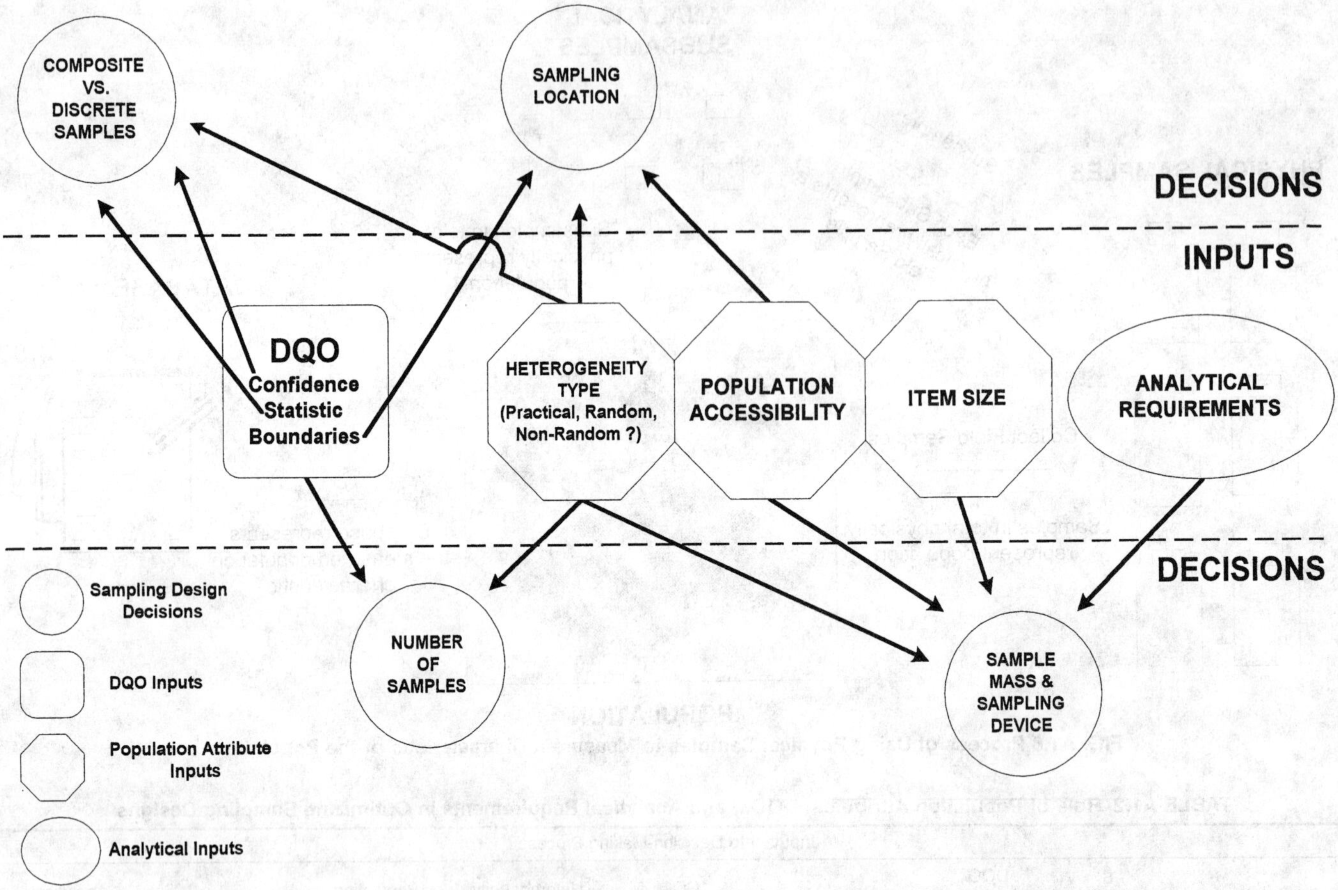

NOTE 1—Regulatory requirements addressed during the DQO process may impact sampling design.

FIG. A1.9 Role of Population Attributes, DQOs, and Analytical Requirements in Optimizing Sampling Designs

APPENDIX

(Nonmandatory Information)

X1. CASE STUDY FOR DESIGN OF A SAMPLING APPROACH FOR A HIGHLY STRATIFIED WASTE

X1.1 The following is a hypothetical scenario of how sample design can be optimized for highly stratified waste.

X1.1.1 A storage area contains 4000 drums of waste generated over a 15-year period. The drum contents are highly stratified and contain a myriad of wastes from process waste; destruction and construction debris such as wood, concrete; laboratory wastes including broken glassware, paper, or empty bottles. The initial stages of the planning process identified beryllium and solvents as the target characteristics and the mean and variance as the statistics of interest. In addition, groundwater modeling indicated that the storage area is the source of a plume contaminated with solvents and beryllium.

X1.2 *Sampling Design:*

X1.2.1 *What are the target characteristics?* Solvents and beryllium are the target characteristics identified during the planning process. The average and variance are the statistics of interest.

X1.2.2 *Are the target characteristics correlated with an identifiable strata or source?* The source of beryllium is traceable to one process, whose waste easily should be identifiable even when drum markings are not legible. The solvents are likewise traceable to a machine shop that disposed of its waste in easily identified drums. Testing will have to be performed to determine if there is any correlation with item size, space, time, or components in the waste.

X1.2.3 *Can any waste components or strata be eliminated from consideration during sampling?* Historical information indicated that 400 drums of construction debris were generated during construction of a new warehouse. The information indicates that the virgin nature of the materials may make these drums candidates for less intensive sampling or no sampling. Likewise, the source of beryllium contamination is a beryllium sludge that exists in drums by itself or in drums commingled with shredded packing material and laboratory wastes that were generated during physical testing of the beryllium product. Since the materials commingled with the beryllium waste are known not to be a source of contamination, the commingled material can be discriminated against during sampling, and only the beryllium sludge sampled and the volume contribution of the commingled material noted.

X1.2.4 *Are contamination levels correlated with item size-?* Some of the older beryllium sludge has dried and formed a cementaceous aggregate of different item sizes. Since the sludge is known to be homogeneous within a batch, by process knowledge and preliminary sampling data, sampling can be restricted to the more easily sampled, smaller item sizes.

X1.2.5 *Is contamination innate or surface adsorbed?* The waste from the machine shop consists of varied material from fine metallic filings to large chunks of metal and out-of-specification metal product. Since the only contamination in the machine shop is solvents and cutting oils and the waste matrix is impervious, the contamination is surface adsorbed in nature. Sampling of these wastes, therefore, will consist of the sampling of fines that will be subjected to extraction, wipe sampling of the large metallic objects, and notation of the volume contributions of the different item sizes. It is essential that all assumptions (that is, any correlations) be verified at least by knowledge of the waste and preferably confirmed by exploratory sampling and analyses.

X1.3 In the preceding hypothetical case, the proposed strategy for characterizing the 4000 drums resulted in the following:

X1.3.1 The identification of two large strata that constitute the majority of the waste (that is, the beryllium sludge and the solvent and cutting oil contaminated machine shop waste),

X1.3.2 The elimination of the need to sample 10 % of the drums (that is, the construction debris) if preliminary testing verifies waste disposal information,

X1.3.3 Simplified sampling of the beryllium commingled waste by restricting sampling to the beryllium sludge and not the other commingled materials if preliminary testing verifies waste disposal information, and

X1.3.4 Simplified sampling of the cementaceous beryllium sludge by limiting sampling to the more easily sampled small items if preliminary testing verifies waste disposal information.

X1.4 Simplifying the sampling of the machine wastes, since the source of contamination is surface adsorbed and not innate to the waste materials if preliminary testing, verifies waste disposal information.

REFERENCES

(**1**) "Characterizing Heterogeneous Wastes," EPA 600/R-92/033, February 1992.

(**2**) "Environmental Monitoring Issues," EPA/600/R-93/033, March 1993.

(**3**) "Guidance for the Data Quality Objectives Process," EPA QA/G-4, September 1994.

(**4**) "Preparation of Soil Sampling Protocols: Sampling Techniques and Strategies," EPA 600/R-92/128, July 1992.

(**5**) "Test Methods for Evaluating Solid Waste Physical/Chemical Methods," EPA SW-846, 3rd Edition, March 1995.

(**6**) Pitard, F. F., *Pierre Gy's Sampling Theory and Sampling Practice*, CRC Press, Ann Harbor, MI, 1993.

(**7**) Gilbert, Richard O., *Statistical Methods for Environmental Pollution Monitoring*, Van Nostrand Reinholt Co., New York, NY, 1987.

(**8**) Keith, L. H., *Environmental Sampling and Analysis*, Lewis Publisher, Inc., Chelsea, MI, 1991.

(**9**) Keith, L. H., *Principles of Environmental Sampling*, ACS, Washington, DC, 1991.

(**10**) Taylor, J. K., *Statistical Techniques for Data Analysis*, Lewis Publishers, Inc, Chelsea, MI, 1990.

Designation: D 5975 – 96

Standard Test Method for Determining the Stability of Compost by Measuring Oxygen Consumption[1]

This standard is issued under the fixed designation D 5975; the number immediately following the designation indicates the year of original adoption or, in the case of revision, the year of last revision. A number in parentheses indicates the year of last reapproval. A superscript epsilon (ε) indicates an editorial change since the last revision or reapproval.

1. Scope

1.1 This test method covers the stability of a compost sample by measuring oxygen consumption after exposure of the test compost to a well-stabilized compost under controlled-composting conditions on a laboratory scale involving active aeration. This test method is designed to yield reproducible and repeatable results under controlled conditions that resemble the end of the active composting phase. The compost samples are exposed to a well-stabilized compost inoculum that is prepared from municipal solid waste or waste similar to the waste from which the test materials are derived. The aerobic composting takes place in an environment where temperature, aeration, and humidity are monitored closely and controlled.

1.2 This test method yields a cumulative amount of oxygen consumed/g of volatile solids in the samples over a four-day period. The rate of oxygen consumption is monitored as well.

1.3 This test method is applicable to different types of compost samples including composts derived from wastes, such as municipal solid waste, yard waste, source-separated organics, biosolids, and other types of organic wastes that do not have toxicity levels that are inhibitory to the microorganisms present in aerobic composting systems.

1.4 The values stated in SI units are to be regarded as the standard.

1.5 There is no similar or equivalent ISO method.

1.6 *This standard does not purport to address all of the safety concerns, if any, associated with its use. It is the responsibility of the user of this standard to establish appropriate safety and health practices and determine the applicability of regulatory limitations prior to use.* Specific hazard statements are given in Section 8.

2. Referenced Documents

2.1 *ASTM Standards:*

D 515 Test Methods for Phosphorus in Water[2]

D 883 Terminology Relating to Plastics[3]

D 1293 Test Methods for pH of Water[2]

D 1888 Test Methods for Particulate and Dissolved Matter in Water[4]

D 2908 Practice for Measuring Volatile Organic Matter in Water by Aqueous-Injection Gas Chromatography[5]

D 3590 Test Methods for Total Kjeldahl Nitrogen in Water[5]

D 4129 Test Method for Total and Organic Carbon in Water by High-Temperature Oxidation and Coulometric Detection[5]

D 5338 Test Method for Determining Aerobic Biodegradation of Plastic Materials Under Controlled Composting Conditions[6]

2.2 *APHA-AWWA-WPCF Standards:*

2540 D Total Suspended Solids Dried at 103°–105°C[7]

2540 E Fixed and Volatile Solids Ignited at 550°C[7]

3. Terminology

3.1 Definitions of terms in this test method appear in Terminology D 883.

4. Summary of Test Method

4.1 This test method consists of the following:

4.1.1 Selecting a compost sample for the determination of the stability.

4.1.2 Producing a fully stabilized compost from a similar waste stream under well-controlled laboratory conditions.

4.1.3 Exposing the compost test samples to the fully stabilized compost under controlled composting conditions.

4.1.4 Measuring the oxygen consumption rate and determining the cumulative oxygen consumption.

4.2 Obtaining the level of stability from the cumulative oxygen consumption.

5. Significance and Use

5.1 A measurement of compost stability is needed for several reasons. It aids in assessing whether the composting

---

[1] This test method is under the jurisdiction of ASTM Committee D-34 on Waste Management and is the direct responsibility of Subcommittee D34.03 on Treatment.

Current edition approved Aug. 10, 1996. Published October 1996.

[2] *Annual Book of ASTM Standards*, Vol 11.01.

[3] *Annual Book of ASTM Standards*, Vol 08.01.

[4] Discontinued; see *1991 Annual Book of ASTM Standards*, Vol 11.01.

[5] *Annual Book of ASTM Standards*, Vol 11.02.

[6] *Annual Book of ASTM Standards*, Vol 08.03.

[7] "Standard Methods for the Examination of Water and Wastewater," 17th Edition, 1989, *American Public Health Association*, 1740 Broadway, New York, NY 19919.

process has proceeded sufficiently far to allow the finished compost to be used for its intended application. A different compost stability may be required for different applications of the compost.

5.2 A measurement of compost stability also is needed to verify whether a composting plant is processing the waste to previously agreed levels of stability. This measurement is useful in the commissioning of composting plants and the verification of whether plant operators are satisfying permit requirements.

5.3 The level of compost stability also will indicate its potential to cause odors if the compost is stored without aeration, as well as the level to which it has been hygienized and how susceptible the compost is to renewed bacterial and possible pathogenic activity. Compost stability is an important parameter with regard to phytotoxicity and plant tolerance of the compost.

5.4 The determination of compost stability will allow the selection of well performing composting technologies, as well as the safe application of compost in its various markets. The method indicates a degree of stability, but does not necessarily indicate that one level is preferable over another level of stability.

6. Apparatus

6.1 *Stabilized-Compost Inoculum Preparation Bin* (see Fig. 1):

6.1.1 A stabilized-compost inoculum preparation bin with a volume of 100 to 200 L, with insulation sufficient to maintain composting temperatures of 50 to 65°C during a period of at least two weeks when composting similar waste as the waste from which the samples were derived, and equipped with air distribution plate, inlet and outlet, and airtight lid.

6.1.2 *Pressurized Air*, provided to the composting bin at a precise and controllable rate up to 200 L/kg waste/day.

6.1.3 *Thermometer*, with temperature measurement up to 80°C (± 2°C).

6.1.4 Suitable devices for measuring oxygen and CO_2 (optional) concentrations in the exhaust air of the composting bin, such as sensors or appropriate gas chromatography.

6.2 *Composting Apparatus* (see Fig. 2):

6.2.1 A series of at least nine composting vessels (one test substance, one blank, one positive reference, all in three replicates) of 2 to 5 L of volume.

6.2.2 *Incubators*, water baths, or other temperature controlling means capable of maintaining the temperature of the composting vessels at 58°C (± 2°C).

6.2.3 *Pressurized-Air System*, providing H_2O-saturated air to each of the composting vessels at the appropriate aeration rates.

6.2.4 Suitable devices for measuring oxygen concentration in the exhaust air of the composting vessels, such as specific sensors or appropriate gas chromatography.

6.3 *Miscellaneous*:

6.3.1 *Balance* (± 1 mg), to weigh sample and stabilized compost.

6.3.2 *Scales* (± 0.1 kg), to weigh composting waste for

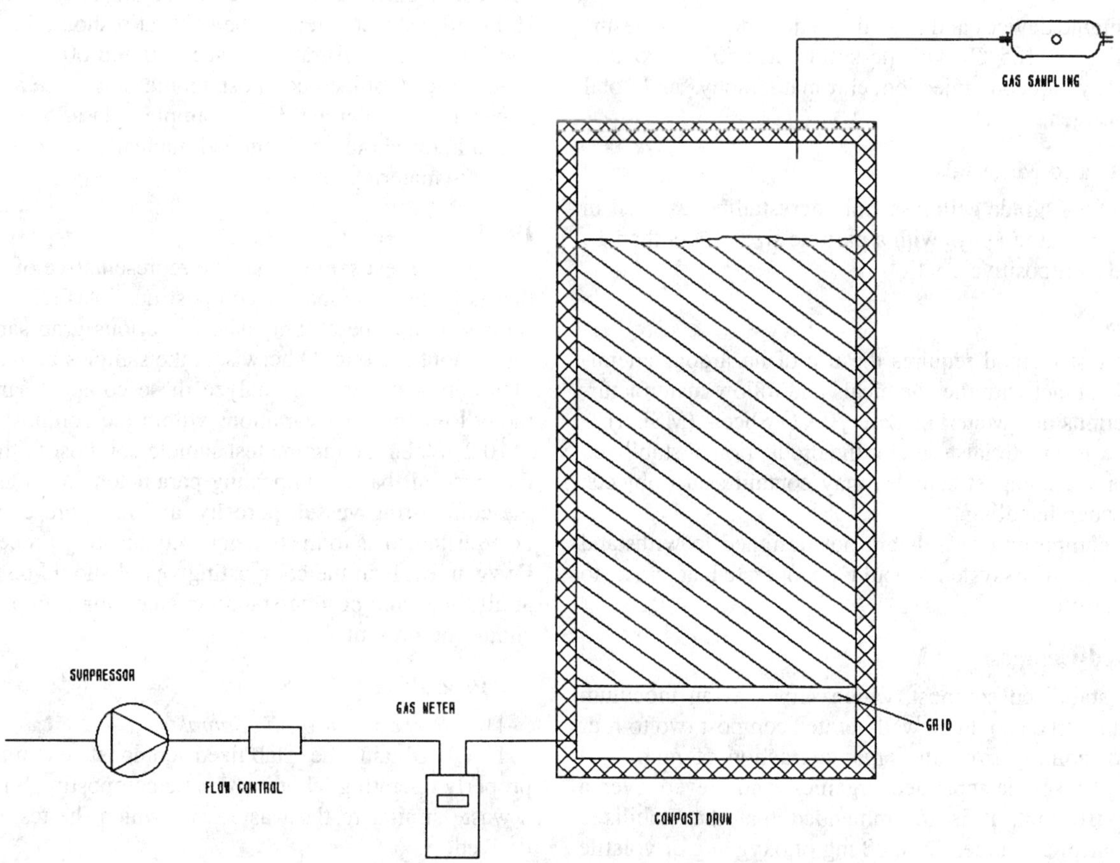

FIG. 1 Optional Set-Up Compost Preparation Bin

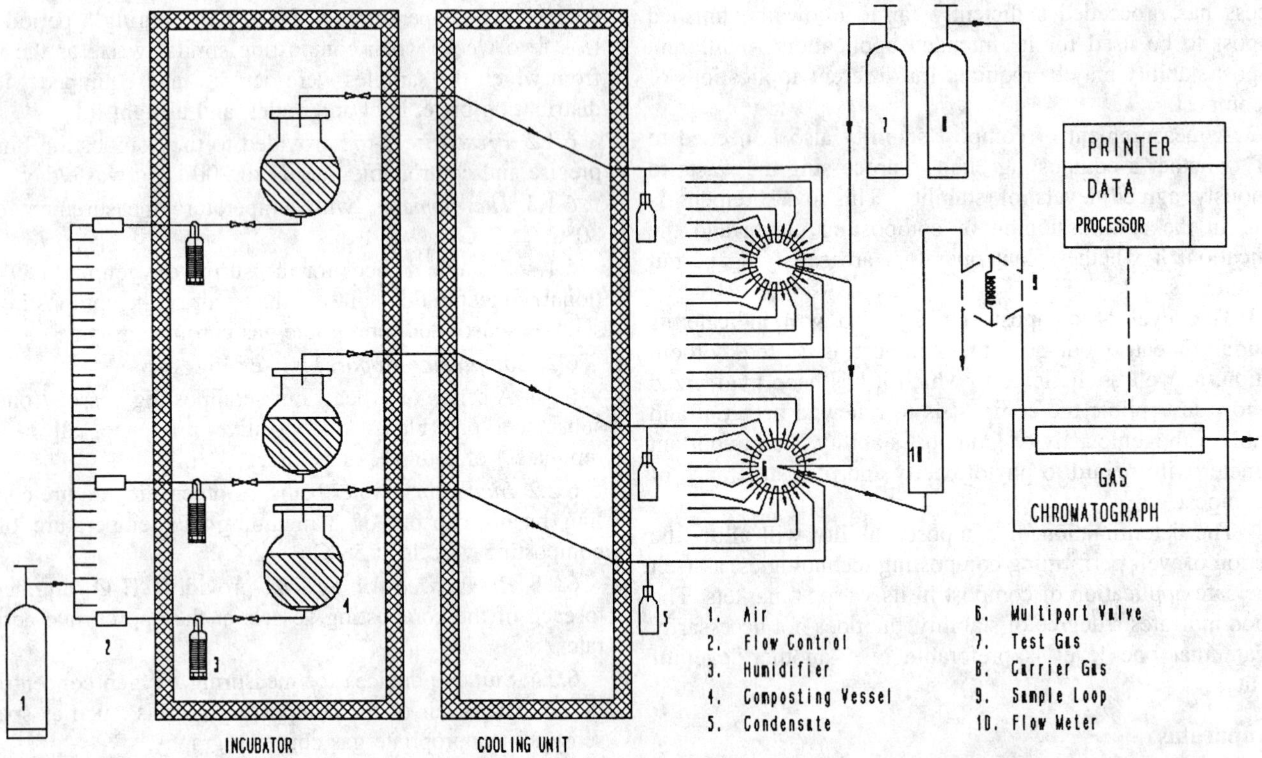

FIG. 2 Optional Set-Up Using Gas Chromatograph (see also Test Method D 5338)

stabilized compost production.

6.3.3 Normal laboratory glassware, equipment, and chemicals.

6.3.4 Suitable devices and analytical equipment for measuring dry solids (at 105°C), volatile solids (at 550°C), volatile fatty acids by aqueous-injection chromatography, and total Kjeldahl nitrogen.

7. Reagents and Materials

7.1 Analytical-grade cellulose (microcrystalline, as used in thin-layer chromatography) with a particle size of less than 10 μm, for use as a positive control.

8. Hazards

8.1 This test method requires the use of hazardous chemicals. Avoid contact with the chemicals and follow manufacturer's instructions and Materials Safety Data Sheets (MSDS).

8.2 The waste materials used for the production of stabilized compost, or the compost samples may contain sharp objects. Take care when handling.

8.3 The composting vessels are not designed to withstand high pressures. The system should be operated at close to ambient pressure.

9. Stabilized Compost

9.1 The stabilized compost, which serves as an inoculum and the test matrix, should be well-aerated compost two to four months old, coming from the organic fraction of municipal solid waste or source-separated organics, and sieved over a screen of <10 mm. It is recommended that the stabilized compost control consumes 15 to 80 mg of oxygen/g of volatile solids over the four-day test period. The stabilized compost must have a total solids content between 50 and 60 % on wet weight, an ash content of less than 70 % on total solids, a pH between 7 and 8 and be free of volatile fatty acids (less than 100 mg/L as acetic acid). The C/N ratio should be between 10 and 20 and the C/P ratio between 30 and 60.

9.2 The stabilized compost should be as free as possible of larger inert materials (for example, glass, stones, metals). These items should be removed manually to produce a homogeneous material.

10. Test Samples

10.1 The test sample must be representative of the compost that is being assessed for compost quality. Preferably the test sample should be a composite of various grab samples taken throughout the pile. Otherwise, take samples at various depths of a compost pile and analyze these compost samples separately to assess any variations within the compost pile itself.

10.2 When adding the test sample compost to the stabilized compost, all basic composting parameters, such as oxygen in the composting vessel, porosity, and moisture content should be adjusted so as to make a good composting process possible. Oxygen levels in the composting vessel should be at least 6 % at all times and no free-standing water nor clumps of material should be present.

11. Procedure

11.1 *Preparation of the Samples*:

11.1.1 Obtain the stabilized compost inoculum from a properly operating laboratory-scale composting bin processing a waste similar to the waste from which the test samples are derived.

11.1.1.1 Screen the stabilized compost to less than 10 mm

and manually remove and discard any large inert items (pieces of glass, stone, wood) that went through the 10 mm screen. Determine on the fraction less than 10 mm the volatile solids, dry solids and carbon, nitrogen and phosphorus contents according to Test Methods D 515, D 1888, D 3590, D 4129, and APHA Test Methods 2540 D and 2540 E and . Also determine pH and volatile fatty acids as described in 11.4.2. Add ammonium chloride if the C/N ratio is more than 20 and adjust to a C/N ratio of 15. Add NaH_2PO_4 if the C/P is more than 60 and adjust to a C/P ratio of 45.

11.1.1.2 Determine the volatile solids and dry solids of all the test sample composts in accordance with APHA Test Methods 2540 D and 2540 E and .

11.1.1.3 Weigh out precise amounts of stabilized compost inoculum and test sample compost (roughly 500 g of each per composting vessel) and mix thoroughly. The relation between the dry weight of the stabilized compost and the dry weight of test material should be about 1:1. Adjust with water the dry solids content of the mixture to approximately 50 %, and add the mixture to three composting vessels. Weigh the vessels with contents.

11.1.1.4 The blank consists of the stabilized compost only, containing about 1000 g wet weight/composting vessel for each of the three replicates. For the positive control, add 50 g of microcrystalline cellulose to 1000 g of stabilized compost for each of the three replicates.

11.2 *Start-Up Procedure*—Initiate aeration of the composting vessels with air-flow rates that are sufficiently high to ensure that oxygen levels do not drop below 6 % in the exhaust air. Oxygen levels should be closely controlled during the first 36 h and measured at least four times daily. Adjust air-flow rates as needed for the remainder of the test.

11.3 *Operating Procedure*:

11.3.1 Incubate the composting vessels in the dark for a period of four days. Keep the temperature at 58°C (± 2°C) for the duration of the test. The incubation time may be extended if the rate of oxygen consumption during the last 24 h is higher than during the previous 24 h in the vessels containing the samples.

11.3.2 Check O_2 concentrations in the outgoing air, four times daily, with a maximum time interval of 5 h.

11.3.3 Check air flow daily at the connections before and after the composting vessels and at the outlets, ensuring that no leaks are present in the complete system. Adjust air flow to maintain an oxygen concentration of at least 6 % v/v in the exhaust air.

11.3.4 Ensure proper composting conditions and shake composting vessels after the first day of testing.

11.4 *End of the Test*:

11.4.1 At the end of the test weigh the vessels with the contents and determine the dry solids content remaining in the compost.

11.4.2 Measure the pH in conformance with Test Methods D 1293. Measure the pH by diluting the sample to a 5:1 w/w mixture of distilled water to compost, mix by shaking manually and measure immediately. If the pH is less than 7, measure the volatile fatty acids in accordance with Practice D 2908 in the liquid phase after centrifugation of the diluted sample at 3000 G forces. The volatile fatty acids must be below 2000 mg/L as acetic acid.

12. Calculation

12.1 Determine the volumetric cumulative oxygen consumption for each composting vessel over the whole test period, using for each reading during the test the following formula:

$$V_1 = (O_i - O_e) \times F \times \Delta t \quad (1)$$

where:

V_1 = cummulative volumetric oxygen consumption (L),
O_i = oxygen concentration in incoming air (volume %),
O_e = oxygen concentration in exhaust air (volume %),
F = air flow rate (in L/h), and
Δt = period of time.

12.2 The cumulative volumetric oxygen consumption in litres, obtained from the previous calculation, is recalculated to standard conditions of temperature and pressure, using the formula:

$$V_2 = V_1 \times \frac{T_2}{T_1} \times \frac{p_1}{p_2} \quad (2)$$

where:

V_1 = cumulative volumetric oxygen consumption (L),
V_2 = cumulative volumetric oxygen consumption under standard conditions (L),
T_2 = standard temperature (273°K)
T_1 = ambient temperature (in °K),
p_1 = ambient pressure (in atm), and
p_2 = standard pressure (1 atm).

Subsequently, the oxygen consumption is recalculated using the formula:

$$C = V_2 \times \frac{32\ \text{g}}{22.414\ \text{L}} \quad (3)$$

where:

C = oxygen consumption in grams,
22.414 L = volume of 1 mole of oxygen under standard conditions of temperature and pressure, and
32 g = weight of 1 mole of oxygen.

12.3 Determine the net cumulative oxygen consumption of the test compost sample by subtracting the oxygen consumption of the stabilized compost inoculum in the test vessel from the total oxygen consumption of the test vessel, (the oxygen consumption of the inoculum in the test vessel is the consumption per gram of the inoculum in the control vessels multiplied by the weight of inoculum in the test vessel).

12.4 Express the oxygen consumption in mg oxygen/g volatile solids and over four days (mg O_2/g VS.4d).

13. Interpretation of the Results

13.1 Results of other analyses of the test compost sample such as volatile fatty acids, pH, ammonia, nitrate and rotting degree and information on the ecotoxicity of the test substance may be useful in the interpretation of inhibitive effects.

13.2 A reference or control substance known to biodegrade is necessary to check the activity of the stabilized compost, which serves as an inoculum. If sufficient oxygen consumption (a minimum of 20 % of the theoretical oxygen demand for cellulose within the duration of the test) is not observed with the positive reference, the test must be regarded as invalid and should be repeated using new stabilized compost inoculum.

14. Report

14.1 Report the following data and information:

14.1.1 Information on the stabilized compost inoculum, including source, percent dry solids, percent volatile solids, pH, volatile fatty acids, carbon, total Kjeldahl nitrogen, phosphorus, C/N, C/P, activity (oxygen consumption), date of collection, storage and handling.

14.2 Information on the compost sample including source, percent dry solids, percent volatile solids, pH, date of collection, storage and handling.

14.3 Apparatus used for carrying out the test method. Weight of vessels with contents before and at the end of the test.

14.4 Cumulative oxygen consumption over time (display graphically).

14.5 Average oxygen consumption for each test compost in mg oxygen/g volatile solids/four days, along with standard deviation.

14.6 Evolution over four-day test period and final percentage of mineralization of positive reference.

14.7 pH of final residues. Volatile fatty acids concentration for vessels with a final pH of less than 7.

14.8 Ratio of oxygen consumption versus CO_2 consumption (optional).

15. Precision and Bias

15.1 The precision and bias of this test method is being determined.

15.2 Preliminary results of a test with six different composts are given in Table 1. The compost stability through oxygen consumption in mg O_2/g VS.4d is compared with the NO_3/NH 4 ratio and with the rotting degree obtained in a respirometry method without inoculation. The evolution of the oxygen consumption for the six different composts is shown in Fig. 3. Analyses were performed in duplicate.

16. Keywords

16.1 compost; oxygen consumption; stability

TABLE 1 Results of NO_3/NH_4 Ratio, Rotting Degree and Compost Stability for Six Different Composts

	NO_3/NH_4 Ratio	Rotting Degree	Stability mg O_2/g VS.4d
Compost 1	0	II	258
Compost 2	0	III	109
Compost 3	0	IV	35
Compost 4	∞	IV	23
Compost 5	∞	IV	20
Compost 6	53	IV	8

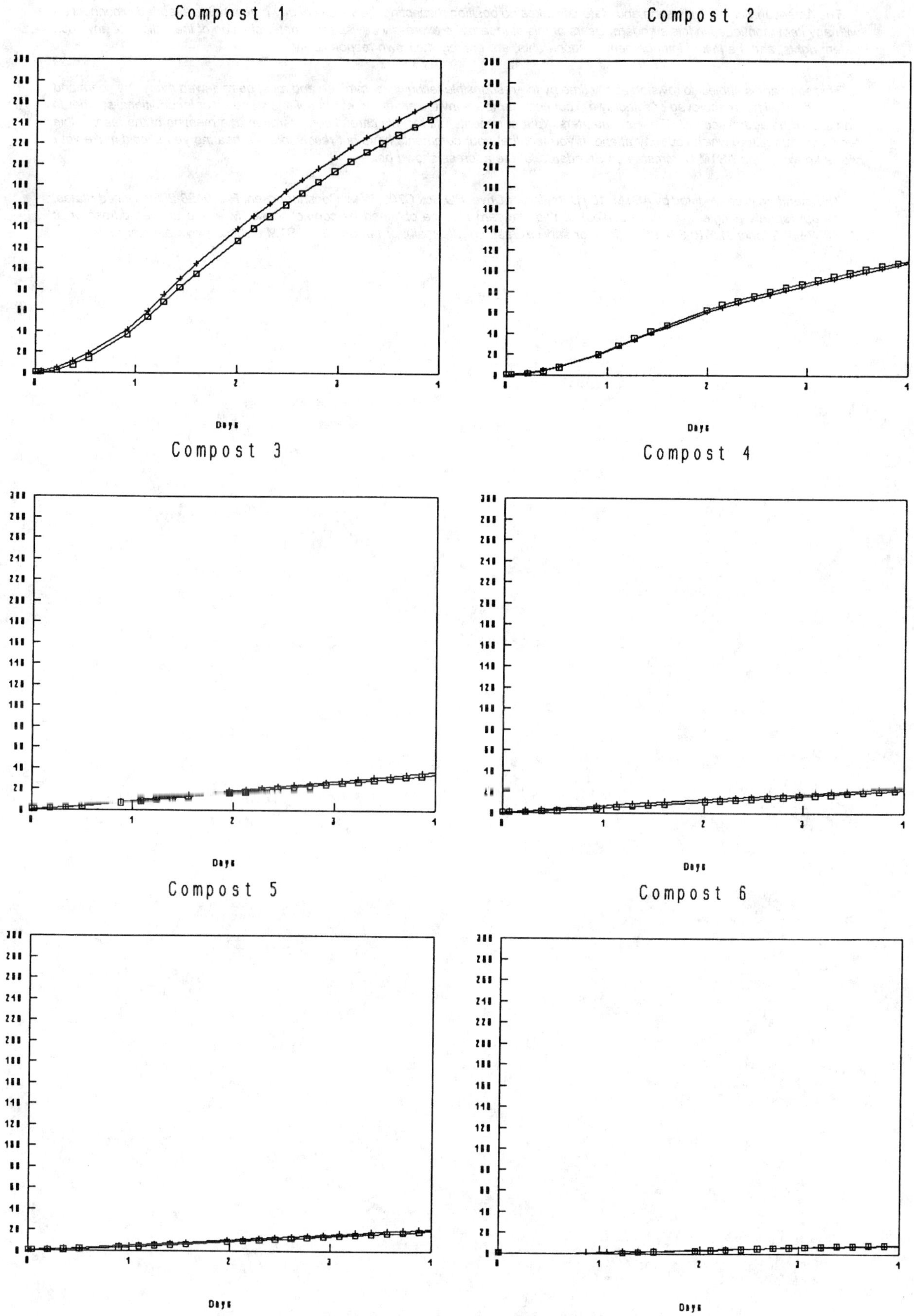

FIG. 3 Oxygen Consumption of Six Different Composts Over Time

Designation: D 6008 – 96

Standard Practice for Conducting Environmental Baseline Surveys[1]

This standard is issued under the fixed designation D 6008; the number immediately following the designation indicates the year of original adoption or, in the case of revision, the year of last revision. A number in parentheses indicates the year of last reapproval. A superscript epsilon (ϵ) indicates an editorial change since the last revision or reapproval.

1. Scope

1.1 *Purpose*—The purpose of this practice is to define good commercial and customary practice in the United States for conducting an environmental baseline survey (EBS) in order to determine certain elements of the environmental condition of federal real property, including excess and surplus property at closing and realigning military installations. This effort is conducted to fulfill certain requirements of the Comprehensive Environmental Response, Compensation and Liability Act of 1980 (CERCLA) section 120(h), as amended by the Community Environmental Response Facilitation Act of 1992 (CERFA). As such, this practice is intended to help a user to gather and analyze data and information in order to classify property into seven environmental condition of property area types (in accordance with the Standard Classification of Environmental Condition of Property Area Types). Once documented, the EBS is used to support Findings of Suitability to Transfer (FOSTs), Findings of Suitability to Lease (FOSLs), or uncontaminated property determinations, or a combination thereof, pursuant to the requirements of CERFA. Users of this practice should note that it does not address (except where explicitly noted) requirements for appropriate and timely regulatory consultation or concurrence, or both, during the conduct of the EBS or during the identification and use of the standard environmental condition of property area types.

1.1.1 *Environmental Baseline Survey*—In accordance with the Department of Defense (DoD) policy, an EBS will be prepared or evaluated for its usefulness (and updated if necessary) for any property to be transferred by deed or leased. The EBS will be based on existing environmental information related to storage, release, treatment, or disposal of hazardous substances or petroleum products on the property to determine or discover the obviousness of the presence or likely presence of a release or threatened release of any hazardous substance or petroleum product. In certain cases, additional data, including sampling, if appropriate under the circumstances, may be needed in the EBS to support the FOST or FOSL. A previously conducted EBS may be updated as necessary and used for making a FOST or FOSL. An EBS also may help to satisfy other environmental requirements (for example, to satisfy the requirements of CERFA or to facilitate the preparation of environmental condition reports). In addition, the EBS provides a useful reference document and assists in compliance with hazard abatement policies related to asbestos and lead-based paint. The EBS process consists of discrete steps. This practice principally addresses EBS-related information gathering and analysis.

1.1.2 *CERCLA Section 120(h) Requirements*—This practice is intended to assist with the identification of installation areas subject to the notification and covenant requirements of CERCLA § 120(h) relating to the deed transfer of contaminated Federal real property (42 USC 9601 *et seq.*).

1.1.3 *CERFA Requirements*—This practice can be used to provide information that can be used to partially fulfill the identification requirements of CERFA [Pub. L. 102-426, 106 Stat. 2174], which amended CERCLA. Property classified as area Type *1*, in accordance with Classification D 5746 is eligible for reporting as "uncontaminated" under the provisions of CERFA. Additionally, certain property classified as area Type *2*, where evidence indicates that storage occurred for less than one year, may also be identified as uncontaminated. At installations listed on the National Priorities List, Environmental Protection Agency (EPA) concurrence must be obtained for the property to be considered "uncontaminated" and therefore transferable under CERCLA § 120(h)(4). The EPA has stated that there may be instances in which it would be appropriate to concur with the DoD Component that certain property can be identified as uncontaminated under CERCLA § 120(h)(4) although some limited quantity of hazardous substances or petroleum products have been stored, released, or disposed of on the property. If the information available indicates that the storage, release, or disposal was associated with activities that would not be expected to pose a threat to human health or the environment (for example, housing areas, petroleum-stained pavement areas, and areas having undergone routine application of pesticides), such property should be eligible for expeditious reuse.

1.1.4 *Petroleum Products*—Petroleum products and their derivatives are included within the scope of this practice. Areas on which petroleum products or their derivatives were stored for one year or more, known to have been released or disposed of [CERCLA§ 120(h)(4)] are not eligible to be reported as "uncontaminated property" under CERFA.

1.1.5 *Other Federal, State, and Local Environmental Laws*—This practice does not address requirements of any

---

[1] This practice is under the jurisdiction of ASTM Committee D-34 on Waste Management and is the direct responsibility of Subcommittee D34.05 on Site Remediation.

Current edition approved Oct. 10, 1996. Published December 1996. Originally published as PS 37. Last previous edition PS 37 – 95.

federal, state, or local laws other than the applicable provisions of CERCLA identified in 1.1.2 and 1.1.3. Users are cautioned that federal, state, and local laws may impose additional EBS or other environmental assessment obligations that are beyond the scope of this practice. Users should also be aware that there are likely to be other legal obligations with regard to hazardous substances or petroleum products discovered on property that are not addressed in this practice and that may pose risks of civil or criminal sanctions, or both, for noncompliance.

1.1.6 *Other Federal, State, and Local Real Property and Natural and Cultural Resources Laws*—This practice does not address requirements of any federal, state or local real property or natural and cultural resources laws. Users are cautioned that numerous federal, state, and local laws may impose additional environmental and other legal requirements that must be satisfied prior to deed transfer of property that are beyond the scope of this practice.

1.2 *Objectives*—Objectives guiding the development of this practice are (*1*) to synthesize and put in writing a standard practice for conducting a high quality EBS, (*2*) to facilitate the development of high quality, standardized environmental condition of property maps to be included in an EBS that can be used to support FOSTs, FOSLs, and other applicable environmental condition reports, (*3*) to facilitate the use of the standard classification of environmental condition of property area types, and (*4*) to facilitate the development of a standard guide for preparing and updating EBS reports.

1.3 *Limitations*— Users of this practice should note that, while many of the elements of an EBS are performed in a manner consistent with other "due diligence" functions, an EBS is not prepared to satisfy a purchaser of real property's duty to conduct an "appropriate inquiry" in order to establish an "innocent landowner defense" to CERCLA § 107 liability. Any such use of any EBS by any party is outside the control of the United States Department of Defense and its components and beyond the scope of any EBS. No warranties or representations are made by the United States Department of Defense, its components, its officers, employees, or contractors that any EBS Report satisfies any such requirement for any party.

1.4 *Organization of This Practice*—This practice has 15 sections. Section 1 is the scope. Section 2 identifies referenced documents. Section 3, Terminology, includes definitions of terms not unique to this practice, descriptions of terms unique to this practice, and acronyms and abbreviations. Section 4 is the significance and use of this practice. Section 5 describes user's responsibilities. Sections 6-13 are the main body of the data gathering analysis steps of the EBS process. Section 14 briefly describes the EBS Step 3 classification of environmental condition of property area types. Section 15 contains a list of keywords.

1.5 *This standard does not purport to address all of the safety concerns, if any, associated with its use. It is the responsibility of the user of this standard to establish appropriate safety and health practices and determine the applicability of regulatory limitations prior to use.*

2. Referenced Documents

2.1 *ASTM Standards:*

E 1527 Practice for Environmental Site Assessments: Phase I Environmental Site Assessment Process[2]

E 1528 Practice for Environmental Site Assessments: Transaction Screen Process[2]

D 5746 Classification of Environmental Condition of Property Area Types[2]

2.2 *Department of Defense Policies:*

DoD Policy on the Environmental Review Process to Reach a Finding of Suitability to Lease (FOSL), September 1993[3]

DoD Policy on the Environmental Review Process to Reach a Finding of Suitability to Transfer (FOST) for Property Where No Release or Disposal Has Occurred, June 1994[3]

DoD Policy on the Environmental Review Process to Reach a Finding of Suitability to Transfer (FOST) for Property Where Release or Disposal Has Occurred, June 1994[3]

DoD Policy on the Implementation of the Community Environmental Response Facilitation Act (CERFA), September 1993[3]

2.3 *Department of Defense Guidance Document:*

BRAC Cleanup Plan Guidebook, Fall 1993[3]

2.4 *Federal Standards:*[4]

Title 40, Code of Federal Regulations (CFR), Part 300, National Oil and Hazardous Substances Pollution Contingency Plan

Title 40, Code of Federal Regulations (CFR), Part 302, Designation Reportable Quantities and Notification

Title 40, Code of Federal Regulations (CFR), Part 355, Emergency Planning and Notification

3. Terminology

3.1 This section provides definitions (of terms not unique to this practice), descriptions of terms specific to this practice, and a list of acronyms and abbreviations used herein. The terms are an integral part of this practice and are critical to its understanding and use. Many of these terms are also found in Practice E 1527.

3.2 *Definitions:*

3.2.1 *asbestos*—six naturally occurring fibrous minerals found in certain types of rock formations. Of the six, the minerals chrysotile, amosite, and crocidolite have been most commonly used in building products. When mined and processed, asbestos is typically separated into very thin fibers. Because asbestos is strong, incombustible, and corrosion-resistant, asbestos was used in many commercial products beginning early in this century and peaking in the period from World War II into the 1970s. When inhaled in sufficient quantities, asbestos fibers can cause serious health problems.

3.2.2 *asbestos-containing material (ACM)*—any material or product that contains more than 1 % asbestos.

3.2.3 *Comprehensive Environmental Response, Compensation, and Liability Information System (CERCLIS)*—the list of sites compiled by EPA that EPA has investigated or is currently

[2] *Annual Book of ASTM Standards*, Vol 11.04.

[3] Available from Department of Defense, Office of Environmental Security, 3400 Defense Pentagon, Washington, DC 20301-3400.

[4] Available from the Superintendent of Documents, U.S. Government Printing Office, Washington, DC 20402.

investigating for potential hazardous substance contamination for possible inclusion on the National Priorities List.

3.2.4 *contaminated public wells*—public wells used for drinking water that have been designated by a government entity as contaminated by toxic substances (for example, chlorinated solvents), or as having water unsafe to drink without treatment.

3.2.5 *drum*—a container (typically, but not necessarily, holding 55 gal [208 L] of liquid) that may have been used to store hazardous substances or petroleum products.

3.2.6 *dwelling*—structure or portion thereof used for residential habitation.

3.2.7 *environmental lien*—a charge, security, or encumbrance upon title to a property to secure the payment of a cost, damage, debt, obligation, or duty arising out of response actions, cleanup, or other remediation of hazardous substances or petroleum products upon a property, including (but not limited to) liens imposed pursuant to CERCLA 42 USC § 9607(1) and similar state or local laws.

3.2.8 *ERNS list*—EPA's Emergency Response Notification System list of reported CERCLA hazardous substance releases or spills in quantities equal to or greater than the reportable quantity, as maintained by the National Response Center. Notification requirements for such releases or spills are codified in 40 CFR Parts 302 and 35515.

3.2.9 *Federal Register (FR)*—publication of the United States government published daily (except for Federal holidays and weekends) containing all proposed and final regulations and some other activities of the Federal government. When regulations become final, they are included in the Code of Federal Regulations (CFR) as well as published in the Federal Register.

3.2.10 *hazardous substance*—a substance defined as a hazardous substance pursuant to CERCLA 42 USC § 9601(14), as interpreted by EPA regulations and the courts: "(*A*) any substance designated pursuant to section 1321(b)(2)(A) of Title 33, (*B*) any element, compound, mixture, solution, or substance designated pursuant to Section 9602 of this title, (*C*) any hazardous waste having the characteristics identified under or listed pursuant to Section 3001 of the Solid Waste Disposal Act (42 USC § 6921) (but not including any waste the regulation of which under the Solid Waste Disposal Act (42 USC§ 6921 *et seq.*) has been suspended by Act of Congress), (*D*) any toxic pollutant listed under Section 1317(a) of Title 33, (*E*) any hazardous air pollutant listed under Section 112 of the Clean Air Act (42 USC § 7412), and (*F*) any imminently hazardous chemical substance or mixture with respect to which the Administrator (of EPA) has taken action pursuant to Section 2606 of Title 15. The term does not include petroleum, including crude oil or any fraction thereof which is not otherwise specifically listed or designated as a hazardous substance under subparagraphs (*A*) through (*F*) of this paragraph, and the term does not include natural gas, natural gas liquids, liquefied natural gas, or synthetic gas usable for fuel (or mixtures of natural gas and such synthetic gas)." Users of this practice should note that certain states may expand this definition to include other substances not meeting the above definition. The user or environmental professional should consider whether the state in which the installation is located has identified such identified substances.

3.2.11 *hazardous waste*—any hazardous waste having the characteristics identified under or listed pursuant to section 3001 of the Solid Waste Disposal Act (42 USC § 6901 *et seq.*) (but not including any waste the regulation of which under the Solid Waste Disposal Act has been suspended by Act of Congress) and so forth.

3.2.12 *landfill*—a place, location, tract of land, area, or premises used for the disposal of solid wastes as defined by state solid waste regulations. The term is synonymous with the term solid waste disposal site and is also known as a garbage dump, trash dump, or similar term.

3.2.13 *local street directories*—directories published by private (or sometimes government) sources that show ownership, occupancy, or use of sites, or combination thereof, by reference to street addresses. Often local street directories are available at libraries of local governments, colleges or universities, or historical societies.

3.2.14 *material safety data sheet (MSDS)*—written or printed material concerning a hazardous substance which is prepared by chemical manufacturers, importers, and employers for hazardous chemicals pursuant to OSHA's Hazard Communication Standard, 29 CFR 1910.1200.

3.2.15 *National Contingency Plan (NCP)*—the National Oil and Hazardous substances Pollution Contingency Plan found at 40 CFR § 300, which is the EPA's regulations for how hazardous substances are to be cleaned up pursuant to CERCLA.

3.2.16 *National Priorities List*—list compiled by EPA pursuant to CERCLA 42 USC § 9605(a)(8)(B) of properties with the highest priority for cleanup pursuant to EPA's Hazard Ranking System. See 40 CFR Part 300.

3.2.17 *occupants*—those tenants, subtenants, or other persons or entities using the property or a portion of the property.

3.2.18 *owner*—generally the fee owner of record of the property.

3.2.19 *petroleum exclusion*—the exclusion from CERCLA liability provided in 42 USC § 9601(14), as interpreted by the courts and EPA: "The term (hazardous substance) does not include petroleum, including crude oil or any fraction thereof which is not otherwise specifically listed or designated as a hazardous substance under subparagraphs (*A*) through (*F*) of this paragraph, and the term does not include natural gas, natural gas liquids, liquefied natural gas, or synthetic gas usable for fuel (or mixtures of natural gas and such synthetic gas)."

3.2.20 *petroleum products*—those substances included within the meaning of the petroleum exclusion to CERCLA, 42 USC § 9601(14) as interpreted by the courts and EPA, that is: "petroleum, including crude oil or any fraction thereof which is not otherwise specifically listed or designated as a hazardous substance under subparagraphs (*A*) through (*F*) of this paragraph, and the term does not include natural gas, natural gas liquids, liquefied natural gas, or synthetic gas usable for fuel (or mixtures of natural gas and such synthetic gas)."

3.2.21 *Phase I Environmental Site Assessment*— the process described in Practice E 1527.

3.2.22 *pits, ponds, or lagoons*—man-made or natural depressions in a ground surface that are likely to hold liquids or sludge containing hazardous substances or petroleum products. The likelihood of such liquids or sludge being present is determined by evidence of factors associated with the pit, pond, or lagoon, including, but not limited to, discolored water, distressed vegetation, or the presence of an obvious wastewater discharge.

3.2.23 *property*—the real property that is the subject of the EBS described in this practice as well as the real property adjacent to the subject property (which may be privately owned). Real property includes buildings and other fixtures and improvements located on the property and affixed to the land.

3.2.24 *property tax files*—the files kept for property tax purposes by the local jurisdiction where the property is located and includes records of past ownership, appraisals, maps, sketches, photos, or other information that is reasonably ascertainable and pertaining to the property.

3.2.25 *RCRA generators*—those persons or entities that generate hazardous wastes, as defined and regulated by RCRA.

3.2.26 *RCRA generators list*—list kept by EPA of those persons or entities that generate hazardous wastes, as defined and regulated by RCRA.

3.2.27 *RCRA TSD facilities*—those facilities on which treatment, storage, or disposal, or a combination thereof, of hazardous wastes takes place, as defined and regulated by RCRA.

3.2.28 *RCRA TSD facilities list*—list kept by EPA of those facilities on which treatment, storage, or disposal, or a combination thereof, of hazardous wastes takes place, as defined and regulated by RCRA.

3.2.29 *recorded land title records*—records of fee ownership, leases, land contracts, easements, liens, and other encumbrances on or of the property recorded in the place where land title records are, by law or custom, recorded for the local jurisdiction in which the property is located. (Commonly, such records are kept by a municipal or county recorder or clerk.) Such records may be obtained from title companies or directly from the local government agency. Information about the title to the property that is recorded in a U.S. district court or any place other than where land title records are, by law or custom, recorded for the local jurisdiction in which the property is located, are not considered part of recorded land title records. See 3.3.33 and 7.2.4.

3.2.30 *records of emergency release notifications* (SARA§ 304)—Section 304 of EPCRA or Title III of SARA requires operators of facilities to notify their local emergency planning committee (as defined in EPCRA) and State emergency response commission (as defined in EPCRA) of any release beyond the facility's boundary of any reportable quantity of any extremely hazardous substance. Often the local fire department is the local emergency planning committee. Records of such notifications are "records of emergency release notifications" (SARA § 304).

3.2.31 *solid waste disposal site*—a place, location, tract of land, area, or premises used for the disposal of solid wastes as defined by state solid waste regulations. The term is synonymous with the term landfill and is also known as a garbage dump, trash dump, or similar term.

3.2.32 *solvent*—a chemical compound that is capable of dissolving another substance and a hazardous substance, used in a number of manufacturing/industrial processes including but not limited to the manufacture of paints and coatings for industrial and household purposes, equipment clean-up, and surface degreasing in metal fabricating industries.

3.2.33 *State registered USTs*—State lists of underground storage tanks required to be registered under Subtitle I, Section 9002 of RCRA.

3.2.34 *sump*—a pit, cistern, cesspool, or similar receptacle where liquids drain, collect, or are stored.

3.2.35 *underground storage tank (UST)*—any tank, including underground piping connected to the tank that is or has been used to contain hazardous substances or petroleum products and the volume of which is 10 % or more beneath the surface of the ground.

3.2.36 *USGS 7.5 Minute Topographic Map*—the map (if any) available from or produced by the United States Geological Survey, entitled "USGS 7.5 Minute Topographic Map" and showing the property.

3.2.37 *wastewater*—water that (*1*) is or has been used in an industrial or manufacturing process, (*2*) conveys or has conveyed sewage, or (*3*) is directly related to manufacturing, processing, or raw materials storage areas at an industrial plant. Wastewater does not include water originating on or passing through or adjacent to a site, such as stormwater flows, that has not been used in industrial or manufacturing processes, has not been combined with sewage, or is not directly related to manufacturing, processing, or raw materials storage areas at an industrial plant.

3.3 *Definitions of Terms Specific to This Standard:*

3.3.1 *adjacent properties*—those properties contiguous or partially contiguous to the boundaries of the property being surveyed during an EBS or other activity intended to classify the property into a standard environmental condition of property area type, or other properties relatively near the installation that could pose significant environmental concern and/or have a significant impact on the results of an EBS or on the classification of installation property into standard environmental condition of property area types.

3.3.2 *aerial photographs*—photographs, taken from an aerial platform, having sufficient resolution to allow identification of development and activities of areas encompassing the property. Aerial photographs are commonly available from government agencies or private collections unique to a local area.

3.3.3 *all remedial action taken*—for the purposes of this practice, all remedial action, as described in CERCLA§ 120(h)(3)(B)(i), has been taken if "the construction and installation of an approved remedial design has been completed, and the remedy has been demonstrated to the administrator [of EPA] to be operating properly and successfully. The carrying out of long-term pumping and treating, or operation and maintenance, after the remedy has been demonstrated to the administrator to be operating properly and successfully does not preclude the transfer of the property." [42 USC §

9620(h)(3)]. Alternatively, in circumstances where a remedy has been constructed, but no ongoing treatment or operation and maintenance is required, for example, "clean closure" or excavation of soil with off-site treatment, all remedial action means that all action required to meet applicable state or federal regulatory standards, including, as required, state or federal regulatory approval, has been taken.

3.3.4 *applicable requirements*—those cleanup standards, standards of control, and other substantive requirements, criteria, or limitations promulgated under federal environmental or State environmental or facility siting laws that specifically address a hazardous substance, pollutant, contaminant, remedial action, location, or other circumstances found at a CERCLA site. Only those state standards that are identified by a state in a timely manner and that are more stringent than federal requirements may be applicable.

3.3.5 *approximate minimum search distance*—the area for which records must be obtained and reviewed pursuant to Section 7 subject to the limitations provided in that section. This may include areas outside the property and shall be measured from the nearest property boundary. This term is used instead of radius to include irregularly shaped properties.

3.3.6 *BRAC statutes*—Title II of the Defense Authorization Amendments and Base Closure and Realignment Act of 1988 (Pub. L. 100-526, 10 USC 2687, note.) and the Defense Base Closure and Realignment Act of 1990 (Part A of Title XXIX of Pub. L. 101-510, 10 USC 2687, note.), collectively.

3.3.7 *closing military installation*—installations identified for closure pursuant to BRAC statutes, or installations previously closed under the authority of 10 USC 2687.

3.3.8 *DoD Component*—collectively, the Office of the Secretary of Defense, the Military Departments, the Chairman of the Joint Chiefs of Staff, the Inspector General of the Department of Defense, the Defense Agencies and the DoD Field Activities.

3.3.9 *disposal*—the discharge, deposit, injection, dumping, spilling, leaking, or placing of any hazardous substances, or petroleum products or their derivatives into or on any land or water so that such hazardous substances, or petroleum products or their derivatives or any constituent thereof may enter the environment or be emitted into the air or discharged into any waters including ground water.

3.3.10 *due diligence*—the process of inquiring into the environmental characteristics of a parcel of commercial real estate or other conditions, usually in connection with a commercial real estate transaction. The degree and kind of due diligence vary for different properties and differing purposes.

3.3.11 *environmental audit*—the investigative process to determine if the operations of an existing facility are in compliance with applicable environmental laws and regulations. This term should not be used to describe Practices E 1527, E 1528, or this practice, although an environmental audit may include an EBS or, if prior audits or EBSs are available, may be part of an EBS.

3.3.12 *environmental baseline survey* (EBS)—a survey of federal real property based on all existing environmental information related to storage, release, treatment, or disposal of hazardous substances or petroleum products or derivatives on the property to determine or discover the obviousness of the presence or likely presence of a release or threatened release of any hazardous substance or petroleum product. In certain cases, additional data, including sampling and analysis, may be needed in the EBS to support the classification of the property into one of the standard environmental condition of property area types. Additionally, an EBS may also satisfy the uncontaminated property identification requirements of CERFA. An EBS will consider all sources of available information concerning environmentally significant current and past uses of the real property, and shall, at a minimum, consist of the following:

3.3.12.1 Detailed search and review of available information and records in the possession of the DoD Components or records made available by the regulatory agencies or other involved federal agencies. The DoD Components are responsible for requesting and making reasonable inquiry into the existence and availability of relevant information and records to include any additional study information (for example, surveys for radioactive materials, asbestos, radon, lead-based paint, transformers containing PCB, RCRA Facility Assessments and Investigations, Underground Storage Tank Cleanup Program) to determine the environmental condition of the property;

3.3.12.2 Review of all reasonably obtainable federal, state, and local government records for each adjacent facility where there has been a release or likely release of any hazardous substance or any petroleum product, and which is likely to cause or contribute to a release or threatened release of any hazardous substance or any petroleum product on the federal real property;

3.3.12.3 Analysis of aerial photographs that may reflect prior uses of the property, which are in the possession of the federal government or are reasonably obtainable through state or local government agencies;

3.3.12.4 Interviews with current or former employees, or both, involved in operations on the real property;

3.3.12.5 Visual inspections of the real property; any buildings, structures, equipment, pipe, pipeline, or other improvements on the real property; and of properties immediately adjacent to the real property, noting sewer lines, runoff patterns, evidence of environmental impacts (for example, stained soil, stressed vegetation, dead or ill wildlife) and other observations which indicate actual or potential release of hazardous substances or petroleum products;

3.3.12.6 Identification of sources of contamination on the installation and on adjacent properties which could migrate to the parcel during federal government ownership;

3.3.12.7 Ongoing response actions or actions that have been taken at or adjacent to the parcel; and

3.3.12.8 A physical inspection of property adjacent to the real property, to the extent permitted by owners or operators of such property.

3.3.13 *environmental baseline survey (EBS) report*—the written record of an EBS that includes the following:

3.3.13.1 An executive summary briefly stating the areas of real property (or parcels) evaluated and the conclusions of the EBS;

3.3.13.2 The property identification (for example, address,

assessor parcel number, legal description);

3.3.13.3 Any relevant information obtained from a detailed search of federal government records pertaining to the property, including available maps;

3.3.13.4 Any relevant information obtained from a review of the recorded chain of title documents regarding the real property. The review should address those prior ownerships/uses that could reasonably have contributed to an environmental concern, and, at a minimum, cover the preceding 60 years;

3.3.13.5 A description of past and current activities, including all past DoD uses to the extent such information is reasonably available, on the property and on adjacent properties,

3.3.13.6 A description of hazardous substances or petroleum products management practices (to include storage, release, treatment, or disposal) at the property and at adjacent properties;

3.3.13.7 Any relevant information obtained from records reviews and visual and physical inspections of adjacent properties;

3.3.13.8 Description of ongoing response actions or actions that have been taken at or adjacent to the property;

3.3.13.9 An evaluation of the environmental suitability of the property for an intended lease or deed transaction, if known, including the basis for the determination of such suitability; and

3.3.13.10 Reference to key documents examined (for example, aerial photographs, spill incident reports, investigation results).

3.3.14 *environmental condition of property area type*—any of the seven standard environmental condition of property area types defined in the Standard Classification of Environmental Condition of Property Area Types.

3.3.15 *environmental condition of property map*— a map, prepared on the basis of all environmental investigation information conducted to date, that shows the environmental condition of a DoD installation's real property in terms of the seven standard environmental condition of property area types as defined in the standard classification.

3.3.16 *environmental investigation*—any investigation intended to determine the nature and extent of environmental contamination or to determine the environmental condition of property at a BRAC installation. Environmental investigations may include, but are not limited to, environmental site assessments, preliminary assessments, site inspections, remedial investigations, EBSs, RCRA facility assessments, and RCRA facility investigations.

3.3.17 *environmental professional*—a person possessing sufficient training and experience necessary to conduct an EBS including all activities related to this practice, and from the information and data gathered by such activities, having the ability to develop conclusions regarding environmental condition of property and recognized environmental conditions in connection with the property being evaluated. An individual's status as an environmental professional may be limited to the type of EBS to be performed or to specific steps of the EBS for which the professional is responsible. The person may be an independent contractor of an employee of the Department of Defense or its components.

3.3.18 *fill dirt*—dirt, soil, sand, or other earth, that is obtained off-site, that is used to fill holes or depressions, create mounds, or otherwise artificially change the grade or elevation of real property. It does not include material that is used in limited quantities for normal landscaping activities.

3.3.19 *innocent landowner defense*—that defense to CERCLA liability provided in 42 USC § 9601(35) and 42 USC § 9607(b)(3). One of the requirements to qualify for this defense is that the party make" all appropriate inquiry into the previous ownership and uses of the property consistent with good commercial or customary practice." There are additional requirements to qualify for this defense.

3.3.20 *installation restoration program (IRP)*— the DoD program, mandated by 10 USC § 2407 to assess and respond to releases of hazardous substances on military property under the control of the military services. Additionally, based upon policy decisions, the IRP serves as an umbrella program for environmental response in all media, including RCRA corrective action, LUST corrective action, as well as CERCLA removals and remedial actions. Generally, where field sampling or intrusive environmental testing is required, the IRP will serve as a vehicle for such testing. The IRP is also known as the Defense Environmental Restoration Program.

3.3.21 *interviews*—sessions with current or former employees involved in operations on the real property, conducted to ascertain if storage, release, treatment, or disposal of hazardous substances, petroleum products or their derivatives occurred or is occurring on the real property.

3.3.22 *local government agencies*—those agencies of municipal or county government having jurisdiction over the property. Municipal and county government agencies include, but are not limited to, cities, parishes, townships, and similar entities. Local government agencies may also include, where appropriate, state agencies with local jurisdiction which perform functions commonly performed in other locations by local government agencies.

3.3.23 *migration*—the movement of contaminant(s) away from a source through permeable subsurface media (such as the movement of a ground water plume of contamination), or movement of contaminant(s) by a combination of surficial and subsurface processes.

3.3.24 *obviousness*—the condition of being plain or evident. A condition or fact which could not be ignored or overlooked by a reasonable observer while conducting a records search or while physically or visually observing the property in conjunction with an EBS.

3.3.25 *other historical sources*—any source or sources other than those designated in 7.2.1-7.2.4 that are credible to a reasonable person and that identify past uses of the property. The term includes, but is not limited to: miscellaneous maps, newspaper archives, and records in the files and/or personal knowledge of the property owner and/or occupants.

3.3.26 *physical setting sources*—sources that provide information about the geologic, hydrogeologic, hydrologic, or topographic characteristics of a property.

3.3.27 *practically reviewable*—information that is practically reviewable is information provided by the source in a manner and in a form that, upon examination, yields information relevant to the property without the need for extraordinary analysis of irrelevant data. The form of the information shall be such that the user can review the records for a limited geographic area. Records that cannot be feasibly retrieved by reference to the location of the property or a geographic area in which the property is located are not generally practically reviewable. Most data bases of public records are practically reviewable if they can be obtained from the source agency by the county, city, zip code, or other geographic area of the facilities listed in the record system. Records that are sorted, filed, organized, or maintained by the source agency only chronologically are not generally practically reviewable. This term has the same meaning as provided in Practice E 1527.

3.3.28 *preliminary assessment (PA)*—review of existing information and an off-site reconnaissance, if appropriate to determine if a release or potential release may require additional investigation or action. A PA may include an on-site reconnaissance, if appropriate.

3.3.29 *publicly available*—information that is publicly available means that the source of the information allows access to the information by anyone upon request.

3.3.30 *reasonably available*—information that is (*1*) publicly available, (*2*) obtainable from its source within reasonable time and cost constraints, and (*3*) practically reviewable. This term has the same meaning as the term "reasonably ascertainable" as provided in Practice E 1527.

3.3.31 *reasonably obtainable*—information that is (*1*) publicly available, (*2*) obtainable from its source within reasonable time and cost constraints, and (*3*) practically reviewable. This term has the same meaning as the term "reasonably ascertainable" as provided in Practice E 1527. Reasonably available and reasonably obtainable are synonyms.

3.3.32 *recognized environmental conditions*—the presence or likely presence of any hazardous substances or petroleum products on any federal real property under conditions that indicate an existing release, a past release, or a material threat of a release of any hazardous substances or petroleum products into the environment. The term includes hazardous substances or petroleum products even under conditions in compliance with laws. The term is not intended to include *de minimis* conditions that generally do not present a material risk of harm to public health or the environment and that generally would not be the subject of an enforcement action if these conditions were brought to the attention of appropriate governmental agencies. This term is introduced in Practice E 1527, and is used herein only in conjunction with EBS Steps *1* and *2* (see 6.2), as an intermediate outcome prior to the Step *3* classification of environmental condition of property area types. A Phase I Site Assessment results in recognized environmental conditions, but not environmental condition of property area types.

3.3.33 *recorded chain of title documents*—this term has the same meaning as recorded land title records.

3.3.34 *records search and/or review*—detailed search and review of available information and records in the possession of the DoD components and records made available by the regulatory agencies or other involved federal agencies, including, but not limited to IRP studies and analyses, surveys for radioactive materials, asbestos, radon, lead-based paint, electrical devices (that is, transformers) containing PCB, RCRA facility assessments and Investigations to determine what, if any, hazardous substances or petroleum products may be present on the property. For the purposes of adjacent facilities, a records search includes the review of all reasonably obtainable federal, state, and local government records for each adjacent facility where there has been a release or likely release of any hazardous substance or any petroleum product, and which is likely to cause or contribute to a release or threatened release of any hazardous substance or any petroleum product on the federal real property.

3.3.35 *release*—any spilling, leaking, pumping, pouring, emitting, emptying, discharging, injecting, escaping, leaching, dumping, or disposing into the environment (including the abandonment or discarding of barrels, containers, and other closed receptacles) of any hazardous chemical, extremely hazardous substance, or CERCLA hazardous substance.

3.3.36 *relevant and appropriate requirements*— those cleanup standards, standards of control, and other substantive requirements, criteria, or limitations promulgated under federal environmental or State environmental or facility siting laws that, while not "applicable" to a hazardous substance, pollutant, contaminant, remedial action, location, or other circumstance at a CERCLA site, address problems or situations sufficiently similar to those encountered at the CERCLA site that their use is well suited to the particular site. Only those state standards that are identified in a timely manner and are more stringent than federal requirements may be relevant and appropriate.

3.3.37 *remedial actions*—those actions consistent with a permanent remedy taken instead of, or in addition to, removal action in the event of a release or threatened release of a hazardous substance into the environment, to prevent or minimize the release of hazardous substances so that they do not migrate to cause substantial danger to present or future public health or welfare or the environment.

3.3.38 *removal*—the cleanup or removal of released hazardous substances from the environment; such actions as may be necessary to take in the event of the threat of release of hazardous substances into the environment; such actions as may be necessary to monitor, assess, and evaluate the release or the threat of release of hazardous substances; the disposal of removed material; or the taking of such other actions as may be necessary to prevent, minimize, or mitigate damage to the public health or welfare or to the environment, which may otherwise result from a release or threat of release.

3.3.39 *required remedial actions*—remedial actions determined necessary to comply with the requirements of CERCLA § 120(h)(3)(B)(i).

3.3.40 *required response actions*—removal and/or remedial actions determined necessary to comply with the requirements of CERCLA§ 120(h)(3)(B)(i).

3.3.41 *significant and significance*—in this practice, significant and significance connote the opposite of trivial or *de minimis*. An event or condition is considered significant if it has the potential to present a nontrivial risk to human health and the environment, using the risk range established by the NCP. A probability is considered significant when an environmental professional estimates the probability as nontrivial. For example, in the hypothetical case of an underground tank that was installed and removed prior to the existence of regulatory requirements for tank closure, the environmental professional must evaluate the possibility of release from the tank in the absence of soil testing results. If such an evaluation, based upon observed site conditions and documented soil corrosivity characteristics were to conclude that the probability of release is trivial or very close to zero, then no soil testing would be undertaken in the absence of a specific regulatory requirement for such testing. On the other hand, there is a significant probability of release if such an evaluation were to determine that the probability of release were greater than the extremely low probability encompassed by the concept of trivial. The evaluation of significance, in this sense, is a matter of professional judgment on the part of the environmental professional and should be so documented.

3.3.42 *site inspection (SI)*—an on-site investigation to determine whether there is a release or potential release and the nature of the associated threats. The purpose is to augment the data collected in the PA and to generate, if necessary, sampling and other field data to determine if further action or investigation is appropriate.

3.3.43 *standard classification*—the Standard Classification of Environmental Condition of Property Area Types.

3.3.44 *standard environmental condition of property area type*—one of the seven environmental condition of property area types defined in the Standard Classification.

3.3.45 *standard practice*—the activities set forth in this practice and Practices E 1527 and E 1528, where referenced.

3.3.46 *storage*—the containment of hazardous substances, petroleum products or their derivatives, either on a temporary basis or for a period of years, in such a manner as not to constitute disposal of such hazardous substances, petroleum products, or their derivatives.

3.3.47 *transaction screen process*—the process described in Practice E 1528.

3.3.48 *transaction screen questionnaire*—the questionnaire provided in Practice E 1528.

3.3.49 *user*—the party seeking to use this practice to perform an EBS of the property. A user may include, without limitation, a DoD component (acting as owner of the property).

3.3.50 *visual and/or physical inspection*—actions taken during an EBS to include observations made by vision while walking through or otherwise traversing a property and structures located on it and observations made by the sense of smell, particularly observations of noxious or foul odors.

3.4 *Acronyms:Acronyms and Abbreviations:*

3.4.1 *ARARs*—applicable or relevant and appropriate requirements.

3.4.2 *ASTM*—American Society for Testing and Materials.

3.4.3 *BRAC*—Base Realignment and Closure.

3.4.4 *CERCLA*—Comprehensive Environmental Response, Compensation and Liability Act of 1980, as amended (42 USC 9620 *et seq.*).

3.4.5 *CERCLIS*—Comprehensive Environmental Response, Compensation and Liability Information System.

3.4.6 *CERFA*—Community Environmental Response Facilitation Act of 1992 (102 Pub. L. 426, 106 Stat. 2174).

3.4.7 *CFR*—Code of Federal Regulations.

3.4.8 *DoD*—Department of Defense.

3.4.9 *EBS*—environmental baseline survey.

3.4.10 *ECP*—environmental condition of property.

3.4.11 *USEPA*—United States Environmental Protection Agency.

3.4.12 *EPCRA*—Emergency Planning and Community Right to Know Act, 42 USC.

3.4.13 *ERNS*—Emergency Response Notification System.

3.4.14 *ESA*—environmental site assessment.

3.4.15 *FOIA*—U.S. Freedom of Information Act (5 USC 552 *et seq.*).

3.4.16 *FOSL*—Finding of Suitability to Lease as described in applicable DoD Policy.

3.4.17 *FOST*—Finding of Suitability to Transfer as described in applicable DoD Policy.

3.4.18 *FR*—Federal Register.

3.4.19 *IRP*—Installation Restoration Program.

3.4.20 *LUST*—leaking underground storage tank.

3.4.21 *MSDS*—material safety data sheet.

3.4.22 *NCP*—National Contingency Plan.

3.4.23 *PA*—preliminary assessment.

3.4.24 *PCBs*—polychlorinated biphenyls.

3.4.25 *RCRA*—The Resource Conservation and Recovery Act, as amended, 42 USC 6901 *et seq.*

3.4.26 *SARA*—Superfund Amendments and Reauthorization Act of 1986.

3.4.27 *SI*—site inspection.

3.4.28 *TSD*—treatment, storage, and disposal.

3.4.29 *USC*—United States Code.

3.4.30 *USGS*—United States Geological Survey.

3.4.31 *UST*—underground storage tank.

4. Significance and Use

4.1 *Uses*—This practice is intended for use by DoD components and environmental professionals in order to facilitate EBS efforts. It is also intended for use by preparers and reviewers of environmental condition of property maps and EBS Reports used to support CERFA uncontaminated property identifications and property suitable for transfer by lease or by deed.

4.2 *Clarifications on Use*:

4.2.1 *Use Not Limited to CERCLA*—This practice is designed to assist the user in developing information about the environmental condition of a property and as such has utility for a wide range of persons, including those who may have no actual or potential CERCLA liability.

4.2.2 *Residential Tenants/Purchasers and Others*—No implication is intended that it is currently customary practice for residential tenants of multifamily residential buildings, tenants of single-family homes or other residential real estate, or purchasers of dwellings for one's own residential use, to conduct an EBS in connection with these transactions. Thus, these transactions are not included in the term commercial real estate transactions. Thus, although such property may be included within the scope of an EBS, their occupants shall not be treated as key site personnel with regard to the housing occupied for the purpose of conducting an EBS.

4.2.3 *Site-Specific*— This practice is site-specific in that it relates to assessment of environmental conditions of federal real property. Consequently, this practice does not address many additional issues raised in transactions such as purchases of business entities; or interests therein, or of their assets, that may well involve environmental liabilities pertaining to properties previously owned or operated or other off-site environmental liabilities.

4.3 *Related Practices*—See Practices E 1527 and E 1528.

4.4 *Principles*—The following principles are an integral part of this practice and all related practices and are intended to be referred to in resolving any ambiguity or exercising such discretion as is accorded the user or environmental professional in performing an EBS or in judging whether a user or environmental professional has conducted appropriate inquiry or has otherwise conducted an adequate EBS.

4.4.1 *Uncertainty Not Eliminated*—No EBS can wholly eliminate uncertainty regarding the potential for recognized environmental conditions in connection with a property. Performance of this practice is intended to reduce uncertainty regarding the potential for recognized environmental conditions in connection with a property to the minimum practicable level, but not eliminate such uncertainty altogether, as well as to recognize reasonable limits of time and cost for property information (see 7.1.3.2).

4.4.2 *Not Exhaustive*— Appropriate inquiry does not mean an exhaustive assessment of an uncontaminated property. There is a point at which the cost of information obtained or the time required to gather it outweighs the usefulness of the information and, in fact, may be a material detriment to the orderly completion of transactions. One of the purposes of this practice is to identify a balance between the competing goals of limiting the costs and time demands inherent in performing an EBS and the reduction of uncertainty about unknown conditions resulting from additional information.

4.4.3 *Level of Inquiry Is Variable*—Not every property will warrant the same level of EBS effort. Consistent with good practice, the appropriate level of EBS will be guided by the type of property subject to EBS and the information developed in its conduct.

4.4.4 *Comparison With Subsequent Inquiry*—It should not be concluded or assumed that an inquiry was not an appropriate inquiry merely because the inquiry did not identify recognized environmental conditions in connection with a property. The EBSs must be evaluated based on the reasonableness of judgments made at the time and under the circumstances in which they were made. Subsequent EBSs should not be considered valid standards to judge the appropriateness of any prior EBS based on hindsight, new information, use of developing technology or analytical techniques, or other factors.

4.5 *Continued Viability of Environmental Baseline Survey*—An EBS meeting or exceeding this practice and completed less than 180 days prior to the date of a subsequent use is presumed to be valid for that use. An EBS not meeting or exceeding this practice or completed more than 180 days previously may be used to the extent allowed by 4.6-4.6.5.

4.6 *Prior EBS Usage*— This practice recognizes that EBSs performed in accordance with this practice or otherwise containing information which was reasonably accurate at the time prepared will include information that subsequent users may want to use to avoid undertaking duplicative EBS procedures. Therefore, this practice describes procedures to be followed to assist users in determining the appropriateness of using information in EBSs performed previously. The system of prior EBS usage is based on the following principles that should be adhered to in addition to the specific procedures set forth elsewhere in this practice:

4.6.1 *Use of Prior Information*—Subject to 4.6.4, users and environmental professionals may use information in prior EBSs provided such information was generated as a result of procedures that meet or exceed the requirements of this practice or accurately state the limitations of the information presented. When using information from an EBS which, as a whole, fails to meet or exceed the requirements of this practice, the use shall be limited to those portions of the EBS which, based upon the limitations and methodology of the EBS Report, the environmental professional finds to be reasonably accurate.

4.6.2 *Prior EBS Meets or Exceeds*—Subject to 4.6.4, a prior EBS may be used in its entirety, without regard to the specific procedures set forth in these practices if, in the reasonable judgment of the user, the prior EBS meets or exceeds the requirements of this practice and the conditions at the property likely to affect environmental condition of property area types in connection with the property are not likely to have changed materially since the prior EBS was conducted. In making this judgment, the user should consider the type of property subject to the EBS and the conditions in the area surrounding the property.

4.6.3 *Current Investigation*—Except as specifically provided in 4.6.2, prior EBSs should not be used without current investigation of conditions likely to affect the environmental condition of property in connection with the property that may have changed materially since the prior EBS was conducted. For an EBS to be consistent with this practice, a new visual inspection, interviews, an update of the records review, and other appropriate activities may have to be performed.

4.6.4 *Actual Knowledge Exception*—If the user or environmental professional(s) conducting an EBS has actual knowledge that the information being used from a prior EBS is not accurate or if it is obvious, based on other information obtained by means of the EBS or known to the person conducting the EBS, that the information being used is not accurate, such information from a prior EBS may not be used.

4.6.5 *Contractual Issues Regarding Prior EBS Usage*—The contractual and legal obligations between prior and subsequent users of EBSs or between environmental professionals who conducted prior EBSs and those who would like to use such prior EBSs are beyond the scope of this practice.

5. User's Responsibilities

5.1 *Scope*—This section is limited to the responsibilities of users of this practice. Users may be either DoD component staff or environmental professionals contractually engaged to perform EBSs. Users of this practice should be familiar with its entire contents before conducting or documenting an EBS, and to use best professional judgment regarding its applicability to a particular situation.

5.1.1 *DoD Component Staff*—DoD component staff who have both the requisite specialized knowledge and experience and appropriate training can use this practice as a starting point for conducting or updating EBSs. Although this practice has been designed to help DoD components meet certain legal and policy requirements, it should not be used as a substitute for meeting environmental, BRAC statute, or health and safety legal requirements that exist under various laws, regulations, and DoD and DoD component policies and guidance.

5.1.2 *Environmental Professionals*—Environmental professionals who have both the requisite specialized knowledge and experience, and appropriate training can use this practice as a starting point for conducting or updating EBSs. Although this practice has been designed to help environmental professionals contractually engaged by DoD components to conduct EBSs, in accordance with applicable legal and policy requirements, it should not be used as a substitute for meeting environmental, BRAC statute, or health and safety legal requirements that exist under various laws, regulations, and DoD and DoD component policies and guidance. Contractually engaged environmental professionals should not use this practice to perform tasks that are inherently governmental functions.

5.2 *Specialized Knowledge or Experience of the User*—Users of this practice are expected to have the requisite environmental and health and safety training necessary to conduct the tasks identified in this practice. The DoD components are responsible for identifying appropriate staff for conducting these functions, and are also responsible for contractually ensuring that environmental professionals engaged to perform EBSs have appropriate qualifications. These qualifications should be identified in the contract or scope of work.

6. Environmental Baseline Survey Process

6.1 *Objective*—In accordance with DoD policy, the purpose of the EBS is to determine or discover and to document the obviousness of the presence or likely presence of a release or threatened release of any hazardous substance or petroleum product. In certain cases, additional data, including sampling and analysis, may be needed in the EBS or EBS supplement to support the classification of the property into one of the standard environmental condition of property area types. Additionally, an EBS may also satisfy the uncontaminated property identification requirements of CERFA. Users are cautioned that elements of this practice pertain to an initial EBS conducted by a DoD component as well as to EBS updates, supplemental EBSs, or site-specific EBSs, or a combination thereof, (however termed by the DoD component). As such, it is anticipated that it will only be necessary to complete all steps and tasks identified in this practice for the initial EBS for the property. The user or environmental professional should obtain the input of the DoD component end user regarding the level of effort to be used during any supplemental EBS efforts.

6.2 *Five Steps*—Within the limitations described in 6.1, it is anticipated that the EBS process will commonly consist of at least four and possibly five discrete steps. These are summarized as follows:

6.2.1 *EBS Step 1*—Gathering of data and information in accordance with the process described in the applicable DoD policy referenced in 2.2 and as further elaborated in Sections 7-13 of this practice.

6.2.2 *EBS Step 2*—Analysis of data and information in accordance with the process described in Sections 7-13 of this practice.

6.2.3 *EBS Step 3*—Determination of the environmental condition of property area type for the real property being evaluated by the EBS, in accordance with the process described in this practice and DoD policy.

6.2.4 *EBS Step 4*—Preparation of an EBS Report in accordance with the format described in the applicable DoD policy.

6.2.5 *EBS Step 5*—Updating and enhancing, as necessary, an EBS Report to support property transfer transactions (for example, FOSLs, FOSTs, or environmental condition reports). This process may require repeating Steps 1 through 3 to incorporate additional information or data, or both, generated between the time an initial EBS report is issued and the time an updated version is used to support a property transfer transaction.

6.3 *Additional Explanation of EBS Steps 1 and 2*—Steps 1 and 2 of the EBS process will consider all sources of available information concerning environmentally significant current and past uses of the real property, and shall, at a minimum, consist of the following eight components, as described in overview as follows (more detailed descriptions of each component are found in Sections 7-13):

6.3.1 *Records Search and Review Scope*—Detailed search and review of available information and records in the possession of the DoD components or records made available by the regulatory agencies or other involved federal agencies. Department of Defense (DoD) components are responsible for requesting and making reasonable inquiry into the existence and availability of relevant information and records to include any additional study information (for example, surveys for radioactive materials, asbestos, radon, lead-based paint, drinking water quality, indoor air quality, transformers containing PCBs, RCRA Facility Assessments and Investigations, and Underground Storage Tank Cleanup Program) to help support the determination of the environmental condition of property area type.

6.3.2 *Adjacent Facility Records Search and Review Scope*—Review of all reasonably obtainable federal, state, and local government records for each adjacent facility where there has been a release or likely release of any hazardous substance or any petroleum product, and which is likely to cause or contribute to a release or threatened release of any hazardous substance or any petroleum product on the real property.

6.3.3 *Aerial Photography Analysis*—Analysis of aerial photographs that are in the possession of the federal government or are reasonably obtainable through state or local government agencies that may reflect prior uses of the property.

6.3.4 *Interviews*—Interviews with key current or former employees, or both, involved in operations on the real property.

6.3.5 *Visual Inspections*—Nonintrusive visual inspections of the real property; any buildings, structures, equipment, pipe, pipeline, or other improvements on the real property; and of properties immediately adjacent to the real property, noting sewer lines, runoff patterns, evidence of environmental impacts (for example, stained soil, stressed vegetation, dead or ill wildlife), and other observations which indicate actual or potential release of hazardous substances or petroleum products.

6.3.6 *Contamination Source Identification*—Identification of sources of contamination on the installation and on adjacent properties which could migrate to the real property.

6.3.7 *Ongoing Response Actions*—Ongoing response actions or actions that have been taken at or adjacent to the property will be identified and documented.

6.3.8 *Physical and Visual Inspection of Adjacent Property*—A physical inspection of property adjacent to the real property, to the extent permitted by owners or operators of such property. A visual inspection will be accomplished from areas of public access if a physical inspection is not authorized by the owners or operators of such property.

7. Records Search and Review

7.1 *Introduction*— Reasonable prudence, CERFA requirements (in the case of an EBS performed to support the identification of uncontaminated property), and DoD guidance mandate that the federal real property be evaluated in order to support real property transactions. One component of this evaluation is the review of all reasonably obtainable federal, state, and local government records to determine where, on the installation, there has been storage, release or likely release of any hazardous substance or any petroleum product, and which is likely to cause or contribute to a release or threatened release of any hazardous substance or any petroleum product on the real property.

7.1.1 *Objective*—The objective of the records review is to perform those parts of Steps 1 and 2 of the EBS process pertaining to obtaining and reviewing adequate and complete records that will help the user or environmental professional make an environmental condition of property area type determination regarding the federal real property.

7.1.2 *Accuracy and Completeness*—Accuracy and completeness of record information varies among information sources, including governmental sources. Record information is often inaccurate or incomplete. The user or environmental professional is not obligated to identify mistakes or insufficiencies in information provided. However, the environmental professional reviewing the records shall make a reasonable effort to compensate for mistakes or insufficiencies in the information reviewed that are obvious in light of other information of which the environmental professional has actual knowledge.

7.1.3 *Reasonably Obtainable/Standard Sources*—Availability of record information varies from information source to information source, including governmental jurisdictions. The user or environmental professional is not obligated to identify, obtain, or review every possible record that might exist with respect to a property. Instead, this practice identifies record information that shall be reviewed from standard sources, and the user or environmental professional is required to review only record information that is reasonably ascertainable from those standard sources. Record information that is reasonably obtainable means: (*1*) information that is publicly available, (*2*) information that is obtainable from its source within reasonable time and cost constraints, and (*3*) information that is practically reviewable.

7.1.3.1 *Publicly Available*—Information that is publicly available means that the source of the information allows access to the information by anyone upon request.

7.1.3.2 *Reasonable Time and Cost*—Information that is obtainable within reasonable time and cost constraints means that the information will be provided by the source within a reasonable amount of time of receiving a written, telephone, or in-person request at no more than a nominal cost intended to cover the source's cost of retrieving and duplicating the information. Information that can only be reviewed by a visit to the source is reasonably ascertainable if the visit is permitted by the source within a reasonable amount of time of the request.

7.1.3.3 *Practically Reviewable*—Information that is practically reviewable means that the information is provided by the source in a manner and in a form that, upon examination, yields information relevant to the property without the need for extraordinary analysis of irrelevant data. The form of the information shall be such that the user can review the records for a limited geographic area. Records that cannot be feasibly retrieved by reference to the location of the property or a geographic area in which the property is located are not generally practically reviewable. Most databases of public records are practically reviewable if they can be obtained from the source agency by the county, city, zip code, or other geographic area of the facilities listed in the record system. Records that are sorted, filed, organized, or maintained by the source agency only chronologically are not generally practically reviewable. For large databases with numerous facility records (such as RCRA generators and registered USTs), the records are not practically reviewable unless they can be obtained from the source agency in the smaller geographic area of zip codes. Even when information is provided by zip code for some large databases, it is common for an unmanageable number of sites to be identified within a given zip code. In these cases, it is not necessary to review the impact of all of the sites that are likely to be listed in any given zip code because that information would not be practically reviewable. In other words, when so much data is generated that it cannot be feasibly reviewed for its impact on the property, it is not required to be reviewed.

7.1.4 *Alternatives to Standard Sources*—Alternative sources may be used instead of standard sources if they are of similar or better reliability and detail, or if a standard source is not reasonably ascertainable.

7.1.5 *Coordination*— If records are not reasonably ascertainable from standard sources or alternative sources, the environmental professional shall attempt to obtain the requested information by other means specified in this practice such as questions posed to the current owner or occupant(s) of the property or appropriate persons available at the source at the time of the request.

7.1.6 *Sources of Standard Source Information*—Standard

source information or other record information from government agencies may be obtained directly from appropriate government agencies or from commercial services. Government information obtained from nongovernmental sources may be considered current if the source updates the information at least every 90 days or, for information that is updated less frequently than quarterly by the government agency, within 90 days of the date the government agency makes the information available to the public.

7.1.7 *Documentation of Sources Checked*—The EBS report shall document each source that was used, even if a source revealed no findings. Sources shall be sufficiently described, including name, date request for information was filled, date information provided was last updated by source, date information was last updated by original source (if provided other than by original source; see 7.1.4) so as to facilitate reconstruction of the research at a later date.

7.1.8 *Significance*— If a standard environmental record source (or other sources in the course of conducting the EBS) identifies the property or another site within the approximate minimum search distance, the EBS report shall include the environmental professional's judgment about the significance of the listing to the analysis of recognized environmental conditions in connection with the property (based on the data retrieved pursuant to this section, additional information from the government source, or other sources of information). In doing so, the environmental professional may make statements applicable to multiple sites (for example, a statement to the effect that none of the sites listed is likely to have a negative impact on the property except ...).

7.2 *EBS Step 1: Records Gathering*—In accordance with 6.2.1, this section specifies the general level of effort required to complete EBS Step 1 tasks associated with records gathering. At a minimum, records to be gathered and reviewed when an EBS is initially conducted include the following:

7.2.1 *Background and Physical Setting Records.*

7.2.1.1 *Physical Setting Sources*—A current USGS 7.5 Minute Topographic Map showing the area on which the property is located shall be reviewed, provided it is reasonably obtainable. If a current USGS 7.5 Minute Topographic Map is not readily obtainable, a current 15 Minute Topographic Map showing the area on which the property is located shall be reviewed, provided it is reasonably obtainable. It is the only standard physical setting source and the only physical setting source that is required to be obtained (and only if it is reasonably obtainable). One or more additional physical setting sources may be obtained in the discretion of the environmental professional. Because such sources provide information about the geologic, hydrogeologic, hydrologic, or topographic characteristics of a site, discretionary physical setting sources shall be sought when: (*1*) conditions have been identified in which hazardous substances or petroleum products are likely to migrate to the property or from or within the property into the ground water or soil and (*2*) more information than is provided in the current USGS 7.5 Minute Topographic Map is generally obtained, pursuant to local good commercial or customary practice in initial environmental site assessments in the type of commercial real estate transaction involved, in order to assess the impact of such migration on recognized environmental conditions in connection with the property.

Standard Physical Setting Source: Current USGS
7.5 Minute Topographic Map

If this map is unavailable, obtain the USGS 15 Minute Map, if available. If neither the USGS 7.5 Minute Topographic Map nor the 15 Minute Map are available, a larger scale (for example, 1:250 000) USGS topographic map should be considered. Where appropriate, a comparable topographic map, prepared by the Defense Mapping Agency, may be used instead of the USGS 7.5 Minute Topographic Map.

Other Physical Setting Sources:

The following sources may also be utilized if requested:

- USGS and/or State Geological Survey—Groundwater Maps
- USGS and/or State Geological Survey—Bedrock Geology Maps
- USGS and/or State Geological Survey—Surficial Geology Maps
- Soil Conservation Service—Soil Maps
- Other physical setting sources that are reasonably credible (as well as reasonably ascertainable)

7.2.2 Department of Defense (DoD) Component records maintained on the property or elsewhere, but reasonably obtainable, which are relevant to classification of environmental condition of property area types, including, but not limited to records of:

7.2.2.1 Ongoing and completed site remediation and environmental response activities, including IRP activities, corrective action programs, LUST responses, and similar activities. This includes all records in the Administrative Record maintained under CERCLA.

7.2.2.2 Records of reported spills of hazardous substances and responses.

7.2.2.3 Records of hazardous waste accumulation, storage, treatment, or disposal, including satellite accumulation records, manifests, and records maintained in connection with permitted hazardous waste activities.

7.2.2.4 Records of hazardous substance and petroleum usage and/or storage.

7.2.2.5 Records of potential hazard surveys, including, but not limited to asbestos surveys, lead-based paint surveys, radioactive materials surveys, mercury surveys, PCB surveys, and radon surveys.

7.2.2.6 Environmental compliance records not specifically included in other required records. This includes, but is not limited to Safe Drinking Water Act reports, Clean Water Act permits and discharge reports, Clean Air Act permits and discharge and emission reports, EPCRA reports, hazardous waste minimization plans and reports, and pollution prevention plans and reports.

7.2.2.7 Additional records to include planning maps, base historian records, the base comprehensive plan or base master plan, military construction records, real property records, fire department records, historical photographs, and facility and utility records.

7.2.3 Federal, state, and local agency records.

7.2.4 Recorded chain of title documents.

7.3 *EBS Step 2: Records Analysis*—Upon review of the required records gathered to complete EBS Step 1, the user or environmental professional shall indicate in the EBS Report whether the search revealed any of the following on the property:

7.3.1 Spills of hazardous substances or petroleum products, or both,

7.3.2 Leaks of hazardous substances or petroleum products, or both,

7.3.3 Discharges of hazardous substances or petroleum products, or both,

7.3.4 Leaching of hazardous substances or petroleum products, or both,

7.3.5 Injection of hazardous substances or petroleum products, or both,

7.3.6 Dumping of hazardous substances or petroleum products, or both,

7.3.7 Abandoned or discarded barrels, containers, or other,

7.3.8 Receptacles containing hazardous substances or petroleum products,

7.3.9 Automotive batteries,

7.3.10 Industrial batteries,

7.3.11 Pesticides in containers, cartons, sacks, storage bins, canisters,

7.3.12 Paints in containers, cartons, sacks, storage bins, canisters,

7.3.13 Drums containing hazardous substances or petroleum products, or both,

7.3.14 Tanks containing hazardous substances or petroleum products, or both,

7.3.15 Fill dirt from a contaminated site,

7.3.16 Fill pipes,

7.3.17 PCBs in transformers or capacitors,

7.3.18 Heavy industrial equipment, including hydraulic equipment in storage or use,

7.3.19 Ditches subject to contaminated runoff or discharges,

7.3.20 Railroad loading/unloading areas,

7.3.21 Ordinance,

7.3.22 Medical/biohazardous waste,

7.3.23 Radioactive materials and mixed wastes, or

7.3.24 Mercury, for example, seals.

7.4 If the records review revealed any of the above, the EBS Report shall document it/them.

8. Adjacent Facility Records Search and Review

8.1 *Introduction*— Reasonable prudence, CERFA requirements (in the case of an EBS performed to support the identification of uncontaminated property), and DoD guidance mandate that the federal real property be evaluated in order to categorize real property into applicable environmental condition of property area types. One component of this evaluation is the review of all reasonably obtainable federal, state, and local government records for each adjacent facility where there has been a release or likely release of any hazardous substance or any petroleum product, and which is likely to cause or contribute to a release or threatened release of any hazardous substance or any petroleum product on the real property or which might migrate to the federal real property. In this connection, adjacent has the meaning provided in 3.3.1 and includes those properties near enough to the federal real property to present a reasonable probability of affecting the environmental condition of property on the federal real property.

8.1.1 *Objective*—The objective of the adjacent facility records search and review is to perform those parts of Steps 1 and 2 of the EBS process pertaining to identifying, obtaining, and reviewing those reasonably available Federal, State, and local agency records that might disclose information which would affect the environmental condition of property area type determination regarding the federal real property.

8.1.2 *Approximate Minimum Search Distance*—Adjacent facility records pertain not only to facilities adjacent to the federal real property, but also pertain to properties within an additional approximate minimum search distance in order to help assess the likelihood of problems from migrating hazardous substances or petroleum products. When the term approximate minimum search distance includes areas outside the property, it shall be measured from the nearest property boundary. The term approximate minimum search distance is used instead of radius in order to include irregularly shaped properties.

8.1.2.1 *Reduction of Approximate Minimum Search Distance*—When allowed by 8.2.1.1, the approximate minimum search distance for a particular record may be reduced at the discretion of the environmental professional. Factors to consider in reducing the approximate minimum search distance include: (*1*) the density (for example, urban, rural, or suburban) of the setting in which the property is located; (*2*) the distance that the hazardous substances or petroleum products are likely to migrate based on local geologic or hydrogeologic conditions; and (*3*) other reasonable factors. The justification for each reduction and the approximate minimum search distance actually used for any particular record shall be explained in the EBS report.

8.1.3 *Accuracy and Completeness*—See 7.1.2.

8.1.4 *Reasonably Obtainable/Standard Sources*—See 7.1.3.

8.1.4.1 *Publicly Available*—See 7.1.3.1.

8.1.4.2 *Reasonable Time and Cost*—See 7.1.3.2.

8.1.4.3 *Practically Reviewable*—See 7.1.3.3.

8.1.5 *Alternatives to Standard Sources*—See 7.1.4.

8.1.6 *Coordination*— See 7.1.5.

8.1.7 *Sources of Standard Source Information*—See 7.1.6.

8.1.8 *Documentation of Sources Checked*—See 7.1.7.

8.1.9 *Significance*—See 7.1.8.

8.2 *EBS Step 1: Adjacent Facility Records Gathering*—In accordance with 6.2.1, this section specifies the level of effort required to complete EBS Step 1 tasks associated with adjacent facility records gathering. At a minimum, the following records are to be searched:

8.2.1 *Standard Environmental Sources*—The following standard environmental record sources shall be reviewed, subject to the conditions of 7.1.1-7.1.8:

8.2.1.1 *Standard Environmental Record Sources: Federal and State*—The approximate minimum search distance should be established for each installation or portion of an installation, based on the physical setting and surrounding land use. Table

1 includes recommended approximate minimum search distances. An approximate minimum search distance for a particular record may be reduced at the discretion of the environmental professional. Factors to consider in reducing the approximate minimum search distance include: (*1*) the density (for example, urban, rural, or suburban) of the setting in which the property is located; (*2*) the distance that the hazardous substances or petroleum products are likely to migrate based on local geologic or hydrogeologic conditions; and (*3*) other reasonable factors. The justification for each reduction and the approximate minimum search distance actually used for any particular record should be explained in the EBS report.

8.2.2 *Additional Environmental Record Sources: State or Local*—One or more additional state sources or local sources of environmental records may be checked, at the discretion of the environmental professional, to enhance and supplement federal and state sources identified in Table 1. Factors to consider in determining which local or additional state records, if any, should be checked include: (*1*) whether they are reasonably ascertainable, (*2*) whether they are sufficiently useful, accurate, and complete in light of the objective of the records review (see 7.1.1), and (*3*) whether they are generally obtained, pursuant to local good commercial or customary practice, in initial environmental site assessments in the type of commercial real estate transaction involved. To the extent additional state sources or local sources are used to supplement the same record types listed in Table 1, approximate minimum search distances should not be less than those specified (adjusted as provided in 8.1.2.1 and 8.2.1.1). Some types of records and sources that may be useful include:

Types of Local Records

Lists of Landfill/Solid Waste Disposal Sites
Lists of Hazardous Waste/Contaminated Sites
Lists of Registered Underground Storage Tanks
Records of Emergency Release Reports (SARA Part 304)
Records of Contaminated Public Wells

Local Sources

Department of Health/Environmental Division
Fire Department
Planning Department
Building Permit/Inspection Department
Local/Regional Pollution Control Agency
Local/Regional Water Quality Agency

TABLE 1 Recommended Approximate Minimum Search Distances

Record(s) Source	Approximate Minimum Search Distance, miles (kilometres)
Federal NPL site list	1.0 (1.6)
Federal CERCLIS list	0.5 (0.8)
Federal RCRA TSD facilities list	1.0 (1.6)
Federal RCRA generators list	property and adjoining properties
Federal ERNS list	property only
State lists of hazardous waste sites Identified for investigation or remediation (NPL and CERCLIS equivalents)	1.0 (1.6)
State landfill and/or solid waste disposal site lists	0.5 (0.8)
State leaking UST lists	0.5 (0.8)
State registered UST lists	property and adjoining properties

Local Electric Utility Companies (for records relating to PCBs)

8.2.3 *Physical Setting Sources*—See 7.2.1.

8.3 *EBS Step 2: Adjacent Facility Records Analysis*—Upon review of adjacent facility records listed in 8.2, the environmental professional shall indicate in the EBS report whether or not the search revealed any of the following on the property:

8.3.1 Potential or actual migration of hazardous substances or petroleum products, or both, into the area in question from sources of these substances.

8.3.2 The presence of actual sources of hazardous substances or petroleum products, or both, on adjacent property and facilities with suspected migration that has not been evaluated or characterized.

8.3.3 Uncontrolled migration of hazardous substances or petroleum products, or both, in the immediate vicinity of a boundary of the area in question.

8.4 If the records review revealed any of the preceding, the EBS report shall document it/them.

9. Aerial Photography Analysis

9.1 *Introduction*— Analysis of aerial photography can provide an extremely useful source of supplemental information regarding both land use and the environmental condition of property area type. This analysis may encompass both the federal real property and the adjacent facilities. Analysis of aerial photography should be focused on patterns of land use and human activities as well as direct and indirect evidence of the potential existence of a recognized environmental condition.

9.1.1 *Objective*—The objective of the aerial photography analysis is to perform those parts of Steps 1 and 2 of the EBS process pertaining to assembling, if reasonably available, an adequately complete set of adjacent properties encompassing both the federal real property and the adjacent facilities, analyzing those photographs for patterns of land use and human activities as well as direct and indirect evidence of the potential existence of a recognized environmental condition, and thereafter, incorporating that information into the overall environmental condition of property area type. Aerial photographs will be sought for the period encompassing the past 60 years of facility use.

9.2 *EBS Step 1: Aerial Photography Gathering*—In accordance with 6.2.1, this section specifies the level of effort required to complete EBS Step 1 tasks associated with identifying and gathering adjacent properties. At a minimum, the following sources are to be searched:

9.2.1 *Standard Sources: Aerial Photographs*—Aerial photographs are commonly available from government agencies or private collections unique to a local area, and may also be obtained from universities, colleges, and history museums.

9.2.2 *Optional Sources*— Other sources of similar information, including, but not limited to satellite reconnaissance may be sought where the environmental professional feels that these records would be reasonably available and would materially enhance the analysis required in this section.

9.3 *EBS Step 2: Aerial Photography Analysis*—The environmental professional shall perform the analysis of photographs using an accepted photo interpretation land use and land cover classification system. Analyzed photographs should preferably be at a 1:24 000 scale or smaller scale (that is, more magnified) to reveal adequate surface detail and necessary spatial coverage. Upon review of adjacent properties in accordance with the previously stated guidelines, the environmental professional shall specify the analysis method used in the EBS Report and also indicate whether any of the following were identified on the property:

9.3.1 Evidence of excavation activities of unknown type, or of industrial operations.

9.3.2 Evidence of dumping or disposing of waste materials.

9.3.3 Evidence of significant storage activities involving drums, tanks, or pipelines containing hazardous substances or petroleum products.

9.3.4 Evidence of staining associated with industrial activities or activities of unknown origin or type.

9.4 If the aerial photography analysis revealed any of the preceding, the EBS report shall document it/them.

10. Interviews

10.1 *Introduction*— Reasonable prudence, CERFA requirements (in the case of an EBS performed to support the identification of uncontaminated property), and DoD guidance mandate that the federal real property be evaluated in order to categorize real property into applicable environmental condition of property area types. One component of this evaluation are interviews of current or former occupants, or both, involved in operations on the property that are conducted to aid in identifying a recognized environmental condition on the federal real property and other information necessary to determine standard environmental condition of property area types. Interviews of current or former personnel, or both, on the federal real property and interviews of appropriate local government officials supplement documented information and may also provide keys to effective interpretation of such information. Where available information indicates the presence of recognized environmental conditions, it may be necessary to conduct interviews relating to those portions of the installation that have been, or are currently being, actively investigated under the IRP or similar investigation effort.

10.1.1 *Objective*—The objective of conducting interviews is to perform those parts of Steps 1 and 2 of the EBS process pertaining to obtaining information indicating recognized environmental conditions in connection with the property, so that environmental condition of property area type determinations can be made.

10.2 *EBS Step 1: Interviews with Site Personnel*—In accordance with 6.2.1, this section specifies the level of effort required to complete EBS Step 1 tasks associated with interviewing current or former personnel, or both, and appropriate local governmental officials. At a minimum, the following should be incorporated into this process:

10.2.1 *Content*—Interviews with site personnel consist of questions to be asked in the manner and of persons as described in this section. The content of questions to be asked shall attempt to obtain information about uses and conditions as described in Section 9, as well as the information described in Sections 11-13.

10.2.2 *Medium*—Questions to be asked pursuant to this section may be asked in person, by telephone, or in writing, at the discretion of the environmental professional.

10.2.3 *Timing*—Except as specified in 9.8 and 9.9, it is at the discretion of the environmental professional whether to ask questions before, during, or after the site visit described in Section 11, or in some combination thereof.

10.2.4 *Who Should be Interviewed*:

10.2.4.1 *Key Site Manager*— Prior to the site visit, the owner can be asked to identify a person with good knowledge of the uses and physical characteristics of the property (the key site manager). Often the key site manager will be the installation or base commander, base civil engineer, public works commander or other property manager, chief physical plant supervisor or head maintenance person. (If the user is the current property owner, the user has an obligation to identify a key site manager, even if it is the user himself or herself.) If a key site manager is identified, the person conducting the site visit shall make at least one reasonable attempt (in writing or by telephone) to arrange a mutually convenient appointment for the site visit when the key site manager agrees to be there. If the attempt is successful, the key site manager shall be interviewed in conjunction with the site visit. If such an attempt is unsuccessful, when conducting the site visit, the environmental professional shall inquire whether an identified key site manager (if any) or if a person with good knowledge of the uses and physical characteristics of the property is available to be interviewed at that time; if so, that person shall be interviewed. In any case, it is within the discretion of the environmental professional to decide which questions to ask before, during, or after the site visit or in some combination thereof.

10.2.4.2 *Occupants, Including Current and Former Employees*—A reasonable attempt shall be made to interview a reasonable number of occupants of, if any, and current employees involved in operations on the property.

(1) Residential Properties—For residential properties, residential occupants do not need to be interviewed, but if the property has nonresidential uses, interviews can be held with the nonresidential occupants based on criteria specified in 10.2.4.2.

(2) Major Occupants—Except as specified in *residential properties*, if the property has five or fewer current occupants, a reasonable attempt shall be made to interview a representative of each one of them. If there are more than five current occupants, a reasonable attempt shall be made to interview the major occupant(s) and those other occupants whose operations are likely to indicate recognized environmental conditions in connection with the property.

(3) Reasonable Attempts to Interview—Examples of reasonable attempts to interview those occupants specified in *major occupants* include (but are not limited to) an attempt to interview such occupants when making the site visit or calling such occupants by telephone. In any case, when there are several occupants to interview, it is not expected that the site visit must be scheduled at a time when they will all be available to be interviewed.

(4) Occupant Identification—The EBS report shall identify the occupants interviewed and the duration of their occupancy.

10.2.4.3 *Local Agency Officials*—A reasonable attempt shall be made to interview at least one staff member of any one of the following types of local government agencies:

(1) Local fire department that serves the property,

(2) Local health agency or local/regional office of state health agency serving the area in which the property is located, or

(3) Local agency or local/regional office of state agency having jurisdiction over hazardous waste disposal or other environmental matters in the area in which the property is located.

10.2.5 *Prior Assessment Usage*—Persons interviewed as part of a prior EBS consistent with this practice do not need to be questioned again about the content of answers they provided at that time. However, they can be questioned about any new information learned since that time, or others can be questioned about conditions since the prior EBS consistent with this practice.

10.2.6 *Quality Of Answers*—The person(s) interviewed should be asked to be as specific as reasonably feasible in answering questions. The person(s) interviewed should be asked to answer in good faith and to the extent of their knowledge.

10.2.7 *Incomplete Answers*—In accordance with 10.2.7.1 and 10.2.7.2, the person conducting the interview(s) has an obligation to ask questions, in certain instances the persons to whom the questions are addressed may not have an obligation to answer them.

10.2.7.1 *User/DoD Component Personnel*—If the person to be interviewed is the user (an employee of a DoD component on whose behalf the EBS is being conducted), the user has an obligation to answer all questions posed by the person conducting the interview, in good faith, to the extent of his or her actual knowledge, or to designate a key site manager to do so. If answers to questions are unknown or partially unknown to the user or such key site manager, this interview section of the EBS shall not thereby be deemed incomplete.

10.2.7.2 *Non-User*—If the person conducting the interview(s) asks questions of a person other than a user but does not receive answers or receives partial answers, this section of the EBS shall not thereby be deemed incomplete, provided that (*1*) the questions have been asked (or attempted to be asked) in person or by telephone and written records have been kept of the person to whom the questions were addressed and the responses, or (*2*) the questions have been asked in writing sent by first class mail or by private, commercial carrier and no answer or incomplete answers have been obtained and at least one reasonable follow-up (telephone call or written request) was made again asking for responses.

10.2.8 *Questions About Helpful Documents*—Prior to the site visit, the property owner, key site manager (if any is identified), and user (if different from the property owner) shall be asked if they know whether any of the documents listed in 10.2.8.1 exist and, if so, whether copies can and will be provided to the environmental professional within reasonable time and cost constraints. Even partial information provided may be useful. If so, the environmental professional conducting the site visit shall review the available documents prior to or at the beginning of the site visit.

10.2.8.1 *Helpful Documents*:

(1) Environment site assessment reports, including PA, SI, or other similar reports.

(2) Environment audit reports.

(3) Environmental permits (for example, solid waste disposal permits, hazardous waste disposal permits, wastewater permits, NPDES permits).

(4) Registrations for underground and above-ground storage tanks.

(5) Material safety data sheets.

(6) Community right-to-know plan.

(7) Safety plans; preparedness and prevention plans; spill prevention, countermeasure, control plans, and so forth.

(8) Reports regarding hydrogeologic conditions of the property or surrounding area.

(9) Notices or other correspondence from any government agency relating to past or current violations of environmental laws with respect to the property or relating to environmental liens encumbering the property.

(10) Hazardous waste—generator notices or reports.

(11) Geotechnical studies.

10.2.9 *Proceedings Involving the Property*—Prior to the site visit, the property owner, key site manager (if any is identified), and user (if different from the property owner) shall be asked whether they know of: (*1*) any pending, threatened, or past litigation relevant to hazardous substances or petroleum products in, on, or from the property, (*2*) any pending, threatened, or past administrative proceedings relevant to hazardous substances or petroleum products in, on, or from the property, and (*3*) any notices from any governmental entity regarding any possible violation of environmental laws or possible liability relating to hazardous substances or petroleum products.

10.2.10 *Interview Questions*—The person conducting each interview shall, at a minimum, ask and document responses to the following questions:

10.2.10.1 Was or is the area in question used as a gasoline station, motor repair facility, dry cleaners, photo developing laboratory, plating shop, medical or dental facility, junkyard or landfill, training area, or as a waste treatment, storage, disposal, processing, or recycling facility?

10.2.10.2 Has there been any damaged or discarded automotive or industrial batteries, or pesticides, paints, or other chemical or individual containers stored or used in the area in question?

10.2.10.3 Are there drums, sacks, cartons, or other containers of chemicals located on the property in question?

10.2.10.4 Was or is the area in question used for any waste generation or disposal activities?

10.2.10.5 Was or is the area in question used as a firing or bombing range, or both?

10.2.10.6 Have there been or are there storage tanks containing hazardous substances or petroleum products located on the property in question?

10.2.10.7 Have spills, leaks, or other releases of hazardous substance or petroleum products occurred to the best of your knowledge?

10.2.10.8 Have unidentified waste materials, tires, automotive or industrial batteries, ordnance or any other waste materials been dumped, buried, or burned, or a combination thereof, in the area in question?

10.3 *EBS Step 2: Interview Analysis*—Upon completion of Step 1 of the EBS process related to interviews, the environmental professional shall refer to 7.3 and indicate in the EBS report whether or not the responses revealed any of the items in 7.3 on the property. If the responses revealed any of these items, the EBS report shall document it/them.

11. Visual and Physical Inspections

11.1 *Introduction*— Reasonable prudence, CERFA requirements (in the case of an EBS performed to support the identification of uncontaminated property), and DoD guidance mandate that the federal real property be evaluated in order to categorize real property into applicable environmental condition of property area types. One element of this evaluation is the requirement that visual inspections be conducted of the installation property and adjacent property. The visual inspection supplements the documentary record, including interviews, records developed during records search, aerial photography analysis, and the other components of the EBS. The terms "visual inspection" and "visual site inspection" as used in this practice are synonymous. For the purpose of this practice, visual inspection includes two similar, but distinct inspections, the visual inspection of the federal real property and the visual inspection of the adjacent property. The visual inspection of the federal real property will be conducted as described in 11.3. The visual and physical inspection of the adjacent property will be conducted as described in 11.4. The environmental professional and the user should exercise reasonable prudence in scoping and executing the tasks described in this section, recognizing that it is not the purpose of the EBS visual inspection to take the place of other regulatory program requirements, for example, asbestos surveys.

11.2 *Objective*—The objective of the visual inspection is to perform those parts of Steps 1 and 2 of the EBS process pertaining to visually obtaining information indicating the likelihood of recognized environmental conditions in connection with the property, so that environmental condition of property area type determinations can be made.

11.3 *EBS Step 1: Visual Inspection*—In accordance with 6.2.1, this section specifies the level of effort required to complete EBS Step 1 tasks associated with making a visual inspection of the installation property. At a minimum, the following should be incorporated into this process:

11.3.1 *Observation*— On a visit to the property (the site visit), the environmental professional shall visually and physically observe the property and any structure(s) located on the property to the extent not obstructed by bodies of water, adjacent buildings, or other obstacles. The environmental professional shall also prepare a visual inspection record of the visit. The visual inspection record, which can be field notes and the like, will be used to prepare the EBS report preserved to supplement the EBS report.

11.3.1.1 *Exterior*—The periphery of the property shall be visually and physically observed, as well as the periphery of all structures on the property, and the property should be viewed from all adjacent public thoroughfares. If roads or paths with no apparent outlet are observed on the property, the use of the road or path should be identified to determine whether it was likely to have been used as an avenue for disposal of hazardous substances or petroleum products. For each road without an apparent outlet, the environmental professional shall travel or overfly the length of the road and observe whether areas adjacent to the road appear to have been used for waste disposal.

11.3.1.2 *Interior*—On the interior of structures on the property, accessible common areas expected to be used by occupants or the public (such as lobbies, hallways, utility rooms, recreation areas, and so forth), maintenance and repair areas, including boiler rooms, and a representative sample of occupant spaces, should be visually and physically observed. It is generally not necessary to look under floors, above ceilings, or behind walls. The environmental professional shall exercise best professional judgment and appropriately coordinate with the DoD component when observing the interior of structures on the property.

11.3.1.3 *Methodology*— The environmental professional shall document, in the EBS report, the method used (for example, grid patterns or other systematic approaches used for large properties, which spaces for owner or occupants were observed) to observe the property. For example, representative inspections may be appropriate for structures: (*1*) that were built using like construction methods, during a fixed time period, (*2*) are geographically co-located, (*3*) were subjected to similar categorical use throughout their history, and (*4*) have no significant differences in their other investigations criteria (for example, records search, interviews, and aerial photos). Common examples would include: military housing units, barracks, officer and enlisted quarters. In consultation with the user, the environmental professional may determine the percentage of structures/properties to be inspected and establish an inspection pattern that provides sufficient representation of the area in question.

11.3.1.4 *Limitations*— The environmental professional shall document, in the EBS report, general limitations and bases of review, including limitations imposed by physical obstructions such as adjacent buildings, bodies of water, asphalt, or other paved areas, and limiting conditions (for example, snow and rain).

11.3.1.5 *Frequency*—It is not expected that more than one visit to the property shall be made by the environmental professional in connection with an EBS. The one visit constituting part of the EBS may be referred to as the site visit.

11.3.2 *Prior EBS Usage*— The information supplied in connection with the visual inspection portion of a prior EBS may be used for guidance but shall not be relied upon without determining through a new visual inspection whether any conditions that are material to recognized environmental conditions in connection with the property have changed since the prior EBS.

11.3.3 *Uses and Conditions*—The environmental professional(s) conducting the visual inspection should note the uses and conditions specified in 11.3.3.1-11.3.3.4 to the extent visually or physically observed during the site visit. The uses and conditions specified in 11.3.3.1-11.3.3.4 should also be the subject of questions asked as part of interviews of owners and occupants (see Section 10). Uses and conditions to be noted shall be recorded in field notes of the environmental professional(s) conducting the visual inspection but are only required to be described in the EBS report to the extent specified in 11.3.3.1-11.3.3.4. The environmental professional(s) performing the EBS are obligated to identify uses and conditions only to the extent that they may be visually and physically observed on a site visit, as described in this practice, or to the extent that they are identified by the interviews (see Sections 9 and 10) or records search (see Section 7) processes described in this practice.

11.3.3.1 *General Site Setting*:

(1) Current Use(s) of the Property—The current use(s) of the property shall be identified in the EBS report. Any current uses likely to involve the use, treatment, storage, disposal, or generation of hazardous substances or petroleum products shall be identified in the EBS report. Unoccupied occupant spaces should be noted. In identifying current uses of the property, more specific information is more helpful than less specific information. (For example, it is more useful to identify uses such as a commissary or base exchange rather than simply retail use.)

(2) Past Use(s) of the Property—To the extent that indications of past uses of the property are visually or physically observed on the site visit, or are identified in the interviews or record review, they shall be identified in the EBS report, and past uses so identified shall be described in the EBS report if they are likely to have involved the use, treatment, storage, disposal, or generation of hazardous substances or petroleum products. (For example, there may be signs indicating a past use or a structure indicating a past use.)

(3) Geologic, Hydrogeologic, Hydrologic, and Topographic Conditions—The topographic conditions of the property shall be noted to the extent visually or physically observed or determined from interviews, as well as the general topography of the area surrounding the property that is visually or physically observed from the periphery of the property. If any information obtained shows there are likely to be hazardous substances or petroleum products on the property or on nearby properties and those hazardous substances or petroleum products are of a type that may migrate, topographic observations shall be analyzed in connection with geologic, hydrogeologic, hydrologic, and topographic information obtained pursuant to records review (see 7.2.3) and interviews to evaluate whether hazardous substances or petroleum products are likely to migrate to the property, or within or from the property, into ground water or soil.

(4) General Description of Structures—Generally, the EBS report shall describe the structures or other improvements on the property for example: number of buildings, number of stories each, approximate age of buildings, ancillary structures (if any), and so forth.

(5) Thoroughfares—Public thoroughfares adjoining the property shall be identified in the EBS Report, and any roads, streets, railroads, and parking facilities on the property shall be described in the EBS Report.

(6) Potable Water Supply—The source of potable water for the property shall be identified in the EBS report.

(7) Sewage Disposal System—The sewage disposal system for the property shall be identified in the EBS report. Inquiry shall be made as to the age of the system as part of the process under Sections 7, 9, or 10.

(8) Storm Drains—The storm drains for the property shall be identified in the EBS report.

(9) Access Vaults—Access vaults occurring on the property shall be identified in the EBS report.

11.3.3.2 *Interior and Exterior Observations*:

(1) Current Use(s) of the Property—The current use(s) of the property shall be identified in the EBS report. Any current uses likely to involve the use, treatment, storage, disposal, or generation of hazardous substances or petroleum products shall be identified in the EBS Report. Unoccupied occupant spaces should be noted. In identifying current uses of the property, more specific information is more helpful than less specific information. (For example, it is more useful to identify uses such as a commissary or base exchange rather than simply retail use.)

(2) Past Use(s) of the Property—To the extent that indications of past uses of the property are visually or physically observed on the site visit, or are identified in the interviews or records review, they shall be identified in the EBS report, and past uses so identified shall be described in the EBS report if they are likely to have involved the use, treatment, storage, disposal, or generation of hazardous substances or petroleum products. (For example, there may be signs indicating a past use or a structure indicating a past use.)

(3) Hazardous Substances and Petroleum Products in Connection with Identified Uses—To the extent that present uses are identified that use, treat, store, dispose of, or generate hazardous substances and petroleum products on the property. The hazardous substances and petroleum products shall be identified or indicated as unidentified in the EBS report, and the approximate quantities involved, types of containers (if any) and storage conditions shall be described in the EBS report. To the extent that past uses are identified that used, treated, stored, disposed of, or generated hazardous substances and petroleum products on the property, the information shall be identified to the extent it is visually or physically observed during the site visit or identified from the interviews or the records review.

(4) Storage Tanks—Above-ground storage tanks, or solvents or vent pipes, fill pipes or access ways indicating solvents shall be identified (for example, content, capacity, and age) to the extent visually or physically observed during the site visit or identified from the interviews or records review.

(5) Odors—Strong, pungent, or noxious odors shall be described in the EBS report and their sources shall be identified in the EBS report to the extent visually or physically observed or identified from the interviews or records review.

(6) Pools of Liquid—Standing surface water (other than common rain puddles) shall be noted. Pools or sumps containing liquids likely to be hazardous substances or petroleum products shall be described in the EBS report to the extent

visually or physically observed or identified from the interviews or records review.

(7) Drums—To the extent visually or physically observed or identified from the interviews or records review, drums shall be described in the EBS report, whether or not they are leaking, unless it is known that their contents are not hazardous substances or petroleum products (in that case, the contents should be described in the EBS report). Drums commonly hold 55 g (208 L) of liquid, but containers as small as 5 g (19 L) should also be described.

(8) Hazardous Substance and Petroleum Products Containers (Not Necessarily in Connection With Identified Uses)—When containers identified as containing hazardous substances or petroleum products are visually or physically observed on the property and are or might be a recognized environmental condition: the hazardous substances or petroleum products shall be identified or indicated as "unidentified" in the EBS report, and the approximate quantities involved, types of containers (for example, cartons, tanks, cans), and storage conditions shall be described in the EBS report.

(9) Unidentified Substance Containers—When open or damaged containers containing unidentified substances suspected of being hazardous substances or petroleum products are visually or physically observed on the property, the approximate quantities involved, types of containers, and storage conditions shall be described in the EBS report.

(10) PCBs—Electrical or hydraulic equipment known to contain PCBs or likely to contain PCBs shall be described in the EBS report to the extent visually or physically observed or identified from the interviews or records review. Fluorescent light ballast likely to contain PCBs does not need to be noted.

11.3.3.3 *Interior Observations*:

(1) Heating/Cooling—The means of heating and cooling the buildings on the property, including the fuel source for heating and cooling, shall be identified in the EBS report (for example, heating oil, gas, electric, radiators from steam boiler fueled by gas).

(2) Stains or Corrosion—To the extent visually or physically observed or identified from the interviews, stains or corrosion on floors, walls, or ceilings shall be described in the EBS report, except for staining from water.

(3) Drains and Sumps—To the extent visually or physically observed or identified from the interviews, floor drains and sumps shall be identified in the EBS report. Floor drains and sumps where a release or suspected release occurred should be further described in the EBS report.

11.3.3.4 *Exterior Observations*:

(1) Pits, Ponds, or Lagoons—To the extent visually or physically observed or identified from the interviews or records review, pits, ponds, or lagoons on the property shall be identified in the EBS report, and further described if they have been used in connection with waste disposal or waste treatment. Pits, ponds, or lagoons on properties adjoining the property shall be described in the EBS report to the extent they are visually or physically observed from the property or identified in the interviews or records review.

(2) Stained Soil or Pavement—To the extent visually or physically observed or identified from the interviews, areas of stained soil or pavement shall be described in the EBS report.

(3) Stressed Vegetation—To the extent visually or physically observed or identified from the interviews, areas of stressed vegetation (from something other than insufficient water) shall be described in the EBS report.

(4) Solid Waste—To the extent visually or physically observed or identified from the interviews or records review, areas that are apparently filled or graded by nonnatural causes (or filled by fill of unknown origin) suggesting trash or other solid waste disposal, or mounds or depressions suggesting trash or other solid waste disposal, shall be described in the EBS report.

(5) Waste Water—To the extent visually or physically observed or identified from the interviews or records review, waste water or other liquid (including storm water) or any discharge into a drain, ditch, or stream on or adjacent to the property shall be described in the EBS report.

(6) Wells—To the extent visually or physically observed or identified from the interviews or records review, all wells (including dry wells, irrigation wells, injection wells, abandoned wells, or other wells) shall be described in the EBS report.

(7) Septic Systems—To the extent visually or physically observed or identified from the interviews or records review, indications of on-site septic systems or cesspools should be described in the EBS report.

11.4 *Visual Inspection of Adjacent Property*—The visual inspection of adjacent property will be accomplished when consent has been obtained to enter the adjacent facility. The user or environmental professional shall visually and physically observe the adjacent property to evaluate and identify, if possible, conditions which could give rise to recognized environmental conditions on the property, through migration or other transport of petroleum products or hazardous substances.

11.4.1 *Consent of Owner/Operator of Adjacent Property*—Access to the adjacent property is required to undertake a visual and physical inspection of the adjacent property. Written consent will be sought by either the DoD Component or the environmental professional. Where voluntary consent is obtained, a visual and a physical inspection will be conducted, as described in this section. If consent is not given, a visual inspection will be conducted from outside the facility boundaries, for example, from public rights of way, or adjacent facilities where consent has been given and from other suitable sites. If voluntary consent is not obtained, no direct access to the adjacent property will be sought through involuntary means, for example, condemnation of an easement or right of entry or obtaining an administrative warrant.

11.4.2 *General Site Setting*:

11.4.2.1 *Current Uses of Adjacent Properties*—To the extent that current uses of adjacent properties are visually or physically observed on the site visit, or are identified in the interviews or records review, they shall be identified in the EBS report, and current uses so identified shall be described in the EBS report if they are likely to indicate recognized environmental conditions in connection with the adjoining properties or the property.

11.4.2.2 *Past Uses of Adjacent Properties*—To the extent

that indications of past uses of adjacent properties are visually or physically observed on the site visit, or are identified in the interviews or record review, they shall be noted by the environmental professional, and past uses so identified shall be described in the EBS report if they are likely to indicate recognized environmental conditions in connection with the adjacent properties or the property.

11.4.3 *Current or Past Uses in the Surrounding Area*—To the extent that the general type of current or past uses (for example, residential, commercial, industrial) of properties surrounding the property are visually or physically observed on the site visit or going to or from the property for the site visit, or are identified in the interviews or record review, they shall be noted by the environmental professional, and uses so identified shall be described in the EBS report if they are likely to indicate recognized environmental conditions in connection with the property.

11.5 *EBS Step 2: Visual Inspection Analysis*—If the visual inspection revealed any conditions indicating the storage, release, or disposal of hazardous substances or petroleum products, the EBS report shall document them.

12. Contamination Source Identification

12.1 *Introduction*— The identification of sources of actual or potential contamination is required by published DoD policy statements relating to each of the actions that require an EBS, particularly the DoD FOST guidance. It is anticipated that, in the ordinary course of events, all relevant records to accomplish this task will be obtained through the records search and review tasks, interview tasks, site visit tasks, and other data collection efforts prescribed in this practice. The principal additional goal of this section is to fulfill Step 1 and Step 2 EBS requirements, by analyzing information and data, where necessary to specifically identify sources of actual or potential contamination (recognized environmental conditions), leading to the identification of an environmental condition of property area type.

12.2 *Objective*—Identify sources of actual or potential contamination so as to evaluate whether they affect or may affect the categorization of the environmental condition of the property, including whether all required remedial action has been taken.

12.3 *EBS Step 1: Contamination Source Identification*—In accordance with 6.2.1, this section specifies the level of effort required to complete EBS Step 1 tasks associated with contamination source identification. At a minimum, this step should be completed by following the procedures described in Sections 7-11. If current records are incomplete regarding the nature and extent of contaminant sources, the user or environmental professional will note the status of records search efforts and which records remain incomplete.

12.4 *EBS Step 2: Contamination Source Identification Analysis*—If records are complete for a contaminant source, its nature and extent will be used to determine the environmental condition of the real property. If current records are incomplete regarding the nature and extent of contaminant sources, the user or environmental professional will note the strategy for completing relevant records so that a determination other than environmental condition of property area Type 7 can be made (for example, "contaminant source currently being characterized through remedial investigation; anticipated completion in month/year").

13. Ongoing Response Actions

13.1 *Introduction*— The identification of ongoing response actions is required by published DoD policy statements relating to each of the actions that require an EBS, particularly the DoD FOST guidance. It is anticipated that, in the ordinary course of events, all relevant records to accomplish this task will be obtained through the records search and review tasks, interview tasks, site visit tasks, and other data collection efforts prescribed in this practice. The principal additional goal of this section is to fulfill Step 1 and Step 2 EBS requirements, by analyzing information and data, where necessary to specifically identify ongoing response actions and determine relevant information regarding effectiveness and completeness of any ongoing response actions, so that an accurate environmental condition of property area type determination can be made.

13.2 *Objective*—Identify ongoing response actions and determine what portion of the property is affected and whether the environmental condition of the property satisfies the requirements of CERCLA § 120(h)(3), including whether all required remedial action has been taken.

13.3 *EBS Step 1: Ongoing Response Actions*—In accordance with 6.2.1, this section specifies the level of effort required to complete EBS Step 1 tasks associated with identification of ongoing response actions at the installation or on properties that would affect environmental condition of property area type determinations for all or parts of the installation. At a minimum, this step should be completed by following the procedures described in Sections 7-11. If current records are incomplete regarding ongoing response actions, the user or environmental professional will note the status of records search efforts and which records remain incomplete.

13.4 *EBS Step 2: IRP and Other Ongoing Response Action Analysis*—The purpose of the Step 2 analysis is to support the objective of 13.2. This analysis can be critical to the differentiation between or among certain environmental condition of property area types (for example, between area Type 5 areas, which are not transferrable by deed, and area Type 4 areas, which are), as performed as EBS Step 3.

14. Determining Environmental Condition of Property Area Type

14.1 *Introduction*— The user or environmental professional will generally take information obtained from the activities described in Sections 7-13 and classify federal real property into the seven standard environmental condition of property area types identified in the Standard Classification. Steps 1 and 2 of the EBS process are intended to collect and analyze information that will provide the basis for assigning, in EBS Step 3, an environmental condition of property (ECP) area type for each portion of the installation. The actual classification of any portion of installation property is based on both the information and analyses at hand and on a variety of site-specific conditions (for example, geologic conditions, hydrologic conditions, nature, and extent of contamination, and so forth) and, with the exception of the general guidelines presented in this section, is beyond the scope of this practice. Ideally, the classification of any property into an ECP area type should reflect a consensus among the Military Department and the federal and state regulators. In general, Steps 1 and 2 of the EBS process will have accomplished their purpose if they provide all information and analyses required to make accurate Step 3 ECP area type determinations.

14.2 *Objective*—The objective of this section is to identify a general process to guide the EBS Step 3 classification of areas of federal real property into one of seven standard ECP area types. The ECP area type classification is a tool that is intended to provide an EBS user with a complete and accurate "snapshot" of relevant aspects of the environmental condition of installation property in support of property transfer and reuse decisions (see also Section 1).

14.3 *General Process for EBS Step 3—Determining ECP Area Types*:

14.3.1 *Process for Determining ECP Area Type 1 Property*—Complete Steps 1 and 2 of the EBS process (see 6.2.1 and 6.2.2). Identify ECP area Type 1 property in accordance with CERFA criteria (see 1.1.3). The ECP area Type 1 property may be identified as either part of the CERFA process or subsequent to the CERFA statutory deadline.

14.3.2 *Process for Determining ECP Area Type 2 Property*—Complete Steps 1 and 2 of the EBS process (see 6.2.1 and 6.2.2). Identify ECP area Type 2 property in accordance with CERFA criteria (see 1.1.3). Identify real property where only storage for less than one year occurred (CERFA uncontaminated) and where only storage occurred for more than one year. Real property may be proposed as ECP area Type 2 through the CERFA process or subsequent to the CERFA statutory deadline.

14.3.3 *Process for Determining ECP Area Type 3 Property*—Complete Steps 1 and 2 of the EBS process (see 6.2.1 and 6.2.2). If a determination can be made that concentrations of hazardous substances or petroleum products are below action levels, in accordance with the criteria contained in the Standard Classification, the real property may be classified as ECP area Type 3. The ECP area Type 3 real property may be identified at any time after completion of Steps 1 and 2 of the EBS process.

14.3.4 *Process for Determining ECP Area Type 4 Property*—Complete Steps 1 and 2 of the EBS process (see 6.2.1 and 6.2.2). If the determination can be made that all required remedial actions have been taken, in accordance with CERCLA and the criteria contained in the Standard Classification, the real property may be classified as ECP area Type 4. The ECP area Type 4 real property may be identified at any time after completion of Steps 1 and 2 of the EBS process.

14.3.5 *Process for Determining ECP Area Type 5 Property*—Complete Steps 1 and 2 of the EBS process (see 6.2.1 and 6.2.2). If the determination can be made that a remedy has been selected but that all required remedial actions have not yet been taken, in accordance with CERCLA and the criteria contained in the Standard Classification, the real property may be classified as ECP area Type 5. The ECP area Type 5 real property may be identified at any time after completion of Steps *1* and *2* of the EBS process.

14.3.6 *Process for Determining ECP Area Type 6 Property*—Complete Steps 1 and 2 of the EBS process (see 6.2.1 and 6.2.2). If the determination can be made that concentrations of hazardous substances or petroleum products are above action levels, in accordance with the criteria contained in federal, state, and local statutes and the Standard Classification, the real property should be designated as ECP area Type 6. The ECP area Type 6 real property may be identified at any time after completion of Steps 1 and 2 of the EBS process.

14.3.7 *Process for Determining ECP Area Type 7 Property*—Complete Steps 1 and 2 of the EBS process (see 6.2.1 and 6.2.2). If the real property cannot be conclusively categorized into area Types 1 through 6, as defined in the Standard Classification, the real property should be designated as ECP area Type 7. The ECP area Type 7 real property may be identified at any time after completion of Steps 1 and 2 of the EBS process.

15. Keywords

15.1 environment; environmental assessment; environmental baseline; environmental condition; environmental condition of property; restoration; site assessment; site characterization; site remediation

Designation: D 6009 – 96 (Reapproved 2001)

Standard Guide for Sampling Waste Piles[1]

This standard is issued under the fixed designation D 6009; the number immediately following the designation indicates the year of original adoption or, in the case of revision, the year of last revision. A number in parentheses indicates the year of last reapproval. A superscript epsilon (ε) indicates an editorial change since the last revision or reapproval.

1. Scope

1.1 This guide provides guidance for obtaining representative samples from waste piles. Guidance is provided for site evaluation, sampling design, selection of equipment, and data interpretation.

1.2 Waste piles include areas used primarily for waste storage or disposal, including above-grade dry land disposal units. This guide can be applied to sampling municipal waste piles.

1.3 This guide addresses how the choice of sampling design and sampling methods depends on specific features of the pile.

1.4 *This standard does not purport to address all of the safety concerns, if any, associated with its use. It is the responsibility of the user of this standard to establish appropriate safety and health practices and determine the applicability of regulatory limitations prior to use.*

2. Referenced Documents

2.1 *ASTM Standards:*

D 1452 Practice for Soil Investigation and Sampling by Auger Borings[2]

D 1586 Test Method for Penetration Test and Split-Barrel Sampling of Soils[2]

D 1587 Practice for Thin-Walled Tube Geotechnical Sampling of Soils[2]

D 4547 Practice for Sampling Waste and Soils for Volatile Organics[3]

D 4687 Guide for General Planning of Waste Sampling[3]

D 4700 Guide for Soil Sampling from the Vadose Zone[2]

D 4823 Guide for Core-Sampling Submerged, Unconsolidated Sediments[4]

D 5088 Practice for Decontamination of Field Equipment Used at Nonradioactive Sites[5]

D 5314 Guide for Soil Gas Monitoring in the Vadose Zone[5]

D 5451 Practice for Sampling Using a Trier Sampler[3]

D 5518 Guide for Acquisition of File Aerial Photography and Imagery for Establishing Historic Site-Use and Surficial Conditions[5]

D 5730 Guide to Site Characterization for Environmental Purposes with Emphasis on Soil, Rock, the Vadose Zone and Ground Water[5]

3. Terminology

3.1 *Definitions of Terms Specific to This Standard:*

3.1.1 *hot spots*—strata that contain high concentrations of the characteristic of interest and are relatively small in size when compared with the total size of the materials being sampled.

3.1.2 *representative sample*—a sample collected such that it reflects one or more characteristics of interest (as defined by the project objectives) of the population from which it was collected.

3.1.2.1 *Discussion*—A representative sample can be a single sample, a set of samples, or one or more composite samples.

3.1.3 *waste pile*—unconfined storage of solid materials in an area of distinct boundaries, above grade and usually uncovered. This includes the following:

3.1.3.1 *chemical manufacturing waste pile*—a pile consisting primarily of discarded chemical products (whether marketable or not), by-products, radioactive wastes, or used or unused feedstocks.

3.1.3.2 *scrap metal or junk pile*—a pile consisting primarily of scrap metal or discarded durable goods such as appliances, automobiles, auto parts, or batteries.

3.1.3.3 *trash pile*—a pile of waste materials from municipal sources, consisting primarily of paper, garbage, or discarded nondurable goods that contain or have contained hazardous substances. It does not include waste destined for recyclers.

4. Significance and Use

4.1 This guide is intended to provide guidance for sampling waste piles. It can be used to obtain samples for waste characterization related to use, treatment, or disposal; to monitor an active pile; to prepare for closure of the waste pile; or to investigate the contents of an abandoned pile.

4.2 Techniques used to sample include both in-place evaluations of the pile and physically removing a sample. In-place evaluations include techniques such as remote sensing, on-site gas analysis, and permeability.

4.3 Sampling strategy for waste piles is dependent on the following:

4.3.1 Project objectives including acceptable levels of error when making decisions;

[1] This guide is under the jurisdiction of ASTM Committee D34 on Waste Management and is the direct responsibility of Subcommittee D34.01.01 on Planning for Sampling.

Current edition approved Oct. 10, 1996. Published December 1996.

[2] *Annual Book of ASTM Standards*, Vol 04.08.

[3] *Annual Book of ASTM Standards*, Vol 11.04.

[4] *Annual Book of ASTM Standards*, Vol 11.02.

[5] *Annual Book of ASTM Standards*, Vol 04.09.

4.3.2 Physical characteristics of the pile, such as its size and configuration, access to all parts of it, and the stability of the pile;

4.3.3 Process that generated the waste and the waste characteristics, such as hazardous chemical or physical properties, whether the waste consists of sludges, dry powders or granules, and the heterogeneity of the wastes;

4.3.4 History of the pile, including dates of generation, methods of handling and transport, and current management methods;

4.3.5 Regulatory considerations, such as regulatory classification and characterization data;

4.3.6 Limits and bias of sampling methods, including bias that may be introduced by waste heterogeneity, sampling design, and sampling equipment.

4.4 It is recommended that this guide be used in conjunction with Guide D 4687, which addresses sampling design, quality assurance, general sampling considerations, preservation and containerization, cleaning equipment, packaging, and chain of custody.

4.5 A case history of the investigation of a waste pile is included in Appendix X1.

5. Site Evaluation

5.1 Site evaluations are performed to assist in designing the most appropriate sampling strategy. An evaluation may consist of on-site surveys and inspections, as well as a review of historical data. Nonintrusive geophysical and remote sensing methods are particularly useful at this stage of the investigation (see Guide D 5518). Table 1 summarizes the effects that various factors associated with the waste pile, such as the history of how the pile was generated, have upon the strategy and design of the sampling plan. The strategic and design considerations are discussed as well.

5.2 *Generation History*—The waste pile may have been created over an extended time period. A remote sensing method that is very useful in establishing historical management practices for waste piles is aerial imagery. Aerial photographs are widely available and may be used to determine the history of a waste pile, sources of waste, and the presence and distribution of different strata. Satellite imagery could be used for larger waste piles.

TABLE 1 Strategy Factors

Waste Pile Factors	Strategic Considerations	Design Considerations
Generation history	Date of generation	Analysis required
	Types of processes	Location of samples
	Characteristics by process	
	Delivery method	
	Current management	
	Regulatory considerations	
Physical characteristics of pile:	Physical variability of pile	Number of samples
– size	Access	Location of samples
– shape	Safety	Equipment selection
– stability		
Waste characteristics	Constituents present	Number of samples
	Constituent distribution	Analysis required
	Heterogeneity	Location of samples
	– physical variability	Representative samples
	– chemical variability	Equipment selection

5.2.1 The date of generation could be important with respect to the types of processes that generated the waste, the characteristics of the waste, the distribution of the constituents, and regulatory concerns.

5.2.2 The type of process that generated the waste will determine the types of constituents that may be present in the waste pile. Chemical variability will influence the number of samples that are required to characterize the waste pile unless a directed (biased) sampling approach is acceptable.

5.2.3 The delivery method of the material to the waste pile could influence the concentrations of the constituents, affect the overall shape of the pile, or create physical dissimilarity within the waste pile through sorting by particle size or density.

5.2.4 If the pile is under current management and use, the variability in constituent types and concentrations may be affected. Current management activities also may influence the regulatory status of the waste pile.

5.2.5 Regulatory considerations will typically focus on waste identification questions, in other words is the material a solid waste that should be regulated and managed as a hazardous waste (**1**).[6] This may involve a limited, directed sampling approach, particularly if a regulatory agency is conducting the investigation. A more comprehensive sampling design may be required to determine if the waste classifies as hazardous. Remediation efforts and questions regarding permits may focus on characterizing the entire pile, possibly as the removal of material is occurring. It should be noted that concentrations of contaminants near regulatory levels may increase the number of samples required to meet the objectives of the investigation. These regulatory levels could be those established to determine if a waste is hazardous, or "cleanup" levels set for a removal or remediation.

5.3 *Physical Characteristics of Pile*— Several physical characteristics of the waste pile must be considered during the site evaluation. Variability in size, shape, and stability of the pile affects access to it to obtain samples as well as safety considerations. Physical variability will influence the number of samples that are required to characterize the waste pile unless a directed (biased) sampling approach is considered to be acceptable. Techniques that might be used include resistivity and seismic refraction (for determining the depth of very large piles).

5.3.1 The size of the waste pile will influence the sampling strategy in that increasing size is often accompanied by increased variability in the physical characteristics of the waste pile. The number of samples, however, that are needed to characterize a waste pile adequately will typically be a function of the study objectives as well as the inherent variability of the pile.

5.3.2 The shape of the waste pile can influence the sampling strategy by limiting access to certain locations within the pile, and if it is topologically complex it is difficult to lay out a sampling grid. Also, a waste pile may extend vertically both above and below grade, making decisions regarding the depth of sample collection difficult.

[6] The boldface numbers in parentheses refer to the list of references at the end of this standard.

5.3.3 The stability of the waste pile also can limit access to both the face and the interior of the pile. The use of certain types of heavier sampling equipment also could be limited by the ability of the pile to bear the weight of the equipment.

5.4 *Waste Characteristics*:

5.4.1 The constituents could include inorganics, volatile organic compounds (VOCs), and semivolatile organic compounds (including pesticides and polychlorinated biphenyls (PCBs)) (see Practice D 4547). Speciality analyses may be warranted, such as leaching tests or analyses for dioxin/furans or explosive compounds. Soil gas sampling is a minimally intrusive technique that may detect the presence and distribution of volatile organic compounds in soils and in porous, unconsolidated materials. Appropriate applications for soil gas monitoring are identified in Guide D 5314.

5.4.2 The distribution of constituents in the waste pile could be influenced by changes in the manufacturing process which resulted in changes in the composition of the waste; the length of time the material has remained in the pile (particularly for VOCs); the mode of delivery of the waste materials to the pile; and management practices, such as mixing together wastes from more than one process.

5.4.3 Physical and chemical variabilities would include variability in the chemical characteristics of the material within the pile, as well as variability in particle size, density, hardness, whether brittle or flexible, moisture content, consolidated, or unconsolidated. The variability may be random or found as strata of materials having different properties or containing different types or concentrations of constituents.

5.4.3.1 Geophysical survey methods may be used on piles to estimate physical homogeneity, which may or may not be related to chemical homogeneity, and to detect buried objects, both of which may need to be considered during the development of the sampling design and the safety plan for the investigation. The most suitable technique for detecting nonmetallic objects is electromagnetics. Ground-penetrating radar, a more sophisticated and complex technique, also may be considered. Electromagnetic techniques are suited particularly to large piles that contain leachate plumes (for example, mine tailings) or for the detection of large discontinuities in a pile (for example, different types of wastes or the transition from a disposal area to background soils). For metallic objects, metal detectors and magnetometers are useful and relatively easy to use in the field.

5.5 *Potential Investigation Errors*:

5.5.1 Equipment selection can bias sampling results even if the equipment is used properly. Bias can result from the incompatibility of the materials that the sampling equipment is made of with the materials being sampled. For example, the equipment could alter the characteristics of the sample. Some equipment will bias against the collection of certain particles sizes, and some equipment cannot penetrate the waste pile adequately.

5.5.2 Equipment, use, and operation can introduce error (bias) into the characterization of a waste pile. Sampling errors typically are caused when certain particle sizes are excluded, when a segment of the waste pile is not sampled, or when a location outside the pile is inadvertently sampled.

5.5.3 When stratification, layering, or solid phasing occurs it may be necessary to obtain and analyze samples of each of the distinct phases separately to minimize sampling bias. Care should be taken when sampling stratified layers to minimize cross contamination. Proper decontamination procedures should be used for all sampling equipment (see Practice D 5088).

5.5.4 Statistical bias includes situations where the data are not normally distributed or when the sampling strategy does not allow the potential for every portion of the pile to be sampled.

6. Sampling Strategy

6.1 Developing a strategy for sampling a waste pile requires a thorough examination of the site evaluation factors listed in Section 5. The location and frequency of sampling (number of samples) should be outlined clearly in the sampling plan, as well as provisions for the use of special sampling equipment, access of heavy equipment to all areas of the pile, if necessary, and so forth.

6.1.1 *Representative Sampling*—The collection of a representative set of samples from a waste pile typically will be complicated by the presence of a number of the site evaluation factors **(2,3)**.

6.1.2 *Heterogeneous Wastes*—Waste piles may be homogeneous, for applied purposes, or may be quite heterogeneous in particle size and contaminant distribution. If the particle sizes of the material in the waste pile and the distribution of contaminants are known, or can be estimated, then less sampling may be necessary to define the properties of interest in the waste pile. An estimate of the variability in contaminant distribution may be based on process knowledge or determined by preliminary sampling **(4)**. The more heterogeneous the waste pile is, the greater the planning and sampling requirements.

6.1.3 *Strata and Hot Spots*—A waste pile also could contain strata that have less internal variation in physical properties or concentrations of chemical constituents than the remainder of the waste pile **(2,5)**. For example, strata may be present in a waste pile due to changes in the process that generated the waste, or if different processes at a facility contribute waste to different parts of the waste pile. A stratified sampling strategy would consider this situation by conducting independent sampling of each stratum, which could reduce the number of samples required. These strata could be in specific areas of the waste pile **(4)**. Also, hot spots may be present in the waste pile that are unique in composition **(2,5)**.

6.2 *Specific Sampling Strategies*:

6.2.1 Although the most appropriate method for evaluating material in waste piles is to sample at or immediately following the point of generation (for example, conveyor belt), most sampling problems involve existing or in-place waste piles. Therefore, the following discussion will focus on in-place waste piles. Sampling strategies available for waste piles include directed or judgmental sampling, simple random sampling, stratified random sampling, systematic grid sampling, and systematic sampling over time **(2,6)**. General concerns about the collection of a representative sample, the existence of potential heterogeneity in the waste pile, the presence of strata within the waste pile, and the existence of distinct hot spots within the waste pile may also influence the selection of an appropriate sampling strategy and development of the sampling plan **(5)**. The following paragraphs provide an introduction to determining the appropriate number of samples to collect and the sampling strategies available for determining sample locations.

6.2.2 *Determining the Frequency or Number of Samples*—The frequency of sampling or the number of samples to collect typically will be based on several factors including the study objectives, properties of wastes in the pile, degree of confidence required, access to sampling points, and budgetary constraints. Practical guidance for determining the number of samples is included in Guide D 4687 and Refs **(2, 3)**.

6.2.3 *Directed Sampling*—Directed sampling (Fig. 1) is based on the judgment of the investigator and will not result necessarily in a sample that reflects the characteristics of the entire waste pile. Directed sampling also is called judgmental sampling, authoritative sampling, or nonprobability sampling. The experience of the investigator often is the basis for sample collection, and, depending on the study objectives, bias should be recognized as a potential problem. For preliminary screening investigations of a waste pile and for certain regulatory investigations, however, directed sampling may be appropriate. A directed sampling strategy could call for the collection of a composite sample from the surface area or the collection of discrete grabs at the surface of the pile (see Fig. 1). Directed sampling would typically focus on worst case conditions in a waste pile, for example, the most visually contaminated area or most recently generated waste.

6.2.4 *Simple Random Sampling*—Simple random sampling (Fig. 2) ensures that each element in the waste pile has an equal chance of being included in the sample **(2)**. This may be the method of choice when, for purposes of the investigation, the waste pile is randomly heterogeneous **(5)**. If the waste pile contains trends or patterns of contamination, a stratified random sampling or systematic grid sampling strategy would be more appropriate **(2)** (see 6.2.5 and 6.2.6).

6.2.4.1 A simple random approach could use a grid with random grids selected for sample collection (see Fig. 2). Note that the grid size could be selected based on the number of samples that are required (some guidance suggests having at least ten times the number of grids as samples required). Once the grid is overlaid and the sampling locations are selected, the decision must be made to collect either a discrete grab sample (surface), a composite of surface samples taken from predesignated locations within the grid cell (based on compass points), a vertical composite to a specified depth, or discrete

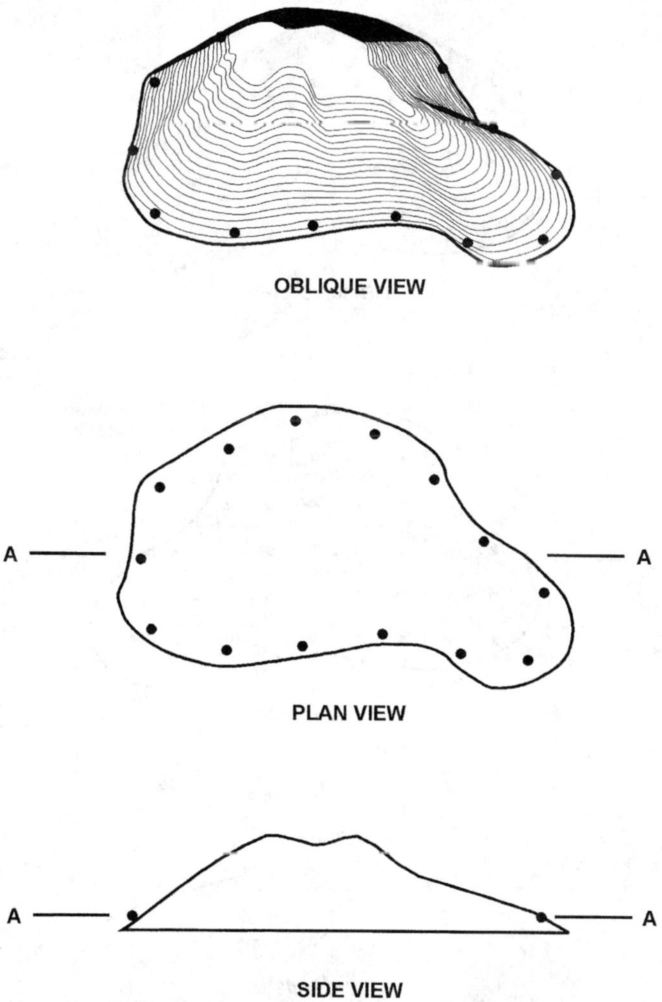

FIG. 1 Waste Pile Sampling Strategy—Directed Sampling

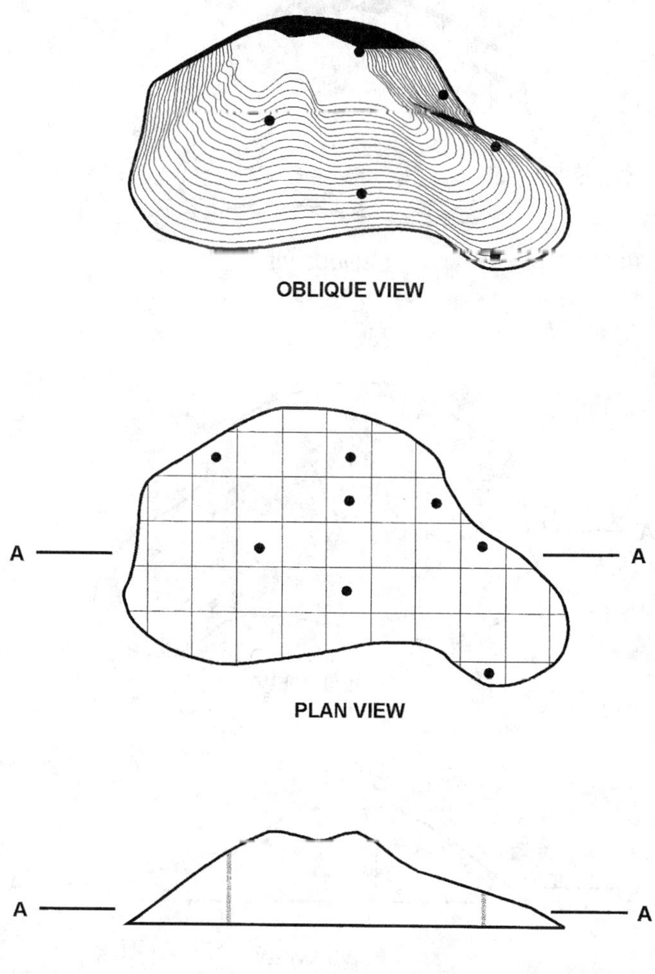

FIG. 2 Waste Pile Sampling Strategy—Simple Random Sampling

grab samples at specified depths. If discrete grab samples are desired at specified depths, they typically would be collected at the same location as the bore hole is advanced into the pile. Fig. 2 illustrates the collection of vertical composites at each of the randomly selected locations.

6.2.5 *Stratified Random Sampling*—Stratified random sampling (see Fig. 3) may be useful when distinct strata or homogeneous subgroups are identified within the waste pile **(2)**. The strata may be located in different areas of the pile or may be comprised of different layers (see Fig. 3). This approach is useful when the individual strata may be considered internally homogeneous or at least have less internal variation in what would otherwise be considered a heterogeneous waste pile **(2)**. Information on the waste pile usually is required to establish the location of individual strata unless process knowledge or changes in the composition of the material is obvious, such as with discoloration or with the type of waste. The grid may be utilized for sampling several horizontal layers if the strata are oriented horizontally **(4)**. A simple random sampling approach then is used within each stratum. The use of a stratified random sampling strategy may result in the collection of fewer samples. Fig. 3 illustrates a scenario where the number of samples collected in each stratum varies (plan view), and discrete grabs are collected in each boring at predesignated depths (side view).

6.2.6 *Systemic Grid Sampling*—Systematic grid sampling (see Fig. 4) involves the collection of samples at fixed intervals and is useful when the contamination is assumed to be distributed randomly **(2)**. This method also is commonly used with waste piles when estimating trends or patterns of contamination or when the objective is to locate hot spots. This approach may not be acceptable if the entire waste pile is not accessible or if the sampling grid locations become phased with variations in the distribution of contaminants within the waste pile **(6)**. It also may be useful for identifying the presence of strata within the pile. The grid and starting points should be laid out randomly over the waste pile, yet the method allows for rather easy location of exact sample locations by means of the grid (see Fig. 4). The same considerations discussed in 6.2.4 concerning the depth of each sample (surface, vertical composite, discrete grabs at depth) also should be considered. Fig. 4 illustrates the collection of vertical composites at each grid, which could be difficult and costly. Also note that the grid size typically would be adjusted according to the number of samples that are required.

6.2.7 *Systematic Sampling Over Time*—Systematic sampling over time at the point of generation is useful if the material is being sampled from a conveyor belt or being

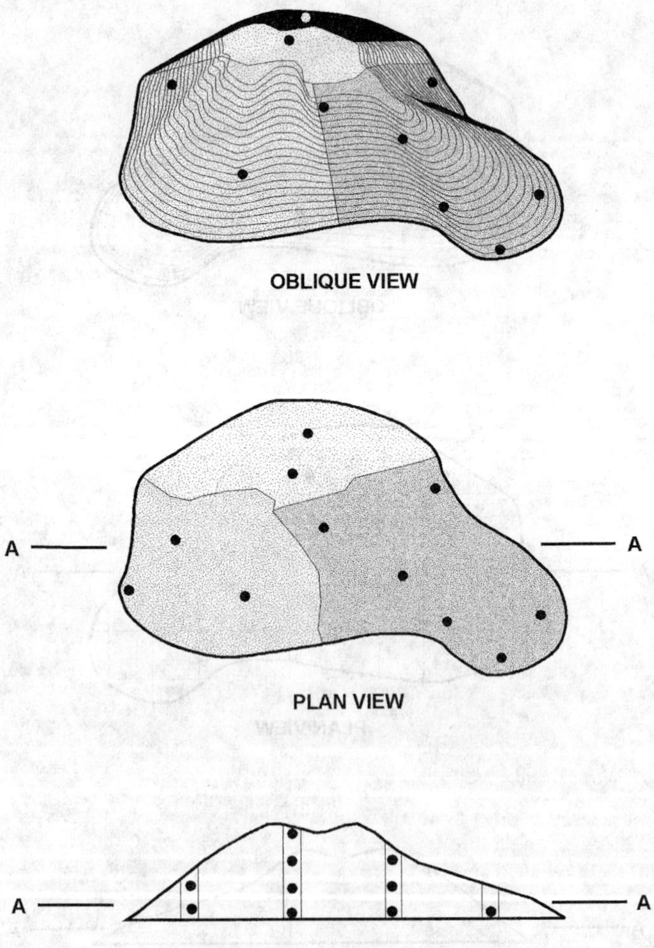

FIG. 3 Waste Pile Sampling Strategy—Stratified Random Sampling

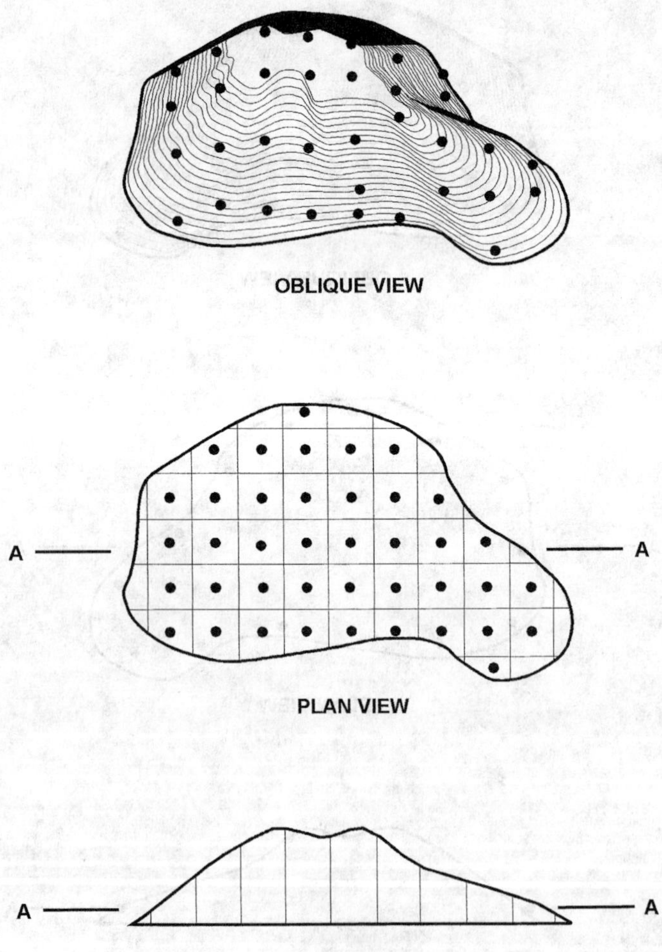

FIG. 4 Waste Pile Sampling Strategy—Systematic Grid Sampling

delivered by means of truck or pipeline to the waste pile. The sampling interval can be determined on a time basis, for example, every hour from a conveyor belt or pipeline discharge, or from every third truck load. The time between intervals is influenced by the factors addressed in 6.2.2.

6.2.8 *Alternative Approach*—In many cases, an objective of waste pile characterization is to determine the impact of the pile on the environment. At times this may be accomplished more easily by sampling the routes by which contaminants are dispersed from the pile than through direct sampling of the pile, especially for piles that are difficult to characterize. For example, ground water up-and-down gradient from the pile could be sampled to check for ground water contamination. The vadose zone below the pile also might be sampled to detect leachate (and potential ground water contamination) through soil sampling, vacuum lysimeters, or soil gas. Surface water and sediment in drainage channels down gradient from the pile also might be sampled. Surface soils, air samples, and contaminants deposited on vegetation can be used as indicators of atmospheric transport of contaminants from the pile, including both particulate and volatile materials. Such approaches will seldom replace pile sampling completely, but they may reduce the number of pile samples needed to make remedial action decisions (see Guide D 5730), also Refs **(7-9)**.

7. Selection of Sampling Equipment

7.1 Wastes in piles are often complex, multiphase mixtures of solids and semisolids. The wastes can range from powders to granules to large, heterogeneous solid fragments and can cover many acres in area. No single type of sampler can be used to collect representative samples of all types of waste from piles. Large, thick piles may require drill rigs to obtain samples from depth. The sampling of gases from within the pile requires other types of equipment. Table 2 lists typical waste types and the corresponding recommended samplers to use.

7.2 Sampling at depth from inside the pile may require heavy equipment designed for excavation or removal of soil or rock. Table 3 lists such equipment and its applications for sampling waste piles **(10)**.

7.3 Sampling equipment should be constructed of materials that are compatible with the waste to be sampled. Compatibility refers to the physical durability, lack of chemical reactivity with the waste, and lack of potential for contamination of the waste with analytes of concern. Typical materials of construction include stainless steel, plastic, and glass.

8. Data Use

8.1 The decisions that will be made based upon the data must be identified early in the planning process since these affect the approach to the problem and how the data will be evaluated. Decisions affecting waste classification, closure, and post-closure issues, are examples of the uses of the data. Methods to determine the volume of contaminated material in a pile or pile strata may be needed. Standard mathematical formulas for calculating the volume of a cone, cylinder, various prisms, and so forth, may be used.

8.2 *Statistical Considerations*:

8.2.1 Data quality assessment (DQA) methods are used to evaluate the data for any anomalies and to evaluate the assumptions for statistical evaluation. The statistician makes use of both subjective judgment (graphical analysis for identification of trends and anomalies) and statistical models and inference (for example, outlier detection, autocorrelation estimation) in the investigation of data for validity of the assumptions needed to make a statistical test. Classical statistical models assume that the samples collected from the population of interest are independent and have an identical probability distribution (that is, normal distribution with constant mean and variance). Random sampling is a method to ensure independence. The probability distributional assumptions are part of DQA that will determine if the classical statistical

TABLE 2 Sampling Devices Suitable for Waste Piles[A]

Location and Waste Type	Sampling Devices	ASTM Standard	Limitations
Subsurface			
Powdered, granular, or soil-like solids; sludges	split-barrel push coring device	D 1586 D 1587 D 4700 D 4823	Limited application for sampling moist and sticky solids, or particles with diameter 0.6 cm (0.25 in.) or more. Depth limitation of about 1 m.
	trier	D 5451	May not retain core sample of very dry granular materials. Not applicable to sampling solid wastes with particle diameter >½ the diameter of the sampling tube.
	auger	D 1452 D 4700	Does not collect undisturbed sample.
	thin-walled tube sampler	D 4823 D 4700	Collects relatively undisturbed core. Difficult to use on gravelly or rocky soils.
	drill rigs		Used for geoenvironmental exploration. To minimize sample contamination, avoid those using a water-based drilling fluid.
	soil gas samplers	D 5314	Used for volatile organic compounds.
Surface			
Powdered, granular, or soil-like solids; sludges	trowel or scoop	D 4700	Not applicable to sampling deeper than 8 cm (3 in.). Difficult to obtain reproducible mass of sample. May exclude certain particle sizes, especially large aggregates.
Slag	hammer/chisel Impact device		Changes particle size.

[A] This table is not all inclusive; other equipment may be used.

TABLE 3 Excavation and Removal Equipment for Waste Piles

Excavation and Removal Equipment	General Excavation	Ability to Excavate Hard and Compacted Material	Soil Hauling	Mixing of Solids, Soil	Spreading Cover	Site Maneuverability
Wheel or crawler Mounted backhoe	A[A]	A	B[B]/O[C]	A	A	A/B
Wheel or crawler Mounted front-end loader	A	A	A/B	A	A	A/B
Skid steer loader	A	B	B	A	B	A
Bulldozer	A	A	O	O	A	B

[A] A = Good choice. Equipment is fully capable of performing function listed.
[B] B = Secondary choice. Equipment is marginally capable of performing function listed.
[C] O = Not applicable or poor choice.

model is appropriate for the collected data. For directed sampling, the sampling is subjective and the sample results are typically judged on a qualitative basis.

8.2.2 Simple random sampling will provide an unbiased estimate of the average waste concentration, that is, an estimate of the mean. This unbiased estimate is independent of the geometry of the pile and of the distribution of the concentration of the contaminants, but it may not have the smallest variance. Other sampling designs, such as systematic grid sampling or stratified random sampling, may provide an average that has a smaller variance. If the waste pile has uneven topography, the calculation of the mean concentration of the pile should be a volume-weighted average, using core volume as the weighting factor to reduce the variance of the estimated mean.

8.2.2.1 For simple random sampling and systematic grid sampling designs, histogram and normal probability plots of the sample data can be used to judge if the data conform to normal distribution. If not, there are several alternatives. First, the classical statistical model may still be considered robust for the decision-making process. Second, a transformation of the data may approximate a normal distribution of the data. For example, logarithmic transformation will normalize data that are lognormal originally. If the data are lognormal, the question of whether to use the arithmetic mean or the geometric mean for decision-making purposes must be decided. Third, an alternative statistical model based on nonparametric methods, but which uses weaker assumptions, may be proposed to analyze the decision-making process. It may be advisable to consult a statistician.

8.2.2.2 For the stratified random sampling design, the test of normality is not straightforward. Generally, it requires a mathematical model to take out the strata effects first, then test for normality using the residuals. A statistician should be consulted.

8.2.2.3 In any of these cases, alternative consequences of the level of uncertainty can be calculated prior to collecting the data. These alternatives can be used by decision-makers to select the best strategy to minimize the environmental risks.

9. Keywords

9.1 piles; sampling; waste

APPENDIX

(Nonmandatory Information)

X1. WASTE PILE—A CASE HISTORY

X1.1 **Background**—The waste pile was generated by a facility that produces brass alloys from scrap metal. The byproduct from this operation was slag, which was generated in the recovery furnace. The slag was ground subsequently in a ball mill prior to being reintroduced into the recovery furnace. A large amount of the ground slag was disposed of in a waste pile which covered about one acre. No active management was occurring with the waste pile. No buried containers or extremely heterogenous material (unground slag) was suspected of being present in the waste pile based on facility records and interviews of personnel.

X1.1.1 Lead and cadmium were the constituents of concern based on process knowledge, and the possibility for the waste being hazardous was the regulatory consideration. The potential for off-site migration of contaminants was also an immediate concern, and this was considered in the development of the Phase 1 study design. Fig. X1.1 shows a site map of the facility and the slag pile. Fig. X1.2 shows a computer enhancement of the slag pile, and Fig. X1.3 shows a topographic view of the pile.

X1.2 Phase 1:

X1.2.1 *Objective*—The primary objective of the initial investigation was to determine if the slag in the waste pile classified as hazardous based on the concentration of lead and cadmium in a leach test. A secondary objective was to provide preliminary information on the potential migration and transport of contaminants from the waste pile off-site. The sampling plan for this initial investigation utilized a directed sampling strategy to provide a preliminary estimate of the lead concentration in the waste, the variability of contaminant concentrations in the pile, and the potential for leaching using the applicable leaching procedure mandated in regulations. Four composite samples were collected from the surface (0 to 15 cm or 0 to 6 in.) of the waste pile at locations within the four

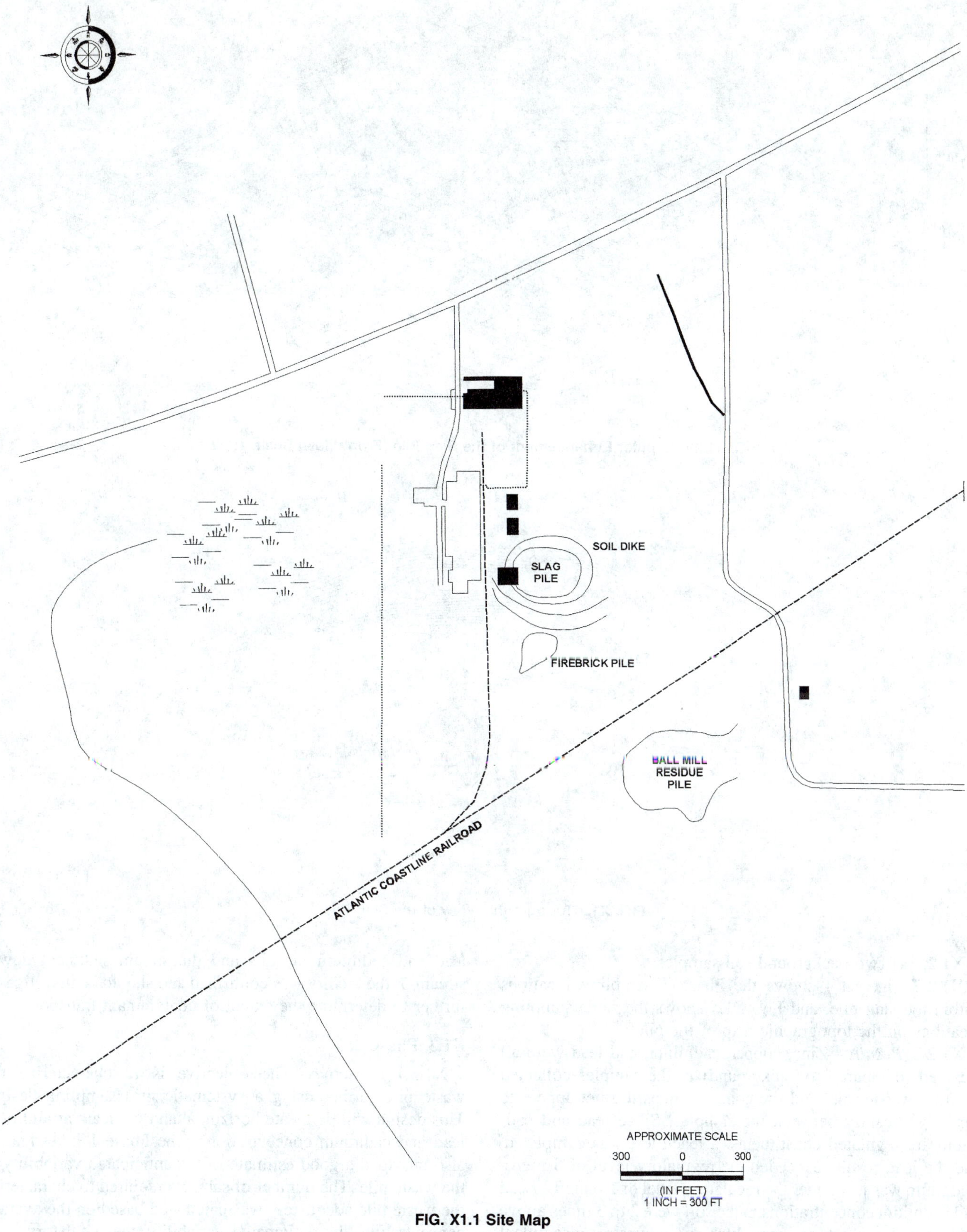

FIG. X1.1 Site Map

quadrants. The following environmental samples were also collected:

X1.2.1.1 Several soil samples in the vicinity of the waste pile,

X1.2.1.2 Sediment upstream and downstream in a stream which borders the facility,

X1.2.1.3 Sediment in a ditch which contained runoff from the pile, and

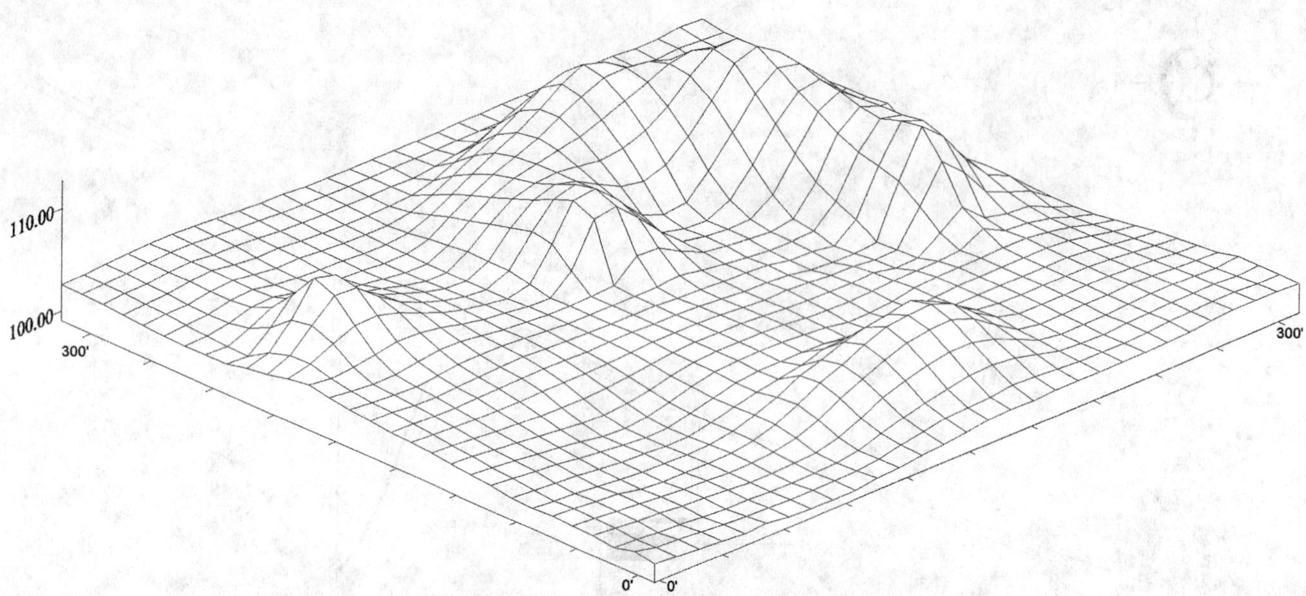

FIG. X1.2 Computer Enhancement of the Slag Pile (Front View) Scale 1:1:2

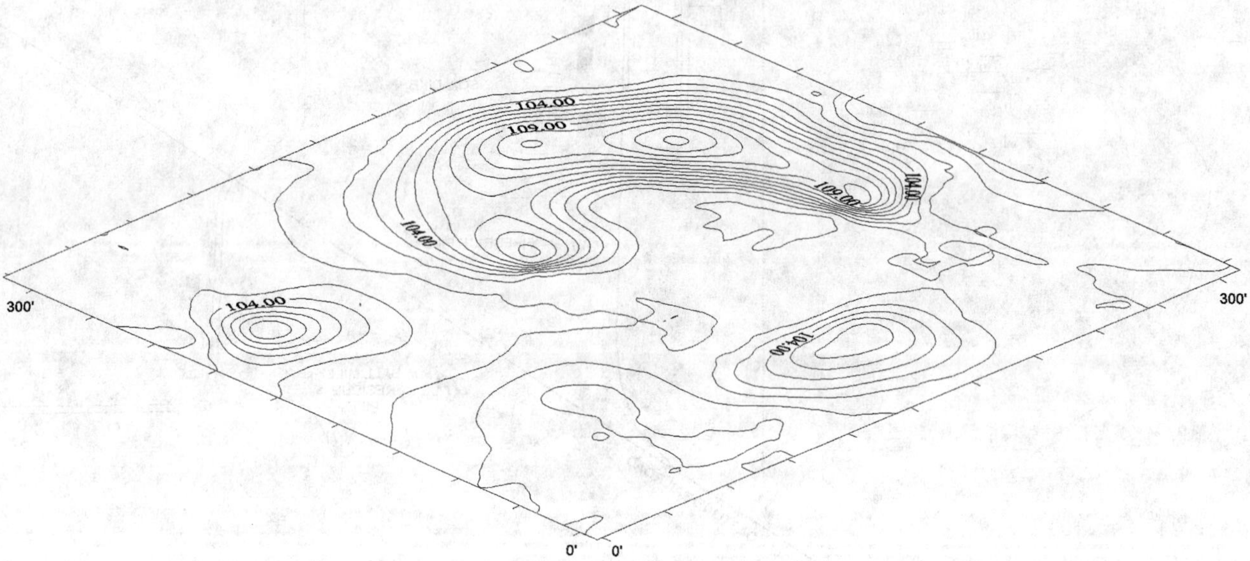

FIG. X1.3 Topographic View of the Slag Pile

X1.2.1.4 Two background soil samples.

X1.2.2 Fig. X1.4 shows the Phase 1 sampling locations within the slag pile, and Fig. X1.5 shows the same sampling locations on the topographic map of the pile.

X1.2.3 *Results*—Zinc, copper, cadmium, and lead were all elevated (compared to background) in the samples collected from the waste pile, and the concentrations did not appear to vary significantly between the samples. Since lead and cadmium are regulated constituents, a leach test was completed, and the lead results exceeded the regulatory level of 5 mg/L. Cadmium was just under the regulatory level of 1.0 mg/L. Lead and cadmium concentrations in the soil were 2 to 3 times above background, and the drainage ditch and downstream sediment sample also had elevated lead and cadmium levels.

X1.2.4 *Conclusion*— The waste pile contained slag that is hazardous for lead. The waste pile required further characterization to determine the variability in the pile. The presence of lead and cadmium in soils and the stream sediment downstream of the facility was confirmed and should be investigated further to determine the extent of contaminant transport.

X1.3 Phase 2:

X1.3.1 *Objective*—The objective is to characterize the waste pile further using a systematic grid sampling design. This design will delineate horizontal and vertical variability in lead and cadmium concentrations. The Phase 1 investigation also provided a good estimate of the anticipated variability in the waste pile. The number of samples required to characterize the waste pile adequately was calculated based on the average concentration, the anticipated variability, the regulatory level of concern, and the specified confidence interval. The grid size then was adjusted to accommodate the projection on the required number of samples. Composite samples were collected within each grid cell based on one center point and eight

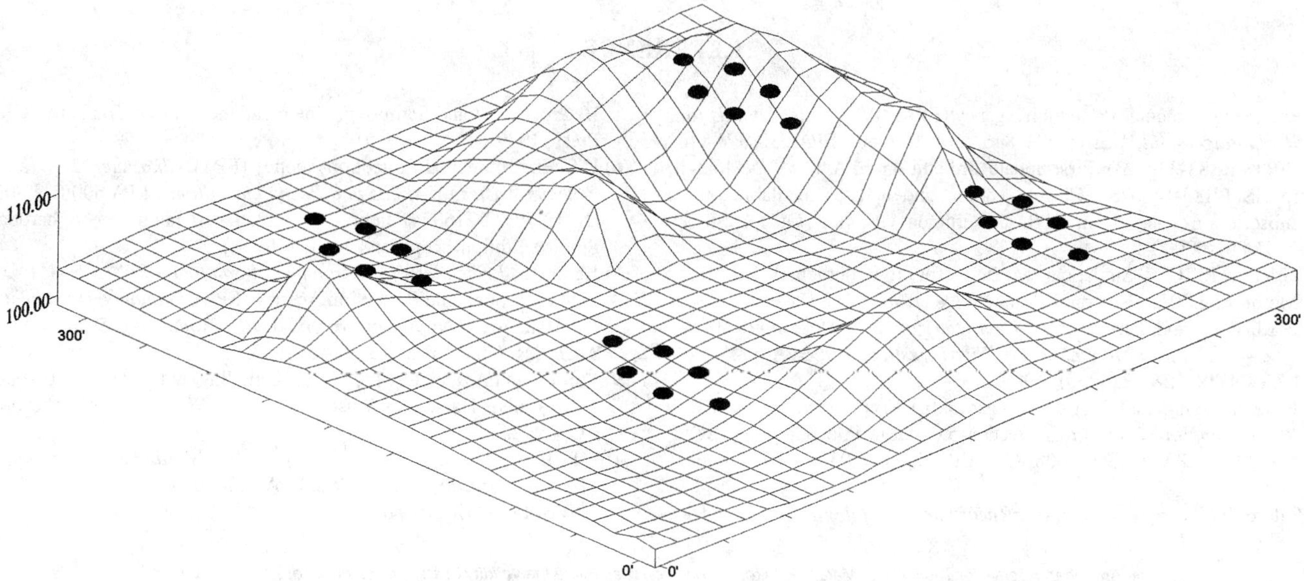

FIG. X1.4 Front View of the Slag Pile Showing Sampling Locations Scale 1:1:2

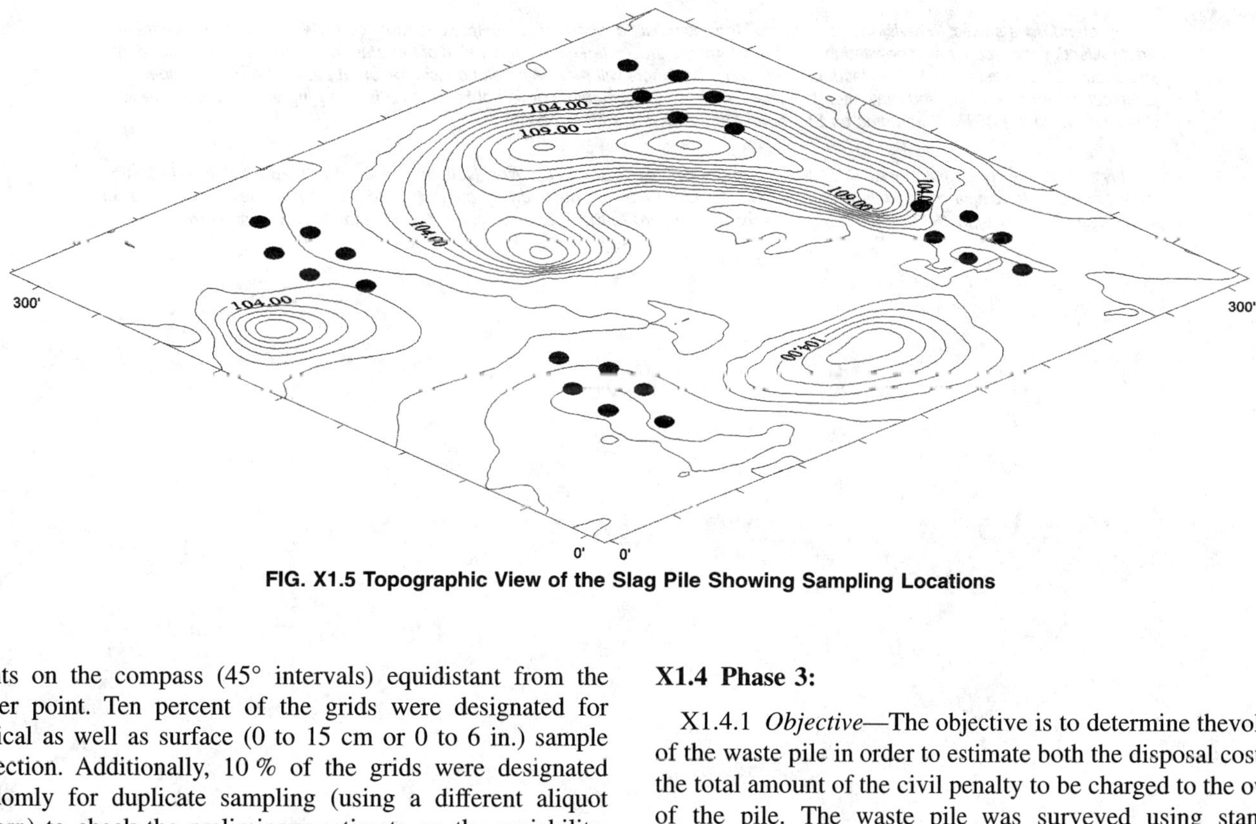

FIG. X1.5 Topographic View of the Slag Pile Showing Sampling Locations

points on the compass (45° intervals) equidistant from the center point. Ten percent of the grids were designated for vertical as well as surface (0 to 15 cm or 0 to 6 in.) sample collection. Additionally, 10 % of the grids were designated randomly for duplicate sampling (using a different aliquot pattern) to check the preliminary estimate on the variability. Additional environmental sampling was conducted but will not be covered in this discussion.

X1.3.2 *Results*—The results supported the initial Phase 1 investigation with lead consistently exceeding the regulatory level. Cadmium consistently was below the regulatory level.

X1.3.3 *Conclusion*— The waste pile was characteristic for lead and classified as hazardous according to the applicable regulations. There was no significant variability with depth, although several gradients were noticed across the grid based on lead concentration (scan) results.

X1.4 Phase 3:

X1.4.1 *Objective*—The objective is to determine thevolume of the waste pile in order to estimate both the disposal cost and the total amount of the civil penalty to be charged to the owner of the pile. The waste pile was surveyed using standard surveying techniques.

X1.4.2 *Results*—The results were used to calculate the volume using geometric principles. Also, a computer program was utilized which constructs contours based on the surveying information. The computer program was used as a check of the manual method, which produced a result that was 10 % higher in volume than the computer program.

X1.4.3 *Conclusion*— For penalty calculation purposes, the smaller estimate was utilized; however, the actual treatment and disposal costs could reflect the larger estimate.

REFERENCES

(1) U.S. Environmental Protection Agency (EPA). 1986. *Test Methods for Evaluating Solid Waste*; SW-846, 3rd Edition. EPA/530/SW-846 (NTIS PB88-239223); First update, 3rd edition. EPA/530/SW-846.3-1 (NTIS PB89-148076). Current edition and updates available on a subscription basis from U.S. Government Printing Office, Stock #955-001-00000-1.

(2) Gilbert, R. O., "Statistical Methods for Environmental Pollution Monitoring," Van Nostrand Reinhold Co., 1987.

(3) Ford, P. J., and Turina, P. J., *Characterization of Hazardous Waste Sites—A Methods Manual, Vol 1: Site Investigations*, EPA 600/4-84/075, (NTIS PB85-215960), 1985.

(4) U.S. Environmental Protection Agency (EPA), *Compendium of ERT Waste Sampling Procedures*, Section 5.0 Waste Pile Sampling, SOP No. 2017, EPA 540/P-91/008, OSWER Directive 9360.4-07, January 1991.

(5) Pitard, F., *Pierre Gy's Sampling Theory and Sampling Practice*, Vol 1: Heterogeneity and Sampling, Chemical and Rubber Company (CRC) Press, 1989.

(6) U.S. Environmental Protection Agency (EPA), *Characterizing Heterogeneous Wastes: Methods and Recommendations*, EPA 600/R-92/033, (NTIS PB92-216894) February 1992. [Also published as hardback Smoley Edition: Rupp and Joens (1993).]

(7) Keith, L., *Principles of Environmental Sampling*, Ed. ACS, 1988.

(8) U.S. Environmental Protection Agency (EPA). *Guidance Document on the Statistical Analysis of Ground Water Monitoring Data at RCRA Facilities*, Office of Solid Waste, 1993.

(9) McCoy and Associates, Inc., "Soil Sampling and Analysis—Practices and Pitfalls," Hazardous Waste Consultant, Vol 10, No. 6, Lakewood, CO, 1992.

(10) PEI Associates, 1991, *Survey of Materials-Handling Technologies Used at Hazardous Waste Sites*, EPA 540/2-91/010, June 1991. (NTIS PB91-186924), 225 pp.

Designation: D 6010 – 96 (Reapproved 2001)

Standard Practice for Closed Vessel Microwave Solvent Extraction of Organic Compounds from Solid Matrices[1]

This standard is issued under the fixed designation D 6010; the number immediately following the designation indicates the year of original adoption or, in the case of revision, the year of last revision. A number in parentheses indicates the year of last reapproval. A superscript epsilon (ε) indicates an editorial change since the last revision or reapproval.

1. Scope

1.1 This practice describes the closed vessel microwave extraction of soils, sediments, sludges, and wastes for subsequent determination of solvent extractable semivolatile and nonvolatile organic compounds by such techniques as gas chromatography and gas chromatography-mass spectrometry.

1.1.1 Compounds listed in Tables 1–5 can be extracted from the preceding materials.

1.2 This test method is applicable to samples that will pass through a 10-mesh (approximately 2-mm opening) screen.

1.3 The detection limit and linear concentration range for each compound is dependent on the gas chromatograph or gas chromatograph-mass spectrometer technique employed and may be found in the manual accompanying the instrument used.

1.4 *This standard does not purport to address all of the safety concerns, if any, associated with its use. It is the responsibility of the user of this standard to establish appropriate safety and health practices and determine the applicability of regulatory limitations prior to use.* See Section 8 for specific hazard statements.

2. Referenced Documents

2.1 *ASTM Standards:*

D 1193 Specification for Reagent Water[2]

D 3976 Practice for Preparation of Sediment Samples for Chemical Analysis[3]

D 5368 Test Method for the Gravimetric Determination of Total Solvent Extractable Content (TSEC) of Solid Waste Samples[4]

2.2 *Other Standards:*

United States Environmental Protection Agency (USEPA), Test Methods for Evaluating Solid Waste Volume 1A: Laboratory Manual Physical/Chemical Methods[5]

Title 21, Code of Federal Regulations (CFR), Part 1030, and Title 47, Part 18[5]

3. Summary of Practice

3.1 This procedure ensures intimate contact of the sample matrix with 115°C extraction solvent.

3.2 A 1 to 5-g portion of a solid sample is extracted in a sealed microwave transparent extraction vessel with 30 mL of acetone-hexane (1 + 1).

3.3 Up to 12 samples may be extracted simultaneously.

3.4 After extraction the vessels are cooled to room temperature, opened, and the solvent and sample are separated by decanting, filtration, or centrifuging.

3.5 This practice provides a sample suitable for analysis by gas chromatography or gas chromatography-mass spectrometry.

4. Significance and Use

4.1 Extraction of organic pollutants from wastes can provide information on the susceptibility of compounds to leeching, water quality changes, or other site conditions.

4.2 Rapid heating, in combination with temperatures in excess of the atmospheric boiling point of organic solvents, reduces sample extraction times.

4.3 Small amounts of solvents (30 mL) are used resulting in reduced sample preparation cost and time.

5. Interferences

5.1 Method interferences may be caused by contaminants in solvents, labware, and other hardware used in sample processing that lead to discrete artifacts or elevated baselines in gas chromatograms. The analyst must demonstrate, through the analysis of reagent blanks, that the system and the materials are free from interferents.

5.2 The use of high-purity solvents helps to minimize interference problems.

5.3 Matrix interferences are caused by contaminants that are coextracted from the sample. The extent of matrix interferences may vary considerably from sample to sample.

5.4 After cleaning, vessel liners and covers should be stored in a clean environment to prevent accumulation of contaminants.

---

[1] This practice is under the jurisdiction of ASTM Committee D34 on Waste Management and is the direct responsibility of Subcommittee D34.01.06 on Analytical Methods.

Current edition approved Oct. 10, 1996. Published December 1996.

[2] *Annual Book of ASTM Standards*, Vol 11.01.

[3] *Annual Book of ASTM Standards*, Vol 11.02.

[4] *Annual Book of ASTM Standards*, Vol 11.04.

[5] Available from the Superintendent of Documents, U.S. Government Printing Office, Washington, DC 20402.

TABLE 1 Semivolatile Analyte Recovery from Freshly Spiked Topsoil

Analyte	Spike Level, mg/kg	Average Recovery, %[A]	RSD, %
Acenaphthene	5.0	97.6	9.8
Acenaphthylene	5.0	100	10
Acetophenone	5.0	92.2	12
4-Aminobiphenyl	5.0	77.3	9.5
Aniline	5.0	68.1	7.5
Anthracene	5.0	108	9.2
Benzidine	5.0	0	
Benzoic acid	5.0	42.3	13
Benzo(a)anthracene	5.0	113	9.4
Benzo(b)fluoranthene	5.0		
Benzo(k)fluoranthene	5.0	116	9.3
Benzo(g,h,i)perylene	5.0	111	4.7
Benzo(a)pyrene	5.0	110	8.6
Benzyl alcohol	5.0	96.1	9.0
Bis(2-chloroethoxy)methane	5.0	92.4	9.8
Bis(2-chloroethyl)ether	5.0	96.0	11
Bis(2-chloroisopropyl)ether	5.0	95.2	12
Bis(2-ethylhexyl)phthalate	5.0	116	9.3
4-Bromophenylphenyl ether	5.0	108	9.0
Butyl benzyl phthalate	5.0	116	9.8
4-Chloroaniline	5.0	97.0	9.2
1-Chloronaphthalene	5.0	104	12
2-Chloronaphthalene	5.0	91.8	7.3
4-Chloro-3-methylphenol	5.0	107	12
2-Chlorophenol	5.0	94.5	7.8
4-Chlorophenyl phenyl ether	5.0	106	9.7
Chrysene	5.0	111	8.8
Dibenzo(a,j)acridine	5.0	10.6	34
Dibenzo(a,h)anthracene	5.0	110	5.9
Dibenzofuran	5.0	98.8	9.9
Di-*n*-butyl phthalate	5.0	113	9.4
1,2-Dichlorobenzene	5.0	89.9	12
1,3-Dichlorobenzene	5.0	87.6	13
1,4-Dichlorobenzene	5.0	87.3	13
3,3-Dichlorobenzidine	5.0	96.8	12
2,4-Dichlorophenol	5.0	97.5	8.0
2,6-Dichlorophenol	5.0	93.1	12
Diethyl phthalate	5.0	111	8.0
Dimethylaminoazobenzene	5.0	116	11
7,12-Dimethylbenz(a)anthracene	5.0	128	7.0
αα-Dimethylphenethylamine	5.0	7.0	4.1
2,4-Dimenthylphenol	5.0	107	9.4
Dimethyl phthalate	5.0	106	8.4
4,6-Dinitro-2-methylphenol	5.0	57.6	9.3
2,4-Dinitrophenol	5.0	17.2	39
2,4-Dinitrotoluene	5.0	98.2	6.2
2,6-Dinitrotoluene	5.0	98.5	9.9
1,2-Diphenylhydrazine[B]	5.0	108	11
Di-*n*-octyl phthalate	5.0	117	12
Ethyl methanesulfonate	5.0	77.9	10
Fluoranthene	5.0	110	8.7
Fluorene	5.0	101	10
Hexachlorobenzene	5.0	108	8.9
Hexachlorobutadiene	5.0	89.5	11
Hexachlorocyclopentadiene	5.0	60.9	14
Hexachloroethane	5.0	83.7	13
Indeno(1,2,3-cd)pyrene	5.0	99.2	6.2
Isophorone	5.0	88.7	8.5
3-Methylcholanthrene	5.0	117	8.6
Methyl methanesulfonate	5.0	48.5	28
2-Methylnaphthalene	5.0	104	9.3
2-Methylphenol	5.0	95.1	8.5
4-Methylphenol	5.0	92.4	11
Naphthalene	5.0	95.0	12
1-Naphthylamine	5.0	57.8	8.7
2-Naphthylamine	5.0	73.5	9.0
2-Nitroaniline	5.0	100	7.7
3-Nitroaniline	5.0	96.8	8.5
4-Nitroaniline	5.0	99.0	8.5
Nitrobenzene	5.0	88.4	19
2-Nitrophenol	5.0	85.3	10
4-Nitrophenol	5.0	104	6.0
N-nitroso-di-*n*-butylamine	5.0	97.5	9.3
N-nitroso-di-*n*-propylamine	5.0	87.5	20
N-nitrosopiperidine	5.0	90.8	7.6
Pentachlorobenzene	5.0	101	9.1
Pentachloronitrobenzene	5.0	109	9.7
Pentachlorophenol	5.0	86.2	8.1
Phenacetin	5.0	97.0	12
Phenanthrene	5.0	109	8.5
Phenol	5.0	97.3	9.2
2-Picoline	5.0	7.7	30
Pronamid	5.0	120	11
Pyrene	5.0	113	8.4
1,2,4,5-Tetrachlorobenzene	5.0	91.2	8.6
2,3,4,6-Tetrachlorophenol	5.0	104	7.3
1,2,4-Trichlorobenzene	5.0	89.3	11
2,4,5-Trichlorophenol	5.0	95.1	12
2,3,6-Trichlorophenol	5.0	96.4	6.7
2-Fluorobipenyl	2.5	92.9	8.0
2-Fluorophenol	5.0	95.4	7.7
Nitrobenzene-d_5	2.5	92.2	9.8
Phenol-d_5	5.0	98.9	9.7
Terphenyl-d_{14}	2.5	112	10
2,4,6-Tribromophenol	5.0	92.3	7.7

[A] The topsoil was dry when spiked. The number of determinations was three. Determinations were made by gas chromatography-mass spectrometry. All recoveries were corrected for analyte losses incurred during blowdown evaporation of solvent to determine, specifically, recoveries by microwave extraction.

[B] Determined as azobenzene.

6. Apparatus

6.1 *Microwave Heating System*—A laboratory microwave heating system capable of delivering a minimum of 900 W of microwave energy. The system should be capable of 1 % power adjustments and 1-s time adjustments. The microwave

TABLE 2 Semivolatile Analyte Recovery from ERA Soil (Lot 324)[A]

Analyte	Certified Concentration mg/kg[B]	Average Recovery, %[C,D]	RSD, %
Anthracene	1.01	68.6	4.7
Benzo(a)anthracene	2.03	103	6.7
Bis(2-ethylhexyl)phthalate	7.12	150	11.2
Butyl benzyl phthalate	10.6	128	10.8
2-Chlorophenol	5.08	76.2	15.7
Chrysene	2.35	114	8.5
Dibenzofuran	6.79	88.8	1.9
2,4-Dinitrotoluene	5.0	83.0	4.2
Fluorene	6.06	72.1	1.0
Naphthalene	1.64	64.3	15.7
Pentachlorophenol	12.2	85.0	6.8
Phenanthrene	1.57	110	6.8
Pyrene	8.03	110	12.8
2,4,5-Trichlorophenol	7.99	96.9	1.3
2,4,6-Trichlorophenol	4.56	71.1	4.7
2-Fluorobiphenyl	6.0	102	8.8
2-Fluorophenol	20.0	99.5	14.1
Nitrobenzene-d_5	5.0	87.4	15.8
Phenol-d_5	20.0	96.0	13.1
Terphenyl-d_{14}	5.0	142	8.4
2,4,6-Tribromophenol	20.0	94.8	3.9

[A] This soil was obtained from Environmental Resources Associates (ERA) in Arvada, CO, and has been certified for the compounds listed in this table.

[B] Reported by ERA.

[C] The number of determinations was four. The recoveries are based on the ERA certified values.

[D] Recoveries corrected for analyte losses incurred during blowdown evaporation of solvent to determine, specifically, recoveries by microwave extraction.

TABLE 3 Chlorinated Pesticide Recovery from Freshly Spiked Topsoil[A]

Pesticide	Spike Level, μg/kg	Average Recovery, %[A]	RSD, %
Alachlor	200	87.6	2.3
Aldrin	20	87.0	2.1
α-BHC	20	94.4	4.1
β-BHC	20	90.4	3.6
γ-BHC	20	89.6	1.3
δ-BHC	20	96.9	2.8
Captafol	200	122	4.7
Captan	200	105	1.7
Chlorobenzilate	100	82.6	5.6
α-Chlordane	20	80.0	3.9
γ-Chlordane	20	86.2	2.9
Chloroneb	100	69.6	2.4
Chloropropylate	100	86.0	5.6
Chlorothalonil	60	83.4	0.9
DBCP	100	55.1	19.8
DCPA	60	93.3	1.5
4,4′-DDD	20	76.9	4.5
4,4′-DDE	20	84.7	3.1
4,4′-DDT	20	116	5.6
Diallate	200	98.6	4.1
Dichlone	200	77.7	2.6
Dichloran	60	110	9.3
Dieldrin	20	85.9	3.8
Endosulfan I	20	86.8	3.1
Endosulfan Ii	20	71.9	6.3
Endosulfan sulfate	20	81.6	6.5
Endrin	20	97.4	1.9
Endrin aldehyde	20	78.3	3.5
Endrin ketone	20	90.6	3.0
Etridiazole	100	92.6	14
Heptachlor	20	110	1.4
Heptachlor epoxide	20	95.3	2.7
Hexachlorobenzene	200	80.8	1.6
Hexachlorocyclopentadiene	100	107	12
Isodrin	20	103	1.3
Methoxychlor	60	90.9	2.7
Mirex	20	93.5	3.4
Nitrofen	100	90.2	3.6
PCNB	100	94.5	1.8
Perthane	200	86.8	4.0
Propachlor	100	91.5	8.4
Trans-Nonachlor	20	81.7	5.4
Cis-Permethrin	60	81.2	4.8
Trans-Permethrin	60	103	32
Trifluralin	40	101	2.4

[A] The number of determinations was three. The recoveries were corrected for analyte losses incurred during blowdown evaporation of solvent to determine, specifically, recoveries by microwave extraction.

unit must be capable of measuring and controlling solvent temperature within an extraction vessel. The microwave cavity should be constructed so as to prevent any possible metal to metal arcing from occurring within the cavity. The oven cavity should be equipped with an exhaust ventilation sufficient to provide ten chamber exchanges/min. The ventilation exhaust should contain an air flow sensor and a solvent sensor capable of detecting no air flow and solvent concentrations below their lower explosive limits and shutting the microwave source off. The cavity shall have a 360° oscillating turntable to ensure even sample heating and be capable of removing contained vessel-vented solvents. Safety interlocks to shut off magnetron power output shall be contained in the cavity door opening mechanism. The system shall comply with the Department of Health and Human Services Standards under the CFR, Part 1030.10, Subpart (c)(1), (c)(2), and (c)(3), for microwave leakage. The system should have Federal Communication Commission (FCC) type approval for operations under FCC Rule Part 18. If the system is used outside of the United States it should comply with health and human service standards for the country in which it is used.

TABLE 4 Organophosphorus Pesticide and Industrial Chemical Compound Recovery from Freshly Spiked Topsoil

Pesticide	Spike Level mg/kg	Average Recovery, %[A]	RSD, %
Aspon	1.0	91.8	3.4
Azinphos-methyl	2.0	103	0.9
Azinphos-ethyl	1.0	122	1.4
Bolstar	1.0	103	1.8
Chlorofenvinphos	2.0	1.02	3.8
Chlorpyrifos	1.0	90.9	3.8
Chlorpyrifos methyl	1.0	95.5	3.6
Coumaphos	2.0	100	0.7
Crotoxyphos	2.0	109	0.8
Demeton-O	1.0	98.7	7.1
Demeton-S	1.0	95.2	7.3
Dichlorofenthion	1.0	73.5	9.6
Dichlorvos	1.0	63.4	14
Dicrotophos	1.0	79.5	2.5
Dimethoate	1.0	85.8	8.8
Dioxathon	2.0	99.0	1.7
Disulfoton	1.0	97.5	3.5
EPN	1.0	102	1.3
Ethion	1.0	102	2.4
Ethoprop	1.0	106	3.6
Famphur	1.0	101	1.7
Fenithrothion	1.0	119	1.1
Fensulfothion	1.0	75.7	34
Fenthion	1.0	103	1.8
Fonophos	1.0	102	3.3
Leptophos	1.0	95.9	2.9
Malathion	1.0	105	1.1
Merphos	1.0	115	0.6
Mevinphos	2.0	66.1	17
Monochrotophos	1.0	9.4	8.3
Naled	2.0	9.4	15
Parathion ethyl	1.0	98.0	3.8
Phorate	1.0	94.4	4.9
Phosmot	1.0	99.1	0.8
Phosphamidon	2.0	16.6	29
Ronnel	1.0	99.5	2.9
Stirophos	2.0	99.0	2.7
Sulfotepp	1.0	93.0	4.3
TEPP	2.0	5.4	13
Thionazin	0.5	104	2.1
Tokuthion	1.0	99.0	3.0
Trichlorfon	1.0	7.8	16
Trichloronate	1.0	96.9	1.7
Industrial Chemicals			
Hexamethylphosphoramide (HMPA)	0.5	5.6	32
Tri-*o*-cresylphosphate (TOCP)	1.0	100	1.6
Triazine Herbicides			
Atrazine	1.0	100	0.9
Simazine	1.0	123	0.8

[A] The number of determinations was three. The analyte recoveries were corrected for losses incurred during blowdown evaporation of solvent to determine specificly recoveries by microwave extraction.

6.2 *Extraction Vessels*—A closed lined extraction vessel of approximately 100-mL capacity. The vessel consists of a fluoropolymer inner liner and cover to contain and isolate a solvent solution from a higher strength polymer outer pressure vessel body. The vessel assembly shall be transparent to microwave energy and capable of withstanding an internal pressure of 200 psi (13.8×10^{6} dynes/cm^2) and a temperature of 200°C. The vessel assembly shall contain a safety pressure relief device that will prevent possible vessel rupture and be connected to a sealed venting system capable of removing vented solvent vapors from the microwave cavity.

TABLE 5 PCB Recovery from Soils

Sample	Certified Value or Spike Level, mg/kg	Aroclor Type	Average Recovery, %[A]	RSD, %
Environmental Resource Associates soil	394[B]	1260	89.9[C]	2.6
Freshly spiked topsoil	0.100[D]	1016	85.9[E]	5.6
		1260	82.5[E]	4.2

[A] Analyte recoveries were corrected for losses incurred during blowdown evaporation of solvent to determine specificly recoveries by microwave extraction.
[B] Certified by the Environmental Resource Associates (ERA).
[C] The number of determinations was four.
[D] Spiked in the laboratory.
[E] The number of determinations was three.

6.3 *Temperature-Pressure Control Vessel*—A closed lined extraction vessel with temperature and pressure monitoring ports capable of withstanding an internal pressure of 200 psi (13.8×10^6 dynes/cm^2) and a temperature of 200°C. The vessel assembly shall contain a safety pressure relief device that will prevent possible vessel rupture and that is connected to a sealed venting system capable of removing vented solvent vapors from the microwave cavity.

7. Reagents

7.1 *Purity of Reagents*—Reagent grade chemicals shall be used in all extractions. Unless otherwise indicated, it is intended that all reagents shall conform to the specifications of the committee on Analytical Reagents of the American Chemical Society, where such specifications are available.[6]

7.2 *Acetone*, pesticide quality or equivalent.

7.3 *Hexane*, pesticide quality or equivalent.

7.4 *Purity of Water*— Unless otherwise indicated, references to water shall be understood to mean reagent water conforming to Type III of Specification D 1193.

8. Hazards

8.1 The microwave system shall be operated and maintained in accordance with the manufacturer's recommended safety precautions.

8.2 The extraction vessels shall be used and maintained in accordance with the manufacturer's recommended safety precautions.

8.3 Use only microwave systems and vessels approved for solvent use.

9. Sampling

9.1 Collect a soil, sediment, sludge, or waste sample using an appropriate sampling technique.

9.2 Prepare the soil, sediment, sludge, or waste sample in accordance with Practice D 3976, Test Method D 5368, and EPA Test Methods for Evaluating Solid Waste, Vol 1A.

9.3 *Determination of Water Content*—In certain cases, sample results are to be reported on a dry-weight basis. When such data are desired, a sample portion for water content determination should be weighed out at the same time as the portion used for organic analytical determination. Water content then is determined on that portion of the sample in accordance with Practice D 3976.

10. Vessel Cleaning

10.1 Follow the manufacturer's recommended cleaning procedure or clean the vessel liners and covers as follows:

10.1.1 Clean each liner and cover as soon as possible after use by rinsing with acetone-hexane (1 + 1), followed by washing with hot water and detergent, followed by rinsing with tap water and then with organic-free reagent water. Rinse with acetone-hexane (1 + 1), dry and store in a clean dust-free environment.

11. Procedure

11.1 Take a 5-g portion of the sample prepared in 9.2, weigh to the nearest 0.1 mg, and transfer into extraction vessels. Include an empty extraction vessel in each set as a method blank. The temperature-pressure control vessel shall contain 5 g of sample material.

11.2 Add 30 mL of acetone-hexane (1 + 1) to each sample and blank extraction vessel.

11.3 Close each extraction vessel according to the manufacturer's recommended procedures.

11.4 Place the closed extraction vessel into the instrument turntable and assemble following the manufacturer's suggested procedure.

11.5 Place the turntable containing evenly spaced vessels into the oven cavity, and make the proper connections to the control vessel following the manufacturers recommended procedures.

NOTE 1—If samples have different water contents, it is recommended that dry samples be extracted separately from wet samples, since dry and wet sample-solvent suspensions will heat at different rates, and thereby reach different temperatures.

11.6 Heat the vessels according to the extraction heating parameters in Table 6.

NOTE 2—For less than 7 vessels, microwave power can be reduced to less than 100 %.

11.7 Allow the vessels to cool to room temperature.

11.8 Open the vessels, decant, and filter or centrifuge to separate the solvent from the sample for further processing and cleanup if necessary before performing analyte determination.

12. Data

12.1 Typical analyte recoveries are contained in Tables 1-5.

12.2 Each laboratory using this practice will operate a formal Quality Control Program. The minimum requirements of this program must consist of an initial demonstration of

TABLE 6 Extraction Heating Parameters

Microwave power	100 %
Vessels pressure	150 psi
Vessels temperature	115°C
Heating time at 115°C	10 min

[6] *Reagent Chemicals, American Chemical Society Specifications*, American Chemical Society, Washington, DC. For suggestions on the testing of reagents not listed by the American Chemical Society, see *Analar Standards for Laboratory Chemicals*, BDH Ltd., Poole, Dorset, U.K., and the *United States Pharmacopeia and National Formulary*, U.S. Pharmacopeial Convention, Inc. (USPC), Rockville, MD.

laboratory capability and the ongoing analysis of method blanks, duplicate and spiked samples to evaluate, and document quality data.

13. Precision and Bias

13.1 It is the responsibility of the user to establish the precision and bias of the extraction method in conjunction with the analytical measuring method.

14. Keywords

14.1 extraction; microwave; organic compounds; safety; solvent; vessel

Designation: D 6044 – 96

Standard Guide for Representative Sampling for Management of Waste and Contaminated Media[1]

This standard is issued under the fixed designation D 6044; the number immediately following the designation indicates the year of original adoption or, in the case of revision, the year of last revision. A number in parentheses indicates the year of last reapproval. A superscript epsilon (ε) indicates an editorial change since the last revision or reapproval.

1. Scope

1.1 This guide covers the definition of representativeness in environmental sampling, identifies sources that can affect representativeness (especially bias), and describes the attributes that a representative sample or a representative set of samples should possess. For convenience, the term" representative sample" is used in this guide to denote both a representative sample and a representative set of samples, unless otherwise qualified in the text.

1.2 This guide outlines a process by which a representative sample may be obtained from a population. The purpose of the representative sample is to provide information about a statistical parameter(s) (such as mean) of the population regarding some characteristic(s) (such as concentration) of its constituent(s) (such as lead). This process includes the following stages: (*1*) minimization of sampling bias and optimization of precision while taking the physical samples, (*2*) minimization of measurement bias and optimization of precision when analyzing the physical samples to obtain data, and (*3*) minimization of statistical bias when making inference from the sample data to the population. While both bias and precision are covered in this guide, major emphasis is given to bias reduction.

1.3 This guide describes the attributes of a representative sample and presents a general methodology for obtaining representative samples. It does not, however, provide specific or comprehensive sampling procedures. It is the user's responsibility to ensure that proper and adequate procedures are used.

1.4 The assessment of the representativeness of a sample is not covered in this guide since it is not possible to ever know the true value of the population.

1.5 Since the purpose of each sampling event is unique, this guide does not attempt to give a step by step account of how to develop a sampling design that results in the collection of representative samples.

1.6 Appendix X1 contains two case studies, which discuss the factors for obtaining representative samples.

1.7 *This standard does not purport to address all of the safety concerns, if any, associated with its use. It is the responsibility of the user of this standard to establish appropriate safety and health practices and determine the applicability of regulatory limitations prior to use.*

2. Referenced Documents

2.1 *ASTM Standards:*

D 3370 Practices for Sampling Water from Closed Conduits[2]

D 4448 Guide for Sampling Groundwater Monitoring Wells[3]

D 4547 Practice for Sampling Waste and Soils for Volatile Organics[3]

D 4700 Guide for Soil Sampling from the Vadose Zone[4]

D 4823 Guide for Core-Sampling Submerged, Unconsolidated Sediments[5]

D 5088 Practice for Decontamination of Field Equipment Used at Nonradioactive Waste Sites[6]

D 5792 Practice for Generation of Environmental Data Related to Waste Management Activities: Development of Data Quality Objectives[3]

D 5956 Guide for Sampling Strategies for Heterogeneous Wastes[3]

D 6051 Guide for Composite Sampling and Field Subsampling for Environmental Waste Management Activities[3]

3. Terminology

3.1 *analytical unit, n*—the actual amount of the sample material analyzed in the laboratory.

3.2 *bias, n*—a systematic positive or negative deviation of the sample or estimated value from the true population value.

3.2.1 *Discussion*—This guide discusses three sources of bias—sampling bias, measurement bias, and statistical bias.

There is a sampling bias when the value inherent in the physical samples is systematically different from what is inherent in the population.

There is a measurement bias when the measurement process produces a sample value systematically different from that inherent in the sample itself, although the physical sample is

[1] This guide is under the jurisdiction of ASTM Committee D34 on Waste Management and is the direct responsibility of Subcommittee D34.01.01 on Planning for Sampling.

Current edition approved Nov. 10, 1996. Published January 1997.

[2] *Annual Book of ASTM Standards*, Vol 11.01.

[3] *Annual Book of ASTM Standards*, Vol 11.04.

[4] *Annual Book of ASTM Standards*, Vol 04.08.

[5] *Annual Book of ASTM Standards*, Vol 11.02.

[6] *Annual Book of ASTM Standards*, Vol 04.09.

itself unbiased. Measurement bias can also include any systematic difference between the original sample and the sample analyzed, when the analyzed sample may have been altered due to improper procedures such as improper sample preservation or preparation, or both.

There is a statistical bias when, in the absence of sampling bias and measurement bias, the statistical procedure produces a biased estimate of the population value.

Sampling bias is considered the most important factor affecting inference from the samples to the population.

3.3 *biased sampling, n*—the taking of a sample(s) with prior knowledge that the sampling result will be biased relative to the true value of the population.

3.3.1 *Discussion*—This is the taking of a sample(s) based on available information or knowledge, especially in terms of visible signs or knowledge of contamination. This kind of sampling is used to detect the presence of localized contamination or to identify the source of a contamination. The sampling results are not intended for generalization to the entire population. This is one form of authoritative sampling (see *judgment sampling*.)

3.4 *characteristic, n*—a property of items in a sample or population that can be measured, counted, or otherwise observed, such as viscosity, flash point, or concentration.

3.5 *composite sample, n*—a combination of two or more samples.

3.6 *constituent, n*— an element, component, or ingredient of the population.

3.6.1 *Discussion*—If a population contains several contaminants (such as acetone, lead, and chromium), these contaminants are called the constituents of the population.

3.7 *Data Quality Objectives, DQOs, n*—qualitative and quantitative statements derived from a DQO process describing the decision rules and the uncertainties of the decision(s) within the context of the problem(s) (see Practice D 5792).

3.8 *Data Quality Objective Process*—a quality management tool based on the Scientific Method and developed by the U.S. Environmental Protection Agency to facilitate the planning of environmental data collection activities. The DQO process enables planners to focus their planning efforts by specifying the use of data (the decision), the decision criteria (action level), and the decision maker's acceptable decision error rates. The products of the DQO process are the DQOs (see Practice D 5792).

3.9 *error, n*—the random or systematic deviation of the observed sample value from its true value (see *bias* and *sampling error*).

3.10 *heterogeneity, n*— the condition or degree of the population under which all items of the population are not identical with respect to the characteristic(s) of interest.

3.10.1 *Discussion*—Although the ultimate interest is in the statistical parameter such as the mean concentration of a constituent of the population, heterogeneity relates to the presence of differences in the characteristics (for example, concentration) of the units in the population. It is due to the presence of fundamental heterogeneity (or fundamental error)[7] in the population that sampling variance arises. Degree of sampling variance defines the degree of precision in estimating the population parameter using the sample data. The smaller the sampling variance is, the more precise the estimate is. See also *sampling error*.

3.11 *homogeneity, n*— the condition of the population under which all items of the population are identical with respect to the characteristic(s) of interest.

3.12 *judgment sampling, n*—taking of a sample(s) based on judgment that it will more or less represent the average condition of the population.

3.12.1 *Discussion*—The sampling location(s) is selected because it is judged to be representative of the average condition of the population. It can be effective when the population is relatively homogeneous or when the professional judgment is good. It may or may not introduce bias. It is a useful sampling approach when precision is not a concern. This is one form of authoritative sampling (see *biased sampling*.)

3.13 *population, n*— the totality of items or units of materials under consideration.

3.14 *representative sample, n*—a sample collected in such a manner that it reflects one or more characteristics of interest (as defined by the project objectives) of a population from which it is collected.

3.14.1 *Discussion*—A representative sample can be a single sample, a collection of samples, or one or more composite samples. A single sample can be representative only when the population is highly homogeneous.

3.15 *representative sampling, n*—the process of obtaining a representative sample or a representative set of samples.

3.16 *representative set of samples, n*—a set of samples that collectively reflect one or more characteristics of interest of a population from which they were collected. See *representative sample*.

3.17 *sample, n*—a portion of material that is taken for testing or for record purposes.

3.17.1 *Discussion*—Sample is a term with numerous meanings. The scientist collecting physical samples (for example, from a landfill, drum, or monitoring well) or analyzing samples considers a sample to be that unit of the population that was collected and placed in a container. A statistician considers a sample to be a subset of the population, and this subset may consist of one or more physical samples. To minimize confusion, the term *sample*, as used in this guide, is a reference to either a physical sample held in a sample container, or that portion of the population that is subjected to in situ measurements, or a set of physical samples. See *representative sample*.

3.17.1.1 The term *sample size* also means different things to the scientist and the statistician. To avoid confusion, terms such as sample mass/sample volume and number of samples are used instead of sample size.

3.18 *sampling error*— the systematic and random deviations of the sample value from that of the population. The

[7] Pitard, F. F., "*Pierre Gy's Sampling Theory and Sampling Practice: Heterogeneity, Sampling Correctness and Statistical Process Control*," 2nd ed., CRC Press Publishers, 1993.

systematic error is the *sampling bias*. The random error is the *sampling variance*.

3.18.1 *Discussion*—Before the physical samples are taken, potential sampling variance comes from the inherent population heterogeneity (sometimes called the "fundamental error," see *heterogeneity*). In the physical sampling stage, additional contributors to sampling variance include random errors in collecting the samples. After the samples are collected, another contributor is the random error in the measurement process. In each of these stages, systematic errors can occur as well, but they are the sources of bias, not sampling variance.

3.18.1.1 Sampling variance is often used to refer to the total variance from the various sources.

3.19 *stratum, n*—a subgroup of the population separated in space or time, or both, from the remainder of the population, being internally similar with respect to a target characteristic of interest, and different from adjacent strata of the population.

3.19.1 *Discussion*—A landfill may display spatially separated strata, such as old cells containing different wastes than new cells. A waste pipe may discharge into temporally separated strata of different constituents or concentrations, or both, if night-shift production varies from the day shift. In this guide, strata refer mostly to the stratification in the concentrations of the same constituent(s).

3.20 *subsample, n*— a portion of the original sample that is taken for testing or for record purposes.

4. Significance and Use

4.1 Representative samples are defined in the context of the study objectives.

4.2 This guide defines the meaning of a representative sample, as well as the attributes the sample(s) needs to have in order to provide a valid inference from the sample data to the population.

4.3 This guide also provides a process to identify the sources of error (both systematic and random) so that an effort can be made to control or minimize these errors. These sources include sampling error, measurement error, and statistical bias.

4.4 When the objective is limited to the taking of a representative (physical) sample or a representative set of (physical) samples, only potential sampling errors need to be considered. When the objective is to make an inference from the sample data to the population, additional measurement error and statistical bias need to be considered.

4.5 This guide does not apply to the cases where the taking of a nonrepresentative sample(s) is prescribed by the study objective. In that case, sampling approaches such as judgment sampling or biased sampling can be taken. These approaches are not within the scope of this guide.

4.6 Following this guide does not guarantee that representative samples will be obtained. But failure to follow this guide will likely result in obtaining sample data that are either biased or imprecise, or both. Following this guide should increase the level of confidence in making the inference from the sample data to the population.

4.7 This guide can be used in conjunction with the DQO process (see Practice D 5792).

4.8 This guide is intended for those who manage, design, and implement sampling and analytical plans for waste management and contaminated media.

5. Representative Samples

5.1 Samples are taken to infer about some statistical parameter(s) of the population regarding some characteristic(s) of its constituent(s) of interest. This is discussed in the following sections.

5.2 *Samples*—When a representative sample consists of a single physical sample, it is a sample that by itself reflects the characteristics of interest of the population. On the other hand, when a representative sample consists of a set of physical samples, the samples collectively reflect some characteristics of the population, though the samples individually may not be representative. In most cases, more than one physical sample is necessary to characterize the population, because the population in environmental sampling is usually heterogeneous.

5.3 *Constituents and Characteristics*—A population can possess many constituents, each with many characteristics. Usually it is only a subset of these constituents and characteristics that are of interest in the context of the stated problem. Therefore, samples need to be representative of the population only in terms of these constituent(s) and characteristic(s) of interest. A sampling plan needs to be designed accordingly.

5.4 *Parameters*—Similarly, samples need to be representative of the population only in the parameter(s) of interest. If the interest is only in estimating a parameter such as the population mean, then composite samples, when taken correctly, will not be biased and therefore constitute a representative sample (regarding bias) for that parameter. On the other hand, if the interest happens to be the estimation of the population variance (of individual sampling units), another parameter, then the variance of the composite samples is a biased estimate of the population variance and therefore is not representative. (It is to be noted that composite samples are often used to increase the precision in estimating the population mean and not to estimate the population variance of individual sampling units.)

5.5 *Population*—Since the samples are intended to be representative of a population, a population must be well defined, especially in its spatial or temporal boundaries, or both, according to the study objective.

5.6 *Representativeness*—The word "reflects" in this guide is used to mean a certain degree of low bias and high precision when comparing the sample value(s) to the population value(s). This is a broad definition of sample representativeness used in this guide. A narrower definition of representativeness is often used to mean simply the absence of bias.

5.6.1 *Bias*—Bias is sometimes mistakenly taken to be "a difference between the observed value of a physical sample and the true population value." The correct definition of bias is "a *systematic* (or consistent) difference between an observed (sample) value and the true population value." The word "systematic" here implies "on the average" over a set of physical samples, and not a single physical sample. Recall that sampling error consists of the random and systematic deviations of a sample (or estimated) value from that of the population. Although random deviations may occur on occasions due to imprecision in the sampling or measurement processes, or both, they balance out on the average and lead to no systematic difference between the sample (or estimated) value and the population value. The random deviation corresponds to the observation of "a random difference between a single physical sample value and the true population value," which can be randomly positive or negative, and is not a bias. On the other hand, a persistent positive or negative difference is a systematic error and is a bias.

5.6.1.1 In order to assess bias, the true population value must be known. Since the true population value is rarely known, bias cannot be quantitatively assessed. However, this guide provides an approach to identifying the potential sources of bias and general considerations for controlling or minimizing these potential biases.

5.6.2 *Precision*—Precision has to do with the level of confidence in estimating the population value using the sample data. If the population is totally homogeneous and the measurement process is flawless, a single sample will provide a completely precise estimate of the population value. When the population is heterogeneous or the measurement process is not totally precise, or both, a larger number of samples will provide a more precise estimate than a smaller number of samples.

5.6.2.1 In the case of bias, the goal in environmental sampling is its absence. In the case of precision, the goal in sampling will depend on factors such as:

(1) The precision level needed to achieve the desired levels of decision errors, both false positive and false negative errors,

(2) If the true value is known or suspected to be well below the regulatory limit, high precision in the samples may not be needed, and

(3) The study budget.

5.6.2.2 Note that the second item applies similarly to bias as well.

5.6.2.3 Since bias, especially during sampling, can be very large when proper procedures are not followed, it is considered to be the first necessary condition for sample representativeness. On the other hand, precision can be more or less controlled, for example, by increasing the number of samples taken or by decreasing the sampling or measurement variabilities, or both.

5.6.2.4 The optimal number of samples to take to achieve a desired level of precision is typically an issue in optimization of a sampling plan. Therefore, the precision issue will be covered only briefly in this guide.

6. A Systematic Approach to Representative Sampling

6.1 A systematic approach is one that first defines the desired end result and then designs a process by which such a result can be obtained. In representative sampling, the desired end result is a sample or a set of samples that achieves desired levels of low bias and high precision.

6.2 A representative sampling process is described in Fig. 1. The key components in the process are described in this section.

6.3 *Study Objective*— A sampling plan is designed according to a defined problem or a stated study objective. The samples are then collected according to the sampling plan. Generally, the study objective dictates that representative samples be taken for the purpose of inference about the population. In that case, these samples will need to be collected according to this guide in order for the inference to be valid.

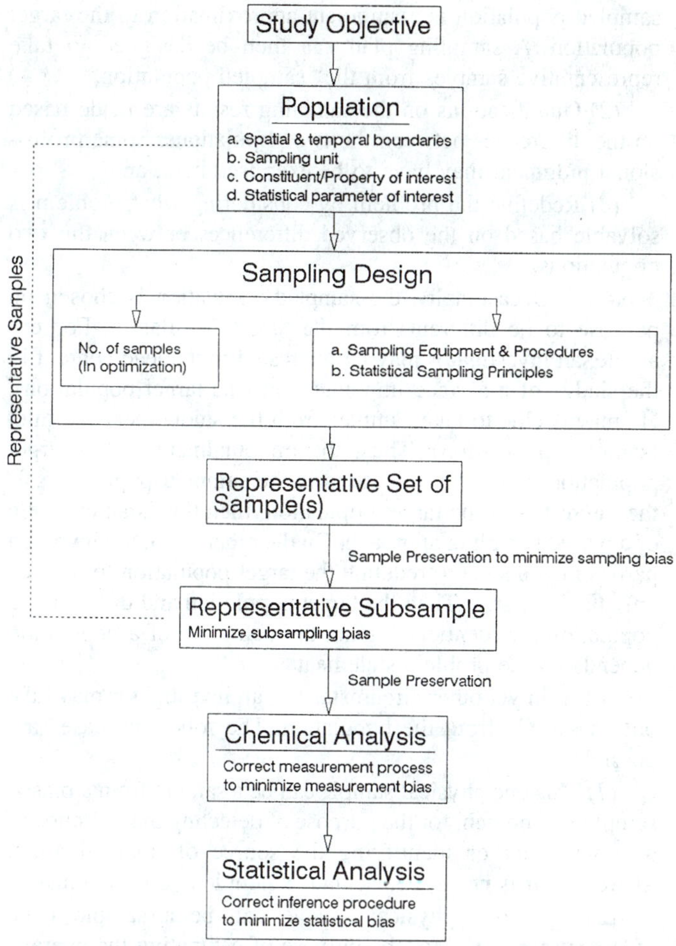

FIG. 1 A Systematic Approach to Representative Sampling

Occasionally, the objective is merely to detect the presence of a contaminant or to obtain a "worst case" sample. In that case, an authoritative sampling approach (biased sampling or judgment sampling) may be taken and this guide does not apply.

6.4 *Population*—A population consists of the totality of items or units of materials under consideration (Compilation of ASTM Standard Definitions, 1990). Its boundaries (spatial or temporal, or both) are defined according to the problem statement. This population is usually called the *target population*. In order to solve the stated problem, samples must be taken from the target population.

6.4.1 *Sampled Population*—Sometimes some parts of the target population may not be amenable to sampling due to factors such as accessibility. The boundaries of the target population actually sampled due to factors such as incomplete accessibility define the sampled population.

6.4.1.1 Although the samples taken from the sampled population may be representative of the sampled population, they may not be representative of the target population. In this case, potential exists that the samples taken from the sampled population may systematically deviate from the true value of the target population, thereby introducing bias when making inference from the samples to the target population.

6.4.1.2 When the boundaries of the target and sampled populations are not identical, some possible solutions are:

(1) The parties to the decision-making may agree that the

sampled population is a sufficient approximation to the target population. A sampling plan can then be designed to take representative samples from the" sampled population,"

(2) Qualifications on the sampling results are made based on the differences between the two populations. Some professional judgment may have to be exercised here, and

(3) Redefine the problem by considering what problem is solvable based on the observed differences between the two populations.

6.4.1.3 Occasionally, the sampled population is chosen on purpose to be different from the target population. For example, an investigator may be interested in the lead content in the sludge of a surface impoundment (the target population). He may decide to take samples from the sludge near the inlet (sampled population). Thus, the impoundment is the target population, while the inlet area is the sampled population. If the interest is in the target population, then this is an example of a biased sampling approach. On the other hand, the involved parties may decide to redefine the target population to include only the inlet area. Then the target population and the sampled population are identical. Again, the definition of a population depends on the problem statement.

6.4.1.4 In yet other circumstances, an investigator may take only a sample from the population. The following cases are possible:

(1) This one physical sample can be a sample from a biased sampling approach, for the purpose of detecting the presence of a contaminant or identifying the source of contamination. Therefore, it is not a representative sample due to its bias,

(2) This one physical sample can be a sample from judgment sampling, for the purpose of estimating the average condition of the population. Bias may or may not exist depending to some degree on the expertise of the sampler,

(3) This sample can be viewed as a population itself if the investigator is interested in the sample alone and a result from this sample is not to be used to infer to areas outside the sample. In this case, no bias exists, and

(4) If this sample is the composite of a few samples taken from the population, bias is likely to be minimal if the original samples are carefully taken.

6.4.2 *Decision Unit*— Often a population may be divided into several exposure units, cleanup units, or strata. If the environmental management decision is to be made for the entire population as a whole, representative samples can be obtained by designs such as a stratified random sampling design. Here the entire population is the decision unit. On the other hand, if the decision is to be made on each unit or stratum, then each unit or stratum is the decision unit. In this case, representative sample(s) need to be taken from each unit or stratum as if the unit or stratum is the population.

6.4.2.1 If the units or strata are relatively small in size or too numerous to take many samples per unit or stratum, composite sample(s) can be taken from each unit or stratum to increase precision without introducing bias. Alternatively, if precision is not a concern and there is sufficient professional expertise to avoid bias, a judgment sample(s) can be taken from each unit or stratum.

6.4.3 *Heterogeneity*— Heterogeneity is discussed in greater detail in Guide D 5956.

6.4.3.1 The degree and extent of population heterogeneity affect potential bias and precision in the samples. Population heterogeneity can be viewed at least in three different ways:

(1) When the population is heterogeneous in a random manner in only the distribution of the concentration, but not in the physical materials such as particle sizes, designs such as a simple random sampling design will generally produce samples with minimal bias. Its precision will then depend on the number of samples taken,

(2) When the population is randomly heterogeneous in concentrations due to large differences in the materials such as particle size, a simple random sampling design may still be effective if the sample volume/weight and sampling equipment are chosen to accommodate the largest particles and thereby prevent introduction of bias, and

(3) If the population is systematically heterogeneous, such as the presence of stratification in concentrations, then a simple random sampling design may not be biased, but will be less precise than an alternative design such as stratified random sampling.

6.4.3.2 Heterogeneity in the population affects the sampling variance. Sampling variance is a function of factors such as the population heterogeneity and the sample volume or weight. It is clear that the more heterogeneous the population is, the larger the inherent sampling variance is. It is also clear that samples of smaller volume or weight will have a higher sampling variance than those with greater volume or weight. However, the reduction in sampling variance due to increased volume or weight may eventually reach a limit. Determination of the optimal sample volume or weight is beyond the scope of this guide.[7]

6.4.3.3 The proper procedure is to first determine the right sample volume or weight, then to determine the number of samples needed for the chosen sample volume or weight.

6.4.3.4 Since stratification as a phenomenon of population heterogeneity is fairly common, it is discussed in greater details as follows.

6.4.4 *Stratification*— There are generally three types of stratification affecting sample representativeness. One is a stratification in the distribution of the contaminant concentration distribution alone. The second is a stratification in sampling materials or matrices alone. The third is a combination of both types. Stratification of any type is not a big problem regarding sample representativeness if each stratum is a decision unit. In that case, the units in a stratum are by definition relatively similar, apart from the random variations in concentrations. A simple random sampling design can be used to obtain representative samples (unbiased) for each stratum. The question of sample representativeness becomes more complicated when a decision is to be made over all the strata in the population.

6.4.4.1 *A Single Representative Sample in A Stratified Population*—When the objective is to obtain a single (physical) representative sample of all the strata, the sample must be a composite of individual samples from the strata (for example, at least one individual sample per stratum). Here the volumes or weights of the individual samples should be proportional to the relative stratum sizes. The composite sample so obtained would be unbiased. However, since there is only one composite sample, precision of the composite sample cannot be estimated. If there are existing data on the precision of the individual samples in the strata, then the precision of the composite sample can be inferred from the precision of the individual samples by theoretical or empirical relationship. See Guide D 6051.

6.4.4.2 *A Representative Set of Samples*—When the population is stratified, a set of samples obtained by statistical designs such as stratified random sampling, where the number of samples to be taken from the strata are proportional to the relative sizes of the strata, is unbiased and more precise than a set of samples taken without considering the stratification.

6.4.5 *Parameter(s) of Interest*—This refers to the statistical parameter such as mean or variance of the population. It is often used with a characteristic such as concentration of a constituent(s) of the population. An example is the mean (parameter) concentration (characteristic) of lead (constituent). Another example is a population of mixture of silt-size calcium carbonate particles and large cobble-size particles of calcium carbonate. The interest here could be in the mean (parameter) particle size or chemical composition (characteristic) of calcium carbonate (constituent), depending on the study objective.

6.5 *Develop A Sampling Design*—The objectives of a sampling design are to minimize bias and achieve a desired level of precision. Precision and bias are an issue at various stages of the process of inferring from the samples to the population. The first stage is the act of obtaining the physical samples. The second stage is the act of analyzing the physical samples and translating them into data. The third stage is the use of statistical method to infer from the sample data to the population. At the first stage, the main concerns are sampling precision and bias. At the second stage, the concerns are measurement of precision and bias. At the third stage, the concern is statistical bias.

6.5.1 At the first stage of obtaining physical samples, the issues of precision and bias are sometimes grouped together as sampling design issues.

6.5.2 Bias at this stage is often called the sampling bias. Sampling bias is the systematic difference between the value inherent in the physical samples and the true population value. The word "inherent" is used because at this point the physical samples have not been translated into data.

6.5.3 The phrase "systematic difference" implies a persistent difference in long-term average or expectation, not the occasional random difference. Representative samples, apart from the issue of precision, are obtained when this long-term expected difference is zero or nearly so.

6.5.4 Since the true population value is typically not known, sampling bias cannot be assessed. However, efforts to minimize sampling bias can be attempted in at least two areas:

6.5.4.1 *Proper Statistical Sampling Design*—Statistical sampling design has to do with where and how samples are to be taken, where equal probability of selecting any of the units or items in the population is often a primary requirement. If the probability of selection is not equal, it is highly likely that bias will have been introduced into the physical samples so obtained. Depending on the layout of the population, designs such as simple random sampling or stratified random sampling can be used.

6.5.4.2 *Proper Sampling Procedures and Sampling Equipment*—This includes proper procedures for compositing, subsampling, sample preparation and preservation, and proper use of the chosen sampling equipment. This is a major source affecting precision and bias, especially bias.

6.5.5 In the case of precision, it can be controlled by things such as the number of samples taken, the use of composite samples, or more precise sampling techniques. Often, the number of samples to take is considered the key design issue. Some considerations regarding precision are:

6.5.5.1 If a population is relatively small compared to the sample mass/volume and the distribution of the characteristic of interest is random, it may be appropriate to collect a smaller number of samples by a random or systematic sampling approach, and

6.5.5.2 If a population is relatively large compared to sample mass/volume and the characteristic of interest is not randomly distributed (for example, stratified), a greater number of samples and a stratified sampling approach may be needed.

6.5.6 *Compositing*— Compositing is the combination of two or more individual physical samples into a single sample. It is often used to reduce the analytical costs, while maintaining or increasing precision relative to the individual samples (see Guide D 6051). Bias may or may not be introduced in compositing, depending on the study objective and the physical means of compositing. For example:

6.5.6.1 If the study calls for the estimation of the population variance (or standard deviation) of individual samples, then composite samples will surely underestimate the population variance, and

6.5.6.2 If the physical means of compositing changes the characteristics of the samples, then bias may have been introduced (unless such changes are part of the study design).

6.6 *Subsampling*— Sampling bias can be introduced in subsampling unless the same proper sampling protocol is followed as in taking samples from the original population.

6.6.1 *Discussion*—After the physical samples have been obtained and before they are measured, bias can be prevented by following proper sample preservation and preparation procedures. It is not important whether these procedures are viewed as part of the sampling process or as part of the measurement process. It is only important in following the proper procedures to prevent bias.

6.7 *Measurement of Precision and Bias*:

6.7.1 The measurement process, like the sampling process, also consists of a random error and a systematic error. The random errors define the degree of measurement precision, and the systematic error defines the degree of measurement bias.

6.7.2 Like sampling precision, measurement precision is controlled by things such as the number of replicate analyses performed per sample and refinements of the analytical method.

6.7.3 Measurement bias is a systematic difference between the sample value produced by the measurement process and the true population value, assuming that the physical samples are unbiased before the analysis. The bias can come from contamination, loss or alteration of the sample materials, systematic errors in the measurement device, or from systematic human errors.

6.7.4 Often the measurement bias can be reasonably estimated in a laboratory testing setting when the true value is known. Laboratory samples spiked with known quantities of a chemical or certified reference standard can often be used to assess potential measurement bias. Minimization or adjustment for such estimable bias in the measurement process is essential in order to obtain data that are unbiased. When estimation of bias is not possible, care in measurement protocol and training is probably the only recourse.

6.7.4.1 *Discussion*—It is important to note that, when inferring from the sample data to the population, all the sources of imprecision, including sampling, subsampling, and measurement, need to be combined. The process of accumulating these sources of variation is sometimes called the "propagation of errors." The determination of the optimal numbers of samples, subsamples, and replicates are an issue of optimization and is not covered in this guide.

6.8 *Statistical Bias*— Statistical bias can result from an inappropriate sampling design or inappropriate estimation procedures, or both:

6.8.1 *Selection Bias from Sampling Design*—In the course of taking the sample, if the population units do not have the same probability of being selected, bias can be introduced. This bias can be prevented or minimized when a statistical sampling design is carefully selected, based on the study objective and the layout of the population. Some possible designs are the simple random sampling design and the stratified random sampling design.

6.8.2 *Estimation of Bias from Estimation Procedures*—This bias occurs when the expected value of the statistical estimator does not equal the true value.

6.8.2.1 Estimation bias can occur when the wrong statistical distribution of the data is used. For example, if the normal distribution assumption is used when the true data distribution is lognormal, the interval estimate of the mean concentration will be an biased estimate against the true interval. Thus, the expected value of the estimator will not be equal to the true value. To avoid this potential bias, it is wise to check the data distribution.

6.8.2.2 Estimation bias can also occur when a wrong statistical estimator is used. For example, if the sum of squares of deviations from the sample mean divided by the number of samples (that is, $\Sigma_{i=1,n}(x_i - \bar{x})^2/n$) is used to estimate the population variance, then this estimator is biased (its mathematical expected value is not equal to the population variance). If its denominator is modified to be $(n-1)$, then it is an unbiased estimator. For an unbiased statistical estimator, the reader is advised to check with a statistician.

7. Attributes of Representative Samples

7.1 The attributes of a representative (physical) sample or a representative set of (physical) samples can be described in the chronological order in which samples are taken. Note that these attributes apply only to how representative the physical samples are of the population. This corresponds to the upper half of Fig. 1.

7.2 *Design Considerations*:

7.2.1 A well-defined target population. The target population includes all the population units as determined from the stated problem.

7.2.2 The sampled population equals the target population in their spatial or temporal boundaries, or both. The sampled population consists of the population units directly available for measurement.[8]

7.2.2.1 When all the population units in the target population are accessible and directly available for measurement, then the sampled population is identical to the target population in its spatial or temporal boundaries, or both.

7.2.2.2 When not all the population units are directly available for measurement, then the inference from the sample is made to the sampled population, not the target population.

7.2.3 Size (weight or volume) of the sampling unit is well defined.

7.2.3.1 The population can be divided into various sizes (weight or volume) of population units. The size of the sampling unit is the size of the population unit most appropriate for the sampling purposes.

7.2.3.2 The appropriate size of the sample is determined by degree of heterogeneity of the materials to be sampled, such as particle size or shape.

7.3 *Sampling and Measurement Considerations*:

7.3.1 Correct sampling procedures are followed to minimize sampling bias.

7.3.1.1 Absence or minimization of bias is a key attribute of representative samples. Sampling bias can be minimized by following correct sampling procedures. Correct sampling procedures have two components.

(1) A sampling procedure that maximizes the potential of population units having equal probability of selection as sampled, and

(2) Correct sampling procedures. This includes the selection of appropriate equipment and proper use of that equipment.

7.3.2 Sample integrity is maintained during sampling and before chemical analysis.

7.3.3 If subsampling is performed, correct sampling procedures are followed to minimize sampling bias.

7.3.4 Sample preparation errors such as contamination and loss or alteration of constituents are prevented or minimized.

7.3.5 The samples, in the end, collectively reflect the target population within the context of the problem.

7.3.6 These attributes can be summarized into three broad categories:

7.3.6.1 A well-defined population,

7.3.6.2 Correct sampling procedures, and

7.3.6.3 Samples collected in the context of the stated problem.

[8] Gilbert, Richard O., *Statistical Methods for Environmental Pollution Monitoring*, Van Nostrand Reinholt Co., New York, NY 1987.

8. Practical Considerations

8.1 *Sampling Equipment*—The choice of appropriate sampling equipment can be crucial to the task of collecting a representative sample or a representative set of samples. Depending on the goals of the sampling activity, the sampling device used should minimize bias by having certain characteristics and capabilities, such as:

8.1.1 The ability to access and extract from every location in the target population,

8.1.2 The ability to collect a sample of proper shape,

8.1.3 The ability to collect a sufficient mass or volume of sample such that the distribution of particle sizes in the population are represented, and

8.1.4 The ability to collect a sample without the addition or loss of contaminants of interest.

8.2 *Equipment Design*— The improper design of sampling equipment may result in the collection of samples that are not representative of the population.

8.2.1 An example of equipment design influencing sampling results is samplers which exclude certain sized particles from a soil matrix or waste pile sample. The shape of some scoops may influence the distribution of particle sizes collected from a sample. Dredges used to collect river or estuarine sediments may also exclude certain sized particles, particularly the fines fraction which may contain a significant percentage of some contaminants such as polynuclear aromatic hydrocarbons (PAHs). Specific considerations in equipment design are outlined as follows.

8.2.1.1 *Sample Volume Capabilities*— Most sampling devices will provide adequate sample volume. However, the sampling equipment volumes should be compared to the volume necessary for all required analyses and the additional amount necessary for quality control (QC), split and repeat samples. Taking more than one aliquot to obtain an adequate sample volume can impact the representativeness of a sample.

8.2.1.2 *Compatibility*— It is important that sampling equipment, other equipment that may come in contact with samples (such as gloves, mixing pans, knives, spatulas, spoons, etc.) and sample containers be constructed of materials that are compatible with the matrices and analytes of interest. Incompatibility may result in the contamination of the sample and the degradation of the sampling equipment.

8.2.1.3 *Decontamination (see Practice D 5088) and Reuse*—Inadequate decontamination of sampling equipment can result in contamination of the sample and affects its representativeness. Due to design, some equipment is very difficult to adequately decontaminate. In some instances, it may even be desirable to either dispose of sampling equipment after use or to dedicate the equipment to a sampling point.

8.3 *Sampling Procedure*—Inappropriate use of sampling equipment is one of the largest sources of sampling bias. While it is beyond the scope of this guide to discuss it in depth, examples of how bias can be introduced during the sampling procedure are discussed in the following paragraphs. This guide does not provide comprehensive sampling procedures. It is the responsibility of the user to ensure that proper and adequate procedures are used.

8.3.1 *Ground Water*— For a more comprehensive discussion of sampling ground water refer to Guide D 4448.

8.3.1.1 Ground-water samples are usually collected through an in-place well, either temporarily or permanently installed. The following is a list of concerns that should be considered when collecting a ground-water sample.

(1) The well should be purged before collecting samples in order to clear the well of stagnant water which is not representative of aquifer conditions. Purging and sampling rates can cause chemical or physical changes in the water.

(2) Purging can be performed in such a way that the entire column of water is not removed. The best method for avoiding this situation is by lowering a pump or bailer into the top of the column of water.

(3) Bailing may stir up sediment in the well if conducted too vigorously. Increased turbidity can result in a higher metal content in the sample than in a non-turbid sample.

(4) Samples for volatile organic analysis should be collected in a fashion that minimizes agitation of the sample.

(5) Wells with in-place plumbing must also be purged. Samples should be collected immediately following purging. In order to collect a sample representative of ground water, samples should be collected before the water travels through any hoses or in-line treatment devices.

8.3.2 *Surface Water and Sediment*—For a more comprehensive discussion of sampling surface water and sediment, refer to Practice D 3370 and Guide D 4823. General and specific sampling concerns for collection of surface water and sediment samples are as follows:

8.3.2.1 *General Considerations*.

(1) Although bridges and piers may provide access for water and sediment sampling, these structures can also alter the nature of water flow and thus influence sediment deposition or scouring. Depending on the construction materials, these structures can contaminate samples collected in the immediate vicinity.

(2) Wading for water samples should be done with caution since bottom deposits are easily disturbed resulting in increased sediment in surface water samples and a removal of fines from the sediment sample.

8.3.2.2 *Rivers, Streams, and Creeks*:

(1) A good location to collect a vertically mixed surface water sample is immediately downstream of a riffle area. This location is also a likely area for deposition of sediment since the greatest deposition occurs where stream velocity slows down.

(2) Horizontal (cross-channel) mixing occurs in constrictions in the channel. However, this is a poor sediment sample collection area because of scouring.

(3) Surface water samples will be affected by point sources, such as tributaries and industrial and municipal effluents.

(4) Locations immediately upstream or downstream from the confluence of two streams or rivers may not immediately mix, and at times, due to possible back flow, can upset the normal flow patterns.

(5) Unless a stream is extremely turbulent, it is nearly impossible to measure the effect of a waste discharge or tributary immediately downstream of the source. Inflow frequently "hugs" the stream bank with very little cross-channel

mixing for some distance. Samples from quarter points across a stream may miss the wastes altogether and reflect only the quality of water upstream from the waste source. Samples collected within the portion of the cross section containing the wastes would indicate excessive effects of the wastes with respect to the river as a whole.

(6) When sampling tributaries, care should be exercised to avoid collecting water from the main stream that may flow into the mouth of the tributary on either the surface or bottom.

8.3.2.3 *Lakes, Ponds, and Impoundments*:

(1) Stratification of surface water is of greater concern in standing water. For example: A turbidity difference may occur vertically where a highly turbid river enters a lake, and each layer of the stratified water column may need to be considered. In addition, stratification may be caused by water temperature difference; cooler, heavier river water is beneath the warmer lake water.

(2) Dredges used to collect sediment samples can displace and miss lighter materials if allowed to drop freely.

(3) Core samplers used to sample vertical columns of sediment are useful when there is a need to know the history of sediment deposition. Coring devices also minimize the disturbance of fines at the sediment-water interface. However, coring devices can only sample a relatively small surface area. Depending on the core diameter, larger particles may be excluded and a single aliquot may not be sufficient for analytical needs.

8.3.3 *Soils*—For more detailed information, refer to Practice D 4547 and Guide D 4700. General areas of concern for sampling soils are as follows:

8.3.3.1 Soil samples for purgeable organic analyses should be collected with a minimum disturbance of the sample.

8.3.3.2 Samples for VOA analysis should not be mixed.

8.3.3.3 Two potential problems are associated with compositing soil samples. Low concentrations of contaminants present in individual aliquots may be diluted to the extent that the total composite concentration is below the minimum quantification limit. In addition, depending on the soil type, it can be very difficult to produce a homogeneous mixture.

8.3.4 *Waste*—Wastes referred to in this section include any liquid, solid, or sludge from pits, ponds, lagoons, waste piles, landfills, and open or closed containers such as drums, tank trucks, and storage tanks.

8.3.4.1 Any of these units may have multiple phases (floating solids, different density liquid phases, and sludge) and one or all of them may need to be sampled.

8.3.4.2 If sampling from access valves or ports on an open or closed container, care should be taken to be sure that the desired layer is sampled. For example, bottom sampling ports would allow only the heavier contents to be sampled while surface or top sampling would allow only sampling of the lighter layers.

8.4 *Subsampling (Field)*:

8.4.1 Different analyses require different types of bottles and preservation. For multiple analyses of the same waste stream, this may require subsampling in the field. Subsampling in the laboratory may require many of the same procedures; however, laboratory subsampling is beyond the scope of this guide.

8.4.1.1 Samples for organic analyses should always be taken from the first material collected. This minimizes loss of volatile organics during handling of the material.

8.4.1.2 If necessary, place the appropriate volume of material in a tray or other suitable container to composite. The volume is dependent on the needed analyses, and should be specified by the analytical laboratory.

8.4.1.3 Transfer the material into the required containers for analyses. If subsampling takes place, then the analytical sample is the final portion of the material subsampled from the original sampling unit and analyzed in the laboratory.

8.4.2 In subsampling, the original sampling unit can be considered as the population and the correct sampling procedures must be followed to ensure a representative subsample.

9. Keywords

9.1 bias; contaminated media; precision; representative; sample; waste; waste management

APPENDIX

(Nonmandatory Information)

X1. TWO CASE STUDIES OF REPRESENTATIVE SAMPLING

X1.1 Case Study One—Waste Pile Investigation

X1.1.1 *Background*— An industrial facility has managed recovery furnace slag and baghouse dust in a waste pile located on the site. No active management was occurring with the waste pile. No buried containers or extremely heterogenous material (debris) was suspected of being present in the waste pile based on facility records and interviews of personnel.

X1.1.1.1 Lead and cadmium were the constituents of concern based on process knowledge, and the possibility for the waste being hazardous by means of the Toxicity Characteristic (TC) Rule was the regulatory consideration. No preliminary information on the variability of lead and cadmium within the piles was available. The potential for off-site migration of contaminants by means of a drainage ditch that leads to a stream adjacent to the facility was an immediate concern.

X1.1.2 *Phase 1: Objective*—The primary objective of the initial investigation was to determine if the slag and baghouse dust in the waste piles were characteristic for lead via the Toxicity Characteristic Rule. A secondary objective was to provide preliminary information on potential migration and transport of contaminants from the waste piles off site.

X1.1.2.1 The sampling design for this initial investigation

utilized a judgmental sampling strategy to provide a preliminary estimate of the lead and cadmium concentrations in the waste pile, the variability of contaminant concentrations in the pile, and the potential for leaching using the TCLP. Four areal composite samples were collected from the surface (0 to 6 in.) at the four quadrants of the waste pile. Borings were completed at the center of each area that was sampled on the surface. Each four-foot interval was analyzed to assess vertical variability.

X1.1.2.2 The following environmental samples were also collected using a judgmental approach:

(1) Several soil samples in the vicinity of the waste pile,

(2) Sediment upstream and downstream in a stream that borders the facility,

(3) Sediment in a ditch which contained run-off from the pile, and

(4) Two background soil samples.

X1.1.2.3 *Results*—Zinc, copper, cadmium, and lead were all elevated (compared to background) in the samples collected from the waste piles. Since lead and cadmium are TC Rule constituents, the TCLP was completed, and the lead results exceeded the regulatory level of 5 mg/L. Cadmium was just under the regulatory level of 1.0 mg/L. Lead and cadmium concentrations in the soil near the waste piles were 2 to 3 times above background, and the drainage ditch and downstream sediment sample also had elevated lead and cadmium levels.

X1.1.2.4 *Conclusion*— The waste piles contain slag and baghouse dust that is hazardous for lead. The waste pile requires further characterization to determine the variability in the pile. The presence of lead and cadmium in soils and the stream sediment downstream of the facility was confirmed and should be further investigated to determine the extent of contaminant transport.

X1.1.3 *Phase 2: Objective*—The sampling design utilized a systematic grid approach. This design will delineate horizontal and vertical variability in lead and cadmium concentrations. The Phase 1 investigation also provided a good estimate of the anticipated variability in the waste pile.

X1.1.3.1 The number of samples required to adequately characterize the waste pile was calculated based on the anticipated variability, the regulatory level of concern, and the specified confidence interval. The grid sizes were then adjusted to accommodate the projection on the required number of samples. Composite samples were collected within each grid cell based on one center point and eight points on the compass (45 deg intervals) equidistant from the center point.

X1.1.3.2 Twenty percent of the grids were designated for vertical characterization (at the grid center) at four-foot intervals, as well as surface (0 to 6 in.) sample collection. Additionally, ten percent of the grids were randomly designated for duplicate sampling (using a different aliquot pattern within the cell) to check the preliminary estimate on the variability.

X1.1.3.3 Additional environmental sampling was conducted that included a systematic sampling design for the stream adjacent to the facility with sediment samples collected at 100-ft intervals. A systematic approach was also used for the drainage ditch (50-ft intervals), with judgmental samples being collected at any location where visible staining was observed.

X1.1.3.4 *Results*—The results supported the initial investigation with lead consistently exceeding the TC Rule regulatory level; cadium was consistently below the regulatory level. Vertical differences in the lead and cadmium concentrations were not significant. Lead and cadmium were detected at elevated concentrations (relative to background) in the adjacent stream at a point downstream of the confluence with the drainage ditch.

X1.1.3.5 *Conclusion*— The waste pile was characteristic for lead and subject to Subtitle C of RCRA. There was no significant variability with depth, although several gradients were noticed across the grid (horizontally) based on lead concentration (scan) results.

X1.2 Case Study Two—Drum Sampling

X1.2.1 *Background*— An industry has two areas where drums of waste have been stored. One area is a warehouse adjacent to an off-line plating process that contains less than 25 drums (55 gal). The drums have manufacturers' labels indicating they contain an acid solution, and all of the drums are similar in appearance. A second area is a covered shed that has an estimated 100 drums from a variety of processes, several of which are no longer in use at the facility. Information on the content of these drums is not available.

X1.2.2 *Objective*—The objective of the initial investigation was to survey both of the storage areas for safety purposes, assess and record information on the drums, and open drums that were candidates for screening. All drums that were opened were surveyed using an organic vapor analyzer (PID, FID), pH paper, halogen detector, cyanide detector, and radiation meter.

X1.2.2.1 A judgmental sampling design was utilized in the warehouse where the anticipated variability was low. Based on the site screening (pH measurement), six samples were collected for pH analysis from the warehouse.

X1.2.2.2 The drums in the shed were screened in a similar fashion. A variety of results were obtained which included elevated pH, high organic vapor readings, and so forth. A simple random sampling design was used which called for the collection of 15 samples, with five from each major group of drums based on the screening (five corrosives, five potential ignitables with no halogens, and five with elevated halogen readings).

X1.2.2.3 *Results*—The warehouse samples were all corrosive with pH values from 1 to 2 S.U. The shed samples resulted in the collection of five corrosive wastes, three that were both ignitable and characteristic for non-halogenated TC Rule constituents, and two that were ignitable and characteristic for halogenated constituents. In summary, of the 15 drums sampled, 10 contained hazardous waste.

X1.2.2.4 *Conclusions*— All of the drums in the warehouse are subject to Subtitle C of RCRA. The drums in the shed require further assessment due to the fact that several of those sampled did not contain hazardous waste.

Designation: D 6050 – 97

Standard Test Method for Determination of Insoluble Solids in Organic Liquid Hazardous Waste[1]

This standard is issued under the fixed designation D 6050; the number immediately following the designation indicates the year of original adoption or, in the case of revision, the year of last revision. A number in parentheses indicates the year of last reapproval. A superscript epsilon (ϵ) indicates an editorial change since the last revision or reapproval.

1. Scope

1.1 This test method covers the determination of the approximate amount of insoluble, suspended solid material in organic liquid hazardous waste (OLHW).

1.2 This test method is intended to be used in approximating the amount of insoluble, suspended solids in determining the material handling characteristics and fuel quality of OLHW. It is not intended to replace more sophisticated procedures for the determination of total solids.

1.3 *This standard does not purport to address all of the safety concerns, if any, associated with its use. It is the responsibility of the user of this standard to establish appropriate safety and health practices and determine the applicability of regulatory limitations prior to use.*

2. Referenced Documents

2.1 *ASTM Standards:*

D 96 Test Method for Water and Sediment in Crude Oil by Centrifuge Method (Field Procedure)[2]

3. Summary of Test Method

3.1 A 10-mL aliquot of OLHW sample is decanted into a 15-mL graduated centrifuge tube and centrifuged for 3 min. The separated liquid phase of the OLHW is decanted into an appropriate waste vessel. The centrifuge tube with the separated solid material is brought back to its original 10-mL volume with a user-selected blend of clean solvents and agitated to mix the solid and liquid phases. The tube is centrifuged for 2 min, and the amount of remaining solid material is read.

4. Significance and Use

4.1 Facilities utilizing bulk liquid hazardous waste fuels are concerned about material handling characteristics and the overall quality of the fuel. A high percentage of insoluble, suspended solid material can create pumping, filtering, or grinding difficulties in the off-loading of bulk shipments of OLHW and can contribute to excessive wear on processing equipment. High solids can also decrease the quality and consistency of commingled fuel by decreasing the effectiveness of agitation in storage tanks.

5. Apparatus

5.1 *Centrifuge*—Capable of spinning two or more centrifuge tubes at a speed controlled to give a relative centrifugal force of between 1200 to 1400. The speed to achieve this is generally between 3100 to 3600 rpm. The rotation speed necessary to achieve the relative centrifugal force can be determined from one of the following equations:

$$rpm = 1335\, rcf/d \qquad (1)$$

$$rpm = 265\, rcf/d \qquad (2)$$

where:

rpm = rotation speed, in revolutions per min,
rcf = relative centrifugal force,
d = diameter of swing, in mm (Eq 1) or in. (Eq 2), measured between the tips of opposite tubes when the tubes are in rotating position.

NOTE 1—Eq 1 and Eq 2 are described in Test Method D 96.

5.2 *Centrifuge Tubes*— Centrifuge tubes shall be cone shaped, made of glass or a solvent resistant plastic or polymer, have a minimum capacity of 15 mL when filled to volume, and graduated with minimum subdivisions of 0.5 mL. Class A centrifuge tubes are recommended. If any grade other than Class A is used, refer to the section on Calibration and Standardization.

6. Reagents and Materials

6.1 *Purity of Reagents*—Reagent grade chemicals shall be used in all tests. Unless otherwise indicated, it is intended that all reagents conform to the specifications of the Committee on Analytical Reagents of the American Chemical Society where such specifications are available.[3] Other grades may be used, providing that it is first ascertained that the reagent is of sufficiently high purity to permit its use without lessening the accuracy of the determination.

[1] This test method is under the jurisdiction of ASTM Committee D34 on Waste Management and is the direct responsibility of Subcommittee D34.01.06 on Analytical Methods.

Current edition approved Jan. 10, 1997. Published March 1997.

[2] *Annual Book of ASTM Standards*, Vol 05.01.

[3] *Reagent Chemicals, American Chemical Society Specifications*, American Chemical Society, Washington, DC. For suggestions on the testing of reagents not listed by the American Chemical Society, see *Analar Standards for Laboratory Chemicals*, BDH Ltd., Poole, Dorset, U.K., and the *United States Pharmacopeia and National Formulary*, U.S. Pharmacopeial Convention, Inc. (USPC), Rockville, MD.

6.2 *Clean Solvent Blend*—The user should formulate the clean solvent blend based on the "average" normal chemical composition of the commingled OLHW. The chemical composition of OLHW is usually determined by gas chromatographic analysis. The clean solvent blend will vary in complexity depending on the specifications of the user.

NOTE 2—The clean solvent blend may be as simple as a single component (for example, toluene) or a complex mixture of aliphatic and aromatic hydrocarbons, alcohols, esters, ketones, ethers, terpenoids, and halogenated compounds. The user should formulate the clean solvent blend based on compositional knowledge of the OLHW that the sample material is to be commingled with. If the commingled OLHW contains water, it may be appropriate to include in the clean solvent blend.

7. Sample

7.1 Because settling of insoluble solid material in liquid samples is probable, the laboratory sample should be thoroughly mixed by shaking prior to withdrawing a portion for testing.

8. Calibration and Standardization

8.1 Centrifuge tubes other than Class A must be periodically checked for accuracy. The frequency of this check will be determined by the user, but should be minimally done for each new box or lot number of tubes.

9. Procedure

9.1 Fill a centrifuge tube to the 10-mL graduation mark with the OLHW sample.

NOTE 3—It will be necessary to use an additional tube with an equal mass of water or other suitable liquid to balance the centrifuge.

9.2 Place the tubes from 8.1 in the trunnion cups or centrifuge tube slots on opposite sides of the centrifuge. Close the centrifuge lid and engage safety lock, if so equipped.

9.3 Set the centrifuge speed (r/min) to provide a minimum relative centrifugal force between 1200 to 1400 (see 5.1).

9.4 Engage centrifuge, and allow to spin for 3 min.

9.5 Remove the tube containing OLHW, and decant the separated liquid phase to an appropriate waste vessel.

9.6 Return the tube with solids to its original 10-mL volume with the clean solvent blend. Cap or stopper the tube and shake vigorously in order to completely mix the centrifuged solid phase with the solvent phase. It may be necessary to use a Vortex mixer for thorough mixing.

9.7 Replace the OLHW tube in the centrifuge, and set centrifuge as described in 9.2 and 9.3.

9.8 Engage centrifuge, and allow to spin for 2 min.

9.9 Remove the OLHW tube, and decant the separated liquid phase of each to an appropriate waste vessel.

9.9.1 Visually estimate and record the amount of centrifuged solids in the tubes to the nearest 0.25 mL.

10. Quality Control

10.1 Each laboratory using this test method will operate a formal quality control program. This program shall include elements that address analyst proficiency through the evaluation of method blanks, duplicates, and reference materials if available.

11. Calculation

11.1 The estimated amount of solid material is calculated as follows:

$$(A/10\ mL)\ 100 = \text{Insoluble solids, percent} \quad (3)$$

where:

A = volume of solids from centrifuge tube.

12. Precision and Bias

12.1 *Precision*—The precision of this test method is being determined.

12.2 *Bias*—The bias of this test method has not been determined because there are no recognized reference materials.

13. Keywords

13.1 centrifuge; hazardous waste; insoluble solids; organic liquid hazardous waste; suspended solids

Standard Guide for Composite Sampling and Field Subsampling for Environmental Waste Management Activities[1]

This standard is issued under the fixed designation D 6051; the number immediately following the designation indicates the year of original adoption or, in the case of revision, the year of last revision. A number in parentheses indicates the year of last reapproval. A superscript epsilon (ϵ) indicates an editorial change since the last revision or reapproval.

1. Scope

1.1 Compositing and subsampling are key links in the chain of sampling and analytical events that must be performed in compliance with project objectives and instructions to ensure that the resulting data are representative. This guide discusses the advantages and appropriate use of composite sampling, field procedures and techniques to mix the composite sample and procedures to collect an unbiased and precise subsample(s) from a larger sample. It discusses the advantages and limitations of using composite samples in designing sampling plans for characterization of wastes (mainly solid) and potentially contaminated media. This guide assumes that an appropriate sampling device is selected to collect an unbiased sample.

1.2 The guide does not address: where samples should be collected (depends on the objectives) (see Guide D 6044), selection of sampling equipment, bias introduced by selection of inappropriate sampling equipment, sample collection procedures or collection of a representative specimen from a sample, or statistical interpretation of resultant data and devices designed to dynamically sample process waste streams. It also does not provide sufficient information to statistically design an optimized sampling plan, or determine the number of samples to collect or calculate the optimum number of samples to composite to achieve specified data quality objectives (see Practice D 5792). Standard procedures for planning waste sampling activities are addressed in Guide D 4687.

1.3 The sample mixing and subsampling procedures described in this guide are considered inappropriate for samples to be analyzed for volatile organic compounds. Volatile organics are typically lost through volatilization during sample collection, handling, shipping and laboratory sample preparation unless specialized procedures are used. The enhanced mixing described in this guide is expected to cause significant losses of volatile constituents. Specialized procedures should be used for compositing samples for determination of volatiles such as combining directly into methanol (see Practice D 4547).

1.4 *This standard does not purport to address all of the safety concerns, if any, associated with its use. It is the responsibility of the user of this standard to establish appropriate safety and health practices and determine the applicability of regulatory limitations prior to use.*

2. Referenced Documents

2.1 *ASTM Standards:*

C 702 Practice for Reducing Samples of Aggregate to Testing Size[2]

D 1129 Terminology Relating to Water[3]

D 4439 Terminology for Geosynthetics[4]

D 4547 Practice for Sampling Waste and Soils for Volatile Organics[5]

D 4687 Guide for General Planning of Waste Sampling[5]

D 5088 Practice for Decontamination of Field Equipment Used at Nonradioactive Waste Sites[4]

D 5792 Practice for Generation of Environmental Data Related to Waste Management Activities: Development of Data Quality Objectives[5]

D 6044 Guide for Representative Sampling for Management of Wastes and Contaminated Media[5]

E 856 Definitions of Terms and Abbreviations Relating to Physical and Chemical Characteristics of Refuse-Derived Fuel[5]

3. Terminology

3.1 *Definitions:*

3.1.1 *composite sample, n*—a combination of two or more samples. **D 1129**

3.1.2 *sample, n*—a portion of material taken from a larger quantity for the purpose of estimating properties or composition of the larger quantity. **E 856**

3.1.3 *specimen, n*—a specific portion of a material or laboratory sample upon which a test is performed or which is taken for that purpose. **D 4439**

3.1.4 *subsample, n*—a portion of a sample taken for the purpose of estimating properties or composition of the whole sample.

3.1.4.1 *Discussion*—a subsample, by definition, is also a sample.

---

[1] This guide is under the jurisdiction of ASTM Committee D34 on Waste Managementand is the direct responsibility of Subcommittee D34.01.01 on Planning for Sampling.

Current edition approved Dec. 10, 1996. Published February 1997.

[2] *Annual Book of ASTM Standards*, Vol 04.02.

[3] *Annual Book of ASTM Standards*, Vol 11.01.

[4] *Annual Book of ASTM Standards*, Vol 04.09.

[5] *Annual Book of ASTM Standards*, Vol 11.04.

4. Summary of Guide

4.1 This guide describes how the collection of composite samples, as opposed to individual samples, may be used to: more precisely estimate the mean concentration of a waste analyte in contaminated media, reduce costs, efficiently determine the absence or possible presence of a hot spot (a highly contaminated local area), and, when coupled with retesting schemes, efficiently locate hot spots. Specific procedures for mixing a sample(s) and collecting subsamples for transport to a laboratory are provided.

5. Significance and Use

5.1 This guide provides guidance to persons managing or responsible for designing sampling and analytical plans for determining whether sample compositing may assist in more efficiently meeting study objectives. Samples must be composited properly, or useful information on contamination distribution and sample variance may be lost.

5.2 The procedures described for mixing samples and obtaining a representative subsample are broadly applicable to waste sampling where it is desired to transport a reduced amount of material to the laboratory. The mixing and subsampling sections provide guidance to persons preparing sampling and analytical plans and field personnel.

5.3 While this guide generally focuses on solid materials, the attributes and limitations of composite sampling apply equally to static liquid samples.

6. Attributes of Composite Sampling for Waste Characterization

6.1 In general, the individual samples to be composited should be of the same mass, however, proportional sampling may be appropriate in some cases depending upon the objective. For example, if the objective is to determine the average drum concentration of a contaminant, compositing equals volumes of waste from each drum would be appropriate. If the objective is to determine average contaminant concentration of the waste contained in a group of drums, the volume of each sample to be composited should be proportional to the amount of waste in each drum. Another example of proportional sampling is estimating the contaminant concentration of soil overlying an impermeable zone. Soil cores should be collected from the surface to the impermeable layer, regardless of core length.

6.2 The principal advantages of sample compositing include: reduction in the variance of an estimated average concentration **(1)**,[6] increasing the efficiency of locating/identifying hot spots **(2)**, and reduction of sampling and analytical costs **(3)**. These main advantages are discussed in the following paragraphs. However, a principle assumption needed to justify compositing is that analytical costs are high relative to sampling costs. In general, appropriate use of sample compositing can:

6.2.1 Reduce inter-sample variance, that is, improve the precision of the mean estimation while reducing the probability of making an incorrect decision,

[6] The boldface numbers in parentheses refer to a list of references at the end of this guide.

6.2.2 Reduce costs for estimating a total or mean value, especially where analytical costs greatly exceed sampling costs (also may be effective when analytical capacity is a limitation),

6.2.3 Efficiently determine the absence or possible presence of hot spots or hot containers and, when combined with retesting schemes, identify hot spots, as long as the probability of hitting a hot spot is low,

6.2.4 Be especially useful for situations, where the nature of contaminant distribution tends to be contiguous and non-random and the majority of analyses are "non-detects" for the contaminant(s) of interest, and

6.2.5 Provide a degree of anonymity where population, rather than individual statistics are needed.

6.3 *Improvement in Sampling Precision*—Samples are always taken to make inferences to a larger volume of material, and a set of composite samples from a heterogeneous population provides a more precise estimate of the mean than a comparable number of discrete samples. This occurs because compositing is a "physical process of averaging." Averages of samples have greater precision than the individual samples. Likewise, a set of composite samples is always more precise than an equal number of individual samples. Decisions based on a set of composite samples will, for practical purposes, always provide greater statistical confidence than for a comparable set of individual samples.

6.3.1 If an estimated precision of a mean is desired, then more than one composite sample is needed; a standard deviation cannot be calculated from one composite sample. However, the precision of a single composite sample may be estimated when there are data to show the relationship between the precision of the individual samples that comprise the composite sample and that of the composite sample. The precision (standard deviation) of the composite sample is approximately the precision of the individual samples divided by the square root of the number of individual samples in the composite.

6.4 *Example 1*—An example of how a single composite sample can be used for decision-making purposes is given here. Assume a regulatory limit of 1 mg/kg and a standard deviation of 0.5 mg/kg for the individual samples. If the concentration of a site is estimated to be around 0.6 mg/kg, how many individual samples should be composited to have relatively high confidence that the true concentration does not exceed the regulatory limit when only one composite sample is used? Assuming the composite is well mixed, then the precision of a composite is a function of the number of samples as follows:

Number of Individual Samples in Composite	Precision (standard deviation ÷ $\sqrt{n}$) of One Composite Sample
2	0.35
3	0.29
4	0.25
5	0.22
6	0.20

Thus, if six samples are included in a composite, the composite concentration of 0.6 mg/kg is two standard deviations below the regulatory limit. Therefore, if the composite concentration is actually observed to be in the neighborhood of 0.6 mg/kg, we can be reasonably confident (approximately

95 %) that the concentration of the site is below the regulatory limit, using only one composite sample.

6.5 *Example 2*—Another example is when the standard deviation of the individual samples in the previous example is relatively small, say 0.1 mg/kg. Then the standard deviation of a composite of 6 individual samples is 0.04 mg/kg (0.1 mg/kg divided by the square root of 6 = 0.04 mg/kg), a very small number relative to the regulatory limit of 1 mg/kg. In this case, simple comparison of the composite concentration to the regulatory limit is often quite adequate for decision-making purposes.

6.5.1 The effectiveness of compositing depends on the relative magnitude of sampling and analytical error. When sampling uncertainty is high relative to analytical error (as is usually assumed to be the case) compositing is very effective in improving precision. If analytical errors are high relative to field errors, sample compositing is much less effective.

6.5.2 Because compositing is a physical averaging process, composite samples tend to be more normally distributed than the individual samples. The normalizing effect is frequently an advantage since calculation of means, standard deviations and confidence intervals generally assume the data are normally distributed. Although environmental residue data are commonly non-normally distributed, compositing often leads to approximate normality and avoids the need to transform the data.

6.5.3 The spatial design of the compositing scheme can be important. Depending upon the locations from which the individual samples are collected and composited, composites can be used to determine spatial variability or improve the precision of the parameter being estimated. Fig. 1 and Fig. 2 represent a site divided into four cells. Composite all samples with the same number together. The sampling approach in Fig. 1 is similar to sample random sampling, except they are now composite samples. Each composite sample in this case is a representative sample of the entire site, eliminates cell-to-cell variability, and leads to increased precision in estimating the mean concentration of the site. If there is a need to estimate the cell-to-cell variability, then the approach in Fig. 2 is suitable. In addition, if the precision of estimating the mean concentration of the cell is needed, multiple composite samples should be collected from that cell.

6.6 *Effect on Cost Reduction*—Because the composite samples yield a more precise mean estimate than the same number of individual samples, there is the potential for substantial cost saving. Given the higher precision associated with composite samples, the number of composite samples required to achieve a specified precision is smaller than that required for individual samples. This cost saving opportunity is

1	2	4	3
4	3	2	1
4	2	1	4
3	1	2	3

FIG. 1 Example of Composing Across a Site

1	1	2	2
1	1	2	2
3	3	4	4
3	3	4	4

FIG. 2 Example of Within Cell Compositing

especially pronounced when the cost of sample analysis is high relative to the cost of sampling, compositing, and analyzing.

6.7 *Hot Container/Hot Spot Identification and Retesting Schemes*—Samples can be combined to determine whether an individual sample exceeds a specified limit as long as the action limit is relatively high compared with the actual detection limit and the average sample concentration. Depending on the difficulty and probability of having to resample, it may be desirable to retain a split of the discrete samples for possible analysis depending on the analytical results from the composite sample.

6.8 *Example 3*—One hundred drums are to be examined to determine whether the concentration of PCBs exceeds 50 mg/kg. Assume the detection limit is 5 mg/kg and most drums have non-detectable levels. Compositing samples from ten drums for analysis would permit determining that none of the drums in the composite exceed 50 mg/kg as long as the concentration of the composite is <5 mg/kg. If the detected concentration is >5 mg/kg, one or more drums may exceed 50 mg/kg and additional analyses of the individual drums are required to identify any hot drum(s). The maximum number of samples that can theoretically be composited and still detect a hot sample is the limit of concern divided by the actual detection limit (for example, 50 mg/kg ÷ 5 mg/kg = 10).

6.9 *Example 4*—Assume background levels of dioxin are non detectable, and the analytical detection limit is 1 µg/kg and the action level is 50 µg/kg. The site is systematically gridded (the most efficient sampling design for detecting randomly distributed hot spots) using an appropriate design, and cores to a depth of 10 cm are collected. Composite samples are collected since analytical costs for dioxin are high. In theory, groups of up to 50 samples could be composited and if the resultant concentration were <1 µg/kg, all samples represented in the composite should be below 50 µg/kg. If the contaminant concentration is >1 µg/kg, one or more spots may exist that exceed 50 µg/kg in the area covered by the composite sample although the precise location and areal extent would not be known without further sampling and analyses. Compositing fewer samples would probably be more practical, however.

6.9.1 The relative efficiency of compositing individual samples to detect a hot spot depends on the probability of a "hot" discrete sample being used to form a composite sample. According to Garner et al. **(1)**, if the probability can be estimated as low, say 1 %, the optimum number of samples to composite is about ten, which would result in a cost saving of about 80 % (assuming there is no detection limit problem). When the probability of collecting a sample from a hot spot rises to 10 %, the optimal number of samples to composite is 4, which results in a 40 % cost savings. By the time the probability of sampling a hot spot rises to 40 %, there is no cost benefit to compositing. Other resampling and testing schemes are possible and may lead to somewhat different cost saving potentials.

7. Limitations of Composite Sampling

7.1 The principal limitations of sample compositing involve the loss of the discrete information contained in a single sample and the potential for dilution of the contaminants in a sample with uncontaminated material; however, in that case, the dilution factor can be used to estimate the maximum number of samples that can be composited. The following situations may not lend themselves to cost-effective sample compositing:

7.1.1 When the integrity of individual sample values change because of compositing, for example, chemical interaction occurs between constituents in the samples being combined or volatiles are lost during mixing,

7.1.2 Where the composite sample cannot be properly mixed and subsampled or the whole composite sample cannot be analyzed,

7.1.3 When the goal is to detect hotspots and a large proportion of the samples are expected to test positive for an attribute, compositing and retesting schemes may not be cost effective,

7.1.4 When analytical costs are low relative to sampling costs (for example, in situ field portable X-ray fluorescence takes only 30 s with no sample preparation so analytical costs/sample are very low), and

7.1.5 When regulations specify that a grab sample must be collected (usually a composite sample covering a limited area is still preferred from a technical standpoint).

8. Sample Mixing Procedures

8.1 Prior to sample mixing, project-specific instructions should be followed regarding sample collection, which may include removal of extraneous sample materials such as twigs, grass, rocks, etc. If samples are sieved or large materials are removed, it may be necessary to record the mass of materials removed for later estimation of contaminant concentration in the original sample. According to particulate sampling theory **(4,5)** the following sample masses are adequate to represent the corresponding maximum size particles in the sample with a relative standard deviation of 15 %.

Sample Mass, g	Maximum Particle Size, cm
5	0.170
50	0.37
100	0.46
500	0.79
1000	1.0
5000	1.7

8.1.1 Frequently it is necessary to mix an individual or composite sample and obtain a representative subsample(s) for transport to the analytical laboratory. This occurs when multiple containers of the identical material are desired (for example, separate sample jars for metals, semivolatile organics, etc. are desired) or when the original sample (or composite sample) size is greater than accepted by the laboratory. Even when the original sample volume is acceptable, it may be desirable to thoroughly mix the sample prior to transport to an analytical laboratory. However, some samples that have been well mixed in the field may segregate during shipment to the laboratory.

8.1.2 A laboratory typically collects a 0.5 to 30 g specimen (100 g for some extraction tests) from the sample for analysis. Specimens are frequently collected from the surface material in the container or after minimal mixing. Such procedures are inadequate to obtain a small representative specimen from a 100 to 300 g sample. Special mixing and subsampling procedures are necessary to obtain a representative subsample unless the sample is already homogenous. Field mixing should be considered essential unless it is known that the sample in the container is homogeneous or it is known that the laboratory will homogenize the sample and collect a representative specimen. To help ensure that an unbiased and precise specimen is collected, the analytical laboratory should be provided instructions (preferably with the sample shipment) on homogenizing and obtaining a specimen for analysis. Few laboratories follow good sample homogenizing and specimen collection practices. To meet both sampling and analytical objectives, field and analytical personnel, and the end-user of the data must be aware of the laboratories standard practices for handling, mixing, and obtaining a specimen or specify such practices with the sample shipment.

8.1.3 To avoid subsampling it may be possible to collect a small sample (or composite samples) directly into the sample container that is delivered to the laboratory (**Caution:** small sample sizes may result in bias by excluding large particles). While no field mixing and subsampling is needed as long as the laboratory homogenizes the sample, it may be advisable to mix such samples anyway (see 8.1.2).

8.1.4 Soil, sediment, sludge and waste samples collected for purgeable/volatile organic compounds' analyses should *not* be mixed and subsampled using procedures described in this guide but other specialized procedures such as combining samples directly into methanol (see Practice D 4547) may be appropriate.

8.1.5 A significant problem with analyzing very small samples is that the smaller the volume of sample actually extracted or analyzed, the less representative that sample may be unless thoroughly mixed/homogenized and subsampled. Therefore, sample compositing without thorough mixing can nullify the potential benefits of compositing.

8.1.6 Methods that may be applicable to field mixing, depending on the matrix, include hand mixing in a pan, sieving, particle size reduction, kneading, etc. For highly heterogeneous waste such as municipal refuse, field comminution (grinding) may be needed. Some of these methods may be inappropriate if trace levels of contamination are a primary concern. The use of disposable equipment for mixing should be considered to minimize field decontamination problems. Field personnel should use care to ensure that samples do not become contaminated during the sampling, mixing and subsampling process.

8.1.7 Once a sample has been collected, it may have to be split into separate containers for different analyses. A true split of soil, sediment, or sludge samples may be difficult to accomplish under field conditions.

8.1.8 The following are some common methods for mixing

soils, sludges, etc. While it is not always possible to determine that a sample is adequately mixed, following standard procedures and observing sample texture, color, and particle distribution are practical methods. While some materials cannot be homogenized, following the subsampling procedures in Section 9 will help ensure that a representative subsample is collected. Under certain conditions, some of the procedures that follow are applicable when trace level contaminants are of concern.

8.1.8.1 *Pan Mixing/Quartering*—One common method of mixing is referred to as quartering. Place the material in a glass or stainless steel sample pan and divide into quarters. Mix each quarter separately, then mix all quarters into the center of the pan. Repeat this procedure several times until the sample is adequately mixed (usually a minimum of three repetitions). If round bowls are used for sample mixing, adequate mixing is achieved by stirring the material in a circular fashion and occasionally turning the material over.

8.1.8.2 *Mixing Square*— Combine samples through a non-contaminating screen into an appropriate clean mixing container. Mix in the container and pour onto a 1 metre square of non-contaminating material such as plastic for metals analyses or polytetrafluoroethylene for organics. Roll the sample backward and forward on the sheet while alternately lifting and releasing opposite side corners of the sheet. This is appropriate for flowable granular materials **(6)**. If polytetrafluoroethylene sheeting is used, this procedure could be acceptable for trace level contaminants.

8.1.8.3 *Kneading*—Place the sample in a non-contaminating bag and knead as in bread making to mix the sample. This may be appropriate for viscous or clay-like materials. If a non-contaminating bag is used, this approach would be acceptable for trace level contaminants.

8.1.8.4 *Sieving and Mixing*—If a laboratory requires a small specimen (1 to 30 g) or if less than a specific particle size is required, disruption of aggregated particles or sieving, or both, followed by mixing may be needed. Sieving allows only those particles below a desired size to pass through the sieve into a mixing pan for subsequent mixing and subsampling into containers. Sieving works best with relatively dry granular materials. Sieving and the exclusion of large particles can result in very biased results and should only be conducted when designed into a sampling plan.

8.1.8.5 *Particle Size Reduction*—When particle size reduction is appropriate and trace contaminants are of concern, non-contaminating materials compatible with objectives should be used (for example, glass, ceramic, stainless steel). Other materials may be acceptable if trace levels of contaminants are not a concern. The reduction method can be as simple as using a hammer to break apart large pieces into smaller pieces that are either acceptable to the laboratory or that can pass through a sieve. This method of reduction creates a great deal of fine material which may or may not be included in the sample container, and could introduce bias. More complex reducers, such as ball mills, ceramic plate grinders, etc., are available, but usually require relatively dry samples and thorough decontamination to avoid cross contamination. Such a process may be more appropriately conducted in a laboratory.

8.1.9 With thorough decontamination (see Practice D 5088) of the particle size reducer, sieve and the mixing pan, these procedures could be acceptable for trace level contaminants.

8.1.10 *Other Mixing Equipment*—Riffle splitters, coning and quartering, etc., involve equipment and materials that are difficult to decontaminate, and awkward to use on a routine basis for waste management sampling. Since these procedures are not routinely used, the devices are not considered in this guide. However, procedures for coning and quartering, and the use of riffle splitters are described in Practice C 702 and could be modified for subsampling contaminated media.

9. Field Subsampling Procedures

9.1 If mixing procedures could ensure a truly homogenous sample, subsampling would be simple. Mixing of various particle sizes may, however, cause the particles to segregate according to size, and improper subsampling could introduce bias. Since homogeneity is frequently not achieved, appropriate subsampling procedures should be used by field personnel to provide representative subsamples. The procedures that follow are appropriate for collecting a representative sample from a larger sample. As noted previously, riffle splitters and coning and quartering procedures can also be used for subsampling as well as mixing (see Practice C 702).

9.1.1 *Rectangular Scoop*—As the final step of mixing, the material is arranged in a pile along the long axis of the rectangular pan. A flat bottomed scoop with vertical sides is moved across the entire width of the short axis of the pile to collect a swath of sample (Fig. 3). Multiple evenly-spaced swaths are collected until the subsample container is full. Multiple containers are filled by rearranging the remaining material and collecting swaths as just described.

9.1.2 *Alternate Scoop*— The volume of material required for filling sample containers is compared to the volume of the mixed sample. Scoops of mixed material are placed in the sample container(s) or are discarded, that is, three scoops are discarded for every scoop saved when collecting a 25 % subsample (Fig. 4). Care should be taken that each scoop of material is of the same size and is collected in a consistent manner to minimize bias **(5)**.

9.1.3 *Slab-cake*—The cohesive or clay-like materials as discussed in 8.1.8.3 on kneading. The sample can be flattened, cut into cubes (Fig. 5) and the cubes randomly or systematically combined into subsample(s) **(5)**. The subsample should be re-kneaded before shipment to the laboratory unless it can be ensured that the laboratory will homogenize the subsample before collecting a specimen.

10. Keywords

10.1 composite; compositing; hot spot; particle size reduction; sample; sampling; subsample; subsampling

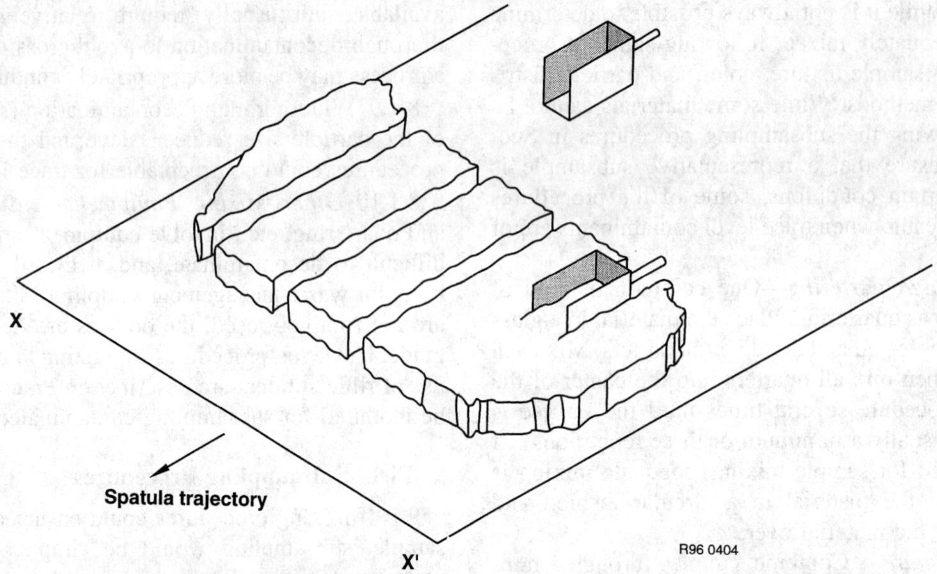

FIG. 3 Rectangular Scoop as Used to Collect Swaths for Subsample

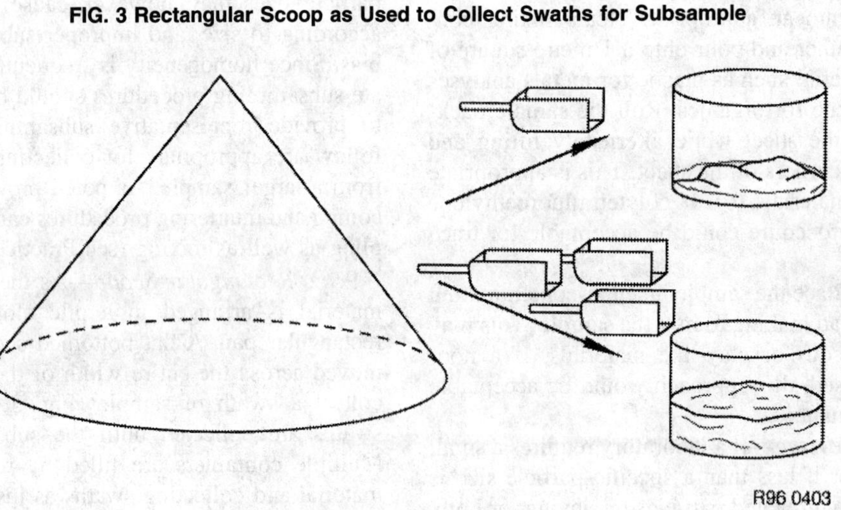

FIG. 4 Alternate Scoop Subsampling Technique

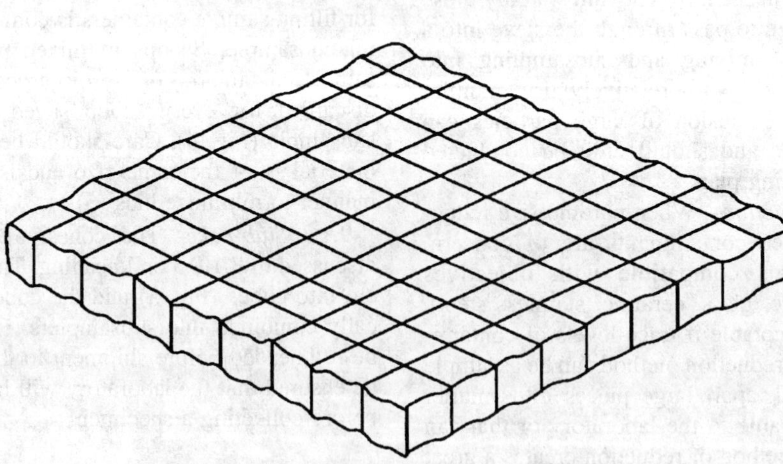

FIG. 5 Slab Cake Subsampling Technique

REFERENCES

(1) Garner et al., "Composite Sampling for Environmental Monitoring in Principals of Environmental Sampling," in *Principles of Environmental Sampling*, American Chemical Society (ACS), Keith, L., ed., 1988, pp. 363–374.

(2) Mack, G. A., and Robinson, P. E., "Use of Composited Samples to Increase the Precision and Probability of Detection of Toxic Chemicals," in *Environmental Applications of Chemometrics*, ACS Symposium Series 292, ACS, Washington, DC, 1985, pp. 174–183.

(3) Rajagopal, R., and Williams, L. R., "Economics of Sample Compositing as a Screening Tool in Ground Water Monitoring," *Ground Water Monitoring Review*, 1989, pp. 186–192.

(4) Ramsey, C. A., Ketterer, M. E., and Lowery, J. H., "Application of Gy's Sampling Theory to the Sampling of Solid Waste Materials," in *Proceedings of the EPA Fifth Annual Waste Testing and Quality Assurance Symposium*, 1989, p. II-494.

(5) Pitard, F. F., "Pierre Gy's Sampling Theory and Practice," *Sampling Correctness and Sampling Practice*, Volume II, CRC Press, Boca Raton, FL, 1989, p. 247.

(6) U.S. Environmental Protection Agency (EPA), *Description and Sampling of Contaminated Soils, A Field Pocket Guide*, EPA/625/12-91/002, 1991.

Designation: D 6052 – 97

Standard Test Method for Preparation and Elemental Analysis of Liquid Hazardous Waste by Energy-Dispersive X-Ray Fluorescence[1]

This standard is issued under the fixed designation D 6052; the number immediately following the designation indicates the year of original adoption or, in the case of revision, the year of last revision. A number in parentheses indicates the year of last reapproval. A superscript epsilon (ε) indicates an editorial change since the last revision or reapproval.

1. Scope

1.1 This test method covers the determination of trace and major element concentrations by energy-dispersive X-ray fluorescence spectrometry (EDXRF) in liquid hazardous waste (LHW).

1.2 This test method has been used successfully on numerous samples of aqueous and organic-based LHW for the determination of the following elements: Ag, As, Ba, Br, Cd, Cl, Cr, Cu, Fe, Hg, I, K, Ni, P, Pb, S, Sb, Se, Sn, Tl, V, and Zn.

1.3 This test method is applicable for other elements (Si-U) not listed in 1.2.

1.4 *This standard does not purport to address all of the safety concerns, if any, associated with its use. It is the responsibility of the user of this standard to establish appropriate safety and health practices and determine the applicability of regulatory limitations prior to use.*

2. Referenced Documents

2.1 *ASTM Standards:*

C 982 Guide for Selecting Components for Energy-Dispersive X-ray Fluorescence (XRF) Systems[2]

D 1193 Specification for Reagent Water[3]

2.2 *Other ASTM Documents:*

ASTM Data Series DS 46 X-Ray Emission Wavelengths and KeV Tables for Nondiffractive Analysis[4]

3. Summary of Test Method

3.1 A weighed portion of activated alumina and sample are combined in a mixing vessel and shaken until well mixed. The sample mixture is transferred into a disposable sample cup and placed in the spectrometer for analysis.

3.2 The *K* spectral emission lines are used for elements Si-Ba.

3.3 The *L* spectral emission lines are used for elements with atomic numbers greater than Ba.

4. Significance and Use

4.1 The elemental analysis of liquid hazardous waste is often important for regulatory and process specific requirements. This test method provides the user an accurate, rapid method for trace and major element determinations.

5. Interferences

5.1 *Spectral Overlaps (Deconvolution)*:

5.1.1 Samples containing a mixture of elements often exhibit X-ray emission line overlap. Modern Si (Li) detectors generally provide adequate resolution to minimize the effects of spectral overlap. In cases where emission line overlap exists, techniques of peak fitting exist for extracting corrected analyte emission line intensities. For example, the PbLα "line overlaps with the AsKα." The PbLβ line can be used to avoid this overlap and the AsK lines can then be resolved from the PbLα overlap. The actual lines used for any particular element should be such that overlaps are minimized. Follow the EDXRF manufacturer's recommendation concerning spectral deconvolution. Reference should be made to ASTM Data Series DS 46 for detailed information on potential line overlaps.

5.2 *Matrix Interferences (Regression)*:

5.2.1 Matrix interference in the measurement of "as received" LHW samples using EDXRF has been the principle limitation in the development and expanding use of this instrumental technique. Using well understood XRF principles for controlling matrix effects, for example, dilution and matrix modification using lithium borate fusion and addition of heavy absorbers, a matrix can be stabilized. Using calcined alumina and the above principles matrices are stabilized for quantitative EDXRF analysis.

5.2.2 The response range of this test method should be linear with respect to the elements of interest and their regulatory or process control, or both, action thresholds. Large concentration variations of element or matrix, or both, components in LHW samples can result in non-linear X-ray intensity response at increasing element concentrations.

6. Apparatus

6.1 *Energy-dispersive X-ray Fluorescence Spectrometer*, capable of measuring the wavelengths of the elements listed in 1.2. Refer to Guide C 982 for system specifications.

---

[1] This test method is under the jurisdiction of ASTM Committee D34 on Waste Management and is the direct responsibility of Subcommittee D34.01.06 on Analytical Methods.

Current edition approved Jan. 10, 1997. Published March 1997.

[2] *Annual Book of ASTM Standards*, Vol 12.01.

[3] *Annual Book of ASTM Standards*, Vol 11.01.

[4] Available from ASTM Headquarters, Customer Service.

6.2 *Analytical Balance*, capable of weighing to 0.001 g.

7. Reagents and Materials

7.1 *Purity of Reagents*—Reagent grade chemicals shall be used in all tests. Unless otherwise indicated, it is intended that all reagents conform to the specifications of the Committee on Analytical Reagents of the American ChemicalSociety, where such specifications are available.[5] Other grades may be used, provided it is first ascertained that the reagent is of sufficiently high purity to permit its use without lessening the accuracy of the determination.

7.2 *Purity of Water*—Unless otherwise indicated, references to water shall be understood to mean meeting the numerical requirements of Type II water as defined by Specification D 1193.

7.3 *Aluminum Oxide, Al_2O_3*—pre-calcined at 1500°C, approximately 100 to 125 mesh.

7.4 *Aqueous or organic-based Atomic Absorption Standards (AAS)*, 1000 mg/L for the elements Ag, As, Ba, Cd, Cr, Cu, Fe, Hg, K, Ni, Pb, Sb, Se, Sn, Tl, V, and Zn. Standard solutions for elements not listed are also available.

NOTE 1—AAS standards are typically presented in mass/vol units. The density of these solutions can be considered as unity (that is, 1) thus they can be considered as % mass/mass (m/m).

7.5 *1-bromonaphthalene, trichlorobenzene, iodobenzoic acid, triethyl phosphate and dithiodiglycol* are the recommended standards for the elements Br, Cl, I, P and S, respectively.

7.6 *Low Molecular Weight Polyethylene Glycol (PEG 400, or equivalent) or Water* is used for producing method blank.

7.7 *High-Density Polyethylene (HDPE) Wide-mouth, Round, Screw-Cap Bottles*, 50 to 60 mL capacity.

7.8 *Mixing Balls*, approximately 1 cm diameter, stainless steel or equivalent.

NOTE 2—Potential low level Cr, Fe or Ni (<20 mg/kg[1]) contamination due to the use of stainless steel may exist. Other suitable materials would be tungsten carbide, Zr or Ta.

7.9 *Thin-film Support*.

NOTE 3—The user should select a thin-film support that provides for maximum transmittance and is resistant to typical components in LHW. The thin-film supports used in the development of this test method were a polypropylene base and a high-purity, 4 µm polyester film.

7.10 *Sample Cups*, vented.

7.11 *Helium, He*—minimum 99.99 purity for use as a chamber purge gas for the analysis of Cl, P and S. This numerical purity is intended to specify a general grade of helium. Ultra-high purity helium is not required for this test method.

8. Sample

8.1 Because of the potential heterogeneous nature of LHW, all possible efforts should be made to ensure that representative samples are taken.

9. Preparation of Apparatus

9.1 Follow the manufacturer's instructions for set-up, conditioning, preparation and maintenance of the XRF spectrometer.

9.2 When required, reference spectra should be obtained from pure element standards for all deconvoluted elements.

9.3 Spectral and matrix interferences as listed in the Interferences section must be addressed per the manufacturer's recommendations.

10. Calibration and Standardization

10.1 The spectrometer must be calibrated using an appropriate reference element(s) at a minimum frequency as recommended by the manufacturer.

10.2 Analytical standards should be prepared gravimetrically by blending the solution or pure element standards with Al_2O_3 to suitable standard concentrations as determined by the user's analytical requirements. Table 1 gives recommended concentration ranges for regression. Standards can be single or multi-element mixtures. Standard solutions are generally mixed with Al_2O_3 at a ratio of 3:1.

NOTE 4—More than one standard element(s) solution can be added to a single 15 g Al_2O_3 mass provided the total mass of standard is 5 g. This will maintain the proper 3:1 ratio while allowing mixtures of potentially incompatible elements to be combined in a single standard.

10.2.1 The number of standards required to produce calibrations is dependent on the number of elements to be determined. Generally, two calibrations are produced, the first is to determine potentially major elements such as halogens, S & P. The second is to determine trace elements, typically toxic metals and heavy elements. The minimum number of standards required can be determined from the following equation: minimum standards required = number elements determined plus two. Both of the above calibrations should use a minimum of ten standards each to cover the element concentration ranges shown in Table 1 and to ensure that adequate data is available to assess spectral overlaps as described in 5.1.

10.3 The Al_2O_3 + element(s) specimen is placed into an XRF sample cup supported by a suitable thin-film. The sample is gently tapped on a flat, hard surface to settle the powder against the thin-film support and ensure there are no air gaps.

TABLE 1 Recommended Standards Ranges

Analyte	Low Concentration Range, mg/kg	High Concentration Range, mg/kg	Analyte	Low Concentration Range, mg/kg	High Concentration Range, mg/kg
Ag	5	600	Zn	5	600
Ba	5	600	As	5	600
P	0.1 %	5 %	Se	5	600
S	0.05 %	5 %	Br	10	5000
Cl	0.05 %	5 %	Cd	5	600
K	0.1 %	5 %	Sb	5	600
V	5	600	Sn	5	600
Cr	5	600	I	5	600
Fe	5	600	Hg	5	600
Ni	5	600	Tl	5	600
Cu	5	600	Pb	5	600

[5] *Reagent Chemicals, American Chemical Society Specifications*, American Chemical Society, Washington, DC. For suggestions on the testing of reagents not listed by the American Chemical Society, see *Analar Standards for Laboratory Chemicals*, BDH Ltd., Poole, Dorset, U.K., and the *United States Pharmacopeia and National Formulary*, U.S. Pharmacopeial Convention, Inc. (USPC), Rockville, MD.

10.3.1 The standard specimen in the sample cups is placed in the spectrometer's sample holder avoiding any contact with the film or rough handling that may disturb the standards.

10.4 Two methods of calibration are available.

10.4.1 *Method A*—Empirical calibration method using a suit of standard concentrations. Standard concentrations are limited to 600 mg/kg for Ag, As, Ba, Br, Cd, Cr, Cu, Fe, Hg, I, K, Ni, Pb, Sb, Se, Sn, Tl, V, and Zn. Standard concentrations are limited to 5 % for Cl, P, S, and other light elements (that is, $<z = 22$ and <0.5 % for Br). The limits ensure staying within the linear range and due to the limited concentration range of available traceable standards. The standards should provide a linear response of element intensity to concentration. Serial dilutions of analyte standards can be used to set up the calibration for each element. Multi-element standards can then be used to assess the deconvolution requirements of the spectrometer and check for calibration linearity.

NOTE 5—Standards may be diluted into the linear range using low molecular weight polyethylene glycol (PEG) or water. The choice of diluent is dependent upon whether the original standard solution is aqueous- or organic-based. For example, a 5000 mg/kg organic-based Pb standard solution can be diluted into the 0–600 mg/kg range by combining and mixing 15 g of Al_2O_5 + 0.5 g of 5000 mg/kg Pb standard solution + 4.5 g PEG. This yields a ten-fold dilution yielding a prepared standard concentration of 500 mg/kg.

10.4.1.1 *Drift Correction Monitors*—To correct for instrumental drift, use physically stable, solid disks or pressed pellets containing at least one element measured under each instrumental condition used.[6] At least two disks are necessary to correct both sensitivity and base-line drifts. One should provide a high net count-rate similar to standards from the upper end of the calibration range and the other should provide a low net count-rate similar to the blank. Measure the net count-rate for each element in the high concentration disk in such a way that the counting statistical error due to random fluctuation of the X-ray flux is less than 0.5 % relative to the net count-rate. Counting times must be long enough to collect 40 000 net counts for each element in the high-concentration disk. Use the same counting times when measuring the low concentration or blank disk.

10.4.2 *Method B*—Fundamental Parameters method. Most EDXRF manufacturers provide software capable of estimating the composition of materials without the use of a suite of standards. The setup of a particular manufacturer's fundamental parameters method may require a high and low concentration or mid-range concentration for each element present to determine the initial sensitivity for the elements in the alumina matrix. Other manufacturers provide the initial sensitivity with the added option to align the sensitivity to a specific matrix type for more accurate determinations using a single similar standard containing the elements of interest. By measuring the X-ray intensity (cps) for each element and using the above determined sensitivity factor for each element plus various equations to account for X-ray absorption and enhancement effects, the concentration of all elements present can be estimated. The exact equations used will differ for each manufacturer.

10.4.2.1 Follow the manufacturer's fundamental parameters set-up recommendations. The stoichiometric set-up of the fundamental parameters method for the analysis of the LHW mixed with alumina should allow for the manual input of a fixed 75 % Al_2O_3 concentration and the use of carbon as a balance estimate of the solvent/aqueous phase with the elements of interest determined directly according to the principles of 10.4.2.

10.4.3 Two control samples are needed for monitoring instrument stability. One control sample is a blank preparation using PEG or the low concentration drift correction monitor used in 10.4.1.1. The other sample is a stable mixture containing a suitable range and number of elements (for example, S, V, Zn, Pb, and Ba) at concentrations near the middle of the calibration ranges. A mixture of leftover samples/standards, spiked with element concentrations as needed and carefully mixed may be used.

10.4.4 Restandardization should be carried out whenever quality control results defined in Section 14 are outside data quality objectives as determined by the user. Method A: The initial linear regressions are performed only once as per 10.4.1. A day zero measurement of the drift correction monitors, 10.4.1.1 during the set-up of the initial regression allows for subsequent re-calibration to be performed using the two standards defined in 10.4.1.1, via a restandardization procedure in order to check the values of the slope and intercept for each regressed element. NOTE: Restandardization using drift correction monitors is often part of instrumental software. Follow the manufacturer's recommendations for the set-up of restandardization using two standards. Method B: Follow the manufacturer's recommendations for the set-up of initial element sensitivities and the appropriate fundamental parameters methods using the principles outlined in 10.4.2.

11. Procedure

11.1 Thoroughly mix the LHW sample before withdrawing a portion with a pipet, or equivalent device for preparation. See the Sample section.

11.2 Weigh 15.00 g $Al_2O_3 \pm$ 0.05 g directly into the HDPE bottle.

11.3 Weigh a 5.00 g ± 0.05 g portion of the sample directly into the blending vial/vessel containing the Al_2O_3 from 11.1.

11.4 Add two mixing balls to the HDPE bottle from 11.3 and seal.

11.5 Shake vigorously for approximately 30 s. Tapping the bottle on a hard surface will aid in proper mixing. The user may feel a mild exothermic reaction.

11.6 Empty the Al_2O_3/sample mixture from the bottle to an XRF sample cup as described in 10.3.

11.7 Place the sample cup into the spectrometer sample holder and initiate data acquisition. If re-analysis of the sample is required, the user should obtain a new portion from the HDPE bottle following the instructions in 11.6.

11.8 After analysis is completed by the instrument, process analytical results according to Section 12.

[6] Pressed aluminum powder doped with the elements of interest has been found to be satisfactory. Preparation guidelines can be found in: Forte, M.," Fabrication and Use of Permanent Monitors and Standards," *X-ray Spectrometry*, 1983, Vol 3, pp. 115.

12. Calculation

12.1 Concentration units normally reported are mg/kg. A number of corrections may be required for samples.

12.1.1 *Conversion of Analysis Units to Mass/Volume*—A density will be required. This is determined by ensuring the sample is well mixed. Record the mass in grams of a 1 cm^3 volume to calculate the density. The conversion calculations are as follows:

$$\text{mass/vol (mg/l)} = \text{Density (g/cm}^3\text{)} \times \text{reported concentration (mg/kg)} \quad (1)$$

12.1.2 *Re-calculation of Concentration if Sample Dilution has been Used*—For samples that have been diluted, that is further than the 15 g alumina: 5 g sample ratio according to 10.4.1. The following calculations are required to give correct sample concentrations:

$$\text{Corrected sample concentration} = 5 \times \frac{\text{reported concentration}}{\text{Actual mass of sample used, g}} \quad (2)$$

13. Quality Control

13.1 Each laboratory using this test method should operate a formal quality control program.

13.2 Before using this test method, it is strongly recommended that the user fully investigate and implement any specific regulatory quality control requirements.

13.3 A measurement of the two control samples defined in 10.4.3 at a frequency of not less than one per day should be carried out.

13.4 Results of quality control sample measurements made in 13.3 must be evaluated.

13.4.1 Should values fall outside data quality objectives when using Method A in 10.4.1 then the regression must be restandardized following the procedure in 10.4.4, Method A.

13.4.2 Should values fall outside data quality objectives when using Method B in 10.4.2 then the instrument must be restandardized following the procedure in 10.4.4, Method B.

14. Precision and Bias

14.1 *Precision*—No statement is made about the precision at this time. Appendix X1 provides limited precision information to the user.

14.2 *Bias*—No statement is made about bias at this time. This information will be acquired in the future.

14.3 Tables X1.1-X1.6 in Appendix X1 contain analytical/quality control results for most of the elements in 1.2 following this test method on typical LHW, check standard, MS/MSD and method blank. Lower limits of detection are also given.

15. Keywords

15.1 EDXRF; halides; liquid hazardous waste; spectrometry; spectroscopy; trace elements; toxic metals; waste-derived fuel; XRF

APPENDIX

(Nonmandatory Information)

X1. BLANK, ACCURACY AND MATRIX SPIKE RESULTS

X1.1 Errors shown in Tables X1.1-X1.3 are taken from the results output of the instrument and nominally ± 2 sigma.[7] These represent the total error attributed to spectrum processing and counting statistics. See Stratham[12] for details of the calculations used in the error calculations.

[7] Stratham, P., *Analytical Chemistry*, 1977, Vol 49, pp. 2149.

TABLE X1.1 Blank

NOTE 1—A sample containing only PEG was used to check for any bias in the calibrations at the zero concentration level.

Element	V	Cr	Fe	Ni	Cu	Zn	As	Se	Cd	Sn	Sb	I	Hg	Tl	Pb
mg kg^{1}	1.2	0.0	17.3	0.0	0.0	1.6	1.9	2.6	0.0	1.3	0.0	3.9	1.9	2.2	0.0
mg kg^{1} error	0.5	0.7	0.2	3.6	0.04	0.4	7.6	2.5	2.5	0.4	0.9	10	2.9	3.9	1.0

TABLE X1.2 Accuracy

NOTE 1—Standards were run against the calibrations to assess accuracy and to check for bias in the calibrations due to either matrix or spectral effects. The % accuracy figure shows the match of given versus calculated concentration for each element.

Element	P % m/m	S % m/m	Cl % m/m	Se mg/kg	As mg/kg	Br mg/kg	Cd mg/kg	Sn mg/kg	Sb mg/kg	I mg/kg
Calculation concentration	0.96	2.09	0.33	93	393	4779	179	47	201	587
Error	0.02	0.02	0.005	4	10	15	4	3	7	13
Given concentration	1.024	2.095	0.598	100	400	4838	200	50	200	600
% accuracy	96	100	89	93	98	99	98	94	100	98

Element	Hg	Tl	Pb	V	Cr	Fe	Ni	Cu	Zn
Calculation concentration, mg kg^{1}	99	96	94	109	6	589	53	55	95
Error	9	13	7	7	3	0.2	8	3	5
Given concentration	100	100	100	100	10	600	50	50	100
% accuracy	99	96	94	109	60	98	106	110	93

TABLE X1.3 Light Elements and Halides

Single Phase Solvent	P	S	Cl	Br	I
MS/MSD recovery (%)	111.3	120	110	95	100
error, % m/m	0.02	0.01	0.007	0.0007	0.0032
spike concentration, % m/m	0.9156	0.926	1.279	0.0599	0.856
Turbid Waste					
MS/MSD recovery,%	98	91	89	104	100
error, % m/m	0.01	0.009	0.005	0.0006	0.0034
spike concentration, % m/m	0.849	0.805	0.858	0.042	0.845
Biphasal Waste					
MS/MSD recovery, %			98		
error, % m/m			0.008		
spike concentration, % m/m			1.654		

X1.2 Table X1.4 gives the lower limit of detection for each analyte. This is based on the following equation:

$$LLD = \frac{3x\sqrt{bg}}{\text{net peak}} \times \frac{1}{\sqrt{T}} \times \text{concentration} \quad \text{(X1.1)}$$

where:

bg = background intensity under analyte peak cps,
net peak = fitted peak intensity of analyte cps,
T = count time, s, and
conc. = concentration of analyte.

X1.3 *Matrix/Matrix Spike Recoveries:*

X1.3.1 A series of experiments was conducted to test the performance of the alumina method. Three types of actual waste solutions were selected from routine test samples taken at an incineration plant. These samples were:

X1.3.1.1 Clear solutions,

X1.3.1.2 Turbid solutions, that is, contained significant solids not in suspension, and

X1.3.1.3 Biphasal solutions, that is, contained two distinctly immiscible liquid phases.

X1.3.2 For each matrix type a sub-sample was spiked with a known concentration of analyte. The spiked sample was prepared using the alumina technique and measured. Using Eq X1.2, a recovery figure for each analyte in each matrix type was determined. The results, referred to as a matrix spike/matrix spike duplicate (MS/MSD), are shown below.

$$\text{\% MS/MSD recovery} - ((C2\ (D1 \times C1))/C3) \times 100 \quad \text{(X1.2)}$$

where:

$D1$ = dilution factor due to matrix spike addition = 1 − (mass of spike)/(total mass of sample + spike),
$C1$ = calculation concentration of matrix without spike,
$C2$ = calculation concentration of matrix + spike, and
$C3$ = given concentration of matrix spike.

X1.4 *Precision of measurement:*

X1.4.1 A number of repeat measurements were made on a waste sample.

X1.4.1.1 A single measurement from each of ten repeat sample preparations was made for Cl content. A repeat of this process was made on newly prepared samples 48 h later.

X1.4.1.2 A single measurement from each of ten repeat sample preparations was made for a waste sample spiked with 54.3 mg kg^{-1} Cd.

X1.4.2 A single Cl analysis of the same waste sample used in X1.4.1.1 was measured by a second laboratory using the alumina sample preparation technique and is shown in the last column of Table X1.6.

TABLE X1.4 3 Sigma Lower Limits of Detection

Element	LLD mg/kg	Element	LLD mg/kg	Element	LLD mg/kg
P	n/a[A]	Cu	17	Se	6
S	n/a[A]	Zn	11	I	7
Cl	n/a[A]	As	5	Hg	7
V	8	Se	5	Tl	4
Cr	8	Br	5	Pb	4
Fe	14	Cd	3		
Ni	16	Sn	5		

[A] n/a = these elements were calibrated at concentrations significantly higher than their respective detection limits.

TABLE X1.5 Toxic Elements

Biphasal Waste	Cd	Hg	Tl	Se	As	Pb	V	Cr	Fe	Ni	Cu	Zn	Sb	Sn
Ms/MSD recovery,%	108	103	98	109	109	99	96	111	84	86	73	105	84	103
error mg kg^{-1}	3	9	10	4	9	7	6	5	1	9	4	5	6	3
spike concentration mg kg^{-1}	47.1	101.8	102.5	104.6	109.7	183.7	97.4	98.3	97.1	98.2	99.3	102.0	33.3	53.5
Turbid Waste	**Cd**	**Hg**	**Tl**	**Se**	**As**	**Pb**	**V**	**Cr**	**Fe**	**Ni**	**Cu**	**Zu**	**Sb**	**Sn**
MS/MSD recovery,%	92	114	43	90	121	108	102	110	104	118	94	105	108	103
error mg kg^{-1}	3	11	13	5	10	10	6	6	2	11	5	5	6	3
spike concentration mg kg^{-1}	44.8	96.8	47.7	93.6	94.8	177	103.4	102	106.4	104.5	103.5	106.6	53.2	53.2
Single Phase Solvent	**Cd**	**Hg**	**Tl**	**Se**	**As**	**Pb**	**V**	**Cr**	**Fe**	**Ni**	**Cu**	**Zn**	**Sb**	**Sn**
Ms/MSD recovery,%	100	98	77	108	108	error^{A}	110	114	109	104	105	90	112	95
error mg kg^{-1}	3	9	10	4	9		6	6	2	10	5	5	6	3
spike concentration mg kg^{-1}	45.5	97.0	96	103.8	105		116.8	109.5	100	105.7	109.9	105.3	54.6	55.5

^{A} Spike not added.

TABLE X1.6 Precision

Element	10 repeats—Day 1			10 repeats—Day 3			second lab
	mean	S.D.^{A}	%R.S.D.^{B}	mean	S.D.^{A}	%R.S.D.^{B}	single result
Cl % m/m	1.45	0.027	1.9	1.43	0.037	2.6	1.38
Cd mg/kg	56	1.3	2.2				

^{A} S.D. = 1 sigma standard deviation.
^{B} % R.S.D. = % relative standard deviation.

The American Society for Testing and Materials takes no position respecting the validity of any patent rights asserted in connection with any item mentioned in this standard. Users of this standard are expressly advised that determination of the validity of any such patent rights, and the risk of infringement of such rights, are entirely their own responsibility.

This standard is subject to revision at any time by the responsible technical committee and must be reviewed every five years and if not revised, either reapproved or withdrawn. Your comments are invited either for revision of this standard or for additional standards and should be addressed to ASTM Headquarters. Your comments will receive careful consideration at a meeting of the responsible technical committee, which you may attend. If you feel that your comments have not received a fair hearing you should make your views known to the ASTM Committee on Standards, at the address shown below.

Designation: D 6063 – 96 (Reapproved 2001)

Standard Guide for Sampling of Drums and Similar Containers by Field Personnel[1]

This standard is issued under the fixed designation D 6063; the number immediately following the designation indicates the year of original adoption or, in the case of revision, the year of last revision. A number in parentheses indicates the year of last reapproval. A superscript epsilon (ε) indicates an editorial change since the last revision or reapproval.

1. Scope

1.1 This guide covers information, including flow charts, for field personnel to follow in order to collect samples from drums and similar containers.

1.2 The purpose of this guide is to help field personnel in planning and obtaining samples from drums and similar containers, using equipment and techniques that will ensure that the objectives of the sampling activity will be met. It can also be used as a training tool.

1.3 *This standard does not purport to address all of the safety concerns, if any, associated with its use. It is the responsibility of the user of this standard to establish appropriate safety and health practices and determine the applicability of regulatory limitations prior to use.*

2. Referenced Documents

2.1 *ASTM Standards:*

C 783 Practice for Core Sampling of Graphite Electrodes[2]

D 1452 Practice for Soil Investigation and Sampling by Auger Borings[3]

D 1586 Test Method for Penetration Test and Split-Barrel Sampling of Soils[3]

D 1587 Practice for Thin-Walled Tube Geotechnical Sampling of Soils[3]

D 2113 Practice for Diamond Core Drilling for Site Investigation[3]

D 4448 Guide for Sampling Groundwater Monitoring Wells[4]

D 4687 Guide for General Planning of Waste Sampling[4]

D 4700 Guide for Soil Sampling from the Vadose Zone[3]

D 4823 Guide for Core-Sampling Submerged, Unconsolidated Sediments[5]

D 4840 Guide for Sample Chain of Custody Procedures[6]

D 5088 Practice for Decontamination of Field Equipment Used at Nonradioactive Waste Sites[7]

D 5283 Practice for Generation of Environmental Data Related to Waste Management Activities: Quality Assurance and Quality Control Planning and Implementation[4]

D 5358 Practice for Sampling with a Dipper or Pond Sampler[4]

D 5451 Practice for Sampling Using a Trier Sampler[4]

D 5495 Practice for Sampling with a Composite Liquid Waste (COLIWASA) Sampler[4]

3. Terminology

3.1 *Definitions:*

3.1.1 *bung, n*—usually a 2-in. (5-cm) or ¾-in. (1.3-cm) diameter threaded plug specifically designed to close a bung hole.

3.1.2 *bung hole, n*—an opening in a barrel or drum through which it can be filled, emptied or vented.

3.1.3 *consolidated solid, n—as used in this guide*, a compact solid not easily compressed or broken into smaller portions.

3.1.4 *drum, n—when used in the flow charts in this guide*, the word implies any drum, barrel or non-bulk container of 5 to 110 gal (19 to 400 L) capacity.

3.1.5 *representative sample, n*—a sample collected such that it reflects one or more characteristics of interest of the lot or population from which it was collected.

3.1.6 *sample, n*—one or more items or portions collected from a lot or population.

3.1.7 *sampler, n*—the device used to obtain a sample.

3.1.8 *sludge, n—as used in this guide*, any mixture of solids that settles out of solution; sludges contain liquids that are not apparent as free liquids.

3.1.9 *unconsolidated solid, n—as used in this guide*, uncemented or uncompacted material that is easily separated into smaller portions.

3.1.10 *work plans, n*—plans that are specific to sampling at a particular site; examples are Health and Safety Plans and Sampling and Analysis Plans.

4. Summary of Guide

4.1 This guide uses a decision-tree format to lead persons intending to sample waste materials from drums and similar containers through a series of questions. The answers to the questions result in recommended actions, including the selection of appropriate sampling equipment. Brief instructions on

[1] This guide is under the jurisdiction of ASTM Committee D34 on Waste Management and is the direct responsibility of Subcommittee D34.01.02 on Sampling Techniques.

Current edition approved Dec. 10, 1996. Published February 1997.

[2] *Annual Book of ASTM Standards*, Vol 15.01.

[3] *Annual Book of ASTM Standards*, Vol 04.08.

[4] *Annual Book of ASTM Standards*, Vol 11.04.

[5] *Annual Book of ASTM Standards*, Vol 11.02.

[6] *Annual Book of ASTM Standards*, Vol 11.01.

[7] *Annual Book of ASTM Standards*, Vol 04.09.

the use of the equipment are included.

4.2 This guide addresses commonly used sampling equipment and devices; it is not intended to cover all that might be purchased or custom made.

5. Significance and Use

5.1 This guide is intended to assist field personnel in obtaining samples from drums and similar containers for laboratory analysis. The costs associated with sampling and analysis make it essential that samples be taken correctly before submitting them for chemical analysis or physical testing, or both. Incorrect sampling can invalidate resulting data.

5.2 This guide may be used by personnel who have no formal workplan. It draws their attention to issues that must be addressed before, during, and after taking a sample. It provides guidance in choosing the sampling technique and equipment suitable for specific situations. It can serve as a training tool for those who are unfamiliar with sampling. It is recommended that this guide be used as a supplement to a written workplan.

5.3 Some sections of this guide contain flow charts (see Figs. 1-5) that must be worked through, starting from the top of each page. By answering the questions in the diamond-shaped boxes, and following the appropriate arrows, the person planning to sample will be guided towards the most suitable procedures and equipment. The numbers at the bottom of some boxes refer to corresponding paragraphs in the text, which provide information to help the person sampling answer the questions.

5.4 Figs. 6-15 are examples of types of equipment. Similar devices that do the same job in the same way are not intended to be excluded.

6. Objectives of Sampling

6.1 The purpose of sampling is to collect a representative sample of all or part of the contents of the drum or similar container, to determine the physical and chemical characteristics of those contents (see Fig. 1). This information may then be used to:

6.1.1 Select suitable methods of treatment and disposal of the contents,

6.1.2 Provide evidence for use in a court of law,

6.1.3 Comply with regulations, such as those for the transportation of hazardous materials,

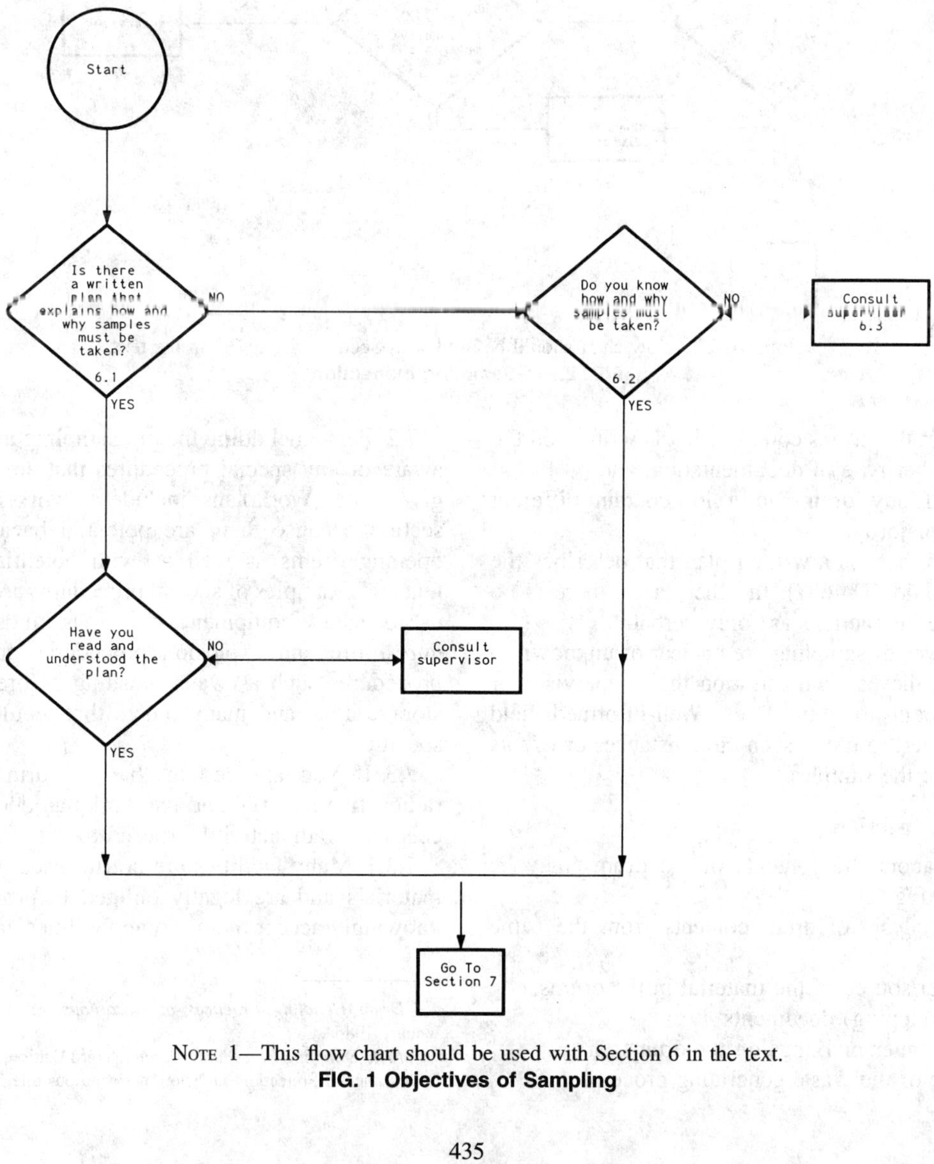

NOTE 1—This flow chart should be used with Section 6 in the text.

FIG. 1 Objectives of Sampling

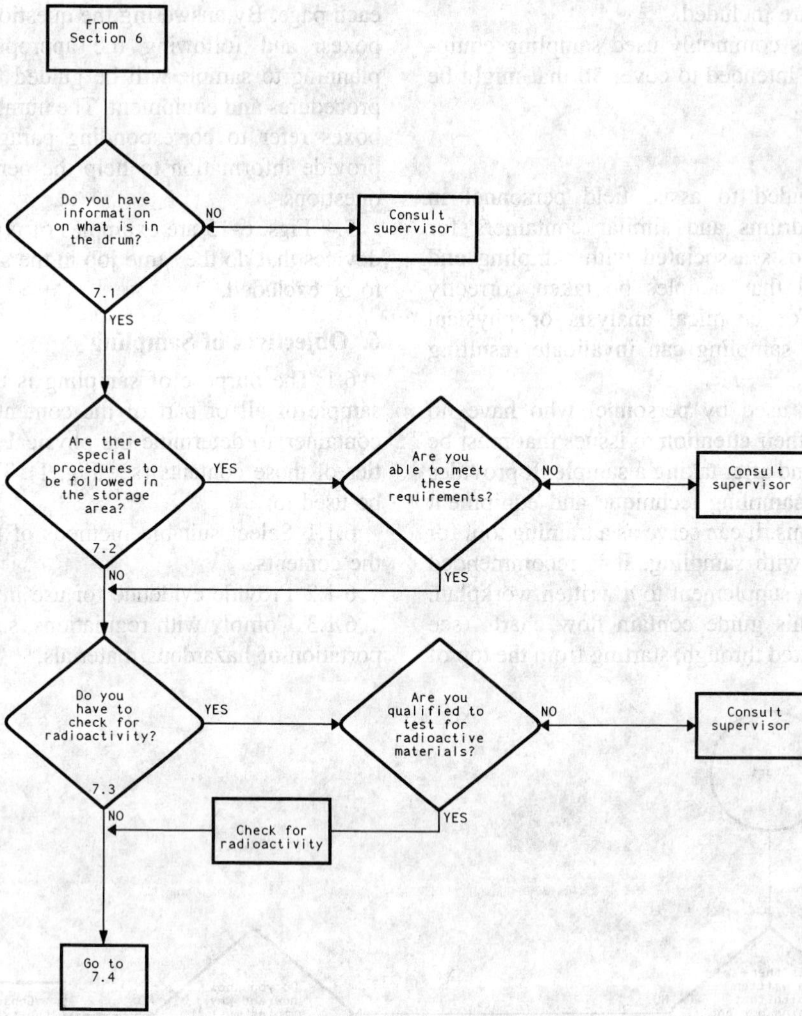

NOTE 1—This flow chart should be used with Sections 7.1 to 7.3 in the text.

FIG. 2 Pre-Sampling Inspection

6.1.4 Confirm that the drums contain what is written on the label, manifest or other type of documentation, and

6.1.5 Find out if any drums in a lot contain different materials from the majority.

6.2 In most cases there is a written plan that describes the work to be done (Guide D 4687). In other cases, there is no written plan and the instructions are only verbal.

6.3 If the objectives of sampling are unclear or unknown to the field personnel, they should question their supervisor or project manager about the objectives. Well-informed field personnel are then alert to unforeseen circumstances or events that might invalidate the samples.

7. Pre-Sampling Inspection

7.1 Information about the contents of the drums may be available from (see Fig. 2):

7.1.1 Previous analysis of drum contents from the same source,

7.1.2 The supplier/source of the material in the drums,

7.1.3 Manifest (shipping) documents,

7.1.4 Labels and other markings on the drums, or

7.1.5 Knowledge of the waste generating process.

7.2 Personnel doing the pre-sampling and sampling must be aware of any special procedures that are to be followed at a given site. Workplans include a worker health and safety section because there are potential hazards associated with opening drums as well as with potentially hazardous contents.[8,9] Examples of special procedures are change of clothing, use of safety equipment of various kinds, evacuation procedures, fire and explosion procedures and vehicle cleaning procedures such as water washing before leaving the site or storage area, and many others that would be site or storage specific.

7.3 If you are certain that the drum does not contain radioactive material and the workplan does not require you to check for radioactivity, proceed to 7.4.

7.3.1 Many facilities are not licensed to handle radioactive materials and are legally obliged to prove that they do not knowingly accept them. Some facilities are licensed to handle

[8] *Drum Handling Practices at Hazardous Waste Sites*, EPA/600/2-86/013, January 1986.

[9] *Field Sampling Procedures Manual*, Third Edition, New Jersey Department of Environmental Protection, Division of Hazardous Site Mitigation, February 1988.

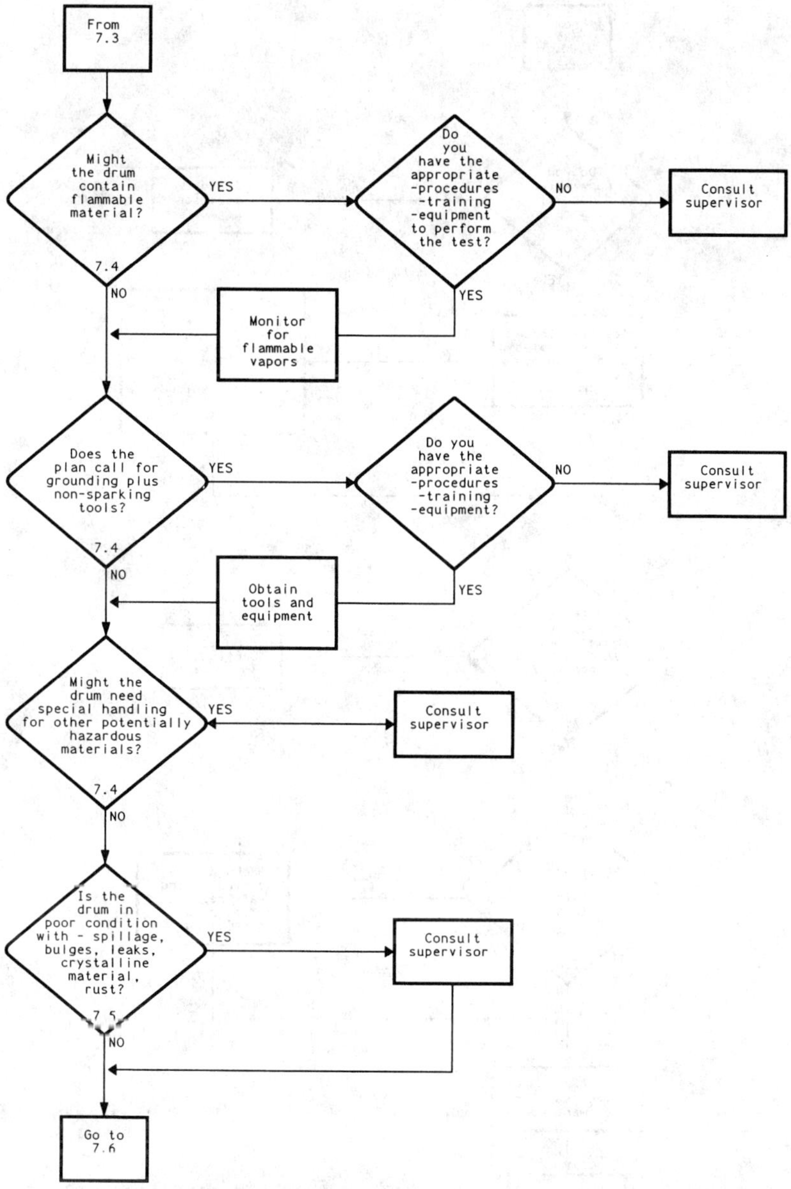

NOTE 1—This flow chart should be used with Sections 7.3 to 7.5 in the text.

FIG. 2 Pre-Sampling Inspection *(continued)*

radioactive materials; they need to have a measure of how radioactive the material is for the safety of their workers.

7.3.2 Hand-held monitors that check for radioactivity should always be used if you suspect that radioactive material might be present or if the workplan requires it. It is important that the monitor has been calibrated correctly, according to the manufacturer's instructions. Monitoring should be done only by those with the appropriate written procedures, training and equipment.

7.3.3 It is prudent to monitor a storage area before entering it. If radioactive material is found to be present when it should not be, leave the area immediately, post warning signs to alert other workers, and consult your supervisor.

7.4 Drums may contain flammable materials, strong oxidizers or reducing materials, light-sensitive materials, corrosive acids or bases, and materials sensitive to moisture. All of these drums require special handling, including segregation.

7.4.1 Many solvents, like benzene, evaporate into air space in and around the drum where the vapour may be easily ignited.

7.4.2 If you are sampling a potentially flammable or unknown material, non-sparking tools should be used and the drums should be grounded.

7.4.3 If the drums are stored in a closed room or confined space, the air in the area should be tested by a hand-held monitor to check for flammable vapors. It is important that the monitor has been calibrated according to the manufacturer's instructions. The monitoring should be done by those with the appropriate training and written procedures.

NOTE 1—**Warning:** Flammable materials should be sampled in a well-ventilated area. There are other safety considerations that must be considered regarding confined spaces. It may be necessary to check for explosivity or oxygen levels.

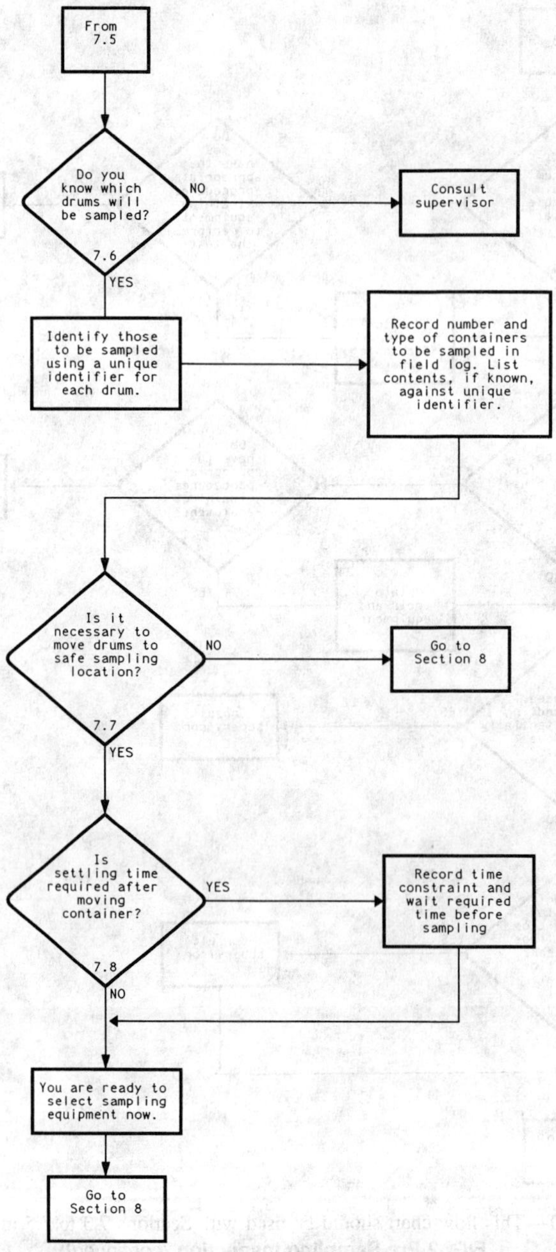

NOTE 1—This flow chart should be used with Sections 7.6 to 7.8 in the text.

FIG. 2 Pre-Sampling Inspection *(continued)*

7.4.4 Labels on drums of waste materials may not be accurate. Unless the drums come from a reliable source, for example, the generator of the material and the process that created the waste are known to you, it is prudent to assume that the labels may not match the contents.

NOTE 2—**Precaution:** Attempting to open a drum that is in poor condition can expose a worker to the possibility of a serious, even fatal, accident. Special precautions should be taken when the bungs are rusted or corroded since the drum top may give way, exposing the worker to vapor or liquid. Overpacking before sampling should be considered for drums in poor condition.

7.5 It is not always necessary to sample every drum in a lot. The workplans provide direction as to how many, and possibly which, drums should be sampled. Each drum that will be sampled must be identified in a unique way in case a second sample has to be taken later. Colored labels, crayons, paint sticks or pens, or stencilled paint can be used to identify drums. Any new identification system should not cover the existing labels or identifiers.

7.6 Sometimes drums have to be moved to another location for sampling; this is known as "drum staging". This is required if:

7.6.1 Sampling the drum in its present location poses a high risk to surrounding property and individuals,

7.6.2 The drum cannot be accessed for sampling in its current location, or

7.6.3 Exposure to climatic conditions alter the sample, for example, formation of ice; or create a health and safety risk, for

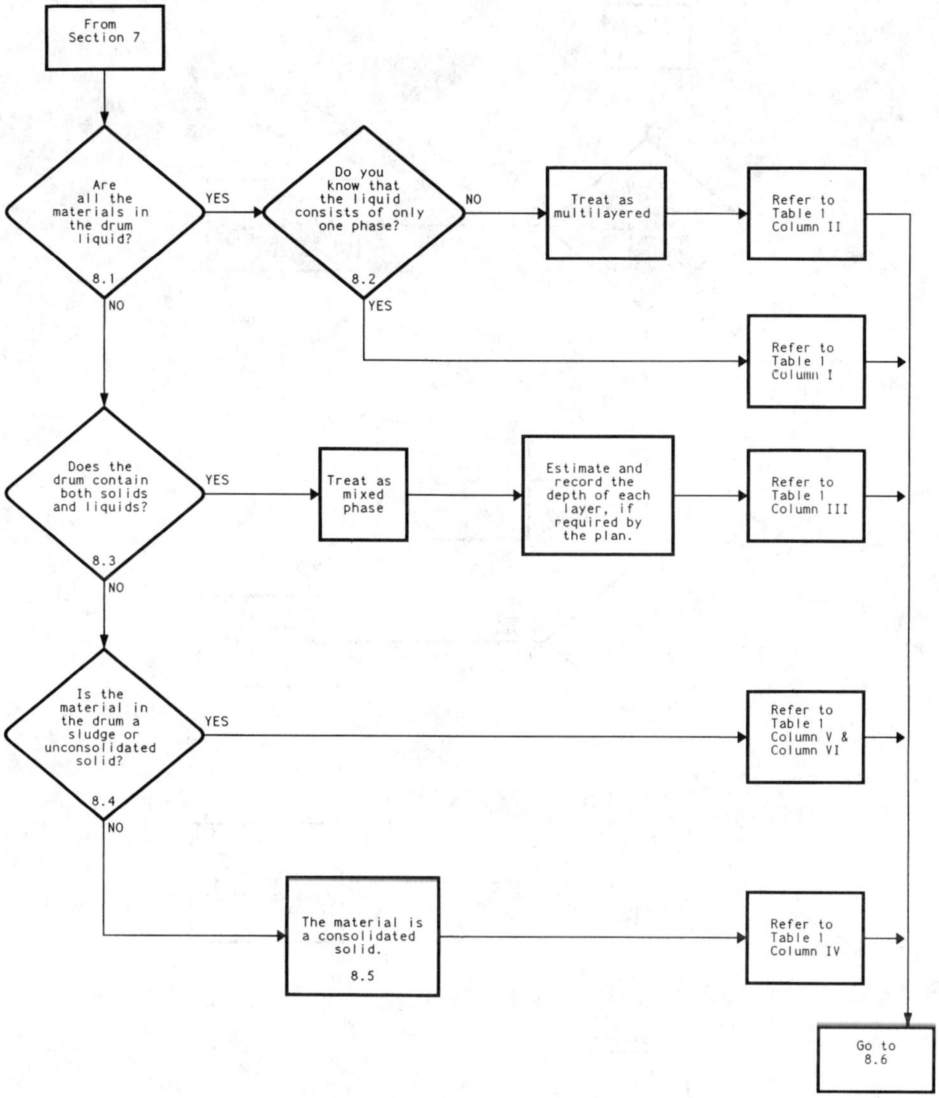

NOTE 1—This flow chart should be used with Sections 8.1 to 8.5 in the text.

FIG. 3 Selection of Suitable Sampling Procedure

example, the sun heating a drum containing solvents.

7.7 The physical condition of drums must be evaluated before attempting to open or move them (see Note 2). Drum carriers, which lock on the drum lip, should not be used to move the drum if the condition is poor.

7.8 Materials in layers, such as oil with water, can become mixed together when moved. If you want to sample each layer separately, the material may need time to settle before opening and sampling the drum.

8. Selection of Suitable Sampling Procedure

8.1 The physical state(s) of the material(s) being sampled is an important criterion when sampling (see Fig. 3).

8.2 A drum containing one liquid, such as water, or a mixture of liquids, such as a stable emulsion like hand cream, that does not separate into two layers regardless of time, is said to contain one phase. A drum containing two liquids, such as oil and water, which form two distinct layers when they are not stirred is said to contain two phases.

8.3 When it is necessary to know the amounts of solid and liquid layers in a drum, a calibrated measuring device or the sampling equipment (for example, a COLIWASA) can be inserted into the opened drum (see Section 12) and the liquid level measured.

8.4 Although sludges behave like sticky solids and are not usually pumped, they can contain quite a high proportion of liquids, such as oil or water, which is not visible as free liquid.

8.5 An unconsolidated solid is a material like sand or a powder. A consolidated solid consists of material, like sandstone or concrete.

8.5.1 A drum containing mixed materials, such as disposable personal protective equipment and laboratory supplies, is treated as one with unconsolidated solids.

8.6 If the waste material is likely to attack the sampling equipment, such as an acid corroding a metal thief:

8.6.1 The equipment may partially dissolve, adding constituents, such as metals to the sample. Faulty conclusions may be drawn about the composition of the sample, leading to

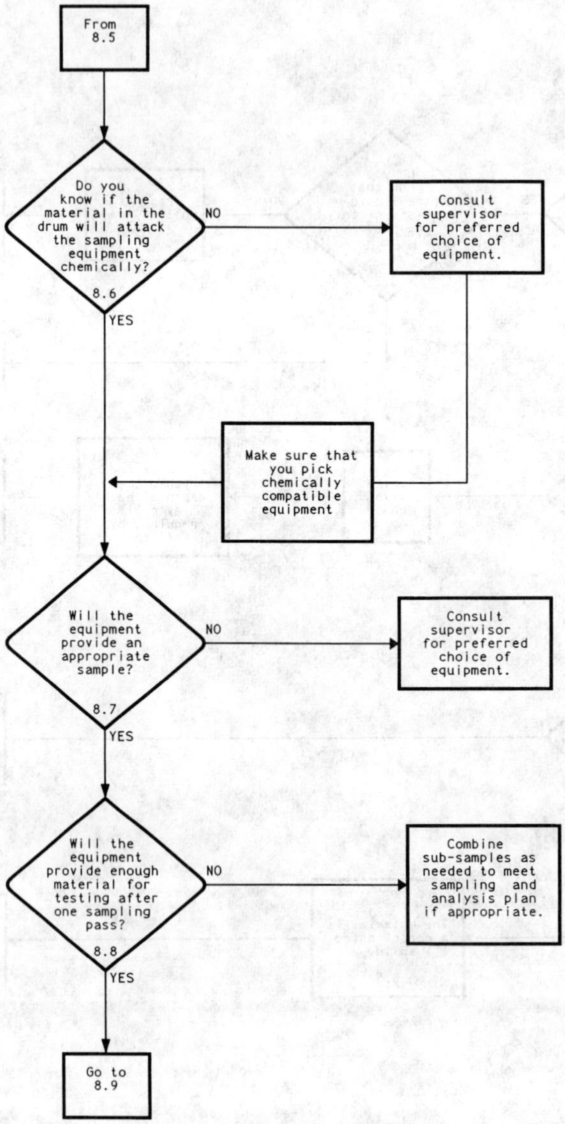

NOTE 1—This flow chart should be used with Sections 8.6 to 8.8 in the text.

FIG. 3 Selection of Suitable Sampling Procedure *(continued)*

costly and unnecessary remedial actions, and

8.6.2 The equipment will have to be replaced frequently, adding costs to the project.

8.7 When selecting equipment, it is important to be aware of the limitations of the tools. The design of some equipment can result in part of the material not being sampled. For example, if the size of the opening(s) that allow the sample to enter a trier is smaller than some of the particles in the drum, the sample will not be representative.

8.7.1 Volatile organic constituents are likely to be lost if the sampling equipment causes a buildup of heat or agitation of the sample, as will exposure to air for more than a very short time or storage in a sample container with a headspace above the sample.

8.7.2 See Table 1 for more information on the limitations of sampling equipment.

8.8 When the quantity of material that is removed from a drum by the equipment will not provide enough for the laboratory to perform all the required tests, a number of subsamples must be taken and combined. In doing this it is critical not to disturb layers; this could result in an unsuitable sample.

8.9 Pumps may require electricity to operate. If the sampling location is outdoors, they may need to be protected from the weather. If flammable vapors were observed in 7.4, consult your supervisor about sources of ignition, such as pumps, electrical connections and switches.

8.10 Consideration should be given to having a separate clean sampling device for each drum since this eliminates cross-contamination and may be more efficient than cleaning one sampling device after each drum. It is generally easier to clean the sampling devices and other equipment in the laboratory or another suitable place where solvent disposal and drying equipment are readily available.

8.11 It is worthwhile to make a comparison of the costs of cleaning or decontaminating equipment, or both, including

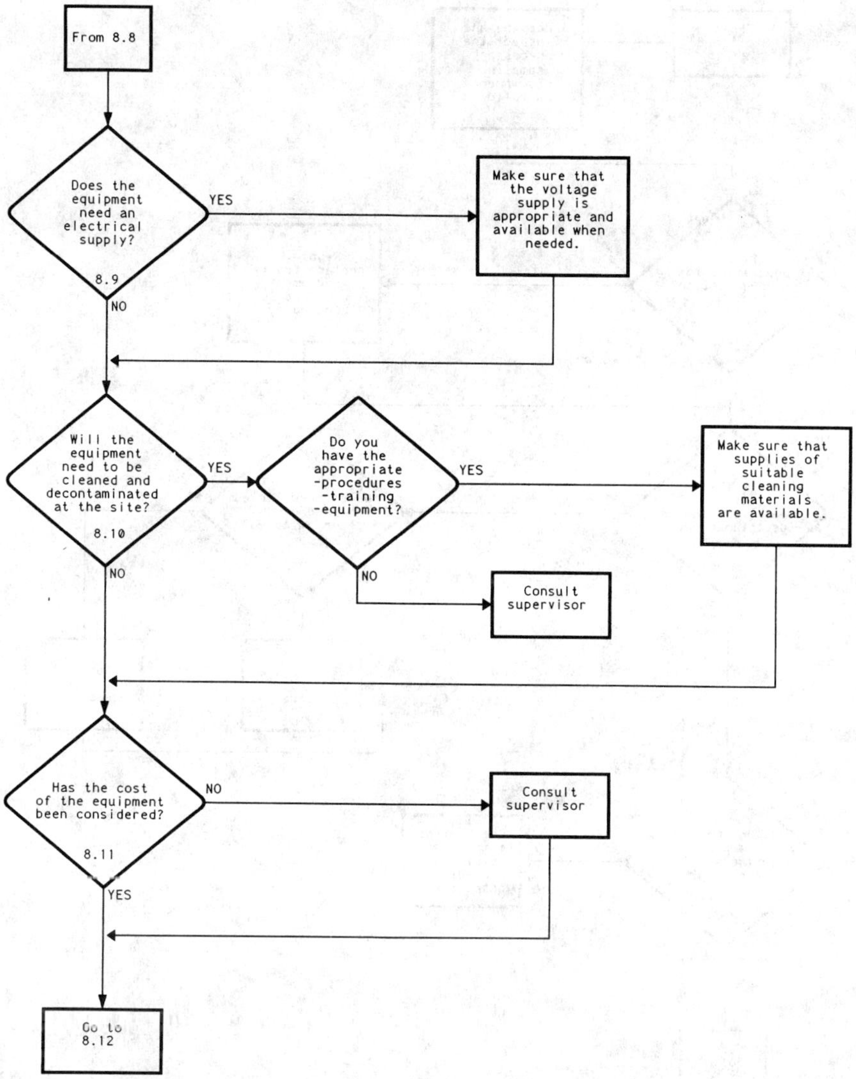

NOTE 1—This flow chart should be used with Sections 8.9 to 8.11 in the text.

FIG. 3 Selection of Suitable Sampling Procedure *(continued)*

disposal of the cleaning agents, versus using disposable equipment when selecting and preparing equipment. The initial purchase price of the equipment may also be a factor in the selection. There may be more than one suitable device (see Table 2).

8.12 Table 1 and Table 2 list equipment that is commonly used for sampling liquid and solid wastes, and their limitations. They contain more than one type of equipment for some types of waste materials but are not intended to include all possible equipment or devices.

8.12.1 Other factors that will guide you in making the final choice follow in Section 13.

9. Preparation of Sampling Equipment

9.1 Damaged sampling equipment can affect the sample, for example, dull cutting edges or chipped glass parts. It may also be a safety hazard for the worker (see Fig. 4).

9.2 Cleaning and decontamination procedures should be identified in the plan (see Practice D 5088).

9.3 Check the workplan for preservatives, refrigeration, holding times, type and size of sample container, packaging, and shipping requirements. If the samples are going to be analyzed for volatile organic compounds, such as gasoline, special bottles must be used.

9.4 If the samples are going to be analyzed for trace amounts of volatile organic compounds, such as solvents, plan to fill and seal the special sample bottles required for this analysis first, before filling the other sample containers.

9.5 Personal safety equipment will vary depending on the hazards associated with the task. It must comply with the health and safety requirements of the organization that employs you.

9.5.1 Basic safety equipment includes:

9.5.1.1 Safety glasses,

9.5.1.2 Synthetic rubber gloves,

9.5.1.3 Safety shoes or boots, and

9.5.1.4 Protective clothing.

9.5.2 More hazardous situations may require equipment such as:

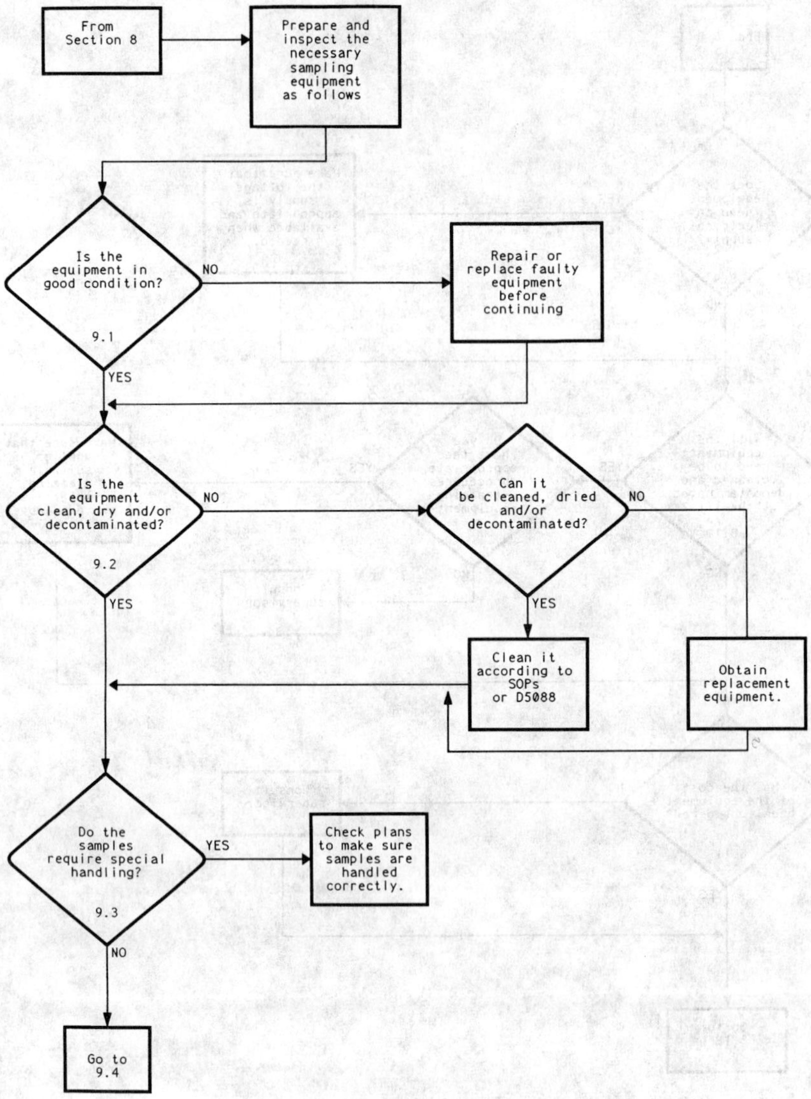

NOTE 1—This flow chart should be used with Sections 9.1 to 9.3 in the text.

FIG. 4 Preparation of Sampling Equipment

9.5.2.1 Hard hat,

9.5.2.2 Respirators,

9.5.2.3 Face shields,

9.5.2.4 Chemically-resistant suits,

9.5.2.5 Self-contained breathing apparatus, and

9.5.2.6 Two way radio communication.

9.5.3 If appropriate, segregate the drums and surrounding area from casual intrusion by using barricades or caution tape.

9.5.4 When sampling inside buildings, ventilation may be desirable.

9.6 Records that associate a sample with a drum are usually required. A field technician's log and chain of custody forms are commonly used (see Section 10).

10. Report

10.1 *Chain of Custody Forms*:

10.1.1 The purpose of chain of custody forms is to show that the samples analyzed are the same ones that were collected. They are required for regulatory purposes. They serve as legal documentation that sample integrity was maintained. When complete, they should show that there were no lapses in accountability. It is not always necessary to use chain of custody procedures but some form of sample tracking is necessary.

10.1.2 A chain of custody form may originate with the laboratory at the time the sample containers are prepared or in the field after the sample has been taken. Each time the samples change hands, the chain of custody form must be signed. The originator of the chain of custody keeps one copy to confirm that the samples were passed on and to whom; they may have been given directly to the laboratory or to a courier. The completed form, with all remaining copies intact, must accompany the samples. Copies of the custody forms will be returned to the parties involved, to confirm that the samples have been received at the laboratory.

10.1.3 Sometimes security seals are placed over the caps of empty, clean sample containers and signed off by the laboratory. The person sampling must break these seals in order to fill

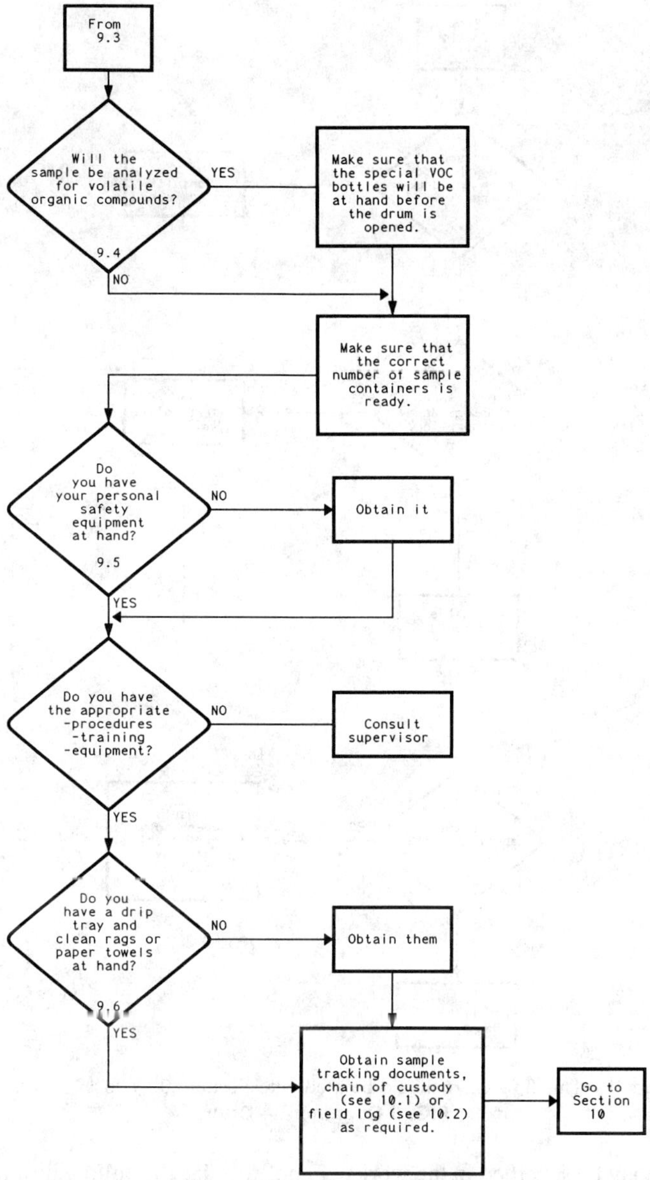

NOTE 1—This flow chart should be used with Sections 9.4 to 10.2 in the text.

FIG. 4 Preparation of Sampling Equipment *(continued)*

the containers. Security seals may also be attached after the sample container has been filled. The date, time of sampling and name of the person sampling are then written on the seals. The purpose of the seal is to indicate possible tampering with the sample.[10] (See Guide D 4840.)

10.2 *Field Log*:

10.2.1 Ideally, a field log is maintained in a bound book with printed page numbers. Information should be recorded in indelible ink. Errors should be crossed out with a single line and initialled. Items normally documented include:

10.2.1.1 Type of waste, for example, sludge, wastewater,

10.2.1.2 Suspected waste composition, including concentrations,

10.2.1.3 Number and volume of sample taken,

10.2.1.4 Description of sampling point and sampling method,

10.2.1.5 Date, time and location of each sample collected,

10.2.1.6 Preservatives used, if any (including ice),

10.2.1.7 Analytical parameters to be measured,

10.2.1.8 Unique sample identification number,

10.2.1.9 Types and number of QC samples,

10.2.1.10 Equipment used to obtain samples,

10.2.1.11 Field observations; unusual events that may have an impact on the sample(s),

10.2.1.12 Field measurements and results (for example, pH),

10.2.1.13 Signature of person who took the sample, and

10.2.1.14 Names of personnel on sampling team.

[10] *Guide to the Collection and Submission of Samples for Laboratory Analysis*, 6th Edition, Ontario Ministry of the Environment and Energy, April 1989.

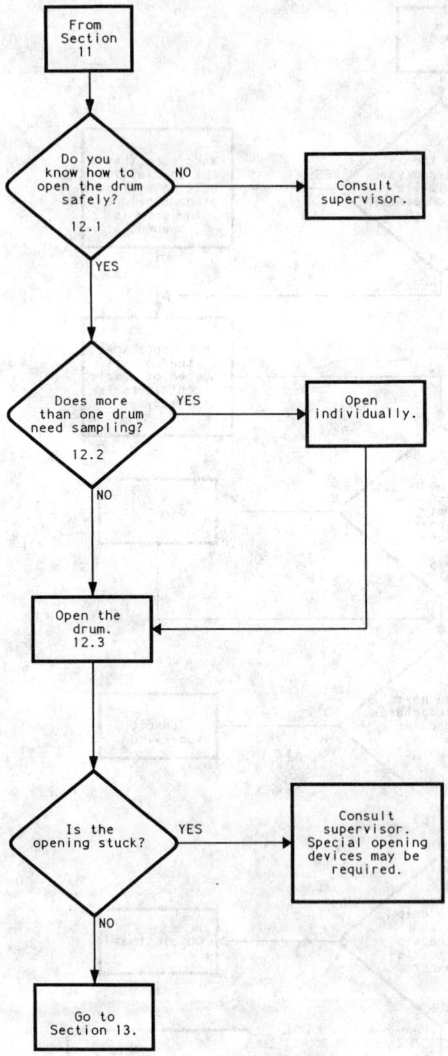

NOTE 1—This flow chart should be used with Sections 12.1 to 12.3 in the text.

FIG. 5 Opening the Drum

10.2.2 Additional information may be specified in the workplan. (See Practice D 5283.)

10.3 *Labels*:

10.3.1 Information written on labels may include:

10.3.1.1 Sample ID number,

10.3.1.2 Name of person sampling,

10.3.1.3 Initials or signature of person sampling,

10.3.1.4 Date and time of sampling,

10.3.1.5 Sample location,

10.3.1.6 Sampling information (for example, grab or composite),

10.3.1.7 Preservative/preservation required,

10.3.1.8 Special instructions, and

10.3.1.9 Analysis request.

10.3.2 Labels may be put on containers before or after sampling. Consult the workplan.

11. Mixing of Liquids, With or Without Solids, Before Sampling

11.1 Thorough mixing of the material prior to sampling is not always possible or necessary. When deciding whether or not to mix, the following points should be considered.

11.1.1 Does the plan require that the materials in the drum be mixed before sampling? Or does the plan prohibit mixing?

11.1.2 If the material is hazardous and mixing it will create a safety problem, such as allowing acidic fumes to escape, do not mix.

11.1.3 The results of the laboratory tests that the sample will undergo may be influenced by mixing. Examples are (*1*) if the material contains volatile compounds that could be lost during the mixing process, do not mix, and (*2*) if a thin layer of solids on the bottom of the drum contain the metals of concern, mixing is necessary.

11.1.4 If the sampling plan has been designed such that several smaller samples will be taken from various locations and combined into one sample, or if numerous discrete samples will be taken, there is less need to mix the material because the sampling plan has addressed variations in the material.

11.1.5 Sampling equipment designed to take a profile of the waste (for example, COLIWASA) does not require the material to be mixed first.

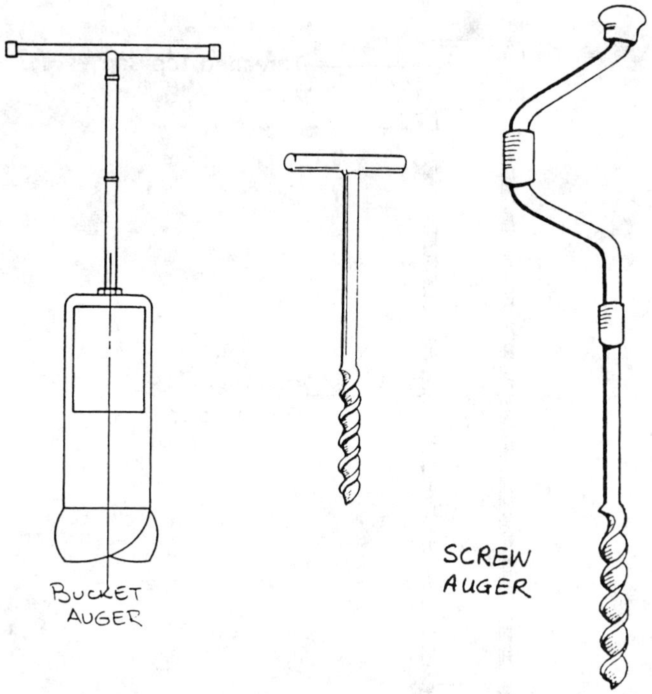

FIG. 6 Augers

11.1.6 If the drum was moved or agitated before sampling, samples may be taken only when the contents of the drum are either completely settled or are thoroughly mixed.

11.2 If the materials are going to be mixed, do so in such a way that no contamination is introduced to the drum and sample container and no material is lost. It should be performed in a safe manner (for example, don't roll a 200-L drum back and forth with the hope of mixing the contents—it rarely works and the potential for injury is high).

11.3 A recirculating pump can be used to mix a liquid material in, for example, a 200-L drum.

12. Opening the Drum

12.1 The bung, ring or other fastening device that secures the lid should be removed slowly, allowing any pressure or vacuum to equalize (see Fig. 5).

12.1.1 Pails with "snap-on" lids may be difficult to open. Be careful to avoid splashing the contents when opening them.

12.1.2 If the top of the drum is dished inward (dimpled), solid or liquid materials, or both, may have accumulated there. The lid may "pop" when equalizing pressure, spraying the person sampling with any material that has accumulated on the lid. To avoid this, material on top of the drum must be removed before opening the drum.

12.1.3 If the top of the drum is bulging, opening it too quickly may cause the top to fly upward, possibly injuring the person sampling.

12.1.4 If there is evidence of a chemical reaction or sudden pressure buildup (for example, fumes escaping, heat build-up, rocking drum), the person sampling should leave the area immediately and decide if remote drum opening equipment should be used.

12.1.5 Drums should be grounded before being opened, if grounding is required.

12.2 Drums should be opened, sampled and closed individually to minimize possible volatilization of organic compounds and also exposure of the person sampling to the material(s).

12.3 When opening a drum with a bung, if the lid is dimpled or bulging, loosen the smaller bung first. This will make it easier to control the release of pressure. Also, cover a manual bung wrench with a cloth to control potential liquid spray. Then remove the large bung.

12.3.1 When opening a drum with a removable lid, slowly loosen the bolt with a manual wrench or an air impact wrench. If the lid is dimpled or bulging, allow the pressure to equalize before completely loosening the bolt. Remove the bolt, the ring and the lid.

12.3.2 When opening a pail with a removable lid (side-lever lock ring), slowly release the lever. If the lid is dimpled or bulging, allow the pressure to equalize before loosening the ring sufficiently for lid removal.

13. How to Use the Equipment

13.1 *Auger*:

13.1.1 *Description*—An auger samples hard or packed solid materials or soil. It consists of sharpened spiral blades attached to a hard metal central shaft (see Fig. 6).

13.1.2 *Steps to Extract a Sample*:

13.1.2.1 Collect the sample by rotating the handle of the auger in a clockwise direction while applying slight downwards pressure.

13.1.2.2 Continue turning until the desired depth has been reached. Pull the auger straight up out of the material.

13.1.2.3 Remove the material that has been withdrawn in the screw thread of the auger, and place it in the container. This is usually done with the auger placed on a clean plastic sheet.

13.1.2.4 Transfer the material to the appropriate container for analysis.

13.2 *Coring Devices*:

13.2.1 *Description*—A coring device samples hard or packed solid waste or soil (see Fig. 7). They are available with coring tips on all types and auger tips on some. The coring tip cuts a sample a little smaller than the inside diameter of the sampler body. The sampler is connected to a metal extension that allows the sampler to be driven into the material either with downward hand pressure or with a slide hammer. When a sampler with an auger tip is used, the sampler is rotated into the material being sampled, using the extension and attached cross handle. The coring sampler with butterfly valve is designed for use in sludges and liquid-saturated materials. The valve closes and prevents loss of sludges and semi-solids when raised from the sampling medium.

13.2.1.1 A stainless steel or plastic liner may be inserted into the hollow tube prior to use.

NOTE 3—Plastic liners may be inappropriate if the sample will be analyzed for organics.

13.2.2 *Steps to Extract a Sample*:

13.2.2.1 Place the coring device on the surface of the material to be sampled.

13.2.2.2 Twist the device slightly, and apply downward

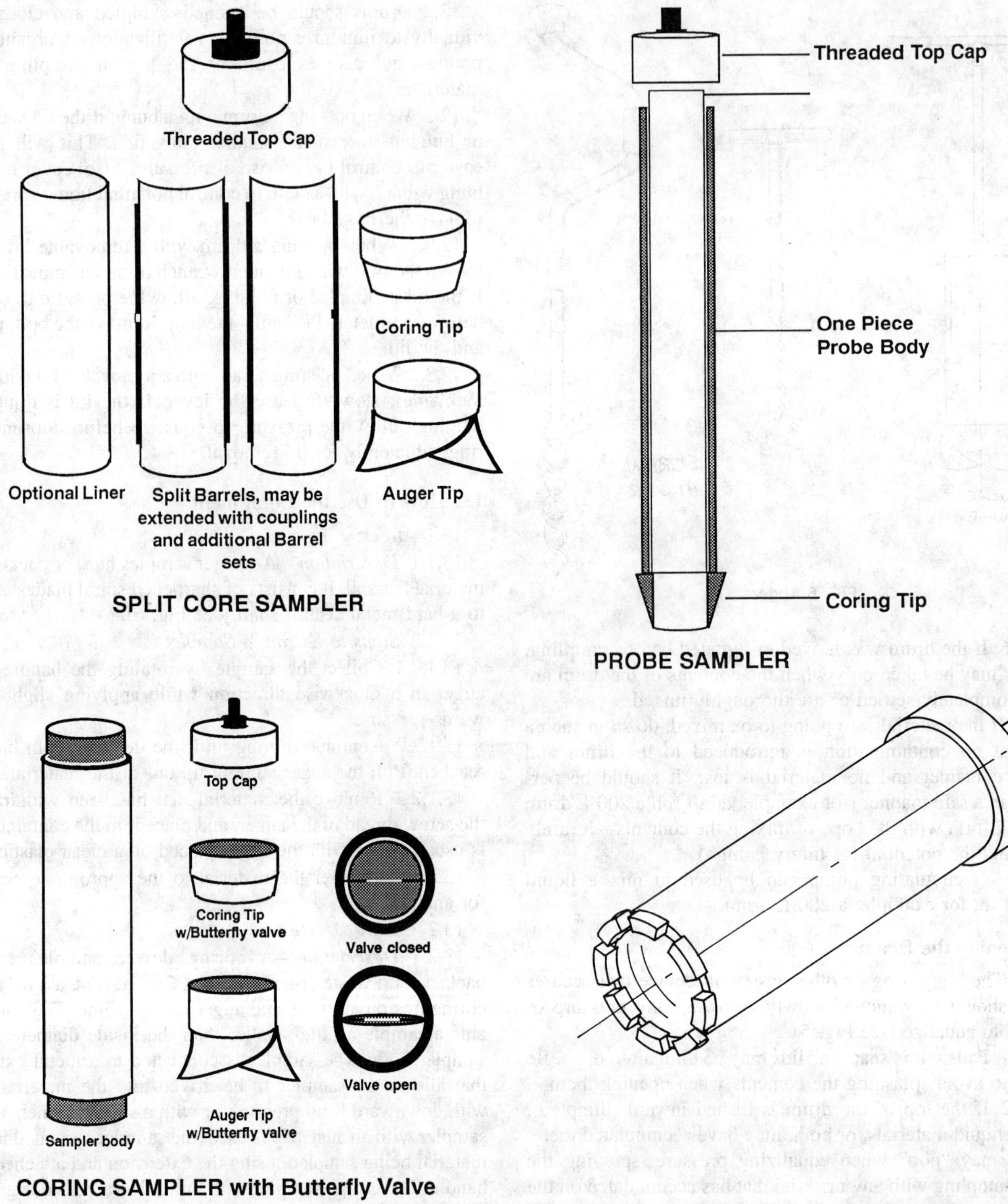

FIG. 7 Coring Devices

pressure to begin penetration. This steadies the device for the next step.

13.2.2.3 Use the slide hammer when manual pressure is no longer effective to drive the coring device into the material, until the desired depth is reached. Always mark the extension with a reference point equivalent to when the tip is near the bottom of the container. Take care not to overdrive the sampler to prevent damage to the bottom of the drum or container.

13.2.2.4 Twist the device again to break the core off from the bottom.

13.2.2.5 Pull the coring device straight up out of the material.

13.2.2.6 Open the device. Either cap off the liner, if present, or remove the material and place it into the appropriate sample container for analysis. This is usually done with the coring device placed on a clean plastic sheet.

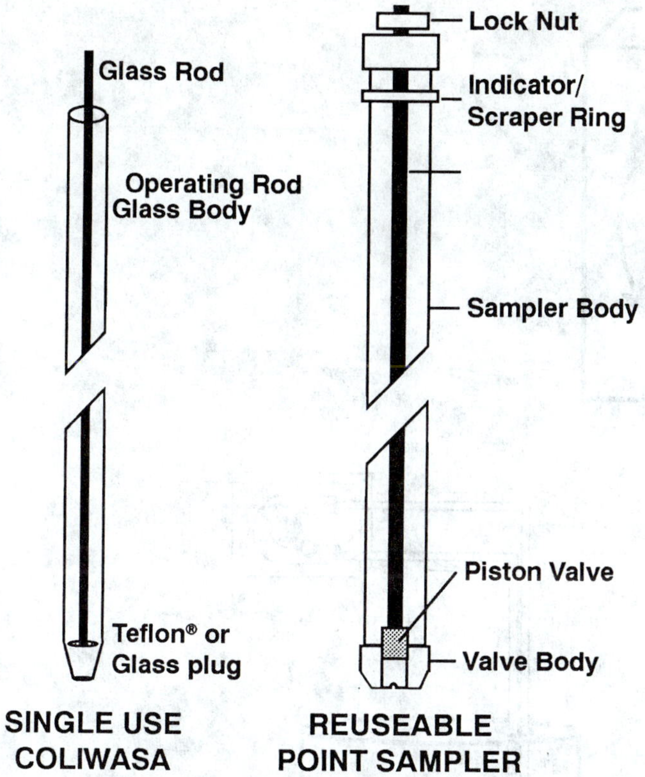

FIG. 8 COLIWASA

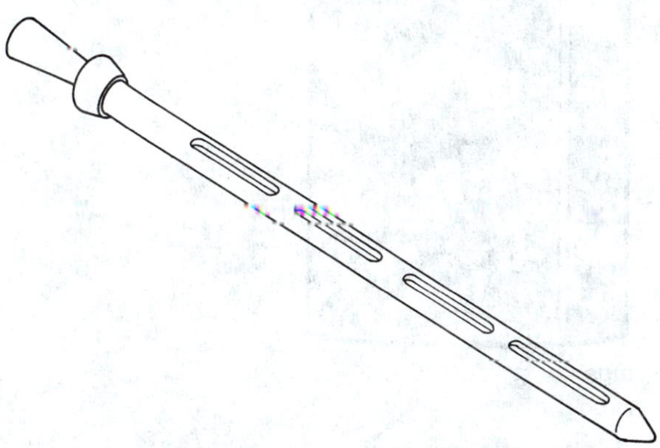
FIG. 9 Concentric Tube Thief

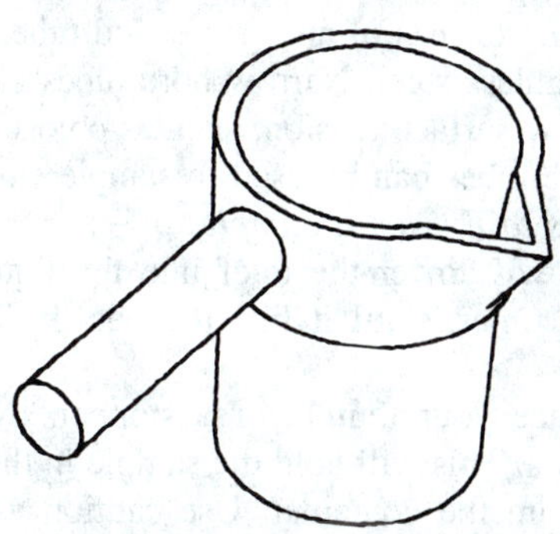
FIG. 10 Dipper

13.3 *Composite Liquid Waste Sampler (COLIWASA)*:

13.3.1 *Description*—COLIWASAs are available commercially with different types of stoppers and mechanisms, but they all operate using the same principle (see Fig. 8). They can

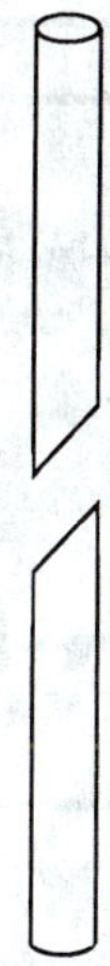
FIG. 11 Drum Thief

be constructed of materials such as polyvinyl chloride (PVC), glass, metal or polytetrafluoroethylene (PTFE). They usually consist of two sections. The outer section is a sleeve that may be tapered at the end. The inner section is a rod with some type of stopper on the end. When the inner section is fitted inside the outer section, a seal is formed, and the unit is locked. There are variations in the mechanisms of opening and closing COLIWASAs. They can be used to obtain samples from specific depths. Sometimes it is necessary to enter the drum a number of times with this sampler to obtain enough sample to meet the needs of the analytical laboratory.

13.3.2 *Steps to Extract a Sample*:

13.3.2.1 Place the COLIWASA in the open position.

13.3.2.2 Lower the COLIWASA slowly into the liquid, keeping it vertical at all times, making sure that the level of the liquid inside and outside the sampler tube remain about the same.

13.3.2.3 When the unit touches the bottom of the container, or has reached the desired depth, close the COLIWASA.

13.3.2.4 Remove the sampler from the liquid with one hand while wiping the outer tube with a disposable cloth or rag with the other hand.

13.3.2.5 Open the COLIWASA over the sample container.

13.4 *Concentric Tube Thief*:

13.4.1 *Description*—A concentric tube thief (grain sampler) (see Fig. 9) is used to sample free-flowing, dry granules or powdery materials whose particle diameter is less than one-third the width of the slots in the sampler. It consists of two slotted concentric tubes, usually made of stainless steel or brass. The outer tube has a conical pointed tip that permits the sampler to penetrate the material being sampled. The inner tube is rotated to open and close the sampler.

13.4.2 *Steps to Extract a Sample*:

13.4.2.1 Ensure the sampler is in the closed position to start.

13.4.2.2 Insert the sampler into the material so that the largest cross-section of material will be sampled.

13.4.2.3 Open the unit by rotating the inner tube.

13.4.2.4 Jiggle the sample to encourage the material to enter the sampler.

13.4.2.5 Close the unit by rotating the inner tube and remove the sampler from the material.

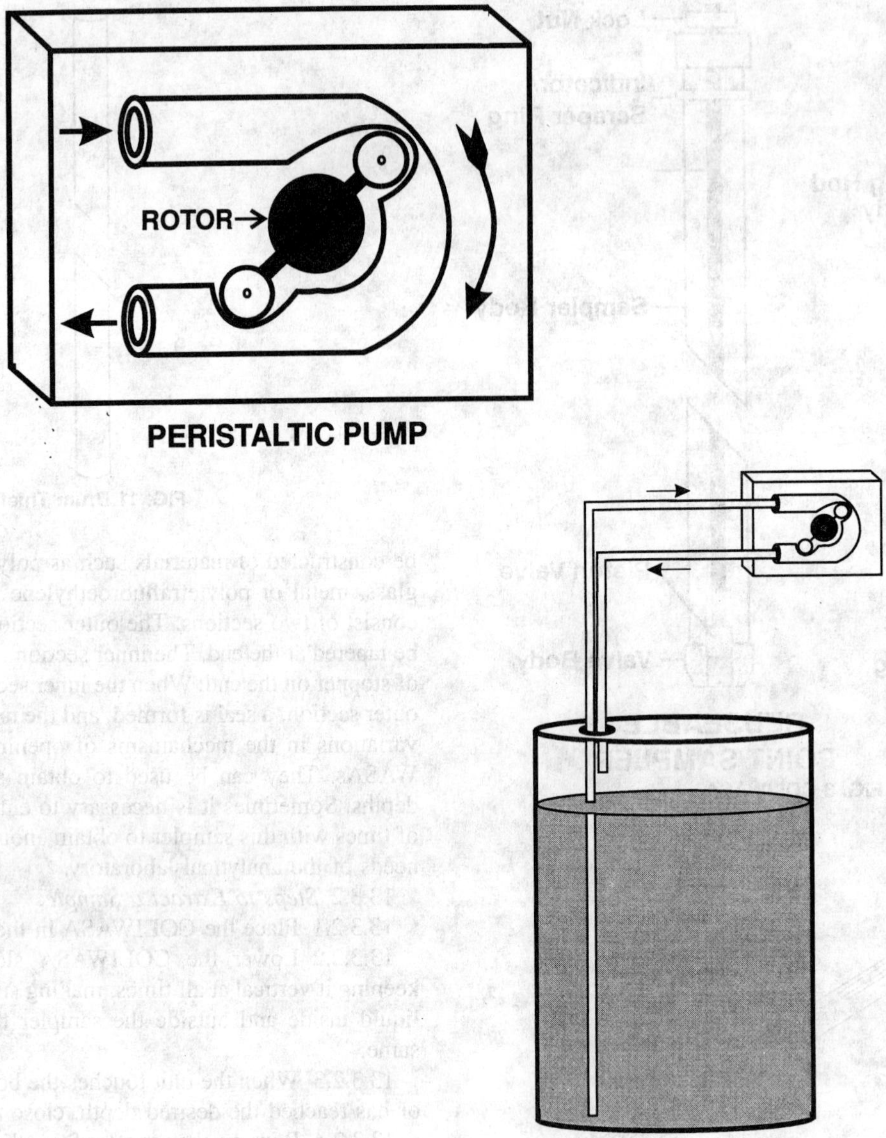

FIG. 12 Peristaltic Pumps

13.4.2.6 Lay the sampler on a clean surface with the slots facing upwards. This is usually done on a clean plastic sheet. Note: some types of grain samplers allow the two tubes to separate from each other, allowing for easier removal of the sample.

13.4.2.7 Remove the sample and place in appropriate container for analysis.

13.5 *Dipper*:

13.5.1 *Description*—A dipper samples single phased liquids (see Fig. 10). It consists of a glass, metal or plastic beaker clamped to the end of a two- or three-piece telescoping aluminum or fiberglass pole that serves as a handle. Samples are taken at, or just below, the surface.

13.5.2 *Steps to Extract a Sample*:

13.5.2.1 Submerge the dipper into the material slowly, to cause minimum surface disturbance.

13.5.2.2 Allow the beaker to fill and slowly bring it to the surface.

13.5.2.3 Slowly pour the contents into the sample container.

13.6 *Drum Thief*:

13.6.1 *Description*—A drum thief is used to sample liquids (see Fig. 11). It consists of an open-ended tube, usually made of glass or stainless steel. Narrow bore tubes can be used for liquids with low surface tension, such as chlorinated solvents, and wider bore tubes can be used to sample sludges.

13.6.2 *Steps to Extract a Sample*:

13.6.2.1 Slowly lower the thief into the liquid, keeping it vertical at all times, until it hits touches the bottom of the container.

13.6.2.2 Place your thumb, or a stopper, over the top to create a vacuum. This will hold the sample in the tube while it is removed from the container. Use caution as this form of vacuum does not always hold the sample in place. If the material releases prematurely from the bottom of the thief, this equipment is unsuitable.

13.6.2.3 Place the thief over the appropriate container for analysis and release the vacuum by removing your thumb or the stopper.

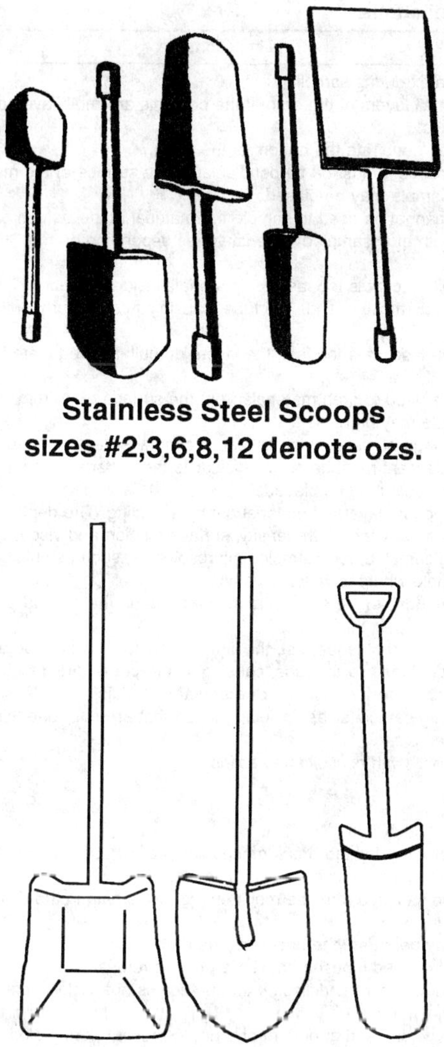

FIG. 13 Scoop, Spoon, Trowel

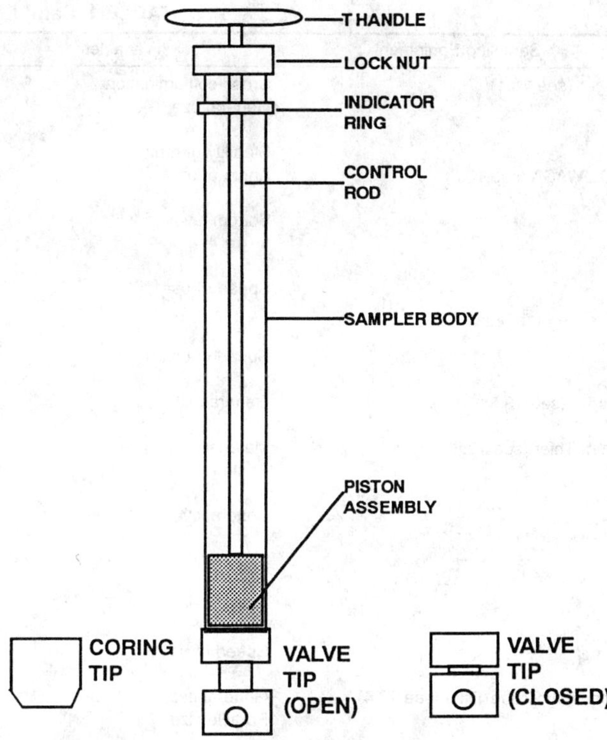

FIG. 14 Syringe Sampler

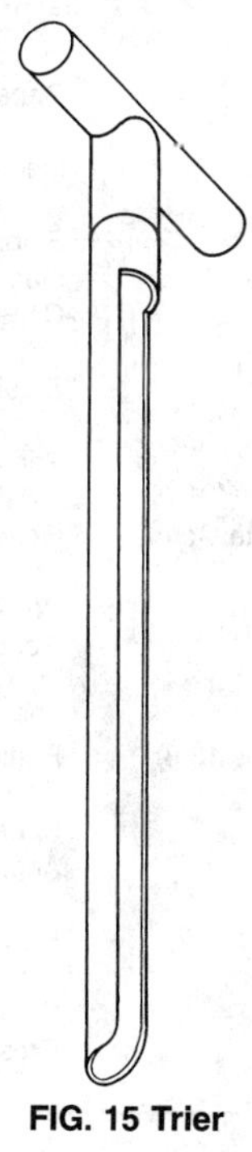

FIG. 15 Trier

13.7 *Peristaltic Pumps*:

13.7.1 *Description*—Peristaltic pumps (see Fig. 12), in which a cam rotates against flexible tubing, thereby moving the material inside the tube through it, are used for sampling liquids and slurries (see Guide D 4448). Hosing should be compatible with the material being sampled. General-use plastic tubing works well for low hazard situations where the sample is water-based, contains few or no organic constituents, and will not be analyzed for organics. High hazard materials and situations where organic compounds are present, and which may be analyzed, should be sampled using fluorocarbon resin tubing.

13.7.2 *Steps to Extract a Sample*:

13.7.2.1 Place the end of the inlet tubing within the container at the depth from which the sample will be extracted.

13.7.2.2 Place the end of the outlet tubing within the container and turn on the pump.

13.7.2.3 Once the sample is adequately mixed, turn off the pump and place the end of the outlet tubing into the appropriate container for analysis.

13.7.2.4 Turn on the pump and collect the sample into the container.

13.8 *Scoop, Spoon, Trowel*:

13.8.1 *Description*—Scoops, spoons and trowels (see Fig. 13) are used to sample a wide variety of materials, including sludge, soil, powder or hard solid waste.

13.8.2 *Steps to Extract a Sample*:

13.8.2.1 Collect sample from the appropriate depth by digging and rotating the sampler. Trowels may not be suitable

TABLE 1 Sampling Equipment Limitations

Sampling Equipment	Variables	Limitations
Auger (see 13.1)	Cross-contamination	Will disturb matrix during sampling.
	General use	May not reach all layers of the drum if the contents are multi-layered, and so may bias results.
	Stability of drum	Equipment may penetrate the bottom of the drum.
COLIWASA (see 13.3)	Angle of use	The angle of descent must be perpendicular to the surface of the material, or the resulting sample may be biased.
	Bottom layer	This sampler cannot be used to sample the material in the bottom of the drum. The actual depth of unsampled material varies depending on the COLIWASA used.
	Speed of use	If the speed of descent is too fast, the materials inside the tube will not be at the same level as those outside the tube, causing incorrect proportions in the sample.
	Speed of use	If the speed of descent is too fast, the layers of multi-layered materials will be disturbed.
Dipper (see 13.5)	General use	This sampler is used to grab materials from the surface of a drum, so the resulting sample may be biased.
Drum Thief (see 13.6)	Material of sampler	If sampler is made of glass, chips or cracks can cause an imperfect seal.
	Angle of use	The angle of descent must be perpendicular to the surface of the material, or the resulting sample may be biased.
	Bottom layer	May lose the bottom layers of material during sampling. The depth of the unsampled material varies with density, surface tension and viscosity of the material being sampled, for example, chlorosolvents, such as chloroform, are difficult to sample whereas water is easy.
	Consistency	With viscous materials, more material may end up on the outside of the tube than inside it.
	Speed of use	If the speed of descent is too fast, the liquid in the tube will not be at the same level as the liquid outside the tube, causing incorrect proportions in the sample.
Concentric Tube Thief (see 13.4)	General use	Can be used only for dry, powdery or granular materials.
	Particle size	Excludes certain particle sizes, including those that are over one third the slot width of the sampler.
Impact Devices Chipper Hammer Chisel	General use	Can be used only with consolidated solids.
	General use	These samplers are used to break off chunks of a consolidated solid from the surface only and so may bias resulting sample.
Push Coring Devices Split Barrel (see 13.2)	Media	Materials to be sampled must be moist enough to remain in the device.
	Stability of drum	Equipment may penetrate the bottom of the drum.
Push Coring Devices Thin-walled	Media	Tube will be damaged if pushed past the point of refusal.
	Consistency	Retrieving an intact core is difficult if the material is not sticky. Cannot be used in coarse, rocky materials.
	Stability of drum	Care required when used at bottom of drum to prevent penetration, particularly when auger tip is used.
Pump, Peristaltic (see 13.7)	Particle size	It is critical to realise that this sampler pumps suspendable solids, not heavy sludge. An example would be rainwater with suspended solids.
Rotating Core Device Concrete Corer (see 13.2)	General use	Used to sample consolidated solids only. May not get all the layers of a multi-layered drum, so may bias the sample.
	Power source	This device must be attached to a power drill.
Scissors and Tongs	General use	Used to cut pieces of unconsolidated solids within a drum, for example, to cut dark spots off gloves or rags to look for "worst case" areas. May not provide sample representative of all "parts" of the drum.
Scoop, Spoon and Trowel (see 13.8)	Particle size	May exclude certain particle sizes, especially large aggregates.
	General use	Used to grab material from the top of a drum and so may bias results.
Syringe Sampler (see 13.9)	Speed of use	The sampler must be lowered slowly into the drum to minimize mixing.
	Angle of use	The angle of descent should be perpendicular to the surface of the material to be sampled.
	Media	Material to be sampled must be viscous enough to remain in the device when the coring tip is used. The valve tip cannot be used with viscous materials.
Trier (see 13.10)	Consistency	Material will not be held in place during removal if it is not sticky.
	Cross-contamination	There may be some contamination of the material within the bounds of the trier by the material outside the bounds.

for collecting solids from depths, or if digging is limited by the sides of the container.

13.8.2.2 Transfer the material to the appropriate container for analysis.

13.9 *Syringe Sampler*:

13.9.1 *Description*—A syringe sampler (see Fig. 14) is used to sample thick liquids, sludges and tar-like substances. It can also draw samples when only a small amount remains at the bottom of a drum or tank. Syringe samplers are available commercially. They usually include a manually operated piston assembly consisting of a T-handle, lock nut, control rod (polytetrafluoroethylene-covered aluminum rod facilitates operation of the piston), piston assembly, sampling tube and two tips for the lower end, one with a closeable valve and one with a coring tip.

13.9.2 *Steps to Extract a Sample*:

TABLE 2 Equipment Selection

Equipment or Device	ASTM Standard	One Liquid Layer	Two or More Layers of Liquids	Both Liquid and Solid Layers	Consolidated Solids^A	Unconsolidated Solids^A	Sludge^A	See Paragraph
Column Number		I	II	III	IV	V	VI	
Auger	D 4700 D 1452	^{B}	—	—	X^C	X	—	13.1
Chipper, Hammer and Chisel		—	—	—	X	N^D	—	—
COLIWASA	D 5495	X	X	N	—	—	N	13.3
Concentric Tube Thief		—	—	—	—	X	—	13.4
Dipper	D 5358	X	—	N	—	—	N	13.5
Drum Thief		X	X	X	—	—	X	13.6
Peristaltic Pump	D 4448	X	X	N	—	—	N	13.7
Push Coring Devices:	D 1586 D 1587							
Split Barrel	D 4700	—	—	—	N	X	N	13.2
Thin-walled Tube	D 4823	—	—	—	N	X	—	
Butterfly Valve		—	—	N	—	N	X	
Rotating Coring Device	C 783 D 2113	—	—	—	X	—	—	13.2
Scissors and Tongs		—	—	—	—	X	—	—
Scoop, Spoon and Trowel		N	—	—	—	X	N	13.8
Syringe Sampler		N	N	X	—	—	X	13.9
Trier	D 5451	—	—	—	—	N	N	13.10

^{A} The suitability of the sampling device depends on the consistency of the material to be sampled.
^{B} Equipment is probably unsuitable.
^{C} Equipment may usually be used with this type of waste.
^{D} Not equipment of choice but may be used at discretion of supervisor.

13.9.2.1 Make sure that the top and bottom fittings are secured to the sampling tube and the valve tip, if used, is open. The piston assembly should be at the lower end of the sampler.

13.9.2.2 Slowly lower the syringe sampler into the drum or tank until it contacts the surface of the material to be sampled.

13.9.2.3 To collect a point sample, hold the sampler body, loosen the lock nut and gradually raise the T-handle to draw the sample into the sampler.

13.9.2.4 To collect a core sample with the coring tip, loosen the lock nut, hold the T-handle and slowly push the sampler body down into the material.

13.9.2.5 When the desired sample is obtained, hand tighten the lock nut to secure the piston rod.

13.9.2.6 Remove the sampler from the drum with one hand while wiping the sampler body with a disposable cloth or rag with the other hand.

13.9.2.7 Hold the syringe sampler over the sample container; open the valve tip, if fitted and closed.

13.9.2.8 Loosen the lock nut and force the sample out of the sampler by depressing the piston rod.

13.10 *Trier*:

13.10.1 *Description*—A trier is used to sample moist or sticky solids with a particle diameter less than one-half the diameter of the trier. It consists of a handle and a tube cut in half lengthwise, with a sharpened tip that allows the sampler to cut into sticky materials and loosen solids.

13.10.2 *Steps to Extract a Sample*:

13.10.2.1 Hold the trier either horizontally or with the handle end tilted slightly downwards.

13.10.2.2 Insert the trier into the material to be sampled, at this angle.

13.10.2.3 Cut a core of the material by rotating the trier once or twice.

13.10.2.4 Stop the rotation with the open face pointing upwards.

13.10.2.5 Slowly remove the trier and empty the contents into the appropriate container for analysis.

14. Keywords

14.1 drum; sampling; waste

Designation: D 6232 – 00

Standard Guide for Selection of Sampling Equipment for Waste and Contaminated Media Data Collection Activities[1]

This standard is issued under the fixed designation D 6232; the number immediately following the designation indicates the year of original adoption or, in the case of revision, the year of last revision. A number in parentheses indicates the year of last reapproval. A superscript epsilon (ε) indicates an editorial change since the last revision or reapproval.

1. Scope

1.1 This guide covers criteria that should be considered when selecting sampling equipment for collecting environmental and waste samples for waste management activities. This guide includes a list of equipment that is used and is readily available. Many specialized sampling devices are not specifically included in this guide. However, the factors that should be weighed when choosing any piece of equipment are covered and remain the same for the selection of any piece of equipment. Sampling equipment described in this guide includes automatic samplers, pumps, bailers, tubes, scoops, spoons, shovels, dredges, coring and augering devices. The selection of sampling locations is outside the scope of this guide.

1.1.1 Table 1 lists selected equipment and its applicability to sampling matrices, including water (surface and ground), sediments, soils, liquids, multi-layered liquids, mixed solid-liquid phases, and consolidated and unconsolidated solids. The guide does not address specifically the collection of samples of any suspended materials from flowing rivers or streams. Refer to Guide D 4411 for more information.

1.2 Table 2 presents the same list of equipment and its applicability for use based on compatibility of sample and equipment; volume of the sample required; physical requirements such as power, size, and weight; ease of operation and decontamination; and whether it is reusable or disposable.

1.3 Table 3 provides the basis for selection of suitable equipment by the use of an Index.

1.4 Lists of advantages and disadvantages of selected sampling devices and line drawings and narratives describing the operation of sampling devices are also provided.

1.5 The values stated in both inch-pound and SI units are to be regarded separately as the standard. The values given in parentheses are for information only.

1.6 This guide offers an organized collection of information or a series of options and does not recommend a specific course of action. This guide cannot replace education or experience and should be used in conjunction with professional judgement. Not all aspects of this guide may be applicable in all circumstances. This guide is not intended to represent or replace the standard of care by which the adequacy of a given professional service must be judged, nor should this document be applied without consideration of a project's many unique aspects. The word "Standard" in the title of this guide means only that it has been approved through the ASTM consensus process.

1.7 *This standard does not purport to address all of the safety concerns, if any, associated with its use. It is the responsibility of the user of this standard to establish appropriate safety and health practices and determine the applicability of regulatory limitations prior to use.*

2. Referenced Documents

2.1 *ASTM Standards:*

D 1452 Practice for Soil Investigation and Sampling by Auger Borings[2]

D 1586 Test Method for Penetration Test and Split-Barrel Sampling of Soils[2]

D 1587 Practice for Thin-Walled Tube Geotechnical Sampling of Soils[2]

D 3550 Practice for Ring-Lined Barrel Sampling of Soils[2]

D 4136 Practice for Sampling Phytoplankton with Water-Sampling Bottles[3]

D 4342 Practice for Collecting of Benthic Macroinvertebrates with Ponar Grab Sampler[3]

D 4343 Practice for Collecting Benthic Macroinvertebrates with Ekman Grab Sampler[3]

D 4348 Practice for Collecting Benthic Macroinvertebrates with Holme (Scoop) Grab Sampler[3]

D 4387 Guide for Selecting Grab Sampling Devices for Collecting Benthic Macroinvertibrates[3]

D 4411 Guide for Sampling Fluvial Sediment in Motion[4]

D 4448 Guide for Sampling Groundwater Monitoring Wells[5]

D 4547 Practice for Sampling Waste and Soils for Volatile Organics[5]

D 4687 Guide for General Planning of Waste Sampling[5]

---

[1] This guide is under the jurisdiction of ASTM Committee D34 on Waste Management and is the direct responsibility of Subcommittee D34.01.01 on Planning for Sampling.

Current edition approved June 10, 2000. Published August 2000. Originally published as D 6232 – 98. Last previous edition D 6232 – 98.

[2] *Annual Book of ASTM Standards*, Vol 04.08.

[3] *Annual Book of ASTM Standards*, Vol 11.05.

[4] *Annual Book of ASTM Standards*, Vol 11.02.

[5] *Annual Book of ASTM Standards*, Vol 11.04.

TABLE 1 Equipment Selection—Matrix Guide

Equipment (May be used for discrete sample collection	Water and Waste Water: Surface Water	Water and Waste Water: Ground Water	Water and Waste Water: Point Discharge	Sediment	Soil	Waste: Liquid	Waste: Multi-Layer Liquid	Waste: Mixed Phase Solid/Liquid	Waste: Consolidated Solid	Waste: Unconsolidated Solid
Pumps and Siphons										
Automatic Sampler—Non volatiles	*		*	-	-	N	N	N	-	-
Automatic Composite Sampler—Volatiles	*		*	-	-	-	-	-	-	-
Air/Gas Displacement Pump		*D 4448^{G}	*	-	-		*	-	-	-
Piston Displacement Pump		*D 4448^{G}	*	-	-		N	-	-	-
Bladder Pumps		*D 4448^{G}	*	-	-	N	N	-	-	-
Peristaltic Pump	*	*D 4448^{G}	*	-	-	*	*	N		
Centrifugal Submersible Pump		*	*	-	-			-	-	-
Dredges										
Ekman Dredge	-	-	-	*D 4387^{G} D 4343^{P}	-	-	-	-	-	-
Petersen Dredge	-	-	-	*D 4387^{G}	-	-	-	-	-	-
Ponar Dredge	-	-	-	*D 4387^{G}	-	-	-	-	-	-
Discrete Depth Samplers										
Bacon Bomb	*	-	-	-	-	*	N	-	-	-
Kemmerer Sampler	*D 4136^{P}	-	-	-	-	*	N	-	-	-
Syringe Sampler	*D 5743^{G}	-	N	-	-	*	*	*	-	-
Peristaltic Pump	*	*D 4448^{G}	*	-	-	*	*	N		
Lidded Sludge/Water Sampler	-	-	-	-	-	N	N	*	-	N
Discrete Level Sampler	*	*	*	-	-	*	*	-	-	-
Push Coring Devices										
Temporary G.W. Sampler	-	*	-	-	-	N	-	-	-	-
Penetrating Probe Sampler	-	-	-	N	*	-	-	N	-	*
Split Barrel Sampler	-	-	-	*	*D 1586^{TM} *D 4700^{G}	-	-	N	-	N
Concentric Tube Thief	-	-	-	-	-	-	-	-	-	*
Trier	-	-	-	-	*	-	-	N	-	*D 5451^{P} *E 300^{P}
Thin Walled Tube	-	-	-	*D 4823^{G}	*D 1587^{P} D 4700^{G}	-	-	-	-	*
Coring Type w/Valve	-	-	-	N	*D 4823^{G}	-	-	*	-	*
Miniature Core Sampler	-	-	-	N	*D 4547^{G} D 6418^{P}	-	-	-	-	N
Modified Syringe Sampler	-	-	-	N	*D 4547^{G}	-	-	-	-	N
Sort Sediment Sampler	-	-	-	*	N	-	-	N		N
Rotating Coring Devices										
Bucket Auger	-	-	-	N	*D 1452^{P} D 4700^{G}	-	-	-	-	*
Screw Auger	-	-	-	-	-	-	-	-	*	-
Rotating Coring Device	-	-	-	*D 4823^{G}	*D 4700^{G}	-	-	-	*	-
Liquid Profile Devices										
COLIWASA	-	-	-	-	-	*D 5495^{P} D 5743^{G}	*D 5495^{P} D 5743^{G}	-	-	
Reuseable Point Sampler	N	-	N	-	-	*	*	*	-	-
Drum Thief	-	-	-	-	-	*	*	*	-	-
Valved Drum Sampler	-	-	-	-	-	*	*	*	-	-
Plunger Type Sampler	N	-	N	-	-	*D 5743^{G}	*D 5743^{G}	*D 5743^{G}	-	-
Liquids Profiler	N	-	N	-	-	*	*	*	-	-
Surface Sampling Devices										
Bailer	N	*D 4448^{G}	-	-	-	N	N	-	-	-
Point Sampling Bailer	N	*D 4448^{G}	-	-	-	N	N	-	-	-
Differential Pressure Bailer	-	*	-	-	-	N	N	-	-	-
Dipper	*D 5358^{P}	-	*D 5013^{P}	-	-	*D 5358^{P}	-	*D 5358^{P}	-	-
Liquid Grab Sampler	*	-	N	-	-	*	*	*	-	-
Swing Jar Sampler	*	-	N	N	-	*	*	N	-	-
Impact Devices	-	-	-	-	-	-	-	-	*	-
Spoon	N	-	N	-	*D 4700^{G}	N	N	-	-	N
Scoops and Trowel	-	-	-	N	*D 4700^{G}	N	-	N	-	*
Shovels	-	-	-	N	*D 4700^{G}	-	-	N	-	*

* Equipment may be used with this matrix N=Not equipment of choice but use is possible -Not recommended
^{G}=ASTM Guide ^{TM}=ASTM Test Method ^{P}= ASTM Practice

TABLE 2 Sampling Equipment Selection Guide

Equipment	Chemical	Physical	Effect on Sample	Volume Range	Physical	Ease of Operation	Decon	Disposal or Reuse
Pumps and Siphon								
Automatic Sampler—Nonvolatiles	•	•	√	U	B/P	•	•	R
Automatic Composite Sampler—Volatiles	•	•	√	U	B/P	•	•	R
Air/Gas Displacement Pump	•	•	•	U	P/S/W	•	•	R
Piston Displacement Pump	•	•	•	U	P/S/W	•	•	R
Bladder Pumps	√	•	√	U	P/S/W	•	•	R
Peristaltic Pump	•	•	√	U	B/P	•	√	R
Centrifugal Submersible Pump	•	•	•	U	P/S/W	√	•	R
Dredges								
Ekman Dredge	√	√	•	0.5-3.0	N	•	•	R
Petersen Dredge	√	√	•	0.5-3.0	W	•	•	R
Ponar Dredge	√	√	•	0.5-3.0	W	•	•	R
Discrete Depth Samplers								
Bacon Bomb	•	√	√	0.1-0.5	N	√	•	R
Kemmerer Sampler	•	√	√	1.0-2.0	N	√	•	R
Syringe Sampler	√	√	√	0.2-0.5	N	√	√	R
Lidded Sludge/Water Sampler	√	•	•	1.0	S/W	•	•	R
Discrete Level Sampler	√	•	√	0.2-0.5	N	√	•	R
Push Coring Devices								
Temporary G.W. Sampler	√	√	√	0.1-0.3	P/S/W	•	•	R
Penetrating Probe Sampler	√	√	√	0.2-2.0	S/W	•	√	R
Split Barrel Sampler	√	√	•	0.5-30.0	S/W	√	√	R
Concentric Tube Theif	√	√	√	0.5-1.0	N	√	√	R
Trier	√	√	√	0.1-0.5	N	√	√	R
Thin Walled Tube	√	√	•	0.5-5.0	S/W	√	√	R
Coring Type w/Valve	√	√	√	0.2-1.5	N	√	√	R
Miniature Core Sampler	√	√	√	0.01-0.05	N	√	√	D
Modified Syringe Sampler	√	√	√	0.01-0.05	N	√	√	D
Soft Sediment Sampler	√	√	√	1.6-7.0	N	√	√	R
Rotating Coring Devices								
Bucket Auger	√	√	•	0.2-1.0	N	•	√	R
Screw Auger	√	√	•	0.1-0.3	N	•	√	R
Rotating Coring Device	√	√	•	0.5-1.0	B/P	√	√	R
Liquid Profile Devices								
COLIWASA	√	•	√	0.5-3.0	N	√	•	D/R
Reuseable Point Sampler	√	√	√	0.2-0.6	N	√	√	R
Drun Thief	√	•	√	0.1-0.5	N	√	•	D/R
Valved Drum Sampler	√	√	√	0.3-1.6	N	√	√	D/R
Plunger Type Sampler	√	•	√	0.2-U	N	√	√	D/R
Liquids Profiler	•	√	√	1.3-4.0	N	√	√	R
Surface Sampling Devices								
Bailer	•	√	•	0.5-2.0	N	√	√	D/R
Point Sampling Bailer	•	√	√	0.5-2.0	N	√	√	R
Differential Pressure Bailer	√	√	√	0.04-1.0	N	√	√	R
Dipper	√	√	√	0.5-1.0	N	√	√	R
Liquid Grab Sampler	√	√	√	0.5-1.0	N	√	√	R
Swing Jar Sampler	•	√	√	0.5-1.0	N	√	√	R
Impact Devices	•	•	•	N/A	B/P	√	√	R
Spoon	√	√	•	N/A	N	√	√	R
Scoops and Trowel	√	√	•	0.1-0.6	N	√	√	R
Shovels	√	√	•	1.0-5.0	N	√	√	R

• = Significant operation consideration
√ = Not a significant operational consideration

Range of Volume (liters)
U = Unlimited
N/A = Not Applicable

Physical Requirements:
B = Battery W = Weight
P = Power S = Size
N = No limitations

Disposal and Reuse:
R = Reusable
D = Single-Use

D 4696 Guide for Pore-Liquid Sampling in the Vadose Zone[2]

D 4700 Guide for Soil Sampling from the Vadose Zone[2]

D 4823 Guide for Core Sampling Submerged, Unconsolidated Sediments[3]

D 5013 Practices for Sampling Wastes from Pipes and Other Point Discharges[3]

D 5079 Practices for Preserving and Transporting Rock Core Samples[6]

D 5088 Practice for Decontamination of Field Equipment Used at Nonradioactive Waste Sites[6]

D 5283 Practice for Generation of Environmental Data Related to Waste Management Activities: Quality Assurance and Quality Control Planning and Implementation[5]

D 5314 Guide for Soil Gas Monitoring in the Vadose Zone[6]

D 5358 Practice for Sampling with a Dipper or Pond Sampler[5]

D 5451 Practice for Sampling Using a Trier Sampler[5]

D 5495 Practice for Sampling with a Composite Liquid Waste Sampler COLIWASA[5]

[6] *Annual Book of ASTM Standards*, Vol 04.09.

D 5633 Practice for Sampling with a Scoop[5]

D 5679 Practice for Sampling Consolidated Solids in Drums or Similar Containers[5]

D 5680 Practice for Sampling Unconsolidated Solids in Drums or Similar Containers[5]

D 5730 Guide for Site Characterization for Environmental Purposes with Emphasis on Soil, Rock, the Vadose Zone and Ground Water[6]

D 5743 Practice for Sampling Single or Multilayered Liquids, With or Without Solids, in Drums or Similar Containers[5]

D 5778 Test Method for Performing Electronic Friction Cone and Piezocone Penetration Testing of Soils[6]

D 5781 Guide for Use of Dual-Wall Reverse-Circulation Drilling for Geoenvironmental Exploration and the Installation of Subsurface Water-Quality Monitoring Devices[6]

D 5782 Guide for Use of Direct Air-Rotary Drilling for Geoenvironmental Exploration and the Use of Subsurface Water-Quality Monitoring Devices[6]

D 5783 Guide for Use of Direct Rotary Drilling with Water-Based Drilling Fluid for Geoenvironmental Exploration and the Installation of Subsurface Water-Quality Monitoring Devices[6]

D 5784 Guide for Use of Hollow-Stem Augers for Geoenvironmental Exploration and the Installation of Subsurface Water-Quality Monitoring Devices[6]

D 5875 Guide for Use of Cable-Tool Drilling and Sampling Methods for Geoenvironmental Exploration and Installation of Subsurface Water-Quality Monitoring Devices[6]

D 5876 Guide for Use of Direct Rotary Wireline Casing Advancement Drilling Methods for Geoenvironmental Exploration and Installation of Subsurface Water-Quality Monitoring Devices[6]

D 6001 Guide for Direct-Push Water Sampling for Geoenvironmental Investigations[6]

D 6044 Guide for Representative Sampling and Management of Waste and Contaminated Media[5]

D 6063 Guide for of Sampling Drums and Similar Containers by Field Personnel[5]

D 6067 Guide for Using the Electronic Cone Penetrometer for Environmental Site Characterization[6]

D 6151 Practice for Using Hollow-Stem Augers for Geotechnical Exploration and Soil Sampling[6]

D 6169 Guide for Selection of Soil and Rock Sampling Devices Used with Drill Rigs for Environmental Investigations[6]

D 6282 Guide for Direct Push Soil Sampling for Environmental Site Characterizations[6]

D 6286 Guide for Selection of Drilling Methods of Environmental Site Characterization[6]

D 6418 Practice for Using the Disposable En Core Sampler for Sampling and Storing Soil for Volatile Organic Analysis[5]

E 300 Practice for Sampling Industrial Chemicals[7]

E 1391 Guide for Collection, Storage, Characterization, and Manipulation of Sediments for Toxicological Testing[3]

3. Terminology

3.1 *Definitions of Terms Specific to This Standard:*

3.1.1 *consolidated, adj*—a compact solid not easily compressed or broken into smaller particles.

3.1.2 *decontamination, n*—the process of removing or reducing to a known level undesirable physical or chemical constituents, or both, from a sampling apparatus to maximize the representativeness of physical or chemical analyses proposed for a given sample.

3.1.3 *data quality objectives (DQOs), n*—qualitative or quantitative statement(s) derived from the DQO process describing the problem(s), the decision rule(s) and the uncertainties of the decision(s) stated in the context of the problem.

3.1.4 *environmental data, n*—defined for use in this document to mean data in support of environmental activities.

3.1.5 *matrix, n*—the principal constituent(s) of a material.

3.1.6 *unconsolidated, adj*—defined for use in this guide to mean uncemented or uncompacted material that is easily separated into smaller portions.

3.1.7 *representative sample, n*—a sample collected in such a manner that it reflects one or more characteristics of interest (as defined by the project objectives) of a population from which it was collected. **(D 6044)**

4. Summary of Guide

4.1 This guide discusses important criteria which should be considered when choosing sampling equipment.

4.1.1 Criteria discussed in this guide include physical and chemical compatibility, sample matrix, sample volume, physical requirements, ease of operation and decontamination. Costs are considered, where appropriate.

4.2 A limited list of sampling equipment is presented in two separate tables. The list attempts to include a variety of different types of equipment. However, this list is in no way all inclusive, as there are many excellent pieces of equipment not included. Table 1 lists matrices (surface and ground water, stationary sediment, soil and mixed phase wastes) and indicates which sampling devices are appropriate for use with these matrices. It also includes ASTM method references (draft standards are not included). Table 2 indicates physical requirements (such as battery), electrical power, and weight; physical and chemical compatibility; effect on matrix; range of volume; ease of operation; decontamination; and reusability. Table 3 provides sampler type selection process based upon the sample type and matrix to be sampled.

---

[7] *Annual Book of ASTM Standards*, Vol 15.05.

TABLE 3 Index of Sampling Equipment

Media Type	Sampler Type	Section	Sample Type
Consolidated Solid	Rotating Corer	(7.6.3)	Surface or Depth, Undisturbed
	Screw Auger	(7.6.2)	Surface, Disturbed
Unconsolidated Solid	Lidded Sludge	(7.4.4)	Discrete, Composite
	Penetrating Probe	(7.5.2)	Discrete, Undisturbed
	Split Barrel	(7.5.3)	Discrete, Undisturbed
	Concentric Tube Thief	(7.5.4.1)	Surface, Disturbed, Selective
	Trier	(7.5.4.2)	Surface, Relatively Undisturbed, Selective
	Thin Walled Tube	(7.5.5)	Surface or Depth, Undisturbed
	Coring Type w/Valve	(7.5.6)	Surface or Depth, Disturbed
	Bucket Auger	(7.6.1.1)	Surface or Depth, Disturbed
	Spoon	(7.9.2)	Surface, Disturbed, Selective
	Scoops/Trowel	(7.9.3)	Surface, Disturbed, Selective
	Shovel	(7.9.4)	Surface, Disturbed
	Miniature Core	(7.5.7)	Surface, Undisturbed
	Modified Syringe	(7.5.8)	Surface, Undisturbed
Soil	Penetrating Probe	(7.5.2)	Discrete, Undisturbed
	Split Barrel	(7.5.3)	Discrete, Undisturbed
	Trier	(7.5.4.2)	Surface, Relatively Undisturbed, Selective
	Thin Walled Tube	(7.5.5)	Surface or Depth, Undisturbed
	Coring Type w/Valve	(7.5.6)	Surface or Depth, Disturbed
	Bucket Auger	(7.6.1.1)	Surface or Depth, Disturbed
	Rotating Corer	(7.9.1)	Surface or Depth, Undisturbed
	Spoon	(7.9.2)	Surface, Disturbed, Selective
	Scoops/Trowel	(7.9.3)	Surface, Disturbed, Selective
	Shovel	(7.9.4)	Surface, Disturbed
	Miniature Core	(7.5.7)	Surface, Undisturbed
	Modified Syringe	(7.5.8)	Surface, Undisturbed
Mixed Solid/Liquid	AutoSampler, Non V.	(7.2.1)	Shallow, Composite-Suspended Solids only
	Peristaltic Pump	(7.2.4)	Shallow, Discrete or Composite-Suspended Solids Only
	Syringe Sampler	(7.4.3)	Shallow, Discrete, Disturbed
	Lidded Sludge/Water	(7.4.4)	Discrete
	Penetrating Probe	(7.5.2)	Depth, Discrete, Undisturbed
	Split Barrel	(7.5.3)	Depth, Discrete, Undisturbed
	Trier	(7.5.4.2)	Surface, Semi-solid only, Selective
	Coring Type w/Valve	(7.5.6)	Depth, Disturbed
	COLIWASA	(7.7.1)	Shallow, Composite, Semi-liquid only
	Reuseable Point	(7.7.1.2)	Shallow, Discrete
	Plunger Type	(7.7.4)	Shallow, Discrete
	Liquids Profiler	(7.7.5)	Depth, Composite-Suspended Solids only
	Drum Thief	(7.7.2)	Shallow, Composite
	Valved	(7.7.3)	Shallow, Composite
	Dipper	(7.8.4)	Shallow, Composite
	Liquid Grab	(7.8.5)	Shallow, Composite-Suspended Solids only
	Swing Jar	(7.8.6)	Shallow, Composite
	Scoops/Trowel	(7.8.9)	Shallow, Composite, Semi-solid only
	Shovel	(7.8.10)	Shallow, Composite, Semi-solid only
Sediments	Ekman Dredge	(7.3.1)	Bottom Surface, Soft only, Disturbed
	Petersen Dredge	(7.3.2)	Bottom Surface, Rocky or Soft, Disturbed
	Ponar	(7.3.3)	Bottom Surface, Rocky or Soft, Disturbed
	Penetrating Probe	(7.5.2)	Bottom Surface or Depth, Undisturbed
	Split Barrel	(7.5.3)	Bottom Surface or Depth, Undisturbed
	Thin Walled Tube	(7.5.5)	Bottom Surface or Depth, Undisturbed
	Coring Type w/Valve	(7.5.6)	Bottom Surface or Depth, Disturbed
	Bucket Auger	(7.6.1.1)	Bottom Surface, Disturbed
	Soft Sediment	(7.5.9)	Bottom, Depth, Undisturbed
	Rotating Corer	(7.9.1)	Bottom Surface, Undisturbed if solid
	Scoops, Trowel	(7.9.3)	Exposed Surface only, Disturbed, Selective
	Shovel	(7.9.4)	Exposed Surface only, Disturbed

TABLE 3 *Continued*

Media Type	Sampler Type	Section	Sample Type
	AutoSplr. -Non Vols.	(7.2.1)	25-ft Lift, Discrete or Composite
	Auto Splr. - Vols.	(7.2.1)	25-ft Lift, Discrete
	Air/Gas Displacement	(7.2.2.1)	Depth, Discrete
	Piston Displacement	(7.2.2.2)	Depth, Discrete
	Bladder Pump	(7.2.3)	Depth, Discrete
	Peristaltic Pump	(7.2.4)	Shallow(25-ft), Discrete
	Centrifugal Sub. Pump	(7.2.5)	Depth, Discrete
	Bacon Bomb	(7.4.1)	Depth, Discrete
	Kemmerer	(7.4.2)	Depth, Discrete
	Discrete Level	(7.4.5)	Depth, Discrete
	Reuseable Point	(7.7.1.2)	Shallow (8-ft), Discrete
	Plunger Type	(7.7.4)	Shallow (12-ft), Discrete
Surface Water	Liquids Profiler	(7.7.5)	Shallow, Composite
	Bailer	(7.8.1)	Depth, Discrete
	Point Sampling Bailer	(7.8.2)	Depth, Discrete
	Diff. Pressure Bailer	(7.8.3)	Depth, Discrete
	Dipper	(7.8.4)	Shallow (10-ft.), Composite
	Liquid Grab	(7.8.5)	Shallow (6-ft), Composite
	Swing Jar	(7.8.6)	Shallow, (10-ft), Composite
	Spoon	(7.8.8)	Shallow (1-in.), Composite
	AutoSplr. -Non Vols.	(7.2.1)	25-ft Lift, Discrete or Composite
	Auto Splr. - Vols.	(7.2.1)	25-ft Lift, Discrete
	Air/Gas Displacement	(7.2.2.1)	Depth, Discrete
	Piston Displacement	(7.2.2.2)	Depth, Discrete
Ground Water	Bladder Pump	(7.2.3)	Depth, Discrete
	Peristaltic Pump	(7.2.4)	25-ft Lift, Discrete
	Centrifugal Sub. Pump	(7.2.5)	Depth, Discrete
	Discrete Level	(7.4.5)	Depth, Discrete
	Temp. Ground Water	(7.5.1.1)	Depth, Discrete
	Bailer	(7.8.1)	Depth, Composite
	Point Sampling Bailer	(7.8.2)	Depth, Discrete
	Diff. Pressure Bailer	(7.8.3)	Depth, Discrete
	AutoSplr. -Non Vols.	(7.2.1)	Shallow (25-ft), Discrete or Composite
	Auto Splr. - Vols.	(7.2.1)	Shallow (25-ft), Discrete
	Air/Gas Displacement	(7.2.2.1)	Depth, Discrete
	Piston Displacement	(7.2.2.2)	Depth, Discrete
	Bladder Pump	(7.2.3)	Depth, Discrete
Liquid Effluent	Peristaltic Pump	(7.2.4)	Shallow (25-ft), Discrete
	Centrifugal Sub. Pump	(7.2.5)	Depth, Discrete
	Syringe Sampler	(7.4.3)	Shallow (8-ft), Discrete
	Discrete Level	(7.4.5)	Depth, Discrete
	Reuseable Point	(7.7.1.2)	Shallow (8-ft), Discrete
	Plunger Type	(7.7.4)	Shallow (12-ft), Discrete
	Liquids Profile	(7.7.5)	Shallow, Composite
	Dipper	(7.8.4)	Shallow (10-ft), Composite
	Liquid Grab	(7.8.5)	Shallow (6-ft), Composite
	Swing Jar	(7.8.6)	Shallow (10-ft), Composite
	Spoon	(7.8.8)	Shallow (1-in.), Composite

TABLE 3 *Continued*

Media Type	Sampler Type	Section	Sample Type
Liquid	Air/Gas Displacement	(7.2.2.1)	Depth, Discrete
	Piston Displacement	(7.2.2.2)	Depth, Discrete
	Bladder Pump	(7.2.3)	Depth, Discrete
	Peristaltic Pump	(7.2.4)	Shallow (25-ft), Discrete
	Centrifugal Sub. Pump	(7.2.5)	Depth, Discrete
	Syringe Sampler	(7.4.3)	Shallow (8-ft), Discrete
	Lidded Sludge/Water	(7.4.4)	Shallow (8-ft), Discrete
	Discrete Level	(7.4.5)	Depth, Discrete
	Temp. Ground Water	(7.5.1.1)	Depth, Discrete
	COLIWASA	(7.7.1)	Shallow (4-ft), Composite
	Reuseable Point	(7.7.1.2)	Shallow (8-ft), Discrete
	Plunger Type	(7.7.4)	Shallow, (12-ft), Discrete
	Liquids Profile	(7.8.5)	Shallow, Composite
	Drum Thief	(7.7.2)	Shallow (3-ft), Composite
	Valved Sampler	(7.7.3)	Shallow (8-ft), Composite
	Bailer	(7.8.1)	Depth, Discrete
	Point Sampling Bailer	(7.8.2)	Depth, Discrete
	Diff. Pressure Bailer	(7.8.3)	Depth, Discrete
	Dipper	(7.8.4)	Shallow (10-ft), Composite
	Liquid Grab	(7.8.5)	Shallow (6-ft), Composite
	Swing Jar	(7.8.6)	Shallow, (10-ft), Composite
	Spoon	(7.8.8)	Shallow (1-in.), Composite
	Scoops & Trowel	(7.8.9)	Shallow, (1-in.), Composite
Multi Layer Liquid	Air/Gas Displacement	(7.2.2.1)	Depth, Discrete
	Piston Displacement	(7.2.2.2)	Depth Discrete
	Bladder Pump	(7.2.3)	Depth, Discrete
	Peristaltic Pump	(7.2.4)	Shallow(25-ft), Discrete
	Centrifugal Sub. Pump	(7.2.5)	Depth, Discrete
	Syringe Sampler	(7.4.3)	Shallow (8-ft), Discrete
	Lidded Sludge/Water	(7.4.4)	Shallow (8-ft), Discrete
	Discrete Level	(7.4.5)	Depth, Discrete
	Temp. Ground Water	(7.5.1.1)	Depth, Discrete
	COLIWASA	(7.7.1)	Shallow (4-ft), Composite
	Reuseable Point	(7.7.1.2)	Shallow (8-ft), Discrete
	Plunger Type	(7.7.4)	Shallow, (12-ft), Discrete
	Liquids Profile	(7.8.5)	Shallow, Composite
	Drum Thief	(7.7.2)	Shallow (3-ft), Composite
	Valved Sampler	(7.7.3)	Shallow (8-ft), Composite
	Bailer	(7.8.1)	Depth, Discrete
	Point Sampling Bailer	(7.8.2)	Depth, Discrete
	Diff. Pressure Bailer	(7.8.3)	Depth, Discrete
	Dipper	(7.8.4)	Shallow (10-ft), Composite
	Liquid Grab	(7.8.5)	Shallow (6-ft), Composite
	Swing Jar	(7.8.6)	Shallow (10-ft), Composite
	Spoon	(7.8.8)	Shallow (1-in.), Composite

5. Significance and Use

5.1 Although many technical papers address topics important to efficient and accurate sampling investigations (DQO's, study design, QA/QC, data assessment), the selection and use of appropriate sampling equipment is assumed or omitted.

5.2 The choice of sampling equipment can be crucial to the task of collecting a sample appropriate for the intended use.

5.3 When a sample is collected, all sources of potential bias should be considered, not only in the selection and use of the sampling device, but also in the interpretation and use of the data generated. Some major considerations in the selection of sampling equipment for the collection of a sample are listed below:

5.3.1 The ability to access and extract from every relevant location in the target population,

5.3.2 The ability to collect a sufficient mass of sample such that the distribution of particle sizes in the population are represented, and

5.3.3 The ability to collect a sample without the addition or loss of constituents of interest.

5.4 The characteristics discussed in 5.3 are particularly important in investigations when the target population is heterogeneous such as when particle sizes vary, liquids are present in distinct phases, a gaseous phase exists or material from different sources are present in the population. The consideration of these characteristics during the equipment selection process will enable the data user to make appropriate statistical inferences about the target population based on the sampling results.

6. Selection Criteria

6.1 Refer to Table 1 and Table 2 for a summary of matrix compatibility and selection criteria. Refer to Table 3 for an index of sampling equipment based upon sample type and matrix to be sampled.

NOTE 1—Information on sample containers and equipment used in sampling that is not used in the actual collection of the sample is not within the scope of this guide.

6.2 *Compatibility*—It is important that sampling equipment, other equipment which may come in contact with samples (such as gloves, mixing pans, knives, spatulas, spoons, etc.) and sample containers be constructed of materials that are compatible with the matrices and analytes of interest. Incompatibility may result in the contamination of the sample and the degradation of the sampling equipment. Appropriate sampling equipment must be compatible chemically and physically.

6.2.1 *Chemical Compatibility*—The effects of a matrix on the sampling equipment is usually considered in the light of the analytes, or groups of analytes of interest. For example, polyvinyl chloride (PVC) has been found to degrade in the presence of many organic compounds; therefore, it would be preferable to collect ground water samples for organic analyses using polytetrafluoroethylene (PTFE), stainless steel, or glass sampling equipment (**1, 2**).[8] Acids, bases, and high chloride ground water in coastal areas, and wastes with high concentrations of solvents may also degrade many types of sampling equipment over time. The residence or contact time, the time the sample is in contact with the sampling equipment, may be significant in terms of chemical interaction between the sampled matrix and the equipment.

6.2.1.1 The choice of materials used in the construction of sampling devices should be based upon a knowledge of what constituents may be present in the sampling environment because the constituents and materials may interact chemically or be incompatible. Consult available chemical compatibility charts.

6.2.2 *Physical Compatibility*—The sampling equipment should also be compatible with the physical characteristics of the matrices to be sampled. Equipment used to dig or core (shovels, augers, coring type samplers) should be constructed of material that will not deform during use, or be abraded by the material being sampled. Equipment abrasion may result in the contribution of contaminants to the sample being collected. For example, plastic or glass would not be appropriate for difficult to access matrices, and stainless steel equipment may contribute small amounts of metals if significantly abraded by the matrix.

6.3 *Equipment Effects on the Matrix*:

6.3.1 *Equipment Design*— Samples collected using inappropriate sampling equipment may not provide *representative samples* (**1, 3**). An example of equipment design influencing sample results is a sampler which excludes certain sized particles from a soil matrix or waste pile sample. The shape of some scoops may influence the distribution of particle sizes collected from a sample (**1**). Dredges used to collect river or estuarine stationary sediments may also exclude certain sized particles, particularly the fines fraction which may contain a significant percentage of some contaminants such as polynuclear aromatic hydrocarbons (PAHs).

6.3.2 *Equipment Use*— Inappropriate use of sampling equipment can influence analytical results. For example, if a displacement pump (bladder, piston or air/gas displacement) is used to purge a well and the intake is placed below the well screen, sediment in the sump can be put into suspension and become part of the water sample (**4**). Excessive vacuum generated by sampling pumps can cause loss of volatile constituents or change valence states of some ions. The use of bailers for well purging and sample collection also may cause increased turbidity levels in ground water samples. When sampling containerized liquids, insertion of a COLIWASA-sampler at too fast a rate may prevent it from collecting a representative, depth integrated sample.

6.4 *Sample Volume Capabilities*—Most sampling devices will provide adequate sample volume. However, the sampling equipment volumes should be compared to the volume necessary for all required analyses including the additional amount necessary for quality control (QC), split and repeat samples (**4, 5**). Sampling devices that may not provide an adequate volume would be small diameter glass tubes and triers. In this case the investigator must consider the following options:

6.4.1 A similar device with an increased capacity,

6.4.2 An alternate device with an increased capacity, or

6.4.3 Modification of an existing device (often difficult or impractical).

6.4.4 If these alternatives are not acceptable or available, then the investigator must consider the collection of multiple aliquots to fulfill the sample volume requirement. The effect of multiple aliquots on the data quality collection objectives should be considered.

6.5 *Physical Requirements*—Sampling equipment selection should always consider factors such as the size and weight of the equipment, power requirements (battery/110V), and ancillary equipment required (drill rig for split barrel samplers). Most sampling equipment used in the collection of environmental samples is relatively easy to transport and use in the field. The use of equipment with significant physical requirements may impede the progress of a sampling investigation.

6.6 *Ease of Operation*—Much of the equipment used for environmental sampling is rather simple to employ. Samples may be collected easily as long as properly selected equipment is used with adequate consideration of the matrix of interest. Sampling errors may occur as a result of inadequate consideration of matrix effects, and poor collection techniques (**1,3**). Training requirements should focus on the proper use of equipment in varying environmental matrices.

6.7 *Decontamination and Reuse of Equipment*:

6.7.1 *Decontamination (Practice D 5088)*—Inadequate decontamination of sampling equipment can result in significant errors in analytical results. When choosing sampling equipment, ease of decontamination must be a consideration. Pumps, automatic samplers, Kemmerer samplers and dredges require more effort to decontaminate than does a bailer or split barrel sampler. The investigator should consider decontamination requirements prior to the study to avoid significant delays.

6.7.2 *Reuse*—Due to the expense of materials associated with modern sampling equipment (stainless steel, PTFE), most equipment is reusable following proper decontamination. Some equipment such as bailers may be disposed of after use or dedicated to a sampling point to save time during extensive field investigations. Drum thieves and COLIWASA samplers

[8] The boldface numbers given in parentheses refer to a list of references at the end of the text.

are typically not reused, particularly when waste samples have been collected.

6.8 *Cost*—Detailed information on the cost of sampling equipment is not contained within this guide. Cost is usually a major consideration in the process of sampling equipment selection. In general, the cost of PTFE and stainless steel equipment will be greater than equipment made of glass, PVC, or other plastics. However, the life expectancy for PTFE or stainless steel equipment is usually longer. In addition, labor costs for decontamination of reusable equipment versus the disposal costs of single use equipment are considerations. Comments on costs are included in the "Advantages and Limitations" tables, where appropriate.

7. Sampling Equipment

7.1 Presented below are brief descriptions of some sampling equipment used in waste management and in the collection of environmental samples as they relate to waste management activities **(6)**. This is by no means an inclusive list of the sampling equipment that is available to investigators. There are many pieces of equipment that have been designed for specific sampling needs. In addition, investigators may design their own pieces of equipment for a specific project. In all these instances, an investigator must keep in mind the criteria for sampling equipment selection which have been discussed previously in this guide.

7.2 *Pumps and Siphons (Guide D 4448)*— Pumps used for the collection of waste and environmental liquid samples for waste management include automatic samplers and displacement, bladder, peristaltic, and centrifugal pumps.

7.2.1 *Automatic Samplers (*Fig. 1 *and* Fig. 2*)* — Automatic samplers may be used when samples are to be collected at frequent intervals. They frequently are used in waste-water collection systems and treatment plants, but they also can be used during stream sampling investigations. They may be used to collect time composite or flow proportional samples. In the flow proportional sampling mode, the samplers are activated by a compatible flow meter. Peristaltic and vacuum pumps commonly are employed as the sampling mechanism. Automatic samplers designed specifically for the collection of samples for volatile organic analyses are available. See Table 4 for advantages and limitations.

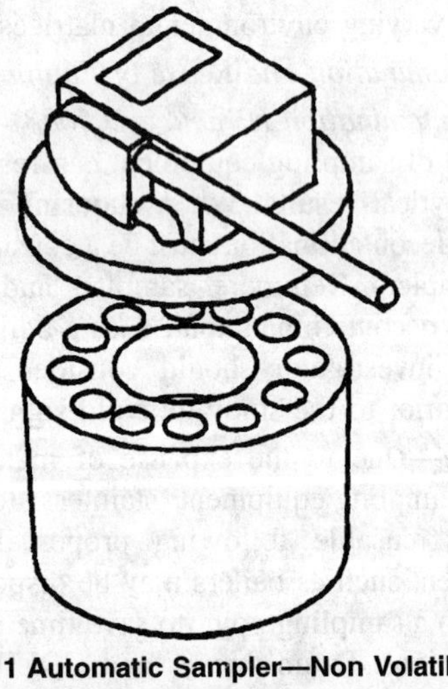

FIG. 1 Automatic Sampler—Non Volatiles

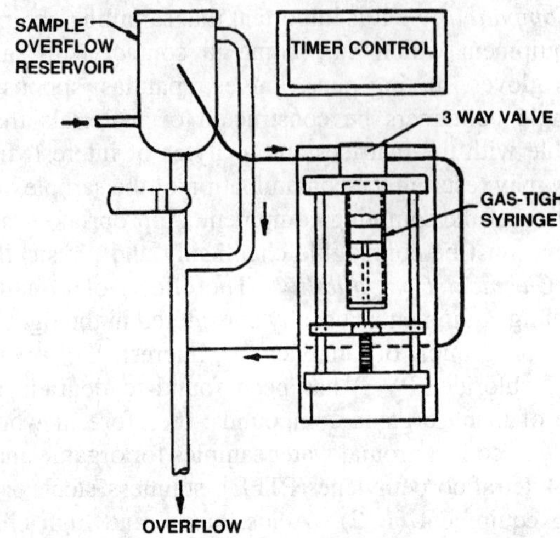

FIG. 2 Automatic Composite Sampler—Volatiles

TABLE 4 Automatic Samplers—Advantages and Limitations

Advantages	Limitations
Can collect either grab samples over time or a composite sample	May be unsuitable for samples requiring volatile organic analysis or samples containing dissolved gases
Will operate unattended	Need power source/battery
Versatile—can be programmed to sample proportional to flow	May be difficult to decontaminate due to design and/or construction materials
	May be incompatible with liquid streams containing a high percentage of solids

NOTE 2—Flow proportional samples also can be collected using a discrete sampler and a flow recorder and manually compositing the individual aliquots in flow proportional amounts.

7.2.2 *Displacement Pumps*—Displacement pumps are designed for ground water sampling and mechanically force a discrete column of water to the surface. The air displacement pump uses compressed air while the piston displacement pump uses an actuating rod powered either from the surface or from a separate sealed air or electric actuator. (See Table 5 for advantages and limitations.)

7.2.2.1 The air displacement pump (Fig. 3) operates by applying a positive pressure to the gas line causing the inlet check valve of the sampling device to close and the sample discharge line check valve to open, forcing the contents to the surface. Cyclical removal of gas pressure will cause the flow to

TABLE 5 Displacement Pumps—Advantages and Limitations

Advantages	Limitations
Commonly constructed on PVC, or stainless steel, or both, but can be constructed of PTFE to reduce risk of contamination when trace levels of organics are of interest	Potential loss of dissolved gases and VOCs from the pumped sample or contamination from the driving gas
	Large gas volume required
Ease to decontaminate (air displacement)	May be difficult to decontaminate (piston displacement)

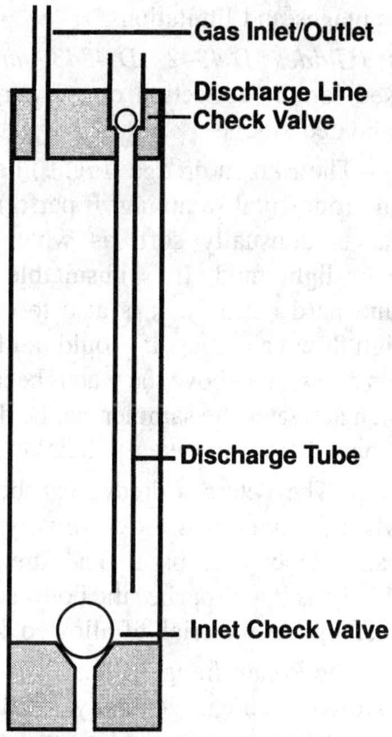

FIG. 3 Air/Gas Displacement Pump

stop, the discharge line check valve to close and the inlet check valve of the sampling device to open, allowing the sampling device to fill.

7.2.2.2 The piston displacement pump (Fig. 4) uses a mechanically operated plunger to deliver the sample to the surface at the same time as the chamber fills. It has a flexible flap valve on the piston and an inlet check valve.

7.2.3 *Bladder Pumps*— Bladder pumps are used for sampling ground water and are constructed with a flexible bladder inside a rigid sample container. There are two types. The squeeze type (Fig. 5) has the bladder connected to the sample discharge line. The chamber between the bladder and the

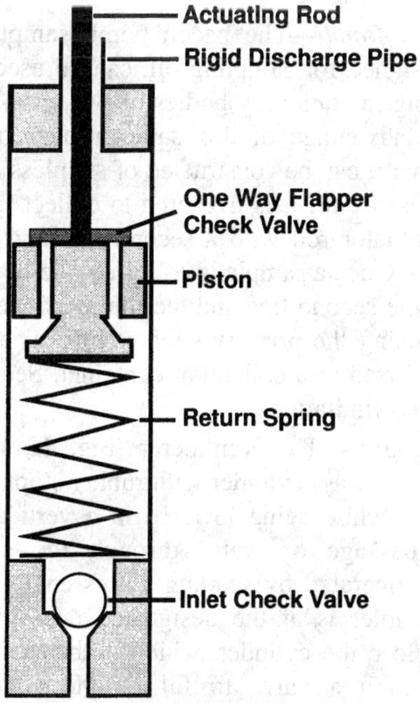

FIG. 4 Piston Displacement Pump

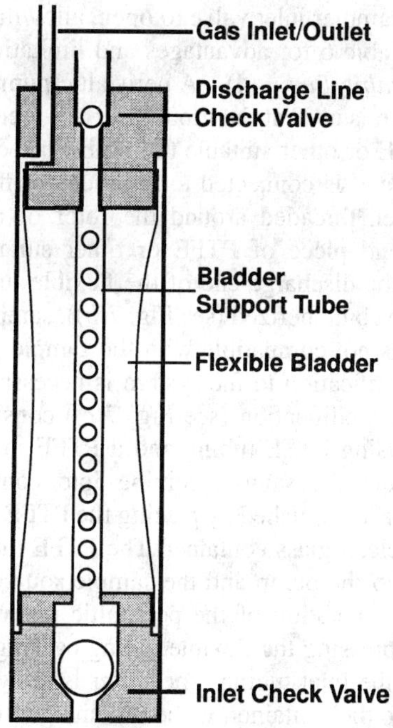

FIG. 5 Bladder Pump—Squeeze Type

sampler body is connected to the gas line. The expanding type (Fig. 6) has the bladder connected to the gas line with the sample discharge line connected to the chamber surrounding the bladder.

7.2.3.1 The pump operates by applying a positive pressure to the gas line causing either the bladder to expand or be compressed, dependant on the type. The sampler inlet valve closes and the sample discharge valve opens forcing the contents of the sampler up the discharge line. Cyclic removal of the gas pressure causes the flow to stop, the sample valve to

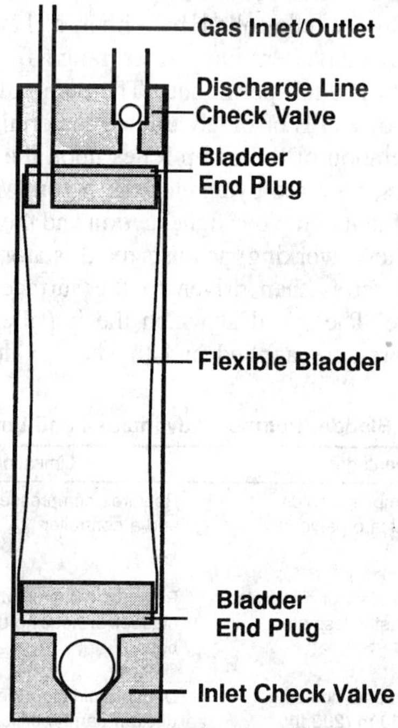

FIG. 6 Bladder Pump—Expanding Type

close and the sampler inlet valve to open, allowing the sampler to refill. See Table 6 for advantages and limitations.

7.2.4 *Peristaltic Pump* (**4**)—A peristaltic pump is a suction lift pump which is used at the ground surface (see Fig. 7*(a)*). A length of PTFE or other suitable tubing is placed in the liquid and the other end is connected to the piece of flexible tubing which has been threaded around the rotor of the peristaltic pump. A second piece of PTFE or other suitable tubing is connected to the discharge end of the flexible tubing to allow the water to be containerized (see Fig. 7*(b)*), sampled etc. If the pump tubing is not compatible with the sample parameters of concern, a modification to the system is necessary.

7.2.4.1 The modification (see Fig. 7*(c)*) consists of a peristaltic pump using PTFE tubing and a PTFE insert to collect samples without the sample coming into contact with the pump. This is accomplished by placing the PTFE insert into the opening of a clean glass container. The PTFE tubing connects the container to the pump and the sample source.

7.2.4.2 The operation of the peristaltic pump results from the rotor compressing the flexible tubing causing a vacuum to be applied to the inlet tubing. The water is drawn up the inlet tubing and into the container, without coming into contact with the pump flexible tubing.

7.2.4.3 Samples for purgeable organic compounds analyses may be collected by attaching the PTFE tubing to the intake side of the peristaltic pump, pumping the tubing full of the liquid, disconnecting the tubing, and allowing the PTFE tube to drain into the sample vials. A peristaltic pump also can be used to mix and sample liquids from drums (see Guide D 6063). See Table 7 for advantages and limitations.

7.2.5 *Centrifugal Submersible Pump (*Fig. 8*)* — Centrifugal submersible pumps may be used for purging and sampling monitoring wells, waste water impoundments or point discharges. Water contacting parts may be made of PTFE and stainless steel. The motor cavity may be either filled with air, deionized, or distilled water that may be replaced as necessary. The pump may be controlled by either a 12v (DC) or a 110/220v (AC) converter. Flow rates range from 9 gal per minute down to 100 mL per minute. The pump discharge hose may be made of PTFE or other suitable material.

7.2.5.1 Operation of the pump relies upon the rotation of a set of impellers, powered by an electric motor. Water is drawn into the centrifugal pump by slight suction and then pressurized by the impellers working against fixed stator plates. The pressurized water is then driven to the surface through the discharge hose. The speed at which the impellers are driven controls the pressure applied and thence the flow rate. See Table 8 for advantages and limitations.

TABLE 6 Bladder Pumps—Advantages and Limitations

Advantages	Limitations
Suitable for sampling liquids containing volatile organic compounds	Requires compressed air or gas and a controller
Available in a variety of materials, such as PTFE, stainless steel, PVC, etc.	Potential contamination from the bladder or housing materials, or both
Have an operational pumping head of up to 60 m (200 ft)	Decontamination (depending on design) can be difficult

7.3 *Dredges (Guides D 4342, D 4343, and D 4387)*—Dredges are used for the collection of submerged sediments and semi-consolidated sludge.

7.3.1 *Ekman*—The Ekman dredge (Fig. 9) has only limited usefulness in environmental sampling. It performs well where bottom material is unusually soft, as when covered with organic sludge or light mud. It is unsuitable, however, for sandy, rocky, and hard bottoms. It is also too light to use in streams with high flow velocities. It should not be used from a bridge more than a few feet above the water, because the spring mechanism which activates the sampler can be damaged by the messenger if dropped from too great a height.

7.3.2 *Petersen*—The Petersen dredge can be used for routine analyses when the bottom is rocky, in very deep water, or when the stream velocity is high. The dredge should be lowered very slowly as it approaches the bottom, because it can displace and miss lighter materials if allowed to drop freely.

7.3.3 *Ponar*—The Ponar dredge (Fig. 10) is a modification of the Petersen dredge and is generally similar in size and weight. Smaller, lighter versions are also available. It has been modified by the addition of side plates and a screen on top of the sample compartment. The screen over the sample compartment permits water to pass through the sampler as it descends thus reducing the "shock wave". The Ponar dredge is easily operated by one person in the same fashion as the Petersen dredge. The Ponar dredge is one of the most effective samplers for general use on all types of substrates. See Table 9 for advantages and limitations.

7.4 *Discrete Depth Samplers (Guide D 4448)*—These samplers are usually used in lakes, ponds, etc., but they also can be used to collect liquid waste samples in large tanks or lagoons. If liquid samples are desired at a specific depth, a Bacon Bomb or Kemmerer sampler may be used. Other types of discrete depth samplers are also available. (For shallow tanks and drums, refer to 7.6).

7.4.1 *Bacon Bomb*—The bacon bomb sampler (Fig. 11), originally designed for sampling oil, can be used for discrete depth sampling in stationary bodies of water, lakes, or waste. The primary advantage of this sampler over other discrete samplers is that it can be constructed of stainless steel and that it remains closed until it is triggered to collect the sample by raising the actuator rod with a second line and allowing the sampler to fill. Once a sample is collected, the device is closed by releasing the second line and the sampler is returned to the surface by raising the primary support line. The sample may then be transferred to a collection container. See Table 10 for advantages and limitations.

7.4.2 *Kemmerer*—The Kemmerer (Fig. 12) sampler is a stainless steel or brass cylinder with rubber stoppers that leave the ends open while being lowered in a vertical position to allow free passage of water through the cylinder. The Kemmerer is operated by sending a messenger down a rope when the sampler is at the designated depth, to cause the stoppers to close the cylinder, which is then raised. Water is removed through a valve to fill sample containers. With multiple depth samples, care should be taken not to stir up

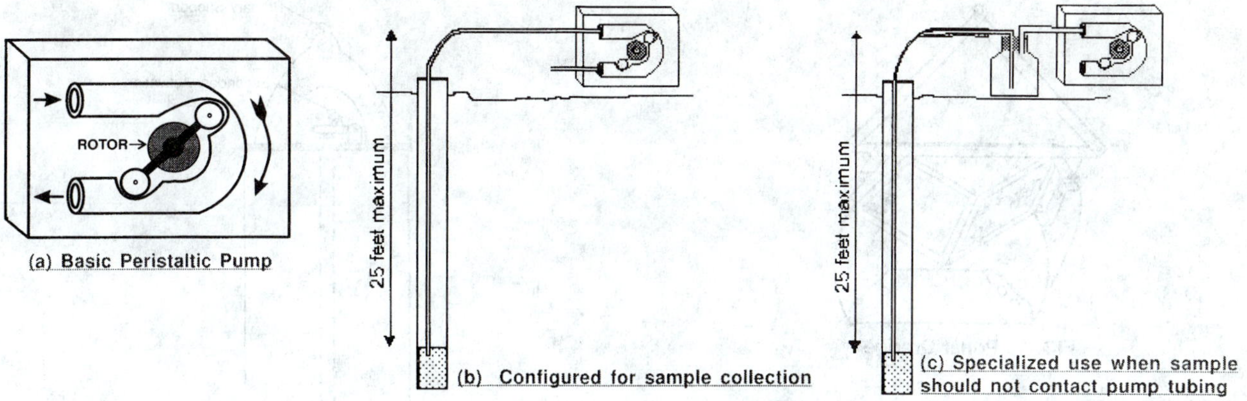

FIG. 7 Peristaltic Pump

TABLE 7 Peristaltic Pumps—Advantages and Limitations

Advantages	Limitations
May be used in small diameter (1 inch) wells	Depth to the liquid surface cannot exceed about 7.6 m (25 ft)
Decontamination of the pump motor is not necessary	May cause a loss of dissolved gases including volatile organic compounds.
Easy to replace the pump tubing without decontamination.	

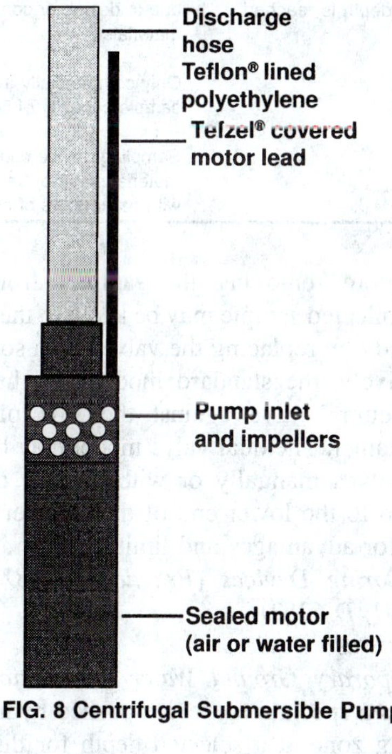

FIG. 8 Centrifugal Submersible Pump

TABLE 8 Centrifugal Submersible Pumps—Advantages and Limitations

Advantages	Limitations
Constructed of materials easily decontaminated, stainless steel and PTFE	Requires an electric power source
May be used to pump liquids up to a 76 m (250 ft) head	May be incompatible with liquids containing a high percentage of solids
Flow rate is adjustable	Portable use may require a winch or reel system
	May not be suitable for collecting samples of liquids containing volatile organic compounds

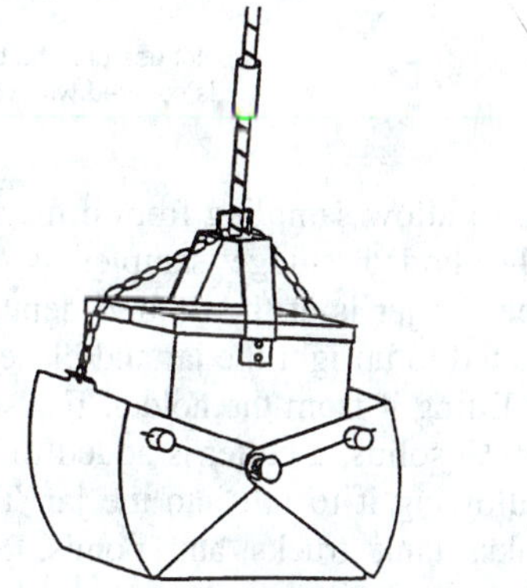

FIG. 9 Ekman Dredge

bottom sediment and thus bias the sample. All PTFE construction is available. See Table 11 for advantages and limitations.

7.4.3 *Syringe Sampler (*Fig. 13*)*—A syringe sampler is used to sample highly viscous liquids, sludges and tar-like substances. It can also draw samples when only a small amount remains at the bottom of a tank or drum. Syringe samplers are available commercially, they usually include a piston assembly consisting of a T-handle, safety locking nut, control rod (PTFE covered aluminum rod facilitates operation of the piston) piston body assembly, sampling tube assembly, and standard bottom valve or coring bottom. The assembled sampler with the bottom valve opened is positioned at the sampling point. By raising the T-handle, the sample is drawn into the sampler. The bottom valve is closed by pressing down on the sampler against the side or bottom of the container. To empty the sampler, open the bottom valve and extrude the sampler into a container by pushing down on the T-handle. See Table 12 for advantages and limitations.

7.4.4 *Lidded Sludge/Water Sampler*—A stainless steel sampling device used to collect sludges or waste fluids in a 1- L glass jar (see Fig. 14). The jar is removed and transported to the laboratory. No transfer of the sample to another container is necessary; this decreases handling and cross contamination. A PTFE insert is placed in the lid and is replaced between collection of samples. Handle extensions with depth markings

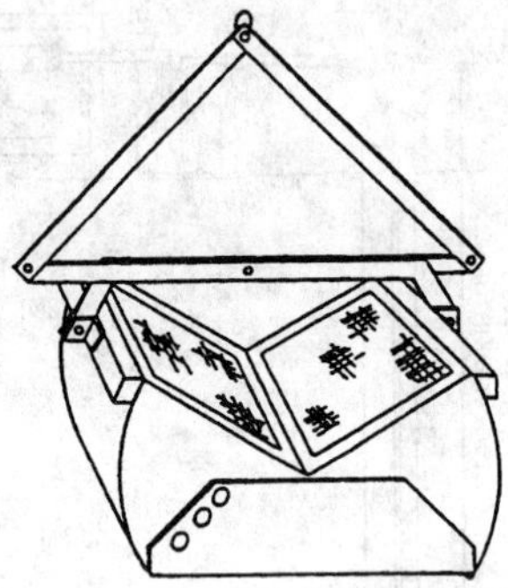

FIG. 10 Ponar Dredge

TABLE 9 Dredges—Advantages and Limitations

Advantages	Limitations
Ability to sample most types of stationary sediments from silt to granular material	Are not capable of collecting undisturbed samples
Light weight Ponar dredges are available	Are not capable of collecting a representative lift or repeatedly sampling to the same depth and position
	Petersen and other dredges with extra weights are very heavy
	Care must be taken to minimize disturbance and sample washout as the dredge is retreived through the liquid column
	May be difficult to decontaminate due to construction or materials
	Not suitable for use in rough waters
	Not useful if the bottom to be sampled is covered with vegetation

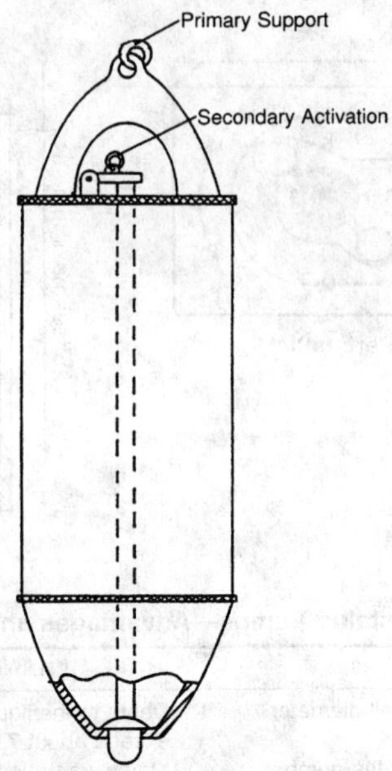

FIG. 11 Bacon Bomb

TABLE 10 Bacon Bomb—Advantages and Limitations

Advantages	Limitations
Sampler is not opened until the desired sample depth is reached	May be difficult to decontaminate due to design or construction materials
	Only commercially available with a sample capacity of 500 mL
	Sampling device construction material may not be compatible with parameters of concern

are available to allow sampling from difficult to access areas.

7.4.4.1 The lidded sludge sampler is lowered into the sludge. When the jar is at the desired depth, the top actuator handle is rotated to upright the jar and close the lid. The jar is removed by lifting it from the holder. For samples containing more than 40 % solids, a cutter is added to the jar which cuts the sludge allowing it to fall into the jar. This device can be used in tanks, tank trucks and ponds. See Table 13 for advantages and limitations.

7.4.5 *Discrete Level Sampler (*Fig. 15*)* —A sampler that can be used to sample liquids in drums, tanks, surface waters or wells. It is fitted with manually operated valve or valves on the ends of the sample collection chamber. Made from PTFE and stainless steel and designed to be reusable. The sampler is assembled with either a rigid control tube and rod or a flexible tube and inner cable attached to the upper end of the sampler. The proximal ends of the controls are provided with a handle and inner rod or cable actuator. The standard model is provided with an upper manually operated valve for filling and a lower spring retained dump valve for emptying. The dual valve model has manually operated valves at each end.

7.4.5.1 The sampler is lowered into the liquid column to the desired sampling level. The valve or valves are opened manually and the liquid sample collected. The valve or valves are closed before removing the sampler from the liquid column. The collected sample may be taken to the laboratory in the sampler body by replacing the valves with solid PTFE end caps. Alternatively, the standard model may be emptied by pressing the dump valve against the side of the sample collection container. The dual valve model may be emptied by opening the valves manually or with the use of a metering device attached to the lower end of the sampler (not shown). See Table 14 for advantages and limitations.

7.5 *Push Coring Devices (Practice D 1587 and Guides D 4700, D 6001, D 6282)*:

7.5.1 *Direct Push Devices*:

7.5.1.1 *Temporary Ground Water Penetration Sampler*— This sampler is a specialized device pushed hydraulically into a water-bearing zone at a selected depth for discrete ground water sampling (see Fig. 16). This device may be used in conjunction with a hollow-stem auger where the center plug is removed and then the device is inserted into the auger and pushed or driven to the required location in the aquifer. The device is then opened to allow entry of water and subsequently closed and retrieved. The device can also be used as a temporary well. Use of this device reduces the volume of soil which may have to be handled as an investigative derived waste. See Table 15 for advantages and limitations.

7.5.2 *Penetrating Probe Samplers (*Fig. 17 *and* Fig. 18*)* —Probe samplers can be used for sampling soil vapor, soil, and

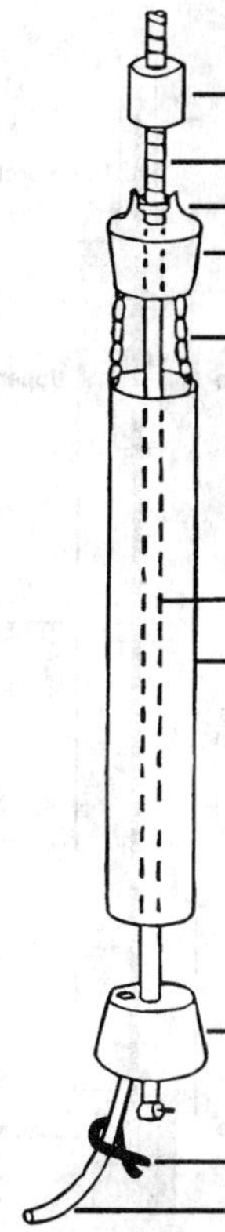

FIG. 12 Kemmerer Sampler

TABLE 11 Kemmerer—Advantages and Limitations

Advantages	Limitations
Able to sample at discrete depths	Sampling container is exposed to medium being sampled while being lowered to sampling point
	May be difficult to decontaminate due to construction or materials

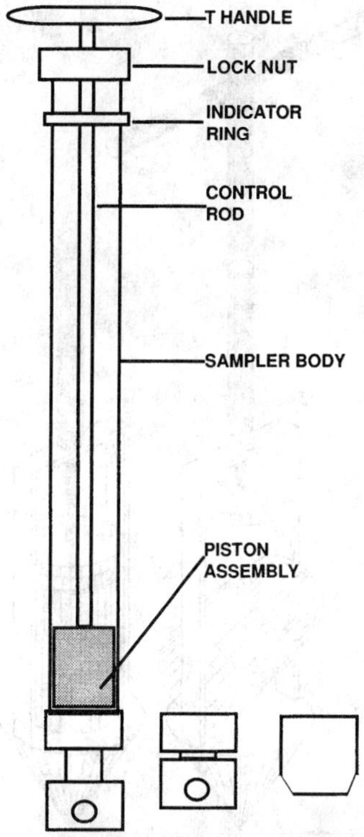

FIG. 13 Syringe Sampler

TABLE 12 Syringe Sampler—Advantages and Limitations

Advantages	Limitations
Simple to use and decontaminate, all sample contacting parts are PTFE	With viscous materials, more materials may end up on the outside of the sampler than inside it
Ability to sample at discrete depths, including the bottom of the container	
May be used to depths of about 1.8 m (6 ft)	

ground water. These samplers range in construction from a simple small diameter tube (usually less than 25 mm (1 in.) in diameter with a hard detachable point that are hand-driven to more complex devices that are rig-driven and can be opened after penetrating the ground surface.

7.5.2.1 The rig-driven probe samplers generally consist of single or multiple threaded steel tubes, 50 mm (2 in.) or more in diameter with detachable hard steel tips and are pushed or hydraulically driven into the subsurface materials. Some probes are equipped with adjustable screens or retractable inner rods to allow for sampling of soil vapor or ground water. Ground water can be retrieved using a peristaltic pump or miniature bailer. (See 7.2.4 on peristaltic pumps.) Soil samples can be collected at discrete intervals using specialized attachments.

7.5.2.2 Samples can be prepared for on-site analysis in a field laboratory or off site depending upon volumes obtained and the use of the data as determined by the data quality objectives process. See Table 16 for advantages and limitations.

7.5.3 *Split-Barrel*— (See Test Method D 1586 and Practice D 3550) A split-barrel sampler is used to collect soil samples at depth. The sampler consists of a length of steel tubing split longitudinally and equipped with a drive shoe and a drive head (see Fig. 19 and Fig. 20). They are available in a variety of diameters and lengths.

7.5.3.1 The sampling tube can be driven manually, or mechanically with a drill rig drive weight assembly or hydraulically pushed using rig hydraulics.

7.5.3.2 Drill and direct-push rigs offer the capability of collecting soil samples from greater depths. For all practical purposes, the depth of investigation achievable by this method is controlled only by the depth of soil overlying bedrock, which may be in excess of 31 m (100 ft).

7.5.3.3 When used in conjunction with drilling, split-barrel samplers are usually driven either inside a hollow-stem auger or inside an open borehole after rotary drilling equipment has been temporarily removed. The barrel is driven with a 140 lb drop hammer through a distance of up to 24 in. (61 cm) and

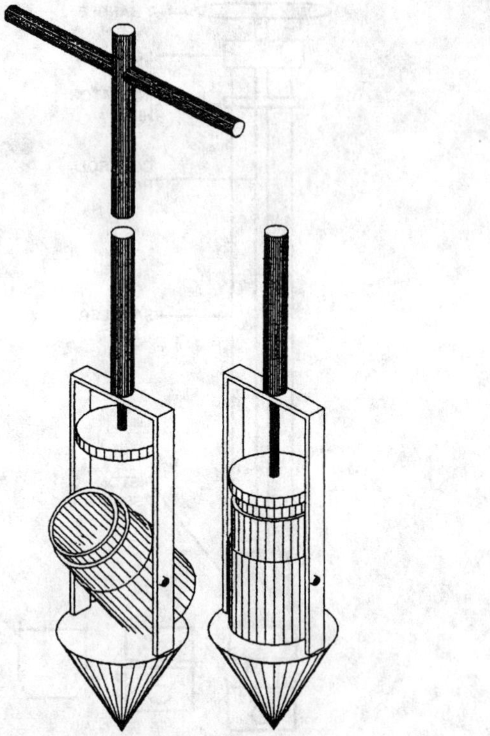

FIG. 14 Lidded Sludge/Water Sampler

TABLE 13 Lidded Sludge/Water Sampler—Advantages and Limitations

Advantages	Limitations
Sampler is not opened until desired depth is reached, allowing collection of samples from discrete depths	Thick sludge is difficult to sample with the device
Sturdy construction, prevents personnel contact with the sample	Equipment is heavy
Bottles and lids are unique to each sample container decontamination of these is not required	Limited to one bottle size

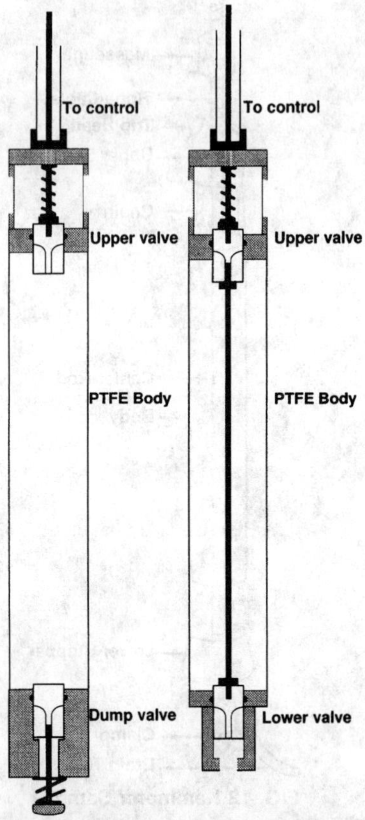

FIG. 15 Discrete Level Sampler

TABLE 14 Discrete Level Sampler—Advantages and Limitations

Advantages	Limitations
May be easily decontaminated	May be unsuitable for sampling liquids containing a high percentage of solids
May be used to sample liquids in most environmental situations	Sample chamber capacities 240 to 475 mL
Sample quality minimally affected by liquids above the sampling point	
Remote operation for hazardous environments.	

removed. If geotechnical data is also required, the number of blows with the hammer for each 6 in. (15.2 cm) driven interval also is recorded (see Test Method D 1586).

7.5.3.4 Split-barrel samplers may be used to obtain 1.52 m (5 ft) foot long, continuous samples approximately 7.6 to 12.7-cm (3 to 5-in.) in diameter. These devices are located inside a five foot hollow stem auger section and advanced with the auger during drilling. As the auger advances, the central core of soil moves into the sampler and is retained until retrieval.

7.5.3.5 Split-barrel samplers are sometimes used with liners. The advantage of a liner is that the sample can be removed from the sampler with a minimum amount of disturbance; and, if used correctly, they can minimize contamination of the samples. Liners are often used in situations where volatile organic compounds are constituents of concern or where there is an interest in trace elements or compounds. It is important that the investigator chooses liners composed of materials that are chemically compatible with the matrix and constituents of concern.

7.5.3.6 Split-barrel samplers may be fitted with a core-catcher immediately behind the drive tip. This will allow the sampler to collect samples of wet or cohesionless soils. See Table 17 for advantages and limitations.

7.5.4 *Concentric Tube Thief and Trier (See Practices D 5451 and E 300)*—These devices can be used for sampling powdered or granular materials or wastes in piles or in bags, drums or similar containers.

7.5.4.1 The concentric tube thief (Fig. 21) consists of two slotted telescoping tubes, constructed of stainless steel, brass or other material. The outer tube has a conical, pointed tip on one end that allows the thief to penetrate the material being

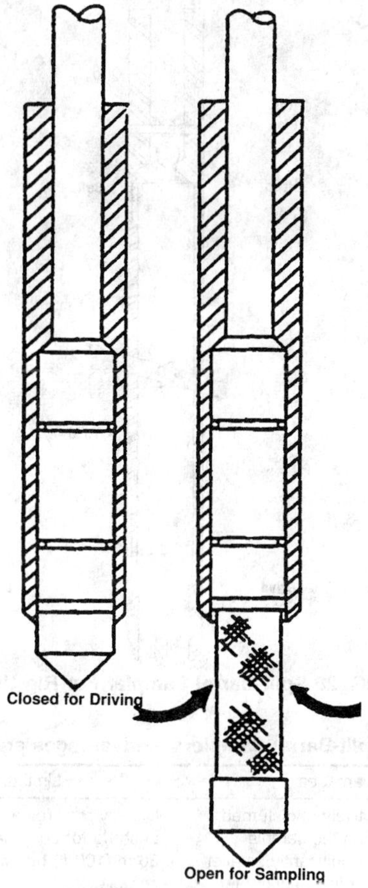

FIG. 16 Temporary Ground Water Penetration Sampler

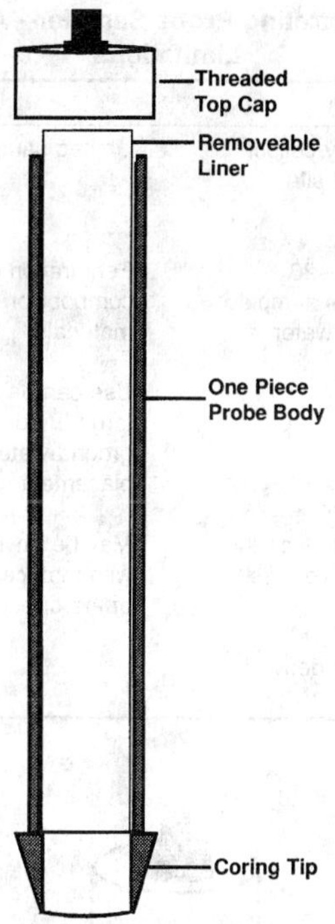

FIG. 17 Penetrating Probe Sampler for Hand Use

TABLE 15 Temporary Ground Water Penetration Sampler—Advantages and Limitations

Advantages	Limitations
Provides discrete depth ground water samples without the installation of a monitoring wall	Volume of sample is limited which may influence the type of analysis possible
Can aid in more appropriate placement of permanent monitoring wells when used with onsite analytical methods	Requires the use of a drill or direct push rig and some specialized training for use
Can be used to rapidly and inexpensively collect samples for expedited site characterization	May be physically incompatible with matrices that result in refusal of the direct push device
Can be used in depths reached by appropriate drilling equipment	
Reduces investigative derived waste	

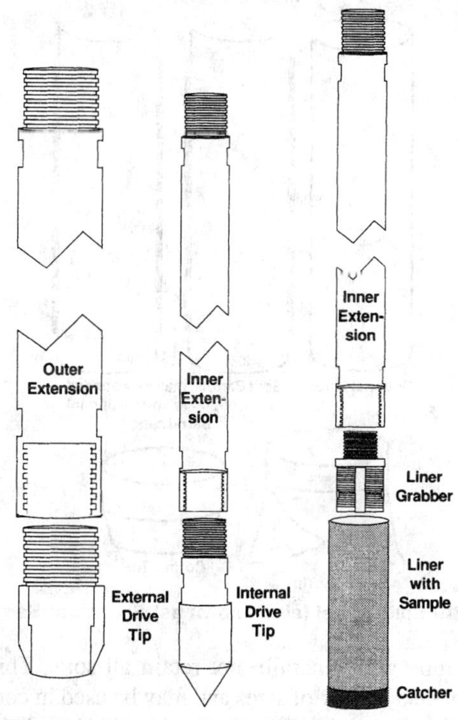

FIG. 18 Penetrating Probe Sampler for Rig Use

sampled. The thief is opened and closed by rotating the inner tube.

7.5.4.2 The trier (Fig. 22) is essentially a tube with a slot that extends along most of its length. The tip and edges of the tube slot are sharpened to allow the trier to cut a core of the material to be sampled when rotated after insertion into the material. Commercially available triers are usually constructed from stainless steel. See Table 18 for advantages and limitations.

7.5.5 *Thin-Walled Tube (See Practice D 1587)*—This is generally constructed of stainless steel and has a beveled leading edge, that is twisted and pushed directly into the soil. This type of sampling device is particularly useful if a relatively undisturbed sample is required (see Fig. 23). The sampling device is removed from the push head, then the sample is extruded from the tube into a pan or sample container with a spoon or special extruder. Even though the push head is equipped with a check valve to help retain samples, the

TABLE 16 Penetrating Probe Sampler—Advantages and Limitations

Advantages	Limitations
Can be used to rapidly collect samples for expedited site characterization	Limited sample volume
Versatile, generally 15–20 locations a day can be sampled for soil vapor, ground water, soil, or any combination	Penetration can be limited by composition of subsurface materials
	Use can be limited by depth to target media such as deeper ground water or accessibility of placement unit
These samplers can reduce the costs of more expensive off-site analyses	May be physically incompatible with matrices that result in refusal of the direct push device
Reduces investigative derived wastes	

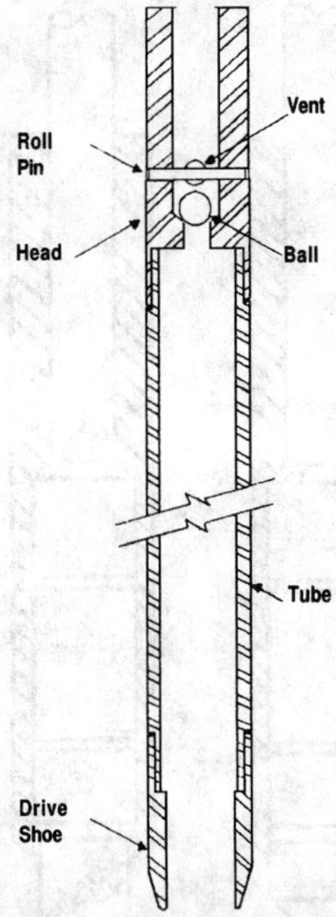

FIG. 20 Split-Barrel Sampler For Rig Use

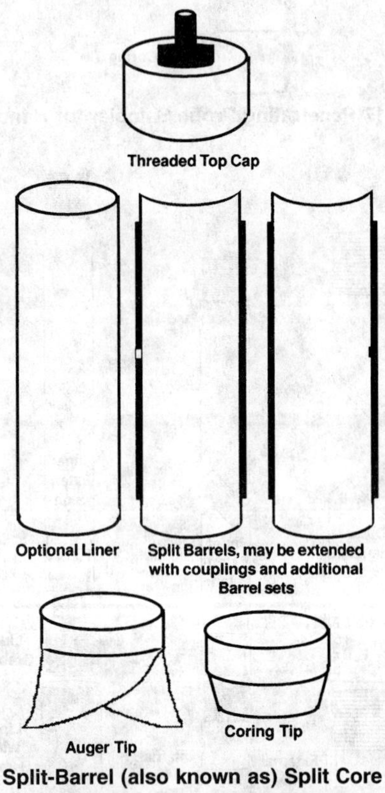

FIG. 19 Split-Barrel (also known as) Split Core Sampler

TABLE 17 Split-Barrel Samplers—Advantages and Limitations

Advantages	Limitations
Provides a relatively undisturbed sample, providing the sample particle size is significantly smaller than the sampler inside diameter	Usually requires a drill or direct push rig for deep samples up to 30 m (100 ft) below the soil surface
Since the sample is not extruded, fewer volatile organic compounds may be lost	The sample is exposed to the atmosphere, potentially allowing loss of volatile organic compounds, unless subsampling is immediately performed
Samples collected in capped liners can be stored for limited times before subsampling	

thin-wall tube will generally not retain all soils. Thin-walled tubes come in a variety of sizes and may be used in conjunction with drills, from hand-held to full sized drill rigs. See Table 19 for advantages and limitations.

7.5.6 *Coring Type Sampler With Valve (see Guide D 4823)*—This is designed for sampling sediments, sludges and free flowing powders (see Fig. 24). It is a stainless steel cylindrical sampler with a non-return valve at the lower end behind a coring or augering tip. The sample will normally be collected in an optional liner. It is operated by attaching a handle or an extension with a handle to the top of the coring device. The corer is lowered to the sampling point, pushed into the material being sampled and then removed. To recover the sample, the top cap is removed and the contents emptied into a sample container. Alternatively, the liner can be removed and capped on both ends for subsequent shipment to a laboratory. See Table 20 for advantages and limitations.

7.5.7 *Miniature Core Sampler (see Practice D 6418)*—This device is designed to collect and store small volume soil samples and allow transportation to a laboratory for subsequent chemical analysis of VOCs. Constructed from an inert composite polymer, the device consists of a coring body, a plunger, and an end cap (Fig. 25). Stainless steel handles are available to assist in collecting and subsequently extruding the sample. The sampler is available in sizes to allow collection of volumetric samples of approximately 5 and 25 g. Air-tight sealing is achieved with viton O-rings placed on the plunger and in the cap. This device may be used to retrieve samples of soil from the ground surface or trench walls. Also, it is used frequently to collect subsamples from soil cores. See Table 21 for advantages and limitations.

7.5.8 *Modified Syringe Sampler (see Guide D 4547)*—This sampler is used for collecting a small volume sample from a

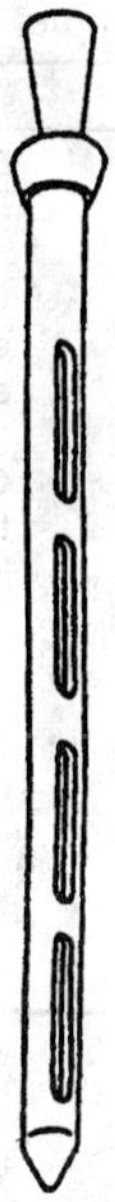

FIG. 21 Concentric Tube Thief

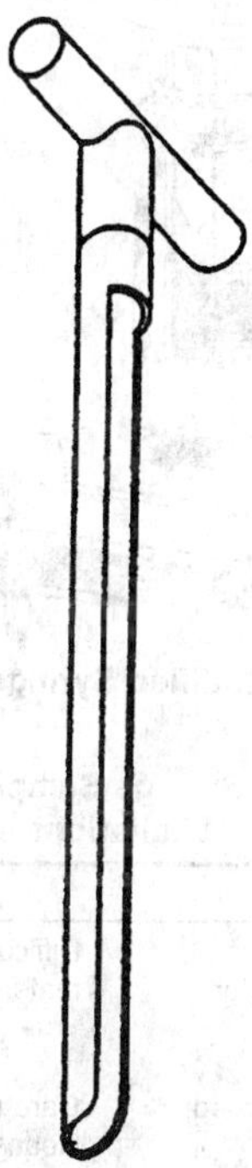

FIG. 22 Trier

TABLE 18 Tube Thief and Trier—Advantages and Limitations

Advantages	Limitations
Concentric tube thief is best used in dry, unconsolidated materials	Does not collect samples containing all particle sizes if the diameter of the largest solid particle is greater than one third of the slot width
The trier is best for moist or sticky materials	Samples may not be representative

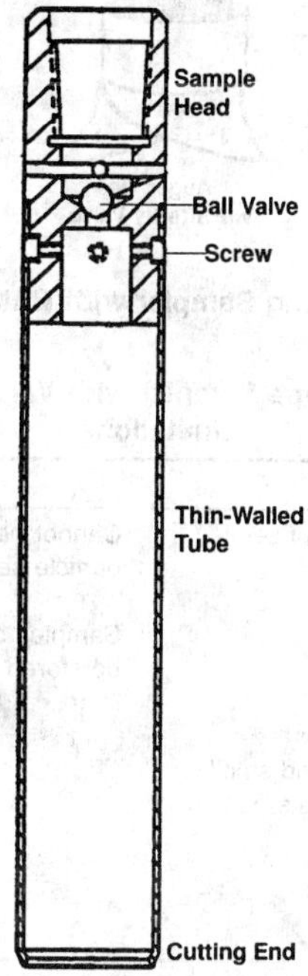

FIG. 23 Thin-Walled Tube

TABLE 19 Thin-Walled Tube Sampler—Advantages and Limitations

Advantages	Limitations
Provides a core sample	Cannot be used in gravel or rocky soils
Collects a relatively undisturbed sample which minimizes loss of volatiles	Loss of volatile organic compounds possible if the sample is extruded
Can be deployed down a bore hole to collect deep samples	Samples containing VOCs cannot be stored in the liner
Inexpensive and easily decontaminated	Not effective in cohesionless soils

material surface or more usually to subsample a core for subsequent VOC analysis. The sample then is transferred immediately to a vial for transportation and analysis. This device is available commercially or made by modifying a plastic, disposable syringe. The lower end and the attachment for a needle and plunger cap are removed (see Fig. 26). The plunger is pushed in until it is flush with the cut end. The syringe sampler is then pushed into the soil core to collect the sample which then should be placed in a prepared, air-tight glass vial for transport to a laboratory until analysed. The vial mouth should have a diameter larger than the syringe barrel. See Table 22 for advantages and limitations.

7.5.9 *Soft Sediment Sampler (Fig. 27)*—Designed for sampling saturated soft materials either on the surface or beneath liquids. It is constructed from stainless steel with aluminum guide rods used for placement. This sampler collects a 2- or 4-in. square core sample up to 6-ft long. The sampler is comprised of two angular blades, one with a sample retaining flap valve, the other with an alignment block used to position the blades correctly. The blades are driven consecutively to collect an undisturbed sample. See Table 23 for advantages and limitations.

7.6 *Rotating Coring Devices*—Includes bucket augers that collect disturbed samples of unconsolidated materials and

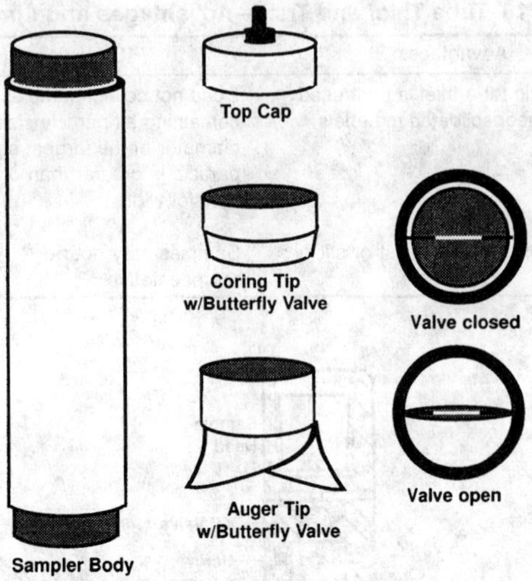

FIG. 24 Coring Sampler with Butterfly Valve

TABLE 20 Coring Type Sampler with Valve—Advantages and Limitations

Advantages	Limitations
Provides a core sample of semi-liquid materials	Cannot be used in gravel or large particle sediments or sludges
Easily decontaminated	Samples containing VOCs cannot be stored and transported in the liner
May be used in drums and small containers as well as tanks, lagoons, and waste impoundments	
Usually hand operated	

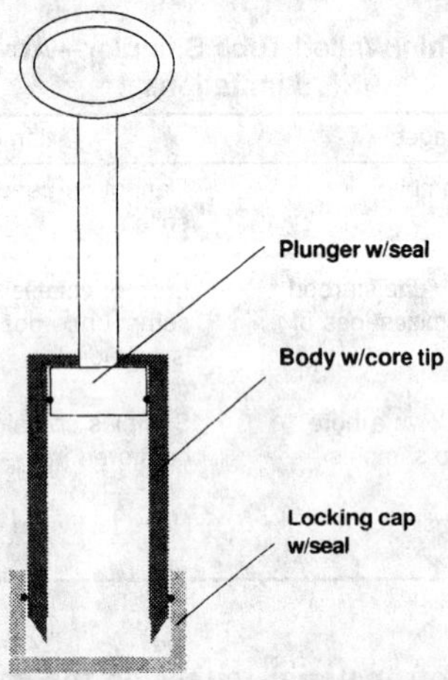

FIG. 25 Miniature Core Sampler

TABLE 21 Miniature Core Sampler—Advantages and Limitations

Advantages	Limitations
Provides a core sample from a soil surface or trench wall	Difficult to use in dry sandy materials
Collects a relatively undisturbed sample	Care required to ensure that soil does not compromise the end cap seals
Sampler is designed as a single use device for collection, storage and transportation of samples containing VOCs.	Cost may be a consideration for this single use device
Collects a sample of suitable size for analysis. Laboratory or field subsampling is not required	
Sliding plunger prevents air entrapment and allows sample extrusion	

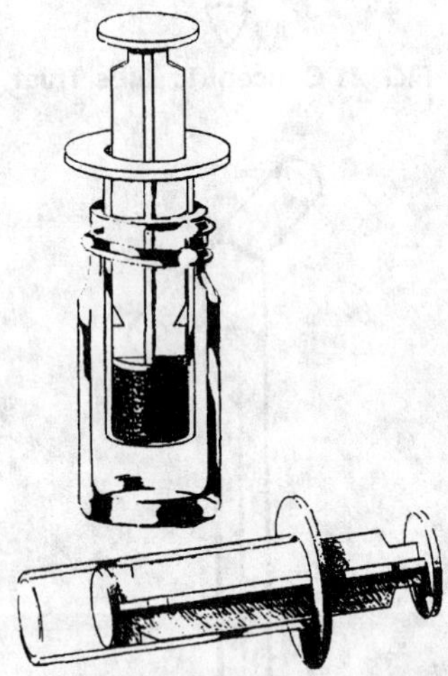

FIG. 26 Modified Syringe Sampler

TABLE 22 Modified Syringe Sampler—Advantages and Limitations

Advantages	Limitations
Provides a core sample if sampled from a soil surface or trench wall	Difficult to use in dry sandy materials
Collects a relatively undisturbed sample	Care required to ensure device is clean before use
Sampler is a low cost single use device	
Collects a sample suitable for VOC analysis, laboratory, or field subsampling is not required	
Sliding plunger prevents air entrapment and allows sample extrusion	

soils, screw augers that collect cuttings of consolidated materials and rocks, and a rotating corer that collects cores of consolidated materials.

7.6.1 *Auger (See Practice D 1452 and Guide D 4700)*—Many different types and designs are available, ranging from the hand-held to portable power-driven to pick-up or van mounted to full-scale drill rigs. Two separate types of augers, one for unconsolidated solids and the other used for consolidated solids, are described below.

7.6.1.1 *Bucket Augers (*Fig. 28*)* —Typically, bucket augers with cutting heads are pushed and twisted into the media and removed as the buckets are filled. The auger holes are advanced one bucket at a time. The practical depth of investigation using a hand auger is related to the material being sampled. In sands,

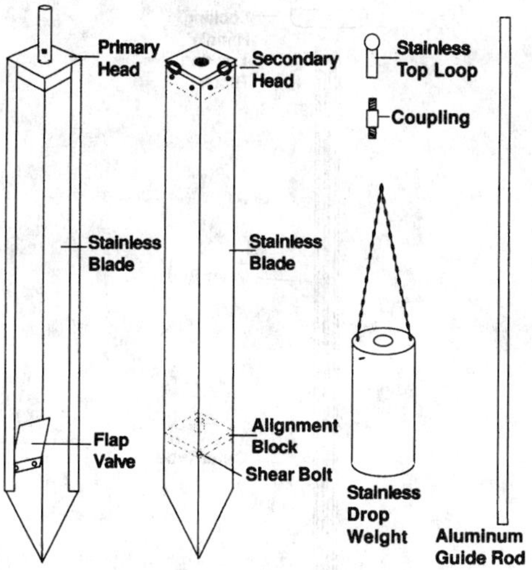

FIG. 27 Soft Sediment Sampler

TABLE 23 Soft Sediment Sampler—Advantages and Limitations

Advantages	Limitations
Provides a core sample	May require a floating platform for use in sampling beneath liquids
Collects a relatively undisturbed profile sample	May require a winch and/or additional personnel for recovery of long samples from beneath a deep liquid layer
May be used to collect samples beneath liquids up to 30 ft deep	
Manually operated, usually with one person	

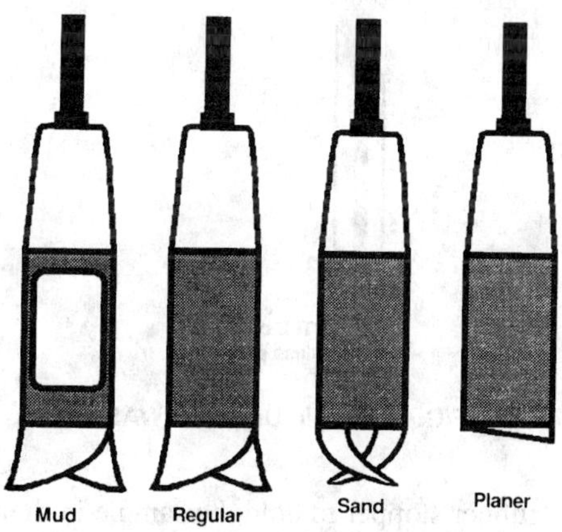

FIG. 28 Bucket Augers

augering is usually easily accomplished, but the depth of investigation is controlled by the depth at which sands begin to cave. At this point, auger holes usually begin to collapse and cannot practically be advanced to lower depths, and further samples, if required, must be collected using some type of pushed or driven device. Hand augering may also become difficult in tight clays or cemented sands. At depths approaching 20 ft (6 m), torquing of hand auger extensions becomes so severe that in resistant materials, powered methods must be used if deeper samples are required.

7.6.1.2 When a vertical sampling interval has been established, one bucket auger is used to advance the auger hole to the first desired sampling depth. If the sample at this location is to be a verticle composite of all intervals, the same bucket may be used to advance the hole, as well as collect subsequent samples in the same hole. However, if discrete grab samples are to be collected to characterize each depth, a clean bucket auger must be used to collect the next sample. The top several inches of material should be removed from the bucket to minimize chances of cross-contamination of the sample from fall-in of material from the upper portions of the hole.

7.6.1.3 The Planer type bucket auger may be used to remove loose material from the bottom of an augered hole, prior to core sampling. It also may be used to collect samples of solid materials from the bottom of drums and tanks. See Table 24 for advantages and limitations.

7.6.2 *Screw Augers*— For sampling consolidated solids, such as construction materials, soft rock, and wood. These augers are similar to drill bits and can be operated by hand (brace and bit) or powered by a portable electric drill (see Fig. 29). As the auger advances into the material being sampled, the cuttings move up the auger stem to the surface where they are collected for the sample. See Table 25 for advantages and limitations.

7.6.3 *Rotating Coring Device*—This device is used to obtain a core of consolidated solid (see Fig. 30). It consists of a diamond or carbide tipped open steel cylinder attached to an electric drill. The drill may be hand held or mounted on a stand placed on the ground surface. Water is usually used to cool and lubricate the cutting edge. The core barrel diameter ranges from 5 to 15 cm (2 to 6 in.). See Table 26 for advantages and limitations.

7.7 *Liquid Profile Devices*:

7.7.1 *COLIWASA (See Practice D 5495 and Practice D 5743)*—The COLIWASA (Composite Liquid Waste Sampler) sampler is used to obtain a vertical column of liquid of the sampled material (see Figs. 31 and 32). It's most common use is for sampling containerized liquids, such as tanks, barrels, and drums. It may also be used for pools and other open bodies of stagnant liquids. They can be constructed of any material that would be compatible with the samples being collected.

7.7.1.1 COLIWASA's are available commercially with different types of stoppers and locking mechanisms, but all operate using the same principle. In use, the device is lowered into the liquid, tapered end first. The COLIWASA should be open at both ends so that the material flows through it as it is lowered to the desired sampling depth. This must be done slowly because the container may contain solid material which might break the tube and injure the sampler, and slowly lowering the tube allows the liquid phases to stay in equilibrium with the COLIWASA sampler.

7.7.1.2 The reuseable point sampler (Fig. 33) is used in the

TABLE 24 Bucket Augers—Advantages and Limitations

Advantages	Limitations
Easy and quick for shallow subsurface samples	Collects only disturbed samples
	Inappropriate for sampling soils for volatile organic compounds

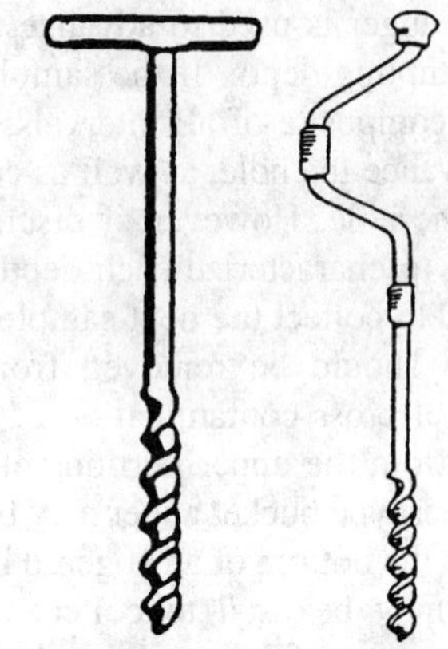

FIG. 29 Screw Augers

TABLE 25 Screw Augers—Advantages and Limitations

Advantages	Limitations
Allows collection of a sample from a solid material	Destroys layers and soil horizons and cannot obtain an undisturbed sample

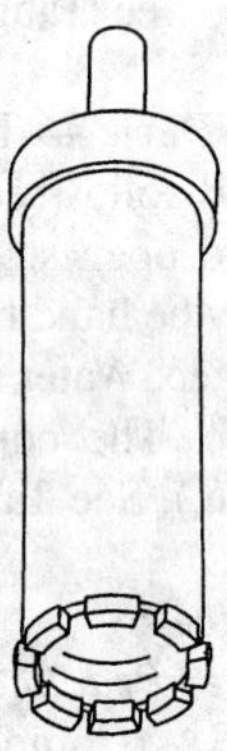

FIG. 30 Rotating Corer

TABLE 26 Rotating Corer—Advantages and Limitations

Advantages	Limitations
Can obtain a solid core	Need power and water source
	Difficult to operate
	May affect integrity of the matrix

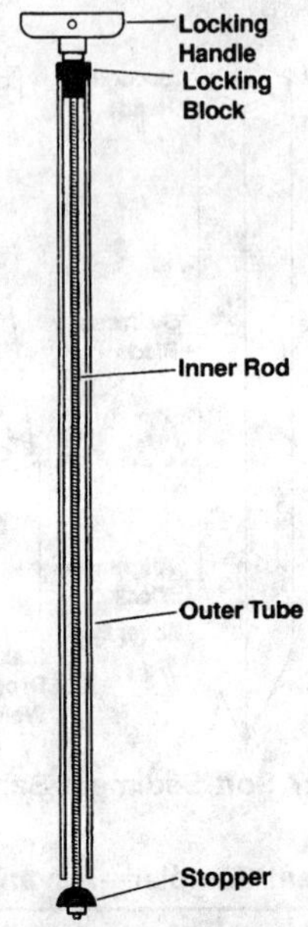

FIG. 31 Original COLIWASA

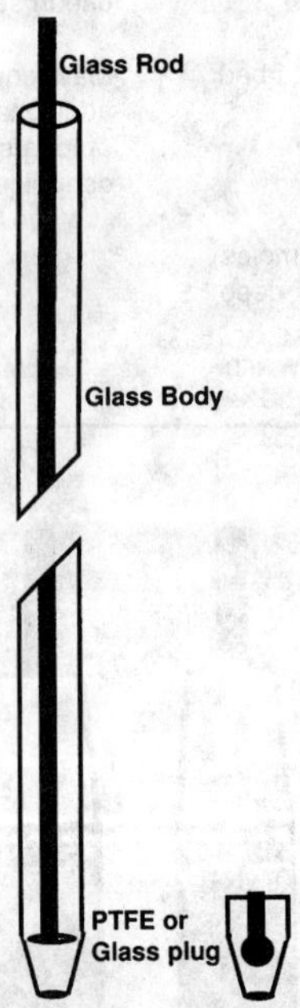

FIG. 32 Single Use COLIWASA

same way as the COLIWASA. In addition it may be used to sample at a specific point in the liquid column. This sampler is usually made of PTFE.

7.7.1.3 Once the COLIWASA has filled, the stopper mechanism is seated and both tubes are withdrawn from the material together. By manipulating the inner tube, the sampler can control the rate of flow of sampled liquid into the sample container. See Table 27 for advantages and limitations.

7.7.2 *Drum Thief (see Practice D 5743)*—A drum thief is a 1.3 m (4 ft) long tube, used to sample liquids in drums and similiar containers, usually made of glass, but can be constructed of other materials (see Fig. 34). In most instances, glass tubes with a 1 cm (½in. or less) inside diameter work best. The tube is inserted into the opening of the drum or barrel as far as possible. The open end is then sealed either with the thumb or a rubber stopper to hold the sample in the tube while removing the tube from the container. The sample is then placed in an appropriate container, and the procedure repeated until an adequate amount of sample is collected. See Table 28 for advantages and limitations.

7.7.3 *Valved Sampler*—This device allows collection of a vertical column of liquid from a drum or tank (see Fig. 35). It may be constructed from PTFE for reuse or polypropylene for single use. The device is operated by first opening the top plug and the bottom valve and then lowering it vertically and slowly into the liquid to allow levels inside and outside to equalize. The top plug is closed manually and the bottom valve is pressed against the side or bottom of the container to close it. To empty the sampler, the contents are poured from the top into

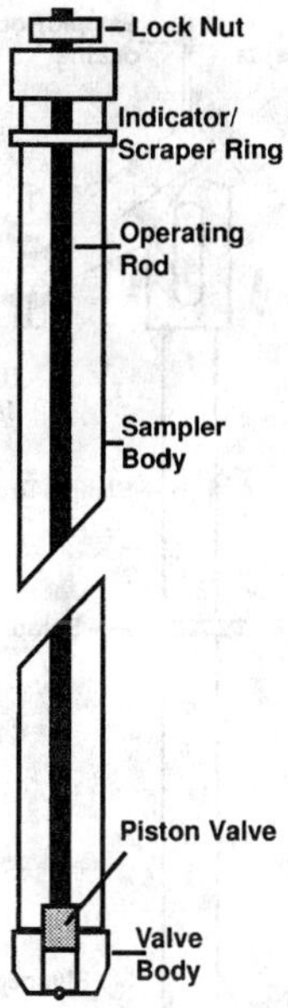

FIG. 33 Reuseable Point Sampler

TABLE 27 COLIWASA—Advantages and Limitations

Advantages	Limitations
Simple to use	Depth to sample limited to length of sampler
Reuseable and single use models available	Stopper mechanism may not allow collection of approximately the bottom inch of material
Inexpensive	High viscosity fluids difficult to sample
	May break if made of glass and used in consolidated matrices
	If constructed of glass and reused, decontamination may be difficult

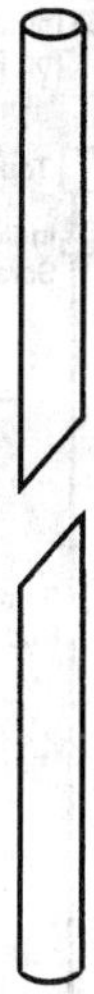

FIG. 34 Drum Thief

TABLE 28 Drum Thief—Advantages and Limitations

Advantages	Limitations
Simple to use	Depth to sample limited to length of sampler
Usually single use	High viscosity fluids difficult to sample
Inexpensive	Drum size tubes have a small volume capability, possibly requiring repeated use to obtain a sample Larger sizes are available, however, two or more people may be required
	May be difficult to hold sample in the tube
	May break if used in consolidated matrices
	If made of glass and reused, decontamination may be difficult

a suitable container. See Table 29 for advantages and limitations.

7.7.4 *Plunger-Type Sampler (see Practice D 5743)*—The plunger-type sampler is used to obtain a vertical column of liquid or slurries from drums, tanks, or similar containers. It is made from high density polyethylene or PTFE with an optional glass sampling tube (see Fig. 36). It has an open lower end and a fixture at the upper end to hold a sampling bottle. The device is lowered into the liquid to be sampled, the plunger is engaged to secure the sample aliquot, and the cord or rod is raised to transfer the sample directly into the sampling bottle or jar. The plunger can be pushed back down the sampling tube to reset the sampler. They are available in lengths suitable for sampling drums, road tankers, and rail cars. See Table 30 for advantages and limitations.

7.7.5 *Liquids Profiler*—The sampler is made from clear PVC and is provided with 1-ft depth markings on the 5-ft sampler body sections, a check valve on the lower section and a cord on the upper section (see Fig. 37). Its primary use is to allow measurement and sampling of settleable solids as would be found in sewage treatment plants, waste settling ponds, and impoundments containing waste materials. In use, it is assembled, using threaded connections to the length needed and lowered into the liquid to allow it to fill. A slight tug on the cord will set the check valve and allow it to be removed. The levels of settleable solids can be measured using the markings. It may be emptied by pressing the protruding pin on the lower end against a hard surface, or it may be pushed in and held manually. See Table 31 for advantages and limitations.

7.8 *Liquid Grab Sampling Devices*:

7.8.1 *Bailer (See Guide D 4448)*—A bailer is essentially a length of PTFE, stainless steel or PVC pipe with a check valve on the bottom (see Fig. 38). Preferably, the top should be closed, except for a pouring opening, to keep matter on the inside of the well casing from falling into the bailer while sampling. The bottom valve allows the bailer to fill with sample and retain it while being brought to the surface. Bailers are available in numerous sizes to accommodate a wide variety of well sizes, as either reusable or single-use sampling devices. See Table 32 for advantages and limitations.

7.8.1.1 When using a top-emptying bailer, samples can be recovered with a minimum of aeration if care is taken to

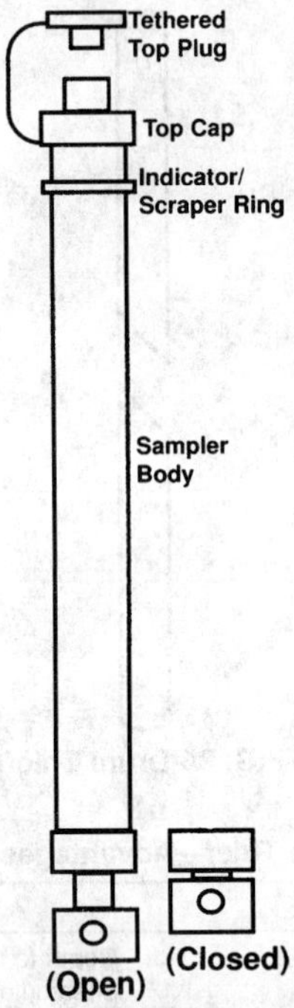

FIG. 35 Valved Sampler

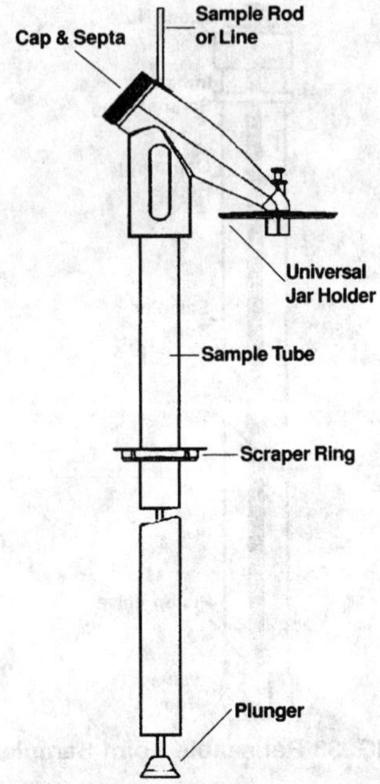

FIG. 36 Plunger Type Sampler

TABLE 29 Valved Sampler—Advantages and Limitations

Advantages	Limitations
Simple to use	Bottom valve prevents collection of the bottom 1.25 cm (½ in.)
Reusable if made from PTFE; single use if made from polypropylene	High viscosity liquids may be difficult to sample
Unbreakable and can sample to depths of about 6.5 m (21 ft), using body extensions	

TABLE 30 Plunger Type Sampler—Advantages and Limitations

Advantages	Limitations
Simple to use	Care needed when using a glass sampling tube
Provides a sealed collection system	Heavy contamination may be difficult to remove, particularly when a glass sampling tube is used
May be used as either a reusable or single use device	
Relatively inexpensive and available in various lengths	

gradually lower the bailer until it contacts the water surface and then is allowed to sink as it fills. The bailer should be raised to the surface slowly. When transferring the bailer contents to a sample container, the bailer should be tipped only enough that a slow discharge from the top of the bailer is allowed to flow into the container.

7.8.1.2 Bottom-emptying bailers with controlled flow valves are also available. This type of bailer is particularly good for collecting samples for volatile organic analyses (VOA) since they minimize agitation of the sample.

7.8.2 *Point Sampling Bailer (see Guide D 4448)*—The point-sampling bailer is similar in construction to the bailer described in the prior section. A point-source bailer has an additional check valve at the top of the body (see Fig. 39). As the bailer is lowered through the liquid column the liquid flows through the bailer. At the sampling point the two check valves will close to contain the sample and prevent mixing with the liquids above as the sampler is retrieved. See Table 33 for advantages and limitations.

7.8.3 *Differential Pressure Bailer*—The differential-pressure bailer comprises a sealed tubular body with two small diameter tubes built in to the removable top (see Fig. 40). Usually, it is made from stainless steel to provide sufficient weight to allow it to sink quickly to the sampling point. Hydrostatic pressure allows the bailer to fill through the lower tube at the same time as displacing air through the upper tube. See Table 34 for advantages and limitations.

7.8.4 *Dipper (See Practice D 5358)*—This sampling device is used to collect liquid samples from ponds, pits, lagoons, etc. (see Fig. 41). The sampler can consist of a variety of pieces of equipment assembled in a manner to obtain a sample. One type consists of an adjustable clamp attached to the end of a piece of metal tubing. The tubing forms the handle; the clamp is used to secure a beaker, sample container, etc. Another device is made using a stainless steel scoop clamped to a movable bracket that is attached to a piece of rigid tube. The scoop may face either toward or away from the person collecting the sample, and the angle of the scoop to the pipe is adjustable. See Table 35 for advantages and limitations.

7.8.5 *Liquid Grab Sampler*—The liquid grab sampler is used to collect liquid or slurry samples at specific depths beneath the liquid surface (see Fig. 42). Usually, it is made from polypropylene or PTFE with an aluminum or stainless steel handle and stainless steel fittings. The sampling jar, usually glass but plastic is available, is threaded into the

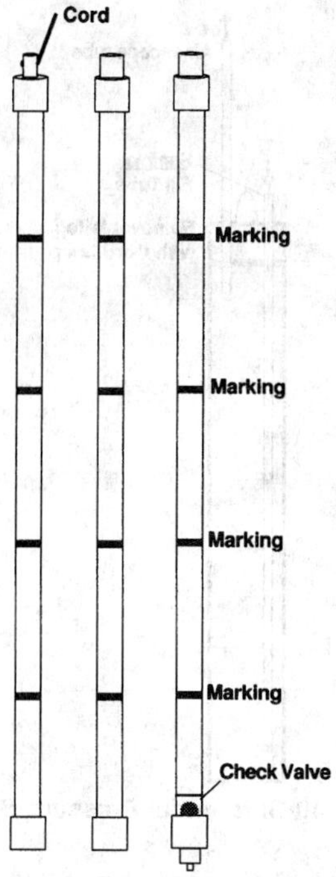

FIG. 37 Liquids Profiler

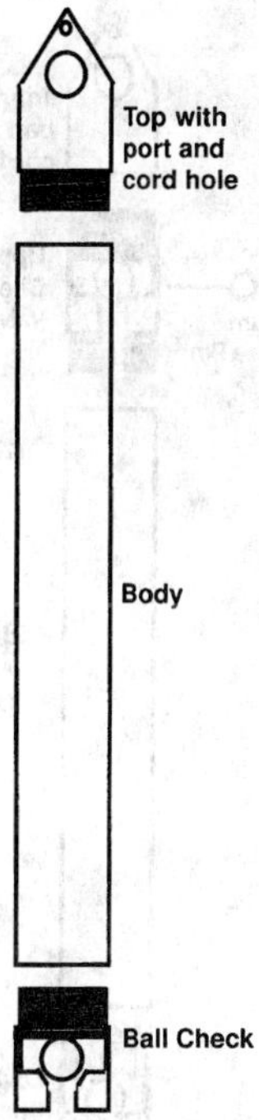

FIG. 38 Bailer

TABLE 31 Liquids Profiler—Advantages and Limitations

Advantages	Limitations
Allows length measurement of liquid/settleable solids columns of any length	Suitable for sampling non-caustic liquids
Easily assembled and used	High viscosity materials may be difficult to sample
Unbreakable in normal use and reusable	

TABLE 32 Bailer—Advantages and Limitations

Advantages	Limitations
Simple to use	Unable to collect samples from specific depths below the liquid surface
External power source not required	Transfer of sample from top-emptying bailer to sample container may aerate sample if not poured carefully
Relatively economical compared to other sampling methods; a separate bailer could be dedicated to each well	May disturb sample in water column if the bailer is lowered too rapidly
Can be made from almost any material that is compatible with the parameters of interest	May be chemically incompatible with certain matrices unless constructed of resistant material

sampler head assembly and lowered to the desired position beneath the liquid surface. The value is opened, by pulling up on the finger ring, to allow the jar to fill and then closed before retrieving the sample. See Table 36 for advantages and limitations.

7.8.6 *Swing Jar Sampler*—The swing jar sampler comprises an extendable aluminum handle attached to a plastic jar holder using pivot (see Fig. 43). The open top jar is held in the holder with an adjustable clamp. The pivot allows samples to be collected at different angles. It may be used to sample liquids, powders or small solids at distances of up to 3½ m (12 ft). Normally used with high density polyethylene sample jars. See Table 37 for advantages and limitations.

7.9 *Surface Sampling Devices*:

7.9.1 *Impact Devices (*Fig. 44*)* —These devices are used for sampling consolidated solids. The most common "device" is a hammer and hand chisel. Another device is the pneumatic chisel where compressed air takes the place of the hammer. See Table 38 for advantages and limitations.

7.9.2 *Spoon (*Fig. 45*)* —A spoon may be used to sample particulate materials on the ground surface or from an open container or waste pile. Small samples of liquid may also be collected with this device, although it is not the preferred method. Made from stainless steel or PTFE they can be easily cleaned for re-use. Plastic spoons may be used as they are inexpensive and can be considered a single use item. See Table 39 for advantages and limitations.

7.9.3 *Scoops and Trowels (See Practice D 5633)*—These have limited application for collecting surface soil samples but may be used for solid waste sampling. These devices come in different sizes and materials (see Fig. 46). Unpainted stainless steel is preferred. Scoops are available from laboratory and field equipment supply houses, trowels can be obtained from hardware stores. See Table 40 for advantages and limitations.

7.9.4 *Shovels*—Shovels used for environmental sample retrieval are usually made from stainless steel or suitable plastic materials (see Fig. 47). Their primary use is collection of surface materials or large samples from waste piles. Their other use is the mixing of large sample volumes as may be required

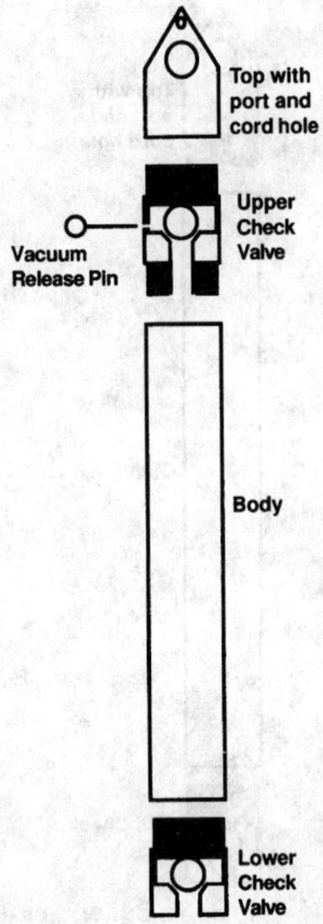

FIG. 39 Point Sampling Bailer

TABLE 33 Point Sampling Bailer—Advantages and Limitations

Advantages	Limitations
Simple to use	Bailer may be compromised as it is lowered through contaminated layers in the liquid column
Allows sample collection at a specific depth in the liquid	Requires a means to unseat the upper check valve to break the vacuum as the sample is eluted through the lower check valve using a bottom emptying device

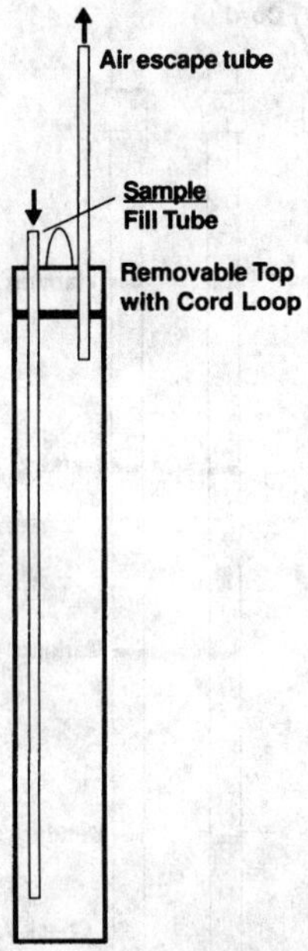

FIG. 40 Differential Pressure Bailer

TABLE 34 Differential Pressure Bailer—Advantages and Limitations

Advantages	Limitations
Simple to use	Decontamination requires care to ensure that all parts of the device, including the air escape and sample entry tubes, are clean
Allows sample collection at a specific depth in the liquid column, without risk of contaminants in upper layers compromising the sample	

for the collection and mixing of composite samples. See Table 41 for advantages and limitations.

8. Keywords

8.1 environmental; liquid; monitoring; sampling; sampling equipment; sediment; soil; waste management; water

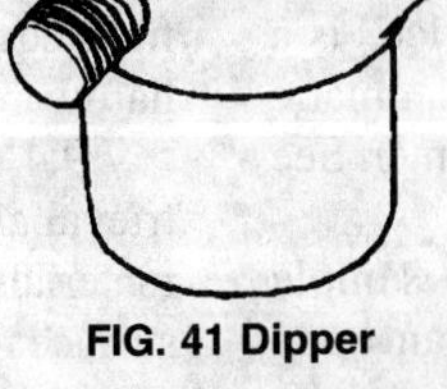

FIG. 41 Dipper

TABLE 35 Dipper—Advantages and Limitations

Advantages	Limitations
Inexpensive	When liquids are stratified, it cannot be used to obtain a sample containing the same proportions of the strata as the location being sampled
When attached to a rigid tube, can reach easily 3-4 m (10 to 13 ft) away from the person collecting samples	Can be used only to obtain surface samples

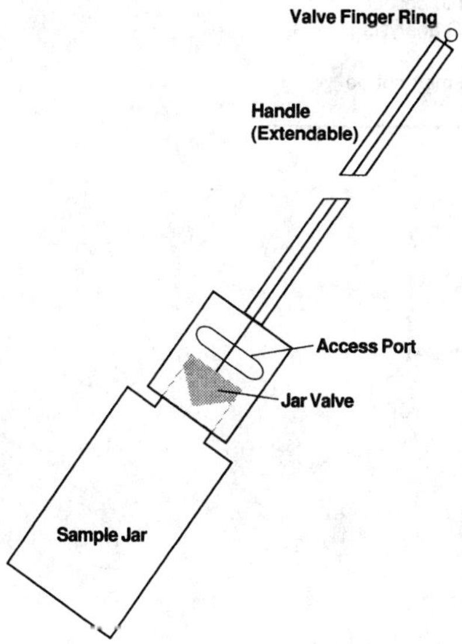

FIG. 42 Liquid Grab Sampler

TABLE 36 Liquid Grab Sampler—Advantages and Limitations

Advantages	Limitations
Simple and easy to use	Care in use is required to prevent breakage of the glass sample jar
May be used to sample ponds, impoundments, tanks, drums and through manholes	Construction materials should be compatible with the media being sampled
The closed sampler prevents contaminants in upper layers compromising the collected sample	
The filled sampling container may be capped, stored, and shipped	

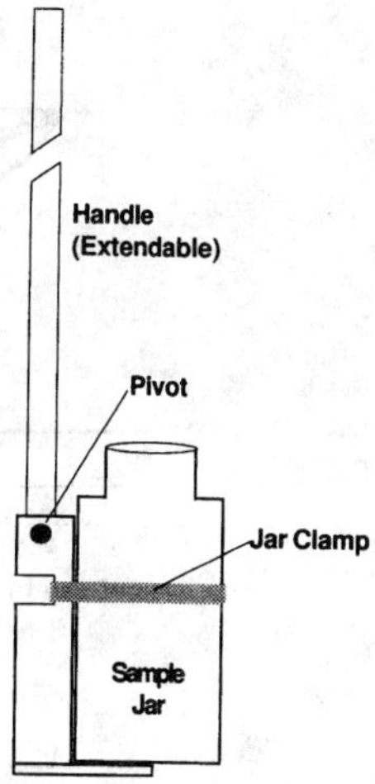

FIG. 43 Swing Jar Sampler

TABLE 37 Swing Jar Sampler—Advantages and Limitations

Advantages	Limitations
Simple and easy to use	Cannot collect discrete samples
Easily adaptable to sample jars of different sizes and materials	Construction materials should be compatible with the media being sampled
The filled sampling container may be capped, stored and shipped	Care required to prevent breakage when using a glass sample jar

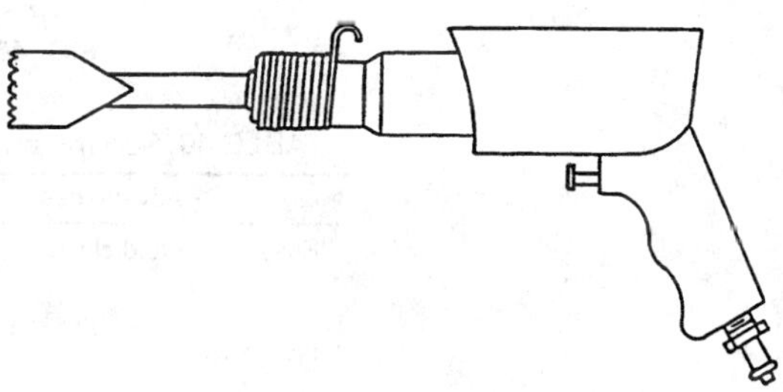

FIG. 44 Impact Device

TABLE 38 Impact Devices—Advantages and Limitations

Advantages	Limitations
Can obtain a sample of a solid material by chipping or flaking at the surface of the material	Pneumatic system needs an air source
	May not collect all layers of a heterogeneous solid

FIG. 45 Spoon

TABLE 39 Spoon—Advantages and Limitations

Advantages	Limitations
Inexpensive	Small sample volume
Easy to use and clean	Cannot be used to collect samples for VOC analysis
	A single sample may not be representative

Sizes #2,3,6,8,12 denote ozs.

FIG. 46 Stainless Steel Scoops

TABLE 40 Scoops and Trowels—Advantages and Limitations

Advantages	Limitations
Easy to use and clean	May affect the matrix during sample collection by selecting certain particle sizes
Inexpensive	May not be constructed in a shape that is compatible with the dimensions of the matrix
	May exacerbate the loss of volatile organic compounds by disturbance

APPENDIX

(Nonmandatory Information)

X1. ADDITIONAL RELATED PUBLICATIONS

US-EPA, *RCRA Ground-Water Monitoring Technical Enforcement Guidance Document* (TEGD), OSWER 9950.1, Office of Solid Waste and Emergency Response (OSWER), Washington, DC, September, 1986.

American Chemical Society, *Principles of Environmental Sampling*, L.H. Keith, Editor, 1988.

McCoy and Associates, Inc., "Soil Sampling and Analysis - Practices and Pitfalls," *Hazardous Waste Consultant*, Volume 10, No. 6, Lakewood, CO, November/December 1992.

US-EPA, *Test Methods for Evaluating Solid Waste, 3rd Edition*, EPA/530/SW-846 (NTIS, PB88-239223), Washington, DC, 1986.

US-EPA, *RCRA Ground Water Monitoring: Draft Technical Guidance*, EPA/530/R-93/001 (NTIS PB93-139350), Washington, DC, 1993.

US-EPA, *Description and Sampling of Contaminated Soils:*

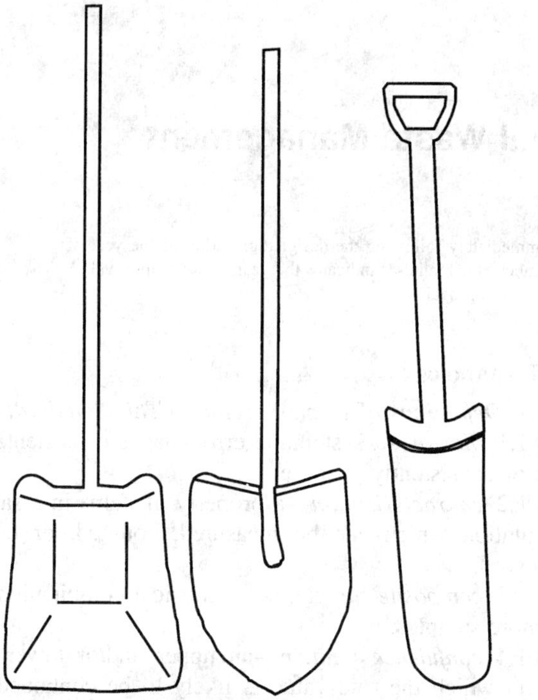

FIG. 47 Stainless Steel Shovels

TABLE 41 Shovels—Advantages and Limitations

Advantages	Limitations
Easy to use and clean	For surface use only
Rugged for use with hard materials	Cannot be easily used to fill sample containers
	Cannot be used to collect samples for VOC analysis

A Field Pocket Guide, EPA/652/2-91/002, Washington, DC, 1991.

US-EPA, *Subsurface Characterization and Monitoring Techniques: A Desk Reference Guide, Volume I: Solids and Ground Water, Volume II, The Vadose Zone, Chemical Field Screening and Analysis*, EPA/625/R-93/003a, EPA/625/R-93/0003b, Washington, DC, 1993.

Boulding, J.R., *Description and Sampling of Contaminated Soils: A Field Guide, Revised and Expanded 2nd Edition*, Lewis Publishers, Chelsea, MI, 1994.

US-EPA, *Soil Sampling and Analysis for Volatile Organic Compounds*, EPA/540/4-91/001, Washington, DC, 1991.

US-EPA, *Emergency Response Team Standard Operating Procedures Compendia: Compendium of ERT Soil Sampling and Surface Geophysics Procedures* (EPA/540/P-91/006; *Compendium of ERT Ground Water Sampling Procedures* (EPA/540/P-91/007); *Compendium of ERT Waste Sampling Procedures* (EPA/540/P-91/008; *Compendium of ERT Toxicity and Testing Procedures* EPA/540/P-91/009), Washington, DC, 1991.

US-EPA, *Field Methods Compendium Draft*, OEER # 9285.2-11, Analytical Operation Branch, Hazardous Site Evaluation Division, Office of Emergency and Remedial Response, Washington, DC, 1993.

US-EPA, *Sediment Sampling Quality Assurance User's Guide, 2nd Edition*, EPA/608/8-89/046, Washington, DC, 1989.

US-EPA, *Preparation of Soil Sampling Protocols: Sampling Techniques and Strategies*, EPA/600/R-92/128, Washington, DC, 1992.

US-EPA, *Methods Manual for Bottom Sediment Sample Collection*, EPA/905/4-85/004, Washington, DC, 1985.

US-EPA, Environmental Investigations Standard Operating Procedures and Quality Assurance Manual. http//www.epa.gov/region04/sfd/eisopqam/eisop9am.html Region 4, Science and Ecosystem Support Division: Athens, GA 1996.

REFERENCES

(1) USEPA, *Environmental Investigations Standard Operating Procedures and Quality Assurance Manual*, Athens, GA, May 1996.

(2) USEPA, *Final RCRA Comprehensive Ground-Water Monitoring Evaluation (CME) Guidance Document*, Final OSWER Directive 9950.2 (NTIS PB91-140194), Washington, DC, 1986.

(3) Pitard, F.F., *Pierre Gy's Sampling Theory and Sampling Practice* - Volumes I and II, CRC Press, Boca Raton, FL, 1989.

(4) USEPA, *RCRA Ground-Water Monitoring: Draft Technical Guidance*, EPA/530K-93-0001, Office of Solid Waste and Emergency Response (OSWER), Washington, DC, November 1992.

(5) USEPA, *A Compendium of Superfund Operation Methods*, EPA/540/P-97/001 (OSWER 9355.0-14), Office of Solid Waste and Emergency Response, Washington, DC, December 1997.

(6) USEPA, *Characterizing Heterogenous Wastes: Methods and Recommendations*, EPA/600/R-92/033, Office of Research and Development, Washington, DC, February 1992.

Designation: D 6233 – 98

Standard Guide for Data Assessment for Environmental Waste Management Activities[1]

This standard is issued under the fixed designation D 6233; the number immediately following the designation indicates the year of original adoption or, in the case of revision, the year of last revision. A number in parentheses indicates the year of last reapproval. A superscript epsilon (ε) indicates an editorial change since the last revision or reapproval.

1. Scope

1.1 This guide covers a practical strategy for examining an environmental project data collection effort and the resulting data to determine if they will support the intended use. It covers the review of project activities to determine conformance with the project plan and impact on data usability. This guide also leads the user through a logical sequence to determine which statistical protocols should be applied to the data.

1.1.1 This guide does not establish criteria for the acceptance or use of data but instructs the assessor/user to use the criteria established by the project team during the planning (data quality objective process), and optimization and implementation (sampling and analysis plan) process.

1.2 *This standard does not purport to address all of the safety concerns, if any, associated with its use. It is the responsibility of the user of this standard to establish appropriate safety and health practices and determine the applicability of regulatory limitations prior to use.*

2. Referenced Documents

2.1 *ASTM Standards:*

D 4687 Guide for General Planning of Waste Sampling[2]

D 5088 Practice for Decontamination of Field Equipment Used at Nonradioactive Waste Sites[3]

D 5283 Practice for Generation of Environmental Data Related to Waste Management Activities: Quality Assurance and Quality Control Planning and Implementation Activities[2]

D 5792 Practice for Generation of Environmental Data Related to Waste Management Activities: Development of Data Quality Objectives[2]

D 5956 Guide for Sampling Strategies for Heterogeneous Wastes[2]

D 6044 Guide for Representative Sampling for Management of Waste and Contaminated Media[2]

3. Terminology

3.1 *Definitions of Terms Specific to This Standard:*

3.1.1 *bias, n*—a systematic error that is consistently negative or consistently positive.

3.1.2 *characteristic, n*—a property of items in a sample or population which can be measured, counted, or otherwise observed.

3.1.3 *composite sample, n*—a physical combination of two or more samples.

3.1.4 *confidence limit, n*—an upper and/or lower limit(s) within which the true value is likely to be contained with a stated probability or confidence.

3.1.5 *continuous data, n*—data where the values of the individual samples may vary from minus infinity to plus infinity.

3.1.6 *data quality objectives (DQOs), n*—DQOs are qualitative and quantitative statements derived from the DQO process describing the decision rules and the uncertainties of the decision(s) within the context of the problem(s).

3.1.7 *data quality objective process, n*—a quality management tool based on the scientific method and developed to facilitate the planning of environmental data collection activities.

3.1.8 *discrete data, n*—data made up of sample results that are expressed as a simple pass/fail, yes/no, or positive/negative.

3.1.9 *heterogeneity, n*—the condition of the population under which all items of the population are not identical with respect to the parameter of interest.

3.1.10 *homogeneity, n*—the condition of the population under which all items of the population are identical with respect to the parameter of interest.

3.1.11 *population, n*—the totality of items or units under consideration.

3.1.12 *representative sample, n*—a sample collected in such a manner that it reflects one or more characteristics of interest (as defined by the project objectives) of a population from which it is collected.

3.1.13 *sample, n*—a portion of material which is taken from a larger quantity for the purpose of estimating properties or composition of the larger quantity.

3.1.14 *sampling design error, n*—error which results from the unavoidable limitations faced when media with inherently

---

[1] This guide is under the jurisdiction of ASTM Committee D34 on Waste Management and is the direct responsibility of Subcommittee D34.01.01 on Planning for Sampling.

Current edition approved Feb. 10, 1998. Published June 1998.

[2] *Annual Book of ASTM Standards*, Vol 11.04.

[3] *Annual Book of ASTM Standards*, Vol 04.09.

variable qualities are measured and incorrect judgement on the part of the project team.

3.1.15 *subsample, n*—a portion of a sample that is taken for testing or for record purposes.

4. Significance and Use

4.1 This guide presents a logical process for determining the usability of environmental data for decision making activities. The process describes a series of steps to determine if the enviromental data were collected as planned by the project team and to determine if the *a priori* expectations/assumptions of the team were met.

4.2 This guide identifies the technical issues pertinent to the integrity of the environmental sample collection and analysis process. It guides the data assessor and data user about the appropriate action to take when data fail to meet acceptable standards of quality and reliability.

4.3 The guide discusses, in practical terms, the proper application of statistical procedures to evaluate the database. It emphasizes the major issues to be considered and provides references to more thorough statistical treatments for those users involved in detailed statistical assessments.

4.4 This guide is intended for those who are responsible for making decisions about environmental waste management projects.

5. General Considerations

5.1 This guide provides general guidance about applying numerical and other techniques to the assessment of data resulting form environmental data collection activities associated with waste management activities.

5.2 The environmental measurement process is a complex process requiring input from a variety of personnel to properly address the numerous issues related to the integrity of the sample collection and measurement process in sufficient detail. Table 1 lists many of the topics that are common to most environmental projects. A well-executed project planning activity (see Guide D 4687, Practices D 5088, D 5283, and D 5792) should consider the impact of each of these issues on the reliability of the final project decision. The data assessment process must then evaluate the actual performance in these areas versus that expected by the project planners. Significant deviations from the *a priori* performance level of any one or combination of these issues may impact the reliability of the project decision and necessitate a reconsideration of the decision criteria by the project decision makers.

5.3 Appropriate professionals must assess the project planning documents and completed project records to determine if the project findings match the conceptual model and decision logic. In areas where the findings don't match, the assessors must document and report their findings and, if possible, the potential impact on the decision process. Items subject to numerical confirmation are compared to the project plan and any discrepancies and their potential impact noted.

5.4 Effective quality control (QC) programs are those that empower the individuals performing the work to evaluate their performance and implement real-time corrections during the sampling or measurement process, or both. When quality control processes (including documentation) are properly implemented, they result in data sets (see Fig. 1) that are generated by in-control processes or out-of control processes that were not amenable to corrective action but whose details are explained by the project staff conducting the work. Good QC programs lead to reliable data that are seldom called into question during the assessment process. However, in cases where the absence of staff responsibility or authority to self-monitor and correct deficiencies at the working level is missing, the burden of assuring data integrity is placed on the quality assurance (QA) function. The data assessment process must determine the location (working level or QA level) where effective quality control occurs (detection of error and execution of corrective action) in the data collection process and focus on how well the QC function was executed. As a general rule, if the QC function is not executed in real-time and thoroughly documented by the staff performing the work, the more likely the data assessor will be to find questionable data.

5.5 In addition to addressing the issues listed in Table 1, the data assessment process must search for unmeasurable factors whose impact cannot be detected by the review of the project records against expectations or numerical techniques. These are the types of things that are controlled by effective quality

TABLE 1 Information Needed to Evaluate the Integrity of the Environmental Sample Collection and Analysis Process

General Project Details	• Site History • Process Description • Waste Generation Records • Waste Handling/Disposal Practices • Sources of Contamination • Conceptual Site Model • Potential Contaminants of Concern • Fate and Transport Mechanisms • Exposure Pathways • Boundaries of the Study Area • Adjacent Properties
Sampling Issues	• Sampling Strategy • Sample Location • Sample Number • Sample Matrix • Sample Volume/Mass • Discrete/Composite Samples • Sample Representativeness • Sampling Equipment, Containers and Preservatives
Analytical Issues	• Laboratory Sub-sampling • Sample Preparation Methods • Analytical Method • Detection Limits • Matrix Interferences • Bias • Holding Times • Calibration • Quality Control Results • Contamination • Reporting Requirements • Reagents/Supplies
Validation and Assessment	• Data Quality Objectives • Chain of Custody • Action Level • Completeness • Laboratory Audit Results • Field and Laboratory Records • Level of Uncertainty in Reported Values

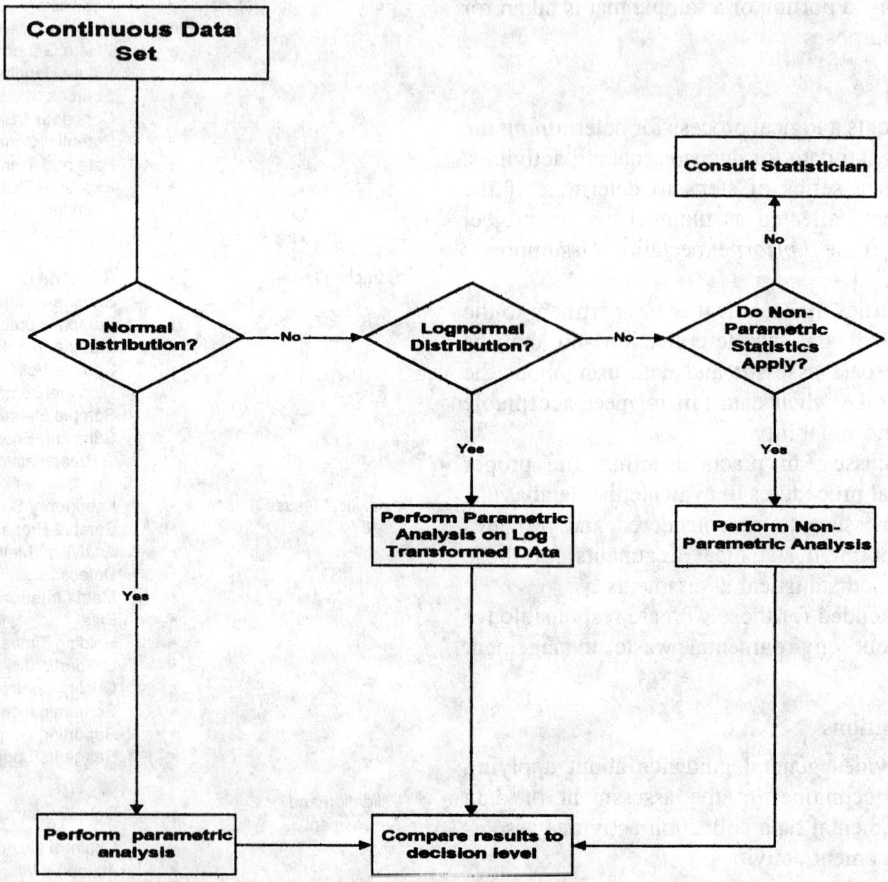

FIG. 1 General Strategy for Assessment of Continuous Data Sets

TABLE 2 Examples of Unmeasurable Factors Affecting the Integrity of Environmental Data Collection Efforts

• Biased Sampling/Subsampling	• Incorrect Dilutions
• Sampling Wrong Area or Material	• Incorrect Documentation
• Sample Switching (Mis-labeling)	• Matrix-Specific Artifacts
• Misweighing/Misaliquoting	

assurance programs, standard operating procedures, documentation practices, and staff training. Historically, efforts have been focused on the control of data collection errors through data review and the quality control process but little emphasis has been placed on the detection and evaluation of immeasurable errors using the quality assurance process. These unmeasurable sources of error are often the greatest source of uncertainty in the data collected for environmental projects. Examples of unmeasurable factors are given in Table 2.

5.6 Once the data assessment process has determined the degree to which the actual data collection effort met the expectations of the planners, the assessment process moves into the next phase to determine if the data generated by the effort can be verified and validated and whether it pass statistical tests for useability. These issues are discussed in the next sections.

6. Sources of Sampling Error

6.1 Sample collection may cause random or systematic errors. Random error affects the data by increasing the imprecision, whereas systemic error biases the data. The data assessment process should examine the available sampling records to determine if errors were introduced by improper sampling. A discussion of some of the more common sources of error follow.

6.1.1 *Random Error*:

6.1.1.1 Flaws in the sampling design which result in too few quality control samples being taken in the field can result in undetected errors in the sampling program. Adequate numbers of field QC samples (for example, field splits, co-located samples, equipment rinsate blanks, and trip blanks) are necessary to assess inconsistencies in sample collection practices, contaminated equipment, and contamination during the shipment process.

6.1.1.2 Variations (heterogeneity) in the media being sampled can cause concentration and property differences between and within samples. Field sampling and laboratory sub-sampling records should be examined to determine if heterogeneity was noted. This can explain wide variations in field and/or laboratory duplicate data.

6.1.1.3 Samples from the same population (including co-located samples) can be very different from each other. For example, one sample might be taken from a hot spot that was not visually obvious while the other was taken outside the perimeter of the hot spot. If data from areas of high concentration is contained in data sets consisting primarily of uncontaminated material, statistical outlier analysis might suggest the sample data should be omitted from consideration when evaluating a site. This can cause serious decision errors. Prior to declaring the data point(s) to be outliers, it is important for the assessor to examine the QC records from the analysis yielding the suspect data. If the QC data indicates the system was in control and review of the raw sample data reveals no handling or calculation errors, the suspect data should be discussed in the assessor's report but it should not be discounted. The site history and operating records may hold clues to the possible existence of hot spots.

6.2 *Systematic Error*:

6.2.1 Flaws in the sampling design that result in sampling of inappropriate locations can result in significant bias in the data. The samples collected from such a flawed plan will not be representative of the population and can result in incorrect decisions. The assessor should review the sampling plan for signs of potential bias and discuss their findings in the final report.

6.2.2 Sampling tools and equipment can deselect certain parts of a sample based on the physical properties (density, particle size, multi-phasic materials, particle geometry, etc.). If the sample is biased because of some physical characteristic, then any constituent that is distributed in the material based on that characteristic, will be incorrectly reported. Both field and laboratory sampling equipment can introduce this type of bias.

6.2.3 Incorrect sampling procedures can cause losses of certain constituents of a sample such as volatile organics. Failure to control the loss of of constituents that exist in the gaseous state often comprises the collection of unsaturated media for volatile compound characterization. Deterioration of the sample can also occur after collection due to improper storage and transportation. For example, samples left standing in sunlight or in a hot vehicle can undergo photochemical reactions or lose volatile constituents.

6.2.4 Interactions between the sample and the material of the sampling equipment or container, or both, are potential sources of positive or negative bias.

6.2.5 Inappropriate preservation of the sample can cause a shift in chemical equilibria, loss of target analytes, or degradation, or all of these. For example, when analyzing a water sample for dissolved metals, addition of nitric acid to a water sample containing suspended solids might dissolve metals from the solids, resulting in an incorrect high concentration being reported. Failure to preserve water samples intended for organic analysis may allow significant biological alteration of the sample.

6.2.6 The time of day and prevailing weather conditions when samples are collected can affect the sample. For example, strong winds can blow dust that can contaminate the samples. Cool mornings or evening can lead to higher retention of volatile components in near-surface soil samples compared to the samples collected in the heat of the day.

6.2.7 The above examples only serve to illustrate the need for an experienced professional to review the sampling activities and to place the resulting analytical data in the proper context of the sampling activity. Such assessments add materially to the usability of the data.

7. Sources of Analytical Error

7.1 Variation in the analytical process may cause random or systematic error. Random error affects the data by increasing the imprecision, whereas systematic error increases the bias of the data. The data assessment process should examine the available analytical records to determine if errors were introduced in the data by the analytical process. Analytical results can also be impacted by sample matrix effects. Discussion of some of the more common sources these types of error follow.

7.1.1 *Random Error*:

7.1.1.1 Random errors in the analytical process are often uncontrollable and unobserved. They are usually distributed between positive and negative error and tend to cancel out and so have little effect. However, for any one measurement, random error can be significant.

7.2 *Systematic Error*—The bias resulting from systematic error can be either positive or negative but it affects all results in a data set(s) the same way. Sources of systematic error are most often associated with sample preparation or analysis. Incomplete digestion or insufficient reaction time during sample preparation are examples that can produce negatively biased results during the preparation process. Improperly calibrated instruments, incorrect standards, dirty detectors, and leaking sample introduction systems are examples of instrumental problems that cause systematic error. They are most often detected when reference samples and laboratory control samples fail to produce the expected results.

7.3 *Sample Matrix Effects*:

7.3.1 The sample matrix can introduce either systematic or random error in analytical results. Consistently high or low results (systematic error) can be obtained when the matrix contains a non-target constituent that interferes with the accurate measurement of the target analyte. The interfering substance must be uniformly distributed in the matrix to produce consistent deviations from the true value. If the interference is non-uniformly distributed in the matrix, the error will appear as a random error.

7.3.2 The relationship between the sample matrix and the analytical method can result in an important class of matrix errors. When the method selected is not appropriate to the matrix, errors may result. One of the most common types of mismatches of method and matrix is using methods designed for water analysis to analyze soils. Another is the use of methods designed for the analysis of naturally occurring materials, such as groundwater or soils, for the analysis of waste materials.

7.3.3 Most sample matrix and method selection errors can be detected by examining the results of matrix spike quality control samples where known amounts of the target analyte(s) are introduced into the sample before analysis. Spike results should be evaluated to determine the presence of any matrix effect. For certain types of analyses, simple dilution of the sample and re-analysis will demonstrate matrix effects when the second result, corrected for the dilution factor, is not consistent with the initial result.

8. Assessment of Environmental Data Sets

8.1 Data are usually verified and validated prior to comparing the results of environmental analysis to some decision level by suitable statistical processes. Data verification determines whether the laboratory carried out all steps required by the sampling and analysis plan or a contract, or both. After data is verified, it is validated. Validation examines the available laboratory data to determine whether an analyte is present or absent in a sample and the degree of overall uncertainty associated with the reported value. After data has been validated, it is normally compared to a decision level using suitable statistical techniques to determine the appropriate course of action.

8.2 The verification process compares the laboratory data package to a list of required data. These requirements are generated by two separate activities. The first is the contract for analytical services between the project and the laboratory and the second is the project sampling and analysis plan with its accompanying quality assurance project plan (QAPP) developed by project and laboratory staff. These two activities determine, *a priori*, the procedures the laboratory must use to produce data of known quality and the content of the analytical data package. Verification compares the material delivered by the laboratory against these requirements and produces a report that identifies those requirements which were not met (called *exceptions*). Verification exceptions normally identify:

8.2.1 Required steps not carried out by the laboratory (that is, incomplete analysis of all samples, lack of proper signatures, etc.),

8.2.2 Procedures not conducted at the required frequency (that is, too few blanks, duplicates, etc.),

8.2.3 Procedures which did not meet pre-set acceptance criteria (poor laboratory control sample recovery, unacceptable duplicate precision, etc).

8.3 The validation process begins with a review of the verification report or the laboratory data package, or both, to rapidly screen the areas of strength and weakness of the data set (tests of quality control). It continues with objective evaluation of sample data to confirm the presence or absence of an analyte (tests of detection) and to establish the statistical uncertainty (precision) of the measurement process for the analyte (test of uncertainty). Each data point is then qualified as to its integrity and dependability in the contest of all available laboratory data.

8.4 Examples of some important data project information that must be examined during the assessment of data are given in Table 1. Examples of some of the shortcomings that can occur are shown in Table 3. Some important characteristics of the data set that are frequently determined when examining quality control sample performance are given in Table 4. Data points not meeting the quality control criteria should be flagged and the magnitude and direction of any bias should be documented and made available for reference during the statistical evaluation processes that follow.

8.5 If project quality requirements are not met, further data assessment should not be undertaken until the data limitations are discussed with the project team. Data assessment cannot overcome basic design/execution flaws in the data collection process. Many times however, the project team can evaluate the problem and establish revised data quality objectives (different project expectations and new data requirements) factoring in the realities of the data collection effort which can then be used as the basis for data assessment.

TABLE 3 Common Data Requirements and Potential Shortcomings

Data Requirement	Potential Shortcomings
Number of samples	• Too few samples may have been collected or analyzed to be representative of the target population. • Too few samples were collected to narrow the estimate of the dispersion (variance, standard deviation, coeffiecient of variation, etc) of the measured results to acceptable levels.
Location of samples	• Samples were collected from the wrong locations due to error or inaccessibility.
Analyte/method	• Incorrect choice of analyte/method for the sample matrix
Quality control	• Measurement system not calibrated • Contamination found in field, trip, or method blanks • Method performance on reference samples unsatisfactory • Calculation errors
Method sensitivity	• Failure to meet minimum detectable limits
Method precision	• Failure to achieve satisfactory duplicate results for analysis of field samples due to sample characteristics or other analytical problems
Method bias	• Failure to demonstrate method performance on reference materials or analytical standards • Failure to demonstrate satisfactory target analyte spike/surrogate recoveries in field sample analysis
Interferences	• Presence of unanticipated materials/analytes in field samples that render accurate analysis suspect
Action level	• Not provided

9. Statistical Evaluation of Data Sets

9.1 The US EPA *Guidance for Data Quality Assessment, QA/G-9*(**1**)[4] is a good source for information on the following statistical approaches to data assessment.

9.1.1 *Continuous Data*:

9.1.1.1 Continuous data are data where the values of the individual samples may vary form zero to any maximum value. Examples of continuous data are the concentration of a constituent in soil or the percent moisture in an environmental sample. This is the type of information most frequently collected in environmental waste management projects. It is normally used to establish a statistical characteristic of the target population which is then compared to a decision level resulting in an action. This is referred to as the "decision rule" and normally takes the form:

[4] The boldface numbers given in parentheses refer to a list of references at the end of the text.

TABLE 4 Information Derived From Quality Control Samples[A]

Type of QC Sample	Type of Information										
	Precision			Bias			Contamination				
	Sampling	Splitting	Preparation and Analysis	Spiking	Field/ Shipping/ Storage	Laboratory	Containers and Preservatives	Field Environment	Equipment	Cross-Contamination	Laboratory
Replicates											
Splits, field		X	X								
Collocated, field	X		X								
Splits, laboratory		X	X								
Spikes											
Field				X	X	X					
Laboratory, matrix				X		X					
Blanks											
Trip							X			X	
Field							X	X		X	
Equipment							X	X	X	X	
Method											X

[A] Can be assessed using numerical techniques.

If (characteristic of the population) (method of comparison) (action level), then (action). Otherwise, (alternate action).

where the items in parentheses are determined by the project team on a project-specific basis. Two examples are:

If (the average concentration of mercury in the top 15 cm of soil over the site) (is greater than) (100 mg/kg), then (excavate the top 30 cm of soil and dispose of in a RCRA landfill). Otherwise, (no remediation is required).

and:

If (less than one half the randomly selected waste oil drums have an average organic halide concentration) (of less than 500 ppm), then (composite the contents of all drums and use it as boiler fuel). Otherwise, (send all drums to a RCRA treatment and disposal facility).

9.1.2 Before beginning the statistical interpretation of a continuous data set, plots of the data should be constructed to guide the statistical interpretation of the data that follows. Examples of the types of plots that can be constructed are:

9.1.2.1 Concentration versus time, and

9.1.2.2 Concentration versus location in two or three dimensions as appropriate

9.1.2.3 These types of plots provide a picture of the distribution of the parameter of interest and permit the identification of strata as a function of time or location. Plots also identify data points which are abnormally high or low with respect to the surrounding data. These are potential outliers and they can be more rigorously evaluated by the verification and validation process to determine whether there is an analytically-related explanation. This information will identify random or stratified data sets and outliers or QC-failed data prior to statistical evaluation.

9.1.3 *Normally Distributed Data*:

9.1.3.1 Once the data evaluation described above have been completed, statistical techniques should be used to evaluate the data against the decision criteria. The key steps in the sequence to evaluate continuous data are shown in Fig. 2.

9.1.3.2 The first step is to determine if the data are normally distributed. That is, are there an approximately equal number of values that are less than and greater than the mean and is the range of values approximately equal on either side of the mean (See Fig. 2). This property of normal distribution is a reasonable model of the behavior of certain random phenomena and can be used to approximate many kinds of data.

9.1.3.3 There are several graphical techniques that can be applied to determine if data are normally distributed. Among them are: stem- and leaf- diagrams, histogram/frequency plots, box and whiskers plots, ranked data plots, quantile plots, and, normal probability plots (quantile-quantile plots).

9.1.3.4 The use of plots to determine if data are normally distributed involves a subjective decision on the part of the individuals making the assessment. This is easy when the data are very non-normal but more difficult as the data approach normal distribution. There are series of formal numerical methods to test for normal distribution. The Shapiro - Wilkes test can be applied to data sets of less than 50 samples. For larger size sample sets (up to 1000 data points), Fillben's Statistic is frequently used. Both methods are difficult to implement by hand because of the large number of calculations required but are readily accomplished by computer programs.

9.1.4 Once the normal distribution of the data is shown, the straightforward calculation of the statistical quantities used in the project decision rule can be performed. For example, the two-sided confidence limits for the mean (that is a parametric population characteristic) can be performed. This allows the data user to determine the interval in which the true mean is expected to be found with specified confidence. The mean lead level, interval and confidence are frequently expressed as:

the level of lead in the soil is $X \pm x$ *at the* 95 % *confidence level*

9.1.4.1 The width of the interval, $\pm 2x$, can be calculated for varying degrees of confidence (selected by the data user) to meet project-specific tolerable error rates for making the

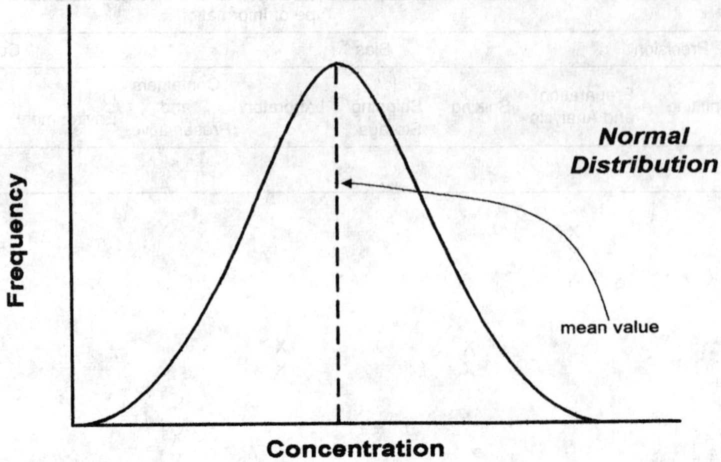

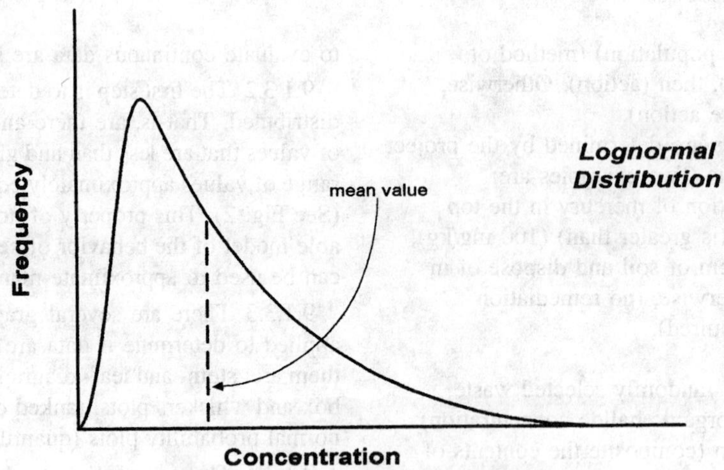

FIG. 2 Two Types of Data Distribution

correct decision. For a two-sided confidence interval, *x* is given by:

$$x = \frac{(t_{97.5,\,n-1})\,(s)}{\sqrt{n}} \quad (1)$$

where:
s = standard deviation,
n = the number of samples, and
t = the student *t*-statistic.

For the one-sided confidence interval, *x* is given by:

$$x = \frac{(t_{0.95,\,n-1})\,(s)}{\sqrt{n}} \quad (2)$$

9.1.5 Some types of statistical quantities which can be calculated from normally distributed data include, but are not limited to: mean, range, variance, standard deviation, coefficient of variation, and, confidence limits.

9.1.6 The choice of which statistics should be calculated is dependent on the characteristic of the population that will be used in the decision rule. After the appropriate statistical quantities are calculated from the field sample data, they should be compared to the assumed values which were the basis of the DQO calculation of tolerable error rate at the decision level.

9.1.6.1 Fig. 3 shows four examples of the comparison of various types of laboratory data against a regulatory decision level. In the examples, the upper confidence limit of the data set is compared to regulatory decision level. The figure represents four common characteristics of continuous data sets. These are: unbiased and precise, unbiased and imprecise, biased and precise, biased and imprecise.

9.1.6.2 To distinguish between the biased and unbiased situations pictured in the figure, one can refer to the verification and validation results described previously. The positive bias displayed in the bottom two examples of Fig. 3 should be reflected in high matrix spike recoveries and higher than normal recoveries on reference samples or laboratory control samples. None of the examples in Fig. 3 reflect sampling bias, they only apply to analytical bias.

9.1.6.3 The results shown in Fig. 3 may also prompt a re-examination of the decision criteria reached in the DQO process. The impact of imprecision and bias on the decision

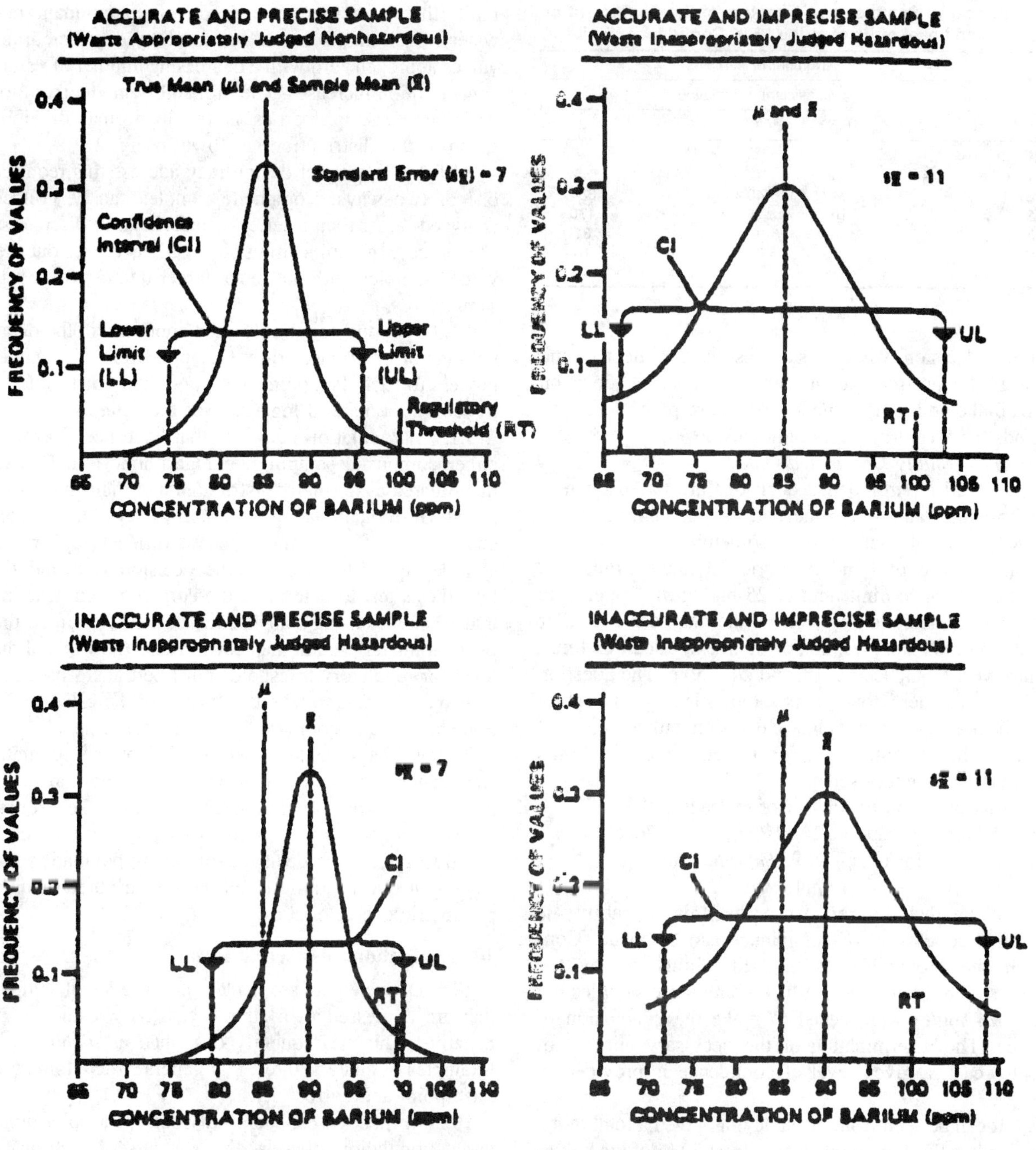

NOTE: In All Cases, Confidence Interval for $\mu = \bar{x} \pm t_{.20}\, s_{\bar{x}}$.

FIG. 3 Four Examples of Laboratory Data

making process increases as the mean approaches the regulatory threshold (or action level) and as imprecision increases. At some mean value, increased variance in the mean will determine that a correct decision cannot be reached with acceptable certainty within the project budget.

9.1.6.4 The opposite of the above example may also occur, the variance may be much less than estimated and the mean value of the parameter of interest may be low compared to the action level. It follows then, that the decision error rate and overall certainty will be improved so that a much higher degree of confidence in the final decision can be attained.

9.1.7 The complex inter-relationship between confidence level, relative error, action levels, variance, number of samples, and population characteristic will determine if an acceptable decision can be reached. Therefore, the data assessor must evaluate if the final outcome meets the data quality objective level of confidence and is within the tolerable error window. If so, the data can be said to be acceptable for making the intended decision. If not, the DQO objectives for the final decision must be changed or additional sampling and analysis must be conducted to meet the objectives. Table 5 (**2**) shows how the two variables, level of error and variance, impact the correct number of samples.

9.1.7.1 The assessor must determine whether the data conform to the project design criteria for level of error and coefficient of variation in the data for the desired confidence

TABLE 5 Number of Samples as a Function of the Coefficient of Variation and Level of Error at the 95 % Confidence Limit

	Number of Samples				
	Coefficient of Variation				
	0.1	0.5	0.65	1	2
Relative Error					
0.1	4	96	162	384	1537
0.2	1	24	40	96	384
0.3	...	10	18	42	170
0.5	...	4	6	15	61
1.0	...	1	2	4	15
2.0	...	...	...	1	4

level in the decision when *n* samples were collected and analyzed. If the design criteria were not met, the decision makers can take additional samples, use more precise analytical methods, or accept a lower confidence level.

9.1.8 *Non-Normally Distributed Data*:

9.1.8.1 Not all environmental decisions are based on normally distributed data. When the data are not normal, non-parametric statistical methods can sometimes be used. An example is the use of non-parametric tolerance limits **(3)**. Suppose a compliance limit limit of 25 mg/kg of copper is in place at a soil remediation project. Further suppose that after the collection of six samples, non of them exceeded this limit. The values were 9, 11, 13, 15, 18, and 24 mg/kg. The question may be asked whether a total of six samples is large enough to allow for the making of a complinace decision with a high level of confidence. In the context of non-parametric tolerance limits the question can be expressed as:

What is the probability of of exceeding the largest measured value of 24 mg/kg at some level of confidence (for example, 95 %) when there are only six sample values?

9.1.8.2 When there is confidence that this probability is small, then a conclusion of compliance can be made. Conversely, if the probability exceeds the limits set by the stakeholders then the question of how many more samples not exceeding 24 mg/kg are needed to make the conclusion of compliance. The determination of the necessary number of samples to reach a given level of confidence is provided in Table 6 **(2)**.

9.1.8.3 It can be seen from the table that, given a total of six samples, the it is 95 % probable that at least 50 % of the values from this site do not exceed 24 mg/kg. If the stakeholders prefer an 80 % degree of confidence, then a total of fourteen samples not exceeding 24/kg are needed.

9.1.9 Many environmental data sets represents populations where the parameter of interest approaches a lower theoretical limit (that is, zero concentration of contaminant in soil or water). Such data sets are not normally distributed, and most values approach zero with a decreasing number of values as the concentration increases. The probability model that most often describes these properties is the lognormal distribution. A graph of this distribution is shown in Fig. 2.

TABLE 6 Relationship of Degree of Confidence to the Percentage of the Population< Maximum Measured Value of 24 (n = 6)

Degree of Confidence	Percentage of Population < Maximum Measured Value
98	50
85	70
80	75
70	80

9.1.9.1 The project design may address this requirement in one of two ways. Composite samples can be collected and analyzed rather than a series of individual or discrete samples. The process of compositing physically averages out the higher valued samples with the much larger number of lower valued samples.

9.1.9.2 In one commonly used approached, the data set can be transformed (changed by the application of a mathematical process to each data point) into a normally distributed data set by taking the natural logarithm of the data.

9.1.9.3 Population data sets that have been normalized by either composite sampling or logarithmic transformation are then treated as normally distributed data. This means the same statistical reductions can be used to yield the population characteristic (for example, mean, variance, upper and lower confidence limits) used in the decision rule and the same hypothesis test to determine the correct action. It is important that whenever using transformed data to determine the statistical parameter used in the decision rule, the value of the action level (or regulatory threshold) must be transformed as well. It is also acceptable to take the antilog of the calculated population statistic and compare that to the action level.

9.1.9.4 If a suitable transformation of the population data base cannot be found which results in a normal distribution, more advanced statistical technique may be required (see statistical texts and monographs). The project team can also be asked to restate the decision rule using population characteristics that don't require a normally distributed data set (non-parametric characteristics).

10. Evaluation of Discrete Data

10.1 Discrete data are made up of a series of sample results that are expressed as a simple pass/fail, yes/no, or positive/negative. This is frequently described as *dichotomous data*. Examples of analytical test that generate dichotomous data are flash point and corrosivity.

10.2 In most cases, each individual data point represents a target population (that is, the contents of a drum) and the decision is made by comparing the individual value with the action level. For example, each drum sample that ignites when tested by a flammability test is determined to have failed the test (the dichotomous response) and the project response is to send the drum to a hazardous waste treatment and disposal facility. In this case (each sample represents the target population) no statistical assessment of the data is possible, and the decision is made on a qualitative basis. In most cases, the test results supported by the data requirements in Table 3 become the foundation for the qualitative determination.

10.3 In cases where a set of discrete samples from a large population (that is, 25 individual drum samples randomly selected from a set of 125 drums) are analyzed by a test producing dichotomous results, the proportion of drums failing the test is the statistic used in the decision rule. Tests of proportion require that the data set be normally distributed or capable of being transformed into a normally distributed set.

11. Outliers

11.1 Individual measurements that are extremely large or small relative to the majority of the data are generally suspected of misrepresenting the target population from which they were collected. These values are generally referred to a outliers. Outliers may be the result of sampling or analytical errors, or both, or may represent true extremes in the population. Before determining the value of population statistic used in the decision rule, statistical outlier tests should be performed to identify outliers which are then carefully investigated to see if the value may be the result of an error.

11.1.1 There are four commonly used techniques to evaluate data for outliers.

11.1.1.1 *Comparison to Historical Data*—If sufficient data exist over time from the same site or population, potential outlying values can be compared to past data from the same sampling point or location. If past data show the same high or low values as the current data, it is reasonable to assume that the current extreme values are valid. The data assessor should describe their findings in the evaluation report and a description of how the extreme values were used in past decision making processes provided. This will allow project decision makers to apply consistent decision logic over time.

11.1.1.2 *Trend Analysis*—As discussed in 9.1.2, plots of data points versus time or three dimensional location will reveal trends in data sets. If a trend exist, it can be used to support the validity of data points at the high or low extremes. Data points that are inconsistent with the trend should be more carefully evaluated to determine if they are real.

11.1.1.3 *Comparison with Companion Data Sets*—The existence of a correlationship between two data sets from a single population (for example, total petroleum hydrocarbons and volatile organic compounds in soil) is strong support for the validity of extreme values. If samples show both high total petroleum hydrocarbon (TPH) and toluene values, then both data points can be considered valid. A clear picture of this type of data relationship can be gained by plotting the two values for a single sample (TPH and toluene) on a set of coordinates keyed to two variables. The points that trend in the same general straight line can be considered valid. Points that lie off the line may do so because one or both of the variables is an outlier.

11.1.1.4 *Checks for Errors*—Errors in interpreting sampling instructions (location, sampling equipment, sample preservation), errors in analysis (incorrect sample size, uncalibrated instruments), errors in calculations or reporting can lead to apparent outliers in 11.1.1.1 through 11.1.1.3 above. The data assessor should follow up on potential outliers identified in the above processes and check for errors. If errors can be found and documented, the corrected data points should be introduced to the data set and the entire set re-evaluated.

11.2 Suspect values that can be documented based on some scientific observation or quality assurance basis should be omitted from the calculation of the statistic. They are reported to the data user in the final report but they are flagged as outlying values not to be used in the project decision or calculation of the population statistic (for example, mean value). If no error can be found, the outlying value should be retained but its presence and impact on the decision statistic clearly reported to the decision maker who may ask for re-sampling and analysis in critical situations.

11.3 There are other more powerful statistical tools that can be applied to the detection and treatment of outliers but they should be applied under the guidance of an experienced statistician.

12. Non-Detect Values

12.1 The treatment of non-detect values in a population data base can greatly affect the final population statistic. There are a variety of ways to treat values that lie below the detection limit of the analytical method. However, there are no general procedures that are applicable in all cases. The choice of data analysis method is dependent on the percentage of samples in the data base that are below the detection limit. The EPA's *Guidance for Data Quality Assessment* (**1**) provides a general discussion of several approaches outlined below. A brief discussion of some generally accepted approaches follows.

12.2 *Less than 15 % Non-Detects*—When less than 15 % of the reported data falls below the detection limit, it is possible to replace the non-detected values with a small number less than the detection limit. The number most frequently chosen is one-half the detection limit. For difficult to analyze matrices, one-half the practical quantitation limit can be used if the PQL has been determined for the analyses in the matrix being analyzed.

12.3 *Between 15 to 50 % Non-Detects*—There are three methods that can be applied when 15 to 50 % of the data are non-detects.

12.3.1 Cohen's method uses the sample mean, the variance, the total number of observations below the detection limit, and a set of statistical tables to determine corrected values for the sample mean and variance. Cohen's method is constrained by the requirement that the data above the detection limit be normally distributed and the detection limit always be the same.

12.3.2 The method of trimmed means determines the number of observations below the detection limit and directs that this number of measurements be deleted from both the high end and low end of the data. The remaining data is used to calculate the trimmed population mean and variance.

12.3.3 The method of winsorized means replaces the set non-detects (*n*-observations) with values equal to the *n*-lowest values above the detection limit. The highest values are adjusted downward in a similar fashion, that is, the *n*-highest values are dropped and replaced by the series of *n*-highest values remaining in the data set.

12.4 At the interval between 50 to 90 % non-detects, a test of proportions is recommended. For sets where less than 10 % of the values are above the detection limit, a Poisson Distribution approach is recommended.

12.5 All the methods mentioned here can be found in more detail in standard statistical texts.

13. Final Use of Data in the Decision Rule

13.1 Once it has been established that the data meet the

assumptions of the DQO process in terms of anticipated values using this assessment process, they are tested against the action level given in the decision rule using hypothesis tests. These tests determine if the data meet or exceed the action level and thereby direct the data user to the agreed upon course of action for the project. A discussion of the types of hypothesis testing is beyond the scope of this standard. Guidance can be found in standard statistical texts and monologues

REFERENCES

(**1**) Environmental Protection Agency, *Guidance for Data Quality Assurance, EPA QA/G-9*, 1996.
(**2**) Gilbert, R. O., *Statistical Methods for Environmental Pollution Monitoring*, Van Nostrand Reinhold, New York, 1987.
(**3**) Owen, D.B., 1962, *Handbook of Statistical Tables*, Addison-Wesley Publishing Company, Reading, Massachusetts.
(**4**) Cochran, W.G., *Sampling Techniques*, 3rd Ed., John Wiley and Sons, Inc., New York, 1977.
(**5**) Desu, M.M. and D. Raghavarao, *Sample Size Methodology*, Academic Press, San Diego, 1990.

Standard Test Method for Shake Extraction of Mining Waste by the Synthetic Precipitation Leaching Procedure[1]

This standard is issued under the fixed designation D 6234; the number immediately following the designation indicates the year of original adoption or, in the case of revision, the year of last revision. A number in parentheses indicates the year of last reapproval. A superscript epsilon (ε) indicates an editorial change since the last revision or reapproval.

1. Scope

1.1 This test method covers a procedure for the shake leaching of mining waste containing at least 80 % dry solids (≤ 20 % moisture) in order to generate a solution to be used to determine the inorganic constituents leached under the specified testing conditions, that conform to the synthetic precipitation leaching procedure (SPLP).

1.2 This test method calls for the shaking of a known weight of mining waste with acidic extraction fluid of a specified composition as well as the separation of the liquid phase for analysis. The pH of the extraction fluid is to reflect the pH of acidic precipitation in the geographic region in which the waste being tested is to be disposed.

NOTE 1—Possible sources of information concerning the pH of the precipitation in the geographic region of interest include state and federal environmental agencies, state universities, libraries, etc. pH values given in USEPA Method 1312, that are 4.2 east of the Mississippi River and 5.0 west of the Mississippi River and are based on acid precipitation maps, are examples of values that can be used. If the pH of the laboratory water is less than the desired pH for the site, do not use this test method, use Test Method D 3987.

1.3 This test method is intended to describe the procedure for performing single batch extractions only. It does not describe all types of sampling and analytical requirements that may be associated with its application.

1.4 The values stated in SI units are to be regarded as the standard.

1.5 *This standard does not purport to address all of the safety concerns, if any, associated with its use. It is the responsibility of the user of this standard to establish appropriate safety and health practices and determine the applicability of regulatory limitations prior to use.*

2. Referenced Documents

2.1 *ASTM Standards:*

D 75 Practice for Sampling Aggregates[2]
D 420 Guide to Site Characterization for Engineering, Design and Construction Purposes[3]
D 653 Terminology Relating to Soil, Rock, and Contained Fluids[3]
D 1129 Terminology Relating to Water[4]
D 1193 Specification for Reagent Water[4]
D 2234 Test Methods for Collection of a Gross Sample of Coal[5]
D 2777 Practice for Determination of Precision and Bias of Methods of Committee D-19 on Water[4]
D 3370 Practices for Sampling Water[4]
D 3987 Test Method for Shake Extraction of Waste with Water[6]
D 5744 Test Method for Accelerated Weathering of Solid Materials Using a Modified Humidity Cell.[6]
E 691 Practice for Conduction an Interlaboratory Test Study to Determine the Precision of Test Methods[7]
E 877 Practice for Sampling and Analysis of Iron Ores and Related Materials[8]

2.2 *EPA Document:*

U.S. Environmental Protection Agency, Synthetic Precipitation Leaching Procedure, Method 1312 in SW-846, Test Methods for Evaluating Solid Waste, Physical/Chemical Methods, Third Edition[9]

3. Terminology

3.1 *Definitions*—For definitions of terms used in this test method, see Terminology D 1129.

3.2 *Definitions of Terms Specific to This Standard*:

3.2.1 *mining waste,n*—overburden or waste rock excavated and disposed of during mining operations.

3.3 *Symbols*:

Variables listed in this test method are defined in the individual sections in which they are discussed.

4. Significance and Use

4.1 This test method is intended as a means for obtaining an extract of mining waste. The extract may be used to estimate the release of certain inorganic constituents of the waste under

[1] This test method is under the jurisdiction of ASTM Committee D34 on Waste Management and is the direct responsibility of Subcommittee D34.01.04 on Waste Leaching Techniques.

Current edition approved Feb. 10, 1998. Published June 1998.

[2] *Annual Book of ASTM Standards,* Vol 04.03.

[3] *Annual Book of ASTM Standards,* Vol 04.08

[4] *Annual Book of ASTM Standards,* Vol 11.01.

[5] *Annual Book of ASTM Standards,* Vol 05.05.

[6] *Annual Book of ASTM Standards,* Vol 11.04.

[7] *Annual Book of ASTM Standards,* Vol 14.02.

[8] *Annual Book of ASTM Standards,* Vol 03.06.

[9] Available from U.S. Government Printing Office, Washington, DC 20402. Request Publication Number 955-001-00000-1.

the laboratory conditions described in this test method. The user is advised to minimize the holding time between sampling and testing if the waste is suspected to contain reactive sulfide minerals.

NOTE 2—This method is not intended to be used as a kinetic test to simulate weathering of mining wastes. For kinetic testing of mining wastes, refer to Test Method D 5744 to determine release rates for constituents of interest.

4.2 The pH of the extraction fluid used in this test method is to reflect the pH of acidic precipitation in the geographic region in which the waste being tested is to be disposed (see 1.2).

4.3 An intent of this test method is for the final pH of the extract to reflect the interaction of the extractant with the buffering capacity of the waste.

4.4 This test method is not intended to provide an extract that is representative of the actual leachate produced from a waste in the field or to produce extracts to be used as the sole basis of engineering design. If the conditions of this test method are not suitable for the test material USEPA Method 1312 may be used.

4.5 This test method has not been demonstrated to simulate actual disposal site leaching conditions.

4.6 This test method produces extracts that are amenable to the determination of both major and minor (trace) inorganic constituents. When minor constituents are being determined, it is especially important that precautions be taken in sample storage and handling to avoid possible contamination of the samples.

4.7 This test method has been tested to determine its applicability to certain inorganic components in the waste. This test method has not been tested for applicability to organic substances, volatile matter (see Note 4), or biologically active samples. This test method has undergone limited testing to determine its reproducibility.

5. Apparatus

5.1 *Straight Edge*, such as a thin-edged yardstick.

5.2 *Impermeable Sheet*, of glazed paper, oil cloth, or other flexible material of a composition suitable to the analytes of interest.

5.3 *Drying Pans or Dishes*, (for example, aluminum tins, porcelain dishes, glass weighing pans), two per waste, suitable to the waste being tested and the instructions given in 9.2.

5.4 *Drying Oven*—Any thermostatically controlled drying oven capable of maintaining a steady temperature of ± 2°C in a range of 100 to 110°C.

5.5 *Desiccator*, having a capacity to hold the drying pans described in 5.3 and the crucibles described in 5.16.

5.6 *Laboratory Balance*, capable of weighing to 0.1 g.

5.7 *Erlenmeyer Flask*, 2-L capacity, equipped with a magnetic stir bar.

5.8 *Magnetic Stir Plate*.

5.9 *Graduated cylinder*, 1 or 2-L capacity.

5.10 *Pipet*, 1-mL capacity.

5.11 *Volumetric Flask*, 1-L capacity.

5.12 *Pipet*, 10-mL capacity. (Various other sized pipets, including micropipets, may be necessary for 9.3.2.)

5.13 *pH Meter*—Any pH meter with a readability of 0.01 units and an accuracy of ± 0.05 units at 25°C.

5.14 *Carboy-Type Container*, with spigot, 20 to 50-L capacity, of a composition suitable to the nature of the analyses to be performed (see Practices D 3370).

5.15 *Large Glass Funnel*.

5.16 *Crucibles*, porcelain, 20-mL capacity each, two per waste.

5.17 *Wash Bottle*, 500-mL capacity.

5.18 *Agitation Equipment*, of any type that rotates the extraction vessel in an end-over-end fashion at a rate of 30 ± 2 r/min such that the axis of rotation is horizontal and it passes through the center of the bottle (see Fig. 1).

5.19 *Pressure Filtration Assembly*—A pressure filtration device using pressure regulated compressed gas of a composition suitable to the nature of the analyses to be performed and equipped with a 0.45 or 0.8-μm pore size filter (see Note 6).

5.20 *Extraction Vessels*, cylindrical, wide-mouth, of a composition suitable to the nature of the waste and analyses to be performed, constructed of materials that will not allow sorption of the constituents of interest, and sturdy enough to withstand the impact of the falling sample fragments. The size of the container should he selected so that the sample, plus extraction fluid occupy approximately 95 % of the container. The containers must have water-tight closures.

5.20.1 Extraction vessels should be cleaned in a manner consistent with the analyses to he performed (see Section 13 of Practice D 3370).

6. Reagents

6.1 *Purity of Reagents*—Reagent grade chemicals shall be used in all tests. Unless otherwise indicated, it is intended that all reagents shall conform to the specifications of the Committee on Analytical Reagents of the American Chemical Society, where such specifications are available[10]. Other grades may be used, provided it is first ascertained that the reagent is of sufficiently high purity to permit its use without lessening the accuracy of the determination.

6.2 *Purity of Water*—Unless otherwise indicated, references to water shall be understood to mean Type IV reagent water at 18 to 27°C conforming to Specification D 1193. The method by which the water is prepared, that is, distillation, ion exchange, reverse osmosis, electrodialysis, or a combination thereof, should remain constant throughout testing.

6.3 *Sulfuric Acid/Nitric Acid Solution*—A 60/40 weight percent (weight %) mixture prepared using 95 to 98 weight % sulfuric acid and 69 to 71 weight % nitric acid. (See 9.3 for instructions on the preparation of this solution.)

7. Sampling

7.1 Obtain a representative sample of the mining waste to be tested by using, where available, ASTM sampling methods developed for the specific industry (see Practice D 75, Guide D 420, Terminology D 653, Test Method D 2234, and Practice E 877).

---

[10] *Reagent Chemicals, American Chemical Society Specifications*, American Chemical Society, Washington, DC. For suggestions on the testing of reagents not listed by the American Chemical Society, see *Analar Standards for Laboratory Chemicals*, BDH Ltd., Poole, Dorset, U.K., and the *United States Pharmacopeia and National Formulary*, U.S. Pharmaceutical Convention, Inc. (USPC) , Rockville, MD.

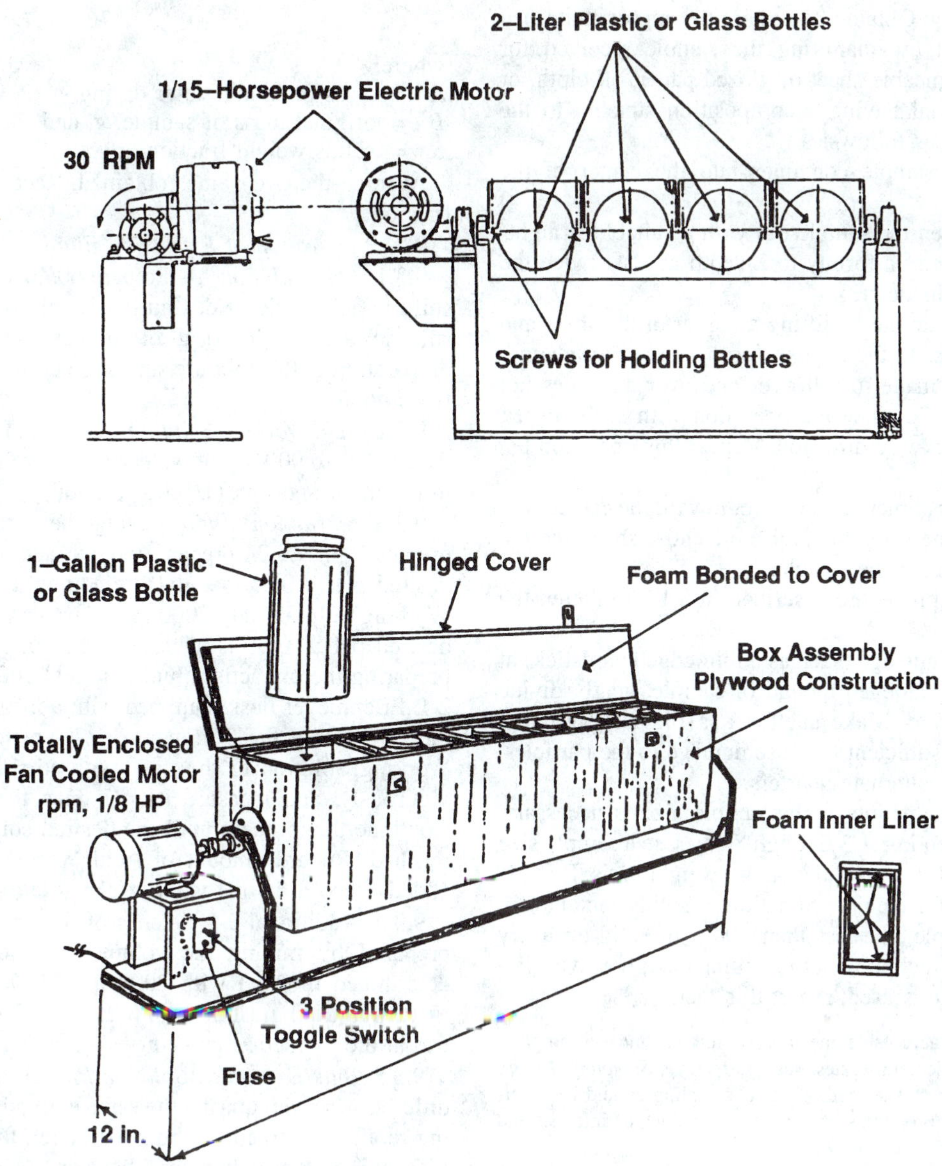

FIG. 1 Extractors

7.2 Sampling methodology for materials of similar physical form shall be used where no specific methods are available.

7.3 The amount of sample to be sent to the laboratory should he sufficient to perform the solids content determination as specified in 9.2, and to provide 100 g of sample on a dry weight basis for extraction.

7.4 It is important that the sample of the mining waste be representative with respect to the inorganic constituents to be determined.

NOTE 3—Information on obtaining representative samples can also be found in Pierre Gy's *Sampling Theory and Sampling Practice*.[11]

7.5 In order to prevent sample contamination or constituent loss prior to extraction, keep the samples in closed containers appropriate to sample type and desired analysis. See Practices D 3370 for guidance. Record the storage conditions and handling procedures in the report.

7.6 The time between collection and extraction of the sample should be determined by the nature of the sample and the information desired. See Practices D 3370 for guidance. Report the length of time between sample collection and extraction. The user is advised to minimize the holding time between sampling and testing if the waste is suspected to contain reactive sulfide minerals. Sample containing sulfide minerals may be preserved by filling the container with nitrogen gas and storing at 10°C.

8. Sample Preparation

8.1 Pass the gross sample through a 9.5-mm (3/8 in.) sieve and stage crush any oversize material no more than necessary to pass it through the sieve. Divide the gross sample in a riffle splitter with 25-mm (1 in.) chutes, repeatedly if necessary, to obtain a representative laboratory sample with a weight in the

[11] Pitard, F., *Pierre Gy's Sampling Theory and Sampling Practice*, 2nd Edition, CRC Press, Boca Raton, FL, 1993.

range of 250 to 500 g. Obtain a sample of the approximate size required in the test by quartering the sample received for testing on an impermeable sheet of glazed paper, oil cloth, or other flexible material having a composition suitable to the analytes of interest, as follows:

8.1.1 Empty the sample container into the center of the sheet.

8.1.2 Gently flatten the sample out with a suitable straightedge until it is spread uniformly to a depth at least twice the maximum particle diameter.

8.1.3 Remix the sample by lifting a corner of the sheet and drawing it low across to the opposite corner in such a manner that the material is made to roll over and over and does not merely slide along. Continue the operation with each corner, proceeding in a clockwise direction. Repeat this operation ten times.

8.1.4 Lift all four corners of the sheet toward the center and, holding all four corners together, raise the entire sheet into the air to form a pocket for the sample.

8.1.5 Repeat the procedure described in 8.1.2 to flatten the sample out.

8.1.6 With a straightedge (such as a thin-edged yardstick) at least as long as the flattened mound of sample, gently divide the sample into quarters. Make an effort to avoid using pressure on the straightedge sufficient to cause damage to the particles.

8.1.7 Discard the alternate quarters.

8.1.8 If further reduction of the sample size is necessary, repeat the steps given in 8.1.3 through 8.1.7. Use a sample size to provide at least 100 g of solid on a dry weight basis. Provide additional samples for the determination of solids content (see 9.2). Use of a sample size less than 100 g of solid on a dry weight basis for extraction is not recommended; however, if a different sample size is used, report this fact.

NOTE 4—For other acceptable methods of mixing and subsampling free-flowing solid particulate wastes, see Pierre Gy's *Sampling Theory and Sampling Practice.*[10] The method of subsampling should be determined by the physical properties of the waste, analytes of interest, and equipment available.

9. Procedure

9.1 Record a physical description of the sample to be tested, including particle size so far as it is known.

9.2 *Solids Content*—Determine the solids weight fraction of two separate portions of the sample as follows:

9.2.1 Dry to a constant weight, at 104 ± 2°C, two dishes or pans of size suitable to the solid waste being tested. Cool in a desiccator and weigh. Record the values to ± 0.1 g.

9.2.2 Place at least 50 g of the waste to be tested into each pan. Record the mass of sample in each pan to ± 0.1 g.

9.2.3 Dry for 16 to 20 h at 104 ± 2°C. Record the temperature and time of the drying period.

9.2.4 Cool to room temperature in a desiccator and reweigh. Record the mass to ± 0.1 g.

9.2.5 Repeat the steps given in 9.2.3 and 9.2.4, if necessary, until constant container-sample masses are obtained. Record the final weights and discard the dried samples following completion of this step.

9.2.6 Calculate the solids weight fraction of the sample from the data obtained in 9.2.1, 9.2.2, and 9.2.4 as follows:

$$S = A/B \tag{1}$$

where:
A = mass of sample after drying, g,
B = original mass of sample, g, and
S = solids weight fraction, g/g.

Average the two values obtained. Record the solids weight fraction.

9.3 *Preparation of Extraction Fluid*:

9.3.1 *Acid Solution A*—Prepare 60/40 weight % mixture of sulfuric acid/nitric acid. Cautiously mix 60 g of concentrated sulfuric acid with 40 g of concentrated nitric acid. The preparation of this mixture should be performed in a laboratory fume hood.

9.3.2 *Acid Working Solution*—Add 1 mL of Acid Solution A (See 9.3.1) to 900 mL of water in a 1-L volumetric flask, dilute to the mark and mix (1/1000 solution).

9.3.3 *Extraction Fluid*—Using the acid working solution prepared in 9.3.2, prepare the extraction fluid having the desired pH ± 0.05 (see 4.2) by adding a volume of the Acid Working Solution into 2000 mL of water with mixing until the desired pH ± 0.05 is achieved. A recommended method for preparing the extraction fluid is to add 2000 mL of water to a 2-L Erlenmeyer flask equipped with a magnetic stir bar. Place the Erlenmeyer flask on magnetic stir plate, and add the Acid Working Solution to the flask with stirring. Stir the mixture vigorously, and measure its pH once the solution is static. Continue this process until the desired solution pH ± 0.05 is reached. Record amount of Acid Working Solution solution added. Record the pH value of the solution.

9.3.4 Additional 2-L batches of the extraction fluid can be prepared by mixing the volume of acid working solution determined in 9.3.3 with 2000 mL of water. The pH of the extraction fluid must he within ± 0.05 of the desired value for use in the extraction procedure.

9.3.5 *Bulk Extraction Solution Preparation*—For extracting different wastes requiring the same extraction fluid or performing replicate extractions, bulk extraction fluid can be prepared in a carboy-type container of a composition suitable to the nature of the analyses to be performed, using the Acid Working Solution volume determined in 9.3.3 for each 2 L volume of extraction fluid desired, and measured for correct pH. If the pH is within ± 0.05 of the desired value, extraction fluid portions may be dispensed from the carboy for use in the extraction procedure.

9.4 *Extraction Procedure*:

9.4.1 *Test Sample*—Add at least 100 g (weighed to ± 0.1 g) of test sample on a dry weight basis to the extraction vessel. Calculate the minimum amount of as-received waste to add as follows:

$$M = 100/S \tag{2}$$

where:
S = solids weight fraction (g/g) determined in 9.2.6, and
M = minimum mass of as-received waste (weighed to ± 0.1 g) to add to the extraction vessel to yield at least 100 g of solid waste on a dry weight basis.

9.4.2 Calculate the dry test sample weight as follows:

$$M_d = MS \tag{3}$$

where:
M_d = mass of dry test sample, g,
S = solids weight fraction (g/g) determined in 9.2.6, and
M = mass of test sample, g.

9.4.3 Calculate the mass of water in the test sample as follows:

$$M_d = M - M_d \quad (4)$$

where:
M_w = mass of water in the test sample g,
M = mass of the test sample g, and
M_d = mass of the dry test sample, g.

9.4.4 The mass of the extraction fluid to add is twenty times the dry weight of the test sample and is calculated as follows:

$$M_e = 20M_d - M_w \quad (5)$$

where:
M_e = mass of extraction fluid, g,
M_d = mass of the dry test sample, g, and
M_w = mass of water in the test sample, g.

This will provide a solid-to-liquid ratio of 1:20 in the extraction vessel.

9.4.5 Agitate at 30 ± 2 r/min, continuously, for 18 ± 0.25 h at 18 to 27°C. Samples which release gasses during agitation must be checked periodically and vented. Record the agitation time and temperature.

NOTE 5—The venting of the container and filtration have the potential to affect the concentration of volatile compounds in the extracts.

9.4.6 Open the extraction vessel. Observe and record any visible changes in the sample and leaching solution. Record the pH of the waste/leaching solution slurry.

9.5 *Filtration*—Quantitatively transfer as much of the waste/leaching solution as possible through a large glass funnel to a pressure filtration device equipped with a 0.45 or 0.8-μm filter. Pressure filter the liquid through the filter using a compressed gas that will not contaminate or change the integrity of the sample in increments of 10 psi, increasing the pressure by 10 psi, only when extract is not expelled for a two minute interval. Do not exceed a maximum pressure of 50 psi. After the extract has passed through the filter, continue running gas through the filtration device at 30 psi for 3 min. The filtrate obtained is the extract mentioned in this test method (see 9.6). Determine the mass of the filtrate collected and report it as M_f for the extraction step. Measure the pH of the extract immediately, and then preserve the extract portions in a manner consistent with the chemical analyses or biological testing procedures to be performed (see Practices D 3370, Section 15).

NOTE 6—Analytical results may be affected by the type of filter used. If a 0.8 μm filter pore size is used, the resulting extract should he digested prior to metals analysis. The composition of the filter should also be considered. If the filter is composed of material that may contaminate the extract during filtration, the filter should be washed in the filtration device in a manner consistent with the chemical analyses or biological testing procedures to he performed on the extract. For example, for elemental analysis of the extract, if a filter composed of borosilicate glass fiber is used, in order to prevent contamination, it should be washed in the filtration device with a dilute acid solution and rinsed with approximately 2 L of water prior to filtration.

NOTE 7—It is recommended that all filtrations be performed in a hood.

9.6 Analyze the extracts for specific constituents or properties, or use the extracts for biological testing procedures as desired, using appropriate ASTM or other appropriate standard test methods. Where no appropriate standard test methods exists, other appropriate test methods may be used and documented in the report. Whether visible phase separation during storage of the extracts occurs or not, appropriate mixing should be used to ensure the homogeneity of the extracts prior to their use in such analyses or testing.

10. Report

10.1 Report the following information:

10.2 Source of information concerning the pH value of the precipitation in the geographic region of interest;

10.3 Source of the waste, date of sampling, methods of sampling and sample preservation, storage conditions, handling procedures, and length of time between sample collection and extraction;

10.4 Description of the mining waste, including its physical characteristics and particle size, if known (See 9.1);

10.5 Solids weight fraction (See 9.2);

10.6 Mass of mining waste on a dry weight basis extracted;

10.7 The pH of the extraction fluid used for the extraction;

10.8 Time and temperature used in the determination of solids weight fraction;

10.9 Agitation temperature and time;

10.10 Filter pore size used and filter composition;

10.11 Use of a prefilter, prefilter pore size, and composition;

10.12 Observations of changes in the test material or leaching solution (see 9.4.6);

10.13 The pH before and after filtration, and the results of specific analyses; and

10.14 Dates on which extraction was started and completed, preservation used for extract portions, and dates of analyses.

NOTE 8—Fig. 2 presents a detailed laboratory worksheet that may be helpful in performing the test method.

11. Precision and Bias

11.1 *Precision*—An interlaboratory study was undertaken to test the precision of the method in accordance with Practice D 2777 on two wastes from six data sets[12]. The results from the mining waste sample tested in the study are summarized in Table 1. The within laboratory repeatability was measured using six analytical standards in triplicate. The between laboratory reproducibility was measured on the mining waste using

TABLE 1 Statistical Summary

Constituents	Mean, μm/mL	Repeatability (*r*, Practice E 691), μg/mL	Reproducibility (*R*, Practice E 691), μg/mL
Barium	1.9	0.6	0.8
Calcium	1815	389	666
Lead	63	8	22
Magnesium	107	20	34
Manganese	36.7	5.6	15.4
Silicon	55	6	58
Zinc	297	87	98

[12] Supporting data for the precision and bias information are available from ASTM Headquarters. Request PNC:33-000004-56.

Sample Description:

Solids Weight Fraction Determination

Dish #1 mass after drying: = ____________

Dish #2 mass after drying: = ____________

Sample #1 mass before drying (B#1) : = ____________

Sample #2 mass before drying (S#2) : = ____________

Drying temperature: = ____________

Drying time: = ____________

Sample #1 mass after drying (A#1) : = ____________

Sample #2 mass after drying (A#2) : = ____________

Solids Weight Fraction #1 (S#1) : S = A#1/B#1 = ____________

Solids Content #2 (S#2) : S = A#2~B#2 = ____________

Average Solids Weight Fraction (S) = ____________

Extraction

Mass of test sample (M) : M - 100/S = ____________

Dry mass of test sample (M_d) : MS = ____________

Mass of water in the test sample: (M_w) : M - M_d= ____________

Mass of extraction fluid (M_a) : 20M_d- M_w = ____________

pH of extraction fluid: = ____________

Extraction time and temperature: = ____________

pH of Slurry: = ____________

Filter type and pore size: = ____________

Mass of filtrate collected (M_f) : = ____________

pH of extract: = ____________

Observations:

FIG. 2 Laboratory Worksheet

a pH 5.0 ± 0.05 synthetic precipitation extraction fluid.

11.2 The data generated in this collaborative study are specific to the test materials used in the study, the elements of interest to the study, the pH values of the extraction fluids used, and 0.45-μm filter filter pore-size filtration. For other materials, elements, pH values, and filter types, these data may not apply.

11.3 *Bias*—Determination of the extraction bias of this test method is not possible, as no standard reference material exists. The analytical bias, however, was evaluated in the research report.[11]

12. Keywords

12.1 extract; extraction fluid; leaching procedure; mining waste; shake extraction; SPLP; synthetic precipitation; waste leaching technique

Designation: D 6249 – 98

Standard Guide for Alkaline Stabilization of Wastewater Treatment Plant Residuals[1]

This standard is issued under the fixed designation D 6249; the number immediately following the designation indicates the year of original adoption or, in the case of revision, the year of last revision. A number in parentheses indicates the year of last reapproval. A superscript epsilon (ε) indicates an editorial change since the last revision or reapproval.

1. Scope

1.1 This document provides guidance for use of reactive alkaline materials (quicklime, hydrated lime, high lime fly ash, or other byproducts) for treating wastewater solids (biosolids) to reduce pathogen levels and achieve compliance with regulatory requirements. Federal (40 CFR, Part 503) regulations for use or disposal of biosolids became effective on March 22, 1993; refer to USEPA regulations and guidance documents for information on other treatment processes or for specific requirements for use or disposal of biosolids.

1.2 Additional requirements may be imposed by individual states, and these are available through state regulatory agencies that issue permits for treatment and use or disposal, or both, of biosolids.

2. Referenced Documents

2.1 *ASTM Standards:*

C 25 Test Methods for Chemical Analysis of Limestone, Quicklime, and Hydrated Lime[2]

C 110 Test Methods for Physical Testing of Quicklime, Hydrated Lime, and Limestone[2]

C 400 Test Method for Quicklime and Hydrated Lime for Neutralization of Waste Acid[2]

2.2 *USEPA Publication:*

Title 40, Code of Federal Regulations (CFR), Part 503, Standards for the Use or Disposal of Sewage Sludge; Final Rules, 58 FR 9248-9404[3]

3. Terminology

3.1 *acronyms*—these are defined by operating parameters (for example, time, temperature) whose values must be met in order for biosolids to be used in various ways as a nutrient source/soil conditioner. Ref. 40 CFR Part 257.

3.1.1 *PFRP*—Processes to Further Reduce Pathogens (equivalent to 503 Class A).

3.1.2 *PSRP*—Processes to Significantly Reduce Pathogens (equivalent to 503 Class B).

4. Significance and Use

4.1 Operators of power and other plants producing alkaline by-products and wastewater treatment plant operators needing to treat and manage wastewater solids will find this guide helpful in dealing with their materials.

4.2 This guide provides the tests, procedures, and parameters that should be considered to significantly reduce pathogens in wastewater treatment plant solids by the addition of manufactured or by-product alkaline materials **(1)**.[4]

5. Alkaline Materials Characteristics

5.1 *Chemical Composition*:

Alkaline materials may be tested for Available Lime Index (ALI) in accordance with the optional chemical test of Table 1. Other chemical components, if required, may be determined in accordance with the appropriate procedure when requested by the purchaser.

5.2 *Reactivity*:

5.2.1 Alkaline materials should be tested for pH and heat of hydration (heat rise or slaking rate) in accordance with the recommended tests of Table 2.

5.3 *Physical Characteristics*:

5.3.1 Alkaline materials should be tested to determine the particle size in accordance with the recommended physical tests of Table 3.

6. Process Performance

6.1 *PFRP (Class A) Alkaline Treatment of Biosolids*:

6.1.1 *Mixing*—Thorough mixing of the biosolids and stabilization reagent must be provided to ensure uniform pH distribution and pathogen reduction throughout the biosolids mass **(2)**. Effective mixing depends upon achieving the appropriate ratio of alkaline material to biosolids cake uniformly distributed throughout the treated biosolids.

6.1.1.1 Biosolids with a high moisture content will require less mixing energy than high-solids biosolids cake.

6.1.1.2 Biosolids characteristics will determine the proper type of equipment or system required for adequate mixing. Incomplete mixing can cause odor release during product storage or application and may lead to failure to meet regulatory requirements for pathogen and vector control.

---

[1] This guide is under the jurisdiction of ASTM Committee D-34 on Biotechnology and is the direct responsibility of Subcommittee D34.03 on Treatment.

Current edition approved April 10, 1998. Published September 1998.

[2] *Annual Book of ASTM Standards*, Vol 04.01.

[3] Available from the Superintendent of Documents, U.S. Government Printing Office, Washington, DC 20402.

[4] The boldface numbers in parentheses refer to the list of references at the end of this guide.

TABLE 1 Optional Chemical Test

Test Method	Component	Specification
C 25	Available Lime Index (ALI), %	[A]

[A] To be specified only as required by purchaser.

TABLE 2 Recommended Reactivity Tests

ASTM Test Method	Component	Specification
C 400	pH	≥ 12.0[A]
C 110[B]	Heat rise, C	[C]

[A] Based on 40 CFR Part 503 for pH >12 for 2 h or more.
[B] Modify Test Method C 110 to proportion alkaline reagent in lieu of quicklime. Alkaline material and water ratio may need to be modified to obtain measurable results. Any modification of Test Method C 110 must be clearly stated on the analysis report.
[C] To be specified only as required by purchaser.

TABLE 3 Recommended Physical Test

ASTM Test Method	Component	Specification
C 110	Amount retained on 600 μm (No. 30 mesh), %	[A]
C 110	Amount retained on 75 μm (No. 200 mesh), %	[A]

[A] To be specified only as required by purchaser.

6.1.2 *Particle Size*:

6.1.2.1 Given an adequate moisture supply using alkaline agents (for example, CaO) with smaller particle sizes will facilitate rapid and efficient mixing of agents with biosolids and increase reaction rates and pH, resulting in higher temperatures and greater pathogen reduction.

6.1.2.2 Since dusts are more easily generated from finely divided particles, precautions should be taken to prevent exposure to eyes and mucous membranes, which may result in irritation.

6.1.2.3 Reactivity and particle size also affect the rate of dust and mist emissions from reactors or mixing devices, or both. Particulate release may require scrubbing, water spray, or other emission controls on reactors or mixing devices for aesthetic reasons or to meet regulatory requirements.

6.1.2.4 Very small particle size may also lead to "air slaking" or recarbonation of active lime particles if the material is exposed to high humidity. Air slaked/recarbonated materials will not achieve the pH necessary to meet regulatory requirements.

6.1.3 *Reactivity (Heat and pH Elevation)*:

6.1.3.1 Reactivity is dependent upon the interaction between the alkaline reagent and the material to be treated. Reactions occur as the alkaline material contacts the biosolids. The finer the alkaline product, the greater the potential for a more rapid pH/temperature elevation.

6.1.3.2 Reagent reactivity affects mixing time and dosage rate and must be considered in process design.

6.1.4 *Moisture Content*:

6.1.4.1 Adequate moisture must be present to react with the free CaO (as measured by Available Lime Index, ALI, as per Test Methods C 25) to generate heat and elevate pH. Generally, dry biosolids cakes (18 to 30 %) require a more intimate mix to ensure proper penetration and reaction than is required by wet biosolids (less than 18 %).

6.1.4.2 The calcium oxide in the reagent must react with the moisture in biosolids (hydration) producing calcium hydroxide and heat. The moisture content in the biosolids mass must be sufficient to allow the hydration reaction to occur between the selected reagent (CaO concentration and fineness) and biosolids mass.

6.1.4.3 Biosolids cakes with a high moisture content will tend to react faster than biosolids with a low moisture content.

6.1.5 *Biosolids Type*:

6.1.5.1 Case-by-case alkaline material demand should be determined for each biosolids type through pilot testing using the actual biosolids cake and proposed reagents for each project. Develop process guidelines for alkaline additions by biosolids type and alkaline additive characteristics.

6.1.5.2 Biosolids with a high moisture content may require a higher dose ratio than drier dewatered biosolids cake when dosage ratios are expressed on a dry weight basis.

6.1.6 *Reaction Time and Curing Time*:

6.1.6.1 Heat will be generated as hydration of calcium oxide occurs. The reaction time will vary depending on reagent composition moisture content of the biosolids mass, and mixer efficiency.

6.1.6.2 Reaction times to effect pathogen reduction are established by applicable federal and state regulations. Reaction/cure times depend upon a number of variables and should be pilot-tested using the actual biosolids cake, alkaline admixture, mix unit, cure vessel, and testing protocol and acceptance criteria to assure compliance with regulatory standards.

6.1.6.3 For alkaline treatment processes, one of the three performance criteria is required.

(1) The time-temperature relation established in 40 CFR 503.32 (a) (3) (Alternative 1). Selected time-temperature values are as follows:

Biosolids	Temp, °C	50	55	60	65	70	75	80
Moisture >7 %	Time, hours	316	63	13	2.5	0.5	0.10	0.020
Moisture <7 %	Time, hours	120	24	4.8	0.95	0.19	0.04	0.008

(2) The pH-time, temperature-time, drying procedure in 40 CFR 503.32 (a) (4) (Alternative 2). Basically, the biosolids are held at a pH above 12 for 72 hours with a 12-hour period in which the temperature exceeds 52°C, followed by air drying to a solids content exceeding 50 %.

(3) Pasteurization (40 CFR 503, App. B, Part B—PFRP Option 7) in which the biosolids are maintained at a minimum temperature of 70°C for 30 minutes.

6.1.6.4 Ammonia or other odors released may require water spray, scrubbing, gas capture, or control of emission.

6.1.7 *Reaction/Cure Vessels or Containers*:

6.1.7.1 Consideration should be given to minimize heat losses through materials management, configuration, and materials of construction for processes that require extended curing times.

6.1.7.2 Proper temperatures can be maintained without an insulated vessel by adding adequate alkaline reagent to compensate for heat loss. However, some situations may benefit from an insulated vessel to efficiently retain the heat to meet PFRP temperature requirements.

6.1.7.3 For processes using a windrow, at a minimum, the mixture should be at least 18 inches thick at all locations in the pile to ensure heat retention throughout the entire mass for the applicable curing time. Thin areas at the pile extremities should be avoided as they will not retain adequate heat and can lead to potential regrowth and recontamination of the entire mass.

6.1.8 *Process Testing Requirements*:

6.1.8.1 Process testing requirements vary with the specific alkaline process selected.

6.1.8.2 Temperature and pH measurements for the requisite time periods should be recorded to comply with pathogen reduction (and vector attraction reduction). Daily testing may include monitoring and documenting the elevated temperature and pH for a predetermined period of time (see 6.2.6 for details). Some methods also require documentation of reduced moisture content and mixing.

6.1.8.3 Procedures to monitor or collect samples for analysis are developed for each project based upon site-specific conditions considering the process selected, equipment utilized, volumes of materials to be processed, local state and federal regulatory requirements, and local conditions **(3)**.

6.1.9 *Process Testing Schedule*:

6.1.9.1 Testing must be conducted in accordance with federal, state, and local regulations. Product testing, to meet end-use requirements, will be site-specific. Under 40 CFR Part 503 regulations, pathogen (salmonella, virus, protozoan, and helminth egg) or indicator organism testing and pollutant concentration (metals) testing requirements are:

Amount of Biosolids (dry metric tons/365 days)	Frequency
> 0 and < 290	Once per year
≥ 290 and < 1500	Once per quarter
≥ 1500 and < 15 000	Once per 60 days
≥ 15 000	Once per month

6.1.9.2 In addition, at a minimum, daily operating records should be maintained documenting compliance with applicable requirements (for example, process time/temperature, pH, solids concentrations, and fecal coliform levels) **(4)**.

6.1.10 *Product Solids Content*:

6.1.10.1 Product utilization may be affected by solids content to aid in control of microbial regrowth during storage to minimize odor potential at application sites and during storage or to improve end-product marketability and physical handling characteristics.

6.1.11 *On-Site Storage*:

6.1.11.1 On-site storage may be required for a land application or marketing program of the tested product. Programs that produce an end-product for sale should consider on-site storage capacity to meet the seasonal fluctuations in market demand, the scheduling needs of the consumer, and production rates of the generator.

6.1.11.2 On-site storage may require odor control for end-products with high moisture content or a low alkaline reagent dosage rate.

6.2 *PSRP (Class B) Alkaline Treatment of Biosolids*:

6.2.1 *Mixing*—Thorough mixing of the biosolids and chemical reagent must be provided to ensure uniform pH distribution and pathogen reduction **(2)**. Mechanical mixing to achieve a homogeneous blend of reagent throughout the biosolids mass depends upon a number of factors, including achieving the proper ratio of alkaline reagent to biosolids, and sufficient moisture to enable the reaction to occur.

6.2.1.1 Incomplete mixing or an inadequate reagent dosage rate can cause odor generation and release during product storage or application and failure to meet regulatory pathogen or vector control requirements.

6.2.2 *Alkaline Reagent Particle Size*:

6.2.2.1 Reactions occur as the alkaline reagents contact the moist biosolids particles. The rate of reaction of the alkaline reagent tends to increase with: (1) finer reagent particle size, and (2) increased free moisture content. Reagents with a finer (smaller) particle size distribution generally are more easily and uniformly blended into the biosolids.

6.2.2.2 When using an alkaline slurry, or treating liquid biosolids, particle size may not be as critical as long as sufficient mixing and reaction time are provided.

6.2.3 *Reactivity (pH elevation)*:

6.2.3.1 Care must be taken to ensure moisture will not come into contact with the reagent prior to entering the mixer. Air slaking of the alkaline reagent can be a problem in long-term storage or pneumatic transfer systems. Guidance for proper storage of reactive alkaline materials can be found in Lime—Handling, Application and Storage, National Lime Association Bulletin 213.

6.2.3.2 Increasing the alkaline reagent oxide concentration may increase potential for reactivity.

6.2.4 *Moisture Content*:

6.2.4.1 Moisture content affects the pH by permitting the alkaline reagents to dissociate, releasing hydroxide ions, increasing the pH. As moisture increases, it is easier for the alkaline material to disperse, penetrate, react, and dissociate, thus generating hydroxide ions throughout the mixture.

6.2.4.2 Care should be exercised when working with biosolids cakes with greater than 30 % solids content to ensure a complete uniform reaction is achievable.

6.2.5 *Biosolids Type*—Biosolids characteristics will impact on the required dosage rates for specific alkaline reagents. Therefore, site-specific process guidelines should be developed through actual pilot studies for alkaline additions by biosolids type and alkaline additive characteristics. Liquid biosolids require less mixing energy to achieve a uniform distribution of alkaline reagent (and therefore uniform pH) throughout the mixture than do cake biosolids.

6.2.6 *Reaction Time*:

6.2.6.1 For dry reagents, after uniform mixing with biosolids is complete, the alkaline metal oxides must hydrate to hydroxides and penetrate the biosolids mass. The presence of hydroxides will increase biosolids pH.

6.2.6.2 When working with biosolids cake with low moisture content or reagents with low reactivity due to particle size or oxide content, longer curing time (contact period) may be required to allow the hydration reaction to occur.

6.2.6.3 At elevated pH values, ammonia compounds in the biosolids or water dissociate to release ammonia gas (NH_3). The biosolids may exhibit a transient pH of 12 or more due to unreleased ammonia in the form of unstable ammonium hydroxide (NH_4OH). Once the NH_4OH dissociates and the ammonia dissipates, the pH may fall. Therefore, the amounts of reagent added should take this factor into account and not be based on the short-term NH_4OH effect on pH. The strong odor of ammonia release is a signal to be alert to this process occurring.

6.2.7 *Process Testing Requirements*—pH > 12 for two hours.

6.2.7.1 When testing a liquid/solid mixture with a system designed for liquids (pH probe), care must be taken to keep flow of sample moving across the pH probe surface for accuracy. Clean the electrode frequently, paying particular attention to the reference electrode junction. Bits of adherent biosolids or alkaline agent can strongly alter the measured pH, resulting in erroneous values not representative of the entire mass of material. pH measurements should be made at 25°C, or corrected for sample temperature differences as shown in Table 4, which demonstrates the temperature/pH relationship as the measured pH approaches the regulatory limit.

NOTE 1—At high pH (above 11), temperature has a significant effect on pH readings. Therefore, at high pH levels, the reading must be corrected to 25°C, standard temperature. Table 4 and following examples may be useful for a better understanding.

NOTE 2—pH meters often have temperature correction controls. These controls correct for probe temperature and conductivity only. Changes in the concentration of the test sample due to solubility that changes with temperature are not compensated by these controls. For example, at 30°C:

$$\text{Correction Factor (CF)} = 0.03 \text{ pH units} \times (T_{meas} - 25°C)$$
$$CF = 0.03 \times (30 - 25); CF = 0.03 \times 5$$
$$CF = +0.15$$
$$\text{Actual pH} = \text{Measured pH} \pm \text{the Correction Factor}$$
$$\text{Actual pH} = 12.304 + 0.15 = 12.454$$

6.2.7.2 pH—Calibrate routinely with laboratory grade standards developed for high pH monitoring.

6.2.8 *On-Site Storage*:

6.2.8.1 PSRP (Class B) biosolids have regulatory restrictions preventing application of material to a site during inclement weather, standing water, ice- or snow-covered fields.

6.2.8.2 On-site storage should be considered to hold biosolids, unless backup options are available when application sites cannot be used. Storage design must address regulatory concerns for runoff or leachate and local site-specific concerns for odor. Storage capacity should be consider worst-case weather conditions.

6.2.8.3 Microbial recontamination and regrowth is possible at pH < 10, with concomitant odor potential.

7. End Use Criteria

7.1 *Matrix of Product Factors*:

TABLE 4 pH Temperature Correction Factors at Varying Temperatures from Standard

Temperature Measured,° C	pH Meter Reading	Correction Factor (CF), +/–		
30	12.304	+	0.15	
29	12.334	+	0.12	
28	12.364	+	0.09	
27	12.394	+	0.06	
26	12.424	+	0.03	
25	12.454		0.00	12.454[A]
24	12.484	–	0.03	
23	12.514	–	0.06	
22	12.544	–	0.09	
21	12.574	–	0.12	
20	12.604	–	0.15	

[A] pH of a saturated lime solution at 25°C.

7.1.1 Table 5 addresses various factors that influence specific product criteria for alkaline-treated biosolids. The relative importance of these criteria will depend on the type of product and its intended use; for example, for agricultural liming material, $CaCO_3$ equivalency and friability for handling will be important, along with metal levels, which must meet Table 3 of EPA's 503 Rule for a product to be sold or distributed.

EPA 40 CFR Part 503.13 (13 (Table 3) Metal Concentration Limits

Metal	Pollutant Concentrations (mg/kg)
Arsenic	41
Cadmium	39
Copper	1,500
Lead	300
Mercury	17
Molybdenum	75[A]
Nickel	420
Selenium	100
Zinc	2,800

[A] This is the ceiling concentration for Molybdenum. EPA is still developing a pollutant concentration value.

8. Federal Regulations

8.1 *Introduction*:

8.1.1 *Pathogen reduction*—Alkaline materials reduce pathogens in municipal wastewater solids by creating a series of stresses on the microbial population in the solids. The addition of an alkaline material raises the pH of the mixture to levels that are toxic to some microbes, resulting in their reduction or elimination (**5**). The exothermic reaction of quicklime with water also releases heat (often intense) and the resultant temperature rise helps inactivate pathogens. Hydration of alkaline oxide materials helps dry the biosolids, further stressing the microorganisms.

8.2 *Class A Pathogen Reduction*:

8.2.1 All 40 CFR Part 503 Class A alternatives must meet the following criteria:

8.2.1.1 Either the density of fecal coliform in the biosolids must be less than 1000 most probable numbers (MPN) per gram total solids (dry-weight basis), or the density of *Salmonella* sp. bacteria in the biosolids must be less than 3 MPN per 4 g of total solids (dry-weight basis).

8.2.1.2 Either of these requirements must be met at one of the following times: *(1)* when the biosolids are used or disposed; *(2)* when the biosolids are prepared for sale or give-away in a bag or other container for land application; or *(3)* when the biosolids or derived materials are prepared to meet the requirements of CFR 503 (**6**).

8.2.1.3 Vector attraction reduction must take place before or at the same time as pathogen reduction except when the pH adjustment, percent solids vector attraction, injection, or incorporation options are met for vector attraction reduction.

8.3 *Alkaline Processes for Meeting Class A Pathogen Reduction Requirements*:

8.3.1 Alternative for Meeting Class A: Biosolids Treated in a High Temperature Process (through Chemically Produced Heat).

8.3.1.1 The time-temperature relation established in 40 CFR

TABLE 5 Matrix of Product Factors

End-Use Criterion	Related Process Parameter: Input Biosolids	Alkaline Additive	Process Heat	Other Parameters
Percent solids	Percent solids	Percent solids	Heat of hydration and evaporative heat loss	
Friability/particle size/plasticity	Percent solids	Amount added		Percent solids of final mix; bulking agent; type and amount of mixing
Nutrient content	Nutrient content	Amount added (dilution)		Amount of ammonia driven off; amount of "mineralization" of nitrogen in chemical reaction; nutrient content of bulking agent
Organic matter content	Percent volatile solids	Amount added (dilution)	Loss of volatile organics	"Mineralization" of organic content in chemical reaction; organic content of bulking agent
Metals content	Metals content	Metals content		Metals content of bulking agent
Liming potential	$CaCO_3$ equivalency	Ca content Mg content		Ca, Mg content of bulking agent
pH level and duration	pH, percent solids, ammonia and alkaline metals content, buffering capacity	CaO $Ca(OH)_2$		Final pH of product; granularity or plasticity of end-product (permeability to air and water allowing reaction to product carbonate)
Pathogen reduction		pH 12 (CaO and $Ca(OH)_2$ content)	Heat level and duration	Dryness of end-product; chemical destruction of pathogens
Odor potential	Percent ammonia, percent volatile solids, digestion process, polymer type and content	pH-driven ammonia release; neutralization of malodorous organics	Volatilization of organics	Thoroughness of mix; odor stripping during process
Bearing strength	Non-volatile solids	Material available for forming carbonates; silicates; aluminates; aluminoferrites	Vaporization of water	Time since product was made; CO_2 reaction producing carbonates; characteristics of bulk agent(s)
Permeability		Cementitious compounds		Time since product was made; CO_2 process producing carbonates; plasticity/friability of bulking agent; characteristics of bulking agents; density

503.32(a)(3) (Alternative 1). (See 6.1.6 for time-temperature examples.)

8.3.2 Alternative 2 for Meeting Class A: Biosolids Treated in a High pH-High Temperature-Air Drying Process 40 CFR 503.32(a)(4).

8.3.2.1 Elevating the pH to greater than 12 (measured at 25°C) for 72 h or longer,

8.3.2.2 Maintaining the temperature above 52°C for at least 12 h during the period that the pH is greater than 12,

8.3.2.3 Air-drying to over 50 % solids after the 72-h period of elevated pH, and

8.3.2.4 Meeting all the requirements in 6.2.

8.3.3 Alternative 5 for Meeting Class A: Treated in a PFRP (Pasteurization) Process Listed in 40 CFR 503, App. B, Part B-Option 7

8.3.3.1 Combined with an Alkaline Method of Vector Attraction Reduction 40 CFR 503.33(b)(6),

raise the temperature to ≥70°C for a minimum of 30 minutes, and

8.3.3.2 Combined naturally with the vector attraction reduction, raise the pH to ≥12 for a minimum of two hours, and maintain the pH ≥11.5 for an additional 22 h.

8.4 *Alkaline Process for Meeting Class B Pathogen Reduction Requirements*:

8.4.1 Alternative 2—PSRP listed in Appendix B, 40 CFR Part 503: Alkaline stabilization—sufficient alkaline reagent is added to the biosolids to raise the pH of the biosolids to 12 after two hours of contact time.

8.4.2 Unlike the comparable Class A requirements, this alternative does not require microbiological monitoring for regrowth of fecal coliform or *Salmonella* sp. bacteria.

8.5 *Alkaline Process for Meeting Vector Attraction Reduction*:

8.5.1 Option 6 Alkali[5], addition under specified conditions:

8.5.2 Biosolids are considered to be adequately reduced in vector attraction if sufficient alkaline material is added to achieve the following:

8.5.2.1 Raise the pH to at least 12, measured at 25°C, and without the addition of more alkaline material, maintain a pH of at least 12 for two hours, and

[5] 40 CFR Part 503 uses the term "alkali" for any alkaline material. For beneficial use options, sodium (Na)-based alkaline materials are usually unsuitable due to the negative effects of sodium on soil physical properties.

8.5.2.2 Maintain a pH of at least 11.5 without addition of more alkaline material for an additional 22 h.

8.5.3 The conditions required under this option are designed to ensure that the biosolids can be stored for at least several days at the treatment works, transported, and then used or disposed without the pH falling to the point where putrefaction occurs and vectors are attracted.

8.6 *Septage Treatment*:

8.6.1 *Option 12–Alkaline Treatment for Domestic Septage:*

8.6.2 This option pertains only to vector attraction reduction for domestic septage. Under this option, the pH of domestic septage must be raised to at least 12 and remain at pH 12 or above for a minimum of 30 min during which no additional alkaline material may be added.

9. Keywords

9.1 alkaline stabilization; biosolids; high-lime byproducts; sewage sludge; stabilization; wastewater treatment plant

REFERENCES

(**1**) Lime—Handling, Application and Storage, *National Lime Association Bulletin 213*, Arlington VA. 1995.

(**2**) Technical Support Document for Reduction of Pathogens and Vector Attraction in Sewage Sludge, *USEPA, Office of Water*, NTIS: PB93-110609.

(**3**) POTW Sludge Sampling and Analysis Guidance Document, *USEPA, Permits Division*, Washington, DC, First edition. August 1989, NTIS: PB93-227957.

(**4**) Preparing Sewage Sludge for Land Application or Surface Disposal—A Guide for Preparers of Sewage Sludge on the Monitoring, Record Keeping and Reporting Requirements of the Federal Standards for Use or Disposal of Sewage Sludge,40 CFR Part 503, *USEPA, Office of Wastewater Enforcement and Compliance*, EPA/813-B-93-002a. NTIS: PB94-102415.

(**5**) Environmental Regulations and Technology: Control of Pathogens and Vector Attraction in Sewage Sludge, *USEPA, Office of Research and Development*, EPA/625/R-92/013.

(**6**) A Plain English Guide to the EPA Part 503 Biosolids Rule,*USEPA, Office of Wastewater Management*, EPA/832/R-93/003.

Standard Practice for Derivation of Decision Point and Confidence Limit for Statistical Testing of Mean Concentration in Waste Management Decisions[1]

This standard is issued under the fixed designation D 6250; the number immediately following the designation indicates the year of original adoption or, in the case of revision, the year of last revision. A number in parentheses indicates the year of last reapproval. A superscript epsilon (ε) indicates an editorial change since the last revision or reapproval.

1. Scope

1.1 This practice covers a logical basis for the derivation of a decision point and confidence limit when mean concentration is used for making environmental waste management decisions. The determination of a decision point or confidence limit should be made in the context of the defined problem. The main focus of this practice is on the determination of a decision point.

1.2 In environmental management decisions, the derivation of a decision point allows a direct comparison of a sample mean against this decision point, where similar decisions can be made by comparing a confidence limit against a concentration limit (for example, a regulatory limit, which will be used as a surrogate term for any concentration limit throughout this practice). This practice focuses on making environmental decisions using this kind of statistical comparison. Other factors, such as any qualitative information that may be important to decision-making, are not considered here.

1.3 A decision point is a concentration level statistically derived based on a specified decision error and is used in a decision rule for the purpose of choosing between alternative actions.

1.4 This practice derives the decision point and confidence limit in the framework of a statistical test of hypothesis under three different presumptions. The relationship between decision point and confidence limit is also described.

1.5 Determination of decision points and confidence limits for statistics other than mean concentration is not covered in this practice. This practice also assumes that the data are normally distributed. When this assumption does not apply, a transformation to normalize the data may be needed. If other statistical tests such as nonparametric methods are used in the decision rule, this practice may not apply. When there are many data points below the detection limit, the methods in this practice may not apply.

2. Referenced Documents

2.1 *ASTM Standards:*

D 4687 Guide for General Planning of Waste Sampling[2]

D 5792 Practice for Generation of Environmental Data Related to Waste Management Activities: Development of Data Quality Objectives[2]

D 4790 Terminology of Aromatic Hydrocarbons and Related Chemicals[3]

E 456 Terminology Relating to Quality and Statistics[4]

E 1138 Terminology of Technical Aspects of Products Liability Litigation[4]

2.2 *Other Documents:*

USEPA (1989a) Statistical Analysis of Ground-Water Monitoring Data at RCRA Facilities. Interim Final Guidance. Office of Solid Waste Management Division, Washington, D.C. (PB89-15-1047)[5]

USEPA (1989b) Methods for Evaluating the Attainment of Cleanup Standards. Vol. 1: Soils and Solid Media. Statistical Policy Branch (PM-223)[5]

USEPA (1992) Statistical Methods for Evaluating the attainment of Superfund Cleanup Standards. Vol. 2: Groundwater. DRAFT, Statistical Policy Branch, Washington, D.C[5]

USEPA (1994) Guidance for the Data Quality Objectives Process. EPA QA/G4, Quality Assurance Management Staff, USEPA, September, 1994[5]

3. Terminology

3.1 *Definitions:*

3.1.1 *decision point, n*—the numerical value which causes the decision maker to choose one of the alternative actions (for example, conclusion of compliance or noncompliance).

3.1.1.1 *Discussion*—In the context of this practice, the numerical value is calculated in the planning stage and prior to the collection of the sample data, using a specified hypothesis, decision error, an estimated standard deviation, and number of samples. In environmental decisions, a concentration limit such as a regulatory limit usually serves as a standard for judging attainment of cleanup, remediation, or compliance objectives. Because of uncertainty in the sample data and other factors,

[1] This practice is under the jurisdiction of ASTM Committee D34 on Waste Management and is the direct responsibility of Subcommittee D34.01.01 on Planning for Sampling.

Current edition approved April 10, 1998. Published December 1998.

[2] *Annual Book of ASTM Standards*, Vol 11.04.

[3] *Annual Book of ASTM Standards*, Vol 06.04.

[4] *Annual Book of ASTM Standards*, Vol 14.02.

[5] Available from the Superintendent of Documents, U.S. Government Printing Office, Washington, DC 20402.

actual cleanup or remediation, for example, may have to go to a level lower or higher than this standard. This new level of concentration serves as a point for decision-making and is, therefore, termed the decision point.

3.1.2 *confidence limits*, *n*—the limits on either side of the mean value of a group of observations which will, in a stated fraction or percent of the cases, include the expected value. Thus the 95 % confidence limits are the values between which the population mean will be situated in 95 out of 100 cases. **D 4790**

3.1.2.1 *Discussion*—A one-sided upper or lower confidence limit can also be used when appropriate. An upper confidence limit is a value below which the population mean is expected to be with the specified confidence. Similarly, a lower confidence limit is a value above which the population mean is expected to be with the specified confidence. It is to be noted that confidence limits are calculated after the collection of sample data.

3.1.3 *decision rule*, *n*—a set of directions in the form of a conditional statement that specify the following: (*1*) how the sample data will be compared to the decision point, (*2*) which decision will be made as a result of that comparison, and (*3*) what subsequent action will be taken based on the decisions. **D 5792**

3.1.3.1 *Discussion*—For this practice, the comparison in (1) in 3.1.3 can be made in two equivalent ways: (*1*) a comparison between the sample mean (calculated from the sample data) and a decision point (calculated during the planning stage), or (*2*) a comparison between a confidence limit(s) (calculated from the sample data) and a regulatory limit.

3.1.4 *false negative error*, *n*—occurs when environmental data mislead decision maker(s) into not taking action specified by a decision rule when action should be taken. **D 5792**

3.1.4.1 *Discussion*—For this practice, this is an error defined in the context of a regulatory decision in waste management. In this context, it is an error in concluding that the true value is smaller than the regulatory limit when in fact it is not. The calculation of the false negative error will depend on how the hypotheses are framed (see Appendix X1).

3.1.5 *false positive error*, *n*—occurs when environmental data mislead decision maker(s) into taking action specified by a decision rule when action should not be taken. **D 5792**

3.1.5.1 *Discussion*—For this practice, this is an error defined in the context of a regulatory decision in waste management. In this context, it is an error in concluding that the true value is equal to or greater than the regulatory limit when in fact it is not. The calculation of the false positive error will depend on how the hypotheses are framed (see Appendix X1).

3.1.6 *hypothesis*, *n*—a supposition or conjecture put forward to account for certain facts and used as a basis for further investigation by which it may be proved or disproved. **E 1138**

3.1.6.1 *Discussion*—For this practice, a hypothesis is a postulation of what the true value is, typically framed for the purpose of making a statistical test of the hypothesis. In a statistical test, there are two competing hypotheses: the null hypothesis and the alternative hypothesis. The null hypothesis is a hypothesis "put up" for consideration and is the presumed hypothesis of choice before the data are collected. The alternative hypothesis is favored only when the data reject the null hypothesis.

3.1.7 *statistic*, *n*—a quantity calculated from a sample of observations, most often to form an estimate of some population parameter. **E 456**

4. Significance and Use

4.1 Environmental decisions often require the comparison of a statistic to a decision point or the comparison of a confidence limit to a regulatory limit to determine which of two alternate actions is the proper one to take.

4.2 This practice provides a logical basis for statistically deriving a decision point, or a confidence limit as an alternative, for different underlying presumptions.

4.3 This practice is useful to users of a planning process generally known as the data quality objectives (DQO) process (see Practice D 5792), in which calculation of a decision point is needed for the decision rule.

5. Overview of Decision Point Determination

5.1 The determination of a decision point is usually a part of an overall planning process. For example, the decision rule in the DQO planning process often includes the specification of a decision point. A brief summary of the steps needed to determine a decision point is given below.

5.1.1 State the problem and the decision rule (see Section 6),

5.1.2 Consider the alternative presumptions in the hypotheses based on the relative consequences of false positive and false negative errors (see 7.6),

5.1.3 Choose the form of the hypotheses to be used in the decision rule based on the chosen presumption (see 7.5 through 7.6 and Fig. 1),

5.1.4 Obtain an estimated standard deviation and the number of samples used in that estimation,

5.1.5 Specify acceptable decision errors (see Section 8), and

5.1.6 Calculate the decision point (see Section 8).

5.2 The following sections discuss in practical terms the topics of decision rule, presumptions and test of hypothesis, calculation of a decision point for specified decision errors, ways to control decision errors, and the use of a confidence limit as an alternative approach in decision-making.

6. Decision Rule in Waste Management Decisions

6.1 A decision rule is constructed according to a problem statement defined and agreed to by all the parties concerned, through a planning process. The decision rule can be carried out in two similar ways.

6.1.1 *When Using A Decision Point*:

6.1.1.1 The general construct of the decision rule in this case is:

If (sample mean) $\geq$ (decision point), then (one action). Otherwise, (alternate acti

6.1.1.2 Because a decision point is needed in the above decision rule, this practice provides a logical basis for developing such a decision point. Because the above decision rule can also be carried out similarly using confidence limits, it is also presented that way in 6.2.

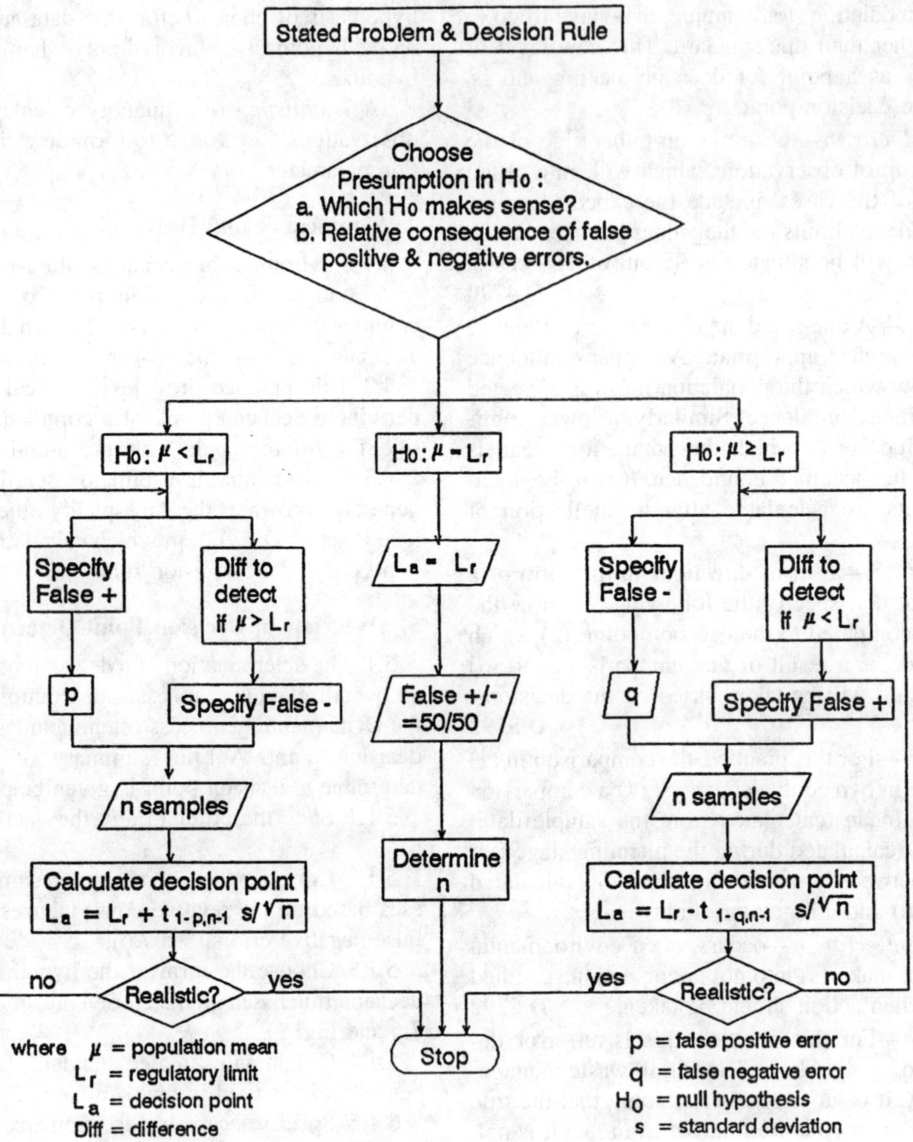

FIG. 1 Decision Point Determination for Mean Concentration

6.1.1.3 Note that when data can be measured with certainty, the regulatory limit defines the decision point. For example, sample data taken from a totally homogeneous population, in the absence of measurement error, have no variability. This means that the standard deviation of the data is zero and the decision point is reduced to the regulatory limit (see 8.6.3).

6.1.1.4 When data cannot be measured precisely or the population is not totally homogeneous, this variability needs to be incorporated to obtain a decision point. The decision point then includes both the original regulatory limit and a margin of uncertainty that is reflected in the standard deviation, which is a component in the calculation of the decision point (see 8.6.3). The way to incorporate this uncertainty depends on how a hypothesis is formulated and which presumption is adopted. This is discussed in Section 7.

6.1.1.5 An example of carrying out the decision rule using a decision point is:

> If (average concentration of cadmium in a truck load) ≥ (decision point), then (dispose of the waste fly ash in an RCRA landfill).
> Otherwise, (dispose the waste fly ash in a sanitary landfill).

6.1.1.6 The inputs needed for the calculation of the decision point in 6.1.1.5 are: form of the hypotheses to be tested, acceptable maximum decision error, number of samples, and estimated standard deviation. The standard deviation should include all the sources of variation in the sampling and measurement processes. Decision errors include the false positive error and false negative error. Details are given in Section 8.

6.2 *When Using Confidence Limit*:

6.2.1 The general construct of the decision rule in this case is:

> If (confidence limit) > or < (regulatory limit), then (one action)
> Otherwise, (alternate action).

where the confidence limit can be the upper confidence limit or the lower confidence limit, depending on the chosen presumption in the null hypothesis (see Section 7). A special case where the confidence limit is replaced by the sample mean in the above decision rule is also discussed in Section 7.

6.2.2 Two examples corresponding to the > and < signs in the decision rule are:

6.2.2.1 If upper confidence limit of mean concentration of cadmium) > (regulatory limit), then (dispose of the waste fly ash in a RCRA landfill). Otherwise, (dispose of the waste fly ash in a sanitary landfill).

6.2.2.2 If lower confidence limit of mean concentration of cadmium) < (background concentration), then (dispose of the waste fly ash in a sanitary landfill). Otherwise, (dispose of the waste fly ash in a RCRA landfill).

6.2.3 The relationship between the decision point approach and the confidence limit approach is described in Appendix X1.

6.2.4 The decision point approach and the confidence limit approach are identical in decision-making if the standard deviation (*s*) and number of samples (*n*) used in the calculations are identical. Since the decision point is calculated during the planning stage and before the sample data are collected, its *s* and *n* may be different from those used in the calculation of the confidence limit (which is calculated after the data are collected).

6.2.5 Note that similar to 6.1.1, when data can be measured with certainty, the confidence limit in the decision rule above is reduced to the sample mean, because the standard deviation of the data is zero (see 8.6.4).

7. Test of Hypothesis

7.1 This section is a brief introduction to the concept of statistical test of hypothesis and how it relates to the determination of a decision point.

7.2 A statistical test of hypothesis is framed by two hypotheses: a null hypothesis and an alternative hypothesis.

7.3 The null hypothesis is a hypothesis "put up for consideration" or "being tested." That is, the null hypothesis is presumed to be the hypothesis of choice before the data are collected. If, after the data are collected, the sample data are consistent with this hypothesis, then the null hypothesis is not rejected and the alternative hypothesis is discarded. If, on the other hand, the data are not consistent with the null hypothesis, then the null hypothesis is rejected in favor of the alternative hypothesis.

7.4 Thus, it is the alternative hypothesis that bears the "burden of proof." That is, the alternative hypothesis is not favored until the data suggest that the null hypothesis is not tenable and cause the rejection of the null hypothesis.

7.5 *Presumptions in Null Hypothesis*—In environmental testing, two presumptions can be postulated for the null hypothesis. A third presumption can be constructed as a compromise between the first two presumptions based on practical considerations.

7.5.1 *Presumption Number 1*—The true (population) mean concentration is presumed to be below the regulatory limit, with an opposite presumption in the alternative hypothesis.

7.5.1.1 This presumption of no exceedance would require" cleanup" down to a concentration level not statistically significantly higher than the regulatory limit. In this case, the decision point will be higher than the regulatory limit.

7.5.2 *Presumption Number 2*—The true (population) mean concentration is presumed to be equal to or greater than the regulatory limit, with an opposite presumption in the alternative hypothesis.

7.5.2.1 This presumption of exceedance would require" cleanup" down to a concentration level statistically significantly lower than the regulatory limit. In this case, the decision point will be lower than the regulatory limit.

7.5.3 *Presumption Number 3*—a neutral presumption that the true mean concentration is neither higher nor lower than the regulatory limit.

7.5.3.1 This presumption would require "cleanup" down to the regulatory limit. In this case, the decision point is identical to the regulatory limit.

7.5.3.2 This presumption is a compromise between the first two presumptions based on practical considerations. See X1.6 of Appendix X1 for details.

7.6 *Choice of Presumption*

7.6.1 The presumption of "exceedance over the regulatory limit" (Presumption 2) is a reasonable choice when the contaminants are highly toxic or when protection of health and environment is of first priority. However, it will tend to incur higher costs in environmental waste management.

7.6.2 The first presumption of "no exceedance" is considered reasonable for the following example situations.

7.6.2.1 When comparing environmental cleanup or remediation against background, cleanup below the background has little technical merit. For example, RCRA groundwater regulations stipulate use of the first presumption in the statistical comparisons (USEPA, 1989a).[5]

7.6.2.2 Frequently, the regulatory limit is arrived at with a series of assumptions. For example, the regulatory limit may be derived from animal data, and extrapolation is made to estimate human risk. Often the extrapolation chooses a set of data with the most severe health effects; uses a conservative model to extrapolate from the high dose response to the low dose response; uses conservative assumptions in extrapolating from animal risk to human risk; and uses conservative assumptions in estimating human exposure.

7.6.2.2.1 When this kind of conservatism is built into the regulatory limit, it is often unnecessary to impose the additional conservatism of the more strict presumption of "exceedance" in the null hypothesis.

7.6.2.3 When the regulatory limit is low or close to zero, use of the second presumption may lead to a decision point being negative or zero, which is impractical.

7.6.2.4 More fundamentally, the choice can be made by considering the decision errors and the severity of their consequences.

7.6.2.5 When the consequence of a false negative decision error is more severe than that of the false positive decision error, then the presumption of "exceedance over the regulatory limit" (Presumption 2) may be reasonable. When the consequence of a false positive decision error is more severe than that of the false negative decision error, then the presumption of "no exceedance over the regulatory limit" (Presumption 1) may be reasonable. When neither decision error seems to be dominant, then a neutral presumption (Presumption 3) may be reasonable. It is by weighing the decision errors and their respective consequences that a reasonable presumption can be reached.

7.6.2.6 At times, choice of a presumption is mandated by regulations.

8. Determination of A Decision Point

8.1 An overview of how a decision point is derived is given in Section 5. This section provides the details.

8.2 Fig. 1 provides a schematic description of how a decision point is derived within the framework of hypothesis testing under different presumptions in the null hypothesis. The left-hand side of Fig. 1 corresponds to 8.6, the right-hand side to 8.7, and the middle part of Fig. 1 to 8.8. Paragraphs 8.6 through 8.8 also show how the decision rules can be carried out similarly using either the decision point or confidence limit approach. Note that all the boxes here apply to the confidence limit approach, while only some of the boxes apply to the decision point approach. Details are discussed in the appropriate following sections.

8.3 The mathematical details for deriving a decision point or a confidence limit are given in Appendix X1. Normal distribution of the data is assumed throughout this practice. When the data are not normally distributed, a transformation to normalize the data may be necessary. Other statistical tests for non-normal data can also be used, but they will not be covered here.

8.4 Note that when formulating the decision rule during the planning process, the decision point is calculated before the data are collected from a formal sampling plan. As can be seen from Eq 4 and 6, values for the standard deviation, *s*, and the number of samples, *n*, need to be provided for this calculation. They can come from previously available data. If existing data are limited or non-existent, a pilot study to obtain this information may be necessary. In any event, if the values of *s* and *n* so obtained are crude, the derived decision point will be only an approximation. If this approximation is sufficient for testing purposes, the test can proceed. If not, either a pilot study needs to be conducted or the confidence limit approach can be used in place of the decision point approach. Since the confidence limit is calculated after the data are collected, *s* and *n* are readily available.

8.5 The previous observations apply to all three cases of presumptions.

8.6 *First Presumption—True Mean Concentration Is Below the Regulatory Limit*:

8.6.1 The statistical test of whether or not the true mean concentration exceeds the regulatory limit can be carried out similarly using either the decision point or the confidence limit. Determination of a decision point under this presumption is schematically given in Fig. 1 (left-hand side of figure).

8.6.2 The relationships between the decision to "accept" a certain hypothesis and the associated decision errors under this presumption are given in Fig. 2.

8.6.3 *Using a Decision Point in the Decision Rule*:

8.6.3.1 Under this presumption, the decision point L_a can be calculated as follows:

$$L_a = L_r + t_{1-p,n-1} s/\sqrt{n} \quad (1)$$

where:

L_a = decision point,
L_r = regulatory limit,
$t_{1-p,n-1}$ = tabled t-value with 100*p* % false positive error and (*n*-1) degrees of freedom,
p = specified maximum false positive error (in fraction) and is typically a number smaller than 0.5,
s = estimated standard deviation, and
n = number of samples in the estimation of *s*.

8.6.3.2 The steps in the statistical test are as follows:

(1) Before the sample data are collected: Specify the statistical comparison in the decision rule. Specify the two alternative actions in the decision rule. Specify the maximum acceptable false positive error p, $p < 0.5$. For given number of samples (*n*), estimated standard deviation (*s*), regulatory limit (L_r) and tabled t-value, calculate the decision point L_a according to Eq 1. If the calculated L_a is realistic, a decision point has been determined. If not, the previous steps can be reiterated for different values of acceptable false positive error (*p*). Note that the number of samples (*n*) and standard deviation (*s*) are given values. However, different values of *s* or *n*, or both, may be tried for scenario analysis. The reiteration may need to go back to earlier steps, including restating the problem.

(2) After the sample data are collected, Calculate the sample mean $\bar{x}$. Compare $\bar{x}$ to L_a. If $\bar{x} \geq L_a$, conclude that the regulatory limit has been exceeded and take one course of action in the decision rule. Otherwise, conclude differently and take the alternate action.

8.6.4 *Using a Confidence Limit in the Decision Rule*:

8.6.4.1 Under this presumption, the 100 (1-*p*) % lower confidence limit (LCL), associated with a 100*p* % false positive error and corresponding to the decision point in Eq 1 is:

$$\text{LCL} = \bar{x} - t_{1-p,n-1} s/\sqrt{n} \quad (2)$$

8.6.4.2 The steps in the statistical test are as follows:

(1) Before the sample data are collected: Specify the statistical comparison in the decision rule. Specify the two alternative actions in the decision rule. Specify the maximum acceptable false positive error *p*, where $p<0.5$.

(2) After the sample data are collected, Calculate the lower confidence limit (LCL) in accordance with Eq 2. Compare the LCL to the regulatory limit L_r. If LCL $\geq L_r$, conclude that the regulatory limit has been exceeded and take one course of action. Otherwise, conclude differently and take the alternate action.

8.6.4.3 Application of the decision rule under Presumption Number 1 using either a decision point or a lower confidence limit is given graphically in Fig. 3.

NOTE 1—Although specification of the false negative error is not needed for the calculation of a decision point or a confidence limit, such specification is necessary to determine the number of samples needed to achieve the desired false negative error. This subject is beyond the scope of this practice and is not covered.

8.6.4.4 In confidence limit approach, the determination of how many samples to collect does require inputs such as false positive error, false negative error, and a difference from the hypothesized value (in the null hypothesis) important to detect. These inputs are described in the left hand side of Fig. 1. Specifics regarding determination of the number of samples is

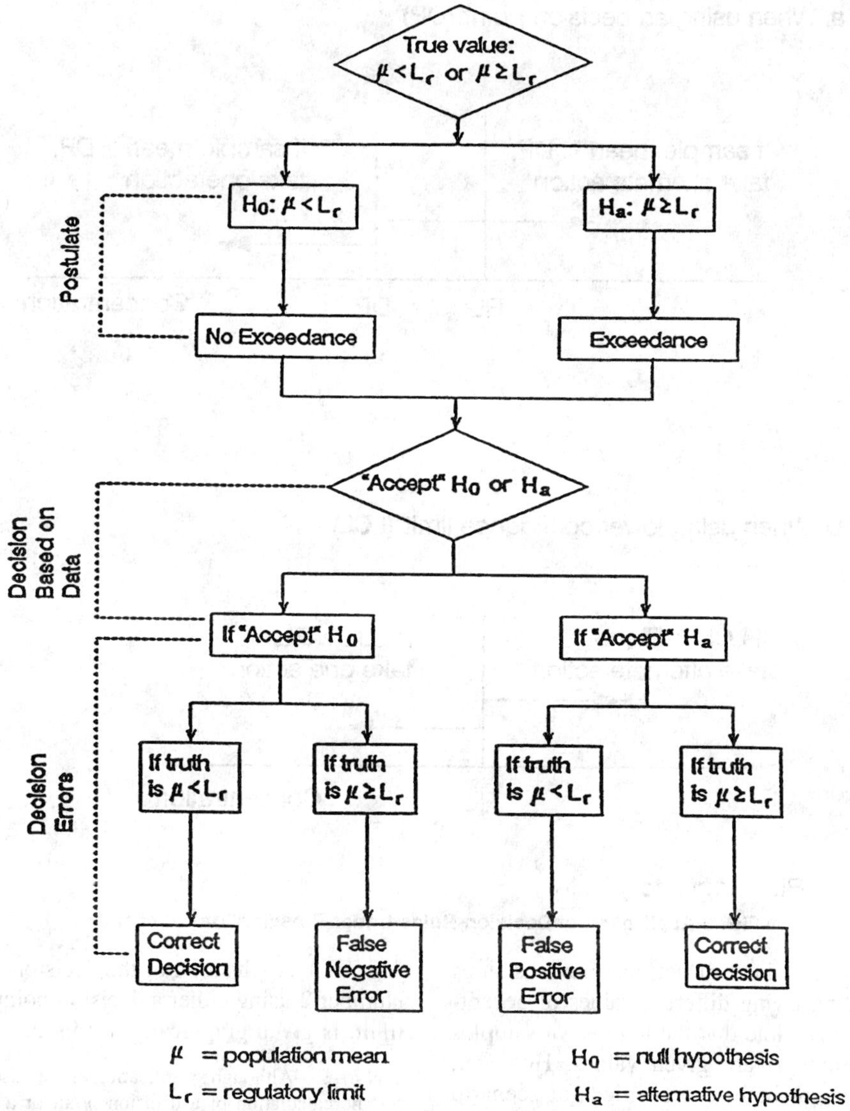

FIG. 2 Relationships Between Hypotheses and Decision Errors Under First Presumption

not covered in this practice.

8.6.4.5 Some example criteria for judging whether a derived decision point is realistic or not could include: (*1*) Does the value of the decision point make sense? A decision point that is below the detection limit or negative is unlikely to be useful. (*2*) Does the decision point reflect a good balance between the potential consequences of the two types of decision errors? A decision point that leads to either unaffordable costs or a consequence of high toxicity or other adverse effects needs to be re-examined carefully.

8.6.4.6 Note that the decision point given in Eq 1 includes both the regulatory limit and the uncertainty in the data. When the data are measured with certainty, the standard deviation, *s*, becomes zero, and the regulatory limit becomes the decision point.

8.7 *Second Presumption—True Mean Concentration Is Equal to or Higher Than the Regulatory Limit*:

8.7.1 Again, the statistical test of whether or not the true mean concentration exceeds the regulatory limit can be carried out similarly using either the decision point or the confidence limit.

8.7.2 Determination of a decision point under this presumption is schematically given in Fig. 1 (right hand side of figure).

8.7.3 The relationships between the decision to "accept" a certain hypothesis and the associated decision errors, under this presumption, are given in Fig. 4.

8.7.4 *Using a Decision Point in the Decision Rule*:

8.7.4.1 Under this presumption, the decision point L_a can be calculated as follows:

$$L_a = L_r - t_{1-q,n-1} s/\sqrt{n} \quad (3)$$

where:

q = specified maximum acceptable false negative error (in fraction), where $q < 0.5$.

8.7.4.2 The steps in the statistical test are as follows:

(*1*) Before the sample data are collected: Specify the statistical comparison in the decision rule. Specify the two alternative actions in the decision rule. Specify acceptable maximum false negative error q, $q < 0.5$. For given number of samples (n), estimated standard deviation (s), regulatory limit (L_r) and tabled t-value, calculate the decision point L_a in accordance with Eq 3. If the calculated L_a is realistic, a

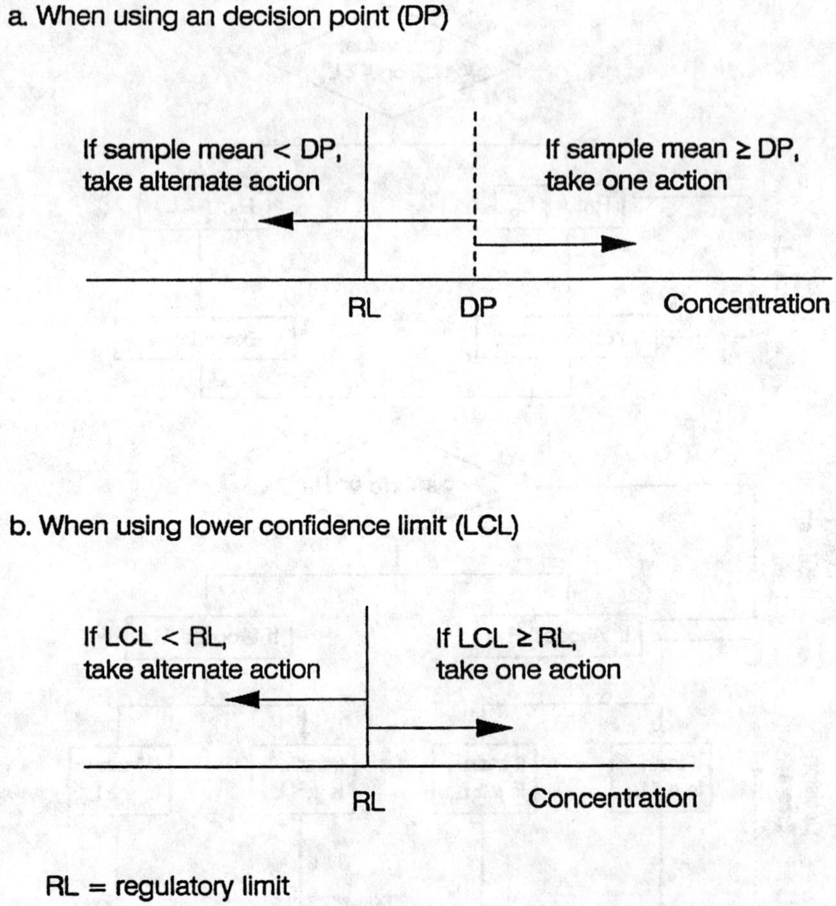

FIG. 3 Application of Decision Rules Under Presumption Number 1

decision point has been determined. If not, the previous steps can be reiterated, including trying different values of acceptable false negative error (q). Note that the number of samples (n) and standard deviation (s) are given values. However, different values of s or n, or both, can be tried for scenario analysis.

(2) After the sample data are collected, Calculate the sample mean x. Compare x to L_a. If $x \geq L_a$, conclude that the regulatory limit has been exceeded and take one course of action. Otherwise, conclude differently and take the alternate action.

8.7.5 *Using a Confidence Limit in the Decision Rule*:

8.7.5.1 The 100 (1-q) % upper confidence limit (UCL), associated with a 100q % false negative error and corresponding to the decision point in Eq 3 is:

$$UCL = \bar{x} + t_{1-q,n-1} s/\sqrt{n} \quad (4)$$

8.7.5.2 The steps in the statistical test are as follows:

(1) Before the sample data are collected: Specify the statistical comparison in the decision rule. Specify the two alternative actions in the decision rule. Specify acceptable maximum false negative error q, $q < 0.5$.

(2) After the sample data are collected, Calculate the upper confidence limit (UCL) in accordance with Eq 4. Compare the UCL to the regulatory limit L_r. If UCL $\geq L_r$, then conclude that the regulatory limit has been exceeded and take one course of action. Otherwise, conclude differently and take the alternate action.

8.7.5.3 Application of the decision rules under Presumption Number 2 using either a decision point or an upper confidence limit, is given graphically in Fig. 5.

NOTE 2—Although specification of the false positive error is not needed for the calculation of a decision point or a confidence limit here, such specification is necessary to determine the number of samples needed to achieve the desired false positive error. This subject is beyond the scope of this document and is not covered.

8.7.5.4 In confidence limit approach, the determination of how many samples to collect does require inputs such as false positive error, false negative error and a difference from the hypothesized value (in the null hypothesis) important to detect. These inputs are described in the right side of Fig. 1. Specifics regarding determination of the number of samples is not covered in this standard.

8.7.5.5 Other comments in Note 1 on judging how realistic a calculated decision point may also apply here.

(1) Note that the decision point given in Eq 2 includes both the regulatory limit and the uncertainty in the data. When the data are measured with certainty, the standard deviation s becomes zero and the regulatory limit becomes the decision point.

8.8 *Third Presumption—True Mean Concentration Is neither Higher nor Lower Than the Regulatory Limit*:

8.8.1 In this case, the statistical test is the same for either the decision point approach or the confidence limit approach.

8.8.2 Under this presumption, the decision point L_a is the

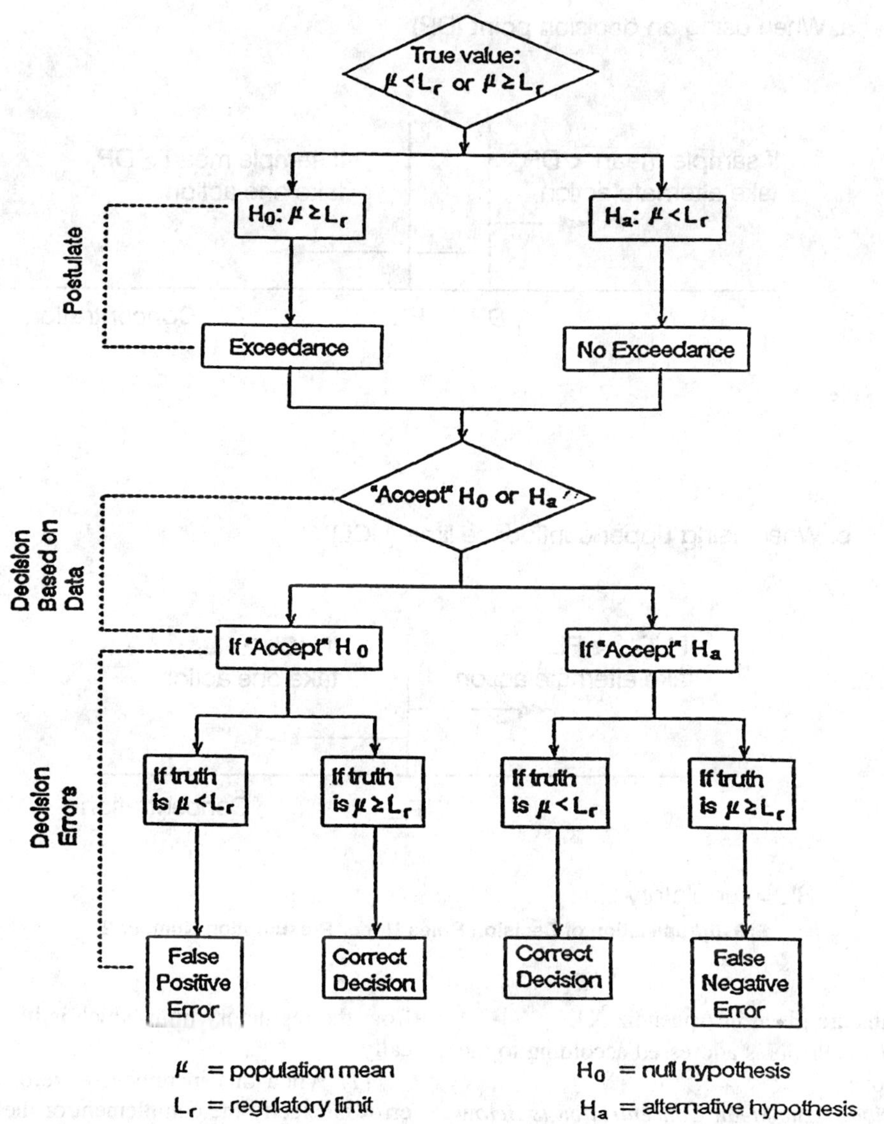

FIG. 4 Relationships Between Hypotheses and Decision Errors Under Second Presumption

regulatory limit L_r (see X1.6 of Appendix X1 for details). Namely,

$$L_a = L_r \tag{5}$$

8.8.3 The steps in the statistical test are as follows:

8.8.3.1 Before the sample data are collected:

(1) Specify the statistical comparison in the decision rule as a comparison between x and L_r.

(2) Specify the two alternative actions in the decision rule.

8.8.3.2 After the sample data are collected:

(1) Calculate the sample mean x.

(2) Compare x to L_r. If $x \geq L_r$, conclude that the regulatory limit has been exceeded and take one course of action. Otherwise, conclude differently and take the alternate action.

8.8.4 Note that the false positive error and the false negative error are equal, at 50 % each, when the true value is at the regulatory limit (see Appendix X1).

8.8.5 Application of the decision rule under Presumption Number 3 is similar to Fig. 3 and Fig. 5. In this case, the decision rule is a simple comparison between the sample mean concentration x and the regulatory limit L_r.

9. Control of Decision Errors

9.1 Note that an environmental decision involves two parts: *(1)* the true relationship (or relative positions) between the population mean concentration (true value) and the regulatory limit, and *(2)* a conclusion on this relationship based on empirical data. A decision error of false positive or false negative is made only when the two parts are in conflict (see Fig. 3 and Fig. 5).

9.2 A false positive error is made when an empirical conclusion of the true mean concentration being higher than the regulatory limit is made when the true relationship is otherwise. Similarly, a false negative error is made when an empirical conclusion of the true mean concentration being lower than the regulatory limit is made when the true relationship is otherwise.

9.3 Because both kinds of decision errors are anchored at the regulatory limit, it is important to know what the decision error rate is when the true mean concentration is around the

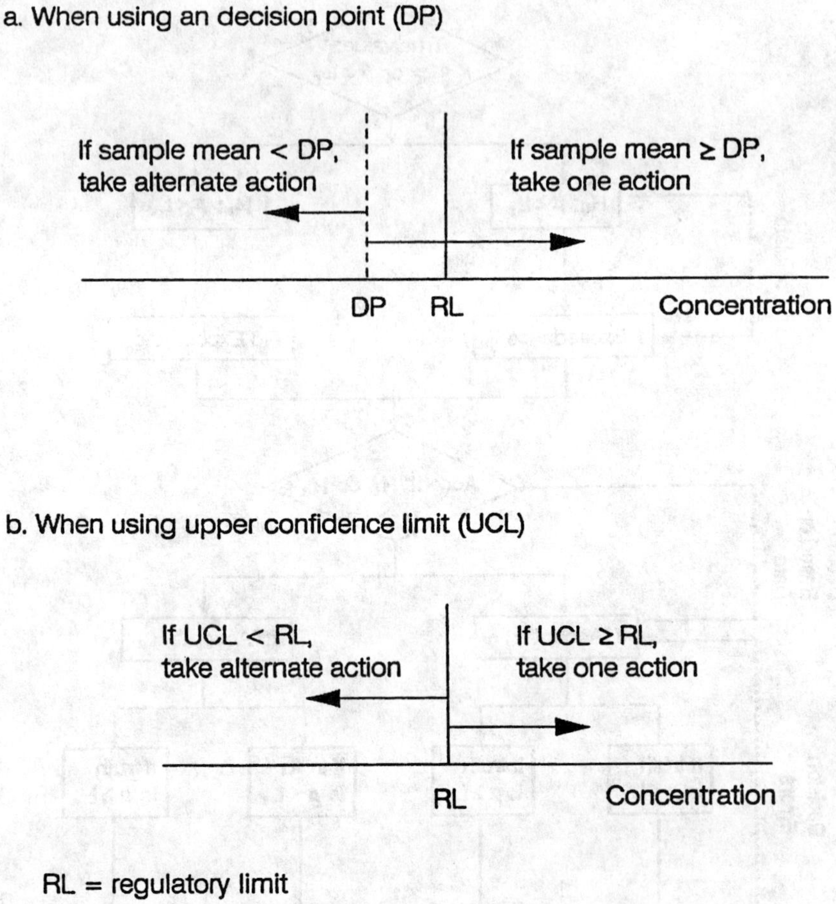

FIG. 5 Application of Decision Rules Under Presumption Number 2

regulatory limit. Details are given in Appendix X1.

9.4 Control of decision errors is addressed according to the underlying presumptions.

9.5 *First Presumption—True Mean Concentration Is Below the Regulatory Limit*:

9.5.1 Under this presumption, the acceptable maximum false positive decision error is specified first. It is usually specified based on negotiated agreement by the concerned parties after considering the risk and consequence of the error and the costs necessary to control the error to the specified level. When the true value is at the regulatory limit, the corresponding false negative error is 100 % minus the specified false positive error (in percent).

9.5.2 The second step is to control the false negative error by any of the following:

9.5.2.1 Reducing the variance of the sample data. The smaller the variance is, the lower the false negative error is, all other things (such as the decision point and regulatory limit) being the same. This can be accomplished, for example, by reducing the variance due to sampling or measurement, or both, and compositing of the samples.

9.5.2.2 Increasing the number of samples. This needs to be weighed against the increase in sampling and analytical costs. If composite samples are used, the increase in costs may be controlled.

9.5.2.3 Specifying an allowable increment in concentration from the regulatory limit which is important to detect statistically.

(1) When this increment is zero, then the false negative error is exactly the complement of the specified false positive error. Namely, they total 100 %.

(2) If, on the other hand, there is some flexibility in the decision rule, then the increment could be defined as an increase above the regulatory limit which would not pose a substantial increase in health or environmental risk, but an increase beyond that would. If such is the case, then a false negative error is defined as the probability of failing to detect such an increase. As a matter of fact, false negative errors associated with different increments from the regulatory limit can be calculated for scenario analysis (see Appendix X1).

9.6 *Second Presumption—True Mean Concentration Is Equal to or Higher Than the Regulatory Limit*:

9.6.1 Under this presumption, the acceptable maximum false negative decision error is specified first. It is usually specified based on negotiated agreement by the concerned parties after considering the risk and consequence of the error and the costs necessary to control the error to the specified level. When the true value is at the regulatory limit, the corresponding false positive error is 100 % minus the specified false negative error (in percent).

9.6.2 False positive error can then be controlled by any of the following:

9.6.2.1 Reducing the variance of the sample data. The smaller the variance is, the lower the false positive error is, all other things (such as the decision point and regulatory limit) being the same. This can be accomplished, for example, by reducing the variance due to sampling or measurement variance, or both, and compositing of the samples.

9.6.2.2 Increasing the number of samples. This needs to be weighed against the increase in sampling and analytical costs. If composite samples are used, the increase in costs may be controlled.

9.6.2.3 Specifying an allowable decrement in concentration from the regulatory limit which is important to detect statistically.

(1) When this decrement is zero, then the false positive error is the exact complement of the specified false negative error. Namely, they total 100 %.

(2) If, on the other hand, there is some flexibility in the decision rule, then the decrement could be defined as a decrease below the regulatory limit which would not pose a substantial increase in cleanup or remediation costs, but a decrease beyond that would. If such is the case, then a false positive error can be defined as the probability of failing to detect such an decrement. As a matter of fact, false positive errors can be calculated for different decrements below the regulatory limit for scenario analysis (see Appendix X1).

9.7 *Third Presumption—True Mean Concentration Is Neither Higher nor Lower Than the Regulatory Limit*:

9.7.1 Under this presumption, the false positive and false negative errors are equal at 50 % each when the true value is at the regulatory limit.

9.7.2 In this case, the reduction in variance and the increase in the number of samples do not actually change the probabilities of false positive and false negative errors. However, they do tighten the distribution of the sample data such that there is less uncertainty in the decision. Namely, a smaller variance will allow statistical detection of a smaller deviation from the regulatory limit.

10. Keywords

10.1 confidence limit; data quality objectives; decision error; decision point; false negative; false positive; hypothesis; mean concentration; presumption; waste management

APPENDIX

(Nonmandatory Information)

X1. MATHEMATICAL DERIVATION OF A DECISION POINT FROM STATISTICAL TEST OF HYPOTHESIS FOR MEAN CONCENTRATION

X1.1 The derivation of a decision point has the purpose of statistically testing if the true mean concentration of a well-defined population exceeds the regulatory limit, using sample data taken from the population.

X1.2 The decision point L_a can be derived from a statistical test of hypothesis under three different presumptions.

X1.2.1 *Decision Point Derivation Under First Presumption*—The true (population) mean concentration is presumed to be below the regulatory limit.

X1.2.1.1 The steps in deriving a decision point under this presumption are given in Fig. 1 (left hand side of figure).

X1.2.1.2 Under this presumption, the null and alternative hypotheses are:

$$H_o: u < L_r \text{ versus } H_a: u \geq L_r \qquad \text{(X1.1)}$$

where:

H_o = null hypothesis,
H_a = alternative hypothesis,
u = true (population) mean concentration, and
L_r = regulatory limit.

X1.2.1.3 Associated with the null hypothesis above, the primary concern is the false positive error. A false positive decision error is the probability of concluding that the true concentration is greater than the regulatory limit when in fact it is not so. Note that in the decision rule, such a conclusion is reached only when $x \geq L_a$, where L_a is the decision point (see 8.6.3). Namely,

False positive error:

= probability of saying that the true mean concentration is not lower than the regulatory limit when it is so, (X1.2)

= probability that the data observe ($\bar{x} \geq L_a$) in favor of H_a when H_o is the correct one,

= Prob(rejecting H_o | when H_o is true), and

= Prob($\bar{x}$ ($\geq L_a$ H_o).

X1.2.1.4 An acceptable maximum false positive error, p, can be specified so that it cannot be exceeded. Given this specification, the above expression becomes:

$$\text{Prob}(\bar{x} \geq L_a \mid H_o) \leq p \qquad \text{(X1.3)}$$

X1.2.1.5 Generally, x in Eq X1.3 can be assumed to follow a normal distribution due to the central limit theorem in statistics (when the number of samples is relatively large). Namely, x can be assumed to follow a normal distribution with variance σ^2/n and a mean concentration below the regulatory limit under the null hypothesis H_o. In some cases, a transformation of the data may be necessary for the normality assumption to apply.

X1.2.1.6 Since the null hypothesis can be rewritten as H_o:$\mu = L_r - \delta_1$, $\delta_1 > 0$, the false positive error in Eq X1.3 becomes:

$$\text{Prob}(\bar{x} \geq L_a) = \text{Prob}\{[\bar{x} - (L_r - \delta_1)] / (\sigma/\sqrt{n}) \geq [L_a - (L_r - \delta_1)] / (\sigma/\sqrt{n})\} \qquad \text{(X1.4)}$$

$$< \text{Prob}[Z \geq (L_a - L_r) / (\sigma/\sqrt{n})] = p, \text{ since } \delta_1 > 0$$

where Z is normally distributed with mean zero and variance 1, and n is the number of samples.

X1.2.1.7 Given Eq X1.4, the decision rule in a statistical test of hypothesis can be used to derive the decision point. The decision rule in test of hypothesis, for $p < 0.5$, is:

(1) If $(L_a - L_r) / (\sigma/\sqrt{n}) \geq Z_{1-p}$, then reject H_o and "accept" the alternative hypothesis of exceedance over the regulatory limit.

(2) If $(L_a - L_r) / (\sigma/\sqrt{n}) < Z_{1-p}$, then "accept" the null hypothesis H_o of no exceedance over the regulatory limit.

X1.2.1.8 The decision point is the cutoff point in this decision. Namely, set:

$$(L_a - L_r) / (\sigma/\sqrt{n}) = Z_{1-p} \quad \text{(X1.5)}$$

X1.2.1.9 Rearranging Eq X1.5, we obtain the decision point L_a for the hypotheses in Eq X1.1. Namely,

$$L_a = L_r + Z_{1-p}\ \sigma/\sqrt{n} \quad \text{(X1.6)}$$

X1.2.1.10 Eq X1.6 can be used to obtain the decision point L_a in the decision rule if the population standard deviation, σ, is known. When the population standard deviation, σ, is not known, its estimate s needs to be used and the decision point is given by Eq X1.7:

$$L_a = L_r + t_{1-p,n-1} s/\sqrt{n} \quad \text{(X1.7)}$$

where s is the estimated standard deviation, and $t_{1-p,n-1}$ a tabled t-value with $(n-1)$ degrees of freedom.

X1.2.1.11 Since the term $(t_{1-p,n-1} s/\sqrt{n})$ in Eq X1.7 is non-negative, the decision point L_a is equal to or greater than the regulatory limit L_r.

X1.2.1.12 The decision rule using the decision point in Eq X1.7 is carried out as follows:

(1) If $x < L_a$, then do not reject H_o and conclude no exceedance over the regulatory limit.

(2) If $x \geq L_a$, then reject the null hypothesis H_o and conclude exceedance over the regulatory limit.

X1.3 *False Negative Decision Error*:

X1.3.1 The corresponding false negative error under this presumption can be similarly calculated. Note that the alternative hypothesis can be rewritten as $H_a : \mu = L_r + \delta_2$, $\delta_2 \geq 0$. Thus,

False negative decision error:

= Probability of saying that the true mean concentration is lower than the regulatory limit when it is not so, (X1.8)

= Probability that the data observe $(\bar{x} < L_a)$ in favor of the null hypothesis H_o when the alternative hypothesis H_a is the correct one,

= $\text{Prob}\ (\bar{x} < L_a \mid H_a)$,

= $\text{Prob}\{Z < [L_a - (L_r + \delta_2)] / (\sigma/\sqrt{n})\}$, and

= $\text{Prob}\ \{Z < Z_{1-p} - \delta_2/(\sigma/\sqrt{n})\}$.

where $Z_{1-p} = (L_a - L_r)/(\sigma/\sqrt{n})$ from Eq X1.5, and $\delta_2 \geq 0$, $p < 0.5$.

X1.3.2 As can be seen above, false negative error is a function of how much the true mean u is higher than the regulatory limit L_r. This difference is δ_2 expressed in the unit of the standard error of the mean $(\sigma/\sqrt{n})$, or altogether $\delta_2/(\sigma/\sqrt{n})$.

X1.3.3 False negative error rates, using Eq X1.8, for different values of $\delta_2/(\sigma/\sqrt{n})$ are given in Table X1.1. When $\delta_2/(\sigma/\sqrt{n}) = 0$, the true value is right where the regulatory limit, L_r, is. As the true value becomes larger than L_r (when $\delta_2/(\sigma/\sqrt{n})>0$), the false negative error becomes smaller.

TABLE X1.1 False Negative Error (*q*) Under Presumption Number 1

NOTE 1—For Some Values of False Positive Error (*p*), and True Values $\delta_2/(\sigma\ \sqrt{n})$.[A]

False Positive Error, *p*	If True Value As Number of Standard Errors Away from Regulatory Limit, or $\delta_2/(\sigma/\sqrt{n})$, Is	False Negative Error, *q*
0.05	0	0.95
	1	0.74
	2	0.36
	3	0.09
0.10	0	0.90
	1	0.61
	2	0.24
	3	0.04
0.20	0	0.80
	1	0.44
	2	0.12
	3	0.02

[A] Where the true value = regulatory limit when $\delta_2/(\sigma\ \sqrt{n}) = 0$.

X1.3.4 It is to be noted that the false negative error is the exact complement of the chosen false positive error when the true value is at L_r; there the two errors total 100 %.

X1.3.5 When the population standard deviation, σ, is replaced by sample standard deviation, s, in Eq X1.8, the z-statistic is replaced by the t-statistic. The calculation of the false negative error now involves a non-centrality parameter and a statistician needs to be consulted.

X1.4 *Equivalency Between Decision Point Approach and Test of Hypothesis Approach*:

X1.4.1 The decision rule using the decision point given in Eq X1.7 can be made to be equivalent to one using the lower 100 (1-*p*) % confidence limit in a statistical test of hypothesis.

X1.4.2 Recall that the decision rule using a decision point is to conclude exceedance of the regulatory limit if $x \geq L_a$.

X1.4.3 But $L_a = L_r + t_{1-p,n-1} s/\sqrt{n}$, from Eq X1.7. Thus $x \geq L_a$ becomes $x \geq L_r + t_{1-p,n-1} s/\sqrt{n}$.

X1.4.3.1 Rearranging:

$$\bar{x} - t_{1-p,n-1} s/\sqrt{n} \geq L_r \quad \text{(X1.9)}$$

X1.4.3.2 Note that the left-hand side of Eq X1.9 is the lower 100 (1-*p*) % confidence limit in the framework of a test of hypothesis. Thus, the decision point approach from Eq X1.7 and the hypothesis test approach from Eq X1.9 are equivalent in decision-making.

X1.4.3.3 The decision rule using the lower confidence limit in Eq X1.9 is carried out as follows:

(1) If $(x - t_{1-p,n-1} s/\sqrt{n}) < L_r$, then do not reject H_o and conclude no exceedance over the regulatory limit.

(2) If $(x - t_{1-p,n-1} s/\sqrt{n}) \geq L_r$, reject the null hypothesis H_o and conclude exceedance over the regulatory limit.

X1.5 *Decision Point Derivation Under Second Presumption*—The true (population) mean concentration is presumed to be equal to or greater than the regulatory limit.

X1.5.1 The steps in deriving a decision point under this presumption are given in Fig. 1 (right hand side of figure).

X1.5.2 Under this presumption, the null and alternative hypotheses are:

$$H_o : u \geq L_r \text{ versus } H_a : u < L_r \quad \text{(X1.10)}$$

X1.5.3 Associated with the null hypothesis in X1.5.2, the primary concern is the false negative error. A false negative error is the probability of concluding that the true concentration is lower than the regulatory limit when in fact it is not so. Note that in the decision rule, such a conclusion is reached only when $x < L_a$, where L_a is the decision point (see 8.7.4). Namely,

False negative decision error:

= probability of saying
that the true mean concentration is lower than
the regulatory limit when it is not so, (X1.11)

= probability that the data observe ($\bar{x} < L_a$) in favor of H_a when H_o is the correct one,

= Prob(rejecting $H_o \mid H_o$), and

= Prob ($\bar{x} < L_a \mid H_o$).

X1.5.4 An acceptable maximum false negative error, q, can be specified so that it cannot be exceeded. Given this specification, the Eq X1.11 becomes:

$$\text{Prob } (\bar{x} < L_a \mid H_o) \leq q \quad \text{(X1.12)}$$

X1.5.5 The left-hand side of Eq X1.12 can be expanded under normal theory or the central limit theorem in statistics. Namely, under the null hypothesis H_o, $\bar{x}$ can be assumed to follow a normal distribution with variance σ^2/n and some mean concentration above the regulatory limit. Namely, the null hypothesis can be rewritten as $H_o : u = L_r + \delta_3$, $\delta_3 \geq 0$.

X1.5.5.1 Thus, the false negative error in Eq X1.12 becomes:

$$\text{Prob } (\bar{x} < L_a) = \text{Prob}\{[\bar{x} - (L_r+\delta_3)]/(\sigma/\sqrt{n}) < [L_a - (L_r+\delta_3)]/(\sigma/\sqrt{n})\} \quad \text{(X1.13)}$$

$$\leq \text{Prob}[Z < (L_a - L_r)/(\sigma/\sqrt{n})] = q$$

where Z is normally distributed with mean zero and variance 1, n, is the number of samples, and $q < 0.5$.

X1.5.6 Given Eq X1.13, the decision rule in statistical test of hypothesis is:

X1.5.6.1 If $(L_a - L_r)/(\sigma/\sqrt{n}) < -Z_{1-q}$, then reject H_o and "accept" the alternative hypothesis of no exceedance over the regulatory limit (for $q < 0.5$).

X1.5.6.2 If $(L_a - L_r)/(\sigma/\sqrt{n}) \geq -Z_{1-q}$, then "accept" the null hypothesis, H_o, of exceedance over the regulatory limit.

X1.5.7 The decision point is the cutoff point in this decision. Namely,

$$(L_a - L_r) / (\sigma/\sqrt{n}) = Z_{1-q} \quad \text{(X1.14)}$$

X1.5.8 Rearranging Eq X1.14, we obtain the decision point L_a for the hypotheses in Eq X1.10. Namely,

$$L_a = L_r - Z_{1-q}\sigma/\sqrt{n} \quad \text{(X1.15)}$$

X1.5.9 Eq X1.15 can be used to obtain the decision point L_a in the decision rule if the population standard deviation, σ, is known. When the population standard deviation, σ, is not known, its estimate s needs to be used and the decision point is given by Eq X1.16:

$$L_a = L_r - t_{1-q,n-1}s/\sqrt{n} \quad \text{(X1.16)}$$

where s is the estimated standard deviation, and $t_{1-q,n-1}$ a tabled t-value with (n-1) degrees of freedom.

X1.5.10 Since the term $(t_{1-q,n-1}s/\sqrt{n})$ in Eq X1.16 is non-negative, the decision point L_a is equal to or smaller than the regulatory limit L_r.

X1.5.11 The decision rule using the decision point in Eq X1.16 is carried out as follows:

X1.5.11.1 If $x < L_a$, then reject H_o and conclude no exceedance over the regulatory limit.

X1.5.11.2 If $x \geq L_a$, then do not reject the null hypothesis H_o and conclude exceedance over the regulatory limit.

X1.5.12 *False Positive Error*:

X1.5.12.1 The corresponding false positive error under this presumption can be calculated as well. Under the alternative hypothesis, the true mean is postulated to be lower than the regulatory limit. Namely, H_a: $u = L_r - \delta_4$, $\delta_4 > 0$. Thus, False positive decision error:

= probability of saying
that the true mean concentration is not lower
than the regulatory limit when it is so, (X1.17)

= probability that the data observe ($\bar{x} \geq L_a$) in favor of the null hypothesis H_o when the alternative hypothesis H_a is the correct one,

= Prob ($\bar{x} \geq L_a \mid H_a$)

= Prob$\{Z \geq [L_a - (L_r - \delta_4)] / (\sigma/\sqrt{n})\}$, and

= Prob$\{Z \geq Z_{1-q} + \delta_4/(\sigma/\sqrt{n})\}$, $\delta_4 > 0$, $q < 0.5$

X1.5.12.2 As can be seen in Eq X1.17, the false positive error is a function of how much the true mean, μ, is lower than the regulatory limit L_r. This difference is δ_4 expressed in the unit of the standard error of the mean $(\sigma/\sqrt{n})$, or altogether $\delta_4/(\sigma/\sqrt{n})$.

X1.5.12.3 False positive error rates for different values of $\delta_4/(\sigma/\sqrt{n})$ are given in Table X1.2. When δ_4 approaches zero (thus, $\delta_4/(\sigma/\sqrt{n}) - 0$), the true value approaches the regulatory limit L_r. As the true value becomes smaller than L_r (when $\delta_4/(\sigma/\sqrt{n}) > 0$), the false positive error becomes larger.

X1.5.12.4 It is to be noted that the false positive error is the exact complement of the chosen false negative error when the true value is at L_r; there the two errors total 100 %.

X1.5.12.5 When the population standard deviation, σ, is

TABLE X1.2 False Positive Error (*p*) Under Presumption Number 2

NOTE 1—For Some Values of False Negative Error (*q*), and True Values $\delta_4/(\sigma \sqrt{n})$.[A]

False Negative Error, *q*	If True Value As Number of Standard Errors Away from Regulatory Limit, or $\delta_4/(\sigma/\sqrt{n})$, Is	False Positive Error, *p*
0.05	0	0.95
	1	0.74
	2	0.36
	3	0.09
0.10	0	0.90
	1	0.61
	2	0.24
	3	0.04
0.20	0	0.80
	1	0.44
	2	0.12
	3	0.02

[A] Where the true value = regulatory limit when $\delta_4/(\sigma \sqrt{n}) = 0$.

replaced by sample standard deviation, s, in Eq X1.17, the z-statistic is replaced by the t-statistic. The calculation of the false positive error now involves a non-centrality parameter and a statistician should be consulted.

X1.5.13 *Equivalency Between Decision Point Approach and Test of Hypothesis Approach*:

X1.5.13.1 The decision rule using the decision point given in Eq X1.15 or Eq X1.16 can be made to be equivalent to one using the upper 100 (1-p) % confidence limit approach in a statistical test of hypothesis.

X1.5.13.2 Recall that the decision rule using a decision point is to conclude exceedance of the regulatory limit if $x \geq L_a$.

X1.5.13.3 But $L_a = L_r - t_{1-q,n-1}s/\sqrt{n}$ from Eq X1.16. Thus, $x \geq L_a$ becomes $x \geq L_r - t_{1-q,n-1}s/\sqrt{n}$.

X1.5.13.4 Rearranging, we can conclude exceedance of the regulatory limit (using confidence limit approach) if:

$$\bar{x} + t_{1-q,n-1}s/\sqrt{n} \geq L_r \quad \text{(X1.18)}$$

X1.5.13.5 Note that the left-hand side of Eq X1.18 is the upper 100 (1-q) % confidence limit in a statistical test of hypothesis. Thus, the decision point approach from Eq X1.16 and the hypothesis test approach from Eq X1.18 are equivalent in decision-making.

X1.5.13.6 The decision rule using the upper confidence limit in Eq X1.18 is carried out as follows:

(1) If $(x + t_{1-q,n-1}s/\sqrt{n}) < L_r$, then reject H_o and conclude no exceedance over the regulatory limit.

(2) If $(x + t_{1-q,n-1}s/\sqrt{n}) \geq L_r$, do not reject the null hypothesis H_o and conclude exceedance over the regulatory limit.

X1.6 *Decision Point Derivation Under Third Presumption*—The true (population) mean concentration is presumed to be neither higher nor lower than the regulatory limit.

X1.6.1 The steps in deriving a decision point under this presumption are given in Fig. 1 (middle part of figure).

X1.6.2 Recall that the decision point derived from the first presumption is Eq X1.7, and that the decision point derived from the second presumption is Eq X1.16.

X1.6.3 Eq X1.7 leads to a decision point higher than the regulatory limit and Eq X1.16 leads to one lower than the regulatory limit. Since the third presumption is a neutral one, it seems reasonable to have a corresponding decision point which is neither higher nor lower than the regulatory limit. This is achieved by setting the decision errors p and q equal to each other at 0.50 (or 50 %). When $p = q = 0.50$, $t_{1-p,n-1} = t_{1-q,n-1} = 0$ and the two equations above become identical and are reduced to Eq X1.19:

$$L_a = L_r \quad \text{(X1.19)}$$

X1.6.4 This is the derived decision point based on practical considerations under the third presumption.

X1.6.5 The decision rule using the decision point in Eq X1.19 is carried out as follows:

X1.6.5.1 If $x < L_a$, then conclude no exceedance over the regulatory limit.

X1.6.5.2 If $x \geq L_a$, then conclude exceedance over the regulatory limit.

X1.6.6 The confidence limit approach under this presumption is identical to Eq X1.19, since the confidence limits in Eq X1.9 and Eq X1.18 are similarly reduced to Eq X1.19, when:

$p = q = 0.50$.

X1.6.7 The false positive and false negative errors, when $L_a = L_r$, are:

False positive error:

$$= \text{Prob}\ (\bar{x} \geq L_a \mid u < L_r), \quad \text{(X1.20)}$$

$= \text{Prob}\ (\bar{x} \geq L_r \mid u < L_r$

$= \text{Prob}\ (Z \geq \delta_5/(\sigma/\sqrt{n}))$, where $u = L_r - \delta_5$, $\delta_5 > 0$

$< \text{Prob}\ (Z \geq 0)$, and

$= 0.5$.

False negative error:

$$= \text{Prob}\ (\bar{x} < L_a \mid u \geq L_r), \quad \text{(X1.21)}$$

$= \text{Prob}\ (\bar{x} < L_r \mid u \geq L_r$

$= \text{Prob}\ (Z < \delta_6/(\sigma/\sqrt{n}))$, where $u = L_r - \delta_6$, $\delta_6 \geq 0$,

$\leq \text{Prob}\ (Z < 0)$, and

$= 0.5$.

When δ_5 and δ_6 are relatively small, the false positive and false negative errors will be approximately equal at 0.5 each. When δ_5 and δ_6 are not small, the errors will deviate from 0.5 substantially, depending on where the true mean u is. When the true mean is exactly at the regulatory limit, L_r, then $\delta_5 = \delta_6 = 0$, and the two errors are equal at 0.5.

Standard Practice for
Use of Scrap Tires in Civil Engineering Applications[1]

This standard is issued under the fixed designation D 6270; the number immediately following the designation indicates the year of original adoption or, in the case of revision, the year of last revision. A number in parentheses indicates the year of last reapproval. A superscript epsilon (ε) indicates an editorial change since the last revision or reapproval.

1. Scope

1.1 This practice provides guidance for testing the physical properties and gives data for assessment of the leachate generation potential of processed or whole scrap tires in lieu of conventional civil engineering materials, such as stone, gravel, soil, sand, or other fill materials. In addition, typical construction practices are outlined.

2. Referenced Documents

2.1 *ASTM Standards:*

C 127 Test Method for Specific Gravity and Absorption of Coarse Aggregate[2]

D 422 Test Method for Particle-Size Analysis of Soils[3]

D 698 Test Method for Laboratory Compaction Characteristics of Soil Using Standard Effort (12,400 ft-lbf/ft^3 (600 kN-m/m^3))[3]

D 1557 Test Method for Laboratory Compaction Characteristics of Soil Using Modified Effort (56,000 ft-lbf/ft^3 (2,700 kN-m/m^3))[3]

D 2434 Test Method for Permeability of Granular Soils (Constant Head)[3]

D 3080 Test Method for Direct Shear Test of Soils Under Consolidated Drained Conditions[3]

D 4253 Test Methods for Maximum Index Density and Unit Weight of Soils Using a Vibratory Table[3]

2.2 *AASHTO Standard:*

T 274 Standard Method of Test for Resilient Modulus of Subgrade Soils[4]

2.3 *USEPA Standard:*

Method 1311 Toxicity Characteristics Leaching Procedure[5]

3. Terminology

3.1 *Definitions:*

3.1.1 *baling, n*—a method of volume reduction whereby tires are compressed into bales.

3.1.2 *bead, n*—the anchoring part of the tire which is shaped to fit the rim and is constructed of bead wire wrapped by the plies.

3.1.3 *bead wire, n*—a high tensile steel wire surrounded by rubber, which forms the bead of a tire that provides a firm contact to the rim.

3.1.4 *belt wire, n*—a brass plated high tensile steel wire cord used in steel belts.

3.1.5 *buffing rubber, n*—vulcanized rubber usually obtained from a worn or used tire in the process of removing the old tread in preparation for retreading.

3.1.6 *carcass, n*—see **casing**.

3.1.7 *casing, n*—the basic tire structure excluding the tread (Syn. *carcass*).

3.1.8 *granulated rubber, n*—particulate rubber composed of mainly nonspherical particles that span a broad range of maximum particle dimension, from below 425 μm (40 mesh) to 12 mm (also refer to **particulate rubber**).[6]

3.1.9 *ground rubber, n*—particulate rubber composed of mainly nonspherical particles that span a range of maximum particle dimensions, from below 425 μm (40 mesh) to 2 mm (also refer to **particulate rubber**).[6]

3.1.10 *nominal size, n*—the average size product (chip) that comprises 50 % or more of the through put in a scrap tire processing operation; scrap tire processing operations generate products (chips) above and below the nominal size.

3.1.11 *particulate rubber, n*—raw, uncured, compounded or vulcanized rubber that has been transformed by means of a mechanical size reduction process into a collection of particles, with or without a coating of a partitioning agent to prevent agglomeration during production, transportation, or storage (also see definition of **buffing rubber, granulated rubber, ground rubber,** and **powdered rubber**).[6]

3.1.12 *passenger car tire, n*—a tire with less than a 457-mm rim diameter for use on cars only.

3.1.13 *powdered rubber, n*—particulate rubber composed of mainly nonspherical particles that have a maximum particle dimension equal to or below 425 μm (40 mesh) (also refer to **particulate rubber**).[6]

3.1.14 *rough shred, n*—a piece of a shredded tire that is larger than 50 mm by 50 mm by 50 mm, but smaller than 762 mm by 50 mm by 100 mm.

3.1.15 *rubber fines, n*—small particles of ground rubber that

---

[1] This practice is under the jurisdiction of ASTM Committee D-34 on Biotechnology and is the direct responsibility of Subcommittee D34.06 on Recovery and Reuse.

Current edition approved June 10, 1998. Published August 1998.

[2] *Annual Book of ASTM Standards*, Vol 04.02.

[3] *Annual Book of ASTM Standards*, Vol 04.08.

[4] *Standard Specifications for Transportation Materials and Methods of Sampling and Testing, Part II: Methods of Sampling and Testing,* American Association of State Highway and Transportation Officials, Washington, D.C.

[5] *Test Methods for Evaluating Solid Waste: Physical/Chemical Methods,* 3rd ed., Report No. EPA 530/SW-846, U.S. Environmental Protection Agency, Washington, D.C.

[6] The defined term is the responsibility of Committee D-11 on Rubber.

result as a by-product of producing shredded rubber.

3.1.16 *scrap tire, n*—a tire, which can no longer be used for its original purpose due to wear or damage.

3.1.17 *shred sizing, n*—a term which generally refers to the process of particles passing through a rated screen opening rather than those which are retained on the screen.

3.1.18 *shredded tire, n*—a size reduced scrap tire where the reduction in size was accomplished by a mechanical processing device, commonly referred to as a shredder.

3.1.19 *shredded rubber, n*—pieces of scrap tires resulting from mechanical processing.

3.1.20 *sidewall, n*—the side of a tire between the tread shoulder and the rim bead.

3.1.21 *single pass shred, n*—a shredded tire that has been processed by one pass through a shear type shredder and the resulting pieces have not been classified by size.

3.1.22 *steel belt, n*—rubber coated steel cords that run diagonally under the tread of steel radial tires and extend across the tire approximately the width of the tread.

3.1.23 *tire chips, n*—Pieces of scrap tires that have a basic geometrical shape and are generally between 12 mm and 50 mm in size and have most of the wire removed (Syn. *chipped tire*).

3.1.24 *tire shreds, n*—Pieces of scrap tires that have a basic geometrical shape and are generally between 50 mm and 305 mm in size.

3.1.25 *tread, n*—that portion of the tire which contacts the road.

3.1.26 *truck tire, n*—a tire with a rim diameter of 500 mm or larger.

3.1.27 *waste tire, n*—a tire which is no longer capable of being used for its original purpose but which has been disposed of in such a manner that it cannot be used for any other purpose.

3.1.28 *whole tire, n*—a scrap tire that has been removed from a rim but which has not been processed.

3.1.29 *x-mm minus, n*—pieces of classified, size reduced scrap tires where the maximum size of 95 % of the pieces is less than x-mm in any dimension (that is, 25-mm minus; 50-mm minus; 75-mm minus, etc).

4. Significance and Use

4.1 This practice is intended for use of scrap tires including tire chips or tire shreds comprised of pieces of scrap tires, tire chip/soil mixtures, tire sidewalls, and whole scarp tires in civil engineering applications. This practice includes the use of tire chips, tire shreds, and tire chip/soil mixtures as lightweight embankment fill, lightweight retaining wall backfill, drainage layers, thermal insulation to limit frost penetration beneath roads, insulating backfill to limit heat loss from buildings, and replacement for soil or rock in other fill applications. Use of whole scrap tires and tire sidewalls includes construction of retaining walls and drainage culverts, as well as use as fill when whole tires have been compressed into bales. It is the responsibility of the design engineer to determine the appropriateness of using scrap tires in a particular application and to select applicable tests and specifications to facilitate construction and environmental protection. This practice is intended to encourage wider utilization of scrap tires in civil engineering applications.

4.2 Three tire shred fills with thicknesses in excess of 7 m have experienced a serious heating reaction; however, more than 70 fills with a thickness less than 3 m have been constructed with no evidence of a deleterious heating reaction (**1**)[7]. Guidelines have been developed to minimize internal heating of tire shred fills (**2**) as discussed in 6.10. The guidelines are applicable to fills less than 3 m thick; thus, this practice should be applied only to tire shred fills less than 3 m thick.

5. Material Characterization

5.1 The specific gravity and water absorption capacity of tire shreds should be determined in accordance with Test Method C 127; however, the specific gravity of tire shreds is less than half the value obtained for common earthen coarse aggregate, so it is permissible to use a minimum weight of test sample that is half of the specified value. The particle density or density of solids of tire shreds (ρ_s) may be determined from the apparent specific gravity using the following equation:

$$\rho_s = S_a\,(\rho_w) \tag{1}$$

where:

S_a = apparent specific gravity, and
ρ_w = density of water.

5.2 The gradation of tires shreds should be determined in accordance with Test Method D 422; however, the specific gravity of tire shreds is less than half the values obtained for common earthen materials so it is permissible to use a minimum weight of test sample that is half of the specified value.

5.3 The laboratory compacted dry density, or bulk density, of tire chips and tire chip/soil mixtures with less than 30 % retained on the 19.0-mm sieve can be determined in accordance with Test Method D 698 or D 1557. Tire Shred and tire shred/soil mixtures used for civil engineering applications, however, almost always have more than 30 % retained on the 19.0-mm sieve, so these methods generally are not applicable. A larger compaction mold should be used to accommodate the larger size of the tire shreds. The sizes of typical compaction molds are summarized in Table 1. The larger mold requires that the number of layers, or the number of blows of the rammer/layer, or both, be increased to produce the desired compactive energy/unit volume. Compactive energies ranging from 60 % of Test Method D 698 (60 % × 600 kN-m/m^3 = 360 kN-m/m^3) to 100 % of Test Method D 1557 (2,700 kN-m/m^3) have been

[7] The boldface numbers in parentheses refer to the list of references at the end of this standard.

TABLE 1 Size of Compaction Molds Used to Determine Dry Density of Tire Shreds

Maximum Particle Size (mm)	Mold Diameter (mm)	Mold Volume (m^3)	Reference
75	254	0.0125	(3)
75	305	0.0146	(4)
51	203 and 305	N.R.[A]	(5)

[A] N.R. = not reported.

used. Compaction energy only has a small effect on the resulting dry density (**3**); thus, for most applications it is permissible to use a compactive energy equivalent to 60 % of Test Method D 698. To achieve this energy with a mold volume of 0.0125 m^3 would require that the sample be compacted in five layers with 44 blows/layer with a 44.5 N rammer falling 457 mm. The water content of the sample only has a small effect on the compacted dry density (**3**) so it is permissible to perform compaction tests on air or oven-dried samples.

5.3.1 The dry densities for tire shreds loosely dumped into a compaction mold and tire shreds compacted by vibratory methods (similar to Test Method D 4253) are about the same (**4, 5, 6**); thus, vibratory compaction of tire shreds in the laboratory (see Test Method D 4253) should not be used.

5.3.2 When estimating an in-place density for use in design, the compression of a tire shred layer under its own self-weight and under the weight of any overlying material must be considered. The dry density determined as discussed in 5.3 are uncompressed values. In addition, short-term time dependent settlement of tire shreds should be accounted for when estimating the final in-place density (**7**).

5.4 The compressibility of tire shreds and tire shred/soil mixtures can be measured by placing tire shreds in a rigid cylinder with a diameter several times greater than the largest particle size and then measuring the vertical strain caused by an increasing vertical stress. If it is desired to calculate the coefficient of lateral earth pressure at rest K_O, the cylinder can be instrumented to measure the horizontal stress of the tire shreds acting on the wall of the cylinder.

5.4.1 The high compressibility of tire shreds necessitates the use of a relatively thick sample. In general, the ratio of the initial specimen thickness to sample diameter should be greater than one. This leads to concerns that a significant portion of the applied vertical stress could be transferred to the walls of the cylinder by friction. If the stress transferred to the walls of the cylinder is not accounted for, the compressibility of the tire shreds will be underestimated. For all compressibility tests, the inside of the container should be lubricated to reduce the portion of the applied load that is transmitted by side friction from the sample to the walls of the cylinder. For testing where a high level of accuracy is desired, the vertical stress at the top and the bottom of the sample should be measured so that the average vertical stress in the sample can be computed. A test apparatus designed for this purpose is illustrated in Fig. 1 (**8**).

5.5 The resilient modulus (M_R) of subgrade soils can be expressed as:

$$M_R = A\Theta^B \quad (2)$$

where:

Θ = first invariant of stress (sum of the three principal stresses),

A = experimentally determined parameter, and

B = experimentally determined parameter.

Tests for the parameters A and B can be conducted according to AASHTO T 274. The maximum particle size typically is limited to 19 mm by the testing apparatus, which precludes the general applicability of this procedure to the larger size tire chips and shreds typically used for civil engineering applications.

5.6 The coefficient of lateral earth pressure at rest K_O and Poisson's ratio μ can be determined from the results of confined compression tests where the horizontal stresses were

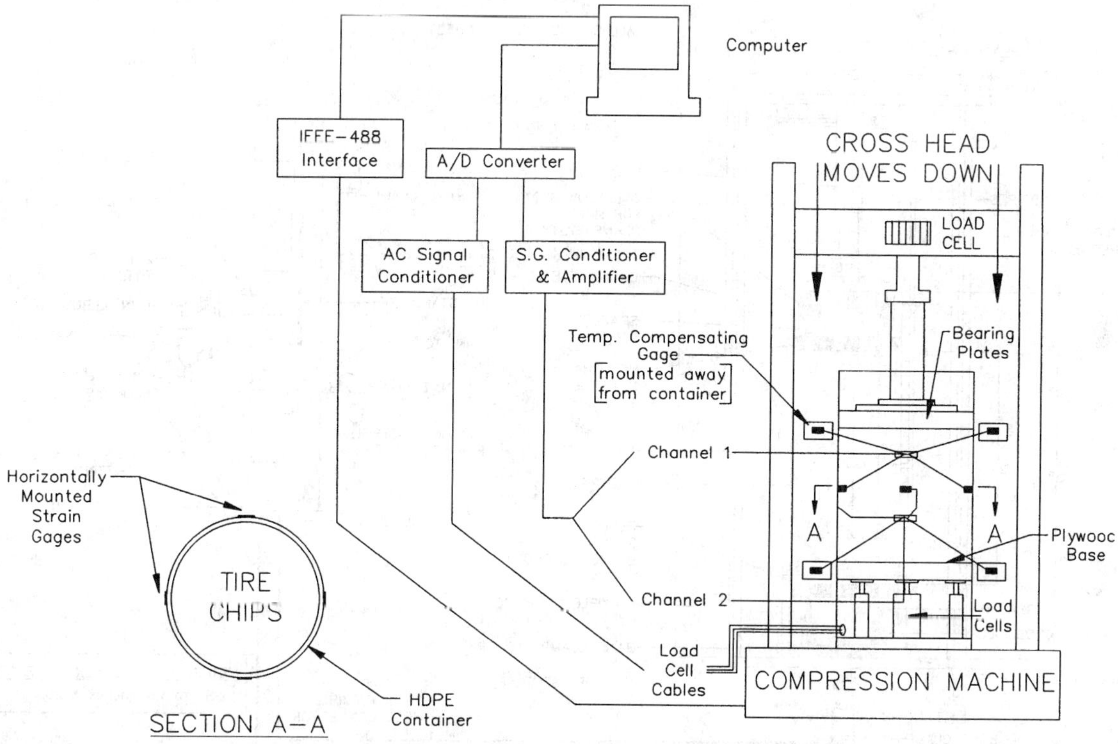

FIG. 1 Compressibility Apparatus for Tire Shreds Designed to Measured Lateral Stress and the Portion of the Vertical Load Transferred by Friction from Tire Shreds to Container (8)

measured. A test apparatus designed for this purpose is shown in Fig. 1 K_O and μ are calculated from:

$$K_O = \sigma_h / \sigma_v \quad (3)$$

$$\mu = K_O / (1 + K_O) \quad (4)$$

where:
σ_h = measured horizontal stress, and
σ_v = measured vertical stress.

5.7 The shear strength of tire shreds may be determined in a direct shear apparatus in accordance with Test Method D 3080 or using a triaxial shear apparatus. The large size of tire shreds typically used for civil engineering applications requires that specimen sizes be several times greater than used for common soils. Because of the limited availability of large triaxial shear apparatus, this method is generally restricted to tire chips 25 mm in size and smaller. Extrapolation of results on small size pieces to the 75-mm and larger size shreds used for civil engineering applications is uncertain since small pieces are nearly equidimensional while larger tire chips and shreds tend to be long and flat. Furthermore, the triaxial shear apparatus generally is not suitable for tire shreds that have steel belts protruding from the cut edges of the shreds since the wires would puncture the membrane used to surround the specimen. The interface strength between tire shreds and geomembrane can be measured in a large scale direct shear test apparatus (**9**).

5.8 The hydraulic conductivity (permeability) of tire shreds and tire shred/soils mixtures should be measured with a constant head permeameter with a diameter several times greater than the maximum particle size. Tire chips with a maximum size smaller than 19 mm can be determined in accordance with Test Method D 2434; however, tire shreds and tire shred/soil mixtures used for civil engineering applications almost always have a majority of their particles larger than 19 mm so this method is generally not applicable. Samples should be tested at a void ratio comparable to the value expected in the field. This may require a permeameter capable of appling a vertical stress to the sample to simulate the compression that would occur under the weight of overlying material. The high hydraulic conductivity of tire shreds should be accounted for in design of the permeameter. The design shall include provisions for an adequate supply of water and measuring the head loss across the sample using standpipes mounted on the body of the permeameter. An apparatus designed taking these factors into account is shown in Fig. 2 (**10**).

5.9 The thermal conductivity of tire shreds is significantly lower than for common soils. For tire chips smaller than 25 mm in size, the thermal conductivity can be measured using commercially available guarded hot plate apparatus. For tire shreds larger than 25 mm, it is necessary to construct a large scale hot plate apparatus (**11**). The thermal conductivity of tire shreds also can be backcalculated from field measurements (**11**).

6. Construction Practices

6.1 Tire shreds have a compacted dry density that is one-third to one-half of the compacted dry density of typical soil. This makes them an attractive lightweight fill for embankment construction on weak, compressible soils where slope stability or excessive settlement are a concern.

6.2 The thermal resistivity of tire shreds is approximately eight times greater than for typical granular soil. For this reason, tire shreds can be used as a 150 to 450-mm thick insulating layer to limit the depth of frost penetration beneath

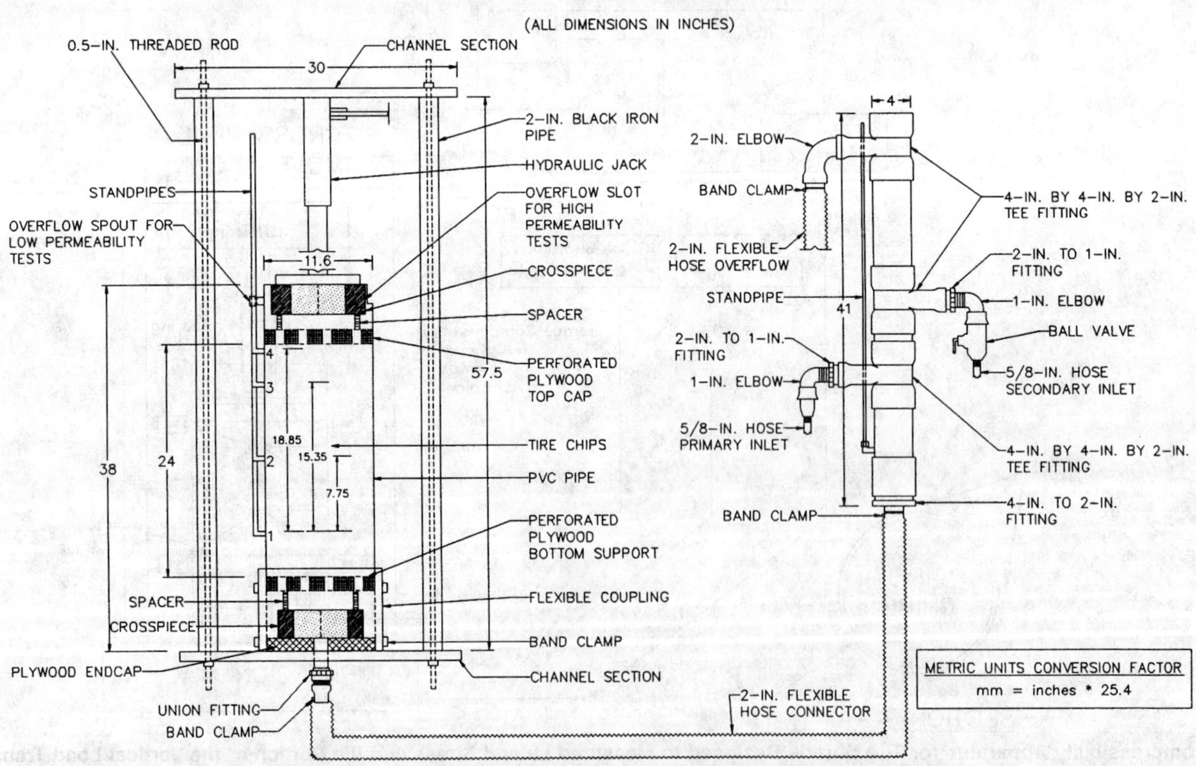

FIG. 2 Hydraulic Conductivity Apparatus for Tire Shreds with Provisions for Application of Vertical Stress (10)

roads. This reduces frost heave in the winter and improves subgrade support during the spring thaw. In addition, tire shreds can be used as backfill around basements to limit heat lost through basement walls, thereby reducing heating costs.

6.3 The low-compacted dry density, high-hydraulic conductivity, and low-thermal conductivity makes tire shreds very attractive for use as retaining wall backfill. Lateral earth pressures for tire shred backfill can be about 50 % of values obtained for soil backfill (**7**). Tire shreds also can be used as backfill for geosynthetic-reinforced retaining walls.

6.4 The high hydraulic conductivity of tire shreds, generally greater than 1 cm/s, makes them suitable for many drainage applications, including French drains, drainage layers in landfill liner and cover systems, and leach fields for on-site sewage disposal systems.

6.5 Two different sizes of tire shreds commonly are used for the applications discussed above. One has a maximum size of 75 mm and the other has a maximum size of 300 mm. Rough shreds also can been used for some applications provided all tires are shredded such that the largest shred is the lesser of one-quarter circle in shape or 600 mm in length. In all cases, at least one side wall should be severed from the tread.

6.6 Tire shreds with a maximum size of 75 mm or 300 mm generally are placed in 300-mm thick lifts and compacted by a tracked bulldozer, sheepsfoot roller, or smooth drum vibratory roller with a minimum operating weight of 90 kN. Rough shreds generally are placed in 900-mm thick lifts and compacted by a tracked bulldozer. For most applications, a minimum of six passes of the compaction equipment should be used.

6.7 Tire shreds should be covered with a sufficient thickness of soil to limit deflections of overlying pavement caused by traffic loading. Soil cover thicknesses as low as 0.8 m may be suitable for roads with light traffic. For roads with heavy traffic, 1 to 2 m of soil cover may be required. For unpaved applications, 0.3 to 0.5 m of soil cover may be suitable depending on the traffic loading. The designer should assess the actual thickness of soil cover needed based on the loading conditions, tire-shred layer thickness, pavement thickness, and other conditions as appropriate for particular project. Regardless of the application, the tire shreds should be covered with soil to prevent contact between the public and the tire shreds, which may have exposed steel belts.

6.8 In applications where pavement will be placed over the tire shred layer and in drainage applications, the tire shred layer should be wrapped completely in a layer of nonwoven or woven geotextile to minimize infiltration of soil particles into the voids between the tire shreds.

6.9 Whole tires and tire sidewalls that have been cut from the tire carcass can be used to construct retaining walls and bound together to form drainage culverts.

6.10 Tire shred fills should be designed to minimize the possibility of an internal heating reaction (**2**). Possible causes of the reaction are oxidation of the exposed steel belts and oxidation of the rubber. Microbes may play a role in both reactions. Although details of the reaction are under study, the following factors are thought to create conditions favorable for oxidation of exposed steel, or rubber, or both; free access to air; free access to water; retention of heat caused by the high insulating value of tire shreds in combination with a large fill thickness; large amounts of exposed steel belts; smaller tire shred sizes and excessive amounts of granulated rubber particles; and, the presence of inorganic and organic nutrients that would enhance microbial action.

6.10.1 The design guidelines given in the following sections have been developed to minimize the possibility for heating of tire shred fills by minimizing factors that could possibly create conditions favorable for this reaction. As more is learned about the causes of the reaction, it may be possible to ease some of the guidelines. In developing these guidelines, the insulating effect caused by increasing fill thickness and the favorable performance of projects with tire shred fills less than 4-m thick have been considered; thus, design guidelines are less stringent for projects with thinner tire shred layers. The guidelines are divided into two classes; Class I Fills with tire shred layers less than 1-m thick, and Class II Fills with tire shred layers in the range of 1-m to 3-m thick. Although there have been no projects with less than 4-m of tire shred fill that have experienced a catastrophic heating reaction, to be conservative, tire shred layers greater than 3-m thick are not recommended. In addition to the guidelines given below, the designer must choose the maximum tire shred size, thickness of overlying soil cover, etc., to meet the requirements imposed by the engineering performance of the project. These guidelines are for use in designing tire shred fills. Design of fills that are mixtures or alternating layers of tire shreds and mineral soil that is free from organic matter should be handled on a case by case basis.

6.10.2 For both Class I and II Fills, the tire shreds shall be free of all contaminants, such as oil, grease, gasoline, diesel fuel, etc., that could create a fire hazard. In no case shall the tire shreds contain the remains of tires that have been subjected to a fire because the heat of a fire may liberate liquid petroleum products from the tire that could create a fire hazard when the shreds are placed in a fill.

6.10.3 For Class I Fills, the tire shreds shall have a maximum of 50 % (by weight) passing the 38-mm sieve and a maximum of 5 % (by weight) passing the 4.75-mm sieve. No special design features are required to minimize heating of Class I Fills.

6.10.4 For Class II Fills, the tire shreds shall have a maximum of 25 % (by weight) passing the 38-mm sieve and a maximum of 1 % (by weight) passing the 4.75-mm sieve. The tire shreds shall be free from fragments of wood, wood chips, and other fibrous organic matter. The tire shreds shall have less than 1 % (by weight) of metal fragments, which are not at least partially encased in rubber. Metal fragments that are encased partially in rubber shall protrude no more than 25 mm from the cut edge of the tire shred on 75 % of the pieces and no more than 50 mm on 100 % of the pieces.

6.10.5 Class II Fills shall be constructed in such a way that infiltration of water and air is minimized; moreover, there shall be no direct contact between tire shreds and soil containing organic matter, such as topsoil. One possible way to accomplish this is to cover the top and sides of the fill will a 0.5-m thick layer of compacted mineral soil with a minimum of 30 % fines. The mineral soil should be free from organic matter and should be separated from the tire shreds with a geotexile. The top of the mineral soil layer should be sloped so that water will drain away from the tire shred fill. Additional fill may be placed on top of the mineral soil layer as needed to meet the overall design of the project. If the project will be paved, it is recommended that the pavement extend to the shoulder of the embankment or that other measures be taken to minimize infiltration at the edge of the pavement.

6.10.6 For Class II Fills, use of drainage features located at the bottom of the fill that could provide free access to air should be avoided. Use of drainage features includes, but is not limited to, open graded drainage layers daylighting on the side of the fill and drainage holes in walls. Under some conditions, it may be possible to use a well graded granular soil as a drainage layer. The thickness of the drainage layer at the point where it daylights on the side of the fill should be minimized. For tire shred fills placed against walls, it is recommended that the drainage holes in the wall be covered with well graded granular soil. The granular soil should be separated from the tire shreds with geotextile.

7. Leachate

7.1 The Toxicity Characteristics Leaching Procedure (TCLP) (USEPA Method 1311) is used to determine if a waste is a hazardous waste, thereby posing a significant hazard to human health due to leaching of toxic compounds. The TCLP test represents the worst case scenario of acid rain percolating through the waste and exiting as leachate. For all regulated metals and organics, the results for tire shreds are well below the TCLP regulatory limits (**12, 13, 14**); therefore, tire shreds are not classified as a hazardous waste.

7.2 In addition to TCLP tests, laboratory leaching studies have been performed following several test protocols. Results show that metals are leached most readily at low pH and that organics are leached most readily at high pH (**14, 15**); thus, it is preferable to use tire shreds in environments with a near neutral pH.

7.3 Field studies of tire shred fills located above the ground water table show that tire shreds tend to leach manganese, and under some circumstances, iron at levels above their secondary drinking water standard (**4, 16**). Since secondary standards are based on aesthetic factors, such as color, odor, and taste, rather than health concerns, release of manganese and iron from tire shreds is not a significant concern. Release of organics from tire shreds placed above the water table generally is below test method detection limits (**16**); thus, release of organics from tire shreds placed above the water table is not a significant concern.

7.4 For tire shreds placed below the water table, tire shreds release levels of manganese and iron that are significantly above their secondary drinking water standards (**14**); thus, tire shreds should be used below the water table only where the aesthetic concerns raised by elevated levels of manganese and iron have been examined. Tire shreds placed below the water table leach low levels of a few organic compounds into the ground water (**14**). Further study is needed to determine if these levels are high enough to be of concern. Pending continued studies of the effect of tire shreds placed below the water table on organic levels, the use of tire shreds should be limited to above water table applications.

8. Keywords

8.1 construction practices; landfills; leachate; lightweight fill; retaining walls; roads; scrap tires

APPENDIX

(Nonmandatory Information)

X1. Typical Material Properties

X1.1 This appendix contains typical properties of tire chips and shreds to aid in the selection of values for preliminary designs and to provide a basis for comparison for test results.

X1.2 Values of specific gravity and water absorption capacity reported in the literature are summarized in Table X1.1. Table X1.2 summarizes the compacted and uncompacted dry density of tire shreds. Compaction results for mixtures of tire shreds and soil also are available (**4, 5, 6, 17**). The results from one study are summarized in Fig. X1.1.

X1.3 Typical compressibility results are summarized in Table X1.3. A measure of compressibility applicable to vehicle loads is resilient modulus. Results determined by Ahmed (**5**) using AASHTO T 274-82 for mixtures of tire chips and soil are summarized in Table X1.4. The parameter A, and therefore M_R, decreases as the percent tire sheds by dry weight of the mix increases. Results determined by Edil and Bosscher (**4, 17**) for mixtures of tire shreds and sand are summarized in Fig. X1.2. Shao et al (**18**) performed resilient modulus tests on crumb rubber (7-mm maximum size) and rubber buffings (1-mm maximum size). The resilient modulus values ranged from 700 to 1700 kPa.

X1.4 Typical values of coefficient of lateral earth pressure at rest and Poisson's ratio, measured as part of vertical compression tests, are presented in Table X1.5.

X1.5 The shear strength of tire shreds has been measured using triaxial shear (**5, 19, 20**) and using direct shear (**9, 21, 23**). Failure envelopes for tests conducted at low stress levels (less than about 100 kPa) are compared in Fig. X1.3. The failure envelopes are non-liner and concave down, so when fitting a linear failure envelope to the data, it is important that this be done over the range of stresses that will occur in the field.

X1.6 The shear strength of tire shred/soil mixtures has been measured using triaxial shear (**4, 24**) and direct shear (**4, 25**).

TABLE X1.1 Summary of Specific Gravity and Water Absorption Capacity

Tire Shred Type	Specific Gravity			Water Absorption Capacity (%)	Reference
	Bulk	Saturated Surface Dry	Apparent		
Glass belted (F and B)	- - - -	- - - -	1.14	3.8	(21)
Glass belted	0.98	1.02	1.02	4	(26)
Steel belted	1.06	1.01	1.10	4	(26)
Mixture	1.06	1.16	1.18	9.5	(19)
Mixture (Pine State)	- - - -	- - - -	1.24	2	(21)
Mixture (Palmer)	- - - -	- - - -	1.27	2	(21)
Mixture (Sawyer)	- - - -	- - - -	1.23	4.3	(21)
Mixture	1.01	1.05	1.05	4	(26)
Mixture (12.7 mm to 50.8 mm)	- - - -	0.88 to 1.13	- - - -	- - - -	(5)

TABLE X1.2 Summary of Laboratory Dry Densities of Tire Shreds

Compaction Method[A]	Particle Size Range (mm)	Tire Shred Type	Source of Tire Shreds	Dry Density (kg/m^3)	(Reference)
Loose	2 to 75	Mixed	Palmer Shredding	341	(21, 22)
Loose	2 to 51	Mixed	Pine State Recycling	482	(21, 22)
Loose	2 to 25	Glass	F and B Enterprises	495	(21, 22)
Loose	2 to 51	Mixed	Sawyer Environmental	409	(3, 26)
Loose	51 max	Mixed	- - - -	466	(5, 6)
Loose	25 max	Mixed	- - - -	489	(5, 6)
Vibration	25 max	Mixed	- - - -	496	(5, 6)
Vibration	13 max	Mixed	- - - -	473	(5, 6)
50 % Standard	51 max	Mixed	- - - -	614	(5, 6)
50 % Standard	25 max	Mixed	- - - -	641	(5, 6)
60 % Standard	2 to 75	Mixed	Palmer Shredding	620	(21, 22)
60 % Standard	2 to 51	Mixed	Pine State Recycling	643	(21, 22)
60 % Standard	2 to 25	Glass	F and B Enterprises	618	(21, 22)
60 % Standard	2 to 51	Mixed	Sawyer Environmental	625	(3, 26)
Standard	2 to 51	Mixed	Sawyer Environmental	640	(3, 26)
Standard	51 max	Mixed	- - - -	635	(5, 6)
Standard	38 max	Mixed	- - - -	645	(5, 6)
Standard	25 max	Mixed	- - - -	653	(5, 6)
Standard	10 max	Mixed	- - - -	633	(5, 6)
Standard	20 to 75	- - - -	Rodefeld	504[B]	(4, 17)
Standard	20 to 75	- - - -	Rodefeld	560[C]	(4, 17)
Modified	2 to 51	Mixed	Sawyer Environmental	660	(3, 26)
Modified	51 max	Mixed	- - - -	668	(5, 6)
Modified	25 max	Mixed	- - - -	685	(5, 6)
- - - -	50.8	Mixed	- - - -	410 to 570	(19)

[A] Compaction methods:

Loose	=	no compaction; tire shreds loosely dumped into compaction mold
Vibration	=	Test Method D 4253
50 % Standard	=	Impact compaction with compaction energy of 296.4 kJ/m^3.
60 % Standard	=	Impact compaction with compaction energy of 355.6 kJ/m^3.
Standard	=	Impact compaction with compaction energy of 296.4 kJ/m^3.
Modified	=	Impact compaction with compaction energy of 2693 kJ/m^3.

[B] 152-mm diameter mold compacted by 4.54 kg rammer falling 305 mm.

[C] 305-mm diameter mold compacted by 27.4 kg rammer falling 457 mm.

Table X1.6 and Table X1.7 summarize the results from Ahmed (**5**). Edil and Bosscher (**4**), and Benson and Khire (**25**) are interested primarily in the reinforcing effect of tire shreds when added to a sand. Under some circumstances, the shear strength is increased by adding tire shreds.

X1.7 Typical hydraulic conductivities for tire shreds and mixtures of tire shreds and soil are reported in Table X1.8 and Table X1.9 and Fig. X1.4.

X1.8 Measured thermal conductivities ranged from 0.0838 Cal/m-hr-°C for 1-mm particles tested in a thawed state with a water content less than 1 % and with low compaction to 0.147 Cal/m-hr-°C for 25 mm tire shreds tested in a frozen state with a water content of 5 % and high compaction (**18**). The thermal conductivity increased with increasing particle size, increased water content, and increased compaction. The thermal conductivity was higher for tire shreds tested under frozen conditions than when tested under thawed conditions. A thermal conductivity of 0.2 Cal/m-hr-°C was backcalculated from a field trial constructed using tire shreds with a maximum size of 51 mm (**11**). It is reasonable that the back calculated thermal conductivity is higher than found by Shao et al (**18**) since the tire

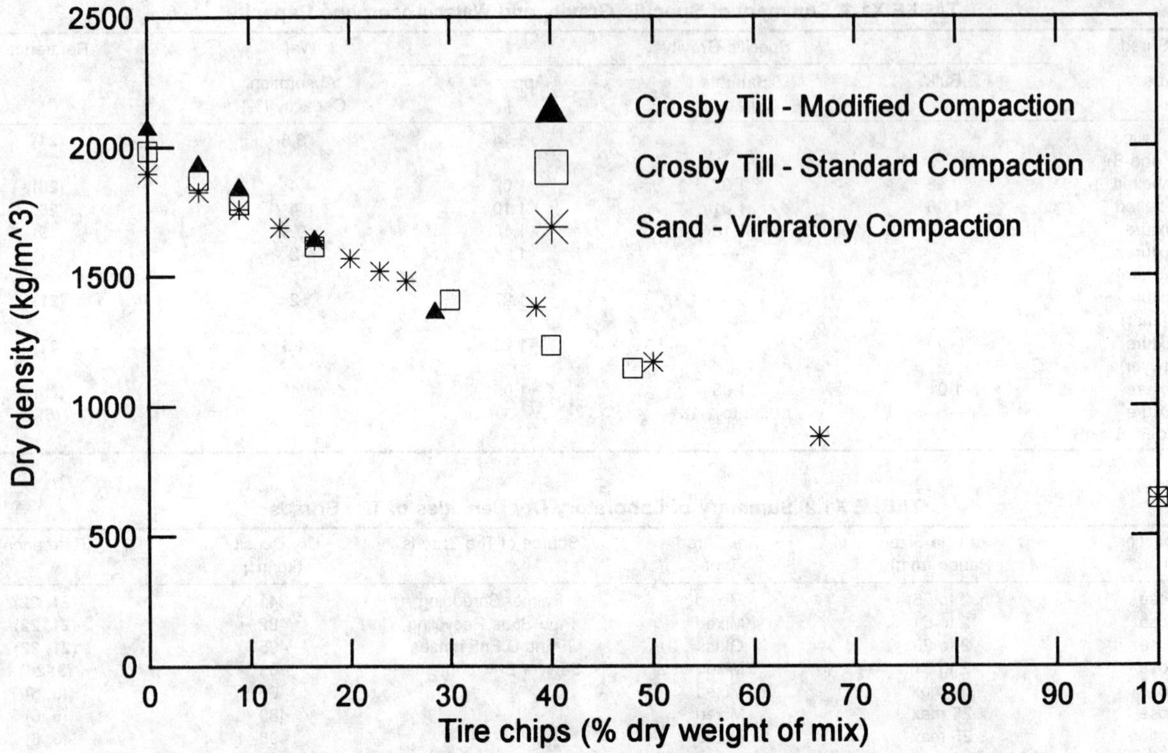

FIG. X1.1 Comparison of Compacted Dry-Density of Mixtures of Tire Shreds with Ottawa Sand and Crosby Till (5)

TABLE X1.3 Compressibility on Initial Loading

Particle Size Range (mm)	Tire Shred Type	Tire Shred Source	Initial Dry Density (kg/m^3)	Vertical Strain (%) at Indicated Vertical Stress (kPa)					Reference
				10	25	50	100	200	
2 to 75	Mixed	Palmer	Compacted	7 to 11	16 to 21	23 to 27	30 to 34	38 to 41	**(26)**
2 to 51	Mixed	Pine State	Compacted	8 to 14	15 to 20	21 to 26	27 to 32	33 to 37	**(21)**
2 to 25	Glass	F and B	Compacted	5 to 10	11 to 16	18 to 22	26 to 28	33 to 35	**(21)**
2 to 51	Mixed	Sawyer	Compacted	5 to 10	13 to 18	17 to 23	22 to 30	29 to 37	**(26)**
	Mixed		Compacted	4 to 5	8 to 11	13 to 16	18 to 23	27	**(5)**
75 max	Mixed	Pine State	510 to 670	12 to 20	18 to 28	- - - -	- - - -	- - - -	**(8)**
2 to 51	Mixed	Pine State	Loose	18	34	41	46	52	**(21)**
2 to 25	Mixed	F and B	Loose	8	18	28	37	45	**(21)**
	- - - -		Loose	9	12 to 17	17 to 24	24 to 31	30 to 38	**(27)**

shreds for the former were larger and contained more steel bead wire and steel belt.

X1.9 The results of TCLP tests for regulated metals are summarized in Table X1.10. Results of field studies of the effect of tire chips on water quality are summarized in Table X1.11 and Table X1.12, as well as Fig. X1.5 and Fig. X1.6.

X1.10 A typical material safety data sheet for whole scrap tires is included in Fig. X1.7.

TABLE X1.4 Resilient Modulus of Tire Shreds and Tire Shred/Soil Mixtures (5)

NOTE 1—Constants A and B are the constants for the regression equation and r^2 is the regression coefficient.
NOTE 2—Standard = Standard Proctor Energy = 296.4 kJ/m^3.
NOTE 3—The constants A and B assume the units for θ and M_R are psi (1 psi = 6.89 kPa).

Test No.	Tire Shred Maximum Size (mm)	Sample Preparation	% tire Shreds Based on Total Weight	Soil Type	Constant A	Constant B	r^2
AH01	No. shreds	Vibratory	No shreds	Sand	1071.5	0.84	0.95
AH02	13	Vibratory	15	Sand	524.8	0.83	0.95
AH03	13	Vibratory	30	Sand	269.2	0.90	0.67
AH04	13	Vibratory	38	Sand	42.7	1.15	0.89
AH05	13	Vibratory	50	Sand	38.9	0.83	0.84
AH06	13	Vibratory	100	Sand	36.3	0.55	0.74
AH07	19	Vibratory	38	Sand	34.7	1.21	0.92
AH08	No shreds	Standard	No shreds	Crosby Till	3162.3	0.49	0.83
AH09	13	Standard	15	Crosby Till	53.7	1.15	0.91
AH10	13	Standard	29	Crosby Till	61.7	0.91	0.94
AH11	13	Standard	38	Crosby Till	55.0	0.67	0.95

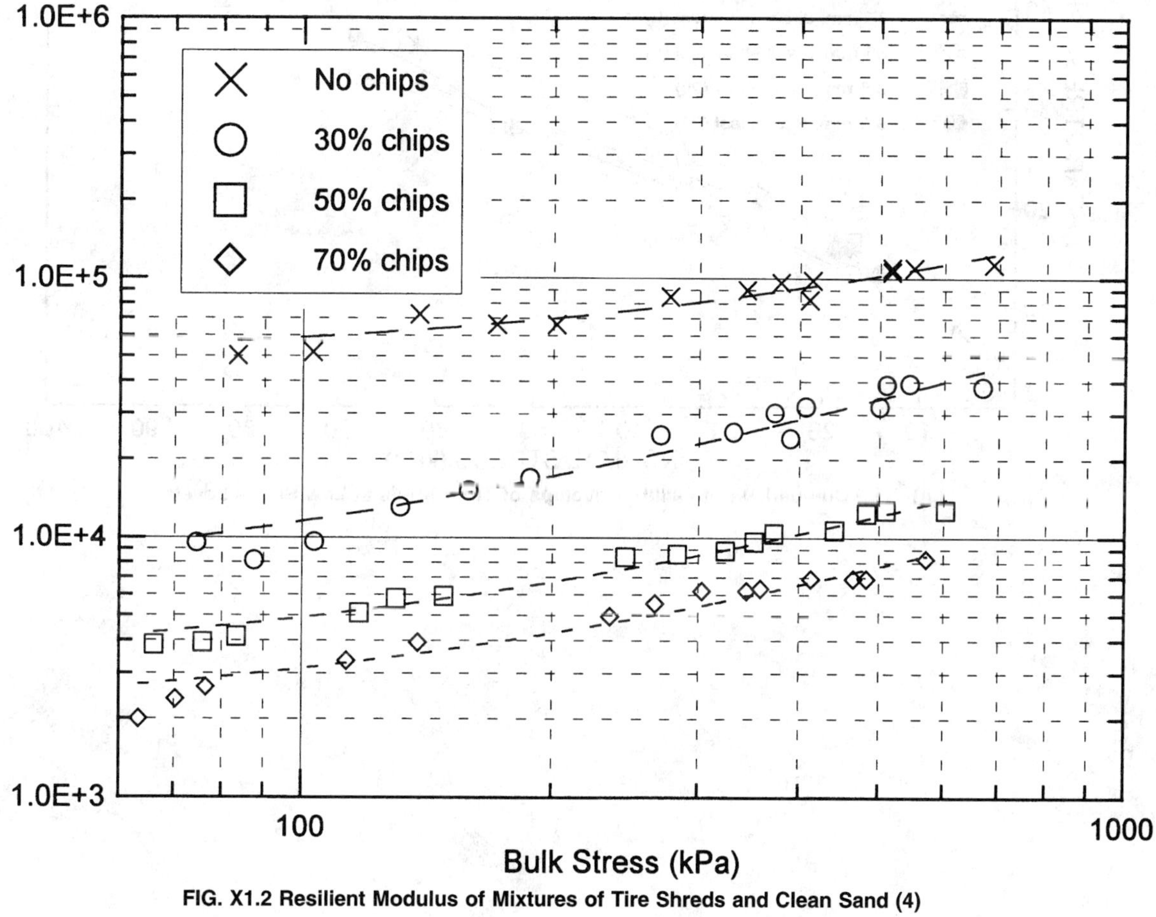

FIG. X1.2 Resilient Modulus of Mixtures of Tire Shreds and Clean Sand (4)

TABLE X1.5 Summary of Coefficient of Lateral Earth Pressure at Rest and Poisson's Ratio

Particle Size Range (mm)	Tire Shred Type	Source of Tire Shreds	K_O	-μ	References
2 to 51	Mixed	Sawyer Environmental	0.44	0.30	**(3, 26)**
2 to 75	Mixed	Palmer Shredding	0.26	0.20	**(21, 22)**
2 to 51	Mixed	Pine State Recycling	0.41	0.28	**(21, 22)**
2 to 25	Glass	F and B Enterprises	0.47	0.32	**(21, 22)**
- - - -	- - - -	- - - -	- - - -	0.3 to 0.17	**(4, 17)**
13 to 51	Mixed	Maust Tire Recyclers	0.4^A	0.3	**(27)**

[A] For vertical stress less than 172 kPa.

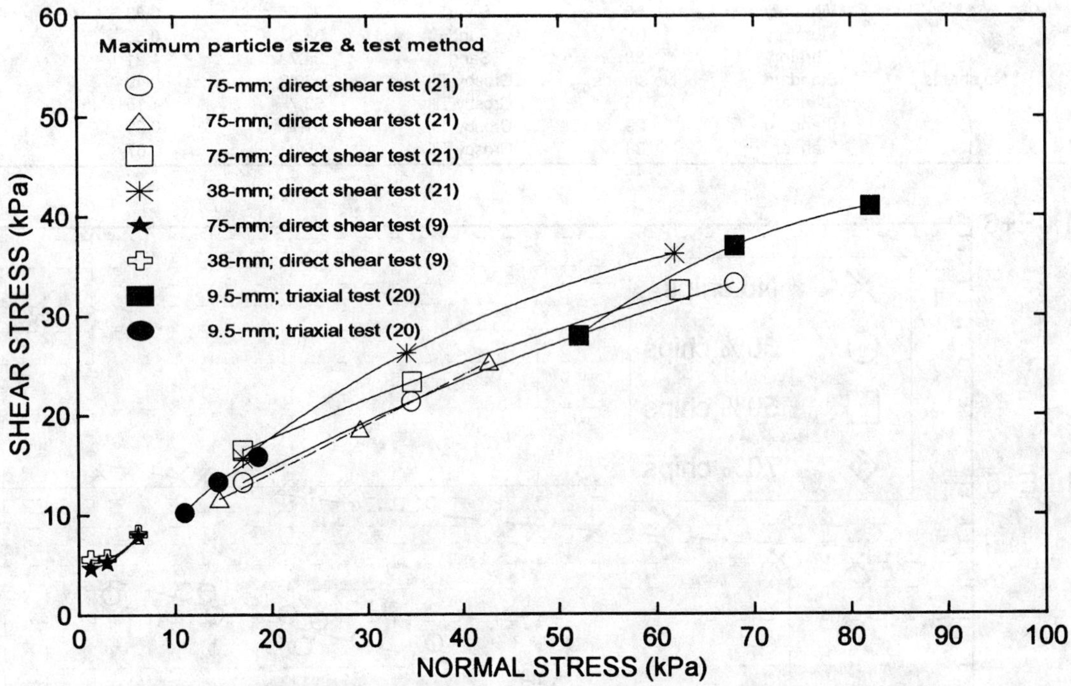

FIG. X1.3 Comparison of Failure Envelops of Tire Shreds at Low Stress Levels

TABLE X1.6 Shear Strength of Mixtures of Tire Shreds and Ottawa Sand (5)

NOTE 1—All samples are prepared by using vibratory compaction.

NOTE 2—Chip ratio is the air dried weight to chips divided by dry weight of mix, expressed in percent.

NOTE 3—$\sin \phi = \tan \alpha$; $c = a/\cos \phi$.

Test No.	Size of Chips (in.)	Chip/Mix Ratio (%)	Confining Pressure (psi)	Strain Levels (%)	a (psi)	tan α	r^2	c (psi)	ϕ (°)
TRS01	No-Chip	0	4.50	5	-0.24	0.6615	0.9998	0	41.41
TRS02	No-Chip	0	14.36	10	-	-	-	-	-
TRS03	No-Chip	0	28.86	15	-	-	-	-	-
TRS04	1.00	16.5	4.64	5	2.17	0.6006	0.9996	2.71	36.91
TRS05	1.00	16.5	14.50	10	1.05	0.6252	0.9998	1.35	38.70
TRS06	1.00	16.5	28.86	15	-	-	-	-	-
TRS07	1.00	29.16	4.50	5	5.52	0.4944	0.9943	6.35	29.63
TRS08	1.00	29.16	14.50	10	3.04	0.6110	0.9992	3.84	37.66
TRS09	1.00	29.16	28.86	15	2.65	0.6286	0.9993	3.41	38.95
TRS10	1.00	40.00	4.64	5	5.15	0.3957	0.9988	5.61	23.31
TRS11	1.00	40.00	14.36	10	5.13	0.5413	0.9972	6.10	32.77
TRS12	1.00	40.00	28.86	15	4.09	0.6013	0.9999	5.12	36.96
TRS13	1.00	50.00	4.64	5	-0.68	0.3562	0.9601	0.00	20.87
TRS14	1.00	50.00	14.36	10	4.54	0.4362	0.9988	5.05	25.86
TRS15	1.00	50.00	28.71	15	3.84	0.5519	0.9986	4.60	33.50
TRS16	1.00	66.54	4.50	5	2.23	0.1699	0.9999	2.26	9.78
TRS17	1.00	66.54	14.36	10	1.89	0.3324	0.9901	2.00	19.41
TRS18	1.00	66.54	28.71	15	4.91	0.3759	0.9992	5.30	22.08
TRS19	0.50	37.85	4.64	5	5.26	0.3891	0.9998	5.71	22.90
TRS20	0.50	37.85	14.50	10	5.48	0.5383	1.0000	6.50	32.57
TRS21	0.50	37.85	28.71	15	4.42	0.6238	0.9998	5.66	38.59
TRS22	1.00	38.78	4.64	5	6.55	0.4299	0.9964	7.25	25.46
TRS23	1.00	39.32	14.36	10	5.17	0.5684	0.9985	6.28	34.64
TRS24	1.00	39.37	28.71	15	4.08	0.617	0.9999	5.18	38.10

TABLE X1.7 Shear Strength of Mixtures of Tire Shreds and Crosby Till (5)

NOTE 1—Chip ratio is the air dried weight of chips divided by dry weight of mix, expressed in percent.

NOTE 2—$\sin \phi = \tan \alpha$; $c = a/\cos \phi$.

Test No.	Size of Chips (in.)	Chip Ratio (%)	Confining Pressure (psi)	Strain Levels (%)	a (psi)	tan α	r^2	c (psi)	ϕ (°)
TRC01	No-Chip	0	4.50	5	6.14	0.4299	0.9970	6.80	25.46
TRC02	No-Chip	0	14.50	10	9.28	0.4914	1.0000	10.66	[illegible]
TRC03	No-Chip	0	28.71	15	9.72	0.5099	0.9996	11.30	30.66
				20	9.58	0.5151	0.9996	11.18	30.00
TRC04	1.00	16.27	4.64	5	7.43	0.3873	0.9979	8.06	22.79
TRC05	1.00	16.27	14.36	10	6.21	0.5810	0.9982	7.63	35.52
TRC06	1.00	16.27	28.71	15	7.77	0.5686	0.9992	9.45	34.65
				20	5.71	0.6232	0.9992	7.30	38.55
TRC07	1.00	30.18	44.52	5	6.82	0.2612	0.9991	7.67	15.14
TRC08	1.00	30.18	14.36	10	9.96	0.3740	0.9997	10.74	21.96
TRC09	1.00	30.18	28.86	15	9.88	0.4748	0.9973	11.23	28.35
				20	8.82	0.5460	0.9971	10.53	33.09
TRC10	1.00	40.05	4.64	5	5.50	0.2205	0.9947	5.64	12.74
TRC11	1.00	40.05	14.36	10	7.65	0.3598	0.9990	8.20	21.09
TRC12	1.00	40.05	28.71	15	8.19	0.4543	0.9991	9.42	27.02
				20	8.44	0.5271	0.9999	9.93	31.81
TRC13	1.00	48.49	4.64	5	4.93	0.2025	0.9985	5.03	11.68
TRC14	1.00	48.49	14.36	10	6.69	0.3472	0.9999	7.13	20.32
TRC15	1.00	48.49	28.86	15	7.81	0.4441	0.9999	8.72	26.37
				20	7.92	0.5208	0.9999	9.28	31.39
TRC16	0.50	39.80	4.64	5	6.17	0.1173	0.9980	6.21	6.74
TRC17	0.50	39.80	14.36	10	9.37	0.2181	0.9875	9.60	12.60
TRC18	0.50	39.80	28.86	15	11.07	0.3130	0.9866	11.66	18.24
TRC19	0.50	39.64	14.36						
TRC20	0.50	39.79	14.36						

TABLE X1.8 Summary of Reported Hydraulic Conductivities of Tire Shreds

Particle Size (mm)	Void Ratio	Dry Density (kg/m³)	Hydraulic Conductivity (cm/s)	Reference
25 to 64		469	5.3 to 23.5	(19)
25 to 64		608	2.9 to 10.9	
5 to 51		470	4.9 to 59.3	
5 to 51		610	3.8 to 22.0	
38	- - - -	- - - -	1.4 to 2.6	(28)
19	- - - -	- - - -	0.8 to 2.6	
10 to 51	0.925	644	7.7	(21, 22)
10 to 51	0.488	833	2.1	
20 to 76	1.114	601	15.4	
20 to 76	0.583	803	4.8	
10 to 38	0.833	622	6.9	
10 to 38	0.414	808	1.5	
10 to 38		653	0.58	(5)

TABLE X1.9 Hydraulic Conductivities of Mixtures of Tire Shreds and Soil (5)

Tire Shred Maximum Size (mm)	Soil Type	% Tire Shreds Based on Total Weight	Dry Density (kg/m³)	Hydraulic Conductivity (cm/s)
- - - -	Ottawa Sand	0	1890	1.6×10^{-4}
25	Ottawa Sand	15.5	1680	1.8×10^{-3}
25	Ottawa Sand	30.1	1530	3.5×10^{-3}
25	Ottawa Sand	37.7	1410	8.7×10^{-3}
- - - -	Crosby till	0	1910	8.9×10^{-7}
25	Crosby till	14.8	1700	1.8×10^{-5}
25	Crosby till	30.1	1390	2.1×10^{-3}
25	Crosby till	40	1200	8.8×10^{-3}
13	Crosby till	40	1190	9.7×10^{-3}

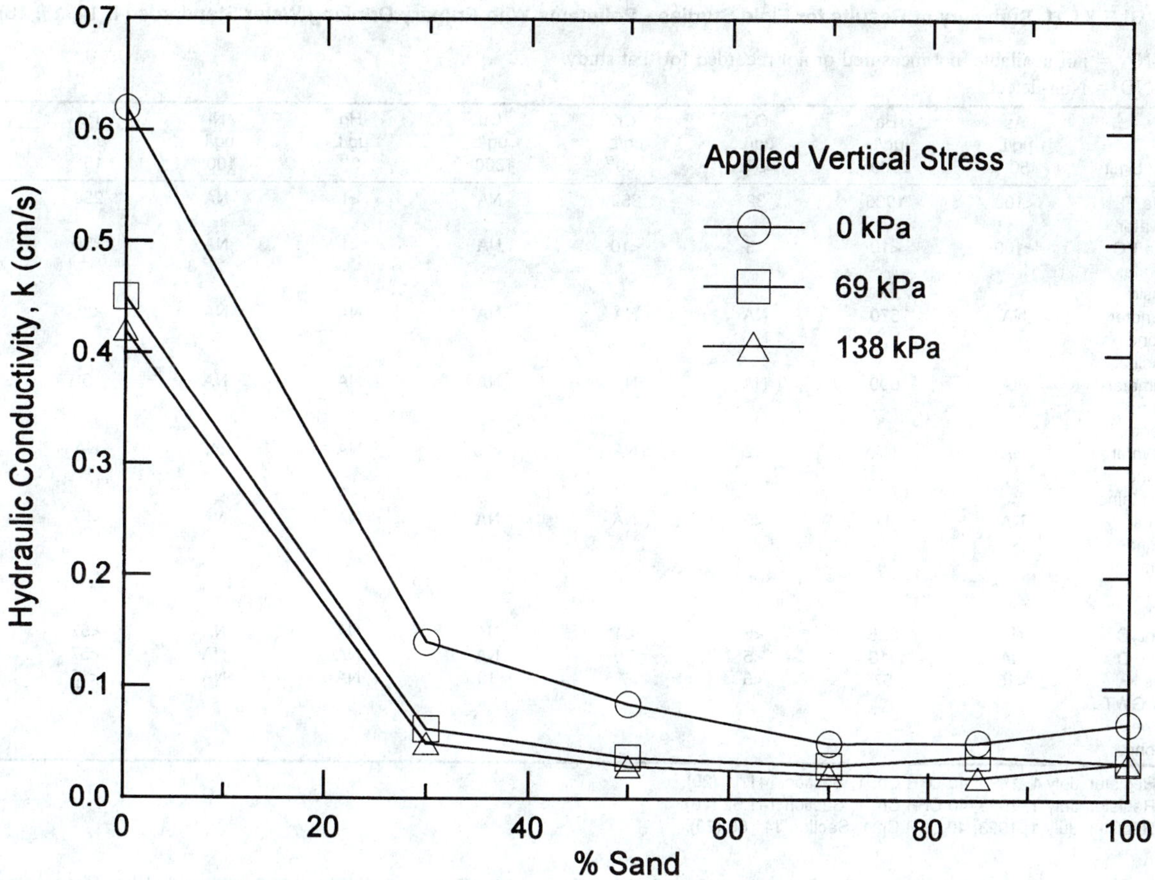

FIG. X1.4 Hydraulic Conductivities of Mixtures of Tire Shreds and Clean Sand (4)

TABLE X1.10 Summary of TCLP Results for Regulated Metals (12, 13, 14)

Concentration in Extract	Ag µg/L (ppb)	As µg/L (ppb)	Ba µg/L (ppb)	Cd µg/L (ppb)	Cr µg/L (ppb)	Hg µg/L (ppb)	Pb µg/L (ppb)	Se µg/L (ppb)
TCLP Regulatory Limit	5000	5000	100 000	1000	5000	200	5000	1000
Virigina DOT	NA[A]	NA	NA	1.55	2.8	NA	19.6	NA
Scrap Tire Management[B]	ND[C]	2	590	ND	48	0.4	16	ND
Maine	ND	ND	357	185	84	ND	216	ND

[A] NA = not available, that is not measured or not reported for that study.
[B] Maximum value reported for the seven tire products that were tested.
[C] ND = non-detect

TABLE X1.11 Summary of Results for Field Studies - Pollutants With Primary Drinking Water Standards (4, 14, 15, 16)

NOTE 1—NA = not available, not measured or not recorded for that study.
NOTE 2—ND = Non-detect

Regulatory Limit	As µg/L 50[A]	Ba µg/L 2000[B]	Cd µg/L 5[B]	Cr µg/L 100[B]	Cu µg/L 1300[C]	Hg µg/L 2[B]	Ni µg/L 100[B]	Pb µg/L 15[C]	Se µg/L 50[B]
Minnesota FL Groundwater	<100	1930	32	350	NA	<1	NA	230	<100
Minnesota PC Groundwater	<100	<10	<3	<10	NA	<1	NA	<20	<100
Wisconsin East Lysimeter Max Conc	NA	570	NA	NA	NA	NA	NA	22	NA
Wisconsin West Lysimeter Max Conc	NA	690	NA	NA	NA	NA	NA	5	NA
Tire Pond Surface Water Max Conc	NA	NA	<2	NA	<25	NA	<40	NA	NA
Richmond, Maine Field Trial Max Conc	NA	45	<5	NA	NA	NA	NA	<57	NA
North Yarmouth Maine Field Trial									
Max Conc C	NA	225	<5	64	NA	NA	NA	<57	NA
Max Conc D	NA	113	<5	70	NA	NA	NA	<57	NA
Maine TC Below GWT Field Trial Max Conc	<15	57	<5	7	19	NA	NA	<15	NA

[A] Federal Register, July 1, 1993; 40 CFR Ch. 1, Section 141.11 **(29)**.
[B] Federal Register, July 1, 1993; 40 CFR Ch. 1, Section 141.62 **(29)**.
[C] Federal Register, July 1, 1993; 40 CFR Ch.1, Section 141.80 **(29)**.

TABLE X1.12 Summary of Results for Field Studies - Metals With Secondary Drinking Water Standards and Other Parameters (4, 14, 15, 16)

NOTE 1—NA = not available, not measured or not recorded for that study.
NOTE 2—ND = non-detect

	Al µg/L	Ca mg/L	Fe mg/L	Mg mg/L	Mn µg/L	Zn µg/L	SO_4^{2-} mg/L
Regulatory Limit	50-200[A]	-	0.3[A]	-	50[A]	5000[A]	250[A]
Minnesota FL Groundwater	180000	1080	298	383	NA	870	NA
Minnesota PC Groundwater	1800	14.4	4.4	2.8	NA	<10	NA
Wisconsin East Lysimeter Max Conc	NA	340	5.3	390	NA	560	450
Wisconsin West Lysimeter Max Conc	NA	300	4	320	NA	750	150
Tire Pond Surface Water Max Conc	NA	NA	1.83	NA	NA	30	17
Richmond, Maine Field Trial Max Conc	170	105	<0.1	18	290	<100	65
North Yarmouth Maine Field Trial							
Max Conc C	150	437	1.5	141	902	1100	58
Max Conc D	<100	481	4.8	157	3200	13	25
Maine TC Below GWT Field Trial Max Conc	362	30	87	13	3430	123	22

[A] Viessman and Hammer **(30)**.

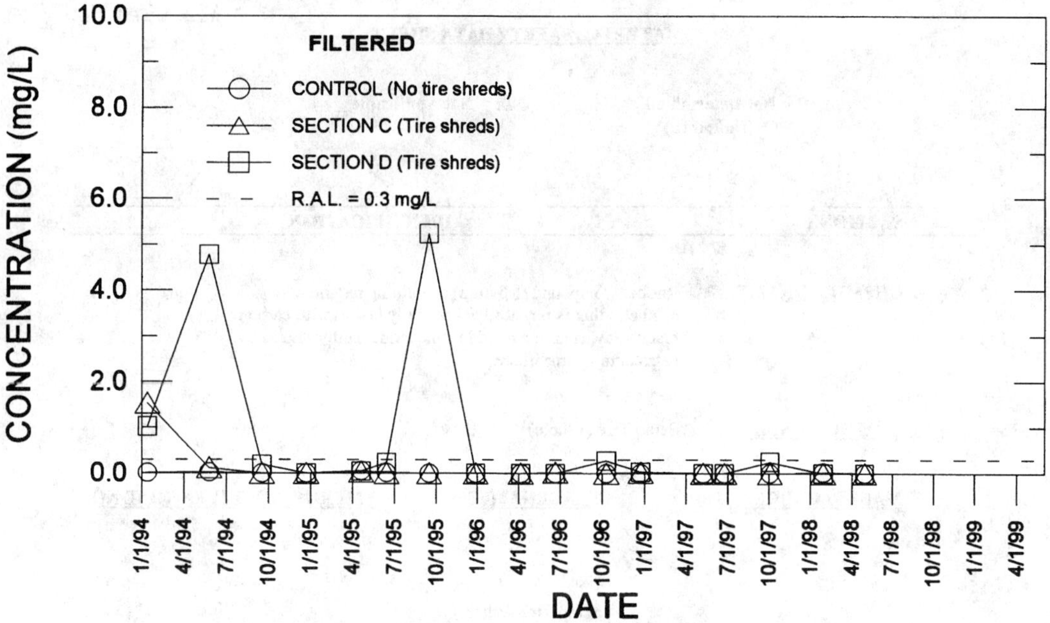

FIG. X1.5 Iron Levels for Filtered Samples at North Yarmouth Field Trial (16)

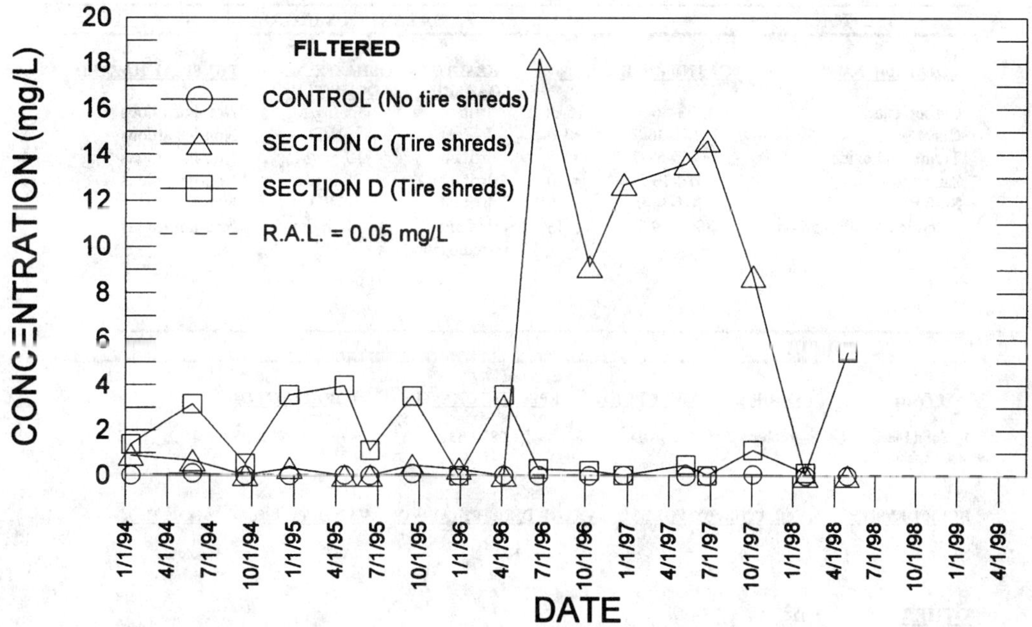

FIG. X1.6 Manganese Levels for Filtered Samples at North Yarmouth Field Trial (16)

Whole Scrap Tire

MATERIAL SAFETY DATA SHEET

N/D = Not Determined (Unknown) **N/A = Not Applicable**

SECTION I	IDENTIFICATION

CHEMICAL NAME Rubber Compound (Mixture) containing natural and synthetic that is physically/chemically bound with carbon black, clay, titanium dioxide, zinc oxide, sulfur and petroleum hydrocarbons.

COMMON NAME Scrap Tire (Whole)

MANUFACTURERS	TRADENAMES	EMERGENCY TELEPHONE NO.
	Scrap Tire (Whole)	

SECTION II	HAZARDOUS INGREDIENTS

CHEMICAL NAME	CAS NUMBER	%	HEALTH HAZARD	OSHA (PEL) AOGIH (TLV)	PHYSICAL HAZARD
Carbon Black	1333-86-4	16-36	Irritant	3.5 mg/M^3	Non-hazardous
Clay	12141-46-7	<1.0	Irritant	N/D	Non-hazardous
Titanium dioxide	13463-67-7	<1.5	Irritant	10 mg/M^3	Non-hazardous
Zinc oxide	1314-13-2	<2.0	Irritant	5.0 mg/m^3	Non-hazardous
Sulfur	7704-34-9	<1.5	Irritant	N/D	Non-hazardous
Peteroleum hydrocarbons	8002-29-7	5-13	Irritant Carcinogen	5.0 mg/m^3	Non-hazardous

SECTION III	PHYSICAL DATA

APPEARENCE	ODOR	MELT POINT	SPECIFIC GRAVITY	BOILING POINT
Solid Black Rubber	Rubber	N/D	1.085-1.331	N/A

BULK DENSITY	%VOLATILE BY VOLUME	VAPOR DENSITY (AIR=1)	VAPOR PRESS.	%SOL. H20
N/A	0	N/A	N/A	Insoluble

OTHER N/D

FIG. X1.7 Material Safety Data Sheet for Whole Scrap Tires

Whole Scrap Tire

MATERIAL SAFETY DATA SHEET

N/D = Not Determined (Unknown) **N/A = Not Applicable**

SECTION IV	**FIRE AND EXPLOSION HAZARD DATA**

FLASH POINT & METHOD	**IGNITION TEMP.**	**FLAMMABLE LIMITS**	
		LOWER	**UPPER**
N/D	N/D	N/D	N/D

FIRE EXTINGUISHING AGENTS AND SPECIAL PROCEDURES

Any of the following extinguishing agents may be used to combat any fires of this material: water (dispersed with fog nozzles), carbon dioxide, dry chemical, Halon or alcohol foam. Water, dispersed with fog nozzles, may be used to cool fire-exposed containers and to prevent pressure build-up.

Full protective clothing and MSHA/NIOSH (Mine Safety and Health Administration/National Institute for Occupatinal Safety and Health) approved, positive pressure, self-contained breathing apparatus should be used while firefighting. Thermal decomposition by-products may present a health hazard.

UNUSUAL EXPLOSIVE HAZARDS NONE

PRODUCTS EVOLVED WHEN SUBJECTED TO HEAT OR COMBUSTION

Potentially carcinogenic materials (including nitrosamines), carbon oixdes (carbon monoxide and carbon dioxide), acrid fumes, and flammable hydrocarbons may be liberated as a result of thermal decomposition or combustion. Avoid the smoke and fumes that result from thermal decomposition or combustion.

SECTION V	**HEALTH EFFECTS - EFFECTS of Exposure**

LD50 ORAL (INGESTION)	**LD50 DERMAL (SKIN CONTACT)**	**LC50 (INHALATION)**
N/D	N/D	N/D

THRESHOLD LIMIT VALUE (TLV)	**PRIMARY ROUTE OF EXPOSURE**
N/D	skin (dermal) contact

EFFECT OF ACUTE (SHORT TERM) EXPOSURE:

No known health effects due to acute (short term) exposure.

EFFECT OF CHRONIC (REPEATED) EXPOSURE:

This material contains untreated naphthenic or aromatic extender oil. This oil could be released from the surface through skin contact. Prolonged contact with these oils has been shown to cause skin cancer in laboratory studies with animals. Untreated naphthenic and aromatic oils are classified as carcinogenic by IARC (International Agency for Research on Cancer). Prolonged or repeated contact may cause skin irritation or sensitization (allergic skin reaction).

MEDICAL CONDITIONS AGGRAVATED BY EXPOSURE: Skin Disorders

FIG. X1.7 Material Safety Data Sheet for Whole Scrap Tires *(continued)*

Whole Scrap Tire

MATERIAL SAFETY DATA SHEET

N/D = Not Determined (Unknown) **N/A = Not Applicable**

SECTION VI **EMERGENCY AND FIRST AID PROCEDURES**

EYES: Not expected to be a problem.

SKIN: Wash thoroughly with soap and water. If reddening or irritation develops, obtain supportive medical attention.

INGESTION: Not expected to be a problem.

INHALATION: Not expected to be a problem.

OTHER INSTRUCTIONS:

Employees who have prolonged contact with material should practice good personal hygiene by frequent washing of hands and arms with soap and water. Remove contaminated clothing and launder before reuse. Shower at the end of each work day.

SECTION VII **CHEMICAL REACTIVITY**

CONDITIONS CAUSING INSTABILITY Stable under normal conditions.

INCOMPATIBILITY (MATERIALS TO AVOID) None.

HAZARDOUS DECOMPOSITION PRODUCTS

Potentially carcinogenic materials (including nitrosamines), carbon oxides (carbon monoxide and carbon dioxide), acid fumes, and flammable hydrocarbons may be liberated as a result of thermal decomposition or combustion. Avoid the smoke and fumes that result from thermal decomposition or combustion.

SECTION VIII **SPILL, LEAK, AND DISPOSAL INFORMATION**

STEPS TO BE TAKEN IN CASE MATERIAL IS SPILLED OR RELEASED

WASTE DISPOSAL: Reclaim or recycle material if possible. Dispose of materials-in accordance with applicable federal, state and local guidelines and regulations.

FIG. X1.7 Material Safety Data Sheet for Whole Scrap Tires *(continued)*

Whole Scrap Tire

MATERIAL SAFETY DATA SHEET

N/D = Not Determined (Unknown) **N/A = Not Applicable**

SECTION IX SPECIAL PROTECTION INFORMATION

All rubber products should be handled so as to prevent eye contact and excessive or repeated skin contact. Appropriate skin protection should be employed. Inhalation of dusts should be avoided

EYES: Not required for normal use.

SKIN: Use of protective gloves is recommended. Wash hands before eating, smoking or using the restroom.

INHALATION: Under normal conditions of use, respiratory protection should not be required.

ADDITIONAL PERSONAL PROTECTION INFORMATION:

Employees who have prolonged contact with material should practice good personal hygiene by frequent washing of hands and arms with soap and water. Remove contaminated clothing and launder before reuse. Shower at the end of each work day.

SECTION X STORAGE INFORMATION

PRECAUTIONS TO BE TAKEN IN HANDLING AND STORAGE

Store indoors in a cool, dry, well ventilated area under ambient conditions. (Temperatures: 32-100^{0}F 0^0-38^{0}C). Do not store in direct sunlight. Store and dispose of material in accordance with applicable federal, state and local guidelines and regulations.

SECTION XI ADDITIONAL COMMENTS

Components of this product are included in the EPA Toxic Substances Control Act (TCSA) Chemical Substances Inventory.

FOR ADDITIONAL INFORMATION

NAME **COMPANY** **TELEPHONE**

FIG. X1.7 Material Safety Data Sheet for Whole Scrap Tires *(continued)*

References

(**1**) Humphrey, D.N., "Investigation of Exothermic Reaction in Tire Shred Fill Located on SR 100 in Ilwaco, Washington," *Report to the Federal Highway Administration,* Washington, DC, March, 1996.

(**2**) Ad Hoc Civil Engineering Committee, "Design Guidelines to Minimize Internal Heating of Tire Shred Fills," *Scrap Tire Management Council,* Washington, DC, 1997.

(**3**) Humphrey, D.N., and Manion, W.P., "Properties of Tire Chips for Lightweight Fill," *Grouting, Soil Improvement, and Geosynthetics,* ASCE, Vol 2, 1992, pp. 1344-1355.

(**4**) Edil, T.B., and Bosscher, P.J., "Development of Engineering Criteria for Shredded or Whole Tires in Highway Applications," *Report No. WI 14-92,* Department of Civil and Environmental Engineering, University of Wisconsin, Madison, WI, November, 1992.

(**5**) Ahmed, I., "Laboratory Study on Properties of Rubber Soils, *"Report No. FHWA/IN/JHRP-93/4,* Purdue University, West Lafayette, IN, 1993.

(**6**) Ahmed, I., and Lovell, C.W., "Rubber Soils as Lightweight Geomaterial," *Transportation Research Record*, No. 1422, 1993, pp. 61-70.

(**7**) Tweedie, J., Humphrey, D.N., and Sandford, T.C., "Tire Chips as Lightweight Backfill for Retaining Walls-Phase II, Field Trial," *A Study for the New England Transportation Consortium*, Department of Civil Engineering, University of Maine, Orono, ME, 1997.

(**8**) Nickels, W.L., Jr., and Humphrey, D.N., "The Effect of Tire Chips as Subgrade Fill on Paved Roads," Technical Services Division, Maine Department of Transportation, Augusta, ME, 1997.

(**9**) Cosgrove, T.A., "Interface Strength Between Tire Chips and Geomembrane for Use as a Drainage Layer in a Landfill Cover," *Proceedings* *of Geosynthetics'95,* Industrial Fabrics Association, St. Paul, MN, Vol 3, 1995, pp. 1157-1168.

(**10**) Humphrey, D.N., Chen, L.H., and Lawrence, B.K., "Use of Tire Chip/Soil Mixtures to Limit Frost Heave and Pavement Damage of Paved Roads," *A Study for the New England Transportation Consortium*, Department of Civil Engineering, University of Maine, Orono, ME, 1998.

(**11**) Humphrey, D.N., Chen, L.H., and Eaton, R.A.," Laboratory and Field Measurement of the Thermal Conductivity of Tire Chips for Use as Subgrade Insulation," Preprint NO. 971289, "Transportation Research Board, Washington, DC, 1997.

(**12**) Zelibor, J.L., "Leachate from Scrap Tires: RMA TCLP Report," *Education Seminar on Scrap Tire Management*, Scrap Tire Management Council, Washington, DC, September, 1991, pp. 381-391.

(**13**) Ealding, W., "Final Report on Leachable Metals in Scrap Tires," Virginia Department of Transportation Materials Division, Virginia Department of Transportation Scrap Tire Task Force, 1992.

(**14**) Downs, L.A., Humphrey, D.N., Katz, L.E., and Rock, C.A., "Water Quality Effects of Using Tire Chips Below the Groundwater Table," A Report Prepared by the Department of Civil and Environmental Engineering, University of Maine, Orono, Maine for the Maine Department of Transportation, 1997.

(**15**) Twin City Testing," Environmental Study of the Use of Shredded Waste Tires for Roadway Sub-Grade Support," Twin City Testing Corp., St. Paul, MN, Waste Tire Management Unit, Site Response Section, Groundwater and Solid Waste Divison, Minnesota Pollution Control Agency, St. Paul, MN, February, 1990.

(**16**) Humphrey, D.N., Katz, L.E., and Blumenthal, M., "Water Quality Effects of Tire Chip Fills Placed Above the Groundwater Table," *Testing Soil Mixed with Waste or Recycled Materials,* ASTM STP 1275, ASTM, 1997. pp. 299-313.

(**17**) Edil, T.B., and Bosscher, P.J., "Engineering Properties of Tire Chips and Soil Mixtures," *Geotechnical Testing Journal*, Vol 17, No. 4, December, 1994, pp. 453-464.

(**18**) Shao, J., Zarling, J.P., and ESCH, D., "Thermal Conductivity of Recycled Tire Rubber for Insulation Beneath Roadways," *Transportation Research Board*, 1995.

(**19**) Bressette, T., "Used Tire Material as an Alternative Permeable Aggregate," *Report No. FHWA/CA/TL-84/07*, Office of Transportation Laboratory, California Department of Transportation, Sacramento, CA, 1984.

(**20**) Benda, C.C., "Engineering Properties of Scrap Tires Used in Geotechnical Applications," *Report 95-1*, Materials and Research Division, Vermont Agency of Transportation, Montpelier, VT, 1995.

(**21**) Humphrey, D.N., Sandford, T.C., Cribbs, M.M., Gharegrat, H., and Manion, W.P., "Tire Chips as Lightweight Backfill for Retaining Walls - Phase I, "A Study for the New England Transportation Consortium, Department of Civil Engineering, University of Maine, Orono, ME, 1992.

(**22**) Humphrey, D.N. and Sandford, T.C., "Tire Chips as Lightweight Subgrade Fill and Retaining Wall Backfill," *Proceedings of the Symposium on Recovery and Effective Reuse of Discarded Materials and By-Products for Construction of Highway Facilities*, Federal Highway Administration, Washington, DC, 1993, pp. 5-87 to 5-99.

(**23**) Humphrey, D.N. Sandford, T.C., Cribbs, M.M., Gharegrat, H., and Manion, W.P., "Shear Strength and Compressibility of Tire Chips for Use as Retaining Wall Backfill," *Transportation Research Record No. 1422*, Transportation Research Board, 1993, pp. 29-35.

(**24**) Hannon, J.B., and Forsyth, R.A., "Fill Stabilization Using Nonbiogradable Waste Products-Phase I," *Report No. CA-DOT-TL-2124-1-72-25,* Transportation Laboratory, California Department of Transportation, Sacramento, CA, August 1973.

(**25**) Benson, C.H., and Khire, M.V., "Closure to: Reinforcing Sand with Strips of Reclaimed High-Density Polyethylene," *Journal of Geotechnical Engineering,* Vol 121, No. 4, April, 1995, pp. 400-401.

(**26**) Manion, W.P., and Humphrey, D.N., "Use of Tire Chips as Lightweight and Conventional Embarkment Fill, Phase I - Laboratory," *Technical Paper 91-1*, Technical Services Division, Maine Department of Transportation, Augusta, ME, 1992.

(**27**) Drescher, A., and Newcomb, D.E., "Development of Design Guidelines for Use of Shredded Tires as a Lightweight Fill in Road Subgrade and Retaining Walls," *Report No. MN/RC-94/04*, Department of Civil and Mineral Engineering, University of Minnesota, Minneapolis, MN, 1994.

(**28**) Hall, T.J., "Reuse of Shredded Waste Tire Material for Leachate Collection Systems at Municipal Solid Waste Landfills," for Iowa Department of Natural Resources Waste Management and Authority Division, by Shive-Hattery Engineers and Architects, Inc., 1990.

(**29**) Federal Register, July 1, 1993, US Office of the Federal Register, Washington, DC.

(**30**) Viessman, W., Jr., and Hammer, M.J., *Water Supply and Pollution Control,* 4th ed., Harper Collins College Publishers, New York, NY, 1985, pp. 213-231.

Designation: D 6311 – 98

Standard Guide for Generation of Environmental Data Related to Waste Management Activities: Selection and Optimization of Sampling Design[1]

This standard is issued under the fixed designation D 6311; the number immediately following the designation indicates the year of original adoption or, in the case of revision, the year of last revision. A number in parentheses indicates the year of last reapproval. A superscript epsilon (ε) indicates an editorial change since the last revision or reapproval.

1. Scope

1.1 This document provides practical guidance on the selection and optimization of sample designs in waste management sampling activities, within the context of the requirements established by the data quality objectives or other planning process.

1.2 This document (*1*) provides guidance for selection of sampling designs; (*2*) outlines techniques to optimize candidate designs; and (*3*) describes the variables that need to be balanced in choosing the final optimized design.

1.3 The contents of this guide are arranged by section as follows:

1.4 *This standard does not purport to address all of the safety concerns, if any, associated with its use. It is the responsibility of the user of this standard to establish appropriate safety and health practices and determine the applicability of regulatory limitations prior to use.*

2. Referenced Documents

2.1 *ASTM Standards:*

D 4687 Guide for General Planning of Waste Sampling[2]

D 5283 Practice for Generation of Environmental Data Related to Waste Management Activities: Quality Assurance and Quality Control Planning and Implementation[2]

D 5792 Practice for Generation of Environmental Data Related to Waste Management Activities: Development of Data Quality Objectives[2]

D 5956 Guide for Sampling Strategies for Heterogeneous Wastes[2]

D 6044 Guide for Representative Sampling for Management of Waste and Contaminated Media[2]

D 6051 Guide for Composite Sampling and Field Subsampling for Environmental Waste Management Activities[2]

D 6232 Guide for Selection of Sampling Equipment for Waste and Contaminated Media Data Collection Activities[2]

D 6233 Guide for Data Assessment for Environmental Waste Management Activities[2]

D 6250 Practice for Derivation of Decision Point and Confidence Limit for Statistical Testing of Mean Concentration in Waste Management Decisions[2]

D 6323 Guide for Laboratory Subsampling of Media Related to Waste Management Activities[2]

E 135 Terminology Relating to Analytical Chemistry for Metals, Ores and Related Materials[3]

---

[1] This guide is under the jurisdiction of ASTM Committee D34 on Waste Management and is the direct responsibility of Subcommittee D34.01.01 on Planning for Sampling.

Current edition approved Sept. 10, 1998. Published November 1998.

[2] *Annual Book of ASTM Standards*, Vol 11.04.

[3] *Annual Book of ASTM Standards*, Vol 03.05.

E 943 Terminology Relating to Biological Effects and Environmental Fate[4]

2.2 *USEPA Documents:*

USEPA, Guidance for the Data Quality Objectives Process, EPA QA/G-4, Quality Assurance Management Staff, Washington, DC, March 1995[5]

USEPA, Data Quality Objectives Process for Superfund - Workbook, EPA 540/R-93/078 (OSWER 9355.9-01A), Office of Emergency and Remedial Response, Washington, D.C., September, 1993[5]

USEPA, Environmental Investigations Branch Standard Operating Procedures and Quality Assurance Manual (EISOPQAM), Region 4 - Science and Ecosystem Support Division, Athens, GA, May 1996[5]

2.3 There are numerous useful references available from ASTM, USEPA, and private sector publishers. Appendix X1 contains a list, which is by no means comprehensive, of additional commonly used references.

3. Terminology

3.1 *accuracy, n*—closeness of a measured value to the true or an accepted reference or standard value. **(E 135)**

3.2 *attribute, n*—a quality of samples or a population. **(D 5956)**

3.3 *characteristic, n*—a property of items in a sample or population that can be measured, counted, or otherwise observed. **(D 5956)**

3.3.1 *Discussion*—A characteristic of interest may be the cadmium concentration or ignitability of a population.

3.4 *composite sample, n*—a combination of two or more samples.

3.5 *confidence interval, n*—a numerical range used to bound the value of a population parameter with a specified degree of confidence (that the interval would include the true parameter value).

3.5.1 *Discussion*—When providing a confidence interval, the number of observations on which the interval is based should be identified.

3.6 *confidence level, n*—the probability, usually expressed as a percent, that a confidence interval will contain the parameter of interest.

3.7 *data quality objectives (DQO), n*—qualitative and quantitative statements derived from the DQO process describing the decision rules and the uncertainties of the decision(s) within the context of the problem(s). **(D 5956)**

3.8 *data quality objective process, n*—a quality management tool based on the scientific method and developed by the U.S. Environmental Protection Agency to facilitate the planning of environmental data collection activities. **(D 5956)**

3.8.1 *Discussion*—The DQO process enables planners to focus their planning efforts by specifying the use of the data (the decision), the decision criteria (action level) and the decision maker's acceptable decision error rates. The products of the DQO Process are the DQOs.

3.9 *decision rule, n*—a set of directions in the form of conditional statements that specifies: (*1*) how the sample data will be compared to the decision point or action level, (*2*) which decision will be made as a result of that comparison, and (*3*) what subsequent action will be taken based on the decisions.

3.10 *false negative error, n*—an error which occurs when (environmental) data misleads the decision maker(s) into not taking action when action should be taken.

3.11 *false positive error, n*—an error which occurs when environmental data misleads the decision maker(s) into taking action when action should not be taken.

3.12 *heterogeneity, n*—the condition of the population under which items of the population are not identical with respect to the characteristic of interest. **(D 5956)**

3.13 *homogeneity, n*—the condition of the population under which all items of the population are identical with respect to the characteristic of interest. **(D 5956)**

3.14 *representative sample, n*—a sample collected such that it reflects one or more characteristics of interest (as defined by the project objectives) of a population from which it was collected. **(D 5956)**

3.15 *risk, n*—the probability or likelihood that an adverse effect will occur. **(E 943)**

3.16 *sample, n*—a portion of material which is collected for testing or for record purposes. **(D 5956)**

3.16.1 *Discussion*—Sample is a term with numerous meanings. The project team member collecting physical samples (for example, from a landfill, drum or waste pipe) or analyzing samples considers a sample to be that unit of the population collected and placed in a container. In statistics, a sample is considered to be a subset of the population and this subset may consist of one or more physical samples. To minimize confusion, the term "physical sample" is a reference to the sample held in a sample container or that portion of the population which is subjected to measurement.

3.17 *sampling design, n*—(*1*) the sampling schemes specifying the point(s) for sample collection; (*2*) the sampling schemes and associated components for implementation of a sampling event.

3.17.1 *Discussion*—Both of the above definitions are commonly used within the environmental community. Therefore, both are used within this document.

4. Significance and Use

4.1 The intended use of this guide is to provide practical assistance in the development of an optimized sampling design. This standard describes or discusses:

4.1.1 Sampling design selection criteria,

4.1.2 Factors impacting the choice of a sampling design,

4.1.3 Selection of a sampling design,

4.1.4 Techniques for optimizing candidate designs, and

4.1.5 The criteria for evaluating an optimized sampling design.

4.2 Within a formal USEPA data generation activity, the planning process or Data Quality Objectives (DQO) development is the first step. The second and third are the implementation of the sampling and analysis design and the data quality assessment. Within the DQO planning process, the selection

[4] *Annual Book of ASTM Standards*, Vol 11.05.

[5] Available from the Superintendent of Documents, U.S. Government Printing Office, Washington, DC 20402.

and optimization of the sampling design is the last step, and therefore, the culmination of the DQO process. The preceding steps in the DQO planning process address:

4.2.1 The problem that needs to be addressed,

4.2.2 The possible decisions,

4.2.3 The data input and associated activities,

4.2.4 The boundaries of the study,

4.2.5 The development of decision rules, and

4.2.6 The specified the limits on decision error.

4.3 This guide is not intended to address the aspects of the planning process for development of the project objectives. However, the project objectives must be outlined and communicated to the design team, prior to the selection and optimization of the sample design.

4.4 This guide references statistical aspects of the planning and implementation process and includes an appendix for the statistical calculation of the optimum number of samples for a given sampling design.

4.5 This guide is intended for those who are responsible for making decisions about environmental waste management activities.

5. Summary of Guide

5.1 The selection and optimization process is an iterative process of selecting and then evaluating the selected design alternatives and determining the most resource-effective design which satisfies the project objectives or DQOs. Fig. 1 illustrates this approach.

5.2 An appropriate sampling design may be implemented without a formal optimization, however, the following steps are recommended. Each evaluation step typically results in fewer design alternatives.

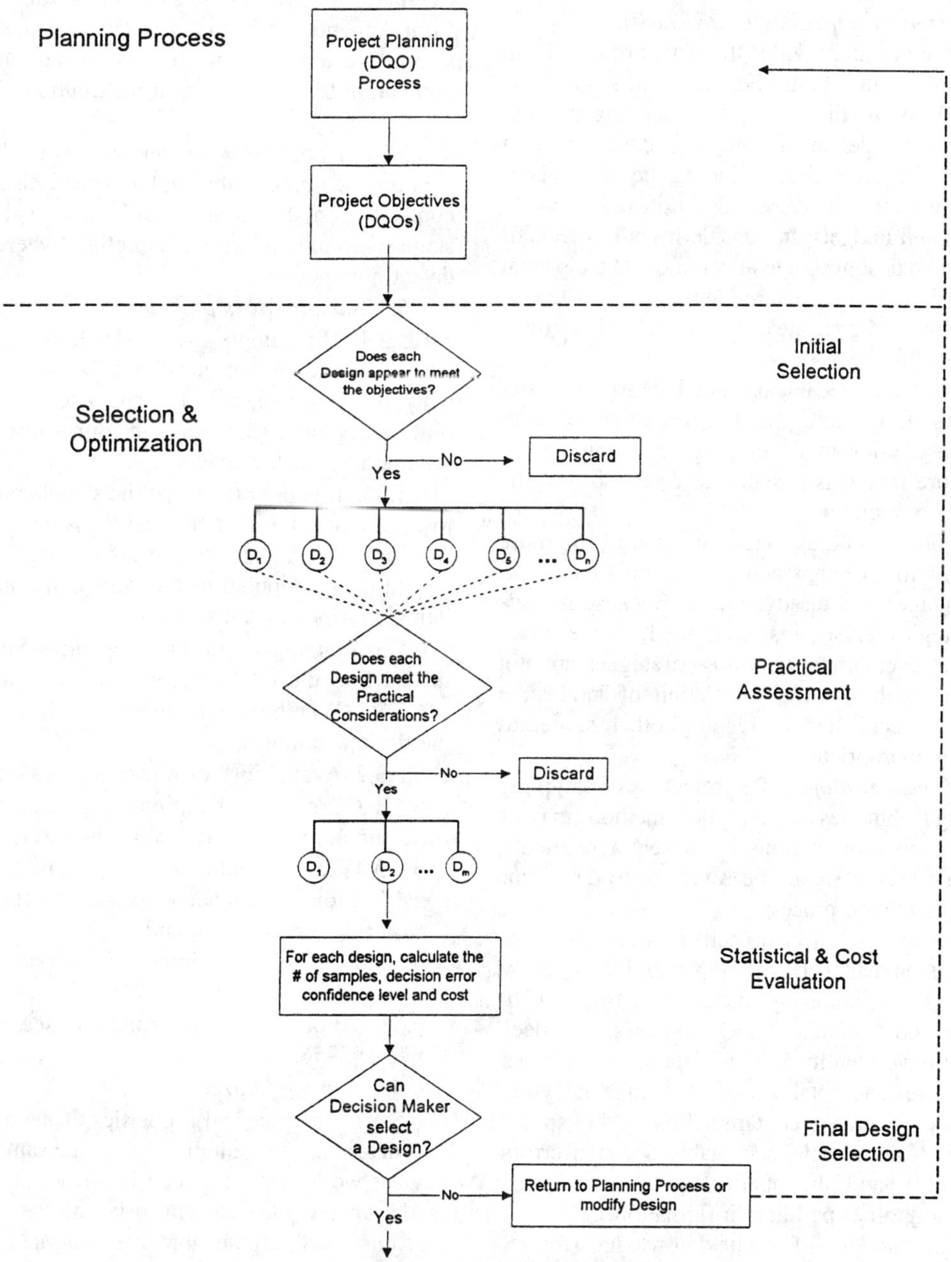

FIG. 1 Implement Sampling Design

5.2.1 Evaluation of the designs against the project's practical considerations (for example, time, personnel, and material resources),

5.2.2 Calculation of the design cost and statistical uncertainty, and

5.2.3 Choice of the sample design decision by the decision makers.

5.3 The process steps for the evaluation can be followed in any order. And for a small project, the entire selection and optimization process may be conducted at the same time. If ultimately, a design meeting the project constraints, for example, schedule and budget, cannot be identified among the candidate sampling designs, it may be necessary to modify the closest candidate design or reevaluate and revise the project objectives.

6. Factors Affecting Sampling Design Selection

6.1 *Sampling Design Performance Characteristics*:

6.1.1 The sampling design provides the structure and detail for the sampling activity and should be chosen in light of the project objectives. Prior to this point, the planning process should have addressed and defined the project needs for each of the sampling design characteristics, including the characteristics of interest, population boundaries, decision rule, acceptable decision errors and budgets. In considering all aspects of the project, the selected design should accommodate the spatial and temporal distribution of contaminants at the site, be practical, cost effective and generate data that allow the project objectives to be met.

6.1.2 Whenever possible, technical guidelines for measurement of the sources of variability and levels of uncertainty should be established prior to developing sampling design alternatives, to ensure that it is possible to establish that the program objectives are met.

6.1.3 Annex A1 presents an overview of some of the more commonly used sampling designs and design tools and summarizes their advantages and disadvantages. Because numerous sampling strategies exist, this is limited to the more common. If the more common sampling strategies are not cost-effective or applicable to the population of interest, a statistician should be consulted to identify other strategies which may be more appropriate.

6.2 *Regulatory Considerations*—The selection of sampling design, the sampling techniques and analytical methods may be dictated by current regulation, permits or consent agreements, applicable to the site. These should be reviewed to determine their impact on the selection process.

6.3 *Project Objectives*—Project objectives are usually determined by the decision makers (for example, regulatory body, consent agreement group, company management) during the initial investigation and planning or DQO process. The decision makers should have identified the population boundaries, characteristics of interest, acceptability of an average analytical value, the need to locate areas of contamination or "hot spots," the statistical needs (for example, acceptable decision errors, levels of uncertainty), and the quality control acceptance criteria, as well as any other pertinent information.

6.4 *Knowledge of the Site*—The site knowledge (for example, geography/topography, utilities, past site use) used to determine project objectives, will also provide for a more resource efficient sampling design, for example, divide a site into separate design areas for sampling or exclude an area from sampling.

6.5 *Physical Sample Issues*—The physical material to be sampled and its location on or within the site will usually impact the sampling design and limit the choices of equipment and methods.

6.5.1 *Number of Samples*:

6.5.1.1 The project objectives should specify the confidence levels for decision making. Using this level of decision error, the proximity to a threshold or action limit and the anticipated population variance, the number of samples can be calculated. The statistical parameter of interest, for example, mean or 95 percentile, and type of frequency of distribution, for example, normal or log normal, will determine which equation is used to calculate the appropriate number of samples. Equation X3.5 from Appendix X3, can be used to calculate the number of samples when the objective is to measure the mean for a population that has a normal distribution for the characteristic of interest.

6.5.1.2 Appendix X3 contains statistical approaches to calculating the number of samples needed for estimating the mean concentration, for simple random, statistical random, multistage sampling and search sampling (where the objective is to detect hot spots).

6.5.2 *Sample Mass or Volume*:

6.5.2.1 The sample mass or volume is determined by the size of the items that constitute the population, the heterogeneity of the population, the characteristics of available sampling equipment (for example, dimensions) and the mass or volume needed for analysis.

6.5.2.2 It is important that the sample mass be large enough to accommodate all item sizes or parts of all items. If items such as fine granular sand or large discarded automobile parts constitute the population, the sample may need to include those items or wipes of those items.

6.5.3 *Sample Access and Logistics*—Site access and logistics such as the following, can alter the sampling design:

6.5.3.1 Whether equipment can maneuver on site to obtain the desired samples,

6.5.3.2 Availability of power and water,

6.5.3.3 Presence of buried, suspended or surficial utilities, for example, power lines, water lines, etc.,

6.5.3.4 Terrain including slope, stability of site (subsidence considerations), presence of brush or trees, and soil condition (hard pack versus mud), and

6.5.3.5 Noise of equipment which could constitute a nuisance.

6.5.3.6 For further information, see Guide D 6232 and Guide D 5956.

6.5.4 *Sample Matrix*:

6.5.4.1 The physical properties of the matrix to be sampled will determine the suitability of some sampling devices. Some devices work best with cohesive material, such as moist soils, while other work best with dry materials. Equipment used to dig, core and sample abrasive materials needs to be strong enough to maintain its integrity during sampling. The sampling program should not be compromised by incompatibilities between the sampling device and the waste.

6.5.4.2 Heterogeneity will impact both the sampling design and the physical means of collecting the samples. Non-uniform distribution of the contaminant(s) of interest and/or varying particle size of the material, for example, soil, concrete, building material, vegetation, will require different sampling equipment and sampling strategies. For further information, see Guide D 6232 and Guide D 5956.

6.6 *Communication with the Laboratory*—Advanced planning with the laboratory should address the sampling schedule, sample preparation techniques, any subsampling instructions, analytical procedures, analytes of interest, matrices to be analyzed, data report format, data to be reported, and any specific requirements for accuracy, precision, quality control, calibration and needed turn around time.

6.7 *Analytical Turn Around Time*—Turn around time is usually the time from sample receipt in the laboratory to analytical data delivery. This usually depends on the analytical considerations and the laboratory capabilities.

6.8 *Analytical Method Constraints*—The analytical methods need to be chosen prior to or in conjunction with the optimization of the sampling design. The selection of appropriate methods needs to take into account at least the following areas.

6.8.1 *Analytical Method Sensitivity*—The analytical method sensitivity, usually expressed as the method detection limit or detection limit, may dictate the mass or volume of sample needed, the selection of the analytical methods, and the accuracy and precision of the data. Analytical method sensitivity is influenced by a number of factors, including sample preparation, sample volume, percent moisture, dilutions, and analytical method used.

6.8.1.1 *Analytical Aliquot Mass or Volume*—In general, for a given method, the larger the analytical aliquot (up to the maximum accommodated by the analytical method), the lower the detection limit and the more representative the data. However, typical aliquots used by most methods range from a few milliliters to 1 L or 100 g. The laboratory instrumentation may not be physically capable of managing a much larger aliquot.

6.8.1.2 *Dilutions*—Any analytical dilution will decrease sensitivity and increase the detection limit, as a multiplier of the dilution factor. When the sample has parameters in high concentrations, the lab may dilute the sample to allow the parameter to fall within the analytical instruments' calibration range.

*(1)*Samples containing parameters of varied concentrations may need to be prepared and analyzed at different dilutions. The manner of reporting these multiple analyses need to be agreed upon with the laboratory prior to analysis.

6.8.1.3 *Action Levels*—Detection levels need to be lower than the decision points or regulatory levels. If the detection limit is at the action or regulatory level, the increased levels of imprecision will increase the uncertainty in the decision. Low detection limit requirements may require special method development. The validation and ruggedization of new methods can be costly and impact schedules.

6.8.2 *Moisture Content of Samples*—Reporting analytical results on a dry weight basis may increase the sample mass requirements. Dry weight reporting may be accomplished in one of two ways.

6.8.2.1 Dry the sample aliquot prior to analysis on the same sample aliquot. This approach usually yields the lower detection limit. However, drying may change the sample. For example, it may affect the results of an analytical extraction, such as oxidizing a constituent, for example, hexavalent chromium.

6.8.2.2 Employ two sample aliquots. One aliquot is used to determine the moisture content, which is then used to calculate a dry weight analytical result, based on an analysis of the second aliquot. This second approach can result is raised detection limits, but it is required for the analysis of volatile analytes, which would be lost during drying.

6.8.3 *Holding Times*—The holding time is usually the time from sample collection to sample extraction or analysis. Most regulatory agencies will not accept or will limit the use of data from a sample analyzed outside the specified holding time. Holding times differ depending on the media, analyte, and regulation.

6.8.3.1 Analyses such as pH, hexavalent chromium, semi-volatile and volatile organics have short holding times and may necessitate special planning. Samples with very short holding times will need to be shipped as soon as possible to allow sufficient processing time or will need on-site analysis.

6.8.4 *Screening Measurements*:

6.8.4.1 *Screening*—Screening methods can implemented in the field or the laboratory and can provide either qualitative or semi-quantitative analyses. Screening methods are faster and generally less costly than traditional laboratory methods, but may be less sensitive and employ less quality control than traditional methods. However, they allow field personnel to define problematic areas quickly and to guide the sampling and verification analyses, using traditional methods, for final compliance determination.

6.8.4.2 *Field and In-Situ Analyses*—When hold or turn around times cannot be met by traditional laboratory analyses or to save time, reduce cost, or increase the number of samples, analytical testing may be performed at the sampling site. Field analytical methods include chemical specific kits and portable instrumentation for various organic and inorganic compounds. Care should be taken that the needed detection limits, regulatory requirements and quality control are achievable and that accuracy and precision criteria, trained staff, and data management practices are in place to produce data to meet the planning objectives or DQOs.

6.9 *Health and Safety*—Personnel safety must be considered. Of particular concern are any potential exposure of field personnel to hazardous materials and any possibility for explosion or fire which might be triggered by sampling equipment. Additionally, intrusive sampling, such as drilling, can result in the release of hazardous materials to the environment and potentially impact off-site personnel.

6.10 *Budget/Cost Considerations*—Budgets are almost always a significant factor. The challenge is to design a cost-effective sampling plan, while still achieving the specified project objectives. The sample design cost estimate comparisons need to include:

6.10.1 Personnel costs including travel and per diem,

6.10.2 Sampling equipment, including purchases/rentals,

6.10.3 Mobilization and demobilization costs,

6.10.4 Decontamination of equipment,

6.10.5 Waste collection and disposal,

6.10.6 Sample analyses and/or field screening,

6.10.7 QA or QC samples, or both,

6.10.8 Consumables, and

6.10.9 Health and safety.

6.11 *Representativeness*—Representativeness is the degree to which samples collected reflect the characteristics of the population. The sampling design must be chosen such that bias is minimized and the other project objectives achieved. For further guidance, see Guide D 6044.

7. Initial Design Selection

7.1 Sampling design options need to be selected consistent with the project objectives. Prior to selecting designs, the parameter(s) of interest (target compounds), population boundaries, decision rule, the spatial and temporal distribution of contaminants at the site (if known), the acceptable decision errors and budgets should have been considered in the planning process. In addition the final design should be practical and cost-effective. See Annex A1 for a listing of commonly used sampling designs.

7.2 *Meeting Project Objectives*—Prior to the selection of the initial set of sampling designs, those responsible for the project planning or DQO process need to establish and communicate the project objectives. Fig. 2 provides a guide to some common sampling designs as they could potentially satisfy some basic project objectives. Fig. 3 gives a guide to some of the attributes of the same designs.

7.2.1 *Estimating Population Parameters*—Waste classification, evaluation of waste treatability or determining compatibility of wastes are types of projects where information of the population parameters, such as the mean and variance, may be required. Estimation of these parameters generally relies on a statistical sampling design and classical inferential statistics.

7.2.2 *Monitoring for Routine Purposes*—Monitoring, such as ground water or monitoring changes in waste streams over time may be useful in determining for example whether a characteristic of a waste stream has exceeded a prespecified quality control or permit limit.

7.2.3 *Describing Spatial or Temporal Distribution, or Both*—Information for corrective action purposes may be used to define spatially or temporally those portions of a waste stream that are to be managed in different ways (for example, disposal versus treatment, etc.) Information may also be used for locating additional sampling units for increased precision in defining boundaries separating wastes to be manage differently.

7.2.4 *Non-Compliance Monitoring*—Identifying hot spots is a common non-compliance monitoring objective. Search sampling is used to locate or detect constituents of interest, objects," hot spots" in the area to be sampled. Authoritative sampling based on site knowledge is frequently used to identify the possibility of the "worst case" scenario non-compliance.

7.3 *Sampling Designs for Complex Sites*—Many sites for environmental sampling are complex and require the selection of multiple sampling designs to address the various suspected problems. For these sites, optimization of several sample designs is needed. Once the specific areas are defined, the process is similar to any other optimization. The following is an example.

7.3.1 *Example*—Fig. 4 illustrates a complex site, one where a multi-design program is appropriate. It represents a source of potential contamination such as a waste lagoon that is leaking contaminated liquids to the subsurface and the ground water. To determine the extent of the problem, it is necessary to collect samples from separate areas of the site and answer the following:

7.3.1.1 What are the potential contaminants present in the lagoon?

7.3.1.2 What are the background levels of contaminants?

7.3.1.3 What are the levels of the potential contaminants in

Objectives \ Sampling Designs	Sampling Designs							
	Authoritative		Statistical				Systematic	
	Judgmental	Biased	Simple Random	Stratified	Search	Unequal Probability	Line	Grid
Estimate population parameters			✓	✓				✓
Routine Monitoring	✓		✓	✓			✓	
Describe spatial and/or temporal distribution				✓	✓	✓	✓	✓
Non-Compliance Monitoring	✓	✓			✓			

FIG. 2 Project Objectives and Sampling Designs Guidance

Sampling Designs / Attributes	Designs: Authoritative: Judgmental	Biased	Statistical: Simple Random	Stratified Random	Search	Unequal Probability	Systematic: Line	Grid
Some Knowledge of Contaminant Distribution	✓	✓		✓		✓		
Usually is less Costly	✓	✓						
Simple to Implement	✓	✓	✓				✓	✓
Estimates Average Conditions			✓	✓		✓	✓	✓
Estimates Variability			✓	✓				✓
Minimizes Personal Bias			✓	✓			✓	✓
Effectively Accommodates and Identifies Strata				✓			✓	✓
Identifies or Locates Hot Spots		✓			✓			✓
Considers Trends or Cycles in Contaminant Distribution						✓		✓
Identifies Trends and Cycles in Contaminant Distribution							✓	✓

FIG. 3 Relationships Between Sampling Designs and Some Attributes Guidance

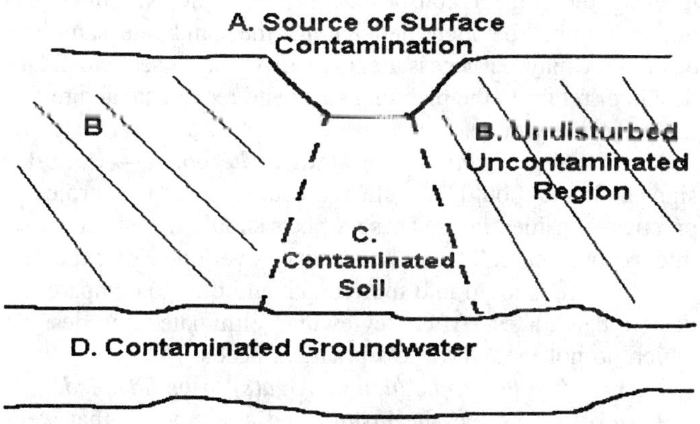

FIG. 4 Complex Site

the soil immediately adjacent to and beneath the lagoon?

7.3.1.4 Has the contamination reached the ground water and to what extent?

7.3.2 Assuming the planning team is familiar with the process waste and the spatial heterogeneity of the lagoon, the first question can be answered by an authoritative sampling. The second question can be answered by a systematic sampling of the areas adjacent to the site. The answer to the third question can be found by a systematic sampling design around and beneath the lagoon. The fourth question can be answered by a systematic sampling of the ground water along a line from the point of origin (the lagoon) in the direction of ground water flow.

7.3.3 The final integrated plan for the entire site should consider all the information needs integrate multiple sampling designs (per area) and stage the field sampling to collect samples in such a manner that it will satisfy more than one area or question. For example, samples from the lagoon can be scheduled first and the results used to determine the analyte list for the soils and ground water. The soil borings used to determine the contamination around and beneath the lagoon and the ground water samples from the ground water beneath the site will provide information about the extent of any plume. This type of integrated planning occurs after the selection of the designs to answer the individual questions. Many times, considerable cost savings can be realized by this type of selection and optimization.

	Area of Site and Description	Candidate Design
A.	Lagoon (source of the contamination)	Authoritative Sampling
B.	Undisturbed soil area (presumed uncontaminated)	Systematic Sampling
C.	Soil directly under the spill (known to be contaminated, need to map extent of contamination)	Systematic Sampling
D.	Ground water (need to know the extent of contamination of ground water plume)	Grid Sampling

7.4 *Statistical versus Non-statistical Designs*:

7.4.1 *Non-Statistical Designs*—Sampling designs can be classified as statistical or non-statistical sampling designs. Non-statistical sampling designs are sometimes referred to as non-probability or authoritative (biased or judgmental) sampling. These strategies rely upon a person's judgment or a pre-arranged decision-rule to decide which portions of a population will be sampled. Non-statistical sampling designs can be the optimum strategy for certain populations or times in a sampling program. Non-statistical sampling may be appropriate under circumstances such as the following: (*1*) pilot studies (preliminary information is needed to facilitate planning); (*2*) spills: a spill of an unknown chemical has been encountered; (*3*) limited access to portions of the population; (*4*) field screening to select a limited number of samples for laboratory analysis; (*5*) historical site knowledge is available; and (*6*) non-compliance determinations.

7.4.1.1 While non-statistical sampling can generate useful data, because of its subjective nature, the logic used to choose the sampling location must be explained and defensible.

7.4.1.2 It is very important not to confuse non-statistical sampling with the use of historical information during sampling design. For example, if one area of a site is known to be heavily contaminated while another area is believed not to be contaminated, this information can be used to defensibly divide the site into strata or de-select an area from sampling. Use of historical information in conjunction with statistical sampling strategies should generate unbiased, representative and defensible data.

7.4.2 *Statistical Designs*—Statistical sampling designs are also referred to as probability, non-biased or non-judgmental designs and rely upon a random selection of sampling locations to minimize any bias in the sample selection process. Statistical sampling strategies allow for large populations to be characterized with a measured degree of confidence. In addition to considering all the other information, the following may apply:

7.4.2.1 Usually, the greater the number of samples, the narrower or tighter the confidence interval for the parameter of interest.

7.4.2.2 Composite samples are useful for locating the hot spot areas, although they may not identify a specific point source contaminant location.

7.4.2.3 For containerized waste, the sampling error for both the within (an individual) container and between multiple containers need to be considered.

7.4.2.4 Because sampling errors are usually larger and may be more difficult to quantify than analytical errors, field QC samples need to be included to help determine the potential errors.

7.4.2.5 Systematic grid sampling is preferred when spatial structure (correlation) is suspected or known. A random factor may be introduced by a random choice of origin. Systematic grid sampling usually provides a more accurate estimate of the mean.

8. Optimization Criteria

8.1 Optimization involves choosing between the initially selected sample designs which may or may not meet the project objectives. The optimum sample design will minimize project variables such as cost, time and risk, the objective being to achieve a balance between the costs of acquiring environmental data and the costs or consequences of incorrect waste management decisions.

8.2 In general the criteria for an optimized sampling design are that the design:

8.2.1 Be resource and cost effective,

8.2.2 Provide data of known quality,

8.2.3 Meet or not to exceed the acceptable level of decision errors,

8.2.4 Be practical or at least possible to implement, appropriately,

8.2.5 Comply with regulatory requirements,

8.2.6 Be implementable within the project schedule,

8.2.7 Have high reliability, and

8.2.8 Meet any other project specific objectives.

8.3 In the optimization process the above criteria will be used to choose the optimum design from the candidate sampling designs.

9. Optimization Process

9.1 The optimization process is an iterative process of evaluating the initially selected design alternatives and determining the most resource-effective design which satisfies the project objectives or DQOs. An appropriate sampling design may be implemented without a formal optimization, however, the following steps are recommended: (*1*) evaluation of the designs against the project's practical considerations (for example, time, personnel, and material resources); (*2*) calculation of the design cost and statistical uncertainty; and (*3*) choice of the sample design decision by the decision makers. Fig. 1 and Section 5, illustrate this approach.

9.1.1 The process steps for the evaluation can be followed in any order. For a small project, the entire selection and optimization process may be conducted simultaneously. Typically as the evaluation continues each evaluation step will result in fewer design alternatives. If, ultimately, a design meeting the project constraints, for example, schedule and budget, cannot be identified among the candidate sampling designs, it may be necessary to modify the closest candidate design or return to the planning stage and reevaluate and revise the project objectives.

9.2 *Practical Evaluation of Design Alternatives*—Each design candidate should be evaluated with respect to the project's practical considerations. These aspects should have been taken into account initially and some may overlap. However, the purpose here is to go into more depth and then to compare the design candidates. After reviewing, eliminate any designs which do not meet the site's practical needs.

9.2.1 *Define the Population or Area(s) to be Sampled*:

9.2.1.1 Review the site history and assumptions that were used to define the population boundaries. This information may allow for stratification of the site, identification of specific areas of interest and an estimate of heterogeneity. Determine which sampling design best accommodates the spatial and temporal boundaries of the population.

9.2.1.2 Subdivision of the site may involve spatial boundaries such as drums, tanks, an area within a grid, a boring location on a grid, a depth interval in a boring, distance along a conveyor belt, or any other appropriate defined physical unit from which material can be obtained. For example, a defined search area may be the answer to locate an 8–ft diameter area of PCB contamination from a 55–gal drum PCB spill.

9.2.2 *Determine Optimum Number of Samples*:

9.2.2.1 Budgets and the acceptable levels of uncertainty as defined by the DQO or project objective planning process are competing factors that affect the number of samples. Statistical techniques for balancing these competing factors are discussed in a number of places in the literature[6] and Appendix X1.

9.2.2.2 The following illustrates a calculation for the commonly used systematic grid sampling and the iterations which may be necessary, if the calculated number of samples should prove, for practical reasons, to be too large to implement.

9.2.2.3 *Example Calculation*—The number of samples to be collected can be calculated based on variance information derived from previous sampling data or estimated based on professional judgment. Usually the contaminants of concern (COCs)s are parameters which are closest to or in excess of an action level. Their presence is normally the driving force behind the need to determine the extent and levels of contamination. The statistic of interest here is the mean and it assumes a normal distribution.

(1) Select a margin of error (p) acceptable for the subsequent use of the data. For soil studies, a margin of error of 0.20 is not unusual. The margin of error may be calculated by dividing the needed precision, in units of concentration, for example, = 10 ppm, by the known or anticipated mean concentration of the COCs. Note, that if the actual precision or mean concentration for the COC differs from those estimated during the planning process, a re-evaluation of the assumed margin of error may be necessary.

(2) A coefficient of variation (CV), which is defined as the standard deviation of a COC divided by the mean of the COC, is either obtained using previous sampling data, or estimated based on anticipated variability. If a CV above 0.65 is obtained, a large number of samples will usually be needed to make a decision with the selected margin of error.

(3) A confidence level 100(1-α) % needs to be established. For work involving hazardous wastes, a confidence level of 95 % is frequently used. For a 95 % confidence level, using a standard Z statistical table, this corresponds to a one-sided statistical factor of $Z_\alpha = 1.645$.

(4) If a 1-sided inference about the population is desired (for example, comparing a mean concentration to a regulatory threshold), the required number of samples is calculated using the following formula:

$$n = \{Z_\alpha\ (CV)/P\}^2 \quad (1)$$

where:

n = number of samples to collect,
Z_α = statistical factor for the desired confidence level,
CV = coefficient of variation, and
p = margin of error.

In a case where no previous sampling data is available, the values used in the above discussion can be used as a starting point.

$$n = \frac{(1.65)^2(0.65)^2}{(0.20)^2} \quad (2)$$

$$n = 29 \text{ samples} \quad (3)$$

If a two-sided inference is desired (for example the mean is equal to 10 ppm), the Z value of 1.96 is used in the formula, instead of 1.65. The result is an n=40.

(5) Upon completion of the calculation the number of samples and the margin of error is reviewed to determine that each is acceptable. If the value of n number of samples is too great, then an adjustment to the margin of error should be considered, or the sampling design may be modified. Alternately, if the population is stratified by concentration, the number of samples required may be reduced by selecting a sampling design for each of the strata. The inter-strata variability would then be removed from the calculation of the needed number of samples.

(6) Table 1 illustrates the number of samples required at a 95 % confidence level (Z-table factor of 1.65) with varying margins of error (p) and coefficients of variation (CV).

(7) Note that as the CV increases at a set margin of error, the number of samples required increases. When the variability is low (as measured by the standard deviation or the square root of the variance) relative to the mean of the data, then the CV is low. However, as the variability in the population begins to increase relative to the mean of the data, then the CV increases and the number of samples required increases if characterization of the site at a 95 % confidence level and a set margin of error is desired.

(8) A similar relationship is observed for the margin of error. When the precision required (say ± 10 ppm lead) is high, relative to the mean of the data (say 100 ppm lead), then the margin of error is low (in this case 0.1). In this case 115 samples would be required with a CV of 0.65. If the investigators could accept a higher margin of error (for example, ± 20 %), and the mean concentration of the data is still 100 ppm lead, then the resulting margin of error (0.2) would result in a lower number of required samples. Note that 29 samples would be required at the same CV of 0.65 and a one-sided inference.

(9) If the confidence level is decreased to 80 %, then the required number of samples reflected in this figure would be lower for each margin of error and CV combination. However, the confidence level may fixed. One alternative to analyzing the larger number of samples may be to use compositing.

9.2.2.4 *Site/Event Considerations*—The site and physical sampling event(s) constitute the majority of the practical aspects to be evaluated. Each design should be evaluated against all practical aspects to determine whether or not a given design will be practical to implement. This evaluation is subject to professional judgment as to whether or not a practical aspect, for example, the level of personnel training needed, is practical or acceptable, or both. If it is not, then the design needs to be modified or eliminated. These aspects include, but are not limited to the following:

TABLE 1 Number of Samples (*n*) for given Coefficient of Variation and Margin of Error

	Coefficient of Variation (CV)				
	0.1	0.5	0.65	1.0	2.0
Margin of Error (*p*)					
0.1	3	68	115	272	1089
0.2	1	17	29	68	272
0.3	-	8	13	30	121
0.5	-	3	5	11	44
1.0	-	1	1	3	11
2.0	-	-	-	1	3

[6] Gilbert, R. O., *Statistical Methods for Environmental Pollution Monitoring*, Van Nostrand Reinhold Co., New York, 1987.

(1) Site Access and Conditions —Site considerations - cross contamination potential; limits on access to sampling locations (for example, buildings, refusals).

(2) Equipment and Personnel —Equipment limitations; experience of the field sampling team; experience of the analysts; field and laboratory resources.

(3) Sampling Event —Special site concerns (for example, unexploded ordnance); special analytical needs (for example, low level analyses, dioxin); special analytical concerns (for example, interferences, multiple phases, incompatibility); health and safety considerations; resistant matrices (for example, solidified material); investigation derived waste (IDW) generation.

(4) Schedule —Transitory events (for example, start-up, shut-downs); potential impacts on project schedule.

(5) Safety Considerations.

9.3 *Statistical and Cost Evaluation*—Following the evaluation of each sampling design for the practical considerations, there should be a reduced set of design candidates. At this point, the statistical aspects (for example, uncertainty) and estimated costs should be calculated in preparation for the final review by the decision makers. The types of calculations and the degree to which statistics are needed will be dependent on the project objectives. The statistical evaluation routinely addresses issues of false positive or negative error, accuracy and precision, representativeness, and objectivity versus subjectivity.

9.3.1 *Statistical Considerations*—Statistical evaluation techniques are numerous and discussed in a number of places in the literature, including ASTM and USEPA guidance documents. The degree to which the statistical evaluation needs to be employed depends on the project objectives. Routinely, the statistical evaluation includes the acceptable limits of the potential sampling and analytical error and any mechanisms established for their controlling. The following are some general guidelines.

9.3.1.1 *Sensitivity Analysis*—A sensitivity analysis will determine how each design performs when the underlying assumptions about the sampling activity are modified. Typically, this involves changing specific parameters within some reasonable range and establishing how each of these changes influences the expected decision error rates. A statistical power curve is a useful statistical tool used to evaluate whether a sampling design has the ability to meet the project objectives.

9.3.1.2 *Hypothesis Test*:

(1) Each statistical sampling design should include a statistical hypothesis test. A statistical model should be developed which describes the relationship of the measured value to the "true" value. This mathematical formulation clarifies how data generated from a design is to be interpreted and processed in testing the hypothesis. A tentative analytic form for analyzing the resulting data (for example, a student's *t*-test or a tolerance interval) should be specified in the project objectives. This information can be used to determine the minimum sample size which satisfies the project objective's limits on decision error.

(2) The objectives of a statistical design are to limit the total error, which is a combination of sampling and measurement error, to acceptable levels. Traditional laboratory methods tend to minimize measurement error, but can be so expensive that only a limited number of samples can be analyzed within budget. The advantage to using less precise methods which are relatively less expensive, is that it allows a significantly larger number of samples to be collected and analyzed. This may trade off an increase in measurement error for a decrease in sampling error. If so, given the natural variability in many environmental studies, this approach may reduce overall costs while limiting the total decision error rates to acceptable levels.

(3) Appendix X2 provides an example approach for a statistical treatment of the choice of an analytical method based on the analytical variance and the analytical cost per sample.

9.3.1.3 *Error Statements*—When authoritative or nonstatistical designs are used, quantitative statements about data quality are limited to the measurement error component of the total study error. A statistical approach would allow a quantitative statement about the sampling error component of the total project error to be made and allow for determining the probability of making a decision error regarding the overall sampling event.

9.3.1.4 *Comparison of Sampling Designs Based on Statistical Considerations*:

(1) When Population Concentration Distribution is Random —When there are no trends, stratification or spatial correlation in the distribution of the contaminant concentration over an area, systematic grid and simple random sampling are usually equally precise.

(2) When Population Concentration Distribution Has Trends —In general, systematic grid sampling is more precise than simple random sampling and is less precise than stratified random sampling (in the estimation of the population mean), assuming strata are appropriately identified.

(3) When Population Concentrations Are Spatially Correlated —Spatial correlation refers to the fact that the concentrations of two samples taken in close proximity tend to be similar or correlated and that this correlation decreases as the distance between the two samples increases. Often, the presence of spatial correlation or clustering can be minimized by taking samples spatially far apart. A grid design and geostatistical data analysis may be used to eliminate the error associated with a random design. However, random sampling is useful in order to avoid human bias and is the design of choice when too little is known to stratify or grid. If the site has known differences due to the historical insult, fate, and/or transport of the pollutants, then this knowledge can best be used by a stratified design. Stratified random sampling will generally be more precise than simple random sampling.

9.3.2 *Cost Estimates*—Development of an estimate of the total cost of sample collection and analysis for each design option enables the decision team to compare the financial aspects of the sampling designs which meet the project objectives. These should include all aspects of the sampling design and analytical event(s). The cost differential between the alternate designs will frequently although not always, be closely related to the number of samples required. Detailed examples of cost estimates are included in USEPA's Guidance on DQO's (Appendix B).

10. Final Selection

10.1 The decision team should now have all the information needed to make a final selection of the most resource-effective design. The information is presented to the decision maker(s). The final comparison or evaluation may take the form of 10.2 or 10.3.

10.2 *Risk versus Cost*—Here, risk is an estimate of the probability of an incorrect decision and its associated degree of hazard. Because usually, the lower the risk, the more the implementation will cost, the acceptable level of risk is at the discretion of the decision maker. Assuming that each design option presented meets the project objectives, the dynamics of the final decision are usually a balance between risk and cost, as displayed in Fig. 5. The objective is to minimize the cost or risk, or both, relative to the project constraints (for example, false positives, false negatives).

10.3 *Confidence Level versus Cost versus Time*—The balancing of the confidence level, cost and time constraints is shown in Fig. 6. Design modifications to achieve the needed balance can include: (*1*) an increase or decrease in the number of samples; (*2*) use of design tools such as compositing; (*3*) satisfying practical limitations; (*4*) a change in the study boundaries; (*5*) an increase in the tolerable design errors; (*6*) a relaxation of project constraints; or (*7*) modifications in the initial hypotheses.

10.3.1 Significant changes to one or more of the alternate designs would require that the optimization process be repeated, with a review of the potential practical constraints, re-calculation of the number of samples, re-development of the statistical and cost estimates, and a final review by the decision maker.

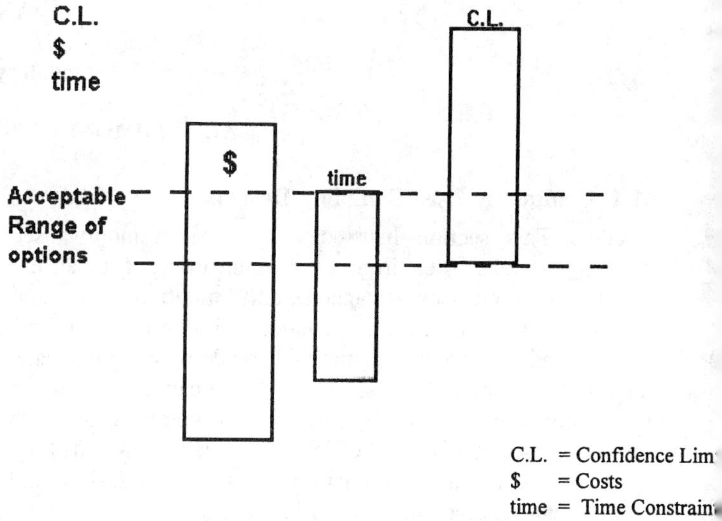

FIG. 6 Decision Variables

10.4 *Final Selection Options*:

10.4.1 Using the information generated to compare cost, risk, confidence level, time constraints and any other pertinent data, the decision maker can either (*1*) choose a sample design, or (*2*) reject the candidate designs.

10.4.2 If no design satisfactorily meets the requirements of the project objectives, including budgetary constraints and acceptable level of risk/error, then the sampling design or the project objectives will need to be modified. The design team should discuss with the decision maker whether modifications to the project objectives or the design option(s) are the most appropriate.

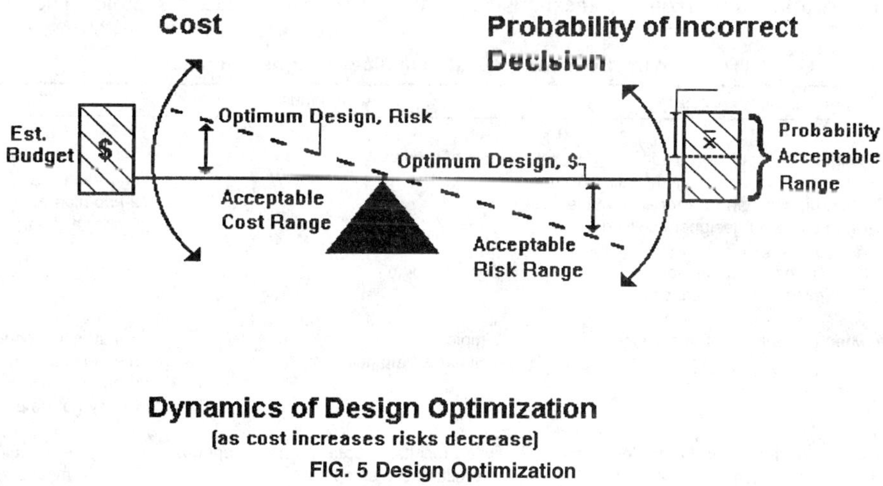

FIG. 5 Design Optimization

ANNEX

(Mandatory Information)

A1. COMMONLY USED SAMPLING DESIGNS

A1.1 Commonly Used Sampling Designs

A1.1.1 This section lists some of the commonly used sampling designs in environmental applications, with an explanation on their uses, advantages and limitations. The sampling designs discussed here include: authoritative sampling, simple random sampling, stratified random sampling, and systematic sampling. Each sampling design can be used alone, in combination with each other, or in combination with sampling design tools (see Section A1.2) such as compositing. Tables A1.1-A1.3 are a summary of the sampling designs and tools for quick reference.

A1.1.1.1 *Authoritative Sampling*—Authoritative sampling implies that some subjective human judgment is involved in the selection of the sampling location(s). There are generally two kinds: judgment sampling and biased sampling. However, these sampling methods are not haphazard sampling, implying that any sampling location is acceptable. When used properly, these sampling schemes can be highly cost-effective. However, in authoritative sampling, no objective probability of selection is assigned to any of the units in the population. As a result, there is usually no way to estimate the sampling variance of the population.

A1.1.1.2 *Judgment Sampling*—In judgment sampling, the sampling location(s) is selected because it is deemed to be representative of the average conditions of the population. The sampling objective here is to obtain a quick, inexpensive sample(s) so that an estimate of the general (average) conditions of the population can be obtained. More than one sample can be taken to increase the confidence in the estimate of the population developed from the sample data.

(1) Uses —*(1)* When an estimate of the population mean is needed quickly and inexpensively, (2) when the population is relatively homogeneous, (3) when there is a high tolerable margin of error in developing a view of the population, or (4) when the sampling individual or design is knowledgeable and a good judgment can be made as to where to take the samples.

(2) Advantages —*(1)* May be quick and inexpensive, particularly if no re-sampling is needed.

(3) Limitations —*(1)* When the population is highly heterogeneous and more statistically based results are needed, its average conditions may not be easily estimated. (2) When the tolerable margin of error is small, this method cannot be used, unless the population is relatively homogeneous or more samples are taken. When too many samples are needed, its advantage in cost may disappear and its disadvantage in accuracy (absence of bias) and precision over other sampling designs may begin to rise. (3) The sampling variance calculated from "judgment" samples may not be a good estimate of the true population sampling variance.

A1.1.1.3 *Biased Sampling*—In biased sampling the objective is usually to detect the presence of contamination or to obtain a "worst case" sample. The sampling location(s) is

TABLE A1.1 Commonly Used Sampling Designs: Summary

Sampling Design	Uses	Advantages	Limitations
Authoritative			
Judgment	1-estimate of population mean 2-when population homogeneous 3-when high margin of error acceptable 4-when sampling designer knowledgeable	1-cost effective	1-if population heterogeneous, mean not easily estimated 2-has high margin of error 3-poor estimate of variance
Biased	1-ID localized contamination 2-Determine non-compliance	1-cost effective	1-cannot generalize to the entire population
Simple Random	1-when population not stratified	1-simple 2-estimates variance	1-if stratified populations, may not estimate mean accurately 2-need more samples 3-may not be easy logistically
Stratified Random	1-when population can be divided into relatively homogeneous strata	1-when resultant strata are homogeneous 2-representative samples 3-estimates variance	1-may be difficult logistically 2-strata must correctly reflect any contaminant stratification
Systematic Line Space Random	1-locate hot spots 2-map trends	1-samples easily identified and collected 2-can define contamination patterns 3-more accurate estimate of mean concentration	1-unrecognized trends or cycles may cause poor accuracy or precision, or both
Search	1-locate hot spots	1-cost effective 2-minimum samples needed 3-easy to implement	1-hot spot may be undetected
Unequal Probability	1-heterogeneous population 2-contaminant expected in specific fraction	1-more precise estimate of the chemical contamination in a heterogeneous material. 2-less costly	1-unrecognized trends or cycles may cause poor accuracy or precision, or both

TABLE A1.2 Sampling Design Tools: Summary

Sampling Design	Uses	Advantages	Disadvantages
Compositing	1-mean estimate of population 2-where optimum number of samples high	1-reduce inter-sample variance 2-reduce costs 3-determine presence/absence of hot spots	1-loss of volatiles during mixing 2-when cannot be properly mixed 3-when analytical costs low relative to sampling costs
Particulate Material Method	1-heterogeneous material mixture of granular material	1-statistical and defensible	1-very large sample size needed (exceeding the limit of the some field and most laboratory equipment)
Geostatistics	1-where need isoplath maps 2-where data collected on a spatial network	1-maximizes use of available data 2-modeling	1-highly specialized math modeling 2-subjective
Sequential: Formal or Wald	1-as a contributing information source 2-subsidiary to other methods	1-use minimum number of data points 2-cost effective	1-decisions restricted by formal rules 2-not allow for alternatives, value judgments, etc.

TABLE A1.3 Combination Sampling Design: Summary

Sampling Design		Advantages	Disadvantages
Phased Sampling	1-where screening and then confirmation appropriate 2-verification of screening	1-cost effective (field screening)	1-increased turn-around time of the project sampling phase
Sequential	1-informal approach is acceptable	1-reduced cost by initially undersampling, and adding additional samples as needed	1-unrecognized trends or cycles may cause poor accuracy or precision, or both 2-action alternatives need to be carefully defined

selected based on available information or knowledge, especially in terms of visually contaminated areas or knowledge of contamination.

(1) Uses —*(1)* Detecting the presence of localized contamination, identifying the source of contamination, or determining non-compliance.

(2) Advantages —*(1)* Cost-effective for the intended purposes.

(3) Limitations —*(1)* Local detection of contamination from this method cannot be generalized to the entire site area or population. (The question of degree or extent of contamination in the entire population cannot be answered with this kind of sampling.)

A1.1.2 *Simple Random Sampling*—Simple random sampling considers all units in the population equally (see also A1.1.3 and A1.1.4). In this sampling scheme, all the units in the population have equal probability of being selected as one of the samples, and the selection of one sample does not affect the selection of the remaining samples. An appropriate number of samples are randomly collected from that population.

A1.1.2.1 *Uses*—*(1)* When potential personal bias needs to be minimized; (2) when the population has no identified patterns of contamination, such as stratification.

A1.1.2.2 *Advantages*—*(1)* This is the most simple random sampling design. Because the sample selection is based on probability, a representative set of samples can be expected and the sample data can be used to obtain an estimate of the population variance.

A1.1.2.3 *Limitations*—*(1)* If patterns or stratification exists, it may not provide an accurate (in both bias and precision) estimate of the population characteristic. (2) A large number of samples may be needed in order to achieve the desired level of precision in estimating the population characteristics (especially when the population is highly heterogeneous). *(3)* Logistically, it may not be easy to take samples at the specified random locations.

A1.1.3 *Stratified Random Sampling*—In stratified random sampling, the population is divided into non-overlapping subgroups (in space or time) called strata, where the contaminant concentration is similar within a stratum and dissimilar between the strata. Sampling locations are selected using simple random sampling within each stratum. One can also use a systematic design in each stratum. An overall estimate of the population characteristic (for example, mean concentration) is the weighted average of the stratum means, the weights being the relative sizes of the strata. The relative size of a stratum is defined as the ratio of the size of a stratum to that of the entire population.

A1.1.3.1 *Uses*—*(1)* When a population can be divided into different strata, especially when the stratification boundaries are consistent with the contaminant concentration, such that each stratum is relatively homogeneous relative to the characteristic of interest.

A1.1.3.2 *Advantages*—*(1)* If the division of the population into strata creates relatively homogeneous strata, this design will obtain a more accurate estimate of the population characteristic(s) than simple random sampling, using the same number of samples. (2) A representative set of samples can be obtained. *(3)* Sampling variance can be estimated.

A1.1.3.3 *Limitations*—*(1)* Collecting samples at random locations within a stratum may not be easy to implement logistically. (2) Greater accuracy/precision is probable only if the known stratification correctly reflects the strata, with maximum between-strata variability and minimum within-strata variability. *(3)* If each stratum is used as a unit upon which decision is to be made, then each stratum can be viewed as a population and a simple random sampling design can be used for each stratum.

NOTE A1.1—It is often desirable to have a more uniform distribution of the sampling locations throughout the population than a simple random sampling design. In such cases (for example, when sampling along a line or over an area), the population can be divided into blocks of equal or unequal sizes. These blocks can be construed to be "arbitrary" strata. These arbitrary strata can be defined by topography, political boundaries, or physical demarcations such as roads and rivers. A set of random or systematic samples can then be taken from within each block.

A1.1.4 *Systematic Sampling*—Systematic sampling involves the collection of samples at predetermined, regular intervals, in space or time. However, this method of sampling does not preclude the incorporation of random elements. Several of the commonly used designs will be discussed here.

A1.1.4.1 *Uses*—*(1)* These designs are applicable when there is a need to obtain samples more uniformly located throughout the population, as when a population is essentially random or contains a minimum of strata. (2) Identification of hot spots. (*3*) Mapping distribution of contaminants.

A1.1.4.2 *Advantages*—*(1)* It is relatively easy to identify the samples for collection and to implement under most field conditions. (*2*) It may be useful for defining patterns of contamination (trends or cycles). (*3*) In many cases, it will produce a more accurate estimate of mean concentration. (*4*) Particularly useful in detecting hot spots.

A1.1.4.3 *Limitations*—*(1)* Care needs to be taken so that the sampling locations do not coincide with unsuspected patterns or periodicity in contaminant concentration. Such coincidence may produce biased estimate. (2) An accurate estimate of the population sampling variance may be difficult.

A1.1.4.4 *Sampling Along A Line*—When sampling over a time or along a physical line, for example, a ditch, samples can be taken on a predetermined, regular interval along the line. Once the number of samples to be collected (*n*) is known, the sampling interval in space or time (*k*) can be determined.

$$k = N/n \quad \text{(A1.1)}$$

where:
N = is the total length of the line, and
k = is determined to the nearest integer.

One way to obtain the n samples is as follows: (*1*) picture the total length of the line (*N*) as extending around a circle (*2*) choose a random number between 1 and *N* as a starting point and (*3*) select every kth sample around the circle, until *n* of them have been obtained. This will provide a set of samples from which an unbiased estimate of the mean concentration can be determined.

A1.1.4.5 *Sampling Over Space (Square Grid)*—The most simple grid sampling design is a square grid design. One step wise procedure is:

(1) Calculated grid size *G* as $G = \sqrt{(A/n)}$, where *A* = area size, *n* = number of samples needed.

(2) Select an initial random point in the area.

(3) Construct coordinate axes going through the initial point.

(4) Construct lines parallel to the vertical axis, separated by grid distance *G*.

(5) Similarly, construct lines parallel to the horizontal axis.

(6) Locate sampling points at grid line intersections or at the center of grid cells within the site boundary.

(7) The total number of samples identified on the grid may be different from the number *n* which are needed. At this point, the several options include: (*a*) Taking all the samples identified on the grid, (*b*) taking only *n* samples out of the total located, *n* being based on best judgment, (*c*) taking samples at a random location in the grid squares, and (*d*) compositing the samples taken from within grid cells.

NOTE A1.2—There are many variations of this design. However, they offer essentially the same advantages and limitations.

A1.1.4.6 *Systematic Random Sampling*—Systematic random sampling is applied to a population by choosing a random starting time or location with the aid of random number tables or random number generator. The initial starting point is where a grid of equally spaced sampling locations is initiated. A second random number can be chosen to determine the orientation of the grid over the population. The grid dimensions or the distance between sampling locations is often determined by the budget which will determine how many samples can be collected and analyzed. The sampling process is greatly facilitated by the fact that field personnel can easily measure to adjacent sampling locations. The fact that the first location and the grid orientation are chosen in a random manner ensures that operator bias does not impact the choice of sampling locations.

(1) Uses —*(1)* Where a uniform coverage of the site is needed.

(2) Advantages —*(1)* Detects concentration trends over time and space and under certain circumstances can yield a more accurate estimate of mean concentration. (*2*) Can be applied to a temporal population (for example, a wastewater effluent).

(3) Limitations —*(1)* Substantial bias can result if unknown concentration cycles coincide with the systematic sampling locations. Therefore historical or waste generation information is recommended to determine the likelihood of cyclic patterns in the population. (*2*) Inaccurate variance estimates occur when the concentration distribution is not random (for example, the systematic collection of samples crosses different strata) or if concentrations are serially correlated. (*3*) Substantial number of samples are required.

A1.1.5 *Search Sampling*—Search sampling is a strategy to increase the likelihood that at least one sample coincides with any existing areas of high concentrations or "hot spots." This involves (*1*) the estimation of the likely shape and size of the hot spot, and (*2*) the determination of the acceptable risk of not detecting the hot spot. With this information, the shape of the sampling grid units (for example, square or triangular) are chosen and the distance between sampling locations is calculated. The sampling locations may form a grid pattern similar to that of systematic random sampling.

A1.1.5.1 *Uses*—*(1)* To locate hot spots.

A1.1.5.2 *Advantages*—*(1)* This design provides for the maximum use of any knowledge of the site or location, or both. (*2*) Cost effective. (*3*) Minimum number of samples needed. (*4*) Ease of implementation.

A1.1.5.3 *Limitations*—*(1)* If the assumptions are incorrect, it is possible that a hot spot will exist between the sampling locations and remain undetected.

A1.1.6 *Unequal Probability Sampling*—Unequal probability sampling is similar to simple random sampling, except that a rule is used to unequally weight the likelihood that a given sample will be included in the data set. The samples to be included or excluded may be chosen based on a specific area of the strata, the size of the group or population, or the particle size of the waste.

A1.1.6.1 *Uses*—*(1)* To sample heterogeneous materials in which the contaminant of interest is expected to be found in a particular particle fraction, for example, the fines. The remaining items, for example, the large particles, can be ignored as part of the sample collected.

NOTE A1.3—It may still be necessary to know the mass of the large particles, depending on the project objectives.

A1.1.6.2 *Advantages*—*(1)* For a given number of samples of specific size, it often provides a more precise estimate of the chemical contamination in a heterogeneous material (than simple random sampling). *(2)* This design is less costly because fewer samples are needed for a given mass of waste.

A1.1.6.3 *Limitations*—*(1)* Poor accuracy and precision can occur when unrecognized trends or cycles occur in the population.

A1.2 Sampling Design Tools

A1.2.1 *Sample Compositing*—The compositing or combining of individual samples to form a composite sample is a mechanism to reduce the analytical costs, where a mean estimate value is either acceptable or the objective. The main physical issues to address are whether the individual sample matrices are compatible and how many samples it is appropriate to composite. See Guide D 6051. For example, if the regulatory limit is 50 ppm and the detection limit is 5 ppm, no more than 10 samples may be composited together in order to detect the possibility of a single sample exceeding the regulatory level of 50 ppm.

A1.2.1.1 *Uses*—*(1)* Where sample collection is inexpensive but analysis is costly. *(2)* Less than 20 % of the results are expected to exceed the blank values. *(3)* The objective is a mean estimation of the population.

A1.2.1.2 *Advantages*—*(1)* Reduce inter-sample variance; *(2)* reduce costs for estimating a total or mean value, especially where analytical costs greatly exceed sampling costs; *(3)* efficiently determine the absence or possible presence of hot spots or hot containers and, when combined with retesting schemes, identifying hot spots, as long as the probability of hitting a hot spot is low; *(4)* situations where the nature of the contaminant distribution tends to be contiguous and non-random and the majority of analyses are "non-detects" for the contaminant of interest; *(5)* provides a degree of anonymity where population, rather than individual statistics are needed.

A1.2.1.3 *Limitations*—*(1)* Loss of volatiles during mixing; *(2)* when the composite sample cannot be properly mixed and subsampled or the whole composite sample cannot be analyzed; *(3)* when the goal is to detect hotspots and a large proportion of the samples are expected to test positive for an attribute; *(5)* when analytical costs are low relative to sampling costs (for example, insitu field portable X-ray fluorescence takes only 30 s with no sample preparation so analytical costs/sample are very low; *(6)* when regulations specify that a grab sample must be collected.

A1.2.2 *Particulate Material Method*—The particulate material method is statistically based and extends into sample preparation. It seeks to provide an overall framework for sampling design, whereas, traditional random sampling schemes address only the statistical aspects of sample identification and data computation, leaving the question of particle size and sample volume to the chemist. This method was developed for the mining industry to sample ore. Characterization of ore is dependent on the sample volume, particle size, the degree of particle size reduction and the information available about the ore.

A1.2.2.1 *Uses*—*(1)* For sampling heterogeneous mixtures of granular material.

A1.2.2.2 *Advantages*—*(1)* The advantage to using this method is the defensibility afforded by the statistical approach.

A1.2.2.3 *Limitations*—*(1)* Extraordinarily large sample volumes may be required. Most laboratories cannot handle such volumes and the aliquots used for environmental analytical methods make the sample volume unrealistic.

A1.2.3 *Geostatistic Sampling*—Geostatistic sampling includes techniques such as kriging and conditional simulation. Preliminary or exploratory study data can be used to optimize a sampling design where geostatistical data analysis methods are to be used. Statistically, conditional simulation techniques assume that the population is treated as though it is a realization of a random process: the population is random and the sample is deterministic. In other words, because the population is truly random, the sample locations may be selected subjectively rather than randomly. First, assumptions are selected and the population is mathematically modeled. Then from data or past experience one estimates a function which describes the spatial interrelation within the population, which is often subjectively selected when there is little available information. As data is obtained, the estimates are updated to provide final answers which are based on the most up-to-date co-variance estimates.

A1.2.3.1 *Kriging* is a technique in which a weighted moving average method is used to interpolate values from a sample data set onto a grid of points for contouring. It can be used to identify areas of high uncertainty and guide additional sampling. Kriging provides an unbiased estimate (minimizing the expected-square-error) of the average of a given attribute over a specified spatial region of the population. The method varies depending on the assumed statistical structure of the population and the attribute(s) to be estimated.

A1.2.3.2 *Conditional simulation* techniques use detailed computer simulations of a population, consistent with available data and assumptions of the population's statistical structure and spatial interrelations. Multiple simulations can be produced to demonstrate the degree of variability following collection of a suite of data points. This provides valuable information in determining future sampling locations or possible responses to remediation action, consistent with the current available data. If this shows too much variation, then more data should be collected.

(1) Uses —*(1)* Site assessment and monitoring situations

where data are collected on a spatial network of sampling locations; (2) where contour maps of pollutant concentration (or other variables) are desired; (3) when sample collection and analysis are costly and it is necessary to maximize the information from each data point.

(2) Advantages —(*1*) These methods maximize use of the available data.

(3) Limitations —(*1*) These techniques are subjective. Judgment, experience, and skill are important factors.

A1.2.4 *Formal or Wald Sequential Testing*—In formal sequential testing, a hypothesis is developed; samples are collected and analyzed; and the data is used in a formal hypothesis test procedure with three possible actions: (*1*) accept the null hypothesis; (*2*) reject the null hypothesis; (*3*) insufficient data, collect additional samples. A formal stopping rule is used to determine which of the three actions is most appropriate.

A1.2.4.1 *Uses*—(*1*) Primarily as a contributing source of information for a practical decision process. (*2*) As a subsidiary to other methods.

A1.2.4.2 *Advantages*—(*1*) Action levels are reached with a minimum number of data points.

A1.2.4.3 *Limitations*—(*1*) The formal procedures restrict the decisions made by not allowing for the alternate courses of action, value judgments, etc.

A1.3 Combination Sample Designs

A1.3.1 The following are frequently used sampling designs which combine any of the above discussed designs, sampling tools and/or employ a multi-step sampling design. The uses, advantages and limitations will be the same as those of the primary design(s) or tools used.

A1.3.1.1 *Phased or Two Stage Sampling*—Phased sampling is a general term for a two step method using one or more of the previously discussed designs. A first set of samples is collected and analyzed. The resulting data is then used to design the collection and analysis of the second set of samples. Most commonly, a large number of samples is selected to be analytically screened, inexpensively. Based on the first sample results a second, smaller set of samples is selected, and analyzed with more precise, costly methods. With *random sampling* models, the first data set is used to obtain estimates of the population variance. This provides guidance for selection of the most appropriate number or location, or both, of the subsequent set of samples. With *geostatistical techniques* the first data set is chosen to estimate or verify the population intercorrelation or other structural population assumptions. The second data set provides the basis for least-cost computation of the estimates of the target precision.

(1) Uses —(*1*) Particularly useful for verifying field screening results using more rigorous methods. The closer the screening procedure correlates with more accepted/rigorous technique, the more acceptable the screening procedure. If the screening measurement is relatively inexpensive, a more accurate characterization of a waste site can be achieved by increasing the number of measurements.

(2) Advantages —(*1*) The principal advantage to this method is the cost reduction by using field screening methods, followed by fewer more precise and expensive, regulatory required sampling. (*2*) Using random sampling methods reduces the variance on the final samples because of the number of samples. (*3*) Geostatistical sampling provides for reduced costs and maximum information from each data point.

(3) Limitations —(*1*) The turn-around time of the sampling phase of the project is relatively long, because the sampling and analyses are performed twice.

A1.3.2 *Sequential Sampling*—This more informal sequential testing is an extension of double sampling and is not restricted to the formal Wald process. It includes aspects of sequential testing, allows decision making with minimal data and may be used with either random sampling or geostatistical techniques.

A1.3.2.1 *Uses*—(*1*) Where the more formal approach is not as critical.

A1.3.2.2 *Advantages*—(*1*) The sequential sampling approach reduces cost by initially under sampling then adding samples, as needed.

A1.3.2.3 *Limitations*—(*1*) The action alternatives, types of data needed at each stage, etc., need to be carefully defined.

APPENDIXES

(Nonmandatory Information)

X1. ADDITIONAL REFERENCES

X1.1 General

USEPA, *Characterizing Heterogeneous Wastes: Methods and Recommendations*, ORD and USDOE, EPA 600/R-92/033, February 1992.

USEPA, Guidance for the Data Quality Assessment, EPA QA/G-9, EPA 600/R-96/055, Quality Assurance Management Staff, Washington, DC, September 1994.

USEPA, *Test Methods for Evaluating Solid Waste;* SW-846, third ed., Washington DC: U.S. Environmental Protection Agency Office of Solid Waste, 1991.

X1.2 Sampling Design

Cochran, W. G., *Sampling Techniques*, 3rd ed., John Wiley and Sons, Inc., New York, 1977.

ACS Committee on Environmental Improvement, "Principles of Environmental Analysis," *Anal. Chem.* 55, 2210-2218, 1983.

Desu, M. M. and D. Raghavarao, *Sample Size Methodology*, Academic Press, San Diego, 1990.

Nelson, L. S., "Economic Balance of Sampling and Testing," *J. of Quality Technology*, 24, 162-163, 1992.

"Preparation of Soil, Sampling Protocols: Sampling Techniques and Strategies," *EPA/600/R-92/128*, EMSL - LV, Benjamin Mason, 1992, Preissue draft.

X1.3 Specific Sampling Device Selection

ASTM D 4547 Practice for Sampling Water and Soils for Volatile Organics

ASTM D 5679 Practice for Sampling Consolidated Solids in Drums or Similar Containers

ASTM D 5680 Practice for Sampling Unconsolidated Solids in Drums or Similar Containers

ASTM D 5743 Practice for Sampling Single or Multilayered Liquids, With or Without Solids, in Drums or Similar Containers

ASTM 6232 Guide for Selection of Sampling Equipment for Waste and Contaminated Media Data Collection Activities

A Compendium of Superfund Field Operations Methods, U.S. EPA, Washington, DC PB88-181557, Dec. 1987.

The Environmental Survey Manual, U.S. Department of Energy, DOE/EH-0053, second ed. Vol 5, Appendix E, January 1989.

Characterization of Hazardous Waste Sites—A Methods Manual: Volume II. Available Sampling Methods, second ed. U.S. Environmental Protection Agency, Las Vegas, Nevada. EPA-600/4-84-076.

Sampling for Hazardous Materials, U.S. Environmental Protection Agency, Office of Emergency and Remedial Response, Hazardous Response Support Division, 1988.

Lewis, T.E., A.B. Crockett, R.L. Sicgrist, and K. Zarrabi, "Soil Sampling and Analysis for Volatile Organic Compounds, Ground-Water Issue," U.S. Environmental Protection Agency, Las Vegas, NV, *EPA/540/4-91/001*, February 1991.

Garner, F.C., M.A. Stapanian, and L.R. Williams, "Composite Sampling for Environmental Monitoring," in *Principles of Environmental Sampling*, L.H. Keith, ed., American Chemical Society, Washington D.C., 1988, pp. 363-374.

Rajagopal, R., Economics of Screening in the Detection of Organics in Ground Water, *Chemometrics and Intelligent Laboratory Systems*, 9:261-272, 1990.

Rajagopal, R., and L.R. Williams, "Economics of Sample Compositing as a Screening Tool in Ground Water Quality Monitoring," *Ground Water Monitoring Review*, Winter 1989, pp. 186-192.

X1.4 Statistics

Gilbert, Richard O., *Statistical Methods for Environmental Pollution Monitoring*, Van Nostrand Reinhold Co., New York, NY, 1987.

Pitard, F.F., *Pierre Gy's Sampling Theory and Sampling Practice, Vol. I: Heterogeneity and Sampling*, CRC Press, Boca Raton, FL, 1989.

Pitard, F. F., *Pierre Gy's Sampling Theory and Sampling Practice. Volume II, Sampling Correctness and Sampling Practice*, CRC Press, Boca Raton, FL, 1989.

Siegmund, D., *Sequential Analysis, Tests and Confidence Intervals*, Springer-Verlag, New York, 272 pp., 1985.

Wetherill, G.B., and K.D. Glazebrook, *Sequential Methods in Statistics*, 3rd ed., Chapman and Hall, New York, 264 pp., 1986.

Garner, F.C., M.A. Stapanian, E.A. Yfantis, and L.R. Williams, "Probability Estimation with Sample Compositing Techniques," *Journal of Official Statistics* 5(4):364-374, 1989.

Borgman L.E., and W.F. Quimby, "Sampling for Test of Hypothesis When Data are Correlated in Space and Time," In *Principles of Environmental Sampling* (Keith, Lawrence H., ed.), p 25-43, The American Chemical Society, 1988.

ASTM C 1215 Guide for Preparing and Interpreting Precision and Bias Statements in Test Method Standards Used in the Nuclear Industry[7]

Borgman, L.E., "New Advances in Methodology for Statistical Test Useful in Geostatistical Studies," *Mathematical Geology* 20(4):383-403, 1988.

Gilbert, R.O., *Statistical Methods for Environmental Pollution Monitoring*, Van Nostrand Reinhold Co., New York, 320 pp., 1987.

Gy, P., L'Enchantillonnage des Minerais en Vrac: Tome I: Theorie Generale, *Revue de L'Industrie Minerale*. Also published in *Memoires du Bureau de Recherches Geologique et Minieres, No. 56*, 1967, 186 pp. (Chapter 4, Theorie de L'Enchantillonnage Equiprobable, gave an excellent presentation of the theoretical foundation for Gy's method).

Borgman, L.E., An exposition of Gy's derivation in L'Enchantillonnage des Minerais en Vrac, Theorie Generale. *Statistics Laboratory Report No. 91-2*, Department of Statistics, University of Wyoming, Laramie, WY, 1991.

Gy, P.M., *Sampling of Particulate Materials, Theory and Practice*, Elsevier Scientific Pub., New York, 1982.

X1.5 Software

Gaugush, R.F., 1990, Sampling Design Software, version 2.0. Environmental Laboratory, U.S.A.E., Waterways Experiment Station, P.O. Box 631, Vicksburg, MS 39180.

Pitard, F.F., Software Programs: QE1, VARIO, and COAL, Francis F. Pitard, 14710 Tejon Street, Broomfield, CO 80020.

[7] *Annual Book of ASTM Standards*, Vol 12.01.

X2. CHOOSING ANALYTICAL METHOD BASED ON VARIANCE AND COST

X2.1 Choosing and analytical method based on the analytical variance and analytical cost per sample can be formalized through the following calculations. These are based on the comparison of two methods at a time.

Let s_i^2 = estimated analytical variance for method *i*,
c_i = analytical cost per sample for method *i*,
C_i = cost of analysis for n_i samples for the *i*th method $= n_i c_i$,
X_i = mean of n_i analyses for the *i*th method, and
i = 1 or 2 (representing the first or second method being considered).

X2.2 Equal precision between X_i and X_2 can be achieved by setting:

$$\text{Var}(X_1) = \text{Var}(X_2) \quad (X2.1)$$

$$s_1^2 / n_1 = s_2^2 / n_2 \quad (X2.2)$$

X2.3 Rearranging these equations yields:

$$n_1 = n_2(s_1^2/s_2^2) \quad (X2.3)$$

X2.4 Substituting for n_1, therefore creates the expression:

$$C_1 = n_1 c_1 = n_2(s_1^2/s_2^2)c_1 \quad (X2.4)$$

and dividing by $C_2 = n_2 c_2$ gives:

$$C_1/C_2 = (s_1^2/s_2^2)(c_1/c_2) \quad (X2.5)$$

X2.5 Given estimates s_1^2, s_2^2, c_1 and c_2, the ratio C_1/C_2 can be calculated. The optimal cost at an equal precision can be established by using the following:

X2.5.1 If $C_1/C_2 < 1$, choose method 1.

X2.5.2 If $C_1/C_2 > 1$, choose method 2.

X3. CALCULATING THE NUMBER OF SAMPLES: A STATISTICAL TREATMENT

X3.1 Summary of Calculations

Sample Design	Assumption	Equation No.
Simple Random Sample	Prescribed Precision	Eq X3.1
	Estimated Variance	Eq X3.2
	Prescribed Margin of Error Prescribed Precision	Eq X3.3
	Prescribed Margin of Error Estimated Variance	Eq X3.4
	Prescribed Margin of Error Prescribed Relative Error	Eq X3.5
	Specified False Positive and False Negative Errors	Eq X3.6 Eq X3.7
Stratified Random Sampling	Optimal Allocation of Samples to Strata	Eq X3.8-X3.11
	Total Number of Samples for Prescribed Fixed Cost	Eq X3.12
	Total Number of Samples for Prescribed Variance	Eq X3.13
	Total Number of Samples for Prescribed Margin of Error	Eq X3.14
Multi-Stage Sampling	Two Sources of Variation	Eq X3.16-X3.18
Search Sampling for Hot Spots	Hot Spots are Small Sizes	Eq X3.20
Grid Sampling for Hot Spots	Hot Spots of Defined Size and Shape	Eq X3.22
	Multiple Hot Spots	Eq X3.26
Composite Sampling	Estimation Problem	Eq X3.27-X3.30
	Classification Problem	Eq X3.33-X3.38

X3.2 How to Determine the Number of Samples

X3.2.1 Many of the following methods are described in footnotes 6 and 8.[6,8] Paragraph X3.2.2 describes the methods for calculating the number of samples needed for estimation of the population mean concentration, for simple random and stratified random sampling designs. Paragraph X3.2.3 describes multi-stage sampling methods when the objective is to statistically detect hot spots. The needed number of samples can vary according to site specific conditions (for example, degree of heterogeneity in the sampling materials) and specifications (for example, acceptable precision, false positive error and false negative error).

X3.2.2 *For Estimation of Mean Concentration*—This section assumes that the population is so large that it can be construed to contain an infinite number of sampling units or items.

X3.2.2.1 The formula for calculating the needed number of samples are often given for two cases: when the population variance is known and when the population variance is estimated. The population variance can be construed to be known when there is history or experience as to what the size of the variance is. When this is the case, the z statistic is used in the formula. On the other hand, if the population variance is estimated from certain data, then the t statistic is used instead. Oftentimes, the difference in the numbers of samples produced by the two different approaches do not differ very substantially.

X3.2.2.2 Since there is generally lack of data to estimate the population variance, the formula with the z statistic can often be used as an approximation. If high precision in the needed number of samples is required, then a pilot study to provide a good estimate of the variance may be called for.

X3.2.2.3 The methods also assumes little or no spatial correlation between the samples. Where appropriate, a one-sided inference about the population is assumed.

X3.2.2.4 *Sample Random Sampling*:

(1) For Prescribed Precision — The variance of the estimated mean concentration is a measure of precision. In many applications, it is often required that this variance be no larger than a prescribed value v. The number of samples needed in this situation is given below.

When the population variance σ^2 is known

$$n = \sigma^2/v \quad \text{(X3.1)}$$

where:

n = needed number of samples,
σ^2 = known population variance, and
v = prescribed size of the variance of sample mean Var (x).

When the population variance is estimated—When population variance is estimated from an initial set of n_1 samples to compute the variance s^2 as an estimate of the population variance σ^2, then:

$$n = (s^2/v)\,(1 + 2/n_1) \quad \text{(X3.2)}$$

(2) For Prescribed Margin of Error — When the absolute deviation of the sample mean from the true mean must be within a prescribed limit d with 100 (1-α) % probability. If the population mean is μ and the sample estimate is x, then $d = \bar{x} - \mu$.

When population variance is known

$$n = (z_{1-\alpha}\sigma/d)^2 \quad \text{(X3.3)}$$

where:

d = specified or acceptable absolute deviation of the estimate from the true value, or the absolute limit of error in estimation,
$1-\alpha$ = confidence level of not exceeding the prescribed margin of error d, and
$z_{1-\alpha}$ = the 100 (1–α) percentile point of standard normal distribution.

(3) The relationships among selected values of α, $z_{1-\alpha}$ and confidence level (for a one-sided inference about the population) are given below:

α	Confidence Level, %	$z_{1-\alpha}$
0.20	80	0.842
0.10	90	1.282
0.05	95	1.645
0.025	97.5	1.960
0.01	99	2.576

For a two-sided z value, simply multiply the α value by 2, re-calculated the confidence level as 100 (1–α), and then look for the corresponding z value. For example, the two-sided z-value with 95 % confidence is 1.96. Often a z value of 2 is used because of the uncertainty in estimating the standard deviation (σ).

When the population variance is not known and is estimated from m samples, then z in Eq X3.3 is replaced by t, resulting in Eq X3.4:

$$n = (t_{1-\alpha,m-1}s/d)^2 \quad \text{(X3.4)}$$

where:

$t_{1-\alpha,m-1}$ = the 100 (1–α) percentile point of the t-distribution with (m–1) degrees of freedom.

(4) For Prescribed Relative Error — Often a reliable value of σ^2 is not available but a coefficient of variation (CV) can be provided. In such cases, the margin of error in *(2)* can be specified as a percentage of the mean such that this relative error is no larger than a prescribed value d_r with 100 (1–α) % confidence. Thus,

$$n = (z_{1-\alpha}CV/d_r)^2 \quad \text{(X3.5)}$$

where:

CV = coefficient of variation in % = 100s/x,
d_r = relative margin of error in % = $100\,|\bar{x}|/\mu$, and
μ = population mean.

Eq X3.5 is useful because it is often easier to estimate CV than the variance σ^2.

(5) For Both False Positive and False Negative Errors —When it is desirable to estimate the number of samples needed for specified false positive error (typically denoted as α) and false negative error (typically denoted as β) for a decision rule, then the formula is:

$$n = (z_{1-\alpha} + z_{1-\beta})^2\sigma^2/\delta^2 \quad \text{(X3.6)}$$

[8] Cochran, William G., *Sampling Techniques*, 3rd ed., John Wiley and Sons, 1977.

where:
α = false positive error,
β = false negative error,
σ = population standard deviation, and
δ = deviation from the true mean important to detect.

Some example values of $z_{1-\alpha}$ and $z_{1-\beta}$ are given in the table in the previous section.

Eq X3.6 can be modified when the population variance is estimated by s^2, an approximate number of samples is:

$$n = (z_{1-\alpha} + z_{1-\beta})^2 s^2/\delta^2 + (z_{1-\alpha})^2/2 \quad \text{(X3.7)}$$

The quantity δ in Eq X3.6 and X3.7 is a quantity which needs to be negotiated and agreed to. It often represents a departure from some reference value (such as regulatory limit) which is important to detect statistically. When it is large, the tolerance for departure from the reference value is high and the resultant number of samples required will be small. When it is small, the tolerance for departure from the reference value is low and the resultant number of samples required will be large. A balance in these considerations would lead to an agreed δ.

NOTE X3.1—It needs to be noted that when this method is used (Eq X3.6 or Eq X3.7), the number of samples required is in general much larger than the previous methods (for example, Eq X3.3-X3.5). The number of samples increases rather rapidly as the specified false negative error decreases.

X3.2.2.5 *Stratified Random Sampling*—This section describes the methods to determine the number of samples to collect from each stratum in stratified random sampling, for the purpose of estimating the mean concentration over all the strata. If the sampling objective is to estimate the mean concentration of each of the strata, then a simple random sampling design can be applied to each stratum.

(1) Optimal Allocation of Samples to Strata —If the sampling cost function is:

$$\text{Cost} = C = c_0 + \Sigma_{i=1,s} c_i n_i \quad \text{(X3.8)}$$

where:
c_0 = fixed overhead cost,
s = number of strata,
c_i = cost of taking a sample from the *i*th stratum, and
n_i = number of samples to be taken from the *i*th stratum.

then the optimal allocation is to allocate n_i samples (of the total number n to be collected) to *i*th stratum as follows:

$$n_i = n \frac{w_i\sigma_i/\sqrt{c_i}}{\Sigma_{i=1,s}(w_i\sigma_i/\sqrt{c_i})} \quad \text{(X3.9)}$$

where:
n = total number of samples to be taken from all the strata, and
w_i = proportion of stratum i to the entire population (N_i/N).

If the cost of sampling is the same among the strata, Eq X3.9 reduces to:

$$n_i = n \frac{w_i s_i}{\Sigma_{i=1,s} w_i s_i} \quad \text{(X3.10)}$$

where:
s_i = standard deviation of samples from stratum *i* estimated from prior data.

A simple alternative to Eq X3.9 and X3.10 is proportional allocation. In this scheme,

$$n_i = w_i n \quad \text{(X3.11)}$$

That is, if a stratum is *p* % of the population, then *p* % of the total samples will come from that stratum. For example, *p*=30 %.

The above equations requires the knowledge of *n*, the total number of samples *n* to take over all the strata. The sections below deals with this issue.

(2) Total Number of Samples for Prescribed Fixed Cost —When the total cost *C* is fixed, the optimal total number of samples is:

$$n = \frac{(C - c_0)\Sigma_{i=1,s}(w_i s_i/\sqrt{c_i})}{\Sigma_{i-1,s}(w_i s_i\sqrt{c_i})} \quad \text{(X3.12)}$$

where:
s_i = estimated standard deviation of samples from stratum i, and
$C-c_0$ = money available for sampling and analysis.

(3) Total Number of Samples for Prescribed Variance — If the variance of the estimated mean cannot exceed a prescribed value *v*, an approximation is:

$$n = \Sigma[(w_i^2 s_i^2)/f_i] / [v + (w_i s_i^2)/N] \quad \text{(X3.13)}$$

where:
f_i = sampling fraction from stratum *i* (or the proportion of total number of samples n to be taken from stratum *i*).

f_i in Eq X3.13 can be made identical to w_i.

(4) Total Number of Samples for Prescribed Margin of Error —If proportional allocation is used, then the optimal number of samples can be obtained for a specified (absolute) deviation of the estimated mean from the true value by no more than d with 100 (1–α) % confidence. An approximation is:

$$n = [Z^2_{1-\alpha}\Sigma_{i=1,s} w_i^2 s_i^2/d^2] / [1 + z^2_{1-\alpha}\Sigma_{i=1,s} w_i s_i^2 / (d^2 N)] \quad \text{(X3.14)}$$

When *N* is very large, Eq X3.14 is reduced to:

$$n = Z^2_{1-\alpha}\Sigma_{i=1,s} w_i^2 s_i^2 / d^2 \quad \text{(X3.15)}$$

X3.2.3 *Multi-Stage Sampling/Measurement Process*—Often a sampling and analytical plan consists of several stages of sampling and analysis, leading to multiple sources of variation. One objective here is to determine the optimal number of samples needed at each of the stages.

X3.2.3.1 The precision of the estimate of the population mean concentration often depends on, for example, the number of field samples, the number of laboratory samples and the number of analyses performed on each of the laboratory samples. In this example, the sampling stages include field, laboratory and analysis, leading to sources of variation from field sample variation, laboratory sample variation and "analytical variation." Note that the "analytical variance" here may not be the true analytical variance, depending on whether the split samples analyzed are identical or not. If the laboratory sample is homogenous, leading to identical splits, then the "anatical variance" component reflects the true analytical variance. If the splits are not identical due to laboratory subsampling, then the Analytical variance "component" above actually includes analytical variance and laboratory subsampling variance.

X3.2.3.2 A simpler example is the one above but without replicate analyses, where each laboratory sample is analyzed in total. In this case, there are two sources of variation: field sample to field sample (or between sample) variation and analytical (or within sample) variation.

X3.2.3.3 In any case, a mathematical model can be laid out to describe the sampling and analytical plan, by which these sources of variation can be estimated (called variance components of the total variation). This is an area where a statistician needs to be consulted, regarding the optimal design of the plan and the proper estimation of the variance components.

X3.2.3.4 In the simplest case of only two sources of variation (between sample and within sample), the standard deviation of the mean concentration from n samples, each analyzed r times, is:

$$\sigma_m = \sqrt{[\sigma_b^2/n + \sigma_w^2/(nr)]} \quad (X3.16)$$

where:

σ_m = standard deviation of mean concentration,
σ_b^2 = between sample variance,
σ_w^2 = within sample variance,
n = number of samples, and
r = number of replicate analyses for each sample.

And, as a result, the total cost of sampling and analysis is:

$$C_t = n(C_b + rC_w) \quad (X3.17)$$

where:

C_b = cost of obtaining a sample, and
C_w = cost of one analysis of a sample.

X3.2.3.5 Given Eq X3.16 and X3.17 and estimates of σ_b^2 and σ_w^2, one way to determine n and r is as follows:

(1) Specify an acceptable value of σ_m.

(2) Then, the optimal number of analyses per sample is:

$$r = (\sigma_w/\sigma_b)\sqrt{(C_b/C_w)} \quad (X3.18)$$

(3) The value of r from Eq X3.18 is substituted into Eq X3.16 to obtain the optimal value of n.

NOTE X3.2—Note that the calculated n and r are rounded up to the nearest integer.

(4) This approach is quite useful as it takes into account all the important sources of variations to determine the optimal combination of numbers of field samples, laboratory samples, laboratory subsamples, and laboratory analyses, and so forth. And this is part of optimization of the sampling plan.

X3.2.4 *Search Sampling: For Detection of Hot Spots*—Localized areas of high concentration are commonly called hot spots. Some examples of determining the numbers of samples for the detection of hot spots are given in this section.

X3.2.4.1 *When Hot Spots are of Small Sizes*:

(1) When hot spots are of small sizes, the number of samples needed for detecting at least one of them can be determined so that if none is detected there is little likelihood that any hot spots exist.

(2) If we let $x_1, x_2, \ldots, x_n$ be a set of n random sample from the population and assume that they are independent, then one way is to find this number of samples n such that we have high confidence (with 100(1–α) % confidence) that at least one hot spot will be detected. If p is the probability of the existence of a hot spot and this p is constant for all hot spots, then the probability of finding at least one hot spot is:

$$1 - (1-p)^n = (1-\alpha) \quad (X3.19)$$

Therefore, the number of samples needed in this case is:

$$n = ln\alpha/ln(1-p) \quad (X3.20)$$

where α is suitably small (for example, 0.05 or 0.10). The number of samples n for some values of p and α is given in Table X3.1.

X3.2.4.2 *Grid Sampling for Detecting Hot Spot of Defined Size and Shape*:

(1) Systematic grid sampling methods have been used in detection of hot spots of defined size and shape. For a defined target population, the intersections of the grid lines determine the number of samples to be taken. In general, the smaller the grid distance (the distance between two adjacent intersections) is, the larger the number of samples is and the higher the chance of detecting a hot spot of a given size and shape.

(2) In general, triangular grid design seems to be superior to various patterns of square grid design, in terms of probability of detection.

(3) The following parameters need to be specified in order to determine the grid distance for detecting a circular or elliptical hot spot: *(a)* L, the length of semi-major axis of the hot spot (this defines the size of the hot spot); *(b)* S, ratio of the length of the short axis to the length of the long axis (this defines the shape of the hot spot); and *(c)* β, the acceptable probability of not finding the hot spot.

(4) A major axis is the long side of an ellipsoid. Semi-major axis is simply half of the major axis. In the case of a circular hot spot, the major axis equals the minor axis, which is the short side of an ellipsoid. The following discussions will focus on ellipitcal hot spot.

(5) Once these parameters have been specified, graphs[6] can be used to determine the grid distance G, which will leads to the numbers of samples needed.

(6) If the concentration defining a hot spot is higher than chemical method detection limit, then an alternative method is to define the size of the hot spot by the detection limit, instead

TABLE X3.1 Number of Samples *n* as a Function of *p* and α (Eq X3.18)

α	p	n
0.05	0.05	59
	0.1	29
	0.2	14
	0.3	9
	0.4	6
	0.5	5
	0.6	4
	0.7	3
	0.8	2
	0.9	2
0.10	0.05	45
	0.1	22
	0.2	10
	0.3	7
	0.4	5
	0.5	4
	0.6	3
	0.7	2
	0.8	2
	0.9	2

of the original cutoff concentration for a hot spot. Since the detection limit is lower than the original cutoff concentration, the size of the hot spot is now larger and therefore a smaller number of samples is required to detect with the same probability of detection.

(7) This is equivalent to re-specifying L, the length of the semi-major axis of the hot spot using the detection limit, instead of the original cut-off.

(8) Namely, if the hot spot concentration limit is c_h and it is higher than the method detection limit c_d, then, for the purpose of detection, we can define the boundary of the hot spot by c_d, instead of c_h.

(9) The next thing to do is to determine the semi-major axis L* based on c_d, instead of L based on c_h. If $c_h > c_d$ and there is a gradient of declining concentration from the center of the ellipsoid outward toward the edge of the hot spot, then two potential solutions for finding L^* are:

(10) Asssuming Change in Concentration Is Inversely Proportional to Change In Semi-Major Axis —This assumption means that, as the semi-major axis moves from L to L^* ($L^*>L$), the decrease in concentration is inversely proportional to the length of the semi-major axis. Namely,

$$L^*/L = c_h/(pc_h) \quad \text{(X3.21)}$$

where:

c_h = hot spot concentration limit at the outer edge of L, and
p = c_d/c_h, $0<p<1$.

Thus,

$$L^* = L/p \quad \text{(X3.22)}$$

For example. Let $L = 50'$ and $p = 0.1$ (that is, c_d is only 10 % c_h). Then

$$L^* = 500' \quad \text{(X3.23)}$$

L^* can then be used to determine the grid distance, leading to a smaller number of samples than when L is used.

(11) Asssuming Change in Log Concentration Is Inversely Proportional to Change In Semi-Major Axis —This is similar to above, except that the assumed relationship is a logarithmic one. Namely,

$$L^*/L = ln[c_h/(pc_h)] = ln(1/p) \quad \text{(X3.24)}$$

Thus,

$$L^* = L[ln(1/p)] \text{ for } 0 < p < 0.36 \quad \text{(X3.25)}$$

where $[1n\ (1/p)]$ is multiple factor of L^* relative to L (called here the multiplier effect). In order for this effect to be greater than 1, p needs to be smaller than 0.36. This effect as a function of p is given in Table X3.2.

(12) Other relationships can be postulated, in addition to those described in Eq X3.22 and X3.25. In place of detection limit, some other quantities such as $(c_h + c_d)/2$ can be considered as well.

TABLE X3.2 Multiplier Effect Between *L* and *L (Eq X3.25)**

p	Multiplier
0.3	1.20
0.1	2.30
0.05	3.00
0.01	4.61

(13) Since $L^*>L$, therefore the grid distance is increased, thereby decreasing the number of samples needed.

(14) The relationships between grid distance G and semi-major axis L are illustrated in Table X3.3.

(15) This method of reducing the number of samples by increasing the semi-major axis of the hot spot could lead to detection of larger number of "hot spots" based on c_d, which may not be hot spots at all based on c_h. Thus, this method is likely to be useful only when the target population is known to be relatively uncontaminated and the chance of false detection based on c_d is likely to be small.

(16) Once a hot spot based on c_d is identified, then either a field method or other sampling approaches can be used to further define the extent or boundary of the true hot spot of concern, namely the hot spot based on the original cut-off concentration c_h.

X3.2.4.3 *Multiple Hot Spots in Grid Sampling*:

(1) The grid distance determination in the section above is for detecting a single hot spot. When there are multiple hot spots, the grid distance can be determined for detecting at least one hot spot with a high degree on confidence. When none is detected, the probability of the existence of any hot spots can be expected to be very small. This is analogous to X3.2.4.1, except here the size and shape of the hot spots are considered in grid sampling.

(2) Again, in the determination of grid distance G, it requires specification of the following parameters:

L = length of semi-major axis of the smallest hot spot,
S = ratio of the short axis to the long axis of the hot spot, and
β = probability of not finding a hot spot of size L or larger when it actually exists.

(3) In the above definitions, β is the probability of not finding a hot spot. When there are multiple hot spots, under the simplifying assumption of equal size and shape, the probability of not finding any hot spots, for given β, is β^h, where h is the

TABLE X3.3 Approximate Relationships Between Grid Distance *G* and Semi-Major Axis *L* in Circular Hot Spots

L (ft)	β	G (ft)
25	0.1	45
	0.2	49
	0.3	53
	0.5	63
50	0.1	89
	0.2	98
	0.3	106
	0.5	125
100	0.1	180
	0.2	196
	0.3	213
	0.5	250
200	0.1	357
	0.2	392
	0.3	426
	0.5	500
300	0.1	536
	0.2	588
	0.3	638
	0.5	750

number of hot spots. As β is typically defined to be very small (0<β<1), β^h can become very close to zero as *h* increases, leading to very short and impractical grid spacing.

(4) One way to overcome this problem in the case of multiple hot spots is to set the probability of detecting at least one out of h hot spots to be the same as the probability of detecting a hot spot in a single hot spot case. From there, the corresponding β for determining the grid distance in the multiple hot spots case can be derived.

(5) Let it be given that $\beta=\beta_0$ for the single hot spot case. Then:

$$\text{Prob(detecting a hot spot } |h=1) = 1 - \beta_0$$

where *h* = number of hot spots.

(6) In the multiple hot spots case,

$$\text{Prob(detecting at least 1 hot spot } |h>1) = 1 - \beta^h$$

Let:

$$\text{Prob(detecting at least 1 hot spot } |h>1) = 1 - \beta_0 = 1 - \beta^h$$

Thus, the β to use to determine the grid distance for detecting at least one out of h hot spots is:

$$\beta^h = \beta_0 \qquad \text{(X3.26)}$$

$$\beta = (\beta_0)^{1/h}$$

Thus, once the acceptable probability of not detecting a single hot spot is specified to be β_0, the value of β for the multiple hot spots case can be determined by Eq X3.26. We can then use this new β to determine *G*. Since $\beta>\beta_0$, the new grid distance will be larger than the one based on β_0, leading to smaller number of samples needed.

(7) For example, the originally specified probability of not finding a hot spot, β_0, may be 0.1. If there are *h*=4 hot spots, the new β is 0.56 (= $0.1^{1/4}$). Because of this higher value of β, the grid distance *G* is increased accordingly.

(8) If this degree of increase in β is too steep, it can be modified, for example, by using $\sqrt{h}$ instead of *h*, in Eq X3.26.

(9) This approach seems reasonable in areas where no known contamination is present. For such areas, the sampling objective is to determine if some potential hot spots may be present. Thus, a sampling plan designed to detect, with high probability, at least one out of several potential hot spots seems to be a reasonable one. If the sampling plan fails to detect any, then the chance of presence of hot spots is small. On the other hand, if it does detect some hot spot(s), then more extensive sampling or field methods may be required to better define the areas of contamination.

(10) There is the issue of how the number of hot spots, *h*, is to be determined in the use of Eq X3.26. A reasonable approach is to assume that hot spots are associated with topographical depressions. In that case, the number of hot spots can be assumed to be the number of identifiable depressions.

X3.2.4.4 *Composite Sampling*—Composite sampling can be used in two broad areas. Both areas have potential of substantially reducing analytical costs. The two areas are: (*1*) Estimation of mean and variance with greater precision (in standard error of the mean). This is viewed as an estimation problem; and (*2*) detection of presence or exceedance in any of the samples in the composite. This is viewed as a classification problem.

NOTE X3.3—Since composite sampling has a wide variety of applications, a statistician can be consulted. This section describes some simple cases.

(1) Estimation Problem — A simple example of composite sampling is given below.

(2) Let *x* be a random variable representing the concentrations of independent and identical (same volume or weight) individual samples and *x* is normally distributed as *N* (μ, σ^2). Let y_k be the composite of *k* individual samples. Then it is known that y_k is an unbiased estimator of the population mean with greater precision because:

$$E(y_k) = \mu \qquad \text{(X3.27)}$$

$$\text{Var}(y_k) = \sigma^2/k$$

if the composite sample is totally homogeneous. The factor *k* is greater than 1 by definition of compositing. In case *m* (*m*>1) composite samples are used for assessing attainment of cleanup standard, then:

y_m = mean concentration of m composite samples, each a composite of *k* individual samples
= $\Sigma y_{k,i}/m$, i=1,2, . . .,*m*

Again, this mean of composite samples is unbiased with even greater precision than that observed by Eq X3.27, because:

$$E(y_m) = \mu \qquad \text{(X3.28)}$$

$$\text{Var}(\bar{y}_m) = \sigma^2/(km)$$

(3) If we have an estimate of the variance of the individual samples σ^2, then we can find combination(s) of *k* and *m* which achieves some desired level of the variance (or the standard deviation) of y_m. Some example reductions in the variance of the mean as a function of *m* and *k*, relative to the variance of individual samples, are given in Table X3.4.

(4) This kind of composite sampling can be applied to many other situations.

TABLE X3.4 Percent Reduction in Standard Deviation of Mean Concentration Estimate, Assuming Total Homogeneity in Composite (Eq X3.28)

m	*k*	*mk*	% Reduction
1	1	1	0
	2	2	29
	3	3	42
	4	4	50
	5	5	55
	10	10	68
2	1	2	29
	2	4	50
	3	6	59
	4	8	65
	5	10	68
	10	20	78
3	1	3	42
	2	6	59
	3	9	67
	4	12	71
	5	15	74
	10	30	82

(5) In estimation, it is to be noted that, if only a subsample of the composite is analyzed chemically, then proper mixing to achieve homogeneity in the composite sample is important. In case when the composite sample is not totally or nearly homogeneous, then subsampling would introduce an additional source of variation, which may be called the heterogeneity variance σ_h^2. When σ_h^2 is not nearly zero, Eq X3.27 becomes:

$$\text{Var}\ (y_k) = \sigma^2/k + \sigma_h^2 \quad \text{(X3.29)}$$

and Eq X3.28 becomes:

$$\text{Var}\ (\bar{y}_m) = [\sigma^2/k + \sigma_h^2]/m \quad \text{(X3.30)}$$

(6) All the correct sampling and subsampling procedures must be followed to prevent biased results and assure reduction in sampling variance.

(7) On some occasions when analytical costs are high (relative to sampling costs), one composite sample may be used for inference purposes, especially when it is a composite of many individual samples. Since it is a composite of many samples, the concentration of the composite can be expected to have a much lower standard error than the individual samples. When this standard error is judged to be small, the concentration of the composite can be used for comparison against a standard or limit. This is especially appropriate when there is a pilot study to provide the relationship between the variances of the individual samples and the composite samples.

(8) Regarding questions such as the optimal number of individual samples needed in a composite, the number of composites, and subsamples of the composite and the like, a statistician needs to be consulted.

(9) *Classification Problem* — Some discussions on classification are given in Principles of Environmental Sampling.[9]

(10) Let $x_1, \ldots, x_n$ be n individual samples and y_k ($k \leq n$) be a composite sample of k individual samples. Classification into two categories (for example, presence or absence of a contaminant) is to be made on each x based on analysis of y_k first, and perhaps some or all of the x's if y_k is positive.

(11) Because compositing may dilute the concentration to a level below detection limit, the number k can be determined to ensure that this does not happen.

(12) Assume that one individual sample has concentration C and all the other (k–1) samples have concentration zero. Then the composite concentration is:

$$y_k = C/k \quad \text{(X3.31)}$$

Given that the detection limit is d, we want a number k such that

$$C/k > d \quad \text{(X3.32)}$$

Namely,

$$k < C/d \quad \text{(X3.33)}$$

(13) If regulatory limit L is the concern, instead of d, then Eq X3.30 can be modified to to find the right k:

$$k < L/d \quad \text{(X3.34)}$$

(14) In the case of using Eq X3.34, if $y_k < L$, then $x_i < L$, i=1, . . ., k. Namely, it can be concluded that all the individual samples are less than the regulatory limit L. And only one analysis of the composite sample is required to reach this conclusion.

(15) If $y_k \geq L$, then the simplest scheme is to analyze all the individual samples to determine if any exceeds the limit. And the total number of analysis is (k+1).

(16) Let q = probability ($y_k < L$), then the expected number of analyses under compositing is:

$$E(N) = q + (1-q)\ (k+1) \quad \text{(X3.35)}$$

(17) Thus, the cost saving potential of composite sampling can be measured by the following relative efficiency index (REI), which is the ratio of the expected number of analyses under composite sampling to the number of analyses for individual samples:

$$\text{REI} = E(N)/k = 1 - q + 1/k \quad \text{(X3.36)}$$
$$= p + 1/k$$

where:

p = $1 - q$
= probability of exceedance.

(18) Savings from compositing is realized when p is small and only when REI<1, which translates into:

$$\text{REI} < 1 \quad \text{(X3.37)}$$
$$1 - q + 1/k < 1$$
$$q > 1/k$$

Or,

$$0 < p < 1 - 1/k \quad \text{(X3.38)}$$

[9] Keith, L. H., *Principles of Environmental Sampling*, L. H. Keith, ed., American Chemical Society, 1988.

(19) The example percent reductions in the total number of analyses as a function of k and p in Eq X3.36 are given in Table X3.5.

TABLE X3.5 Relative Efficiency of Composite Sampling versus Individual Tests (Eq X3.36)

k	REI<1	p	REI	% Savings Relative to Individual Tests
2	p<0.50	0.4	0.9	10
		0.3	0.8	20
		0.2	0.7	30
		0.1	0.6	40
		0.05	0.55	45
		0.01	0.51	49
3	p<0.67	0.4	0.73	27
		0.3	0.63	37
		0.2	0.53	47
		0.1	0.43	59
		0.05	0.38	62
		0.01	0.34	66
4	p<0.75	0.4	0.65	35
		0.3	0.55	45
		0.2	0.45	55
		0.1	0.35	65
		0.05	0.30	70
		0.01	0.26	74
5	p<0.80	0.4	0.6	40
		0.3	0.5	50
		0.2	0.4	60
		0.1	0.3	70
		0.05	0.25	75
		0.01	0.21	79
10	p<0.90	0.4	0.5	50
		0.3	0.4	60
		0.2	0.3	70
		0.1	0.2	80
		0.05	0.15	85
		0.01	0.11	89

Designation: D 6323 – 98

Standard Guide for Laboratory Subsampling of Media Related to Waste Management Activities[1]

This standard is issued under the fixed designation D 6323; the number immediately following the designation indicates the year of original adoption or, in the case of revision, the year of last revision. A number in parentheses indicates the year of last reapproval. A superscript epsilon (ε) indicates an editorial change since the last revision or reapproval.

1. Scope

1.1 This guide covers common techniques for obtaining representative subsamples from a sample received at a laboratory for analysis. These samples may include solids, sludges, liquids, or multilayered liquids (with or without solids).

1.2 The procedures and techniques discussed in this guide depend upon the sample matrix, the type of sample preparation and analysis performed, the characteristic(s) of interest, and the project specific instructions or data quality objectives.

1.3 This guide includes several sample homogenization techniques, including mixing and grinding, as well as information on how to obtain a specimen or split laboratory samples.

1.4 This guide does not apply to air or gas sampling.

1.5 The values stated in SI units are to be regarded as the standard.

1.6 *This standard does not purport to address all of the safety concerns, if any, associated with its use. It is the responsibility of the user of this standard to establish appropriate safety and health practices and determine the applicability of regulatory limitations prior to use.*

2. Referenced Documents

2.1 *ASTM Standards:*

C 702 Practice for Reducing Samples of Aggregate to Testing Size[2]

C 859 Terminology Relating to Nuclear Materials[3]

D 346 Practice for the Collection and Preparation of Coke Samples for Laboratory Analysis[4]

D 2234 Test Methods for Collection of a Gross Sample of Coal[5]

D 4547 Guide for Sampling Waste and Soils for Volatile Organic Compounds[5]

D 4823 Guide for Core Sampling Submerged Unconsolidated Sediments[5]

D 5743 Practice for Sampling Single or Multilayered Liquids, With or Without Solids, in Drums or Similar Containers[6]

D 5792 Practice for Generation of Environmental Data Related to Waste Management Activities: Development of Data Quality Objectives[6]

D 5956 Guide for Sampling Strategies for Hetergeneous Wastes[6]

D 6051 Guide for Composite Sampling and Field Subsampling for Environmental Waste Management Activities[6]

3. Terminology

3.1 *contaminant unit, n*—the largest particle size that contains the contaminant of interest

3.1.1 *Discussion*—The contaminant of concern, as defined by the project objectives, may be associated with all the particle sizes or associated with only a certain particle size or sizes. At the time of waste generation, discharge or spill, the particle size of this contaminant of concern may be on the atomic or molecular scale, such as solvent spill into sand, or a macro scale, such as lead acid batteries at a dump site. The contaminant unit may also be in-between these scales, such as lead particles encapsulated in coal. In practice, the contaminant unit may change if the contaminant unit becomes absorbed or adsorbed to particles larger than the contaminant unit. It is the size of the contaminant unit at the time of subsampling, not at the time of generation, that is referred to as the contaminant unit.

3.2 *maximum allowable particle size, n*—the largest lineal dimension of a sample's individual particles accepted for a given sample mass.

3.2.1 *Discussion*—The maximum allowable particle size is sometimes referred to as the allowable particle size. A simple method of measurement is a sieve.

3.3 *multilayered sample, n*—a sample consisting of two or more clearly differentiated components.

3.3.1 *Discussion*—Multilayered samples are those with two or more distinct visual layers of material. These layers may be the result of differences in density, such as liquid/liquid layers (for example, chlorinated solvents and water, water and oil), liquid/solid layers (for example, sludge), solid/solid layers (for example, small rocks and large rocks), or combinations of these layers (for example water, oil, and soil). These layers

---

[1] This guide is under the jurisdiction of ASTM Committee D34 on Waste Management and is the direct responsibility of Subcommittee D34.01.01 on Planning for Sampling.

Current edition approved Sept. 10, 1998. Published December 1998.

[2] *Annual Book of ASTM Standards*, Vol 04.02.

[3] *Annual Book of ASTM Standards*, Vol 12.01.

[4] *Annual Book of ASTM Standards*, Vol 05.05.

[5] *Annual Book of ASTM Standards*, Vol 11.02.

[6] *Annual Book of ASTM Standards*, Vol 11.04.

may also be the result of depositional layering, such as green clay and silty sand from a coring sample.

3.4 *particle size, n*—the controlling lineal dimension of individual particles (see Terminology C 859).

3.5 *representative subsample, n*—a subsample collected in such a manner that it reflects one or more characteristics of interest (as defined by the project objectives) of the laboratory sample from which it was collected.

3.5.1 *Discussion*—A representative subsample can apply to a single sample, or a composite sample.

3.6 *sample, n*—a portion of material taken from a larger quantity for the purpose of estimating properties or composition of the larger quantity, (see Guide D 6051).

3.7 *sludge, n*—Any mixture of solids that settles out of solution. Sludges contain liquids that are not apparent as free liquids, (see Practice D 5743).

3.8 *subsample, n*—a portion of a sample taken for the purpose of estimating properties or composition of the whole sample, (see Guide D 6051).

3.8.1 *Discussion*—A subsample is also commonly referred to by such terms as specimen, aliquot, subsample, or analytical subsample. To avoid confusion, the term specimen is used to refer to the final portion of the sample used during analysis. The material received at the laboratory will be referred to as the laboratory sample. All portions subsampled in-between receipt at the laboratory and the specimen are referred to as subsamples. The subsample may be the specimen.

4. Significance and Use

4.1 This guide discusses options for taking a subsample from a sample submitted to a laboratory. If followed, it will minimize the bias and variance of the characteristic of interest of the laboratory sample prior to analysis.

4.2 The guide will describe appropriate instructions to be submitted to the laboratory with the field sample.

4.3 This guide is intended for use in the laboratory to take a representative subsample or specimen of the whole field sample for direct analysis or sample preparation for analysis. It is intended for field personnel, data users, laboratory sample reception personnel, analysts, and managers.

4.4 To obtain a representative subsample, layer analysis, grinding, mixing, and changing the physical state such as digesting, drying, melting or freezing may be required. This guide considers cone and quartering, riffle splitting, and particle size reduction.

5. General Considerations

5.1 Successful implementation of this standard depends on effective communication between the data user and the laboratory staff. The selection of optimal subsampling procedures, techniques, and strategy by the laboratory depends on the intended use of the data. The data user should submit appropriate instructions with all samples and, when necessary, the laboratory staff should contact the data user for confirmation or further clarification of these instructions.

5.1.1 The appropriate instructions must be reviewed by both the laboratory receiving personnel and the analyst(s) or supervisor. If there are no instructions, the appropriate laboratory personnel should contact the data user. Options should be discussed and clarified prior to initiating any subsampling procedure. These instructions may include such options as those found in Table 1. The limitations and advantages of these methods are also found in this table. The data user should be informed about the limitations and advantages of all subsampling procedures prior to deciding which one to use.

5.1.2 If the data user still provides no instructions upon being contacted, laboratory personnel should explain to the data user that the laboratory's standard operating procedures, which reflect the concerns and issues discussed in this standard, will be used. The sample should be treated as if the scale of the contamination is on the micro level, and no artifacts can be removed. Since sample matrices and types and mechanisms of contamination are infinitely variable and require judgments to be made, it is advisable that experienced analysts decide which subsampling techniques be employed. These procedures must be discussed and clarified with the data user prior to initiating any subsampling procedure.

5.2 If the sample integrity or composition is not as anticipated, the data user must be contacted to confirm or clarify the instructions. An example of when this would be necessary would be a case where a coring sleeve was received at the laboratory. On opening the container, the analyst notices clay in one end of the sleeve, and sand at the other end. Before the analyst can proceed, the appropriate instructions from the data user must be obtained.

5.2.1 Field samples should be collected in appropriate containers for the analyses requested. If the submitted sample is improperly collected, the data user should be contacted by laboratory personnel. If the data user authorizes the laboratory to continue with the analysis, a note should be made in the receival documentation, and also in the case narrative in the final report.

5.3 Documentation during the subsampling process is critical. Since subsampling techniques may bias the results, the subsampling method used must be noted in the analytical logbook.

5.3.1 Anytime the analytical result will be biased, it must be documented, and the data user should be notified prior to beginning any subsampling technique. For example, if headspace exists in a container arriving at the laboratory, some volatile components will have partitioned into that headspace. However, if the data user decides to proceed with the analysis, the analytical logbook and the case narrative on the final report should indicate this condition.

5.4 Particle size is the physical dimension of an object's pieces or parts. The maximum particle size contained within a laboratory sample is the largest of these pieces. The contaminant of concern, as defined by the project objectives, may be associated with all particle sizes or associated with only a certain particle size or sizes. The largest of these particle sizes, that contain the contaminant of interest, would be the contaminant unit. The contaminant unit, at the time of waste generation, discharge or spill, may be on the atomic or molecular scale, such as a solvent spill into sand, or a macro scale, such as lead acid batteries at a dump site. The contaminant unit may also be in-between these scales, such as lead particles encapsulated in coal. In practice, the contaminant unit may change if

TABLE 1 Limitations and Advantages of Sample Preparation Options

Instruction	Limitations	Advantages
Remove artifacts, such as rocks and twigs, from the sample prior to subsampling	(*1*) May bias analytical results by altering contaminant concentration, (*2*) May bias sample if results are not properly weight averaged.	(*1*) May be easier to subsample, (*2*) May be easier to analyze, (*3*) Appropriate if the target population is material minus artifacts.
Digest or extract the contaminant from the outside of the large particles	May bias sample if contaminant is within the large particles.[A]	(*1*) May be easier to analyze, (*2*) May prevent need for weight average calculation.
Digest or extract particle sizes separately	(*1*) Separation of particle sizes may be difficult,[A] (*2*) May bias sample if results are not properly weight averaged, (*3*) Higher cost.	(*1*) Allows some particle size consistency during analysis, (*2*) May be easier to subsample within portions after separation.
Form an emulsion layer so that the material may be treated as homogeneous liquid	May bias the sample if a complete emulsion is not achieved.[A]	(*1*) May be easier to subsample as a homogeneous liquid.
Separate liquid layers	(*1*) Separation of layers may be difficult, especially at the interface, (*2*) May bias sample if results are not properly weight averaged.	(*1*) May be easier to analyze, (*2*) May be easier to subsample within portions after separation, (*3*) Allows different preparation methods within each layer.
Dry sample	May alter chemistry or change stability of some compounds.[A]	(*1*) Allows for consistency of subsampling for liquid/solid mixtures, (*2*) Analytes reported unbiased by moisture content.
Change the physical state, such as freeze the material so that it may be treated as a solid, or melt the material so that it may be treated as a liquid	May be difficult to achieve complete freezing or melting and maintain it long enough to get a subsample.[A]	Allows for consistency of subsampling.
Analyze only one layer of multilayered samples, such as analyze only the oil portion of an oil/water mixture for PCBs	May bias the sample if there is cross-contamination between layers	(*1*) Possible cost savings to customer, (*2*) May be easier to subsample from a single layer, (*3*) May be easier to analyze.
Composite portions of the sample for volatile analysis directly in a purge unit vs. individual analysis of these portions	(*1*) May overload gas chromatography columns if sample has high amounts of solvents in each portion, (*2*) Separation of portions may be difficult.	(*1*) May prevent losses of volatile because the sample is handled only once, (*2*) Possible cost savings to customer, (*3*) May prevent need for weight average calculation.
Reduce particle size	(*1*) Increasing surface area may effect data in some procedure with particle digestion or extraction, such as TCLP,[A] (*2*) May be difficult, depending on matrix.	Allows for consistency of subsampling.
Use standard methods for solids (for example, cone and quarter, grind, riffle, sieve)	Dependent on method.[A] See 7.1.5 for more information.	Dependent on method. See 7.1.5 for more information.

[A] May be unsuitable for samples to be analyzed for volatile constituents.

the contaminant becomes absorbed or adsorbed to particles larger than the contaminant unit.

5.4.1 Knowledge of the contaminant unit may be used to determine the preliminary steps to subsampling. For example, if the contaminant unit is on a molecular scale and was adsorbed to soil particles and rocks, removal of large rocks with their relatively small surface area may not affect the data as long as the results are weight averaged. If one is unsure of the mechanism of contamination, that determines how the contamination is dispersed within the sample matrix, one may not be able to discard any particles during subsampling.

5.5 Sampling theory requires that subsample *mass* should increase as the size of the largest particle in the sample increases. If the subsample mass recommended by sampling theory is larger than that normally used in the sample preparation method, the subsample mass may be increased and the extraction/digestion procedure scaled accordingly. The standard volume of digestate/extract is then submitted for analysis. However, if the subsample mass is too large to be accommodated by the sample preparation procedure, multiple subsamples (of equal mass) can be extracted/digested, and the extracts/digests combined and mixed prior to removal of the standard volume specimen needed for analysis. Another alternative is to reduce the particle size of the entire sample or subsample as specified by sampling theory. If the particle size is reduced enough, a subsample of the mass recommended by sampling theory and the extraction/digestion method can be obtained.

5.6 All subsampling should be performed in an area which is free from contamination, easily decontaminated, and vented to control dust and remove fumes.

5.7 Prior to subsampling liquids, the analyst must consider the property or characteristic requested, and the container size received. If the analyte has the ability to adsorb onto the container, the field sample should arrive at the laboratory in an appropriately sized bottle, such that the whole sample will be used. The sides of the container should then be rinsed properly to assure that all the contaminants are transferred into the analytical vessel.

5.8 Subsampling techniques are different when analyzing for volatile compounds than non-volatile compounds. The differences are discussed for each sample matrix.

6. Matrix Specific Subsampling

6.1 *Solids*:

6.1.1 Table 2 gives the relationship between the maximum

TABLE 2 Recommended Minimum Subsample Mass for Particulate Materials to Achieve a Fundamental Error of 15 %

Recommended Minimum Subsample Mass, g	Maximum Allowable Particle Size[A], cm	Size Class[B]	U.S. Standard Sieve Mesh Size
0.1	0.05	Clay to coarse sand	35
1	0.1	Coarse to very coarse sand	18
2	0.13	Very coarse sand	13
5	0.17	Very coarse sand	12
10	0.21	Granule gravel	10
30	0.31	Granule gravel	7
50	0.37	Granule gravel	6
100	0.46	Pebble gravel	5

[A]The maximum allowable particle size allowed can be approximated for other sample masses by using the equation, allowable particle size (cm) = the cube root of 0.001 × sample mass in grams

[B]Wentworth Size Class (See **(3)** and **(4)** These soil descriptions are added for those readers who equate particle size with the Wentworth descriptions and are not meant to indicate that the application of this table is limited to soil particles. The table can also be used for non-soil particles such as waste.

particle size and sample mass to achieve a fundamental error of 15 %. For information about fundamental error, see Ref **(1)**[7]. If the maximum allowable particle size is greater than that listed in the table, then sampling theory would suggest particle size reduction, such as grinding, or use of a larger sample mass. For example, if granule gravel has an allowable particle size of 0.21 cm, according to this table, a subsample mass of 10 g is needed. Many digestions/extractions commonly use 1 g. The 10 g subsample would be ground to a size of 0.1 cm, and the specimen would be taken from the ground material. Approximations were employed to construct this table. The actual fundamental error could be much larger. More information on calculating the maximum allowable particle size for various fundamental errors, and particles shapes is included in Annex A1.

6.1.1.1 Particle size reduction can be avoided by modifying the sample digestion/extraction methods to use large specimen masses. Following digestion/extraction of the large specimen, the homogeneous digestate/extract is analyzed. For example, if the contaminant of concern was a chromium solution spilled into a mixture of gravel and soil, knowing that the chromium coats the medium allows a choice of sample preparation. The analyst may totally dissolve the specimen. He/she may digest the chromium from the outside of the particles, analyzing the digestate. Or, he/she may extract the material, and analyze the extractant. If the analyst had no information about the mechanism of contamination, he/she would have to assume that the contamination is throughout all particles of the laboratory sample.

6.1.1.2 If information is available regarding the type and mechanism of contamination, then it may be possible to avoid particle size reduction and the use of large specimen masses. For example, if contamination occurred on a molecular level (for example, a solvent is discharged to soil), an argument can be made to preclude large particles (for example, rocks), since insignificant amounts of contamination will be adsorbed to the small surface of the large particles. If the large particles are excluded, the appropriate mass can be based on the smaller particle sizes, which have a much larger surface area. If the decision is made to remove the large particles, the mass of both the large and small particles would be used to calculate the final concentration of the contaminant of interest.

6.1.2 If larger sample masses are required, two ASTM Standards may be helpful. Test Method D 2234 discusses test methods and procedures for the collection of a sample under various conditions of sampling. It describes general and special purpose sampling procedures for coal by size, condition, and characteristics. The document also includes in the annex, a test method for determining the variance of components of a coal and a test method for estimating the overall variance for increments of one fixed weight of a given coal.

6.1.2.1 A second method which is often referenced by other ASTM standards dealing with coals is Practice D 346. The practice is designed to provide a representative sample of the coke from which it is collected. It considers the variability of coke and the wide variety of sampling equipment.

6.1.3 If the desired subsample mass is equal to or less than the maximum allowable particle size, subsampling can proceed without any need for particle size reduction. If the particle size is greater than the allowable particle size in Table 2, then the sample matrix, the type and mechanism of contamination, the scale of the contaminant unit, as well as particle size reduction should be considered. These allowable particle sizes are based on the assumption that the contaminant exists as a particle or is adsorbed to particles. In those cases where the contaminant unit is on the atomic or molecular scale or is a particle much smaller than the allowable particle size of the sample, then alternative approaches can be employed with acceptable bias.

6.1.4 *Solid Samples for Volatile Analysis*:

6.1.4.1 Due to the volatility of this class of compounds, laboratory subsampling procedures addressing the sample's allowable particle size could result in large losses of the parameter of interest, rendering the results unrepresentative. The possible exception is when cryogenic freezing conditions are maintained throughout the handling process.

NOTE 1—If cryogenic techniques (working with frozen samples) or other special techniques are used, the procedure should be validated on laboratory control samples (LCS) which have been spiked with known concentrates of the parameter of interest. The nature of the test matrix for the LCS and the spiking solution should be chosen to allow uniform distribution of the spike in the matrix prior to subsampling. Regardless of the particle size reduction techniques chosen, it is advisable to carry an LCS through each preparation batch as a quality control sample. A subsample of the field sample without particle size reduction through the validation procedure will demonstrate the amount of volatile loss (from the LCS results) and increased volatiles recovery (from a comparison of the unreduced field sample with the size reduced sample) using particle size reduction.)

6.1.4.2 When analyzing samples for low levels of volatile constituents (0.5 to 200 µg/kg range), disctete samples should be placed into a container that, once sealed, is analyzed without opening (see Guide D 4547). For this reason, and because sample mass is often limited by the analysis system, low level samples typically are not amenable to either particle size reduction procedures or the analysis of larger sample mass.

[7] The boldface numbers given in parentheses refer to a list of references at the end of the text.

When contaminant levels exceed 200 µg/kg, discrete samples can be transferred to vessels containing an appropriate solvent, such as methanol. Because the solvent retains the volatiles in solution, and only a small portion of the solvent is removed for analysis, particle and contaminant unit size with respect to subsample mass requirements can be considered.

NOTE 2—Current analytical guidance recommends solvent extraction only for volatile organic concentrations at greater than 200 µg/kg. However, solvent extraction may be used for solvents of lower concentration levels if the solvent/solid ratio is reduced, or by using an analysis system with lower detection limits.

6.1.4.3 The following subsections discuss different subsampling procedures that can be used as particle size and contaminant unit vary in relationship to the allowable particle sizes listed in 6.1.1.

6.1.5 *Laboratory Sample Particle Size and Contaminant Unit Less Than the Allowable Particle Size*—Prepare sample vessel with the appropriate solvent and solvent volume as specified in the method. Add the appropriate mass of sample to the prepared vessel with a pre-cleaned coring total (see Guide D 4547) or spatula as quickly as possible. The coring tool must be large enough to accommodate the largest sample matrix particle. After volatile extraction, remove the appropriate volume of the solvent for analysis from the liquid portion of the specimen.

6.1.6 *Laboratory Sample Particle Size Greater Than the Allowable Particle Size and the Contaminant Unit Less Than the Allowable Particle Size*—Particle size reduction cannot generally be used, since it may result in a loss of volatile contaminants. There are cases where the chemistry of the contaminant and the support will not allow the volatile to be lost. For these types of contaminants, see the methods in 6.1.5. Alternate approaches to subsampling may be viable if knowledge exists regarding the sample matrix, type and mechanism of contamination, and the contaminant unit.

6.1.6.1 If the contaminant unit is on a molecular scale, then the distribution of the contaminant will be a function of the absorptive properties, and the larger surface area of the smaller particles results in more contaminants being adsorbed to the smaller particles. If the data user is willing to accept a biased-high concentration for the contaminant of interest, a subsample of only the smaller particles may be made.

6.1.6.2 Another method is to use a solvent appropriate for analysis, such as methanol, to extract the material. Use a 4 g specimen of the original laboratory sample in 10 mL of solvent. Increase the volumes proportionally to accommodate the specified sample mass. An appropriate portion of the solvent is analyzed once the contaminant is dissolved in the solvent.

6.1.7 *Laboratory Sample Particle Size and Contaminant Unit Larger than the Allowable Particle Size*—When sample masses become prohibitive with regard to logistical constraints, sample extracts can be compositied. An example of this is the case where soil of particle size 0.46 cm is contaminated with volatile organic compounds. The suggested subsample mass in this case from Table 2 is 100 g, therefore requiring a vessel with 250 mL of solvent. As an alternate to a single large sample, five 20 g subsamples could be extracted. Microliter amounts of each of the five resulting extracts could be composited into the same purge chamber. Once the contaminants are dissolved in the solvent, the concern of contaminant heterogeneity is decreased and the use of small subsamples of the extracts can be justified.

6.1.8 *Solid Sample for Non-Volatile Analysis*—The method of subsampling is dependent on the size of the allowable particle size and the size of the contaminant unit. The following subsections discuss different subsampling procedures that can be used as laboratory sample particle size and contaminant unit vary in relationship to the allowable particle sizes listed in Table 2. If the laboratory sample requires mixing prior to subsampling, the methods in section 6.1.6 should be used.

6.1.9 *Laboratory Sample Particle Size and Contaminant Unit Less Than the Allowable Particle Size*:

6.1.9.1 *Transversal Subsampling (Rectangular Scoop) (see Fig. 1)*—One method of subsampling the materials is to use transversal subsampling (see Guide D 6051). The entire laboratory sample is emptied onto a non-contaminating smooth surface. The sample is shaped into an elongated pile with a flattened top surface (**1**). Complete top-to-bottom transversal cuts are made across the pile and the extracted material is transferred into a tared container. The transversal cuts are repeated until; the appropriate mass is obtained. Cuts are made with a rectangular scoop. The width of the transversal section should be wide enough to accommodate the largest particle size, but narrow enough to require multiple transversal sections be used to obtain the appropriate mass. Two thin sheets of Teflon, metal, or polyethylene may be used to facilitate the removal of material within the transversal cut. After the removal of a few transversal sections, the pile may be reshaped to facilitate further cuts. The entire unused portion of the laboratory sample should be returned to the original sample container.

6.1.9.2 *Cone and Quartering (see Fig. 2)*—Another method of subsampling materials involves cone and quartering (**2**). This method may introduce large biases from the grouping and segregation error. It is discussed here due to it's widespread use, but the user should take note of its limitations (see Table

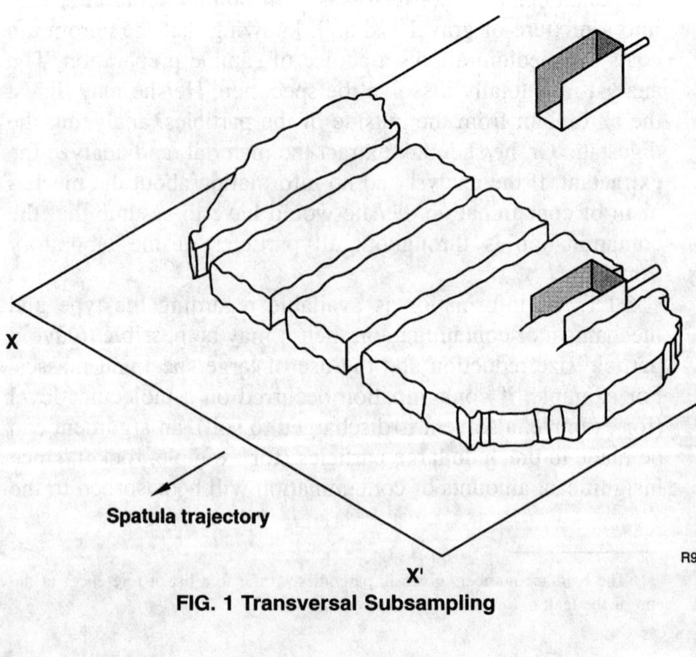

FIG. 1 Transversal Subsampling

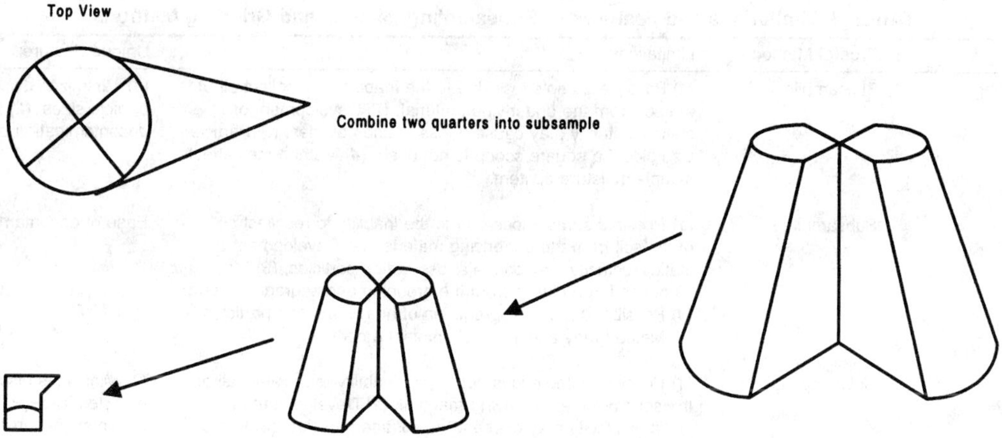

FIG. 2 Cone and Quartering

TABLE 3 Commonly Used Particle Size Reduction Devices for Solids

Device	Manufacturer's Recommended Particle Size or Sample Mass	Reduced to	Uses of
Cutting Mill	50 mm or less	Almost any size depending on mill	Foliage, grains, soils
Micro-Mill	20 to 50 mg	Almost any size depending on mill	Can be temperature controlled
Centrifugal cutting mill	3 to 100 g	40 μm	Chemicals, coal, seeds, plant stems, polypropylene or polyester.
Jar mill	pint to 2 gal	Almost any size depending on material and time	Can be used wet or dry
Horizontal and Vertical Disc Mills	5 to 20 mm particle size	250 to 1000 μm	Chemicals, clay, foods, plants, leather, cardboard
Dish and puck mill	10 g to 5 kg	10 to 100 μm	Brick, cement, chemicals, coal, coke, glass, minerals, ore, plants, rock, metals, sand, slag, soil
Mortar grinders	1 to 8 mm particle size	5 μm to 8 mm	Can be used wet or dry, organic or inorganic substances
Jar Crushers	1 to 100 mm particle size	1 to 40 mm	Medium hard, brittle, and tough materials, for example, chemicals, coke, concrete, glass, coal, gravel, quartz, slag.

4). The entire laboratory sample is emptied out onto a non-contaminating smooth surface. Material is piled into a cone with a flattened top surface. Two top-to-bottom cuts are made through the cone at perpendicular angles to form four equal portions, or quarters. Two opposite quarters are compiled into a new cone, and the process is repeated until the proper sample mass is achieved. See C 702 for more information.

6.1.9.3 *Riffle Splitter (see Fig. 3)*—A riffle splitter is a mechanical device with an equal number of narrow sloping chutes (**2**). Alternate chutes discharge the sample into opposite directions into two collection pans. Subsampling is achieved by pouring the entire laboratory sample into the riffle splitter. One portion is removed, and the collection pan is replaced with a clean pan. This portion is poured back through the riffle splitter so that half of the subsample will combine with the original material, and the other half will drop into the clean collection pan. This process is repeated until the proper sample mass is achieved. There are many types of riffle splitters available. The analyst must select the one that meets the data user's needs. The riffle splitter opening should be about three times the largest particle diameter, and have an even number of chutes.

6.1.9.4 *Alternate Scoop*—This method is discussed in Guide 6051. Scoops of the mixed laboratory sample are either placed in the analytical vessel, or discarded. In the example in Fig. 4, three scoops are discarded for every scoop saved. Care should be taken that each scoop of material is of the same size and is taken in a consistent manner to minimize bias.

6.1.9.5 *Sectorial Splitter (Sample Divider)—(See Fig. 5)*—This is a spinning riffler-type sample divider. The sample is placed in the opening, and the material is vibrated through shoots to the spinning bottles. The specimen may be taken directly from the bottle, or it may be further subsampled by discarding every other bottle, placing the remaining sample back in the opening, and starting the process over again.

6.1.10 *Laboratory Sample Particle Size Greater Than the Allowable Particle Size and the Contaminant Unit Less Than the Allowable Particle Size*:

TABLE 4 Limitations and Features of Subsampling, Mixing, and Grinding Methods

Method	Type of Method	Limitations	Unique Features
Transversal Subsampling	Subsampling	(*1*) Possible sample loss due to the inability to recollect all of the soil from the underlying material, (*2*) Development of static electricity may cause a loss of fine particles, (*3*) Particle size bias if a square scoop is not used, (*4*) Method may affect sample moisture content.	(*1*) Minimizes discrimination between particle sizes, (*2*) Ease of decontamination of equipment.
Cone and Quartering	Subsampling	(*1*) Possible sample loss due to the inability to recollect all of the soil from the underlying material, (*2*) Development of static electricity may cause a loss of fine particles, (*3*) May introduce large biases from the grouping and segregation error, (*4*) Possible unequal segregation of heavy and fine particles, (*5*) Method may affect sample moisture content.	Ease of decontamination of equipment.
Alternate Scoop	Subsampling	(*1*) Possible sample loss due to the inability to recollect all of the soil from the underlying material, (*2*) Development of static electricity may cause a loss of fine particles, (*3*) Particle size bias if a square scoop is not used, (*4*) Bias may occur if each scoop is not of the same size or if not taken in a consistent manner, (*5*) Method may affect sample moisture content.	(*1*) Minimizes discrimination between particle sizes, (*2*) Ease of decontamination of equipment, (*3*) Simpler than Transversal Subsampling, since there is no need to shape the pile.
Sectorial Splitter (Sample Divider)	Subsampling	(*1*) Limited to use with small particle sizes (about 6 mm), (*2*) May be difficult to decontaminate, (*3*) Method may affect sample moisture content.	(*1*) Minimizes discrimination between particle sizes, (*2*) Proper use eliminates most biases, (*3*) Simple to use, mechanical device.
Riffle Splitting	Subsampling or Mixing	(*1*) Possible loss of fine particles due to dusting when the sample is introduced, (*2*) May introduce biases from the grouping and segregation error, (*3*) Some riffle splitters are difficult to decontaminate.	(*1*) Proper use eliminates most biases, (*2*) Many choices in size available from manufacturers, including closed or open systems.
Grinding and Sieving/Mills/Jaw Crushers	Grinding	(*1*) Development of static electricity may cause a loss of fine particles, (*2*) Possible loss of fine particles due to dusting when the sample is introduced into the grinder, or the sieve, (*3*) Increasing the surface-to-mass ratio, which may effect the leachability during partial digestion or extraction, (*4*) Grinding equipment may be difficult to decontaminate, (*5*) Grinding and sieving may be labor intensive.	(*1*) Simple method to reduce particle size, (*2*) Many choices in size and type available from manufacturers.
Mixing Squares	Mixing	(*1*) Mixing sheet must not be made of material that will contaminate samples, (*2*) Development of static electricity may cause a loss of fine particles.	(*1*) Non-mechanical method of mixing, (*2*) Cost efficient, (*3*) Ease of decontamination of equipment.
Mechanical Mixers	Mixing	(*1*) Possible segregation of heavy versus light, or large versus small particles, (*2*) Some mixers are difficult to decontaminate.	Simple method to evenly distribute particles in solids that are difficult to mix by hand.

6.1.10.1 If the data user can accommodate worst-case, such as biased-high concentration of the parameter of interest, then subsampling can be performed without the need for large sample masses or particle size reduction. The procedure described in 6.1.11.1 will work, except that the particles larger than the allowable particle size are removed from the transversal section prior to addition to the subsample.

6.1.10.2 If the data user cannot accommodate worst-case, refer to 6.1.11.

6.1.11 *Laboratory Sample Particle Size and Contaminant Unit Larger than the Allowable Particle Size*:

6.1.11.1 Since the contaminant is associated with the large particles in this type of laboratory sample, the option of selectively removing large particles is not viable. Optimum solutions would be to increase the sample mass, or reduce the particle size. If these methods are not practical, a multiple subsample approach may be employed. Ideally, enough of these subsamples will be analyzed or extracted so that the desired sample mass is achieved. The subsamples will then be subjected to preparation for analysis, such as solvent extraction or acid digestion. The resulting extracts or digestates can be analyzed separately and the results can be weight-averaged, or the extracts may be combined prior to analysis.

6.1.11.2 *Grinding or Milling*—A alternate method to increasing the sample mass is particle size reduction. This includes such methods as grinding or milling. Grinding can be achieved by hand methods, such as a mortar and pestle, and by using such mechanical devices as mills, grinders, or jaw crushers. Table 3 contains a list of commonly used devices that are commercially available.

6.1.11.3 Each device can be constructed out of different materials to prevent contamination of the laboratory sample.

6.1.11.4 Once particle size reduction has stopped, the laboratory sample should be poured through a sieve into a pan to demonstrate that the desired particle size has been achieved.

6.1.11.5 The sample should be reground if it doesn't go through the sieve. The subsample is then obtained from the sieved material using one of the methods outlined in section 6.1.9

6.1.12 The limitations of each method discussed in 6.1.11

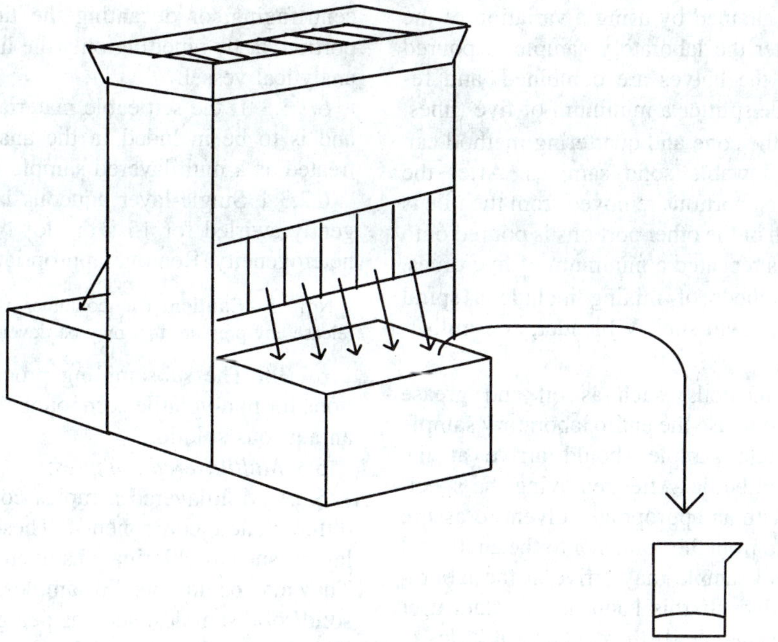

FIG. 3 Riffle Splitter

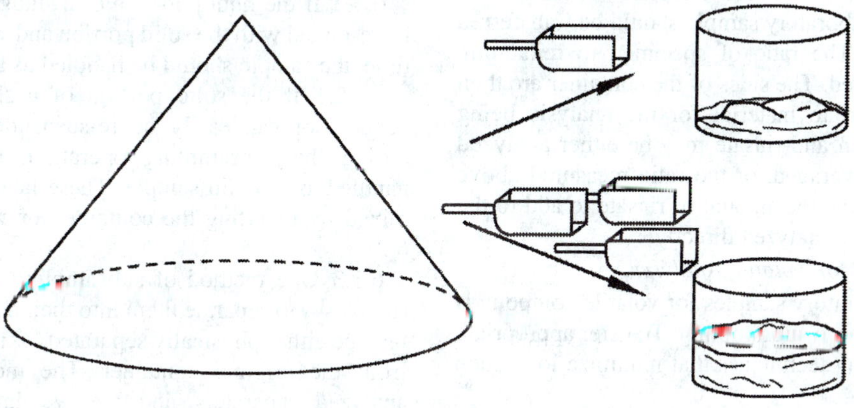

FIG. 4 Alternate Scoop

must be considered before subsampling begins. Table 4 discusses the limitations of each of these methods.

6.1.13 *Mixing*—To reduce heterogeneity, a sample or subsample must be mixed prior to removing the specimen. However, care should be taken since mixing may actually increase the variance. For example, if a subsample had lead shot in soil, differences in size, shape and density may increase heterogeneity further by the development of different strata upon agitation or vibration of the sample.

6.1.14 Mixing techniques are included in the following section.

6.1.14.1 Mixing of a subsample may be achieved manually, or by the use of a mechanical mixer. Manual methods can include stirring with a spatula for 2 to 3 min, until the sample appears uniform, or use of a mixing square. A mixing square is a non-contaminating square sheet, such as Teflon. The laboratory sample is poured into the middle of the sheet, and a rolling pin is used to roll the sample backward and forward while alternately lifting and releasing opposite side corners of the sheet. (See Refs (**3**) and (**4**)).

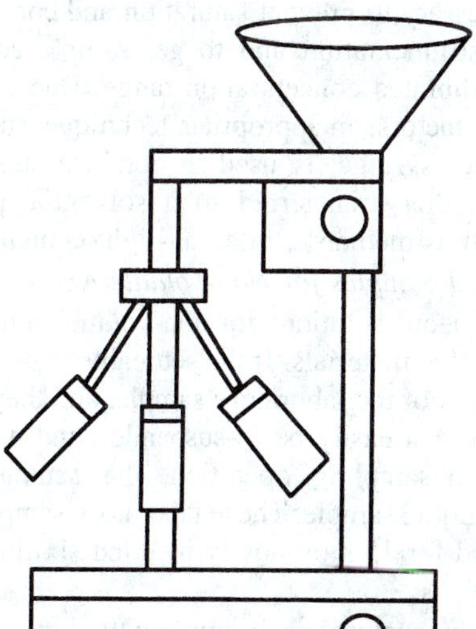

FIG. 5 Sectorial Splitter

6.1.14.2 Mixing can be attained by using a variation of the riffle splitting method. After the laboratory sample is poured through the riffle splitter, the halves are combined, and re-introduced through the riffle splitter a minimum of five times.

6.1.14.3 A variation of the cone and quartering method can be used to homogenize flowable solid samples. After the material is quartered, the first portions removed from the pile is re-formed into a cone. Each of the other portions is poured onto the cone, and the process is repeated a minimum of five times.

6.1.14.4 Mechanical methods of mixing include a spiral mixer, a cement mixer, or a twin-shell V blender, and mills.

6.2 *Liquids*:

6.2.1 Some analytical methods, such as oil and grease analyses, require the analyst to use the entire laboratory sample as the specimen. This field sample should arrive at the laboratory in a proper sized bottle. After removing the specimen, the bottle is rinsed with an appropriate solvent to assure transfer of the entire sample from the container to the analytical vessel. Occasionally, a field sample may arrive at the laboratory in an improper container. If this happens, the data user should be contacted for a resample. If the data user decides to go ahead with the inappropriately sized field sample, the analyst should remove the specimen from the container, and the rest of the original laboratory sample should be transferred into another container. The ratio of specimen to remaining material must be measured. The sides of the container are then rinsed with the appropriate material for the analysis being performed. The specimen and rinsate may be either analyzed separately and weight averaged, or the ratio measured above could be used to determine the amount of rinsate to add to the specimen and it could be analyzed directly.

6.2.2 *Liquid Samples for Volatile Analyses*:

6.2.2.1 Aqueous laboratory samples for volatile compounds should not be mixed prior to subsampling. Transfer appropriate volume for analysis using techniques that minimize loss, such as a syringe.

6.2.2.2 Subsampling of an non-aqueous solution for volatile compounds is often the first step in a sample dilution process which is necessary to prevent saturation and contamination of analytical instrumentation and to get sample concentrations within the calibrated concentration range. Due to the usually large dilution factors, an appropriate technique, such as the use of a microliter syringe, is used to obtain a subsample. The subsample is then transferred to a solvent appropriate for analysis, such as methanol, in an air-tight container.

6.2.3 *Liquid Samples for Non-Volatile Analyses*:

6.2.3.1 Aqueous solutions for non-volatile compounds may contain settleable materials. If the settleable materials are to be included as part of the laboratory sample, and they will remain suspended, or can easily be re-suspended and will remain so during the subsampling operation, the sample should be handled as a liquid sample. These laboratory samples should be gently swirled for 15 s or slowly inverted six times to reduce heterogeneity.

6.2.3.2 If the analyst is subsampling the liquid portion only, the settleable material must be allowed to sink to the bottom before withdrawing the subsample from the liquid portion. The liquid subsample or specimen may be obtained by filtering, centrifuging, or decanting the liquid portion from the solid portion, or by pipetting only the liquid portion directly into the analytical vessel.

6.2.3.3 If the settleable material will not remain suspended, and is to be included in the analysis, the sample should be treated as a multilayered sample (see 6.3).

6.2.3.4 Single-layer aqueous laboratory samples should be gently swirled for 15 s or slowly inverted 6 times to reduce heterogeneity. Remove appropriate subsample for analysis.

NOTE 3—**Caution:** The container should be vented periodically to release any pressure that may be developed during mixing.

6.2.3.5 The subsampling process for a non-aqueous solutions for non-volatile compounds is the same as the process for an aqueous solution.

6.3 *Multilayered Samples*:

6.3.1 Multilayered samples consist of two or more clearly differentiated components. These may include liquid/liquid layers, such as chlorinated solvents and water, or water and oil. They may be liquid/solid samples, such as sludge. They may be solid/solid samples, such as pea gravel and sand. Or they may include combinations of these materials, such as a mixture of oil, water, and sludge.

6.3.2 If the liquid portion of a sludge is so small that it can be re-mixed with the solid portion and will only reseparate over time, the sample should be handled as a solid sample (see 6.1).

6.3.2.1 If the solid portion of a sludge will remain suspended, or can easily be re-suspended and will remain so during the subsampling operation, the sample should be handled as a liquid sample. These laboratory samples may be mixed by inverting the container, or with slight shaking (see 6.2).

6.3.3 One method of subsampling multilayered laboratory samples is to separate them into their component layers. Layers may be either physically separated, or individually subsampled from the original container. The individual layers can be analyzed separately and the weighted average calculated. Portions of the separate phases can be proportionally recombined, either in the analytical vessel for non-volatile analyses, or directly on a purge unit during volatile analysis to form a representative specimen.

6.3.4 *Multilayered Samples for Volatile Analysis/Separation of Layers*:

6.3.4.1 For liquid/liquid layers, transfer appropriate volume for analysis using techniques that minimize loss, such as a syringe or a pipette, from each layer.

6.3.4.2 If laboratory samples have significant solid and liquid layers that cannot be mixed vigorously enough to suspend a solid phase without the potential of losing volatile components, materials may be separated into their component phases. The solid/liquid phase separation may be achieved by gently centrifuging the unopened container, or by allowing it to sit undisturbed until the solid portion is settled. The percentage of each phase should be determined, either by volume or weight.

6.3.4.3 Transfer appropriate mass for analysis using techniques that minimize loss, such as syringe or a pipette. Decant the remaining liquid into another container.

6.3.4.4 A specimen of the remaining solid phase is weighed

for analysis. See 6.1.4 for more information.

6.3.5 *Multilayered Samples for Non-Volatile Analyses/ Separation of Layers*:

6.3.5.1 For liquid/liquid layers, separate the layers using techniques such as a separatory funnel or pipette. Each layer of the material may either be transferred into another container, or directly into the analytical vessel. Each separated portion is then handled as a homogeneous liquid sample. See 6.2.3 for more information.

6.3.5.2 For liquid/solid layers, the liquid subsample or specimen may be obtained by filtering, centrifuging, or decanting the liquid portion from the solid portion, or by pipetting the liquid portion directly into the analytical vessel. The liquid portion is then handled as a homogeneous liquid sample. The solid portion is handled as a solid laboratory sample.

6.3.6 *Alternative Methods for Multilayered Samples for Non-Volatile Analyses*—An alternative method of subsampling liquid/solid samples is to freeze the material (see Guide D 4823).

6.3.6.1 The laboratory sample is transferred into a rectangular box and the solids are allowed to settle. Dry ice or liquid nitrogen can be used to freeze the material. The container may also be placed in a freezer until frozen. Subsamples are sliced out of the box through all the layers. This subsample or specimen is then treated as a solid.

6.3.6.2 An alternate freezing method is to place the laboratory sample into an appropriate container. The whole sample is then frozen using dry ice of liquid nitrogen, and the subsample is obtained using methods described in 6.1.11.

6.3.6.3 An important technique in sampling liquid/solid samples is continuous agitation during the subsampling process. This can be applied to the subsampling of non-volatile compounds for laboratory samples that are pre-dominantly liquid or have liquid characteristics. This is particularly useful for laboratory samples that can form a semi-emulsified form, but only with continuous agitation. During the mixing process, remove an appropriate specimen from the original laboratory sample for analysis.

6.3.7 Solid/solid samples may be subsampled according to 6.1 of this guide.

6.3.8 One or more of the previous methods can be utilized to subsample multilayered samples, depending on how many layers are encountered in the sample.

7. Keywords

7.1 sample preparation; subsampling; waste

ANNEX

(Mandatory Information)

A1. CONSIDERATION OF FUNDAMENTAL ERROR IN THE DETERMINATION OF APPROPRIATE SUBSAMPLE MASS

A1.1 The Table A1.1 is a guide for the maximum allowable particle sizes (centimetres) that can be accommodated by a given subsample mass (grams) with a specified percent relative standard deviation accrued from fundamental error alone. This table is based on the equation for fundamental error as described in Ramsey et al. **(5)**. The tabularized data were generated using the following equation:

$$S^2 = 18*f*e*d^3/M_S \quad (A1.1)$$

where:

S^2 = the relative variance of the contaminant concentration due to the fundamental error,

S = the relative standard deviation of the contaminant concentration due to the fundamental error,

f = a dimensionless factor related to particle shape (a value of 0.5 can be taken as typical, Gy, 1982 **(6)**,

e = the population's average density (g/cm^3). An average density is assumed to be 2.5 g/cm^3,

d = the diameter of the largest particle in centimetres, and

M_S = the mass of the sample in grams.

TABLE A1.1 Maximum Allowable Particle Size (Centimetres) that can be Accommodated by a Given Subsample Mass at Various % Relative Standard Deviation (RSD)

Recommended Subsample Mass, (g)	U.S. Standard Sieve Mesh Size	5 % RSD	10 % RSD	15 % RSD
0.1	35	0.02	0.04	0.05
1	18	0.05	0.08	0.10
2	13	0.06	0.10	0.13
5	12	0.08	0.13	0.17
10	10	0.10	0.16	0.22
30	7	0.15	0.24	0.31
50	6	0.18	0.28	0.37
100	5	0.22	0.35	0.46

A1.2 Using the above assumptions for density and the shape factor this equation can be rearranged to calculate the largest particle size that can be representatively accommodated by a given subsample mass and given fundamental error.

$$d = \sqrt[1/3]{(M_S/22.5)*S^2} \quad (A1.2)$$

A1.3 If the density of the material being sampled varies significantly from 2.5 g/cm^3, then the actual density should be employed. If the material being sampled does not have a typical shape, the factors in Table A1.2 can be substituted for the 0.5 value used to calculate the tabularized data (**1**).

TABLE A1.2 *"f"* Values for Different Particle Shapes

Particle Shape	*f*
Cubic	1
Spheres	0.5
Flakes	0.1
Soft solids shaped by mechanical stress	0.2
Needles	> 1 to ≤ 10

A1.4 Table A1.1 summarizes the maximum allowable particle size in centimetres that can be accommodated by a given subsample mass at varying fundamental errors. The tighter the required precision, the larger the sample mass. This table assumes that the particle is the typical shape described above by Gy, and serves as an example of how the maximum allowable particle size changes as a function of these parameters. It is important to note that the fundamental error should be no larger than 15 % (**1**).

REFERENCES

(**1**) Pitcard, Francis F., *Pierre Gy's Sampling Theory and Practice*; 2nd Edition, CRC Press, Inc., Boca Raton, FL, 1993.

(**2**) Simmons, Milagros S., *Hazardous Waste Measurements*, Lewis Publishers, 200 Corporate Blvd., NW, Boca Raton, FL, 1991.

(**3**) US-EPA, *Description and Sampling of Contaminated Soils, A Field Pocket Guide*, EPA/625/12-91/002, 1991.

(**4**) Boulding, J.R., *Description and Sampling of Contaminated Soils: A Field Guide*, Lewis Publishers, 2000 Corporate Blvd., NW, Boca Raton, FL, 1994.

(**5**) Ramsey, Charles A., Ketterer, Michael E., and Lowry, Joe H., "Application of Gy's Sampling Theory to the Sampling of Solid Waste Materials," In *Proceedings of the EPA Fifth Annual Waste Testing and Quality Assurance Symposium*, Vol 11, Washington, DC, July 24-28, 1989, pp. 11-494.

(**6**) Gy, P.M., *Sampling of Particulate Material*, Elsevier, Amsterdam, 1982.

Designation: D 6346 – 98

Standard Guide for Accepting, Segregating and Packaging Materials Collected Through Household Hazardous Waste Programs[1]

This standard is issued under the fixed designation D 6346; the number immediately following the designation indicates the year of original adoption or, in the case of revision, the year of last revision. A number in parentheses indicates the year of last reapproval. A superscript epsilon (ϵ) indicates an editorial change since the last revision or reapproval.

1. Scope

1.1 This guide covers Household Hazardous Waste (HHW) programs for accepting, segregating and packaging materials collected through HHW programs to minimize the risk associated with managing these materials and to identify applicable regulations. This guide does not address storing, vehicle loading or transporting collected and packaged materials. This guide does not attempt to define the hazardous nature of materials.

1.2 Certain existing local, state and federal regulations apply to HHW program operations. This guide does not replace these existing regulations, and is not intended to be used as the basis for regulations for HHW Programs. This guide does not reference all applicable applications, since applicable regulations will vary. HHW Programs should research all applicable regulations before establishing a materials collection program.

1.3 *This standard does not purport to address all of the safety concerns, if any, associated with its use. It is the responsibility of the user of this guide to establish appropriate safety and health practices and determine the applicability of regulatory limitations prior to use.*

2. Referenced Documents

2.1 *Code of Federal Regulations (CFR):*

Title 49, Transportation, Parts 171–178[2]

CFR Title 40, Protection of Environment, Part 261.5[2]

3. Terminology

3.1 *Definitions:*

3.1.1 *bulking*—the act of emptying multiple containers of compatible materials and mixing those materials together in a single package unit destined for shipment. This would also include material placed in storage tanks to be packaged for shipment at a later date, or pumped into a bulk tank truck for shipment.

3.1.2 *consolidation*—the act of combining two or more materials to make a single package unit. Common types of consolidation packaging used by HHW programs include: bulking, lab packaging, and composite packaging.

3.1.3 *package or outside package*—a package plus its contents.

3.1.4 *packaging*—a receptacle and any other components or materials (drums, boxes, liners, absorbents, etc.) necessary for the receptacle to perform its containment function in conformance with the minimum packing requirements of 40 CFR 171, 172, 173.

4. Significance and Use

4.1 This guide is intended to provide general guidance to HHW programs for accepting and segregating materials, selecting a management method, and packaging materials collected by HHW programs.

5. Procedure

5.1 *Accepting Materials*—HHW Programs should establish a list of materials that will be accepted at its facilities or event collections. The list should be distributed to the public prior to the facility opening or event collection. Types of materials commonly collected by HHW Programs include:

5.1.1 Materials that exhibit a hazardous characteristic, as defined in 40 CFR 261.5, such as ignitability, toxicity, corrosivity, oxidizer, reactivity, or are a listed hazardous waste;

5.1.2 Materials prohibited from local municipal waste management systems; and

5.1.3 Materials collected to provide a convenience or public service to a community.

5.1.4 HHW Programs will need to establish traffic flow, vehicle unloading and safety procedures. All personnel involved with material acceptance should receive training in material identification, proper lifting procedures, safety and personnel protective equipment.

5.1.5 Leaking containers should be repackaged immediately into a container that is compatible with the material in the container.

5.1.6 Unlabeled containers should be segregated from other wastes. The individual who brought the container should be interviewed to assist in identification. Any further categorization and identification of the material should be completed as soon as possible. A hazardous materials contractor or trained staff could perform tests to identify the materials.

5.2 *Segregating Materials*—As the materials are collected they should be segregated into general categories based on DOT incompatibility and packaging regulations.

---

[1] This guide is under the jurisdiction of ASTM Committee D-34 on Waste Management and is the direct responsibility of Subcommittee D34.04 on Waste Minimization.

Current edition approved Dec. 10, 1998. Published February 1999.

[2] Available from U.S. Government Printing Office, Washington, DC.

5.2.1 *Commonly Collected Materials*—DOT classifies materials generated from households as a consumer commodity, which exempts them from DOT regulations. The materials listed below are sub-categories of DOT classes and are not intended to be all inclusive, but a guide to the most common materials collected by HHW programs.

Oil Based Paint	Water Based (latex) Paint/Sealer
Fuels	Antifreeze
Corrosive Acids	Corrosive Bases
Cyanides	Dioxins
Pesticide Liquids (including herbicides)	Pesticide Solids (including herbicides)
Flammables and Chlorinated Products	Organic Peroxides
Oxidizers	Reactive Solids
Arsenic/Heavy Metals	Mercury Compounds
PCB Containing Devices	Motor Oil
Compressed gas cylinders	Aerosols
Batteries	Fluorescent and HID lamps

5.2.1.1 Once a material is collected by a HHW program, it should be handled in an appropriate manner based on whether it exhibits a hazardous characteristic. A material that exhibits a hazardous characteristic or meets the definition of a listed hazardous waste or material should be transported in accordance with DOT regulations for its specific hazard class.

5.2.2 *Materials Requiring Special Handling*—The materials listed below require special handling. HHW Programs do not typically promote accepting these materials through their programs. However, these materials are brought to collection facilities and it is wise to be prepared to refer people to the appropriate local management option or to handle the materials at the facility. Management options are suggested below for these materials. The handling of these materials should be addressed in the site safety and emergency response plan for the facility.

5.2.2.1 *Explosives, Ammunition or Fireworks*—Contact local law enforcement or bomb squad.

5.2.2.2 *Radioactive Materials*—Refer to appropriate management option or contact Nuclear Regulatory Commission.

5.2.2.3 *Friable Asbestos*—Refer to local management facility or accept in accordance with local, state and federal regulations.

5.2.2.4 *Medical/Infectious Materials*—Refer to local management option or accept in accordance with local, state and federal regulations.

5.2.2.5 *Consumer Electronics and Appliances*—These items may have hazardous components and should be handled by a local handler, or may be managed through the HHW program.

5.2.3 *Incompatibility and Segregation of Materials*—Incompatible materials, when mixed, generate chemical reactions which may result in serious safety and health concerns. These include toxic gas generation, heat generation, flammable gas generation, fire, explosion, or polymerization of materials. Containers of incompatible materials should be adequately separated using bins, drums, dikes, walls or other devices, such that mixing of incompatible materials is prevented in the event of a spill, or a break or leak in a container.

5.2.3.1 Material and container segregation procedures should be established before waste is accepted. Procedures should be based on the type of materials accepted, facility design and facility operation. Personnel responsible for segregating wastes should be properly trained in chemical management and safety. The following list gives only general guidance on incompatible categories of materials.

Always Segregate:	From:
Acids	Bases, cyanides
Oxidizers	Flammables
Cyanides	Acids, oxidizers

5.3 *Selecting a Management Method for Collected Materials*—Prior to consolidating materials, the ultimate management method for the material should be considered. The management method selected may affect how the materials will be packaged. Common management practices include: reuse, recycling, energy recovery, on-site treatment, incineration, land disposal, and waste water treatment.

5.3.1 When selecting the most appropriate material management method, HHW Programs should consider several factors, including:

5.3.1.1 Compliance with permitting requirements and local, state and federal regulations (especially when performing on-site treatment of materials);

5.3.1.2 Operational or structural limitations at facilities or event collections (that is, no product reuse available on-site, no connection to sewer treatment facility, limited space);

5.3.1.3 Personnel training and expertise;

5.3.1.4 Consideration of a materials management hierarchy that favors reuse, recycling, and energy or material recovery over incineration or land disposal;

5.3.1.5 Cost;

5.3.1.6 Potential liability related to each management method; and

5.3.1.7 Contractual or contractor restrictions or preferences.

5.4 *Packaging Collected Materials*—Collected materials need to be packaged or handled according to facility standards, local, state and federal regulations. Materials management contractors may also have specific packaging protocol that must be followed so that they can properly and efficiently process materials at their facilities. Unlabeled containers or questionable materials should not be packaged until they are properly identified.

5.4.1 *On-site Management Options and Packaging*:

5.4.1.1 *Reuse*—Usable materials in original containers that are offered to local residents or organizations during an event collection or at a permanent site should be left in their original containers. This allows users to reference original label information for proper use, and preserves the manufacturers' product liability.

5.4.1.2 Paints or other materials may be bulked into larger containers. However, the HHW Program may be considered to be engaging in a manufacturing operation when this is done.

5.4.1.3 *Recycle*—If materials are to be recycled on-site, they do not need to be packaged according to DOT protocol. Some items may be bulked or packaged into larger containers according to facility protocol and local regulations.

5.4.1.4 *Treatment or Disposal on-Site*—Many water-based cleaners and non-hazardous liquid materials may be sewered if the site is connected to the local wastewater treatment plant and has permission from the plant to do so. Some acids and bases may also be neutralized and sewered on-site in accordance with local or state regulations. Materials do not need to be packaged according to DOT protocol if treated or sewered on-site. A site safety or operational plan should be developed to describe in detail the process used to treat or dispose of wastes on-site.

5.4.1.5 Non-hazardous solids may be disposed of using the facility's solid waste management service and container.

5.4.1.6 *Energy Recovery*—If an energy recovery option is available on-site (that is, used oil furnace, municipal solid waste incinerator), the material should be handled and packaged according to the facility's standard protocol and all applicable regulations.

5.4.2 *Off-Site Management and Packaging*—Materials that will be shipped off-site for further management at recycling, energy recovery, incineration or land disposal facilities should conform to DOT regulations. Therefore, it makes sense to initially package all materials which will leave the site in accordance with DOT regulations. These packaging methods are fully detailed in DOT regulations, 49CFR Parts 171–178. These regulations describe the specific container types, packaging procedures, and container labeling required for each sub-category of material identified in 5.2.1. HHW Programs that intend to package their own materials need to become familiar with DOT regulations in order to properly package materials. Programs may also hire trained contractors to perform packaging on-site.

6. Keywords

6.1 compatibility; household hazardous waste; incompatibility; packaging; segregation

Standard Practice for Using the Disposable En Core Sampler for Sampling and Storing Soil for Volatile Organic Analysis[1]

This standard is issued under the fixed designation D 6418; the number immediately following the designation indicates the year of original adoption or, in the case of revision, the year of last revision. A number in parentheses indicates the year of last reapproval. A superscript epsilon (ε) indicates an editorial change since the last revision or reapproval.

ε1 NOTE—The title was corrected editorially in August 1999.

1. Scope

1.1 This practice is intended for application to soils that may contain volatile waste constituents.

1.2 This practice provides a procedure for using the disposable En Core[2] sampler to collect and store a soil sample for volatile organic analysis.

1.3 It is recommended that this standard be used in conjunction with Practice D 4547 and Guide D 4687.

1.4 *This standard does not purport to address all of the safety concerns, if any, associated with its use. It is the responsibility of the user of this standard to establish appropriate safety and health practices and determine the applicability of regulatory limitations prior to use.*

NOTE 1—ASTM takes no position respecting the validity of any patent rights asserted in connection with any item mentioned in this standard. Users of this standard are expressly advised that determination of the validity of any such patent rights, and the risk of infringement of such rights, are entirely their own responsibility.

2. Referenced Documents

2.1 *ASTM Standards:*

D 2777 Practice for Determination of Precision and Bias of Applicable Methods of Committee D-19 on Water[3]

D 4547 Practice for Sampling Waste and Soils for Volatile Organics[4]

D 4687 Guide for General Planning of Waste Sampling[4]

D 5792 Practice for Generation of Environmental Data Related to Waste Management Activities: Development of Data Quality Objectives[4]

3. Summary of Practice

3.1 This practice provides a procedure for using the disposable En Core sampler to collect and store a soil sample of approximately 5 or 25 g for volatile organic analysis. The En Core sampler, which consists of a coring body/storage chamber, O-ring sealed plunger, and O-ring sealed cap, is designed to collect and hold a soil sample during shipment to the laboratory for analysis.

3.2 After the sample is collected in the En Core sampler, the coring body/storage chamber is sealed with a slide-on cap and immediately becomes a sample storage chamber. To minimize loss of volatile compounds due to volatilization, or biodegradation, or both, from the time of collection until analysis or preservation in the laboratory (see Practice D 4547), sample storage in the device currently is specified at 4 ±2°C for up to 48 h. For cases where holding times beyond 48 h are necessary, storage of samples in the En Core sampler at 4 ±2°C or –12 ± 2°C for longer than 48 h is an option, provided it can be shown that the longer storage time used does not influence the concentrations of the volatile organic compounds (VOCs) of interest in the samples, or that the data generated by analysis of the samples meet the data quality objectives (DQOs), (see Practice D 5792). This practice does not use methanol preservation or other chemical preservatives in the field.

4. Significance and Use

4.1 This practice is for use in collecting and storing 5- or 25-g soil samples for volatile organic analysis in a manner that minimizes loss of contaminants due to volatilization, or biodegradation, or both. The En Core sampler serves as both the sampling device and sample storage chamber.

4.2 In performing the practice, the integrity of the soil sample structure is maintained during sample collection, storage, and transfer in the laboratory for analysis or preservation.

4.3 During sample collection, storage, and transfer, there is very limited exposure of the sample to the atmosphere.

4.4 Laboratory subsampling is not required when performing this practice. The sample is expelled directly from the coring body/storage chamber into the appropriate container for analysis or preservation without disrupting the integrity of the sample. Subsampling from the En Core device should not be performed to obtain smaller sample sizes for analysis.

4.5 This practice specifies sample storage in the En Core sampler at 4 ±2°C for up to 48 h. For cases where holding

[1] This practice is under the jurisdiction of ASTM Committee D34 on Waste Management and is the direct responsibility of Subcommittee D34.01.03 on Sampling Equipment.

Current edition approved June 10, 1999. Published August 1999.

[2] En Core is a registered trademark of En Novative Technologies, Inc., Green Bay, WI. The En Core sampler is covered by a patent. Interested parties are invited to submit information regarding the identification of an alternative(s) to this patented item to the ASTM Headquarters. Your comments will receive careful consideration at a meeting of the responsible technical committee, which you may attend.

[3] *Annual Book of ASTM Standards*, Vol 11.01.

[4] *Annual Book of ASTM Standards*, Vol 11.04.

times beyond 48 h are necessary, storage of samples in the En Core sampler at 4 ±2°C or –12 ±2°C for longer than 48 h is an option, provided it can be shown that the longer storage time used does not influence the concentrations of the VOCs of interest in the samples, or that the data generated by analysis of the samples meet the DQOs (see Practice D 5792).

4.6 This practice does not use methanol preservation or other chemical preservatives in the field. As a result, there are no problems associated with flammability hazards, shipping restrictions, or dilution of samples containing low volatile concentrations due to solvents being added to samples in the field.

4.7 The En Core sampler is a single-use device. It should not be cleaned, or reused, or both.

4.8 This practice cannot be used for sampling cemented material, consolidated material, or material having fragments coarse enough to interfere with proper coring techniques.

5. Apparatus

5.1 *En Core Sampler*—The En Core sampler has three components: the coring body/storage chamber, which is volumetrically designed to collect and store a soil sample of approximately 5 or 25 g; an O-ring sealed plunger for nondisruptive extrusion of the sample into an appropriate container for analysis or preservation; and, a slide-on cap having an O-ring seal and locking arm mechanism (see Fig. 1). The coring body/storage chamber of the 5-g sampler has a volume of 3 cm^3 to give a sample size of approximately 5 g, assuming a soil density of 1.7 g/cm^3. The coring body/storage chamber of the 25-g sampler has a volume of 14.5 cm^3 to give a sample size of approximately 25 g, assuming a soil density of 1.7 g/cm^3. The seals of the sampler are provided by three polytetrafluoroethylene-coated fluoroelastomer O-rings (see Fig. 1). The coring body/storage chamber, plunger, and cap of the En Core sampler are constructed of an inert composite polymer. The En Core sampler is certified as clean and should not be reused. There are two stainless steel reusable attachments that are used with the En Core sampler. These are a T-handle, which is used to push the sampler into the soil for sample collection; and an extrusion tool, which attaches to the plunger for extrusion of the sample from the coring body/storage chamber (see Fig. 2). Each En Core sampler is supplied with a protective moisture-proof bag for shipment to the laboratory.

5.2 *Minimum/Maximum Thermometer*—This is any thermometer that registers the minimum and maximum temperatures reached during any given period of time in °C and has a range that includes the specified storage temperature in divisions of 1°C.

6. Procedure

6.1 The size of the En Core sampler used is determined by the size of the sample required by the laboratory procedure that will be used to analyze the sample. If a sample size of approximately 5 g is required for analysis, the 5-g En Core sampler, and not the 25-g device, should be used to collect and

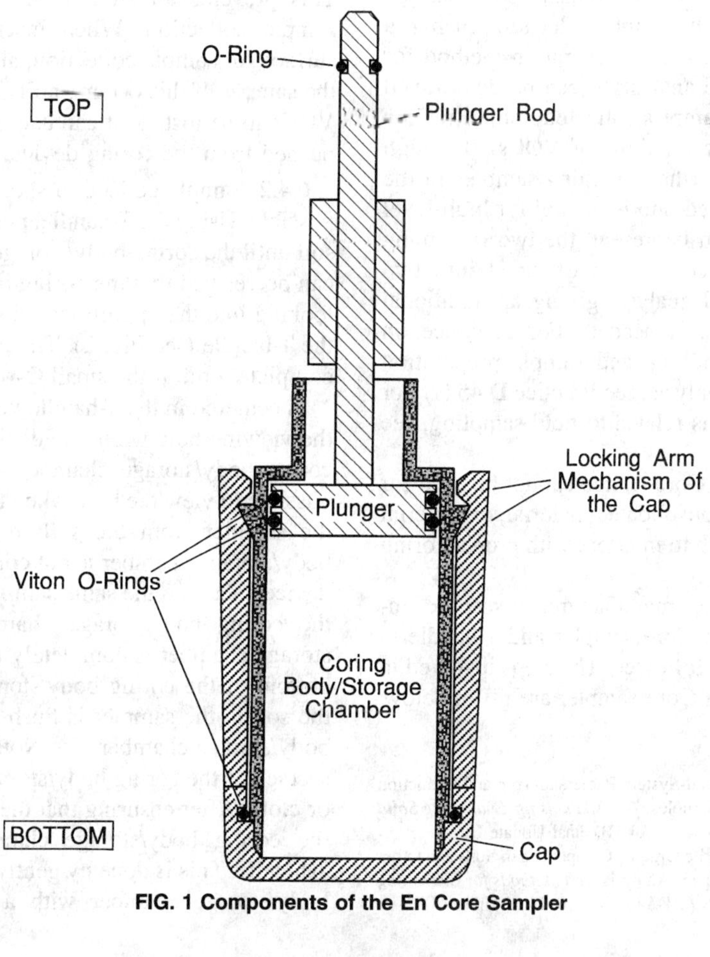

FIG. 1 Components of the En Core Sampler

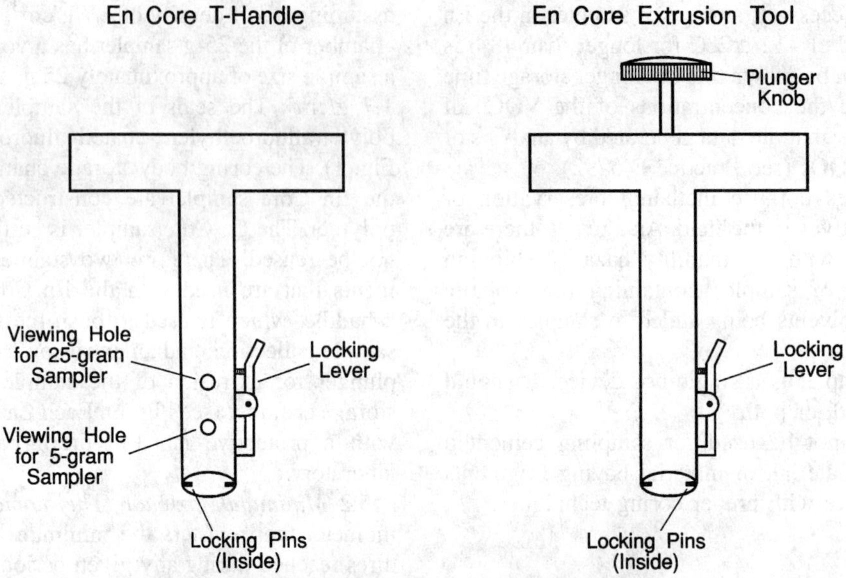

FIG. 2 Reusable Attachments to the En Core Sampler

store the sample. Subsampling from the En Core devices should not be performed to obtain smaller sample sizes for analysis.

6.2 If volatile contaminant levels in the soil being sampled are not known, it is recommended that three samples be collected at each sampling location using three En Core samplers. If determination of moisture content is required for reporting sample results on a dry weight basis, a fourth sample should be collected from each sampling location using a suitable container. The proper sample preparation method for low-level or high-level volatile analysis[5,6] can be determined by screening one of the three samples collected in the En Core samplers for high or low concentrations of VOCs. If a high concentration is present, one of the remaining samples in the En Core devices can be extruded into methanol for high-level analysis; if a low concentration is present, the two remaining samples in the En Core devices can be extruded into two appropriate vials for low-level analysis giving an additional low-level sample for analysis, if needed. For guidance on sample collection, sample handling, and sample preparation methods for volatile organic analysis, see Practice D 4547. For quality assurance considerations related to field sampling, see Guide D 4687.

6.3 As stated in 4.8, the En Core device cannot be used for sampling cemented material, consolidated material, or material having fragments coarse enough to interfere with proper coring techniques.

6.4 *Sample Collection*—The manufacturer's specific instructions for operating the En Core sampler and T-handle to collect a soil sample should be followed. The steps involved in sample collection using the En Core sampler are given below.

[5] U.S. EPA, 1996, Method 5035: Closed-System Purge-and-Trap and Extraction for Volatile Organics in Soil and Waste Samples. *Test Methods for Evaluating Solid Waste: Physical/Chemical Methods (SW-846)*, Vol 1B, Final Update III.

[6] U.S. EPA, 1996, Method 5021: Volatile Organic Compounds in Soils and Other Solid Matrices Using Equilibrium Headspace Analysis. *Test Methods for Evaluating Solid Waste: Physical/Chemical Methods (SW-846)*, Vol 1B, Final Update III.

6.4.1 Before collecting a sample in the En Core sampler, the plunger rod (see Fig. 1) should be positioned so that the plunger can be moved freely from the top to the bottom of the coring body/storage chamber. For sample collection, the T-handle (see Fig. 2) should be attached to the sampler, and the plunger should be positioned so that the bottom of the plunger is flush with the bottom of the coring body/storage chamber. This prevents air from being trapped in the device during sample collection. When inserting a coring device into a surface for sample collection, air should not be trapped behind the sample. If this occurs, air can pass through the soil causing VOCs to be lost, or it can cause the sample to be prematurely pushed from the coring device.

6.4.2 Sample collection should be performed as quickly as possible. Using the T-handle, push the En Core sampler into the soil until the coring body/storage chamber is completely full. It can be verified that the coring body/storage chamber is full by looking into the appropriate viewing hole (either 5 or 25-g) in the T-handle (see Fig. 2). The coring body/storage chamber is completely full if the small O-ring on the plunger rod (see Fig. 1) is centered in the T-handle viewing hole. If possible, look at the viewing hole while the device is in the soil to check if the coring body/storage chamber is completely full. If it is not possible to view the hole when the device is in the soil, remove the sampler from the soil to view the hole. If the coring body/storage chamber is not completely full, quickly insert the device back into the same sampling spot and push harder to fill the coring body/storage chamber. When the coring body/storage chamber is completely full, scrape a spatula across the bottom of the coring body/storage chamber so the surface of the soil in the sampler is flush with the opening of the coring body/storage chamber (see Note 2). Quickly wipe the external surface of the coring body/storage chamber with a clean tissue or cloth. After ensuring that the sealing surfaces are clean, cap the coring body/storage chamber while it is still on the T-handle. This is done by gently sliding the cap onto the coring body/storage chamber with a twisting motion. The cap is locked into position when the grooves on the locking arms seat over the ridge on the coring body/storage chamber (see Fig. 1). If the cap of the En Core sampler is difficult to lock into position, the O-ring in the cap (see Fig. 1) may be bent out of position, and a new cap should be used to seal the device. A bent O-ring in the cap may result in loss of VOCs from the stored sample (see X1.1.8.2 and X1.1.9.2). After the cap is locked into position, the T-handle is removed from the sampler, and the plunger is locked into position by rotating the plunger rod (see manufacturer's specific operating instructions).

NOTE 2—For drier soils that are difficult to compact in the sampler, scraping a spatula across the surface of the soil to make it flush with the opening of the coring body/storage chamber may loosen particles of the soil in the coring body/storage chamber. These particles may scatter when the device is capped and may affect the seal between the cap and coring body/storage chamber (see X1.1.9.3). Caution should be used when working with these types of soils to prevent scattering from happening.

6.4.3 A sample label showing the sample number is attached to the cap of the En Core sampler, and the sampler is placed in its protective moisture-proof shipping bag (provided with each En Core sampler). This bag has a label attached for recording sample name, date collected, comments, and sample number.

6.5 *Sample Storage*—The capped coring body immediately becomes a sealed sample storage chamber. A sample can be stored in the sealed coring body/storage chamber at 4 ±2°C for up to 48 h. For cases where holding times beyond 48 h are necessary, storage of samples in the En Core sampler at 4 ±2°C or –12 ±2°C for longer than 48 h is an option, provided it can be shown that the longer storage time used does not influence the concentrations of the VOCs of interest in the samples, or that the data generated by analysis of the samples meet the DQOs (see Practice D 5792). Data showing performance of the En Core sampler as a storage chamber for samples of three different soil types stored at 4 ±2°C for 48 h; 4 ±2°C for four days; 4 ±2°C for 48 h followed by storage for 5 days at –12± 2°C; and, –12 ±2°C for seven days are given in Appendix X1.

6.5.1 Storage of samples at 4 ±2°C can be achieved by placing the protective shipping bag, in which the En Core sampler has been placed, in a cooler with ice, cold packs, or in a refrigerated compartment regulated at 4 ±2°C. If the optional storage temperature of –12± 2°C is used as discussed in 6.5, the En Core sampler should be stored in a freezer compartment regulated at –12 ±2°C. A minimum/maximum thermometer should be placed with the samples during storage to verify that the temperature requirement is maintained.

6.6 Arrangements with the receiving laboratory for sample log in, sample handling, required storage conditions, and analysis should be made.

6.7 *Extrusion of the Soil Sample from the En Core Sampler*—The manufacturer's specific instructions for operating the En Core sampler and extrusion tool to extrude the soil sample from the coring body/storage chamber should be followed. The steps involved in sample extrusion from the En Core sampler are outlined in 6.7.1.

6.7.1 At the laboratory, the soil is transferred into the appropriate container without disturbing the integrity of the sample by removing the cap from the coring body/storage chamber and using the plunger to expel the soil into the receiving container (see Note 3). This is done by performing the following steps according to the manufacturer's specific operating instructions: attach the extrusion tool (see Fig. 2) to the En Core sampler; rotate the extrusion tool plunger knob into position for sample extrusion; unlock the locking arms of the cap; carefully remove the cap from the sampler; and, push down on the plunger knob of the extrusion tool (see Fig. 2) to expel the sample from the coring body/storage chamber directly into the appropriate container for analysis or preservation (see Practice D 4547). After the sample is expelled from the En Core sampler, the O-ring in the sampler cap should be inspected to make sure that the O-ring was not bent when the cap was placed on the coring body/storage chamber. A bent O-ring in the sampler cap can result in loss of VOCs from the stored sample and should be documented (see X1.1.8.2 and X1.1.9.2).

NOTE 3—Samples that have been stored at –12 ±2°C should be allowed to sit at room temperature for 2 to 3 min before extrusion from the En Core sampler, which will facilitate extrusion of the frozen sample from the device.

7. Report

7.1 For guidance on information that should be recorded for a sampling activity, see Guide D 4687.

8. Precision and Bias

8.1 *Precision*—Data on the precision of this practice are to be generated in the near future.

8.2 *Bias*—No information can be given on the bias of the sampling/storage procedure described in this practice because there is no standard reference material for sampling soil for VOCs in the field; however, data have been generated to provide information on the performance of the En Core sampler for storage of soil samples at 4± 2°C for 48 h, at 4 ±2°C for four days, at 4 ±2°C for 48 hours followed by storage for five days at –12 ±2°C, and at –12 ±2°C for seven days[7]. These data are shown in Appendix X1.

9. Keywords

9.1 En Core sampler; soil sampling; storage of soil samples; volatile organic compounds (VOCs); VOC analysis

[7] A copy of the research report on the study described in Appendix X1 is available from ASTM Headquarters. Request RR:D34-1012.

APPENDIX

(Nonmandatory Information)

X1. PERFORMANCE OF THE EN CORE SAMPLER TO STORE VOC-SPIKED SOIL SAMPLES

X1.1 A study was conducted to evaluate the performance of the 5- and 25-g En Core samplers to store three different soil types spiked with an aqueous solution containing nine volatile compounds. The En Core samplers used in this study were manufactured in 1998. The 5-g devices that were used represent lot numbers K018166, K028132, and K038132. The 25-g devices that were used represent lot numbers K218266, K018140, P018138, 018134, 018135, 018126, and 018128. The manufacturer can be contacted for information on these lot numbers.

X1.1.1 The soils used in the study are representative of different environments and contained native microbial populations. They are a river bank soil having 49 % sand, 26 % silt, 24 % clay, 5.3 % organic material, approximately 14 % moisture, and a dehydrogenase (microbial) activity of 22-mg total product formed (TPF)/g/24 h; a mountain soil having 75 % sand, 13 % silt, 12 % clay, 4.3 % organic material, approximately 12 % moisture, and a dehydrogenase activity of 11 mg TPF/g/24 h; and, a prairie soil having 67 % sand, 17 % silt, 16 % clay, 1.5 % organic material, approximately 8 % moisture, and a dehydrogenase activity of 17 mg TPF/g/24 h.

X1.1.2 The VOCs used in the study are *cis*-dichloroethylene (CDCE), benzene, trichloroethylene (TCE), toluene, perchloroethylene (PCE), ethylbenzene, m/p-xylene, o-xylene, and methylethylketone (MEK). These compounds were selected as the analytes of interest because they are representative of halogenated, aromatic, and polar VOCs typically found in contaminated soils.

X1.1.3 In the study, soil samples were collected in the En Core samplers from a large container of loose soil and then spiked with an aqueous solution containing the compounds listed in X1.1.2 at concentrations of approximately 50 mg/L. The spiking solution was injected into the middle of the soil plug in the sampler, after which, the sampler was immediately capped. The 5-g samples were spiked with 250 µL of the spiking solution, and the 25-g samples were spiked with 1250 µL of the spiking solution to give an approximate concentration of 2.5 µg/g of each analyte of interest in the samples. This analyte concentration in the soil was selected to limit the influence of the analytical method on the data.

X1.1.4 After all samples were spiked and capped, five random samples for each soil type were extruded from each size of En Core sampler into methanol for analysis to give time-zero concentrations of the analytes of interest. The remaining samples were stored under the storage conditions shown in Table X1.1. Storage temperatures were monitored to make sure they were at the specified temperature. After the samples were held for the appropriate times, they were extruded into methanol for extraction and analysis. The methanol extracts of the samples were analyzed using EPA Method 8021B.[8]

X1.1.5 To evaluate the data, the mean values of the analytes of interest in the stored samples were compared to their mean values in the time-zero samples by calculating average percent recovery. Before average percent recovery was calculated, the data sets were evaluated for outlier data points as described in Practice D 2777.

X1.1.6 The data generated by this testing are specific to the experimental design of the study. The data give information on the performance of the En Core samplers for storing soil samples collected from loose soil, spiked with an aqueous solution of selected analytes of interest, and then stored under specific storage conditions. The data generated by this testing also are specific to the soils used in the study, the analytes of interest, the analyte concentrations, and the storage conditions that were evaluated. For other soil types, analytes, analyte concentrations, and storage conditions, these data may not apply.

X1.1.7 The average percent recoveries of the VOCs of interest from samples of the river bank soil stored in the 5-g En Core samplers are shown in Table X1.2; and, the average percent recoveries of the VOCs of interest from samples of the river bank soil stored in the 25-g En Core samplers are shown in Table X1.3. In each of these tables, the percent relative standard deviation of the concentration values in the stored samples is given in parentheses next to the corresponding average percent recovery value. The percent relative standard deviation of the concentration values in the time-zero samples is given in a footnote to each of the tables.

X1.1.7.1 As shown in Table X1.2, the VOCs of interest in this study have average percent recovery values ranging from

[8] U.S. EPA, 1996, Method 8021B: Aromatic and Halogenated Volatiles by Gas Chromatography Photoionization and/or Electrolytic Conductivity Detectors. *Test Methods for Evaluating Solid Waste: Physical/Chemical Methods (SW-846)*, Vol 1B, Final Update III.

TABLE X1.1 Storage Conditions for Testing the Disposable En Core Samplers

Storage Times	Storage Conditions	No Storage	4 ±2°C on Ice	4 ±2°C on Ice for 48 h then Refrigerated at 4 ±2°C	4 ±2°C on Ice for 48 h then Frozen at −12 ±2°C	Frozen at −12 ±2°C
Time-zero		5 samples				
48 h			5 samples			
4 days				5 samples		
7 days					5 samples	5 samples

TABLE X1.2 Average Percent Recoveries of VOCs from Samples of Soil 1[A] Stored in 5-G En Core Samplers

VOCs	Storage Conditions	4 ±2°C/48 h	4 ±2°C/4 Days	4 ±2°C/48 h then −12± 2°C/5 Days	−12± 2°C/7 Days
CDCE		91 %[B] (15 %)[C]	96 %[B] (4 %)[C]	89 %[B] (3 %)[C]	98 %[B] (7 %)[C]
Benzene		93 % (3 %)	94 % (4 %)	87 % (5 %)	93 % (10 %)
TCE		97 % (1 %)	100 % (3 %)	95 % (4 %)	101 % (8 %)
Toluene		99 % (1 %)	96 % (4 %)	96 % (3 %)	99 % (6 %)
PCE		100 % (1 %)	100 % (1 %)	99 % (3 %)	99 % (7 %)
Ethylbenzene		101 % (3 %)	100 % (4 %)	101 % (3 %)	101 % (6 %)
m\p-Xylene		102 % (2 %)	100 % (3 %)	96 % (9 %)	99 % (6 %)
o-Xylene		99 % (1 %)	99 % (4 %)	91 % (7 %)	100 % (7 %)
MEK		100 % (0 %)	100 % (0 %)	92 % (6 %)	91 % (6 %)

[A] Soil 1 is a river bank soil (49 % sand, 26 % silt, 24 % clay, 5.3 % organic material, and approximately 14 % moisture).
[B] Average percent recovery based on mean values determined for five time-zero samples and five stored samples
[C] Value in parentheses is the percent relative standard deviation of the concentration values in the five stored samples. The percent relative standard deviation of the concentration values in the time-zero samples ranged from 0 to 6 % for all tests using soil 1 in the 5-g device.

TABLE X1.3 Average Percent Recoveries of VOCs from Samples of Soil 1[A] Stored in 25-G En Core Samplers

VOCs	Storage Conditions	4 ±2°C/48 h	4 ±2°C/4 Days	4 ±2°C/48 h then −12± 2°C/5 Days	−12± 2°C/7 Days
CDCE		91 %[B] (1 %)[C]	94 %[B] (3 %)[C]	81 %[B] (8 %)[C]	90 %[B] (8 %)[C]
Benzene		90 % (4 %)	94 % (4 %)	79 % (11 %)	87 % (12 %)
TCE		92 % (3 %)	96 % (8 %)	87 % (6 %)	100 % (6 %)
Toluene		94 % (3 %)	94 % (6 %)	91 % (4 %)	98 % (6 %)
PCE		96 % (3 %)	100 % (5 %)	92 % (5 %)	102 % (6 %)
Ethylbenzene		98 % (3 %)	100 % (1 %)	98 % (3 %)	102 % (4 %)
m\p-Xylene		98 % (1 %)	98 % (2 %)	95 % (6 %)	100 % (3 %)
o-Xylene		98 % (3 %)	100 % (2 %)	97 % (5 %)	102 % (4 %)
MEK		96 % (1 %)	96 % (3 %)	90 % (4 %)	93 % (4 %)

[A] Soil 1 is a river bank soil (49 % sand, 26 % silt, 24 % clay, 5.3 % organic material, and approximately 14 % moisture).
[B] Average percent recovery based on mean values determined for five time-zero samples and five stored samples
[C] Value in parentheses is the percent relative standard deviation of the concentration values in the five stored samples. The percent relative standard deviation of the concentration values in the time-zero samples ranged from 1 to 4 % for all tests using soil 1 in the 25-g device.

87–102 % for storage of the river bank soil samples in the 5-g En Core samplers for all of the storage conditions used in this study. Taking into account the error introduced in preparation and analysis of the samples, these data show little loss of the VOCs of interest from the river bank soil during storage in the 5-g En Core samplers. The mean of the nine average percent recovery values listed in Table X1.2 for the analytes of interest in the samples of river bank soil stored at 4 ±2°C for 48 h is 98 % recovery with a standard deviation of 4 % and a percent relative standard deviation of 4 %. The overall mean of the 36 average percent recovery values listed in Table X1.2 for the analytes of interest and various storage conditions is 97 % recovery with a standard deviation of 4 % and a percent relative standard deviation of 4 %.

X1.1.7.2 As shown in Table X1.3, the VOCs of interest in this study have average percent recovery values ranging from 79–102 % for storage of the river bank soil samples in the 25-g En Core samplers for all of the storage conditions used in this study. Taking into account the error introduced in preparation and analysis of the samples, these data show little loss of the VOCs of interest from the river bank soil during storage in the 25-g En Core samplers. The mean of the nine average percent recovery values listed in Table X1.3 for the analytes of interest in the samples of river bank soil stored at 4 ±2°C for 48 h is 95 % recovery with a standard deviation of 3 % and a percent relative standard deviation of 3 %. The overall mean of the 36 average percent recovery values listed in Table X1.3 for the analytes of interest and various storage conditions is 95 % recovery with a standard deviation of 6 % and a percent relative standard deviation of 6 %.

X1.1.8 The average percent recoveries of the VOCs of interest from samples of the mountain soil stored in the 5-g En Core samplers are shown in Table X1.4; and, the average percent recoveries of the VOCs of interest from samples of the mountain soil stored in the 25-g En Core samplers are shown in Table X1.5. In each of these tables, the percent relative standard deviation of the concentration values in the stored samples is given in parentheses next to the corresponding average percent recovery value. The percent relative standard deviation of the concentration values in the time-zero samples is given in a footnote to each of the tables.

X1.1.8.1 As shown in Table X1.4, TCE, toluene, PCE, ethylbenzene, m/p-xylene, and o-xylene have average percent recovery values for storage of the mountain soil at 4 ±2°C for 48 h in the 5-g En Core samplers that range from 90–97 %. Average percent recovery values of these compounds from the mountain soil samples stored for 7 days at −12 ±2°C are similar, ranging from 91–101 %. Taking into account the error introduced in preparation and analysis of the samples, these data show very little or no loss of the six VOCs from the mountain soil during storage in the 5-g samplers under the two different storage conditions. The average percent recovery values for TCE, toluene, PCE, ethylbenzene, m/p-xylene, and o-xylene from the mountain soil samples stored for four days at 4 ±2°C and for 48 h at 4 ±2°C followed by five-day storage at −12± 2°C in the 5-g En Core samplers are slightly lower, ranging from 80–92 %. The same trend is shown in Table X1.4 for CDCE and benzene. These compounds have average

TABLE X1.4 Average Percent Recoveries of VOCs from Samples of Soil 2[A] Stored in 5-G En Core Samplers

VOCs	Storage Conditions	4 ±2°C/48 h	4 ±2°C/4 Days	4 ±2°C/48 h then −12± 2°C/5 Days	−12± 2°C/7 Days
CDCE		87 %[B] (10 %)[C]	77 %[B] (10 %)[C]	74 %[B] (7 %)[C]	84 %[B] (11 %)[C]
Benzene		86 % (11 %)	74 % (12 %)	69 % (9 %)	84 % (12 %)
TCE		91 % (8 %)	83 % (9 %)	80 % (5 %)	91 % (9 %)
Toluene		90 % (5 %)	83 % (7 %)	86 % (2 %)	97 % (8 %)
PCE		96 % (4 %)	89 % (6 %)	90 % (5 %)	99 % (6 %)
Ethylbenzene		92 % (7 %)	90 % (6 %)	90 % (10 %)	96 % (16 %)
m\p-Xylene		92 % (2 %)	86 % (6 %)	89 % (2 %)	101 % (5 %)
o-Xylene		97 % (2 %)	92 % (6 %)	92 % (5 %)	97 % (10 %)
MEK		83 % (0 %)	76 % (7 %)	74 % (4 %)	71 % (9 %)

[A] Soil 2 is a mountain soil (75 % sand, 13 % silt, 12 % clay, 4.3 % organic material, and approximately 12 % moisture).
[B] Average percent recovery based on mean values determined for five time-zero samples and five stored samples
[C] Value in parentheses is the percent relative standard deviation of the concentration values in the five stored samples. The percent relative standard deviation of the concentration values in the time-zero samples ranged from 3 to 8 % for all tests using soil 2 in the 5-g device.

TABLE X1.5 Average Percent Recoveries of VOCs from Samples of Soil 2[A] Stored in 25-G En Core Samplers

VOCs	Storage Conditions	4 ±2°C/48 h	4 ±2°C/4 Days	4 ±2°C/48 h then − 12 ±2°C/5 Days	−12 ±2°C/7 Days
CDCE		87 %[B] (8 %)[C]	82 %[B] (11 %)[C]	59 %[D] (17 %)[C]	71 %[B] (25 %)[C]
Benzene		90 % (11 %)	81 % (14 %)	56 % (18 %)	71 % (30 %)
TCE		94 % (6 %)	88 % (9 %)	69 % (9 %)	82 % (17 %)
Toluene		93 % (6 %)	87 % (5 %)	78 % (5 %)	89 % (12 %)
PCE		98 % (6 %)	92 % (4 %)	85 % (4 %)	92 % (11 %)
Ethylbenzene		96 % (4 %)	94 % (4 %)	90 % (2 %)	98 % (7 %)
m\p-Xylene		93 % (4 %)	91 % (3 %)	92 % (2 %)	98 % (7 %)
o-Xylene		96 % (4 %)	96 % (5 %)	93 % (2 %)	96 % (6 %)
MEK		86 % (2 %)	83 % (6 %)	75 % (2 %)	73 % (2 %)

[A] Soil 2 is a mountain soil (75 % sand, 13 % silt, 12 % clay, 4.3 % organic material, and approximately 12 % moisture).
[B] Average percent recovery based on mean values determined for five time-zero samples and five stored samples
[C] Value in parentheses is the percent relative standard deviation of the concentration values in the stored samples. The percent relative standard deviation of the concentration values in the time-zero samples ranged from 1 to 8 % for all tests using soil 2 in the 25-g device.
[D] Average percent recovery based on mean values determined for five time-zero samples and three stored samples.

percent recovery values ranging from 84–87 % for the samples stored at 4 ±2°C for 48 h and –12 ±2°C for seven days. While the percent recovery values of CDCE and benzene from the mountain soil samples stored for four days at 4 ±2°C and for 48 h at 4 ±2°C followed by five-day storage at –12 ±2°C are slightly lower, ranging from 69 to 77 %. The average percent recovery values of MEK from the mountain soil samples stored under the four storage conditions in the 5-g En Core samplers range from 71–83 %. These values are lower than would be expected for this less volatile compound, especially when compared to the average percent recovery values of the other VOCs listed in Table X1.4. The reason for the lower average percent recovery values for MEK from the mountain soil may be that the ketone reacts with this soil causing a decrease in MEK concentrations with time.

X1.1.8.2 As shown in Table X1.5, TCE, toluene, PCE, ethylbenzene, m/p-xylene, and o-xylene have average percent recovery values ranging from 93–98 % for storage of the mountain soil samples in the 25-g En Core samplers at 4 ±2°C for 48 h. Average percent recovery values of CDCE and benzene from samples of the mountain soil stored in the 25-g En Core samplers for 48 h at 4 ±2°C are 87 and 90 %, respectively. For four-day storage at 4 ±2°C in the 25-g En Core samplers, TCE, toluene, PCE, ethylbenzene, m/p-xylene, and o-xylene have average percent recovery values from the mountain soil samples ranging from 87–96 %; and average percent recoveries of CDCE and benzene from these samples are 82 and 81 %, respectively. In testing storage of the spiked mountain soil samples in the 25-g En Core samplers for 48 h at 4 ±2°C followed by five-day storage at –12 ±2°C, two of the devices used to store the samples had O-rings in their caps that were bent out of position as discussed in 6.4.2 and 6.7.1. Because of this error in capping these devices, the seal between the cap and coring body/storage chamber was compromised. As a result, data for only three of the stored samples were used to calculate the average percent recovery values for this storage condition. As shown in Table X1.5, these values range from 56–69 % for CDCE, benzene, and TCE; and 78–93 % for toluene, PCE, ethylbenzene, m/p-xylene, and o-xylene. For seven-day storage at –12 ±2°C in the 25-g En Core samplers, TCE, toluene, PCE, ethylbenzene, m/p-xylene, and o-xylene have average percent recovery values from the mountain soil samples ranging from 82–98 %; and average percent recovery values of CDCE and benzene from these samples are both 71 %. The average percent recovery values of MEK from the mountain soil samples stored under the four storage conditions in the 25-g En Core samplers range from 73–86 %. As discussed in X1.1.8.1, these values are lower than would be expected and may be due to a chemical reaction between MEK and the mountain soil.

X1.1.8.3 Differences in the average percent recovery values of the VOCs of interest in this study from the river bank soil (Tables X1.2 and X1.3) and mountain soil (Tables X1.4 and X1.5) may be due to the difference in the composition of these soils. As described in X1.1.1, the river bank soil is 49 % sand, 26 % silt, and 24 % clay, while the mountain soil contains

75 % sand, 13 % silt, and 12 % clay.

X1.1.9 The average percent recoveries of the VOCs of interest from samples of the prairie soil stored in the 5-g En Core samplers are shown in Table X1.6; and the average percent recoveries of the VOCs of interest from samples of the prairie soil stored in the 25-g En Core samplers are shown in Table X1.7. In each of these tables, the percent relative standard deviation of the concentration values in the stored samples is given in parentheses next to the corresponding average percent recovery value. The percent relative standard deviation of the concentration values in the time-zero samples is given in a footnote to each of the tables.

X1.1.9.1 As shown in Table X1.6, PCE, ethylbenzene, m/p-xylene, o-xylene, and MEK have average percent recovery values ranging from 90–92 % for storage of the prairie soil samples in the 5-g En Core samplers at 4 ±2°C for 48 h. Under these same storage conditions, average percent recovery values of CDCE, benzene, TCE, and toluene from samples of the prairie soil stored in the 5-g En Core samplers range from 75–82 %. For four-day storage at 4 ±2°C in the 5-g En Core samplers, ethylbenzene, m/p-xylene, o-xylene, and MEK have average percent recovery values from the prairie soil samples ranging from 85–92 %. Under these same storage conditions, average percent recovery values of CDCE, benzene, TCE, toluene, and PCE from samples of the prairie soil stored in the 5-g En Core samplers range from 42–75 %. For 48-h storage at 4 ±2°C followed by five-day storage at –12 ±2°C in the 5-g samplers, PCE, ethylbenzene, m/p-xylene, o-xylene, and MEK have average percent recovery values from the stored prairie soil samples ranging from 64–85 %. Under these same storage conditions, CDCE, benzene, TCE, and toluene have average percent recovery values ranging from 30–52 %. For seven-day storage at –12 ±2°C in the 5-g En Core samplers, PCE, ethylbenzene, m/p-xylene, o-xylene, and MEK have average percent recovery values from the prairie soil samples ranging from 83–99 %. Under these same storage conditions, average percent recovery values of CDCE, benzene, and TCE from samples of the prairie soil stored in the 5-g En Core samplers range from 36–61 %; and, the average percent recovery value of toluene is 77 %.

X1.1.9.2 As shown in Table X1.7, PCE, ethylbenzene, m/p-xylene, o-xylene, and MEK have average percent recovery values ranging from 86–97 % for storage of the prairie soil samples in the 25-g En Core samplers at 4 ±2°C for 48 h. Under these same storage conditions, average percent recovery values of CDCE, benzene, TCE, and toluene from samples of the prairie soil stored in the 25-g En Core samplers range from 69–79 %. For four-day storage at 4 ±2°C in the 25-g En Core samplers, ethylbenzene, m/p-xylene, o-xylene, and MEK have average percent recovery values from the prairie soil samples ranging from 79–93 %. Under these same storage conditions, average percent recovery values of CDCE, benzene, TCE, and toluene from samples of the prairie soil stored in the 25-g En Core samplers range from 50–64 %; and, the average percent recovery value of PCE is 73 %. In testing storage of the spiked prairie soil samples in the 25-g En Core samplers for 48-h at 4 ±2°C followed by five-day storage at –12 ±2°C, two of the devices used to store the samples had O-rings in their caps that were bent out of position as discussed in 6.4.2 and 6.7.1. Because of this error in capping these devices, the seal between the cap and coring body/storage chamber was compromised. As a result, data for only three of the stored samples were used to calculate the average percent recovery values for this storage condition. As shown in Table X1.7, ethylbenzene, m/p-xylene, o-xylene, and MEK have average percent recovery values that range from 71–85 %. Under these same storage conditions, percent recovery values for toluene and PCE are 46 and 57 %, respectively; and, average percent recovery values for CDCE, benzene, and TCE range from 22–33 %. For seven-day storage at –12 ±2°C in the 25-g En Core samplers, PCE, ethylbenzene, m/p-xylene, o-xylene, and MEK have average percent recovery values from the prairie soil samples ranging from 86–98 %. Under these same storage conditions, average percent recovery values of CDCE, benzene, and TCE range from 54–63 %; and, the average percent recovery value of toluene is 75 %.

X1.1.9.3 The average percent recovery values for the VOC-spiked prairie soil samples stored in the 5-g and 25-g En Core samplers are lower than the values reported for the river bank soil (Tables X1.2 and X1.3) and mountain soil (Tables X1.4 and X1.5). As described in section X1.1.1, the river bank and mountain soils contain approximately 12–14 % moisture, while the prairie soil contains approximately 8 % moisture. Samples of this drier soil collected from the large container of loose soil used in this study were difficult to compact in the En Core samplers. In addition, scraping the surface of the prairie soil to make it flush with the opening of the coring body/storage chamber, so the sampler could be capped, loosened particles of the dry soil. It is believed that for many of the

TABLE X1.6 Average Percent Recoveries of VOCs from Samples of Soil 3[A] Stored in 5-G En Core Samplers

VOCs	Storage Conditions	4 ±2°C/48 h	4 ±2°C/4 Days	4 ±2°C/48 h then –12± 2°C/5 Days	–12± 2°C/7 Days
CDCE		82 %[B] (9 %)[C]	64 %[B] (40 %)[C]	36 %[B] (6 %)[C]	51 %[B] (15 %)[C]
Benzene		75 % (13 %)	42 % (23 %)	30 % (8 %)	36 % (58 %)
TCE		79 % (10 %)	51 % (16 %)	42 % (5 %)	61 % (17 %)
Toluene		82 % (8 %)	75 % (9 %)	52 % (3 %)	77 % (11 %)
PCE		91 % (5 %)	58 % (8 %)	64 % (3 %)	89 % (11 %)
Ethylbenzene		91 % (3 %)	85 % (16 %)	70 % (4 %)	93 % (8 %)
m\p-Xylene		90 % (1 %)	87 % (13 %)	71 % (3 %)	96 % (7 %)
o-Xylene		92 % (2 %)	86 % (4 %)	79 % (3 %)	99 % (7 %)
MEK		92 % (0 %)	92 % (6 %)	85 % (0 %)	83 % (13 %)

[A] Soil 3 is a prairie soil (67 % sand, 17 % silt, 16 % clay, 1.5 % organic material, and approximately 8 % moisture).
[B] Average percent recovery based on mean values determined for five time-zero samples and five stored samples
[C] Value in parentheses is the percent relative standard deviation of the concentration values in the five stored samples. The percent relative standard deviation of the concentration values in the time-zero samples ranged from 0 to 11 % for all tests using soil 3 in the 5-g device.

TABLE X1.7 Average Percent Recoveries of VOCs from Samples of Soil 3[A] Stored in 25-G En Core Samplers

VOCs	Storage Conditions	4 ±2°C/48 h	4 ±2°C/4 Days	4 ±2°C/48 h then −12± 2°C/5 Days	−12± 2°C/7 Days
CDCE		76 %[B] (17 %)[C]	59 %[B] (34 %)[C]	29 %[D] (42 %)[C]	59 %[B] (47 %)[C]
Benzene		69 % (25 %)	50 % (50 %)	22 % (51 %)	54 % (57 %)
TCE		72 % (22 %)	55 % (39 %)	33 % (36 %)	63 % (44 %)
Toluene		79 % (14 %)	64 % (26 %)	46 % (20 %)	75 % (29 %)
PCE		86 % (14 %)	73 % (24 %)	57 % (15 %)	86 % (21 %)
Ethylbenzene		89 % (6 %)	79 % (11 %)	71 % (7 %)	91 % (12 %)
m\p-Xylene		88 % (6 %)	81 % (9 %)	73 % (5 %)	94 % (10 %)
o-Xylene		94 % (4 %)	89 % (6 %)	79 % (3 %)	98 % (8 %)
MEK		97 % (3 %)	93 % (0 %)	85 % (2 %)	92 % (5 %)

[A] Soil 3 is a prairie soil (67 % sand, 17 % silt, 16 % clay, 1.5 % organic material, and approximately 8 % moisture).
[B] Average percent recovery based on mean values determined for five time-zero samples and five stored samples
[C] Value in parentheses is the percent relative standard deviation of the concentration values in the stored samples. The percent relative standard deviation of the concentration values in the time-zero samples ranged from 1 to 6 % for all tests using soil 3 in the 25-g device.
[D] Average percent recovery based on mean values determined for five time-zero samples and three stored samples.

prairie soil samples prepared in this study, the loose soil particles scattered when the samplers were capped affecting the seals of the samplers. Many of the percent relative standard deviation values shown in Tables X1.6 and X1.7 for the stored prairie soil samples are higher than those reported for the river bank soil samples (Tables X1.2 and X1.3) and mountain soil samples (Tables X1.4 and X1.5). These values reflect the variability of analyte concentrations in the stored samples of the spiked prairie soil. This variability is believed to be due to scattered soil particles affecting the seals of the En Core devices to varying degrees during capping and storage. A cautionary note to 6.4.2 of this practice has been included to alert the user to use caution when working with loose, dry soils, (see Note 2).

Designation: D 6498 – 99

Standard Guide for Household Hazardous Waste Training Outline for Household Hazardous Waste Collection Operations[1]

This standard is issued under the fixed designation D 6498; the number immediately following the designation indicates the year of original adoption or, in the case of revision, the year of last revision. A number in parentheses indicates the year of last reapproval. A superscript epsilon (ε) indicates an editorial change since the last revision or reapproval.

1. Scope

1.1 This guide covers recommended health and safety training topics for workers at operations for the collection of household hazardous waste or conditionally exempt small quantity generator waste, or both, regardless of the type of collection. Although this guide is intended to protect the worker, public health, and the environment, it is not intended to satisfy all the health and safety training requirements under the Occupational Safety and Health Act of 1970 or the Resource Conservation and Recovery Act of 1976. Additionally, local and state requirements may also vary. Therefore, it is recommended that the operator of a household hazardous waste collection operation also check federal, state, and local regulations for additional requirements.

1.2 *This standard does not purport to address all of the safety concerns, if any, associated with its use. It is the responsibility of the user of this guide to establish appropriate safety and health practices and determine the applicability of regulatory limitations prior to use.*

2. Referenced Documents

2.1 *ANSI Standards:*

ANSI Z88.2 Practices for Respiratory Protection[2]

ANSI Z358.1 Emergency Eyewash and Shower Equipment[2]

3. Terminology

3.1 *Definitions of Terms Specific to This Standard:*

3.1.1 *conditionally exempt small quantity generator*—a generator that generates not more than 100 Kg of hazardous waste in a calendar month and generates not more than 1 Kg of acute hazardous waste in a calendar month, and stores not more than 1000 Kg of hazardous waste on site at any one time during the month.

3.1.2 *household hazardous waste collection*—a permanent site, temporary location, or mobile or residential operation for the collection of household hazardous waste. Some household hazardous waste collections also accept/collect conditionally exempt small quantity generator waste in addition to household hazardous waste.

3.1.3 *state OSHA program*—a state which has been authorized by the United States Assistant Secretary of Labor for Occupational Safety and Health to implement the Williams-Steiger Occupational Safety and Health Act of 1970.

3.1.4 *worker*—an employee or volunteer of a household hazardous waste collection or an employee of a contractor who has been contracted to perform services at a household hazardous waste collection.

3.2 *Acronyms:Acronyms:*

3.2.1 *CESQG*—conditionally exempt small quantity generator.

3.2.2 *CFR*—Code of Federal Regulations.

3.2.3 *D.O.T.*—United States Department of Transportation.

3.2.4 *HHW*—household hazardous waste.

3.2.5 *HHWC*—household hazardous waste collection.

3.2.6 *OSHA*—United States Department of Labor, Occupational Safety and Health Administration.

3.2.7 *RCRA*—Resource Conservation and Recovery Act.

3.2.8 *USEPA*—United States Environmental Protection Agency.

4. Significance and Use

4.1 This guide is written for all persons involved with HHWCs, but especially for those primarily responsible for establishing and providing training to workers at HHWCs. This guide is intended to provide recommended training topics which should be covered during the initial baseline and annual refresher training. The actual topics and depth of training for each worker must be assessed by the HHWC operator on a case-by-case basis with emphasis on the particular job assignment for each worker. The major factors the operator should consider are what are the anticipated hazards to which each worker may be exposed and what are the controls/work practices which the worker must know in order to do his or her job assignments safely.

4.2 Another factor which the operator must consider is the areas in which each worker is expected to work or the areas to which each worker has access to as part of job assignments. For example, if an administrative assistant only has access to the office area, and never enters the active HHWC area, then training may be limited to required actions as part of the

---

[1] This guide is under the jurisdiction of ASTM Committee D34 on Waste Management and is the direct responsibility of Subcommittee D34.07 on Land Management.

Current edition approved Dec. 10, 1999. Published February 2000.

[2] Available from American National Standards Institute, 11 W. 42nd St., 13th Floor, New York, NY 10036.

contingency plan. If, however, the administrative assistant is periodically required to enter into the active HHWC area, more detailed training is required based on anticipated hazards which the administrative assistant may be exposed to while in the HHWC area.

4.3 When deciding on the training topics for the annual refresher training, the HHWC operator must decide on topics in Section 7 which are relevant to the HHWC workers attending the refresher training class. Factors which should be considered are:

4.3.1 Hazards to which the workers may be exposed to,

4.3.2 Changes in the standard operating procedures, contingency plan since the previous training,

4.3.3 Length of time from which a worker has received training in the topics listed in Section 7, and

4.3.4 Identification of incidents, hazards, unsafe conditions, or any other situations which indicate a need to retrain workers in a particular topic to ensure a safe and healthful workplace.

5. Initial and Periodic Training

5.1 Listed below is the recommended initial baseline training and frequency of periodic and ongoing training for HHWCF workers.

5.1.1 *Initial*—Initial baseline training shall be provided to the worker prior to assignment of job duties at a HHWC. The initial baseline training should cover all the applicable modules in Section 7 which are specific for each worker's job duties/assignments. It is recommended that the initial baseline training consist of a combination of classroom instruction and supervised on-the-job type training. HHWC workers should not be allowed to work in unsupervised positions until they have completed all the required baseline training. An effective baseline training must also include instruction in site-specific hazards and site-specific procedures/protocols. Therefore, if a worker receives his or her baseline training from an outside source, such as a university extension or at a conference, the HHWC operator must then supplement the outside training with on-the-job-training and site/job-specific information, hazards, procedures and work practices.

5.1.2 *Periodic*—Periodic refresher training shall be provided at least annually.

5.1.3 *"Tailgate" Safety Meetings*—Supplementary to the periodic/annual training, it is recommended that the HHWC operator provide ongoing training via short "tailgate" safety meetings in which health and safety topics are discussed with workers. It is also recommended that tailgate meetings should be held prior to collections and whenever any of the following occur:

5.1.3.1 Change in standard operating procedures,

5.1.3.2 Change in the contingency plan,

5.1.3.3 Introduction of new equipment, supplies, or machinery which may result in worker exposure to new hazards,

5.1.3.4 Identification of new or previously unidentified hazards at the HHWC.

5.1.3.5 Worker assignment to a new/different job/task for which the worker has not previously received training within the past six months,

5.1.3.6 Identification of inadequacies in worker's knowledge to indicate that the worker(s) has/have not retained the requisite understanding or knowledge from the previous training classes,

5.1.3.7 Introduction of a new or previously unidentified waste stream,

5.1.3.8 Any other situation arises in which retraining is necessary to ensure a safe and healthful workplace,

5.1.3.9 Problems or incidents, or both, which have occurred at other household hazardous waste collections.

6. Qualification of Trainers

6.1 *Initial and Periodic Training*—Trainers shall have sufficient health and safety experience/knowledge in the topics outlined below.

6.2 *Tailgate Meetings*—Tailgate training can be facilitated by managers, supervisors, and/or lead persons, provided such persons have sufficient knowledge and experience in the health and safety topics being discussed.

7. Training Modules

7.1 *Chemical and Physical Hazards*—This module should discuss the basic chemical terms that participants need to know to understand chemical hazards. The module should discuss the criteria/characteristics that make a chemical or waste hazardous, including:

7.1.1 Chemical and physical properties,

7.1.2 Reactivity,

7.1.3 Flammability,

7.1.4 Corrosivity,

7.1.5 Toxicity,

7.1.6 Chemical nomenclature, and

7.1.7 Chemical compatibility.

7.2 *Hazard Awareness*—This module should discuss potential chemical, physical and biological hazards which may be encountered at the HHWC.

7.2.1 *Introduction*:

7.2.2 *Chemical Hazards*:

7.2.2.1 Chemical exposure,

7.2.2.2 Fire and explosion,

7.2.2.3 Chemical reactions,

7.2.2.4 Explosive or flammable atmospheres,

7.2.2.5 Oxygen enriched atmospheres,

7.2.2.6 Reactives,

7.2.2.7 Compressed gases,

7.2.2.8 Oxygen deficiency,

7.2.2.9 Bulking of volatiles,

7.2.2.10 Carbon monoxide from vehicle exhaust,

7.2.2.11 Asbestos,

7.2.2.12 Dusts and particulates,

7.2.2.13 Pesticides, herbicides, and rodenticides, and

7.2.2.14 Hazards of common types of HHW.

7.2.3 *Biological Hazards*—Potential biological hazards which may be encountered at HHWCs should be discussed. These hazardous should include but not be limited to: medical and infectious waste; bites or stings from animals or insects; and toxic effects from plants.

7.2.4 *Physical Hazards*:

7.2.4.1 Ionizing radiation,

7.2.4.2 Electrical hazards,

7.2.4.3 Vehicles and machinery,

7.2.4.4 Noise,

7.2.4.5 Slips, trips, and falls,

7.2.4.6 Ergonomics, safe lifting, repetitive motion injuries, and

7.2.4.7 Thermal stress (heat stress/heat strain and cold stress).

7.3 *Toxicology*—This module should discuss basic toxicological principles in order for the worker to understand and recognize hazards which he or she may be exposed to.

7.3.1 Introduction,

7.3.2 Hazard versus risk,

7.3.3 Toxicology,

7.3.4 Classification of toxic materials,

7.3.5 Routes of exposure,

7.3.6 Dose - response relationship,

7.3.7 Length of exposure, and

7.3.8 Signs and symptoms of exposure.

7.4 *Industrial Hygiene*—This module should discuss basic industrial hygiene principles such as recognition, evaluation, and control of hazards.

7.4.1 *Recognition*:

7.4.1.1 Personal and air monitoring equipment,

7.4.1.2 Monitoring equipment chart,

7.4.1.3 Oxygen deficient atmospheres,

7.4.1.4 Explosive/flammable atmospheres, and

7.4.1.5 Toxic atmospheres.

7.4.2 *Evaluation*:

7.4.2.1 Calculation of dose and exposure levels, and

7.4.2.2 Evaluation of hazards.

7.4.3 *Control*:

7.4.3.1 *Selection of Worker Health and Safety Protective Measures*—Engineering controls, administrative controls; or personal protective equipment, or all of these, and

7.4.3.2 Selection and monitoring of personal protective clothing and equipment.

7.5 *Rights and Responsibilities of Workers Under OSHA*—This module should discuss the rights and responsibilities of workers under the OSHA Act or the applicable State OSHA Program.

7.5.1 *Employer Responsibility (OSHA or Applicable Equivalent State Program; General Duty Clause)*:

7.5.1.1 Manager and supervisor responsibilities, and

7.5.1.2 Primary responsibilities of the HHWC operator over contractors.

7.5.2 *Applicable Provisions Under Title 29 CFR (Or Equivalent State Regulations)*:

7.5.2.1 Hazard communication regulations,

7.5.2.2 Access to employee exposure and medical records, and

7.5.2.3 Injury illness log (OSHA Log 200).

7.5.3 *Worker Rights*:

7.5.3.1 Worker representation during OSHA inspection,

7.5.3.2 File a complaint,

7.5.3.3 Contest time allowed for abatement of citation, and

7.5.3.4 Refuse unsafe work.

7.6 *Site Safety*—This module should discuss the site safety procedures, practices, and protocols which have been established for the safe operation of the HHWC.

7.6.1 Preparedness and prevention,

7.6.2 Safe practices,

7.6.3 Safety briefings and meetings,

7.6.4 Standard operating procedures,

7.6.5 Site safety maps,

7.6.6 Safety committee, and

7.6.7 Reporting of observed hazards, unsafe conditions, or unsafe work practices.

7.7 *Emergencies*—This module should discuss the HHWC's established emergency response plan, contingency plan, along with any coordination which may be required with other local, state, or federal agencies. Each worker should be trained and instructed in their role/responsibilities in the event of an emergency. The module should also discuss the required emergency response equipment which is located at the HHWC, along with a discussion/training in the use of the emergency equipment which each worker may be required to use. Lastly, the module should discuss follow-up investigations, critique, reports, or documentation which may be required following any emergency, spills, or release.

7.7.1 Emergency equipment,

7.7.2 Emergency response plan,

7.7.3 Contingency plan,

7.7.4 Releases,

7.7.5 Initial response,

7.7.6 Emergency help and self rescue,

7.7.7 Emergency decontamination,

7.7.8 Follow-up investigation and documentation, and

7.7.9 Use of fire extinguishers.

7.8 *Standard Operating Procedures (SOPs)* This module should discuss the established SOPs for operation of the HHWC. Each worker must be instructed as to the safe work practices for each job/task which he or she must perform.

7.8.1 Waste acceptance,

7.8.2 Receiving,

7.8.3 Unloading and containment of leaking and open containers,

7.8.4 Ergonomics,

7.8.5 Waste identification and classification,

7.8.6 Timely testing and identification of unknowns,

7.8.7 Segregation of incompatible chemicals,

7.8.8 Handling,

7.8.9 Bulking, and

7.8.10 Lab pack/loose pack.

7.8.11 *Preparation for Transportation*:

7.8.11.1 Manifesting,

7.8.11.2 Hazard class divisions,

7.8.11.3 Shipping descriptions,

7.8.11.4 Emergency response information,

7.8.11.5 Marking requirements,

7.8.11.6 Labeling requirements,

7.8.11.7 Performance oriented packaging,

7.8.11.8 Placarding,

7.8.11.9 Segregation during transportation,

7.8.12 Safe loading practices,

7.8.13 Storage,

7.8.14 Other onsite processing

7.8.15 Equipment use and lockout/tagout, and

7.8.16 Decontamination procedures.

7.9 *Control of Exposures*—This module should discuss the use of engineering controls, administrative controls, or personal protective equipment, or all of these, to control harmful exposures to workers and/or the general public. Engineering controls should be discussed as the preferred method, followed by administrative controls and personal protective equipment as the second and third choices respectively to control exposures.

7.9.1 *Engineering Controls*:

7.9.1.1 Ventilation,

7.9.1.2 Use of barriers,

7.9.1.3 Process isolation (eating area or bulking area),

7.9.1.4 Substituting a less toxic chemical (maintenance operations), and

7.9.1.5 Mechanized processing equipment.

7.9.2 *Administrative Controls*:

7.9.2.1 Minimize worker exposure time,

7.9.2.2 Rotation of workers, and

7.9.2.3 Rotation of work hours.

7.9.3 Personal protective equipment.

7.10 *Personal Protective Equipment*—This module should discuss the HHWC's established personal protective equipment program and the types of personal protective equipment each worker is required to use.

7.10.1 *Personal Protective Equipment Program*:

7.10.1.1 General protection,

7.10.1.2 Uses and limitations,

7.10.1.3 Inspection, cleaning, maintenance, storage, and sanitation,

7.10.1.4 Required personal protective equipment for each task,

7.10.1.5 Required personal protective equipment in designated areas,

7.10.1.6 Potential hazards from wearing personal protective equipment,

7.10.1.7 Personal protective equipment SOPs,

7.10.1.8 Required marking/identification of personal protective equipment,

7.10.1.9 Donning and doffing, and

7.10.1.10 Plan to ensure the effectiveness of the personal protective equipment program.

7.10.2 *Type of Personal Protective Equipment*:

7.10.2.1 Head protection,

7.10.2.2 Eye and face protection,

7.10.2.3 Hearing protection,

7.10.2.4 Hand and arm protection,

7.10.2.5 Foot protection,

7.10.2.6 Body protection, and

7.10.2.7 Respiratory protection.

7.10.3 *USEPA Levels of Protection*:

7.10.3.1 Level A,

7.10.3.2 Level B,

7.10.3.3 Level C, and

7.10.3.4 Level D.

7.11 *Medical Program*—This module should discuss the medical surveillance program which the operator of the HHWC has implemented to ensure the satisfactory maintenance of worker health and ascertain the effectiveness of the control methods. The module should also discuss the first aid materials available at the HHWC along with the required first aid/cardiopulmonary resuscitation (CPR) training for designated workers. In the event of an injury/emergency, the module should discuss each worker's responsibility.

7.11.1 Design, planning, and implementation of the medical monitoring program,

7.11.2 Respirator use certification,

7.11.3 Audiometric examinations,

7.11.4 First aid,

7.11.5 Stress recognition,

7.11.6 First aid for chemical exposures,

7.11.7 CPR, and

7.11.8 Emergency drills.

7.12 *Legal and Regulatory Aspects*—This module should discuss the environmental and occupational laws, regulations and agencies applicable to the operation of the HHWC.

7.12.1 Introduction to RCRA,

7.12.2 D.O.T.,

7.12.3 OSHA,

7.12.4 State and local regulations, and

7.12.5 Fire codes.

7.13 *Hygiene and Sanitation*—This module should discuss hygiene and sanitation practices to minimize potential inhalation, ingestion, or absorption of toxic materials, prevent fire hazards, and minimize slip/trip/fall injuries.

7.13.1 General housekeeping,

7.13.2 Consumption and storage of food and beverages,

7.13.3 Change rooms,

7.13.4 Washing facilities,

7.13.5 Potable drinking water, and

7.13.6 Sanitary facilities.

7.14 *Reference Materials*—This module should discuss reference materials which are available to the workers and operators at the HHWC. Recommended reference material are **(1-12)**[3]:

7.14.1 ANSI Z88.2, and

7.14.2 ANSI Z358.1.

8. Keywords

8.1 CESQG; conditionally exempt small quantity generator; HHW; home chemicals; household hazardous waste; pollution prevention; waste disposal; worker health and safety training

[3] The boldface number given in parentheses refer to a list of references at the end of the text.

APPENDIXES

X1. OSHA REGULATIONS INFORMATION SOURCE (ADVISORY)

X1.1 The following is a list of OSHA regulatory sections in 29 CFR which may be applicable to HHWCs. Applicability of a regulation would depend on the operation(s) or procedure(s) which are specific to each individual HHWC operation. In states which are authorized state OSHA programs, the HHWC operator should consult the applicable equivalent state regulations.

OSHA Regulation (29 CFR)[4]	Subject
1903.2	Posting of Notice; Availability of the Act, Regulations and
1904.1 - 1904.16	Recordkeeping and Reporting Occupational Injuries and Illnesses
1910.21 - 1910.30	Walking - Working Surfaces
1910.35 - 1910.37	Means of Egress
1910.38	Employee Emergency Plans and Fire Prevention Plan
1910.66 - 1910.68	Powered Platforms, Manlifts, and Vehicle-Mounted Work
1910.94	Ventilation
1910.95	Occupational Noise Exposure
1910.97	Non-Ionizing Radiation
1910.101	Compressed Gases
1910.106	Flammable and Combustible Liquids
1910.120	Hazardous Waste Operations and Emergency Response
1910.132 - 1910-139	Personal Protective Equipment
1910.141	Sanitation
1910.144	Safety Color Code for Marking Physical Hazards
1910.145	Specifications for Accident Prevention Signs and Tags
1910.146 - 1910.147	Confined Spaces
1910.151	Medical Services and First Aid
1910.157	Portable Fire Extinguishers
1910.158	Standpipe and Hose Systems
1910.159 - 1910.163	Fixed Fire Suppression Equipment
1910.164	Fire Detection Systems
1910.165	Employee Alarm Systems
1910.169	Compressed Gas and Compressed Air Equipment
1910.176	Handling Materials - General
1910.178	Powered Industrial Trucks
1910.211 - 1910.212	Machinery and Machine Guarding
1910.241 - 1910-244	Hand and Portable Powered Tools and Other Hand-Held
1910.303	Electrical Safety - General Requirements
1910.304	Electrical - Wiring Design and Protection
1910.305	Electrical - Wiring Methods, Components and Equipment for
1910.306	Electrical - Specific Purpose Equipment and Installations
1910.307	Electrical - Hazardous (Classified) Locations
1910.1000	Air Contaminants
1910.1001	Asbestos
1910.1003 - 1910.1016	13 Carcinogens
1910.1017	Vinyl Chloride
1910.1018	Inorganic Arsenic
1910.1020	Access to Employee Exposure and Medical Records
1910.1025	Lead
1910.1027	Cadmium
1910.1028	Benzene
1910.1030	Bloodborne Pathogens
1910.1044	1,2-Dibromo-3-chloropropane
1910.1045	Acrylonitrile
1910.1047	Ethylene Oxide
1910.1048	Formaldehyde
1910.1050	Methylenedianiline
1910.1051	1,3-Butadine
1910.1052	Methylene Chloride
1910.1096	Ionizing Radiation
1910.1200	Hazard Communication
1910.1450	Occupational Exposure to Hazardous Chemicals in Laboratories

[4] Available from Superintendent of Documents, US Government Printing Office, Washington, DC 20402.

X2. DOT REGULATIONS INFORMATION SOURCE (ADVISORY)

X2.1 The following is a list of DOT regulatory sections in 49 CFR which may be applicable to HHWCs. Applicability of a regulation would depend on the operation(s) and procedure(s) which are specific to each individual HHWC operation. THE HHWC operator should also consult if there are additional state or local transportation regulations.

Dot Regulation (49 CFR)[4]	Subject
Part 172 Subpart H	Training
172.704	Training Requirements
Part 177 Subpart A	General Information and Regulations
177.800	Responsibility for Compliance and Training
177.816	Driver Training

X3. CANADIAN REGULATIONS INFORMATION SOURCE (ADVISORY)

X3.1 In Canada, the HHWC operator should consult the applicable Canada Labour regulations or the equivalent provincial or territorial regulations. Listed below are acts and regulations which may be applicable to HHWCs in Canada:

Federal:
Transportation of Dangerous Good Act, 1992
Canadian Environmental Protection Act
Canada Labour Code
Provincial and Territorial Acts and Regulations:
Prince Edward Island:
As of January 1, 1999 there were no provincial acts or regulations which directly apply to HHWC.
Newfoundland:
As of January 1, 1999 there were no provincial acts or regulations which directly apply to HHWC. The "Waste Materials Act" was passed on December 15, 1998.
Nova Scotia:
The Environment Act
New Brunswick:
Transportation of Dangerous Goods Regulation
Quebec:
Loi sur la qualité de l'environment (L.R.Q., chapitre Q-2)
Règlement sur les matières dangereuses (Q-2, r. 15.2)
Règlement sur le transport des matières dangereuses (pratiquement identique au TDG) les normes d'hygiènes au travail tel qu'édictées par la Commission de la santé et sécurité au travail (CSST).
Ontario:
Occupational Health and Safety Act
Ontario Environmental Protection Act
Environmental Assessment Act
Workplace Hazardous Materials Information System (WHMIS) Regulation: R.R.O. 1990, Regulation 860 under OHSA (Provincial)
Ontario Regulation 347 under the Provincial EPA (to be replaced soon by a new Waste Management Regulation)
Environmental Bill of Rights under the Provincial EPA

X3.2 Occupational Health and Safety Act, Revised Statutes of Ontario, 1990, Chapter 0.1, (Distributed February 1998):

X3.2.1 List of sections which may be applicable to HHWC's. Please note that it seems that the OHSA of Ontario has a slightly different role from the U.S. OSHA.

1.(1) Definitions
Part I: Application
2.(2) Application of other Acts
Part II: Administration
8.(1)-(5) Health and Safety Representative
8.(6)-(10) Inspections
8.(11)-(16) Powers of Representatives
9. Joint Health and Safety Committee
11. Industrial Hygiene
12. Summary of Lost Workdays, etc.
Part III: Duties of Employers and Other Persons
25. Duties of Employers
26. Additional Duties of Employers
27. Duties of Supervisor
28. Duties of Workers
29. Duties of Owners
32. Duties of Directors & Officers of a Corporation
Part IV: Toxic Substances
33. Orders of Director
36. Hazardous Materials Inventory
37. Hazardous Material Identification and Data Sheets
38. Inventory and Material Safety Data Sheets to be Made Available
40.1 Information Privileged
42. Instruction and Training
Part V: Right to Refuse to Stop Work where Health or Safety in Danger
43. -49.
Part VI: Reprisals by Employer Prohibited
all subsections
Part VII: Notices
all subsections
Part VII: Enforcement
54.(5) Entitlement to Time From Work
59. Notice of Compliance
61. Appeals from Order of an Inspector
62. Obstruction of/Assistance to Inspector
63. Information Confidential
Part IX: Offences & Penalties
66. -69.

Manitoba:
As of January 1, 1999 there were no provincial acts or regulations which directly apply to HHWC.
Saskatchewan:
Hazardous Substances and Waste Dangerous Good Regulation
Transportation of Dangerous Goods Act (Saskatchewan)
Alberta:
Alberta Environmental Protection and Enhancement Act
Occupational Health and Safety Act
Transportation of Dangerous Goods Control Act
British Columbia:
Waste Management Act
Special Waste Regulation
Return of Used Lubricating Oil Regulation
Post-Consumer Residual Stewardship Program Regulation
Post-Consumer Paint Stewardship Program Regulation
Yukon:
Special Waste Regulations
ODS Regulations
Dangerous Goods Transportation Act (Yukon)
Northwest Territories:
As of January 1, 1999, the Northwest Territories are splitting into two territories. There are no territorial acts or regulations which are specific to HHWC.

X4. USEPA REGULATIONS HHW AND CESQG EXEMPTIONS

X4.1 *HHW*—Household waste for the purposes of regulation are excluded from 40 CFR,[4] Chapter I (Parts 260 -265) if the resource recovery facility is within the exclusions of 40 CFR 261.4(b)(1).

X4.2 *CESQG*—A generator is a CESQG if the generator meets the requirements in 40 CFR 261.5.

REFERENCES

(1) *The Condensed Chemical Dictionary*, Van Nostrand Reinhold Publications.

(2) Sax, N.I., *Dangerous Properties of Industrial Materials.*

(3) *Farm Chemicals Handbook*, Meister Publishing.

(4) Federal, State, and Local Regulations and Statutes.

(5) *First Aid Manual for Chemical Accidents.*

(6) *Handbook of Toxic and Hazardous Chemicals and Carcinogens*, Sittig, Noyes Publications.

(7) International Air Transport Association (IATA) Dangerous Good Regulations.

(8) Material Safety Data Sheets (MSDSs).

(9) National Fire Protection Association (NFPA) Fire Codes.

(10) National Institute for Occupational Safety and Health (NIOSH) Pocket Guide to Chemical Hazards.

(11) *North American Emergency Response Guidebook.*

(12) Personal Protective Equipment Selection Guidance Documents

Designation: D 6538 – 00

Standard Guide for Sampling Wastewater With Automatic Samplers[1]

This standard is issued under the fixed designation D 6538; the number immediately following the designation indicates the year of original adoption or, in the case of revision, the year of last revision. A number in parentheses indicates the year of last reapproval. A superscript epsilon (ε) indicates an editorial change since the last revision or reapproval.

1. Scope

1.1 This guide covers the selection and use of automatic wastewater samplers including procedures for their use in obtaining representative samples. Automatic wastewater samplers are intended for the unattended collection of samples that are representative of the parameters of interest in the wastewater body. While this guide primarily addresses the sampling of wastewater, the same automatic samplers may be used to sample process streams and natural water bodies.

1.2 The guide does not address general guidelines for planning waste sampling activities (see Guide D 4687), development of data quality objectives (see Practice D 5792), the design of monitoring systems and determination of the number of samples to collect (see Practice D 6311), operational details of any specific type of sampler, in-situ measurement of parameters of interest, data assessment and statistical interpretation of resultant data (see Guide D 6233), or sampling and field quality assurance (see Guide D 5612). It also does not address sampling groundwater.

1.3 The values stated in SI units are to be regarded as the standard. The inch-pound units given in parentheses are for information only.

1.4 *This standard does not purport to address all of the safety concerns, if any, associated with its use. It is the responsibility of the user of this standard to establish appropriate safety and health practices and determine the applicability of regulatory limitations prior to use.*

2. Referenced Documents

2.1 *ASTM Standards:*

D 1129 Terminology Relating to Water[2]

D 3694 Practices for Preparation of Sample Containers and for Preservation of Organic Constituents[3]

D 3856 Guide for Good Laboratory Practices in Laboratories Engaged in Sampling and Analysis of Water[2]

D 4687 Guide for General Planning of Waste Sampling[4]

D 4840 Guide for Sampling Chain-of-Custody Procedures[2]

D 5088 Practice for Decontamination of Field Equipment Used at Nonradioactive Waste Sites[5]

D 5283 Practice for Generation of Environmental Data Related to Waste Management Activities Quality Assurance and Quality Control Planning and Implementation[4]

D 5612 Guide for Quality Planning and Field Implementation of a Water Quality Measurement Program[2]

D 5792 Practice for Generation of Environmental Data Related to Waste Management Activities: Development of Data Quality Objectives[4]

D 5851 Guide for Planning and Implementing a Water Monitoring Program[3]

D 5956 Guide for Sampling Strategies for Heterogeneous Wastes[4]

D 6233 Guide for Data Assessment for Environmental Waste Management Activities[4]

D 6311 Practice for Generation of Environmental Data Related to Waste Management Activities: Optimizing Sampling Design[4]

E 856 Definitions of Terms and Abbreviateions Relating to Physical and Chemical Characteristics of Refuse Derived Fuel[4]

3. Terminology

3.1 *composite sample, n*—a combination of two or more samples. **(D 1129)**

3.2 *representative sample, n*—a sample collected such that it reflects one or more characteristics of interest (as defined by the project objectives) of a population from which it was collected. **(D 5956)**

3.3 *sample, n*—a portion of material taken from a larger quantity for the purpose of estimating properties or composition of the larger quantity. **(E 856)**

4. Significance and Use

4.1 This guide provides persons responsible for designing and implementing wastewater sampling programs with a summary of the types of automatic wastewater samplers, discusses the advantages and disadvantages of the different types of samplers and addresses recommended procedures for their use.

---

[1] This guide is under the jurisdiction of ASTM Committee D34 on Waste Management and is the direct responsibility of Subcommittee D34.01.01 on Planning for Sampling.

Current edition approved May 10, 2000. Published July 2000.

[2] *Annual Book of ASTM Standards*, Vol 11.01.

[3] *Annual Book of ASTM Standards*, Vol 11.02.

[4] *Annual Book of ASTM Standards*, Vol 11.04.

[5] *Annual Book of ASTM Standards*, Vol 04.09.

TABLE 1 Advantages and Disadvantages of Manual versus Automatic Sampling of Wastewater (3)

Type	Advantages	Disadvantages
Manual	Low capital cost	Increased variability due to sample handling
	Personnel can compensate for various situations	Inconsistency in collection
	Personnel can document unusual conditions	High cost of labor assuming composite or multiple grab samples are collected
	No maintenance	Repetitious and monotonous task for personnel
	Extra samples can be collected in a short time if necessary	
Automatic	Consistent samples	Considerable maintenance for batteries and cleaning; susceptible to plugging by solids
	Decreased variability caused by sample handling	Restricted in size to the general specifications
	Minimal labor requirement for sampling	Greater potential for sample contamination
	Capable of collecting multiple grab and multiple aliquot composite samples	May be subject to damage by vandals
		High capital cost

5. Automatic Versus Manual Sampling (1, 2)[6]

5.1 The advantages and disadvantages of manual and automatic sampling are summarized in Table 1. The decision as to whether to use manual or automatic sampling involves many considerations in addition to equipment costs. In general, manual sampling is indicated when infrequent samples are required from a site, when biological or sediment samples, or both, are also required, when investigating special incidents, where sites will not allow the use of automatic devices, for most bacteriological sampling, where concentrations remain relatively constant, etc. The use of automatic samplers is indicated where frequent sampling is required at a given site, where long-term compositing is desired, where simultaneous sampling at many sites is necessary, etc. Automatic sampling is often the method of choice for storm-generated discharge studies, for longer outfall monitoring, for treatment plant efficiency studies, where 24-h composite samples are required, etc. The user should review 7.1.22 before selecting manual or automatic sampling.

6. Types of Samples Collected by Automatic Samplers

6.1 *Grab Samples*—As defined under the U.S. Environmental Protection Agency's (EPA) National Pollutant Discharge Elimination Program, grab samples are individual samples collected over a period of time not exceeding 15 min and are representative of conditions at the time of sampling **(4)**. Grab samples are sometimes also called individual or discrete samples **(5)**. Sequential grab samples are a series of grab samples collected at constant increments of either time or flow and provide a history of variation. Grab samples are appropriate when samples are needed to:

6.1.1 Characterize an effluent that is not continuous,

6.1.2 Provide information about instantaneous concentrations of pollutants,

6.1.3 Allow collection of samples of varied volume,

6.1.4 Corroborate composite samples,

6.1.5 Monitor parameters not amenable to compositing (for example, pH, temperature, dissolved oxygen, chlorine, purgeable organics (unless a specialized sampler is used), oil and grease and others specified by a permit which may include phenols, sulfites and hexavalent chromium).

6.1.6 Characterize a waste stream in detail where rapid fluctuations of parameters occur (sequential grabs).

6.2 *Composite Samples*—Composite samples are collected over time, either by continuous sampling or by mixing discrete samples, and represent the average characteristics of the waste stream during the compositing period. Composite samples are collected when stipulated in a permit, when average pollutant concentration during the compositing period is to be determined, and when wastewater characteristics are highly variable. There are four types of composite samples.

6.2.1 *Time Composite Samples*—This method requires discrete sample aliquots be collected in one container at constant time intervals. The method is appropriate when the flow of the stream is constant (flow rate does not vary more than ±10 % of the average flow rate **(4)**) or when flow monitoring equipment is not available. The EPA allows time-proportional sampling and requires samples be collected every 15-min, on average, over a 24-h period.

6.2.2 *Flow-Proportional Composite Samples*—There are two methods used for this type of sample **(4)**. The most commonly used method with automatic samplers collects a constant sample volume at varying time intervals proportional to stream flow based on input from a flow monitor (for example, a 200-mL aliquot is collected for every 5000 L of flow). In the other flow-proportional compositing method, the sample is collected by varying the volume of each aliquot as the flow varies, while maintaining a constant time interval between the aliquots.

6.2.3 *Sequential Composite Samples*—A sequential composite sample is composed of a series of short-period composites, each of which is held in an individual container, for example, four sample aliquots are composited (one every 15 min) to form hourly composites **(4)**. The 24-h sequential composite is then manually made by compositing the individual 1-h composite sample.

6.2.4 *Continuous Composite Samples*—This method requires that the sample be collected continuously at a constant rate or proportional to flow **(4)**. This method is seldom used with automatic samplers.

7. Attributes of Automatic Samplers

7.1 The EPA **(6)** developed a list of attributes of the ideal automatic sampler for their use and EPA Region 4 **(7)** and others **(3)** have noted other important attributes. These attributes and requirements may be specific to EPA's use and were primarily directed at suction lift type automatic samplers. Not all these sampler characteristics will be important to all users but their consideration may guide persons selecting automatic samplers. The desirable features of automatic samplers listed below have been summarized and combined from

[6] The boldface numbers given in parentheses refer to a list of references at the end of the standard.

the referenced documents.

7.1.1 Capable of AC/DC operation with adequate dry battery energy storage for 120-h operation at 1-h sampling intervals.

7.1.2 Suitable for suspension in a standard manhole yet still accessible for inspection and sample removal. A secure harness or mounting device if the sampler is placed in a sewer.

7.1.3 Total weight, including batteries, less than 18 kg. Compact and portable enough for one-person installation.

7.1.4 Sample collection interval adjustable from 10 min to 4 h.

7.1.5 Capable of collecting a single 9.5-L (2.5-gal) sample and/or collecting 500-mL (0.13-gal) discrete samples in a minimum of 24 containers. The individual sample aliquot must be at least 100 mL.

7.1.6 Capable of multiplexing repeated aliquots into discrete bottles.

7.1.7 One intake hose with a minimum inner diameter of 0.64 cm (0.25 in) and a weighted, streamlined intake screen which will prevent accumulation of solids.

7.1.8 Intake hose liquid velocity adjustable from 0.6 to 3 m/s (2.0 to 10 ft/s) with dial setting.

7.1.9 Minimum lift capacity of 6.1 m (20 ft).

7.1.10 Explosion proof construction.

7.1.11 Watertight exterior case to protect components in the event of rain or submersion.

7.1.12 Exterior case capable of being locked, including lugs for attaching steel cable to prevent tampering and to provide security.

7.1.13 An integral sample container compartment capable of maintaining samples at 4 to 6°C for a period of 24 h at ambient temperatures up to 38°C.

7.1.14 Capable of operating in a temperature range from–10 to 40°C with the exception of the intake hose.

7.1.15 A purge cycle to flush the sample intake tubing before and after each collection interval, and a mechanism to sense and clear a plugged sample line and then collect the complete sample.

7.1.16 Capable of collecting flow-proportional and time-composite samples.

7.1.17 Materials of construction that contact the sample must not compromise the integrity of the sample for the intended use.

NOTE 1—Some references prohibited sample contact with metal **(6)** and contact with plastic or metal parts when parameters to be analyzed could be impacted by these materials **(7)**.

7.1.18 Water velocity in intake hose (greater than 0.6 m/s [2.0 ft/s]) and aliquot volume are independent of lift heights experienced during the sampling event.

7.1.19 Overall construction, including casing, of materials resistant to corrosion (plastics, fiberglass, stainless steel).

7.1.20 Exterior surface a light color to reflect sunlight.

7.1.21 Low cost, availability of spare parts, warranty, ease of maintenance, reliability and ruggedness of construction, and capable of being repaired in the field.

7.1.22 *Other Factors*—Other factors **(3)** that should be considered in selecting an automatic sampler are the:

7.1.22.1 Expected variation in water or wastewater composition with time,

7.1.22.2 Variation of flow rate with time,

7.1.22.3 Specific gravity of the liquid

7.1.22.4 Concentration and density of suspended solids of interest,

7.1.22.5 Presence of floating materials,

7.1.22.6 Characteristics of the site where the sampler will be placed,

7.1.22.7 Range of intended use (a permanent site or traveling sampler),

7.1.22.8 Skill level required for installation and operation of the sampler, and

7.1.22.9 The level of accuracy desired.

8. Types of Automatic Samplers (1,2,3)

8.1 There are three main types of automatic samplers, suction lift, pressure or forced flow, and mechanical. Each has its advantages and limitations and all types are available in models designed to preserve samples via cooling (iced or refrigerated). While all automatic samplers can collect samples through time, some samplers are designed to be triggered by inputs from online devices measuring flow, pH, temperature, conductance, etc., and collect samples under specific conditions (for example, pH >9.0).

8.2 *Suction Lift*—Suction lift devices can be further subdivided into peristaltic and vacuum type samplers. Peristaltic pump devices are the most commonly used type in the United States and use a rotating head to pinch a flexible hose creating a vacuum to transport the sample to the container. Vacuum devices **(8)** are more popular in Europe and use a vacuum pump to transport the sample to the sample container. Suction lift samplers are portable, versatile due to their light weight and can purge the transport line between samples. Their main limitation is that their lift capacity which is claimed to range up to 9 m but may be significantly less. Also, since suction lift devices use a vacuum to transport samples, they cause some degassing of the sample and are generally unsuitable for sampling dissolved gases and volatile organic compounds (VOC). Peristaltic pump samplers can be used to collect pressurized line samples if the pressure is less than 100 kPa (15 psi). Vacuum systems must use a flow-through cell which adds cost to the system.

8.3 *Pressure or Forced Flow*—With pressure or forced flow systems, a submersible pump or pneumatic pressure is used to force the sample from the source to the collection container. Because the sample is under positive pressure during collection, gases and VOC are less likely to be lost than in suction lift systems. Pressure systems can be used to lift samples considerable heights.

8.4 *Mechanical*—Mechanical systems use scoops, or cups on a chain or paddle wheel to remove a sample from the source and transfer it to the collection container. Some mechanical systems use a plunger to trap a small volume sample and transfer it to the collection container. This latter type is particularly suited for sampling pressurized pipes. Mechanical systems are prone to fouling. The scoop and cup mechanical types are often large, and unable to be moved but can lift samples to great heights and can collect samples integrated across the entire flow path.

8.5 *Samplers for Volatile Organic Compounds*—Most of the previously mentioned automatic samplers are not designed to retain VOC in the collection container, even if they are not volatilized during sample collection. Automatic samplers specifically designed for the collection of VOCs are now commercially available from several sources (**2**). One system uses a bladder pump to collect up to 24 discrete 40-mL vial samples with zero headspace over a 24-h period. Another device collects a 100-mL composite sample using a motor driven syringe pump. The syringe is then transported to the analytical laboratory for sample analysis. A third device used a hydropneumatic sample system to collect a composite sample which is then transferred to volatile organic analysis vials for shipment to the analytical laboratory.

9. Using Automatic Samplers

9.1 *General*—Sampling is conducted to achieve specific objectives that must be well defined at the outset. Those objectives may be to demonstrate or determine compliance with regulations, check on industrial processes, check on the performance of industrial or waste processes, or acquire specific data needed to achieve certain data quality objectives. The design of a liquid waste monitoring program is beyond the scope of this standard, but in general, there should be a written sampling plan or procedures defining:

9.1.1 Overall and sampling objectives,

9.1.2 A monitoring system design (sampling location, types of samples, frequency of sampling, parameters of interest) (see Practice D 6311),

9.1.3 Detailed sampling procedures (sampling equipment and cleaning procedures, exact sampling location, sample containers and volumes required, sample preservation and shipping procedures, documentation requirements including chain-of-custody procedures if needed, etc.) or references to them (see Guide D 4687),

9.1.4 Analytical procedures,

9.1.5 Field and laboratory quality assurance procedures (see Guide D 5612),

9.1.6 Plans for data assessment and interpretation (see Practice D 5792 and Guide D 6233), and

9.1.7 Safety procedures.

9.2 *Selection of Sampling Locations*—In many cases, sampling locations may be specified in a permit but samples should be collected from representative sites. "Representative point" is defined (**3**) as a location in surface or ground waters at which specific conditions or parameters may be measured in such a manner as to characterize or approximate the quality or condition of the water body; or a location in process waters or wastewaters where specific conditions or parameters are measured that adequately reflect the actual conditions of those waters or wastewaters. Some general criteria that can aid in selecting the best specific sampling sites include:

9.2.1 *Accessability, Convenience and Practicality*—These factors are important but they must be secondary to the representativeness of sampling. Manholes on busy streets should be avoided if possible. Sites with a history of surcharging or submergence by surface water, or both, should also be avoided if possible. Sites with AC power may be preferred, especially if the samples require preservation by cooling to 4°C. Sites where the automatic sampler can be protected from vandalism are preferred.

9.2.2 *Safety*—Hazardous sampling locations should be avoided if at all possible, especially those requiring explosion proof samplers, respiratory protection or confined space entry.

9.2.3 *Flow Characteristics*—The homogeneity or heterogeneity of the water or wastewater is vitally important. Turbulence and good mixing resulting from a hydraulic jump, enhance the homogeneity or uniform distribution of the constituents in the body of water and samples should be collected in such areas. Influent wastewater to sewage treatment plants is preferably sampled at highly turbulent flow locations but in many instances the most preferable location is not accessible. Preferable influent locations include: the upflow siphon following a comminutor (in the absence of a grit chamber); the upflow distribution box following pumping from the main plant wet well; aerated grit chamber; flume throat; pump wet well when the pump is operating; or downstream of the preliminary screening (**1**). An exception to this guidance is that when sampling for volatile organic compounds, a site with low turbulence should be selected to reduce the amount of entrapped air in the sample (**5**). The sampling site should be far enough downstream of tributary inflow to ensure mixing of the tributary with the main stream. Sampling sites should be in a straight section of channel, at least 20 channel widths below bends if heavier suspended, floating solids or particles with a specific gravity much less than 1 are of interest (**1**). A dye such as Rhodamine WT can be used as an aid in determining the most representative sampling location.

9.3 *Sample Intake Design and Placement*—The appropriate location of the sampler intake depends on the characteristics of the body of water or wastewater being sampled and the parameters of interest. Particles that rapidly sink or float are difficult to sample representatively and if such particles are of interest more detailed guidance should be sought (**1**). Sample intake velocity should be at least equivalent to the flowrate of the wastewater being sampled but not less than 0.6 m/s (2 ft/s). The following recommendations on sample inlet placement are based on an assumed need to representatively sample particulate material with a hydraulic size (hydraulic size is based on particle settling velocity) no greater than fine sand. When particulate matter is not of concern, or the particles of interest have a small hydraulic size (silt, clay, neutrally buoyant particles, etc), the inlet placement recommendations below are much less important.

9.3.1 In general, samples should be collected where the sample stream is well mixed, but not at the point of maximum turbulence. Samples should be collected at a point midway between the point of maximum turbulence and the edge, at ⅔ to ¾ (**1**) the depth of the liquid (**5**).

9.3.2 Under circumstances where low flows need to be sampled in an intermittent stream such as a surface runoff channel, the intake should be mounted horizontally and at a right angle to the flow, in the middle of the stream and with its lowest surface about 5 cm (2 in.) above the bottom (**1**).

9.3.3 For continuously flowing, small natural streams, such as might be encountered in an urban runoff study, the intake should be mounted near the low flow mid-depth, in a vertical position with the intake opening facing up.

9.3.4 In man-made channels and conduits, there is no concern for bottom scour and the intake screen can be allowed to rest on the bottom unless significant bedload depths are anticipated. Trailing the intake downstream by the intake tubing is a common practice but this is not recommended because the depth of sampling is never known. An alternate approach is to fix the sample intake tubing in the stream and let a short section of tubing trail downstream. This allows the intake depth to fluctuate over a limited range and accommodate changes in flow velocity. Initial studies also objected to trailing the intake downstream because it was thought not to gather representative solids due to momentum effects **(1)**. However, subsequent study has shown that pointing the nozzle directly downstream appears to have a higher particulate sampling efficiency than when the nozzle is oriented at a right angle to the flow **(9)**.

9.3.5 Where flow is continuous but variable, the intake should be firmly mounted near the low flow mid-depth point. As opposed to natural streams, in many man-made conduits it will be more convenient to dangle the weighted intake from above with the suction tube pointing down. Although the vertical orientation is preferable (intake facing up), the dangling approach is acceptable **(1)**.

9.3.6 For the rare case where relatively steady flow is anticipated in either natural or man-made conduits, the intake should be mounted at about ⅔ **(5)** to ¾depth unless site specific information indicates otherwise **(1)**.

9.3.7 Due to the nature of most wastewater flows, a fine mesh screen over the intake to prevent blockage and exclude debris is not desirable. A screen made up of a number of rather large holes, about 0.3 to 0.6 cm (⅛ to ¼ in.) in diameter is preferred **(1)**.

9.3.8 The sample intake hose should be as short as possible, sample lift should be minimized, there should be no bends or kinks in the inlet hose and the hose should be as vertical as possible or at least on a constant slope to aid in drainage **(1)**.

9.4 *Sampling Procedures*:

9.4.1 All equipment should be properly maintained, and materials contacting samples should be thoroughly cleaned and chemically compatible with the parameters to be measured. Sample collection containers are usually glass or chemically resistant plastic (polyethylene or polypropylene). For most analytes, either type of container is acceptable but glass is specified for samples being collected for determination of: oil and grease, and organic priority pollutants including phenols, pesticides, and polychlorinated biphenyls. Depending on the analytes of interest sample tubing should be poly vinyl chloride or TFE-fluorocarbon (for priority pollutant organics) and on peristaltic pumps, pump head tubing should be silicone or specialized flexible tubing that is chemically resistant and minimally leachable (for priority pollutant organics). All tubing should be new and precleaned using procedures appropriate for the analytes of interest. For more information on tubing selection, see Selection of Equipment for Water Sampling **(10)** and for EPA procedures on cleaning automatic wastewater samplers and tubing (see Appendix C of Ref **(7)**) (also see Practice D 5088).

9.4.2 The automatic sampler should be installed as close to the sampling point as practical to minimize sample lift while avoiding submergence. The equipment should be firmly secured at the sampling site to help prevent vandalism.

9.4.3 When collecting samples or installing sampling equipment, personnel should always wear a new pair of appropriate protective gloves (for example, disposable latex gloves, rubber gloves, etc.) to prevent contamination of the sample and reduce exposure to hazardous substances.

9.4.4 The automatic sampler should be programmed according to the manufacturers instructions and consistent with the permit or sampling plan. The automatic sampler should initially be operated manually to ensure correct operation and that sample volumes are correct. It may also be necessary to collect a rinsate sample prior to formal sampling. The sample line should be flushed prior to and immediately after a sample aliquot has been collected and the sample flow rate in tubing should be a minimum of 0.61 m/s (2 ft/s).

9.4.5 As appropriate, samples should be preserved by chilling during sampling, preserved after sampling and during containerization and sample splitting, and sent to the analytical laboratory as soon as possible to minimize sample holding times.

9.4.6 Standard documentation should be recorded in field logbooks on sample labels, chain-of-custody logs, etc. (see Guide D 4687).

10. Keywords

10.1 automatic sampler; composite; sample; wastewater; water sampling

REFERENCES

(1) Shelly, P. E, *Sampling of Water and Wastewater*, EPA-600/4-77-039, PB272664, US Environmental Protection Agency, Washington, DC, 1977.

(2) Dick, E.M., *Automatic Water and Wastewater Sampling, In: Principles of Environmental Sampling* (L.H. Keith, ed.), *American Chemical Society*, Chapter 13, 1996, pp. 237–258.

(3) U.S. Environmental Protection Agency (EPA), *Handbook for Sampling and Sample Preservation of Water and Wastewater*, EPA-600/4-82-029, EPA Environmental Monitoring and Support Laboratory, Cincinnati, OH, 1982.

(4) U.S. Environmental Protection Agency (EPA), 1994, *NPDES Compliance Inspection Manual*, EPA 300/B-94-014, Washington, DC, 1994.

(5) Water Environment Federation, *Wastewater Sampling for Process and Quality Control*, Manual or Practice No. OM-1, Alexandria, VA, 1996.

(6) Harris, D. J. and Keffer, W. J., *Wastewater Sampling Methodologies and Flow Measurement Techniques*, EPA 907/9-74-005, U.S. Environmental Protection Agency, 1974.

(7) U.S. Environmental Protection Agency (EPA), *Environmental Investigations Standard Operating Procedures and Quality Assurance Manual*, Region 4, EPA, Athens, GA, 1996. http://www.epa.gov/region04/sesd/eisopqam/eisopqam.html

(8) Carroll, C. M and Carroll, D. M., "Selecting Wastewater Sampling Pumps, Revisited," Environmental Technology, Jan/Feb, 1997, pp 37-38.

(9) Edwards, T. K. and Glysson, G. D., *Field Methods for Measurement of Fluvial Sediment*, TWRI Book 3, Chapter C2, United States Geological Survey, 1999.

(10) Wilde, F. D., Radtke, D. B., Gibs, J and Iwatsubo, R. T., editors, *Selection of Equipment for Water Sampling*, TWRI Book 9, Chapter A2, United States Geological Survey, 1998.

Designation: D 6582 – 00

Standard Guide for
Ranked Set Sampling: Efficient Estimation of a Mean Concentration in Environmental Sampling[1]

This standard is issued under the fixed designation D 6582; the number immediately following the designation indicates the year of original adoption or, in the case of revision, the year of last revision. A number in parentheses indicates the year of last reapproval. A superscript epsilon (ε) indicates an editorial change since the last revision or reapproval.

1. Scope

1.1 This guide describes ranked set sampling, discusses its relative advantages over simple random sampling, and provides examples of potential applications in environmental sampling.

1.2 Ranked set sampling is useful and cost-effective when there is an auxiliary variable, which can be inexpensively measured relative to the primary variable, and when the auxiliary variable has correlation with the primary variable. The resultant estimation of the mean concentration is unbiased, more precise than simple random sampling, and more representative of the population under a wide variety of conditions.

1.3 *This standard does not purport to address all of the safety concerns, if any, associated with its use. It is the responsibility of the user of this standard to establish appropriate safety and health practices and determine the applicability of regulatory limitations prior to use.*

2. Referenced Documents

2.1 *ASTM Standards:*

D 5792 Practice for Generation of Environmental Data Related to Waste Management Activities: Development of Data Quality Objectives[2]

D 6044 Guide for Representative Sampling for Management of Waste and Contaminated Media[2]

3. Terminology

3.1 *Definitions:*

3.1.1 *auxiliary variable, n*—the secondary characteristic or measurement of interest.

3.1.1.1 *Discussion*—In ranked set sampling, information contained in an auxiliary variable is useful for ranking the samples. This ranking may mimic the rankings of the samples with respect to the values of the primary variable when there is correlation between the auxiliary variable and the primary variable. Auxiliary information may include visual inspection, inexpensive quick measurement, knowledge of operational history, previous site data, or any other similar information.

3.1.2 *data quality objectives (DQO) process, n*—a quality management tool based on the scientific method and developed by the U.S. Environmental Protection Agency (EPA) to facilitate the planning of environmental data collection activities. **(D 5792)**

3.1.3 *equal allocation, n*—this occurs when the number of sets in ranked set sampling is an integer multiple of the size of the set.

3.1.4 *primary variable, n*—the primary characteristic or measurement of interest.

3.1.5 *ranked set sampling, n*—a sampling method in which samples are ranked by the use of auxiliary information on the samples and only a subset of the samples are selected for the measurement of the primary variable.

3.1.6 *representative sample, n*—a sample collected in such a manner that it reflects one or more characteristics of interest (as defined by the project objectives) of a population from which it is collected. **(D 6044)**

3.1.6.1 *Discussion*—A representative sample can be a single sample, a collection of samples, or one or more composite samples. A single sample can be representative only when the population is highly homogeneous. **(D 6044)**

4. Significance and Use

4.1 Ranked set sampling is cost-effective, unbiased, more precise and more representative of the population than simple random sampling under a variety of conditions **(1)**.[3]

4.2 Ranked set sampling (RSS) can be used when:

4.2.1 The population is likely to have stratification in concentrations of contaminant.

4.2.2 There is an auxiliary variable.

4.2.3 The auxiliary variable has strong correlation with the primary variable.

4.2.4 The auxiliary variable is either quick or inexpensive to measure, relative to the primary variable.

4.3 This guide provides a ranked set sampling method only under the rule of equal allocation. This guide is intended for those who manage, design, and implement sampling and analysis plans for management of wastes and contaminated media. This guide can be used in conjunction with the DQO process (see Practice D 5792).

---

[1] This guide is under the jurisdiction of ASTM Committee D34 on Waste Management and is the direct responsibility of Subcommittee D34.01.01 on Planning for Sampling.

Current edition approved Aug. 10, 2000. Published October 2000.

[2] *Annual Book of ASTM Standards*, Vol 11.04.

[3] The boldface numbers in parentheses refer to the list of references at the end of this standard.

5. Ranked Set Sampling (RSS)

5.1 Environmental sampling typically requires the identification of the locations where the samples are to collected. Subsequent analyses of these samples to quantify the characteristics of interest allow inference on the population mean concentration from the sample data.

5.2 A simple random sampling (SRS) approach is one sampling design that can be used. In this case, a set of random samples is identified and collected from a population and all of these samples are analyzed (for the primary variable).

5.3 Ranked set sampling (RSS) is similar to SRS in the identification and collection of the samples, but only a subset of the samples are selected for analysis. The selection is done by ranking the samples using auxiliary information on the samples and selecting a subset based on the rankings of the samples.

5.4 As can be seen from the steps described below, RSS is in fact a "stratified random sampling at the sample level," meaning that stratification of the population is induced after sampling and no construction of the strata is needed before sampling. Increased precision of stratified random sampling in the estimation of the population mean, relative to SRS, is well known, especially when the population is stratified by concentrations.

5.5 Increased precision of RSS relative to SRS means that the same precision can be achieved with fewer samples analyzed for the primary variable under RSS. RSS is therefore more cost-effective than SRS. When the objective is to minimize sampling and analytical costs, the number of samples can be determined so that RSS has precision equal to that of SRS at a lower cost.

5.6 The actual steps to conduct RSS are given below.

5.7 *Steps in Ranked Set Sampling (RSS)*:

5.7.1 Determine the total number of sample analyses (n) agreed to by the stakeholders. A planning process, such as a data quality objectives (DQO) process (Practice D 5792), may be used to determine this number.

5.7.2 Determine the primary variable and the auxiliary variable of interest.

5.7.3 Determine the size of the set, m. Study the auxiliary measurement and determine its capability in ranking the samples. For example, if the auxiliary measurement is visual inspection, its capability in ranking the samples may be somewhat limited. Namely, it may be capable of ranking 3–4 samples, but may have difficulty in ranking greater than 5 or 6 samples based on visual inspection; thus, the preferred size of the set (m) in ranked set sampling is about 3 or 4. On the other hand, an instrument-based quick-test may be capable of a larger m (see 5.14 for ranking criteria).

5.7.4 Calculate the needed number of replicates, r (the number of times the ranked sets are to be repeated). Divide n by m and round it up to whole number to obtain the needed r. Namely, $r = n/m$ and round up to whole number.

5.7.5 Randomly select a total of m^2r samples from the population, for example, by simple random sampling design, and randomly divide them into r replicates, with m^2 (m times m) samples in each replicate.

NOTE 1—In practice, the m^2r samples may not be taken all at once. More often, m random samples may be taken from a geographical sub-area of the population and are then ranked according to the auxiliary variable. This is repeated m times to obtain the first replicate of m^2 (m times m) samples. This entire process is repeated r times to obtain the needed r replicates.

5.7.6 Start with the first replicate of m^2 (m times m) samples. Arrange these samples into m sets of size m (an m by m matrix).

5.7.7 For each of the m sets in this replicate, rank the samples within each set by using the auxiliary measurement on the samples. When the observations on the auxiliary variable cannot be distinguished from each other, these observations are called "ties." Ties can be broken arbitrarily (namely, arbitrarily assigning one rank to one sample and a succeeding rank to the other).

5.7.8 Select samples for the measurement on the primary variable as follows. In set i, select and measure the sample with rank i, $i = 1, 2, ..., m$. Completion of this step leads to a total of m samples to be analyzed for the primary variable, out of a total of m^2 samples collected.

5.7.9 Repeat steps 5.7.6 through 5.7.8 for r times to obtain a total of $m \times r = n$ samples to be analyzed and measured for the primary variable.

5.8 Since the number of sets (m) in step 5.7.6 equals the size of the set (m), this is called equal allocation. RSS under unequal allocation tends to have additional gains in precision, relative to equal allocation; but, this gain is, in general, not large compared to the gain against SRS, and is not covered in this guide.

5.9 The value of n can be the total number of samples for which the budget can afford to analyze.

5.10 The rounding up in step 5.7.4 may cause the total number of analyses for the primary variable to exceed n. When this is the case, there are two options:

5.10.1 Obtain buy-in from the stakeholders to accept the slightly higher total number of sample analyses, or

5.10.2 Try different values of m and r to get the total number of analyses as close to n as possible.

5.11 *Estimation of Mean and Standard Error of the Mean*:

5.11.1 In 5.7, if $n = 12$, $m = 3$, and $r = 4$, the data on the primary variable obtained from the steps in that section may be summarized as in Table 1. The true mean concentration of the characteristic of interest is estimated by the arithmetic sample mean of the measured samples. For the hypothetical example

TABLE 1 Sample Values on Primary Variable

Set	Replicate	Value
1	1	X_{11}
	2	X_{12}
	3	X_{13}
	4	X_{14}
2	1	X_{21}
	2	X_{22}
	3	X_{23}
	4	X_{24}
3	1	X_{31}
	2	X_{32}
	3	X_{33}
	4	X_{34}

in Table 1 (and assuming normal distribution of the data), the mean (M) is estimated as follows:

$$M = (X_{11} + X_{12} + X_{13} + X_{14} + X_{21} + X_{22} + ... + X_{34})/12 .... \quad (1)$$

The standard error of the mean (S_M) is estimated as follows:

$$S_M = \{[(X_{11} - X_{1.})^2 + (X_{12} - X_{1.})^2 + (X_{13} - X_{1.})^2 + (X_{14} - X_{1.})^2 + (X_{21} - X_{2.})^2 + (X_{22} - X_{2.})^2 + (X_{23} - X_{2.})^2 + ... + (X_{34} - X_{3.})^2]/(m^2r(r-1)\}^{1/2} \quad (2)$$

where:

X_{ij} = the value of the primary variable from the ith set and the jth replicate, and

X_i = the average of set i.

NOTE 2—The numerator of Eq 2 represents the squared differences between a value and its set average.

5.11.2 Given these estimates, inference about the population mean concentration can be made from the sample data (with some typical assumptions about the underlying statistical distribution of the data). This includes the use of confidence limits to estimate the population mean.

5.12 *An Illustration of RSS*:

5.12.1 An illustration of the steps in 5.7 and 5.11 is given in 5.12.2.1 through 5.12.2.9. The objective of this example is to estimate the mean of total petroleum hydrocarbon (TPH) concentration in the soil of a 1-acre site, down to the depth of one inch from the surface. Assume that the stakeholders agree that, due to cost and other considerations, a total of 12 analyses are the limit. Further assume that coloration on the surface of the soil is positively correlated with TPH concentration, with darker color indicating higher concentration.

5.12.2 The steps to carry out ranked set sampling are as follows:

5.12.2.1 $n = 12$, the desired total number of analyses for the primary variable.

5.12.2.2 The primary variable = TPH concentration and the auxiliary variable = soil coloration, where coloration is observed in-situ.

5.12.2.3 Professional judgment may indicate that visual inspection of soil coloration is capable of ranking three samples; thus, $m = 3$.

5.12.2.4 $r = n/m = 12/3 = 4$, the number of replicates.

5.12.2.5 Randomly select a total of m^2r samples = (3)(3)(4) = 36 samples. Divide these samples into 4 replicates with 9 samples in each replicate.

5.12.2.6 Take the first replicate of 9 samples and arrange them into 3 by 3 matrix; each row is called a set with set size of 3 (three samples in that set).

5.12.2.7 Rank the three samples within each of the three sets (namely, assigning ranks to the samples) according to the ranking of the soil coloration, giving a rank of 1 for lightest coloration, a rank of 3 for the darkest coloration.

5.12.2.8 In the first set, select the sample with rank 1 for the measurement of the primary variable. In set 2, select the sample with rank 2 for the measurement of the primary variable. And so forth for the third set. After this step, a total of m = 3 samples have been chosen for the measurement of the primary variable.

5.12.2.9 Repeat steps 5.12.2.6 through 5.12.2.8 for four times to obtain a total of $m \times r = 3 \times 4 = 12$ samples.

5.12.3 After steps 5.12.2.1 through 5.12.2.5, the 36 samples to be taken from the population may appear graphically as in Fig. 1. The samples in Fig. 1 are arbitrarily numbered from 1 through 36.

5.12.4 After steps 5.12.2.1 through 5.12.2.9, the rankings on the auxiliary variable and the measured values on the primary variable may appear as in Table 2. Each row of three samples can be called a cluster, and they are so designated in Table 2. These clusters can be marked in Fig. 1.

5.12.5 Table 3 is a summary of the data on the primary variable in Table 2. Note that the sample values of the primary variable and the bold-faced rank data in Table 2 happen to have the same ordering in all the replicates, except replicate 4, implying good correlation between the auxiliary variable and the primary variable.

5.12.6 When the data of the primary variable in Table 3 follow a normal distribution, the sample mean (M) and standard error of the mean (S_M) can be calculated as follows:

$$M = (9+10+12+15+15+16+20+14+17+18+23+20)/12 = 15.75, \quad (3)$$

$$S_M = \{[(9-11.5)^2 + (10-11.5)^2 + (12-11.5)^2 + (15-11.5)^2 + (15-16.25)^2 + (16-16.25)^2 + (20-16.25)^2 + (14-16.25)^2 + (17-19.5)^2 + (18-19.5)^2 + (23-19.5)^2 + (20-19.5)^2]/[(3^2)(4)(4-1)]\}^{1/2} = (62.75/81)^{1/2} = 0.76. \quad (4)$$

5.12.7 One- or two-sided confidence limits (CL) can be calculated from the sample mean and sample standard error of the mean (under the assumption that the data normally are distributed). For example, two-sided 95 % confidence limits are as follows:

$$CL = M \pm t_{0.95,11}S_M \quad (5)$$
$$= 15.75 \pm (2.201)(0.76)$$
$$= 14.08 \text{ to } 17.42,$$

where the degrees of freedom for the t-statistic is (12-1) = 11.

NOTE 3—The standard error of the mean of the 12 sample values would have been 1.21 under SRS, much higher than the value of 0.76 under RSS. This improvement in precision in RSS is due to stratification "using an auxiliary variable."

5.12.7.1 In this illustration, the standard deviation under RSS is ($\sqrt{12}$) (0.76) = 2.64. If it is desired to determine the

1 2 3 4 5 6 7 8 9 10 11

12 13 14 15 16 17 18 19 20

21 22 23 24 25 26 27 28

29 30 31 32 33 34 35 36

FIG. 1 Thirty-Six Sample Locations in the Population

TABLE 2 Auxiliary Rankings[A] and Primary Variable of Thirty-Six Soil Samples in Four Replicates

Cluster	Replicate	Set	Sample Number[A]			Color Rank[A,B]			TPH
A	1	1	**1**	2	3	**1**	3	2	9
B	1	2	4	5	**6**	3	1	**2**	15
C	1	3	7	**8**	9	1	**3**	2	17
D	2	1	**10**	11	12	**1**	3	2	10
E	2	2	13	**14**	15	1	**2**	3	16
F	2	3	16	17	**18**	2	1	**3**	18
G	3	1	**19**	20	21	**1**	3	2	12
H	3	2	**22**	23	24	**2**	3	1	20
I	3	3	**25**	26	27	**3**	1	2	23
J	4	1	28	**29**	30	2	**1**	3	15
K	4	2	31	32	**33**	1	3	**2**	14
L	4	3	**34**	35	36	**3**	1	2	20

[A] Samples in bold-type are the ones to be analyzed for the primary variable.
[B] Samples are ranked within each set of three samples using soil coloration, where 1 = light, 2 = medium, and 3 = dark.

TABLE 3 Sample Values on Primary Variable (TPH, ppm)

Replicate	Sample Value for Set: 1	2	3
1	9	15	17
2	10	16	18
3	12	20	23
4	15	14	20
Set mean	11.50	16.25	19.50

number of samples needed for RSS so that its precision (standard error) is equal to that of SRS, we can derive it by setting:

$$2.64/\sqrt{n} = 1.21 \quad (6)$$

Thus,

$$n = 5. \quad (7)$$

5.12.7.2 Thus, RSS can be designed so that it yields a total of five samples (for the primary variable). For example, a combination of $m = 3$ and $r = 2$ will yield a total of six samples. This will lead to the approximate level of desired precision. It can be seen that the six samples under RSS are only 50 % of the 12 samples needed under SRS for equal precision. The cost savings are obvious.

5.13 *Relative Advantages of Ranked Set Sampling*:

5.13.1 Ranked set sampling has the following advantages:

5.13.1.1 It provides an unbiased estimate of the population mean.

5.13.1.2 It is a more efficient estimator of the population mean than that from simple random sampling.

5.13.1.3 It has higher probability of obtaining representative samples relative to simple random sampling.

5.13.1.4 It exhibits superior performance in estimating the population mean for various underlying statistical distributions.

5.13.2 *Unbiased Estimate of the Mean*—According to Takahashi and Wakimoto **(2)**, the ranked set sample mean is an unbiased estimate of the population mean. Dell and Clutter **(3)** show that it is an unbiased estimator regardless of imprecision in ranking using the auxiliary information.

5.13.3 *Relative Precision (RP) of RSS*:

5.13.3.1 The relative precision (RP) of RSS, relative to SRS, is the ratio of the variance of the SRS sample mean to the variance of the RSS sample mean. Namely,

$$RP = \frac{\text{Var(SRS sample mean)}}{\text{Var(RSS sample mean)}} \quad (8)$$

McIntyre **(4)** and Dell & Clutter **(3)** show that:

$$1 \leq RP \leq (m + 1)/2. \quad (9)$$

5.13.3.2 RP = 1 when ranking using auxiliary information has no value, meaning that the precision of RSS equals that of SRS. When ranking is near perfect, RP is very close to and not much below $(m + 1)/2$.

5.13.3.3 When RP is greater than 1, it means that the precision of RSS is better than that of SRS. Since the number of sets, m, in RSS is 2 or greater, RP is equal to 1.5 or greater. This means that the standard error of an RSS sample mean always is smaller than that of an SRS sample mean, given the same number of samples analyzed for the primary variable.

NOTE 4—When m=1, RSS is reduced to SRS and no auxiliary variable is needed.

5.13.3.4 This increased precision of RSS relative to SRS means that the same precision can be achieved with fewer samples (analyzed for the primary variable) under RSS. RSS is therefore more cost-effective than SRS under the stated conditions. For more details on the potential cost advantages of RSS over SRS, see 5.12.1 through 5.12.7.2 and Refs **(5 and 6)**.

5.13.4 *Greater Representativeness of the Population*:

5.13.4.1 Lacayo and Neerchal **(7)** simulated elephant herds with different proportions of calves in the herd populations and find that RSS has smaller probabilities of obtaining nonrepresentative samples than SRS. This advantage should hold in environmental sampling, as well.

5.13.4.2 Sample representativeness of the population is an important environmental objective (Guide D 6044).

5.13.5 *Relative Precision Under Different Statistical Distributions*:

5.13.5.1 Patil, et al. **(1)** show that the variance of the sample mean under RSS is smaller than that obtained under SRS under a variety of statistical distributions for the population (including normal, lognormal, rectangular, exponential, gamma, Weibul, and triangular).

5.13.5.2 Some example relative precision (RP) results are given in Table 4.

5.14 *Ranking Criteria*:

5.14.1 Auxiliary variables that can be used to rank the samples within a set include the following:

5.14.1.1 Visual inspection.

5.14.1.2 Inexpensive quick measurement.

TABLE 4 Relative Precision (RP) Under Different Population Distributions

Distribution		m in RSS	RP
Normal N (0, 1)		3	1.91
		5	2.77
Lognormal,	σ = 0.1	3	1.90
	σ = 0.5	3	1.70
	σ = 1.0	3	1.34
Lognormal,	σ = 0.1	10	4.73
	σ = 0.5	10	3.58
	σ = 1.0	10	2.06

where σ is the standard deviation for the underlying distribution.

5.14.1.3 Knowledge of operational history.

5.14.1.4 Previous site data.

5.14.2 A quick method could be an in-situ measurement with an instrument. Operational history or previous site data may provide information about which areas are more highly contaminated than the others. In fact, stratified random sampling typically uses this kind of information to define the strata. When information on the stratification of the population is available in advance for the entire site, stratified random sampling may be more efficient than RSS, depending on how precise the knowledge of the population stratification is.

5.14.3 Near perfect ranking is achieved when the correlation between the auxiliary and primary variables approaches +1.0 or −1.0; however, perfect ranking is not needed for increased precision of RSS relative to SRS. Correlation close to +1.0 means that the ordering of the rank data on the auxiliary variable would be similar to the ordering of the data on the primary variable. Correlation close to −1.0 means that the ordering of the rank data on the auxiliary variable would be almost in reverse order of the primary variable. In this case of −1.0 correlation, the purpose of stratification "at the sample level" using the auxiliary variable also is achieved, thus leading to increased precision for RSS relative to SRS.

5.14.4 The relative efficiency (increase in precision) of RSS over SRS depends on ranking accuracy, the number of samples, and the underlying population distribution **(1)**. Details on the last item are given in 5.13.2.

5.14.5 When there are ties and the ties are broken arbitrarily, this can reduce effectiveness of ranking, and thus, reduces the advantage of RSS to some extent. Tie-breaking, however, does not totally negate its advantage, nor does it introduce bias.

6. Uses of RSS in Environmental Sampling

6.1 RSS can be used when:

6.1.1 There is an auxiliary variable.

6.1.2 The auxiliary variable is either quick or inexpensive to measure, relative to the primary variable.

6.1.3 Cost of ranking the auxiliary variable is low relative to measurement of the primary variable.

6.1.4 The auxiliary variable has good correlation with the primary variable.

6.2 The advantages of RSS would be reduced or become negligible if any of the conditions listed in 6.1.1 through 6.1.4 are absent. For example, if the cost of measuring the auxiliary variable is large relative to that of measuring the primary variable, there may be an increase in total measurement costs. There would still be an increase in precision, however, relative to SRS (due to the effect of stratification using the auxiliary variable). In these cases when the advantages are not clear-cut, the stakeholder will need to weigh the pros and cons.

6.3 Nussbaum, et al. **(5)** describe the use of field RVP (Reid Vapor Pressure) as the auxiliary variable and the laboratory RVP as the primary variable in testing regulatory compliance of emissions at gas pumps, using RSS.

6.4 Pacific Northwest Laboratory **(8)** suggests the following potential applications:

6.4.1 Use of *in situ* readings (for example, readings of exterior drum conditions) to rank drums before a subset is selected for detailed inspection (on the assumption that the objective is to estimate the average condition of the drums).

6.4.2 In deciding which archived samples should be measured for a new contaminant, previous analytical results of these samples on other contaminants can be used to rank and select the samples for analysis of the new contaminant.

7. Keywords

7.1 auxiliary variable; bias; mean concentration; precision; primary variable; ranked set sampling; representativeness; sampling

REFERENCES

(1) Patil, G. P. et al. (1994), "Ranked Set Sampling," *In Handbook of Statistics 12*, G. P. Patil and N. P. Rao, eds, pp. 167–200, North-Holland, NY.

(2) Takahashi, K. and Wakimoto, K. (1968), "On Unbiased Estimates of the Population Mean Based on the Sample Stratified by Means of Ordering," Ann. of Institute of Stat. Math, 20:1–31.

(3) Dell, T. R. and Clutter, J. L. (1972), "Ranked Set Sampling Theory with Order Statistics Background," *Biometrics 28*, 545–555, June 1972.

(4) McIntyre, G. A. (1952), "A Method for Unbiased Selective Sampling, Using Ranked Sets," *Australian J. of Agricultural Research*, Vol 3, pp. 395–390.

(5) Nussbaum, B. D. et al (1997), "Cost Effective Gasoline Sampling Using Ranked Set Sampling," *1997 Proceedings of the Sec. on Statistics and the Environment*, American Statistical Association.

(6) Mode, N. A. et al. (1999), "Ranked Set Sampling for Ecological Research: Accounting for the Total Costs of Sampling," *Environmetrics*, Vol 10, pp. 179–194.

(7) Lacayo and Neerchal (1996), cited by Steven Millard and Nagaraj Neerchal in "Introduction to Environmental Statistics Using Environmental Stats for S-Plus," August 10, 1997, at *ASA Joint Statistical Meetings*, Anaheim, CA.

(8) Pacific Northwest Laboratory (1995), "DQO Statistics Bulletin," Vol 1, No. 1, June 1995, PNL-SA-26377.

Designation: D 6597 – 00

Standard Practice for
Assessment of Attaining Clean Up Level for Site Closure[1]

This standard is issued under the fixed designation D 6597; the number immediately following the designation indicates the year of original adoption or, in the case of revision, the year of last revision. A number in parentheses indicates the year of last reapproval. A superscript epsilon (ϵ) indicates an editorial change since the last revision or reapproval.

1. Scope

1.1 This practice covers the steps necessary to determine if a site is ready for closure with emphasis on general sampling techniques and how to determine if the agreed upon level of remediation has been achieved.

1.2 A minimum of statistical procedures is used in this practice.

1.3 The standard practice does not address radioactive waste sites nor does it cover every situation that can occur in the field. It does cover the following general sampling techniques: soil sampling, swipe sampling of buildings and equipment, surface impoundments, waste piles, and layered tanks. It does not cover drum sampling, general water sampling and monitoring wells and their construction.

1.4 *This standard does not purport to address all safety concerns, if any, associated with its use. It is the responsibility of the user of this standard to establish appropriate safety and health practices to determine the applicability of regulatory limitations prior to use.*

2. Referenced Documents

2.1 *ASTM Standards:*

D 1452 Practice for Soil Investigation and Sampling by Auger Borings[2]

D 1586 Test Method for Penetration Test and Split-Barrel Sampling of Soils[2]

D 1587 Practice for Thin-Walled Tube Geotechnical Sampling of Soils[2]

D 3550 Practice for Ring-Lined Barrel Sampling of Soils[2]

D 4547 Guide for Sampling Waste and Soils for Volatile Organic Compounds[3]

D 4687 Guide for General Planning of Waste Sampling[3]

D 5088 Practice for Decontamination of Field Equipment Used at Nonradioactive Waste Sites[2]

D 5283 Practice for Generation of Environmental Data Related to Waste Management Activities: Quality Assurance and Quality Control Planning and Implementation[3]

D 5434 Guide for Field Logging of Subsurface Explorations of Soil and Rock[2]

D 5451 Practice for Sampling Using a Trier Sampler[3]

D 5730 Guide for Site Characterization for Environmental Purposes With Emphasis on Soil, Rock, the Vadose Zone and Ground Water[2]

D 5792 Practice for Generation of Environmental Data Related to Waste Management Activities: Development of Data Quality Objectives[3]

D 5956 Guide for Sampling Strategies for Heterogeneous Wastes[3]

D 6009 Guide for Sampling Waste Piles[3]

D 6051 Guide for Composite Sampling and Field Subsampling for Environmental Waste Management Activities[3]

D 6232 Guide for Selection of Sampling Equipment for Waste and Contaminated Media Data Collection Activities[3]

D 6250 Practice for Derivation of Decision Point and Confidence Limit for Statistical Testing of Mean Concentration in Waste Management Decisions[3]

D 6311 Guide for Generation of Environmental Data Related to Waste Management Activities: Selection and Optimization of Sampling Design[3]

E 1728 Practice for Field Collection of Settled Dust Samples Using Wipe Sampling Methods for Lead Determination by Atomic Spectrometry Techniques[4]

2.2 *API Standard:*

API Standard 2015 Cleaning Petroleum Storage Tanks[5]

2.3 *Federal Standards:*

Title 40, Code of Federal Regulation (CFR), Part 245, [6]

Title 40, Code of Federal Regulation (CFR), Part 265, Guidance on Demonstrating Equivalence [6]

3. Terminology

3.1 *Definitions of Terms Specific to This Standard:*

3.1.1 *clean up level, n*—a concentration of a compound or element, in a defined volume of material, that is allowed to remain on a site.

3.1.1.1 *Discussion*—The numerical level usually is determined during the DQO process using a mean, median, or percentage value based upon health risk, fate and transport consideration, site location and intended use.

---

[1] This practice is under the jurisdiction of ASTM Committee D34 on Waste Management and is the direct responsibility of Subcommittee D34.01.01 on Planning for Sampling.

Current edition approved Oct. 10, 2000. Published December 2000.

[2] *Annual Book of ASTM Standards*, Vol 04.08.

[3] *Annual Book of ASTM Standards*, Vol 11.04.

[4] *Annual Book of ASTM Standards*, Vol 04.11.

[5] Available from American Petroleum Institute, 1801 K Street NW, Washington, DC 20226.

[6] Available from Superintendent of Documents, US Government Printing Office, Washington, DC 20402.

3.1.2 *closure, n*—all waste and manufactured residue are remediated to an acceptable clean up level.

4. Summary of Practice

4.1 The site is determined to be remediated and ready to be closed when site contaminant concentrations meet the agreed upon cleanup level for the site for each media of concern and sampling and analysis procedures meet data quality objectives (DQOs).

4.2 Optimization of sampling design is discussed in brief or by reference.

4.3 The site closure document assumes that the site has been characterized, the characterization was found to be adequate, and DQOs were met.

4.4 A copy of the table of contents of a closure plan submitted to a government agency is included to give an overall view of some items that should be considered. This example should be used only for guidance (see Appendix X1.2).

4.5 This site closure practice contains lists of equipment and sampling methods used in sampling of soils and buildings to confirm the cleanup level(s) has/have been reached.

5. Significance and Use

5.1 This practice contains a mechanism to determine that the hazardous constituents have been remediated to an acceptable level.

5.2 This practice assists both the regulator and the owner/operator in establishing cost-effective steps and procedures necessary to determine if the site has achieved its objective for site closure.

5.3 This practice can be applied to hazardous and nonhazardous spills or leaks, surface impoundments, waste piles, buildings, and tanks. The general nature of this practice allows its use in a large variety of situations.

6. Planning Using the DQO Process

6.1 The Practice D 5792 is a companion to this practice. A thorough study of this practice will prevent many costly mistakes later in the remediation program. The Code of Federal Regulations document 40 CFR 245 covers the DQo process).

6.2 The DQO process should be completed and agreed to by the stakeholders.

6.2.1 The DQO process must specify some numerical value or values that the stakeholders agree beforehand that the decision for site closure shall depend on (see Note 1). For example, the decision rule may state that if the lower 95 % confidence limit is less than the regulatory limit, then the site is safe to close (see Practice D 6250). In that case, no additional sampling is needed and the site may be closed. The conditions for site closure must be stated clearly.

NOTE 1—At some sites, the cleanup goal may be to eliminate an exposure pathway, as is done for installation of slurry walls or caps around or on a site. In these cases no numerical cleanup level is specified.

6.3 After the site has been remediated, the question becomes whether the effort has met the clean up level. The verification process is established during the DQO process. The process may require additional sampling or simply a visual confirmation.

6.3.1 The number of samples and where they will be taken will depend upon the values and acceptable risk established during the DQO process.

NOTE 2—The number of samples may change if the sampling results indicate additional remediation is required.

7. General Sampling Considerations

7.1 A written sampling plan should be prepared before the sampling begins. There are several ASTM documents the stakeholders may want to review during the planning process, and they include: Guides D 4687, D 5434, and D 6051 and Practice D 5283.

7.2 Complete records of all sampling activities must be maintained.

7.3 Before the final sampling commences, the project manager or other responsible party should confirm that the site is ready for the final sampling in preparation for closure. Any differences between planned and actual remediation activities should be documented.

8. Sample Design

8.1 The sampling design is established and optimized during the DQO process.

8.1.1 The sampling design may change if the data from the samples do not confirm clean closure has been achieved. The sampling design should be reevaluated because the site conditions have changed due to the remediation effort. Fewer samples may be required for the second confirmation try.

8.1.2 After further remediation has occurred, the site again is sampled for confirmation that the site meets the clean closure requirements and the DQOs. If the cleanup level(s) has/have been reached and the DQOs have been met, the site is ready for closure.

9. Determination of Attainment of Clean Closure

9.1 The decision rule for closure attainment typically is specified in Step 5 of a DQO process (see Practice D 5792).

9.2 Clean closure value or values usually are developed and agreed to by the stakeholders during the DQO process. The values often are mandated by regulations.

9.3 Examples of closure attainment in the decision rule can include the following:

9.3.1 When the upper 95 % confidence limit of the mean is less than the regulatory limit,

9.3.2 When the lower 95 % confidence limit of the mean is less than the regulatory limit,

9.3.3 When the sample mean is less than the regulatory limit, and

9.3.4 When there is a 90 % confidence that no more than 5 % of the samples are expected to exceed the regulatory limit (in statistical terms, when the upper 95 % tolerance limit is less than the regulatory limit with 90 % confidence).

9.4 The first three examples (9.3.1, 9.3.2 and 9.3.3) have to do with closure decisions based on the mean concentration of a site. Details of how these decision rules can be formulated statistically are given in Practice D 6250.

9.5 At other times, the decision rule can be a qualitative one. For example, a decision-maker can conclude that closure standard has been achieved when all the sample values are below detection limit.

9.6 A review of the US Government guidance document to help the user demonstrating clean colsure equivalence will be helpful. It is found in Code of Federal Regulation, 40 CFR 265.

9.7 Other documents that contain more detailed statistical methods are "Statistical Methods for Evaluating the Attainment of Cleanup Standards"[7] and USEPA document "Methods for Evaluating the Attainment of Cleanup Standards.[8]

10. Sampling Soils

10.1 Selecting appropriate sampling equipment for a site investigation or closure can be a challenging task. Sampling equipment should be selected to accommodate all of the known physical characteristics of concern or chosen, such that the effect of any sampling bias is minimized. In general, the same types of equipment used during the investigation phase of sampling can be used during the confirmation phase.

10.1.1 The sampling equipment should not discriminate against certain physical characteristics, for example, phase or particle size.

10.1.2 Other considerations in selecting the sampling equipment is the ability to access and extract the sample from the relevant location in the target population, ability to collect a sufficient mass of sample so that the target population is represented, ease of operation, cost of the equipment, and the ability to properly decontaminate the sampling apparatus. Guide D 6232 provides criteria for selecting sampling equipment. Also consider Guide D 5730.

10.2 A list of samplers that are in use to sample soils are found in Appendix X1.

11. Sampling Waste Piles and Surface Impoundments

11.1 It is assumed that the waste pile or surface impoundment has been remediated. The question is whether the site has been remediated to the condition specified in the DQOs. In most cases, automated soil boring is most appropriate for taking samples from the sides and bottom of the remediated site. Hand equipment can be used when sampling locations are based on knowledge gained during the site assessment stage, such as locations where the highest concentration of the compounds of interest was located.

NOTE 3—Guides D 6009 and D 6311 will help the user in the determination of how to sample waste piles and set up a sampling plan.

11.2 If the clean-up level(s) has/have been reached and the DQOs have been met, then site closure can proceed. If it is determined that the clean up level has not been reached, further excavation and additional sampling may be required. If unexpected exceedances are detected, the conceptual model may have to be revisited to determine if the model is still valid.

12. Sampling Tanks

12.1 The tank should be inspected visually before any sampling is done to ensure that the contents have been removed and the tank is clean.

12.2 After the visual check has been completed from the top access panel and the oxygen content is adequate for entry, the tank may be opened and a swipe sample or other appropriate samples taken to determine if level of remaining material is at a level below the clean up level (see API Standard 2015). The number of samples is dependent upon the level of confidence required (see Guide D 6311).

13. Buildings and Equipment

13.1 A site may contain buildings and equipment used in conjunction with hazardous materials resulting in the need for decontamination before being removed from the site. Procedures such as washing with detergents and water or solvents can be used to clean up the building or equipment (see Practice D 5088). After the cleaning procedure has been completed, the results may require verification before the building or equipment is removed or dismantled.

13.2 Wet-wipe sampling (see Practice E 1728) or cyclone vacuuming procedures are used frequently and the results compared with a fixed standard. Generally, authoritative sampling is used; however, systematic random sampling can be used if a more accurate estimate of the mean concentration is required.

14. Keywords

14.1 clean; closure; decontaminate; impoundment; site; soils; tank; waste

[7] Gilbert, R. O. and Simpson, J. C., "Statistical Methods for Evaluating the Attainment of Cleanup Standards", Vol 3, "Reference-Based Standards for Soils and Solid Media", Prepared for the USEPA by Pacific Northwest Laboratory, Richland, WA, PNL-7409 Vol 3, Rev. 1, 1992.

[8] USEPA, "Methods for Evaluating the Attainment of Cleanup Standards", Vol 1: Soils and Solid Media," Statistical Policy Branch (PM-223), 1989b.

APPENDIX

(Nonmandatory Information)

X1. SOIL SAMPLING DEVICES

X1.1 Augers (see Practice D 1452) and trier samplers (see Practice D 5451) are used primarily for disturbed or grab samples used in chemical analyses. Hand augers will reach surface soils to depths of 3 to 15 ft (1 to 5 m).

X1.2 Split barrel samplers (see Test Method D 1586) provide samples that allow description of lithology and other subsurface features and laboratory measurements of soil parameters unaffected by sample disturbance.

X1.3 Thin-wall tube (see Practice D 1587) and ring-lined barrel samplers (see Practice D 3550) collect undisturbed samples of unconsolidated material that are suitable for laboratory testing.

X1.4 Surface sampling techniques such as scoop, spoon, trawl, hand-corers, and shovel can be used to verify whether a site has met the site objectives. The hand corer is the best to minimize bias.

X1.5 Special handling is required for sampling for volatile organics to minimize the lost of contaminants (see Guide D 4547).

Designation: D 6640 – 01

Standard Practice for Collection and Handling of Soils Obtained in Core Barrel Samplers for Environmental Investigations[1]

This standard is issued under the fixed designation D 6640; the number immediately following the designation indicates the year of original adoption or, in the case of revision, the year of last revision. A number in parentheses indicates the year of last reapproval. A superscript epsilon (ε) indicates an editorial change since the last revision or reapproval.

1. Scope

1.1 This practice covers procedures for obtaining soils from core barrel samplers for chemical and physical analysis, with an emphasis on the collection and handling procedures that maintain the representativeness of the chemical contaminants of concern. Core barrel samplers are initially empty (hollow) until they are pushed into the ground to collect and retrieve a cylindrical soil sample with minimal disturbance. The selection of equipment and the sample handling procedures are dependent on the soil properties, the depth of sampling, and the general properties of the chemical contaminants of concern, that is, volatile organic compounds, semi-volatile organic compounds, and inorganic constituents. The sampling procedures described are designed to maintain representative concentrations of the contaminants regardless of their physical state(s), that is, solid, liquid or gas.

1.2 Four general types of core barrel samplers are discussed in this practice: split-barrel, ring-lined barrel, thin-walled tube, and solid-barrel samplers.

1.3 This document does not cover all the core barrel devices that are available for the collection of soil samples.

1.4 The procedures described may or may not be applicable to handling of samples for assessing certain geotechnical properties, for example, soil porosity.

1.5 *This standard does not purport to address all of the safety concerns, if any, associated with its use. It is the responsibility of the user of this standard to establish appropriate safety and health practices and determine the applicability of regulatory limitations prior to use.*

2. Referenced Documents

2.1 *ASTM Standards:*

D 653 Terminology Relating to Soil, Rock, and Contained Fluids[2]

D 1586 Standard Test Method for Penetration Test and Split-Barrel Sampling of Soils[3]

D 1587 Standard Practice for Thin-Walled Tube Geotechnical Sampling of Soils[3]

D 3550 Standard Practice for Ring-Lined Barrel Sampling of Soils[3]

D 3694 Standard Practices for Preparation of Sample Containers and for Preservation of Organic Constituents[4]

D 4547 Standard Guide for Sampling Waste and Soils for Volatile Organic Compounds[5]

D 4687 Guide for General Planning of Sampling[5]

D 4700 Standard Guide for Soil Sampling from the Vadose Zone[3]

D 5088 Standard Practice for Decontamination of Field Equipment Used at Nonradioactive Waste Sites[3]

D 5784 Standard Guide for Use of Hollow-Stem Augers for Geoenvironmental Exploration and the Installation of Subsurface Water-Quality Monitoring Devices[2]

D 5792 Standard Practice for Generation of Environmental Data Related to Waste Management Activities: Development of Data Quality Objectives[5]

D 5875 Standard Guide for Use of Cable-Tool Drilling and Sampling Methods for Geoenvironmental Exploration and the Installation of Subsurface Water-Quality Monitoring Devices[2]

D 5876 Standard Guide for Use of Direct Rotary Wireline Casing Advancement Drilling Methods for Geoenvironmental Exploration and the Installation of Subsurface Water-Quality Monitoring Devices[2]

D 6051 Standard Guide for Composite Sampling and Field Subsampling for Environmental Waste Management Activities[5]

D 6151 Standard Practice for Using Hollow-Stem Augers for Geotechnical Exploration and Soil Sampling[2]

D 6169 Standard Guide for Selection of Soil and Rock Sampling Used With Drill Rigs for Environmental Investigations[2]

D 6232 Standard Guide for Selection of Sampling Equipment for Waste and Contaminated Media Data Collection Activities[5]

D 6282 Standard Guide for Direct Push Soil Samplings for Environmental Site Characterization[2]

D 6286 Standard Guide for Selection of Drilling Methods

---

[1] This practice is under the jurisdiction of ASTM Committee D34 on Waste Management and is the direct responsibility of Subcommittee D34.01.03 on Sampling Equipment.

Current edition approved Feb. 10, 2001. Published April 2001.

[2] *Annual Book of ASTM Standards*, Vol 04.09.

[3] *Annual Book of ASTM Standards*, Vol 04.08.

[4] *Annual Book of ASTM Standards*, Vol 11.02.

[5] *Annual Book of ASTM Standards*, Vol 11.04.

for Environmental Site Characterization[2]

3. Summary of Practice

3.1 Obtaining soil samples from the surface and subsurface for chemical and physical analysis often involves the advancement of a core barrel sampler into the ground. A core barrel sampler can be operated by hand or mechanically, and it may be of a closed or open design (D 6282). Once the core barrel has been filled, the sampler is recovered from the bore hole and the soil sample is handled appropriately for the chemical constituents of concern.

3.2 This practice describes collection and handling procedures used with four types of core barrel samplers. The standards related to data quality objectives (D 5792), equipment specifications (D 6232, D 6169 and D 4700), their limitations and advantages (D 6282), and the site-specific geological and hydrological data should be reviewed to determine the soil coring equipment that is best suited for a specific project.

4. Significance and Use

4.1 Often during environmental investigations, soils are analyzed after being collected from the surface, the vadose zone (D 653) and sometimes from below the ground water table to identify and quantify the presence of a chemical contaminant. A contaminant is a substance that is typically hazardous and either is not normally present or that occurs naturally but is of an uncharacteristically high concentration (D 4687). A three-dimensional spatial array of samples can often provide information as to the source and route(s) of migration of the contaminant. The resultant information is used to direct remedial and corrective actions or can be used for monitoring purposes. Obtaining a soil sample with a core barrel sampler involves driving this device into the ground and then retrieving it for sample processing. Several methods for advancing a core barrel are generally acceptable (e.g., D 1586, D 1587, D 3550, D 4700, D 5784, D 5875, D 5876, D 6151, D 6282, and D 6286). Drilling methods that use drilling fluids (liquids or air) should be avoided because they are more susceptible to cross-contamination (See section 5.1.6).

5. Equipment Selection Criteria

5.1 Important criteria to consider when selecting a core barrel sampler for soil sampling are:

5.1.1 The materials that come into direct contact with the soil sample (barrel or barrel liner) should be compatible with the chemical or physical properties of the contaminant(s) of concern and the chemical properties of the soil. As a general rule samples obtained for semi-volatile organic compound analysis can be obtained within a core barrel or core barrel liner that is composed of stainless steel, steel, or brass. When only inorganic constituents are of concern, a plastic core barrel liner would be more appropriate than the previously cited materials. All of these materials are suitable for volatile organic compounds as long as the contact time is minimized. Often all of the above samples (semi-volatile organic, inorganic, and volatile organic compounds) are obtained from a single soil core. In this situation, soils should be taken from the interior of the soil core to avoid potential interferences between the contaminants of concern and the surface of the core barrel that is in direct contact with the sample.

5.1.2 The design of the core barrel sampler should allow for easy access to the sampled substrate for all subsequent handling procedures.

5.1.3 Core barrel size requirements depend on the type and number of chemical constituents of concern. For example, more soil is needed for the collection of samples intended for semi-volatile organic compound analysis than for the analysis of volatile organic compounds or inorganic constituents or both. Typically, a 250-mL (8-oz) bottle is filled for the analysis of semi-volatile organic compounds, a 125-mL (4-oz) bottle for inorganic constituents, and only 5-g subsamples are taken for volatile organic compounds.

5.1.4 Sampling depth (see D 6286, for selection of drilling method).

5.1.5 Suitability for soil type, that is, grain size, cohesion properties, and moisture content. For example, when sampling non-cohesive materials or when sampling below the water table, a core catcher (basket) should be used to limit the loss of sample and ground water during retrieval (D 4700 and D 6282).

5.1.6 The spreading of contamination between sampling depths should be minimized. Sealed hollow-barrel samplers (section 6.1.4) or cased bore holes and proper advancement techniques should be used to limit cross-contamination between sampling depths in the vadose zone. Although a cased bore hole does not guarantee that contamination will not be spread from one sampling depth to another, it lowers the possibility and eliminates the potential for bore hole collapse and side wall slough that can compromise the integrity of the samples removed from uncased bore holes. When there is standing ground water in the bore hole, a sealed sampler system should be used.

5.1.7 Site accessibility.

5.1.8 Time constraints.

5.1.9 Appropriate equipment availability.

6. Sampling Equipment

6.1 The types of core barrel samplers discussed in this practice are the split-barrel, ring-lined barrel, thin-walled tube, and solid-barrel samplers. These samplers are part of equipment that is either manually or mechanically powered, are designed to excavate a bore hole to the sampling depth, and may remain in the bore hole during the sample collection activity (i.e., cased bore hole or dual-walled casing).

6.1.1 *Split-Barrel Sampler*—As described in Methods D 1586 and D 4700, the split-barrel sampler is composed of a core barrel cut in half along the length of the barrel, a hardened metal drive shoe (cutting tip) and a sample head that vents (e.g., a ball check valve) to allow air to be displaced as it is filled (see Fig. 1). The shoe and the head thread onto opposite ends of the split barrel and hold the two halves together. A common barrel size is 5.08 cm (2 in.) outside diameter and 3.81 cm (1.5 in.) inside diameter. The drive shoe used with this particular barrel size has an inside diameter of 3.49 cm (1.375 in.). If fitted with a liner for encasing the sample, the inside diameter of the core barrel liner should not be less than the drive shoe. Several other sizes of split-barrel samplers are available, with inside diameters ranging from 2.5 to 10 cm. A

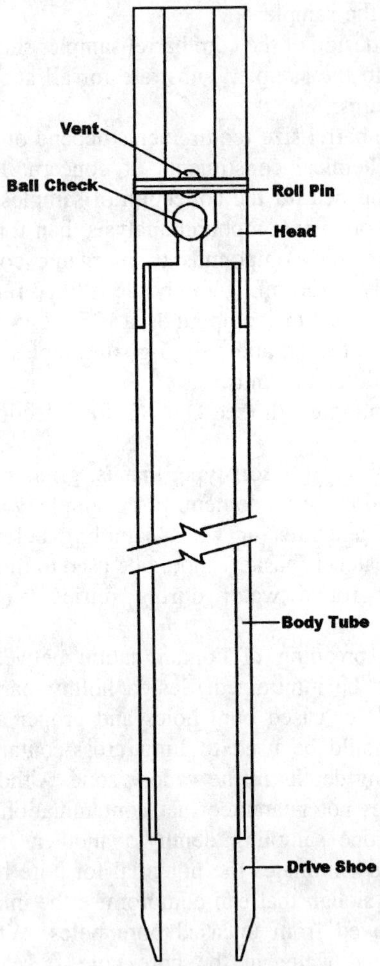

FIG. 1 Split Barrel Sampler

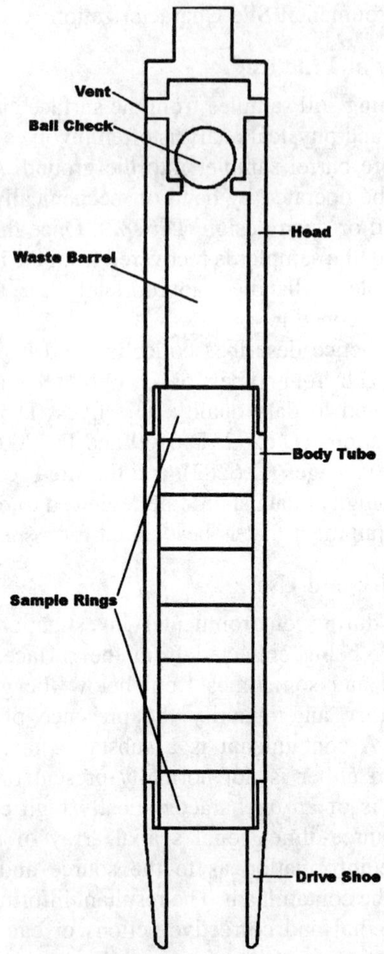

FIG. 2 Ring-Lined Barrel Sampler

core barrel catcher (basket) can be used with this type of sampler to help retain non-cohesive materials. The split-barrel sampler is often used in conjunction with hollow-stem continuous-flight augers or an equivalent drilling or direct-push probe system (D 4700 and D 6282).

6.1.2 *Ring-Lined Barrel Sampler*—As described in Methods D 3550 and D 4700, the ring-lined barrel sampler consists of an intact barrel or two split-barrel halves, a drive shoe, rings, a waste barrel, and a sampler head that vents (e.g., a ball check valve) to allow air to be displaced as it is filled (Fig. 2). The rings, which come in various lengths and are made of several different materials, should fit snugly within the barrel. The inside diameter of the rings should not be less than that of the drive shoe opening. The lengths of the rings will depend on the sampling plan so as to either allow quick access to a cross section of the soil core at a depth of interest or allow the appropriate size sample to be collected (see section 8.1.2). The waste barrel section allows the rings to be filled with undisturbed soil by leaving space to contain the disturbed soil that often is present at the bottom of a hole. A core catcher (basket) can be used with this type of sampler to help retain non-cohesive materials. This sampler is often used in conjunction with hollow-stem continuous flight augers, or an equivalent drilling or direct-push probe system (D 4700 and D 6282).

6.1.3 *Thin-Walled Tube Sampler*—As described in Methods D 1587 and D 4700, the thin-walled tube sampler consists of a core barrel attached to a head (Fig. 3). The head connects to standard drill rods and contains a vent (e.g., a ball check valve) to allow air to be displaced as it is filled. This sampler description is unique to the Shelby tube, which is available with outside diameters of 5.08, 7.62, and 12.70 cm (2, 3, and 5 in.), and a length of approximately 0.91 m (36 in.). In all cases the drive end of the thin-walled tube sampler has a slightly smaller diameter than the inside of the tube. A core barrel catcher (basket) cannot be used with this type of sampler; therefore, non-cohesive materials may be lost during retrieval. These samplers are often used in conjunction with hollow-stem continuous-flight augers or an equivalent drilling or direct-push probe system (D 4700 and D 6282).

6.1.4 *Solid Barrel Sampler*—The solid-barrel sampler is similar in design to the thin-tube sampler, with some important exceptions; the walls are thicker and they often can be equipped with a liner(s) and a core catcher (Method D 6282, see Figs. 4 and 5). Most of the core barrel samplers in D 6282 are designed to be a closed chamber until the depth of interest is reached, then either the entire sampler or an inner barrel is driven to a greater depth, without advancing or after retracting the drive point or piston tip. The open chamber (no drive point or piston tip) sampler design is often used for near-surface sampling equipment (Fig. 5).

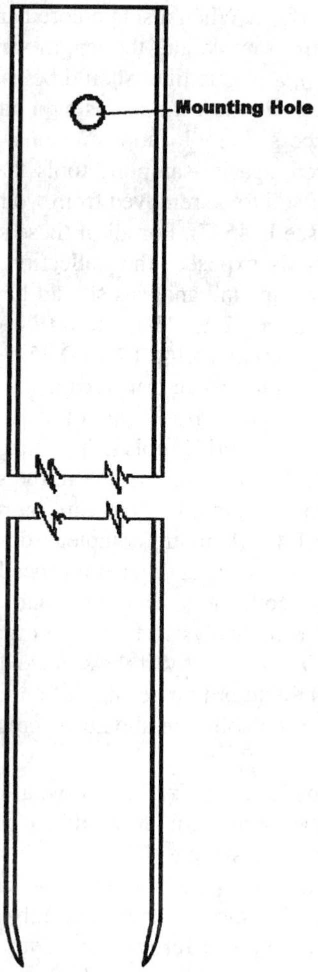

FIG. 3 Thin-Walled Tube Sampler

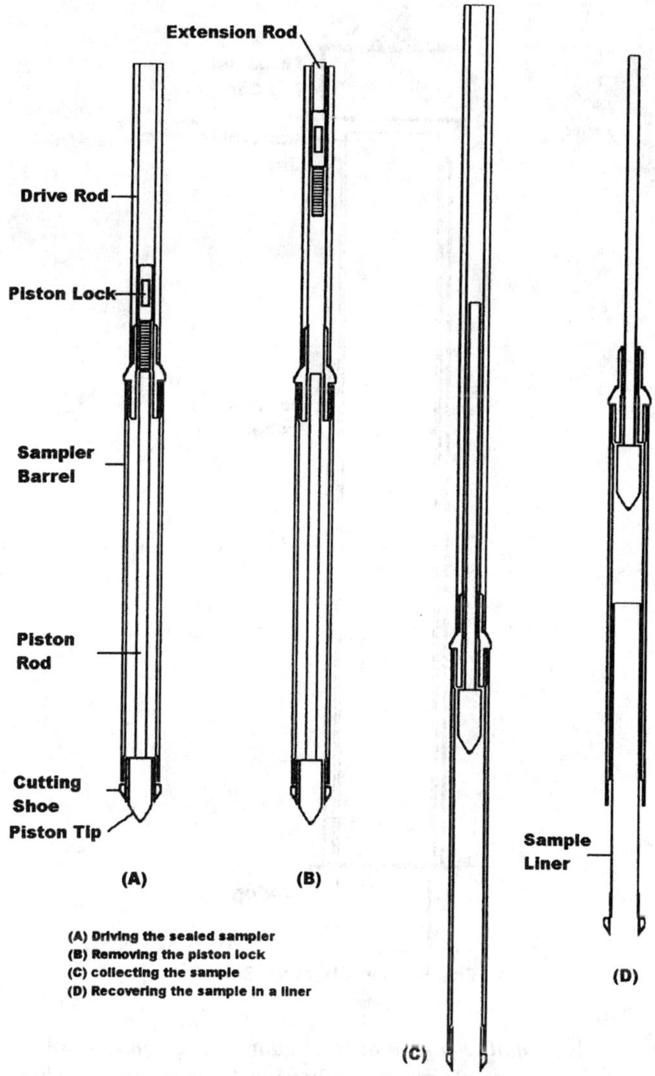

FIG. 4 Direct Push, Single Tube Sampler

7. Pre-Sampling

7.1 The pre-sampling activities are:

7.1.1 Equipment that comes into direct contact with the sample should be steam cleaned, washed with a detergent solution, and rinsed with contaminant-free water, desorbing agents, and deionized water (D 5088). The other pieces of equipment should be either steam-cleaned or washed with a detergent solution and rinsed with contaminant-free water.

7.1.2 The cutting tip or shoe should have a sharp edge. Once dulled, the cutting tip or shoe should be sharpened or replaced.

7.1.3 A work station should be set up for capping the core barrel sampler or when opening to remove subsamples. This work station should consist of a flat clean surface and be set up in a location which provides protection from extreme weather conditions and that is upwind of any local emission sources.

7.1.4 When sampling for volatile organic compounds or when subsampling on-site for constituents that are subject to oxidation or rapid biodegradation, the vessels used for transportation, the on-site sample storage chamber, and the equipment used for obtaining subsamples should be prepared for immediate use. For example, when using an on-site chemical or physical method of sample preservation, it is necessary to perform all of the steps leading up to the transfer of soil subsample from the sampler to the sample container. This may involve adding a solvent or acid solution to a vessel and recording a tared weight or having a chamber available that is capable of maintaining the appropriate refrigeration temperature.

8. Sampling

8.1 *Sampling Method*—A core barrel sampler can be used at the surface or threaded onto extension rods and lowered to the bottom of a cased or uncased bore hole or pushed as a single unit (direct push D 6282) to the depth of interest. The core barrel sampler is then driven into the soil manually or by some mechanical means to a depth that does not exceed the length of the core barrel. The filled sampler is then pulled out as smoothly (minimizing vibration) as possible to retain a maximum amount of soil in the sample chamber. Depending on the chemical and physical characteristics of the contaminants of concern, either the collection of subsamples should be initiated immediately or the sampler or barrel liner(s) should be capped and preserved in preparation for storage and transportation. The core barrel sampler should be handled by personnel wearing a fresh pair of gloves, and if opened on site, this operation should be performed on a clean surface that is positioned upwind of any local emission sources.

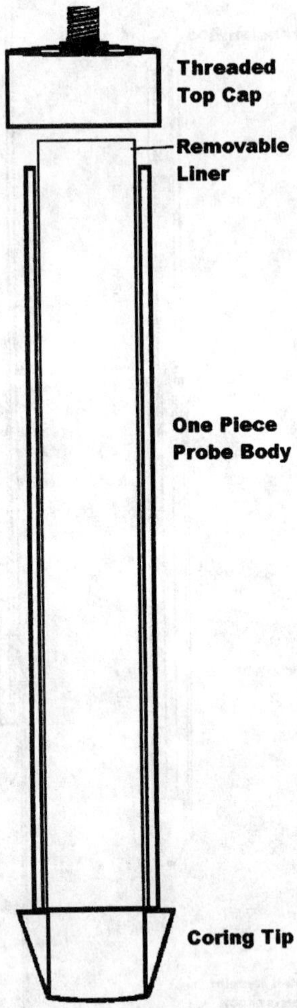

FIG. 5 Open Chamber Sampler

8.1.1 *Sampling for Volatile Organic Compounds*—Volatile organic compounds often simultaneously exist in three physical phases, that is, gas, liquid and solid. In general, the loss of volatile organic compounds from the gas phase presents the greatest challenge to intrusive sampling protocols. To minimize losses of volatile organic compounds, subsurface soils should be retrieved and subsampled as quickly as possible, taking special care to limit exposure and disaggregration of the soil's physical structure. Therefore, the core barrel should not be removed from the borehole until the sampling personnel are prepared, and the collection of subsamples for volatile organic compounds analysis should be the first activity performed (i.e., pre-empting even the logging of core barrel sample recovery and other soil core characteristics) once a core barrel has been opened.

8.1.1.1 When removing the drive shoe or drive head or both, care should be taken not to disaggregate (vibrate) the sample. The ends should be removed using the appropriate wrenches while the sampler is held in a vise. In the case of a split-barrel sampler without a liner, once the two ends (shoe and head, Fig. 1) have been removed, the sample is exposed by removing one half of the barrel. For a ring-lined barrel sampler, after removing the ends the rings should be gently pushed out of the barrel as a single piece, if possible. Once removed, the rings should be separated using a flat-bladed knife to create a flat cross-sectional surface. When a single core barrel liner is used to encase the entire sample and the length sampled is greater than 15 cm (6 in.), a plastic liner should be used. Thin-walled plastic liners can easily be cut open using a hand saw or razor knife to allow access at any location. When a 15-cm or longer metal liner is used, special sampling tools that allow smaller diameter cores of soil to be removed from well within the liner should be used (see D 4547). For all of these samplers, once a fresh soil surface is exposed, the collection of samples for volatile organic compound analysis should be initiated immediately. Subsamples of 5 to 25 g should be obtained with a rigid, thin-walled, small coring tool (D 4547). For example, coring tools for the purpose of transferring a subsample can be made by cutting off the front end of a disposable plastic syringe. These smaller coring tools help to maintain the sample structure during collection and transfer to the sample bottle, as do their larger counterparts, i.e., the core barrel samplers (see sections 6.1.1-6.1.4). A more complete description of the sampling tools, containers, and general procedures that should be used for the collection, handling, and preservation of samples taken for the analysis of volatile organic compounds can be found in D 4547. Lastly, if the surface has been exposed for more than a minute or two because of an unforeseen delay, a fresh surface in the soil core should be created just prior to sampling.

8.1.1.2 For the characterization of volatile organic compounds in soil, the sampling plan should specify the collection of either a discrete subsample (i.e., approximately 5 g of soil, D 4547) or one that is representative of the sampling interval. When a discrete subsample is specified, only a single location in the soil core is sampled for analysis. Typically, the highest reading established by a field-portable instrument capable of continuously measuring the total concentration of detectable volatile organic compounds, in the vapor phase adjacent to the surface of the soil core, is used to select the sampling location. Traditionally, only a single subsample has been taken from a soil core for the analysis of volatile organic compounds because most homogenization techniques fail to maintain representative concentrations for this class of constituents. A sample that is more representative of the concentration of volatile organic compounds over the sampling interval can be obtained by composite sampling (D 6051). For volatile organic compounds this could be performed by placing several (at least 3 per 30 cm [1 ft] of vertical distance, evenly spaced) subsamples into a vessel containing a solvent (D 4547) or by obtaining several subsamples and averaging the concentration for the soil core by either combining extracts when an organic solvent is used for extraction or after performing the analysis of several discrete subsamples.

8.1.1.3 To limit loss of soil gases, small (< 2.5-cm) diameter hollow-barrel samplers (large sample surface area to volume ratio) or ones that require that the sample be extruded from the core barrel (compaction) are not recommended for obtaining soils for volatile organic compound characterization.

8.1.1.4 Core barrel and core barrel liners are not recommended for storage and transportation since sheets of polytetrafluoroethylene or aluminum foil do not serve as a hermetic barrier for volatile organic compounds, sealing waxes readily sorb volatile organic compounds, and many adhesives contain volatile organic compounds of interest (D 4547).

8.1.2 *Sampling for Semi-Volatile Organic Compounds and Inorganic Constituents*—When sampling for contaminants that have low vapor pressures and that are not rapidly affected by oxidation or biodegradation, the sample handling does not have to be as careful or as expeditious as for volatile organic compounds. In general the same handling procedures stated above for the different soil core barrel samplers should be used (see section 8.1.1.1). One major difference is that samples can be removed from the core barrel or core barrel liner using a clean stainless steel spatula or similar plastic device, depending on the constituents of concern. However, since many semi-volatile organic compounds are susceptible to biodegradation, they should be preserved if not analyzed immediately. Preservation methods and the preparation of sample containers for many of the compounds that fall under the general classification of organic constituents can be found in D 3694.

8.1.2.1 For the characterization of semi-volatile and inorganic constituents in soil cores, sampling plans often specify that the subsample be representative of the sampling interval. Therefore, sample compositing and mixing schemes should be used prior to removing a representative sample for analysis. D 6051 describes procedures that could be used to homogenize the soil obtained from a core barrel that could either be used in the field or in a laboratory setting prior to removing a subsample for analysis. When oxidation or rapid biodegradation is a concern, practices similar to those cited above (see 8.1.1.2) for handling several discrete or composite samples for the characterization of volatile organic compounds should be used. That is, the entire sampling process should be performed as quickly as possible from the point of filling the core barrel sampler to the point where subsampling is performed. When removing subsamples for chemical analysis, care should be taken to limit exposure to the atmosphere and, if not analyzed immediately, they should be appropriately preserved. If subsampling is performed off-site, the ends of the core barrel or core barrel liner should be sealed with the appropriate materials and the entire soil core preserved by placing it in a refrigerated container.

8.1.2.2 Larger samples are usually collected for the analysis of semi-volatile organic compound than for either volatile organic compounds or inorganic constituents (see section 5.1.3). When sampling plans specify an 240-mL (8-oz; approximately 400-g) bottle be filled to capacity for the collection of a sample for semi-volatile organic compound analysis, a small-diameter (< 2.5-cm) hollow-barrel sampler would not be recommended.

9. Post-Sampling

9.1 The core barrel sampler and metal core barrel liners and the rest of the down-hole equipment should be decontaminated in accordance with D 5088 (see 7.1.1).

9.2 Nonreusable equipment (that is, plastic core barrel liners and baskets) should be disposed of properly.

9.3 The sample handling table should be properly decontaminated between subsampling activities.

10. Keywords

10.1 contaminants; core barrel samplers; split-barrel, ring-lined barrel; subsurface soil sampling; thin-walled tube

Designation: D 6661 – 01

Standard Practice for Field Collection of Organic Compounds from Surfaces Using Wipe Sampling[1]

This standard is issued under the fixed designation D 6661; the number immediately following the designation indicates the year of original adoption or, in the case of revision, the year of last revision. A number in parentheses indicates the year of last reapproval. A superscript epsilon (ϵ) indicates an editorial change since the last revision or reapproval.

1. Scope

1.1 This practice addresses sampling of organic compounds (i.e., PCBs, dioxins, many pesticides and similar compounds) from smooth nonporous surfaces using a solvent-wetted wipe sampling method. Samples are collected in a manner that permits the solvent extraction of the organic compound(s) of interest from the wipes and subsequent determination using a laboratory analysis technique such as gas chromatography with a suitable detector. This practice is, however, unsuitable for the collection of volatile organic compounds.

1.2 This practice should only be used to collect samples for the determination of organic compound(s) on a loading basis (e.g., mass per unit area). It cannot be used to collect samples for the determination of organic compounds on a concentration basis (e.g., mass per unit mass).

1.3 This wipe sampling practice is not recommended for collecting samples of organic compounds from rough or porous surfaces such as upholstery, carpeting, brick, rough concrete, ceiling tiles, and bare wood. It is also not intended for the collection of dust samples (see Practice E 1728) or sampling to estimating human exposure to contaminated surfaces.

1.4 To ensure valid conclusions are reached, a sufficient number of samples must be obtained as directed by a sampling design (the number and location of samples including quality control samples) and a quality assurance/quality control plan. This practice does not address the sampling designs used to achieve the data quality objectives (see Practice D 5792).

1.5 The values stated in SI units are to be regarded as the standard.

1.6 *This standard does not purport to address all of the safety concerns, if any, associated with its use. It is the responsibility of the user of this standard to establish appropriate safety and health practices and determine the applicability of regulatory limitations prior to use.*

2. Referenced Documents

2.1 *ASTM Standards:*

D 4687 Guide for General Planning of Waste Sampling[2]

D 5792 Practice for Generation of Environmental Data Related to Waste Management Activities: Development of Data Quality Objectives[2]

E 1728 Practice for Field Collection of Settled Dust Samples Using Wipe Sampling Methods for Lead Determination by Atomic Spectrometry Techniques[3]

3. Terminology

3.1 *Definitions:*

3.1.1 *wipe*, *n*—sorbent material (e.g., cotton gauze) that is rubbed on a surface to collect a sample for chemical analysis.

4. Summary of Practice

4.1 A wipe sample is collected from a smooth nonporous surface with a solvent-wetted wipe following a specified pattern of wiping to ensure complete coverage of an area of specified dimensions. The wipe is then extracted and analyzed to detect and quantify (at least semiquantitatively) the presence of organic compounds on surfaces.

5. Significance and Use

5.1 Wipe sampling is typically used by persons involved in hazardous waste site investigations to characterize the areal extent and the level of contamination on walls, floors, equipment, etc. Wipe sampling is also used to determine compliance with regulations.

5.2 There are many factors that contribute to variation in sampling results during wipe sampling including, the use of different pressures applied to the wipe, different kinds of wipes, different wiping patterns, the texture of the surface being wiped, and perhaps even the duration of wiping. The significance of this practice is that it standardizes wiping procedures to reduce sampling variability in the collection of samples from smooth, nonporous surfaces such as metal, glass, painted or sealed surfaces, tile, etc., in and around buildings, and from pipes, tanks, decontaminated equipment, etc.

6. Sampling Equipment and Supplies

6.1 *Sample Containers*—Airtight amber glass sample containers with PTFE-lined caps such as 40-mL volatile organic analysis vials are recommended. Larger 125-mL wide mouth bottles may also be used which eliminate the need for forceps to place or remove wipes from the sample container. To

---

[1] This practice is under the jurisdiction of ASTM Committee D34 on Waste Management and is the direct responsibility of Subcommittee D34.01 on Sampling and Monitoring.

Current edition approved April 10, 2001. Published May 2001.

[2] *Annual Book of ASTM Standards*, Vol 11.04.

[3] *Annual Book of ASTM Standards*, Vol 04.11.

TABLE 1 Contaminant Recovery Data Using Common Solvents and TLC Pads[4]

Compound	Solvent	Percent Recovery
Chlordane	Acetone	71
	Isooctane	54
Chlorpyrifos	Acetone	72
	Isooctane	56
Malathion	Dichloromethane	81
	Isooctane	80
Diazinon	Isooctane	70
Aroclor 1260	Isooctane	80
	Acetone	76
Bendiocarb	Acetone	85
	Isopropanol	84
Propoxur	Isopropanol	90
	Acetone	90

minimize solvent handling in the field, wipes may be wetted with solvent in the laboratory and shipped to the field in the sample container.

6.2 *Wipes*—Cotton gauze pads 7.6-cm square are to be used. Sterile surgical gauze pads are typically used without precleaning however, samples of the pads should be analyzed or otherwise determined to be free of the target compounds and substances that could interfere with the analytical method. If necessary, pads should be precleaned by solvent extraction in a laboratory prior to field use.

6.3 *Solvent*—A high purity solvent (one which is free of contaminants that might interfere with analysis), capable of solublizing the target organic compound and compatible with the surface being wiped, should be used. For collecting PCBs and most pesticides (e.g., chlordane, chlorpyrifos and malathion) isooctane is an effective solvent. For carbamates or known polar pesticides, isopropanol is more effective. Some guidance on solvent selection (Table 1) was generated by EPA[4] using thin layer chromatography (TLC) saturation pads (essentially a heavy filter paper) which generally performs similarly to cotton gauze pads. Hexane is another commonly used solvent to consider for PCB sampling. Some effective solvents such as acetone are not the most desirable because interfering compounds from some surfaces can also be recovered.[4] The analytical laboratory should be able to assist in selecting a proper wiping solvent compatible with the surface to be sampled and with the analytical procedures.

6.4 *Disposable Gloves*—Powderless gloves which protect the sampler's hands from the solvent and do not contribute any possibly interfering contaminants should be used. A new pair of gloves should be used for each wipe.

6.5 *Sampling Template (Optional)*—Templates made of stainless steel, aluminum, disposable heavy-duty aluminum foil or other inert material can be used to expose a 10-cm by 10-cm surface area to be wiped.

6.6 *Other*—Standard field sampling supplies are discussed in Guide D 4687 and may include a copy of the sampling plan, chain-of-custody forms, custody seals, logbook, camera, field data sheets, sample labels, forceps, noncontaminating marker (e.g., pencil, scribe), decontamination supplies, and solvent dispenser. Additional detailed equipment lists are included in EPA publications.[5, 6]

7. Procedure

7.1 Review the sampling plan and sampling procedures, assemble sampling equipment, ensure personnel are adequately trained for their tasks, arrange logistics, and ensure supplies will be available at the site when needed. Since sampling results can vary between operators sampling identical surfaces, the same person should collect all wipe samples at a given site[5] to minimize variability and enhance comparison of results from various locations.

7.2 Locate the sampling points as specified in, or according to the guidance of, the sampling plan.

7.3 Install the sample template or otherwise delineate the area to be sampled, normally a 10-cm by 10-cm area. This can be achieved by either taping a template in place (caution, tape used to secure a template should not be wiped since this may contaminate the sample), or by drawing the boundary of the area to be sampled with a noncontaminating marker. Although a 10-cm by 10-cm area is the standard-size template, the area does not have to be square as long as a 100-cm^2 area is being sampled. If contaminant levels are expected to be low, greater sensitivity may be achieved by sampling a larger area. In all cases, the location and dimensions (e.g., length, width, diameter) of the area sampled must be recorded and possibly photographed.

7.4 Don a new glove and obtain a clean wipe. If precleaned wipes were shipped in sample containers, forceps can be used to remove the wipe from the container.

7.5 If solvent-wetted wipes were not shipped to the field, dispense solvent (2 mL recommended although somewhat more may be desired if shipping solvent-wetted wipes to the field in 125-mL bottles) onto the cotton gauze pad. A repeating dispenser can be used for dispensing the same amount of solvent to each wipe. The EPA's PCB program specifies the use of a saturated, but not dripping, wipe (~5 mL of solvent),[5] which may slightly increase contaminant recovery, but research has shown more consistent results using 2 mL of solvent.[4]

7.6 Wipe the entire surface to be sampled using firm strokes by pressing with the fingertips. Wipe vertically and then horizontally to ensure there is complete coverage in both directions with minimal overlap of the previous stroke. The objective is to systematically, thoroughly, and consistently wipe the entire target area twice, each time from a different direction.[5] Excess wiping (e.g., more than single coverage in each direction) has been shown to reduce organic contaminant recovery.[4]

7.7 Fold the wipe with the sampled side inward, place it in the sample container, and cap the container. EPA's PCB program specifies air drying the saturated cotton gauze pad,

[4] Carr, B. L. and Hill, D. F., *Sampling of Common Pesticides and PCBs from Inert Surfaces*, EPA 330/1-90-001, National Enforcement Investigations Center, Denver, CO, 1989.

[5] Smith, J. H., *Wipe Sampling and Double Wash/Rinse Cleanup as Recommended by the Environmental Protection Agency PCB Spill Cleanup Policy*, U.S. Environmental Protection Agency, Washington, DC, 1991.

[6] U.S. Environmental Protection Agency, *Chip, Wipe and Sweep Sampling*, SOP#: 2011, Environmental Response Team, NJ, 1994.

either in the laboratory or the field. Field drying can be accomplished by placing wipes on clean aluminum foil or in the sample container with the lid off (ensure no liquid solvent is lost when placing wet wipes in the sample jar).

7.8 Label the sample container and complete standard documentation procedures.

7.9 Store the sample out of direct sunlight, cool to 4°C and ship or transport the sample(s) to the laboratory.

7.10 Quality control samples should be collected as specified in the sampling or quality assurance/quality control plan. The types of quality control samples may include blank, second wipe, duplicate, and spiked samples[5] as described below but may include other types as needed to achieve the objectives.

7.10.1 The first type of blank sample is a wipe in an unopened sample container (if provided to the field in this manner) or a clean wipe (with or without solvent) placed in a sample container. This type of blank is useful in determining whether the wipes and possibly the solvent are contaminated. The second type of blank is a wipe sample collected from a control area for each type of surface sampled. This type of blank is useful in determining whether contaminants may have been extracted from the surface sampled (e.g., target or interfering contaminants from painted, plastic, tile, etc. surfaces).

7.10.2 Two other types of quality control samples are second wipes and duplicate samples. A second wipe sample is collected from exactly the same area and immediately after the regular sample is collected. These sample results are used to estimate the residual contamination remaining after regular sampling. The data can be used to calculate an estimate of total surface contamination (this calculation is beyond the scope of this standard). This type of wipe is important for sampling relatively more porous surfaces such as vinyl tile. A duplicate sample is collected immediately adjacent to the regular sample and can be used to estimate sample collection precision (assuming the immediately adjacent area has the same level of contamination).

8. Keywords

8.1 sample collection; surface sample; wipe; wipe sample

Designation: E 688 – 94 (Reapproved 1999)

Standard Test Methods for Waste Glass as a Raw Material for Glass Manufacturing[1]

This standard is issued under the fixed designation E 688; the number immediately following the designation indicates the year of original adoption or, in the case of revision, the year of last revision. A number in parentheses indicates the year of last reapproval. A superscript epsilon (ϵ) indicates an editorial change since the last revision or reapproval.

1. Scope

1.1 These test methods give the various tests for assessing the compliance of glass recovered from wastes for use as a raw material for glass manufacturing.

1.2 The test methods combine visual examinations with both chemical and physical tests. A flow chart of the testing sequence is included in this test method (see Fig. 1).

1.3 The values stated in SI units are to be regarded as the standard. The values given in parentheses are for information only.

1.4 *This standard does not purport to address all of the safety concerns, if any, associated with its use. It is the responsibility of the user of this standard to establish appropriate safety and health practices and determine the applicability of regulatory limitations prior to use.* For hazard statements, see 5.3, and Section 6.

2. Referenced Documents

2.1 *ASTM Standards:*

C 169 Test Methods for Chemical Analysis of Soda-Lime and Borosilicate Glass[2]

C 566 Test Method for Total Moisture Content of Aggregate by Drying[3]

C 702 Practice for Reducing Samples of Aggregate to Testing Size[3]

C 729 Test Method for Density of Glass by the Sink-Float Comparator[2]

D 1068 Test Methods for Iron in Water[4]

D 1193 Specification for Reagent Water[4]

D 2576 Test Method for Metals in Water and Waste Water by Atomic Absorption Spectrophotometry[5]

D 4129 Test Method for Total and Organic Carbon in Water by High Temperature Oxidation and by Coulometric Detection[6]

E 11 Specification for Wire-Cloth Sieves for Testing Purposes[7]

E 105 Practice for Probability Sampling of Materials[7]

E 122 Practice for Choice of Sample Size to Estimate a Measure of Quality for a Lot or Process[7]

3. Significance and Use

3.1 These test methods provide a means for determining whether waste glass is suitable for use as a raw material for glass manufacturing.

4. Apparatus

4.1 The following various items of equipment required may be purchased from most laboratory supply houses:

4.1.1 *Aspirator.*

4.1.2 *Balance.*

4.1.3 *Burner*, Fisher (Meker) type.

4.1.4 *Crucible*, porcelain or other ceramic.

4.1.5 *Crucible*, platinum.

4.1.6 *Flask*, filtering, with side tube, 2000-ml.

4.1.7 *Funnel*, Büchner, approximately 171 mm in diameter.

4.1.8 *Funnel*, approximately 150 mm in diameter, filtering.

4.1.9 *Furnace*, 540°C or 1000°F.

4.1.10 *Clamps*, tubing, screw compressor.

4.1.11 *Magnet*, C-shaped, Alnico.

4.1.12 *Magnifier*, 5×, 10×.

4.1.13 *Oven*, 110°C or 230°F.

4.1.14 *Scales*, triple-beam.

4.1.15 *Sieves*, U.S. Standard Series—50 mm (2 in.), 6.3 mm (¼ in.), 1.18 mm (No. 16), 850 µm (No. 20), 425 µm (No. 40), 250 µm (No. 60), 106 µm (No. 140), conforming to Specification E 11.

4.1.16 *Sink-Float Standard*, sp gr 2.65.

4.1.17 *Triangle*, platinum.

4.1.18 *Tubing*, vinyl.[8]

4.1.19 Other ancillary laboratory equipment.

5. Reagents and Materials

5.1 *Purity of Reagents*—Reagent grade chemicals shall be used in all tests. Unless otherwise indicated, it is intended that

[1] These test methods are under the jurisdiction of ASTM Committee D-34 on Waste Management and are the direct responsibility of Subcommittee D34.06 on Recovery and Reuse.

Current edition approved Nov. 15, 1994. Published January 1995. Originally published as E 688 – 79. Last previous edition E 688 – 79 (1988) $^{\epsilon 1}$

[2] *Annual Book of ASTM Standards*, Vol 15.02.

[3] *Annual Book of ASTM Standards*, Vol 04.02.

[4] *Annual Book of ASTM Standards*, Vol 11.01.

[5] Discontinued—See *1980 Annual Book of ASTM Standards*, Part 31.

[6] *Annual Book of ASTM Standards*, Vol 11.02.

[7] *Annual Book of ASTM Standards*, Vol 14.02.

[8] Tygon plastic tubing, available from Norton Co., Plastics and Synthetics Div., Dept TR2, 12 East Ave., Tallmadge, OH 44278, or equivalent, has been found suitable for this purpose.

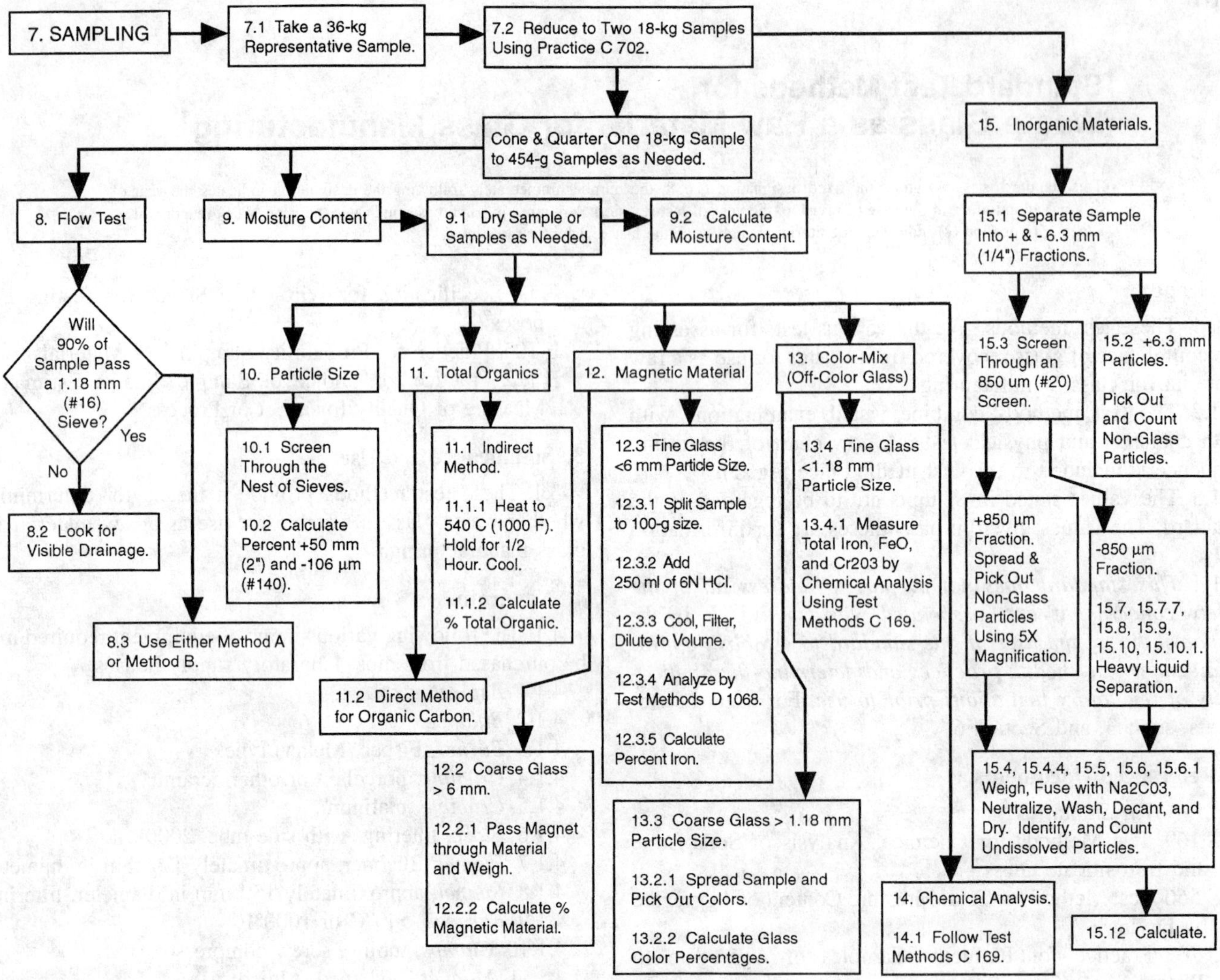

FIG. 1 Simplified Testing Flow Chart

all reagents shall conform to the specifications of the Committee on Analytical Reagents of the American Chemical Society, where such specifications are available.[9] Other grades may be used, provided it is first ascertained that the reagent is of sufficiently high purity to permit its use without lessening the accuracy of the determination.

5.2 *Purity of Water*— Unless otherwise indicated, references to water shall be understood to mean reagent water, Type II, as defined in Specification D 1193.

5.3 **Warning:** *Acetone* (*see 15.5*)—This substance is highly flammable (Class B) and must not be used in the vicinity of open flames or other ignition sources. Vapors should not be inhaled, since they can cause skin and membrane irritation.

5.4 *Ethyl Alcohol*, denatured.

5.5 *Hydrochloric Acid (3 N and 6 N) (see 12.3.2/15.11)*—Prepare 3 *N* acid by diluting 1 part of concentrated hydrochloric acid (HCl, sp gr 1.19) with 3 parts of water. Prepare 6 *N* acid by diluting 1 part of concentrated HCl with 1 part of water.

NOTE 1—**Caution:** These materials are corrosive and injurious to the skin as well as irritating to the eyes and mucous membranes.

5.6 *Potassium Hydroxide, Saturated Solution*—Add 100 g of potassium hydroxide (KOH) slowly, while stirring, to 100 ml water. Store this solution in a polyethylene bottle. This solution is corrosive and injurious to the skin.

5.7 *Sodium Carbonate* (Na_2CO_3).

5.8 *sym-Tetrabromoethane (Acetylene Tetrabromide)* (*sp gr 2.964*)—This substance has a threshold limit value (8 h time-weighted average exposure) of 1 ppm and a short time exposure limit (15 min) of 1.25 ppm. It must be used in a hood or under conditions of ensured ventilation.

6. Hazards

6.1 The analyst should be aware of good laboratory practices. Adequate ventilation is necessary, particularly for handling sym-tetrabromoethane and the density liquids used in 15.7.1-15.10.1. The flammability of acetone must be considered when it is used in 15.5.

6.2 Due to the origins of glass samples in municipal wastes

[9] *Reagent Chemicals, American Chemical Society Specifications*, American Chemical Society, Washington, DC. For suggestions on the testing of reagents not listed by the American Chemical Society, see *Analar Standards for Laboratory Chemicals*, BDH Ltd., Poole, Dorset, U.K., and the *United States Pharmacopeia and National Formulary*, U.S. Pharmaceutical Convention, Inc. (USPC), Rockville, MD.

destined for disposal, common sense dictates that some precautions should be observed when conducting tests on the samples. Recommended hygienic practices include using gloves when handling waste glass and washing hands before eating or smoking.

7. Sampling

7.1 *Gross Sample*— Take a grab sample weighing 36 kg (80 lb) in such a manner that it will be representative of the lot as described in Practices E 105 and E 122.

7.2 *Test Sample Preparation*—Reduce the gross sample to two samples, each weighing 18 kg (40 lb) by a method as described in Practice C 702. Use one sample for testing for inorganic material. Reduce the other sample by coning and quartering to produce at least seven samples, each weighing 454 g (1 lb), to be used for the remaining tests described in Sections 8-14.

7.3 *Sample Preservation*—Store the gross sample and subsequent samples in such a manner as to prevent not only the loss of contaminants but to prevent further contamination, until the necessary tests can be performed. It is recommended that samples be stored in sealed plastic bags (75 µm or 3 mil) or other airtight containers in order to avoid gain or loss of moisture.

8. Flow Test[10]

8.1 This test is used to ensure that the sample of glass shows no drainage, is noncaking, and free flowing.

8.2 If the sample is 90 % larger than will pass through a 1.18-mm (No. 16) sieve, the requirements will be met if the sample shows no drainage of water.

8.3 *Method A*—If the sample contains in excess of 10 % through a 1.18-mm (No. 16) sieve, screen a sample through the 1.18-mm sieve, 454 g (1 lb) of this material must flow out of the funnel as follows: Fit a powder funnel having a uniform internal stem diameter of 18 mm, a stem length of 30 mm, and a top of 80 mm, in a rack, and close off the bottom by the palm of a hand. Pour the sample to be tested into the funnel until the funnel is full. When the hand over the opening is removed, glass meeting the requirements will flow out of the funnel. One or two light taps on the funnel may be used to begin or aid flow. If the glass fails to flow out of the specified funnel and both producers and users agree, a second funnel test (Method B) may be performed; this test will measure the flow properties of a somewhat more sticky glass.

8.4 *Method B*—Use the same procedure as in Method A except that the dimensions of the second funnel shall be: stem internal diameter, 38 mm; stem length, 48 mm; top diameter, 170 mm.

9. Moisture Content

9.1 *Procedure*—Dry five 454-g (1-lb) test samples (see 7.2) to constant weight in accordance with Test Method C 566. Drying time may be in the order of 2 h at 110°C (230°F). Record the dry weight of at least one sample for use in 9.2. Use the remaining dry samples for testing according to Sections 10-14.

9.2 *Calculation*— Calculate the moisture content as follows:

$$\text{Moisture, \%} = \frac{(\text{original weight} - \text{dry weight}) \times 100}{\text{original weight}} \quad (1)$$

10. Particle Size

10.1 *Procedure*—Screen a sample from 9.1 on a 50-mm (2-in.) sieve. Weigh the material remaining on the sieve. Screen the material passing the sieve on a 106-µm (No. 140) sieve. An intervening sieve, such as a 212 µm (No. 70) may be used merely to reduce the amount presented to the test sieve. Weigh material passing through the 106-µm sieve. Shake all sieves mechanically for 10 min or by hand to achieve equivalent results. Other intervening screen sizes may be utilized.

10.2 *Calculations*— Calculate the percent of plus 50-mm and minus 106-µm material as follows:

$$\text{Plus 50-mm material, \%} = (A/W) \times 100 \quad (2)$$

$$\text{Minus 106-µm material, \%} = (B/W) \times 100 \quad (3)$$

where:

A = weight of material on 50-mm sieve,
B = weight of material through 106-µm sieve, and
W = dry weight of sample.

11. Total Organics (Paper, Plastic, and Other Combustibles)

11.1 *Indirect Method*:

11.1.1 *Procedure*—Using a sample from 9.1, place the sample in an uncovered ceramic crucible(s) and heat to 540°C (1000°F). Maintain this temperature for ½ h or until all flame and smoke have ceased.

NOTE 2—**Caution:** Overheating can cause the glass particles to fuse together. Allow the sample to cool to room temperature, weigh, and calculate the percent total organics. Reserve the sample or use an alternative dry sample for subsequent tests.

11.1.2 *Calculation*— Calculate the percent total organics as follows:

$$\text{Total organics, \%} = 100 - \frac{\text{weight after ignition} \times 100}{\text{dry weight of sample}} \quad (4)$$

11.2 *Direct Method for Organic Carbon*—Organic carbon can only be inferred from the method in 11.1. Organic carbon can be determined directly by the method in this section.

11.2.1 *Procedure*—Organic carbon can be determined directly by an instrumental method such as coulometrics.[11] Test Method D 4129 uses this instrumentation for total and organic carbon in water. The instrument can be readily adapted to solid materials such as waste glass.

12. Magnetic Material (Iron Contamination)

12.1 The method of testing for iron contamination will depend on the particle size of the sample.

[10] More detailed testing procedures and the effect of compaction are discussed by Carson, J. W., in *International Journal of Powder Metallurgy and Powder Technology*, vol. 11, 1975, pp. 233–239.

[11] Model 5010 Coulometer, available from Coulometrics Inc., a subsidiary of UIC Inc., P.O. Box 563, Joliet, IL 60434.

12.2 *Determination of Iron Contamination in Coarse Glass (Larger Than 6 mm in Size)*:

12.2.1 *Procedure*—Using the sample from 9.1 or from a replicate dry sample, spread a portion of the sample over a clean, dry surface in such a manner as not to have any piles over three particles deep. Cover the poles of a C-shaped Alnico magnet or its equivalent with a piece of paper or plastic film. Draw the magnet slowly through the glass particles so as to collect any magnetic material present. Transfer the magnetic material to an appropriate container by removing the cover from the magnet, while holding it over the container. Continue this process until all of the sample has been exposed to the magnet. Screen at 6.3 mm (¼ in.); then combine the magnetic material from all portions of the sample, weigh, and calculate the percent magnetic material. Reserve all portions of the sample, or from a replicate sample for subsequent tests.

12.2.2 *Calculation*— Calculate the percent magnetic material as follows:

$$\text{Magnetic material, \%} = \frac{\text{weight of tramp iron} \times 100}{\text{dry weight of sample}} \quad (5)$$

12.3 *Determination of Iron Contamination in Fine Glass (Smaller than 6 mm Size)*:

12.3.1 Weigh 100 g from the sample from 11.1 or from a replicate.

12.3.2 Add 250 ml of 6 *N* hydrochloric acid (HCl) and slowly heat to a vigorous boil. Stir while boiling.

NOTE 3—**Caution:** Rapid heating may cause "bumping."

12.3.3 Cool the sample and filter through a medium filter paper. Catch the filtrate in a 500-ml volumetric flask and wash the insolvable sample at least five times with warm distilled water. Wash water should not exceed 500 ml. Allow the flask and its contents to cool to room temperature and dilute to volume with distilled water.

12.3.4 Transfer an appropriate sized aliquot to a 200-ml volumetric flask and analyze for iron, by either Test Methods D 1068 or Test Method D 2576.

12.3.5 *Calculation*— Calculate the percent iron as follows:

$$\text{Iron, \%} = \frac{\text{weight of iron } (Fe^{++}) \times 100}{\text{dry weight of sample}} \quad (6)$$

To report results as % Fe_2O_3, multiply by 1.43.

13. Off-Color Glass

13.1 This test is used to test for off-color glass in color-sorted glass used by the container industry. Glass samples that are plus 1.18-mm (No. 16 sieve) particle size can be tested by visual means. Glass samples that are minus 1.18-mm particle size cannot be tested by visual means but must be chemically analyzed.

13.2 *Coarse Material*— Plus 1.18-mm.

13.2.1 *Procedure*—Spread the dry sample from 9.1 on a clean, dry surface that is illuminated to at least 2150 lx (200 fc). Visually pick out the off-color glass particles. The off-color particles should be segregated into amber, green (emerald green and Georgia green should be segregated when examining a flint glass because emerald green has a chromium content ten times that of Georgia green), flint, and other glass fractions. Georgia green can be visually distinguished from emerald green, since Georgia green is lighter in color. Other glass colors are those other than the colors listed. Save the fractions.

13.2.2 *Calculation*— Calculate the percent off-color glass as follows:

$$\text{Off-color glass, \%} = \frac{\text{weight of off-color glass} \times 100}{\text{dry weight of sample}} \quad (7)$$

13.3 *Examples*—The following examples are used to demonstrate the method of calculating the off-color fraction:

13.3.1 *Example 1*:

Given: A flint glass analytical sample with a dry weight of 447.8 g.

Off-Color Fractions:

Amber = 13.4 g

Georgia green = 21.9 g

Emerald green = 2.2 g

Other = 2.2 g

Then:

$$\text{Amber, \%} = \frac{13.4 \times 100}{447.8} = 3.0$$

$$\text{Georgia green, \%} = \frac{21.9 \times 100}{447.8} = 4.9$$

$$\text{Emerald green, \%} = \frac{2.2 \times 100}{447.8} = 0.5$$

$$\text{Equivalent emerald green, \%} = 0.5 + \frac{4.9}{10} = 1.0$$

$$\text{Other, \%} = \frac{2.2 \times 100}{447.8} = 0.5$$

$$\text{Flint, \%} = 100 - (3.0 + 1.0 + 0.5) = 95.5$$

13.3.2 *Example 2*:

Given: An amber glass analytical sample with a dry weight of 446 g.

Off-Color Fractions:

[A] Emerald and Georgia green need not be segregated for amber glass determination.

Then:

$$\text{Flint, \%} = \frac{19.9 \times 100}{446} = 4.5$$

$$\text{Green, \%} = \frac{17.8 \times 100}{446} = 4.0$$

$$\text{Other, \%} = \frac{4.5 \times 100}{446} = 1.0$$

$$\text{Amber, \%} = 100 - (4.5\ \% + 4.0 + 1.0) = 90.5$$

13.4 *Fine Material*— Minus 1.18 mm.

13.4.1 *Procedure*—Chemically analyze a representative sample of the glass in accordance with procedures found in Test Method C 169. Test for: (*1*) Cr_2O_3; (*2*) Fe_2O_3; (*3*) FeO, and (*4*) any other coloring species of interest.

13.4.2 Color is an indicator of the oxidation state of cullet. The buyers and sellers of cullet may agree to limits on the concentration of coloring species because changes in oxidation state can affect heat transfer in the melt, cause the release of dissolved gases, and thus change furnace efficiency and glass quality. Limits on coloring species may be expressed through the REDOX NUMBER as follows, where

$$\text{Redox number} = \frac{Fe^{++}}{Fe^{+++}} = \frac{\%FeO \times 0.7779}{(\%Fe_2O_3 \times 0.6994) - (\%FeO \times 0.7779)} \quad (8)$$

An increasing Redox number indicates a more reduced state of the material.

14. Chemical Analysis

14.1 A chemical analysis may be necessary to establish the acceptability of a glass sample. Chemical analysis of glass samples shall be done in accordance with procedures for routine analysis found in Test Methods C 169 or to procedures mutually agreeable to the parties involved.

15. Inorganic Material (Nonmagnetic Metal, Refractories, and Other Solid Inorganics)

15.1 *Sample Preparation*—Using the 18-kg (40-lb) sample from 7.2, screen through a 6.3-mm (¼-in.) sieve or screen. If the sample is not dry, the material on the screen or sieve must be washed on the screen or sieve with a flowing stream of water to remove adhering particles. Reserve the minus 6.3-mm material for 15.3 and dry if necessary.

15.2 *For Material Remaining on the Screen*—Pick out non-glass particles and determine if any are of magnetic metal, nonmagnetic metal or other inorganic materials. Count the number of such particles. Discard any of the remaining particles and reserve the rest for calculating total inorganic materials. From the remaining glass material, pick any pieces that appear to be coated with a painted or foil paper label or other coating. Save these for subsequent test for the presence of free metals. Discard the remaining plus 6.3-mm material.

15.3 Screen the dry minus 6.3-mm material from 15.1 through a 850-μm (No. 20) sieve. Reserve the minus 850 μm material for 15.7. Spread the plus 850-μm material on a clean dry surface that is illuminated to 2150 lx (200 fc) and visually pick out the non-glass particles using 5× magnification. Discard any particles attracted to a magnet and all organic material. Weigh the remaining inorganic particles and reserve the weight for calculating total inorganic material. Pick out and count the nonmagnetic metal particles. In addition, from the remaining glass material, pick any pieces that appear to be coated with a painted or foil paper label or other coating. Save these for subsequent test for the presence of free metals. Reserve the remaining inorganic particles for 15.4.

15.4 Weigh the inorganic particles from 15.3. If the sample weighs more than 1 g, split the sample into aliquots so that not more than 1 g will be fused at one time in a single crucible. Weigh out seven times as much sodium carbonate (Na_2CO_3) as the weight of the sample(s). Mix the sample(s) with its charge of sodium carbonate and add to the crucible(s). Platinum crucibles are preferred but certain metals may destroy the crucible. Porcelain crucibles with covers may be substituted. Heat the mixture on a Meker-type laboratory burner and fuse the mixture for 10 min at the maximum heat attainable.

15.4.1 The purpose of the fusion is to eliminate particles that are obviously not refractory so as to facilitate the inspection of the remaining particles. The particles remaining after fusion may not necessarily be refractory in glass furnaces so must be evaluated individually.

15.5 Place the cooled crucible in a 250-ml beaker. Neutralize the fused material with 3 *N* hydrochloric acid (HCl). Rinse the particles from the crucible. Dilute the solution with water, decant and dispose of the spent liquid. Wash the residue with water. If the presence of a residue obscures the particles, add a few millilitres of concentrated potassium hydroxide (KOH) solution and warm to dissolve. Again dilute the solution with water, decant, discard the spent liquid, and wash the residue with water. Rinse the particles with acetone and allow to air dry.

15.6 Examine the particles remaining. Particles that remain nondriable when probed with a needle and appear unaltered except for possible reduction in size will be reserved for routine examination by a petrographic microscope or subjected to a standard glass melt test, or both. Questionable particles can be examined by other means (for example, X-ray diffraction, emission spectroscopy, or microchemistry). Record the number of particles.

15.6.1 These materials are shown to be refractory in glass furnaces when they exist as particles of 250 μm or larger: corundum, mullite, zircon, chromite, spinel, sillimanite, andalusite, kyanite, and cassiterite.

15.7 Weigh the minus 250-μm material from 15.3. If the sample weight exceeds 450 g, take a representative sample weighing 450 g; otherwise use the whole sample. Separate the heavy inorganic particles from the glass using a dense liquid of 2.65 sp gr.

15.7.1 A dense liquid of 2.65 sp gr may be prepared by slowly adding denatured ethyl alcohol while stirring to *sym*-tetrabromoethane until a 2.65 sp gr density standard is no longer buoyant. Other liquids of 2.65 sp gr may be used.

15.8 Construct a separatory apparatus by slipping a 100-mm piece of vinyl tubing about 15 mm in diameter on the stem of a large filtering funnel. Install two screw compression-type tubing clamps on the tubing about 50 mm apart. Tighten the lower clamp to completely close off the tubing and adjust the upper clamp merely tight enough to remain in place. Add enough dense liquid to fill the separatory apparatus about half full.

15.9 Add the sample from 15.7 and enough dense liquid to nearly fill the separatory apparatus. Stir the sample periodically, and allow sufficient intervals between the stirring to permit the materials to separate. Should the accumulation of heavy particles in the tubing tend to fill up the upper clamp, close off the upper clamp, place a 100-ml beaker under the tubing, open the lower clamp, and flush the particles into the beaker. Close off the lower clamp and open the upper clamp as before. Continue stirring for a total of ten times. It may be necessary to repeat the removal of particles between stirrings and after the final stirring. Allow the particles to settle in the beaker, and decant the excess heavy liquid back into the separatory apparatus.

15.10 After the final decantation into the separatory apparatus, dilute the remaining heavy liquid in the beaker with acetone, allow the particles to settle, and decant the dilute heavy liquid into the evaporating dish. (See 15.10.1) Wash the heavy particles with acetone, decant, and discard the liquid.

Allow the heavy particles to dry and weigh. Reserve this material for calculating total inorganic materials. If the heavy particles exceed 1 g, screen through a 250-μm sieve and reserve the plus 250-μm heavy particles for 15.11. Discard the minus 250-μm particles.

15.10.1 The heavy liquid containing glass and other light particles may be conserved by filtering the liquid through a Büchner funnel, using a rapid filter paper and vacuum. Transfer the filtered liquid to a storage reservoir, which should be covered to retard evaporation of alcohol. Rinse the particles remaining on the filter with acetone to remove additional heavy liquid. Transfer the dilute heavy liquid to the evaporating reservoir, and allow the acetone to evaporate. Periodically test the density of the dilute liquid with a 2.65 sp gr sink-float density standard (see Test Method C 729). When the liquid exceeds the 2.65 sp gr, filter and transfer to the storage reservoir for reuse. Discard the glass and other light particles.

15.11 Test the heavy particle sample from 15.10 in the same manner as found in 15.4 and 15.5 by fusion with sodium carbonate and dissolution of the fusion with 3 *N* HCl. If a platinum crucible is used, be sure all metals have been removed. After drying, screen the particles through the 425-μm and 250-μm sieves. Particles that remain nonfriable when probed with a needle and appear unaltered except for possible reduction in size will be considered potential refractories. Record the number of potential refractory particles found on each sieve. As in 15.6 these particles may be examined further.

15.12 *Calculations*:

15.12.1 Calculate the percent weight of inorganic materials larger than 6.3 mm (¼ in.), W_i (+ 6.3 mm), as follows:

$$W_i\ (+6.3\ \text{mm}) = \frac{100\ W}{18} \tag{9}$$

where W is the weight of +6.3 mm inorganic material reserved from 15.2.

15.12.2 Calculate the percent weight of inorganic materials, between 6.3 mm and 850 μm (¼in. and No. 20) W_i (+850 μm), as follows:

$$W_i\ (+850\ \mu\text{m}) = \frac{100\ W_2}{18} \tag{10}$$

where W_2 is the weight of 6.3 mm − 850 μm inorganic material reserved from 15.3.

15.12.3 Calculate the percent weight of inorganic materials, smaller than 850 μm (No. 20), W_i (−850 μm), as follows:

$$W_i\ (-850\ \mu\text{m}) = \frac{100\ W_3}{18} \tag{11}$$

where W_3 is the weight of − 850 μm inorganic material reserved from 15.10.

15.12.4 Calculate the percent weight of total inorganic materials, W_t, as follows:

$$W_t = W_i\ (+6.3\ \text{mm}) + W_i\ (+850\ \mu\text{m}) + W_i\ (-850\ \mu\text{m}). \tag{12}$$

15.13 *Test for Presence of Free-Metals*—After following the calculation procedures in 15.12, gather the inorganic material particles from 15.2, 15.3, 15.6, and 15.10 along with the pieces of coated glass from 15.3 and 15.4.

15.13.1 *Procedure*—Place the inorganic particles in a 15-mL or other appropriate sized test tube. Add a sufficient amount of a 1 % alcoholic solution of o-dinitrobenzene to half cover the particles in each test tube. Add 0.5 N NaOH to each test tube until the particles are covered (this will constitute a 1:1 mixture). Suspend the test tubes in boiling water. The development of a violet color indicates the presence of aluminum, zinc, lead, or tin metal, or a combination thereof.

NOTE 4—This test is effective even if these metals have an oxide surface coating.

15.14 *Chemical Analysis*—This procedure does not separate or identify any inorganic materials other than glass smaller than 850 μm in size and having a specific gravity less than 2.65. If such material is suspected to be present, the methods cited in Section 14 should be used.

16. Precision and Bias

16.1 The precision and bias of these methods have not yet been established.

17. Flow Chart

17.1 The accompanying flow chart (see Fig. 1) is included to clarify the complex sequence of tests to be carried out in accordance with Sections 7-15.

18. Keywords

18.1 chemical tests; glass; glass cullet; glass manufacturing; glass recovery; inorganic contaminants; metallic contaminants; off-color glass; organic contaminants; physical tests; refractory contaminants; test methods; waste glass

Standard Test Methods for Municipal Ferrous Scrap[1]

This standard is issued under the fixed designation E 701; the number immediately following the designation indicates the year of original adoption or, in the case of revision, the year of last revision. A number in parentheses indicates the year of last reapproval. A superscript epsilon (ε) indicates an editorial change since the last revision or reapproval.

1. Scope

1.1 These test methods cover various tests for assessing the usefulness of a ferrous fraction recovered from municipal wastes.

1.2 These test methods comprise both chemical and physical tests, as follows:

	Section
Sampling	5
Bulk Density	6
Total Combustibles	7
Chemical Analysis (for Industries Other Than the Detinning Industry)	8
Magnetic Fraction (for the Detinning Industry)	9
Chemical Analysis for Tin (for the Detinning Industry)	10
Metallic Yield for All Industries Other Than the Copper Industry and the Detinning Industry	11

1.3 The values stated in inch-pound units are to be regarded as the standard. The values given in parentheses are for information only.

1.4 *This standard does not purport to address all of the safety concerns, if any, associated with its use. It is the responsibility of the user of this standard to establish appropriate safety and health practices and determine the applicability of regulatory limitations prior to use.*

2. Referenced Documents

2.1 *ASTM Standards:*

C 29 Test Method for Unit Weight and Voids in Aggregate[2]

C 702 Practice for Reducing Field Samples of Aggregate to Testing Size[2]

D 2234 Test Methods for Collection of a Gross Sample of Coal[3]

E 30 Test Methods for Chemical Analysis of Steel, Cast Iron, Open-Hearth Iron, and Wrought Iron[4]

E 122 Practice for Choice of Sample Size to Estimate a Measure of Quality for a Lot or Process[5]

E 350 Test Methods for Chemical Analysis of Carbon Steel, Low-Alloy Steel, Silicon Electrical Steel, Ingot Iron, and Wrought Iron[4]

E 351 Test Methods for Chemical Analysis of Cast Iron—All Types[4]

E 415 Test Method for Optical Emission Vacuum Spectrometric Analysis of Carbon and Low-Alloy Steel[6]

E 702 Specification for Municipal Ferrous Scrap[7]

3. Significance and Use

3.1 The establishment of these test methods for municipal ferrous scrap as a raw material for certain industries (see Specification E 702) will aid commerce in such scrap by providing the chemical and physical tests for the characterization of the scrap needed as a basis for communication between the purchaser and supplier.

4. Hazards

4.1 Due to the origins of municipal ferrous scrap in waste destined for disposal, common sense dictates that some precautions should be observed when conducting tests on the samples. Recommended hygienic practices include using gloves when handling municipal ferrous scrap and washing hands before eating or smoking.

5. Sampling

5.1 *Gross Sample of Loose Ferrous Scrap*:

5.1.1 Take a minimum of one gross sample having a volume of 7 ft^3 (0.2 m^3) (approximately equal to a 55-gal drum). Guidance for determining the number of gross samples needed to characterize a given lot of material and methods for accumulating a gross sample can be found in Practice E 122 and Test Method D 2234, respectively. In all cases, the actual sampling procedures to be used and the number of gross samples required to obtain a representative sample of the lot shall be established in accordance with an agreement between the purchaser and supplier.

5.1.2 Air-dry the gross sample at ambient temperature for a period of 24 h by spreading the sample on a clean, dry surface to one-layer thickness. Protect the sample from contamination by falling dust and debris. Reduce the gross sample to four samples by the method of coning and quartering, as described in Method B of Practice C 702.

5.2 *Gross Sample of Baled Ferrous Scrap*—Take a minimum of two bales. Guidance for determining the number of bales needed to characterize a given lot of material and

[1] These test methods are under the jurisdiction of ASTM Committee D-34 on Waste Management and are the direct responsibility of Subcommittee D34.06 on Recovery and Reuse.

Current edition approved Jan. 22, 1980. Published March 1980.

[2] *Annual Book of ASTM Standards*, Vol 04.02.

[3] *Annual Book of ASTM Standards*, Vol 05.05.

[4] *Annual Book of ASTM Standards*, Vol 03.05.

[5] *Annual Book of ASTM Standards*, Vol 14.02.

[6] *Annual Book of ASTM Standards*, Vol 03.06.

[7] *Annual Book of ASTM Standards*, Vol 11.04.

methods for selecting the bales can be found in Practice E 122. In all cases, the actual sampling procedures to be used and the number of gross samples required to obtain a representative sample of the lot shall be established in accordance with an agreement between the purchaser and supplier.

6. Bulk Density

6.1 *Loose Ferrous Scrap*:

6.1.1 *Apparatus*:

6.1.1.1 *Container*, constructed of suitable materials, for example, plywood, having the following approximate internal dimensions: base of 1 by 1 ft (300 by 300 mm) and a height of at least 2 ft (600 mm). Measure the internal dimensions of the box to the nearest 0.1 in. (3 mm). Suitable handles may be attached to the exterior of the container to aid in subsequent handling. Alternatively, containers of other geometries, agreeable to the purchaser and supplier, may be employed provided the area of the base is at least 1 ft^2 (0.09 m^2).

NOTE 1—The operator should be aware that this test method is not intended for those occasional pieces whose size is of the order of the dimensions of the box. As a guide, the maximum length of a single piece should not exceed three fourths of the maximum dimension of the base.

6.1.1.2 *Balance or Scale*, accurate within 0.1 % of the test load within the range of use. The range of use shall be considered to extend from the weight of the container empty to the weight of the container plus its contents at 100 lb/ft^3 (1600 kg/m^3).

6.1.1.3 *Measuring Rod*, calibrated in 0.1-in. (3-mm) intervals having a blunt end with an area of 4 $in.^2$ (26 cm^2).

6.1.2 *Procedure*:

6.1.2.1 Use each of the four samples from 5.1.2 to determine the bulk density.

6.1.2.2 Before each determination, weigh the empty container to the nearest 0.1 lb (0.05 kg).

6.1.2.3 Place oversize pieces, likely to protrude above the surface of the material in the container, at the bottom of the container prior to filling with the remainder of the sample.

6.1.2.4 Fill the container in three approximately equal layers. After each layer, place the container on a firm base, for example, a concrete floor, raising the opposite sides alternately about 2 in. (50 mm) and allowing the container to drop in such a manner as to hit with a sharp, resounding impact. Do this settling step ten times, five times on each side, in the manner described. Level the surface of the material manually to minimize surface irregularities.

6.1.2.5 Using the measuring rod described in 6.1.1.3, measure the distance from the top of the container to the surface of the material to the nearest 0.1 in. (3 mm) in each of the four corners of the container. Subtract the average of the four measurements from the inside height of the container to determine the height of the material.

6.1.2.6 Weigh the filled container to the nearest 0.1 lb (0.05 kg).

6.1.3 *Calculation*— Calculate the bulk density as follows:

$$\text{Bulk density, lb/ft}^3\ (\text{kg/m}^3) = \frac{a - b}{c \times d \times e} \times f \qquad (1)$$

where:

a = weight of container plus material, lb (or kg),
b = weight of container, lb (or kg),
c = inside length of container base, in. (or m),
d = inside depth of container base, in. (or m),
e = height of material in container, in. (or m),
f = 1 for container dimensions measured in metres, or 1728 for container dimensions measured in inches.

6.1.4 *Report*—Report each bulk density determination and the average of the four determinations.

6.2 *Baled Ferrous Scrap*:

6.2.1 *Procedure*:

6.2.1.1 Determine the weight of each bale from 5.2 to the nearest 0.1 lb (0.05 kg) using a scale described in 6.1.1.2.

6.2.1.2 Measure individually the length, width, and height of the bale to the nearest 0.1 in. (3 mm).

6.2.2 *Calculations*— Calculate the bulk density as follows:

$$\text{Bulk density, lb/ft}^3\ (\text{kg/m}^3) = \frac{g}{h \times i \times j} \times k \qquad (2)$$

where:

g = weight of bale, lb (or kg),
h = length of bale, in. (or m),
i = width of bale, in. (or m),
j = height of bale, in. (or m),
k = 1 for bale dimensions measured in metres, or 1728 for bale dimensions measured in inches.

6.2.3 *Report*—Report each bulk density determination and the average of all of the determinations.

7. Total Combustibles

7.1 *Procedure*:

7.1.1 Use two of the four bulk density volumes from 6.1.2.1 for the total combustibles determination. Reduce the size of each sample, if necessary, to approximately 20 lb (9.1 kg) by the method of coning and quartering, as described in Method B of Practice C 702. Determine the weight of each of the two samples to the nearest 0.1 lb (0.05 kg) before heating.

7.1.2 Heat each of the two samples in excess air at 750°F (400°C) for 60 min. An external source of air at low flow rates and pressures can be introduced at several locations within the sample to provide for combustion and excess air. The sample may be stirred every 15 min to expose fresh surface. Determine the weight of each of the two samples after heating to the nearest 0.1 lb (0.05 kg).

NOTE 2—For example, the amount of air needed can be estimated as in the following example:

Assume a 20-lb (9-kg) sample containing 10 % combustibles that are 40 % carbon. For complete combustion, the amount of carbon to be removed is:

$$20 \times 0.1 \times 0.4 = 0.8 \text{ lb} \qquad (3)$$

For the combustion reaction $C + O_2 = CO_2$, 0.8 lb of carbon requires 0.8 × (32 lb/lb·mol)/(12 lb/lb·mol) = 2.13 lb of oxygen or 2.13 × (359 ft^3 /lb·mol)/(32 lb/lb·mol) of O_2 = 23.9 ft^3 of oxygen at standard temperature and pressure (STP). Assuming the oxygen contribution from the sample is zero, and since air is 21 % oxygen by volume, 23.9/0.21 = 114 ft^3 of air at STP is required to react with the carbon. For air at 25°C (77°F), the volume of air required is 114 × (273 + 25)/273 = 124 ft^3, and assuming a 50 % excess air requirement, the total air necessary is 124 + 0.5 = 186 ft^3. For a combustion time of 60 min, the flow rate of air needed is 186/60 = 3.1 ft^3/min.

7.2 *Calculation*— Calculate the total combustibles as follows:

$$\text{Total combustibles, weight \%} = [1 - (w_1/w_2)] \times 100 \quad (4)$$

where:
w_1 = sample weight after heating, and
w_2 = sample weight before heating.

7.3 *Report*—Report each determination of total combustibles and the average of the two determinations.

8. Chemical Analysis (for Industries Other Than the Detinning Industry)

8.1 Reduce the two bulk density volumes remaining after Section 7 to two 30-lb (13.6-kg) samples, if necessary, and melt each in an induction furnace under a blanket of argon gas.

8.2 Take a sample of each melt and prepare for chemical analysis in accordance with one of the following test methods: E 30, E 350, E 351, E 415, or to procedures mutually agreed upon by the purchaser and the supplier.

8.3 Report the chemical composition of each melt and the average composition of the two melts.

9. Magnetic Fraction (for the Detinning Industry)

9.1 *Procedure*:

9.1.1 Weigh each of the two bulk density volumes remaining after Section 7 to 0.1 lb (0.05 kg) and manually separate using a hand magnet into two fractions: magnetic and nonmagnetic.

9.1.2 Wash the magnetic fraction in a galvanized tub of approximately 20-gal (0.08-m^3) capacity for 2 min with 180°F (82°C) water. Locate a 2-in. quick-drain valve, or equivalent, at the base of the tub to drain the water and wash the residue. When the drain valve is opened, use water from a garden hose for approximately 1 min to wash off any remaining residue. Place a ¼-in. (6.3-mm) hardware cloth with sufficient screening area at the exit of the drain valve to collect any of the magnetic fraction that may be washed out through the drain valve during draining of the tub. Next, repeat the previously described wash cycle. Manually remove the magnetic fraction from the tub and drain, if necessary, any residue or retained water, or both, from the individual pieces. After draining the water, air-dry the magnetic fraction at ambient temperature for a period of 24 h by spreading the sample onto a clean, dry surface to one-layer thickness, or as required by Test Method C 29, and weigh to the nearest 0.1 lb (0.05 kg). While drying, protect the sample from contamination by falling dust and debris.

9.2 *Calculation*— Calculate the magnetic fraction as follows:

$$\text{Magnetic fraction, weight \%} = \frac{w_3}{w_4} \times 100 \quad (5)$$

where:
w_3 = weight of magnetic fraction, and
w_4 = weight of as-received sample (from 9.1.1).

9.3 *Report*—Report each determination of the magnetic fraction and the average of the two determinations.

10. Chemical Analysis for Tin (for the Detinning Industry)

10.1 *Procedure*:

10.1.1 Separate manually each dried magnetic portion from Section 9 into "cans and other." Weigh the can and other fractions to the nearest 0.1 lb (0.05 kg). Prepare the can fraction for sampling by compacting it to sufficient density to maintain its integrity during subsequent drilling. The cylindrical compact should have a volume of approximately 10 $in.^3$ (160 cm^3).

10.1.2 Drill two ¼-in. (6-mm) holes through the cylinder from top to bottom. Locate the holes on the base of the cylinder, midway between the cylinder axis and the cylinder edge on a common diameter.

10.1.3 Combine the drillings from the two holes for the chemical analysis described in 10.1.5. Exercise caution to ensure the collection of all drillings.

NOTE 3—Experience has shown that approximately 20 g of drillings is a sufficient sample for the tin analysis.

10.1.4 Alternatively, the can fraction can be sampled by any other procedure mutually agreed upon between the purchaser and the supplier.

10.1.5 Prepare the sample for tin analysis in accordance with Sections IIIA and IIIB1 of the *Treatise on Analytical Chemistry*[8] or to procedures mutually agreeable to the purchaser and the supplier. The analysis result is the weight percent tin in the can fraction.

10.2 *Calculation*— Calculate the tin content as follows:

$$\text{Tin content of as-received sample, weight \%} = \frac{w_5}{w_6} \times w_7 \quad (6)$$

where:
w_5 = weight of can fraction,
w_6 = weight of as-received sample (from 9.1.1), and
w_7 = weight % of tin in can fraction.

10.3 *Report*—Report the tin content as percent tin by weight of the as-received sample.

11. Metallic Yield for All Industries Other Than the Copper Industry and the Detinning Industry

11.1 *Procedure*—Determine the metallic yield from each of the samples used for the chemical analysis in Section 8.

11.2 *Calculation*— Calculate the metallic yield as follows:

$$\text{Metallic yield, weight \%} = \frac{w_8}{w_9} \times 100 \quad (7)$$

where:
w_8 = weight of metal after melting (Note 4), and
w_9 = weight of total sample before melting.

NOTE 4—Weight includes portion removed for chemical analysis and excludes weight of slag formed.

11.3 *Report*—Report each determination of metallic yield and the average of the two determinations.

[8] *Treatise on Analytical Chemistry*, edited by Kolthoff, Elving, and Sandell, Part II, Vol 3, Interscience Publishers, New York, N. Y., 1961.

12. Precision and Bias

12.1 The precision and bias of these test methods have not yet been established.

13. Keywords

13.1 bulk density; chemical analysis; magnetic fraction; metallic yield; municipal ferrous scrap; sampling; test methods; total combustibles

Designation: E 702 – 85 (Reapproved 1999)

Standard Specification for Municipal Ferrous Scrap[1]

This standard is issued under the fixed designation E 702; the number immediately following the designation indicates the year of original adoption or, in the case of revision, the year of last revision. A number in parentheses indicates the year of last reapproval. A superscript epsilon (ε) indicates an editorial change since the last revision or reapproval.

1. Scope

1.1 This specification covers the chemical and physical requirements of municipal ferrous scrap that is intended for use by such industries listed as follows:

1.1.1 Copper industry (precipitation process),

1.1.2 Iron and steel foundries,

1.1.3 Iron and steel production,

1.1.4 Detinning industry, and

1.1.5 Ferroalloy industry.

1.2 Questions concerning material rejection, downgrading, and retesting based on failure to meet the requirements of this specification shall be dealt with through contractual arrangements between the purchaser and the supplier.

2. Referenced Documents

2.1 *ASTM Standards:*

E 701 Test Methods for Municipal Ferrous Scrap[2]

3. Terminology

3.1 *Definitions:*

3.1.1 *metallic yield*—the weight percent of the municipal ferrous scrap that is generally recoverable as metal or alloy.

3.1.2 *municipal ferrous scrap*—ferrous waste that is collected from industrial, commercial, or household sources and destined for disposal facilities. Typically, municipal ferrous scrap consists of a metal or alloy fraction, a combustible fraction, and an inorganic noncombustible fraction that includes metal oxides.

3.1.3 *total combustibles*—materials that include paints, lacquers, coatings, plastics, etc., associated with the original ferrous product, as well as combustible materials (paper, plastic, textiles, etc.) which become associated with the ferrous product after it is manufactured.

4. Chemical Composition

4.1 Municipal ferrous scrap shall conform to the requirements as to chemical composition for the respective end uses prescribed in Table 1.

4.2 The chemical requirements listed in Table 1 are based on melt analyses except where noted.

5. Physical Properties

5.1 Municipal ferrous scrap shall conform to the physical properties for the respective end uses prescribed in Table 2.

6. Test Methods

6.1 Determine the physical and chemical requirements of municipal ferrous scrap in accordance with Test Methods E 701.

7. Keywords

7.1 chemical requirements; copper industry; detinning industry; ferroalloy production; iron and steel foundries; iron and steel production; municipal ferrous scrap; physical requirements

---

[1] This specification is under the jurisdiction of ASTM Committee D-34 on Waste Disposal and is the direct responsibility of Subcommittee D34.06 on Recovery and Reuse.

Current edition approved July 26, 1985. Published February 1986. Originally published as E 702 – 79. Last previous edition E 702 – 79.

[2] *Annual Book of ASTM Standards*, Vol 11.04.

TABLE 1 Chemical Requirements

Element	Composition, %				
	Copper Industry (Precipitation Process)	Iron and Steel Foundries	Iron and Steel Production[A]	Detinning Industry[B]	Ferroalloy Production
Phosphorus, max	...	0.03	0.03	...	0.03
Sulfur, max	...	0.04	0.04	...	...
Nickel, max	...	0.12	0.08	...	...
Chromium, max	...	0.15	0.10	...	0.15
Molybdenum, max	...	0.04	0.025	...	...
Copper, max	...	0.20	0.10	...	0.20
Aluminum, max	...	0.50	0.50	4.00[C]	0.15
Tin	...	0.30 max[D]	0.30 max	0.15 min[E]	0.30
Lead, max	...	0.03	0.15	...	...
Zinc, max	...	0.06	0.06	...	...
Iron (metallic), min	96.0	...	...	...	...
Silicon, max	...	...	0.10	...	...
Manganese, max	...	...	...	...	0.35
Carbon, max	...	...	...	...	0.6
Titanium, max	...	...	...	...	0.025
Total combustibles, max	0.2[F]	4.0	4.0	...	0.5[G]
Metallic yield, min	...	90.0	90.0	...	90.0

[A] Experience has shown that material which has been incinerated probably will not meet these requirements.
[B] A minimum of 95 weight % of the material delivered shall be magnetic. Nonmagnetic material attached to the original magnetic article may be included in the minimum requirement.
[C] Not based on melt analyses due to aluminum losses during melting; to be determined by a method mutually agreed upon between the purchaser and supplier.
[D] For steel castings, the requirement for tin content is 0.10 max %.
[E] Refer to sections on magnetic fraction and chemical analysis of tin in Test Methods E 701. Normal separation of white goods and heavy iron yields tin contents equal to or greater than 0.15 weight %. Lesser tin contents would impact severely the value of the scrap to detinners.
[F] The scrap shall be appropriately processed (for example, by burning, chemical detinning, etc.) to be virtually free of combustibles.
[G] The scrap shall be appropriately processed (for example, by burning, chemical detinning, etc.) to be virtually free of combustibles.

TABLE 2 Physical Requirements

End-Use	Property	
	Bulk Density, lb/ft^3(kg/m^3)	Form
Copper Industry (Precipitation Process)	30 (480) max	loose, shredded as agreed upon between purchaser and supplier; shall not be balled or baled[A]
Iron and Steel Foundries	50 (800) min	loose, balled, or baled[B] as agreed upon between purchaser and supplier
Iron and Steel Production	75 (1200) min	loose[C] or baled[B] as agreed upon between purchaser and supplier
Detinning Industry	30 (480) max	shredded, 95 weight % shall be – 6, + ½ in. (–152, + 12.5 mm); shall not be balled, baled, burned, incinerated, or pyrolyzed
Ferroalloy Production	50 (800) min	loose, as agreed upon between purchaser and supplier

[A] Various consumers may establish gage limitations on the material they purchase.
[B] Industry practice is to specify a maximum bale size that may vary among users.
[C] Experience has shown that if the size range is 95 weight %, – 2, + ¼ in. (–50, + 6.3 mm), the bulk density requirement can be met and the material will be loose and free flowing.

The American Society for Testing and Materials takes no position respecting the validity of any patent rights asserted in connection with any item mentioned in this standard. Users of this standard are expressly advised that determination of the validity of any such patent rights, and the risk of infringement of such rights, are entirely their own responsibility.

This standard is subject to revision at any time by the responsible technical committee and must be reviewed every five years and if not revised, either reapproved or withdrawn. Your comments are invited either for revision of this standard or for additional standards and should be addressed to ASTM Headquarters. Your comments will receive careful consideration at a meeting of the responsible technical committee, which you may attend. If you feel that your comments have not received a fair hearing you should make your views known to the ASTM Committee on Standards, 100 Barr Harbor Drive, West Conshohocken, PA 19428.

Designation: E 708 – 79 (Reapproved 1999)

Standard Specification for Waste Glass as a Raw Material for the Manufacture of Glass Containers[1]

This standard is issued under the fixed designation E 708; the number immediately following the designation indicates the year of original adoption or, in the case of revision, the year of last revision. A number in parentheses indicates the year of last reapproval. A superscript epsilon (ε) indicates an editorial change since the last revision or reapproval.

1. Scope

1.1 This specification covers particulate glass (cullet material, recovered from waste destined for disposal, smaller than 6 mm intended for reuse as a raw material in the manufacture of glass containers.

1.2 The values stated in SI units are to be regarded as the standard. The values given in parentheses are for information only.

2. Referenced Documents

2.1 *ASTM Standards:*

C 162 Terminology of Glass and Glass Products[2]

C 169 Test Methods for Chemical Analysis of Soda-Lime and Borosilicate Glass[2]

C 429 Test Method for Sieve Analysis of Raw Materials for Glass Manufacture[2]

E 688 Test Methods for Waste Glass as a Raw Material for Glass Manufacturing[3]

3. Terminology

3.1 *Definitions:*

3.1.1 *flint glass cullet*—a particulate glass material that contains no more than 0.1 weight % Fe_2O_3, or 0.0015 weight % Cr_2O_3, as determined by chemical analysis.

3.1.2 For definitions of other terms used in this specification, refer to Terminology C 162.

4. Representative Sample

4.1 The following requirements qualify the glass lot to be used for direct use in soda-lime glass container manufacturing. Sample should be prepared and examined in accordance with Test Methods E 688.

NOTE 1—A preponderant proportion of glass cullet will be soda-lime bottle glass, the glass cullet having a composition as follows, as determined by Test Methods C 169.

Oxide	Composition, Weight %
SiO_2	66 to 75
Al_2O_3	1 to 7
CaO + MgO	9 to 13
Na_2O	12 to 16

NOTE 2—All percents referred to in this specification are weight percents.

5. General Requirements

5.1 The sample shall show no drainage of liquid and be noncaking and free flowing. A moisture content of less than 0.5 weight % is required to meet the free-flowing characteristics of a cullet that is predominantly of smaller particle size, 1.18-mm (No 16) sieve or smaller.

5.2 *Screen Size*—No material shall be retained on a 6-mm (¼-in.) screen. Material not exceeding 15 weight % shall pass through a 106-µm (No. 140) screen.

5.3 *Organic Materials*—The total content of organic materials, as measured in accordance with Section 6 shall not exceed 0.2 weight % of dry sample, except for color-mixed glass where the content of organic material may exceed 0.2 weight %. However, a content of organic material greater than 0.2 weight % must be held within a tolerance of ±0.05 weight %, with a maximum organic limit of 0.4 weight %.

5.4 *Magnetic Materials*—The total magnetic materials shall not exceed 0.05 weight % of dry sample weight for flint glass and 0.14 weight % for colored glass of dry sample weight in accordance with Section 6.

5.5 *Permissible Color Mix for Color Sorted Glass Cullet by Weight*:

5.5.1 *Amber Glass Cullet*:

90 to 100 % amber
0 to 10 % flint
0 to 10 % green
0 to 5 % other colors

5.5.2 *Green Glass Cullet*:

50 to 100 % green
0 to 35 % amber
0 to 15 % flint
0 to 4 % other colors

5.5.3 *Flint Glass Cullet*:

95 to 100 % flint
0 to 5 % amber
0 to 1 % green
0 to 0.5 % other colors

---

[1] This specification is under the jurisdiction of ASTM Committee D-34 on Waste Management and is the direct responsibility of Subcommittee D34.06 on Recovery and Reuse.

Current edition approved Nov. 30, 1979. Published January 1980.

[2] *Annual Book of ASTM Standards*, Vol 15.02.

[3] *Annual Book of ASTM Standards*, Vol 11.04.

5.5.3.1 Percents above 0.1 weight % of Fe_2O_3 or 0.0015 weight % of Cr_2O_3, or both, as determined by chemical analysis shall be considered mixed color glass. These limits are consistent with industry experience on raw material.

5.5.3.2 Flint glass cullet may contain up to 1 weight % emerald green or 10 weight % Georgia green, or a combination within the limits: 1 % Georgia green = 0.1 % emerald green.

5.6 *Other Inorganic Material* (such as nonmagnetic metals or refractories)—As measured, material larger than 850-μm (No. 20) screen size shall not exceed 0.1 % of the dry sample weight. Material smaller than 850-μm screen size shall not exceed 0.5 % of the dry sample weight.

5.6.1 *Refractories*—Based upon U.S. series screen size and sample weight, the following refractory particle limits shall apply for each screen fraction as stated below.

+20 mesh	1 particle per 18-kg (40-lb) sample
−20, +40 mesh	2 particles per 450-g (1-lb) sample
−40, +60 mesh	20 refractory particles per 450-g (1-lb) sample

5.6.2 *Nonmagnetic Metals:*—

+20 mesh	1 particle per 18-kg (40-lb) sample

Upon failure to meet the previously stated specification limits, retesting is permissible.

6. Sampling and Testing

6.1 Sampling and testing shall be in accordance with Test Methods E 688.

The American Society for Testing and Materials takes no position respecting the validity of any patent rights asserted in connection with any item mentioned in this standard. Users of this standard are expressly advised that determination of the validity of any such patent rights, and the risk of infringement of such rights, are entirely their own responsibility.

This standard is subject to revision at any time by the responsible technical committee and must be reviewed every five years and if not revised, either reapproved or withdrawn. Your comments are invited either for revision of this standard or for additional standards and should be addressed to ASTM Headquarters. Your comments will receive careful consideration at a meeting of the responsible technical committee, which you may attend. If you feel that your comments have not received a fair hearing you should make your views known to the ASTM Committee on Standards, 100 Barr Harbor Drive, West Conshohocken, PA 19428.

Designation: E 711 – 87 (Reapproved 1996)

Standard Test Method for Gross Calorific Value of Refuse-Derived Fuel by the Bomb Calorimeter[1]

This standard is issued under the fixed designation E 711; the number immediately following the designation indicates the year of original adoption or, in the case of revision, the year of last revision. A number in parentheses indicates the year of last reapproval. A superscript epsilon (ϵ) indicates an editorial change since the last revision or reapproval.

1. Scope

1.1 This test method covers the determination of the gross calorific value of a prepared analysis sample of solid forms of refuse-derived fuel (RDF) by the bomb calorimeter method.

1.2 *This standard does not purport to address all of the safety concerns, if any, associated with its use. It is the responsibility of the user of this standard to establish appropriate safety and health practices and determine the applicability of regulatory limitations prior to use.* For specific cautionary and precautionary statements see 6.10 and Section 8.

2. Referenced Documents

2.1 *ASTM Standards:*

D 1193 Specification for Reagent Water[2]

D 3177 Test Method for Total Sulfur in the Analysis Sample of Coal and Coke[3]

E 1 Specification for ASTM Thermometers[4]

E 180 Practice for Determining the Precision of ASTM Methods for Analysis and Testing of Industrial Chemicals[5]

E 775 Test Methods for Total Sulfur in the Analysis Sample of Refuse-Derived Fuel[6]

E 790 Test Method for Residual Moisture in a Refuse-Derived Fuel Analysis Sample[6]

E 829 Practice for Preparing Refuse-Derived Fuels (RDF) Laboratory Samples for Analysis[6]

3. Terminology

3.1 *Definitions:*

3.1.1 *calorific value*—the heat of combustion of a unit quantity of a substance. It may be expressed in joules per gram (J/g), British thermal units per pound (Btu/lb), or calories per gram (cal/g) when required.

NOTE 1—The unit equivalents are as follows:

1 Btu (International Table) = 1055.06 absolute joules
1 Calorie (International Table) = 4.1868 absolute joules
1 Btu/lb = 2.326 J/g
1.8 Btu/lb = 1.0 cal/g

3.1.2 *gross calorific value*—the heat produced by combustion of a unit quantity of solid fuel, at constant volume, in an oxygen bomb calorimeter under specified conditions such that all water in the products remains in liquid form.

3.1.3 *net calorific value*—a lower value calculated from the gross calorific value. It is equivalent to the heat produced by combustion of a unit quantity of solid fuel at a constant pressure of one atmosphere, under the assumption that all water in the products remains in the form of vapor.

3.2 *Definitions of Terms Specific to This Standard:*

3.2.1 *calorimeter*—describes the bomb, the vessel with stirrer, and the water in which the bomb is immersed.

3.2.2 *energy equivalent*—the energy required to raise the temperature (Note 2) of the calorimeter system 1°C (or 1°F) per gram of sample. This is the number that is multiplied by the corrected temperature rise in degrees and divided by the sample weight in grams to give the gross calorific value after thermochemical corrections have been applied.

NOTE 2—Temperature change is measured in thermal units. Temperature changes may also be recorded in electromotive force, ohms, or other units when other types of temperature sensors are used. Consistent units must be used in both the standardization and actual calorific determination. Time is expressed in minutes. Weights are measured in grams.

3.2.3 *refuse-derived fuels*—solid forms of refuse-derived fuels from which appropriate analytical samples may be prepared are defined as follows in *ASTM STP 832:*[7]

RDF-1—Wastes used as a fuel in as-discarded form with only bulky wastes removed.

RDF-2—Wastes processed to coarse particle size with or without ferrous metal separation.

RDF-3—Combustible waste fraction processed to particle sizes, 95 % passing 2-in. square screening.

RDF-4—Combustible waste fraction processed into powder form, 95 % passing 10-mesh screening.

RDF-5—Combustible waste fraction densified (compressed) into the form of pellets, slugs, cubettes, or briquettes.

---

[1] This test method is under the jurisdiction of ASTM Committee D-34 on Waste Management and is the direct responsibility of Subcommittee D34.08 on Thermal Treatment.

Current edition approved Aug. 28, 1987. Published October 1987.

[2] *Annual Book of ASTM Standards*, Vol 11.01.

[3] *Annual Book of ASTM Standards*, Vol 05.05.

[4] *Annual Book of ASTM Standards*, Vol 14.03.

[5] *Annual Book of ASTM Standards*, Vol 15.05.

[6] *Annual Book of ASTM Standards*, Vol 11.04.

[7] *Thesaurus on Resource Recovery Terminology, ASTM STP 832*, ASTM, 1983, p. 72.

4. Summary of Test Method

4.1 Calorific value is determined in this method by burning a weighed analysis sample in an oxygen bomb calorimeter under controlled conditions. The calorific value is computed from temperature observations made before and after combustion, taking proper allowance for thermometer and thermochemical corrections. Either isothermal or adiabatic calorimeter jackets may be used.

5. Significance and Use

5.1 The calorific value, or heat of combustion, is a measure of the energy available from a fuel. Knowledge of this value is essential in assessing the commercial worth of the fuel and to provide the basis of contract between producer and user.

6. Apparatus

6.1 *Test Room*—The apparatus should be operated in a room or area free of drafts that can be kept at a reasonably uniform temperature and humidity for the time required for the determination. The apparatus should be shielded from direct sunlight and radiation from other sources. Controlled room temperature and humidity are desirable.

6.2 *Oxygen Bomb*, constructed of materials that are not affected by the combustion process or products sufficiently to introduce measurable heat input or alteration of end products. If the bomb is lined with platinum or gold, all openings shall be sealed to prevent combustion products from reaching the base metal. The bomb shall be designed so that all liquid combustion products can be completely recovered by washing the inner surfaces. There shall be no gas leakage during a test. The bomb shall be capable of withstanding a hydrostatic pressure test to 21 MPa (3000 psig) at room temperature without stressing any part beyond its elastic limit.

6.3 *Calorimeter*, made of metal (preferably copper or brass) with a tarnish-resistant coating and with all outer surfaces highly polished. Its size shall be such that the bomb will be completely immersed in water when the calorimeter is assembled. It shall have a device for stirring the water thoroughly and at a uniform rate, but with minimum heat input. Continuous stirring for 10 min shall not raise the calorimeter temperature more than 0.01°C (0.02°F) starting with identical temperatures in the calorimeter, room, and jacket. The immersed portion of the stirrer shall be coupled to the outside through a material of low heat conductivity.

6.4 *Jacket*—The calorimeter shall be completely enclosed within a stirred water jacket and supported so that its sides, top, and bottom are approximately 10 mm from the jacket walls. The jacket may be arranged so as to remain at constant temperature or with provisions for rapidly adjusting the jacket temperature to equal that of the calorimeter for adiabatic operation. It shall be constructed so that any water evaporating from the jacket will not condense on the calorimeter.

6.5 *Thermometers*—Temperatures in the calorimeter and jacket shall be measured with the following thermometers or combinations thereof:

6.5.1 *Mercury-in-Glass Thermometers*, conforming to the requirements for Thermometers 116°C or 117°C (56°F or 57°F) as prescribed in Specification E 1. Other thermometers of equal or better accuracy are satisfactory. These thermometers shall be tested for accuracy against a known standard (preferably by the National Bureau of Standards) at intervals no greater than 2.0°C (3.6°F) over the entire graduated scale. The maximum difference in correction between any two test points shall not be more than 0.02°C (0.04°F).

6.5.2 *Beckmann Differential Thermometer*, having a range of approximately 6°C in 0.01°C subdivisions reading upward and conforming to the requirements for Thermometer 115°C, as prescribed in Specification E 1. Each of these thermometers shall be tested for accuracy against a known standard at intervals no larger than 1°C over the entire graduated scale. The maximum difference between any two test points shall not be more than 0.02°C.

6.5.3 *Calorimetric-Type Platinum Resistance Thermometer, 25-*, tested for accuracy against a known standard.

6.5.4 *Other Thermometers*—A high precision electronic thermometer employing balanced thermistors or a quartz thermometer may be used, provided the temperature rise indication is accurate within ± 0.003°C per 1°C rise.

6.6 *Thermometer Accessories*—A magnifier is required for reading mercury-in-glass thermometers to one tenth of the smallest scale division. This shall have a lens and holder designed so as to introduce no significant errors due to parallax. A Wheatstone bridge and galvanometer capable of measuring resistance to 0.0001 Ω are necessary for use with resistance thermometers.

6.7 *Sample Holder*—Samples shall be burned in an open crucible of platinum, quartz, or acceptable base-metal alloy. Base-metal alloy crucibles are acceptable if after a few preliminary firings the weight does not change significantly between tasks.

6.8 *Firing Wire* shall be 100 mm of No. 34 B & S nickel-chromium alloy wire or 100 mm of No. 34 B & S iron wire. Equivalent platinum or palladium wire may be used provided constant ignition energy is supplied, or measured, and appropriate corrections made.

6.9 *Firing Circuit*—A 6 to 16-V alternating or direct current is required for ignition purposes with an ammeter or pilot light in the circuit to indicate when current is flowing. A stepdown transformer connected to an alternating current lighting circuit or batteries may be used.

6.10 **CAUTION:** The ignition circuit switch shall be of momentary double-contact type, normally open, except when held closed by the operator. The switch should be depressed only long enough to fire the bomb.

7. Reagents

7.1 *Purity of Reagents*—Reagent grade chemicals shall be used in all tests. Unless otherwise indicated, it is intended that all reagents shall conform to the specifications of the Committee on Analytical Reagents of the American Chemical Society, where such specifications are available.[8] Other grades may be used, provided it is first ascertained that the reagent is of sufficiently high purity to permit its use without lessening the

[8] "Reagent Chemicals, American Chemical Society Specifications," Am. Chemical Soc., Washington, DC. For suggestions on the testing of reagents not listed by the American Chemical Society, see "Analar Standards for Laboratory U.K. Chemicals," BDH Ltd., Poole, Dorset, and the "United States Pharmacopeia."

accuracy of the determination.

7.2 *Purity of Water*—Unless otherwise indicated, references to water shall be understood to mean reagent water, Type III, conforming to Specification D 1193.

7.3 *Benzoic Acid, Standard (C_6H_5COOH)*—Use National Bureau of Standards SRM (Standard Reference Material) benzoic acid. The crystals shall be pelletized before use. Commercially prepared pellets may be used provided they are made from National Bureau of Standards benzoic acid. The value of heat of combustion of benzoic acid, for use in the calibration calculations, shall be in accordance with the value listed in the National Bureau of Standards certificate issued with the standard.

7.4 *Methyl Orange, Methyl Red, or Methyl Purple Indicator*—may be used to titrate the acid formed in the combustion. The indicator selected shall be used consistently in both calibrations and calorific determinations.

7.5 *Oxygen*, free of combustible matter. Oxygen manufactured from liquid air, guaranteed to be greater than 99.5 % pure, will meet this requirement. Oxygen made by the electrolytic process may contain a small amount of hydrogen rendering it unfit without purification.

7.6 *Sodium Carbonate, Standard Solution* (0.34 *N*)—One millilitre of this solution should be equivalent to 20.0 J in the nitric acid (HNO_3) titration. Dissolve 18.02 g of anhydrous sodium carbonate (Na_2CO_3) in water and dilute to 1 L. The Na_2CO_3 should be previously dried for 24 h at 105°C. The buret used for the HNO_3 titration shall be of such accuracy that estimations to 0.1 mL can be made. A more dilute standard solution may be used for higher sensitivity.

8. Precautions

8.1 Due to the origins of RDF in municipal waste, common sense dictates that some precautions should be observed when conducting tests on the samples. Recommended hygienic practices include use of gloves when handling RDF and washing hands before eating or smoking.

8.2 The following precautions are recommended for safe calorimeter operation:

8.2.1 The weight of solid fuel sample and the pressure of the oxygen admitted to the bomb must not exceed the bomb manufacturer's recommendations.

8.2.2 Bomb parts should be inspected carefully after each use. Threads on the main closure should be checked frequently for wear. The bomb should be returned to the manufacturer occasionally for inspection and possibly proof of firing.

8.2.3 The oxygen supply cylinder should be equipped with an approved type of safety device, such as a reducing valve, in addition to the needle valve and pressure gage used in regulating the oxygen feed to the bomb. Valves, gages, and gaskets must meet industry safety codes. Suitable reducing valves and adaptors for 2 to 3.5-MPa (300 to 500-psig) discharge pressure are obtainable from commercial sources of compressed gas equipment. The pressure gage shall be checked periodically for accuracy.

8.2.4 During ignition of a sample, the operator shall not permit any portion of his body to extend over the calorimeter.

9. Sampling [9]

9.1 RDF products are frequently nonhomogeneous. For this reason significant care should be exercised to obtain a representative laboratory sample for the RDF lot to be characterized.

9.2 The sampling method for this procedure should be based on agreement between the involved parties.

9.3 The laboratory sample must be air-dried and particle size reduced to pass a 0.5-mm screen as described in Practice E 829.

10. Standardization

10.1 Determine the energy equivalent of the calorimeter as the average of a series of ten individual runs, made over a period of not less than 3 days or more than 5 days. To be acceptable, the standard deviation of the series shall be 6.9 kJ/°C (6.5 Btu/°C) or less (see Appendix X1, Table X1.1). For this purpose, any individual run may be discarded only if there is evidence indicating incomplete combustion. If this limit is not met, repeat the entire series until a series is obtained with a standard deviation below the acceptable limit.

10.2 The weights of the pellets of benzoic acid in each series should be regulated to yield the same temperature rise as that obtained with the various samples tested in the individual laboratories. The usual range of weight is 0.9 to 1.3 g. Make each determination in accordance with the procedure described in Section 11, and compute the corrected temperature rise, *T*, as described in 12.1. Determine the corrections for HNO_3 and firing wire as described in 12.2 and substitute into the following equation:

$$E = [(H)(g) + e_1 + e_3 + e_4] \times t \quad (1)$$

where:

E = energy equivalent, J/°C,
H = heat of combustion of benzoic acid, as stated in the National Bureau of Standards certificate, J/g,
g = weight of benzoic acid, g,
t = corrected temperature rise, °C,
e_1 = titration correction, J,
e_3 = fuse wire correction, J, and
e_4 = correction for ignition energy if measured and corrected for, J.

10.3 Standardization tests should be repeated after changing any part of the calorimeter and occasionally as a check on both calorimeter and operating technique.

11. Procedure

11.1 *Weight of Sample*—Thoroughly mix the analysis sample of solid fuel in the sample bottle, taking care that the heavies and lights (fluff) are distributed in the sample (Note 3). Carefully weigh approximately 1 g of the sample directly into the crucible in which it is to be burned or into a tared weighing scoop from which the sample is transferred to the crucible. Weigh the sample to the nearest 0.1 mg. Some form of compaction may be necessary to ensure satisfactory ignition and complete combustion.

---

[9] ASTM Subcommittee E38.01 is currently in the process of developing procedures for sampling RDF-3 and the preparation of an analysis sample. The chairman of E38.01 should be contacted for details.

NOTE 3—In the event segregation of the heavies and lights cannot be avoided, attempt to remove sample from the bottle in such a way that a representative sample is transferred.

NOTE 4—Perform the residual moisture determination of the sample simultaneously using Test Method E 790.

11.2 *Water in Bomb*—Add 1.0 mL of water to the bomb by a pipet. Before adding this water, rinse the bomb, and drain the excess water, and leave undried.

11.3 *Firing Wire*—Connect a measured length of firing wire to the ignition terminals with enough slack to allow the firing wire to maintain contact with the sample.

11.4 *Oxygen*—Charge the bomb with oxygen to a consistent pressure between 20 and 30 atm (2.03 and 3.04 MPa). This pressure must remain the same for each calibration and for each calorific determination. If, by accident, the oxygen introduced into the bomb should exceed the specified pressure, do not proceed with the combustion. Detach the filling connection and exhaust the bomb in the usual manner. Discard this sample.

11.5 *Calorimeter Water*—It is recommended that calorimeter water temperature be adjusted before weighing as follows:

11.5.1 *Isothermal Jacket Method*, 1.6 to 2.0°C (3.0 to 3.5°F) below jacket temperature (Note 4).

11.5.2 *Adiabatic Jacket Method*, 1.0 to 1.4°C (2.0 to 2.5°F) below room temperature.

NOTE 5—This initial adjustment will ensure a final temperature slightly above that of the jacket for calorimeters having an energy equivalent of approximately 10 200 J/K (2450 cal/°C). Some operators prefer a lower initial temperature so that the final temperature is slightly below that of the jacket. This procedure is acceptable, provided it is used in all tests, including standardization. Use the same amount (± 0.5 g) of water in the calorimeter vessel for each test and for calibration. The amount of water (2000 g is usual) can be most satisfactorily determined by weighing the calorimeter vessel and water together on a balance. The water may be measured volumetrically if it is always measured at the same temperature. Tap water may be satisfactory for use in calorimeter bucket.

11.6 *Observations, Isothermal Jacket Method*—Assemble the calorimeter in the jacket and start the stirrer. Allow 5 min for attainment of equilibrium; then record the calorimeter temperatures (Note 6) at 1-min intervals for 5 min. Fire the charge at the start of the sixth minute and record the time and temperature, T^a. Add to this temperature 60 % of the expected temperature rise, and record the time at which the 60 % point is reached (Note 5). After the rapid-rise period (about 4 to 5 min), record temperatures at 1-min intervals on the minute until the difference between successive readings has been constant for 5 min.

NOTE 6—Use a magnifier and estimate all readings (except those during the rapid rise period) to the nearest 0.002°C (0.005°F) when using ASTM Bomb Calorimeter Thermometer 56C (56F). Estimate Beckmann thermometer readings to the nearest 0.001°C. Tap mercurial thermometers with a pencil just before reading to avoid errors caused by mercury sticking to the walls of the capillary.

NOTE 7—When the approximate expected rise is unknown, the time at which the temperature reaches 60 % of the total can be determined by recording temperatures at 45, 60, 75, 90, and 105 s after firing and interpolating.

11.7 *Observations, Adiabatic Jacket Method*—Assemble the calorimeter in the jacket and start the stirrer. Adjust the jacket temperature to be equal to or slightly lower than the calorimeter, and run for 5 min to obtain equilibrium. Adjust the jacket temperature to match the calorimeter with ± 0.01°C (0.02°F) and hold for 3 min. Record the initial temperature (Note 6) and fire the charge. Adjust the jacket temperature to match that of the calorimeter during the period of rise, keeping the two temperatures as nearly equal as possible during the rapid rise, and adjusting to within ± 0.01°C (0.02°F) when approaching the final equilibrium temperature. Take calorimeter readings at 1-min intervals until the same temperature is observed in three successive readings. Record this as the final temperature. Do not record time intervals since they are not critical in the adiabatic method.

11.8 *Analysis of Bomb Contents*—Remove the bomb and release the pressure at a uniform rate, in such a way that the operation will require not less than 1 min. Examine the bomb interior and discard the test if unburned sample or sooty deposits are found. Carefully wash the interior of the bomb including the capsule with distilled or deionized water containing the titration indicator until the washings are free of acid. Collect the washings in a beaker and titrate the washings with standard carbonate solution. Remove and measure or weigh the combined pieces of unburned firing wire, and subtract from the original length or weight to determine the wire consumed in firing. Determine the sulfur content of the sample by any of the procedures described in Test Methods E 775.

12. Calculation

12.1 *Temperature Rise in Isothermal Jacket Calorimeter*—Using data obtained as prescribed in 11.6, compute the temperature rise, T, in an isothermal jacket calorimeter as follows:

$$T = T_c - T_a - r_1(b - a) - r_2(c - b) \quad (2)$$

where:

T = corrected temperature rise,
a = time of firing,
b = time (to nearest 0.1 min) when the temperature rise reaches 60 % of total,
c = time at beginning of period in which the rate of temperature change with time has become constant (after combustion),
T_a = temperature at time of firing, corrected for thermometer error (Note 7),
T_c = temperature at time c, corrected for thermometer error (Note 7),
r_1 = rate (temperature units per minute) at which temperature was rising during 5-min period before firing, and
r_2 = rate (temperature units per minute) at which temperature was rising during the 5-min period after time c. If the temperature is falling, r_2 is negative and the quantity $r_2\ (c - b)$ is positive.

12.2 *Temperature Rise in Adiabatic Jacket Calorimeter*—Using data obtained as prescribed in 11.7 compute the corrected temperature rise, T, as follows:

$$T = T_f - T_a \quad (3)$$

where:

T = corrected temperature rise, °C or° F,

T_a = initial temperature when charge was fired, corrected for thermometer error (Note 8), and
T_f = final temperature corrected for thermometer error.

NOTE 8—With all mercury-in-glass thermometers, it is necessary to make the following corrections if the total heat value is altered by 12 J/g or more. This represents a change of 0.001°C (0.002°F) in a calorimeter using approximately 2000 g of water. The corrections include the calibration correction as stated on the calibration certificate, the setting correction for Beckman thermometers according to the directions furnished by the calibration authority, and the correction for emergent stem. Directions for these corrections are given in Appendix X2.

12.3 *Thermochemical Corrections* (Appendix X3)—Compute the following for each test:

e_1 = correction for the heat of formation of HNO_3, J. Each millilitre of standard alkali is equivalent to 20.0 J.
e_2 = correction for heat of formation of H_2SO_4, J
= 55.2 × percent of sulfur in sample × weight of sample, g.
e_3 = correction for heat of combustion of firing wire, J (Note 10)
= 9.6 J/cm or 5980 J/g for No. 34 B & S gage Chromel C
= 11.3 J/cm or 7330 J/g for No. 34 B & S iron wire.
e_4 = correction for ignition energy of platinum or palladium if measured and corrected for.

NOTE 9—There is no correction for platinum or palladium wire, provided the ignition energy is constant.

12.4 *Calorific Value*:

12.4.1 Calculate the gross calorific value (gross heat of combustion) as follows:

$$H_s = [(T)(E) - e_1 - e_2 - e_3 - e_4]/g \quad (4)$$

where:
H_s = gross calorific value, J/g,
T = corrected temperature rise as calculated in 12.1 or 12.2, °C or °F, consistent with the water equivalent value,
E = energy equivalent (see Section 10), e_1, e_2, e_3, e_4 = corrections as prescribed in 12.3, and
g = weight of sample, g.

12.4.2 Calculate the net calorific value (net heat of combustion) as follows:

$$H_i = H_s - 23.96\,(H \times 9) \quad (5)$$

where:
H_i = net calorific value (net heat of combustion), J/g,
H_s = gross calorific value (gross heat of combustion), J/g, and
H = total hydrogen, %.

13. Precision and Bias [10]

13.1 *Precision*—The standard deviations of individual determinations, in Btu/lb, are:

Average	Within-laboratory	Between-laboratories
HHV-1:		
6400	27.1	135.5
5200	48.8	239.6
HHV-2:		
7900	32.3	118.0
7400	38.1	227.8
HHV-3:		
9700	111.3	290.4
9500	99.2	249.2
9300	40.3	67.6

13.2 These precision estimates are based on an interlaboratory study conducted in accordance with Practice E 180.

[10] Supporting data are available on loan from ASTM Headquarters. Request RR:E38-1000.

APPENDIXES

(Nonmandatory Information)

X1. CALCULATION OF STANDARD DEVIATIONS FOR CALORIMETER STANDARDIZATION

X1.1 The example given in Table X1.1 illustrates the method of calculating standard deviations for calorimeter standardizations.

X2. THERMOMETER CORRECTIONS

X2.1 It is necessary to make the following corrections in the event they result in an equivalent change of 0.001°C or more.

X2.1.1 *Calibration Correction* shall be made in accordance with the calibration certificate furnished by the calibration authority.

X2.1.2 *Setting Correction* is necessary for the Beckmann thermometer. It shall be made in accordance with the directions furnished by the calibration authority.

X2.1.3 *Differential Emergent Stem Correction*—The calculation depends upon the way the thermometer was calibrated and how it is used. The following two conditions are possible:

(*a*) *Thermometers Calibrated in Total Immersion and Used in Partial Immersion*—This emergent stem correction is made as follows:

$$\text{Correction} = K(t_c - t_a)\,(t_c + t_a - L - T) \quad (X2.1)$$

TABLE X1.1 Standard Deviations for Calorimeter Standardization[A]

Standardization Number	Column A Water Equivalent, (Btu/lb) × (g/°C)	Column B Code to 4400 (Column A-4400)	Column C (Column B)^2
1	4412	12	144
2	4407	7	49
3	4415	15	225
4	4408	8	64
5	4404	4	16
6	4406	6	36
7	4409	9	81
8	4410	10	100
9	4412	12	144
10	4409	9	81
Sum		92	940

Average = $\bar{x}$ = x/10 = (92/10) + 4400 = 4409
Variance = s^2 = Column C – (Column B)^2/n/n– 1 = 940 – (92)^2/10/9 = 10.4
Standard deviation, s = Variance = 10.4 = 3.22

[A] In this example the values of water equivalent are typical for a calorimeter calibrated such that the water equivalent multiplied by the temperature rise in °C/g of sample will give the calorific value of the sample in Btu/lb.

where:

K = 0.00016 for thermometers calibrated in° C,
0.00009 for thermometers calibrated in °F,
L = scale reading to which the thermometer was immersed,
T = mean temperature of emergent stem,
t_a = initial temperature reading, and
t_c = final temperature reading.

NOTE X2.1— *Example*—Suppose the point L, to which the thermometer was immersed was 16°C; its initial reading, t_a, was 24.127°C, its final reading, t_c, was 27.876°C, the mean temperature of the emergent stem, T, was 26°C,

then:

Differential stem correction

$$= + 0.00016\ (28 - 24)\ (28 + 24 - 16 - 26) \qquad (X2.2)$$

$$= + 0.006°C$$

(*b*) *Thermometers Calibrated and Used in Partial Immersion but at a Different Temperature than the Calibration Temperature*—This emergent stem correction is made as follows:

$$\text{Correction} = K\ (t_c - t_a)\ (t_1 - t°) \qquad (X2.3)$$

where:

K = 0.00016 for thermometers calibrated in° C,
0.00009 for thermometers calibrated in °F,
t_a = initial temperature reading,
t_c = final temperature reading,
t_1 = observed stem temperature, and
$t°$ = stem temperature at which the thermometer was calibrated.

NOTE X2.2—*Example*—Suppose the initial reading, t_a, was 80°F, the final reading, t_c, was 86°F, and that the observed stem temperature, t_1, was 82°F, and the calibration temperature, $t°$, was 72°F; then:

Differential stem correction

$$= 0.00009\ (86 - 90)(82 - 72) \qquad (X2.4)$$

$$= 0.005°F$$

X3. THERMOCHEMICAL CORRECTIONS

X3.1 *Heat of Formation of Nitric Acid*—A correction (e^1, in 12.3) of 20 J is applied for each 1 mL of standard Na_2CO_3 solution used in the acid titration. The standard solution (0.34 *N*) contains 18.02 g of Na_2CO_3/L. This correction is based on assumption that all the acid titrated is HNO_3 formed by the following reaction: ½ N_2 (g + 5⁄4 O_2 (g) + ½ H_2O (l) = HNO_3 (in 500 mol H_2O), and (2) the energy of formation of 1 mol of HNO_3 is approximately 500 mol of water under bomb conditions is 14.1 kcal/mol.[6] When H_2SO_4 is also present part of the correction for H_2SO_4 is contained in the e_1 correction and the remainder in the e_2 correction.

X3.2 *Heat of Formation of Sulfuric Acid*—By definition the gross calorific value is obtained when the product of the combustion of sulfur in the sample is SO_2 (g). However, in actual bomb combustion processes, the sulfur is found as H_2SO_4 in the bomb washings. A correction (e_2 in 12.4.1) of 55.2 J is applied for each percent of sulfur in the 1-g sample, that is converted to H_2SO_4. This correction is based upon the energy of formation of H_2SO_4 in solutions such as will be present in the bomb at the end of a combustion. This energy is taken as – 70.5 kcal/mol.[11] A correction, of 2 × 14.1 kcal/mol of sulfur was applied in the e_1 correction, so the additional correction necessary is 70.5 – (2 × 14.1) = 42.3 kcal/mol or 5520 J of sulfur in the sample (55.2 J × weight of sample in grams × % sulfur in sample).

X3.2.1 The value of 5520 J/g of sulfur is based on a coal containing about 5 % sulfur and about 5 % hydrogen. The assumption is also made that the H_2SO_4 is dissolved entirely in water condensed during combustion of the sample.[12] If a 1-g sample of such a fuel is burned, the resulting H_2SO_4 condensed with water formed on the walls of the bomb will have a ratio of about 15 mol of water to 1 mol of H_2SO_4. For this concentration the energy of the reaction.

$$SO_2\ (g) + \frac{1}{2} O_2\ (g) + H_2O\ (l) = H_2SO_4\ (\text{in } 15 \text{ mol } H_2O) \qquad (X3.1)$$

under the conditions of the bomb process is – 70.5 kcal/mol.

X3.2.2 Basing the calculation upon a sample of comparatively large sulfur content reduces the overall possible errors, because for smaller percentages of sulfur the correction is smaller.

X3.3 *Fuse Wire*—Calculate the heat in SI units contributed by burning the fuse wire in accordance with the directions

[11] Calculated from data in National Bureau of Standards Circular 500.

[12] Mott, R. A., and Parker, C., "Studies in Bomb Calorimetry IX-Formation of Sulfuric Acid," *Fuel*, Vol 37, 1958, p. 371.

furnished by the supplier of the wire. For example, the heat of combustion of No. 34 B & S gage Chromel C wire is equivalent to 9.6 J/cm or 5980 J/g and that of No. 34 B & S gage iron wire is equivalent to 11.3 J/cm or 7330 J/g. There is no correction for platinum or palladium wire provided the ignition energy is constant.

X4. REPORTING RESULTS IN OTHER UNITS

X4.1 *Reporting Results in British Thermal Units (Btu) per Pound*—The gross calorific value can be expressed in British thermal units by using the thermochemical correction factors in Table X4.1 and the water equivalent expressed in (Btu/lb) × (g/°C).

TABLE X4.1 Thermochemical Correction Factors (Units in BTU)

Correction	Multiplication Factor	Multiply by
e_1 (HNO_3)	10.0	mL of 0.394 *N* Na_2CO_3 solution
e_2 (H_2SO_4)	23.7	% of sulfur in sample times weight of sample in grams
e_3 (fuse wire)	4.1 or	cm of No. 34 B & S gage Chromel C wire
	2570	weight (g) of Chromel C wire
e_3 (fuse wire)	4.9 or	cm of No. 34 B & S gage iron wire
	3150	weight (g) of iron wire

Designation: E 753 – 80 (Reapproved 1999)

Standard Specification for Municipal Aluminum Scrap (MAS)[1]

This standard is issued under the fixed designation E 753; the number immediately following the designation indicates the year of original adoption or, in the case of revision, the year of last revision. A number in parentheses indicates the year of last reapproval. A superscript epsilon (ε) indicates an editorial change since the last revision or reapproval.

1. Scope

1.1 This specification covers municipal-refuse originated aluminum alloy scrap (MAS), not source-separated, that is recovered from industrial, commercial, or household wastes destined for disposal facilities.

1.2 Municipal aluminum scrap (MAS) covered by this specification is suitable for use by the following industries:

1.2.1 Secondary aluminum smelters,

1.2.2 Primary aluminum producers,

1.2.3 Aluminum scrap dealers,

1.2.4 Iron and steel industry,

1.2.5 Foundries,

1.2.6 Nonintegrated aluminum producers, and

1.2.7 Independent aluminum fabricators.

1.3 The values stated in inch-pound units are to be regarded as the standard. The values given in parentheses are for information only.

1.4 *This standard does not purport to address all of the safety concerns, if any, associated with its use. It is the responsibility of the user of this standard to establish appropriate safety and health practices and determine the applicability of regulatory limitations prior to use.* The hazard statement applies only to the test method sections in the annexes of this specification.

2. Referenced Documents

2.1 *ASTM Standards:*

D 2013 Test Method for Preparing Coal Samples for Analysis[2]

E 11 Specification for Wire-Cloth Sieves for Testing Purposes[3]

E 34 Test Methods for Chemical Analysis of Aluminum and Aluminum Alloys[4]

E 101 Test Method for Spectrographic Analysis of Aluminum and Aluminum Alloys by the Point-to-Plane Technique[4]

E 122 Practice for Choice of Sample Size to Estimate a Measure of Quality for a Lot or Process[3]

E 227 Test Method for Optical Emission Spectrometric Analysis of Aluminum and Aluminum Alloys by the Point-to-Plane Technique[4]

E 276 Test Method for Particle Size or Screen Analysis at No. 4 (4.75-mm) Sieve and Finer for Metal Bearing Ores and Related Materials[4]

3. Terminology

3.1 *Definitions of Terms Specific to This Standard:*

3.1.1 *combustible material (organic)*—material that is measured by weight loss of a dried sample input after heating to red heat in an open crucible in a vented furnace. Combustibles include both loose organics and organic coatings.

3.1.2 *loose combustible material (organic)*—loose combustible organics (LCO) that consist of, but are not limited to, nonmetallic materials such as paper, rags, plastic, rubber, wood, food wastes, and yard or lawn wastes, etc., which are not permanently attached to noncombustible objects. The LCOs are defined as material larger than 12 mesh (U.S. Standard Sieve). A determination of LCOs is best done by sampling the material and handpicking, handcleaning, and visually identifying the materials described previously.

3.1.3 *moisture percent*—liquid content, as determined by weight loss when sample material is dried to a constant weight at 110° ± 5°C.

3.1.4 *recovery*—the percent material recovered after an assay using the procedures prescribed in this specification.

4. Classification

4.1 This specification covers two classes based on fines content (7.2) and six grades based on chemical composition of MAS material, as listed in Table 1 (see also 7.2).

5. Ordering Information

5.1 It is recognized that variations in the MAS may occur due to the heterogeneous nature of the solid waste stream. The grades indicated are intended as a means for the purchaser and the seller to establish the value and quality of the MAS.

5.2 MAS shall be considered to be of a particular grade if the value for each element specified, as obtained by the test method agreed upon between the purchaser and seller, does not exceed any of the limits for that grade.

6. Chemical Composition

6.1 The MAS shall conform to the requirements as to

---

[1] This specification is under the jurisdiction of ASTM Committee D-34 on Waste Disposal and is the direct responsibility of Subcommittee D 34.06 on Recovery and Reuse.

Current edition approved Sept. 2, 1980. Published November 1980.

[2] *Annual Book of ASTM Standards*, Vol 05.05.

[3] *Annual Book of ASTM Standards*, Vol 14.02.

[4] *Annual Book of ASTM Standards*, Vol 03.05.

TABLE 1 Chemical Requirements

Element[A]	Composition, Maximum % Allowable					
	Grade 1	Grade 2	Grade 3	Grade 4	Grade 5	Grade 6
Silicon	0.30	0.30	0.50	1.00	9.00	9.00
Iron	0.60	0.70	1.00	1.00	0.80	1.00
Copper	0.25	0.40	1.00	2.00	3.00	4.00
Manganese	1.25	1.50	1.50	1.50	0.60	0.80
Magnesium	2.00	2.00	2.00	2.00	2.00	2.00
Chromium	0.05	0.10	0.30	0.30	0.30	0.30
Nickel	0.04	0.04	0.30	0.30	0.30	0.30
Zinc	0.25	0.25	1.00	2.00	1.00	3.00
Lead	0.02	0.04	0.30	0.50	0.10	0.25
Tin	0.02	0.04	0.30	0.30	0.10	0.25
Bismuth	0.02	0.04	0.30	0.30	0.10	0.25
Titanium	0.05	0.05	0.05	0.05	0.10	0.25
Others (each)	0.04	0.05	0.05	0.08	0.10	0.10
Others (total)	0.12	0.15	0.15	0.20	0.30	0.30
Aluminum	balance	balance	balance	balance	balance	balance

[A] By agreement between the purchaser and the seller, analysis may be required, and limits established for elements or compounds not specified in this table.

chemical composition prescribed in Table 1 (see Test Methods E 34, E 101, and E 227).

7. Physical Properties

7.1 *Density*—The density for MAS is not specified and shall be agreed upon between the purchaser and the seller.

7.2 *Fineness*—MAS shall contain not more than the amount of minus 12-mesh (U.S. Standard Sieve) (see Specification E 11) material, described in 7.2.1 and 7.2.2.

7.2.1 *Class A* material shall contain not more than 1 weight % fines.

7.2.2 *Class B* material shall contain not more than 3 weight % fines.

7.3 *Loose Combustibles*—MAS shall contain not more than 2.0 weight % of loose combustible material.

7.4 *Moisture*—MAS shall contain not more than 0.5 weight % of moisture.

7.5 *Metal Recovery*— A minimum metal recovery of 85 % is required.

7.6 *Magnetics*—The presence of free magnetic material is not specified and shall be as agreed upon between the purchaser and seller as part of the purchase contract.

8. Sampling

8.1 Sampling shall be in accordance with the procedures described in Annex A1 or Annex A2. Either procedure may be used, as determined by agreement between the purchaser and the seller.

8.1.1 Annex A1 covers sampling at the point of origin.

8.1.2 Annex A2 covers sampling at the point of receipt.

9. Test Methods

9.1 Determine the properties enumerated in this specification in accordance with the following:

9.1.1 *Fineness*—Annex A3.

9.1.2 *Moisture*—Annex A3.

9.1.3 *Metal Recovery*—Annex A3.

10. Rejection and Rehearing

10.1 Material that fails to conform to the requirements of this specification may be rejected. Rejection should be reported to the producer or supplier promptly and in writing. In case of dissatisfaction with the results of the test, the producer or supplier may make claim for a rehearing.

11. Shipping

11.1 MAS shall be shipped in rail cars, trailers, or other containers as agreed upon between the purchaser and the seller. The shipping equipment shall be sufficiently watertight to prevent the MAS from becoming wet during shipment.

12. Keywords

12.1 assay; chemical composition; classification system; gross sample; municipal aluminum scrap; physical properties; sampling; test methods; user industries

ANNEXES

(Mandatory Information)

A1. TEST METHOD FOR COLLECTION OF A SAMPLE OF ALUMINUM SCRAP, RECOVERED FROM MUNICIPAL SOLID WASTE, AND ITS PREPARATION FOR ANALYSIS

A1.1 Scope

A1.1.1 This test method describes procedures for collection of a sample of shredded aluminum metal scrap recovered from municipal refuse, and the shredding, mixing, and secondary sampling of the metal for analysis.

A1.2 Summary of Test Method

A1.2.1 A selected size, gross sample of shredded, nonferrous metal scrap is taken from the metal recovery system conveyor belt in increments. Sample increments are taken at timed intervals from a full cross section of the conveyor while it is stopped or by briefly taking the total flow at the discharge of the conveyor while it is moving.

A1.2.2 The quantity of gross sample may be further reduced by mixing, cone-and-quarter sampling, and riffling.

A1.3 Hazards

A1.3.1 In solids sampling, each step must be designed to eliminate accidental classification by size or gravity. Different sizes usually have different analyses.

A1.3.2 The increments obtained during the sampling period shall be protected from changes in composition due to exposure to the weather.

A1.3.3 Plan the sampling arrangement to avoid contamination of the increments with foreign material.

A1.3.4 A satisfactory sampling arrangement is one that takes an unbiased sample at the desired degree of precision of the constituent for which the sample is to be analyzed. The weight or volume of the collected sample is compared with that of the total lot to assure a constant sampling ratio.

A1.3.5 It is preferable that the nonferrous metal scrap be weighed and sampled at about the same time. If there is a long lapse in time between these two events, both the purchaser and seller should give consideration to changes in moisture during this interval and the consequent shift in relationship of moisture to the true content at the instant when ownership of the nonferrous metal scrap transfers from one to the other.

A1.3.6 Samples and subsamples shall be collected in such a manner that there is no unmeasured loss of moisture of significant amount. The samples shall be weighed before and after drying or other operations to measure all significant weight loss. The material balances shall be adjusted accordingly.

A1.4 Selection of Gross Sample Size

A1.4.1 Choose the gross sample size by methods given in Practice E 122, whenever practicable. The chief difficulty for implementing this practice can be that insufficient information concerning possible variation is available. This information should be gathered with practice. Due to the heterogeneity in size and type of material comprising municipal solid waste, the choice of a large sample is desirable.

A1.5 Taking a Gross Sample

A1.5.1 In order to obtain complete representation of materials in a gross sample, it is desirable that the sample increments be withdrawn from the full cross section of the stream. The best possible increment is either a full cross section removed from a stopped conveyor belt or the total flow at the discharge of the moving conveyor taken during a suitable interval of time.

A1.5.2 The choice of sample size can be estimated using Practice E 122. It is imperative for a given degree of precision that not less than the minimum size and number of sample increments be collected from a lot (see Table A1.1).

A1.5.3 *Number of Gross Samples*—For quantities up to approximately 20 tons (18 000 kg), it is recommended that one gross sample represent the lot. Take this sample in accordance with the requirements prescribed in Table A1.1.

A1.5.4 *Distribution of Increments*—It is essential that the increments be distributed throughout the lot to be sampled. The taking of increments shall be at regularly spaced intervals.

A1.6 Sample Preparation

A1.6.1 Cone and quarter the sample until approximately 2 ft^3 (0.06 m^3) remains. Pile the material to be sampled into a conical heap and then spread out into a circular cake. Divide the cake into quarters, take two of the diagonally opposite quarters as the sample, and reject the two remaining quarters. Collect the two quarters taken as the sample and repeat the procedure of coning and quartering until the desired size is obtained.

A1.6.2 Divide the sample into approximately equal parts. Take one half (1 ft^3) (0.03 m^3) for use in the melt test (see Annex A3). Divide the sample by riffling until the analytical sample is obtained. (Typical riffles can be found in the apparatus section of Test Method D 2013.)

A1.6.3 Store the prepared analytical sample in a covered, labeled, corrosion-resistant metal can or plastic container until needed for chemical analysis.

TABLE A1.1 Number and Weight of Increments for Sampling

	Top Size, in. (mm)		
	⅝ (15)	2 (50)	6 (150)
Minimum number of increments	15	15	15
Minimum weight of increments, lb (kg)	2 (1)	6 (3)	18 (9)

A2. SAMPLING AT POINT OF RECEIPT

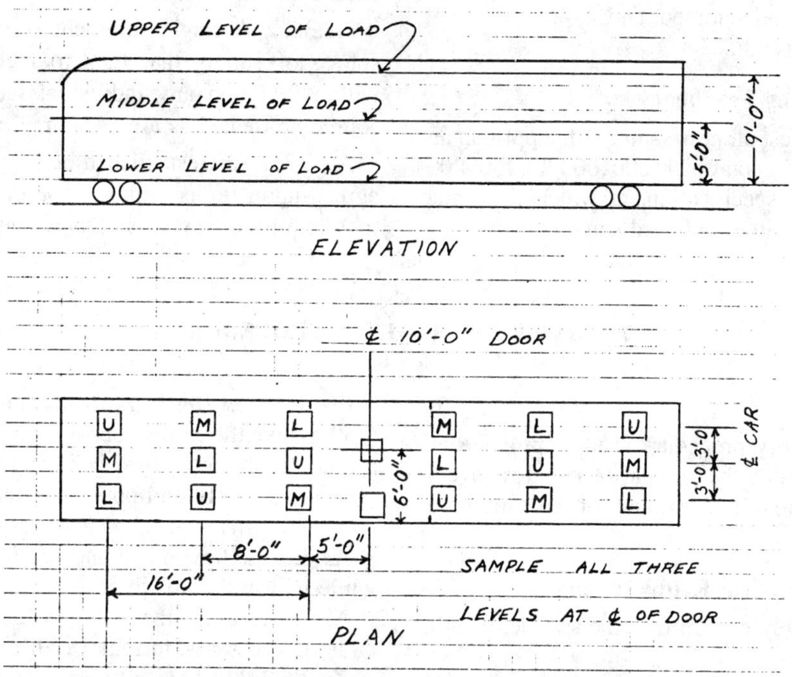

NOTE 1—All samples consist of two No. 2 shovels from each location sampled.

FIG. A2.1 Sample Location Chart for 50-ft Railway Car

A2.1 Sampling During Unloading

A2.1.1 *Sample Size*— Take a representative quantity of approximately 1 yd^3 (0.76 m^3) from each car or truck of shredded aluminum scrap received.

A2.1.1.1 *Car Sample*— Take two shovels (No. 2 size) of aluminum scrap from the top, middle, and bottom of opened doorway area of car in two locations of doorway area as shown in Fig. A2.1 to obtain six samples. Take an additional 18 samples as shown in Fig. A2.1. Place all material sampled into a suitable receptacle for the total sample from the car. Adequately identify the sample container. Weigh and record the sample from the car.

A2.1.1.2 *Truck Sample*— Take two shovels (No. 2 size) of aluminum scrap from upper, middle, and lower areas of shredded material starting at the rear of the truck. Starting at the truck rear, take samples every 8 ft (2.4 m) at locations one third the distance from the side of the truck. Place all material sampled into a suitable receptacle for the total sample of the truck. Adequately identify the sample container. Weigh and record the sample from the truck.

A2.1.2 *Reduction of Sample Size*—Using riffle sampling equipment[5] or coning and quartering (or equivalent method), reduce the size of sample to approximately 1 ft^3 (0.03 m^3) by putting through the riffle sampler five times. Retain one half of the split sample on each pass through the riffle sampler until a fine sample of approximately 1 ft^3 is attained. Carefully bag and identify the sample for the assay and retain a duplicate sample. Retain the duplicate sampleuntil assay is completed and accepted. If assay is not accepted, then the duplicate sample may be used for settling the claim.

A2.1.3 *Identity Ticket*— It is recommended that the following information be included on the ticket:

A2.1.3.1 Supplier,

A2.1.3.2 Car number or truck identification,

A2.1.3.3 Net weight of car or truck,

A2.1.3.4 Date unloaded and sampled, and

A2.1.3.5 Initial sample weight.

A2.2 Off-Specification Shipments

A2.2.1 *Sampling*—If shipments are received as off-specification material or are suspected of being off-specification, sample the shipment in the doorway of the car or truck using core sampling equipment,[5] or other acceptable procedure, from at least ten different locations prior to car unloading. Take at least 2 ft^3 (0.06 m^3) of representative sample. Weigh and record the sample weight using a scale accurate to within ± 0.1 lb (0.05 kg) or ± 0.5 %, whichever is the more precise.

A2.2.2 *Reduction of Sample Size*—Using riffle sampling equipment or coning and quartering (or equivalent procedure), split the sample into equal parts 1 ft^3 (0.03 m^3) remains in each of the last split fractions. Weigh each split fraction, identify, and retain duplicate sample until acceptance or rejection of shipment.

A2.2.3 *Identity Ticket*— It is recommended that the following information be included on the ticket:

[5] Core sampling equipment, available from W. S. Tyler Co., or equivalent, has been found suitable for this purpose.

A2.2.3.1 Supplier,

A2.2.3.2 Car number or truck identification,

A2.2.3.3 Net weight of car or truck,

A2.2.3.4 Date unloaded and sampled, and

A2.2.3.5 Initial sample weight.

A2.3 Recommended Testing Frequency

A2.3.1 *New Sources*— Test all new sources of supply on an "every shipment" basis until a total of 1 000 000 lb (450 000 kg) have been received on-specification, at which time, the supplier is considered to be an established source.

A2.3.2 *Established Sources*:

A2.3.2.1 Randomly sample shipments from established source and test a minimum of 20 % of all shipments.

A2.3.2.2 A single shipment from any established or new source of supply that fails to meet the agreed-upon limits within 20 % on any individual factor may require that the source be treated as a new source (see A2.3.1).

A2.3.2.3 Any established or new source failing to meet the agreed-upon limits within 10 % on any two or more factors may require treatment as a new source (see A2.3.1).

A3. ASSAY AND ANALYSIS PROCEDURE

A3.1 Significance

A3.1.1 This is a laboratory procedure used to produce an analytical specimen, determine the percent of metal recovery by remelting, and determine the amount of contaminants present.

A3.2 Determination of Physical Requirements

A3.2.1 Take approximately one third of the sample, weigh, and record the weight.

A3.2.2 Dry this one-third sample for 2 h at 110 ± 5°C and record the weight of the dried material.

A3.2.3 Remove the magnetic materials and separately record the weight of the magnetic materials removed and the remaining nonmagnetic materials.

A3.2.4 Screen the sample over a 12-mesh (U.S. Standard) sieve using the Dry Screening Procedure of Test Method E 276. Record the weight of the sieved material and the material remaining on the screen.

A3.3 Determination by Assay and Chemical Analysis

A3.3.1 Take the remaining two thirds of the original sample weight and place in one or more crucibles using a minimum sample size of about 8 oz (225 g).

A3.3.2 Prepare a salt flux comprised of 47.5 % NaCl, 47.5 % KCl, and 5 % cryolite. (If cryolite is not available, sodium fluoride may be substituted.)

A3.3.3 Mix a minimum of 2 parts by weight of salt flux to 1 part metal (a salt-to-metal ratio of 3:1 is preferred).

A3.3.4 Place the crucible containing the sample in a gas-fired furnace to obtain a sample temperature of 815 ± 55°C (1500 ± 100°F). Stir occasionally with a graphite rod until all metal is melted. Hold for 15 min.

A3.3.5 *Alternative Procedure A for Pouring Sample*:

A3.3.5.1 Decant salt and sludge and save for removal of metal fines.

A3.3.5.2 When entire sample is melted in one crucible, pour remaining sample into a mold.

(a) *Samples over 1 lb*—If sample weighs more than 1 lb (450 g), pour analytical sample into appropriate mold.

(b) *Samples Under 1 lb*—If sample weighs less than 1 lb (450 g) take the entire sample of metal recovered and remelt in a suitable laboratory furnace.[6] Melt the sample at 675 to 735°C (1250 to 1350°F) and pour into an analytical sample mold.

A3.3.6 *Alternative Procedure B for Pouring Sample*:

A3.3.6.1 Allow metal and salt to cool to room temperature while still in the crucible.

A3.3.6.2 Invert the crucible to remove metal; break crucible, if necessary, to remove all metallics.

A3.3.6.3 Mechanically separate salt from metal and pick out any metallics that may be entrapped in the flux.

A3.3.7 Weigh all metal. Wash with water and then dilute nitric acid (0.1 *N*), rinse, dry, and weigh.

A3.4 Calculations

A3.4.1 Calculate the percent recovery on the basis of the as-received sample weight, before drying, screening, and magnetic separation as follows:

$$\% \text{ recovery} = \frac{R}{S} \times 100 \quad \text{(A3.1)}$$

where:

R = weight of metal recovered in A3.3.7, oz, and
S = weight of sample in A3.3.1, oz.

A3.5 Spectrochemical Analysis

A3.5.1 Determine the composition of the sample using an appropriate spectrochemical technique.

NOTE A3.1—It should be recognized that if the metal recovered in the assay is analyzed, the results for magnesium will be lower than the actual magnesium content of the MAS. This is due to a reaction of the assay flux with magnesium. If there is reason to suspect that the magnesium content of the MAS is higher than the maximum specified, analysis should be made on a duplicate sample of the MAS, which is remelted either under a salt flux mixture or in a small induction furnace.

[6] Laboratory furnaces, available from Jelrus and Lindberg, or equivalent, have been found suitable for this purpose.

Designation: E 775 – 87 (Reapproved 1996)

Standard Test Methods for Total Sulfur in the Analysis Sample of Refuse-Derived Fuel[1]

This standard is issued under the fixed designation E 775; the number immediately following the designation indicates the year of original adoption or, in the case of revision, the year of last revision. A number in parentheses indicates the year of last reapproval. A superscript epsilon (ϵ) indicates an editorial change since the last revision or reapproval.

1. Scope

1.1 These test methods cover two alternative procedures for the determination of total sulfur in prepared analysis samples of solid forms of refuse-derived fuel (RDF). Sulfur is included in the ultimate analysis of RDF.

1.2 The test methods appear in the following order:

Test	Sections
Eschka Method	8-11
Bomb Washing Method	12 and 13

1.3 These test methods may be applicable to any waste material from which a laboratory analysis sample can be prepared.

1.4 The values stated in inch-pound units are to be regarded as standard. The values given in parentheses are for information only.

1.5 *This standard does not purport to address all of the safety concerns, if any, associated with its use. It is the responsibility of the user of this standard to establish appropriate safety and health practices and determine the applicability of regulatory limitations prior to use.* For specific precautionary statements see Section 6.

2. Referenced Documents

2.1 *ASTM Standards:*

D 1193 Specification for Reagent Water[2]

E 180 Practice for Determining the Precision of ASTM Methods for Analysis and Testing of Industrial Chemicals[3]

E 711 Test Method for Gross Calorific Value of Refuse-Derived Fuel by the Bomb Calorimeter[4]

E 829 Practice for Preparing Refuse-Derived Fuels (RDF) Laboratory Samples for Analysis[4]

3. Terminology

3.1 *Definitions of Term Specific to This Standard:*

3.1.1 *refuse-derived fuels*—solid forms of refuse-derived fuels from which appropriate analytical samples may be prepared are defined as follows in *ASTM STP 832*:[5]

RDF-1—Wastes used as a fuel in as-discarded form with only bulky wastes removed.

RDF-2—Wastes processed to coarse particle size with or without ferrous metal separation.

RDF-3—Combustible waste fraction processed to particle sizes, 95 % passing 2-in. square screening.

RDF-4—Combustible waste fraction processed into powder form, 95 % passing 10-mesh screening.

RDF-5—Combustible waste fraction densified (compressed) into the form of pellets, slugs, cubettes, or briquettes.

4. Summary of Test Methods

4.1 *Eschka Method*— A weighed sample and Eschka mixture are ignited together and the sulfur is precipitated from the resulting solution as barium sulfate ($BaSO_4$). The precipitate is filtered, ashed, and weighed.

4.2 *Bomb Washing Method*—Sulfur is precipitated as $BaSO_4$ from the oxygen-bomb calorimeter washings and the precipitate is filtered, ashed, and weighed.

5. Significance and Use

5.1 The standards are available to producers and users of RDF for determining the total sulfur content of the fuel.

6. Precautions

6.1 Due to the origins of RDF in municipal waste, common sense dictates that some precautions should be observed when conducting tests on the samples. Recommended hygienic practices include use of gloves when handling RDF; wearing dust masks (NIOSH-approved type), especially while milling RDF samples; conducting tests under negative pressure hood when possible; and washing hands before eating or smoking.

7. Sampling

7.1 RDF products are frequently nonhomogeneous. For this reason significant care should be exercised to obtain a representative laboratory sample from the RDF lot to be characterized.

7.2 The sampling method for this procedure should be based on agreement between the involved parties.

7.3 The laboratory sample must be air-dried and particle size reduced to pass a 0.5-mm screen as described in Practice

---

[1] These test methods are under the jurisdiction of ASTM Committee D-34 on Waste Management and is the direct responsibility of Subcommittee D34.03 on Treatment.

Current edition approved Aug. 28, 1987. Published October 1987. Originally published as E 775 – 81. Last previous edition E 775 – 81.

[2] *Annual Book of ASTM Standards*, Vol 11.01.

[3] *Annual Book of ASTM Standards*, Vol 15.05.

[4] *Annual Book of ASTM Standards*, Vol 11.04.

[5] *Thesaurus on Resource Recovery Terminology, ASTM STP 832*, ASTM, 1983, p.72.

E 829. This procedure must be performed carefully to preserve the sample's representative characteristics (other than particle size) while preparing the analysis sample to be used in the procedures.

TEST METHOD A—ESCHKA METHOD

8. Apparatus

8.1 *Gas (*Note 1*) or Electric Muffle Furnace or Burners*, for igniting the sample with Eschka mixture and for igniting the barium sulfate ($BaSO_4$).

NOTE 1—Gas may contain sulfur compounds in sufficient quantities to affect the results.

8.2 *Crucibles or Capsules*—Porcelain capsules, ⅞ in. (22 mm) in depth and 1¾ in. (44 mm) in diameter, or porcelain crucibles of 30-mL capacity, high or low-form, or platinum crucibles of similar size shall be used for igniting the sample with the Eschka mixture. Porcelain, platinum, Alundum, or silica crucibles of 10 to 15-mL capacity shall be used for the final ignition step (see 10.3.8).

9. Reagents

9.1 *Purity of Reagents*—Reagent grade chemicals shall be used in all tests. Unless otherwise indicated, it is intended that all reagents shall conform to the specifications of the American Chemical Society, where such specifications are available. Other grades[6] may be used, provided it is first ascertained that the reagent is of sufficiently high purity to permit its use without lessening the accuracy of the determination.

9.2 *Purity of Water*— Unless otherwise indicated, references to water shall be understood to mean reagent water, Type III conforming to Specification D 1193.

9.3 *Barium Chloride Solution* (100 g/L)—Dissolve 100 g of barium chloride ($BaCl_2{\cdot}2H_2O$) and dilute to 1 L with water.

9.4 *Bromine Water* (saturated)—Add an excess of bromine to 1 L of water.

9.5 *Eschka Mixture*— Thoroughly mix 2 parts by weight of light calcined magnesium oxide (MgO) with 1 part of anhydrous sodium carbonate (Na_2CO_3). Both materials should be as free as possible from sulfur.

9.6 *Hydrochloric Acid* (1 + 1)—Mix equal volumes of concentrated HCl (sp gr 1.19) and water.

9.7 *Hydrochloric Acid* (1 + 9)—Mix 1 volume of concentrated HCl (sp gr 1.19) with 9 volumes of water.

9.8 *Methyl Orange Indicator Solution* (0.2 g/L)—Dissolve 0.2 g of methyl orange in 1000 mL of hot water and filter.

9.9 *Sodium Carbonate* (saturated solution)—Dissolve approximately 60 g of crystallized sodium carbonate ($Na_2CO_3{\cdot}10H_2O$) or 20 g of anhydrous sodium carbonate (Na_2CO_3) in 100 mL of water, using a sufficient excess of Na_2CO_3 to ensure a saturated solution.

9.10 *Sodium Hydroxide Solution* (100 g/L)—Dissolve 100 g of sodium hydroxide (NaOH) in 1 L of water. This solution may be used in place of Na_2CO_3 solution.

[6] "Reagent Chemicals, American Chemical Society Specification," *American Chemical Society,* Washington, DC. For suggestions on testing of reagents not listed by the American Chemical Society, see "Analar Standards for Laboratory U. K. Chemicals," BDH Ltd., Poole, Dorset, and the "United States Pharmacopeia."

10. Procedure

10.1 *Preparation of Sample and Eschka Mixture*—Weigh to the nearest 0.1 mg about 1 g of mixed air-dried analysis sample and 3 g of Eschka mixture on glazed paper. Mix thoroughly. The amount of sample to be taken will depend on the amount of $BaCl_2$ solution required (see 10.3.5 and Note 2).

10.1.1 Quantitatively transfer the mixture to a porcelain capsule or porcelain crucible or platinum crucible, and cover with about 1 g of Eschka mixture.

10.2 *Ignition*—Heat the crucible over a gas flame as described in 10.2.1, or in a gas- or electrically heated muffle furnace as described in 10.2.2. The use of artificial gas for heating the sample and Eschka mixture is permissible only when the crucibles are heated in a muffle (see Note 2).

10.2.1 Heat the crucible, placed in a slanting position partially covered on a triangle, over a very low flame. This prevents rapid expulsion of the volatile matter and affords more complete oxidation of the sulfur. After 30 min of low flame heating, gradually increase the temperature and occasionally stir the mixture until all black particles have disappeared, which is an indication of complete combustion.

10.2.2 Place the crucible in a cold muffle furnace and gradually raise the temperature to 800 ± 25°C in about 1 h. Maintain this maximum temperature until upon stirring all black particles have disappeared (about 1½ h).

10.3 *Subsequent Treatment*:

10.3.1 Remove the crucible, cool, and empty the contents quantitatively into a 200-mL beaker. Digest with 100 mL of hot water for ½ to ¾ h with occasional stirring.

10.3.2 Decant the supernatant liquid through a filter into a 600-mL beaker. Wash the insoluble matter with hot water several times using 25 mL of water at each washing and filter the washings through the filter paper into the 600-mL beaker. After washing, transfer the insoluble matter to the filter and wash five times with hot water, keeping the mixture well agitated, collecting the wash waters in the 600-mL beaker.

10.3.3 Treat the filtrate with 10 to 20 mL of saturated bromine water. Make slightly acid with HCl and boil to expel the liberated bromine.

10.3.4 Neutralize using methyl orange indicator with NaOH or Na_2CO_3 solution; then add 1 mL of HCl solution (1 + 9).

10.3.5 Boil again and then, while stirring constantly, add slowly from a pipet 10 mL or more of $BaCl_2$ solution.

NOTE 2—Barium chloride solution must be added in excess. If more than 10 mL of $BaCl_2$ solution is required, reduce the weight of sample to about 0.5 g and repeat the ignition and digestion.

10.3.6 Continue boiling for 15 min and allow to stand for at least 2 h, or preferrably overnight, at a temperature just below boiling.

NOTE 3—When standing overnight at a temperature slightly less than boiling, cover the flask with a watchglass to prevent the solution from evaporating to dryness and to protect it from external contamination.

10.3.7 Filter the solution and the precipitate of barium sulfate ($BaSO_4$) through an ashless paper and wash the $BaSO_4$ residue with hot water until 1 drop of silver nitrate ($AgNO_3$) solution produces no more than a slight opalescence when added to 8 to 10 mL of filtrate.

10.3.8 Place the wet filter containing the precipitate of barium sulfate ($BaSO_4$) in a preweighed platinum, porcelain, silica, or Alundum crucible, allowing a free access of air by folding the paper over the precipitate loosely to prevent spattering.

10.3.9 Smoke the paper off gradually over a gas burner or in a gas or electrically heated muffle furnace. At no time allow it to burn with a flame (see Note 4). After charring of the paper, raise the temperature to approximately 925°C and heat to constant weight.

NOTE 4—Partially covering the crucible while smoking and smoldering will aid in the prevention of the flaming of the paper.

10.3.10 Transfer the crucibles to a desiccator and weigh when cooled to room temperature.

10.4 *Blanks and Corrections*—In all cases a correction must be applied. Either a reagent blank may be run exactly as described above, using the same amount of all reagents that were employed in the routine determination, or a more accurate correction may be made by analyzing a weighed portion of a standard sulfate using the prescribed reagents and operations.

NOTE 5—If the latter procedure is carried out once a week, or whenever a new supply of a reagent is used for a series of solutions covering the approximate range of sulfur concentrations in the samples, it is only necessary to add or subtract from the weight of $BaSO_4$ determined for the sample the deficiency or excess found by the appropriate "check" determination. This procedure is more accurate than the simple reagent blank because, for the amounts of sulfur in question and condition of precipitation prescribed, the solubility error for $BaSO_4$ is probably the largest one considered. Barium sulfate is soluble in acids[7] and in pure water, and the solubility limit is reached almost immediately on contact with the solvent. Hence, if very high-purity reagents are used or extra precaution is exercised, there may be no sulfate apparent in the "blank." In other words, the solubility limit for $BaSO_4$ has not been reached or at any rate not exceeded; consequently, some sulfate in the sample may remain in solution or has redissolved.

11. Calculation

11.1 Calculate the sulfur content as follows:

$$\text{Sulfur, \%} = (A - B) \times 13.738/W \quad (1)$$

where:

A = grams of $BaSO_4$ precipitated,
B = grams of $BaSO_4$ correction,
W = grams of sample used, and
13.738 = percentage of sulfur in $BaSO_4$.

TEST METHOD B—BOMB WASHING METHOD[8]

12. Reagents

12.1 *Purity of Reagents*—see 9.1.

12.2 *Purity of Water*— see 9.2.

12.3 *Ammonium Hydroxide* (sp gr 0.90)—Concentrated ammonium hydroxide (NH_4OH).

12.4 *Bromine Water* (saturated)—see 9.4.

12.5 *Hydrochloric Acid* (1 + 1)—see 9.6.

12.6 *Sodium Carbonate Solution*—Dissolve 18.02 g of anhydrous sodium carbonate (Na_2CO_3) in water and dilute to 1 L. The Na_2CO_3 should be previously dried for 24 h at 105°C.

NOTE 6—Other concentrations of sodium carbonate solution may be used.

12.7 *Wash Solution*— Dilute 1 mL of a saturated solution of methyl orange to 1 L of water.

13. Procedure

13.1 *Ignition*—Sulfur is determined in the washings from the oxygen-bomb calorimeter following calorimetric determination in accordance with Test Method E 711. The type of bomb, amount of water in the bomb, oxygen pressure, and amount of sample taken shall be the same as specified under the calorimetric determination. The bomb shall stand in the calorimeter water for not less than 5 min after firing.

13.2 *Subsequent Treatment*:

13.2.1 Remove the bomb from the calorimeter water and open the valve carefully so as to allow the gases to escape at an approximately even rate so the pressure is reduced to atmospheric in not less than 1 min. Bombs equipped with valves other than needle valves, such as compression valves, shall be provided with a device so the valve can be controlled to permit a slow and uniform release of the gases.

13.2.2 Open the bomb and examine the inside for traces of unburned material or sooty deposit. If these are found, discard the determination. Wash carefully all parts of the interior of the bomb, including the capsule with a fine jet of water containing methyl orange (12.7) until no acid reaction is observed. It is essential to wash through the valve opening in the case of bombs equipped with compression valves, or other types of valves with large openings, as considerable spray may collect in such valve openings.

13.2.3 Collect the washings in a 250-mL beaker and titrate with standard sodium carbonate solution (12.6) to obtain the acid correction for the heating value, as specified in the calorimetric determination of Test Method E 711.

13.2.4 Adjust the pH to between 5.5 and 7.0 with dilute NH_4OH and heat the solution to boiling.

13.2.5 Filter through a qualitative filter paper and wash the residue and paper thoroughly five or six times with hot water collecting the filtrate and washings.

13.2.6 To the filtrate and washings add 1 mL of saturated bromine water (12.4) and sufficient HCl (12.5) to make it slightly acid. Boil the solution to expel the excess bromine.

13.2.7 Adjust the acidity, precipitate and determine the sulfur as described in 10.3.4-11.1 inclusive.

14. Precision and Bias[9]

14.1 *Precision*:

---

[7] Allen, E. T., and Johnston, John, "The Exact Determination of Sulfur in Soluble Sulfates," *Journal of the American Chemical Society*, Vol 32, No. 5, 1910, pp. 588-617; and Johnston, John, and Adams, L. H., "The Phenomenon of Acclusion in Precipitates of Barium Sulfate, and its Relation to the Exact Determination of Sulfate," *Journal of the American Chemical Society*, Vol 33, No. 6, 1911, pp. 829-845.

[8] Selvie, W. A. and Fieldner, A. C., "Check Determinations of Sulfir in Coal and Coke by the Eschka, Bomb Washing and Sodium Peroxide Fusion Methods," *Industrial and Engineering Chemistry*, Vol 29, 1927, pp. 729-733.

[9] Data from preliminary testing and round-robin tests are on file in ASTM Research Report RR-E38-1000. A copy of the report is available on loan from ASTM Headquarters, 1916 Race St., Philadelphia, PA 19103.

14.1.1 The standard deviation of individual determinations in percent absolute, are as follows:

Typical average value	0.35 %
Within-laboratory	0.03 %
Between-laboratory	0.06 %

14.1.2 The above precision estimates are based on an interlaboratory study conducted in accordance with Practice E 180.[9]

14.2 *Bias*—The bias of this test method has not been determined because of the lack of a recognized standard reference material.

Designation: E 776 – 87 (Reapproved 1996)

Standard Test Method for Forms of Chlorine in Refuse-Derived Fuel[1]

This standard is issued under the fixed designation E 776; the number immediately following the designation indicates the year of original adoption or, in the case of revision, the year of last revision. A number in parentheses indicates the year of last reapproval. A superscript epsilon (ε) indicates an editorial change since the last revision or reapproval.

1. Scope

1.1 This test method covers the determination of the forms of chlorine in refuse-derived fuel-three (RDF): total chlorine, water-soluble chloride, and water-insoluble chlorine.

1.2 This test method may be applicable to any waste material from which a laboratory analysis sample can be prepared.

1.3 *This standard does not purport to address all of the safety concerns, if any, associated with its use. It is the responsibility of the user of this standard to establish appropriate safety and health practices and determine the applicability of regulatory limitations prior to use.* For specific precautionary statements see Section 6 and 11.2.1.

2. Referenced Documents

2.1 *ASTM Standards:*

D 1193 Specification for Reagent Water[2]

E 144 Practice for Safe Use of Oxygen Combustion Bombs[3]

E 180 Practice for Determining the Precision of ASTM Methods for Analysis and Testing of Industrial Chemicals[4]

E 287 Specification for Burets[3]

E 711 Test Method for Gross Calorific Value of Refuse-Derived Fuel by the Bomb Calorimeter[5]

E 829 Practice for Preparing Refuse-Derived Fuel (RDF) Laboratory Samples for Analysis[5]

3. Terminology

3.1 *Definitions of Terms Specific to This Standard:*

3.1.1 *refuse-derived fuels*—solid forms of refuse-derived fuels from which appropriate analytical samples may be prepared are defined as follows in *ASTM STP 832:*[6]

RDF-1—Wastes used as a fuel in as-discarded form with only bulky wastes removed.

RDF-2—Wastes processed to coarse particle size with or without ferrous metal separation.

RDF-3—Combustible waste fraction processed to particle sizes, 95 % passing 2-in. square screening.

RDF-4—Combustible waste fraction processed into powder form, 95 % passing 10-mesh screening.

RDF-5—Combustible waste fraction densified (compressed) into the form of pellets, slugs, cubettes, or briquettes.

3.1.2 *total chlorine*—all chlorine as determined in the refuse-derived fuel.

3.1.3 *water-insoluble chlorine*—water-insoluble chlorides and chlorine in the refuse-derived fuel.

3.1.4 *water-soluble chloride*—those chlorides which are water-solubilized by water extraction as determined in the refuse-derived fuel.

4. Summary of Test Method

4.1 The forms of chloride and chlorine are determined. The various procedures in the method convert the forms of chlorine into a water-soluble chloride form that can be quantitated by titration.

4.1.1 *Total Chlorine*—The sample is combusted in an oxygen atmosphere. The chlorine is converted to chloride and absorbed in an alkaline solution.

4.1.2 *Water-Soluble Chlorides*—A portion of the analysis sample is successively extracted with hot chloride-free water.

4.1.3 *Water-Insoluble Chlorine*—Water-insoluble chlorine is calculated from the results of the total chlorine and the water-soluble chloride determination where:

$$\text{water-insoluble chlorine} = \text{total chlorine} - \text{water-soluble chlorides} \quad (1)$$

4.2 The chlorides contained in the alkaline solution (4.1.1) and the extraction solution (4.1.2) are determined by potentiometric (see Section 13) or modified Volhard titration (see Section 14).

5. Significance and Use

5.1 The standard is available to producers and users of RDF for determining the content and forms of chlorine present in the fuel.

6. Precautions

6.1 Due to the origins of RDF in municipal waste, common sense dictates that some precautions should be observed when conducting tests on the samples. Recommended hygienic practices include use of gloves when handling RDF; wearing

[1] This test method is under the jurisdiction of ASTM Committee D-34 on Waste Management and is the direct responsibility of Subcommittee D34.08 on Thermal Treatment.

Current edition approved Aug. 28, 1987. Published October 1987. Originally published as E 776 – 81. Last previous edition E 776 – 81.

[2] *Annual Book of ASTM Standards*, Vol 11.01.

[3] *Annual Book of ASTM Standards*, Vol 14.02.

[4] *Annual Book of ASTM Standards*, Vol 15.05.

[5] *Annual Book of ASTM Standards*, Vol 11.04.

[6] *Thesaurus on Resource Recovery Terminology, ASTM STP 832*, ASTM, 1983, p. 72.

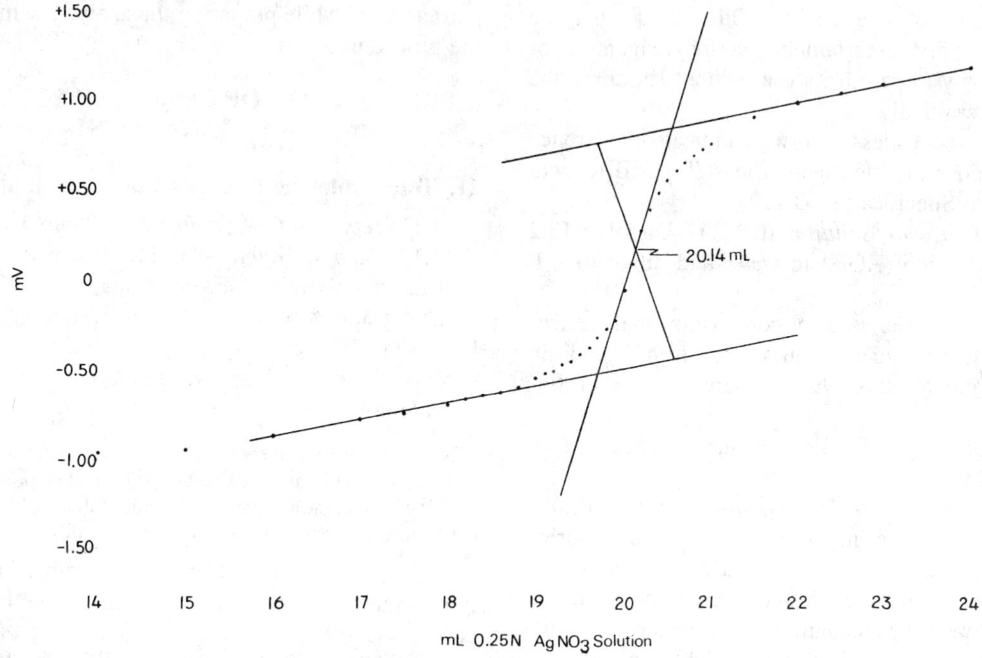

FIG. 1 Graph From a Potentiometric Titration of Chloride

dust masks (NIOSH-approved type), especially while milling RDF samples; conducting tests under negative pressure hood when possible; and washing hands before eating or smoking.

7. Interferences

7.1 *Potentiometric Titration Method A*:

7.1.1 Iodide and bromide are also titrated as chloride. Ferricyanide causes high results and must be removed. Chromate and dichromate interfere and should be reduced to the chromic state or be removed. Ferric iron interferes if present in an amount substantially higher than the amount of chloride. Chromic ions, ferrous ions, and phosphates do not interfere.

7.1.2 Grossly contaminated sample solutions usually require pretreatment. Where contamination is minor, some contaminants can be destroyed simply by the addition of nitric acid.

7.2 *Volhard Titration Method B*:

7.2.1 Compounds that have a strong oxidizing action interfere by reacting with thiocyanate. These compounds should be reduced beforehand by treatment with ferrous sulfate or a similar reducing agent.

7.2.2 Salts of mercury and palladium interfere by reacting with thiocyanate. They may be removed by precipitation with hydrogen sulfide before the addition of silver nitrate. The excess of sulfide is easily removed by gently boiling the acid solution for a few minutes. Sulfite can be eliminated in the same way.

7.2.3 Cyanide is also precipitated by silver nitrate. It is usually determined separately by the Liebig-Deniges method and a correction is applied to the results of the Volhard titration.[7]

7.2.4 The Volhard method, as with the potentiometric method, directly applied to a mixture of halides can determine only total halide content excluding fluoride. Preliminary treatment is necessary for the determination of chloride alone in a mixture.[8]

8. Apparatus

8.1 *Balance*, having a sensitivity of 0.1 mg.

8.2 *Apparatus for Bomb Combustion of the Sample*.

8.2.1 *Oxygen Bomb*, similar to that used in the determination of the calorific value of refuse-derived fuels as described in Test Method E 711.

8.2.2 *Capsule*, for holding the sample, approximately 25 mm in diameter at the top, approximately 12 mm deep, and conforming to Test Method E 711.

8.2.3 *Firing Wire*, as specified in Test Method E 711.

8.2.4 *Firing Circuit*, as specified in Test Method E 711.

8.2.5 *Metal Vessel*, cylindrical, such that the bomb will be fully immersed when approximately 2 L of water are added.

8.3 *Magnetic Stirrer and Stirring Bars*.

8.4 *Apparatus for Potentiometric Titration*:

8.4.1 *Potentiometric Titration Assembly*, using a silver indicator electrode and a calomel reference electrode containing a saturated sodium nitrate solution as a bridge.

NOTE 1—All glassware and graduated apparatus should be Class A or equivalent as described in Specification E 287.

9. Reagents

9.1 *Purity of Reagents*—Reagent grade chemicals shall be used in all tests. Unless otherwise indicated, it is intended that all reagents shall conform to the specifications of the Committee on Analytical Reagents of the American Chemical Society

[7] Scott's *Standard Method of Chemical Analysis*, edited by M. H. Furman, D. Van Nostrand Co., Inc., New York, NY.

[8] Koltoff, I. M., and Stenger, V. A., *Volumetric Analysis II*, Interscience Publishers, Inc., New York, NY.

where such specifications are available.[9] Other grades may be used, provided it is first ascertained that the reagent is of sufficiently high purity to permit its use without lessening the accuracy of the determination.

9.2 *Purity of Water*—Unless otherwise indicated, reference to water shall be understood to mean at least Type III reagent water conforming to Specification D 1193.

9.3 *Potassium Hydroxide Solution* (0.2 *N*)—Dissolve 13.2 g of potassium hydroxide (KOH) in water and dilute to 1 L with water.

9.4 *Oxygen*—The oxygen used for combustion shall be free of combustible matter. Oxygen manufactured from liquid air, guaranteed to be greater than 99.5 % pure, will meet this requirement.

9.5 *Sodium Chloride* (NaCl)—Primary standard quality (purity of 100 ± 0.02 %).

9.6 *Sodium Chloride, Primary Standard Solution* (0.025 *N*)—Crush 10 to 20 g of primary standard sodium chloride (NaCl) to 100-mesh fineness and dry in a glass container at 120°C for 2 h. Stopper and keep desiccated. Dissolve 5.844 g ± 0.1 mg of dried primary standard NaCl in water and dilute to 1 L. Dilute 25.00 mL of this solution to 100.0 mL.

9.7 *Methanol.*

9.8 *Nitric Acid* (1 + 1)—Mix equal volumes of concentrated nitric acid (HNO_3, sp, gr, 1.42) and water.

9.9 *Silver Nitrate, Standard Solution* (0.025 *N*)—Dissolve 4.247 g of silver nitrate ($AgNO_3$) in water and dilute to 1 L. Store in an amber glass bottle. Standardize against 0.025 *N* sodium chloride solution as directed in 13.1.1 and 14.1.1.

9.10 *Potassium Chromate – Potassium Dichromate Indicator*—(K_2CrO_4 – $K_2Cr_2O_7$)—Dissolve 4.2 g of K_2CrO_4 and 0.7 g of $K_2Cr_2O_7$ in 100 mL of water.

9.11 *Nitrobenzene.*

9.12 *Ferric Ammonium Sulfate Indicator Solution*—Add sufficient concentrated HNO_3 (sp gr 1.42) to a cold saturated solution of ferric ammonium sulfate [$FeNH_4(SO_4)_2 \cdot 12H_2O$] to remove the brown color.

9.13 *Potassium Thiocyanate, Standard Solution* (0.025 *N*)—Dissolve 3 g of potassium thiocyanate (KCNS) in freshly distilled or boiled water, dilute to 1 L, and standardize against 0.025 *N* $AgNO_3$ solution as directed in 14.1.2.

10. Sampling

10.1 RDF products are frequently nonhomogeneous. For this reason significant care should be exercised to obtain a representative laboratory sample from the RDF lot to be characterized.

10.2 The sampling method for this procedure should be based on agreement between the involved parties.

10.3 The laboratory sample must be air-dried and the particle size reduced to pass a 0.5-mm screen as described in Practice E 829. This procedure must be performed carefully to preserve the sample's representative characteristics (other than particle size) while preparing the analysis sample to be used in the procedures.

PREPARATION OF CHLORINE SOLUTIONS FOR ANALYSIS

11. Total Chlorine (Oxygen Bomb Method)

11.1 *Preparation of Sample and Bomb* (see Note 3):

11.1.1 *Sample Weight*—Weigh to the nearest 0.1 mg about 1 g of thoroughly mixed air-dried analysis RDF sample into the bomb capsule. A pellet may be made from the air-dried analysis RDF sample, accurately weighed, and placed into the bomb capsule. Place the capsule containing the sample into the capsule holder.

NOTE 2—There is a tendency for chlorine to adhere to the bomb walls, especially if the bomb is pitted or has been used previously to determine high levels of chlorine. Unless the bomb is thoroughly cleaned before use, the blanks may have values in excess of reality.

11.1.2 *Firing Wire*—Connect a length of firing wire to the ignition terminals in such a manner that the loop of firing wire is in contact with the sample.

11.1.3 *Bomb Preparation*—Add 20 to 25 mL of 0.02 *N* potassium hydroxide solution to the bomb and wet the entire internal surface of the bomb with this solution (see Note 3). Assemble the bomb.

NOTE 3—Sodium hydroxide solution at appropriate concentration may be used.

11.2 *Addition of Oxygen*—Admit oxygen to the bomb slowly to avoid blowing the sample from the capsule until a pressure of 25 atm is reached.

11.2.1 **Caution**—The following precautions are recommended for safe calorimeter operation. Additional precautions are given in Practice E 144.

11.2.1.1 The weight of RDF sample and the pressure of the oxygen admitted to the bomb must not exceed the bomb manufacturer's recommendations.

11.2.1.2 Bomb parts should be inspected carefully after each use. Threads on the main closure should be checked frequently for wear. Cracked or significantly worn parts should be replaced. The bomb should be returned to the manufacturer occasionally for inspection and possibly proof firing.

11.2.1.3 The oxygen supply cylinder should be equipped with an approved type of safety device, such as a reducing valve, in addition to the needle valve and pressure gage used in regulating the oxygen feed to the bomb. Valves, gages, and gaskets must meet industry safety codes. Suitable reducing valves and adaptors for 300 to 500 psi (2070 to 3450 kPa) discharge pressures are obtainable from commercial sources of compressed gas equipment. The pressure gage shall be checked periodically for accuracy.

11.2.1.4 During ignition of a sample, the operator must not permit any portion of his body to extend over the calorimeter.

11.3 *Preparation of the Chlorine Solution*—Immerse the bomb in a cold-water bath, connect it to the firing circuit, and close the circuit to ignite the sample. Allow the bomb to stand in the water bath for not less than 10 min after firing. Remove the bomb from the water bath, invert the bomb and shake for about 10 min (Note 4). Release the pressure at a slow, uniform

[9] Reagent Chemicals, American Chemical Society Specifications, Am. Chemical Soc., Washington, DC. For suggestions on the testing of reagents not listed by the American Chemical Society, see "Analar Standards for Laboratory U.K. Chemicals," BDH Ltd., Poole, Dorset, and the "United States Pharmacopeia."

rate so that the pressure is reduced to atmospheric in not less than 1 min. Open the bomb and examine the inside for traces of unburned material or sooty deposits. If any are found, discard the determination and thoroughly wash all parts of the bomb interior before using it again. If no unburned material or sooty deposits are present, rinse the interior of the bomb, the sample capsule, and the interior surface of the bomb cover with a fine jet of hot water and collect the contents and washings in a beaker or flask.

NOTE 4—Inverting and shaking a heavy, wet, and possibly slippery bomb may present a hazard. Precautions should be taken when performing this function.

12. Extraction of Water-Soluble Chlorides

12.1 *Preparation of the Chloride Solution*—Weigh to the nearest 0.1 mg about 5 g of thoroughly mixed, air-dried RDF analysis sample. Transfer to a 500-mL beaker and add 100 mL of hot water. Heat to close to boiling for ½ h with occasional stirring to ensure thorough wetting of the sample. Decant the supernatant liquid through a fast, qualitative filter paper collecting the filtrate in a beaker. Repeat the extraction two more times, each time using 100 mL of water and pooling the filtrates. After the third extraction, wash the sample thoroughly using 100 mL of hot water, adding the wash water to the pooled filtrates.

12.2 *Preparation of the Chloride Solution* (Alternative Method)—Weigh to the nearest 0.1 mg about 5 g of thoroughly mixed air-dried analysis sample (Note 5). Transfer to a 500-mL beaker and add 100 mL of hot water. Heat to close-to-boiling for ½ h with occasional stirring to ensure thorough wetting of the sample. Remove the beaker from the hot plate and allow the sample to settle. Decant the supernatant to a 100-mL centrifuge tube and centrifuge for ½ h at approximately 1200 rpm. Decant the supernatant through a fast qualitative filter paper. Repeat the extraction and centrifuging two more times, each time using 60 mL of water and pooling the filtrates. After the third extraction, wash the centrifuge tube and filter paper thoroughly using 60 mL of hot water, adding the wash water to the pooled filtrates (Notes 6 and 7).

NOTE 5—The weight of the sample used can be adjusted when higher or lower concentrations of water-soluble chlorides are anticipated.

NOTE 6—When high concentrations of chlorides are anticipated, dilute the cooled extraction solution to volume in a suitably sized volumetric flask and continue the analysis with a suitably sized aliquot.

NOTE 7—If the chloride solution is highly colored or turbid the potentiometric method (Section 13) should be used. If the Volhard method is to be used (Section 14), the color and turbidity must first be removed by aluminum hydroxide [$Al(OH)_3$] coagulation or by some other suitable method of turbidity or color removal without interfering with the chloride content of the solution.[10]

ALTERNATIVE PROCEDURES

13. Potentiometric Titration (Method A)

13.1 *Standardization of Reagents*—Prepare a standard solution of silver nitrate by adding 10.0 mL of standard 0.025 *N* NaCl solution to a 250-mL beaker. Add 40 to 50 mL of water and 50 mL of methanol (Note 9). Add 2 mL of $HNO_3(1 + 1)$ and titrate potentiometrically with the $AgNO_3$ solution as directed in 13.3. Calculate the normality of the $AgNO_3$ solution as follows:

$$\text{Normality of } AgNO_3 = \text{mL of NaCl solution used} \times \text{normality of NaCl solution/mL of } AgNO_3 \text{ solution required} \quad (2)$$

NOTE 8—Methanol increases the sensitivity of the inflection point. However, it may be omitted if it is found to interfere with or damage certain electrode systems.

13.2 *Blank Determination*— Prepare a solution for blank determination in the same manner as the preparation of chlorine solutions for the respective type of chlorine desired as described in Sections 11 or 12 using all the reagents in the same quantities and eliminating the RDF sample. Add to that blank solution 0.025 N NaCl solution as used in the standardization of $AgNO_3$ solution (13.1.1). Proceed as directed in 13.3. The difference in the volume of $AgNO_3$ solution required in this titration and that volume required for the standardization titration (13.1.1) is the true titer blank.

$$A - B = \text{true titer blank} \quad (3)$$

where:

A = millilitres of $AgNO_3$ required for the blank determination and
B = millilitres of $AgNO_3$ required for the standardization determination (13.1).

13.3 *Procedure*:

13.3.1 Concentrate the chlorides in the prepared solution (11.3, 12.1, or 12.2, respectively) for the type of chlorine desired by evaporation to about 50 mL. Add 50 mL of methanol and neutralize with $HNO_3(1 + 1)$ using phenolphthalein as an indicator. Add 2 mL of $HNO_3(1 + 1)$ in excess. Immerse the electrodes into the sample solution. While gently stirring on a magnetic stirrer, titrate by adding 0.10-mL increments of standard $AgNO_3$ solution and record the millivolt reading after each addition (Notes 9 and 10). Titrate well beyond the end point.

NOTE 9—Larger increments of $AgNO_3$ solution may be added before and after the end point. However, as the end point is reached, the 0.10-mL increment will give the most accurate end point.

NOTE 10—The end point is reached when the change of voltage is greatest upon the addition of the given increment of standard $AgNO_3$.

13.3.2 Plot the millivolt readings versus the millilitres of standard $AgNO_3$ solution added. Determine the end point (point of inflection) graphically from the titration curve or mathematically (second derivative) from the tabulated data. (See Table 1 and Fig. 2 for examples.)

13.4 *Calculations*—Calculate the weight percent of chloride as follows:

$$\text{Chloride, weight \%} = [C - (A - B)] \times N \times 0.03545 \times 100/W \quad (4)$$

[10] APHA, 14th ed., Method 408-A, Section 3a.

TABLE 1 Typical Data Sheet and Calculation of End Point from a Potentiometric Titration of Chloride

Volume of 0.025 *N* $AgNO_3$ solution, mL	Millivolt Reading	Volume of 0.025 *N* $AgNO_3$ solution, mL	Millivolt Reading	Volume of 0.025*N* $AgNO_3$ solution, mL	Millivolt Reading
0.0	−1.38	18.4	−0.65	20.2	+ 0.26
1.0	−1.29	18.6	−0.63	20.3	+ 0.38
2.0	−1.26	18.8	−0.59	20.4	+ 0.47
4.0	−1.21	19.0	−0.55	20.5	+ 0.55
6.0	−1.18	19.1	−0.53	20.6	+ 0.60
8.0	−1.14	19.2	−0.51	20.7	+ 0.65
10.0	−1.10	19.3	−0.48	20.8	+ 0.69
12.0	−1.04	19.4	−0.46	20.9	+ 0.72
14.0	−0.97	19.5	−0.42	21.0	+ 0.75
15.0	−0.95	19.6	−0.38	21.5	+ 0.90
16.0	−0.86	19.7	−0.33	22.0	+ 0.98
17.0	−0.79	19.8	−0.27	22.5	+ 1.04
17.5	−0.75	19.9	−0.19	23.0	+ 1.09
18.0	−0.70	20.0	−0.06	24.0	+ 1.17
18.2	−0.67	20.1	+ 0.08	25.0	+ 1.24

The greatest change of voltage per unit increment addition of $AgNO_3$ solution occurs between 19.8 and 20.5 mL. Determine the end point and titer as follows:

mL	mV	1st Derivative	2nd Derivative
19.8	−0.27		
19.9	−0.19	+ 0.08	+ 0.05
20.0	−0.06	+ 0.13	+ 0.01
20.1	+ 0.08	+ 0.14	+ 0.04
20.2	+ 0.26	+ 0.18	−0.06
20.3	+ 0.38	+ 0.12	−0.03
20.4	+ 0.47	+ 0.09	−0.01
20.5	+ 0.55	+ 0.08	

Interpolation:

$$20.1\ mL + \left(\frac{0.04}{0.10} \times 0.1\right) = 20.14\ mL$$

where:

A = millilitres of $AgNO_3$ solution required for the titration of the blank (13.2),
B = millilitres of $AgNO_3$ solution required for the titration of the NaCl standard (13.1.1),
C = millilitres of $AgNO_3$ solution required for the titration of the sample (13.3.2),
N = normality of the $AgNO_3$ solution (13.1),
W = grams of the analysis sample used, and
0.03545 = milliequivalent weight of chloride.

14. Volhard Titration (Method B)

14.1 *Standardization of Reagents*:

14.1.1 *Silver Nitrate, Standard Solution*—Add 20.0 mL of standard 0.025 *N* NaCl solution to a 250-mL Erlenmeyer flask. Add 1 mL of K_2CrO_4– $K_2Cr_2O_7$ indicator. Dilute to about 100 mL with water. Titrate with $AgNO_3$ solution to a faint red-brown end point. Calculate the normality of the $AgNO_3$ solution as follows:

$$\text{Normality of } AgNO_3 = \text{mL of NaCl solution used} \times \text{normality of NaCl solution/mL of } AgNO_3 \text{ solution required} \quad (5)$$

14.1.2 *Potassium Thiocyanate, Standard Solution*—Add 20.0 mL of 0.025 *N* $AgNO_3$ solution to a 250-mL glass-stoppered Erlenmeyer flask. Add 10 mL of $HNO_3(1 + 1)$ and make up to approximately 100 mL with water. Add 5 to 10 mL of nitrobenzene, stopper, and shake for 1 min. Add 8 to 10 drops of $FeNH_4(SO_4)_2$ indicator solution and titrate with KCNS solution. The end point is reached when the solution becomes faintly orange-pink in color. Calculate the normality of the KCNS solution as follows:

$$\text{Normality of KCNS} = \text{mL of } AgNO_3 \text{ solution used} \times \text{normality of } AgNO_3 \text{ solution/mL of KCNS solution required} \quad (6)$$

14.2 *Blank Determination*—Prepare a solution for the blank determination in the same manner as the preparation of chlorine solutions for the respective type of chlorine desired as described in Sections 11 and 12, using all the reagents in the same quantities and eliminating the RDF sample. Proceed as described in 14.3.

14.3 *Procedure*—Transfer quantitatively the chloride solution (11.3, 12.1, or 12.2, respectively) for the type of chlorine desired into a 250-mL glass-stoppered Erlenmeyer flask. Add 10 mL of $HNO_3(1 + 1)$, stirring gently during the acid addition. Add 20.0 mL of standard $AgNO_3$ solution. Stopper, mix, and let stand in the dark for 15 min. Cool to room temperature. Add 5 to 10 mL of nitrobenzene and shake for 1 min. Add 8 to 10 drops of $FeNH_4(SO_4)_2$ solution and titrate with standard KCNS solution against a white background. The end point of the determination is reached when the solution becomes faintly orange-pink in color.

14.4 *Calculations*—Calculate the weight percent of chloride as follows:

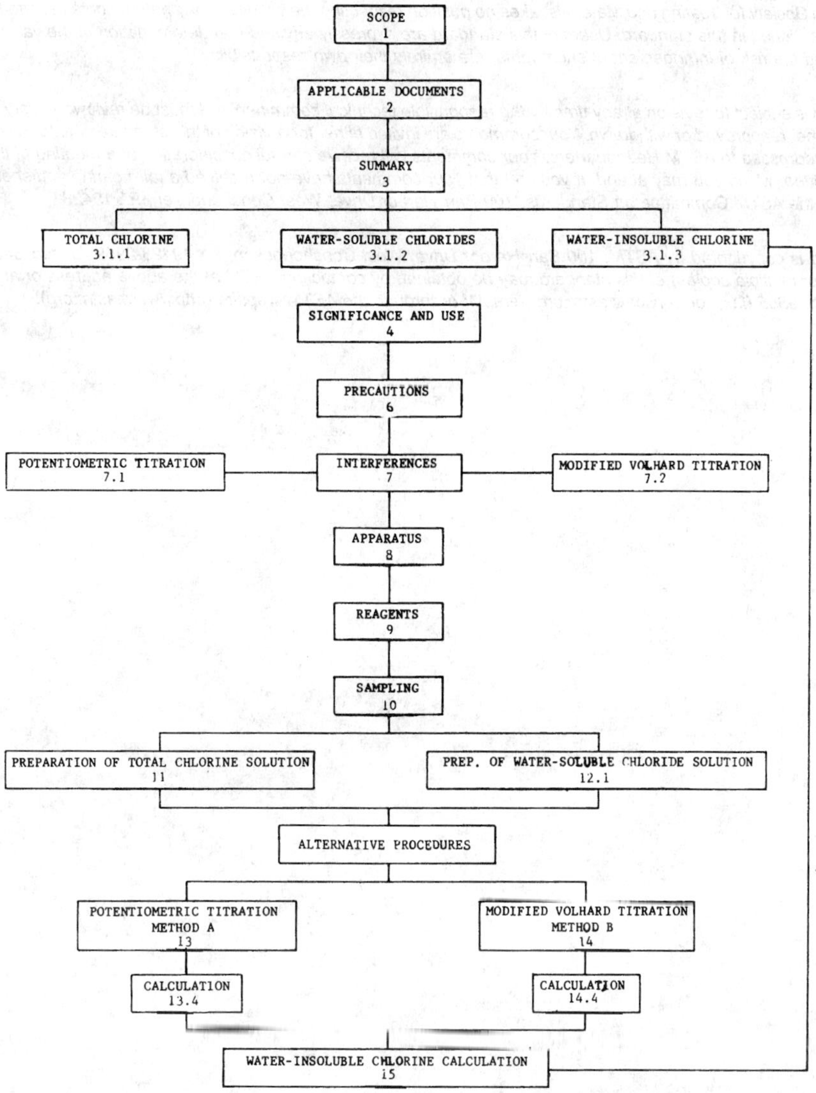

FIG. 2 Forms of Chlorine in Refused-Derived Fuel

$$\text{Chloride (chlorine), weight \%} = \frac{(B - A) \times N \times 0.03545 \times 100}{W} \qquad (7)$$

where:

A = millilitres of KCNS solution required for the titration of the sample (14.3),
B = millilitres of KCNS solution required for the titration of the blank (14.2),
N = normality of the KCNS solution (14.1.2),
W = grams of the analysis sample used, and
0.03545 = milliequivalent weight of chloride.

15. Water-Insoluble Chlorine

15.1 Calculate the percent of water-insoluble chlorine as follows: percent of total chlorine minus percent of water-insoluble chloride.

16. Precision and Bias [11]

16.1 *Precision:*

16.1.1 The standard deviations of individual determinations, in percent absolute, are as follows:

Typical Average Value,%	Within-Laboratory, %	Between Laboratories, %
Total Chlorine:		
0.49	0.03	0.11
Water-Soluble Chlorides:		
0.37	0.02	0.05

16.1.2 These precision estimates are based on an interlaboratory study conducted in accordance with Practice E 180.

16.2 *Bias*—The bias of this test has not been determined because of the lack of a recognized standard reference material.

[11] Supporting data are available on loan from ASTM Headquarters. Request RR: E38-1000.

Standard Test Method for Carbon and Hydrogen in the Analysis Sample of Refuse-Derived Fuel[1]

This standard is issued under the fixed designation E 777; the number immediately following the designation indicates the year of original adoption or, in the case of revision, the year of last revision. A number in parentheses indicates the year of last reapproval. A superscript epsilon (ϵ) indicates an editorial change since the last revision or reapproval.

1. Scope

1.1 This test method covers the determination of total carbon and hydrogen in a sample of refuse-derived fuel (RDF). Both carbon and hydrogen are determined in one operation. This test method yields the total percentages of carbon and hydrogen in RDF as analyzed and the results include not only the carbon and hydrogen in the organic matter, but also the carbon present in mineral carbonates and the hydrogen present in the free moisture accompanying the analysis sample as well as hydrogen present as water of hydration.

NOTE 1—It is recognized that certain technical applications of the data derived from this test procedure may justify additional corrections. These corrections could involve compensation for the carbon present as carbonates, the hydrogen of free moisture accompanying the analysis sample, and the calculated hydrogen present as water of hydration.

1.2 This test method may be applicable to any waste material from which a laboratory analysis sample can be prepared.

1.3 *This standard does not purport to address all of the safety concerns, if any, associated with its use. It is the responsibility of the user of this standard to establish appropriate safety and health practices and determine the applicability of regulatory limitations prior to use.* For specific precautionary statements see Section 8.

2. Referenced Documents

2.1 *ASTM Standards:*

D 1193 Specification for Reagent Water[2]

E 180 Practice for Determining the Precision of ASTM Methods for Analysis and Testing of Industrial Chemicals[3]

E 790 Test Method for Residual Moisture in a Refuse-Derived Fuel Analysis Sample[4]

E 791 Test Method for Calculating Refuse-Derived Fuel Analysis Data from As-Determined to Different Bases[4]

E 829 Practice for Preparing Refuse-Derived Fuel (RDF) Laboratory Samples for Analysis[4]

3. Terminology

3.1 *Definitions of Terms Specific to This Standard:*

3.1.1 *refuse-derived fuels*—solid forms of refuse-derived fuels from which appropriate analytical samples may be prepared are defined as follows in *ASTM STP 832:*[5]

RDF-1— Wastes used as a fuel in as-discarded form with only bulky wastes removed.

RDF-2—Wastes processed to coarse particle size with or without ferrous metal separation.

RDF-3—Combustible waste fraction processed to particle sizes, 95 % passing 2-inch square screening.

RDF-4—Combustible waste fraction processed into powder form, 95 % passing 10-mesh screening.

RDF-5—Combustible waste fraction densified (compressed) into the form of pellets, slugs, cubettes, or briquettes.

4. Summary of Test Method

4.1 The determination is made by burning the sample to convert all of the carbon to carbon dioxide and all of the hydrogen to water. The combustion is carried out by high purity oxygen that has been passed through a purifying train. The carbon dioxide and water are recovered in an absorption train. The combustion tube packing is used to remove any interfering substances. This test method gives the total percentages of carbon and hydrogen in the RDF as analyzed, including the carbon in carbonates and the hydrogen in any form of water.

5. Significance and Use

5.1 The standard sample is available to producers and users of RDF as a method of determining the weight percent of carbon and hydrogen in the analysis sample.

5.2 Carbon and hydrogen are part of the ultimate analysis of a fuel and can be used for calculations of combustion parameters.

6. Apparatus

6.1 *Oxygen-Purifying Train*—The high-purity oxygen is passed through water and carbon dioxide absorbers prior to use for combustion. The oxygen-purifying train consists of the following three units in order of passage of oxygen (see Fig. 1):

[1] This test method is under the jurisdiction of ASTM Committee D-34 on Waste Management and is the direct responsibility of Subcommittee D34.08 on Thermal Treatment.

Current edition approved Aug. 28, 1987. Published October 1987. Originally published as E 777 – 81. Last previous edition E 777 – 81.

[2] *Annual Book of ASTM Standards*, Vol 11.01.

[3] *Annual Book of ASTM Standards*, Vol 15.05.

[4] *Annual Book of ASTM Standards*, Vol 11.04.

[5] *Thesaurus on Resource Recovery Terminology, ASTM STP 832*, ASTM, 1983, p. 72.

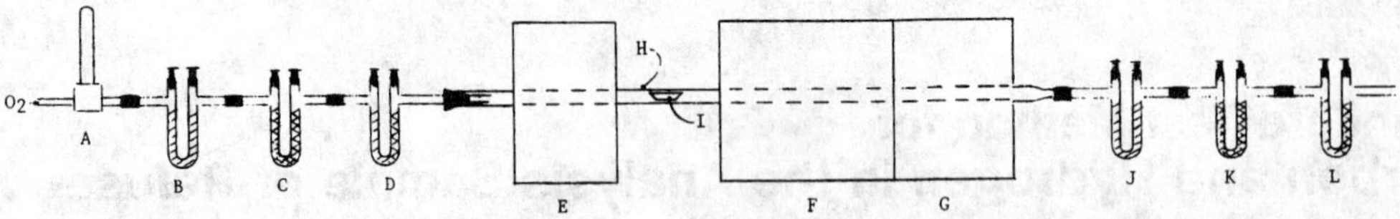

A = flowmeter (6.2)
B, C, D = oxygen purifying train (6.1)
B = first water absorber (6.1.1)
C = carbon dioxide absorber (6.1.2)
D = second water absorber (6.1.3)

E, F, G, H, I = combustion unit (6.3)
E = furnace 1 (6.3.1)
F = furnace 2 (6.3.2)
G = furnace 3 (6.3.3)
H = combustion tube (6.3.4)
I = combustion boat (6.3.5)

J, K, L = absorption train (6.4)
J = first water absorber (6.4.1)
K = carbon dioxide absorber (6.4.2)
L = guard tube (6.4.3)

NOTE 1—*C* and *K* can substitute a Vanier bulb if liquid absorbent is used.

FIG. 1 Set-Up of Apparatus

6.1.1 *First Water Absorber*—A container constructed so that the oxygen must pass through a column of reagent. The container shall have a capacity for at least 45 cm^3 of solid reagent, and the minimum gas travel through the reagent shall be at least 80 mm. A container of large volume and long path of oxygen travel through the reagent will be found to be advantageous where many carbon and hydrogen determinations are made.

6.1.2 *Carbon Dioxide Absorber*—If solid reagents are used for carbon dioxide absorption, the container shall be as described in 6.1.1. If a solution is used, the container shall be a Vanier bulb. It shall provide a column of reagent adequate to remove the carbon dioxide completely.

6.1.3 *Second Water Absorber*—Same as specified in 6.1.1.

6.2 *Flowmeter*, used to permit volumetric measurement of the rate of flow of oxygen during the determination. It shall be suitable for measuring flow rates within the range from 50 to 100 mL/min (standard temperature and pressure). The use of a double-stage pressure-reducing regulator with gage and needle valve is recommended to permit easy and accurate adjustment to the rate of flow.

6.3 *Combustion Unit*, consisting of three electrically heated furnace sections, individually controlled, which may be mounted on rails for easy movement. The upper part of each furnace may be hinged so that it can be opened for inspection of the combustion tube. The three furnace sections shall be as follows (see Fig. 1):

6.3.1 *Furnace Section 1*—Furnace 1 is nearest the oxygen inlet end of the combustion tube, approximately 130 mm long and used to heat the inlet end of the combustion tube and the sample. It shall be capable of rapidly attaining an operating temperature of 875 ± 25°C.

NOTE 2—Combustion tube temperature shall be measured by means of a thermocouple placed immediately adjacent to the tube near the center of the appropriate tube section.

6.3.2 *Furnace Section 2*—Furnace 2 shall be approximately 330 mm in length and used to heat that portion of the tube filled with cupric oxide. The operating temperature shall be 850 ± 20°C (see Note 2).

6.3.3 *Furnace Section 3*—Furnace 3 shall be approximately 230 mm long, and used to heat that portion of the tube filled with lead chromate or silver. The operating temperature shall be 500 ± 50°C.

6.3.4 *Combustion Tube*, made of fused quartz, or high-silica glass and having a nominal inside diameter which may vary within the limits of 19 to 22 mm and a minimum total length of 970 mm. The exit end shall be tapered down to provide a tubulated section for connection to the absorption train. The tubulated section shall have a length of 20 to 25 mm, an internal diameter of not less than 3 mm, and an external diameter of approximately 7 mm. The total length of the reduced end shall not exceed 60 mm. If a translucent fused quartz tube is used, a transparent section 190 mm long, located 250 mm from the oxygen inlet end of the tube, will be found convenient (see Fig. 2).

6.3.5 *Combustion Boat*, made of glazed porcelain, fused silica, or platinum. Boats with internal dimensions of approximately 70 by 8 by 8 mm have been found convenient.

6.4 *Absorption Train*, identical to the oxygen absorption train indicated in 5.1 to obtain system equilibrium. Therefore, the absorption train shall consist of the following units arranged as listed in the order of passage of oxygen (see Fig. 1):

6.4.1 *First Water Absorber*, as described in 5.1.1.

6.4.2 *Carbon Dioxide Absorber*, as described in 6.1.2.

6.4.3 *Second Water Absorber*, as described in 6.1.3. The second water absorber is also known as a guard tube.

7. Reagents

7.1 *Purity of Reagents*—Reagent grade chemicals shall be used in all tests. Unless otherwise indicated, it is intended that all reagents shall conform to the specifications of the American Chemical Society, where such specifications are available.[6] Other grades may be used, provided it is first ascertained that the reagent is of sufficiently high purity to permit its use without lessening the accuracy of the determination.

7.2 *Purity of Water*—Unless otherwise indicated, references to water shall be understood to mean reagent water, Type III, conforming to Specification D 1193.

[6] "Reagent Chemicals, American Chemical Society Specifications," Am. Chemical Soc., Washington, DC. For suggestions on the testing of reagents not listed by the American Chemical Society, see "Analar Standards for Laboratory U.K. Chemicals," BDH Ltd., Poole, Dorset, and the "United States Pharmacopeia".

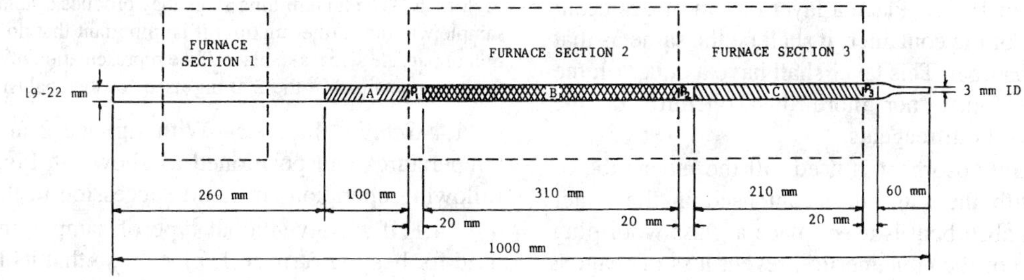

A =clear fused quartz section (optional when a translucent quartz tube is used)
B =cupric oxide filling
C =lead chromate or silver gauze filling
P_1, P_2, P_3 =oxidized copper gauze plugs

NOTE 1—When furnace sections longer than those specified in 6.3 are to be used, changes in the above dimensions shall be in accordance with provisions of Note 5.

FIG. 2 Arrangement of Tube Filling for Combustion Tube

7.3 *Oxygen*, with minimum acceptable purity 99.5 %.

NOTE 3—If the blank tests for flow (see 10.3.2) indicate interfering impurities in the oxygen supply by consistent weight-gain in the absorbers, eliminate these impurities by using a preheater furnace and tube, filled with cupric oxide. Operate this preheater at 850 ± 20°C and insert in series between the supply tank of oxygen and the purification train.

7.4 *Combustion Tube Reagents*:

7.4.1 *Cupric Oxide* (CuO), wire form, dust-free.

7.4.2 *Fused Lead Chromate* ($PbCrO_4$), approximately 2.38 to 0.84 mm in size.

7.4.3 *Silver Gauze*, 99.9 % silver minimum purity, 0.84 mm, made from approximately No. 27 B&S gage wire.

7.4.4 *Copper Gauze*, 99.0 % copper minimum purity, 0.84 mm, made from approximately No. 26 B&S gage wire.

7.5 *Purification and Absorption Train Reagents*:

7.5.1 *Water Absorbent*—Anhydrous magnesium perchlorate ($Mg(ClO_4)_2$) approximately 2.38 to 0.35 mm in size.[7]

7.5.2 *Carbon Dioxide Absorbent*—If a solid reagent is used, it shall be sodium or potassium hydroxide (NaOH or KOH) impregnated in an inert carrier approximately 2.38 to 0.84 mm in size. Use of soda lime in place of the above or in admixture with them is permissible (Note 4). If a solution is used, it shall be 30 weight % potassium hydroxide (KOH).

NOTE 4—Acceptable carbon dioxide absorbing reagents using sodium or potassium hydroxide are sold under the tradenames: Ascarite, Caroxite, and Mikohbite. If soda lime is used in admixture with any of the foregoing, it should not exceed 30 weight % of the total reagent. In using Ascarite it may be necessary to add a few drops of water to this reagent to assure complete absorption of carbon dioxide.

8. Precautions

8.1 Due to the origins of RDF in municipal waste, common sense dictates that some precautions should be observed when conducting tests on the samples. Recommended hygienic practices include use of gloves when handling RDF; wearing dust masks (NIOSH-approved type), especially while milling RDF samples; conducting tests under a negative pressure hood when possible; and washing hands before eating or smoking.

9. Sampling

9.1 RDF products are frequently nonhomogeneous. For this reason significant care should be exercised to obtain a representative laboratory sample from the RDF lot to be characterized.

9.2 The sampling method for this procedure should be based on agreement between the involved parties.

9.3 The laboratory sample must be air-dried and particle size reduced to pass a 0.5-mm screen as described in Practice E 829. This procedure must be performed carefully to preserve the sample's representative characteristics (other than particle size) while preparing the analysis sample to be used in the procedures.

10. Preparation of Apparatus

10.1 *Combustion Tube Packing*—To ensure complete oxidation of combustion products and complete removal of interfering substances such as oxides of sulfur, the combustion tube shall be packed with cupric oxide and lead chromate or silver gauze. The arrangement and lengths of the tube fillings and separating plugs shall be as shown in Fig. 2 (see Note 5). It is recommended that the tube be placed in a vertical position (constricted end downward) for packing. When filling the tube with lead chromate, any residual reagent adhering to the walls of the empty portion of the tube must be removed. When silver gauze is used as a tube filling, the required length of filling may be prepared conveniently from three or four strips 150 to 200 mm in length, by rolling each strip into a cylindrical plug and inserting the strips end-to-end in the tube.

NOTE 5—Longer furnaces with appropriate lengths of tube packing will be satisfactory.

10.2 *Purification and Absorption Trains*:

10.2.1 *Water Absorbers*—Fill a container, described in 6.1.1, with a permissible solid desiccant, as described in 7.5.1, by adding the required amount in small portions and settling each portion by gently tapping between additions. Place a glass wool plug between the reagent and absorber outlet to prevent loss of reagent dust.

10.2.2 *Carbon Dioxide Absorbers*—If a solid reagent is used for the retention of carbon dioxide, 7.5.2, fill the absorber,

[7] Tradenames of this reagent are Anhydrone or Dehydrite.

6.1.2, as described in 10.2.1. Place a layer or cap of desiccant in the outlet section of the container; it shall be the same as that used in the water absorber. This layer shall have a bulk volume not less than one fourth nor more than onethird of the combined volume of both reagents.

10.2.2.1 If a liquid absorbent is used, fill the inner tube of the Vanier bulb with the same desiccant used in the water absorber. If a solid absorbent is used, place a glass wool plug in the outlet section of the container to prevent loss of reagent dust.

10.2.3 *Guard Tube*—Pack a container, as described in 5.1.1, with equal volumes of the water absorbent and a solid carbon dioxide absorbent.

10.2.4 *Connections*—To ensure a closed system from the supply tank of oxygen to the guard tube at the end of the absorption train, it is recommended that all connections be glass-to-glass or glass-to-quartz butt joints with short lengths of flexible tubing as seals. The connection between the purification train and the combustion tube may be made by means of a rubber stopper or other suitable device. All connections shall be gas tight. No lubricant shall be used for making tubing connections in the absorption train.

10.3 *Conditioning of Apparatus.*

10.3.1 *Newly Packed Combustion Tube*—Burn a sample of RDF as described in 11.4 except that the products of combustion need not be fixed in a weighed absorption train.

10.3.2 *Used Combustion Tube*—After any extended shut down (one day or more) test the combustion train under procedure conditions, but without burning a sample, for 40 min with weighed absorbers connected. A variation of not more than 0.5 mg of both water and carbon dioxide absorbers shall be considered satisfactory (see Note 3).

10.3.3 *Absorption Train*—Condition freshly packed absorbers and guard tubes by burning a sample of RDF, as described in 11.4, except that the absorber weights need not be determined.

10.3.4 Make standard checks frequently, particularly when intermittent use of the combustion train is common or when any changes have been made in the system. Burn a standard substance of certified analysis, such as benzoic acid or sucrose as furnished by the National Bureau of Standards, as described in Section 11. A variation from the theoretical of not more than 0.07 % for hydrogen nor more than 0.30 % for carbon shall be considered satisfactory.

11. Procedure

11.1 After the combustion tube and absorbers have been conditioned as prescribed in Section 10, conduct the test as follows:

11.2 *Absorption Train*—Bring the water and carbon dioxide absorbers to room temperature near the balance for 15 to 20 min, vent momentarily to the atmosphere, wipe with a chamois or lint-free cloth in the areas where handled, and weigh the water and carbon dioxide absorbers to the nearest 0.1 mg.

11.3 *Sample*—After thoroughly mixing the RDF analysis sample to provide the best possible mix of heavy fines with milled fluff, weigh approximately 0.2 g to the nearest 0.1 mg of sample into a combustion boat.

NOTE 6—The final milling stage may produce a suitably mixed analysis sample without further mixing. It is important that no unwanted segregation of particle sizes take place. True representation of the analysis sample must be contained in the 200 mg used for the analysis.

11.4 *Sample Analysis*—With furnace 2 and 3 at specified temperatures and positioned as shown in Fig. 2, perform the following operations in rapid succession in the order listed:

11.4.1 If a conventional type of sample heating furnace is used for heating furnace 1, place it so that its left-hand edge is about 100 mm from the oxygen inlet end of the combustion tube.

11.4.2 Attach the weighed absorption train to the combustion tube.

11.4.3 Push the sample boat containing the weighed sample into the combustion tube to a point within approximately 20 mm from plug P_1 (see Fig. 2).

11.4.4 Close the tube and adjust the oxygen flow to a rate of 50 to 100 mL/min (standard temperature and pressure), being the same as used in blanking (see 10.3.2).

11.4.5 Apply full heat to heating Furnace 1 to bring it to an operating temperature of 850 to 900°C as rapidly as possible. Move the heater slowly toward the boat so that it completely covers the boat and is brought into contact with furnace 2 over a period of 10 to 20 min (Note 7). Allow it to remain in this position for an additional 5 to 10 min, and then shut off the heat and return the sample heater to its original position. Continue the flow of oxygen through the system for 10 min (Note 9), close the absorbers under a positive pressure of oxygen and detach them from the train. Remove the absorbers to the vicinity of the balance, allow them to cool to room temperature for 15 to 20 min, vent momentarily to the atmosphere, wipe them with a chamois or lint-free cloth in the areas handled, and finally weigh them to the nearest 0.1 mg. While the absorbers are cooling, it is recommended that the ash remaining in the combustion boat be examined for tracers of unburned carbon which, if present, will nullify the determination.

NOTE 7—Some variation in operating technique and heater manipulation may be permitted at the discretion of the analyst, provided that it is conducive to a gradual and controlled release of volatile matter. Conditions that lead to visible burning (flame combustion) of the sample shall be avoided.

NOTE 8—Since water may condense in the cooler outlet end of the combustion tube or in the inlet arm of the water absorber, the use of an external or internal heat-conducting device (a metal heat bridge) is recommended to prevent such condensation or promote reevaporation during this flushing period.

12. Calculation

12.1 Calculate the percentage of carbon in the analysis sample as follows:

$$\text{Carbon, \%} = \frac{A}{B} \times 27.289 \quad (1)$$

where:

A = increase in weight of CO_2 absorber, g,
B = grams of sample used, and
27.289 = percentage of carbon in CO_2.

12.2 Calculate the percentage of hydrogen in the analysis sample as follows (see Note 7).

$$\text{Hydrogen, \%} = \frac{C}{B} \times 11.19 \quad (2)$$

where:

C = increase in weight of water absorber, g,
B = grams of sample used, and
11.19 = percentage of hydrogen in water.

NOTE 9—The water absorbed in the water absorber includes not only water formed as a product of combustion, but also free water (moisture) in the sample and water of hydration of materials that may be contained in RDF.

13. Report

13.1 The results of the carbon and hydrogen analysis may be reported on any of a number of bases, differing from each other in the manner by which moisture is treated.

13.2 The numerical moisture value established by Test Method E 790 shall be used for converting carbon and hydrogen data on the as-determined basis to the dry basis as in Test Method E 791.

14. Precision and Bias [8]

14.1 *Precision*:

14.1.1 The standard deviations of individual determinations, in percent absolute, are as follows:

Typical Average Value, %	Within-Laboratory, %	Between-Laboratories, %
Carbon:		
40	0.5	1.6
Hydrogen:		
5.4	0.2	0.5

14.1.2 These precision estimates are based on an interlaboratory study conducted in accordance with Practice E 180.

14.2 *Bias*—The bias of this test method has not been determined because of the lack of a recognized standard reference material.

[8] Supporting data are available on loan from ASTM Headquarters. Request RR:E38-1000.

Designation: E 778 – 87 (Reapproved 1996)

Standard Test Methods for Nitrogen in the Analysis Sample of Refuse-Derived Fuel[1]

This standard is issued under the fixed designation E 778; the number immediately following the designation indicates the year of original adoption or, in the case of revision, the year of last revision. A number in parentheses indicates the year of last reapproval. A superscript epsilon (ϵ) indicates an editorial change since the last revision or reapproval.

1. Scope

1.1 These test methods cover the determination of total Kjeldahl nitrogen in (prepared analysis) samples of (solid forms) of refuse-derived fuel (RDF). The procedures measure free ammonia or ammonia formed from the conversion of organic nitrogenous compounds such as amino acids and proteins. However, the procedures may not convert the nitrogenous compounds of some wastes to ammonia. Examples of such compounds that may not be measured are nitro compounds, hydrozones, oxines, nitrates, semicarbazones, pyridines, and some refractory tertiary amines.

1.2 Two alternatives are described for the final determination of the ammonia, the Kjeldahl-Gunning Test Method and the Acid-Titration Test Method.

1.3 The analytical data from these test methods are to be reported as part of the ultimate analysis where ultimate analysis is requested.

1.4 These test methods may be applicable to any waste material from which a laboratory analysis sample can be prepared.

1.5 *This standard does not purport to address all of the safety concerns, if any, associated with its use. It is the responsibility of the user of this standard to establish appropriate safety and health practices and determine the applicability of regulatory limitations prior to use.* For specific precautionary statements see 8.4.1 and Section 9.

2. Referenced Documents

2.1 *ASTM Standards:*

D 1193 Specification for Reagent Water[2]

E 180 Practice for Determining the Precision of ASTM Methods for Analysis and Testing of Industrial Chemicals[3]

E 200 Practice for Preparation, Standardization, and Storage of Standard and Reagent Solutions for Chemical Analysis[3]

E 790 Test Method for Residual Moisture in Refuse-Derived Fuel Analysis Sample[4]

E 791 Test Method for Calculating Refuse-Derived Fuel Analysis Data from As-Determined to Different Bases[4]

E 829 Practice for Preparing Refuse-Derived Fuel (RDF) Laboratory Samples for Analysis[4]

3. Terminology

3.1 *Definitions of Terms Specific to This Standard:*

3.1.1 *refuse-derived fuel*—solid forms of refuse-derived fuels from which appropriate analytical samples may be prepared are defined as follows in *ASTM STP 832:*[5]

RDF-1—Wastes used as a fuel in as-discarded form with only bulky wastes removed.

RDF-2—Wastes processed to coarse particle size with or without ferrous metal separation.

RDF-3—Combustible waste fraction processed to particle sizes, 95 % passing 2-in. square screening.

RDF-4—Combustible waste fraction processed into powder form, 95 % passing 10-mesh screening.

RDF-5—Combustible waste fraction densified (compressed) into the form of pellets, slugs, cubettes, or briquettes.

4. Summary of Test Methods

4.1 The determination of nitrogen is made by either the Kjeldahl-Gunning Test Method (Section 11) or the Acid-Titration Test Method (Section 12). In both these methods the nitrogen in the sample is converted into ammonium salts by destructive digestion of the sample with a hot, catalyzed mixture of concentrated sulfuric acid and potassium sulfate. The salts are subsequently decomposed in a hot alkaline solution from which the ammonia is recovered by distillation, and finally determined by alkalimetric or acidimetric titration.

5. Significance and Use

5.1 The standard sample is available to producers and users of RDF as a method for determining the weight percent of nitrogen in the analysis sample.

5.2 Nitrogen is part of the ultimate analysis and can be used for calculation of combustion parameters.

6. Interferences and Limitations

6.1 Because of the nature of RDF, nitrogenous compounds may be present which will not readily be converted to ammonia by this test method (1.1). Modifications to the digestion of the

[1] These test methods are under the jurisdiction of ASTM Committee D-34 on Waste Management and are the direct responsibility of Subcommittee D34.08 on Thermal Treatment.

Current edition approved Aug. 28, 1987. Published October 1987. Originally published as E 778 – 81. Last previous edition E 778 – 81.

[2] *Annual Book of ASTM Standards*, Vol 11.01.

[3] *Annual Book of ASTM Standards*, Vol 15.05.

[4] *Annual Book of ASTM Standards*, Vol 11.04.

[5] *Thesaurus on Resource Recovery Terminology, ASTM STP 832*, ASTM, 1983, p. 72.

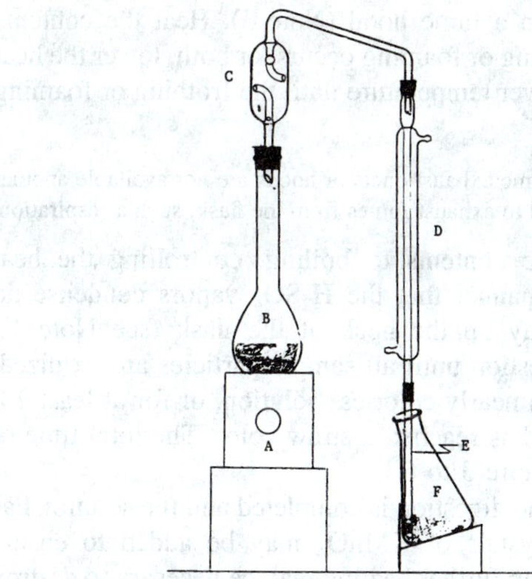

A = electric heater
B = Kjeldahl digestion flask
C = Kjeldahl connecting bulb
D = condenser
E = connecting tube
F = receiving flask

FIG. 1 Kjeldahl Distillation Apparatus

waste may enhance the conversion of these nitrogenous compounds to the ammonium salts.[6]

7. Apparatus

7.1 *Digestion Unit*—An electrical heater of approximately 500-W minimum rating or a gas burner of comparable capacity. Either type of heater shall be provided with adequate means of control to maintain digestion rates as described in 11.1 (Note 1). Commercially made, multiple-unit digestion racks provided with fume exhaust ducts may be used.

NOTE 1—If commercially made electrical heaters are used, auxiliary voltage control equipment, such as an autotransformer, may be needed to maintain the specified rates of digestion and distillation.

7.2 *Distillation Unit* (Fig. 1)—An electrical heater or gas burner as described in 7.1. Either type shall be provided with adequate means of control to maintain rates as described in 11.2. Commercially made, multiple-unit distillation racks provided with water-cooled glass or block tin condensers may be used.

7.3 *Condenser*, glass, water-cooled, having a minimum jacket length of 500 mm.

7.4 *Kjeldahl Digestion Flask*, of heat-resistant glass, having a capacity of 500 or 800 mL. Borosilicate glass has been found satisfactory for this purpose.

7.5 *Kjeldahl Connecting Bulb*, cylindrical type, 45 mm in diameter by 100 mm long, or larger, with curved inlet and outlet tubes.

7.6 *Receiving Flasks*—Erlenmeyer flask having a capacity of 250 or 300 mL.

7.7 *Connecting Tube*—Glass tubing approximately 10 mm in outside diameter by 200 mm in length.

7.8 *Pure Gum Rubber Tubing*.

8. Reagents

8.1 *Purity of Reagents*—Reagent grade chemicals shall be used in all tests. Unless otherwise indicated, it is intended that all reagents shall conform to the specifications of the Committee on Analytical Reagents of the American Chemical Society, where such specifications are available.[7] Other grades may be used, provided it is first ascertained that the reagent is of sufficiently high purity to permit its use without lessening the accuracy of the determination.

8.2 *Purity of Water*—Unless otherwise indicated, reference to water shall be understood to mean reagent water, Type II, conforming to Specification D 1193, prepared by the passage through an ion-exchange column containing a strongly acid cation resin in the hydrogen form.

8.3 *Potassium Sulfate* (K_2SO_4), crystals.

NOTE 2—Other satisfactory and permissible catalysts for the digestion, together with the quantities of K_2SO_4 required in their use, are as follows:

(1) Five grams of a mixture containing 32 parts by weight of K_2SO_4, 5 parts by weight of mercuric sulfate ($HgSO_4$), and 1 part by weight of selenium.

(2) Three-tenths gram of mercuric selenite ($HgSeO_3$) with 7 to 10 g of K_2SO_4.

(3) Three-tenths gram of cupric selenite dihydrate ($CuSeO_2{\cdot}2H_2O$) with 7 to 10 g of K_2SO_4. When this mixture is used, the addition of a sulfide to the alkali solution is not necessary.

8.4 *Mercury*, metal (see Note 2).

8.4.1 **Precaution**—Appropriate safety precautions should be used when handling and disposing of mercury and selenium compounds.

8.5 *Sulfuric Acid* (sp gr 1.84)—Concentrated sulfuric acid (H_2SO_4).

8.6 *Potassium Permanganate* ($KMnO_4$), crystals.

8.7 *Zinc*, mossy or granular.

8.8 *Alkali Solution*—Dissolve 8.0 g of potassium sulfide (K_2S) and 500 g of sodium hydroxide (NaOH) in water and dilute to 1 L. The use of appropriate amounts of sodium sulfide (Na_2S) or potassium hydroxide (KOH) may be substituted for the above, if desired (Note 2 (*c*)).

8.9 *Ethyl Alcohol* (95 %)—Ethyl alcohol conforming to Formula No. 30 or 2A of the U.S. Bureau of Internal Revenue. Methyl alcohol may be substituted.

8.10 *Sucrose*—National Bureau of Standards primary standard grade.

8.11 *Reagents Required for Kjeldahl-Gunning Test Method*:

8.11.1 *Methyl Red Indicator Solution* (0.4 to 1 g/L)—Dissolve 0.04 to 0.1 g of methyl red in 50 mL of 95 % ethyl alcohol or methyl alcohol and add 50 mL of water. Bromcresol green indicator solution of equal concentration may be used.

8.11.2 *Sodium Hydroxide, Standard Solution* (0.1 to 0.2 *N*)—Prepare and standardize a 0.1 to 0.2 *N* sodium hydroxide (NaOH) solution against a primary standard, as described in Practice E 200.

[6] Kolthoff, I. M., and Stenger, V. A., *Volumetric Analysis II*, Intersciences Publishers, Inc., New York, NY, pp. 173–176.

[7] "Reagent Chemicals, American Chemical Society Specifications," Am. Chemical Soc., Washington, DC. For suggestions on the testing of reagents not listed by the American Chemical Society, see "Analar Standards for Laboratory U.K. Chemicals," BDH Ltd., Poole, Dorset, and the "United States Pharmacopoeia".

8.11.3 *Sulfuric Acid, Standard Solution* (0.2 *N*)—Prepare and standardize a 0.2 *N* sulfuric acid (H_2SO_4) solution as described in Practice E 200.

8.12 *Reagents Required Only for Acid-Titration Test Method*:

8.12.1 *Boric Acid Solution* (50 g/L)—Dissolve 5 g of boric acid (H_3BO_3) in 100 mL of boiling water. Allow to cool to room temperature before use.

8.12.2 *Mixed Indicator Solution*—Prepare a solution containing 0.125 % methyl red and 0.083 % methylene blue in 95 % ethyl alcohol or in methyl alcohol. Prepare fresh solution at bimonthly intervals.

8.12.3 *Sulfuric Acid, Standard Solution* (0.1 to 0.2 *N*)—Prepare and standardize a 0.1 to 0.2 *N* sulfuric acid (H_2SO_4) solution. Hydrochloric acid (HCl) of similar concentration, as described in Practice E 200, may be substituted.

9. Precautions

9.1 Due to the origins of RDF in municipal waste, common sense dictates that precautions should be observed when conducting tests on the samples. Recommended hygienic practices include use of gloves when handling RDF; wearing dust masks (NIOSH-approved type), especially while milling RDF samples; conducting tests under a negative pressure hood when possible; and washing hands before eating or smoking.

9.2 The hot acidic and basic solutions in this procedure pose a significant potential hazard. Proper laboratory safety practices and equipment should be employed throughout this procedure.

10. Sampling

10.1 RDF products are frequently nonhomogeneous. For this reason significant care should be exercised to obtain a representative laboratory sample from the RDF lot to be characterized.

10.2 The sampling method for this procedure should be based on agreement between the involved parties.

10.3 The laboratory sample must be air-dried and particle size reduced to pass a 0.5-mm screen as described in Practice E 829. This procedure must be performed carefully to preserve the sample's representative characteristics, other than particle size, while preparing the analysis sample to be used in the procedures.

11. Procedure for Kjeldahl-Gunning Test Method

11.1 *Digestion of Sample.*

11.1.1 After thoroughly mixing the RDF analysis sample to provide the best possible mix of heavy fines with milled fluff, weigh approximately 1 g to the nearest 1 mg of sample into a weighing scoop.

11.1.2 Carefully transfer the sample into a 500 or 800-mL Kjeldahl flask containing 7 to 10 g of K_2SO_4 and 0.6 to 0.8 g of mercury (see Note 3).

11.1.3 Add 30 mL of H_2SO_4(sp gr 1.84) to the mixture by pouring it down the neck of the flask while rotating the flask to wash any sample adhering to the walls into the mixture. Swirl the contents of the flask several times to ensure thorough mixing and wetting of the sample.

11.1.4 Incline the flask at an angle of 45 to 60° on the digestion heater in a fume hood (Note 3). Heat the contents gradually. If frothing or foaming occurs, or both, lower the heat and digest at a lower temperature until the frothing or foaming ceases.

NOTE 3—When fume exhaust ducts or hoods are not available another method must be used to exhaust fumes from the flask, such as aspiration.

11.1.5 Heat the contents to boiling, controlling the heat input in such a manner that the H_2SO_4 vapors condense no more than halfway up the neck of the flask (see Note 1). Continue the digestion until all sample particles are oxidized, as evidenced by a nearly colorless solution, or for at least 2 h after the solution has reached a straw color. The total time of digestion will require 3 to 6 h.

11.1.6 When the digestion is completed and the solution has cooled, a few crystals of $KMnO_4$ may be added to ensure complete oxidation; further heating may be necesary to destroy the excess permanganate and decolorize the solution.

11.2 *Distillation of Digestate* (see Fig. 1).

11.2.1 Dilute the cooled digestion mixture to about 300 mL with water and remove any heat of dilution by cooling the flask under running water or by allowing it to stand until cool.

11.2.2 Accurately pipet 20.0 mL of 0.2 *N* H_2SO_4 into a 250 or 300-mL Erlenmeyer flask. Add 6 drops of methyl red or bromcresol green indicator solution.

11.2.3 Attach the glass connecting tube to the discharge end of the condenser, using a short piece of rubber tubing as a seal.

11.2.4 Incline the Erlenmeyer flask at a suitable angle and insert this tube so that the end is immersed well below the surface of the acid solution (see Fig. 1).

11.2.5 Add 1 to 2 g of granular zinc to the digestion mixture in the Kjeldahl flask (two or three small pieces, if mossy zinc is used), and slowly add 100 mL of alkali solution so that it forms a distinct layer under the acid solution. This may be accomplished by inclining the flask at an angle of 45 to 60° and pouring the alkali solution slowly down the neck of the flask. Failure to maintain discrete layers during the operation may lead to a fairly fast exothermic reaction and loss of ammonia.

11.2.6 Quickly connect the flask to the distilling condenser through the Kjeldahl connecting bulb and swirl the contents to promote thorough mixing.

NOTE 4—All connections must be air-tight so no loss of ammonia will be experienced.

11.2.7 Bring the contents of the Kjeldahl flask to a boil carefully in order to avoid bumping or foaming, or both, and distill the ammonia over into the acid solution in the Erlenmeyer flask.

11.2.8 Continue the distillation at a maximum rate of approximately 350 mL/h until 100 to 150 mL of distillate have been collected.

11.2.9 Discontinue the boiling, and remove the glass connecting tube from the condenser and Erlenmeyer flask. Rinse the tube with water, collecting the washings in the Erlenmeyer flask.

11.2.10 Titrate the excess acid in the Erlenmeyer flask to a methyl red or bromcresol green end point, using 0.1 to 0.2 *N* NaOH solution as the titrant.

11.3 *Blank Determination*—Run a blank determination in

the same manner as described in 11.1 and 11.2, using approximately 1 g of sucrose (weighed to the nearest 1 mg) as the sample material.

NOTE 5—A blank determination must be made with every series of analyses performed. The blank determination serves two principal functions:

(1) Since the principle of the method is ascertaining the amount of standardized acid being consumed in a reaction and a back-titration is necessary, the blank is a check on the concentration of the standard solutions.

(2) The blank serves as a correction for nitrogen from sources other than the sample.

12. Calculation

12.1 Calculate the percentage of nitrogen in the analysis sample as follows:

$$\text{Nitrogen, \%} = \frac{(B - A)\,N \times 0.014}{C} \times 100 \quad (1)$$

where:

A	=	millilitres of NaOH solution required for titration of the sample,
B	=	millilitres of NaOH solution required for titration of the blank,
N	=	normality of the NaOH solution,
C	=	grams of sample used, and
0.014	=	milliequivalent weight of nitrogen.

13. Procedure for Acid-Titration Test Method

13.1 *Digestion of Sample.*

13.1.1 Digest the sample as described in 11.1.

13.2 *Distillation of Digestate* (see Fig. 1).

13.2.1 Dilute the cooled digestion mixture to about 300 mL with water and remove any heat of dilution by cooling the flask under running water or by allowing it to stand until cool.

13.2.2 Add 20 mL of H_3BO_3 solution into a 250 or 300-mL Erlenmeyer flask and add 6 drops of mixed indicator solution.

13.2.3 Set up the distillation apparatus and distill as described in 11.2.3-11.2.9.

13.2.4 Titrate the ammonia collected in the Erlenmeyer flask containing the H_3BO_3 to the mixed indicator end point using 0.2 *N* H_2SO_4 as the titrant.

13.3 *Blank*—Run a blank determination in the same manner as described in 11.1 and 11.2, using approximately 1 g (weighed to the nearest 1 mg) of sucrose as the sample material (Note 5 (*2*)).

14. Calculation

14.1 Calculate the percent of nitrogen in the analysis sample as follows:

$$\text{Nitrogen, \%} = \frac{(A - B)\,N \times 0.014}{C} \times 100 \quad (2)$$

where:

A	=	millilitres of H_2SO_4 required for titration of the sample,
B	=	millilitres of H_2SO_4 required for titration of the blank,
N	=	normality of H_2SO_4,
C	=	grams of sample used, and
0.014	=	milliequivalent weight of nitrogen.

15. Report

15.1 The results of the nitrogen analysis may be reported on any number of bases, differing from each other in the manner by which moisture is treated.

15.2 The numerical moisture value established by Test Method E 790, shall be used for converting nitrogen data from the as-determined basis to the dry basis as in Test Method E 791.

16. Precision and Bias [8]

16.1 *Precision*:

16.1.1 The standard deviations of individual determinations, in percent absolute, are as follows:

Typical Average Value, %	Within-Laboratory, %	Between-Laboratories, %
0.6	0.04	0.05

16.1.2 These precision estimates are based on an interlaboratory study conducted in accordance with Practice E 180.

16.2 *Bias*—The bias of this test method can not be determined due to the lack of a recognized standard reference material.

[8] Supporting data are available on loan from ASTM Headquarters. Request RR:E38-1000.

Designation: E 790 – 87 (Reapproved 1996)

Standard Test Method for Residual Moisture in a Refuse-Derived Fuel Analysis Sample[1]

This standard is issued under the fixed designation E 790; the number immediately following the designation indicates the year of original adoption or, in the case of revision, the year of last revision. A number in parentheses indicates the year of last reapproval. A superscript epsilon (ε) indicates an editorial change since the last revision or reapproval.

1. Scope

1.1 This test method covers the measurement of the residual moisture in an analysis sample of RDF. It is used to calculate to the dry basis other determinations performed on the analysis sample. It is used with the air-dry moisture results to calculate total moisture (Note 1). The total moisture is used to calculate as-received values or other analyses performed on the sample.

NOTE 1—In some instances RDF moisture may change during the size-reduction steps of the analysis sample preparation procedure. This moisture change, unless suitable corrections are made, will affect the accuracy of the total moisture value as calculated from the air-dry and residual moisture results.

1.2 The values stated in acceptable metric units are to be regarded as standard. The values given in parentheses are for information only.

1.3 *This standard does not purport to address all of the safety concerns, if any, associated with its use. It is the responsibility of the user of this standard to establish appropriate safety and health practices and determine the applicability of regulatory limitations prior to use.* For more specific precautionary information see Section 7.

2. Referenced Documents

2.1 *ASTM Standards:*

D 3173 Test Method for Moisture in the Analysis Sample of Coal and Coke[2]

E 180 Practice for Developing Precision of ASTM Methods for Analysis and Testing of Industrial Chemicals[3]

3. Terminology

3.1 *Definitions of Terms Specific to This Standard:*

3.1.1 *air drying*—a process of partial drying of RDF to bring its moisture content near to equilibrium with the atmosphere in which further reduction, division, and characterization of the sample are to take place. In order to bring about the equilibrium, the RDF is usually subjected to drying under controlled temperature conditions ranging from 30 to 40°C.

3.1.2 *analysis sample*—the final subsample prepared from the air-dried laboratory sample but reduced by passing through a mill with a 0.5-mm (0.02-in.) size or smaller final screen.

3.1.3 *bias*—a systematic error that is consistently negative or consistently positive. The mean of errors resulting from a series of observations that does not tend towards zero.

3.1.4 *gross sample*—a sample representing one lot and composed of a number of increments on which neither reduction nor division has been performed.

3.1.5 *laboratory sample*—a representative portion of the gross sample received by the laboratory for analysis.

3.1.6 *lot*—a large designated quantity (greater than the quantity of the final sample) of RDF which can be represented by a properly selected gross sample.

3.1.7 *precision*—a term used to indicate the capability of a person, an instrument, or a method to obtain reproducible results; specifically, a measure of the random error as expressed by the variance, the standard error, or a multiple of the standard error.

3.1.8 *refuse-derived fuels*—solid forms of refuse-derived fuels from which appropriate analytical samples may be prepared are defined as follows in *ASTM STP 832*.[4]

RDF-1—Wastes used as a fuel in as-discarded form with only bulky wastes removed.

RDF-2—Wastes processed to coarse particle size with or without ferrous metal separation.

RDF-3—Combustible waste fraction processed to particle sizes, 95 % passing 2-in. square screening.

RDF-4—Combustible waste fraction processed into powder form, 95 % passing 10-mesh screening.

RDF-5—Combustible waste fraction densified (compressed) into the form of pellets, slugs, cubettes, or briquettes.

3.1.9 *representative sample*—a sample collected in such a manner that it has characteristics equivalent to the lot sample.

3.1.10 *sample division*—the process of extracting a smaller sample from a sample so that the representative properties of the larger sample are retained. During this process it is assumed that no change in particle size or other characteristics occurs.

3.1.11 *sample preparation*—the process that includes drying, size reduction, division, and mixing of a laboratory sample for the purpose of obtaining an unbiased analysis sample.

---

[1] This test method is under the jurisdiction of ASTM Committee D-34 on Waste Management and is the direct responsibility of Subcommittee D34.08 on Thermal Treatment.

Current edition approved Aug. 28, 1987. Published October 1987. Orginally published as E 790 – 81. Last previous edition E 790 – 81.

[2] *Annual Book of ASTM Standards*, Vol 05.05.

[3] *Annual Book of ASTM Standards*, Vol 15.05.

[4] *Thesaurus on Resource Recovery Terminology, ASTM STP 832*, ASTM, 1983, p. 72.

3.1.12 *sample reduction*—the process whereby sample particle size is reduced without change in sample weight.

3.1.13 *significant loss*—any loss that introduces a bias in final results that is of appreciable importance to concerned parties.

4. Summary of Test Method

4.1 This test method is based on the loss in weight of an air-dried analysis sample of RDF under rigidly controlled conditions of temperature, time, and air flow.

4.2 The total moisture is calculated from the loss or gain in air drying and the residual moisture as determined by this test method.

5. Significance and Use

5.1 The treatment of the sample as specified herein is intended for the purpose of determining the residual moisture present in an analysis sample of RDF.

5.2 The residual moisture value is used to convert as-determined analyses such as gross heating value, sulfur, and ash to a dry sample basis.

6. Apparatus

6.1 *Drying Oven*:

6.1.1 *Referee Type*—The oven shall be so constructed as to have a uniform temperature within the specimen chamber, have a minimum excess air volume, and be capable of constant temperature regulation at 107 ± 3°C. Provision shall be made for renewing the preheated air in the oven at the rate of two to four times a minute, with the intake air dried by passing it through a desiccant. An oven similar to the one illustrated in Fig. 1 of Test Method D 3173 is suitable.

6.1.2 *Routine Type*—A drying oven of either the mechanical or natural circulation type which is capable of constant uniform temperature within the specimen chamber regulated at 107 ± 3°C.

NOTE 2—Either type of oven may be used for routine determinations. However, the referee-type oven shall be used to resolve differences between determinations.

6.2 *Containers*—A convenient form that allows the ash determination to be made on the same sample is a porcelain capsule 22 mm in depth and 44 mm in diameter or a fused silica capsule of similar shape. This shall be used with a well-fitting flat aluminum cover. Platinum crucibles or glass capsules with ground-glass caps may also be used. They should be as shallow as possible consistent with convenient handleability.

6.3 *Analytical Balance*, with 0.1 mg sensitivity.

6.4 *Analysis Sample Containers*—Heavy (minimum 4 mil), vapor-impervious bags, properly sealed; or noncorroding cans, glass jars, or plastic bottles with air-tight sealing covers to store RDF samples for analysis. Containers shall be checked for suitability by measuring weight loss or gain of the sample and container stored for 1 week under ambient laboratory conditions. The weight loss or gain should be less than 0.5 % of the sample weight stored in the container.

7. Precautions

7.1 Due to the origins of RDF in municipal waste, common sense dictates that some precautions should be observed when conducting tests on the samples. Recommended hygienic practices include use of gloves when handling RDF; wearing dust masks (NIOSH-approved type), especially when shredding RDF samples; conducting tests under negative pressure hood when possible; and washing hands before eating or smoking.

7.2 Laboratory sample handling shall be performed by trained personnel. All operations shall be done rapidly as possible to avoid sample moisture changes due to atmospheric exposure.

7.3 Since heavy fine particles tend to segregate rapidly in the RDF analysis sample, the analyst should exercise care to assure that the analysis sample is well-mixed prior to performing this determination.

7.4 When the residual moisture is to be used for the determination of total moisture, special care shall be taken to avoid any change in sample moisture between the completion of air drying and analysis for residual moisture. It is recommended that the delay between sample preparation and the determination of residual moisture be a maximum of 72 h.

8. Sampling [5]

8.1 RDF products are frequently nonhomogeneous. For this reason, significant care should be exercised to obtain a representative laboratory sample from the RDF lot to be characterized.

8.2 The sample method for this procedure should be based on agreement between the involved parties.

8.3 The laboratory sample must be air-dried and particle size reduced to pass a 0.5-mm screen for this analysis. This procedure must be performed carefully to preserve the sample's representative characteristics (other than particle size) while preparing the analysis sample to be used in this procedure.

9. Procedure

9.1 Heat the empty containers and covers under the conditions at which the sample is to be dried, place the stopper or cover on the container, cool over a desiccant for about 15 to 20 min, and weigh. Mix the sample, if necessary, and dip out with a spoon or spatula from the sample bottle approximately 1 g of the sample. Put the sample quickly into the container, cover, and weigh at once.

NOTE 3—If weighing bottles with air-tight covers are used, it may not be necessary to preheat the moisture analysis container nor to desiccate it after drying.

9.2 Remove the cover and place in a desiccator. Quickly place the uncovered container into an oven preheated to 107 ± 3°C through which is passed a current of dry air. Close the oven at once and heat for 1 hr. Open the oven, remove, cover the container quickly, and cool in a desiccator over desiccant. Weigh the sample and container as soon as cooled to room temperature.

[5] ASTM Subcommittee E38.01 is currently in the process of developing procedures for sampling RDF and the preparation of an analysis sample. The chairman of E38.01 should be contacted for details.

10. Calculation (see Note 1)

10.1 Calculate the percent residual moisture, R, in the analysis sample as follows:

$$R = \frac{S - B}{S} \times 100 \quad (1)$$

where:
S = grams of analysis sample used, and
B = grams of sample after heating.

10.2 Calculate the percent total moisture in the laboratory sample, as follows:

$$M = \frac{R\,(100 - A)}{100} + A \quad (2)$$

where:
R = residual moisture, %, and
A = air dry loss determined during preparation of the analysis sample, %.

10.3 To convert other parameters determined on the analysis sample, such as ash, sulfur, and gross calorific value, to a dry sample basis, the following equation can be used:

$$P_{dry} = \frac{P_{ad}\,(100)}{100 - R} \quad (3)$$

where:
P_{ad} = parameter, %" as-determined" on the analysis sample,
R = residual moisture, % (see 10.1), and
P_{dry} = parameter, % expressed on a dry sample basis.

11. Precision and Bias

11.1 *Precision*:

11.1.1 The standard deviations of individual determinations, in percent absolute are as follows:

Typical Average Value	Within Laboratories	Between Laboratories
2.5–4.5 %	0.15 %	0.50 %

11.1.2 The above precision estimates are based on an interlaboratory study conducted in accordance with Practice E 180.

11.2 *Bias*—The bias of this test method has not been determined due to a lack of a recognized standard reference material.

Designation: E 791 – 90 (Reapproved 1996)

Standard Test Method for Calculating Refuse-Derived Fuel Analysis Data from As-Determined to Different Bases[1]

This standard is issued under the fixed designation E 791; the number immediately following the designation indicates the year of original adoption or, in the case of revision, the year of last revision. A number in parentheses indicates the year of last reapproval. A superscript epsilon (ϵ) indicates an editorial change since the last revision or reapproval.

1. Scope

1.1 This test method gives equations to enable analytical data from the application of RDF analyses procedures to be expressed on various different bases in common use. Such bases are: as-received; dry; dry, ash-free; and others (see 2.1.12).

1.2 *This standard does not purport to address all of the safety concerns, if any, associated with its use. It is the responsibility of the user of this standard to establish appropriate safety and health practices and determine the applicability of regulatory limitations prior to use.*

2. Terminology

2.1 *Definitions of Terms Specific to This Standard:*

2.1.1 *ash*—inorganic residue remaining after ignition of combustible substances, determined by definite prescribed methods. Ash may not be identical, in composition or quantity, with the inorganic substances present in the material before ignition.

2.1.2 *as-determined basis*—experimental data obtained from the analysis sample of RDF. These data represent the numerical values obtained for a particular moisture or ash content, or both, in the sample at the time of measurement.

2.1.3 *air drying*—a process of partial drying of RDF to bring its moisture content near to equilibrium with the atmosphere in which further reduction, division, and characterization of the sample are to take place. In order to bring about this equilibrium, the RDF is usually subjected to drying under controlled temperature conditions ranging from 30 to 40°C.

2.1.4 *air dry loss*—the decrease in mass presumed to be moisture of a sample due to air drying.

2.1.5 *as-received basis*—experimental data calculated to the moisture condition of the sample as it arrived in the laboratory and before any laboratory processing or conditioning. The total moisture value that is calculated from the air dry loss and residual moisture value is used to convert data from the dry basis to the as-received basis (see *dry ash-free basis*).

2.1.6 *dry ash-free basis*—the experimental data calculated to a theoretical base of no moisture or ash associated with the sample. Numerical values (air-dry loss, residual moisture values, and ash content) are used for converting the as-determined data to a moisture and ash-free basis.

2.1.7 *dry basis*—the experimental data calculated to a theoretical base of no moisture associated with the sample. The numerical value (residual moisture value) is used for converting the as-determined data to a dry basis.

2.1.8 *fixed carbon*—the ash-free carbonous material that remains after volatile matter is driven off during the proximate analysis of a dry sample.

2.1.9 *gross calorific value* (gross heat of combustion at constant volume) Q_v (gross)—the heat produced by combustion of a unit quantity of solid fuel, at constant volume, in an oxygen bomb calorimeter under specified conditions such that all water in the products remains in liquid form.

NOTE 1—The conditions are: initial oxygen pressure of 20 to 40 atm (2 to 4 MPa), initial and final temperatures between 68 and 95°F (20 and 35°C).

NOTE 2—The gross calorific value is closely related to the internal energy of combustion for the same reaction at constant standard temperature and pressure (Δυ comb.). It is of opposite sign and differs by a small amount due to energy effects resulting from compression, temperature differences, and solution effects that vary with combustion conditions. Because of the variation of conditions allowed in Note 1, the definition of gross calorific value does not lead to a unique value for any given fuel. However, the specified conditions limit the possible value to a narrow range for which approximate limits can be calculated for a given fuel.

2.1.10 *higher heating value*—synonym for *gross calorific value*.

2.1.11 *proximate analysis*—the determination, by prescribed methods, of moisture, volatile matter, fixed carbon (by difference), and ash. Unless otherwise specified, the term proximate analysis does not include determinations of chemical elements or any determinations other than those named.

2.1.12 *refuse-derived fuels*—solid forms of refuse-derived fuels from which appropriate analytical samples may be prepared are defined as follows in *ASTM STP 832*:[2]

RDF-1—Wastes used as a fuel in as-discarded form with only bulky wastes removed.

RDF-2—Wastes processed to coarse particle size with or without ferrous separation.

RDF-3—Combustible waste fraction reduced to particle

[1] This test method is under the jurisdiction of ASTM Committee D-34 on Waste Management and is the direct responsibility of Subcommittee D34.08 on Thermal Treatment.

Current edition approved June 1, 1990. Published August 1990. Originally published as E 791 – 81. Last previous edition E 791 – 87.

[2] *Thesaurus on Resource Recovery Terminology, ASTM STP 832*, ASTM, 1983, p. 72.

sizes, 95 % passing 2 in. square screening.

RDF-4—Combustible waste fraction processed into powder form, 95 % passing 10-mesh screening.

RDF-5—Combustible waste fraction densified (compressed) into the form of pellets, slugs, cubettes, or briquettes.

2.1.13 *residual moisture*—the moisture content remaining in an RDF sample after it has been milled down to an analysis sample. Prior to milling, the RDF sample should have been subjected to either a total moisture determination (single stage), or an air drying procedure.

2.1.14 *total moisture*—the weight loss resulting from drying a sample to constant weight in an oven usually maintained between 103 and 107°C.

2.1.15 *ultimate analysis*—the determination of the percentages of carbon, hydrogen, sulfur, nitrogen, chlorine, ash, and oxygen in a dry sample. The percentage of oxygen may be obtained by calculating the difference between 100 % and the other determined elemental analyses.

2.1.16 *volatile matter*—those products, exclusive of moisture, given off by a material as gas or vapor, determined by definite prescribed methods which may vary according to the nature of the material.

2.2 *Symbols:Symbols:*

2.2.1 The symbols used in this test method are as follows:

M = moisture, weight %,
M_{ar} = moisture as-received (total moisture), weight %,
M_{ad} = moisture as-determined (residual moisture, weight %,
ADL = air-dry loss, weight %,
P = any analysis parameter listed in 4.1, weight % (except gross calorific value in Btu/lb),
H = hydroxygen, weight %,
O = oxygen, weight %, and
A = ash, weight %.

2.2.2 Subscripts used in this test method are as follows:

ad = as-determined,
ar = as-received,
d = dry, and
daf = dry, ash-free (equivalent to moisture and ash-free, maf).

3. Significance and Use

3.1 This test method is available to producers and users of RDF to use in converting laboratory data from one basis to another.

4. Applicable Parameters

4.1 The calculation procedures defined in 6.1.3, 6.2.2, and 6.3.2 are applicable to the following analysis parameters when expressed as a weight percent (except gross calorific value as Btu/lb).

4.1.1 Ash,
4.1.2 Carbon,
4.1.3 Chloride, water-soluble,
4.1.4 Chlorine, total,
4.1.5 Calorific value (gross),
4.1.6 Fixed carbon,
4.1.7 Nitrogen,
4.1.8 Sulfur, and
4.1.9 Volatile matter.

5. Sampling

5.1 Sampling techniques are not directly applicable to this test method. However, sampling procedures are identified in the respective methods of analyses.

6. Methods for Calculating Data

6.1 Converting from the as-determined analysis sample basis to the as-received basis:

6.1.1 *Moisture*:

$$M_{ar} = [M_{ad} \times (100 - ADL)/100] + ADL \quad (1)$$

6.1.2 *Hydrogen and Oxygen*—Inasmuch as hydrogen and oxygen values may be reported on the basis of containing or not containing the hydrogen and oxygen in water (moisture) associated with the sample, alternate conversion procedures are defined as follows:

6.1.2.1 Hydrogen and oxygen reported include hydrogen and oxygen in water:

$$H_{ar} = \left[(H_{ad} - 0.1119\, M_{ad}) \times \frac{100 - M_{ar}}{100 - M_{ad}}\right] + 0.119\, M_{ar} \quad (2)$$

$$O_{ar} = \left[(O_{ad} - 0.8881\, M_{ad}) \times \frac{100 - M_{ar}}{100 - M_{ar}}\right] + 0.881\, M_{ar} \quad (3)$$

where:
0.1119 = ratio of the weight of hydrogen in water to the molecular weight of water, and
0.8881 = ratio of the weight of oxygen in water to the molecular weight of water.

6.1.2.2 Hydrogen and oxygen reported do not include hydrogen and oxygen in water:

$$H_{ar} = (H_{ad} - 0.1119\, M_{ad}) \times \frac{100 - M_{ar}}{100 - M_{ad}} \quad (4)$$

$$O_{ar} = (O_{ad} - 0.881\, M_{ad}) \times \frac{100 - M_{ar}}{100 - M_{ad}} \quad (5)$$

6.1.3 *Other Parameters*—The following equation is applicable to all parameters, *P*, listed in 4.1:

$$P_{ar} = P_{ad} \times \frac{100 - M_{ar}}{100 - M_{ad}} \quad (6)$$

6.2 Converting from the as-determined analysis sample basis to the dry basis:

6.2.1 *Hydrogen and Oxygen*:

$$H_d = (H_{ad} - 0.1119\, M_{ad}) \times \frac{100}{100 - M_{ad}} \quad (7)$$

$$O_d = (O_{ad} - 0.881\, M_{ad}) \times \frac{100}{100 - M_{ad}} \quad (8)$$

6.2.2 *Other Parameters*—The following equation is applicable to all parameters, *P*, listed in 4.1:

$$P_d = P_{ad} \times 100/(100 - M_{ad}) \quad (9)$$

6.3 Converting from the as-determined analysis sample basis to the dry, ash-free basis (see Note Note 3).

6.3.1 *Hydrogen and Oxygen*:

$$H_{daf} = (H_{ad} - 0.1119\,M_{ad}) \times \frac{100}{100 - M_{ad} - A_{ad}} \qquad (10)$$

$$O_{daf} = (O_{ad} - 0.8881\,M_{ad}) \times \frac{100}{100 - M_{ad} - A_{ad}} \qquad (11)$$

6.3.2 *Other Parameters*—The following equation is applicable to all parameters, P, listed in 4.1:

$$P_{daf} = P_{ad} \times \frac{100}{100 - M_{ad} - A_{ad}} \qquad (12)$$

NOTE 3—It may not always be practical to report some parameters on a dry, ash-free basis due to oxidation of constituents as well as some elements being retained in the ash. The data calculated are to a theoretical base of no moisture or ash present in the sample.

7. Conversion Chart

7.1 To convert any of the analysis values for the parameters listed in 4.1 from one basis to another, multiply the given value by the value shown in the appropriate column in Table 1.

7.2 Hydrogen and oxygen on the as-determined basis include hydrogen and oxygen in free water (moisture) associated with the analysis sample. However, hydrogen and oxygen values reported on other moisture-containing bases may be reported either as containing or not containing the hydrogen and oxygen in water (moisture) reported on that basis. Conversion procedures are shown in Table 2.

8. Sample Calculations

8.1 An example of analysis data tabulated for a hypothetical RDF on various bases is shown in Table 3.

9. Reporting of Results

9.1 Report the results of the proximate and ultimate analysis based on the example shown in Table 3.

9.2 To avoid ambiguity and to provide a means for conversion of data to other than the reported basis, it is essential that an appropriate moisture and ash content be given in the data report.

9.3 If hydrogen or oxygen data, or both, are reported on an as-received basis (or any other moisture basis), a footnote or some other means shall be employed in the report to indicate whether the values reported do or do not include hydrogen and oxygen in the moisture associated with the sample.

TABLE 1 Conversion Chart

NOTE 1— To convert any of analyses values for the parameters listed in 4.1 from one basis to another, multiply the given value by the value shown in the appropriate column.

To convert from / Multiply by	Conversion Factor: As-Determined (ad)	As-Received (ar)	Dry (d)	Dry, Ash-Free (daf)
As-determined (ad)	. . .	$\frac{100 - M_{ar}}{100 - M_{ad}}$	$\frac{100}{100 - M_{ad}}$	$\frac{100}{100 - (M_{ad} - A_{ad})}$
As-received (ar)	$\frac{100 - M_{ad}}{100 - M_{ar}}$	. . .	$\frac{100}{100 - M_{ar}}$	$\frac{100}{100 - (M_{ar} - A_{ar})}$
Dry (d)	$\frac{100 - M_{ad}}{100}$	$\frac{100 - M_{ar}}{100}$	. . .	$\frac{100}{100 - A_d}$
Dry, ash-free (daf)	$\frac{100 - (M_{ad} - A_{ad})}{100}$	$\frac{100 - (M_{ar} - A_{ar})}{100}$	$\frac{100 - A_d}{100}$	. . .

TABLE 2 Procedures for Converting As-Determined Values to Other Bases

Parameter	As-Determined	As Received: H_{ar} and O_{ar} include H and O in M_{ar}	As Received: H_{ar} and O_{ar} do not include H and O in M_{ar}	Dry Basis	Dry, Ash-Free Basis
P (ash, carbon, nitrogen, sulfur, chlorine)	no corrections (see standard method)	$P_{ar} = P_{ad} \times \left(\frac{100 - M_{ar}}{100 - M_{ad}}\right)$	. . .	$P_d = P_{ad} \times \left(\frac{100}{100 - M_{ad}}\right)$	$P_{daf} = P_{ad} \times \left(\frac{100}{100 - M_{ad} - A_{ad}}\right)$
H (hydrogen)	no corrections (see standard method)	$H_{ar} = \left[(H_{ad} - 0.1119\, M_{ad}) \times \left(\frac{100 - M_{ar}}{100 - M_{ad}}\right)\right] + 0.1119\, M_{ar}$	$H_{ar} = (H_{ad} - 0.1119\, M_{ad}) \times \left(\frac{100 - M_{ar}}{100 - M_{ad}}\right)$	$H_d = (H_{ad} - 0.1119\, M_{ad}) \times \left(\frac{100}{100 - M_{ad}}\right)$	$H_{daf} = (H_{ad} - 0.1119\, M_{ad}) \times \left(\frac{100}{100 - M_{ad} - A_{ad}}\right)$
O (oxygen)	$O_{ad} = 100 - (A_{ad} + C_{ad} + H_{ad} + N_{ad} + S_{ad} + Cl_{ad})$	$O_{ar} = \left[(O_{ad} - 0.8881\, M_{ad}) \times \left(\frac{100 - M_{ar}}{100 - M_{ad}}\right)\right] + 0.8881\, M_{ar}$	$O_{ar} = (O_{ad} - 0.8881\, M_{ad}) \times \left(\frac{100\text{ b} - M_{ar}}{100 - M_{ad}}\right)$	$O_d = (O_{ad} - 0.8881\, M_{ad}) \times \left(\frac{100}{100 - M_{ad}}\right)$ or $O_d = 100 - (A_d + C_d + H_d + N_d + S_d + Cl_d)$	$O_{daf} = (O_{ad} - 0.8881\, M_{ad}) \times \left(\frac{100}{100 - M_{ad} - A_{ad}}\right)$

TABLE 3 Standard Reporting Form Analyses Data of RDF

	As-Determined	As-Received: H_{ar} and O_{ar} include H and O in sample M_{ar}	As-Received: H_{ar} and O_{ar} do not include H and O in sample M_{ar}	Dry Basis	Dry, Ash-Free Basis
Air dry loss, %	18.69	...	...	...	...
Residual moisture (M_{ad}), %	2.15	...	...	...	...
Total moisture (M_{ar}), %	...	20.44	20.44	...	...
Ash, %	20.68	16.82	16.82	21.14	...
Volatile matter, %	62.29	50.65	50.65	63.66	80.73
Fixed carbon, %	14.88	12.09	12.09	15.20	19.27
Total, proximate	100.00	100.00	100.00	100.00	100.00
Total, moisture, %	...	...	20.44	...	...
Ash, %	20.68	16.82	16.82	21.14	...
Carbon, %	39.66	32.25	32.25	40.53	51.40
Hydrogen, %	5.12	6.25	3.97	4.99	6.32
Nitrogen, %0.60	0.60	0.49	0.49	0.61	0.78
Chlorine, %	0.30	0.24	0.24	0.31	0.39
Sulfur, %0.15	0.15	0.12	0.12	0.15	0.19
Oxygen, %	33.49	43.83	25.68	32.28	40.93
Total, ultimate	100.00	100.00	100.00	100.00	100.00
Gross heating value, Btu/lb	6868	5584	5584	7019	8900
Chloride, water soluble, %	0.19	0.15	0.15	0.19	0.25

Designation: E 828 – 81 (Reapproved 1997)

Standard Test Method for Designating the Size of RDF-3 From its Sieve Analysis[1]

This standard is issued under the fixed designation E 828; the number immediately following the designation indicates the year of original adoption or, in the case of revision, the year of last revision. A number in parentheses indicates the year of last reapproval. A superscript epsilon (ε) indicates an editorial change since the last revision or reapproval.

1. Scope

1.1 This test method of designating the size of refuse-derived fuel from its sieve analysis is applicable to the classified light fraction (RDF-3) of shredded municipal or industrial waste materials less than 0.15 m (6 in.) in size.

1.2 The values stated in acceptable metric units are to be regarded as standard. The values given in parentheses are for information only.

1.3 *This standard does not purport to address all of the safety concerns, if any, associated with its use. It is the responsibility of the user of this standard to establish appropriate safety and health practices and determine the applicability of regulatory limitations prior to use.* For more specific precautionary information see Section 7.

2. Referenced Documents

2.1 *ASTM Standards:*

D 2234 Test Method for Collection of a Gross Sample of Coal[2]

E 11 Specification for Wire-Cloth Sieves for Testing Purposes[3]

E 177 Practice for Use of the Terms Precision and Bias in ASTM Test Methods[3]

3. Terminology

3.1 *Definitions:*

3.1.1 *air drying*—a process of partial drying of RDF-3 to bring its moisture content near to equilibrium with the atmosphere in the room in which the sieving is to take place.

3.1.2 *gross sample*—a sample representing a lot of RDF and composed of a number of increments on which neither reduction nor division has been performed.

3.1.3 *laboratory sample*—a representative portion of the gross sample delivered to the laboratory for further analysis.

3.1.4 *lot*—a large designated quantity of RDF-3.

3.1.5 *representative sample*—a sample collected in such a manner that it has characteristics equivalent to the material being sampled.

3.1.6 *sample division*—the process of extracting a smaller sample from a gross sample wherein the representative properties of the large sample are retained.

3.2 *Definitions of Terms Specific to This Standard:*

3.2.1 *refuse-derived fuel (RDF-3)*—a shredded fuel derived from municipal solid waste (MSW) that has been processed to remove metal, glass, and other inorganics. This material has a particle size such that 95 weight % passes through a 2-in. square-mesh screen.

NOTE 1—Other refuse-derived fuel may be classified as follows:

RDF-1—Wastes used as a fuel in as-discarded form.
RDF-2—Wastes processed to coarse particle size with or without ferrous metal separation.
RDF-4—Combustible waste processed into powder form, 95 weight % passing 10-mesh screening.
RDF-5—Combustible waste densified (compressed) into the form of pellets, slugs, cubettes, or briquettes.
RDF-6—Combustible waste processed into liquid fuel.
RDF-7—Combustible waste processed into gaseous fuel.

4. Summary of Test Method

4.1 This test method covers the separation of an RDF sample into defined size fractions and expressing those fractions as a weight percent of an air-dried sample.

5. Significance and Use

5.1 The purpose of this test method is to provide a means of designating the size classification of RDF-3 for use by consumers and producers of RDF-3.

6. Apparatus

6.1 *Sieves*:

6.1.1 Use sieves conforming to Specification E 11. For recommended sizes see Table 1.

6.1.1.1 For RDF-3 and larger than 50 mm (2 in.) screens having rectangular frames 0.6 to 0.7 m^2 (6 to 8 ft^2) sieve area are satisfactory.

6.1.1.2 For RDF-3 50 mm (2 in.) or smaller, rectangular frames having 2 to 4 ft^2 (0.2 to 0.4 m^2) sieve area are satisfactory.

6.1.1.3 For RDF-3 smaller than 0.01 m (0.5 in.), circular sieves 0.3 m (12 in.) or 0.2 m (8 in.) in diameter are satisfactory.

6.2 *Sieving Devices*:

6.2.1 Hand sieving is permissible.

6.2.2 Sieving machines that provide the necessary agitation and tumbling action may be used. See Annex A1 for recommended screen sizes and machines.

---

[1] This test method is under the jurisdiction of ASTM Committee D-34 on Waste Management and is the direct responsibility of Subcommittee D34.08 on Thermal Treatment.

Current edition approved July 31, 1981. Published February 1982.

[2] *Annual Book of ASTM Standards*, Vol 05.05.

[3] *Annual Book of ASTM Standards*, Vol 14.02.

TABLE 1 Recommended Sieve Sizes (ASTM E11 – 70)

Standard (mm)	Alternative (in. or mesh)
For screening RDF-3 the following screen series is recommended:	
100 mm	4 in.
50 mm	2 in.
25 mm	1 in.
12.5 mm	½in.
6.3 mm	¼in.
3.35 mm	No. 6
1.70 mm	No. 12
850 μm	No. 20
425 μm	No. 40
The following intermediate screen sizes may be used as needed:	
Standard (mm)	Alternative (in. or mesh)
75 mm	3 in.
37.5 mm	1.5 in.
19.0 mm	¾in.
9.5 mm	⅜in.
4.75 mm	No. 4
2.36 mm	No. 8
1.18 mm	No. 16
600μ m	No. 30

6.3 *Balance* (laboratory sample), having sufficient capacity to weigh the sample and container with a sensitivity of 0.5 g in 1000 g.

7. Precautions

7.1 Due to the origins of RDF-3 in municipal waste, common sense dictates that some precautions be observed when conducting tests on the samples. Recommended hygienic practices include use of gloves when handling RDF-3; wearing masks (NIOSH-approved type), especially while shredding RDF-3 samples; conducting tests under a negative pressure hood when possible; and washing hands before eating or smoking.

8. Sampling

8.1 Collect increments regularly and systematically so that the entire quantity of RDF sampled will be representative proportionately in the gross sample, and with such frequency that a gross sample of the required amount shall be collected. No sampling procedure shall be used that alters the particle size distribution.

8.2 Establish the sampling procedures to be used, the number and size of samples required to obtain a representative sample, and the method of division of the gross sample into the laboratory sample in accordance with an agreement between purchaser and supplier.

NOTE 2—The statistical methods described in Test Method D 2234 may be applicable in determining the number and size of samples.

8.3 Division of the gross sample into the laboratory sample may be done by coning and quartering, riffling, or by other appropriate method.

8.4 The sample shall be approximately 2 kg (4.4 lb) in weight.

8.5 Air-dry the sample in a ventilated drying oven to constant weight at 10 to 15°C above the ambient temperature. (Calculate the loss in weight to percentage of moisture that shall constitute the air-dry loss in the sieve analysis sample.)

9. Procedure

9.1 Weigh the air-dried sample.

9.2 *Hand Sieving*:

9.2.1 Starting with the sieve having the largest opening, sieve a portion of the RDF-3 in such an increment as will allow the individual pieces to be in direct contact with the meshes of the screen after the completion of shaking of each increment. In shaking, apply a vertical as well as horizontal motion in order to allow all small particles to pass through the openings, until no more material will pass. Hand fitting is not permitted.

9.2.2 Pass the material through successively smaller sieves in increments small enough to avoid matting of the material to the extent that the undersized material cannot reach and pass the screen.

9.2.3 Continue to shake the sieve after each increment is added until no significant amount of material passes through the screen.

9.3 *Machine Sieving*:

9.3.1 When sieving machines are used, test their thoroughness of sieving by comparison with hand methods as described in 9.2.

9.3.2 Stack the sieves progressively starting with the smallest aperture size, above the pan, to the largest aperture size at the top.

9.3.3 Introduce the air-dried sample above the largest screen in small enough increments such that matting of the material does not occur to an extent that prevents the undersize material from reaching and passing the screen. The amount of RDF-3 added to the top screen in any increment must not exceed one third of the volume of the screen, in order to prevent matting or blinding.

9.3.4 After adding each increment, assemble the pans or trays in the machine and turn on agitation for 10 min, or up to 15 min if necessary, to complete screening.

9.3.5 Inspect each screen for evidence of matting. If a screen is mostly or entirely covered with a mat, decrease the size of the initial increments such that no mat forms on any sieve, and repeat the tests.

9.3.6 When sieving of each increment is complete, promptly determine the weight of material remaining on each screen to the nearest 0.5 g. If more than one increment is sieved to pass the entire sample, add the incremental weights remaining on each sieve. If the sum of the weights show a loss of 2 % or more, reject the analysis and make another test using a second sample.

NOTE 3—In order to obtain a complete characterization of the size range of an RDF-3 sample, it is necessary that the number of sieves be such that no more than 25 % of the gross sample weight will be retained on any given sieve. The recommended screen series are listed in Table 1.

NOTE 4—The sand and glass contained in a sample of RDF-3 has a strong tendency to segregate from the light fraction. For this reason great care must be taken to include the entire sample in the sieve analysis. When a sample is divided, the sand will probably not divide equally into the sample portions. Samples may be divided for convenience in feeding the sieving apparatus, but the weights of all portions of the sample must be properly summed so that the entire sample has been included in the sieve analysis.

NOTE 5—Some abrasion and physical degradation of the sample by the screen can occur during the sieving operation. The analyst shall monitor and report his observations of any sample degradation.

10. Report

10.1 Report the weights of the size fractions as a percentage of the weight of the air-dried laboratory sample of RDF-3. Calculate to the nearest 0.5 % the percentages of the size fractions remaining on each sieve, and the percentage passing through the smallest aperture sieve. See sample report form in Fig. 1.

10.2 Record the results starting with the largest aperture size. If desirable, the percentage can also be reported on a cumulative basis, as cumulative percentage greater than size or cumulative percentage less than size where size refers to sieve aperture size or mesh number.

10.3 The sieve aperture defining the upper particle size limit shall be that sieve of the series with the smallest aperture size that will retain less than 5 % of the sample weight. This sieve size is the nominal top particle size (see Annex A2 for definition).

10.4 The sieve aperture size defining the lower particle size limit shall be the smallest aperture sieve of the series which passes less than 15 % of the sample weight. This sieve size is the nominal bottom particle size (see Annex A2 for definition).

10.5 The term defining particle sizes shall be written with the nominal top size first, followed by the nominal bottom size.

10.6 International Standard sieve sizes shall be expressed as millimetre (mm), or micrometre, (μm), representing the actual size of the sieve opening. U.S. Standard sieve sizes shall be

REPORT OF SIEVE ANALYSIS OF RDF-3

All weights in grams

Sample Identification no. ____________________

Date sampled: ____________________

Sample rec'd. ____________________

Source: ____________________

____

AIR DRY LOSS

Weight Before ____________

Weight After ____________

Weight Loss ____________

% A-D Loss ____________

Sieve No. (mm)	Alternative (in. mesh)	Weight retained	% retained	% cumulative greater than size
100 mm	4 in.			
50 mm	2 in.			
25 mm	1 in.			
12.5 mm	½ in.			
6.3 mm	¼ in.			
3.35 mm	No. 6			
1.70 mm	No. 12			
850 μm	No. 20			
425 μm	No. 40			
In pan through 425 μm	No. 40			
(Total)				

NOTE 1—Calculate percentages to the nearest 0.5 %.
NOTE 2—If the sum of the weights shows a loss of over 2 %, reject the analysis and make another test.
NOTE 3—Nominal top size sieve shall retain 1 % to 5 % of A-D sample weight. The nominal bottom size sieve shall allow no more than 15 % of A-D sample to pass.

Analyst: ____________

Date: ____________

FIG. 1 Sample Report Form

expressed as inches (in.) or by mesh numbers representing the number of mesh wires per inch. For sieves No. 4 and smaller sieve sizes, the abbreviation No., shall be used each time a sieve is indicated by a mesh number.

11. Precision and Bias

11.1 The precision and bias of this test method are yet to be determined.

ANNEXES

(Mandatory Information)

A1. SIEVING DEVICES

A1.1 Horizontal Rotating Cylindrical Screens

A1.1.1 Horizontal rotating cylindrical screens are preferable for screening flat materials such as RDF-3, because they readily provide the lifting and tumbling action required to bring all materials to the screen surface. However, no screens of this sort are commercially available at this time.

A1.2 Rectangular Testing Screen[4] (Fig. A1.1)

A1.2.1 Trays have 0.46 by 0.66-m (18 by 26-in.) clear screen area, designed primarily for the 0.1 m (4 in.) to No. 4 mesh size, but will handle small amounts of finer material down to No. 200 mesh. The screens handle samples up to a maximum of 1 ft^3(0.03 m^3). Screening motion is essentially a vertical variation, which is factory set for the type of material to be tested. Up to six screen trays can be held in the vibrating unit.

A1.3 Rotary Pan Sieve[5] (Fig. A1.2)

A1.3.1 This device can be operated with up to nine full height 200 mm (8-in.) or 300 mm (12-in.) sieving screens and a pan, assembled together in a set, and held at an angle of 45° while rapping and rotating the assembly. A timer is provided to stop the mechanical action after time intervals up to 15 min.

A1.4 Testing Sieve Shaker[6]

A1.4.1 This device reproduces the circular and tapping motion given testing sieves in hand-sieving, and can hold a series of size full-height, 200-mm (8-in.) sieves in one operation of the machine.

[4] Gilson testing screen model TS-1, having six screens and a pan 0.46 by 0.66 m (18 by 26 in.), has been found to be satisfactory for RDF-3 under 0.05 mm (2 in.), when equipped with a special low-amplitude drive shaft. Available from Gilson Screen Co., P.O. Box 99-T, Malinta, OH 43535.

[5] Rainhart Rotary Pan Sieve using 300-mm (12-in.) or 200-mm (8-in.) circular sieves has been found to be satisfactory for RDF-3 under 12.5 mm (0.5 in.). Available from Rainhart Co. (Testing Equipment), 604-T Williams, Austin, TX 78752.

[6] A Rotap screening machine with 200-mm (8-in.) circular sieve has been found to be satisfactory for RDF-3 under 12.5 mm (0.5 in.). Available from W. S. Tyler Co., Inc., 8200 Tyler Blvd., Mentor, OH 44060.

FIG. A1.1 Rectangular Testing Screen

FIG. A1.2 Rotary Pan Sieve

A2. METHOD FOR DETERMINATION OF NOMINAL AND MEAN PARTICLE SIZES

A2.1 Graphical Form

A2.1.1 The graphical form (see Fig. A2.1) is suitable for recording the sieve analysis data, determining the percentage retained, the cumulative percentage, and for plotting the cumulative percentage on the Rosin-Rammler graph.

A2.1.2 The characteristics of the size distribution can be determined from the plotted cumulative percentage, resulting in a distribution coefficient *n* and an absolute constant, or mean particle size *x* in accordance with the techniques of Rosin-Rammler analysis.

A2.2 Procedure for Determining Coefficients

A2.2.1 Plot "Percent Cumulative Greater than Size" against size on the Graphical Form.

A2.2.2 Note the size retaining 36.79 %. This is the mean particle size. Use the nearest-standard screen opening or mesh designation to describe the mean particle size.

A2.2.3 Draw a straight line through the plotted points, passing through the mean particle size, and extend this line to the upper axis (1 % oversize).

A2.2.4 Read the size at 5 % oversize. Determine the next

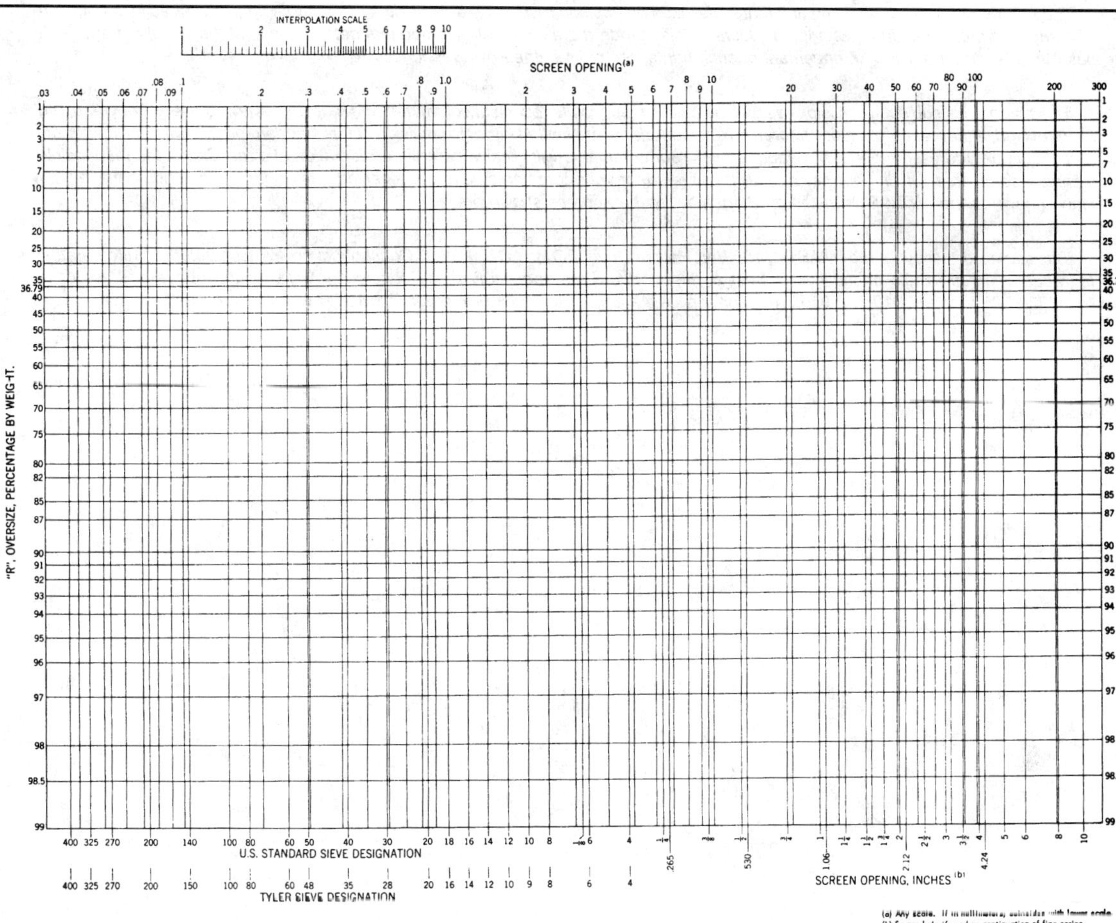

GRAPHICAL FORM FOR REPRESENTING DISTRIBUTION OF SIZES OF BROKEN COAL.

(From: Landers, W. S., and Reid, W. T., A Graphical Form for Applying the Rosin and Rammler Equation to the Size Distribution of Broken Coal; Bureau of Mines Inf. Circular 7346, 1946.)

Average particle size (intersection of size distribution line with 63.21 % passing line)	$\bar{X}$
Slope of size distribution line (tangent of angle)	*N*

[A] Any scale, if in millimetres, coincides with lower scale.

[B] Square hole if used as continuation of fine series.

[C] From: Landers, W. S., and Reid, W. T., A Graphical Form for Applying the Rosin and Rammler Equation to the Size Distribution of Broken Coal; Bureau of Mines Inf. Circular 7346, 1946. Pristine forms are available from ASTM Headquarters. Order PCN 12-508280-4.

FIG. A2.1 Graphical Form for Representing Distribution of Sizes[C]

larger standard screen opening. This is the "Nominal Top Particle Size" of the sample, defined as the size retaining less than 5 % of the sample.

A2.2.5 Read the size at 85 % oversize. Determine the next smaller standard screen opening. This is the "Nominal Bottom Particle Size."

A2.2.6 Select two points, "A" one screen size less than and "B" one size greater than the mean particle size, lying on the straight line drawn through the plotted points.

A2.2.6.1 Measure the horizontal distance of points A and B from the left axis in mm (or inches), and enter them in the table "characteristics" along with the percentage retained. Take the difference between x and y .

A2.2.6.2 The distribution coefficient, n, is the slope of the line, $\frac{\Delta y}{\Delta x}$.

A2.2.6.3 Measure the horizontal distance, x, of points A and B from the left vertical axis, and record in the table "characteristics."

A2.2.6.4 Measure the vertical distance, y, of points A and B from the bottom axis. Record in the table and subtract the x and y measurements to obtain the differences, Δx and Δy.

Designation: E 829 – 94

Standard Practice for Preparing Refuse-Derived Fuel (RDF) Laboratory Samples for Analysis[1]

This standard is issued under the fixed designation E 829; the number immediately following the designation indicates the year of original adoption or, in the case of revision, the year of last revision. A number in parentheses indicates the year of last reapproval. A superscript epsilon (ε) indicates an editorial change since the last revision or reapproval.

1. Scope

1.1 This practice covers the preparation of RDF laboratory samples for analysis, the laboratory samples having been previously obtained from representative RDF samples.

1.2 The determination of the air-dry loss of the RDF is part of this preparation procedure and must be performed prior to the particle size reduction.

1.3 The practice given may also be used for other RDF types but additional sample preparation steps may be necessary prior to the application of this method.

1.4 *This standard does not purport to address all of the safety concerns, if any, associated with its use. It is the responsibility of the user of this standard to establish appropriate safety and health practices and determine the applicability of regulatory limitations prior to use.* Specific hazard statements are given in Section 7.

2. Referenced Documents

2.1 *ASTM Standards:*

E 180 Practice for Determining the Precision of ASTM Methods for Analysis and Testing of Industrial Chemicals[2]

E 790 Test Method for Residual Moisture in a Refuse-Derived Fuel Analysis Sample[3]

E 791 Test Method for Calculating Refuse-Derived Fuel Analysis Data from As-Determined to Different Bases[3]

3. Terminology

3.1 *Definitions:*

3.1.1 *air-drying*—a process of partially drying RDF to bring its moisture content to near equilibrium with the ambient atmosphere in which further reduction, division, and characterization of the sample are to take place. In order to bring about this equilibrium, the RDF is usually subjected to drying under controlled temperature conditions ranging from 30 to 40°C.

3.1.2 *analysis sample*—final subsample prepared from the air-dried laboratory sample but reduced by passing through a mill with a 0.5-mm (0.02-in.) size or smaller final screen.

3.1.3 *bias*—a constant or systematic error in the test results. The error is a persistent positive or negative deviation from the accepted reference value.

3.1.4 *gross sample*—a sample representing one lot and composed of a number of increments on which neither reduction nor division has been performed.

3.1.5 *laboratory sample*—a representative portion of the gross sample received by the laboratory for further analysis.

3.1.6 *lot*—a large designated quantity (greater than the quantity of the final sample) of RDF that can be represented by a properly selected gross sample.

3.1.7 *precision*—a term used to indicate the capability of a person, an instrument, or a method to obtain reproducible results; specifically, a measure of the random error as expressed by the variance, the standard error, or a multiple of the standard error.

3.1.8 *refuse-derived fuel*—solid forms of refuse-derived fuels from which appropriate analytical samples may be prepared defined as follows in *ASTM STP 832:*[4]

RDF-1—wastes used as a fuel in as-discarded form with only bulky wastes removed.

RDF-2—wastes processed to coarse particle size with or without ferrous metal separation.

RDF-3—combustible waste fraction processed to particle sizes, 95 % weight passing 2-in. [2] screening.

RDF-4—combustible waste fraction processed into powder form, 95 % weight passing 10-mesh screening.

RDF-5—combustible waste fraction densified (compressed) into the form of pellets, slugs, cubettes, or briquettes.

3.1.9 *representative sample*—a sample collected in such a manner that it has characteristics equivalent to the lot sample.

3.1.10 *sample division*—the process of extracting a smaller sample from a sample so that the representative properties of the larger sample are retained. During this process it is assumed that no change in particle size or other characteristics occurs.

3.1.11 *sample preparation*—the process that includes drying, size reduction, division, and mixing of a laboratory sample for the purpose of obtaining an unbiased analysis sample.

3.1.12 *sample reduction*—the process whereby sample particle size is reduced without change in sample weight.

3.1.13 *significant loss*—any loss that introduces a bias in

[1] This practice is under the jurisdiction of ASTM Committee D-34 on Waste Management and is the direct responsibility of Subcommittee D34.03 on Treatment.

Current edition approved Aug. 15, 1994. Published October 1994. Originally published as E 829 – 81. Last previous edition E 829 – 88.

[2] *Annual Book of ASTM Standards*, Vol 15.05.

[3] *Annual Book of ASTM Standards*, Vol 11.04.

[4] *Thesaurus on Resource Recovery Terminology, ASTM STP 832*, ASTM, 1983.

final results that is of appreciable importance to concerned parties.

4. Summary of Practice

4.1 Sample moisture is reduced by air-drying to allow the mechanical reduction of the sample without significant change to the sample's fuel properties. The final sample is in a form suitable for further analysis.

5. Significance and Use

5.1 Using this procedure a sample of RDF can be converted into a physical form suitable for laboratory fuel analysis.

5.2 As indicated in Test Method E 791, air-dry moisture, which is determined by this procedure, is essential to the calculation of other laboratory results on an as-received basis. The air-dry moisture value is used in conjunction with the results of the residual moisture determination in Test Method E 790 to calculate total sample moisture.

6. Apparatus

6.1 *Air-Drying:*

6.1.1 *Drying Oven*—A large chamber mechanical draft oven capable of maintaining a controlled temperature in the range from 25 to 40°C. Air changes should be at the rate of 1 to 4 changes per minute. Air flow should be baffled to prevent samples from being blown out of the sample containers.

6.1.2 *Drying Pan*—A noncorroding pan or mesh basket to be used for holding the sample during air-drying operations.

6.1.3 *Balance* (Laboratory Sample)—A balance of sufficient capacity to weigh the sample and container with a sensitivity of 0.5 g.

6.2 *Sample Reduction:*

6.2.1 *Mill*—A mill operating on the principle of cutting or shearing action shall be used for sample particle size reduction. It shall have the capability to regulate the particle size of the final product by means of either interchangeable screens or mill adjustments. The mill shall be enclosed and should generate a minimum amount of heat during the milling process to minimize the potential for loss of moisture. The final product shall pass through a 0.5-mm or smaller screen into a receiver integral with the mill. Access should be provided so that the mill can be quickly and easily cleaned between samples.

6.3 *Analysis Sample Containers*—Heavy, vapor impervious bags, properly sealed; or noncorroding cans, glass jars, or plastic bottles with airtight sealing covers may be used to store RDF samples for analysis. Containers shall be checked for suitability by measuring weight loss or gain of the sample and container stored for 1 week under ambient laboratory conditions. The weight loss or gain should be less than 0.5 % of the sample weight stored in the container.

6.4 *Drying Oven*—A drying oven of either the mechanical or natural circulation type which is capable of constant uniform temperature within the specimen chamber regulated at 107 ± 3°C.

6.5 *Shredder*—A laboratory shredder capable of shredding or cutting larger particle sizes of solid waste. The final product shall pass through a 2-in. or smaller screen into a receiver integral with the shredder.

NOTE 1—A garden-type shrubbery shredder equipped with a screen and bag for collection of shredded samples is satisfactory.

7. Hazards

7.1 Due to the origins of RDF in municipal waste, common sense dictates that some precautions should be observed when conducting tests on the samples. Recommended hygienic practices include use of gloves when handling RDF; wearing dust masks (NIOSH-approved type), especially while shredding RDF samples; conducting tests under a negative pressure hood when possible; and washing hands before eating or smoking.

7.2 Laboratory sample handling and reduction shall be performed by trained personnel. If all precautions regarding sample preparations are not followed, the error in the preparation may bias some or all of the analyses performed on the sample.

7.2.1 All preparation steps shall be done rapidly and in as few operations as possible, since moisture loss depends on several factors other than total moisture content, such as time required for milling, atmospheric temperature and humidity, and the type of laboratory sample reduction equipment.

7.2.2 At all times RDF samples should be protected from moisture change due to exposure to rain, snow, and sun, or contact with absorbent materials.

7.2.3 Samples should be transported to the laboratory and analyzed as soon as possible. If any sample-handling step involved an extended time period, the sample and container should be weighed before and after the process to determine any weight gain or loss. This weight gain or loss shall be included in the calculation of moisture content.

7.2.4 Force-feeding of the sample through the mill can overload the motor. An overload can cause rapid heating of the rotor and mill chamber with possible loss of residual moisture.

8. Sampling [5]

8.1 RDF products are frequently nonhomogeneous. For this reason significant care should be exercised to obtain a representative sample from the RDF lot to be characterized.

8.2 The sampling method for this procedure should be based on agreement between the involved parties.

8.3 For this procedure the laboratory sample size will normally not exceed 2 kg with some variation possible depending on the laboratory equipment available.

8.3.1 Due to the heterogeneous nature of RDF, dividing a laboratory sample to a very small size analyses sample may result in nonrepresentative results. Since milling operations mix the sample as well as reduce particle size, laboratory samples should not be divided before the initial preparation steps have been completed.

9. Procedure

9.1 Weigh the entire laboratory sample into a tared air-drying pan. Sample depth in the drying pan shall be no greater than 100 mm (4 in.) and any lumps of sample should be broken

[5] ASTM Subcommittee E38.01 is currently in the process of developing a procedure for sampling RDF. The chairman of E38.01 should be contacted for details.

up. Use more than one pan if necessary. If a very fine mesh-type drying pan is used, size the mesh such that the sample will not be lost.

9.2 Air-dry the sample at 10 to 15°C above ambient, but not greater than 40°C until the weight loss is less than 0.1 % of the sample weight per hour. Samples can normally be allowed to air-dry for a set time period such as overnight or 24 h. To speed the drying stage the sample may be carefully stirred avoiding loss of sample.

NOTE 2—The air discharge of the forced draft air-drying oven should be filtered prior to discharge to minimize laboratory contamination by air-entrained RDF dust.

9.3 Separate and weigh the millables and nonmillables for classification and use, or analysis, if necessary. Calculate the millables and nonmillables as described in 10.2.

9.4 Dry a representative portion of the air-dried millable fraction at 107 ± 3°C to constant weight as follows:

9.4.1 Heat a clean, empty drying pan at a minimum temperature of 107 ± 3°C for at least 1 h. When cool, transfer to a desiccator and tare weigh to an accuracy of 0.5 g.

9.4.2 Place the laboratory sample of RDF in the drying pan(s). A maximum sample depth of 50 to 100 mm is recommended. Weigh the pan and sample to an accuracy of 0.5 g. More than one drying pan may be necessary.

NOTE 3—If a mesh-type pan is used, place a clean sheet of aluminum foil under the pan to check for any sample fall through. If any occurs, a smaller-mesh drying pan is required.

9.4.3 Place the pan and sample in the drying oven at 107 ± 3°C for a minimum of 1 h.

NOTE 4—Observe the sample periodically to make certain that the sample does not decompose or ignite at this temperature.

9.4.4 After an appropriate drying time, remove the pan and sample from the oven and place in a desiccator to cool. When cool, weigh the sample and pan to the nearest 0.5 g.

9.4.5 Place the sample and pan in the oven for an additional 1 h at 107 ± 3°C.

9.4.6 Remove the sample and pan and place in the desiccator to cool. When cool, weigh the sample and pan to the nearest 0.5 g. If the sample weight loss was less than 0.1 %/h of the original sample weight, the determination is complete; if not, repeat 9.4.5 and 9.4.6.

NOTE 5—At this point the dried sample can be used for further analysis if desired.

9.4.7 Calculate the moisture of the nonmilled millable fraction of the laboratory sample as described in 10.3.

9.5 Reduce the air-dried sample to a smaller particle size by using a cutting- or shearing-type shredder or mill. The final product should pass through a 0.5-mm or smaller screen. Depending on the specific RDF product, this step may involve more than one stage or reduction, that is, passing the sample through a shredder or mill with larger size screens first and then milling to pass the final screen. Even though the sample has been air dried, minimum atmospheric exposure is recommended and the milling process should be conducted to avoid significant moisture change. If necessary, the milled sample should be well mixed by either manual or mechanical means to ensure thorough intermix of heavy fines and milled fluff.

9.6 The mixed, air-dried, finely ground laboratory sample can then be further subdivided to an analysis-size sample (Note 6). Retain a minimum of 50 g as the analysis sample. Any division method used shall ensure that the retained analysis sample is representative of the original laboratory sample.

NOTE 6—If it is possible to riffle the product, a small laboratory riffle can be used to divide the sample. If it is not possible to riffle the sample, use some other valid method to divide the sample.

9.7 Keep the analysis sample in a labeled sample container having a moisture-tight seal.

9.8 Determine the residual moisture of the analysis sample as described in Test Method E 790.

10. Calculation

10.1 Calculate the air-dry moisture as follows:

$$ADL = [(G - L)/G] \times 100 \quad (1)$$

where:

ADL = air-dry loss, %,
G = weight of the laboratory sample before air-drying, and
L = weight of the laboratory sample after air-drying.

10.2 Calculate the millables and the nonmillables of the air-dried sample as follows:

$$M = [W_M/(W_M + W_{NM})] \times 100 \quad (2)$$

where:

M = millables, % ,
W_M = weight of the millables in the air-dried sample,
NM = nonmillables, % , and
W_{NM} = weight of the nonmillables in the air-dried sample.

10.3 Calculate the moisture in the nonmilled millable fraction of the air-dried fraction of the laboratory sample as follows:

$$MAD = [(W_{bd} - W_{ad})/W_{bd}] \times 100 \quad (3)$$

where:

MAD = moisture of nonmilled millable fraction of the air-dried sample, % ,
W_{bd} = weight of the nonmilled millable fraction of the air-dried sample before drying at 107 ± 3°C, and
W_{ad} = weight of the nonmilled millable fraction of the air-dried sample after drying at 107 ± 3°C.

NOTE 7—Moisture loss or gain during particle-size reduction (shredding/milling) can be determined by comparing the moisture of the nonmilled millable fraction of the air-dried sample with the residual moisture-content value and used accordingly.

NOTE 8—Nonmillables usually are noncombustible and can be used in contributing to the ash value or noncombustible-value portion of the sample.

11. Precision and Bias

11.1 The standard deviations of the individual determinations, in percent absolute, are:

Typical Average Value	Within Laboratory	Between Laboratories
16	1.8	2.9

11.1.1 These precision estimates are based on an interlaboratory study conducted in accordance with Practice E 180.

11.2 *Bias*—Due to the variability in what is considered to be "in equilibrium with ambient conditions" it is not possible to describe any measure of bias regarding this practice.

Designation: E 830 – 87 (Reapproved 1996)

Standard Test Method for Ash in the Analysis Sample of Refuse-Derived Fuel[1]

This standard is issued under the fixed designation E 830; the number immediately following the designation indicates the year of original adoption or, in the case of revision, the year of last revision. A number in parentheses indicates the year of last reapproval. A superscript epsilon (ϵ) indicates an editorial change since the last revision or reapproval.

1. Scope

1.1 This test method covers determination of the ash content in the analysis sample of refuse-derived fuel (RDF). The results obtained can be applied as the weight percent ash in the proximate analysis and in the ultimate analysis.

1.2 The values stated in acceptable metric units are to be regarded as standard. The values given in parentheses are for information only.

1.3 *This standard does not purport to address all of the safety concerns, if any, associated with its use. It is the responsibility of the user of this standard to establish appropriate safety and health practices and determine the applicability of regulatory limitations prior to use.* For specific precautionary statements see Section 6.

2. Referenced Documents

2.1 *ASTM Standards:*

E 180 Practice for Determining the Precision of ASTM Methods for Analysis and Testing of Industrial Chemicals[2]

E 790 Test Method for Residual Moisture in a Refuse-Derived Fuel Analysis Sample[3]

E 829 Practice for Preparing Refuse-Derived Fuel (RDF) Laboratory Samples for Analysis[3]

3. Terminology

3.1 *Definitions of Terms Specific to This Standard:*

3.1.1 *refuse-derived fuel*—Solid forms of refuse-derived fuels from which appropriate analytical samples may be prepared are defined as follows in *ASTM STP 832:*[4]

RDF-1—Wastes used as a fuel in as-discarded form with only bulky wastes removed.

RDF-2—Wastes processed to coarse particle size with or without ferrous metal separation.

RDF-3—Combustible waste fraction processed to particle sizes, 95 % passing 2-in. square screening.

RDF-4—Combustible waste fraction processed into powder form, 95 % passing 10-mesh screening.

RDF-5—Combustible waste fraction densified (compressed) into the form of pellets, slugs, cubettes, or briquettes.

4. Summary of Test Method

4.1 Ash is determined by weighing the residue remaining after burning the prepared analysis sample under rigidly controlled conditions of sample weight, temperature, and furnace atmosphere.

5. Significance and Use

5.1 This test method is available to producers and users of RDF as a method of determining the weight percent of ash in the analysis sample.

6. Apparatus

6.1 *Electric Furnace*—For determination of the ash content of RDF, the furnace shall have adequate air ventilation and shall be capable of temperature regulation up to at least 750 ± 25°C. An air change rate of 1 to 4 furnace volumes of air per minute has been found adequate.

NOTE 1—It may be possible to reduce the rate of air flow below the suggested minimum without adversely affecting results of the ash determination.

6.2 *Porcelain Capsules*, about 22 mm (⅞ in.) in depth, and 44 mm (1¾ in.) in diameter, or similar containers.

NOTE 2—Weighing bottles of borosilicate glass may be safely used without deformation or softening at temperatures of 600°C or less.

7. Precautions

7.1 Due to the origins of RDF in municipal waste, common sense dictates that some precautions should be observed when conducting tests on the samples. Recommended hygienic practices include use of gloves when handling RDF; wearing dust masks (NIOSH-approved type), especially while milling RDF samples; conducting tests under a negative pressure hood when possible; and washing hands before eating or smoking.

8. Sampling

8.1 The laboratory sample shall be obtained in accordance with sampling methods developed for materials of similar physical form.

8.2 The laboratory sample must be air-dried and particle

---

[1] This test method is under the jurisdiction of ASTM Committee D-34 on Waste Management and is the direct responsibility of Subcommittee D34.08 on Thermal Treatment.

Current edition approved Aug. 28, 1987. Published October 1987. Orginally published as E 830 – 81. Last revised E 830 – 81.

[2] *Annual Book of ASTM Standards*, 15.05.

[3] *Annual Book of ASTM Standards*, 11.04.

[4] *Thesaurus on Resource Recovery Terminology, ASTM STP 832*, ASTM, 1983, p. 72.

size reduced to pass a 0.5-mm screen as described in Practice E 829.

9. Procedure

9.1 After thoroughly mixing the analysis sample analysis sample to provide the best possible mix of heavy fines with the milled fluff, transfer approximately 1 g of the sample to a tared, previously fired container (weighed to the nearest 0.1 mg) with a scoop or spatula. Quickly weigh sample and container to the nearest 0.1 mg. As an alternate method use the dried analysis sample from the residual moisture determination. See Test Method E 790.

9.2 Place the uncovered container containing the sample in the furnace at low temperature and gradually heat to ignition at such a rate as to avoid mechanical loss from too rapid expulsion of volatile matter.

9.3 Finish the ignition to constant weight 9±0.001 g/h) at 575 ± 25°C. It may be determined that a constant weight can be routinely established by allowing a sample to ash within the prescribed temperature range for a set period of time.

NOTE 3—Experience has shown that particles of glass and sand tend to sinter to each other and also to porcelain crucibles at temperatures close to 675°C. If laboratory conditions necessitate maintaining consistency in the maximum furnace temperature used for ash tests of other fuels, the ignition may be finished to constant weight (± 0.001 g/h) at a temperature of 725 ± 25°C. If this option is invoked, it should be also noted that prolonged exposure to high temperatures may actually result in changes in weight due to possible chemical reactions.

9.4 Cool in a desiccator over desiccant and weigh as soon as possible after the container and ash reach the temperature of the area in which weighing is performed.

10. Calculation

10.1 Calculate the ash percent in the analysis sample as follows:

$$\text{Ash as-determined, \%} = [(A - B)/C] \times 100 \quad (1)$$

where:
A = weight of container and ash residue, g,
B = weight of empty container, g, and
C = weight of ash analysis sample, g (includes residual moisture).

10.2 Use the numerical moisture value established by Test Method E 790 for converting ash data on the as-determined basis to the dry basis.

11. Report

11.1 Difficulty may be experienced in securing satisfactory check determinations of ash in the same or different laboratories for RDF rich in heavy fines. This is caused by siliceous matter such as glass and sand as well as a wide variety of other particles of different densities entrained in the milled RDF in nonuniform strata. When such a condition is anticipated or encountered, a paired set of determinations should be made, and the results reported as an average. If one determination of a paired set is accidentally ruined, another pair must be run. An off or unusual value does not constitute a ruined determination. In such cases, an additional set of duplicate determinations should be run and all values reported as an average of the two sets.

12. Precision and Bias

12.1 *Precision*:

12.1.1 The standard deviations of individual determinations in percent absolute are as follows:

Typical Average Value, %	Within-Laboratory, %	Between-Laboratories, %
20.0	0.6	1.3

12.1.2 These precision estimates are based on an interlaboratory study conducted in accordance with Practice E 180.

12.2 *Bias*—The bias of this test method can not be determined due to the lack of a recognized standard reference material.

The American Society for Testing and Materials takes no position respecting the validity of any patent rights asserted in connection with any item mentioned in this standard. Users of this standard are expressly advised that determination of the validity of any such patent rights, and the risk of infringement of such rights, are entirely their own responsibility.

This standard is subject to revision at any time by the responsible technical committee and must be reviewed every five years and if not revised, either reapproved or withdrawn. Your comments are invited either for revision of this standard or for additional standards and should be addressed to ASTM Headquarters. Your comments will receive careful consideration at a meeting of the responsible technical committee, which you may attend. If you feel that your comments have not received a fair hearing you should make your views known to the ASTM Committee on Standards, 100 Barr Harbor Drive, West Conshohocken, PA 19428.

Designation: E 850 – 95

Standard Practice for Use of Inorganic Process Wastes as Structural Fill[1]

This standard is issued under the fixed designation E 850; the number immediately following the designation indicates the year of original adoption or, in the case of revision, the year of last revision. A number in parentheses indicates the year of last reapproval. A superscript epsilon (ε) indicates an editorial change since the last revision or reapproval.

1. Scope

1.1 This practice provides guidance for use of selected process wastes as structural fills by listing representative test methods for predicting and evaluating those physical characteristics of waste that are related to the integrity of fills and to protection of ground and surface waters.

1.2 Table 1 lists references which provide engineering practices and test procedures that may be applied to process waste for use as structural fill.

1.3 Some process wastes may produce leachate that exceeds environmentally acceptable limits. Special provisions are included to accommodate this class of materials (see 7.2).

2. Referenced Documents

2.1 *ASTM Standards:*

C 294 Descriptive Nomenclature for Constituents of Natural Mineral Aggregate[2]

C 295 Guide for Petrographic Examination of Aggregates for Concrete[2]

C 593 Specification for Fly Ash and Other Pozzolans for Use with Lime[3]

C 821 Specification for Lime for Use with Pozzolans[3]

D 420 Guide to Site Characterization for Engineering, Design, and Construction Purposes[4]

D 421 Practice for Dry Preparation of Soil Samples for Particle-Size Analysis and Determination of Soil Constants[4]

D 422 Test Method for Particle-Size Analysis of Soils[4]

D 4318 Test Method for Liquid Limit, Plastic Limit, and Plasticity Index of Soils[4]

D 559 Test Methods for Wetting and Drying Compacted Soil-Cement Mixtures[4]

D 560 Test Methods for Freezing and Thawing Compacted Soil-Cement Mixtures[4]

D 653 Terminology Relating to Soil, Rock, and Contained Fluids[4]

D 698 Test Method for Laboratory Compaction Characteristics of Soil Using Standard Effort (12,400 ft-lbf/ft^3 (600 kN-m/m^3))[4]

D 854 Test Method for Specific Gravity of Soils[4]

D 1140 Test Method for Amount of Material in Soils Finer Than the No. 200 (75-μm) Sieve[4]

D 1452 Practice for Soil Investigation and Sampling by Auger Borings[4]

D 1556 Test Method for Density and Unit Weight of Soil in Place by the Sand-Cone Method[4]

D 1557 Test Method for Laboratory Compaction Characteristics of Soil Using Modified Effort (56,000 ft-lbf/ft^3 (2,700 kN-m/m^3))[4]

D 1586 Test Method for Penetration Test and Split-Barrel Sampling of Soils[4]

D 1587 Practice for Thin-Walled Tube Sampling of Soils[4]

D 1633 Test Method for Compressive Strength of Molded Soil-Cement Cylinders[4]

D 2049 Test Method for Relative Density of Cohesionless Soils[5]

D 2166 Test Method for Unconfined Compressive Strength of Cohesive Soil[4]

D 2167 Test Method for Density and Unit Weight of Soil in Place by the Rubber Balloon Method[4]

D 2216 Test Method for Laboratory Determination of Water (Moisture) Content of Soil and Rock[4]

D 2217 Practice for Wet Preparation of Soil Samples for Particle-Size Analysis and Determination of Soil Constants[4]

D 2434 Test Method for Permeability of Granular Soils (Constant Head)[4]

D 2487 Classification of Soils for Engineering Purposes (Unified Soil Classification System)[4]

D 2488 Practice for Description and Identification of Soils (Visual-Manual Procedure)[4]

D 2573 Test Method for Field Vane Shear Test in Cohesive Soil[4]

D 2664 Test Method for Triaxial Compressive Strength of Undrained Rock Core Specimens Without Pore Pressure Measurements[4]

D 2850 Test Method for Unconsolidated, Undrained Compressive Strength of Cohesive Soils in Triaxial Compression[4]

---

[1] This practice is under the jurisdiction of ASTM Committee D-34 on Waste Management and is the direct responsibility of Subcommittee D34.06 on Recovery and Reuse.

Current edition approved Oct. 10, 1995. Published December 1995. Originally published as E 850 – 82. Last previous edition E 850 – 88.

[2] *Annual Book of ASTM Standards*, Vol 04.02.

[3] *Annual Book of ASTM Standards*, Vol 04.01.

[4] *Annual Book of ASTM Standards*, Vol 04.08.

[5] Discontinued—See *1983 Annual Book of ASTM Standards*. Vol 04.08. Replaced by D4253.

D 2922 Test Methods for Density of Soil and Soil-Aggregate in Place by Nuclear Methods (Shallow Depth)[4]
D 2937 Test Method for Density of Soil in Place by the Drive-Cylinder Method[4]
D 3017 Test Method for Water Content of Soil and Rock in Place by Nuclear Methods (Shallow Depth)[4]
D 3080 Test Method for Direct Shear Test of Soils Under Consolidated Drained Conditions[4]
D 3974 Practice for Extraction of Trace Elements from Sediments[6]
D 3987 Test Method for Shake Extraction of Solid Waste with Water[7]
D 5084 Test Method for Measurement of Hydraulic Conductivity of Saturated Porous Materials Using a Flexible Wall Permeameter[8]

3. Terminology

3.1 *cemented materials*—materials consisting of one or more substances that develop hardness by chemical reaction after placement of the material in a fill.

3.2 *coarse material*— material coarser than a No. 200 (75-μm) U.S. standard sieve.

3.3 *effective coefficient of permeability*—the coefficient of permeability that characterizes a fill and is the result of combined materials characteristics and construction techniques including compaction, capping, placement of impermeable layers, etc.

3.4 *fill material*— material used in the construction of a structural fill.

3.5 *fine material*— material finer than No. 200 (75-μm) U.S. standard sieve.

3.6 *leachate*—liquid that has percolated through or passed over a solid waste or other medium and contains dissolved or suspended materials, or both, from the medium.

3.7 *process waste*— inorganic by-product materials such as mine tailings, culm piles, coal processing conversion and combustion wastes, cement and limekiln dust, by-product gypsum, and chemically treated compositions made from these wastes or waste mixtures.

3.8 *structural fill*— man-made deposits of solid materials. Examples include backfills, landfills, embankments, earth dams, linings and blankets, foundations, canals, road base, footings, and trenches.

4. Significance and Use

4.1 This practice is intended for inorganic process wastes that can be used as replacements for natural material such as soil or rock suitable for construction applications. Selection of appropriate and feasible fill materials and selection of applicable materials, tests, and specifications to facilitate construction and environmental protection are the responsibility of the design engineer. This practice is intended to encourage wider utilization of waste materials.

5. Determination of Material Characteristics

5.1 Table 1 contains representative test methods recommended for determining and evaluating characteristics of process wastes, of either candidate or in situ fill materials. Appropriate numerical values of materials characteristics will vary depending on design requirements and are selected on the basis of accepted engineering practice and regulatory requirements. Testing of process wastes that may result in chemical reactions or contain cementitious materials should be performed on specimens that have been cured and aged to duplicate in situ conditions as closely as possible. Examples of such test procedures are listed in Table 1 under Cemented Materials. Cured specimens carefully removed from the fill may be used in carrying out the laboratory or field procedures (Table 1). Solubility of the waste material must be suitable for the intended use.

[6] *Annual Book of ASTM Standards*, Vol 11.02.
[7] *Annual Book of ASTM Standards*, Vol 11.04.
[8] *Annual Book of ASTM Standards*, Vol 04.09

TABLE 1 Representative Test Methods Recommended for Determining and Evaluating Characteristics of Process Wastes Suitable for Fill Construction

Characteristics	Test Methods[A]
General:	
Laboratory Procedures:	
Dry Preparation of Soil Samples for Particle-Size Analysis and Determination of Soils Constants	D 421
Particle-Size Analysis of Soils	D 422
Liquid Limit of Soils	D 4318
Plastic Limit and Plasticity Index of Soils	D 4318
Terminology Relating to Soil, Rock, and Contained Fluids	D 653
Moisture-Density Relations of Soils Using 5.5 lb Rammer and 12-in. Drop	D 698
Specific Gravity of Soils	D 854
Amount of Materials in Soils Finer than the No. 200 Sieve	D 1140
Moisture-Density Relations of Soils Using 10 lb. Rammer and 18-in. Drop	D 1557
Laboratory Determination of Moisture Content of Soil	D 2216
Wet Preparation of Soil Samples for Particle-Size Analysis and Determination of Soil Constants	D 2217
Classification of Soils for Engineering Purposes	D 2487
Description of Soils	D 2488
Field Procedures:	

TABLE 1 *Continued*

Characteristics	Test Methods[A]
Investigating and Sampling Soil and Rock for Engineering Purposes	D 420
Soil Investigation and Sampling by Auger Borings	D 1452
Density of Soil In Place by Sand-Cone Method	D 1556
Penetration Test and Split-Barrel Sampling of Soils	D 1586
Thin-Walled Tube Sampling of Soils	D 1587
Density of Soil in Place by Rubber-Balloon Method	D 2167
Density of Soil and Soil-Aggregate in Place by Nuclear Methods	D 2922
Density of Soil In Place by Drive-Cylinder Method	D 2937
Moisture Content of Soil and Soil-Aggregate In Place by Nuclear Methods	D 3017
Field Permeability Tests in Boreholes	**(7)**
Materials:	
Coarse Materials:	
Relative Density of Cohesionless Soils	D 2049
Permeability of Granular Soils, Constant Head	D 2434
Direct Shear Test of Soils under Consolidated Drained Conditions	D 3080
Fine Materials:	
Unconfined Compressive Strength of Cohesive Soil	D 2166
Permeability of Fine Materials, Falling Head	**(5)**
Permeability of Fine Materials, Flexible Wall	D 5084
Field Vane Shear Test in Cohesive Soil	D 2573
Unconsolidated, Undrained Compressive Strength of Cohesive Soils in Triaxial Compression	D 2850
Cemented Materials:	
Wetting-and-Drying Tests of Compacted Soil-Cement Mixtures	D 559
Freezing-and-Thawing Tests of Compacted Soil-Cement Mixtures	D 560560
Compressive Strength of Molded Soil-Cement Cylinders	D 1633
Permeability of Fine Materials, Falling Head	**(5)**
Triaxial Compressive Strength of Undrained Rock Core Specimens Without Pore Pressure Measurements	D 2664
Fly Ash and Other Pozzolans for Use with Lime	C 593
Lime for Use with Pozzolans	C 821
Descriptive Nomenclature for Constituents of Natural Mineral Aggregates	C 294
Petrographic Examination of Aggregates for Concrete	C 295
Anhydrous Minerals and Organic Materials as Sources of Distress in Concrete, W. C. Hansen	**(1)**
Reactions of Aggregates Involving Solubility, Oxidation, Sulfates, or Sulfides, Richard C. Mielenz	**(2)**
The Handbook of Concrete Aggregates—A Petrographic and Technological Evaluation, L. Dolar Mantuani	**(3)**
Environment	
Extraction Procedures:	
Extraction of Trace Elements with Sediments	D 3974
Shake Extraction of Solid Waste with Water	D 3987
EPA Extraction Procedure	**(6)**

[A] Test Methods are ASTM procedures except as noted.

5.2 Some by-product and waste materials contain constituents that may produce volume changes that would not be detected by the standard procedures listed in Table 1. Volume increases may be caused by chemical reactions such as oxidation or hydration and by long-time increases in moisture content. Decreases in volume and loss of stability can result from drying over a long period of time or from the dissolution of soluble constituents. Problems may also result from other effects of chemical and physical weathering processes or from attack by fill constituents on structures with which it is in contact. Wastes that have no prior performance record should be subjected to chemical, geotechnical, and petrographic studies to determine the presence of constituents known to produce such effects. Information helpful in such investigations may be found in Descriptive Nomenclature C 294, Guide C 295, and in the literature by Hansen **(1)**,[9] Mielenz **(2)**, and Mantuani **(3)**.

6. Construction Practice

6.1 Construction of a fill with process waste should conform to standard practices employed with conventional fill materials. Methods such as those described in the *U.S. Department of the Interior Earth Manual* **(4)** and the *Corps of Engineers Soil Testing Manual* **(5)** are suitable construction practices. Inspection of the fill should be made during construction to ensure that fill construction specifications are fulfilled.

[9] The boldface numbers in parentheses refer to the list of references at the end of this standard.

7. Special Provisions

7.1 Materials having predicted in situ coefficients of permeability less than 1×10^{-7} cm/s and adequate engineering properties, including shear strength, compressibility, soluble salts, etc. (see Table 1, Materials) may be suitable for fill construction when placed a minimum of 5 ft (1.5 m) above the historical high-water table.

7.2 The following provisions apply when process wastes used as fill materials produce concentrations of constituents in the leachate that exceed acceptable limits as determined by Test Method D 3987 or the EPA method **(6)**.

7.2.1 Fills having predicted effective coefficients of permeability greater than 1×10^{-5} cm/s shall have environmentally acceptable underdrain and leachate collection and disposal systems.

7.2.2 Fills having predicted effective coefficients of permeability less than 1×10^{-5} cm/s do not require leachate collection and treatment systems provided that surfaces exposed to the environment are sloped so that the minimum predicted runoff from the surfaces is 90 % of the incident precipitation and provided the materials are placed a minimum of 5 ft (1.5 m) above the historical high-water table where the coefficient of permeability of the subsoil is greater than 1×10^{-5} cm/s.

7.3 In situ monitoring of fill permeability and integrity may be required during and following fill construction to ensure that the performance of the fill as constructed is adequate. The design, implementation, and interpretation of in situ permeability tests shall be accomplished under the direction of the engineer.

NOTE 1—These special provisions are industry standards. The user is advised to contact local, state, and federal environmental agencies about the regulations dealing with the placement of regulated materials outside the limits of a permitted landfill.

REFERENCES

(1) Hansen, W. C., "Anhydrous Minerals and Organic Materials as Sources of Distress in Concrete," *Highway Research Record No. 43*, National Research Council/Transportation Research Board, 1983, pp. 1–7.

(2) Mielenz, R. C., "Reactions of Aggregates Involving Solubility, Oxidation, Sulfates, or Sulfides," *Highway Research Record No. 43*, National Research Council/Transportation Research Board, 1983, pp. 8–18.

(3) Mantuani, L. D., *The Handbook of Concrete Aggregates—A Petrographic and Technological Evaluation*, Noyes Publication, Mill Rd., Park Ridge, NJ, 07656, 1983.

(4) *U.S. Department of the Interior, Earth Manual*, (Second Edition), 1974.

(5) Corps of Engineers Soil Testing Manual, 1110-2, 1906.

(6) *Resource Conservation and Recovery Act, Federal Register*, Environmental Protection Agency, EPA Publication No. SW-846, Sept. 13, 1979.

(7) Field Permeability Tests in Boreholes, U.S. Dept. of Interior, Federal Registar, U.S. Government Printing Office, Superintendent of Documents, Washington, DC.

Designation: E 856 – 83 (Reapproved 1998)

Standard Definitions of Terms and Abbreviations Relating to Physical and Chemical Characteristics of Refuse Derived Fuel[1]

This standard is issued under the fixed designation E 856; the number immediately following the designation indicates the year of original adoption or, in the case of revision, the year of last revision. A number in parentheses indicates the year of last reapproval. A superscript epsilon (ε) indicates an editorial change since the last revision or reapproval.

accuracy—generally, a term used to indicate the reliability of a measurement, or an observation. Specifically, a measure of closeness of agreement between a test result and a recognized standard value. For example, the difference between the observed and a standard value for the sulfur content of a sample of refuse-derived fuel. This difference includes both random and systematic errors.

air-drying—a process of partial drying of refuse-derived fuel (RDF) to bring its moisture content near to equilibrium with the atmosphere in which further reduction, division, and characterization of the sample are to take place. In order to bring about this equilibrium, the RDF is usually subjected to drying under controlled temperature conditions ranging from 30 to 40°C.

air-dry loss—the decrease in sample mass due to air-drying. This decrease is presumed to be moisture.

analysis sample—final subsample prepared from the air-dried laboratory sample but reduced by passing through a mill with a 0.5 mm (0.02 in.) size or smaller final screen.

as-determined basis—analytical data obtained from the analysis sample after conditioning and preparation in accordance with Method E 829. As-determined basis data represent the numerical values obtained at the particular moisture and ash level in the sample at the time of analysis.

ash—the residue remaining after ignition of refuse-derived fuel determined by definite prescribed methods. The ash may not be identical in composition or quantity with the inorganic substances present in the refuse-derived fuel before ignition.

as-received basis—the test data calculated to the moisture condition of the sample as it arrived in the laboratory and before any laboratory processing or conditioning.

bias—a systematic error that is consistently negative or consistently positive. The mean of errors resulting from a series of observations which does not tend toward zero.

calorific value—the energy of combustion of a unit quantity of refuse-derived fuel. It may be expressed in kilojoules per kilogram (kJ/kg), British thermal units per pound (Btu/lb), or calories per gram (cal/g) (obsolete).

combustible—that portion of the refuse-derived fuel sample which is consumed upon ignition exclusive of the moisture present in the sample.

d-RDF—abbreviation for densified refuse-derived fuel.

dry, ash-free—the test data calculated to a theoretical base of no moisture or ash associated with the sample.

dry basis—the test data calculated to a theoretical base of no moisture associated with the sample.

duplicate analysis—paired determinations on the same sample performed by one analyst at essentially the same time.

gross calorific value—the energy released by combustion of a unit quantity of refuse-derived fuel at constant volume or constant pressure in a suitable calorimeter under specified conditions such that all water in the products is in the liquid form. This is the measure of calorific value predominantly used in the United States. **Synonym:** higher heating value.

gross sample—a sample representing one lot, normally composed of a number of increments, on which neither reduction nor division has been performed.

higher heating value (HHV)—a synonym for gross calorific value.

increment—a portion of the lot collected by one individual manual or mechanical sampling operation and normally combined with other increments from the lot to make a gross sample.

laboratory sample—a representative portion of the gross sample received by the laboratory for analysis.

lot—a designated quantity of refuse-derived fuel that can be represented by a properly selected gross sample.

lower heating value (LHV)—a synonym for net calorific value.

milling—a reduction in particle size of refuse-derived fuel by shearing, cutting, or grinding to a suitable particle size for analysis and characterization (see **sample reduction**).

net calorific value—the energy released by combustion of a unit quantity of refuse-derived fuel at constant volume or constant pressure in a suitable calorimeter under specified conditions such that all water in the products remains in the gaseous form. The net calorific value is the lower heating value that can be calculated from the gross calorific value by making a correction for the heat of vaporization of the water. This is the measure of calorific value in Europe but is rarely used in the United States. **Synonym:** lower heating value.

noncombustible—that fraction of a macrosample remaining after moisture and combustibles are driven off by heat and

[1] These definitions are under the jurisdiction of ASTM Committee D-34 on Waste Management and are the direct responsibility of Subcommittee D34.03 on Treatment.

Current edition approved Oct. 26, 1983. Published December 1983.

combustion. It is composed of metallic and glass particles in addition to the residue from the combustion of organic substances.

precision—a term used to indicate the capability of a person, an instrument, or a method to obtain reproducible results; specifically, a measure of the random error as expressed by the variance, the standard error, or a multiple of the standard error.

proximate analysis—the determination, by prescribed methods, of moisture, volatile matter, fixed carbon (by difference), and ash. Unless otherwise specified, the term proximate analysis does not include determinations of chemical elements or any determinations other than those named.

refuse-derived fuel (RDF-3)—refuse-derived fuel-3 (RDF-3) is defined as a shredded fuel derived from municipal solid waste (MSW) which has been processed to remove metal, glass, and other inorganic materials. This material has a particle size such that 95 weight % passes through a 2-in. square mesh screen.

NOTE 1—Other refuse-derived fuel may be classified as follows:
RDF-1—Wastes used in as-discarded form.
RDF-2—Wastes processed to coarse particle size with or without ferrous metal separation.
RDF-4—Combustible waste processed into powder form, 95 weight % passing 10-mesh screening.
RDF-5—Combustible waste densified (compressed) into the form of pellets, slugs, cubettes, or briquetes.
RDF-6—Combustible waste processed into liquid fuels.
RDF-7—Combustible waste processed into gaseous fuel.

residual moisture—the moisture content remaining in an RDF sample after it has been air-dried and milled down to an analysis sample.

representative sample—a sample collected in such a manner that it has characteristics equivalent to the lot.

sample—a portion of material taken from a larger quantity for the purpose of estimating properties or composition of the larger quantity. See **analysis sample, gross sample, laboratory sample, and representative sample.**

sample division—the process of obtaining a smaller sample from a larger sample so that the representative properties of the larger sample are retained. During this process it is assumed that no change in particle size or other characteristics occurs.

sample preparation—the process that includes drying, size reduction, division, and mixing of a laboratory sample for the purpose of obtaining an unbiased analysis sample.

sample reduction—the process whereby sample particle size is reduced without change in sample weight (see milling).

standard deviation—the square root of the variance.

systematic error—an error caused by undetected fundamental flaws in the test equipment by inadequate understanding of the theory underlying the measurement or by repetitious errors on the part of the analyst.

total moisture—the water contained in a sample. The determination of the total moisture is made by drying a sample under controlled conditions of temperature, time, and air flow. The determination may consist of a single-stage or a two-stage drying process.

ultimate analysis—the determination of the percentages of carbon, hydrogen, sulfur, nitrogen, chlorine, ash, and oxygen in a dry refuse-derived fuel sample. The percentage of oxygen may be obtained by difference between 100 % and the sum of the other determined elemental analyses.

variance—the mean square of deviations (or errors) of a set of observations; the sum of square deviations (or errors) of individual observations with respect to their arithmetic mean divided by the number of observations less one (degrees of freedom); the square of the standard deviation (or standard error).

volatile matter—those products, exclusive of moisture, given off by a material as gas or vapor, determined by definite prescribed methods.

Standard Test Methods for Conducting Performance Tests on Mechanical Conveying Equipment Used in Resource Recovery Systems[1]

This standard is issued under the fixed designation E 868; the number immediately following the designation indicates the year of original adoption or, in the case of revision, the year of last revision. A number in parentheses indicates the year of last reapproval. A superscript epsilon (ε) indicates an editorial change since the last revision or reapproval.

1. Scope

1.1 These test methods include descriptions for conducting and reporting throughput and electrical power tests on mechanical conveying equipment for municipal solid waste and recovered products from resource recovery systems. Other aspects of performance testing such as spillage, conveyor tracking, dusting, slippage, transfer points, etc., should be considered in the interpretation of the results. These test methods can be used on equipment handling raw refuse, processed refuse, magnetic scrap metals, nonferrous scrap metals, mixed glass, and residues or tailings. These test methods may also be used for materials in other industries.

1.2 These test methods cover mechanical conveying equipment including apron, belt, drag, flight, screw, slat, and vibrating conveyors and bucket elevators.

1.3 These test methods are applicable specifically to the resource recovery industry since municipal solid wastes are heterogeneous mixtures and the composition and bulk densities vary considerably depending on many factors. Because of the varying composition of municipal solid waste, a number of samples must be taken to determine accurately the performance of the mechanical conveying equipment.

1.4 Test Methods for determining the approximate as-conveyed bulk density of the material and for determining the electrical horsepower input of the equipment motors are also included.

1.5 It is intended that the tests be made and reported by personnel trained in the proper application and use of the various instruments and methods involved.

1.6 The values stated in SI units are to be regarded as the standard. The values given in parentheses are for information only.

1.7 *This standard does not purport to address all of the safety concerns, if any, associated with its use. It is the responsibility of the user of this standard to establish appropriate safety and health practices and determine the applicability of regulatory limitations prior to use.* For hazard statement, see Section 7.

2. Referenced Documents

2.1 *ASTM Standards:*

E 380 Practice for Use of the International System of Units (SI) (the Modernized Metric System)[2]

E 856 Definitions of Terms and Abbreviations Relating to Physical and Chemical Characteristics of Refuse-Derived Fuel[3]

2.2 *Other Standard:*

No. 550 Classification and Definitions of Bulk Materials[4]

3. Terminology

3.1 *Definitions:*

3.1.1 *oversize bulky waste (OBW)*—items whose large size precludes or complicates processing or sampling.

3.1.2 *performance test*—a test devised to permit observation and measurement of the performance of a system or unit of equipment operating under prescribed load conditions.

3.2 For definitions of other terms used in these test methods, refer to Definitions E 856. For an explanation of the metric system including symbols and conversion factors, refer to Practice E 380.

4. Summary of Test Methods

4.1 The conveying equipment performance can be calculated by determining the volume or weight of a representative sample of material on the conveying equipment and measuring its speed. Another method for calculating the conveying equipment performance is to measure the infeed or discharge weight or volume in a given length of time. The minimum recommended number of test runs and size of samples are provided for various types of materials (see Table 1).

4.2 The material flow rate may be reported in any unit; sample calculations are given only for selected (common) units.

[1] These test methods are under the jurisdiction of ASTM Committee D-34 on Waste Management and are the direct responsibility of Subcommittee D34.06 on Recovery and Reuse.

Current edition approved March 26, 1982. Published December 1982.

[2] *Annual Book of ASTM Standards*, Vol 14.02.

[3] *Annual Book of ASTM Standards*, Vol 11.04.

[4] Available from Conveyor Equipment Manufacturers Association, 1000 Vermont Ave., N.W. Washington, DC 20005.

4.3 Motor wattage (or amperage and voltage) may be measured and used to calculate the electrical power consumption.

5. Significance and Use

5.1 These test methods may be used to measure the equipment performance.

5.2 These test methods are applicable when the conveying equipment is of sufficient length and is accessible for taking the samples and measuring the speed, or when the discharge is accessible to collect a sample in a given length of time. Not all pieces of equipment in a processing plant may be accessible; therefore, the input or total of inputs to adjacent upstream equipment/output or total of outputs of adjacent downstream equipment may be used to determine the throughput of the conveying equipment in question. Judgement must be used to determine any loss of material or changes in bulk density.

6. Apparatus

6.1 *Ammeter/Voltmeter*—Multimeter or individual meters to permit reading the maximum current and voltage anticipated. Meters may be the snap-on type with analog or digital readout.

6.2 *Bulk Density-Measuring Container*—An open-top container constructed of suitable materials such as plywood or plastic and having the following internal dimensions: 300 mm wide by 300 mm long (1 by 1 ft) and 600 mm (2 ft) high may be used for material normally smaller than 150 mm (6 in.) in size. Suitable handles may be attached to the exterior of the container to aid in subsequent handling.

NOTE 1—Alternatively, containers of other dimensions may be employed provided the base area is known and sides are perpendicular. Dimensions of the container shall be a minimum of two times the largest particle size.

6.3 *Bulk Density Measuring Rod*—A round or square rod, approximately 50 mm (2 in.) in diameter or square by 600 mm (2 ft) long, calibrated in 5-mm (0.1-in.) intervals starting from one end. The end should be cut off square to prevent sinking into sample.

6.4 *Tachometer or Speed Indicator*— A tachometer with linear speed indicator or surface speed indicator. Indicator may be hand type with digital readout.

6.5 *Wattmeter*—Industrial analyzer or individual wattmeter to provide two wattmeter indications for three-phase power. Meters may be analog or digital type.

7. Safety Hazards

7.1 These test methods may involve the use of hazardous materials, operations, and equipment. It is the responsibility of whomever uses this standard to establish appropriate safety practices and to determine the applicability of regulatory limitations prior to use.

7.2 Due to the origins of municipal solid waste, common sense dictates that some precautions should be observed when conducting tests. Recommended hygienic practices include use of gloves when handling the waste, wearing dust masks (NIOSH-approved type), and washing hands before eating or smoking.

7.3 Safety precautions should be taken when measuring conveyor speeds or collecting samples near open, moving equipment and taking electrical measurements.

NOTE 2—**Precaution:** Include use of eye protection and hard hats, avoidance of loose-fitting clothing that could become entangled in machinery, and adopt the use of a" buddy system" in which the person conducting the test is always within sight and hearing of a coworker.

8. Sampling

8.1 *Minimum Number of Tests and Size of Sample*—The minimum recommended number of tests and minimum size of samples to be collected are shown in Table 1, based on the type of material and particle sizes.

NOTE 3—The quantity and size of samples have to be statistically verified. ASTM is conducting additional sampling experiments to verify this information.

8.2 *Frequency of Sample Collection*—No more than one sample shall be taken at a time. After collecting a representative sample, a minimum of 15 min processing time shall elapse before taking the next sample.

9. Procedures

9.1 *General*—Install, lubricate, and align the equipment to be tested in accordance with the manufacturer's recommendations. It is advisable to make one or more preliminary tests for the purpose of determining the adequacy of the instruments and apparatus, and the training of the personnel, if required. Before the tests are begun, run the equipment under stable conditions for sufficient length of time to bring about equilibrium and steady readings.

9.2 *Recording Data*— Keep complete records of all information relevant to the tests. A suggested form for recording the data and calculating the results is given in Fig. 1. Additional observations such as material wetness, particle size variations, unusual constituents in the waste or unusually high concentrations of a particular constituent, and conveying equipment spillage, rollback of material, dusting, etc., should be recorded on the back of Fig. 1 or on a separate sheet. Before removing test equipment, compute the results to determine if they are reasonable. If so, the test can be considered terminated and the test equipment removed.

9.3 *Calibration of Instruments*—Properly calibrate all instruments in accordance with the manufacturer's instructions. Confirm that the instruments are in good condition and are being used under conditions corresponding to those existing at the time of their calibration.

9.4 *Sample Collection on Open Conveyors*—On open conveyors such as apron, belt, drag, and flight conveyors that can be stopped, cut two bulkheads from plywood or similar material to fit the contour of the conveying surface to prevent material rollback. Place these contoured bulkheads on the conveyor to establish sample (gathering) boundary. Stop the conveyor. Place one bulkhead, perpendicular to the length of the conveyor and perpendicular to the conveying surface, making sure the material is separated uniformly. Place the second bulkhead in a similar manner either upstream or downstream from the first, a sufficient distance to obtain one of the recommended size samples from the conveyor. Measure the length between the bulkheads (L_s) to the nearest 10 mm (0.25 in.) and record. Carefully remove all of the material, including the fine, particulate materials, from between the two bulkheads.

Owner's Name ______ Sheet ______ of ______
Plant location ______ Material description ______
Type of Equipment ______ Weather ______
Item. No. ______ Name ______
Arrangement: Horizontal length ______ Height ______ Incline ______ Width ______
Rated capacity ______ Rated bulk density ______ Rated volume ______
Motor Type ______ Frame ______
Manufacturer ______ Serial No. ______
Rated horsepower ______ Rated speed ______ Full load current ______
Characteristics: Voltage ______ Phase ______ Frequency ______
Gear Reducer Type ______ Belt or Chain ______
Manufacturer ______ Serial No. ______
Speed ratio ______ Class ______ Horsepower ______ Service Factor ______

TEST INFORMATION:

(A) *Open Conveying Equipment Feed Performance Determinations:*

Test No.	Sample Weight (W_s)	Sample Volume	Conveyor Speed (N)	Sample Length (L_s)	Calculated Capacity (C)	Percent Difference from Rated
1						
2						
3						
4						
5						
				Average:		

Make and serial number of instruments/date calibrated ______

(B) *Closed Conveying Equipment Performance Determinations:*

Test No.	Sample Weight (W_s)	Sample Volume	Collection Time (T)	Calculated Capacity (C)	Percent Difference from Rated
1					
2					
3					
4					
5					
			Average:		

Make and serial number of instruments/date calibrated ______

(C) *Bulk Density Determinations:*

Test No.	Filled Container Weight (W_f)	Empty Container Weight (W_e)	Area of Base (a)	Height of Material (h)	Calculated Bulk Density (BD)	Percent Difference from Rated
1						
2						
3						
4						
5						
				Average:		

Make and serial number of instruments/date calibrated ______

(D) *Electrical Horsepower Input Determinations:*

Test No.	Motor Watts (W)	Motor Voltage (E)	Motor Amperage (I)	Calculated Horsepower (Ehp)	Percent Difference from Rated
1					
2					
3					
4					
5					
			Average:		

Make and serial number of instruments/date calibrated ______

Date ______ Test made by ______ Witnessed by ______

FIG. 1 Mechanical Conveying Equipment Suggested Performance Test Data Sheet

Place the sample in adequate container(s) or plastic bag(s) for material where moisture is a factor. If the length of time before measuring the bulk density is to be more than 2 h, it is recommended to use double plastic bags with ends sealed separately to prevent moisture loss or gain. Do not squeeze or compress the sample. Determine the bulk density as soon as possible after collection to prevent drying, moisture gain, or settling.

9.5 *Sample Collection at Equipment Discharge*—On screw conveyors, vibrating conveyors, enclosed conveyors (or otherwise nonaccessible), and bucket elevators or conveyors that cannot be stopped, collect the sample from the discharge. Collect the entire cross section of the discharge in suitable size container(s) or plastic bag(s) for a time period. Do not squeeze or compress the sample. Determine the bulk density as soon as possible after collection. Record the length of time (T)

required to fill each container to the nearest 1 s.

9.6 *Weighing Sample*—Carefully weigh each sample (W_s) collected above to the nearest 0.05 kg (0.1 lb) using a suitable weighing scale.

9.7 *Measure Conveyor Speed*—Measure the speed of smooth belt conveyors (N) in m/min or ft/min by using a tachometer with surface speed indicator. One alternative method may be by measuring the distance a mark on the conveyor moves in a given length of time. The latter method should be performed a minimum of three times and the average of the three measurements may be recorded as N. (Three measurements is arbitrary to minimize human or other errors.) On metal apron conveyors, the r/min of the headshaft, pitch, and number of teeth in the headshaft sprocket or diameter of drive pulley may be used.

9.8 *Truck Scale Weight/Unit of Time*—As an alternative method of calculating throughput, weigh a quantity of raw refuse using truck weighing scales, if available. Measure the length of time required for processing the weighed quantity to determine the throughput rate.

9.9 *Measuring Bulk Density of Material*[5]—Determine the approximate as-conveyed bulk density for each of the samples collected in accordance with 8.1. Determine the empty weight of the bulk density-measuring container (W_e) by weighing the empty container before each determination to the nearest 0.5 kg (0.1 lb). Carefully fill the measuring container with material that would be representative of the sample, that is, approximately one third from the top, one third from the middle, and one third from the bottom, including fines. Carefully level the surface of the material manually to minimize surface irregularities. Take care not to tamp the material or cause settling so as to maintain as nearly as possible the as-conveyed density. Carefully measure the distance from the top of the container to the surface of the material to the nearest 5 mm (0.1 in.) in each of the four corners of the container using the bulk density measuring rod described in 6.3. Subtract the average of the four measurements from the inside height of the container to determine the height of the material (h). Weigh the filled container to the nearest 0.05 kg (0.1 lb) to determine the filled weight (W_f) of the container plus contents.

9.10 *Measuring Electrical Power*[6]—The electrical power may be measured by using one or two wattmeters to measure the wattage (W) or an ammeter to measure the motor current (I) and a voltmeter to measure the motor voltage (E). These measurements may be made at the motor control center (or starter) terminals. The average measurement of each leg should be used.

NOTE 4—On direct current type motor drives, the power of the controller and blower will be measured as well. Also, power factors are included in the wattmeter methods and are not included in the current and voltage method.

10. Calculation

10.1 Calculate apron, belt, drag, or flight conveyor performance (C) in megagrams per hour using the sample weights and belt speeds in accordance with the following equation:

$$C = [(W_s \times N)/(L_s)] \times 0.06 \quad (1)$$

where:

C = conveyor performance, Mg/h,
W_s = mass of sample collected, kg,
N = speed of conveyor, m/min,
L_s = length of conveyor between bulkheads over which the sample was distributed, m, and
0.06 = conversion factor of min to h and kg to Mg.

10.1.1 To calculate the conveyor performance in short tons per hours, use Eq 1 with pounds and inch measurements and use 0.0025 as the conversion factor.

10.1.2 To calculate the conveyor performance in cubic metres per hours or cubic feet per hour, use Eq 1 with the mass of the sample divided by the bulk density and use 60 (5.0) as the conversion factor.

10.2 Calculate screw conveyor, vibrating conveyors or bucket elevator performance (C) in megagrams per hour using the collected sample weights and measured times of collection in accordance with the following equation:

$$C = (W_s)/(T) \times 3.6 \quad (2)$$

where:

C = conveyor or bucket elevator performance, Mg/h
W_s = mass of collected sample, kg,
T = length of time in seconds required to collect sample, and
3.6 = conversion factor of s to h and kg to Mg.

10.2.1 To calculate the conveyor performance in short tons per h, use Eq 2 with W_s lb and use 1.8 as the conversion factor in place of 3.6.

10.2.2 To calculate the conveyor performance in m^3/h or ft^3/h, use Eq 2 with the volume measurement to the nearest 0.1 m^3(0.1 ft^3) in place of the sample weight and use 3600 as the conversion factor.

10.3 Calculate the bulk density of the sample using the following equation:

$$BD = (W_f - W_e)/(a \times b) \quad (3)$$

where:

BD = bulk density of the material, kg/m^3 (or lb/ft^3),
W_f = weight of measuring container plus material, kg (lb),
W_e = weight of empty measuring container, kg (lb),
a = inside area of container base, m^2 (ft^2), and
h = average inside height of material in container, m (ft).

10.4 Calculate the electrical horsepower input to the drive using either of the following three equations:

10.4.1 Industrial analyzer method of measurement:

$$E_{hp} = W/746 \quad (4)$$

where:

E_{hp} = electrical horsepower input to the drive, and
W = analyzer wattmeter reading, W.

10.4.2 Two-wattmeters method of measurement:

$$E_{hp} = (W_1 + W_2)/(746) \quad (5)$$

[5] A separate method for measuring bulk density is in preparation. When approved, this section will be revised to refer to the new standard.

[6] A separate method for measuring electric power consumption is in preparation. When approved, this section will be revised to refer to the new standard.

TABLE 1 Minimum Number of Test Runs and Sample Sizes for Performance Determination

Type	Description	Normal Top Particle Size, mm (in.)	Number of Test Runs	Sample Size, m^3 (ft^3)
Raw refuse (as discarded)	municipal solid waste; residential, commer- cial, and industrial (excludes oversize bulky waste)	900 (36) max in any one dimension	8	1.5 (50)
Processed Refuse	primary screen undersize	200 (8)	8	0.1 (2)
	primary screen oversize	900 (36) max in any one dimension	8	1.5 (50)
	first-stage shredded refuse	300 (12)	8	0.25 (8)
	air-classified heavy fraction	300 (12)	8	0.25 (8)
	air-classified light fraction	300 (12)	8	0.25 (8)
	second-stage shredded light fraction	100 (4)	8	0.25 (1)
	screened air-classified light fraction	50 (2)	8	0.25 (1)
	densified refuse-derived fuel	40 (1½)	8	0.25 (1)
	pulverized or powdered RDF	10 (⅜)	8	0.25 (½)
Magnetic scrap	first-stage shredded metal	300 (12)	8	0.25 (1)
	light-gage iron scrap after air classification	300 (12)	8	0.25 (1)
	shredded or nuggetized iron scrap	25 (1)	8	0.25 (1)
Nonferrous scrap metal	aluminum can stock	100 (4)	8	0.25 (1)
	shredded aluminum scrap	50 (2)	8	0.25 (1)
	other nonferrous metals	50 (2)	8	0.25 (1)
Mixed glass	coarse cullet glass	75 (3)	8	0.25 (1)
	fine cullet glass	6 (¼)	8	0.25 (1)

where:

E_{hp} = electrical horsepower input to the drive,
W_1 = reading of first wattmeter, and
W_2 = reading of second wattmeter.

10.4.3 Voltage and ammeter method for single phase measurements:

$$E_{hp} = [(E \times I)/746] \times (PF/100) \quad (6)$$

where:

E_{hp} = electrical horsepower input to the drive,
E = average motor voltage, V,
I = average motor amperage, A and,
PF = power factor, %.

10.4.4 For three phase measurements use:

$$E_{hp} = (1.732\, E \times I \times PF)/(746 \times 100) \quad (7)$$

10.5 The calculated feed rates, bulk densities, and electrical horsepower inputs may be compared to the specification values to evaluate equipment performance in the given application. Electrical horsepower input calculation should be used for reference purposes only since drive efficiencies are not included in the above horsepower calculations.

11. Precision and Bias

11.1 Precision and bias have not yet been developed.

Designation: E 884 – 82 (Reapproved 2001)

Standard Practice for Sampling Airborne Microorganisms at Municipal Solid-Waste Processing Facilities[1]

This standard is issued under the fixed designation E 884; the number immediately following the designation indicates the year of original adoption or, in the case of revision, the year of last revision. A number in parentheses indicates the year of last reapproval. A superscript epsilon (ε) indicates an editorial change since the last revision or reapproval.

1. Scope

1.1 This practice covers sampling of airborne microorganisms at municipal solid-waste processing facilities, hereafter referred to as facilities. Investigators should consult Practice D 1357 for the general principles of conducting an air-sampling program.

1.2 This practice applies only to sampling airborne bacteria and fungi, not viruses. Since sampling airborne viruses is significantly more difficult than sampling bacteria and fungi, reliable methods of sampling viruses are not yet available.

2. Referenced Documents

2.1 *ASTM Standards:*

D 1356 Terminology Relating to Atmospheric Sampling and Analysis[2]

D 1357 Practice for Planning the Sampling of the Ambient Atmosphere[2]

2.2 *Other Standards:*

Microbiological Methods for Monitoring the Environment, Water and Wastes[3]

Air Sampling Instruments for the Evaluation of Atmospheric Contaminants[4]

3. Definitions

3.1 *microbiological aerosol*—an airborne particle partially or exclusively composed of microorganisms including bacteria and fungi.

3.2 For definitions of other terms used in this practice, refer to Terminology D 1356.

4. Summary of Practice

4.1 Concentrations of selected airborne bacteria and fungi are determined using both liquid impinger and multi-stage impactor samplers.

4.2 Procedures are included for selecting sampling locations; determining numbers of samples, types of microorganisms to be sampled, intervals between sample collection and analysis; choosing sampling equipment; preserving samples; and reporting results.

5. Significance and Use

5.1 Bacteria and fungi present in municipal solid wastes (as well as in other forms of waste) may become airborne as dusts during waste processing. Several investigations to determine the health significance of these microbiological aerosols have been hindered by the lack of standardized procedures for sampling airborne bacteria and fungi in an industrial environment and by the absence of standards for assessing their health significance. Because it is difficult to correlate airborne levels of bacteria and fungi with epidemiological data, this standard is designed to permit the formation of a data base to aid in the assessment of the health significance of airborne microorganisms. It is intended that the use of this practice will improve sampling precision and thereby facilitate comparisons between sampling results.

6. Apparatus

6.1 Two types of samplers are used in each sampling program for microbiological aerosols at waste processing facilities (**5**).[5]

6.1.1 *Multi-Stage Impactor*, for collection of airborne microbes on agar plates. It is recommended that an impactor be used for sampling all of the types of bacteria and fungi listed in 10.6.1.[6]

6.1.2 *All-Glass Impinger*, for collection of airborne microbes in a liquid medium. It is recommended that an impinger be used for sampling fecal coliforms and for determination of total plate count.[7]

6.2 *Air Sampling Pumps*, providing approximately 40 L per min (1.4 CFM) free-flow capacity.

6.3 Additional equipment such as carts, stands, and tool

---

[1] This practice is under the jurisdiction of ASTM Committee D34 on Waste Management and is the direct responsibility of Subcommittee D34.01.02 on Sampling Techniques.

Current edition approved July 30, 1982. Published May 1983.

[2] *Annual Book of ASTM Standards*, Vol 11.03.

[3] Available from the National Technical Information Service, 5285 Port Royal Road, Springfield, Va. 22161. Request EPA-600/8-78-017.

[4] Available from American Conference of Governmental and Industrial Hygienists, 6500 Glenway Avenue, Building D-5, Cincinnati, Ohio 45211.

[5] The boldface numbers in the parentheses refer to the list of references at the end of the method.

[6] The six-stage and two-stage microbiological samplers manufactured by Anderson Samplers, Inc. have been found to be satisfactory.

[7] Air sampling impinger No. 7540 manufactured by Ace Glass, Inc. (AGI 30) has been found to be satisfactory.

boxes are routinely used during dust-sampling programs.

7. Reagents and Materials

7.1 *Agars for Use with the Multi-Stage Impactor*:

7.1.1 *Littman Oxgall*, for total number of fungi present and for identification of the following species of fungi: (*a*) *Aspergillus flavus* and (*b*) *A. fumigatus*.

7.1.2 *Vogel and Johnson*, selective for *Staphylococcus aureus*.

NOTE 1—A fungicide such as nystatin should be used with these agars.

7.1.3 *Levine eosin methylene blue*, specific for enterics including *Klebsiella* spp. (Note 1)

7.1.4 *Trypticase soy*, for total bacteria count. (Note 1)

7.2 *Liquid Media for Use in Impingers*:

7.2.1 *Lactose Broth with Antifoam A*, for analysis of fecal coliform and total plate count.

7.2.2 The exact amount of Antifoam A to be added should be determined prior to field sampling. Sufficient antifoam should be added to prevent loss of fluid from the impinger, but excess should be avoided.

7.3 *Media Preparation*:

7.3.1 Conduct the following according to *Microbiological Methods for Monitoring the Environment, Water and Wastes* **(14)**: (*a*) laboratory quality assurance, (*b*) selection and use of laboratory apparatus, (*c*) washing and sterilization, and (*d*) preparation of culture media.

7.3.2 Preincubate all sampling media to determine if contamination has occurred and to dry the agar surface. Excessive evaporation from the media or excessive contamination of the exterior surfaces of the petri dishes must be guarded against during this preliminary incubation.

7.3.3 Media level in the sampling container is critical to collection efficiency.

7.3.3.1 *Impactor*—The petri dishes must be of such a size that the agar surface is at the manufacturer's specified distance below each stage. The manufacturer of the Andersen impactor specifies 27 mL of agar per standard Andersen petri dish. The agar surface must be smooth and free of bubbles to ensure an even air flow.

7.3.3.2 *Impinger*—For the all glass impinger, 20 mL of broth is recommended **(17)**. Autoclave impingers, and then aseptically add 20 mL of sterile broth. Mark its level on the impinger, and record any significant loss during sampling. After sampling, the volume must be reconstituted to the original or the actual volume carefully calculated because a known volume must be used for quantitative work.

8. Precautions

8.1 Due to the nature of municipal refuse, common sense dictates that some precautions should be observed when sampling dusts at municipal solid-waste processing facilities. Recommended safety practices include wearing hard hats, safety shoes, safety glasses, gloves, and respirators as well as washing hands before eating or smoking.

9. Sampling

9.1 *Location and Number of Sampling Sites*:

9.1.1 All sampling shall be carried out during normal plant operations.

9.1.2 Use not less than two sampling locations inside the facility at work sites or zones where employees are most likely to be exposed to airborne dust concentrations **(7)**. (Note 2) Among these locations, those where sampling equipment can be located without interfering with facility operations shall be preferred.

NOTE 2—Examples of potential sampling locations are (*a*) on a tipping floor near or on a front end loader; (*b*) at a hand-picking station along a conveyor belt; and (*c*) along catwalks or platforms in frequent use by employees.

9.1.3 Outside the facility, locate at least one sampling site 300 m (1000 ft) upwind from the facility and at least one sampling site 100 m (330 ft) downwind from the facility. Measure the distances upwind and downwind from the same point, the point at which the emissions leave the facility or, in the case of multiple discharge points, from a central point equidistant from the discharge points.

9.1.4 Carefully measure and record the actual distances of the sampling sites from the points of emission and wind direction and velocity.

9.2 *Position of Sampling Inlet*—Locate the sampling inlet(s) 1.5 m (5 ft) above the floor level to approximate the breathing zone of a worker or other person exposed to the dusts. Locate the vacuum pumps where they will not disturb the air flow patterns around the sampling inlet(s).

9.3 *Number of Samples*:

9.3.1 Inside the facility, collect not less than 5 replicate samples at each sampling site.

9.3.2 Outside the facility, collect not less than 3 replicate samples at the upwind site(s) and not less than 5 replicate samples at the downwind site(s).

9.3.3 Wide variations in reported microbiological aerosol levels within facilities make it unlikely that the collection of five samples will yield a tight distribution of results; therefore, where economically feasible, it is recommended that the sample size be increased to more than five.

9.4 *Air Temperature*:

9.4.1 Collect samples when the air temperature at the sampling site is above 5°C (40°F).

9.4.2 At temperatures below 5°C (40°F), the sampling medium may crystallize, thus affecting recovery of microorganisms.

10. Procedure

10.1 Record air temperature and relative humidity for each location sampled.

10.2 Label all impingers to denote sampling run and location. Label all petri dishes to denote sampling run, location, and stage of impactor.

10.3 *Air-Flow Rates*:

10.3.1 Determine the air-flow rate by an in-line flow meter. Where this is not possible, calibrate air-flow rate with a gas-flow meter according to the procedure described in Ref **(16)**. The recommended flow rate for the Andersen impactor is 28.3 L/min. The optimum flow rate for the all-glass impinger is 12.5 L/min.

10.3.2 Maintain a constant air-flow rate through the sampler during the sampling time. Before sampling, allow the vacuum pump to warm up for not less than 1 min. Use clamps, T-shaped connectors, and in-line membrane filters with 1-mm pore size to pull filtered air through the pump during the warmup without pulling air through the sampler. Select clamps and T-shaped connectors that will not alter the flow rate through the samplers.

10.3.3 Secure all connections to keep the air loss less than 4 % of the average sampling rate or less than 0.00057 m^3/min (0.02 ft^3/min), whichever is smaller. Measure the leakage-flow rate with a suitable dry-gas meter connected to the discharge side of the vacuum pump while the inlet to the sampling apparatus is plugged and a 380-mm (15-in. Hg) vacuum is drawn. A lower vacuum may be used provided it is not exceeded during sampling.

NOTE 3—Many of the vane-type air sampling pumps (including the one furnished for use with the Andersen sampler) use a needle valve to control the air flow through the sampler by bleeding in air that bypasses the sampler. The air flow through the pump is therefore constant, and a meaningful measure of the flow through the sampler can only be made at this location in the sample stream.

10.4 *Sampling time*—The length of time needed to collect each sample is dependent upon the type of sampler used and the concentration of microbiological aerosols present in the air. Trial sampling runs may be necessary to determine if a satisfactory plate loading can be obtained within the limitations of the equipment used.

10.4.1 For the all-glass impinger operating at a flow rate of 12.5 L/min, the normal sampling time is 20 min.

10.4.2 When using a multistage impactor, choose the sampling time to avoid overloading the impaction plates, that is, the loading on any of the plates should not exceed 300 colonies per plate. The sampling time for the multistage impactors will vary depending on the medium used for sampling collection and the concentration of airborne dust. Suggested initial sampling times for the various media are in Table 1.

10.5 *Care During Sampling and Transport*—Collect, pack, transport, and manipulate the sample prior to analysis in a manner that safeguards against any change in the microbial activity in the sample, such as, extreme heat and cold and radiation, including sunlight. Use the proper media to ensure preservation of the sample until its identification. If samples must be shipped prior to analysis, positive controls should be included with each shipment. Federal regulations must be followed when they apply to these shipments.

10.5.1 *Care During Sampling with the Impactor*:

10.5.1.1 Carry out impactor loading and unloading in an atmosphere of minimal microbial activity, preferably in a portable polyethylene glove bag or a similar container. Invert the petri dishes immediately when the sampler is unloaded. Sanitize the impactor with a 70 % alcohol solution and dry thoroughly between samplings. Do not sanitize in the glove bag. To provide a control check for contamination, load and unload the impactor without sampling using a set of trypticase soy agar petri dishes, and then subject these petri dishes to the same processing steps and analytical procedures applied to the samples.

TABLE 1 1 Suggested Initial Sampling Times

Type of Media	Suggested Initial Sampling Time, min.[A]
Littman-Oxgall	1.5
Vogel and Johnson	8
Levine eosin methylene blue	10
Trypticase soy	0.5

[A] The initial sampling times suggested above are based on reported concentration levels in an enclosed facility. These times are subject to adjustment based upon the initial test results.

10.5.1.2 Minimize uneven distribution of colonies on the plates by centering the plates on the three pegs in each stage of the impactor and, once loaded, handling the impactor carefully to maintain this position.

10.5.2 *Care During Sampling with the Impinger*:

10.5.2.1 Include a negative (sterile) control with the impingers to determine whether the samples become contaminated while in transit or at the test site.

10.5.3 Preserve all samples by placing each one in a closed container at 4 ± 2°C immediately after taking them. Protect the plates from direct contact with the ice to prevent contamination.

NOTE 4—Sealed ice packets have been found to be satisfactory and convenient for this purpose.

10.5.4 Return the samples to the laboratory as soon as possible and not later than 6 h after sampling. Process the samples and place in a incubator as soon as possible.

10.5.5 For impinger samples, rinse the neck of the impinger and add this material to the sample. The volume of the rinse solution must be measured so that the final sample volume is known.

10.6 *Identification of Colonies*:

10.6.1 Analyze for the types of bacteria and fungi listed in 10.6.1.1-10.6.1.4. This is a minimum list of bacteria and fungi recommended for identification and quantification. Individual investigators may wish to sample for additional organisms. Among the other microorganisms that have been sampled at facilities are *Aspergillus niger*, *Mycobacterium* species and other members of the *Actinomycetales* order.

10.6.1.1 Total plate count (impactor) (Note 5),

10.6.1.2 Total plate count (impinger) (Note 5),

10.6.1.3 Bacteria, (*a*) Fecal coliforms, (*b*) *Klebsiella* species, (*c*) *Staphylococcus aureus*, and

10.6.1.4 *Fungi*, (*a*) *Aspergillus fumigatus* and (*b*) *Aspergillus flavus*.

NOTE 5—The sum of total bacteria counts and total fungi counts should be reported separately.

10.6.2 Agar plates and impinger solutions are to be processed by standard microbiological procedures such as those described in Refs. **1-3**, **9**, **13**, and **14**.

10.6.3 Identify representatives of both typical and atypical colony morphology. The microorganisms are being recovered from a stressed environment, consequently the colony morphology on the various media may not be typical.

11. Enumeration of the Microorganisms

11.1 *Impactor Samples*:

11.1.1 Count impactor plates as specified by the manufacturer of the impactor. Correct impactor plate counts using the positive hold conversion table supplied by the manufacturer.

11.1.2 When multistage impactors are used, report results for each stage.

11.1.3 Because the loading varies among the impactor plates, report which stages have less than 30 and more than 300 colonies.

11.1.4 If none of the plates had more than 30 colonies, report the computed counts as an estimated total count.

11.1.5 If at least one of the plates had more than 300 colonies, report the computed counts as greater than (>) the calculated concentration of total count.

11.1.6 Plates containing more than 399 colonies are reported as too numerous to count (TNTC).

11.1.7 Plates containing spreading colonies must be so reported on the data sheet. If "spreaders" exceed one-half of the total plate area, the plate should not be used and the results for the plate should be reported as "No results, spreaders."

11.1.8 Colonies can be counted on representative portions of plates if spreading colonies constitute less than one-half of the total plate area following the procedure in *Microbiological Methods for Monitoring the Environment, Water and Wastes* **(14)**.

11.2 *Impinger Samples*:

11.2.1 Serial dilution procedures should be used to adjust the sample concentration prior to further processing. Choose a dilution that will provide 30 to 300 colonies per plate.

11.2.2 Standard plate counts should be determined and recorded as specified in Paragraph 5.6, Part III, of Ref **(14)**. Report all results as the number of colony-forming units per cubic metre of air.

12. Report

12.1 Report all concentrations as the number of colony-forming units (that is, particles that contain at least one living organism) per cubic metre of air.

12.2 For each site and each type of microorganism analyzed, report the mean and standard deviation of the logtransform of the concentrations according to the procedure described in Part IV (Quality Assurance) of *Microbiological Methods for Monitoring the Environment, Water and Wastes* **(14)**.

NOTE 6—It is recognized that the standard deviation may be quite large, particularly, if minimum sample sizes are used, but the reporting of standard deviations will at least indicate the spread of the data and will be useful in putting perspective on the sampling results.

12.3 The report of results should include the following:

12.3.1 Concentrations at each sampling site,

12.3.2 Descriptions of the number and location of sampling sites,

12.3.3 Activities at sampling sites,

12.3.4 The temperature and relative humidity of the collected air at the sampling site,

12.3.5 The flow rates,

12.3.6 The sampling times,

12.3.7 The holding time and temperature between the time of sampling and the time of assay,

12.3.8 Descriptions of the microbiological procedures used in the identification of organisms,

12.3.9 The suppliers and lot numbers of the various medias used, and

12.3.10 The colony counts, if any, recovered on "control" petri dishes described in 10.5.1.1.

12.4 Additional meteorological parameters such as solar radiation, precipitation, wind speed, and wind direction may also influence recoveries and should be reported where possible.

13. Precision and Accuracy

13.1 The information in this section is derived from the data collected at a facility and analyzed by L. Lembke and R.N. Kniseley at Ames Laboratory, U.S. Department of Energy **(12)**.

13.2 *Repeatability* (single-laboratory)—The coefficient of variation of total plate counts obtained with gelatin milk broth **(16)** in the all-glass impinger was estimated to be 0.38 for an observed aerosol concentration of 10^4 colony-forming units per cubic metre of air in an operating municipal solid-waste processing facility. The coefficient of variation of counts obtained with 5 % sheep blood agar plates in the six-stage Andersen impactor was estimated to be 0.23 for an observed aerosol concentration of 10^4 colony-forming units per cubic metre of air in an operating municipal solid-waste processing facility.

13.3 *Accuracy*—No procedure presently exists to unequivocably determine the accuracy of this method. At Ames Laboratory, simultaneously operated all-glass impingers yielded colony counts that indicated a strong linear relationship between samplers over an observed aerosol concentration of 1.1×10^3 to 2.8×10^7 colony-forming units per cubicmetre of air in an operating municipal solid-waste processing facility. Simultaneously operated six-stage impactors also yielded colony counts that indicated a strong linear relationship between samplers over an observed aerosol concentration of 3.9×10^3 to 1.9×10^5colony-forming units per cubic metre of air in an operating municipal solid-waste processing facility.

REFERENCES

(1) *Compendium of Methods for the Microbiological Examination of Foods*, American Public Health Assn. (APHA), Section 31.93, Washington, D.C., 1976.

(2) *Standard Methods for the Examination of Water and Wastewater*, American Public Health Assn. (APHA), Fourteenth Edition, Washington, D.C., 1976.

(3) *Standard Methods for the Examination of Dairy Products*, American Public Health Assn. (APHA), Eleventh Edition, Washington, D.C., 1960.

(4) Andersen, A. A., *New Sampler for the Collection, Sizing and Enumeration of Viable Airborne Particles.* J. Bacteriol., Vol 76, 1958, pp. 471–484.

(5) Brachman, P. S.; Erlich, R.; Eichenwald, H. F.; Cabelli, V. J.; Kethley, C. W.; Madin, S. H.; Maltman, J. R.; Middlebrook, G.; Morton, J. D.; Silver, I. H.; and Wolfe, E. K., "Standard Sampler for Assay of Airborne Microorganisms," *Science*, Vol 144, 1964, (3642): p. 1295.

(6) Chatigny, Mark A., "Sampling Airborne Microorganisms", *Air Sampling Instruments for Evaluation of Atmospheric Contaminants*, Fifth Edition, American Conference of Government of Industrial Hygienists, Cincinnati, Ohio, 1978.

(7) Corn, M. and Esman, N. A., "Workplace Exposure Zones for Classification of Employee Exposures to Physical and Chemical Agents," *Amer. Industrial Hygiene Assoc. Journal*, Vol 40, 1979, pp. 47–56.

(8) Dimmick, R. L. and Akers, A. B., *An Introduction to Experimental Aerobiology*, Wiley-Interscience, New York, N.Y., 1969.

(9) Emmons, C. W., Binford, C. H., and Utz, J. P., *Medical Mycology*, Lea and Feibiger, Philadelphia, Pa., 1963.

(10) Gregory, P. H., *The Microbiology of the Atmosphere*, John Wiley & Sons, New York, N.Y., 1973.

(11) Larson, E. W., Demchak, P., and Dominik, J. W., "Comparisons Between the Single-Stage Impaction Device and Multistage Liquid Impinger Systems for Sampling and Sizing Biological Aerosols," *Airborne Transmission and Airborne Infection*, ed. by Hers, J. F., Ph. and Winkeler, K. C., John Wiley & Sons, New York, N.Y., 1973.

(12) Lembke, L. L., Kniseley, R. N., Van Nostrand, R. C., and Hale, M. D. "Precision of the All-Glass Impinger and the Andersen Microbial Impactor for Air Sampling in Solid-Waste Handling Facilities," *Applied and Environmental Microbiology*, Vol 42, 1981, pp. 222–225.

(13) Raper, K. B. and Fennell, D. I., *The Genus Aspergillus*, Robert E. Krieger, Huntington, New York, 1977.

(14) *Microbiological Methods for Monitoring the Environment, Water and Wastes*, EPA-600/8-78-017, U. S. Environmental Protection Agency, December 1978. Available from National Technical Information Service.[3]

(15) Wells, W. F., *Airborne Contagion and Air Hygiene*, Harvard University Press, Cambridge, Mass., 1955.

(16) White, L. A., Hadley, D. J., Davids, D. E., and Naylor, R., "Improved Large-Volume Sampler for the Collection of Bacterial Cells from Aerosol," *Applied Microbiology*, Vol 29, 1975, pp. 335–339.

(17) Wolf, H. W., Skaliy, P., Hall, L. B., Harris, M. M., Decker, H. M., Buchanan, L. M., and Dahlgren, C. M., "Sampling Microbiological Aerosols," *Public Health Monograph*, No. 60, U. S. Department of Health, Education, and Welfare, Public Health Service Publication No. 686, Washington, D.C., 1959.

Standard Test Methods for Analyses of Metals in Refuse-Derived Fuel by Atomic Absorption Spectroscopy[1]

This standard is issued under the fixed designation E 885; the number immediately following the designation indicates the year of original adoption or, in the case of revision, the year of last revision. A number in parentheses indicates the year of last reapproval. A superscript epsilon (ϵ) indicates an editorial change since the last revision or reapproval.

1. Scope

1.1 These test methods cover the determination of metals in solution by atomic absorption spectroscopy (AAS).

1.2 The following sections outline the operating parameters for the individual metals:

	Sections
Aluminum, Direct Aspiration	12
Aluminum, Furnace Technique	13
Antimony, Direct Aspiration	14
Antimony, Furnace Technique	15
Arsenic, Furnace Technique	16
Arsenic, Gaseous Hydride Method	17
Barium, Direct Aspiration	18
Barium, Furnace Technique	19
Beryllium, Direct Aspiration	20
Beryllium, Furnace Technique	21
Cadmium, Direct Aspiration	22
Cadmium, Furnace Technique	23
Calcium, Direct Aspiration	24
Chromium, Direct Aspiration	25
Chromium, Furnace Technique	26
Chromium, Chelation-Extraction	27
Chromium, Hexavalent, Chelation-Extraction	28
Cobalt, Direct Aspiration	29
Cobalt, Furnace Technique	30
Copper, Direct Aspiration	31
Copper, Furnace Technique	32
Iron, Direct Aspiration	33
Iron, Furnace Technique	34
Lead, Direct Aspiration	35
Lead, Furnace Technique	36
Lithium, Direct Aspiration	37
Magnesium, Direct Aspiration	38
Manganese, Direct Aspiration	39
Manganese, Furnace Technique	40
Mercury, Cold Vapor Technique	41
Molybdenum, Direct Aspiration	42
Molybdenum, Furnace Technique	43
Nickel, Direct Aspiration	44
Nickel, Furnace Technique	45
Potassium, Direct Aspiration	46
Selenium, Furnace Technique	47
Selenium, Gaseous Hydride	48
Silver, Direct Aspiration	49
Silver, Furnace Technique	50
Sodium, Direct Aspiration	51
Tin, Direct Aspiration	52
Tin, Furnace Technique	53
Titanium, Direct Aspiration	54
Titanium, Furnace Technique	55
Vanadium, Direct Aspiration	56
Vanadium, Furnace Technique	57
Zinc, Direct Aspiration	58
Zinc, Furnace Technique	59

1.3 Detection limits, sensitivity, and optimum ranges of the test methods will vary with the various makes and models of atomic absorption spectrophotometers. The data shown in Table 1 provide some indication of the actual concentration ranges measurable by direct aspiration and using furnace techniques. In the majority of instances, the concentration range shown in the table by direct aspiration may be extended much lower with scale expansion and conversely extended upwards by using a less sensitive wavelength or by rotating the burner head. Detection limits by direct aspiration may also be extended through concentration of the sample or through solvent extraction techniques, or both. Lower concentrations may also be determined using the furnace techniques. The concentration ranges given in Table 1 are somewhat dependent on equipment such as the type of spectrophotometer and furnace accessory, the energy source, and the degree of electrical expansion of the output signal.

1.4 When using the furnace techniques, the analyst should be cautioned as to possible chemical reactions occurring at elevated temperatures that may result in either suppression or enhancement of the analysis element. To ensure valid data with furnace techniques, the analyst must examine each matrix for interference effects (see 6.2) and if detected, treat accordingly using either successive dilution, matrix modification or method of standard additions (see 10.5).

1.5 Where direct aspiration atomic absorption techniques do not provide adequate sensitivity, in addition to the furnace procedure, reference is made to specialized procedures such as gaseous hydride method for arsenic and selenium, the cold-vapor technique for mercury and the chelation-extraction procedure for selected metals.

1.6 *This standard does not purport to address all of the safety concerns, if any, associated with its use. It is the responsibility of the user of this standard to establish appropriate safety and health practices and determine the applicability of regulatory limitations prior to use.* For hazard statement, see 8.4 and 17.2.3.

[1] These test methods are under the jurisdiction of ASTM Committee D-34 on Waste Management and are the direct responsibility of Subcommittee D34.08 on Thermal Treatment.

Current edition approved March 25, 1988. Published May 1988. Originally published as E 885 – 82. Last previous edition E 885 – 82.

TABLE 1 Atomic Absorption Concentrations[A]

Metal	Direct Aspiration			Furnace Technique[B,C]	
	Detection Limit, mg/L	Sensitivity, mg/L	Optimum Concentration Range, mg/L	Detection Limit, µg/L	Optimum Concentration Range, µg/L
Aluminum	0.1	1	5 to 50	3	20 to 200
Antimony	0.2	0.5	1 to 40	3	20 to 300
Arsenic[D]	0.002	...	0.002 to 0.02	1	5 to 100
Barium (*P*)	0.1	0.4	1 to 20	2	10 to 200
Beryllium	0.005	0.025	0.05 to 2	0.2	1 to 30
Cadmium	0.005	0.025	0.05 to 2	0.1	0.5 to 10
Calcium	0.01	0.08	0.2 to 7	...	...
Chromium	0.05	0.25	0.5 to 10	1	5 to 100
Cobalt	0.05	0.2	0.5 to 5	1	5 to 100
Copper	0.02	0.1	0.2 to 5	1	5 to 100
Iron	0.03	0.12	0.3 to 5	1	5 to 100
Lead	0.1	0.5	1 to 20	1	5 to 100
Lithium	...	0.035	...	...	...
Magnesium	0.001	0.007	0.02 to 0.5	...	...
Manganese	0.01	0.05	0.1 to 3	0.2	1 to 30
Mercury[E]	0.0002	...	0.0002 to 0.01	...	...
Molybdenum (*P*)	0.1	0.4	1 to 40	1	3 to 60
Nickel (*P*)	0.04	0.15	0.3 to 5	1	5 to 100
Potassium	0.01	0.04	0.1 to 2	...	...
Selenium[D]	0.002	...	0.002 to 0.02	2	5 to 100
Silver	0.01	0.06	0.1 to 4	0.2	1 to 25
Sodium	0.002	0.015	0.03 to 1	...	...
Tin	0.8	4	10 to 300	5	20 to 300
Titanium (*P*)	0.4	2	5 to 100	10	50 to 500
Vanadium (*P*)	0.2	0.8	2 to 100	4	10 to 200
Zinc	0.005	0.02	0.05 to 1	0.05	0.2 to 4

[A] The concentrations shown are not contrived values and should be obtainable with any satisfactory atomic absorption spectrophotometer.
[B] For furnace sensitivity values consult instrument operating manual.
[C] The listed furnace values are those expected when using a 20µ L injection and normal gas flow except in the case of arsenic and selenium where gas interrupt is used. The symbol (*p*) indicates the use of pyrolytic graphite with the furnace procedure.
[D] Gaseous hydride method.
[E] Cold vapor technique.

2. Referenced Documents

2.1 *ASTM Standards:*

D 1193 Specification for Reagent Water[2]

D 3223 Test Method for Total Mercury in Water[2]

E 926 Test Methods of Preparing Refuse-Derived Fuel (RDF) Samples for Analyses of Metals[3]

3. Terminology

3.1 *Definitions of Terms Specific to This Standard:*

3.1.1 *detection limit*—detection limits can be expressed as either an instrumental or method parameter. The limiting factor of the former using acid water standards would be the signal to noise ratio and degree of scale expansion used; while the latter would be more affected by the sample matrix and preparation procedure used.

3.1.1.1 The Scientific Apparatus Makers Association (SAMA) has approved the following definition: The detection limit is that concentration of an element which would yield an absorbance equal to twice the standard deviation of a series of measurements of a solution, the concentration of which is distinctly detectable above, but close to blank absorbance measurement.

3.1.1.2 The detection limit values listed in Table 1 and on individual metal methods are to be considered minimum working limits achievable with the procedures outlined in these test methods.

3.1.2 *optimum concentration range*—a range defined by limits expressed in concentration, below which scale expansion must be used and above which curve correction should be considered. The range will vary with the sensitivity of the instrument and the operating condition employed.

3.1.3 *sensitivity*—the concentration in milligrams of metal per litre that produces an absorption of 1 %.

4. Summary of Test Methods

4.1 In direct aspiration atomic absorption spectroscopy, a sample is aspirated and atomized in a flame. The light beam from a hollow cathode lamp whose cathode is made of the element to be determined is directed through the flame into a monochromator, and into a detector that measures the amount of light absorbed. Absorption depends upon the presence of free unexcited ground state atoms in the flame. Since the wavelength of the light beam is characteristic of only the metal being determined, the light energy absorbed by the flame is a measure of the concentration of that metal in the sample. This principle is the basis of atomic absorption spectroscopy.

4.2 Pretreatment of a solid sample is necessary for complete dissolution of the metals and complete breakdown of organic material prior to analysis (see Methods E 926). This process may vary because of the metals to be determined and the nature of the sample being analyzed.

4.3 When using the furnace technique in conjunction with an atomic absorption spectrophotometer, a representative aliquot of the sample is placed in the graphite tube in the furnace, evaporated to dryness, charred, and atomized. As a greater

[2] *Annual Book of ASTM Standards,* Vol 11.01.
[3] *Annual Book of ASTM Standards,* Vol 11.04.

percentage of available atoms are vaporized and dissociated for absorption in the tube than the flame, the use of small sample volumes or detection of low concentrations of elements is possible. The principle is essentially the same as with direct aspiration atomic absorption except a furnace, rather than a flame, is used to atomize the sample. Radiation from a given excited element is passed through the vapor containing ground state atoms of that element. The intensity of the transmitted radiation decreases in proportion to the amount of the ground state element in the vapor. The metal atoms to be measured are placed in the beam of radiation by increasing the temperature of the furnace, thereby causing the injected specimen to be volatilized. A monochromator isolates the characteristic radiation from the hollow cathode lamp, and a photosensitive device measures the attenuated transmittal radiation.

5. Significance and Use

5.1 Metals in solution may be readily determined by atomic absorption spectroscopy (AAS). The method is simple, rapid, and applicable to a large number of metals in solution. Solid type samples may be analyzed after proper treatment.

6. Interferences

6.1 *Direct Aspiration*:

6.1.1 The most troublesome type of interference in atomic absorption spectrophotometry is usually termed "chemical" and is caused by lack of absorption of atoms bound in molecular combination to the flame. This phenomenon can occur when the flame is not sufficiently hot to dissociate the molecule, as in the case of phosphate interference with magnesium, or because the dissociated atom is immediately oxidized to a compound that will not dissociate further at the temperature of the flame. The addition of lanthanum will overcome the phosphate interference in the magnesium, calcium, and barium determinations. Similarly, silica interference in the determination of manganese can be eliminated by the addition of calcium.

6.1.2 Chemical interferences may also be eliminated by separating the metal from the interfering material. While complexing agents are primarily employed to increase the sensitivity of the analysis, they may also be used to eliminate or reduce interferences.

6.1.3 Highly dissolved solids in the sample being aspirated may result in an interference from nonatomic absorbance such as light scattering. If background correction is not available, a nonabsorbing wavelength should be checked. Preferably, high solid content solutions should be extracted (see 6.1.1 and 11.2).

6.1.4 Ionization interferences occur where the flame temperature is sufficiently high to generate the removal of an electron from a neutral atom, giving a positive charged ion. This type of interference can generally be controlled by the addition, to both standard and sample solutions, of a large excess of an easily ionized element.

6.1.5 Although quite rare, spectral interference can occur when an absorbing wavelength of an element present in the sample but not being determined falls within the width of the absorption line of the element of interest. The results of the determination will then be erroneously high, due to the contribution of the interfering element to the atomic absorption signal. Also, interference can occur when resonant energy from another element in a multi-element lamp or a metal impurity in the lamp cathode falls within the bandpass of the slit setting and that metal is present in the sample. This type of interference may sometimes be reduced by narrowing the slit width.

6.2 *Flameless Atomization*:

6.2.1 Although the problem of oxide formation is greatly reduced with furnace procedures because atomization occurs in an inert atmosphere, the technique is still subject to chemical and matrix interferences. The composition of the sample matrix can have a major effect on the analysis. It is those effects that must be determined and taken into consideration in the analysis of each different matrix encountered. To help verify the absence of matrix or chemical interference, use the following procedure. Withdraw from the sample two equal aliquots. To one of the aliquots, add a known amount of analyte and dilute both aliquots to the same predetermined volume. (The dilution volume should be based on the analysis of the undiluted sample. Preferably, the dilution should be 1:4 while keeping in mind the optimum concentration range of the analysis. Under no circumstances should the dilution be less than 1:1). The diluted aliquots should then be analyzed and the unspiked results multiplied by the dilution factor should be compared to the original determination. Agreement of the results (within ± 10 %) indicates the absence of interference. Comparison of the actual signal from the spike to the expected response from the analyte in an aqueous standard should help confirm the finding from the dilution analysis. Those samples that indicate the presence of interference should be treated in one or more of the following ways.

6.2.1.1 The samples should be successively diluted and reanalyzed to determine if the interference can be eliminated.

6.2.1.2 The matrix of the sample should be modified in the furnace. Examples are the addition of ammonium nitrate to remove alkali chlorides, ammonium phosphate to retain cadmium, and nickel nitrate for arsenic and selenium analysis **(1)**.[4] The mixing of hydrogen with the inert purge gas has also been used to suppress chemical interference. The hydrogen acts as a reducing agent and aids in molecular dissociation.

6.2.1.3 Analyze the sample by method of standard additions while noting the precautions and limitations of its use (see 10.5).

6.2.2 Gases generated in the furnace during atomization may have molecular absorption bands encompassing the analytical wavelength. When this occurs, either the use of background correction or choosing an alternate wavelength outside the absorption band should eliminate this interference. Background correction can also compensate for nonspecific broad band absorption interference.

6.2.3 Interference from a smoke-producing sample matrix can sometimes be reduced by extending the charring time at a higher temperature or using an ashing cycle in the presence of air. Care must be taken, however, to prevent loss of the analysis element.

6.2.4 Samples containing large amounts of organic materials should be oxidized by conventional acid digestion prior to

[4] The boldface numbers in parentheses refer to the list of references at the end of these test methods.

being placed in the furnace. In this way, broad-band absorption will be minimized.

6.2.5 From anion-interference studies in the graphite furnace it is generally accepted that nitrate is the preferred anion. Therefore, nitric acid is preferable for any digestion or solubilization step. If another acid in addition to HNO_3 is required, a minimum amount should be used. This applies particularly to hydrochloric and to a lesser extent to sulfuric and phosphoric acids.

6.2.6 Carbide formation resulting from the chemical environment of the furnace has been observed with certain elements that form carbides at high temperatures. Molybdenum may be cited as an example. When this takes place, the metal will be released very slowly from the carbide as atomization continues. For molybdenum, one may be required to atomize for 30 s or more before the signal returns to baseline levels. This problem is greatly reduced, and the sensitivity increased with the use of pyrolytically-coated graphite.

6.2.7 Ionization interferences have to date not been reported with furnace techniques.

6.2.8 For comments on spectral interference see 6.1.5.

6.2.9 Contamination of the sample can be a major source of error because of the extreme sensitivities achieved with the furnace. The sample-preparation work area should be kept scrupulously clean. All glassware should be cleaned as directed. Pipet tips have been known to be a source of contamination. If suspected, they should be acid soaked with 1:5 HNO_3 and rinsed thoroughly with tap and deionized water. The use of a better grade pipet tip can greatly reduce this problem. It is very important that special attention be given to reagent blanks in both analysis and the correction of analytical results. Lastly, pyrolytic graphite, because of the production process and handling, can become contaminated. As many as five to possibly ten high temperature burns may be required to clean the tube before use.

7. Apparatus

7.1 *Atomic Absorption Spectrophotometer*—Single or dual channel, single- or double-beam instrument having a grating monochromator, photomultiplier detector, adjustable slits, a wavelength range from 190 to 800 nm, and provisions for interfacing with a strip-chart recorder.

7.2 *Burner*—The burner recommended by the particular instrument manufacturer should be used. For certain elements the nitrous oxide burner is required.

7.3 *Hollow Cathode Lamps*—Single-element lamps are to be preferred but multi-element lamps may be used. Electrodeless discharge lamps may also be used when available.

7.4 *Graphite Furnace*—Any furnace device capable of reaching the specified temperatures is satisfactory.

7.5 *Strip Chart Recorder*—A recorder is strongly recommended for furnace work so that there will be a permanent record, and any problems with the analysis such as drift, incomplete atomization, losses during charring, changes in sensitivity, etc., can be easily recognized.

7.6 *Pipets*—Microliter with disposable tips. Sizes can range from 5 to 100 µL as required.

7.7 *Pressure-reducing Valves*—The supplies of fuel and oxidant shall be maintained at pressures somewhat higher than the controlled operating pressure of the instrument by suitable valves.

7.8 *Separatory Flasks*—250 mL, or larger, for extraction with organic solvents.

7.9 *Glassware*—All glassware, linear polyethylene, polypropylene or Teflon containers, including sampling bottles, should be washed and rinsed in the following order: washed with detergent; rinsed with tap water, 1:1 nitric acid, tap water, 1:1 hydrochloride acid, tap water, and deionized distilled water.

7.10 *Borosilicate Glass Distillation Apparatus.*

8. Reagents and Materials

8.1 *Purity of Reagents*—Reagent grade chemicals shall be used in all tests. Unless otherwise indicated, it is intended that all reagents shall conform to the specifications of the Committee on Analytical Reagents of the American Chemical Society, where such specifications are available.[5] Other grades may be used, provided it is first ascertained that the reagent is of sufficiently high purity to permit its use without lessening the accuracy of the determination.

8.2 *Purity of Water*—Unless otherwise indicated, references to water shall be understood to mean reagent water as defined by Type II of Specification D 1193.

8.3 *Deionized Distilled Water*—Prepare by passing distilled water through a mixed bed of cation and anion exchange resins. Use deionized distilled water for the preparation of all reagents, calibration standards, and as dilution water.

8.4 *Nitric Acid* (concentrated)—If metal impurities are found to be present, distill reagent grade nitric acid in a borosilicate glass distillation apparatus, or use a spectrograde acid.

NOTE 1—**Precaution:** Perform distillation in hood with protective sash in place.

8.4.1 *Nitric Acid* (1:1)—Prepare a 1:1 dilution with deionized, distilled water by adding the concentrated acid to an equal volume of water.

8.5 *Hydrochloric Acid* (1:1)—Prepare a 1:1 solution of reagent grade hydrochloric acid and deionized distilled water. If metal impurities are found to be present, distill this mixture from a borosilicate glass distillation apparatus or use a spectrograde acid.

8.6 *Stock Standard Metal Solutions*—Prepare as directed in 10.1 and under the individual metal procedures. Commercially available stock standard solutions may be used.

8.7 *Calibration Standards*—Prepare a series of standards of the metal by dilution of the appropriate stock metal solution to cover the concentration range desired.

8.8 *Fuel and Oxidant*—Commercial grade acetylene is generally acceptable. Air may be supplied from a compressed air line, a laboratory compressor, or from a cylinder of compressed air. Reagent grade nitrous oxide is also required for certain

[5] *Reagent Chemicals, American Chemical Society Specifications*, American Chemical Society, Washington, DC. For suggestions on the testing of reagents not listed by the American Chemical Society, see *Analar Standards for Laboratory Chemicals*, BDH Ltd., Poole, Dorset, U.K., and the *United States Pharmacopeia and National Formulary*, U.S. Pharmaceutical Convention, Inc. (USPC), Rockville, MD.

determinations. Standard, commercially available argon and nitrogen are required for furnace work.

8.9 *Special Reagents for the Extraction Procedure*:

8.9.1 *Pyrrolidine Dithiocarbamic Acid (PDCA)*[6]—Prepare by adding 18 mL of analytical reagent grade pyrrolidine to 500 mL of chloroform in a litre flask.[7] Cool and add 15 mL of carbon disulfide in small portions and with swirling. Dilute to 1 L with chloroform. The solution can be used for several months if stored in a brown bottle in a refrigerator.

8.9.2 *Ammonium Hydroxide*, 2*N*—Dilute 3 mL concentrated NH_4OH to 100 mL with deionized distilled water.

8.9.3 *Bromphenol Blue Indicator* (1 g/L)—Dissolve 0.1 g bromphenol blue in 100 mL of 500 % ethanol or isopropanol.

8.9.4 *HCL* (2.5 % v/v)—Dilute 2 mL redistilled HCl to 40 mL with deionized distilled water.

9. Sample Handling and Preservation

9.1 See Test Methods E 926 for sample handling and preservation procedures.

10. Preparation of Standards and Calibration

10.1 *Stock Standard Solutions*, are prepared from high purity metals, oxides, or nonhygroscopic reagent grade salts using deionized distilled water and redistilled nitric or hydrochloric acids. (See individual analysis sheets for specific instruction.) Sulfuric or phosphoric acids should be avoided as they produce an adverse effect on many elements. The stock solutions are prepared at concentrations of 1000 mg of the metal per litre. Commercially available standard solutions may also be used.

10.2 *Calibration Standards*, are prepared by diluting the stock metal solutions at the time of analysis. For best results, calibration standards should be prepared fresh each time an analysis is to be made and discarded after use. Prepare a blank and at least four calibration standards in graduated amounts in the appropriate range. The *calibration standards* should be prepared using the same type of acid or combination of acids and at the same concentration as will result in the samples following processing. Beginning with the blank and working toward the highest standard, aspirate the solutions and record the readings. Repeat the operation with both the calibration standards and the samples a sufficient number of times to secure a reliable average reading for each solution. Calibration standards for furnace procedures should be prepared as described on the individual sheets for that metal.

10.3 Where the sample matrix is so complex that viscosity, surface tension, and components cannot be accurately matched with standards, the method of standard addition must be used. This technique relies on the addition of small, known amounts of the analysis element to portions of the sample, the absorbance difference between those, and the original solution giving the slope of the calibration curve. The method of standard addition is described in greater detail in 10.5.

10.4 For those instruments that do not read out directly in concentration, a calibration curve is prepared to cover the appropriate concentration range. Usually, this means the preparation of standards that produce an absorption of 0 to 80 %. The correct method is to convert the percent absorption readings to absorbance and plot that value against concentration. The following relationship is used to convert absorption values to absorbance:

$$\text{absorbance} = \log(100/\%\ T) = 2 - \log(\%\ T) \qquad (1)$$

where:

$\%\ T$ = 100 − % absorption

As the curves are frequently nonlinear, especially at high absorption values, the number of standards should be increased in that portion of the curve.

10.5 *Method of Standard Additions*:

10.5.1 In this test method, equal volumes of sample are added to a deionized distilled water blank and to three standards containing different known amounts of the test element. The volume of the blank and the standards must be the same. The absorbance of each solution is determined and then plotted on the vertical axis of a graph, with the concentrations of the known standards plotted on the horizontal axis. When the resulting line is extrapolated back to zero absorbance, the point of interception of the abscissa is the concentration of the unknown. The abscissa on the left of the ordinate is scaled the same as on the right side, but in the opposite direction from the ordinate. An example of a plot so obtained is shown in Fig. 1.

10.5.2 The method of standard additions can be very useful. For the results to be valid, the following limitations must be taken into consideration:

10.5.2.1 The absorbance plot of sample and standards must be linear over the concentration range of concern. For best results the slope of the plot should be nearly the same as the slope of the aqueous standard curve. If the slope is significantly different (more than 20 %) caution should be exercised.

10.5.2.2 The effect of the interference should not vary as the ratio of analyte concentration to sample matrix changes, and the standard addition should respond in a similar manner as the analyte.

10.5.2.3 The determination must be free of spectral interference and corrected for nonspecific background interference.

11. General Procedure for Analysis by Atomic Absorption

11.1 *Direct Aspiration*—Differences between the various makes and models of satisfactory atomic absorption spectrophotometers prevent the formulation of detailed instructions applicable to every instrument. The analyst should follow the manufacturer's operating instructions for his particular instrument. In general, after choosing the proper hollow cathode lamp for the analysis, allow the lamp to warm up for a minimum of 15 min unless operated in a double beam mode. During this period, align the instrument, position the monochromator at the correct wavelength, select the proper monochromator slit width, and adjust the hollow cathode current according to the manufacturer's recommendation. Subsequently, light the flame and regulate the flow of fuel and oxidant, adjust the burner and nebulizer flow rate for maximum

[6] The name pyrrolidine dithiocarbamic acid (PDCA), although commonly referenced in the scientific literature is ambiguous. From the chemical reaction of pyrrolidine and carbon disulfide a more proper name would be 1-pyrrolidine carbodithioic acid, PCDA (CAS Registry No. 25769-03-3).

[7] An acceptable grade of pyrrolidine may be obtained from the Aldrich Chemical Co., 940 West St. Paul Ave., Milwaukee, WI 53233.

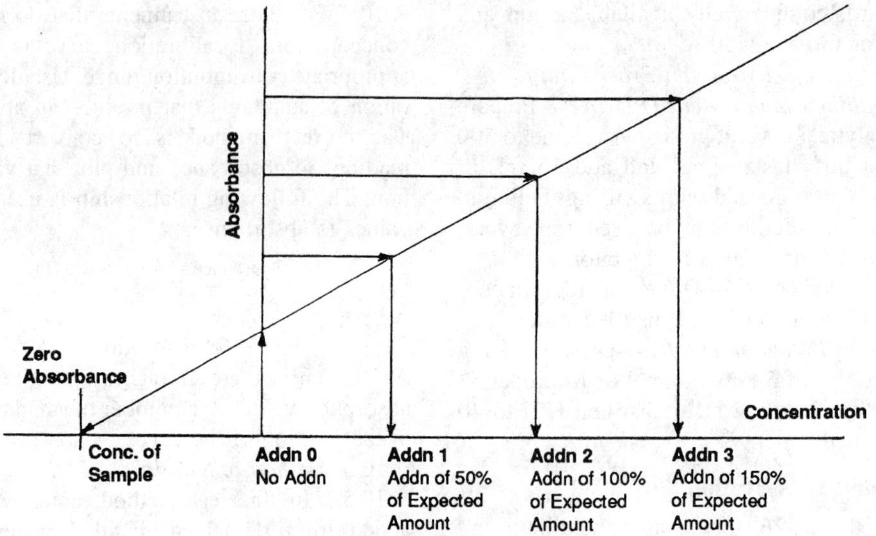

FIG. 1 Standard Addition Plot

percent absorption and stability, and balance the photometer. Run a series of standards of the element under analysis and construct a calibration curve by plotting the concentrations of the standards against the absorbance. For those instruments which read directly in concentration set the curve corrector to read out the proper concentration. Aspirate the samples and determine the concentrations either directly or from the calibration curve. Standards must be run each time a sample or series of samples are run.

11.1.1 *Calculation for Direct Determination of Liquid Samples*—Read the metal value in mg/L from the calibration curve or directly from the readout system of the instrument.

11.1.1.1 If dilution of sample was required:

$$\text{mg/L metal in sample} = A\left(\frac{C+B}{C}\right) \quad (2)$$

where:
A = mg/L of metal in diluted aliquot from calibration curve,
B = mL of deionized distilled water used for dilution, and
C = mL of sample aliquot.

11.1.2 For solid samples: report all concentrations as mg/kg dry weight.

11.1.2.1 Dry sample:

$$\text{mg metal/kg sample} = \frac{A \times V}{D} \quad (3)$$

where:
A = mg/L of metal in processed sample from calibration curve,
V = final volume of the processed sample in mL, and
D = weight of dry sample in grams.

11.1.2.2 Wet sample:

$$\text{mg metal/kg sample} = \frac{A \times V}{W \times P} \quad (4)$$

where:
A = mg/L of metal in processed sample from calibration curve,
V = final volume of the processed sample in mL,
W = weight of wet sample in grams, and
P = percent solids.

11.2 *Special Extraction Procedure*—When the concentration of the metal is not sufficiently high to determine directly, or when considerable dissolved solids are present in the sample, certain metals may be chelated and extracted with organic solvents. Ammonium pyrrolidine dithiocarbamate (APDC)[8] in methyl isobutyl ketone (MIBK) is widely used for this purpose and is particularly useful for zinc, cadmium, iron, manganese, copper, silver, lead and chromium^{+6}. Trivalent chromium does not react with APDC unless it has first been converted to the hexavalent form **(2)**. This procedure is described under method for chromium (chelation extraction). Aluminum, beryllium, barium and strontium also do not react with APDC. While the APDC-MIBK chelating-solvent system can be used satisfactorily, it is possible to experience difficulties.

NOTE 2—Certain metal chelates, manganese-APDC in particular, are not stable in MIBK and will redissolve into the aqueous phase on standing. The extraction of other metals is sensitive to both shaking rate and time. As with cadmium, prolonged extraction beyond 1 min, will reduce the extraction efficiency, whereas 3 min of vigorous shaking is required for chromium. Also, when multiple metals are to be determined either larger sample volumes must be extracted or individual extractions made for each metal being determined. The acid form of APDC-pyrrolidine dithiocarbamic acid prepared directly in chloroform as described by Lakanen has been found to be most advantageous **(3)**. In this procedure the more dense chloroform layer allows for easy combination of multiple extractions which are carried out over a broader pH range favorable to multielement extraction. Pyrrolidine dithiocarbamic acid in chloroform is very stable and may be stored in a brown bottle in the refrigerator for months. Because chloroform is used as the solvent, it may not be aspirated into the flame. The procedure described in 11.2.1 is suggested.

11.2.1 *Extraction Procedure with Pyrrolidine Dithiocarbamic Acid (PDCA) in Chloroform*:

[8] The name ammonium pyrrolidine dithiocarbamate (APDC) is somewhat ambiguous and should more properly be called ammonium, 1-pyrollidine carbodithioate (APCD), CAS Registry No. 5108-96-3.

11.2.1.1 Transfer 200 mL of sample into a 250-mL separatory funnel, add 2 drops bromphenol blue indicator solution (8.9.3) and mix.

11.2.1.2 Prepare a blank and sufficient standards in the same manner and adjust the volume of each to approximately 200 mL with deionized distilled water. All of the metals to be determined may be combined into single solutions at the appropriate concentration levels.

11.2.1.3 Adjust the pH by addition of 2N NH_4OH solution (8.9.2) until a blue color persists. Add HCl (8.9.4) dropwise until the blue color just disappears; then add 2.0 mL HCl (8.9.4) in excess. The pH at this point should be 2.3. (The pH adjustment may be made with a pH meter instead of using indicator.)

11.2.1.4 Add 5 mL of PDCA-chloroform reagent (8.9.1) and shake vigorously for 2 min. Allow the phases to separate and drain the chloroform layer into a 100-mL beaker.

NOTE 3—If hexavalent chromium is to be extracted, the aqueous phase must be readjusted back to a pH of 2.3 after the addition of PDCA-chloroform and maintained at that pH throughout the extraction. For multielement extraction, the pH may be adjusted upward after the chromium has been extracted.

11.2.1.5 Add a second portion of 5 mL PDCA-chloroform reagent (8.7.1) and shake vigorously for 2 min. Allow the phases to separate and combine the chloroform phase with that obtained in 11.2.1.4.

11.2.1.6 Determine the pH of the aqueous phase and adjust to 4.5.

11.2.1.7 Repeat 11.2.1.4 again combining the solvent extracts.

11.2.1.8 Readjust the pH to 5.5, and extract a fourth time. Combine all extracts and evaporate to dryness on a steam bath.

11.2.1.9 Hold the beaker at a 45° angle, and slowly add 2 mL of concentrated distilled nitric acid, rotating the beaker to effect thorough contact of the acid with the residue.

11.2.1.10 Place the beaker on a low temperature hotplate or steam bath and evaporate just to dryness.

11.2.1.11 Add 2 mL of nitric acid (1:1) to the beaker and heat for 1 min. Cool, quantitatively transfer the solution to a 10-mL volumetric flask and bring to volume with distilled water. The sample is now ready for analysis.

11.2.2 Prepare a calibration curve by plotting absorbance versus the concentration of the metal standard (μg/L) in the 200-mL extracted standard solution. To calculate sample concentration read the metal value in μg/L from the calibration curve or directly from the readout system of the instrument. If dilution of the sample was required use the following equation:

$$\text{mg/L metal in sample} = Z\left(\frac{C + B}{C}\right) \quad (5)$$

where:

Z = μg/L of metal in diluted aliquot from calibration curve,
B = mL of deionized distilled water used for dilution, and
C = mL of sample aliquot.

11.3 *Furnace Procedure*—Furnace devices (flameless atomization) are a most useful means of extending detection limits. Because of differences between various makes and models of satisfactory instruments, no detailed operating instructions can be given for each instrument. Instead, the analyst should follow the instructions provided by the manufacturer of his particular instrument and use as a guide the temperature settings and other instrument conditions listed on the individual analysis sheets which are recommended for the Perkin-Elmer HGA-2100.[9] In addition, the following points may be helpful.

11.3.1 With flameless atomization, background correction becomes of high importance especially below 350 nm. This is because certain samples, when atomized, may absorb or scatter light from the hollow cathode lamp. It can be caused by the presence of gaseous molecular species, salt particles, or smoke in the sample beam. If no correction is made, sample absorbance will be greater than it should be, and the analytical result will be erroneously high.

11.3.2 If during atomization all the analyte is not volatilized and removed from the furnace, memory effects will occur. This condition is dependent on several factors such as the volatility of the element and its chemical form, whether pyrolytic graphite is used, the rate of atomization and furnace design. If this situation is detected through blank burns, the tube should be cleaned by operating the furnace at full power for the required time period as needed at regular intervals in the analytical scheme.

11.3.3 Some of the smaller size furnace devices, or newer furnaces equipped with feedback temperature control employing faster rates of atomization, can be operated using lower atomization temperatures for shorter time periods than those listed in this manual.[10]

11.3.4 Although prior digestion of the sample in many cases is not required providing a representative aliquot of sample can be pipeted into the furnace, it will provide for a more uniform matrix and possibly lessen matrix effects.

11.3.5 Inject a measured microlitre aliquot of sample into the furnace and atomize. If the concentration found is greater than the highest standard, the sample should be diluted in the same acid matrix and reanalyzed. The use of multiple injections can improve accuracy and help detect furnace pipetting errors.

11.3.6 To verify the absence of interference, follow the procedure as given in part 6.2.1.

11.3.7 A check standard should be run approximately after every 10 sample injections. Standards are run in part to monitor the life and performance of the graphite tube. Lack of reproducibility or significant change in the signal for the standard indicates that the tube should be replaced. Even though tube life depends on sample matrix and atomization temperature, a conservative estimate would be that a tube will last at least 50 firings. A pyrolytic-coating would extend that estimate by a factor of 3.

11.3.8 *Calculation*—For determination of metal concentration by the furnace: Read the metal value in μg/L from the calibration curve or directly from the readout system of the instrument.

[9] The Perkin-Elmer HGA-2100 available from Perkin-Elmer Corp., Instruments Division, Main Ave., Norwalk, CT 06858 has been found suitable.

[10] Instrumentation Laboratories Model 555 available from Instrumentation Laboratory, Inc., Analytical Instrumentation Division, Jonspin Road, Wilmington, MA 01887; Perkin-Elmer Models HGA2200 and HGA7613; and Varian Model CRA-90 available from Varian Associates, Inc., 611 Hansen Way, Palo Alto, CA 94303 have been found suitable.

11.3.8.1 If different size furnace injection volumes are used for samples rather than for standards, calculate as follows:

$$\mu g/L \text{ of metal in sample} = Z\left(\frac{S}{U}\right) \quad (6)$$

where:
Z = µg/L of metal read from calibration curve or readout system,
S = µL volume standard injected into furnace for calibration curve, and
U = µL volume of sample injected for analysis.

11.3.8.2 If dilution of sample was required but sample injection volume was the same as for the following standard:

$$\mu g/L \text{ of metal in sample} = Z\left(\frac{C + B}{C}\right) \quad (7)$$

where:
Z = µg/L metal in diluted aliquot from calibration curve,
B = mL of deionized distilled water used for dilution, and
C = mL of sample aliquot.

11.3.9 For solid samples, report all concentrations as mg/kg dry weight.

11.3.9.1 *Dry sample*:

$$\text{mg metal/kg sample} = \frac{\left(\frac{Z}{1\,000}\right)V}{D} \quad (8)$$

where:
Z = µg/L of metal in processed sample from calibration curve (see 11.3.8.1),
V = final volume of processed sample in millilitres, and
D = weight of dry sample in grams.

11.3.10 *Wet sample*:

$$\text{mg metal/kg sample} = \frac{\left(\frac{Z}{1\,000}\right)V}{W \times P} \quad (9)$$

where:
Z = µg/L of metal in processed sample from calibration curve (see 11.3.8.1),
V = final volume of processed sample in millilitres,
W = weight of wet sample in grams, and
P = percent solids.

12. Aluminum—Direct Aspiration

12.1 *Requirements*:

12.1.1 *Optimum Concentration Range*, 5 to 50 mg/L using a wavelength of 309.3 nm (see Note 4 and Note 5).

12.1.2 *Sensitivity*, 1 mg/L.

12.1.3 *Detection Limit*, 0.1 mg/L.

NOTE 4—The following lines may also be used:
308.2 nm Relative Sensitivity 1
396.2 nm Relative Sensitivity 2
394.4 nm Relative Sensitivity 2.5

NOTE 5—For concentrations of aluminum below 0.3 mg/L, the furnace procedure is recommended.

12.2 *Preparation of Standard Solution*:

12.2.1 *Stock Solution*—Carefully weigh 1.000 g of aluminum metal (analytical reagent grade). Add 15 mL of concentrated HCl to the metal, cover the beaker, and warm gently. When solution is complete, transfer quantitatively to a 1 L-volumetric flask and make up to volume with deionized distilled water. 1 mL = 1 mg Al (1000 mg/L).

12.2.2 *Potassium Chloride Solution*—Dissolve 95 g potassium chloride (KCl) in deionized distilled water and make up to 1 L.

12.2.3 Prepare dilutions of the stock solution to be used as calibration standards at the time of analysis. The *calibration standards* should be prepared using the same type of acid and at the same concentration as that of the sample being analyzed either directly or after processing. To each 100 mL of standard and sample alike add 2.0 mL potassium chloride solution.

12.3 *General Instrumental Parameters*:

12.3.1 *Aluminum Hollow Cathode Lamp*.

12.3.2 *Wavelength*—309.3 nm.

12.3.3 *Fuel*—Acetylene.

12.3.4 *Oxidant*—Nitrous oxide.

12.3.5 *Type of flame*—Fuel rich.

12.4 *Analysis Procedure*—For analysis procedure and calculation, see "Direct Aspiration," 11.1.

12.5 *Interferences*—Aluminum is partially ionized in the nitrous oxide-acetylene flame. This problem may be controlled by the addition of an alkali metal (potassium, 1000 µg/mL) to both sample and standard solutions.

13. Aluminum—Furnace Technique

13.1 *Requirements*:

13.1.1 *Optimum Concentration Range*, 20–200 µg/L (see Note 6).

13.1.2 *Detection Limit*, 3 µg/L.

NOTE 6—The above concentration values and instrument conditions are for a Perkin-Elmer HGA-2100, based on the use of a 20-µL injection, continuous flow purge gas and nonpyrolytic graphite.

13.2 *Preparation of Standard Solution*:

13.2.1 *Stock Solution*—Prepare as described under "direct aspiration method."

13.2.2 Prepare dilutions of the stock solution to be used as calibration standards at the time of analysis. Also use these solutions for "standard additions."

13.2.3 Dilute the calibration standard to contain 0.5 % (v/v) HNO_3.

13.3 *General Instrument Parameters*:

13.3.1 *Drying Time and Temperature*—30 s at 125°C.

13.3.2 *Ashing Time and Temperature*—30 s at 1300°C.

13.3.3 *Atomizing Time and Temperature*—10 s at 2700°C.

13.3.4 *Purge Gas Atmosphere*—Argon.

13.3.5 *Wavelength*—309.3 nm.

13.3.6 Other operating parameters should be set as specified by the particular instrument manufacturer (see Note 7 and Note 8).

NOTE 7—Background correction may be required if the sample contains high dissolved solids.

NOTE 8—It has been reported that chloride ion and that nitrogen used as a purge gas suppress the aluminum signal. Therefore, the use of halide acids and nitrogen as a purge gas should be avoided.

13.4 *Analysis Procedure*—For the analysis procedure and the calculation, see "Furnace Procedure" 11.3 (see Note 9 and Note 10).

NOTE 9—For every sample matrix analyzed, verification is necessary to determine that the method of standard additions is not required (see 6.2.1).

NOTE 10—If the method of standard additions is required, follow the procedure given earlier in 10.5.

14. Antimony—Direct Aspiration

14.1 *Requirements*:

14.1.1 *Optimum Concentration Range*, 1 to 40 mg/L using a wavelength of 217.6 nm (see Note 11).

14.1.2 *Sensitivity*, 0.5 mg/L.

14.1.3 *Detection Limit*, 0.2 mg/L.

NOTE 11—For concentrations of antimony below 0.35 mg/L, the furnace procedure is recommended.

14.2 *Preparation of Standard Solution*:

14.2.1 *Stock Solution*—Carefully weigh 2.7426 g of antimony potassium tartrate (analytical reagent grade) and dissolve in deionized distilled water. Dilute to 1 L with deionized distilled water. 1 mL = 1 mg Sb (1000 mg/L).

14.2.2 Prepare dilutions of the stock solution to be used as calibration standards at the time of analysis. The *calibration standards* should be prepared using the same type of acid and at the same concentration as that of the sample being analyzed either directly or after processing.

14.3 *General Instrumental Parameters*:

14.3.1 *Antimony Hollow Cathode Lamp*.

14.3.2 *Wavelength*—217.6 nm.

14.3.3 *Fuel*—Acetylene.

14.3.4 *Oxidant*—Air.

14.3.5 *Type of Flame*—Fuel Lean.

14.4 *Analysis Procedure*—For analysis procedure and calculation, see "Direct Aspiration," 11.1.

14.5 *Interferences*:

14.5.1 In the presence of lead (1000 mg/L), a spectral interference may occur at the 217.6-nm resonance line. In this case the 231.1-nm antimony line should be used.

14.5.2 Increasing acid concentrations decrease antimony absorption. To avoid this effect, the acid concentration in the samples and in the standards should be matched.

15. Antimony—Furnace Technique

15.1 *Requirements*:

15.1.1 *Optimum Concentration Range*, 20–300 μg/L (see Note 12).

15.1.2 *Detection Limit*, 3 μg/L.

NOTE 12—The concentration values and instrument conditions are for a Perkin-Elmer HGA-2100, based on the use of a 20-μL injection, continuous flow purge gas and non-pyrolytic graphite. Smaller sized furnace devices or those employing faster rates of atomization can be operated using lower atomization temperatures for shorter time periods than the recommended settings.

15.2 *Preparation of Standard Solution*:

15.2.1 *Stock Solution*—Prepare as described under "direct aspiration method."

15.2.2 Prepare dilutions of the stock solution to be used as calibration standards at the time of analysis. Also use these solutions for "standard additions."

15.2.3 Dilute the calibration standard to contain 0.2 % (v/v) HNO_3.

15.3 *General Instrument Parameters*:

15.3.1 *Drying Time and Temperature*—30 s at 125°C.

15.3.2 *Ashing Time and Temperature*—30 s at 800°C.

15.3.3 *Atomizing Time and Temperature*—10 s at 2700°C.

15.3.4 *Purge Gas Atmosphere*—Argon.

15.3.5 *Wavelength*—217.6 nm.

15.3.6 Other operating parameters should be set as specified by the particular instrument manufacturer (see Note 13 and Note 14).

NOTE 13—The use of background correction is recommended.

NOTE 14—Nitrogen may also be used as the purge gas.

15.4 *Analysis Procedure*—For the analysis procedure and the calculation, see "Furnace Procedure" 11.3 (see Note 15, Note 16, and Note 17).

NOTE 15—If chloride concentration presents a matrix problem or causes a loss previous to atomization, add an excess of 5 mg of ammonium nitrate to the furnace and ash using a ramp accessory or with incremental steps until the recommended ashing temperature is reached.

NOTE 16—For every sample matrix analyzed, verification is necessary to determine that the method of standard additions is not required (see 6.2.1).

NOTE 17—If the method of standard additions is required, follow the procedure given in 10.5.

16. Arsenic—Furnace Technique

16.1 *Requirements*:

16.1.1 *Optimum Concentration Range*, 5–100 μg/L (see Note 18).

16.1.2 *Detection Limit*, 1 μg/L.

NOTE 18—The above concentration values and instrument conditions are for a Perkin-Elmer HGA 2100, based on the use of a 20-μL injection, continuous flow purge gas and non-pyrolytic graphite. Smaller size furnace devices or those employing faster rates of atomization can be operated using lower atomization temperatures for shorter time periods than the above recommended settings.

16.2 *Preparation of Standard Solution*:

16.2.1 *Stock Solution*—Dissolve 1.320 g of arsenic trioxide, As_2O_3 (analytical reagent grade) in 100 mL of deionized distilled water containing 4 g NaOH. Acidify the solution with 20 mL concentrated HNO_3 and dilute to 1 L. 1 mL = 1 mg As (1000 mg/L).

16.2.2 *Nickel Nitrate Solution*, 5 %—Dissolve 24.780 g of ACS reagent grade $Ni(NO_3)_2{\cdot}6H_2O$ in deionized distilled water and make up to 100 mL.

16.2.3 *Nickel Nitrate Solution*, 1 %—Dilute 20 mL of the 5 % nickel nitrate to 100 mL with deionized distilled water.

16.2.4 *Working Arsenic Solutions*—Prepare dilutions of the stock solution to be used as calibration standards at the time of analysis. Withdraw appropriate aliquots of the stock solution, add 1 mL of concentrated HNO_3, 2 mL of 30 % H_2O_2 and 2 mL of the 5 % nickel nitrate solution. Dilute to 100 mL with deionized distilled water.

16.3 *Sample Preparation*:

16.3.1 Transfer 100 mL of well-mixed sample to a 250-mL Griffin beaker. Add 2 mL of 30 % H_2O_2 and sufficient concentrated HNO_3 to result in an acid concentration of 1 % (v/v). Heat for 1 h at 95°C or until the volume is slightly less than 50 mL.

16.3.2 Cool and bring back to 50 mL with deionized distilled water.

16.3.3 Pipet 5 mL of this digested solution into a 10 mL volumetric flask, add 1 mL of the 1 % nickel nitrate solution and dilute to 10-mL with deionized distilled water. The sample is now ready for injection into the furnace.

NOTE 19—If solubilization or digestion is not required, adjust the HNO_3 concentration of the sample to 1 % (v/v) and add 2 mL of 30 % H_2O_2 and 2 mL of 5 % nickel nitrate to each 100 mL of sample. The volume of the calibration standard should be adjusted with deionized distilled water to match the volume change of the sample.

16.4 *General Instrument Parameters*:

16.4.1 *Drying Time and Temperature*—30 s at 125°C.

16.4.2 *Ashing Time and Temperature*—30 s at 1100°C.

16.4.3 *Atomizing Time and Temperature*—10 s at 2700°C.

16.4.4 *Purge Gas Atmosphere*—Argon.

16.4.5 *Wavelength*—193.7 nm.

16.4.6 Other operating parameters should be set as specified by the particular instrument manufacturer.

NOTE 20—The use of background correction is recommended.

16.5 *Analysis Procedure*—For the analysis procedure and the calculation, see "Furnace Procedure" 11.3.

NOTE 21—For every sample matrix analyzed, verification is necessary to determine that the method of standard additions is not required (see 6.2.1).

NOTE 22—If the method of standard additions is required, follow the procedure given in 10.5.

17. Arsenic Gaseous Hydride Method

17.1 *Scope and Application:* The gaseous hydride method determines inorganic arsenic when present in concentrations at or about 2 µg/L. The method is applicable to drinking water and most fresh and saline waters in the absence of high concentrations of chromium, cobalt, copper, mercury, molybdenum, nickel, and silver.

17.2 *Summary of Test Method*:

17.2.1 Arsenic in the sample is first reduced to the trivalent form using $SnCl_2$ and converted to arsine, AsH_3, using zinc metal. The gaseous hydride is swept into an argon-hydrogen flame of an atomic absorption spectrophotometer. The working range of the method is 2 to 20 µg/L. The 193.7 nm wavelength is used.

17.2.2 Organic arsenic must be converted to inorganic compounds.

NOTE 23—**Precaution:** Arsine is a toxic gas. Precautions should be made to keep the system closed to the atmosphere.

17.3 Except for the perchloric acid step, the procedure to be used for this determination is found in Standard Methods for the Examination of Water and Wastewater **(4)**.

18. Barium—Direct Aspiration

18.1 *Requirements*:

18.1.1 *Optimum Concentration Range*, 1–20 mg/L using a wavelength of 553.6 nm (see Note 24).

18.1.2 *Sensitivity*, 0.4 mg/L.

18.1.3 *Detection Limit*, 0.1 mg/L.

NOTE 24—For concentrations of barium below 0.2 mg/L, the furnace procedure is recommended.

18.2 *Preparation of Standard Solution*:

18.2.1 *Stock Solution*—Dissolve 1.7787 g barium chloride ($BaCl_2 \cdot 2H_2O$, analytical reagent grade) in deionized distilled water and dilute to 1 L. 1 mL = 1 mg Ba (1000 mg/L).

18.2.2 *Potassium Chloride Solution*—Dissolve 95 g potassium chloride, KCl, in deionized distilled water and make up to 1 L.

18.2.3 Prepare dilutions of the stock barium solution to be used as calibration standards at the time of analysis. To each 100 mL of standard and sample alike, add 2.0 mL potassium chloride solution. The *calibration standards* should be prepared using the same type of acid and the same concentration as that of the sample being analyzed either directly or after processing.

18.3 *General Instrumental Parameters*:

18.3.1 *Barium hollow cathode lamp.*

18.3.2 *Wavelength*—553.6 nm.

18.3.3 *Fuel*—Acetylene.

18.3.4 *Oxidant*—Nitrous oxide.

18.3.5 *Type of Flame*—Fuel rich.

18.4 *Analysis of Procedure*—For analysis procedure and calculation, see "Direct Aspiration," 11.1.

18.5 *Interferences*:

18.5.1 The use of nitrous oxide-acetylene flame virtually eliminates chemical interference. However, barium is easily ionized in this flame, and potassium must be added (1000 mg/L) to standards and samples alike to control this effect.

18.5.2 If the nitrous oxide flame is not available and acetylene-air is used, phosphate, silicon and aluminum will severely depress the barium absorbance. This may be overcome by the addition of 2000 mg/L lanthanum.

19. Barium—Furnace Technique

19.1 *Requirements*:

19.1.1 *Optimum Concentration Range*, 10 to 200 µg/L (see Note 25).

19.1.2 *Detection Limit*, 2 µg/L.

NOTE 25—The above concentration values and instrument conditions are for a Perkin-Elmer HGA-2100, based on the use of a 20-µL injection, continuous flow purge gas and pyrolytic graphite.

19.2 *Preparation of Standard Solution*:

19.2.1 *Stock Solution*—Prepare as described under "direct aspiration method."

19.2.2 Prepare dilutions of the stock solution to be used as calibration standards at the time of analysis. These solutions are also to be used for "standard additions."

19.2.3 The calibration standard should be diluted to contain 0.5 % (v/v) HNO_3.

NOTE 26—The use of halide acid should be avoided.

19.3 *General Instrument Parameters*:

19.3.1 *Drying Time and Temperature*—30 s at 125°C.

19.3.2 *Ashing Time and Temperature*—30 s at 1200°C.

19.3.3 *Atomizing Time and Temperature*—10 s at 2800°C.

19.3.4 *Purge Gas Atmosphere*—Argon.

NOTE 27—Because of possible chemical interaction, nitrogen should not be used as a purge gas.

19.3.5 *Wavelength*—553.6 nm.

19.3.6 *Other Operating Parameters*, should be set as specified by the particular instrument manufacturer.

19.4 *Analysis Procedure*—For the analysis procedure and the calculation, see "Furnace Procedure" 11.3.

NOTE 28—For every sample matrix analyzed, verification is necessary to determine that the method of standard additions is not required (see 6.2.1).

NOTE 29—If the method of standard additions is required, follow the procedure given in 10.5.

20. Beryllium—Direct Aspiration

20.1 *Requirements*:

20.1.1 *Optimum Concentration Range*, 0.05 to 2 mg/L using a wavelength of 234.9 nm (see Note 30 and Note 31).

20.1.2 *Sensitivity*, 0.025 mg/L.

20.1.3 *Detection Limit*, 0.005 mg/L.

NOTE 30—The "aluminon colorimetric method" may also be used **(6)**. The minimum detectable concentration by this method is 5 µg/L.

NOTE 31—For concentrations of beryllium below 0.02 mg/L, the furnace procedure is recommended.

20.2 *Preparation of Standard Solution*:

20.2.1 *Stock solution*—Dissolve 11.6586 g beryllium sulfate, $BeSO_4$, in deionized distilled water containing 2 mL concentrated nitric acid and dilute to 1 L. 1 mL = 1 mg Be (1000 mg/L).

20.2.2 Prepare dilutions of the stock solution to be used as calibration standards at the time of analysis. The *calibration standards* should be prepared using the same type of acid and at the same concentration as that of the sample being analyzed either directly or after processing.

20.3 *General Instrumental Parameters*:

20.3.1 *Beryllium Hollow Cathode Lamp*.

20.3.2 *Wavelength*—234.9 nm.

20.3.3 *Fuel*—Acetylene.

20.3.4 *Oxidant*—Nitrous oxide.

20.3.5 *Type of Flame*—Fuel-rich.

20.4 *Analysis Procedure*:

20.4.1 For analysis procedure and calculation, see "Direct Aspiration," 11.3.

20.5 *Interferences*:

20.5.1 Sodium and silicon at concentrations in excess of 1000 mg/L have been found to severely depress the beryllium absorbance.

20.5.2 Bicarbonate ion is reported to interfere; however, its effect is eliminated when samples are acidified to a pH of 1.5.

20.5.3 Aluminum at concentrations of 500 µg/L is reported to depress the sensitivity of beryllium **(5)**.

21. Beryllium—Furnace Technique

21.1 *Requirements*:

21.1.1 *Optimum Concentration Range*, 1 to 30 µg/L (see Note 32).

21.1.2 *Detection Limit*, 0.2 µg/L.

NOTE 32—The above concentration values and instrument conditions are for a Perkin-Elmer HGA-2100, based on the use of a 20-µL injection, continuous flow purge gas and non-pyrolytic graphite. Smaller sized furnace devices or those employing faster rates of atomization can be operated using lower atomization temperatures for shorter time periods than the above recommended settings.

21.2 *Preparation of Standard Solution*:

21.2.1 *Stock Solution*—Prepare as described under "direct aspiration method."

21.2.2 Prepare dilutions of the stock solution to be used as calibration standards at the time of analysis. Also use these solutions for "standard additions."

21.2.3 The calibration standard should be diluted to contain 0.5 % (v/v) HNO_3.

21.3 *General Instrumental Parameters*:

21.3.1 *Drying Time and Temperature*—30 s at 125°C.

21.3.2 *Ashing Time and Temperature*—30 s at 1000°C.

21.3.3 *Atomizing Time and Temperature*—10 s at 2800°C.

21.3.4 *Purge Gas Atmosphere*—Argon.

21.3.5 *Wavelength*—234.9 nm.

21.3.6 The operating parameters should be set as specified by the particular instrument manufacturer.

NOTE 33—The use of background correction is recommended.

NOTE 34—Because of possible chemical interaction and reported lower sensitivity, nitrogen should not be used as the purge gas.

21.4 *Analysis Procedure*—For the analysis procedure and the calculation see "Furnace Procedure," 11.3.

NOTE 35—For every sample matrix analyzed, verification is necessary to determine that the method of standard additions is not required (see 6.2.1).

NOTE 36—If the method of standard additions is required, follow the procedure given in 10.5.

22. Cadmium—Direct Aspiration

22.1 *Requirements*:

22.1.1 *Optimum Concentration Range*, 0.05 to 2 mg/L using a wavelength of 228.8 nm (see Note 37).

22.1.2 *Sensitivity*, 0.025 mg/L.

22.1.3 *Detection Limit*, 0.005 mg/L.

NOTE 37—For levels of cadmium below 20 µg/L, either the Special Extraction Procedure given in 11.2 or the furnace technique is recommended.

22.2 *Preparation of Standard Solution*:

22.2.1 *Stock Solution*—Carefully weigh 2.282 g of cadmium sulfate ($3CdSO_4 \cdot 8H_2O$, analytical reagent grade) and dissolve in deionized distilled water. 1 mL = 1 mg Cd (1000 mg/L).

22.2.2 Prepare dilutions of the stock solution to be used as calibration standards at the time of analysis. The *calibration standards* should be prepared using the same type of acid and at the same concentration as that of the sample being analyzed either directly or after processing.

22.3 *General Instrumental Parameters*:

22.3.1 *Cadmium Hollow Cathode Lamp*.

22.3.2 *Wavelength*—228.8 nm.

22.3.3 *Fuel*—Acetylene.

22.3.4 *Oxidant*—Air.

22.3.5 *Type of Flame*—Oxidizing.

22.4 *Analysis Procedure*—For analysis procedure and calculation, see "Direct Aspiration," 11.1.

23. Cadmium—Furnace Technique

23.1 *Requirements*:

23.1.1 *Optimum Concentration Range*, 0.5 to 10 µg/L (see Note 38).

23.1.2 *Detection Limit*, 0.1 μg/L.

NOTE 38—The above concentration values and instrument conditions are for a Perkin-Elmer HGA-2100, based on the use of a 20-μL injection, continuous flow purge gas and non-pyrolytic graphite. Smaller sized furnace devices or those employing faster rates of atomization can be operated using lower atomization temperatures for shorter time periods than the above recommended settings.

23.2 *Preparation of Standard Solution*:

23.2.1 *Stock Solution:* Prepare as described under "direct aspiration method."

23.2.2 *Ammonium Phosphate Solution* (40 %)—Dissolve 40 g of ammonium phosphate, $(NH_4)2HPO_4$ (analytical reagent grade) in deionized distilled water and dilute to 100 mL.

23.2.3 Prepare dilutions of the stock cadmium solution to be used as calibration standards at the time of analysis. To each 100 mL of standard and sample alike add 2.0 mL of the ammonium phosphate solution. The calibration standards should be prepared to contain 0.5 % (v/v) HNO_3.

23.3 *General Instrument Parameters*:

23.3.1 *Drying Time and Temperature*—30 s at 125°C.

23.3.2 *Ashing Time and Temperature*—30 s at 500°C.

23.3.3 *Atomizing Time and Temperature*—10 s at 1900°C.

23.3.4 *Purge Gas Atmosphere*—Argon.

23.3.5 *Wavelength*—228.8 nm.

23.3.6 Other operating parameters should be set as specified by the particular instrument manufacturer.

NOTE 39—The use of background correction is recommended.

23.4 *Analysis Procedure*—For the analysis procedure and the calculation see "Furnace Procedure," 11.3.

NOTE 40—Contamination from the work area is critical in cadmium analysis. Use of pipet tips which are free of cadmium is of particular importance.

NOTE 41—For every sample matrix analyzed, verification is necessary to determine that the method of standard additions is not required (see 6.2.1).

NOTE 42—If the method of standard additions is required, follow the procedure given in 10.5.

24. Cadmium—Direct Aspiration

24.1 *Requirements*:

24.1.1 *Optimum Concentration Range*, 0.2 to 7 mg/L using a wavelength of 422.7 nm (see Note 43 and Note 44).

24.1.2 *Sensitivity*, 0.08 mg/L.

24.1.3 *Detection Limit*, 0.01 mg/L.

NOTE 43—Phosphate, sulfate and aluminum interfere but are masked by the addition of lanthanum. Since low calcium values result if the pH of the sample is above 7, both standards and samples are prepared in dilute hydrochloric acid solution. Concentrations of magnesium greater than 1000 mg/L also cause low calcium values. Concentrations of up to 500 mg/L each of sodium, potassium and nitrate cause no interference.

NOTE 44—The 239.9 nm line may also be used. This line has a relative sensitivity of 120.

24.2 *Preparation of Standard Solution*:

24.2.1 *Stock Solution*—Suspend 1.250 g of $CaCO_3$ (analytical reagent grade) dried at 180°C for 1 h before weighing, in deionized distilled water, and dissolve cautiously with a minimum of dilute HCl. Dilute to 1000 mL with deionized distilled water. 1 mL = 0.5 mg Ca (500 mg/L).

24.2.2 *Lanthanum Chloride Solution*—Dissolve 29 g of La_2O_3, slowly and in small portions, in 250 mL concentrated HCl (**Caution**—Reaction is violent). Dilute to 500 mL with deionized distilled water.

24.2.3 Prepare dilutions of the stock calcium solutions to be used as calibration standards at the time of analysis. To each 10 mL volume of calibration standard and sample alike, add 1.0 mL of the lanthanum chloride solution, that is, 20 mL of standard or sample + 2 mL $LaCl_3$ = 22 mL.

24.3 *General Instrumental Parameters*:

24.3.1 *Cadmium Hollow Cathode Lamp.*

24.3.2 *Wavelength*—422.7 nm.

24.3.3 *Fuel*—Acetylene.

24.3.4 *Oxidant*—Air.

24.3.5 *Type of Flame*—Reducing.

24.4 *Analysis Procedure*—For analysis procedure and calculation, see "Direct Aspiration," 11.1.

NOTE 45—Anionic chemical interferences can be expected if lanthanum is not used in samples and standards.

NOTE 46—The nitrous oxide-acetylene flame will provide two to five times greater sensitivity and freedom from chemical interferences. Ionization interferences should be controlled by adding a large amount of alkali to the sample and standards. The analysis appears to be free from chemical suppressions in the nitrous oxide-acetylene flame **(7)**.

25. Chromium—Direct Aspiration

25.1 *Requirements*:

25.1.1 *Optimum Concentration Range:* 0.5 to 10 mg/L using a wavelength of 357.9 nm (see Note 47 and Note 48).

25.1.2 *Sensitivity*—0.25 mg/L.

25.1.3 *Detection Limit*—0.05 mg/L.

NOTE 47—The following wavelengths may also be used:
359.3 nm Relative Sensitivity 1.4,
425.4 nm Relative Sensitivity 2,
427.5 nm Relative Sensitivity 3, and
428.9 nm Relative Sensitivity 4.

NOTE 48—For levels of chromium between 50 and 200μ g/L, where the air-acetylene flame cannot be used or for levels below 50 μg/L, either the furnace procedure or the extraction procedure is recommended.

25.2 *Preparation of Standard Solution*:

25.2.1 *Stock Solution:* Dissolve 1.923 g of chromium trioxide (CrO_3, reagent grade) in deionized distilled water. When solution is complete, acidify with redistilled HNO_3 and dilute to 1 L with deionized distilled water. 1 mL = 1 mg Cr (1000 mg/L).

25.2.2 Prepare dilutions of the stock solution to be used as calibration standards at the time of analysis. The *calibration standards* should be prepared using the same type of acid and at the same concentration as that of the sample being analyzed either directly or after processing.

25.3 *General Instrumental Parameters*:

25.3.1 *Chromium Hollow Cathode Lamp.*

25.3.2 *Wavelength*—357.9 nm.

25.3.3 *Fuel*—Acetylene.

NOTE 49—The fuel-rich air-acetylene flame provides greater sensitivity but is subject to chemical and matrix interference from iron, nickel, and other metals. If the analysis is performed in a lean flame the interference can be lessened but the sensitivity will also be reduced.

NOTE 50—The suppression of both Cr (III) and Cr (VI) absorption by most interfering ions in fuel rich air-acetylene flames is reportedly controlled by the addition of 1 % ammonium bifluoride in 0.2 % sodium sulfate (**8**). A 1 % oxine solution is also reported to be useful.

25.3.4 *Oxidant*—Nitrous oxide.

25.3.5 *Type of Flame*—Fuel rich.

25.4 *Analysis Procedure*—For analysis procedure and calculation, see "Direct Aspiration," 11.1.

26. Chromium—Furnace Technique

26.1 *Requirements*:

26.1.1 *Optimum Concentration Range*, 5 to 100 μg/L (see Note 51).

26.1.2 *Detection Limit*, 1 μg/L.

NOTE 51—The concentration values and instrument conditions are for a Perkin-Elmer HGA-2100, based on the use of 20 μL injection, continuous flow purge gas and non-pyrolytic graphite.

26.2 *Preparation of Standard Solution*:

26.2.1 *Stock Solution*—Prepare as described under "direct aspiration method."

26.2.2 *Calcium Nitrate Solution*—Dissolve 11.8 g of calcium nitrate, $Ca(NO_3)_2 \cdot 4H_2O$ (analytical reagent grade) in deionized distilled water and dilute to 100 mL. 1 mL = 20 mg Ca.

26.2.3 Prepare dilutions of the stock chromium solution to be used as calibration standards at the time of analysis. The calibration standards should be prepared to contain 0.5 % (v/v) HNO_3. To each 100 mL of standard and sample alike, add 1 mL of 30 % H_2O_2 and 1 mL of the calcium nitrate solution.

NOTE 52—Hydrogen peroxide is added to the acidified solution to convert all chromium to the trivalent state. Calcium is added to a level above 200 mg/L where its suppressive effect becomes constant up to 1000 mg/L.

26.3 *General Instrument Parameters*:

26.3.1 *Drying Time and Temperature*—30 s at 125°C.

26.3.2 *Ashing Time and Temperature*—30 s at 1000°C.

26.3.3 *Atomizing Time and Temperature*—10 s at 2700°C.

26.3.4 *Purge Gas Atmosphere*—Argon.

26.3.5 *Wavelength*—357.9 nm.

26.3.6 Other operating parameters should be set as specified by the particular instrument manufacturer.

NOTE 53—Background correction may be required if the sample contains high dissolved solids.

NOTE 54—Nitrogen should not be used as a purge gas because of possible CN band interference.

26.4 *Analysis Procedure*—For the analysis procedure and the calculation, see "Furnace Procedure" 11.3.

NOTE 55—Pipet tips have been reported to be a possible source of contamination.

NOTE 56—For every sample matrix analyzed, verification is necessary to determine that the method of standard additions is not required (see 6.2.1).

NOTE 57—If the method of standard additions is required, follow the procedure given in 10.5.

27. Chromium—Chelation-Extraction

27.1 *Scope*—This test method may be used to analyze samples containing from 1.0 to 25 μg of chromium per litre of solution.

27.2 *Summary of the Test Method*:

27.2.1 This test method is based on the chelation of hexavalent chromium with ammonium pyrrolidine dithiocarbamate (APDC) following oxidation of trivalent chromium. The chelate is extracted with methyl isobutyl ketone (MIBK) and aspirated into the flame of the atomic absorption spectrophotometer.

27.2.2 Hexavalent chromium may also be chelated with pyrrolidine dithiocarbamic acid in chloroform as described in 11.2.

27.3 *Interferences*—High concentrations of other reactive metals, as may be found in wastewaters, may interfere. The method is free from interferences from elements normally occurring in fresh water.

27.4 *General Instrumental Parameters*:

27.4.1 *Chromium Hollow Cathode Lamp.*

27.4.2 *Wavelength*—357.9 nm.

27.4.3 *Fuel*—Acetylene.

27.4.4 *Oxidant*—Air.

27.4.5 *Type of Flame*—Fuel rich (adjust for organic solvent).

27.5 *Reagents*:

27.5.1 *Ammonium Pyrrolidine Dithiocarbamate* (APDC) *Solution*—Dissolve 1.0 g APDC in dimineralized water and dilute to 100 mL. Prepare fresh daily.

27.5.2 *Bromphenol Blue Indicator Solution*—Dissolve 0.1 g bromphenol blue in 100 mL 50 % ethanol.

27.5.3 *Potassium Dichromate Standard Solution* (1.0 mL = 0.08 mg Cr)—Dissolve 0.2263 g dried analytical reagent grade $K_2Cr_2O_7$ in demineralized water, and make up to 1000 mL.

27.5.4 *Trivalent Chromium Stock Solution* (1.0 mL = 0.002 mg $Cr^{+}3$)—Pipet 5.00 mL of the potassium dichromate standard solution (5.3) into an Erlenmeyer flask. Add approximately 15 mg Na_2SO_3 and 0.5 mL concentrated HNO_3. Gently evaporate to dryness; strong heating reoxidizes the chromium. Add 0.5 mL concentrated HNO_3 and again evaporate to dryness to destroy any excess sulfite. Take up in 1 mL concentrated HNO_3 with warming and dilute to 200.00 mL with demineralized water.

27.5.5 *Trivalent Chromium Working Solution* (1.0 mL = 0.005 mg Cr^{+3})—Immediately before use, dilute 25.0 mL of trivalent chromium stock solution (27.5.4) to 100.0 mL with demineralized water.

27.5.6 *Potassium Permanganate* (0.1 N)—Dissolve 0.32 g potassium permanganate in 100 mL demineralized water.

27.5.7 *Sodium Azide* (0.1 %)—Dissolve 100 mg sodium azide in demineralized water and dilute to 100 mL.

27.5.8 *Methyl Isobutyl Ketone* (MIBK).

27.5.9 *Sodium Hydroxide Solution* (1 M)—Dissolve 40 g NaOH in demineralized water and dilute to 1 L.

27.5.10 *Sulfuric Acid* (0.12 M)—Slowly add 6.5 mL concentrated H_2SO_4 (sp gr 1.84) to demineralized water and dilute to 1 L.

27.6 *Procedure*:

27.6.1 Pipet a volume of sample containing less than 2.5 μg chromium (100 mL maximum) into a 200 mL volumetric flask, and adjust the volume to approximately 100 mL. The pH must be 2.0 or less. Add concentrated HNO_3 if necessary.

27.6.2 Acidify a litre of demineralized water with 1.5 mL concentrated HNO_3. Prepare a blank and sufficient standards using trivalent chromium, and adjust volumes to approximately 100 mL with the acidified demineralized water.

27.6.3 Add 0.1 N $KMnO_4$ dropwise to both standards and samples until a faint pink color persists.

27.6.4 Heat on a steam bath for 20 min. If the color disappears, add additional $KMnO_4$ solution dropwise to maintain a slight excess.

27.6.5 While still on the steam bath, add sodium azide solution dropwise until the $KMnO_4$ color just disappears. Heat for about 2 min between each addition and avoid adding any excess. Continue heating for 5 min after adding the last drop of sodium azide solution.

27.6.6 Transfer the flasks to a water bath and cool to room temperature.

27.6.7 Remove from the water bath and filter (through Whatman No. 40 filter paper or equivalent) any sample that has a brownish precipitate or coloration which may interfere with the pH adjustment.

27.6.8 Add 2.0 mL of 1 M NaOH and 2 drops bromphenol blue indicator solution. Continue the addition of 1 M NaOH dropwise to all samples and standards in which the indicator change from yellow to blue has not occurred. Add 0.12 M H_2SO_4 dropwise until the blue color just disappears, then add 2.0 mL in excess. The pH at this point will be 2.4.

27.6.9 The pH adjustment to 2.4 may also be made with a pH meter instead of using an indicator.

27.6.10 Add 5.0 mL APDC solution and mix. The pH should then be approximately 2.8.

27.6.11 Add 10.0 mL MIBK and shake vigorously for 3 min.

27.6.12 Allow the layers to separate and add demineralized water until the ketone layer is completely in the neck of the flask.

27.6.13 Aspirate the ketone layer, record the instrument reading for each sample and standard against the blank. Repeat, and average the duplicate results.

27.7 *Calculations*:

27.7.1 Determine the µg/L Cr in each sample from a plot of the instrument readings of standards. A working curve must be prepared with each set of samples. Report Cr concentrations as follows:

27.7.1.1 Less than 10 µg/L, nearest µg/L, and

27.7.1.2 10 µg/L and above, two significant figures.

27.7.2 Calculate the mg metal per kg of samples as outlined in 11.2.

28. Chromium, Hexavalent—Chelation–Extraction

28.1 *Scope*:

28.1.1 This test method may be used to analyze samples containing from 1.0 to 25 µg of chromium per litre of solution.

28.2 *Summary of Test Method*:

28.2.1 This method is based on the chelation of hexavalent chromium with ammonium pyrrolidine dithiocarbamate (APDC) and extraction with methyl isobutyl ketone (MIBK). The extract is aspirated into the flame of the atomic absorption spectrophotometer.

28.2.2 Hexavalent chromium may also be chelated with pyrrolidine dithiocarbamic acid in chloroform as described in 11.2. A pH of 2.3 must be maintained throughout the extraction.

28.2.3 The diphenylcarbazide colorimetric procedure as found in "Standard Methods for the Examination of Water and Wastewater" may also be used **(9)**.

28.3 *Sample Handling and Preservation*:

28.3.1 Stability of hexavalent chromium is not completely understood at this time. Therefore, the chelation and extraction should be carried out as soon as possible.

28.3.2 To retard the chemical activity of hexavalent chromium, the sample should be transported and stored until time of analysis at 4°C.

28.4 *Interferences*:

28.4.1 High concentrations of other reactive metals, as may be found in wastewaters, may interfere. The method is free from interferences from elements normally occurring in fresh water.

28.5 *General Instrumental Parameters*:

28.5.1 *Chromium hollow cathode lamp.*

28.5.2 *Wavelength*—357.9 nm.

28.5.3 *Fuel*—Acetylene.

28.5.4 *Oxidant*—Air.

28.5.5 *Type of Flame:* Fuel-rich (adjust for organic solvent).

28.6 *Reagents*:

28.6.1 *Ammonium Pyrrolidine Dithiocarbamate (APDC) Solution*—Dissolve 1.0 g APDC in demineralized water and dilute to 100 mL. Prepare fresh daily.

28.6.2 *Bromphenol Blue Indicator Solution*—Dissolve 0.1 g bromphenol blue in 100 mL 50 % ethanol.

28.6.3 *Chromium Standard Solution I* (1.0 mL = 100 µg Cr)—Dissolve 0.2829 g pure, dried $K_2Cr_2O_7$ in demineralized water and dilute to 1000 mL.

28.6.4 *Chromium Standard Solution II* (1.0 mL = 10.0 µg Cr)—Dilute 100 mL chromium standard solution I to 1000 mL with demineralized water.

28.6.5 *Chromium Standard Solution III* (1.0 mL = 0.10 µg Cr)—Dilute 10.0 mL chromium standard solution II to 1000 mL with demineralized water.

28.6.6 *Methyl Isobutyl Ketone* (MIBK).

28.6.7 *Sodium Hydroxide Solution* (1 M)—Dissolve 40 g NaOH in demineralized water and dilute to 1 L.

28.6.8 *Sulfuric Acid* (0.12 M)—Slowly add 6.5 mL concentrated H_2SO_4 (sp gr 1.84) to demineralized water and dilute to 1 L.

28.7 *Procedure*:

28.7.1 Pipet a volume of sample containing less than 2.5 µg chromium (100 mL maximum) into a 200-mL volumetric flask, and adjust the volume to approximately 100 mL.

28.7.2 Prepare a blank and sufficient standards, and adjust the volume of each to approximately 100 mL.

28.7.3 Add 2 drops bromphenol blue indicator solution. (The pH adjustment to 2.4 may also be made with a pH meter instead of using an indicator.)

28.7.4 Adjust the pH by addition of 1 M NaOH solution dropwise until a blue color persists. Add 0.12 M H_2SO_4 dropwise until the blue color just disappears in both the standards and sample. Then add 2.0 mL of 0.12 M H_2SO_4 in excess. The pH at this point should be 2.4.

28.7.5 Add 5.0 mL APDC solution and mix. The pH should then be approximately 2.8.

28.7.6 Add 10.0 mL MIBK and shake vigorously for 3 min.

28.7.7 Allow the layers to separate and add demineralized water until the ketone layer is completely in the neck of the flask.

28.7.8 Aspirate the ketone layer, and record the scale reading for each sample and standard against the blank. Repeat and average the duplicate results.

28.8 *Calculations*:

28.8.1 Determine the µg/L Cr^{+6} in each sample from a plot of scale readings of standards. A working curve must be prepared with each set of samples. Report Cr^{+6} concentrations as follows: Less than 10 µg/L, nearest µg/L; 10 µg/L and above, two significant figures.

28.8.2 Calculate the mg metal per kg of sample as outlined in 11.2.

29. Cobalt—Direct Aspiration

29.1 *Requirements*:

29.1.1 *Optimum Concentration Range*—0.5 to 5 mg/L using a wavelength of 240.7 nm (see Note 58).

29.1.2 *Sensitivity*—0.2 mg/L.

29.1.3 *Detection Limit*—0.05 mg/L.

NOTE 58—For levels of cobalt below 100 µg/L, either the special extraction procedure (11.2), or the furnace technique is recommended.

29.2 *Preparation of Standard Solution*:

29.2.1 *Stock Solution*—Dissolve 4.307 g of cobaltous chloride, $CoCl_2 \cdot 6H_2O$ (analytical reagent grade), in deionized distilled water. Add 10 mL of concentrated nitric acid and dilute to 1 L with deionized distilled water. 1 mL = 1 mg Co (1000 mg/L).

29.2.2 Prepare dilutions of the stock cobalt solution to be used as calibration standards at the time of analysis. The *calibration standards* should be prepared using the same type of acid and at the same concentration as that of the sample being analyzed either directly or after processing.

29.3 *General Instrumental Parameters*:

29.3.1 *Cobalt Hollow Cathode Lamp.*

29.3.2 *Wavelength*—240.7 nm.

29.3.3 *Fuel*—Acetylene.

29.3.4 *Oxidant*—Air.

29.3.5 *Type of Flame*—Oxidizing.

29.4 *Analysis Procedure*—For analysis procedure and calculation, see "Direct Aspiration," 11.1.

30. Cobalt—Furnace Technique

30.1 *Requirements*:

30.1.1 *Optimum Concentration Range*, 5–100 µg/L (see Note 59).

30.1.2 *Detection Limit*, 1 µg/L.

NOTE 59—The above concentration values and instrument conditions are for a Perkin-Elmer HGA-2100, based on the use of a 20 µL injection, continuous flow purge gas and nonpyrolytic graphite. Smaller sized furnace devices or those employing faster rates of atomization can be operated using lower atomization temperatures for shorter time periods than the above recommended settings.

30.2 *Preparation of Standard Solution*:

30.2.1 *Stock Solution*—Prepare as described under "direct aspiration method."

30.2.2 Prepare dilutions of the stock solution to be used as calibration standards at the time of analysis. These solutions are also to be used for "standard additions."

30.2.3 The calibration standard should be diluted to contain 0.5 % (v/v) HNO_3.

30.3 *General Instrument Parameters*:

30.3.1 *Drying Time and Temperature*—30 s at 125°C.

30.3.2 *Ashing Time and Temperature*—30 s at 900°C.

30.3.3 *Atomizing Time and Temperature*—10 s at 2700°C.

30.3.4 *Purge Gas Atmosphere*—Argon.

30.3.5 *Wavelength*—240.7 nm.

30.3.6 Other operating parameters should be set as specified by the particular instrument manufacturer.

NOTE 60—The use of background correction is recommended.

NOTE 61—Nitrogen may also be used as the purge gas but with reported lower sensitivity.

30.4 *Analysis Procedure*—For the analysis procedure and the calculation, see "Furnace Procedure" 11.3.

NOTE 62—For every sample matrix analyzed, verification is necessary to determine that the method of standard additions is not required (see 6.2.1).

NOTE 63—If the method of standard additions is required, follow the procedure given earlier in 10.5.

31. Copper—Direct Aspiration

31.1 *Requirements*:

31.1.1 *Optimum Concentration Range*, 0.2–5 mg/L using a wavelength of 324.7 nm (see Note 64 and Note 65).

31.1.2 *Sensitivity*, 0.1 mg/L.

31.1.3 *Detection Limit*, 0.02 mg/L.

NOTE 64—For levels of copper below 50 µg/L, either the Special Extraction Procedure, given in 11.2 or the furnace technique is recommended.

NOTE 65—Numerous absorption lines are available for the determination of copper. By selecting a suitable absorption wavelength, copper samples may be analyzed over a very wide range of concentration. The following lines may be used:

327.4 nm Relative Sensitivity 2,
216.5 nm Relative Sensitivity 7, and
222.5 nm Relative Sensitivity 20.

31.2 *Preparation of Standard Solution*:

31.2.1 *Stock Solution*—Carefully weigh 1.00 g of electrolyte copper (analytical reagent grade). Dissolve in 5 mL redistilled HNO_3, and make up to 1 L with deionized distilled water. Final concentration is 1 mg Cu per mL (1000 mg/L).

31.2.2 Prepare dilutions of the stock solution to be used as calibration standards at the time of analysis. The *calibration standards* should be prepared using the same type of acid and at the same concentration as that of the sample being analyzed either directly or after processing.

31.3 *General Instrumental Parameters*:

31.3.1 *Copper Hollow Cathode Lamp.*

31.3.2 *Wavelength*—324.7 nm.

31.3.3 *Fuel*—Acetylene.

31.3.4 *Oxidant*—Air.

31.3.5 *Type of Flame*—Oxidizing.

31.4 *Analysis Procedure*—For analysis procedure and calculation, see "Direct Aspiration", 11.1.

32. Copper—Furnace Technique

32.1 *Requirements*:

32.1.1 *Optimum Concentration Range*, 5 to 100 µg/L (see Note 66).

32.1.2 *Detection Limit*, 1 µg/L.

NOTE 66—The above concentration values and instrument conditions are for a Perkin-Elmer HGA-2100, based on the use of a 20-µL injection, continuous flow purge gas and non-pyrolytic graphite. Smaller sized furnace devices or those employing faster rates of atomization can be operated using lower atomization temperatures for shorter time periods than the above recommended settings.

32.2 *Preparation of Standard Solution*:

32.2.1 *Stock Solution*—Prepare as described under "direct aspiration method."

32.2.2 Prepare dilutions of the stock solution to be used as calibration standards at the time of analysis. These solutions are also to be used for "standard additions."

32.2.3 The calibration standard should be diluted to contain 0.5 % (v/v) HNO_3.

32.3 *General Instrument Parameters*:

32.3.1 *Drying Time and Temperature*—30 s at 125°C.

32.3.2 *Ashing Time and Temperature*—30 s at 900°C.

32.3.3 *Atomizing Time and Temperature*—10 s at 2700°C.

32.3.4 *Purge Gas Atmosphere*—Argon.

32.3.5 *Wavelength*—324.7 nm.

32.3.6 Other operating parameters should be set as specified by the particular instrument manufacturer.

NOTE 67—Background correction may be required if the sample contains high dissolved solids.

NOTE 68—Nitrogen may also be used as the purge gas.

32.4 *Analysis Procedure*—For the analysis procedure and the calculation, see "Furnace Procedure" 11.3.

NOTE 69—For every sample matrix analyzed, verification is necessary to determine that the method of standard additions is not required (see 6.2.1).

NOTE 70—If the method of standard additions is required, follow the procedure given in 10.5.

33. Iron—Direct Aspiration

33.1 *Requirements*:

33.1.1 *Optimum Concentration Range*, 0.3 to 5 mg/L using a wavelength of 248.3 nm (see Note 71 and Note 72).

NOTE 71—The following lines may also be used:
248.8 nm relative sensitivity 2,
271.9 nm relative sensitivity 4,
302.1 nm relative sensitivity 5,
252.7 nm relative sensitivity 6, and
372.0 nm relative sensitivity 10.

NOTE 72—For concentrations of iron below 0.05 mg/L, either the Special Extraction Procedure given in 11.2 or the furnace procedure, is recommended.

33.1.2 *Sensitivity*, 0.12 mg/L.

33.1.3 *Detection Limit*, 0.03 mg/L.

33.2 *Preparation of Standard Solution*:

33.2.1 *Stock Solution*—Carefully weigh 1.000 g of pure iron wire (analytical reagent grade) and dissolve in 5 mL redistilled HNO_3, warming if necessary. When solution is complete, make up to 1 L with deionized distilled water. 1 mL = 1 mg Fe (1000 mg/L).

33.2.2 Prepare dilutions of the stock solution to be used as calibration standards at the time of analysis. The *calibration standards* should be prepared using the same type of acid and at the same concentration as will result in the sample to be analyzed either directly or after processing.

33.3 *General Instrumental Parameters*:

33.3.1 *Iron Hollow Cathode Lamp.*

33.3.2 *Wavelength*—248.3 nm.

33.3.3 *Fuel*—Acetylene.

33.3.4 *Oxidant*—Air.

33.3.5 *Type of Flame*—Oxidizing.

33.4 *Analysis Procedure*—For analysis procedure and calculation, see "Direct Aspiration," 11.1.

34. Iron—Furnace Technique

34.1 *Requirements*:

34.1.1 *Optimum Concentration Range*—5 to 100 µg/L (see Note 73).

34.1.2 *Detection Limit*—1 µg/L.

NOTE 73—The concentration values and instrument conditions are for a Perkin-Elmer HGA-2100, based on the use of a 20 µL-injection, continuous flow purge gas, and nonpyrolytic graphite. Smaller sized furnace devices or those employing faster rates of atomization can be operated using lower atomization temperatures for shorter time periods than the recommended settings.

34.2 *Preparation of Standard Solution*:

34.2.1 *Stock Solution*—Prepare as described under "direct aspiration method."

34.2.2 Prepare dilutions of the stock solution to be used as calibration standards at the time of analysis. These solutions are also to be used for "standard additions."

34.2.3 The calibration standard should be diluted to contain 0.5 % (v/v) HNO_3.

34.3 *General Instrument Parameters*:

34.3.1 *Drying Time and Temperature*—30 s at 125°C.

34.3.2 *Ashing Time and Temperature*—30 s at 1000°C.

34.3.3 *Atomizing Time and Temperature*—10 s at 2700°C.

34.3.4 *Purge Gas Atmosphere*—Argon.

34.3.5 *Wavelength*—248.3 nm.

34.3.6 Other operating parameters should be set as specified by the particular instrument manufacturer.

NOTE 74—The use of background correction is recommended.

NOTE 75—Nitrogen may also be used as the purge gas.

34.4 *Analysis Procedure*—For the analysis procedure and the calculation, see "Furnace Procedure," 11.3.

NOTE 76—For every sample matrix analyzed, verification is necessary to determine that the method of standard additions is not required (see 6.2.1).

NOTE 77—If the method of standard additions is required, follow the procedure given in 10.5.

35. Lead—Direct Aspiration

35.1 *Requirements*:

35.1.1 *Optimum Concentration Range*, 1–20 mg/L using a wavelength of 283.3 nm (see Note 78 and Note 79).

35.1.2 *Sensitivity*, 0.5 mg/L.

35.1.3 *Detection Limit*, 0.1 mg/L.

NOTE 78—For levels of lead below 200 µg/L, either the Special Extraction Procedure given in 11.2 or the furnace technique is recommended.

NOTE 79—The following lines may also be used:
217.0 nm Relative Sensitivity 0.4, and
261.4 nm Relative Sensitivity 10.

35.2 *Preparation of Standard Solution*:

35.2.1 *Stock Solution*—Carefully weigh 1.599 g of lead nitrate, $Pb(NO_3)_2$ (analytical reagent grade), and dissolve in deionized distilled water. When solution is complete, acidify with 10 mL redistilled HNO_3 and dilute to 1 L with deionized distilled water. 1 mL = 1 mg Pb (1000 mg/L).

35.2.2 Prepare dilutions of the stock solution to be used as calibration standards at the time of analysis. The *calibration standards* should be prepared using the same type of acid and at the same concentration as that of the sample being analyzed either directly or after processing.

35.3 *General Instrumental Parameters*:

35.3.1 *Iron Hollow Cathode Lamp.*

35.3.2 *Wavelength*—283.3 nm.

35.3.3 *Fuel*—Acetylene.

35.3.4 *Oxidant*—Air.

35.3.5 *Type of Flame*—Oxidizing.

35.4 *Analysis Procedure*—For analysis procedure and calculation, see "Direct Aspiration," 11.1.

NOTE 80—The analysis of this metal is exceptionally sensitive to turbulence and absorption bands in the flame. Therefore, some care should be taken to position the light beam in the most stable, center portion of the flame. To do this, first adjust the burner to maximize the absorbance reading with a lead standard. Then, aspirate a water blank and make minute adjustments in the burner alignment to minimize the signal.

36. Lead—Furnace Technique

36.1 *Requirements*:

36.1.1 *Optimum Concentration Range*, 5 to 100 µg/L (see Note 81).

36.1.2 *Detection Limit*, 1 µg/L.

NOTE 81—The above concentration values and instrument conditions are for a Perkin-Elmer HGA-2100, based on the use of a 20-µL injection, continuous flow purge gas and nonpyrolytic graphite. Smaller sized furnace devices or those employing faster rates of atomization can be operated using lower atomization temperatures for shorter time periods than the above recommended settings.

36.2 *Preparation of Standard Solution*:

36.2.1 *Stock Solution*—Prepare as described under "direct aspiration method."

36.2.2 *Lanthanum Nitrate Solution*—Dissolve 58.64 g of ACS reagent grade La_2O_3 in 100 mL concentrated HNO_3, and dilute to 1000 mL with deionized distilled water. 1 mL = 50 mg La.

36.2.3 *Working Lead Solution*—Prepare dilutions of the stock lead solution to be used as calibration standards at the time of analysis. Each calibration standard should contain 0.5 % (v/v) HNO_3. To each 100 mL of diluted standard, add 10 mL of the lanthanum nitrate solution.

36.3 *General Instrument Parameters*:

36.3.1 *Drying Time and Temperature*—30 s at 125°C.

36.3.2 *Ashing Time and Temperature*—30 s at 500°C.

36.3.3 *Atomizing Time and Temperature*—10 s at 2700°C.

36.3.4 *Purge Gas Atmosphere*—Argon.

36.3.5 *Wavelength*—283.3 nm.

NOTE 82—Greater sensitivity can be achieved using the 217.0-nm line, but the optimum concentration range is reduced. The use of a lead electrodeless discharge lamp at this lower wavelength has been found to be advantageous. Also a lower atomization temperature (2400°C) may be preferred.

36.3.6 Other operating parameters should be set as specified by the particular instrument manufacturer.

NOTE 83—The use of background correction is recommended.

36.4 *Analysis Procedure*—For the analysis procedure in the calculation see "Furnace Procedure," 11.3.

NOTE 84—To suppress sulfate interference (up to 1500 ppm) lanthanum is added as the nitrate to both samples and calibration standards (**10**).

NOTE 85—Since glassware contamination is a severe problem in lead analysis, all glassware should be cleaned immediately prior to use, and once cleaned, should not be open to the atmosphere except when necessary.

NOTE 86—For every sample matrix analyzed, verification is necessary to determine that the method of standard additions is not required (see 6.2.1).

NOTE 87—If the method of standard additions is required, follow the procedure given in 10.5.

37. Lithium—Direct Aspiration

37.1 *Requirements*:

37.1.1 *Optimum Concentration Range*—to 0.2 mg/L using a wavelength of 670.8 nm (see Note 88).

37.1.2 *Sensitivity*—0.035 mg/L.

NOTE 88—The following lines may also be used:
323.3 nm relative sensitivity 235, and
610.4 nm relative sensitivity 3600.

37.2 *Preparation of Standard Solution*:

37.2.1 *Stock Solution*—Dissolve 5.324 g of lithium carbonate, Li_2CO_3 in a minimum volume of (1 + 1) HCl and dilute to 1 L with deionized water. 1 mL = 1.00 mg Li (1000 mg/L).

37.2.2 Prepare dilutions of the stock lithium solution to be used as calibration standards at the time of analysis. The *calibration standards* should be prepared using the same type of acid and at the same concentration as that of the sample being analyzed either directly or after processing.

37.3 *General Instrumental Parameters*:

37.3.1 *Lithium Hollow Cathode Lamp.*

37.3.2 *Wavelength*—670.8 nm.

37.3.3 *Fuel*—Acetylene.

37.3.4 *Oxidant*—Air.

37.4 *Analysis Procedure*—For analysis procedure and calculations, see "Direct Aspiration," 11.1.

38. Magnesium—Direct Aspiration

38.1 *Requirements*:

38.1.1 *Optimum Concentration Range*, 0.02 to 0.5 mg/L using a wavelength of 285.2 nm (see Note 89 and Note 90).

38.1.2 *Sensitivity*, 0.007 mg/L.

38.1.3 *Detection Limit*, 0.001 mg/L.

NOTE 89—The following line may also be used: 202.5 nm relative sensitivity 25.

NOTE 90—To cover the range of magnesium values normally observed in surface waters (0.1 to 20 mg/L), it is suggested that either the 202.5 nm line be used or the burner head be rotated. A90° rotation of the burner head will produce approximately one-eighth the normal sensitivity.

38.2 *Preparation of Standard Solution*:

38.2.1 *Stock Solution*—Dissolve 0.829 g of magnesium oxide, MgO (analytical reagent grade), in 10 mL of redistilled HNO_3, and dilute to 1 L with deionized distilled water. 1 mL = 0.50 mg Mg (500 mg/L).

38.2.2 *Lanthanum Chloride Solution*—Dissolve 29 g of La_2O_3, slowly and in small portions in 250 mL concentrated HCl, (**Caution**—Reaction is violent), and dilute to 500 mL with deionized distilled water.

38.2.3 Prepare dilutions of the stock magnesium solution to be used as calibration standards at the time of analysis. These *calibration standards* should be prepared using the same type of acid and at the same concentration as that of the sample being analyzed directly or after processing. To each 10-mL volume of *calibration standard* and sample alike add 1.0 mL of the lanthanum chloride solution, that is, 20 mL of standard or sample + 2 mL $LaCl_3$ = 22 mL.

38.3 *General Instrumental Parameters*:

38.3.1 *Magnesium Hollow Cathode Lamp.*

38.3.2 *Wavelength*—285.2 nm.

38.3.3 *Fuel*—Acetylene.

38.3.4 *Oxidant*—Air.

38.3.5 *Type of Flame*—Oxidizing.

38.4 *Analysis Procedure*—For analysis procedure and calculation, see "Direct Aspiration," 11.1.

NOTE 91—The interference caused by aluminum at concentrations greater than 2 mg/L is masked by addition of lanthanum. Sodium, potassium and calcium cause no interference at concentrations less than 400 mg/L.

39. Manganese—Direct Aspiration

39.1 *Requirements*:

39.1.1 *Optimum Concentration Range*, 0.1 to 3 mg/L using a wavelength of 279.5 nm (see Note 92).

39.1.2 *Sensitivity*, 0.05 mg/L.

39.1.3 *Detection Limit*, 0.01 mg/L.

NOTE 92—The following line may also be used: 403.1 nm Relative Sensitivity 10.

39.2 *Preparation of Standard Solution*:

39.2.1 *Stock Solution*—Carefully weigh 1.000 g of manganese metal (analytical reagent grade), and dissolve in 10 mL of redistilled HNO_3. When solution is complete, dilute to 1 L with 1 % (v/v) HCl. 1 mL = 1 mg Mn (1000 mg/L).

39.2.2 Prepare dilutions of the stock solution to be used as calibration standards at the time of analysis. The *calibration standards* should be prepared using the same type of acid and at the same concentration as that of the sample being analyzed either directly or after processing.

39.3 *General Instrumental Parameters*:

39.3.1 *Manganese hollow cathode lamp.*

39.3.2 *Wavelength*—279.5 nm.

39.3.3 *Fuel*—Acetylene.

39.3.4 *Oxidant*—air.

39.3.5 *Type of Flame*—oxidizing.

39.4 *Analysis Procedure*—For analysis procedure and calculation, see "Direct Aspiration," 11.1.

NOTE 93—For levels of manganese below 25 µg/L, either the furnace procedure or the Special Extraction Procedure given in 10.2 is recommended. The extraction is carried out at a pH of 4.5 to 5. The manganese chelate is very unstable and the analysis must be made without delay to prevent its solution in the aqueous phase.

40. Manganese—Furnace Technique

40.1 *Requirements*:

40.1.1 *Optimum Concentration Range*, 1 to 30 µg/L (see Note 94).

40.1.2 *Detection Limit*, 0.2 µg/L.

NOTE 94—The above concentration values and instrument conditions are for a Perkin-Elmer HGA-2100, based on the use of a 20-uL injection, continuous flow purge gas and nonpyrolytic graphite. Smaller sized furnace devices or those employing faster rates of atomization can be operated using lower atomization temperatures for shorter time periods than the above recommended settings.

40.2 *Preparation of Standard Solution*:

40.2.1 *Stock Solution*—Prepare as described under "direct aspiration method."

40.2.2 Prepare dilutions of the stock solution to be used as calibration standards at the time of analysis. These solutions are also to be used for "standard additions."

40.2.3 The calibration standard should be diluted to contain 0.5 % (v/v) HNO_3.

40.3 *General Instrument Parameters*:

40.3.1 *Drying Time and Temperature*—30 s at 125°C.

40.3.2 *Ashing Time and Temperature*—30 s at 1000°C.

40.3.3 *Atomizing Time and Temperature*—10 s at 2700°C.

40.3.4 *Purge Gas Atmosphere*—Argon.

40.3.5 *Wavelength*—279.5 nm.

40.3.6 Other operating parameters should be set as specified by the particular instrument manufacturer.

NOTE 95—The use of background correction is recommended.

NOTE 96—Nitrogen may also be used as the purge gas.

40.4 *Analysis Procedure*—For the analysis procedure and the calculation, see "Furnace Procedure" 11.3.

NOTE 97—For every sample matrix analyzed, verification is necessary to determine that the method of standard additions is not required (see 6.2.1).

NOTE 98—If the method of standard additions is required, follow the procedure given earlier in 10.5.

41. Mercury Cold Vapor Technique

41.1 *Scope and Application* (**11**):

41.1.1 In addition to inorganic forms of mercury, organic mercurials may also be present. These organo-mercury compounds will not respond to the cold vapor atomic absorption technique unless they are first broken down and converted to mercuric ions. Potassium permanganate oxidizes many of these compounds, but recent studies have shown that a number of organic mercurials, including phenyl mercuric acetate and methyl mercuric chloride, are only partially oxidized by this reagent. Potassium persulfate has been found to give approximately 100 % recovery when used as the oxidant with these compounds. Therefore, a persulfate oxidation step following the addition of the permanganate has been included to ensure that organo-mercury compounds, if present, will be oxidized to the mercuric ion before measurement. A heat step is required for methyl mercuric chloride when present in or spiked to a natural system. For distilled water the heat step is not necessary.

41.1.2 The range of the test method may be varied through instrument or recorder expansion, or both. Using a 100-mL sample, a detection limit of 0.2 µg Hg/L can be achieved. Concentrations below this level should be reported as <0.2.

NOTE 99—If additional sensitivity is required, a 200 mL sample with recorder expansion may be used provided the instrument does not produce undue noise. Using a Coleman MAS-50 with a drying tube of magnesium perchlorate and a variable recorder, 2 mv was set to read full scale. With these conditions, and distilled water solutions of mercuric chloride at concentrations of 0.15, 0.10, 0.05 and 0.025 µg/L, the standard deviations were ±0.027, ±0.006,± 0.01, and ±0.004. Percent recoveries at these levels were 107, 83, 84, and 96 %, respectively.

41.2 *Summary of Test Method*:

41.2.1 The flameless AA procedure is a physical method based on the absorption of radiation at 253.7 nm by mercury vapor. The mercury vapor passes through a cell positioned in the light path of an atomic absorption spectrophotometer. Absorbance (peak height) is measured as a function of mercury concentration and recorded in the usual manner.

41.3 *Sample Handling and Preservation*:

41.3.1 Until more conclusive data are obtained, samples should be preserved by acidification with nitric acid to a pH of 2 or lower immediately at the time of collection. If only dissolved mercury is to be determined, the sample should be filtered through an all glass apparatus before the acid is added. For total mercury, the filtration is omitted.

41.4 *Interference*:

41.4.1 Possible interference from sulfide is eliminated by the addition of potassium permanganate. Concentrations as high as 20 mg/L of sulfide as sodium sulfide do not interfere with the recovery of added inorganic mercury from distilled water.

41.4.2 Copper has also been reported to interfere; however, copper concentrations as high as 10 mg/L had no effect on recovery of mercury from spiked samples.

41.4.3 High chloride concentrations require additional permanganate (as much as 25 mL). During the oxidation step, chlorides are converted to free chlorine which will also absorb radiation of 253 nm. Care must be taken to ensure that free chlorine is absent before the mercury is reduced and swept into the cell. This may be accomplished by using an excess of hydroxylamine sulfate reagent (25 mL). In addition, the dead air space in the BOD bottle must be purged before the addition of stannous sulfate. Both inorganic and organic mercury spikes have been quantitatively recovered from sea water using this technique.

41.4.4 Interference from certain volatile organic materials that will absorb at this wavelength is also possible. A preliminary run without reagents should determine if this type of interference is present.

NOTE 100—The possibility of absorption from certain organic substances actually being present in the sample does exist. This is mentioned only to caution the analyst of the possibility. A simple correction that may be used is as follows: If an interference has been found to be present, the sample should be analyzed both by using the regular procedure and again under oxidizing conditions only, that is without the reducing reagents. The true mercury value can then be obtained by subtracting the two values.

41.5 *Apparatus*:

41.5.1 Atomic Absorption Spectrophotometer; (See Note 101) Any atomic absorption unit having an open sample presentation area in which to mount the absorption cell is suitable. Instrument settings recommended by the particular manufacturer should be followed.

NOTE 101—Instruments designated specifically for the measurement of mercury using the cold vapor technique are commercially available and may be substituted for the atomic absorption spectrophotometer.

41.5.2 *Mercury Hollow Cathode Lamp*—argon filled, or equivalent.[11]

41.5.3 *Recorder*—Any multi-range variable speed recorder that is compatible with the UV detection system is suitable.

41.5.4 *Absorption Cell*—Standard spectrophotometer cells 10 cm long, having quartz end windows may be used. Suitable cells may be constructed from plexiglass tubing, 1 outside diameter by 4½ in. The ends are ground perpendicular to the longitudinal axis and quartz windows (1-in. diameter by 1/16-in. thickness) are cemented in place. The cell is strapped to a burner for support and aligned in the light beam by use of two 2 by 2 in. cards. One-inch diameter holes are cut in the middle of each card; the cards are then placed over each end of the cell. The cell is then positioned and adjusted vertically and horizontally to give the maximum transmittance.

41.5.5 *Air Pump*—Any peristaltic pump capable of delivering 1 L of air per minute may be used. A Masterflex pump with electronic speed control has been found to be satisfactory.

41.5.6 *Flowmeter*—Capable of measuring an air flow of 1 L per minute.

41.5.7 *Aeration Tubing*—A straight-glass frit having a coarse porosity. Tygon tubing is used for passage of the mercury vapor from the sample bottle to the absorption cell and return.

41.5.8 *Drying Tube*—6 by ¾ in. diameter tube containing 20 g of magnesium perchlorate (see Note 102). The apparatus is assembled as shown in Fig. 2.

NOTE 102—In place of the magnesium perchlorate drying tube, a small reading lamp with 60 W bulb may be used to prevent condensation of moisture inside the cell. The lamp is positioned to shine on the absorption cell maintaining the air temperature in the cell about 10.25°C above ambient.

41.6 *Reagents*:

41.6.1 *Sulfuric Acid, Concentrated*—Reagent grade.

41.6.1.1 *Sulfuric Acid,* 0.5 N—Dilute 14.0 mL of concentrated sulfuric acid to 1.0 L.

41.6.2 *Nitric Acid,* Concentrate—Reagent grade of low mercury content.

NOTE 103—If a high reagent blank is obtained, it may be necessary to distill the nitric acid.

41.6.3 *Stannous Sulfate*—Add 25 g stannous sulfate to 250 mL of 0.5 *N* sulfuric acid. This mixture is a suspension and

[11] Westinghouse WL-22847 available from any laboratory supply house has been found suitable.

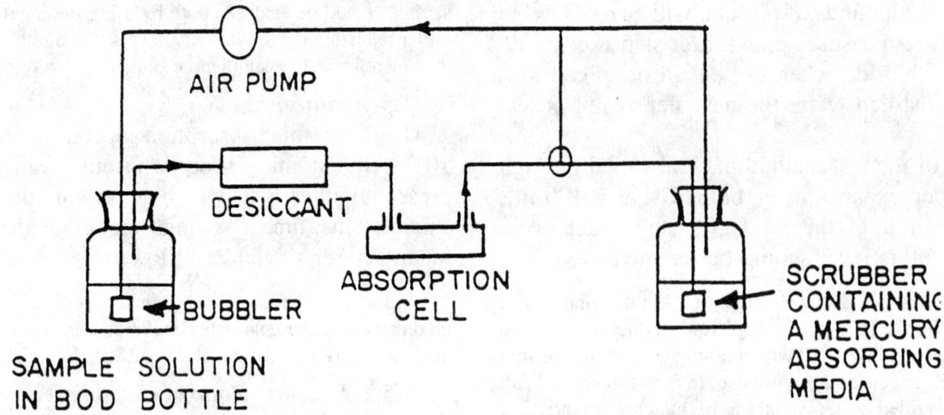

FIG. 2 Apparatus for Flameless Mercury Determination

should be stirred continuously during use. (Stannous chloride may be used in place of stannous sulfate.)

41.6.4 *Sodium Chloride-Hydroxylamine Sulfate Solution*—Dissolve 12 g of sodium chloride and 12 g of hydroxylamine sulfate in distilled water and dilute to 100 mL. (Hydroxylamine hydrochloride may be used in place of hydroxylamine sulfate.)

41.6.5 *Potassium Permanganate* (5 % solution, w/v)—Dissolve 5 g of potassium permanganate in 100 mL of distilled water.

41.6.6 *Potassium Persulfate* (5 % solution, w/v)—Dissolve 5 g of potassium persulfate in 100 mL of distilled water.

41.6.7 *Stock Mercury Solution*—Dissolve 0.1354 g of mercuric chloride in 75 mL of distilled water. Add 10 mL of concentrated nitric acid and adjust the volume to 100.0 mL. 1 mL = 1 mg Hg.

41.6.8 *Working Mercury Solution*—Make successive dilutions of the stock mercury solution to obtain a working standard containing 0.1 µg per mL. This working standard and the dilutions of the stock mercury solution should be prepared fresh daily. Acidity of the working standard should be maintained at 0.15 % nitric acid. This acid should be added to the flask as needed before the addition of the aliquot.

41.7 *Calibration*:

41.7.1 Transfer 0, 0.5, 1.0, 2.0, 5.0, and 10.0 mL aliquots of the working mercury solution containing 0 to 1.0 µg of mercury to a series of 300 mL BOD bottles. Add enough distilled water to each bottle to make a total volume of 100 mL. Mix thoroughly and add 5 mL of concentrated sulfuric acid and 2.5 mL of concentrated nitric acid to each bottle. Add 15 mL of $KMnO_4$ solution to each bottle, and allow to stand at least 15 min. Add 8 mL of potassium persulfate to each bottle and heat for 2 h in a water bath maintained at 95°C. Cool and add 6 mL of sodium chloride-hydroxylamine sulfate solution to reduce the excess permanganate. When the solution has been decolorized wait 30 s, add 5 mL of the stannous sulfate solution and immediately attach the bottle to the aeration apparatus forming a closed system. At this point the sample is allowed to stand quietly without manual agitation. The circulating pump, which has previously been adjusted to a rate of 1 L per minute, is allowed to run continuously. (An open system where the mercury vapor is passed through the absorption cell only once may be used instead of the closed system.)

41.7.2 The absorbance will increase and reach maximum within 30 s. As soon as the recorder pen levels off, approximately 1 min, open the bypass valve and continue the aeration until the absorbance returns to its minimum value. Because of the toxic nature of mercury vapor, precaution must be taken to avoid its inhalation. Therefore, a bypass has been included in the system to either vent the mercury vapor into an exhaust hood or pass the vapor through some absorbing media such as follows:

(a) equal volumes of 0.1 M $KMnO_4$ and 10 % H_2SO_4, and
(b) 0.25 % iodine in a 3 % KI solution.

NOTE 104—Directions for the disposal of mercury-containing wastes are given in Test Method D 3223.[12]

41.7.3 Close the bypass valve, remove the stopper and frit from the BOD bottle and continue the aeration. Proceed with the standards and construct a standard curve by plotting peak height versus micrograms of mercury.

41.8 *Procedure*:

41.8.1 Transfer 100 mL, or an aliquot diluted to 100 mL, containing not more than 1.0 µg of mercury, to a 300-mL BOD bottle. Add 5 mL of sulfuric acid and 2.5 mL of concentrated nitric acid mixing after each addition. Add 15 mL of potassium permanganate solution to each sample bottle. For sewage, sample additional permanganate solution, if necessary, until the purple color persists for at least 15 min. Add 8 mL of potassium persulfate to each bottle and heat for 2 h in a water bath at 95°C. Cool and add 6 mL of sodium chloride-hydroxylamine sulfate to reduce the excess permanganate. After a delay of at least 30 s, add 5 mL of stannous sulfate and immediately attach the bottle to the aeration apparatus. Continue as described in 41.7.

41.9 *Calculation*:

41.9.1 Determine the peak height of the unknown from the chart, and read the mercury value from the standard curve.

41.9.2 Calculate the mercury concentration in the sample by the formula:

$$\mu g\ Hg/L = (\mu g\ Hg\ in\ aliquot)\left(\frac{1{,}000}{volume\ of\ aliquot\ in\ mL}\right) \quad (10)$$

41.9.3 Report mercury concentrations as follows: Below 0.2

---

[12] A specially treated charcoal that will absorb mercury vapor is also available from Barneby and Cheney, East 8th Ave. and North Cassidy St., Columbus, OH 43219, Catalog No. E580-13 or 580-22.

μg/L, <0.2; between 1 and 10 μg/L, one decimal; above 10 μg/L, whole numbers.

42. Molybdenum—Direct Aspiration

42.1 *Requirements*:

42.1.1 *Optimum Concentration Range*, 1 to 40 mg/L using a wavelength of 313.3 nm (see Note 105).

42.1.2 *Sensitivity*, 0.4 mg/L.

42.1.3 *Detection Limit*, 0.1 mg/L.

NOTE 105—For concentrations of molybdenum below 0.2 mg/L, the furnace procedure is recommended.

42.2 *Preparation of Standard Solution*:

42.2.1 *Stock Solution*—Dissolve 1.840 g of ammonium molybdate $(NH_4)_6Mo_7O_{24}\cdot 4H_2O$ (analytical reagent grade) in deionized distilled water and dilute to 1 L. 1 mL = 1 mg Mo (1000 mg/L).

42.2.2 *Aluminum Nitrate Solution*—Dissolve 139 g aluminum nitrate, $Al(NO_3)_3\cdot 9H_2O$, in 150 mL of deionized distilled water; heat to effect solution. Allow to cool and make up to 200 mL.

42.2.3 Prepare dilutions of the stock molybdenum solution to be used as calibration standards at the time of analysis. The *calibration standards* should be prepared using the same type of acid and at the same concentration as that of the sample being analyzed either directly or after processing. To each 100 mL of standard and sample alike, add 2 mL of the aluminum nitrate solution.

42.3 *General Instrumental Parameters*:

42.3.1 *Molybdenum Hollow Cathode Lamp*.

42.3.2 *Wavelength*—313.3 nm.

42.3.3 *Fuel*—Acetylene.

42.3.4 *Oxidant*—Nitrous Oxide.

42.3.5 *Type of Flame*—Fuel rich.

42.4 *Analysis Procedure*—For analysis procedure and calculation, see "Direct Aspiration," 11.1.

42.5 *Interferences*:

42.5.1 With the recommended nitrous oxide-acetylene flame, interferences of calcium and other ions may be controlled by adding 1000 mg/L of a refractory metal such as aluminum **(12)**. This should be done to both samples and standards alike.

43. Molybdenum—Furnace Technique

43.1 *Requirements*:

43.1.1 *Optimum Concentration Range*, 3 to 60 μg/L (see Note 106).

43.1.2 *Detection Limit*, 1 μg/L.

NOTE 106—The concentration values and instrument conditions are for a Perkin-Elmer HGA-2100, based on the use of a 20-μL injection, continuous flow purge gas and pyrolytic graphite.

43.2 *Preparation of Standard Solution*:

43.2.1 *Stock Solution*—Prepare as described under "direct aspiration method".

43.2.2 Prepare dilutions of the stock solution to be used as calibration standards at the time of analysis. These solutions are also to be used for "standard additions".

43.2.3 The calibration standard should be diluted to contain 0.5 % (v/v) HNO_3.

43.3 *General Instrument Parameters*:

43.3.1 *Drying Time and Temperature*—30 s at 125°C.

43.3.2 *Ashing Time and Temperature*—30 s at 1400°C.

43.3.3 *Atomizing Time and Temperature*—15 s at 2800°C.

43.3.4 *Purge Gas Atmosphere*—Argon.

43.3.5 *Wavelength*—313.3 nm.

43.3.6 Other operating parameters should be set as specified by the particular instrument manufacturer.

NOTE 107—Background correction may be required if the sample contains high dissolved solids.

NOTE 108—The use of nitrogen as a purge gas is not recommended.

43.4 *Analysis Procedure*—For the analysis procedure and the calculation, see "Furnace Procedure," 11.3.

NOTE 109—For every sample matrix analyzed, verification is necessary to determine that the method of standard additions is not required (see 6.2.1).

NOTE 110—If the method of standard additions is required, follow the procedure given in 10.5.

44. Nickel—Direct Aspiration

44.1 *Requirements*:

44.1.1 *Optimum Concentration Range*, 0.3 to 5 mg/L using a wavelength of 232.0 nm (see Note 111).

44.1.2 *Sensitivity*, 0.15 mg/L.

44.1.3 *Detection Limit*, 0.04 mg/L.

NOTE 111—For levels of nickel below 100 μg/L, either the Special Extraction Procedure, given in 11.2, or the furnace technique is recommended.

44.2 *Preparation of Standard Solution*:

44.2.1 *Stock Solution*—Dissolve 4.953 g of nickel nitrate, $Ni(NO_3)_2\cdot 6H_2O$ (analytical reagent grade) in deionized distilled water. Add 10 mL of concentrated nitric acid and dilute to 1 L with deionized distilled water. 1 mL = 1 mg Ni (1000 mg/L).

44.2.2 Prepare dilutions of the stock nickel solution to be used as calibration standards at the time of analysis. The *calibration standards* should be prepared using the same type of acid and at the same concentration as that of the sample being analyzed either directly or after processing.

44.3 *General Instrumental Parameters*:

44.3.1 *Nickel Hollow Cathode Lamp*.

44.3.2 *Wavelength*—232.0 nm.

44.3.3 *Fuel*—Acetylene.

44.3.4 *Oxidant*—Air.

44.3.5 *Type of Flame*—Oxidizing.

44.4 *Analysis Procedure*—For analysis procedure and calculation, see "Direct Aspiration", 11.1.

44.5 *Interferences*—The 352.4-nm wavelength is less susceptible to interference and may be used. The calibration curve is more linear at this wavelength; however, there is some loss of sensitivity.

45. Nickel—Furnace Technique

45.1 *Requirements*:

45.1.1 *Optimum Concentration Range*, 5 to 100 μg/L (see Note 112).

45.1.2 *Detection Limit*, 1 μg/L.

NOTE 112—The concentration values and instrument conditions are for a Perkin-Elmer HGA-2100, based on the use of a 20-µL injection, continuous flow purge gas and pyrolytic graphite. Smaller sized furnace devices or those employing faster rates of atomization can be operated using lower atomization temperatures for shorter time periods than the recommended settings.

45.2 *Preparation of Standard Solution*:

45.2.1 *Stock Solution*—Prepare as described under "direct aspiration method."

45.2.2 Prepare dilutions of the stock solution to be used as calibration standards at the time of analysis. These solutions are also to be used for "standard additions."

45.2.3 The calibration standard should be diluted to contain 0.5 % (v/v) HNO_3.

45.3 *General Instrument Parameters*:

45.3.1 *Drying Time and Temperature*—30 s at 125°C.

45.3.2 *Ashing Time and Temperature*—30 s at 900°C.

45.3.3 *Atomizing Time and Temperature*—10 s at 2700°C.

45.3.4 *Purge Gas Atmosphere*—Argon.

45.3.5 *Wavelength*—232.0 nm.

45.3.6 Other operating parameters should be set as specified by the particular instrument manufacturer.

NOTE 113—The use of background correction is recommended.

NOTE 114—Nitrogen may also be used as the purge gas.

45.4 *Analysis Procedure*—For the analysis procedure and the calculation, see "Furnace Procedure", 11.3.

NOTE 115—For every sample matrix analyzed, verification is necessary to determine that the method of standard additions is not required (see 6.2.1).

NOTE 116—If the method of standard additions is required, follow the procedure given in 10.5.

46. Potassium—Direct Aspiration

46.1 *Requirements*:

46.1.1 *Optimum Concentration Range*—0.1 to 2 mg/L using a wavelength of 766.5 nm (see Note 117).

46.1.2 *Sensitivity*—0.04 mg/L.

46.1.3 *Detection Limit*—0.01 mg/L.

NOTE 117—The 404.4-nm line may also be used. This line has a relative sensitivity of 500.

46.2 *Preparation of Standard Solution*:

46.2.1 *Stock Solution*—Dissolve 0.1907 g of KCl (analytical reagent grade), dried at 110°C, in deionized distilled water and make up to 1 L. 1 mL = 0.10 mg K (100 mg/L).

46.2.2 Prepare dilutions of the stock solution to be used as calibration standards at the time of analysis. The *calibration standards* should be prepared using the same type of acid and at the same concentration as that of the sample being analyzed either directly or after processing.

46.3 *General Instrumental Parameters*:

46.3.1 *Potassium Hollow Cathode Lamp.*

46.3.2 *Wavelength*—766.5 nm.

46.3.3 *Fuel*—Acetylene.

46.3.4 *Oxidant*—Air.

46.3.5 *Type of Flame*—Slightly oxidizing.

46.4 *Analysis Procedure*—For analysis procedure and calculation, see "Direct Aspiration", 11.1.

NOTE 118—In air-acetylene or other high temperature flames (>2800°C), potassium can experience partial ionization which indirectly affects absorption sensitivity. The presence of other alkali salts in the sample can reduce this ionization and thereby enhance analytical results. The ionization suppressive effect of sodium is small if the ratio of Na to K is under 10. Any enhancement due to sodium can be stabilized by adding excess sodium (1000 µg/mL) to both sample and standard solutions. If more stringent control of ionization is required, the addition of cesium should be considered. Reagent blanks should be analyzed to correct for potassium impurities in the buffer stock.

NOTE 119—To cover the range of potassium values normally observed in surface waters (0.1 to 20 mg/L), it is suggested that the burner head be rotated. A90° rotation of the burner head provides approximately one-eighth the normal sensitivity.

47. Selenium—Furnace Technique

47.1 *Requirements*:

47.1.1 *Optimum Concentration Range*, 5 to 100 µg/L (see Note 120).

47.1.2 *Detection Limit*, 2 µg/L.

NOTE 120—The above concentration values and instrument conditions are for a Perkin-Elmer HGA-2100, based on the use of a 20-µL injection, continuous flow purge gas and non-pyrolytic graphite. Smaller sized furnace devices or those employing faster rates of atomization can be operated using lower atomization temperatures for shorter time periods than the above recommended settings.

47.2 *Preparation of Standard Solution*:

47.2.1 *Stock Selenium Solution*—Dissolve 0.3453 g of selenous acid (actual assay 94.6 % H_2SeO_3) in deionized distilled water and make up to 200 mL. 1 mL = 1 mg Se (1000 mg/L).

47.2.2 *Nickel Nitrate Solution, 5 %*—Dissolve 24.780 g of ACS reagent grade Ni $(NO_3)_2 \cdot 6H_2O$ in deionized distilled water and make up to 100 mL.

47.2.3 *Nickel Nitrate Solution, 1 %*—Dilute 20 mL of the 5 % nickel nitrate to 100 mL with deionized distilled water.

47.2.4 *Working Selenium Solution*—Prepare dilutions of the stock solution to be used as calibration standards at the time of analysis. Withdraw appropriate aliquots of the stock solution, add 1 mL of concentrated HNO_3, 2 mL of 30 % H_2O_2 and 2 mL of the 5 % nickel nitrate solution. Dilute to 100 mL with deionized distilled water.

47.3 *Sample Preparation*:

47.3.1 Transfer 100 mL of well-mixed sample to a 250-mL Griffin beaker, add 2 mL of 30 % H_2O_2 and sufficient concentrated HNO_3 to result in an acid concentration of 1 % (v/v). Heat for 1 h at 95°C or until the volume is slightly less than 50 mL.

47.3.2 Cool and bring back to 50 mL with deionized distilled water.

47.3.3 Pipet 5 mL of this digested solution into a 10 mL volumetric flask, add 1 mL of the 1 % nickel nitrate solution and dilute to 10 mL with deionized distilled water. The sample is now ready for injection into the furnace.

NOTE 121—If solubilization or digestion is not required, adjust the HNO_3 concentration of the sample to 1 % (v/v) and add 2 mL of 30 % H_2O_2 and 2 mL of 5 % nickel nitrate to each 100 mL of sample. The volume of the calibration standard should be adjusted with deionized distilled water to match the volume change of the sample.

47.4 *Instrument Parameters*:

47.4.1 *Drying Time and Temperature*—30 s at 125°C.

47.4.2 Charring Time and Temperature—30 s at 1200°C.

47.4.3 *Atomizing Time and Temperature*—10 s at 2700°C.

47.4.4 *Purge Gas Atmosphere*—Argon.

47.4.5 *Wavelength*—196.0 nm.

47.4.6 Other operating parameters should be set as specified by the particular instrument manufacturer.

NOTE 122—The use of background correction is recommended.

47.5 *Analysis Procedure*—For the analysis procedure and the calculation see "Furnace Procedure," 11.3.

NOTE 123—Selenium analysis suffers interference from chlorides (>800 mg/L) and sulfate (>200 mg/L). For the analysis of industrial effluents and samples with concentrations of sulfate from 200 to 2000 mg/L, both samples and standards should be prepared to contain 1 % nickel.

NOTE 124—For every sample matrix analyzed, verification is necessary to determine that the method of standard additions is not required (see 6.2.1).

NOTE 125—If the method of standard additions is required, follow the procedure given in 10.5.

48. Selenium—Gaseous Hydride

48.1 *Scope and Application*—The gaseous hydride method determines inorganic selenium when present in concentrations at or above 2 µg/L. The test method is applicable in the absence of high concentrations of chromium, cobalt, copper, mercury, molybdenum, nickel, and silver.

48.2 *Summary of Test Method*—Selenium in the sample is reduced from the + 6 oxidation state to the + 4 oxidation state by the addition of $SnCl_2$. Zinc is added to the acidified sample, producing hydrogen and converting the selenium to the hydride, SeH_2. The gaseous selenium hydride is swept into an argon-hydrogen flame of an atomic absorption spectrophotometer. The working range of the test method is 2 to 20 µg/L using the 196.0-nm wavelength.

48.3 *Discussion*:

48.3.1 In analyzing drinking water and most surface and ground waters, interferences are rarely encountered. Industrial waste samples should be spiked with a known amount of selenium to establish adequate recovery.

48.3.2 Organic forms of selenium must be converted to an inorganic form and organic matter must be oxidized before beginning the analysis. The oxidation procedure given in Procedure 4.1 of Method 206.5 should be used **(13)**.

48.4 References

48.4.1 Except for the perchloric acid step, the procedure to be used for this determination is found in *Standard Methods for the Examination of Water and Wastewater* **(14)**.

49. Silver—Direct Aspiration

49.1 *Requirements*:

49.1.1 Optimum Concentration Range, 0.1 to 4 mg/L using a wavelength of 328.1 nm (see Note 126 and Note 127).

49.1.2 *Sensitivity*, 0.06 mg/L.

49.1.3 *Detection Limit*, 0.01 mg/L.

NOTE 126—For levels of silver below 30 µg/L, either the Special Extraction Procedure, given in 11.2 or the furnace procedure is recommended.

NOTE 127—The 338.2-nm wavelength may also be used. This has a relative sensitivity of 2.

49.2 *Preparation of Standard Solution*:

49.2.1 *Stock Solution*—Dissolve 1.575 g of $AgNO_3$ (analytical reagent grade) in deionized distilled water, add 10 mL concentrated HNO_3 and make up to 1 L. 1 mL = 1 mg Ag (1000 mg/L).

NOTE 128—Silver nitrate standards are light sensitive. Dilutions of the stock should be discarded after use as concentrations below 10 mg/L are not stable over long periods of time.

49.2.2 Prepare dilutions of the stock solution to be used as calibration standards at the time of analysis. The *calibration standards* should be prepared using nitric acid and at the same concentration as that of the sample being analyzed either directly or after processing.

49.2.3 *Iodine Solution* (1 *N*)—Dissolve 20 g of potassium iodide, KI (analytical reagent grade) in 50 mL of deionized distilled water, add 12.7 g of iodine, I_2 (analytical reagent grade) and dilute to 100 mL. Store in a brown bottle.

49.2.4 *Cyanogen Iodide* (CNI) *Solution*—To 50 mL of deionized distilled water add 4.0 mL concentrated NH_4OH, 6.5 g KCN, and 5.0 mL of 1.0 *N* I_2 solution. Mix and dilute to 100 mL with deionized distilled water. Fresh solution should be prepared every 2 weeks.

49.3 *General Instrumental Parameters*:

49.3.1 *Silver Hollow Cathode Lamp.*

49.3.2 Wavelength—328.1 nm.

49.3.3 *Fuel*—Acetylene.

49.3.4 *Oxidant*—Air.

49.3.5 *Type of Flame*—Oxidizing.

49.4 *Analysis Procedure*—For the analysis procedure and the calculation, see "Direct Aspiration," 11.1.

NOTE 129—If absorption to container walls or the formation of AgCl is suspected, make the sample basic using concentrated NH_4OH and add 1 mL of (CNI) solution per 100 mL of sample. Mix the sample and allow to stand for 1 h before proceeding with the analysis.

50. Silver—Furnace Technique

50.1 *Requirements*:

50.1.1 *Optimum Concentration Range*—1 to 25 µg/L (see Note 130).

50.1.2 *Detection Limit*—0.2 µg/L.

NOTE 130—The concentration values and instrument conditions are for a Perkin-Elmer HGA-2100, based on the use of a 20-µL injection, continuous flow purge gas and nonpyrolytic graphite. Smaller sized furnace devices or those employing faster rates of atomization can be operated using lower atomization temperatures for shorter time periods than the above recommended settings.

50.2 *Preparation of Standard Solution* :

50.2.1 *Stock Solution*—Prepare as described under "direct aspiration method."

50.2.2 Prepare dilutions of the stock solution to be used as calibration standards at the time of analysis. These solutions are also to be used for "standard additions."

50.2.3 The calibration standard should be diluted to contain 0.5 % (v/v) HNO_3.

NOTE 131—The use of halide acids should be avoided.

50.3 *General Instrument Parameters*:

50.3.1 *Drying Time and Temperature*—30 s at 125°C.

50.3.2 *Ashing Time and Temperature*—30 s at 400°C.

50.3.3 *Atomizing Time and Temperature*—10 s at 2700°C.

50.3.4 *Purge Gas Atmosphere*—Argon.

50.3.5 *Wavelength*—328.1 nm.

50.3.6 Other operating parameters should be set as specified by the particular instrument manufacturer.

NOTE 132—Background correction may be required if the sample contains high dissolved solids.

50.4 *Analysis Procedure*—For the analysis procedure and the calculation, see "Furnace Procedure," 11.3 (see Note 3, Note 5, Note 6, and Note 7).

NOTE 133—If adsorption to container walls or formation of AgCl is suspected, see the Direct Aspiration Method.

NOTE 134—For every sample matrix analyzed, verification is necessary to determine that the method of standard additions is not required (see 6.2.1).

NOTE 135—If the method of standard additions is required, follow the procedure given in 10.5.

51. Sodium—Direct Aspiration

51.1 *Requirements*:

51.1.1 *Optimum Concentration Range*, 0.03 to 1 mg/L using a wavelength of 589.6 nm (see Note 136).

51.1.2 *Sensitivity*, 0.015 mg/L.

51.1.3 *Detection Limit*, 0.002 mg/L.

NOTE 136—The 330.2 nm resonance line of sodium, which has a relative sensitivity of 185, provides a convenient way to avoid the need to dilute more concentrated solutions of sodium.

51.2 *Preparation of Standard Solution*:

51.2.1 *Stock Solution*—Dissolve 2.542 g of NaCl (analytical reagent grade), dried at 140°C, in deionized distilled water and make up to 1 L. 1 mL = 1 mg Na (1000 mg/L).

51.2.2 Prepare dilutions of the stock solution to be used as calibration standards at the time of analysis. The *calibration standards* should be prepared using the same type of acid and at the same concentration as that of the sample being analyzed either directly or after processing.

51.3 *General Instrumental Parameters*:

51.3.1 *Sodium Hollow Cathode Lamp*.

51.3.2 *Wavelength*—589.6 nm.

51.3.3 *Fuel*—Acetylene.

51.3.4 *Oxidant*—Air.

51.3.5 *Type of Flame*—Oxidizing.

NOTE 137—Low-temperature flames increase sensitivity by reducing the extent of ionization of this easily ionized metal. Ionization may also be controlled by adding potassium (1000 mg/L) to both standards and samples.

51.4 *Analysis Procedure*—For analysis procedure and calculation, see "Direct Aspiration," 11.1.

52. Tin—Direct Aspiration

52.1 *Requirements*:

52.1.1 Optimum Concentration Range, 10 to 300 mg/L using a wavelength of 286.3 nm (see Note 138).

52.1.2 *Sensitivity*, 4 mg/L.

52.1.3 *Detection Limit*, 0.8 mg/L.

NOTE 138—For concentrations of tin below 2 mg/L, the furnace procedure is recommended.

52.2 *Preparation of Standard Solution*:

52.2.1 *Stock Solution*—Dissolve 1.000 g of tin metal (analytical reagent grade) in 100 mL of concentrated HCl and dilute to 1 L with deionized distilled water. 1 mL = 1 mg Sn (1000 mg/L).

52.2.2 Prepare dilutions of the stock tin solution to be used as calibration standards at the time of analysis. The *calibration standards* should be prepared using the same type of acid and at the same concentration as that of the sample being analyzed either directly or after processing.

52.3 *General Instrumental Parameters*:

52.3.1 *Tin Hollow Cathode Lamp*.

52.3.2 *Wavelength*—286.3 nm.

52.3.3 *Fuel*—Acetylene.

52.3.4 *Oxidant*—Nitrous oxide.

52.3.5 *Type of Flame*—Fuel rich.

52.4 *Analysis Procedure*—For the analysis procedure and the calculation, see "Direct Aspiration," 11.1.

53. Tin—Furnace Technique

53.1 *Requirements*:

53.1.1 *Optimum Concentration Range*, 20 to 300 μg/L (see Note 139).

53.1.2 *Detection Limit*, 5 μg/L.

NOTE 139—The concentration values and instrument conditions are for a Perkin-Elmer HGA-2100, based on the use of a 20 μL injection, continuous flow purge gas and non-pyrolytic graphite. Smaller sized furnace devices or those employing faster rates of atomization can be operated using lower atomization temperatures for shorter time periods than the recommended settings.

53.2 *Preparation of Standard Solution*:

53.2.1 *Stock Solution*—Prepare as described under "direct aspiration method".

53.2.2 Prepare dilutions of the stock solution to be used as calibration standards at the time of analysis. These solutions are also to be used for "standard additions".

53.2.3 The calibration standard should be diluted to contain 2 % (v/v) HNO_3.

53.3 *Instrument Parameters*:

53.3.1 *Drying Time and Temperature*—30 s at 125°C.

53.3.2 *Ashing Time and Temperature*—30 s at 600°C.

53.3.3 *Atomizing Time and Temperature*—10 s at 2700°C.

53.3.4 *Purge Gas Atmosphere*—Argon.

53.3.5 *Wavelength*—224.6 nm.

53.3.6 Other operating parameters should be set as specified by the particular instrument manufacturer.

NOTE 140—The use of background correction is recommended.

NOTE 141—Nitrogen may also be used as the purge gas.

53.4 *Analysis Procedure*—For the analysis procedure and the calculation see "Furnace Procedure," 11.3.

NOTE 142—Tin analysis is sensitive to chloride concentration. If chloride concentration presents a matrix problem or causes a loss previous to atomization, add an excess of 5 mg of ammonium nitrate to the furnace and ash using a ramp necessary or with incremental steps until the recommended ashing temperature is reached. Extended ashing times have been reported to improve precision.

NOTE 143—For every sample matrix analyzed, verification is necessary to determine that the method of standard additions is not required (see 6.2.1).

NOTE 144—If the method of standard additions is required, follow the procedure given in 10.5.

54. Titanium—Direct Aspiration

54.1 *Requirements*:

54.1.1 *Optimum Concentration Range*, 5 to 100 mg/L using a wavelength of 365.3 nm (see Note 145).

54.1.2 *Sensitivity*, 2 mg/L.

54.1.3 *Detection Limit*, 0.4 mg/L.

NOTE 145—For concentrations of titanium below 1.0 mg/L, the furnace procedure is recommended.

54.2 *Preparation of Standard Solution*:

54.2.1 *Stock Solution*—Dissolve 4.008 g of titanium sulfate, $Ti_2(SO_4)_3$, in dilute HCl and make up to 1 L with deionized distilled water. 1 mL = 1 mg Ti (1000 mg/L).

54.2.2 *Potassium Chloride Solution*—Dissolve 95 g potassium chloride, KCl, in distilled water and make up to 1 L.

54.2.3 Prepare dilutions of the stock titanium solution to be used as calibration standards at the time of analysis. The *calibration standards* should be prepared using the same type of acid and at the same concentration as that of the sample being analyzed either directly or after processing. To each 100 mL of standard and sample alike, add 2 mL of potassium chloride solution.

54.3 *General Instrumental Parameters*:

54.3.1 *Titanium Hollow Cathode Lamp*.

54.3.2 *Wavelength*—365.3 nm.

54.3.3 *Fuel*—Acetylene.

54.3.4 *Oxidant*—Nitrous oxide.

54.3.5 *Type of Flame*—Fuel rich.

54.4 *Analysis Procedure*—For the analysis procedure and the calculation, see "Direct Aspiration," 11.1.

54.5 *Interferences*—A number of elements increase the sensitivity of titanium. To control this problem, potassium (1000 mg/L) must be added to standards and samples alike **(15)**.

55. Titanium Furnace Technique

55.1 *Requirements*:

55.1.1 *Optimum Concentration Range*, 50 to 500 μg/L (see Note 146).

55.1.2 *Detection Limit*, 10 μg/L.

NOTE 146—The above concentration values and instrument conditions are for a Perkin-Elmer HGA-2100, based on the use of a 20-μL injection, continuous flow purge gas and pyrolytic graphite.

55.2 *Preparation of Standard Solution*:

55.2.1 *Stock Solution*—Prepare as described under "direct aspiration method."

55.2.2 Prepare dilutions of the stock solution to be used as calibration standards at the time of analysis.

55.2.3 The *calibration standards* should be prepared using the same type of acid and at the same concentration as that of the sample being analyzed either directly or after processing.

55.3 *General Instrument Parameters*:

55.3.1 *Drying Time and Temperature*—30 s at 125°C.

55.3.2 *Ashing Time and Temperature*—30 s at 1400°C.

55.3.3 *Atomizing Time and Temperature*—15 s at 2800°C.

55.3.4 *Purge Gas Atmosphere*—Argon.

55.3.5 *Wavelength*—365.4 nm.

55.3.6 Other operating parameters should be set as specified by the particular instrument manufacturer.

NOTE 147—Background correction may be required if the sample contains high dissolved solids.

NOTE 148—Because of possible chemical interactions, nitrogen should not be used as the purge gas.

55.4 *Analysis Procedure*—For the analysis procedure and the calculation, see "Furnace Procedure," 11.3.

NOTE 149—For every sample matrix analyzed, verification is necessary to determine that the method of standard additions is not required (see 6.2.1).

NOTE 150—If the method of standard additions is required, follow the procedure given in 10.5.

56. Vanadium—Direct Aspiration

56.1 *Requirements*:

56.1.1 *Optimum Concentration Range*—2 to 100 mg/L using a wavelength of 318.4 nm (see Note 151).

56.1.2 *Sensitivity*—0.8 mg/L.

56.1.3 *Detection Limit*—0.2 mg/L.

NOTE 151—For concentrations of vanadium below 0.5 mg/L, the furnace procedure is recommended.

56.2 *Preparation of Standard Solution*:

56.2.1 *Stock Solution*—Dissolve 1.7854 g of vanadium pentoxide, V_2O_5 (analytical reagent grade) in 10 mL of concentrated nitric acid and dilute to 1 L with deionized distilled water. 1 mL = 1 mg V (1000 mg/L).

56.2.2 *Aluminum Nitrate Solution*—Dissolve 139 g aluminum nitrate, $Al(NO_3)_3 \cdot 9H_2O$, in 150 mL of deionized distilled water; heat to effect solution. Allow to cool and make up to 200 mL.

56.2.3 Prepare dilutions of the stock solution to be used as calibration standards at the time of analysis. The *calibration standards* should be prepared using the same type of acid and at the same concentration as that of the sample being analyzed either directly or after processing. To each 100 mL of standard and sample alike, add 2 mL of the aluminum nitrate solution.

56.3 *General Instrumental Parameters*:

56.3.1 *Vanadium Hollow Cathode Lamp* .

56.3.2 *Wavelength*—318.4 nm.

56.3.3 *Fuel*—Acetylene.

56.3.4 *Oxidant*—Nitrous oxide.

56.3.5 *Type of Flame*—Fuel rich.

56.4 *Analysis Procedure*—For the analysis procedure and the calculation, see "Direct Aspiration," 11.1.

56.5 *Interferences*—It has been reported that high concentrations of aluminum and titanium increase the sensitivity of vanadium. This interference can be controlled by adding excess aluminum (1000 ppm) to both samples and standards **(16)**.

57. Vanadium—Furnace Technique

57.1 *Requirements*:

57.1.1 *Optimum Concentration Range*, 10 to 200 μg/L (see Note 152).

57.1.2 *Detection Limit*, 4 μg/L.

NOTE 152—The above concentration values and instrument conditions are for a Perkin-Elmer HGA-2100, based on the use of a 20-μL injection, continuous flow purge gas and pyrolytic graphite. Smaller sized furnace devices or those employing faster rates of atomization can be operated using lower atomization temperatures for shorter time periods than the above recommended settings.

57.2 *Preparation of Standard Solution*:

57.2.1 *Stock Solution*—Prepare as described under "direct aspiration method."

57.2.2 Prepare dilutions of the stock solution to be used as calibration standards at the time of analysis. These solutions are also to be used for "standard additions."

57.2.3 The calibration standard should be diluted to contain 0.5 % (v/v) HNO_3.

57.3 *General Instrument Parameters*:

57.3.1 *Drying Time and Temperature*—30 s at 125°C.

57.3.2 *Ashing Time and Temperature*—30 s at 1400°C.

57.3.3 *Atomizing Time and Temperature*—15 s at 2800°C.

57.3.4 *Purge Gas Atmosphere*—Argon.

57.3.5 *Wavelength*—318.4 nm.

57.3.6 Other operating parameters should be set as specified by the particular instrument manufacturer.

NOTE 153—Background correction may be required if the sample contains high dissolved solids.

NOTE 154—Because of possible chemical interaction, nitrogen should not be used as the purge gas.

57.4 *Analysis Procedure*—For the analysis procedure and the calculation, see "Furnace Procedure," 11.3.

NOTE 155—For every sample matrix analyzed, verification is necessary to determine that the method of standard additions is not required (see 6.2.1).

NOTE 156—If the method of standard additions is required, follow the procedure given in 10.5.

58. Zinc—Direct Aspiration

58.1 *Requirements*:

58.1.1 *Optimum Concentration Range*, 0.05 to 1 mg/L using a wavelength of 213.9 nm (see Note 157).

58.1.2 *Sensitivity*, 0.02 mg/L.

58.1.3 *Detection Limit*, 0.005 mg/L.

NOTE 157—For concentrations of zinc below 0.01 mg/L, either the Special Extraction Procedure or the Furnace Procedure, is recommended.

58.2 *Preparation of Standard Solution*:

58.2.1 *Stock Solution*—Carefully weigh 1.00 g of zinc metal (analytical reagent grade) and dissolve cautiously in 10 mL HNO_3. When solution is complete make up to 1 L with deionized distilled water. 1 mL = 1 mg Zn (1000 mg/L).

58.2.2 Prepare dilutions of the stock solution to be used as calibration standards at the time of analysis. The *calibration standards* should be prepared using the same type of acid and at the same concentration as that of the sample being analyzed either directly or after processing.

58.3 *Instrumental Parameters*:

58.3.1 *Zinc Hollow Cathode Lamp.*

58.3.2 *Wavelength*—13.9 nm.

58.3.3 *Fuel*—Acetylene.

58.3.4 *Oxidant*—Air.

58.3.5 *Type of Flame*—Oxidizing.

58.4 *Analysis Procedure*—For analysis procedure and calculation, see "Direct Aspiration," 11.1.

NOTE 158—High levels of silicon may interfere.

NOTE 159—The air-acetylene flame absorbs about 25 % of the energy at the 213.9 nm line.

NOTE 160—The sensitivity may be increased by the use of low-temperature flames.

NOTE 161—Some sample container cap liners can be a source of zinc contamination. To circumvent or avoid this problem, the use of polypropylene caps is recommended.

59. Zinc—Furnace Technique

59.1 *Requirements*:

59.1.1 *Optimum Concentration Range*—0.2 to 4 µg/L (see Note 162).

59.1.2 *Detection Limit*—0.05 µg/L.

NOTE 162—The concentration values and instrument conditions are for a Perkin-Elmer HGA-2100, based on the use of a 20-µL injection, continuous flow purge gas and non-pyrolytic graphite. Smaller sized furnace devices or those employing faster rates of atomization can be operated using lower atomization temperatures for shorter time periods than the recommended settings.

59.2 *Preparation of Standard Solution*:

59.2.1 *Stock Solution*—Prepare as described under "Direct Aspiration Method."

59.2.2 Prepare dilutions of the stock solution to be used as calibration standards at the time of analysis. These solutions are also to be used for "standard additions."

59.2.3 The calibration standard should be diluted to contain 0.5 % (v/v) HNO_3.

59.3 *General Instrument Parameters*:

59.3.1 *Drying Time and Temperature*—30 s at 125°C.

59.3.2 *Ashing Time and Temperature*—30 s at 400°C.

59.3.3 *Atomizing Time and Temperature*—10 s at 2500°C.

59.3.4 *Purge Gas Atmosphere*—Argon.

59.3.5 *Wavelength*—213.9 nm.

59.3.6 Other operating parameters should be set as specified by the particular instrument manufacturer.

NOTE 163—The use of background correction is recommended.

NOTE 164—Nitrogen may also be used as a purge gas.

59.4 *Analysis Procedure*—For the analysis procedure andthe calculation, see "Furnace Procedure," 11.3.

NOTE 165—The analysis of zinc by the graphite furnace is extremely sensitive and very subject to contamination from the work area, reagents, and pipet tips. Since all these factors affect the precision and accuracy, zinc should be analyzed by the direct aspiration procedure whenever possible.

NOTE 166—For every sample matrix analyzed, verification is necessary to determine that the method of standard additions is not required (see 6.2.1).

NOTE 167—If the method of standard additions is required, follow the procedure given in 10.5.

60. Precision and Bias

60.1 Precision and bias data for the methods described on RDF are not available at this time.

REFERENCES

(1) *Atomic Absorption Newsletter*, Vol 14, No. 5, Sept. and Oct., Perkin-Elmer Corp., Norwalk, CT, 1975, p. 127.
(2) *Atomic Absorption Newsletter*, Vol 6, 1967, p. 128.
(3) *Atomic Absorption Newsletter*, Vol 5, 1966, p. 17.
(4) *Standard Methods for the Examination of Water and Wastewater*, 14th Ed. Method 301A (VII), American Public Health Assn., American Water Works Assn. and Water Pollution Control Federation, 1975, p. 159.
(5) *Spectrochimica Acta*, Vol 22, Pergamon Press, 1966, p. 1325.
(6) *Standard Methods for the Examination of Water and Wastewater*, 14th Ed., 1975, p. 177.
(7) *Atomic Absorption Newsletter*, Vol 14, 1975, p. 29.
(8) *Talanta*, Vol 20, An International Journal of Analytical Chemistry, Pergamon Press, Inc. Journals Division, Maxwell House, Fairview Park, Elmford, NY 10523, 1973, p. 631.
(9) *Standard Methods for the Examination of Water and Wastewater*, 14th Ed., 1975, p. 192.
(10) *Atomic Absorption Newsletter*, Vol 15, No. 3, May and June 1976, p. 71.
(11) Kopp, J. F., Longbottom, M. C., and Lobring, L. B., "Cold Vapor Method for Determining Mercury," American Water Works Assoc., Vol 64, Jan. 1971, p. 20.
(12) *Analytical Chimica Acta*, Vol 44, Elsevier Scientific Publishing Co., Box 211, 1000 AE, Amsterdam, Netherlands, 1969, p. 437.
(13) *Standard Methods for the Examination of Water and Wastewater*, 14th Ed., 404B, 1975, p. 285.
(14) *Standard Methods for the Examination of Water and Wastewater*, 14th Ed., Method 301A (VII), 1975, p. 159.
(15) *Atomic Absorption Newsletter*, Vol 6, 1967, p. 86.
(16) *Talanta*, Oxford/NY, 1968, p. 871.

Standard Practices for Dissolution of Refuse-Derived Fuel (RDF) Ash Samples for Analyses of Metals[1]

This standard is issued under the fixed designation E 886; the number immediately following the designation indicates the year of original adoption or, in the case of revision, the year of last revision. A number in parentheses indicates the year of last reapproval. A superscript epsilon (ϵ) indicates an editorial change since the last revision or reapproval.

1. Scope

1.1 These practices described herein cover the preparation of RDF ash, fly ash, bottom ash, or slag for analyses of metals by atomic absorption spectroscopy or inductively coupled plasma spectroscopy, or both (see Appendix X1).

1.2 These practices may be applicable to any waste material from which a laboratory analysis sample can be prepared.

1.3 Three practices are described in this standard:

1.3.1 *Practice A*—Lithium tetraborate ($Li_2B_4O_7$) fusion,

1.3.2 *Practice B*—Aqua regia dissolution, and

1.3.3 *Practice C*—Bomb, acid digestion method.

1.4 *This standard does not purport to address all of the safety concerns, if any, associated with its use. It is the responsibility of the user of this standard to establish appropriate safety and health practices and determine the applicability of regulatory limitations prior to use.* For hazard statements, see Sections 4, 20.3, and 20.6.

2. Referenced Documents

2.1 *ASTM Standards:*

D 1193 Specification for Reagent Water[2]

E 885 Test Methods for Analyses of Metals in Refuse-Derived Fuel by Atomic Absorption Spectroscopy[3]

3. Significance and Use

3.1 These practices are available to producers and users of RDF as methods of preparing RDF ashes for subsequent analyses of metals.

3.2 The chemical composition of laboratory prepared RDF ashes do not exactly represent the composition of mineral matter in the RDF or the composition of fly ash, bottom ash, and slag resulting from commercial-scale burning of RDF. These practices are for analyzing the various ash samples as received in the laboratory.

4. Hazards

4.1 The hot acidic solutions in these practices pose a significant potential hazard. Proper laboratory safety practices and equipment should be employed throughout these procedures.

5. Sampling

5.1 The method of sampling for these procedures should be based on agreement between involved parties.

PRACTICE A—LITHIUM TETRABORATE FUSION

6. Summary of Practice

6.1 The RDF ash, fly ash, bottom ash, or slag is fused with lithium tetraborate ($Li_2B_4O_7$) followed by a final dissolution of the melt in dilute hydrochloric acid (HCl). The solution is analyzed for metals by atomic absorption spectroscopy or inductively coupled plasma spectroscopy.

7. Apparatus

7.1 *Analytical Balance*, capable of weighing to 0.0001 g.

7.2 *Muffle Furnace*— The furnace shall have adequate air ventilation and shall be capable of temperature regulation up to at least 1000°C. An air change rate of 1 to 4 furnace volumes of air per minute has been found adequate.

7.3 *Platinum Dish*, 35 to 85-mL capacity, or

7.3.1 *Graphite Crucibles*, 35 to 85-mL capacity.

7.4 *Stirring Hotplate and Stirring Bars*, operating temperature of 200°C.

7.5 *Volumetric Flasks*, 50, 100, and 200-mL.

8. Reagents and Materials

8.1 *Purity of Reagents*—Reagent grade chemicals shall be used in this test. Unless otherwise indicated, it is intended that all reagents shall conform to the specifications of the Committee on Analytical Reagents of the American Chemical Society where such specifications are available.[4] Other grades may be used, provided it is first ascertained that the reagent is of sufficiently high purity to permit its use without lessening the accuracy of the determination.

---

[1] These practices are under the jurisdiction of ASTM Committee D-34 on Waste Management and are the direct responsibility of Subcommittee D34.03 on Treatment.

Current edition approved Aug. 15, 1994. Published October 1994. Originally published as E 886 – 82. Last previous edition E 886 – 88.

[2] *Annual Book of ASTM Standards*, Vol 11.01.

[3] *Annual Book of ASTM Standards*, Vol 11.04.

[4] *Reagent Chemicals, American Chemical Society Specifications*, American Chemical Society, Washington, DC. For suggestions on the testing of reagents not listed by the American Chemical Society, see *Analar Standards for Laboratory Chemicals*, BDH Ltd., Poole, Dorset, U.K., and the *United States Pharmacopeia and National Formulary*, U.S. Pharmaceutical Convention, Inc. (USPC), Rockville, MD.

8.2 *Purity of Water*— Unless otherwise indicated, reference to water shall be understood to mean at least Type III reagent water conforming to Specification D 1193.

8.3 *Lithium Tetraborate* —($Li_2B_4O_7$), powder.

8.4 *Hydrochloric Acid* (HCl), (5+95)—Dilute 50 mL of concentrated hydrochloric acid (HCl, sp gr 1.19) to 1000 mL with water.

9. Preparation of Sample

9.1 Prepare the RDF ash, fly ash, bottom ash, or slag by grinding the sample in an agate mortar to a particle size to pass a No. 200 (74-μm) sieve.

10. Procedure

10.1 *Sample Fusion*:

NOTE 1—Stronger or weaker concentrations may be required to establish a suitable range for those metals of varying concentrations outside the range of the typical sample. Each analyst must determine the sensitivity and linear range of his own equipment and choose concentration ranges from standards, sample weight, and dilution factors all compatible with the instrument specific for his work. For very trace concentrations of any metal in the ash, multiple fusions may be made and the dissolved fusion beads pooled into one sample.

10.1.1 Weigh 0.1 ± 0.0001 g of sample as prepared in Section 9 into a platinum dish or graphite crucible.

10.1.2 Add 0.5 g of $Li_2B_4O_7$. Mix the sample and $Li_2B_4O_7$ powder well. Then add an additional 0.5 g of $Li_2B_4O_7$ to cover the mixture.

10.1.3 Place the dish in a clean silica or refractory tray and place in a muffle furnace preheated to 1000°C; 15 min at 1000°C is sufficient to fuse the mixture completely.

10.1.4 Remove the tray and dish from the muffle furnace, and cool to room temperature.

10.2 *Dissolution*:

10.2.1 Carefully rinse the bottom and outside of the platinum dish or graphite crucible to remove possible contamination. Then place it in a clean 250 or 400-mL beaker.

10.2.2 Place a clean TFE-fluorocarbon-coated stirring magnet inside the dish and add 150 mL of HCl (5+95) to the beaker and dish, and place immediately on the stirring hotplate.

10.2.3 Heat the solution to just below boiling temperature and maintain for not more than 30 min with constant stirring. This time and temperature is sufficient to completely dissolve the metal. If stirring is not maintained constantly, some of the ash constituents are apt to precipitate and the analysis must be repeated.

10.2.4 Remove the beaker from the hotplate, and permit to cool to room temperature.

10.2.5 Quantitatively transfer the solution to a 200-mL volumetric flask, wash the platinum dish or graphite crucible and beaker with small amounts of HCl (5+95), and dilute to the 200-mL mark with the HCl solution. This solution (200 mL) contains those metals as were contained in 0.1 g of the ash (500 ppm) and 5 grams per litre of $Li_2B_4O_7$ solution.

10.2.6 This solution (10.2.5) can be either concentrated or diluted depending upon the magnitude of concentration of the metal in question (see Note 1).

10.3 Analyze this solution for the metal analyte in question in accordance with Test Methods E 885.

10.4 Prepare reagent blanks using the procedure outlined in 10.1.2 through 10.3.

PRACTICE B—AQUA REGIA DISSOLUTION

11. Summary of Practice

11.1 The RDF ash, fly ash, bottom ash, or slag is dissolved in mineral acids. The resulting solution is analyzed for metals by atomic absorption spectroscopy or inductively coupled plasma spectroscopy.

12. Apparatus

12.1 *Analytical Balance*, capable of weighing to 0.0001 g.

12.2 *Bottles*, polyethylene or polytetrafluoroethylene, 125-mL capacity with screw-cap lids, capable of withstanding 130°C.

12.3 *Steam Bath.*

12.4 *Volumetric Flasks*, 50, 100, and 200-mL.

12.5 *Crucibles*, 50-mL quartz or high silica.

13. Reagents

13.1 *Purity of Reagents*—Reagent grade chemicals shall be used in this test. Unless otherwise indicated, it is intended that all reagents shall conform to the specifications of the Committee on Analytical Reagents of the American Chemical Society where such specifications are available.[4] Other grades may be used, provided it is first ascertained that the reagent is of sufficiently high purity to permit its use without lessening the accuracy of the determination.

13.2 *Purity of Water*— Unless otherwise indicated, reference to water shall be understood to mean at least Type III reagent water conforming to Specification D 1193.

13.3 *Aqua Regia Solution*—Mix 1 part concentrated nitric acid (HNO_3, sp gr 1.42), 3 parts concentrated hydrochloric acid (HCl, sp gr 1.19), and 1 part water.

13.4 *Hydrofluoric Acid* (HF)—Concentrated sp gr 1.15.

13.5 *Boric Acid* (H_3BO_3)—Saturated solution.

14. Preparation of Sample

14.1 Prepare the RDF ash, fly ash, bottom ash, or slag by grinding the sample in an agate mortar to a particle size to pass a No. 200 (74-μm) sieve.

15. Procedure

15.1 *Dissolution*:

15.1.1 Weigh accurately about 0.2 g of ash sample as prepared in 14.1, and add to a 125-mL plastic bottle with a screw cap.

15.1.2 Add 3 mL of aqua regia solution and 5 mL of HF to the sample.

15.1.3 Tighten the screw-cap, and place bottle on a steam bath for at least 2 h.

15.1.4 Add 50 mL of saturated H_3BO_3 solution to the resultant solution (see Note 2). If a residue remains, the mixture may be reheated for about 1 h to help dissolve it (see Note 3).

NOTE 2—Boric acid (H_3BO_3) not only complexes fluoride (F^-), but it also has been shown to have good flame properties and acts as a flame buffer.

NOTE 3—If a residue persists, repeat 15.1.2 through 15.1.4.

15.1.5 Cool the solution to room temperature, quantitatively transfer the contents of the plastic bottle to a 100-mL volumetric flask and dilute to volume with water. This solution (100 mL) contains those metals as were contained in 0.2 g of ash (2000 ppm).

15.1.6 Prepare blanks by using the above procedure. Store samples and blanks in polyethylene bottles.

15.1.7 The sample solution (15.1.5) can be either concentrated or diluted depending upon the magnitude of concentration of the metal in question.

NOTE 4—Stronger or weaker concentrations may be required to establish a suitable range for those metals of varying concentrations outside the range of the typical sample. Each analyst must determine the sensitivity and linear range of calibration of his own equipment and choose concentration ranges for standards, sample weights used, and dilution factors all compatible with the instrument specific for his work.

15.2 Analyze the sample solutions and blank solutions for the metal analyte in question in accordance with Test Methods E 885.

PRACTICE C—BOMB, ACID DIGESTION METHOD

16. Summary of Practice

16.1 The RDF ash, fly ash, bottom ash, or slag is oxidized and dissolved in an acid digestion bomb containing strong mineral acid. The resulting solution is analyzed for metals by atomic absorption spectroscopy or inductively coupled plasma spectroscopy.

17. Apparatus

17.1 *Analytical Balance*, capable of weighing to 0.0001 g.

17.2 *Bomb*, acid digestion, polymer of tetrafluoroethylene lined, 25-mL capacity.

17.3 *Oven*, circulating air capable of maintaining a temperature of 100 ± 0.5°C.

17.4 *Beakers*, polymer of tetrafluoroethylene, 125-mL capacity.

17.5 *Steam Bath*.

17.6 *Volumetric Flasks*, 50, 100, and 250-mL.

18. Reagents

18.1 *Purity of Reagents*—Reagent grade chemical shall be used in this test. Unless otherwise indicated, it is intended that all reagents shall conform to the specifications of the Committee on Analytical Reagents of the American Chemical Society, where such specifications are available.[4] Other grades may be used, provided it is first ascertained that the reagent is of sufficiently high purity to permit its use without lessening the accuracy of the determination.

18.2 *Purity of Water*— Unless otherwise indicated, reference to water shall be understood to mean at least Type III reagent water conforming to Specification D 1193.

18.3 *Nitric Acid* (HNO_3)—Concentrated, sp gr 1.42.

18.4 *Hydrochloric Acid* (HCl)—Concentrated, sp gr 1.19.

18.5 *Aqua Regia*, 3 parts concentrated HNO_3 and 1 part concentrated HCl.

18.6 *Hydrofluoric Acid* (HF)—48 %.

18.7 *Boric Acid* (H_3BO_3), saturated solution.

19. Preparation of Sample

19.1 Prepare the RDF ash, fly ash, bottom ash, or slag by grinding the sample in an agate mortar to a particle size to pass a No. 200 (74-μm) sieve.

20. Procedure

20.1 Weigh accurately about 0.1 g of the ash sample as prepared in 19.1 into the polymer of tetrafluoroethylene cup of the acid digestion bomb.

20.2 Add no less than 2.5 nor more than 3.0 mL of aqua regia to the sample in the polymer of tetrafluoroethylene cup.

20.3 Assemble the polymer of tetrafluoroethylene cup and the acid digestion bomb (see Note 6).

NOTE 5—**Precaution:** Acid digestion bombs must be used according to the manufacturer's specifications. That is, the limits on the amounts and nature of the material that can be safely digested, the volume of acid added and the temperatures of digestion cannot be exceeded.

NOTE 6—Most acid digestion bombs have sockets for applying a spanner wrench to the cap and for holding the bomb in a rigid clamp during opening and closing operations. Normally these bombs can be sealed tightly by simply turning the knurled cap down firmly by hand, but sometimes additional tightening with a wrench may be required. Some bombs have an accessory spanner-jack available for holding the bomb during the opening and closing operations. Spring loaded pins on the arm of the spanner hold the body of the bomb firm while applying a flat face spanner to the screw-cap. The spanner-jack serves also as a valuable tool for pushing the polymer of tetrafluoroethylene cup out of the body if it becomes distorted after long exposure to high temperatures and pressures or locked in the body by corrosive products on the metal wall.

20.4 Place the bomb and contents into a circulating hot air oven at 100°C for 2 h.

20.5 Remove the bomb and contents from the oven and cool to room temperature.

20.6 Open the bomb and remove the polymer of tetrafluoroethylene cup (see Note 6).

NOTE 7—**Caution:** When removing the polymer of tetrafluoroethylene cup from the bomb, care must be exercised to keep from spilling the contents of the cup.

20.7 Quantitatively transfer the contents of the polymer of tetrafluoroethylene cup to a 100-mL volumetric flask, and dilute to mark with water. This solution (100 mL) contains those metals as were contained in 0.1 g of ash (1000 ppm).

NOTE 8—Should any undissolved residue be contained in the bomb's polymer of tetrafluoroethylene cup, quantitatively transfer the cup's contents to a 125-mL polymer of tetrafluoroethylene beaker, add 5 mL of HF and heat on a steam bath for 2 h. Add 50 mL of saturated H_3BO_3 solution, cool, and proceed as described in 20.7.

20.8 Prepare a reagent blank in the same manner as described in 20.2 through 20.7.

20.9 The sample solution (20.7) can be either concentrated or diluted depending upon the magnitude of concentration of metal in question.

NOTE 9—Stronger or weaker concentrations may be required to establish a suitable range for those metals of varying concentrations outside the range of the typical sample. Each analyst must determine the sensitivity and linear range of calibration of his own equipment and choose concentration ranges for standards, sample weights used, and dilution factors all compatible with the instrument specific for his work.

20.10 Analyze the sample solutions and blank solutions for

the metal analyte in question in accordance with Test Methods E 885.

CALCULATION AND PRECISION AND BIAS

21. Calculation

21.1 The general calculations for the metal in ash are as follows:

$$\% \text{ metals in ash} = \left[\frac{(A-B)}{(C-B)}\right] \times \frac{N}{M} \times 100 \quad (1)$$

$$\text{mg metal/kg ash} = \left[\frac{(A-B)}{(C-B)}\right] \times \frac{N}{M} \times 10^6 \quad (2)$$

$$\% \text{ metal oxide in ash} = \left[\frac{(A-B)}{(C-B)}\right] \times \frac{N}{M} \times F \times 100 \quad (3)$$

$$\text{mg metal oxide/kg ash} = \left[\frac{(A-B)}{(C-B)}\right] \times \frac{N}{M} \times F \times 10^6 \quad (4)$$

where:
A = absorbance of sample solution,
B = absorbance of blank solution,
C = absorbance of standard,
N = ppm metal in standard,
M = ppm sample in solution,
F = conversion factor from element to oxide, and
F = molecular weight of metal oxide/atomic weight of metal

A and C absorbances must be in the linear range of the calibration curve for the metal analyzed. See Test Methods E 885.

22. Precision and Bias

22.1 Precision and bias statements cannot be made on these test methods at this time. Precision and bias statements will have to be determined on each metal analyzed by a given method of analysis and will be a function of both the sample preparation and the method of analysis.

APPENDIX

(Nonmandatory Information)

X1. FACTORS FOR CHOOSING A METHOD FOR DISSOLUTION OF RDF ASH SAMPLES FOR ANALYSIS OF METALS

X1.1 Test Method E 886 consists of three possible methods of preparing RDF ash, fly ash, bottom ash, or slag for metals by atomic absorption spectroscopy or inductively coupled plasma spectroscopy, or both. No one method has been ascertained to be superior or inferior to any other method. Many factors must be considered when choosing the method to be used. Some of these factors include the following:

X1.1.1 The metal or metals to be analyzed.

X1.1.2 The proper laboratory facilities available.

X1.1.3 The training and experience of the analyst.

X1.1.4 The amount of sample to be treated.

X1.1.5 The magnitude of concentration of metal in the sample.

X1.1.6 The detection limits required.

X1.2 Metals may show variations in concentrations in RDF ash, fly ash, bottom ash, and slag samples. After dissolution of the RDF ash samples, the dilution to which the dissolved ash is taken depends upon the magnitude of concentration of the metal or metals in question. The range of metal concentrations shown in Table X1.1 can be used as a guide in determining the amount of RDF ash sample to be used and the final dilution factor for subsequent analysis.

TABLE X1.1 RANGE OF METAL CONCENTRATIONS

Metal (M)		Concentration Range, (mg M/kg Ash)
Aluminum	(Al)	49 000 ± 800
Antimony	(Sb)	120 ± 90
Barium	(Ba)	1 400 ± 600
Cadmium	(Cd)	41 ± 15
Calcium	(Ca)	40 000 ± 18 000
Chromium	(Cr)	520 ± 240
Cobalt	(Co)	70 ± 10
Copper	(Cu)	450 ± 190
Iron	(Fe)	16 000 ± 6 000
Lead	(Pb)	1 700 ± 800
Lithium	(Li)	19 ± 3
Magnesium	(Mg)	12 500 ± 2 600
Manganese	(Mn)	3 100 ± 1 700
Nickel	(Ni)	210 ± 250
Potassium	(K)	6 300 ± 1 400
Silver	(Ag)	38 ± 8
Sodium	(Na)	8 200 ± 1 400
Tin	(Sn)	400
Zinc	(Zn)	5 500 ± 1 500

Standard Test Method for Silica in Refuse-Derived Fuel (RDF) and RDF Ash[1]

This standard is issued under the fixed designation E 887; the number immediately following the designation indicates the year of original adoption or, in the case of revision, the year of last revision. A number in parentheses indicates the year of last reapproval. A superscript epsilon (ϵ) indicates an editorial change since the last revision or reapproval.

1. Scope

1.1 This test method covers the determination of silica in RDF, RDF ash, fly ash, bottom ash, or slag.

1.2 The test method is an acid dehydration gravimetric procedure and is independent of interferences.

1.3 The values stated in SI units are to be regarded as the standard. The values given in parentheses are for information only.

1.4 *This standard does not purport to address all of the safety concerns, if any, associated with its use. It is the responsibility of the user of this standard to establish appropriate safety and health practices and determine the applicability of regulatory limitations prior to use.* For hazard statement, see Section 6.

2. Referenced Documents

2.1 *ASTM Standards:*

D 1193 Specification for Reagent Water[2]

E 791 Test Method for Calculating Refuse-Derived Fuel Analysis Data from As-Determined to Different Bases[3]

E 829 Practice for Preparing Refuse-Derived-Fuel (RDF) Laboratory Samples for Analysis[3]

E 830 Test Method for Ash in the Analysis Sample of Refuse-Derived Fuel[3]

E 856 Definitions of Terms and Abbreviations Relating to Physical and Chemical Characteristics of Refuse-Derived Fuel[3]

3. Summary of Test Method

3.1 Silicon compounds in RDF ash, fly ash, bottom ash, or slag are dissolved by alkali fusion and dehydrated with hydrochloric acid (HCl). Dehydration is completed by ignition, and the silica is volatilized as silicon tetrafluoride.

4. Apparatus

4.1 *Analytical Balance*, capable of weighing to 0.0001 g.

4.2 *Muffle Furnace*—The furnace shall have an operating temperature of up to 1200°C.

4.3 *Hot Plate or Steam Bath.*

4.4 *Platinum Crucibles*, 35 to 85-mL capacity.

4.5 *Graphite Crucibles*, 35 to 85-mL capacity.

4.6 *Fused Quartz Dishes*, 35 to 85-mL capacity.

5. Reagents and Materials

5.1 *Purity of Reagents*—Reagent grade chemicals shall be used in this test. Unless otherwise indicated, it is intended that all reagents shall conform to the specifications of the Committee on Analytical Reagents of the American Chemical Society where such specifications are available.[4] Other grades may be used, provided it is first ascertained that the reagent is of sufficiently high purity to permit its use without lessening the accuracy of the determination.

5.2 *Purity of Water*—Unless otherwise indicated, reference to water shall be understood to mean at least Type III reagent water conforming to Specification D 1193.

5.3 *Sodium Carbonate* (Na_2CO_3), anhydrous powder.

5.4 *Hydrochloric Acid* (HCl), concentrated, sp gr 1.19.

5.5 *Hydrochloric Acid* (1 + 3), Mix 1 volume of concentrated HCl with 3 volumes of water.

5.6 *Hydrochloric Acid* (1 + 1)—Mix 1 volume of concentrated HCl with 1 volume of water.

5.7 *Hydrochloric Acid* (1 + 99)—Mix 1 volume of concentrated HCl with 99 volumes of water.

5.8 *Sulfuric Acid* (1 + 1)—Mix 1 volume of concentrated sulfuric acid (H_2SO_4, sp gr 1.84) with 1 volume of water.

5.9 *Hydrofluric Acid* (HF), concentrated 48 to 51 %.

6. Hazards

6.1 Due to the origins of RDF in municipal waste, common sense dictates that precautions should be observed when conducting tests on the samples. Recommended hygienic practices include use of gloves when handling RDF; wearing dust masks (NIOSH-approved type), especially while milling RDF samples; conducting tests under negative pressure hoods when possible; and washing hands before eating or smoking.

7. Sampling

7.1 *Refuse-Derived Fuel (RDF)*:

NOTE 1—ASTM Subcommittee E38.01 is currently in the process of

[1] This test method is under the jurisdiction of ASTM Committee D-34 on Waste Management and is the direct responsibility of Subcommittee D34.08 on Thermal Treatment.

Current edition approved March 25, 1988. Published May 1988. Originally published as E 887 – 82. Last previous edition E 887 – 82.

[2] *Annual Book of ASTM Standards*, Vol 11.01.

[3] *Annual Book of ASTM Standards*, Vol 11.04.

[4] "Reagent Chemicals, American Chemical Society Specifications," Am. Chemical Soc., Washington, DC. For suggestions on the testing of reagents not listed by the American Chemical Society, see "Analar Standards for Laboratory U.K. Chemicals," BDH Ltd., Poole, Dorset, and the "United States Pharmacopeia."

developing procedures for sampling RDF.

7.1.1 RDF products are frequently nonhomogeneous. For this reason, significant care should be exercised to obtain a representative laboratory sample from the RDF lot to be characterized.

7.1.2 The sampling method for these procedures should be based on agreement between involved parties.

7.1.3 The laboratory sample must be air-dried and particle size reduced to pass a 0.5-mm screen for analysis. This procedure must be performed carefully to preserve the sample's representative characteristics (other than particle size) while preparing the analysis sample to be used in this procedure (see Practice E 829).

7.2 *Refuse-Derived Fuel Ash, Fly Ash, Bottom Ash, or Slag*—The method of sampling for this procedure should be based on agreement between involved parties.

8. Sample Preparation

8.1 *Refuse-Derived Fuel*:

8.1.1 Weigh accurately 30 to 50 g of RDF analysis sample as prepared in 7.16.1 into a conditioned and preweighed fused quartz dish.

8.1.2 Spread out the analysis sample of RDF in a layer not over 38.1 mm (1½ in.) in depth.

8.1.3 Place the dish in the muffle at a low temperature (not greater than 100°C) and gradually heat to redness at such a rate as to avoid mechanical loss from too rapid expulsion of volatile matter.

8.1.4 Complete the conversion to ash at a temperature of 800 to 900°C (1470 to 1650°F).

8.1.5 Cool in a desiccator and stir the ash to ensure homogeneity of particle sizes. Be careful not to lose any ash from the dish during this stirring.

8.1.6 Spread the ash in a thin layer in the dish, and ignite in a stream of oxygen for 1½ h at 800 to 850°C (1470 to 1560°F) to ensure complete and uniform oxidation of the ash.

8.1.7 Cool the ash to room temperature in a desiccator.

8.1.8 Weigh the dish and the ash.

8.1.9 Calculate the percent ash as follows:

$$\%\ \text{Ash} = (C - A)/(B - A) \times 100 \tag{1}$$

where:

A = weight of fused quartz, g,

B = weight of fused quartz dish and sample, g, and

C = weight of fused quartz dish and ash, g.

8.2 *Refuse-Derived Fuel Ash, Fly Ash, Bottom Ash, or Slag*:

8.2.1 Prepare the RDF ash, fly ash, bottom ash, or slag by grinding the sample in an agate mortar to a particle size to pass a No. 200 (75-µm) sieve.

8.2.2 Weigh accurately 6 to 10 g of RDF ash, fly ash, bottom ash, or slag as prepared in 8.2.1 into a conditioned preweighed fused quartz dish.

8.2.3 Spread out the analysis sample of the ash to be analyzed in a layer not over 6.4 mm (¼ in.) in depth.

8.2.4 Place the dish in the muffle at a low temperature (not greater than 100°C), and gradually heat to redness at such a rate as to avoid mechanical loss from too rapid expulsion of volatile matter.

8.2.5 Complete the conversion to ash at a temperature of 800 to 900°C (1470 to 1659°F).

8.2.6 Cool in a desiccator and stir the ash to ensure homogeneity of particle sizes. Be careful not to lose any ash from the dish during this stirring.

8.2.7 Spread the ash in a thin layer in the dish and ignite in a stream of oxygen for 1½ h at 800 to 850°C (1470 to 1560°F) to ensure complete and uniform oxidation of the ash.

8.2.8 Cool the ash to room temperature in a desiccator.

8.2.9 Weigh the dish and the ash.

8.2.10 Calculate the percent residue after ignition as follows:

$$\%\ \text{Residue after ignition} = (F - D)/(E - D) \times 100 \tag{2}$$

where:

D = weight of fused quartz dish, g,

E = weight of fused quartz dish and sample, of RDF ash, fly ash, bottom ash or slag, g, and

F = weight of fused quartz dish and residue after ignition of RDF ash, fly ash, bottom ash, or slag, g.

9. Procedure

9.1 *Sample Fusion*:

9.1.1 Weigh accurately 0.010 to 0.100 g of the prepared RDF ash as prepared in 8.1 or the residue of RDF ash, fly ash, bottom ash, or slag as prepared in 8.2 into a platinum or graphite crucible.

9.1.2 Add 1.0 g of Na_2CO_3. Mix the ash and Na_2CO_3 well, then add an additional 0.5 g of Na_2CO_3 to cover the mixture.

9.1.3 Place the crucible into a clean silica or refractory tray, and place in a muffle furnace preheated to 1000°C and maintain until the mass is quiescent (about 45 min).

9.1.4 Set the crucible aside to cool.

9.1.5 Rinse off the outside of the crucible, and place it on its side in a 300-mL casserole or beaker about one-third full of water. Warm and stir until the cake disintegrates and can be easily removed.

9.1.6 By means of platinum tipped tongs, lift the crucible out of the liquid, rinsing it thoroughly with water followed by rinsing with dilute hydrochloric acid (HCl, 1 + 3); adding the rinse to the casserole or beaker.

9.2 *Acid Dehydration*:

9.2.1 Very slowly and cautiously add 20 mL of concentrated hydrochloric acid (HCl, sp gr 1.19) to the covered casserole or beaker (see Note 2). Remove the cover and rinse. If any gritty particles are present, the fusion is incomplete and must be repeated, using a new sample and a larger amount of Na_2CO_3.

NOTE 2—This solution will tend to "creep" over the rim of the casserole unless the rim is kept hot and dry. This can be accomplished by applying heat from above by means of infrared lamps.

9.2.2 Evaporate the solution to dryness on a steam bath.

9.2.3 Without heating the residue any further, treat it with 5 to 10 mL of HCl (concentrated), wait at least 2 min, then add an equal amount of water.

9.2.4 Cover the casserole or beaker, and digest for 10 min on the steam bath or hot plate.

9.2.5 Dilute the solution with an equal volume of hot water, immediately filter through medium-textured ashless paper, and wash the residue thoroughly with hot HCl (1 + 99), then with hot water.

9.2.6 Reserve the residue.

9.2.7 Again, evaporate the filtrate to dryness and bake the residue in an oven for 1 h at 105 to 110°C.

9.2.8 Cool, add 10 to 15 mL HCl (1 + 1) and digest on the steam bath or hot plate for 10 min.

9.2.9 Dilute with an equal volume of water, filter immediately on a fresh filter paper, and wash the small amount of residue thoroughly with hot HCl (1 + 99), then with hot water.

9.2.10 Transfer the papers containing the residues (from 9.2.6 and 9.2.9) to a weighed conditioned platinum crucible.

9.2.11 Dry and ignite the papers, first at a low heat until the carbon of the filter paper is completely consumed without flaming, and finally ignite at 1100 to 1200°C until the weight becomes constant. Record weight of residue after ignition.

9.3 *Volatilization of Silicon Tetrafluoride*:

9.3.1 Treat the silica (SiO_2) thus obtained, which will contain impurities, in the crucible with 0.5 to 1 mL of water, 2 drops of H_2SO_4 (1 + 1) and 10 mL of HF.

9.3.2 Cautiously evaporate to dryness on a hot plate or hot sand bath.

9.3.3 Finally, ignite the small residue at 1050 to 1100°C (1922 to 2012°F) for 5 min.

9.3.4 Cool in a desiccator and weigh.

9.3.5 The difference between this weight and the weight previously obtained in 9.2.11 represents the amount of SiO_2.

9.4 *Blank*—Make a blank determination, following the same procedures as used in 9.1.2 through 9.3.5 using the same amounts of reagents and correct the obtained in the analysis accordingly.

10. Calculation

10.1 Calculate the concentration of SiO_2 as follows:

10.1.1 Percent SiO_2 on prepared sample as used in 9.1.1.

$$\% \, SiO_2 = (W_1 - W_2) - (W_3 - W_4)/W_5 \times 100 \quad (3)$$

where:

W_1 = weight of residue after first ignition as in 9.2.11, g,
W_2 = weight of blank after first ignition, g,
W_3 = weight of residue after treatment with HF and second ignition as in 9.3.4, g,
W_4 = weight of blank after treatment with HF and second ignition, g, and
W_5 = weight of sample of prepared RDF, RDF ash, fly ash, bottom ash, or slag as used in 9.1.1, g.

10.1.2 Percent SiO_2 in RDF (6.1).

$$\% \, SiO_2 \text{ in } RDF = (G \times H)/(100) \quad (4)$$

where:

G = ash as found in 8.1.9, %, and
H = SiO_2 as found in 10.1.1, %.

10.1.3 Percent SiO_2 in RDF ash, fly ash, bottom ash, or slag (8.2).

$$\% \, SiO_2 \textit{ in RDF Ash, fly ash, bottom ash, or slag} = (I \times H)/(100) \quad (5)$$

where:

I = residue after ignition as found in 8.2.10, %, and
H = SiO_2 as found in 10.1.1, %.

See Method E 791 for procedures to convert values to other bases.

11. Precision and Bias

11.1 Precision and bias statements cannot be made at this time for the measurement of SiO_2 by this test method.

Designation: E 889 – 82 (Reapproved 1996)

Standard Test Method for Composition or Purity of a Solid Waste Materials Stream[1]

This standard is issued under the fixed designation E 889; the number immediately following the designation indicates the year of original adoption or, in the case of revision, the year of last revision. A number in parentheses indicates the year of last reapproval. A superscript epsilon (ε) indicates an editorial change since the last revision or reapproval.

1. Scope

1.1 This test method covers the determination of the composition of a materials stream in a solid waste resource recovery processing facility. The composition is determined with respect to one or more defined components. The results are used for determining the purity resulting from the operation of one or more separators, and in conjunction with Test Method E 1108 used to measure the efficiency of a materials separation device.

1.2 The values stated in SI units are to be regarded as the standard. The values given in parentheses are for information only.

1.3 *This standard does not purport to address all of the safety concerns, if any, associated with its use. It is the responsibility of the user of this standard to establish appropriate safety and health practices and determine the applicability of regulatory limitations prior to use.* For hazard statements, see Section 7.

2. Referenced Documents

2.1 *ASTM Standards:*

C 566 Test Method for Total Moisture Content of Aggregate by Drying[2]

C 702 Practice for Reducing Field Samples of Aggregates to Testing Size[2]

D 75 Practice for Sampling Aggregates[3]

D 644 Test Method for Moisture Content of Paper and Paperboard by Oven Drying[4]

D 2013 Method of Preparing Coal Samples for Analysis[5]

D 2234 Test Methods for Collection of a Gross Sample of Coal[5]

E 380 Practice for Use of the International System of Units (SI) (the Modernized Metric System)[6]

E 1107 Method for Measuring the Throughput of Resource-Recovery Unit Operations[7]

E 1108 Test Method for Determination of the Recovery of a Product in a Materials Separation Device[7]

3. Terminology Definitions

3.1 *binary separator*—a device that separates a single input feed stream into two output or product streams.

3.2 *gross sample*—a sample representing one lot and composed of a number of increments on which neither reduction nor division has been performed.

3.3 *laboratory sample or analysis sample*—a portion of one gross sample representative of a lot and taken at random from the gross sample.

3.4 *polynary separator*—a device that separates a single input feed stream into three or more output product streams.

3.5 *purity*—The purity of a stream is defined in terms of one or more identifiable components, x, y, z, etc. The purity for any component such as x is the mass of x in a stream divided by the total mass of that stream. In some cases, the mass of x must be defined in practical terms that relate to the origin of the feed. For example, the purity of a ferrous product magnetically recovered from refuse can be expressed as the purity of ferrous by proximate analysis. Alternatively, it can be expressed as the purity by manual sorting, with all nonferrous materials that cannot readily be removed by hand as the contaminants. In any case, purity must be defined for each application.

4. Summary of Test Method

4.1 A gross sample of a preselected process stream is taken, and subdivided into four laboratory samples. Two of the laboratory samples are analyzed for composition of the component(s) or item(s) of interest by hand-picking and weighing. The third sample is used if the results of the first two do not agree within specified limits. The composition is computed as the weight ratio of the component(s) of interest to the weight of the sample.

4.2 The composition is expressed as the purity of the stream with respect to the component(s) indicated.

5. Significance and Use

5.1 This method is used to document the ability of solid waste resource recovery separators to concentrate or classify a particular component (or components) present in solid waste.

5.2 The purity determined in this way is used to calculate the recovery achieved by a separator as another measure of its performance, according to Test Method E 1108.

---

[1] This test method is under the jurisdiction of ASTM Committee D-34 on Waste Management and is the direct responsibility of Subcommittee D34.08 on Thermal Treatment.

Current edition approved July 30, 1982. Published May 1983.

[2] *Annual Book of ASTM Standards*, Vol 04.02.

[3] *Annual Book of ASTM Standards*, Vol 04.03.

[4] *Annual Book of ASTM Standards*, Vol 15.09.

[5] *Annual Book of ASTM Standards*, Vol 05.05.

[6] *Annual Book of ASTM Standards*, Vol 14.02.

[7] *Annual Book of ASTM Standards*, Vol 11.04.

6. Apparatus

6.1 *Scales*—Several size scales must be available to weigh gross samples and laboratory samples that will range in size from less than 1 kg (2.2 lb) to more than 100 kg (220 lb). All scales should have a precision and accuracy of ±0.1 %.

6.2 *Sorting Apparatus*—This may be any convenient work surface, such as a table and bins as decribed by Kaiser *et al.*[8]

6.3 *Polyethylene Bags*, used to store laboratory and sometimes gross samples as the means of preserving their moisture content. These bags should be 0.10 to 0.15 mm (0.004 to 0.006 in.) thick and supplied with metal ties.

6.4 *Laboratory Drying Oven*, and general associated equipment are required.

7. Hazards

7.1 This procedure calls for the hand-picking of solid waste and its processed fractions. Because the origin of all of the materials is generally unknown, workers must use proper safety precautions when handling samples. Workers shall wear gloves and safety glasses. When appropriate, dust masks shall be worn. Workers must be cautioned to wash their hands thoroughly before eating or smoking.

7.2 Particular caution shall be exercised when collecting samples near moving equipment.

8. Sampling

8.1 Samples are taken from processor streams, or separator output or input streams, according to the procedures outlined in Method E 1107.

8.2 The purity is determined with respect to a particular component in the stream, such as steel cans or glass. Hence, sampling and containment of samples must be done in a manner to preserve the integrity of the components being analyzed and the mass of the entire sample. As an example, if a stream is to be analyzed for the purity with respect to pieces of glass larger than 10 mm (0.4 in.) in size, the sample must be handled and stored in a manner in which pieces of glass larger than 10 mm are not broken.

9. Test Specimen and Samples

9.1 The size of sample is determined by the particle size of the material in accordance with Practice D 75. The size of sample specified constitutes the gross sample.

9.1.1 For particle sizes greater than 90 mm (3½ in.), not included in Table 1, Size of Samples, of Practice D 75, the size of sample shall be 250 kg (550 lb).

9.1.2 Gross samples shall be weighed, without subdivision.

9.1.3 A gross sample may be sized in accordance with the section on Test Specimen and Samples of Method E 1107 in circumstances where Practice D 75 cannot be practiced.

9.2 Gross samples shall be subdivided in accordance with Practice C 702 to form four laboratory samples. Each is weighed and labeled. This is recorded as "as-received weight."

9.3 Samples from wet processing steps, for example, wet screening, spiral classification, jigging, etc., are to be weighed after draining the water, in accordance with 11.3. The weight of such samples is recorded as the "drained weight."

10. Conditioning

10.1 Weigh the samples immediately after being taken with due precaution so that they neither gain nor lose weight from natural drying or from being left uncovered in wet or dusty areas. This is known as" as-received weight."

10.2 Condition wet samples by draining in accordance with 11.8. This is known as "drained weight."

10.3 The weight of dry samples, either of gross or laboratory samples, must exclude the contained moisture. Dry samples in accordance with 11.10. This is known as "dry weight."

11. Procedure

11.1 Take a gross sample in accordance with the section on Procedures, in Method E 1107.

11.1.1 Contain the gross sample in a moisture barrier container in accordance with 11.8.1 if the moisture content is to be preserved.

11.1.2 Record the weight of the gross sample.

11.2 Subdivide the gross sample into four near equal parts in accordance with Practice C 702 to form four laboratory samples.

11.2.1 Place the laboratory samples in moisture barrier containers, in accordance with 11.10 if the moisture content is to be preserved.

11.2.2 Record the weight of each laboratory sample.

11.3 Choose two of the laboratory samples at random for analysis; retain the other two, if needed, in accordance with 11.5.1.

11.4 Perform the analysis by spreading a laboratory sample on a clean, flat surface, and hand-picking the component(s) of interest.

11.4.1 Place the component(s) of interest in a clean, tared container. If the moisture content is to be preserved, this must be a moisture barrier container in accordance with 11.8.1.

11.4.2 Carefully remove adhering tramp substances from the component(s) of interest. Do not remove tramp substances that are physically attached so as not to be readily removed by hand and classify the entire piece as either components of interest (accept) or not of interest (reject). Record this decision on the data sheet.

11.4.3 Record the weights of component(s) picked from each laboratory sample as "as-received weight."

11.5 Compute the purity of the stream in regard to the component(s) of interest in accordance with Section 12.

11.5.1 The computed values of the purity of the two laboratory samples must agree within 10 % to be accepted. If they do not so agree, repeat the analysis procedure for a third laboratory sample chosen at random from the remaining two. Discard the fourth sample. If the value of the purity of the third sample does not agree within 10 % of the value of either of the first two, discard all samples, repeat the entire determination using a larger gross sample.

11.6 Hand-picking and weighing must be done without spillage and loss of material. Remove adhering tramp materials on the component(s) of interest that can be removed before the

[8] Kaiser, E. R., "Sampling and Analysis of Solid-Incinerator Refuse and Residue," *Proceedings of 1970 National Incineration Conference*, Am. Soc. Mechanical Engrs., pp. 25–31.

component(s) are placed in the tared container.

11.7 Record within 0.1 % all weights and in accordance with the precautions of Section 10 and 11.8.

11.8 Handle samples and weigh them in a manner to preserve the moisture content, except for samples taken from a wet process or separator, as described in 11.9.

11.8.1 Store laboratory samples in polyethylene bags, 0.10 to 0.15 mm (0.004 to 0.006 in.) in wall thickness. Secure the bags with two metal ties at the neck, and label.

11.8.2 In transferring the contents of material from a polyethylene bag, if the weight is to be preserved (that is, the contents are to be transferred quantitatively), take note of any condensed moisture or fine particulate solid materials adhering to the inside of the bag. Shake out any adherent solid material into the sample before or after the step outlined in 11.8.3.

11.8.3 If the bag contains condensed moisture, the weight of this moisture must be included in the weight of the sample transferred from this bag. After the bag is emptied, weigh it and place it in a forced-air drying oven at 55°C (130°F) and dry it to constant weight. Shake out the remaining solid materials (see 11.8.1). Add the weight loss of the bag to the weight of the sample as "contained moisture." Add the weight loss, after shaking out solid materials, to the weight of the sample as part of the solids. Determine moisture content in accordance with 11.10.

11.8.4 If the contents of a laboratory sample do not have to be transferred quantitatively, 11.8.2 and 11.8.3 are not applicable.

11.9 As mentioned in 9.3, samples from wet separators are to be weighed after draining the water. Record the weight as "drained weight."

11.9.1 Draining is accomplished by having the laboratory sample remain in a sieve until water is no longer observed dripping from the sieve but not longer than 1 h.

11.9.2 If drained weight is used, report purity on a" wet basis."

11.9.3 Alternatively, samples after draining may be dried to constant weight to determine the moisture content in accordance with 11.10. Subsequently, hand-pick the dried sample for the component(s) of interest. In this case, report the purity on a "dry basis."

11.10 The moisture content of paper or paper-like streams or components is to be determined by Test Method D 644. Determine the moisture content of inorganic or aggregate-like streams or components by Test Method C 566.

11.10.1 The temperature of drying in these methods may have to be lowered to avoid damaging any of the components in the sample. Melting pieces of plastic beyond recognition is an example of such damage.

12. Calculation

12.1 Calculate the purity as a mass fraction with respect to component x as follows:

$$\text{Purity of } x = P_x = \frac{\text{mass of x in the stream}}{\text{total mass of the stream}} \tag{1}$$

12.2 Calculate the mass of the stream, M_s, as follows:

$$M_s = C - T \tag{2}$$

where:

C = weight of the container filled with the laboratory sample, and

T = weight of that container alone.

12.3 Calculate the mass of the component of interest, M_x, as follows:

$$M_x = C' - T' \tag{3}$$

where:

C' = weight of the container filled with the hand-picked component x, and

T' = the weight of that container alone.

13. Report

13.1 Report purity, P_x, of component x for each of the laboratory samples analyzed. Report the average of the two measurements as the P_x for the separator stream analyzed.

13.2 The location of the separator stream as well as the component(s) of interest are identified.

13.3 The report must distinguish if the purity is based on as-received weight, drained weight, or dry weight. In anycase, note the amount of moisture in the component(s) of interest and in the process stream.

13.4 The recommended report form is shown in Fig. 1. Complete all items.

14. Precision and Bias

14.1 There are not yet sufficient data available to compute the expected precision and bias of this test method.

Date: ______________________ Location: ______________________
Time: ______________________ Operator: ______________________

Process Stream[A] ______________________

Component ______________________

Gross Sample + Container Weight			Indicate if Dry Weight ☐
Tare ____			Drained Weight ☐
Mass of Gross Sample			As-received Weight ☐
	No. 1	No. 2	
Laboratory Sample + Container, C			Dry Weight ☐
Tare, T	___	___	Drained Weight ☐
$M_s = C - T$			As-received Weight ☐
Component + Container, C'			Dry Weight ☐
Tare, T'	___	___	Drained Weight ☐
$M_x = C' - T'$			As-received Weight ☐

Purity, $P_x = M_s/M_x \times 100$ ___% ___%

Average P_x =

based on dry wt. ☐, drained ☐, or as-received ☐ NOTE: All weights

Moisture Contents, weight %

Gross Sample: ______, M_s ______, M_x ______

Observations:

[A] Repeat calculations, using a separate sheet, for each process stream or separation output. Use a separate sheet for each component of interest.

FIG. 1 Data Compilation and Calculation Sheet for the Determination of the Composition or Purity of a Solid Waste Materials Stream

Designation: E 897 – 88 (Reapproved 1998)

Standard Test Method for Volatile Matter in the Analysis Sample of Refuse-Derived Fuel[1]

This standard is issued under the fixed designation E 897; the number immediately following the designation indicates the year of original adoption or, in the case of revision, the year of last revision. A number in parentheses indicates the year of last reapproval. A superscript epsilon (ε) indicates an editorial change since the last revision or reapproval.

1. Scope

1.1 This test method covers the determination of the percentage of gaseous products, exclusive of moisture vapor, in the analysis sample which is released under specific conditions of the test. The knowledge of the volatile matter content assists in predicting burning characteristics of RDF.

1.2 This test method may be applicable to any waste material from which a laboratory analysis sample can be prepared.

1.3 *This standard does not purport to address all of the safety concerns, if any, associated with its use. It is the responsibility of the user of this standard to establish appropriate safety and health practices and determine the applicability of regulatory limitations prior to use.*

2. Referenced Documents

2.1 *ASTM Standards:*

E 180 Practice for Determining the Precision Data of ASTM Methods for Analysis and Testing of Industrial Chemicals[2]

E 790 Test Method for Residual Moisture in a Refuse-Derived Fuel Analysis Sample[3]

E 829 Practice for Preparing Refuse-Derived Fuel (RDF) Laboratory Samples for Analysis[3]

3. Terminology

3.1 *Definition:*

3.2 *refuse-derived fuel* (RDF):

RDF–1—Waste used as a fuel in as-discarded form.

RDF–2—Waste processed to coarse particle size with or without ferrous metal separation.

RDF–3—shredded fuel derived from municipal solid waste (MSW) that has been processed to remove metal, glass, and other inorganics. This material has a particle size such that 95 weight % passes through a 2-in. square mesh screen.

RDF–4—Combustible waste processed into powder form—95 weight % passing a 10–mesh screen.

RDF–5—Combustible waste densified (compressed) into the form of pellets, slugs, cubettes or briquettes.

RDF–6—Combustible waste processed into liquid fuel.

RDF–7—Combustible waste processed into gaseous fuel.

4. Summary of Test Method

4.1 Volatile matter is determined by establishing the loss in weight resulting from heating refuse-derived fuel under rigidly-controlled conditions. The measured weight loss, corrected for moisture as determined in Test Method E 790, establishes the volatile matter content.

5. Apparatus

5.1 *Platinum or Fused Quartz Crucible*, with closely fitting cover. The crucible shall be of not less than 10 nor more than 20 mL capacity, not less than 25 nor more than 35 mm in diameter, and not less than 30 nor more than 35 mm in height.

5.2 *Vertical Electric Tube Furnace*— It shall be regulated to maintain a temperature of 950 ± 20°C in the crucible, as measured by a thermocouple positioned in the furnace.

6. Hazards

6.1 Due to the origins of RDF in municipal waste, common sense dictates that precautions should be observed when conducting tests on the samples. Recommended hygienic practices include use of gloves when handling RDF, wearing dust masks (NIOSH-approved type), especially while milling RDF samples, conducting tests under a negative-pressure hood when possible, and washing hands before eating or smoking.

NOTE 1—**Caution**—Exercise care when placing the sample into the volatile furnace. The possibility of an explosion always exists when heating samples of unknown origin.

7. Procedure

7.1 Weigh to the nearest 0.1 mg about 1 g of thoroughly mixed air-dried analysis RDF sample in a weighed crucible. Close with a cover (Note 2), place on a platinum or Nichrome-wire support and insert directly into the furnace chamber, which shall be maintained at a temperature of 950 ± 20°C. Lower the crucible immediately to the 950°C zone. Regulation of the temperature to within the prescribed limits is critical. After the more rapid discharge of volatile matter has subsided as shown by disappearance of the luminous flame, inspect the crucible to verify that the lid is still seated. If necessary, reseat the lid to guard against the admission of air into the crucible.

[1] This test method is under the jurisdiction of ASTM Committee D-34 on Waste Management and is the direct responsibility of Subcommittee D34.03 on Treatment.

Current edition approved March 25, 1988. Published May 1988. Originally published as E 897 – 82. Last previous edition E 897 – 82.

[2] *Annual Book of ASTM Standards,* Vol 15.05.

[3] *Annual Book of ASTM Standards,* Vol 11.04.

Do this as rapidly as possible by raising the crucible to the top of the furnace chamber, reposition the lid to more perfectly seal the crucible, then lower immediately back to the 950°C zone.

NOTE 2—The cover should fit closely enough so that the carbon deposit from the refuse-derived fuel does not burn away from the underside.

7.2 After heating for a total of exactly 7 min, remove the crucible from the furnace and, without disturbing the cover, allow it to cool on a metal cooling block. Weigh as soon as cold (Note 3). The percentage loss of weight minus the percentage moisture in accordance with Test Method E 790 is the volatile matter.

NOTE 3—To ensure uniformity of results, the cooling period should be kept constant and should not be prolonged beyond 15 min.

8. Calculation

8.1 Calculate the percentage of volatile matter on an" as–determined" basis, V_{ad} as follows:

$$V_{ad} = \left[\frac{A - B}{A} \times 100\right] - M_{ad} \quad (1)$$

where:

A = weight of sample used, g,
B = weight of sample after heating, g, and
$M(_{ad})$ = moisture (as–determined), %.

9. Precision and Bias

9.1 *Precision*:

9.1.1 The standard deviation of individual determinations, in percent absolute, is as follows:
Typical Average Value, 69 %
Within-Laboratory, 0.7 %
Between-Laboratories, 2.1 %

9.1.2 The precision estimates in 9.1.1 are based on an interlaboratory study conducted in accordance with Practice E 180.

9.2 *Bias*:

9.2.1 The bias of this test method has not been determined.

9.2.2 Precision estimates are based on ASTM Report No. RR:E 38–1000 which describes the preliminary testing and round–robin tests.[4]

[4] Supporting data are available on loan from ASTM Headquarters. Request RR: E 38–1000.

Designation: E 926 – 94

Standard Practices for Preparing Refuse-Derived Fuel (RDF) Samples for Analyses of Metals[1]

This standard is issued under the fixed designation E 926; the number immediately following the designation indicates the year of original adoption or, in the case of revision, the year of last revision. A number in parentheses indicates the year of last reapproval. A superscript epsilon (ε) indicates an editorial change since the last revision or reapproval.

1. Scope

1.1 These practices described herein cover the preparation of milled refuse-derived fuel (RDF) sample for analyses of metals by atomic absorption spectroscopy or inductively coupled plasma spectroscopy (see Appendix X1), or both.

1.2 These practices may be applicable to any waste material from which a laboratory analysis sample can be prepared.

1.3 The following four practices are described in this standard:

1.3.1 *Practice A*—Nitric-Hydrofluoric-Perchloric Acid Digestion.

1.3.2 *Practice B*—Nitric-Sulfuric-Hydrofluoric Acid Digestion.

1.3.3 *Practice C*—Bomb, Acid Digestion Method.

1.3.4 *Practice D*—Oxygen Bomb Combustion Method.

1.4 The four practices will cover the scope of preparing most RDF samples for subsequent analyses of most metals. Many times the nature of the sample and the metal in question will dictate the sample preparation technique. In such cases, modifications of these practices may be made with care as to not alter the integrity of the analyte in question.

1.5 *This standard does not purport to address all of the safety concerns, if any, associated with its use. It is the responsibility of the user of this standard to establish appropriate safety and health practices and determine the applicability of regulatory limitations prior to use.* For hazard statements, see Sections 9, 10.6, 14, 19, and 24.

2. Referenced Documents

2.1 *ASTM Standards:*

D 1193 Specification for Reagent Water[2]

E 711 Test Method for Gross Calorific Value of Refuse-Derived Fuel by the Bomb Calorimeter[3]

E 829 Practice for Preparing Refuse-Derived Fuel (RDF) Laboratory Samples for Analysis[3]

3. Summary of Practice

3.1 The organic matter of the analysis sample is completely destroyed by strong acid digestion and oxidation. The inorganic constituents are solubilized and the resulting solution analyzed for metals by either atomic absorption spectroscopy or inductively coupled plasma spectroscopy.

4. Significance and Use

4.1 These practices are available to producers and users of RDF for preparing RDF samples for subsequent analyses of metals.

5. Apparatus

5.1 *Analytical Balance*, capable of weighing to 0.0001 g.

5.2 *TFE-fluorocarbon Beakers,*[4] 500-mL capacity.

5.3 *Hot Plate*, capable of heating to at least 510°C.

5.4 *Volumetric Flasks*, 50, 100, and 250-mL.

6. Reagents and Materials

6.1 *Purity of Reagents*—Reagent grade chemicals shall be used in this test. Unless otherwise indicated, it is intended that all reagents shall conform to the specifications of the Committee on Analytical Reagents of the American Chemical Society where such specifications are available.[5] Other grades may be used, provided it is first ascertained that the reagent is of sufficiently high purity to permit its use without lessening the accuracy of the determination.

6.2 *Purity of Water*—Unless otherwise indicated, reference to water shall be understood to mean at least Type III reagent water conforming to Specification D 1193.

6.3 *Nitric Acid* (HNO_3), concentrated sp gr 1.42.

6.4 *Hydrofluoric Acid* (HF), 48 %.

---

[1] These practices are under the jurisdiction of ASTM Committee D-34 on Waste Management and is the direct responsibility of Subcommittee D34.08 on Thermal Treatment.

Current edition approved Aug. 15, 1994. Published October 1994. Originally published as E 926 – 83. Last previous edition E 926 – 88.

[2] *Annual Book of ASTM Standards*, Vol 11.01.

[3] *Annual Book of ASTM Standards*, Vol 11.04.

[4] ®Tradename for tetrafluoroethylene (TFE) fluorocarbon polymer.

[5] *Reagent Chemicals, American Chemical Society Specifications*, American Chemical Society, Washington, DC. For suggestions on the testing of reagents not listed by the American Chemical Society, see *Analar Standards for Laboratory Chemicals*, BDH Ltd., Poole, Dorset, U.K., and the *United States Pharmacopeia and National Formulary*, U.S. Pharmaceutical Convention, Inc. (USPC), Rockville, MD.

6.5 *Perchloric Acid* ($HClO_4$), concentrated 70 to 72 %.

6.6 *Hydrochloric Acid* (HCl), concentrated sp gr 1.19.

7. Hazards

7.1 Due to the origins of RDF in municipal waste, common sense dictates that precautions should be observed when conducting tests on samples. Recommended hygienic practices include use of gloves when handling RDF; wearing dust masks (NIOSH-approved type), especially while milling RDF samples; conducting tests under a negative pressure hood when possible; and washing hands before eating or smoking.

7.2 The hot acidic solutions in these procedures pose a significant potential hazard. Proper laboratory safety practices and equipment should be employed throughout these procedures.

7.3 All wet ashing must be performed in a fume hood. Where perchloric acid is used a perchlorate hood must be used.

8. Sampling [6]

8.1 RDF products are frequently non-homogeneous. For this reason significant care should be exercised to obtain a representative laboratory sample from the RDF lot to be characterized.

8.2 The sampling method for these procedures should be based upon agreement between involved parties.

8.3 The laboratory sample is prepared in accordance with Practice E 829.

PRACTICE A—NITRIC-HYDROFLUORIC-PERCHLORIC ACID DIGESTION

9. Hazards

9.1 Hot acidic solutions in this procedure pose a significant potential hazard. Proper laboratory safety practices and equipment should be employed throughout.

9.2 All wet ashing must be performed in a perchlorate fume hood.

9.3 Perchloric acid is a colorless, hygroscopic, strong oxidizing agent. It can explode in contact with organic materials, or by shock or heat. Controlled redox potential of perchloric acid is imperative when used for oxidation of organic material (see Appendix X2).

10. Procedure

10.1 Weigh accurately approximately 5 g of the analysis RDF sample into a 500-mL TFE-fluorocarbon beaker.

10.2 Add 50 mL of concentrated HNO_3. Cover the beaker with a loose fitting cover.

10.3 Heat on low heat for several hours. Add HNO_3 as needed. Continue heating until most of the organic matter is decomposed.

NOTE 1—Additions of HNO_3 may be made to hot solutions without fear of violent splattering if the acid is added by dropwise additions, or as a thin stream from a wash bottle down the sides of the beaker.

10.4 Slowly add 10 mL of 48 % HF, cover, and continue digesting the sample over low heat. Frequently rinse the inside walls of the beaker with deionized water to ensure that all material is in contact and can react with the acid mixtures.

10.5 Continue digesting until the solution becomes clear. Add additional HNO_3 as needed.

NOTE 2—Some samples will require several days for complete digestion to occur.

10.6 In a perchlorate hood add 10 mL of $HClO_4$ to the digestion mixture, cover, and continue digestion on low heat for several hours. At frequent intervals rinse the inside walls of the beaker with HF. Approximately 20 to 25 mL of additional HF can be added during this digestion (see Appendix X2).

NOTE 3—**Warning:** An excessively high temperature may cause an explosion if carbonaceous material is present. The addition of a cold solution to hot acid solutions may cause splattering.

10.7 Remove the cover, elevate the hot plate temperature, and slowly evaporate to dryness.

10.8 Cool the contents and beaker to room temperature.

10.9 Dissolve the solids in 35 to 50 mL of 10 % V/V solution of HCl using moderate heat. If all material is not soluble, add 20 to 25 mL of HF and digest on low heat for an additional hour. Continue as described in 10.7, 10.8, and 10.9.

10.10 Quantitatively transfer the sample solution to a 100-mL volumetric flask and dilute to mark with 1 % HCl solution.

10.11 This solution will contain those metals as contained in 5 g of sample RDF in 100 mL of solution. Should a more concentrated or a more dilute solution be required, the sample size or the final dilution, or both, can be adjusted accordingly. This adjustment will be a function of the magnitude of concentration of the metal in question and the sensitivity of the method of analysis.

10.12 Prepare a reagent blank in the same manner as described in 10.2-10.10 using the same volumes of acids used for the digestion of the sample.

PRACTICE B—NITRIC-SULFURIC-HYDROFLUORIC ACID DIGESTION

11. Summary of Practice

11.1 The organic matter of the analysis sample is completely destroyed by strong acid digestion and oxidation. The inorganic constituents are solubilized and the resulting solution analyzed for metals by either atomic absorption spectroscopy or inductively coupled plasma spectroscopy.

12. Apparatus

12.1 *Analytical Balance*, capable of weighing to 0.0001 g.

12.2 *Berzelius Beakers*, 600-mL capacity.

12.3 *Hot Plate*, capable of heating to at least 510°C.

12.4 *Volumetric Flasks*, 50, 100, and 250-mL.

13. Reagents and Materials

13.1 *Purity of Reagents*—See 6.1.

13.2 *Purity of Water*—See 6.2.

13.3 *Nitric Acid* (NHO_3), concentrated sp gr 1.42.

13.4 *Sulfuric Acid* (H_2SO_4), concentrated sp gr 1.84.

13.5 *Hydrofluoric Acid* (HF), 48 %.

14. Hazards

14.1 Hot acidic solutions in this procedure pose a significant

[6] ASTM Subcommittee E38.01 is currently in the process of developing procedures for sampling RDF.

potential hazard. Proper laboratory safety practices and equipment should be employed throughout.

14.2 All wet ashing must be performed in a fume hood.

15. Procedure

15.1 Weigh a suitable amount of RDF analysis sample and place into a 600-mL Berzelius beaker.

NOTE 4—Should very low or high concentrations of the element in question be anticipated, the amount of sample used and the dilution of the final solution can be adjusted accordingly.

NOTE 5—Recommended acid additions are given in Table 1.

15.2 Add a suitable volume of concentrated HNO_3 and cautiously digest over low to medium heat (see Note 6).

NOTE 6—**Warning:** On prolonged digestion with nitric acid, some materials from highly reactive compounds that erupt violently, result in critical loss of sample and possible danger to the operator.

15.3 When the initial vigorous reaction ceases, gradually increase the temperature and evaporate to a 10 to 15-mL volume.

15.4 Cool slightly and add H_2SO_4 slowly in small increments (5 to 10 mL).

15.5 Return the beaker to high heat and evaporate to dense white fumes of sulfur trioxide (SO_3).

15.6 Allow to fume until the solution in the beaker darkens, then fume for 10 min more.

15.7 Remove from the heat, cool until SO_3 fumes are no longer observed, then add HNO_3 slowly and cautiously down the inside of the beaker.

NOTE 7—Additions of HNO_3 may be made to hot (not fuming) H_2SO_4 without fear of violent splattering if the acid is added by dropwise additions, or as a thin stream from a wash bottle down the sides of the beaker.

15.8 Repeat 15.5, 15.6, and 15.7 until no further darkening of the solution is noted on subsequent fumings.

15.9 If the solution contains gelatinious silica (SiO_2) precipitate, transfer the solution quantitatively to a 250-mL TFE-fluorocarbon beaker with a minimum amount of water.

15.10 Add 5 mL of HF and evaporate to fumes of SO_3.

NOTE 8—When any additions of HF or HCl are to be made to an H_2SO_4 solution, the H_2SO_4 solution must first be diluted with water and be near room temperature. If highly accurate metals analysis (particularly Si, Al, and Fe) is not required, the HF may be added directly to the sample in the glass beaker.

15.11 Allow the solution to cool to room temperature, add 5 mL of water and 5 mL of HCl, and again fume until a volume of approximately 2 mL is reached.

NOTE 9—Samples with appreciable salt content will render judgement of this final volume nearly impossible. In such instances, evaporate the H_2SO_4 to the lowest liquid volume discernible, then cease fuming when salts form a wet sludge.

TABLE 1 Recommended Acid Additions

Sample Size, g	Initial Additions HNO_3, mL	Initial Additions HSO_4, mL	Second Additions, HNO_3, mL	Third Additions, HNO_3, mL	Subsequent Additions, HNO_3, mL
1	25	15	10	5	5
2	50	20	10	5	5
5	50	20	10	10	10
10	50	25	10	10	10

15.12 Remove from heat and cool to room temperature.

15.13 Transfer the solution quantitatively to a 100-mL volumetric flask, and dilute to mark with water.

15.14 This solution will contain those metals as contained in the given amount of sample of RDF used in 100 mL of solution. Should a more concentrated or a more dilute solution be required, the sample size or the final dilution, or both, can be adjusted accordingly. This adjustment will be a function of the magnitude of concentration of the metal in question and the sensitivity of the method of analysis.

15.15 Prepare a reagent blank in the same manner as described in 15.2-15.13, using the same volumes of acids used for the digestion of the sample.

PRACTICE C—BOMB, ACID DIGESTION METHOD

16. Summary of Practice

The organic matter of the analysis sample is completely destroyed by strong acid digestion and oxidation in a closed acid digestion bomb at low temperature. The inorganic constituents are solubilized and the resulting solution analyzed for metals by either atomic absorption spectroscopy or inductively coupled plasma spectroscopy.

17. Apparatus

17.1 *Analytical Balance*, capable of weighing to 0.0001 g.

17.2 *Bomb, Acid Digestion*, TFE-fluorocarbon-lined, 25-mL capacity, designed for high pressure decompositions.

17.3 *Oven*, circulating air capable of maintaining a temperature of 110 ± 0.5°C.

17.4 *Volumetric Flasks*, 50, 100, and 250-mL.

18. Reagents and Materials

18.1 *Purity of Reagents*—See 6.1.

18.2 *Purity of Water*—See 6.2.

18.3 *Nitric Acid* (HNO_3), concentrated sp gr 1.42.

19. Hazards

19.1 Acid digestion bombs must be used according to the manufacturers' specifications. That is, the limits on the amounts of material that can be safely digested, the volume of acid added, and the temperatures of digestion cannot be exceeded.

20. Procedure

20.1 Weigh accurately about 0.1 g of RDF analysis sample into the TFE-fluorocarbon cup of the acid digestion bomb.

20.2 Add no less than 2.5 nor more than 3.0 mL of concentrated HNO_3 to the sample in the TFE-fluorocarbon cup.

20.3 Without delay, assemble the TFE-fluorocarbon cup and the acid digestion bomb and place the assembly into an oven controlled to 110°C for 2 h (see Note 12).

NOTE 10—**Caution:** The possibility of the cellulose and nitric acid forming nitrocellulose is present.

NOTE 11—Most acid digestion bombs have sockets for applying a spanner wrench to the cap and for holding the bomb in a rigid clamp during opening and closing operations. Normally these bombs can be sealed tightly by simply turning the knurled cap down firmly by hand but sometimes additional tightening with a wrench may be required. Some bombs have an accessory spanner-jack available for holding the bomb during the closing and opening operations.

20.4 Remove the bomb and contents from the oven and cool to room temperature.

20.5 Open the bomb and remove TFE-fluorocarbon cup (see Note 12).

NOTE 12—**Caution:** When removing the TFE-fluorocarbon cup from the bomb, care must be exercised to keep from spilling contents of the cup.

20.6 Quantitatively transfer the contents of the TFE-fluorocarbon cup to a 100-mL volumetric flask and dilute to mark with water.

20.7 This solution will contain those metals as contained in the given amount of RDF used (20.1) in 100 mL of solution. Should a more concentrated or a more dilute solution be required, the sample size or the final dilution, or both, can be adjusted accordingly. This adjustment will be a function of the magnitude of concentration of the metal in question and the sensitivity of the method of analysis.

20.8 Prepare a reagent blank in the same manner as described in 20.2-20.8 using the same volumes of acids used for the digestion of the sample.

PRACTICE D—OXYGEN BOMB COMBUSTION METHOD

21. Summary of Practice

21.1 The analysis sample of RDF is combusted in an oxygen combustion bomb. The oxidized form of the metals are trapped in an acid solution and the resulting solution analyzed for metals by either atomic absorption spectroscopy or inductively coupled plasma spectroscopy.

22. Apparatus

22.1 *Analytical Balance*, capable of weighing to 0.0001 g.

22.2 *Apparatus for Bomb Combustion of the Sample* (see Note 14):

22.2.1 *Oxygen Bomb*, similar to that used in the determination of calorific value of RDF as described in Test Method E 711.

22.2.2 *Capsule*, for holding the sample, approximately 25 mm in diameter at the top, approximately 12-mm deep of fused quartz or silica.

22.2.3 *Quartz or Platinum Liner for Oxygen Bomb.*

22.2.4 *Firing Wire*, platinum as specified in Test Method E 711.

22.2.5 *Firing Circuit*, as specified in Test Method E 711.

22.2.6 *Electrodes in Oxygen Combustion Bomb*, platinum.

22.2.7 *Metal Vessel*, cylindrical such that the bomb will be fully immersed when approximately 2 L of water are added.

22.3 *Water Bath*, contain water at room temperature.

22.4 *Hot Plate.*

22.5 *TFE-fluorocarbon Beakers*, 150-mL capacity.

22.6 *Volumetric Flasks*, 25, 50, 100, and 250-mL capacity.

NOTE 13—The apparatus for oxygen bomb combustion is different from the conventional oxygen bombs used for calorific values. The type of bomb and accessories as well as to its proper use can be obtained from the manufacturer of oxygen combustion bombs. The manufacturer's specifications should be followed.

23. Reagents and Materials

23.1 *Purity of Reagents*—See 6.1.

23.2 *Purity of Water*—See 6.2.

23.3 *Nitric Acid* (HNO_3), concentrated, sp gr 1.42.

23.4 *Nitric Acid* (1 + 10 HNO_3), one part concentrated HNO_3 and nine parts water.

23.5 *Hydrochloric Acid* (HCl), concentrated, sp gr 1.19.

23.6 *Aqua Regia*, three parts concentrated HNO_3 and one part concentrated HCl.

23.7 *Hydrofluoric Acid* (HF), 48 %.

23.8 *Oxygen*—The oxygen used for combustion shall be free of combustible matter, guaranteed to be greater than 99.5 % pure, will meet this requirement.

24. Hazards

24.1 Hot acidic solutions in this procedure pose a significant potential hazard. Proper laboratory safety practices and equipment should be employed throughout.

24.2 Oxygen combustion bombs must be used according to the manufacturer's specifications.

25. Procedure

25.1 *Preparation of Sample and Bomb*:

25.1.1 *Sample Weight*—Weigh to the nearest 0.1 mg about 1 g of thoroughly mixed analysis RDF sample in the bomb capsule. A pellet may be made from the sample, accurately weighed, and placed into the bomb capsule. Place the capsule containing the sample into the capsule holder.

25.1.2 *Firing Wire*—Connect a length of firing wire to the ignition terminals in such a manner that the loop of firing wire is in contact with the sample.

25.1.3 *Bomb Preparation*—Add 10 mL of 1:10 HNO_3 to either the bomb or to the quartz liner, wetting the entire internal surface of the bomb or liner with the solution.

NOTE 14—To prevent contamination of the combusted material by metallic elements from the bomb walls either the use of a quartz liner or a platinum-lined bomb is recommended.

25.1.4 Assemble the bomb.

25.2 *Addition of Oxygen*—Admit oxygen to the bomb slowly to avoid blowing the sample from the capsule until a pressure of 15 atm is reached.

25.3 *Bomb Combustion*—Immerse the bomb in a cold water bath, connect it to the firing circuit, and close the circuit to ignite the sample. Allow the bomb to stand in the water for not less than 10 min after firing. Remove the bomb from the water bath. Release the pressure at a slow, uniform rate so that the pressure is reduced to atmospheric in not less than 1 min. Open the bomb and examine the inside for traces of unburned material or sooty deposits. If any are found, discard the determination and thoroughly wash all parts of the bomb interior before using it again. If no unburned material or sooty deposits are present, rinse the interior of the bomb or the liner, the capsule, electrodes, and the interior surface of the bomb or liner. Cover with a fine jet of hot water and collect the contents of the bomb and washings into a 150-mL TFE-fluorocarbon beaker.

NOTE 15—Inverting and shaking a heavy, wet, and possible slippery bomb may present a hazard. Precautions should be taken when performing this function.

25.4 *Dissolution*:

25.4.1 Evaporate the contents of the beaker to dryness on a hotplate.

25.4.2 Cool and add 5 mL of aqua regia and 5 mL of HF (48 %). Heat to dryness.

25.4.3 Dissolve the residue in 1 % HNO_3.

25.4.4 Quantitatively transfer the sample solution to a volumetric flask and dilute to mark with 1 % HNO_3. For samples having ash contents of up to 5 %, use a 25-mL volumetric flask. Samples with ash contents up to 10 %, use a 50-mL volumetric flask, and those samples containing between 10 and 20 % ash, use a 100-mL volumetric flask (see X1.2). Should a more concentrated or more dilute solution be required, the sample size or the final dilution, or both, can be adjusted accordingly.

25.5 *Reagent Blank*—Prepare a reagent blank in the same manner as described in 25.1-25.3 using a combustible material known to be free of metals in place of the sample.

26. Precision and Bias

26.1 Precision and bias statements cannot be made on these test methods at this time. Precision and bias statements will have to be determined on each metal analyzed by a given method of analyses and will be a function of both the sample preparation and the method of analysis.

APPENDIXES

(Nonmandatory Information)

X1. CONSIDERATIONS FOR SELECTION OF PRACTICES

X1.1 These practices consist of four methods of preparing a milled RDF sample for analyses of metals by atomic absorption spectroscopy or inductively coupled plasma spectroscopy, or both.

X1.1.1 No one practice has been ascertained to be superior or inferior to any other practice. Many factors must be considered when choosing the practice to be used. Some of these factors are as follows:

X1.1.1.1 The metal or metals to be analyzed.

X1.1.1.2 The proper laboratory facilities available.

X1.1.1.3 The training and experience of the analyst.

X1.1.1.4 The amount of sample to be treated.

X1.1.1.5 The magnitude of concentration of metal in the sample.

X1.1.1.6 The detection limits required.

X1.2 Metals may show variations in concentration in RDF samples. After dissolution of the RDF analysis sample, the concentration or dilution to which the digestate is taken depends upon the magnitude of concentration of the metal or metals in question. The range of metal concentrations (see Table X1.1) can be used as a guide in determining the amount of analysis RDF sample to be used and the final concentration or dilution factor for subsequent analysis.

TABLE X1.1 Range of Metal Concentrations

Metal	(M)	Concentration Range, mg M/kg sample
Aluminum	(Al)	5400–12000
Antimony	(Sb)	20–80
Arsenic	(As)	<1
Barium	(Ba)	47–450
Beryllium	(Be)	<1
Cadmium	(Cd)	2–22
Calcium	(Ca)	5900–17000
Chromium	(Cr)	20–100
Cobalt	(Co)	<3–5
Copper	(Cu)	80–900
Iron	(Fe)	1000–3500
Lead	(Pb)	110–1500
Lithium	(Li)	<2–7
Magnesium	(Mg)	880–2100
Manganese	(Mn)	50–240
Mercury	(Hg)	1–2
Nickel	(Ni)	9–90
Potassium	(K)	920–1900
Selenium	(Se)	<1
Silver	(Ag)	<3
Sodium	(Na)	1800–7400
Tin	(Sn)	<20–40
Titanium	(Ti)	<1
Vanadium	(V)	<1
Zinc	(Zn)	200–2500

X2. PROPERTIES OF PERCHLORIC ACID

X2.1 Among the many valuable properties of perchloric acid as applied to analytical reactions of a wide variety of applications, there are two which prove of great value in the wet oxidation of organic composition. First of these provides for the dissolution of a wide variety of organic materials. Cellulose, starch, proteins, and sugars are readily soluble in perchloric acid. Unlike the sulfuric acid oxidation of such entities at higher temperatures there is no carbonization. The solvent promoting properties of perchloric acid provides mostly clear solutions throughout their oxidation. Such condition is continuous over the temperature range up to the boiling point of the perchloric acid-water azeotropic composition of approximately 200°C. The value of the dissolution of organic compositions, provided by perchloric acid as oxygen donor, cannot be over-estimated. It is of outstanding significance.[7]

X2.2 The second and most valuable property of perchloric acid as oxygen donor is not duplicated by any of the other oxygen donor acids, chloric, nitric, periodic, chlorous acid ($HClO_2$), hypochlorous acid (HClO), or their salts in acid media. This perchloric acid property of relinquishing its high oxygen content at increasing oxidation potential only in small increasing increments that increase in concentration and boiling point is not duplicated by any of these other cited oxygen donors. The availability of its oxygen is a fixed function of acid concentration and the temperature at which each acid concentration is controlled. By regulation, the acid concentration at its boiling point (under reflux to keep acid concentration constant) and oxidation potential magnitude between 0.85 to 2.0 V can be realized and statically controlled. This property is alien to all the oxygen donor acids previously cited.[7]

[7] "The Wet Chemical Oxidation of Organic Compositions Employing Perchloric Acid," G. Frederick Smith Chemical Co., Inc., P.O. Box 23214, Columbus, OH 43223.

Standard Test Method for Measuring Electrical Energy Requirements of Processing Equipment[1]

This standard is issued under the fixed designation E 929; the number immediately following the designation indicates the year of original adoption or, in the case of revision, the year of last revision. A number in parentheses indicates the year of last reapproval. A superscript epsilon (ε) indicates an editorial change since the last revision or reapproval.

1. Scope

1.1 This test method covers the determination of the energy and power requirements of processing equipment using an electrical metering system.

1.2 This test method can be used to measure energy and power requirements of processing equipment driven by an electrical motor operating on alternating current.

1.3 This test method includes instructions for installation and checkout of the energy metering system, procedures for measuring and recording energy usage, and methods for calculating the average gross power, average freewheeling power, and average net power requirements of processing equipment.

1.4 *This standard does not purport to address all of the safety concerns, if any, associated with its use. It is the responsibility of the user of this standard to establish appropriate safety and health practices and determine the applicability of regulatory limitations prior to use.* For hazard statements, see Section 6.

2. Terminology Definitions

2.1 *electrical metering system*—a system composed of current and potential transformers and a wattmeter electrically connected in such a manner so as to measure the energy usage of a piece of equipment driven by an electric motor.

2.2 *freewheeling condition*—a piece of equipment under an unloaded condition wherein the electrical energy is dissipated due to friction and windage.

2.3 *freewheeling power*—power requirement of a piece of equipment under unloaded, or freewheeling, conditions.

2.4 *gross energy*— energy usage of a piece of equipment operating under loaded conditions as measured using an electrical metering system.

2.5 *gross power*— power requirement of a piece of equipment under loaded conditions.

2.6 *loaded condition*— equipment doing processing work on solids, liquids, or gases, or all of these, (for example, moving material, changing its characteristics, or separating it into different streams).

2.7 *net power*—the difference between gross power and freewheeling power; net power is the power required for processing.

2.8 *specific energy*— energy consumption expressed on the basis of unit mass of throughput.

2.9 *unloaded condition*—equipment not doing processing work (for example, moving, changing the characteristics of, or separating materials), but operating in a freewheeling, or idling, condition.

3. Summary of Test Method

3.1 An electrical metering system is installed and checked.

3.2 The metering instrumentation and processing equipment is allowed to warmup.

3.3 Using the electrical metering system, the energy used by the processing equipment under no-load and loaded conditions is measured and recorded.

3.4 The average gross power, average freewheeling power, and average net power required by the equipment is calculated.

4. Significance and Use

4.1 Energy usage and power requirements of processing equipment are important from the standpoint of determining if equipment is operating within specification and meeting performance criteria.

4.2 Having determined the energy usage and power requirements of the processing equipment using this method, specific energy may be calculated, with the use of system throughput, and used as one criterion to compare the performance of similar pieces of equipment operating under similar operating conditions.

4.3 Measurements of energy usage can be used for the purpose of identifying inefficient electrical motors and processing equipment.

5. Apparatus

5.1 *Calibrated Watthour Meter.*

5.2 *Volt-Ammeter.*

5.3 *Stopwatch*, accurate to 0.1 s.

5.4 *Incandescent Lamps*, for use as a known load.

5.5 *Current Transformers* (CTs).

5.6 *Potential (Voltage) Transformers* (PTs).

---

[1] This test method is under the jurisdiction of ASTM Committee D-34 on Waste Management and is the direct responsibility of Subcommittee D34.06 on Recovery and Reuse.

Current edition approved Feb. 25, 1983. Published May 1983.

Watthour Meter	Serial No.	Type	Class	K_h	Accuracy	Date Calibrated

Current Transformer	Serial No.	Type	Ratio	Accuracy Class	Date Calibrated

Potential Transformer	Serial No.	Type	Ratio	Accuracy Class	Date Calibrated

FIG. 1 Electrical Metering System Installation Form

6. Hazards

6.1 When installing metering equipment always de-energize the load side of the processing equipment by locking out the main switch on the electrical control panel.

6.2 Dangerous high voltage results from open current transformer secondaries. Therefore, to avoid equipment damage and electrical shock, use circuit-closing devices or equipment to short circuit the secondaries of current transformers.

6.3 Always observe the polarity markings of current and potential transformers during their installations to ensure proper connection of the metering equipment. These polarity markings are usually denoted on the transformers as white dots, blocks, or "HX" marks.

6.4 Closely observe polarities, and check connections of instrument transformers to the watthour meter.

7. Equipment Calibration

7.1 Calibrate all meters and instrument transformers used for energy measurements in accordance with standard practice of calibration.[2,3,4 ,5] The accuracy of the meters and trans formers shall be duly noted on the Electrical Metering Service Installation Form, see Fig. 1.

8. Procedure

8.1 *Meter Installation*:

8.1.1 For the piece of equipment to be tested, determine the type of electrical service (for example, single-phase two-wire, three-phase three-wire), voltage requirements, full load power, and current rating of the motor from the motor nameplate or manufacturer's specifications. For the purpose of meter selection and installation, it can be assumed that 1 hp = 1 kW = 1 kVA. Select the metering system that is compatible with the type of electrical service and with the load on the motor.

8.1.1.1 Self-contained single phase watthour meter can be used when the load is less than 48 kVA.

8.1.1.2 Self-contained polyphase meters can be used when the load is less than 96 kVA (except 480 V delta).

8.1.1.3 Above 48 or 96 kVA, respectively, for single and polyphase loads, use transformer type watthour meters.

8.1.2 For any meter installation, do not exceed the meter's overload capability listed as follows:

8.1.2.1 *Class 10*—Nominal 2.5-A meter, 10-A overload capability.

8.1.2.2 *Class 20*—Nominal 2.5-A meter, 20-A overload capability.

8.1.2.3 *Class 60*—Nominal 15-A meter, 60-A overload capability.

8.1.2.4 *Class 100*—Nominal 15-A meter, 100-A overload capability.

8.1.2.5 *Class 200*—Nominal 30-A meter, 200-A overload capability.

8.1.2.6 *Class 320*—Nominal 50-A meter, 320-A overload capability.

8.1.3 *Instrument Transformers*—For meter installations requiring instrument transformers (that is, when the primary current or voltage, or both, exceed the operating specifications of the watthour meter), use current and potential (voltage) transformers. Select current and potential transformers with an accuracy class rating of 0.3 (0.3 %) and compatibility with the primary electrical service. If transformers with an accuracy class of 0.3 are not available, substitute another accuracy class and note on the Electrical Metering System Installation Form (Fig. 1).

8.1.4 *Current Transformers*—Calculate the current transformer ratio (CTR) using the following definition.

$$\text{CTR} = \text{Primary Current/Watthour Meter Nominal Current Rating} \quad (1)$$

Generally, current transformer ratios are denoted such that the secondary current will be 5 amperes when rated amperes are flowing in the primary circuit.

8.1.5 *Potential Transformers* — Potential transformers are used with watthour meters where the primary circuit voltage exceeds the rating of the meter, generally above 480 V and frequently above 240 V. The potential transformer ratio (PTR) can be calculated using the following definition.

$$\text{PTR} = \text{Primary Voltage/Watthour Meter Nominal Voltage Rating} \quad (2)$$

---

[2] *Meter and Instrument Transformer Application Guide*, 5th Edition, Westinghouse Electric Company, Raleigh, NC.

[3] *Metermen's Handbook*, Duncan Electric Company, Lafayette, IN, No. 5M, April 1976.

[4] *Electrical Metermen's Handbook*, Edison Electric Institute, New York, NY.

[5] *Guide for Installing General Electric Watthour Meters*, General Electric Company, Somersworth, NH, April 1976.

8.1.6 Phase relations will be retained if the polarity markings are observed and the current in the potential circuit is considered to flow in on the primary terminal polarity mark and out on the corresponding secondary terminal polarity mark.

8.1.7 The Electrical Metering Service Installation Form (Fig. 1) is recommended for documenting the equipment used for the test.

8.1.8 Mount instrument transformers and watthour meters in an upright position and in a area free from heavy vibration.

8.2 *Checking Meter Installation*:

8.2.1 Check meter installations for correct connections as soon as the wiring is completed. For installation of self-contained watthour meters this is comparatively simple. It is only necessary to see that line and load wires, and potential taps where required, are connected to the proper points. A quick check on the operation under load conditions may be made to see that the meter is rotating in the proper direction and at approximately the right speed.

8.2.2 Where instrument transformers are used, the installation is more liable to incorrect connections and should, therefore, be checked carefully. It is possible to have incorrect registration even with proper connections, due to a wrong transformer polarity marking, a reversed meter coil, incorrect transformer ratio marking, etc. It is generally not possible to completely check all of these items in the field; however, by making several of the tests listed in Annex A1, it will be possible to determine most of the inconsistencies or incorrect connections that might occur.

8.3 *Measurements*:

8.3.1 After installation and check-out of the energy metering equipment, measure and record the energy used by the equipment under no-load and loaded conditions in order to determine the average gross and freewheel power requirements of the equipment.

8.3.2 Determine the average freewheeling power of the equipment to be tested by measuring the energy usage of the motor under no-load conditions over a specified time interval. After a suitable warm-up period, time ten disk revolutions to establish the freewheel energy usage at the beginning and end of the test. Prior to taking the first measurement for determining the freewheel energy usage, take two preliminary freewheel energy measurements (10 disk revolutions) approximately 5 min apart. If the preliminary readings differ by more than 10 % or more, extend the warm-up period until two consecutive preliminary measurements fall within 10 % of one another.

8.3.3 After the suitable warm-up period, take and record three initial disk timings of ten revolutions each. Likewise, after the conclusion of the load tests, take and record three final disk timings of ten revolutions each. An Energy Measurement Data Sheet for recording the freewheeling energy measurements is given in Fig. 2. The freewheeling power calculations are described in Section 9.

8.3.4 Determine the gross energy usage (E_g) of the equipment undergoing testing by calculating the difference in the register readings or by counting the number of disk revolutions of the watthour meter while operating the processing equipment under loaded conditions for a suitable measuring period. A suitable measuring period consists of a time span that is long enough to attain at least one disk revolution or at least one complete rotation of the least significant register dial, whichever applies to the particular test situation.

8.3.5 An Energy Measurement Data Sheet for recording the measured data from the tests conducted under loaded conditions is given in Fig. 2. The calculations for determining power demand under loaded conditions are described in Section 9.

8.3.6 *Alternative Procedure for Constant Load Power Measurements*—If the processing equipment exhibits a constant load as evidenced by power fluctuations of less than ±10 % of the average reading (that is, as may be the case for a conveyor or blower, etc.), a clamp-on wattmeter, an analog wattmeter, or recording wattmeter can be used to measure power if the metering equipment and electrical service can be made compatible with one another. For this procedure, power measurements for both unloaded (freewheeling) and loaded conditions should be made in sufficient numbers so that a reliable average reading can be calculated. The power requirement is read directly from the instrument. The measurements may be recorded on the Gross and Net Power Data Sheet, Fig. 3.

9. Calculation

9.1 *Average Freewheeling Power Requirements*:

9.1.1 Calculate freewheeling power (P_{fw}), in kilowatts, as follows:

$$P_{fw} = 600\,(Kh)(CTR)(PTR)/t \tag{3}$$

where:

Kh = disk constant of the watthour meter (kWh/disk revolution),
CTR = current transformer ratio,
PTR = potential transformer ratio, and
t = time duration for 10 disk revolutions (minutes).

9.1.2 Average the three initial freewheeling power measurements and the three final measurements to give the average initial freewheeling power ($\bar{P}_{fw_i}$) and final freewheeling power ($\bar{P}_{fw_f}$). Then average the average values ($\bar{P}_{fw_i}$ and $\bar{P}_{fw_f}$) and use as the average freewheeling power requirement $\bar{P}_{fw}$ of the equipment corresponding to the interval of gross energy measurement. Record the average value for the freewheeling power in the column titled "Average Freewheel Power" of the Gross and Net Power Data Sheet (see Fig. 3).

9.1.3 Calculate average initial freewheel power ($\bar{P}_{fw_i}$) as follows:

$$\bar{P}_{fw_i} = (P_{fw_a} + P_{fw_b} + P_{fw_c})/3 \tag{4}$$

where:

P_{fw_a}, P_{fw_b}, and P_{fw_c} are the three initial freewheeling power measurements.

9.1.4 Calculate average final freewheel power ($\bar{P}_{fw_f}$) as follows:

$$\bar{P}_{fw_f} = (P_{fw_x} + P_{fw_y} + P_{fw_z})/3 \tag{5}$$

where:

P_{fw_x}, P_{fw_y}, and P_{fw_z} are the three final freewheeling power measurements.

9.1.5 Calculate average freewheel power as follows:

CTR: ___ PTR: ___

Test No.	Watthour Meter Reading Initial (kWh)	Watthour Meter Reading Final (kWh)	Time Interval (Min)	Disk Revolutions (n)	Kh (kWh/rev.)	Gross Energy (kWh)
Loaded Condition 1						
2						
3						
4						
5						
6						
7						
No-load Condition (Freewheel)						
Initial (P_{fwi}) 1						
2						
3						
Average Initial ($\bar{P}_{fw)i}$						
Final (P_{fwf}) 1						
2						
3						
Average Final ($\bar{P}_{fwf}$)						

Average of Average Initial and Average Final Freewheeling Power: ___

FIG. 2 Energy Measurement Data Sheet

$$\bar{P}_{fw} = (\bar{P}_{fw_i} + \bar{P}_{fw_f})/2 \tag{6}$$

9.2 Determine the gross energy (E_g) usage of the equipment undergoing testing by calculating the difference in the register readings (Case 1) or by counting the number of disk revolutions of the watthour meter after a suitable measuring period (Case 2).

9.2.1 For Case 1, calculate the gross energy (E_g) as follows:

$$E_g = (R_f - R_i)(PTR)(CTR) \tag{7}$$

where:

R_f = final meter reading (kWh),
R_i = initial meter reading (kWh),
PTR = potential transformer ratio, and
CTR = current transformer ratio.

9.2.2 For Case 2, calculate the gross energy (E_g) as follows:

$$E_g = 60\, n\, (Kh)(PTR)(CTR)/t \tag{8}$$

where:

n = number of disk revolutions (rev.),
Kh = disk constant of the watthour meter (kWh/r),
PTR = potential transformer ratio,
CTR = current transformer ratio, and
t = time in minutes.

9.3 *Average Gross Power Requirement*—Calculate the average gross power requirement ($\bar{P}_g$) for the equipment under load using the calculated value of gross energy and the length of the time interval over which the gross energy usage was measured:

$$\bar{P}_g = 60E_g/t \tag{9}$$

where:
$\bar{P}_g$ has units of kW,

E_g = gross energy (kWh) measured during the timing interval, and
t = time in minutes.

9.4 *Average Net Power Requirement*—Calculate the average net power requirement ($\bar{P}_n$) as follows:

$$\bar{P}_n = \bar{P}_g - \bar{P}_{fw} \tag{10}$$

where:
$\bar{P}_n$, $\bar{P}_g$, and $\bar{P}_{fw}$ have units of kW.

9.5 A gross and net power data sheet for recording the data and results is included in Fig. 3.

10. Precision and Bias

10.1 *Precision*—The precision of this test method has not been established.

10.2 *Bias*—The bias of this test method has not been established. As a guideline, the accuracy of watthour meters is estimated to be 98.0 to 99.5 %. The accuracy of potential and current transformers (0.3 accuracy class) is 99.7 %. The

CTR:___
PTR:___

Test No.	Gross Energy (E_g) (kWh)	Time Interval (t) (Mins.)	Average Gross Power ($\bar{P}_g$) (kW)	Average Freewheel Power ($\bar{P}_{fw}$) (kW)	Average Net Power ($\bar{P}_n$)

FIG. 3 Gross and Net Power Data Sheet

accuracy of clamp-on wattmeters is typically 5 % of full-scale deflection.

ANNEX

(Mandatory Information)

A1. PROCEDURES FOR CHECKING METER INSTALLATIONS

A1.1 Single Phase

A1.1.1 For single phase meter installation, it is sufficient to see that the line and load conductors are connected to the proper terminals of the meters. A very rough but quick check is to turn on a load and see that the disk rotates in the correct direction. A more careful check (usually unnecessary on single-phase) is to connect a known load and time the revolution of the disk.

A1.2 Three-Phase, Three-Wire

A1.2.1 *Cross-phase Check*—Interchange top and bottom line side potentials. Meter should stop on balanced load regardless of power factor.

A1.2.2 Open common or number two potential. Meter should run at half-speed on balanced load.

A1.2.3 Read current in each lead. On three-phase, three-wire installations using two current transformers, all leads should read the same on balanced load. This will give a check on connections and comparative ratio of transformers.

A1.2.4 Determine relative speed and direction of stators when power factor is known.

A1.3 Three-Phase, Four-Wire Delta (Two Stators)

A1.3.1 Check one stator at a time. Speed of meter should be

the same for each stator on balanced load regardless of power factor.

A1.3.2 On unbalanced loads, when the approximate three phase and single-phase loads are known, check each stator at a time. The top stator measures half of the three-phase load.

A1.4 Three-Phase, Four-Wire "Y" (Two Stators)

A1.4.1 *Cross Phase Check*—Interchange top and bottom line potentials. Meter should stop on balanced load regardless of power factor.

A1.4.2 Open common potential. Meter should run at half-speed on balanced load. The No. 2 current does not affect this check.

A1.4.3 Check one current at a time with normal potential on each stator. Meter should run at same speed at balanced load for each current regardless of power factor.

A1.4.4 Determine relative speed and direction of stators when approximate power factor is known.

A1.4.5 Read current in each lead. On balanced load, the three line currents should be the same and the neutral current should be zero.

A1.5 Time Load Method (Not to be used for calibration)

A1.5.1 This method of checking the accuracy of a watthour meter consists of connecting a known load to a two-wire watthour meter in the conventional manner, and timing the disk for a desired number of revolutions. (Be sure that loads other than those being considered are not connected to the meter at the time of the check.) One of the most consistent and readily available loads is a standard incandescent lamp. Lamp wattage can, if desired, be pre-checked in the shop before use in the field. However, the actual load is usually close enough to the watt rating at rated voltage for making approximate field checks of meter accuracy. The accuracy of the field check can be materially improved by measuring the service voltage in each case and adjusting the "known" wattage accordingly. For voltages within plus or minus 10 volts of lamp rating (which is usually well within the variation found in service), the watt load of the lamp will increase or decrease 1½ % for each 1 % of voltage above or below the voltage rating of the lamps.

A1.5.2 The disk should be timed for a convenient number of revolutions depending on the rating of the meter and the load used. It is usually desirable to run meters for about one minute or over to minimize errors in reading time. It is preferable when timing the disk to use a stop watch or timing device with a second hand. When a pocket or wrist watch without a stop second hand is used, hold the watch in front of the disk so that both the second hand on the watch and the meter disk can be seen at the same time and repeat the time trials until consistent results are obtained.

A1.5.3 The required number of seconds for a given num ber of revolutions of the disk in an accurately calibrated meter measuring a known watt load is given by the equation:

$$\mathrm{Kh} \times 3600 \times R/W = t \tag{A1.1}$$

where:

Kh	=	watthour (or disk) constant (watthours for one revolution),
3600	=	60 min × 60 s = 1 h (used in equation to convert watt-hours to watt-seconds),
R	=	revolutions of meter disk for time of test,
W	=	watt load on meter, and
t	=	time of test in seconds.

A1.5.4 The watthour or disk constant (Kh) will be found on the nameplate of all modern meters; for some older types it is marked on the disk. In any cases of doubt, constants of any meter may be obtained from the manufacturer if the rating, type, and serial number of the meter is given.

A1.5.5 In checking three-wire meters, load should be applied to both current coils. This may be done by dividing the load between the two live conductors and the neutral, or connecting to first one side and then the other.

A1.5.6 For checking 15 ampere (TA 15) three-wire meters, it is usually desirable to use at least 600 watts or about 20 % of rated capacity. A light load check may be made with 300 watts or 10 % load. In some cases, particularly when the normal load is small (below the light load calibration point of 10 percent) it is desirable to also make a check below the 10 % point. There are some rare cases where a meter is improperly calibrated on light load and may be out considerably more at extremely light loads. If such a case is found the meter should, of course, be tested with a rotating standard or equivalent method and recalibrated.

A1.5.7 A Time-Load chart for use in this method of checking is given in Table A1.1.

A1.5.8 The accuracy of the Time-Load method of checking should not be expected to be better than two percent.

TABLE A1.1 Time-Load Chart for Watthour Meter

Meter Rating	Manufacturer-Type	Disk Constant (Kh)	Rev. (R)	Required Time for Known Load		Load Ws = 3600 Kh (R)
				Watts (W)	Seconds (S)	
5-ampere	Duncan MF, MD; Westinghouse C,	1/3	15	300	60	18000
2-wire	OC, OB; Sangamo HF, HC	1/3	3	60	60	3600
120-volt	G.E. I-50, I-30, I-20, I-16	0.6	10	300	72	21600
	Sangamo J	0.6	2	60	72	4320
15-ampere	Duncan MF, MD; Westinghouse C-OC-OB; Sangamo HF-HC	1	10	600	60	36000
2-wire	G.E. I-30	1.5	8	600	72	43200
120-volt	G.E. I-50, I-20, I-16, Sangamo J	1.8	6	600	64.8	38880
	Duncan MK, MQ Westinghouse D, D-2 Sangamo J-2, J-3	3.0	4	600	72	43200
15-ampere	Duncan MF, MD; Westinghouse	2	5	600	60	36000
3-wire	C-OC-OB; Sangamo HF-HC					
	G.E. I-30	3	4	600	72	43200
240-volt	G.E. I-55, I-50, I-20, I-16;	3.6	3	600	64.8	38880
	Sangamo J. Duncan MK, MQ, Westinghouse D, D-2 Sangamo J-2, J-3	6.0	2	600	72	43200
30-ampere	Duncan MF, MD; Westinghouse	6–2/1	5	2000	60	120000
3-wire	C-OC-OB; Sangamo HF-HC					
240-volt	G.E. I-50, I-30, I-20, I-16;	12	3	2000	64.8	129600
	Duncan MK, MQ, Sangamo J, J-2, J-3					
50-ampere	Duncan MK, MQ	12	3	2000	64.8	129600
3-wire	G.E. I-60	7.2	5	1800	72	129600
240-volt	Westinghouse D, D-2;	7.2	5	1800	72	129600
	Sangamo J-2, J-3	12	3	2000	64.8	129600

Designation: E 949 – 88 (Reapproved 1996)

Standard Test Method for Total Moisture in a Refuse-Derived Fuel Laboratory Sample[1]

This standard is issued under the fixed designation E 949; the number immediately following the designation indicates the year of original adoption or, in the case of revision, the year of last revision. A number in parentheses indicates the year of last reapproval. A superscript epsilon (ε) indicates an editorial change since the last revision or reapproval.

1. Scope

1.1 This test method covers the measurement of the total moisture in RDF as it exists at the time it is sampled. Because of its empirical nature, strict adherence to test procedures are required for valid results. The standard is available to producers, vendors, and consumers as a total, two-stage moisture method.

1.2 Since RDF can vary from extremely wet (water saturated) to relatively dry, special emphasis must be placed on sampling, sample preparation, and the method of determination.

1.3 *This standard does not purport to address all of the safety concerns, if any, associated with its use. It is the responsibility of the user of this standard to establish appropriate safety and health practices and determine the applicability of regulatory limitations prior to use.* For hazard statement, see Section 7.

2. Referenced Documents

2.1 *ASTM Standards:*

D 3173 Test Method for Moisture in the Analysis Sample of Coal and Coke[2]

E 790 Test Method for Residual Moisture in Refuse-Derived Fuel Analysis Sample[3]

E 829 Practice for Preparing Refuse-Derived Fuel (RDF) Laboratory Samples for Analysis[3]

3. Terminology Definitions

3.1 *air drying*—a process of partial drying of RDF to bring its moisture content near to equilibrium with the atmosphere in which further reduction, division, and characterization of the sample are to take place. In order to bring about the equilibrium, the RDF is usually subjected to drying under controlled temperature conditions ranging from 30 to 40°C.

3.2 *analysis sample*—the final subsample prepared from the air-dried laboratory sample but reduced by passing through a mill with a 0.5 mm (0.02 in.) size or smaller final screen.

3.3 *bias*—a systematic error that is consistently negative or consistently positive. The mean of errors resulting from a series of observations that does not tend towards zero.

3.4 *gross sample*—a sample representing one lot and composed of a number of increments on which neither reduction nor division has been performed.

3.5 *laboratory sample*—a representative portion of the gross sample received by the laboratory for analysis.

3.6 *lot*—a large designated quantity (greater than the quantity of the final sample) of RDF which can be represented by a properly selected gross sample.

3.7 *precision*—a term used to indicate the capability of a person, an instrument, or a method to obtain reproducible results; specifically, a measure of the random error as expressed by the variance, the standard error, or a multiple of the standard error.

3.8 *forms of refuse-derived fuel (RDF)*:

RDF-1—Wastes used as a fuel in as-discarded form.

RDF-2—Wastes processed to coarse particle size with or without ferrous metal separation.

RDF-3—shredded fuel derived from municipal solid waste (MSW) that has been processed to remove metal, glass, and other inorganics. This material has a particle size such that 95 weight % passes through a 2-in. (50-mm) square mesh screen.

RDF-4—Combustible waste processed into powder form, 95 weight % passing 10-mesh screening.

RDF-5—Combustible waste densified (compressed) into the form of pellets, slugs, cubettes, or briquettes.

RDF-6—Combustible waste processed into liquid fuel.

RDF-7—Combustible waste processed into gaseous fuel.

3.9 *representative sample*—a sample collected in such a manner that it has characteristics equivalent to the lot sampled.

3.10 *sample division*—the process of extracting a smaller sample from a sample so that the representative properties of the larger sample are retained. During this process it is assumed that no change in particle size or other characteristics occurs.

3.11 *sample preparation*—the process that includes drying, size reduction, division, and mixing of a laboratory sample for the purpose of obtaining an unbiased analysis sample.

3.12 *sample reduction*—the process whereby sample particle size is reduced without change in sample weight.

3.13 *significant loss*—any loss that introduces a bias in final results that is of appreciable importance to concerned parties.

4. Summary of Test Method

4.1 This test method is based on the loss in weight of RDF

---

[1] This test method is under the jurisdiction of ASTM Committee D-34 on Waste Management and is the direct responsibility of Subcommittee D34.08 on Thermal Treatment.

Current edition approved March 25, 1988. Published May 1988. Originally published as E 949 – 83. Last previous edition E 949 – 83.

[2] *Annual Book of ASTM Standards,* Vol 05.05.

[3] *Annual Book of ASTM Standards,* Vol 11.04.

in an air atmosphere under controlled conditions of temperature, time, and air flow.

4.2 The laboratory sample is air-dried to near equilibrium with the atmosphere in the area where division and reduction will take place. The residual moisture determination is made in a heated, forced-circulation oven, under rigidly defined conditions.

4.3 The total moisture is calculated from losses in air-drying and the residual moisture as shown in 11.

5. Significance and Use

5.1 The collection and treatment of the sample as specified herein is intended for the specific purpose of determining the total moisture in a laboratory sample of RDF.

5.2 This test method is available as the method for the determination of total moisture unless alternative techniques or modifications have been agreed upon by involved parties.

6. Apparatus

6.1 *Air Dry Moisture*:

6.1.1 *Drying Oven*—A large chamber mechanical draft oven capable of maintaining a controlled temperature in the range of 25 to 40°C. Air changes should be at the rate of 1 to 4 changes per minute. Air flow should be baffled to prevent samples from being blown out of the sample containers.

6.1.2 *Drying Pan*—A non-corroding pan or mesh basket to be used for holding the sample during air drying operations.

6.1.3 *Balance (Laboratory Sample)*—A balance of sufficient capacity to weigh the sample and container with a sensitivity of 0.5 g.

6.2 *Sample Reduction*:

6.2.1 *Mill*—A mill operating on the principle of cutting or shearing action shall be used for sample particle size reduction. It shall have the capability to regulate the particle size of the final product by means of either interchangeable screens or mill adjustments. The mill shall be enclosed and should generate a minimum amount of heat during the milling process to minimize the potential for loss of moisture. The final product shall pass through a 0.5 mm or smaller screen into a receiver integral with the mill. Access should be provided so that the mill can be cleaned quickly and easily between samples.

6.3 *Residual Moisture*:

6.3.1 *Drying Oven*:

6.3.1.1 *Referee Type*—The oven shall be so constructed as to have a uniform temperature within the specimen chamber, have a minimum excess air volume, and be capable of constant temperature regulation at 107 ± 3°C. Provision shall be made for renewing the preheated air in the oven at the rate of two to four times a minute, with the intake air dried by passing it through a desiccant. An oven similar to the one illustrated in Fig. 1, Moisture Oven, of Test Method D 3173 is suitable.

6.3.1.2 *Routine Type*—A drying oven of either the mechanical or natural circulation type which is capable of constant uniform temperature within the specimen chamber regulated at 107 ± 3°C.

NOTE 1—Either type of oven may be used for routine determinations. However, the referee-type oven shall be used to resolve differences between determinations.

6.3.2 *Containers*—A convenient form that allows the ash determination to be made on the same sample is a porcelain capsule 22 mm in depth and 44 mm in diameter or a fused silica capsule of similar shape. This shall be used with a well-fitting flat aluminum cover. Platinum crucibles or glass capsules with ground glass caps may also be used. They should be as shallow as possible consistent with convenient handleability.

6.3.3 *Analytical Balance*, with 0.1 mg sensitivity.

6.3.4 *Analysis Sample Containers*—Heavy (minimum 4 mil), vapor-impervious bags, properly sealed; or noncorroding cans, glass jars, or plastic bottles with air-tight sealing covers to store RDF samples for analysis. Containers shall be checked for suitability by measuring weight loss or gain of the sample and container stored for 1 week under ambient laboratory conditions. The weight loss or gain should be less than 0.5 % of the sample weight stored in container.

7. Hazards

7.1 Due to the origins of RDF in municipal waste, common sense dictates that some precautions should be observed when conducting tests on the samples. Recommended hygienic practices include use of gloves when handling RDF; wearing dust masks (NIOSH-approved type), especially when shredding RDF samples; conducting tests under negative pressure hood when possible; and when washing hands before eating or smoking.

7.2 Laboratory sample handling shall be performed by trained personnel. All operations shall be done as rapidly as possible to avoid sample moisture changes due to atmospheric exposure.

7.3 At all times RDF samples should be protected from moisture change due to exposure to rain, snow and sun, or contact with absorbent materials.

7.4 Since heavy fine particles tend to segregate rapidly in the RDF analysis sample, the analyst should exercise care to assure that the analysis sample is well mixed prior to performing the residual moisture determination.

7.5 When the residual moisture is to be used for the determination of total moisture, special care shall be taken to avoid any change in sample moisture between the completion of air drying and analysis for residual moisture. It is recommended that the delay between sample preparation and the determination of residual moisture be a maximum of 72 h.

7.6 Samples should be transported to the laboratory and analyzed as soon as possible. If any sample handling step involves an extended time period, the sample and container should be weighed before and after the process to determine any weight gain or loss. This weight gain or loss shall be included in the calculation of moisture content.

7.7 Force-feeding of the sample through the mill can overload the motor. An overload can cause rapid heating of the rotor and mill chamber with possible loss of residual moisture.

8. Sampling (Note 2)

8.1 RDF products are frequently nonhomogeneous. For this reason, care should be exercised to obtain a representative sample from the RDF lot to be characterized.

8.2 The sampling method for this procedure should be based on agreement between the involved parties.

8.3 For this procedure the laboratory sample size will normally not exceed 2 kg with some variation possible depending on the laboratory equipment available.

8.3.1 Due to the heterogeneous nature of RDF, dividing a laboratory sample to a very small size analysis sample may result in non-representative results. Since milling operations mix the sample as well as reduce particle size, laboratory samples should not be divided before the initial preparation steps have been completed.

NOTE 2—ASTM Subcommittee E38.01 is currently in the process of developing a procedure for sampling RDF. The chairman of E38.01 should be contacted for details.

9. Sample Preparation

9.1 The principles, terms, organization and preparation procedures as established in Practice E 829 shall apply to the handling and preparation of RDF for determination of total moisture by the two-stage method.

9.2 This procedure provides for using an air-drying oven to equilibrate laboratory sample moisture prior to reduction in size or amount and a moisture oven for determination of residual moisture on the air-dried analysis sample.

9.3 The laboratory sample must be air dried and particle size reduced to pass a 0.5 mm screen as described in Practice E 829, for the residual moisture (second stage) of the total moisture determination.

10. Procedure

10.1 *Air Drying Laboratory Sample*:

10.1.1 Weigh the entire laboratory sample into a tared air-drying pan. Use more than one pan if necessary. If a very fine mesh type of drying pan is used, size the mesh such that the sample will not be lost through it. Sample depth in the drying pan shall be no greater than 10 cm (4 in) and any lumps of sample shall be broken up.

10.1.2 Air dry the sample at 10 to 15°C above ambient, but not greater than 40°C until the weight loss is less than 0.1 % of the sample weight per hour. Normally, allow thc sample to air dry for a set time period such as overnight or 24 h. To speed the drying stage, stir the sample carefully while avoiding loss of sample (Note 3).

NOTE 3—The air discharge of the forced draft air drying oven should be filtered prior to discharge to minimize laboratory contamination by air entrained RDF dust.

10.1.3 At the end of the air drying period, cool the sample to room temperature and weigh. Protect the sample from contamination and loss during the cool-down process but do not place in a desiccator. Calculate air dry moisture loss percent in accordance with Section 11.

10.1.4 Separate, weigh, and hold non-millables for further classification and use for analysis if necessary (Note 4). Mill the remainder of the sample in accordance with Practice E 829.

10.1.5 The calculation for the decimal percent of non-millables (NM) is:

$$NM = \frac{\text{Weight in grams of non-millables}}{\text{Weight in grams of air-dried sample}} \quad (1)$$

NOTE 4—Non-millables are those materials which will not pass through the milling screen, or may damage the milling apparatus, or both.

10.2 *Residual Moisture on Air-Dried Analysis Sample*:

10.2.1 Heat the empty containers and covers under the conditions at which the sample is to be dried, place the stopper or cover on the container, cool over a desiccant for about 15 to 20 min, and weigh. Mix the sample, if necessary, and dip out with a spoon or spatula from the sample bottle approximately 1 g of the sample. Put the sample quickly into the container, cover and weigh at once (Note 5).

NOTE 5—If weighing bottles with air-tight covers are used, it may not be necessary to preheat the moisture analysis container nor to desiccate it after drying.

10.2.2 Remove the cover and place in a desiccator. Quickly place the uncovered container into an oven preheated to 107 ± 3°C through which is passed a current of dry air. Close the oven at once and heat for 1 h. Open the oven, remove, cover the container quickly, and cool in a desiccator over desiccant. Weigh the sample and container as soon as cooled to room temperature.

11. Calculation

11.1 The air dry moisture, A, is calculated as follows:

$$A = \frac{G - L}{G} \times 100 \quad (2)$$

where:

A = air dry moisture, %,
G = weight, in grams, of laboratory sample before air drying, and
L = weight, in grams, of laboratory sample after air drying.

11.2 Calculate the percent residual moisture, R, in the analysis sample as follows:

$$R = \frac{S - B}{S} \times 100\,(1 - NM) \quad (3)$$

where:

R = residual moisture, %,
S = grams of analysis sample used,
B = grams of sample after heating at 107°C, and
NM = decimal percent of non-millables as determined in 10.1.5.

11.3 Calculate the percent total moisture, M, in the laboratory sample, as follows:

$$M = \frac{R(100 - A)}{100} + A \quad (4)$$

where:

M = total moisture, %,
R = residual moisture, %, and
A = air dry moisture, %

12. Precision and Bias

12.1 Precision and bias has not been determined.

Designation: E 953 – 88 (Reapproved 1998)

Standard Test Method for Fusibility of Refuse-Derived Fuel (RDF) Ash[1]

This standard is issued under the fixed designation E 953; the number immediately following the designation indicates the year of original adoption or, in the case of revision, the year of last revision. A number in parentheses indicates the year of last reapproval. A superscript epsilon (ε) indicates an editorial change since the last revision or reapproval.

1. Scope

1.1 This test method covers the observation of the temperatures at which triangular pyramids (cones) prepared from RDF ash attain and pass through certain stages of fusing and flow when heated at a specific rate in controlled, mildly-reducing, and oxidizing atmospheres.

1.2 The test method is empirical, and strict observance of the requirements and conditions is necessary to obtain reproducible temperatures and enable different laboratories to obtain concordant results.

1.3 The values stated in SI units are to be regarded as the standard. The values given in parentheses are for information only.

1.4 *This standard does not purport to address all of the safety problems, if any, associated with its use. It is the responsibility of the user of this standard to establish appropriate safety and health practices and determine the applicability of regulatory limitations prior to use.* See Section 6 for additional hazard information.

2. Referenced Documents

2.1 *ASTM Standards:*

E 180 Practice for Determining the Precision Data of ASTM Methods for Analysis and Testing of Industrial Chemicals[2]

E 829 Practice for Preparing Refuse-Derived Fuel (RDF) Laboratory Samples for Analysis[3]

3. Terminology

3.1 *Definitions and Symbols:*

3.1.1 The critical temperature points to be observed are as follows, denoting the atmosphere used:

3.2 *initial deformation temperature, IT*—the temperature at which the first rounding of the apex of the cone occurs. Shrinking or warping of the cone is ignored if the tip remains sharp. In Fig. 1, the first cone shown is an unheated one; the second cone, IT, is a typical cone at the initial deformation stage.

3.3 *softening temperature, ST*—the temperature at which

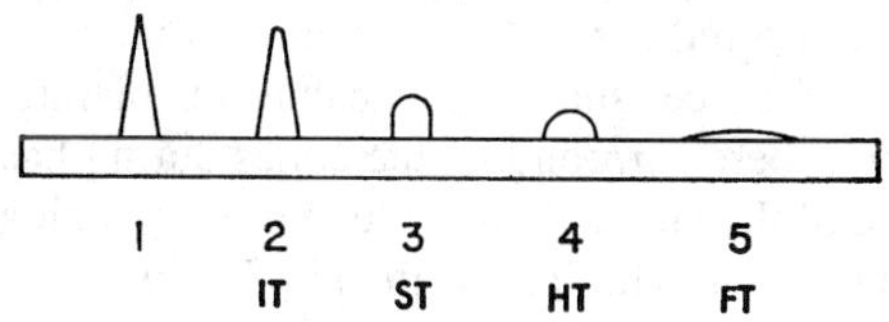

FIG. 1 Critical Temperature Points

the cone has fused down to a spherical lump in which the height is equal to the width at the base as shown by the third cone, ST, in Fig. 1.

3.4 *hemispherical temperature, HT*—the temperature at which the cone has fused down to a hemispherical lump at which point the height is one half the width of the base as shown by the fourth cone, HT, in Fig. 1.

3.5 *fluid temperature, FT*—the temperature at which the fused mass has spread out in a nearly flat layer with a maximum height of 1.6 mm (1⁄16 in.) as shown in the fifth cone, FT, in Fig. 1.

3.6 *forms of refuse-derived fuel* (RDF):

RDF-1—Waste used as a fuel in as-discarded form.

RDF-2—Waste processed to coarse particle size with or without ferrous metal separation.

RDF-3—Shredded fuel derived from municipal solid waste (MSW) that has been processed to remove metal, glass, and other inorganics. This material has a particle size such that 95 weight % passes through a 2-in. square mesh screen.

RDF-4—Combustible waste processed into powder form, 95 weight % passing a 10-mesh screening.

RDF-5—Combustible waste densified (compressed) into the form of pellets, slugs, cubettes, or briquettes.

RDF-6—Combustible waste processed into liquid fuel.

RDF-7—Combustible waste processed into gaseous fuel.

4. Significance and Use

4.1 The standard is available to producers and users of RDF to use in determining the fusibility of ash produced from RDF.

4.2 *Limitations of Ash Fusibility Data*—Ash fusibility data are too often over-interpreted. In practice, types of burning equipment, rate of burning, temperature and thickness of fire bed or ball, distribution of ash forming mineral matter in the RDF, and viscosity of the molten ash may influence ash behavior more than the ash fusibility characteristics determined by the laboratory test. Furthermore, conditions existing during applied combustion of RDF are so complex that they are impossible to duplicate completely in a small-scale laboratory test. Therefore, the test should be considered an empirical one

---

[1] This test method is under the jurisdiction of ASTM Committee D-34 on Waste Management and is the direct responsibility of Subcommittee D34.03 on Treatment.

Current edition approved March 25, 1988. Published May 1988. Originally published as E 953 – 83. Last previous edition E 953 – 83.

[2] *Annual Book of ASTM Standards*, Vol 15.05.

[3] *Annual Book of ASTM Standards*, Vol 11.04.

and the data, at best, only qualitative.

5. Apparatus

5.1 *Furnace*—Any gas-fired or electric furnace conforming to the following requirements may be used.

5.1.1 The furnace shall be capable of maintaining a uniform temperature zone in which to heat the ash cones. This zone shall be such that the difference in the melting point of 12.7 mm (½ in.) pieces of pure gold wire when mounted in place of the ash cones on the cone support shall be not greater than 11°C (20°F) in a reducing atmosphere test run.

5.1.2 The furnace shall be capable of maintaining the desired atmosphere surrounding the cones during heating. The composition of the atmosphere, reducing or oxidizing, shall be maintained within the limits specified in Section 7. The desired atmosphere in the gas-fired furnace surrounding the cones shall be obtained by regulation of the ratio of gas to air in the combustion mixture. The desired atmosphere in the electric furnace shall be obtained by means of gases introduced into the heating chamber. The muffle shall be gas-impervious, free from cracks, and the closure-plug tight fitting. Since state-of-the-art furnaces vary somewhat in design, the gas supply inlet tube shall be installed per instructions of furnace manufacturer.

5.1.3 The furnace shall be capable of regulation so that the rate of temperature rise shall be 8 ± 3°C (15 ± 5°F) per minute.

5.1.4 The furnace shall provide a means of observing the ash cones during the heating. Observation shall be on the same horizontal plane as the cone-support surface.

5.2 *Cone Mold*—A commercially available cone mold as shown in Fig. 2 shall be used. The cone shall be 19 mm (¾ in.) in height and 6.4 mm (¼ in.) in width at each side of the base which is an equilateral triangle.

5.2.1 A steel spatula with a pointed tip, ground off to fit the cone depression in the mold, is suitable for removal of the ash cone.

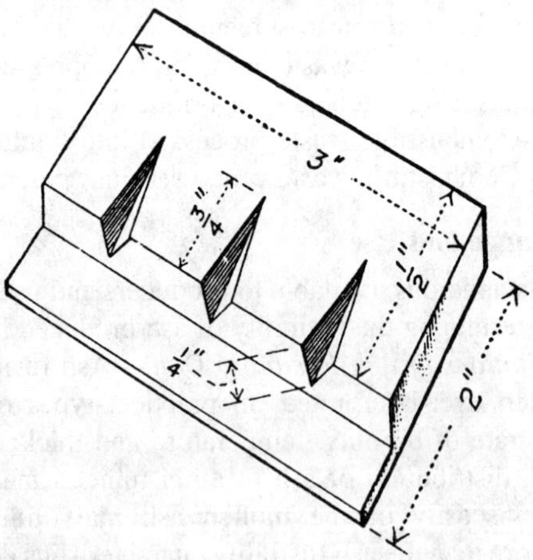

U.S. Customary Units. in.	SI Units, mm
¼	6.4
¾	19.1
1½	38.1
2	50.2
3	76.2

FIG. 2 Brass Cone Mold

5.3 *Optical Pyrometer or Thermocouple*, for temperature measurements, conforming to the following requirements:

5.3.1 *Optical Pyrometer*—An optical pyrometer of the disappearing filament type shall be used. The instrument shall have been calibrated to be accurate within 11°C (20°F) up to 1400°C (2550°F) and within 16°C (30°F) from 1400 to 1600°C (2550°F to 2900°F) (Note 2). The pyrometer filament shall be sighted on the cones until the softening point temperature (Fig. 1) has been passed, and then sighted on the cone support. The pyrometer shall have readable graduations not larger than 5.5°C (10°F).

Note 1—The pyrometer equipment shall be standardized periodically by a suitably equipped standardizing laboratory such as that of the National Bureau of Standards, or checked periodically against equipment certified by the National Bureau of Standards.

5.3.2 *Thermocouple*—A thermocouple of platinum and platinum-rhodium, protected from the furnace gases by a glazed porcelain sheath, shall be used with a high-resistance millivoltmeter or potentiometer accurate and readable to within 5.5°C (10°F). The sheath shall be sealed to the furnace wall by alundum cement. The hot junction of the thermocouple shall touch the end of the sheath and shall be located in the center of the muffle and immediately to the rear of the cones. The thermocouple protective sheath shall be checked periodically for cracks. The thermocouple and its meter shall meet the requirements of Section 10. The potentiometer or millivoltmeter shall be located or shielded adequately as to prevent radiant or convection heating of the cold junction. The room temperature compensator shall be adjusted to the existing temperature.

5.4 *Ash-Cone Refractory Support*—The ash cone shall be mounted on a refractory base composed of a mixture of equal parts by weight of kaolin and alumina conforming to the following requirements:

5.4.1 *Kaolin*—NF-Grade powder passing a No. 200 (0.074 mm) sieve.

5.4.2 *Aluminum Oxide*—Reagent grade powder passing a No. 100 (0.149 mm) sieve.

5.5 *Refractory Support Mold*—A mold with flat top and bottom surfaces to provide a refractory support of suitable thickness to minimize warping shall be used. A side mold not over ¼ in. (6.4 mm) high of any convenient shape, placed on an iron plate so that the top surface of the refractory mix can be struck off flat and parallel to the base by means of a straightedge, is satisfactory. For electric furnace use, legs not over 3 mm (⅛ in.) long may be provided on the corners of the cone support by suitable holes bored in the iron base plate of the mold.

5.6 *Gold Wire*, [4] Twenty-four gage or larger round wire of 99.98 % purity, but drawn from metal of 99.99 % purity, and having a melting point of 1063°C (1945°F).

5.7 *Nickel Wire*, [5] Twenty-four gage or larger round wire of CP nickel 99.98 % pure, fully annealed, and having a melting

[4] Gold wire of this purity can be purchased from the Baker Dental Division of Engelhard Industries, Inc., 850 Passaic Ave., East Newark, NJ 07029.

[5] Nickel wire of this purity can be purchased with the additional specifications of having a minimum coefficient of resist for 0 to 100°C of 0.00673 Ohm/Ohm° C, from the Baker Dental Division of Engelhard Industries, Inc., 850 Passaic Ave., East Newark, NJ 07029.

point of 1452°C (2645°F).

6. Hazards

6.1 Due to the origins of RDF in municipal waste, common sense dictates that some precautions should be observed when conducting tests on the samples. Recommended hygenic practices include use of gloves when handling RDF; wearing dust masks (NIOSH-approved type); especially while milling RDF samples; conducting tests under a negative pressure hood when possible; and washing hands before eating or smoking.

6.2 Good laboratory practices dictate the precautions that should be observed when using compressed gases such as hydrogen or carbon monoxide as reducing gases.

7. Test Atmosphere[6]

7.1 *Gas Fired Furnace*:

7.1.1 *Reducing Atmosphere Test*—A mildly reducing atmosphere surrounding the cones shall be maintained during the test in the gas-fired furnace. Hydrogen, hydrocarbons, and carbon monoxide shall be considered as reducing gases; oxygen, carbon dioxide, and water vapor shall be considered as oxidizing gases. Nitrogen is inert. The ratio by volume of reducing gases to oxidizing gases in the atmosphere shall be between the limits of 20 to 80 %, that is, on a nitrogen-free basis, the total amount of reducing gases present shall be between the limits of 20 and 80 volume %. A flame 150 to 200 mm (6 to 8 in.) in height and tinged with yellow above the furnace outlet has been found to provide an atmosphere within the specified limits.

7.1.2 *Oxidizing Atmosphere Test*—An atmosphere containing a minimum amount of reducing gases shall be maintained surrounding the cones during the test in the gas-fired furnace. On a nitrogen-free basis, the volume of the reducing gases present in the atmosphere will not exceed 10 volume %. Combustion with the maximum possible quantity of air with preservation of the specified rate of temperature increase has been found to provide an atmosphere within the specified limits. A completely blue flame, not over 50 mm (2 in.) in height above the outlet at the beginning of the test, provides the desired atmosphere; and, by regulation of the combustion gas-air ratio, the specified atmosphere and temperature rise can be maintained.

7.2 *Electric Furnace*:

7.2.1 *Reducing Atmosphere Test*—A regulated flow of gas of the nominal composition, 60 % carbon monoxide and 40 ± 5 volume % carbon dioxide,[7] shall be maintained in the heating chamber throughout the test (Note 2) in the electric furnace. The gas stream shall be regulated by any convenient means to provide a measured flow of 1.3 to 1.5 furnace volumes per minute.

NOTE 2—New cylinders of the mixed gas for which a certified analysis is not available should be mixed before use by laying the cylinder on its side with the protective screw cap in place. The cylinder should then be rolled back and forth, 6 to 10 ft in each direction, approximately 15 times. Certified analyses of each cylinder or batch can be obtained for a small extra charge.

7.2.2 *Oxidizing Atmosphere Test*—A regulated stream of air shall be maintained throughout the test in the electric furnace. The gas stream shall be regulated by any convenient means to provide a measured flow of 1.3 to 1.5 furnace volumes per minute.

8. Preparation of Ash

8.1 Use RDF milled to passing a 0.5-mm (0.02-in.) sieve prepared in accordance with Practice E 829 to obtain the ash by incineration in a well-ventilated muffle furnace. The quantity of RDF required will vary with the ash content; usually 3 to 5 g of ash will suffice for cones for several check determinations. Spread out the analysis sample of RDF in a layer approximately 6.4 mm (¼ in.) in depth in a fireclay or porcelain roasting dish. Place the dish in the muffle at a low temperature, and gradually heat to redness at such a rate as to avoid mechanical loss from too rapid expulsion of volatile matter. The rate of temperature rise of 500°C (932°F) in 1 h was found to be satisfactory. Complete the conversion to ash at a temperature of 800 to 900°C (1470 to 1650°F). Transfer the ash to an agate mortar (Note 3), and grind so it will pass a No. 200 (0.074 mm) sieve.

NOTE 3—A mechanical agate mortar grinder will save time where many determinations are made. An iron mortar or pestle is not recommended because of metallic contamination.

8.2 Spread the ash in a thin layer in a fireclay, silica, or porcelain dish and ignite it in a stream of oxygen for 1½ h at 800 to 850°C (1470 to 1560°F) to ensure complete and uniform oxidation of the ash. Any tube or muffle-type furnace which, when supplied with an oxygen flow of not less than one furnace volume in 5 min will maintain a highly oxidizing atmosphere, is suitable.

NOTE 4—It has been found that in most samples, the initial ignition outlined in 8.1 is sufficient to convert the RDF to ash and the reignition step in 8.2 is not necessary. Reignition of the ash should be made only if an observable amount of noncombustible matter or carbon is present.

9. Preparation of Cones

9.1 Thoroughly mix the ignited ash in a mechanical mixer or on a sheet of glazed paper or oil cloth by raising first one corner to roll the ash over, and then raising each of the other corners in rotation in the same manner until each corner has been raised five times or more.

9.2 Take sufficient ash for the number of cones desired from various parts of the bulk ash. Moisten the ash with a few drops of a clear, filtered (if necessary) 10 % solution of dextrin containing 0:1 % salicylic acid as a preservative, and work it into a stiff plastic mass with a spatula. Press the plastic material firmly with a spatula into the cone mold to form the triangular pyramids. Strike off the exposed surfaces of the material smooth and remove the cones from the mold by applying pressure at the base with a suitably pointed spatula. Previous coating of the mold with a thin layer of petroleum jelly, thinned with kerosine (if necessary), aids in preventing adherence of the cones to the mold and in providing the sharp point and edges desired in the cone.

[6] For information concerning the effect of various atmospheres, see U.S. Bureau of Mines Bulletin 129, 1918.

[7] This gas is available from the Matheson Division of Searle Medical Products, USA, Inc.

9.3 Place the cones in a suitable location to dry sufficiently to permit handling without deformation. Mount the dried cone vertically on a freshly prepared refractory base. Moisten a portion of the well-mixed kaolin-alumina mixture with a minimum amount of water to make a workable, but stiff, plastic mass and firmly press it into the support mold. Strike off the surface of the mass flat and smooth with a steel spatula, moistening with one or two drops of water is necessary to obtain a smooth surface. A number of cones may be mounted on one base. Make shallow triangular depressions, not over 0.8 mm (1⁄32 in.) in depth, with a triangular file ground to the correct size to produce a depression to fit the base of the cone, and locate the cones sufficiently distant from adjacent cones so that no merging of the fusing material of the cones shall occur during the test. Mount the cones vertically in the depressions while the base is still wet without the use of ash or refractory as mounting aid.

NOTE 5—The intent of the triangular depression is to enable the cones to be mounted in a sufficiently stable manner to permit handling of the prepared support with cones.

NOTE 6—Gold wires can be mounted on each cone support beside the ash cones, and the gold melting point observed concurrently with the ash cones in both oxidizing and reducing atmospheres.

9.4 For the calibration check run of Section 10, mount 12.7 mm (1⁄2in.) lengths of gold and nickel wire as described in 5.6 and 5.7 vertically on a support similar to those prepared as described in 9.3, but on which no ash cones are mounted. Insert the pieces of wire into the support before drying. Two or more pieces of each kind of wire should be mounted on this support in locations which correspond to ash cone positions. Dry the support as described in 9.5 but do not ignite. At least one such support with mounted pieces of gold and nickel wire shall be prepared for each week of observation of ash fusibility.

9.5 Dry the mounted cones at 110°C (230°F), and ignite at 750°C (1382°F) to remove all carbonaceous material.

10. Calibration

10.1 At least once during each week of operation, check the optical pyrometer or the thermocouple and its meter for calibration in place under routine test conditions with the reducing atmosphere by observing the melting point of gold. At the same time, check the adequacy of the furnace atmosphere by observing the melting point of nickel.

10.1.1 Insert a support with mounted pieces of gold and nickel wire into the test furnace. Locate at the position used for tests of ash cones.

10.1.2 Establish the reducing gas atmosphere, and heat the furnace chamber in accordance with 11.1.

10.1.3 Observe the temperatures shown on the meter when the pieces of wire melt.

10.1.4 The indicated melting points should be within± 10°F (5.5°C) of the following:

Gold	1945°F (1063°C)
Nickel	2645°F (1452°C)

10.1.5 If the indicated melting point of gold wire frequently falls outside of the desired range, readjust or calibrate the meter so that the average temperature from several observations of the gold melting point is within the specified range.

10.1.6 If the indicated melting point for nickel wire frequently falls outside the desired range after applying the corrections in 10.1.5, the consistency of indicated temperatures and the subsequent appearance of the specimen should be examined closely. Erratic readings of failures to obtain melting at 2645°F (1452°C) can be due to nickel oxidation caused by an insufficient reducing atmosphere. If a consistent error of more than 25°F (14°C) is found, the furnace atmosphere and the temperature measurement equipment should be suspected of having faults; necessary corrections should be made.

11. Procedure

11.1 *Reducing Atmosphere Test*:

11.1.1 Place the mounted test cones in the furnace at a temperature of not over 1470°F (800°C) for the gas-fired furnace and not over 750°F (400°C) for the electric furnace in order to provide sufficient time to purge the air from the uniform temperature zone and establish the desired atmosphere.

11.1.2 If the furnace temperature is below the respective temperature specified in 11.1.1, raise it rapidly to the specified temperature; then control the rate of heating to give a rate of temperature increase of 15 ± 5°F (8 ± 3°C) per minute. Maintain this rate throughout the test.

11.1.3 Establish the mildly reducing atmosphere surrounding the cones, as specified in 7.1.1 and 7.2.1, at the temperature specified in 11.1.1 for the respective furnace type. Maintain this atmosphere throughout the test.

NOTE 7—At temperatures of the order of 2500 to 2600°F (1370 to 1430°C) and above in the gas-fired furnace, it may not be possible to maintain the reducing gases above the 20-volume % limit specified while also preserving the specified rate of temperature increase. At such temperatures, the effect of the atmosphere is not so critical as the maintenance of the specified heating rate. Make every effort to maintain the reducing gases as near the 20- volume % level as possible at such temperature.

11.2 *Oxidizing Atmosphere Test*:

11.2.1 Place the mounted test cones in the furnace at a temperature of not over 800°C (1470°F) for the gas-fired furnace, and not over 400°C (750°F) for the electric furnace. If the furnace temperature is below the respective temperature specified, raise it rapidly to the specified temperature, then control the rate of heating to give a rate of temperature increase of 8 ± 3°C (15 ± 5°F) per minute. Maintain this rate throughout the test.

11.2.2 Establish the oxidizing atmosphere surrounding the cones, as specified in 7.1.2 and 7.2.2 at the temperature specified in 11.2.1 for the respective furnace type. Maintain this atmosphere throughout the test.

12. Report

12.1 Report the following information:

12.1.1 Type of atmosphere,

12.1.2 Initial deformation temperature, IT,

12.1.3 Softening temperature, ST,

12.1.4 Hemispherical temperature, HT, and

12.1.5 Fluid temperature, FT.

13. Precision and Bias

13.1 *Precision:*[8]

13.1.1 The standard deviations of individual determinations are given in Table 1:

13.1.2 The precision estimates in Table 1 apply to both reducing and oxidizing atmospheres. The standard deviation for the 1049–1177°C (1950–2150°F) range include initial deformation (IT), softening temperature (ST), and hemispherical temperature (HT). The standard deviations for the 1204–1260°C (2200–2300°F) range cover fluid temperature (FT).

TABLE 1 Standard Deviations of Individual Determinations

Temperature Range		Standard Deviation			
°C	(°F)	Within-Laboratory		Between-Laboratories	
		Sw		Sb	
1049–1177	(1920–2150)	8.9°C	(16°F)	36.7°C	(66°F)
1204–1260	(2200–2300)	11.0°C	(20°F)	92.2°C	(166°F)

13.1.3 The precision estimates in Table 1 are based on an interlaboratory study conducted in accordance with Practice E 180.

13.2 *Bias*—The bias of this test method has not been determined.

[8] Supporting data describing the preliminary testing and round-robin tests on which precision estimates are based is available on loan from ASTM Headquarters. Request RR: E38 – 1000.

Designation: E 954 – 94

Standard Practice for Packaging and Shipping of Laboratory Samples of Refuse-Derived Fuel[1]

This standard is issued under the fixed designation E 954; the number immediately following the designation indicates the year of original adoption or, in the case of revision, the year of last revision. A number in parentheses indicates the year of last reapproval. A superscript epsilon (ε) indicates an editorial change since the last revision or reapproval.

1. Scope

1.1 This practice covers a procedure for packaging a refuse-derived fuel sample at its point of origin for shipping this sample to the laboratory for subsequent analyses.[2]

1.2 This practice may be applicable to any waste material from which a laboratory analysis sample can be prepared.

1.3 *This standard does not purport to address all of the safety concerns, if any, associated with its use. It is the responsibility of the user of this standard to establish appropriate safety and health practices and determine the applicability of regulatory limitations prior to use.* Additional hazard information is found in Section 6.

2. Terminology

2.1 *Definitions:*

2.1.1 *gross sample*—a sample representing one lot and composed of a number of increments on which neither reduction nor division has been performed.

2.1.2 *laboratory sample*—a representative portion of the gross sample (or lot) received by the laboratory for analysis.

2.1.3 *lot*—a large designated quantity (greater than the quantity of the final sample) of RDF which can be represented by a properly selected gross sample.

2.1.4 *forms of refuse-derived fuel (RDF)*:

RDF-1 Waste used as a fuel in as-discarded form.

RDF-2—Waste processed to coarse particle size with or without ferrous metal separation.

RDF–3—shredded fuel derived from municipal solid waste (MSW) which has been processed to remove metal, glass, and other inorganics. This material has a particle size such that 95 weight % passes through a 2-in. square mesh screen.

RDF-4—Combustible waste processed into powder form, 95 weight % passing a 10-mesh screen.

RDF-5—Combustible waste densified (compressed) into the form of pellets, slugs, subettes, and briquettes.

RDF-6—Combustible waste processed into liquid fuel.

RDF-7—Combustible waste processed into gaseous fuel.

3. Summary of Practice

3.1 A representative portion of a gross sample of RDF is packaged in such a manner that all physical and chemical properties of the material are retained during shipment to the laboratory for analyses. Labeling and chain-of-custody procedures are also included in this practice.

4. Significance and Use

4.1 The packaging of a laboratory sample as specified herein is intended to ensure that the physical and chemical characteristics of the sample as received in the laboratory are not changed during shipment.

5. Apparatus

5.1 *Polyethylene Bags*, 3 mL, having a capacity to hold 2 kg (1 to 2 ft^3) of RDF-3.

5.2 *Polyethylene Bag Seal*, twist tie or equivalent.

5.3 *Box*, corrugated.

5.4 *Box Seal*, strapping tape or equivalent.

6. Hazards

6.1 Due to the origins of RDF in municipal waste, common sense dictates that some precautions should be observed when handling samples. Recommended hygienic practices include use of gloves when handling RDF; wearing a dust mask (NIOSH—approved type), especially when shredding samples; conducting tests under a negative pressure hood when possible; and washing hands before eating and smoking.

6.2 Sample handling shall be performed by trained personnel. All operations shall be done as rapidly as possible to avoid sample moisture changes due to atmospheric exposure.

6.3 At all times, RDF samples should be protected from moisture, sun or contact with absorbent materials.

6.4 Packaging must be done at the sampling site without delay between sampling and bagging.

7. Sampling

7.1 Refuse-Derived Fuel products are frequently nonhomogeneous. For this reason, significant care should be exercised to obtain a representative laboratory sample from the RDF lot to be characterized.

7.2 The sampling method should be based on an agreement between involved parties.

---

[1] This practice is under the jurisdiction of ASTM Committee D-34 on Waste Waste Management and is the direct responsibility of Subcommittee D34.03on Treatment.

Current edition approved Aug. 15,1994. Published October 1994. Originally published as E 954 – 83. Last previous edition E 954 – 88.

[2] Supporting data for this test method are available on loan from ASTM Headquarters, 1916 Race St., Philadelphia, PA 19103. Request RR: E38-1001.

8. Packaging and Package Marking

8.1 Place the sample as it is obtained into a preweighed 3-mil polyethylene bag. Secure the bag with a seal to prevent moisture loss.

8.2 Place the polyethylene bag containing the sample in 8.1 into a second preweighed 3-mil polyethylene bag. Secure the bag with a seal to prevent moisture loss.

8.3 Weigh the double-bagged sample, and record the weight of the bags and sample on a chain-of-custody tag. Attach the tag to the outside of the sample bag (see Fig. 1).

$$D = C - (A + B) \tag{1}$$

where:

A = weight in grams of polyethylene bag including seal in 8.1,

B = weight in grams of polyethylene bag including seal in 8.2,

C = weight in grams of sample, polyethylene bags, and seals in 8.3, and

D = weight in grams of sample as shipped.

8.4 Place the double-bagged sample into a corrugated container, seal the container, and ship to the testing laboratory. Any appropriate means of shipment may be used.

8.5 A chain-of-custody tag shall accompany the sample (see Fig. 1).

Client ____________ Sampler ____________
Sample Identification ____________
Date sampled ________ Date packaged ________ weight of gross sample ____________
Weight of sample, bags, and seals as shipped ____________
Weight of bags and seals ____________
Weight of sample as shipped ____________
Weight of samples as received ____________

Analyses requested ____________
____

Method of shipment ____________
____

Relinquished by ________ Received by ________ Date ________ Time ________
Relinquished by ________ Received by ________ Date ________ Time ________
Relinquished by ________ Received by ________ Date ________ Time ________
Other Information ____________
____
____

NOTE—Remove tag before weighing.

FIG. 1 Typical Chain-of-Custody Tag

Standard Test Method for Thermal Characteristics of Refuse-Derived Fuel Macrosamples[1]

This standard is issued under the fixed designation E 955; the number immediately following the designation indicates the year of original adoption or, in the case of revision, the year of last revision. A number in parentheses indicates the year of last reapproval. A superscript epsilon (ε) indicates an editorial change since the last revision or reapproval.

1. Scope

1.1 This test method covers the determination of moisture, noncombustibles and combustibles, and the calculation of higher heating value content of a large mass of refuse-derived fuel-three (RDF).

1.2 This test method may be applicable to any waste material, including residues from combustion, from which a representative sample can be prepared.

1.3 The values stated in SI units are to be regarded as the standard. The values given in parentheses are for information only.

1.4 *This standard does not purport to address all of the safety concerns, if any, associated with its use. It is the responsibility of the user of this standard to establish appropriate safety and health practices and determine the applicability of regulatory limitations prior to use.* See Section 7 for additional hazard information.

2. Referenced Documents

2.1 *ASTM Standards:*

E 711 Test Method for Gross Calorific Value of Refuse-Derived Fuel by the Bomb Calorimeter[2]

E 791 Test Method for Calculating Refuse-Derived Fuel Analysis Data from As-Determined to Different Bases[2]

3. Terminology

3.1 *Definitions of Terms Specific to This Standard:*

3.1.1 *combustibles*—that portion of the RDF sample which is consumed upon ignition exclusive of the moisture present in the sample.

3.1.2 *macrosample*—a representative sample in the order of 1 kg mass is used to determine moisture, combustible, and noncombustible content without further processing or size reduction.

3.1.3 *noncombustibles*—that fraction of a macrosample remaining after moisture and combustibles are driven off by heat and combustion. It is composed of metallic and glass particles in addition to the residue from combustion of organic substances.

3.1.4 *forms of refuse-derived fuel (RDF):* —

RDF-1—Wastes used as a fuel in as-discarded form.

RDF-2—Wastes processed to coarse particle size with or without ferrous metal separation.

RDF-3—shredded fuel derived from municipal solid waste (MSW) that has been processed to remove metal, glass, and other inorganics. This material has a particle size such that 95 weight % passes through a 2-in. (50 mm) square mesh screen.

RDF-4—Combustible waste processed into powder form, 95 weight % passing 10-mesh screening.

RDF-5—Combustible waste densified (compressed) into the form of pellets, slugs, cubettes, or briquettes.

RDF-6—Combustible waste processed into liquid fuel.

RDF-7—Combustible waste processed into gaseous fuel.

4. Summary of Test Method

4.1 A macrosample of RDF is dried and ashed successively. The moisture, combustibles, and noncombustibles content are determined gravimetrically.

4.2 Heating value of a macrosample of RDF is calculated using an established moisture and noncombustible free heating value.

4.2.1 Normal practice is for contracting practices to rate RDF on a higher heating value basis.

4.2.2 If contracting parties choose to rate RDF on a lower heating value basis, provision is made using an established moisture and non-combustible free lower heating value.

5. Significance and Use

5.1 This test method is available to producers and users of RDF as a means of determining thermal characteristics of a large sample of RDF without extensive processing of the laboratory sample. It is intended that the results obtained be used to monitor changes in the fuel characteristics of RDF over a period of time.

6. Apparatus

6.1 *Oven*—A large chamber mechanical draft oven (approximately 508 by 508 by 508 mm (20 by 20 by 20 in.) inside dimensions) capable of maintaining a controlled temperature between 100 and 500°C may be used. A minimum of one air

---

[1] This test method is under the jurisdiction of ASTM Committee D-34 on Waste Management and is the direct responsibility of Subcommittee D34.08 on Thermal Treatment.

Current edition approved March 25, 1988. Published May 1988. Originally published as E 955-83. Last previous edition E 955-83.

[2] *Annual Book of ASTM Standards*, Vol 11.04.

change per minute is satisfactory. Air flow should be baffled to prevent any sample loss due to air currents.

NOTE 1—A home electric oven with self-cleaning mode of operation, modified with an air inlet has been found satisfactory.[3]

6.2 *Balance*, having a sensitivity of 0.5 g and a capacity of at least 2000 g.

6.3 *Sample Container*—A noncorroding pan (stainless steel or aluminum) approximately 15 dm^3(0.5 ft^3).

NOTE 2—Disposable aluminum roasting pans (381 × 508 × 82.6 mm) (12 by 20 by 3¼)) have been suitable for this purpose. (Two pans will probably be required for a 1 kg sample).[3]

7. Hazards

7.1 Due to the origins of RDF in municipal waste, common sense dictates that some precautions should be observed when conducting tests on samples. Recommended hygienic practices include use of gloves when handling RDF; wearing dust masks (NIOSH approved type); conducting tests under negative pressure hood; and washing hands before eating or smoking.

8. Sampling

8.1 RDF products are heterogeneous. For this reason, significant care should be exercised to obtain a representative laboratory sample from the RDF lot to be characterized.

8.2 The sampling method for this procedure should be based on agreements between both involved parties.[4]

9. Procedure

9.1 *Moisture Determination*:

9.1.1 Place a sample of approximately 1 kg into a tared container(s) weighed to the nearest 0.5 g. A maximum depth of 10 cm is recommended. Quickly weigh the container, and sample to the nearest 0.5 g.

9.1.2 Place the container and sample in the oven at 105°C (220°F) until the sample has attained constant weight. (Constant weight is reached when the loss in sample weight is less than 0.1 % per hour of the original sample weight).

NOTE 3—After several runs, a period of time can be established for the material to reach constant weight. Overnight drying has been found to be convenient and sufficient for some RDF products.[3]

9.1.3 After a sufficient drying time, remove the samples from the oven and immediately weigh while hot to the nearest 0.5 g.

NOTE 4—The practicality of this procedure step is demonstrated in Footnote 3.

9.2 *Noncombustible Determination*:

9.2.1 Place the container with the dried sample into a laboratory hood, and ignite the sample with a flame. Stirring is required to maintain the flame. Take care so no sample is lost when stirring.

NOTE 5—**Caution:** Because of the unknown nature of the material, use caution during ignition and stirring.

9.2.2 When a flame can no longer be maintained, place the container with the partially combusted sample into the oven at 500°C until constant weight is reached.

NOTE 6—It may be determined that constant weight can be routinely established by allowing samples to ash at the prescribed temperature for a set period of time. (Two hours have been found to be sufficient).

NOTE 7—A home electric oven in the self cleaning mode modified to permit air circulation has been found to achieve sufficient ashing during a 3 h clean cycle.[3]

9.2.3 Remove the container and noncombustibles from the oven. As soon as the container is cool enough to handle, weigh to the nearest 0.5 g.

10. Calculation

10.1 Calculate the percent moisture as follows:

$$A = \frac{B - C}{B} \times 100 \quad (1)$$

where:
A = mass percent moisture,
B = "as received" sample mass in grams, and
C = "dry" sample mass in grams.

Round value to nearest percent.

10.2 Calculate the percent non-combustibles as follows:

$$D = \frac{E}{B} \times 100 \quad (2)$$

where:
D = mass percent noncombustibles,
E = non-combustibles mass in grams, and
B = "as received" sample mass in grams.

Round value to nearest percent.

10.3 Calculate the percent combustibles as follows:

$$F = 100 - (A + D) \quad (3)$$

where:
F = mass percent combustibles,
A = mass percent moisture, and
D = mass percent non-combustibles.

10.4 Calculate the higher heating value as follows:

$$G = \frac{F \times H}{100} \text{ or } G = \frac{[100 - (A + D)]H}{100} \quad (4)$$

where:
G = higher heating value "as received,"
F = mass percent combustibles,
H = higher heating value, moisture and non-combustibles free, in MJ/kg (Btu/lb) established by the contracting parties,
A = mass percent moisture, and
D = mass percent non-combustibles free.

NOTE 8—Round robin testing sponsored by Committee E-38 on Resource Recovery has found that the mean moisture/ash free heating value on six different samples of RDF analyzed in replicate (4 times) by as many as twelve different laboratories is 21.92 MJ/kg (9423 Btu/lb) with a

[3] Hecklinger, R. S., and Large, R. M., "Determination of the Fuel Characteristics of Refuse-Derived Fuels by Macroanalysis," *Proceedings of the Seventh Mineral Waste Utilization Symposium*, U.S. Bureau of Mines, 1980, pp. 84–90.

[4] ASTM Subcommittee E38.01 on Energy is currently in the process of developing procedures for sampling RDF-3.

standard deviation of 772 kJ/kg (332 Btu/lb).[5] The relative constancy of RDF higher heating value, moisture and ash free, is corroborated by research conducted by the National Bureau of Standards.[6] Therefore, those using this method may agree to establish a higher heating value, moisture and ash free, of 22 MJ/kg (9400 Btu/lb). As an alternative, those using this test method may agree to establish or adjust the higher heating value, moisture and ash free, for a particular plant by means of Test Methods E 711 and E 791.

10.5 *Optional Calculation*—Calculate the lower heating value as follows:

$$I = \frac{[100 - (A + D)]J}{100} - (A \times 10.3) \quad (5)$$

where:

I = lower heating value "as received,"
A = mass percent moisture,
D = mass percent non-combustibles,
J = lower heating value moisture and non-combustibles, and free, in MJ/kg (Btu/lb) established by the contacting parties.

NOTE 9—The lower heating value of 20 MJ/kg (8600 Btu/lb) is equivalent to 22 MJ/kg (9400 Btu/lb) higher heating value from Note 8.

NOTE 10—If the analysis is performed on residue from combustion of municipal solid waste, the combustibles in the residue should be largely fixed carbon. If bomb calorimetry is not available to establish a moisture and ash free heating value, an estimated moisture and ash free heating value of 28 MJ/kg (12 000 Btu/lb) may be used for residue.

11. Report

11.1 The report shall include the following RDF Analysis" as received"

Moisture	*A* %
Combustibles	*F*
Noncombustibles	*D*
Total	100 %
Either	
Higher heating value—MJ/kg (Btu/lb)	
Or	
Lower heating value—MJ/kg (Btu/lb)	

12. Precision and Bias

12.1 The precision and bias of this test method has yet to be determined.

[5] Supporting data (Kieffer, John K., "An Approach for Determining the Heating Value of Municipal Waste From its Combustible Content") are available from ASTM Headquarters, 1916 Race St., Philadelphia, PA 19103. Request RR: E-38-1002.

[6] Kirklin, D. R., et al, "The Variability of Municipal Solid Waste and Its Relationship to the Determination of the Calorific Value of Refuse-Derived Fuels," *Resources and Conservation*, Vol 9, 1982, pp. 281–300.

Designation: E 956 – 83 (Reapproved 1999)

Standard Classification for Municipal-Mixed Nonferrous Metals (MNM)[1]

This standard is issued under the fixed designation E 956; the number immediately following the designation indicates the year of original adoption or, in the case of revision, the year of last revision. A number in parentheses indicates the year of last reapproval. A superscript epsilon (ε) indicates an editorial change since the last revision or reapproval.

1. Scope

1.1 This classification covers municipal mixed nonferrous metals (MNM), not source-separated, that are recovered from municipal waste destined for disposal.

1.2 The mixed nonferrous metals (MNM) have been subdivided according to processing history, nonferrous metal content, size, and moisture content.

1.3 The values stated in inch-pound units are to be regarded as the standard. The values given in parentheses are for information only.

1.4 *This standard does not purport to address all of the safety concerns, if any, associated with its use. It is the responsibility of the user of this standard to establish appropriate safety and health practices and determine the applicability of regulatory limitations prior to use.*

2. Referenced Documents

2.1 *ASTM Standards:*

D 2013 Method for Preparing Coal Samples for Analysis[2]

E 11 Specification for Wire-Cloth Sieves for Testing Purposes[3]

E 34 Test Methods for Chemical Analysis of Aluminum and Aluminum Base Alloys[4]

E 35 Test Methods for Chemical Analysis of Magnesium and Magnesium Alloys[4]

E 47 Test Methods for Chemical Analysis of Zinc Die-Casting Alloys[4]

E 54 Test Methods for Chemical Analysis of Special Brasses and Bronzes[4]

E 101 Test Method for Spectrographic Analysis of Aluminum and Aluminum Alloys By the Point-to-Plane Technique[5]

E 122 Practice for Choice of Sample Size to Estimate a Measure of Quality for a Lot or Process[3]

E 227 Test Method for Optical Emission Spectrometric Analysis of Aluminum and Aluminum Alloys By the Point-to-Plane Technique[5]

E 276 Test Method for Particle Size or Screen Analysis at No. 4 (4.75-mm) Sieve and Finer for Metal Bearing Ores and Related Materials[4]

E 478 Test Methods for Chemical Analysis of Copper Alloys[4]

E 753 Specification for Municipal Aluminum Scrap (MAS)[6]

3. Terminology

3.1 *Definitions of Terms Specific to This Standard:*

3.1.1 *loose combustible material (organic)*—loose combustible organics (LCO) that consist of, but are not limited to, nonmetallic materials such as paper, rags, plastic, rubber, wood, food wastes, and yard or lawn wastes, etc., which are not permanently attached to noncombustible objects. The LCOs are defined as material larger than No. 12 Mesh (U.S. Standard Sieve) as stated in Specification E 11. A determination of LCOs is best done by sampling the material and handpicking, handcleaning, and visually identifying the materials described previously.

3.1.2 *mixed nonferrous metal content*—mixed nonferrous metals remaining after removal of magnetics, combustibles, and other nonmetals (for example, glass, rock, etc.). Further methods of separation and identification may be agreed upon between purchaser and seller and can include picking or dense-media separation.

3.1.3 *moisture percent—liquid content,* as determined by weight loss when sample material is dried to a constant weight at 110° ± 5°C.

3.1.4 *total combustibles*—materials that include paints, lacquers, coatings, plastics, etc., associated with the original nonferrous products, as well as combustible materials (paper, plastic, textile, etc.) which become associated with the nonferrous product after it is manufactured.

4. Significance and Use

4.1 This classification is intended for use in the marketing of mixed nonferrous metals.

4.2 Mixed nonferrous metals covered by this classification are suitable for use by one or more of the following industries:

4.2.1 Secondary aluminum smelters,

4.2.2 Primary aluminum producers,

4.2.3 Scrap dealers and processors,

[1] This classification is under the jurisdiction of ASTM Committee D-34 on Waste Management and is the direct responsibility of Subcommittee D34.06 on Recovery and Reuse.

Current edition approved July 29, 1983. Published December 1983.

[2] *Annual Book of ASTM Standards,* Vol 05.05.

[3] *Annual Book of ASTM Standards,* Vol 14.02.

[4] *Annual Book of ASTM Standards,* Vol 03.05.

[5] *Annual Book of ASTM Standards,* Vol 03.06.

[6] *Annual Book of ASTM Standards,* Vol 11.04.

4.2.4 Zinc refiners, and
4.2.5 Copper refiners.

5. Basis of Classification (refer to Table 1)

5.1 This classification covers material, either processed to recover aluminum or not processed to recover aluminum.

5.2 This classification consists of four types of MNM, based on nonferrous metal content. The four types of MNM are further subdivided into three grades, based on size. The three grades of MNM are further subdivided into two classes, based on moisture content.

6. Ordering Information

6.1 Mixed nonferrous metals may be identified using the criteria in Table 1. This is an unusual material, and the table is included for information purposes rather than to establish limits. The unit operations used to recover MNM may aid the purchaser and seller in establishing a classification. Some of the more common unit operations used to recover MNM include the following:

6.1.1 *Wet Processing*—water elutriators, dense media, and mineral jig.

6.1.2 *Dry Processing*—air classifier, air knife, gravity or concentrating table, Humphrey Spiral, eddy-current separator, electrostatic separator, and handpicking.

6.2 It is recognized that variations in the MNM may occur due to the heterogeneous nature of the solid waste stream. The criteria indicated are intended as a means for the purchaser and the seller to establish the value and quality of the MNM.

6.3 Mixed nonferrous metals shall be considered to be of a particular classification if the value for each component specified, as obtained by the test method agreed upon between the purchaser and seller, shall not exceed any of the limits for that grade.

TABLE 1 Classification of Municipal Mixed Nonferrous Metals (MNM)[A]

	Type I	Type II	Type III	Type IV
Classification	Over 90 % MNM	50 to 90 % MNM	30 to 50 % MNM	Less Than 30 % MNM
Grades	1 over 6 in. (150 mm)	2 2 to 6 in. (50 to 150 mm) incl		3 under 2 in. (50 mm)
Class	A high (5 % moisture and over)		B low (under 5 % moisture)	

[A]It is important for both purchaser and seller to note whether material *had been* processed to recover aluminum or whether it *had not been* processed to recover aluminum and what, if any, procedures where used.

7. Physical Requirements

7.1 In addition to Table 1, the MNM physical requirements include the following:

7.1.1 *Bulk Density*—The density for MNM is not specified and shall be agreed upon between the purchaser and the seller.

7.1.2 *Fineness*—Acceptability of contained fines shall be determined by the purchaser and seller.

7.1.3 *Loose Combustibles*—As agreed upon between purchaser and seller.

7.1.4 *Magnetics*—The presence of free magnetic material is not specified and shall be as agreed upon between the purchaser and seller as part of the purchase contract.

8. Sampling

8.1 Sampling shall be in accordance with the procedures described in Annex A1 or Annex A2. Either procedure may be used, as determined by agreement between the purchaser and the seller.

8.1.1 Annex A1 covers sampling at the point of origin.

8.1.2 Annex A2 covers sampling at the point of receipt.

9. Test Methods

9.1 Determine the properties of fineness, moisture, and metal recovery in accordance with the procedures described in Annex A3.

10. Rejection and Rehearing

10.1 Material that fails to conform to the requirements of this classification may be rejected. Rejection should be reported to the seller promptly and in writing. In case of dissatisfaction with the results of the test, the seller may make claim for a rehearing.

11. Shipping

11.1 Mixed nonferrous metals shall be shipped in rail cars, trailers, or other containers as agreed upon between the purchaser and the seller. The shipping equipment shall be sufficiently water-tight to prevent the MNM from becoming wet during shipment.

ANNEXES

(Mandatory Information)

A1. TEST METHOD FOR COLLECTION OF A SAMPLE OF MNM SCRAP RECOVERED FROM MUNICIPAL SOLID WASTE AND ITS PREPARATION FOR ANALYSIS

A1.1 Scope

A1.1.1 This test method describes procedures for collection of a sample of MNM scrap recovered from municipal refuse, and the preparation and secondary sampling of the metal for analysis.

A1.2 Summary of Test Method

A1.2.1 A selected size, gross sample of MNM scrap, is taken from the metal recovery system conveyor in increments. Sample increments are taken at timed intervals from a full cross section of the conveyor while it is stopped, or by briefly taking the total flow at the discharge of the conveyor while it is moving.

A1.2.2 The quantity of gross sample may be further reduced by mixing, cone-and-quarter sampling, and riffling.

A1.3 General Precautions

A1.3.1 In solids sampling, each step must be designed to eliminate accidental classification by size or gravity. Different sizes usually have different analyses.

A1.3.2 The increments obtained during the sampling period shall be protected from changes in composition due to exposure to the weather.

A1.3.3 Plan the sampling arrangement to avoid contamination of the increments with foreign material.

A1.3.4 A satisfactory sampling arrangement is one that takes an unbiased sample at the desired degree of precision of the constituent for which the sample is to be analyzed. The weight or volume of the collected sample is compared with that of the total lot to ensure a constant sampling ratio.

A1.3.5 It is preferable that the MNM scrap be weighed and sampled at about the same time. If there is a long lapse in time between these two events, both the purchaser and seller should give consideration to changes in moisture during this interval and the consequent shift in relationship of moisture to the true content at the instant when ownership of the nonferrous metal scrap transfers from one to the other.

A1.3.6 Samples and subsamples shall be collected in such a manner that there is no unmeasured loss of moisture of significant amount. The samples shall be weighed before and after drying or other operations to measure all significant weight loss. The material balances shall be adjusted accordingly.

A1.4 Selection of Gross Sample Size

A1.4.1 Choose the gross sample size by methods given in Practice E 122, whenever practicable. The chief difficulty for implementing this practice can be that insufficient information concerning possible variation is available. This information should be gathered with practice. Due to the heterogeneity in size and type of material comprising municipal solid waste, the choice of a large sample is desirable.

A1.4.2 Shredding is recommended for metals larger than 4 to 6 in. (100 to 150 mm) for ease in sampling and handling.

A1.5 Taking A Gross Sample

A1.5.1 In order to obtain complete representation of materials in a gross sample, it is desirable that the sample increments be withdrawn from the full cross-section of the stream. The best possible increment is either a full cross section removed from a stopped conveyor belt or the total flow at the discharge of the moving conveyor taken during a suitable interval of time.

A1.5.2 The choice of sample size can be estimated using Practice E 122. It is imperative for a given degree of precision that not less than the minimum size and number of sample increments be collected from a lot (see Table A1.1).

A1.5.3 *Number of Gross Samples*—For quantities up to approximately 20 tons, it is recommended that one gross sample represent the lot. Take this sample in accordance with the requirements prescribed in Table A1.1.

A1.5.4 *Distribution of Increments*—It is essential that the increments be distributed throughout the lot to be sampled. The taking of increments shall be at regularly spaced intervals.

A1.6 Sample Preparation

A1.6.1 Cone and quarter the sample until approximately 2 ft^3 (0.06 m^3) remains. Pile the material to be sampled into a conical heap and then spread out into circular cake. Divide the cake into quarters, take two of the diagonally opposite quarters as the sample, and reject the two remaining quarters. Collect the two quarters taken as the sample and repeat the procedure of coning and quartering until the desired size is obtained.

A1.6.2 Divide the sample into approximately equal parts. Take one half 1 ft^3 (0.03 m^3) for use in the melt test (see Annex A3.). Divide the sample by riffling until the analytical sample is obtained. (Typical rifflers can be found in the apparatus section of Method D 2013.)

A1.6.3 Store the prepared analytical sample in a covered, labeled, corrosion-resistant metal can or plastic container until needed for chemical analysis.

TABLE A1.1 Number and Weight of Increments for Sampling

Top Size, in. (mm)	⅝ (15)	2 (50)	6 (150)	12 (300)
Minimum number of increments	15	15	15	15
Minimum weight of increments, lb (kg)	2 (1)	6 (3)	18 (9)	36 (18)

A2. SAMPLING AT POINT OF RECEIPT

A2.1 Sampling During Unloading

A2.1.1 *Sample Size*—Take a representative quantity of approximately 1 yd^3 (0.76 m^3) from each car or truck of MNM scrap received.

A2.1.1.1 *Car Sample*—Take two shovels (No. 2 size) of MNM scrap from the top, middle, and bottom of opened doorway area of car in two locations of doorway as shown in Fig. A2.1 to obtain six samples. Take an additional 18 samples as shown in Fig. A2.1. Place all material sampled into a suitable receptacle for the total sample from the car. Adequately identify the sample container. Weigh and record the sample from the car.

A2.1.1.2 *Truck Sample*—Take two shovels (No. 2 size) of MNM scrap from upper, middle, and lower areas of material starting at the rear of the truck. Starting at the truck rear, take samples every 8 ft (2.4 m) at locations one third the distance from the side of the truck. Place all material sampled into a suitable receptacle for the total sample of the truck. Adequately identify the sample container. Weigh and record the sample from the truck.

A2.1.2 *Reduction of Sample Size*—In the event of material larger than 4 to 6 in., (100 to 150 mm) it is recommended that it be shredded to facilitate further processing. Using riffle sampling equipment[7] or coning and quartering (or equivalent method), reduce the size of sample to approximately 1 ft^3 (0.03 m^3) by putting it through the riffle sampler five times. Retain one half of the split sample on each pass through the riffle sampler until a fine sample of approximately 1 ft^3 (0.03 m^3) is attained. Carefully bag and identify the sample for assay, and retain a duplicate sample. Retain the duplicate sample until assay is completed and accepted. If the assay is not accepted, then the duplicate sample may be used for settling the claim.

A2.1.3 *Identity Ticket*—It is recommended that the following information be included on the ticket:

A2.1.3.1 Supplier,

A2.1.3.2 Car number or truck identification,

A2.1.3.3 Net weight of car or truck,

A2.1.3.4 Date unloaded and sampled, and

A2.1.3.5 Initial sample weight.

A2.2 Off-Specification Shipments

A2.2.1 *Sampling*—If shipments are received as off-specification material or are suspected of being off-specification, sample the shipment in the doorway of the car or truck using core-sampling equipment,[7] or other acceptable procedure, from at least ten different locations prior to car unloading. Take at least 2 ft^3 (0.06 m^3) of representative sample. Weigh and record the sample weight using a scale accurate to within ±0.1 lb (0.05 kg) or ±0.05 %, whichever is the more precise.

A2.2.2 *Reduction of Sample Size*—In the event of material larger than 4 to 6 in., (100 to 150 mm) it is recommended that it be shredded to facilitate further processing. Using riffle-sampling equipment or coning and quartering (or equivalent procedure), split the sample into equal parts until 1 ft^3 (0.03 m^3) remains in each of the last split fractions. Weigh each split

[7] Core and riffle sampling equipment, available from W. S. Tyler Co., or equivalent, has been found suitable for this purpose.

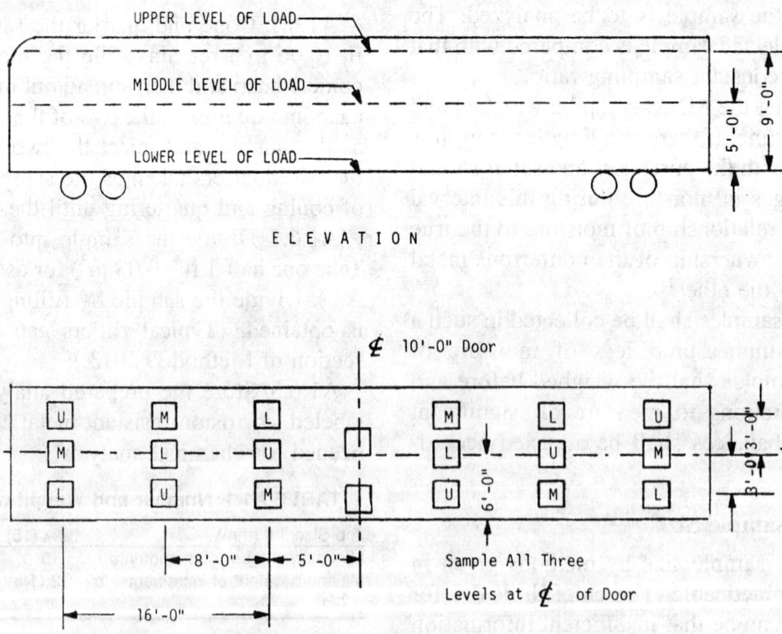

NOTE 1—All samples consist of two No. 2 shovels from each location sampled.

FIG. A2.1 Sample Location Chart for 50-ft Railway Car

fraction, identify, and retain duplicate sample until acceptance or rejection of shipment.

A2.2.3 *Identity Ticket*—It is recommended that the following information be included on the ticket:

A2.2.3.1 Supplier,

A2.2.3.2 Car number or truck identification,

A2.2.3.3 Net weight of car or truck,

A2.2.3.4 Date unloaded and sampled, and

A2.2.3.5 Initial sample weight.

A2.3 Recommended Testing Frequency

A2.3.1 *New Sources*—Test all new sources of supply on an "every shipment" basis until a total of 1 000 000 lb (450 000 kg) have been received on-specification, at which time, the supplier is considered to be an established source.

A2.3.2 *Established Sources*:

A2.3.2.1 Randomly sample shipments from established source, and test a minimum of 20 % of all shipments.

A2.3.2.2 A single shipment from any established or new source of supply that fails to meet the agreed-upon limits within 20 % of any individual factor shown in Table 1 may require that the source be treated as a new source (see A2.3.1).

A2.3.2.3 Any established or new source failing to meet the agreed-upon limits within 10 % on any two or more factors shown in Table 1 may require treatment as a new source (see A2.3.1).

A3. ASSAY AND ANALYSIS PROCEDURE

A3.1 Significance

A3.1.1 This is a laboratory procedure used to produce an analytical specimen, determine the percent of metal recovery by remelting, and determine the amount of contaminants present.

A3.2 Determination of Physical Requirements

A3.2.1 Take approximately one third of the sample, weigh, and record the weight.

A3.2.2 Dry this one-third sample for 2 h at 110°C± 5°C and record the weight of the dried material.

A3.2.3 Remove the magnetic materials and separate, record the weight of the nonmagnetic materials.

A3.2.4 Screen the sample over a 12-mesh (U.S. Standard) sieve using the Dry Screening Procedure of Test Method E 276. Record the weight of the sieved material and the material remaining on the screen. This section (A3.2.4) does not apply when the material is shredded at the receiving point.

A3.3 Determination by Chemical Analysis

A3.3.1 For material with high aluminum content, based on 4.1, refer to Specification E 753 for assay procedure.

A3.3.2 For material with low aluminum content based on 4.1, use wet analytical procedure in accordance with the ASTM standards referenced in 2.1.

Designation: E 959 – 83 (Reapproved 1999)

Standard Test Method for Characterizing the Performance of Refuse Size-Reduction Equipment[1]

This standard is issued under the fixed designation E 959; the number immediately following the designation indicates the year of original adoption or, in the case of revision, the year of last revision. A number in parentheses indicates the year of last reapproval. A superscript epsilon (ε) indicates an editorial change since the last revision or reapproval.

1. Scope

1.1 This test method covers measuring the performance of solid waste size reduction equipment.

1.2 This test method can be used to measure the flow (that is, throughput) of solid waste through the size-reduction equipment, energy usage of the size-reduction device, and particle size of the shredded product.

1.3 This test method includes instructions for measuring energy usage, solid waste throughput, net processing time, and particle size distribution.

1.4 This test method applies only to size reduction equipment that produces a shredded product with a size corresponding to 90 % cumulative passing in the range of 0.5 to 15 cm (0.2–6 in.) on an air-dry weight basis. For material with nominal sizes outside of this range, the precision and bias statements for particle size designation (Section 14) may not apply.

1.5 This test method can be applied to size reduction equipment located anywhere within a processing line.

1.6 The values stated in SI units are to be regarded as the standard. The values given in parentheses are for information only.

1.7 *This standard does not purport to address all of the safety concerns, if any, associated with its use. It is the responsibility of the user of this standard to establish appropriate safety and health practices and determine the applicability of regulatory limitations prior to use.* See Section 7 for specific hazard information.

2. Referenced Documents

2.1 *ASTM Standards:*

E 828 Test Method for Designating the Size of RDF-3 From Its Sieve Analysis[2]

E 929 Test Method for Measuring Electrical Energy Requirements of Processing Equipment[2]

3. Terminology *Definitions:*

3.1 *characteristic product size*—the screen size corresponding to 63.2 % cumulative passing by weight.

3.2 *discrete throughput method*—the method whereby average throughput is calculated as the average of a number of discrete throughput measurements conducted during a test period.

3.3 *idling time*— time periods during which a size reduction device is freewheeling, that is, not processing refuse.

3.4 *net processing time*—the time during which refuse is processed through the size reduction device.

3.5 *nominal product size*—the screen size corresponding to 90 % cumulative passing by weight.

3.6 *size reduction device or equipment*—a device which size reduces (Synonyms: shredder, grinder, pulverizer, and mill).

3.7 *stationary belt method*—a method of gross sample collection in which the conveyor belt is stopped and the sample of material is removed manually.

3.8 *time-averaged throughput method*—the method whereby the average throughput is calculated by dividing the total mass size reduced by the net processing time.

3.9 *test interval*— a test interval is equal to one-quarter of the test period.

3.10 *test period*— the test period is two to four continuous h of net-processing time.

4. Summary of Test Method

4.1 The duration of the test period is established and refuse is prepared for processing.

4.2 An energy measuring system is installed.

4.3 Solid waste is processed through the size reduction equipment, energy usage and throughput is measured, and samples for analysis of product particle size distribution are collected.

4.4 Average throughput, power requirements, specific energy, and particle size of the shredded product are calculated.

4.5 Two methods (Time-Averaged Throughput Method and Discrete Throughput Method [Section 10]) for measuring the performance of size reduction equipment are described. The selection of a particular method is governed by the layout of the processing equipment, the location of the size-reduction equipment relative to the other processing equipment, and the preference of the parties conducting the test.

5. Significance and Use

5.1 Throughput, power and energy requirements, and product size are key parameters that describe the operation and

[1] This test method is under the jurisdiction of ASTM Committee D-34 on Waste Management and is the direct responsibility of Subcommittee D34.06 on Recovery and Reuse.

Current edition approved Aug. 26, 1983. Published February 1984.

[2] *Annual Book of ASTM Standards*, Vol 11.04.

performance of solid waste size-reduction equipment.

5.2 This test method can be used to determine if the size-reduction equipment is operating within specifications and meeting performance criteria.

5.3 Having determined the parameters given in 5.1, the equipment that has been subjected to the test may be compared to other equipment similarly tested in order to establish relative levels of performance among equipment.

5.4 The basic test period is a continuous two to four h duration. The use of several test periods may be warranted to assess adequately the performance of size reduction equipment.

6. Apparatus

6.1 *Hand Broom.*

6.2 *Dust Pan.*

6.3 *Wide-mouthed Shovel.*

6.4 *Clock or Stopwatch*, accurate to 0.1 s.

6.5 *Plastic Bags*, large containers, or both.

6.6 *Push-broom.*

6.7 *Ties and Labels.*

6.8 *Electrical Metering System.*

6.9 *Sieving Equipment*, manual or mechanical.

7. Hazards

7.1 The test procedure described in 11.4 requires the removal of shredded material from a stopped conveyor belt by test personnel. Precautions should be taken to ensure that the belt cannot be started while occupied. These precautions consist of lockout of the electrical power to the conveyor, ready access to a safety "stop" cord located on the conveyor, or both.

7.2 This test method requires installation of electrical metering equipment. Consequently, the precautions described in Test Method E 929 should be observed.

7.3 Gross samples should be collected sufficiently far from the size reduction equipment such that test personnel are protected from potential explosions and flying objects from the equipment.

8. Equipment Calibration

8.1 All electrical metering equipment used for energy measurement shall be calibrated in accordance with Test Method E 929.

8.2 All weight-measuring equipment shall be calibrated according to the manufacturer's instructions.

9. Preparation for Test

9.1 *Refuse Preparation and Establishment of Test Intervals*—The duration of the test period is to be a minimum of 2 h and a maximum of 4 of net-processing time. During the test period, collect four gross samples of shredded product from which subsamples for particle size distribution analysis will be taken subsequently. The test period is divided into four equal test intervals (that is, test intervals 1, 2, 3, and 4). Calculate the approximate duration of the test intervals using the following relation:

$$t_i^* \simeq \frac{t_p^*}{4} \quad (1)$$

where:

t_p^* = estimate of the duration of the test interval (h), and

t_p^* = estimated duration of the test period (h), subject to the condition $2\ h \leq t_p^* \leq 4\ h$.

Weigh refuse, uniformly mixed as much as possible, and form into four discrete piles, each of which has an approximate (nominal) weight as calculated by the following relation:

$$M_i^* \simeq \frac{\dot{m}^* t_i^*}{4} \quad (2)$$

where:

M_i^* = approximate weight of the refuse pile in Mg,

$\dot{m}^*$ = nominal throughput value (Mg/h) established for the test, and

t_i^* = estimated duration of the test interval (h) derived from Eq 1.

The measured weight of each pile (M_i) is to be within ± 5 % of the nominal weight (M_i^*). Record the weight of each pile on the sample data form shown in Fig. 1.

9.2 *Time Measurements and Logbook*—Keep a time log during the conduct of the test program, the primary purpose of which is to allow the calculation of net-processing time. A sample format for the log is shown in Fig. 2.

9.2.1 The key time recordings for each time interval are as follows:

9.2.1.1 Starting time of the time interval,

9.2.1.2 Starting time of idling periods in which the size reduction device is electrically energized but in which no size reduction of refuse is occurring,

Pile No.	Weight of Pile, M_i (Mg)
1	
2	
4	
Total, M:	
5	
6	
7	
8	
Total, M:	
9	
10	
11	
12	
Total, M:	
13	
14	
15	
16	
Total, M:	
17	
18	
19	
20	
Total, M:	

FIG. 1 Sample Data Sheet for Throughput Measurement Using the Time-Averaged Throughput Method

Site: ______________________ Date: ______________________

Type of Size Reduction Device: ______________________ Recorded By: ______________________

Model No.: ______________________ Test Period No.: ______________________

Serial No.: ______________________ Test Interval No.: ______________________

Time	Description of Activity/Reason for Shutdown	(A) Shredding Time, Δt_s(h)	(B) Idling Time, Δt_x^A (h)	(C) Shredder Shutdown, Δt_y^B (h)
Totals				

^{A} Power on to size reduction equipment, but no processing of material.
^{B} Power off to size reduction equipment.

FIG. 2 Time Log for Testing Size Reduction Equipment

9.2.1.3 Starting time of any periods in which the size reduction device is electrically shut down (de-energized), and

9.2.1.4 Finishing time of the test interval.

9.2.2 In order to obtain representative test data, it is recommended that the net-processing time be a minimum of 75 % of the duration of the test period. For example, if a four-h test period is chosen, the net processing time should be equal to or greater than three h.

9.3 *Setup and Use of the Energy Measuring Equipment*—Measure energy usage of the size reduction device during the test period using Test Method E 929. Use a rotating disk-type wattmeter or equivalent as the measuring instrument. Install and test the energy measuring equipment prior to initiating the test period.

10. Time-Averaged Throughput Method Procedure:

10.1 The Time-Averaged Throughput Method may be used in those instances where there is no stream-splitting apparatus upstream of the size-reduction device, for example, there is no pre-trommel screen upstream of the size-reduction device.

10.2 After an initial one-half hour warmup period during which refuse is shredded and the power measuring equipment is functioning, allow the shreading device to empty. Subsequent to its emptying, measure the initial freewheeling power draw while the machine is idling using a rotating disk-type wattmeter, as described in Test Method E 929. Record measurements in accordance with Fig. 2, Energy Measurement Data Sheet of Test Method E 929.

10.3 After completion of the initial freewheeling power measurements and at the onset of the first time interval, note the starting time and record on the time log (Fig. 2). Record the initial wattmeter reading in accordance with Fig. 2, Energy Measurement Data Sheet of Test Method E 929. Simultaneously, initiate the processing of one of the four pre-weighed piles of refuse. Make every reasonable effort to supply a constant flow of refuse into the size-reduction device. The measured duration of the test interval is to be within ±10 % of that estimated for t_i^* in Eq 1.

10.4 Approximately midway through the first time interval of the test period, collect a representative gross sample for product particle size analysis downstream of the shredder discharge. The appropriate weight of the gross sample is as indicated in Fig. 3.

10.5 The preferred method of collection for the gross sample is through diversion of the entire cross section of the shredded refuse stream into a collection container or through collection of the entire cross section of the stream at a conveying transition point. Where neither of the two preferred methods of collection can be employed, the collection of a partial stream sample may be substituted. If partial stream sampling is used, make a notation on the data sheet used for recording the weights of gross and laboratory samples (Fig. 4) and the data sheets used to record particle size distribution data (Fig. 5).

10.6 Weigh the gross sample and store in a waterproof container or bag until the representative laboratory samples are chosen. Record weight data on the data sheet shown in Fig. 4.

10.7 At the conclusion of the time interval, note the time and record the reading on the time log.

10.8 Collect the second, third, and fourth gross samples for product particle size analysis approximately midway into the second, third, and fourth test intervals, respectively, of the test period, following the procedures in 10.4 through 10.7 . Note and record on the time logs the starting times of the subsequent test intervals.

10.9 Following the fourth time interval and immediately upon size reducing the remainder of the fourth pile of refuse, note the time and record the final wattmeter reading on Table X4 of Test Method E 929. After the final wattmeter reading has been noted, measure the final freewheeling power draw and record the data in Fig. 2, Energy Measurement Data Sheet of Test Method E 929.

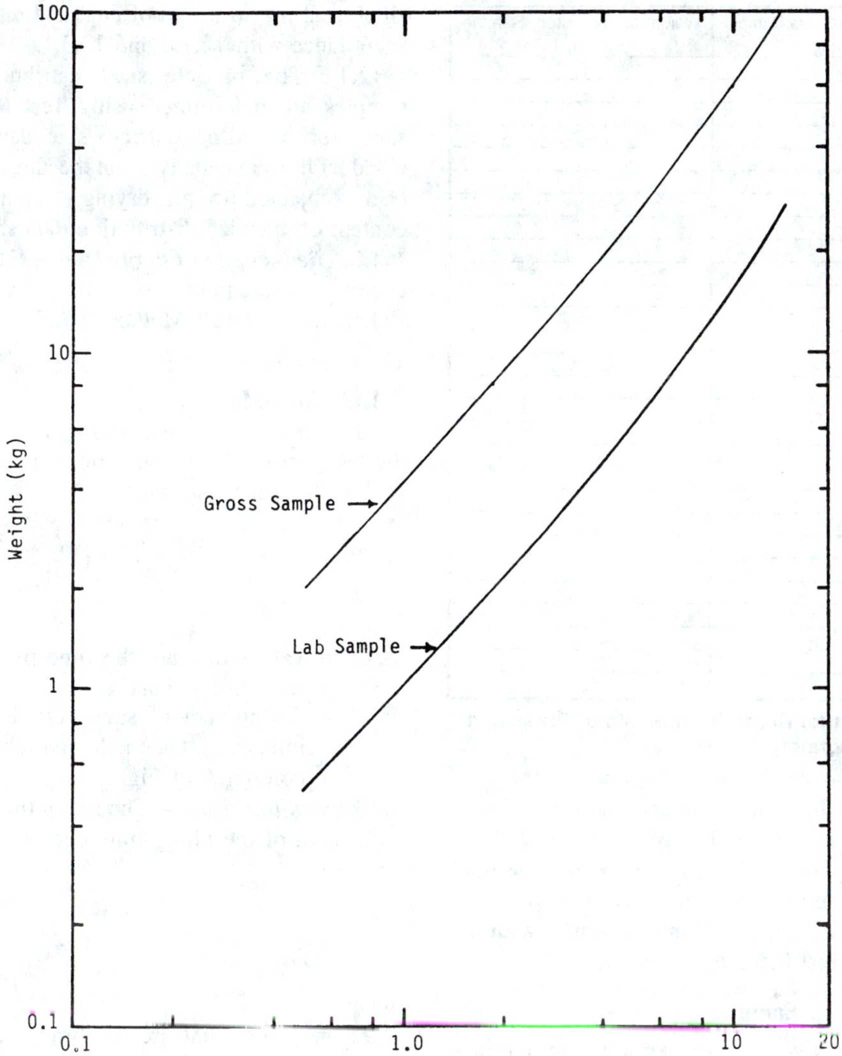

FIG. 3 Weight Requirements for Gross and Lab Samples as a Function of Nominal Product Size

11. Discrete Throughput Method Procedure

11.1 The Discrete Throughput Method is used in those instances where splitting of the raw refuse stream occurs prior to its entering the size-reduction device, for example, in those systems where a pre-trommel screen is located upstream of the size-reduction device.

11.2 Follow the procedures in 10.2 through 10.4 . The preferred method of collection of the gross sample is by diversion of the entire cross section of the shredded refuse stream into a collection container or through collection of the entire cross section of the stream in free fall at a conveying transition point. Use a stopwatch to measure the time during which the throughput sample is being collected.

11.3 Weigh and store the gross sample in a waterproof container or bag until the representative laboratory samples are chosen. Record the weight of the throughput sample and the elapsed time of sample collection on the data sheet shown in Fig. 6.

11.4 Where neither of the methods of 11.2 (that is, diversion of the entire cross section of the process stream) can be employed, collect throughput samples from a suitable length of conveyor belt downstream of the shredder discharge, using the Stationary Belt Method. Simultaneously stop both the shredder infeed conveyor and the belt from which the throughput sample is to be taken. After the conveyors are stopped, collect and remove the shredded material from a measured length of the belt. The weight of material to be removed will be approximately as indicated in Fig. 3.

11.5 Immediately prior to stopping the belt for the purpose of collecting the throughput sample, note the time and record the reading on the time log, Fig. 2.

11.6 Weigh the gross sample and store in a waterproof container or bag until the representative laboratory samples are chosen. Record the weights and conveyor information on the data sheet shown in Fig. 7.

11.7 After removal of the gross sample, start the conveyor and begin shredding refuse. Note the time and record the reading on the time log.

11.8 Repeat the procedures in 11.5 through 11.7 for the second, third, and fourth test intervals.

11.9 Immediately at the conclusion of the fourth test interval, note the final time reading and record it on the time log. In

Test Interval No.	Weight of Gross Sample (kg)	Weight of Laboratory Sample (kg)
1		
2		
3		
4		
5		
6		
7		
8		
9		
10		
11		
12		
13		
14		
15		
16		
17		
18		
19		
20		
21		
22		
23		
24		
25		
26		
27		
28		

FIG. 4 Sample Data Sheet for Recording Weights of Gross and Laboratory Samples

addition, note and record the final wattmeter reading in Fig. 2, Energy Measurement Data Sheet of Test Method E 929. After the final meter reading is recorded, make and record the final freewheeling power measurements in Fig. 2, Energy Measurement Data Sheet of Test Method E 929 in accordance with the procedures described in Test Method E 929.

12. Analyzing Laboratory Samples

12.1 Take the laboratory samples for particle size determinations from the gross samples using the following procedures for cone-and-quartering of the material:

12.1.1 Empty the contents of the container or bag containing the gross sample onto a clean, smooth, and level surface.

12.1.2 Using a wide-mouthed shovel, form the gross sample into a symmetrical cone, uniformly mixing the material as the cone is formed.

12.1.3 Using the blade of the shovel, carefully partition the cone of material into one-quarter segments. Use a vertical as well as a sideways motion of the blade to promote the separation of the one-quarter segments. Cut the cone completely to the bottom of the pile.

12.1.4 Select two one-quarter segments that are 180° opposite each other, weigh and bag each in a waterproof bag, and label them. In collecting the one-quarter segments, take care to gather all of the material, including dirt and glass fines. The weight of the laboratory samples should be approximately as shown in Fig. 3.

12.1.5 Two laboratory samples (that is, twin samples) are subsampled from each gross sample. Subject at least one representative laboratory sample from each test interval to the procedures for particle size analysis. The twin laboratory sample may also be analyzed for particle size distribution. Subject all laboratory samples to air drying to constant weight, label, seal in a waterproof bag, and retain for later analysis in accordance with 12.1.6 and 13.3.1.

12.1.6 The particle size distribution of the laboratory samples are determined using Test Method E 828. A data sheet for recording particle size data is shown in Fig. 5. Conduct the sieve analysis on the shredded material after it has been subjected to air drying. Report the air-dry moisture content on the size distribution data sheet.

12.2 *Reduction of Energy Data*—Calculate the results of the energy measurements using the procedures in the section on Calculations of Test Method E 929.

13. Calculation

13.1 *Time Measurements*:

13.1.1 *Net-Processing Time*—The net-processing time for the test period (T_n) is the sum of the net-processing times for each of the four test intervals:

$$T_n = \sum_{j=1}^{4} \left(\sum_{i=1}^{n} (\Delta t_s)_i \right)_j \quad (3)$$

where:

$\Delta\, t_s$ = values that are the time periods during which size reduction occurs, and

n = the number of such periods during any given test interval, j. Calculate and tabulate the $\Delta\, t_s$ values in column A of Fig. 2.

13.1.2 *Idling Time*— The idling time for the test period (T_x) is the sum of the idling time periods for each of the four test intervals,

$$T_x = \sum_{j=1}^{4} \left(\sum_{i=1}^{n} (\Delta t_x)_i \right)_j \quad (4)$$

where:

$\Delta\, t_x$ = values that are the time periods during which the size reduction device is idling (that is freewheeling), and

n = the number of such periods during any given test interval, j. The Δt_x values are calculated and tabulated in column B of Fig. 2.

13.2 *Throughput*:

13.2.1 *Time-Averaged Throughput Method* (to be used with Section 10): Compute the average throughput ($\bar{\dot{m}}$) using the following relation:

$$\bar{\dot{m}} = \frac{M}{T_n} \quad (5)$$

where:

M = total as-received weight of the refuse processed during the test period, Mg and

T = the net-processing time of the test period in hours.

13.2.2 *Discrete Throughput Method* (to be used with Section 11)— Compute the average throughput ($\bar{\dot{m}}$) for the test period using the following relation:

$$\bar{\dot{m}} = \frac{\sum_{i=1}^{4} \bar{\dot{m}}_i}{4} \quad (6)$$

13.3 *Particle Size Distribution*:

13.3.1 Plot the particle size distribution data for each

Size Distribution Data Sheet

Date: ______ Sample Wet Weight: ______

Site: ______ Sample Dry Weight: ______

Test No.: ______ Water Content: ______

Grinder: ______ Moisture Content: ______

Material: ______ Screening Time: ______

Screen Size	Bottom Screen Size ( )	Gross Weight Retained by Bottom Screen ( )	Tare Weight ( )	Net Weight Retained by Bottom Screen ( )	% of Feed on Bottom Screen	Cumulative Weight % Passing Bottom Screen
Pan						

Total Sample Weight:

NOTES: ______

FIG. 5 Size Distribution Data Sheet

Test Interval No.	Weight of Throughput Sample, m (kg)	Collection Time, t_c (s)	Calculated Throughput, $\dot{m}_i^A$ (Mg/h)
1			
2			
3			
4			

Average Throughput $(\bar{\dot{m}})^B$ = ______

$^A\ \dot{m}_i = 3.6\frac{m}{t_c}$

$^B\ \bar{\dot{m}} = \frac{\sum_{i=1}^{4}\dot{m}_i}{4}$

FIG. 6 Sample Data Sheet for Throughput Measurements Using Procedure 9.3.2

laboratory sample plotted on Rosin-Rammler coordinates (Fig. 8). Any or all of the twin laboratory samples (see 12.1.5) may be screened and used as additional data. Draw a smooth curve for each of the particle size distributions.

13.3.2 Determine the nominal and characteristic product sizes (corresponding to 90 % and 63.2 % cumulative % passing, respectively) from the Rosin-Rammler plots for each sample and record on the sample data sheet shown in Fig. 9. Calculate the average nominal size ($\bar{X}_{90}$) in centimetres using the following relation:

$$\bar{X}_{90} = \frac{1}{n}\sum_{i=1}^{n}(X_{90})_i \tag{7}$$

Calculate the average characteristic size ($\bar{X}_o$), cm, using the following relation:

$$\bar{X}_o = \frac{1}{n}\sum_{i=1}^{n}(X_o)i \tag{8}$$

13.4 *Energy Usage*:

13.4.1 *Gross Energy Usage*—Calculate the gross energy (E_g) in kWh used for size reduction during the test period using the following relation:

$$E_g = E_{tot} - \bar{P}_{fw}T_x \tag{9}$$

where:

E_{tot} = the energy measured by the wattmeter (kWh),
$\bar{P}_{fw}$ = the average freewheeling power draw (kW), and
T_x = the idling time expressed in hours.

Calculate the average freewheeling power draw, $\bar{P}_{fw}$, in accordance with the procedures given in the section on Calculations of Test Method E 929.

13.4.2 *Net Energy Usage*— Calculate the net energy (E_n) in kWh used the size reduction during the test period using the following relation:

$$E_n = E_g - \bar{P}_{fw}T_n \tag{10}$$

13.5 *Power Requirements*:

13.5.1 *Gross Power Requirements*—Calculate the gross average power requirements ($\bar{P}_g$) of the size reduction device in kW as the quotient of the gross energy (E_g) measured during the test period and the net-processing time (T_n):

$$\bar{P}_g = \frac{E_g}{T_n} \tag{11}$$

The units of E_g and T_n are kW and h, respectively.

13.5.2 *Average Net Power Requirement*—Calculate the average net power requirement ($\bar{P}_n$) of the size reduction device in kW as follows:

$$\bar{P}_n = \frac{E_n}{T_n} \tag{12}$$

The units of E_n and T_n are kWh and h, respectively.

Test Interval No.	Weight Gross Sample, m (kg)	Length of Conveyor Belt Section (l) (m)	Belt Speed of the Conveyor (s) (m/s)	Calculated Throughput ($\dot{m}_i$)^A (Mg/h)
1				
2				
3				
4				

Average Throughput $(\bar{\dot{m}})^B$ = ______

^{A} $\dot{m} = 3.6 \frac{ms}{l}$

^{B} $\bar{\dot{m}} = \frac{1}{4} \sum_{i=1}^{4} \dot{m}_i$

FIG. 7 Measured Parameters for Gross Samples Collected Using the Stopped Belt Method (11.4)

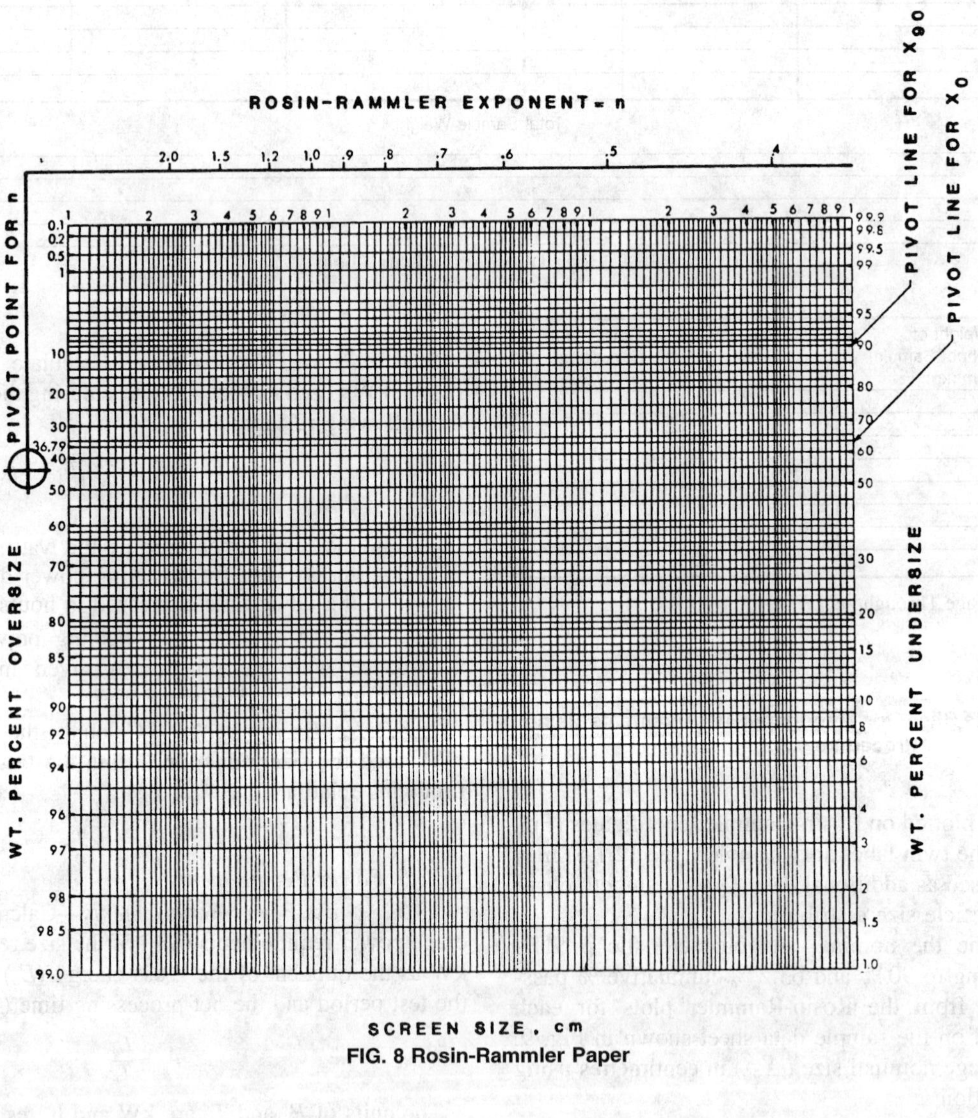

FIG. 8 Rosin-Rammler Paper

13.6 *Specific Energy Requirements*:

13.6.1 *Gross Specific Energy*—Calculate the gross specific energy requirement, $(E_o)_g$, in kWh/Mg using the following relation:

$$(E_o)_g = \frac{E_g}{\bar{\dot{m}}} \tag{13}$$

where:

E_g = the gross energy usage in kWh, and

$\dot{m}$ = the average throughput in Mg/h.

The gross specific energy requirement includes the free-wheeling component.

Test Interval	Product Size (cm)	
	Nominal (X_{90}) (90 %)	Characteristic (X_o) (63.2 %)
1		
2		
3		
4		
Alternative Samples		
1		
2		
3		
4		
Average[A,B]		

[A] Average nominal size ($\bar{X}_{90}$) =

$$1/n \sum_{i=1}^{n} (X_{90})_i$$

[B] Average characteristic size ($\bar{X}_o$) =

$$1/n \sum_{i=1}^{n} (X_o)_i$$

FIG. 9 Summary of Product Size Distribution Data

13.6.2 *Net Specific Energy*—Calculate the net specific energy requirement, E_o, in kWh/Mg using the following relation:

$$E_o = \frac{E_n}{\dot{\bar{m}}} \quad (14)$$

where:

E_n = the net energy usage in kWh, and
$\dot{m}$ = the average throughput in Mg/h.

13.7 *Recording of Results*:

13.7.1 The calculated results for average throughput ($\overline{\dot{m}}$), average gross ($\bar{P}_g$) and net ($\bar{P}_n$) power requirements, and average nominal ($\bar{X}_{90}$) and characteristic ($\bar{X}_o$) product sizes, and average gross ($(E_o)_g$) and net (E_o) specific energy requirements may be recorded on the sample summary data sheet shown in Fig. 10.

14. Precision and Bias

14.1 The bias of this method has not been established. The following estimates are given as guidelines:

14.1.1 The bias of watthour metres is estimated to be 98.0 to 99.5 %.

14.1.2 The bias of potential and current transformers (0.3 accuracy class) is 99.7 %.

14.1.3 The bias of the particle size designation (X_{90} and X_o) is a function of the number of samples analyzed and the degree of confidence; for example at a 90 % confidence level the following estimates apply:

Number of Samples	Precision (± %)
4	20 to 35
8	10 to 20
16	7 to 15

14.2 The precision of this test method has not been established.

Date: ______________________
Site: ______________________

Type of Size Reduction Device: ______________________
Model No.: ______________________
Serial No.: ______________________
Type of Solid Waste: ______________________

Test Period No.	Average Throughput (Mg/h)	Average Gross Power Requirement, $\bar{P}_g$(kW)	Average Net Power Requirement, $\bar{P}_n$(kW)	Average Product Size (cm)		Specific Energy Requirement (kWh/Mg)	
				Nominal X_{90}	Characteristic X_o	Gross $(E_o)_g$	Net (E_o)

FIG. 10 Summary of Test Results

Designation: E 1037 – 84 (Reapproved 1996)$^{\epsilon1}$

Standard Test Method for Measuring Particle Size Distribution of RDF-5[1]

This standard is issued under the fixed designation E 1037; the number immediately following the designation indicates the year of original adoption or, in the case of revision, the year of last revision. A number in parentheses indicates the year of last reapproval. A superscript epsilon (ϵ) indicates an editorial change since the last revision or reapproval.

ϵ^1 NOTE—Editorial changes were made throughout in October 1996.

1. Scope

1.1 This test method is used to determine the size distribution of a RDF-5 sample. Size is defined as the maximum length of the particle, where length is determined by the RDF-5 manufacturing process. That is, a pellet, cubette, or briquette all have a recognizable length. Fig. 1 displays the sizes and shapes of some RDF-5 particles.

1.2 An air dried RDF-5 sample is separated into categories of differing particle sizes. The size distribution is measured as the weight percentage of each size category. A graph of a function of the cumulative fraction of material by weight finer than particle size versus particle size is plotted. From this plot are taken values which describe the size distribution—the uniformity constant and the characteristic particle size.

1.3 *This standard does not purport to address all of the safety concerns, if any, associated with its use. It is the responsibility of the user of this standard to establish appropriate safety and health practices and determine the applicability of regulatory limitations prior to use.*

2. Terminology

2.1 *Definition:*

2.1.1 *RDF-5*—solid fuel derived from municipal solid waste in which the processed combustible fraction is densified (compressed) into the form of pellets, cubettes or briquettes.

3. Significance and Use

3.1 The particle size distribution of RDF-5 strongly influences the storage and handling characteristics of the fuel. Small particles tend to block flow through storage bins and feed hoppers, although correct bin and hopper designs will alleviate this problem of blockage.

3.2 This test method of measuring size by hand allows accurate description of RDF-5 particle size distribution. Measurement by hand is superior to sieving techniques, wherein particles may be broken by the size separation technique itself. However, hand measurement is more time-consuming than sieving techniques.

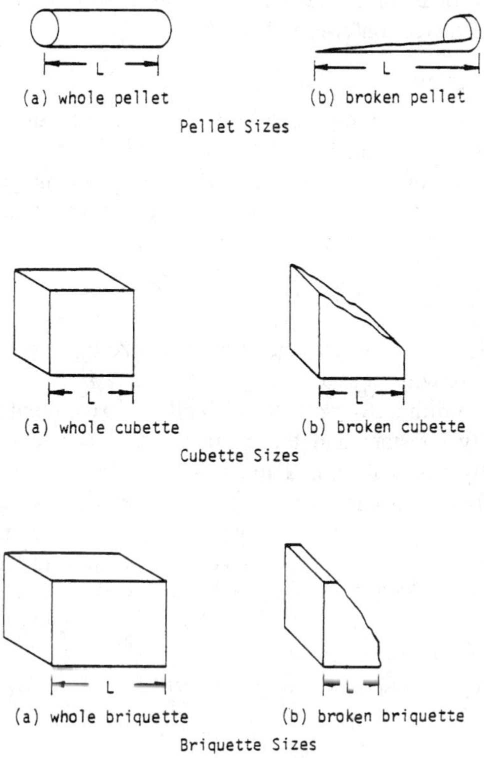

FIG. 1 RDF-5 Sizes

4. Apparatus

4.1 *Labelled Containers*, used to hold the particles which are separated by size. Appropriate containers are beakers or pans labelled "≥70 mm", "≥60 mm – <70 mm", etc. The tare weight of each container shall be recorded to 0.1 g.

4.2 *Scale*, capable of weighing the sample and container with an accuracy of 0.1 g.

4.3 *Vernier Calipers*, a length-measuring instrument having an accuracy of 0.1 mm.

5. Procedure

5.1 The sample shall weigh 1.0 ± 0.1 kg (2.2 ± 0.2 lb) unless otherwise specified. Record the weight of the sample to the nearest 0.1 g.

5.2 Beginning with the largest particles, measure the length of each particle in the sample. Separate the particles into containers labelled as the size categories of less than 10 mm,

[1] This test method is under the jurisdiction of ASTM Committee D-34 on Waste Management and is the direct responsibility of Subcommittee D34.06 on Recovery and Reuse.

Current edition approved Dec. 28, 1984. Published June 1985.

10 to less than 20 mm, 20 to less than 30 mm, 30 to less than 40 mm, etc. as needed.

5.3 Record the weight of each size category to 0.1 g.

5.4 Add the weights of the size components. If this sum differs by more than 2 % from the sample weight recorded initially, then reject the analysis and begin another test.

5.5 Use the sum of the separate size categories as the total weight to determine weight percentages of each size fraction. If the weight percentage of any fraction is greater than or equal to 25 %, then separate that fraction into two portions of 5-mm size categories. Continue separating the size fraction into categories of 2.5 mm, 1.25 mm, etc., until the weight percentage of each fraction is less than 25 %.

6. Calculation

6.1 With each particle size for which there is data, calculate *y*, the cumulative fraction of the sample less than that particle size. This is found by summing the weight percentages of all fractions less than that size. Also calculate the value of the Rosin-Rammler function for each particle size, as follows:

$$ln\left(\frac{1}{1-y}\right) \quad (1)$$

6.2 Plot ln[1/(1 − *y*)] versus particle size on log-log paper. Data points will yield a straight line.

6.3 Determine two values which characterize this line—the uniformity constant and the characteristic particle size. The uniformity constant, *n*, is defined as the slope of the line. The characteristic particle size, x_0, is defined as the size at which 63.2 % of the particles (by weight) are smaller. The characteristic particle size is the size corresponding to $ln[1/(1-y)] = 1.0$. Fig. 2 is an example of this plot.

7. Report

7.1 Report the uniformity constant, *n* (a dimensionless number), and the characteristic particle, x_0 (millimeters).

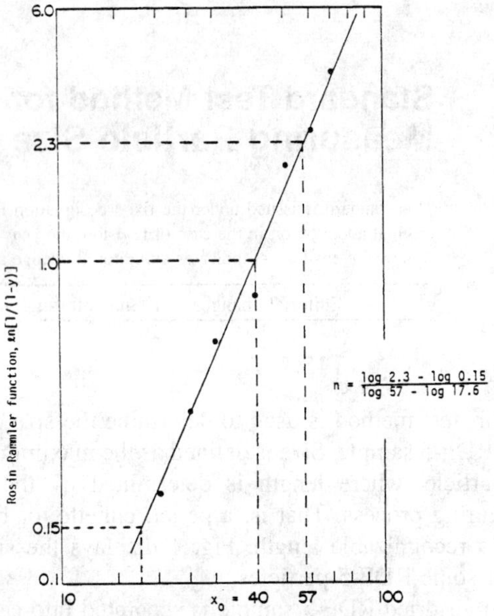

FIG. 2 Size Distribution Example—Test Number 1

8. Precision

8.1 Table 1 presents the results of intra- and inter-laboratory precision testing. Intra-laboratory and inter-laboratory tests were performed on different populations of RDF-5.

TABLE 1 Results of Precision Testing

		Uniformity Constant	Characteristic Size
Intra-laboratory	Sample mean	2.42	3 g
	Standard deviation	0.16	2 mm
Inter-laboratory	Sample mean	2.65	36 mm
	Standard deviation	0.18	2 mm

Designation: E 1076 – 85 (Reapproved 1999)

Standard Practice for Maintaining Health and Safety Records at Solid Waste Processing Facilities[1]

This standard is issued under the fixed designation E 1076; the number immediately following the designation indicates the year of original adoption or, in the case of revision, the year of last revision. A number in parentheses indicates the year of last reapproval. A superscript epsilon (ε) indicates an editorial change since the last revision or reapproval.

1. Scope

1.1 The purpose of this practice is to provide guidance to solid waste processing facility managers responsible for maintaining records of the health and safety experience of their employees. This practice describes general principles for establishing a procedure to collect and document health and safety data within a solid waste processing facility and provides specific information on the forms and procedures to be used in recording illnesses among employees.

1.2 *This standard does not purport to address all of the safety concerns, if any, associated with its use. It is the responsibility of the user of this standard to establish appropriate safety and health practices and determine the applicability of regulatory limitations prior to use.*

2. Referenced Documents

2.1 *OSHA Standard:*

29 CFR 1910.20 Access to Records[2]

3. Significance and Use

3.1 This practice is intended for use in resource recovery and other types of solid waste processing facilities. The occupational health risks in such facilities are not well documented and the purpose of this practice is to facilitate recording of information that can be useful in assessing the occupational health significance of working in these facilities.

3.2 The records developed and maintained in accordance with this practice can be used as a basis for spotting trends (if any) in the types or frequency, or both, of injuries and illnesses. This information may, in combination with data on environmental health conditions within a waste processing facility, be useful in identifying possible cause/effect relationships.

3.3 This practice is not intended as a design guide for an occupational health and safety program but rather is intended to build upon existing occupational health programs and to utilize currently available health and environmental records.

4. Records to Be Maintained

4.1 *Personal and Work History Records*—These data are essential to a recordkeeping system because they describe the workers' previous work experience and identify the worker with the work site.

4.1.1 For each employee the following information should be recorded:

4.1.1.1 Name,

4.1.1.2 Employer identification number,

4.1.1.3 Birthdate,

4.1.1.4 Social security number,

4.1.1.5 Sex,

4.1.1.6 Race,

4.1.1.7 Date of employment,

4.1.1.8 Past employment history, and

4.1.1.9 Hobbies or recreational activities.

4.1.2 *Work History Data*—For each assignment at the facility, record the following information:

4.1.2.1 Job classification and description,

4.1.2.2 Department,

4.1.2.3 Work location, and

4.1.2.4 Time spent at each work location.

4.2 *Health Records*:

4.2.1 *Physical Examinations*—All employees should receive entry and yearly examinations. The former should include a comprehensive medical history questionnaire indicating the following:

4.2.1.1 Pre-existing medical conditions,

4.2.1.2 Immunizations,

4.2.1.3 Alcohol, drug, and tobacco use habits, and

4.2.1.4 Family history data, indicating previous respiratory, allergy, neurological or cardiovascular conditions.

4.2.2 Clinical tests should be conducted yearly, including CBC, audiometry, pulmonary function and others as suggested by the physician. New employees should receive baseline audiograms during their entry examinations. If a threshold shift is observed, in a subsequent audiometric examination, the test should be repeated after the individual has been in a low noise environment for at least 14 h.

4.2.3 Results of these examinations should be kept under the control of the employer's physician or another designated physician.

[1] This practice is under the jurisdiction of ASTM Committee D-34 on Waste Management and is the direct responsibility of Subcommittee D34.06 on Recovery and Reuse.

Current edition approved Nov. 14, 1985. Published February 1986.

[2] Available from Superintendent of Documents, U.S. Government Printing Office, Washington, DC 20402.

4.2.4 *Illness Reports*— An employee experiencing an illness (occupational or nonoccupational) should report that illness to the physician referred to in 3.2.1. The report shall include at least the following information:

4.2.4.1 The nature of the illness,

4.2.4.2 The data of onset and duration,

4.2.4.3 The number of days off from work,

4.2.4.4 Treatment, and

4.2.4.5 Name and address of any physician consulted.

4.2.5 See Fig. 1 for an example of an illness report used by the U.S. Air Force.

4.3 *Accident Records*— Employers should maintain records of industrial injuries resulting in lost time on the job. These reports should be filed according to employee classification and include copies of the following information, where applicable:

4.3.1 *Supervisors Accident Investigation Report*, describing the following:

4.3.1.1 Who was involved,

4.3.1.2 Time,

4.3.1.3 Place,

4.3.1.4 Nature and circumstances of the incident,

4.3.1.5 Bodily injury involved, and

4.3.1.6 Other information required by locally applicable Workman's Compensation Laws.

4.3.2 *Employee's Report of the Accident to the Workman's Compensation Board.*

4.3.3 *Report/Findings of the Workman's Compensation Board.*

4.3.4 Any other illness and injury-related records, especially recommended work restrictions and records of any hospitalizations, lost time, compensation claims and payments.

4.4 *Work Environmental Records*—Records should be maintained for all industrial hygiene surveys. These data may provide insight into the relationship between exposure to

TOXICOLOGICAL EXPOSURE INCIDENT/OCCUPATIONAL ILLNESS REPORT
(THIS FORM IS AFFECTED BY THE PRIVACY ACT OF 1974 (Use Blanket PAS - DD Form 2005)

I. PATIENT IDENTIFICATION			
1. NAME	2. SSAN	3. GRADE: MIL / CIV	
4. SEX	5. AGE	6. ORGANIZATION AND SYMBOL	7. INSTALLATION
8. WORK LOCATION		9. OCCUPATION *(Job Title/AFSC)*	
10. DUTY PHONE	11. SUPERVISOR		12. SUPERVISOR'S DUTY PHONE
II. INCIDENT/ILLNESS DATA			
13. DATE AND TIME OF INCIDENT/ONSET OF ILLNESS			
14. STATUS AT TIME SPECIFIED ABOVE	☐ ON DUTY ☐ OFF DUTY	☐ LEAVE ☐ TDY	☐ OTHER *(Explain)*
15. DESCRIPTION OF INCIDENT OR ILLNESS ONSET SYMPTOMS *(Specify any hazardous material/agent)*			
16. WITNESS *(If exposure incident)*		17. PHONE	
III. MEDICAL DATA			
18. DIAGNOSIS AND RELEVANT MEDICAL DATA *(Indicate affected body parts)*		19. CLASSIFICATION 2	OSHA CODE
		OCCUPATIONAL SKIN DISEASE	21
		DUST DISEASE OF LUNGS	22
		RESPIRATORY CONDITION DUE TO TOXIC AGENT	23
		SYSTEMATIC EFFECT OF TOXIC MATERIAL *(Poisoning)*	24
		DISORDER DUE TO PHYSICAL AGENT *(Other than toxic material)*	25
		DISORDER DUE TO REPEATED TRAUMA *(Exclude hearing loss)*	26
FATALITY	RESULTED IN UNCONSCIOUSNESS	OTHER OCCUPATIONAL ILLNESS	29
20. DATE/TIME OF INITIAL TREATMENT/DIAGNOSIS		21. MEDICAL FACILITY	
22. TREATMENT ADMINISTERED *(Check One)*	☐ FIRST AID 1	☐ DEFINITIVE CARE *(Specify in Remarks)*	
23. DISPOSITION OF PATIENTS			

YES	NO		NO. OF DAYS	
		RETURN TO NORMAL DUTY		ADMITTED TO HOSPITAL 2
		REFER TO PRIVATE PHYSICIAL		PLACED ON QUARTERS 2
		OTHER *(Specify in Remarks)*		RETURN TO LIMITED DUTY 2

24. NAME OF MEDICAL OFFICER

25. REMARKS

IV. ENVIRONMENTAL DATA

26. DESCRIBE JOB TASKS THAT RESULTED IN EXPOSURE TO HAZARDOUS MATERIAL/AGENTS *(Specify the material/agent)*

27. DURATION OF EXPOSURE DESCRIBED ABOVE

1. One-time treatment of minor scratches, cuts, burns, and splinters which do not require professional medical care.
2. See AFR 127-12.

AF FORM JAN 77 190

NOTE 1—This is a standard governmental form used by the U.S. Air Force.

FIG. 1 An Example of an Illness Report Form

contaminants and reported illnesses in future epidemiologic studies.[3] Data should be maintained on the following information:

4.4.1 *Airborne Contaminants*:

4.4.1.1 *Dust Concentraions*—Records of total dust and respirable dust levels that have been measured should be maintained.

4.4.1.2 *Microbiological Organisms*—Records of sampling airborne microbiological organisms should be maintained.

4.4.1.3 *Vapors and Other Airborne Contaminants*—Any measurements of concentrations of chemical vapors, fumes from welding operations or other airborne contaminants should be recorded. Records should indicate sampling location, sampling method employed, employee(s) potentially exposed, duration of sample, and analytical method. Any incident involving chemical exposure should be recorded with all available information such as identity of the substance, level of exposure and identity of people exposed.

4.4.2 *Other data*—Results of other industrial hygiene surveys such as those for general housekeeping conditions, rodent and fly problems, and use of personal protective equipment should be recorded.

4.5 *Occupational Safety Records*—Employers should maintain records of periodic occupational safety training provided to employees and of safety surveys conducted at the facility.

4.5.1 *Safety Training*— Records should be kept of fire prevention and other occupational safety training provided to employees. Such records may be maintained as part of the personnel file of each individual employee or in a central file that lists the participants in all safety training programs conducted at the facility.

4.5.2 *Safety Surveys*— Records of safety surveys conducted within the facility should be maintained in a central file. Safety survey records should include information such as the following:

4.5.2.1 Unsafe conditions or practices noted during periodic surveys or reported by employees;

4.5.2.2 Safety equipment inventories;

4.5.2.3 Maintenance records for safety equipment; and

4.5.2.4 Records of actions taken to correct unsafe conditions.

5. Preservation of Records

5.1 Each employer shall ensure the preservation and retention of records as follows:

NOTE 1—These recommendations for retention of records are consistent with the regulations of the U.S. Occupational Safety and Health Administration (OSHA) governing access to employee exposure and medical records as described in 29 CFR 1910.20.

5.1.1 *Employee Medical Records*—for at least the duration of employment plus 30 years.

5.1.2 *Employee Exposure Records*—at least 30 years, except for background data and worksheets (1 year) as long as sampling results, sampling plan, analytical method and summary of data relative to interpretation is retained for at least 30 years.

5.1.3 *Analyses Using Exposure or Medical Data*—30 years.

[3] Examples of the forms used in recording these data can be seen in Chapter 3 of *Patty's Industrial Hygiene and Toxicology*, (Third Edition, 1978) entitled "Industrial Hygiene Records and Reports."

Designation: E 1107 – 86 (Reapproved 1996)$^{\epsilon 1}$

Standard Test Method for Measuring the Throughput of Resource-Recovery Unit Operations[1]

This standard is issued under the fixed designation E 1107; the number immediately following the designation indicates the year of original adoption or, in the case of revision, the year of last revision. A number in parentheses indicates the year of last reapproval. A superscript epsilon (ϵ) indicates an editorial change since the last revision or reapproval.

ϵ[1] NOTE—Editorial changes were made throughout in October 1996.

1. Scope

1.1 This test method is for measuring the throughput, or mass flowrate, of a resource-recovery unit operation, or series of unit operations.

1.2 *This standard does not purport to address all of the safety concerns, if any, associated with its use. It is the responsibility of the user of this standard to establish appropriate safety and health practices and determine the applicability of regulatory limitations prior to use.* Specific precautionary information is given in Section 7.

2. Referenced Documents

2.1 *ASTM Standards:*

C 702 Practice for Reducing Field Samples of Aggregate to Testing Size[2]

D 75 Practice for Sampling Aggregates[3]

E 380 Practice for Use of the International System of Units (SI) (the Modernized Metric System)[4]

E 868 Test Methods for Conducting Performance Tests on Mechanical Conveying Equipment Used in Resource Recovery Systems[5]

3. Terminology

3.1 *Descriptions of Terms Specific to This Standard:*

3.1.1 *binary separator*—a mechanical device that separates single input feed stream into two output feed streams.

3.1.2 *polynary separator*—a mechanical device that separates single input feed stream into three or more output feed streams.

3.1.3 *processor*—a type of resource recovery unit operation with a single input feed stream and single output stream. Its function is to alter the physical or physico-chemical properties of the input feed stream. The mass flow rates of input and output streams should be equal unless moisture is lost.

3.1.4 *throughput*—the mass flowrate through a unit operation, expressed, preferably, in units of kilograms per hour (kg/h) or alternatively in units of pounds per hour (lb/h).

4. Summary of Test Method

4.1 The output streams of a separator or processor are collected over a measured period of time and weighed. Collection of the output stream is either in containers or by stopping, then clearing, portions of conveyor belts or chutes. For processing equipment in which materials separation is not accomplished, the input stream may be sampled if this is more convenient.

5. Significance and Use

5.1 This test method is used to document the mass flowrate of resource recovery unit operations as a means of documenting the operating of a plant (or portions of a plant) and as a means of relating operation to design objectives.

5.2 This test method is also used in conjunction with measurements of the performance of materials separators (particularly recovery and purity). As such, throughput should not generally be measured by sampling the feed since this may change its performance. Processing equipment that does not perform separations can be sampled at either the feed or product streams.

6. Apparatus

6.1 *Collection Bins*— Several size collection bins are required. The size is determined by the size of sample, which in turn, is determined by the throughput of the plant. Some streams can be sampled into drums or barrels.

6.1.1 All containers must be clean and in good mechanical condition, and not have rusting, flaking, or mechanically weakened sections. Containers should be cleaned with water or an air hose prior to use. (**CAUTION**: Air hoses must be used with appropriate safety equipment to avoid personal injury.)

6.1.2 The tare weight of the containers should be clearly marked with paint and checked periodically. Provision may be made for bin covers for the sampling and containment of materials that can be blown or spilled. Covers must be marked to indicate whether or not they are included in the tare weight of the container.

---

[1] This test method is under the jurisdiction of ASTM Committee D-34 on Waste Management and is the direct responsibility of Subcommittee D34.06 on Recovery and Reuse.

Current edition approved April 25, 1986. Published June 1986.

[2] *Annual Book of ASTM Standards*, Vol 04.02.

[3] *Annual Book of ASTM Standards*, Vol 04.03.

[4] *Annual Book of ASTM Standards*, Vol 14.02.

[5] *Annual Book of ASTM Standards*, Vol 11.04.

6.2 *Scales*—The type and size of scales varies with the size of the samples obtained. Containers as small as perhaps a drum to as large as perhaps a tote bin, roll-off container, or even full truck may be used. All scales should have a precision and accuracy of ±0.1% .

6.3 *Stopwatch*—Flow times are determined with a stopwatch capable of measuring to the nearest 0.1 s.

6.4 *Miscellaneous*— A variety of scoops, shovels, brushes, and similar tools are required to transfer materials.

7. Precautions

7.1 If samples are taken by transferring materials from a conveyor belt, it is essential to measure the belt's speed and use appropriate tools to be certain that all of the material, especially including fine particulate materials, are transferred.

7.2 Because the origin of all of the materials in solid waste is generally unknown, workers must use proper safety precautions when handling samples. Workers shall wear gloves and safety glasses. When appropriate, dust masks shall be worn. Workers shall be cautioned to wash their hands thoroughly before eating or smoking.

7.3 Safety precautions shall be taken when collecting samples or working near moving equipment.

8. Sampling

8.1 Samples are taken after the equipment has reached a steady-state operation. A steady-state operation is arbitrarily assumed after the equipment is operating for at least 30 min under what are considered to be normal conditions, or as otherwise agreed. The composition and type of feed may not be changed during this time.

8.2 After steady-state, samples are taken at agreed intervals.

8.3 The sample is taken by whatever method in Section 11 suits the separator or processor being sampled.

9. Test Specimen and Samples

9.1 The size of sample is taken in relation to the particle size of the material or estimated throughput of the process, or both.

9.1.1 The minimum size of sample is determined by its particle size in accordance with Practice D 75, or by 9.1.2 or 9.1.3, whichever is greater.

9.1.2 For particle sizes greater than 90 mm (not included in Table 1 of Practice D 75), the size of sample is 250 kg (550 lb).

9.1.3 The minimum weight of sample shall correspond to the estimated throughput for 1.0 min or the minimum weight will be determined by the procedure in Test Methods E 868.

9.2 Test samples corresponding to 9.1 are weighed without subdivision.

9.3 Three test samples shall be taken for each randomly chosen sampling time. Two of the samples will be weighed; the third shall be retained and weighed if the calculated throughputs based on the first two differ by more than 10 %, as described in 12.4.

9.4 If possible, both binary and polynary separators should normally be sampled at the output sides.

10. Conditioning

10.1 Weigh the samples immediately after they are taken. Take precautions to ensure that they neither gain nor lose weight from natural drying or drainage or from ambient moisture or dirt.

11. Procedure

11.1 Use separate procedures for sampling conveyors, chutes, or discharge containers.

11.2 *Conveyors*:

11.2.1 Conveyors are most conveniently sampled by catching the discharge at the end of the conveyor in the tared container. Take care that the bin is wider than the width of the conveyor and the entire contents of the belt width is collected.

11.2.2 Conveyors can also be sampled by stopping the belt and removing a portion of the belt load. In this method, determine the belt speed by timing the movement of a mark on the belt as it passes between two marks on the sides of the conveyor. Measure the distance between the latter two marks and use the measurement to calculate the belt speed. Alternatively, use a tachometer for determining the speed of the belt. Stop the conveyor and shovel the material lying on a predetermined length of the conveyor into a suitable tared container. Take care to include all fine particulate materials. Also, take care to ensure that the belt load moves at the same speed as the belt and is not hindered by the sides of the conveyor, thus causing slippage.

11.3 *Chutes*—Sample material falling through a chute by placing a tared container of suitable size under the chute and collecting the material for a predetermined length of time, measured with a stopwatch to the nearest 0.1 s. If it is not possible to sample the discharge of a chute, a diverting chute member may have to be added along with a gate. Exercise care at the discharge ends of chutes to ensure that all of the material flowing is collected in the container. Flexible spouts may be fastened (even if temporarily) on the ends of the chutes and directed into the containers.

11.4 *Discharge Containers*—Sample discharge points by using a tared container to collect the material. Preferably, fasten a flexible spout or a diverter to the discharge point to the usual collection bin so that the flow can be suddenly diverted to the tared container without significant spillage. Without such a flexible spout, it may not be possible to time accurately the discharge.

11.5 Sampling and weighing must be done without spilling any of the material. Note any spillage more than dusting. Spillage of more than 1% of the collected sample, visually estimated, is reason to discard the sample and start over.

11.6 Weigh the tared containers containing the samples immediately and record the filled gross weights. Record weights within 0.1 % of the total filled weight in accordance with the precautions of 10.1.

12. Calculation

12.1 Record the following information:

12.1.1 Method of sampling (11.2.1, 11.2.2, 11.4.1, or 11.3.1);

12.1.2 Location;

12.1.3 Time of day and date;

12.1.4 Tare of container;

12.1.5 Weight of filled container; and

12.1.6 Special observations.

12.2 Calculate the sampling time for the procedure in 11.2.2 as follows:

$$\text{Sampling time} = t = L/Y \quad (1)$$

where:
L = length of conveyor swept, and
Y = conveyor speed, calculated as $Y = C/T$;
C = measured distance between the two marks on the conveyor sidewall, m.
T = measured time for the mark on the conveyor belt to move between these two marks, s.

12.2.1 The conveyor speed is in units of metres per second (m/s) and the sampling time t is in units of seconds (s).

12.3 Measure the sampling time for other procedures (11.2.1, 11.3.1, and 11.4.1) with the stopwatch.

12.4 Calculate the throughput, Q, on a wet-weight basis for a processor, sampling either the input or output stream as follows:

$$Q = \frac{(W - A)\,3600}{t} \quad (2)$$

where:
W = weight of the filled container,
A = tare (empty) weight of the container, and
t = time of collection of the sample, s.

12.4.1 If W and A are in kilograms, Q is in kilograms per hour.

12.5 Paragraph 12.4 can be used to calculate the throughput of a processor if the input stream is sampled. However, note the information in 9.4.

12.6 A binary or polynary separator may be sampled at each of its output streams with all samples taken at the same time and for equal time intervals. Calculate the throughput for each stream in accordance with 12.4 and sum to obtain the total throughput.

13. Report

13.1 The report shall include the information on the data compilation and calculation sheet (see Fig. 1).

14. Precision and Bias

14.1 There are not yet sufficient data available to compute the precision and bias of this test method.

Data Compilation and Calculation Sheet[A]

Date: __________ Location: __________
Time of Day: __________ Operator: __________

Conveyor Sampling:

Speed = __________ = Y
Length of belt swept __________ = L
Sampling time $t = L/Y$ = __________

Throughput (wet weight basis):

Container No. _____, Tar Weight A = __________
Filled Weight W = __________
Throughput $Q = (W\text{-}A)/t$ = __________

Special Observations:

[A] Repeat calculations using a separate sheet for each processor or separator stream sample.

FIG. 1 Sample Data Compilation and Calculation Sheet

Designation: E 1108 – 86 (Reapproved 1996)

Standard Test Method for Determination of the Recovery of a Product in a Materials Separation Device[1]

This standard is issued under the fixed designation E 1108; the number immediately following the designation indicates the year of original adoption or, in the case of revision, the year of last revision. A number in parentheses indicates the year of last reapproval. A superscript epsilon (ε) indicates an editorial change since the last revision or reapproval.

1. Scope

1.1 This test method covers the determination of the recovery of a desired product in a device processing solid waste for the purpose of concentrating a component of interest. The recovery is determined with respect to the amount of the desired component in one output stream (accepts) as opposed to another output stream (rejects). The results of this calculation determine the effectiveness of component separation when coupled with a measure of product purity as described in Test Method E 889.

1.2 *This standard does not purport to address all of the safety concerns, if any, associated with its use. It is the responsibility of the user of this standard to establish appropriate safety and health practices and determine the applicability of regulatory limitations prior to use.* Specific precautionary information is given in Section 7.

2. Referenced Documents

2.1 *ASTM Standards:*

E 889 Test Method for Composition or Purity of a Solid Waste Materials Stream[2]

E 1107 Test Method for Measuring the Throughput of Resource-Recovery Unit Operations[2]

3. Terminology

3.1 *Definitions of Terms Specific to This Standard:*

3.1.1 *accepts*—the output stream from the materials separation device that contains the highest concentration (purity) of the component(s) that the device is designed to separate.

3.1.2 *binary separator*—a device that separates a single input feed stream into two output or product streams.

3.1.3 *component*—any identifiable and defined fraction of solid waste.

3.1.4 *feed*—the input stream.

3.1.5 *gross sample*—a sample representing one lot and composed of a number of increments on which neither reduction nor division has been performed.

3.1.6 *laboratory sample (or analysis sample)*—a portion of one gross sample representative of a lot and taken at random from the gross sample.

3.1.7 *polynary separator*—a device that separates a single input feed stream into three or more output product streams.

3.1.8 *purity*—the purity of a stream is defined in terms of one or more identifiable components, *x, y, z*, etc. as separated by hand sorting. The composition for any component such as *x* is the mass of *x* in a stream divided by the total mass of that stream. Purity is expressed as a fraction, decimal fraction or percent.

3.1.9 *recovery*—the fraction (on a percentage basis) of a specific component, *x*, in an input process stream (feed) captured or recovered into an output process stream (accepts). Recovery is calculated as a mass of *x* in the output stream per unit time divided by the mass of *x* in the input stream per the same time, and the fraction is multiplied by 100. Alternatively, the concentration of the component(s) of interest (defined as purity in Test Method E 889) are measured in all input and exit streams and used to calculate recovery.

3.1.10 *rejects*—the output stream(s) that have low concentrations of the component(s) of interest.

NOTE 1—In mass per time terms, Feed = Accepts + Rejects for any conservation component if there is no storage term or loss of mass, for example, moisture.

3.1.11 *steady state operation*—a condition in which the materials separator is achieving a constant recovery performance over a given time.

4. Summary of Test Method

4.1 Percent recovery of a component can be calculated either by measuring the throughput of component *x* (mass per unit time) in the input (feed) and the throughput of component *x* in the output (accepts) process streams or by measuring the purity of the input and output streams.

4.1.1 The throughputs are measured as described in Test Method E 1107.

4.1.2 The purity analyses are described in Test Method E 889.

5. Significance and Use

5.1 This test method is used to document the ability of a solid waste resource recovery separator to capture the component(s) of interest in the desired process stream.

---

[1] This test method is under the jurisdiction of ASTM Committee D-34 on Waste Management and is the direct responsibility of Subcommittee D34.06 on Recovery and Reuse.

Current edition approved April 25, 1986. Published July 1986.

[2] *Annual Book of ASTM Standards*, Vol 11.04.

5.2 The recovery determined in this way is used in conjunction with the purity of the product, as described in Test Method E 889.

6. Precautions

6.1 This procedure calls for the hand-picking of solid waste and its processed fractions. Because the origin of all of the materials is generally unknown, workers must use proper safety precautions when handling samples. Workers shall wear gloves and safety glasses. When appropriate, dust masks shall be worn. Workers must be cautioned to wash their hands thoroughly before eating or smoking.

6.2 Safety precautions must be taken when collecting samples or working near moving equipment.

7. Sampling

7.1 If throughputs are to be measured, samples are taken from input and output process streams, in accordance with procedures included in Test Method E 1107.

7.2 For a binary separator, only the input and one output or both output streams are sampled. If only one output stream is sampled, it should be the one in which the component(s) of interest is concentrated (the accept). The output stream is always sampled before the input stream in order not to disturb the steady state operation of the separator.

7.3 For a polynary separator, only the throughput method is used. Input plus the output containing the component(s) of interest are sampled.

7.4 Sampling and containment must be done in a manner to preserve the integrity of the component(s) of interest and the mass of the entire sample.

8. Test Specimens and Samples

8.1 If throughput is to be measured, follow the procedures in Test Method E 1107.

8.2 If purity is to be measured, follow the procedures in Test Method E 889.

9. Conditioning

9.1 If the throughput method is to be used, follow the procedure in Test Method E 1107.

9.2 If the purity method is to be used, follow the procedure in Test Method E 889.

10. Procedure

10.1 It is necessary that the separator attain steady state operation before sampling is done. Performing only one recovery analysis does not ensure steady state conditions. The separator must be operating without changes in feed quantity or composition, and without changes in operation mode, for at least 30 min before sampling. The results of a series of recovery determinations will confirm the presence of steady state operation.

10.2 If the throughput method is to be used, follow the procedures in Test Method E 1107.

10.2.1 Sample the input stream and the output stream that contains the concentrated fraction of the component(s) of interest.

10.3 If the purity method is to be used, follow the procedures in Test Method E 889.

10.3.1 Sample the input stream and both output streams.

11. Calculations

11.1 If throughputs are measured, calculate the percent recovery of a component x as follows:

$$\text{Percent recovery of } x = R_x = Q_A \times (100)/Q_F \tag{1}$$

where:

Q_A = the throughput of x in the concentrated output stream (accepts) as kg/s, and

Q_F = the throughput of x in the input stream, as kg/s.

11.2 If purity is measured, compute the percent recovery of a component x in a binary separator as follows:

$$\text{Percent recovery of } x = R_x = P_A[P_F - P_B] \times (100)/P_F[P_A - P_B] \tag{2}$$

where:

P_F = the purity of the input stream, as mass of x in the stream divided by the total mass of the stream,

P_A = the purity of the concentrated output stream (accepts), and

P_B = the purity of the dilute process stream (rejects), with purity expressed as before.

11.3 If a polynary separator is employed, calculate the recovery of x in any one output stream as in 11.1.

12. Report

12.1 If throughput is used, calculate the recovery and record on the report form shown as Fig. 1.

12.2 If purity is used, calculate the recovery and record on the report form shown as Fig. 2.

12.2.1 A modified form is needed for polynary separators.

13. Precision and Bias

13.1 Since recovery is a calculation based on other standard sampling methods, the precision and bias is a function of their standard methods.

Determination of the Recovery of a Product in a Materials Separation Device

NOTE—This calculation sheet is to be used when recovery is based on throughput.

Throughput of input (feed stream (Q_F)	________	kg/s
Throughput of concentrated (accepts) output stream (Q_A)	________	kg/s
Recovery = $\frac{(Q_A \times (100)}{Q_F}$ =	________	%

FIG. 1 Sample Throughput Report Form

Determination of the Recovery of a Product in a Materials Separation Device

NOTE—This calculation is to be used when recovery is based on purity.

Purity of input (feed) stream (P_F)	________
Purity of concentrated (accepts) output stream (P_A)	________
Purity of dilute (rejects) output stream (P_B)[A]	________
$\text{Recovery} = \dfrac{P_A(P_F - P_B) \times (100)}{P_F(P_A - P_B)} =$	________ %

[A] If a polynary separator is employed, the purity of the rejects is a weighed average as described in 12.3.

FIG. 2 Sample Purity Report Form

Designation: E 1109 – 86 (Reapproved 1996)

Standard Test Method for Determining the Bulk Density of Solid Waste Fractions[1]

This standard is issued under the fixed designation E 1109; the number immediately following the designation indicates the year of original adoption or, in the case of revision, the year of last revision. A number in parentheses indicates the year of last reapproval. A superscript epsilon (ε) indicates an editorial change since the last revision or reapproval.

1. Scope

1.1 This test method determines the bulk density of various fractions from the resource recovery processing of municipal solid waste. It is intended as a means of characterizing such fractions and for providing data useful to designers of solid waste processing plants.

1.2 *This standard does not purport to address all of the safety concerns, if any, associated with its use. It is the responsibility of the user of this standard to establish appropriate safety and health practices and determine the applicability of regulatory limitations prior to use.* Specific precautionary information is given in Section 7.

2. Referenced Documents

2.1 *ASTM Standards:*

C 702 Practice for Reducing Field Samples of Aggregate to Testing Size[2]

D 75 Practice for Sampling Aggregates[3]

E 689 Reference Radiographs for Ductile Iron Castings[4]

E 1107 Test Method for Measuring the Throughput of Resource-Recovery Unit Operations[5]

3. Terminology

3.1 *Definitions of Terms Specific to This Standard:*

3.1.1 *bulk density*—the mass per unit volume of particulate matter. Bulk density is not an absolute material property as is the density of individual particles of a material. The bulk density depends on the size of the container and how the material is loaded into the container. For example, the bulk density of material placed loosely in a container will be less than that of material tamped into a container. Also, some materials placed loosely in a container will settle with time due to its own weight; thus, its bulk density will increase.

4. Summary of Test Method

4.1 A sample of a solid waste stream is tamped in a cubical container of known volume and then weighed. Bulk density is calculated from the weight of the contents and volume of the container.

5. Significance and Use

5.1 This test method describes a physical property of solid waste in processing facilities; a property that characterizes the solid waste streams and hence the operation of resource recovery separators and processors.

5.2 The bulk density is an important property for the design of materials handling equipment, separators, and processors.

6. Apparatus

6.1 *Balance*—A balance or scale accurate within 0.1% of the test load at any point within the range of use. The range of use shall be considered to extend from the weight of the measure empty to the weight of the measure plus its contents.

6.2 *Cubical Measure*, with internal dimensions approximately 60 by 60 by 60 cm (24 by 24 by 24 in.) provided with handles. The interior surfaces of the measure shall be nonabsorbent to moisture.

NOTE 1—A satisfactory weigh box has been constructed of ¾ in. (1.9 cm) exterior grade plywood with the finished surface on the inside. Reinforcing strips, approximately 5 cm (2 in.) wide and 1.3 cm (½ in.) thick were fastened to the outside walls; two of them in a way to provide handles.

6.3 Larger cubical weigh boxes shall be used as necessary to accommodate large particles of material. No particle larger than two-thirds the length, height, or width of the weigh box shall be put into the weigh box in determining the bulk density.

7. Precautions

7.1 This procedure calls for the handling of solid waste and its processed fractions. Because the origin of all the materials is generally unknown, workers must use proper safety precautions when handling samples. Workers shall wear gloves and safety glasses. When appropriate, dust masks shall be worn. Workers must be cautioned to wash their hands thoroughly before eating or smoking.

8. Sampling

8.1 Sample shall be obtained from the processor streams or separator output streams in accordance with 11.1, Test Method E 1107, and shall constitute the gross sample.

8.2 The amount of material in each sample shall conform to

---

[1] This test method is under the jurisdiction of ASTM Committee D-34 on Waste Management and is the direct responsibility of Subcommittee D34.06 on Recovery and Reuse.

Current edition approved April 25, 1986. Published June 1986.

[2] *Annual Book of ASTM Standards*, Vol 04.02.

[3] *Annual Book of ASTM Standards*, Vol 04.03.

[4] *Annual Book of ASTM Standards*, Vol 03.03.

[5] *Annual Book of ASTM Standards*, Vol 11.04.

the requirements of Table 1, Practice D 75, except that samples of material with greater than 90 mm (3½ in.) maximum nominal size shall weigh not less than 250 kg (550 lb).

9. Sample Preparation

9.1 Prepare and test the samples as soon as practicable with due precaution that they neither gain nor lose weight from natural drying or being left uncovered in wet or dusty areas, as described in Section 11, Test Method E 1107.

9.2 Weigh the gross samples before being subdivided. Drain wet samples in accordance with 11.g of Method E 689 before weighing and report the test results as "per drained weight."

9.3 Subdivide gross samples in accordance with Practice C 702 to form four test specimens. If the cone and quartering Method of Practice C 702 is used, take special care to include the fine particulate material that may sift to the bottom of the pile. Keep the test specimens in moisture barrier containers until tested or discarded.

10. Procedure

10.1 Measure the internal dimensions of the weight box to ±1 mm (0.05 in). Then calculate and record the volume to ±0.1%.

10.2 Determine the weight of the weigh box to ±0.1%.

10.3 Select a test specimen at random from the four and fill the weigh box to overflowing, taking special care to avoid segregation, compaction or spillage; especially loss of the fine particulate material. Then tamp the box three times by lifting it 6 cm (2½ in.) above the ground and dropping squarely. After tamping, draw a rigid straightedge across the top of the box to level the contents. Some agitation of the straightedge may be necessary to force particulate either in or out of the box. If the contents remain below the top of the box, empty the box and start the test again using another specimen.

10.4 Weigh the filled box ±0.1 % and calculate the bulk density to three significant figures.

10.5 Repeat the procedures of 10.3 and 10.4 with another test specimen selected at random from the remaining three.

10.6 Calculate the average of the two bulk densities. If the densities are more than 10 % different, select at random another specimen from the remaining two specimens and make a third bulk density determination. If the third density is more than 5 % above or below the average of the first two, discard all four test specimens, obtain another gross sample and repeat the test procedure using a larger weigh box. If the third determination is within 5 % of the average of the first two, report the average of the three determinations.

11. Calculation

11.1 Calculate the bulk density as follows:

$$\text{Bulk Density} = (W - W_T)/V$$

where:

W = the weight of the box full,
W_T = the weight of the box empty, and
V = the volume of the box as determined in 10.1.

11.2 If the box is weighed in kilograms, the volume of the box shall be expressed in cubic metres. If the box is weighed in pounds, the volume of the box shall be expressed in cubic feet and the bulk density in pounds per cubic feet.

12. Report

12.1 The bulk density is reported as the average of the two or three determinations.

12.2 The report shall indicate whether the density is based on as-received weight, drained weight, or other.

12.3 The report form is shown in Fig. 1. All items shall be completed.

13. Precision and Bias

13.1 There are not yet sufficient data available to determine the precision of this test method. There are no known standard materials for determining the bias of this test method.

Computation and Reporting of Bulk Density

Date ____________________ Location ____________________
Time ____________________ Operation ____________________
Pressure Stream ____________________

Weight Box Weight = W_T = ____________________

Box + Sample = W=	______ Sample 1	______ Sample 2	______ Sample 3*
Bulk Density − $(W - W_T)/v$ =	______ Sample 1	______ Sample 2	______ Sample 3*
Average			
Average +5 %			
Average −5 %			

Indicate if: Dry weight ______
Drained weight ______
As-received weight ______

FIG. 1 Sample Bulk Density Computation and Report Sheet

Standard Specification for Source-Separated Steel Cans[1]

This standard is issued under the fixed designation E 1134; the number immediately following the designation indicates the year of original adoption or, in the case of revision, the year of last revision. A number in parentheses indicates the year of last reapproval. A superscript epsilon (ε) indicates an editorial change since the last revision or reapproval.

1. Scope

1.1 This specification covers the chemical and physical requirements of source-separated steel cans that are intended for use by such industries as follows:

1.1.1 Copper industry (precipitation process),

1.1.2 Iron and steel foundry industry,

1.1.3 Iron and steel production industry,

1.1.4 Detinning industry, and

1.1.5 Ferroalloy industry.

1.2 Questions concerning material rejection, downgrading, and retesting based on failure to meet the requirements of this specification shall be dealt with through contractural arrangements between purchaser and supplier.

1.3 Source-separated steel cans in this specification covers two types of steel cans:

1.3.1 Bi-metal food and beverage cans, and

1.3.2 All other steel cans.

2. Referenced Documents

2.1 *ASTM Standards:*

E 701 Test Methods of Testing Municipal Ferrous Scrap[2]

E 702 Specification for Municipal Ferrous Scrap[2]

3. Descriptions of Terms Specific to This Standard

3.1 *source-separated steel cans*—post-consumer products that are generated as separated can fractions by commercial or household sources.

3.2 *bi-metal beverage cans*—steel cans with nonferrous metal convenience ends (normally made of aluminum), originally containing beer or carbonated beverages, but not including other contaminants.

3.3 *bi-metal food cans*—steel cans with nonferrous metal (usually aluminum) convenience ends, originally containing snack foods, but not including other contaminants.

3.4 *all other steel cans*—include containers for food products or liquids with a maximum capacity of 5 gal.

3.5 *total combustibles*—materials that include paints, lacquers, coatings, plastics, etc. associated with the original container (ferrous product), as well as combustible materials (paper, plastic, textiles, etc.) which become associated with the container (ferrous product).

3.6 *metallic yield*— the weight percent of the source-separated steel cans that is generally recovered as metal or alloy.

4. Chemical Composition

4.1 Source-separated steel cans shall conform to the requirements as to chemical composition for the respective end uses prescribed in Table 1.

4.2 The chemical requirements listed in Table 1 are based on melt analysis except where noted.

5. Physical Requirements

5.1 Source-separated steel cans shall conform to the physical properties for the respective end uses prescribed in Table 2.

6. Test Methods

6.1 Determine the physical and chemical requirements of source-separated steel cans in accordance with Test Methods E 701.

[1] This specification is under the jurisdiction of ASTM Committee D-34 on Waste Management and is the direct responsibility of Subcommittee D34.06 on Recovery and Reuse.

Current edition approved Oct. 31, 1986. Published December 1986.

[2] *Annual Book of ASTM Standards*, Vol 11.04.

TABLE 1 Chemical Requirements

Element	Composition, Weight %					
	Cans, as Manufactured[A]	Copper Industry (Precipitation Process)	Iron and Steel Foundry Industry	Iron and Steel Production Industry	Detinning Industry	Ferroalloy Industry
Bi-metal Cans:						
Aluminum, max	11.0	...	0.50	11.0	[B]	0.15
Tin	0.17–0.20	...	0.3 max[C]	0.3 max	...	0.3 max
Iron (metallic), min	87.0	96.0	...	...	...	...
Total combustibles, max	1.75	0.2	1.75	...	...	0.5
Metallic yield, min	...	...	90.0	90.0	...	90.0
Further processing required to meet industry requirements	...	yes	yes	no	yes	yes
All Other Steel Cans:[D]						
Aluminum, max	0.0	...	0.50	11.0	[B]	0.15
Tin	0.23–0.27	...	0.3 max[C]	0.3 max	[E]	0.3 max
Iron (metallic), min	98.7	96.0	...	...	...	...
Total combustibles, max	1.0	0.2	1.75	...	...	0.5
Metallic yield, min	...	...	90.0	90.0	...	90.0
Further processing required to meet industry requirements	...	yes	no	no	no	yes

[A] Values in this first column are typical values shown for reference.
[B] The presence of aluminum is not acceptable to the detinning industry.
[C] For steel foundries, the requirement for tin content is 0.10 %, maximum.
[D] Washed, free of visible foodstuffs and paper labels.
[E] Tin content of the container, as manufactured, is acceptable to the detinning industry.

TABLE 2 Physical Requirements

End-use	Property	
	Bulk Density, kg/m^3(lb/ft^3)	Form[A]
Copper industry (precipitation process)	480 (30) max	loose, shredded; shall not be nuggetized or baled[B]
Iron and steel foundry industry	800 (50) min	loose, nuggetized, or baled[C]
Iron and steel production industry	1200 (75) min[D]	loose[E] or baled[C]
Detinning industry	480 (30) max	loose, flattened, shredded, or baled[F]; 95 % by weight shall be in size range from − 152 to + 12.5 mm (−6 to + ½ in.); shall not be nuggetized, burned, incinerated, or pyrolyzed
Ferroalloy industry	800 (50) min	loose and free flowing.

[A] As agreed between purchaser and supplier.
[B] Various purchasers may establish gage limitations on the material they purchase.
[C] Industry practice is to specify a maximum bale size that may vary among purchasers.
[D] Some purchasers prefer that bale density be as high as 1600 (100).
[E] Experience has shown that if the size range is 95 % by weight, from − 50 to + 6.3 mm (−2 to + ¼ in.) the bulk density requirement can be met and the material will be loose and free flowing.
[F] Bales must conform to bulk density requirements and fall apart when dropped.

The American Society for Testing and Materials takes no position respecting the validity of any patent rights asserted in connection with any item mentioned in this standard. Users of this standard are expressly advised that determination of the validity of any such patent rights, and the risk of infringement of such rights, are entirely their own responsibility.

This standard is subject to revision at any time by the responsible technical committee and must be reviewed every five years and if not revised, either reapproved or withdrawn. Your comments are invited either for revision of this standard or for additional standards and should be addressed to ASTM Headquarters. Your comments will receive careful consideration at a meeting of the responsible technical committee, which you may attend. If you feel that your comments have not received a fair hearing you should make your views known to the ASTM Committee on Standards, at the address shown below.

Designation: E 1183 – 87 (Reapproved 1996)

Standard Test Method for Air Drying RDF-5 for Further Analysis[1]

This standard is issued under the fixed designation E 1183; the number immediately following the designation indicates the year of original adoption or, in the case of revision, the year of last revision. A number in parentheses indicates the year of last reapproval. A superscript epsilon (ε) indicates an editorial change since the last revision or reapproval.

1. Scope

1.1 This test method covers the process of air drying a gross or laboratory sample of RDF-5. The air dry loss is determined by air drying on a drying floor.

1.2 The values given in SI units are to be regarded as the standard. The values in parentheses are for information only.

1.3 *This standard does not purport to address all of the safety concerns, if any, associated with its use. It is the responsibility of the user of this standard to establish appropriate safety and health practices and determine the applicability of regulatory limitations prior to use.* For specific precautionary statements, see Section 5.

2. Terminology

2.1 *RDF-5*—solid fuel derived from municipal solid waste in which the processed combustible fraction is densified (compressed) into the form of pellets, cubettes or briquettes.

3. Significance and Use

3.1 Air drying is performed on a gross sample if it cannot be handled without loss of moisture. A laboratory sample is air dried in order to yield a RDF-5 sample with moisture content near to equilibrium with the atmosphere. Such equilibrium is necessary to ensure uniform conditions throughout subsequent analysis of the RDF-5 laboratory sample for density, size distribution, hydrophilia, breakage, and similar tests. The test is not applicable if the sample is to be tested for total moisture.

4. Apparatus

4.1 *Drying Floor*, a smooth, clean floor area in a room free of contamination by dust or other material and that permits air circulation without excessive heat or air currents.

4.2 *Drying Pans*, clean, dry, non-corroding pans used to contain the sample during the drying process.

4.3 *Balance or Scale*, a device capable of weighing the sample and container with a sensitivity of 0.1 g in 1000 g (0.0002 lb in 2.2 lb.)

4.4 *Glass Stirring Rod*, a clean, smooth glass rod used to stir the sample to facilitate drying.

4.5 *Shovel*, used to stir the sample to facilitate drying.

5. Precautions

5.1 All operations shall be done rapidly and in as few operations as possible because moisture loss depends on several factors other than total moisture content, such as atmospheric temperature and humidity.

5.2 At all times RDF-5 samples should be protected from moisture change due to exposure to rain, snow, wind, and sun or contact with absorbant materials.

5.3 Samples should be transported to the laboratory and analyzed as soon as possible. If any sample handling step involves an extended time period, the sample and container should be weighed before and after the process to determine any weight gain or loss. This weight gain or loss shall be included in the calculation of moisture content.

6. Procedure

6.1 Weigh the sample to the nearest 0.5 g (0.001 lb). Spread the sample on the drying floor to a depth of not more than 15 cm (6 in.). Mix or stir the RDF-5 with a shovel from time to time, being careful not to lose or break particles.

6.1.1 Continue air drying and mixing until the surface of the sample appears dry. Weigh the entire sample and redistribute over the floor for further drying. Continue the drying, mixing, and weighing until the weight loss of the total sample is ≤0.1 %/h. Refrain from excessive drying and rough handling of the RDF-5 in order to avoid particle breakage.

7. Report

7.1 This test method does not yield a numerical quantity that is reported.

---

[1] This test method is under the jurisdiction of ASTM Committee D-34 on Waste Management and is the direct responsibility of Subcommittee D34.06 on Recovery and Reuse.

Current edition approved Aug. 28, 1987. Published October 1987.

Designation: E 1248 – 90 (Reapproved 1997)

Standard Practice for Shredder Explosion Protection[1]

This standard is issued under the fixed designation E 1248; the number immediately following the designation indicates the year of original adoption or, in the case of revision, the year of last revision. A number in parentheses indicates the year of last reapproval. A superscript epsilon (ε) indicates an editorial change since the last revision or reapproval.

1. Scope

1.1 This practice covers general recommended design features and operating practices for shredder explosion protection in resource recovery plants and other refuse processing facilities.

1.2 Hammermills and other types of size reduction equipment (collectively termed shredders) are employed at many facilities that mechanically process solid wastes for resource recovery. Flammable or explosive materials (for example, gases, vapors, powders, and commercial and military explosives) may be present in the as-received waste stream. There is potential for these materials to be released, dispersed, and ignited within or near a shredder. Therefore, explosion prevention and damage amelioration provisions are required.

2. Referenced Documents

2.1 *National Fire Protection Association Standards:*
National Electrical Code
NFPA 13 Sprinkler Systems
NFPA 68 Guide for Explosion Venting
NFPA 69 Explosion Prevention Systems
NFPA 497A Classification of Class I Hazardous (Classified) Locations for Electrical Installations in Chemical Process Areas

3. Terminology

3.1 *Definitions:*

3.1.1 *deflagration*—an explosion in which the flame or reaction front propagates at a speed well below the speed of sound in the unburned medium, such that the pressure is virtually uniform throughout the enclosure (shredder) at any time during the explosion.

3.1.2 *detonation*—an explosion in which the flame or reaction front propagates at a supersonic speed into the unburned medium, such that pressure increases occur in the form of shock waves.

3.1.3 *explosion*—a rapid release of energy (usually by means of combustion) with a corresponding pressure buildup capable of damaging equipment and building structures.

3.1.4 *explosion venting*—the provision of an opening(s) in the shredder enclosure and contiguous enclosed areas to allow gases to escape during a deflagration and thus prevent pressures from reaching the damage threshold.

3.1.5 *explosion suppression*—the technique of detecting and extinguishing incipient explosions in the shredder enclosure and contiguous enclosed areas before pressures exceed the damage threshold.

3.1.6 *inerting*—the technique by which a combustible mixture is rendered nonflammable by addition of a gas incapable of supporting combustion.

3.1.7 *shredder*—a size-reduction machine that tears or grinds materials to a smaller and more uniform particle size.

4. Significance and Use

4.1 Shredder explosions have occurred in most refuse processing plants with shredding facilities. Lessons learned in these incidents have been incorporated into this practice along with results of relevant test programs and general industrial explosion protection recommended practices. Recommendations in this practice cover explosion protection aspects of the design and operation of shredding facilities and equipment used therein.

4.2 This practice is not intended to be a substitute for an operating manual or a detailed set of design specifications. Rather, it represents general principles and guidelines to be addressed in detail in generating the operating manual and design specifications.

5. Design Practices

5.1 *Design Rationale*:

5.1.1 Each of the following design features is better suited for some types of combustible/explosive materials and shredders than for others. The selection of a particular combination of explosion prevention features or damage control features, or both, should be made with an understanding of the types of refuse entering the shredder, shredder operating conditions, the inherent strength of the shredder and surrounding structures, and the operating controls for screening input materials and restricting personnel access during shredding operations.

5.1.2 Several of the following explosion protection design practices are effective for deflagrations but not for detonations. Deflagrations usually result from accumulations of flammable gas-air, vapor-air, or powder(dust) air mixtures in or around the shredder. However, commercial explosives and military ordnance usually generate detonations. A few flammable gases (for example, acetylene and hydrogen) are also prone to

[1] This practice is under the jurisdiction of ASTM Committee D-34 on Waste Management and is the direct responsibility of Subcommittee D34.16 on Processing Equipment and Unit Operations.

Current edition approved Dec. 28, 1990. Published February 1991.

detonate when dispersed in highly turbulent, strong ignition source environments such as exist inside a shredder. Because many explosion protection design practices are not applicable to detonations, rigorous visual detection and removal of detonable material before it enters the shredder is particularly important (6.1).

5.1.3 In view of the difficulties in preventing and controlling all types of shredder explosions, it is important to isolate the shredder and surrounding enclosure from vulnerable equipment and occupied areas in the plant. This is best achieved by locating the shredder outdoors or, if indoors, in a location suitable for explosion venting directly outside. Locations in or near the center of a processing building are not desirable. If the shredder is situated in an isolated, explosion resistant structure, the structure should be designed to withstand the explosion pressures specified in NFPA 68.

5.1.4 The shredder and all contiguous enclosures should be equipped with an explosion protection system consisting of one or more of the following: inerting system (5.2); explosion vents (5.3); explosion suppression system (5.4). Water spray systems (5.5), combustible gas detectors (5.6), and industrial fire protection systems (5.7) should also be installed for additional protection. Adjacent structures and personnel should be protected (5.8).

5.2 *Inerting Systems*:

5.2.1 An inerting system is intended to prevent combustion explosions within a shredder (and contiguous enclosures) by maintaining oxygen concentrations below the level required to support combustion.

5.2.2 The following factors must be accounted for in designing a shredder inerting system: inert gas source and distribution; operating controls and associated instrumentation; leakage of inert gas from and entry of air into enclosures; maintenance and inspection constraints in an oxygen deficient atmosphere during normal operations; effect of inert gas on shredder materials and waste throughput; and contingency plans for inert gas source supply interruption.

5.2.3 Flue gas from an on-site furnace or boiler can be a suitable inert gas providing there is a reliable means to prevent flame propagation into the shredding system and providing flue gas conditioning is installed to maintain suitable temperature (to prevent steam condensation or spontaneous ignition) and flue gas composition (including dew point, oxygen, carbon monoxide, soot, and contaminant concentrations).

5.2.4 Steam from an on-site boiler can be a suitable inert gas providing the temperatures of the shredder and contiguous enclosures are sufficiently high (at least 180°F (82°C)) to prevent steam condensation and the associated increase in oxygen and flammable gas concentrations.

5.2.5 Oxygen concentrations in the shredder and all contig uous enclosures should be no higher than 10 % by volume, unless test data for the particular inert gas employed and the variety of combustibles expected in the shredder demonstrate that a higher oxygen concentration can be tolerated without generating a flammable mixture. Test data for maximum oxygen concentrations for nitrogen and carbon dioxide inerting are as listed in Appendix C of NFPA 69.

5.2.6 Reliable oxygen concentration monitors should be installed, calibrated, and maintained to verify that the maximum oxygen concentration is not being exceeded in the shredder and contiguous enclosures. This will require multiple monitors and sampling points depending on the extent and uniformity of flow in the enclosed volume. Provision for cleaning and clearing sample lines, as recommended in 5.4.5 are needed.

5.2.7 The inert gas distribution system should be designed in accordance with the provisions of Chapter 2 of NFPA 69.

5.3 *Explosion Venting*:

5.3.1 Explosion venting is intended to limit structural damage incurred during deflagrations by allowing unburned gas and combustion products to be discharged from the shredder or contiguous enclosures, or both, before combustion and the associated potentially destructive pressure rise is completed. The effectiveness of explosion venting for a particular explosion depends on the rate of combustion versus the rate of discharge of gases through the explosion vents. The rate of combustion in the shredder or adjacent enclosure depends upon the composition of the combustible gas-air, vapor-air, or dust-air mixture, the size of the shredder/enclosure, and the turbulence level as determined by air flow rates and hammer tip speed.

5.3.2 In general, explosion venting is most effective with large vent areas, low vent deployment pressures, low vent panel weight, and vent locations near the expected ignition source (which is often hammer impact sparks within the shredder). The following quantitative guidelines for these factors are intended to protect against near worst-case flammable gas-air mixtures occupying the entire shredder internal volume.

5.3.3 Explosion vent areas should be sufficiently large to maintain explosion pressures under the damage threshold value for the particular shredder installation. Previously published guidelines relating peak pressure to vent area are not directly applicable to Municipal Solid Waste (MSW) shredders because shredder hammer velocities can increase the combustion rate well above that considered in establishing previous guidelines. The following recommended relationship is based on propane-air explosion tests conducted in a full-scale large shredder mock-up, including rotating hammers **(1)**.[2]

5.3.3.1 The vent area, A_v, required to maintain explosion pressures under the shredder damage threshold (in units of psig), P_M, is given by the equation:

$$A_v = 0.13V^{2/3}P_M^{-0.435}(5 + 0.034v_H) \quad (1)$$

where:

V = shredder internal volume, and

v_H = hammer tip velocity, ft/s.

The calculated vent area will be in the same units as $V^{2/3}$. The metric equivalent, if P_M is in bar, and v_H is in m/s, is

$$A_v = 0.041V^{2/3}P_M^{-0.435}(5 + 0.112v_H) \quad (2)$$

5.3.3.2 If the shredder discharge is at least 3 ft (0.91 m)

[2] The boldface numbers in parentheses refer to the list of references at the end of this practice.

above an unenclosed discharge conveyor, half the discharge area can be credited toward attaining the required vent area, A_v. The difference should be made up with unobstructed explosion vents. No credit should be taken for the inlet area which is usually too obstructed to be an effective vent.

5.3.3.3 To illustrate the use of Eq 1 and 2, consider a hypothetical shredder with an internal volume of 1000 ft^3 (28.3 m^3), including the portion of the inlet hood directly above the hammermill. Let us suppose that structural calculations indicate that the weakest structural member can withstand an applied load equivalent to a hydrostatic pressure of 10 psig (0.70 bar). At the design shaft speed in this shredder, the hammer tip speed is 250 ft/s (76.2 m/s). Substitution of these values into Eq 1 and 2 results in a calculated required vent area of 64 ft^2 (5.95 m^2). If the shredder discharge area is 20 ft^2 (1.9 m^2), an explosion vent of at least 54 ft^2 (5.0 m^2) area should be installed on the shredder.

5.3.4 The explosion vent opening should discharge combustion gases and flame into an unoccupied outdoor area. If the shredder is situated inside a building, vent ducting will be needed to channel gases and flame out of the building. This ducting, which should have a strength at least equal to the shredder itself, should be kept as short as possible in order to avoid further burning and gas compression during venting.

5.3.4.1 Vent ducting of any length will cause the pressure to increase significantly above the value expected for unrestricted venting. The increased pressure can be related to the unrestricted (no duct) vented explosion pressure through Fig. 1. The parameter in Fig. 1 that determines this relationship is the ratio of vent duct volume to shredder volume. In the example in 5.3.3.3, the use of only a 5.5-ft (1.7-m) long duct attached to the 54-ft^2 (5.0-m^2) vent area would represent a duct volume of 300 ft^3 (8.5 m^3), corresponding to a duct/shredder volume ratio of 0 to 3. According to Fig. 1, an explosion pressure of 10 psig (0.7 bar) without the duct would be increased to about 21 psig (1.5 bar) with a duct/shredder volume ratio of 0 to 3.

5.3.4.2 If the pressure increases shown in Fig. 1 are intolerable, a duct with a diverging cross-section area should be used. Apparently, there have not been any published test data on how much divergence is required to prevent significant pressure increases above the unrestricted vent values given by

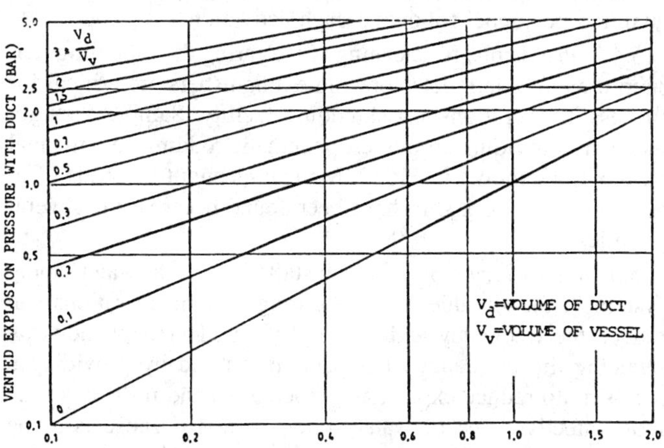

FIG. 1 Influence of Vent Duct Volume on Vented Explosion Peak Pressures (Ref. 2)

Eq 1 and 2. Even with large divergence angles, the vent duct should be designed to withstand a pressure equal to the shredder damage threshold pressure.

5.3.4.3 It is desirable to prevent flammable gas from entering and accumulating in a vent duct during normal shredder operation. Although this is difficult to achieve, two possible approaches are use of a sturdy vent cover (5.3.5), or vent cover and projectile deflector to separate the shredder from the vent duct; or, as a less desirable alternative, use of air sweeping of the vent duct by the induced draft of the shredder or by a high-capacity dust collection or pneumatic transport system, or both. These systems should be equipped with their own explosion protection systems.

5.3.5 Vent covers are usually needed either (preferably) directly on the shredder, or at the far end of the vent duct. Without these covers, dust and debris generated during the shredding process would be ejected and would possibly create a health and safety hazard to nearby personnel. Since impact forces from large ejected debris could prematurely open the vent cover, deflection gratings, heavy chain links, or wire rope are often employed to rebound these missiles back into the shredder.

5.3.5.1 The opening pressure of the vent cover should be low in comparison to the shredder structural damage threshold, P_M. Based on the explosion tests described in EPA Report M2052 **(1)**, it is recommended that the static deployment pressure be no more than $P_M/5$ since the cover will open at a somewhat higher pressure under rapid explosion loads than under static test conditions. The vent opening pressure should also be higher than pressures developed by air motions and waste throughput during normal shredding operations.

5.3.5.2 Several different types of vent covers can be used. Some of the simplest covers consist of rubber flaps on the vent duct outlet. Rain hoods situated at least one vent duct diameter from the end of the duct have also been used in conjunction with a deflector grating inside the duct, but the rain hoods must be restrained from being blown off during the explosion. Vent cover construction and release mechanisms are described in NFPA 68.

5.3.5.3 If explosion vent panels or doors are used, they should satisfy several criteria in addition to the vent area and release pressure criteria cited-previously. First, vent panel/door inertia should be as low as possible, so as not to obstruct the open vent. NFPA 68 specifies that the vent panel area density be less than 2.5 lb/ft^2 (12.5 kg/m^2). Another important factor is that a rugged and shock-resistant hinge or steel cable (of at least 1-cm diameter) is required to prevent the panel/door from flying off during explosions. Finally, there is a need to periodically inspect the vent actuation mechanism so that it will function when needed. Some explosion doors have not deployed at all during shredder explosions involving major damage. Therefore, an inspection and preventive maintenance program is recommended.

5.3.6 During an explosion, flame and potentially damaging pressure waves propagate out from the shredder explosion vent and from the inlet and discharge openings. Blast wave pressures can be estimated from the following hypothesized generalization of test data. Thus, the peak overpressure, P_d, at a distance, d, directly ahead of the vent opening is given by the equation:

$$\frac{P_d}{P_M} = \frac{k}{k + d\sqrt{A_v}} \tag{3}$$

where:

P_M = maximum pressure in shredder (same units as P_d) (as used in Eq 1 and 2, also),

A_v = area of shredder vent or other opening closest to target structure (same units as d^2), and

k = constant Z1.7.

The value of k is probably dependent upon gas composition and shredder volume. The suggested value of 1.7 is based on test data for methane-air explosions in 28-m^3(1000-ft^3) enclosure.

5.3.6.1 Blast pressures exerted on a target located at some angular displacement from a line perpendicular to the vent opening will probably be somewhat less than given by Eq 3 at the same distance, d. However, test data are inconsistent as to how much of a reduction occurs. Therefore, a conservative approach would be to use Eq 3 to estimate pressures at off-axis locations as well as in-line targets.

5.3.6.2 Adjacent structures vulnerable to blast waves include the shredder feed conveyor enclosure and picking station (for removing oversized hazardous materials from the waste streams), the discharge conveyor, dust collection ducting, and the nearest building wall. As an example of the use of Eq 3, consider the blast wave loading on the discharge conveyor for the same hypothetical 1000-ft^3 (28.3-m^3) shredder described in 5.3.3.3. If the discharge conveyor is 4.5-ft (1.4-m) below the 20-ft^2(1.9-m^2) discharge opening, the calculated value of P_d/P_M is 0.63. Thus, the discharge conveyor could be subjected to a 6.3-psig (0.43-bar) blast loading and should be strengthened accordingly to avoid damage. If the discharge conveyor is enclosed, the enclosure may also be subject to internal pressure loads during an explosion and it should be strengthened or explosion vented, or both, in accordance with 5.3.7.

5.3.7 *Discharge Conveyor and Other Downstream Enclosures*—Discharge conveyors, dust collectors, and other enclosures downstream of the shredder require their own explosion vents to cope with explosions initiated either in the shredder or downstream of the shredder. Discharge conveyor and dust collector ducts and other elongated enclosures should have vents installed at regular intervals along the duct, with each vent having an area at least equal-to the duct cross-sectional area. The distance between vents depends on the strength of the duct, but should be no further than 15 vent diameters. Detailed vent area and spacing requirements for elongated pressure-resistant enclosures can be found in NFPA 68. If discharge conveyors are constructed of thin sheet metal without any significant pressure resistance, adjacent equipment and personnel should be protected from metal sections torn asunder during the explosion.

5.3.8 *Exposure to Vented Flames*—Flames emerging from shredder explosion vents can seriously injure nearby personnel and can ignite other combustibles in the vicinity causing secondary fires and explosions. Therefore, personnel and vulnerable equipment should be restricted from the area around the shredder explosion vents during shredder operation. The restricted area should also include the outlet of any explosion vent ducting.

5.4 *Explosion Suppression Systems*:

5.4.1 Explosion suppression systems are intended to detect and suppress an incipient deflagration before pressures reach the damage threshold and before flame is vented from the shredder. Detection of incipient shredder explosions is accomplished with diaphragmatic pressure sensors. Suppression is achieved by the rapid discharge of an extinguishing agent from pressurized containers mounted on the shredder walls. Suppression systems are effective only for deflagrations that require shredder air for oxidation. Deflagrations involving combustible gases or dust/powders with exceptionally rapid burning velocities (for example, hydrogen, acetylene, and aluminum) are not amenable to explosion suppression.

5.4.2 To be effective in a shredding facility, suppression system detectors and agent distribution should cover the entire shredder and all contiguous enclosed areas including inlet hoods, discharge conveyors, reject chutes, dust collection systems, and so forth.

5.4.3 Halogenated and dry chemical suppression agents have been shown to be effective for most deflagration-type explosions. Most agents are equally effective for explosions with relatively slow or moderate rates of pressure rise, but some of the dry chemicals (particularly monoammonium phosphate) may be more effective for explosions with a rapid rate of pressure rise. Halon 1011 (chlorobromomethane) and a hybrid combination of dry chemical and Halon 1301 (bromotrifluoromethane) have been used extensively in refuse shredder suppression systems. The success of a particular agent in a particular installation depends upon the pressure detector setting and location, agent distribution uniformity, and retention of minimum agent concentrations in the shredder/enclosure until combustion is terminated. The latter requirement implies that explosion vent deployment pressures should be greater than suppression system activation pressures.

5.4.4 Explosion suppression system components should be tested and accepted by a nationally accredited approval testing organization. Component system design, installation, inspection, and maintenance should be in accordance with Chapter 4 of NFPA 69. This includes system recharging and testing by qualified personnel following each activation.

5.4.5 Provisions for cleaning and clearing the mountings for suppression system detectors and extinguishers are critical for successful operation in a shredding facility. Sample arrangements for providing air purges, pneumatic rodding, and manual cleanout are shown in Fig. 2. Manual cleanout intervals of at least once per work shift have been found necessary in several facilities.

5.5 *Water Spray Systems*—Installation of a water spray system in the shredder for use during normal operations can reduce the frequency and severity of shredder explosions by reducing the frequency of impact sparks and by providing a heat sink to reduce explosion temperatures and pressures. The most effective type of water spray for this purpose is a fog composed of small droplets in high concentrations, generated from high pressure or atomizing spray nozzles (**5**). An alternative to use during normal operations is to actuate the spray

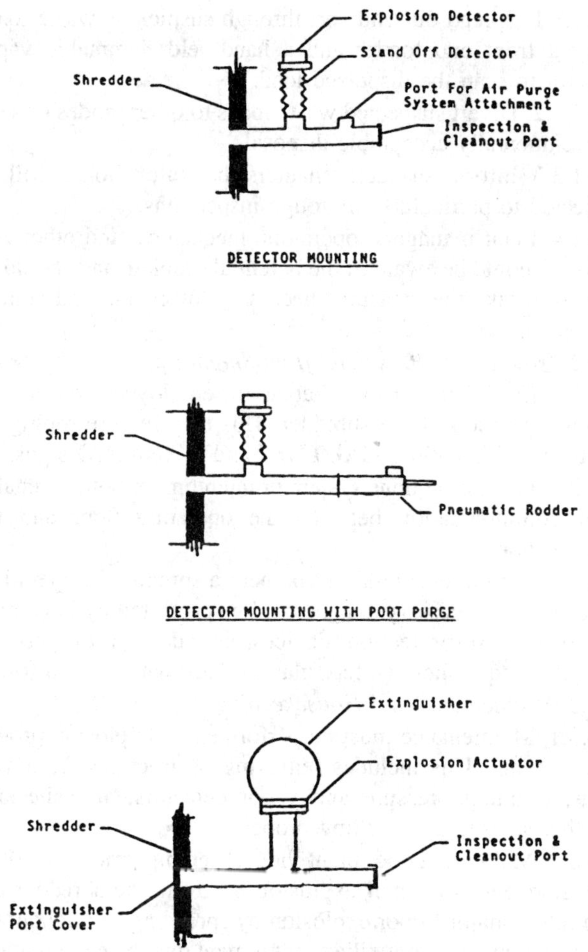

FIG. 2 Conceptual Schematic of Extinguisher Mounting on Side of Shredder (Ref. 3)

nozzles by gas detectors or by any other indication of flammable vapor release in the shredder. A spray system is also useful in extinguishing post-explosion fires and fires ignited during shredder coast-down.

5.6 *Combustible Gas Detectors*:

5.6.1 A combustible gas detection system is an important supplemental explosion protection measure. For use in a refuse shredding facility, a typical system would consist of one or more sampling lines, a sampling pump, a flammable gas sensor, an audible and visual alarm and possibly electrical relays to control other explosion protection or refuse processing equipment. Several types of gas-sensing units and calibration schemes are available. The selected unit should be listed with a nationally recognized approval organization and should be responsive to a wide variety of flammable vapors including hydrocarbons, alcohols, ketones and ethers. Periodic thorough inspection and maintenance (including calibration) is critical if the system is to remain operational in a refuse shredding environment.

5.6.2 The locations of the sampling lines should be based on site specific considerations. For example, if there is a larger air flow rate out of the shredder towards the shredder inlet hood, it would be desirable to locate a sample line in the hood. If there is very little air flow, it would be desirable to locate a sample line in the shredder itself. One possible location in the shredder where vapors have been reported (**1**) to accumulate is at the end of the shaft in a horizontal hammermill. It is also desirable to locate a detector downstream of the shredder. Periodic automated purging and rodding of sensor ports, as recommended for explosion suppression system components, should also be used on sampling parts and sensors outfitted with flame arrestors.

5.6.3 For most catalytic combustion-type sensors, calibration with heptane provides the most sensitive indicator of common hydrocarbon vapors. Catalytic sensors must be tested and replaced periodically (monthly intervals are reported in Ref. **4**) because of contamination/poisoning by a variety of vapors including compounds containing lead, silicon, or halogens. The sensor should be tested following activation of a suppression system using a halogenated agent such as Halon 1011.

5.6.4 Sensor set points commonly used in other processing industries include alarm annunciation at 25 % of the calibrated lower explosive limit (LEL) and relay tripping at 50 % of the LEL. In a refuse shredding facility, the relay might signal building evacuation, shut down the shredder and the conveyor that feeds the shredder, and trigger a water spray system in the shredder.

5.6.5 Battery-operated, hand-held, flammable vapor detectors are useful for screening incoming waste and for checking for the presence of flammable vapor in other areas of the shredding facility following an installed vapor detector alarm or an explosion, or both.

5.7 *Fire Protection*:

5.7.1 Standard industrial fire protection equipment, including automatic sprinklers, applicable to plants with ordinary combustible process materials, should be provided throughout the plant in order to fight a fire generated following a shredder explosion. Water flow rates, sprinkler size and spacing, and so forth, should be as specified in NFPA 13. Sprinkler piping and supports should be located so as to provide protection against damage from explosions.

5.7.2 A water spray on the shredder inlet and outlet conveyors is useful for putting out fires caused by shredder explosions and for wetting materials in the event the liquids or gases are detected. Spray nozzles should be able to be actuated from the control room and other work stations. Hose stations near the shredder and at intervals along the feed and discharge conveyors should also be provided.

5.8 *Protective Barriers*:

5.8.1 Shredders should be located away from control rooms, offices, restrooms, lounges, and so forth. The potential effects of an explosion on critical plant systems such as electrical cables, control cables, or air lines should also be taken into account when choosing shredder location. If a control room is located near a shredder, the room should be designed to withstand blast pressures (5.3.6) and the impact from projectiles. Control room doors should be designed to open outward, control room windows should be of small cross-sectional area (for example, 10 by 10 in. is the NFPA recommendation for the maximum size of fire-rated windows) and constructed of pressure and impact resistant materials (for example, polycarbonate plastic or wired glass).

5.8.2 Blast mats or similar blast resistant barricades should be installed between the shredder and plant areas normally occupied by personnel. Picking stations represent one such area sometimes located near the shredder inlet. It should be noted that blast mats stop missiles and shrapnel, but only deflect pressure fronts. The blast mat supports should be designed to carry the blast load (specified in 5.3.6) exerted on the mat during an explosion.

5.9 *External Ignition Sources*—Flammable vapors generated and released in the shredder have sometimes been ignited by equipment located downstream and, in at least one incident, upstream from the shredder. This can occur if additional vaporization or vapor accumulation, or both, occurs downstream, or if the flammable vapor does not encounter an ignition source in the shredder, but is exposed to a downstream ignition source. Potential downstream ignition sources include furnaces, various electrical equipment such as motors, lights, relays, and so forth, and conveyor transfer points (where scrap metal can be lodged against moving parts and subjected to frictional heating or sparking). Electrical equipment on or near the shredder discharge conveyor and other contiguous equipment should be approved for use in hazardous (classified) areas as defined by the National Electrical Code and NFPA Standard 497A.

6. Operating Practices

6.1 *Employee Training and Visual Inspection*—Refuse haulers as well as plant personnel should be trained to identify potentially exposive materials and instructed in regard to safe removal and disposal. Visual detection and removal of potentially explosive or damaging materials is one of the most effective methods of reducing the size and frequency of shredder explosions. However, it cannot be relied upon to prevent all, or even most, explosions.

6.1.1 All plant personnel should be made aware that, by looking for potentially dangerous materials, they can play an important part in reducing the hazard to themselves and the equipment. Among the types of materials to be visually detected and removed are gasoline cans or tanks, lawnmower engines, compressed gas cylinders, cases of bullets or other ammunition, commercial and military explosives, combustible powders (for example, aluminum powder), paint thinner, and cartons of aerosol spray cans.

6.1.2 Visual detection is a tedious job and provision must be made to relieve monotony among plant personnel (for example, by rotating job assignments). The effectiveness of the visual inspection increases as the depth of the waste stream on the conveyor decreases. Visual inspection of the input stream to the shredder may be done from the control room, preferably through the use of remote TV cameras.

6.1.3 Incoming vehicles should be considered suspect and subjected to special screening in the following situations: they have previously been identified as hauling potentially explosive waste; they are hauling industrial or commercial waste from facilities known to contain large quantities of combustible liquids or powders; or, they trigger a positive response from a hand-held flammable vapor detector. Once the suspected haulers have been identified, the measures described in 6.1.3.1-6.1.3.3 can be taken to reduce potential explosions.

6.1.3.1 Spread out and sort through suspicious waste loads using a front end loader and a hand-held flammable vapor detector to scan the dispersed load.

6.1.3.2 Divert suspected waste loads to other modes of safe, environmentally acceptable disposal.

6.1.3.3 Inform suspected haulers that future loads will be subjected to particularly thorough inspections.

6.1.4 Plant managers, operators, mechanics, and other employees should be aware of the potential problem and should be trained to take the measures necessary to prevent and control explosions.

6.2 *Isolation of Personnel from Shredder*:

6.2.1 Establish "off limit" areas for employees and visitors during operation of the shredder. This may require roping off the area and installing **CAUTION** and **WARNING** signs.

6.2.2 Provide a phone system or intercom, or both, to enable clear communication between the operating floor and the control room.

6.2.3 Personnel should not be near a shredder in operation. If absolutely necessary, they should be protected by blast mats or other blast protection devices, in addition to personal protective equipment (safety glasses, hard hats, and so forth).

6.3 *Maintenance and Housekeeping*:

6.3.1 Maintenance must be performed on explosion protective systems. This includes removing obstructions from vent areas, cleaning pressure and vapor detectors, and checking whether sensors are working properly.

6.3.2 Maintain good plant housekeeping practices. What may start out as a small explosion or fire in the shredder can turn into a major fire or explosion by spreading throughout the conveyors and other auxiliary equipment due to accumulations of dust and shredder refuse.

6.4 *Post-Explosion Procedures*—The post-explosion procedures described in 6.4.1-6.4.8 through are recommended to protect personnel, reduce fire damage, and facilitate recovery from the explosion.

6.4.1 Evacuate the building and assemble all personnel in a predesignated place for a head count. Toxic gases may be generated and secondary explosions may occur following a shredder explosion. Therefore, personnel should not reenter the facility immediately. A portable combustible gas detector should verify the dispersal of residual flammable gas concentrations throughout the facility before personnel reenter to begin cleanup and repair operations.

6.4.2 Following an actuation of an explosion suppression system that uses a halogenated hydrocarbon such as bromochloromethane (Halon 1011) as the suppression agent, operating equipment must be purged before permitting access to personnel or personnel must use auxiliary breathing equipment when reentering the shredder area. Eye protection must be worn at all times.

6.4.3 Stop the conveyor that feeds the shredder.

6.4.4 The decision whether to keep the dust collection system running is site-specific and will be dependent primarily upon the type of dust cleaning devices employed, and the type of fire protection used within the dust collection system.

6.4.5 If any nonautomatic sprinkler systems are used in the areas affected by the explosion, these should be activated as soon as possible after the explosion, consistent with personnel safety.

6.4.6 Call the fire department. An automatic or semiautomatic alarm to the nearest fire department should be considered.

6.4.7 If operable, the shredder outlet conveyor may be kept running after an explosion in order to clear burning material from the shredder. This is especially advisable where the burning material can be diverted outside the facility to facilitate fire-fighting. The decision to keep outlet conveyors running is site-specific, however, because for some facilities it may be easier to fight a fire in the shredder itself than in areas fed by the outlet conveyors.

6.4.8 At the discretion of the facility operator, the shredder itself may either be kept running or stopped. Keeping the shredder running has the advantage of clearing the shredder of burning material. Stopping the shredder may reduce windage that may fan the flames but may also leave a *plug* of burning material inside the shredder to be extinguished and later removed (usually manually). The shredder will continue running (by momentum) for a period of several minutes even after the motor is turned off, but this period may not be sufficient to clear the shredder. For each facility, the decision of whether to keep the shredder running will be more dependent upon operator preferences for fire-fighting and material cleanup than upon explosion safety considerations.

6.5 *Miscellaneous Plant Specific Operating Practices*—A detailed listing of miscellaneous pre-explosion and post-explosion operating practices that have evolved in one plant in which the operators have experienced numerous documented shredder explosions is reported in Ref. **4.**

REFERENCES

(1) Zalosh, R. G., and Coll, J. P., "Determination of Explosion Venting Requirements for Municipal Solid Waste Shredders," *EPA Report M2052*, July 1982.

(2) van Laar, G. F. M., "Limitations in the Use of Explosion Reliefs," *A Practical Introduction to Gas and Dust Explosion Venting*, 1985 Seminar Proceedings published by European Information Centre for Explosion Protection, Hove-Antwerp, Belgium. Original figure from Aellig, A., and Gramlich, R., paper in VDI-Berichte No. 494, pp. 176–183, 1984

(3) Christiensen, H., Director, Department of Solid Waste, Monroe County, New York, Presentation at ASTM E-38 Committee Seminar on Shredder Explosions, Bal Harbor, FL, November 1985. Available from ASTM Headquarters.

(4) Nollet, A. R., Robinson, D. F., and Greeley, R. H., "Uptake on Shredder Explosions," ASME Solid Waste Processing Division Conference, 1986.

(5) Nollet, A. R., Sherwin, E. T., and Madora, A. W.," An Approach to Energy Attenuation of Explosive Wastes in Processing Equipment," U.S. Bureau of Mines/I.I.T. Research Institute, Proceeds of the Sixth Mineral Waste Utilization Symposium, Chicago, IL, May 1978.

Designation: E 1266 – 88 (Reapproved 1999)

Standard Practice for Processing Mixtures of Lime, Fly Ash, and Heavy Metal Wastes in Structural Fills and Other Construction Applications[1]

This standard is issued under the fixed designation E 1266; the number immediately following the designation indicates the year of original adoption or, in the case of revision, the year of last revision. A number in parentheses indicates the year of last reapproval. A superscript epsilon (ϵ) indicates an editorial change since the last revision or reapproval.

1. Scope

1.1 This practice provides descriptions and references of existing test methods and commercial practices relating to the processing of lime, fly ash, and heavy metal wastes in construction applications.

1.2 *This standard does not purport to address all of the safety concerns, if any, associated with its use. It is the responsibility of the user of this standard to establish appropriate safety and health practices and determine the applicability of regulatory limitations prior to use.*

2. Referenced Documents

2.1 *ASTM Standards:*

C 5 Specification for Quicklime for Structural Purposes[2]

C 25 Test Method for Chemical Analysis of Limestone, Quicklime, and Hydrated Lime[2]

C 109 Test Method for Compressive Strength of Hydraulic Cement Mortars (Using 2-in. or 50-mm Cube Specimens)[2]

C 110 Test Methods for Physical Testing of Quicklime, Hydrated Lime, and Limestone[2]

C 206 Specification for Finishing Hydrated Lime[2]

C 207 Specification for Hydrated Lime for Masonry Purposes[2]

C 311 Test Method for Sampling and Testing Fly Ash or Natural Pozzolans for Use as a Mineral Admixture in Portland Cement Concrete[3]

C 400 Test Method for Quicklime and Hydrated Lime for Neutralization of Waste Acid[2]

C 593 Specification for Fly Ash and Other Pozzolans for Use with Lime[2]

C 618 Specification for Fly Ash and Raw or Calcined Natural Pozzolan for Use as a Mineral Admixture in Portland Cement Concrete[3]

C 821 Specification for Lime for Use with Pozzolans[2]

C 911 Specification for Quicklime, Hydrated Lime, and Limestone for Chemical Uses[2]

C 977 Specification for Quicklime and Hydrated Lime for Soil Stabilization[2]

D 559 Test Methods for Wetting and Drying Compacted Soil-Cement Mixtures[4]

D 560 Test Methods for Freezing and Thawing Compacted Soil-Cement Mixtures[4]

D 1557 Test Method for Laboratory Compaction Characteristics of Soil Using Modified Effort (56,000 ft-lbf/ft^3(2,700 kN-m/m^3))[4]

D 1633 Test Method for Compressive Strength of Molded Soil-Cement Cylinders[4]

D 2434 Test Method for Permeability of Granular Soils (Constant Head)[4]

D 2435 Test Method for One-Dimensional Consolidation Properties of Soils[4]

D 3877 Test Methods for One-Dimensional Expansion, Shrinkage, and Uplift Pressure of Soil-Lime Mixtures[4]

D 3987 Test Method for Shake Extraction of Solid Waste with Water[5]

D 4318 Test Method for Liquid Limit, Plastic Limit, and Plasticity Index of Soils[4]

E 850 Practice for Use of Inorganic Process Wastes as Structural Fill[5]

2.2 *Environmental Protection Agency Documents:*

EPA Resource Conservation and Recovery Act (RCRA)[6]

EPA/SW-846 Test Methods for Evaluation of Solid Waste[7]

EPA Method 1310 Extraction Procedure (EP) Toxicity Test Method and Structural Integrity Test[7]

EPA/SW-872 Properties of Stabilized/Solidified Waste[7]

RCRA Document EPA-IAG-D4-0569 Guide to the Disposal of Chemically Stabilized and Solidified Waste[7]

Solvents[8]

Hazardous and Solid Waste Amendments (HSWA)[8]

Method 9095 Paint Filter Liquid Test (PFLT)[7]

EPA/530-SW-85-0031 Petitions to Delist Hazardous Waste:

---

[1] This practice is under the jurisdiction of ASTM Committee D-34 on Waste Management and is the direct responsibility of Subcommittee D34.06 on Recovery and Reuse.

Current edition approved Oct. 12, 1988. Published November 1988.

[2] *Annual Book of ASTM Standards*, Vol 04.01.

[3] *Annual Book of ASTM Standards*, Vol 04.02.

[4] *Annual Book of ASTM Standards*, Vol 04.08.

[5] *Annual Book of ASTM Standards*, Vol 11.04.

[6] Documents 12/18/78, 9/13/79, 5/26/82, 7/26/82, and 4/4/83, available from *Federal Register* U.S. Government Printing Office, Superintendent of Documents, Washington, DC 20402.

[7] Available from Environmental Protection Agency, U.S. Government Printing Office.

[8] Amendments available from Federal Register.

A Guidance Manual, NTIS: PB 85-194488[7]

EPA/530-SW-86-008 OWSER Policy Directive No. 9527.00-1A, Guidance Manual for Research, Development, and Demonstration Permits Under 40 CFR 270.65[7]

EPA/530-SW-86-016 OWSER Policy Directive No. 9487.00-2A, Prohibition on the Placement of Bulk Liquid Hazardous Waste in Landfills[7]

EPA/540-2-86-001 Handbook for Stabilization/ Solidification of Hazardous Waste, Superfund Document[7]

2.3 *Code of Federal Regulations:*

40 CFR 264 Subpart B, section 264.13, Hazardous Waste Management System, Land Disposal Restrictions, Proposed Rule, Dec. 11, 1988[8]

40 CFR 268 Hazardous Waste Management System; Land Disposal Restrictions; and California List Constituents

2.4 *Department of the Interior Document:*

U.S. Department of the Interior Earth Manual (Section Edition), 1974[9]

2.5 *Corps of Engineers Document:*

1110-2-1906 Permeability of Fine Materials, Falling Head Aug. 12, 1987.[10]

3. Terminology

3.1 *Definitions:*

3.1.1 *fly ash*—finely sized ash generated from combustion of pulverized coal. Descriptions and types are listed in Specifications C 593 and C 618.

3.1.2 *heavy metal wastes*—industrial wastes containing heavy metals such as arsenic, cadmium, chromium, barium, lead, silver, selenium, and mercury; these wastes are generally liquids, sludges, or filter cakes.

3.1.3 Heavy metal wastes may also contain small amounts of organic compounds. Special provisions are referenced to accommodate this class of material as stated in 8.4.

3.1.4 *lime*—a commercial product derived from the calcination of high calcium or dolomitic limestone. A number of ASTM standards relating to lime are given in 2.1.

3.1.5 *monolithic mass*—a mass that has good dimensional stability, to freezing and thawing resistance, low permeability, a high bearing capacity, and resistance to attack by biological agents. The EPA states that an end product such as this could be used as a foundation for buildings or roads, or simply buried and covered over in a landfill (EPA/SW-872).

3.1.6 *resource application*—use of stabilized products in specific areas such as earth liners, foundations, road base, backfills, embankments, earth dams, etc.

3.1.7 *resource structural products*—structural products produced by lime, fly ash, and heavy metal waste; examples are block, brick, aggregates, gabions, and miscellaneous structural shapes.

3.1.8 *solidification*—a binding physical and chemical treatment process that transforms materials containing free liquids into a solid, soil-like, or clayey material. This solid material can be a monolithic block with structural integrity.

3.1.9 *stabilization*—a treatment process that involves both a physical and chemical reaction for treating heavy metal waste. Heavy metal wastes are considered stabilized when they meet current applicable regulatory requirements.

3.1.10 *structural landfill*—man-made earth work meeting engineered practices and structural requirements. The fill must also be environmentally acceptable and meet EPA requirements. (See 40 CFR 268.)

4. Significance and Use

4.1 This practice provides users with current methods for preconditioning, handling, processing, and means of characterizing the materials that are produced.

4.2 Lime and fly ash, and mixtures of lime and fly ash can be useful for treating hazardous and nonhazardous waste as follows:

4.2.1 Treating hazardous waste for potential resource recovery application,

4.2.2 Solidifying liquids and sludges that are banned from land disposal because of excess free liquid content,

4.2.3 Treating hazardous waste that may require treatment because of hazardous constituents prior to land disposal, and,

4.2.4 Treating hazardous waste for potential delisting to a nonhazardous waste status. Each one of these applications, however, must comply with requirements of the Resource Recovery and Conservation Act and the Hazardous and Solid Waste Amendments .

5. Properties and Uses of Materials Applicable to the Practice

5.1 *Commercial Lime*— The following are properties and uses of commercial lime.

5.1.1 Neutralizes acids;

5.1.2 Precipitates and reduces the solubilities of heavy metals;

5.1.3 Provides high absorption rates of aqueous and non-aqueous liquids;

5.1.4 Solidifies and hardens a number of inorganic waste sludges;

5.1.5 Reacts chemically with soils, particularly clays, and thereby reduces plasticity; improves dimensional stability; and develops and controls structural applications;

5.1.6 Develops cements when mixed with natural pozzolans, such as diatomaceous earth, cherts, shales, volcanic ash, and also fly ash formed in the combustion of pulverized coal; and

5.1.7 Capable of increasing pH of heavy metal waste.

5.2 *Pulverized Coal Fly Ash*—The following are properties and uses of pulverized coal fly ash.

5.2.1 Serves as a filler in the treatment of liquid waste;

5.2.2 Provides siliceous glass that reacts with lime to form cementitious compounds (tobermorites);

5.2.3 Provides aluminous glass which reacts with lime and sulfates to form cementitious compounds (ettringites); and

5.2.4 Contributes to stabilizing heavy metals that are insolubilized with lime.

5.2.5 Fly ash is available in different classes depending on the type of coal. These classes are described in Specification C 618 and in Test Method C 311. Class C contains some free

[9] Available from Bureau of Reclamation, Department of the Interior, Code D/7923A, P.O. Box 25007, Denver, CO 80225.

[10] Available from Department of the Army, U.S. Army Corps of Engineers, Public Depot, 2803 52nd Ave., Hyattsville, MD 20781.

calcium oxide that can generate considerable heat when mixed with water. In some applications, this type of fly ash may need to be preconditioned as described in 7.1.1. Standards pertaining to lime and lime/fly ash are Test Methods C 25, C 110, C 311, and C 400, Specifications C 5, C 206, C 207, C 593, C 618, C 821, C 911, and C 977.

NOTE 1—Additional information may be found in Test Methods C 109, D 1557, D 1633, D 2434, D 2435, D 3877, D 3987, and D 4318.

6. Applications Pertaining to Hazardous Wastes

6.1 *Resource Recovery Application*—Lime fly ash mixtures can be used to solidify and stabilize the heavy metal waste and render these treated wastes suitable for use as a resource structural product. In this application, the lime and fly ash mixtures solidify the waste and stabilize the heavy metals contained in the waste.

6.2 *Solidifying Waste Liquids and Sludges*—Lime/fly ash mixtures may be useful for stabilizing/solidifying liquids and sludges that are banned from land disposal because they contain free liquids. Mixtures of lime/fly ash can be used to react with the aqueous portion of the waste, thereby solidifying it so that the treated waste will pass the EPA paint filter liquid test (PFLT) (Method 9095) and other RCRA regulatory requirements as specified under SW-846 and thus be acceptable for disposal into hazardous waste landfills.[11] In some cases, the liquid waste treated by the lime/fly ash mixtures may be required to also pass an unconfined compressive strength test, in conjunction with the PFLT. Requirements and guidance for the PFLT and compression test can be found in EPA/530-SW-86-016.

6.3 *Treating of Hazardous Waste Prior to Land Disposal*—Lime and fly ash may be acceptable materials for treating selected heavy metal waste by stabilization/solidification when such waste requires treatment prior to land disposal because of specific hazardous constituents. The EPA is evaluating stabilization and solidification for treating heavy metal waste materials. Specific requirements are listed in EPA/530-SW-86-016.

6.4 *Delisting of Hazardous Waste*—In some cases, lime/fly ash mixtures may be useful in treating hazardous waste to render them nonhazardous and, therefore, potentially applicable for delisting. Appropriate mixtures of lime and fly ash for treating a waste for delisting will need to be determined on a case by case basis. Procedures and requirements for petitioning for delisting of a hazardous waste could require a research development and demonstration project permit (see EPA/530-SW-85-0031 and EPA/530-SW-86-008).

6.5 The appropriate mixtures of the lime/fly ash that will treat the waste to meet the requirements will need to be determined on a case by case basis. Presence of organics may interfere in the treatment process, and appreciable amounts can obviate the use of the lime/fly ash systems.

7. Laboratory Procedures to Determine Design of Mixtures

7.1 Hydrated lime/fly ash mixtures and proportions are prepared and tested using the following ASTM standards:

Unconfined compressive strength	Test Method C 109
Lime for use with pozzolans	Specification C 821
Lime for chemical uses	Specification C 911
Moisture density	Test Method D 1557
Confined compressive strength	Test Method D 1633

7.1.1 The results of these tests may serve as a basis for establishing mixtures appropriate for the structural applications under consideration. Compressive strength may range from a high strength value for applications as listed in (3.6)3.1.7 to low strengths for products as listed in (3.4)3.1.5. Sufficient lime is added to obtain the desired strength at optimum moisture content.

7.2 Lime/sludge mixtures are run to determine quantity of lime necessary to neutralize acid and precipitate the heavy metals to an appropriate pH. The EPA provides the solubilities of metal hydroxides as a function of pH (40 CFR 268). Methods C 400 is also helpful in this item.

7.3 The lime/fly ash blend is added to the lime-treated heavy metal waste in sufficient quantities to comply with the necessary requirements for the contemplated use.

7.3.1 Compressive strength tests of the final mixture may be compared with the previous results (7.1)7.1. If major changes such as loss in strength occur, determine if additional curing time or an increase in the lime dosage is needed.

NOTE 2—Quicklime can be used as an alternate to hydrated lime. This may reduce the amount of water in a heavy metal sludge because of the heat of hydration when quicklime is used in place of the hydrate. Since quicklime consumes considerable water in hydration, the quicklime/fly ash blend may be added dry to the wet, heavy-metal waste sludge as an alternate procedure that may reduce the lime/fly ash requirement.

7.4 To complete the laboratory tests, the following test methods may be useful, depending on the particular application:

Wet/dry weathering	Test Methods D 559
Freezing and thawing	Test Methods D 560
Falling head permeability	Corps of Engineers 1110-2-1906
EP Toxicity Test	EPA Method 1310

NOTE 3—If Extraction Procedure, (EP) Toxicity Test shows excessive concentration of soluble ingredients, additional curing may be beneficial.

8. Construction Practice

8.1 Lime and fly ash are usually stored in closed bins such as employed at plants that are designed to provide lime/fly ash/aggregate mixtures for use in construction of roads. These plants frequently employ conventional equipment for blending lime/fly ash and soil, and are adaptable for weighing and mixing lime and fly ash with wet sludges. This equipment is frequently portable and can be located at the construction sites.

8.1.1 Class F fly ash can be stockpiled wet for a maximum of two weeks. Longer periods of stockpiling may affect the reactivity of the ash. Class C fly ash should not be stored wet.

8.1.2 When a dry Class C fly ash is used, adding water to the lime/Class C fly ash mixture will usually generate considerable heat. After cooling the freshly formed mixture, the sludge should be added within a few hours. If the sludge cannot be used within 24 h, it is generally necessary to precondition the lime/Class C fly ash and water mixture by rerunning the blend through a pug mill to avoid formulations of solid slabs. After remixing, the lime/fly ash can be kept in a stockpiled condition until the heavy metal waste is available for preparing the final

[11] When the Liquids Release Test (Method 9096) is promulgated, it will be required under RCRA and specified under this practice.

mixture for the field project. This also can be accomplished in a pug mill.

8.2 Where structural shapes are formed, it is generally acceptable to supply the lime and fly ash mixture in a moistcondition and use separate containers for the waste. These materials can be fed through the plant equipment using a variety of mixers where the blending and addition of water is accomplished. Examples are found in block or brick plants. The equipment and practice is found in commercial plants which produce items such as briquettes, concrete block, or brick. Aggregate can be formed by crushing and screening the shapes to desired size.

8.3 Storage of the processed waste is a major factor when the material is intended to be used as a resource structural product. This is useful in providing curing time for the shapes. It is also necessary to build stockpiles in order to meet scheduling required by the contractors. Storage of the material is carried out in several ways, such as open warehouse, open piles using tarps, and open or closed bins.

8.4 Construction of monolithic fill should conform to standard practices employed with conventional materials. The U. S. Department of the Interior Earth Manual provides suitable construction practices. Inspection of the fill should be carried out during construction to ensure compliance with specifications. Practice E 850 contains special provisions that are related to this section. The EPA has developed requirements to restrict the disposal of untreated industrial waste containing heavy metal wastes and organic materials in the landfills such as EPA/SW-872, EPA/530-SW-85-0031, EPA/530-SW-86-008, EPA/530-SW-86-016; and EPA/540-2-86-001.

Last ASTM Designation: E 1430 – 91

Standard Guide for Using Release Detection Devices with Underground Storage Tanks

This guide provides tank owners/operators, regulators, equipment manufacturers/suppliers, and others concerned with tanks, with information relevant to the selection and use of release detection devices for underground storage tanks (USTs) that contain petroleum products or other regulated substances. As used in the phrase "release detection devices," the meaning of release includes leak or spill. The "state-of-the-art" in release detection technology and techniques is rapidly evolving; therefore, this guidance cannot and will not provide design-related answers to when, where, or how to use or select appropriate detection devices. Rather, this guide introduces a common approach for using and selecting detectors as part of the design and use of an overall detection system.

Formerly under the jurisdiction of ASTM Committee E50 on Environmental Assessment, this guide was discontinued in February 2000.

Designation: E 1526 – 93

Standard Practice for Evaluating the Performance of Release Detection Systems for Underground Storage Tank Systems[1]

This standard is issued under the fixed designation E 1526; the number immediately following the designation indicates the year of original adoption or, in the case of revision, the year of last revision. A number in parentheses indicates the year of last reapproval. A superscript epsilon (ε) indicates an editorial change since the last revision or reapproval.

INTRODUCTION

This practice provides the general procedure for evaluating and reporting the evaluation results of a system designed to detect releases from underground storage tanks containing regulated substances. It includes release detection systems that are designed to detect leaks from tanks, pipelines, and leaks into the interstitial area of double-walled tanks and secondary containment systems, and releases and spills in the backfill and native soil outside of the tank system. Internal release detection systems include but are not limited to volumetric leak detection systems (also known as precision test and tank integrity tests), automatic tank gaging systems, and pipeline leak detectors. External release detection systems include but are not limited to vapor monitors and liquid monitors.

Intended Use—This practice is intended to provide minimum requirements for evaluating all types of release detection systems, and a standard reporting procedure for the results. The procedure is general so that all types of systems and applications of these systems will be included.

Two Levels of Evaluation—A procedure is provided in this practice for evaluating performance characteristics of the release detection systems, and the performance of the systems under actual field test conditions. This practice requires that the performance characteristics of all systems be evaluated, and that an estimate be made of the capability of all interstitial and in-tank release detection systems to detect leaks in the tank walls, piping, or connections in the field.

Disclosure of Experimental Conditions—The evaluation of the release detection method must be done experimentally, and a full disclosure must be made of the experimental conditions under which the evaluation was performed, and the method was recommended for use.

ASTM Designation or Marking—In accordance with Section 6, the manufacturer may mark on its literature and the detection system, that the system was evaluated and the evaluation data reported in accordance with this practice.

1. Scope

1.1 This practice covers a uniform protocol for evaluating the performance of systems used to detect releases of regulated substances from underground storage tank systems or facilities. This practice applies to users, manufacturers, vendors, government regulators, and others using or concerned with release detection systems.

1.2 This practice covers release detection systems that are used internally, externally, or within the interstitial area of an underground storage tank system that has secondary containment.

1.3 This practice does not specify minimum performance standards.

1.4 The release detection system may consist of an individual or multiple tests. The release detection system may operate in a continuous or intermittent mode, and may produce either quantitative output (measured value) or qualitative-output (on/off) results.

1.5 The release detection system may be applicable to any part of a storage tank system, a single storage tank system, or combination of systems. The manufacturer may further limit the use of a release detection system to specific test conditions (for example, tank sizes, geographical regions, etc.).

1.6 The evaluation given in this practice provides two separate performance descriptions: an estimate of the performance characteristics of the release detection system, and an estimate of probability of release detection and probability of false alarm.

1.6.1 This practice requires that the system performance characteristics must be determined for all release detection systems as described in 1.2.

1.6.2 Further, this practice currently requires the determination of an estimate of probability of false alarm.

1.6.3 Estimates of probability of detecting releases and false

[1] This practice is under the jurisdiction of ASTM Committee E-50 on Environmental Assessment and is the direct responsibility of Subcommittee E50.01 on Storage Tanks.

Current edition approved March 15, 1993. Published May 1993.

alarms for external release detection systems are not required.

1.7 The values stated in inch-pound units are to be regarded as the standard. The SI units given in parentheses are for information only.

1.8 *This standard does not purport to address all of the safety problems, if any, associated with its use. It is the responsibility of the user of this standard to establish appropriate safety and health practices and determine the applicability of regulatory limitations prior to use.* A full disclosure is required if hazardous materials are released into the environment as part of the test procedure.

1.9 The sections and subsections in this practice are arranged in the following order:

2. Terminology

2.1 *Definitions:*

2.1.1 Definitions of pertinent terms are presented in Appendix X1 of this practice.

3. Release Detection System Description

3.1 The manufacturer must briefly describe the release detection system, its intended use, application, operational conditions, recommended calibration schedule, expected lifetime, and safety considerations.

3.1.1 *General Description*—A brief description of the release detection system must include the following:

3.1.1.1 The type of release detection system (internal, interstitial, or external) must be stated.

3.1.1.2 A clear statement of what the release detection system is designed to detect must be given.

3.1.1.3 A brief description of the scientific principle of operation of the release detection system must be given.

3.1.1.4 The frequency of the measurement (that is, continuous, periodic, or discrete) must be given.

3.1.1.5 A brief description of the apparatus and materials, test protocol, data analysis, and detection criteria must be given.

3.2 *Intended Use*—The intended use of the system must be given.

3.2.1 The applicability of the release detection system with different types and sizes of underground storage tank systems must be stated.

3.2.2 The part of the underground storage tank system to which the release detection system is applicable must be specified.

3.2.3 The applicability of the release detection system with specific stored products must be stated.

3.3 *Operational Conditions*—The range of applicable environmental conditions under which the release detection system is expected to operate successfully must be stated, and the conditions, practices, and materials for which the system is not expected to operate successfully should also be stated.

3.4 *Recommended Calibration Schedule*—Calibration procedures and recommended frequency for calibration of the release detection system or system components must be clearly stated.

3.5 *Expected Lifetime*—The expected lifetime (under specified conditions) of the release detection system and its components must be stated.

3.6 *Safety Considerations*—Known safety considerations, including compliance with OSHA regulations, special chemical concerns, and specific safety procedures must be documented.

4. Development of Performance Data

4.1 The performance evaluation in this practice requires the determination of performance characteristics for all release detection systems; it further requires an estimate of probability to detect release and false alarm for internal and interstitial release detection systems.

4.1.1 *Performance Characteristics*—An evaluation of the performance characteristics of all release detection systems, at a minimum, requires determination of the specificity of the release detection system to the chemical substance or physical property to be detected (that is, signal); measurement of the detection limit, dynamic range, precision and bias, response time for the signal; and quantification of those physical properties that significantly affect the magnitude of the signal; and identification and quantification of those chemical substances or physical properties that the system also responds to and that interferes with the detection of the signal (that is, interferences or ambient noise).

4.1.2 The performance results will be summarized in tables. Technical guidance on how to make the calculations is presented in Appendix X2 of this practice. This technical guidance is taken from reports that are intended to estimate these quantities for external release detection systems. Any standard engineering estimate of these quantities will suffice.

4.2 *Estimate of Probability to Detect a Release and False Alarm*:

4.2.1 The performance of internal and interstitial release detection systems to detect releases is specified in terms of the probability of detection and the probability of false alarm for a given release. The performance estimate includes an estimate of the uncertainty (that is, confidence intervals) on the probability of detection and probability of false alarm, and a statement of the experimental conditions used to generate the performance estimate (that is, the noise and the signal-plus-noise).

4.2.2 The performance results will be presented as specified in 4.2.2.1-4.2.2.3. Technical guidance on how to calculate these results is presented in Appendix X3 of this practice.

4.2.2.1 The manufacturer must specify the probability of false alarm of the release detection system using the manufacturer's detection criterion. In addition, the manufacturer must tabulate the release detection thresholds that are associated with the following: 0.1, 0.075, and 0.05. If the data are sufficient, estimates can also be reported for any of the following probabilities: 0.01, 0.005, or 0.001.

4.2.2.2 The manufacturer must specify and tabulate the release measurements associated with the following probabilities of detection using the manufacturer's detection threshold: 0.95, 0.90, 0.80, and 0.50.

4.2.2.3 For comparison to other release detection systems, the manufacturer must specify the release that can be detected with the release detection system for the combinations of probabilities of detection and false alarm given in Table 1.

5. Evaluation of Performance Data

5.1 The manufacturer's performance claim must be supported and documented by an engineering evaluation that can be made available for review upon request by the purchaser or user of the release detection system. The evaluation must contain at a minimum the information specified below.

5.1.1 *General*—All performance evaluations will contain the following general information:

5.1.1.1 The manufacturer will briefly describe the important features of the release detection system that will be evaluated. The description will include the measurement systems, sensor geometry, test protocol, data analysis, and detection criterion of the system that will be evaluated.

5.1.1.2 The manufacturer will describe the protocol that will be used to evaluate the release detection system. The protocol should be specified and include the theoretical calculations, experiments, data to be collected, test conditions, and analyses to be performed.

5.2 *Performance Characteristics*:

5.2.1 *System Response to the Signal*—Under this practice, system response to a signal (the chemical substance or physical property to be detected) must be experimentally determined for the release detection system performance parameters of specificity, lower detection limit, the dynamic range of detection, accuracy and bias, precision, and response time for the signal. Evaluation of system response to the signal also requires an estimation of the effects of physical properties that significantly affect the magnitude of the signal.

TABLE 1 Performance of the Release Detection System is Required at the Specified Probability of Detection and Probability of False Alarm

Probability of Detection	Probability of False Alarm
0.95	0.10
0.95	0.075
0.95	0.05
0.90	0.05
0.80	0.05
0.50	0.05

5.2.2 *Specificity*:

5.2.2.1 The specificity of the release detection system must be experimentally estimated for each chemical substance or physical property to be detected by the system. Determination of specificity is dependent on the type of output that a release detection system produces.

5.2.2.2 For quantitative-output systems, specificity is reported as a response factor. For these systems, specificity is mathematically defined as the system response factor and reported as the ratio of the chemical substance or physical property measured at the output of the release detection system to the actual test value expressed as a percentage.

5.2.2.3 For qualitative-output systems, specificity is determined by whether the system activates or not in response to a known signal level (that is, a true or false response).

5.2.3 *Lower Detection Limit*—The mean and standard deviation of the lower detection limit for the designated signal sources as determined in 5.2.2 must be experimentally estimated.

5.2.4 *Dynamic Range*—An experimental estimate of the operating range of the release detection system must be made for the designated signal sources as determined in 5.2.2.

5.2.5 *Precision and Bias*:

5.2.5.1 The precision and bias of the release detection system must be made experimentally over the operating range for the designated signal sources as determined in 5.2.2.

5.2.5.2 For quantitative-output detectors, the precision and bias are estimated from the standard deviation and mean obtained from respective measurements at multiple representative known signal levels within the operating range of the release detection system.

5.2.5.3 For a qualitative-output system the precision is calculated as the number of positive responses expressed as a percentage of the total number of tests at a known signal level. An estimate of bias cannot be made for qualitative-output systems.

5.2.6 *Response Time*:

5.2.6.1 The response time of the release detection system must be experimentally determined for the signal sources designated in 5.2.2. Response time is a general term that refers to the more specific terms of lag time and rise time, that together constitute detection time and fall time of a release detection system.

5.2.6.2 The lag time is the elapsed time from a detector's first contact with the measured signal (chemical substance or physical property) to the first detectable signal.

5.2.6.3 The rise time is the elapsed time from a detector's first detectable signal in response to a product or measured physical property to an output that is 95 % of full scale.

5.2.6.4 The fall time is the elapsed time after a signal is displaced by a total absence of signal until its output returns to 95 % of its original baseline level or there is no detectable signal output.

5.2.7 *Effects of Physical Properties*—An estimate of the change in signal strength at the release detection threshold must be experimentally made for each physical property that has a significant effect.

5.2.8 *System Response to Individual Sources of Interference or Ambient Noise*—Significant sources of interference or ambient noise that may produce false alarms and missed detections must be identified and quantified.

5.3 *Estimate of Probability to Detect a Release and a False Alarm*—It is assumed that the evaluation requirements specified in 5.1.1 and 5.2 have already been completed, and that the following requirements will be satisfied:

5.3.1 The manufacturer will estimate the performance of the release detection system to be achieved during the applicable field test conditions. The evaluation may include both theoretical and experimental analyses. A solely theoretical analysis without any experimental verification is not considered an acceptable evaluation procedure. The manufacturer must specify that sources of ambient noise (that is, interferences) are included in the evaluation and the magnitude of these interferences.

5.3.1.1 The evaluation will include an estimate of the ambient noise histogram. This is usually obtained from release detection measurements made over the operational conditions in the absence of a release. If possible, the raw data used to generate the histogram should be tabulated.

5.3.1.2 The evaluation will include an estimate of the signal-plus-noise histogram. This is usually estimated by developing a relationship between the signal and the noise.

5.3.2 The manufacturer will generate a table or a set of curves that present the performance results as follows:

5.3.2.1 Probability of detection versus the probability of false alarm for a minimum of three release quantities,

5.3.2.2 Probability of detection versus threshold for a minimum of three release quantities, and

5.3.2.3 Probability of false alarm versus threshold for a minimum of three release quantities.

5.3.3 At the completion of the evaluation, a report will be prepared that describes the protocol used to evaluate the release detection system, the applicable test operational conditions, the theoretical calculations, the experiments conducted, the experimental data, the analyses, the performance results. The report should be prepared with sufficient detail that it can be critically reviewed.

6. Marking with ASTM Designation

6.1 Release detection systems that comply with this practice may be marked with the following: "The results of the release detection system performance evaluation have been developed and reported in accordance with ASTM Practice E 1526."

6.2 This marking may be placed on the evaluation report, sales literature, and the release detection system.

APPENDIXES

(Nonmandatory Information)

X1. DEFINITIONS

X1.1 *Bias*—A systematic error that contributes to the difference between a population mean of the measurements of test results and an accepted reference as true value.[2]

X1.2 *Practice*—A definitive procedure for performing one or more specific operations or functions that does not produce a test result.[3]

X1.3 *Precision*—The closeness of agreement between randomly selected individual measurements or test results.[2]

[2] *Form and Style for ASTM Standards*, ASTM, 1989.

[3] *Regulations Governing ASTM Technical Committees*, ASTM, 1988.

X2. TECHNICAL GUIDANCE TO CALCULATE PERFORMANCE CHARACTERISTICS OF RELEASE DETECTION SYSTEMS

X2.1 Specific forms of equations for data treatment will always be related to the experimental design under which data are developed. Therefore, no specific forms of equations can be given until the methods for specific performance parameters are defined. However, the following definitions should offer some goals for the types of mathematical observations that should be made for the performance criteria in this practice. The definitions and calculations were derived from publications designed to evaluate the performance characteristics of external release detection systems:

X2.1.1 *Bias and Precision*—The closeness of agreement between randomly selected individual measurements or test results. It may be calculated as the sum of the absolute value of the arithmetic mean of the differences and the absolute value of the confidence coefficient. Bias may be evaluated by comparing the precision of a single set of values to the arithmetic mean of the difference of a data set. Precision can be calculated as the absolute mean difference between a group of measured values and the true value, plus the error confidence coefficient, divided by the true value.

X2.1.2 *Detection Time*—Detection time is the sum of lag time and rise time. Lag time is the elapsed time from a detector's first contact with a leak condition until the first detectable signal. Rise time is the elapsed time required to respond to a leak condition after the first detectable signal.

X2.1.3 *Fall Time*—The elapsed time after a test simulating leaking conditions is displaced by a "no leak" (or "no contamination") condition.

X2.1.4 *Interferences*—Interferences can be quantified using response factors. The ratio of the measured to a signal plus the interference divided by the response without the introduction of interference indicates the relative effect of an interference.

X2.1.5 *Response Time*—The elapsed time it takes a detector to respond to a leak condition or the return time to baseline conditions, or both, when the leak condition is replaced with a no leak environment.

X2.1.6 *Specificity*—Specificity can be calculated using response factors. The response factor to the chemical substance or physical properties is equal to the ratio of the measured response divided by the known response time.

X3. TECHNICAL GUIDANCE TO CALCULATE THE PERFORMANCE OF A RELEASE DETECTION SYSTEM BY DETERMINING PROBABILITY OF DETECTION AND PROBABILITY OF FALSE ALARM

X3.1 The performance of a detection system can be assessed from the statistical characteristics of the signal and the noise. The statistical characteristics can be determined from a compilation of the output of the system with and without the signal present. The output of the system without the signal present represents the noise. The noise is estimated from a histogram, that is a graphical display of the number of times a test result occurs in a certain interval. The signal produced by a volumetric leak detection system, for example, is the flow rate (that is, volume change) of the product, defined at a constant pressure, through a hole in the tank or pipe. The noise produced by a volumetric system is the sum of the volume changes in a nonleaking tank that could be mistaken for the signal. An example of the histogram and the frequency distribution for a generic volumetric leak detection system is shown in Fig. X3.1. These figures were generated from the sample test results that are tabulated in Table X3.1 and Table X3.2. The frequency distribution is essentially the same as the histogram, except that the values on the *y* axis of the histogram have been divided by the total number of tests to arrive at a fractional representation of results. The frequency distribution thus describes the fraction of the total number of test results in a defined interval or bin; the data in Fig. X3.1 are plotted in 0.01-gal/h (0.038-L/h) bins centered on a volume rate of 0.0 gal/h (0 L/h) (−0.005 gal/h (0.0189 L/h) < volume rate ≤ + 0.005 gal/h (L/h)). The likelihood of exceeding a specified noise level is described by the integral of the frequency distribution. The resulting cumulative frequency distribution, that corresponds to Fig. X3.1, is shown in Fig. X3.2, and is obtained by summing the fractional frequencies in the frequency distribution. If the signal is constant over time and is independent and additive with the noise, the signal-plus-noise distribution can be estimated directly from the distribution of the noise only. For this signal, the signal-plus-noise distribution has the same shape as the distribution of the noise only, but the mean of the noise distribution is shifted by an amount equal to the signal strength. An example of a signal-plus noise cumulative frequency distribution for a − 0.10-gal/h (0.378 L/h) leak detected by a volumetric system is shown in Fig. X3.3.

X3.2 Statistical models of the noise and signal-plus-noise can also be developed from the histograms or frequency distributions by means of standard statistical models. The probability density and cumulative distribution functions, that are continuous representations of the noise, can then be used in place of the frequency and cumulative frequency distributions, that are discrete representations. A statistical model that uses a normal distribution to estimate the central tendency of the probability density function and an exponential distribution, or another extreme-value distribution, to estimate the tails of the probability density function has been shown historically to be a good model of the noise.

X3.3 Fig. X3.4 presents one statistical model, based on the cumulative frequency distributions shown in Fig. X3.2 and Fig. X3.3, that can be used to estimate the performance of a detection system in terms of probability of detection (P_D) and probability of false alarm (P_{FA}) (that is, a false positive). The cumulative frequency distribution shown in this figure is centered about zero, and shows volume fluctuation level during tests in nonleaking tanks. When a tank has a leak, LR, the cumulative frequency distribution is reflected by the dashed curve. This curve has been shifted to the left by an amount equal to the leak rate. The model shown in Fig. X3.4 can be used to determine the performance of the detection system against a − 0.10-gal/h (0.378 L/h) leak; we can estimate the performance against other leaks by shifting the signal-plus-noise cumulative frequency distribution accordingly. A leak is declared whenever the measured volume rate exceeds the threshold. For a specified detection threshold, *T*, the P_{FA} is the fractional time that the noise will exceed the threshold; the P_{FA} is represented by the solid box on the cumulative frequency distribution of the noise, and is always based upon the noise distribution *only*. The P_D is the fractional time that the measured volume rate *with the signal present* will exceed the threshold; the P_D is represented by the solid box on the signal-plus-noise (dashed) cumulative frequency distribution. The probability of a missed detection (that is, false negative) is $1.0 - P_D$.

X3.4 The P_D, P_{FA}, *T*, and LR are all interrelated; changing one parameter affects the value of one or more of the other parameters. The choice of parameters affects the conclusions to be drawn from leak detection tests (that is, the reliability of the test result). For example, once the threshold has been selected, the P_{FA} is determined and does not change, regardless of the leak rate to be detected. The P_D, however, does change with leak rate if the threshold is kept constant. Since the P_D increases as the detectable leak rate increases, there is a better chance of finding large leaks than small leaks. The threshold is usually chosen to balance the P_D and P_{FA} in such a way that

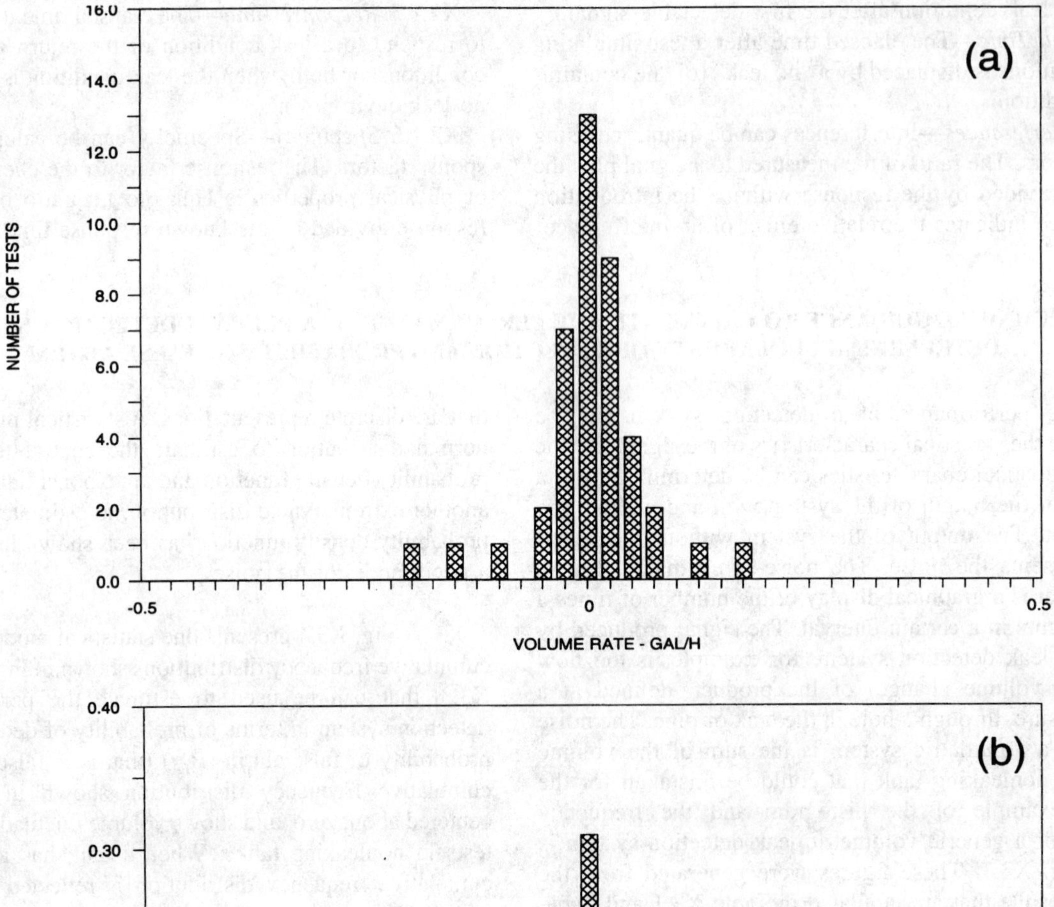

NOTE 1—The mean and standard deviation are 0.0036 and 0.0610 gal/h (0.0136 and 0.2308 L/h), respectively.

FIG. X3.1 Histogram (*a*) and Frequency Distribution (*b*) of the Noise Compiled from 42 Leak Detection Tests on Nonleaking Tanks by a Volumetric Leak Detection System

there is an acceptable balance between economic and environmental risks. If the threshold is low (that is, if very small leaks are to be detected), the P_D will be high, but so will the P_{FA}. On the other hand, if the threshold is high, there exists less chance of false alarm but also a greater probability of missed detections (because the P_D is lower). Thus, any adjustment made to the threshold for the purpose of improving the P_D carries with it an increased risk of false alarm. Conversely, any adjustment made to the threshold for the purpose of lowering the P_{FA} automatically implies an increased risk of missed detections.

X3.5 An estimate of performance for a given leak detection method is developed from the results of tests conducted with that method. A summary of the basic steps required to develop this estimate is given below:

X3.5.1 Tabulate the available results of tests on a nonleaking tank, arranging them in order from the lowest value to the highest and numbering them sequentially (1 being the lowest).

X3.5.2 Assign an individual frequency to each test result equal to $1/(n + 1)$ where n is the total number of test results.

TABLE X3.1 Values of the Cumulative Frequency Distribution of the Noise Shown in Fig. X3.2

Cumulative Frequency	Test Result, gal/h (L/h)	Cumulative Frequency	Test Result, gal/h (L/h)
0.0233	−0.200 (0.757)	0.5116	0.007 (0.026)
0.0465	−0.140 (0.530)	0.5349	0.008 (0.030)
0.0698	−0.100 (0.378)	0.5581	0.010 (0.038)
0.0930	−0.060 (0.227)	0.5814	0.012 (0.045)
0.1163	−0.045 (0.169)	0.6047	0.015 (0.057)
0.1395	−0.035 (0.132)	0.6279	0.018 (0.068)
0.1628	−0.032 (0.121)	0.6512	0.020 (0.076)
0.1860	−0.030 (0.114)	0.6744	0.022 (0.083)
0.2093	−0.030 (0.114)	0.6977	0.026 (0.098)
0.2326	−0.025 (0.095)	0.7209	0.028 (0.106)
0.2558	−0.025 (0.095)	0.7442	0.029 (0.109)
0.2791	−0.015 (0.057)	0.7674	0.030 (0.113)
0.3023	−0.011 (0.042)	0.7907	0.032 (0.121)
0.3256	−0.010 (0.038)	0.8140	0.048 (0.182)
0.3488	−0.007 (0.026)	0.8372	0.050 (0.189)
0.3721	−0.0055 (0.0206)	0.8605	0.059 (0.223)
0.3953	−0.004 (0.015)	0.8837	0.060 (0.227)
0.4186	−0.002 (0.008)	0.9070	0.071 (0.269)
0.4419	0.000 (•)	0.9302	0.080 (0.303)
0.4651	0.000 (•)	0.9535	0.125 (0.473)
0.4884	0.003 (0.011)	0.9767	0.175 (0.662)

TABLE X3.2 Values of the Cumulative Frequency Distribution of the Signal-Plus-Noise Shown in Fig. X3.3 Generated for Leak Rate (Signal) of − 0.10 gal/h (0.378 L/h)

Cumulative Frequency	Test Result, gal/h (L/h)	Cumulative Frequency	Test Result, gal/h (L/h)
0.0233	−0.300 (1.135)	0.5116	−0.093 (0.352)
0.0465	−0.240 (0.908)	0.5349	−0.092 (0.348)
0.0608	−0.200 (0.757)	0.5581	−0.090 (0.340)
0.0930	−0.160 (0.606)	0.5814	−0.088 (0.333)
0.1163	−0.145 (0.549)	0.6047	−0.085 (0.322)
0.1395	−0.135 (0.511)	0.6279	−0.082 (0.310)
0.1628	−0.132 (0.499)	0.6512	−0.080 (0.303)
0.1860	−0.130 (0.492)	0.6744	−0.078 (0.295)
0.2093	−0.130 (0.492)	0.6977	−0.074 (0.280)
0.2326	−0.125 (0.473)	0.7209	−0.072 (0.272)
0.2558	−0.125 (0.473)	0.7442	−0.071 (0.268)
0.2791	−0.115 (0.435)	0.7674	−0.070 (0.265)
0.3023	−0.111 (0.420)	0.7907	−0.068 (0.257)
0.3256	−0.110 (0.416)	0.8140	−0.052 (0.197)
0.3488	−0.107 (0.405)	0.8372	−0.050 (0.189)
0.3721	−0.1055 (0.3993)	0.8605	−0.041 (0.155)
0.3953	−0.104 (0.393)	0.8837	−0.040 (0.151)
0.4186	−0.102 (0.386)	0.9070	−0.029 (0.109)
0.4419	−0.100 (0.378)	0.9302	−0.020 (0.075)
0.4651	−0.100 (0.378)	0.9535	0.025 (0.094)
0.4884	−0.097 (0.367)	0.9767	0.075 (0.284)

X3.5.3 Develop the cumulative frequency for each test result by multiplying the individual frequency of each result by the sequential number of each test. For example, the fifth test result would have a cumulative frequency of 0.1163, that is equal to five times the individual frequency (that is, $5/(n + 1)$), and would have the values shown in Table X3.1.

X3.5.4 Generate a curve by plotting the test result on the abscissa (x axis) versus the cumulative frequency on the ordinate (y axis). This curve is the cumulative frequency distribution of the noise (the distribution of test results from nonleaking tanks). This curve corresponds to Fig. X3.2.

X3.5.5 Locate the threshold on the abscissa of the curve generated in X3.5.4.

X3.5.6 Estimate the P_{FA} from the intersection of the threshold and the cumulative distribution curve. This value is read from the ordinate at the intersection point. For a threshold of − 0.05 gal/h (0.189 L/h), the P_{FA} equals 0.1085 for the data plotted in Fig. X3.4. This value can be estimated by interpolation of the data in Table X3.1.

X3.5.7 Generate a cumulative distribution curve of the signal-plus-noise. This is obtained by adding the leak rate of interest (−0.10 gal/h (0.378 L/h), for example) to each of the tabulated test results from X3.5.3. The resulting table would correspond to Table X3.2. Negative values mean that product is flowing *out* of the tank or pipeline.

X3.5.8 Plot the data from this second table against the same cumulative frequency values. The resulting curve represents, in this case, the cumulative distribution of the signal-plus-noise for a leak rate of − 0.10 gal/h (0.378 L/h). This curve corresponds to Fig. X3.3.

X3.5.9 Estimate the P_D from the intersection of the threshold and the signal-plus-noise cumulative frequency curve generated in X3.5.7 and shown in Fig. X3.3. The P_D is estimated for a leak rate of − 0.10 gal/h (0.378 L/h). Changing the threshold has an effect on the P_{FA}. This effect can be assessed directly from Fig. X3.2. The effects of varying the leak rate can be gaged by generating a new curve for Fig. X3.3 for the leak rate of interest. For each new leak rate (signal plus-noise) curve, the effects on the P_D of changing the threshold can be estimated directly from the intersection of the threshold with the curve.

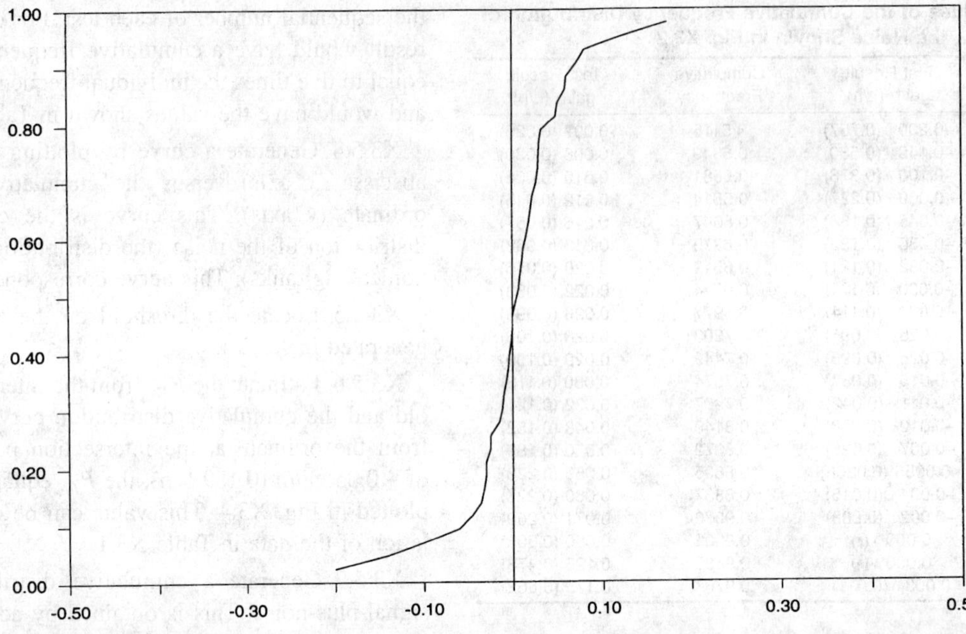

FIG. X3.2 Cumulative Frequency Distribution of the Noise Derived from the Frequency Distribution in Fig. X3.1

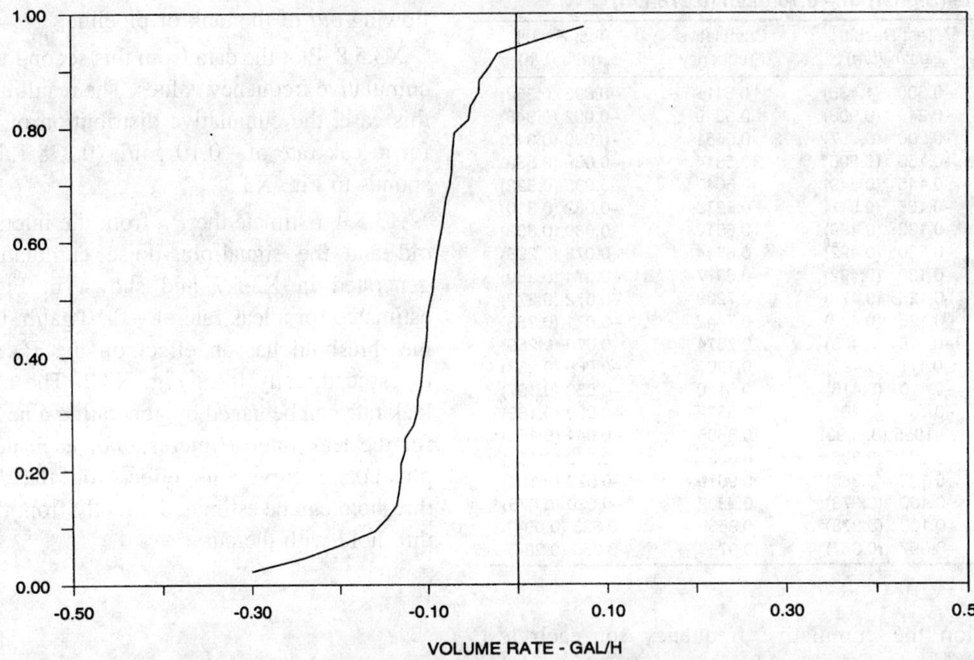

FIG. X3.3 Cumulative Frequency Distribution of the Signal-Plus-Noise Generated for a Leak Rate (Signal) of − 0.10 gal/h (0.378 L/h) Using the Cumulative Frequency Distribution of the Noise Shown in Fig. X3.2

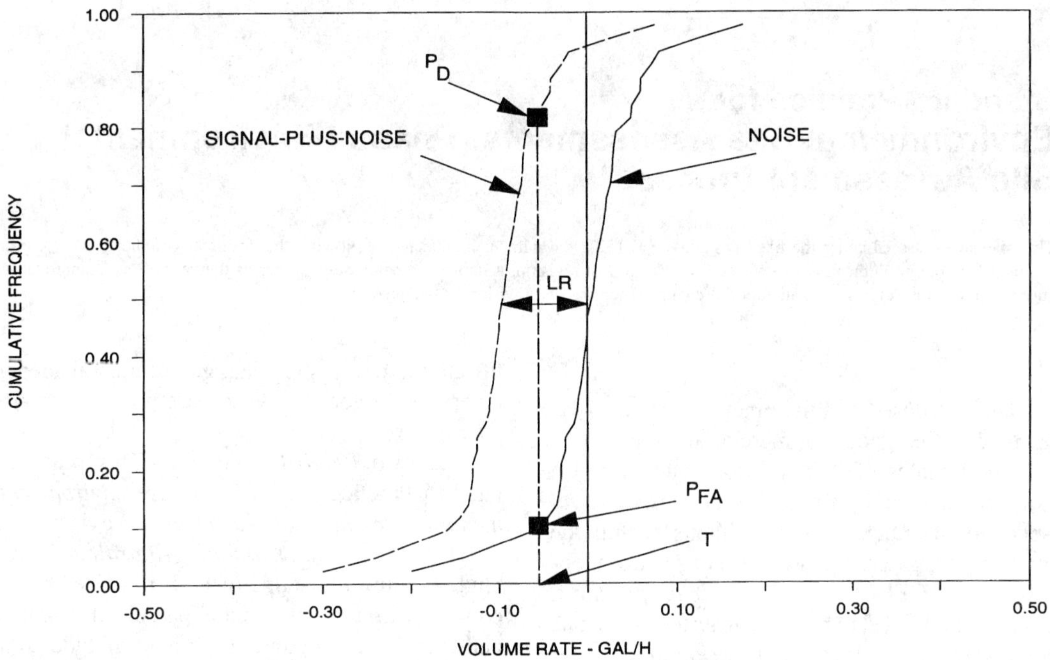

NOTE 1—The model is set up to calculate the performance of the leak detection system against a – 0.10-gal/h (0.378 L/h) leak rate. For a threshold of – 0.05 gal/h (0.189 L/h), the P_{FA} = 0.1085 and the P_D = 0.8372.

FIG. X3.4 Performance Model Used to Calculate the P_D and P_{FA} of a Volumetric Leak Detection System

Designation: E 1527 – 00

Standard Practice for Environmental Site Assessments: Phase I Environmental Site Assessment Process[1]

This standard is issued under the fixed designation E 1527; the number immediately following the designation indicates the year of original adoption or, in the case of revision, the year of last revision. A number in parentheses indicates the year of last reapproval. A superscript epsilon (ε) indicates an editorial change since the last revision or reapproval.

1. Scope

1.1 *Purpose*—The purpose of this practice, as well as Practice E 1528, is to define good commercial and customary practice in the United States of America for conducting an *environmental site assessment*[2] of a parcel of *commercial real estate* with respect to the range of contaminants within the scope of Comprehensive Environmental Response, Compensation and Liability Act (CERCLA) and *petroleum products*. As such, this practice is intended to permit a *user* to satisfy one of the requirements to qualify for the *innocent landowner defense* to CERCLA liability: that is, the practices that constitute "all appropriate inquiry into the previous ownership and uses of the property consistent with good commercial or customary practice" as defined in 42 USC § 9601(35)(B). (See Appendix X1 for an outline of CERCLA's liability and defense provisions.) An evaluation of *business environmental risk* associated with a parcel of commercial real estate may necessitate investigation beyond that identified in this practice (see Sections 1.3 and 12).

1.1.1 *Recognized Environmental Conditions*—In defining a standard of good commercial and customary practice for conducting an *environmental site assessment* of a parcel of *property*, the goal of the processes established by this practice is to identify *recognized environmental conditions*. The term *recognized environmental conditions* means the presence or likely presence of any *hazardous substances* or *petroleum products* on a *property* under conditions that indicate an existing release, a past release, or a material threat of a release of any *hazardous substances* or *petroleum products* into structures on the *property* or into the ground, ground water, or surface water of the *property*. The term includes *hazardous substances* or *petroleum products* even under conditions in compliance with laws. The term is not intended to include *de minimis* conditions that generally do not present a material risk of harm to public health or the environment and that generally would not be the subject of an enforcement action if brought to the attention of appropriate governmental agencies. Conditions determined to be *de minimis* are not *recognized environmental conditions*.

1.1.2 *Two Related Practices*—This practice is closely related to Practice E 1528. Both are *environmental site assessments* for *commercial real estate*. See 4.3.

1.1.3 *Petroleum Products*—*Petroleum products* are included within the scope of both practices because they are of concern with respect to many parcels of *commercial real estate* and current custom and usage is to include an inquiry into the presence of *petroleum products* when doing an *environmental site assessment* of *commercial real estate*. Inclusion of *petroleum products* within the scope of this practice and Practice E 1528 is not based upon the applicability, if any, of CERCLA to *petroleum products*. (See Appendix X1 for discussion of *petroleum exclusion* to CERCLA liability.)

1.1.4 *CERCLA Requirements Other Than Appropriate Inquiry*—This practice does not address whether requirements in addition to *appropriate inquiry* have been met in order to qualify for CERCLA's *innocent landowner defense* (for example, the duties specified in 42 USC § 9607(b)(3)(a) and (b) and cited in Appendix X1).

1.1.5 *Other Federal, State, and Local Environmental Laws*—This practice does not address requirements of any state or local laws or of any federal laws other than the appropriate inquiry provisions of CERCLA's *innocent landowner defense*. *Users* are cautioned that federal, state, and local laws may impose environmental assessment obligations that are beyond the scope of this practice. *Users* should also be aware that there are likely to be other legal obligations with regard to *hazardous substances* or *petroleum products* discovered on *property* that are not addressed in this practice and that may pose risks of civil and/or criminal sanctions for noncompliance.

1.1.6 *Documentation*—The scope of this practice includes research and reporting requirements that support the user's ability to qualify for the *innocent landowner defense*. As such, sufficient documentation of all sources, records, and resources utilized in conducting the inquiry required by this practice must be provided in the written report (refer to 7.1.8 and 11.2).

1.2 *Objectives*—Objectives guiding the development of this practice and Practice E 1528 are (1) to synthesize and put in writing good commercial and customary practice for *environmental site assessments* for *commercial real estate*, (2) to

---

[1] This practice is under the jurisdiction of ASTM Committee E-50 on Environmental Assessment and is the direct responsibility of Subcommittee E50.02 on Commercial Real Estate Transactions.

Current edition approved May 10, 2000. Published July 2000. Originally published as E 1527 – 93. Last previous edition E 1527 – 97.

[2] All definitions, descriptions of terms, and acronyms are defined in Section3. Whenever terms defined in 3.2 or described in 3.3 are used in this practice, they are in *italics*.

facilitate high quality, standardized *environmental site assessments*, (3) to ensure that the standard of *appropriate inquiry* is practical and reasonable, and (4) to clarify an industry standard for *appropriate inquiry* in an effort to guide legal interpretation of CERCLA's *innocent landowner defense*.

1.3 *Considerations Beyond Scope*—The use of this practice is strictly limited to the scope set forth in this section. Section 12 of this practice, identifies, for informational purposes, certain environmental conditions (not an all-inclusive list) that may exist on a *property* that are beyond the scope of this practice but may warrant consideration by parties to a *commercial real estate* transaction. The need to include an investigation of any such conditions in the environmental professional's scope of services should be evaluated based upon, among other factors, the nature of the property and the reasons for performing the assessment (for example, a more comprehensive evaluation of *business environmental risk*) and should be agreed upon between the *user* and *environmental professional* as additional services beyond the scope of this practice prior to initiation of the *Environmental Site Assessment* process.

1.4 *Organization of This Practice*—This practice has several parts and two appendixes. Section 1 is the Scope. Section 2 is Referenced Documents. Section 3, Terminology, has definitions of terms not unique to this practice and descriptions of terms unique to this practice and acronyms. Section 4 is Significance and Use of this practice. Section 5 describes User's Responsibilities. Sections 6-11 are the main body of the Phase I Environmental Site Assessment, including evaluation and report preparation. Section 12 provides additional information regarding non-scope considerations (see 1.3). The appendixes are included for information and are not part of the procedures prescribed in either this practice or Practice E 1528. Appendix X1 explains the liability and defense provisions of CERCLA that will assist the user in understanding the user's responsibilities under CERCLA; it also contains other important information regarding CERCLA and this practice. Appendix X2 provides a recommended table of contents and report format for a Phase I Environmental Site Assessment Report. Appendix X3 includes guidance to assist users in the preparation for, and selection of, an *environmental professional* to conduct a Phase I *Environmental Site Assessment*.

1.5 *This standard does not purport to address all of the safety concerns, if any, associated with its use. It is the responsibility of the user of this standard to establish appropriate safety and health practices and determine the applicability of regulatory limitations prior to use.*

1.6 *This practice offers a set of instructions for performing one or more specific operations and should be supplemented by education, experience and professional judgment. Not all aspects of this practice may be applicable in all circumstances. This ASTM standard practice does not necessarily represent the standard of care by which the adequacy of a given professional service must be judged, nor should this document be applied without consideration of a project's unique aspects. The word "standard" in the title means only that the document has been approved through the ASTM consensus process.*

2. Referenced Documents

2.1 *ASTM Standards:*

E 1528 Practice for Environmental Site Assessments: Transaction Screen Process[3]

3. Terminology

3.1 This section provides definitions, descriptions of terms, and a list of acronyms for many of the words used in this practice and Practice E 1528. The terms are an integral part of both practices and are critical to an understanding of the practices and their use.

3.2 *Definitions:*

3.2.1 *activity and use limitations*—legal or physical restrictions or limitations on the use of, or access to, a site or facility: (*1*) to reduce or eliminate potential exposure to hazardous substances in the soil or ground water on the property, or (*2*) to prevent activities that could interfere with the effectiveness of a response action, in order to ensure maintenance of a condition of no significant risk to public health or the environment. These legal or physical restrictions, which may include institutional and/or engineering controls, are intended to prevent adverse impacts to individuals or populations that may be exposed to hazardous substances in the soil or ground water on the property.

3.2.2 *Comprehensive Environmental Response, Compensation, and Liability Information System (CERCLIS)*—the list of sites compiled by EPA that EPA has investigated or is currently investigating for potential hazardous substance contamination for possible inclusion on the National Priorities List.

3.2.3 *construction debris*—concrete, brick, asphalt, and other such building materials discarded in the construction of a building or other improvement to property.

3.2.4 *contaminated public wells*—public wells used for drinking water that have been designated by a government entity as contaminated by toxic substances (for example, chlorinated solvents), or as having water unsafe to drink without treatment.

3.2.5 *CORRACTS list*—list of hazardous waste treatment, storage, or disposal facilities and other RCRIS facilities (due to past interim status or storage of hazardous waste beyond 90 days) who have been notified by the U.S. Environmental Protection Agency to undertake corrective action under RCRA.

3.2.6 *demolition debris*—concrete, brick, asphalt, and other such building materials discarded in the demolition of a building or other improvement to property.

3.2.7 *drum*—a container (typically, but not necessarily, holding 55 gal (208 L) of liquid) that may be used to store *hazardous substances* or *petroleum products*.

3.2.8 *dry wells*—underground areas where soil has been removed and replaced with pea gravel, coarse sand, or large rocks. Dry wells are used for drainage, to control storm runoff, for the collection of spilled liquids (intentional and non-intentional) and wastewater disposal (often illegal).

3.2.9 *dwelling*—structure or portion thereof used for residential habitation.

3.2.10 *engineering controls*—physical modifications to a

[3] *Annual Book of ASTM Standards*, Vol 11.04.

site or facility (for example, capping, slurry walls, or point of use water treatment) to reduce or eliminate the potential for exposure to hazardous substances in the soil or ground water on the property.

3.2.11 *environmental lien*—a charge, security, or encumbrance upon title to a *property* to secure the payment of a cost, damage, debt, obligation, or duty arising out of response actions, cleanup, or other remediation of *hazardous substances* or *petroleum products* upon a *property*, including (but not limited to) liens imposed pursuant to CERCLA 42 USC § 9607(1) and similar state or local laws.

3.2.12 *ERNS list*—EPA's emergency response notification system list of reported CERCLA hazardous substance releases or spills in quantities greater than the reportable quantity, as maintained at the National Response Center. Notification requirements for such releases or spills are codified in 40 CFR Parts 302 and 355.

3.2.13 *Federal Register, (FR)*—publication of the United States government published daily (except for federal holidays and weekends) containing all proposed and final regulations and some other activities of the federal government. When regulations become final, they are included in the Code of Federal Regulations (CFR), as well as published in the Federal Register.

3.2.14 *fire insurance maps*—maps produced for private fire insurance map companies that indicate uses of properties at specified dates and that encompass the property. These maps are often available at local libraries, historical societies, private resellers, or from the map companies who produced them. See Question 23 of the transaction screen process in Practice E 1528 and 7.3.4.2 of this practice.

3.2.15 *hazardous substance*—A substance defined as a hazardous substance pursuant to CERCLA 42 USC§ 9601(14), as interpreted by EPA regulations and the courts:" (A) any substance designated pursuant to section 1321(b)(2)(A) of Title 33, (B) any element, compound, mixture, solution, or substance designated pursuant to section 9602 of this title, (C) any hazardous waste having the characteristics identified under or listed pursuant to section 3001 of the Solid Waste Disposal Act (42 USC § 6921) (but not including any waste the regulation of which under the Solid Waste Disposal Act (42 USC§ 6901 *et seq.*) has been suspended by Act of Congress), (D) any toxic pollutant listed under section 1317(a) of Title 33, (E) any hazardous air pollutant listed under section 112 of the Clean Air Act (42 USC § 7412), and (F) any imminently hazardous chemical substance or mixture with respect to which the Administrator (of EPA) has taken action pursuant to section 2606 of Title 15. The term does not include petroleum, including crude oil or any fraction thereof which is not otherwise specifically listed or designated as a hazardous substance under subparagraphs (A) through (F) of this paragraph, and the term does not include natural gas, natural gas liquids, liquefied natural gas, or synthetic gas usable for fuel (or mixtures of natural gas and such synthetic gas)." (See Appendix X1.)

3.2.16 *hazardous waste*—any hazardous waste having the characteristics identified under or listed pursuant to section 3001 of the Solid Waste Disposal Act (42 USC § 6921) (but not including any waste the regulation of which under the Solid Waste Disposal Act (42 USC § 6901 *et seq.*) has been suspended by Act of Congress). The Solid Waste Disposal Act of 1980 amended RCRA. RCRA defines a hazardous waste, in 42 USC § 6903, as: "a solid waste, or combination of solid wastes, which because of its quantity, concentration, or physical, chemical, or infectious characteristics may—(A) cause, or significantly contribute to an increase in mortality or an increase in serious irreversible, or incapacitating reversible, illness; or (B) pose a substantial present or potential hazard to human health or the environment when improperly treated, stored, transported, or disposed of, or otherwise managed."

3.2.17 *institutional controls*—a legal or administrative restriction (for example, deed restriction, restrictive zoning) on the use of, or access to, a site or facility to reduce or eliminate potential exposure to hazardous substances in the soil or ground water on the property.

3.2.18 *landfill*—a place, location, tract of land, area, or premises used for the disposal of solid wastes as defined by state solid waste regulations. The term is synonymous with the term *solid waste disposal site* and is also known as a garbage dump, trash dump, or similar term.

3.2.19 *local street directories*—directories published by private (or sometimes government) sources that show ownership, occupancy, and/or use of sites by reference to street addresses. Often local street directories are available at libraries of local governments, colleges or universities, or historical societies. See 7.3.4.6 of this practice.

3.2.20 *material safety data sheet (MSDS)*—written or printed material concerning a hazardous substance which is prepared by chemical manufacturers, importers, and employers for hazardous chemicals pursuant to OSHA's Hazard Communication Standard, 29 CFR 1910.1200.

3.2.21 *National Contingency Plan (NCP)*—the National Oil and Hazardous Substances Pollution Contingency Plan, found at 40 CFR § 300, that is the EPA's blueprint on how hazardous substances are to be cleaned up pursuant to CERCLA.

3.2.22 *National Priorities List (NPL)*—list compiled by EPA pursuant to CERCLA 42 USC § 9605(a)(8)(B) of properties with the highest priority for cleanup pursuant to EPA's Hazard Ranking System. See 40 CFR Part 300.

3.2.23 *occupants*—those tenants, subtenants, or other persons or entities using the *property* or a portion of the *property*.

3.2.24 *owner*—generally the fee owner of record of the *property*.

3.2.25 *petroleum exclusion*—The exclusion from CERCLA liability provided in 42 USC § 9601(14), as interpreted by the courts and EPA: "The term (hazardous substance) does not include petroleum, including crude oil or any fraction thereof which is not otherwise specifically listed or designated as a hazardous substance under subparagraphs (A) through (F) of this paragraph, and the term does not include natural gas, natural gas liquids, liquefied natural gas, or synthetic gas usable for fuel (or mixtures of natural gas and such synthetic gas)."

3.2.26 *petroleum products*—those substances included within the meaning of the *petroleum exclusion* to CERCLA, 42 USC § 9601(14), as interpreted by the courts and EPA, that is:

petroleum, including crude oil or any fraction thereof which is not otherwise specifically listed or designated as a hazardous substance under Subparagraphs (A) through (F) of 42 USC § 9601(14), natural gas, natural gas liquids, liquefied natural gas, and synthetic gas usable for fuel (or mixtures of natural gas and such synthetic gas). (The word fraction refers to certain distillates of crude oil, including gasoline, kerosene, diesel oil, jet fuels, and fuel oil, pursuant to *Standard Definitions of Petroleum Statistics*.[4]

3.2.27 *Phase I Environmental Site Assessment*— the process described in this practice.

3.2.28 *pits, ponds, or lagoons*—man-made or natural depressions in a ground surface that are likely to hold liquids or sludge containing *hazardous substances* or *petroleum products*. The likelihood of such liquids or sludge being present is determined by evidence of factors associated with the pit, pond, or lagoon, including, but not limited to, discolored water, distressed vegetation, or the presence of an obvious wastewater discharge.

3.2.29 *property*—the real property that is the subject of the *environmental site assessment* described in this practice. Real property includes buildings and other fixtures and improvements located on the property and affixed to the land.

3.2.30 *property tax files*—the files kept for property tax purposes by the local jurisdiction where the property is located and includes records of past ownership, appraisals, maps, sketches, photos, or other information that is reasonably ascertainable and pertaining to the property. See 7.3.4.3.

3.2.31 *RCRA generators*— those persons or entities that generate hazardous wastes, as defined and regulated by RCRA.

3.2.32 *RCRA generators list*—list kept by EPA of those persons or entities that generate hazardous wastes as defined and regulated by RCRA.

3.2.33 *RCRA TSD facilities*—those facilities on which treatment, storage, and/or disposal of hazardous wastes takes place, as defined and regulated by RCRA.

3.2.34 *RCRA TSD facilities list*—list kept by EPA of those facilities on which treatment, storage, and/or disposal of hazardous wastes takes place, as defined and regulated by RCRA.

3.2.35 *recorded land title records*—records of fee ownership, leases, land contracts, easements, liens, and other encumbrances on or of the property recorded in the place where land title records are, by law or custom, recorded for the local jurisdiction in which the *property* is located. (Often such records are kept by a municipal or county recorder or clerk.) Such records may be obtained from title companies or directly from the local government agency. Information about the title to the property that is recorded in a U.S. district court or any place other than where land title records are, by law or custom, recorded for the local jurisdiction in which the property is located, are not considered part of recorded land title records. See 7.3.4.4.

3.2.36 *records of emergency release notifications (SARA § 304)*—Section 304 of EPCRA or Title III of SARA requires operators of facilities to notify their local emergency planning committee (as defined in EPCRA) and state emergency response commission (as defined in EPCRA) of any release beyond the facility's boundary of any reportable quantity of any extremely hazardous substance. Often the local fire department is the local emergency planning committee. Records of such notifications are "Records of Emergency Release Notifications" (SARA § 304).

3.2.37 *report*—the written record of a transaction screen process as required by Practice E 1528 or the written report prepared by the environmental professional and constituting part of a "Phase I Environmental Site Assessment," as required by this practice.

3.2.38 *solid waste disposal site*—a place, location, tract of land, area, or premises used for the disposal of solid wastes as defined by state solid waste regulations. The term is synonymous with the term *landfill* and is also known as a garbage dump, trash dump, or similar term.

3.2.39 *solvent*—a chemical compound that is capable of dissolving another substance and may itself be a *hazardous substance*, used in a number of manufacturing/industrial processes including but not limited to the manufacture of paints and coatings for industrial and household purposes, equipment clean-up, and surface degreasing in metal fabricating industries.

3.2.40 *state registered USTs*—state lists of underground storage tanks required to be registered under Subtitle I, Section 9002 of RCRA.

3.2.41 *sump*—a pit, cistern, cesspool, or similar receptacle where liquids drain, collect, or are stored.

3.2.42 *TSD facility*—treatment, storage, or disposal facility (see *RCRA TSD facilities*).

3.2.43 *underground storage tank (UST)*—any tank, including underground piping connected to the tank, that is or has been used to contain *hazardous substances* or *petroleum products* and the volume of which is 10 % or more beneath the surface of the ground.

3.2.44 *USGS 7.5 Minute Topographic Map*—the map (if any) available from or produced by the United States Geological Survey, entitled "USGS 7.5 Minute Topographic Map," and showing the property. See 7.3.4.5.

3.2.45 *wastewater*—water that (*1*) is or has been used in an industrial or manufacturing process, (*2*) conveys or has conveyed sewage, or (*3*) is directly related to manufacturing, processing, or raw materials storage areas at an industrial plant. Wastewater does not include water originating on or passing through or adjacent to a site, such as stormwater flows, that has not been used in industrial or manufacturing processes, has not been combined with sewage, or is not directly related to manufacturing, processing, or raw materials storage areas at an industrial plant.

3.2.46 *zoning/land use records*—those records of the local government in which the *property* is located indicating the uses permitted by the local government in particular zones within its jurisdiction. The records may consist of maps and/or written records. They are often located in the planning department of a municipality or county. See 7.3.4.8.

3.3 *Definitions of Terms Specific to This Standard:*

[4] *Standard Definitions of Petroleum Statistics*, American Petroleum Institute, Fourth Edition, 1988.

3.3.1 *actual knowledge*—the knowledge actually possessed by an individual who is a real person, rather than an entity. Actual knowledge is to be distinguished from constructive knowledge that is knowledge imputed to an individual or entity.

3.3.2 *adjoining properties*—any real property or properties the border of which is contiguous or partially contiguous with that of the property, or that would be contiguous or partially contiguous with that of the property but for a street, road, or other public thoroughfare separating them.

3.3.3 *aerial photographs*—photographs taken from an airplane or helicopter (from a low enough altitude to allow identification of development and activities) of areas encompassing the property. Aerial photographs are often available from government agencies or private collections unique to a local area. See 7.3.4.1 of this practice.

3.3.4 *appropriate inquiry*—that inquiry constituting "all appropriate inquiry into the previous ownership and uses of the property consistent with good commercial or customary practice" as defined in CERCLA, 42 USC § 9601(35)(B), that will give a party to a *commercial real estate* transaction the *innocent landowner defense* to CERCLA liability (42 USC§ 9601(A) and (B) and § 9607(b)(3)), assuming compliance with other elements of the defense. See Appendix X1.

3.3.5 *approximate minimum search distance*—the area for which records must be obtained and reviewed pursuant to Section 7 subject to the limitations provided in that section. This may include areas outside the *property* and shall be measured from the nearest *property* boundary. This term is used in lieu of radius to include irregularly shaped properties.

3.3.6 *building department records*—those records of the local government in which the property is located indicating permission of the local government to construct, alter, or demolish improvements on the property. Often building department records are located in the building department of a municipality or county. See 7.3.4.7.

3.3.7 *business environmental risk*—a risk which can have a material environmental or environmentally-driven impact on the business associated with the current or planned use of a parcel of commercial real estate, not necessarily limited to those environmental issues required to be investigated in this practice. Consideration of *business environmental risk* issues may involve addressing one or more non-scope considerations, some of which are identified in Section 12.

3.3.8 *commercial real estate*—any real property except a dwelling or property with no more than four dwelling units exclusively for residential use (except that a dwelling or property with no more than four dwelling units exclusively for residential use is included in this term when it has a commercial function, as in the building of such dwellings for profit). This term includes but is not limited to undeveloped real property and real property used for industrial, retail, office, agricultural, other commercial, medical, or educational purposes; property used for residential purposes that has more than four residential dwelling units; and property with no more than four dwelling units for residential use when it has a commercial function, as in the building of such dwellings for profit.

3.3.9 *commercial real estate transaction*—a transfer of title to or possession of real property or receipt of a security interest in real property, except that it does not include transfer of title to or possession of real property or the receipt of a security interest in real property with respect to an individual dwelling or building containing fewer than five dwelling units, nor does it include the purchase of a lot or lots to construct a dwelling for occupancy by a purchaser, but a commercial real estate transaction does include real property purchased or leased by persons or entities in the business of building or developing dwelling units.

3.3.10 *due diligence*—the process of inquiring into the environmental characteristics of a parcel of *commercial real estate* or other conditions, usually in connection with a commercial real estate transaction. The degree and kind of due diligence vary for different properties and differing purposes. See Appendix X1.

3.3.11 *environmental audit*—the investigative process to determine if the operations of an existing facility are in compliance with applicable environmental laws and regulations. This term should not be used to describe Practice E 1528 or this practice, although an environmental audit may include an *environmental site assessment* or, if prior audits are available, may be part of an environmental site assessment. See Appendix X1.

3.3.12 *environmental professional*—a person possessing sufficient training and experience necessary to conduct a *site reconnaissance*, *interviews*, and other activities in accordance with this practice, and from the information generated by such activities, having the ability to develop opinions and conclusions regarding *recognized environmental conditions* in connection with the *property* in question. An individual's status as an environmental professional may be limited to the type of assessment to be performed or to specific segments of the assessment for which the professional is responsible. The person may be an independent contractor or an employee of the *user*.

3.3.13 *environmental site assessment (ESA)*—the process by which a person or entity seeks to determine if a particular parcel of real *property* (including improvements) is subject to *recognized environmental conditions*. At the option of the user, an environmental site assessment may include more inquiry than that constituting *appropriate inquiry* or, if the user is not concerned about qualifying for the *innocent landowner defense*, less inquiry than that constituting *appropriate inquiry*. See Appendix X1. An environmental site assessment is both different from and less rigorous than an *environmental audit*.

3.3.14 *fill dirt*—dirt, soil, sand, or other earth, that is obtained off-site, that is used to fill holes or depressions, create mounds, or otherwise artificially change the grade or elevation of real property. It does not include material that is used in limited quantities for normal landscaping activities.

3.3.15 *hazardous waste/contaminated sites*—sites on which a release has occurred, or is suspected to have occurred, of any *hazardous substance*, *hazardous waste*, or *petroleum products*, and that release or suspected release has been reported to a government entity.

3.3.16 *historical recognized environmental condition*—environmental condition which in the past would have been considered a *recognized environmental condition,* but which may or may not be considered a *recognized environmental condition* currently. The final decision rests with the *environmental professional* and will be influenced by the current impact of the *historical recognized environmental condition* on the property. If a past release of any hazardous substances or petroleum products has occurred in connection with the property and has been remediated, with such remediation accepted by the responsible regulatory agency (for example, as evidenced by the issuance of a no further action letter or equivalent), this condition shall be considered an *historical recognized environmental condition* and included in the findings section of the Phase I *Environmental Site Assessment* report. The *environmental professional* shall provide an opinion of the current impact on the property of this *historical recognized environmental condition* in the opinion section of the report. If this *historical recognized environmental condition* is determined to be a *recognized environmental condition* at the time the Phase I *Environmental Site Assessment* is conducted, the condition shall be identified as such and listed in the conclusions section of the report.

3.3.17 *innocent landowner defense*—that defense to CERCLA liability provided in 42 USC § 9601(35) and § 9607(b)(3). One of the requirements to qualify for this defense is that the party make "all appropriate inquiry into the previous ownership and uses of the property consistent with good commercial or customary practice." There are additional requirements to qualify for this defense. See Appendix X1.

3.3.18 *interviews*—those portions of this practice that are contained in Section 9 and 10 thereof and address questions to be asked of *owners* and *occupants* of the *property* and questions to be asked of local government officials.

3.3.19 *key site manager*—the person identified by the *owner* of a *property* as having good knowledge of the uses and physical characteristics of the property. See 9.5.1.

3.3.20 *local government agencies*—those agencies of municipal or county government having jurisdiction over the *property.* Municipal and county government agencies include but are not limited to cities, parishes, townships, and similar entities.

3.3.21 *LUST sites*—state lists of leaking underground storage tank sites. Section 9003 (h) of Subtitle I of RCRA gives EPA and states, under cooperative agreements with EPA, authority to clean up releases from UST systems or require owners and operators to do so.

3.3.22 *major occupants*—those tenants, subtenants, or other persons or entities each of which uses at least 40 % of the leasable area of the *property* or any anchor tenant when the *property* is a shopping center.

3.3.23 *material threat*—a physically observable or obvious threat which is reasonably likely to lead to a release that, in the opinion of the *environmental professional,* is threatening and might result in impact to public health or the environment. An example might include an aboveground storage tank that contains a hazardous substance and which shows evidence of damage. The damage would represent a material threat if it is deemed serious enough that it may cause or contribute to tank integrity failure with a release of contents to the environment.

3.3.24 *obvious*—that which is plain or evident; a condition or fact that could not be ignored or overlooked by a reasonable observer while *visually or physically observing the property.*

3.3.25 *other historical sources*—any source or sources other than those designated in 7.3.4.1 through 7.3.4.8 that are credible to a reasonable person and that identify past uses of the property. The term includes, but is not limited to: miscellaneous maps, newspaper archives, and records in the files and/or personal knowledge of the *property owner* and/or *occupants.* See 7.3.4.9.

3.3.26 *physical setting sources*—sources that provide information about the geologic, hydrogeologic, hydrologic, or topographic characteristics of a *property.* See 7.2.3.

3.3.27 *practically reviewable*—information that is practically reviewable means that the information is provided by the source in a manner and in a form that, upon examination, yields information relevant to the *property* without the need for extraordinary analysis of irrelevant data. The form of the information shall be such that the user can review the records for a limited geographic area. Records that cannot be feasibly retrieved by reference to the location of the *property* or a geographic area in which the *property* is located are not generally *practically reviewable.* Most databases of public records are *practically reviewable* if they can be obtained from the source agency by the county, city, zip code, or other geographic area of the facilities listed in the record system. Records that are sorted, filed, organized, or maintained by the source agency only chronologically are not generally practically reviewable. Listings in publicly available records which do not have adequate address information to be located geographically are not generally considered practically reviewable. For large databases with numerous facility records (such as RCRA hazardous waste generators and registered underground storage tanks), the records are not *practically reviewable* unless they can be obtained from the source agency in the smaller geographic area of zip codes. Even when information is provided by zip code for some large databases, it is common for an unmanageable number of sites to be identified within a given zip code. In these cases, it is not necessary to review the impact of all of the sites that are likely to be listed in any given zip code because that information would not be *practically reviewable.* In other words, when so much data is generated that it cannot be feasibly reviewed for its impact on the *property,* it is not *practically reviewable.*

3.3.28 *preparer*—the person preparing the *transaction screen questionnaire* pursuant to Practice E 1528, who may be either the user or the person to whom the user has delegated the preparation of the *transaction screen questionnaire.*

3.3.29 *publicly available*—information that is publicly available means that the source of the information allows access to the information by anyone upon request.

3.3.30 *reasonably ascertainable*—for purposes of both this practice and Practice E 1528, information that is (*1*) *publicly available,* (*2*) obtainable from its source within reasonable time and cost constraints, and (*3*) *practically reviewable.*

3.3.31 *recognized environmental conditions*—the presence or likely presence of any *hazardous substances* or *petroleum products* on a *property* under conditions that indicate an existing release, a past release, or a material threat of a release of any *hazardous substances* or *petroleum products* into structures on the *property* or into the ground, ground water, or surface water of the *property*. The term includes *hazardous substances* or *petroleum products* even under conditions in compliance with laws. The term is not intended to include *de minimis* conditions that generally do not present a material risk of harm to public health or the environment and that generally would not be the subject of an enforcement action if brought to the attention of appropriate governmental agencies. Conditions determined to be *de minimis* are not *recognized environmental conditions.*

3.3.32 *records review*—that part that is contained in Section 7 of this practice addresses which records shall or may be reviewed.

3.3.33 *site reconnaissance*—that part that is contained in Section 8 of this practice and addresses what should be done in connection with the *site visit*. The site reconnaissance includes, but is not limited to, the *site visit* done in connection with such a Phase I Environmental Site Assessment.

3.3.34 *site visit*—the visit to the property during which observations are made constituting the *site reconnaissance* section of this practice and the *site visit* requirement of Practice E 1528.

3.3.35 *standard environmental record sources*— those records specified in 7.2.1.1.

3.3.36 *standard historical sources*—those sources of information about the history of uses of property specified in 7.3.4.

3.3.37 *standard physical setting source*—a current USGS 7.5 minute topographic map (if any) showing the area on which the property is located. See 7.2.3.

3.3.38 *standard practice(s)*—the activities set forth in either and both this practice and Practice E 1528.

3.3.39 *standard sources*—sources of environmental, physical setting, or historical records specified in Section 7 of this practice.

3.3.40 *transaction screen process*—the process described in Practice E 1528.

3.3.41 *transaction screen questionnaire*—the questionnaire provided in Section 6 of Practice E 1528.

3.3.42 *user*—the party seeking to use Practices E 1527 or E 1528 to perform an *environmental site assessment* of the *property*. A user may include, without limitation, a purchaser of *property*, a potential tenant of property, an *owner* of *property*, a lender, or a property manager.

3.3.43 *visually and/or physically observed*—during a *site visit* pursuant to Practice E 1528, or pursuant to this practice, this term means observations made by vision while walking through a *property* and the structures located on it and observations made by the sense of smell, particularly observations of noxious or foul odors. The term "walking through" is not meant to imply that disabled persons who cannot physically walk may not conduct a *site visit*; they may do so by the means at their disposal for moving through the *property* and the structures located on it.

3.4 *Acronyms:Acronyms:*

3.4.1 *CERCLA*—Comprehensive Environmental Response, Compensation and Liability Act of 1980 (as amended, 42 USC § 9601 *et seq.*).

3.4.2 *CERCLIS*—Comprehensive Environmental Response, Compensation and Liability Information System (maintained by EPA).

3.4.3 *CFR*—Code of Federal Regulations.

3.4.4 *CORRACTS*—facilities subject to Corrective Action under RCRA.

3.4.5 *EPA*—United States Environmental Protection Agency.

3.4.6 *EPCRA*—Emergency Planning and Community Right to Know Act ((also known as SARA Title III), 42 USC § 11001 *et seq.*).

3.4.7 *ERNS*—emergency response notification system.

3.4.8 *ESA*—environmental site assessment (different than an *environmental audit*; see 3.3.13).

3.4.9 *FOIA*—U.S. Freedom of Information Act (5 USC 552 *et seq.*).

3.4.10 *FR*—Federal Register.

3.4.11 *LUST*—leaking underground storage tank.

3.4.12 *MSDS*—material safety data sheet.

3.4.13 *NCP*—National Contingency Plan.

3.4.14 *NFRAP*—former CERCLIS sites where no further remedial action is planned under CERCLA.

3.4.15 *NPDES*—national pollutant discharge elimination system.

3.4.16 *NPL*—national priorities list.

3.4.17 *PCBs*—polychlorinated biphenyls.

3.4.18 *PRP*—potentially responsible party (pursuant to CERCLA 42 USC § 9607(a)).

3.4.19 *RCRA*—Resource Conservation and Recovery Act (as amended, 42 USC§ 6901 *et seq.*).

3.4.20 *SARA*—Superfund Amendments and Reauthorization Act of 1986 (amendment to CERCLA).

3.4.21 *TSDF*—hazardous waste treatment, storage or disposal facility.

3.4.22 *USC*—United States Code.

3.4.23 *USGS*—United States Geological Survey.

3.4.24 *UST*—underground storage tank.

4. Significance and Use

4.1 *Uses*—This practice is intended for use on a voluntary basis by parties who wish to assess the environmental condition of *commercial real estate*. While use of this practice is intended to constitute *appropriate inquiry* for purposes of CERCLA's *innocent landowner defense*, it is not intended that its use be limited to that purpose. This practice is intended primarily as an approach to conducting an inquiry designed to identify *recognized environmental conditions* in connection with a *property*, and *environmental site assessments* that are both more and less comprehensive than this practice (including, in some instances, no *environmental site assessment*) may be appropriate in some circumstances. Further, no implication is intended that a person must use this practice in order to be deemed to have conducted inquiry in a commercially prudent or reasonable manner in any particular transaction. Nevertheless, this practice is intended to reflect a commercially prudent and reasonable inquiry.

4.2 *Clarifications on Use*:

4.2.1 *Use Not Limited to CERCLA*—This practice and Practice E 1528 are designed to assist the *user* in developing information about the environmental condition of a *property* and as such has utility for a wide range of persons, including those who may have no actual or potential CERCLA liability and/or may not be seeking the *innocent landowner defense*.

4.2.2 *Residential Tenants/Purchasers and Others*—No implication is intended that it is currently customary practice for residential tenants of multifamily residential buildings, tenants of single-family homes or other residential real estate, or purchasers of dwellings for one's own residential use, to conduct an *environmental site assessment* in connection with these transactions. Thus, these transactions are not included in the term commercial real estate transactions, and it is not intended to imply that such persons are obligated to conduct an *environmental site assessment* in connection with these transactions for purposes of *appropriate inquiry* or for any other purpose. In addition, no implication is intended that it is currently customary practice for *environmental site assessments* to be conducted in other unenumerated instances (including but not limited to many commercial leasing transactions, many acquisitions of easements, and many loan transactions in which the lender has multiple remedies). On the other hand, anyone who elects to do an *environmental site assessment* of any *property* or portion of a property may, in such person's judgment, use either this practice or Practice E 1528.

4.2.3 *Site-Specific*— This practice is site-specific in that it relates to assessment of environmental conditions on a specific parcel of *commercial real estate*. Consequently, this practice does not address many additional issues raised in transactions such as purchases of business entities, or interests therein, or of their assets, that may well involve environmental liabilities pertaining to properties previously owned or operated or other off-site environmental liabilities.

4.3 *Two Related Practices*—This practice sets forth one procedure for an *environmental site assessment* known as a "Phase I Environmental Site Assessment" or a "Phase I ESA" or simply a "Phase I." This practice is a companion to Practice E 1528. These practices are each intended to meet the standard of *appropriate inquiry* necessary to qualify for the *innocent landowner defense*. It is essential to consider that these practices, taken together, provide for two alternative practices of *appropriate inquiry*.

4.3.1 *Election to Commence with Either Practice*—The *user* may commence inquiry to identify *recognized environmental conditions* in connection with a *property* by performing either the *transaction screen process*, or the Phase I Environmental Site Assessment given in this practice.

4.3.2 *Who May Conduct*— The *transaction screen process* may be conducted either by the *user* (including an agent or employee of the user) or wholly or partially by an *environmental professional*. The *transaction screen process* does not require the judgment of an *environmental professional*. Whenever a Phase I Environmental Site Assessment is conducted, it must be performed by an *environmental professional* to the extent specified in 6.5.1. Further, at the Phase I Environmental Site Assessment level, no practical standard can be designed to eliminate the role of judgment and the value and need for experience in the party performing the inquiry. The professional judgment of an *environmental professional* is, consequently, vital to the performance of appropriate inquiry at the Phase I Environmental Site Assessment level.

4.4 *Additional Services*—As set forth in 11.9, additional services may be contracted for between the *user* and the *environmental professional*. Such additional services may include *business environmental risk* issues not included within the scope of this practice, examples of which are identified in 12.1.4 under Non-Scope Considerations.

4.5 *Principles*—The following principles are an integral part of both practices and are intended to be referred to in resolving any ambiguity or exercising such discretion as is accorded the *user* or *environmental professional* in performing an *environmental site assessment* or in judging whether a *user* or *environmental professional* has conducted *appropriate inquiry* or has otherwise conducted an adequate *environmental site assessment*.

4.5.1 *Uncertainty Not Eliminated*—No *environmental site assessment* can wholly eliminate uncertainty regarding the potential for *recognized environmental conditions* in connection with a *property*. Performance of this practice or E 1528 is intended to reduce, but not eliminate, uncertainty regarding the potential for *recognized environmental conditions* in connection with a property, and both practices recognize reasonable limits of time and cost.

4.5.2 *Not Exhaustive*—*Appropriate inquiry* does not mean an exhaustive assessment of a clean *property*. There is a point at which the cost of information obtained or the time required to gather it outweighs the usefulness of the information and, in fact, may be a material detriment to the orderly completion of transactions. One of the purposes of this practice is to identify a balance between the competing goals of limiting the costs and time demands inherent in performing an *environmental site assessment* and the reduction of uncertainty about unknown conditions resulting from additional information.

4.5.3 *Level of Inquiry Is Variable*—Not every *property* will warrant the same level of assessment. Consistent with good commercial or customary practice, the appropriate level of *environmental site assessment* will be guided by the type of *property* subject to assessment, the expertise and risk tolerance of the *user*, and the information developed in the course of the inquiry.

4.5.4 *Comparison With Subsequent Inquiry*—It should not be concluded or assumed that an inquiry was not *appropriate inquiry* merely because the inquiry did not identify *recognized environmental conditions* in connection with a *property*. *Environmental site assessments* must be evaluated based on the reasonableness of judgments made at the time and under the circumstances in which they were made. Subsequent *environmental site assessments* should not be considered valid standards to judge the appropriateness of any prior assessment based on hindsight, new information, use of developing technology or analytical techniques, or other factors.

4.6 *Continued Viability of Environmental Site Assessment*—An *environmental site assessment* meeting or exceeding either Practice E 1528 or this practice and completed less than 180 days previously is presumed to be valid. An *environmental site assessment* meeting or exceeding either practice and completed more than 180 days previously may be used to the extent allowed by 4.7-4.7.5.

4.7 *Prior Assessment Usage*—Both Practice E 1528 and this practice recognize that *environmental site assessments* performed in accordance with these practices will include information that subsequent *users* may want to use to avoid undertaking duplicative assessment procedures. Therefore, the practices describe procedures to be followed to assist users in determining the appropriateness of using information in *environmental site assessments* performed previously. The system of prior assessment usage is based on the following principles that should be adhered to in addition to the specific procedures set forth elsewhere in these practices:

4.7.1 *Use of Prior Information*—Subject to 4.7.4, *users* and *environmental professionals* may use information in prior *environmental site assessments* provided such information was generated as a result of procedures that meet or exceed the requirements of this practice or Practice E 1528 and then only provided that the specific procedures set forth in the appropriate practice are met.

4.7.2 *Prior Assessment Meets or Exceeds*—Subject to 4.7.4, a prior *environmental site assessment* may be used in its entirety, without regard to the specific procedures set forth in these practices, if, in the reasonable judgment of the *user*: the prior *environmental site assessment* meets or exceeds the requirements of Practice E 1528 or this practice and the conditions at the *property* likely to affect *recognized environmental conditions* in connection with the *property* are not likely to have changed materially since the prior *environmental site assessment* was conducted. In making this judgment, the *user* should consider the type of *property* assessed and the conditions in the area surrounding the *property*.

4.7.3 *Current Investigation*—Except as provided in 4.7.2 and 4.7.2 of Practice E 1528 prior *environmental site assessments* should not be used without current investigation of conditions likely to affect *recognized environmental conditions* in connection with the *property* that may have changed materially since the prior *environmental site assessment* was conducted. At a minimum, for a Phase I Environmental Site Assessment consistent with this practice, a new *site reconnaissance*, interviews, and an update of the *records review* should be performed.

4.7.4 *Actual Knowledge Exception*—If the *user* or *environmental professional(s)* conducting an *environmental site assessment* has *actual knowledge* that the information being used from a prior *environmental site assessment* is not accurate or if it is *obvious*, based on other information obtained by means of the *environmental site assessment* or known to the person conducting the *environmental site assessment*, that the information being used is not accurate, such information from a prior *environmental site assessment* may not be used.

4.7.5 *Contractual Issues Regarding Prior Assessment Usage*—The contractual and legal obligations between prior and subsequent *users* of *environmental site assessments* or between *environmental professionals* who conducted prior *environmental site assessments* and those who would like to use such prior *environmental site assessments* are beyond the scope of this practice.

4.8 *Rules of Engagement*—The contractual and legal obligations between an *environmental professional* and a *user* (and other parties, if any) are outside the scope of this practice. No specific legal relationship between the *environmental professional* and the user is necessary for the *user* to meet the requirements of this practice.

5. User's Responsibilities

5.1 *Scope*—The purpose of this section is to describe tasks that will help identify the possibility of *recognized environmental conditions* in connection with the *property*. These tasks do not require the technical expertise of an *environmental professional* and are generally not performed by *environmental professionals* performing a Phase I Environmental Site Assessment. They may be performed by the *user*.

5.2 *Checking Title Records for Environmental Liens or Activity and Land Use Limitations—Reasonably ascertainable recorded land title records* (see 7.3.4.4) should be checked to identify *environmental liens* or *activity and use limitations*, if any, that are currently recorded against the *property*. Any *environmental liens* or *activity and use limitations* so identified shall be reported to the *environmental professional* conducting a Phase I Environmental Site Assessment. This practice does not impose on the *environmental professional* the responsibility to check for recorded *environmental liens* or *activity and use limitations*. Rather the *user* should check or engage a title company or title professional to check *reasonably ascertainable recorded land title records* for *environmental liens* or *activity and use limitations* currently recorded against the *property*.

5.2.1 *Reasonably Ascertainable—Environmental liens* or *activity and use limitations* that are unrecorded or are recorded any place other than *recorded land title records* are not considered to be in *recorded land title records* that are *reasonably ascertainable*. *Recorded land title records* need not be checked if they otherwise do not meet the definition of the term *reasonably ascertainable*.

5.3 *Specialized Knowledge or Experience of the User*—If the *user* is aware of any specialized knowledge or experience that is material to *recognized environmental conditions* in connection with the *property*, it is the user's responsibility to communicate any information based on such specialized knowledge or experience to the *environmental professional*. The *user* should do so before the *environmental professional* does the *site reconnaissance*.

5.4 *Reason for Significantly Lower Purchase Price*—In a transaction involving the purchase of a parcel of *commercial real estate*, if a user has *actual knowledge* that the purchase price of the *property* is significantly less than the purchase price of comparable properties, the *user* should try to identify an explanation for the lower price and to make a written record of such explanation. Among the factors to consider will be the information that becomes known to the *user* pursuant to the Phase I Environmental Site Assessment.

5.5 *Other*—Either the *user* shall make known to the *environmental professional* the reason why the *user* wants to have the Phase I *Environmental Site Assessment* performed or, if the *user* does not identify the purpose of the Phase I *Environmental Site Assessment*, the *environmental professional* shall assume the purpose is to qualify for the innocent landowner defense to CERCLA liability and state this in the report. In addition to satisfying one of the requirements to qualify for the innocent landowner defense to CERCLA liability, another reason for performing a Phase I *Environmental Site Assessment* might include the need to understand potential environmental conditions that could materially impact the operation of the business associated with the parcel of *commercial real estate*. The *user* and the *environmental professional* may also need to modify the scope of services performed under this practice for special circumstances, including, but not limited to, operating industrial facilities or large tracts of land (large areas or corridors).

6. Phase I Environmental Site Assessment

6.1 *Objective*—The purpose of this Phase I Environmental Site Assessment is to identify, to the extent feasible pursuant to the processes prescribed herein, recognized environmental conditions in connection with the property. (See 1.1.1.)

6.2 *Four Components*— A Phase I Environmental Site Assessment shall have four components, as described as follows:

6.2.1 *Records Review*— Review of records; see Section 7,

6.2.2 *Site Reconnaissance*, A visit to the property; see Section 8,

6.2.3 *Interviews*:

6.2.3.1 Interviews with current *owners* and *occupants* of the property; see Section 9, and

6.2.3.2 Interviews with local government officials; see Section 10, and

6.2.4 *Report*—Evaluation and report; see Section 11.

6.3 *Coordination of Parts*:

6.3.1 *Parts Used in Concert*—The *records review*, *site reconnaissance*, and *interviews* are intended to be used in concert with each other. If information from one source indicates the need for more information, other sources may be available to provide information. For example, if a previous use of the *property* as a gasoline station is identified through the *records review*, but the present *owner* and *occupants* interviewed report no knowledge of an underground storage tank, the person conducting the *site reconnaissance* should be alert for signs of the presence of an underground storage tank.

6.3.2 *User Obligations*— The *environmental professional* shall note in the report whether or not the *user* has reported to the *environmental professional* any *environmental liens* encumbering the *property* or any specialized knowledge or experience of the *user* that would provide important information about previous ownership or uses of the *property* that may be material to identifying *recognized environmental conditions*.

6.4 *No Sampling*— This practice does not include any testing or sampling of materials (for example, soil, water, air, building materials).

6.5 *Who May Conduct a Phase I*:

6.5.1 *Environmental Professional's Duties*—The *interviews* and *site reconnaissance*, as well as review and interpretation of information upon which the report is based and overseeing the writing of the report, are all portions of a Phase I Environmental Site Assessment that shall be performed by an *environmental professional* or *environmental professionals*. If more than one environmental professional is involved in these tasks, they shall coordinate their efforts.

6.5.2 *Environmental Professional Supervision*—Information for the *records review* needed for completion of a Phase I Environmental Site Assessment may be provided by a number of parties including government agencies, third-party vendors, the *user*, and present and past *owners* and *occupants* of the *property*, provided that the information is obtained by or under the supervision of an *environmental professional* or is obtained by a third-party vendor specializing in retrieval of the information specified in Section 7. Prior assessments may also contain information that will be appropriate for usage in a current *environmental site assessment* provided the prior usage procedures set forth in Sections 7, 8, and 9 are followed. The *environmental professional(s)* participating in the *site reconnaissance* and responsible for the report shall review all of the information provided.

6.5.2.1 *Reliance*—An environmental professional is not required to verify independently the information provided but may rely on information provided unless he or she has *actual knowledge* that certain information is incorrect or unless it is *obvious* that certain information is incorrect based on other information obtained in the Phase I Environmental Site Assessment or otherwise actually known to the *environmental professional*.

7. Records Review

7.1 *Introduction*:

7.1.1 *Objective*—The purpose of the *records review* is to obtain and review records that will help identify *recognized environmental conditions* in connection with the *property*.

7.1.2 *Approximate Minimum Search Distance*—Some records to be reviewed pertain not just to the *property* but also pertain to properties within an additional *approximate minimum search distance* in order to help assess the likelihood of problems from migrating *hazardous substances* or *petroleum products*. When the term *approximate minimum search distance* includes areas outside the *property*, it shall be measured from the nearest *property* boundary. The term *approximate minimum search distance* is used in lieu of radius in order to include irregularly shaped properties.

7.1.2.1 *Reduction of Approximate Minimum Search Distance*—When allowed by 7.2.1.1, the *approximate minimum search distance* for a particular record may be reduced in the discretion of the environmental professional. Factors to consider in reducing the *approximate minimum search distance* include: (*1*) the density (for example, urban, rural, or suburban) of the setting in which the *property* is located; (*2*) the distance that the *hazardous substances* or *petroleum products* are likely to migrate based on local geologic or hydrogeologic conditions; and (*3*) other reasonable factors. The justification for each reduction and the *approximate minimum search distance* actually used for any particular record shall be explained in the report. If the *approximate minimum search distance* is specified as "property only," then the search shall be limited to the *property* and may not be reduced unless the particular record is not *reasonably ascertainable*.

7.1.3 *Accuracy and Completeness*—Accuracy and completeness of record information varies among information sources, including governmental sources. Record information is often inaccurate or incomplete. The *user* or *environmental professional* is not obligated to identify mistakes or insufficiencies in information provided. However, the *environmental professional* reviewing records shall make a reasonable effort to compensate for mistakes or insufficiencies in the information reviewed that are *obvious* in light of other information of which the *environmental professional* has *actual knowledge*.

7.1.4 *Reasonably Ascertainable/Standard Sources*—Availability of record information varies from information source to information source, including governmental jurisdictions. The *user* or *environmental professional* is not obligated to identify, obtain, or review every possible record that might exist with respect to a *property*. Instead, this practice identifies record information that shall be reviewed from standard sources, and the *user* or *environmental professional* is required to review only record information that is reasonably ascertainable from those standard sources. Record information that is *reasonably ascertainable* means (*1*) information that is publicly available, (*2*) information that is obtainable from its source within reasonable time and cost constraints, and (*3*) information that is *practically reviewable*.

7.1.4.1 *Publicly Available*—Information that is *publicly available* means that the source of the information allows access to the information by anyone upon request.

7.1.4.2 *Reasonable Time and Cost*—Information that is obtainable within reasonable time and cost constraints means that the information will be provided by the source within 20 calendar days of receiving a written, telephone, or in-person request at no more than a nominal cost intended to cover the source's cost of retrieving and duplicating the information. Information that can only be reviewed by a visit to the source is *reasonably ascertainable* if the visit is permitted by the source within 20 days of request.

7.1.4.3 *Practically Reviewable*—Information that is *practically reviewable* means that the information is provided by the source in a manner and in a form that, upon examination, yields information relevant to the *property* without the need for extraordinary analysis of irrelevant data. The form of the information shall be such that the *user* can review the records for a limited geographic area. Records that cannot be feasibly retrieved by reference to the location of the *property* or a geographic area in which the *property* is located are not generally *practically reviewable*. Most databases of public records are practically reviewable if they can be obtained from the source agency by the county, city, zip code, or other geographic area of the facilities listed in the record system. Records that are sorted, filed, organized, or maintained by the source agency only chronologically are not generally *practically reviewable*. Listings in publicly available records which do not have adequate address information to be located geographically are not generally considered practically reviewable. For large databases with numerous facility records (such as RCRA generators and registered USTs), the records are not *practically reviewable* unless they can be obtained from the source agency in the smaller geographic area of zip codes. Even when information is provided by zip code for some large databases, it is common for an unmanageable number of sites to be identified within a given zip code. In these cases, it is not necessary to review the impact of all of the sites that are likely to be listed in any given zip code because that information would not be *practically reviewable*. In other words, when so much data is generated that it cannot be feasibly reviewed for its impact on the *property*, it is not required to be reviewed.

7.1.5 *Alternatives to Standard Sources*—Alternative sources may be used instead of standard sources if they are of similar or better reliability and detail, or if a *standard source* is not *reasonably ascertainable*.

7.1.6 *Coordination*— If records are not *reasonably ascertainable* from *standard sources* or alternative sources, the *environmental professional* shall attempt to obtain the requested information by other means specified in this practice such as questions posed to the current *owner* or *occupant(s)* of the *property* or appropriate persons available at the source at the time of the request.

7.1.7 *Sources of Standard Source Information*—*Standard source* information or other record information from government agencies may be obtained directly from appropriate government agencies or from commercial services. Government information obtained from nongovernmental sources may be considered current if the source updates the information at least every 90 days or, for information that is updated less frequently than quarterly by the government agency, within 90 days of the date the government agency makes the information available to the public.

7.1.8 *Documentation of Sources Checked*—The report shall document each source that was used, even if a source revealed no findings. Sources shall be sufficiently documented, including name, date request for information was filled, date information provided was last updated by source, date information was last updated by original source (if provided other than by original source; see 7.1.7). Supporting documentation shall be included in the report or adequately referenced to facilitate reconstruction of the assessment by an environmental professional other than the *environmental professional* who conducted it.

7.1.9 *Significance*—If a standard *environmental record source* (or other sources in the course of conducting the *Phase I Environmental Site Assessment*) identifies the *property* or another site within the *approximate minimum search distance*, the report shall include the *environmental professional's* judgment about the significance of the listing to the analysis of *recognized environmental conditions* in connection with the *property* (based on the data retrieved pursuant to 7.2, additional information from the government source, or other sources of information). In doing so, the *environmental professional* may make statements applicable to multiple sites (for example, a statement to the effect that none of the sites listed is likely to have a negative impact on the property except ...).

7.2 *Environmental Information*:

7.2.1 *Standard Environmental Sources*— The following *standard environmental record sources* shall be reviewed, subject to the conditions of 7.1.1 through 7.1.7:

7.2.1.1 *Standard Environmental Record Sources: Federal and State*—The *approximate minimum search distance* may be reduced, pursuant to 7.1.2.1, for any of these *standard environmental record sources* except the Federal NPL site list and Federal RCRA TSD list.

	Approximate Minimum Search Distance, miles (kilometres)
Federal NPL site list	1.0 (1.6)
Federal CERCLIS list	0.5 (0.8)
Federal CERCLIS NFRAP site list	property and adjoining properties
Federal RCRA CORRACTS facilities list	1.0 (1.6)
Federal RCRA non-CORRACTS TSD facilities list	0.5 (0.8)
Federal RCRA generators list	property and adjoining properties
Federal ERNS list	property only
State lists of hazardous waste sites identified for investigation or remediation:	
State-equivalent NPL	1.0 (1.6)
State-equivalent CERCLIS	0.5 (0.8)
State landfill and/or solid waste disposal site lists	0.5 (0.8)
State leaking UST lists	0.5 (0.8)
State registered UST lists	property and adjoining properties

7.2.2 *Additional Environmental Record Sources: State or Local*—One or more additional state sources or local sources of environmental records may be checked, in the discretion of the *environmental professional*, to enhance and supplement federal and state sources identified above. Factors to consider in determining which local or additional state records, if any, should be checked include (*1*) whether they are *reasonably ascertainable*, (*2*) whether they are sufficiently useful, accurate, and complete in light of the objective of the *records review* (see 7.1.1), and (*3*) whether they are generally obtained, pursuant to local good commercial or customary practice, in initial *environmental site assessments* in the type of *commercial real estate* transaction involved. To the extent additional state sources or local sources are used to supplement the same record types listed above, *approximate minimum search distances* should not be less than those specified above (adjusted as provided in 7.2.1.1 and 7.1.2.1). Some types of records and sources that may be useful include:

Types of Local Records

Brownfield Sites
Lists of Landfill/Solid Waste Disposal Sites
Lists of Hazardous Waste/Contaminated Sites
Lists of Registered Underground Storage Tanks
Local Land Records (*for activity and use limitations*)
Records of Emergency Release Reports (SARA § 304)
Records of Contaminated Public Wells

Local Sources

Department of Health/Environmental Division
Fire Department
Planning Department
Building Permit/Inspection Department
Local/Regional Pollution Control Agency
Local/Regional Water Quality Agency
Local Electric Utility Companies (for records relating to PCBs)

7.2.3 *Physical Setting Sources*—A current USGS 7.5 Minute Topographic Map (or equivalent) showing the area on which the *property* is located shall be reviewed, provided it is *reasonably ascertainable*. It is the only *standard physical setting source* and the only *physical setting source* that is required to be obtained (and only if it is *reasonably ascertainable*). One or more additional *physical setting sources* may be obtained in the discretion of the *environmental professional*. Because such sources provide information about the geologic, hydrogeologic, hydrologic, or topographic characteristics of a site, discretionary *physical setting sources* shall be sought when (*1*) conditions have been identified in which *hazardous substances* or *petroleum products* are likely to migrate to the *property* or from or within the *property* into the ground water or soil and (*2*) more information than is provided in the current USGS 7.5 Minute Topographic Map (or equivalent) is generally obtained, pursuant to local good commercial or customary practice in initial *environmental site assessments* in the type of *commercial real estate transaction* involved, in order to assess the impact of such migration on *recognized environmental conditions* in connection with the *property*.

Mandatory Standard Physical Setting Source

USGS—Current 7.5 Minute Topographic Map (or equivalent)

Discretionary and Non-Standard Physical Setting Sources

USGS and/or State Geological Survey—Groundwater Maps
USGS and/or State Geological Survey—Bedrock Geology Maps
USGS and/or State Geological Survey—Surficial Geology Maps
Soil Conservation Service—Soil Maps
Other Physical Setting Sources that are reasonably credible (as well as reasonably ascertainable)

7.3 *Historical Use Information*:

7.3.1 *Objective*—The objective of consulting historical sources is to develop a history of the previous uses of the *property* and surrounding area, in order to help identify the likelihood of past uses having led to *recognized environmental conditions* in connection with the *property*.

7.3.2 *Uses of the Property*—All *obvious* uses of the *property* shall be identified from the present, back to the *property's obvious* first developed use, or back to 1940, whichever is earlier. This task requires reviewing only as many of the *standard historical sources* in 7.3.4.1 through 7.3.4.8 as are necessary and both *reasonably ascertainable* and likely to be useful (as defined under *Data Failure* in 7.3.2.3). For example, if the *property* was developed in the 1700's, it might be feasible to identify uses back to the early 1900's, using sources such as *fire insurance maps* or *USGS 7.5 minute topographic maps* (or equivalent). Although other sources such as *recorded land title records* might go back to the 1700's, it would not be required to review them unless they were both *reasonably ascertainable* and likely to be useful. As another example, if the *property* was reportedly not developed until 1960, it would still be necessary to confirm that it was undeveloped back to 1940. Such confirmation may come from one or more of the *standard historical sources* specified in 7.3.4.1 through 7.3.4.8, or it may come from *other historical sources* (such as someone with personal knowledge of the *property*; see 7.3.4.9). However, checking other historical sources (see 7.3.4.9) would not be required. For purposes of 7.3.2, the term "developed use" includes agricultural uses and placement of fill. The report shall describe all identified uses, justify the earliest date identified (for example, records showed no development of the *property* prior to the specific date), and explain the reason for any gaps in the history of use (for example, *data failure*).

7.3.2.1 *Intervals*—Review of *standard historical sources* at less than approximately five year intervals is not required by this practice (for example, if the *property* had one use in 1950 and another use in 1955, it is not required to check for a third use in the intervening period). If the specific use of the *property* appears unchanged over a period longer than five years, then it is not required by this practice to research the use during that period (for example, if *fire insurance maps* show the same apartment building in 1940 and 1960, then the period in between need not be researched).

7.3.2.2 *General Type of Use*—In identifying previous uses, more specific information about uses is more helpful than less specific information, but it is sufficient, for purposes of 7.3.2, to identify the general type of use (for example: office, retail, and residential) unless it is *obvious* from the source(s) consulted that the use may be more specifically identified. However, if the general type of use is industrial or manufacturing (for example, *zoning/land use records* show industrial zoning), then additional *standard historical sources* should be reviewed if they are likely to identify a more specific use and are *reasonably ascertainable*, subject to the constraints of *data failure* (see 7.3.2.3).

7.3.2.3 *Data Failure*—A *standard historical source* may be excluded *(1)* if the source is not *reasonably ascertainable*, or *(2)* if past experience indicates that the source is not likely to be sufficiently useful, accurate, or complete in terms of satisfying 7.3.2. *Other historical sources* specified in 7.3.4.9 may be used to satisfy 7.3.2, 7.3.2.1, and 7.3.2.2, but are not required to comply with this practice. Whatever history of previous uses is derived from checking the *standard historical sources* specified in 7.3.4.1 through 7.3.4.8 (except those excluded by *(1)* and *(2)* of 7.3.2.3) shall be deemed sufficient historical use information to comply with this practice.

7.3.3 *Uses of Properties in Surrounding Area*—Uses in the area surrounding the *property* shall be identified in the report, but this task is required only to the extent that this information is revealed in the course of researching the *property* itself (for example, an *aerial photograph* or *fire insurance map* of the *property* will usually show the surrounding area). If the *environmental professional* uses sources that include the surrounding area, surrounding uses should be identified to a distance determined at the discretion of the *environmental professional* (for example, if an aerial photo shows the area surrounding the *property*, then the *environmental professional* shall determine how far out from the *property* the photo should be analyzed). Factors to consider in making this determination include, but are not limited to: the extent to which information is *reasonably ascertainable*; the time and cost involved in reviewing surrounding uses (for example, analyzing *aerial photographs* is relatively quick, but reviewing *property tax files* for adjacent properties or reviewing *local street directories* for more than the few streets that surround the site is typically too time-consuming); the extent to which information is useful, accurate, and complete in light of the purpose of the records review (see 7.1.1); the likelihood of the information being significant to *recognized environmental conditions* in connection with the *property*; the extent to which potential concerns are *obvious*; known hydrogeologic/geologic conditions that may indicate a high probability of *hazardous substances* or *petroleum products* migration to the property; how recently local development has taken place; information obtained from *interviews* and other sources; and local good commercial or customary practice.

7.3.4 *Standard Historical Sources*:

7.3.4.1 *Aerial Photographs*—The term "aerial photographs" means photographs taken from an airplane or helicopter (from a low enough altitude to allow identification of development and activities) of areas encompassing the *property*. Aerial photographs are often available from government agencies or private collections unique to a local area.

7.3.4.2 *Fire Insurance Maps*—The term *fire insurance maps* means maps produced by private fire insurance map companies that indicate uses of properties at specified dates and that encompass the *property*. These maps are often available at local libraries, historical societies, private resellers, or from the map companies who produced them.

7.3.4.3 *Property Tax Files*—The term *property tax files* means the files kept for property tax purposes by the local jurisdiction where the *property* is located and includes records of past ownership, appraisals, maps, sketches, photos, or other information that is *reasonably ascertainable* and pertaining to the *property*.

7.3.4.4 *Recorded Land Title Records*—The term *recorded land title records* means records of fee ownership, leases, land contracts, easements, liens, and other encumbrances on or of the *property* recorded in the place where land title records are, by law or custom, recorded for the local jurisdiction in which the *property* is located. (Often such records are kept by a municipal or county recorder or clerk.) Such records may be obtained from title companies or directly from the local government agency. Information about the title to the *property* that is recorded in a U.S. district court or any place other than where land title records are, by law or custom, recorded for the local jurisdiction in which the *property* is located, are not considered part of recorded land title records, because often this source will provide only names of previous owners, lessees, easement holders, etc. and little or no information about uses or occupancies of the *property*, but when employed in combination with another source recorded land title records may provide helpful information about uses of the *property*. *This source cannot be the sole historical source consulted. If this source is consulted, at least one additional standard historical source must also be consulted.*

7.3.4.5 *USGS 7.5 Minute Topographic Maps*— The term *USGS 7.5 Minute Topographic Maps* means the map (if any) available from or produced by the United States Geological Survey, entitled "USGS 7.5 minute topographic map," and showing the property.

7.3.4.6 *Local Street Directories*—The term *local street directories* means directories published by private (or sometimes government) sources and showing ownership and/or use of sites by reference to street addresses. Often local street directories are available at libraries of local governments, colleges or universities, or historical societies.

7.3.4.7 *Building Department Records*—The term *building department records* means those records of the local government in which the property is located indicating permission of the local government to construct, alter, or demolish improvements on the property. Often building department records are located in the *building department* of a municipality or county.

7.3.4.8 *Zoning/Land Use Records*—The term *zoning/land use records* means those records of the local government in which the property is located indicating the uses permitted by the local government in particular zones within its jurisdiction. The records may consist of maps and/or written records. They are often located in the *planning department* of a municipality or county.

7.3.4.9 *Other Historical Sources*—The term *other historical sources* means any source or sources other than those designated in 7.3.4.1 through 7.3.4.8 that are credible to a reasonable person and that identify past uses of the *property*. This category includes, but is not limited to: miscellaneous maps, newspaper archives, and records in the files and/or personal knowledge of the property *owner* and/or *occupants*.

7.4 *Prior Assessment Usage— Standard historical sources* reviewed as part of a prior *environmental site assessment* do not need to be searched for or reviewed again, but uses of the *property* since the prior *environmental site assessment* should be identified either through *standard historical sources* (as specified in 7.3) or by alternatives to *standard historical sources*, to the extent such information is reasonably ascertainable. (See 4.7.)

8. Site Reconnaissance

8.1 *Objective*—The objective of the *site reconnaissance* is to obtain information indicating the likelihood of identifying *recognized environmental* conditions in connection with the property.

8.2 *Observation*—On a visit to the *property* (the *site visit*), the *environmental professional* shall *visually and physically* observe the *property* and any structure(s) located on the *property* to the extent not obstructed by bodies of water, adjacent buildings, or other obstacles.

8.2.1 *Exterior*—The periphery of the *property* shall be *visually and/or physically observed*, as well as the periphery of all structures on the property, and the *property* should be viewed from all adjacent public thoroughfares. If roads or paths with no apparent outlet are observed on the *property*, the use of the road or path should be identified to determine whether it was likely to have been used as an avenue for disposal of *hazardous substances* or *petroleum products*.

8.2.2 *Interior*—On the interior of structures on the *property*, accessible common areas expected to be used by *occupants* or the public (such as lobbies, hallways, utility rooms, recreation areas, etc.), maintenance and repair areas, including boiler rooms, and a representative sample of *occupant* spaces, should be *visually and/or physically observed*. It is not necessary to look under floors, above ceilings, or behind walls.

8.2.3 *Methodology*—The *environmental professional* shall document, in the report, the method used (for example, grid patterns or other systematic approaches used for large properties, which spaces for *owner* or *occupants* were observed) to observe the *property*.

8.2.4 *Limitations*—The *environmental professional* shall document, in the report, general limitations and bases of review, including limitations imposed by physical obstructions such as adjacent buildings, bodies of water, asphalt, or other paved areas, and limiting conditions (for example, snow, rain).

8.2.5 *Frequency*—It is not expected that more than one visit to the *property* shall be made by the *environmental professional* in connection with a *Phase I Environmental Site Assessment*. The one visit constituting part of the *Phase I Environmental Site Assessment* may be referred to as the *site visit*.

8.3 *Prior Assessment Usage*—The information supplied in connection with the *site reconnaissance* portion of a prior *environmental site assessment* may be used for guidance but shall not be relied upon without determining through a new *site reconnaissance* whether any conditions that are material to *recognized environmental conditions* in connection with the *property* have changed since the prior *environmental site assessment*.

8.4 *Uses and Conditions*—The *environmental professional(s)* conducting the *site reconnaissance* should note the uses and conditions specified in 8.4.1 through 8.4.4.8 to the extent *visually and/or physically observed* during the *site visit*. The uses and conditions specified in 8.4.4 through 8.4.4.8 should also be the subject of questions asked as part of *interviews* of *owners* and *occupants* (see Section 9). Uses and conditions to be noted shall be recorded in field notes of the *environmental professional(s)* conducting the *site reconnaissance* but are only required to be described in the report to the extent specified in 8.4.1 through 8.4.4.8. The *environmental professional(s)* performing the *Phase I Environmental Site Assessment* are obligated to identify uses and conditions only to the extent that they may be *visually and/or physically observed* on a *site visit*, as described in this practice, or to the extent that they are identified by the *interviews* (see Sections 9 and 10) or *record review* (see Section 7) processes described in this practice.

8.4.1 *General Site Setting*.

8.4.1.1 *Current Use(s) of the Property*— The current use(s) of the *property* shall be identified in the report. Any current uses likely to involve the use, treatment, storage, disposal, or generation of *hazardous substances* or *petroleum products* shall be identified in the report. Unoccupied occupant spaces should be noted. In identifying current uses of the *property*, more specific information is more helpful than less specific information. (For example, it is more useful to identify uses such as a hardware store, a grocery store, or a bakery rather than simply retail use.)

8.4.1.2 *Past Use(s) of the Property*—To the extent that indications of past uses of the *property* are *visually and/or physically observed* on the *site visit*, or are identified in the *interviews* or *record review*, they shall be identified in the report, and past uses so identified shall be described in the report if they are likely to have involved the use, treatment, storage, disposal, or generation of *hazardous substances* or *petroleum products*. (For example, there may be signs indicating a past use or a structure indicating a past use.)

8.4.1.3 *Current Uses of Adjoining Properties*—To the extent that current uses of *adjoining properties* are *visually and/or physically observed* on the *site visit*, or are identified in the *interviews* or *records review*, they shall be identified in the report, and current uses so identified shall be described in the report if they are likely to indicate *recognized environmental conditions* in connection with the *adjoining properties* or the *property*.

8.4.1.4 *Past Uses of Adjoining Properties*— To the extent that indications of past uses of *adjoining properties* are *visually and/or physically observed* on the *site visit*, or are identified in the *interviews* or *record review*, they shall be noted by the *environmental professional*, and past uses so identified shall be described in the report if they are likely to indicate *recognized environmental conditions* in connection with the *adjoining properties* or the *property*.

8.4.1.5 *Current or Past Uses in the Surrounding Area*—To the extent that the general type of current or past uses (for example, residential, commercial, industrial) of properties surrounding the *property* are *visually and/or physically observed* on the *site visit* or going to or from the *property* for the *site visit*, or are identified in the *interviews* or *record review*, they shall be noted by the *environmental professional*, and uses so identified shall be described in the report if they are likely to indicate *recognized environmental conditions* in connection with the *property*.

8.4.1.6 *Geologic, Hydrogeologic, Hydrologic, and Topographic Conditions*—The topographic conditions of the *property* shall be noted to the extent *visually and/or physically observed* or determined from *interviews*, as well as the general topography of the area surrounding the *property* that is *visually and/or physically observed* from the periphery of the *property*. If any information obtained shows there are likely to be *hazardous substances* or *petroleum products* on the property or on nearby properties and those *hazardous substances* or *petroleum products* are of a type that may migrate, topographic observations shall be analyzed in connection with geologic, hydrogeologic, hydrologic, and topographic information obtained pursuant to *records review* (see 7.2.3) and *interviews* to evaluate whether *hazardous substances* or *petroleum products* are likely to migrate to the *property*, or within or from the *property*, into ground water or soil.

8.4.1.7 *General Description of Structures*— The report shall generally describe the structures or other improvements on the *property*, for example: number of buildings, number of stories each, approximate age of buildings, ancillary structures (if any), etc.

8.4.1.8 *Roads*—Public thoroughfares adjoining the *property* shall be identified in the report and any roads, streets, and parking facilities on the *property* shall be described in the report.

8.4.1.9 *Potable Water Supply*—The source of potable water for the *property* shall be identified in the report.

8.4.1.10 *Sewage Disposal System*—The sewage disposal system for the *property* shall be identified in the report. Inquiry shall be made as to the age of the system as part of the process under Sections 7, 9, or 10.

8.4.2 *Interior and Exterior Observations*:

8.4.2.1 *Current Use(s) of the Property*— The current use(s) of the *property* shall be identified in the report. Any current uses likely to involve the use, treatment, storage, disposal, or generation of *hazardous substances* or *petroleum products* shall be identified in the report. Unoccupied *occupant* spaces should be noted. In identifying current uses of the *property*, more specific information is more helpful than less specific information. (For example, it is more useful to identify uses such as a hardware store, a grocery store, or a bakery rather than simply retail use.)

8.4.2.2 *Past Use(s) of the Property*—To the extent that indications of past uses of the *property* are *visually and/or physically observed* on the *site visit*, or are identified in the *interviews* or *records review*, they shall be identified in the report, and past uses so identified shall be described in the report if they are likely to have involved the use, treatment, storage, disposal, or generation of *hazardous substances* or *petroleum products*. (For example, there may be signs indicating a past use or a structure indicating a past use.)

8.4.2.3 *Hazardous Substances and Petroleum Products in Connection with Identified Uses*—To the extent that present uses are identified that use, treat, store, dispose of, or generate *hazardous substances* and *petroleum products* on the *property*: (*1*) the *hazardous substances* and *petroleum products* shall be identified or indicated as unidentified in the report, and (*2*) the approximate quantities involved, types of containers (if any) and storage conditions shall be described in the report. To the extent that past uses are identified that used, treated, stored, disposed of, or generated *hazardous substances* and *petroleum products* on the *property*, the information shall be identified to the extent it is *visually and/or physically observed* during the *site visit* or identified from the *interviews* or the *records review*.

8.4.2.4 *Storage Tanks*—Above ground storage tanks, or underground storage tanks or vent pipes, fill pipes or access ways indicating underground storage tanks shall be identified (for example, content, capacity, and age) to the extent *visually and/or physically observed* during the *site visit* or identified from the *interviews* or *records review*.

8.4.2.5 *Odors*—Strong, pungent, or noxious odors shall be described in the report and their sources shall be identified in the report to the extent *visually and/or physically observed* or identified from the *interviews* or *records review*.

8.4.2.6 *Pools of Liquid*—Standing surface water shall be noted. Pools or *sumps* containing liquids likely to be *hazardous substances* or *petroleum products* shall be described in the report to the extent *visually and/or physically observed* or identified from the *interviews* or *records review*.

8.4.2.7 *Drums*—To the extent *visually and/or physically observed* or identified from the *interviews* or *records review*, *drums* shall be described in the report, whether or not they are leaking, unless it is known that their contents are not *hazardous substances* or *petroleum products* (in that case the contents should be described in the report). *Drums* often hold 55 gal (208 L) of liquid, but containers as small as 5 gal (19 L) should also be described.

8.4.2.8 *Hazardous Substance and Petroleum Products Containers (Not Necessarily in Connection With Identified Uses)*— When containers identified as containing *hazardous substances* or *petroleum products* are *visually and/or physically observed* on the *property* and are or might be a *recognized environmental condition*: the *hazardous substances* or *petroleum products* shall be identified or indicated as unidentified in the report, and the approximate quantities involved, types of containers, and storage conditions shall be described in the report.

8.4.2.9 *Unidentified Substance Containers*— When open or damaged containers containing unidentified substances suspected of being *hazardous substances* or *petroleum products* are *visually and/or physically observed* on the *property*, the approximate quantities involved, types of containers, and storage conditions shall be described in the report.

8.4.2.10 *PCBs*—Electrical or hydraulic equipment known to contain PCBs or likely to contain PCBs shall be described in the report to the extent *visually and/or physically observed* or identified from the *interviews* or *records review*. Fluorescent light ballast likely to contain PCBs does not need to be noted.

8.4.3 *Interior Observations*:

8.4.3.1 *Heating/Cooling*—The means of heating and cooling the buildings on the property, including the fuel source for heating and cooling, shall be identified in the report (for example, heating oil, gas, electric, radiators from steam boiler fueled by gas).

8.4.3.2 *Stains or Corrosion*—To the extent *visually and/or physically observed* or identified from the *interviews*, stains or corrosion on floors, walls, or ceilings shall be described in the report, except for staining from water.

8.4.3.3 *Drains and Sumps*—To the extent *visually and/or physically observed* or identified from the *interviews*, floor drains and *sumps* shall be described in the report.

8.4.4 *Exterior Observations*:

8.4.4.1 *Pits, Ponds, or Lagoons*—To the extent *visually and/or physically observed* or identified from the *interviews* or *records review*, *pits, ponds, or lagoons* on the *property* shall be described in the report, particularly if they have been used in connection with waste disposal or waste treatment. *Pits, ponds, or lagoons* on *properties* adjoining the *property* shall be described in the report to the extent they are *visually and/or physically observed* from the *property* or identified in the *interviews* or *records review*.

8.4.4.2 *Stained Soil or Pavement*—To the extent *visually and/or physically observed* or identified from the *interviews*, areas of stained soil or pavement shall be described in the report.

8.4.4.3 *Stressed Vegetation*—To the extent *visually and/or physically observed* or identified from the *interviews*, areas of stressed vegetation (from something other than insufficient water) shall be described in the report.

8.4.4.4 *Solid Waste*—To the extent *visually and/or physically observed* or identified from the *interviews* or *records review*, areas that are apparently filled or graded by non-natural causes (or filled by fill of unknown origin) suggesting trash or other solid waste disposal, or mounds or depressions suggesting trash or other solid waste disposal, shall be described in the report.

8.4.4.5 *Waste Water*—To the extent *visually and/or physically observed* or identified from the *interviews* or *records review*, waste water or other liquid (including storm water) or any discharge into a drain, ditch, or stream on or adjacent to the *property* shall be described in the report.

8.4.4.6 *Wells*—To the extent *visually and/or physically observed* or identified from the *interviews* or *records review*, all wells (including dry wells, irrigation wells, injection wells, abandoned wells, or other wells) shall be described in the report.

8.4.4.7 *Septic Systems*—To the extent *visually and/or physically observed* or identified from the *interviews* or *records review*, indications of on-site septic systems or cesspools should be described in the report.

9. Interviews With Owners and Occupants

9.1 *Objective*—The objective of *interviews* is to obtain information indicating *recognized environmental conditions* in connection with the *property*.

9.2 *Content*—Interviews with *owners* and *occupants* consist of questions to be asked in the manner and of persons as described in this section. The content of questions to be asked shall attempt to obtain information about uses and conditions as described in Section 8, as well as information described in 9.8 and 9.9.

9.3 *Medium*—Questions to be asked pursuant to this section may be asked in person, by telephone, or in writing, in the discretion of the *environmental professional*.

9.4 *Timing*—Except as specified in 9.8 and 9.9, it is in the discretion of the *environmental professional* whether to ask questions before, during, or after the *site visit* described in Section 8, or in some combination thereof.

9.5 *Who Should be Interviewed*:

9.5.1 *Key Site Manager*—Prior to the *site visit*, the *owner* should be asked to identify a person with good knowledge of the uses and physical characteristics of the *property* (the *key site manager*). Often the *key site manager* will be the property manager, the chief physical plant supervisor, or head maintenance person. (If the *user* is the current *property* owner, the user has an obligation to identify a *key site manager*, even if it is the *user* himself or herself.) If a *key site manager* is identified, the person conducting the *site visit* shall make at least one reasonable attempt (in writing or by telephone) to arrange a mutually convenient appointment for the *site visit* when the *key site manager* agrees to be there. If the attempt is successful, the *key site manager* shall be interviewed in conjunction with the *site visit*. If such an attempt is unsuccessful, when conducting the *site visit*, the *environmental professional* shall inquire whether an identified *key site manager* (if any) or if a person with good knowledge of the uses and physical characteristics of the *property* is available to be interviewed at that time; if so, that person shall be interviewed. In any case, it is within the discretion of the *environmental professional* to decide which questions to ask before, during, or after the *site visit* or in some combination thereof.

9.5.2 *Occupants*—A reasonable attempt shall be made to interview a reasonable number of *occupants* of the *property*.

9.5.2.1 *Multi-Family Properties*—For multi-family residential properties, residential *occupants* do not need to be interviewed, but if the *property* has nonresidential uses, interviews should be held with the nonresidential *occupants* based on criteria specified in 9.5.2.2.

9.5.2.2 *Major Occupants*—Except as specified in 9.5.2.1, if the *property* has five or fewer current *occupants*, a reasonable attempt shall be made to interview a representative of each one of them. If there are more than five current *occupants*, a reasonable attempt shall be made to interview the *major occupant(s)* and those other *occupants* whose operations are likely to indicate *recognized environmental conditions* in connection with the *property*.

9.5.2.3 *Reasonable Attempts to Interview*— Examples of reasonable attempts to interview those *occupants* specified in 9.5.2.2 include (but are not limited to) an attempt to interview such *occupants* when making the *site visit* or calling such *occupants* by telephone. In any case, when there are several *occupants* to interview, it is not expected that the *site visit* must be scheduled at a time when they will all be available to be interviewed.

9.5.2.4 *Occupant Identification*—The report shall identify the occupants interviewed and the duration of their occupancy.

9.5.3 *Prior Assessment Usage*—Persons interviewed as part of a prior *Phase I Environmental Site Assessment* consistent with this practice do not need to be questioned again about the content of answers they provided at that time. However, they should be questioned about any new information learned since that time, or others should be questioned about conditions since the prior *Phase I Environmental Site Assessment* consistent with this practice.

9.6 *Quality of Answers*—The person(s) interviewed should be asked to be as specific as reasonably feasible in answering questions. The person(s) interviewed should be asked to answer in good faith and to the extent of their knowledge.

9.7 *Incomplete Answers*—While the person conducting the interview(s) has an obligation to ask questions, in many instances the persons to whom the questions are addressed will have no obligation to answer them.

9.7.1 *User*—If the person to be interviewed is the user (the person on whose behalf the *Phase I Environmental Site Assessment* is being conducted), the *user* has an obligation to answer all questions posed by the person conducting the interview, in good faith, to the extent of his or her *actual knowledge* or to designate a *key site manager* to do so. If answers to questions are unknown or partially unknown to the *user* or such *key site manager*, this *interview* section of the *Phase I Environmental Site Assessment* shall not thereby be deemed incomplete.

9.7.2 *Non-user*—If the person conducting the interview(s) asks questions of a person other than a user but does not receive answers or receives partial answers, this section of the *Phase I Environmental Site Assessment* shall not thereby be deemed incomplete, provided that (*1*) the questions have been asked (or attempted to be asked) in person or by telephone and written records have been kept of the person to whom the questions were addressed and the responses, or (*2*) the questions have been asked in writing sent by first class mail or by private, commercial carrier and no answer or incomplete answers have been obtained and at least one reasonable follow up (telephone call or written request) was made again asking for responses.

9.8 *Questions About Helpful Documents*— Prior to the *site visit*, the *property owner*, *key site manager* (if any is identified), and *user* (if different from the *property owner*) shall be asked if they know whether any of the documents listed in 9.8.1 exists and, if so, whether copies can and will be provided to the *environmental professional* within reasonable time and cost constraints. Even partial information provided may be useful. If so, the *environmental professional* conducting the *site visit* shall review the available documents prior to or at the beginning of the *site visit*.

9.8.1 *Helpful Documents*:

9.8.1.1 Environment site assessment reports,

9.8.1.2 Environment audit reports,

9.8.1.3 Environmental permits (for example, solid waste disposal permits, hazardous waste disposal permits, wastewater permits, NPDES permits),

9.8.1.4 Registrations for underground and above-ground storage tanks,

9.8.1.5 Material safety data sheets,

9.8.1.6 Community right-to-know plan,

9.8.1.7 Safety plans; preparedness and prevention plans; spill prevention, countermeasure, and control plans; etc.,

9.8.1.8 Reports regarding hydrogeologic conditions on the property or surrounding area,

9.8.1.9 Notices or other correspondence from any government agency relating to past or current violations of environmental laws with respect to the property or relating to environmental liens encumbering the property,

9.8.1.10 Hazardous waste generator notices or reports, and

9.8.1.11 Geotechnical studies.

9.9 *Proceedings Involving the Property*— Prior to the *site visit*, the *property owner*, *key site manager* (if any is identified), and *user* (if different from the *property owner)* shall be asked whether they know of: (*1*) any pending, threatened, or past litigation relevant to *hazardous substances* or *petroleum products* in, on, or from the *property*; (*2*) any pending, threatened, or past administrative proceedings relevant to *hazardous substances* or *petroleum products* in, on or from the *property*; and (*3*) any notices from any governmental entity regarding any possible violation of environmental laws or possible liability relating to *hazardous substances* or *petroleum products*.

10. Interviews With Local Government Officials

10.1 *Objective*—The objective of *interviews* with *local government officials* is to obtain information indicating *recognized environmental conditions* in connection with the *property*.

10.2 *Content*—Interviews with *local government officials* consist of questions to be asked in the manner and of persons as described in this section. The content of questions to be asked shall be decided in the discretion of the *environmental professional(s)* conducting the *Phase I Environmental Site Assessment*, provided that the questions shall generally be directed towards identifying *recognized environmental conditions* in connection with the *property*.

10.3 *Medium*—Questions to be asked may be asked in person or by telephone, in the discretion of the *environmental professional*.

10.4 *Timing*—It is in the discretion of the *environmental professional* whether to ask questions before or after the *site visit* described in Section 8, or in some combination thereof.

10.5 *Who Should Be Interviewed*:

10.5.1 *Local Agency Officials*—A reasonable attempt shall be made to interview at least one staff member of any one of the following types of local government agencies:

10.5.1.1 Local fire department that serves the *property*,

10.5.1.2 Local health agency or local/regional office of state health agency serving the area in which the *property* is located, or

10.5.1.3 Local agency or local/regional office of state agency having jurisdiction over hazardous waste disposal or other environmental matters in the area in which the *property* is located.

10.6 *Prior Assessment Usage*—Persons interviewed as part of a prior *Phase I Environmental Site Assessment* consistent with this practice do not need to be questioned again about the content of answers they provided at that time. However, they should be questioned about any new information learned since that time, or others should be questioned about conditions since the prior *Phase I Environmental Site Assessment* consistent with this practice.

10.7 *Quality of Answers*—The person(s) interviewed should be asked to be as specific as reasonably feasible in answering questions. The person(s) interviewed should be asked to answer in good faith and to the extent of their knowledge.

10.8 *Incomplete Answers*—While the person conducting the *interview(s)* has an obligation to ask questions, in many instances the persons to whom the questions are addressed will have no obligation to answer them. If the person conducting the *interview(s)* asks questions but does not receive answers or receives partial answers, this section shall not thereby be deemed incomplete, provided that questions have been asked (or attempted to be asked) in person or by telephone and written records have been kept of the person to whom the questions were addressed and their responses.

11. Evaluation and Report Preparation

11.1 *Report Format*—The report of findings for the *Phase I Environmental Site Assessment* should generally follow the recommended report format attached as Appendix X2 unless otherwise required by the user.

11.2 *Documentation*—The findings, opinions and conclusions in the Phase I *Environmental Site Assessment* report shall be supported by documentation. If the *environmental professional* has chosen to exclude certain documentation from the report, the *environmental professional* shall identify in the report the reasons for doing so (for example, a confidentiality agreement). Supporting documentation shall be included in the report or adequately referenced to facilitate reconstruction of the assessment by an environmental professional other than the *environmental professional* who conducted it. Sources that revealed no findings also shall be documented.

11.3 *Contents of Report*—The report shall include those matters required to be included in the report pursuant to various provisions of this practice. In addition, the report shall state whether the *user* reported to the *environmental professional* any information pursuant to the user's responsibilities described in Section 5 of this practice (for example, an *environmental lien* encumbering the *property* or any relevant specialized knowledge or experience of the user).

11.4 *Scope of Services*—The report shall describe all services performed in sufficient detail to permit another party to reconstruct the work performed.

11.5 *Findings*—The report shall have a findings section which summarizes known or suspect environmental conditions associated with the property, and which may include *recognized environmental conditions*, *historical recognized environmental conditions* and *de minimis* conditions, among other environmental conditions.

11.6 *Opinion*—The report shall include the *environmental professional's* opinion(s) of the impact on the property of known or suspect environmental conditions identified in the findings section. The logic and reasoning used by the *environmental professional* in evaluating information collected during the course of the investigation related to known or suspect environmental conditions shall be discussed. The opinion shall specifically include the *environmental professional's* rationale for concluding that a known or suspect environmental condition is or is not currently a *recognized environmental condition*. Known or suspect environmental conditions identified by the *environmental professional* as *recognized environmental conditions* currently shall be listed in the conclusions section of the report.

11.7 *Conclusions*—The report shall include a conclusions section that summarizes all *recognized environmental conditions* connected with the property and the impact of these *recognized environmental conditions* on the property. The report shall include one of the following statements:

11.7.1 "We have performed a Phase I *Environmental Site Assessment* in conformance with the scope and limitations of ASTM Practice E 1527 of [insert address or legal description], the *property*. Any exceptions to, or deletions from, this practice are described in Section [ ] of this report. This assessment has revealed no evidence of *recognized environmental conditions* in connection with the property," or

11.7.2 "We have performed a Phase I *Environmental Site Assessment* in conformance with the scope and limitations of ASTM Practice E 1527 of [insert address or legal description], the property. Any exceptions to, or deletions from, this practice are described in Section [ ] of this report. This assessment has revealed no evidence of *recognized environmental conditions* in connection with the *property* except for the following: (list)."

11.8 *Deviations*—All deletions and deviations from this practice (if any) shall be listed individually and in detail and all additions should be listed.

11.9 *Additional Services*—Any additional services contracted for between the *user* and the *environmental professional(s)*, including a broader scope of assessment, more detailed conclusions, liability/risk evaluations, recommendation for Phase II testing, remediation techniques, etc., are beyond the scope of this practice, and should only be included in the report if so specified in the terms of engagement between the *user* and the *environmental professional*.

11.10 *References*—The report shall include a references section to identify published referenced sources relied upon in preparing the Phase I *Environmental Site Assessment*. Each referenced source shall be adequately annotated to facilitate retrieval by another party.

11.11 *Signature*—The *environmental professional(s)* responsible for the Phase I *Environmental Site Assessment* shall sign the report.

11.12 *Qualifications*—The report shall include a qualifications statement of the *environmental professional(s)* responsible for conducting the Phase I *Environmental Site Assessment* and preparation of the report.

11.13 *Appendices*—The report shall include an appendix section containing supporting documentation.

12. Non-Scope Considerations

12.1 *General*:

12.1.1 *Additional Issues*—There may be environmental issues or conditions at a property that parties may wish to assess in connection with *commercial real estate* that are outside the scope of this practice (the non-scope considerations). As noted by the legal analysis in Appendix X1 of this practice, some substances may be present on a property in quantities and under conditions that may lead to contamination of the property or of nearby properties but are not included in CERCLA's definition of hazardous substances (42 USC § 9601(14)) or do not otherwise present potential CERCLA liability. In any case, they are beyond the scope of this practice.

12.1.2 *Outside Standard Practices*—Whether or not a *user* elects to inquire into non-scope considerations in connection with this practice or any other environmental site assessment, no assessment of such non-scope considerations is required for appropriate inquiry as defined by this practice.

12.1.3 *Other Standards*—There may be standards or protocols for assessment of potential hazards and conditions associated with non-scope conditions developed by governmental entities, professional organizations, or other private entities.

12.1.4 *List of Additional Issues*—Following are several non-scope considerations that persons may want to assess in connection with commercial real estate. No implication is intended as to the relative importance of inquiry into such non-scope considerations, and this list of non-scope considerations is not intended to be all-inclusive:

12.1.4.1 Asbestos-Containing Materials,

12.1.4.2 Radon,

12.1.4.3 Lead-Based Paint,

12.1.4.4 Lead in Drinking Water,

12.1.4.5 Wetlands,

12.1.4.6 Regulatory compliance,

12.1.4.7 Cultural and historic resources,

12.1.4.8 Industrial hygiene,

12.1.4.9 Health and safety,

12.1.4.10 Ecological resources,

12.1.4.11 Endangered species,

12.1.4.12 Indoor air quality, and

12.1.4.13 High voltage powerlines.

APPENDIXES

(Nonmandatory Information)

X1. LEGAL BACKGROUND TO FEDERAL LAW AND THE PRACTICES ON ENVIRONMENTAL ASSESSMENTS IN COMMERCIAL REAL ESTATE TRANSACTIONS

INTRODUCTION

The legal section of Subcommittee E50.02 on Environmental Assessments In Commercial Real Estate Transactions provides the following background to the Comprehensive Environmental Response, Compensation, and Liability Act (CERCLA), as amended including amended by the Superfund Amendments and Reauthorization Act (SARA), 42 USC § 9601 *et seq.* The background to CERCLA, commonly known as the Superfund law, outlines the potential liability for the cleanup of hazardous substances, available defenses to such liability, appropriate inquiry under Superfund, statutory definition of hazardous substances, petroleum products and petroleum exclusion to CERCLA, and reasons why certain environmental-hazards are excluded from the scope of Superfund and this practice and Practice E 1528.

There are several elements of Superfund liability and the commonly termed "innocent purchaser" defense, that arises out of the statutory third-party defense, that may impact on the development and understanding of this practice and Practice E 1528.

X1.1 *Superfund Liability:*

X1.1.1 All of the following elements of liability under Superfund must be established by a plaintiff before a defendant will be held liable under superfund for a *government's* response costs:[5]

X1.1.1.1 The site is a facility, as defined at § 9601(9),

X1.1.1.2 A release or threatened release of a hazardous substance from the site occurred (release is defined at

[5] 42 USC §9607(a). (All statutory references are to Title 42 of the United States Code, unless otherwise specified.) See *United States versus Aceto Agricultural Chemicals Corp.*, 872 F.2d 1373 (8th Cir. 1989). Private plaintiffs, as well as the government, may seek response costs under Superfund from defendants. While many users of these ASTM practices or other private parties may think in terms of how to defend against Superfund liability, they should recognize that they may decide to conduct cleanup actions and seek response costs from other parties.

§ 9601(22) as *any*amount of any hazardous substance;" hazardous substance" is defined at § 9601(14) (see statutory definition of "hazardous substance"),

X1.1.1.3 A release or threatened release caused the plaintiff to incur response costs. Response costs are defined at § 9601(25) to mean costs related to both removal actions (§ 9601(23)) and remedial actions (§ 9601(24)), and

X1.1.1.4 Defendants fall within at least one of the four classes of responsible parties.[6]

X1.1.2 In order to recover response costs, a government plaintiff must prove that the costs were not inconsistent with the National Oil and Hazardous Substances Pollution Contingency Plan (commonly referred to as the National Contingency Plan or NCP), 40 CFR § 300.[7] A private plaintiff must prove its costs were necessary costs of response and that the response action was consistent with the NCP. 42 USC § 9607 (a).[8]

X1.1.3 If there is a release or threatened release of hazardous substances on a site, private parties, even if they are not PRPs, may decide to incur response costs and seek recovery from other private parties, and PRPs may seek contribution from other PRPs.

X1.1.4 There is an important difference between government's burden to show that its response costs are "not inconsistent with the NCP" and the burden a private party bears to show that its response costs are "consistent with the NCP". See § 9607 (a)(4)(A) and (B). Courts have interpreted this statutory difference to give the government a rebuttable presumption that its response costs are consistent with the NCP, whereas a private party who undertakes response costs and seeks recovery from responsible parties bears the burden of proving its response was consistent with the NCP.[9] The EPA takes the position that a private party who undertakes a response action must be only in "substantial compliance", rather than strict technical compliance, with the NCP, as long as a CERCLA-quality cleanup is achieved. The NCP requirements for a private party response-action are set forth at 40 CFR § 300.700.

X1.2 *Defenses to Liability:*

X1.2.1 Assuming all the elements of liability exist, a party may avoid liability only by meeting one of the defenses listed in§ 9607(b). These listed defenses are exclusive of all others.[10] Section 9607(b) states (*emphasis added*):

"There shall be no liability under subsection (a) for a person otherwise liable who can establish by a preponderance of the evidence [the lowest evidentiary standard available, meaning more probable than not] that the release or threat of release of a hazardous substance and the damages resulting therefrom were *caused solely by*—

1) an act of God;

2) an act of war;

3) *an act or omission of a third party other*than an employee or agent of the defendant, or*than one whose act or omission occurs in connection with a contractual relationship*[see the definition of "contractual relationship" in X1.2.2], existing directly or indirectly, with the defendant ..., if the defendant establishes by a preponderance of the evidence that (a) he exercised due care with respect to the hazardous substance concerned ..., and (b) he took precautions against foreseeable acts or omissions of any such third party and the consequences that could foreseeably result from such actions or omissions."

X1.2.2 Under § 9601(35)(A), a contractual relationship "includes, but is not limited to, land contracts, deeds, or other instruments transferring title or possession ...". These contractual relationships with third parties eliminate the defense to liability unless the defendant is an innocent purchaser. Or as stated by the statute at § 9601(35) (A)(emphasis added), a contractual relationship with the third party defeats the defense.

"unless the real property on which the facility is located was acquired by the defendant after disposal or placement of the hazardous substance and one or more of the following circumstances is also established by the defendant by a preponderance of the evidence:

(i) At the time the defendant acquired the facility the defendant *did not know and had no reason to know*that any hazardous substance which is the subject of the release or threatened release was disposed of on, in, or at the facility.

(ii) The defendant is the government ...

(iii) The defendant acquired the facility by inheritance or bequest."

X1.2.3 Therefore, the so-called innocent purchaser defense arises out of the third-party defense of § 9607(b)(3). Restated, this defense to Superfund liability is available only if the defendant shows the following:

X1.2.3.1 The release or threat of release was caused solely by a third party,

X1.2.3.2 The third party is not an employee or agent of the defendant,

X1.2.3.3 The acts or omissions of the third party did not occur in connection with a direct or indirect contractual relationship to the defendant, or if there was a contractual relationship, the defendant acquired the property after disposal or placement of the hazardous substance, and at the time the defendant acquired the facility the defendant *did not know and had no reason to know* that any hazardous substance that is the subject of the release or threatened release was disposed of on, in, or at the facility, and

X1.2.3.4 The defendant exercised due care with respect to

[6] The four classes of potentially responsible parties (PRPs) are listed as § 9607(a) as follows:

(*1*) Owner and operator of a facility (See § 9601 (20)(A): the term "owner or operator" does not include a person, who, without participating in the management of a facility, holds indicia of ownership primarily to protect his security interest in the facility. In *re: Bergsoe Metal Corporation*, 910 F.2d 668 (9th Cir. Aug. 9, 1990); *Guidice versus BFG Electroplating and Manufacturing Co.*, 732 F. Supp. 556 (W.D.Pa. 1989); *United States v. Mirabile*, 23 ERC 1511 (E.D.Pa. 1985). But see *United States versus Fleet Factors Corp.*, 901 F.2d 1550 (11th Cir. 1990); *United States versus Maryland Bank and Trust Co.*, 632 F. Supp. 573 (D. Md. 1986). For clarification of the security interest exclusion, see EPA's rule on lender liability under CERCLA, 57 Federal Register 18344 (April 29, 1992),

(*2*) Any person who at the time of disposal of any hazardous substance owned or operated any facility at which such hazardous substances were disposed of,

(*3*) Any person who by contract, agreement, or otherwise arranged for disposal or treatment or transport of hazardous substances, and

(*4*) Any person who accepts hazardous substances for transport to a facility selected by such person.

[7] The National Contingency Plan is the federal government's blueprint on how hazardous substances are to be cleaned up pursuant to CERCLA.

[8] See *Dedham Water Co. versus Cumberland Farms Dairy, Inc.*, 889 F.2d 1146 (1st Cir. 1989); other cases cited at ABA, *Natural Resources, Energy, and Environmental Law: 1989 The Year In Review*, p. 215, Note 155.

[9] *Amland Properties Corp. versus Aluminum Co. of America*, 711 F. Supp. 784, 794 (D. N.J. 1989); *Artesian Water Co. versus New Castle County*, 659 F. Supp. 1269, 1291 (D. Del. 1987); *United States versus Northeastern Pharmaceutical and Chemical Co.*, 579 F. Supp. 823 (W.D. Mo. 1984), *aff'd in part, rev'd on other grounds*, 810 F.2d 726 (8th Cir. 1986).

[10] *United States versus Aceto Agricultural Chemicals Corp.*, 872 F.2d 1373 (8th Cir. 1989). But see *United States versus Marisol, Inc.*, 725 F. Supp. 833 (M.D. Pa. 1989) (equitable defenses under CERCLA may be available after the development of a factual record).

the hazardous substances and took precautions against foreseeable acts or omissions of the third party.

X1.2.4 The statute then states at § 9601(35)(B) (*emphasis added*):

"To establish that the defendant had no reason to know, as provided [above], *the defendant must have undertaken, at the time of acquisition, all appropriate inquiry into the previous ownership and uses of the property consistent with good commercial or customary practice*in an effort to minimize liability... . [T]he court shall take into account any specialized knowledge or experience on the part of the defendant, the relationship of the purchase price to the value of the property if uncontaminated, commonly known or reasonably ascertainable information about the property, the obviousness of the presence or likely presence of contamination at the property, and the ability to detect such contamination by appropriate inspection."

X1.3 *Appropriate Inquiry in Commercial Real Estate Transactions:*

X1.3.1 One of the major questions that parties to commercial real estate transactions face when considering their potential Superfund liability is, "What level of inquiry into the previous ownership and uses of the property is appropriate to establish the innocent purchaser defense to Superfund liability?" These practices are structured to articulate the level of inquiry under Superfund that is appropriate for different situations.

X1.3.2 *The Appropriate Level of Inquiry*:

X1.3.2.1 The level of environmental inquiry that is appropriate under Superfund cannot be the same for every property or every party to a real estate transaction. The level of inquiry, in fact, will change depending on the particular property or party involved in a transaction. The statutory language, Congressional history, and common sense support this conclusion.

X1.3.2.2 First, it must be noted that little case law exists to serve as guidance about the minimum level of inquiry that will be deemed appropriate for the innocent purchaser defense. See, for example, *United States versus Serafini*, 706 F. Supp. 346 (M.D. Pa. 1988), and 1990 U.S. Dist. LEXIS 18466 (M.D. Pa. 1990) (By entertaining disputed facts as to the custom and practice of viewing land prior to purchase, the court implied that appropriate inquiry necessarily varies on a site-by-site basis); *United States versus Pacific Hide and Fur Depot, Inc.*, 716 F. Supp. 1341 (D. Idaho 1989) (No inquiry was required by those who received an ownership interest in property via corporate stock transfer and warranty deed under the facts of this case); *International Clinical Laboratories, Inc. versus Stevens*, 30 ERC 2066, 20 ELR 20,560 (E.D.N.Y. 1990). (Despite a long history of toxic wastewater disposal and presence of the site on the state's hazardous waste disposal site list, the purchaser established the innocent purchaser defense since there were no visible environmental problems at the site, the defendant had no knowledge of environmental problems at the site and the purchase price did not reflect a reduction on account of the problem.)

X1.3.2.3 While the statute does not specifically distinguish certain types of properties and uses from others, or certain types of parties from others, it does list certain factors courts should consider in determining whether one's inquiry under the circumstances is appropriate. The statute, as explained in X1.2, requires a court to consider a party's specialized knowledge or experience. The statute further mandates a court to consider what is "*reasonably* ascertainable information about the property", what contamination is *obviously* present, and the *party's* "ability to detect such contamination". The very use of terms such as "appropriate" and "reasonably", and the use of "specialized knowledge and experience" and "ability" in conjunction with the specific person attempting to utilize the defense signifies that Congress did not intend the appropriateness of the inquiry be judged by a bright line standard. If it so intended, Congress would have stated, but did not, that the same inquiry should be made in every case.

X1.3.2.4 What is reasonable and obvious to one party may not be so to other parties, and ability, by necessity, varies among all parties. The statute, therefore, recognizes that different properties and parties must be treated differently. That is, different parties may conduct different levels of inquiry appropriate to their circumstances.

X1.3.2.5 The statutory standard of "appropriate inquiry" suggests the level of inquiry will depend on the circumstances and the underlying facts. Since the facts are almost always different, the level of inquiry must change with them. The legislative history on this particular issue demonstrates that Congress intended that the level of inquiry changes with the type of property and party:[11]

The duty to inquire under this provision shall be judged as of the time of acquisition. Defendants shall be held to a higher standard as public awareness of the hazards associated with hazardous releases has grown, as reflected by this Act, the 1980 Act [CERCLA] and other Federal and State statutes. Moreover, good commercial or customary practice with respect to inquiry in an effort to minimize liability shall mean that a reasonable inquiry must have been made in all circumstances, in light of best business and land transfer principles. Those engaged in commercial transactions should, however, be held to a higher standard than those who are engaged in private residential transactions.

X1.3.2.6 Because few cases address the standard of inquiry in the innocent purchaser defense and the legislative history describing Congressional intent is sparse, common sense is a useful guide in interpreting statutory language. If CERCLA mandated that the level of inquiry be the same for every property or potential defendant, then a lay consumer (renter or buyer) of a home, a purchaser of a small environmentally benign business, and a multinational corporate buyer of an industrial complex would have to conduct the same environmental site assessment (ESA) of the different properties in question. Additionally, the statute makes no mention of a Phase I ESA or any other specific type of inquiry one is to conduct in order for the inquiry to be deemed appropriate. If all inquires had to be at the same level to be "appropriate", it would be illogical to stop at a Phase I ESA since some commercial or industrial properties routinely undergo, in the exercise of good commercial and customary practices, intrusive sampling (typically a Phase II ESA activity). Therefore, since routinely some properties undergo sampling, an inflexible standard would require sampling of all properties, no matter what its use. This could not have been the intended result of SARA.

X1.3.3 *The Minimum Inquiries to Satisfy All Appropriate Inquiry*:

X1.3.3.1 Recognizing that inquiry changes with the underlying circumstances, the next question concerns that level of

[11] H.R. Rep. No. 962, 99th Cong., 2d Sess. 187 (1986), *reprinted at* 1986 U.S. Code Cong. and Admin. News 3276, 3280.

inquiry, if any, that Superfund requires to utilize the innocent landowner defense.

X1.3.3.2 As noted above, in some real estate transactions a Phase II ESA is routinely conducted. A Phase I ESA is conducted in these transactions only as a necessary prerequisite to outline the scope of the Phase II ESA. A Phase II ESA typically involves taking soil, water, and air samples to determine their contaminant content or verify that no contaminants are present or likely present. Note, however, that this simplistic outline of the Phase II ESA is misleading since the party can always dig down one foot deeper, take one more sample, or conduct one more test. The problem of how much inquiry is conducted, or at what level a party should begin, involves proving a negative, that is, that no contamination is present.[12] Since, according to the statute, inquiries should be judged by the circumstances existing at the time of acquisition, then there could be some properties and parties to real estate transactions where it may be appropriate to begin the inquiry with an intrusive Phase II ESA in order to invoke the innocent purchaser defense to liability.

X1.3.3.3 At the other extreme, the minimum level of inquiry that a party would be expected to conduct is found by looking at the least environmentally obtrusive class of property and party from a CERCLA perspective. This transaction likely involves the lay buyer of a home or the renter of an apartment. Assuming these parties meet the other prerequisites for the innocent purchaser defense, what level of environmental inquiry must they conduct to avoid Superfund liability? While there are no recorded court cases on this issue, the answer is probably none, unless a particular residential purchaser or renter has some specialized knowledge about or experience with the property in question that would lead a court to conclude that some questions should have been asked. Beyond these rare situations, it is highly unlikely that Congress intended to saddle housing consumers with the burden of investigating or cleaning up contaminated sites. In fact, EPA has issued a statement of enforcement policy to the effect that it will not generally pursue owners of single family residences pursuant to CERCLA.[13] Therefore, for some properties and parties to real estate transactions, it is appropriate to conduct no environmental inquiry in order to meet one's innocent purchaser defense to liability.

X1.3.3.4 The minimum level of appropriate inquiry under Superfund, therefore, ranges from no specialized inquiry to conducting an intrusive Phase II ESA. In order to satisfy these practices, to do no specialized inquiry, such as the Transaction Screen or Phase I ESA, is not enough for commercial real estate transactions. Under current commercial and customary practice and in light of best business and land transfer principles, however, no environmental site assessments are conducted in many real estate transactions, particularly those involving smaller properties, vacant land, or transactions of low monetary value. This practice and Practice E 1528 and the minimum level of inquiry under these practices, actually raises the average level of inquiry that should be performed where the parties want to come within the protection of the innocent landowner defense.

X1.3.3.5 The burden of proof is on the defendant to sustain by a preponderance of the evidence, the innocent purchaser defense. This is the least onerous burden of proof available to a party in litigation. The defendant must show only that the evidence offered to support the level of inquiry that was taken at the time of acquisition is of greater weight or more convincing than the evidence offered in opposition to it. In other words, the evidence on the inquiry issue taken as a whole shows that the fact sought to be proved is more probable than not. There may be technical or business judgments on whether the inquiry conducted or any other fact in a particular case is sufficient to meet the needs or concerns of a party to the real estate transaction. The bottom line, however, is that the judgment on whether the specific facts of a case, in light of statutory language, are sufficient to produce liability or a viable defense to liability is a legal one and such judgments constitute the practice of law.

X1.3.3.6 Practice E 1528 is designed as the minimum level of inquiry to satisfy the practice from which a party to a commercial real estate transaction should proceed, recognizing that some parties to some commercial real estate transactions may wish to proceed by beginning with a Phase I or a Phase II ESA.

X1.4 *Statutory Definition of Hazardous Substance:*

X1.4.1 The statute at 42 USC § 9601(14)(A – F) defines hazardous substance by referring to five other statutes as well as to Superfund's own § 9602. The following is a description of the relevant portions of the other statutes and § 9602 of Superfund:

42 USC § 9601(14)(A):" [A]ny substance designated pursuant to section 1321(b)(2)(A) of Title 33." Title 33 USC § 1321 lies within the Clean Water Act and refers to, among other things, hazardous substance liability. 33 USC§ 1321(b)(2)(A) states that the EPA shall develop, "as may be appropriate, regulations designating as hazardous substances, other than oil as defined in this section, such elements and compounds which, when discharged in any quantity into" the navigable waters of the United States ..., present an imminent and substantial danger to the public ... health or welfare, including, but not limited to, fish, shellfish, wildlife, shorelines, and beaches."

[12] The inability to prove a negative creates a dilemma for the potential defendant. If the party's inquiry discovers contamination, then under the statute the party will not be able to avail itself of the innocent purchaser defense. If the inquiry does not discover contamination, EPA or another private party can argue in a response action that the inquiry was not "appropriate" and, therefore, the defendant can have no defense. This dilemma is explicitly recognized by the Subcommittee E50.02 as beyond any reasonable interpretation of Congressional intent. The scope of the E50.02 Standard Practices resolves the party's dilemma in the only reasonable way by stating: "It should not be concluded or assumed that the inquiry was not appropriate inquiry merely because the inquiry did not identify existing recognized environmental conditions in connection with a property. Environmental site assessments must be evaluated based on the reasonableness of the judgments made at the time and under the circumstances in which they were made." See 4.5.4.

[13] EPA, *Policy Towards Owners of Residential Property at Superfund Sites*, OSWER Directive No. 9834.6, July 3, 1991.

42 USC § 9601(14)(B):" [A]ny element, compound, mixture, solution, or substance designated pursuant to section 9602 of this title." Section 9602 gives EPA the authority to designate as a hazardous substance "such elements, compounds, mixtures, solutions, and substances which, when released into the environment may present substantial danger to the public health or welfare or the environment... ."

42 USC § 9601(14)(C):" [A]ny hazardous waste having the characteristics identified under or listed pursuant to section 3001 of the Solid Waste Disposal Act [42 USCA§ 6921] (but not including any waste the regulation of which under the Solid Waste Disposal Act [42 USCA § 6901 et seq.] has been suspended by Act of Congress)." The Solid Waste Disposal Act of 1980 amended the Resource Conservation and Recovery Act (RCRA). 42 USCA § 6921 of RCRA provides authority to the EPA to develop criteria for identifying characteristics of hazardous waste and for listing particular hazardous wastes within the meaning of 42 USCA § 6903 (5) of RCRA. RCRA, § 6903(5), defines hazardous waste to mean

"a solid waste, or combination of solid wastes, which because of its quantity, concentration, or physical, chemical, or infectious characteristics may—

(A) cause, or significantly contribute to an increase in mortality or an increase in serious irreversible, or incapacitating reversible, illness; or

(B) pose a substantial present or potential hazard to human health or the environment when improperly treated, stored, transported, or disposed of, or otherwise managed."

For the identification and listing of hazardous wastes under RCRA, see 40 CFR §§ 261.1 *et seq.*

42 USC § 9601(14)(D):" [A]ny toxic pollutant listed under Section 1317(a) of Title 33." Section 1317(a) of Title 33 refers to toxic and pretreatment effluent standards under the Clean Water Act. The EPA is charged in this section with publishing and revising from time to time a list of toxic pollutants, taking "into account toxicity of the pollutant, its persistence, degradability, the usual or potential presence of the affected organisms in any waters, the importance of the affected organisms, and the nature and extent of the effect of the toxic pollutant on such organisms." Each toxic pollutant listed according to this section shall be subject to effluent limitations. For toxic pollutant effluent standards, see 40 CFR §§ 129.1 et seq.

42 USC § 9601(14)(E):" [A]ny hazardous air pollutant listed under Section 112 of the Clean Air Act [42 USCA § 7412]." Section 7412 of Title 42 deals with national emission standards for hazardous air pollutants. The EPA is charged here with publishing and revising from time to time "a list which includes each hazardous air pollutant for which [it] intends to establish an emission standard under this section." The term "hazardous air pollutant" means an air pollutant that in EPA's judgment "causes, or contributes to, air pollution which may reasonably be anticipated to result in an increase in mortality or an increase in serious irreversible, or incapacitating reversible, illness." For emission standards for hazardous pollutants, see 40 CFR § 61.01 et seq.

42 USC § 9601(14)(F):" [A]ny imminently hazardous chemical substance or mixture with respect to which the [EPA] has taken action pursuant to Section 2606 of Title 15." Section 2606 of Title 15 deals with imminent hazards under the Toxic Substances Control Act (TSCA). The EPA is authorized under 15 USC § 2606 to seize an imminently hazardous chemical substance or mixture or seek other relief, such as requiring notice to users of the chemical substance or public notice of the risk associated with the substance or mixture. The term " 'imminently hazardous chemical substance or mixture' means a chemical substance or mixture which presents an imminent and unreasonable risk of serious or widespread injury to health or the environment."

X1.4.2 After Subsections A–F, outlined above, the Superfund definition of "hazardous substance" in § 9601(14) then goes on to state:

"The term does not include petroleum, including crude oil or any fraction thereof which is not otherwise specifically listed or designated as a hazardous substance under Subparagraphs (A) through (F) of this paragraph, and the term does not include natural gas, natural gas liquids, liquefied natural gas, or synthetic gas usable for fuel (or mixtures of natural gas and such synthetic gas)."

X1.4.3 The EPA has collected a list of "those substances in the statutes referred to in Section 101(14) of the Act [42 USC § 9601(14)]" 40 CFR § 302.1 (1989) ("List of Hazardous Substances And Reportable Quantities," 40 CFR Part 302). This list changes with notices in the Federal Register. Also, any time a new hazardous waste is listed, the waste automatically becomes a hazardous substance.

X1.5 *Petroleum Products:*

X1.5.1 Under the petroleum exclusion of CERCLA (42 USC§ 9601(14)), petroleum and crude oil have been explicitly excluded from the definition of hazardous substances under CERCLA. Nevertheless, petroleum products are included within the scope of both practices because they are of concern in many commercial real estate transactions and current custom and usage is to include an inquiry into the presence of petroleum products in an environmental site assessment. Inclusion of petroleum products within the scope of the practices is not based upon the applicability, if any, of CERCLA to petroleum products.

X1.5.2 One reason to include petroleum products within the scope of the practices is because to do so reflects custom and usage: when environmental assessments are conducted in connection with commercial real estate transactions, they customarily include an assessment of the presence or likely presence of petroleum products under conditions that may lead to contamination. For example, environmental assessments ordinarily seek to assess whether there may be underground or above-ground storage tanks that may be leaking, whether those tanks contain petroleum products or some other product.

X1.5.3 In addition, although CERCLA may exclude petroleum products, other laws require cleanup of releases or spills of petroleum products. For example, petroleum products sometimes (for example, when they cannot be reclaimed from soil) become hazardous wastes subject to RCRA Subtitle C (42 USC § 6921 *et seq.*), must be cleaned up if released from underground storage tanks pursuant to RCRA Subtitle I (42 USC § 6991 *et seq.*), must be cleaned up pursuant to the Oil Pollution Act of 1990 (33 USC § 1321 *et seq.*), and must be cleaned up if released into the navigable waters of the United States pursuant to the Clean Water Act (33 USC § 1251 *et seq.*).

X1.5.4 Moreover, case law and EPA interpretations of the petroleum exclusion require an analysis of the facts of each case to determine whether a particular petroleum product is included in CERCLA's petroleum exclusion. The exclusion has been broadly interpreted to exclude gasoline and leaded gasoline from CERCLA's definition of hazardous substances regardless of the fact that gasoline and leaded gasoline contain certain indigenous components and additives which have themselves been designated as hazardous pursuant to CERCLA. See *Wilshire Westwood Associates versus Atlantic Richfield Corporation*, 881 F.2d 801 (9th Cir. 1989). The interpretation was narrowed when a judicial distinction was made between petroleum fractions produced by distillation processes and waste products resulting from contaminated tank scale. See *United States versus Western Processing Co.*, 761 F. Supp. 713 (W.D. Wash. 1991). Another decision narrowly interpreted CERCLA's petroleum exclusion to be inapplicable to oil-related wastes containing hazardous substances because the primary purpose of the exclusion is to remove "spills or other releases strictly of oil" from the scope of CERCLA response and liability (not releases of hazardous substances mixed with oil). See *City of New York versus Exxon Corporation*, 744 F. Supp. 474 (S.D.N.Y. 1990). For additional discussion, see EPA Memorandum entitled, "The Petroleum Exclusion Under the Comprehensive Environmental Response Compensation and

Liability Act," issued by EPA's General Counsel, Francis S. Blake, July 31, 1987.

X1.6 *Exclusion of Certain Hazards From Superfund:*

X1.6.1 The information that follows is provided to explain why these potential environmental hazards are not covered by Superfund's appropriate inquiry responsibilities:

X1.6.2 As a preliminary matter, it should be noted that an environmental site assessment that does not address substances excluded from CERCLA (whether those substances are excluded because they are petroleum products or by virtue of other characteristics) but that otherwise constitutes" all appropriate inquiry into the previous ownership and uses of the property consistent with good commercial or customary practice" should nevertheless entitle the user to the innocent purchaser defense, assuming that other requirements of the defense are met.

X1.6.3 *Radon*:

X1.6.3.1 A case discussing Superfund and radon is *Amoco Oil Company versus Borden, Inc.*, 889 F.2d 664 (5th Cir. 1989). This case dealt with a private cost recovery action by the buyer of a site against the seller for response costs relating to radiation from phosphogypsum wastes left on the site. Radon emanated from these radioactive wastes. The case points out that the "EPA has designated radionuclides as hazardous substances under § 9602(a) of CERCLA... . Additionally, the ... EPA under § 112 of the Clean Air Act ... list radionuclides as a hazardous air pollutant. Radon and its daughter products are considered radionuclides, which are defined as 'any nuclide that emits radiation.' " Therefore, radon is a CERCLA hazardous substance. Also, when discussing what constitutes a release of a hazardous substance under the statute, the statute is plain that there is no quantitative requirement and that a release, broadly defined at 42 USC § 9601(22), of *any* amount constitutes a CERCLA release.

X1.6.3.2 Liability under Superfund depends on several factors, as noted in X1.1. Only one of four factors is the release or threatened release of a hazardous substance. The other three factors are (*1*) the site is a facility, (*2*) the defendant falls within at least one of four classes of potentially responsible parties (PRPs), and (*3*) the release or threatened release *caused* the plaintiff (that can be the government or another private party) to incur response costs. Further, response costs must not be inconsistent with the National Contingency Plan (NCP), *and must not be limited by § 9604(a)(3)*. And, of course, there is no need to raise the innocent purchaser defense and its appropriate inquiry requirements unless the elements of liability will be met.

X1.6.3.3 Where radon from any source occurs in a building, three of the liability elements under CERCLA are met. There is a release of a hazardous substance, the building is a facility, and we can assume the defendant is a PRP. However, under 42 USC § 9604(a)(3)(A), "[r]emedial actions taken in response to hazardous substances as they occur naturally are specifically excluded from the NCP and are therefore not recoverable." *Amoco Oil Company versus Borden, Inc.*, 889 F.2d at 570. The statute is plain.[14]

"(3) Limitations on response

The President shall not provide for a removal or remedial action under this section in response to a release or threat of release—

(A) of a naturally occurring substance in its unaltered form, or altered solely through naturally occurring processes or phenomena, from a location where it is naturally found;

(B) from products which are part of the structure of, and result in exposure within, residential buildings or business or community structures;[15] or

(C) into public or private drinking water supplies due to deterioration of the system through ordinary use.[16]

(4) Exception to limitations

Notwithstanding paragraph of this subsection, to the extent authorized by this section, the President may respond to any release or threat of release if in the President's discretion, it constitutes a public health or environmental emergency and no other person with the authority and capability to respond to the emergency will do so in a timely manner."

X1.6.3.4 Therefore, no liability under CERCLA attaches for naturally occurring radon. If a party to a real estate transaction wants to look for radon within a building, no amount of radon investigation will have any bearing on one's innocent purchaser defense under Superfund. Investigation of naturally occurring radon would be included, if at all, in the portion of the practice that deals with non-scope issues.

X1.6.4 *Asbestos*:

X1.6.4.1 The analysis of asbestos is similar to that involving radon. Before considering appropriate inquiry responsibilities, the four elements of CERCLA liability must be satisfied. Once again, as with radon, they are not met.

X1.6.4.2 Section 9604(a)(3)(B) of CERCLA prohibits response actions involving a release or threat of release "from products which are part of the structure of, and result in exposure within, residential buildings or business or community structures." There are a number of cases dealing with asbestos that interpret this statutory language. One such case is *First United Methodist Church of Hyattsville versus United States Gypsum Co.* that cites to other relevant cases.

X1.6.4.3 In *First United* the church brought a private cost recovery action against the manufacturer of asbestos-containing acoustical plaster. In holding that the action was barred by a state statute of repose (a certain time allowed by statute for bringing litigation) and that CERCLA did not preempt the state statute of repose, the court stated that § 9604(a)(3)(B) of CERCLA "represents much more than a procedural limitation on the President's authority; it is a substantive limitation of the breadth of CERCLA itself."[17] Therefore, the limitations of § 9604(a)(3) apply to private parties as well.

X1.6.4.4 Citing to the legislative history, the *First United* court concluded, "[i]n view of this clear expression of Congressional intent, we wil[l] not expand CERCLA to encompass asbestos-removal actions." The court further explained:[18]

---

[14] 42 USC § 9604(a)(3) and (4) (*emphasis added*).

[15] This provision has implications for asbestos and lead-based paint. See X1.6.4 and X1.6.5.

[16] This provision has implications for lead from lead pipes and solder. See X1.6.5.

[17] One such case is *First United Methodist Church of Hyattsville versus United States Gypsum Co.*, 882 F.2d 862 (4th Cir. 1989), that cites to other relevant cases.

[18] The same at 869; See also *3550 Stevens Creek Associates versus Barclays Bank of California*, 915 F.2d 1355 (9th Cir. 1990).

"In closing, we note that this interpretation of CERCLA fully comports with the most fundamental guide to statutory construction—common sense. To extend CERCLA's strict liability scheme to all past and present owners of buildings containing asbestos as well as to all persons who manufactured, transported, and installed asbestos products into buildings, would be to shift literally billions of dollars of removal cost liability based on nothing more than an improvident interpretation of a statute that Congress never intended to apply in this context. [FN12] ... Certainly, if Congress had intended for CERCLA to address the monumental asbestos problem, it would have said so more directly when it passed SARA... .

FN12—It is for this reason, that Congress simply did not intend for CERCLA to remedy the asbestos-removal problem, that we decline to follow the reasoning of *Prudential, Knox and Covalt* in rejecting First United's preemption argument. Instead of recognizing the fact that CERCLA is out of context in this situation, these courts rejected similar attempts to invoke the statute by construing CERCLA's key terms in a way to exclude asbestos-removal actions. *Covalt*, 860 F.2d [1434] at 1438-39 (defining" environment" to exclude the interior of a workplace); *Knox*, 690 F. Supp at 756-57 (defining "release" in terms of "spills" or "disposal"); *Prudential*, [711 F. Supp 1244] at 1254-55 (defining "disposal" to exclude the sale of a product for consumer use). We find this analysis unsatisfactory because it runs the risk of unnecessarily restricting the scope of CERCLA merely to dispose of claims that the statute was never intended to encompass in the first place. It is far better to simply acknowledge the inapplicability of CERCLA to asbestos-removal claims than to restrict its operative terms."

X1.6.4.5 Since asbestos that is a part of the structure of, and results in exposure within, residential buildings or business or community structures is excluded from CERCLA liability, it should not be investigated pursuant to a party's innocent purchaser appropriate inquiry requirements. Like naturally occurring radon, investigation of asbestos-containing materials that are part of the structure of buildings should be included, if at all, in the portion of this practice that deals with non-scope issues. Note, however, if asbestos is disposed of on a site and, therefore, is no longer part of the structure of a building, the cleanup of the disposed asbestos is subject to Superfund response actions. Likewise, if a building is sold with the knowledge that it will be demolished, one court ruled that the sale constitutes a disposal falling under CERCLA's liability provisions.[19]

X1.6.5 *Lead in Drinking Water and Lead-Based Paint*

These hazards can be evaluated in terms of the exclusions of 42 USC § 9604(a)(3)(B) and (C), in an analysis similar to the analysis applied above to radon and asbestos. While there is no reported case law on these environmental issues as they relate to Superfund, the statutory language seems clear that these environmental hazards are not encompassed by Superfund's appropriate inquiry responsibilities. Note, however, like asbestos, where there is a disposal of these substances on the site or in a facility, CERCLA liability may arise.

[19] *CP Holdings, Inc. versus Goldberg-Zoino & Associates, Inc.*, 769 F. Supp. 432 (D.N.H. 1991).

X2. RECOMMENDED TABLE OF CONTENTS AND REPORT FORMAT

X2.1 Summary

X2.2 Introduction

X2.2.1 Purpose
X2.2.2 Detailed Scope-of-Services
X2.2.3 Significant Assumptions
X2.2.4 Limitations and Exceptions
X2.2.5 Special Terms and Conditions
X2.2.6 User Reliance

X2.3 Site Description

X2.3.1 Location and Legal Description
X2.3.2 Site and Vicinity General Characteristics
X2.3.3 Current Use of the Property
X2.3.4 Descriptions of Structures, Roads, Other Improvements on the Site (including heating/cooling system, sewage disposal, source of potable water)
X2.3.5 Current Uses of the Adjoining Properties

X2.4 User Provided Information

X2.4.1 Title Records
X2.4.2 Environmental Liens or Activity and Use Limitations
X2.4.3 Specialized Knowledge
X2.4.4 Valuation Reduction for Environmental Issues
X2.4.5 Owner, Property Manager, and Occupant Information
X2.4.6 Reason for Performing Phase I
X2.4.7 Other

X2.5 Records Review

X2.5.1 Standard Environmental Record Sources
X2.5.2 Additional Environmental Record Sources
X2.5.3 Physical Setting Source(s)
X2.5.4 Historical Use Information on the Property
X2.5.5 Historical Use Information on Adjoining Properties

X2.6 Site Reconnaissance

X2.6.1 Methodology and Limiting Conditions
X2.6.2 General Site Setting
X2.6.3 Exterior Observations
X2.6.4 Interior Observations

X2.7 Interviews

X2.7.1 Interview with Owner
X2.7.2 Interview with Site Manager
X2.7.3 Interviews with Occupants
X2.7.4 Interviews with Local Government Officials
X2.7.5 Interviews with Others

X2.8 Findings

X2.9 Opinion

X2.10 Conclusions

X2.11 Deviations

X2.12 Additional Services

X2.13 References

X2.14 Signature(s) of Environmental Professional(s)

X2.15 Qualification(s) of Environmental Professional(s)

X2.16 Appendices

X2.16.1 Site (Vicinity) Map

X2.16.2 Site Plan

X2.16.3 Site Photographs

X2.16.4 Historical Research Documentation (aerial photographs, fire insurance maps, historical topographical maps, etc.)

X2.16.5 Regulatory Records Documentation

X2.16.6 Interview Documentation

X2.16.7 Special Contractual Conditions between User and Environmental Professional

X3. GUIDANCE TO ASSIST USERS IN THE PREPARATION FOR, AND SELECTION OF, AN ENVIRONMENTAL PROFESSIONAL TO CONDUCT A PHASE I ENVIRONMENTAL SITE ASSESSMENT

X3.1 User Preparation

X3.1.1 Once a property is identified for which obtaining a Phase I *Environmental Site Assessment* is being considered, certain information should be collected, if available, prior to the selection of an *environmental professional*. This information should be provided to the *environmental professional* selected to conduct the Phase I. The information includes:

X3.1.1.1 The reason why the Phase I is required,

X3.1.1.2 The type property and type property transaction, for example, sale, purchase, exchange, etc.,

X3.1.1.3 The complete and correct address for the property (a map or other documentation showing property location and boundaries is helpful),

X3.1.1.4 Whether or not there are any environmental liens on the property, or *activity and use limitations* (such as a deed restriction),

X3.1.1.5 Whether the value of the property has been reduced below comparable properties due at least in part to environmental conditions associated with the property (the *environmental professional* will be concerned with the environmental condition causing the value reduction),

X3.1.1.6 The scope of services desired for the Phase I (including whether any parties to the property transaction may have a required standard scope of services on whether any considerations beyond the requirements of Practice E 1527 are to be considered),

X3.1.1.7 Identification of all parties who will rely on the Phase I report,

X3.1.1.8 Identification of the site contact and how the contact can be reached,

X3.1.1.9 Any special terms and conditions which must be agreed upon by the *environmental professional*,

X3.1.1.10 Any other knowledge or experience with the property that may be pertinent to the *environmental professional* (for example, copies of any available prior environmental site assessment reports, documents, correspondence, etc., concerning the property and its environmental condition).

X3.2 Environmental Professional Selection

X3.2.1 Investigating past uses of a property and the surrounding area, which may result in identifying *recognized environmental conditions*, requires expertise that can only come with training and experience. Selecting an unqualified or inexperienced *environmental professional* to perform the work may present unnecessary risk. In the selection of a qualified *environmental professional* to perform the work, consideration should be given to both the individual(s) conducting the Phase I and the firm (assuming the individual(s) works for a firm).

X3.2.2 *Individual*—Before retaining an *environmental professional* to conduct a Phase I, the technical competence of the individual(s) retained to perform the required scope of services should be evaluated. Consideration should be given to the following:

X3.2.2.1 The formal education of the individual,

X3.2.2.2 Any environmental site assessment training the individual received in the classroom or in the field,

X3.2.2.3 Length of time the individual has been conducting Phase I Environmental Site Assessments,

X3.2.2.4 Experience of the individual in performing the scope of services required for the type of property,

X3.2.2.5 Familiarity of the individual with the current Practice E 1527,

X3.2.2.6 Sample reports prepared by the individual,

X3.2.2.7 References who used the individual's services.

X3.2.3 *Firm*—Assuming the individual works for a firm, consideration should be given to the following in the evaluation process:

X3.2.3.1 Quality assurance/quality control used by the firm in the review of Phase I Environmental Site Assessments conducted by individuals employed by the firm,

X3.2.3.2 The firm's internal risk management program to manage the risk associated with the conduct of Phase I Environmental Site Assessments,

X3.2.3.3 The firm's standard terms and conditions (including any limitations to liability),

X3.2.3.4 The firms errors and omissions (E&O) professional liability insurance policy (including the amount of coverage, limits, deductibles, and exclusions).

Designation: E 1528 – 00

Standard Practice for Environmental Site Assessments: Transaction Screen Process[1]

This standard is issued under the fixed designation E 1528; the number immediately following the designation indicates the year of original adoption or, in the case of revision, the year of last revision. A number in parentheses indicates the year of last reapproval. A superscript epsilon (ε) indicates an editorial change since the last revision or reapproval.

1. Scope

1.1 *Purpose*—The purpose of this practice, as well as Practice E 1527, is to define good commercial and customary practice in the United States of America for conducting an *environmental site assessment*[2] of a parcel of *commercial real estate* with respect to the range of contaminants within the scope of the Comprehensive Environmental Response Compensation and Liability Act (CERCLA) and *petroleum products*. As such, this practice is intended to permit a user to satisfy one of the requirements to qualify for the *innocent landowner defense* to CERCLA liability: that is, the practices that constitute "all appropriate inquiry into the previous ownership and uses of the property consistent with good commercial or customary practice" as defined in 42 USC § 9601(35)(B). (See Appendix X1 for an outline of CERCLA's liability and defense provisions.) An evaluation of *business environmental risk* associated with a parcel of commercial real estate may necessitate investigation beyond that identified in this practice (see Sections 1.3 and 11).

1.1.1 *Recognized Environmental Conditions*—In defining a standard of good commercial and customary practice for conducting an *environmental site assessment* of a parcel of *property*, the goal of the processes established by this practice is to identify *recognized environmental conditions*. The term *recognized environmental conditions* means the presence or likely presence of any *hazardous substances* or *petroleum products* on a *property* under conditions that indicate an existing release, a past release, or a material threat of a release of any *hazardous substances* or *petroleum products* into structures on the *property* or into the ground, ground water, or surface water of the *property*. The term includes *hazardous substances* or *petroleum products* even under conditions in compliance with laws. The term is not intended to include *de minimis* conditions that generally do not present a material risk of harm to public health or the environment and that generally would not be the subject of an enforcement action if brought to the attention of appropriate governmental agencies. Conditions determined to be *de minimis* are not *recognized environmental conditions*.

1.1.2 *Two Related Practices*—This practice is closely related to Practice E 1527, a *Phase I Environmental Site Assessment*. Both are *environmental site assessments* for *commercial real estate*. See 4.3.

1.1.3 *Petroleum Products*—*Petroleum products* are included within the scope of both practices because they are of concern on many parcels of *commercial real estate* and current custom and usage is to include an inquiry into the presence of *petroleum products* when doing an *environmental site assessment* of *commercial real estate*. Inclusion of *petroleum products* within the scope of this practice is not based upon the applicability, if any, of CERCLA to *petroleum products*.

1.1.4 *CERCLA Requirements Other Than Appropriate Inquiry*—This practice does not address whether requirements in addition to *appropriate inquiry* have been met in order to qualify for CERCLA's *innocent landowner defense* (for example, the duties specified in 42 USC § 9607(b)(3)(a) and (b) and cited in Appendix X1).

1.1.5 *Other Federal, State, and Local Environmental Laws*—This practice does not address requirements of any state or local laws or of any federal laws other than the appropriate inquiry provisions of CERCLA's *innocent landowner defense*. Users are cautioned that federal, state, and local laws may impose environmental assessment obligations that are beyond the scope of this practice. *Users* should also be aware that there are likely to be other legal obligations with regard to *hazardous substances* or *petroleum products* discovered on *property* that are not addressed in this practice and may pose risks of civil and/or criminal sanctions for non-compliance.

1.2 *Objectives*—Objectives guiding the development of this practice and Practice E 1527 are (1) to synthesize and put in writing good commercial and customary practice for *environmental site assessments* for *commercial real estate*, (2) to facilitate highquality, standardized *environmental site assessments*, (3) to ensure that the standard of *appropriate inquiry* is practical and reasonable, and (4) to clarify an industry standard for *appropriate inquiry* in an effort to guide legal interpretation of CERCLA's *innocent landowner defense*.

1.3 *Considerations Beyond the Scope*—The use of this practice is strictly limited to the scope set forth in this section.

---

[1] This practice is under the jurisdiction of ASTM Committee E-50 on Environmental Assessment and is the direct responsibility of Subcommittee E50.02 on Commercial Real Estate Transactions.

Current edition approved May 10, 2000. Published July 2000. Originally published as E 1528 – 93. Last previous edition E 1528 – 96.

[2] All definitions, descriptions of terms, and acronyms are defined in Section 3 of Practice E 1527. Whenever terms defined in 3.2 or described in 3.3 are used in this practice, they are in *italics*.

Section 11 of this practice identifies, for informational purposes, certain environmental conditions (not an all-inclusive list) that may exist on a *property* that are beyond the scope of this practice but may warrant consideration by parties to a *commercial real estate* transaction. The need to include an investigation of any such conditions in the scope of services should be evaluated based upon, among other factors, the nature of the property and the reasons for performing the assessment (for example, a more comprehensive evaluation of *business environmental risk*) and should be agreed upon as additional services beyond the scope of this practice prior to initiation of the *Environmental Site Assessment* process.

1.4 *Organization of This Practice*—This practice has several parts and two appendixes. Section 1 is the Scope. Section 2 refers to other ASTM standards in the Referenced Documents. Section 3, Terminology, has definitions of terms not unique to this practice, descriptions of terms unique to this practice, and acronyms. Section 4 is Significance and Use of this practice. Section 5 is the Introduction to the Transaction Screen Questionnaire. Section 6 sets forth the Transaction Screen Questionnaire itself. Sections 7-10 contain the Guide to the Transaction Screen Questionnaire and its various parts. Section 11 provides additional information regarding non-scope considerations (see 1.3). The appendixes are included for information and are not part of the procedures prescribed in either Practice E 1527 or this practice. Appendix X1 and Practice E 1527 explain the liability and defense provisions of CERCLA that will assist the user in understanding the user's responsibilities under CERCLA; it also contains other important information regarding CERCLA and this practice. Appendix X2 provides information referred to in the guide to the transaction screen questionnaire.

1.5 *This standard does not purport to address all of the safety concerns, if any, associated with its use. It is the responsibility of the user of this standard to establish appropriate safety and health practices and determine the applicability of regulatory limitations prior to use.*

2. Referenced Documents

2.1 *ASTM Standards:*

E 1527 Practice for Environmental Site Assessments: Phase I Environmental Site Assessment Process[3]

3. Terminology

3.1 *Scope*—This section provides definitions, descriptions of terms, and a list of acronyms for many of the words used in this practice and Practice E 1527. The terms are an integral part of both practices and are critical to an understanding of the written practices and their use.

3.2 *Definitions:*

3.2.1 *activity and use limitations*—legal or physical restrictions or limitations on the use of, or access to, a site or facility: *(1)* to reduce or eliminate potential exposure to hazardous substances in the soil or ground water on the property, or *(2)* to prevent activities that could interfere with the effectiveness of a response action, in order to ensure maintenance of a condition of no significant risk to public health or the environment. These legal or physical restrictions, which may include institutional and/or engineering controls, are intended to prevent adverse impacts to individuals or populations that may be exposed to hazardous substances in the soil or ground water on the property.

3.2.2 *Comprehensive Environmental Response, Compensation and Liability Information System (CERCLIS)*—the list of sites compiled by EPA that EPA has investigated or is currently investigating for potential hazardous substance contamination for possible inclusion on the National Priorities list.

3.2.3 *construction debris*—concrete, brick, asphalt, and other such building materials discarded in the construction of a building or other improvement to property.

3.2.4 *contaminated public wells*—public wells used for drinking water that have been designated by a government entity as contaminated by toxic substances, (for example, chlorinated solvents), or as having water unsafe to drink without treatment.

3.2.5 *CORRACTS list*—list of hazardous waste treatment, storage or disposal facilities and other RCRIS facilities (due to past interim status or storage of hazardous waste beyond 90 days) who have been notified by the U.S. Environmental Protection Agency to undertake corrective action under RCRA.

3.2.6 *demolition debris*—concrete, brick, asphalt, and other such building materials discarded in the demolition of a building or other improvement to property.

3.2.7 *drum*—a container (typically, but not necessarily, holding 55 gal (208 L) of liquid) that may be used to store *hazardous substances* or *petroleum products*.

3.2.8 *dry wells*—underground areas where soil has been removed and replaced with pea gravel, coarse sand, or large rocks. Dry wells are used for drainage, to control storm runoff, for the collection of spilled liquids (intentional and non-intentional) and wastewater disposal (often illegal).

3.2.9 *dwelling*—structure or portion thereof used for residential habitation.

3.2.10 *engineering controls*—physical modifications to a site or facility (for example, capping, slurry walls, or point of use water treatment) to reduce or eliminate the potential for exposure to hazardous substances in the soil or ground water on the property.

3.2.11 *environmental lien*—a charge, security, or encumbrance upon title to a *property* to secure the payment of a cost, damage, debt, obligation, or duty arising out of response actions, cleanup, or other remediation of *hazardous substances* or *petroleum products* upon a *property*, including (but not limited to) liens imposed pursuant to CERCLA 42 USC § 9607(1) and similar state or local laws.

3.2.12 *ERNS list*—EPA's Emergency Response Notification System list of reported CERCLA hazardous substance releases or spills in quantities greater than the reportable quantity, as maintained at the National Response Center. Notification requirements for such releases or spills are codified in 40 CFR Parts 302 and 355.

3.2.13 *Federal Register (FR)*—publication of the United States government published daily (except for federal holidays and weekends) containing all proposed and final regulations

[3] *Annual Book of ASTM Standards*, Vol 11.04.

and some other activities of the federal government. When regulations become final, they are included in the Code of Federal Regulations (CFR), as well as published in the Federal Register.

3.2.14 *fire insurance maps*—maps produced for private fire insurance map companies that indicate uses of properties at specified dates and that encompass the property. These maps are often available at local libraries, historical societies, private resellers, or from the map companies who produced them. See Question 2310.3.1.3 of the transaction screen process in this practice and 7.3.4.2 of Practice E 1527.

3.2.15 *hazardous substance*—a substance defined as a hazardous substance pursuant to CERCLA 42 USC § 9601(14), as interpreted by EPA regulations and the courts: "(A) any substance designated pursuant to Section 1321(b)(2)(A) of Title 33, (B) any element, compound, mixture, solution, or substance designated pursuant to Section 9602 of this title, (C) any hazardous waste having the characteristics identified under or listed pursuant to Section 3001 of the Solid Waste Disposal Act (42 USC § 6921) (but not including any waste the regulation of which under the Solid Waste Disposal Act (42 USC § 6901 *et seq.*) has been suspended by Act of Congress), (D) any toxic pollutant listed under Section 1317(a) of Title 33, (E) any hazardous air pollutant listed under Section 112 of the Clean Air Act (42 USC § 7412), and (F) any imminently hazardous chemical substance or mixture with respect to which the administrator (of EPA) has taken action pursuant to Section 2606 of Title 15. The term does not include petroleum, including crude oil or any fraction thereof which is not otherwise specifically listed or designated as a hazardous substance under Subparagraphs (A) through (F) of this paragraph, and the term does not include natural gas, natural gas liquids, liquefied natural gas, or synthetic gas usable for fuel (or mixtures of natural gas and such synthetic gas)," (See Appendix X1.)

3.2.16 *hazardous waste*—any hazardous waste having the characteristics identified under or listed pursuant to Section 3001 of the Solid Waste Disposal Act (42 USC § 6921) (but not including any waste the regulation of which under the Solid Waste Disposal Act (42 USC § 6901 *et seq.*) has been suspended by Act of Congress). The Solid Waste Disposal Act of 1980 amended RCRA. The RCRA defines a hazardous waste, in 42 USC § 6903, as:

"a solid waste, or combination of solid wastes, which because of its quantity, concentration, or physical, chemical, or infectious characteristics may—(A) cause, or significantly contribute to an increase in mortality or an increase in serious irreversible, or incapacitating reversible, illness; or (B) pose a substantial present or potential hazard to human health or the environment when improperly treated, stored, transported, or disposed of, or otherwise managed."

3.2.17 *institutional controls*—a legal or administrative restriction (for example, deed restriction, restrictive zoning) on the use of, or access to, a site or facility to reduce or eliminate potential exposure to hazardous substances in the soil or ground water on the property.

3.2.18 *landfill*—a place, location, tract of land, area, or premises used for the disposal of solid wastes as defined by state solid waste regulations. The term is synonymous with the term *solid waste disposal site* and is also known as a garbage dump, trash dump, or similar term.

3.2.19 *local street directories*—directories published by private (or sometimes government) sources that show ownership, occupancy, use of sites and/or by reference to street addresses. Often local street directories are available at libraries of local governments, colleges or universities, or historical societies. See 7.3.4.6 of the Records Review Section of Practice E 1527.

3.2.20 *material safety data sheet (MSDS)*—written or printed material concerning a *hazardous substance* which is prepared by chemical manufacturers, importers, and employers for hazardous chemicals pursuant to OSHA's Hazard Communication Standard, 29 CFR 1910.1200(g).

3.2.21 *National Contingency Plan (NCP)*—the National Oil and Hazardous Substances Pollution Contingency Plan, found at 40 CFR § 300, that is the EPA's blueprint on how hazardous substances are to be cleaned up pursuant to CERCLA.

3.2.22 *National Priorities List (NPL)*—list compiled by EPA pursuant to CERCLA 42 USC § 9605(a)(8)(B) of properties with the highest priority for cleanup pursuant to EPA's hazard ranking system. See 40 CFR Part 300.

3.2.23 *occupants*—those tenants, subtenants, or other persons or entities using the *property* or a portion of the *property*.

3.2.24 *owner*—generally the fee owner of record of the *property*.

3.2.25 *petroleum exclusion*—the exclusion from CERCLA liability provided in 42 USC § 9601(14), as interpreted by the courts and EPA: "The term (hazardous substance) does not include petroleum, including crude oil or any fraction thereof which is not otherwise specifically listed or designated as a hazardous substance under Subparagraphs (A) through (F) of this paragraph, and the term does not include natural gas, natural gas liquids, liquefied natural gas, or synthetic gas usable for fuel (or mixtures of natural gas and such synthetic gas)."

3.2.26 *petroleum products*—those substances included within the meaning of the terms within the *petroleum exclusion* to CERCLA, 42 USC § 9601(14), as interpreted by the courts and EPA, that is: petroleum, including crude oil or any fraction thereof that is not otherwise specifically listed or designated as a *hazardous substance* under Subparagraphs (A) through (F) of 42 USC § 9601(14), natural gas, natural gas liquids, liquefied natural gas, and synthetic gas usable for fuel (or mixtures of natural gas and such synthetic gas). (The word fraction refers to certain distillates of crude oil, including gasoline, kerosene, diesel oil, jet fuels, and fuel oil, pursuant to *Standard Definitions of Petroleum Statistics*.[4]

3.2.27 *Phase I Environmental Site Assessment*— the process described in Practice E 1527.

3.2.28 *pits, ponds, or lagoons*—man-made or natural depressions in a ground surface that are likely to hold liquids or sludge containing *hazardous substances* or *petroleum products*. The likelihood of such liquids or sludge being present is

[4] *Standard Definitions of Petroleum Statistics*, American Petroleum Institute, Fourth Edition, 1988.

determined by evidence of factors associated with the pit, pond, or lagoon, including, but not limited to, discolored water, distressed vegetation, or the presence of an obvious wastewater discharge.

3.2.29 *property*—the real property that is the subject of the environmental site assessment described in this practice. Real property includes buildings and other fixtures and improvements located on the property and affixed to the land.

3.2.30 *property tax files*—the files kept for property tax purposes by the local jurisdiction where the property is located and includes records of past ownership, appraisals, maps, sketches, photos, or other information that is reasonably ascertainable and pertaining to the property. See 7.3.4.3 of the Records Review Section of Practice E 1527.

3.2.31 *RCRA generators*—those persons or entities which generate *hazardous wastes*, as defined and regulated by RCRA.

3.2.32 *RCRA generators list*—list kept by EPA of those persons or entities that generate *hazardous wastes* as defined and regulated by RCRA.

3.2.33 *RCRA TSD facilities*—those facilities on which treatment, storage, and/or disposal of *hazardous wastes* takes place, as defined and regulated by RCRA.

3.2.34 *RCRA TSD facilities list*—list kept by EPA of those facilities on which treatment, storage, and/or disposal of *hazardous wastes* takes place, as defined and regulated by RCRA.

3.2.35 *recorded land title records*—records of fee ownership, leases, land contracts, easements, liens, and other encumbrances on or of the *property* recorded in the place where land title records are, by law or custom, recorded for the local jurisdiction in which the property is located. (Often such records are kept by a municipal or county recorder or clerk.) Such records may be obtained from title companies or directly from the local government agency. Information about the title to the *property* that is recorded in a U.S. district court or any place other than where land title records are, by law or custom, recorded for the local jurisdiction in which the property is located, are not considered part of recorded land title records. See 7.3.4.4 of the Records Review Section of Practice E 1527.

3.2.36 *records of emergency release notifications (SARA § 304)*—Section 304 of EPCRA or Title III of SARA requires operators of facilities to notify their local emergency planning committee (as defined in EPCRA) and State Emergency Response Commission (as defined in EPCRA) of any release beyond the facility's boundary of any reportable quantity of any extremely *hazardous substance*. Often the local fire department is the local emergency planning committee. Records of such notifications are" Records of Emergency Release Notifications" (SARA §304).

3.2.37 *report*—the written record of a transaction screen process as required by Practice E 1527 or the written report prepared by the environmental professional and constituting part of a *Phase I Environmental Site Assessment*, as required by Practice E 1527.

3.2.38 *solid waste disposal site*—a place, location, tract of land, area, or premises used for the landfill disposal of solid wastes as defined by state solid waste regulations. The term is synonymous with the term *landfill* and is also known as a garbage dump, trash dump, or similar term.

3.2.39 *solvent*—a chemical compound that is capable of dissolving another substance and may itself be a *hazardous substance* used in a number of manufacturing/industrial processes including, but not limited to, the manufacture of paints and coatings for industrial and household purposes, equipment clean-up, and surface degreasing in metal fabricating industries.

3.2.40 *state registered USTs*—state lists of underground storage tanks required to be registered under Subtitle I, Section 9002 of RCRA.

3.2.41 *sump*—a pit, cistern, cesspool, or similar receptacle where liquids drain, collect, or are stored.

3.2.42 *TSD Facility*—treatment, storage, or disposal facility (see definition of *RCRA TSD facilities*).

3.2.43 *underground storage tank (UST)*—any tank, including underground piping connected to the tank, that is or has been used to contain *hazardous substances* or *petroleum products* and the volume of which is 10 % or more beneath the surface of the ground.

3.2.44 *USGS 7.5 Minute Topographic Map*—the map (if any) available from or produced by the United States Geological Survey, entitled "USGS 7.5 Minute Topographic Map," and showing the *property*. See 7.3.4.5 of Practice E 1527.

3.2.45 *wastewater*—water that is or has been used in an industrial or manufacturing process, conveys or has conveyed sewage, or is directly related to manufacturing, processing, or raw materials storage areas at an industrial plant. Wastewater does not include water originating on or passing through or adjacent to a site, such as stormwater flows, that has not been used in industrial or manufacturing processes, has not been combined with sewage, or is not directly related to manufacturing, processing, or raw materials storage areas at an industrial plant.

3.2.46 *zoning/land use records*—those records of the local government in which the *property* is located indicating the uses permitted by the local government in particular zones within its jurisdiction. The records may consist of maps and/or written records. They are often located in the planning department of a municipality or county. See 7.3.4.8 of the Records Review Section of Practice E 1527.

3.3 *Definitions of Terms Specific to This Standard:*

3.3.1 *actual knowledge*—the knowledge actually possessed by an individual who is a real person, rather than an entity. Actual knowledge is to be distinguished from constructive knowledge, that is, knowledge imputed to an individual or entity. (See 5.5.3.)

3.3.2 *adjoining properties*—any real property or properties the border of which is contiguous or partially contiguous with that of the property, or that would be contiguous or partially contiguous with that of the property but for a street, road, or other public thoroughfare separating them.

3.3.3 *aerial photographs*—photographs taken from an airplane or helicopter of areas encompassing the property. Aerial photographs are often available from government agencies or private collections unique to a local area. See 7.3.4.1 of Practice E 1527.

3.3.4 *appropriate inquiry*—that inquiry constituting "all

appropriate inquiry into the previous ownership and uses of the property consistent with good commercial or customary practice" as defined in CERCLA, 42 USC § 9601(35)(B), that will give a party to a *commercial real estate* transaction the *innocent landowner defense* to CERCLA liability (42 USC § 9601(A) and (B) and 9607(b)(3)), assuming compliance with other elements of the defense. See Appendix X1 and Practice E 1527.

3.3.5 *approximate minimum search distance*—the area for which records must be obtained and reviewed pursuant to the Records Review Section of Practice E 1527, subject to the limitations provided in that section. The term *approximate minimum search distance* may include areas outside the *property* and shall be measured from the nearest *property* boundary. The term *approximate minimum search distance* is used instead of radius to include irregularly shaped properties.

3.3.6 *building department records*—those records of the local government in which the property is located indicating permission of the local government to construct, alter, or demolish improvements on the property. Often building department records are located in the building department of a municipality or county. See 7.3.4.7 of Practice E 1527.

3.3.7 *business environmental risk*—a risk which can have a material environmental or environmentally-driven impact on the business associated with the current or planned use of a parcel of commercial real estate, not necessarily limited to those environmental issues required to be investigated in this practice. Consideration of *business environmental risk* issues may involve addressing one or more non-scope considerations, some of which are identified in Section 11.

3.3.8 *commercial real estate*—any real property except a dwelling or property with no more than four dwelling units exclusively for residential use (except that a dwelling or property with no more than four dwelling units exclusively for residential use is included in the term commercial real estate when it has a commercial function, as in the building of such dwellings for profit). The term *commercial real estate* includes but is not limited to undeveloped real property and real property used for industrial, retail, office, agricultural, other commercial, medical, or educational purposes; *property* used for residential purposes that has more than four residential dwelling units; and *property* with no more than four dwelling units for residential use when it has a commercial function, as in the building of such dwellings for profit.

3.3.9 *commercial real estate transaction*—a transfer of title to or possession of real property or receipt of a security interest in real property, except that it does not include transfer of title to or possession of real property or the receipt of a security interest in real property with respect to an individual dwelling or building containing fewer than five dwelling units, nor does it include the purchase of a lot or lots to construct a dwelling for occupancy by a purchaser, but a commercial real estate transaction does include real property purchased or leased by persons or entities in the business of building or developing dwelling units.

3.3.10 *due diligence*—the process of inquiring into the environmental characteristics of a parcel of *commercial real estate* or other conditions, usually in connection with a commercial real estate transaction. The degree and kind of due diligence vary for different properties and differing purposes. See Appendix X1 and Practice E 1527.

3.3.11 *environmental audit*—the investigative process to determine if the operations of an existing facility are in compliance with applicable environmental laws and regulations. The term *environmental audit* should not be used to describe this practice or Practice E 1527 although an environmental audit may include an *environmental site assessment* or, if prior audits are available, may be part of an environmental site assessment. See Appendix X1 and Practice E 1527.

3.3.12 *environmental professional*—a person possessing sufficient training and experience necessary to conduct a *site reconnaissance, interviews*, and other activities in accordance with Practice E 1527, and from the information generated by such activities, having the ability to develop opinions and conclusions regarding *recognized environmental conditions* in connection with the *property* in question. An individual's status as an environmental professional may be limited to the type of assessment to be performed or to specific segments of the assessment for which the professional is responsible. The person may be an independent contractor or an employee of the *user*.

3.3.13 *environmental site assessment (ESA)*—the process by which a person or entity seeks to determine if a particular parcel of real *property* (including improvements) is subject to *recognized environmental conditions*. At the option of the user, an environmental site assessment may include more inquiry than that constituting *appropriate inquiry* or, if the user is not concerned about qualifying for the *innocent landowner defense*, less inquiry than that constituting *appropriate inquiry*. See Appendix X1 and Practice E 1527. An environmental site assessment is both different from and less rigorous than an *environmental audit*.

3.3.14 *fill dirt*—dirt, soil, sand, or other earth, that is obtained off-site, that is used to fill holes or depressions, create mounds, or otherwise artificially change the grade or elevation of real property. It does not include material that is used in limited quantities for normal landscaping activities.

3.3.15 *hazardous waste/contaminated sites*—sites on which a release has occurred, or is suspected to have occurred, of any *hazardous substance, hazardous waste*, or *petroleum products*, and on which release or suspected release has been reported to a government entity.

3.3.16 *historical recognized environmental condition*—environmental condition which in the past would have been considered a *recognized environmental condition*, but which may or may not be considered a *recognized environmental condition* currently. The final decision will be influenced by the current impact of the *historical recognized environmental condition* on the property.

3.3.17 *innocent landowner defense*—that defense to CERCLA liability provided in 42 USC § 9601(35) and § 9607(b)(3). One of the requirements to qualify for this defense is that the party make "all appropriate inquiry into the previous ownership and uses of the property consistent with good commercial or customary practice." There are additional requirements to qualify for this defense. See discussion in

Appendix X1 and Practice E 1527.

3.3.18 *interviews*—those portions of the *Phase I Environmental Site Assessment* in Practice E 1527 that are contained in Sections 9 and 10 thereof and addresses questions to be asked of *owners* and *occupants* of the *property* and questions to be asked of local government officials.

3.3.19 *key site manager*—the *key site manager* is the person identified by the *owner* of a *property* as having good knowledge of the uses and physical characteristics of the property. See 9.5.1 of Practice E 1527.

3.3.20 *local government agencies*—those agencies of municipal or county government having jurisdiction over the *property*. Municipal and county government agencies include but are not limited to cities, parishes, townships, and similar entities.

3.3.21 *LUST sites*—state lists of leaking underground storage tank sites. Section 9003 (h) of Subtitle I of RCRA gives EPA and states, under cooperative agreements with EPA, authority to clean up releases from UST systems or require owners and operators to do so.

3.3.22 *major occupants*—those tenants, subtenants, or other persons or entities each of which uses at least 40 % of the leasable area of the *property* or any anchor tenant when the *property* is a shopping center.

3.3.23 *material threat*—a physically observable or obvious threat which is reasonably likely to lead to a release that, in the opinion of the *environmental professional*, is threatening and might result in impact to public health or the environment. An example might include an aboveground storage tank that contains a hazardous substance and which shows evidence of damage. The damage would represent a material threat if it is deemed serious enough that it may cause or contribute to tank integrity failure with a release of contents to the environment.

3.3.24 *obvious*—that which is plain or evident; a condition or fact which could not be ignored or overlooked by a reasonable observer while *visually or physically observing* the *property*.

3.3.25 *other historical sources*—any source or sources other than those designated in 7.3.4.1 through 7.3.4.8 of Practice E 1527 that are credible to a reasonable person and that identify past uses or occupancies of the *property*. The term includes records in the files, and/or personal knowledge of the *property owner* and/or *occupants*. See 7.3.4.9 of the Records Review Section of Practice E 1527.

3.3.26 *physical setting sources*—sources that provide information about the geologic, hydrogeologic, hydrologic, or topographic characteristics of a *property*. See 7.2.3 of Practice E 1527.

3.3.27 *practically reviewable*—information that is practically reviewable means that the information is provided by the source in a manner and in a form that, upon examination, yields information relevant to the *property* without the need for extraordinary analysis of irrelevant data. The form of the information shall be such that the user can review the records for a limited geographic area. Records that cannot be feasibly retrieved by reference to the location of the *property* or a geographic area in which the *property* is located are not generally *practically reviewable*. Listings in publicly available records that do not have adequate address information to be located geographically are not generally considered *practically reviewable*. Most databases of public records are *practically reviewable* if they can be obtained from the source agency by the county, city, zip code, or other geographic area of the facilities listed in the record system. Records that are sorted, filed, organized, or maintained by the source agency only chronologically are not generally practically reviewable. For large databases with numerous facility records (such as RCRA hazardous waste generators and registered underground storage tanks), the records are not *practically reviewable* unless they can be obtained from the source agency in the smaller geographic area of zip codes. Even when information is provided by zip code for some large databases, it is common for an unmanageable number of sites to be identified within a given zip code. In these cases, it is not necessary to review the impact of all of the sites that are likely to be listed in any given zip code because that information would not be practically reviewable. In other words, when so much data is generated that it cannot be feasibly reviewed for its impact on the property, it is not *practically reviewable*.

3.3.28 *preparer*—the person preparing the *transaction screen questionnaire* pursuant to this practice, who may be either the *user* or the person to whom the *user* has delegated the preparation.

3.3.29 *publicly available*—information that is publicly available means that the source of the information allows access to the information by anyone upon request.

3.3.30 *reasonably ascertainable*—for purposes of both this practice and Practice E 1527 information that is *publicly available*, obtainable from its source within reasonable time and cost constraints, and *practically reviewable*.

3.3.31 *recognized environmental conditions*—the presence or likely presence of any *hazardous substances* or *petroleum products* on a *property* under conditions that indicate an existing release, a past release, or a material threat of a release of any *hazardous substances* or *petroleum products* into structures on the *property* or into the ground, ground water, or surface water of the *property*. The term includes *hazardous substances* or *petroleum products* even under conditions in compliance with laws. The term is not intended to include *de minimis* conditions that generally do not present a material risk of harm to public health or the environment and that generally would not be the subject of an enforcement action if brought to the attention of appropriate governmental agencies. Conditions determined to be *de minimis* are not *recognized environmental conditions*.

3.3.32 *records review*—that part of the *Phase I Environmental Site Assessment* in Practice E 1527 that is contained in Section 7 thereof and addresses which records shall or may be reviewed.

3.3.33 *site reconnaissance*—that part of the *Phase I Environmental Site Assessment* in Practice E 1527 that is contained in Section 8 thereof and addresses what should be done in connection with the *site visit*. The site reconnaissance includes, but is not limited to, the *site visit* done in connection with such as *Phase I Environmental Site Assessment*.

3.3.34 *site visit*—the visit to the property during which

observations are made constituting the *site reconnaissance* section of the *Phase I Environmental Site Assessment* in Practice E 1527 and the *site visit* requirement of the transaction screen process in this practice.

3.3.35 *standard environmental record sources*— those records specified in 7.2.1.1 of the Records Review Section of the *Phase I Environmental Site Assessment* of Practice E 1527.

3.3.36 *standard historical sources*—those sources of information about the history of uses of property specified in 7.3.4 of the Records Review Section of the *Phase I Environmental Site Assessment* of Practice E 1527.

3.3.37 *standard physical setting source*—a current USGS 7.5 minute topographic map (if any) showing the area on which the property is located. See 7.2.3 of Practice E 1527.

3.3.38 *standard practice(s)*—the activities set forth in either this practice or Practice E 1527, or both, for the conduct of environmental site assessments.

3.3.39 *standard sources*—sources of environmental, physical setting, or historical records specified in the Records Review Section (Section 7) of the *Phase I Environmental Site Assessment* of Practice E 1527.

3.3.40 *transaction screen questionnaire*—the questionnaire provided in Section 6 of Practice E 1527.

3.3.41 *transaction screen process*—the process described in Practice E 1527.

3.3.42 *user*—the party seeking to use the transaction screen process of this practice or the *Phase I Environmental Site Assessment* of Practices E 1527 or E 1528 to perform an *environmental assessment* of the *property*. A *user* may include, without limitation, a purchaser of *property*, a potential tenant of *property*, an *owner* of *property*, a lender, or a property manager.

3.3.43 *visually and/or physically observed*— during a *site visit* pursuant to the *transaction screen process* of this practice or pursuant to a *Phase I Environmental Site Assessment* of Practice E 1527, the term *visually and physically observed* means observations made by vision upon walking through a *property* and the structures located on it and observations made by the sense of smell, particularly observations of noxious or foul odors. The term *walking through* is not meant to imply that disabled persons who cannot physically walk may not conduct a *site visit*; they may do so by the means at their disposal for moving through the *property* and the structures located on it.

3.4 *Acronyms:*

3.4.1 *CERCLA*—Comprehensive Environmental Response, Compensation and Liability of 1980 Act (as amended, 42 USC § 9601 *et seq.*).

3.4.2 *CERCLIS*—Comprehensive Environmental Response, Compensation and Liability Information System maintained by EPA.

3.4.3 *CFR*—Code of Federal Regulations.

3.4.4 *CORRACTS*—facilities subject to Corrective Action under RCRA.

3.4.5 *EPA*—United States Environmental Protection Agency.

3.4.6 *EPCRA*—Emergency Planning and Community Right to Know Act (also known as SARA Title III), (42 USC § 11001 *et seq.*).

3.4.7 *ERNS*—Emergency Response Notification System.

3.4.8 *ESA*—environmental site assessment (different than an environmental audit; see 3.3.13).

3.4.9 *FOIA*—U.S. Freedom of Information Act (5 USC § 552 *et seq.*).

3.4.10 *FR*—Federal Register.

3.4.11 *LUST*—leaking underground storage tank.

3.4.12 *MSDS*—material safety data sheet.

3.4.13 *NCP*—National Contingency Plan.

3.4.14 *NFRAP*—former CERCLIS sites where no further remedial action is planned under CERCLA.

3.4.15 *NPDES*—National Pollutant Discharge Elimination System.

3.4.16 *NPL*—National Priorities List.

3.4.17 *PCBs*—polychlorinated biphenyls.

3.4.18 *PRP*—potentially responsible party (pursuant to CERCLA 42 USC § 9607(a).

3.4.19 *RCRA*—Resource Conservation and Recovery Act (as amended, 42 USC§ 6901 *et seq.*).

3.4.20 *SARA*—Superfund Amendments and Reauthorization Act of 1986 (amendment to CERCLA).

3.4.21 *TSDF*—hazardous waste treatment, storage or disposal facility.

3.4.22 *USC*—United States Code.

3.4.23 *USGS*—United States Geological Survey.

3.4.24 *UST*—underground storage tank.

4. Significance and Use

4.1 *Uses*—This practice is intended for use on a voluntary basis by parties who wish to assess the environmental condition of *commercial real estate*. While use of this practice is intended to constitute *appropriate inquiry* for purposes of CERCLA's *innocent landowner defense*, it is not intended that its use be limited to that purpose. This practice is intended primarily as an approach to conducting an inquiry designed to identify *recognized environmental conditions* in connection with a *property*, and *environmental site assessments* that are both more and less comprehensive than this practice (including, in some instances, no *environmental site assessment*) may be appropriate in some circumstances. Further, no implication is intended that a person must use this practice in order to be deemed to have conducted inquiry in a commercially prudent or reasonable manner in any particular transaction. Nevertheless, this practice is intended to reflect a commercially prudent and reasonable inquiry.

4.2 *Clarifications on Use*:

4.2.1 *Use Not Limited to CERCLA*—This practice and Practice E 1527 are designed to assist the *user* in developing information about the environmental condition of a *property* and as such has utility for a wide range of persons, including those who may have no actual or potential CERCLA liability and/or may not be seeking the *innocent landowner defense*.

4.2.2 *Residential Tenants/Purchasers and Others*—No implication is intended that it is currently customary practice for residential tenants of multifamily residential buildings, tenants of single-family homes or other residential real estate, or purchasers of dwellings for residential use, to conduct an *environmental site assessment* in connection with these transactions. Thus, these transactions are not included in the term commercial real estate transactions, and it is not intended to imply that such persons are obligated to conduct an *environmental site assessment* in connection with these transactions for purposes of *appropriate inquiry* or for any other purpose. In addition, no implication is intended that it is currently customary practice for *environmental site assessments* to be conducted in other unenumerated instances (including but not limited to many commercial leasing transactions, many acquisitions of easements, and many loan transactions in which the lender has multiple remedies). On the other hand, anyone who elects to do an *environmental site assessment* of any *property* or portion of a property may, in such person's judgment, use either this practice or Practice E 1527.

4.2.3 *Site-Specific*— This practice is site-specific in that it relates to assessment of environmental conditions on a specific parcel of *commercial real estate*. Consequently, this practice does not address many additional issues raised in transactions such as purchases of business entities, or interests therein, or of their assets, that may well involve environmental liabilities pertaining to properties previously owned or operated or other off-site environmental liabilities.

4.3 *Two Related Practices*—This practice sets forth one procedure for an environmental site assessment for purposes of appropriate inquiry necessary to qualify for CERCLA's *innocent landowner defense*, known as a "transaction screen process" or a transaction screen." This practice is a companion to Practice E 1527 that is called the "Phase I Environmental Site Assessment Process" or a "Phase I ESA" or simply a "Phase I." These practices are each intended to meet the standard of *appropriate inquiry* necessary to qualify for the *innocent landowner defense* of CERCLA. It is essential to consider that these two practices, taken together, provide for two alternative practices of appropriate inquiry: the *transaction screen process* described in this practice and the *Phase I Environmental Site Assessment* described in Practice E 1527.

4.3.1 *Election to Commence With Either Practice*—The *user* may commence inquiry to identify recognized *environmental conditions* in connection with a *property* by performing either the *transaction screen process* or the *Phase I Environmental Site Assessment*.

4.3.2 *Who May Conduct*— The *transaction screen process* may be conducted either by the *user* (including an agent, independent contractor or employee of the *user*) or wholly or partially by an *environmental professional*. The *transaction screen process* does not require the judgment of an *environmental professional*. Whenever a *Phase I Environmental Site Assessment* is conducted, it must be performed by an *environmental professional* to the extent specified in 6.5.1 through 6.5.2.1 of Practice E 1527. Further, at the *Phase I Environmental Site Assessment* level, no practical standard should be designed to eliminate the role of judgment and the value and need for experience in the party performing the inquiry. The professional judgment of an *environmental professional* is, consequently, vital to the performance of appropriate inquiry at the *Phase I Environmental Site Assessment* level.

4.3.3 *Completion of Transaction Screen Process*—Performance of the *transaction screen process* may allow the *user* to conclude that no further inquiry is needed to assess the potential for identifying any *recognized environmental condition* at the property and hence that performance of the *transaction screen process* constitutes *appropriate inquiry* without undertaking the *Phase I Environmental Site Assessment*. Upon completion of the *transaction screen process*, the *user* should conclude either (1) no further inquiry into recognized environmental conditions at the property is needed for purposes of *appropriate inquiry*, or (2) further inquiry is needed to assess *recognized environmental conditions* appropriately for purposes of *appropriate inquiry*. If no further inquiry is needed, the *user* has completed his or her *appropriate inquiry* of the *property*.

4.3.4 *Inquiry Beyond the Transaction Screen Process*—If further inquiry is needed after performance of the *transaction screen process* (as described in 4.3.3), the *user* must determine, in the exercise of the *user's* reasonable business judgment, whether further inquiry may be limited to those specific issues identified as of concern or should proceed to a full *Phase I Environmental Site Assessment*.

4.4 *Additional Services*—As set forth in 11.9 of Practice E 1527, additional services may be contracted for between the *user* and the *environmental professional(s)*.

4.5 *Principles*—The following principles are an integral part of this practice and are intended to be referred to in resolving any ambiguity or exercising such discretion as is accorded the *user* or *environmental professional* in performing an *environmental site assessment* or in judging whether a *user* or *environmental professional* has conducted *appropriate inquiry* or has otherwise conducted an adequate *environmental site assessment*.

4.5.1 No *environmental site assessment* can wholly eliminate uncertainty regarding the potential for *recognized environmental conditions* in connection with a *property*. Performance of either this practice or Practice E 1527 is intended to reduce but not eliminate uncertainty regarding the existence of *recognized environmental conditions* in connection with a property, and both practices recognize reasonable limits of time and cost.

4.5.2 *Appropriate inquiry* does not mean an exhaustive assessment of a clean *property*. There is a point at which the cost of information obtained or the time required to gather it outweighs the usefulness of the information and, in fact, may be a material detriment to the orderly completion of transactions. One of the purposes of this practice is to identify a balance between the competing goals of limiting the costs and time demands inherent in performing an *environmental site assessment* and the reduction of uncertainty about unknown conditions resulting from additional information.

4.5.3 Not every *property* will warrant the same level of assessment. Consistent with good commercial or customary practice, the appropriate level of *environmental site assessment* will be guided by the type of property subject to assessment, the expertise and risk tolerance of the *user*, and the information developed in the course of the inquiry.

4.5.4 It should not be concluded or assumed that an inquiry was not an *appropriate inquiry* merely because the inquiry did not identify *recognized environmental conditions* in connection with a *property*. *Environmental site assessments* must be evaluated based on the reasonableness of judgments made at the time and under the circumstances in which they were made. Subsequent *environmental site assessments* should not be considered valid standards to judge the appropriateness of any prior assessment based on hindsight, new information, use of developing technology or analytical techniques, or other factors.

4.6 *Continued Viability of Environmental Site Assessment*—An *environmental site assessment* meeting or exceeding either this practice or Practice E 1527 and completed less than 180 days previously is presumed to be valid. An *environmental site assessment* meeting or exceeding either practice and completed more than 180 days previously may be used to the extent allowed by 4.7 through 4.7.5.

4.7 *Prior Assessment Usage*—Both this practice and Practice E 1527 recognize that *environmental site assessments* performed in accordance with these practices will include information which *users* or subsequent users may want to use to avoid undertaking duplicative assessment procedures. Therefore, the practices describe procedures to be followed to assist users in determining the appropriateness of using information in *environmental site assessments* performed previously. The system of prior assessment usage is based on the following principles that should be adhered to in addition to the specific procedures set forth elsewhere in these practices:

4.7.1 Subject to 4.7.4, *users* and *environmental professionals* may use information in prior *environmental site assessments* provided such information was generated as a result of procedures that meet or exceed the requirements of this practice or Practice E 1527 and then only provided that the specific procedures set forth in each practice are met.

4.7.2 Subject to 4.7.4, a prior *environmental site assessment* may be used in its entirety, without regard to the specific procedures set forth in these practices, if, in the reasonable judgment of the *user*: the prior *environmental site assessment* meets or exceeds the requirements of this practice or Practice E 1527, and the conditions at the *property* likely to affect *recognized environmental conditions* in connection with the *property* are not likely to have changed materially since the prior *environmental site assessment* was conducted. In making this judgment, the *user* should consider the type of *property* assessed and the conditions in the area surrounding the *property*.

4.7.3 Except as provided in 4.7.2 and 4.7.2 of Practice E 1527, prior *environmental site assessments* should not be used without current investigation of conditions likely to affect *recognized environmental conditions* in connection with the *property* that may have changed materially since the prior *environmental site assessment* was conducted. At a minimum, for a *transaction screen process* consistent with this practice, a new *site visit* should be performed.

4.7.4 If the *user* or *environmental professional(s)* conducting an *environmental site assessment* has *actual knowledge* that the information being used from a prior *environmental site assessment* is not accurate or if it is *obvious*, based on other information obtained by means of the *environmental site assessment* or known to the person conducting the *environmental site assessment*, that the information being used is not accurate, such information from a prior *environmental site assessment* may not be used.

4.7.5 The contractual and legal obligations between prior and subsequent *users* of *environmental site assessments* or between *environmental professionals* who conducted prior *environmental site assessments* and those who would like to use such prior *environmental site assessments* are beyond the scope of this practice.

4.8 The contractual and legal obligations between an *environmental professional* and a *user* (and other parties, if any) are beyond the scope of this practice. No specific legal relationship between the *environmental professional* and the *user* is necessary for the *user* to meet the requirements of this practice.

4.9 If the *user* is aware of any specialized knowledge or experience that is material to *recognized environmental conditions* in connection with the *property*, and the *preparer* is not the *user*, it is the *user*'s responsibility to communicate any information based on such specialized knowledge or experience to the *preparer*. The *user* should do so before the preparer makes the site visit.

4.10 In a transaction involving the purchase of a parcel of *commercial real estate*, if a *user* has *actual knowledge* that the purchase price of the property is significantly less than the purchase price of comparable properties, the *user* should try to identify an explanation for the lower price and to make a written record of such explanation. Among the factors to consider will be the information that becomes known to the *user* pursuant to the *transaction screen environmental site assessment*.

5. Introduction to Transaction Screen Questionnaire

5.1 *Process*—The *transaction screen process* consists of asking questions contained within the *transaction screen questionnaire* of *owners* and *occupants* of the *property*, observing site conditions at the *property* with direction provided by the *transaction screen questionnaire*, and, to the extent *reasonably ascertainable*, conducting limited research regarding certain *government records* and certain *standard historical sources*. The questions asked of *owners* are the same questions as those asked of *occupants*.

5.2 *Guide*—The *transaction screen questionnaire* is followed by a guide designed to assist the person completing the *transaction screen questionnaire*. The guide to the *transaction screen questionnaire* is set out in Sections 7-10 of this practice. The guide is divided into three sections: Guide for Owner/Occupant Inquiry, Guide to Site Visit, and Guide to Government Records/Historical Sources Inquiry.

5.2.1 To assist the *user*, its employee or agent, or the *environmental professional* in preparing a report, the guide repeats each of the questions set out in the *transaction screen questionnaire* in both the guide for *owner/occupant inquiry* and the guide to *site visit*. The questions regarding *government records/historical sources inquiry* are also repeated in the guide to that section.

5.2.2 The guide also describes the procedures to be followed to determine if reliance upon the information in a prior *environmental site assessment* is appropriate under this practice.

5.2.3 A *user*, his employee or agent, or *environmental*

professional conducting the *transaction screen process* should not use the *transaction screen questionnaire* without reference to, or familiarity from prior usage with, the guide.

5.3 *User and Preparer*—The *user* conducting the *transaction screen process* is the party seeking to perform *appropriate inquiry* with respect to the *property*. The *user* may delegate the preparation of the *transaction screen questionnaire* to an employee or agent of the *user* or may contract with a third party to prepare the questionnaire on behalf of the *user*. The person preparing the questionnaire is the *preparer*, who may be either the *user* or the person to whom the *user* has delegated the preparation of the *transaction screen questionnaire*.

5.4 *Exercise of Care*— The *preparer* conducting the transaction screen process should use good faith efforts in determining answers to the questions set forth in the transaction screen questionnaire. The *user* should take time and care to check whatever records are in the user's possession. The *preparer* should ask all persons to whom questions are directed to give answers to the best of the respondent's knowledge. As required by Section 9601(35)(B) of CERCLA, the *user* or *preparer* should discuss with a responsible person in authority in the user's organization (if any) any specialized knowledge or experience relating to *hazardous substances* on the *property* and the *preparer* should understand such information.

5.5 *Knowledge*—The owner or occupant of the *property* to which portions of the transaction screen questionnaire are directed should have sufficient knowledge and experience with respect to the property or in the owner's or occupant's particular business to understand the purpose and use of the transaction screen questionnaire. All answers should be given to the best of the owner's or occupant's actual knowledge.

5.5.1 While the person conducting the *transaction screen process* has an obligation to ask the questions set forth in the *transaction screen questionnaire*, in many instances the parties to whom the questions are addressed will have no obligation to answer them. The *user* is only required to obtain information to the extent it is *reasonably ascertainable*.

5.5.2 If the preparer asks the questions set forth in the *transaction screen questionnaire*, but does not receive any response or receives partial responses, the questions will be deemed to have been answered provided the questions have been asked, or were attempted to be asked, in person or by telephone and written records have been kept of the person to whom the questions were addressed and their responses, or the questions have been asked in writing sent by certified or registered mail, return receipt requested, postage prepaid, or by private, commercial overnight carrier and no responses have been obtained after at least two follow-up telephone calls were made or written request was sent again asking for responses.

5.5.3 The *transaction screen questionnaire* and the *transaction screen guide* sometimes include the phrase "to the best of your knowledge." Use of this phrase shall not be interpreted as imposing a constructive knowledge standard when it is not included or as imposing anything other than an *actual knowledge* standard for the person answering the questions, regardless of whether it is used. It is sometimes included as an assurance to the person being questioned that he or she is not obligated to search out information he or she does not currently have in order to answer the particular question.

5.6 *Conclusions Regarding Affirmative or Unknown Answers*—If any of the questions set forth in the transaction screen questionnaire are answered in the affirmative, the user must document the reason for the affirmative answer. If any of the questions are not answered or the answer is unknown, the user should document such nonresponse or answer of unknown and evaluate it in light of the other information obtained in the transaction screen process, including, in particular, the *site visit* and the government records/historical sources inquiry. If the *user* decides no further inquiry is warranted after receiving no response, an answer of unknown or an affirmative answer, the *user* must document the reasons for any such conclusion.

5.6.1 Upon obtaining an affirmative answer, an answer of unknown or no response, the *user* should first refer to the guide. The guide may provide sufficient explanation to allow a *user* to conclude that no further inquiry is appropriate with respect to the particular question.

5.6.2 If the guide to a particular question does not, in itself, permit a *user* to conclude that no further inquiry is appropriate, then the *user* should consider other information obtained from the *transaction screen process* relating to this question. For example, while on the site performing a *site visit*, a person may find a storage tank on the *property* and therefore answer Question 10 of the *transaction screen questionnaire* in the affirmative. However, during or subsequent to the *owner/occupant inquiry*, the *owner* may produce evidence that substances now or historically contained in the tank (for example, water) are not likely to cause contamination.

5.6.3 If either the guide to the question or other information obtained during the *transaction screen process* does not permit a *user* to conclude no further inquiry is appropriate with respect to such question, then the user must determine, in the exercise of the *user's* reasonable business judgment, based upon the totality of unresolved affirmative answers or answers of unknown received during the *transaction screen process*, whether further inquiry may be limited to those specific issues identified as of concern or should proceed with a full *Phase I Environmental Site Assessment*.

5.7 *Presumption*— A presumption exists that further inquiry is necessary if an affirmative answer is given to a question or because the answer was unknown or no response was given. In rebutting this presumption, the *user* should evaluate information obtained from each component of the transaction screen process and consider whether sufficient information has been obtained to conclude that no further inquiry is necessary. The user must determine, in the exercise of the user's reasonable business judgment, the scope of such further inquiry: whether to proceed with a *Phase I Environmental Site Assessment* prepared in accordance with Practice E 1527 or a lesser inquiry directed at specific issues raised by the questionnaire.

5.8 *Further Inquiry Under Practice E 1527*—Upon completing the *transaction screen questionnaire*, if the *user* concludes that a *Phase I Environmental Site Assessment* is needed, the *user* should proceed with such inquiry with the advice and guidance of an *environmental professional*. Such further inquiry should be undertaken in accordance with Practice E 1527.

5.9 *Signature*—The *user* and the *preparer* of the *transaction screen questionnaire* must complete and sign the questionnaire as provided at the end of the questionnaire.

6. Transaction Screen Questionnaire

6.1 *Persons to Be Questioned*—The following questions should be asked of (*1*) the current *owner* of the *property*, (*2*) any major *occupant* of the *property* or, if the *property* does not have any major *occupants*, at least 10 % of the *occupants* of the *property*, and (*3*) in addition to the current *owner* and the *occupants* identified in (*2*), any *occupant* likely to be using, treating, generating, storing, or disposing of *hazardous substances* or *petroleum products* on or from the *property*. A major *occupant* is any *occupant* using at least 40 % of the leasable area of the property or any anchor tenant when the *property* is a shopping center. In a multifamily property containing both residential and commercial uses, the *preparer* does not need to ask questions of the residential *occupants*. The preparer should ask each person to answer all questions to the best of the respondent's *actual knowledge* and in good faith. When completing the *site visit* column, the *preparer* should be sure to observe the *property* and any buildings and other structures on the *property*. The guide provides further details on the appropriate use of this questionnaire.[5]

[5] Unk = "unknown" or "no response."

Description of Site: Address:

____

____

____

Question	Owner[7]			Occupants (if applicable)			Observed During Site Visit	
1a. Is the *property* used for an industrial use?	Yes	No	Unk	Yes	No	Unk	Yes	No
1b. Is any *adjoining property* used for an industrial use?	Yes	No	Unk	Yes	No	Unk	Yes	No
2a. Did you observe evidence or do you have any prior knowledge that the *property* has been used for an industrial use in the past?	Yes	No	Unk	Yes	No	Unk	Yes	No
2b. Did you observe evidence or do you have any prior knowledge that any *adjoining property* has been used for an industrial use in the past?	Yes	No	Unk	Yes	No	Unk	Yes	No
3a. Is the *property* used as a gasoline station, motor repair facility, commercial printing facility, dry cleaners, photo developing laboratory, junkyard or landfill, or as a waste treatment, storage, disposal, processing, or recycling facility (if applicable, identify which)?	Yes	No	Unk	Yes	No	Unk	Yes	No
3b. Is any *adjoining property* used as a gasoline station, motor repair facility, commercial printing facility, dry cleaners, photo developing laboratory, junkyard or landfill, or as a waste treatment, storage, disposal, processing, or recycling facility (if applicable, identify which)?	Yes	No	Unk	Yes	No	Unk	Yes	No
4a. Did you observe evidence or do you have any prior knowledge that the *property* has been used as a gasoline station, motor repair facility, commercial printing facility, dry cleaners, photo developing laboratory, junkyard or landfill, or as a waste treatment, storage, disposal, processing, or recycling facility (if applicable, identify which)?	Yes	No	Unk	Yes	No	Unk	Yes	No
4b. Did you observe evidence or do you have any prior knowledge that any *adjoining property* has been used as a gasoline station, motor repair facility, commercial printing facility, dry cleaners, photo developing laboratory, junkyard or landfill, or as a waste treatment, storage, disposal, processing, or recycling facility (if applicable, identify which)?	Yes	No	Unk	Yes	No	Unk	Yes	No
5a. Are there currently any damaged or discarded automotive or industrial batteries, pesticides, paints, or other chemicals in individual containers of >5 gal (19 L) in volume or 50 gal (190 L) in the aggregate, stored on or used at the *property* or at the facility?	Yes	No	Unk	Yes	No	Unk	Yes	No

Question	Owner[7]			Occupants (if applicable)			Observed During Site Visit	
5b. Did you observe evidence or do you have any prior knowledge that there have been previously any damaged or discarded automotive or industrial batteries, or pesticides, paints, or other chemicals in individual containers of >5 gal (19 L) in volume or 50 gal (190 L) in the aggregate, stored on or used at the *property* or at the facility?	Yes	No	Unk	Yes	No	Unk	Yes	No
6a. Are there currently any industrial *drums* (typically 55 gal (208 L)) or sacks of chemicals located on the property or at the facility?	Yes	No	Unk	Yes	No	Unk	Yes	No
6b. Did you observe evidence or do you have any prior knowledge that there have been previously any industrial *drums* (typically 55 gal (208 L)) or sacks of chemicals located on the property or at the facility?	Yes	No	Unk	Yes	No	Unk	Yes	No
7a. Did you observe evidence or do you have any prior knowledge that *fill dirt* has been brought onto the property that originated from a contaminated site?	Yes	No	Unk	Yes	No	Unk	Yes	No
7b. Did you observe evidence or do you have any prior knowledge that *fill dirt* has been brought onto the property that is of an unknown origin?	Yes	No	Unk	Yes	No	Unk	Yes	No
8a. Are there currently any *pits, ponds*, or *lagoons* located on the *property* in connection with waste treatment or waste disposal?	Yes	No	Unk	Yes	No	Unk	Yes	No
8b. Did you observe evidence or do you have any prior knowledge that there have been previously, any *pits, ponds*, or *lagoons* located on the *property* in connection with waste treatment or waste disposal?	Yes	No	Unk	Yes	No	Unk	Yes	No
9a. Is there currently any stained soil on the *property*?	Yes	No	Unk	Yes	No	Unk	Yes	No
9b. Did you observe evidence or do you have any prior knowledge that there has been previously, any stained soil on the *property*?	Yes	No	Unk	Yes	No	Unk	Yes	No
10a. Are there currently any registered or unregistered storage tanks (above or underground) located on the *property*?	Yes	No	Unk	Yes	No	Unk	Yes	No
10b. Did you observe evidence or do you have any prior knowledge that there have been previously, any registered or unregistered storage tanks (above or underground) located on the *property*?	Yes	No	Unk	Yes	No	Unk	Yes	No
11a. Are there currently any vent pipes, fill pipes, or access ways indicating a fill pipe protruding from the ground on the *property* or adjacent to any structure located on the *property*?	Yes	No	Unk	Yes	No	Unk	Yes	No
11b. Did you observe evidence or do you have any prior knowledge that there have been previously, any vent pipes, fill pipes, or access ways indicating a fill pipe protruding from the ground on the *property* or adjacent to any structure located on the *property*?	Yes	No	Unk	Yes	No	Unk	Yes	No
12a. Is there currently evidence of leaks, spills or staining by substances other than water, or foul odors, associated with any flooring, drains, walls, ceilings, or exposed grounds on the property?	Yes	No	Unk	Yes	No	Unk	Yes	No

Question	Owner[7]			Occupants (if applicable)			Observed During Site Visit	
12b. Did you observe evidence or do you have any prior knowledge that there have been previously any leaks, spills, or staining by substances other than water, or foul odors, associated with any flooring drains, walls, ceilings or exposed grounds on the property?	Yes	No	Unk	Yes	No	Unk	Yes	No
13a. If the property is served by a private well or non-public water system, is there evidence or do you have prior knowledge that contaminants have been identified in the well or system that exceed guidelines applicable to the water system?	Yes	No	Unk	Yes	No	Unk	Yes	No
13b. If the property is served by a private well or non-public water system, is there evidence or do you have prior knowledge that the well has been designated as contaminated by any government environmental/health agency?	Yes	No	Unk	Yes	No	Unk	Yes	No
14. Does the *owner* or *occupant* of the *property* have any knowledge of *environmental liens* or governmental notification relating to past or recurrent violations of environmental laws with respect to the *property* or any facility located on the *property*?	Yes	No	Unk	Yes	No	Unk		
15a. Has the *owner* or *occupant* of the *property* been informed of the past existence of *hazardous substances* or *petroleum products* with respect to the *property* or any facility located on the *property*?	Yes	No	Unk	Yes	No	Unk		
15b. Has the *owner* or *occupant* of the *property* been informed of the current existence of *hazardous substances* or *petroleum products* with respect to the *property* or any facility located on the *property*?	Yes	No	Unk	Yes	No	Unk		
15c. Has the *owner* or *occupant* of the *property* been informed of the past existence of environmental violations with respect to the *property* or any facility located on the *property*?	Yes	No	Unk	Yes	No	Unk		
15d. Has the *owner* or *occupant* of the *property* been informed of the current existence of environmental violations with respect to the *property* or any facility located on the *property*?	Yes	No	Unk	Yes	No	Unk		
16. Does the *owner* or *occupant* of the *property* have any knowledge of any *environmental site assessment* of the *property* or facility that indicated the presence of *hazardous substances* or *petroleum products* on, or contamination of, the *property* or recommended further assessment of the *property*?	Yes	No	Unk	Yes	No	Unk		
17. Does the *owner* or *occupant* of the *property* know of any past, threatened, or pending lawsuits or administrative proceedings concerning a release or threatened release of any *hazardous substance* or *petroleum products* involving the *property* by any owner or occupant of the *property*?	Yes	No	Unk	Yes	No	Unk		
18a. Does the *property* discharge waste water (not including sanitary waste or storm water) onto or adjacent to the *property* and/or into a storm water system?	Yes	No	Unk	Yes	No	Unk	Yes	No

Question	Owner[7]			Occupants (if applicable)			Observed During Site Visit	
18b. Does the *property* discharge waste water (not including sanitary waste or storm water) onto or adjacent to the *property* and/or into a sanitary sewer system?	Yes	No	Unk	Yes	No	Unk	Yes	No
19. Did you observe evidence or do you have any prior knowledge that any *hazardous substances* or *petroleum products*, unidentified waste materials, tires, automotive or industrial batteries, or any other waste materials have been dumped above grade, buried and/or burned on the *property*?	Yes	No	Unk	Yes	No	Unk	Yes	No
20. Is there a transformer, capacitor, or any hydraulic equipment for which there are any records indicating the presence of PCBs?	Yes	No	Unk	Yes	No	Unk	Yes	No

Government Records/Historical Sources Inquiry

(See guide, Section 10)

21. Do any of the following Federal government record systems list the property or any property within the search distance noted below:

	Approximate Minimum Search Distance, miles (kilometers)		
Federal NPL site list	1.0 (1.6)	Yes	No
Federal CERCLIS list	0.5 (0.8)	Yes	No
Federal CERCLIS NFRAP site list	property and adjoining properties	Yes	No
Federal RCRA CORRACTS facilities list	1.0 (1.6)	Yes	No
Federal RCRA non-CORRACTS TSD facilities list	0.5 (0.8)	Yes	No
Federal RCRA generators list	property and adjoining properties	Yes	No
Federal ERNS list	property only	Yes	No

22. Do any of the following state record systems list the property or any property within the search distance noted below:

	Approximate Minimum Search Distance, miles (kilometers)		
State lists of hazardous waste sites identified for Investigation or remediation:			
State — Equivalent NPL	1.0 (1.6)	Yes	No
State — Equivalent CERCLIS	0.5 (0.8)	Yes	No
State landfill and/or solid waste disposal site lists	0.5 (0.8)	Yes	No
State leaking UST lists	0.5 (0.8)	Yes	No
State registered UST lists	property and adjoining properties	Yes	No

23. Based upon a review of *fire insurance maps* 10.3.1.3 or consultation with the local fire department serving the *property*, all as specified in the guide, are any buildings or other improvements on the *property* or on an *adjoining property* identified as having been used for an industrial use or uses likely to lead to contamination of the *property*? Yes No

The *preparer* of the *transaction screen questionnaire* must complete and sign the following. (For definition of "preparer" and "user," see 5.3 or 3.3.28.)

The *Owner* questionnaire was completed by:

Name
Title
Firm
Address

Phone number
Date
Preparer's relationship to site
Preparer's relationship to user (for example, principal, employee, agent, consultant)

The *Occupant* questionnaire was completed by:

Name
Title
Firm
Address

Phone number
Date
Preparer's relationship to site
Preparer's relationship to user (for example, principal, employee, agent, consultant)

The *Site Visit* questionnaire was completed by:

Name
Title
Firm
Address

Phone number
Date
Preparer's relationship to site
Preparer's relationship to user (for example, principal, employee, agent, consultant)

The *Government Records* and *Historical Sources Inquiry* questionnaire was completed by:

Name
Title
Firm
Address

Phone number
Date
Preparer's relationship to site
Preparer's relationship to user (for example, principal, employee, agent, consultant)

User's relationship to the site (for example, owner, prospective purchaser, lender, etc.)

If the preparer(s) is different from the user, complete the following:

Name of User
User's address

User's phone number

Copies of the completed questionnaires have been filed at:

Copies of the completed questionnaires have been mailed or delivered to:

Preparer represents that to the best of the preparer's knowledge the above statements and facts are true and correct and to the best of the preparer's actual knowledge no material facts have been suppressed or misstated.

Signature ________________ Date________________
Signature ________________ Date________________
Signature ________________ Date________________

7. Guide to Transaction Screen Questionnaire

7.1 The following sets forth the guide to the *transaction screen questionnaire*. The guide accompanies the *transaction screen questionnaire* to assist the *preparer* in completing the questionnaire. Questions found in the *transaction screen questionnaire* are repeated in the guide.

7.2 If the *preparer* completing the *transaction screen questionnaire* is familiar with the guide from prior usage, the questionnaire may be completed without reference to the guide.

7.3 The *site visit* portion of the guide considers most of the same questions set forth in the guide to *owner/occupant inquiry* because the *transaction screen process* requires both questions of *owners* and *occupants* of the *property* and observations of the *property* by the *preparer*.

7.4 Prior *environmental site assessment* usage procedures

are contained in the guide to *owner/occupant inquiry* and the guide to *government records/historical sources inquiry*. The information supplied in connection with the *site visit* portion of a prior *environmental site assessment* may be used for guidance, but may not be relied upon without determining through a new *site visit* whether any conditions that are material to recognized environmental conditions in connection with the *property* have changed since the prior *environmental site assessment*. Therefore, the guide to the *site visit* does not contain any prior assessment procedures.

7.5 In performing the *site visit* portion of the *transaction screen process*, the *preparer* should *visually and physically observe* the *property* and any structure located on the *property* to the extent not obstructed by bodies of water, cliffs, adjacent buildings, or other impassable obstacles.

7.5.1 The periphery of the *property* should be *visually and physically observed*, as well as the periphery of all structures on the *property*, and the *property* should be viewed from all adjacent public thoroughfares. Any overgrown areas should be inspected, including roads or paths with no apparent outlet that should be *visually and physically* observed to their ends.

7.5.2 On the interior of structures on the *property*, accessible common areas expected to be used by building *occupants* or the public (such as lobbies, hallways, utility rooms, and recreation areas), a representative sample of *owner* and *occupant* spaces, and maintenance and repair areas, including boiler rooms, should be *visually and physically* observed. It is not necessary to look under floors, above ceilings, or behind walls.

7.5.3 After completing the *site visit*, the *preparer* of the *transaction screen questionnaire* may obtain "yes" answers that require the *preparer* once again to ask questions of the *owner* of the *property* or *occupants* of the *property* to satisfy the *user* that no further inquiry is necessary.

7.6 In addition to asking questions of the *owner* of the *property* and *occupants* of the *property* (Section 8) and *visually and physically* observing the *property* (Section 9), the *user* completing the *transaction screen process* should determine, either from governmental agencies or through commercial services providing government environmental records, whether certain known or suspected contaminated sites or activities involving the release of *hazardous substances* or *petroleum products* occur on or near the *property*. See Section 10.

7.6.1 These records may be obtained either directly from the government agencies or from commercial services that provide the records for a fee. Because of the numerous sources that must be searched and the response time of government agencies, commercial services are available that provide a single source for federal and state records. These services may provide a quicker response than the government agencies but fees will be charged for the information.

7.6.2 If government information is obtained from a commercial service, the firm should provide assurances that its records stay current with the government agency record sources. Government information obtained from non-government sources may be considered current if the source updates the information at least every 90 days, or, for information that is updated less frequently than quarterly by the government agency, within 90 days of the date the government agency makes the updated information available to the public.

7.6.3 The identity of firms providing this type of government information may be obtained through local telephone directories or through an inquiry of environmental professionals in the area of the preparer completing the transaction screen questionnaire.

8. Guide for Owner/Occupant Inquiry

8.1 Is the property used for an industrial use?

___Yes ___No ___Unknown

8.1.1 Is any adjoining property used for an industrial use?

___Yes ___No ___Unknown

	Land Use
Property:	________________
Adjoining properties north:	________________
Adjoining properties south:	________________
Adjoining properties east:	________________
Adjoining properties west:	________________

8.1.2 *Guide*:

8.1.2.1 It is recommended that the *preparer* describe the use of the *property* and *adjoining properties*.

8.1.2.2 Certain industrial uses on the *property* may raise concerns regarding the possibility of contamination affecting the *property*. For purposes of the *transaction screen questionnaire*, an industrial use is an activity requiring the application of labor and capital for the production or distribution of a product or article, including, without limitation, manufacturing, processing, extraction, refining, warehousing, transportation, and utilities. Manufacturing is defined as a process or operation of producing by hand, machinery, or other means a finished product or article from raw material. Industrial uses may be categorized as light or heavy industrial uses, depending upon the scale of the operations and the impact upon surrounding property in terms of smoke, fumes, and noise. Regardless of such categorization, the concern for purposes of the transaction screen process is whether the use involves the processing, storage, manufacture, or transportation of *hazardous substances* or *petroleum products*. For example, further inquiry would be necessary if the industrial use concerned the manufacture of paints, oils, solvents, and other chemical products but not if the use concerned the storage of inert goods in containers.

8.1.2.3 The term *adjoining properties* means any real property or properties the border of which is contiguous or partially contiguous with that of the property, or that would be contiguous or partially contiguous with that of the property but for a street, road, or other public thoroughfare separating them. *Adjoining properties* means the *property* and include *properties* across the street or any right of way from the property.

8.1.2.4 To use the information supplied in response to this question in a prior *environmental site assessment*, the *preparer* must determine if there were changes in the use of the *property* or any *adjoining property* since the prior *environmental site assessment* that are material to *recognized environmental conditions* in connection with the *property*. If not, using information in the prior *environmental site assessment* is appropriate. If so, the information requested must be supplied for each *property* for which the use has so changed.

8.2 Did you observe evidence or do you have any prior knowledge that the *property* has been used for an industrial use in the past?

___Yes ___No ___Unknown

8.2.1 Did you observe evidence or do you have any prior knowledge that any *adjoining property* has been used for an industrial use in the past?

___Yes ___No ___Unknown

8.2.2 *Guide*—See guide for question 8.1.

	Owner	Use	Dates
Previous use of property			
Previous use of properties to north			
Previous use of properties to south			
Previous use of properties to east			
Previous use of properties to west			

8.3 Is the *property* used as a gasoline station, motor repair facility, commercial printing facility, dry cleaners, photo developing laboratory, junkyard, or landfill, or as a waste treatment, storage, disposal, processing, or recycling facility (if applicable, identify which)?

___Yes ___No ___Unknown

8.3.1 Is any *adjoining property* used as a gasoline station, motor repair facility, commercial printing facility, dry cleaners, photo developing laboratory, junkyard or landfill, or as a waste treatment, storage, disposal, processing, or recycling facility (if applicable, identify which)?

___Yes ___No ___Unknown

	Land Use
Property:	
Adjoining properties north:	
Adjoining properties south:	
Adjoining properties east:	
Adjoining properties west:	

8.3.2 *Guide*:

8.3.2.1 It is recommended that the *preparer* describe the uses of the *property* and *adjoining properties*.

8.3.2.2 Gasoline stations, motor vehicle repair facilities (with or without supplying gas for the motor vehicles), dry cleaners, photo developing laboratories, commercial printing facilities, junkyards or landfills, and waste treatment, storage, disposal, processing, or recycling facilities all involve the use of *hazardous substances* or *petroleum products* and therefore require further inquiry concerning the possible release of such substances.

8.3.2.3 The term *adjoining properties* means any real property or properties the border of which is contiguous or partially contiguous with that of the *property*, or that would be contiguous or partially contiguous with that of the *property* but for a street, road, or other public thoroughfare separating them.

8.3.2.4 To rely on the information supplied in response to this question in a prior *environmental site assessment*, the *preparer* must determine if there were changes in the use of the property or any *adjoining property* since the prior *environmental site assessment* that are material to *recognized environmental conditions* in connection with the *property*. If not, then use of information in the prior *environmental site assessment* is appropriate. If so, the information requested must be supplied for each *property* for which the use has so changed.

8.4 Did you observe evidence or do you have any prior knowledge that the *property* has been used as a gasoline station, motor repair facility, commercial printing facility, dry cleaners, photo developing laboratory, junkyard or landfill, or as a waste treatment, storage, disposal, processing, or recycling facility (if applicable, identify which)?

___Yes ___No ___Unknown

8.4.1 Did you observe evidence or do you have any prior knowledge that any *adjoining property* has been used as a gasoline station, motor repair facility, commercial printing facility, dry cleaners, photo developing laboratory, junkyard or landfill, or as a waste treatment, storage, disposal, processing, or recyling facility (if applicable, identify which)?

___Yes ___No ___Unknown

8.4.2 *Guide*—See guide for question 8.3.

	Owner	Use	Dates
Previous use of property			
Previous use of properties to north			
Previous use of properties to south			
Previous use of properties to east			
Previous use of properties to west			

LAND ISSUES

8.5 Are there currently any damaged or discarded automotive or industrial batteries, pesticides, paints, or other chemicals in individual containers of >5 gal (19 L) in volume or 50 gal (190 L) in the aggregate, stored on or used at the property or at the facility?

___Yes ___No ___Unknown

8.5.1 Did you observe evidence or do you have any prior knowledge that there have been previously any damaged or discarded automotive or industrial batteries, or pesticides, paints, or other chemicals in individual containers of >5 gal (19 L) in volume or 50 gal (190 L) in the aggregate, stored on or used at the *property* or at the facility?

___Yes ___No ___Unknown

8.5.2 *Guide*:

8.5.2.1 Are there any containers on the site that may contain any of these items? Is there any reason to suspect that chemicals or *hazardous substances* in such quantities may be stored on the site? Sheltered areas, cartons, sacks, storage bins, large canisters, sheds, or cellars of existing improvements are examples of containers and areas where chemicals or *hazardous substances* may be stored. If the answer to this question is "yes," list the items and the location(s) where they are stored. If unfamiliar with the contents of any container located on the site, the question must be answered "yes" until the materials are identified.

8.5.2.2 *Hazardous substances* may often be unmarked. The *preparer* should never open any containers that are unmarked because they may contain explosive materials or acids.

8.5.2.3 Consumer products in undamaged containers used for routine office maintenance or business, such as copy toner, should not create a need for further inquiry unless the quantity of such products is in excess of what would be customary for such use. The Environmental Protection Agency has published a guidance document that identifies hazardous substances that must be reported under Sections 311 and 312 of the Emergency Planning and Community Right to Know Act ("EPCRA").[6] This document lists in tabular form the CERCLA Section 103 chemicals. If a preparer has a question regarding whether the substance is a hazardous substance under CERCLA, the preparer may refer to the list of lists or 40 CFR Part 302. In addition, the Environmental Protection Agency has also published a guidance document.[7] This document sets forth the *hazardous substances* found in many common consumer products listed by trade name.

8.5.2.4 A *preparer* should not rely exclusively upon a prior *environmental site assessment* in supplying this information.

8.6 Are there currently any industrial drums (typically, 55 gal (208 L)) or sacks of chemicals located on the property or at the facility?

___Yes ___No ___Unknown

8.6.1 Did you observe evidence or do you have any prior knowledge that there have been previously any industrial *drums* (typically 55 gal (208 L)) or sacks of chemicals located on the property or at the facility?

___Yes ___No ___Unknown

8.6.2 *Guide*:

8.6.2.1 Chemicals are frequently stored in large 55 gal (208–L) drums and dry chemicals are often stored in 20 lb (9 kg) sacks. See Appendix X2 for examples of 55-gal (208–L) drums and for surface staining resulting from improper drum storage.

8.6.2.2 A *preparer* should not rely exclusively upon a prior *environmental site assessment* in supplying this information.

8.7 Did you observe evidence or do you have any prior knowledge that *fill dirt* has been brought onto the property that originated from a contaminated site?

___Yes ___No ___Unknown

8.7.1 Did you observe evidence or do you have any prior knowledge that *fill dirt* has been brought onto the property that is of an unknown origin?

___Yes ___No ___Unknown

8.7.2 *Guide*:

8.7.2.1 The origin of *fill dirt* brought onto the *property* should be investigated to determine whether such dirt originated from a contaminated site. The term *fill dirt* is defined in the definitions and the *preparer* should refer to the definitions if the *preparer* has any question concerning the meaning of the term.

8.7.2.2 If any structures have been demolished on the property, the *preparer* should investigate whether the structures were demolished in place and *fill dirt* compacted over them because such demolition debris may contain asbestos or *hazardous substances*.

8.7.2.3 To use the information supplied in response to this question in a prior *environmental site assessment*, the *preparer* must determine if there has been any filling at the site since the prior *environmental site assessment*. If not, then using information in the prior *environmental site assessment* is appropri ate. If so, the information requested must be supplied for any *fill dirt* brought on the property since the prior *environmental site assessment*.

8.8 Are there currently any *pits, ponds, or lagoons* located on the *property* in connection with waste treatment or waste disposal?

___Yes ___No ___Unknown

8.8.1 Did you observe evidence or do you have any prior knowledge that there have been previously, any *pits, ponds*, or *lagoons* located on the *property* in connection with waste treatment or waste disposal?

___Yes ___No ___Unknown

8.8.2 *Guide*:

8.8.2.1 The presence of *pits, ponds, or lagoons*, together with waste treatment or waste disposal may indicate contaminated property. See the definitions with respect to the definition of *pits, ponds, or lagoons* in 3.2.28.

8.8.2.2 A *preparer* should not rely exclusively upon a prior *environmental site assessment* in supplying this information.

8.9 Is there currently any stained soil on the *property*?

___Yes ___No ___Unknown

8.9.1 Did you observe evidence or do you have any prior knowledge that there has been previously, any stained soil on the *property*?

___Yes ___No ___Unknown

8.9.2 *Guide*:

8.9.2.1 Stained soils are frequently associated with contamination and often are an indication of either current or previous leakage associated with piping and liquid storage containers. Soils that are stained show a marked discoloration as compared to other soils in the immediate vicinity.

8.9.2.2 A *preparer* should not rely exclusively upon a prior *environmental site assessment* in supplying this information.

8.10 Are there currently any registered or unregistered storage tanks (above or underground) located on the *property*?

___Yes ___No ___Unknown

8.10.1 Did you observe evidence or do you have any prior knowledge that there have been previously, any registered or unregistered storage tanks (above or underground) located on the *property*?

___Yes ___No ___Unknown

8.10.2 *Guide*:

8.10.2.1 Tanks are often used to store heating fuels, chemicals, and petroleum products; while tanks may be associated with the storage of chemicals, they are most often associated

[6] "Title III List of Lists, Consolidated List of Chemicals Subject to Reporting Under Title III of the Superfund Amendments and Reauthorization Act (SARA) of 1986," U.S. EPA, Office of Toxic Substances, January 1989.

[7] "Common Synonyms for Chemicals Listed Under Section 313 of the Emergency Planning and Community Right to Know Act" Office of Toxic Substances, U.S. EPA, January 1988.

with liquid fuel heating systems (for example, oil furnaces).

8.10.2.2 To use the information supplied in response to this question in a prior *environmental site assessment*, the *user* must determine if there were storage tanks installed on the site since the prior *environmental site assessment*. If not, then using information in the prior *environmental site assessment* is appropriate. If so, the information requested must be supplied on all storage tanks installed on the site since the prior *environmental site assessment*.

8.11 Are there currently any vent pipes, fill pipes, or access ways indicating a fill pipe protruding from the ground on the *property* or adjacent to any structure located on the *property*?

___Yes ___No ___Unknown

8.11.1 Did you observe evidence or do you have any prior knowledge that there have been previously, any vent pipes, fill pipes, or access ways indicating a fill pipe protruding from the ground on the *property* or adjacent to any structure located on the *property*?

___Yes ___No ___Unknown

8.11.2 *Guide*:

8.11.2.1 Vent or fill pipes often signal the current or previous existence of underground storage tanks.

8.11.2.2 Additionally, in answering this question the owner and occupant should consider any asphalt or concrete patching that would indicate the possibility of previous underground storage tank removal. Examples of vent and fill pipes are illustrated in Appendix X2.

8.11.2.3 A *preparer* should not rely exclusively upon a prior *environmental site assessment* in supplying this information.

STRUCTURE ISSUES

8.12 Is there currently evidence of leaks, spills or staining by substances other than water, or foul odors, associated with any flooring, drains, walls, ceilings or exposed grounds on the property?

___Yes ___No ___Unknown

8.12.1 Did you observe evidence or do you have any prior knowledge that there have been previously any leaks, spills or staining by substances other than water, or foul odors, associated with any flooring, drains, walls, ceilings or exposed grounds on the property?

___Yes ___No ___Unknown

8.12.2 *Guide*:

8.12.2.1 Stains (other than water stains) or foul odors may indicate leaks or spills of hazardous substances of contaminants. Floor drains located within a building adjacent to hazardous substance storage areas or connected to an on-site disposal system (for example, septic system) present a potential source of subsurface discharge of contaminants.

8.12.2.2 A *preparer* should not rely exclusively upon a prior *environmental site assessment* in supplying this information.

OTHER ISSUES

8.13 If the *property* is served by a private well or non-public water system, is there evidence or do you have prior knowledge that contaminants have been identified in the well or system that exceed guidelines applicable to the water system?

___Yes ___No ___Unknown

8.13.1 If the property is served by a private well or non-public water system, is there evidence or do you have prior knowledge that the well has been designated as contaminated by any government environmental/health agency?

___Yes ___No ___Unknown

8.13.2 *Guide*:

8.13.2.1 Private wells and non-public water systems are not monitored daily for water quality as municipal systems are monitored. If the system is private, it probably has been tested for contamination or evidence that it is free from contamination, and the results of any such tests should be produced by the *owner* or *occupant* of the well. The *preparer* is not required to test the water system to conduct the *transaction screen*.

8.13.2.2 A *preparer* should not rely exclusively upon a prior *environmental site assessment* in supplying this information.

8.14 Does the *owner* or *occupant* of the property have any knowledge of *environmental liens* or governmental notification relating to past or recurrent violations of environmental laws with respect to the *property* or any facility located on the property?

___Yes ___No ___Unknown

8.14.1 *Guide*:

8.14.1.1 In most cases, the federal or state government will notify the property owner prior to filing a lien on the *property*. Sections 302, 311, 312, and 313 of The Emergency Planning and Community Right-to-Know Act and other provisions of federal and state environmental laws establish reporting requirements with respect to businesses storing or using *hazardous substances* in excess of certain quantities. These businesses should be making periodic reports to a federal, state, or local environmental department, agency, or bureau. The government may periodically inspect such facilities to ensure compliance with environmental laws. In the event of a release of a reportable quantity within a 24-h period (as defined in CERCLA and the regulations promulgated pursuant to CERCLA), the person in charge of the facility is obligated to notify the U.S. EPA of the release. Any notification or response by any governmental entity will be in writing.

8.14.1.2 The information supplied in response to this question in a prior *environmental site assessment* may be used provided it is updated to the present time.

8.15 Has the *owner* or *occupant* of the *property* been informed of the past existence of *hazardous substances* or *petroleum products* with respect to the *property* or any facility located on the *property*?

___Yes ___No ___Unknown

8.15.1 Has the *owner* or *occupant* of the *property* been informed of the current existence of *hazardous substances* or *petroleum products* with respect to the *property* or any facility located on the *property*?

___Yes ___No ___Unknown

8.15.2 Has the *owner* or *occupant* of the *property* been informed of the past existence of environmental violations with respect to the *property* or any facility located on the *property*?

___Yes ___No ___Unknown

8.15.3 Has the *owner* or *occupant* of the *property* been informed of the current existence of environmental violations with respect to the *property* or any facility located on the *property*?

___Yes ___No ___Unknown

8.15.4 *Guide*:

8.15.4.1 Consider whether any *environmental professionals* familiar with *hazardous substances* or *petroleum products* have observed or determined that contamination existed on the property. *Hazardous substances* or *petroleum products* from the *property* may have affected soils, air quality, water quality, or otherwise affected structures located on the *property*.

8.15.4.2 The information supplied in response to this question in a prior *environmental site assessment* may be used provided it is updated to the present time.

8.16 Does the *owner* or *occupant* of the *property* have any knowledge of any *environmental site assessment* of the *property* or facility that indicated the presence of *hazardous substances* or *petroleum products* on, or contamination of, the *property* or recommended further assessment of the *property*?

___Yes ___No ___Unknown

8.16.1 *Guide*:

8.16.1.1 Copies of *reasonably ascertainable* prior *environmental site assessments* of the *property* or any portion thereof should be obtained and examined to determine whether further action or inquiry is necessary in connection with any environmental problems raised by a prior *environmental site assessment*.

8.16.1.2 The information supplied in response to this question in a prior *environmental site assessment* may be used provided it is updated to the present time.

8.17 Does the *owner* or *occupant* of the *property* know of any past, threatened, or pending lawsuits or administrative proceedings concerning a release or threatened release of any *hazardous substance* or *petroleum products* involving the *property* by any owner or occupant of the *property*?

___Yes ___No ___Unknown

8.17.1 *Guide*:

8.17.1.1 The *user* is not required to make an independent investigation or search of records on file with a court or public agency in answering this question; this question is to be answered by the *owner* or *occupant* based upon their respective *actual knowledge* and review of *reasonably ascertainable* records in their possession.

8.17.1.2 The information supplied in response to this question in a prior *environmental site assessment* may be used provided it is updated to the present time.

8.18 Does the *property* discharge wastewater (not including sanitary waste or storm water) onto or adjacent to the property and/or into a storm water system?

___Yes ___No ___Unknown

8.18.1 Does the property discharge wastewater (not including sanitary waste or storm water) onto or adjacent to the property and/or into a sanitary sewer system?

___Yes ___No ___Unknown

8.18.2 *Guide*:

8.18.2.1 The *owner* and each *occupant* should be asked where drain traps lead and the purpose of drainage pipes at the facility. Domestic sewage is not a CERCLA issue and the reference to *wastewater* does not include domestic sewage.

8.18.2.2 To use the information supplied in response to this question in a prior *environmental site assessment*, the *preparer* must determine if there was any change in discharge practices at the facility since the prior *environmental site assessment*. If not, using information in the prior *environmental site assessment* is appropriate. If so, the information requested must be supplied for all new or changed discharge practices.

8.18.2.3 Some jurisdictions require facilities with large roof or paved areas and construction sites to collect and divert runoff through a treatment process prior to discharging the stormwater runoff to municipal, separate storm sewer systems, or the waters of the United States. Such units are often called stormwater treatment systems. Oil-water separators are most often found outside a building under a manhole and require routine servicing to remove oil. Oil-water separators are usually in restaurants, repair garages, and service stations. An example of an oil-water separator is shown in Appendix X2. If any such oil-water separators or treatment systems have been installed at the *property* since a prior *environmental site assessment*, the requested information must be supplied for each new installation.

8.19 Did you observe evidence or do you have any prior knowledge that any *hazardous substances* or *petroleum products*, unidentified waste materials, tires, automotive or industrial batteries, or any other waste materials have been dumped above grade, buried and/or burned on the *property*?

___Yes ___No ___Unknown

8.19.1 *Guide*:

8.19.1.1 Past waste disposal practices should be examined because these may have resulted in *hazardous substances* or *petroleum products* being released on the *property*. Does the *property* evidence any mounds or depressions that suggest a disposal site?

8.19.1.2 To use the information supplied in response to this question in a prior *environmental site assessment*, the *preparer* must determine if there was any dumping, burying, or burning of such materials at the site since the prior *environmental site assessment*. If not, then using information in the prior *environmental site assessment* is appropriate. If so, the information requested must be supplied for all such events since the prior *environmental site assessment*.

8.20 Is there a transformer, capacitor, or any hydraulic equipment for which there are any records indicating the presence of PCBs?

___Yes ___No ___Unknown

8.20.1 *Guide*:

8.20.1.1 The PCBs are regulated by the Toxic Substances Control Act 15 USC. Section 2601 *et seq.* and, in the absence of a release, are not regulated by CERCLA. The provisions of CERCLA do apply if there is a release of PCBs. Accordingly, if an affirmative answer is obtained to this question, the further focus should be on whether there have been any instances of insulating oil leakage and, if so, whether these are suspected of being PCB or PCB-contaminated.

8.20.1.2 Transformers containing PCBs may have many different sizes and shapes. Some of the more commonly used transformers are set forth in Appendix X2. Transformers are to be registered pursuant to 40 CFR § 761.30.

8.20.1.3 Elevators and auto lifts are often run by hydraulically controlled systems containing PCBs. If inspection or maintenance records for the elevator, capacitor, or other hydraulic equipment indicate no release has occurred or that regular, scheduled maintenance has taken place and the machinery does not appear to be damaged or leaking, no further inquiry is required.

8.20.1.4 To use the information supplied in response to this question in a prior *environmental site assessment*, the *preparer* must determine if there were any transformers installed at the site since the prior *environmental site assessment* that are not owned by a utility, cooperative, or association. If not, then using information in the prior *environmental site assessment* is appropriate, except that for any transformer identified in the prior *environmental site assessment*, the PCB status should be updated. If new transformers have been installed, their PCB status should also be verified.

9. Guide to Site Visit

9.1 Is the *property* used for an industrial use?

___Yes ___No

9.1.1 Is any *adjoining property* used for an industrial use?

___Yes ___No

	Land Use
Property:	
Adjoining properties north:	
Adjoining properties south:	
Adjoining properties east:	
Adjoining properties west:	

9.1.2 *Guide*:

9.1.2.1 It is recommended that the *preparer* describe the uses of the *property* and *adjoining properties*.

9.1.2.2 Certain industrial uses on the *property* may raise concerns regarding the possibility of contamination affecting the *property*. For purposes of the *transaction screen questionnaire*, an industrial use is an activity requiring the application of labor and capital for the production or distribution of a product or article, including, without limitation, manufacturing, processing, extraction, refining, warehousing, transportation, and utilities. Manufacturing is defined as a process or operation of producing by hand, machinery, or other means, a finished product or article from raw material. Industrial uses may be categorized as light or heavy industrial uses, depending upon the scale of the operations and the impact upon surrounding property in terms of smoke, fumes, and noise. Regardless of such categorization, the concern for purposes of the transaction screen process is whether the use involves the processing, storage, manufacture, or transportation of *hazardous substances* or *petroleum products*. For example, further inquiry would be necessary if the industrial use concerned the manufacture of paints, oils, solvents, and other chemical products but not if the use concerned the storage of inert goods in containers.

9.1.2.3 The term *adjoining properties* means any real property or properties the border of which is contiguous or partially contiguous with that of the *property*, or that would be contiguous or partially contiguous with that of the *property* but for a street, road, or other public thoroughfare separating them.

9.2 Did you observe evidence or do you have any prior knowledge that the *property* has been used for an industrial use in the past?

___Yes ___No

9.2.1 Did you observe evidence or do you have any prior knowledge that any *adjoining property* has been used for an industrial use in the past?

___Yes ___No

9.2.2 *Guide:*

9.2.2.1 The *user* should inspect for any indications present on the *property* that would cause the *user* to suspect an industrial facility may once have existed on the site. Old buildings, pipes, containers, or other debris are indicators of previous industrial use of the site.

9.2.2.2 See guide for 9.1.

	Owner	Use	Dates
Previous use of property			
Previous use of properties to north			
Previous use of properties to south			
Previous use of properties to east			
Previous use of properties to west			

9.3 Is the property used as a gasoline station, motor repair facility, commercial printing facility, dry cleaners, photo developing laboratory, junkyard or landfill, or as a waste treatment, storage, disposal, processing, or recycling facility (if applicable, identify which)?

___Yes ___No

9.3.1 Is any *adjoining property* used as a gasoline station, motor repair facility, commercial printing facility, dry cleaners, photo developing laboratory, junkyard or landfill, or as a waste treatment, storage, disposal, processing, or recycling facility (if applicable, identify which)?

___Yes ___No

	Land Use
Property:	
Adjoining properties north:	
Adjoining properties south:	
Adjoining properties east:	
Adjoining properties west:	

9.3.2 *Guide*:

9.3.2.1 It is recommended that the *preparer* describe the uses of the *property* and *adjoining properties*.

9.3.2.2 Gasoline stations, motor vehicle repair facilities (with or without supplying gas for the motor vehicles), dry cleaners, photo developing laboratories, commercial printing facilities, junkyards or landfills, and waste treatment, storage, disposal, processing, or recycling facilities all involve the use of *hazardous substances* or *petroleum products* and therefore require further inquiry concerning the possible release of such substances.

9.3.2.3 The term *adjoining properties* means any real property or properties the border of which is contiguous or partially contiguous with that of the *property*, or that would be contiguous or partially contiguous with that of the *property* but for a street, road, or other public thoroughfare separating them. Adjoining properties include those that border the property and include properties across the street or any right of way from the property.

9.4 Did you observe evidence or do you have any knowledge that the *property* has been used as a gasoline station, motor repair facility, commercial printing facility, dry cleaners, photo developing laboratory, junkyard or landfill, or as a waste treatment, storage, disposal, processing, or recycling facility (if applicable, identify which)?

___Yes ___No

9.4.1 Did you observe evidence or do you have any prior knowledge that any *adjoining property* has been used as a gasoline station, motor repair facility, commercial printing facility, dry cleaners, photo developing laboratory, junkyard or landfill, or as a waste treatment, storage, disposal, processing, or recyling facility (if applicable, identify which)?

___Yes ___No

9.4.2 *Guide*—See guide for 9.2 and 9.3.

	Owner	Use	Dates
Previous use of property			
Previous use of properties to north			
Previous use of properties to south			
Previous use of properties to east			
Previous use of properties to west			

LAND ISSUES

9.5 Are there currently any damaged or discarded automotive or industrial batteries, pesticides, paints, or other chemicals in individual containers of >5 gal (19 L) in volume or 50 gal (190 L) in the aggregate, stored on or used at the property or at the facility?

___Yes ___No

9.5.1 Did you observe evidence or do you have any prior knowledge that there have been previously any damaged or discarded automotive or industrial batteries, or pesticides, paints, or other chemicals in individual containers of >5 gal (19 L) in volume or 50 gal (190 L) in the aggregate, stored on or used at the *property* or at the facility?

___Yes ___No

9.5.2 *Guide*:

9.5.2.1 Are there any containers on the site that may contain any one of these items? Is there any reason to suspect that chemicals or *hazardous substances* or *petroleum products* in such quantities may be stored on the site? Sheltered areas, cartons, sacks, storage bins, large canisters, sheds, or cellars of existing improvements should be investigated because these are areas where chemicals or *hazardous substances* or *petroleum products* may be stored. If the answer to this question is "yes," list the items and the location(s) where they are stored. If you are unfamiliar with the contents of any container located on the site, the question must be answered "yes" until the materials are identified. The existence of any damaged or opened containers identified as containing *hazardous substances* or *petroleum products* requires further investigation.

9.5.2.2 *Hazardous substances* or *petroleum products* may often be unmarked. The *preparer* should never open any unmarked containers at the facility because they may contain explosive materials or acids.

9.5.2.3 Consumer products in undamaged containers used for routine office maintenance or business, such as copy toner, should not create a need for further inquiry unless the quantity of such products is in excess of what would be customary for such use. The Environmental Protection Agency has published a guidance document that identifies hazardous substances or petroleum products that must be reported under Section 311 and 312 of EPCRA.[6] This document lists in tabular form the CERCLA Section 103 chemicals. If a preparer has a question regarding whether the substance is a hazardous substance under CERCLA, the preparer may refer to the list of lists or 40 CFR Part 302. In addition, the Environmental Protection Agency has also published a guidance document.[7] This document sets forth the hazardous substances or petroleum products found in many common consumer products listed by trade name.

9.6 Are there currently any industrial *drums* (typically, 55 gal (208 L)) or sacks of chemicals located on the *property* or at the facility?

___Yes ___No

9.6.1 Did you observe evidence or do you have any prior knowledge that there have been previously any industrial *drums* (typically 55 gal (208 L)) or sacks of chemicals located on the property or at the facility?

___Yes ___No

9.6.2 *Guide*—If found, they will require further examination with respect to any *hazardous substance* associated with them.

9.7 Did you observe evidence or do you have any prior knowledge that *fill dirt* has been brought onto the *property* that originated from a contaminated site?

___Yes ___No

9.7.1 Did you observe evidence or do you have any prior knowledge that *fill dirt* has been brought onto the property that is of an unknown origin?

___Yes ___No

9.7.2 *Guide*— *Fill dirt* brought onto the *property* may appear as mounds or depressions that do not appear to be naturally occurring. *Fill dirt* may be added in construction of a facility. The term *fill dirt* is defined in the definitions, and the *preparer* should refer to the definitions if the *preparer* has any question concerning the meaning of the term.

9.8 Are there currently any *pits, ponds, or lagoons* located on the *property* in connection with waste treatment or waste disposal?

___Yes ___No

9.8.1 Did you observe evidence or do you have any prior knowledge that there have been previously, any *pits, ponds*, or *lagoons* located on the *property* in connection with waste treatment or waste disposal?

___Yes ___No

9.8.2 *Guide*—The presence of *pits, ponds, or lagoons*, together with waste treatment or waste disposal may indicate contaminated property. See the definitions with respect to the definition of *pits, ponds, or lagoons* in 3.2.28.

9.9 Is there currently any stained soil on the *property*?

___Yes ___No

9.9.1 Did you observe evidence or do you have any prior knowledge that there has been previously, any stained soil on the *property*?

___Yes ___No

9.9.2 *Guide*—Stained soils are frequently associated with contamination and often are an indication of either current or previous leakage associated with piping and liquid storage containers. Soils that are stained show a marked discoloration as compared to other soils in the immediate vicinity.

9.10 Are there currently any registered or unregistered storage tanks (above or underground) located on the *property*?

___Yes ___No

9.10.1 Did you observe evidence or do you have any prior knowledge that there have been previously, any registered or unregistered storage tanks (above or underground) located on the *property*?

___Yes ___No

9.10.2 *Guide*—Tanks are often used to store heating fuels, chemicals, and *petroleum products*; while tanks may be associated with storage of chemicals, they are most often associated with liquid fuel heating systems (that is, oil furnaces). Examples of tanks are illustrated in Appendix X2.

9.11 Are there currently any vent pipes, fill pipes, or access ways indicating a fill pipe protruding from the ground on the property or adjacent to any structure located on the *property*?

___Yes ___No

9.11.1 Did you observe evidence or do you have any prior knowledge that there have been previously, any vent pipes, fill pipes, or access ways indicating a fill pipe protruding from the ground on the *property* or adjacent to any structure located on the *property*?

___Yes ___No

9.11.2 *Guide*—Vent or fill pipes often signal the current or previous existence of underground storage tanks. Additionally, observations should be made regarding any asphalt or concrete patching that would indicate the possibility of previous underground storage tank removal. Examples of vent and fill pipes are illustrated in Appendix X2.

STRUCTURE ISSUES

9.12 Is there currently evidence of leaks, spills or staining by substances other than water, or foul odors, associated with any flooring, drains, walls, ceilings or exposed grounds on the property?

___Yes ___No

9.12.1 Did you observe evidence or do you have any prior knowledge that there have been previously any leaks, spills, or staining by substances other than water, or foul odors, associated with any flooring, drains, walls, ceilings or exposed grounds on the property?

___Yes ___No

9.12.2 *Guide*—Stains (other than water stains) or foul odors may indicate leaks of *hazardous substances* or *petroleum products* or contaminants. Floor drains located within a building adjacent to *hazardous substance* storage areas or connected to an on-site disposal system (for example, septic system) present a potential source of subsurface discharge of contaminants.

OTHER ISSUES

9.13 If the *property* is served by a private well or non-public water system, is there evidence or do you have prior knowledge that contaminants have been identified in the well or system that exceed guidelines applicable to the water system?

___Yes ___No

9.13.1 If the property is served by a private well or non-public water system, is there evidence or do you have prior knowledge that the well has been designated as contaminated by any government environmental/health agency?

___Yes ___No

9.13.2 *Guide*—Evidence of well water generally consists of a 4- to 12-in. (102- to 305-mm) diameter low level pipe protruding from the ground that is capped, as illustrated in Appendix X2.

9.14 Does the *property* discharge *wastewater* (not including sanitary waste or storm water) onto or adjacent to the property and/or into a storm water system?

___Yes ___No

9.14.1 Does the *property* discharge *wastewater* (not including sanitary waste or storm water) onto or adjacent to the property and/or into a sanitary sewer system?

___Yes ___No

9.14.2 *Guide*:

9.14.2.1 All drain traps and pipes should be examined and their end points should be determined. Any ditches or streams on or adjacent to the site should be *visually and physically observed* for *wastewater* flow.

9.14.2.2 Some jurisdictions require facilities with large roof or paved areas and construction sites to collect and divert such runoff through a treatment process prior to discharging the stormwater runoff to municipal, separate storm sewer systems, or the waters of the United States. Such units are often called stormwater treatment systems. Oil-water separators are most often found outside a building under a manhole and require routine servicing to remove oil. Oil-water separators are usually in restaurants, repair garages, and service stations. An example of an oil-water separator is shown in Appendix X2.

9.15 Did you observe evidence or do you have any prior knowledge that any *hazardous substances* or *petroleum products*, unidentified waste materials, tires, automotive or industrial batteries, or any other waste materials have been dumped above grade, buried and/or burned, on the *property*?

___Yes ___No

9.15.1 *Guide*—Past waste disposal practices should be examined because these may have resulted in *hazardous substances* being released on the *property*. Does the site evidence any mounds or depressions that suggest a disposal site?

9.16 Is there a transformer, capacitor, or any hydraulic equipment for which there are any records indicating the presence of PCBs?

___Yes ___No

9.16.1 *Guide*:

9.16.1.1 The PCBs are regulated by the Toxic Substances Control Act 15 USC Section 2601 *et seq.* and, in the absence of a release, are not regulated by CERCLA. The provisions of CERCLA do apply if there is a release of PCBs. Accordingly, if an affirmative answer is obtained to this question, the further focus should be on whether there have been any instances of insulating oil leakage and, if so, whether these are suspected of being PCB or PCB-contaminated.

9.16.1.2 Elevators and auto lifts are often operated by hydraulically controlled systems containing PCBs. If inspection or maintenance records for the elevator, capacitor, or other hydraulic equipment indicate no release has occurred and the machinery does not appear to be damaged or leaking, no further inquiry is required.

9.16.1.3 Transformers containing PCBs may have many different sizes and shapes. Some of the more commonly used transformers are set forth on Appendix X2. Transformers are to be registered pursuant to 40 CFR § 761.30.

10. Guide to Government Records/Historical Sources Inquiry

10.1 Do any of the following federal government record systems list the *property* or any *property* within the search distance noted below:

National Priorities List—within 1.0 mile (1.6 Km)?	___Yes	___No
CERCLIS List—within 0.5 mile (0.8 Km)?	___Yes	___No
RCRA CORRACTS Facilities—within 1.0 mile (1.6 Km)?	___Yes	___No
RCRA non-CORRACTS TSD Facilities—within 0.5 mile (0.8 Km)?	___Yes	___No

10.1.1 *Guide*:

10.1.1.1 The NPL or National Priorities List is a list compiled by EPA pursuant to CERCLA 42 USC § 9605(a)(8)(B) of properties with the highest priority for cleanup pursuant to EPA's Hazard Ranking System. See 40 CFR Part 300.

10.1.1.2 The Comprehensive Environmental Response Compensation and Liability Information System (CERCLIS) is the list of sites compiled by EPA that EPA has investigated or is currently investigating for potential hazardous substance contamination for possible inclusion on the national Priorities List.

10.1.1.3 RCRA CORRACTS Facilities are those facilities which treat, store and/or dispose of hazardous wastes on-site and at which corrective remedial action is underway, as defined and regulated by RCRA. The RCRA non-CORRACTS TSD Facilities List are those facilities on which treatment, storage, and/or disposal of hazardous wastes takes place and at which corrective remedial action has not been required by EPA, as defined and regulated by RCRA.

10.1.1.4 If the preparer elects to obtain the records directly from government agencies, those records typically must be obtained through a formal written request to the office within each agency that is responsible for maintaining the records or for responding to public requests for records. At the federal level, these requests are governed by the Freedom of Information Act (FOIA). FOIA requires a written request and the request should identify the records the preparer requires and should identify the site and geographic area for which the preparer needs the records (for example, the address of the site and the appropriate city, county, or zip code to be searched). The request should be directed to the FOIA officer for the regional EPA office responsible for the region in which the site is located. A list of the FOIA offices for each of the EPA regions may be obtained from the federal government or local library. From the federal EPA offices, the *preparer* should anticipate a response no sooner than four to eight weeks.

10.1.1.5 If government information is obtained from a commercial service, the firm should provide ensurances that its records stay current with the government agency record sources. Government information obtained from commercial sources may be considered current if the source updates the information at least every 90 days, or for information that is updated less frequently than quarterly by the government agency, within 90 days of the date the government agency makes the updated information available to the public.

10.1.1.6 The information supplied in response to this question in a prior *environmental site assessment* may be used provided it is updated to the present time.

10.2 Do any of the following state record systems list the *property* or any *property* within the search distance noted below:

List maintained by state environmental agency of hazardous waste sites identified for investigation or remediation that is the state agency equivalent to NPL—within 1.0 mile (1.6 Km)?

___Yes ___No

List maintained by state environmental agency of sites identified for investigation or remediation that is the state equivalent to CERCLIS—within 0.5 mile (0.8 Km)?

___Yes ___No

Leaking Underground Storage Tank (LUST) List—within 0.5 mile (0.8 Km)?

___Yes ___No

Solid Waste/Landfill Facilities—within 0.5 mile (0.8 Km)?

___Yes ___No

10.2.1 *Guide*:

10.2.1.1 The LUST list is a list of sites containing one or more underground storage tanks that have been identified as having leaked or are potentially leaking their contents into the ground or ground water; these sites may be involved in a state cleanup program.

10.2.1.2 The solid waste/landfill facilities list is a list of sites that currently accept, or have accepted in the past, waste of any kind for disposal on site. Solid waste/landfill facilities lists typically are obtained through a state office of solid waste management that is often a division of the primary state environmental agency.

10.2.1.3 Although many states do not have specific Freedom of Information laws, if the preparer elects to obtain the records directly from government agencies, a similar written request for state records should be made to the primary state agency responsible for environmental regulation in that state. Typically, the office responsible for maintaining the records and for responding to requests for records are the same. Once again, the written request should identify the specific records requested and identify the site and geographic area for which the preparer needs the records. The state agency response will vary from state to state and agency to agency, but the *preparer* should anticipate a minimum of four weeks for a response.

10.2.1.4 In some cases, the request should be directed to a specific state office. For example, leaking underground storage tank requests should be made through either the state agency's ground water management division, the state Fire Marshall's office, or the state Emergency Planning and Management Agency.

10.2.1.5 The identity of the state office to which the request should be made can be obtained by contacting the primary state environmental agency. Also, there are publications listing agency sources for each state. The local public library may contain these publications.

10.3 Based upon a review of *fire insurance maps* or consultation with the local fire department serving the property all as specified in the guide, are any buildings or other improvements on the property or on an adjoining property identified as having been used for an industrial use or uses likely to lead to contamination of the property.

___Yes ___No ___Not Applicable

10.3.1 *Guide*:

10.3.1.1 The focus of this research is to determine whether any past use of the *property* would suggest the presence of contamination associated with the *property*. If *reasonably ascertainable*, one of two sources of data should be examined in the following order of preference: *fire insurance maps* showing the *property* or the local fire department serving the property. However, if the user has first-hand knowledge of the use of the *property* from the present back to 1940 or if the *preparer* interviewed disinterested people with such knowledge, then the *preparer* may eliminate this research and answer "not applicable" to the questions above. In addition, the preparer may eliminate this research and answer "not applicable" to the question if the *preparer* is unable to find appropriate sources of *fire insurance maps* or individuals at the local fire department for the property with knowledge of the property's past use, after making a reasonable effort in good faith to locate such information or if the information is otherwise not *reasonably ascertainable*.

10.3.1.2 Subject to the previous paragraph, the *preparer* should obtain *fire insurance maps* from the period(s) not covered by the first-hand knowledge of the *user* or of those interviewed, beginning with when the maps are first available for the area or when the area was first thought to be developed. At least two maps should be ordered at points in time separated by at least ten years.

10.3.1.3 *Fire insurance maps* are defined in 3.2.14 and may be available for review from public libraries, colleges, and local historical societies, or from commercial services.

10.3.1.4 In examining a *fire insurance map*, the *user* is only required to review those areas shown in the given source. For example, if the *property* is at the edge of a map sheet, the user need not order the adjoining sheet. If a source covers a large area, the *user* need only review the area within approximately ⅛ mile (200 m) of the *property*.

10.3.1.5 *Fire insurance maps* reviewed as part of a prior *environmental site assessment* do not need to be searched for or reviewed again, but the *preparer* should make a reasonable effort to determine the uses of the *property* since the last use identified in a prior *environmental site assessment*.

11. Non-Scope Considerations

11.1 *General*:

11.1.1 There may be environmental issues or conditions at a *property* that parties may wish to assess in connection with *commercial real estate* that are outside the scope of this practice (the non-scope considerations). As noted by the legal analysis in Appendix X1 of this practice, some substances may be present on a property in quantities and under conditions that may lead to contamination of the property or of nearby properties but are not included in CERCLA's definition of *hazardous substances* (42 USC 9601(14)) or do not otherwise present potential CERCLA liability. In any case, they are beyond the scope of this practice.

11.1.2 Whether or not a *user* elects to inquire into non-scope considerations in connection with this practice or any other *environmental site assessment*, no assessment of such non-scope considerations is required for appropriate inquiry as defined by this practice.

11.1.3 There may be standards or protocols for assessment of potential hazards and conditions associated with non-scope conditions developed by governmental entities, professional organizations, or other private entities.

11.1.4 Following are several non-scope considerations that persons may want to assess in connection with commercial real estate. No implication is intended as the relative importance of inquiry into such non-scope considerations, and this list of non-scope considerations is not intended to be all-inclusive:

11.1.4.1 Asbestos-containing materials,

11.1.4.2 Radon,

11.1.4.3 Lead-based paint,

11.1.4.4 Lead in drinking water,

11.1.4.5 Wetlands,

11.1.4.6 Regulatory compliance,

11.1.4.7 Cultural and historic resources,

11.1.4.8 Industrial hygiene,

11.1.4.9 Health and safety,

11.1.4.10 Ecological resources,

11.1.4.11 Endangered species,

11.1.4.12 Indoor air quality, and

11.1.4.13 High voltage powerlines.

APPENDIXES

(Nonmandatory Information)

X1. LEGAL BACKGROUND TO FEDERAL LAW AND THE PRACTICES ON ENVIRONMENTAL ASSESSMENTS IN COMMERCIAL REAL ESTATE TRANSACTIONS

INTRODUCTION

The legal section of Subcommittee E.50.02 on Environmental Assessments in Commercial Real Estate Transactions provides the following background to the Comprehensive Environmental Response, Compensation, and Liability Act (CERCLA), as amended including amended by the Superfund Amendments and Reauthorization Act (SARA), 42 USC §§ 9601 *et seq.* The background to CERCLA, commonly known as the Superfund Law, outlines the potential liability for the cleanup of hazardous substances, available defenses to such liability, appropriate inquiry under Superfund, statutory definition of hazardous substances, petroleum products and petroleum exclusion to CERCLA, and reasons why certain environmental hazards are excluded from the scope of Superfund and this practice and Practice E 1527.

There are several elements of Superfund liability and the commonly termed "innocent purchaser" defense, that arises out of the statutory third-party defense, that may impact on the development and understanding of this practice and Practice E 1527.

X1.1 *Superfund Liability:*

X1.1.1 All of the following elements of liability under Superfund must be established by a plaintiff before a defendant will be held liable under Superfund for a *government's* response costs:[8]

X1.1.1.1 The site is a facility, as defined as § 9601(9),

X1.1.1.2 A release or threatened release of a hazardous substance from the site occurred (release is defined at § 9601(22) as *any* amount of any hazardous substance; "hazardous substance" is defined at § 9601(14) (see statutory definition of hazardous substance below),

X1.1.1.3 A release or threatened release caused the plaintiff to incur response costs. Response costs are defined at § 9601(25) to mean costs related to both removal actions (§ 9601(23)) and remedial actions (§ 9601(24)), and

X1.1.1.4 Defendants fall within at least one of the four classes of responsible parties.[9]

X1.1.2 In order to recover response costs, a government plaintiff must prove that the costs were not inconsistent with the National Oil and Hazardous Substances Pollution Contingency Plan (commonly referred to as the National Contingency Plan or NCP), 40 CFR § 300.[10] A private plaintiff must prove its costs were necessary costs of response and that the response action was consistent with the NCP 42 USC§ 9607 (a).[11]

X1.1.3 If there is a release or threatened release of hazardous substances on a site, private parties, even if they are not PRPs, may decide to incur response costs and seek recovery from other private parties. The PRPs may seek contribution from other PRPs.

X1.1.4 There is an important difference between the government's burden to show that its response costs are "not inconsistent with the NCP" and the burden a private party bears to show that its response costs are "consistent with the NCP."

[8] USC § 9607(a). (All statutory references are to Title 42 of the United States Code, unless otherwise specified.) See *United States versus Aceto Agricultural Chemicals Corp.*, 872 F.2d 1373 (8th Cir. 1989). Private plaintiffs, as well as the government, may seek response costs under Superfund from defendants. While many users of these practices or other private parties may think in terms of how to defend against Superfund liability, they should recognize that they may decide to conduct cleanup actions and seek response costs from other parties.

[9] The four classes of potentially responsible parties (PRPs) are listed at § 9607(a) as follows:

(1) Owner and operator of a facility (see § 9601 (20)(A): the term "owner or operator" does not include a person, who, without participating in the management of a facility, holds indicia of ownership primarily to protect his security interest in the facility. In *re: Bergsoe Metal Corporation*, 910 F.2d 668 (9th Cir. Aug. 9, 1990); *Guidice versus BFG Electroplating and Manufacturing Co.*, 732 F. Supp. 556 (W.D. Pa. 1989); *United States versus Mirabile*, 23 ERC 1511 (E.D. Pa. 1985). But see *United States versus Fleet Factors Corp.*, 901 F.2d 1550 (11th Cir. 1990); *United States versus Maryland Bank and Trust Co.*, 632 F. Supp. 573 (D. Md. 1986). For clarification of the security interest exclusion, see EPA's rule on lender liability under CERCLA, 57 Federal Register 18344 (April 29, 1992);

(2) Any person who at the time of disposal of any hazardous substance owned or operated any facility at which such hazardous substances were disposed of;

(3) Any person who by contract, agreement, or otherwise arranged for disposal or treatment or transport of hazardous substances; and

(4) Any person who accepts hazardous substances for transport to a facility selected by such person.

[10] The National Contingency Plan is the federal government's blueprint on how hazardous substances are to be cleaned up pursuant to CERCLA.

[11] See *Dedham Water Co. versus Cumberland Farms Dairy, Inc.*, 889 F.2d 1146 (1st Cir. 1989); other cases cited at ABA, *National Resources, Energy, and Environmental Law: 1989 The Year In Review*, p. 215, Note 155.

See § 9607 (a)(4)(A) and (B). Courts have interpreted this statutory difference to give the government a rebuttable presumption that its response costs are consistent with the NCP, whereas a private party who undertakes response costs and seeks recovery from responsible parties bears the burden of proving its response was consistent with the NCP.[12] The EPA takes the position that a private party who undertakes a response action must be only in "substantial compliance," rather than strict technical compliance, with the NCP, as long as a CERCLA-quality cleanup is achieved. The NCP requirements for a private party response-action are set forth at 40 CFR § 300.700.

X1.2 *Defenses to Liability:*

X1.2.1 Assuming all the elements of liability exist, a party may avoid liability only by meeting one of the defenses listed in§ 9607(b). These listed defenses are exclusive of all others.[13] Section 9607(b) states (*emphasis added*):
"There shall be no liability under subsection (a) for a person otherwise liable who can establish by a preponderance of the evidence (the lowest evidentiary standard available, meaning more probable than not) that the release or threat of release of a hazardous substance and the damages resulting therefrom were *caused solely by*—
1) an act of God;
2) an act of war;
3) *an act or omission of a third party other* than an employee or agent of the defendant, or *than one whose act or omission occurs in connection with a contractual relationship* (see the definition of "contractual relationship" below), existing directly or indirectly, with the defendant ..., if the defendant establishes by a preponderance of the evidence that (a) he exercised due care with respect to the hazardous substance concerned ..., and (b) he took precautions against foreseeable acts or omissions of any such third party and the consequences that could foreseeably result from such actions or omissions."

X1.2.2 Under § 9601(35)(A), a contractual relationship "includes, but is not limited to, land contracts, deeds, or other instruments transferring title or possession ...". These contractual relationships with third parties eliminate the defense to liability unless the defendant is an innocent purchaser. Or as stated by the status of § 9601(35)(A) (*emphasis added*), a contractual relationship with the third party defeats the defense "unless the real property on which the facility is located was acquired by the defendant after disposal or placement of the hazardous substance ... and one or more of the following circumstances is also established by the defendant by a preponderance of the evidence:
(i) At the time the defendant acquired the facility the defendant did not know and had no reason to know that any hazardous substance which is the subject of the release or threatened release was disposed of on, in, or at the facility.
(ii) The defendant is the government ...
(iii) The defendant acquired the facility by inheritance or bequest."

X1.2.3 Therefore, the so-called innocent purchaser defense arises out of the third party defense of § 9607(b)(3). Restated, this defense to Superfund liability is available only if the defendant shows the following:

X1.2.3.1 The release or threat of release was caused solely by a third party,

X1.2.3.2 The third part is not an employee or agent of the defendant,

X1.2.3.3 The acts or omissions of the third party did not occur in connection with a direct or indirect contractual relationship to the defendant, or if there was a contractual relationship, the defendant acquired the property after disposal or placement of the hazardous substance, and at the time the defendant acquired the facility the defendant *did not know and had no reason to know* that any hazardous substance which is the subject of the release or threatened release was disposed of on, in, or at the facility, and

X1.2.3.4 The defendant exercised due care with respect to the hazardous substances and took precautions against foreseeable acts or omissions of the third party.

X1.2.4 The statute then states at § 9601(35)(B)(*emphasis added*):
"To establish that the defendant had no reason to know, as provided (above), *the defendant must have undertaken, at the time of acquisition, all appropriate inquiry into the previous ownership and uses of the property consistent with good commercial or customary practice* in an effort to minimize liability... . [T]he court shall take into account any specialized knowledge or experience on the part of the defendant, the relationship of the purchase price to the value of the property if uncontaminated, commonly known or reasonably ascertainable information about the property, the obviousness of the presence or likely presence of contamination at the property, and the ability to detect such contamination by appropriate inspection."

X1.3 *Appropriate Inquiry in Commercial Real Estate Transactions:*

X1.3.1 One of the major questions that parties to commercial real estate transactions face when considering their potential Superfund liability is, "What level of inquiry into the previous ownership and uses of the property is appropriate to establish the innocent purchaser defense to Superfund liability?" These practices are structured to articulate the level of inquiry under Superfund that is appropriate for different situations.

X1.3.2 *The Appropriate Level of Inquiry*:

X1.3.2.1 The level of environmental inquiry that is appropriate under Superfund cannot be the same for every property or every party to a real estate transaction. The level of inquiry, in fact, will change depending on the particular property or party involved in a transaction. The statutory language, Congressional history, and common sense support this conclusion.

X1.3.2.2 First, it must be noted that little case law exists to

[12] *Amland Properties Corp. versus Aluminum Co. of America*, 711 F. Supp. 784, 794 (D. N.J. 1989); *Artesian Water Co. versus New Castle County*, 659 F. Supp. 1269, 1291 (D. Del. 1987); *United States versus Northeastern Pharmaceutical Chemical Co.*, 579 F. Supp. 823 (W.D. Mo. 1984), *aff'd in part, rev'd on other grounds*, 810 F.2d 726 (8th Cir. 1986).

[13] *United States versus Aceto Agricultural Chemicals Corp.*, 872 F.2d 1373 (8th Cir. 1989). But see *United States versus Marisol, Inc.*, 725 F. Supp. 833 (M.D. Pa. 1989) (equitable defenses under CERCLA may be available after the development of a factual record).

serve as guidance about the minimum level of inquiry that will be deemed appropriate for the innocent purchaser defense. See, for example, *United States versus Serafini*, 706 F. Supp. 346 (M.D. Pa. 1988), and 1990 U.S. Dist. LEXIS 18466 (M.D. Pa. 1990) (By entertaining disputed facts as to the custom and practice of viewing land prior to purchase, the court implied that appropriate inquiry necessarily varies on a site-by-site basis); *United States versus Pacific Hide and Fur Depot, Inc.*, 716 F. Supp. 1341 (D. Idaho 1989) (No inquiry was required by those who received an ownership interest in property by means of corporate stock transfer and warranty deed under the facts of this case.); *International Clinical Laboratories, Inc. versus Stevens*, 30 ERC 2066, 20 ELR 20,560 (E.D.N.Y. 1990) (Despite a long history of toxic wastewater disposal and presence of the site on the state's hazardous waste disposal site list, the purchaser established the innocent purchaser defense since there were no visible environmental problems at the site, the defendant had no knowledge of environmental problems at the site, and the purchase price did not reflect a reduction on account of the problem.)

X1.3.2.3 While the statute does not specifically distinguish certain types of properties and uses from others, or certain types of parties from others, it does list certain factors courts should consider in determining whether one's inquiry under the circumstances is appropriate. The statute, as explained above in X1.2 requires a court to consider a party's specialized knowledge or experience. The statute further mandates a court to consider what is "*reasonably* ascertainable information about the property," what contamination is *obviously* present, and the *party's "ability* to detect such contamination". The very use of terms such as" appropriate" and "reasonably", and the use of "specialized knowledge and experience" and "ability" in conjunction with the specific person attempting to utilize the defense signifies that Congress did not intend the appropriateness of the inquiry be judged by a bright line standard. If it is so intended, Congress would have stated, but did not, that the same inquiry should be made in every case.

X1.3.2.4 What is reasonable and obvious to one party may not be so to other parties, and ability, by necessity, varies among all parties. The statute, therefore, recognizes that different properties and parties must be treated differently. That is, different parties may conduct different levels of inquiry appropriate to their circumstances.

X1.3.2.5 The statutory standard of "appropriate inquiry" suggests the level of inquiry will depend on the circumstances and the underlying facts. Since the facts are almost always different, the level of inquiry must change with them. The legislative history on this particular issue demonstrates that Congress intended that the level of inquiry changes with the type of property and party.[14]

The duty to inquire under this provision shall be judged as of the time of acquisition. Defendants shall be held to a higher standard as public awareness of the hazards associated with hazardous releases has grown, as reflected by this Act, the 1980 Act [CERCLA] and other Federal and State statutes.

Moreover, good commercial or customary practice with respect to inquiry in an effort to minimize liability shall mean that a reasonable inquiry must have been made in all circumstances, in light of best business and land transfer principles.

Those engaged in commercial transactions should, however, be held to a higher standard than those who are engaged in private residential transactions.

X1.3.2.6 Because few cases address the standard of inquiry in the innocent purchaser defense and the legislative history describing Congressional intent is sparse, common sense is a useful guide in interpreting statutory language. If CERCLA mandated that the level of inquiry be the same for every property or potential defendant, then a lay consumer (renter or buyer) of a home, a purchaser of a small environmentally benign business, and a multinational corporate buyer of an industrial complex would have to conduct the same environmental site assessment (ESA) of the different properties in question. Additionally, the statute makes no mention of a Phase I ESA or any other specific type of inquiry one is to conduct in order for the inquiry to be deemed appropriate. If all inquiries had to be at the same level to be "appropriate," it would be illogical to stop at a Phase I ESA since some commercial or industrial properties routinely undergo, in the exercise of good commercial and customary practices, intrusive sampling (typically a Phase II ESA activity). Therefore, since routinely some properties undergo sampling, an inflexible standard would require sampling of all properties no matter what its use. This could not have been the intended result of SARA.

X1.3.3 *The Minimum Inquiries to Satisfy All Appropriate Inquiry*:

X1.3.3.1 Recognizing that inquiry changes with the underlying circumstances, the next question concerns that level of inquiry, if any, that Superfund requires to utilize the innocent landowner defense.

X1.3.3.2 As noted above, in some real estate transactions a Phase II ESA is routinely conducted. A Phase I ESA is conducted in these transactions only as a necessary prerequisite to outline the scope of the Phase II ESA. A Phase II ESA typically involves taking soil, water, and air samples to determine their contaminant content or verify that no contaminants are present or likely present. Note, however, that this simplistic outline of the Phase II ESA is misleading since the party can always dig down one foot deeper, take one more sample, or conduct one more test. The problem of how much inquiry is conducted, or at what level a party should begin, involves proving a negative—that is, that no contamination is present.[15] Since, according to the statute, inquiries should be

[14] H.R. Rep. No. 962, 99th Cong., 2d Sess. 187 (1986), *reprint at* 1986 U.S. Code Cong. and Admin. News 3276, 3280.

[15] The inability to prove a negative creates a dilemma for the potential defendant. If the party's inquiry discovers contamination, then under the statute the party will not be able to avail itself of the innocent purchaser defense. If the inquiry does not discover contamination, EPA or another private party can argue in a response action that the inquiry was not "appropriate" and, therefore, the defendant can have no defense. The dilemma is explicitly recognized by Subcommittee E50.02 as beyond any reasonable interpretation of Congressional intent. The scope of these practices resolves the party's dilemma in the only reasonable way by stating: "It should not be concluded or assumed that the inquiry was not appropriate inquiry merely because the inquiry did not identify existing recognized environmental conditions in connection with a property. Environmental site assessments must be evaluated based on the reasonableness of the judgments made at the time and under the circumstances in which they were made". See 4.5.4.

judged by the circumstances existing at the time of acquisition, then there could be some properties and parties to real estate transactions where it may be appropriate to begin the inquiry with an intrusive Phase II ESA in order to invoke the innocent purchaser defense to liability.

X1.3.3.3 At the other extreme, the minimum level of inquiry that a party would be expected to conduct is found by looking at the least environmentally obtrusive class of property and party from a CERCLA perspective. This transaction likely involves the lay buyer of a home or the renter of an apartment. Assuming these parties meet the other prerequisites for the innocent purchaser defense, what level of environmental inquiry must they conduct to avoid Superfund liability? While there are no recorded court cases on this issue, the answer is probably none, unless a particular residential purchaser or renter has some specialized knowledge about or experience with the property in question that would lead a court to conclude that some questions should have been asked. Beyond these rare situations, it is highly unlikely that Congress intended to saddle housing consumers with the burden of investigating or cleaning up contaminated sites. In fact, EPA has issued a statement of enforcement policy to the effect that it will not generally pursue owners of single family residences pursuant to CERCLA.[16] Therefore, for some properties and parties to real estate transactions, it is appropriate to conduct no environmental inquiry in order to meet the innocent purchaser defense to liability.

X1.3.3.4 The minimum level of appropriate inquiry under Superfund, therefore, ranges from no specialized inquiry to conducting an intrusive Phase II ESA. In order to satisfy the practices, to do no specialized inquiry, such as the transaction screen or Phase I ESA, is not enough for commercial real estate transactions. Under current commercial and customary practice and in light of best business and land transfer principles, however, no environmental site assessments are conducted in many real estate transactions, particularly those involving smaller properties, vacant land, or transactions of low monetary value. This practice and Practice E 1527 and the minimum level of inquiry under these practices, actually raises the average level of inquiry that should be performed where the parties want to come within the protection of the innocent landowner defense.

X1.3.3.5 The burden of proof is on the defendant to sustain by a preponderance of the evidence, the innocent purchaser defense. This is the least onerous burden of proof available to a party in litigation. The defendant must show only that the evidence offered to support the level of inquiry that was taken at the time of acquisition is of greater weight or more convincing than the evidence offered in opposition to it. In other words, the evidence on the inquiry issue taken as a whole shows that the fact sought to be proved is more probable than not. There may be technical or business judgments on whether the inquiry conducted or any other fact in a particular case is sufficient to meet the needs or concerns of a party to the real estate transaction. The bottom line, however, is that the judgment on whether the specific facts of a case, in light of statutory language, are sufficient to produce liability or a viable defense to liability is a legal one and such judgments constitute the practice of law.

X1.3.3.6 This practice is designed as the minimum level of inquiry to satisfy the practice from which a party to a commercial real estate transaction should proceed, recognizing that some parties to some commercial real estate transactions may wish to proceed by beginning with a Phase I or a Phase II ESA.

X1.4 *Statutory Definition of Hazardous Substance:*

X1.4.1 The statute at 42 USC § 9601(14)(A–F) defines hazardous substance by referring to five other statutes as well as to Superfund's own § 9602. The following is a description of the relevant portions of the other statutes and § 9602 of Superfund:

42 U.S.C. § 9601(14)(A): "[A]ny substance designated pursuant to section 1321(b)(2)(A) of Title 33." Title 33 USC § 1321 lies within the Clean Water Act and refers to, among other things, hazardous substance liability. 33 USC§ 1321(b)(2)(A) states that the EPA shall develop, "as may be appropriate, regulations designating as hazardous substances, other than oil as defined in this section, such elements and compounds which, when discharged in any quantity into" the navigable waters of the United States ..., present an imminent and substantial danger to the public ... health or welfare, including, but not limited to, fish, shellfish, wildlife, shorelines, and beaches."

42 U.S.C.§ 9601(14)(B): "[A]ny element, compound, mixture, solution, or substance designated pursuant to section 9602 of this title." Section 9602 gives EPA the authority to designate as a hazardous substance "such elements, compounds, mixtures, solutions, and substances which, when released into the environment may present substantial danger to the public health or welfare or the environment ..."

42 U.S.C.§ 9601(14)(C): "[A]ny hazardous waste having the characteristics identified under or listed pursuant to section 3001 of the Solid Waste Disposal Act [42 USCA § 6921](but not including any waste the regulation of which under the Solid Waste Disposal Act [42 USCA § 6901 et seq.] has been suspended by Act of Congress)." The Solid Waste Disposal Act of 1980 amended the Resource Conservation and Recovery Act (RCRA). 42 USCA § 6921 of RCRA provides authority to the EPA to develop criteria for identifying characteristics of hazardous waste and for listing particular hazardous wastes within the meaning of 42 USCA § 6903(5) of RCRA. RCRA, § 6903(5), defines hazardous waste to mean

"a solid waste, or combination of solid wastes, which because of its quantity, concentration, or physical, chemical, or infectious characteristics may—

(A) cause, or significantly contribute to an increase in mortality or an increase in serious irreversible, or incapacitating reversible, illness; or

(B) pose a substantial present or potential hazard to human health or the environment when improperly treated, stored, transported, or disposed of, or otherwise managed."

For the identification and listing of hazardous wastes under RCRA, see 40 CFR. §§ 261.1 *et seq.*

[16] EPA, *Policy Towards Owners of Residential Property at Superfund Sites*, OSWER Directive No. 9834.6, July 3, 1991.

42 USC § 9601(14)(D): "[A]ny toxic pollutant listed under section 1317(a) of Title 33." Section 1317(a) of Title 33 refers to toxic and pretreatment effluent standards under the Clean Water Act. The EPA is charged in this section with publishing and revising from time to time a list of toxic pollutants, taking "into account toxicity of the pollutant, its persistence, degradability, the usual or potential presence of the affected organisms in any waters, the importance of the affected organisms, and the nature and extent of the effect of the toxic pollutant on such organisms." Each toxic pollutant listed according to this section shall be subject to effluent limitations. For toxic pollutant effluent standards, see 40 CFR§§ 129.1 *et seq.*

42 USC § 9601(14)(E): "[A]ny hazardous air pollutant listed under section 112 of the Clean Air Act [42 USCA§ 7412]." Section 7412 of the Title 42 deals with national emission standards for hazardous air pollutants. The EPA is charged here with publishing and revising from time to time "a list which includes each hazardous air pollutant for which [it] intends to establish an emission standard under this section." The term "hazardous air pollutant" means an air pollutant that in EPA's judgment "causes, or contributes to, air pollution which may reasonably be anticipated to result in an increase in mortality or an increase in serious irreversible, or incapacitating reversible, illness." For emission standards for hazardous pollutants, see 40 CFR§ 61.01 *et seq.*

42 USC § 9601(14)(F):"[A]ny imminently hazardous chemical substance or mixture with respect to which the [EPA] has taken action pursuant to section 2606 of Title 15." Section 2606 of Title 15 deals with imminent hazards under the Toxic Substances Control Act (TSCA). The EPA is authorized under 15 USC § 2606 to seize an imminently hazardous chemical substance or mixture or seek other relief, such as requiring notice to users of the chemical substance or public notice of the risk associated with the substance or mixture. The term "'imminently hazardous chemical substance or mixture' means a chemical substance or mixture which presents an imminent and unreasonable risk of serious or widespread injury to health or the environment."

X1.4.2 After Subsections A through F, the Superfund definition of "hazardous substance" in § 9601(14) then goes on to state:

"The term does not include petroleum, including crude oil or any fraction thereof which is not otherwise specifically listed or designated as a hazardous substance under Subparagraphs (A) through (F) of this paragraph, and the term does not include natural gas, natural gas liquids, liquefied natural gas, or synthetic gas usable for fuel (or mixtures of natural gas and such synthetic gas)."

X1.4.3 The EPA has collected a list of "those substances in the statutes referred to in section 101(14) of the Act [42 USC § 9601(14)]." 40 CFR § 302.1 (1989) ("List of Hazardous Substances and Reportable Quantities," 40 CFR Part 302). This list changes with notices in the Federal Register. Also, any time a new hazardous waste is listed, the waste automatically becomes a hazardous substance.

X1.5 *Petroleum Products:*

X1.5.1 Under the petroleum exclusion of CERCLA (42 USC§ 9601(14)), petroleum and crude oil have been explicitly excluded from the definition of hazardous substances under CERCLA. Nevertheless, petroleum products are included within the scope of both practices because such they are of concern in many commercial real estate transactions and current custom and usage is to include an inquiry into the presence of petroleum products in an environmental site assessment. Inclusion of petroleum products within the scope of the practices is not based upon the applicability, if any, of CERCLA to petroleum products.

X1.5.2 One reason to include petroleum products within the scope of the practice is because to do so reflects custom and usage: when environmental assessments are conducted in connection with commercial real estate transactions, they customarily include an assessment of the presence or likely presence of petroleum products under conditions that may lead to contamination. For example, environmental assessments ordinarily seek to assess whether there may be underground or above-ground storage tanks that may be leaking, whether those tanks contain petroleum products or some other product.

X1.5.3 In addition, although CERCLA may exclude petroleum products, other laws require cleanup of releases or spills of petroleum products. For example, petroleum products sometimes (for example, when they cannot be reclaimed from soil) become hazardous wastes subject to RCRA Subtitle C (42 USC § 6921 *et seq.*), must be cleaned up if released from underground storage tanks pursuant to RCRA Subtitle I (42 USC § 6991 *et seq.*), must be cleaned up pursuant to the Oil Pollution Act of 1990 (33 USC § 1321 *et seq.*), and must be cleaned up if released into the navigable waters of the United States pursuant to the Clean Water Act (33 USC § 1251 *et seq.*).

X1.5.4 Moreover, case law and EPA interpretations of the petroleum exclusion require an analysis of the facts of each case to determine whether a particular petroleum product is included in CERCLA's petroleum exclusion. The exclusion has been broadly interpreted to exclude gasoline and leaded gasoline from CERCLA's definition of hazardous substances regardless of the fact that gasoline and leaded gasoline contain certain indigenous components and additives which have themselves been designated as hazardous pursuant to CERCLA. See *Wilshire Westwood Associates versus Atlantic Richfield Corporation*, 881 F.2d 801 (9th Cir. 1989). The interpretation was narrowed when a judicial distinction was made between petroleum fractions produced by distillation processes and waste products resulting from contaminated tank scale. *United States versus Western Processing Co.*, 761 F. Supp. 713 (W.D. Wash. 1991). Another decision narrowly interpreted CERCLA's petroleum exclusion to be inapplicable to oil-related wastes containing hazardous substances because the primary purpose of the exclusion is to remove "spills or other releases strictly of oil" from the scope of CERCLA response and liability (not releases of hazardous substances mixed with oil). See *City of New York versus Exxon Corporation*, 744 F. Supp. 474 (S.D. N.Y. 1990). For additional discussion, see EPA memorandum entitled "The Petroleum Exclusion under the Comprehensive Environmental Response Compensation and Liability Act," issued by EPA's General Counsel, July 31, 1987.

X1.6 *Exclusion of Certain Hazards from Superfund:*

X1.6.1 The information that follows is provided to explain why these potential environmental hazards are not covered by Superfund's appropriate inquiry responsibilities.

X1.6.2 As a preliminary matter, it should be noted that an environmental site assessment that does not address substances excluded from CERCLA (whether those substance are excluded because they are petroleum products or by virtue of other characteristics) but that otherwise constitutes" all appropriate inquiry into the previous ownership and uses of the property consistent with good commercial or customary practice" should nevertheless entitle the user to the innocent purchaser defense, assuming that other requirements of the defense are met.

X1.6.3 *Radon*:

X1.6.3.1 A case discussing Superfund and radon is *Amoco Oil Company versus Borden, Inc.* This case dealt with a private cost recovery action by the buyer of a site against the seller for response costs relating to radiation from phosphogypsum wastes left on the site. Radon emanated from these radioactive wastes. The case points out that the "EPA has designated radionuclides as hazardous substances under§ 9602(a) of CERCLA.... Additionally, the ... EPA under§ 112 of the Clean Air Act ... list radionuclides as a hazardous air pollutant." "Radon and its daughter products are considered radionuclides, which are defined as 'any nuclide that emits radiation.'" Therefore, radon is a CERCLA hazardous substance. Also, when discussing what constitutes a release of a hazardous substance under the statute, the statute is plain that there is no quantitative requirement and that a release, broadly defined at 42 USC § 9601(22), of *any* amount constitutes a CERCLA release.

X1.6.3.2 Liability under Superfund depends on several factors, as noted in X1.1. Only one of four factors is the release or threatened release of a hazardous substance. The other three factors are the site is a facility, the defendant falls within at least one of four classes of potentially responsible parties (PRPs), and the release or threatened release *caused* the plaintiff (which can be the government or another private party) to incur response costs. Further, response costs must not be inconsistent with the National Contingency Plan (NCP), *and must not be limited by § 9604(a)(3)*. And, of course, there is no need to raise the innocent purchaser defense and its appropriate inquiry requirements unless the elements of liability will be met.

X1.6.3.3 Where radon from any source occurs in a building, three of the liability elements under CERCLA are met. There is a release of a hazardous substance, the building is a facility, and we can assume the defendant is a PRP. However, under 42 USC § 9604(a)(3)(A), "[r]emedial actions taken in response to hazardous substances as they occur naturally are specifically excluded from the NCP and are therefore not recoverable." A case discussing superfund and radon is *Amoco Oil Company versus Borden, Inc.*, 889 F.2d 664 (5th Cir. 1989). The statute is plain (42 USC § 9604(a)(3) and (4)(*emphasis added*):

"(3) Limitations on response

The President shall not provide for a removal or remedial action under this section in response to a release or threat of release—

(A) *of a naturally occurring substance in its unaltered form, or altered solely through naturally occurring processes or phenomena, from a location where it is naturally found;*

(B) from products which are part of the structure of, and result in exposure within, residential buildings or business or community structures;[17] or

(C) into public or private drinking water supplies due to deterioration of the system through ordinary use.[18]

(4) Exception to limitations

Notwithstanding Paragraph (3) of this subsection, to the extent authorized by this section, the President may respond to any release or threat of release if in the President's direction, it constitutes a public health or environmental emergency and no other person with the authority and capability to respond to the emergency will do so in a timely manner."

X1.6.3.4 Therefore, no liability under CERCLA attaches for naturally occurring radon. If a party to a real estate transaction wants to look for radon within a building, no amount of radon investigation will have any bearing on one's innocent purchaser defense under Superfund. Investigation of naturally occurring radon would be included, if at all, in the portion of the practice that deals with non-scope issues.

X1.6.4 *Asbestos*:

X1.6.4.1 The analysis of asbestos is similar to that involving radon. Before considering appropriate inquiry responsibilities, the four elements of CERCLA liability must be satisfied. Once again, as with radon, they are not met.

X1.6.4.2 Section 9604(a)(3)(B) of CERCLA prohibits response actions involving a release or threat of release "from products which are part of the structure of, and result in exposure within, residential buildings or business or community structures." There are a number of cases dealing with asbestos that interpret this statutory language. One such case is *First United Methodist Church of Hyattsville versus United States Gypsum Co.*,[19] that cites to other relevant cases.

X1.6.4.3 In *First United* the church brought a private cost recovery action against the manufacturer of asbestos-containing acoustical plaster. In holding that the action was barred by a state statute of repose (a certain time allowed by statute for bringing litigation) and that CERCLA did not preempt the state statute of repose, the court stated that § 9604(a)(3)(B) of CERCLA "represents much more than a procedural limitation on the President's authority; it is a substantive limitation of the breadth of CERCLA itself.[20] Therefore, the limitations of § 9604(a)(3) apply to private parties as well.

X1.6.4.4 Citing to the legislative history, the *First United* court concluded, "[i]n view of this clear expression of Congressional intent, we will not expand CERCLA to encompass asbestos-removal actions." The court further explained:[20]

[17] This provision has implications for asbestos and lead-based paint. See X1.6.4.

[18] This provision has implications for lead from lead pipes and solder. See X1.6.5.

[19] One such case is *First United Methodist Church of Hyattsville versus United States Gypsum Co.*, 882 F.2d 862 (4th Cir. 1989), that cites to other relevant cases.

[20] See also *3550 Stevens Creek Associates versus Barclays Bank of California*, 915 F.2d 1355 (9th Cir. 1990).

"In closing, we note that this interpretation of CERCLA fully comports with the most fundamental guide to statutory construction—common sense. To extend CERCLA's strict liability scheme to all past and present owners of buildings containing asbestos as well as to all persons who manufactured, transported, and installed asbestos products into buildings, would be to shift literally billions of dollars of removal cost liability based on nothing more than an improvident interpretation of a statute that Congress never intended to apply in this context. [FN12] ... Certainly, if Congress had intended for CERCLA to address the monumental asbestos problem, it would have said so more directly when it passed SARA... .

FN12—It is for this reason, that Congress simply did not intend for CERCLA to remedy the asbestos-removal problem, that we decline to follow the reasoning of *Prudential, Knox* and *Covalt* in rejecting First United's preemption argument. Instead of recognizing the fact that CERCLA is out of context in this situation, these courts rejected similar attempts to invoke the statute by construing CERCLA's key terms in a way to exclude asbestos-removal actions. *Covalt*, 860 F.2d [1434] at 1438-39 (defining "environment" to exclude the interior of a workplace); *Knox*, 690 F. Supp at 756-57 (defining "release" in terms of "spills" or "disposal"); *Prudential*, [711 F. Supp 1244] at 1254-55 (defining "disposal" to exclude the sale of a product for consumer use). We find this analysis unsatisfactory because it runs the risk of unnecessarily restricting the scope of CERCLA merely to dispose of claims that the statute was never intended to encompass in the first place. It is far better to simply acknowledge the inapplicability of CERCLA to asbestos-removal claims than to restrict its operative terms."

X1.6.4.5 Since asbestos that is a part of the structure of, and results in exposure within, residential buildings or business or community structures is excluded from CERCLA liability, it should not be investigated pursuant to a party's innocent purchaser appropriate inquiry requirements. Like naturally occurring radon, investigation of asbestos-containing materials that are part of the structure of buildings should be included, if at all, in the portion of the practice that deals with non-scope issues. Note, however, if asbestos is disposed of on a site and, therefore, is no longer part of the structure of a building, the cleanup of the disposed asbestos is subject to Superfund response actions. Likewise, if a building is sold with the knowledge that it will be demolished, one court ruled that the sale constitutes a disposal falling under CERCLA's liability provisions.[21]

X1.6.5 *Lead in Drinking Water and Lead-Based Paint*—These hazards can be evaluated in terms of the exclusions of 42 USC § 9604(a)(3)(B) and (C), in an analysis similar to the analysis applied above to radon and asbestos. While there is no reported case law on these environmental issues as they relate to Superfund, the statutory language seems clear that these environmental hazards are not encompassed by Superfund's appropriate inquiry responsibilities. Note, however, like asbestos, where there is a disposal of these substances on the site or in a facility, CERCLA liability may arise.

[21] *CP Holdings, Inc. versus Goldberg-Zoino and Associates, Inc.*, 769 F. Supp. 432 (D.N.H. 1991).

X2. SUPPLEMENTAL INFORMATION FOR USE IN CONNECTION WITH THE GUIDE ON ENVIRONMENTAL SITE ASSESSMENTS OF COMMERCIAL REAL ESTATE

FIG. X2.1 Chemical Storage in 55-gal (208-L) Steel Drums

FIG. X2.2 Chemical Storage in 55-gal (208-L) Plastic Drums

FIG. X2.3 Typical Pole-Mounted Transformer

NOTE 1—Oil-water separators are often located under manholes outside repair garages, or at any location where it is necessary to separate oil from water prior to discharge.

FIG. X2.4 Manhole Cover Outside Repair Garage

NOTE 1—Floor drains come in various shapes and sizes. Shown here is one type of floor drain. It is important to know the point of discharge of any floor drain.

FIG. X2.5 Example of Floor Drain

NOTE 1—Floor drains come in various shapes and sizes. Shown here is one type of floor drain. It is important to know the point of discharge of any floor drain.

FIG. X2.6 Example of Floor Drain

FIG. X2.7 Two Fill and Vent Pipes Leading to Two Underground Storage Tanks

FIG. X2.8 Single Tall Vent Pipe (Arrow) for Underground Storage Tank on Side of Building

NOTE 1—Approximately 2½-in. (64-mm) diameter with screw cap.

FIG. X2.9 Fill Pipe for Residential Underground Fuel Oil Storage Tank

NOTE 1—Approximately 8-in. (203-mm) diameter.

FIG. X2.10 Water Supply Well for Residential Property

NOTE 1—Approximately 8-in. (203-mm) diameter

FIG. X2.11 Water Supply Well for Residential Property

FIG. X2.12 Surface Staining from Improper Drum Storage

Designation: E 1599 – 94

Standard Guide for Corrective Action for Petroleum Releases[1]

This standard is issued under the fixed designation E 1599; the number immediately following the designation indicates the year of original adoption or, in the case of revision, the year of last revision. A number in parentheses indicates the year of last reapproval. A superscript epsilon (ε) indicates an editorial change since the last revision or reapproval.

1. Scope

1.1 This guide covers corrective action for petroleum releases. It describes the approach for assessment and remediation of releases to protect human health, safety, and the environment. It is intended to complement but not supersede federal, state, and local regulations, as well as complement other recommended practices on this subject (for example, NFPA 329 and API 1628).

1.2 The approach described in this guide is not the only way that a corrective action could be conducted, but experience has shown that following these guidelines will help ensure cost effective and timely remediation.

1.3 This guide is not intended to address field and site specific contractor health and safety issues. For guidance concerning contractor health and safety issues appropriate OSHA and other industry standards should be consulted. This guide does not address specific details of sample preparation or preservation or sampling quality assurance/quality control practices. For guidance concerning sampling practices see Appendix X1.

1.4 As shown in Fig. 1, assessment and remedial activities occur at many points in the corrective action process. Each round of assessment and remediation may result in additional steps until the corrective action goal has been achieved. The precise sequence and timing of these activities will depend on the site and the techniques that are used. However, the assessment and remedial activities shown in Fig. 1 may be conducted concurrently.

1.5 Once sufficient information has been gathered, remedial action can begin prior to defining the full extent of contamination. In many cases, an interim remedial action may be appropriate when contaminants are mobile. The ultimate effectiveness and the cost of remediation are often related to the migration of the contamination. Timely action will improve the effectiveness of the remediation and minimize its cost.

1.6 Regulators, consultants, contractors, owners, operators, insurance companies, and the public all need to have good communication throughout the corrective action process. Some of the forms that this communication can take are:

1.6.1 Site visits,

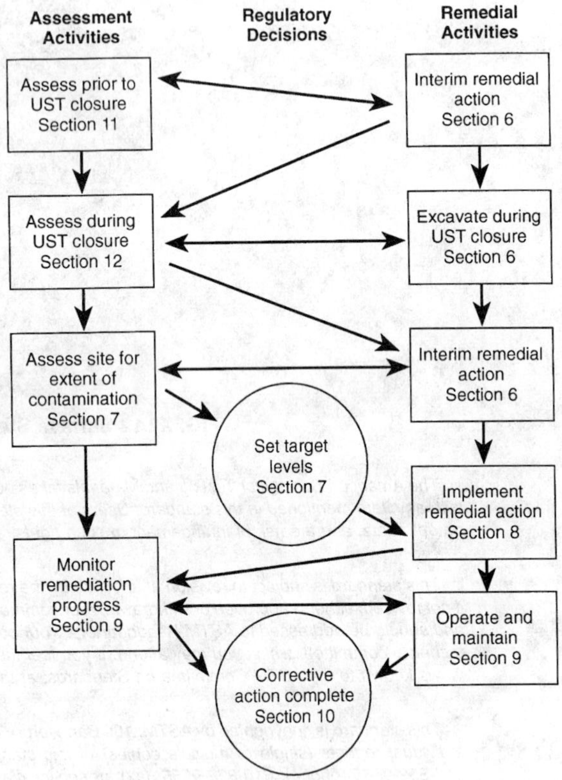

FIG. 1 Corrective Action Activities

1.6.2 Telephone conversations,

1.6.3 Notification forms,

1.6.4 Progress reports, and

1.6.5 Project plans.

1.7 It is important to note that a report in and of itself is not communication; someone has to read and understand it for there to be communication. Reports must be complete, presenting pertinent information that is necessary to lead to an appropriate corrective action decision.

1.8 Progress reports play a key role in the communication. These reports should be clear and sufficient so that all parties involved in the remediation can understand them.

1.9 This guide is organized as follows: Section 2 lists referenced documents, Section 3 defines terminology used in this guide, Section 5 discusses how indicator compounds can be used in the corrective action process, Section 6 discusses interim remedial actions, Section 7 describes site assessments, Section 8 discusses remedial actions, Section 9 describes

---

[1] This guide is under the jurisdiction of ASTM Committee E-50 on Environmental Assessment and is the direct responsibility of Subcommittee E50.01 on Storage Tanks.

Current edition approved March 15, 1994. Published May 1994.

operation, maintenance, and monitoring requirements for remedial actions, Section 10 discusses completion of the corrective action process, Section 11 discusses a pre-excavation evaluation (PEE) option that can help identify and plan for contaminated materials that may be encountered during construction activities at UST sites, Section 12 discusses assessments associated with tank removal or abandonment, Sections 11 and 12 are specific to underground storage tank (UST) system closures. When a release is discovered and confirmed to have been caused by other means, the activities or portions of the activities described in Sections 11 and 12 may not be needed. Finally, Appendix X1 identifies additional documents related to assessment and remediation activities.

1.10 The values stated in inch-pound units are to be regarded as the standard. The SI units given in parentheses are for information only.

1.11 *This standard does not purport to address all of the safety concerns, if any, associated with its use. It is the responsibility of the user of this standard to establish appropriate safety and health practices and determine the applicability of regulatory limitations prior to use.*

2. Referenced Documents

2.1 *EPA Standards:*

SW 846, USEPA Recommended Analytical Procedures, Test Methods for Evaluating Solid Waste—Physical/Chemical Methods[2]

USEPA Publication No. USGPO 055-000-00368-8, Field Measurement Technics: Dependable Data When You Need It[2]

2.2 *API Standards:*

RP 1628, A Guide to the Assessment and Remediation of Underground Petroleum Releases[3]

RP 1629, A Guide for Assessing and Remediating Petroleum Hydrocarbons in Soil[3]

2.3 *NFPA Standard:*

NFPA 329, Leakage and Repair Safeguards for Flammable and Combustible Liquids[4]

3. Terminology

3.1 *Definitions of Terms Specific to This Standard:*

3.1.1 *corrective action*—actions taken to identify and clean up a release of petroleum. These activities include site assessment, interim remedial action, remedial action, operation and maintenance of equipment, monitoring of progress, and termination of the remedial action.

3.1.2 *corrective action goal*—the corrective action goal is to reduce levels of contamination to protect human health, safety, and the environment.

3.1.3 *natural cycle*—normally one annual fluctuation of the ground water levels. This time may differ depending on site specific and climatic conditions.

3.1.4 *pre-excavation evaluation (PEE)*—an assessment of the potential for contamination and its relative extent prior to an excavation at an UST site. A typical PEE could include the sampling of soil and ground water in the area of the UST excavation and the product dispensers.

3.1.5 *petroleum*—including crude oil or any fraction thereof that is liquid at standard conditions of temperature and pressure (60°F (16°C) and 14.7 psia (101.3 kPa)). The term includes petroleum-based substances comprised of a complex blend of hydrocarbons derived from crude oil through processes of separation, conversion, upgrading, and finishing, such as motor fuels, jet oils, lubricants, petroleum solvents, and used oils.

3.1.6 *receptors*—persons, structures, utilities, surface waters, and water supply wells that are or may be adversely affected by a release.

3.1.7 *regulatory agency*—USEPA or the designated state and local agencies responsible for carrying out the UST or other corrective action program.

3.1.8 *release*—a discharge of petroleum to the environment.

3.1.9 *remediation/remedial action*—activities conducted to protect human health, safety, and the environment. These activities include evaluating risk, making no further action determinations, monitoring, and designing and operating cleanup equipment.

3.1.10 *site assessment*—an evaluation of subsurface geology, hydrology, and surface characteristics to determine if a release has occurred, the levels of contamination, and the extent of contaminant migration. The site assessment generates information to support remedial action decisions.

3.1.11 *source area*—the source area is defined as either the location of liquid hydrocarbons or the location of highest soil and ground water contamination levels.

3.1.12 *UST closure*—the removal from the ground or decommissioning in place of an UST system, including the evaluation of the surrounding soil to determine if a release has occurred.

3.1.13 *UST system*—a storage tank and underground piping connected to the tank, that has at least 10 % of its volume below the ground.

4. Significance and Use

4.1 The purpose of this guide is to provide a logical, timely, economical framework and general sequence for site assessment and remediation activities for petroleum releases that contaminate the subsurface. However, this guide does not recommend particular techniques. Where state and local regulations exist, the intent is to provide a model to enable streamlining of the regulatory processes and to allow the corrective action to proceed in an effective manner. The corrective action goal is to reduce levels of contamination to protect human health, safety, and the environment, and to demonstrate that the impacts of the contamination have been addressed.

NOTE 1—Activities described in this guide should be conducted by a person familiar with assessment and remediation techniques.

5. Corrective Action Indicators

5.1 *Selection and Use of Indicator Compounds*:

5.1.1 Indicator compounds for sampling and analysis are

[2] Available from the Superintendent of Documents, U.S. Government Printing Office, Washington, DC 20402.

[3] Available from American Petroleum Institute, 1801 K Street N.W., Washington, DC 20226.

[4] Available from National Fire Protection Assoc., Batterymarch Park, Quincy, MA 02269.

easy to select when the released product is known. If, however, the type of product is unknown or more than one type of hydrocarbon product is suspected to have been released, the initial sampling and analysis should include indicator compounds for all suspected products. Once the compounds of concern have been identified, then further analysis can be limited to the identified compounds. When gasoline is the suspected release, benzene, toluene, ethylbenzene, and xylenes (BTEX) may be indicator compounds of concern. Other possible indicator compounds may be methyl tertiary butyl ether (MTBE) and tertiary butyl alcohol (TBA). When diesel, other distillates, or an unknown mixture of petroleum products is the suspected released hydrocarbon, indicator compounds may include naphthalenes and other semi-volatiles.

5.1.2 Indicator compounds in ground water and soil can be used to confirm the extent of contamination, defining the remedial action target levels discussed in 8.5, monitoring progress of the remedial action, and identifying the termination point of the remedial action.

5.2 *Field Screening Indicators*—Field screening techniques may be a cost-effective and timely assessment methodology. Field screening utilized during the assessment process may use one or more of a wide variety of qualitative or quantitative measurement techniques. The screening process includes defining the likely sources of contamination, the possible direction of contamination movement, and the likely extent of contamination. Some examples of field screening indicators are dissolved oxygen anomalies (O_2), carbon dioxide anomalies (CO_2), and volatile organics. For further information on field measurement techniques see USEPA Publication No. 055-000-00368-8.

5.3 *Indicator Compound Analysis*—The analysis of specific indicator compounds can occur in both soil and ground water. In general, analysis in soil should be limited to those compounds that are adversely affecting or are expected to adversely affect the ground water or other receptors. Unless specifically outlined by the regulatory agency, when investigating a petroleum release, the following analytical methodologies in Table 1 are commonly used and are recommended. Other methodologies or protocols that provide comparable results may be used.

6. Interim Remedial Action

6.1 *Introduction*:

6.1.1 The primary goals of interim remedial action are to mitigate fire and safety hazards and to prevent further migration of hydrocarbons in their vapor, dissolved, or liquid phase. Interim remedial action is most effective when the regulatory agency limits its oversight to being notified of the activities taken. From initial assessment through actual remediation, interim remedial action may be warranted or desired. Situations that warrant interim remedial action include the following:

6.1.1.1 Hydrocarbon vapors in occupied buildings or subsurface structures,

6.1.1.2 Dissolved hydrocarbons in drinking water wells,

6.1.1.3 Liquid hydrocarbons floating on ground water, and

6.1.1.4 Hydrocarbons apparently confined to the soils immediately adjacent to a recent release.

6.1.1.5 In addition, interim remedial action should be used in situations where it will be timely and cost effective and will not adversely affect the final remedial action plan.

6.1.2 *General Methods*—The following methods are the most common alternatives used in handling hydrocarbon-contaminated soils. Other methods may be locally competitive in both cost effectiveness and environmental compatibility. The methods may be used alone or together. (**Warning**—See Note 2.)

6.1.2.1 Liquid hydrocarbon recovery can be accomplished either by control of the ground water (ground water depression through pumping) or by passive recovery methods not requiring ground water pumping.

NOTE 2—**Warning:** Pumping ground water, pumping free product from the ground water, or sparging air into the ground water should only be used when sufficient understanding of the hydrogeologic impact of a method has been acquired or in an emergency situation. If done improperly, the plume may spread into previously uncontaminated areas.

6.1.2.2 Hydrocarbon vapor abatement can be accomplished through vapor extraction or limited source excavation.

6.1.2.3 Dissolved hydrocarbon recovery can be accomplished through ground water pump-and-treat methods or air sparging.

7. Site Assessment

7.1 *Introduction*:

7.1.1 The goals of site assessment are to determine the source of the contamination, the extent of contamination, and the potential impact of the contamination on human health, safety, and the environment. The data collected during site assessment become the base for determining the need for remedial action and for evaluating remedial-action alternatives. Consequently, the site assessment should be designed to collect

TABLE 1 Recommended Analytical Methodologies

	Soil	Water
Gasoline	volatile organic aromatics using SW846 Method 8020	volatile organic aromatics using SW846 Method 8020 modified to detect MTBE and TBA
Middle distillates (for example, No. 2 fuel oil, JP4, diesel)	volatile organic aromatics using SW846 Method 8020	volatile organic aromatics using SW846 Method 8020
	poly-nuclear aromatics (PNAs) using SW846 8100 (Naphthalenes)	poly-nuclear aromatics (PNAs) using SW846 8100 (Naphthalenes)
Heavier fuel oils and lubricating oil or unknown (for example, motor oil, used oil, No. 6 oil)	volatile organic aromatics using SW846 Method 8240	volatile organic aromatics using SW846 Method 8240
	semi-volatile organics using SW846 Method 8270	semi-volatile organics using SW846 Method 8270

data on the extent and movement of the contamination, and the impact to structures and utilities which will assist in the selection of remedial alternatives.

7.1.2 The site assessment should begin as soon as practical after a release has been confirmed. Site-assessment activities can and should coincide with any initial abatement or interim free-product removal activities.

7.1.3 Regardless of the approach and methods used, site assessment should include all of the following components:

7.1.3.1 Initial site evaluation,

7.1.3.2 On-site delineation, and

7.1.3.3 Off-site delineation.

7.1.4 A variety of technologies and methods can be used to conduct a site assessment. The use of some technologies or methodologies may be limited by the geology of the site, the hydrology of the site, surface and subsurface structures, the availability of remediation equipment, characteristics of the substance released, and requirements of the regulatory agency. Ground water monitoring wells may need to be installed. The discussion here is limited to a general approach rather than to specific technologies or methodologies.

7.2 *Initial Site Evaluation*—Prior to drilling or other subsurface site assessment activities, a review of the local and regional geologic and hydrogeologic characteristics and a survey and identification of potential receptors should be conducted. Information collected during this phase will be used to direct subsurface investigation activities and to identify potential initial abatement activities.

7.2.1 The following activities should be performed during the initial site evaluation:

7.2.1.1 Review local and regional geology and hydrogeology through the use of USGS maps, local well logs, environmental agency information, and the like,

7.2.1.2 Estimate the depth to ground water,

7.2.1.3 Identify subsurface structures that may promote contaminant transport (for example, water, sewer, and utility lines),

7.2.1.4 Identify and locate private and public water supply wells within a minimum of a 0.5-mile (0.8-km) radius of the site,

7.2.1.5 Identify surface waters and streams within a minimum of a 0.5-mile (0.8-km) radius of the site,

7.2.1.6 Identify surrounding land use, and

7.2.1.7 Identify other potential contaminant sources.

7.2.2 A suitable base map and surrounding land-use map should be developed summarizing the results of the initial site evaluation. The base map should include the property lines of the site, all structures and underground utilities on the site, and all known underground storage tanks, related piping, and UST excavations. The surrounding land-use map should include the property lines of the site, streets, alleys, utilities, neighboring structures (including structures across the streets from the site), and water supply wells within 0.5 miles (0.8 km) of the site. The base map is also used for summarizing future site assessment data.

7.3 *On-Site Delineation*—Utilizing the results from the initial site evaluation, a subsurface investigation should be initiated to determine site-specific geologic and hydrogeologic characteristics and to identify the extent of contamination on the release site (within the property boundaries). The first objective of the on-site delineation is to identify the source area(s) of the contamination. The second objective is to identify the extent of the contamination in the release site and determine if off-site migration has occurred.

7.3.1 Subsurface investigation activities should begin in the area of the release or in the area of the underground storage tanks and piping. Subsurface investigation activities should be expanded from these areas until the source area has been identified. After the source area(s) has been identified, the vertical and horizontal extent of contamination should be investigated. Variations in contamination levels below and above the water table as well as across the site are important to quantifying the release, its mobility, and potential remedial action alternatives. In addition, the depth to ground water, the soil characteristics, the hydraulic gradient, and the thickness of liquid hydrocarbon should be determined as part of the subsurface investigation. Based on these data, a determination of contaminant migration can be made and further investigative activities can be conducted to determine the extent of contamination within the property boundaries of the release site. Investigative activities for the on-site delineation should continue until the full extent of contaminant migration has been determined or until the investigation reaches the release site property lines.

7.3.2 If after the completion of the on-site delineation, there are data sufficient to design and implement a remedial action, an on-site delineation report describing the on-site activity should be prepared and submitted to the regulatory agency. Utilizing the data collected during the on-site delineation phase, this report should include at the minimum the following:

7.3.2.1 Data collected during the initial site evaluation,

7.3.2.2 Logs of any installed boring(s) or other subsurface investigation and schematics of any installed monitoring well(s) with ground water levels indicated showing screened intervals and the geologic unit where the screen intercepts,

7.3.2.3 Depth-to-fluid, depth-to-water, top-of-casing elevation, water table elevations, and product thickness measurements summarized in tabular and map form,

7.3.2.4 Water table elevations in map form and analytical results tabulated on a site map,

7.3.2.5 Analytical results presented in tabular form,

7.3.2.6 Laboratory reports including chain of custody forms, quality assurance/quality control (QA/QC) procedures, data and chromatograph results, and

7.3.2.7 A conceptual design of the remedial action approach.

7.4 *Off-Site Delineation*—If the extent of contamination extends beyond the property boundaries of the site on which the release is located, investigative activities on neighboring properties may be necessary. Off-site work will require negotiation and may require execution of off-site access agreements that may be difficult to obtain and time-consuming to complete. As a result, it is difficult to anticipate the time required to complete off-site delineation. However, if access is not obtained within a reasonable period of time (for example, 90 days), the regulatory agency should be notified and its assistance solicited.

7.4.1 Off-site investigative activities are a continuation of the activities conducted on-site; they should continue until the extent of contaminant migration has been determined. It is important to recognize that definition of full extent of contamination may not be possible because of structures, access problems, or insufficient data. Assessment activities should continue until enough data have been accumulated to determine the need for remediation, the design of the remedial action, and the method for monitoring its progress and measuring success.

7.4.2 During the preparation of the on-site delineation report, any suspected off-site contamination should be explored. Once the off-site delineation has been completed, a supplemental off-site delineation report may be prepared and submitted to the regulatory agency for approval. The supplemental off-site delineation report should contain the remaining data to complete the requirements of the site assessment report.

7.5 *Sampling/Analysis*—Indicator compounds should be analyzed as noted in Section 5. Every effort should be made to achieve the method detection limits for analysis of compounds in soil and ground water. The initial analytical results may be used to modify subsequent soil and ground water sampling and analysis. If initial testing does not reveal the presence of an indicator compound or it has been determined that its presence is the result of background concentrations, then subsequent testing for those compounds is not required.

7.6 *Site Assessment Methodologies*:

7.6.1 There are a number of methodologies that can be used to conduct site assessment activities. Application of these methods varies according to the site being assessed. For further information see API RP 1628 and API RP 1629. These methods include the following:

7.6.1.1 Soil borings,

7.6.1.2 Direct push technologies (for example, cone penetrometers, formation water sampling devices),

7.6.1.3 Geophysical surveys,

7.6.1.4 Soil gas surveys,

7.6.1.5 Ground penetrating radar,

7.6.1.6 Trenching or test pits,

7.6.1.7 Monitoring wells, and

7.6.1.8 Aerial photographic analysis.

7.7 *Reports*:

7.7.1 *Site Assessment Report*—If an on-site delineation report had not been submitted, then after completion of the site assessment activities a site assessment report should be prepared and submitted to the regulatory agency. The site assessment report should at a minimum include the following:

7.7.1.1 Data collected during the initial site evaluation,

7.7.1.2 Map of facility indicating boring and well locations, structures, property lines, tanks, piping, pump islands, roads, adjacent and neighboring property owners, and other significant features,

7.7.1.3 Logs of any borings or other subsurface investigation and schematics of any installed monitoring well with ground water levels indicated and results of field screening of soils,

7.7.1.4 Depth-to-fluid, depth-to-water, top-of-casing elevations, water table elevations, and product thickness measurements in tabular form,

7.7.1.5 Water table elevations in map form and analytical results tabulated on a site map,

7.7.1.6 Analytical results,

7.7.1.7 Chain of custody forms, and

7.7.1.8 Recommendations for any further actions.

7.7.2 *Progress Reports*—A progress report should be prepared and submitted on a quarterly basis or other interval specified by the regulatory agency. The initial progress report should include the results of the initial site evaluation and any interim remedial action activities. Subsequent progress reports should include at a minimum the following:

7.7.2.1 A base map of the site indicating the location of structures, boring or other subsurface investigations and well locations, property lines, and UST systems and excavations,

7.7.2.2 A surrounding-use map indicating the location of roads, adjacent and neighboring property owners, and other significant features,

7.7.2.3 Logs of borings or any subsurface investigation and schematics of any monitoring well installed since the last report,

7.7.2.4 Results of laboratory analysis received since the last report and data (in tabular or graphic form) from previous sampling events,

7.7.2.5 Depth-to-fluid, depth-to-water, top-of-casing elevations, water table elevations, and product thickness measurements for each well,

7.7.2.6 Water table elevations in map form and analytical results in tabular form on a site map,

7.7.2.7 A brief description of further proposed activities, and

7.7.2.8 Description of any interim remedial action.

8. Remedial Action

8.1 *Remedial Action Determination*:

8.1.1 Prior to the development of a remedial action approach, a determination of the need and extent for remedial action must be made. The need for remedial action will be determined subsequent to an evaluation of the contaminant levels, exposure pathways, contaminant source(s), and potential receptors for contaminants in both the soil and the ground water. The evaluation must address the types of hydrocarbon contamination (for example, vapor, dissolved, floating liquid, or residuals bound in the soils). The following three possible courses of action can result from this evaluation:

8.1.1.1 No further action,

8.1.1.2 Passive remediation (monitoring only), or

8.1.1.3 Active remediation.

8.1.2 The course of action taken will depend upon the site-specific risks to human health and the environment associated with the contaminant levels identified during the site assessment. Remedial action should be a performance-based activity that is focused on target levels, methods of measuring progress, and completion. The following discussions contain general criteria for determining when each course of action would be appropriate. These criteria may not be appropriate for all circumstances. Site-specific circumstances may warrant selecting an option not indicated by these criteria.

8.2 *No Further Action*—The no-further-action course of action could be appropriate when the levels of contaminants identified at the site pose no threat to human health and the environment. Examples of when no further action may be appropriate include the following:

8.2.1 Floating liquid hydrocarbon is not present,

8.2.2 Drinking water supplies are not adversely affected and not likely to be adversely affected,

8.2.3 Underground structures or utilities are not adversely affected and not likely to be adversely affected by vapors,

8.2.4 Surface waters are not adversely affected and not likely to be adversely affected,

8.2.5 Useable portions of USEPA or state-designated sole source aquifers are not adversely affected and not likely to be adversely affected,

8.2.6 The aquifer is unusuable for drinking water,

8.2.7 The contaminants are immobile, or otherwise separated from potential receptors,

8.2.8 State standards for soil or ground water quality are not exceeded, or

8.2.9 Other areas of local environmental concern are not adversely affected and not likely to be adversely affected.

8.3 *Passive Remediation (Monitoring Only)*:

8.3.1 Passive remediation or a monitoring-only action could be appropriate when the levels of contaminants resulting from a site pose a minimal risk to human health, safety, and the environment. Examples of when passive remediation may be appropriate include the following:

8.3.1.1 An exposure evaluation determines the potential for the impacts from the contaminants to be low,

8.3.1.2 Migration is minimal (for example, concentrations of contaminants at the perimeter of the plume are stable or decreasing), or

8.3.1.3 The source of ground water contamination has been eliminated.

8.3.2 If passive remediation is selected, then a monitoring plan should be developed. The goal of the monitoring is to determine if contaminant concentrations are stable or decreasing. The monitoring plan should include the following information:

8.3.2.1 Where ground water is adversely affected, a minimum of three monitoring points is recommended; one down-gradient from the area of contamination, one in the area of highest contamination, and one up-gradient from the area of contamination. Additional monitoring points may be necessary based on site-specific conditions (for example, multiple-contaminant plumes, multiple aquifers, or multiple sources) and to monitor ground water flow direction.

8.3.2.2 The monitoring period should usually be a minimum of one year or other reasonable period of time depending on site-specific conditions. Data evaluation should emphasize identification of significant increases in contaminant concentrations at the monitoring points. If monitoring results are erratic or increasing, the period should be extended or active remediation begun.

8.3.2.3 Monitoring points should be sampled for identified indicator compounds on a regular basis agreed upon with the regulatory agency.

8.4 *Active Remediation*—Active remediation should be proposed when concentrations of contaminants identified at the site pose a potential risk to human health, safety, and the environment. Active remediation could be implemented when no further action and passive remediation courses of action are not appropriate.

8.5 *Target Levels*:

8.5.1 Once a remedial action has been determined to be necessary, the objective or target level for that action must be defined. The target level must be either an achievable numeric value or other performance criteria that protect human health, safety, and the environment. For passive remediation, the target level will typically be a performance-based criteria (for example, monitoring of selected wells for increases in contaminant levels). For active remediation, specific levels are usually chosen as both design criteria and goals for remediation. Termination of the remedial activity may occur at levels other than the target levels as discussed in Section 10. Target levels could also include the following:

8.5.1.1 Numeric values based on an evaluation of the risk of the contaminant levels at the site,

8.5.1.2 EPA maximum contaminant levels (MCL) when drinking water supplies are adversely affected or other health-based levels, such as recommended allowable limits (RAL) outlined by the regulatory agency when private drinking water wells are adversely affected, or

8.5.1.3 Minimization or elimination of exposure to potential receptors.

8.5.2 *Contaminated Media*—There are four types of hydrocarbon contamination (vapor, dissolved, floating liquid, and residuals bound in the soil) that need to be addressed when determining target levels and identifying remedial alternatives. In many cases, the impact of one type of contamination on another may be the driving force in determining the need for remediation. For example, contamination in soil may only be significant when it migrates to ground water; the contamination in soil itself may not pose a risk. In other cases, the potential for dermal contact or breathing vapors resulting from the contamination in soil may warrant remedial action of the soil.

8.6 *Technology Selection*:

8.6.1 The technology selection should be based upon the type of contamination to be addressed, the target levels to be achieved, site-specific conditions, and regulatory requirements. In most cases, more than one technology may be required to address the various media and achieve the defined target levels. Technology selection should include the following:

8.6.1.1 An evaluation of the effectiveness of the technologies in achieving the target levels,

8.6.1.2 An evaluation of the ability of the technologies to address the four types of hydrocarbon contamination discussed in 8.5.2,

8.6.1.3 A cost performance and cash flow analysis comparing remedial action alternatives to include total project costs to design, implement, operate, maintain, and complete the remedial activity,

8.6.1.4 An evaluation of the maintainability and reliability of the technologies,

8.6.1.5 An estimate of the time required to complete the remediation,

8.6.1.6 An identification of permitting requirements,

8.6.1.7 A determination of handling and disposition of effluent and off-gasses,

8.6.1.8 A determination of handling and disposition of waste streams, and

8.6.1.9 An evaluation of contractor availability and qualifications including a demonstration of knowledge and capability to implement the selected technology.

8.6.2 *Technologies*—A detailed discussion of the available technologies is beyond the scope or intent of this guide. New and innovative technologies continue to be developed. In general, commonly utilized technologies include soil vapor extraction; air sparging and vapor extraction; bioremediation; hydraulic methods; excavation/treatment; thermal treatment; and soil flushing. For further information see API RP 1628 and API RP 1629.

8.6.3 *Combination Technologies*—The majority of sites may best be approached using more than one technology if more than one type of hydrocarbon contamination is present. Integrated remediation will allow a more effective response at the site. Examples of effective combination technologies include air sparging of the saturated zone while soil venting from the unsaturated zone and in-situ biorestoration with both ground water extraction and vacuum extraction methods.

8.7 *Remedial Action Plan*:

8.7.1 One of the primary goals of the remedial action plan is to establish with the regulatory agency the performance criteria for the remedial activity. The remedial action plan should include the following:

8.7.1.1 Identification of target levels to be achieved,

8.7.1.2 A method for evaluating remediation progress specific to the technology selected (for example, changes in indicators in monitoring wells or effluent). Proper and timely progress monitoring can identify problems in the design or implementation of the selected technology which may require additional investigation or modifications to the design of the remediation. Monitoring and maintenance are as important to remedial action as the technology selection and design, and

8.7.1.3 The criteria to be used for closure determination. Once the target levels have been achieved, it should be demonstrated that they can be maintained. In general, monitoring should occur for at least one natural cycle of hydrologic events.

9. Operation, Maintenance, and Monitoring

9.1 *Introduction*:

9.1.1 Once the remedial action method has been selected, a plan for monitoring the operation and maintaining the remediation equipment should be developed. In addition, a plan for periodic monitoring of the progress of the remediation effort should be defined.

9.1.2 The remediation equipment should be inspected for proper operation. Equipment should be inspected for wear, leakage, and other signs that indicate a potential failure. Preventive maintenance should be performed on the equipment.

9.1.3 A monitoring plan should be developed to track the progress of the remediation activity. This plan should include the procedures and schedule for sampling and analysis. Sampling locations can include influent and effluent points as well as designated monitoring wells. Monitoring results should be submitted in accordance with the remedial action plan.

9.2 *Modification of the Remedial Action*—If results from the periodic monitoring indicate that the remediation is not proceeding as planned, changes in approach should be considered. These changes could include modifying the existing equipment and procedures or implementing a different remedial action technology.

9.3 *Partial Site Activity Termination*—Modifications to monitoring and operating procedures may be made based on periodic monitoring results. One example is the termination of monitoring of designated wells once the target levels have been reached and maintained for a reasonable period of time. At sites where the remedial action uses several technologies or recovery points, portions of the remedial action may be terminated when target levels have been reached.

10. Completion of the Corrective Action Process

10.1 Once a decision has been made to terminate remediation, a period of monitoring will typically be required. Subsequent to the monitoring period, a no-furtheraction determination can be made. Termination of remedial action can occur under several circumstances:

10.1.1 Target levels achieved,

10.1.2 Technology limits reached,

10.1.3 Alternate target levels achieved, or

10.1.4 Isolation from potential receptors.

10.2 *Target Levels Achieved*—Remedial action can be terminated when no floating liquid hydrocarbon is present in a recovery well or monitoring well, and when the target levels identified have been achieved. Monitoring in accordance with the passive remediation approach (see 8.3) should be conducted once the target levels have been achieved.

10.3 *Technology Limits Reached*:

10.3.1 Remedial action may be terminated if after a minimum of one natural cycle of hydrologic events, concentrations of contaminants have leveled off or reached an asymptotic level. Appropriate methods should be applied to demonstrate this conclusion. This demonstration can be made using a visual graph of concentration over time, an exponential formula demonstrating a near-zero slope, a regression analysis that matches the concentrations over time to a near flat line or other appropriate method. The one-year minimum treatment period may be shortened if it can be demonstrated to the regulatory agency that a shorter time period is appropriate (that is, soil venting in highly permeable soils frequently requires less than one year to achieve target levels). Prior to termination of the remedial action:

10.3.1.1 The technical feasibility and costs of other proven treatment techniques to further reduce the contaminant levels at the site should be evaluated, and

10.3.1.2 The effects of the remaining contamination on the use of the water resource, and other issues of human health, safety, and the environment should be evaluated.

10.3.2 Monitoring in accordance with the passive remediation approach described in 8.3 should be conducted once the asymptotic level has been achieved. Designated monitoring wells should be sampled on a regular basis for the target compounds for a minimum of one natural cycle of hydrologic events. If during this period, concentrations of the discharged substance increase significantly for two consecutive monitoring rounds, remedial action should be re-instituted; otherwise, remedial action should be deemed complete when the "levelling off" level has remained constant or decreased for a period of one year.

10.4 *Alternate Target Levels Achieved*:

10.4.1 Remedial action may be terminated based on an assessment of the risk to human health, safety, and the environment posed by remaining contaminant levels. Monitoring in accordance with the passive remediation approach should be conducted after termination of the remediation activity.

10.4.2 Remedial action may be terminated if modified land use, modified ground water classification, or other changes occur during the remediation process that affect the parameters of the risk evaluation such that the existing levels of contaminants no longer present a risk to human health, safety, and the environment. Monitoring in accordance with the passive remediation approach should be conducted after termination of the remediation activity.

10.5 *Isolation from Potential Receptors*—Remedial action may be terminated once it has been determined that the contamination is confined, immobilized, or otherwise isolated from potential receptors. After termination of the remedial activity, monitoring should be conducted in accordance with the passive remediation approach.

10.6 *Partial Termination of Remedial Action*—At sites where the remedial action uses several technologies or recovery points it may be appropriate to terminate portions of the remedial activity. For instance, if soil venting and ground water recovery are being used together, it may be appropriate to terminate the soil venting once its target levels have been achieved even though ground water recovery may need to continue. Another example would be a situation in which multiple ground water recovery wells are in use and concentrations in some of the wells have met the target levels. Abandonment of some of the wells may be appropriate.

10.7 *Notice of Residual Contamination*—Once remedial action has been completed and a no further action status has been achieved, residual contamination may still remain on the site. Based on the current use, this contamination may not pose any potential for exposure. However, future activities on the site may be affected by that residual contamination. For example, if soil remediation was terminated because the contaminated soil was isolated under the pavement and a new use would require excavation of the pavement, then potential exposure may exist. As a result, it is recommended that potential purchasers and future property owners be notified of the existence of residual contamination on the site.

11. Pre-Excavation Evaluation

11.1 *Uses in the Corrective Action Process*:

11.1.1 Conducting an assessment or pre-excavation evaluation (PEE) prior to UST removal or excavation is an optional step and may be beneficial in many circumstances. Generally, contamination is not solely the result of leaks in the USTs or their piping systems. The most common causes of UST-related contamination are residuals from past releases, overfills, and spills associated with equipment maintenance. Often this type of contamination is minor and is limited to the backfill, and immediate excavation area; typically, it can be remediated with minimal action.

11.1.2 The following three basic issues that can be addressed by assessing the site prior to removing an UST system:

11.1.2.1 The potential for and general extent of contamination can be estimated.

11.1.2.2 Handling, storage, and disposal or remediation options for contaminated soil, vapors, and water generated during an UST removal can be identified.

11.1.2.3 Both short-term and long-term remediation options can be identified.

11.1.3 The information gathered during the assessment can be used to identify the personnel, equipment, permits, and the like that will be required to perform the tank removal. The benefits of this assessment include identification of worker and public safety issues; minimization of unnecessary project delays; and potential cost savings.

11.2 *PEE Design*:

11.2.1 Prior to an excavation at an UST site, an assessment of the potential for contamination and its relative extent can be determined by conducting a PEE. A typical PEE includes the installation of borings or wells in the area of the excavation and product dispensing systems. Generally, four or five borings or test points (two on opposite sides of the UST excavation and one between each dispensing island and the USTs) are necessary for the typical UST installation. Placement of the borings should also consider actual or anticipated ground water flow patterns. Samples of ground water and soil may be collected and analyzed for the suspected petroleum compounds. Soil type data and ground water level information can be used to evaluate the subsurface characteristics. Other analysis may be appropriate to evaluate treatment and disposal options. As part of the PEE, accessible areas such as pump pits, dispenser pits, and curb box areas should be inspected for signs of contamination.

11.2.2 The PEE is not designed or intended to determine the full extent of contamination, but rather to:

11.2.2.1 Identify whether or not contamination exists outside the UST system area,

11.2.2.2 Indicate the relative levels or concentrations of contaminants in the soil or ground water,

11.2.2.3 Indicate the hydrogeologic and geologic characteristics of the site, and

11.2.2.4 Provide an opportunity to plan and develop contingencies for interim remedial actions.

11.2.3 The data generated from the PEE can be used to evaluate the amount and extent of contamination to be handled, the potential remedial action alternatives, and the disposal or remediation options for contaminated water and soil generated during the UST removal. For example, if the PEE identifies a high water table, treatment and disposal alternatives for this water may have to be addressed. Given prior knowledge of the need for water handling, a temporary treatment system could be designed and approved prior to the excavation, thereby eliminating a costly emergency response involving storage and transportation for disposal. Also, preliminary landfill approval for soil disposal may be received, or other treatment options evaluated based on soil data.

11.2.4 The most important result of the PEE will be to provide data to determine whether excavation or other remedial action option at the time of the UST removal is appropriate and economical. If the PEE reveals high levels of contamination outside the UST excavation, extensive excavation may not be the most efficient or cost-effective option. However, if the PEE reveals minor or no contamination outside the excavation area, then excavation may be an appropriate and economical option. Regardless of the results of the PEE, the backfill will probably contain levels of contamination that will necessitate special handling.

11.2.5 If contamination is identified and confirmed during the PEE, it may need to be reported in accordance with regulatory requirements.

12. Assessment During Tank Removal or Closure in Place

12.1 *Purpose*—UST owners who remove or close their tanks in place are required to assess the soil or ground water, or both, for the presence of a release. The objective of this activity is to collect sufficient data to determine the existence of a release or to confirm that no further action is needed. If a field measurement or laboratory analyses indicates that a release has occurred, site assessment activities as described in Section 7 may be required. The UST system removal should be conducted in accordance with regulatory requirements and applicable standards.

12.2 *General*:

12.2.1 The information gathered during UST removal or closure in place may be useful in evaluating interim remedial actions. For instance, knowledge of contaminants in soil samples, depth to ground water, and soil types may allow for an estimate of the potential for ground water impact. The following actions should be performed during tank removal or closure in place.

12.2.1.1 Use personnel who are familiar with the required assessment activities and follows accepted health and safety plans,

12.2.1.2 Arrange for a PEE prior to the tank removal if deemed necessary as described in Section 11,

12.2.1.3 If contamination is identified, it may need to be reported in accordance with regulatory requirements, and

12.2.1.4 Plan for the temporary storage of petroleumcontaminated soil that may be excavated (that is, prevent infiltration and run-off while the soil is being stored).

12.3 *Soil and Ground Water Assessment for UST Removal*:

12.3.1 *Soil Samples*—Unless specifically directed by the regulatory agency, the soil and ground water assessment should include a minimum of two soil samples for each tank removed, having a nominal capacity of less than 10 000 gal (37850 L), and a minimum of three soil samples for each tank removed, having a nominal capacity greater than or equal to 10 000 gal (37850 L). For the tanks with a capacity of less than 10 000 gal (37850 L), one soil sample should be collected from directly below each end of the tank. For the tanks with a capacity of greater than or equal to 10 000 gal (37850 L), one soil sample should be collected from directly below each end and the center of each tank (see Note 3). The intent of this section is to provide minimum sampling criteria for most UST removal. Alternate sample locations may be appropriate based on site specific conditions, that is, sampling sidewalls where lateral migration has occurred. Additional sampling and analysis may be appropriate in highly sensitive areas to document that no further action is required (for example, near public or private water supplies and a geologic condition that lends itself to increases in contaminant migration such as karst terrains, glacial deposits, alluvial and colluvial deposits, or highly fractured strata). Soil samples should be collected within 2 ft (0.6 m) below the bottom of the excavation or above the water table if encountered. All samples should be collected from freshly exposed soil. In addition, a minimum of one soil sample should be collected from beneath each dispenser island that is being removed and a minimum of one soil sample should be collected from below every 20 linear ft (6 m) of underground piping. Samples should be evaluated using field measurement techniques capable of detecting the identified field screening indicators. One sample with the highest field measurement reading from each of three areas, the tank excavation (regardless of the number of tanks in the excavation), the line excavation (regardless of the number of lines in the excavation), and the dispenser islands (regardless of the number of dispenser islands) should be analyzed for the indicator compounds discussed in Section 5. If field measurement techniques do not detect the field screening indicators, then one sample from each of the three areas previously described should be analyzed.

Note 3—The intent of this guide is to provide a minimum number of soil samples for typical situations, alternate sampling locations based on-site specific conditions may be utilized, that is, sidewall samples.

12.3.2 *Water Samples*—If water in the excavation has not been determined to be ground water, then it should be completely evacuated and disposed of in accordance with all applicable regulations. If within 24 h, the water recharges into the excavation to a level sufficient for sample collection, then a sediment-free sample should be collected as soon as practical. However, if water exists in the excavation and site conditions warrant immediate backfilling (that is, collapsing side walls or other safety issues), a water sample should be collected. Water sampled directly from inside an excavation or from a soil boring is not necessarily representative of normal ground water conditions and should not be evaluated as a ground water sample. Such samples may, however, be used to document the existence of a release.

12.3.3 *Analyses*:

12.3.3.1 On-site field measurement techniques can provide immediate results that are useful and may be less expensive than those obtained from laboratory analyses. These may be used to indicate a release. For further information on field measurement techniques see USEPA Publication No. 055-000-00368-8.

12.3.3.2 Samples identified for analysis should be analyzed for indicator compounds as described in Section 5. The regulatory agency may require additional or other analyses. If contamination is identified in the samples, it may need to be reported to the regulatory agency as a release.

12.4 *Soil and Ground Water Assessment for Closure in Place*:

12.4.1 Assessment of petroleum contamination is required if the USTs are permanently closed in place.

12.4.2 *Sampling*—Unless specifically outlined by the regulatory agency, the soil and ground water assessment should include a minimum of two soil borings per tank within the tank backfill area, regardless of the size of the tanks. Soil borings should extend to a maximum of 5 ft (1.5 m) below the bottom of the tank or to the water table if encountered. Samples should be evaluated using field measurement techniques capable of detecting the identified field screening indicators. The sample with the highest field measurement reading from each of the borings should be analyzed for the indicator compounds discussed in 5.7.3.3. If the field measurement reading does not indicate contamination, a sample from the bottom of each boring should be submitted for laboratory analysis. If ground water is encountered within 5 ft (1.5 m) of the tank bottom, a soil sample from the soil/water interface should be submitted for laboratory analysis.

12.5 *Recordkeeping*—The owner and operator of the tank should keep a permanent record of the closure assessment activities.

13. Keywords

13.1 closure; corrective action; hydrocarbons; indicators; interim remedial action; no further action; passive remediation; remediation; site assessment; target levels; termination

APPENDIX

(Nonmandatory Information)

X1. RELATED DOCUMENTS

X1.1 *ASTM Standards:*

D 653 Terminology Relating to Soil, Rock, and Contained Fluids (contains many terms and definitions related to ground water and ground water contamination)

D 1125 Test Methods for Electrical Conductivity and Resistivity of Water

D 1293 Test Methods for pH of Water

D 1452 Practice for Soil Investigation and Sampling by Auger Borings

D 1586 Test Method for Penetration Test and Split-Barrel Sampling of Soils

D 1587 Practice for Thin-Walled Tube Sampling of Soils

D 2113 Practice for Diamond Core Drilling for Site Investigation

D 3370 Practices for Sampling Water

D 3550 Practice for Ring-Lines Barrel Sampling of Soils

D 3584 Practice for Indexing Papers and Reports on Soil and Rock for Engineering Purposes

D 4043 Guide for Selection of Aquifer-Test Field and Analytical Procedures in Determination of Hydraulic Properties by Well Techniques

D 4220 Practice for Preserving and Transporting Soil Samples

D 4448 Guide for Sampling Groundwater Monitoring Wells

D 4700 Guide for Soil Sampling from the Vadose Zone

D 4750 Test Method for Determining Subsurface Liquid Levels in a Borehole or Monitoring Well (Observation Well)

D 5092 Recommended Practice for Design and Installation of Ground Water Monitoring Wells in Aquifers

X1.2 *EPA Standards:*

40 CFR Part 280, Technical Standards and Corrective Action Requirements for Owners and Operators of Under-ground Storage Tanks

SW 846, USEPA Recommended Analytical Procedures, Test Methods for Evaluating Solid Waste—Physical/Chemical Methods

USEPA Publication USGPO 055-000-00368-8, Field Measurement Technics: Dependable Data When You Need It

X1.3 *API Standards:*

RP 1628, A Guide to the Assessment and Remediation of Underground Petroleum Releases

RP 1629, A Guide for Assessing and Remediating Petroleum Hydrocarbons in Soil

Publication 4510, Technological Limits of Ground Water Remediation: A Statistical Evaluation Method

Publication 4367, Ground Water Monitoring and Sample Bias

Publication 4394, Detection of Hydrocarbons in Ground Water by Analysis of Shallow Soil Gas/Vapor

Publication 4440, Manual of Sampling and Analytical Methods for Petroleum Hydrocarbons in Ground Water and Soil

Publication 4447, Phased Separated Hydrocarbon Contaminant Modeling for Corrective Action

Publication 4476, Hydrogeologic Data Base for Ground Water Modeling

Publication 4509, An Evaluation of Soil Gas and Geophysical Techniques for Detection of Hydrocarbons

Publication 4631, Solubility of BTEX from Gasoline-Oxygenate Mixtures

Publication 4414, Literature Survey: Hydrocarbon Solubilities and Attenuation Mechanisms

X1.4 *NFPA Standards:*

NFPA 30, Flammable and Combustible Liquids Code

NFPA 329, Leakage and Repair Safeguards for Flammable and Combustible Liquids

Designation: E 1609 – 94

Standard Guide for Development and Implementation of a Pollution Prevention Program[1]

This standard is issued under the fixed designation E 1609; the number immediately following the designation indicates the year of original adoption or, in the case of revision, the year of last revision. A number in parentheses indicates the year of last reapproval. A superscript epsilon (ϵ) indicates an editorial change since the last revision or reapproval.

1. Scope

1.1 This guide covers guidance on a logical progression of tasks and procedures to be followed in a pollution prevention program to reduce or eliminate the generation of waste, the loss of natural resources, and process emissions through source reduction, reuse, recycling, and reclamation.

1.2 *Summary*—The basic components of a pollution prevention program should include the following seven activities:

1.2.1 Develop an organizational commitment to pollution prevention (see Section 4).

1.2.2 Establish goals, objectives, and an implementation schedule (see Section 5).

1.2.3 Generate baseline information (see Section 6).

1.2.4 Develop a resource, emissions, and waste measurement and tracking system (see Section 7).

1.2.5 Analyze pollution prevention opportunities (see Section 8).

1.2.6 Prioritize pollution prevention opportunities (see Section 9).

1.2.7 Implement and maintain the progress of a pollution prevention program (see Section 10).

1.3 *Organization of Text*—This guide is organized based on the activities previously enumerated. Each section of the guide describes the manner in which the specified activity may be conducted to implement a program of pollution prevention at a facility.

1.4 *This standard does not purport to address all of the safety concerns, if any, associated with its use. It is the responsibility of the user of this standard to establish appropriate safety and health practices and determine the applicability of regulatory limitations prior to use.*

2. Terminology

2.1 *Descriptions of Terms Specific to This Standard:*

2.1.1 *by-product*—material, other than principal product, that is generated as a consequence of a process and that has productive use.

2.1.2 *capital budget*—a statement of the firm's planned financial expenditures, generally based on estimates of future sales, costs, production, and research and development (R&D) needs and the availability of capital.

2.1.3 *cost accounting system*—the internal procedure used to track and allocate production costs and revenues to a product or process.

2.1.4 *cost allocation*—a process within an internal cost accounting system of assigning costs and revenues to cost and profit centers for the purposes of product pricing, cost tracking, and performance evaluation.

2.1.5 *fugitives*—emissions or releases that leave a system or process without containment or capture.

2.1.6 *full-cost accounting*—a method of managerial accounting that accounts for both the direct and indirect costs of an item. Full-cost accounting uses historical data to assign all costs to a process, product, or product line, most often for the purposes of pricing.

2.1.7 *materials*—physical substances that are used, applied, produced, formed, or processed.

2.1.8 *media*—any or all specific physical components of the environment, such as air, water (surface water and ground water), and soil.

2.1.9 *minimization*—the process of determining and achieving the optimal amount of resources necessary to perform a particular function. In addition, minimization is the process of determining and achieving the least practicable harmful effects of a particular function or activity.

2.1.10 *performance measurements*—a means by which a system or process can be evaluated for effectiveness or efficiency, or both. It must have a quantitative and consistent basis for evaluating the effectiveness or efficiency, or both, of a particular function or activity.

2.1.11 *point source*—the discrete location of emissions or releases that leave a system or process through a duct, pipe, or other containment device.

2.1.12 *pollutant*—any substance that directly or indirectly creates an adverse human health or environmental effect when introduced into any environmental media.

2.1.13 *pollution*—the introduction of a pollutant into the environment.

2.1.14 *pollution control*—a control device, mechanism, or system that is used to reduce the quantity or toxicity of a pollutant, or both, before it is introduced into the environment,

---

[1] This guide is under the jurisdiction of ASTM Committee E-50 on Environmental Assessment and is the direct responsibility of Subcommittee E50.03on Pollution Prevention, Reuse, Recycling and Environmental Efficiency.

Current edition approved Sept. 15, 1994. Published November 1994.

or to reduce the probability of release of a pollutant to the environment.

2.1.15 *pollution prevention*—the act of reducing or eliminating the use, release, or generation of a pollutant or potential pollutant through source reduction, recycling, reuse, reclamation, or modification of operating practices.[2]

2.1.16 *pollution prevention program*—a comprehensive management, planning, capital budgeting, and monitoring program to promote and support the development and implementation of pollution prevention throughout an organization or at a specific facility. Such a program should have a statement of policy and goals; plan for measuring performance; and specified time frames for implementation, measurement of progress, and reevaluation.

2.1.17 *process*—a method, or collection of methods, or series of progressive and interdependent steps for generating a product or performing a service.

2.1.18 *product*—the intended output of a process or a facility operation, other than a waste.

2.1.19 *reclaim*—a procedure of either regenerating or processing a material, either used or unused, to recover or make a usable product.

2.1.20 *recycle*—the act of using the output of a process as an ingredient to produce a product in the same, similar, or related process from which that output was generated.

2.1.21 *release*—any spilling, leaking, pumping, pouring, emitting, discharging, injecting, escaping, leaching, dumping, or disposing any material or pollutant into the environment.

2.1.22 *resource*—a material that has a recoverable value.

2.1.23 *reuse*—the act of using the output of a process as an ingredient to produce a product in a process other than the same, similar, or related process from which that output was generated originally.

2.1.24 *source reduction*—any activity that eliminates or decreases wastes by avoiding their creation, typically by materials substitution, process design, or product redesign.

2.1.25 *total cost assessment (TCA)*—a comprehensive financial analysis of the lifecycle costs and savings of a pollution prevention project. A TCA approach includes the following:

(*1*) Internal allocation of environmental costs to product lines or processes through full-cost accounting;

(*2*) Inclusion in a project financial analysis of direct and indirect costs and short- and long-term costs, liability costs, and less tangible benefits of an investment;

(*3*) Evaluation of project costs and savings over a long time period (for example, 10 to 15 years); and

(*4*) Use measures of profitability that capture the long-term profitability of the project (for example, net present value and internal rate of return).

2.1.26 *total quality management (TQM)*—a management philosophy involving continuous process improvement activities involving all personnel in an organization in an integrated effort toward improving performance at every level.

2.1.27 *treatment*—any mechanism used for reducing the quantity or toxicity of a waste after its generation in a process.

2.1.28 *waste*—any output from a process or facility operation that is not used, reused, reclaimed, or recycled productively, and that is placed directly into the environment or treated through pollution control.

2.1.29 *waste minimization*—to eliminate or decrease, to the maximum extent practicable, the generation of waste by any method of source reduction, reuse, reclamation, or recycling.

2.1.30 *waste reduction*—to decrease or eliminate the generation of waste by any method of source reduction, reuse, reclamation, or recycle.

3. Significance and Use

3.1 This guide for development and implementation of a pollution prevention program is applicable to any organization or facility that releases materials to any of the three environmental media (such as the air, water, or land) and that wishes to reduce those releases, without using treatment or transferring them to one of the other two media primarily for the purpose of disposal. Incentives for applying this standard of practice include concern for the environment, conservation of natural resources, economic considerations, and current and future regulatory compliance. Effective pollution prevention can also increase the efficiency of operations and use of resources, employee morale, and profitability while reducing liability.

3.2 A successful pollution prevention program can save money by reducing waste management costs and raw material purchases, reduce potential emissions and disposal liabilities, protect public health and worker health and safety, and protect the environment. It will also position an organization to compete domestically and internationally through both long-term cost reductions and participation in green marketing opportunities.

4. Development of an Organizational Commitment to Pollution Prevention

4.1 *Introduction*—The purpose of this activity is to develop an organizational commitment to the pollution prevention process; establish program goals; assign strategic and tactical responsibilities; and identify and procure requisite financial, material, and human resources necessary for successful pollution prevention.

4.1.1 The inclusion of a policy statement on pollution prevention in an organization's overall environmental policy, or as an independent policy, is necessary to convey, to all employees, customers, suppliers, shareholders, and the public, the organization's commitment to pollution prevention.

4.1.2 The statement should express clearly the reasons that a pollution prevention program is being implemented, how it will be accomplished, and who will be involved. The policy should be relevant to the organization's activities, products, and services and the associated environmental effects. The policy statement can be general in nature, allowing for the specific pollution prevention methods to be developed by organization employees. For example, the policy statement might be stated simply as follows:

4.1.3 It is this organization's policy for all employees to reduce or eliminate the use of toxic materials, to reduce the

[2] It should be noted that ASTM's definition of "pollution prevention" is different from some definitions used by the Environmental Protection Agency. See, for example, 58 Fed. Reg. 6478 (Jan. 29, 1993, Council on Environmental Quality), and 58 Fed. Reg. 41,981 (Aug. 6, 1993, Executive Order).

volume and toxicity of all waste generated, and to recycle, reuse, or reclaim materials whenever possible.

4.1.4 Alternatively, the policy statement may cover specific components of the pollution prevention program.

4.1.5 Commitment throughout an organization is necessary in order to motivate employees and to encourage cooperation and participation in achieving overall success in the program. The appointment of a team, including members from all business units in an organization, to assist in the development of this policy can enhance the overall success of the program. Approval of the policy by the board of directors of a corporation and the executive officer lends credibility to the policy and demonstrates commitment at the highest level of the organization.

4.1.6 Organizational commitment is the key to success of the pollution prevention process. Without sufficient commitment, planning, and organization, any changes recommended by the following activities may not be implemented properly, and the goals of the pollution prevention program may not be attained fully.

4.1.7 *Procedure*—Suggested components for the development of a pollution prevention commitment include the following:

4.1.8 Develop a philosophical and resource commitment to the process among key members of the organization.

4.1.9 Determine long- and short-range strategies.

4.1.10 Define and obtain infrastructure, human, and financial resources.

4.1.10.1 Infrastructure includes the necessary office space and information management systems.

4.1.10.2 Human resources include a structure for implementing a pollution prevention program, such as pollution prevention teams or committees.

4.1.10.3 Financial resources include a budget and funds for pollution prevention program startup, human resources, and process changes or raw material substitutions.

4.1.11 Organize and mobilize the process.

4.2 *Organization Structure and Responsibilities*—Development of an organizational structure and responsibilities that emphasize and support the importance of a pollution prevention program are paramount to the success of a program. This could include appointment of a pollution prevention manager, as well as production area or group managers who could work together in order to meet the goals of the pollution prevention program.

4.2.1 *Integration*—Pollution prevention must become part of the everyday work ethic. The following activities should be performed to establish an organizational structure that promotes pollution prevention:

4.2.1.1 Define the organization of the program and individual responsibilities.

4.2.1.2 Assign a lead responsibility with authority.

4.2.1.3 Define the program responsibilities across all parts of the organization.

4.2.2 *Responsibilities*—Once the program organization is defined, specific job responsibilities can be defined for the program lead, as well as all others involved throughout the organization. The defined responsibilities should include accountability both internally and externally. Program responsibilities should be as consistent as possible with other policies and commitments in order to provide a solid foundation for the pollution prevention program.

4.2.3 *Level of Effort*—There should be an ongoing commitment to allocate human resources commensurate with the scope of the program. Human resources should be adjusted by number and skill as demanded by the program objectives.

5. Establishment of Goals, Objectives, and Implementation Schedule

5.1 *Introduction*—When implementing a pollution prevention program, it is important to establish goals, objectives, and schedules that are consistent with the organization's policy statement and organizational structure.

5.2 *Development of General Goals*:

5.2.1 It is important that the goals of the pollution prevention program reflect the management style and operation of the organization. An organization that operates by total quality management (TQM) may focus on goals that allow and encourage participation by all employees, whereas an organization that is more traditionally managed may focus on specific pollution prevention projects identified by management.

5.2.2 Since each organization's culture, management structure, and operational procedures can differ dramatically, this activity may be accomplished in a variety of ways. For example, pollution prevention goals may be qualitative and strive to meet "a significant reduction" in waste generation. A company that is TQM based may direct its goals toward continuous improvement or zero discharges. However, a company that focuses on yearly productivity goals may focus on a specific waste reduction goal, such as "90 % waste reduction."

5.2.3 It is important that the organization's overall goals be incorporated within each department's goals in order to allow the opportunity to develop an acceptable and achievable program directed toward its specific processes. An effective program will find its way into all aspects of the operations of an organization, including marketing, product design, procurement, materials control, production, and environmental affairs.

5.2.4 Goals can cover a number of areas, such as costs, waste quantities generated, waste disposed of, or waste toxicity. At minimum, the pollution prevention program should incorporate the goal of reducing the generation of all wastes and the release of pollutants. The following are other examples of program goals:

5.2.5 Reduce risks to human health and the environment.

5.2.6 Reduce the costs of waste management, including costs of potential long-term liability.

5.2.7 Improve worker health and safety.

5.2.8 Enhance the reliability of service and productivity when reducing wastes.

5.2.9 Eliminate the land disposal of hazardous wastes.

5.2.10 Reduce or eliminate hazardous waste generation from specific plant sources.

5.3 *Development of Specific Objectives*:

5.3.1 In addition to general goals, a pollution prevention program should incorporate specific objectives in order to make progress toward the goals. The objectives should cover a variety of areas such as risk, costs, safety, and reliability. Some examples of objectives include the following:

5.3.2 Seek a specific significant (such as percent) reduction in the generation of hazardous and industrial wastes over a specific time period.

5.3.3 Demonstrate a continuous reduction in the rate of waste generation per unit of output.

5.3.4 Save sufficient resources through pollution prevention programs to offset half the costs of environmental management.

5.3.5 Replace oil-based paints with water-based paints.

5.4 *Measurement of Objectives*—The pollution prevention program objectives should be sufficiently quantitative to allow measurement of the program's success. Quantifiable objectives provide a clear and understandable guide to the program expectations.

5.5 *Implementation Schedule*:

5.5.1 A schedule should be an integral part of the goals and objectives and should be used as a feedback mechanism for determining whether they have been met. The goals, objectives, and implementation schedule should be sufficiently flexible to encourage employee involvement in implementation of the pollution prevention program.

5.5.2 The initial development of the pollution prevention program should include specific time frames for all aspects of implementation of the program. This includes employee awareness and participation, completion of baseline measurements, and submission of the first and subsequent reports.

5.5.3 The implementation schedule should include a periodic evaluation of the program goals and objectives. Changes in technology, process input material supplies, and environmental regulations may mandate adjustments to the goals.

6. Generation of Baseline Information

6.1 *Introduction*—The objective of this step of the pollution prevention process is to identify the baseline facility or process information that will serve as the means for identifying pollution prevention opportunities and the basis for demonstrating the success, progress, and accomplishments of the program. A baseline assessment should be made of a facility's current processes and operations, including the process limitations and the materials use, products, byproducts, and wastes of each process and operation. The development of baseline information is important to providing adequate information to highlight pollution prevention options and frame them in a proper perspective.

6.2 *Impact of Organizational Structure*—The amount and complexity of the data that are required is highly dependent on the organization's management and decision-making structure. In an organization in which individuals at all levels of the organization are empowered to take appropriate actions in order to reduce or eliminate waste, an individual, within his or her level of authority, would take action on waste reduction based on a goal of continuous reduction of waste and his or her professional knowledge and evaluation of the available data. The baseline under this structure may tend to be more general rather than an extensive, highly detailed, and centrally located database.

6.3 *Examples*—The following is a list of possible information sources that may be used to develop the baseline information. The list is not all inclusive, and it is not necessary to use all of the information sources in every case.

6.3.1 *Design Information*:

6.3.1.1 Process flow diagrams and schematics.

6.3.1.2 Material balances for production and pollution control processes.

6.3.1.3 Operating manuals and process descriptions.

6.3.1.4 Standard batch sheets.

6.3.1.5 Quality control manuals.

6.3.1.6 Equipment lists, specifications, and data sheets.

6.3.1.7 Piping and instrument diagrams.

6.3.1.8 Site plans, including equipment layouts.

6.3.2 *Environmental Records*:

6.3.2.1 Permits and permit applications.

6.3.2.2 Periodic hazardous waste generation reports.

6.3.2.3 Waste manifests.

6.3.2.4 Air emission inventories, including fugitives.

6.3.2.5 Emergency planning and community right-to-know act (EPCRA) submittals.

6.3.2.6 National pollutant discharge elimination system (NPDES) discharge monitoring reports.

6.3.2.7 Previous environmental assessments.

6.3.2.8 Spill and release reports.

6.3.2.9 Media reports.

6.3.3 *Raw Material and Production Data*:

6.3.3.1 Material safety data sheets.

6.3.3.2 Product data (quantities and specifications).

6.3.3.3 Product and raw material inventory records.

6.3.3.4 Purchasing records.

6.3.3.5 Operator procedures and data logs.

6.3.3.6 Production schedules.

6.3.4 *Economic Information*:

6.3.4.1 *Direct Costs or Benefits*:

(*1*) Changes in product revenues;

(*2*) Capital expenditures (buildings, equipment, utility connections, equipment installation, and project engineering);

(*3*) Operation and maintenance expenses and revenues (raw materials and supplies, labor (maintenance and operations), waste disposal, utilities (for example, water, sewer, electricity, and fuels), and revenue from recovered materials); and

(*4*) Environmental impairment and liability insurance premium.

6.3.4.2 *Indirect Costs or Benefits*:

(*1*) Compliance costs (permitting, reporting, tracking, monitoring, manifesting, training, waste handling, recordkeeping, labeling, testing, emergency preparedness, and medical surveillance);

(*2*) Waste storage;

(*3*) Operation of on-site pollution control equipment;

(*4*) Replacement of non-product output;

(*5*) Marginal change in costs due to the reduced use or generation of hazardous materials; and

(*6*) Legal fees.

6.3.4.3 *Liability Costs*:

(*1*) Penalties and fines;

(*2*) Damage to property and casualty loss;

(*3*) Legal fees; and

(*4*) Insurance premiums.

6.3.5 *Less Tangible Benefits*:

6.3.5.1 Revenue from enhanced product quality.

6.3.5.2 Revenue from an increased market share of green products.

6.3.5.3 Reduced worker compensation claims and absenteeism costs from improved employee health.

6.3.5.4 Increased productivity from improved employee relations.

6.3.5.5 Reduced staff time in dealing with community concerns.

7. Development of Resource, Emissions, and Waste Measurement and Tracking System

7.1 *Introduction*—The resource, emissions, and waste measurement and tracking system will be used to set pollution prevention goals and objectives, quantify and document progress in reducing wastes, prepare progress reports, and identify and solve problems. An important aspect of the program is the development and maintenance of accurate records on existing resource use and waste generation. However, it is recognized that estimates may be appropriate for developing a new pollution prevention program.

7.2 *Measurement and Tracking*—The measurement and tracking of resource use and waste generation must be specific to the individual organization or facility. There are several specific attributes, as follow, that are necessary for any system to be effective.

7.2.1 *The system must be accurate.* Any system for measuring resource use and waste generation must be accurate and reproducible. The ability to demonstrate progress in pollution prevention will be impeded seriously if it is an arbitrary or estimated value.

7.2.2 *It must fit the organization's goals and objectives.* The goals and objectives that have been established must follow through to the measurement and tracking methodology. The measurement system must be capable of measuring all wastes if the goal is to reduce all wastes. The measurement system must include all costs if the objective is to reduce disposal costs.

7.2.3 *The system should be simple.* In order to ensure the buy-in of all employees, the measurement system must be simple and user friendly. When possible, the use of data already generated for production or accounting purposes is preferred. A system that would require extensive additional measurement or recordkeeping by facility personnel is less likely to succeed than a simple system.

7.2.4 *The system should be integrated with the corporate culture and operational methods.* If an organization is focused heavily on TQM, the measurement system should fit into a TQM system. A TQM system may include all wastes since TQM involves all employees working toward a common goal. On the other hand, a company that does not use TQM, but instead establishes short-range goals, could focus on specific waste streams and consequently adapt its measurement system to that method of operation.

7.2.5 *The system should be flexible.* The measurement system should allow flexibility in reporting. It should be easy to correct or update. Furthermore, the capacity to normalize data should be assessed.

7.2.6 *The system should be continuous.* Once the measurement and tracking system is in place, it should be monitored and modified as required to ensure the continuous accuracy and value of the information.

7.3 *Measurement Methods*—A variety of approaches customized to suit each process should be considered. Measurement methods are listed as follows:

7.4 *Measurement Techniques*:

7.4.1 Material balance.

7.4.2 Materials accounting.

7.4.2.1 Process specific.

7.4.2.2 Facility wide.

7.4.3 Hybrid.

8. Analysis of Pollution Prevention Opportunities

8.1 *Introduction*—The purpose of the opportunity analysis (OA) activity in the pollution prevention program is to develop data with which to prioritize and identify the various pollution prevention opportunities available. The success of the overall pollution prevention process frequently hinges on the success of the first projects undertaken. Success can be defined in many ways: volumetric reductions, avoided liabilities, cost savings, improved communications, etc., according to the organization and its demeanor. Whatever the prevailing factors, pollution prevention priorities should be established according to the financial, physical, and human resources available to the process. It is therefore critical to match opportunities with available resources in order to determine a hierarchy of activities. This prioritization does not ensure, but certainly improves, the probability for success of the pollution prevention program. An example of a checklist for a pollution prevention OA is given in Appendix X2.

8.2 *Source Reduction*—This section of the guide provides direction for identifying opportunities for pollution prevention as a part of the program strategy.

8.2.1 *Inventory Management*:

8.2.1.1 Purchasing practices.

8.2.1.2 Just-in-time manufacturing.

8.2.2 *Product Redesign*—Material substitution.

8.2.3 *Modification of Production Processes*—Equipment modifications.

8.2.4 *Reduction of Waste Volume*—Segregation.

8.3 *Recycling*:

8.3.1 Internal recycling.

8.3.2 Recovery of waste.

8.3.3 External recycling.

8.4 *Reuse/Reclamation.*

8.5 *Operating Practices.*

9. Prioritization of Pollution Prevention Opportunities

9.1 *Introduction*:

9.1.1 A priority setting system for pollution prevention opportunities should be developed based on the management structure of the organization or facility. The part of the organization or level of management charged with identifying and implementing pollution prevention opportunities will determine the degree of formality required for setting priorities in the organization's pollution prevention program.

9.1.2 For example, within a TQM-oriented organization,

each employee would prioritize the opportunities available within his or her level of authority. However, some opportunities that are beyond the expenditure authority of the individuals at a specific facility may be subject to prioritization at a higher level of management. The organization or facility should evaluate the overall impact of proposed opportunities within specific activities.

9.1.3 Within an organizational structure that requires a formal priority setting system prior to program implementation, it will be possible to identify pollution prevention opportunities following the collection of all necessary facility and process information. Once the opportunities have been identified, each must be examined carefully to decide which can be implemented. Flexibility is required in setting priorities.

9.2 *Feasibility Analysis*:

9.2.1 *Technical Feasibility Analysis*—The first step is to determine the technical and economic feasibility of each pollution prevention opportunity. The outcome of the these analyses is the decision whether a specific opportunity will work in the proposed application and a summary of the economic impact. The following factors should be considered in the technical feasibility analysis:

9.2.1.1 *Inventory Management Option*:

(*1*) Affected departments,

(*2*) Impact on production,

(*3*) Training or procedural changes, and

(*4*) Ease of implementing.

9.2.1.2 *Modification of Production Processes Option*:

(*1*) Affected products and production processes,

(*2*) Affected departments,

(*3*) Availability of materials,

(*4*) Availability of equipment,

(*5*) Testing or scale-up requirements,

(*6*) Impact on production,

(*7*) Impact on product quality,

(*8*) Safety considerations,

(*9*) Additional storage or handling requirements,

(*10*) Additional laboratory and analytical requirements,

(*11*) Training or procedural changes,

(*12*) Material testing and quality control requirements, and

(*13*) Ease of implementation.

9.2.1.3 *Waste Reduction and Recycling Option*:

(*1*) Affected departments,

(*2*) Impact on production,

(*3*) Training or procedural changes,

(*4*) Ease of implementation, and

(*5*) Safety considerations.

9.2.1.4 *Waste Recovery, Reuse, and Reclamation Option*:

(*1*) Affected departments,

(*2*) Availability of equipment,

(*3*) Testing or scale-up requirements,

(*4*) Impact on product quality,

(*5*) Additional storage or handling requirements,

(*6*) Additional laboratory and analytical requirements,

(*7*) Training or procedural changes,

(*8*) Ease of implementation, and

(*9*) Safety considerations.

9.2.2 *Economic Feasibility Analysis*—In the economic feasibility analysis, an estimate of the total costs and benefits expected from each option is obtained, based on an evaluation of the following inputs:

9.2.2.1 *Direct Costs or Benefits*:

(*1*) Changes in product revenues;

(*2*) Capital expenditures (buildings, equipment, utility connections, equipment installation, and project engineering);

(*3*) Operation and maintenance expenses and revenues (raw materials and supplies, labor (maintenance and operations), waste disposal, utilities (for example, water, sewer, electricity, and fuels), revenue from recovered materials); and

(*4*) Environmental impairment and liability insurance premium.

9.2.2.2 *Indirect Costs or Benefits*:

(*1*) Compliance costs (permitting, reporting, tracking, monitoring, manifesting, training, waste handling, recordkeeping, labeling, testing, emergency preparedness, and medical surveillance);

(*2*) Waste storage;

(*3*) Operation of on-site pollution control equipment;

(*4*) Replacement of non-product output;

(*5*) Marginal change in costs due to the reduced use or generation of hazardous materials; and

(*6*) Legal fees.

9.2.2.3 *Liability Costs*:

(*1*) Penalties and fines;

(*2*) Damage to property and casualty loss;

(*3*) Legal fees; and

(*4*) Insurance premiums.

9.2.2.4 *Intangibles*:

(*1*) Revenue from enhanced product quality;

(*2*) Revenue from an increased market share of green products;

(*3*) Worker compensation claim and absenteeism costs from improved employee health;

(*4*) Productivity from improved employee relations; and

(*5*) Staff time in dealing with community concerns.

9.3 *Prioritization*:

9.3.1 Upon completion of the technical and economic feasibility analyses, other factors should be considered in order to make the final decisions in setting priorities. These include the ability of the organization to maintain compliance with environmental regulations, overall environmental protection benefits to employees and the community, and overall reduction of the environmental and safety risks of the facility.

9.3.2 Any prioritization method selected is a dynamic process. As the capital and operating costs of individual opportunities change, it may become necessary to reevaluate the opportunities. See Appendix X3 for an example of a method of prioritizing pollution prevention opportunities. As new information becomes available on different opportunities, some may become technically feasible while others become economically impractical.

9.3.3 An effective priority-setting system should do the following:

9.3.3.1 Evaluate the performance of potential projects in terms of goals and objectives such as the following:

(*1*) Total cost savings, such as process savings, treatment and disposal cost savings, and avoidance of estimated future liability. (Savings estimates must be consistent with facility level tracking data and must include estimates of uncertainty.)

(*2*) Waste elimination, such as volume reduction and reduction in waste toxicity. (Estimates must once again be consistent with facility level tracking data.)

(*3*) Response to directives (for example, Montreal Protocol on Substances that Deplete the Ozone Layer or the Basel Convention on the Control of Transboundary Movements of Hazardous Wastes and Their Disposal) to curtail the use of specific chemicals.

(*4*) Response to public and national concerns, such as facility-specific environmental issues, nationwide phase-outs of specific products, and the protection of endangered species.

9.3.3.2 Incorporate feedback on actual plant performance over time in order to adjust annual goals on the basis of objective data in the tracking system.

9.3.3.3 Become more refined over time as better data become available and the program focuses on the more difficult pollution prevention projects.

9.3.3.4 Remain dynamic as new information is obtained.

10. Implementation and Maintenance of Progress of a Pollution Prevention Program

10.1 *Introduction*—After pollution prevention opportunities have been prioritized, it is appropriate to begin to implement and maintain the pollution prevention program. A large part of this step involves developing awareness, training, and motivating the participants. These aspects of the program become increasingly important as the program matures.

10.2 *Publication and Promotion of Pollution Prevention Program*:

10.2.1 *Communications*—It is important for an organization to develop and maintain a continuous process of internal communications addressing the development and milestones of the pollution prevention program. The first step will require the development of an internal communications plan that identifies the goals, objectives, and methods of internal communications.

10.2.2 *Objectives*—Some of the objectives that may be identified include the following:

10.2.2.1 Promote a corporate culture that embraces pollution prevention,

10.2.2.2 Introduce awareness of pollution prevention to all employees,

10.2.2.3 Obtain a commitment to pollution prevention at all management levels,

10.2.2.4 Bring concepts of pollution prevention into daily tasks associated with facility operation and maintenance,

10.2.2.5 Encourage employees to extend pollution prevention concepts into their homes and communities,

10.2.2.6 Commit to an ongoing organizational dialogue on pollution prevention, and

10.2.2.7 Encourage a team approach.

10.2.3 *Tools*:

10.2.3.1 A variety of tools may be used for internal communications. Newsletters or stories can be used as educational tools or for selling the success of the program. The newsletter can initially be used to report the organization's pollution prevention goals. When the program is underway, it can then be used to report on progress toward achieving those goals. Employees may be encouraged to use the newsletter as a forum for suggestions. Furthermore, successful pollution prevention efforts can be spotlighted to motivate employees toward continued success. Organizations may choose to distribute a brochure periodically that identifies the pollution prevention goals for the upcoming year.

10.2.3.2 The communications may also include an employee suggestion box for ideas relating to pollution prevention. A system for understanding the suggestions, as well as monitoring and acting on the employee suggestions, should be developed.

10.2.3.3 Finally, a free-standing or mounted display may be produced that provides a continuous report on that status toward achieving the program goals. The display may also be used to recognize individual, departmental, or team pollution prevention efforts periodically.

10.2.3.4 The following are examples of activities that can be used for publicizing and promoting a pollution prevention program:

(*1*) Conducting open houses and facility tours;

(*2*) Establishing a pollution prevention speakers' bureau;

(*3*) Preparing educational materials;

(*4*) Encouraging the formation of community advisory panels;

(*5*) Sponsorship of community activities and pollution prevention workshops;

(*6*) Providing monetary awards and financial support to employees and community members for pollution prevention activities;

(*7*) Promoting media communications; and

(*8*) Communicating with customers, vendors, and suppliers.

10.2.4 *Reports*—An effective communications program should flow up and down. Periodic reports to upper management that identify program successes and failures are essential. Of equal importance would be senior management's appraisal of the program to the rank-and-file. The latter reports could provide employees with much needed encouragement, even in the face of some failures, to continue their commitment. The organization may choose to distribute a brochure annually that identifies the pollution prevention goals for the upcoming year. Employees may be able to display this in the work areas.

10.3 *Integration of Pollution Prevention Concepts into Ongoing Training Programs*:

10.3.1 *Introduction*:

10.3.1.1 Training is an essential component of a corporate pollution prevention program. A high level of commitment to training is a prerequisite for a viable program. The degree and level of sophistication of training afforded may be tailored to employees or functions, or both, in an organization. However, at a minimum, all employees should receive training that articulates the philosophy of the program and the particular activities that can be taken by the employees involved.

10.3.1.2 Internal training is a critical means for the organization to stress its dedication to, and the importance of, the pollution prevention program. Adding an introduction to the pollution prevention program at new employee orientation programs will emphasize the importance of the programs. In addition, adding elements of the pollution prevention program to the routine training programs will also emphasize the importance of the program to the employees.

10.3.2 *Resources*:

10.3.2.1 Employees must be provided with training in order to be able to plan, develop, implement, and maintain a pollution prevention program. It is therefore important that organizations allocate adequate time and resources for training programs.

10.3.2.2 A training plan developed early in the process should identify the training objectives, scope of training, and audience. The complexity of the training program will be determined by the number of tasks to be covered, as well as the different audiences to be involved. It may be appropriate to develop the training program in modules in order to allow maximum flexibility. For example, employees evaluating and setting priorities for pollution prevention opportunities may benefit more from training in life-cycle costing, while employees operating a production process may benefit more from training in specific pollution prevention option implementation.

10.3.3 *Delivery*—Training might involve discussions in group meetings or may require the delivery of a detailed training course. In addition to serving as reference materials, training manuals help to minimize the most common causes of lack of consistency in performance, namely, a lack of accountability and poor understanding of the requirements.

10.4 *Personnel or Group Awards and Incentive Programs*:

10.4.1 *Introduction*—The development of individual or group incentive programs may also be useful for training employees in the importance of the pollution prevention program and their role in such a program. This may include monthly recognition awards for achieving certain levels of performance or an annual award for overall performance.

10.4.2 *Recognition*:

10.4.2.1 Recognition and incentive programs that acknowledge the successful efforts of employees can be significant motivators. Programs should be structured to identify the types of activities that deserve recognition. For example, an organization may recognize activities that are keyed to measurable goals or linked to the accomplishment of these goals.

10.4.2.2 While the incentives may take the form of special recognition or monetary awards, it is also important to recognize the potential for overall improvement in the quality of the work environment due to pollution prevention efforts.

10.5 *Quality Assurance and Quality-Control Procedures*:

10.5.1 *Introduction*:

10.5.1.1 The development and implementation of quality assurance and quality control (QA/QC) procedures may be useful for determining the baseline levels of operations and establishing performance goals, while accomplishing quality goals. Whenever possible, it is important that the pollution prevention QA/QC program be integrated with the organization's other QA/QC programs.

10.5.1.2 The development and implementation of QA/QC procedures can be critical to the effectiveness and success of the pollution prevention program. Commitment to quality builds a pollution prevention program that encourages continuous improvement.

10.5.2 *Process*—A quality improvement process generally includes the following activities:

10.5.2.1 Measurement of existing conditions using quantifiable parameters.

10.5.2.2 Identification of improvement opportunities, that is, specific projects.

10.5.2.3 Establishment of agreed upon goals.

10.5.2.4 Development of a detailed plan for improvement for each project.

10.5.2.5 Identification and implementation of training requirements.

10.5.2.6 Measurement, monitoring, and documentation of results of the plan.

10.5.2.7 Evaluation of results, analysis of failures, and communication of the lessons learned.

10.5.2.8 Go to Step 1 (see 10.5.2.1) and repeat.

10.6 *Contingency Plan*—The development of a contingency plan will allow the organization to cope with unforeseen factors that may impact the development and implementation of the organization's pollution prevention program positively or negatively. Contingency planning should be built into every component of the pollution prevention program. Environmental and safety factors or equipment deficiencies, or both, caused by factors beyond those recognized as part of prudent planning and operating practices, could hinder pollution prevention efforts. This planning should establish an effective means of anticipating, assessing, and responding, in an appropriate and coordinated manner, to a long- or short-term deficiency. Contingency planning may range from developing an organizational plan to creating a public relations program.

10.7 *Public Relations Program*:

10.7.1 *Introduction*—Developing a sound and responsive public relations program will provide the basis for enhancing an organization's credibility within its community. To some extent, the pollution prevention public relations program can dovetail with the organization's risk communication efforts to address the public's concerns about health, safety, environmental protection, and public policy. The public relations program must have clearly defined objectives and must identify specific action plans for communicating with all targeted sectors of the public, including the local community, the public and private sector organizations, and the media.

10.7.2 *Objectives*—The objectives of the public relations program may include the following:

10.7.2.1 Projecting a positive public image.

10.7.2.2 Establishing good relationships within the community.

10.7.2.3 Establishing and maintaining an ongoing dialogue with the public.

10.7.2.4 Communicating responsible pollution prevention practices as good corporate citizens.

10.7.2.5 Creating a continuum of daily practices promoting pollution prevention as a framework for daily life within the community.

10.7.2.6 Promoting and coordinating efforts with local public and private sector organizations.

10.7.3 *Action Plans*—Action plans must be tailored specifically to the targeted audience, reflecting its perspective, level of technical knowledge, and concerns. The following activities can be used for meeting the selected objectives:

10.7.3.1 Conducting open houses and facility tours.

10.7.3.2 Establishing a speakers' bureau.

10.7.3.3 Preparing educational materials.

10.7.3.4 Encouraging the formation of community advisory panels.

10.7.3.5 Sponsorship of community activities and pollution prevention workshops.

10.7.3.6 Providing monetary awards and financial support to community members' pollution prevention efforts.

10.7.3.7 Promoting media communications.

10.8 *Review of Program*—A periodic review of the program should be performed as a method of evaluating the past and ensuring the future effectiveness of the pollution prevention program. This should include review of the following:

10.8.1 Initial program goals and baseline information.

10.8.2 Results of program activities.

10.8.3 Program priorities.

10.8.4 Implementation of contingencies.

10.9 The best methods for evaluating the success of the pollution program will depend on the goals and objectives of the program. In cases in which economic criteria are important, it might be appropriate to use a method of economic evaluation.

10.10 The review should be announced and supported by senior management, facilitated by the pollution prevention task force members, and implemented by members of the general work force. Review should be performed frequently enough that it provides the organization and program with information necessary to function in the most efficient and effective manner, normally every one to three years.

10.11 Failures, as well as successes, should be communicated throughout the organization, in objective terms. Individuals deserving credit should be recognized formally. Unsuccessful efforts should be evaluated for their cause and corrective actions taken and communicated.

APPENDIXES

(Nonmandatory Information)

X1. SUGGESTED ADDITIONAL PUBLICATIONS

Achievements in Source Reduction and Recycling for Ten Industries in the United States, U.S. Environmental Protection Agency, EPA/600/2-91/052, 1991.

Approaches to Source Reduction: Practical Guidelines from Existing Policies and Programs, Environmental Defense Fund, 1986.

Campbell, John M., Jr., "Market-Based Incentives for Pollution Prevention Projects," presented at the *Hazardous Materials Management Conference & Exhibition International*, Atlantic City, NJ, June 1988.

Cebon, Peter B., "Organizational Behavior as a Key Element in Waste Management Decision Making," presented at the *Second International Conference on Pollution Prevention*, Washington, DC, July 1990.

Cunningham, V., et al., *ECAS-Environmental Cost Analysis System*, Van Nostrand Reinhold, New York, NY, 1986.

Dharmavaram, Seshasayi, J. Mount, Brian, and Donahue, Bernard A., "Automated Economic Analysis Model for Hazardous Waste Minimization," *Journal of Air and Waste Management Association*, Vol 40, No. 7, July 1990, pp. 1004–1011.

Economic Implications of Waste Reduction, Recycling, Treatment and Disposal of Hazardous Wastes, The Fourth Biennial Report, California Department of Health Services, Toxic Substances Control Division, Alternative Technology Section, Sacramento, CA, 1988.

Economic Incentives: Options for Environmental Protection, Report of the U.S. EPA Economic Incentives Task Force, Office of Policy, Planning, and Evaluation, U.S. Environmental Protection Agency, 1991.

Financial Analysis of Waste Management Alternatives, General Electric Company, Corporate Environmental Programs, 1987.

Fromm, Carl H., and Butler, David," Practical Guidelines for Estimating the Profitability of Waste Minimization Measures," Jacobs Engineering Group, Inc., Pasadena, CA.

Gulledge, William P., "Developing a Pollution Liability Insurance Underwriting Model: Managing for the Potential Exposure from Toxic Releases," *The Environmental Professsional*, Vol 11, 1989, pp. 447–453.

Karem, Joseph, Verderese, James, and Bailey, Paul, "How Full Cost Accounting Can Drive the Economic Viability of a Pollution Prevention Project," presented at the *Hazardous Materials Management Conference & Exhibit/International*, Atlantic City, NJ, June 1990.

MacLean, Richard W., "Financial Analysis of Waste Minimization Options—Considerations in Developing a Corporate Program," presented at the *1987 Waste Minimization Workshop*, Chemical Manufacturer's Association, Washington, DC, Nov. 11–13, 1987.

Pollution Prevention Benefits Manual: Vol 1: The Manual, Phase II, U.S. Environmental Protection Agency, 1989.

Report on Barriers to Pollution Prevention, Minnesota Office of Waste Management, 1991.

Wigglesworth, David, *Profiting From Waste Reduction in Your Small Business*, Alaska Health Project, Anchorage, AK, 1988.

X2. EXAMPLE OF A POLLUTION PREVENTION OPPORTUNITY ANALYSIS CHECKLIST

X2.1 *Documentation Review*—The following data sources may be accessed (if applicable and available) in order to understand facility operations thoroughly and in order to begin identifying source reduction opportunities:

X2.1.1 *Product Design Criteria*.

X2.1.2 *Process Flow Diagrams for all Solid Waste, Wastewater, and Air Emissions Production Sources*—Ideally, these diagrams will include, or be accompanied by, the following:

X2.1.2.1 Flow data.

X2.1.2.2 Waste concentration or physical and chemical characterization data, or both; raw material input data.

X2.1.2.3 Typical production rates for end products.

X2.1.2.4 Points of waste generation.

X2.1.2.5 Continuous versus batch nature of process.

X2.1.2.6 Wastewater and hazardous waste treatment facilities.

X2.1.2.7 Air pollution control devices.

X2.1.2.8 Machinery lubricating systems.

X2.1.2.9 Hydraulic oil circulating systems.

X2.1.3 *Site Maps, Showing the Locations of All Pertinent Units and Process Flow Diagrams as Described Above*—Information including the following:

X2.1.3.1 Cooling towers.

X2.1.3.2 Cooling water users.

X2.1.3.3 Wastewater treatment systems.

X2.1.3.4 Air pollution control devices.

X2.1.3.5 Points of discharge to the sewer system.

X2.1.3.6 Land treatment or disposal units, or both.

X2.1.3.7 Bulk raw and waste material storage components, systems, and facilities.

X2.1.4 *Additional Documentation*—Information such as follows:

X2.1.4.1 Material safety data sheets (MSDS) for all raw materials and process equipment cleaning chemicals.

X2.1.4.2 Any MILSPECs (military specifications), FDA, or other customer specifications that would preclude or hamper the process modifications (for example, if cyanide zinc plating were required by customer specifications).

X2.1.4.3 Publicly owned treatment works (POTW) permits, a national pollutant discharge elimination system (NPDES) permit, and any reports of exceedence of those permits.

X2.1.4.4 Resource conservation and recovery act (RCRA) permits and relevant requirements.

X2.1.4.5 Hazardous waste manifests and annual and biennial reports.

X2.1.4.6 SARA Title III reports of chemical usage, discharge, and waste minimization for the latest years.

X2.1.4.7 State and local air, waste, solid, and hazardous waste permits and applications.

X2.1.5 *Economic Data*—Information such as follows:

X2.1.5.1 Cost of raw materials management, including material cost, storage, and health and safety programs.

X2.1.5.2 Cost of air, wastewater, and hazardous waste treatment chemicals.

X2.1.5.3 Waste management operating and maintenance costs, including labor, transportation, sewer use fees, and solid and hazardous waste disposal costs.

X2.1.6 *Managerial Information*—Information including the following may be reviewed:

X2.1.6.1 Environmental policies and procedures at all relevant levels, including organization charts.

X2.1.6.2 Prioritization of waste management concerns (such as whether any environmental issues take precedence over other issues).

X2.1.6.3 Automated waste tracking system databases or other relevant computerized waste management systems.

X2.1.6.4 Inventory and distribution procedures including procurement policies and routine and non-routine maintenance scheduling practices.

X2.1.6.5 The basis of source reduction decisions and policies (for example, economics, environmental and safety liability concerns, capital costs, potential operational benefits, and ease of implementation).

X2.1.6.6 Any planned modifications or revisions to existing operations that would impact waste generation activities.

X2.2 *Facility Tour*:

X2.2.1 *Overview*—A tour of the facility should be scheduled after review of all documentation. Department managers should be available to describe and discuss their area of the facility as it relates to pollution prevention. In addition to production department managers, maintenance, purchasing, R&D, and engineering management information systems (MIS) managers should be contacted as appropriate.

X2.2.2 *Purpose of Plant Tour*—The purpose of the plant tour is to verify the information contained in the documentation and to gain a better understanding of the points of view of each of the individuals involved. These personnel typically have unique insights into potential projects that may not be identified from the documentation. (For example, a production department manager may be aware that they already have plans to make the required source reductions or perhaps discontinue the production process.)

X2.2.3 *Identification of Materials*—The material needs to be defined clearly before going to the marketplace. Each component material needs to be identified by type, origin, supply manufacturer, each number, and trade name. The basic parameters of the material should be identified (such as viscosity of the liquid, durometer of the rubber, hardness of the metal, weight of the paper, flow rate of the plastic, etc.).

X2.2.3.1 *Sources of Information*:

(*1*) Vendor product data.

(*2*) Material safety data sheets (MSDS).

(*3*) Waste stream profile sheet.

(*4*) Purchase specifications.

(*5*) Modifications during processing (chemical reactions and mechanical modifications and additive introduced).

X2.2.3.2 *Quantification*—To determine the volume and weight generated on a time basis, both the steady-state and periodic state of generation must be reviewed. For purposes of reuse and reclamation, a sufficient quantity must be available to justify the expense of qualification for use.

X2.2.3.3 *Marketing Plan Development*—After it is understood that the material is to be reused and reclaimed, a market analysis must be completed. The following sources should be reviewed.

(*1*) Survey of existing literature.

(*2*) Discussion with brokers, dealers, and recyclers.

(*3*) Discussion with internal technical personnel.

(*4*) Discussion with universities and consultants.

X2.2.3.4 *Implementation*—Sample both internal and external users and recyclers.

X2.2.3.5 *Other Considerations*:

(*1*) Transportation to accumulation point (method of transportation, who will pay transportation charges, whether the material is hazardous, and whether special handling is required).

(*2*) Accumulation (who will develop the accumulation system; where the accumulation system will be located; whether the accumulation point will have to be volume reduced, for either confidential reasons or freight savings; and who will pay the accumulation costs).

(*3*) Transportation to the recycling location (what the method of transportation is, who will pay the transportation charges, whether the material is hazardous, and whether special handling is required).

(*4*) Company(s) to reuse and reclaim the material (whether the reclaimer credit is worthy (Has a credit check been done?), whether the material is hazardous (Has the facility been inspected?), ownership of the facility, and contractual arrangements).

X3. EXAMPLE OF A METHOD OF EVALUATION OF POLLUTION PREVENTION OPPORTUNITIES

X3.1 *Summary*—Each waste stream or operation is assigned a priority rank using a weighted sum method based on a ranking criteria, the relative weight of each criteria, and the rating of the waste stream or operation for each criteria.

X3.2 *Procedure*:

X3.2.1 *Ranking Criteria*—Establish the ranking criteria on the basis of corporate or facility pollution prevention policies and objectives, or both. Typical criteria, or factors used in establishing the criteria, may include the following:

X3.2.1.1 Compliance with current regulations.

X3.2.1.2 Compliance with future regulations.

X3.2.1.3 Total cost of materials management, including waste management, and including all labor costs.

X3.2.1.4 Potential environmental and safety liability.

X3.2.1.5 Quantity of waste.

X3.2.1.6 Hazardous properties of input materials or wastes.

X3.2.1.7 Ease of implementation potential for the pollution prevention method.

X3.2.1.8 Potential operational benefits.

X3.2.1.9 Pollution prevention budget.

X3.2.2 *Weight Criteria*—Establish the relative weight of each ranking criteria, and assign a numerical value to each criteria that reflects its importance (for example, 1 through 10).

X3.2.3 *Rating Scale*—Establish a rating scale (for example, 1 through 10) to be applied uniformly to each criteria.

X3.2.4 *Ratings*—Assign a rating to each waste stream for each criteria, calculate the product of each rating and the criteria weight, and sum the products to determine the waste stream priority rating score.

X3.2.5 *Order*—Rank order each waste stream or operation based on the priority rating score.

X3.3 *Report*—Brief descriptions of each waste stream operation should be prepared and listed in accordance with rank order from the most important to least important.

Standard Guide for Risk-Based Corrective Action Applied at Petroleum Release Sites[1]

ϵ^1 NOTE—Editorial changes were made throughout in December 1996.

1. Scope

1.1 This is a guide to risk-based corrective action (RBCA), which is a consistent decision-making process for the assessment and response to a petroleum release, based on the protection of human health and the environment. Sites with petroleum release vary greatly in terms of complexity, physical and chemical characteristics, and in the risk that they may pose to human health and the environment. The RBCA process recognizes this diversity, and uses a tiered approach where corrective action activities are tailored to site-specific conditions and risks. While the RBCA process is not limited to a particular class of compounds, this guide emphasizes the application of RBCA to petroleum product releases through the use of the examples. Ecological risk assessment, as discussed in this guide, is a qualitative evaluation of the actual or potential impacts to environmental (nonhuman) receptors. *There may be circumstances under which a more detailed ecological risk assessment is necessary* (see Ref **(1)**.[2]

1.2 The decision process described in this guide integrates risk and exposure assessment practices, as suggested by the United States Environmental Protection Agency (USEPA), with site assessment activities and remedial measure selection to ensure that the chosen action is protective of human health and the environment. The following general sequence of events is prescribed in RBCA, once the process is triggered by the suspicion or confirmation of petroleum release:

1.2.1 Performance of a site assessment;

1.2.2 Classification of the site by the urgency of initial response;

1.2.3 Implementation of an initial response action appropriate for the selected site classification;

1.2.4 Comparison of concentrations of chemical(s) of concern at the site with Tier 1 Risk Based Screening Levels (RBSLs) given in a look-up table;

1.2.5 Deciding whether further tier evaluation is warranted, if implementation of interim remedial action is warranted or if RBSLs may be applied as remediation target levels;

1.2.6 Collection of additional site-specific information as necessary, if further tier evaluation is warranted;

1.2.7 Development of site-specific target levels (SSTLs) and point(s) of compliance (Tier 2 evaluation);

1.2.8 Comparison of the concentrations of chemical(s) of concern at the site with the Tier 2 evaluation SSTL at the determined point(s) of compliance or source area(s);

1.2.9 Deciding whether further tier evaluation is warranted, if implementation of interim remedial action is warranted, or if Tier 2 SSTLs may be applied as remediation target levels;

1.2.10 Collection of additional site-specific information as necessary, if further tier evaluation is warranted;

1.2.11 Development of SSTL and point(s) of compliance (Tier 3 evaluation);

1.2.12 Comparison of the concentrations of chemical(s) of concern at the site at the determined point(s) of compliance or source area(s) with the Tier 3 evaluation SSTL; and

1.2.13 Development of a remedial action plan to achieve the SSTL, as applicable.

1.3 The guide is organized as follows:

1.3.1 Section 2 lists referenced documents,

1.3.2 Section 3 defines terminology used in this guide,

1.3.3 Section 4 describes the significance and use of this guide,

1.3.4 Section 5 is a summary of the tiered approach,

1.3.5 Section 6 presents the RBCA procedures in a step-by-step process,

1.3.6 Appendix X1 details physical/chemical and toxicological characteristics of petroleum products,

1.3.7 Appendix X2 discusses the derivation of a Tier 1 RBSL Look-Up Table and provides an example,

1.3.8 Appendix X3 describes the uses of predictive modeling relative to the RBCA process,

1.3.9 Appendix X4 discusses considerations for institutional controls, and

1.3.10 Appendix X5 provides examples of RBCA applications.

1.4 This guide describes an approach for RBCA. It is intended to compliment but not supersede federal, state, and local regulations. Federal, state, or local agency approval may

---

[1] This guide is under the jurisdiction of ASTM Committee E-50 on Environmental Assessment and is the direct responsibility of Subcommittee E50.01 on Storage Tanks.

Current edition approved Sept. 10, 1995. Published November 1995. Originally published as ES 38 – 94. Last previous edition ES 38 – 94.

[2] The boldface numbers in parentheses refer to the list of references at the end of this guide.

be required to implement the processes outlined in this guide.

1.5 The values stated in either inch-pound or SI units are to be regarded as the standard. The values given in parentheses are for information only.

1.6 *This standard does not purport to address all of the safety concerns, if any, associated with its use. It is the responsibility of the user of this standard to establish appropriate safety and health practices and determine the applicability of regulatory limitations prior to use.*

2. Referenced Documents

2.1 *ASTM Standards:*

E 1599 Guide for Corrective Action for Petroleum Releases[3]

2.2 *NFPA Standard:*

NFPA 329 Handling Underground Releases of Flammable and Combustible Liquids[4]

3. Terminology

3.1 *Definitions of Terms Specific to This Standard:*

3.1.1 *active remediation*—actions taken to reduce the concentrations of chemical(s) of concern. Active remediation could be implemented when the no-further-action and passive remediation courses of action are not appropriate.

3.1.2 *attenuation*—the reduction in concentrations of chemical(s) of concern in the environment with distance and time due to processes such as diffusion, dispersion, absorption, chemical degradation, biodegradation, and so forth.

3.1.3 *chemical(s) of concern*—specific constituents that are identified for evaluation in the risk assessment process.

3.1.4 *corrective action*—the sequence of actions that include site assessment, interim remedial action, remedial action, operation and maintenance of equipment, monitoring of progress, and termination of the remedial action.

3.1.5 *direct exposure pathways*—an exposure pathway where the point of exposure is at the source, without a release to any other medium.

3.1.6 *ecological assessment*—a qualitative appraisal of the actual or potential effects of chemical(s) of concern on plants and animals other than people and domestic species.

3.1.7 *engineering controls*—modifications to a site or facility (for example, slurry walls, capping, and point of use water treatment) to reduce or eliminate the potential for exposure to a chemical(s) of concern.

3.1.8 *exposure*—contact of an organism with chemical(s) of concern at the exchange boundaries (for example, skin, lungs, and liver) and available for absorption.

3.1.9 *exposure assessment*—the determination or estimation (qualitative or quantitative) of the magnitude, frequency, duration, and route of exposure.

3.1.10 *exposure pathway*—the course a chemical(s) of concern takes from the source area(s) to an exposed organism. An exposure pathway describes a unique mechanism by which an individual or population is exposed to a chemical(s) of concern originating from a site. Each exposure pathway includes a source or release from a source, a point of exposure, and an exposure route. If the exposure point differs from the source, a transport/exposure medium (for example, air) or media also is included.

3.1.11 *exposure route*—the manner in which a chemical(s) of concern comes in contact with an organism (for example, ingestion, inhalation, and dermal contact).

3.1.12 *facility*—the property containing the source of the chemical(s) of concern where a release has occurred.

3.1.13 *hazard index*—the sum of two or more hazard quotients for multiple chemical(s) of concern or multiple exposure pathways, or both.

3.1.14 *hazard quotients*—the ratio of the level of exposure of a chemical(s) of concern over a specified time period to a reference dose for that chemical(s) of concern derived for a similar exposure period.

3.1.15 *incremental carcinogenic risk levels*—the potential for incremental carcinogenic human health effects due to exposure to the chemical(s) of concern.

3.1.16 *indirect exposure pathways*—an exposure pathway with at least one intermediate release to any media between the source and the point(s) of exposure (for example, chemicals of concern from soil through ground water to the point(s) of exposure).

3.1.17 *institutional controls*—the restriction on use or access (for example, fences, deed restrictions, restrictive zoning) to a site or facility to eliminate or minimize potential exposure to a chemical(s) of concern.

3.1.18 *interim remedial action*—the course of action to mitigate fire and safety hazards and to prevent further migration of hydrocarbons in their vapor, dissolved, or liquid phase.

3.1.19 *maximum contaminant level (MCL)*—a standard for drinking water established by USEPA under the Safe Drinking Water Act, which is the maximum permissible level of chemical(s) of concern in water that is delivered to any user of a public water supply.

3.1.20 *Monte Carlo simulation*—a procedure to estimate the value and uncertainty of the result of a calculation when the result depends on a number of factors, each of which is also uncertain.

3.1.21 *natural biodegradation*—the reduction in concentration of chemical(s) of concern through naturally occurring microbial activity.

3.1.22 *petroleum*—including crude oil or any fraction thereof that is liquid at standard conditions of temperature and pressure (60°F and 14.7 lb/in.^{2} absolute; (15.5°C and 10 335.6 kg/m^2)). The term includes petroleum-based substances comprised of a complex blend of hydrocarbons derived from crude oil through processes of separation, conversion, upgrading, and finishing, such as motor fuels, jet oils, lubricants, petroleum solvents, and used oils.

3.1.23 *point(s) of compliance*—a location(s) selected between the source area(s) and the potential point(s) of exposure where concentrations of chemical(s) of concern must be at or below the determined target levels in media (for example, ground water, soil, or air).

[3] *Annual Book of ASTM Standards*, Vol 11.04.

[4] Available from National Fire Protection Association, 1 Batterymarch Park, P.O. Box 9101, Quincy, MA 02269.

3.1.24 *point(s) of exposure*—the point(s) at which an individual or population may come in contact with a chemical(s) of concern originating from a site.

3.1.25 *qualitative risk analysis*—a nonnumeric evaluation of a site to determine potential exposure pathways and receptors based on known or readily available information.

3.1.26 *reasonable maximum exposure (RME)*—the highest exposure that is reasonably expected to occur at a site. RMEs are estimated for individual pathways or a combination of exposure pathways.

3.1.27 *reasonable potential exposure scenario*— a situation with a credible chance of occurence where a receptor may become directly or indirectly exposed to the chemical(s) of concern without considering extreme or essentially impossible circumstances.

3.1.28 *reasonably anticipated future use*—future use of a site or facility that can be predicted with a high degree of certainty given current use, local government planning, and zoning.

3.1.29 *receptors*—persons, structures, utilities, surface waters, and water supply wells that are or may be adversely affected by a release.

3.1.30 *reference dose*—a preferred toxicity value for evaluating potential noncarcinogenic effects in humans resulting from exposure to a chemical(s) of concern.

3.1.31 *remediation/remedial action*—activities conducted to protect human health, safety, and the environment. These activities include evaluating risk, making no-further-action determinations, monitoring institutional controls, engineering controls, and designing and operating cleanup equipment.

3.1.32 *risk assessment*—an analysis of the potential for adverse health effects caused by a chemical(s) of concern from a site to determine the need for remedial action or the development of target levels where remedial action is required.

3.1.33 *risk reduction*—the lowering or elimination of the level of risk posed to human health or the environment through interim remedial action, remedial action, or institutional or engineering controls.

3.1.34 *risk-based screening level/screening levels (RBSLs)*—risk-based site-specific corrective action target levels for chemical(s) of concern developed under the Tier 1 evaluation.

3.1.35 *site*—the area(s) defined by the extent of migration of the chemical(s) of concern.

3.1.36 *site assessment*—an evaluation of subsurface geology, hydrology, and surface characteristics to determine if a release has occurred, the levels of the chemical(s) of concern, and the extent of the migration of the chemical(s) of concern. The site assessment collects data on ground water quality and potential receptors and generates information to support remedial action decisions.

3.1.37 *site classification*—a qualitative evaluation of a site based on known or readily available information to identify the need for interim remedial actions and further information gathering. Site classification is intended to specifically prioritize sites.

3.1.38 *site-specific target level (SSTL)*—risk-based remedial action target level for chemical(s) of concern developed for a particular site under the Tier 2 and Tier 3 evaluations.

3.1.39 *site-specific*—activities, information, and data unique to a particular site.

3.1.40 *source area(s)*—either the location of liquid hydrocarbons or the location of highest soil and ground water concentrations of the chemical(s) of concern.

3.1.41 *target levels*—numeric values or other performance criteria that are protective of human health, safety, and the environment.

3.1.42 *Tier 1 evaluation*—a risk-based analysis to develop non-site-specific values for direct and indirect exposure pathways utilizing conservative exposure factors and fate and transport for potential pathways and various property use categories (for example, residential, commercial, and industrial uses). Values established under Tier 1 will apply to all sites that fall into a particular category.

3.1.43 *Tier 2 evaluation*—a risk-based analysis applying the direct exposure values established under a Tier 1 evaluation at the point(s) of exposure developed for a specific site and development of values for potential indirect exposure pathways at the point(s) of exposure based on site-specific conditions.

3.1.44 *Tier 3 evaluation*—a risk-based analysis to develop values for potential direct and indirect exposure pathways at the point(s) of exposure based on site-specific conditions.

3.1.45 *user*—an individual or group involved in the RBCA process including owners, operators, regulators, underground storage tank (UST) fund managers, attorneys, consultants, legislators, and so forth.

4. Significance and Use

4.1 The allocation of limited resources (for example, time, money, regulatory oversight, qualified professionals) to any one petroleum release site necessarily influences corrective action decisions at other sites. This has spurred the search for innovative approaches to corrective action decision making, which still ensures that human health and the environment are protected.

4.2 The RBCA process presented in this guide is a consistent, streamlined decision process for selecting corrective actions at petroleum release sites. Advantages of the RBCA approach are as follows:

4.2.1 Decisions are based on reducing the risk of adverse human or environmental impacts,

4.2.2 Site assessment activities are focussed on collecting only that information that is necessary to making risk-based corrective action decisions,

4.2.3 Limited resources are focussed on those sites that pose the greatest risk to human health and the environment at any time,

4.2.4 The remedial action achieves an acceptable degree of exposure and risk reduction,

4.2.5 Compliance can be evaluated relative to site-specific standards applied at site-specific point(s) of compliance,

4.2.6 Higher quality, and in some cases faster, cleanups than are currently realized, and

4.2.7 A documentation and demonstration that the remedial action is protective of human health, safety, and the environment.

4.3 Risk assessment is a developing science. The scientific

approach used to develop the RBSL and SSTL may vary by state and user due to regulatory requirements and the use of alternative scientifically based methods.

4.4 Activities described in this guide should be conducted by a person familiar with current risk and exposure assessment methodologies.

4.5 In order to properly apply the RBCA process, the user should avoid the following:

4.5.1 Use of Tier 1 RBSLs as mandated remediation standards rather than screening levels,

4.5.2 Restriction of the RBCA process to Tier 1 evaluation only and not allowing Tier 2 or Tier 3 analyses,

4.5.3 Placing arbitrary time constraints on the corrective action process; for example, requiring that Tiers 1, 2, and 3 be completed within 30-day time periods that do not reflect the actual urgency of and risks posed by the site,

4.5.4 Use of the RBCA process only when active remediation is not technically feasible, rather than a process that is applicable during all phases of corrective action,

4.5.5 Requiring the user to achieve technology-based remedial limits (for example, asymptotic levels) prior to requesting the approval for the RBSL or SSTL,

4.5.6 The use of predictive modelling that is not supported by available data or knowledge of site conditions,

4.5.7 Dictating that corrective action goals can only be achieved through source removal and treatment actions, thereby restricting the use of exposure reduction options, such as engineering and institutional controls,

4.5.8 The use of unjustified or inappropriate exposure factors,

4.5.9 The use of unjustified or inappropriate toxicity parameters,

4.5.10 Neglecting aesthetic and other criteria when determining RBSLs or SSTLs,

4.5.11 Not considering the effects of additivity when screening multiple chemicals,

4.5.12 Not evaluating options for engineering or institutional controls, exposure point(s), compliance point(s), and carcinogenic risk levels before submitting remedial action plans,

4.5.13 Not maintaining engineering or institutional controls, and

4.5.14 Requiring continuing monitoring or remedial action at sites that have achieved the RBSL or SSTL.

5. Tiered Approach to Risk-Based Corrective Action (RBCA) at Petroleum Release Sites

5.1 RBCA is the integration of site assessment, remedial action selection, and monitoring with USEPA-recommended risk and exposure assessment practices. This creates a process by which corrective action decisions are made in a consistent manner that is protective of human health and the environment.

5.2 The RBCA process is implemented in a tiered approach, involving increasingly sophisticated levels of data collection and analysis. The assumptions of earlier tiers are replaced with site-specific data and information. Upon evaluation of each tier, the user reviews the results and recommendations and decides whether more site-specific analysis is warranted.

5.3 *Site Assessment*— The user is required to identify the sources of the chemical(s) of concern, obvious environmental impacts (if any), any potentially impacted humans and environmental receptors (for example, workers, residents, water bodies, and so forth), and potentially significant transport pathways (for example, ground water flow, utilities, atmospheric dispersion, and so forth). The site assessment will also include information collected from historical records and a visual inspection of the site.

5.4 *Site Classification*—Sites are classified by the urgency of need for initial response action, based on information collected during the site assessment. Associated with site classifications are initial response actions that are to be implemented simultaneously with the RBCA process. Sites should be reclassified as actions are taken to resolve concerns or as better information becomes available.

5.5 *Tier 1 Evaluation*—A look-up table containing screening level concentrations is used to determine whether site conditions satisfy the criteria for a quick regulatory closure or warrant a more site-specific evaluation. Ground water, soil, and vapor concentrations may be presented in this table for a range of site descriptions and types of petroleum products ((for example, gasoline, crude oil, and so forth). The look-up table of RBSL is developed in Tier 1 or, if a look-up table has been previously developed and determined to be applicable to the site by the user, then the existing RBSLs are used in the Tier 1 process. Tier 1 RBSLs are typically derived for standard exposure scenarios using current RME and toxicological parameters as recommended by the USEPA. These values may change as new methodologies and parameters are developed. Tier 1 RBSLs may be presented as a range of values, corresponding to a range of risks or property uses.

5.6 *Tier 2 Evaluation*—Tier 2 provides the user with an option to determine SSTLs and point(s) of compliance. It is important to note that both Tier 1 RBSL and Tier 2 SSTLs are based on achieving similar levels of protection of human health and the environment (for example, 10^{-4} to 10^{-6} risk levels). However, in Tier 2 the non-site-specific assumptions and point(s) of exposure used in Tier 1 are replaced with site-specific data and information. Additional site-assessment data may be needed. For example, the Tier 2 SSTL can be derived from the same equations used to calculate the Tier 1 RBSL, except that site-specific parameters are used in the calculations. The additional site-specific data may support alternate fate and transport analysis. At other sites, the Tier 2 analysis may involve applying Tier 1 RBSLs at more probable point(s) of exposure. Tier 2 SSTLs are consistent with USEPA-recommended practices.

5.7 *Tier 3 Evaluation*—Tier 3 provides the user with an option to determine SSTLs for both direct and indirect pathways using site-specific parameters and point(s) of exposure and compliance when it is judged that Tier 2 SSTLs should not be used as target levels. Tier 3, in general, can be a substantial incremental effort relative to Tiers 1 and 2, as the evaluation is much more complex and may include additional site assessment, probabilistic evaluations, and sophisticated chemical fate/transport models.

5.8 *Remedial Action*— If the concentrations of chemical(s) of concern at a site are above the RBSL or SSTL at the point(s)

of compliance or source area, or both, and the user determines that the RBSL or SSTL should be used as remedial action target levels, the user develops a remedial action plan in order to reduce the potential for adverse impacts. The user may use remediation processes to reduce concentrations of the chemical(s) of concern to levels below or equal to the target levels or to achieve exposure reduction (or elimination) through institutional controls discussed in Appendix X4, or through the use of engineering controls, such as capping and hydraulic control.

6. Risk-Based Corrective Action (RBCA) Procedures

6.1 The sequence of principal tasks and decisions associated with the RBCA process are outlined on the flowchart shown in Fig. 1. Each of these actions and decisions is discussed as follows.

6.2 *Site Assessment*— Gather the information necessary for site classification, initial response action, comparison to the RBSL, and determining the SSTL. Site assessment may be conducted in accordance with Guide E 1599. Each successive tier will require additional site-specific data and information that must be collected as the RBCA process proceeds. The user may generate site-specific data and information or estimate reasonable values for key physical characteristics using soil survey data and other readily available information. The site characterization data should be summarized in a clear and concise format.

6.2.1 The site assessment information for Tier 1 evaluation may include the following:

6.2.1.1 A review of historical records of site activities and past releases;

6.2.1.2 Identification of chemical(s) of concern;

6.2.1.3 Location of major sources of the chemical(s) of concern;

6.2.1.4 Location of maximum concentrations of chemical(s) of concern in soil and ground water;

6.2.1.5 Location of humans and the environmental receptors that could be impacted (point(s) of exposure);

6.2.1.6 Identification of potential significant transport and exposure pathways (ground water transport, vapor migration through soils and utilities, and so forth);

6.2.1.7 Determination of current or potential future use of the site and surrounding land, ground water, surface water, and sensitive habitats;

6.2.1.8 Determination of regional hydrogeologic and geologic characteristics (for example, depth to ground water, aquifer thickness, flow direction, gradient, description of confining units, and ground water quality); and

6.2.1.9 A qualitative evaluation of impacts to environmental receptors.

6.2.2 In addition to the information gathered in 6.2.1, the site assessment information for Tier 2 evaluation may include the following:

6.2.2.1 Determination of site-specific hydrogeologic and geologic characteristics (for example, depth to ground water, aquifer thickness, flow direction, gradient, description of confining units, and ground water quality);

6.2.2.2 Determination of extent of chemical(s) of concern relative to the RBSL or SSTL, as appropriate;

6.2.2.3 Determination of changes in concentrations of chemical(s) of concern over time (for example, stable, increasing, and decreasing); and

6.2.2.4 Determination of concentrations of chemical(s) of concern measured at point(s) of exposure (for example, dissolved concentrations in nearby drinking water wells or vapor concentrations in nearby conduits or sewers).

6.2.3 In addition to the information gathered in 6.2.1 and 6.2.2, the site assessment information for Tier 3 evaluation includes additional information that is required for site-specific modeling efforts.

6.3 *Site Classification and Initial Response Action*—As the user gathers data, site conditions should be evaluated and an initial response action should be implemented, consistent with site conditions. This process is repeated when new data indicate a significant change in site conditions. Site urgency classifications are presented in Table 1, along with example classification scenarios and potential initial responses. *Note that the initial response actions given in Table 1 may not be applicable for all sites. The user should select an option that best addresses the short-term health and safety concerns of the site while implementing the RBCA process.*

6.3.1 The classification and initial response action scheme given in Table 1 is an example. It is based on the current and projected degree of hazard to human health and the environment. This is a feature of the process that can be customized by the user. "Classification 1" sites are associated with immediate threats to human health and the environment; "Classification 2" sites are associated with short-term (0 to 2-year) threats to human health, safety, and the environment; "Classification 3" sites are associated with long-term (greater than 2-year) threats to human health, safety, and the environment; "Classification 4" sites are associated with no reasonable potential threat to human health or to the environment.

6.3.2 Associated with each classification scenario in Table 1 is an initial response action; the initial response actions are implemented in order to eliminate any potential immediate impacts to human health and the environment as well as to minimize the potential for future impacts that may occur as the user proceeds with the RBCA process. Note that initial response actions do not always require active remediation; in many cases the initial response action is to monitor or further assess site conditions to ensure that risks posed by the site do not increase above acceptable levels over time. The initial response actions given in Table 1 are examples, and the user is free to implement other alternatives.

6.3.3 The need to reclassify the site should be evaluated when additional site information is collected that indicates a significant change in site conditions or when implementation of an interim response action causes a significant change in site conditions.

6.4 *Development of a Tier 1 Look-Up Table of RBSL*—If a look-up table is not available, the user is responsible for developing the look-up table. If a look-up table is available, the user is responsible for determining that the RBSLs in the look-up table are based on currently acceptable methodologies and parameters. The look-up table is a tabulation for potential exposure pathways, media (for example, soil, water, and air), a range of incremental carcinogenic risk levels (10E-4 to 10E-6

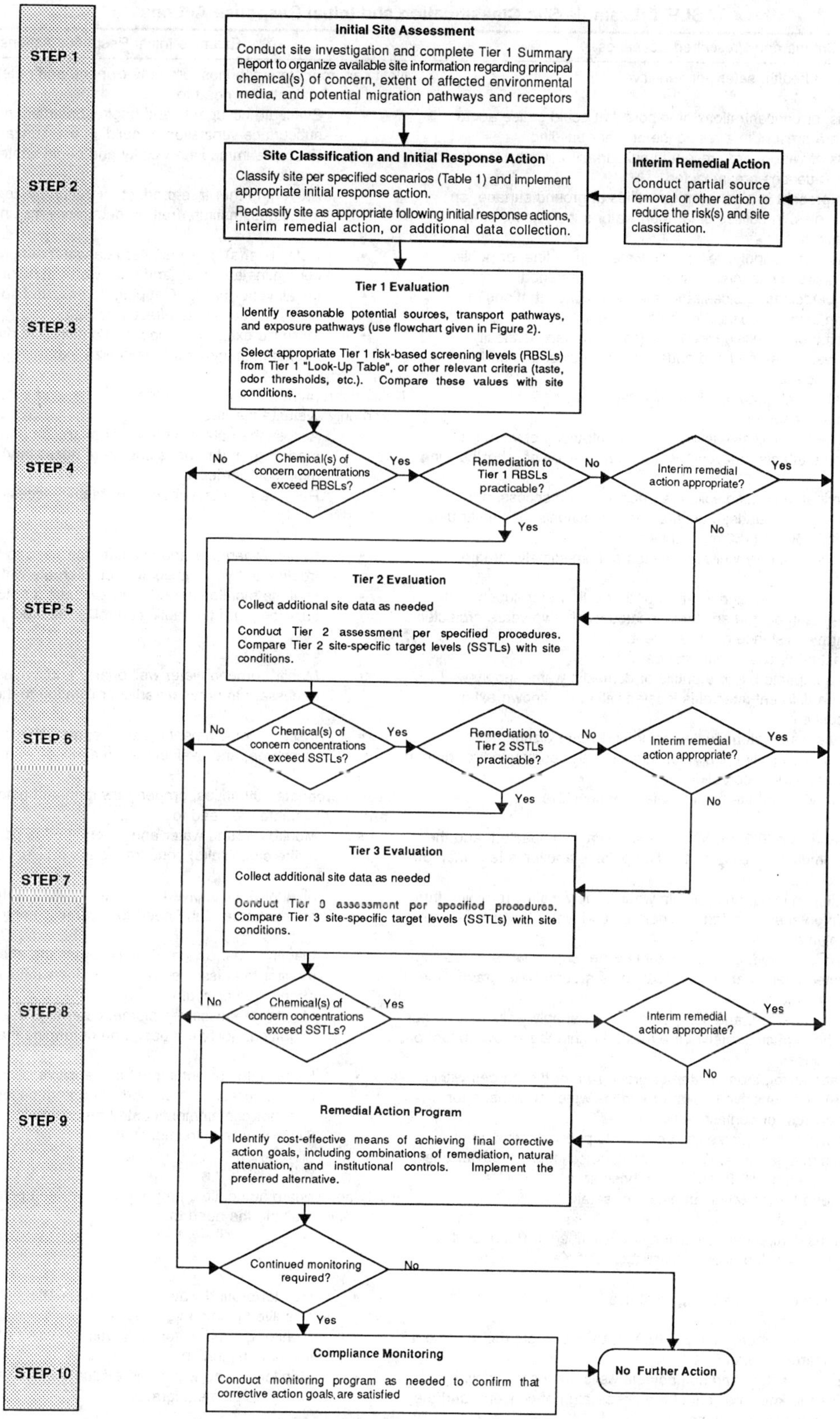

FIG. 1 Risk-Based Corrective Action Process Flowchart

are often evaluated as discussed in Appendix X1 paragraph X1.7, Discussion of Acceptable Risk) and hazard quotients equal to unity, and potential exposure scenarios (for example, residential, commercial, industrial, and agricultural) for each chemical(s) of concern.

6.4.1 The RBSLs are determined using typical, non-sitespecific values for exposure parameters and physical parameters for media. The RBSLs are calculated according to methodology suggested by the USEPA. For each exposure scenario, the RBSLs are based on current USEPA RME

TABLE 1 Example Site Classification and Initial Response Actions[A]

Criteria and Prescribed Scenarios	Example Initial Response Actions[B]
1. Immediate threat to human health, safety, or sensitive environmental receptors	Notify appropriate authorities, property owners, and potentially affected parties, and only evaluate the need to
• Explosive levels, or concentrations of vapors that could cause acute health effects, are present in a residence or other building.	• Evacuate occupants and begin abatement measures such as subsurface ventilation or building pressurization.
• Explosive levels of vapors are present in subsurface utility system(s), but no building or residences are impacted.	• Evacuate immediate vicinity and begin abatement measures such as ventilation.
• Free-product is present in significant quantities at ground surface, on surface water bodies, in utilities other than water supply lines, or in surface water runoff.	• Prevent further free-product migration by appropriate containment measures, institute free-product recovery, and restrict area access.
• An active public water supply well, public water supply line, or public surface water intake is impacted or immediately threatened.	• Notify user(s), provide alternate water supply, hydraulically control contaminated water, and treat water at point-of-use.
• Ambient vapor/particulate concentrations exceed concentrations of concern from an acute exposure or safety viewpoint.	• Install vapor barrier (capping, foams, and so forth), remove source, or restrict access to affected area.
• A sensitive habitat or sensitive resources (sport fish, economically important species, threatened and endangered species, and so forth) are impacted and affected.	• Minimize extent of impact by containment measures and implement habitat management to minimize exposure.
2. Short-term (0 to 2 years) threat to human health, safety, or sensitive environmental receptors	Notify appropriate authorities, property owners, and potentially affected parties, and only evaluate the need to
• There is potential for explosive levels, or concentrations of vapors that could cause acute effects, to accumulate in a residence or other building.	• Assess the potential for vapor migration (through monitoring/modeling) and remove source (if necessary), or install vapor migration barrier.
• Shallow contaminated surface soils are open to public access, and dwellings, parks, playgrounds, day-care centers, schools, or similar use facilities are within 500 ft (152 m) of those soils.	• Remove soils, cover soils, or restrict access.
• A non-potable water supply well is impacted or immediately threatened.	• Notify owner/user and evaluate the need to install point-of-use water treatment, hydraulic control, or alternate water supply.
• Ground water is impacted, and a public or domestic water supply well producing from the impacted aquifer is located within two-years projected ground water travel distance down gradient of the known extent of chemical(s) concern.	• Institute monitoring and then evaluate if natural attenuation is sufficient, or if hydraulic control is required.
• Ground water is impacted, and a public or domestic water supply well producing from a different interval is located within the known extent of chemicals of concern.	• Monitor ground water well quality and evaluate if control is necessary to prevent vertical migration to the supply well.
• Impacted surface water, storm water, or ground water discharges within 500 ft (152 m) of a sensitive habitat or surface water body used for human drinking water or contact recreation.	• Institute containment measures, restrict access to areas near discharge, and evaluate the magnitude and impact of the discharge.
3. Long-term (>2 years) threat to human health, safety, or sensitive environmental receptors	Notify appropriate authorities, property owners, and potentially affected parties, and only evaluate the need to
• Subsurface soils (>3 ft (0.9 m) BGS) are significantly impacted, and the depth between impacted soils and the first potable aquifer is less than 50 ft (15 m).	• Monitor ground water and determine the potential for future migration of the chemical(s) concerns to the aquifer.
• Ground water is impacted, and potable water supply wells producing from the impacted interval are located >2 years ground water travel time from the dissolved plume.	• Monitor the dissolved plume and evaluate the potential for natural attenuation and the need for hydraulic control.
• Ground water is impacted, and non-potable water supply wells producing from the impacted interval are located >2 years ground water travel time from the dissolved plume.	• Identify water usage of well, assess the effect of potential impact, monitor the dissolved plume, and evaluate whether natural attenuation or hydraulic control are appropriate control measures.
• Ground water is impacted, and non-potable water supply wells that do not produce from the impacted interval are located within the known extent of chemical(s) of concern.	• Monitor the dissolved plume, determine the potential for vertical migration, notify the user, and determine if any impact is likely.
• Impacted surface water, storm water, or ground water discharges within 1500 ft (457 m) of a sensitive habitat or surface water body used for human drinking water or contact recreation.	• Investigate current impact on sensitive habitat or surface water body, restrict access to area of discharge (if necessary), and evaluate the need for containment/control measures.
• Shallow contaminated surface soils are open to public access, and dwellings, parks, playgrounds, day-care centers, schools, or similar use facilities are more than 500 ft (152 m) of those soils.	• Restrict access to impact soils.
4. No demonstrable long-term threat to human health or safety or sensitive environmental receptors	Notify appropriate authorities, property owners, and potentially affected parties, and only evaluate the need to
Priority 4 scenarios encompass all other conditions not described in Priorities 1, 2, and 3 and that are consistent with the priority description given above. Some examples are as follows:	
• Non-potable aquifer with no existing local use impacted.	• Monitor ground water and evaluate effect of natural attenuation on dissolved plume migration.
• Impacted soils located more than 3 ft (0.9 m) BGS and greater than 50 ft (15 m) above nearest aquifer.	• Monitor ground water and evaluate effect of natural attenuation on leachate migration.
• Ground water is impacted, and non-potable wells are located down gradient outside the known extent of the chemical(s) of concern, and they produce from a nonimpacted zone.	• Monitor ground water and evaluate effect of natural attenuation on dissolved plume migration.

[A] Johnson, P. C., DeVaull, G. E., Ettinger, R. A., MacDonald, R. L. M., Stanley, C. C., Westby, T. S., and Conner, J., "Risk-Based Corrective Action: Tier 1 Guidance Manual," Shell Oil Co., July 1993.

[B] Note that these are potential initial response actions that may not be appropriate for all sites. The user is encouraged to select options that best address the short-term health and safety concerns of the site, while the RBCA process progresses.

parameters and current toxicological information given in Refs **(2, 3)** or peer-reviewed source(s). Consequently, the RBSL look-up table is updated when new methodologies and parameters are developed. For indirect pathways, fate and transport models can be used to predict RBSLs at a source area that corresponds to exposure point concentrations. An example of the development of a Tier 1 Look-Up Table and RBSL is given in Appendix X2. *Fig. 2 and Appendix X2 are presented solely for the purpose of providing an example development of the RBSL, and the values should not be viewed as proposed RBSLs.*

6.4.2 Appendix X2 is an example of an abbreviated Tier 1 RBSL Look-Up Table for compounds of concern associated with petroleum releases. The exposure scenarios selected in the example case are for residential and industrial/commercial scenarios characterized by USEPA RME parameters for adult males. The assumptions and methodology used in deriving the example are discussed in Appendix X2. Note that not all possible exposure pathways are considered in the derivation of the example. *The user should always review the assumptions and methodology used to derive values in a look-up table to make sure that they are consistent with reasonable exposure scenarios for the site being considered as well as currently accepted methodologies.* The value of creating a look-up table is that users do not have to repeat the exposure calculations for each site encountered. The look-up table is only altered when RME parameters, toxicological information, or recommended methodologies are updated. Some states have compiled such tables for direct exposure pathways that, for the most part, contain identical values (as they are based on the same assumptions). Values for the cross-media pathways (for example, volatilization and leaching), when available, often differ because these involve coupling exposure calculations with predictive equations for the fate and transport of chemicals in the environment. As yet, there is little agreement in the technical community concerning non-site-specific values for the transport and fate model parameters, or the choice of the models themselves. *Again, the reader should note that the example is presented here only as an abbreviated example of a Tier 1 RBSL Look-Up Table for typical compounds of concern associated with petroleum products.*

6.4.3 *Use of Total Petroleum Hydrocarbon Measurements*—Various chemical analysis methods commonly referred to as total petroleum hydrocarbons (TPHs) are often used in site assessments. These methods usually determine the total amount of hydrocarbons present as a single number and give no information on the types of hydrocarbon present. The TPHs should not be used for risk assessment because the general measure of TPH provides insufficient information about the amounts of individual chemical(s) of concern present.

6.5 *Comparison of Site Conditions with Tier 1 Risk-Based Screening Levels (RBSL)*—In Tier 1, the point(s) of exposure and point(s) of compliance are assumed to be located within close proximity to the source area(s) or the area where the highest concentrations of the chemical(s) of concern have been identified. Concentrations of the chemical(s) of concern measured at the source area(s) identified at the site should be compared to the look-up table RBSL. If there is sufficient site assessment data, the user may opt to compare RBSLs with statistical limits (for example, upper confidence levels) rather than maximum values detected. Background concentrations should be considered when comparing the RBSLs, to the site concentrations as the RBSLs may sometimes be less than background concentrations. Note that additivity of risks is not explicitly considered in the Tier 1 evaluation, as it is expected that the RBSLs are typically for a limited number of chemical(s) of concern considered at most sites. Additivity may be addressed in Tier 2 and Tier 3 analyses. To accomplish the Tier 1 comparison:

6.5.1 Select the potential exposure scenario(s) (if any) for the site. Exposure scenarios are determined based on the site assessment information described in 6.2;

6.5.2 Based on the impacted media identified, determine the primary sources, secondary sources, transport mechanisms, and exposure pathways;

6.5.3 Select the receptors (if any) based on current and anticipated future use. Consider land use restrictions and surrounding land use when making this selection.

6.5.4 Identify the exposure scenarios where the measured concentrations of the chemical(s) of concern are above the RBSL.

6.6 *Exposure Evaluation Flowchart*—During a Tier 1 evaluation, the risk evaluation flowchart presented in Fig. 2 may be used as a tool to guide the user in selecting appropriate exposure scenarios based on site assessment information. This worksheet may also be used in the evaluation of remedial action alternatives. To complete this flowchart:

6.6.1 Characterize site sources and exposure pathways, using the data summarized from Tier 1 to customize the risk evaluation flowchart for the site by checking the small check-box for every relevant source, transport mechanism, and exposure pathway.

6.6.2 Identify receptors, and compare site conditions with Tier 1 levels: For each exposure pathway selected, check the receptor characterization (residential, commercial, and so forth) where the concentrations of the chemical(s) of concern are above the RBSL. Consider land use restrictions and surrounding land use when making this selection. Do not check any boxes if there are no receptors present, or likely to be present, or if institutional controls prevent exposure from occurring and are likely to stay in place.

6.6.3 Identify potential remedial action measures. Select remedial action options to reduce or eliminate exposure to the chemical(s) of concern.

6.6.4 The exposure evaluation flowchart (Fig. 2) can be used to graphically portray the effect of the Tier 1 remedial action. Select the Tier 1 remedial action measure or measures (shown as valve symbols) that will break the lines linking sources, transport mechanisms, and pathways leading to the chemical(s) of concern above the RBSL. Adjust the mix of remedial action measures until no potential receptors have concentrations of chemical(s) of concerns above the RBSL with the remedial action measures in place. Show the most likely Tier 1 remedial action measure(s) selected for this site by marking the appropriate valve symbols on the flowchart and recording a remedial action measure on the right-hand-side of this figure.

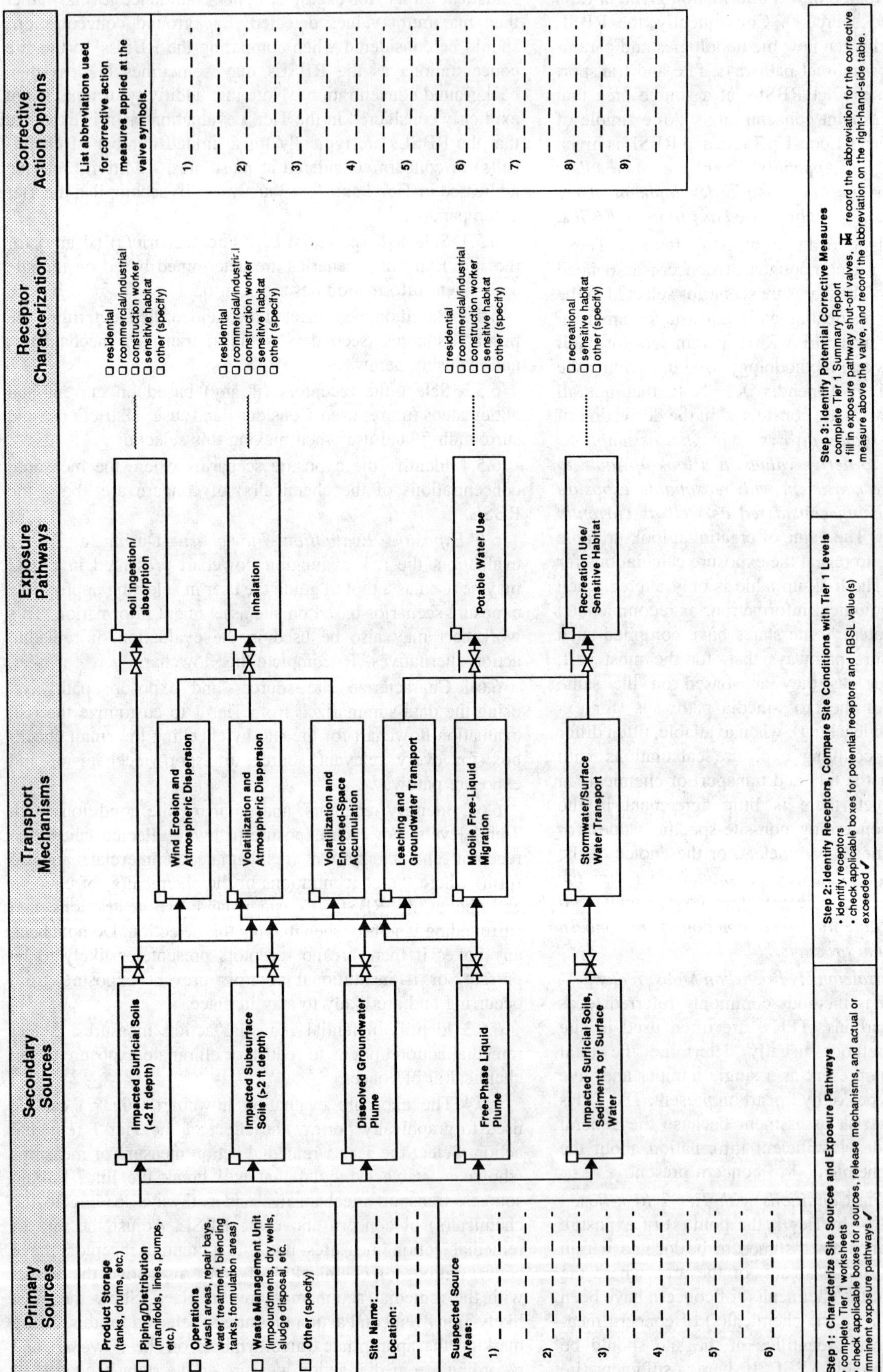

FIG. 2 Exposure Scenario Evaluation Flowchart

6.7 *Evaluation of Tier Results*—At the conclusion of each tier evaluation, the user compares the target levels (RBSLs or SSTLs) to the concentrations of the chemical(s) of concern at the point(s) of compliance.

6.7.1 If the concentrations of the chemical(s) of concern exceed the target levels at the point(s) of compliance, then either remedial action, interim remedial action, or further tier evaluation should be conducted.

6.7.1.1 *Remedial Action*— A remedial action program is designed and implemented. This program may include some combination of source removal, treatment, and containment technologies, as well as engineering and institutional controls. Examples of these include the following: soil venting, bioventing, air sparging, pump and treat, and natural attenuation/passive remediation. When concentrations of chemical(s) of concern no longer exceed the target levels at the point of compliance, then the user may elect to move to 6.7.3.

6.7.1.2 *Interim Remedial Action*—If achieving the desired risk reduction is impracticable due to technology or resource limitations, an interim remedial action, such as removal or treatment of "hot spots," may be conducted to address the most significant concerns, change the site classification, and facilitate reassessment of the tier evaluation.

6.7.1.3 *Further Tier Evaluation*—If further tier evaluation is warranted, additional site assessment information may be collected to develop SSTLs under a Tier 2 or Tier 3 evaluation. Further tier evaluation is warranted when:

(1) The basis for the RBSL values (for example, geology, exposure parameters, point(s) of exposure, and so forth) are not representative of the site-specific conditions; or

(2) The SSTL developed under further tier evaluation will be significantly different from the Tier 1 RBSL or will significantly modify the remedial action activities; or

(3) Cost of remedial action to RBSLs will likely be greater than further tier evaluation and subsequent remedial action.

6.7.2 If the concentrations of chemicals of concern at the point of compliance are less than the target levels, but the user is not confident that data supports the conclusion that concentrations will not exceed target levels in the future, then the user institutes a monitoring plan to collect data sufficient to confidently conclude that concentrations will not exceed target levels in the future. When this data is collected, the user moves to 6.7.3.

6.7.3 If the concentrations of chemicals of concern at the point of compliance are less than target levels, and the user is confident that data supports the conclusion that concentrations will not exceed target levels in the future, then no additional corrective action activities are necessary, and the user has completed the RBCA process. In practice, this is often accompanied by the issuing of a no-further-action letter by the oversight regulatory agency.

6.8 *Tier 2*—Tier 2 provides the user with an option to determine the site-specific point(s) of compliance and corresponding SSTL for the chemical(s) of concern applicable at the point(s) of compliance and source area(s). Additional site assessment data may be required; however, the incremental effort is typically minimal relative to Tier 1. If the user completes a Tier 1 evaluation, in most cases, only a limited number of pathways, exposure scenarios, and chemical(s) of concern are considered in the Tier 2 evaluation since many are eliminated from consideration during the Tier 1 evaluation.

6.8.1 In Tier 2, the user:

6.8.1.1 Identifies the indirect exposure scenarios to be addressed and the appropriate site-specific point(s) of compliance. A combination of assessment data and predictive modeling results are used to determine the SSTL at the source area(s) or the point(s) of compliance, or both; or

6.8.1.2 Applies Tier 1 RBSL Look-Up Table values for the direct exposure scenarios at reasonable point(s) of exposure (as opposed to the source area(s) as is done in Tier 1). The SSTLs for source area(s) and point(s) of compliance can be determined based on the demonstrated and predicted attenuation (reduction in concentration with distance) of compounds that migrate away from the source area(s).

6.8.1.3 An example of a Tier 2 application is illustrated in Appendix X5.

6.8.2 Tier 2 of the RBCA process involves the development of SSTL based on the measured and predicted attenuation of the chemical(s) of concern away from the source area(s) using relatively simplistic mathematical models. The SSTLs for the source area(s) are generally not equal to the SSTL for the point(s) of compliance. The predictive equations are characterized by the following:

6.8.2.1 The models are relatively simplistic and are often algebraic or semianalytical expressions;

6.8.2.2 Model input is limited to practicably attainable site-specific data or easily estimated quantities (for example, total porosity, soil bulk density); and

6.8.2.3 The models are based on descriptions of relevant physical/chemical phenomena. Most mechanisms that are neglected result in predicted concentrations that are greater than those likely to occur (for example, assuming constant concentrations in source area(s)). Appendix X3 discusses the use of predictive models and presents models that might be considered for Tier 2 evaluation.

6.8.3 *Tier 2 Evaluation*—Identify the exposure scenarios where the measured concentrations of the chemical(s) of concern are above the SSTL at the point(s) of compliance, and evaluate the tier results in accordance with 6.7.

6.9 *Tier 3*—In a Tier 3 evaluation, SSTLs for the source area(s) and the point(s) of compliance are developed on the basis of more sophisticated statistical and contaminant fate and transport analyses, using site-specific input parameters for both direct and indirect exposure scenarios. Source area(s) and the point(s) of compliance SSTLs are developed to correspond to concentrations of chemical(s) of concern at the point(s) of exposure that are protective of human health and the environment. Tier 3 evaluations commonly involve collection of significant additional site information and completion of more extensive modeling efforts than is required for either a Tier 1 or Tier 2 evaluation.

6.9.1 Examples of Tier 3 analyses include the following:

6.9.1.1 The use of numerical ground water modeling codes that predict time-dependent dissolved contaminant transport under conditions of spatially varying permeability fields to predict exposure point(s) of concentrations;

6.9.1.2 The use of site-specific data, mathematical models, and Monte Carlo analyses to predict a statistical distribution of exposures and risks for a given site; and

6.9.1.3 The gathering of sufficient data to refine site-specific parameter estimates (for example, biodegradation rates) and improve model accuracy in order to minimize future monitoring requirements.

6.9.2 *Tier 3 Evaluation*—Identify the exposure scenarios where the measured concentrations of the chemical(s) of concern are above the SSTL at the point(s) of compliance, and evaluate the tier results in accordance with 6.7 except that a tier upgrade (6.7.5) is not available.

6.10 *Implementing the Selected Remedial Action Program*—When it is judged by the user that no further assessment is necessary, or practicable, a remedial alternatives evaluation should be conducted to confirm the most cost-effective option for achieving the final remedial action target levels (RBSLs or SSTLs, as appropriate). Detailed design specifications may then be developed for installation and operation of the selected measure. The remedial action must continue until such time as monitoring indicates that concentrations of the chemical(s) of concern are not above the RBSL or SSTL, as appropriate, at the points of compliance or source area(s), or both.

6.11 *RBCA Report*— After completion of the RBCA activities, a RBCA report should be prepared and submitted to the regulatory agency. The RBCA report should, at a minimum, include the following:

6.11.1 An executive summary;

6.11.2 A site description;

6.11.3 A summary of the site ownership and use;

6.11.4 A summary of past releases or potential source areas;

6.11.5 A summary of the current and completed site activities;

6.11.6 A description of regional hydrogeologic conditions;

6.11.7 A description of site-specific hydrogeologic conditions;

6.11.8 A summary of beneficial use;

6.11.9 A summary and discussion of the risk assessment (hazard identification, dose response assessment, exposure assessment, and risk characterization), including the methods and assumptions used to calculate the RBSL or SSTL, or both;

6.11.10 A summary of the tier evaluation;

6.11.11 A summary of the analytical data and the appropriate RBSL or SSTL used;

6.11.12 A summary of the ecological assessment;

6.11.13 A site map of the location;

6.11.14 An extended site map to include local land use and ground water supply wells;

6.11.15 Site plan view showing location of structures, aboveground storage tanks, underground storage tanks, buried utilities and conduits, suspected/confirmed sources, and so forth;

6.11.16 Site photos, if available;

6.11.17 A ground water elevation map;

6.11.18 Geologic cross section(s); and

6.11.19 Dissolved plume map(s) of the chemical(s) of concern.

6.12 *Monitoring and Site Maintenance*—In many cases, monitoring is necessary to demonstrate the effectiveness of implemented remedial action measures or to confirm that current conditions persist or improve with time. Upon completion of this monitoring effort (if required), no further action is required. In addition, some measures (for example, physical barriers such as capping, hydraulic control, and so forth) require maintenance to ensure integrity and continued performance.

6.13 *No Further Action and Remedial Action Closure*—When RBCA RBSLs or SSTLs have been demonstrated to be achieved at the point(s) of compliance or source area(s), or both, as appropriate, and monitoring and site maintenance are no longer required to ensure that conditions persist, then no further action is necessary, except to ensure that institutional controls (if any) remain in place.

APPENDIXES

(Nonmandatory Information)

X1. PETROLEUM PRODUCTS CHARACTERISTICS: COMPOSITION, PHYSICAL AND CHEMICAL PROPERTIES, AND TOXICOLOGICAL ASSESSMENT SUMMARY

X1.1 *Introduction:*

X1.1.1 Petroleum products originating from crude oil are complex mixtures of hundreds to thousands of chemicals; however, practical limitations allow us to focus only on a limited subset of key components when assessing the impact of petroleum fuel releases to the environment. Thus, it is important to have a basic understanding of petroleum properties, compositions, and the physical, chemical, and toxicological properties of some compounds most often identified as the key chemicals or chemicals of concern.

X1.1.2 This appendix provides a basic introduction to the physical, chemical, and toxicological characteristics of petroleum products (gasoline, diesel fuel, jet fuel, and so forth)[5] and other products focussed primarily towards that information which is most relevant to assessing potential impacts due to releases of these products into the subsurface. Much of the information presented is summarized from the references listed at the end of this guide. For specific topics, the reader is referred to the following sections of this appendix:

X1.1.2.1 *Composition of Petroleum Fuels*—See X1.2.

[5] "Alternative products," or those products not based on petroleum hydrocarbons (or containing them in small amounts), such as methanol or M85, are beyond the scope of the discussion in this appendix.

X1.1.2.2 *Physical, Chemical, and Toxicological Properties of Petroleum Fuels*—See X1.3.

X1.1.2.3 *Chemical of Concern*—See X1.4.

X1.1.2.4 *Toxicity of Petroleum Hydrocarbons*—See X1.5.

X1.1.2.5 *Profiles of Select Compounds*—See X1.6.

X1.2 *Composition of Petroleum Products:*

X1.2.1 Most petroleum products are derived from crude oil by distillation, which is a process that separates compounds by volatility. Crude oils are variable mixtures of thousands of chemical compounds, primarily hydrocarbons; consequently, the petroleum products themselves are also variable mixtures of large numbers of components. The biggest variations in composition are from one type of product to another (for example, gasoline to motor oil); however, there are even significant variations within different samples of the same product type. For example, samples of gasoline taken from the same fuel dispenser on different days, or samples taken from different service stations, will have different compositions. These variations are the natural result of differing crude oil sources, refining processes and conditions, and kinds and amount of additives used.

X1.2.2 *Components of Petroleum Products*—The components of petroleum products can be generally classified as either hydrocarbons (organic compounds composed of hydrogen and carbon only) or as non-hydrocarbons (compounds containing other elements, such as oxygen, sulfur, or nitrogen). Hydrocarbons make up the vast majority of the composition of petroleum products. The non-hydrocarbon compounds in petroleum products are mostly hydrocarbon-like compounds containing minor amounts of oxygen, sulfur, or nitrogen. Most of the trace levels of metals found in crude oil are removed by refining processes for the lighter petroleum products.

X1.2.3 *Descriptions and Physical Properties of Petroleum Products*—In order to simplify the description of various petroleum products, boiling point ranges and carbon number (number of carbon atoms per molecule) ranges are commonly used to describe and compare the compositions of various petroleum products. Table X1.1 summarizes these characteristics for a range of petroleum products. Moving down the list from gasoline, increases in carbon number range and boiling range and decreases in volatility (denoted by increasing flash point) indicate the transition to "heavier products." Additional descriptions of each of these petroleum products are provided as follows.

TABLE X1.1 Generalized Chemical and Physical Characterization of Petroleum Products

	Predominant Carbon No. Range	Boiling Range, °C	Flash Point,[A] °C
Gasoline	C4 to C12	25 to 215	−40
Kerosene and Jet Fuels	C11 to C13	150 to 250	<21,[B] 21 to 55,[C] >55[D]
Diesel Fuel and Light Fuel Oils	C10 to C20	160 to 400	>35
Heavy Fuel Oils	C19 to C25	315 to 540	>50
Motor Oils and Other Lubricating Oils	C20 to C45	425 to 540	>175

[A] Typical values.
[B] Jet-B, AVTAG and JP-4.
[C] Kerosene, Jet A, Jet A-1, JP-8 and AVTUR.
[D] AVCAT and JP-5.

X1.2.4 *Gasoline*—Gasoline is composed of hydrocarbons and "additives" that are blended with the fuel to improve fuel performance and engine longevity. The hydrocarbons fall primarily in the C4 to C12 range. The lightest of these are highly volatile and rapidly evaporate from spilled gasoline. The C4 and C5 aliphatic hydrocarbons rapidly evaporate from spilled gasoline (hours to months, depending primarily on the temperature and degree of contact with air). Substantial portions of the C6 and heavier hydrocarbons also evaporate, but at lower rates than for the lighter hydrocarbons.

X1.2.4.1 Fig. X1.1 shows gas chromatograms of a fresh gasoline and the same gasoline after simulated weathering; air was bubbled through the gasoline until 60 % of its initial volume was evaporated. In gas chromatography, the mixture is separated into its components, with each peak representing different compounds. Higher molecular weight components appear further to the right along the *x*-axis. For reference, positions of the *n*-aliphatic hydrocarbons are indicated in Fig. X1.1. The height of, and area under, each peak are measures of how much of that component is present in the mixture. As would be expected by their higher volatilities, the lighter hydrocarbons (up to about C7) evaporate first and are greatly reduced in the weathered gasoline. The gas chromatogram of a fuel oil is also shown for comparison.

X1.2.4.2 The aromatic hydrocarbons in gasoline are primarily benzene (C_6H_6), toluene (C_7H_8), ethylbenzene (C_8H_{10}), and xylenes (C_8H_{10}); these are collectively referred to as "BTEX." Some heavier aromatics are present also, including low amounts of polyaromatic hydrocarbons (PAHs). Aromatics typically comprise about 10 to 40 % of gasoline.

X1.2.4.3 Oxygenated compounds ("oxygenates") such as alcohols (for example, methanol or ethanol) and ethers (for example, methyl tertiarybutyl ether—MTBE) are sometimes added to gasoline as octane boosters and to reduce carbon monoxide exhaust emissions. Methyl tertiarbutyl ether has been a common additive only since about 1980.

X1.2.4.4 Leaded gasoline, which was more common in the past, contained lead compounds added as octane boosters. Tetraethyl lead (TEL) is one lead compound that was commonly used as a gasoline additive. Other similar compounds were also used. Sometimes mixtures of several such compounds were added. Because of concerns over atmospheric emissions of lead from vehicle exhaust, the EPA has reduced the use of leaded gasolines. Leaded gasolines were phased out of most markets by 1989.

X1.2.4.5 In order to reduce atmospheric emissions of lead, lead "scavengers" were sometimes added to leaded gasolines. Ethylene dibromide (EDB) and ethylene dichloride (EDC) were commonly used for this purpose.

X1.2.5 *Kerosene and Jet Fuel*—The hydrocarbons in kerosene commonly fall into the C11 to C13 range, and distill at approximately 150 to 250°C. Special wide-cut (that is, having broader boiling range) kerosenes and low-flash kerosenes are also marketed. Both aliphatic and aromatic hydrocarbons are present, including more multi-ring compounds and kerosene.

X1.2.5.1 Commercial jet fuels JP-8 and Jet A have similar

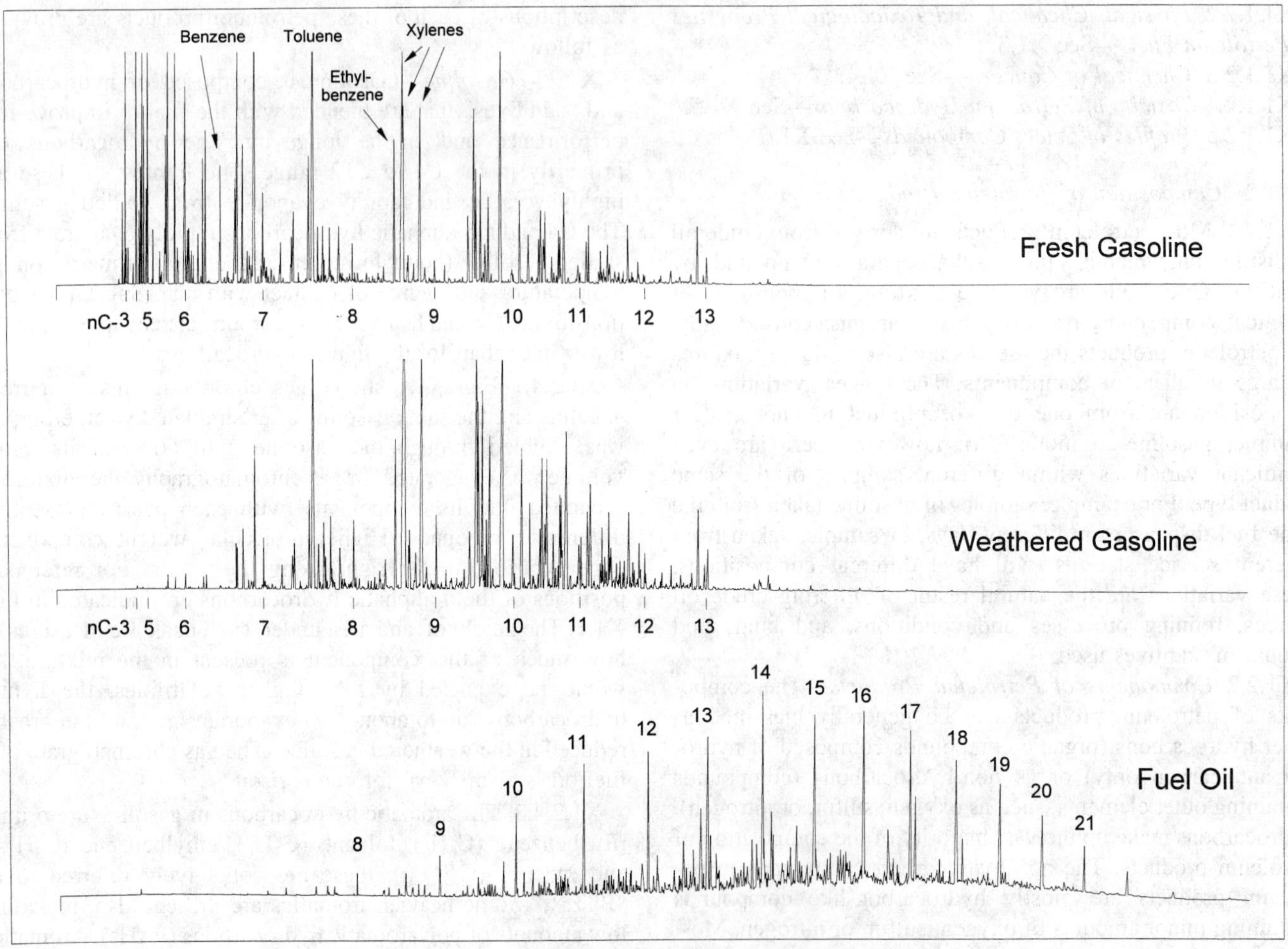

FIG. X1.1 Gas Chromatograms of Some Petroleum Fuels

compositions to kerosene. Jet fuels JP-4 and JP-5 are wider cuts used by the military. They contain lighter distillates and have some characteristics of both gasoline and kerosene.

X1.2.5.2 Aromatic hydrocarbons comprise about 10 to 20 % of kerosene and jet fuels.

X1.2.6 *Diesel Fuel and Light Fuel Oils*—Light fuel oils include No. 1 and No. 2 fuel oils, and boil in the range from 160 to 400°C. Hydrocarbons in light fuel oils and diesel fuel typically fall in the C10 to C20 range. Because of their higher molecular weights, constituents in these products are less volatile, less water soluble, and less mobile than gasoline- or kerosene-range hydrocarbons.

X1.2.6.1 About 25 to 35 % of No. 2 fuel oil is composed of aromatic hydrocarbons, primarily alkylated benzenes and naphthalenes. The BTEX concentrations are generally low.

X1.2.6.2 No. 1 fuel oil is typically a straight run distillate.

X1.2.6.3 No. 2 fuel oil can be either a straight run distillate, or else is produced by catalytic cracking (a process in which larger molecules are broken down into smaller ones). Straight run distillate No. 2 is commonly used for home heating fuel, while the cracked product is often used for industrial furnaces and boilers. Both No. 1 and No. 2 fuel oils are sometimes used as blending components for jet fuel or diesel fuel formulations.

X1.2.7 *Heavy Fuel Oils*— The heavy fuel oils include Nos. 4, 5, and 6 fuel oils. They are sometimes referred to as "gas oils" or "residual fuel oils." These are composed of hydrocarbons ranging from about C19 to C25 and have a boiling range from about 315 to 540°C. They are dark in color and considerably more viscous than water. They typically contain 15 to 40 % aromatic hydrocarbons, dominated by alkylated phenanthrenes and naphthalenes. Polar compounds containing nitrogen, sulfur, or oxygen may comprise 15 to 30 % of the oil.

X1.2.7.1 No. 6 fuel oil, also called "Bunker Fuel" or "Bunker C," is a gummy black product used in heavy industrial applications where high temperatures are available to fluidize the oil. Its density is greater than that of water.

X1.2.7.2 Nos. 4 and 5 fuel oils are commonly produced by blending No. 6 fuel oil with lighter distillates.

X1.2.8 *Motor Oils and Other Lubricating Oils*—Lubricating oils and motor oils are predominately comprised of compounds in the C20 to C45 range and boil at approximately 425 to 540°C. They are enriched in the most complex molecular fractions found in crude oil, such as cycloparaffins and PNAs having up to three rings or more. Aromatics may make up to 10 to 30 % of the oil. Molecules containing nitrogen, sulfur, or oxygen are also common. In addition, used automative crankcase oils become enriched with PNAs and certain metals.

X1.2.8.1 These oils are relatively viscous and insoluble in ground water and relatively immobile in the subsurface.

X1.2.8.2 Waste oil compositions are even more difficult to predict. Depending on how they are managed, waste oils may contain some portion of the lighter products in addition to heavy oils. Used crankcase oil may contain wear metals from engines. Degreasing solvents (gasoline, naphtha, or light chlorinated solvents, or a combination thereof) may be present in some wastes.

X1.3 *Physical, Chemical, and Toxicological Characteristics of Petroleum Products:*

X1.3.1 *Trends in Physical/Chemical Properties of Hydrocarbons*—In order to better understand the subsurface behavior of hydrocarbons it is helpful to be able to recognize trends in important physical properties with increasing number of carbon atoms. These trends are most closely followed by compounds with similar molecular structures, such as the straight-chained, single-bonded aliphatic hydrocarbons. In general, as the carbon number (or molecule size) increases, the following trends are observed:

X1.3.1.1 Higher boiling points (and melting points),

X1.3.1.2 Lower vapor pressure (volatility),

X1.3.1.3 Greater density,

X1.3.1.4 Lower water solubility, and

X1.3.1.5 Stronger adhesion to soils and less mobility in the subsurface.

X1.3.2 Table X1.2 lists physical, chemical, and toxicological properties for a number of hydrocarbons found in petroleum products. In general:

X1.3.2.1 Aliphatic petroleum hydrocarbons with more than ten carbon atoms are expected to be immobile in the subsurface, except when dissolved in nonaqueous phase liquids (NAPLs), due to their low water solubilities, low vapor pressures, and strong tendency to adsorb to soil surfaces.

X1.3.2.2 Aromatic hydrocarbons are more water soluble and mobile in water than aliphatic hydrocarbons of similar molecular weight.

X1.3.2.3 Oxygenates generally have much greater water

TABLE X1.2 Chemical and Toxicological Properties of Selected Hydrocarbons

Compounds	Weight of Evidence Class[A]	Oral RfD, mg/kg-day	Inhalation RfC, mg/m^3	Oral Slope Factor,[A] mg/kg-day^{-1}	Drinking Water MCL,[A] mg/L	Solubility,[B] mg/L	Octanol/Water Partition Coefficient,[B] log K_{ow}	Organic Carbon Adsorption Coefficient,[B] log K_{oc}
Benzene	A	[C]	[C,D]	0.029[D]	0.005	1750	2.13	1.58
Toluene	D	0.2[A]	0.4[A]	...	1	535	2.65	2.13
Ethylbenzene	D	0.1[A]	1[A]	...	0.7	152	3.13	1.98
Xylenes	D	2[A]	0.3[C,E]	...	10.0	198	3.26	2.38
n-Hexane	[C]	0.06[E], 0.6[F]	0.2[F]	...	...	13[G]	...	...
MTBE	...	[C]	3[A]	...	...	18 000[H]	1.06–1.30[I]	1.08[I]
MEK	D	0.6[A]	1[A]	...	[K]	268 000	0.26	0.65
MIBK	...	0.05[E], 0.5[F]	0.08[C,E], 0.8[F]	...	...	...	...	...
Methanol	...	0.5[A]	[C]	...	...	...	...	...
Ethanol	...	...	...	...	...	1 000 000	−0.032	0.34
TBA	...	...	...	...	...	...	...	...
Lead	B2	...	...	...	0.015[L]			...
EDC	B2	...	...	0.091	0.006	8 520	1.48	1.15
EDB	B2	...	[C]	85	0.00006	4 300	1.76	1.64
PNAs:								
Pyrene	D	0.03[A]	...	...	...	0.132	4.88	4.58
Benzo(a)pyrene	B2	...	...	7.3	0.0002[M]	0.00120	5.98	5.59
Anthracene	D	0.3[A]	...	...	...	0.0450	4.45	4.15
Phenanthrene	D	...	...	...	...	1.00	4.46	4.15
Naphthalene	D[C]	0.004[E], 0.04[F]	...	...	...	31.0[G]	3.28[G]	3.11[N]
Chrysene	B2	...	...	1.15[O]	0.0002	0.00180	5.61	5.30
Benzo(k)fluoranthene	B2	...	...	...	0.0002[M]	0.430	6.06	5.74
Fluorene	D	0.04[A]	...	...	...	1.69	4.20	3.86
Fluoranthene	D	0.04[A]	...	...	...	0.206	4.90	4.58
Benzo(g,h,i)perylene	D	...	...	...	...	0.000700	6.51	6.20
Benzo(b)fluoranthene	B2	...	...	...	0.0002[M]	0.0140	6.06	5.74
Benz(a)anthracene	B2	...	...	...	0.0002[M]	0.00670	5.60	6.14

[A] See Ref **(2)**.

[B] See Ref **(4)**.

[C] The data is pending in the EPA-IRIS database.

[D] The inhalation unit risk for benzene is $8.3 \times 10^{-3}(\text{mg/m}^3)^{-1}$. The drinking water unit is 8.3×10^{-4}(mg/L).

[E] Chronic effect. See Ref **(5)**.

[F] Subchronic effect. See Ref **(5)**.

[G] See Ref **(7)**.

[H] See Ref **(8)**.

[I] See Ref **(9)**.

[J] Estimation Equation (from **(10)**):

(1) $\log K_{oc} = -0.55 \log S + 3.64$, where S = water solubility (mg/L)

(2) $\log K_{oc} = 0.544 \log P + 1.377$

[K] Listed in the January 1991 Drinking Water Priority List and may be subject to future regulation (56 FR 1470, 01/14/91).

[L] USEPA. May 1993. Office of Drinking Water. 15 μg/L is an action level; standard for tap water.

[M] Proposed standard.

[N] See Ref **(11)**.

[O] See Ref **(6)**. Health-based criteria for carcinogenic polycyclic aromatic compounds (PAHs) with the exception of dibenzo(a,h)anthracene are set at one tenth of the level of benzo(a)pyrene due to their recognized lesser potency.

solubilities than hydrocarbons of similar molecular weight, and hence are likely to be the most mobile of petroleum fuel constituents in leachate and ground water. The light alcohols, including methanol and ethanol, are completely miscible with water in all proportions.

X1.3.3 *Properties of Mixtures*—It is important to note that the partitioning behavior of individual compounds is affected by the presence of other hydrocarbons in the subsurface. The maximum dissolved and vapor concentrations achieved in the subsurface are always less than that of any pure compound, when it is present as one of many constituents of a petroleum fuel. For example, dissolved benzene concentrations in ground water contacting gasoline-impacted soils rarely exceed 1 to 3 % of the ~1800-mg/L pure component solubility of benzene.

X1.3.4 *Trends in Toxicological Properties of Hydrocarbons*—A more detailed discussion of toxicological assessment is given in X1.5 (see also Appendix X3), followed by profiles for select chemicals found in petroleum products given in X1.6. Of the large number of compounds present in petroleum products, aromatic hydrocarbons (BTEX, PAHs, and so forth) are the constituents that human and aquatic organisms tend to be most sensitive to (relative to producing adverse health impacts).

X1.4 *Chemicals of Concern for Risk Assessments:*

X1.4.1 It is not practicable to evaluate every compound present in a petroleum product to assess the human health or environmental risk from a spill of that product. For this reason, risk management decisions are generally based on assessing the potential impacts from a select group of "indicator" compounds. It is inherently assumed in this approach that a significant fraction of the total potential impact from all chemicals is due to the chemicals of concern. The selection of chemicals of concern is based on the consideration of exposure routes, concentrations, mobilities, toxicological properties, and aesthetic characteristics (taste, odor, and so forth). Historically, the relatively low toxicities and dissolved-phase mobilities of aliphatic hydrocarbons have made these chemicals of concern of less concern relative to aromatic hydrocarbons. When additives are present in significant quantities, consideration should also be given to including these as chemicals of concern.

X1.4.2 Table X1.3 identifies chemicals of concern most often considered when assessing impacts of petroleum products, based on knowledge of their concentration in the specific fuel, as well as their toxicity, water solubility, subsurface mobility, aesthetic characteristics, and the availability of sufficient information to conduct risk assessments. The chemicals of concern are identified by an "X" in the appropriate column.

TABLE X1.3 Commonly Selected Chemicals of Concern for Petroleum Products

	Unleaded Gasoline	Leaded Gasoline	Kerosene/ Jet Fuels	Diesel/ Light Fuel Oils	Heavy Fuel Oils
Benzene	X	X	X	...	...
Toluene	X	X	X	...	...
Ethylbenzene	X	X	X	...	...
Xylene	X	X	X	...	...
MTBE, TBA, MEK, MIBK, methanol, ethanol	when suspected[A]	when suspected[A]	...	...	...
Lead, EDC, EDB	...	X	...	...	...
PNAs[B]	...	...	X	X	X

[A] For example, when these compounds may have been present in the spilled gasoline. These additives are not present in all gasolines.

[B] A list of selected PNAs for consideration is presented in Table X1.2.

X1.5 *Toxicity of Petroleum Hydrocarbons:*

X1.5.1 The following discussion gives a brief overview of origin of the toxicity parameters (reference doses (RfDs)), and slope factors (SFs), a justification for common choices of chemicals of concern and then, in X1.6, a brief summary of the toxicological, physical, and chemical parameters associated with these chemicals of concern.

X1.5.2 *How Toxicity Is Assessed: Individual Chemicals Versus Mixtures*—The toxicity of an individual chemical is typically established based on dose-response studies that estimate the relationship between different dose levels and the magnitude of their adverse effects (that is, toxicity). The dose-response data is used to identify a "safe dose" or a toxic level for a particular adverse effect. For a complex mixture of chemicals, the same approach can be used. For example, to evaluate the toxicity of gasoline, a "pure" reference gasoline would be evaluated instead of the individual chemical. This "whole-product" approach to toxicity assessment is strictly applicable only to mixtures identical to the evaluated mixture; gasolines with compositions different from the reference gasoline might have toxicities similar to the reference, but some differences would be expected. In addition, as the composition of gasoline released to the environment changes through natural processes (volatilization, leaching, biodegradation), the toxicity of the remaining portion may change also.

X1.5.3 An alternative to the "whole-product" approach for assessing the toxicity of mixtures is the "individual-constituent" approach. In this approach, the toxicity of each individual constituent (or a selected subset of the few most toxic constituents, so-called chemicals of concern) is separately assessed and the toxicity of the mixture is assumed to be the sum of the individual toxicities using a hazard index approach. This approach is often used by the USEPA; however, it is inappropriate to sum hazard indices unless the toxicological endpoints and mechanisms of action are the same for the individual compounds. In addition, the compounds to be assessed must be carefully selected based on their concentrations in the mixture, their toxicities, how well their toxicities are known, and how mobile they are in the subsurface. Lack of sufficient toxicological information is often an impediment to this procedure.

X1.5.4 *Use of TPH Measurements in Risk Assessments*—Various chemical analysis methods commonly referred to as TPH are often used in site assessments. These methods usually determine the total amount of hydrocarbons present as a single number, and give no information on the types of hydrocarbon present. Such TPH methods may be useful for risk assessments where the whole product toxicity approach is appropriate. However in general, *TPH should not be used for "individual constituent" risk assessments because the general measure of TPH provides insufficient information about the amounts of individual compounds present.*

X1.5.5 *Toxicity Assessment Process*—Dose-response data

are used to identify a "safe dose" or toxic level for a particular observed adverse effect. Observed adverse effects can include whole body effects (for example, weight loss, neurological observations), effects on specific body organs, including the central nervous system, teratogenic effects (defined by the ability to produce birth defects), mutagenic effects (defined by the ability to alter the genes of a cell), and carcinogenic effects (defined by the ability to produce malignant tumors in living tissues). Because of the great concern over risk agents which may produce incremental carcinogenic effects, the USEPA has developed weight-of-evidence criteria for determining whether a risk agent should be considered carcinogenic (see Table X1.4).

X1.5.6 Most estimates of a "safe dose" or toxic level are based on animal studies. In rare instances, human epidemiological information is available on a chemical. Toxicity studies can generally be broken into three categories based on the number of exposures to the risk agent and the length of time the study group was exposed to the risk agent. These studies can be described as follows:

X1.5.6.1 *Acute Studies*— Acute studies typically use one dose or multiple doses over a short time frame (24 h). Symptoms are usually observed within a short time frame and can vary from weight loss to death.

X1.5.6.2 *Chronic Studies*— Chronic studies use multiple exposures over an extended period of time, or a significant fraction of the animal's (typically two years) or the individual's lifetime. The chronic effects of major concern are carcinogenic, mutagenic, and teratogenic effects. Other chronic health effects such as liver and kidney damage are also important.

X1.5.6.3 *Subchronic Studies*—Subchronic studies use multiple or continuous exposures over an extended period (three months is the usual time frame in animal studies). Observed effects include those given for acute and chronic studies.

X1.5.6.4 Ideally, safe or acceptable doses are calculated from chronic studies, although, due to the frequent paucity of chronic data, subchronic studies are used.

X1.5.6.5 For noncarcinogens, safe doses are based on no observed adverse effect levels (NOAELs) or lowest observed adverse effect levels (LOAELs) from the studies.

X1.5.6.6 Acceptable doses for carcinogens are determined from mathematical models used to generate dose-response curves in the low-dose region from experimentally determined dose-response curves in the high-dose region.

TABLE X1.4 Weight of Evidence Criteria for Carcinogens

Category	Criterion
A	Human carcinogen, with sufficient evidence from epidemiological studies
B1	Probable human carcinogen, with limited evidence from epidemiological studies
B2	Probable human carcinogen, with sufficient evidence from animal studies and inadequate evidence or no data from epidemiological studies
C	Possible human carcinogen, with limited evidence from animal studies in the absence of human data
D	Not classifiable as to human carcinogenicity, owing to inadequate human and animal evidence
E	Evidence of noncarcinogenicity for humans, with no evidence of carcinogenicity in at least two adequate animal tests in different species, or in both adequate animal and epidemiological studies

X1.5.7 Data from the preceding studies are used to generate reference doses (RfDs), reference concentrations (RfCs), and slope factors (SFs) and are also used in generating drinking water maximum concentration levels (MCLs) and goals (MCLGs), health advisories (HAs), and water quality criteria. These terms are defined in Table X1.5 and further discussed in X3.8.

X1.5.8 *Selection of Chemicals of Concern*—The impact on human health and the environment in cases of gasoline and middle distillate contamination of soils and ground water can be assessed based on potential receptor (that is, aquatic organisms, human) exposure to three groups of materials: light aromatic hydrocarbons, PAHs, and in older spills, lead. Although not one of the primary contaminants previously described, EDB and EDC were used as lead scavengers in some leaded gasolines and may be considered chemicals of concern, when present.

X1.5.9 The light aromatics, benzene, toluene, xylenes, and ethylbenzene have relatively high water solubility and sorb poorly to soils. Thus, they have high mobility in the environment, moving readily through the subsurface. When released into surface bodies of water, these materials exhibit moderate to high acute toxicity to aquatic organisms. Although environmental media are rarely contaminated to the extent that acute human toxicity is an issue, benzene is listed by the USEPA as a Group A Carcinogen (known human carcinogen) and, thus, exposure to even trace levels of this material is considered significant.

X1.5.10 Polycyclic aromatics can be broken into two categories: naphthalenes and methylnaphthalenes (diaromatics) have moderate water solubility and soil sorption potential and, thus, their movement through the subsurface tends to be less than monoaromatics, but substantial movement can still occur. When released into surface bodies of water, these materials have moderate to high toxicity to aquatic organisms. The PAHs with three or more condensed rings have very low solubility (typically less than 1 mg/L) and sorb strongly to soils. Thus, their movement in the subsurface is minimal. Several members in the group of three to six-ring PAHs are known or suspected carcinogens and, thus, exposure to low concentrations in drinking water or through the consumption of contaminated soil by children is significant. In addition, materials containing four to six-ring PAHs are poorly biodegradable and, coupled with the potential to bioaccumulate in tissues of aquatic organisms, these materials have the potential to bioconcentrate (be found at levels in living tissue far higher than present in the general surroundings) in the environment.

X1.5.11 Although almost totally eliminated from use in gasolines in the United States, lead is found associated with older spills. Lead was typically added to gasoline either as tetraethyl or tetramethyl lead and may still be found in its original form in areas containing free product. Typically outside the free product zones, these materials have decomposed into inorganic forms of lead. Lead is a neurotoxin and lead in the blood of children has been associated with reduced intellectual development. The ingestion by children of lead-contaminated soils is an exposure route of great concern, as is

TABLE X1.5 Definitions of Important Toxicological Characteristics

Reference Dose—A reference dose is an estimate (with an uncertainty typically spanning an order of magnitude) of a daily exposure (mg/kg/day) to the general human population (including sensitive subgroups) that is likely to be without an appreciable risk of deleterious effects during a lifetime of exposure.
Reference Concentration—A reference concentration is an estimate (with an uncertainty spanning perhaps an order of magnitude) of a continuous exposure to the human population (including sensitive subgroups) that is likely to be without appreciable deleterious effects during a lifetime.
Slope Factor—The slope of the dose-response curve in the low-dose region. When low-dose linearity cannot be assumed, the slope factor is the slope of the straight line from zero dose to the dose at 1 % excess risk. An upper bound on this slope is usually used instead of the slope itself. The units of the slope factor are usually expressed as $(mg/kg/day)^{-1}$
Drinking Water MCLs and MCLGs—Maximum contaminant levels (MCLs) are drinking water standards established by the EPA that are protective of human health. However, these standards take into account the technological capability of attaining these standards. The EPA has, therefore, also established MCL goals (MCLGs) which are based only on the protection of human health. The MCL standards are often used as clean-up criteria.
Drinking Water Health Advisories—The Office of Drinking Water provides health advisories (HAs) as technical guidance for the protection of human health. They are not enforceable federal standards. The HA's are the concentration of a substance in drinking water estimated to have negligible deleterious effects in humans, when ingested for specified time periods.
Water Quality Criteria—These criteria are not rules and they do not have regulatory impact. Rather, these criteria present scientific data and guidance of the environmental effects of pollutants which can be useful to derive regulatory requirements based on considerations of water quality impacts.

the consumption of lead-contaminated drinking water. Ethylene dibromide and ethylene dichloride, used as lead scavengers in gasolines, are of concern because of their high toxicity (potential carcinogens) and their high mobility in the environment.

X1.5.12 In summary, benzene and benzo(a)pyrene (and in some cases EDB and EDC) are chemicals of concern because of their carcinogenicity. Other PAHs may also be grouped with B(a)P because of uncertainties in their carcinogenicity and because they may accumulate (bioconcentrate) in living tissue.

X1.5.13 *Toxicity and Physical/Chemical Properties for Chemicals of Concern*—A summary of health effects and physical/chemical properties for a number of chemicals of concern is provided in Table X1.2. This table provides toxicological data from a variety of sources, regardless of data quality. A refined discussion for selected chemicals of concern is given as follows. The reader is cautioned that this information is only current as of the dates quoted, and the sources quoted may have been updated, or more recent information may be available in the peer-reviewed literature.

X1.5.13.1 The RfD or SF values are generally obtained from a standard set of reference tables (for example, Integrated Risk Information System, IRIS (**2**), or the Health Effects Assessment Summary Tables, HEAST (**3**)). Except as noted, the toxicity evaluations that follow were taken from IRIS (**2**) because these are EPA-sanctioned evaluations. The information in IRIS (**2**), however, has typically only been peer-reviewed within the EPA and may not always have support from the external scientific community. The information in IRIS may also be subject to error (as exampled by recent revisions in the slope factor for B(a)P and RfC for MTBE).

X1.5.13.2 HEAST (**3**) is a larger database than IRIS (**2**) and is often used as a source of health effects information. Whereas the information in IRIS (**2**) has been subject to data quality review, however, the information in the HEAST (**3**) tables has not. The user is expected to consult the original assessment documents to appreciate the strengths and limitations of the data in HEAST (**3**). Thus, care should be exercised in using the values in HEAST (**3**).

X1.5.13.3 References for the physical/chemical properties are provided in Table X1.2. All Henry's law constants quoted in text are from Ref (**11**) except MTBE which is from estimation: $H = (V_p)(MW)/760(S)$, where MW is the molecular weight, V_p = 414 mmHg at 100°F, and S = 48 000 mg/L.

X1.6 *Profiles of Select Compounds:*

X1.6.1 *Benzene:*

X1.6.1.1 *Toxicity Summary*—Based on human epidemiological studies, benzene has been found to be a human carcinogen (classified as a Group A carcinogen, known human carcinogen by the USEPA). An oral slope factor of $2.9 \times 10^{-2} (mg/kg/day)^{-1}$ has been derived for benzene based on the observance of leukemia from occupational exposure by inhalation. The USEPA has set a drinking water maximum contaminant level (MCL) at 5 µg/L. The maximum contaminant level goal (MCLG) for benzene is set at zero.

X1.6.1.2 Although the EPA does not usually set long-term drinking water advisories for carcinogenic materials (no exposure to carcinogens is considered acceptable), a ten-day drinking water health advisory for a child has been set at 0.235 mg/L based on hematological impairment in animals. The EPA is in the process of evaluating noncancer effects and an oral RfD for benzene is pending.

X1.6.1.3 In situations in which both aquatic life and water are consumed from a particular body of water, a recommended EPA water quality criterion is set at 0.66 µg/L. When only aquatic organisms are consumed, the criterion is 40 µg/L. These criteria were established at the one-in-one-million risk level (that is, the criteria represent a one-in-one-million estimated incremental increase in cancer risk over a lifetime).

X1.6.1.4 *Physical/Chemical Parameter Summary*—Benzene is subject to rapid volatilization (Henry's law constant = 5.5×10^{-3} m^3-atm/mol) under common above-ground environmental conditions. Benzene will be mobile in soils due to its high water solubility (2.75×10^{6}µg/L) and relatively low sorption to soil particles (log K_{oc} = 1.92) and, thus, has the potential to leach into ground water. Benzene has a relatively low log K_{ow}value (2.12) and is biodegradable. Therefore, it is not expected to bioaccumulate. In laboratory tests, when a free gasoline phase was in equilibrium with water, typical benzene concentrations in water ranged from 2.42×10^4 to 1.11×10^5µg/L.

X1.6.2 *Toluene:*

X1.6.2.1 *Toxicity Summary*—Using data from animal studies, the USEPA has set an oral RfD for toluene at 0.2 mg/kg/day. In converting a NOAEL from an animal study, in which the critical effect observed was changes in liver and kidney weights, an uncertainty factor of 1000 and a modifying factor of 1 were used. The EPA has assigned an overall medium level of confidence in the RfD because, although the principal study was well performed, the length of the study corresponded to only subchronic rather than a chronic evaluation, and reproductive aspects were lacking. Based on the RfD and assuming 20 % exposure from drinking water, the EPA has set both drinking water MCL and MCLG of 1000 µg/L. Drinking water health advisories range from 1 mg/L (lifetime equivalent to the RfD) to 20 mg/L (one-day advisory for a child).

X1.6.2.2 In situations in which both aquatic life and water are consumed from a particular body of water, the recommended water quality criterion is set at 1.43×10^4µg/L. When only aquatic organisms are consumed, the criterion is 4.24×10^5 µg/L.

X1.6.2.3 An inhalation RfC of 0.4 mg/m^3 was derived based on neurological effects observed in a small worker population. An uncertainty factor of 300 and a modifying factor of 1 were used to convert the lowest observed adverse effect level (LOAEL) to the RfC. The overall confidence in the RfC was established as medium because of the use of a LOAEL and because of the paucity of exposure information.

X1.6.2.4 *Physical/Chemical Parameter Summary*—Toluene is expected to volatilize rapidly, under common above-ground environmental conditions, due to its relatively high Henry's law constant (6.6×10^{-3} m^3-atm/mol). It will be mobile in soils based on an aqueous solubility of 5.35×10^5µg/L and relatively poor sorption to soils (estimated log K_{oc} = 2.48) and, hence, has a potential to leach into ground water. Toluene has a relatively low log K_{ow}(2.73) and is biodegradable. Bioaccumulation of toluene is, therefore, expected to be negligible. In laboratory tests, when a free gasoline phase was in equilibrium with water, typical toluene concentrations in water ranged from 3.48×10^4 to 8.30×10^4µg/L.

X1.6.3 *Xylenes*:

X1.6.3.1 *Toxicity Summary*—Using data from animal studies, the USEPA has set an oral RfD for xylenes at 2.0 mg/kg/day. In converting a NOAEL from the animal study, in which the critical effects observed were hyperactivity, decreased body weight, and increased mortality (among male rats), an uncertainty factor of 100 and a modifying factor of 1 were used. The EPA has assigned an overall medium level of confidence in the RfD because, although the principal study was well designed and performed, supporting chemistry was not performed. A medium level of confidence was also assigned to the database. Based on the RfD and assuming 20 % exposure from drinking water, the EPA has set both drinking water MCL and MCLG of 10 mg/L. Drinking water health advisories of 10 mg/L (lifetime, adult) and 40 mg/L (one-day, ten-day, and long-term child) are quoted by the EPA's Office of Drinking Water. No USEPA ambient water criteria are available for xylenes at this time. Evaluation of an inhalation RfC is pending.

X1.6.3.2 *Physical/Chemical Parameter Summary*—Xylenes are expected to rapidly volatilize under common above-ground environmental conditions based on their Henry's law constants (for *o*-xylene, $H = 5.1 \times 10^{-3}$ m^3-atm/mol). Xylenes have a moderate water solubility (1.46 to 1.98×10^5µg/L) (pure compound) as well as moderate capacities to sorb to soils (estimated log K_{oc} 2.38 to 2.79) and, therefore, they will be mobile in soils and may leach into ground water. Xylenes are biodegradable, and with log K_{ow} values in the range from 2.8 to 3.3, they are not expected to bioaccumulate.

X1.6.4 *Ethylbenzene*:

X1.6.4.1 *Toxicity Summary*—Using data from animal studies, the USEPA has set an oral RfD for ethylbenzene at 0.1 mg/kg/day. In converting a NOAEL from the animal study, in which the critical effects observed were liver and kidney toxicity, an uncertainty factor of 1000 and a modifying factor of 1 were used. The EPA has assigned an overall low level of confidence in the RfD because the study was poorly designed and confidence in the supporting database is also low. Based on the RfD and assuming 20 % exposure from drinking water, the EPA has set both drinking water MCL and MCLG of 700 µg/L. Drinking water health advisories range from 700 µg/L (lifetime equivalent to the RfD) to 32 mg/L (one-day advisory for a child). In situations in which both aquatic life and water are consumed from a particular body of water, a recommended ambient water criterion is set at 1400 µg/L. When only aquatic organisms are consumed, the criterion is 3280 µg/L. An inhalation RfC of 1 mg/m^3 was derived based on developmental toxicity effects observed in rats and rabbits. An uncertainty factor of 300 and a modifying factor of 1 were used to convert the NOAEL to the RfC. Both the study design and database were rated low and, thus, the overall confidence in the RfC was established as low.

X1.6.4.2 *Physical/Chemical Parameter Summary*—Ethylbenzene has a relatively high Henry's law constant (8.7×10^{-3} m^3-atm/mol) and, therefore, can rapidly volatilize under common above-ground environmental conditions. Based on its moderate water solubility (1.52×10^5µg/L) and moderate capacity to sorb to soils (estimated log K_{oc} = 3.04), it will have moderate mobility in soil and may leach into ground water. In laboratory tests, when a free gasoline phase was in equilibrium with water, typical combined ethylbenzene and xylenes concentrations in water ranged from 1.08×10^4 to 2.39×10^4µg/L, due to partitioning effects. Ethylbenzene has a moderate low K_{ow} value (3.15) and is biodegradable. Therefore, it is not expected to bioaccumulate. In laboratory tests, when a free gasoline phase was in equilibrium with water, typical combined ethylbenzene and xylenes concentrations in water ranged from 1.08×10^4 to 2.39×10^4µg/L.

X1.6.5 *Naphthalenes*:

X1.6.5.1 *Toxicity Summary*—In general, poisoning may occur by ingestion of large doses, inhalation, or skin adsorption of naphthalene. It can cause nausea, headache, diaphoresis, hematuria, fever, anemia, liver damage, vomiting, convulsions, and coma. Methylnaphthalenes are presumably less acutely toxic than naphthalene. Skin irritation and skin photosensitization are the only effects reported in man. Inhalation of the vapor may cause headache, confusion, nausea, and sometimes vomiting. The environmental concerns with naphthalenes are primarily attributed to effects on aquatic organisms. As a consequence, the EPA has not set any human health criteria for these materials (that is, there is no RfD or RfC, no drinking water MCL or MCLG or ambient water quality criteria). A risk assessment to define a RfD for these materials is presently under review by the EPA. Drinking water health advisories range from 20 µg/L (lifetime, adult) to 500 µg/L (one-day advisory for a child).[6]

X1.6.5.2 *Physical/Chemical Parameter Summary: Naphthalene*—Naphthalene has a relatively high Henry's law constant (1.15×10^{-3} m^3-atm/mol) and, thus, has the capacity to volatilize rapidly under common above-ground environmental conditions. It has a moderate water solubility (3.10×10^4µg/L) and log K_{oc} (3.11) and has the potential to leach to ground water. A moderate log K_{ow} value of 3.01 has been reported, but because naphthalene is very biodegradable, it is unlikely to bioconcentrate to a significant degree.

X1.6.5.3 *Methylnaphthalenes*—Henry's law constants (2.60×10^{-4} m^3-atm/mol and 5.18×10^{-4} m^3-atm/mol for 1- and 2-methylnaphthalene, respectively) suggest that these materials have the potential to volatilize under common above-ground environmental conditions. 1-Methylnaphthalene exhibits a water solubility similar to naphthalene (2.60×10^4µg/L to 2.8×10^4µg/L). However, solubility decreases with increasing alkylation (dimethylnaphthalenes: 2.0×10^3µg/L to 1.1×10^4µg/L, 1,4,5-trimethylnaphthalene: 2.0×10^3µg/L). These materials are, therefore, expected to be slightly mobile to relatively immobile in soil (for example, log $K_{oc}$is in the range from 2.86 to 3.93 for 1- and 2-methylnaphthalenes). In aquatic systems, methylnaphthalenes may partition from the water column to organic matter contained in sediments and suspended solids. Methylnaphthalenes have high log K_{ow}values (greater than 3.5) and have the potential to bioaccumulate. They do, however, exhibit a moderate degree of biodegradation, which typically decreases with increased alkylation.

X1.6.6 *Three to Six-Ringed PAHs*—The most significant health effect for this class of compounds is their carcinogenicity, which is structure-dependent. Anthracene and phenanthrene have not been shown to cause cancer in laboratory animals. The available data does not prove pyrene to be carcinogenic to experimental animals. On the other hand, benz[a]-anthracene, benzo[a]pyrene, dibenz[a,h]anthracene, and 7,12-dimethylbenz[a]-anthracene have been shown to be carcinogenic in laboratory animals. B(a)P and pyrene are discussed in X1.6.7 and X1.6.8 as representatives of carcinogenic and noncarcinogenic effects of this class.

X1.6.7 *Benzo(a)pyrene (BaP)*:

X1.6.7.1 *Toxicity Summary*—Based on animal data, B(a)P has been classified as a probable human carcinogen (B2 carcinogen) by the USEPA. A range of oral slope factors from 4.5 to 11.7 $(mg/kg/day)^{-1}$ with a geometric mean of 7.3 $(mg/kg/day)^{-1}$ has been derived for B(a)P based on the observance of tumors of the forestomach and squamous cell carcinomas in mice. The data was considered less than optimal but acceptable (note that the carcinogenicity assessment for B(a)P may change in the near future pending the outcome of an on-going EPA review). The EPA has proposed a drinking water MCL at 0.2 µg/L (based on the analytical detection limits). The MCLG for B(a)P is set at zero. In situations in which both aquatic life and water are consumed from a particular body of water, a recommended EPA water quality criterion is set at 2.8×10^{-3} µg/L. When only aquatic organisms are consumed, the criterion is 3.11×10^{-2}µg/L.

X1.6.7.2 *Physical/Chemical Parameter Summary*—When released to water, PAHs are not subject to rapid volatilization (Henry's law constants are on the order of 1.0×10^{-4} m^3-atm/mol or less) under common environmental conditions. They have low aqueous solubility values and tend to sorb to soils and sediments and remain fixed in the environment. Three ring members of this group such as anthracene and phenanthrene have water solubilities on the order of 1000 µg/L. The water solubilities decrease substantially for larger molecules in the group, for example, benzo[a]pyrene has a water solubility of 1.2 µg/L. The log K_{oc} values for PAHs are on the order of 4.3 and greater, which suggests that PAHs will be expected to adsorb very strongly to soil. The PAHs with more than three rings generally have high log K_{ow} values (6.06 for benzo[a]pyrene), have poor biodegradability characteristics and may bioaccumulate.

X1.6.8 *Pyrene*:

X1.6.8.1 *Toxicity Summary*—Using data from animal studies, the USEPA has set an oral RfD for pyrene at 3×10^{-2} mg/kg/day. In converting a NOAEL from the animal study, in which the critical effects observed were kidney toxicity, an uncertainty factor of 3000 and a modifying factor of 1 were used. The EPA has assigned an overall low level of confidence in the RfD because although the study was well-designed, confidence in the supporting database is low. No drinking water MCLs or health advisories have been set. In situations in which both aquatic life and water are consumed from a particular body of water, a recommended EPA water quality criterion is set at 2.8×10^{-3}µg/L. When only aquatic organisms are consumed, the criterion is 3.11×10^{-2}µg/L.

X1.6.8.2 *Physical/Chemical Parameter Summary*—Refer to X1.6.7.2 for BaP. Also see Table X1.2.

X1.6.9 *MTBE*:

X1.6.9.1 *Toxicity Summary*—Using data from animal studies, the USEPA has set an inhalation RfC for MTBE at 3 mg/m^3. In converting a NOAEL from the animal study, in which the critical effects observed included increased liver and kidney weight and increased severity of spontaneous renal lesions (females), increased prostration (females) and swollen pericolar tissue, an uncertainty factor of 100 and a modifying factor of 1 were used. The EPA has assigned an overall medium level of confidence in the RfC because although the study was well-designed, some information on the chemistry was lacking. The confidence in the supporting database is medium to high. No drinking water MCLs or ambient water quality criteria have been set. However, a risk assessment, which may define a RfD for this material, is presently under review by EPA. Drinking water health advisories range from 40µ g/L (lifetime, adult) to 3000 µg/L (one-day advisory for a child).[6]

X1.6.9.2 *Physical/Chemical Parameter Summary*—The Henry's law constant for MTBE is estimated to be approximately 1.0×10^{-3} m^3-atm/mol. It is, therefore, expected to have the potential to rapidly volatilize under common above-ground environmental conditions. It is very water soluble (water solubility is 4.8×10^7µg/L), and with a relatively low

[6] Office of Water, USEPA, Washington, DC.

capacity to sorb to soils (estimated log K_{oc} = 1.08), MTBE will migrate at the same velocity as the water in which it is dissolved in the subsurface. The log K_{ow} value has been estimated to be between 1.06 and 1.30, indicating MTBE's low bioaccumulative potential. It is expected to have a low potential to biodegrade, but no definitive studies are available.

X1.6.10 *Lead*:

X1.6.10.1 *Toxicity Summary—(The following discussion is for inorganic lead—not the organic forms of lead (tetraethyllead, tetramethyllead) that were present in petroleum products.)* A significant amount of toxicological information is available on the health effects of lead. Lead produces neurotoxic and behavioral effects particularly in children. However, the EPA believes that it is inappropriate to set an RfD for lead and its inorganic compounds because the agency believes that some of the effects may occur at such low concentrations as to suggest no threshold. The EPA has also determined that lead is a probable human carcinogen (classified as B2). The agency has chosen not to set a numeric slope factor at this time, however, because it is believed that standard procedures for doing so may not be appropriate for lead. At present, the EPA has set an MCLG of zero but has set no drinking water (MCL) or health advisories because of the observance of low-level effects, the overall Agency goal of reducing total lead exposure and because of its classification as a B2 carcinogen. An action of level of 15 µg/L has been set for water distribution systems (standard at the tap). The recommended EPA water quality criterion for consumption of both aquatic life and water is set at 50 µg/L.

X1.6.10.2 *Physical/Chemical Parameter Summary*—Organic lead additive compounds are volatile (estimated Henry's law constant for tetraethyl lead = 7.98×10^{-2} m^3-atm/mol) and may also sorb to particulate matter in the air. Tetraethyl lead has an aqueous solubility of 800 µg/L and an estimated log K_{oc} of 3.69 and, therefore, should not be very mobile in the soil. It decomposes to inorganic lead in dilute aqueous solutions and in contact with other environmental media. In free product (gasoline) plumes, however, it may remain unchanged. Inorganic lead compounds tightly bind to most soils with minimal leaching under natural conditions. Aqueous solubility varies depending on the species involved. The soil's capacity to sorb lead is correlated with soil pH, cation exchange capacity, and organic matter. Lead does not appear to bioconcentrate significantly in fish but does in some shellfish, such as mussels. Lead is not biodegradable.

X1.7 *Discussion of Acceptable Risk* **(12)**—Beginning in the late 1970s and early 1980s, regulatory agencies in the United States and abroad frequently adopted a cancer risk criteria of one-in-one-million as a negligible (that is, of no concern) risk when fairly large populations might be exposed to a suspect carcinogen. Unfortunately, theoretical increased cancer risks of one-in-one-million are often incorrectly portrayed as serious public health risks. As recently discussed by Dr. Frank Young **(13)**, the current commissioner of the Food and Drug Administration (FDA), this was not the intent of such estimates:

X1.7.1 In applying the de minimis concept and in setting other safety standards, the FDA has been guided by the figure of "one-in-one-million." Other Federal agencies have also used a one-in-one-million increased risk over a lifetime as a reasonable criterion for separating high-risk problems warranting agency attention from negligible risk problems that do not.

X1.7.2 The risk level of one-in-one-million is often misunderstood by the public and the media. It is not an actual risk, that is, we do not expect one out of every million people to get cancer if they drink decaffeinated coffee. Rather, it is a mathematical risk based on scientific assumptions used in risk assessment. The FDA uses a conservative estimate to ensure that the risk is not understated. We interpret animal test results conservatively, and we are extremely careful when we extrapolate risks to humans. When the FDA uses the risk level of one-in-one-million, it is confident that the risk to humans is virtually nonexistent.

X1.7.3 In short, a "one-in-one-million" cancer risk estimate, which is often tacitly assumed by some policy-makers to represent a trigger level for regulatory action, actually represents a level of risk that is so small as to be of negligible concern.

X1.7.4 Another misperception within the risk assessment arena is that all occupational and environmental regulations have as their goal a theoretical maximum cancer risk of 1 in 1 000 000. Travis, et al **(14)** recently conducted a retrospective examination of the level of risk that triggered regulatory action in 132 decisions. Three variables were considered: (*1*) individual risk (an upper-bound estimate of the probability at the highest exposure), (*2*) population risk (an upper-limit estimate of the number of additional incidences of cancer in the exposed population), and (*3*) population size. The findings of Travis, et al **(14)** can be summarized as follows:

X1.7.4.1 Every chemical with an individual lifetime risk above 4×10^{-3} received regulation. Those with values below 1×10^{-6} remained unregulated.

X1.7.4.2 For small populations, regulatory action never resulted for individual risks below 1×10^{-4}.

X1.7.4.3 For potential effects resulting from exposures to the entire United States population, a risk level below 1×10^{-6} never triggered action; above 3×10^{-4} always triggered action.

X1.7.5 Rodricks, et al **(15)** also evaluated regulatory decisions and reached similar conclusions. In decisions relating to promulgation of National Emission Standards for Hazardous Air Pollutants (NESHAPS), the USEPA has found the maximum individual risks and total population risks from a number of radionuclide and benzene sources too low to be judged significant. Maximum individual risks were in the range from 3.6×10^{-5} to 1.0×10^{-3}. In view of the risks deemed insignificant by USEPA, Rodricks, et al **(15)** noted that 1×10^{-5} (1 in 100 000) appears to be in the range of what USEPA might consider an insignificant average lifetime risk, at least where aggregate population risk is no greater than a fraction of a cancer yearly.

X1.7.6 Recently, final revisions to the National Contingency Plan **(16)** have set the acceptable risk range between 10^{-4} and 10^{-6} at hazardous waste sites regulated under CERCLA. In the recently promulgated *Hazardous Waste Management System Toxicity Characteristics Revisions* **(17)**, the USEPA has stated that:

"For drinking water contaminants, EPA sets a reference risk range for carcinogens at 10^{-6} excess individual cancer risk from lifetime exposure. Most regulatory actions in a variety of EPA programs have generally targeted this range using conservative models which are not likely to underestimate the risk."

X1.7.7 Interestingly, the USEPA has selected and promulgated a single risk level of 1 in 100 000 (1×10^{-5}) in the *Hazardous Waste Management System Toxicity Characteristics Revisions* (**17**). In their justification, the USEPA cited the following rationale:

The chosen risk level of 10^{-5} is at the midpoint of the reference risk range for carcinogens (10^{-4} to 10^{-6}) generally used to evaluate CERCLA actions. Furthermore, by setting the risk level at 10^{-5} for TC carcinogens, EPA believes that this is the highest risk level that is likely to be experienced, and most if not all risks will be below this level due to the generally conservative nature of the exposure scenario and the underlying health criteria. For these reasons, the Agency regards a 10^{-5} risk level for Group A, B, and C carcinogens as adequate to delineate, under the Toxicity Characteristics, wastes that clearly pose a hazard when mismanaged."

X1.7.8 When considering these limits it is interesting to note that many common human activities entail annual risks greatly in excess of one-in-one-million. These have been discussed by Grover Wrenn, former director of Federal Compliance and State Programs at OSHA, as follows:

X1.7.9 State regulatory agencies have not uniformly adopted a one-in-one-million (1×10^{-6}) risk criterion in making environmental and occupational decisions. The states of Virginia, Maryland, Minnesota, Ohio, and Wisconsin have employed or proposed to use the one-in-onehundred-thousand (1×10^{-5}) level of risk in their risk management decisions (**18**). The State of Maine Department of Human Services (DHS) uses a lifetime risk of one in one hundred thousand as a reference for non-threshold (carcinogenic) effects in its risk management decisions regarding exposures to environmental contaminants (**19**). Similarly, a lifetime incremental cancer risk of one in one hundred thousand is used by the Commonwealth of Massachusetts as a cancer risk limit for exposures to substances in more than one medium at hazardous waste disposal sites (**20**). This risk limit represents the total cancer risk at the site associated with exposure to multiple chemicals in all contaminated media. The State of California has also established a level of risk of one in one hundred thousand for use in determining levels of chemicals and exposures that pose no significant risks of cancer under the Safe Drinking Water and Toxic Enforcement Act of 1986 (Proposition 65) (**21**). Workplace air standards developed by the Occupational Safety and Health Administration (OSHA) typically reflect theoretical risks of one in one thousand (1×10^{-3}) or greater (**15**).

X1.7.10 Ultimately, the selection of an acceptable and de minimis risk level is a policy decision in which both costs and benefits of anticipated courses of action should be thoroughly evaluated. However, actuarial data and risk estimates of common human activities, regulatory precedents, and the relationship between the magnitude and variance of background and incremental risk estimates all provide compelling support for the adoption of the de minimis risk level of 1×10^{-5} for regulatory purposes.

X1.7.11 In summary, U.S. Federal and state regulatory agencies have adopted a one-in-one-million cancer risk as being of negligible concern in situations where large populations (for example, 200 million people) are involuntarily exposed to suspect carcinogens (for example, food additives). When smaller populations are exposed (for example, in occupational settings), theoretical cancer risks of up to 10^{-4} (1 in 10 000) have been considered acceptable.

X2. DEVELOPMENT OF RISK-BASED SCREENING LEVELS (RBSLs) APPEARING IN SAMPLE LOOK-UP Table X2.1

Introduction: X2.1

X2.1.1 This appendix contains the equations and parameters used to construct the example "Look-Up" (Table X2.1). This table was prepared solely for the purpose of presenting an example Tier 1 matrix of RBSLs, and these values should not be viewed, or misused, as proposed remediation "standards." The reader should note that not all possible pathways have been considered and a number of assumptions concerning exposure scenarios and parameter values have been made. These should be reviewed for appropriateness before using the listed RBSLs as Tier 1 screening values.

X2.1.2 The approaches used to calculate RBSLs appearing in Table X2.1 are briefly discussed as follows for exposure to vapors, ground water, surficial soils, and subsurface soils by means of the following pathways:

X2.1.2.1 Inhalation of vapors,

X2.1.2.2 Ingestion of ground water,

X2.1.2.3 Inhalation of outdoor vapors originating from dissolved hydrocarbons in ground water,

X2.1.2.4 Inhalation of indoor vapors originating from dissolved hydrocarbons in ground water,

X2.1.2.5 Ingestion of surficial soil, inhalation of outdoor vapors and particulates emanating from surficial soils, and dermal absorption resulting from surficial soil contact with skin,

X2.1.2.6 Inhalation of outdoor vapors originating from hydrocarbons in subsurface soils,

X2.1.2.7 Inhalation of indoor vapors originating from subsurface hydrocarbons, and

X2.1.2.8 Ingestion of ground water impacted by leaching of dissolved hydrocarbons from subsurface soils.

X2.1.3 For the pathways considered, approaches used in this appendix are consistent with guidelines contained in Ref (**26**).

X2.1.4 The development presented as follows focuses only on human-health RBSLs for chronic (long-term) exposures.

TABLE X2.1 Example Tier 1 Risk-Based Screening Level (RBSL) Look-up Table[A]

NOTE 1—This table is presented here only as an example set of Tier 1 RBSLs. It is not a list of proposed standards. The user should review all assumptions prior to using any values. Appendix X2 describes the basis of these values.

Exposure Pathway	Receptor Scenario	Target Level	Benzene	Ethylbenzene	Toluene	Xylenes (Mixed)	Napthalenes	Benzo (a)pyrene
				Air				
Indoor air screening levels for inhalation exposure, μ/ m^3	residential	cancer risk = 1E-06	3.92E-01					1.86E-03
		cancer risk = 1E-04	3.92E + 01					1.86E-01
		chronic HQ = 1		1.39E + 03	5.56E + 02	9.73E + 03	1.95E + 01	
	commercial/ industrial	cancer risk = 1E-06	4.93E-01					2.35E-03
		cancer risk = 1E-04	4.93E + 01					2.35E-01
		chronic HQ = 1		1.46E + 03	5.84E + 02	1.02E + 04	2.04E + 01	
Outdoor air screening levels for inhalation exposure, μg/m^3	residential	cancer risk = 1E-06	2.94E-01					1.40E-03
		cancer risk = 1E-04	2.94E + 01					1.40E-01
		chronic HQ = 1		1.04E + 03	4.17E + 02	7.30E + 03	1.46E + 01	
	commercial/ industrial	cancer risk = 1E-06	4.93E-01					2.35E-03
		cancer risk = 1E-04	4.93E + 01					2.35E-01
		chronic HQ = 1		1.46E + 03	5.84E + 02	1.02E + 04	2.04E + 01	
OSHA TWA PEL,μ g/m^3			3.20E + 03	4.35E + 05	7.53E + 05	4.35E + 06	5.00E + 04	2.00E + 02[A]
Mean odor detection threshold,μ g/m^3 [B]			1.95E + 05		6.00E + 03	8.70E + 04	2.00E + 02	
National indoor background concentration range,μ g/m^3 [C]			3.25E + 00 to 2.15E + 01	2.20E + 00 to 9.70E + 00	9.60E-01 to 2.91E + 01	4.85E + 00 to 4.76E + 01		
				Soil				
Soil volatilization to outdoor air, mg/kg	residential	cancer risk = 1E-06	2.72E-01					RES[D]
		cancer risk = 1E-04	2.73E + 01					RES
		chronic HQ = 1		RES	RES	RES	RES	
	commercial industrial	cancer risk = 1E-06	4.57E-01					RES
		cancer risk = 1E-04	4.57E + 01					RES
		chronic HQ = 1		RES	RES	RES	RES	
Soil-vapor intrusion from soil to buildings, mg/kg	residential	cancer risk = 1E-06	5.37E-03					RES
		cancer risk = 1E-04	5.37E-01					RES
		chronic HQ = 1		4.27E + 02	2.00E + 01	RES	4.07E + 01	
	commercial/ industrial	cancer risk = 1E-06	1.69E-02					RES
		cancer risk = 1E-04	1.69E + 00					RES
		chronic HQ = 1		1.10E + 03	5.45E + 01	RES	1.07E + 02	
Surficial soil (0 to 3 ft) (0 to 0.9 m) ingestion/ dermal/ inhalation, mg/kg	residential	cancer risk = 1E-06	5.82E + 00					1.30E-01
		cancer risk = 1E-04	5.82E + 02					1.30E + 01
		chronic HQ = 1		7.83E + 03	1.33E + 04	1.45E + 06	9.77E + 02	
	commercial/ industrial	cancer risk = 1E-06	1.00E + 01					3.04E-01
		cancer risk = 1E-04	1.00E + 03					3.04E + 01
		chronic HQ = 1		1.15E + 04	1.87E + 04	2.08E + 05	1.50E + 03	
Soil-leachate to protect ground water ingestion target level, mg/kg	MCLs		2.93E-02	1.10E + 02	1.77E + 01	3.05E + 02	N/A	9.42E + 00
	residential	cancer risk = 1E-06	1.72E-02					5.50E-01
		cancer risk = 1E-04	1.72E + 00					RES
		chronic HQ = 1		5.75E + 02	1.29E + 02	RES	2.29E + 01	
	commercial/ industrial	cancer risk = 1E-06	5.78E-02					1.85E + 00
		cancer risk = 1E-04	5.78E + 00					RES
		chronic HQ = 1		1.61E + 03	3.61E + 02	RES	6.42E + 01	

TABLE X2.1 *Continued*

Exposure Pathway	Receptor Scenario	Target Level	Benzene	Ethylbenzene	Toluene	Xylenes (Mixed)	Napthalenes	Benzo (a)pyrene
			Ground Water					
Ground water volatilization to outdoor air, mg/L	residential	cancer risk = 1E-06	1.10E + 01					>S[E]
		cancer risk = 1E-04	1.10E + 03					>S
		chronic HQ = 1		>S	>S	>S	>S	
	commercial/ industrial	cancer risk = 1E-06	1.84E + 01					>S
		cancer risk = 1E-04	>S					>S
		chronic HQ = 1		>S	>S	>S	>S	
Ground water ingestion, mg/L	MCLs		5.00E-03	7.00E-01	1.00E + 00	1.00E + 01	N/A	2.00E-04
	residential	cancer risk = 1E-06	2.94E-03					1.17E-05
		cancer risk = 1E-04	2.94E-01					1.17E-03
		chronic HQ = 1		3.65E + 00	7.30E + 00	7.30E + 01	1.46E-01	
	commercial/ industrial	cancer risk = 1E-06	9.87E-03					3.92E-05
		cancer risk = 1E-04	9.87E-01					>S
		chronic HQ = 1		1.02E + 01	2.04E + 01	>S	4.09E-01	
Ground water—vapor intrusion from ground water to buildings, mg/L	residential	cancer risk = 1E-06	2.38E-02					>S
		cancer risk = 1E-04	2.38E + 00					>S
		chronic HQ = 1		7.75E + 01	3.28E + 01	>S	4.74E + 00	
	commercial/ industrial	cancer risk = 1E-06	7.39E-02					>S
		cancer risk = 1E-04	7.39E + 00					>S
		chronic HQ = 1		>S	8.50E + 01	>S	1.23E + 01	

[A] As benzene soluble coal tar pitch volatiles.
[B] See Ref **(22)**.
[C] See Refs **(23-25)**.
[D] RES—Selected risk level is not exceeded for pure compound present at any concentration.
[E] >S—Selected risk level is not exceeded for all possible dissolved levels ($\leq$ pure component solubility).

X2.1.4.1 In the case of compounds that have been classified as carcinogens, the RBSLs are based on the general equation:

$$risk = average\ lifetime\ intake[\mathrm{mg/kg-day}]\ [ml]P5\ potency\ factor[\mathrm{mg/kg-day}]^{-1} \quad (X2.1)$$

where the intake depends on exposure parameters (ingestion rate, exposure duration, and so forth), the source concentration, and transport rates between the source and receptor. The potency factor is selected after reviewing a number of sources, including the USEPA Integrated Risk Information System (IRIS) **(2)** database, USEPA Health Effects Assessment Summary Tables (HEAST) **(3)**, and peer-reviewed sources. The RBSL values appearing in Table X2.1 correspond to probabilities of adverse health effects ("risks") in the range from 10^{-6} to 10^{-4} resulting from the specified exposure. Note that this risk value does not reflect the probability for the specified exposure scenario to occur. Therefore, the actual potential risk to a population for these RBSLs is lower than the 10^{-6} to 10^{-4} range.

X2.1.4.2 In the case of compounds that have not been classified as carcinogens, the RBSLs are based on the general equation:

$$hazard\ quotient = average\ intake\ [\mathrm{mg/kg-day}] / reference\ dose\ [\mathrm{mg/kg-day}] \quad (X2.2)$$

where the intake depends on exposure parameters (ingestion rate, exposure duration, and so forth), the source concentration, and transport rates between the source and receptor. The reference dose is selected after reviewing a number of sources, including the USEPA Integrated Risk Information System (IRIS) **(2)** database, USEPA Health Effects Assessment Summary Tables (HEAST) **(3)**, and peer-reviewed sources. The RBSL values appearing in Table X2.1 correspond to hazard quotients of unity resulting from the specified exposure. Note that this hazard quotient value does not reflect the probability for the specified exposure scenario to occur. Therefore, the actual potential impact to a population for these RBSLs is lower than a hazard quotient of unity.

X2.1.5 Tables X2.2-X2.7 summarize the equations and parameters used to prepare the example look-up Table X2.1. The basis for each of these equations is discussed in X2.2-X2.10.

X2.2 *Air—Inhalation of Vapors (Outdoors/Indoors)* —In this case chemical intake results from the inhalation of vapors. It is assumed that vapor concentrations remain constant over the duration of exposure, and all inhaled chemicals are absorbed. Equations appearing in Tables X2.2 and X2.3 for estimating RBSLs for vapor concentrations in the breathing zone follow guidance given in Ref **(26)**. Should the calculated RBSL exceed the saturated vapor concentration for any individual component, "$>P_{vap}$" is entered in the table to indicate that the selected risk level or hazard quotient cannot be reached or exceeded for that compound and the specified exposure scenario.

X2.3 *Ground Water—Ingestion of Ground Water*— In this case chemical intake results from ingestion of ground water. It is assumed that the dissolved hydrocarbon concentrations remain constant over the duration of exposure. Equations appearing in Tables X2.2 and X2.3 for estimating RBSLs for drinking water concentrations follow guidance given in Ref **(26)** for ingestion of chemicals in drinking water. Should the calculated RBSL exceed the pure component solubility for any individual component, ">S" is entered in the table to indicate that the selected risk level or hazard quotient cannot be reached or exceeded for that compound and the specified exposure scenario (unless free-phase product is mixed with the ingested water).

X2.4 *Ground Water—Inhalation of Outdoor Vapors:*

X2.4.1 In this case chemical intake is a result of inhalation

TABLE X2.2 Equations Used to Develop Example Tier 1 Risk-Based Screening Level (RBSLs) Appearing in "Look-Up" Table X2.1—Carcinagenic Effects[A]

NOTE—See Tables X2.4 through X2.7 for definition of parameters.

Medium	Exposure Route	Risk-Based Screening Level (RBSL)
Air	inhalation[B]	$RBSL_{air}\left[\frac{\mu g}{m^3\text{-}air}\right] = \frac{TR \times BW \times AT_c \times 365\,\frac{days}{years} \times 10^3\,\frac{\mu g}{mg}}{SF_i \times IR_{air} \times EF \times ED}$
Ground water	ingestion (potable ground water supply only)[B]	$RBSL_w\left[\frac{mg}{L\text{-}H_2O}\right] = \frac{TR \times BW \times AT_c \times 365\,\frac{days}{years}}{SF_o \times IR_w \times EF \times ED}$
Ground water[C]	enclosed-space (indoor) vapor inhalation[D]	$RBSL_w\left[\frac{mg}{L\text{-}H_2O}\right] = \frac{RBSL_{air}\left[\frac{\mu g}{m^3\text{-}air}\right]}{VF_{wesp}} \times 10^{-3}\,\frac{mg}{\mu g}$
Ground water[C]	ambient (outdoor) vapor inhalation[D]	$RBSL_w\left[\frac{mg}{L\text{-}H_2O}\right] = \frac{RBSL_{air}\left[\frac{\mu g}{m^3\text{-}air}\right]}{VF_{wamb}} \times 10^{-3}\,\frac{mg}{\mu g}$
Surficial soil	ingestion of soil, inhalation of vapors and particulates, and dermal contact[B]	$RBSL_s\left[\frac{\mu g}{kg\text{-}soil}\right] = \frac{TR \times BW \times AT_c \times 365\,\frac{days}{years}}{EF \times ED\left[\left(SF_o \times 10^{-6}\,\frac{kg}{mg} \times (IR_{soil} \times RAF_o + SA \times M \times RAF_d)\right) + (SF_i \times IR_{air} \times (VF_{ss} + VF_p))\right]}$ For surficial and excavated soils (0 to 1 m)
Subsurface soil[C]	ambient (outdoor) vapor inhalation[D]	$RBSL_s\left[\frac{mg}{kg\text{-}soil}\right] = \frac{RBSL_{air}\left[\frac{\mu g}{m^3\text{-}air}\right]}{VF_{samb}} \times 10^{-3}\,\frac{mg}{\mu g}$
Subsurface soil[C]	enclosed space (indoor) vapor inhalation[D]	$RBSL_s\left[\frac{mg}{kg\text{-}soil}\right] = \frac{RBSL_{air}\left[\frac{\mu g}{m^3\text{-}air}\right]}{VF_{sesp}} \times 10^{-3}\,\frac{mg}{\mu g}$
Subsurface soil[C]	leaching to ground water[D]	$RBSL_s\left[\frac{mg}{kg\text{-}soil}\right] = \frac{RBSL_w\left[\frac{mg}{L\text{-}H_2O}\right]}{LF_{sw}}$

[A] Note that all RBSL values should be compared with thermodynamic partitioning limits, such as solubility levels, maximum vapor concentrations, and so forth. If a RBSL exceeds the relevant partitioning limit, this is an indication that the selected risk or hazard level will never be reached or exceeded for that chemical and the selected exposure scenario.

[B] Screening levels for these media based on other considerations (for example, aesthetic, background levels, environmental resource protection, and so forth) can be derived with these equations by substituting the selected target level for $RBSL_{air}$ or $RBSL_w$ appearing in these equations.

[C] These equations are based on Ref **(26)**.

[D] These equations simply define the "cross-media partitioning factors," VF_{ij} and LF_{sw}.

of outdoor vapors which originate from dissolved hydrocarbons in ground water located some distance below ground surface. Here the goal is to determine the dissolved hydrocarbon RBSL that corresponds to the target RBSL for outdoor vapors in the breathing zone, as given in Tables X2.2 and X2.3. If the selected target vapor concentration is some value other than the RBSL for inhalation (that is, odor threshold or ecological criterion), this value can be substituted for the $RBSL_{air}$ parameter appearing in the equations given in Tables X2.2 and X2.3.

X2.4.2 A conceptual model for the transport of chemicals from ground water to ambient air is depicted in Fig. X2.1. For simplicity, the relationship between outdoor air and dissolved ground water concentrations is represented in Tables X2.2 and X2.3 by the "volatilization factor," VF_{wamb} [(mg/m³-air)/(mg/L-H_2O)], defined in Table X2.5. It is based on the following assumptions:

X2.4.2.1 A constant dissolved chemical concentration in ground water,

X2.4.2.2 Linear equilibrium partitioning between dissolved chemicals in ground water and chemical vapors at the ground water table,

X2.4.2.3 Steady-state vapor- and liquid-phase diffusion through the capillary fringe and vadose zones to ground surface,

X2.4.2.4 No loss of chemical as it diffuses towards ground surface (that is, no biodegradation), and

X2.4.2.5 Steady well-mixed atmospheric dispersion of the

TABLE X2.3 Equations Used to Develop Example Tier 1 Risk-Based Screening Level (RBSLs) Appearing in "Look-Up" Table X2.1—Noncarcinogenic Effects[A]

NOTE—See Tables X2.4 through X2.7 for definition of parameters.

Medium	Exposure Route	Risk-Based Screening Level (RBSL)
Air	inhalation[B]	$RBSL_{air}\left[\frac{\mu g}{m^3\text{-}air}\right] = \frac{THQ \times RfD_i \times BW \times AT_n \times 365 \frac{days}{years} \times 10^3 \frac{\mu g}{mg}}{IR_{air} \times EF \times ED}$
Ground water	ingestion (potable ground water supply only)[B]	$RBSL_w\left[\frac{mg}{L\text{-}H_2O}\right] = \frac{THQ \times RfD_o \times BW \times AT_n \times 365 \frac{days}{years}}{IR_w \times EF \times ED}$
Ground water[C]	enclosed-space (indoor) vapor inhalation[D]	$RBSL_w\left[\frac{mg}{L\text{-}H_2O}\right] = \frac{RBSL_{air}\left[\frac{\mu g}{m^3\text{-}air}\right]}{VF_{wesp}} \times 10^{-3}\frac{mg}{\mu g}$
Ground water[C]	ambient (outdoor) vapor inhalation[D]	$RBSL_w\left[\frac{mg}{L\text{-}H_2O}\right] = \frac{RBSL_{air}\left[\frac{\mu g}{m^3\text{-}air}\right]}{VF_{wamb}} \times 10^{-3}\frac{mg}{\mu g}$
Surficial soil	ingestion of soil, inhalation of vapors and particulates, and dermal contact[B]	$RBSL_s\left[\frac{\mu g}{kg\text{-}soil}\right] = \frac{THQ \times BW \times AT_n \times 365 \frac{days}{years}}{EF \times ED\left(\frac{\left(10^{-6}\frac{kg}{mg} \times (IR_{soil} \times RAF_o + SA \times M \times RAF_d)\right)}{RfD_o} + \frac{(IR_{air} \times (VF_{ss} + VF_p))}{RfD_i}\right)}$ For surficial and excavated soils (0 to 1 m)
Subsurface soil[C]	ambient (outdoor) vapor inhalation[D]	$RBSL_s\left[\frac{mg}{kg\text{-}soil}\right] = \frac{RBSL_{air}\left[\frac{\mu g}{m^3\text{-}air}\right]}{VF_{samb}} \times 10^{-3}\frac{mg}{\mu g}$
Subsurface soil[C]	enclosed space (indoor) vapor inhalation[D]	$RBSL_s\left[\frac{mg}{kg\text{-}soil}\right] = \frac{RBSL_{air}\left[\frac{\mu g}{m^3\text{-}air}\right]}{VF_{sesp}} \times 10^{-3}\frac{mg}{\mu g}$
Subsurface soil[C]	leaching to ground water[D]	$RBSL_s\left[\frac{mg}{kg\text{-}soil}\right] = \frac{RBSL_w\left[\frac{mg}{L\text{-}H_2O}\right]}{LF_{sw}}$

[A] Note that all RBSL values should be compared with thermodynamic partitioning limits, such as solubility levels, maximum vapor concentrations, and so forth. If a RBSL exceeds the relevant partitioning limit, this is an indication that the selected risk or hazard level will never be reached or exceeded for that chemical and the selected exposure scenario.

[B] Screening levels for these media based on other considerations (for example, aesthetic, background levels, environmental resource protection, and so forth) can be derived with these equations by substituting the selected target level for $RBSL_{air}$ or $RBSL_w$ appearing in these equations.

[C] These equations are based on Ref **(26)**.

[D] These equations simply define the "cross-media partitioning factors," VF_{ij} and LF_{sw}.

emanating vapors within the breathing zone as modeled by a "box model" for air dispersion.

X2.4.3 Should the calculated $RBSL_w$ exceed the pure component solubility for any individual component, ">S" is entered in the table to indicate that the selected risk level or hazard quotient cannot be reached or exceeded for that compound and the specified exposure scenario.

X2.5 *Ground Water—Inhalation of Enclosed-Space (Indoor) Vapors:*

X2.5.1 In this case chemical intake results from the inhalation of vapors in enclosed spaces. The chemical vapors originate from dissolved hydrocarbons in ground water located some distance below ground surface. Here the goal is to determine the dissolved hydrocarbon RBSL that corresponds to the target RBSL for vapors in the breathing zone, as given in Tables X2.2 and X2.3. If the selected target vapor concentration is some value other than the RBSL for inhalation (that is, odor threshold or ecological criterion), this value can be substituted for the $RBSL_{air}$ parameter appearing in the equations given in Tables X2.2 and X2.3.

X2.5.2 A conceptual model for the transport of chemicals from ground water to indoor air is depicted in Fig. X2.2. For simplicity, the relationship between enclosed-space air and

TABLE X2.4 Exposure Parameters Appearing in Tables X2.2 and X2.3

Parameters	Definitions, Units	Residential	Commercial/Industrial
AT_c	averaging time for carcinogens, years	70 years	70 years[A]
AT_n	averaging time for noncarcinogens, years	30 years	25 years[A]
BW	adult body weight, kg	70 kg	70 kg[A]
ED	exposure duration, years	30 years	25 years[A]
EF	exposure frequency, days/years	350 days/year	250 days/year[A]
IR_{soil}	soil ingestion rate, mg/day	100 mg/day	50 mg/day[A]
IR_{air}-*indoor*	daily indoor inhalation rate, m^3/day	15 m^3/day	20 m^3/day[A]
IR_{air}-*outdoor*	daily outdoor inhalation rate, m^3/day	20 m^3/day	20 m^3/day[A]
IR_w	daily water ingestion rate, L/day	2 L/day	1 L/day[A]
LF_{sw}	leaching factor, (mg/L-H_2O)/(mg/kg-soil)—see Table X2.5	chemical-specific	chemical-specific
M	soil to skin adherence factor, mg/cm^2	0.5	0.5[B]
RAF_d	dermal relative absorption factor, volatiles/PAHs	0.5/0.05	0.5/0.05[B]
RAF_o	oral relative absorption factor	1.0	1.0
$RBSL_i$	risk-based screening level for media i, mg/kg-soil, mg/L-H_2O, orμ g/m^3-air	chemical-, media-, and exposure route-specific	chemical-, media-, and exposure route-specific
RfD_i	inhalation chronic reference dose, mg/kg-day	chemical-specific	chemical-specific
RfD_o	oral chronic reference dose, mg/kg-day	chemical-specific	chemical-specific
SA	skin surface area, cm^2/day	3160	3160[A]
SF_i	inhalation cancer slope factor, $(mg/kg\text{-}day)^{-1}$	chemical-specific	chemical-specific
SF_o	oral cancer slope factor, $(mg/kg\text{-}day)^{-1}$	chemical-specific	chemical-specific
THQ	target hazard quotient for individual constituents, unitless	1.0	1.0
TR	target excess individual lifetime cancer risk, unitless	for example, 10^{-6} or 10^{-4}	for example, 10^{-6} or 10^{-4}
VF_i	volatilization factor, (mg/m^3-air)/(mg/kg-soil) or (mg/m^3-air)/(mg/L-H_2O)—see Table X2.5	chemical- and media-specific	chemical- and media-specific

[A] See Ref **(27)**.
[B] See Ref **(28)**.

dissolved ground water concentrations is represented in Tables X2.2 and X2.3 by the "volatilization factor" VF_{wesp} [(mg/m^3-air)/(mg/L-H_2O)] defined in Table X2.5. It is based on the following assumptions:

X2.5.2.1 A constant dissolved chemical concentration in ground water,

X2.5.2.2 Equilibrium partitioning between dissolved chemicals in ground water and chemical vapors at the ground water table,

X2.5.2.3 Steady-state vapor- and liquid-phase diffusion through the capillary fringe, vadose zone, and foundation cracks,

X2.5.2.4 No loss of chemical as it diffuses towards ground surface (that is, no biodegradation), and

X2.5.2.5 Steady, well-mixed atmospheric dispersion of the emanating vapors within the enclosed space, where the convective transport into the building through foundation cracks or openings is negligible in comparison with diffusive transport.

X2.5.3 Should the calculated $RBSL_w$ exceed the pure component solubility for any individual component, ">S" is entered in the table to indicate that the selected risk level or hazard quotient cannot be reached or exceeded for that compound and the specified exposure scenario.

X2.6 *Surficial Soils—Ingestion, Dermal Contact, and Vapor and Particulate Inhalation:*

X2.6.1 In this case it is assumed that chemical intake results from a combination of intake routes, including: ingestion, dermal absorption, and inhalation of both particulates and vapors emanating from surficial soil.

X2.6.2 Equations used to estimate intake resulting from ingestion follow guidance given in Ref **(26)** for ingestion of chemicals in soil. For this route, it has been assumed that surficial soil chemical concentrations and intake rates remain constant over the exposure duration.

X2.6.3 Equations used to estimate intake resulting from dermal absorption follow guidance given in Ref **(26)** for dermal contact with chemicals in soil. For this route, it has been assumed that surficial soil chemical concentrations and absorption rates remain constant over the exposure duration.

X2.6.4 Equations used to estimate intake resulting from the inhalation of particulates follow guidance given in Ref **(26)** for inhalation of airborne chemicals. For this route, it has been assumed that surficial soil chemical concentrations, intake rates, and atmospheric particulate concentrations remain constant over the exposure duration.

X2.6.5 Equations used to estimate intake resulting from the inhalation of airborne chemicals resulting from the volatilization of chemicals from surficial soils follow guidance given in Ref **(26)** for inhalation of airborne chemicals.

X2.6.6 A conceptual model for the volatilization of chemicals from surficial soils to outdoor air is depicted in Fig. X2.3. For simplicity, the relationship between outdoor air and surficial soil concentrations is represented in Tables X2.2 and X2.3 by the "volatilization factor" VF_{ss} [(mg/m^3-air)/(mg/kg-soil)] defined in Table X2.5. It is based on the following assumptions:

X2.6.6.1 Uniformly distributed chemical throughout the depth 0—*d* (cm) below ground surface,

X2.6.6.2 Linear equilibrium partitioning within the soil matrix between sorbed, dissolved, and vapor phases, where the partitioning is a function of constant chemical- and soil-specific parameters,

X2.6.6.3 Diffusion through the vadose zone,

X2.6.6.4 No loss of chemical as it diffuses towards ground surface (that is, no biodegradation), and

X2.6.6.5 Steady well-mixed atmospheric dispersion of the emanating vapors within the breathing zone as modeled by a "box model" for air dispersion.

X2.6.7 In the event that the time-averaged flux exceeds that

TABLE X2.5 Volatilization Factors (VF_i), Leaching Factor (LF_{sw}), and Effective Diffusion Coefficients (D_i^{eff})

Symbol	Cross-Media Route (or Definition)	Equation
VF_{wesp}	Ground water → enclosed-space vapors	$VF_{wesp}\left[\frac{(mg/m^3\text{-}air)}{(mg/L\text{-}H_2O)}\right] = \frac{H\left[\frac{D_{ws}^{eff}/L_{GW}}{ER\,L_B}\right]}{1+\left[\frac{D_{ws}^{eff}/L_{GW}}{ER\,L_B}\right]+\left[\frac{D_{ws}^{eff}/L_{GW}}{(D_{crack}^{eff}/L_{crack})\eta}\right]} \times 10^3\,\frac{L}{m^3}$ [A]
VF_{wamb}	Ground water → ambient (outdoor) vapors	$VF_{wamb}\left[\frac{(mg/m^3\text{-}air)}{(mg/L\text{-}H_2O)}\right] = \frac{H}{1+\left[\frac{U_{air}\delta_{air}L_{GW}}{WD_{ws}^{eff}}\right]} \times 10^3\,\frac{L}{m^3}$ [B]
VF_{ss}	Surficial soils → ambient air (vapors)	$VF_{ss}\left[\frac{(mg/m^3\text{-}air)}{(mg/kg\text{-}soil)}\right] = \frac{2W\rho_s}{U_{air}\delta_{air}}\sqrt{\frac{D_s^{eff}H}{\pi[\theta_{ws}+k_s\rho_s+H\theta_{as}]\tau}} \times 10^3\,\frac{cm^3\text{-}kg}{m^3\text{-}g}$ [C] or: $VF_{ss}\left[\frac{(mg/m^3\text{-}air)}{(mg/kg\text{-}soil)}\right] = \frac{W\rho_s d}{U_{air}\delta_{air}\tau} \times 10^3\,\frac{cm^3\text{-}kg}{m^3\text{-}g}$; whichever is less [D]
VF_p	Surficial soils → ambient air (particulates)	$VF_p\left[\frac{(mg/m^3\text{-}air)}{(mg/kg\text{-}soil)}\right] = \frac{P_e W}{U_{air}\delta_{air}} \times 10^3\,\frac{cm^3\text{-}kg}{m^3\text{-}g}$ [E]
VF_{samb}	Subsurface soils → ambient air	$VF_{samb}\left[\frac{(mg/m^3\text{-}air)}{(mg/kg\text{-}soil)}\right] = \frac{H\rho_s}{[\theta_{ws}+k_s\rho_s+H\theta_{as}]\left(1+\frac{U_{air}\delta_{air}L_S}{D_s^{eff}W}\right)} \times 10^3\,\frac{cm^3\text{-}kg}{m^3\text{-}g}$ [F]
VF_{sesp}	Subsurface soil → enclosed-space vapors	$VF_{sesp}\left[\frac{(mg/m^3\text{-}air)}{(mg/kg\text{-}soil)}\right] = \frac{\frac{H\rho_s}{[\theta_{ws}+k_s\rho_s+H\theta_{as}]}\left[\frac{D_s^{eff}/L_S}{ER\,L_B}\right]}{1+\left[\frac{D_s^{eff}/L_S}{ER\,L_B}\right]+\left[\frac{D_S^{eff}/L_s}{(D_{crack}^{eff}/L_{crack})\eta}\right]} \times 10^3\,\frac{cm^3\text{-}kg}{m^3\text{-}g}$ [A]
LF_{sw}	Subsurface soils → ground water	$LF_{sw}\left[\frac{(mg/L\text{-}H_2O)}{(mg/kg\text{-}soil)}\right] = \frac{\rho_s}{[\theta_{ws}+k_s\rho_s+H\theta_{as}]\left(1+\frac{U_{gw}\delta_{gw}}{IW}\right)} \times 10^0\,\frac{cm^3\text{-}kg}{L\text{-}g}$ [B]
D_s^{eff}	Effective diffusion coefficient in soil based on vapor-phase concentration	$D_s^{eff}\left[\frac{cm^2}{s}\right] = D^{air}\frac{\theta_{as}^{3.33}}{\theta_T^2} + D^{wat}\frac{1}{H}\frac{\theta_{ws}^{3.33}}{\theta_T^2}$ [A]
D_{crack}^{eff}	Effective diffusion coefficient through foundation cracks	$D_{crack}^{eff}\left[\frac{cm^2}{s}\right] = D^{air}\frac{\theta_{acrack}^{3.33}}{\theta_T^2} + D^{wat}\frac{1}{H}\frac{\theta_{wcrack}^{3.33}}{\theta_T^2}$ [A]
D_{cap}^{eff}	Effective diffusion coefficient through capillary fringe	$D_{cap}^{eff}\left[\frac{cm^2}{s}\right] = D^{air}\frac{\theta_{acap}^{3.33}}{\theta_T^2} + D^{wat}\frac{1}{H}\frac{\theta_{wcap}^{3.33}}{\theta_T^2}$ [A]
D_{ws}^{eff}	Effective diffusion coefficient between ground water and soil surface	$D_{ws}^{eff}\left[\frac{cm^2}{s}\right] = (h_{cap}+h_v)\left[\frac{h_{cap}}{D_{cap}^{eff}}+\frac{h_v}{D_s^{eff}}\right]^{-1}$ [A]
C_s^{sat}	Soil concentration at which dissolved pore-water and vapor phases become saturated	$C_s^{sat}\left[\frac{mg}{kg\text{-}soil}\right] = \frac{S}{\rho_s} \times [H\theta_{as}+\theta_{ws}+k_s\rho_s] \times 10^0\,\frac{L\text{-}g}{cm^3\text{-}kg}$ [F]

[A] See Ref **(29)**.
[B] See Ref **(30)**.
[C] See Ref **(31)**.
[D] Based on mass balance.
[E] See Ref **(32)**.
[F] See Ref **(33)**.

which would occur if all chemical initially present in the surficial soil zone volatilized during the exposure period, then the volatilization factor is determined from a mass balance assuming that all chemical initially present in the surficial soil zone volatilizes during the exposure period.

X2.7 *Subsurface Soils—Inhalation of Outdoor Vapors:*

X2.7.1 In this case chemical intake is a result of inhalation of outdoor vapors which originate from hydrocarbons contained in subsurface soils located some distance below ground surface. Here the goal is to determine the RBSL for subsurface soils that corresponds to the target RBSL for outdoor vapors in the breathing zone, as given in Table X2.1. If the selected target

TABLE X2.6 Soil, Building, Surface, and Subsurface Parameters Used in Generating Example Tier 1 RBSLs

NOTE 1—See X2.10 for justification of parameter selection.

Parameters	Definitions, Units	Residential	Commercial/Industrial
d	lower depth of surficial soil zone, cm	100 cm	100 cm
D^{air}	diffusion coefficient in air, cm^2/s	chemical-specific	chemical-specific
D^{wat}	diffusion coefficient in water, cm^2/s	chemical-specific	chemical-specific
ER	enclosed-space air exchange rate, 1/sec	$0.00014\ s^{-1}$	$0.00023\ s^{-1}$
f_{oc}	fraction of organic carbon in soil, g-C/g-soil	0.01	0.01
H	henry's law constant, (cm^3-H_2O)/(cm^3-air)	chemical-specific	chemical-specific
h_{cap}	thickness of capillary fringe, cm	5 cm	5 cm
h_v	thickness of vadose zone, cm	295 cm	295 cm
I	infiltration rate of water through soil, cm/years	30 cm/year	30 cm/year
k_{oc}	carbon-water sorption coefficient, cm^3-H_2O/g-C	chemical-specific	chemical-specific
k_s	soil-water sorption coefficient, cm^3-H_2O/g-soil	$f_{oc} \times k_{oc}$	$f_{oc} \times k_{oc}$
L_B	enclosed-space volume/infiltration area ratio, cm	200 cm	300 cm
L_{crack}	enclosed-space foundation or wall thickness, cm	15 cm	15 cm
L_{GW}	depth to ground water = $h_{cap} + h_v$, cm	300 cm	300 cm
L_S	depth to subsurface soil sources, cm	100 cm	100 cm
P_e	particulate emission rate, g/cm^2-s	6.9×10^{-14}	6.9×10^{-14}
S	pure component solubility in water, mg/L-H_2O	chemical-specific	chemical-specific
U_{air}	wind speed above ground surface in ambient mixing zone, cm/s	225 cm/s	225 cm/s
U_{gw}	ground water Darcy velocity, cm/year	2500 cm/year	2500 cm/year
W	width of source area parallel to wind, or ground water flow direction, cm	1500 cm	1500 cm
δ_{air}	ambient air mixing zone height, cm	200 cm	200 cm
δ_{gw}	ground water mixing zone thickness, cm	200 cm	200 cm
η	areal fraction of cracks in foundations/walls, cm^2-cracks/cm^2-total area	0.01 cm^2-cracks/cm^2-total area	0.01 cm^2-cracks/cm^2-total area
θ_{acap}	volumetric air content in capillary fringe soils, cm^3-air/cm^3-soil	0.038 cm^3-air/cm^3-soil	0.038 cm^3-air/cm^3-soil
θ_{acrack}	volumetric air content in foundation/wall cracks, cm^3-air/cm^3 total volume	0.26 cm^3-air/cm^3 total volume	0.26 cm^3-air/cm^3 total volume
θ_{as}	volumetric air content in vadose zone soils, cm^3-air/cm^3-soil	0.26 cm^3-air/cm^3-soil	0.26 cm^3-air/cm^3-soil
θ_T	total soil porosity, cm^3/cm^3-soil	0.38 cm^3/cm^3-soil	0.38 cm^3/cm^3-soil
θ_{wcap}	volumetric water content in capillary fringe soils, cm^3-H_2O/cm^3-soil	0.342 cm^3-H_2O/cm^3-soil	0.342 cm^3-H_2O/cm^3-soil
θ_{wcrack}	volumetric water content in foundation/wall cracks, cm^3-H_2O/cm^3 total volume	0.12 cm^3-H_2O/cm^3 total volume	0.12 cm^3-H_2O/cm^3 total volume
θ_{ws}	volumetric water content in vadose zone soils, cm^3-H_2O/cm^3-soil	0.12 cm^3-H_2O/cm^3-soil	0.12 cm^3-H_2O/cm^3-soil
ρ_s	soil bulk density, g-soil/cm^3-soil	1.7 g/cm^3	1.7 g/cm^3
τ	averaging time for vapor flux, s	9.46×10^8 s	7.88×10^8 s

TABLE X2.7 Chemical-Specific Properties Used in the Derivation Example Tier 1 RBSLs

Chemical	CAS Number	M_w, g/mol	H, L-H_2O/L-air	D^{air}, cm^2/s	D^w, cm^2/s	log(K_{oc}), L/kg	log(K_{ow}), L/kg
Benzene	71-43-2	78^A	0.22^A	0.093^A	1.1×10^{-5A}	1.58^A	2.13^A
Toluene	108-88-3	92^A	0.26^A	0.085^A	9.4×10^{-6B}	2.13^A	2.65^A
Ethyl benzene	100-41-4	106^A	0.32^A	0.076^A	8.5×10^{-6B}	3.11^A	3.13^A
Mixed xylenes	1330-20-7	106^A	0.29^A	0.072^B	8.5×10^{-6B}	2.38^A	3.26^A
Naphthalene	91-20-3	128^A	0.049^A	0.072^B	9.4×10^{-6A}	3.11^A	3.28^A
Benzo(a)pyrene	50-32-8	252^C	5.8×10^{-8D}	0.050^B	5.8×10^{-6B}	5.59^E	5.98^D

Chemical	CAS Number	SF_o, kg-day/mg	SF_i, kg-day/mg	RfD_o, mg/kg-day	RfD_i, mg/kg-day
Benzene	71-43-2	0.029^F	0.029^F	...	...
Toluene	108-88-3	...	...	0.2^F	0.11^F
Ethyl benzene	100-41-4	...	...	0.1^F	0.29^F
Mixed xylenes	1330-20-7	...	...	2.0^F	2.0^F
Naphthalene	91-20-3	...	...	0.004^G	0.004^G
Benzo(a)pyrene	50-32-8	7.3^F	6.1^F	...	...

^{A} See Ref **(34)**.
^{B} Diffusion coefficient calculated using the method of Fuller, Schettler, and Giddings, from Ref **(11)**.
^{C} See Ref **(7)**.
^{D} See Ref **(35)**.
^{E} Calculated from K_{ow}/K_{oc} correlation: $\log(K_{oc}) = 0.937 \log(K_{ow}) - 0.006$, from Ref **(11)**.
^{F} See Ref **(2)**.
^{G} See Ref **(3)**.

vapor concentration is some value other than the RBSL for inhalation (that is, odor threshold or ecological criterion), this value can be substituted for the $RBSL_{air}$ parameter appearing in the equations given in Tables X2.2 and X2.3.

X2.7.2 A conceptual model for the transport of chemicals from subsurface soils to ambient air is depicted in Fig. X2.4. For simplicity, the relationship between outdoor air and soil concentrations is represented in Tables X2.2 and X2.3 by the "volatilization factor," VF_{samb} [(mg/m^3-air)/(mg/kg-soil)], defined in Table X2.5. It is based on the following assumptions:

X2.7.2.1 A constant chemical concentration in subsurface soils,

X2.7.2.2 Linear equilibrium partitioning within the soil matrix between sorbed, dissolved, and vapor phases, where the partitioning is a function of constant chemical- and soil-specific parameters,

X2.7.2.3 Steady-state vapor- and liquid-phase diffusion through the vadose zone to ground surface,

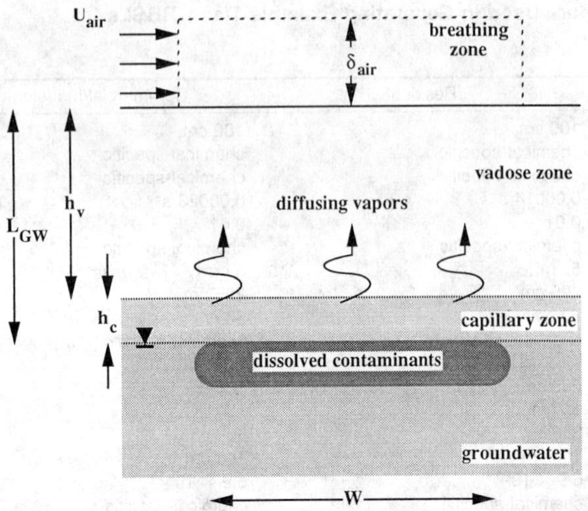

FIG. X2.1 Volatilization from Ground Water to Ambient Air

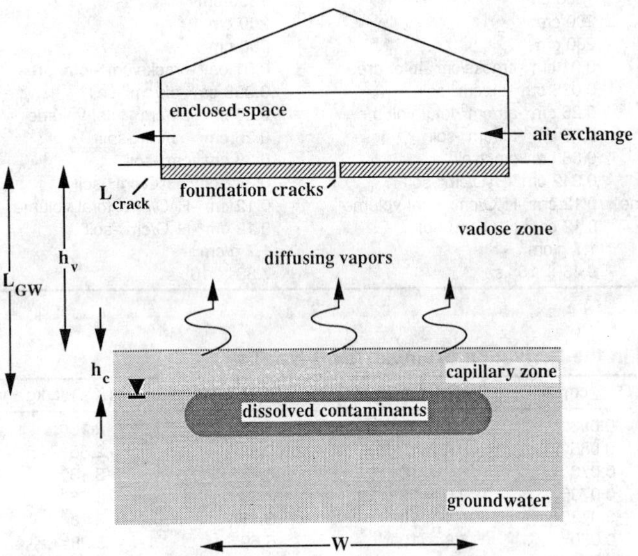

FIG. X2.2 Volatilization from Ground Water to Enclosed-Space Air

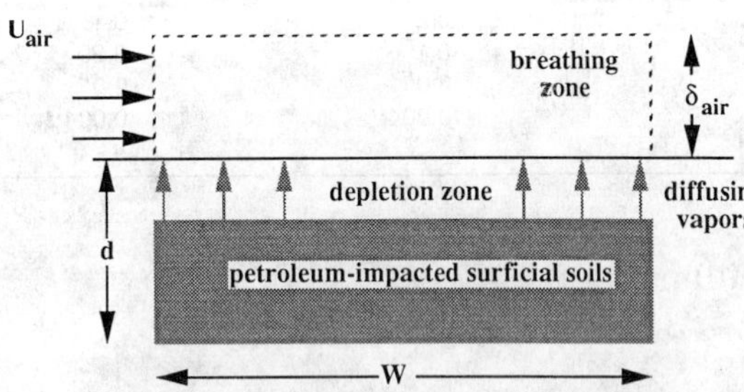

FIG. X2.3 Volatilization from Surficial Soils

X2.7.2.4 No loss of chemical as it diffuses towards ground surface (that is, no biodegradation), and

X2.7.2.5 Steady well-mixed atmospheric dispersion of the emanating vapors within the breathing zone as modeled by a "box model" for air dispersion.

X2.7.3 Should the calculated $RBSL_s$ exceed the value for which the equilibrated vapor and dissolved pore-water phases become saturated, C_s^{sat} [mg/kg-soil] (see Table X2.5 for calculation of this value), "RES" is entered in the table to

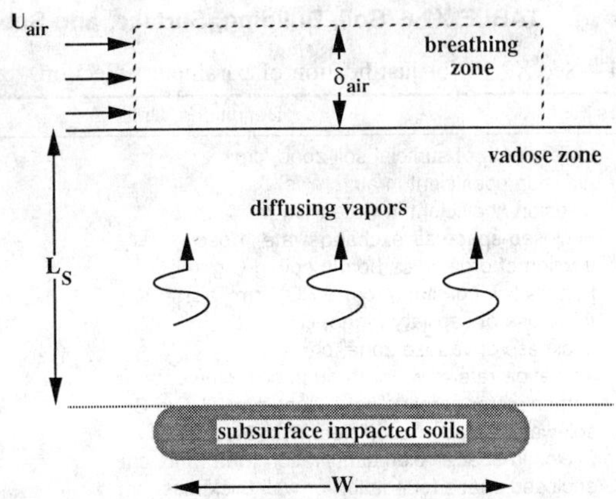

FIG. X2.4 Volatilization from Subsurface Soils to Ambient Air

indicate that the selected risk level or hazard quotient cannot be reached or exceeded for that compound and the specified exposure scenario (even if free-phase product or precipitate is present in the soil).

X2.8 *Subsurface Soils—Inhalation of Enclosed-Space (Indoor) Vapors:*

X2.8.1 In this case chemical intake is a result of inhalation of enclosed-space vapors which originate from hydrocarbons contained in subsurface soils located some distance below ground surface. Here the goal is to determine the RBSL for subsurface soils that corresponds to the target RBSL for indoor vapors, as given in Tables X2.2 and X2.3. If the selected target vapor concentration is some value other than the RBSL for inhalation (that is, odor threshold or ecological criterion), this value can be substituted for the $RBSL_{air}$ parameter appearing in the equations given in Tables X2.2 and X2.3.

X2.8.2 A conceptual model for the transport of chemicals from subsurface soils to enclosed spaces is depicted in Fig. X2.5. For simplicity, the relationship between indoor air and soil concentrations is represented in Tables X2.2 and X2.3 by the "volatilization factor," VF_{sesp} [(mg/m^3-air)/(kg-soil)], defined in Table X2.5. It is based on the following assumptions:

X2.8.2.1 A constant chemical concentration in subsurface soils,

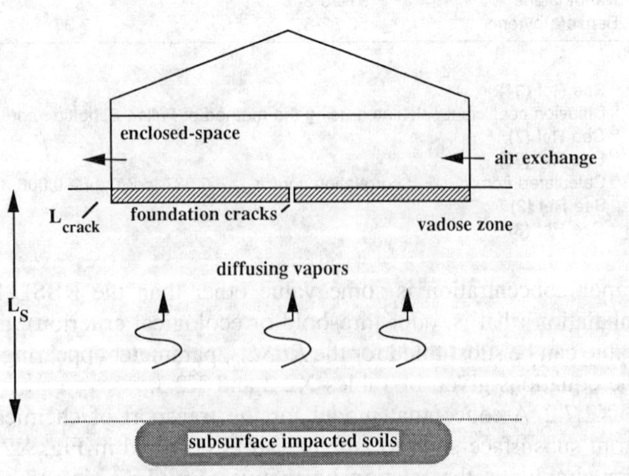

FIG. X2.5 Volatilization from Subsurface Soils to Enclosed-Space Air

X2.8.2.2 Linear equilibrium partitioning within the soil matrix between sorbed, dissolved, and vapor phases, where the partitioning is a function of constant chemical- and soil-specific parameters,

X2.8.2.3 Steady-state vapor- and liquid-phase diffusion through the vadose zone and foundation cracks,

X2.8.2.4 No loss of chemical as it diffuses towards ground surface (that is, no biodegradation), and

X2.8.2.5 Well-mixed atmospheric dispersion of the emanating vapors within the enclosed space.

X2.8.3 Should the calculated $RBSL_s$ exceed the value C_s^{sat} [mg/kg-soil] for which the equilibrated vapor and dissolved pore-water phases become saturated (see Table X2.5 for calculation of this value), "RES" is entered in the table to indicate that the selected risk level or hazard quotient cannot be reached or exceeded for that compound and the specified exposure scenario (even if free-phase product or precipitate is present in the soil).

X2.9 *Subsurface Soils—Leaching to Ground Water:*

X2.9.1 In this case chemical intake is a result of chemicals leaching from subsurface soils, followed by inhalation of enclosed-space vapors, inhalation of outdoor vapors, or ingestion of ground water as discussed in X2.1-X2.3. Here the goal is to determine the RBSL for subsurface soils that corresponds to the target RBSLs for the inhalation or ingestion routes. If the selected target ground water concentration is some value other than an RBSL for ground water (that is, odor threshold or ecological criterion), this value can be substituted for the $RBSL_w$ parameter appearing in the equations given in Tables X2.2 and X2.3.

X2.9.2 A conceptual model for the leaching of chemicals from subsurface soils to ground water is depicted in Fig. X2.6. For simplicity, the relationship between ground water and soil concentrations is represented in Tables X2.2 and X2.3 by the "leaching factor," LF_{sw} [(mg/L-H_2O)/ (mg/kg-soil)], defined in Table X2.5. It is based on the following assumptions:

X2.9.2.1 A constant chemical concentration in subsurface soils,

X2.9.2.2 Linear equilibrium partitioning within the soil matrix between sorbed, dissolved, and vapor phases, where the partitioning is a function of constant chemical- and soil-specific parameters,

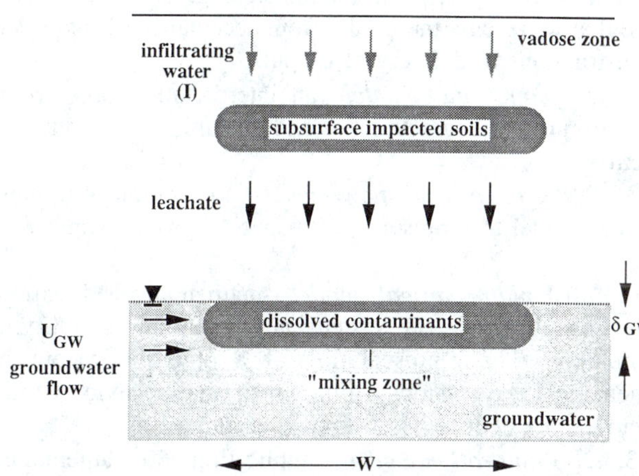

FIG. X2.6 Leaching from Subsurface Soils to Ground Water

X2.9.2.3 Steady-state leaching from the vadose zone to ground water resulting from the constant leaching rate I [cm/s],

X2.9.2.4 No loss of chemical as it leaches towards ground water (that is, no biodegradation), and

X2.9.2.5 Steady well-mixed dispersion of the leachate within a ground water "mixing zone."

X2.9.3 Should the calculated $RBSL_s$ exceed the value C_s^{sat}, for which the equilibrated vapor and dissolved pore-water phases become saturated (see Table X2.5 for calculation of this value), "RES" is entered in the table to indicate that the selected risk level or hazard quotient cannot be reached or exceeded for that compound and the specified exposure scenario (even if free-phase product or precipitate is present in the soil).

X2.9.4 In some regulatory programs, "dilution attenuation factors" (DAFs) are currently being proposed based on fate and transport modeling results. A DAF is typically defined as the ratio of a target ground water concentration divided by the source leachate concentration, and is inherently very similar to the leachate factor, LF_{sw}, discussed here. The difference between these two terms is that LF_{sw} represents the ratio of the target ground water concentration divided by the source area soil concentration. Should a regulatory program already have a technically defensible DAF value, it can be equated to a leachate factor by the following expression:

$$LF_{sw} = \frac{DAF\,[ml]P5\,\rho_s}{[\theta_{ws} + k_s\rho_s + H\theta_{as}]}\,[ml]P5\,10^0 \qquad \text{(X2.3)}$$

where the parameters are defined in Table X2.6.

X2.10 *Parameter Values:*

X2.10.1 Table X2.4 lists exposure parameters used to calculate the RBSLs appearing in sample Look-Up Table X2.1. All values given are based on adult exposures only. With the exception of the dermal exposure parameters (SA, M, and RAF_d), the values given are reasonable maximum exposure (RME) values presented in Ref **(27)** and are regarded as upper bound estimates for each individual exposure parameter.

X2.10.2 The skin surface area, SA = 3160 cm^2/day, is based on the average surface area of the head, hands, and forearms for adult males given in Ref **(27)**. The soil-to-skin adherence factor, M [mg/cm^2], and dermal relative absorption factor, RAF_d[mg-absorbed/mg-applied], are based on guidance issued by Ref **(28)**.

X2.10.3 Soil properties are based on typical values for sandy soils and are consistent with values given in Ref **(30)**.

X2.10.4 Physical dimensions are consistent with the scale of typical underground fuel tank releases.

X2.10.5 Particulate emission rates were estimated by the approach presented by Cowherd, et al **(32)**. It was assumed that the mode of the surficial soil size distribution was 2 mm, the erosion potential was unlimited, there was no vegetative cover, and the mean average annual wind speed was 4 m/s.

X2.10.6 The chemical-specific parameters used are defined in Table X2.7.

X2.10.7 In this development, surficial soils are defined as those soils present within 1 m of ground surface. Subsurface soil RBSLs are based on assumed source depths of 1 m.

Ground water is assumed to be located 3 m below ground surface.

X2.10.8 Once again, the reader is reminded that the parameter (and corresponding RBSL) values are presented here as examples only, and are not intended to be used as standards. At best, the parameters presented are reasonable values based on current information and professional judgment. The reader should review and verify all assumptions prior to using any of the example RBSLs as screening level values.

X3. USE OF PREDICTIVE MODELING IN THE RISK-BASED CORRECTIVE ACTION PROCESS

X3.1 *Scope:*

X3.1.1 Predictive modeling is a valuable tool that can provide information to the risk management process. In a RBCA, modeling is used to predict the location and concentration contaminants and to interpret, or extrapolate, site characterization data, historical monitoring data, and toxicological information. In addition, predictive modeling may be used in evaluation of remedial alternatives and in evaluating compliance targets in monitoring plans. This appendix discusses the following:

X3.1.1.1 Significance and use of predictive modeling in the RBCA process;

X3.1.1.2 Interpretation of predictive modeling results;

X3.1.1.3 Procedures for predictive migration models; and

X3.1.1.4 Procedures for exposure, risk, and dose-response assessment.

X3.1.2 This appendix is not intended to be all inclusive. Each predictive model is unique and may require additional procedures in its development and application. All such additional analyses should be documented in the RBCA process.

X3.2 *Referenced Documents:*

X3.2.1 *ASTM Standards*:

D 653 Terminology Relating to Soil, Rock, and Contained Fluids[7]

D 5447 Guide for Application of a Ground-Water Flow Model to a Site-Specific Problem[8]

D 5490 Guide for Comparing Ground-Water Flow Model Simulations to Site-Specific Information[8]

E 943 Terminology Relating to Biological Effects and Environmental Fate[9]

E 978 Practice for Evaluating Environmental Fate Models of Chemicals[9]

D 5609 Guide for Defining Boundary Conditions in Ground-Water Flow Modeling[8]

D 5610 Guide for Defining Initial Conditions in Ground-Water Flow Modeling[8]

D 5611 Guide for Conducting a Sensitivity Analysis for a Ground-Water Flow Model Application[8]

X3.3 *Terminology:*

X3.3.1 *Definitions*— For definitions of terms used in this appendix, see Terminologies D 653 and E 943.

X3.3.2 *Descriptions of Terms Specific to This Appendix*:

X3.3.2.1 *analytical model*—a model that uses mathematical solutions to governing equations that are continuous in space and time and applicable to the flow and transport process.

X3.3.2.2 *application verification*—using the set of parameter values and boundary conditions from a calibrated model to approximate acceptably a second set of field data measured under similar conditions.

DISCUSSION—Application verification is to be distinguished from code verification, which refers to software testing, comparison with analytical solutions, and comparison with other similar codes to demonstrate that the code represents its mathematical foundation.

X3.3.2.3 *boundary condition*—a mathematical expression of a state of the physical system that constrains the equations of the mathematical model.

X3.3.2.4 *calibration (model application)*—the process of refining the model representation of the fluid and media properties and boundary conditions to achieve a desired degree of correspondence between the model simulation and observations of the real system.

X3.3.2.5 *code validation*— the process of determining how well a modeling code's theoretical foundation and computer implementation describe actual system behavior in terms of the "degree of correlation" between calculated and independently observed cause-and-effect responses of the prototype fluid flow system (for example, research site or laboratory experiment) for which the code has been developed.

X3.3.2.6 *code verification*—the procedure aimed at establishing the completeness, consistency, correctness, and accuracy of modeling software with respect to its design criteria by evaluating the functionality and operational characteristics of the code and testing embedded algorithms and data transfers through execution of problems for which independent benchmarks are available.

X3.3.2.7 *computer code (computer program)*—the assembly of numerical techniques, bookkeeping, and control language that represents the model from acceptance of input data and instructions to delivery of output.

X3.3.2.8 *conceptual model*—an interpretation or working description of the characteristics and dynamics of the physical system.

X3.3.2.9 *ground water flow model*—application of a mathematical model to represent a site-specific ground water flow system.

X3.3.2.10 *mathematical model*—mathematical equations expressing the physical system and including simplifying assumptions. The representation of a physical system by mathematical expressions from which the behavior of the system can be deduced with known accuracy.

X3.3.2.11 *migration model*— application of a mathematical model to represent a site-specific fluid flow system.

[7] *Annual Book of ASTM Standards*, Vol 04.08.

[8] *Annual Book of ASTM Standards*, Vol 04.09.

[9] *Annual Book of ASTM Standards*, Vol 11.04.

X3.3.2.12 *model*—an assembly of concepts in the form of mathematical equations that portray understanding of a natural phenomenon.

X3.3.2.13 *sensitivity (model application)*—the degree to which the model result is affected by changes in a selected model input representing fluid and media properties and boundary conditions.

X3.3.2.14 *simulation*— in migration modeling, one complete execution of a fluid flow modeling computer program, including input and output.

DISCUSSION—for the purposes of this appendix, a simulation refers to an individual modeling run. However, simulation is sometimes also used broadly to refer to the process of modeling in general.

X3.4 *Significance and Use:*

X3.4.1 Predictive modeling is significant in many phases of RBCA, including the following:

X3.4.1.1 Determining the potential urgency of response based on estimated migration and attenuation rates of compounds of concern,

X3.4.1.2 Determining the extent of corrective action based on estimated migration and attenuation rates of compounds of concern,

X3.4.1.3 Establishing relationships between administered doses and adverse impacts to humans and sensitive environmental receptors, and

X3.4.1.4 Determining RBSLs concentrations at points of exposure.

X3.4.2 Examples of predictive modeling uses in the RBCA process include the following:

X3.4.2.1 The prediction of contaminant concentration distributions for future times based on historical trend data, as in the case of ground water transport modeling,

X3.4.2.2 The recommendation of sampling locations and sampling frequency based on current interpretation and future expectations of contaminant distributions, as in the design of ground water monitoring networks,

X3.4.2.3 The design of corrective action measures, as in the case of hydraulic control systems, and

X3.4.2.4 The calculation of site-specific exposure point concentrations based on assumed exposure scenarios, as in the case of direct exposure to surficial soils.

X3.4.3 Predictive modeling is not used in the RBCA process as a substitute for validation of site-specific data.

X3.5 *Interpretation of Predictive Modeling Results:*

X3.5.1 Predictive models are mathematical approximations of real processes, such as the movement of chemicals in the subsurface, the ingestion of chemicals contained in drinking water, and adverse impacts to human health and environmental resources resulting from significant exposures. One key step towards evaluating model results is to assess the accuracy and uncertainty, and to verify the model used.

X3.5.2 The accuracy of modeling-based predictions is evaluated using a post audit and is dependent upon a number of factors, including the following:

X3.5.2.1 The approximations used when describing the real system by mathematical expressions,

X3.5.2.2 The model setup, that is, the input parameters (for example, boundary conditions) used to generate the results, and

X3.5.2.3 The mathematical methods used to solve the governing equations (for example, user selection of numerical solution methods, expansion approximations, numerical parameters, and so forth).

X3.5.3 Predictive modeling results are always subject to some degree of uncertainty. It is important to quantify this uncertainty to properly interpret the results. Many times this is done with a sensitivity analysis in which the user identifies those parameters that most significantly influence the results. If most of all of the parameters do not produce "sensitivity," then the model may need to be reevaluated because it is possible that the key parameters are missing from the model.

X3.5.4 A postaudit may be performed to determine the accuracy of the predictions. While model calibration and verification demonstrate that the model accurately simulates past behavior of the system, the postaudit tests whether the model can predict future system behavior. Postaudits are normally performed several years after the initial assessment and corrective action.

X3.5.5 In the RBCA process, "conservative" is an important criterion of predictive modeling. In the initial evaluation, Tier 1, the most conservative approach, is used, which provides a worst case scenario for potential exposure and risk. Models that, because of their simplicity, neglect factors that yield conservative results are used. Input may include conservative values such as the USEPA RME values. When a more rigorous approach is warranted, such as in Tier 2 of the RBCA process, conservative values are often used, but in conjunction with a more reasonable case scenario. This level requires more specific information about the site and may involve the use of either simple or moderately complex mathematical models. It may involve the use of most likely exposure scenario (that is, USEPA MLE values). This information is used to set conservative corrective action objectives that are still regarded as overly protective. At some sites a comprehensive assessment is required (Tier 3) where SSTLs are determined using a site-specific transport and exposure model and, in some cases, parameter distributions. Tier 3 provides the most realistic evaluation of potential exposure and risk.

X3.6 *Types of Predictive Migration and Risk Assessment Models:*

X3.6.1 Predictive models typically used in the RBCA process can be grouped into two broad categories:

X3.6.1.1 Migration models, and

X3.6.1.2 Exposure, risk, and dose-response assessment models.

X3.6.2 The determination of Tier 1 RBSLs or Tiers 2 and 3 SSTLs generally involves the use of combinations of both types of models. A more detailed description of each type of model is given in X3.7 and X3.8.

X3.7 *Procedures for Predictive Migration Models:*

X3.7.1 Migration (fate and transport) models predict the movement of a petroleum release through soil, ground water, or air, or combination thereof, over time. Most models focus on

TABLE X3.1 Example Screening Level Transport Models

Description	Mathematical Approximation	Parameters
Dissolved Phase Transport:		
Maximum transport rate $u_{d,max}$ [cm/day] of dissolved plume	$u_{d,max} = \frac{K_s i}{\theta_s R_c}$	$C(x)$ = dissolved hydrocarbon concentration along centerline (x, y = 0, z = 0) of dissolved plume [g/cm^3-H_2O] C_{source} = dissolved hydrocarbon concentration in dissolved plume source area [g/cm^3-H_2O] i = ground water gradient [cm/cm] K_s = saturated hydraulic conductivity [cm/day] k_s = sorption coefficient [(g/g-soil)/(g/cm^3-H_2O)]
Minimum time $\tau_{d,min}$ [d] for leading edge of dissolved plume to travel distance, L [cm]	$\tau_{d,min} = \frac{L}{u_{d,max}}$	L = distance downgradient [cm] R_c = retardation factor = $[1 + k_s\rho_s/\theta_s]$ S_w = source width (perpendicular to flow in the horizontal plane) [cm] S_d = source width (perpendicular to flow in the vertical plane) [cm]
Steady-state attenuation [(g/cm^3-H_2O)/(g/cm^3-H_2O)] along the centerline (x, y = 0, z = 0) of a dissolved plume	$\frac{C(x)}{C_{source}} = \exp\left\{\frac{x}{2\alpha_x}\left[1 - \sqrt{\left(1 + \frac{4\lambda\alpha_x}{u}\right)}\right]\right\} \cdot \left(erf\left[\frac{S_w}{4\sqrt{\alpha_y x}}\right]\right)\left(erf\left[\frac{S_d}{4\sqrt{\alpha_z x}}\right]\right)$	u = specific discharge [cm/day] $u_{d,max}$ = maximum transport rate of dissolved plume [cm/day][A] x = distance along centerline from downgradient edge of dissolved plume source zone [cm] y = depth below water table [cm] z = lateral distance away from dissolved plume centerline [cm]
	where: $u = K_s i/\theta_s$	α_x = longitudinal dispersivity [cm] ≈ 0.10 x α_y = transverse dispersivity [cm] ≈ $\alpha_x/3$ α_z = vertical dispersivity [cm] ≈ $\alpha_x/20$ λ = first-order degradation constant [d^{-1}] θ_s = volumetric water content of saturated zone [cm^3-H_2O/cm^3-soil] ρ_s = soil bulk density [g-soil/cm^3-soil] $\tau_{d,min}$ = minimum convective travel time of dissolved hydrocarbons to distance L [d][A] $erf(\eta)$ = error function evaluated for value η
Immiscible Phase Transport: Maximum depth D_{max} [cm] of immiscible phase penetration	$D_{max} = \frac{V_{spill}}{\theta_R \pi R_{spill}^2}$	C_{soil} = total soil hydrocarbon concentration [g/g-soil] $C_{v,eq}$ = equilibrium vapor concentration [g/cm^3-vapor][A] $C_{w,eq}$ = equilibrium dissolved concentration [g/cm^3-H_2O][A]
Equilibrium Partitioning: *Vapor Concentration:* $C_{v,eq}$ [g/cm^3-*vapor*]		D_{max} = maximum depth of immiscible phase penetration [cm][A] H = Henry's Law Constant [(g/cm^3-vapor)/(g/cm^3-H_2O)] k_s = sorption coefficient [(g/g-soil)/(g/cm^3-H_2O)]
Maximum vapor concentration above dissolved hydrocarbons	$C_{v,eq} = HC_{w,eq}$	M_w = molecular weight [g/mol] P_v^i = vapor pressure of compound i [atm]
Maximum vapor concentration when immiscible hydrocarbon is present	$C_{v,eq} = \frac{x_i P_v^i M_w}{RT}$	R = gas constant = 82 cm^3-atm/mol-K R_{spill} = radial extent of hydrocarbon impact [cm] S_i = pure component solubility [g/cm^3-H_2O]
Maximum vapor concentrations in soil pores (no immiscible phase present)	$C_{v,eq} = \frac{HC_{soil}\,\rho_s}{[\theta_w + k_s\rho_s + H\theta_v]}$	T = absolute temperature [K] V_{spill} = volume of hydrocarbon released [cm^3] x_i = mol fraction of component i θ_R = volumetric residual content of hydrocarbon under drainage conditions [cm^3-hydrocarbon/cm^3-soil]
Dissolved Concentration: $C_{w,eq}$ [g/cm^3-H_2O]		θ_w = volumetric content of soil pore water [cm^3-H_2O/cm^3-soil] θ_v = volumetric content of soil vapor [cm^3-vapor/cm^3-soil]
Maximum dissolved concentration when immiscible hydrocarbon is present	$C_{w,eq} = x_i S_i$	π = 3.1416 ρ_s = soil bulk density [g-soil/cm^3-soil]
Maximum dissolved concentration in soil pores (no immiscible phase present)	$C_{w,eq} = \frac{C_{soil}\rho_s}{[\theta_w + k_s\rho_s + H\theta_v]}$	(C_{soil}) = concentration at which immiscible phase forms in soil [g/g-soil][A] D^{air} = pure component diffusion coefficient in air [cm^2/day]
Equilibrium Partioning: *Soil Concentrations [g/g-soil]:* Soil concentration [C_{soil}] [g/g-soil] at which immiscible hydrocarbon phase forms in soil matrix	$(C_{soil}) = \frac{S_i}{\rho_s}[\theta_w + k_s\rho_s + H\theta_v]$	D^{eff} = effective diffusion coefficient for combined vapor and solute transport, expressed as a vapor phase diffusion coefficient (no immiscible hydrocarbon present outside of source area) [cm^2/day][A]
Vapor Phase Transport: Effective porous media diffusion coefficient D^{eff} [cm^2/day] for combined vapor and solute transport, expressed as a vapor phase diffusion coefficient (no immiscible hydrocarbon present outside of source area)	$D^{eff} = \frac{\theta_v^{3.33}}{\theta_T} D^{air} + \frac{1}{H}\frac{\theta_w^{3.33}}{\theta_T} D^w$	D^w = pure component diffusion coefficient in water [cm^2/day] H = Henry's Law Constant [(g/cm^3-vapor)/(g/cm^3-H_2O)] k_s = sorption coefficient [(g/g-soil)/(g/cm^3-H_2O)] k_v = permeability to vapor flow [cm^2] L = distance [cm] R_v = porous media "retardation" factor (no immiscible hydrocarbon present outside of source area) S_i = pure component solubility [g/cm^3-H_2O] $u_{v,max}$ = maximum convective transport rate of vapors [cm/day][A]
Porous media "retardation" factor R_v (no immiscible hydrocarbon present outside of source area)	$R_v = \left[\frac{\theta_w}{H} + \frac{k_s\rho_s}{H} + \theta_v\right]$	∇P = vapor phase pressure gradient [g/cm^2-s^2] θ_w = volumetric content of soil pore water [cm^3-H_2O/cm^3-soil]

TABLE X3.1 *Continued*

Description	Mathematical Approximation	Parameters
Maximum convective transport rate $u_{v,max}$ [cm/day] of vapors	$u_{v,max} = \frac{1}{R_v}\frac{k_v}{\mu_v}\nabla P$	θ_v = volumetric content of soil vapor [cm^3-vapor/cm^3-soil] θ_T = total volumetric content of pore space in soil matrix [cm^3/cm^3-soil]
Minimum time $\tau_{c,min}$ [*d*] for vapors to travel a distance *L* [cm] from source area by convection[A]	$\tau_{c,min} = \frac{L}{u_{v,max}}$	μ_v = vapor viscosity [g/cm-s] ρ_s = soil bulk density [g-soil/cm^3-soil] $\tau_{c,min}$ = minimum time for vapors to travel a distance *L* [cm] by convection [day][A]
Minimum time $\tau_{v,min}$ [*d*] for vapors to travel a distance *L* [cm] from source area by diffusion	$\tau_{d,min} = \frac{L^2}{(D^{eff}/R_v)}$	$\tau_{d,min}$ = minimum time for vapors to travel a distance *L* [cm] by diffusion [day][A] C_{soil} = total soil hydrocarbon concentration [g/g-soil]
Vapor Emissions from Subsurface Vapor Sources to Open Surfaces:		$C_{v,eq}$ = equilibrium vapor concentration [g/cm^3-vapor][A] *d* = distance below ground surface to top of hydrocarbon vapor source [cm]
Maximum diffusive vapor flux F_{max} [g/cm^2-day] from subsurface vapor source located a distance *d* [cm] below ground surface (steady-state, constant source)	$F_{max} = D^{eff}\frac{C_{v,eq}}{d}$	D^{eff} = effective diffusion coefficient for combined vapor and solute transport, expressed as a vapor phase diffusion coefficient (no immiscible hydrocarbon present outside of source area) [cm^2/day][A] R_v = porous media "retardation" factor (no immiscible hydrocarbon present outside of source area)[A]
Maximum time-averaged diffusive vapor flux $\langle F_{max}\rangle$ [g/cm^2-day] from subsurface soils over period from time = 0 to time = τ, single-component immiscible phase present	$\langle F_{max}\rangle = \frac{\rho_s C_{soil}}{\tau}\left\{\sqrt{\left[d^2 + \frac{2C_{v,eq}D^{eff}\tau}{\rho_s C_{soil}}\right]} - d\right\}$	$u_{v,max}$ = maximum convective transport rate $u_{v,max}$ of vapors [cm/day][A] ρ_s = soil bulk density [g-soil/cm^3-soil] τ = averaging time [s] A_B = total area of enclosed space exposed to vapor intrusion (area of foundation) [cm^2]
Maximum combined convective and diffusive vapor flux F_{max} [g/cm^2-day] from subsurface vapor source located a distance *d* [cm] below ground surface	$F_{max} = R_v u_{v,max} C_{v,eq} - \frac{R_v u_{v,max} C_{v,eq}}{\left[1 - \exp\left(\frac{R_v u_{v,max} d}{D^{eff}}\right)\right]}$	A_{crack} = area of foundation through which vapors are transported (area of cracks, open seams, and so forth) [cm^2] C_{soil} = total soil hydrocarbon concentration [g/g-soil] $C_{v,eq}$ = equilibrium vapor concentration [g/cm^3-vapor][A] *d* = distance between foundation/walls and hydrocarbon vapor source [cm] D^{eff} = effective diffusion coefficient through soil for combined vapor and solute transport, expressed as a vapor phase diffusion coefficient (no immiscible hydrocarbon present outside of source area) [cm^2/day][A]
Vapor Emissions from Surface Soils to Open Spaces:		D^{crack} = effective diffusion coefficient through foundation cracks [cm^2/day][A]
Maximum time-averaged diffusive vapor flux $\langle F_{max}\rangle$ [g/cm^2-day] from surface soils over period from time = 0 to time = τ, single-component immiscible phase present	$\langle F_{max}\rangle = \rho_s C_{soil}\sqrt{\frac{2C_{v,eq}D^{eff}}{\rho_s C_{soil}\tau}}$	L_{crack} = thickness of foundation/wall [cm] $M_{w,i}$ = molecular weight of *i* [g/mol] $M_{w,T}$ = average molecular weight of the hydrocarbon mixture [g/mol] P_i^v = vapor pressure of pure component *i* [atm]
Maximum time-averaged diffusive vapor flux $\langle F_{max}\rangle$ [g/cm^2-day] from surface soils over period from time = 0 to time = τ, no immiscible phase present	$\langle F_{max}\rangle = 2\rho_s C_{soil}\sqrt{\frac{D^{eff}}{\pi R_v \tau}}$	Q_B = volumetric flow rate of air within enclosed space [cm^3/s] Q_{soil} = volumetric infiltration flow rate of soil gas into enclosed space [cm^3/s] *R* = gas constant = 82 atm-cm^3/mol-K R_v = porous media "retardation" factor[A] *T* = absolute temperature [K] x_i = mol fraction of component *i*
Maximum time-averaged diffusive vapor flux $\langle F_{max}\rangle$ [g/cm^2-day] from surface soils over period from time = 0 to time = τ, volatile components from relatively nonvolatile immiscible phase (for example, benzene from gasoline)	$\langle F_{max}\rangle = \frac{2D^{eff}\left(\frac{x_i P_i^v M_{w,i}}{RT}\right)}{\sqrt{\pi\alpha\tau}}$ where: $\alpha = \frac{D^{eff}}{\theta_v + \frac{\rho_s RT(C_{soil}/M_{w,T})}{P_i^v}}$	θ_v = volumetric content of soil vapor [cm^3-vapor/cm^3-soil] ρ_s = soil bulk density [g-soil/cm^3-soil] π = 3.1416 τ = averaging time [s] $C_{w,eq}$ = equilibrium dissolved concentration in leachate source area [g/cm^3-H_2O][A] E_B = enclosed space air exchange rate [l/*d*] E_{max} = vapor emission rate into enclosed space [g/day][A] *F* = vapor flux [g/cm^2-day][A] *i* = ground water gradient [cm/cm] K_s = saturated hydraulic conductivity [cm/day] *L* = downwind length of vapor emissions source area [cm] *M* = ground water mixing zone thickness [cm] q_I = water infiltration rate [cm/day]

TABLE X3.1 *Continued*

Description	Mathematical Approximation	Parameters
Vapor Emissions to Enclosed Spaces: Maximum vapor emission rate E_{max} [g/cm²-*d*] to enclosed spaces from subsurface vapor sources located a distance *d* [cm] away from the enclosed spaces	$E_{max} = Q_B C_{v,eq} \left(\frac{D^{eff} A_B}{Q_B d}\right) \exp\left(\frac{Q_{soil} L_{crack}}{D^{crack} A_{crack}}\right) \Big/ \left[\exp\left(\frac{Q_{soil} L_{crack}}{D^{crack} A_{crack}}\right) + \left(\frac{D^{eff} A_B}{Q_{soil}{}^{d}}\right)\left(\exp\left(\frac{Q_{soil} L_{crack}}{D^{crack} A_{crack}}\right) - 1\right)\right]$	u_w = wind speed [cm/day] V_B = volume of enclosed space [cm³] W = width of impacted soil zone [cm] δ = height of breathing zone [cm]
Hydrocarbon Vapor Dispersion:		
Ambient hydrocarbon vapor concentration resulting from area vapor source $C_{outdoor}$ [g/cm³]	$C_{outdoor} = \frac{FL}{u_w \delta}$	
Enclosed space vapor concentration C_{indoor} [g/cm³]	$C_{indoor} = \frac{E_{max}}{V_B E_B}$	
Leachate Transport: *Leaching Impact on Ground Water:* Ground water source area concentration C_{source} [g/cm³-H_2O] resulting from leaching through vadose zone hydrocarbon-impacted soils	$C_{source} = C_{w,eq} \frac{q_i W}{(K_s i M + q_i W)}$	
Ground water source area concentraiton C_{source} [g/cm³-H_2O] resulting from hydrocarbon-impacted soils in direct contact with ground water	$C_{source} = C_{w,eq}$	

[A] Equation for this parameter given in this table.

specific phenomena (for example, ground water transport) and vary in complexity, depending on assumptions made during model development. In RBCA, simplistic screening-level migration models are utilized in Tiers 1 and 2, while more complex models are utilized in Tier 3.

X3.7.2 References to many simplistic models suitable for screening-level evaluations for a number of pathways relevant to hydrocarbon contaminant releases are listed in Table X3.1. Most of the screening-level migration models have a simple mathematical form and are based on multiple limiting assumptions rather than on actual phenomena. For example, a simple model is the use of estimated ground water flow velocity to assess the travel time between the leading edge of a dissolved hydrocarbon plume and a ground water well. The travel time is approximated by the following:

$$[\text{distance to well (ft)/flow velocity (ft/years)}] = \text{travel time (years)} \quad \text{(X3.1)}$$

X3.7.2.1 In the case of a relatively light compound such as benzene dissolved in ground water, the flow velocity may best be equated with the ground water flow velocity. Heavier compounds such as naphthalene may be retarded so that a flow velocity lower than the ground water velocity may be used. If miscible liquids are present on the ground water surface, such as gasoline, the liquid flow velocity may actually exceed the ground water velocity.

X3.7.3 The use of more complex models is not precluded in the RBCA process; however, given limited data and assumptions that must be made, many complex numerical models reduce to the analytical expressions given in Table X3.1.

X3.7.4 *Migration Model Data Requirements*—Predictive migration models require input of site-specific characteristics. Those most commonly required for various simplistic models include the following:

X3.7.4.1 Soil bulk density (for a typical soil: ≈ 1.7 g/cm³),

X3.7.4.2 Total soil porosity (for a typical soil: ≈ 0.38 cm³/cm³),

X3.7.4.3 Soil moisture content can be conservatively estimated in many cases. It is approximately equal to the total soil porosity beneath the water table, and typically >0.05 cm³-H_2O/cm³-soil in the vadose zone; this can be a critical input parameter in the case of diffusion models and may require site-specific determination unless conservative values are used,

X3.7.4.4 Fraction organic matter in soil particles (= 0.00d – 0.01: sandy soil is often conservatively assumed); this can also be a critical parameter requiring site-specific determination unless conservative values are used),

X3.7.4.5 Hydraulic conductivity (generally site-specific determination required),

X3.7.4.6 Ground water gradient and flow direction (requires site-specific determination), and

X3.7.4.7 First-order decay-rate (generally requires site-specific calibration as models are very sensitive to this parameter); see Tables X3.2 and X3.3 and Ref **(41)** for a summary of measured values currently available from the literature. The data in Table X3.3 include retardation and dispersion as well as natural biodegradation in attenuation rates measured. However, sensitivity studies indicate that natural biodegradation is the dominant factor. The sensitivity studies use Ref **(42)**. According to these sensitivity studies, an order of magnitude increase in natural biodegradation rate is 3.5 times as effective as an order of magnitude increase in retardation and 12 times as effective as an order of magnitude increase in dispersion in

TABLE X3.2 Reported Degradation Rates for Petroleum Hydrocarbons

Reference	Source of Data	Chemical Decay Rates (day^{-1}, [half-life days])							
		Benzene	Toluene	Ethyl-Benzene	Xylenes	O-Xylene	MTBE	Naphthalene	Benzo (a)Pyrene
Barker, et al[A]	Borden Aquifer, Canada	0.007 [99]	0.011 [63]	...	...	0.014 [50]	...	...	...
Kemblowski[B]	Eastern Florida Aquifer	0.0085 [82]	...	...	...	...	...	...	...
Chiang, et al[C]	Northern Michigan Aquifer	0.095 [7]	...	...	...	...	...	...	...
Wilson, et al[D]	Traverse City, MI Aquifer	0.007 to 0.024 [99] to [29]	0.067 [10]	...	0.004 to 0.014 [173] to [50]	...	...	...	...
Howard, et al[E]	Literature	0.0009 [730] to 0.069 [10]	0.025 [28] to 0.099 [7]	0.003 [228] to 0.116 [6]	0.0019 [365] to 0.0495 [14]	...	0.0019 [365] to 0.0866 [8]	0.0027 [258]	0.0007 [1058] to 0.0061 [114]

[A] See Ref **(36)**.
[B] See Ref **(37)**.
[C] See Ref **(38)**.
[D] See Ref **(39)**.
[E] See Ref **(40)**.

TABLE X3.3 Results of Exponential Regression for Concentration Versus Time[A]

Site	Compound	k, % per day
Campbell, CA	benzene	1.20
	ethylbenzene	0.67
	xylene	1.12
	benzene	0.42
Palo Alto, CA	benzene	0.30
Virginia Beach, VA	PCE	0.46
	TCE	0.30
Montrose County, CO	benzene	0.42
Provo, UT	benzene	0.23
San Jose, CA	benzene	0.16
	benzene	0.10
Chemical facility	toluene	0.39
	PCE	0.34
	TCE	0.26

[A] Source: Ref **(41)**.

attenuating concentration over distance. Therefore, approximately 80 % of the attenuation shown in the Ref **(41)** data can be attributed to natural biodegradation.

X3.7.4.8 A similar analysis of the sensitivity of attenuation parameters for the vapor transport pathway also indicates that natural biodegradation is the predominant attenuation mechanism **(43)**. Soil geology is not considered an attenuation mechanism directly, but is a stronger determinant of how far contamination travels than even natural biodegradation. Gasoline contamination does not travel very far in clay (less than 30 ft (9 m)) according to the vapor transport model **(43)**.

X3.7.5 Depending on the models selected, other information may be required, such as meteorological information (wind speed, precipitation, temperature), soil particle size distributions, and nearby building characteristics.

X3.7.6 In most cases, measurements of the attenuation (decrease in concentrations) of compounds with distance away from the contaminant source area will be required to calibrate and verify the predictive capabilities of the selected models. The amount of data required varies depending on the following:

X3.7.6.1 The model code used,

X3.7.6.2 The model's sensitivity to changes in input parameters, and

X3.7.6.3 The contribution of the pathway of concern to the total incremental exposure and risk.

X3.7.7 Generally, site-specific physical and chemical properties for the most sensitive parameters are required for migration models to obtain accurate results. However, instead of site-specific data, conservative values selected from the literature may be used with appropriate caution.

X3.7.8 *Migration Modeling Procedure*

The procedure for applying a migration model includes the following steps: definition of study objectives, development of a conceptual model, selection of a computer code or algorithm, construction of the model, calibration of the model and performance of sensitivity analysis, making predictive simulations, documentation of the modeling process, and performing a postaudit. These steps are generally followed in order; however, there is substantial overlap between steps, and previous steps are often revisited as new concepts are explored or as new data are obtained. The iterative modeling approach may also require the reconceptualization of the problem. The basic modeling steps are discussed as follows.

X3.7.8.1 *Modeling Objectives*—Modeling objectives must first be identified (that is, the questions to be answered by the model). The objectives aid in determining the level of detail and accuracy required in the model simulation. Complete and detailed objectives would ideally be specified prior to any modeling activities. Objectives may include interpreting site characterization and monitoring data, predicting future migration, determining corrective action requirements, or predicting the effect of proposed corrective action measures.

X3.7.8.2 *Conceptual Model*—A conceptual model of a subsurface contaminant release, such as a hydrocarbon release from an underground tank, is an interpretation or working description of the characteristics and dynamics of the physical system. The purpose of the conceptual model is to consolidate site and regional data into a set of assumptions and concepts that can be evaluated quantitatively. Development of the conceptual model requires the collection and analysis of physical data pertinent to the system under investigation.

(*1*) The conceptual model identifies and describes important aspects of the physical system, including the following: geologic and hydrologic framework; media type (for example, fractured or porous); physical and chemical processes; and hydraulic, climatic, and vapor properties. The conceptual model is described in more detail for ground water flow systems in Guide D 5447.

(*2*) Provide an analysis of data deficiencies and potential sources of error with the conceptual model. The conceptual model usually contains areas of uncertainty due to the lack of field data. Identify these areas and their significance to the conceptual model evaluated with respect to project objectives.

X3.7.8.3 *Computer Code Selection*—Computer code selection is the process of choosing the appropriate software algorithm, or other analysis technique, capable of simulating the characteristics of the physical system, as identified in the conceptual model. The types of codes generally used in the RBCA process are analytical and numerical models. The selected code should be appropriate to fit the available data and meet the modeling objectives. The computer code must also be tested for the intended use and be well documented.

(*1*) Analytical models are generally based on assumptions of uniform properties and regular geometries. Advantages include quick setup and execution. Disadvantages include, in many cases, that analytical models are so simplistic that important aspects of a given system are neglected.

(*2*) Numerical models allow for more complex heterogeneous systems with distributed properties and irregular geometries. Advantages include the flexibility to simulate more complex physical systems and natural parameter variability. Disadvantages include that the approach is often very time-intensive and may require much more data and information to be collected.

(*3*) Other factors may also be considered in the decision-making process, such as the model analyst's experience and those described as follows for model construction process; factors such as dimensionality will determine the capabilities of the computer code required for the model.

X3.7.8.4 *Model Construction*—Model construction is the process of transforming the conceptual model into a mathematical form. The model typically consists of two parts, the data set and the computer code. The model construction process includes building the data set used by the computer code. Fundamental components of a migration model are dimensionality, discretization, boundary and initial conditions, contaminant, and media properties.

X3.7.8.5 *Model Calibration*—Calibration of a model is the process of adjusting input for which data are not available within reasonable ranges to obtain a match between observed and simulated values. The range over which model parameters and boundary conditions may be varied is determined by data presented in the conceptual model. In the case where parameters are well characterized by field measurements, the range over which that parameter is varied in the model should be consistent with the range observed in the field. The degree of fit between model simulations and field measurements can be quantified using statistical techniques.

(*1*) In practice, model calibration is frequently accomplished through trial-and-error adjustment of the model's input data to match field observations. The calibration process continues until the degree of correspondence between the simulation and the physical system is consistent with the objectives of the project.

(*2*) Calibration of a model is evaluated through analysis of residuals. A residual is the difference between the observed and simulated variable. Statistical tests and illustrations showing the distribution of residuals are described for ground water flow models in Guide D 5490.

(*3*) Calibration of a model to a single set of field measurements does not guarantee a unique solution. To minimize the likelihood of nonuniqueness, the model should be tested to a different set of boundary conditions or stresses. This process is referred to as application verification. If there is poor correspondence to a second set of field data, then additional calibration or data collection are required. Successful verification of an application results in a higher degree of confidence in model predictions. A calibrated but unverified model may still be used to perform predictive simulations when coupled with a sensitivity analysis.

X3.7.8.6 *Sensitivity Analysis*—Sensitivity analysis is a quantitative method of determining the effect of parameter variation on model results. Two purposes of a sensitivity analysis are (*1*) to quantify the uncertainty in the calibrated model caused by uncertainty in the estimates of parameters, stresses, and boundary conditions, and (*2*) to identify the model inputs that have the most influence on model calibration and predictions.

(*1*) Sensitivity of a model parameter is often expressed as the relative rate of change of a selected model calculation during calibration with respect to that parameter. If a small change in the input parameter or boundary condition causes a significant change in the output, the model is sensitive to that parameter or boundary condition.

(*2*) Whether a given change in the model calibration is considered significant or insignificant is a matter of judgment. However, changes in the model's conclusions are usually able to be characterized objectively. For example, if a model is used to determine whether a contaminant is captured by a potable supply well, then the computed concentration is either detectable or not at the location. If, for some value of the input that is being varied, the model's conclusions are changed but the change in model calibration is insignificant, then the model results may be invalid because, over the range of that parameter in which the model can be considered calibrated, the conclusions of the model change. More information regarding conducting a sensitivity analysis for a ground water flow model application is presented in Guide D 5611.

X3.7.8.7 *Model Predictions*—Once these steps have been conducted, the model is used to satisfy the modeling objectives. Predictive simulations should be documented with appropriate illustrations, as necessary, in the model report.

X3.8 *Procedures for Risk, Exposure, and Dose-Response Assessment Models:*

X3.8.1 "Exposure models" are used to estimate the chemical uptake, or dose, while "risk assessment models" are used to relate human health or ecological impacts to the uptake of a chemical. Risk and exposure assessment models are often combined to calculate a target exposure point concentration of a compound in air, water, or soil.

X3.8.1.1 In the case of compounds that have been classified as carcinogens, exposure and risk assessment models are generally linked by the expression:

$$\mathit{risk} = \mathit{average\ lifetime\ intake}[\mathrm{mg/kg-day}]\ [ml]P5\ \mathit{slope\ factor}[\mathrm{mg/kg-day}]^{-1} \qquad (X3.2)$$

where the intake depends on exposure parameters (ingestion rate, exposure duration, and so forth) and the concentration at point-of-exposure. The slope factor (sometimes called the "potency factor") is itself based on a model and set of underlying assumptions, which are discussed as follows.

X3.8.1.2 In the case of compounds that have not been classified as carcinogens, exposure and risk assessment models are generally :

$$\mathit{hazard\ quotient} = \frac{\mathit{average\ intake}\ [\mathrm{mg/kg-day}]}{\mathit{reference\ dose}\ [\mathrm{mg/kg-day}]} \qquad (X3.3)$$

where the intake depends on exposure parameters (ingestion rate, exposure duration, and so forth) and the concentration at point-of-exposure. The reference dose is itself based on a model and set of underlying assumptions, which are discussed as follows.

X3.8.2 *Toxicity Assessment: Dose-Response Models*—Toxicity assessments use dose-estimates of a "safe dose" or toxic level based on animal studies. In some instances, human epidemiological information is available on a chemical. Toxicologists generally make two assumptions about the effects of risk agents at the low concentrations typical of environmental exposures:

X3.8.2.1 Thresholds exist for most biological effects; in other words, for noncarcinogenic, nongenetic toxic effects, there are doses below which no adverse effects are observed in a population of exposed individuals, and

X3.8.2.2 No thresholds exist for genetic damage or incremental carcinogenic effects. Any level of exposure to the genotoxic or carcinogenic risk agent corresponds to some non-zero increase in the likelihood of inducing genotoxic or incremental carcinogenic effects.

X3.8.3 The first assumption is widely accepted in the scientific community and is supported by empirical evidence. The threshold value for a chemical is often called the NOAEL. Scientists usually estimate NOAELs from animal studies. An important value that typically results from a NOAEL or LOAEL value is the RfD. A reference dose is an estimate (with an uncertainty typically spanning an order of magnitude) of a daily exposure (mg/kg/day) to the general human population (including sensitive subgroups) that is likely to be without an appreciable risk of deleterious effects during a lifetime of exposure. The RfD value is derived from the NOAEL or LOAEL by application of uncertainty factors (UF) that reflect various types of data used to estimate RfDs and an additional modifying factor (MF), which is based on a professional judgment of the quality of the entire database of the chemical. The oral RfD, for example, is calculated from the following equation:

$$\mathit{RfD} = \frac{\mathit{NOAEL}}{(\mathit{UF}\ [ml]P5\ \mathit{MF})} \qquad (X3.4)$$

X3.8.4 The second assumption regarding no threshold effects for genotoxic or carcinogenic agents is more controversial but has been adopted by the USEPA. For genotoxic and carcinogenic agents, extrapolations from high experimental doses to low doses of environmental significance require the use of mathematical models to general low dose-response curves. It should be noted that although the EPA uses the linear multi-state model to describe incremental carcinogenic effect, there is no general agreement in the scientific community that this is the appropriate model to use.

X3.8.5 The critical factor determined from the dose-response curve is the slope factor (SF), which is the slope of the dose-response curve in the low-dose region. The units of the slope factor are expressed as $(\mathrm{mg/kg\text{-}day})^{-1}$ and relate a given environmental intake to the risk of additional incidence of cancer above background.

X3.8.6 The RfD or SF values are generally obtained from a standard set of reference tables (for example, Ref **(2)** or Ref **(3)**). It is important to note that the information in IRIS has typically only been peer-reviewed within the EPA and may not always have support from the external scientific community. Whereas the information in IRIS has been subject to agency-wide data quality review, the information in the HEAST tables has not. The user is expected to consult the original assessment documents to appreciate the strengths and limitations of the data in HEAST. Thus, care should be exercised in using the values in HEAST. Some state and local agencies have toxicity factors they have derived themselves or preferences for factors to use if neither IRIS nor HEAST lists a value. Values for a range of hydrocarbons typically of interest are presented in Table X3.1.

X3.8.7 It is important to note that in extrapolating the information obtained in animal studies to humans, a number of conservative assumptions are made.

X3.8.7.1 For noncarcinogens, an arbitrary system of default safety and uncertainty factors, as discussed (in multiples of ten), is used to convert observations, in animals to estimates in humans.

X3.8.7.2 For carcinogens, some of the most important assumptions include: (*1*) the results of the most sensitive animal study are used to extrapolate to humans, (*2*) in general, chemicals with any incremental carcinogenic activity in animals are assumed to be potential human carcinogens, and (*3*) no threshold exists for carcinogens.

X3.8.8 The uncertainty in the RfD and SF values are often neglected in deference to single point values which are then typically summarized in databases such as IRIS and HEAST and assumptions described are risk management policy decisions made by the USEPA. These assumptions are not explicitly defined and further obscure the conservatism in the safe dose estimate. Thus, care must be exercised in interpreting results which have as a basis these conservative toxicity evaluations.

X3.8.9 *Exposure Assessment Modeling*—The goal of exposure assessment modeling is to estimate the chemical uptake that occurs when a receptor is exposed to compounds present in their environment. In principal, the process for developing and using migration models presented in X3.7 is directly applicable to exposure assessment modeling. In this case the user:

X3.8.9.1 Develops a conceptual model by identifying significant exposure pathways and receptors,

X3.8.9.2 Selects a model to describe the contact rate and subsequent uptake of the chemical(s),

X3.8.9.3 Performs a sensitivity analysis to identify critical parameters,

X3.8.9.4 Selects appropriate exposure parameters (breathing rates, and so forth),

X3.8.9.5 Generates estimates of exposure and uptake, and

X3.8.9.6 Assesses the uncertainty in the estimates.

X3.8.10 There are differences between the process outlined in X3.7 and that which can be practically applied to exposure assessment modeling. For example, with the exception of exposures and impacts to environmental resources, it is difficult to calibrate exposure assessment models unless very expensive epidemiological studies are conducted.

X3.8.11 Typically, the models used to estimate uptake are simplistic algebraic expressions, such as those contained in Ref **(27)**. Application of these equations is illustrated in Appendix X2.

X3.8.12 In many cases, exposure parameter values are available in Ref **(27)**, but other more recent information is also available in peer-reviewed publications, and all sources should be carefully reviewed. While point values are often selected for simplicity, statistical distributions for many of the exposure parameters are readily available for Tier 3 analyses.

X3.8.13 It is common for USEPA RME values to be used in exposure assessment calculation, as is done for the example Tier 1 Look-Up Table discussed in Appendix X2. The RME value is generally defined as a statistical upper limit of available data (generally 85 to 90 % of all values are less than the RME value). Therefore, by consistently selecting and multiplying conservative RME values the user models a scenario that is very improbable and always more conservative than the "true" RME exposure scenario. Thus, great care must be exercised, when using combinations of these default values in risk assessments, to avoid a gross overestimation of exposure for a specific site.

X3.9 *Report*—The purpose of the model report is to communicate findings, to document the procedures and assumptions inherent in the study, and to provide detailed information for peer review. The report should be a complete document allowing reviewers and decision makers to formulate their own opinion as to the credibility of the model. The report should describe all aspects of the modeling study outlined in this appendix.

X4. INSTITUTIONAL CONTROLS

X4.1 *Introduction:*

X4.1.1 The purpose of this appendix is to provide a review of generally used institutional controls. For purposes of this appendix, "institutional controls" are those controls that can be used by responsible parties and regulatory agencies in remedial programs where, as a part of the program, certain concentrations of the chemical(s) of concern will remain on site in soil or ground water, or both. Referenced in this appendix are examples of programs from California, Connecticut, Illinois, Indiana, Iowa, Massachusetts, Michigan, Missouri, and New Jersey. In addition, federal programs, such as Superfund settlements and RCRA closure plans have used the following techniques described for some years as a mechanism to ensure that exposure to remaining concentrations of chemical(s) of concern is reduced to the degree necessary.

X4.1.2 The types of institutional controls discussed in this appendix are as follows:

X4.1.2.1 Deed restrictions, or restrictive covenants,

X4.1.2.2 Use restrictions (including well restriction areas),

X4.1.2.3 Access controls,

X4.1.2.4 Notice, including record notice, actual notice, and notice to government authorities,

X4.1.2.5 Registry act requirements,

X4.1.2.6 Transfer act requirements, and

X4.1.2.7 Contractual obligations.

X4.1.3 Institutional controls for environmental remedial programs vary in both form and content. Agencies and landowners can invoke various authorities and enforcement mechanisms, both public and private, to implement any one or a combination of the controls. For example, a state could adopt a statutory mandate (see Appendix X4.2) requiring the use of deed restrictions (see Appendix X4.3) as a way of enforcing use restrictions (see Appendix X4.4) and posting signage (a type of access control, see X4.5). Thus, the institutional controls listed as follows are often used as overlapping strategies, and this blurs the distinctions between them.

X4.2 *Statutory Mandates*—Some states' emergency response programs mandate post-remediation institutional controls and impose civil penalties for noncompliance. The schemes vary from state to state, but all impose obligations on landowners to use one or more institutional controls listed in this appendix.

X4.3 *Deed Restrictions:*

X4.3.1 Deed restrictions place limits and conditions on the use and conveyance of land. They serve two purposes: (*1*) informing prospective owners and tenants of the environmental status of the property and (*2*) ensuring long-term compliance with the institutional controls that are necessary to maintain the integrity of the remedial action over time. Restraining the way someone can use their land runs counter to the basic assumptions of real estate law, so certain legal rules must be satisfied in order to make a deed restriction binding and enforceable.

X4.3.2 There are four requirements for a promise in a deed restriction (also called a "restrictive covenant") to be held against current and subsequent landowners: (*1*) a writing, (*2*) intention by both original parties that particular restrictions be placed on the land in perpetuity, (*3*) "privity of estate," and (*4*) that the restrictions "touch and concern the land."

X4.3.2.1 The first requirement is that of a writing. It is a rule of law that conveyances of land must be documented in a writing. The same rule holds for deed restrictions affecting land. Ideally, a deed restriction used as an institutional control would be written down with particularity and then recorded in the local land records office, in much the same fashion as the documentation and recordation of a sale of land. Parties may also encounter the requirement that the deed restriction be executed "under seal," a legal formality that has been abandoned in most states.

X4.3.2.2 The second requirement is that the deed restriction should precisely reflect what the parties' intentions are in regard to the scope and the duration of the restrictions. Explicitly stating in the deed restriction that the parties intend the restriction to "run with the land" (that is, last forever and bind subsequent owners) is strongly recommended.

X4.3.2.3 The third requirement, privity of estate, arises from a concern that only persons with a certain relationship to the land should be able to enforce a deed restriction. Normally, deed restrictions are promises between the buyer and the seller or between neighbors; therefore, the state or a third party may not enforce a deed restriction. However, even in states that require privity of estate, this concern is addressed if the landowner took the land with knowledge that the restrictions existed and might be enforced by these third parties. Thus, it is also strongly recommended that the deed restriction explicitly state that the state environmental authority may enforce the restriction. Recording of the deed restriction serves as notice to anyone who later purchases or acquires an interest in the land. Therefore, privity of estate should not be a barrier to state enforcement of the deed restriction if the proper steps are taken.

X4.3.2.4 Finally, a deed restriction is only enforceable if the promise "touches and concerns the land." A rough rule of thumb to decide this point is whether the landowner's legal interest in the land is decreased in value due to the deed restriction. If the land is devalued in this way, then the restriction could be said to "touch and concern the land." Note that the focus of the inquiry is on the land itself; promises that are personal in nature and merely concern human activities that happen to take place on the land are least likely to be enforceable. Thus, any deed restriction used as an institutional control should be written so that it centers on the land and the use of the land.

X4.3.3 Due to the potential enforcement hurdles encountered by a governmental agency in enforcing a deed restriction, it may be appropriate for an individual state to seek statutory and regulatory amendments to ensure that such authority exits in regard to all deed restrictions for environmental purposes.

X4.3.4 Remedies for noncompliance with deed restrictions comes in two forms: (*1*) persons or agencies may sue to obtain a court order (injunction) requiring compliance or (*2*) if the state statute allows for it, the state's attorney general can seek enforcement of civil penalties, such as fines, for noncompliance.

X4.3.5 A state program can require a landowner to continue monitoring activities and to allow state environmental officials access to the site to monitor compliance with institutional controls. These arrangements may have to be put in a deed restriction in order to run with the land from owner to owner, but responsible parties can also be required to sign a contract making these promises. Of course, almost every state has authority to issue administrative orders to accomplish some or all of these arrangements.

X4.3.6 The preceding arrangements can also set out procedures that will be followed if some emergency requires that the remediation site be disturbed. If, for example, underground utility lines must be repaired, the landowner would follow this protocol for handling the soil and alerting the state authority.

X4.4 *Use Restrictions:*

X4.4.1 Use restrictions are usually the heart of what is in a deed restriction. Use restrictions describe appropriate and inappropriate uses of the property in an effort to perpetuate the benefits of the remedial action and ensure property use that is consistent with the applicable cleanup standard. Such techniques also prohibit any person from making any use of the site in a manner that creates an unacceptable risk of human or environmental exposure to the residual concentrations of chemical(s) of concern.

X4.4.2 Use restrictions address uses that may disturb a containment cap or any unremediated soils under the surface or below a building. A prohibition on drinking on-site (or off-site by means of well restriction areas discussed as follows) ground water may also be appropriate.

X4.4.3 As an example, a program may allow a restriction of record to include one or more of the following:

X4.4.3.1 Restriction on property use,

X4.4.3.2 Conditioning the change of use from nonresidential on compliance with all applicable cleanup standards for a residential property;

X4.4.3.3 Restricting access; or

X4.4.3.4 Restricting disturbance of department-approved remedial effects.

X4.4.4 Well restriction areas can be a form of institutional control by providing notice of the existence of chemical(s) of concern in ground water, and by prohibiting or conditioning the construction of wells in that area.

X4.4.4.1 This technique preserves the integrity of any ground water remedial action by prohibiting or conditioning the placement and use of any or all types of wells within the area.

X4.4.4.2 Well restrictions of this nature would be subject to agency approval and public notice, and may include the restriction on constructing or locating any wells within a particular designated area. Notice of the well restriction is recorded on the land records and with various health officials and municipal officials. The restrictions can only be released upon a showing that the concentrations of the chemical(s) of concern in the well restriction area is remediated in accordance with state standards.

X4.5 *Access Controls:*

X4.5.1 Another subset of institutional controls is the control of access to any particular site. The state uses the following criteria to determine the appropriate level and means of access control:

X4.5.1.1 Whether the site is located in a residential or

mixed use neighborhood;

X4.5.1.2 Proximity to sensitive land-use areas including day-care centers, playgrounds, and schools; and

X4.5.1.3 Whether the site is frequently traversed by neighbors.

X4.5.2 Access can be controlled by any of the following: fencing and gates, security, or postings or warnings.

X4.6 *Notice*—Regulations of this type generally provide notice of specific location of chemical(s) of concern on the site, and disclose any restrictions on access, use, and development of part or all of the contaminated site to preserve the integrity of the remedial action.

X4.6.1 *Record Notice*:

X4.6.1.1 Some states require that sites having releases of hazardous waste file a notice on the land records providing to any subsequent purchaser of the property information regarding the past or current activities on the site.

X4.6.1.2 The record notice requirement may be broad; the program may require any property subject to a response action to obtain a professional opinion and then prepare and record a Grant of Environmental Restriction that is supported by that opinion.

X4.6.1.3 The record notice requirement can be ancillary to a transfer act (see Appendix X4.8), in which case recording of an environmental statement is only required in conjunction with a land transaction.

X4.6.2 *Actual Notice*:

X4.6.2.1 States may require direct notice of environmental information to other parties to a land transaction. These laws protect potential buyers and tenants, and they also help ensure that use restrictions and other institutional controls are perpetuated.

X4.6.2.2 Actual notice of an environmental defect or failure to provide notice may give a party the right to cancel the transaction and result in civil penalties. For example, landlords and sellers who do not give notice as required by the state may be liable for actual damages plus fines. Nonresidential tenants who fail to notify landowners of suspected or actual hazardous substance releases can have their leases canceled and are subject to fines.

X4.6.3 *Notice to Government Authorities*—Parties to a land transaction may also be required to file the environmental statement with various environmental authorities. Notice to the government may be required before the transaction takes place.

X4.7 *Registry Act Requirements:*

X4.7.1 Some states have registry act programs that provide for the maintenance of a registry of hazardous waste disposal sites and the restriction of the use and transfer of listed sites.

X4.7.2 A typical registry act provides that the state environmental agency establish and maintain a registry of all real property which has been used for hazardous substance disposal either illegally or before regulation of hazardous waste disposal began in that state.

X4.7.3 The state agency is responsible for investigating potential sites for inclusion on the registry. The registry includes the location of the site and a listing of the hazardous wastes on the property, and may also include a classification of the level of health or environmental danger presented by the conditions on the property. The state agency may be required to perform detailed inspections of the site to determine its priority relative to other registered sites.

X4.7.4 Owners of sites proposed for inclusion on the registry have rights of hearing and appeal, and owners of sites on the registry have rights to modify or terminate their listing. In some cases, the owner of a site proposed for inclusion on the registry may obtain the withdrawal of the proposed registration by entering into a consent agreement with the state. Such a consent agreement establishes a timetable and responsibility for remedial action.

X4.7.5 When a site appears on the state registry, the owner must comply with regulatory requirements in regard to use and transfer of the site. The use of a site listed on the registry may not be changed without permission of the state agency. In negotiations for a conveyance of a registered site, the owner may be obligated to disclose the registration early in the process, and permission of the state agency may be required to convey a registered property. Under other schemes, permission to convey is not required, but the seller must notify the state agency of the transaction.

X4.7.6 Finally, registry acts require that the listing of a property on a hazardous materials site registry be recorded in the records of the appropriate locality so that the registration will appear in the chain of title.

X4.8 *Transfer Act Requirements:*

X4.8.1 Some states have transfer act programs that require full evaluation of all environmental issues before or after the transfer occurs. It may be that within such program, institutional controls can be established by way of consent order, administrative order, or some other technique that establishes implementation and continued responsibility for institutional controls.

X4.8.2 A typical transfer act imposes obligations and confers rights on parties to a land transaction arising out of the environmental status of the property to be conveyed. Transfer acts impose information obligations on the seller or lessor of a property (see Appendix X4.6.3). That party must disclose general information about strict liability for cleanup costs as well as property-specific information, such as presence of hazardous substances, permitting requirements and status, releases, and enforcement actions and variances.

X4.8.3 Compliance with transfer act obligations in the manner prescribed is crucial for ensuring a successful conveyance. Sometimes the transfer act operates to render a transaction voidable before the transfer occurs. Failure to give notice in the required form and within the time period required or the revelation of an environmental violation or unremediated condition will relieve the transferee and the lender of any obligation to close the transaction, even if a contract has already been executed. Moreover, violation of the transfer act can be the basis for a lawsuit to recover consequential damages.

X4.9 *Contractual Obligations:*

X4.9.1 One system for ensuring the future restriction on use of a site, or the obligation to remediate a site, is to require private parties to restrict use by contract. While this method is often negotiated among private parties, it will be difficult, if not impossible, to institutionalize some control over that process without interfering with the abilities and rights of private parties to freely negotiate these liabilities.

X4.9.2 Another avenue is for the landowner or responsible party to obligate itself to the state by contract. The state may require a contractual commitment from the party to provide long-term monitoring of the site, use restrictions, and means of continued funding for remediation.

X4.10 *Continued Financial Responsibility*—Another aspect of institutional controls is the establishment of financial mechanisms by which a responsible party ensures continued funding of remediation measures and assurance to the satisfaction of the state.

X4.11 *References:*

X4.11.1 The following references serve as examples and are current as of the fourth quarter of 1993:

X4.11.1.1 *References for Deed Restrictions*:

24 New Jersey Regulations 400 (1992) (New Jersey Administration Code § 7.26D-8.2 (e) (2))
24 New Jersey Regulations 400-02 (1992) (New Jersey Administration Code §§ 7.26D-8.1–8.4)
24 New Jersey Regulations 401 (1992) (New Jersey Administration Code § 7.26D Appendix A, Model Document, Declaration of Environmental Restrictions and Grant of Ease ment, Item 8)
Illinois Responsible Property Transfer Act § 7(c) (1985)
Massachusetts Regulations Code Title § 40.1071 (2) (1) & (k)
Massachusetts Regulations Code, Title § 40.1071(4)
Michigan Administration Code 299.5719 (3) (e) (1990)
Michigan Rules 299.5719 (2), (3) (d)

X4.11.1.2 *References for Use Restrictions*:

24 New Jersey Regulations 400 (New Jersey Administration Code § 7.26D-8.2 (d))
Michigan Administration Code 299.5719 (3) (a), (b), (g)
New Jersey Regulation 7.26D-8.4

X4.11.1.3 *References for Access Controls*:

Iowa Administration Code r. 133.4 (2) (b)
Michigan Rule 299.4719 (3) (f)
New Jersey Regulations § 7.26D-8.2

X4.11.1.4 *References for Notice*:

California Health and Safety Code § 25359.7 (1981)
Illinois Responsible Property Transfer Act (1985)
Indiana Code §§13-7-22.5-1–22 (1989) ("Indiana Environmental Hazardous Disclosure and Responsible Party Transfer Law")
Massachusetts Regulations Code Title §§ 40.1071-1090 (1993)
Michigan Rule 299.5719 (3) (c)

X4.11.1.5 *References for Registry Act Requirements*:

Iowa Code Ann. §§ 455B.426–455B.432, 455B.411 (1) (1990)
Missouri Code Regulations Title 10, §§ 25-10.010, 25-3.260 (1993)

X4.11.1.6 *References for Transfer Act Requirements*:

Connecticut General Stat.§ 22a-134 *et seg*
Illinois Responsible Property Transfer Act (1985)
Indiana Code §§ 13-7-22.5-1–22 (1989) ("Indiana Environmental Hazardous Disclosure and Responsible Party Transfer Law")
New Jersey Senate Bill No. 1070, the Industrial Site Recovery Act, amending the environmental cleanup Responsibility Act, N.J.S.A. 13:1K-6 *et seg*
New Jersey Spill Compensation and Control Act, N.J.S.A. 58:10-23.11 *et seg*

X4.11.1.7 *Reference for Contractual Obligations*:

Michigan Rule 299.5719 (2)

X4.11.1.8 *Reference for Continued Financial Responsibility*:

Michigan Rule 299.5719 (2)

X5. EXAMPLE APPLICATIONS OF RISK-BASED CORRECTIVE ACTION

X5.1 *Introduction*—The following examples illustrate the use of RBCA at petroleum release sites. The examples are hypothetical and have been simplified in order to illustrate that RBCA leads to reasonable and protective decisions; nevertheless, they do reflect conditions commonly encountered in practice.

X5.2 *Example 1—Corrective Action Based on Tier 1 Risk-Based Screening Levels:*

X5.2.1 *Scenario*—A release from the underground storage tank (UST), piping, and dispenser system at a service station is discovered during a real estate divestment assessment. It is known that there are petroleum-impacted surficial soils in the area of the tank fill ports; however, the extent to which the soils are impacted is unknown. In the past, both gasoline and diesel have been sold at the facility. The new owner plans to continue operating the service station facility.

X5.2.2 *Site Assessment*— The responsible party completes an initial site assessment focussed on potential source areas (for example, tanks, lines, dispensers) and receptors. Based on historical knowledge that gasoline and diesel have been dispensed at this facility, chemical analyses of soil and ground water are limited to benzene, toluene, ethylbenzene, xylenes, and naphthalene. Site assessment results are summarized as follows:

X5.2.2.1 Field screening instruments and laboratory analyses indicate that the extent of petroleum-impacted soils is confined to the vicinity of the fill ports for the tanks. A tank and line test reveals no leaks; therefore, evidence suggests that soils are impacted due to spills and overfills associated with filling the storage tank,

X5.2.2.2 The current tanks and piping were installed five years ago,

X5.2.2.3 The concrete driveway is highly fractured,

X5.2.2.4 No other sources are present,

X5.2.2.5 The site is underlain by layers of fine to silty sands,

X5.2.2.6 Ground water, which is first encountered at 32 ft (9.7 m) below ground surface, is not impacted,

X5.2.2.7 Maximum depth at which hydrocarbons are detected is 13 ft (3.9 m). Maximum detected soil concentrations are as follows:

Compound	Depth Below Ground Surface, ft (m)	Concentration, mg/kg
Benzene	8 (2.4)	10
Ethylbenzene	4 (1.2)	4

Toluene	6.5 (1.9)	55
Xylenes	3.5 (1.01)	38
Naphthalene	2 (0.6)	17

X5.2.2.8 A receptor survey indicates that two domestic water wells are located within 900 ft (273.6 m) of the source area. One well is located 500 ft (152.4 m) hydraulically down-gradient from the impacted soil zone, the other well is hydraulically up-gradient. Both wells produce water from the first encountered ground water zone.

X5.2.3 *Site Classification and Initial Response Action*—Based on classification scenarios given in Table 1, this site is classified as a Class 3 site because conditions are such that, at worst, it is a long-term threat to human health and environmental resources. The appropriate initial response is to evaluate the need for a ground water monitoring program (see Table X5.1). At most, this would consist of a single well located immediately down-gradient of the impacted petroleum soils. The responsible party recommends deferring the decision to install a ground water monitoring system until the Tier 1 evaluation is complete, and justifies this recommendation based on no detected ground water impact, the limited extent of impacted soils, and the separation between impacted soils and first-encountered ground water. The regulatory agency concurs with this decision.

X5.2.4 *Development of Tier 1 Look-Up Table of Risk-Based Screening Level (RBSL)*—Assumptions used to derive example Tier 1 RBSL Look-Up Table X2.1 in Appendix X2 are reviewed and presumed valid for this site. A comparison of RBSLs for both pathways of concern indicates that RBSLs associated with the leaching pathway are the most restrictive of the two. As this aquifer is currently being used as a drinking water supply, RBSL values based on meeting drinking water MCLs are selected. In the case of naphthalene, for which there is no MCL, the RBSL value corresponding to a residential scenario and a hazard quotient of unity is used.

X5.2.5 *Exposure Pathway Evaluation*—Based on current and projected future use, the only two potential complete exposure pathways at this site are: (*1*) the inhalation of ambient vapors by on-site workers, or (*2*) the leaching to ground water, ground water transport to the down-gradient drinking-water well, and ingestion of ground water (see Fig. X5.1).

X5.2.6 *Comparison of Site Conditions With Tier 1 RBSLs*—Based on the data given in X5.2.2.7 and the RBSLs given in Look-Up Table X2.1 in Appendix X2, exceedences of Tier 1 RBSLs are noted only for benzene and toluene.

X5.2.7 *Evaluation of Tier 1 Results*—The responsible party decides to devise a corrective action plan to meet Tier 1 standards after considering the following factors:

X5.2.7.1 The shallow aquifer is not yet affected,

X5.2.7.2 Quick (relative to rate of chemical migration) removal of the source will eliminate the need for ground water monitoring,

X5.2.7.3 The new owner plans to install new tanks within six months,

X5.2.7.4 Limited excavation of soils to meet Tier 1 criteria could be performed quickly and inexpensively when the tanks are removed, relative to the cost of proceeding to a Tier 2 analysis, and

X5.2.7.5 An excavation proposal will facilitate the real estate deal.

X5.2.8 *Tier 1 Remedial Action Evaluation*—Excavate all impacted soils with concentrations above the Tier 1 RBSLs when the current tanks are replaced. Subsequently resurface the area with new concrete pavement to reduce future infiltration and leaching potential through any remaining impacted soils. It is agreed that ground water monitoring is not necessary and the governing regulatory agency agrees to issue a No Further Action and Closure letter following implementation of the corrective action plan.

X5.3 *Example 2—RBCA Based on Tier 2 Evaluation:*

X5.3.1 *Scenario*—During the installation of new double-contained product transfer lines, petroleum-impacted soils are discovered in the vicinity of a gasoline dispenser at a service station located close to downtown Metropolis. In the past, both gasoline and diesel have been sold at this facility, which has been operating as a service station for more than twenty years.

X5.3.2 *Site Assessment*— The owner completes an initial

TABLE X5.1 Example 1—Site Classification and Initial Response Actions

Criteria and Prescribed Scenarios	Example Initial Response Actions
3. Long-term (>2 years) threat to human health, safety, or sensitive environmental receptors	Notify appropriate authorities, property owners, and potentially affected parties, and evaluate the need to
• Subsurface soils (>3 ft (0.9 m) BGS) are significantly impacted, and the depth between impacted soils and the first potable aquifer is less than 50 ft (15 m).	• Monitor ground water and determine the potential for future migration of the chemical(s) of concern to the aquifer.
• Ground water is impacted, and potable water supply wells producing from the impacted interval are located >2 years ground water travel time from the dissolved plume.	• Monitor the dissolved plume and evaluate the potential for natural attenuation and the need for hydraulic control.
• Ground water is impacted, and non-potable water supply wells producing from the impacted interval are located >2 years ground water travel time from the dissolved plume.	• Identify water usage of well, assess the effect of potential impact, monitor the dissolved plume, and evaluate whether natural attenuation or hydraulic control are appropriate control measures.
• Ground water is impacted, and non-potable water supply wells that do not produce from the impacted interval are located within the known extent of chemical(s) of concern.	• Monitor the dissolved plume, determine the potential for vertical migration, notify the user, and determine if any impact is likely.
• Impacted surface water, storm water, or ground water discharges within 1500 ft (457 m) of a sensitive habitat or surface water body used for human drinking water or contact recreation.	• Investigate current impact on sensitive habitat or surface water body, restrict access to area of discharge (if necessary), and evaluate the need for containment/control measures.
• Shallow contaminated surface soils are open to public access, and dwellings, parks, playgrounds, day-care centers, schools, or similar-use facilities are more than 500 ft (152 m) of those soils.	• Restrict access to impact soils.

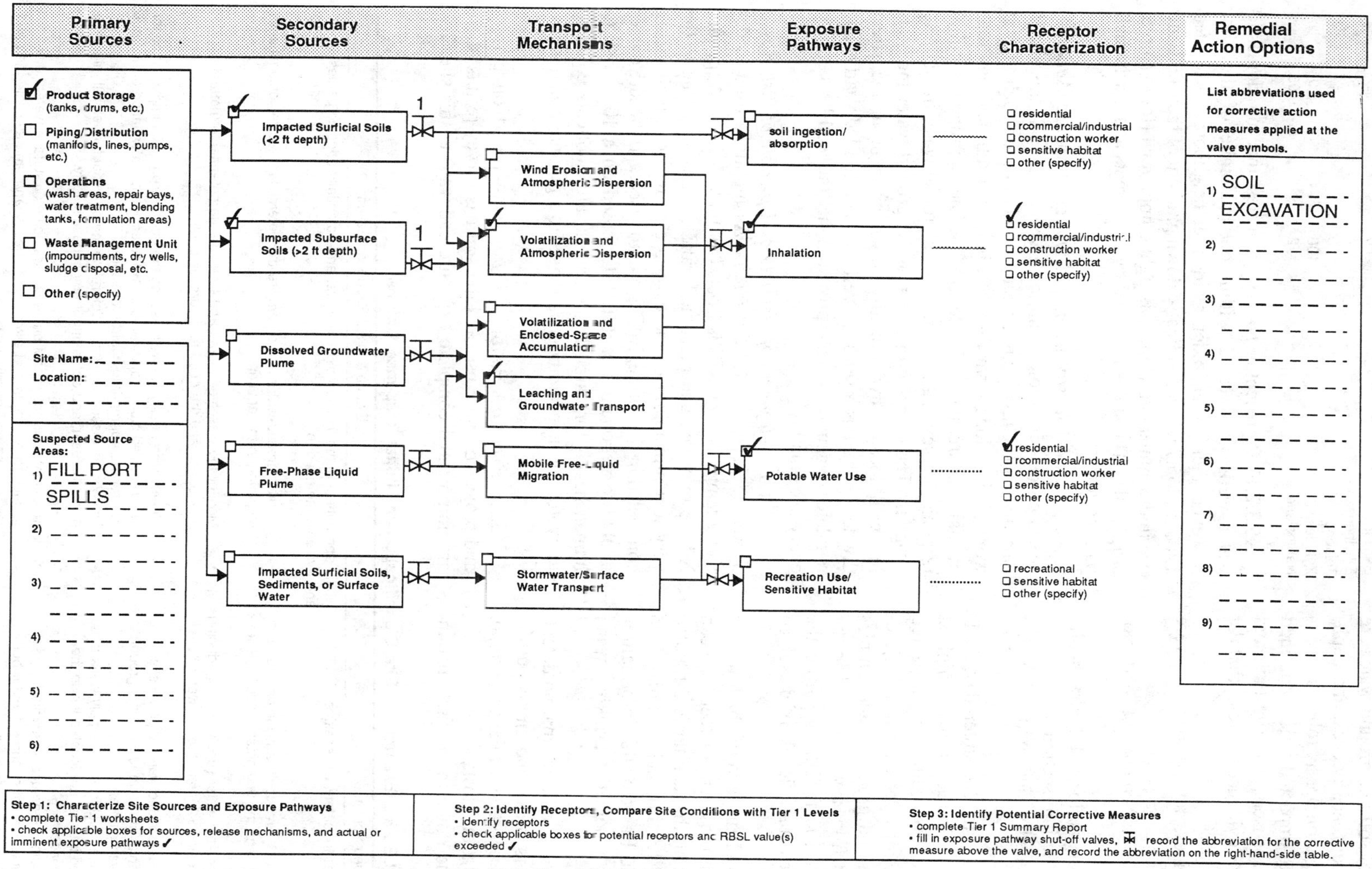

FIG. X5.1 Example 1—Exposure Evaluation Flowchart

site assessment focussed on potential source areas (for example, tanks, lines, dispensers) and receptors. Based on historical knowledge that gasoline and diesel have been dispensed at this facility, chemical analyses of soil and ground water are limited to benzene, toluene, ethylbenzene, xylenes, and naphthalene. Results of the site investigation are as follows:

X5.3.2.1 The extent of petroleum-impacted soils is confined to the vicinity of the tanks and dispensers. A recent tank and line test revealed no leaks; therefore, evidence suggests that the releases occurred sometime in the past,

X5.3.2.2 The current tanks, lines, and dispensers were installed three years ago,

X5.3.2.3 The asphalt driveway is competent and not cracked,

X5.3.2.4 Another service station is located hydraulically down gradient, diagonally across the intersection,

X5.3.2.5 The site is underlain by silty sands with a few thin discontinuous clay layers,

X5.3.2.6 Ground water, which is first encountered at 32 ft (9.7 m) below ground surface, is impacted, with highest dissolved concentrations observed beneath the suspected source areas. Dissolved concentrations decrease in all directions away from the source areas, and ground water samples taken hydraulically down gradient from a well located in the center divider of the street (about 100 ft (30.4 m) from the source area) do not contain any detectable levels of dissolved hydrocarbons,

X5.3.2.7 Ground water flow gradient is very shallow, and ground water flow velocities are at most tens of feet per year,

X5.3.2.8 Ground water yield from this aquifer is estimated to be in excess of 5 gal/min (18.9 L/min), and total dissolved solids levels are less than 700 mg/L. Based on this information, this aquifer is considered to be a potential drinking water supply,

X5.3.2.9 A shallow soil gas survey indicates that no detectable levels of hydrocarbon vapors are found in the utility easement running along the southern border of the property, or in soils surrounding the service station kiosk,

X5.3.2.10 Impacted soils extend down to the first encountered ground water. Maximum concentrations detected in soil and ground water are as follows:

Compound	Soil, mg/kg	Ground water, mg/L
Benzene	20	2
Ethylbenzene	4	0.5
Toluene	120	5
Xylenes	100	5.0
Napthalene	2	0.05

X5.3.2.11 A receptor survey indicates that no domestic water wells are located within one-half mile of the site; however, there is an older residential neighborhood located 1200 ft (365.7 m) hydraulically down gradient of the site. Land use in the immediate vicinity is light commercial (for example, strip malls). The site is bordered by two streets and a strip mall parking lot.

X5.3.3 *Site Classification and Initial Response Action*—Based on classification scenarios given in Table 1, this site is classified as a Class 3 site because conditions are such that, at worst, it is a long-term threat to human health and environmental resources (see Table X5.2). The appropriate initial response is to evaluate the need for a ground water monitoring program. The owner proposes that the ground water monitoring well located hydraulically down gradient in the street divider be used as a sentinel well, and be sampled yearly. The regulatory agency concurs, provided that the well be sampled every six months.

X5.3.4 *Development of Tier 1 Look-Up Table of Risk-Based Screening Level (RBSL) Selection*—Assumptions used to derive example Tier 1 RBSL Look-Up Table X2.1 in Appendix X2 are reviewed and presumed valid for this site. Due to the very low probability of the exposure pathway actually being completed in the future, MCLs are not used and the site owner is able to negotiate Tier 1 RBSLs based on a 10^{-5} risk to human health for carcinogens and hazard quotients equal to unity for the noncarcinogens (based on ground water ingestion).

X5.3.5 *Exposure Pathway Evaluation*—Based on current and projected future use, and the soil gas survey results, there are no potential complete exposure pathways at this site. The

TABLE X5.2 Example 2—Site Classification and Initial Response Actions

Criteria and Prescribed Scenarios	Example Initial Response Actions
3. Long-term (>2 years) threat to human health, safety, or sensitive environmental receptors	Notify appropriate authorities, property owners, and potentially affected parties, and evaluate the need to
• Subsurface soils (>3 ft (0.9 m) BGS) are significantly impacted, and the depth between impacted soils and the first potable aquifer is less than 50 ft (15 m).	• Monitor ground water and determine the potential for future contaminant migration to the aquifer.
• Ground water is impacted, and potable water supply wells producing from the impacted interval are located >2 years ground water travel time from the dissolved plume.	• Monitor the dissolved plume and evaluate the potential for natural attenuation and the need for hydraulic control.
• Ground water is impacted, and non-potable water supply wells producing from the impacted interval are located >2 years ground water travel time from the dissolved plume.	• Identify water usage of well, assess the effect of potential impact, monitor the dissolved plume, and evaluate whether natural attenuation or hydraulic control are appropriate control measures.
• Ground water is impacted, and non-potable water supply wells that do not produce from the impacted interval are located within the known extent of chemical(s) of concern.	• Monitor the dissolved plume, determine the potential for vertical migration, notify the user, and determine if any impact is likely.
• Impacted surface water, storm water, or ground water discharges within 1500 ft (457 m) of a sensitive habitat or surface water body used for human drinking water or contact recreation.	• Investigate current impact on sensitive habitat or surface water body, restrict access to area of discharge (if necessary), and evaluate the need for containment/control measures.
• Shallow contaminated surface soils are open to public access, and dwellings, parks, playgrounds, day-care centers, schools, or similar-use facilities are more than 500 ft (152 m) of those soils.	• Restrict access to impact soils.

down gradient residential neighborhood is connected to a public water supply system, and there is no local use of the impacted aquifer. However, being concerned about future uncontrolled use of the aquifer, the regulatory agency requests that the owner evaluate the ground water transport to residential drinking water ingestion pathway, recognizing that there is a low potential for this to occur (see Fig. X5.2).

X5.3.6 *Comparison of Site Conditions With Tier 1 RBSLs*—Based on the data given in X5.3.2.10 and the RBSLs given in example Look-Up Table X2.1 in Appendix X2, exceedences of Tier 1 soil and ground water RBSLs are noted only for benzene.

X5.3.7 *Evaluation of Tier 1 Results*—The responsible party decides to proceed to a Tier 2 evaluation for benzene and the pathway of concern, rather than devise a corrective action plan to meet Tier 1 standards after considering the following factors:

X5.3.7.1 The shallow aquifer is impacted, but the dissolved plume appears to be stable and ground water movement is very slow,

X5.3.7.2 Excavation of soils to meet Tier 1 criteria would be expensive, due to the depth of impacted soils. Excavation would shut down the facility, and require all tanks and new lines to be removed and reinstalled,

X5.3.7.3 Costs for application of other conventional treatment methods, such as vapor extraction and pump and treat, are estimated to exceed $300 000 over the life of the remediation, and

X5.3.7.4 A tier 2 analysis for this site is estimated to require minimal additional data, and is anticipated to result in equally protective, but less costly corrective action.

X5.3.8 *Tier 2 Evaluation*—The owner collects additional ground water monitoring data and verifies that:

X5.3.8.1 No mobile free-phase product is present,

X5.3.8.2 The dissolved plume is stable and ground water concentrations appear to be decreasing with time,

X5.3.8.3 Extent of the dissolved plume is limited to within 50 ft (15.2 m) of the property boundaries,

X5.3.8.4 Dissolved oxygen concentrations are higher outside of the dissolved plume, indicating some level of aerobic biodegradation,

X5.3.8.5 Ground water movement is less than 50 ft/year (15.2 m), and

X5.3.8.6 Simple ground water transport modeling indicates that observations are consistent with expectations for the site conditions.

X5.3.9 *Remedial Action Evaluation*—Based on the demonstration of dissolved plume attenuation with distance, the owner negotiates a corrective action plan based on the following: (*1*) compliance with the Tier 1 RBSLs at the monitoring well located in the street center divider, provided that deed restrictions are enacted to prevent the use of ground water within that zone until dissolved levels decrease below drinking water MCLs, (*2*) deed restrictions are enacted to ensure that site land use will not change significantly, (*3*) continued sampling of the sentinel/compliance ground water monitoring well on a yearly basis, (*4*) should levels exceed Tier 1 RBSLs at that point for any time in the future, the corrective action plan will have to be revised, and (*5*) closure will be granted if dissolved conditions remain stable or decrease for the next two years.

X5.4 *Example 3—RBCA With Emergency Response and In Situ Remediation:*

X5.4.1 *Scenario*—A 5 000-gal (18 925-L) release of super unleaded gasoline occurs from a single-walled tank after repeated manual gaging with a gage stick. Soils are sandy at this site, ground water is shallow, and free-product is observed in a nearby monitoring well within 24 h. The site is located next to an apartment building that has a basement where coin-operated washers and dryers are located for use by the tenants.

X5.4.2 *Site Assessment*— In this case the initial site assessment is conducted rapidly and is focussed towards identifying if immediately hazardous conditions exist. It is known from local geological assessments that the first encountered ground water is not potable, as it is only about 2 ft (0.6 m) thick and is perched on a clay aquitard. Ground water monitoring wells in the area (from previous assessment work) are periodically inspected for the appearance of floating product, and vapor concentrations in the on-site utility corridors are analyzed with an explosimeter. While this flurry of activity begins, a tenant of the apartment building next door informs the station operator that her laundry room/basement has a strong gasoline odor. Explosimeter readings indicate vapor concentrations are still lower than explosive levels, but the investigation team notes that "strong gasoline odors" are present.

X5.4.3 *Site Classification and Initial Response Action*—This limited information is sufficient to classify this site as a Class 2 site (strong potential for conditions to escalate to immediately hazardous conditions in the short term), based on the observed vapor concentrations, size of the release, and geological conditions (see Table X5.3). The initial response implemented is as follows:

X5.4.3.1 Periodic monitoring of the apartment basement begins to ensure that levels do not increase to the point where evacuation is necessary (either due to explosion or acute health effects). In addition, the fire marshall is notified and building tenants are informed of the activities at the site, potential hazards, and abatement measures being implemented,

X5.4.3.2 A free-product recovery/hydraulic control system is installed to prevent further migration of the mobile liquid gasoline, and

X5.4.3.3 A subsurface vapor extraction system is installed to prevent vapor intrusion to the building.

X5.4.4 *Development of Tier 1 Look-Up Table of Risk-Based Screening Level (RBSL) Selection*—Assumptions used to derive example Tier 1 RBSL Look-Up Table X2.1 in Appendix X2 are reviewed and presumed valid for this site. Target soil and ground water concentrations are determined based on the vapor intrusion scenario. After considering health-based, OSHA PEL, national ambient background, and aesthetic vapor concentrations, target soil levels are based on achieving a 10^{-4} chronic inhalation risk for benzene, and hazard quotients of unity for all other compounds. The agency agrees to base compliance on the volatile monoaromatic compounds in gasoline (benzene, toluene, xylenes, and ethylbenzene), but reserves the right to alter the target levels if aesthetic effects

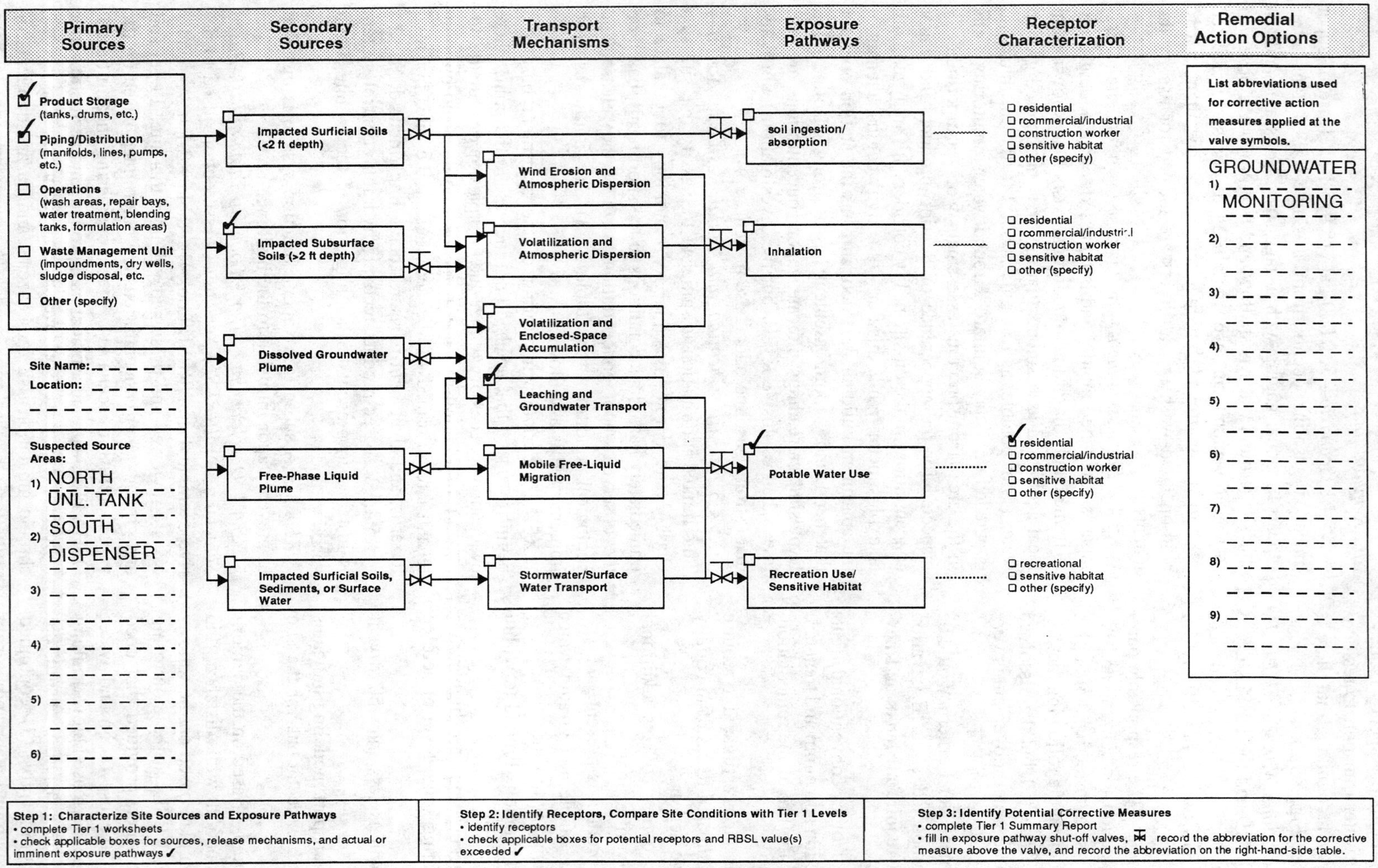

FIG. X5.2 Example 2—Exposure Evaluation Flowchart

TABLE X5.3 Example 3—Site Classification and Initial Response Actions

Criteria and Prescribed Scenarios	Example Initial Response Actions
2. Short-term (0 to 2 years) threat to human health, safety, or sensitive environmental receptors	Notify appropriate authorities, property owners, and potentially affected parties, and evaluate the need to
• There is potential for explosive levels, or concentrations of vapors that could cause acute effects, to accumulate in a residence or other building.	• Assess the potential for vapor migration (through monitoring/ modeling) and remove source (if necessary), or install vapor migration barrier.
• Shallow contaminated surface soils are open to public access, and dwellings, parks, playgrounds, day-care centers, schools, or similar use facilities are within 500 ft (152 m) of those soils.	• Remove soils, cover soils, or restrict access.
• A non-potable water supply well is impacted or immediately threatened.	• Notify owner/user and evaluate the need to install point-of-use water treatment, hydraulic control, or alternate water supply.
• Ground water is impacted, and a public or domestic water supply well producing from the impacted aquifer is located within two-years projected ground water travel distance down gradient of the known extent of chemical(s) of concern.	• Institute monitoring and then evaluate if natural attenuation is sufficient, or if hydraulic control is required.
• Ground water is impacted, and a public or domestic water supply well producing from a different interval is located within the known extent of chemicals of concern.	• Monitor ground water well quality and evaluate if control is necessary to prevent vertical migration to the supply well.
• Impacted surface water, storm water, or ground water discharges within 500 ft (152 m) of a sensitive habitat or surface water body used for human drinking water or contact recreation.	• Institute containment measures, restrict access to areas near discharge, and evaluate the magnitude and impact of the discharge.

persist in the building basement at the negotiated levels.

X5.4.5 *Exposure Pathway Evaluation*—Given that: (*1*) there is a very low potential for ground water usage, (*2*) a 20-ft (6.1-m) thick aquitard separates the upper perched water from any potential drinking water supplies, and (*3*) the close proximity of the apartment building, the owner proposes focusing on the vapor intrusion—residential inhalation scenario (see Fig. X5.3). The agency concurs, but in order to eliminate potential ground water users as receptors of concern, requests that a down-gradient piezometer be installed in the lower aquifer. The owner concurs.

X5.4.6 *Comparison of Site Conditions With Tier 1 RBSLs*—While a complete initial site investigation has yet to be conducted, all parties agree that currently the RBSLs are likely to be exceeded.

X5.4.7 *Evaluation of Tier 1 Results*—The owner decides to implement an interim corrective action plan based on Tier 1 RBSLs, but reserves the right to propose a Tier 2 evaluation in the future.

X5.4.8 *Tier 1 Remedial Action Evaluation*—The owner proposes expanding the vapor extraction system to remediate source area soils. In addition he proposes continuing to operate the free-product recovery/hydraulic control system until product recovery ceases. Monitoring of the piezometer placed in the lower aquifer will continue, as well as periodic monitoring of the apartment building basement. Additional assessments will be conducted to ensure that building vapors are not the result of other sources. After some period of operation, when hydrocarbon removal rates decline, a soil and ground water assessment plan will be instituted to collect data to support a Tier 2 evaluation.

X5.5 *Example 4—RBCA Based on Use of a Tier 2 Table Evaluation*—In circumstances where site specific data are similar among several sites, a table of Tier 2 SSTL values can be created. The following example uses such a table.

X5.5.1 *Scenario*—Petroleum-impacted ground water is discovered in monitoring wells at a former service station. The underground tanks and piping were removed, and the site is ow occupied by an auto repair shop.

X5.5.2 *Site Assessment*— The responsible party completes an initial site assessment to determine the extent of hydrocarbon-impacted soil and ground water. Because gasoline was the only fuel dispensed at the site, the assessment focussed on benzene, toluene, ethylene benzene, and xylenes (BTEX) as the chemicals of concern. Site assessment results are summarized as follows:

X5.5.2.1 The area of hydrocarbon-impacted soil is approximately 18 000 ft^2 (1672 m^2) and the depth of soil impaction is less than 5 ft (1.5 m); The plume is off site,

X5.5.2.2 The site is covered by asphalt or concrete,

X5.5.2.3 The site is underlain by clay,

X5.5.2.4 Hydrocarbon-impacted perched ground water is encountered at 1 to 3 ft (0.3 to 0.9 m) below grade. This water is non-potable. The first potable aquifer is located over 100 ft (30 m) below grade and is not impacted. There is no free product,

X5.5.2.5 Maximum detected concentrations are as follows:

Compound	Soil, mg/kg	Ground water, mg/L
Benzene	39	1.8
Toluene	15	4.0
Ethylbenzene	12	0.5
Xylenes	140	9.0

X5.5.2.6 Ground water velocity is 0.008 ft/day (0.0024 m/day) based on slug tests and ground water elevation survey and assumed soil porosity of 50 %,

X5.5.2.7 A receptor survey indicates that the nearest down gradient water well is greater than 1.0 mile (1.6 km) away and the nearest surface water body is 0.5 miles (0.8 km). The distance to the nearest sensitive habitat is greater than 1.0 mile; however, there is a forest preserve frequented by day hikers and picnickers next to the site. The nearest home is 1000 ft (305 m) away. The commercial building on site is 25 ft (7.6 m) from the area of hydrocarbon-impacted soil.

X5.5.3 *Site Classification and Initial Response Action*—Based on the classification scenarios given in Table 1, this site is classified as a Class 4 site, with no demonstrable long-term threat to human health, safety, or sensitive environmental receptors, because the hydrocarbon-impacted soils are covered

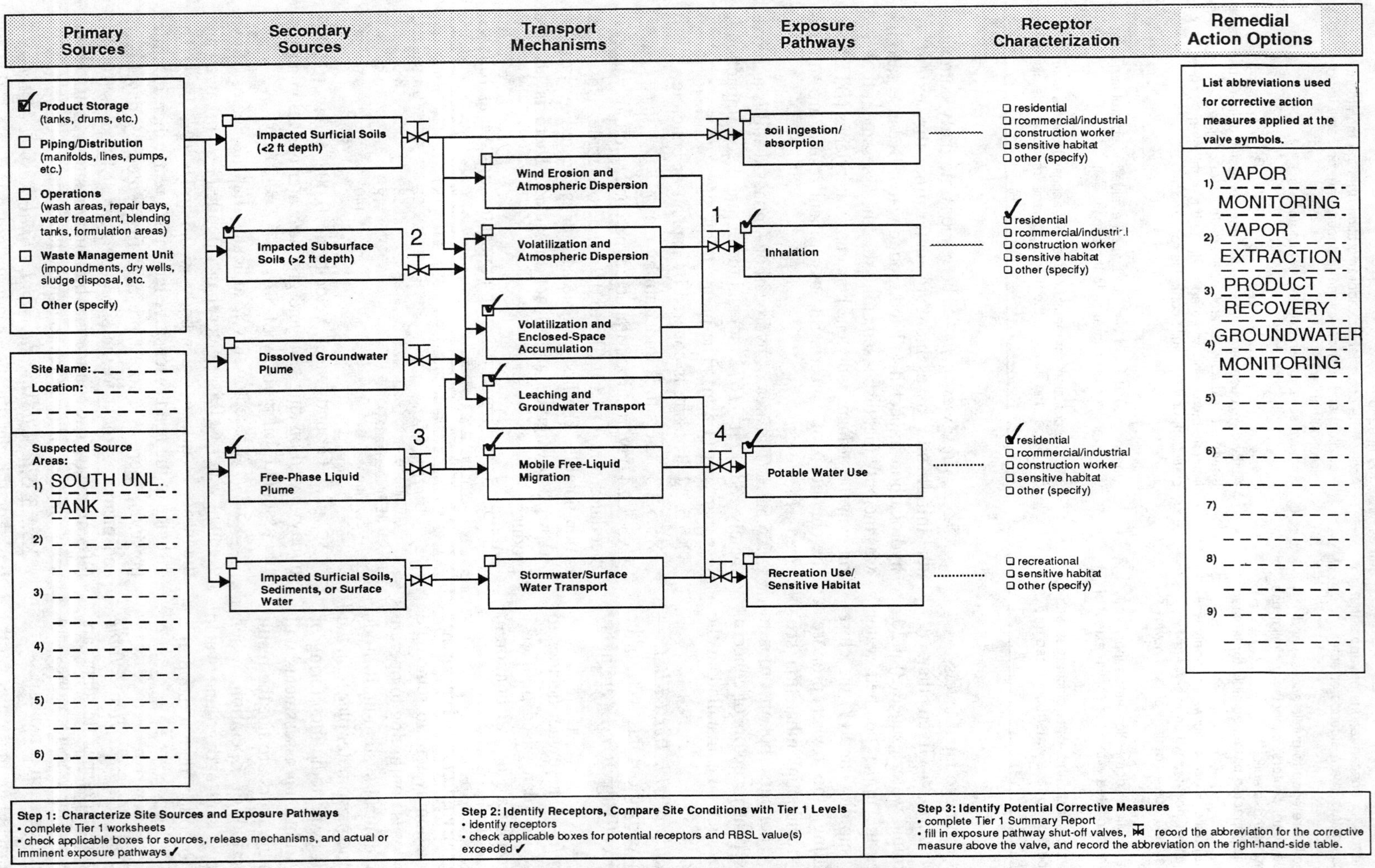

FIG. X5.3 Example 3—Exposure Evaluation Flowchart

by asphalt or concrete and cannot be contacted, only non-potable perched water with no existing local use is impacted, and there is no potential for explosive levels or concentrations that could cause acute effects in nearby buildings. The appropriate initial response is to evaluate the need for a ground water monitoring program.

X5.5.4 *Development of Tier 1 Look-Up Table of Risked-Based Screening Level (RBSL)*—The assumptions used to derive the example Tier 1 RBSL Look-Up Table are presumed valid for this site.

X5.5.5 *Exposure Pathway Evaluation*—The complete pathways are ground water and soil volatilization to enclosed spaces and to ambient air, and direct exposure to impacted soil or ground water by construction workers. A comparison of RBSLs for these pathways of concern indicates that RBSLs associated with soil volatilization to an enclosed space are the most restrictive RBSLs.

X5.5.6 *Comparison of Site Conditions with Tier 1 RBSLs*—Based on the data given in X5.5.2 and the RBSLs given in Table X2.1, exceedances of Tier 1 RBSLs are noted for benzene in soil and ground water and toluene for ground water.

X5.5.7 *Evaluation of Tier 1 Results*—The responsible party decided to proceed to a Tier 2 evaluation for the pathways of concern rather than develop a corrective action plan for the following reasons:

X5.5.7.1 Only shallow perched water is impacted, and the dissolved plume is moving very slowly in tight clay,

X5.5.7.2 Excavation of soils to meet Tier 1 criteria would be expensive and would disrupt activities of the on-site business. Off-site excavation would be impractical and may not be able to clean up ground water to Tier 1 criteria,

X5.5.7.3 Other conventional treatment methods, such as pump and treat and vapor extraction, would be relatively ineffective in the heavy clay, and

X5.5.7.4 A Tier 2 evaluation for this site requires no additional data and is expected to be an equally protective but less costly corrective action.

X5.5.8 *Development of a Tier 2 Table of Site-Specific Target Levels (SSTLs)*—The Tier 2 table is similar to the Tier 1 Look-Up Table with the exception that SSTLs for the pathways of concern are presented as functions of both the distance from the source to the receptor and the soil type.

X5.5.8.1 For the pathways considered, approaches for the Tier 2 table are consistent with guidelines contained in Ref **(26)**.

X5.5.8.2 The equations, assumptions, and parameters used to construct the Tier 1 Look-Up Table and Tier 2 table are similar, except as noted as follows:

(1) Ground Water: Ingestion of Ground Water—A one-dimensional analytical mass balance equation with attenuation mechanisms of retardation, dispersivity, and first-order biological decay (in sandy soil only) was applied in conjunction with the equations in Tables X2.2 and X2.3 to calculate SSTLs. The analytical model is limited to steady-state conditions and longitudinal dispersion. The analytical solution to the mass balance equation is presented in Ref **(44)**.

(2) Ground Water: Inhalation of Outdoor Vapors—This pathway was not considered because exposure concentrations were very low.

(3) Ground Water: Inhalation of Enclosed-Space (Indoor) Vapors—A one-dimensional mass balance equation following Jury, et al **(31)** has been used to model vapor transport **(43)**. This model was used in conjunction with the equations in Tables X2.2 and X2.3 to calculate SSTLs. The model includes concentration attenuation between the source and the building by partitioning into immobile pore water, adsorption onto soil, and biological degradation (in sandy soil only).

(4) Subsurface Soils: Inhalation of Outdoor Vapors—This pathway was not considered because exposure concentrations were very low.

(5) Subsurface Soils: Inhalation of Enclosed-Space (Indoor) Vapors—The SSTLs were calculated using the Jury model **(31)** as discussed in Paragraph (*3*) of X5.5.8.2.

(6) Subsurface Soils: Leaching to Ground Water—The SSTLs were calculated using the one-dimensional mass-balance equation described in Paragraph (*1*) of X5.5.8.2, in conjunction with the lechate factor, LF_{SW}, as discussed in X2.9.4.1.

(7) All exposure parameter values listed in Table X2.4, soil, building surface, and subsurface parameter values listed in Table X2.6, and chemical-specific properties listed in Table X2.7 have not been changed.

(8) First-order decay rates in sandy soil were assumed to be 0.2 % per day for all BTEX compounds. These rates are considered conservative. Chiang, et al **(38)** determined that a DO of 2.0 mg/L is required for rapid and complete biodegradation of benzene. Chiang, et al **(38)** measured a biodegradation rate of 0.95 % per day, and Barker, et al **(36)** measured a biodegradation rate of 0.6 % per day for benzene. In general, published biodegradation rates range from 0.6 to 1.25 % per day. Chiang, et al **(38)** also determined that biodegradation rates may be slower and incomplete at DO concentrations below 2.0 mg/L. This is a conservative value since aerobic biodegradation continues at DO concentrations as low as 0.7 mg/L **(44)**.

(9) Clay properties are as follows:

Total soil porosity, cm^3/cm^3	0.05
Volumetric water content, cm^3/cm^3	0.40
Ground water Darcy velocity, cm/s	25

X5.5.8.3 Assumptions used to derive the example Tier 2 SSTL table are reviewed and presumed valid for this site. Due to the very conservative assumptions used to calculate exposure and the small number of people potentially exposed, the Tier 2 SSTLs are based on a 10^{-5} risk to human health for carcinogens and hazard quotients equal to unity for noncarcinogens.

X5.5.9 *Comparison of Site Conditions with Tier 2 Table SSTLs*—Based on the data given in X5.5.2 and the SSTLs given in the example of Table X5.4, no exceedances of Tier 2 soil or ground water SSTLs are noted.

X5.5.10 *Tier 2 Remedial Action Evaluation*—Based on the fact that Tier 2 soil or ground water SSTLs are not exceeded, the responsible party negotiates a corrective action plan based on the following:

X5.5.10.1 Annual compliance monitoring of ground water

TABLE X5.4 Example Tier 2 Site-Specific Target Level (SSTL) Table—Soil and Ground Water

	Exposure Pathway	Receptor Scenario	Distance to Source, ft (m)	SSTLs at Source Sandy Soil, Natural Biodegradation Carcinogenic Risk = 1×10^{-5}, HQ = 1				SSTLs at Source Clay Soil, No Natural Biodegradation Carcinogenic Risk = 1×10^{-5}, HQ = 1			
				Benzene	Ethylbenzene	Toluene	Xylene	Benzene	Ethylbenzene	Toluene	Xylene
Soil	Soil vapor intrusion from soil to buildings, mg/kg	residential	10 (3)	0.052	18	11	450	1.7	570	300	9500
			25 (7.6)	0.47	160	160	1.7[A]	65	11[A]	10[A]	RES[B]
			100 (30)	3.1[A]	RES	RES	RES	RES	RES	RES	RES
		commercial/ industrial	10 (3)	0.13	39	24	980	4.3	1200	650	2.0[A]
			25 (7.6)	1.2	340	340	3.6[A]	950	24[A]	22.5[A]	RES
			100 (30)	8.0[A]	RES	RES	RES	RES	RES	RES	RES
	Surficial soil ingestion and dermal, mg/kg	residential		22	5100	5400	280	22	5100	5400	280
		commercial/ industrial		120	9600	1.7[A]	1500	117	9600	1.7[A]	1500
	Soil lechate to protect ground water ingestion target level, mg/kg	residential	0 (0)	0.17	47	130	2200	0.17	47	130	2200
			100 (30)	0.32	88	250	4200	0.20	130	760	RES
			500 (152)	4.0	1200	6300	RES	RES	RES	RES	RES
		commercial/ industrial	0 (0)	0.58	130	350	6200	0.58	130	350	6200
			100 (30)	1.1	250	670	1.2[A]	0.70	380	2100	RES
			500 (152)	13	3300	1.75[A]	RES	RES	RES	RES	RES
Ground Water	Ground water ingestion, mg/L	residential	0	0.029	3.6	7.3	73	0.029	3.6	7.3	73
			100	0.054	6.8	14	140	0.035	10	43	>S[C]
			500	0.68	90	350	>S	>S	>S	>S	>S
		commercial/ industrial	0	0.099	10	20	200	0.099	10	20	200
			100	0.185	19	38	>S	0.12	29	120	>S
			500	2.3	250	>S	>S	>S	>S	>S	>S
	Ground water vapor intrusion from ground water to buildings, mg/L	residential	10	0.11	32	17	510	5.0	>S	>S	>S
			25	0.72	210	160	>S	1200	>S	>S	>S
			100	>S	>S	>S	>S	>S	>S	>S	>S
		commercial/ industrial	10	0.28	70	36	>S	13	>S	>S	>S
			25	1.9	>S	350	>S	>S	>S	>S	>S
			100	>S	>S	>S	>S	>S	>S	>S	>S

[A] Weight percent.

[B] RES—Selected risk level is not exceeded for pure compound present at any concentration.

[C] >S—Selected risk level is not exceeded for all possible dissolved levels.

at down gradient monitoring wells will be performed to demonstrate decreasing concentrations,

X5.5.10.2 Should levels exceed Tier 2 SSTLs at any of these monitoring points at any future time, the corrective action plan will be reevaluated, and

X5.5.10.3 Closure will be granted if dissolved concentrations remain stable or decrease for the next two years.

REFERENCES

(**1**) *Ecological Assessment of Hazardous Waste Sites: A Field and Laboratory Reference Document*, EPA/600/3-89/013, NTIS No. PB-89205967, Environmental Protection Agency, Washington, DC, March 1989.

(**2**) *Integrated Risk Information System (IRIS)*, Environmental Protection Agency, Washington, DC, October 1993.

(**3**) *Health Effects Assessment Summary Tables (HEAST)*, OSWER OS-230, Environmental Protection Agency, Washington, DC, March 1992.

(**4**) *Superfund Public Health Evaluation Manuals*, NTIS No. PB87-183125, Environmental Protection Agency, Washington, DC, October 1986.

(**5**) *Health Effects Assessment Summary Tables (HEAST)*, USEPA/OERR 9200.6-303(91.1), NTIS No. PB91-921199, Environmental Protection Agency, Washington, DC, January 1991.

(**6**) *Technical Basis and Background for Cleanup Standards for Contaminated Sites*, New Jersey, 1993.

(**7**) Verschueren, K., *Handbook of Environmental Data on Organic Chemicals*, 2nd Edition, Van Nostrand Reinhold Co., Inc., New York, NY, 1983.

(**8**) *CHEM-BANK, Hazardous Chemical Databank on Compact Disk—HSDB*, U.S. National Library of Medicine.

(**9**) *Information Review Tert-Butyl Methyl Ether*, EPA Contract No. 68-01-6650, CRCS., Inc., Environmental Protection Agency, Washington, DC, March 1986.

(**10**) Dragun, J., *The Soil Chemistry of Hazardous Materials*, Hazardous Materials Control Research Institute, Silver Springs, MD, 1988.

(**11**) Lyman, W. J., Reehl, W. F., Rosenblatt, D. H., *Handbook of Chemical Property Estimation Methods*, McGraw Hill, New York, NY, 1982.

(**12**) Paustenbach, D. J., Jernigan, J. D., Bass, R. Kalmes, R., and Scott, P., "A Proposed Approach to Regulating Contaminated Soil: Identify Safe Concentrations for Seven of the Most Frequently Encountered Exposure Scenarios," *Regulatory Toxicology and Pharmacology*, Vol 16, 1992, pp. 21–56.

(**13**) Young, F. A., "Risk Assessment: The Convergence of Science and Law," *Regulatory Toxicology and Pharmacology*, Vol 7, 1987, pp. 179–184.

(**14**) Travis, C. C., Richter, S. A., Crouch, E. A., Wilson, R., and Wilson, E., "Cancer Risk Management: A Review of 132 Federal Regulatory Decisions," *Environmental Science and Technology*, Vol 21, No. 5, 1987, pp 415–420.

(15) Rodricks, J. V., Brett, S. M., and Wrenn, G. C., "Significant Risk Decisions in Federal Regulatory Agencies," *Toxicology Pharmacology*, Vol 7, 1987, pp. 307–320.
(16) *National Oil and Hazardous Substances Pollution Contingency Plan*, 40 CRF Part 300, Environmental Protection Agency, Washington, DC, 1990.
(17) *Hazardous Waste Management System Toxicity Characteristics Revisions* 55 FR 11798-11863, Environmental Protection Agency, Washington, DC.
(18) Personal communications, State Agencies, 1990.
(19) *Policy for Identifying and Assessing the Health Risks of Toxic Substances*, Environmental Toxicology Program, Division of Disease Control, Bureau of Health, Maine Department of Human Services (DHS), February 1988.
(20) *Draft Interim Guidance for Disposal Site Risk Characterization—In Support of the Massachusetts Contingency Plan*, Office of Research and Standards, Massachusetts Department of Environmental Quality Engineering (DEQE), October 1988.
(21) *Safe Drinking Water and Toxic Enforcement Act of 1986 (Proposition 65)*, Health and Welfare Agency, Office of the Secretary, Sacramento, CA, 1986.
(22) *Odor Thresholds for Chemicals with Established Occupational Health Standards*, American Industrial Hygiene Association, 1989.
(23) Shah and Singh, *Environmental Science and Technology*, Vol 22, No. 12, 1988.
(24) *Toxicological Profiles*, ATSDR, U.S. Public Health Services, 1988.
(25) Wallace, L. A., *Journal of Occupational Medicine*, Vol 28, No. 5, 1986.
(26) *Risk Assessment Guidance for Superfund*, Vol 1, *Human Health Evaluation Manual, Part A*, EPA/540/1-89/002, Environmental Protection Agency, Washington, DC, December 1989.
(27) *Exposure Factors Handbook*, EPA 600/8-89/043, Environmental Protection Agency, Washington, DC, July 1989.
(28) *Supplemental Risk Assessment Guidance for the Superfund Program*, EPA/901/5-89/001, Environmental Protection Agency Region I, Washington, DC, 1989.
(29) Johnson, P. C., and Ettinger, R. A., "Heuristic Model for Predicting the Intrusion Rate of Contaminant Vapors into Buildings," *Environmental Science and Technology*, Vol 25, No. 8, 1991, pp. 1445–1452.
(30) *Superfund Exposure Assessment Manual*, EPA/540/1-88/001, Environmental Protection Agency, Washington, DC, 1988.
(31) Jury, W. A., Spencer, W. F., and Farmer, W. J., "Behavior Assessment Model for Trace Organics in Soil: I, Model Description," *Journal of Environmental Quality*, Vol 12, 1983, pp. 558–564.
(32) Cowherd, C., Muleski, G. E., Englehart, P. J., and Gillett, D. A., *Rapid Assessment of Exposure to Partulate Emissions from Surface Contamination Sites*, Midwest Research Institute, PB85-192219, 1985.
(33) Johnson, P., Hertz, M. B., and Byers, D. I., "Estimates for Hydrocarbon Vapor Emissions Resulting from Service Stations Remediations and Buried Gasoline-Contaminated Soils," *Petroleum Contaminated Soils*, Vol III, Kostecki, P. T., and Calabrese, E. J., eds., Lewis Publishers, Chelsea, MI, 1990, pp. 295–326.
(34) Mullens, M., and Rogers, T., *AIECHE/DIPPR Environmental, Safety, and Health Data*, Design Institute for Physical Property Research—Research Project 911, American Institute for Chemical Engineers, June 1, 1993.
(35) *Hazardous Waste Treatment, Storage, and Disposal Facilities (TSDF)*, OAQPS, Air Emissions Models, EPA/450/3-87/026, Environmental Protection Agency, Washington, DC, 1989.
(36) Barker, J. F., Patrick, G. C., and Major, D., "Natural Attenuation of Aromatic Hydrocarbons in a Shallow Sand Aquifer," *Ground Water Monitoring Review*, Vol 7, 1987, pp. 64–71.
(37) Kemblowski, M. W., Salanitro, J. P., Deeley, G. M., and Stanley, C. C., "Fate and Transport of Residual Hydrocarbons in Ground Water: A Case Study," *Proceedings of the Petroleum Hydrocarbons and Organic Chemicals in Ground Water Conference*, National Well Water Association and American Petroleum Institute, Houston, TX, 1987, pp. 207–231.
(38) Chiang, C. Y., Salanitro, J. P., Chai, E. Y., Colthart, J. D., and Klein, C. L., "Aerobic Biodegradation of Benzene, Toluene, and Xylene in a Sandy Aquifer—Data Analysis and Computer Modeling," *Ground Water*, Vol 27, No. 6, 1989, pp. 823–834.
(39) Wilson, B. H., Wilson, J. T., Kampbell, D. H., Bledsoe, B. E., and Armstrong, J. M., "Biotransformation of Monoaromatic and Chlorinated Hydrocarbons at an Aviation Gasoline Spill Site," *Geomicrobiology Journal*, Vol 8, 1991, pp. 225–240.
(40) Howard, P., et al, *Handbook of Environmental Degradation Rates*, Lewis Publishers Inc., Chelsea, MI, 1991.
(41) Chevron Research and Technology Company, "Evaluation of Intrinsic Bioremediation at Field Sites," *Proceedings of the 1993 Petroleum Hydrocarbons and Organic Chemicals in Ground Water: Prevention, Retention, and Restoration*, Westing Galleria, Houston, TX, Nov. 10–12, 1993.
(42) Van Genuchten and Alves, *Analytical Solutions of the One-Dimensional Convective-Dispersive Solute Transport Equation* , Technical Bulletin No. 1661, U.S. Department of Agriculture, 1982.
(43) Jeng, C. Y., Kremesec, V. J., Primack, H. S., and Olson, C. B. (Amoco Oil Company), "Predicting the Risk in Buildings Posed by Vapor Transport of Hydrocarbon Contaminants," *Hydrocarbon Contaminated Soil and Ground Water, Proceedings for the 5th West Coast Conference: Contaminated Soils and Ground Water*, Vol 5, Association for Environmental Health of Soils, 1994.
(44) Wilson, J. T., "Natural Bioattenuation of Hazardous Organic Compounds in the Subsurface," R. S. Kerr Laboratory, Environmental Protection Agency, Draft Paper, 1993.

Standard Guide for Collection of Multi-Media Field Emission and Discharge Data Associated with Glycol Dehydration Units[1]

This standard is issued under the fixed designation E 1752; the number immediately following the designation indicates the year of original adoption or, in the case of revision, the year of last revision. A number in parentheses indicates the year of last reapproval. A superscript epsilon (ϵ) indicates an editorial change since the last revision or reapproval.

ϵ^1 NOTE—Subsections 6.3.4 and 7.2.5 were corrected editorially in November 2000.

1. Scope

1.1 *Purpose*—This guide covers the proper collection of field emission and discharge data associated with glycol dehydration units used in the natural gas production, processing, transmission, storage, and distribution industries.

1.2 *Background*:

1.2.1 Increasing regulatory pressure has made emissions of benzene, toluene, ethylbenzene, and xylene isomers (collectively known as BTEX) and volatile organic compounds (VOCs) from the still vent of glycol dehydration units a major concern of the natural gas industry. The Clean Air Act Amendments (CAAA) of 1990 have been the impetus for air toxics regulations, and several states are regulating or are considering regulating emissions from glycol units (**1**).[2] Liquid and solid waste discharges are exempt from Subtitle C (hazardous waste) regulation under the Resource Conservation and Recovery Act (RCRA), but may be regulated in the future (**2**).

1.2.2 Measurement of the waste streams from dehydrators is important to determine which units may have emissions above levels of regulatory concern. Measurements of air emissions from glycol dehydration units have been made from a variety of sampling points using different sampling protocols and analytical techniques since no standard methods have been developed by the United States Environmental Protection Agency (USEPA) or state regulatory agencies. Standard sampling methods do not exist for the liquid and solid waste streams since they are exempt from RCRA Subtitle C. The lack of standard protocols has meant that variations of this approach can result in very different emissions measurements (**3**).

1.2.3 Providing guidance on the collection of field emission and discharge data will allow the natural gas industry to quantify emissions and apply appropriate controls to comply with regulations.

1.3 *Summary*—This guide has several parts and an annex. Section 1 is Scope. Section 2 is Terminology that has definitions of terms commonly used with relation to glycol dehydration units in the natural gas industry. Section 3 is Significance and Use of this guide. Section 4 is a process description of glycol dehydration units. Section 5 is a discussion of the waste streams associated with glycol dehydrators. Section 6 presents the Approaches for Collecting Air Emission Data, while Sections 7 and 8 present the approaches for collecting liquid and solid waste discharge data, respectively. The annex includes a standard operating procedure (SOP) for the rich/lean glycol sampling method discussed in this guide.

1.4 The values stated in either inch-pound or SI units are to be regarded separately as the standard. The values given in parentheses are for information only.

1.5 *This standard does not purport to address all of the safety concerns, if any, associated with its use. It is the responsibility of the user of this standard to establish appropriate safety and health practices and determine the applicability of regulatory limitations prior to use.*

2. Terminology

2.1 *Definitions of Terms Specific to This Standard:*

2.1.1 *BTEX*—the compounds benzene, toluene, ethylbenzene, and xylene isomers.

2.1.2 *Clean Air Act Amendments (CAAA)*—the 1990 federal amendments to the Clean Air Act. These amendments address nonattainment areas, hazardous air pollutants, permitting, and enforcement.

2.1.3 *condensate*—water or light hydrocarbons that have changed from a vapor to a condensed liquid state.

2.1.4 *contactor (or absorber)*—a vertical pressure vessel where gas and glycol are intermingled countercurrently to remove water vapor from the gas.

2.1.5 *dehydration*—removal of water vapor from natural gas. Maximum water content of the dehydrated gas is normally 7 lbs (3175 g) H_2O/MMSCF.

2.1.6 *flash tank*—a two- or three-phase separator that is used in the rich glycol stream to remove entrained gas and hydrocarbon liquid.

2.1.7 *gas absorption*—an operation in which a gas mixture is contacted with a liquid for the purposes of preferentially dissolving one or more components of the gas and to provide a solution of these in the liquid.

---

[1] This guide is under the jurisdiction of ASTM Committee E50 on Environmental Assessment and is the direct responsibility of Subcommittee E50.04 on Performance Standards Related to Environmental Regulatory Programs.

Current edition approved Oct. 10, 1995. Published December 1995.

[2] The boldface numbers in parentheses refer to the list of references at the end of the text.

2.1.8 *gas/glycol heat exchanger*—a shell-and-tube, pipe-in-pipe, or other type heat exchanger employed to cool the lean glycol with the gas leaving the contactor before the glycol enters the contactor.

2.1.9 *glycol*—a liquid desiccant used to absorb water vapor from the gas. Triethylene glycol is the most common glycol used in natural gas dehydration.

2.1.10 *lean glycol*—glycol that has been regenerated and is essentially free of water.

2.1.11 *rich glycol*—glycol that has absorbed water.

2.1.12 *reboiler*—the vessel where water is boiled out of the glycol.

2.1.13 *resource conservation and recovery act (RCRA)*—a federal regulatory measure for monitoring liquid and solid waste discharges. Subtitle C of the RCRA, as amended by the 1984 Hazardous and Solid Waste Amendments (HSWA) regulations, provides EPA the statutory authority to adopt a complex set of regulations for identification and management of hazardous waste from "cradle to grave," that is, from the initial generation through the treatment, transportation, storage, and disposal of wastes.

2.1.14 *still column*—vertically mounted fractionation column on top of the reboiler.

3. Significance and Use

3.1 This guide on the proper collection of emission and discharge wastes from glycol dehydrators is applicable to any natural gas industry and supplier that operates glycol dehydration units and that needs to identify which glycol units may have emissions above regulatory levels.

3.2 The emission and discharge sampling methods discussed in this guide are not regulatory standards. Standard protocols have been developed by the Gas Research Institute **(3)** and other gas associations **(4)** and some state regulatory agencies such as the Louisiana Department of Environmental Quality (LDEQ) **(5)** and the Texas Natural Resource Conservation Commission (TNRCC) **(6)** are accepting these data. This guide is **not** intended to instruct the user on how to perform the sampling using these protocols, but to make the user aware of certain practical considerations generally associated with sampling these waste streams.

4. Process Description

4.1 *Introduction*— Glycol dehydration units are commonly used in the natural gas industry to remove water from natural gas streams to prevent corrosion and hydrate formation in pipelines. Up to 40 000 glycol dehydration units may be in operation in the United States **(7)**. TEG is commonly used as the dehydrant since it is more stable than other glycols (for example, ethylene glycol) at high temperatures. A brief description of a typical glycol dehydration unit is presented as follows.

4.2 Fig. 1 is a simplified flow diagram for a glycol dehydration unit. A glycol dehydration unit usually consists of an absorber, a flash tank, heat exchanger(s), filter(s), a glycol reboiler still, and associated pumping and piping equipment. The moist natural gas enters the bottom of the absorber, where it is contacted countercurrently with the cool, lean glycol that absorbs the water. The dry gas exits the top of the absorber.

4.3 The rich glycol (containing water) leaves the bottom of the absorber and often goes through a gas-driven balance pump to a flash tank, which is normally present only on larger units. The flask tank separates much of the natural gas captured under high pressure from the glycol; the flash gas may be used as a fuel source. After a series of heat exchangers and filters, the rich glycol enters the still and reboiler, where water is distilled and stripped from the glycol (some units inject a stripping gas to produce higher-purity glycol at normal reboiler temperatures). The lean glycol is pumped back to the top of the absorber from the surge tank.

4.4 Since some dehydrators may produce emissions that are above the regulatory limits even with optimum operating

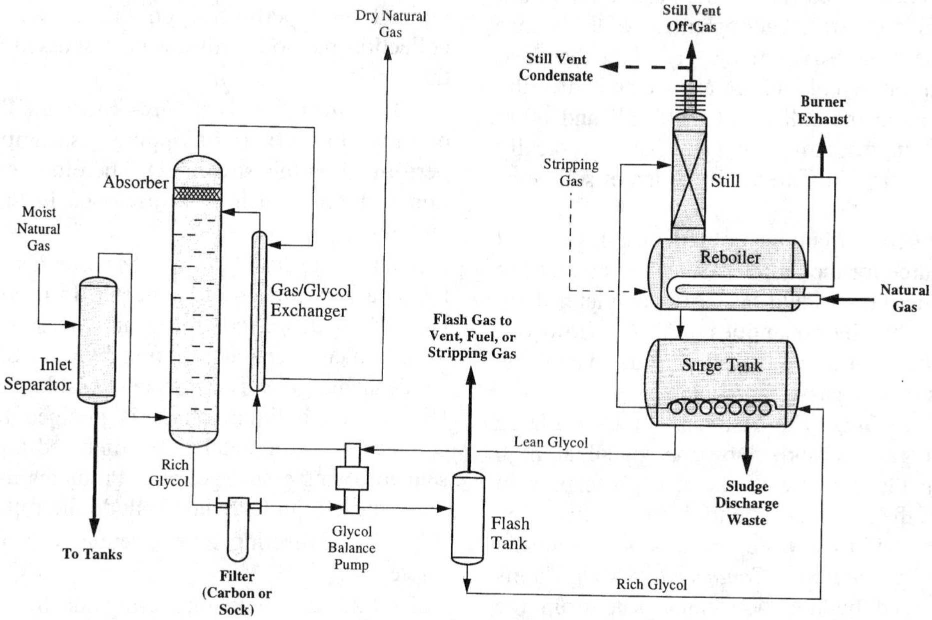

FIG. 1 Typical TEG Dehydrator Waste Streams

conditions, several technologies have been developed to control emissions from glycol dehydrators. The two most common methods for controlling emissions are condensation and combustion (incineration).

4.4.1 Combustion of the vent stream is difficult, however, due to the possibility of condensation of hydrocarbons and water in the flare. Condensation of the flare results in smoking and incomplete combustion of the vapors. Operating costs for flares and incinerators are relatively high because no hydrocarbons are recovered for sale, and supplemental fuel is often required to sustain an adequate flame temperature **(8)**.

4.4.2 Because of the difficulties associated with combustion, condensation has become the most widely used method of controlling emissions from glycol dehydrators. Condensers convert condensable components in the vapor phase to the liquid phase by reducing the temperature of the process stream. Condensers not only reduce emissions but also recover condensable hydrocarbon vapors that may increase production. The main types of condensers currently used by the gas industry are: air-cooled, glycol-cooled, water-cooled, and direct-contact quench condensers **(8)**.

5. Waste Streams

5.1 *Introduction*— Possible waste streams generated by glycol units are shown in Fig. 1. Seven of the waste streams occur on many glycol units: still vent gas, burner flue gas, flash gas, water condensate, waste glycol, reclaimer waste, and spent filters. These waste streams and associated sampling considerations are presented as follows.

5.2 *Waste Streams and Sources*:

5.2.1 *Still Vent Gas*— During the absorption process, the glycol, that has a high affinity for aromatic compounds, removes BTEX and, to a lesser extent, other VOCs from the natural gas. (The BTEX compounds are usually present in the natural gas at levels of less than 1000 parts per million by volume, although concentrations can be higher in some cases.) Since the boiling points of BTEX range from 80 to 140°C, most of these compounds are not lost in the flash tank but are boiled from the glycol in the still. Although many of the lighter hydrocarbons may be removed from the glycol in the flash tank, some remain in the glycol and are also separated in the still. These separations in the still result in BTEX and other VOCs being present in the still vent gas, that is typically released to the atmosphere. This stream is the major source of air emissions.

5.2.2 *Burner Flue Gas*— Many glycol units use a gas-fired burner as the heat source for the reboiler. Since the natural gas combustion in these burners should be fairly complete, these burners should emit only minor quantities of VOCs. However, there will be some NO_x emissions as well as emissions of the combustion products of any gas contaminant.

5.2.3 *Flash Tank Vent Gas*—The flash tank removes a large fraction of the light gases absorbed by the glycol at high pressure by reducing the pressure of the rich glycol stream leaving the bottom of the absorber. The flash gas is often used as fuel, but may be vented as a waste stream at some units.

5.2.4 *Water and Hydrocarbon Condensate*—Many units also collect a water and hydrocarbon condensate from the reboiler still vent gas. As the still vent gas passes through the stack above the distillation section of the column, some condensation may occur on the stack walls. The condensate may be collected through a drain leg off the stack and will likely be contaminated with hydrocarbon species. Still other units install condenser control technologies to recover additional liquid hydrocarbons from the still vent gas.

5.2.5 *Waste Glycol/Reclaimer Waste*—The glycol solution is occasionally replaced in units when it becomes fouled with contaminants such as heavy hydrocarbons, well-treating chemicals, corrosion products, degradation products, and salts. The contaminated glycol may be sent to a vacuum reclaimer for recovery.

5.2.6 *Spent Filters*— Filtration is used to remove low levels of contaminants from the solution and maintain solution quality. Sock filters are typically used in glycol units to remove particulates, while carbon filters are used to remove heavy hydrocarbons and surfactants from solution. These filters must be replaced on a regular basis.

5.3 *Sampling Considerations*:

5.3.1 *Still Vent Gas*— It is difficult to directly measure BTEX and VOC emissions by sampling the still vent stream. The stream temperature is in the range from 200 to 225°F (93.3 to 107.2°C) and may contain 90 + % water, as well as levels of organic compounds that may often be in the percent range. Problems associated with sampling and analyzing this stream include the following **(3)**:

5.3.1.1 Measuring organic compounds in the presence of high levels of water or removing the water without biasing the organic compound measurements,

5.3.1.2 Collecting a representative, homogeneous sample from a low flow rate stream that may have condensing material, and

5.3.1.3 Accurately measuring the low, fluctuating flow rate of this stream.

5.3.2 *Burner Flue Gas*— The USEPA has characterized emissions from natural gas combustion and established factors for estimating these emissions **(2)**. Therefore, sampling is typically not performed on this stream and emission data collection methods will not be discussed in the remainder of this guide.

5.3.3 *Flash Tank Vent Gas*—Since the flash tank vent gas is often used as a fuel or stripping gas, sampling is typically not performed on this stream **(2)**. Therefore, emission data collection methods will not be discussed in the remainder of this guide.

5.3.4 *Water and Hydrocarbon Condensate*—The water and hydrocarbon condensate collected from the reboiler still vent gas and condenser control technologies are typically routed to a three-phase separator (liquid hydrocarbon, water, and noncondensable gas). The noncondensable gas is recycled to the reboiler. Each liquid phase is pumped to a separate waste storage tank. Alternately, the combined liquid streams may be sent to a single storage tank. Problems associated with sampling these liquid streams include the following:

5.3.4.1 Collecting a representative sample of each liquid phase,

5.3.4.2 Accurately measuring the flow rates of the liquid hydrocarbon and water streams since the separator discharge pumps may not run continuously. This is especially a problem when the liquids are not segregated in separate tanks.

5.3.5 *Waste Glycol/Reclaimer Waste*—Waste glycol is usually pumped to storage drums or tanks and then shipped off site for reclamation. The main problem associated with sampling this waste stream is sample homogeneity.

5.3.6 *Spent Filters*— Spent filters are typically stored in drums or dumpsters. The main problem associated with sampling this waste stream is obtaining a representative sample.

6. Approaches for Collecting Air Emission Data

6.1 *Introduction*— There are two often-used approaches for measuring the air emissions from the still vent stack: a rich/lean glycol mass balance (RL) and total capture stack condensation (TCC). Using the RL approach, emissions are calculated from the mass differences between the analytes in the rich and lean samples and the glycol flow rate. The glycol samples may be collected at atmospheric or pressurized operating conditions. Using the TCC approach, the total emissions to the atmosphere from the still vent are collected and separated into three phases (hydrocarbon liquid, water, and noncondensable gas) by chilling the still vent stream with an ice bath. Emission modeling programs such as GRI-GLYCalc[3] can also be used to estimate emissions of BTEX and other VOC from the regenerator still vent gas. Some of the practical considerations recommended when using these sampling approaches are listed below as follows.

6.2 *Practical Considerations for RL Methods*—Glycol dehydration units have different types of hardware (for example, gas-driven or electric pumps, flash tanks, charcoal filters, heat exchangers) and variations in their glycol piping configurations that must be considered before sampling. Gas-driven pumps (for example, Kimray) inject gas into the rich glycol, that results in a larger noncondensable gas fraction in the rich glycol sample line. Some units are equipped with flash tanks, that will remove a portion of the noncondensable gas from the rich glycol line. Charcoal filters may be present and may absorb hydrocarbons from the glycol. Glycol dehydration units may have none, one, or a combination of several heat exchange systems found at different locations (for example, before or after flash tanks) on different units. (Several heat exchange systems found on glycol dehydration units include: concentric pipe rich/lean glycol; rich glycol passed through a coil in the surge tank; rich glycol preheated in the still; and concentric pipe lean glycol/outlet natural gas.) All of these factors play a role in the temperature, pressure, and the amount of gas/foam encountered during glycol sampling **(3, 4)**.

6.2.1 The rich and lean glycol samples should be collected at locations where the glycol stream will be at the lowest possible temperature and pressure. A cooling coil should be used when collecting the rich glycol samples to minimize the loss of volatiles. Sampling points should not be at the top or bottom of filters or anywhere in the glycol stream where gas or liquid can accumulate.

6.2.2 The rich glycol should always be sampled at a point between the glycol pump and the reboiler. The sample point can be selected based on the flow chart in Fig. 2.

6.2.2.1 Sampling the rich glycol stream at units that have sampling points at high pressure (greater than 100 psig (689.5 kPa)), a high noncondensable gas flow rate (no flash tank), and a Kimray gas-driven pump results in a glycol sample that spews (especially in conjunction with the strokes of the Kimray pump) from the line as an aerosol.

6.2.2.2 Units that have rich glycol sample points at low pressure (less than 5 psig (34.5 kPa)), a low noncondensable gas flow rate (flash tank), an electric pump, and a cool glycol stream (glycol temperature less than 70°F (21.1°C)) can result in a viscous foamy glycol sample that flows very slowly through the cooling coil (a cooling coil is not necessary or practical in these situations).

6.2.3 If a charcoal filter is in line between the reboiler and the contactor, the lean glycol should be sampled at a point between the reboiler and charcoal filter. If a charcoal filter is not in line between the reboiler and the contactor, the lean glycol should be sampled at a point between the reboiler and the contactor. In both instances, the sample point should be selected on the low pressure side of the glycol pump. Also, sample points for the lean glycol may be very hot (>300°F (149°C)), especially if collected before any heat exchange device.

6.2.4 Kimray gas-driven pumps and electric pumps are used to determine the glycol flow rate. When Kimray glycol pumps are set up to run at or below their minimum recommended stroke rates, wide variations in glycol circulation rates may

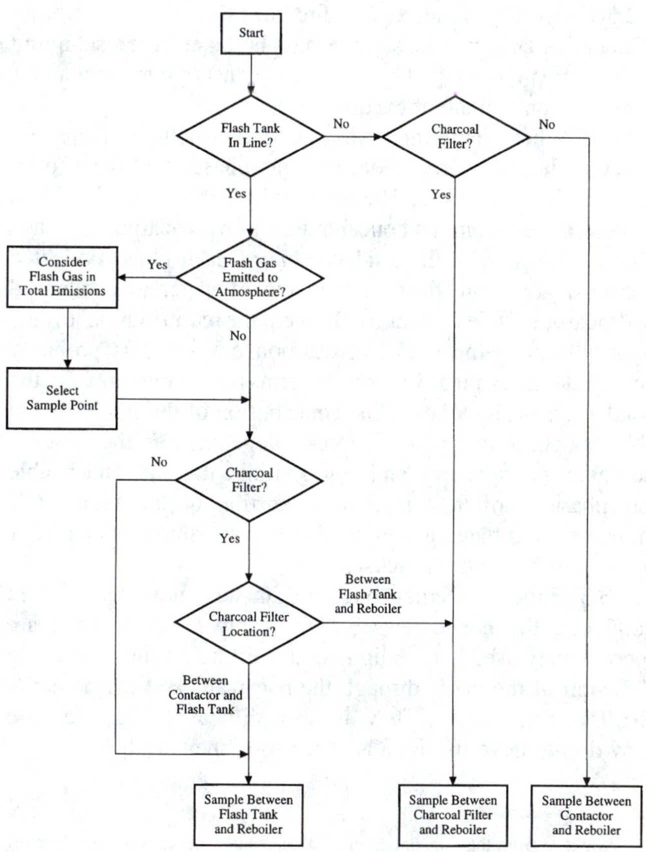

FIG. 2 Selection of the Rich Glycol Sample Port

[3] GRI-GLY Calc is a registered trademark of Gas Research Institute, available from Radian International LLC, P. O. Box 201088, Austin, TX 78720-1088.

occur. Sampling and flow rate measurement should be conducted under conditions in which the glycol pumps are running within their recommended flow rate ranges.

6.2.5 Samples should be analyzed by GC/FID using EPA Methods 8020 and 8260[4] or the direct injection approach outlined in Appendix X1 of this guide.

6.3 *Practical Considerations for TCC Methods*—There are many equipment variations to glycol dehydration units. The specific unit configuration and the process parameters must be considered before sampling to determine emissions from a specific unit. Units that do not have flash tanks or that have flash tanks maintained at low temperatures and high pressures may have condensable gas flow rates that could result in backpressure on the reboiler when the TCC apparatus is attached. When condensing the still vent stream on a glycol unit, one should make sure to minimize backpressure (preferably less than six inches of water column) on the reboiler. Depending on its size, the total capture device can normally only be used to condense the still vent stream on dehydration units drying less than 10 MMSCFD (**3**).

6.3.1 The TCC method should be carried out during both on and off reboiler firing cycles. Glycol reboiler firing cycles create positive and negative still vent flow. During reboiler firing, water and hydrocarbons are distilled from the glycol and their vapors create positive still vent flow. When the reboiler is off, the glycol begins to cool, and no more vapors are separated from the glycol. The existing vapors cool and their volume is reduced, which can result in reverse (negative) flow in the still vent line. Cyclic flow is common in glycol dehydration units and can be minimized or eliminated by maintaining a control of ±2.5°F (1.4°C) on the reboiler fire box. Due to the possibility of negative still vent flow, care must be taken when sampling the gas stream (noncondensable gas) to ensure that ambient air does not contaminate the sample.

6.3.2 When sampling with the TCC method, there are aqueous, liquid hydrocarbon, and gas phases that have to be sampled and measured. The total hydrocarbon emissions will be determined from the concentration of hydrocarbons in each phase coupled with the total volume of each phase collected during a given run time. If the ratio of water to condensed hydrocarbon is less than 16:1, then the contribution of the water phase to the total hydrocarbon emissions is probably negligible (less than 1 % of the emissions contained in the condensed hydrocarbon). The contribution of the noncondensable gas phase to total emissions is dependent on the noncondensable gas flow rate and composition; the noncondensable gas phase is often a significant portion of the total VOC emissions and may also contain significant amounts of BTEX on units without flash tanks.

6.3.3 Either a Tedlar bag or canister may be used in collecting the noncondensable gas sample. Tedlar bags are more widely used but are limited to short hold times due to the diffusion of the gases through the bag walls and the potential for light degradation. They are also difficult to ship because they do not have fixed walls. Therefore, they are better suited for sites that are relatively close so that the Tedlar bags may be driven to the laboratory. By contrast, canisters are not used as widely and tend to be more expensive because of the cleaning procedures needed to reuse them. However, they are easily shipped and are well-suited for remote locations.

6.3.4 Liquid samples should be analyzed using EPA Methods 8020 and 8260[4] or the analytical methods outlined in Ref (**3**). The gas samples should be analyzed using modified EPA compendium Method TO-14 or GPA Methods 2261 and 2286 as outlined in Ref (**3**).

6.4 The BTEX and other VOC emissions may be estimated using GRI-GLYCalc[3] or other commercial software packages. The GRI-GLYCalc uses the inlet natural gas composition and process data, along with fundamental chemical engineering thermodynamics, to make the emissions estimates. (The natural gas may be collected and analyzed using the procedures outlined in GPA Methods 2166 and 2286 or modified EPA TO-14.) The GRI-GLYCalc[3] tends to overpredict BTEX and VOC emissions, rather than underpredict, so emission estimates are conservatively high. Although this approach is less rigorous than the RL and TCC methods described above, it may be used as a rapid, inexpensive screening tool for determining which dehydration units are likely to be in compliance with applicable regulatory emissions levels.

7. Approaches for Collecting Liquid Waste Data

7.1 *Introduction*— The methods and practical considerations recommended when collecting samples from the liquid waste streams associated with glycol dehydrators are discussed as follows.

7.2 *Water and Hydrocarbon Condensate*—Direct pour/grab and scoop sampling techniques may be used to collect water and hydrocarbon condensate samples from waste storage tanks (**9**).

7.2.1 Direct pour sampling involves the transfer of a waste sample from the source directly into the sample container. This collection method should be used when a discharge port is available on the tank, and when there is no means of compositing the waste stream. Prior to collecting the sample, the port and related piping should be purged to ensure that the sample is representative of the material in the vessel. Small containers may be sampled by directly pouring from the container into the sample bottles. Sample homogeneity is a potential concern.

7.2.2 Direct grab sampling consists of dipping a borosilicate beaker or weighted sample bottle into a process or storage vessel. The weighted sample bottle typically consists of a glass or plastic bottle with a sinker and stopper, in addition to a line that is used to lower, raise, and open the bottle. The weighted bottle should be lowered into a tank to the desired level or layer and then opened and filled. Multiple random samples should be composited into a 5-gal (18.9 L) carboy and distributed to the appropriate sampling containers.

7.2.3 Disposable scoops should be used to collect samples of high-viscosity liquid wastes. These samples should be composited directly into wide-mouth bottles by collecting several scoops of material.

7.2.4 Field blank samples should be collected on site when possible.

7.2.5 Water and hydrocarbon condensate samples should be

[4] Available from the Superintendent of Documents, U.S. Government Printing Office, Washington, DC 20036.

analyzed using EPA Methods 8020 and 8260 or the analytical methods outlined in Ref (**3**). Toxicity Characteristic Leaching Procedure (TCLP) testing, as described in 40 CFR Part 261 Appendix II, should be conducted if RCRA toxicity data are needed.

7.3 *Waste Glycol/Reclaimer Waste*—Direct pour/grab and composite liquid sampling techniques (COLIWASA) may be used to collect the waste glycol/reclaimer waste samples from waste storage tanks or drums (**9**).

7.3.1 Waste glycol stored in tanks should be sampled using the direct pour/grab techniques discussed in 8.2.1 and 8.2.2.

7.3.2 Waste glycol stored in drums should be sampled using a composite liquid sampler (COLIWASA). This device typically consists of a 4-ft (1.22-m)-long glass or PVC tube having an inside diameter of approximately 1.25 in. (31.75 mm), with a 0.25-in. (6.35 mm) diameter stainless steel (or TFE-fluorocarbon) rod with attached stoppers passing down the center through the length of the tube. The COLIWASA should be used by opening the stoppered lower end and slowly lowering the tube into the liquid. As the tube passes through the drum, a composite sample is collected over the length of the drum. When the bottom of the drum is reached, the stopper should be pressed back into the tube openings, thereby trapping the sample within the tube. At this point, the COLIWASA contains a composite sample of the drum. When the COLIWASA is withdrawn, it may be emptied into the sample container. When compositing from multiple drums, each sample container should be proportionately filled from each drum.

7.3.3 Field blank samples should be collected on site when possible.

7.3.4 These waste samples should be analyzed using the TCLP procedures described in 40 CFR, Part 261, Appendix II.[4]

8. Approach for Collecting Solid Waste Data

8.1 *Introduction*— The methods and practical considerations recommended when collecting samples from the solid waste stream associated with glycol dehydrators are discussed as follows.

8.2 Filters should not be removed from a unit until differential pressure indicates that the filter should be removed. Care should be used when removing a filter from a high-pressure application.

8.3 The spent filters should be cut into pieces using a hack saw and scissors and composited. If possible, the composite should be made from three different sample filters (**9**).

8.4 These samples should be analyzed using the TCLP procedures described in 40 CFR, Part 261, Appendix II.[4]

APPENDIX

(Nonmandatory Information)

X1. ATMOSPHERIC RICH/LEAN GLYCOL METHOD STANDARD PROCEDURE

X1.1 Purpose

X1.1.1 The purpose is to determine still vent emissions from triethylene glycol (TEG) dehydration units. Still vent emissions are calculated from the mass differences between the analytes in the atmospheric rich and lean glycol samples and the glycol flow rate.

X1.2 Scope and Applicability

X1.2.1 *Scope*—A procedure for collecting and analyzing samples taken from natural gas dehdyration units that use TEG as the desiccant. Determining the glycol flow rate, calculating still vent emissions from a dehydration unit, and using VOC emission correction factors for units with high gaseous non-condensable still vent flow rates are also covered.

X1.2.2 *Applicability*— Still vent emissions may be estimated for benzene, toluene, ethylbenzene, and the xylenes (BTEX), as well as *n*-hexane, 2,2,4-trimethylpentane, C_5 to C_{10} normal alkanes, cyclohexane, and total hydrocarbons.

X1.3 Safety

X1.3.1 Glycol samples may be under high pressure (potentially greater than 1000 psig (6895 KPa)) and may be hot (380°F (193°C)). Samples may have high levels of benzene and other aromatic and aliphatic hydrocarbons that pose exposure hazards. Natural gas present in the glycol sample lines is a potential fire or explosion hazard.

X1.4 Glycol Sampling Procedure

X1.4.1 Obtain atmospheric rich and lean glycol samples from natural gas glycol dehydration units. Collect these samples in 40-mL VOA vials at identified points on a glycol dehydration unit. The results of the rich and lean sample analysis, along with the glycol circulation rate, are used to perform a mass balance of air toxic emissions (*n*-hexane, benzene, toluene, ethylbenzene, xylenes, and other volatile organic compounds VOCs) from the glycol dehydrator still vent.

X1.4.2 *Identification of a Glycol Sampling Location*—Collect both rich and lean glycol samples at locations where the glycol stream will be at the lowest possible temperature and pressure. Sampling points should not be at the top or bottom of filters or anywhere in the glycol stream where gas or liquid can accumulate.

X1.4.3 *Rich Glycol:*

X1.4.3.1 *Step 1*—Always select a sample point between the glycol pump and the reboiler.

X1.4.3.2 *Step 2*—Select a sample point based on the flow chart in Fig. X1.1.

X1.4.4 *Lean Glycol:*

X1.4.4.1 *Step 1*—Determine if a charcoal filter is in line between the reboiler and the contactor. Proceed to Step 2 (X1.4.4.2) if a charcoal filter is in line between the reboiler and the contactor. Proceed to Step 3 (X1.4.4.3) if a charcoal filter

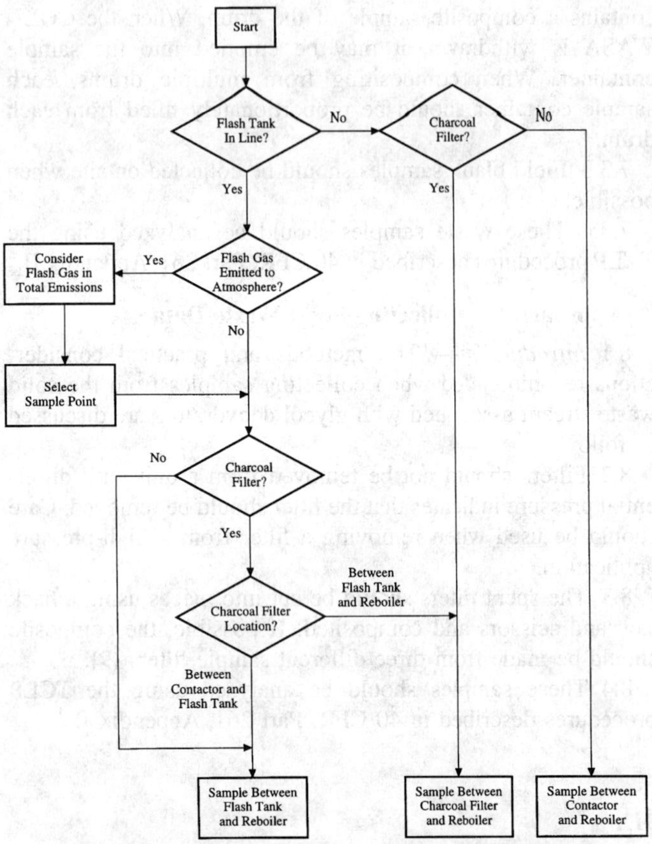

FIG. X1.1 Selection of the Rich Glycol Sample Port

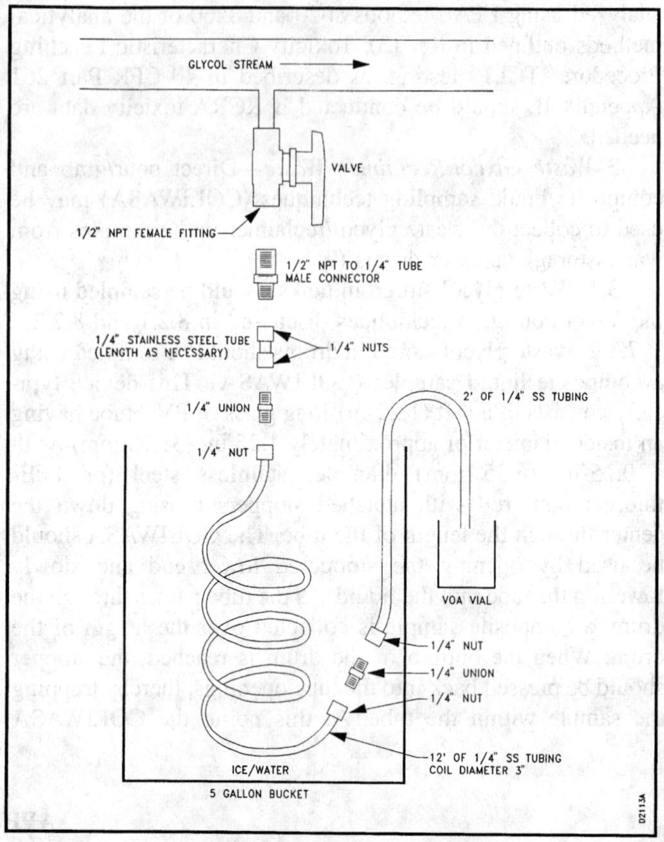

FIG. X1.2 ARL Sampling System

is not in line between the reboiler and the contactor.

X1.4.4.2 *Step 2*—Select a lean glycol sample point between the reboiler and the charcoal filter. Preferably select a sample point on the low-pressure side of the glycol pump.

X1.4.4.3 *Step 3*—Select a lean glycol sample point between the reboiler and the contactor. Preferably select a sample point on the low-pressure side of the glycol pump.

X1.4.5 *Sampling Valves, Lines, and Connections*—After identifying the appropriate sample points, modify them so that the glycol stream may be cooled and sampled at atmospheric pressure. Cooling the glycol sample is not necessary if the temperature of the glycol is less than 70°F (21.1°C).

X1.4.5.1 *Step 1*—The glycol stream lines on most glycol dehydration units are generally 1-in. (25.4-mm) pipe and have National Pipe Thread (NPT) fittings. Adapt the identified sample port from the NPT fitting to a ¼-in. (6.35-mm)-Swagelok male fitting.

X1.4.5.2 *Step 2*—Connect a sampling coil made from approximately 12 ft (3.66 m) of ¼-in. (6.35-mm)-copper or stainless steel tubing to the ¼-in. (6.35-mm)-Swagelok sample valve. See Fig. X1.2.

X1.4.6 *Label the VOA Vials For Glycol Sample Collection*—Before collecting the samples, affix a label to the VOA vial that states the date, sample point, sample type (rich or lean glycol sample), and a unique identification number. Once the label is attached to the VOA vial, cover it with a clear tape. Select the writing utensil carefully, since glycol dissolves ink and felt markings. Pencil, or preferably a tested permanent marker (one that does not leave a glycol soluble mark), works best. The label should be completely filled out, affixed to the vial, and covered with clear tape before sampling begins.

X1.4.7 *Collection of Glycol Samples*:

X1.4.7.1 *Step 1*—Immerse the ¼-in. (6.35-mm)-sampling coil in an ice/water bath contained in a 5-gal (18.9-L) bucket.

X1.4.7.2 *Step 2*—Open the glycol sampling port and purge a stream of glycol through the sampling coil for two minutes. Rich glycol sample generally sprays from the sample line as a foamy aerosol.

X1.4.7.3 *Step 3*—Place a 40-mL glass VOA vial under the flow of glycol and fill the vial to overflowing.

X1.4.7.4 *Step 4*—Place the cap, with the septum TFE-fluorocarbon side down, on the VOA vial immediately after the glycol begins to overflow the vial. Do not overtighten the cap. Trapped gases (headspace) can generally be observed in the vials. Do not try to eliminate the trapped gases in the vials because this will result in loss of additional target analytes from the glycol.

X1.4.8 *Shipment of VOA Vials Containing Glycol Samples*—Place the glycol samples in a cooler on ice and keep them at about 39°F (4°C). Complete a chain of custody form and ship the samples under a chain of custody to a laboratory for analysis.

X1.4.9 *Glycol Circulation Rate*:

X1.4.9.1 *Kimray Gas-Driven Pumps*—Determine the glycol circulation rate by counting pump strokes for a set period of time and converting the stroke rate to a flow rate using Table X1.1. The time required for 25 strokes (strokes should be counted for at least 2 min) should be recorded and then the

TABLE X1.1 Glycol Flow Rates

Pump	Strokes/Minute	Glycol Output Strokes/Gallon (Litres)		Glycol Output Gallons (Litres)/Stroke	
315PV	20 to 100	435	(114.9)	0.0023	(0.0087)
1715PV	8 to 40	59	(15.59)	0.017	(0.0642)
4015PV	12 to 40	59	(15.59)	0.017	(0.642)
9015PV	12 to 40	26.3	(6.948)	0.038	(0.144)
21015PV	10 to 32	9	(2.378)	0.111	(0.421)
45015PV	10 to 28	3.5	(0.925)	0.283	(1.082)
2015SC	5 to 55	147	(38.83)	0.0068	(0.0258)
5015SC	10 to 50	52	(13.74)	0.0193	(0.0728)
10015SC	10 to 48	25	(6.604)	0.040	(0.151)
20015SC	10 to 40	8.8	(2.325)	0.114	(0.430)

TABLE X1.2 Analytical Conditions for the Analysis of Liquid Samples

Instrument	Hewlett-Packard 5880
Detector	FID
Injection system	capillary; split 50 to 1
Column	60-m × 0.32-mm inside diameter 3.0-μ film Restek RTx_{-1}
Carrier gas	helium at 20 psig
Carrier gas flow rate	~3 mL/min
Make-up gas	helium at 30 mL/min
Temperature program	40°C for 2 min; programmed at 8°C/min to 225°C, final hold 10 min
Injector temperature	275°C
Detector temperature	300°C
Injection volume	1 μL

counted number of strokes per minute converted to a glycol circulation rate. Use this flow rate in all calculations requiring the glycol circulation rate.

X1.4.9.2 *Example*—Calculate the glycol flow rate in gallons (litres) per hour for a Kimray Model 20015SC pump in which 22 strokes were counted in 2 min 15 s as follows:

2 min 15 s	=	2.25 min
22 strokes/2.25 min	=	9.8 strokes/min
(9.8 strokes/min) × (60 min/h)	=	588 strokes/h
(588 strokes/h) × (0.114 gal/stroke (L/stroke))	=	67 gal/h (253.6 L/h)

X1.4.10 *Electric Pumps*—Obtain the glycol circulation rate from the manufacturer's data or calculate it as follows.

X1.4.10.1 *Step 1*—Determine the number of displacement pistons (*ndp*).

X1.4.10.2 *Step 2*—Determine the radius (*r*) of the pistons in inches.

X1.4.10.3 *Step 3*—Determine the stroke length (*h*) of the pistons in inches.

X1.4.10.4 *Step 4*—Determine the revolutions per minute (*rpm*) of the pump drive wheel.

X1.4.10.5 *Step 5*—Calculate the pumps glycol displacement in gallons (litres) per revolution (*gpr*):

$$(gpr) = (ndp)(\pi r^2 h)(4.329 \times 10^{-3} \text{ gal/in.}^3 (10^{-6} \text{ L/mm}^3)) \quad (X1.1)$$

X1.4.10.6 *Step 6*—Calculate the glycol flow rate in gallons (litres) per hour (gph):

$$(gph) = (rpm)(gpr)(60 \text{ min/h})(pump\ efficiency) \quad (X1.2)$$

If the pump efficiency is not known use 95 %.

X1.4.10.7 *Example*—Calculate the glycol flow rate in gallons per hour for an electric pump with two pistons. Each piston has a radius of 0.5 in. and a stroke length of 2.25 in. (57.15 mm). The pump speed is 160 rpm's:

$$2 \times (\pi \times (0.5 \text{ in.})^2 \times 2.25 \text{ in. } (57.15 \text{ mm})) \times 4.329 \times 10^3 \text{gal/in.}^3 \quad (X1.3)$$

$$(10^{-6} \text{ L/mm}^3) = 0.015 \text{ gpr}$$

$$160 \text{ rpm} \times 0.015 \text{ gpr} \times 60 \text{ min/h} \times 0.95 \text{ efficiency} = 137 \text{ gph}$$

X1.5 Analytical Conditions and Glycol Analysis

X1.5.1 Use gas chromatographic systems with flame ionization detectors to quantitate the BTEX target analytes, as well as cyclohexane, 2,2,4-trimethylpentane (*iso*octane), C_5 to C_{10} normal alkanes, and total hydrocarbons (see Table X1.2). Use a gas chromatograph equipped with a split/splitless injector and a capillary column to analyze the liquid samples. Quantitation of the target and nontarget analytes is based on individual response factors. Calibration procedures at the end of this section. Use a response factor for *n*-hexane to quantify the unidentified hydrocarbons. Method interferences are discussed in X1.6.

X1.5.2 *Preparation of Stock Solution*:

X1.5.2.1 *Step 1*—Add approximately 10 mL of dichloromethane to a 25-mL volumetric flask.

X1.5.2.2 *Step 2*—Add 500 μL of each target analyte (analytes need to be at room temperature), in order of descending boiling point, to the 25-mL volumetric flask wth a 1-mL gas-tight syringe keeping the tip of the syringe needle below the surface of the diluent.

X1.5.2.3 *Step 3*—Dilute the solution in the 25-mL volumetric flask to volume with dichloromethane, and cap and shake the flask.

X1.5.2.4 *Step 4*—Verify the quality of the stock solution by comparison of the standards made from the stock solution with a second reference standard (see X1.5.4.4).

X1.5.3 *Preparation of Primary Standards*:

X1.5.3.1 *Step 1*—Prepare five standards, each at a different concentration, by removing (with a gas-tight syringe) an aliquot from the stock solution and injecting the aliquot below the surface of the dichloromethane in 10-mL volumetric flasks:

Use 1000 μL of stock solution in Flask 1;
Use 400 μL of stock solution in Flask 2;
Use 100 μL of stock solution in Flask 3;
Use 50 μL of stock solution in Flask 4; and
Use 10 μL of stock solution in Flask 5.

X1.5.3.2 *Step 2*—Dilute the contents in each flask to volume; cap them; shake them to ensure mixing.

X1.5.3.3 *Step 3*—Remove approximately 1-mL aliquots from the flasks and place each in labeled GC crimp-top vials. Cap the crimp-top vials, and use them in X1.5.4 and X1.5.5.

X1.5.4 *Calibration Procedures*—Use the primary standards prepared in X1.5.3 to obtain a calibration factor for each target analyte. The concentrations of the primary standards should bracket the range of concentrations found in glycol samples.

X1.5.4.1 *Step 1*—Inject 1 μL of each primary standard into a GC/FID system.

X1.5.4.2 *Step 2*—Calculate a calibration factor (*CF*) for each analyte at each standard concentration:

$$(CF) = (amount\ injected\ in\ mg/L)/(total\ area\ of\ peak) \quad (X1.4)$$

X1.5.4.3 *Step 3*—Calculate average calibration factor ($\overline{CF}$) for each analyte. If the percent relative standard deviation (%

RSD) of the $\overline{CF}$ is less than 20 % over the working range, linearity through the origin can be assumed, and the average calibration factor can be used in place of a calibration curve:

$$(\% \ RSD) = [(standard\ deviation)/(\overline{CF})] \times 100 \quad \text{(X1.5)}$$

X1.5.4.4 *Step 4*—Verify the validity of the calibration standards by injecting an independent mid-level standard (purchased from a second-source) once following each multipoint calibration using a subset of the target analytes. If the response for the independent calibration standard varies more than ±10 % of the expected value, take corrective action to verify the accuracy of the source materials.

X1.5.5 *Calibration Control Checks*:

X1.5.5.1 Verify the working calibration factor by injecting a mid-level calibration standard at the beginning and end of each working day or analysis sequence. If the response for any target analyte varies from the predicted response by more than ±15 % (relative percent difference (RPD)), conduct a second analysis of the calibration standard. If two calibration checks in a row vary from the predicted response by more than ±15 %, conduct a system check.

X1.5.5.2 Make sure gas flows have not changed, and if they have, reset to initial conditions. Place a new septum in the injector port. Cleaning the injector and detector may also be necessary. After preforming the system check, rerun the calibration check, and if the response still varies by more than 15 %, prepare a new calibration factor (or curve).

$$RPD = [(\overline{CF} - CF\ for\ calibration\ check)/(\overline{CF})] \times \quad \text{(X1.6)}$$

X1.5.6 *Preparing, Analyzing, and Screening Glycol Samples*—Dilute glycol samples with dichloromethane before analyzing them by direct injection into a GC/FID system. Do a preliminary analysis on one rich and one lean glycol sample to make sure that the concentrations of the target analytes in the glycol sample are bracketed by the range of standard concentrations used to generate the standrd curve. If the concentrations of the target analytes fall outside of the concentrations used to generate the standard curve, either use a different dilution ratio for the sample or add another point to the standard curve. Spike at least 10 % of the glycol samples analyzed with a subset of the target analytes, such as *n*-hexane, benzene, toluene, and *n*-octane. The concentrations of the analytes determined by the preliminary analysis can be used to calculate the spike solution (see X1.5.8). The concentrations of the analytes determined by the preliminary analysis can be used to calculate the spike solution.

X1.5.7 *Dilute and Analyze the Glycol Samples*:

X1.5.7.1 *Step 1*—Allow the glycol sample to warm to room temperature before dilution.

X1.5.7.2 *Step 2*—Rinse a 1-mL volumetric flask five times with 1-mL aliquots of dichloromethane. Add approximately 250 µL of dichloromethane to a 1-mL volumetric flask.

X1.5.7.3 *Step 3*—Rinse a 1-mL gas-tight syringe with 3 aliquots of the sample by inserting the needle of the syringe through the septum on a VOA vial containing the glycol sample. Withdraw about 600 µL of sample into the syringe and remove the syringe from the VOA vial. Discharge the sample from the syringe into a waste container.

X1.5.7.4 *Step 4*—Insert the needle of the 1-mL gas-tight syringe through the septum on a VOA vial containing the glycol sample. Withdraw about 600 µL of sample into the syringe and remove the syringe from the VOA vial. Hold the syringe upright at eye level and adjust the plunger so that 500 µL of glycol sample remains in the syringe.

X1.5.7.5 *Step 5*—Place the needle of the syringe that contains the glycol sample under the surface of the 250 µL of dichloromethane in the 1-mL volumetric flask. Slowly introduce the sample into the solvent.

X1.5.7.6 *Step 6*—Repeat X1.5.7.3 and X1.5.7.4 on either 10 % of the samples or at least on one sample per set of analysis. Also repeat X1.5.7.3 and X1.5.7.4 on either 10 % of the samples or at least on one sample per set of analysis for use in the spike dilution process (see X1.5.8).

X1.5.7.7 *Step 7*—After introduction of the sample into the volumetric flask, add dichloromethane to fill the volumetric flask to volume.

X1.5.7.8 *Step 8*—Cap the volumetric flask and shake it briefly.

X1.5.7.9 *Step 9*—Remove the diluted sample from the volumetric flask with a disposable pipet and place it in a labeled crimp top vial. Cap the crimp top vial.

X1.5.7.10 *Step 10*—Analyze 1 µL of the diluted sample by GC/FID. See X1.7 for interferences.

X1.5.7.11 *Step 11*—Rinse the syringe and volumetric flask five times with 1 mL of dichloromethane after each use. After every ten samples, analyze (by GC/FID) the last dichloromethane rinse from washing the volumetric flask and syringe to see if cross contamination is occurring.

X1.5.7.12 *Step 12*—Calculate the concentration of the analytes in the diluted glycol sample. Multiply the average calibration factor for each analyte by the peak area for that analyte.

X1.5.7.13 *Step 13*—Calculate the concentration of the analytes in the original glycol sample. Multiply the concentration of each analyte in the diluted glycol sample by the dilution factor (for example, two if a 1:1 dilution was used).

X1.5.7.14 *Step 14*—Calculate the relative percent difference (RPD) for the analytical duplicates. The RPD for the analytical duplicates should be ±10 %.

$$RPD = \frac{|(C1 - C2)|}{(C1 + C2)/2} \times 100 \quad \text{(X1.7)}$$

where:

$C1$ = concentration of the first sample, and
$C2$ = concentration of the duplicate sample.

X1.5.8 *Glycol Spike Dilution Process*—Spike the samples during the dilution process so that the concentration of the target analytes is approximately twice that of the original sample.

X1.5.9 *Preparing Spike Solution*:

X1.5.9.1 *Step 1*—Prepare the spike solution based on adding 100 µL of a spike solution into the 1-mL volumetric flask used during glycol sample dilution.

X1.5.9.2 *Step 2*—Use *n*-hexane, benzene, toluene, and *n*-octane as the target analytes to prepare the spike solution.

X1.5.9.3 *Step 3*—Calculate the concentration of the spike solution based on the concentrations of target analytes in the diluted glycol sample (see X1.5.7.12).

$$Cs = \frac{(Cd \times Vd)}{Vs} \quad (X1.8)$$

where:

Cs = concentration of the spike solution, mg/L;
Cd = concentration of the analytes in the diluted sample,
Vd = total volume of the diluted sample, mL, and
Vs = volume of the spike solution to be added during the dilution process, mL.

X1.5.9.4 *Step 4*—For example: A diluted glycol sample (see X1.5.7.12) that contained the following four analytes, *n* hexane at 48 ppm, benzene at 1854 ppm, toluene at 1462 ppm, and *n*-octane at 36 ppm, prepare the following spike solution. In the following order, add 3 µL of *n*-octane, 90 µL of toluene, 110 µL of benzene, and 4 µL of *n*-hexane to approximately 5 mL of dichloromethane in a 10-mL volumetric flask. Dilute the resulting solution to a total volume of 10 mL with dichloromethane.

X1.5.10 *Spike Dilution Process*:

X1.5.10.1 *Step 1*—Add approximately 250 µL of dichloromethane to a 1-mL volumetric flask.

X1.5.10.2 *Step 2*—Rinse a 1-mL gas-tight syringe with three aliquots of the sample by inserting the needle of the syringe through the septum on a VOA vial containing the glycol sample. Withdraw about 600 µL of sample into the syringe and remove the syringe from the VOA vial. Discharge the sample from the syringe into a waste container.

X1.5.10.3 *Step 3*—Remove a 500-µL aliquot of the glycol sample from the VOA vial by taking the aliquot through the septum with the 1-mL gas-tight syringe.

X1.5.10.4 *Step 4*—Place the tip of the syringe needle under the surface of the 250 µL of dichloromethane in the 1-mL volumetric flask. Slowly introduce the sample into the solvent.

X1.5.10.5 *Step 5*—Remove a 100-µL aliquot of the spike solution from the 10-mL volumetric flask prepared above.

X1.5.10.6 *Step 6*—Place the needle from the syringe that contains the spike solution under the surface of the dichloromethane in the 1-mL volumetric flask from X1.5.10.3. Slowly introduce the sample into the solvent.

X1.5.10.7 *Step 7*—Dilute the resulting solution in the 1-mL volumetric flask to volume. Place it in a crimp top vial, and cap the vial.

X1.5.10.8 *Step 8*—Analyze 1 µL of the spiked sample by GC/FID. See X1.6 for interferences.

X1.5.10.9 *Step 9*—Calculate the concentration of the spiked analytes in the diluted spiked sample. Multiply the average calibration factor for each analyte by the peak area for that analyte.

X1.5.10.10 *Step 10*—Calculate the percent spike recovery.

$$\% \, rec = \frac{(Ss - Sd)}{Sf} \times 100 \quad (X1.9)$$

where:

$\% \, rec$ = percent recovery;
Ss = diluted spiked sample result, mg/L,
Sd = diluted glycol sample result, mg/L, and
Sf = amount of spike added, mg/L.

X1.5.11 *Spike Solution Analytical Checks*—Perform analytical checks by using the spike solution prepared above.

X1.5.11.1 *Step 1*—Add approximately 500 µL of dichloromethane to a 1-mL volumetric flask.

X1.5.11.2 *Step 2*—Add 100 µL of the spike solution to the volumetric flask with the syringe needle tip under the surface of the solvent.

X1.5.11.3 *Step 3*—Dilute the resulting solution to volume with dichloromethane.

X1.5.11.4 *Step 4*—Analyze 1 µL of the solution by GC/FID.

X1.5.11.5 *Step 5*—Calculate the concentration of the analytes in the diluted spike solution. Multiply the average calibration factor for each analyte by the peak area for that analyte.

X1.5.11.6 *Step 6*—Calculate the concentration of the analytes in the spike solution. Multiply the concentration of each analyte in the diluted sample by the dilution factor (ten if a 1:10 dilution was used).

X1.5.11.7 *Step 7*—Calculate the relative percent difference (RPD) for the spike solution checks:

$$RPD = \frac{|(Cs - Ct)|}{(Cs + Ct)/2} \times 100 \quad (X1.10)$$

where:

Cs = concentration of spike solution, mg/L, and
Ct = theoretical concentration of the spike solution, mg/L.

X1.6 Interferences

X1.6.1 *Autosamplers*— Set autosamplers used for glycol injections into the GC to account for the viscosity of the glycol. When autosamplers are set improperly, two problems are encountered: boiling point discrimination when low boiling diluents are used, and incomplete filling of the syringe when samples are withdrawn from the vial. Both of these problems can be detected by spiking a glycol sample with hexane and benzene at levels that will approximately double their concentrations, followed by analyzing the spiked sample.

X1.6.2 *Ethylene and Diethylene Glycol*—Triethylene glycol (TEG) samples from some glycol units contain diethylene glycol or ethylene glycol, or both, that can be at a high enough concentration to interfere with the quantitation of analytes of interest under certain GC conditions (interferences caused by coelution with analytes of interest). Note that ethylene glycol by itself is not soluble in dichloromethane, that is the diluent used for TEG analysis.

X1.7 Still Vent Emission Calculations

X1.7.1 *Still Vent Emissions Based on Rich/Lean Glycol Samples*—Calculate still vent emissions in tons per year (tpy) from glycol dehydration units by using both field and analytical data.

X1.7.1.1 *Step 1*—Get the average glycol flow rate in gallons (litres) per hour (gph).

X1.7.1.2 *Step 2*—Get the analyte concentrations in the rich and lean glycol samples in milligrams/Litre.

X1.7.1.3 *Step 3*—Average the results from the duplicate analyses to determine a single value for the sample concentrations. Average the duplicate sample results to report a single value for each run.

X1.7.1.4 *Step 4*—Determine compound emissions directly from the concentrations in rich and lean glycol samples and the glycol flow rate:

$$\textit{compound emissions, tpy} = \qquad (X1.11)$$
$$(\textit{analyte concentration in rich glycol, mg/L} -$$
$$\textit{analyte concentration in lean glycol, mg/L}) \times$$
$$(\textit{average glycol flow, gph}) \times \left(\frac{3.7856\ \text{L}}{1\ \text{gal}}\right) \times \left(\frac{1\ \text{lb}}{453\ 600\ \text{mg}}\right) \times$$
$$\left(\frac{1\ \text{ton}}{2000\ \text{lb}}\right) \times \left(\frac{8760\ \text{h}}{1\ \text{year}}\right)$$

X1.7.2 *VOC Still Vent Emissions Correction Factors*—Calculate the VOC still vent emissions for glycol dehydration units that utilize stripping gas or that do not have flash tanks.

X1.7.2.1 *Step 1*—Determine the stripping gas (SG) losses in:

$$SG, tpy = (0.0001156) \times (flow, \text{SCF/h})\ \Sigma\ C_i MW_i \qquad (X1.12)$$

where:

flow = stripping gas flow, SCF/h,
C_i = inlet natural gas concentration of Compound *i*, mole percent; and
MW_i = molecular weight of the compound.

X1.7.2.2 *Step 2*—Determine the flashing (*F*) and pump (*P*) gas losses in tpy :

$$FP, tpy = GCirc\ \Sigma\ C_i CF_i \qquad (X1.13)$$

where:

GCirc = glycol circulation rate in gal (L)/hour,
C_i = inlet natural gas concentration of Compound *i*, mole percent, and
CF_i = correction factor for Compound *i* taken from Tables X1.3-X1.10. Use Table X1.3 for electric glycol pumps and Tables Tables X1.7-X1.10 for Kimray glycol pumps.

X1.7.2.3 *Step 3*—Calculate still vent VOC emissions in tpy:

$$VOC, tpy = ARL, tpy + SG, tpy + FP, tpy \qquad (X1.14)$$

X1.8 Materials and Apparatus

X1.8.1 *Sampling Glycol Streams*:

X1.8.1.1 *Various Swagelok Fittings*—¼-in. (6.35 mm)-unions, ¼-in. (6.35 mm)-front and back ferrules, ¼-in. (6.35 mm)-union elbows, ¼-in. (6.35 mm)-nuts, ¼-in. (6.35 mm)-port connectors, ½-in. (12.7 mm) and ⅜-in. (9.525 mm)-reducing port connectors to ¼ in. (6.35 mm), ¼-in. (6.35 mm)-tube Cajon fractional tube adaptor to male NPT (⅜ in. (9.525 mm), ½ in. (12.7 mm), ¾ in. (19.05 mm), ⅝in. (15.875 mm),

X1.8.1.2 Twenty-five feet (3.66 m) of ¼-in. (6.35 mm)-stainless steel tubing,

X1.8.1.3 Ten-feet (3.05 m) coil of ¼-in. (6.35 mm)-stainless steel tubing (easily prepared by coiling a 12-ft (3.66 m) piece of ¼-in. (6.35 mm)-tubing around a 3-in. (76.2 mm)-piece of pipe),

X1.8.1.4 VOA vials, 40 mL,

X1.8.1.5 Five-gallon (18.9 L) bucket for immersing cooling coil,

X1.8.1.6 Sample labels,

X1.8.1.7 Clear tape,

X1.8.1.8 Gallon-size plastic storage bags,

X1.8.1.9 Hard-shell coolers,

X1.8.1.10 Pens and markers, and

X1.8.1.11 Chain of custodies/paperwork for samples.

X1.8.2 *Sample Preparation*:

X1.8.2.1 *Gas-Tight Syringes*—1 mL, 500, 250, 100, and 10 µL.

X1.8.2.2 *Volumetric Flask*—50, 25, 10, and 1 mL.

X1.8.2.3 *Solvents*—Dichloromethane.

X1.8.2.4 *Standards*—99 + % pure *n*-pentane, *n*-hexane, *n*-heptane, cyclohexane, *iso*octane, *n*-octane, *n*-nonane, *n*-decane, benzene, toluene, ethylbenzene, *m*-xylene, *p*-xylene, 98 % + pure *o*-xylene.

TABLE X1.3 Pentane (C3) Correction Factors (CF) for ARL Total VOC Emissions for Units with Kimray Pump[A,B]

Contactor Temperature, °F (°C)	Contactor Pressure, psig (KPa)								
	200	300	400	500	600	700	800	900	1000
40 (4.4)	0.0073	0.0095	0.0112	0.0142	0.0151	0.0164	0.0163	0.0166	0.0161
50 (10.0)	0.0065	0.0086	0.0101	0.0130	0.0138	0.0151	0.0152	0.0156	0.0153
60 (15.6)	0.0058	0.0078	0.0092	0.0119	0.0127	0.0140	0.0143	0.0147	0.0146
70 (21.1)	0.0053	0.0070	0.0084	0.0109	0.0118	0.0131	0.0134	0.0139	0.0139
80 (26.7)	0.0048	0.0065	0.0078	0.0101	0.0109	0.0122	0.0126	0.0132	0.0133
90 (32.2)	0.0044	0.0060	0.0072	0.0094	0.0102	0.0114	0.0119	0.0125	0.0127
100 (37.8)	0.0041	0.0055	0.0066	0.0088	0.0096	0.0108	0.0113	0.0119	0.0121
110 (43.3)	0.0038	0.0051	0.0062	0.0082	0.0090	0.0102	0.0107	0.0113	0.0116
120 (48.9)	0.0035	0.0048	0.0058	0.0077	0.0085	0.0096	0.0102	0.0108	0.0111
130 (54.4)	0.0033	0.0045	0.0055	0.0072	0.0080	0.0092	0.0097	0.0104	0.0107
140 (60.0)	0.0031	0.0042	0.0052	0.0069	0.0076	0.0088	0.0093	0.0100	0.0103

[A] *Correction = (mole % C3 in gas) × (glycol circulation, gph) × CF* .
[B] If propane mole % is not known, use 4.2 %.

TABLE X1.4 Pentane (C4) Correction Factors (CF) for ARL Total VOC Emissions for Units with Kimray Pump[A,B]

Contactor Temperature, °F (°C)	Contactor Pressure, psig (KPa)								
	200	300	400	500	600	700	800	900	1000
40 (4.4)	0.0119	0.0152	0.0172	0.0220	0.0224	0.0239	0.0228	0.0226	0.0211
50 (10.0)	0.0105	0.0135	0.0154	0.0198	0.0204	0.0220	0.0213	0.0213	0.0202
60 (15.6)	0.0093	0.0120	0.0139	0.0180	0.0187	0.0204	0.0201	0.0202	0.0193
70 (21.1)	0.0083	0.0109	0.0126	0.0165	0.0172	0.0189	0.0189	0.0193	0.0184
80 (26.7)	0.0075	0.0098	0.0115	0.0151	0.0160	0.0177	0.0177	0.0183	0.0178
90 (32.2)	0.0068	0.0090	0.0106	0.0140	0.0148	0.0165	0.0167	0.0173	0.0171
100 (37.8)	0.0062	0.0083	0.0097	0.0130	0.0138	0.0155	0.0158	0.0165	0.0164
110 (43.3)	0.0057	0.0076	0.0091	0.0121	0.0130	0.0146	0.0150	0.0158	0.0157
120 (48.9)	0.0053	0.0070	0.0084	0.0113	0.0122	0.0139	0.0143	0.0151	0.0151
130 (54.4)	0.0049	0.0066	0.0079	0.0107	0.0115	0.0132	0.0205	0.0145	0.0146
140 (60.0)	0.0046	0.0062	0.0074	0.0101	0.0110	0.0125	0.0130	0.0139	0.0141

[A] *Correction = (mole % C4 in gas) × (glycol circulation, gph) × CF .*
[B] If butanes mole % is not known, use 2.1 %.

TABLE X1.5 Pentane (C5) Correction Factors (CF) for ARL Total VOC Emissions for Units with Kimray Pump[A,B]

Contactor Temperature, °F (°C)	Contactor Pressure, psig (KPa)								
	200	300	400	500	600	700	800	900	1000
40 (4.4)	0.0337	0.0411	0.0449	0.0666	0.0652	0.0756	0.0697	0.0713	0.0644
50 (10.0)	0.0280	0.0346	0.0379	0.0571	0.0566	0.0665	0.0618	0.0650	0.0592
60 (15.6)	0.0235	0.0292	0.0326	0.0493	0.0493	0.0587	0.0559	0.0588	0.0544
70 (21.1)	0.0199	0.0250	0.0280	0.0431	0.0434	0.0522	0.0503	0.0540	0.0503
80 (26.7)	0.0171	0.0215	0.0244	0.0377	0.0385	0.0467	0.0454	0.0493	0.0467
90 (32.2)	0.0147	0.0188	0.0214	0.0332	0.0341	0.0419	0.0412	0.0451	0.0432
100 (37.8)	0.0128	0.0164	0.0188	0.0294	0.0307	0.0377	0.0375	0.0414	0.0400
110 (43.3)	0.0112	0.0145	0.0167	0.0264	0.0276	0.0342	0.0342	0.0380	0.0370
120 (48.9)	0.0099	0.0128	0.0149	0.0237	0.0249	0.0312	0.0313	0.0352	0.0344
130 (54.4)	0.0088	0.0115	0.0134	0.0226	0.0214	0.0285	0.0288	0.0326	0.0321
140 (60.0)	0.0079	0.0103	0.0121	0.0207	0.0194	0.0262	0.0266	0.0303	0.0300

[A] *Correction = (mole % C5 in gas) × (glycol circulation, gph) × CF .*
[B] If pentanes mole % is not known, use 0.80 %.

TABLE X1.6 Hexanes + (C6 +) Correction Factors (CF) for ARL Total VOC Emissions for Units with Kimray Pump[A,B]

Contactor Temperature, °F (°C)	Contactor Pressure, psig (KPa)								
	200	300	400	500	600	700	800	900	1000
40 (4.4)	0.0516	0.0614	0.0652	0.1012	0.0962	0.1160	0.1034	0.1079	0.0940
50 (10.0)	0.0419	0.0505	0.0541	0.0850	0.0821	0.1004	0.0909	0.0968	0.0859
60 (15.6)	0.0344	0.0418	0.0454	0.0720	0.0704	0.0872	0.0811	0.0868	0.0783
70 (21.1)	0.0285	0.0349	0.0384	0.0616	0.0609	0.0764	0.0718	0.0787	0.0713
80 (26.7)	0.0239	0.0295	0.0327	0.0530	0.0529	0.0671	0.0639	0.0708	0.0656
90 (32.2)	0.0202	0.0251	0.0280	0.0459	0.0463	0.0592	0.0569	0.0638	0.0599
100 (37.8)	0.0172	0.0215	0.0243	0.0400	0.0407	0.0526	0.0510	0.0578	0.0547
110 (43.3)	0.0147	0.0186	0.0211	0.0351	0.0360	0.0469	0.0460	0.0525	0.0501
120 (48.9)	0.0127	0.0162	0.0185	0.0310	0.0321	0.0421	0.0415	0.0478	0.0460
130 (54.4)	0.0111	0.0142	0.0163	0.0275	0.0287	0.0379	0.0377	0.0437	0.0424
140 (60.0)	0.0099	0.0125	0.0145	0.0246	0.0258	0.0343	0.0343	0.0401	0.0391

[A] *Correction = (mole % C6 in gas) × (glycol circulation, gph) × CF .*
[B] If hexanes plus heavies mole % is not known, use 0.65 %.

TABLE X1.7 Pentane (C3) Correction Factors (CF) for ARL Total VOC Emissions for Units with Kimray Pump[A,B]

Contactor Temperature, °F (°C)	Contactor Pressure, psig (KPa)								
	200	300	400	500	600	700	800	900	1000
40 (4.4)	0.0152	0.0208	0.0257	0.0218	0.0237	0.0252	0.0262	0.0274	0.0279
50 (10.0)	0.0141	0.0194	0.0241	0.0202	0.0222	0.0237	0.0248	0.0261	0.0269
60 (15.6)	0.0131	0.0182	0.0227	0.0188	0.0208	0.0224	0.0236	0.0250	0.0259
70 (21.1)	0.0123	0.0171	0.0214	0.0176	0.0195	0.0212	0.0225	0.0240	0.0249
80 (26.7)	0.0116	0.0162	0.0203	0.0165	0.0185	0.0201	0.0215	0.0230	0.0241
90 (32.2)	0.0110	0.0154	0.0194	0.0156	0.0175	0.0191	0.0205	0.0222	0.0232
100 (37.8)	0.0104	0.0146	0.0185	0.0148	0.0167	0.0183	0.0197	0.0213	0.0225
110 (43.3)	0.0099	0.0140	0.0178	0.0140	0.0159	0.0175	0.0189	0.0206	0.0217
120 (48.9)	0.0095	0.0134	0.0171	0.0134	0.0153	0.0168	0.0183	0.0199	0.0211
130 (54.4)	0.0091	0.0129	0.0165	0.0128	0.0146	0.0162	0.0176	0.0192	0.0204
140 (60.0)	0.0088	0.0124	0.0159	0.0123	0.0140	0.0157	0.0170	0.0187	0.0199

[A] *Correction = (mole % C3 in gas) × (glycol circulation, gph) × CF*.
[B] If propane mole % is not known, use 4.2 %.

TABLE X1.8 Butane (C4) Corrrection Factors (CF) for ARL Total VOC Emissions for Units with Kimray Pump[A,B]

Contactor Temperature, °F (°C)	Contactor Pressure, psig (KPa)								
	200	300	400	500	600	700	800	900	1000
40 (4.4)	0.0179	0.0293	0.0356	0.0327	0.0345	0.0356	0.0356	0.0363	0.0359
50 (10.0)	0.0158	0.0267	0.0328	0.0300	0.0319	0.0333	0.0337	0.0347	0.0347
60 (15.6)	0.0141	0.0244	0.0303	0.0276	0.0297	0.0312	0.0321	0.0332	0.0335
70 (21.1)	0.0126	0.0226	0.0282	0.0255	0.0277	0.0294	0.0305	0.0320	0.0323
80 (26.7)	0.0114	0.0210	0.0265	0.0238	0.0260	0.0277	0.0290	0.0306	0.0313
90 (32.2)	0.0104	0.0196	0.0249	0.0223	0.0245	0.0263	0.0277	0.0294	0.0303
100 (37.8)	0.0095	0.0184	0.0235	0.0209	0.0231	0.0250	0.0265	0.0283	0.0293
110 (43.3)	0.0088	0.0173	0.0224	0.0197	0.0220	0.0239	0.0254	0.0273	0.0284
120 (48.9)	0.0081	0.0164	0.0213	0.0187	0.0209	0.0229	0.0244	0.0264	0.0276
130 (54.4)	0.0076	0.0157	0.0203	0.0178	0.0200	0.0219	0.0315	0.0255	0.0268
140 (60.0)	0.0072	0.0149	0.0195	0.0170	0.0192	0.0211	0.0227	0.0247	0.0260

[A] *Correction = (mole % C4 in gas) × (glycol circulation, gph) × CF*.
[B] If butanes mole % is not known, use 2.1 %.

TABLE X1.9 Pentane (C5) Correction Factors (CF) for ARL Total VOC Emissions for Units with Kimray Pump[A,B]

Contactor Temperature, °F (°C)	Contactor Pressure, psig (KPa)								
	200	300	400	500	600	700	800	900	1000
40 (4.4)	0.0955	0.1178	0.1305	0.1379	0.1362	0.1312	0.1227	0.1240	0.1142
50 (10.0)	0.0802	0.1003	0.1116	0.1190	0.1189	0.1161	0.1095	0.1137	0.1056
60 (15.6)	0.0682	0.0859	0.0973	0.1034	0.1044	0.1033	0.0997	0.1036	0.0977
70 (21.1)	0.0583	0.0745	0.0849	0.0908	0.0924	0.0925	0.0904	0.0958	0.0910
80 (26.7)	0.0507	0.0649	0.0751	0.0801	0.0825	0.0833	0.0823	0.0882	0.0852
90 (32.2)	0.0444	0.0575	0.0668	0.0709	0.0738	0.0754	0.0752	0.0814	0.0794
100 (37.8)	0.0391	0.0510	0.0599	0.0634	0.0669	0.0683	0.0690	0.0753	0.0740
110 (43.3)	0.0348	0.0457	0.0540	0.0572	0.0606	0.0625	0.0635	0.0698	0.0691
120 (48.9)	0.0312	0.0413	0.0490	0.0517	0.0551	0.0575	0.0586	0.0651	0.0649
130 (54.4)	0.0282	0.0375	0.0448	0.0471	0.0505	0.0530	0.0543	0.0608	0.0610
140 (60.0)	0.0258	0.0343	0.0412	0.0432	0.0465	0.0491	0.0506	0.0570	0.0576

[A] *Correction = (mole % C5 in gas) × (glycol circulation, gph) × CF*.
[B] If pentanes mole % is not known, use 0.80 %.

TABLE X1.10 Hexanes + (C6 +) Correction Factors (CF) for ARL Total VOC Emissions for Units with Kimray Pump[A,B]

Contactor Temperature, °F (°C)	Contactor Pressure, psig (KPa)								
	200	300	400	500	600	700	800	900	1000
40 (4.4)	0.2040	0.2434	0.2596	0.2855	0.2720	0.2532	0.2268	0.2407	0.2111
50 (10.0)	0.1662	0.2008	0.2161	0.2402	0.2325	0.2197	0.1997	0.2164	0.1933
60 (15.6)	0.1368	0.1669	0.1823	0.2039	0.1998	0.1191	0.1785	0.1946	0.1767
70 (21.1)	0.1138	0.1401	0.1547	0.1747	0.1732	0.1678	0.1586	0.1768	0.1614
80 (26.7)	0.0956	0.1188	0.1324	0.1506	0.1508	0.1479	0.1414	0.1597	0.1489
90 (32.2)	0.0811	0.1016	0.1142	0.1308	0.1324	0.1309	0.1264	0.1443	0.1363
100 (37.8)	0.0694	0.0876	0.0993	0.1143	0.1167	0.1166	0.1137	0.1310	0.1249
110 (43.3)	0.0599	0.0762	0.0871	0.1006	0.1305	0.1042	0.1027	0.1194	0.1148
120 (48.9)	0.0521	0.0667	0.0768	0.0891	0.0924	0.0938	0.0930	0.1092	0.1058
130 (54.4)	0.0456	0.0589	0.0682	0.0793	0.0829	0.0847	0.0847	0.1001	0.0978
140 (60.0)	0.0408	0.0521	0.0609	0.0710	0.0747	0.0769	0.0774	0.0921	0.0906

[A] *Correction = (mole % C6 in gas) × (glycol circulation, gph) × CF .*

[B] If hexanes plus heavies mole % is not known, use 0.65 %.

REFERENCES

(1) Falzone, Priscilla J., "Impacts of the Clean Air Act Amendments of 1990 on Glycol Dehydration Units," *1992 Glycol Dehydrator Air Emissions Conference*, Radian Corp., New Orleans, LA, September 1992.

(2) Rueter, Curtis O., et al, "Results of the Glycol Waste Survey," Revised Draft Report, *Gas Research Institute*, Chicago, IL, GRI Contract No. 660-016-02-01, November 1994.

(3) Rueter, Curtis O., et al, "Development of Measurement Methods and Emission Estimation Techniques for Determining BTEX and VOC Emissions from Glycol Dehydrators," Topical Report, *Gas Research Institute*, Chicago, IL, GRI Contract No. 5091-254-22-93, March 1995.

(4) Texas Mid-Continent Oil and Gas Association and the Gas Processors Association, Proposed Glycol Dehydration Sampling and Analytical Protocol, December 1990.

(5) Starrett, Tracey, "New Issues in Louisiana Glycol Dehydrator Regulations," *1994 Glycol Dehydrator/Gas Processing Air Toxics Conference*, Louisiana Department of Environmental Quality (LDEQ), San Antonio, TX, June 1994.

(6) Texas Natural Resource Conservation Commission (TNRCC), "Control of Air Pollution from Volatile Organic Compounds, Chapter 115 Proposal," Austin, TX, July 1994.

(7) Wright Killen and Co., "Natural Gas Dehydration: Status and Trends," Final Report, *Gas Research Institute*, Chicago, IL, GRI-94/009, January 1994.

(8) Rueter, Curtis O., et al, "Condensation Systems for Glycol Dehydrator Emissions: Modeling and Design," Revised Draft Report, U.S. EPA, Research Triangle Park, NC, August 1994.

(9) Myerski, F., et al, "Project Work Plan for the Sampling and Analysis of Associated Wastes from Natural Gas Industry Operations," *Gas Research Institute*, Chicago, IL, GRI Contract No. 5084-253-1117, May 1990.

Standard Guide for Use of Coal Combustion By-Products in Structural Fills[1]

This standard is issued under the fixed designation E 1861; the number immediately following the designation indicates the year of original adoption or, in the case of revision, the year of last revision. A number in parentheses indicates the year of last reapproval. A superscript epsilon (ϵ) indicates an editorial change since the last revision or reapproval.

1. Scope

1.1 This guide covers procedures for the design and construction of engineered structural fills using coal combustion by-products (CCBs).

1.2 The utilization of CCBs under this guide is a component of a pollution prevention program; Guide E 1609 describes pollution prevention activities in more detail. Utilization of CCBs in this manner conserves land, natural resources, and energy.

1.3 This guide applies only to CCBs produced solely by the combustion of coal. It does not apply to ash or other combustion products derived from the burning of waste; municipal, industrial, or commercial garbage; sewage sludge or other refuse, or both; derived fuels; wood; wood waste products; rice hulls; agricultural waste; or other non-coal fuels or other such fuels blended with coal, or some combination thereof.

1.4 The testing, engineering, and construction practices for CCB fills are similar to generally accepted practices for natural soil fills. The CCB structural fills should be designed with generally accepted engineering practices.

1.5 Laws and regulations governing the use of CCBs vary by state. The user of this guide has the responsibility to determine and comply with applicable requirements.

1.6 The values stated in inch-pound units are to be regarded as the standard. The SI units given in parentheses are for information only.

1.7 *This standard does not purport to address all of the safety concerns, if any, associated with its use. It is the responsibility of the user of this standard to establish appropriate safety and health practices and determine the applicability of regulatory limitations prior to use.*

2. Referenced Documents

2.1 *ASTM Standards:*

C 150 Specification for Portland Cement[2]
C 188 Test Method for Density of Hydraulic Cement[2]
C 311 Test Methods for Sampling and Testing Fly Ash or Natural Pozzolans for Use as a Mineral Admixture in Portland-Cement Concrete[3]
C 595/C595M Specification for Blended Hydraulic Cements[2]
D 75 Practice for Sampling Aggregates[4]
D 420 Guide to Site Characterization for Engineering, Design, and Construction Purposes[5]
D 422 Test Method for Particle-Size Analysis of Soils[5]
D 653 Terminology Relating to Soil, Rock, and Contained Fluids[5]
D 698 Test Method for Laboratory Compaction Characteristics of Soil Using Standard Effort (12 400 ft-lbf/ft^3 (600 kN-m/m^3))[5]
D 854 Test Method for Specific Gravity of Soils[5]
D 1195 Test Method for Repetitive Static Plate Load Tests of Soils and Flexible Pavement Components, for Use in Evaluation and Design of Airport and Highway Pavements[5]
D 1196 Test Method for Nonrepetitive Static Plate Load Tests of Soils and Flexible Pavement Components, for Use in Evaluation and Design of Airport and Highway Pavements[5]
D 1452 Practice for Soil Investigation and Sampling by Auger Borings[5]
D 1556 Test Method for Density and Unit Weight of Soil In Place by the Sand-Cone Method[5]
D 1557 Test Methods for Laboratory Compaction Characteristics of Soil Using Modified Effort (56 000 ft-lbf/ft^3 (2700 kN-m/m^3))[5]
D 1586 Test Method for Penetration Test and Split-Barrel Sampling of Soils[5]
D 1883 Test Method for CBR (California Bearing Ratio) of Laboratory-Compacted Soils[5]
D 2166 Test Method for Unconfined Compressive Strength of Cohesive Soil[5]
D 2167 Test Method for Density and Unit Weight of Soil in Place by the Rubber Balloon Method[5]
D 2216 Test Method for Laboratory Determination of Water (Moisture) Content of Soil and Rock[5]
D 2435 Test Method for One-Dimensional Consolidation Properties of Soils[5]
D 2850 Test Method for Unconsolidated, Undrained Compressive Strength of Cohesive Soils in Triaxial Compression[5]

---

[1] This guide is under the jurisdiction of ASTM Committee E-50 on Environmental Assessment and is the direct responsibility of Subcommittee E50.03on Pollution Prevention, Reuse, Recycling, and Environmental Efficiency.

Current edition approved March 10, 1997. Published May 1997.

[2] *Annual Book of ASTM Standards*, Vol 04.01.

[3] *Annual Book of ASTM Standards*, Vol 04.02.

[4] *Annual Book of ASTM Standards*, Vol 04.03.

[5] *Annual Book of ASTM Standards*, Vol 04.08.

D 2922 Test Methods for Density of Soil and Soil-Aggregate in Place by Nuclear Methods (Shallow Depth)[5]
D 3080 Test Method for Direct Shear Test of Soils Under Consolidated Drained Conditions[5]
D 3550 Practice for Ring-Lined Barrel Sampling of Soils[5]
D 3877 Test Methods for One-Dimensional Expansion, Shrinkage, and Uplift Pressure of Soil-Lime Mixtures[5]
D 4253 Test Methods for Maximum Index Density and Unit Weight of Soils Using a Vibratory Table[5]
D 4254 Test Method for Minimum Index Density and Unit Weight of Soils and Calculation of Relative Density[5]
D 4318 Test Method for Liquid Limit, Plastic Limit, and Plasticity Index of Soils[5]
D 4429 Test Method for CBR (California Bearing Ratio) of Soils in Place[5]
D 4643 Test Method for Determination of Water (Moisture) Content of Soil by the Microwave Oven Method[5]
D 4959 Test Method for Determination of Water (Moisture) Content of Soil by Direct Heating Method[6]
D 4972 Test Method for pH of Soils[6]
D 5084 Test Method for Measurement of Hydraulic Conductivity of Saturated Porous Materials Using a Flexible Wall Permeameter[6]
D 5239 Practice for Characterizing Fly Ash for Use in Soil Stabilization[6]
E 1527 Practice for Environmental Site Assessments: Phase I Environmental Site Assessment Process[7]
E 1528 Practice for Environmental Site Assessments: Transaction Screen Process[7]
E 1609 Guide for Development and Implementation of a Pollution Prevention Program[7]
G 51 Test Method for pH of Soil for Use in Corrosion Testing[8]
G 57 Test Method for Field Measurement of Soil Resistivity Using the Wenner Four-Electrode Method[8]

2.2 *AASHTO Standards:*[9]
T 288 Determining Minimum Laboratory Soil Resistivity
T 289 Determining pH of Soil for Use in Corrosion Testing
T 290 Determining Water Soluble Sulfate Ion Content in Soil
T 291 Determining Water Soluble Chloride Ion Content in Soil

3. Terminology

3.1 *Definitions*—Definitions are in accordance with Terminology D 653.

3.2 *Definitions of Terms Specific to This Standard:*

3.2.1 *baghouse, n*—a facility constructed at some coal-fired power plants consisting of fabric filter bags that mechanically trap particulate (fly ash) carried in the flue gases.

3.2.2 *beneficial use, n*—projects promoting public health and environmental protection, offering equivalent success relative to other alternatives, and preserving natural resources.

3.2.3 *boiler slag, n*—a molten ash collected at the base of slag tap and cyclone boilers that is quenched in a water-filled hopper and shatters into black, angular particles having a smooth, glassy appearance.

3.2.4 *bottom ash, n*—agglomerated ash particles formed in pulverized coal boilers that are too large to be carried in the flue gases and impinge on the boiler walls or fall through open grates to an ash hopper at the bottom of the boiler. Bottom ash is typically grey to black in color, is quite angular and has a porous surface texture.

3.2.5 *coal combustion by-products, n*—fly ash, bottom ash, boiler slag, or flue gas desulfurization (FGD) material resulting from the combustion of coal.

3.2.6 *drainage blanket, n*—a uniform layer of permeable material (such as sand, crushed stone, or bottom ash) installed with properly designed filter media at the base of a structural fill to maintain the fill in a drained condition.

3.2.7 *electrostatic precipitator, n*—a facility constructed at some coal-fired power plants to remove particulate matter (fly ash) from the flue gas by producing an electric charge on the particles to be collected and then propelling the charged particles by electrostatic forces to collecting curtains.

3.2.8 *flue gas desulfurization (FGD) material, n*—by-products of FGD systems that have been processed using forced oxidation or chemical fixation and stabilization to produce a soil-like material.

3.2.9 *fly ash, n*—coal ash that exits a combustion chamber in the flue gas. Fly ash is typically a pozzolan. Some fly ash also exhibits self-hardening properties in the presence of moisture.

3.2.10 *internal erosion, n*—piping; the progressive removal of soil particles from a mass by percolating water, leading to the development of channels.

3.2.11 *pozzolans, n*—siliceous or siliceous and aluminous materials that in themselves possess little or no cementitious value but will, in finely divided form and in the presence of moisture, chemically react with calcium hydroxide at ordinary temperatures to form compounds possessing cementitious properties.

3.2.12 *stabilized CCBs, n*—CCBs that are blended with a cementitious binder to induce or enhance a pozzolanic reaction.

3.2.13 *structural fill, n*—an engineered fill with a projected beneficial end use that is typically constructed in layers of uniform thickness and compacted to a desired unit weight (density) in a manner to control the compressibility, strength, and hydraulic conductivity of the fill.

3.2.14 *thixotropic, adj*—the property of a material that enables it to stiffen in a relatively short time on standing, but upon agitation or manipulation to change to a very soft consistency or to a fluid of high viscosity, the process being completely reversible.

4. Significance and Use

4.1 *General*—The CCBs can be effective materials for the construction of engineered, structural fills. The CCBs may be used as: structural fill for building sites and foundations; embankments for highways and railroads, dikes, and levees;

---

[6] *Annual Book of ASTM Standards*, Vol 04.09.
[7] *Annual Book of ASTM Standards*, Vol 11.04.
[8] *Annual Book of ASTM Standards*, Vol 03.02.
[9] Available from the American Association of State Highway and Transportation Officials, 444 N. Capitol St., NW, Washington, DC 20001.

and in any other application requiring a compacted fill material. Their low unit weight, relatively high shear strength, ease of handling, and compaction all make CCBs useful as fill material. The CCBs may be a cost-effective fill material in many areas because they are available in bulk quantities and reduce the expenditures required for the purchase, permits, and operation of a soil borrow pit. Also, because CCBs are abundantly produced by-products, their use in large-volume applications such as in the construction of structural fills provides an outlet for material, provided that the CCB is environmentally and geotechnically suitable for the desired use.

4.1.1 Fly ash is typically a pozzolan, however, certain fly ash and FGD by-products may exhibit self-hardening properties as well. The laboratory testing and design considerations for self-hardening and non-self-hardening CCBs vary in some instances. Guidance is provided in Sections 6-9.

4.2 *Engineering Properties and Behavior*—The CCB structural fills may be constructed with fly ash, bottom ash, boiler slag, FGD material, or combinations thereof. Each of these materials typically exhibits general engineering properties that must be considered in the design of a CCB structural fill. These general engineering properties are discussed in the following sections; however, it should be noted that the specific engineering properties of these materials can vary greatly from source to source and must be evaluated for each material, or combination of materials, to be utilized for a structural fill.

4.2.1 *Unit Weight*—Many CCBs have relatively low unit weights. The low unit weight of these materials can be advantageous for some structural fill applications. The lighter weight material will reduce the load on weak layers or zones of soft foundation soils such as poorly consolidated or landslide-prone soils. Additionally, the low unit weight of these materials will reduce transportation costs since less tonnage of material is hauled to fill a given volume.

4.2.1.1 Fly ash is typically lighter than the fill soils it replaces with unit weight ranging from about 50 to 100 pcf (8 to 16 kN/m^3).

4.2.1.2 Bottom ash is also typically lighter than coarse-grained soils of similar gradation.

4.2.1.3 Boiler slag is typically as heavy, if not heavier, than natural soils of similar gradation.

4.2.1.4 The FGD materials are also relatively lightweight, with unit weights ranging from about 50 to 100 pcf (8 to 16 kN/m^3).

4.2.2 *Compaction Characteristics*—Most CCBs can be placed and compacted in a manner very similar to soil and aggregate fill materials. In fact, most CCBs exhibit very little cohesion and are not as sensitive to variations in moisture content as natural soils.

4.2.2.1 Fly ash and FGD material are typically placed and compacted in a manner similar to non-cohesive fine-grained soils. Smooth drum vibratory rollers typically compact these materials most effectively. Although not always, fly ash and FGD material typically exhibit a measurable moisture-density relationship that can be utilized for compaction quality control. It should be noted that fly ash and FGD material that exhibit self-hardening properties must be compacted soon after the addition of water. Compaction criteria are usually not specified for FGD material that exhibits thixotropic properties.

4.2.2.2 Bottom ash is generally placed and compacted in a manner similar to non-cohesive coarse-grained soils or fine aggregate. Smooth drum vibratory rollers typically are most effective for the compaction of these materials. Bottom ash may or may not exhibit consistent moisture-density relationships.

4.2.2.3 Boiler slag is generally placed and compacted in a manner similar to non-cohesive coarse-grained soils or fine aggregate. Smooth drum vibratory rollers typically are most effective for the compaction of these materials. As with bottom ash, boiler slag may or may not exhibit consistent moisture-density relationships.

4.2.3 *Strength*:

4.2.3.1 *Shear Strength*—For non-self-hardening fly ash and bottom ash, shear strength is derived primarily from internal friction. Typical values for angles of internal friction for non-self-hardening fly ash are higher than many natural soils. These ashes are non-cohesive and although the ash may appear cohesive in a partially saturated state, this effect is completely lost when the material is either completely dried or saturated.

(1) Due to its angular shape, the shear strength of bottom ash is typically greater than fly ash and is similar to the shear strength of natural materials of similar gradation. However, friable bottom ash may exhibit lower shear strength than natural materials of similar gradation.

(2) The shear strength of boiler slag may be higher than that of natural materials of similar gradation, in part due to the typically angular shape and hardness of the particles.

4.2.3.2 *Compressive Strength*—Self-hardening and stabilized CCBs and FGD material undergo a cementing process that increases with time. Hydration of dry self-hardening CCBs commences immediately upon exposure to water and can cement the CCB particles in a loose state, reducing the compacted density and strength. Compressive strengths in excess of 1000 psi can be achieved if the CCBs are compacted immediately after incorporation of water.

4.2.4 *Consolidation Characteristics*—Structural fills constructed of fly ash or FGD material typically exhibit small amounts of time-dependent, post-construction consolidation. This is because excess pore water pressures dissipate relatively rapidly, and thus, most of the embankment settlement or deformation occurs due to elastic deformation of the material, rather than by classical consolidation. Most deformation due to the mass of the fill or structure thereon generally occurs during construction.

4.2.4.1 Bottom ash and boiler slag are free-draining materials that can be compacted into a relatively dense, incompressible mass. For these reasons, structural fills constructed of bottom ash or boiler slag also typically exhibit small amounts of time-dependent, post-construction consolidation or deformation, with most deformation occurring during construction.

4.2.4.2 Self-hardening fly ash and FGD material typically exhibit minimal post-construction consolidation or deformation due to cementing and solidification of the CCBs.

4.2.5 *Permeability*—The permeability of non-self-hardening fly ash is similar to values observed for natural silty soils.

4.2.5.1 Self-hardening fly ash and FGD material are relatively impermeable, with permeability values similar to natural clays.

4.2.5.2 Bottom ash and boiler slag are typically as permeable as granular soils of similar gradation.

4.2.6 *Erosion Characteristics*:

4.2.6.1 *Internal Erosion (piping)*—Non-self-hardening fly ash is subject to internal erosion due to its fine-grained, non-cohesive nature. Internal erosion can be controlled by providing adequate surface water controls to minimize infiltration and by providing internal drainage when warranted.

(1) Bottom ash and boiler slag typically are well graded and capable of being compacted to a stable mass. These attributes usually preclude any problems arising from internal piping of material.

(2) Self-hardening fly ash and FGD material are usually not subject to internal erosion.

4.2.6.2 *Surface Erosion*—All CCBs may be eroded by wind or water and require use of erosion controls similar to those commonly used on earthwork construction projects.

4.2.7 *Swelling*—Some self-hardening CCBs may swell with time. Paragraph 6.3.9 provides guidance on evaluating the swelling potential of CCBs.

4.2.8 *Liquefaction and Frost Heave*—Although fine-grained and non-cohesive materials such as fly ash are susceptible to liquefaction and frost heave when saturated, these problems are readily controlled by design practices that allow for drainage away from the ash fill. Because of its sensitivity to moisture, it is standard practice to design fly ash fills to be well drained. Typically, drainage blankets to provide internal drainage and serve as a capillary barrier are included at the base of fills. Also, locating fills in areas where they are not subject to saturation or infiltration by surface or ground water is normally considered in design. Self-hardening and stabilized fly ash and FGD material are not susceptible to liquefaction.

4.2.8.1 Well-compacted bottom ash and boiler slag are not typically susceptible to either liquefaction or frost heave. However, some of the finer bottom ash materials may behave quite similarly to that of fly ash and would require the same consideration for design as fly ash embankments.

4.3 *Environmental Considerations*

4.3.1 *Regulatory Framework*:

4.3.1.1 *Federal*—The U.S. Environmental Protection Agency (USEPA) has completed a study of CCBs for the U.S. Congress and has issued a formal regulatory determination **(1,2)**.[10] The EPA "encourages the utilization of coal combustion by-products and supports State efforts to promote utilization in an environmentally beneficial manner" **(3)**. There is currently no regulatory program at the Federal level that regulates the use of CCBs in structural fills.

4.3.1.2 *State and Local*—There is considerable variation in state-mandated permitting and other regulatory requirements. Some states have specific beneficial use provisions, while other states have no regulations addressing beneficial use.

4.3.2 *Water Quality*—The design and construction of CCB structural fills should consider the potential impacts on ground water and surface water to ensure protection of human health and the environment.

4.3.2.1 The CCB structural fills should typically be designed to provide separation between the CCB fill material and the seasonal high ground water table. Measures taken to protect both ground water and surface water include proper compaction to reduce permeability and erosion and placement of suitable cover material to minimize infiltration and erosion. Structural fills should also be sloped to encourage runoff while minimizing erosion concerns.

4.3.3 *Air Quality*:

4.3.3.1 *Dust Control*—Dusting must be controlled during placement of CCBs in structural fills in order to avoid fugitive dust and to protect workers. See 9.2.1 for recommended dust control measures.

4.3.3.2 *Radionuclides*—As with other structural fill materials, certain radioactive elements are known to occur naturally in CCBs. The model standards and techniques for controlling radon in accordance with Ref **(4)** are recommended for new building construction, where needed.

4.4 *Economic Benefits*—The use of CCBs in structural fill can have economic benefits. These benefits are affected by local and regional factors including production rates, processing and handling costs, transportation costs, availability of competing materials, environmental concerns, and the experience of materials specifiers, design engineers, purchasing agents, contractors, legislators, regulators, and other professionals.

4.4.1 The CCBs can be a cost-effective fill material. In many areas, it is available in bulk quantities. Use of CCBs reduces the expenditures for the purchase, permits, and operation of a soil borrow pit.

5. Site Characterization

5.1 *General*—The siting and design of a CCB structural fill requires the same characterization of site conditions that is typically required of earthwork construction projects of similar size. An investigation of the geologic and hydrologic conditions at the site is required to determine design parameters for the structural fill. In addition, consideration of environmental resources at or near the site is required to avoid or minimize negative environmental consequences. Practices E 1527 and E 1528 may be applied whenever a real estate transaction is involved.

5.2 *Geologic and Hydrologic Investigation*—Subsurface conditions at the site must be understood. This typically involves a review of available information about the site, a site reconnaissance by a geologist or geotechnical engineer, and extraction of soil and rock samples from the subsurface for classification and testing. Guide D 420 provides guidance for conducting subsurface investigations.

5.3 *Environmental Resources*—Many sensitive environmental resources such as wetlands, floodplains, rare and endangered species, and cultural resource areas are afforded

[10] The boldface numbers in parentheses refer to a list of references at the end of this guide.

protection by federal, state, and local regulations and ordinances. Appropriate action should be taken to comply with the requirements of the regulatory agencies having jurisdiction at the structural fill site.

6. Laboratory Test Procedures

6.1 *General*—Laboratory testing of the proposed fill materials is needed to determine and confirm material properties for design. Test results also provide documentation that may be requested or required by site owners and regulatory agencies. The tests to be conducted should be determined based on site conditions, knowledge of the CCBs, end use, and local environmental considerations.

6.2 *Sampling and Handling*—Sampling CCBs for testing purposes should conform to Practice D 75 or Test Methods C 311 as appropriate. Guide D 420 with sample extraction conducted in accordance with Practice D 1452, Test Method D 1586, or Practice D 3550, as appropriate, should be considered. Proper laboratory protocols for handling finegrained material should be followed.

6.3 *Physical and Engineering Characteristics*—Several standard test methods developed for soils may be used to determine CCB properties for use in structural fills. These test methods define physical and engineering parameters for use in design, construction control, and for comparison to other materials.

6.3.1 *Grain-Size Distribution*—Test Method D 422 is commonly used for determining the grain size distribution of CCBs. For fly ash and FGD material, a substantial portion of the material will be finer than the No. 200 sieve, and hydrometer analyses will also be required. Distilled water is used in the hydrometer test with a deflocculating agent added to prevent fly ash or FGD material from forming flocs. Self-hardening fly ash(es) and FGD material may require use of alcohol or other nonreactive solution in place of the standard solution used. Fly ash often has a relatively uniform particle size and precautions against overloading sieves are warranted. Specimen loss through dusting can also be a problem. Specific gravity may vary with particle size. Specific gravity values used in hydrometer analyses should be appropriate to the portion of the sample being tested.

6.3.2 *Specific Gravity*—Test Method D 854 is normally used for CCBs. For some fly ash and FGD samples, a significant portion of the particles may have a density less than water and float. Agitation of the slurry may be needed to keep the particles in suspension so that the average specific gravity can be obtained. Alternately for this ash, self-hardening fly ash, and FGD material, Test Method C 188, which uses kerosene as the fluid, may be used.

6.3.3 *Water Content*—Test Method D 2216 is normally used for CCBs. For self-hardening fly ash and FGD material, lowering the drying temperature to 140°F (60°C) may be considered to avoid driving off the water of hydration.

6.3.4 *Compaction*:

6.3.4.1 *Fly Ash and FGD Material*—Test Methods D 698 or D 1557 may be used depending on end use. For dry self-hardening fly ash and FGD material, the time interval between wetting and compaction in the laboratory should be similar to that anticipated during construction to account for the influence of the rate of hydration on compaction characteristics. Compaction criteria are not typically developed for FGD material that exhibits thixotropic properties because excessive compaction may cause the material to liquefy.

6.3.4.2 *Bottom Ash and Boiler Slag*—Test Methods D 4253 and D 4254 may be used for the determination of maximum and minimum density of coarse-grained CCBs that do not exhibit a moisture-density relationship.

6.3.5 *Strength*—Material strength is defined by shear strength and compressive strength.

6.3.5.1 *Shear Strength*—Test Method D 3080 can be used to determine the shear strength parameters of compacted CCB specimens under drained conditions. This test is preferred because it models the drained conditions that typically exist in a structural fill constructed of CCBs. When using Test Method D 3080, the method is modified in that the shear box is not to be filled with water as required by Test Method D 3080.

6.3.5.2 *Compressive Strength of Non-Self-Hardening CCBs*—Test Method D 2850 can be used to predict the as-constructed compressive strength of the CCB fill and to design for specific site conditions, loading conditions, and fill height. Specimens tested for strength parameters shall be compacted to the unit weights and water contents required by the project compaction specifications.

6.3.5.3 *Compressive Strength of Self-Hardening Fly Ash and FGD Material*—Test Method D 2166 can be used to determine the unconfined compressive strength at various ages to evaluate short-term and long-term strength development.

6.3.6 *Hydraulic Conductivity*—Test Method D 5084 is commonly used to determine the hydraulic conductivity of saturated CCBs. Hydraulic conductivity is used to estimate the quantity of infiltration for designing underdrains.

6.3.7 *Compressibility*—Samples should be prepared at the degree of compaction specified for construction and at the optimum water content determined by the compaction test. This is because fly ash and FGD material tend to lose surface stability in the field when compacted at water contents greater than the optimum for compaction. Test Method D 2435 can be used to determine the compressibility of saturated or unsaturated samples. The CCBs consolidate rapidly, therefore, compressibility, typically, is not a design concern. Because of the non-cohesive nature of some CCBs, extra care in sample handling is needed.

6.3.8 *Plasticity Index*—Test Method D 4318 is a commonly used test for classifying and comparing fine-grained soils. Because fly ash is nonplastic and is vitrified, correlations developed for cohesive soils with clay particles may not be applicable to fly ash. The plastic limit determination may be performed to ensure that the fly ash is nonplastic.

6.3.9 *Swelling*—Test Methods D 3877 can be used to determine the swelling potential of self-hardening fly ash and FGD material. Reactions producing the expansive properties do not commence for a period of more than 30 days after initial ash hydration. The test procedures must address this delayed reaction. The procedure should be modified to extend the wetting and drying cycles to a frequency determined by a qualified design engineer.

6.4 *Chemical Characteristics*—Chemical analyses are routinely conducted by many CCB producers as a means of determining material variation. The results of these analyses should be communicated to users of this material by means of a Material Safety Data Sheet (MSDS) or some similar communication. For the structural fill designer, these results provide information on characteristics that may need to be considered in design, particularly with regard to assessing chemical interaction between fill and other materials or structures. Tests for soluble species may also be required by local regulatory agencies.

6.4.1 *Chemical Composition*—Test Methods C 311 is often used to determine the major chemical constituents of CCB samples.

6.4.2 *pH*—Test Method D 4972 or Practice D 5239 may be used to determine CCB pH. In assessing the test results, consideration should be given to the possibility that the pH of the CCB may vary with age, water content, and other conditions.

6.4.3 *Resistivity*—Test Method G 57, a field test, is used to measure CCB resistivity as an indicator of possible corrosion potential for embedded metals. An alternate laboratory procedure is AASHTO Interim Method of Test T288. Likely, field water contents should be considered in assessing test conditions and results. Field water contents in drained CCB fills are likely to be close to the optimum water content for compaction. AASHTO Interim Methods of Test T289, T290, and T291 provide measurements of the pH, water-soluble sulfate ion content, and water soluble chloride ion content of the CCBs that are useful in evaluating corrosion potential. Test Method G 51 is also used to determine the pH of soil for use in corrosion testing.

6.4.4 *Sulfate*—Sulfate content as determined from the CCBs chemical analysis by Test Methods C 311, or other method is used in a preliminary assessment of the potential for sulfate attack on concrete. As with corrosivity, likely field water conditions and variations in concentrations with time should be considered.

7. Design Considerations

7.1 *General*—Design involves developing a plan that satisfies the site-specific design requirements within the physical and engineering constraints of the fill material and the proposed function of the completed project. The underlying material and CCB must support its own mass and that of the load to be placed on it without excessive settlement, and require no long-term maintenance beyond that typically exercised for the intended use. If applicable, the CCB must have sufficient shear strength to provide stable slopes. The design process for CCBs is similar to that normally followed for cohesionless natural soil materials. Cohesion developed by self-hardening CCBs can also be considered in the design of fill slopes and determination of bearing capacity. References **(5-9)** provide additional information regarding laboratory testing, design, and construction procedures.

7.2 *Design Process*—The design process is an iterative procedure whereby information concerning site and material constraints are balanced against project goals. Information is developed in increasing detail and analyzed to evaluate whether the site development plan is satisfactory. Adjustments to the development plan or material properties are made to accomplish the project goals.

7.2.1 *Conceptual Site Model*—Initially a conceptual site model is developed that identifies specific site characteristics regarding geology, hydrology, and topography. The pertinent material characteristics such as shear strength, load-bearing capability, and other properties regarding compaction, particle (grain) size, hydraulic conductivity, and general chemical properties should be determined for design use. The model should address the changes in CCB material properties that occur with age, such as strength gain. Site and material characteristics are determined by experience, literature search, site reconnaissance, field testing and sampling, and laboratory testing.

7.2.2 *Conceptual Design*—Conceptual design involves preparing a plan for site development that meets project goals within the constraints of site and material characteristics and project finances. A general site layout that balances desired final configuration against current topography, material properties and volumes, and site features is prepared. Since the engineering properties (hydraulic conductivity, strength, and compressibility) of CCBs are a function of the degree of compaction, a study must be performed to establish the degree of compaction that will satisfy the project goals. A general assessment of project feasibility is normally conducted as part of conceptual design.

7.2.3 *Detailed Design*—The detailed design involves drainage design for surface water and ground water, planning of site preparation, and determination of final cover. Analyses of structural performance are performed. If needed, corrosion protection for buried metal or concrete is specified.

7.2.4 *Other Design Considerations*—Consideration must be given to potential water quality and air quality impacts in accordance with 4.3.2 and 4.3.3, respectively.

7.2.4.1 During the design process it is also appropriate to resolve any questions and approvals needed from local or state environmental agencies. While requirements vary from state to state, a thorough geologic and hydrologic survey of the site is commonly required.

7.2.4.2 The ultimate end use of the site can present special design considerations. For example, fly ash or FGD material is not an appropriate medium for septic systems. A thicker soil cover may be appropriate depending on the planned end use of the site. Deed restrictions may be warranted in some instances.

7.3 *Site Preparation and Internal Drainage*—Some structural fills constructed of non-self-hardening fly ash or FGD material must be well drained because of the sensitivity of the material to the flow of water (that is, piping). Problems such as slope stability, liquefaction, and frost heave that may result from saturation of the CCBs are thus avoided. When necessary, a drainage blanket can be used to provide internal drainage and serve as a capillary barrier. Also, locating CCBs in areas where they are not subject to inundation by surface or ground water is normally considered in design.

7.3.1 *Site Preparation*—Site preparation involves developing the site in a suitable condition to facilitate construction of the structural fill. Surface drainage is diverted and controlled.

Erosion and sedimentation controls are installed. If needed, wet areas are allowed to drain and dry. Unsuitable materials such as vegetation and topsoil are removed and the subgrade is prepared. Provisions to stockpile any soil needed for final cover are included.

7.3.2 *Site Drainage*—Provisions for positive site drainage are essential if the structural fill is to be reliably maintained in an unsaturated condition. Drainage of seeps and springs encountered during construction should be provided for in design of a site drainage system. A series of perforated pipe drains or aggregate-filled trenches are commonly used for this purpose. These systems are flexible and can be expanded in areal extent as needed to accommodate conditions encountered during construction. Adequate filter protection of drains to ensure long-term, maintenance-free performance should be included. Any provisions needed to control site ground-water levels through collection and drainage should be included in the design.

7.3.3 *Drainage Blanket*—For non-self-hardening fly ash and FGD material, a drainage blanket of free-draining material may be used. The drainage blanket also serves as a barrier to capillary saturation. Coarser CCBs such as bottom ash often have a suitable particle size range to serve as a drainage blanket. Sand, gravel, or other aggregate can also be used depending upon the gradation of these materials. Adequate filter protection such as a geotextile between the fill and drainage blanket must be considered and included to ensure satisfactory long-term performance. The drainage blanket should be designed so that the outlets will remain freely drained. Including outlet pipes with rodent screens is one method that is often satisfactory.

7.4 *Surface Cover and Drainage*—Provisions must be made for controlling erosion of CCB fills. Due to its fine-grained, non-cohesive nature, non-self-hardening fly ash is readily eroded. Unprotected, compacted CCBs are erodible when exposed to surface runoff or high winds. Erosion control is normally accomplished by controlling surface drainage and establishing permanent cover with pavement or soil and vegetation.

7.4.1 *Cover*—Effective cover to control erosion can be either pavement or soil depending upon the final use of the surface. Surface configuration should include provisions for controlled, positive drainage of surface runoff. Minimum slopes to prevent ponding both on surfaces and in drainage ways of approximately 1 to 3 % are desirable so that settlement and minor surface variations can be accommodated.

7.4.2 *Soil Thickness/Vegetation*—The required thickness of soil cover varies and will depend upon site use, climate, and the type of vegetation to be established. The most important consideration is to control wind and water erosion of the surface. On sites where erosion potential is small, 6 in. (150 mm) of cover may provide protection, but 1 ft is probably a practical minimum thickness in most cases. Where erosion potential is greater, or deeper rooted vegetation is planned, greater thicknesses may be warranted. In some cases CCB soil blends are used as part of the cover to reduce the need for soil borrow. In these applications, testing of the blend to determine its suitability as a growing medium should be conducted.

7.4.3 *Surface Drainage*—Positive surface drainage is needed to prevent ponding that can lead to erosion problems. Suitable channel linings designed to accommodate storm flows without damage are needed. Slopes on surface areas and in drainage channels should be sufficient to prevent ponding and avoid long-term maintenance problems.

7.5 *Structural Performance*—In order to perform satisfactorily, any fill material must support its own mass, that of the loads to be placed on it, and have acceptable settlement. Each of these aspects is analyzed as part of the design process.

7.5.1 *Slope Stability*—Embankment slopes should be stable and able to stand without slumping or sliding. Stability analyses should consider static, dynamic and seismic loadings, and seepage forces, as appropriate. Desired factors of safety typically range from 1.2 (seismic and dynamic) to 1.5 (static). Stability of exterior slopes, foundation soils and embankment combined, and cover soils should be analyzed.

7.5.2 *Bearing Capacity*—Structures located on or within the fill should be stable and function without excessive settlement or tilting. The bearing capacity of foundations supported by the fill and underlying materials should be analyzed. The ability of the fill to support slabs and pavements to be located on the fill surface should be assessed.

7.5.3 *Settlement*—As with any fill material, settlement due to consolidation and compression of the fill and the underlying materials should be considered in design. Settlement may adversely affect project performance if not considered in design. Alternately, consideration of settlement magnitude and duration is commonly compensated for in the design process without difficulty.

7.6 *Compaction*—Proper and uniform compaction (including control of molding water content) of CCBs placed in the structural fill increases the strength of the material, reduces the compressibility, and produces a relatively uniform structural fill. The CCBs are readily spread and compacted by conventional construction equipment; vibratory compactors operated at or near resonant frequency are particularly effective.

7.6.1 *Fly Ash and Non-Thixotropic FGD Material*—Because they are fine-grained, fly ash and non-thixotropic FGD material exhibit compaction behavior under static compaction similar to natural soils in that compaction is sensitive to molding water content. Most fly ash and non-thixotropic FGD material have well-defined compaction relationships, that is, for a given static compactive energy, there exists an optimum water content at which compaction of the CCB will achieve the maximum dry unit weight. Attempting to compact fly ash or FGD material above the optimum water content results in displacement of the fly ash or FGD material and limited densification is attained. Using static compaction, the compaction of fly ash or FGD material with water contents below the optimum water content requires more compactive effort to achieve desired results. However, the compaction of fly ash is not especially sensitive to variations in water content when using vibratory compactors operated at the resonant frequency. Thus, fly ash that is several percent below the optimum water content can be readily compacted using vibratory compactors operated at the resonant frequency. Compaction characteristics of dry self-hardening ash or FGD material change rapidly with time after exposure to water.

7.6.1.1 This property is a result of the rapid rate of hydration that produces a cementitious reaction. A reduction in maximum density of more than 30 lb/ft^3 can occur and must be addressed by the design and compaction procedures.

7.6.2 *Thixotropic FGD Material*—It is not appropriate to specify compaction criteria for thixotropic FGD material as excessive compaction may cause the material to liquefy. For this material, strength is not derived from the internal friction between particles, but from the cementitious reaction that occurs as the material cures. For this reason, temperature effects should be considered in developing the placement procedure.

7.6.3 *Bottom Ash and Boiler Slag*—These CCBs are typically free-draining, therefore, unless saturated, the moisture content of these materials has little influence on their compaction characteristics. Simply wetting the bottom ash or slag sufficiently to prevent bulking will promote adequate compaction.

7.6.4 *Placement of CCBs*—The CCBs should be placed in loose layers of uniform thickness. Each layer should be compacted to the required density because strength is derived from internal friction, and this value is dependent on the relative compaction/unit weight of the CCB. A maximum layer thickness is usually specified to ensure that the required density is achieved through the full depth of the layer. Control of layer thickness is not as important for self-hardening CCBs because additional strength is derived from the cementitious products formed during the hydration process.

7.6.5 *Degree of Compaction*:

7.6.5.1 *Fly Ash and FGD Material*—A typical requirement is that the fill be compacted to a minimum of 95 to 100 % of the maximum dry unit weight, in accordance with Test Method D 698, or 90 to 95 % of the maximum dry unit weight in accordance with Test Method D 1557. Similar requirements are usually applied for the subgrade. In contrast to some soils, variation in absolute values of unit weights determined by the range in percentage of maximum dry unit weights previously specified for the two methods are usually relatively small with fly ash. Therefore either method is suitable. However, the desired performance of the site in terms of safe slopes and adequate performance of foundations, structures, roadways, and so forth, will dictate the degree of compaction needed. Compaction criteria should not be established for FGD material that exhibits thixotropic properties.

7.6.5.2 *Bottom Ash and Boiler Slag*—Granular bottom ash and boiler slag are typically compacted to 70 % relative density, in accordance with Test Method D 4254. Compaction and water content specifications may dictate either the construction method to be used or the performance standard to be attained.

7.6.6 *Method Specifications*—Method specifications specify the type of compaction equipment, the fill material placement methods, and the number of equipment passes to be used in compaction. Method specifications are based on the results of field compaction tests on trial test strips. The test strips are normally conducted at the construction site using the equipment proposed for use and materials or sources that will supply fill material for the project. Method specifications have the advantage of providing continuous quality control by monitoring the ongoing construction activities. If the material source changes or the material itself changes during construction, then the field testing should be repeated on the new material. Method specifications may also be useful for situations where variations in material properties make determination of the appropriate compaction curve difficult.

7.6.7 *Performance Specifications*:

7.6.7.1 *Fly Ash and FGD Material*—The compaction criteria are typically expressed as a percentage of the maximum dry unit weight, in accordance with Test Method D 698 or D 1557 and at molding water contents that do not exceed the optimum water content plus a given percentage and that prevent dusting during placement and compaction. When using static-type compaction, an allowable range of water contents is also usually specified so that the material will be in the range where the required unit weight can be readily achieved. Fly ash and FGD material have a tendency to be displaced under the mass of the compactor when placed above the optimum water content. Specifications requiring placement over a range of water content less than the optimum water content will control this phenomenon. Experience has shown that vibratory compactors operating at the resonant frequency can achieve the required degree of compaction in a minimum of passes over a wide range of water contents, but not excessively wet of the optimum water content. For FGD material that exhibits thixotropic properties, the performance specification will typically specify the allowable layer thickness and strength requirements of samples tested in unconfined compression at specified time intervals.

7.6.7.2 *Bottom Ash and Boiler Slag*—Performance specifications for bottom ash or boiler slag typically specify the compaction criteria as a percentage of the relative density in accordance with Test Method D 4254, and may require use of vibratory compaction equipment.

7.6.8 *Dust Control*—Dusting does not occur during placement and compaction of CCBs when the molding water content of the CCB is sufficient to achieve the desired degree of compaction. The CCB surfaces exposed to the sun and wind can dry out and become susceptible to dusting. Dusting can be controlled by wetting the CCB, applying a dust suppressant, constructing wind screens, or by placing the final soil cover.

7.7 *Protection of Embedded Materials*—When materials are to be embedded in the structural fill, it is prudent during design to assess whether any deleterious reactions are likely to occur. Specifically, the potential for corrosion of pipes, conduits, and other metal structures should be evaluated. Concrete structures such as culverts, footings, and retaining walls should be evaluated for sulfate attack.

7.7.1 *Corrosion Protection*—Low resistivity is commonly used as an indicator of the corrosion potential of soil or aggregates. Field tests with CCBs have shown that additional contributing factors are high or low pH, high soluble sulfate and soluble chlorides, and partially saturated field moisture conditions. It is appropriate to check all of these factors and consider the lifetime and sensitivity of the embedded material. Appropriate test methods are described in 6.4.3. The standards used by the local state transportation agency for evaluating corrosion potential of soil fill may be used as a reference.

7.7.2 *Sulfate Attack on Concrete*—Sulfate attack on concrete in CCB fills has received attention because of the sulfate content in some CCBs. The sulfate exposure is considered severe when the water-soluble sulfate in soil (or CCB) exceeds 0.20 % by weight, or when sulfate in water exceeds 1500 ppm. As with corrosion, other factors such as moisture will be contributing factors. Also as with corrosion, there is a need to assess sensitivity and lifetime of the structure, and the difficulty of replacement or, repair. If sulfate exposure is a concern, the use of blended or sulfate-resistant cements such as those described in Specifications C 595/C 595M and C150, or application of polymer or bituminous coatings may provide protection.

8. Design Methods

8.1 *General*—The underlying materials and the CCB fill must support its own mass and the loads to be placed on it without excessive settlement, and require no long-term maintenance beyond that typically exercised for the intended use. In addition, settlement due to the consolidation of the soils that lie beneath the fill must be evaluated and maintained within tolerable limits considering the intended use of the site. The process of analyzing these conditions for CCBs is similar to that normally followed for conventional natural soil materials. The procedure entails developing an analytical model of the fill and underlying soils, the relevant site conditions, and determining whether expected physical behavior is within allowable limits. All design work and materials testing should be performed in accordance with established engineering practices and in accordance with applicable laws and regulations.

8.2 *Slope Stability*—As is done with conventional fill materials, analysis of structural fill slopes should consider possible failure of the CCB fill as well as failure of the foundation soils resulting from the load of the fill.

8.2.1 *Seepage and Drainage*—Consideration of high water tables, seepage forces, seismic loadings, and excess pore pressures in foundation soils should be considered, as appropriate. Adequacy of drainage provisions to maintain the fill in a drained condition should be considered.

8.2.2 *Material Properties*—Material properties for CCBs should be as determined by laboratory testing. Characterization of site materials and conditions should be in accordance with Guide D 420 with sampling, laboratory, and field testing conducted as appropriate.

8.2.3 *Stability Analyses*—Stability analyses are typically conducted for circular failure surfaces using the friction circle method that is conservative for most cases. For situations where noncircular failure surfaces are to be analyzed, complex conditions are to be assessed, or more precise estimates are required, other appropriate procedures may be used.

8.3 *Settlement*—Settlement analyses should consider compression of the fill resulting from foundations and other loads placed on the structural fill as well as compression of the foundation soils beneath the fill due to the combined mass of the fill and the superimposed loads. Conventional methods of analyses are used as with natural soils.

8.4 *Bearing Capacity*—The ability of the fill to support structures bearing on or within the fill can be calculated by conventional procedures used for natural soils.

8.4.1 *Footings*—Ultimate bearing capacity analysis is appropriate for footings bearing on compacted CCB structural fills. The analysis is simplified by the drained, non-cohesive nature of the fill (except for self-hardening fly ash and FGD material). The relatively low unit weight of CCBs as compared to natural soils should be considered in the analyses. Footings that are wider than the thickness of the fill below the footing or that are located near the edge of slopes are cases that may require special consideration.

8.4.2 *Slabs and Pavements*—The ability of the fill to support slabs and pavements to be located on the fill surface can be assessed by standard pavement design procedures and by determining the modulus of subgrade reaction by Test Methods D 1195 or D 1196, or bearing ratio by Test Methods D 1883 or D 4429, as appropriate.

8.5 *Lateral Earth Pressure*—Conventional methods of analysis of lateral earth pressure can be used for CCBs considering that the material is cohesionless (except for self-hardening CCBs) and has a lower unit weight than many natural soils. For structures that are fixed and unable to yield, earth pressure at rest coefficients of 0.5 are typically used in estimating loads. For most yielding retaining walls, active earth pressures are determined by Rankine's method. Coulomb's method is generally used for walls over 20 ft (6.1 m) in height.

9. Construction

9.1 *General*—Construction procedures for CCB structural fills are similar to conventional earthwork operations. Routine methods employed with soil fills to control dusting, erosion, and sedimentation are similarly required.

9.2 *Weather Restrictions*—Construction should be suspended during severe weather conditions. Operations may proceed during moderately wet periods by reducing the amount of water added at the plant or job site to compensate for precipitation. Dry CCBs can also be disked into excessively wet CCBs to reduce the water content to an acceptable level. Because fly ash obtained directly from silos or hoppers dissipates heat slowly, it may be placed during cold weather. If frost penetrates the surface a few inches, it can be removed from the surface or recompacted upon thawing and drying.

9.2.1 *Dust Control*—When exposed to the elements and allowed to dry out, CCB surfaces are quite susceptible to produce dust. Dust control measures routinely used on earthwork projects are effective in minimizing airborne particulate at CCB fill sites. Typical controls include appropriate hauling methods, use of wind breaks, moisture-conditioning of the CCB, wetting or covering of exposed CCB surfaces, chemically treating CCB surfaces and paving, wetting, and covering of high-traffic haul roads with coarse materials.

9.2.2 *Erosion Control*—The CCBs typically do not require additional sedimentation and erosion control measures beyond those normally employed for soil fills in accordance with state and local requirements.

9.3 *Source and Delivery*—The CCBs are typically supplied from sources containing little or no extraneous or deleterious material. Non-self-hardening fly ash, bottom ash, boiler slag and FGD material are usually hauled in covered dump trucks with tightly sealed tailgates. These CCBs may be conditioned with water at the plant, if necessary. Self-hardening fly ash is hauled in pneumatic tank trucks and conditioned with water at the project site or may be partially conditioned and hauled in covered dump trucks to the project site. Care should be taken to not overfill the trucks so that spillage does not occur. Adequate measures must be taken to ensure proper water content when using fly ash or bottom ash that has been stored in landfills, ponds, and lagoons. Trucks should be spray-cleaned with water at the plant to reduce spillage and dust during transport. Provisions should be made for cleaning of public roads in the event spillage does occur.

9.4 *On-site Storage*—Limit on-site storage of CCBs to the minimum quantity required to maintain the construction schedule. For stockpiles, provide sedimentation and erosion controls in accordance with state and local requirements. Self-hardening fly ash that is not partially conditioned should be stored dry in pneumatic tank trucks or in suitably protected storage silos. Precautions normally taken for bulk storage of cement and lime may be required.

9.5 *Site Preparation*—The base of the fill should be stripped of vegetation and organic soils. The subgrade should be compacted to the desired dry unit weight and underdrains installed, when required.

9.6 *Placement and Compaction*—Place CCBs in uniform layers not exceeding the thickness specified. The CCB must be spread uniformly; otherwise, the compaction equipment will ride on uneven hard spots in the fill, resulting in softer areas between the high spots. Tracking the CCBs with a dozer or truck prior to compaction will facilitate compaction to the required density. Typically a CCB fill is compacted with a vibratory or pneumatic-tired roller. Fill should not be placed on saturated or frozen material. If water must be added to obtain optimum water content condition, allow adequate time for the entire lift to equilibrate, yet compact before the surface dries out. Water should be sprayed uniformly.

9.7 *Cover*—Structural fill slopes should be covered with soil and revegetated as soon as practicable following the fill placement operations. Top surfaces should also be covered promptly to reduce infiltration of precipitation and runoff into the fill and to minimize surface erosion.

9.8 *Quality Control*—Quality control programs for CCB structural fills are similar to quality control programs for earthwork projects. These programs typically include visual observation of CCB placement operations, supplemented with laboratory and field testing to confirm that the structural fill is constructed as designed. The testing requirements will vary depending on whether a method specification or performance specification is used.

9.8.1 Visual observations are typically made to verify lift thickness, the number of passes of the compactor on each lift, and the behavior of the CCB under the weight of the compaction equipment. Laboratory compaction tests (Test Methods D 698, D 1557, D 4253, and D 4254) are performed to establish baseline data needed to control compaction in the field. Field unit weight and water content tests are conducted regularly on compacted lifts to verify that the required degree of compaction is achieved. Test Methods D 1556, D 2167, or D 2922 may be used to determine the field unit weight. Test Methods D 2216, D 4643, or D 4959 may be used to estimate the water content.

9.8.2 It is prudent to maintain daily job logs documenting site conditions, weather, and work activities. Water content and unit weight tests should be taken as specified by the design engineer and whenever visual observations indicate the desired degree of compaction is possibly not being achieved. As a guide in performance specifications, one test for every 1000 to 2000 yd^3 of fill is suggested.

10. Keywords

10.1 boiler slag; bottom ash; coal ash; coal combustion by-products; embankment; flue gas desulfurization materials; fly ash; pollution prevention; resource conservation; structural fill; utilization

REFERENCES

(1) Report to Congress, *Wastes from the Combustion of Coal by Electric Utility Power Plants*, U.S. Environmental Protection Agency, EPA/530-SW-88-002, February 1988.

(2) "Final Regulatory Determination on Four Large Volume Wastes from the Combustion of Coal by Electric Utility Power Plants," *Federal Register*, Aug. 9, 1993, pp. 42466–42482.

(3) Ibid, p. 42480.

(4) "Model Standards and Techniques for Control of Radon in New Residential Buildings," *Federal Register*, March 21, 1994, pp. 13402–13416.

(5) American Coal Ash Association, "Structural Fill Applications of Coal Ash," Alexandria, VA, 1993.

(6) GAI Consultants, Inc., *Fly Ash Design Manual for Road and Site Applications (Vol 1)*, Electric Power Research Institute, EPRI TR-100472, Palo Alto, CA, 1992.

(7) GAI Consultants, Inc., *Fly Ash Construction Manual for Road and Site Applications (2 Vols)*, Electric Power Research Institute, EPRI CS-5981, Palo Alto, CA, 1988.

(8) U.S. Department of Transportation, *Fly Ash as a Construction Material for Highways, A Manual*, Federal Highway Administration, PB-259302, Washington, DC, 1976.

(9) Baker Environmental Consulting Engineers, *FGD By-Product Disposal Manual, Fourth Edition*, Electric Power Research Institute, EPRI TR-104731, Palo Alto, CA, 1995.

Designation: E 1903 – 97

Standard Guide for Environmental Site Assessments: Phase II Environmental Site Assessment Process[1]

This standard is issued under the fixed designation E 1903; the number immediately following the designation indicates the year of original adoption or, in the case of revision, the year of last revision. A number in parentheses indicates the year of last reapproval. A superscript epsilon (ε) indicates an editorial change since the last revision or reapproval.

1. Scope

1.1 This guide[2] covers a framework for employing good commercial and customary practices in conducting a Phase II environmental site assessment (ESA) of a parcel of commercial property with respect to the potential presence of a range of contaminants which are within the scope of CERCLA as well as petroleum products.

1.1.1 This guide is intended to provide practical procedural guidance for the continuation of an assessment conducted in accordance with the most recent edition of Practice E 1527 or E 1528, or both. Practice E 1527 is the practice for conducting Phase I ESAs for a parcel of commercial property and Practice E 1528 is the transaction screen practice. Both practices define a process that is intended to constitute "all appropriate inquiry into the previous ownership and uses of a property" to determine whether hazardous substances or petroleum products have been disposed or released there in order to satisfy one element of the innocent purchaser defense to CERCLA liability.

1.1.2 Because this guide for conducting Phase II ESAs describes a process for further evaluating a parcel of commercial property with recognized environmental conditions, as defined in Practices E 1527 and E 1528, users of this guide should understand the requirements and limitations of those practices. It is strongly recommended that the user refer to and apply the guide in concert with Practices E 1527 and E 1528.

1.1.3 This guide has multiple purposes. It is intended to provide assistance to users in satisfying the appropriate inquiry element of CERCLA's innocent purchaser defense, as defined in 42 U.S.C. § 9601(35)(B), where a previous assessment satisfying that element identified recognized environmental conditions. This guide also is intended to assist a user in gathering reliable information about a property's environmental conditions to guide the user's business decisions. However, this guide does not purport to include the level of specificity required of technical standards that govern full characterization of a site's environmental conditions.

1.2 *Objectives*—The primary objectives of conducting a Phase II ESA are to evaluate the recognized environmental conditions identified in the Phase I ESA or transaction screen process for the purpose of providing sufficient information regarding the nature and extent of contamination to assist in making informed business decisions about the property; and where applicable, providing the level of knowledge necessary to satisfy the innocent purchaser defense under CERCLA.

1.2.1 To achieve these objectives, it may be appropriate to perform more than a single iteration of assessment. The guide fosters an iterative approach to Phase II assessments and allows the user to terminate the Phase II ESA at the point where sufficient data have been generated to meet the user's objectives.

1.2.2 At the completion of a Phase II ESA, the environmental professional should be able to conclude, at a minimum, that either *(a)* the ESA has provided sufficient information to render a professional opinion that there is no reasonable basis to suspect the presence of hazardous substances or petroleum products at the property associated with the recognized environmental conditions under assessment, or *(b)* the ESA has confirmed the presence of hazardous substances or petroleum products at the property under conditions that indicate disposal or release. If the information developed in the ESA is insufficient for the environmental professional to reach either of these conclusions, the environmental professional may recommend additional iterations of assessment if warranted to meet the objectives of the user. If the environmental professional reasonably suspects that unconfirmed hazardous substance or petroleum releases remain but concludes that further reasonable assessment is not expected to provide additional information of significant value, he may recommend that further assessment is not warranted. In such circumstances, the recommendation for no further assessment should be accompanied by an explanation why a reasonable suspicion of releases remains and why further reasonable assessment is not warranted. Depending upon the work scope, the environmental professional may also be able to provide guidance on the nature and extent of contamination in order to assist the user in making business decisions regarding the property.

---

[1] This guide is under the jurisdiction of ASTM Committee E-50 on Environmental Assessment and is the direct responsibility of Subcommittee E50.02 on Commercial Real Estate Transactions.

Current edition approved Dec. 10, 1997. Published February 1998.

[2] As used herein, a "Standard" is a document that has been developed and established within the consensus principles of the Society and that meets the approval requirements of ASTM procedures and regulations. A" Guide" is a compendium of information or a series of options that does not recommend a specific course of action. A guide increases the awareness of information and approaches in a given subject area. A "Practice," in contrast, is a definitive set of instructions for performing one or more specific operations that does not produce a test result. See Form and Style for ASTM Standards, 10th ed., 1996.

1.2.3 This guide is intended to provide guidance for assessing recognized environmental conditions and developing technically sound data. It is not intended to satisfy the level of inquiry that may be necessary to support remedial solutions for a site. For further discussion of the use of this guide, refer to Section 4 on Significance and Use.

1.3 *Needs of the User*—Establishing the innocent purchaser defense may not be a realistic objective in some instances. Accordingly, the extent of assessment is based on the business objectives of the user as well as the degree of uncertainty acceptable to the user. In either case, the primary purpose of a Phase II ESA conducted in accordance with this guide is to assess and evaluate the recognized environmental conditions identified in the Phase I ESA or Transaction Screen Process.

1.3.1 The mere confirmation of contamination or the preliminary indication of the extent and magnitude of contamination may be sufficient for the purposes of many users. If a user desires a more complete characterization of the environmental condition of the property, further assessment may be undertaken. However, this guide should not be construed to require multiple iterations of assessments in all cases, either to establish the innocent purchaser defense or to meet other objectives. Many Phase II ESAs may in fact be restricted to only a single round of assessment, whatever the extent of contamination, if any, that might be revealed.

1.4 *Limitations*—The use of this guide is related to the scope as set forth in Section 1. For information purposes, Section 12 of this guide contains a non-exhaustive list of certain environmental conditions that are beyond the scope of this guide but that may warrant consideration by parties to a commercial property transaction. This guide provides an approach that may be employed to assess the environmental conditions listed in Section 12. Reference also should be made to 4.1.

1.5 *Organization of This Guide*—This guide has twelve sections and one appendix. Section 1 is the Scope section. Section 2 is Referenced Documents. Section 3, Terminology, contains definitions of terms and acronyms used in this guide. Section 4 is Significance and Use of this guide. Section 5 is Contracting Considerations. Sections 6-11 constitute the main body of the Phase II Environmental Site Assessment guide and include objectives (see Section 6), developing the scope of work (see Section 7), assessment activities (see Section 8), evaluation of data (see Section 9), interpretation of results (see Section 10) and recommended report preparation (see Section 11). Section 12 provides additional information regarding non-scope considerations. Appendix X1 provides a sample table of contents and report format for a written Phase II Environmental Site Assessment Report.

2. Referenced Documents

2.1 The references in this Section are for informational purposes. Although Phase II ESAs should utilize government- and industry-accepted practices and methods, this guide does not recommend the use of specific practices in the implementation of a Phase II ESA.

2.2 *ASTM Standards:*

D 5730 Guide to Site Characteristics for Environmental Purposes With Emphasis on Soil, Rock, The Vadose Zone and Ground Water[3,4]

D 653 Terminology Relating to Soil, Rock and Contained Fluids[3]

D 4750 Test Method for Determining Subsurface Liquid Levels in a Borehole or Monitoring Well. (Observation Well)[3]

E 1527 Practice for Environmental Site Assessments: Phase I[5]

E 1528 Practice for Environmental Site Assessments: Transaction Screen Process[3]

3. Terminology

3.1 *Definitions:*

3.1.1 *aquifer*—as defined in Terminology D 653, a geologic formation that is capable of providing a significant quantity of water.

3.1.2 *ARARs*—an acronym for "applicable or relevant and appropriate requirements," a term used in CERCLA and interpreted by EPA regulations. Applicable requirements means "those cleanup standards, standards of control, and other substantive requirements, criteria, or limitations promulgated under federal environmental or state environmental or facility siting laws that specifically address a hazardous substance, pollutant, contaminant, remedial action, location, or other circumstance found at a CERCLA site", 40 C.F.R. § 300.5. Relevant and appropriate requirements means "those cleanup standards, standards of control, and other substantive requirements, criteria, or limitations promulgated under federal environmental or state environmental or facility siting laws that, while not 'applicable' to a hazardous substance, pollutant, contaminant, remedial action, location, or other circumstance at a CERCLA site, address problems or situations sufficiently similar to those encountered at the CERCLA site that their use is well suited to the particular site" 40 C.F.R. § 300.5.

3.1.3 *business decision*—a decision based on various business risk management considerations relating to a specific property, such as a transfer of title or change in financing. Such considerations may also include the potential financial exposure associated with environmental risks, the value of the property compared with the cost of environmental assessment, and the participation and motivations of specific parties to the transaction (that is, owner, seller, buyer, lender, etc.).

3.1.4 *CERCLA*—the acronym for the Comprehensive Environmental Response, Compensation and Liability Act, 42 U.S.C.§ 9601, *et seq.*, the primary federal statute that governs the imposition of liability for environmental cleanups.

3.1.5 *C.F.R.*—Code of Federal Regulations.

3.1.6 *chain of custody*—a written or printed form which is used to document sample possession, condition and responsibility. This custody can include the time from sample container acquisition through transportation, sample collection and laboratory analysis.

3.1.7 *chemical phases*—the physical state of the chemical, (that is; solid, liquid, vapor/gaseous). A chemical's physical

[3] *Annual Book of ASTM Standards*, Vol 04.08.

[4] Guide D 5730 covers the selection of the various ASTM Standards that are available for the investigation of soil, rock, the vadose zone, ground water, and other media where investigations have an environmental purpose.

[5] *Annual Book of ASTM Standards*, Vol 11.04.

state can change from the one phase to another based on its physical environment.

3.1.8 *common law*—as distinguished from law created by the enactment of legislatures, the common law comprises the body of those principles and rules of action, relating to the government and security of persons and property, which derive their authority solely from usages and customs of immemorial antiquity, or from the judgments and decrees of the courts recognizing, affirming and enforcing such usages and customs; and, in this sense, particularly the ancient unwritten law of England.[6]

3.1.9 *cuttings*—soil, rock chips, fragments or other material that is brought to the surface by drilling or sampling, or both.

3.1.10 *disposal*—as defined by CERCLA and interpreted by EPA regulations, the discharge, deposit, injection, dumping, spilling, leaking, or placing of any solid waste or hazardous waste into or on any land or water so that such solid waste or hazardous waste or any constituent thereof may enter the environment or be emitted into the air or discharged into any waters, including ground waters. By judicial construction, the term also may include hazardous substances in repose in the environment, with the result that hazardous substances that have migrated to a site from an off-site source may nonetheless be considered to have been disposed at the site. See, for example, *United States v. Waste Industries*, 734 F.2d 159 (4th Cir. 1984). This construction is not universally accepted. See, for example, *United States v. Petersen Sand and Gravel, Inc.*, 806 F. Supp. 1346 (D.Ill. 1992).

3.1.11 *downgradient*—with respect to a given reference point, it is all points through which ground water originating at the reference point flows until reaching a surface discharge point or a distance or depth that ceases to be a matter of potential environmental concern; the direction of decreasing hydrostatic head.

3.1.12 *environmental media*—soil, rock, water, air, sediment.

3.1.13 *environmental professional*—a person or group of persons possessing sufficient training and experience necessary to prepare and implement a Phase II environmental site assessment in accordance with this guide, and from the information generated by such activities, having the ability to develop sound opinions and conclusions regarding recognized environmental conditions in connection with the property in question. An individual's status as an environmental professional may be limited to the type of assessment to be performed or to specific segments of the assessment for which the professional is responsible. The person may be an independent contractor or an employee of the user.

3.1.13.1 *Discussion*—Some jurisdictions may have licensing requirements for individuals who perform certain activities included in a Phase II ESA.

3.1.14 *EPA*—the United States Environmental Protection Agency.

3.1.15 *ESA*—environmental site assessment.

3.1.16 *fate and transport characteristics*—natural consequences that can be predicted based on the distinguishing characteristics of a substance and the media which carry the substance.

3.1.17 *ground water*—as defined by Terminology D 653, the part of the subsurface that is in the saturated zone. Loosely, all subsurface water as distinct from surface water.

3.1.18 *ground water flow*—as defined by Terminology D 653, the movement of water in the zone of saturation.

3.1.19 *ground water flow direction*—the compass bearing of the movement of water in the zone of saturation (vertical and horizontal components of ground water flow).

3.1.20 *hazardous substance*—a substance defined as a hazardous substance pursuant to CERCLA, 42 U.S.C. § 9601(14), as interpreted by EPA regulations and the courts. The term includes any substance designated pursuant to section 311(b)(2)(A) of the Clean Water Act, 33 U.S.C.§ 1321(b)(2)(A); any element, compound, mixture, solution or substance designated pursuant to section 102 of CERCLA, 42 U.S.C. § 9602; any hazardous waste having the characteristics identified under or listed pursuant to section 3001 of the Solid Waste Disposal Act, 42 U.S.C. § 6921, (but not including any waste the regulation of which under the Solid Waste Disposal Act, 42 U.S.C.§ 6901 *et seq.*, has been suspended by Act of Congress); any toxic pollutant listed under section 307(a) of the Clean Water Act, 33 U.S.C. § 1317(a); any hazardous air pollutant listed under section 112 of the Clean Air Act, 42 U.S.C. § 7412; and any imminently hazardous chemical substance or mixture with respect to which the EPA administrator has taken action pursuant to section 7 of the Toxic Substances Control Act, 15 U.S.C. § 2606. The term does not include petroleum, including crude oil or any fraction thereof which is not otherwise specifically listed or designated as a hazardous substance by EPA and the term does not include natural gas, natural gas liquids, liquefied natural gas, or synthetic gas usable for fuel (or mixtures of natural gas and such synthetic gas). Hazardous substances designated by EPA are listed in 40 C.F.R. § 302.4.

3.1.21 *petroleum products*—those substances included within the petroleum exclusion to CERCLA, 42 U.S.C. § 9601(14), as interpreted by the courts and EPA; that is, petroleum, including crude oil or any fraction thereof which is not otherwise specifically listed or designated as a hazardous substance under Subparagraphs (A) through (F) of 42 U.S.C. § 9601(14), natural gas, natural gas liquids, liquefied natural gas, and synthetic gas usable for fuel (or mixtures of natural gas and such synthetic gas). (The word fraction refers to certain distillates of crude oil, including gasoline, kerosene, diesel oil, jet fuels, and fuel oil, pursuant to Standard Definitions of Petroleum Statistics).[7]

3.1.22 *Phase I environmental site assessments (ESAs)*—the process described in Practice E 1527.

3.1.23 *Phase II environmental site assessments (ESAs)*—the process described in this guide.

3.1.24 *property*—the property that is the subject of the ESA described in this guide. The term includes buildings and other fixtures and improvements located on the property and affixed to the land.

[6] *Black's Law Dictionary 250-251*, 5th ed., 1979.

[7] *Standard Definitions of Petroleum Statistics*, American Petroleum Institute, Fourth Edition 1988.

3.1.25 *purged ground water*—water that is removed from a monitoring well prior to collecting a representative sample of the aquifer, including water generated during well installation and well development.

3.1.26 *quality assurance/quality control (QA/QC)*—the use of standards and procedures to provide reasonable assurance that samples collected and data generated are reliable, reproducible and verifiable.

3.1.27 *recognized environmental condition (REC)*—the presence or likely presence of any hazardous substances or petroleum products on property under conditions that indicate an existing release, a past release, or a material threat of a release of any hazardous substances or petroleum products into structures on the property or into the ground, ground water or surface water of the property. The term includes hazardous substances or petroleum products even under conditions in compliance with laws. The term is not intended to include de minimis conditions that generally do not present a material risk of harm to public health or the environment and that generally would not be the subject of an enforcement action if brought to the attention of appropriate governmental agencies.

3.1.28 *release*—as defined by § 101(22) of CERCLA, 42 U.S.C. § 9601(22), "any spilling, leaking, pumping, pouring, emitting, emptying, discharging, injecting, escaping, leaching, dumping, or disposing into the environment (including the abandonment or discarding of barrels, containers, and other closed receptacles containing any hazardous substance or pollutant or contaminant), but exclud[ing] (A) any release which results in exposure to persons solely within a workplace, with respect to a claim which such persons may assert against the employer of such persons, (B) emissions from the engine exhaust of a motor vehicle, rolling stock, aircraft, vessel, or pipeline pumping station engine, (C) release of source, byproduct, or special nuclear material from a nuclear incident, as those terms are defined in the Atomic Energy Act of 1954 [42 U.S.C. § 2011 *et seq.*], if such release is subject to requirements with respect to financial protection established by the Nuclear Regulatory Commission under section 170 of such Act [42 U.S.C. § 2210], or, for the purposes of section 104 of CERCLA or any other response action, any release of source, byproduct, or special nuclear material from any processing site designated under section 102(a)(1) or 302(a) of the Uranium Mill Tailings Radiation Control Act of 1978; [42 U.S.C. § 7912(a)(1) or 7942(a)], and (D) the normal application of fertilizer."

3.1.29 *sample medium*—the particular matrix that is sampled for laboratory analysis. Examples of commonly encountered matrices include: soil, rock, sludge, sediment, air, wastes, pure or mixtures of chemicals, ground water, drinking water, surface water, etc.

3.1.30 *self-evaluation privilege*—a privilege invoked to protect from disclosure confidential information resulting from a critical self-analysis undertaken by the party seeking to invoke the privilege.

3.1.31 *Superfund*—the common name for the Comprehensive Environmental Response, Compensation and Liability Act, 42 U.S.C. § 9601, *et seq.*, as amended by the Superfund Amendments and Reauthorization Act of 1986 ("SARA"), the primary federal statute that governs the imposition of liability for environmental cleanups.

3.1.32 *surface water*—water above the surface of the ground, including but not limited to lakes, ponds, reservoirs, artificial impoundments, streams, rivers, springs, seeps and wetlands.

3.1.33 *target analyte*—a chemical substance or class of chemicals that is selected for laboratory analysis or field analysis.

3.1.34 *Transaction Screen Process*—the process described in Practice E 1528.

3.1.35 *U.S.C.*—United States Code.

3.1.36 *user*—the party seeking to use this guide to guide performance of an ESA of the property. A user may include, without limitation, a purchaser of property, a potential tenant of property, an owner of property, a lender, or a property manager.

3.1.37 *water table*—as defined in Test Method D 4750, the surface of a ground water body at which the water pressure equals atmospheric pressure. Earth material below the ground water table is saturated with water.

4. Significance and Use

4.1 *Uses:*

4.1.1 This guide is intended for use on a voluntary basis by parties who wish to assess the recognized environmental conditions ("RECs") identified in a Phase I ESA or transaction screen process for the purpose of establishing the innocent purchaser defense, where applicable, or to support business decisions about the property. The level of assessment can, and most likely will, vary depending on the needs of the user and the business situation driving the assessment, as described in 1.3. No implication is intended that a person must use this guide in order to be deemed to have conducted inquiry in a commercially prudent or reasonable manner in any particular transaction. Nevertheless, this guide is intended to reflect good commercial and customary practice for appropriate inquiry sufficient to satisfy the innocent purchaser defense.

4.1.1.1 CERCLA provides that a purchaser of real property qualifies for the innocent purchaser defense if the purchaser acquires a contaminated site after the disposal or placement of hazardous substances at the site, and the purchaser neither knew nor had reason to know that any hazardous substance which is the subject of the release was disposed at the site. 42 U.S.C. § 9601(35)(A)(i). To establish that a purchaser had no reason to know of the disposal or release of any hazardous substance at the site, CERCLA requires the purchaser to show that at the time of acquisition the purchaser undertook all appropriate inquiry into the previous ownership and uses of a property consistent with "good commercial or customary practice". 42 U.S.C.§ 9601(35)(B). To determine whether a purchaser made "all appropriate inquiry", the courts are directed by statute to consider any specialized knowledge or experience on the part of the purchaser, the relationship of the purchase price to the value of the property, if uncontaminated, commonly known or ascertainable information about the property, the obviousness of the presence or likely presence of contamination at the property, and the ability to detect such contamination by appropriate inspection.

4.1.1.2 This guide is intended to meet the business community's need for a written, practical reference describing procedures that constitute" all appropriate inquiry" once recognized environmental conditions are identified. This guide is intended to be a practical reference document which can be used to assess whether good commercial or customary practice (that is, appropriate procedure) has been followed in a particular environmental assessment.

4.1.1.3 This guide does not address the business judgments that follow from the interpretation of data collected through the Phase II process. Assessment of the business risks reflected by such data is dependent upon numerous site- and transaction-specific variables which cannot be adequately anticipated in this guide. Proper analysis of the legal, business and environmental risks associated with a particular site and business transaction is best conducted on a case-by-case basis by legal, business and environmental professionals assessing the known data relating to the particular site and transaction.

4.1.1.4 Likewise, this guide does not assess the legal risks associated with a transaction involving property where hazardous substances or petroleum products are present at levels above those occurring naturally but below those generally considered to be of regulatory concern, that is, *de minimis* levels. What constitutes a *de minimis* level of contamination is an evolving regulatory concept which may vary from jurisdiction to jurisdiction and may turn on site-specific factors.

4.1.1.5 If a Phase II ESA conducted in accordance with this guide provides sufficient information from which the environmental professional can render a professional opinion that there is no reasonable basis for suspecting the disposal or release of hazardous substances at the site, then further assessment is not warranted to establish the innocent purchaser defense. If the Phase II ESA confirms the presence of *de minimis* quantities of hazardous substances on the site under circumstances that suggest these hazardous substances have been released or disposed, the availability of the innocent purchaser defense is not assured as a matter of law. A review of reported case law as of January 1996 revealed no judicial decision concerning the availability of the innocent purchaser defense when *de minimis* quantities of hazardous substances are present on a site. This guide, therefore, does not address the availability of the innocent purchaser defense in those circumstances. Because petroleum products are excluded from the hazardous substances subject to CERCLA, the innocent purchaser defense is inapplicable to petroleum product releases.

4.1.2 *Use Not Limited to CERCLA*—This guide is designed to assist a user in developing information about the environmental condition of the property and has utility for a wide range of users including those who may have no actual or potential CERCLA concerns.

4.1.2.1 Where fewer than all recognized environmental conditions identified in a Phase I ESA or transaction screen are assessed in the Phase II ESA, the conclusion that no further inquiry is warranted applies only to the recognized environmental conditions assessed. However, this guide should not be construed to require Phase II assessment of all recognized environmental conditions identified in a Phase I ESA or a transaction screen. Where the user's objectives may be otherwise satisfied (for example, establishing the innocent purchaser defense is not the user's objective), a recognized environmental condition identified in a Phase I ESA or transaction screen may be eliminated from further assessment in the Phase II ESA.

4.1.3 *Site- and Transaction-Specific*—A Phase II ESA is site-specific in that it relates to assessing recognized environmental conditions as identified in a Phase I ESA or transaction screen. A Phase II ESA is also transaction-specific. This guide is intended to set forth a level of diligence that will satisfy the innocent purchaser defense where it is sought. The level of diligence set forth in this guide is also intended to generate information of sufficient quality to satisfy different or additional objectives of the user.

4.1.4 *Use by Other Parties*—Transaction-specific assumptions and limitations, based on the user's objectives, will be built into the implementation of this guide for any specific site to the extent the user's objectives go beyond establishing the innocent purchaser defense. Any person who seeks to rely on a Phase II ESA conducted for someone else is cautioned to understand these assumptions and objectives prior to review and use of the findings of the Phase II ESA.

4.2 *Principles*—The following principles are an integral part of this guide and are intended to be referred to in resolving any ambiguity or exercising such discretion as is accorded the user or environmental professional.

4.2.1 *Elimination of Uncertainty*—No ESA can eliminate all uncertainty. Furthermore, any sample, either surface or subsurface, taken for chemical analysis may or may not be representative of a larger population. Professional judgment and interpretation are inherent in the process and uncertainty is inevitable. Additional assessment may be able to reduce the uncertainty.

4.2.1.1 *Failure to Detect*—Even when Phase II work is executed with an appropriate site-specific standard of care, certain conditions present especially difficult detection problems. Such conditions may include, but are not limited to, complex geological settings, the fate and transport characteristics of certain hazardous substances and petroleum products, the distribution of existing contamination, physical limitations imposed by the location of utilities and other man-made objects, and the limitations of assessment technologies.

4.2.1.2 *Inadequate Phase I ESA or Transaction Screen*—If Phase II work is based on an inadequate Phase I ESA or an inadequate transaction screen (for example, one that fails to identify all recognized environmental conditions) there will be an increased risk that the disposal or release of hazardous substances or petroleum products on the property may go undetected and jeopardize the availability of the innocent purchaser defense and the quality of data on which judgments are made. A Phase I ESA or transaction screen, or both, conducted in accordance with the latest edition of Practice E 1527 and Practice E 1528 should provide an adequate basis for a Phase II ESA.

4.2.1.3 *Chemical Analysis Error*—Chemical testing methods have inherent uncertainties and limitations. The environmental professional should build quality control and quality assurance tests into the work description, as outlined in 7.1.

The environmental professional should require the laboratory to report any potential or actual problems experienced, or non-routine events which may have occurred during the testing, so that such problems can be considered in evaluating the data. The environmental professional should subsequently identify such problems in any requested reports or documentation provided to the user.

4.2.2 *Level of Assessment*—Phase II ESAs do not generally require an exhaustive assessment of environmental conditions on property. There is a point at which the cost of information obtained and the time required to obtain it outweigh the usefulness of the information and, in fact, may be a material detriment to the orderly completion of transactions. If hazardous substance or petroleum releases are confirmed on a parcel of property, the extent of further assessment is related to the degree of uncertainty that is acceptable to the user with respect to the real estate transaction.

4.2.3 *Comparison With Subsequent Inquiry*—For purposes of the innocent purchaser defense, Phase II ESAs should be evaluated based on the reasonableness of judgments made at the time and under the circumstances in which they were made.

4.2.4 *Data Usability*—Measurements and sampling data only represent the site conditions at the time of data collection. Therefore, the usability of data collected as part of a Phase II ESA may have a finite lifetime depending on the application and use being made of the data. An environmental professional should evaluate whether previously generated data are appropriate for any subsequent use beyond the original purpose for which it was collected.

5. Contracting Considerations

5.1 *Contracts*—The content and form of the contractual relationship between an environmental professional and user are outside the scope of this guide. The outcome of a Phase II ESA may be adversely affected if the environmental professional and the user do not adequately define their relationship. This guide identifies some of the issues that should be resolved prior to initiation of a Phase II ESA. This guide does not purport to identify all issues that may arise in the course of implementing a Phase II ESA. Nor does it specify how these issues should be resolved. Where the user utilizes the services of an in-house environmental professional, contracting considerations may or may not arise.

5.1.1 *Reporting Obligations*—A requirement to report observations from a Phase II ESA to a governmental entity or third party may be imposed by various authorities, including statutes, regulations, common law, and professional standards. In most cases, statutory, regulatory and common law requirements impose reporting obligations only on the owner, operator, or person in charge or control of the facility or property being assessed. In some circumstances, however, reporting obligations may be legally or voluntarily imposed upon a broader group, including the environmental professional. To avoid disagreement, misunderstanding, or unexpected reporting, the contract between the user and the environmental professional should clearly define the obligations of and protocol for both the user and environmental professional to report to governmental entities or third parties.

5.1.2 *Production of Written Reports and Documentation*—The production of written documentation reflecting the findings of a Phase II ESA raises issues of concern to the user and the environmental professional. The user may be concerned, for example, about the potential for disclosure of sensitive information to the government or third parties and the conflicting interest of ensuring documentation to support an innocent landowner claim. The environmental professional may be concerned that the assessment is well documented to minimize misinterpretation, document uncertainty, and clearly present findings to the user. As a result, the agreement between the user and environmental professional should address the type and scope of written documentation that will be developed to reflect the findings of the Phase II ESA. In this regard, consideration should be given to issues such as the attorney-client, work product and self-evaluation privileges, whether recommendations should be provided separately from the Phase II report, whether the report should be written or oral, and the extent to which the user wants to review a report prior to its becoming "final". An example format is attached in Appendix X1.

5.1.3 *Confidentiality*—Agreements for confidential treatment of the Phase II ESA, if any, should be included in the contract. This agreement should include any subcontractors used in performance of the assessment.

5.1.4 *Work Performed by Others*—During the implementation of the Phase II assessment, the environmental professional may employ others (for example, drillers, laboratories) to carry out portions of the work. The contract between the environmental professional and the user may specify whether the environmental professional or the user is to be responsible for selecting subcontractors. The contract also may specify that only qualified subcontractors with current and appropriate certifications and licenses may be employed. The contract also may specify the qualifications required of subcontractors.

5.1.5 *Limitation on Scope of Work, Data, Information, or Time*—Any limitations on the information, data collected or the work to be performed during the Phase II ESA, including time allowed for completing the work, and their effect on the results of the assessment, should be clearly understood by the environmental professional and user. Such limitations may be made part of the contract.

5.1.6 *Third Party Reliance On Reports and Other Documentation*—Responsibility for the use of Phase II ESAs by third parties may be governed by the contractual relationship between the user and environmental professional.

5.1.7 *Generation of Waste*—Wastes may be generated during the assessment implemented as part of the Phase II ESA (for example, drill cuttings and purged ground water). The contract between the environmental professional and the user should clearly address the manner in which such wastes are to be handled and disposed. Techniques that minimize the generation of waste should be utilized to the extent feasible, consistent with the information and data quality objectives of the planned assessment and applicable regulatory requirements.

5.1.8 *Damages Caused by Explorations*—Exploration activities risk damaging structures such as utility lines and underground storage tanks when such are present. Intrusive explorations may also create additional pathways for pollutant migration. Responsibility for identification of subsurface structures may be governed by the contractual relationship between the user and environmental professional. Where the user is not the owner of the property, the owner should be consulted about the location of such structures.

5.1.9 Many states have statutory obligations for contacting utilities through "one-call" programs. Coordination with utility companies or locator services should also be addressed in the contract documents.

5.2 *Responsibility of User*—The following list of responsibilities of the user is not intended to be exhaustive:

5.2.1 The user should provide access to appropriate areas of the site for the environmental professional. The user should provide the environmental professional a site contact name and phone number.

5.2.2 The user should provide to the environmental professional all pertinent documentation and information regarding the site that is reasonably and practicably available, including but not limited to: previously prepared ESAs, other environmental studies, reports, or permits; property appraisals; site plans that identify site features, including buildings and other above ground structures, property boundaries, and underground structures and buried utilities (for example, sanitary and storm sewers, water, gas, telephone, and electric systems, underground storage tanks, oil/water separators, sumps and drywells); site specific health and safety information; and known or suspected environmental conditions.

5.3 *Responsibility of Environmental Professional*—The following list of responsibilities of the environmental professional is not intended to be exhaustive.

5.3.1 The environmental professional should conform to the precepts of this guide and accepted industry practice. The environmental professional should document and explain significant deviations.

5.3.2 The environmental professional should provide the user prompt notice of environmental conditions observed.

5.3.3 The environmental professional should communicate to the user limitations resulting from any time and cost constraints imposed by the user.

5.3.4 The environmental professional should verify with the user, prior to implementation, any substantive deviations from the original scope of work.

5.3.5 The environmental professional should ascertain and observe all site health and safety considerations and regulations applicable to the activities of the environmental professional.

5.3.6 The environmental professional should provide the user a written statement of qualifications, including the qualifications of the individual environmental professional(s) responsible for the Phase II ESA on request.

5.3.7 The environmental professional should not undertake any activity that he or she is not qualified or licensed (where applicable) to perform.

6. Phase II ESA

6.1 *Objective and Purpose*—The objective of conducting a Phase II ESA is to evaluate the recognized environmental conditions identified in the Phase I ESA or transaction screen process. The purpose of conducting a Phase II ESA depends on the objective(s) of the user. Typically, the purposes of a Phase II ESA are to:

6.1.1 Develop sufficient information from which the environmental professional reasonably can render a professional opinion that, with respect to the recognized environmental conditions assessed, hazardous substances have not been disposed or released at the property, thereby satisfying the innocent purchaser defense under CERCLA as to those recognized environmental conditions; or

6.1.2 Develop sufficient information about the presence of a recognized environmental condition at a site to meet the business objectives of the user and to provide sufficient data to assist the user in making informed business decisions, or both.

6.1.3 The Phase II work scope may consist of several iterations and may be terminated at any point, once the objectives of the user have been satisfied.

6.2 *Components*—The components of a Phase II ESA are as follows:

6.2.1 Developing the scope of work,

6.2.2 Assessment activities,

6.2.3 Evaluation and presentation of data, and

6.2.4 Presentation of findings and conclusions.

7. Developing the Scope of Work

7.1 A description of the work to be performed should be developed by the environmental professional to establish the methods and work tasks that achieve the user's Phase II objectives. The description should provide the rationale for planned sampling locations and testing parameters along with identification of selected methodologies and appropriate QA/QC measures. The description may provide for minor field modifications where appropriate. The description should be organized to facilitate the orderly, objective-focused implementation of the plan in the field. A formal written description may not be required or appropriate in all circumstances and may be substituted by another document that contains the same elements, such as proposal or scope of work. Likewise, the sampling, health and safety and chemical testing programs need not be formal, stand-alone documents. In general, there are seven main tasks in developing a description of the work, as described in 7.2 through 7.8

7.2 *Site Limitations*—The environmental professional is responsible for anticipating foreseeable physical and logistical impediments which interfere with, or limit the ability to conduct explorations, analyses and sampling, for example, low ceilings, narrow alleyways, soft ground, steep slopes, known subsurface structures, such as utilities, etc.

7.3 *Review Existing Information*—Readily available existing information should be reviewed to identify or establish those characteristics of the site and its vicinity which:

7.3.1 Constitute the recognized environmental condition(s) being assessed in the Phase II;

7.3.2 Affect the potential distribution and mobility of hazardous substances or petroleum products in structures, on the ground, in ground water, in soils, or in surface water and environs at the site; and

7.3.3 Affect the manner in which the presence of hazardous substances or petroleum products may be assessed.

7.3.4 The review of readily available existing information

should serve as the basis for evaluating the likely distribution of hazardous substances or petroleum products at the site, identifying appropriate sampling locations, and selecting appropriate sampling and analytical methodologies. Readily available information may be gained in review of Phase I ESAs conducted in accordance with Practice E 1527 and transaction screens conducted in accordance with Practice E 1528. Other ESAs or reports may be used where the information, in the judgment of the environmental professional, can be reasonably relied upon.

7.3.5 The environmental professional may review additional sources of information as is appropriate to the property, or as required by the judgment of the environmental professional. Information developed at this stage of the assessment may eliminate recognized environmental conditions identified in a Phase I ESA or transaction screen from further Phase II assessment where, in the judgment of the environmental professional, the additional information adequately demonstrates that hazardous substances or petroleum products associated with a previously identified recognized environmental condition are unlikely to be present at the property. See 10.1.

7.4 *Potential Distributions of Contaminants*—The environmental professional should consider the likely distribution of potential contaminants. This requires that the environmental professional consider the properties, behaviors, and fate and transport characteristics of the contaminant in a setting like that being assessed.

7.5 *Sampling*—A sampling program should be designed to provide collection of potentially contaminated environmental media, if they occur, at locations and depths where the highest concentrations are likely to occur, as determined pursuant to the provisions of this section.

7.6 *Health and Safety*—The work description should include personnel health and safety precautions to be followed in accordance with applicable federal law or state or local equivalents and any requirements imposed by the owner or occupant of the property or by the user.

7.7 *Chemical Testing*—The chemical testing program should be designed to detect the contaminants suspected to be present in the samples collected in implementing 7.5 above. The chemical tests should be appropriate for detecting the indicator constituents of the hazardous substances and petroleum products identified in the review performed under 7.3 above (for example, target analytes), consistent with 7.4 above. The chemical testing program should include tests which provide quality assurance (QA) and techniques that provide quality control (QC) over the chemical analyses. QA/QC, including detection limits of the chemical testing plan, should be consistent with the intended use of the data. The chemical testing program may include both field analytical and laboratory analytical techniques.

7.7.1 The environmental professional should specify in advance the analytical methods to be used as well as the requirements for data deliverables (for example, tabulated results, individual analysis sheets, chromatograms, spike recoveries, instrument calibration data, etc.) and quality control information (that is, project-specific or only laboratory batch-specific QC sample analysis).

7.7.2 An extensive body of methods is available for the analysis of environmental samples gathered in the course of a Phase II ESA. Several distinct compilations of methods for the analysis of environmental samples have been developed, principally by EPA and ASTM, for different purposes, primarily for hazardous waste management, industrial and municipal wastewater monitoring, assessment and cleanup of potential Superfund sites, and monitoring of the public water supply. An example is the compilation titled *Test Methods for Evaluating Solid Waste* (SW-846) prepared by EPA.

7.7.3 For each specific medium and target analyte, a variety of methods may be appropriate, depending upon the particular circumstances. The environmental professional should select methods for the analysis of samples giving consideration to the intended end-use of the resulting data, the user's data quality objectives, the statutory, regulatory, or policy framework applicable to the sample in question, and any limitations or allowances associated with the level of effort as understood by the environmental professional and user. The selected analytical method should be appropriate for the sample medium, target analyte, and detection limits. The choice of analytical methods will influence conclusions regarding recognized environmental conditions.

7.8 *Quality Assurance/Quality Control Procedures*:

7.8.1 Appropriate quality assurance/quality control (QA/QC) procedures should be included in the Phase II ESA to allow for assessment of the quality of the data collected. The QA/QC measures may include, for example, written field sampling protocol, decontamination procedures, instrument calibration, the preparation and analysis of trip blanks, equipment blanks, duplicate samples, and holding times for sample analysis. The data quality objectives of the user will influence the decision whether to include QA/QC samples and the selection of particular QA/QC procedures.

7.8.2 Any analytical laboratory used in the course of the Phase II ESA should have an established QA/QC program sufficient to allow assessment of the precision and accuracy of the data it generates. The environmental professional should verify the certification or licensing of the analytical laboratory where applicable, or otherwise evaluate the appropriateness of the QA/QC program of the analytical laboratory prior to engaging its services. Environmental professionals should use laboratories that hold appropriate licenses and certifications for the analyses to be performed.

8. Assessment Activities

8.1 The assessment activities to be conducted under the Phase II ESA may range from field screening methods to intrusive multi-media sampling and laboratory analysis. This section references an overview of recognized intrusive and non-intrusive site assessment methodologies and common field screening activities and analytical protocols. The selection of one or a combination of these methods will be based on the stated objectives of the user and the judgment of the environmental professional. Appropriate QA/QC measures should be implemented according to the work plan.

8.2 *Field Screening and Field Analytical Techniques*:

8.2.1 Field methods may be used for the purpose of characterizing certain physical attributes of a site and for the assessment of the presence and distribution of a specific chemical or chemical classes at a sampling location. Field screening and field analytical techniques include methods of data gathering which may result in real-time acquisition of data. Field screening and field analytical techniques generally allow rapid, multiple measurements at relatively low cost. Because the accuracy and reliability of data produced by field instruments varies, the selection of particular field instruments and methods is influenced by the data quality objectives of the assessment. The quality of the data generated should be consistent with the intent of the assessment.

8.2.1.1 Some common field screening instruments, such as portable flame and photoionization detectors, have higher detection limits and lower precision and accuracy compared to field or laboratory analytical instruments and are most suited to general qualitative analyses and health and safety monitoring.

8.2.1.2 Field analytical methods include all chemical analysis methods capable of providing chemical-specific quantitative data in the field or non-laboratory setting. More reliable data than obtained through field screening may be obtained in the field or laboratory through the use of portable gas chromatographs or mobile field laboratories. The highest degree of data accuracy can generally be obtained by use of EPA and ASTM analytical methods.

8.2.2 Field screening and field analytical techniques may be used in Phase II ESAs for: qualitative and quantitative confirmation of the presence of contaminants constituting a recognized environmental condition, preliminary mapping of the areal and vertical distribution of contaminants, guiding the collection of samples for more rigorous analysis, and guiding the placement of monitoring wells. The environmental professional should establish standard procedures and protocols prior to the implementation of field screening and field analytical efforts.

8.2.3 Common field methods may assess soil, soil gas, surface water and ground water. Field screening and field analytical methods include use of ion-selective electrodes, detection of volatile compounds using field gas chromatographs, and detection of inorganic constituents and semi-volatile organic chemicals using colorimetric wet chemistry methods, including enzyme immunoassay tests. Section 5.1.5 and Section 12 in Guide D 5730, further address the use of field screening and field analytical methods (see 2.1).

8.3 *Environmental Media Sampling*:

8.3.1 Phase II ESAs generally require environmental sampling and field or laboratory analysis, or both, to determine whether hazardous substances or petroleum products are present on a property. The selection of sampling locations and analytical parameters should be consistent with the work description developed at the onset of the project. Deviations from the description of work to be performed should be noted and justified in the final report. Sampling locations should be documented to allow the sampling event to be reproduced later. The need for surveyed locations is subject to the agreement between the user and the environmental professional.

8.3.2 The Phase II sampling activities should be conducted to avoid the introduction of contaminants into the sample from the sampling equipment, and from environmental factors and to avoid, if feasible, and minimize the spread of contaminants in the environment.

8.3.3 Sampling methods should be selected based on requirements of the local jurisdiction, where applicable. Sampling methods that may be employed during a Phase II site assessment are among the standards identified in Guide D 5730. See 2.1. New technologies are frequently introduced.

8.4 *Sampling Handling*:

8.4.1 Sample handling includes sample collection and containerization, filtration, and preservation. Sample handling procedures should be selected based on requirements of local jurisdictions, where applicable. Sample containerization and preservation techniques are among the standards identified in Guide D 5730. See 2.1. Sample handling methods approved or recommended in the applicable analytical protocol should be used.

8.4.2 A completed chain of custody record should accompany each shipment of samples to the analytical laboratory, and should be included in the written report of the assessment. Chain of custody records provide written documentation regarding sample collection and handling and identify the persons involved in the chain of sample possession. Chain of custody records also provide a written record of requested analytical parameters.

8.4.3 The environmental professional should take appropriate measures to deliver the samples such that the analysis can be completed within the appropriate sample holding times.

9. Evaluation of Data

9.1 The implementation of the work that was planned under Section 7 will generate information and data on environmental conditions at the property that must be evaluated. To complete the Phase II ESA, this information and data must be analyzed to qualify the adequacy of the work performed, and evaluate whether a release or disposal of hazardous substances or petroleum products has occurred.

9.2 *Verification of Assumptions*[8]—The first step in analysis of data is to consider whether the assumptions upon which the work description was based were valid. That is, one asks whether samples were collected of the appropriate media (for example, soil, ground water) at the right location and depth (that is, where the highest concentrations of contaminants should be, if the potential release had actually occurred).

9.2.1 A work description is typically based upon assumptions as to subsurface physical conditions, for example, relative soil or aquifer permeabilities, depth to the water table, ground water flow direction and characteristics of potential contaminants. Actual information on these conditions is usually gained through the explorations, sampling, and observations completed under the work description (at least for those studies where subsurface conditions are of concern). The environmental professional should evaluate whether the assumptions were valid, in light of the actual conditions encountered. If an assumption was not valid, then the work performed may not

[8] In some jurisdictions, the evaluation and analysis of the subsurface conditions may require a licensed individual (for example, registered geologist, professional engineer).

have verified the recognized environmental conditions and additional iterations may be appropriate.

9.2.2 For example, the work description may have called for the installation of a number of observation wells, some of which were intended to be downgradient of areas where releases may have occurred. However, the water levels measured in the wells might have indicated that ground water flowed in a different direction than that inferred for development of the description of work to be performed. Hence the wells were not downgradient of the potential release areas and their purpose was not met. This would indicate that additional site work would have to be done to successfully complete the Phase II ESA. On the other hand, water levels may have confirmed that the wells were downgradient, as intended. They are then qualified as meeting their purpose, and the analysis and interpretation of data relating to them can proceed.

9.3 *Verification of Data*—The second step is to evaluate whether any target analytes detected in the samples are in fact attributable to the site. This requires the environmental professional to determine whether the analytical results should be questioned based on the QA/QC procedures described in 7.8. When data quality is determined to be acceptable, test results should be interpreted.

10. Interpretation of Results

10.1 The results of the Phase II explorations, sampling and testing should be interpreted to determine the significance of the data as they relate to the user's objectives. The data should also be evaluated to determine if other sources of contamination or more highly contaminated media may exist at the site but were not assessed in accordance with this guide. If chemicals that occur in nature are detected, the data must be interpreted to determine whether they are naturally occurring or are present as a result of human activity. For example, lead occurs naturally in soils and waters, at a range of concentrations. Mere detection of lead does not necessarily reflect lead disposal or release. The environmental professional must judge the likelihood that the lead level detected exceeded the naturally occurring lead level in the site vicinity and must evaluate potential sources (for example, fill materials present on the property in an area where use of mine tailings as fill was prevalent) to determine whether disposal or a release has occurred.

10.2 *Elimination of Recognized Environmental Conditions*—If hazardous substances or petroleum products were not detected, and the Phase II ESA information and data has been evaluated and found by the environmental professional to provide sufficient information from which the environmental professional can render a professional opinion that there is no reasonable basis for suspecting the disposal or release of hazardous substances at the site, then no further work is necessary to satisfy the "all appropriate inquiry" element of the innocent purchaser provision of CERCLA with respect to the recognized environmental conditions assessed.

10.2.1 Where fewer than all recognized environmental conditions identified in a Phase I ESA or transaction screen are assessed in the Phase II ESA, the conclusion that no further inquiry is warranted applies only to the recognized environmental conditions assessed. However, this guide should not be construed to require Phase II assessment of all recognized environmental conditions identified in a Phase I ESA or a transaction screen. Where the user's objectives may be otherwise satisfied (for example, establishing the innocent purchaser defense is not the user's objective), a recognized environmental condition identified in a Phase I ESA or Transaction Screen may be eliminated from further assessment in the Phase II ESA.

10.2.2 A recognized environmental condition identified in a Phase I ESA or transaction screen may also be eliminated from the full scope of Phase II ESA procedures by developing further information not included in the Phase I or transaction screen. For example, the user may obtain a tank closure report not included in the Phase I report that confirms there were no releases from the tank that was identified as a recognized environmental condition in the Phase I or transaction screen. This new information may eliminate the need for additional Phase II tasks with respect to that recognized environmental condition. Similarly, a recognized environmental condition identified in the Phase I or transaction screen due to off-site conditions (for example, a nearby CERCLIS or LUST site) may be eliminated from a Phase II ESA if the records associated with such sites confirm that hazardous substances or petroleum products have not migrated or are not likely to migrate onto the property.

10.3 *Confirmation of Recognized Environmental Conditions*—If the Phase II ESA was conducted in accordance with this guide and the presence of hazardous substances or petroleum products at the property is confirmed under conditions that indicate disposal or a release, then the innocent purchaser defense may not be supportable with respect to these confirmed substances. See 4.1.1. If hazardous substances or petroleum products were detected, and if requested by the user, the Phase II information and data may be interpreted to determine what further work is required or desirable to assess the nature and extent of the release. This guide should not be construed to require additional assessment.

10.3.1 Additional assessment may be requested by the user if further information is desired. The objectives of the user will determine the scope of additional assessment, if any.

10.3.2 In order to guide the decision whether to conduct an additional assessment and to form the objectives of that assessment, data which confirms the presence of hazardous substances or petroleum products may be compared to applicable or relevant and appropriate requirements ("ARARs") to determine whether such standards are exceeded. The identification and interpretation of ARARs may be assisted by consultation with legal counsel having environmental expertise and other environmental professionals.[9]

10.3.3 This guide does not address the multitude of factors, in addition to environmental data, that may be considered by a user in determining whether additional assessment is warranted where the presence of hazardous substances or petroleum products under conditions that indicate disposal or release has

---

[9] *ASTM Book Reference No. DS 64* is a compilation of cleanup criteria for contaminated soil and ground water in the fifty states and in selected countries outside the United States. Because such criteria continue to evolve, care should be taken to determine the current status of such criteria.

been confirmed. This guide should not be construed to elevate the consideration of environmental data above economic or other business considerations in business decisions. Such considerations and their impact on transactional decisions are properly within the purview of the parties to the transaction and are beyond the scope of this guide.

11. Phase II ESA Report Preparation

11.1 *Purpose*—The purpose of a written report, when one is prepared, is to describe the work performed and provide documentation of the data and evaluation that *(a)* constitutes the factual findings, *(b)* supports the conclusions of the environmental professional, *(c)* facilitates decisions about the transaction, and *(d)* documents the basis for the decision whether further assessment is warranted. Although generally prepared, a written report is not required of all Phase II ESAs and is subject to the agreement between the user and the environmental professional (see 5.1.2).

11.2 *Characteristics*—A written Phase II ESA report generally has three characteristics: the components of a scientific report, good technical writing, and clear and accurate presentation of the results and conclusions, and, if requested, recommendations to the user. Scientific reports generally contain the following sections: an introduction, a summary of background information, a description of site conditions, a description of work performed and methods used, data presentation and evaluation, and findings and conclusions. The report is typically supported by appropriate tables, figures and appendices. Good technical writing is consistent, accurate, concise, clear and complete. The whole report needs to stand on its own.

11.3 *Format of Report*—An example report format for a written Phase II ESA report is presented in Appendix X1.

11.4 *Content of Report*:

11.4.1 *Introductory Components*—The introductory components of a Phase II ESA usually include a transmittal letter, executive summary, cover page, and table of contents.

11.4.1.1 *Signatures*—The environmental professional(s) responsible for the Phase II ESA should sign the report. Signatures, along with typed names and titles may appear in the transmittal letter, on the cover page, or on a signature page at the end of the main text of the report. Where required by the local jurisdiction, professional seals, license type and license number, as applicable, should be affixed.

11.4.1.2 *Transmittal Letter*—A transmittal letter documents the date of a report's delivery and identifies the intended recipients of the report. It also may contain other important information including a description of the report as draft or final and a designation that the report is confidential or subject to attorney-client privilege, or both. The transmittal letter also may identify the staff responsible for the work.

11.4.1.3 *Cover Page*—The cover or title page identifies the following items: the subject property name and address, the preparer of the report, the person for whom the report was prepared and the date of the report. It may also include a statement of confidentiality. If appropriate, it also may include a statement identifying the document as subject to certain legal privileges, such as attorney-work product and attorney/client privileges.

11.4.1.4 *Table of Contents*—The table of contents refers the reader to the location of the major sections of the report and to the figures, tables and appendices. All section, figure, table and appendix titles should match those of the text. See Appendix X1 for a sample table of contents.

11.4.1.5 *Executive Summary*—An executive summary provides a concise overview of the findings of the ESA and is considered optional. As a summary, it should generally be brief and should not discuss material not included in the text of the report.

11.5 *Main Text*—The main text of the report generally contains an introduction, a discussion of background information, a description of work performed during the Phase II ESA and methods used, data presentation and evaluation, and a discussion of findings and conclusions.

11.5.1 *Introduction*—The introduction should generally include: the purpose and scope of the Phase II ESA; any special terms, conditions, or limitations; the date of authorization of the project and any changes. The introduction also should identify items not included within the scope of the Phase II ESA that might have been expected to be included, for example, recognized environmental conditions not included and the reasons for eliminating them.

11.5.1.1 The introduction also references the guiding standards including contracts/service agreements, regulatory requirements, and the description of work to be performed. The introduction generally identifies the specific tasks comprising the work performed in the order in which they will be described later.

11.5.2 *Background Information*—This section should include a description of the site and its features, a description of the physical site setting, a brief discussion of the site's history and use and the use of adjacent properties, and a summary of previous environmental site assessments. Reference to existing ESA reports may convey this information.

11.5.2.1 This section should identify the data and conclusions that led to the performance of the Phase II ESA. If the Phase II ESA is based on earlier environmental site assessments, the earlier ESAs should be precisely identified (for example, by reference to report title, project number, author and date of any report). Where appropriate, prior ESAs may be included in an appendix. If recommendations for further work were included in a referenced prior ESA but were not followed in the Phase II ESA, an explanation in this section is appropriate.

11.5.3 *Phase II ESA Activities*—This section should summarize the work conducted. This section should describe what was done, identify the methods that were used, identify the location of the Phase II ESA activities and pertinent site conditions. This section also should explain the rationale for the work performed, including the review of supplemental records and the selection of sampling locations, analytical parameters and methodologies. Deviations from the work description and deviations from standard methodologies also should be explained.

11.5.3.1 Methods used should be identified by reference to a specific published standard or method such as ASTM, state protocols, or EPA methodologies. Other methods used (for example, standard operating procedures for a particular company) should be described in detail. Where appropriate, the details of methods and procedures used can be included in an appendix.

11.5.4 *Evaluation and Presentation of Results*—Results of the Phase II ESA should be presented in a logical order and should be summarized to allow the reader to more readily follow the discussion of findings and conclusions which appears in the next section of the report. Results may be presented in text, in tables, and/or in figures. Reference to the location of data in the appendices is appropriate. Results should be presented consistently and accurately.

11.5.4.1 *Evaluation*—Evaluation of data involves verifying assumptions underlying the scope of work as they relate to meeting objectives and determining whether chemical constituents detected in samples were attributable to site conditions or to QA/QC failures. See Section 9 for a detailed discussion of data evaluation.

11.5.4.2 *Presentation*—The presentation of results should be organized to assist the reader's understanding of the interpretation of the chemical analysis data that follows in the next section of the report. Organizational divisions may be by media (for example, soil data presented separately from ground water data), contaminant type (for example, organic vs. inorganic contaminants), source (for example, contamination from leaking underground storage tank versus contamination from drum storage area), or other factors.

11.5.4.3 The specific information presented will depend on the activities performed. Although Phase II ESAs can involve the full range of environmental media, they most typically involve sampling and chemical analysis or screening of soil, vapors and/or ground water.

11.5.5 For purposes of illustration, the following information should generally be presented for the referenced activities:

11.5.5.1 *Soil Sampling*—The locations of sampling points in relation to potential sources; the depths of sampling; the depth to pertinent horizons, such as the water table and the fill/natural soil contact; the sampling and analytical methods used; and the results of chemical analyses.

11.5.5.2 *Soil Gas Sampling*—The locations of sampling points in relation to potential sources; the depths of sampling; the depth to pertinent horizons, such as the water table and the fill/natural soil contact; the sampling and analytical methods used; and the results of chemical analyses.

11.5.5.3 *Ground Water Sampling*—The locations of sampling points in relation to potential sources; water level data (for example, depth to water or water table elevation, screened interval elevation, filtered interval elevation); the sampling and analytical methods used; and the results of chemical analyses.

11.5.6 *Discussion of Findings and Conclusions*—Where a written report of a Phase II ESA is prepared, it should have a findings and conclusions section that includes:

11.5.6.1 A statement that the Phase II ESA was conducted in substantive accordance with the guidance contained in this guide and in accordance with the description of work to be performed, and a description of any deviations from the assessment activities that were planned;

11.5.6.2 A summary description of: the nature of the recognized environmental conditions identified in previous work, the assessments performed to evaluate the identified recognized environmental conditions, the data generated from the assessments, the geologic and hydrogeologic conditions encountered, and any limitations of the data produced and their impact on conclusions;

11.5.6.3 The environmental professional's interpretation of the significance of the data as they relate to the user's objectives. Appropriate citations and references to standards, guidelines and sources for background concentrations should be included. See Section 10 for a detailed discussion of interpretation of results;

11.5.6.4 A statement whether, in the judgment of the environmental professional, the data provides sufficient information to support a professional opinion that there is no reasonable basis for suspecting the disposal or release of hazardous substances or petroleum products at the site with respect to the recognized environmental conditions assessed, and that no further assessment is necessary; or that with respect to the recognized environmental conditions assessed, hazardous substances or petroleum products have been released or disposed at the property; and

11.5.6.5 Where applicable, a specific explanation of data insufficiencies that prevent a conclusion that hazardous substances or petroleum products were or were not released or disposed at the property.

11.5.7 *Recommendations*—Recommendations for further work may be included at the option of the user.

11.5.8 *Concluding Components*—The report should contain tables, figures and appendices as necessary or appropriate to explain and support the main text of the report.

11.5.8.1 *Tables and Figures*—Tables and figures may be used as a summary presentation of data. Guidelines for data presentation in 11.5.4 apply to tables and figures as well as to text.

11.5.8.2 Figures that show potential sources of contamination, sampling points, locations of other activities and surface and subsurface structures can significantly aid the understanding of the reader. Figures should be dated and should include a north arrow, a scale as appropriate to the level of accuracy of the drawing (or notation that the figure is not to scale), a legend, a title, and other appropriate identification. If figures are based on the work of others, the source and its date should be referenced.

11.5.8.3 *Appendices*—The following is a non-exhaustive list of materials that may be appropriate for inclusion in appendices to a Phase II ESA report: photographs, subsurface exploration logs, laboratory reports, with quality control information, chain of custody forms, identification of specific methods, and reports of previous environmental site assessments.

11.6 *Additional Services*—Any additional services including a broader scope of assessment, more detailed conclusions, liability/risk evaluations, recommendations for additional Phase II assessment, remediation techniques, etc., are beyond the scope of this guide, and should only be included in the report if so agreed by the user and the environmental professional.

12. Non-Scope Considerations

12.1 *General*:

12.1.1 *Additional Issues*—There may be environmental issues or conditions at a property that parties may wish to assess in connection with commercial property transactions that are outside the scope of this guide (the non-scope considerations). As noted by the legal analysis in Appendix X1 of Practices E 1527 and E 1528, some substances may be present on property in quantities and under conditions that may lead to contamination of the property or of nearby properties but are not included in CERCLA's definition of hazardous substances (42 U.S.C. § 9601(14)) or do not otherwise present potential CERCLA liability. In any case, they are beyond the scope of this guide.

12.1.2 *Outside Standard Guide*—Whether or not a user elects to inquire into non-scope considerations in connection with an ESA implemented in accordance with this guide or any other ESA, no assessment of such non-scope considerations is required for appropriate inquiry as described by this guide.

12.1.3 *Other Standards*—There may be standards or protocols for assessment of potential hazards and conditions associated with non-scope conditions developed by governmental entities, professional organizations, or other private entities, including ASTM.

12.1.4 *List of Additional Issues*—Several considerations that may be important in property transactions but that are outside the scope of this guide include asbestos-containing materials, radon, lead-based paint, lead in drinking water, and wetlands. No implication is intended as to the relative importance of such considerations, nor is this list intended to be all-inclusive.

APPENDIX

(Nonmandatory Information)

X1. EXAMPLE TABLE OF CONTENTS AND REPORT FORMAT

Where a written report is desired by the user, a Phase II ESA report that includes the sections described below, is suggested:

Transmittal Letter
Cover Page
Table of Contents

X1.1 Executive Summary

X1.2 Introduction
- X1.2.1 Purpose
- X1.2.2 Special Terms and Conditions
- X1.2.3 Limitations and Exceptions of Assessment
- X1.2.4 Limiting Conditions and Methodology Used

X1.3 Background (may be by reference to prior environmental reports):
- X1.3.1 Site Description and Features
- X1.3.2 Physical Setting
- X1.3.3 Site History and Land Use
- X1.3.4 Adjacent Property Land Use
- X1.3.5 Summary of Previous Assessments

X1.4 Phase II Activities
- X1.4.1 Scope of Assessment:
  - X1.4.1.1 Supplemental Record Review
  - X1.4.1.2 Conceptual Site Model and Sampling Plan
  - X1.4.1.3 Chemical Testing Plan
  - X1.4.1.4 Deviations from the Work Plan
- X1.4.2 Field Explorations and Methods
  - X1.4.2.1 Test Pits
  - X1.4.2.2 Test Borings
  - X1.4.2.3 Monitoring Well Installations
    - 1) Ground water elevation measurement
    - 2) Ground water flow direction
  - X1.4.2.4 Other
- X1.4.3 Sampling and Chemical Analyses and Methods
  - X1.4.3.1 Soil
  - X1.4.3.2 Ground water
  - X1.4.3.3 Other

X1.5 Evaluation and Presentation of Results
- X1.5.1 Subsurface Conditions
  - X1.5.1.1 Geologic Setting

List of Tables

List of Figures

List of Appendixes

Designation: E 1912 – 98

Standard Guide for Accelerated Site Characterization for Confirmed or Suspected Petroleum Releases[1]

This standard is issued under the fixed designation E 1912; the number immediately following the designation indicates the year of original adoption or, in the case of revision, the year of last revision. A number in parentheses indicates the year of last reapproval. A superscript epsilon (ϵ) indicates an editorial change since the last revision or reapproval.

1. Scope

1.1 This guide covers a process to rapidly and accurately characterize a confirmed or suspected petroleum release site. This guide is intended to provide a framework for responsible parties, contractors, consultants, and regulators to streamline and accelerate the site characterization process or supplement incomplete characterization data. The accelerated site characterization (ASC) approach may be incorporated in state and local regulations as a cost-effective method of making informed corrective action decisions sooner.

1.2 This guide describes a process for collecting site characterization information in one mobilization, using rapid sampling techniques; field analytical methods; and on-site interpretation and iteration of field data to refine the conceptual model for understanding site conditions as the characterization proceeds. This information can be used to determine the need for interim remedial actions; site classification or prioritization, or both; further corrective actions; and active remediation. The process outlined in this guide can be incorporated into existing corrective action programs, and is organized to be used in conjunction with Guides E 1599 and E 1739.

1.3 For guidance concerning contractor health and safety issues, appropriate federal, state, and local regulations (for example, Occupational Safety and Health Administration) and industry standards should be consulted. For sampling quality assurance/quality control (QA/QC) practices, see references in Section 2. Considerations for field analytical method quality assurance/quality control are discussed in Section 5.

1.4 This guide is organized as follows:

1.4.1 Section 1 describes the scope,

1.4.2 Section 2 lists Referenced Documents,

1.4.3 Section 3 defines Terminology,

1.4.4 Section 4 identifies the Significance and Use,

1.4.5 Section 5 describes the Accelerated Site Characterization Process,

1.4.6 Appendix X1 identifies Additional Referenced Documents,

1.4.7 Appendix X2 provides an Example of a Data Quality Classification System,

1.4.8 Appendix X3 contains a list of physical and chemical properties and hydrogeologic characteristics applicable to site characterizations, and a list of input parameters and methodologies for ASTM RBCA Tier 1 and Tier 2 evaluations, and

1.4.9 Appendix X4 contains a case study example of the ASC process, including a RBCA Tier 1 and Tier 2 evaluation.

1.5 The values stated in inch-pound units are to be regarded as the standard. The SI units given in parentheses are for information only.

1.6 *This standard does not purport to address all of the safety concerns, if any, associated with its use. It is the responsibility of the user of this standard to establish appropriate safety and health practices and determine the applicability of regulatory limitations prior to use.*

2. Referenced Documents

2.1 *ASTM Standards:*

D 5730 Guide to Site Characterization for Environmental Purposes With Emphasis on Soil, Rock, the Vadose Zone, and Ground Water[2]

E 1599 Guide for Corrective Action for Petroleum Releases[3]

E 1689 Guide for Developing Conceptual Site Models for Contaminated Sites[4]

E 1739 Guide for Risk-Based Corrective Action Applied at Petroleum Release Sites[3]

2.2 *EPA Documents:*

USEPA SW 846, Recommended Analytical Procedures, Test Methods for Evaluating Solid Waste-Physical/Chemical Methods[5]

USEPA, Draft Field Methods Compendium, OER 9285.2-11[5]

USEPA, Subsurface Characterization and Monitoring Techniques: A Desk Reference Guide-Vols I and II, EPA 625/R-93/003a and b[5]

USEPA, Description and Sampling of Contaminated Soils: A Field Pocket Guide, EPA 625/12-91/002[5]

USEPA, Environmental Investigations Standard Operating

---

[1] This guide is under the jurisdiction of ASTM Committee E-50 on Environmental Assessment and is the direct responsibility of Subcommittee E50.01 on Storage Tanks.

Current edition approved Jan. 10, 1998. Published June 1998.

[2] *Annual Book of ASTM Standards*, Vol 04.09.

[3] *Annual Book of ASTM Standards*, Vol 11.04.

[4] *Annual Book of ASTM Standards*, Vol 11.05.

[5] Available from Superintendent of Documents, U.S. Government Printing Office, Washington, DC 20402.

Procedures and Quality Assurance Manual, May 1996, USEPA Region[5]

USEPA, Expedited Site Assessment Tools for UST Sites: A Guide for Regulators, EPA 510-B-97-001[5]

3. Terminology

3.1 *Definitions of Terms Specific to This Standard:*

3.1.1 *accelerated site characterization (ASC)*—a process for collecting and evaluating information pertaining to site geology/hydrogeology, nature and distribution of the chemical(s) of concern, source areas, potential exposure pathways and points of exposure in one mobilization. The ASC employs rapid sampling techniques, on-site chemical analysis and hydrogeological evaluation, and field decision making to provide a comprehensive "snap-shot" of subsurface conditions.

3.1.2 *active remediation*—actions taken to reduce the concentrations of chemical(s) of concern. Active remediation could be implemented when the no further action and passive remediation courses of action are not appropriate.

3.1.3 *chemical(s) of concern*—specific constituents that are identified for evaluation in the site characterization process.

3.1.4 *conceptual model*—a summary of information that is known about a site. Available site information is compiled onto one or more simple graphics to develop an understanding of the site conditions. The conceptual model is not an analytical or numerical computer model, but may utilize these tools in developing a conceptual understanding of site conditions.

3.1.5 *corrective action*—activities performed in response to a suspected or confirmed release, which include one or more of the following: site characterization, interim remedial action, remedial action, operation and maintenance of equipment, monitoring of progress, monitoring of natural attenuation, and termination of remedial action.

3.1.6 *exposure pathway*—the course a chemical(s) of concern takes from the source area(s) to an exposed organism. An exposure pathway describes a unique mechanism by which an individual or population is exposed to a chemical(s) of concern originating from a site. Each exposure pathway includes a source or release from a source, a point of exposure, and an exposure route. If the exposure point differs from the source, a transport/exposure medium (for example, air) or media also is included.

3.1.7 *facility*—the property containing the source of the chemical(s) of concern where a release has occurred.

3.1.8 *field analytical methods*—methods or techniques that measure physical properties or chemical presence in soil, soil vapor, and ground water immediately or within a relatively short period of time to be used during a site characterization. Measurement capabilities range from qualitative (positive/negative) response to below parts per billion (sub-ppb) quantitation. Accuracy and precision of data from these methods depends on the method detection limits and QA/QC procedures.

3.1.9 *field-generated analytical data*—information generated on site soon after sample acquisition that is used to direct the site characterization process. These data include: concentrations of chemical(s) of concern in air; soil; soil vapor or ground water, or both; and hydrogeologic conditions.

3.1.10 *indicator compounds*—compounds in ground water, soil, or air, specific to the petroleum product released, used to confirm the existence of the petroleum product, define the distribution of the chemical(s) of concern, define the target levels, monitor progress of the remedial action, and identify the termination point of the remedial action.

3.1.11 *interim remedial action*—the course of action to mitigate fire and safety hazards and to prevent further migration of hydrocarbons in their vapor, dissolved, or liquid phase.

3.1.12 *mobilization*—the movement of equipment and personnel to the site, to prepare for, collect, and evaluate site characterization data. These activities, when conducted as one continuous event (from one day to several weeks), are referred to as a single mobilization. Activities that are not conducted continuously are referred to as multiple-site mobilizations.

3.1.13 *on-site manager*—an individual who is on site and is responsible for directing field activities and decision-making during the site characterization. The on-site manager should be familiar with the purpose of the site characterization, pertinent existing data, and the data collection and analysis program. When conducting an ASC, it is necessary for the on-site manager to also be the principal investigator, developing and refining the conceptual model of site conditions. This individual must have the necessary experience and background to perform the required site characterization activities and to accurately interpret the results and direct the investigation. For the purposes of this guide, sufficient qualification criteria for the on-site manager includes knowledge and experience in the following areas:

3.1.13.1 Soil and ground water sampling and analytical methods to be used at the site;

3.1.13.2 Fate and transport of petroleum hydrocarbons in the subsurface;

3.1.13.3 Local geology/hydrogeology;

3.1.13.4 Local regulations and ordinances, including knowledge of state-specific certification requirements;

3.1.13.5 Personal health and safety requirements; and

3.1.13.6 Evaluation and interpretation of site characterization results.

3.1.14 *petroleum*—including crude oil or any fraction thereof that is liquid at standard conditions of temperature and pressure (60°F (16°C) at 14.7 psia). The term includes petroleum-based substances comprised of a complex blend of hydrocarbons derived from crude oil through processes of separation, conversion, upgrading, and finishing, such as motor fuels, jet oils, lubricants, petroleum solvents, and used oils.

3.1.15 *point(s) of exposure*—the point(s) at which an individual or population may come in contact with a chemical(s) of concern originating from a site.

3.1.16 *quality assurance/quality control (QA/QC)*—the use of standards and procedures to ensure that samples collected and data generated are reliable, reproducible, and verifiable.

3.1.17 *rapid sampling tools*—equipment and techniques that allow personnel to collect samples from different media, in a relatively short period of time, for on-site chemical analysis and hydrogeologic evaluation within the same mobilization.

3.1.18 *receptors*—persons, structures, utilities, surface waters, and water supply wells that are or may be adversely affected by a release.

3.1.19 *regulatory agency*—any state or local program responsible for overseeing underground storage tank (or other petroleum/hazardous material source) site characterization and corrective action.

3.1.20 *release*—any spilling, leaking, emitting, discharging, escaping, leaching, or disposing of petroleum products into ground water, surface water, soils, or air.

3.1.21 *remediation/remedial action*—activities conducted to protect human health, safety, and the environment. These activities include evaluating risk, making no-further-action determinations, monitoring, institutional controls, engineering controls, and designing and operating cleanup systems.

3.1.22 *site characterization*—an evaluation of subsurface geology/hydrogeology, and surface characteristics to determine if a release has occurred, the levels of the chemical(s) of concern, and the distribution of the chemical(s) of concern. The data collected on soil, soil vapor and ground water, potential exposure pathways and location of receptors and point(s) of exposure is used to generate information to support remedial action decisions.

3.1.23 *source area(s)*—the location(s) of liquid hydrocarbons or the zone(s) of highest soil or ground water concentrations, or both, of the chemical(s) of concern.

3.1.24 *user*—an individual or group involved in the ASC process including owners, operators, regulators, petroleum fund managers, attorneys, consultants, legislators, and so forth.

4. Significance and Use

4.1 An ASC is a process for collecting and evaluating information on site geology/hydrogeology, nature and distribution of chemicals of concern, source areas, potential exposure pathways, and points of exposure. The unique goal of an ASC is to complete a site characterization in one mobilization. This can be accomplished by developing and refining a conceptual site model, utilizing rapid sampling tools and techniques, obtaining field-generated analytical data, and on-site interpretation of results. Evaluation of data concurrent with the investigation allows the on-site manager to select subsequent sampling points based on actual subsurface conditions, resulting in a more comprehensive and cost-effective "snapshot" of subsurface conditions.

4.2 The ASC process has the following advantages:

4.2.1 Immediate identification of potential risks to human or environmental receptors or potential liabilities, or both;

4.2.2 Rapid determination of the need for interim remedial actions, site classification, and prioritization;

4.2.3 Rapid sample collection and analysis, near contemporaneous analytical results, and maximum data comparability;

4.2.4 Optimization of sample point locations and analytical methods;

4.2.5 Greater number of data points for resources expended;

4.2.6 Near immediate data availability for accelerating corrective action decisions; and

4.2.7 Collection of vertical and horizontal data, allowing for three-dimensional delineation of chemical(s) of concern in soil, soil vapor, or ground water.

4.3 The ASC process described in this guide is intended for use in situations where the potential exists that petroleum has been released. The same principles may be applicable to other indicator compounds or chemical(s) of concern, and sources (for example, chlorinated solvent releases). If the ASC process is used for chemical(s) of concern, other than petroleum, the user must consider the physical and chemical characteristics of the chemical(s) of concern and the media in which they are present to ensure that the sampling tools and analytical methods are capable of measuring and detecting the chemical(s) of concern.

4.4 A conventional site characterization approach most often involves several mobilizations. Each mobilization typically includes a predefined sampling and analysis plan, where analysis and interpretation of results are performed off-site after demobilization. A conventional site characterization can provide high-quality data; however, multiple mobilizations often prolong the process required to adequately characterize subsurface conditions.

4.5 The ASC process requires an on-site manager to make decisions to guide the characterization. Without an individual on site who is able to interpret data as it is generated, and is authorized to adjust sample locations or scope of the investigation, or both, an ASC has little chance of meeting its stated objective of full characterization in one mobilization. Levels of communication and authority between the on-site manager and the user should be established prior to beginning the characterization.

5. Accelerated Site Characterization Process

5.1 The unique feature of the ASC process is the collection, analysis, and evaluation of hydrogeologic and chemical data while on-site. A flowchart of the ASC process is presented in Fig. 1, and a discussion of each activity begins in 5.2. While many of the steps in an ASC are similar to those in a conventional characterization, the following activities, as illustrated in the area labeled "Field Activities" in Fig. 1, are performed on-site during an ASC:

5.1.1 Interpretation and evaluation of field-generated data as it is collected;

5.1.2 Continuous refinement of the conceptual model, and the understanding of site conditions;

5.1.3 Modification of the sampling and analysis program to address any necessary adjustments in the scope of work; and

5.1.4 Collection of additional data necessary to complete the characterization.

5.2 *Step 1—Identify Site Characterization Purpose*:

5.2.1 *Purpose*—The objectives of any environmental site characterization, as noted previously, are to understand the site geology/hydrogeology, the nature and distribution of the chemicals of concern, the migration pathways and location of potential receptors and point(s) of exposure. The scope of work, however, will vary depending upon the purpose of the specific characterization. Typical purposes include one or more of the following: hazard determination, initial response action, release confirmation, risk determination, remedial action evaluation, regulatory compliance, or real estate transaction. For example, an ASC for an initial response action might focus on defining imminent hazards, potential migration of chemical(s) of concern and the location of receptors and point(s) of exposure, while a characterization for a real estate transaction focuses on identifying the presence of chemicals of concern. A

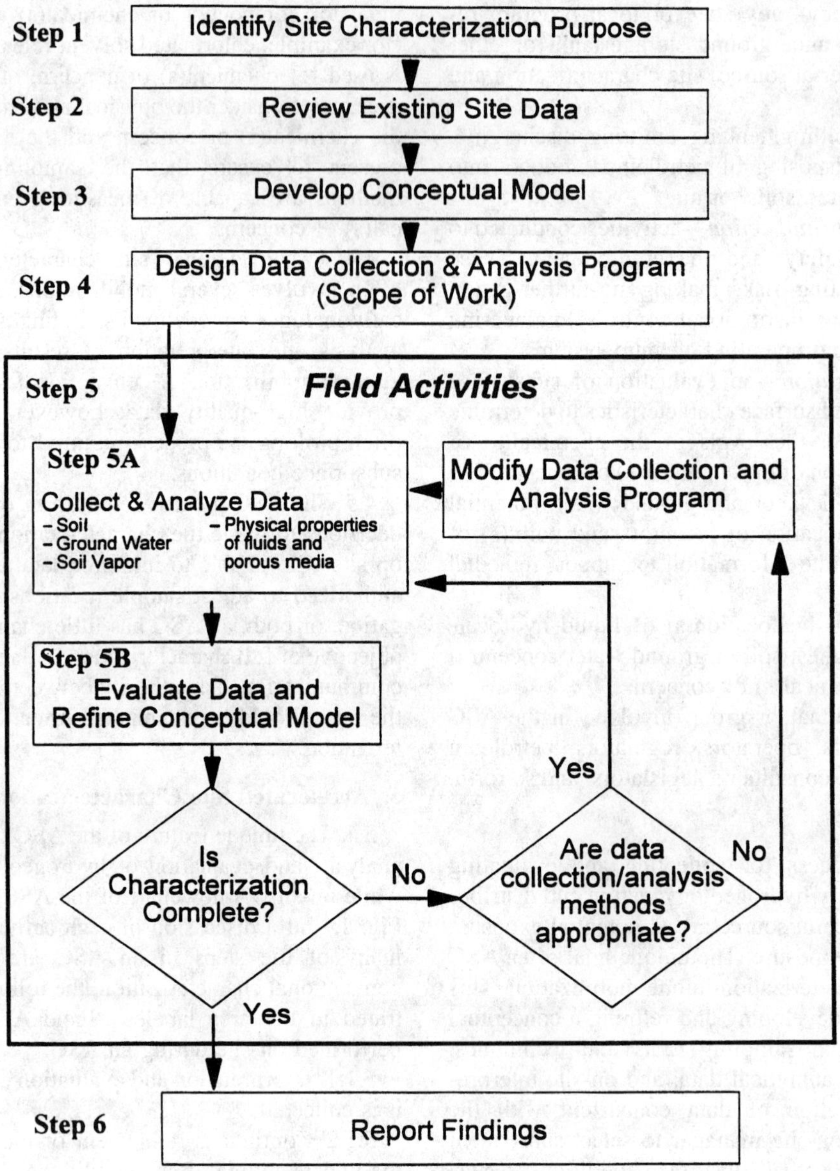

FIG. 1 ASC Process Flowchart

corrective action evaluation will require a higher priority be placed on understanding subsurface hydrogeologic conditions, whereas a risk determination will focus first on receptors, exposure pathways and points of exposure, in addition to levels of chemical(s) of concern.

5.2.2 The scope of the ASC is determined prior to mobilization, but will often be revised based on interpretation of the field-generated data.

5.3 *Step 2—Review Existing Site Information*:

5.3.1 A variety of regional and site-specific information should be obtained prior to mobilization. A review of existing information, and a site visit, are important in the design of a data collection and analysis program, and in the development of the conceptual model. Information obtained through the site visit, interviews, and records search include the following:

5.3.1.1 Local and regional hydrogeologic maps to identify general soil types/regional depth to bedrock, rock type, depth to ground water, aquifer properties, and so forth;

5.3.1.2 Past and current land use history of the site and adjacent properties (including future land use if known);

5.3.1.3 Location of potential sources (for example, current and former storage tank systems);

5.3.1.4 Releases, spills, and overfill incidents on the site and adjacent properties;

5.3.1.5 Previous or on-going corrective action activities, or both, on-site and on nearby properties (that is, existing monitoring wells);

5.3.1.6 Potential receptors and point(s) of exposure including private and public water supply wells, surface waters, utility conduits, basements, sensitive ecosystems, and other sensitive land uses within a given proximity of the site;

5.3.1.7 Potential migration pathways and transport mechanisms to the points of exposure (ground water flow, vapor migration through soils and utilities, and so forth);

5.3.1.8 Other potential off-site sources of chemical(s) of concern; and

5.3.1.9 Site conditions that may affect the health and safety plan.

5.3.2 If the review of existing data does not provide adequate information regarding UST or subsurface structure locations, the use of surface geophysical survey techniques may be appropriate.

5.4 *Step 3—Develop Conceptual Model*:

5.4.1 The initial conceptual model is the starting point of the characterization, and is used as a basis for planning field activities. The model is developed by compiling and interpreting all information obtained from the existing site information review, and may include the following:

5.4.1.1 Anticipated locations and depths of subsurface geologic units;

5.4.1.2 Anticipated ground water depth and flow direction(s) and possible interaction with surface water bodies;

5.4.1.3 Layout of the site, including areas and depths of artificial fill (tank and trench backfill), subsurface utility lines, and subsurfacing piping;

5.4.1.4 Existing soil and ground water analytical data and information regarding the location and volume of the release;

5.4.1.5 Potential releases in the vicinity of the site (especially upgradient from the site);

5.4.1.6 Location of potential receptors, point(s) of exposure, and migration pathways; and

5.4.1.7 Topographic conditions.

5.4.2 The on-site manager should summarize this information onto simple graphics such as a large-scale base map, structure contour maps, ground water elevation contour maps, isoconcentration contour maps, and geologic/hydrogeologic cross sections. These graphics can easily be hand drawn or can be generated using computerized graphics programs before actual field work begins. These documents should be used on-site and updated as the characterization progresses.

5.4.3 The initial conceptual model, developed before beginning any field work, focuses on specific features that are relevant to the characterization objectives. For example, the features of a conceptual model of a leaking underground storage tank site may include preliminary estimates of: source areas; three dimensional distribution of chemical(s) of concern; chemical(s) of concern impact to and distribution in the ground water; geologic units or structures that influence migration of chemical(s) of concern; and ground water depth, flow direction and velocity. The components of the initial conceptual model that are emphasized depends on the purpose of the characterization, and assists the investigator in focusing on the most salient site features. For more information regarding developing conceptual models, see Guide E 1689.

5.5 *Step 4—Design Data Collection and Analysis Program*:

5.5.1 The data collection and analysis program is developed based on the initial conceptual model, prior to mobilization. This program does not need to be a formalized document, but should be agreed upon between the on-site manager and the responsible party prior to initiation of field activities (in some cases, the regulatory agency is involved as well). The exact number and location of data collection points are left somewhat flexible, and are determined in the field based on the actual site conditions. Levels of communication and authority between the on-site manager and the responsible party will keep all parties informed as the ASC progresses.

5.5.2 Proper implementation of the data collection program requires that the on-site manager be familiar with the capabilities and limitations of the sampling tools and field analytical methods, and that he or she interpret the field-generated data as it becomes available.

5.5.3 The design of the data collection and analysis program should consider the following:

5.5.3.1 Purpose of the ASC;

5.5.3.2 Initial conceptual model, including site historical information, hydrogeologic characteristics of the site, and physical properties of fluids and porous media;

5.5.3.3 Methods to collect and analyze data;

5.5.3.4 General location and number of initial samples and the decision process for locating additional samples;

5.5.3.5 Media to be analyzed;

5.5.3.6 Sample collection and analysis criteria (depth, interval, sampling protocol, chemical(s) of concern, data quality levels, analytical methods, and data validation);

5.5.3.7 Specific qualifications of the on-site manager(s);

5.5.3.8 Site constraints (for example, USTs, structures, canopy, limited space, utilities, property boundaries, depth to bedrock, and access constraints);

5.5.3.9 Data for fate and transport modeling, risk evaluations, or corrective action design (for example, soil properties, air permeability, natural attenuation indicators);

5.5.3.10 Level of communication between the on-site manager and the responsible party (for example, agreement on changes to the scope of work or the data collection and analysis program);

5.5.3.11 Contingencies based on reasonably anticipated deviations from expected site conditions, such as shallow bedrock, depth to ground water, disposal of investigatory wastes, change in equipment requirements, and the appearance or detection of unanticipated chemical(s) of concern; and

5.5.3.12 Determination of the possible need for off-site access.

5.5.4 *Data Collection Methods*—The selection of sampling tools should be based on the following:

5.5.4.1 Purpose and anticipated scope of the ASC;

5.5.4.2 Capabilities, limitations, and cost of each tool;

5.5.4.3 Speed by which samples can be obtained;

5.5.4.4 Advantages of using a combination of tools;

5.5.4.5 Site features and layout;

5.5.4.6 Anticipated geologic site conditions;

5.5.4.7 Anticipated chemical(s) of concern and concentrations;

5.5.4.8 Disturbance to site operations and neighboring properties; and

5.5.4.9 Anticipated next steps.

5.5.4.10 Table 1 presents several common tools and devices that can be used to obtain samples. This guide recognizes that additional tools and techniques exist and continue to be developed, and sample collection during an ASC is not limited to those tools listed in Table 1. The sample collection tools are an integral, though not an exclusive part of an accelerated site characterization in defining subsurface structures, potential

TABLE 1 Example Sample Collection Tools[A]

Method	Access[C]	Suitable Media			Sample Depth (m)[B]	Comments
		Soil	Soil Vapor	Ground Water		
Grab samplers (trowels, scoops, shovel, post-hole digger)	M, B	X			< 1	Low cost. Loss of volatiles. Ease of use.
Hand augers	M	X			< 3	Slow. Labor intensive. Shallow depth. Can be used near located utility/product lines.
Slam Bar & Tubing						
Split spoon	DP, DR	X			< 100	Minimal sample disturbance. Difficult to use below water table w/o auger.
Sample sleeve	DP	X			< 100	Difficult in cobbles or hardpan. Visual obs of sample. Can be used below water table. Minimal sample dist.
Other core samplers[D]	M DP DR	X X X			< 2 < 100 < 100	Equipment-specific capabilities and limitations.
Active gas samplers (vacuum pumps & tubing)	OH, DP, DR		X		< 100	Larger sample volume. Loss of volatiles. Low $
Passive gas samplers	M		X		< 1	Time intensive.
Pneumatic depth-specific samplers	OH		X	X	< 100	
Check valve and tubing	OH			X	< 100	Limited sample volume. Low cost.
Exposed-screen sampler	DP			X	< 100	
Bailer	OH			X	< 100	Labor-intensive.
Sheathed Wellpoint	DP, DR			X	< 100	
Peristaltic pump	OH			X	< 10	
Gas-drive/displacement pump	OH			X	< 100	
Gas-drive/piston pump	OH			X	< 100	
Bladder pump	OH			X	< 100	
Helical rotor pump	OH			X	< 100	

[A] Some commonly-used tools for shallow and intermediate depth investigations (generally < 50 meters) are listed. Many other tools are available. Refer to "Subsurface Characterization Monitoring Techniques: A Desk Reference Guide, Vols. I and II," (EPA/625/R-93/003a&b), USEPA, May 1993, for additional information about these and other methods.

[B] Sample depth refers to practical depth limitation range, depending upon the sampling device used and the lithologic conditions.

[C] Access to the sample for collection or installation of sample tool via the listed approaches.
M = manual (hand-operated equipment).
B = backhoe (mechanical excavating equipment).
OH = open hole (unobstructed access to the sample medium via a pit or cavity, a cased well, or narrow-diameter sampling point).
DR = drill rig (mechanical boring equipment, such as hollow-stem auger, mud/air rotary).
DP = direct-push (mechanical, hydraulic, pneumatic or vibratory devices which push or drive narrow diameter sampling points into the subsurface).

[D] Numerous types and sizes available for different soil conditions. Drill rig is the only sample access equipment listed in this table which can be used readily to sample consolidated material.

migration pathways, or barriers at a site, and in selecting areas for further investigation. The case study example in Appendix X4 identifies information and data collection methods which may be necessary to complete an ASC and perform an ASTM Tier 1 or Tier 2 evaluation.

5.5.4.11 In addition to the tools listed in Table 1 to obtain samples, surface and downhole geophysical techniques (for example, ground penetrating radar, electromagnetic induction, electrical resistivity) may assist in obtaining information regarding subsurface features such as undocumented USTs, utility lines or other unknown features. This information can assist the on-site manager in determining locations of intrusive sampling points.

5.5.5 *Sample Analysis*:

5.5.5.1 *Hydrogeologic Conditions and Physical Properties*—Information on the geology/hydrogeology and physical characteristics of the subsurface is essential to refine the conceptual model, evaluate potential migration pathways and transport mechanisms, and to develop an appropriate corrective action plan. A list of characterization properties and parameters, including physical properties, chemical properties, hydrogeologic characteristics and input parameters/methodologies for an ASTM RBCA Tier 1 and Tier 2 evaluations are located in Appendix X3.

5.5.5.2 *Chemical Analysis*—Field analytical methods are used in an ASC to analyze soil, soil vapor, ground water or air, or a combination thereof. On-site analysis for indicator compounds/chemicals of concern allows the on-site manager to determine the location of, or need for additional samples. Field analytical methods can typically provide more data at lower cost with minimal sample disturbance than sending samples to an off-site laboratory. Key considerations in selecting field analytical methods are as follows:

5.5.5.3 *Analyte*—The analytical method(s) selected will depend on the chemical(s) of concern or indicator compound(s) of interest. For example, when gasoline is the suspected release, the indicator compound may be total volatile organics. Therefore, a method that measures total organic vapors may be used. In many cases, specific chemicals of concern, such as benzene, may need to be measured. Depending on the chemical(s) of concern, it may be necessary to use either field analytical method capable of providing chemical-specific results, or a combination of methods capable of analyzing a broader suite of compounds. Table 2 is a summary of commonly used field screening and analytical techniques. For a discussion of the level of data quality produced by each method, see the data quality level discussion below.

5.5.5.4 *Media*—Consideration must be given to the targeted sample media (soil, soil vapor, ground water, air) and the

TABLE 2 Example Sample Field Screening and Analytical Techniques[A]

Method	Analyte	Media: Soil Vapor	Media: Soil	Media: Ground Water	Detection Range: Soil Vapor	Detection Range: Soil	Detection Range: Ground Water	Limitations	Result Time
PID- or FID- headspace	TOV[B]	X	X	X	ppmv	ppmv	ppmv	Temperature. Humidity.	Immediate.
Indicator tube	Specified compound	X	X		ppmv	ppmv		Instrument flowrate. Cross	
O_2	Oxygen	X			%			Sensitivity Issues.	
CO_2	Carbon dioxide	X			ppmv				
pH meter	pH			X			1-14	None.	
DO meter	Dissolved oxygen			X			mg/l	Temperature. Active fouling	
REDOX meter	REDOX potential			X				by materials that react, coat,	
Conductivity meter	Electrical conductivity			X				or clog.	
Ion-specific meter	Indicator compounds			X			mg/l		
Infrared (IR) spectrometer	Indicator compounds		X	X		mg/kg	mg/l	Low bias for aromatics.	Minutes.
Turbidimetric test kit	Indicator compounds		X			mg/kg		Organic rich soils may cause bias.	
Colorimetric methods	Indicator compounds		X	X		mg/kg	mg/l		
Immunoassay kits	Indicator and specific compounds		X	X		mg/kg	ug/l	Cross-reactivity.	
Portable GC	Specific compounds	X	X	X	ppbv	ug/kg	ug/l	Moderate peak resolution.	
Laboratory grade GC (on-site)	Specific compounds	X	X	X	ppbv	ug/kg	ug/l	Negligible.	Minutes to hours.
Laboratory grade mass spectrometer (on-site)	Specific compounds	X	X	X	ppbv	ug/kg	ug/l	Negligible.	
Laboratory grade GC (off-site)	Specific compounds	X	X	X	ppbv	ug/kg	ug/l	Negligible.	Days to weeks.
Laboratory grade mass spectrometer (off-site)	Specific compounds	X	X	X	ppbv	ug/kg	ug/l	Negligible.	

[A] Some commonly-used techniques for analyzing environmental media are listed. Many other techniques are available. This list was generated using "Field Analysis Manual," New Jersey Department of Environmental Protection and Energy, May 1994, and "Subsurface Characterization and Monitoring Techniques: A Desk Reference Guide, Vols. I and II," (EPA/625/R-93/003a&b), USEPA, May 1993.

[B] TOV refers to Total Organic Vapors.

method's capability of measuring concentrations in that medium. The performance of field analytical methods will vary depending on the sample preparation required for the media being analyzed (especially for soil analyses).

5.5.5.5 *Data Quality Level*—The reliability of results is related to the data quality level of the method used. An example of a data quality classification system for commonly used analytical methods is presented in Appendix X2. As shown in the example, several of the field analytical methods are capable of measuring chemical(s) of concern and/or indicator compounds at differing data quality levels. Selection of field analytical methods should be based in part on the chemical of concern or indicator compounds of interest, the intended use of the data, and the capability of the method. For example, lower quality methods (often called field screening methods) may be used for source identification, while higher data quality methods should be used to delineate chemicals of concern at lower detection limits. Both quantitative and qualitative field analytical methods should be used to acquire data necessary to perform a risk evaluation, or to develop future action plans. When determining what level of data quality is most appropriate, the following is considered:

5.5.5.6 The quality level selected should be consistent with the purpose and scope of the ASC and the intended use of the data.

5.5.5.7 Many points containing lower quality level data can provide a better understanding of site conditions than fewer data points at a higher data quality level.

5.5.5.8 Regulatory requirements should be considered with respect to the detection limit of the selected field analytical method.

5.5.5.9 *Limitations*—All analytical methods and instruments have limitations that may affect results. These include affects of temperature or humidity, cross-sensitivity issues, and masking of certain constituents. In addition, the operational expertise of the person performing the analysis may also effect results. These limitations should be considered when selecting analytical methods or instruments.

5.5.5.10 *Regulatory Acceptance*—Field analytical methods are changing rapidly and the appropriate regulatory authority should be consulted in advance of collecting and analyzing data for accepted methods and procedures when an ASC is performed for regulatory purposes.

5.5.5.11 *Method Protocol and QA/QC Considerations*—Each analytical method has a standard protocol established either by the United States Environmental Protection Agency (USEPA), a state regulatory agency, an industry consensus group or manufacturer, or has a protocol specifically developed for use on-site. Prior to performing the analysis, method protocol and quality control procedures should be developed and documented in a quality control plan. A method quality control plan should specify the following: instrument calibration procedures; generation of calibration curves; preparation and analysis of field standards; analysis of matrix spikes, matrix spike duplicates, blanks and control samples; frequency for instrument calibration and quality control sample analysis; and acceptable criteria for results of instrument calibration and quality control samples.

5.5.5.12 The on-site manager must be familiar with the quality control plan and must ensure that the methods are being performed and the samples are being analyzed in accordance with the plan. The results of the quality control sample analysis should be recorded and reviewed as the data is being generated as well as during data evaluation and refining of the conceptual model. Quality control procedures and analytical results should be included in the final site characterization report.

5.5.5.13 Table 2 presents several analytical methods that can be used to analyze soil vapor, soil and ground water samples. The methods are listed in the order of increasing capabilities and time required for analysis. Both field analytical and off-site laboratory methods are listed. This guide recognizes that additional methods continue to be developed and sample analysis during an ASC is not limited to those methods listed in Table 2. The case study in Appendix X4 demonstrates the use of on-site analytical methods which may be used in completing an ASC and in performing an ASTM RBCA Tier 1 and Tier 2 evaluation.

5.6 *Step 5—Field Activities*:

5.6.1 *Step 5A—Collect and Analyze Data*—The established data collection and analysis program is implemented to perform an intensive, short-term field investigation. As samples are acquired, it is important to observe physical appearance and conditions such as lithology, structure, soil staining, color and moisture content (see Appendix X3). Flexibility is a key component for a successful ASC, therefore, the data collection and analysis program should be used to guide the site characterization to completion. As data is collected and analyzed, it may be necessary to adjust the data collection and analysis program to refine the conceptual model and satisfy the purpose of the site characterization.

5.7 *Step 5B—Evaluate Data and Refine Conceptual Model*:

5.7.1 Hydrogeologic, and analytical data collected during the field investigation are periodically interpreted on-site by the field manager. As shown in the flowchart in Fig. 1, the conceptual model is refined in an iterative process of data collection and evaluation. Compilation of the data onto simple graphics is essential for on-site data interpretation. This is best done by updating the maps and cross sections prepared to develop the initial conceptual model. As the investigation proceeds, the maps and cross sections are continually revised (geologic contacts are erased and moved, borehole lithologic data are plotted on cross sections, new isoconcentration contour lines are drawn, and so forth), by incorporating the new data. Using the field-generated graphics, the on-site manager directs the investigation to fill in data gaps or resolve differences between anticipated and actual results, or both. As new data are collected and the investigation proceeds, variances between the initial conceptual model and the data obtained during the characterization are used to adjust the sampling and analysis program in an iterative, scientific manner, until the site geology/hydrogeology, and nature and distribution of the chemical(s) of concern in soil and ground water are accurately defined.

5.7.2 The degree of detail and accuracy of the graphical representation of site conditions varies according to the purpose of the characterization, complexity of the site geology/hydrogeology, and the type and volume of the chemical(s) of concern. As multiple measurements are made and the amount of information that describes more complex subsurface conditions increases, the site data can be compiled on graphical software that is commercially available for laptop computers.

5.7.3 *Data Validation*—To ensure that it is useful, field-generated data must be validated. Considerations for data validation include the following:

5.7.3.1 Quality assurance/quality control (QA/QC) results (for example, duplicates, multi-point calibration curves, calibration checks, blanks, and so forth);

5.7.3.2 Comparison of higher quality level data to check lower quality level data;

5.7.3.3 Consistency of results among analytical methods and sampling techniques;

5.7.3.4 Comparison with results from other media;

5.7.3.5 Comparison with other chemical(s) of concern or indicator compounds;

5.7.3.6 Comparison against previous data, if available; and

5.7.3.7 The data should make sense in the context of the site conditions and previously generated data.

5.7.4 Once the validity of the data has been assessed, it can be used to determine whether data quality requirements have been satisfied.

5.8 *Termination of Data Collection*:

5.8.1 The data collection and evaluation should continue until the on-site manager has determined that the purpose of the site characterization has been met or that constraints prevent complete characterization. Typically, the ASC is complete and no further data collection is required when the following have been satisfied:

5.8.1.1 The conceptual model of the site geology/hydrogeology, the nature and distribution of chemicals of concern, and indicator compounds fit the regional hydrogeologic setting; and

5.8.1.2 The conceptual model of the site generally incorporates/fits all of the site data; and

5.8.1.3 The conceptual model can be used to make accurate predictions of subsurface conditions, and

5.8.1.4 Sufficient detail and delineation of the chemicals of concern have been achieved to fulfill the requirements of the user; or

5.8.1.5 Constraints prevent collection of any additional data.

5.9 *Step 6—Report Findings*:

5.9.1 Upon completion of the field work, a report of findings is provided to the user. The report should contain at a minimum: the purpose of the characterization, a statement of objectives, the background data, a description of the data collection and analysis program, a presentation or summary of the data, and quality assurance/quality control measures. The report may be used to identify the appropriate course of action, which may include the following:

5.9.1.1 No further action;

5.9.1.2 Compliance monitoring;

5.9.1.3 Further risk evaluation under the RBCA process Tier 2 or Tier 3 analysis (data collection during the ASC should be sufficient to meet the requirements of a Tier 1 and Tier 2 analysis); or

5.9.1.4 Evaluation of remedial action alternatives, and subsequent selection of technologies, or combination thereof.

5.9.2 For further information on these courses of action,

please refer to Guides E 1599 and E 1739.

5.9.3 The steps of an ASC process presented in 5.1 to 5.8 are illustrated in the example in Appendix X4. In addition, the example uses the results of the ASC to perform a RBCA Tier 1 and Tier 2 evaluation.

6. Keywords

6.1 accelerated; analytical methods; borings; characterization; chemicals of concern; corrective action; data quality; exposure pathways; field methods; ground water; LUST; mobilization; parameters; petroleum; risk based approach; sampling tools

APPENDIXES

(Nonmandatory Information)

X1. OTHER REFERENCES

X1.1 *ASTM Standards*:

D 1452 Practice for Soil Investigation and Sampling by Auger Borings[6]

D 1586 Test Method for Penetration Test and Split-Barrel Sampling of Soils[6]

D 1587 Practice for Thin-Walled Tube Geotechnical Sampling of Soils[6]

D 2488 Practice for Description and Identification of Soils (Visual-Manual Procedure)[6]

D 3550 Practice for Ring-Lined Barrel Sampling of Soils[6]

D 4447 Guide for Disposal of Laboratory Chemicals and Samples[3]

D 4448 Guide for Sampling Ground Water Monitoring Wells[3]

D 4700 Guide for Soil Sampling from the Vadose Zone[6]

D 4750 Test Method for Determining Subsurface Liquid Levels in a Borehole or Monitoring Well (Observation Well)[6]

D 4823 Guide for Core-Sampling Submerged, Unconsolidated Sediments[7]

D 5092 Practice for Design and Installation of Ground Water Monitoring Wells in Aquifers[2]

D 5299 Guide for the Decommissioning of Ground Water Wells, Vadose Zone Monitoring Devices, Boreholes and Other Devices for Environmental Activities[2]

D 5314 Guide for Soil Gas Monitoring in the Vadose Zone[2]

[6] *Annual Book of ASTM Standards*, Vol 04.08.

[7] *Annual Book of ASTM Standards*, Vol 11.02.

X2. AN EXAMPLE OF A DATA QUALITY CLASSIFICATION SYSTEM

X2.1 *Introduction*:

X2.1.1 This appendix describes an example of a four tiered data quality hierarchy modified from *New Jersey Department of Environmental Protection Field Analysis Manual.*[8] Two significant modifications to the New Jersey Department of Environmental Protection and Energy (NJDEPE) Manual[9] have been incorporated into the example data quality level hierarchy. First, the applications are for petroleum products only. The second modification designates Level 1 as screening levels, either qualitative or semiquantitative, that may require confirmatory analyses with higher data quality methods. Levels 2, 3, and 4 are considered to be essentially quantitative, with Level 2 being less quantitative than Levels 3 or 4. These levels can produce data of sufficient quality that does not necessarily need laboratory confirmation on a routine basis. An overview of these data quality levels are presented in this appendix.

X2.1.2 The USEPA utilizes a two tiered approach to data quality. The first category "Screening Data With Definitive Confirmation" would include data quality Levels 1 and 2. The second category "Definitive Data" would include data quality Levels 3 and 4.

X2.1.3 State regulatory programs may develop their own definitions for data quality for the methods listed in this appendix, and may have specific reporting requirements when using these methods. Details on data quality levels, use of field analytical methods, and specific reporting requirements can be obtained by contacting the appropriate state environmental regulatory agency, fire marshal, or other local jurisdictions.

X2.2 *Data Quality Level 1*:

X2.2.1 Level 1A methods are intended to be used for health and safety evaluations, initial screening of soil and ground water for chemical(s) of concern. The measurements made with these methods (1A) are qualitative and only provide an indication of the presence of contamination above a specified value (for example, pass or fail, positive or negative). Because measurements made with these methods may not always be consistent, the data shall only be used as an initial screening for sample locations for analysis using higher level methods. Clean samples cannot be determined from these methods at this level.

[8] New Jersey Department of Environmental Protection and Energy, *Field Analysis Manual*, July 1994.

[9] New Jersey Department of Environmental Protection and Energy, *Alternative Groundwater Sampling Techniques Guide*.

X2.2.1.1 Instruments used for data quality Level 1 include: photoionization detector (PID) survey instruments, flameionization detector (FID) survey instruments, colorimetric analysis, and headspace analysis.

X2.2.1.2 Quality control procedures are limited primarily to instrument calibration, consistency in method procedure, and background level checks. Since relatively few quality control procedures are employed compared to higher-level field methods, data quality is very much a function of sample handling techniques and analyst skill.

X2.2.2 Level 1B methods can be used for qualitative and semiquantitative screening and defining the location of known types of contamination (that is, orders of magnitude or ranges). Level 1B data can be generated when PIDs and FIDs are used with controlled sample preparation and analysis procedures that include additional QA/QC such as that used with polyethylene bag headspace.

X2.2.2.1 Quality Assurance (QA) procedures include multipoint calibration curves using matrix-spiked field standards, a calibration check using matrix spike duplicates, and a field blank/background sample.

X2.2.2.2 Depending on regulatory requirements, laboratory confirmation may be needed for establishing laboratory-field correlation over the concentration ranges measured for confirming the achievable lower detection limit.

X2.3 *Data Quality Level 2*:

X2.3.1 Level 2 methods are intended to be used for delineation of chemical(s) of concern. These methods can achieve a high degree of reproducibility when required QA/QC procedures are employed.

X2.3.2 Level 2 methods are typically laboratory methods that have been adapted for field use (that is, field gas chromatograph (GC), portable infrared (IR)) or are EPA-derived methods (for example, immunoassay). These methods may not be as rigorous because field extraction's are not directly comparable to laboratory extraction methods.

X2.3.3 Quality assurance (QA) requirements include initial multi-point calibration curves, continuing calibration checks, matrix spike duplicates, background/blank samples, laboratory confirmation of clean samples, and possibly contaminated samples depending on the objective. A matrix spike recovery should be performed on a site-specific basis.

X2.3.4 Level 2 methods that provide a direct numerical value for the indicator measured but do not definitively identify the chemical(s) of concern present (for example, immunoassay, portable IR) are considered semiquantitative. Level 2 methods that measure specific constituents (for example, transportable GC's) are considered quantitative.

X2.3.5 Depending on regulatory requirements, laboratory confirmation of a portion of the samples may be needed for establishing laboratory-field correlation over the concentration ranges measured for confirming the achievable lower detection limit.

X2.3.6 Level 2 methods also include EPA field screening and laboratory methods. The laboratory methods considered to be Level 2 have limited QA information documented. The quality of the data generated using Level 2 laboratory methods depends on the sample handling, storage, and preservation procedures, and analytical procedure and QC used.

X2.4 *Data Quality Level 3*—Level 3 methods are approved laboratory methods with complete QA/QC (for example, EPA Laboratory Methods [see USEPA SW846], third or more recent edition). Level 3 analyses can be performed at off-site laboratories or at on-site mobile laboratories that perform EPA methods. Certain regulatory agencies may require these laboratories to be certified.

X2.5 *Data Quality Level 4*:

X2.5.1 Level 4 methods are generally "state of the art" methods developed specifically for a particular site or chemical(s) of concern. Level 4 methods are used when standard laboratory methods are either unavailable or impractical.

X2.5.2 Generation of Level 4 data may necessitate the use of a laboratory that specializes in methods development.

X3. CHARACTERIZATION PROPERTIES AND PARAMETERS

X3.1 Two sets of parameters are presented in this Appendix. See Table X3.1 for a list of physical and chemical properties and hydrogeologic characteristics and Table X3.2 for a list of input parameters and methodologies for ASTM RBCA Tier 1 and Tier 2 evaluation. These lists are provided as an example of parameters that may be collected and evaluated during an ASC.

X3.2 *List of Physical and Chemical Properties and Hydrogeologic Characteristics*:

X3.2.1 This list is intended to provide an example of a broad range of information that may be collected during a site characterization. It is not comprehensive nor does it imply that all of this information should be collected for every site characterization. A user applying the ASC approach would consider this list, when determining the benefits of collecting information before and during the mobilization.

X3.2.2 The footnoted parameters (see Table X3.1) are listed in Guide D 5730. There are additional ASTM standards and references for methods that may apply but have not been listed in this guide.

TABLE X3.1 List of Physical and Chemical Properties and Hydrogeologic Characteristics

Fluid Properties (Liquid, Dissolved and Vapor-Phase Contaminants):
Density[A]
Viscosity
Interfacial tension[A]
Solubility
Sorptive properties[A]
Vapor transport properties[A]
Chemical composition
Fluid-Media Properties:
Wettability[A]
Capillary pressure-saturation relations
Moisture content[A]
Relative permeabilities (includes air permeability)[A]
Porous Media Properties:
Intrinsic permeabilities
Porosities (total and effective)
Bulk density[A]
Pore volume[A]
Hydraulic conductivity[A]
Grain size distribution[A]
Organic carbon content[A]
Clay content (soil classification)[A]
Infiltration rate[A]
Oxygen and carbon dioxide content
Soil pH[A]
Storativity[A]
Local Geology/Hydrogeology:
Heterogeneities
Stratigraphy/lithology/soil type
Presence, type, and relative abundance of consolidated media
Preferential migration pathways (for example, utilities, fractures, and so forth)
Depth to ground water[A]
Depth to bedrock
Aquifer thickness
Hydraulic gradient[A]
Ground water flow direction[A]
Dissolved oxygen[A]
REDOX potential[A]
Dissolved metals
Total dissolved solids[A]
Ground water pH[A]
Distribution of Chemical(s) of Concern:
Presence of nonaqueous phase liquid.
Depth to impacted soil in the unsaturated or saturated zones.
Spatial distribution of chemical(s) of concern in soils.
Spatial distribution of chemical(s) of concern in ground water.

[A] The parameters listed are in Guide D 5730.

TABLE X3.2 Input Parameters and Methodologies for ASTM RBCA Tier 1 and Tier 2 Evaluations

NOTE 1—Collection of data should not be limited to a specific tier and should be collected based on site specific conditions and potential impacts to receptors.

RBCA Input Parameters	Methodology/Source	Tier
Physical and Chemical Characteristics of the Chemicals of Concern (COC)		
Identify Potential Source, Nature of Release and COC: time, location of potential release, substance released (gasoline, diesel, etc.)	Evaluate historical data, release reports, site specific information	1
Identify Characteristic of COC: most prevalent, mobile, toxic constituents	Review literature to evaluate chemical/toxicological properties of COC	1
Maximum COC Concentrations: maximum concentrations in affected media (source areas)	Obtain samples, conduct chemical analysis of soil, ground water, and/or vapor samples, identify presence/absence of NAPLs	1
Extent of COC in affected media: lateral and vertical extent of COC affected media	Collect soil/ground water/vapor samples, perform chemical analysis to determine concentration and migration pathways of COC	1,2
Representative COC Concentrations: representative COC concentrations in source area	Perform statistical evaluation of sampling analytical results	1,2
Regional and Site Specific Hydrogeologic Conditions		
Regional Surface and Subsurface Geology/Hydrogeology: ground water recharge areas, ground water to surface water discharge rates, topography, location of surface waters, local drainage. General/published soil and aquifer properties (for example, permeability, porosity, hydraulic conductivity)	Review published information: topographic maps, local geologic mapping information. Based on site specific sampling observations, determine general soil and aquifer properties from published literature	1
Site Specific Soil Characteristics: stratigraphy, soil heterogeneity, presence, type, and abundance of consolidated media, permeability, moisture content, bulk density	Collect and observe soil samples in both impacted and non-impacted areas	1,2
Site Specific Aquifer Properties: depth to capillary fringe, depth to ground water, gradient and direction, effective porosity, hydraulic conductivity.	Collect site specific measurements in both impacted or non-impacted areas	1,2
Site Specific Ground water Attenuation Factors: soil fraction organic carbon (f_{oc}), determination of COC decay rate	Collect and analyze samples for f_{oc}, COC, and ground water natural attenuation parameters. Calculate site specific COC decay rate	1,2
Potential Receptors, Current and Future Land Use		
Receptor Survey: private and public potable and non-potable water supply wells, surface waters, public water supply points of intake, wetlands, basements, residential properties	Review state and local health department water supply maps and records. Perform a site specific well canvas to locate unrecorded private wells. Identify nearby subsurface structures. Determine existing or potential receptors	1
Current and Reasonably Potential Future Land Use Survey: residential, commercial and industrial land use, zoning ordinances and restrictions, long term city land use plans	Perform site specific survey to identify land use including residential, commercial and industrial areas, locations of schools and day care facilities. Review local zoning ordinances, city land use plans, determine reasonably potential future land use	1,2
Exposure Assessment		
Transport Media	Identify potential transport media through which COC will migrate. Media based on site specific data	1
Transport Mechanism	Based on transport media, determine the soil to ground water leaching potential, volatilization, dispersion and ground water migration	1,2
Exposure Pathways	Review site specific information, transport media and mechanisms to determine possible exposure pathways such as ingestion of soil and/or ground water, vapor inhalation, dermal contact with soils and/or ground water. (Evaluate preferential migration pathways.)	1,2
Points of Exposure	Review receptor survey and site specific information, identify points of exposure such as drinking water, surface waters, surface soils and basements	1,2

X4. EXAMPLE OF THE USE OF AN ACCELERATED SITE CHARACTERIZATION PROCESS IN A RBCA TIER 1 AND TIER 2 EVALUATION

X4.1 Introduction

X4.1.1 The following example illustrates the ASC process presented in Fig. 1 at a petroleum release site. A hypothetical site with relatively complex geologic conditions is presented in this example to show that the ASC process works for complex sites as well as simple ones. For complex sites, the ASC process can be used to quickly identify gaps in the subsurface data and then fill those gaps while the subsurface sampling tools are still on site.

X4.1.2 In this example, the site is being characterized to provide data necessary to make corrective action decisions following a method similar to Guide E 1739. An ASC can be used at any UST (or other petroleum source) site for accurate and rapid site characterization.

X4.2 Release Scenario

X4.2.1 *Discovery and Notification*—In 1995, a release of

petroleum hydrocarbons at a closed service station was suspected after a contractor noticed a strong gasoline odor in the sanitary sewer adjacent to the station (see Fig. X4.1(*a*)). The contractor contacted the local fire department who determined that the gas vapors did not constitute an immediate explosion hazard. The fire department filed an inspection report with the State Environmental Department (SED). The SED subsequently sent a letter to the current property owners and owners of nearby gas stations, requesting an investigation of the source of the gasoline release.

X4.3 Step 1—Identify Site Characterization Purpose

X4.3.1 The site was investigated because of the letter from the SED to investigate the suspected release. The property owner was also interested in selling the property and needed to define the residual hydrocarbons identified during the tank closure. The site owner retained a consultant who proposed conducting an accelerated site characterization in order to collect the information necessary to perform an ASTM Risk-Based Corrective Action (RBCA) evaluation and determine the appropriate corrective action, an approach that is supported by the SED. The guidelines for risk-based corrective action are outlined in a new state UST corrective action manual, which closely follows the ASTM three-tiered Guide E 1739.

X4.4 Step 2—Review Existing Information

X4.4.1 Before the field activities were defined, available information about the site geology, history, ground water conditions, and nature and distribution of chemical(s) of concern was collected and reviewed. A summary of the review included the following information.

X4.4.2 *Site History*—The former station began dispensing gasoline in the mid-1950's. Gasoline was stored in two 10 000-gal tanks in the southern portion of the property. Two pump islands were located adjacent to the UST area. The station was closed in the early 1980s, and, according to a note in the fire department's file, the tanks and associated piping were removed and the excavation was filled with clean soil. The dispensing pumps were removed, and the site is now operated as a tune up shop and auto repair garage. No records are available regarding the construction of the tanks, the tank removal, or the location of underground piping. No inventory records were found that might have helped define the type and magnitude of the subsurface petroleum hydrocarbon release. From the 1984 source investigation report, the site well installation report, boring logs, and previous analytical results were carefully reviewed. Geologic data shown on the boring logs was compared to published data from USGS and Soil

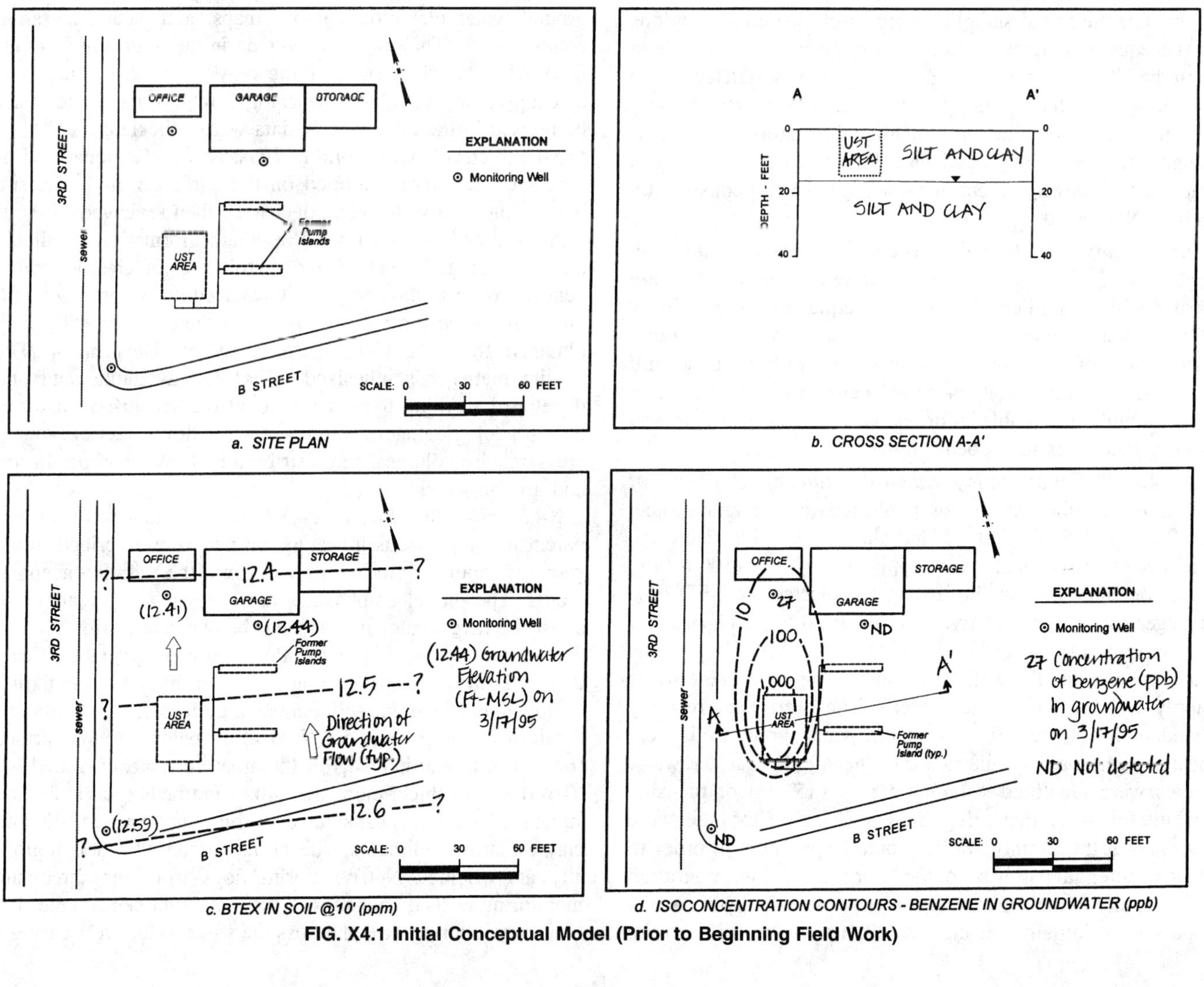

FIG. X4.1 Initial Conceptual Model (Prior to Beginning Field Work)

Conservation Service maps.

X4.4.3 *Regional and Local Hydrogeologic Reports*—Reports published by the United States Geological Survey (USGS), State Geologic Survey, Soil Conservation Service, and the local water service provided valuable information regarding the subsurface geology and ground water conditions (for example, water table elevation, hydraulic properties, general flow direction) near the site. Regional information was also obtained from subsurface investigation reports of nearby sites.

X4.4.4 *Previous Investigations*—An initial source investigation was performed at the site in 1984 by the contractor who removed the tanks. The investigation consisted of installing two ground water monitoring wells along the northern boundary of the site, that was presumed to be downgradient of the UST area and the pump islands. A third well was installed upgradient from the UST area near the sewer line.

X4.4.5 *Ground Water Analysis*—Analyses of ground water samples collected soon after the wells were installed detected 80 mg/L (ppm) of benzene in the well downgradient from the UST area. No benzene was detected in the well located downgradient of the pump islands. No chemical analyses of soil samples from the borings was performed. Additional ground water samples were collected infrequently from the wells after the initial sampling event. Benzene concentrations in subsequent samples collected from the well downgradient from the UST area were erratic, ranging from non-detect (ND) to 150 ppb. The erratic range of analytical results was attributed to several factors including laboratory error, sampling bias, or seasonal ground water level fluctuations. Samples from the well downgradient from the pump islands consistently yielded ND results.

X4.4.6 *Adjacent Property*—The site is located in an urban area. Several other gas stations and industrial facilities are located within ½ mile of the site. Consequently, the possibility of upgradient sources of petroleum hydrocarbons (and other chemical(s) of concern) certainly exists. Moreover, several environmental investigations have been performed at nearby sites, providing valuable information about the geologic and ground water conditions beneath the site.

X4.4.7 *Receptor Survey/Identification of Point(s) of Exposure*—Further review of published reports and project files at the USEPA, SED, and the County indicated the presence of a municipal water supply well with in 1500 ft (457 m) of the site. The on-site manager contacted the SED case manager and a representative from the local water department to determine the current status of the well. According to the water department, the well is no longer being used in the water supply system due to low yield. The well was recently abandoned. No other private or public water supply wells were located with in a one mile radius of the site. No surface water bodies were identified with in 1500 ft (457 m) of the site. As-built plans of the utility lines adjacent to the site were reviewed at the County Public Works Department in order to identify potential pathways of vapor or ground water migration from the site to the sanitary sewer line where the hydrocarbon vapors were initially detected. Additionally, according to the tenants, the on-site tune-up/repair facility has a small basement.

X4.4.8 *Nearby Releases*—Review of USEPA, SED and County files indicated several petroleum release sites and one Superfund site upgradient from the subject property.

X4.4.9 *Interview*—A former employee who worked at the site during its operation was interviewed by the on-site manager. The former employee provided valuable anecdotal information regarding the location of the USTs and piping systems, unreported inventory losses, tank and piping upgrades, and the removal of the underground tanks. Information provided by the former employee was summarized on a scaled base map of the site (see Fig. X4.1(*a*)).

X4.5 Step 3—Develop Conceptual Model

X4.5.1 Based on the review of existing regional and site data, the on-site manager began to formulate an initial conceptual model of the site geology/hydrogeology, and nature and distribution of chemical(s) of concern.

X4.5.2 *Developing Site Maps*—A large amount of data was compiled from various sources. To keep the data organized and accessible, the on-site manager summarized and compiled the information (representing the conceptual model) onto some simple, hand-drawn graphics (see Fig. X4.1). These graphics included a large-scale base map, hydrogeologic cross sections, ground water elevation contour maps, and isoconcentration contour maps. The graphics depicted in the figures are simplified and reduced in size for the purpose of this guide. The actual graphics were larger working drawings that could easily be revised in the field as new data were collected.

X4.5.3 *Developing Working Hypotheses*—Regional hydrogeologic data were compiled on the site base map. The site boring logs and water level data were then reviewed to see if they were consistent with the regional information. Available data regarding the nature and distribution of chemical(s) of concern were compared with the anecdotal information obtained from the former employee and the as-built utility map obtained from the County Public Works Department. The on-site manager synthesized all of the available data and developed working hypotheses about the subsurface distribution of geologic materials, ground water flow direction, source areas, release volumes, and distribution of hydrocarbons in soil and ground water.

X4.5.4 *Key Site Features*—Site features that were not apparent in the previous investigation but were described in the geologic reports were incorporated into the initial conceptual model. The model emphasizes the following key features to assist the investigator in focusing the characterization.

X4.5.4.1 *Fluvial Deposits*—Regional geologic data indicate that the shallow subsurface materials near the site are a mixture of fluvial and estuarine sediments. Granitic bedrock occurs at a depth of approximately 500 ft (152 m) below ground surface (bgs). The fluvial deposits in the area are typically 2 to 15-ft (0.6 to 4.6 m) thick, elongated silty sand beds (that is, buried stream channels) encapsulated within finer-grained silt and clay estuarine sediments. The buried stream channels regionally are oriented N40W. Boring logs from the three site monitoring wells do not show a silty sand unit beneath the site. However, a boring drilled across the street to the west during a previous investigation encountered a 12-ft (3.7 m) thick silty sand bed.

X4.5.4.2 *Aquifer Characteristics*—Unconfined ground water occurs regionally within the unconsolidated sediments at depths ranging from 25 to 30 ft (7.6 to 9.1 m) bgs. This is consistent with water levels measured in the site ground water monitoring wells. Ground water flows regionally to the north, but localized ground water flow patterns exist due to preferential ground water flow within the more permeable buried stream channels. The inferred direction of ground water flow beneath the site is toward the north.

X4.5.4.3 *Source Area and Distribution of Chemical(s) of Concern*—Locations of the former tanks, subsurface piping, on-site utility lines, and areas of artificial fill were compiled onto the site map by the on-site manager. A former employee indicated that strong petroleum odors and discolored soil were evident beneath the supply lines leading to the southern pump island. Several scenarios were developed to estimate the distribution and relative magnitude (volume and concentration) of residual petroleum hydrocarbons resulting from the presumed piping release. Based on the conceptual model of the site geology/hydrogeology, the likely extent of a dissolved hydrocarbon plume was estimated.

X4.6 Step 4—Design Data Collection and Analysis Program

X4.6.1 Before beginning the field investigation, the consultant prepared a "Data Collection and Analysis Program" for the field work. The program included a short discussion of the following:

X4.6.1.1 Methods that would be used to collect subsurface samples,

X4.6.1.2 The media to be analyzed,

X4.6.1.3 The field analytical program,

X4.6.1.4 Protocol for communicating project status to client and SED,

X4.6.1.5 Contingency plans, including plans to procure off-site access, and

X4.6.1.6 Safety program.

X4.6.2 On most projects, a data collection and analysis program is prepared as an internal guide for the use of the project team members. Highlights of the data collection and analysis program for the subject investigation were as follows:

X4.6.3 *Geophysical Survey*—Before a sampling strategy was fully developed, a magnetometer survey was used to confirm the location of the utilities and other potentially buried metallic objects or structures. This information was compared with the other existing information from the previous investigation, as-built plans, and anecdotal information provided by the former employee.

X4.6.4 *Methods to Collect Subsurface Samples*—Because of the unconsolidated nature of the subsurface materials, the relatively shallow depth of the investigation, and the need to sample multiple media (that is, soil, vapor, and ground water) a direct-push (DP) method of sample collection was selected. The DP sampling tools are small-diameter steel probes that are pushed, pounded, or both, into the ground. These sampling tools can be used to collect soil, ground water, and soil vapor samples. The DP sampling tools collect a greater number of depth-discrete samples per day than conventional drilling methods. In addition, small-diameter monitoring wells (monitoring points or microwells) can be installed with most DP rigs, which do not generate drill cuttings, eliminating the cost of soil disposal.

X4.6.5 *Alternative Sampling Methods*—If the site was underlain by consolidated sediments or if the sampling depths were much greater, another method of collecting samples, such as conventional hollow stem auger or rotary drilling, would have been necessary. Accelerated site characterization (ASC) is an approach (not a set of sampling tools) that is fully compatible with conventional sampling methods.

X4.6.6 *The Field Analytical Program*—After discussions with the SED, a mobile laboratory was contracted to perform the analytical testing. The mobile laboratory performs the analyses of soil and ground water on site, providing real-time analytical data to the on-site manager. The mobile laboratory selected was able to process up to 35 samples per day, and was certified by the state to perform analyses for petroleum hydrocarbons. Soil and ground water samples are analyzed for the constituent of concern or indicator compound. Physical properties of the porous media and chemical(s) of concern are observed, measured, or estimated.

X4.6.6.1 *Soil Screening*—All soil samples were to be screened in the field with a portable flame ionization detector (FID). An FID was used because of the high sensitivity to gasoline vapors (ppmv) and especially to alkanes present in weathered gasoline. The FID has a linear response throughout its operating range and can be used to detect a wide range of hydrocarbon concentrations.

X4.6.6.2 *Soil and Ground Water Analyses*—Soil samples between depths of 10 and 30 ft (3.1 and 9.1 m) were to be analyzed by the mobile laboratory every 5 ft (1.5 m), and at intervals where significant hydrocarbon concentrations were indicated by the portable FID. Soil samples were to be analyzed for benzene, toluene, ethylbenzene, and xylenes (BTEX) by EPA Method 8020 and total petroleum hydrocarbons as gasoline (TPH-G) by Modified EPA Method 8015 in accordance with SED requirements. Ground water samples were to be analyzed for BTEX by EPA Method 8020, and for pH, Eh, and dissolved oxygen, using portable field instruments.

X4.6.6.3 *Physical Properties*—In addition to chemical analyses, physical properties of the porous media, ground water, and hydrocarbons present were observed, measured, or estimated as part of the data collection and analysis program. The vadose zone thickness, soil type and porosity, structure, stratigraphy, heterogeneities, moisture content, and location of chemical(s) of concern were documented on soil logs. In addition, soil samples were collected for measuring total organic carbon (TOC), bulk density, porosity and moisture content to make more accurate predictions of the fate and transport of chemical(s) of concern. The aquifer thickness, flow direction, and gradient were determined from water level measurements, and hydraulic conductivity will be determined from slug tests. Ground water quality indicators (for example, pH, total dissolved solids) were measured using portable meters. Dissolved oxygen levels were also measured using portable meters to gain an understanding on whether natural biodegradation is occurring.

X4.6.6.4 *Protocol for Communicating Project Status*—The consultant agreed to update the client and SED with the status of the ASC at the end of each field day. The on-site manager had a pager and portable telephone to communicate with all project participants whenever necessary.

X4.6.6.5 *Permits*—Before beginning the field program, the on-site manager obtained permits for drilling borings (DP probes are considered to be borings by the SED) and installing monitoring wells at the site. One permit was sufficient for any number of borings and wells. Based on the data review, it seemed likely that off-site exploration beneath Third Street would be necessary. Therefore, the on-site manager obtained an encroachment permit from the city and filed a traffic plan with the county public works department. As part of this permit, the on-site manager also included B Street in the permit and plan.

X4.6.6.6 *Utility Clearance*—A private underground utility locating company was hired to locate subsurface utility lines beneath the property and Third and B Streets. This was done in order to avoid penetrating the utility lines with the DP sampling equipment. The utility locating was directly supervised by the on-site manager, because information regarding the location of subsurface utility lines gave the on-site manager valuable information about the location of potential migration pathways for chemical(s) of concern.

X4.7 Step 5—Field Activities

X4.7.1 The field investigation was conducted in three days. The on-site iterative process of collecting information, refining the conceptual model, and guiding the investigation is represented in this example.

X4.7.2 *Day 1: Initial Investigation*—On the first day of the field investigation, samples were collected at the locations shown in Fig. X4.2(*a*). Soil samples were collected at a minimum of every 5 ft (1.5 m). The assistant geologist logged the soil samples in detail. Soil and ground water samples were screened with a portable FID. Select samples were submitted to the on-site mobile laboratory for chemical analysis. Several soil samples were collected and preserved for later analysis of total organic carbon (TOC), bulk density, and moisture content. Site-specific values of those parameters would be necessary for making more accurate estimates of fate and transport of chemical(s) of concern during the RBCA evaluation (see X4.9). Geologic information, depth to ground water, and soil and ground water analytical results were compiled throughout the day onto the field drawings shown in Fig. X4.2.

X4.7.2.1 *Source Area*—To investigate the suspected source

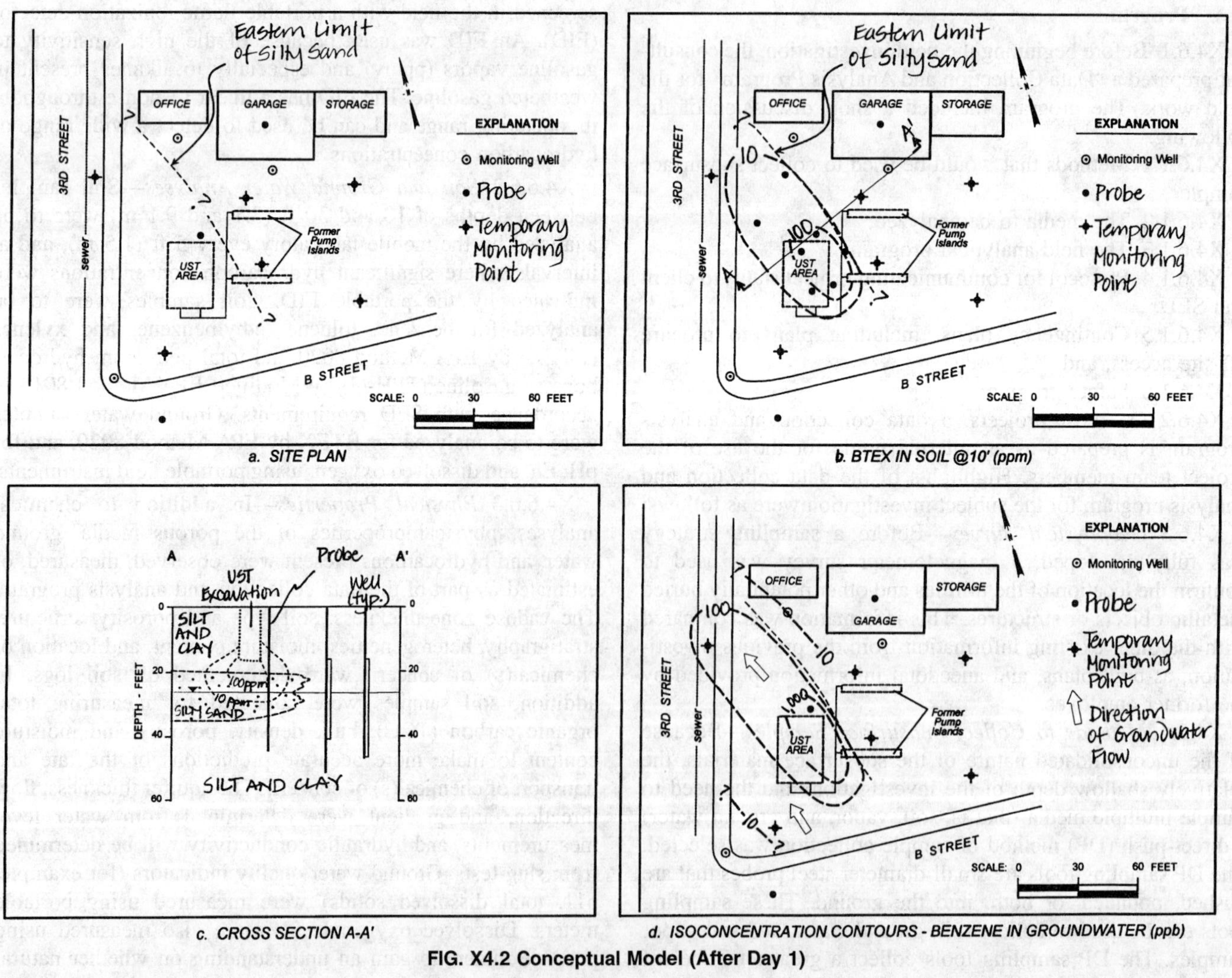

FIG. X4.2 Conceptual Model (After Day 1)

area, soil and ground water samples were collected from locations immediately north of the former UST area and former tank islands. Soil and ground water immediately north of the UST area contained high concentrations of chemical(s) of concern as anticipated from the initial conceptual model, however, the samples adjacent to the pump islands contained little or no concentrations of chemical(s) of concern. Dissolved oxygen levels were also significantly lower than levels measured in the upgradient location. The UST area, therefore, seemed to be the likely source of the release. This is consistent with the initial conceptual model.

X4.7.2.2 *Site Geology/Hydrogeology*—The boring drilled near the former UST area penetrated a native silty sand bed between the depths of 20 and 33 ft (6.1 and 10.1 m) bgs, that was thought to extend beneath the neighboring property to the west (a northwest orientation of this sand bed was consistent with the regional geologic setting). This finding indicated a revision to the initial conceptual model. Soil and ground water samples were then collected from other portions of the site to identify significant geologic units or obvious zones of contamination. Because the subsurface geology was more heterogeneous than initially thought, continuous soil cores were collected below a depth of 15 ft (4.6 m) bgs. (Continuous soil cores are necessary to accurately identify geologic contacts and thin beds.) During the course of the day, the silty sand bed was penetrated in three additional locations, confirming its northwest-southeast orientation. Unconfined ground water was encountered within the silty sand bed at a depth of approximately 26 ft (7.9 m) bgs.

X4.7.2.3 *Distribution of Chemical(s) of Concern in Ground Water*—Once the presence of the buried stream channel was identified, the on-site manager suspected that it may control the movement of ground water, and hence migration of chemical(s) of concern, beneath the site. Indeed, isoconcentration contours of benzene in the ground water samples clearly indicated a northwest alignment of dissolved chemical(s) of concern within the unit (see Fig. X4.2(*d*)). Benzene concentrations in ground water in the southwest portion of the site upgradient from the former UST area, however, were anomalously high (see Fig. X4.3(*b*)). Chromatograms of water samples collected in the southwestern area had a different "fingerprint" than chromatograms of water samples in the UST area, leading the on-site manager to speculate about the likelihood of an upgradient source(s) of dissolved petroleum hydrocarbons. Also, background dissolved oxygen levels were significantly higher than those encountered in the plume.

X4.7.2.4 *Temporary Monitoring Points*—Six small-diameter temporary monitoring points, consisting of ¾-in.

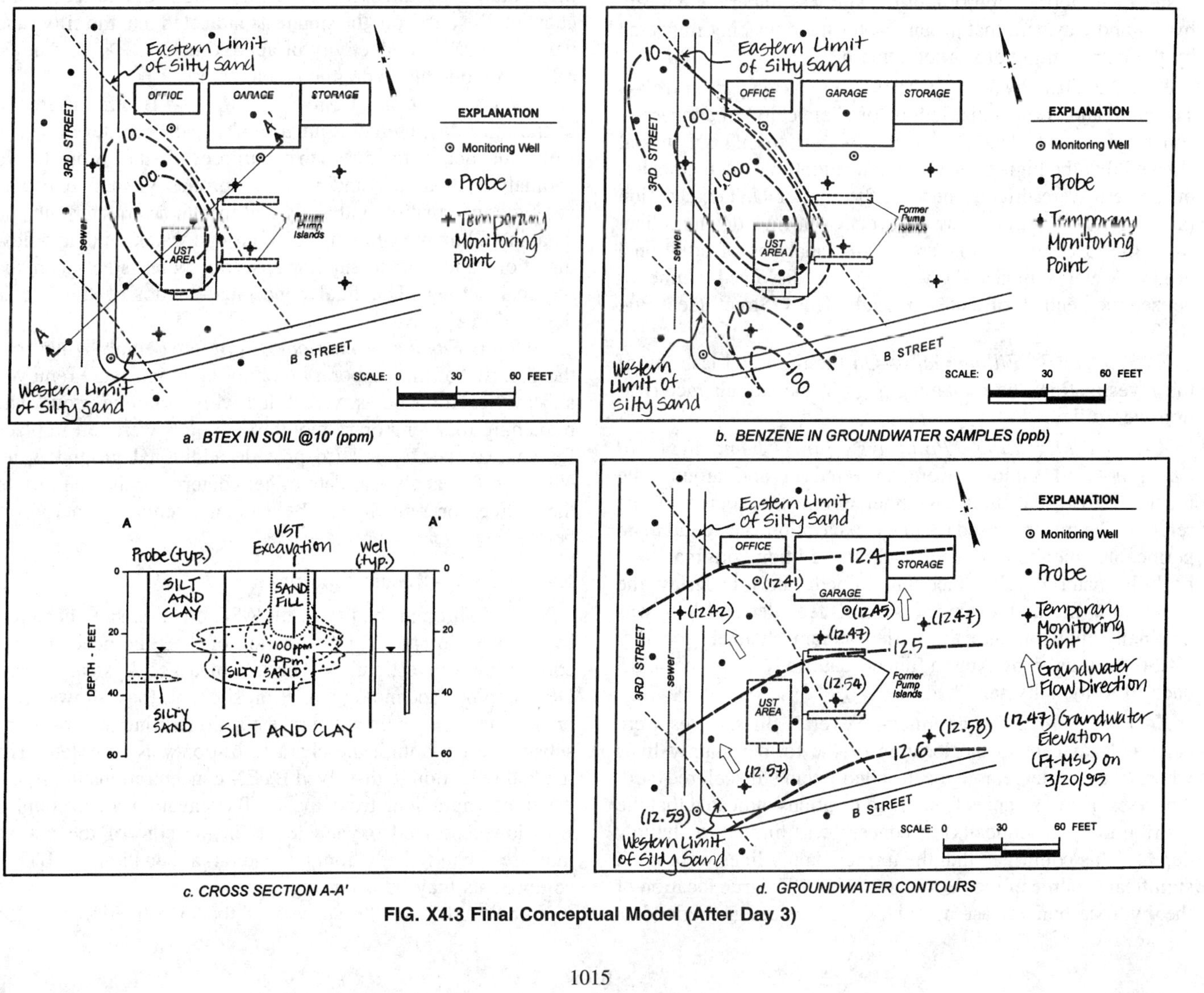

a. BTEX IN SOIL @10' (ppm)

b. BENZENE IN GROUNDWATER SAMPLES (ppb)

c. CROSS SECTION A-A'

d. GROUNDWATER CONTOURS

FIG. X4.3 Final Conceptual Model (After Day 3)

(19.05 mm) diameter slotted PVC, were installed in the DP probe holes shown in Fig. X4.2(*a*). The temporary monitoring points were installed to provide a way to measure the ground water elevation at many locations beneath the site. Installing numerous temporary monitoring points allowed complete definition of ground water flow direction and hydraulic gradient beneath the site. Holes that were not converted to temporary monitoring points were filled with bentonite grout.

X4.7.3 *Day 2—Refining Conceptual Model*—Characterization of the site continued on Day 2, with the on-site manager collecting additional subsurface data to refine the conceptual model of the site geology/hydrogeology, and nature and distribution of chemical(s) of concern. The elevation of the tops of the temporary monitoring points were surveyed relative to the site datum (mean sea level) in order to convert depth-to-water measurements to ground water elevations.

X4.7.3.1 *Defining Buried Stream Channel*—The eastern and western limits of the buried stream channel were refined by collecting additional soil samples from locations in between DP probe holes advanced during Day 1. The western limit of the buried stream channel was found to coincide with the southwestern corner of the site (see Fig. X4.3(*a*) and (*b*)). Additional DP probes were advanced to confirm that the eastern and northern portion of the site was underlain entirely by silt and clay, and that ground water there was not impacted by the petroleum hydrocarbon release.

X4.7.3.2 *Distribution of Chemical(s) of Concern in Soil*—The areal and vertical distribution of chemical(s) of concern in soil began to be clearly defined. Contours of BTEX in soil showed that the highest levels of concentrations of chemical(s) of concern were directly beneath the former UST excavation (see Fig. X4.3(*a*)), and lower levels occurring downgradient from the UST excavation area. Also, analyses of additional ground water samples showed that the dissolved plume of benzene extended off site, beneath Third Street (see Fig. X4.3(*b*)).

X4.7.4 *Day 3—Filling Gaps*—On the third and last day of the investigation, the remaining gaps in the site characterization were filled.

X4.7.4.1 *Migration Pathways for Chemical(s) of Concern*—Soil samples were collected in and around the former UST excavation to further define the source of the release. The material used to fill the excavation was found to be permeable, medium-grained sand. In addition, original tank backfill material (also medium-grained sand) underlay the excavation fill and extended several feet deeper than was originally thought, into the buried stream channel (see Fig. X4.3(*c*)), thus providing a direct pathway from the UST backfill to the silty sand bed.

X4.7.4.2 *Source Confirmation*—Several soil samples were collected adjacent to the location of the former supply lines where the former employee recalled seeing discolored soil. Analyses of soil samples from these locations indicated that the distribution of chemical(s) of concern was limited to shallow depths. This confirmed that the former supply lines were not a significant source of the petroleum release. The true location of the hydrocarbon release was identified using geologic data from two DP probes that penetrated the tank excavation. The floor of the original tank excavation slopes towards the north, therefore, petroleum hydrocarbons that leaked from the USTs accumulated and seeped from the northern end of the original tank excavation. This hypothesis is supported by the distribution of residual petroleum hydrocarbons in soil (see Fig. X4.3(*b*)).

X4.7.4.3 *Potential Off-Site Source*—Several soil and ground water samples were collected in the southwestern portion of the site and beneath B Street to investigate the anomalous analytical data collected during Day 1. The presence of relatively high benzene concentrations in ground water beneath B Street, different chromatographic fingerprint of samples collected in that area, and pattern of decreasing benzene concentrations in ground water samples collected closer to the former UST area indicates that a plume of dissolved petroleum hydrocarbons has migrated onto the subject property from an upgradient source.

X4.7.4.4 *Ground Water Flow Direction/Slug Test*—Water elevations measured in the temporary monitoring points and the existing monitoring wells indicate that ground water within the silt and clay flows toward the north, consistent with the regional ground water flow direction. Not surprisingly, within the buried stream channel, ground water flow toward the northwest is indicated. A slug test was conducted to estimate the hydraulic conductivity of the silty sand in the buried stream channel. Results from the slug tests indicated that the silty sand has a hydraulic conductivity of approximately 2×10^{-3} cm/s, a low to moderate hydraulic conductivity.

X4.7.4.5 *Finalizing Conceptual Model*—By the afternoon of the third day, the conceptual model had been developed in sufficient detail to meet the purpose of the project. No anomalies remained, and new DP probes yielded expected geologic information and analytical results. Moreover, the site data, including the geologic units, ground water depth and flow direction, and upgradient impacts, were consistent with the regional setting. The final conceptual model of the site is depicted in Fig. X4.3.

X4.7.4.6 *Site Decommissioning*—Before demobilizing from the site, two of the temporary monitoring points were removed and the resulting holes were filled with bentonite grout. The remaining four temporary monitoring points were left in place for an extended period (to provide additional ground water elevation and analytical data to help determine the stability of the plume for remediation by natural attenuation purposes) before they were removed.

X4.8 Step 6—Report Findings

X4.8.1 Through the use of the ASC process, site characterization was completed in a fraction of the time needed for a conventional investigation. The use of a geophysical survey, DP sampling technology and on-site analysis allowed the on-site manager to direct the investigation, filling in gaps in the subsurface data, until the characterization was complete. The marked reduction in dissolved BTEX concentrations in ground water downgradient from the UST excavation area, coupled with low dissolved oxygen levels in the core of the plume, indicates that remediation by natural attenuation of BTEX components may be occurring.

X4.8.2 A key to the success of the investigation was the

up-front, pre-field review of available data and development of the initial conceptual model. Because of this work, the presence of the buried stream channel and upgradient petroleum release was not unexpected, and the investigation was not delayed when these anomalies were encountered. The significance of these discoveries should not be minimized. The presence of an upgradient source of dissolved hydrocarbons would certainly complicate efforts to remediate the subject property (if required). The results of this investigation would be more than sufficient for the SED to request an environmental investigation by the owner of the upgradient property.

X4.8.3 The primary migration pathway for chemical(s) of concern beneath the site is certainly different than what was initially thought. The ground water velocity and adsorptive properties of the buried stream channel are significantly different than the values from previous investigations. Moreover, because of the localized northwest ground water flow direction, the probable receptors and point(s) of exposure for chemical(s) of concern in ground water are altogether different than those identified in the previous investigations.

X4.8.4 In this example, the ASC provided accurate data to perform a RBCA Tier 1 and Tier 2 evaluation. Section X4.9 summarizes the details of this RBCA evaluation. If active remediation was deemed necessary, the thorough understanding of the site conditions ensures that effective remedial measures will be undertaken. Ongoing ground water monitoring will likely be required at the example site. The location of permanent monitoring wells can be selected based on the clear understanding of the site ground water flow patterns. In the report which was submitted to the SED within two weeks of the mobilization, the on-site manager summarized the major findings from the accelerated site characterization as follows:

X4.8.4.1 The USTs were the primary source of chemical(s) of concern. Chemical(s) of concern around associated piping was not continuous to ground water;

X4.8.4.2 The areal and vertical distribution and concentrations of chemical(s) of concern in soil and ground water had been defined;

X4.8.4.3 A buried stream channel was the primary migration pathway for the petroleum release;

X4.8.4.4 A potential upgradient source of dissolved hydrocarbons was identified; and

X4.8.4.5 Remediation by Natural Attenuation of petroleum hydrocarbons may be occurring beneath the site.

X4.9 RBCA Tier 1 and Tier 2 Evaluation Example

X4.9.1 *Initial Site Assessment, Site Classification and Initial Response*—Based upon the results from the review of existing information and the data collected during the site characterization, the on-site manager classified the site per specified scenarios (Table 3 of Guide E 1739). The site was classified as 3 - Ground water is impacted and non-potable water supply wells producing from the impacted interval are located > 2 years ground water travel time from the dissolved plume. The initial response action was to identify water usage of wells, assess the effect of potential impact, monitor the dissolved plume, and evaluate whether natural attenuation or hydraulic control are appropriate control measures.

X4.9.2 *Tier 1 Evaluation*—The site characterization information provided in the ASC example was used to perform a RBCA Tier 1 evaluation. Data collected during the site characterization defined the distribution of chemical(s) of concern in soil and ground water, location of potential receptors and point(s) of exposure, migration pathways and potentially complete exposure pathways.

X4.9.2.1 *Exposure Pathway Analysis*—Based upon the results of the receptor survey (see X4.4.7), the following exposure pathways were selected for a Tier 1 Evaluation:

X4.9.2.2 Soil-vapor intrusion from soil to enclosed spaces, and

X4.9.2.3 Ground water - vapor intrusion from ground water to enclosed spaces.

X4.9.2.4 *RBSL Comparison*—The on-site manager then compared the maximum site specific concentrations of chemical(s) of concern to the appropriate Tier 1 risk-based screening levels (RBSL's) from the Guide E 1739 example Tier 1 Look-Up Table (see tables for both soil and ground water below). The following criteria were used in the selecting the appropriate RBSL:

X4.9.2.5 Commercial/Industrial receptor,

X4.9.2.6 Target cancer risk of 10^{-4}, and

X4.9.2.7 Chronic hazard quotient = 1.

X4.9.2.8 The RBSL concentrations were selected from the table for the pathways listed above. A comparison of RBSL's to maximum concentrations of chemical(s) of concern detected during the ASC are listed below:

Soil RBSL's

Chemical of Concern	RBSL Concentration (mg/kg)	Maximum Concentration (mg/kg)	Maximum > RBSL?
Benzene	1.09	13.23	Yes
Toluene	54.50	66.15	Yes
Ethylbenzene	1100.00	19.84	No
Xylenes	> RES^A	112.45	No

Ground Water RBSL's

Chemical of Concern	RBSL Concentration (mg/l)	Maximum Concentration (mg/l)	Maximum > RBSL?
Benzene	7.39	1.13	No
Toluene	85.00	0.08	No
Ethylbenzene	> S^B	1.81	No
Xylenes	> S^B	1.93	No

^A RES - Selected risk level is not exceeded for pure compound present at any concentration.

^B > S - Selected risk level is not exceeded for all possible dissolved levels.

X4.9.2.9 The Tier 1 analysis indicated that the maximum benzene and toluene concentrations in soil exceeded the RBSL's. The investigator therefore performed a Tier 2 analysis.

X4.9.3 *Tier 2 Evaluation*—Before Tier 2 Site Specific Target Levels (SSTL's) were calculated, the on-site manager performed a statistical analysis of the soil analytical data to establish representative soil concentrations at the site. The analysis determined the 95th upper confidence limit (UCL) of the mean of the soil concentrations. The on-site manager compared the 95th UCL concentrations for benzene and toluene in soil to the Tier 1 soil RBSL's to determine if the SSTL's calculation was still necessary. The comparison showed that the 95th UCL concentration for toluene was below the Tier 1

RBSL, however the 95th UCL concentration for benzene exceeded the Tier 1 RBSL. In order to perform the SSTL calculation, the on-site manager summarized the site specific/physical properties data (for example, TOC, bulk density, porosity) for use in the SSTL equations.

X4.9.3.1 *Exposure Pathway Analysis*—The SSTL for benzene was calculated for the following exposure pathway: soil-vapor intrusion from soil to enclosed spaces.

X4.9.3.2 *SSTL Calculation*—The SSTL equations for the remaining pathway were developed using the examples outlined in Appendix X2 of Guide E 1739. The SSTL was calculated using the following criteria: commercial/industrial receptor scenario, target cancer risk of 10^{-4}, chronic hazard quotient = 1, and site specific data collected during the ASC.

X4.9.3.3 *SSTL Comparison*—The on-site manager compared the SSTL concentration to the 95th UCL:

	Soil SSTL's		
Chemical of Concern	SSTL Conc. (mg/kg)	95th UCL Conc. (mg/kg)	95th UCL > SSTL
Benzene	12.60	3.30	No

X4.9.4 *Corrective Action Evaluation*—The results of the Tier 1 evaluation indicated that the maximum concentrations of benzene and toluene in soil and ground water were above the Tier 1 RBSL's for the soil-vapor and ground water-vapor intrusion to enclosed space pathways. Further analysis in the Tier 2 evaluation determined that the 95th UCL concentration for toluene was below the Tier 1 RBSL and the 95th UCL concentration for benzene was below the Tier 2 SSTL. Because the 95th UCL concentrations for toluene and benzene did not exceed the RBSL's or SSTL's, active remediation was not warranted. In order to achieve the final corrective action goals, the on-site manager recommended that a ground water monitoring program be implemented. The monitoring program included collecting and analyzing ground water samples from on-site monitoring points for both concentrations of chemical(s) of concern and natural attenuation indicator parameters (for example, dissolved oxygen, nitrate, sulfate). Although not all natural attenuation parameters were analyzed during the ASC, the on-site manager believed that including natural attenuation parameters in the monitoring program would provide additional evidence that natural attenuation of the chemical(s) of concern was occurring and that the final corrective action goals would be achieved.

Standard Practice for Assessment of Certification Programs for Environmental Professionals: Accreditation Criteria[1]

This standard is issued under the fixed designation E 1929; the number immediately following the designation indicates the year of original adoption or, in the case of revision, the year of last revision. A number in parentheses indicates the year of last reapproval. A superscript epsilon (ε) indicates an editorial change since the last revision or reapproval.

1. Scope

1.1 This practice covers minimum acceptable accreditation criteria to assist in the assessment of certification programs for environmental professionals.

1.2 The practice will assist a user of environmental services in identifying credible certification programs for environmental professionals.

1.3 *This practice does not propose to address any specific certification program for environmental professionals. It is the responsibility of the user of this guide to determine whether or not a specific certification program meets the criteria contained herein.*

1.4 *This standard does not purport to address all of the safety concerns, if any, associated with its use. It is the responsibility of the user of this standard to establish appropriate safety and health practices and determine the applicability of regulatory limitations prior to use.*

2. Referenced Documents

2.1 *CESB Standard:*

Guidelines for Engineering and Related Specialty Certification Programs[2]

2.2 *NCCA Standard:*

Standards for Accreditation of National Certification Organizations[3]

3. Terminology

3.1 *Definitions:*

3.1.1 *accreditation*—the recognition of a certification program as conforming to certain standards that qualify individuals receiving the certification for practice in a specific area. Accreditation applies to institutions and programs, not individuals.

3.1.2 *certification*—recognition by a body that an individual or entity has met specified criteria and is thereby qualified for a designated function.

3.1.3 *environmental professional*—a person possessing sufficient education, training and experience necessary to qualify for practice in a specialty area of the environmental profession.

3.1.4 *licensing*—the authorization granted by a government to an individual or entity to perform a function or service.

3.1.5 *registration*—the listing of an individual with and by a certifying or licensing body. The body may be either governmental or non-governmental. Registration grants no authority nor does it address qualifications.

3.1.6 *specialty area*—a specified, distinctly identifiable area of practice within a profession.

4. Significance and Use

4.1 This practice may be employed by users of environmental services to assess the credibility of certification programs for environmental professionals.

4.2 Environmental professionals seeking certification may use this practice to assess the credibility of certification programs for environmental professionals.

4.3 Certifying bodies for environmental professionals may use this practice to demonstrate that their specific certification program for environmental professionals meets minimum standards for credibility.

4.4 *This practice is not meant to apply to, replace or otherwise interfere with any existing government licensing board or registration program.*

5. Types of Certifying Bodies

5.1 Certifying bodies may be either governmental or non-governmental.

5.2 Governmental bodies such as state licensing boards may have established certification programs to meet their individual requirements, including state laws and programs. This standard is not mandatory for such bodies, however it may be used as a means to improve their programs.

6. Structure of Certifying Body

6.1 *The Certifying Body Shall*:

6.1.1 Be independent from sponsoring professional organizations in matters pertaining to certification;

6.1.2 Consist of a majority of certified individuals (new certifying bodies should refer to 8.1.1.7); and

---

[1] This practice is under the jurisdiction of ASTM Committee E-50 on Environmental Assessment and is the direct responsibility of Subcommittee E50.02 on Commercial Real Estate Transactions.

Current edition approved Jan. 10, 1998. Published June 1998.

[2] Available from Council of Engineering and Scientific Specialty Boards (CESB), 130 Holiday Court, Suite 100, Annapolis, MD 21401–7032.

[3] Available from National Commission for Certifying Agencies (NCCA), 1200 19th Street, N.W., Suite 300, Washington, DC 20036–2401.

6.1.3 Maintain formal written procedures defining the criteria and process used to select or appoint members to the certifying body.

7. Resources of Certifying Body

7.1 *The Certifying Body Shall*:

7.1.1 Possess the expertise, knowledge and skill to implement the certification program.

8. Certification Program Operation

8.1 *The Certifying Body Shall*:

8.1.1 Have a program to evaluate individual competence for certification that is objective and based on the knowledge, skills, and abilities needed to function in the specialty area. Such programs shall include the following as minimum criteria:

8.1.1.1 Have a developed document specifying the "minimum level of skills and knowledge" (referred to as the "body of knowledge") required of a practitioner in the specialty area. The "body of knowledge" document shall be subjected to peer review by practicing professionals in the field of the specialty area. Provision shall be made to update the document to remain current and relevant. Such document shall be made available upon request.

8.1.1.2 Require applicants for certification to meet minimum educational and experience requirements for the specialty area.

8.1.1.3 Require applicants for certification to document education. The certifying body shall verify that minimum educational requirements have been met.

8.1.1.4 Require applicants for certification to supply references supporting relevant experience in the specialty area. References shall be contacted by the certifying body to verify relevant experience.

8.1.1.5 Have a professional code of ethics that applicants requesting certification are provided with and that certificants are required to adhere to in their professional practice.

8.1.1.6 Award certification only if the applicant passes a comprehensive examination based on the requirements of the document specified in 8.1.1.1; and meets all other stipulated certification requirements.

8.1.1.7 New certification programs may grant certification to applicants on the basis of education and demonstrated experience, without examination, for a limited period of time not exceeding approximately two years, provided all other certification requirements are met. After that time, no applicant shall be certified other than by examination.

8.1.1.8 Award certification for a predefined and limited period of time, not exceeding approximately five years, after which recertification shall be required.

8.1.1.9 Establish re-certification requirements.

8.2 Provide any sponsoring professional organization(s), and those who employ or otherwise use environmental services within the specialty area an opportunity for input into the policies of the certifying body;

8.3 Have written mechanisms in place within the organization to re-evaluate on a periodic basis the knowledge, skills, and abilities used to define the specialty area;

8.4 Develop a comprehensive examination designed to test the body of knowledge relevant to the specialty area;

8.5 Take reasonable precautions to protect the integrity of all examinations before, during and after the examination period;

8.6 Have written mechanisms in place to avoid any real or perceived conflict of interest, (for example, an applicant and the evaluator shall not be from the same company); and

8.7 Provide a means for differentiating certifications issued after a substantive revision to the certification requirements of the certifying body from those certifications issued prior to the revision.

9. Public Disclosure of Certification by Certifying Body

9.1 *The Certifying Body Shall*:

9.1.1 Publish and make publicly available a document that clearly defines the certification responsibilities of the certifying body and outlines any other activities of the certifying body which are not related to certification;

9.1.2 Publish and make publicly available a general description of the procedures used in examination construction and validation, administration, and results reporting;

9.1.3 Publish and make publicly available a comprehensive summary of the information, knowledge, or functions covered by any examination required; and

9.1.4 Publish and make publicly available, at least annually, a summary of certification activities, including number of applicants, and number and names of those certified.

10. Certifying Body Responsibilities to Applicants

10.1 *The Certifying Body Shall*:

10.1.1 Not give preferential or prejudicial treatment to any applicant;

10.1.2 Provide all applicants with written copies of procedures to apply for and attain certification;

10.1.3 Maintain a written policy for the periodic review of application and testing procedures to ensure consistency and equitableness;

10.1.4 Report examination results to applicants in a timely manner;

10.1.5 Provide, upon request, to all applicants who fail the examination or fail to meet the requirements of the certification, a written response indicating the general areas of deficiency;

10.1.6 Consider and treat the specific and detailed examination results of each applicant as confidential; and

10.1.7 Publish and maintain a written policy detailing appeal procedures.

11. Certifying Body Responsibilities to the Public and Consumers

11.1 *The Certifying Body Shall*:

11.1.1 Ensure that the examination measures the knowledge, skill, and abilities required for practice in the specialty area;

11.1.2 Award certification only after the examination has been successfully passed and all other certification requirements have been met;

11.1.3 Maintain a publicly accessible roster of individuals awarded certification;

11.1.4 Maintain written due process policies and procedures for the discipline of certificants, for cause, including revocation of certification;

11.1.5 Maintain written provisions for sanctions to certificants resulting from misuse, abuse, or fraudulent use of certification; and

11.1.6 Ensure that any title or credential awarded by the certifying body accurately reflects the specialty area.

12. Independent Accreditation of Certifying Body Certification Programs

12.1 Independent (third party) verification of conformance with the accreditation criteria in this standard brings to any certification program a higher level of credibility.

12.2 Certification bodies are encouraged to obtain independent (third party) verification of conformance with this standard through a nationally recognized body experienced in conducting accreditation. Said acreditation experience would be demonstrated by accreditation of a minimum of three separate certification bodies.

12.3 Accreditation criteria established by either the Council of Engineering and Scientific Specialty Boards (CESB) or the National Commission for Certifying Agencies (NCCA) shall be considered in conformance with this practice.

12.4 A certifying body that has received independent (third party) verification that their certification program is in conformance with this practice may state:

"(name of certifying body and specialty certification) conforms with the accreditation criteria in ASTM Practice E 1929, as independently verified by (state the accreditation organization which conducted the accreditation and provide the address and phone number)."

12.5 Independent (third party) accreditation of conformance with this standard shall be re-evaluated by the accrediting body within a period not exceeding approximately five years.

12.6 Independent (third party) accreditation shall require that the certifying body conform at a minimum with the requirements of this standard.

13. Interim Self-Declaration Option by Certifying Bodies

13.1 A certification body desiring to demonstrate conformity with the accreditation criteria in this practice may, after self-evaluation, declare itself to be in conformance with the accreditation criteria for a period not exceeding three years from the date this practice is first officially published, after which time conformance with this practice shall only be demonstrated by independent (third party) accreditation consistent with Section 12. This interim option is made available in order for the certification body to obtain independent (third party) accreditation. Certification bodies using this interim option shall apply for independent (third party) accreditation within one year from the date this practice is first officially published. Self-declaration shall be supported by written documentation including execution of the Conformity with Standard Form included in Appendix X1.

13.2 A certifying body that, after self-evaluation, is able to demonstrate conformance with the accreditation criteria in this practice may state:

"(name of certifying body and specialty certification) conforms with the accreditation criteria in ASTM Practice E 1929, as determined by self-evaluation. Documentation will be provided upon request."

13.3 A certifying body that, after self-evaluation, declares itself to be in conformance with this practice shall provide a user, upon request, with a copy of the completed Conformity with Standard Form included in Appendix X1.

APPENDIX

(Nonmandatory Information)

X1. DEMONSTRATION OF CONFORMITY WITH PRACTICE E 1929

I. Structure of Certifying Body

(a) Independent in matters pertaining to certification ☐ Yes ☐ No*

(b) Consist of a majority of certified individuals ☐ Yes ☐ No

(c) Written procedures exist for selection of members ☐ Yes ☐ No

II. Resources of Certifying Body

(a) Possess expertise, knowledge and skill to implement the certification program ☐ Yes ☐ No

III. Certification Program Operation

(a) "Body of knowledge" documented ☐ Yes ☐ No

(b) "Body of knowledge" peer reviewed by practicing professionals in the field of the specialty area ☐ Yes ☐ No

(c) Provision to update "body of knowledge" ☐ Yes ☐ No

(d) Minimum educational requirements established ☐ Yes ☐ No
Description of requirements:____________________

(e) Minimum experience requirements established ☐ Yes ☐ No
Description of requirements:____________________

(f) Applicant required to document education ☐ Yes ☐ No

(g) Applicant required to document experience ☐ Yes ☐ No

(h) Applicant required to supply references supporting relevant experience ☐ Yes ☐ No

(i) Verification that applicant meets educational requirements ☐ Yes ☐ No

(j) Verification of applicant's experience ☐ Yes ☐ No

(k) Professional code of ethics exists and required to be adhered to ☐ Yes ☐ No

(l) Examination required ☐ Yes ☐ No
Description:____________________

(m) Examination tests the body of knowledge relevant to specialty area ☐ Yes ☐ No

(n) Precautionary measures taken to protect the integrity of the examination process ☐ Yes ☐ No

(o) Recertification program exists ☐ Yes ☐ No

(p) Written mechanisms in place within the organization to re-evaluate on a periodic basis the knowledge, skills and abilities which define the specialty area ☐ Yes ☐ No

(q) Opportunity for outside input into policies of the certifying body ☐ Yes ☐ No

(r) Written mechanisms exist to avoid conflict of interest ☐ Yes ☐ No

(s) Policy in place to differentiate certifications after any substantive revision(s) to requirements made ☐ Yes ☐ No

IV. Public Disclosure of Certification by Certifying Body

(a) Publish and make publicly available a document defining certification responsibilities and activities ☐ Yes ☐ No

(b) Publish and make publicly available descriptions of procedures used in examination construction and validation, administration, and results reporting ☐ Yes ☐ No

(c) Publish and make publicly available description of information, knowledge, and functions covered by examination ☐ Yes ☐ No

(d) Publish and make publicly available, at least annually, a summary of certification activities ☐ Yes ☐ No

V. Certifying Body Responsibilities to Applicants

(a) Non-discrimination policy in place ☐ Yes ☐ No

(b) Applicants provided with written application procedures ☐ Yes ☐ No

(c) Periodic review of application and testing procedures ☐ Yes ☐ No

(d) Timely reporting of examination results ☐ Yes ☐ No

(e) Deficiency information made available to applicants who fail to be certified ☐ Yes ☐ No

(f) Maintain confidentiality of examination results ☐ Yes ☐ No

(g) Published policy detailing appeal procedures ☐ Yes ☐ No

VI. Certifying Body Responsibilities to the Public and Consumers

(a) Examination measures knowledge, skill and abilities required for practice in the specialty area ☐ Yes ☐ No

(b) Award certification only after successful examination and all certification requirements met ☐ Yes ☐ No

(c) Maintain publicly accessible roster of certified individuals ☐ Yes ☐ No

(d) Written due process policies and procedures exist for disciplining certificants for cause ☐ Yes ☐ No

(e) Provisions exist for sanctions to certificants resulting from misuse, abuse or fraud ☐Yes ☐ No

(f) Name of credential accurately reflects specialty area ☐ Yes ☐ No

** Any negative answer constitutes non-conformance with the practice. An explanation may be attached to this document with the reason for non-conformance.*

*Name of Certifying Body*______________________________

*Address of Certifying Body*______________________________

____

Specialty Certification Description ______________________________

____

☐ New Certification ☐ Existing Certification

*Scope of Practice*______________________________

____

Conformity with ASTM Practice E 1929
for Specialty Certification

☐ Self-Declaration

☐ Applied for Independent Accreditation

Date:______________________________

Accreditation Organization:______________________________

____

☐ Independent Accreditation not yet applied for

☐ Independent Accreditation

Accreditation Organization:

____

____

____

Phone ______________________________

Date Accreditation Received:______________________________

I SWEAR UNDER OATH THAT THE INFORMATION CONTAINED IN THIS DEMONSTRATION OF CONFORMITY WITH ASTM PRACTICE E 1929 DOCUMENT IS TRUE AND CORRECT.

Signature of Certifying Body Officer:

____

Signature

____

Name

____

Title

____

Date

FIG. X1.1 Demonstration of Conformity With ASTM Practice E 1929

Designation: E 1943 – 98

Standard Guide for Remediation of Ground Water by Natural Attenuation at Petroleum Release Sites[1]

This standard is issued under the fixed designation E 1943; the number immediately following the designation indicates the year of original adoption or, in the case of revision, the year of last revision. A number in parentheses indicates the year of last reapproval. A superscript epsilon (ϵ) indicates an editorial change since the last revision or reapproval.

1. Scope

1.1 This is a guide for determining the appropriateness of remediation by natural attenuation and implementing remediation by natural attenuation at a given petroleum release site, either as a stand alone remedial action or in combination with other remedial actions.

1.2 Natural attenuation is a potential remediation alternative for containment and reduction of the mass and concentration of petroleum hydrocarbons in the environment to protect human health and the environment. Remediation by natural attenuation depends upon natural processes such as biodegradation, dispersion, dilution, volatilization, hydrolysis, and sorption to attenuate petroleum constituents of concern to achieve remedial goals.

NOTE 1—Remedial goals must be established through another process as determined by the appropriate regulatory agency.

1.3 In general, remediation by natural attenuation should not be considered a presumptive remedy. A determination of whether remediation by natural attenuation is appropriate for an individual petroleum release site, relative to site-specific remedial goals, requires site characterization, assessment of potential risks, evaluation of the need for source area control, and evaluation of potential effectiveness similar to other remedial action technologies. Application and implementation of remediation by natural attenuation requires demonstration of remedial progress and attainment of remedial goals by use of converging lines of evidence obtained through monitoring and evaluation of resulting data. When properly applied to a site, remediation by natural attenuation is a process for risk management and achieving remedial goals. Monitoring should be conducted until it has been demonstrated that natural attenuation will continue and eventually meet remedial goals.

1.3.1 The primary line of evidence for remediation by natural attenuation is provided by observed reductions in plume geometry and observed reductions in concentrations of the constituents of concern at the site.

1.3.2 Secondary lines of evidence for remediation by natural attenuation are provided by geochemical indicators of naturally occurring degradation and estimates of attenuation rates.

1.3.3 Additional optional lines of evidence can be provided by microbiological information and further analysis of primary and secondary lines of evidence such as through solute transport modeling or estimates of assimilative capacity.

1.4 The emphasis in this guide is on the use of remediation by natural attenuation for petroleum hydrocarbon constituents where ground water is impacted. Though soil and ground water impacts are often linked, this guide does not address natural attenuation in soils separate from ground water or in situations where soils containing constituents of concern exist without an associated ground water impact. Even if natural attenuation is selected as the remedial action for ground water, additional remedial action may be necessary to address other completed exposure pathways at the site.

1.5 This guide does not address enhanced bioremediation or enhanced attenuation.

1.6 Also, while much of what is discussed is relevant to other organic chemicals or constituents of concern, these situations will involve additional considerations not addressed in this guide.

1.7 The guide is organized as follows:

1.7.1 Section 2 lists referenced documents.

1.7.2 Section 3 defines terminology used in this guide.

1.7.3 Section 4 describes the significance and use of this guide.

1.7.4 Section 5 provides an overview of the use of natural attenuation as a remedial action alternative, including;

1.7.4.1 Advantages of remediation by natural attenuation as a remedial alternative;

1.7.4.2 Limitations of remediation by natural attenuation as a remedial alternative; and

1.7.4.3 Using multiple lines of evidence to demonstrate the appropriateness of remediation by natural remediation.

1.7.5 Section 6 describes the decision process for appropriate application and implementation of remediation by natural attenuation including;

1.7.5.1 Initial response, site characterization, selection of chemicals of concern, and establishment of remedial goals;

1.7.5.2 Evaluation of plume status;

1.7.5.3 Collection and evaluation of additional data;

[1] This guide is under the jurisdiction of ASTM Committee E-50 on Environmental Assessment and is the direct responsibility of Subcommittee E50.01 on Storage Tanks.

Current edition approved April 10, 1998. Published August 1998.

1.7.5.4 Comparing remediation by natural attenuation performance to remedial goals;

1.7.5.5 Comparing remediation by natural attenuation to other remedial options;

1.7.5.6 Implementation of a continued monitoring program;

1.7.5.7 Evaluation of progress of remediation by natural attenuation; and

1.7.5.8 No further action.

1.7.6 Section 7 lists keywords relevant to this guide.

1.7.7 Appendix X1 describes natural attenuation processes;

1.7.8 Appendix X2 describes site characterization requirements for evaluating remediation by natural attenuation;

1.7.9 Appendix X3 describes considerations for designing and implementing monitoring for remediation by natural attenuation;

1.7.10 Appendix X4 describes sampling considerations and analytical methods for determining indicator parameters for remediation by natural attenuation;

1.7.11 Appendix X5 describes the interpretation of different lines of evidence as indicators of natural attenuation;

1.7.12 Appendix X6 describes methods for evaluation and quantification of natural attenuation rates; and

1.7.13 Appendix X7 describes example problems illustrating the application and implementation of remediation by natural attenuation.

1.8 *This standard does not purport to address all of the safety concerns, if any, associated with its use. It is the responsibility of the user of this standard to establish appropriate safety and health practices and determine the applicability of any regulatory limitations prior to use.*

2. Referenced Documents

2.1 *ASTM Standards:*

D 888 Test Methods for Dissolved Oxygen in Water[2]

D 1125 Test Methods for Electrical Conductivity and Resistivity of Water[2]

D 1293 Test Methods for ph of Water[2]

D 1452 Practice for Soil Investigation and Sampling by Auger Borings[3]

D 1498 Practice for Oxidation-Reduction Potential of Water[2]

D 1586 Test Method for Penetration Test and Split-Barrel Sampling of Soils[3]

D 4043 Guide for Selection of Aquifer-Test Method in Determining of Hydraulic Properties by Well Techniques[3]

D 4044 Test Method (Field Procedure) for Instantaneous Change in Head (Slug) Tests for Determining Hydraulic Properties of Aquifer Systems[3]

D 4050 Test Method (Field Procedure) for Withdrawal and Injection Well Tests for Determining Hydraulic Properties of Aquifer Systems[3]

D 4104 Test Method (Analytical Procedure) for Determining Transmissivity of Nonleaky Confined Aquifers by Overdamped Well Response to Instantaneous Change in Head (Slug Tests)[3]

D 4105 Test Method (Analytical Procedure) for Determining Transmissivity and Storage Coefficient of Nonleaky Confined Aquifers by the Modified Theis Nonequilibrium Method[3]

D 4106 Test Method (Analytical Procedure) for Determining Transmissivity and Storage Coefficient of Nonleaky Confined Aquifers by the Theis Nonequilibrium Method[3]

D 4372 Specification for Flame-Resistant Materials Used in Camping Tentage[4]

D 4448 Guide for Sampling Groundwater Monitoring Wells[5]

D 4658 Test Method for Sulfide Ion in Water[2]

D 4700 Guide for Soil Sampling From the Vadose Zone[3]

D 4750 Test Method for Determining Subsurface Liquid Levels in a Borehole or Monitoring Well (Observation Well)[3]

D 5092 Practice for Design and Installation of Ground Water Monitoring Wells in Aquifers[6]

D 5269 Test Method for Determining Transmissivity of Nonleaky Confined Aquifers by the Theis Recovery Method[6]

D 5270 Test Method for Determining Transmissivity and Storage Coefficient of Bounded, Nonleaky, Confined Aquifers[6]

D 5434 Guide for Field Logging of Subsurface Explorations of Soil and Rock[6]

D 5473 Test Method for (Analytical Procedure For) Analyzing the Effects of Partial Penetration of Control Well and Determining the Horizontal and Vertical Hydraulic Conductivity in a Nonleaky Confined Aquifer[6]

E 1599 Guide for Corrective Action for Petroleum Releases[5]

E 1689 Guide for Developing Conceptual Site Models for Contaminated Sites[7]

E 1739 Guide for Risk-Based Corrective Action Applied at Petroleum Release Sites[5]

E 1912 Guide for Accelerated Site Characterization for Confirmed or Suspected Petroleum Releases[5]

3. Terminology

3.1 *Definitions of Terms Specific to This Standard:*

3.1.1 *assimilative capacity*—a semi-quantitative estimate of the potential mass of hydrocarbons per unit volume of ground water that can be metabolized by aerobic and anaerobic biodegradation under existing site conditions.

3.1.2 *attenuation rate*—measured reduction in concentration or mass of a compound with time or distance expressed as an amount of reduction per unit time or per unit distance.

3.1.3 *conceptual site model*—a written or pictorial representation of an environmental system and the biological, physical, and chemical processes that determine the transport of constituents of concern from sources through environmental media to environmental receptors within the system.

[2] *Annual Book of ASTM Standards*, Vol 11.01.

[3] *Annual Book of ASTM Standards*, Vol 04.08.

[4] *Annual Book of ASTM Standards*, Vol 07.01.

[5] *Annual Book of ASTM Standards*, Vol 11.04.

[6] *Annual Book of ASTM Standards*, Vol 04.09.

[7] *Annual Book of ASTM Standards*, Vol 11.05.

3.1.4 *constituents of concern*—specific petroleum constituents that are identified as posing a potential risk to human health or the environment.

3.1.5 *corrective action*—actions taken to identify and clean up a release of petroleum. These activities include site assessment, interim remedial action, remedial action, operation and maintenance of equipment, monitoring of progress, and termination of the remedial action.

3.1.6 *electron acceptors*—elements or compounds that are reduced by receiving electrons produced by the oxidation of organic compounds through microbial metabolism or abiotic chemical oxidation processes.

3.1.7 *expanding plume*—configuration where the solute plume margin is continuing to move outward or down gradient from the source area.

3.1.8 *institutional controls*—the restriction on use or access (for example, fences, deed restrictions, restrictive zoning) to a site or facility to eliminate or minimize potential exposure to a constituent(s) of concern.

3.1.9 *monitoring points*—a monitoring well or other monitoring device placed in a selected location for observing parameters such as liquid levels or pressure changes, or for collecting liquid samples. The monitoring point may be cased or uncased, but if cased the casing should have openings to allow flow of borehole liquid into or out of the casing (modified from Test Method D 4750).

3.1.10 *natural attenuation*—reduction in mass or concentration of a compound in ground water over time or distance from the source of constituents of concern due to naturally occurring physical, chemical, and biological processes, such as; biodegradation, dispersion, dilution, sorption, and volatilization.

3.1.11 *optional lines of evidence*—solute transport modeling, estimates of assimilative capacity (to estimate the mass of BTEX and other constituents of concern degraded), and microbiological studies.

3.1.12 *plume*—volume of ground water where constituents of concern are present.

3.1.13 *point of compliance*—a location(s) selected between the source area(s) and potential point(s) of exposure where concentrations of constituents of concern must be at or below the determined ground water target levels.

3.1.14 *primary lines of evidence*—historical concentration data are the primary line of evidence for natural attenuation and are based on measured petroleum hydrocarbon constituent concentrations over time to define the plume as shrinking, stable, or expanding similar to the first line of evidence suggested by NRC (1993).[8]

3.1.15 *receptor*—persons, structures, utilities, ecological receptors, and water supply wells that are or may be adversely affected by a release.

3.1.16 *remedial goals*—remediation objectives established to protect human health and the environment. Remedial goals may be concentration-based target levels applied at specific points throughout the plume or performance-based criteria, such as demonstrated containment of the solute plume or demonstrated reduction in concentrations of constituents of concern over time within the plume or with distance from the source area.

3.1.17 *remediation/remedial action*—activities conducted to protect human health, safety, and the environment. These activities include evaluating risk, making no further action determinations, monitoring, and designing and operating cleanup equipment.

3.1.18 *remediation by natural attenuation*—a remedy where naturally occurring physical, chemical, and biological processes will achieve remedial goals. The use of natural attenuation processes as a remedial action also has been described by a variety of other terms, such as intrinsic remediation, intrinsic bioremediation, passive remediation, natural biodegradation, passive bioremediation, etc. Remediation by natural attenuation does not include remediation methods that require human intervention beyond monitoring.

3.1.19 *secondary lines of evidence*—geochemical indicators of naturally occurring biodegradation and estimates of natural attenuation rate.

3.1.20 *sentinel well*—monitoring points established at a location(s) between the leading edge of the solute plume and a sensitive receptor (for example, drinking water well) to ensure that there will be time for other remedial actions to be taken, if the plume does migrate beyond predicted boundaries.

3.1.21 *shrinking plume*—configuration where the solute plume margin is receding back toward the source area over time and the concentrations at points within the plume are decreasing over time.

3.1.22 *source area*—the location of free phase liquid hydrocarbons or the location of highest soil and ground water concentrations of constituents of concern.

3.1.23 *stable plume*—configuration where the solute plume margin is stationary over time and concentrations at points within the plume are relatively uniform over time or may decrease over time.

3.1.24 *user*—an individual or group involved in the corrective action process at petroleum release sites, which may include environmental consultants, industry, and state, local, and federal regulators.

4. Significance and Use

4.1 The approach presented in this guide is a practical and streamlined process for determining the appropriateness of remediation by natural attenuation and implementing remediation by natural attenuation at a given petroleum release site. This information can be used to evaluate remediation by natural attenuation along with other remedial options for each site.

4.2 In general, remediation by natural attenuation may be used in the following instances:

4.2.1 As the sole remedial action at sites where immediate threats to human health, safety and the environment do not exist or have been mitigated, and constituents of concern are unlikely to impact a receptor;

4.2.2 As a subsequent phase of remediation after another remedial action has sufficiently reduced concentrations/mass in

[8] National Research Council (NRC), 1993, In Situ Bioremediation: When Does It Work? National Academy Press, Washington, DC.

the source area so that plume impacts on receptors are unlikely; or

4.2.3 As a part of a multi-component remediation plan.

4.3 This guide is intended to be used by environmental consultants, industry, and state and federal regulators involved in response actions at petroleum release sites. Activities described in this guide should be performed by a person appropriately trained to conduct the corrective action process.

4.4 The implementation of remediation by natural attenuation requires that the user exercise the same care and professional judgement as with any other remedial alternative by:

4.4.1 Ensuring that site characterization activities focus on collecting information required to evaluate and implement remediation by natural attenuation;

4.4.2 Evaluating information to understand natural attenuation processes present at the site;

4.4.3 Determining whether remediation by natural attenuation is the most appropriate and cost-effective remedial alternative with a reasonable probability of achieving remedial goals; and

4.4.4 Monitoring remedial progress.

4.5 Application and implementation of remediation by natural attenuation is intended to be compatible with Guide E 1739 or other risk-based corrective action programs.

4.6 This guide does not address specific technical details of remediation by natural attenuation implementation such as site characterization (see Guide E 1912), sampling, data interpretation, or quantifying rates. For additional discussion and guidance concerning these technical issues for remediation by natural attenuation see Appendix X1 through Appendix X7.

4.7 This guide does not specifically address considerations and concerns associated with natural attenuation of non-petroleum constituents, such as chlorinated solvents. Care must be taken to ensure that degradation by-products will not cause harm to human health or the environment. In addition, if constituents are present which do not readily attenuate, such as methyl-t-butyl ether (MTBE), remediation by natural attenuation may not be a suitable remedial alternative or may need to be supplemented with other remedial technologies.

4.8 This guide is intended to be consistent with Guide E 1599 and U.S. EPA guidance for implementation of remediation by natural attenuation (U.S. EPA, 1995, Chapter 9).[9]

5. Natural Attenuation as a Remediation Alternative

5.1 At petroleum release sites petroleum migrates outward from a source area through the environment creating a plume of petroleum constituents. The configuration of a solute plume is controlled by the source mass-loading rate relative to the removal rate of natural attenuation processes. Typically, the plume will expand until it reaches steady-state where the rate of petroleum constituents contributed from the source is balanced with the rate of natural attenuation. At steady-state the plume stabilizes. The time scale over which this steady-state condition is reached can vary depending on specific site conditions. When the source area is depleted to the point that the rate of natural attenuation exceeds the source input the result will be a shrinking plume over time.

5.2 Remediation by natural attenuation relies on natural attenuation mechanisms to degrade and reduce concentrations of constituents of concern in ground water. The natural processes involved are physical, chemical, and biological in nature such as dispersion, dilution, volatilization, sorption, and biodegradation. Biodegradation is the process which accounts for the majority of mass removal and associated concentration reduction for constituents of concern. Biodegradation actually reduces the mass of constituents through microbial metabolization of constituents of concern. The ultimate products of this reaction are carbon dioxide, water, and biomass. These mechanisms are described in Appendix X1.

5.3 The processes which contribute to remediation by natural attenuation occur to some extent at all sites. remediation by natural attenuation is effective when these naturally occurring attenuation mechanisms achieve remedial goals. Depending on site conditions, remediation by natural attenuation may be a long-term remedial option. Remediation by natural attenuation is a remedial action approach that is compatible with existing remedy selection processes. It is not exclusive of other options and should be evaluated in the same manner as other remedial action options for a site.

5.4 Remediation by natural attenuation should not be considered to be a presumptive remedy.

5.5 *Advantages of Remediation by Natural Attenuation as a Remediation Alternative*:

5.5.1 Petroleum hydrocarbon constituents of concern which undergo biodegradation can be ultimately transformed to innocuous products (for example, carbon dioxide and water), not just transferred to another phase or location within the environment.

5.5.2 Remediation by natural attenuation is less intrusive; it results in minimal disturbance to the site operations and allows continuing use of the site's infrastructure during remediation.

5.5.3 More conventional remedial technologies can pose greater risk to potential receptors than natural attenuation due to site disruption and/or an inability to properly control these engineered remedial processes (for example, risk to on-site workers, releases to atmosphere, fugitive vapors, induced migration, etc.).

5.5.4 Remediation by natural attenuation can be used in conjunction with conventional remedial technologies such as excavation, pump and treat, soil vapor extraction, bioventing, and dual-phase extraction. It can also be used at sites where other remedial technologies are not technically feasible to use to achieve required cleanup target levels.

5.5.5 Remediation by natural attenuation can be less costly than other currently available remedial technologies when implemented with an appropriate monitoring program.

5.5.6 Remediation by natural attenuation can be evaluated by collecting adequate and appropriate geologic and hydrogeologic data during the site characterization phase. Data can be collected using relatively inexpensive field and laboratory analytical methods (see Appendix X2 and Appendix X4). If it is shown that remediation by natural attenuation is not solely

[9] U.S. EPA, 1995, Evaluating Alternative Cleanup Technologies for Underground Storage Tanks: A Guide for Corrective Action Plan Reviewers. U.S. Environmental Protection Agency, Office of Underground Storage Tanks, Washington, DC, EPA 510-B-95-007, May 1995.

sufficient to provide adequate protection of potential receptors, the data collected for the remediation by natural attenuation study can be used to design supplemental remedial alternatives.

5.5.7 Use of remediation by natural attenuation can help to focus funds and efforts on sites which require active remediation.

5.5.8 Remediation by natural attenuation is not subject to the limitations imposed by the use of mechanized remediation equipment (that is, no equipment down-time) and can be employed for constituents of concern below buildings and other areas that are not accessible.

5.5.9 Constituents such as benzene, toluene, ethyl benzene, and xylenes (BTEX) that typically pose the greatest risk and are commonly the major constituents of regulatory concern at petroleum release sites are generally the most susceptible to biodegradation.

5.6 *Limitations of Remediation by Natural Attenuation as a Remediation Alternative*:

5.6.1 The ability of remediation by natural attenuation to achieve remedial goals can be sensitive to natural and human-induced changes in local hydrogeologic conditions and site operations. Potentially important effects include changes in ground water gradients/velocity, rainfall, temperature, pH, electron acceptor concentrations, exposures not previously anticipated, or potential future releases. Such changes could be brought about by alterations in land use, changes in the local pumping regime, removal of an asphalt cap, or third party impacts, or a change in the location of receptors.

5.6.2 Time frames for achieving remedial goals may be relatively long, particularly for heavier petroleum constituents, compounds which attenuate slowly, and sites with a large source mass. Remediation by natural attenuation may take longer to mitigate constituents of concern than for more aggressive remedial measures. Remediation by natural attenuation may not always achieve the desired cleanup levels within a manageable time-frame.

5.6.3 In the public perception, remediation by natural attenuation may be viewed as a "do-nothing" remedial alternative.

5.6.4 Long-term monitoring for remediation by natural attenuation can represent significant cost and a continued funding commitment.

5.6.5 Application of remediation by natural attenuation may require supplemental source area removal or more active remediation when exposure pathways are completed or receptors are potentially impacted.

5.6.6 Technical limitations may obstruct the implementation or progress of remediation by natural attenuation and require the consideration or use of other remediation alternatives. Such limitations can include constraints associated with inadequate data used to construct the site conceptual model, the inability to implement the monitoring program, insufficient data to perform predictive solute transport modeling, and changes in site conditions.

5.6.6.1 The implementation of remediation by natural attenuation fundamentally requires adequate definition of the solute plume and understanding of site hydrogeology. The lack of necessary site data or inability to obtain representative, or otherwise requisite samples, necessary to construct an acceptable site conceptual model (for example, aquifer parameters, ground water and soil chemistry, etc.) and design an adequate long-term monitoring plan can preclude appropriate implementation of remediation by natural attenuation.

5.6.6.2 Remediation by natural attenuation relies on empirical data generated by ground water monitoring. The inability to place monitoring points and collect ground water samples in appropriate locations due to surface obstructions or other impediments, changes in aquifer water levels rendering monitoring points unusable, and monitoring where the sampling and analytical protocols are not observed can preclude appropriate implementation of remediation by natural attenuation. Also, the inherent variability of the ground water monitoring data may preclude effective evaluation of plume behavior.

5.6.6.3 Remediation by natural attenuation requires that site conditions persist or do not change adversely. Actual or proposed land use changes may result in the site being reclassified to a higher risk level. A new source may introduce additional petroleum product to the system at the site or another up gradient plume may reduce available electron acceptors for biodegradation. Changes in aquifer conditions may alter the long-term ground water transport rates and direction or produce short-term changes that are unacceptable.

5.7 *Multiple Lines of Evidence to Demonstrate Appropriateness of Remediation by Natural Attenuation*:

5.7.1 The National Research Council (1993)[8] suggests a strategy to demonstrate in situ bioremediation which includes three types of evidence:

5.7.1.1 Documented loss of constituents of concern from the site;

5.7.1.2 Evidence showing bioremediation is actually realized in the field; and

5.7.1.3 Laboratory assays showing that microorganisms in site samples have the potential to transform constituents of concern.

5.7.2 This guide suggests the demonstration of remediation by natural attenuation may include primary, secondary, and optional lines of evidence. At a minimum, primary lines of evidence are required to demonstrate the effectiveness remediation by natural attenuation. The decision to collect secondary and optional lines of evidence should be based on the intended use of the data. The cost benefit of obtaining these lines of evidence should also be considered. The primary lines of evidence include constituent of concern data, used to define the plume as shrinking, stable, or expanding, similar to the first line of evidence suggested by NRC (1993).[8] For sites which have sufficient historical monitoring data, the primary lines of evidence will often be adequate to demonstrate remediation by natural attenuation.

5.7.3 Secondary lines of evidence include geochemical indicators of naturally occurring biodegradation and estimates of natural attenuation rate. If the primary lines of evidence are inconclusive, it may be necessary to obtain secondary lines of evidence. For those sites where assessment efforts have recently been initiated, it may be appropriate to supplement the primary lines of evidence by measuring indicators of naturally occurring biodegradation, consistent with the second line of evidence suggested by NRC (1993).[8] Estimates of attenuation rate are based on temporal and/or spatial trends for constituents of concern. Once this secondary line of evidence has been established, the user must continue to monitor and collect data to substantiate the primary line of evidence.

5.7.4 Optional lines of evidence may be used to more rigorously interpret data developed as secondary lines of evidence, particularly if the primary and secondary lines of evidence are inconclusive to demonstrate remediation by natural attenuation. Optional lines of evidence include solute transport modeling, estimates of assimilative capacity (to estimate the mass of BTEX and other constituents of concern degraded), and microbiological studies. Attenuation rates can be used in modeling transport of constituents of concern. Indicators of naturally occurring biodegradation can be used to estimate assimilative capacity. Microbiological studies, as suggested in the third line of evidence by NRC (1993),[8] confirm the presence of microorganisms in the subsurface. Once optional lines of evidence have been established, the user must continue to monitor and collect data to substantiate the primary line of evidence.

6. Decision Process for Appropriate Application and Implementation of Remediation by Natural Attenuation

6.1 The key components of the remediation by natural attenuation process are described in the following sections. The major decisions and actions required to determine the appropriateness of applying and implementing remediation by natural attenuation at a given site been are summarized in the flowchart shown in Fig. 1.

6.1.1 Site characterization and establishment of remedial goals;

6.1.2 Evaluation of plume status;

6.1.3 Comparing RNA performance to remedial goals;

6.1.4 Comparing RNA to other remedial options; and

6.1.5 Development and implementation of an appropriate monitoring program.

6.2 *Initial Response, Site Characterization, Determine Constituents of Concern, and Establish Remediation Goals*:

6.2.1 Initial response should be taken in accordance with implementing agency requirements to report any release of petroleum products; prevent any further release of, or exposure to hydrocarbons in vapor, dissolved, or liquid phase; and mitigate fire and safety hazards. Table 1 in Guide E 1739 provides example site classification and initial response actions.

6.2.2 The site characterization must provide the user with adequate information necessary to determine if remediation by natural attenuation is a viable remedial option for the site, either used by itself or in conjunction with other technologies. Site characterizations may be conducted in accordance with Section 7 of Guide E 1599, and Guide E 1912 taking into consideration evaluation of sources, pathways, and receptors as discussed in 6.2 of Guide E 1739. The types of site characterization information that may be necessary for remediation by natural attenuation are detailed in Appendix X2. Not all the data listed in Appendix X2 may be needed for each site and considerations for when and how this data can and should be used is explained in 6.3 and 6.4.

6.2.2.1 As part of the site characterization process an initial conceptual model should be developed before beginning any field work. The conceptual model should focus on specific features that are relevant to the assessment objectives. For example, the features of a conceptual model of a leaking underground storage tank site may include preliminary estimates of: (*1*) source areas; (*2*) three dimensional distribution of constituents of concern; (*3*) distribution of constituents of concern and impacts to ground water; (*4*) geologic units or structures that influence migration of constituents of concern; (*5*) ground water depth, flow direction and velocity; and (*6*) location of potential receptors and migration pathways. Hydrogeologic, and analytical data collected during the field investigation should be periodically interpreted and used to refine the conceptual model in an iterative process. The components of the conceptual model that are emphasized depends on the purpose of the assessment (See Guide E 1689, and Guide E 1912 5.4 and 5.7).

6.2.3 The determination of constituents of concern is based on the site specific consideration of exposure routes, concentrations, mobilities, toxicological properties, and aesthetic characteristics (taste, odor, etc.). In addition, regulatory requirements may dictate certain constituents of concern. Appendix X1 in Guide E 1739 contains additional discussion regarding determination of constituents of concern.

6.2.4 Remedial goals for the site should be determined by applying the risk-based corrective action process in Guide E 1739 or other accepted state-approved method. Remedial goals may take the form of concentration target levels or performance criteria, including demonstration of containment of the petroleum hydrocarbon plume. Remedial goals may also have some time frame associated with them. An evaluation of the need for source area control measures should be integrated into remedial decision-making at all sites where natural attenuation is under consideration. Source area control measures include physical removal, treatment, and stabilization.

6.2.4.1 Remedial goals may be concentration-based target levels applied at specific points throughout the plume or performance-based criteria, such as demonstrated containment of the solute plume or demonstrated reduction in concentrations of constituents of concern over time within the plume or with distance from the source area. Both must be protective of human health and the environment. In general, remediation by natural attenuation is more amenable to achieving performance-based remedial goals. Also, remediation by natural attenuation performance can provide verification of natural attenuation rates used to determine risk-based target cleanup levels developed through predictive solute transport modeling. When using remediation by natural attenuation as a containment option, institutional controls may be required to manage and prevent on- and off-site exposures.

6.2.5 Once remedial goals have been established, site conditions should be examined to see if these goals have already been met. If remedial goals have already been met at the site, the site may be deemed to require no further action. In some cases continued monitoring may be needed to confirm compliance with remedial goals prior to a determination of no further

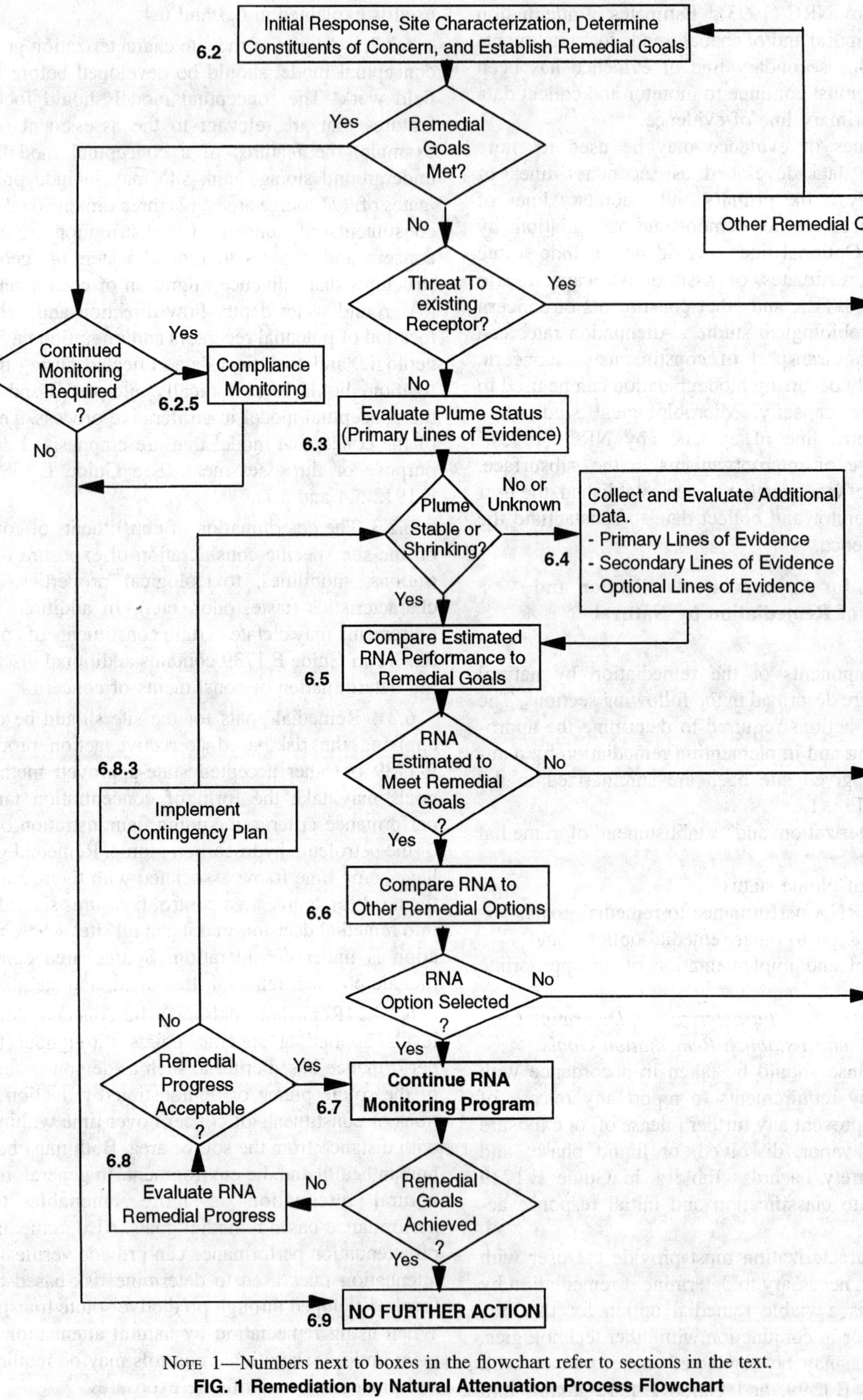

NOTE 1—Numbers next to boxes in the flowchart refer to sections in the text.

FIG. 1 Remediation by Natural Attenuation Process Flowchart

action. If remedial goals have not been met at the site, then additional remedial action will be required.

6.2.6 The potential for impacts to human health and the environment must be determined by conducting surveys of primary and secondary sources, transport mechanisms, viable exposure pathways and potential receptors. Guide E 1739 provides a standardized approach to this type of analysis.

6.2.6.1 If the potential exists for immediate impacts to an identified receptor (for example, see Guide E 1739 Table 1), then other remedial actions or risk-management strategies may be required at the site. If risk-management strategies are not sufficient to prevent impacts to an identified receptor, then remediation by natural attenuation is inappropriate as a stand-alone option.

6.2.6.2 If the potential for a near-term impact to an existing receptor is determined to be low, then remediation by natural attenuation may be used as a stand-alone option for meeting remedial goals within the ground water.

6.3 *Evaluate Plume Status (Primary Lines of Evidence)*:

6.3.1 The dissolved petroleum constituent plume is categorized based on historical constituent of concern concentrations obtained from monitoring points. These historical data are the primary line of evidence for natural attenuation and are based on measured petroleum hydrocarbon concentrations over time to define the plume as shrinking, stable, or expanding. Evidence of reductions of constituents of concern is also the first line of evidence suggested by NRC (1993).[8] The implications of the three plume categories are as follows:

6.3.1.1 A shrinking plume is evidence of natural attenuation. The natural attenuation rate of a shrinking plume necessarily exceeds the mass loading rate of constituents of concern to ground water.

6.3.1.2 A stable plume is evidence of natural attenuation. The source of constituents of concern may persist in soils at the water table but the natural attenuation rate approximately equals the mass loading rate for constituents of concern to ground water.

6.3.1.3 In the case of an expanding plume the mass loading rate of constituents of concern to ground water exceeds the natural attenuation rate. An expanding plume will become stable when the mass loading rate of constituents of concern to ground water is balanced by the natural attenuation rate.

6.3.2 For sites which have sufficient historical monitoring data, the primary lines of evidence will often be adequate to demonstrate remediation by natural attenuation. For sites which have insufficient historical monitoring data, collection and evaluation of geochemical data may be appropriate to expedite the demonstration of remediation by natural attenuation. Paragraph 6.7 and Appendix X3 describe monitoring considerations.

6.3.3 The evaluation of plume status can be accomplished by either of the following methods, which are described in detail in Appendix X3.2.1 and Appendix X5. The effects of historical source removal and remediation efforts should be incorporated into the evaluation of plume status.

6.3.3.1 Monitoring points or other sampling devices should be located to allow the construction of contour maps for BTEX and other constituents of concern concentrations. Ideally, the map will include a non-detect or compliance level contour. Based on changes (or lack of changes) over time, the plume can be characterized as shrinking, stable, or expanding. The example problem in X7.1 illustrates this method.

6.3.3.2 Concentrations of BTEX and other constituents of concern can be determined over time at appropriately located monitoring points down gradient of the source and oriented along the direction of ground water flow (see 6.7 and Appendix X3 for important considerations regarding placement of monitoring points). The trend in BTEX and other constituents of concern concentrations at these points will determine whether the plume is shrinking, stable, or expanding (for example, if the plume is shrinking, concentrations will decrease over time or space; if the plume is stable, concentrations will remain relatively constant over time and space).

6.4 *Collect and Evaluate Additional Data*:

6.4.1 It may be necessary to obtain additional monitoring data before a plume can be defined as stable or shrinking. In the case of a newly discovered petroleum release site, the historical monitoring data necessary to evaluate plume status discussed in 6.3 will not be available. Therefore, one of the methods described in 6.3.3 may be used following additional monitoring events. For newly discovered sites, collection and evaluation of geochemical data may be appropriate to expedite the demonstration of remediation by natural attenuation.

6.4.2 Secondary lines of evidence may be required if the primary line of evidence, the evaluation of plume status, is inadequate or inconclusive to demonstrate remediation by natural attenuation. This may be the case for sites where only one or two monitoring events have been performed.

6.4.3 One secondary line of evidence is to estimate the natural attenuation rate. This estimate is based on the same data used in the evaluation of plume status (see 6.3). Another secondary line of evidence includes geochemical data which serve as indicators of naturally occurring biodegradation. Geochemical parameters are measured in ground water samples.

6.4.3.1 The estimate of attenuation rate can be performed by several methods. A mass balance approach is described in X6.1. The technique includes a calculation for the constituent of concern source rate (mass loading to ground water). This method yields an estimate for attenuation rate depending on whether the plume is shrinking, stable, or expanding.

6.4.3.2 Appendix X6.2 presents graphical and regression techniques to estimate the attenuation rate. These techniques include plots of (*1*) concentration versus time for individual monitoring points and (*2*) concentration versus distance for three or more monitoring points approximately oriented with ground water flow direction. Attenuation rates can be estimated by regression of concentration versus time or distance, or both. By plotting the log of concentration versus time or distance as a straight line (semi-log paper), the assumption of first-order decay can be demonstrated. The attenuation rate is graphically determined by the slope of the straight line. These calculations are described in X6.2. An example problem for concentration versus distance appears in X7.2.

6.4.3.3 Indicators of naturally occurring biodegradation are useful because biological transformation of petroleum hydrocarbons is the single most important process contributing to natural attenuation of petroleum constituents. Other attenuation processes (dispersion, sorption, dilution, volatilization) also contribute to reductions in concentrations of constituents of concern in ground water to a lesser extent. One line of evidence to demonstrate naturally occurring biodegradation, as suggested by the NRC (1993),[8] includes data which show that predicted biodegradation potential is actually realized in the field. To this end ground water monitoring points can be sampled for geochemical parameters to demonstrate naturally occurring biodegradation potential at field sites. These indicator parameters are summarized in Table 1.

NOTE 2—These are the most common parameters, other methods or parameters may also be useful in certain cases.

6.4.3.4 Temperature, pH, and conductivity are standard measurements for ground water sampling. Dissolved oxygen (DO) concentrations define aerobic versus anaerobic conditions. Oxidation reduction potential identifies oxidizing versus reducing conditions in ground water. Nitrate and sulfate may serve as electron acceptors after DO is consumed. Carbon dioxide, methane, ferrous iron, and manganese are the primary products of aerobic or anaerobic biodegradation of petroleum hydrocarbons. Paragraph X5.3.3 of Appendix X5 describes the significance of these indicator parameters. Sampling considerations and analytical methods for the indicator parameters are provided in Appendix X4.

6.4.4 Additional optional lines of evidence may be useful for the a small percentage of sites where the primary and secondary lines of evidence are inconclusive to demonstrate remediation by natural attenuation. These optional lines of evidence may include solute transport modeling, estimates of assimilative capacity, and microbiological studies.

6.4.4.1 Solute transport models may be used for several purposes. Transient analytical solutions can estimate the time required for a shrinking or expanding plume to reach a particular configuration. Steady-state solutions can be used to estimate the extent of a stable plume and aid in selection of locations of down gradient monitoring points. Appendix X6.3 describes the use of a steady-state solution which is coupled to the regression of concentration versus distance (see X6.2.2), for a stable plume.

6.4.4.2 One, two, and three-dimensional analytical solutions are presented in X6.4. The justification for two or three-dimensional analytical models should be based on the availability of data. Two of the more sensitive input parameters are the decay rate and source term. Site-specific attenuation or decay rates, as determined by one of the Appendix X6 methods, can be used in the analytical solution. A source of constituents of concern can be defined as a constant or decaying term.

TABLE 1 Indicator Parameters for Biodegradation

Dissolved oxygen	oxidation/reduction potential
pH	manganese
Temperature	alkalinity
Conductivity	methane
Nitrate	carbon dioxide
Sulfate	ferrous iron

6.4.4.3 Numerical models are appropriate where site characterization data are available to describe a complex hydrogeologic system. Numerical models require input parameters similar to those used for analytical models, but their spatial distributions must be known to warrant the use of these models **(1)**.[10]

6.4.4.4 The estimate of assimilative capacity uses the indicator parameters for naturally occurring biodegradation, presented 6.4.3.3 and described in X5.3. These indicator parameters can be used to estimate the potential mass of BTEX and other constituents of concern degraded per unit volume by aerobic and anaerobic respiration. The qualitative estimate determines the assimilative capacity of the measured electron acceptors to completely metabolize BTEX and other constituents of concern dissolved in ground water. This approach and its limitations are described in X5.3.2.

6.4.4.5 Microbiological studies are another line of evidence to demonstrate naturally occurring biodegradation. The NRC (1993) suggests the use of laboratory assays showing that microorganisms in site samples have the potential to transform the constituents of concern under the expected site conditions. There are at least two techniques to demonstrate the availability of microorganisms, microbial counts and microcosm studies, described in X5.3.5.2 and X5.3.5.3, respectively, of Appendix X5. Naturally occurring biodegradation of petroleum hydrocarbons is rarely limited by the availability of bacteria. For this reason, microbial counts and microcosm studies are not typically performed at petroleum release sites.

6.5 *Compare Estimated Remediation by Natural Attenuation Performance to Remedial Goals*:

6.5.1 Remediation by natural attenuation performance at a given site can be assessed by the following:

6.5.1.1 Plume behavior and containment due to remediation by natural attenuation;

6.5.1.2 Constituents of concern attenuation rates; and

6.5.1.3 Indicators of favorable biological conditions.

6.5.2 The performance of remediation by natural attenuation is generally acceptable if a plume is shrinking or stable (primary line of evidence) and there are no impacts to receptors. Risk reduction, containment, and performance goals are generally met if a plume is shrinking or stable. Secondary lines of evidence, such as estimates of natural attenuation rate and favorable biological conditions may also be used to demonstrate remediation by natural attenuation performance.

6.5.3 If a plume is expanding but at a rate lower than the ground water velocity, the risk reduction and performance goals may still be met depending on the presence and location of receptors. Further investigation and assessment may be necessary to more accurately predict the potential extent of plume migration and ensure protection of receptors.

6.5.4 A concentration-based goal may be achieved within a certain time frame if a plume is already shrinking. Remediation by natural attenuation is a viable option for achieving concentration-based goals if the concentration is applied at an alternate point of compliance some distance from the source or

[10] The boldface numbers in parentheses refer to the list of references at the end of this standard.

the extent of natural attenuation between the source area and potential receptors is considered in setting concentration-based goals for source remediation. However, remediation by natural attenuation is unlikely to meet concentration-based remedial goals which require relatively low concentrations (for example, 5 ppb benzene) at or near the source of a petroleum release in short time frames. For a stable or expanding plume it is more difficult to estimate the time required to meet concentration-based goals at a given site with confidence. However, where a plume is stable and the primary source (for example, tank) is removed and no additional release adds to the source area, mass loading rate will eventually be reduced. An evaluation of the need for source area control measures should be integrated into remedial decision-making at all sites where natural attenuation is under consideration. Any source removal efforts undertaken should focus on those measures that effectively reduce mass loading rates to ground water.

NOTE 3—Source removal may be governed by technical feasibility as well as federal, state, and local guidelines.

6.6 *Compare Remediation by Natural Attenuation to Other Remedial Options*:

6.6.1 The purpose of this subsection is to describe the key considerations for comparing remediation by natural attenuation to other remedial options. The decision to implement remediation by natural attenuation over other alternatives should consider remedial goals, remedial time frame, risk reduction and exposure prevention, cost effectiveness, technical limitations, regulatory constraints, and land use. Each of these comparison criteria is discussed in more detail in the following paragraphs.

6.6.2 *Remedial Goals and Time Frame*—A major consideration when comparing remedial action alternatives is the probability that individual alternatives will meet the established remedial goals. As discussed in 6.5, if remediation by natural attenuation is likely to meet the remedial goals within the desired time frame, then it is a viable alternative. However, if the probability of remediation by natural attenuation meeting remedial goals within the desired time frame is low or uncertain, then supplementary or alternative remedial action measures may be appropriate (see 5.6.2). The time frame for achieving remedial goals is an important criterion for comparison of remediation by natural attenuation with other remedial options. Remediation by natural attenuation is generally a long-term option. However, care should be exercised in estimating remediation time frames for other remedial options so as to not bias the comparison with overly optimistic representations of cleanup time frames. If conflicts arise, time frame considerations are secondary to the goal of receptor protection.

6.6.3 *Risk Reduction and Exposure Prevention*—As part of a risk-based approach to corrective action, remedial options, including remediation by natural attenuation, should be compared to determine which alternative(s) are required to achieve an acceptable level of risk or exposure prevention. Remediation by natural attenuation should be considered a viable option if it provides the adequate level of risk reduction and exposure prevention. Another consideration may be the relative reduction in risk provided by remediation by natural attenuation versus other options and the expense required for the additional risk reduction provided by other remedial options. Additionally, the risks associated with other corrective action measures remedial technologies, such as direct exposure to impacted soils, releases to the atmosphere, and diversion of limited resources from high risk sites, should be considered.

6.6.4 *Cost Effectiveness*—In order to determine if remediation by natural attenuation is a cost effective remedial option, the costs of remediation by natural attenuation implementation need to be understood. Important costs associated with the implementation of remediation by natural attenuation include long-term monitoring and analytical expenses, costs to collect data and evaluate the lines of evidence supporting remediation by natural attenuation, and the potential costs of implementing institutional controls. In some cases, higher cost alternatives in the short term may be considered due to reduced long-term liability and monitoring costs.

6.6.5 *Regulatory Considerations*—The remediation by natural attenuation option, as with other remediation alternatives, is subject to approval by the regulatory agency which is responsible for the oversight of the cleanup of the petroleum release. Issues of regulatory concern may include requirements associated with the delineation of the plume; the degree to which free product needs be removed from the source area; whether performance-based (vis-a-vis concentration-based) remedial goals are acceptable; whether a time constraint is placed on achievement of the remedial goal; offsite migration; and length of time monitoring may be required.

6.6.5.1 Since each state has its own individual requirements regarding the application of remediation by natural attenuation, the user should consult with the appropriate regulatory agency to determine its current policy.

6.6.6 *Land Use*—Remediation by natural attenuation should be considered a viable option at locations where the reasonable potential land use is well defined and changes in land use which could cause exposure to constituents of concern are unlikely to occur without notice (for example, a retail service station to be operated for the foreseeable future in an area zoned commercial industrial). However, if the current land use is expected to change or is not restricted then reasonable potential future land uses should be considered prior to selecting remediation by natural attenuation as the preferred remedial option. In some areas, institutional controls such as restrictions on installation of water supply wells may need to be implemented to ensure that site uses which could create exposure to constituents of concern do not occur.

6.7 *Continue Monitoring Program for Remediation by Natural Attenuation*:

6.7.1 If the remediation by natural attenuation option is selected, it is necessary to develop and implement a monitoring program that both yields adequate information to evaluate the progress of remediation by natural attenuation in meeting remedial goals and is cost-effective. The cost associated with monitoring may well be the most expensive part of a natural attenuation remediation project. The objectives of the monitoring program are:

6.7.1.1 To evaluate performance and progress of remediation by natural attenuation toward meeting remedial goals; and

6.7.1.2 To ensure that the plume is not migrating to an

extent greater than expected or in unexpected directions.

6.7.2 The monitoring program should include appropriate sampling locations, adequate sampling frequency, and meaningful sampling parameters. Monitoring considerations are discussed in Appendix X3. In some cases, the results of a solute transport model can be useful to aid in determining locations of monitoring points and appropriate sampling frequency.

6.7.3 *Sampling Locations*—The monitoring plan should include sufficient ground water monitoring points, both in number and location, to determine changes in ground water flow directions and velocities, trends in concentrations of constituents of concern within the plume (over time or distance, or both), and any further migration of the plume (Appendix X3).

6.7.3.1 For the evaluation of remediation by natural attenuation performance, monitoring point locations must include as a minimum, an up gradient monitoring point, two or more monitoring points within the plume, but outside any free product zone, and a down gradient monitoring point. An up gradient monitoring point will be required to establish the quality of ground water entering the site, both in terms of regulated constituents of concern and in terms of the secondary line of evidence if needed. A down gradient monitoring point, near the edge of the plume, will be necessary to establish the maximum extent of the plume in the direction of ground water flow. Consideration should be given to ground water flow rate and estimated solute transport velocities when selecting well spacing. In addition, monitoring points can be situated in a manner that will allow the gathering of data to determine plume behavior and remediation by natural attenuation progress, as discussed in 6.3, 6.4, 6.5, and Appendix X3.

NOTE 4—The previous discussion only addresses monitoring requirements directly related to evaluating the lines of evidence for natural attenuation. Other monitoring points and monitoring requirements may be necessary to fully evaluate ground water flow direction and seepage velocity.

6.7.3.2 Sentinel wells are monitoring points established at a location(s) between the leading edge of the solute plume and a sensitive receptor (for example, drinking water well) to ensure that there will be time for other remedial actions to be taken, if the plume does migrate beyond predicted boundaries. Sentinel wells always are required where a real use of ground water is threatened, or where entry into a surface water could occur. Sentinel wells may be appropriate but not required where a plume is suspected of expanding and neither of the two above conditions (existing use and potential entry into surface water) exist. Sentinel wells are optional to unnecessary for plumes where natural attenuation is apparent in the existing ground water monitoring network, where no real use of ground water is threatened, and where no entry into a surface water could occur. An adequate amount of site characterization must occur to document which, if any, of these conditions exists and to make the determination that a sentinel well is or is not appropriate.

6.7.4 *Monitoring Frequency*—Monitoring frequency is a site-specific consideration. The frequency with which long-term monitoring should be conducted based on plume status, water table fluctuations, ground water seepage velocity and the distance to receptors. For example, if the initial monitoring indicates that concentrations of constituents of concern fluctuate significantly over time, such as on a seasonal basis, a higher frequency of (shorter interval between) monitoring events will be necessary in order to establish (resolve) a significant trend. Alternatively, if concentrations of constituents of concern are relatively stable on a seasonal basis, a longer interval between monitoring events may be appropriate.

6.7.4.1 Monitoring frequency should be at least quarterly for a minimum of one year in order to define seasonal fluctuations in concentrations of constituents of concern, water table elevations, and hydraulic gradients. The lack of these data could make it very difficult or impossible to adequately resolve concentration trends in subsequent data sets. Subsequent monitoring should be conducted at a frequency appropriate to detect additional plume migration and changes in concentrations of constituents of concern. The length and frequency of monitoring will need to be determined on a site-specific basis and will depend on the present status of the plume, water-table fluctuations, ground water velocity, monitoring point spacing, and the distance from the plume to any sensitive receptor (see Appendix X3).

6.7.5 *Sampling Parameters*—Sampling parameters will include constituents of concern and may also include geochemical parameters as discussed in 6.5.

6.8 *Evaluate Remediation by Natural Attenuation Remedial Progress*:

6.8.1 Monitoring results should be evaluated to determine progress toward meeting remedial goals. As discussed in 6.5, remedial goals may be different depending on site specific conditions and regulatory requirements. If remedial goals are met, then no further action or a site closure plan may be implemented, as discussed in 6.9.

6.8.2 If remedial goals are not met, remediation by natural attenuation remedial progress needs to be evaluated. The evaluation is to determine the plume status and/or to demonstrate that natural attenuation is continuing to occur. This evaluation can be performed using the methodology described in 6.4 and 6.5. If historical data demonstrates that the solute plume has stabilized or is shrinking, then natural attenuation is occurring. If the solute plume is migrating at a rate significantly lower than expected based on the groundwater velocity, then remediation by natural attenuation is occurring to the extent that assumptions about the geology and groundwater conditions are correct."

6.8.3 If remedial progress matches estimates, remediation by natural attenuation monitoring program shall continue. If remedial progress does not match estimates, remediation by natural attenuation should be re-evaluated as to whether it is an appropriate remediation option for the site. If at any point during the long-term monitoring program, data indicates that natural attenuation is not adequate to achieve remedial goals, the contingency plan should be implemented. This plan could include considerations for changes in remedial approach including additional source removal, containment measures, more rigorous institutional controls, and augmenting remediation by natural attenuation with other remedial actions.

6.9 *No Further Action*:

6.9.1 When it can be demonstrated that target cleanup levels or performance-based criteria for the site have been achieved and further monitoring is no longer required to ensure that conditions persist, then no further action is necessary. Mechanisms or procedures must be implemented to ensure that institutional controls (if any) remain in place. Regulatory concurrence should be pursued on a determination of no further action.

6.9.1.1 If natural attenuation is demonstrated to be effective at a site and site conditions will not change, natural attenuation will continue to serve as an ongoing remedial action whether it is monitored or not.

6.9.2 *Key Criteria for No Further Action*—The key criteria for no further action at a site which has undergone remediation by natural attenuation are as follows:

6.9.2.1 There are no existing or potential receptor impacts (see, for example, Guide E 1739).

6.9.2.2 Remedial goals have been met, or it has been demonstrated that natural attenuation will continue and ultimately meet remedial goals (see 6.2.4).

6.9.2.3 The plume is stable or shrinking.

6.9.2.4 If needed, institutional controls are in place and maintained.

7. Keywords

7.1 attenuation; bioremediation; ground water; intrinsic remediation; natural attenuation; passive remediation; remediation; remedial action

APPENDIXES

(Nonmandatory Information)

X1. WHAT IS REMEDIATION BY NATURAL ATTENUATION?

X1.1 *Introduction*:

X1.1.1 Remediation by natural attenuation is the reduction in concentration, mass or mobility of chemical(s) of concern with distance and time due to naturally occurring processes in the environment. These processes can be classified as physical (such as dispersion, diffusion, dilution by recharge, and volatilization), chemical (sorption and chemical or abiotic reaction), and biological (biodegradation). The physical and chemical sorption processes result in the reduction of concentration and/or mobility of a chemical but not the total mass, and are referred to as "nondestructive" mechanisms. The chemical and biological reactions result in the reduction of the total contaminant mass in the system, and are referred to as "destructive" mechanisms. For petroleum hydrocarbons in the subsurface, biological degradation is often the most important process in the reduction of mass because the hydrocarbons are destroyed rather than phase partitioned.

X1.1.2 This appendix provides an overview of the processes of natural attenuation and their significance in the subsurface. It is divided into the following sections:

X1.1.2.1 Physical Processes,

X1.1.2.2 Chemical Processes, and

X1.1.2.3 Biological Processes.

X1.1.3 Much of the information presented is summarized from the references listed at the end of this appendix.

X1.2 *Physical Processes*—The physical processes of natural attenuation include hydrodynamic dispersion (diffusion and mechanical dispersion), dilution by recharge, and volatilization. These non-destructive mechanisms result in a reduction in the concentration of a chemical, but not the total mass in the system.

X1.2.1 Hydrodynamic dispersion, which includes molecular diffusion and mechanical dispersion, is the process whereby a contaminant plume spreads out in directions that are longitudinal and transverse to the direction of groundwater flow. It is generally the primary process causing dilution of dissolved constituents of concern.

X1.2.1.1 Mechanical dispersion describes the spreading of molecules due to interactions between advective movement of the chemical and the porous structure of the medium. It has two components, longitudinal dispersion which is the spreading of a solute in the direction of the ground water flow, and transverse dispersion which is the spreading in the direction perpendicular to the ground water flow. Longitudinal dispersion occurs because of variations in pore size, tortuosity (flow path length), and friction in the pore. Transverse dispersion is caused by the tortuosity of the porous medium which causes flow paths to branch out from the plume centerline **(2)**. Mechanical dispersion is the dominant mechanism causing hydrodynamic dispersion under normal advective flow systems.

X1.2.1.2 Molecular diffusion is the molecular movement of a chemical in response to concentration gradients, even in the absence of ground water flow. The molecular diffusion of a dissolved chemical in ground water is described by Fick's law. It is an attenuation mechanism that is negligible in normal advective flow systems because it is insignificant relative to mechanical dispersion, but is important in no-flow and very low ground water velocity situations **(3)**.

X1.2.1.3 The overall result of hydrodynamic dispersion is spreading and mixing of the contaminant plume with ground water. In addition to the dilution effect, which reduces the plume concentration, dispersion can facilitate biodegradation of organic compounds by spreading the plume into areas with more electron acceptors.

X1.2.2 Recharge due to precipitation, leakage from waterbodies or leakage from underlying aquifers introduces additional water into the solute plume. This results in dilution of the solute plume. An additional effect on natural attenuation is that supplemental dissolved oxygen can be furnished to the contaminant plume to increase the overall electron acceptor capacity for biodegradation of organic hydrocarbons. A similar phenomenon can be observed with the presence of trapped gas due to water level fluctuations caused by seasonal changes in aquifer recharge and discharge **(4)**.

X1.2.3 Volatilization causes a mass loss of volatile organic compounds from the subsurface to the atmosphere under the appropriate conditions. While not a destructive attenuation mechanism, it does remove contaminants from ground water. It is important for fresh spills of petroleum products, for older but highly volatile petroleum constituents (that is, jet fuel), and for free phase product and high levels of dissolved constituents of concern. As time passes after a discharge, this mechanism becomes less important.

X1.2.3.1 The volatilization process usually contains two steps: (*a*) the chemical from the impacted media (soil or ground water) volatilizes to the soil gas, that is, air in the soil pores, and (*b*) when ground water is shallow, the volatilized chemical in the gas phase moves into the atmosphere above the soil surface. The volatilization rate and the amount of mass loss depend on chemical-specific factors (Henry's law constant and diffusion coefficient) as well as media-specific conditions such as climate, depth to water, sorption, temperature, effective porosity, and soil type **(5, 6)**.

X1.2.3.2 Henry's law governs the tendency of a dissolved chemical to volatilize from ground water into the soil gas and is given by:

$$C_a = HC_l \quad \text{(X1.1)}$$

where:
H = Henry's law constant (dimensionless),
C_a = Concentration in air (g/cm^3-vapor), and
C_l = Concentration in water (g/cm^3-water).

The Henry's law constants of hydrocarbons range over several orders of magnitude **(7)**. Values of Henry's law constants for the BTEX compounds are given in Table X1.1. Volatilization can also affect NAPLs in the subsurface but this process is not described by Henry's law. In the study of Chiang et al. **(8)**, volatilization was calculated to account for no more than 5 % of the total mass loss of benzene. At sites where the water table is shallow or highly fluctuating, volatilization may contribute more to the total mass loss.

X1.3 *Chemical Processes*—The chemical processes that cause natural attenuation include sorption and chemical (abiotic) reactions. Sorption is controlled by chemical forces and interactions in nature. It not only affects the chemical's mobility but also the rates of other transport and transformation reactions. Nonetheless, sorption can also be considered as a physical process because the mechanism itself does not result in irreversible changes of the compound. Abiotic chemical reactions refer to those reactions that do not involve metabolically active microorganisms or extracellular enzymes. While many organic chemicals in soil systems can be transformed through one or more chemical reactions, there is no evidence of abiotic chemical transformation of BTEX in the subsurface environment **(9)**.

X1.3.1 The term sorption is used to describe the interphase interaction of a chemical between the water and the soil. The sorption process retards the migration of mainly hydrophobic chemicals in the aquifer relative to the ground water flow velocity, and reduces the dissolved concentration in ground water, but usually results in higher concentrations in the soil phase. Thus, it is an attenuation mechanism that does not reduce the total mass, but affects the mobility and concentration of the chemical.

X1.3.1.1 Sorption is a complicated phenomenon caused by several chemical forces and interactions between chemicals and the aquifer matrix. For hydrocarbons of nonpolar molecular structure, sorption occurs mainly through the process of hydrophobic bonding **(10)**. Two components of the aquifer matrix have the most effect on sorption: organic matter and clay minerals. In most aquifers, sorption is controlled by organic matter. However, if the aquifer matrix has little organic matter, then sorption to the mineral surface may become important.

X1.3.1.2 The kinetics of sorption are also complicated as the result of the nonlinear and nonequilibrium behavior of specific compounds. A sorption isotherm describes the relationship between the concentration of chemical sorbed onto the soil and the concentration remaining in solution at equilibrium. The commonly used isotherms include the Langmuir isotherm, the Freundlich isotherm, and the linear isotherm (a special case of the Freundlich isotherm) **(11)**. The sorptive behavior of multi-compound systems are usually difficult to predict from those of individual compounds **(12)**.

X1.3.1.3 The most commonly used method of representing the distribution of an organic chemical between the soil and the ground water is the distribution coefficient (also called the partition coefficient), K_d:

$$K_d = C_a/C_l \quad \text{(X1.2)}$$

where:
K_d = distribution coefficient (cm^3/g),
C_a = sorbed concentration (g/g-soil), and
C_l = dissolved concentration (g/cm^3-solution).

The distribution coefficient is the slope of the sorption isotherm. The greater the distribution coefficient, the greater the sorption of the chemical. For systems described by a linear isotherm, the distribution coefficient is a constant. The distribution coefficient of an organic chemical also depends on the properties of the soil in the system. In general, the distribution coefficient increases as the fraction of organic carbon (f_{oc}) increases in the soil **(13)**. The linear relationship is likely to fail when the soil contains a low organic carbon fraction (less than

TABLE X1.1 Henry's Law Constants and Organic Carbon Partition Coefficients for Common Fuel Hydrocarbon Compounds[A]

Chemical	Henry's Constant, H (dimensionless)	Organic Carbon Partition Coefficient, log (K_{oc}), L/kg
Benzene	0.22	1.58
Toluene	0.26	2.13
Ethyl Benzene	0.32	1.98
Xylenes (mixed)	0.29	2.38

[A] Guide E 1739 (Standard for Risk-Based Corrective Action (RBCA) at Petroleum Release Sites).

0.001) or large amounts of mineral clays, or when the organic compound is polar **(11)**.

X1.3.1.4 Two approaches are generally used to quantify the distribution coefficient for a given system **(11)**. The first method involves conducting batch or column sorption tests to determine K_d. The second method involves estimating K_d by using the organic carbon partition coefficient (K_{oc}) and the f_{oc} of the soil, where $K_d = (K_{oc})(f_{oc})$. A correlation has been developed to estimate K_{oc} based on readily available chemical properties such as water solubility or the octanol-water partition coefficient. Typical values of K_{oc} used for BTEX compounds are given in Table X1.1. Applicability of the second method has the same limitations as the linear relationship between K_d and f_{oc} discussed earlier. For systems where inorganic matter dominates the sorption, the surface area of the soil particles becomes important in determining K_d **(14)**.

X1.3.1.5 The retardation coefficient *R*, which is determined from the distribution coefficient, is often used to describe the attenuation of the plume due to sorption:

$$R = 1 + \rho_b K_d / n \qquad (X1.3)$$

where:

ρ_b = bulk density (g/cm^3), and
n = total porosity (cm^3/cm^3).

The retarded transport velocity of a sorbing chemical, v_c, is then given by:

$$v_c = v_x / R \qquad (X1.4)$$

where:

v_x = ground water seepage velocity, given by $v_x = ki/n_a$ (k = hydraulic conductivity, i = hydraulic gradient, and n_a = effective porosity).

X1.3.1.6 In addition to the retardation effect on natural attenuation of a chemical, sorption can also influence the relative importance of other attenuation processes. For example, the rates of volatilization, chemical reactions, and biodegradation of many organic chemicals are directly dependent upon the extent of adsorption **(7)**.

X1.3.2 The terms chemical and abiotic reactions in this text refer to non-biological transformation and degradation. Different from all the non-destructive attenuation mechanisms discussed above, they are classified as destructive. The results of an abiotic reaction may produce a compound of more or less environmental concern and of enhanced or decreased biological degradability. For petroleum hydrocarbons such as BTEX, the abiotic reactions are not important in the subsurface environment **(15)**.

X1.3.2.1 For organic compounds, the following chemical reactions are known to occur in the subsurface environment **(5,11)**: *(a)* hydrolysis/substitution occurs when an organic chemical reacts with either water or a hydroxide ion to produce an alcohol (hydrolysis) or with another anion as the nucleophilic agent (substitution); *(b)* elimination is the loss of two adjacent groups within a molecule and results in the formation of a double bond; *(c)* oxidation/reduction involves the transfer of electrons from one compound to another. However, in natural subsurface environments, biological systems tend to be predominant in these redox reactions **(16)**.

X1.4 *Biological Processes*—Like the abiotic chemical reactions, biodegradation is a destructive attenuation mechanism. Many studies have demonstrated that biodegradation by indigenous microbes can contribute significantly to the destruction of organic compounds (see Ref **(1)** for a list of references). For petroleum hydrocarbons, biodegradation is considered to be the primary mechanism by which the solute mass is removed. A brief description of the biological process is given here, Appendix X5 and Ref **(1)** provide further discussion of the metabolism of hydrocarbons involving the various electron acceptors.

X1.4.1 Biodegradation is an electron transfer process. The energy required for cell synthesis and maintenance is obtained through the oxidation of hydrocarbon compounds. Electrons are removed from organic substrates (electron donors) to capture the energy available through the oxidation process. Terminal electron acceptors are needed for the electrons moving through electron transfer or respiratory chains.

X1.4.2 Aerobic respiration is the most facile of the terminal electron accepting reactions used for biodegradation of petroleum hydrocarbons. This degradation process takes place in the presence of aerobic microbes. The ultimate products from the mineralization of organic compounds are carbon dioxide, water and cell mass. Facultative anaerobic organisms can use oxygen when it is present or can switch to alternative electron acceptors

X1.4.3 Other electron acceptors including nitrate, sulfate, ferric ion, manganese, and so forth, will be used when oxygen is not available. Obligate anaerobic organisms become dominant in the absence of oxygen. The rate of degradation may be relatively slow under anaerobic conditions. The metabolic products of anaerobic biodegradation include simple organic acids, CO_2, H_2O, CH_4, H_2, N_2, and cell mass.

X2. SITE CHARACTERIZATION FOR EVALUATION OF NATURAL ATTENUATION

X2.1 The purpose of this appendix is to discuss the site characterization data typically required to evaluate natural attenuation of constituents of concern in ground water. This appendix does not cover all site characterization information which may be required in a site assessment, but does include most of the site characterization data which are used directly in evaluating natural attenuation. Table X2.1 includes a brief description of each piece of site characterization information, the use of the data in evaluating natural attenuation, and a reference containing detailed methods for data collection.

TABLE X2.1 Site Characterization Data for Natural Attenuation at Petroleum Release Sites

NOTE 1—This table describes the typical site characterization information recommended to evaluate natural attenuation. Only the characterization information which is considered directly relevant to most evaluations of natural attenuation is discussed. Additional information may be necessary based on site specific considerations or some of the listed items may not be necessary.

Site Characterization Data	Application	Sources/Methods/Protocols[A,B]
Type of petroleum hydrocarbon released. gasoline, diesel, waste oil, etc.	The type of petroleum hydrocarbon released is needed to evaluate the amount of specific contaminants present, effective solubilities, and long term behavior of the source.	Directly from knowledge of type released. Analytical Methods. Gasoline, jet fuel, kerosene in soil or ground water - EPA 5030/8015. Diesel in soil - EPA 3550/8015. Diesel in ground water - EPA 3550/8015. It is recommended that results of these analyses be reported by carbon number range and chromatograms be examined to identify type of petroleum hydrocarbon. Waste Oil - EPA 418.1.
Date of release and date discovered.	Needed for calculation of expected plume travel distance and evaluation of time required to reach stability.	Inventory records. Release reports. Leak detection monitoring data.
Volume released.	Estimation of mass of each contaminant present and rate of decrease in source concentrations	Inventory records, size of tank. Estimates based on free product thickness in monitoring wells should be considered to have a high degree of uncertainty.
Regional Hydrogeology	Identify drinking water aquifers and regional confining units, and regional flow patterns.	State and Federal government agencies and technical publications.
Definition of site lithology and stratigraphy.	Assess preferential directions of ground water flow and contaminant transport. Guide proper location of monitoring points. Determine appropriate screened intervals for monitoring points. Identify hydrogeologic features which may prevent impact to potential receptors.	Guide D 5434 Field Logging of Subsurface Explorations of Soil and Rock
Survey of nearby water supply wells.	Identify locations of nearest potential ground water receptors.	Local health department and water agency records.
Locations of potential receptors	Identify potential exposure pathways of concern.	Guide E 1739 Risk Based Corrective Action Applied at Petroleum Release Sites
Identify high conductivity pathways. Natural or Anthropogenic.	Determine if natural geologic conditions or utility trenches, ditches, or subsurface utilities may impact groundwater flow direction and velocity. Determine if immediate action is necessary. Properly locate monitoring points.	As-built facility drawings from current or previous property owner. City or locality records on subsurface utilities.
Depth to ground water.	Evaluate impact of unsaturated soil contamination on ground water. Evaluate volatilization of contaminants from ground water.	Test Method D 4750 Determining Subsurface Liquid Levels in a Borehole or Monitoring Well(Observation Well)
Water Level Elevations	Determine direction of ground water flow. Determine horizontal and vertical hydraulic gradients to enable calculation of ground water flow velocities.	Test Method D 4750 Determining Subsurface Liquid Levels in a Borehole or Monitoring Well(Observation Well)
Direction of ground water flow.	Evaluate the predominant direction of flow and any seasonal variations. Proper placement of monitoring points. Identification of potential receptors.	
Range of seasonal water level fluctuations.	Estimate source smear zone. Estimation of mixing zone for leaching. Variability of contaminant concentrations.	Periodic monitoring of water levels.
Hydraulic conductivity	Estimate ground water velocity and rate of transport of contaminants.	Pump tests or slug tests (D 4043-91, ASTM Test Methods D 5270-92, D 4105-91, D 5473-93, D 4104-91, D 4106-91, D 5269-92, D 4044-91, D 4050-91)
Delineation of dissolved phase contaminant plume.	Evaluate transport of contaminants and estimate attenuation rates.	ASTM D 4448 Guide for Sampling Ground water Monitoring Wells ASTM D 5092 Practice for Design and Installation of Ground Water Monitoring Wells in Aquifers.

TABLE X2.1 *Continued*

Site Characterization Data	Application	Sources/Methods/Protocols[A,B]
Potential for vertical migration	Vertical gradients and possible vertical contaminant migration should be assessed if regional hydrogeology or site specific information indicate a reasonable potential for vertical migration.	Multilevel piezometers and sampling points with relatively short screened intervals are recommended for evaluating vertical gradients and extent.
Porosity	Calculate ground water seepage velocity and retardation factors.	Typically determined from the literature based on soil type(s) or calculated based on bulk soil density measurements.
Aquifer thickness	Identify confining layers which prevent impact to deeper aquifers. Proper design of monitoring points.	Soil borings, cone penetrometer, regional hydrogeology.
Collect biological indicator data. Terminal electron acceptors. Degradation byproducts. Environmental conditions.	Described in detail in Appendix X4 and Appendix X5.	Described in detail in Appendix X4 and Appendix X5.
Extent of unsaturated soil impact. Vertical and lateral extent. Residual soil content.	Evaluate expected long term behavior of source.	ASTM D 1452 Practice for Soil Investigation and Sampling by Auger Borings. ASTM D 1586 Test Method for Penetration Test and Split-Barrel Sampling of Soils. ASTM D 4700 Guide for Soil Sampling from the Vadose Zone.
Extent of residual separation phase and sorbed phase hydrocarbon. Soil Only. Soil and Ground Water. (includes both mobile and immobile hydrocarbon)	The extent of residual hydrocarbon impact is needed for evaluation of soluble plume attenuation. Wells at the downgradient edge of the source should be used to represent source concentrations, and distances to other wells should be determined from the downgradient edge of the source. The extent of the source area data on the presence of free product are necessary to evaluate free product transport.	ASTM D 1452 Practice for Soil Investigation and Sampling by Auger Borings. ASTM D 1586 Test Method for Penetration Test and Split-Barrel Sampling of Soils. ASTM D 4700 Guide for Soil Sampling from the Vadose Zone.
Chemical analysis of separate phase hydrocarbon for BTEX.	Estimate source equilibrium concentrations. Evaluate long term leaching behavior of source.	For gasoline, jet fuel, kerosene or diesel use EPA 3580 for dilution/extraction and then analyze for BTEX by EPA 602, 8020, 8240, or 8260.
Locations of ground water recharge and discharge areas.	Identify areas where ground water aeration occurs or where water containing electron acceptors (particularly oxygen) enters the subsurface.	Visual observation of site.
Organic carbon content of saturated zone soils.	Determine retardation factor for components of soluble plume to estimate rates of transport.	Modified Mebius Procedure (heated rapid potassium dichromate oxidation) (Page 1982).

[A] EPA Methods - Test Methods for Evaluating Solid Waste, 1992. USEPA, Office of Solid Waste and Emergency Response, Washington D.C., SW-846.

[B] "Methods of Soil Analysis, Part 2, Chemical and Microbiological Properties, Second Edition," A.L. Page, editor, Agronomy, No. 9 (Part 2), American Society of Agronomy, Inc./Soil Science of America, Inc., 1982. (ISBN 0-89118-072-9 (pt. 2)) pp. 571-573.

X2.2 Table X2.1 lists the site characterization data which can be used as the primary lines or secondary lines of evidence. In characterizing a site for the primary lines of evidence, it is necessary to understand the site lithology, the direction and velocity of ground water flow, the type of hydrocarbons released, the extent of the soluble contaminant plume, and the locations of potential receptors. In the case of a newly discovered petroleum release site, the historical monitoring data necessary to evaluate plume status will not be available. For newly discovered sites, collection and evaluation of geochemical data in addition to the primary lines of evidence may be appropriate to expedite the demonstration of remediation by natural attenuation.

NOTE X2.1—This appendix is not intended to be an exhaustive list of methods and procedures. Other methods which provide equivalent data may be used as well.

X3. MONITORING CONSIDERATIONS

X3.1 *Introduction*:

X3.1.1 In any natural attenuation program, the placement of monitoring wells and the development of a monitoring protocol will be site specific.

NOTE X3.1—The following discussion only addresses monitoring requirements directly related to evaluating the lines of evidence for natural attenuation. Other wells and monitoring requirements may be necessary to fully evaluate groundwater flow direction and seepage velocity. In addition, other methods and requirements may be necessary in fractured rock systems.

X3.1.2 For the evaluation of remediation by natural attenuation performance, monitoring point locations must include as a minimum, an up gradient monitoring point, two or more monitoring points within the plume, but outside any free product zone, and a down gradient monitoring point. Other monitoring points and monitoring requirements may be necessary to fully evaluate ground water flow direction and seepage velocity.

X3.2 *Monitoring Well Placement*—An upgradient well will be required to establish the quality of groundwater entering the site, both in terms of regulated contaminants and in terms of the concentrations of terminal electron acceptors and their reduction products. A downgradient well, outside the plume, will be necessary to establish the maximum extent of the plume in the direction of groundwater flow. Consideration should be given to ground water flow rate and estimated solute transport velocities when selecting well spacing. In addition, monitoring wells should be situated in a manner that will allow the gathering of data to determine one or more of the following:

Plume type (stable, shrinking, expanding)
Natural attenuation rates
Warning of impact on sensitive receptor(s)

X3.2.1 *Characterization of Plume Type*—The primary line of evidence for natural attenuation is to define the plume as stable, shrinking or expanding. Wells must be positioned in a manner which will allow plume characterization. This can be accomplished in one of the two following ways: (*1*) wells should be positioned in a manner that will allow the drawing of BTEX contour maps. Ideally, the map will include a nondetect or compliance level (for example, drinking water criterion) contour. Based on changes (or lack of changes) in the contours over time, the plume can be characterized as expanding, shrinking or stable. (*2*) Alternatively, the BTEX concentration(s) in two or more wells located within the plume and downgradient of the source and oriented along the direction of groundwater flow can be monitored over time. The wells must contain measurable BTEX levels. The trend in BTEX concentrations will determine whether the plume is stable, shrinking or expanding (for example, if the plume is shrinking, concentrations will decrease over time or space; if the plume is stable, concentrations will remain relatively constant over time and space). The positioning of new wells may be based on data from previously installed wells, soil gas surveys, Geoprobe-type investigations or other approaches discussed in Appendix X2. Paragraph X3.2.4 covers other issues related to well placement and groundwater monitoring.

X3.2.1.1 Using secondary lines of evidence, terminal electron acceptors and their reduction products are monitored. This information may be presented by drawing terminal electron acceptor/reduction product contours alongside BTEX contours. As an alternative, electron acceptor/reduction product and BTEX concentrations versus distance from the source area may be plotted. Either approach will yield the information necessary to evaluate the relationship between BTEX and electron acceptor/reduction product concentrations.

X3.2.2 *Determining Natural Attenuation Rates*—Secondary lines of evidence includes evaluating natural attenuation rates. Several methods exist for measuring natural attenuation or biodegradation rates, or both, at a site, including the mass-balance (for any plume type), the concentration-versus-time (for a shrinking plume) and concentration-versus-distance (for a stable plume) approaches. These methods are discussed in detail in Appendix X6. Generally, the positioning of new wells may be based on data from previously installed wells, soil gas surveys, Geoprobe-type investigations or other approaches discussed in Appendix X2. Considerations specific to positioning wells for optimizing rate measurements are as follows:

X3.2.2.1 *Mass-balance Approach*—Well positioning is not a critical issue for the mass-balance approach. This approach will require: (*1*) a reasonable estimate of the source area perpendicular to groundwater flow, which can be determined through site investigation techniques discussed in Appendix X2 and (*2*) a reasonable estimate of site hydraulic conductivity and gradient.

X3.2.2.2 *Concentration-Versus-Time Approach*—If the plume is shrinking and a concentration-versus-time approach is to be used, one or more wells containing measurable BTEX levels will need to be positioned outside any free-product zone. It will be best to evaluate the change in BTEX concentration with time in more than one well.

X3.2.2.3 *Concentration-Versus-Distance Approach*—In a situation where the plume is stable and a concentration-versus-distance approach is used, it is appropriate to locate three wells (although two wells can be used) downgradient of the source and oriented along the direction of groundwater flow. Well placement will be satisfactory if at least two wells have measurable BTEX concentrations (the third, if used, can be nondetect) and each well differs in BTEX concentration by several fold from its upgradient neighbor.

X3.2.3 *Warning of Impact on Sensitive Receptor(s)*—One or more sentinel wells should be established at a compliance point(s) between the impacted groundwater and a sensitive receptor (for example, drinking water well). Such wells will call for longer-term monitoring and will ensure that there will be time for other remediation steps to be taken, if the plume does spread beyond predicted boundaries.

X3.2.4 *Other Considerations for Well Placement*—At any site, the need for additional wells can be evaluated based on a number of factors. Additional wells may be required to demonstrate that no "holes" exist in the monitoring network that would allow contamination to escape undetected. This issue is of particular concern in heterogeneous hydrogeological environments or fractured systems, where contaminants could migrate along discreet (but unmonitored) high-permeability pathways. No rules of thumb exist for the number of wells required; common sense and experience are the best guides. Additional wells may also be required to determine the depth (that is, thickness) of the plume, or the thickness, or both, of the upper-most water-bearing zone. At a particular site, the bottom of the upper-most water-bearing zone may not be defined by site data, and deeper portions of the zone may not be monitored. In this case, the three-dimensional extent of the plume is not characterized, and at least one deeper well may be required to demonstrate that the "bottom" of the plume is defined and/or geologically contained. Finally, it is important to note that care must be taken such that groundwater monitoring wells do not puncture an aquiclude (that is, confining layers between a contaminated and uncontaminated aquifer).

X3.3 *Monitoring Frequency*—The length and frequency of monitoring will need to be determined on a site-specific basis and will depend on the present status of the plume, water-table fluctuations, groundwater velocity, solute transport velocities and the distance from the plume to any sensitive receptor. As a general rule, though, enough information can be gathered in one year of quarterly monitoring to establish the relationship between BTEX and terminal electron acceptor (and their reduction products) concentrations. More than one year of monitoring may be necessary to establish whether the plume is stable, shrinking or expanding.

X3.4 *Monitoring Natural Attenuation at Pre-existing Sites*—It should be noted that at sites with historical contamination, large numbers of wells may have been already installed and several years of monitoring data may exist. In such cases, further installation of wells will not likely be necessary if the extent of the constituents of concern is defined. In addition, because of the already existing data base, monitoring frequency can often be substantially reduced compared to that required at a new site.

X3.5 *Well Installation Procedures*—Appropriate monitoring well installation procedures, as well as procedures for determining hydraulic conductivity and gradient, are covered in various ASTM documents and in the Air Force Technical Protocol for Implementing Intrinsic Remediation with Long-Term Monitoring For Natural Attenuation of Fuel Contamination Dissolved in Groundwater, as well as in State-specific guidelines.

X4. SAMPLING CONSIDERATIONS ND ANALYTICAL METHODS FOR NATURAL ATTENUATION INDICATORS

X4.1 Table X4.1 serves as a guide for selection of natural attenuation indicator parameters to be evaluated, the appropriate analytical methods, special sampling considerations, and the potential use of the data. This table is intended to provide guidance only, site specific considerations should be used when selecting the relevant natural attenuation indicator parameters.

NOTE X4.1—Not all of the parameters listed may be required.

TABLE X4.1 Sampling Parameters, Methods, and Special Considerations for Evaluating Remediation by Natural Attenuation in Groundwater for Petroleum Release Sites

Primary Lines of Evidence (Contaminant plume behavior):
The most reliable and direct demonstration of remediation by natural attenuation is based on the actual historical behavior of the contaminant plume. If historical data demonstrate that a plume has stabilized or is shrinking, then significant remediation by natural attenuation is occurring. If monitoring data indicate a plume is migrating or historical data are not available, then it is recommended that the secondary lines of evidence be evaluated in addition to the primary lines of evidence.

Parameter	Field or Laboratory Method	Analytical Method(s) (or equivalent method)	Comments	Use of Data
Benzene Toluene Ethylbenzene Xylenes Methyl Tertiary Butyl Ether (MTBE)	Laboratory	Purge and Trap GC Soil EPA 5030/8020 EPA 8240/8260 Ground water EPA 602 EPA 5030/8020 EPA 5030/8240 EPA 8260	Standard method - collect samples in 40 ml VOA vials, preserve on ice and acidify with HCl to pH = 2. Sampling frequency should be based on site specific conditions. During first one to two years of monitoring quarterly to semiannual sampling is recommended to understand seasonal variability. Frequency of long term monitoring should depend on plume stability and potential travel time to receptors.	For the primary lines of evidence, historical contaminant plume data is used directly to determine if a plume is expanding, stable, or shrinking. In some cases, historical data showing plume stability and no impact to potential receptors will be sufficient to support application of remediation by natural attenuation.
Conductivity	Field	EPA method 120.1 or SW-9050 direct-reading meter ASTM D 1125[4]	Analyze in flow through cell or collect 100 to 250 mL of water and analyze immediately.	Conductivity can be used as an indicator that samples taken from separate sampling points are from the same hydrogeological zone.
Other potential parmaters/contaminants of interest.				

TABLE X4.1 *Continued*

Secondary Lines of Evidence (Contaminant plume behavior and indicators of biodegradation)

These parameters are collected as apparent indicators of biodegradation. In some cases, sufficient historical contaminant plume data may exist to evaluate remediation by natural attenuation and no further information is necessary. However in some cases it may be necessary or beneficial to better understand the potential for intrinsic biodegradation at a site. This may be the case for new releases with no historical plume data or sites with insufficient historical plume. These parameters may be used as indicators of potential aerobic and anaerobic biodegradation. It is recommended that sampling and analysis of these parameters be conducted in the initial site characterization and investigation. This entire list of parameters is not intended to be mandatory as knowledge of the geologic setting and site specific considerations may be used to select the parameters which are likely to provide important information. After the initial sampling event, only those parameters which may be significant should be included in future sampling events as needed. The cost and benefits of collecting secondary indicator data should be weighed when determining what parameters to collect in future monitoring events and in determining the frequency of sample collection.

Parameter	Field or Laboratory Method	Analytical Method(s)	Comments	Use of Data
pH	Field	EPA Method 150.1 or SW-9040 direct reading meter ASTM D 1293[A]	Can be analyzed in flow through cell or collect 100-250 mL of water in glass or plastic container and analyze immediately. Calibration should be conducted using manufacturers standard solutions.	Difference in pH between contaminated and uncontaminated groundwater may indicate biological activity is occurring.
Temperature	Field	EPA Method 170.1	Available from some DO, ORP, pH, or EC probes. Can be analyzed in flow through cell or from 100-250 mL of water collected and analyzed immediately.	Oxygen solubility is dependent on groundwater temperature. Biodegradation rates may depend on temperature. An increase in temperature may be seen within the solute plume.
Dissolved Oxygen (DO)	Field		With all DO methods extra care must be taken to avoid aeration during all steps of the analysis including well purging and sample collection.	An inverse correlation of DO to BTEX concentrations indicates aerobic biodegradation is occurring. This relationship may also be expressed as depressed or non-detectable levels of DO throughout the plume.

Parameter	Field or Laboratory Method	Analytical Method(s) (or equivalent method)	Comments	Use of Data
		Meter and Probe Flow through cell	Utilizes a flow thru cell with a dissolved oxygen electrode. Other parameters such as temperature, pH, oxidation reduction potential and conductivity can be measured simultaneously. If an oxygen consuming probe is used, then care must be taken to ensure sufficient and continuous flow from the well through the cell.	
		ASTM D 888[A] Winkler titration	Field kits for performing Winkler titrations can be used as the primary method of DO measurement or to confirm meter measurements. A combination of both methods can be used to ensure data quality.	
		Down hole probe	If an oxygen consuming probe is used down hole, then gentile agitation of the probe is required. Vigorous agitation should be avoided to prevent aeration. This technique is recommended only in low permeability conditions where continuous well purging is not possible.	
Ferrous Iron	Field	Colorimetric Std. Methods 18th[B] Edition. Method 3500-Fe D Hach[C] 25140-25	Collect 100 mL of water in glass container. Filter sample with 0.2 μ filter.	Increased concentrations of Fe(II) may indicate Fe(III) is being used as an electron acceptor during anaerobic biodegradation of petroleum hydrocarbons. Also useful in assessing feasibility of other corrective action alternatives.
Total Dissolved Iron	Lab	Atomic adsorption SW-846 Method[D] 7380/7381 ICP SW-846 Method[D] 6010A	filter, 0.2 μ; acidity; refrigerate.	Increased dissolved Fe may indicate Fe(III) is being used as an electron acceptor during anaerobic biodegradation of petroleum hydrocarbons. Also useful in assessing feasibility of other corrective action alternatives.
Oxidation Reduction Potential	Field	Ion Selective Electrode direct reading meter ASTM D 1498[A]	ORP probe can be inserted into flow thru cell and reading obtained simultaneously with DO, pH, T, and EC. Can be taken downhole if necessary.	Define regions of the plume under oxidizing and reducing conditions. Evaluate potential for biologically mediated redox reactions to occur. Helps validate DO measurements. Determine Eh values.

TABLE X4.1 *Continued*

Nitrate	Field Lab	Colorimetric field kit EPA method 353.2 SW9056 ASTM D 4327[A]	Collect 100 mL of water in a glass container. Collect 100 mL of water in a glass or plastic container, cool to 4°C and analyze within 48 hours.	Decreased nitrate concentrations in anaerobic portion of the plume may indicate use of nitrate as an electron acceptor for anaerobic biodegradation of petroleum hydrocarbons.
Sulfate	Field Lab	Colorimetric field kit. EPA Method 375.4 or SW-9056 ASTM D 4327[A]	Collect 100 mL of water in a glass or plastic container, cool to 4°C, analyze immediately. Collect 100 mL water in a glass or plastic container, cool to 4°C, analyze within 28 days.	Decreased sulfate concentrations in anaerobic portion of the plume may indicate use of sulfate as an electron acceptor for anaerobic biodegradation of petroleum hydrocarbons.
Manganese	Field Lab	Colorimetric EPA Method 3010/200	Collect 100 mL of water in a glass container. Filter sample with 0.2 μ filter.	Increased Mn(II) may indicate Mn(IV) is being used during anaerobic biodegradation as a terminal electron acceptor.
Alkalinity	Field Lab	Field test kit EPA method 310.2 Colorimetric	Collect 100 to 250 mL of water in a glass container. Collect 100 mL of water in a glass container cool to 4°C	Typically, total alkalinity is primarily due to carbonate alkalinity. Thus, alkalinity is a measure of dissolved carbonate and bicarbonate. A zone of increased alkalinity indicates biodegradation is either producing organic acids which lower the pH and solubilize carbonate from the soil or CO2 is being produced.
Additional Parameters:	In some cases, it may be necessary to evaluate additional indicators of biodegradation if the primary indicators are insufficient, or if it is desired to further understand the potential mechanisms of biodegradation. These parameters may increase the complexity of interpretation and/or the analytical effort.			
Parameter	Field or Laboratory Method	Analytical Method	Comments	Use of Data
Methane	Field or Lab GC	Head Space; GC/FID. IJEAC[E]	Collect water samples in 40 mL VOA vials with butyl gray/Teflon-lined caps. Head space is analyzed by GC equipped with thermal conductivity and/or flame ionization detector.	Elevated concentrations may indicate anaerobic degradation using carbon dioxide as an electron acceptor.
Carbon Dioxide	Lab GC or Field Method	Head Space; GC/TCD. IJEAC[E] Field titration kit	Collect water samples in 40 mL VOA vials with butyl gray/Teflon-lined caps. Head space is analyzed by GC equipped with thermal conductivity and/or flame ionization detector.	Elevated carbon dioxide levels may indicate aerobic biodegradation or depleted levels may indicate methanogenesis is occurring. Carbon dioxide data must be collected and reviewed carefully due to complex geochemical interactions.
Dissolved sulfides	Field	Colorimetric ASTM D 4658[A]	Collect 100 mL of water in a glass container and analyze immediately.	Increased levels above background may indicate sulfate-based anaerobic respiration.

[A] ASTM - American Society for Testing of Materials, standard methods.
[B] Std. Methods - Standard Methods for the Examination of Water and Wastewater, 1992. American Public Health Assoc., American Water Works Assoc., Water Environment Assoc., 18th Edition.
[C] Hach - HACH Company, P.O. Box 389, Loveland, Colorado.
[D] EPA, SW-846 - Test Methods for Evaluating Solid Waste, 1992. USEPA, Office of Solid Waste and Emergency Response, Washington D.C., SW-846.
[E] IJEAC - Kampbell, D.H., J.T. Wilson, and S. A. Vandegrift, 1989. Dissolved oxygen and methane in water by a gas chromatography headspace equilibrium technique. International Journal of Environmental Analytical Chemistry, 36:249-257.

X4.2 Knowledge of regional hydrogeology and geochemistry may be used to define a more focused list appropriate for a specific geologic setting. The methods recommended here are generally recognized to meet data quality objectives, however this is not meant to be an all encompassing list of acceptable methods. Other methods, which can be demonstrated to meet data quality objectives may be used as well.

X5. INTERPRETATION OF NATURAL ATTENUATION INDICATORS

X5.1 *Introduction*—There are a number of parameters that can be used as indicators of the natural attenuation of petroleum hydrocarbons in ground water. The following sections summarize some of the direct indicators that demonstrate the occurrence of natural attenuation, including indicators of biodegradative processes. Biodegradative processes (often referred to as intrinsic bioremediation) are typically considered the primary mechanism for BTEX attenuation in ground water **(8,17-23)**.

X5.2 *Primary Line of Evidence for Natural Attenuation - Plume Status*—The primary line of evidence for natural attenuation is to define the plume as stable, shrinking or expanding. Knowledge of the distribution and concentration of hydrocarbons of regulatory concern (for example, BTEX) over time is used to assess plume status. In general, a shrinking or stable plume is strong evidence that natural attenuation is effectively remediating a site. It should be noted that even in the case of an expanding plume, the influence of natural attenuation can be demonstrated in many cases.

X5.2.1 *Expanding Plume*—An expanding plume occurs when the margin of the solute plume continues to move outward and/or concentrations at points within the plume continue to increase over time. A total retardation factor can be used to estimate the influence of natural attenuation processes on an expanding plume. This factor is determined by dividing the distance that groundwater has migrated since the release (calculated from a knowledge of ground water velocity and the age of the release) by the distance from the source area to the actual plume edge (based on monitoring well data). If the retardation of plume expansion is due to sorptive processes only, then the natural attenuation factor for a BTEX plume should typically range from 1.5 to 3.0 **(24)**. If the retardation factor is greater than this range, then other processes in addition to adsorption are likely. If a plume is expanding but at a rate lower than the ground water velocity, the risk reduction and performance goals may still be met depending on the presence and location of receptors. Further investigation and assessment may be necessary to more accurately predict the potential extent of plume migration and ensure protection of receptors.

X5.2.2 *Stable Plume*—A stable plume is a configuration where the margin of the solute plume is stable over time and/or the concentrations at points within the plume are relatively uniform over time. If a plume is stable, while ground water continues to move, then remediation through natural attenuation is evident. Simply, for the plume to be stable, the rate of natural attenuation of BTEX in the plume must be equal to the rate that BTEX mass is entering the plume from the source area. The rate of natural attenuation in a stable plume can be calculated through a mass balance approach or through a concentration versus distance approach. Both are discussed in detail in Appendix X6. In addition, the approach for estimating the influence of natural attenuation on an expanding plume given in X5.2.1 is applicable to a stable plume as well.

X5.2.3 *Shrinking Plume*—A shrinking plume is a configuration where the groundwater plume margin is shrinking over time and/or the concentrations at points within the plume are decreasing over time. If a plume is shrinking, while ground water continues to move, then remediation through natural attenuation is evident. Simply, for the plume to be shrinking, the rate of natural attenuation of BTEX in the plume must be greater than the rate that BTEX mass is entering the plume from the source area. The rate of natural attenuation in a shrinking plume can be calculated through a mass balance approach or through a concentration versus time approach. Both are discussed in detail in Appendix X6. In addition, the approach for estimating the influence of natural attenuation on an expanding plume given in X6.2.1 is applicable to a shrinking plume as well.

X5.3 *Secondary Line of Evidence - Indicators of Biodegradation*—Biodegradation is the process in which naturally-occurring subsurface microorganisms biodegrade contaminants, often completely degrading hydrocarbons to carbon dioxide and water. Evaluating indicators specific to the biodegradation process is of critical importance when presenting secondary lines of evidence for natural attenuation. The discussion presented below is confined to an assessment of the biodegradation of benzene, toluene, ethylbenzene and xylenes (BTEX). Because of their toxicity and especially their environmental mobility, these chemicals are typically the constituents of concern in groundwater at petroleum release sites. Other compounds (for example, oxygenated additives, naphthalene) may be of concern on a site specific basis depending on the age and nature of the petroleum release. A discussion of the bioremediation of such compounds is beyond the scope of the following presentation.

X5.3.1 *Introduction to Indicators of Biodegradation*—Microorganisms transform organic molecules, such as benzene, toluene, ethylbenzene and xylenes (BTEX), and obtain carbon and energy from these substrates for survival, growth and reproduction. Initially, the microbes "attack" these hydrocarbons through a series of enzyme-catalyzed oxidative-reduction reactions. The resulting metabolic intermediates are then either divered to biomass-producing pathways or completely oxidized to carbon dioxide and water. The oxidative reactions yield electrons which through a series of enzyme-catalyzed electron transport steps produce the energy the cell needs for maintenance and growth. In order for the electrons to pass through the energy generating steps, though, an electron sink is required. Typically, this electron acceptor (sink) is molecular oxygen (O_2) and the process is called aerobic respiration. In the absence or near absence of molecular oxygen (O_2) and in the order presented, nitrate (NO_3^-), manganese (Mn^{+4}), ferric iron (Fe^{+3}), sulfate (SO_4^{-2}) or carbon dioxide (CO_2) may serve, if available, as terminal electron-acceptors in a process called anaerobic respiration.

X5.3.1.1 The *aerobic* metabolism of benzene, toluene, ethylbenzene and xylenes has been widely documented. In subsurface soils and aquifer materials, both through *ex-situ* microcosm studies and through field studies **(8,17-18,21)**. It is likely that microorganisms capable of aerobically degrading BTEX constituents are nearly ubiquitous in subsurface environments.

X5.3.1.2 The *anaerobic* metabolism of toluene, ethylbenzene and xylenes has been widely demonstrated under nitrate-reducing **(19,25-28)**, sulfate-reducing **(27-30)**, iron- and manganese-reducing **(31-35)** and methanogenic **(36-38)** conditions. The anaerobic metabolism of benzene has not been well documented. Nevertheless, an increasing number of microcosm and field studies are demonstrating the metabolism of benzene under nitrate-, sulfate-, iron-reducing and methanogenic conditions **(32,34,36-42)**.

X5.3.1.3 The equations given in Fig. X5.1 show the reactants and products involved in the aerobic and anaerobic metabolism of BTEX. The reactions below are balanced based on toluene, but the reactions apply to all BTEX constituents.

X5.3.1.4 As shown in previous equations, under aerobic conditions, the metabolism of BTEX results in the consumption of oxygen and the production of carbon dioxide and water. Under anaerobic conditions, the metabolism of BTEX consumes the particular electron acceptor and yields the reduced electron acceptor, carbon dioxide and water (water is not produced in methanogenesis). Thus, coupled with BTEX attenuation, the disappearance in ground water of dissolved oxygen and/or anaerobic electron acceptors, or the appearance of electron-acceptor reduction products, (all relative to upgradient concentrations) is clear evidence of biodegradation.

X5.3.2 *A Quantitative Approach to Evaluating Indicators of Biodegradation*—Quantitatively, the balanced equations given in Fig. X5.1 allow a calculation of: (*1*) the mass of a terminal electron acceptor necessary to completely metabolize a given mass of BTEX; and (*2*) the yield of reduced electron acceptor per mg of BTEX metabolized. Based on the previous equations, 3.1, 4.6 and 4.8 mg/L of oxygen, sulfate and nitrate, respectively, are necessary to completely metabolize 1 mg/L BTEX, while 0.8, 11 and 22 mg/L of methane, Mn^{+2} and Fe^{+2}, respectively, are produced from the complete metabolism of 1 mg/L BTEX.

AEROBIC RESPIRATION

$C_6H_5CH_3 + 9\ O_2 \rightarrow 7\ CO_2 + 4\ H_2O$

ANAEROBIC RESPIRATION

Nitrate Reduction

$5C_6H_5CH_3 + 36NO_3^- + 36H^+ \rightarrow 18N_2 + 35CO_2 + 38H_2O$

Manganese Reduction

$C_6H_5CH_3 + 18MnO_2 + 36H^+ \rightarrow 18Mn{+}2 + 7CO_2 + 22H_2O$

Iron Reduction

$C_6H_5CH_3 + 36Fe(OH)_3 + 72H^+ \rightarrow 36Fe^{+2} + 7CO_2 + 94H_2O$

Sulfate Reduction

$8C_6H_5CH_3 + 36SO_4^{-2} + 72H{+} \rightarrow 36\ H_2S + 56CO_2 + 32H_2O$

Methanogenesis

$8C_6H_5CH_3 + 40H_2O \rightarrow 36CH_4 + 20CO_2$

NOTE 1—Equations are balanced using toluene and would slightly differ for other BTEX constituents. The equations balanced using benzene can be found in Wiedemeier et al. 1995 Ref. **(1)**.

FIG. X5.1 Aerobic and Anaerobic Pathways of BTEX Biodegradation

NOTE X5.1—These are values specific to toluene, but values for other BTEX constituents will vary only slightly.

These are maximum values, since not all of the carbon will be completely oxidized (a significant portion of the carbon will be used for biomass). While a quantitative approach (for example, a decrease in BTEX concentration of 2 mg/L yielded a decrease in sulfate concentration of 10 mg/L) is valuable and has been utilized **(22,23)** the possible environmental sources and sinks for electron acceptors and metabolites can make quantification very difficult. Therefore, a qualitative approach is usually followed, with a direct or inverse correlation between electron acceptors/reduction products and BTEX concentrations providing clear evidence for biodegradation.

X5.3.2.1 Many of the electron acceptors and their reduction products given in Fig. X5.1 can be readily and accurately measured in ground water samples. These include O_2, Fe(II). Mn (II), NO_3, SO_4, CH_4 and CO_2. Measurement of these parameters is discussed in detail in Appendix X4. Thus, these parameters can be used to evaluate the influence of biodegradation on a solute plume. Other means of assessing biodegradation include measuring oxidation-reduction potentials, microcosm studies and microbial counts. Electron acceptors and their reduction products, as well as other methods for assessing biodegradation, are discussed in more detail below.

X5.3.3 *Electron Acceptors and Reduction Products (Secondary Lines of Evidence)*:

X5.3.3.1 *Oxygen*—In zones of high BTEX concentration, dissolved oxygen is depleted because the naturally-occurring microorganisms have utilized the available oxygen as they biodegrade BTEX and other constituents and any oxygen entering this zone is rapidly depleted (due to the high oxygen demand generated by the high BTEX concentrations and other constituents such as Fe(II), Mn(II), etc.). The inverse relationship of high hydrocarbon concentration and low dissolved oxygen concentration can be used as a key indicator of bioremediation.

X5.3.3.2 *Nitrate*—After dissolved oxygen has been depleted (typically considered <0.5 ppm) in a given ground water zone, nitrate, if available, may be used as an electron acceptor for anaerobic biodegradation. In anaerobic ground water zones with high BTEX concentrations, nitrate demand (by nitrate-reducing microorganisms capable of biodegrading BTEX) may be high and nitrate concentrations may be depleted relative to concentrations upgradient and outside the plume. Overall, an inverse relationship between BTEX concentrations and nitrate concentration should be expected.

X5.3.3.3 *Manganese*—The use of manganese (Mn^{+4}) as a terminal electron acceptor by microorganisms yields a reduced water-soluble manganese (Mn^{+2}). In anaerobic ground water zones where BTEX and a source of Mn^{+4} (MnO_2) are present, Mn^{+2} can be used as an indicator of biodegradation. Overall, a positive correlation between BTEX concentrations and Mn^{+2} concentration should be expected.

X5.3.3.4 *Iron*—The use of ferric (Fe^{+3}) iron as a terminal electron acceptor by microorganisms yields water-soluble ferrous (Fe^{+2}) iron. In anaerobic ground water zones where BTEX and a source of ferric iron are present, ferrous iron can be used as an indicator of biodegradation. Overall, a positive correlation between BTEX concentrations and ferrous iron concentration should be expected.

X5.3.3.5 *Sulfate*—Under strongly reducing conditions, after available oxygen, nitrate and ferric iron have been depleted, sulfate can be used as an electron acceptor for anaerobic biodegradation of BTEX. The process results in the production of sulfide, which may precipitate from solution as ferrous sulfide. Under sulfate-reducing conditions and in the presence of high BTEX concentrations, sulfate demand (by sulfate-reducing BTEX degraders) will be high and sulfate concentrations will be depleted relative to concentrations upgradient and outside the plume. Overall, in these sulfate-reducing zones, an inverse relationship between BTEX concentrations and sulfate concentration should be expected.

X5.3.3.6 *Methane*—Methane is produced only under strongly reducing conditions by a group of strict anaerobes. Methanogens either use CO_2 as a terminal electron acceptor, producing methane, or cleave acetate to CO_2 and methane. Because methane is not present in fuels, it can be used as an indicator of biodegradation. Under methanogenic conditions and in the presence of high BTEX concentrations, the rate of methane production can be significant, with the concentration of methane in this zone high relative to areas upgradient and outside the plume (where the methane concentration is typically nondetect). Overall, under methanogenic conditions, a positive correlation between BTEX concentrations and methane concentration should be expected.

X5.3.3.7 *Carbon dioxide*—Both aerobic and anaerobic biodegradative processes can yield large quantities of CO_2, as the BTEX constituents are completely oxidized. An accurate measurement of the CO_2 produced through biodegradation, though, is difficult because the carbonate-buffering system in ground water (measured as alkalinity) serves as both a sink and source of CO_2. Nevertheless, in many circumstances, a positive correlation between BTEX and CO_2 concentrations can be expected and can be used as a qualitative indicator of biodegradation.

X5.3.4 *Summary of the Relationship Between BTEX and Electron Acceptor/Reduction Product Concentrations*—The expected relationship between BTEX concentration and the concentration of a particular electron acceptor or its reduction product is summarized in Table X5.1.

X5.3.5 *Additional Indicators of Biodegradation*:

X5.3.5.1 *Oxidation-Reduction (Redox) Potential*—The oxidation-reduction potential of ground water is a measure of the relative tendency of a solution to accept or donate electrons. Importantly, redox reactions in ground water are usually mediated by microorganisms. Therefore, the redox potential can strongly depend on biodegradative processes and, in turn, the redox potential can strongly influence such processes. While the redox potential of ground water can range from –400 to 800 mv, certain biodegradative processes can only operate within a prescribed range of redox conditions (Fig. X5.2). By measuring upgradient redox values and values within the plume, zones where biodegradation (especially anaerobic processes) is lowering the redox potential can be identified.

TABLE X5.1 Expected Relationship Between BTEX and Electron Acceptor/Reduction Product Concentrations Within a Plume

BTEX	OXYGEN	NITRATE	Mn (II)	FE(II)	SULFATE	METHANE
HIGH	LOW	LOW	HIGH	HIGH	LOW	HIGH
LOW	HIGH	HIGH	LOW	LOW	HIGH	LOW

X5.3.5.2 *Microbial Counts*—Microbial counts, using both counts of total heterotrophic bacteria and hydrocarbon degraders, can be used as an indicator of biodegradation. The ratio of hydrocarbon degraders to total heterotrophic bacteria should increase in aquifer zones where biodegradation of BTEX is occurring. It should be noted, though, that microbial counts are often unreliable indicators of biodegradation (**21**).

X5.3.5.3 *Microcosm Studies*—In microcosms studies, small quantities of ground water and aquifer solids are placed into serum bottles, shake flasks or into columns. The disappearance of a carbon substrate, sometimes along with the disappearance of a terminal electron acceptor or the appearance of the appropriate reduction product, is then measured over time. Microcosm studies are often used to clearly demonstrate the ability of native microorganism to biodegrade a given compound. As discussed, the biodegradation of BTEX constituents is well documented and, therefore, microcosm studies are not typically necessary. In addition, although microcosms studies can be used to estimate *in-situ* biodegradation rates, field methods detailed in Appendix X6 are preferred.

X5.4 *Measuring Nutrient Levels*—Measuring nutrient levels can provide an optional line of evidence. Measuring the concentration of electron acceptors or their reduction products should not be confused with measuring the level of microbial nutrients. Nutrients are incorporated into microbial biomass and are necessary for the formation of proteins, DNA, cell membranes and other components of microbial cells. Microbial nutrients are usually divided into two categories: macronutrients (for example, nitrogen and phosphorus), for which microorganisms require relatively large amounts, and micronutrients (for example, sulfur, manganese, magnesium and many others), for which only a trace amount is required. In contrast to nutrients, electron acceptors and their reduction products are not incorporated into microbial biomass, but the reduced compounds are "excreted" into the environment (for example, the microbes take in sulfate and release H_2S). Macronutrient (nitrogen and phosphorus) levels are often assessed in surface and subsurface environments by measuring ammonium (NH_4^+), nitrate (NO_3^-), organic (Kjeldahl) nitrogen, available phosphorus or phosphate (PO_4^{-3}) and total phosphorus (mostly organic phosphorus compounds + phosphate). Certain molecules, such as nitrate and sulfate, can serve either as nutrients or terminal electron acceptors. While the availability of electron acceptors in the subsurface is a critical factor in assessing the rate and extent of biodegradation (**24**), as previously discussed, nutrient levels are rarely limiting microbial biodegradation activity in the subsurface, and, thus, measuring subsurface nutrient levels is unnecessary.

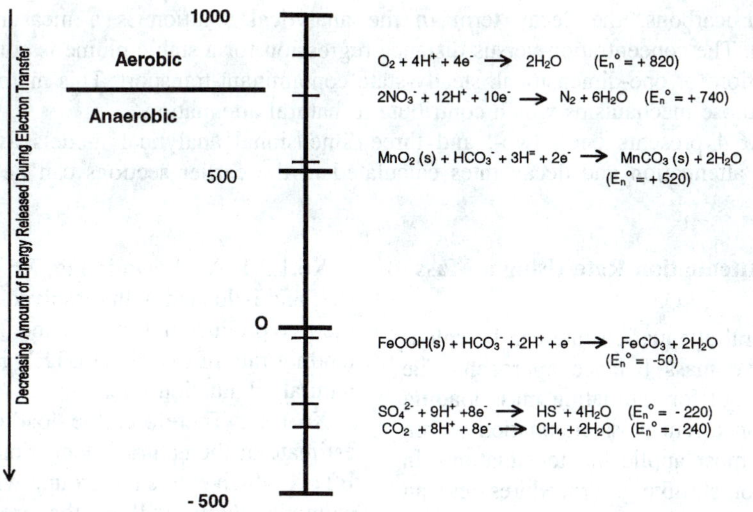

FIG. X5.2 Redox Potentials in Millivolts, pH 7, 25°C (from Wiedemeier et al. 1995)

X6. QUANTIFYING RNA RATES

INTRODUCTION

The primary line of evidence for natural attenuation is to define the plume as stable, shrinking, or expanding and efforts to quantify RNA rate are not necessary. In the secondary line of evidence, the natural attenuation rate may be estimated. The selection of a method to estimate RNA rate is based on the availability of contaminant data. Appendix X6 describes several methods to estimate RNA rate. The effort of calculating a natural attenuation rate achieves the following:

It strengthens the qualitative and semi-quantitative arguments for natural attenuation and;

It lends credence to the argument that natural attenuation can be an effective stand-alone remediation measure. As part of a demonstration of natural attenuation, it may be important to estimate a specific rate of destruction for dissolved BTEX.

This Appendix is divided into the following four sections:

X6.1	Mass Balance Approach,
X6.2	Graphical and Regression Techniques,
X6.3	One-Dimensional Analytical Solution for a Stable Plume, and
X6.4	One-, Two-, and Three-Dimensional Analytical Models.

Appendix X6.1 describes a mass balance approach which calculates the contaminant mass loading to the water table. Depending on the type of plume, the attenuation rate can be estimated based on the contaminant mass loading rate. The attenuation rate for a stable plume is approximately equal to the contaminant mass loading rate. The attenuation rate for a shrinking table plume exceeds the contaminant mass loading rate. In the case of an expanding plume, the attenuation rate is less than the contaminant mass loading rate.

Appendix X6.2 describes graphical and regression techniques useful to determine attenuation rates. Two of the conditions for which natural attenuation contributes to the configuration of a contaminant plume are a shrinking or stable plume. These techniques require data for contaminant concentration versus time and/or distance. Regression of contaminant concentration versus time for individual monitoring wells is an appropriate analysis technique for a shrinking plume. Regression of contaminant concentration versus distance is an appropriate method for either a shrinking or stable plume. Estimates of attenuation rate should be used with caution. The attenuation rate based on concentration versus time for a single monitoring well may not be representative of the entire dissolved plume. On the other hand, if multiple wells display exponential contaminant decay, estimates of attenuation rate may be appropriate to model the plume. Concentration versus distance techniques may yield attenuation rates more representative of the entire plume.

The graphical/regression techniques described in X6.2 yield an attenuation rate which does not distinguish between sorption, dispersion, and biodegradation. The objective of X6.3 is to evaluate the contribution of biodegradation to the overall attenuation rate for stable plumes. This is accomplished by using an analytical solution which includes advection, dispersion, sorption, and decay. For petroleum hydrocarbons, the decay term in the analytical solution is a measure of intrinsic bioremediation. The concentration versus distance regression for a stable plume is coupled with this analytical solution for one-dimensional, steady-state contaminant transport. This method is intended to distinguish those mechanisms which contribute to natural attenuation.

Appendix X6.4 presents one-, two-, and three-dimensional analytical models available in the literature. The attenuation and decay rates calculated in the earlier sections can be used in these models.

X6.1 Estimating Natural Attenuation Rate Using a Mass Balance Approach

X6.1.1 An estimated natural attenuation rate can be calculated by using a quantitative mass balance approach. The mass-balance approach is a tool for estimating mass loading from the source zone and concurrent mass attenuation within the plume. This method is most applicable to situations in which the plume is stable or shrinking. Procedures and an example are presented in this section.

X6.1.2 The additional step of calculating a natural attenuation rate has the following advantages:

X6.1.2.1 It strengthens the qualitative and semi-quantitative arguments given above and;

X6.1.2.2 Calculation of a natural attenuation rate lends credence to the argument that natural attenuation can be an effective stand-alone remediation measure. As part of a demonstration of natural attenuation, it may be important to quote a specific rate of destruction for dissolved BTEX.

X6.1.3 *Site Scenario*:

X6.1.3.1 As shown in Fig. X6.1, ground water flows onto a site, and is loaded with dissolved BTEX, through contact with the free product and smear zone. For a steady-state plume, this loading rate of dissolved BTEX is approximately equal to the natural attenuation rate.

X6.1.3.2 Therefore, the loading rate can be used as an estimate of the natural attenuation rate. The mass of dissolved BTEX which enters the ground water system per unit time (for example, daily) will be the product of the average BTEX concentration in and below the source area and the flow of contaminated ground water in and below the source area.

NOTE X6.1—This calculation is based on the steady state plume configuration. For an expanding plume, the actual natural attenuation rate would be somewhat less than that calculated for a steady state plume, and for a shrinking plume, it would be somewhat more.

X6.1.4 *Procedure and Example*:

X6.1.4.1 The procedure to calculate the dissolved BTEX loading rate (and, therefore, the estimated natural attenuation rate) is presented in this section. An example is used and a

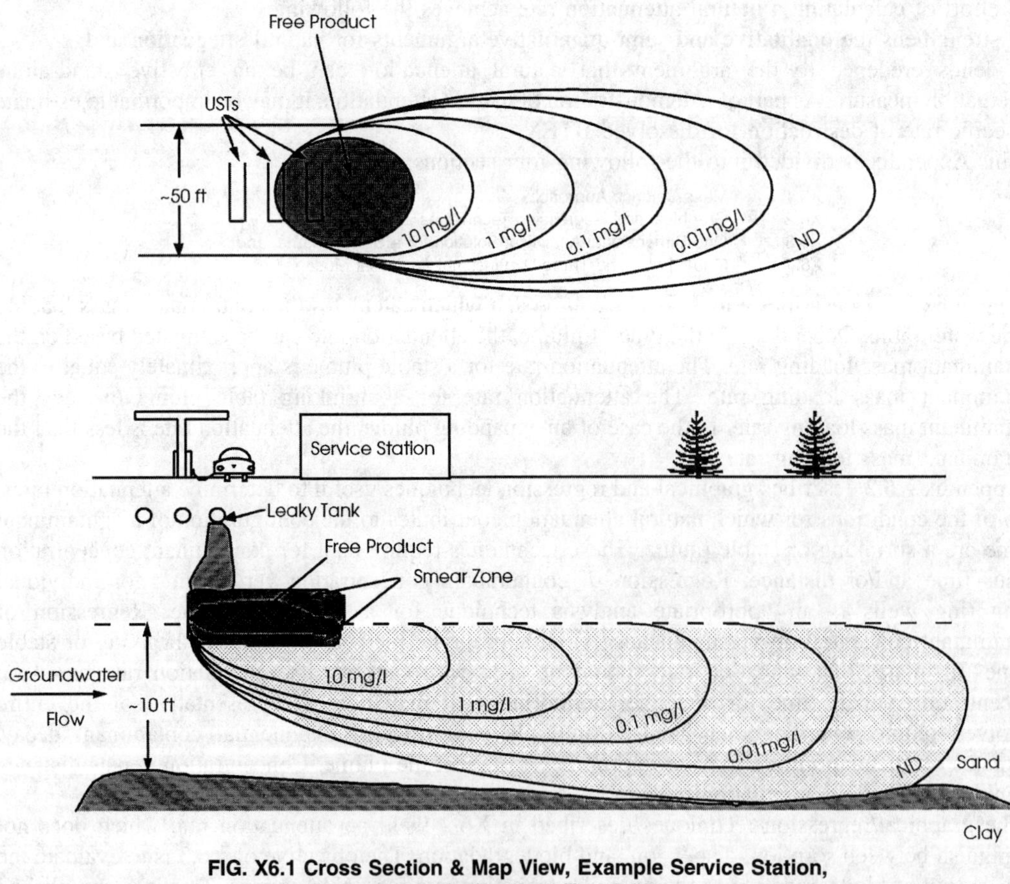

FIG. X6.1 Cross Section & Map View, Example Service Station,
Hydraulic Gradient (i) = 0.005
Hydraulic Conductivity (K) = 100 ft/day

dissolved BTEX loading rate calculated, based on the typical service station site presented in Fig. X6.1.

X6.1.4.2 The dissolved BTEX loading rate has two components: (*1*) the dissolved BTEX added to the ground water that flows horizontally through the smear zone below the water table, and (*2*) the dissolved BTEX added to the ground water beneath the smear zone, caused primarily by vertical dispersion and vertical advection.

X6.1.4.3 For both of these zones, an average concentration is estimated, and multiplied by the ground water flow rate through the zone. The calculation (using appropriate conversion factors, as discussed below) results in the mass of dissolved BTEX contributed for each zone.

X6.1.5 Calculation of BTEX mass added to ground water flowing through the smear zone involves the following steps.

X6.1.5.1 Estimate the cross-sectional area of the smear zone below the water table (A_{sz}), perpendicular to the direction of ground water flow. The average thickness of this zone will be equal to one-half the width of the zone of water table fluctuations. For the example site, historical records indicate maximum water table fluctuations equal to 6′. Therefore, the average thickness of the smear zone below the water table is 3′. The width of the smear zone is 50′ (see Fig. X6.1). Therefore:

$$A_{sz} = 3' \times 50' = 150 \text{ ft}^2.$$

X6.1.5.2 Estimate the hydraulic conductivity of the smear zone (K_{sz}). The value is less than the estimated hydraulic conductivity (*K*) based on slug tests and/or pumping tests conducted in waters containing no free product. This reduction in hydraulic conductivity is caused by the presence of product trapped in soil pores, and can be estimated using Table X6.1 (based on the data and equations of Parker *et al.* 1987 **(43)**: The hydraulic conductivity of the smear zone K_{sz} = the measured *K* value times the appropriate value from the table (or an interpolated value). For the example site (with a medium sand aquifer *K* = 100 ft/day), $K_{sz} = 100 \times 0.3 = 30$ ft/day.

X6.1.5.3 Calculate the ground water flow through the smear zone (Q_{sz}), as follows:

$$Q_{sz} = K_{sz} \times i \times A_{sz} \quad \text{(X6.1)}$$

Q_{sz}, K_{sz} an A_{sz} are as defined above, and
i is the hydraulic gradient.

For the example site, $Q_{sz} = 30 \times 0.005 \times 150 = 22.5 \text{ ft}^3\text{/day}$

X6.1.5.4 The mass flux of dissolved BTEX entering the ground water through the smear zone ($BTEX_{sz}$) is calculated using the following expression:

$$BTEX_{sz} = C_{sz} \times Q_{sz} \times CF \quad \text{(X6.2)}$$

$BTEX_{sz}$ and Q_{sz} are as defined above,
C_{sz} is equal to the theoretical solubility for BTEX in ground water in contact with gasoline components. This value is approximately 100 mg/L based on immiscible partitioning **(44)**. CF is a conversion factor to convert mg/L to lbs/ft^3,
$CF = 6.245 \times 10^{-5}$ $\text{lbs-L/ft}^3\text{-mg}$.

TABLE X6.1 Estimated Hydraulic Conductivities

Soil Type	Ratio of Smear Zone K to Measured K
Silty clay	0.8
Sandy silt/fine sand	0.5
Medium sand	0.3

For the example site, $BTEX_{sz} = 100 \times 22.5 \times 6.245 \times 10^{-5}$ = 0.14 lbs/day

X6.1.6 Calculation of BTEX mass added to ground water flowing below the smear zone involves an estimate of both the thickness of the dissolved plume below the smear zone and the average concentration of this plume.

X6.1.6.1 These parameters are not generally known. For most sites, water quality data in the source area involves measurements of concentration at a single depth, or perhaps no data is available at all (for example, if free product is present, water quality samples are typically not collected). In addition, while the thickness of the shallow water-bearing formation might be known, the vertical extent and distribution of contamination below the source area might not.

X6.1.6.2 However, observations have shown that the vertical extent of contamination below the smear zone is no more than 10 ft. This is based on field observations and theoretical considerations **(8,45-48)**. and personal communication, Gary Robbins, University of Connecticut, Storrs, 1995). The vertical spread of contaminants is dominated by dispersion, and the concentrations can be assumed to have a Guassian distribution (the theoretical distribution resulting from dispersion).

NOTE X6.2—The plume thicknesses discussed above are for relatively small smear zones, that is, measured in tens of feet in the direction of ground water flow. For larger smear zones, actual measurements of plume thicknesses at the downgradient edge of the smear zone are necessary, due to dispersion and/or vertical flow components as ground water travels the length of the smear zone.

X6.1.6.3 Of course, site physical conditions may dictate that the thickness of the plume below the smear zone is less than 10 ft; for example, a sand unit may be underlain by clay at a depth of less than 10 ft below the smear zone. In this case, the vertical concentration profile is the same as for 10 ft; however, the concentration profile is truncated at the lesser thickness, resulting in a higher average concentration.

X6.1.6.4 Assuming a Guassian concentration distribution from 100 mg/L at the smear zone to 0.001 mg/L at 10 ft below the smear zone, the Table X6.2 presents average BTEX concentrations for those cases where the thickness of the dissolved plume is physically constrained by a low permeability unit to less than 10 ft.

X6.1.6.5 The calculation of BTEX mass distributed to the ground water that flows below the smear zone involves the following steps:

(*1*) Estimate the cross sectional area of the dissolved plume below the smear zone (Ab_{sz}), perpendicular to the direction of ground water flow. For the example site, the thickness of this zone is 10 ft, and the width of the smear zone is 50 ft (see Fig. X6.2). Therefore, $A_{bsz} = 10 \times 50 = 500 \text{ ft}^2$. Dissolved hydrocarbon plumes may decrease in size, as observed by

TABLE X6.2 BTEX Concentrations as a Function of Plume Thickness

Plume thickness below smear zone	Average BTEX concentration (mg/L)
2′	87
4′	62
6′	43
8′	33
10′[A]	26

[A] (use 10′ even if hydraulic zone > 10′)

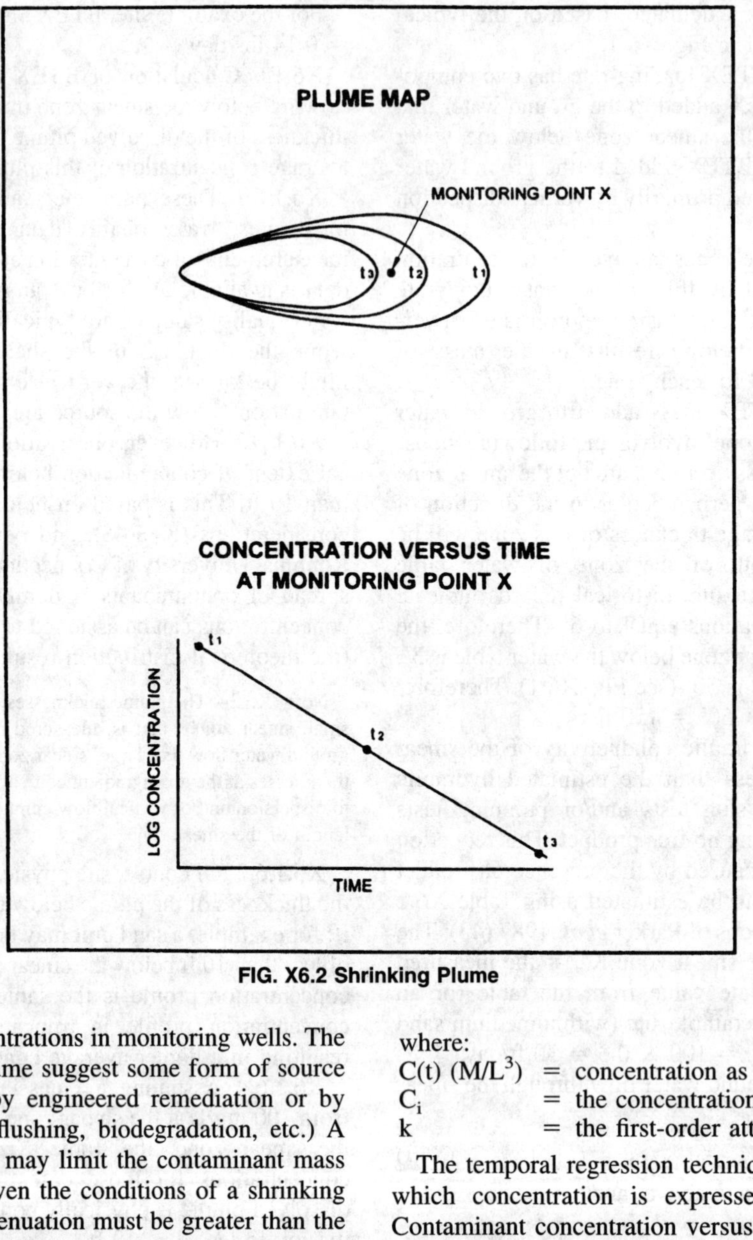

FIG. X6.2 Shrinking Plume

declining contaminant concentrations in monitoring wells. The conditions of a shrinking plume suggest some form of source control or removal, either by engineered remediation or by natural source attenuation (flushing, biodegradation, etc.) A sufficiently deep water table may limit the contaminant mass loading to ground water. Given the conditions of a shrinking plume, the rate of natural attenuation must be greater than the rate of contaminant addition **(24)**. Salanitro **(21)** indicates a shrinking plume has significant reinfiltration of oxygenated water into the aquifer (precipitation). Wells on the periphery of these types of plumes have higher dissolved oxygen (DO) and lower BTEX concentrations, consistent with BTEX biodegradation. Analysis of shallow and deep monitoring wells show that the soluble plume "shrinks" in the longitudinal direction with little vertical dispersion **(8)**. In the three examples described by Salanitro **(21)**, mass balance analysis of the hydrocarbon in the aquifer show that natural biodegradation by soil biota must be responsible for most (80 to 100 %) of the apparent reduction and retardation of the BTEX plume. Exponential regression methods can be used to evaluate whether concentration versus time data fit a first-order decay observed for petroleum hydrocarbons under certain conditions. The solution to the first-order decay is given by the following equation:

$$C(t) = C_i\ e^{-(kt)} \quad (X6.3)$$

where:

$C(t)$ (M/L^3) = concentration as a function of time, t (T),
C_i = the concentration at t = 0, and
k = the first-order attenuation rate (T^{-1}).

The temporal regression technique is based on Eq X6.3, in which concentration is expressed as a function of time. Contaminant concentration versus time is plotted for a hypothetical monitor well in Fig. X6.2. Concentration is plotted on a log scale, consistent with the assumption of first-order degradation. If a first-order decay describes the apparent contaminant concentration decline, the data plot as a straight line. Exponential regression of log concentration versus time yields a value proportional to the slope of the regressed line (*k*), the y-intercept (C_i), and the goodness of fit (R^2 value). A more general form of Eq X6.3 may be appropriate to fit late-time data which are asymptotic. This equation is given as follows:

$$C(t) = (C_i - C_a)e^{-(kt)} + C_a \quad (X6.4)$$

where:

C_a = the asymptotic concentration.

As C_a approaches zero, Eq X6.4 reduces to Eq X6.3. Eq X6.4 is useful to fit data to an asymptotic concentration. It can also be used to confirm the long-term, asymptotic concentration **(8,49)**. The transport processes contributing to these first-order attenuation rates include volatilization, dispersion, sorption, dilution by recharge, and biological degradation. It is difficult to distinguish these mechanisms using the temporal method. However, as partitioning approaches equilibrium, sorption is less significant in reducing solute concentrations with time. Wilson et al **(50)** suggest the use of a recalcitrant, organic compound, having similar sorptive properties as the biodegradable compound of interest, to serve as a tracer.

NOTE X6.3—If the thickness of the hydrogeologic unit below the smear zone exceeds 10 ft, the thickness of this zone is assumed to be 10 ft.

(2) Calculate the ground water flow rate below the smear zone (Q_{bsz}) as follows:

$$Q_{bsz} = K \times i \times A_{bsz} \tag{X6.5}$$

where:
K = hydraulic conductivity estimated by slug or pumping tests,
i = hydraulic gradient, and
A_{bsz} = cross-sectional area below the smear zone.

For example site, $Q_{bsz} = 100 \times 0.005 \times 500 = 250$ ft^3/day.

(3) Estimate the average BTEX concentration (C_{bsz}) from the above table. For the example site, $C_{bsz} = 26$ mg/L.

(4) The mass flux of dissolved BTEX contributed to the ground water flowing below the smear zone ($BTEX_{bsz}$) is calculated using the following expression:

$$BTEX_{bsz} = C_{bsz} \times Q_{bsz} \times CF \tag{X6.6}$$

Q_{bsz} and C_{bsz} and CF are previously defined

For the example site, $BTEX_{bsz} = 26 \times 250 \times 6.245 \times 10^{-5}$ = 0.41 lbs/day.

The estimation of the total dissolved BTEX loading rate ($BTEX_{tot}$) and, therefore, the estimation of the overall natural attenuation rate, involves the addition of the mass flux of dissolved BTEX entering ground water flowing through the smear zone ($BTEX_{sz}$) to the mass flux of dissolved BTEX contributed to ground water below the smear zone ($BTEX_{bsz}$).

$$BTEX_{tot} = \text{Natural Attenuation Rate} = BTEX_{sz} + BTEX_{bsz} \tag{X6.7}$$

$BTEX_{tot}$, $BTEX_{sz}$ and $BTEX_{bsz}$ as previously

For the example site, $BTEX_{tot} = 0.14 + 0.41 = 0.55$ lbs/day = Natural Attenuation Rate

NOTE X6.4—To convert lbs to gallons of gasoline divide by 6.2.

This removal rate is comparable to what might be achieved by a successful pump and treat system. For example, a pump and treat system extracting 50 gpm containing 1 mg/L BTEX removes 0.5 lbs/day.

X6.2 Graphical and Regression Techniques

X6.2.1 *Concentration versus Time*—Shrinking Plume.

X6.2.2 *Concentration versus Distance*—Stable Plume.

X6.2.2.1 A stable plume is characterized by dissolved contaminant concentrations remaining constant over time in individual monitoring wells. Short-term variations in monitoring well concentrations due to water table fluctuation, variability in ground water flow direction, sampling variability, and analytical uncertainty should be distinguished from statistically significant concentration changes. In order for a plume to reach stable conditions, the rate of natural attenuation must be equal to the rate of contaminant addition to the aquifer from the source **(25)**. The contaminant source or influx rate is limited by the compound's effective solubility and the flow rate of water through the source area (infiltration, fluctuating water table, etc.).

X6.2.2.2 Time, *t*, can be described in terms of seepage velocity, *v* (*L/T*) and distance traveled, *x* (*L*).

$$t = \frac{x}{v} \tag{X6.8}$$

X6.2.2.3 The term "x/v" is the residence time for ground water to move some distance, *x*, from the source. Kemblowski *et al.* **(18)** substituted Eq X6.8 into Eq X6.3, and the first-order attenuation rate can be recast for concentration as a function of distance:

$$C(x) = C_o e^{-(k\frac{x}{v})} \tag{X6.9}$$

where:
$C(x)$ = concentration as a function of distance, x (L).
C_o = concentration at x = 0, and
k = first-order attenuation rate (T^{-1}).

X6.2.2.4 The spatial regression technique is based on Eq X6.9. The regression of concentration versus distance is appropriate for the stable plume. It can also be used for a shrinking plume for individual monitoring events. The ground water flow direction is defined based on multiple monitoring events covering the hydrologic cycle. A minimum of three monitoring wells are selected along the ground water trajectory (see Fig. X6.3). Fig. X6.3 is a plot of compound concentration (log scale) versus distance. If data from multiple sampling events are available for a stable plume, the concentrations plotted should be the average concentration over time for each well. If concentrations decline with time, concentration versus distance can be plotted for individual monitoring events. From the exponent in Eq X6.9, the slope of the line in Fig. X6.3 is *k/v* (L^{-1}), the reciprocal of the attenuation distance. If this slope is multiplied by ground water velocity, *v* (*L/T*), we obtain the attenuation rate, *k* (T^{-1}). In the absence of a reliable estimate of ground water velocity, the *k/v* term is useful, particularly for selecting the downgradient monitoring well locations.

X6.2.3 *Expanding Plume*—The calculation of RNA rate is not as straightforward for an expanding plume as it is for a shrinking or stable plume. It is possible to compare the actual rate of migration to the expected rate of migration, assuming sorption only (contaminant velocity). Using the Mass Balance Approach described in X6.1, it is also possible to estimate the attenuation rate (lbs/day), as less than the contaminant mass loading to the water table. Transient analytical solutions can be used to predict the migration distance of an expanding plume or the time required to reach a particular configuration. These analytical models are presented in X6.4.

X6.3 One-Dimensional Analytical Solution for a Stable Plume

X6.3.1 The graphical/regression techniques described in X6.2 yield an attenuation rate which does not distinguish between sorption, dispersion, and biodegradation. The objective of X6.3 is to evaluate the contribution of biodegradation to the overall attenuation rate. This is accomplished by using an

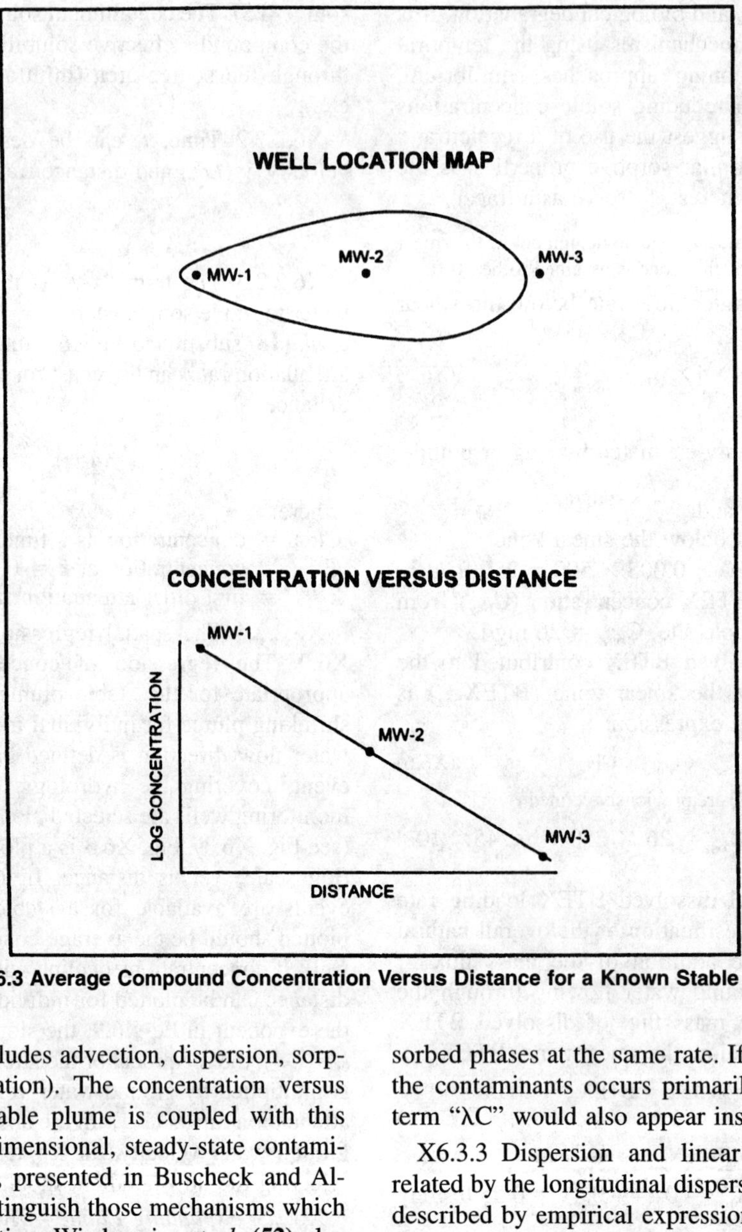

FIG. X6.3 Average Compound Concentration Versus Distance for a Known Stable Plume

analytical solution which includes advection, dispersion, sorption, and decay (biodegradation). The concentration versus distance regression for a stable plume is coupled with this analytical solution for one-dimensional, steady-state contaminant transport. This method, presented in Buscheck and Alcantar **(51)** is intended to distinguish those mechanisms which contribute to natural attenuation. Wiedemeier *et al.* **(52)** also demonstrate the use of this method.

X6.3.2 The general one-dimensional transport equation, with first-order decay of the contaminant, is given by the following equation:

$$\frac{\delta C}{\delta t} = \frac{1}{R}\left[D_x \frac{\delta^2 C}{\delta x^2} - v_x \frac{\delta C}{\delta x}\right] - \lambda C \qquad \text{(X6.10)}$$

Where D_x (L^2/T) is the dispersion of coefficient, v_x (L/T) the seepage velocity in the x direction, R (–) the retardation coefficient and λ (T^{-1}) is the total decay rate. The form of Eq X6.10 assumes D_x is constant and independent of distance, x. While the terms in brackets describe the mass transport by dispersion and advection, respectively, the retardation coefficient characterizes the contribution of sorption. The form of this equation assumes degradation occurs in the aqueous and sorbed phases at the same rate. If biological transformation of the contaminants occurs primarily in the aqueous phase, the term "λC" would also appear inside the brackets.

X6.3.3 Dispersion and linear ground water velocity are related by the longitudinal dispersivity, α_x (L), which has been described by empirical expressions:

$$D_x = \alpha_x v_x \qquad \text{(X6.11)}$$

X6.3.4 The retardation coefficient (*R*) accounts for partitioning between the solid and aqueous phases. *R* describes the relationship between the seepage velocity, v_x, and contaminant velocity, v_c (L/T):

$$R = \frac{v_x}{v_c} \qquad \text{(X6.12)}$$

X6.3.5 Chiang et al. **(8)** demonstrated that the contribution of volatilization to the dissolved contaminant attenuation was only 5 % at one site. Except in the case of very shallow ground water, volatilization is not expected to contribute significantly to the overall attenuation. Therefore, volatilization is neglected and the decay rate is assumed to be a measure of biodegradation of BTEX compounds.

X6.3.6 Bear **(53)** obtained the following solution for one-dimensional transport with a first-order decay:

$$C(x,t) = \left(\frac{C_o}{2}\right) \exp\left[\left(\frac{x}{2\alpha_x}\right)\left[1 - \left(1 + \frac{4\lambda\alpha_x}{v_c}\right)^{\frac{1}{2}}\right]\right] erfc\left[\frac{x - v_c t\left(1 + \frac{4\lambda\alpha_x}{v_c}\right)^{\frac{1}{2}}}{(4\,\alpha_x t v_c)^{\frac{1}{2}}}\right] \quad \text{(X6.13)}$$

This solution is based on the assumptions that the contaminant concentration is initially zero, a continuous dissolved hydrocarbon source exists at C_o, and the transport is contained within a semi-infinite medium. The respective initial and boundary conditions are as follows:

$$C(x,0) = 0$$
$$C(0,t) = C_o$$
$$C(\infty,t) = 0$$

Eq X6.11 reflects a steady-state when the complementary error function (erfc) argument approaches +2, such that the point of observation, *x*, is behind the advective front $vt(1+4\lambda\ \alpha_x/v_c)^{1/2}$ **(54)**. As such, the steady-state solution is:

$$C(x) = C_o \exp\left[\left(\frac{x}{2\alpha_x}\right)\left[1 - \left(1 + \frac{4\lambda\alpha_x}{v_c}\right)^{\frac{1}{2}}\right]\right] \quad \text{(X6.14)}$$

For the case in which decay occurs only in the aqueous phase, the contaminant velocity, v_c, is replaced by the seepage velocity, v_x.

X6.3.7 As the decay rate, λ, increases with respect to the other transport mechanisms, the extent of the plume at steady-state will decrease. As decay rate increases relative to the other mechanisms, the concentration away from the source (x > 0), approaches zero since the material is decaying at a greater rate than it is being transported through the medium. Similarly, as the contaminant velocity increases, the decay becomes ineffective in reducing concentrations as a function of distance. Retarded contaminants therefore have a greater opportunity to decay because retarded velocities favor biodegradation kinetics over transport **(54)**.

X6.3.8 The regression of contaminant concentration versus distance was developed in X6.2.2. This equation yields the reciprocal of the attenuation distance, k/v_x (L^{-1}):

$$C(x) = C_o e^{-(k\frac{x}{v_x})} \quad \text{(X6.15)}$$

Eq X6.14 and Eq X6.15 are of the same form:

$$C(x) = C_o \exp\,(mx) \quad \text{(X6.16)}$$

The slope of the log-linear data is given by m. The one-dimensional, steady-state transport solution also describes the slope, m, of the log-linear data:

$$m = \left(\frac{l}{2\alpha_x}\right)\left[l - \left(l + \frac{4\lambda\alpha_x}{v_c}\right)^{\frac{1}{2}}\right] \quad \text{(X6.17)}$$

Therefore, the term k/v_x and Eq X6.17 both describe the slope of the log-linear data and can be equated to solve for the total decay rate, λ, a measure of intrinsic bioremediation of petroleum hydrocarbons. Dispersivity (α_x), contaminant velocity (v_c), and k/v_x are input to following equation to calculate the decay rate **(51)**.

$$\lambda = \left(\frac{v_c}{4\alpha_x}\right)\left(\left[l + 2\alpha_x\left(\frac{k}{v_x}\right)\right]^2 - l\right) \quad \text{(X6.18)}$$

For the case in which decay occurs only in the aqueous phase, v_c is replaced by v_x in Eq X6.18.

X6.3.9 The Example Problem No. 2 in X7.2 presents a case study which couples the one-dimensional analytical solution to the regression of concentration versus distance for a steady-state or stable plume. This example problem illustrates the contribution of biodecay to the overall attenuation rate.

X6.4 One, Two, and Three-Dimensional Analytical Models

X6.4.1 Analytical models may be used for several purposes. Transient solutions can estimate the time required for an expanding or shrinking plume to reach a particular configuration. Steady-state solutions can be used to estimate the extent of a plume and select the locations of downgradient monitoring wells. The justification for a two or three-dimensional model should be based on the availability of data. Two of the more sensitive input parameters are the decay rate and the ground water seepage velocity. When using any of these models it is recommended that the model be calibrated to actual site data to confirm predicted results. While many of the models described in previous sections may yield similar first order rate constants, the rate constants are not necessarily interchangeable due to the differences in model assumptions and the lumped nature of the parameter. The analytical solutions which follow were derived in three dimensions, but can be easily simplified to the two or one dimensional cases.

X6.4.2 *Application of a Steady State Analytical Model*:

X6.4.2.1 The steady state analytical solution to the transport equation derived by Domenico **(55)** can be used to estimate the plume extent and to determine first order attenuation rate constants. Eq X6.19 represents the steady state, centerline solution of the transport equation assuming a finite sized, but constant source and unidirectional ground water flow. The solution accounts for ground water flow in the x direction, three-dimensional dispersion, and first order solute degradation.

$$\frac{C(x,0,0,\infty)}{C(0,0,0,0)} = \exp\left(\frac{x}{2\alpha_x}\left[1 - \left\{1 + \frac{4\lambda\alpha_x}{v_s}\right\}^{\frac{1}{2}}\right]\right) erf\left(\frac{Y}{4[\alpha_y x]^{\frac{1}{2}}}\right) erf\left(\frac{Z}{4[\alpha_z x]^{\frac{1}{2}}}\right) \quad \text{(X6.19)}$$

Where, $C(x,0.0,\infty)$ is the contaminant concentration along the primary flow direction of the plume, C(0,0,0,0) is the steady state source concentration, and x is the coordinate in the direction of ground water flow. a_x, a_y, and a_z are longitudinal, transverse, and vertical dispersivities, respectively. 1 is the first order degradation rate constant, v_x is the ground water seepage velocity, Y is the width of the source zone in the transverse direction, and Z is the thickness of the source zone in the vertical direction.

X6.4.2.2 To facilitate the use of Eq X6.19 in site specific evaluations, a computer spreadsheet was used to generate solution curves for various parameter values. A nomograph was then generated for the parameters listed in Table X6.3.

X6.4.2.3 An expanded view of Fig. X6.4 from 0 to 300 ft

TABLE X6.3 Typical Values for Parameters in Domenico Analytical Model[A]

Parameter	Value	Reference
Dispersivity longitudinal transverse vertical	α_x= 4 ft α_y= 0.33 α_x α_z= 0.035 α_x	Conservative values. Typical range for α_x is 4 to 20 ft. (Newell et al. (1996)).
Source Dimensions Transverse Vertical	Y = 20 ft Z = 5 ft	Typical width at UST site. Typical smear zone thickness. Site specific data can be used but these are not highly sensitive parameters.
First order degradation constant	λ = 0.001 to 0.01 day^{-1} Typical range for λ to be used as a check on realism of site specific value.	Salanitro (1993)
Ground water velocity	v_s= site specific value	

[A] Fig. X6.4 gives the steady state plume centerline concentration for values of l/v_s from 0.0001 to 1.0 with a longitudinal dispersivity value of 4 ft. As shown in Fig. X6.4, the centerline concentration is very sensitive to the ratio of l/v_s. This ratio provides a measure of the rate of contaminant degradation to the rate of ground water travel. Thus, as the value of l/v_s increases, the predicted extent of the plume decreases.

downgradient from the source is given in Fig. X6.5.

X6.4.3 *Estimation of Plume Extent*—Fig. X6.4 can be used to estimate the probable extent of a plume at steady state with two qualifications. First, a range of values for l/v_s should be considered and secondly, actual monitoring data should be collected to verify plume attenuation estimates. Here the plume extent is defined as the distance from the downgradient edge of the residual source along the plume centerline to the point at which the contaminant concentrations drop below a specified target concentration. To use Fig. 1 to estimate the extent of a plume at steady state, follow these steps:

X6.4.3.1 Select the appropriate range of values for l (the values in Table X6.2 are recommended for BTEX),

X6.4.3.2 Estimate the site specific ground water seepage velocity,

X6.4.3.3 Calculate the range of l/v_s values for the site,

X6.4.3.4 Determine the target concentration for the compound of concern and divide the target level by the source area concentration to get the normalized target concentration (for example, the source concentration is 10 000 ppb benzene and the downgradient target is 5 ppb then the normalized target concentration is 5×10^{-4}), and

X6.4.3.5 Read the range of estimated plume extent from Fig. X6.4 or Fig. X6.5 by determining the point at which the curves corresponding to the l/v_s values calculated in X6.4.3.2 cross the normalized target concentration calculated in X6.4.3.4.

X6.4.4 *Determination of First Order Decay Constant*—As mentioned above, estimates of plume extent should be followed by calibration of the model with site specific data to verify the assumptions and refine the estimates. To calibrate this modeling approach to site specific data and to determine first order rate constants, multiple monitoring points along the primary flowpath of the plume are necessary. The procedure for calibration with site specific data to determine l is outlined below:

X6.4.4.1 Normalize plume centerline concentrations by dividing by the source concentration,

X6.4.4.2 Determine the distance of each monitoring point from the downgradient edge of the source area,

X6.4.4.3 Plot the normalized centerline concentrations on Fig. X6.4 or Fig. X6.5,

X6.4.4.4 Read the value of l/v_s from the nomograph which corresponds to the curve which provides the best fit to the data, and

X6.4.4.5 Multiply the value of l/v_s by the site specific seepage velocity to determine l.

X6.4.5 The proper application of Fig. X6.4 can be demonstrated for an actual UST site (McAllister (**56**)). Historical monitoring data at this site indicate a stable plume and ground water seepage velocity of 0.03 ft/day. Centerline monitoring wells identified at the site and used in this example were MW-1 (source well), MW-4, and MW-8. The data on these wells is summarized in Table X6.4.

X6.4.6 The normalized benzene concentrations in Table X6.4 were plotted versus distance from the source on Fig. X6.6 and the data clearly fall in the expected range of l/vs values, 0.033 to 0.33 ft^{-1}. The calibrated site specific value for l/vs was determined to be 0.1 ft^{-1}. Multiplying by the site ground water seepage velocity, the first order degradation rate constant was determined to be 0.003 day^{-1}.

X6.4.7 *Application of a Transient Analytical Model*:

X6.4.7.1 Consideration of transient plume behavior is important when establishing remedial goals, setting monitoring frequencies, evaluating the plume status, and determining when steady state models may be appropriately applied.

X6.4.7.2 To evaluate transient plume behavior, the transient, centerline analytical solution derived by Domenico (**55**) has been applied. Eq X6.20 represents the transient solution for the centerline concentration as a function of time.

$$\frac{C(x,0,0,t)}{C(0,0,0,0)} = \frac{1}{2}\exp\left(\frac{x}{2\alpha_x}\left[1 - \left\{1 + \frac{4\lambda\alpha_x}{v_s}\right\}^{\frac{1}{2}}\right]\right)\mathrm{erfc}\left(\frac{x - v_s t\left[1 + \frac{4\lambda\alpha_x}{v_s}\right]^{\frac{1}{2}}}{2(\alpha_x x)^{\frac{1}{2}}}\right) \cdot \mathrm{erf}\left(\frac{Y}{4[\alpha_y x]^{\frac{1}{2}}}\right)\mathrm{erf}\left(\frac{Z}{4[\alpha_z x]^{\frac{1}{2}}}\right) \quad (X6.20)$$

By comparison of Eq X6.19 and Eq X6.20 it can be seen that one-half of the steady state concentration is achieved when the erfc term in Eq X6.20 equals unity.

$$\mathrm{erfc}\left(\frac{x - v_s t\left[1 + \frac{4\lambda\alpha_x}{v_s}\right]^{\frac{1}{2}}}{2(\alpha_x x)^{\frac{1}{2}}}\right) = 1\,. \quad (X6.21)$$

Solving Eq X6.21 for *t* gives the following expression for the time required to reach one-half the steady state concentration at any point along the primary flow path of the plume.

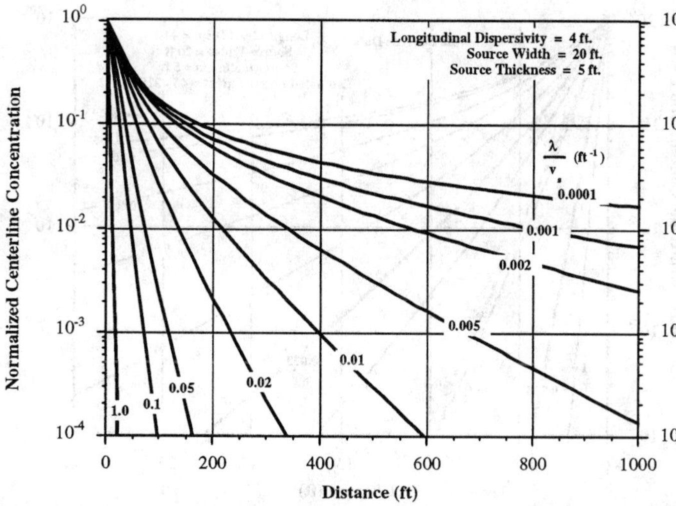

FIG. X6.4 Steady State Nomograph of Domenico (1987) Solution From 0 to 1000 ft Downgradient

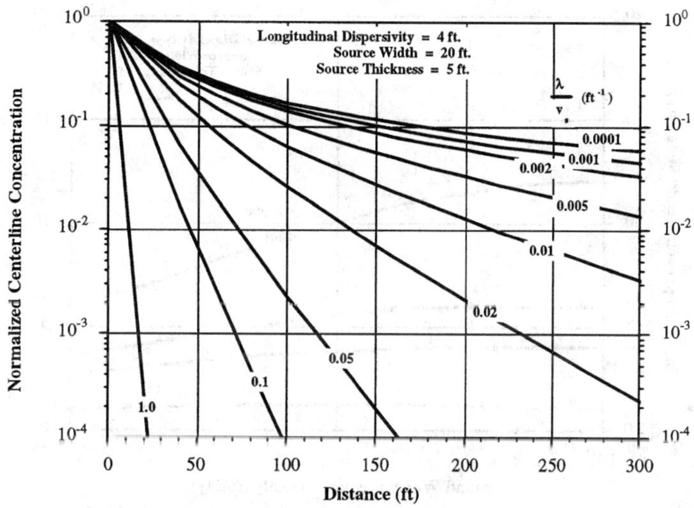

FIG. X6.5 Steady State Nomograph of Domenico (1987) Solution From 0 to 300 ft Downgradient

TABLE X6.4 Summary of UST Site Monitoring Well Data

Well ID	Distance from Source (ft)	Benzene Concentration (µg/L)	Normalized Benzene Concentration
MW-1	0	10 896	1.00
MW-4	23	1 207	0.11
MW-8	90	3	0.00028

$$t_{ss} = \frac{x}{\left(1 + \frac{4\lambda\alpha_x}{v_s}\right)^{\frac{1}{2}} v_s} \quad \text{(X6.22)}$$

Since one-half the steady state concentration is essentially the same as the steady state concentration given the typical variability in actual site data, the time t_{ss}, given in Eq X6.22 will be considered the time to approach steady state.

X6.4.7.3 Eq X6.22 was used to generate a nomograph by the following steps.

(*1*) Parameter values for l, v_s, and a_x were specified.

(*2*) Eq X6.20 and Fig. X6.4 were used to determine the steady state plume extent for a normalized target concentration of 10^{-3}.

(*3*) The value of t_{ss} was calculated from Eq X6.22.

(*4*) The procedure was repeated for various values of l and v_s.

X6.4.8 Shown in Fig. X6.7 are the time to approach steady state and the plume extent at steady state as a function of ground water seepage velocity for various values of l and a normalized target concentration of 10^{-3}(the normalized target concentration is the target concentration divided by the source concentration). The predicted behavior evident in these figures indicates that for high first order degradation rate constants (~0.01 day^{-1}), the time to steady state will become relatively independent of ground water seepage velocity and will be on the order of two years. As the value of l decreases, the predicted time to steady state increases and becomes more dependent on the seepage velocity.

X6.4.9 The use of Fig. X6.7, is demonstrated with data from the example site in X6.4.2. The site specific seepage velocity was 0.03 ft/day and the first order degradation rate constant was 0.003 day^{-1}. Thus, reading from Fig. X6.7 the time to approach steady state is approximately 4 years. The time to approach steady state for this site was consistent with the fact

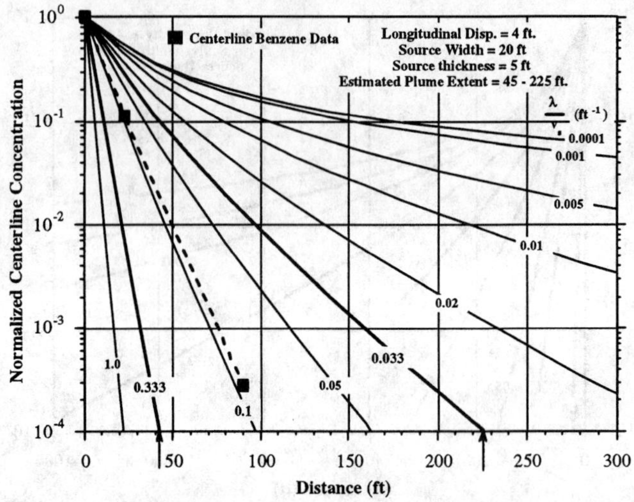

FIG. X6.6 Application of Domenico (1987) Steady State Solution to Estimate Plume Extent and First Order Degradation Rate Constant

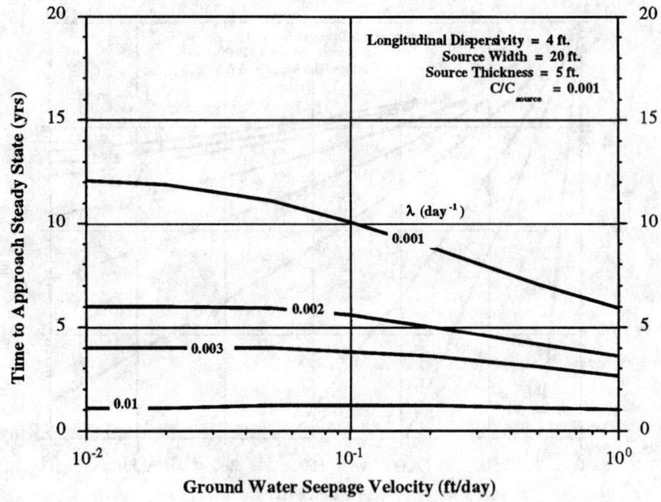

FIG. X6.7 Time to Approach Steady State Using the Domenico (1987) Transient Solution

that the release was over six years old and monitoring data indicated the plume was stable.

X7. EXAMPLE PROBLEMS

X7.1 Example Problem No. 1—Primary Lines of Evidence

X7.1.1 *Introduction*:

X7.1.1.1 The subject site is located on the east coast and is an active retail petroleum service station with nine monitoring wells on-site and one monitoring well offsite to the Northwest (Fig. X7.1). During a service station upgrade conducted in 1991, subsurface hydrocarbon impact was noted during UST removal activities. New USTs were installed along the western side of the property, with the former UST area located just south of MW-1 (Fig. X7.1). Depth to water is approximately 7 ft below grade and groundwater flow is to the northwest. Subsurface soils consist primarily of silts, with minor amounts of silty sands and clayey silts. No measurable free product has been found at the site. There are no private or public wells within a ½ mile radius of the site and city ordinances require hookup to the municipal water supply.

X7.1.1.2 A series of slug tests were conducted in monitoring wells MW-1, MW-4, MW-5, MW-8 and MW-9. The results of the analyses indicated that the site had an average hydraulic conductivity of 3.28 ft/day. With a site gradient of 0.01 ft/ft and a soil porosity of 0.35 (default value based on aquifer texture), the resulting groundwater seepage velocity was 0.1 ft/day (37 ft/yr).

NOTE X7.1—This example does not address methods to properly evaluate groundwater flow direction or velocity. Such information can be found in other documents. For the purpose of this example, groundwater velocity and flow direction are assumed to be as stated.

X7.1.2 *Primary Lines of Evidence*:

X7.1.2.1 The primary lines of evidence require a review of

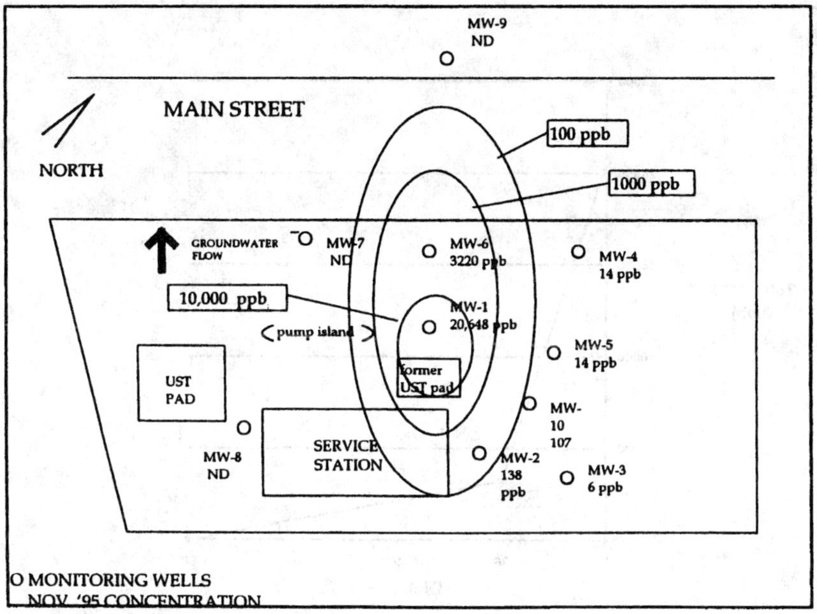

NOTE 1—Total BTEX concentrations and water table elevations in each monitoring well over nine consecutive quarters are given in Table X7.1. 1 in. = 40 ft.

FIG. X7.1 Service Station Site Map

historical analytical data for dissolved BTEX in ground water (see Table X7.1). These data are used first to classify the plume as stable, shrinking or expanding.

X7.1.2.2 In this example, the natural log of BTEX concentrations versus time was plotted (Fig. X7.2) for four wells, in order to characterize the plume. The natural log of well concentrations was used in the evaluation so that increasing, stable or decreasing trends could be clearly differentiated from random concentration fluctuations. Linear regression analysis of the data was also used to determine the equation of the best fit line to the data for each of the four wells (Fig. X7.2). It is clear from this analysis that concentrations in site wells have remained relatively stable for over two years, with the slope of the best fit line for each well approaching zero. In addition, the site map (Fig. X7.1) contains the 100, 1000, and 10 000 ppb total BTEX contours, based on the November 1995 data set (57). Review of the data over the last nine sampling quarters illustrates that these contours have remained relatively stable.

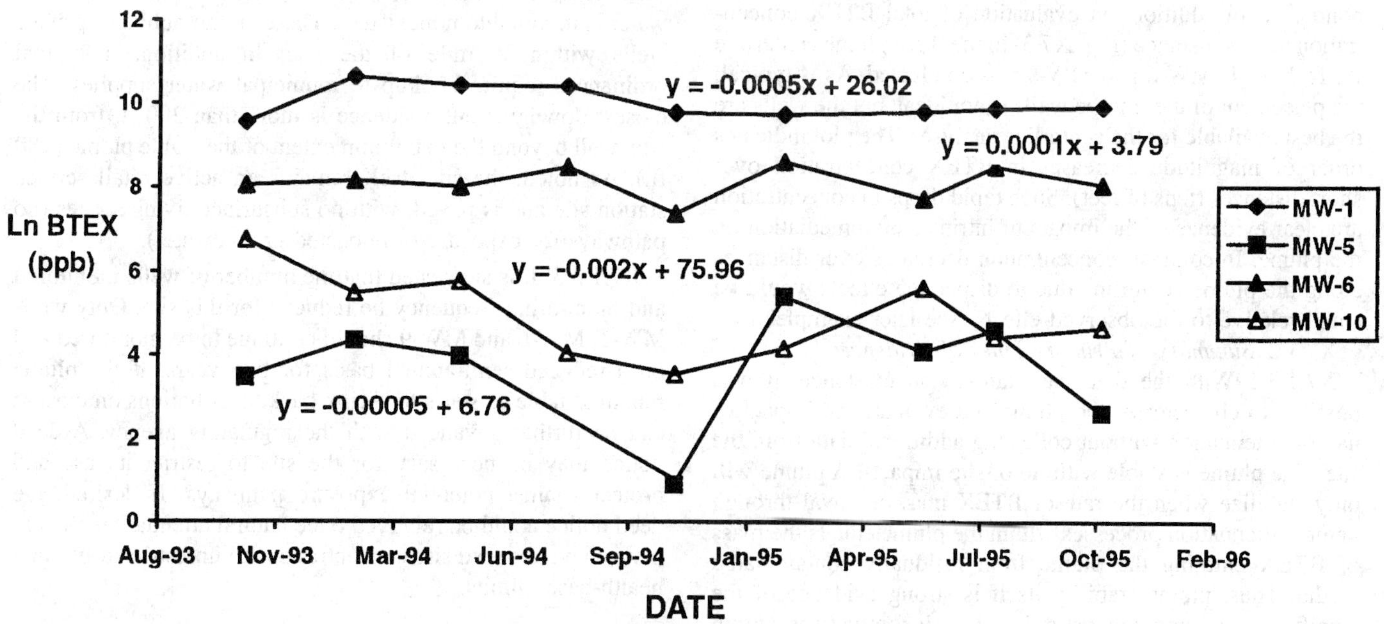

NOTE 1—Equations associated with each data set represents equation of best fit line.

FIG. X7.2 Plot of Natural Log of BTEX Concentrations Versus Time in Four Site Wells

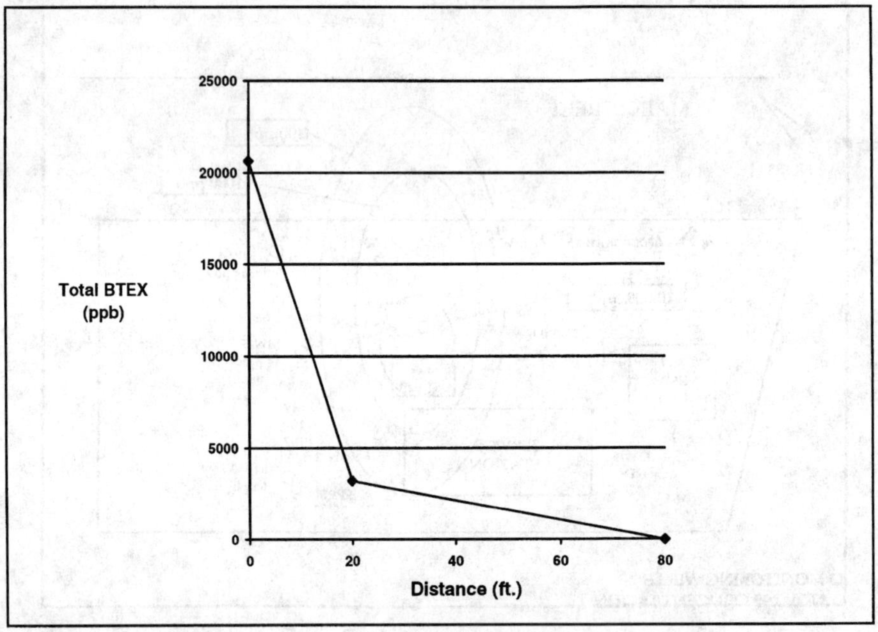

NOTE 1—Assuming MW-1 is in the source area (x = 0 ft.). MW-6 and MW-9 are 20 and 80 ft. from MW-1, respectively

FIG. X7.3 A Plot of Total BTEX Concentration Versus Distance, Using 3 Wells That Run Along the Plume Centerline (that is, MW-1, MW-6, MW-9)

Thus, both the concentration versus time plots for site wells and an evaluation of concentration contours indicates that the plume is stable.

X7.1.2.3 As previously noted, the spill has existed since at least 1991. Based on a ground water velocity of 0.1 ft per day and a very conservative estimate of spill age of 5 years, site ground water has migrated over 185 ft. (0.1 ft/day × 365 days/year × 5 years). In contrast, in June 1995, site data indicated that the maximum extent of plume migration was less than 80 ft, with the downgradient well at 80 ft. remaining nondetect. In addition, an evaluation of total BTEX concentration versus distance (Fig. X7.3) in the three plume centerline wells, MW-1, MW-6 and MW-9 was conducted. As is typical, the placement of these three wells is not ideal, but the wells are the best available for the centerline analyses. The plot indicates order of magnitude decreases in BTEX concentration over short distances (tens of feet). Such rapid drops in concentration are clear evidence of the impact of intrinsic bioremediation on the plume. In contrast, concentration decreases over distances along the plume centerline due to dispersive effects would be small relative to the observed effects (see later examples).

X7.1.3 *Summary of Primary Lines of Evidence*:

X7.1.3.1 With the data base already in existence, it was possible to characterize the plume and evaluate the impact of natural attenuation without collecting additional data from the site. The plume is stable with no offsite impacts. A plume will only stabilize when the rate of BTEX mass removal through natural attenuation processes within the plume equals the mass of BTEX entering the plume from residually contaminated media. Thus, plume stability itself is strong evidence of the significance of ongoing natural attenuation processes within the plume. In addition, as previously noted, an evaluation of total BTEX concentration versus distance in the three plume centerline wells MW-1, MW-6 and MW-9, indicated order of magnitude decreases in BTEX concentration over short distances (tens of feet). Such rapid drops in concentration are clear evidence of the impacts of intrinsic bioremediations on the plume. In contrast, concentration decreases over distance in the plume due to dispersive effects would be small relative to the observed effects.

X7.1.3.2 The risk associated with contact with site groundwater is minimal to nonexistent. There are no public or private wells within ½ mile of the site. In addition, municipal ordinances require hookup to municipal water supplies. The closest downgradient residence is more than 200 ft. from the site, well beyond the maximum extent of the stable plume (<80 ft.). As noted, the site itself remains an active retail service station site and is paved, with no subsurface living spaces (no pathways for exposure to impacted groundwater).

X7.1.3.3 It is suggested that the number of wells monitored and monitoring frequency be reduced for this site. Only wells MW-1, MW-6 and MW-9 should continue to be monitored and on a reduced semi-annual basis for two years. If the plume remains stable or starts to shrink back, negotiations on closure can be further advanced with the regulating agency. A deed notice may be necessary for the site to restrict its use and protect against potential exposure pathways. If desired, the deed notice could be removed once natural attenuation mechanisms have reduced site concentrations to unrestricted-use and health-based limits.

TABLE X7.1 Total BTEX Concentrations and Water Table Elevations (WTE) in Site Monitoring Wells Over Nine Consecutive Quarters

WELL #	DATE	WTE	TOTAL BTEX (ppb)	WELL #	DATE	WTE	TOTAL BTEX (ppb)
MW-1	Nov 93	92.14	13,950	MW-6	Nov 93	91.75	3,180
	Feb 94	93.84	42,080		Feb 94	93.25	3,455
	May 94	94.17	33,899		May 94	93.84	3,022
	Aug 94	92.61	33,678		Aug 94	91.97	4,876
	Nov 94	91.85	18,051		Nov 94	91.55	1,635
	Feb 95	92.89	17,331		Feb 95	92.70	5,569
	Jun 95	92.00	18,474		Jun 95	91.84	2,329
	Aug 95	91.67	20,193		Aug 95	91.59	5,082
	Nov 95	93.16	20,648		Nov 95	92.99	3,220
MW-2	Nov 93	91.80	96	MW-7	Nov 93	91.62	18
	Feb 94	94.01	12		Feb 94	93.10	50
	May 94	94.16	<1		May 94	93.46	107
	Aug 94	92.45	16		Aug 94	92.10	<1
	Nov 94	91.36	8		Nov 94	91.32	<1
	Feb 95	93.27	10		Feb 95	92.67	9
	Jun 95	92.13	394		Jun 95	91.81	<1
	Aug 95	91.51	362		Aug 95	91.56	2
	Nov 95	93.28	138		Nov 95	93.06	3
MW-3	Nov 93	91.70	14	MW-8	Nov 93	92.05	<1
	Feb 94	94.05	<1		Feb 94	93.50	<1
	May 94	94.40	<1		May 94	93.88	<1
	Aug 94	92.40	<1		Aug 94	92.44	<1
	Nov 94	91.47	<1		Nov 94	90.54	<1
	Feb 95	93.27	<1		Feb 95	93.10	<1
	Jun 95	92.07	<1		Jun 95	92.24	<1
	Aug 95	91.42	20		Aug 95	91.77	<1
	Nov 95	93.43	6		Nov 95	93.50	<1
MW-4	Nov 93	90.78	14	MW-9	Nov 93	90.54	<1
	Feb 94	NS	NS		Feb 94	NS	NS
	May 94	93.76	10		May 94	93.19	<1
	Aug 94	92.16	<1		Aug 94	91.43	<1
	Nov 94	90.68	7		Nov 94	90.71	<1
	Feb 95	92.55	6		Feb 95	91.61	<1
	Jun 95	91.54	26		Jun 95	90.44	<1
	Aug 95	90.85	10		Aug 95	90.03	<1
	Nov 95	91.49	14		Nov 95	91.26	<1
MW-5	Nov 93	91.45	32	MW-10	Nov 93	91.17	874
	Feb 94	93.66	78		Feb 94	93.76	244
	May 94	94.11	54		May 94	94.21	321
	Aug 94	92.59	<1		Aug 94	92.48	58
	Nov 94	91.20	2		Nov 94	91.30	34
	Feb 95	92.76	229		Feb 95	92.94	65
	Jun 95	91.78	61		Jun 95	91.84	276
	Aug 95	91.28	102		Aug 95	91.36	86
	Nov 95	92.97	14		Nov 95	92.89	107

POTENTIOMETRIC SURFACE ELEVATIONS (ft)

	April, 1993	July, 1993	November, 1993	February 1994	June, 1994	September, 1994
MW-32	385.53	383.04	379.73	381.16	388.67	381.39
MW-17	379.16	378.25	376.14	377.19	378.44	377.26
MW-15	375.50	374.35	372.96	373.68	374.00	373.21
MW-2	373.37	372.36	371.26	371.87	371.74	371.35
MW-4	366.70	365.55	364.68	365.22	365.70	365.05
MW-37	351.62	349.36	351.76	349.09	349.52	348.02

BENZENE CONCENTRATIONS (mg/L)

	April, 1993	July, 1993	November, 1993	February, 1994	June, 1994	September, 1994
MW-32	2.2	3.1	2.3	2.4	1.9	2.9
MW-17	93	27	12	17	76	10
MW-15	4,700	2,400	1,500	2,100	2,000	1,500
MW-2	3,100	3,700	2,700	2,800	4,200	3,900
MW-4	79	120	210	74	140	310
MW-37	0.1	0.05	0.25	0.25	0.25	0.25

TOLUENE CONCENTRATIONS (mg/L)

	April, 1993	July, 1993	November, 1993	February, 1994	June, 1994	September, 1994
MW-32	<0.5	7.5	<1	<1	<1	<1
MW-17	53	5.4	<1	<1	30	<1

TABLE X7.1 *Continued*

MW-15	3,300	690	86	210	580	360
MW-2	460	700	200	160	390	170
MW-4	1.3	22	2.2	<1	1.6	2.9
MW-37	<0.5	<0.5	<1	<1	<1	<1
ETHYL-BENZENE CONCENTRATIONS (mg/L)						
	April, 1993	July, 1993	November, 1993	February, 1994	June, 1994	September, 1994
MW-32	<0.8	1.5	<1	<1	<1	<1
MW-17	7.8	0.9	<1	<1	2.4	<1
MW-15	400	130	33	73	130	57
MW-2	220	250	180	130	240	160
MW-4	2.5	6.6	6.6	2.4	4.7	9.9
MW-37	<0.8	<0.8	<1	<1	<1	<1
XYLENE CONCENTRATIONS (mg/L)						
	April, 1993	July, 1993	November, 1993	February, 1994	June, 1994	September, 1994
MW-32	<1.7	9.7	<2	<1	<2	<2
MW-17	43	6.2	<2	<2	70	<2
MW-15	2,600	940	610	650	960	480
MW-2	910	1,100	610	510	1,000	630
MW-4	26	54	42	18	37	59
MW-37	<1.7	<1.7	<2	<2	<2	<2

X7.2 Example Problem No. 2—Primary and Secondary Lines of Evidence

X7.2.1 *Introduction*—Example Problem No. 2 demonstrates the application of primary and secondary lines of evidence. Fig. X7.4 is a well location and ground water contour map for a terminal in Fairfax, VA. Ground water flow is to the northwest, as shown on the figure. Ground water elevation data for the past few years indicate flow conditions remain fairly constant over time. Six monitoring wells are highlighted, oriented in the direction of ground water flow. BTEX concentrations suggest a contaminant source in the vicinity of MW-15.

X7.2.2 *Primary Lines of Evidence*—The primary lines of evidence entail a review of historical analytical data for dissolved BTEX in ground water. Table X7.1 summarizes six quarters (4/93-9/94) of ground water elevations and BTEX

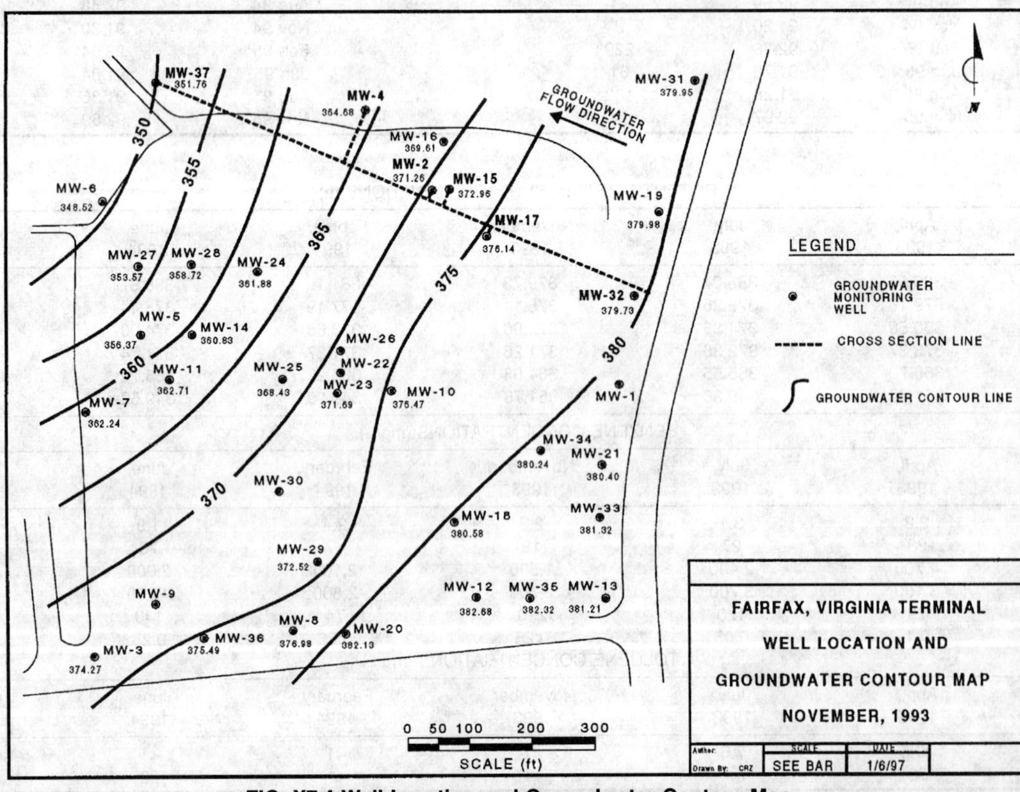

FIG. X7.4 Well Location and Groundwater Contour Map

concentrations for the six monitoring wells oriented along the direction of ground water flow and shown in Fig. X7.4. The ground water flow direction and dissolved concentrations remain fairly constant over time. BTEX concentrations in the downgradient monitoring well, MW-37, are non-detectable for the period of record. These trends suggest a stable dissolved plume. Fig. X7.5 is a plot of concentration versus distance for four monitoring wells, MW-15, MW-2, MW-4, and MW-37. These are average concentrations for the six monitoring events. MW-15 is assumed to be in the vicinity of the contaminant source and is assigned to the origin. The distances separating the three downgradient wells from MW-15 are provided in Table X7.2. The plot in Fig. X7.5 illustrates the exponential decline in concentration versus distance. This relationship of concentration versus distance is further evaluated to obtain the secondary lines of evidence for natural attenuation.

X7.2.3 *Secondary Lines of Evidence*—The secondary lines of evidence include the following tasks: (*1*) regression of concentration versus distance to determine an attenuation rate and (*2*) evaluation of indicator parameters for intrinsic bioremediation.

X7.2.3.1 Log concentration versus distance for monitoring wells MW-15, MW-2, MW-4, and MW-37 is plotted in Fig. X7.6. The concentrations plotted in the figure are the average concentrations for each monitoring well for six quarters, as shown in Table X7.3. The data plot as a straight line, with a goodness of fit, $R^2 = 0.993$, suggesting first-order decay of concentration with distance. The slope of this line, k/v_x, is 0.02 ft^{-1} (attenuation rate divided by seepage velocity). With an estimate of seepage velocity, the attenuation rate can be determined. Pump test data indicate an average seepage velocity of 0.2 ft/day. Therefore, the attenuation rate is estimated as follows:

$$k = (k/v_x)\ v_x \quad \text{(X7.1)}$$

$$k = (0.02\ ft^{-1})\ (0.2\ ft/day)$$

$$k = 0.004\ day^{-1}\ \text{or}\ 0.4\ \%/day$$

X7.2.3.2 The following intrinsic bioremediation indicator parameters were measured in ground water: dissolved oxygen (DO), oxidation-reduction potential, nitrate, sulfate, ferrous iron (Fe^{2+}), pH, biological oxygen demand (BOD), chemical oxygen demand (COD), and others. Fig. X7.7 illustrates plots for DO, oxidation-reduction potential, and ferrous iron versus distance from MW-15 (origin). Benzene concentrations are also shown on these plots. Dissolved oxygen (DO) concentrations are depleted in the middle of the dissolved plume, suggesting aerobic biodegradation is contributing to the configuration of the dissolved plume. The oxidation-reduction potential in the middle of the plume indicates reducing conditions where DO has been consumed. Ferrous iron is elevated in the middle of the plume, suggesting reduction of ferric iron (Fe^{3+}) as an alternative electron acceptor.

X7.2.4 *Optional Lines of Evidence*—An optional line of evidence applies a one-dimensional analytical solution for the stable plume. This solution is coupled with the concentration versus distance regression. The method calculates a decay rate which distinguishes between sorption, dispersion, and biodegradation.

X7.2.4.1 The slope of the line for concentration versus distance was determined by a secondary line of evidence (see X7.2.3.1). The slope, $k/v_x = 0.02\ ft^{-1}$, is used in the one-dimensional analytical solution to calculate a decay rate (Fig. X7.6). The results of equating the spatial regression with the steady-state analytical solution are presented in Table X7.4. The values for source concentration (C_o) and k/v_x were regressed using the data in Table X7.3. Table X7.4 includes k and the ratio, λ/k, the contribution of biodecay to the overall attenuation rate (expressed as %). In Case 1, seepage velocity is 0.2 ft/day based on aquifer pump tests. Retardation is

TABLE X7.2 Concentration Versus Distance

Monitor Well No.	Distance From MW-15 (ft)	Benzene Concentration (μg/L)[A]
MW-15	0	2367
MW-2	25	3400
MW-4	175	156
MW-37	505	0.2

[A]Average of data from 6 quarters (4/93 - 4/94).

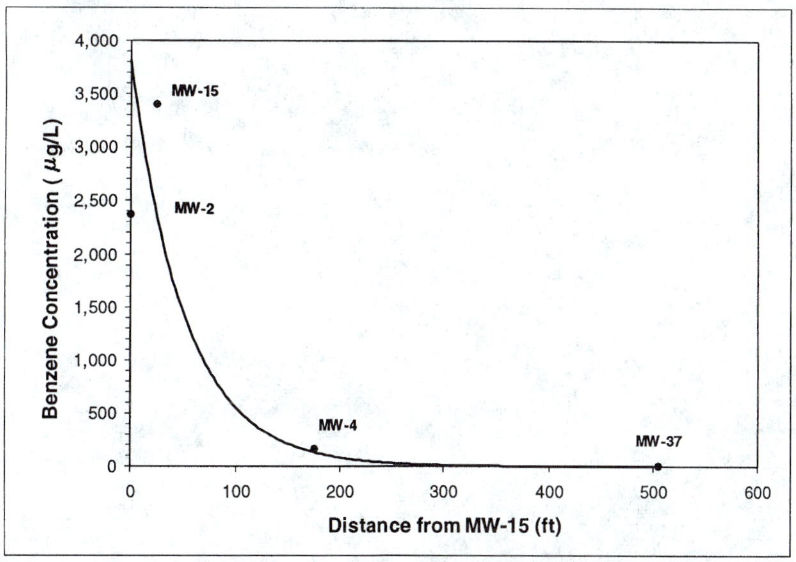

FIG. X7.5 Concentration Versus Distance

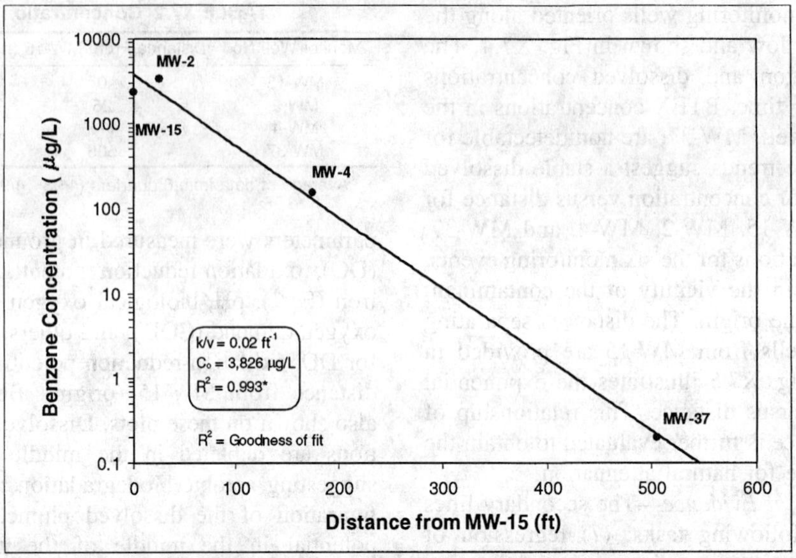

FIG. X7.6 Regression of Concentration Versus Distance

TABLE X7.3 Concentration Versus Distance

Monitor Well No.	Distance From MW-15 (ft)	Benzene Concentration (µg/L)[A]
MW-15	0	2367
MW-2	25	3400
MW-4	175	156
MW-37	505	0.2

[A] Average of data from 6 quarters (4/93 - 4/94).

estimated as 2 and dispersivity is estimated as 25 ft, approximately 5 % of the flow field (distance separating the two furthest wells). In Case 1, λ = 0.30 %/day (0.0030 days^{-1}); λ is 75 % of k for this case. The next three cases are performed to evaluate the sensitivity of changing various input parameters. In each of these cases C_o and k/v_x remain constant. In Case 2, seepage velocity is reduced by a factor of two (v_x = 0.10 ft/day), which reduces the decay rate by the same factor (λ = 0.15 %/day). In this case, only half the decay rate is required to maintain the Case 1 concentration decline with distance; as in Case 1, λ is 75 % of k in Case 2. In Case 3, the dispersivity is increased by a factor of two (α = 50 ft) and λ = 0.40 %/day. More decay is required with a larger dispersivity because more spreading of the contaminant occurs in the direction of ground water flow; λ is equivalent to k in Case 3. In Case 4, the retardation coefficient is 1, (v_c = v_x) and the dispersivity is reduced to 1 ft. In this case, biodecay is the only mechanism contributing to attenuation, therefore λ = k. These four cases demonstrate biodecay contributes significantly to the overall attenuation rate.

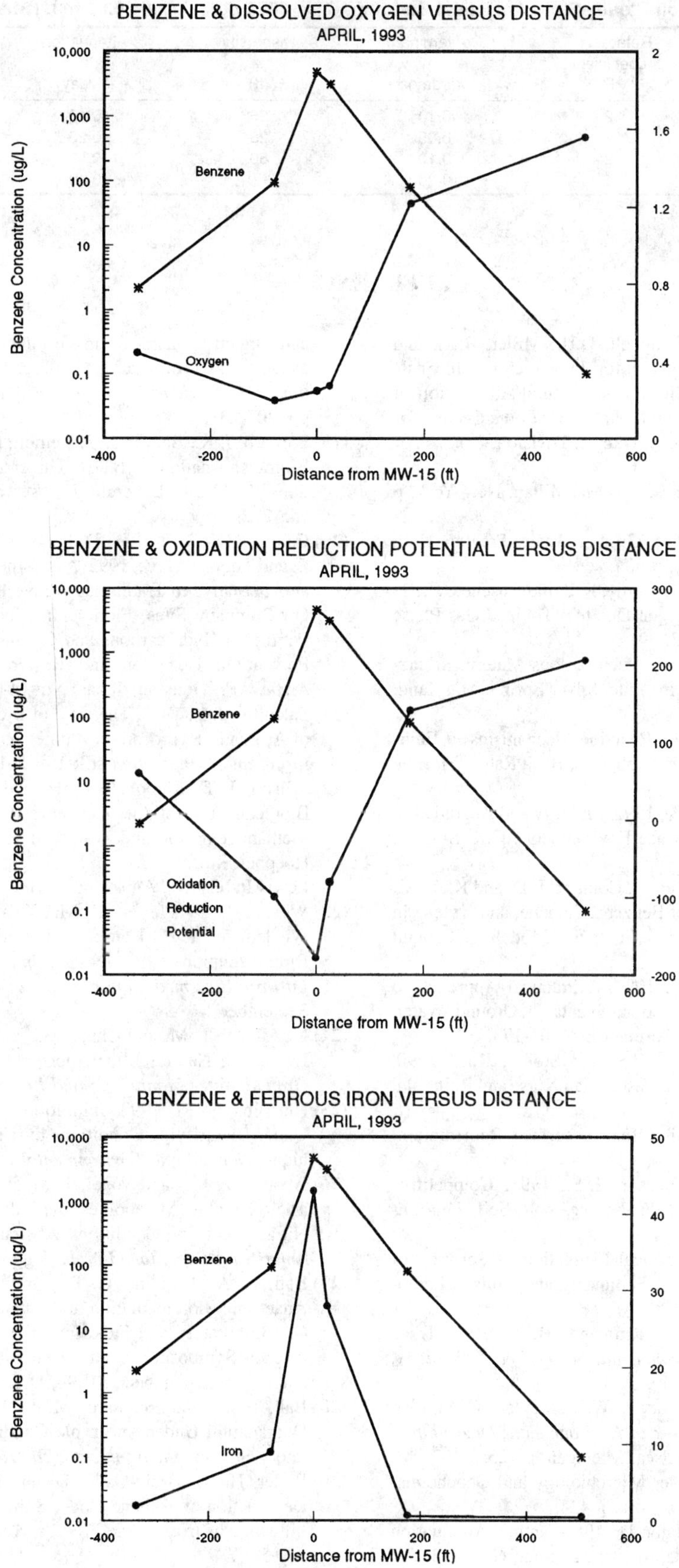

FIG. X7.7 Indicator Parameters

TABLE X7.4 Decay Rates Based on Steady-State Analytical Solution, C_o = 3826 μg/L, k/v_x = 0.02 ft^{-1}. (Buscheck and Alcantar (1995)

	Seepage velocity, v_x (ft/day)	Retardation coefficient R	Contaminant velocity, v_c (ft/day)	Dispersity α (ft)	Attenuation rate, k (%/day)	Decay rate λ (%/day)	λ/k (%)
Case 1	0.20	2	0.10	25	0.40	0.30	75
Case 2	0.10	2	0.05	25	0.20	0.15	75
Case 3	0.20	2	0.10	50	0.40	0.40	100
Case 4	0.20	1	0.20	1.0	0.40	0.40	100

REFERENCES

(**1**) Wiedemeier, T.H., Wilson, J.T., Kampbell, D.H., Miller, R.N., and Hansen, J.E., 1995, Technical Protocol for Implementing Intrinsic Remediation with Long-Term Monitoring for Natural Attenuation of Fuel Contamination Dissolved in Ground Water, *Air Force Center For Environmental Excellence*, Technology Transfer Division, Brooks Air Force Base, San Antonio, TX.

(**2**) Fetter, C.W., 1988. *Applied Hydrogeology:* MacMillan, New York, p 458.

(**3**) Freeze, R.A. and Cherry, J.A., 1979, Ground Water: Prentice-Hall, Inc., Englewood Cliffs, New Jersey, p 604.

(**4**) Fry, V.A., Istok, J.D., Semprini, L., O'Reilly, K.T. and Buscheck, T.E., 1995, Retardation of Dissolved Oxygen Due to a Trapped Gas Phase in Porous Media, Ground Water, 33, 391-398.

(**5**) Dragun, J., 1988, The Soil Chemistry of Hazardous Materials: Hazardous Materials Control Research Institute, Silver Spring, Maryland, p 458.

(**6**) Larson, R.A. and Weber, W.J., 1994, Reaction Mechanisms in Environmental Organic Chemistry: Lewis Publishers, Boca Raton, Florida, p 433.

(**7**) Lyman, W.J., Reidy, P.J. and Levy, B., 1992, Mobility and Degradation of Organic Contaminants in Subsurface Environments: C.K. Smoley, Inc., Chelsea, Michigan, p 395.

(**8**) Chiang, C. Y., Salanitro, J. P., Chai, E. Y., Colthart, J. D. and Klein, C. L., 1989, Aerobic Biodegradation of Benzene, Toluene, and Xylene in a Sandy Aquifer - Data Analysis and Computer Modeling, Ground Water, 27: 823-834.

(**9**) McAllister, P.M. and Chiang, C.Y., 1994, A Practical Approach to Evaluating Natural Attenuation of Contaminants in Ground Water, *Ground Water Monitoring Review*, Spring 1994, 161-173.

(**10**) Devinny, J.S., Everett, L.G., Lu, J.C.S. and Stollar, R.L., 1990, Subsurface Migration of Hazardous Wastes: Van Norstrand Reinhold, New York, p 387.

(**11**) USEPA, 1991 Site Characterization for Subsurface Remediation: EPA/625/4-91/026, p 259.

(**12**) Stuart, B.J., Bowlen, G.F. and Kosson, D.S., 1991, Competitive Sorption of Benzene, Toluene and the Xylenes onto Soil, *Environmental Progress*, 10(2), 104-109.

(**13**) Karickhoff, S.W., 1981, Semi-Empirical Estimation of Sorption of Hydrophobic Pollutants on Natural Sediments and Soils. Chemosphere, 10, 833-846.

(**14**) McCarty, P.L., Reinhard, M. and Rittmann, B.E., 1981, Trace Organics in Ground Water, *Environmental Science & Technology*, 15(1), 40-51.

(**15**) Howard, P.H., Boethling, R.S., Jarvis, W.F., Meylan, W.M. and Michalenko, E.M., 1991, *Handbook of Environmental Degradation Rates:* Lewis Publishers, Inc., Chelsea, Michigan, p 725.

(**16**) Chapelle, F.H., 1993, Ground Water Microbiology and Geochemistry: John Wiley & Sons, Inc., New York, p 424.

(**17**) Barker, J. F., Patrick, G. C. and Major, D., 1987, Natural Attenuation of Aromatic Hydrocarbons in a Shallow Sand Aquifer, *Ground Water Monitoring Review*, 7: 64-71.

(**18**) Kemblowski, M. W., Salanitro, J. P., Deeley, G. M. and Stanley, C. C., 1987, Fate and Transport of Residual Hydrocarbon in Ground Water - A Case Study, In Proceedings of the Petroleum Hydrocarbons and Organic Chemicals in Ground Water Conference: Prevention, Detection and Restoration Conference, *National Water Well Association and American Petroleum Institute*, Houston, TX., Nov., 17-22, pp. 207-231.

(**19**) Barbaro, J. R., Barker, J. F., Lemon, L. A. and Mayfield, C. I., 1992, Biotransformation of BTEX Under Anaerobic, Denitrifying Conditions: Field and Laboratory Observations, *J. Cont. Hydrology* 11: 245-272.

(**20**) Caldwell, K. R., Tarbox, D. L., Barr, K. D., Florenza, S., Dunlap, L. E. and Thomas, S. B., 1992, Assessment of Natural Bioremediation as an Alternative to Traditional Active Remediation at Selected Amoco Oil Company Sites, Florida, In Proceedings of the Conference on Petroleum Hydrocarbons and Organic Chemicals in Ground Water: Prevention, Detection and Restoration, *National Ground Water Assoc./API*, Houston, Texas, Nov. 4-6, pp. 509-515.

(**21**) Salanitro, J. P., 1993, The Role of Bioattenuation in the Management of Aromatic Hydrocarbon Plumes in Aquifers, *Ground Water Monitoring and Remediation*, 13: 150-161.

(**22**) Wilson, J. T., Kampbell, D. H. and Armstrong, J., 1994a, Natural Bioreclamation of Alkylbenzenes (BTEX) from a Gasoline Spill in Methanogenic Ground Water, In Hinchee, R. E., Alleman, B. C., Hoeppel, R. E. and Miller, R. N. (eds) *Hydrocarbon Bioremediation*, Lewis Publishers, Ann Arbor, MI.

(**23**) Wilson, J. T., Pfeffer, F. M., Weaver, J. W., Kampbell, D. H., Wiedemeier, T. H., Hansen, J. E. and Miller, R. N., 1994b, Intrinsic Bioremediation of JP-4 Jet Fuel, *In Proceedings of the Symposium on Intrinsic Bioremediation of Ground Water*, Denver, CO. August 30 - September 1.

(**24**) McAllister, P. M. and Chiang, C. Y., 1994, A Practical Approach to Evaluating Natural Attenuation of Contaminants in Ground Water, *Ground Water Monitoring and Remediation*, 14: 161-173.

(**25**) Hutchins, S. R., 1992, Inhibition of Alkylbenzene Biodegradation Under Denitrifying Conditions by Using the Acetylene Block Technique, *Applied and Environmental Microbiology*, 58:3395-3398.

(**26**) Alvarez, P. J. J. and Vogel, T. M., 1994, Degradation of BTEX and Their Aerobic Metabolites by Indigenous Microorganisms Under Nitrate-Reducing Conditions, Abstracts of the 94th Annual Meeting, *American Society for Microbiology*, Q-339.

(**27**) Ball, H. A., Hopkins, G. D., Orwin, E. and Reinhard, M., 1994, Anaerobic Bioremediation of Aromatic Hydrocarbons in Seal Beach, CA: Laboratory and Field Investigations, Environmental Protection Agency Symposium on Intrinsic Bioremediation of Ground Water, August 30-September 1, 1994, Denver, Colorado.

(**28**) Ball, H. A. and Reinhard, M., 1995, Monoaromatic Hydrocarbon Degradation Under Anaerobic Conditions at Seal Beach, CA:Laboratory Studies, *Environmental Toxicology and Chemistry*, (in press).

(**29**) Beller, H. R., Grbic-Galic, D. and Reinhard, M., 1992, Microbial Degradation of Toluene Under Sulfate-Reducing Conditions and the Influence of Iron on the Process, *Applied Environmental Microbiology* 58: 789-793.

(**30**) Edwards, E. A. and Grbic-Galic, D., 1992, Complete Mineralization of Benzene by Aquifer Microorganisms Under Strictly Anaerobic Conditions, *Applied Environmental Microbiology* 58: 2663-2666.

(**31**) Cozzarelli, I. M., Eganhouse, R. P. and Baedecker, M. J. 1990,

Transformation of Monoaromatic Hydrocarbons to Organic Acids in Anoxic Ground Water Environment, *Environmental Geology and Water Science*, 16:135-141.

(32) Baedecker, M. J., Cozzarelli, I. M., Eganhouse, R. P., Siegel, D. I. and Bennett, P. E., 1993, Crude Oil in a Shallow Sand and Gravel Aquifer - III, Biogeochemical Reactions and Mass Balance Modeling in Anoxic Ground Water, *Applied Geochemistry*, 8:569-586.

(33) Cozzarelli, I. M., Baedecker, M. J., Eganhouse, M. J. and Goerlitz, D. F. 1994, The Geochemical Evolution of Low-Molecular-Weight Organic Acids Derived from the Degradation of Petroleum Contaminants in Ground Water, *Geochimica et Cosmochimica Acta*, 58(2), 883-877.

(34) Lovley, D. R., Woodward, J. C. and Chapelle, F. H. 1994, Stimulated Anaoxic Biodegradation of Aromatic Hydrocarbons Using Fe(III) Ligands, *Nature*, 370:128-131.

(35) Borden, R. C., Gomez, C. A. and Becker, M. T. 1994, Natural Bioremediation of a Gasoline Spill, In *Hydrocarbon Bioremediation*, ed. Hinchee, R. E., Alleman, B. C., Hoeppel, R. E. and Miller, R. N., pp. 290-295, Boca Raton, Lewis Publishers.

(36) Wilson, B. H., Smith, G. B. and Rees, J. F., 1986, Biotransformation of Selected Alkylbenzenes and Halogenated Aliphatic Hydrocarbons in Methanogenic Aquifer Material: A Microcosm Study, *Env. Science Technol*. 20: 997-1002.

(37) Wilson, B. H., Wilson, J. T., Kampbell, D. H., Bledsoe, B. E. and Armstrong, J. M. 1990, Biotransformation of Monoaromatic and Chlorinated Hydrocarbons at an Aviation Gasoline Spill Site, *Geomicrobiology Journal*, 8:235-240.

(38) Edwards, E. A. and Grbic-Galic, D. 1994, Anaerobic Degradation of Toluene and O-Xylene by a Methanogenic Consortium, *Applied and Environmental Microbiology*, 60:313-322.

(39) Major, D. W., Mayfield, C. J. and Barker, J. F. 1988, Biotransformation of Benzene by Denitrification in Aquifer Sand, Ground Water, 26: 8-14.

(40) Lovley, D. R., Coates, J. D., Woodward, J. C. and Phillips, E. J. P. 1995, Benzene Oxidation Coupled to Sulfate Reduction, *Applied and Environmental Microbiology*, 61:953-958.

(41) Caldwell, K. R. and Suffita, J. M. 1995, Anaerobic and Microaerophilic Degradation of Benzene in Aquifer Sediments, Abstracts of the Third International Symposium on In-situ and On Site Bioreclamation. San Diego, California, April 24-27, 1995, E(.6.

(42) Newell, C. J., McLeod, R. K., Gonzales, J. R. 1996, BIOSCREEN Intrinsic Remediation Decision Support System User's Manual Version 1.2.

(43) Parker, J. C., Lenhard, R. J. and Kuppusamy, T., 1987, A Parametric Model for Constitutive Properties Governing Multiphase Flow in Porous Media, *Water Resources Research*, 23: 618-624.

(44) Shiu, W. Y., Maijanen, A., Ng, A. L. Y. and Mackay, D., 1988, Preparation of Aqueous Solutions of Sparingly Soluble Organic Substances: II. Multicomponent Systems - Hydrocarbon Mixtures and Petroleum Products, *Environmental Toxicology Chemistry* 7:125-137.

(45) Martin-Hayden, J. M., Robbins, G. A. and Bristol, R. D., 1991, Mass Balance Evaluation of Monitoring Well Purging, Part II, Field Tests at a Gasoline Contamination Site, Journal of Contaminant Hydrology, 8:225-241.

(46) Robbins, G. A. and Martin-Hayden, J. M. 1991, Mass-Balance Evaluation of Monitoring Well Purging, Part I. *Journal of Contaminant Hydrology*, v. 8, 203-224.

(47) Garbedian, S. P., Leblanc, D. R., Hess, K. M. and Quadri, R. D. 1987, Natural-Gradient Tracer Test in Sand and Gravels: Result of Special Moments Analysis. In B. J. Franks, (ed.), U. S. Geol. Survey Open File Report, 87-109, U. S. Geological Survey Program on Toxic Waste - Ground Water Contamination: Proceedings of the Third Technical Meeting, Pensacola, FL. March 23-27, pp. B13-B16.

(48) Sudicky, E. A., Cherry, L. A., and Frind, E. O., 1983, Migration of Contaminants in Ground Water at a Landfill: A Case Study, *Journal of Hydrology*, 63: 81-108.

(49) American Petroleum Institute (API), 1991, Technological Limits of Groundwater Remediation: A Statistical Evaluation Method, API Publication No. 4510.

(50) Wilson, J. T., Kampbell, D. H. and Armstrong, J., 1993, Natural Bioreclamation of Alkyl Benzenes (BTEX) From a Gasoline Spill in Methanogenic Groundwater, *Proceedings of the Environmental Restoration Technology Transfer Symposium*, San Antonio, TX.

(51) Buscheck, T. E. and Alcantar, C. M., 1995, Regression Techniques and Analytical Solutions to Demonstrate Intrinsic Bioremediation, In Hinchee, R. E., Wilson, J. T. and Downey, D. C. (Eds.), Intrinsic Bioremediation, pp. 109-116.

(52) Wiedemeier, T. H., Swanson, M. A., Wilson, J. T., Kampbell, D. H., Miller, R. N. and Hansen, J. E., 1996, Approximation of Biodegradation Rate Constants for Monoaromatic Hydrocarbons (BTEX) in Ground Water, Ground Water Monitoring and Remediation 16(3): 186-194.

(53) Bear, J. 1979, Hydraulics of Groundwater, McGraw-Hill, New York, NY.

(54) Domenico, P. A. and Schwartz, F. W., 1990, Physical and Chemical Hydrogeology, John Wiley & Sons, New York, NY.

(55) Domenico, P. A., 1987, An Analytical Model for Multidimensional Transport of a Decaying Contaminant Species, *Journal of Hydrology*, 91, 49-58.

(56) McAllister, P. M., 1996, Application of Screening Model Approaches for Evaluation of BTEX Natural Attenuation in Ground Water, Proceedings of Petroleum Hydrocarbons & Organic Chemicals in Groundwater: Prevention, Detection, and Remediation, Nov. 13-15, Houston, Texas.

(57) Buscheck, T. E. and Alcantar, C. M., 1995, Regression Techniques and Analytical Solutions to Demonstrate Intrinsic Bioremediation, In Hinchee, R. E., Wilson, J. T., and Downey, D. C. (Eds.), Intrinsic Biremediation, pp. 109-116.

Standard Guide for Stewardship for the Cleaning of Commercial and Institutional Buildings[1]

This standard is issued under the fixed designation E 1971; the number immediately following the designation indicates the year of original adoption or, in the case of revision, the year of last revision. A number in parentheses indicates the year of last reapproval. A superscript epsilon (ε) indicates an editorial change since the last revision or reapproval.

1. Scope

1.1 This guide covers a procedure to assist owners and operators of commercial and institutional buildings in the stewardship of cleaning and housekeeping operations. The focus of this guide is to address appropriate cleaning activities and processes, to maximize eco-efficiency and to minimize adverse impacts on the building occupants, cleaning personnel, the building structure itself, and the environment. Adherence to the principles set forth in this guide can lead to greater tenant/occupant satisfaction, reduced operational costs and greater productivity (of occupants and cleaning personnel).

1.2 This guide will focus on the development of a stewardship plan and will include the assessment of cleaning processes, product selection, storage, usage, disposal, equipment, training of cleaning personnel and communication throughout the chain-of-commerce.

1.3 This guide addresses issues relating to the operation and maintenance of the heating, ventilating and air conditioning (HVAC) systems which can have a major impact on indoor air quality (IAQ) only to the extent that the HVAC system provides adequate ventilation to lower risk to cleaning personnel, building occupants and the environment during or as a result of the cleaning process.

1.4 This guide is for use in a building that is maintained by either in-house cleaning personnel or an outside cleaning contractor.

1.5 This guide is not intended for construction related activities, but may be appropriate for post construction clean-up.

1.6 This guide is not intended as a procedural guide for cleaning personnel.

1.7 This guide is not intended for use in residential buildings.

1.8 *This standard does not purport to address all of the safety concerns, if any, associated with its use. It is the responsibility of the user of this standard to establish appropriate safety and health practices and determine the applicability of regulatory limitations prior to use.*

2. Referenced Documents

2.1 *ASTM Standards:*

D 5116 Guide for Small-Scale Environmental Chamber Determinations of Organic Emissions From Indoor Materials/Products[2]

D 5157 Guide for Statistical Evaluation of Indoor Air Quality Models[2]

D 5791 Guide for Using Probability Sampling Methods of Indoor Air Quality in Buildings[2]

E 1609 Guide for Development and Implementation of a Pollution Prevention Program[3]

PS 40 Using Indoor Carbon Dioxide Concentrations to Evaluate Indoor Air Quality and Ventilation[2]

3. Terminology

3.1 *Definitions of Terms Specific to This Standard:*

3.1.1 *baseline*—a set of critical observations of data used for comparison or a control.

3.1.2 *benchmarking*—a series of quantitative measurements of performance.

3.1.3 *chain-of-commerce*—manufacturers and suppliers of raw materials; manufacturers, marketers, and distributors of building cleaning products (and systems/equipment); building owners and managers; building cleaning contractors; and cleaning personnel.

3.1.4 *commercial and institutional buildings*—indoor or enclosed workspaces such as, but not limited to, office buildings, educational facilities, health care facilities, retail establishments, and other similar facilities. The term shall not include manufacturing and production facilities, warehouses, residences, and agricultural operations.

3.1.5 *eco-efficiency*—the delivery of competitively priced goods and services that satisfy human needs and bring quality of life, while progressively reducing ecological impacts and resource intensity throughout the life cycle, to a level at least in line with the Earth's carrying capacity.

3.1.6 *extended product responsibility*—a life cycle approach to identifying pollution prevention and resource conservation opportunities. It includes the underlying influence of consumer needs and preferences, government procurement, and the role played by those in the chain of production and distribution.

---

[1] This guide is under the jurisdiction of ASTM Committee E-50 on Environmental Assessment and is the direct responsibility of Subcommittee E50.03 on Global Sustainability/Pollution Prevention.

Current edition approved Oct. 10, 1998. Published January 1999.

[2] *Annual Book of ASTM Standards*, Vol 11.03.

[3] *Annual Book of ASTM Standards*, Vol 11.04.

Importantly, under this principle, manufacturers, suppliers, users and disposers of products share the responsibility for the environmental effects of products and waste streams.

3.1.7 *guide*—a series of options or instructions that do not recommend a specific course of action.

3.1.8 *hazard*—the potential health or physical effect(s) attributable to a specific chemical, mixture or physical agent.

3.1.9 *life cycle*—the stages of a product, process or activity, which encompass raw materials extraction and acquisition, processing, materials manufacture, product fabrication, packaging and distribution, product use/reuse, maintenance, recycling and final disposition.

3.1.10 *material safety data sheet (MSDS)*—a written or printed material concerning a hazardous chemical which contains the information set forth in the OSHA Hazard Communication Standard (see paragraph (g) of 29 CFR 1910.1200) **(1)**.[4]

3.1.11 *pollutant*—any substance that directly or indirectly creates an adverse human health or environmental effect when introduced into any environmental media.[5]

3.1.12 *pollution prevention*—the act of reducing or eliminating the use, release, or generation of a pollutant or potential pollutant through source reduction, recycling, reuse, reclamation, or modification of operating practices.

3.1.13 *risk*—the probability of deleterious health or environmental effects, **(2)**.

3.1.14 *standard*—as used in ASTM, a document that has been developed and established within the consensus principles of the Society and that meets the approval requirements of ASTM procedures and regulations.

3.1.15 *stewardship*—the responsibility for managing, conducting or supervising the quality, state or condition of a commercial or institutional building.

3.1.16 *worker participation*—the involvement of cleaning personnel or their representatives, or both, in all aspects of the cleaning process; including product selection, evaluation and appropriate work practices, training and communication of hazards, and "the process of work."

4. Significance and Use

4.1 Cleaning provides well documented benefits in terms of creating cleaner, safer and healthier surroundings by extracting harmful pollutants from the indoor environment (see Ref **(3)**). An improperly maintained indoor environment could give rise to biological contaminants, and buildup of particulate matter and gases which can have serious health effects. These negative impacts may have adverse affects on worker productivity affecting both cleaning personnel and tenants through increased complaints, absenteeism, injuries, asthmatic incidents or other symptoms. Inappropriate or improper use or selection of cleaning products and processes, along with failure to follow label directions could result in injury or illness to cleaning personnel or building occupants. In addition, it may be detrimental to the physical structure and systems of the building, or to the environment. Moreover, owners and operators maintain the liability for the proper function of the building and its impacts on the occupants and cleaning personnel.

4.1.1 This guide provides a basic reference for the development and preservation of a building environment that is considered safe and healthy for occupants, while reducing the stress on the overall environment as a result of routine maintenance. The anticipated users of this guide include building managers, cleaning personnel, product suppliers and distributors, union representatives and building occupants, who serve together in a stewardship role regarding the maintenance of the building. This guide is intended to raise pertinent questions regarding specific building environments in order that an appropriate stewardship strategy may be developed, for example:

4.1.1.1 How is the building used?

4.1.1.2 Are there any special cleaning requirements?

4.1.1.3 Are there any at-risk populations that need to be considered, such as children, asthmatics, or pregnant woman?

4.1.1.4 How are cleaning materials used?

4.1.1.5 Are there any special issues relevant to construction and furnishings?

4.1.1.6 Are there any issues relating to building age/architectural, such as historic preservation requirements?

4.1.1.7 are there any engineering concerns, such as HVAC systems and natural ventilation?

4.1.1.8 How is the quality of cleaning being evaluated/measured?

4.1.2 Regardless of the specific requirements, this guide will help in the formulation of a comprehensive plan resulting in reduced risk to cleaning personnel, building occupants and the environment.

4.2 This guide will help the building owner and operator understand the cleaning process through the following:

4.2.1 The development of a stewardship plan (see Section 6), will clarify the level of cleanliness that is required or expected, and will ensure that the cleaning process is carried out in a consistent manner with adequate communication feedback to promote success of the plan.

4.2.2 An understanding of extended product responsibility (see Section 7) and the importance of shared responsibility. This section includes task identification and performance requirements, process and product selection, use, storage and disposal.

4.2.3 An identification of the training and communications issues (see Section 8) that will encourage involvement with the entire chain-of-commerce in the cleaning process. These issues are related to both procedural training and feedback opportunities for cleaning personnel, as well as information sharing with building tenants to inform them of possible cleaning process impacts.

5. Stewardship Principles

5.1 *Stewardship Principles*—A building owner or operator should manage the cleaning process according to the following stewardship principles:

[4] The boldface numbers given in parentheses refer to a list of references at the end of the text.

[5] It should be noted that ASTM's definition of Pollution Prevention is different from some definitions used by the Environmental Protection Agency. See, for example, 58Fed.Reg.6478(Jan. 29, 1993, Council on Environmental Quality), and 58Fed.Reg.41,981(Aug. 6, 1993, Executive Order).

5.1.1 Take a comprehensive process approach to cleaning. This includes:

5.1.1.1 Identifying the cleaning task and performance requirements,

5.1.1.2 Defining or outlining steps of the cleaning procedure,

5.1.1.3 Selecting the correct products and equipment,

5.1.1.4 Training cleaning personnel to use correct procedures and to understand the potential health, safety and environmental impacts of the cleaning products and processes,

5.1.1.5 Ongoing inspection and monitoring programs,

5.1.1.6 Communicating clearly with building occupants, and

5.1.1.7 Practicing appropriate storage and disposal methods.

5.1.2 Focusing on only a single area can create unanticipated adverse impacts in other areas.

5.1.3 Foster participation of cleaning personnel and building occupants. A successful cleaning program must encourage participation and input along the entire chain-of-commerce. Cleaning personnel need to participate in the planning, implementation, and continual improvement of the cleaning program. Building occupants should be given the opportunity to participate.

5.1.4 Clean to protect health and safety. Appearances may be deceptive. Even buildings that appear to be clean can be unhealthy. Focus on cleaning for health and safety by controlling microorganisms, spills, gases, dust particles, etc. In most cases the appearance will be addressed at the same time.

5.1.5 Clean and maintain the building as a whole, not just separate components. Cleaning and maintenance in one area of a building can have a major impact on other areas. Appropriate actions must take place to ensure the health and safety throughout the entire building, in addition to the area where the work is being performed.

5.1.6 Recognize occupant impacts on the indoor environment. Occupants must share responsibility for maintaining health and safety. Occupants should quickly and clearly communicate with cleaning personnel to facilitate the rapid solution to problems. Furthermore, occupants must recognize how their behavior, such as food debris resulting from eating in their work areas, can contribute to additional cleaning and pest management requirements, which in turn have further impacts.

5.1.7 Ensure cleaning personnel and building occupant safety at all times. All products and processes may pose some risks, thus safety should always be considered. Safety precautions should be used during all cleaning processes, such as proper ventilation, personal protective equipment, and safety signage where necessary. Safety precautions should also apply to outside contractors, such as pest management or roofing contractors to protect building occupants from the impacts of those products and processes.

5.1.8 Be aware of the impacts to the outdoor environment. Impacts to the outdoor environment can include energy requirements, water usage, disposal of products in terms of solid waste, down-the-drain to the appropriate water treatment works, and emissions to the air during storage or use of products.

6. Developing A Stewardship Plan

6.1 *Plan Development*—Building owners and operators need to oversee the development of a written stewardship plan and its periodic review. This plan will help to ensure that expectations are being met and that the building is being cleaned and protected as required.

6.1.1 Components of the stewardship plan should include both scheduled routine cleaning, as well as responses to building occupant and worker related problems. Furthermore, specific sections need to be developed to address accidents and preventative maintenance programs. These sections of the plan should address common problems, as well as slip and fall accidents, weather-related problems, water leaks, smoke or obnoxious odors.

6.1.1.1 *Building Policy and Goals*—Goals and policies for cleaning activities should be laid out clearly. The building plan should include a definition of the building's primary mission which will assist managers and staff in understanding the scope and priority of stewardship activities. For example, because the primary mission of a health care facility and that of a retail facility are different, the scope and priority of stewardship activities may be substantially different.

6.1.1.2 By carefully considering the traffic level, time to perform the task, types of soil, soil load, safety and hygiene performance requirements, etc. for each part of the building the cleaning process can be optimized. This should result in adequate cleanliness for the function of the building, optimal occupant/tenant satisfaction, improved productivity, optimal cleaning cost effectiveness, appropriate choice of cleaning processes and products, and the greatest achievable eco-efficiency.

6.1.2 *Management Commitment*—The success of the stewardship plan is contingent upon top management commitment. The building owner/operator must be personally committed to success and ensure that senior building management follow through on the stewardship program. Management commitment may be demonstrated through:

6.1.2.1 Selecting a stewardship coordinator and outlining his/her responsibilities,

6.1.2.2 Establishing a stewardship task force with adequate representation by staff from all key areas of the building, such as cleaning personnel, tenants, parents of students, safety and health professionals,

6.1.2.3 Committing staff and resources,

6.1.2.4 Committing funding,

6.1.2.5 Purchasing equipment (such as new vacuums, if required) and ensuring their maintenance,

6.1.2.6 Training programs, and

6.1.2.7 Ongoing communications with cleaning personnel and occupants.

6.1.2.8 The stewardship plan should discuss progress in each of these areas and plan for the future.

6.1.3 *Baseline Study/Benchmarking*—A baseline study or benchmarking will promote the development and implementation of a stewardship plan that may reduce the environmental, health and safety impacts of cleaning activities. This is a key element in the stewardship process. Two primary reasons for benchmarking are goal setting and process development. Baseline study or benchmarking can provide the building owner/manager with measurements to control and manage his/her operations. The key is that the building owner/manager must first define the goal to be achieved, whether it be a desired level of cleanliness, occupant satisfaction, or addressing a particular problem. A baseline study or benchmarking can be accomplished through facility auditing and should focus on the entire cleaning process from which one can identify the greatest risks and opportunities for risk reduction. In scoping the need, cleaning/maintenance management and the building owner/manager should be encouraged to evaluate or audit their operations, procurement, and processes to identify, prioritize, and focus on the greatest opportunities to reduce or eliminate the volume of environmental or human health impacts, while adequately performing the cleaning task.

6.1.4 The study should include a review of the:

6.1.4.1 Use of space within the building, including any potential changes in use or occupancy,

6.1.4.2 Occupant habits, such as smoking or eating at work stations,

6.1.4.3 Occupant or cleaning personnel complaints,

6.1.4.4 Existing information from occupants and cleaning personnel regarding cleaning products or processes,

6.1.4.5 Indoor air quality (IAQ) assessment or test results, if available,

6.1.4.6 Existing cleaning processes/methodologies,

6.1.4.7 Current cleaning schedules,

6.1.4.8 Utilization of cleaning personnel,

6.1.4.9 Training programs,

6.1.4.10 Communications,

6.1.4.11 Document handling,

6.1.4.12 Current product usage, handling, storage, and environmental attributes,

6.1.4.13 Hazard Communications, including Material Safety Data Sheets for cleaning products, and

6.1.4.14 Relevant OSHA injury/illness records and/or workers compensation claims.

6.1.4.15 Some examples of existing forms, outlines and communication tools (letters) to assist in the development of the baseline study can be found in Refs **(4)**, **(5)** and **(6)** The stewardship plan should summarize the information.

6.1.5 *Identification of Needs, Opportunities and Options*—Based on the results of the baseline study, brainstorming sessions should be held among staff and managers to identify systematically needs, opportunities and options for instituting stewardship measures. The plan should summarize the results of these efforts. Paragraphs 6.1.5.1 through 6.1.5.3 identify some of the key issues that can be considered.

6.1.5.1 *Cost*—The major cost element in cleaning is labor and thus will have a critical impact when identifying opportunities and options. In reviewing the overall cost of cleaning it is important to ensure that sufficient provisions are made for initial and continuing worker training and that appropriate staffing levels are allocated to achieve the desired performance requirements. Inadequate training of cleaning personnel and inadequate staffing levels may lead to improper handling of cleaning products and the failure to meet required performance criteria. This, in turn, may lead to increased risk of worker and occupant exposure and the possible improper use of cleaning materials.

6.1.5.2 *Performance*—The question building owner/operator should ask is: "How clean is clean enough?" Building owners/operators should clearly understand the use of the building and the expectations of occupants/tenants. These expectations for cleaning performance should be clearly defined. The cleaning requirements should be agreed upon with cleaning personnel and translated into the size of the cleaning crew, time required on the job and choice of cleaning processes and products.

6.1.5.3 *Time Available to do the Task*—The time available to do a task affects the frequency and duration of the task. Processes and products must be compatible with any unique challenges this poses. For example, there will be greater flexibility in scheduling and completing a cleaning task in a building that is not occupied at night, than a facility that is used 24 h a day. Sometimes, cleaning processes and products may need to address the requirement for rapid execution of a cleaning task in order to avoid interruption of the use of the building by occupants. If the allotted time is shortened to the point where only the most aggressive cleaning processes must be use to successfully accomplish the task, this demand may place cleaning personnel and building occupants at higher risk requiring additional training and more rigorous risk management steps to protect cleaning personnel and occupants when compared to using additional time with less aggressive products/processes. It should also be noted that the need for personnel resources has its own environmental impacts in the amount of employee hours necessary to accomplish any cleaning task.

6.1.6 *Ranking of Options*—Criteria should be developed for prioritizing the needs and opportunities identified and for ranking the options developed. The stewardship plan should explain the criteria used and present the results of the ranking. Typical criteria include (not in order of importance):

6.1.6.1 Worker safety,

6.1.6.2 Tenant and occupant requirements,

6.1.6.3 Costs (life cycle costs should be developed to the degree possible),

6.1.6.4 Liability,

6.1.6.5 Regulatory compliance,

6.1.6.6 Implementation feasibility,

6.1.6.7 Time and staff limitations,

6.1.6.8 Appearance and performance requirements,

6.1.6.9 Environmental impacts, and

6.1.6.10 Staff experience.

6.1.7 *Implementation and Evaluation*—The implementation section of the plan should set schedules for completion of major milestones, such as the completion of the stewardship plan, identify roles and responsibilities, identify barriers encountered or expected, outline communication and training needs for both cleaning personnel and tenants, indicate how success will be measured and evaluated, and outline priorities for future activities.

6.1.8 *New Goals*—After the stewardship plan has been developed, implemented and fully evaluated, the plan should be periodically re-evaluated to identify opportunities for improvement.

7. Extended Product Responsibility

7.1 *Impacts of Extended Product Responsibility*—Extended product responsibility is an approach to identifying pollution prevention and resource conservation opportunities. It identifies the underlying theme of shared responsibility, which includes the role played by those throughout the chain-of-commerce. In this case, the responsibility for reducing impacts is shared among product manufacturers and distributors, cleaning personnel, building owners/managers, and occupants.

7.2 *Process and Product Selection*—By evaluating cleaning procedures, a determination can be made as to which processes are truly necessary, which can be eliminated, and which can be replaced by other technologies. Cleaning procedures should be reviewed to identify and manage hazards.

7.2.1 *General Considerations*—The cleaning process (that includes equipment) and product selection should consider performance, cost, workplace health and safety, and environmental impacts. There is a role for both cleaning personnel and suppliers in ensuring proper process and product selection.

7.2.1.1 Employees responsible for procurement can influence the reduction of environmental impacts and maximize the environmental benefits across many stages of the product's life cycle. It is important to keep in mind that an environmental improvement in one stage of a life cycle may sometime occur at the expense of another life cycle stage. For example, some products that are packaged in a concentrated form may reduce packaging, but could also increase the potential that the product user may be exposed to the concentrate. Exposure to the concentrate may place the product user at greater risk than exposure to the ready-to-use product. Thus, in order to reduce negative environmental impact across both packaging and use stage of the product's life cycle, it is preferable in this example, that products shipped as concentrates are provided in packaging that minimizes concentrate exposure or are accompanied by additional appropriate precautionary measures, or both, and instructions or include the use of portion control equipment designed to minimize exposure to the concentrate, or both. Furthermore, even products that are considered "safe" when used as directed, can pose problems if misused. Risks associated with the use of products classified as "hazardous" can be minimized when all specified directions are followed.

7.2.1.2 It is important that process and product selections be made that consider the capability of the manufacturer/distributor to provide, in addition to fulfilling price and performance requirements, the necessary training and technical support. With proper training, cleaning process and product selection that take into account safety, health and environmental aspects during storage, handling, use, maintenance and disposal can be assured. Manufacturers/distributors should also be able to demonstrate their capability to receive and act upon feedback from cleaning personnel regarding product performance, use and disposal, as well as health, safety and environmental related matters.

7.2.2 *Workplace Health And Safety Considerations*:

7.2.2.1 *Product Labels*—The labels of products designed for the workplace are a valuable source of information which contain the product identity, appropriate hazard warnings, and the name and address of the manufacturer or other responsible party (see 29CFR1910.1200(f)) **(1)**. Product labels also may contain manufacturers' instructions for proper use, as well as storage and disposal requirements for cleaning products. All products should be properly labeled.

7.2.2.2 *Material Safety Data Sheets*—Building owners/operators can use the product Material Safety Data Sheet (MSDS) as a tool in the cleaning process/product selection review. The purpose of the MSDS is to alert employees to the potential risks associated with a product and to provide the employee with appropriate information (that is, precautionary measures) necessary to avoid those risks. In addition, the MSDS will provide information including the physical and health hazards of the product, any generally applicable work practices or personal protective equipment, emergency and first aid procedures, and procedures for clean up of spills and leaks (see 29CFR1910.1200(g)) **(1)**.

7.2.2.3 The MSDS is limited in its content to information of importance to the safety and health of employees and therefore, may not include environmental related information. Furthermore, the MSDS is often written for the concentrated product and may not accurately reflect the product that will be used after appropriately diluted. Additional information can be obtained from product suppliers and through OSHA's Consultation Service.

7.2.3 *Environmental Considerations*:

7.2.3.1 *Source of Environmental Information*—Many suppliers provide environmental information with their product literature which can be used during product or process selection. Often, additional information is available on request from the supplier.

7.2.3.2 *Environmental Marketing Statements*—The Federal Trade Commission (FTC) has published Guides for Environmental Marketing Claims which are administrative interpretations of laws administered by the FTC and which offer guidance in the use of environmental claims in marketing and advertising **(7)**. The Guides specifically address the application of Section 5 of the Federal Trade Commission Act, which prohibits unlawful deceptive advertising acts and practices, to environmental advertising and marketing practices.

7.2.3.3 The FTC has concluded that it is deceptive to misrepresent, directly or by implication, that a product or package offers a general environmental benefit. Unqualified use of terms such as "eco-safe," "environmentally friendly", "environmentally safe", "essentially non-toxic" or "environmentally preferable" is recognized by the FTC to be deceptive.

7.2.3.4 *Evaluating Environmental Impacts*—Cleaning provides well documented (see Ref **(3)**) benefits in terms of creating cleaner, safer, healthier surroundings, but, like all processes, has some environmental impact. One important step in managing the environmental impact is to choose the correct product and process for the job and use only what is needed of the product.

7.2.3.5 Information on environmental impacts is most useful in product/process selection when the use situations and local environmental conditions are considered. For example, some cleaning products may require the use of organic solvents in their formulation to satisfactorily and efficiently perform the task. In such cases, the volatile organic compound (VOC) content does not necessarily indicate high VOC emissions. Differences in VOC content can be overshadowed by in-use amounts needed to do the job. That is to say, using 0.1 oz of a 100 % VOC product may have a lower impact than using 1 oz of a 25 % VOC product.

7.2.3.6 At this time there is no scientifically agreed upon method for making product comparisons. However, efforts are being expended to develop methods at the international, federal, state, and local levels, as well as by private organizations.

7.2.4 *Supplier Support*—A building owner or operator should take into account the ability of a product supplier to support the proposed cleaning program when evaluating which products to use. Indicators of support would include:

7.2.4.1 A commitment to train cleaning personnel (see 7.3.2),

7.2.4.2 The availability of product labeling and other communication systems designed for non-English speaking and illiterate personnel,

7.2.4.3 Open communication channels to ensure two-way information flow to/from product users to encourage product development that meets user needs,

7.2.4.4 A willingness to discuss specific environmental attributes of processes and products on an in-use performance basis, and

7.2.4.5 Access to information to substantiate process and product claims.

7.3 *Use Stage*:

7.3.1 *Guidelines for Proper Use*—A successful cleaning program carried out correctly, will maximize worker/tenant safety, tenant satisfaction, and cost efficiency. Cleaning procedures and product use training are essential components of a stewardship plan.

7.3.1.1 It is important that cleaning products be used in accordance with the manufacturer's recommendations or directions, or both. Common recommendations include, but are not limited to, the proper use concentration for dilutable products and utilization of ventilation appropriate to the product and process being used. The MSDS and the label of cleaning products should also be reviewed for any applicable precautions and control measures for safe handling and use, such as the use of appropriate personal protective equipment, adequate ventilation, and appropriate work practices. A review of the MSDS and product label will also indicate whether additional precautions may be necessary to protect building occupants.

7.3.1.2 *Manufacturer's Instructions*—Cleaning personnel must always follow manufacturer's instructions for proper storage, use and disposal as detailed on the product label. Following those instructions is essential to ensure worker health & safety.

7.3.1.3 *Workplace Health And Safety Requirements*—Compliance with all applicable Occupational, Safety, and Health Administration (OSHA), and other federal, state and local workplace health and safety requirements is mandatory. If products are transferred to other containers, ensure that OSHA-required labeling information accompanies the product. The label must include product identity, appropriate hazard warnings, and the name and address of he manufacturer or other responsible party. In addition, labels often contain information related to the product's use, dilution, proper application, and storage and disposal instructions. Care should be taken to prevent the spread of vapors and spills when chemicals are transferred to other containers.

7.3.1.4 Material Safety Data Sheets (MSDS) must be kept on file for all current products and be readily available for review by cleaning personnel and other employees who may be exposed to such products. The MSDS must include details of hygiene practices, personal protective equipment and clean-up procedures for spills and leaks. Regular safety training sessions, as well as specific training for introduction of new cleaning processes and hazards in the workplace should be provided to cleaning personnel. Personnel protective equipment, if specified on the product's label for specific cleaning tasks should be worn. And finally, appropriate safety signs, such as "Caution—Wet Floor" should be utilized.

7.3.1.5 *Cleaning Schedule*—Cleaning should be scheduled to minimize risks to tenants/occupants. Work should be scheduled in such a way that it does not impact other areas, for example inadvertently via the HVAC system. Building occupants should be informed of cleaning activities that may have an affect on their health and welfare.

7.3.1.6 *Ventilation*—Adequate ventilation (see for additional information on ventilation see Ref **(8)**) should be available for cleaning personnel during normal occupancy hours, as well as other times if cleaning is taking place.

7.3.2 *Training in Product Use*:

7.3.2.1 *Accountability for Training*—Building owners/operators should ensure that appropriate procedural and product training for cleaning staff and supervisors is taking place. Adherence to manufacturer's label instructions, and workplace health and safety requirements are necessary to ensure a safe, healthy environment.

7.3.2.2 *Recognition of Workforce Dynamics*—Where turnover of cleaning personnel is frequent, the building owner or operator will need to be especially diligent with respect to routine process and product use training. Use of multi-lingual labels, pictograms and color coded process/product systems may help reduce the potential for misuse. When training employees, consideration should be given to the employees' education and literacy level, as well as any potential language barriers.

7.4 *Storage and Disposal*:

7.4.1 *Storage*—The stewardship plan should ensure that cleaning equipment and products are stored in a controlled environment. Equipment and products must be stored in accordance with manufacturer's instructions and all federal, state and local requirements to facilitate product identification and safe use. An inventory control system could be used so that only the required amounts of a product are purchased for the job, and that products are rotated so that the oldest product is used first.

7.4.1.1 Storage areas should be secured to allow access only to those qualified and trained in proper use. This may require a lockable facility for buildings with large "at risk" populations, for example school children. Special requirements for equipment and products that pose particular hazards, such as electrical shock, flammability, reactivity, etc. shall be followed. All wastes, including used packaging and wash solutions should be stored or disposed of in a manner consistent with the label instructions on the cleaning product, and with federal, state and local regulations.

7.4.1.2 *Waste Management/Recycling*—The waste management system instituted by building owners and their cleaning personnel/contractors should fully utilize recycling programs available in their communities. This action may keep significant amounts of packaging out of the disposal waste stream. To this end, products should be used completely, so that their packaging can be placed in community recycling programs. It is recommended that purchases of cleaning products be limited to the amount necessary to perform the task. This will minimize the potential problems that can result from the need to dispose of excess product.

7.4.2 *Disposal*—Extended product responsibility requires a life cycle approach. As such, emphasis should be placed on completely using up the product, which is the best method to minimize disposal needs and to facilitate container recycling. If there is product remaining that requires disposal, the manufacturer's instructions provide the starting point for ensuring proper disposal. Compliance with federal, state and local regulations, (including those relating to the nature, quantity and maximum length of time that wastes can be stored on a site without a permit) is mandatory. Rinse containers prior to recycling/disposal and use the rinse water, if appropriate, or dispose properly. If excess product is to be exchanged or given away, it should be in its original container with its label intact and its MSDS provided.

7.4.2.1 Liquid cleaning residues/wastes are typically designed to be disposed of to a sanitary wastewater treatment system and must not be disposed of down an untreated storm drain. Importantly, even though the cleaning product in liquid waste may be safe for disposal, depending on the building being cleaned, substances removed during cleaning may have adverse environmental impact.

7.5 *Occupant Responsibilities*—Occupants must share responsibility for maintaining health and safety. This begins with the recognition of how their own activities may impact the building in total, or the occupants working in nearby areas. For example, food debris resulting from eating in work areas, can impact the building's requirements for cleaning and pest management.

7.5.1 Furthermore, in the event of a spill or accident, such as coffee or toner cartridge spill, occupants must take care of the problem themselves, or communicate quickly and clearly with cleaning personnel. Often the failure to quickly address these problems, results in the need for more rigorous cleaning later.

8. Training and Communications

8.1 *Importance and Legal Requirements*—While training and communications are addressed in almost every section of this guide, the importance of these issues cannot be over emphasized in the successful implementation of a stewardship program. Furthermore, training is a legal requirement for users of hazardous products as part of OSHA's Hazard Communication Standard (see Ref (**1**)) or relevant state regulations.

8.1.1 *Training and Communication On Infectious Disease*—OSHA also requires training for certain employees who are determined to have occupational exposure to blood or other potentially infectious material to reduce the risk of exposure to blood and certain body fluids containing bloodborne pathogens.

8.1.1.1 The OSHA standard requires the employer to identify every worker who performs tasks and procedures in the workplace that may involve exposure to blood or infectious agents. Once occupationally exposed employees have been identified, the next step is to communicate the hazards of the exposure to the employees.

8.1.1.2 Employees who are determined to have occupational exposure to blood and certain body fluids containing blood, must be provided comprehensive training on bloodborne pathogens, the OSHA regulations and the employer's exposure control plan.

8.1.1.3 For those with the potential for occupational exposure to airborne infectious agents, such as tuberculosis (TB), training should be provided that includes signs and symptoms, modes of transmission, the difference between infection and disease, method for diagnosis, ways of preventing transmission and current treatment methods.

8.2 *Training and Communication Plans*—Written training and communication plans should be developed addressing the needs of both cleaning personnel and building occupants.

8.2.1 *Safety Training*—OSHA requires that employers provide employees with effective information and training on hazardous chemicals in their work area at the time of their initial assignment and whenever a new physical or health hazard the employees have not previously been trained about is introduced into their work area (see Ref (**1**)). Implement the safety program with regular meetings to discuss safety related topics. Training should emphasize the necessity of always following the manufacturer's recommendations for use, storage, disposal, safety precautions and first aid.

8.2.2 *Correct Product Usage, Storage and Disposal*—Training on correct product usage, storage and disposal will help to ensure minimal adverse health and environmental impacts, as well as the most efficient utilization of labor.

8.2.3 *Cleaning Personnel/Worker Participation*—Building owners/operators should ensure that cleaning personnel participate on an equal basis, and are provided with the necessary training and education to participate effectively in the stewardship program. Specific attention should be made to individuals susceptible to language and literacy barriers, and for any special situations.

8.2.3.1 Cleaning personnel should select their own representatives to provide input into and participate in all essential aspects of the stewardship program, including process and product selection, usage and evaluation. Building owners/operators should monitor feedback from cleaning personnel who should be assured that communications are open and that retribution will take place, if complaints are made.

8.2.4 *Communicating with Building Occupants*—Communications are essential within a building to inform occupants of cleaning activities. This is especially important in areas where building occupants may have pre-existing health conditions.

8.2.4.1 Furthermore, for the stewardship plan to succeed, occupants must share responsibility for their building environment and should be informed of activities that contribute to building problems, such as storing and consuming of food in their work areas.

8.2.5 *Feedback Loops*—Communication plans should provide feedback loops and a log should be kept to register suggestions, and to track response to problems and complaints. Information should be provided to building occupants and cleaning personnel about how and with whom to log suggestions and complaints. A communications schedule should be established to assure regular and timely interactions.

9. Keywords

9.1 cleaning; commercial buildings; custodial services; extended product responsibility; green buildings; housekeeping; indoor air quality; janitorial services; loss prevention; operations and maintenance; regulatory compliance; risk minimization; stewardship

REFERENCES

(**1**) *Hazard Communication Standard*, Occupational Safety and Health Administration, 29 CFR Part 1900.1200, Washington, DC.

(**2**) Federal Register 57, 104, Washington, DC, May 29, 1992.

(**3**) *Indoor Environmental Characterization of a Non-Problem Building: Assessment of Cleaning Effectiveness*, Research Triangle Institute, CR-815509-02-1, Research Triangle Park, NC, March 1994.

(**4**) *Building Air Quality: A Guide for Building Owners and Facility Managers*, U.S. Environmental Protection Agency, EPA/400/1-91/033, Washington, DC 1991.

(**5**) *Indoor Air Quality: Tools For Schools Action Kit*, U.S. Environmental Protection Agency, 402-K-95-001, Washington, DC, 1995.

(**6**) *Cleaning Makes Cents: Benchmarking for Managing Your Cleaning Operations*, Building Owners and Managers Association (BOMA) International, Washington, DC, 1997.

(**7**) *Guides For The Use Of Environmental Marketing Claims*, Federal Trade Commission (FTC), 16 CFR Part 260.

(**8**) American Society for Heating, Refrigeration and Air-Conditioning Engineers (ASHRAE) Standard 62.

Designation: E 1983 – 98

Standard Guide for Assessment of Wetland Functions[1]

This standard is issued under the fixed designation E 1983; the number immediately following the designation indicates the year of original adoption or, in the case of revision, the year of last revision. A number in parentheses indicates the year of last reapproval. A superscript epsilon (ϵ) indicates an editorial change since the last revision or reapproval.

1. Scope

1.1 This guide covers assisting wetland managers by prescribing a sequence of steps for defining and assessing wetland functions. This guide also identifies properties that must be considered in the selection of a wetland assessment procedure to determine whether it will assist in satisfying the requirements of wetland regulatory programs or produce valid design criteria for planned wetlands, or both. This guide can help wetland managers use existing assessment procedures more effectively during the decision-making process. The outcome of the assessment is dependent on many factors including the selected procedure, the sampling design, and assumptions; therefore, decisions and assumptions made should be documented throughout the process. While this guide is developed to assist in satisfying the requirements of wetland regulatory programs, it can also be used in a variety of planning, management, and educational situations.

1.2 The guide is not intended for use in assigning values to wetland functions in terms of economic (for example, dollars) or other value units. However, the information that is gathered while assessing wetland functions may be useful in meeting this objective when used in conjunction with other information (for example, see Refs **(1)**[2] and **(2)**).

1.3 This guide applies to assessment procedures designed for application at the ecosystem scale. It does not address the less commonly used landscape level models or the use of wetland assessment procedures for cumulative impacts analysis **(3-5)**.

1.4 *Limitations*—This guide does not include a standard wetland assessment procedure or models for assessing function. This guide has been written primarily to complement and to aid in the selection of current procedures. There are several procedures for quantifying wetland functions and each has been developed for specific purposes. The suitability of a procedure depends on assessment objectives, wetland type, availability of applicable models given the wetland type and objectives, and policy of local decision makers. There are continuous efforts to develop new and improved methods that could override any one recommended standard procedure.

1.5 The values stated in inch-pound units are to be regarded as the standard. The SI units given in parentheses are for information only.

1.6 *This standard does not purport to address all of the safety concerns, if any, associated with its use. It is the responsibility of the user of this standard to establish appropriate safety and health practices and determine the applicability of regulatory limitations prior to use.*

2. Terminology

2.1 *Definitions:*

2.1.1 *wetland assessment procedure, n*—a definitive procedure for identifying, characterizing, or measuring the functions that a wetland performs, or a combination thereof.

2.1.2 *wetland functions, n*—the physical, chemical, and biological processes or attributes that contribute to the self-maintenance of wetland ecosystems **(6)** and **(7)**. Wetland functions result directly from the characteristics of a wetland ecosystem and the surrounding landscape and their interaction.

2.1.2.1 *Discussion*—A wetland function is distinguished from wetland value. Wetland functions are a direct result of the characteristics of a wetland and the surrounding landscape. Examples of functions include the removal of dissolved substances, cycling of nutrients, maintenance of plant and animal communities, and short-term storage of surface water. These functions provide benefits, goods, and services which may be assigned a value (economic or noneconomic) describing the relative importance of a wetland function to an individual or group of people. The values of wetlands are estimates, usually subjective, of the worth, merit, quality, or importance of wetland functions **(8)**.

2.2 *Definitions of Terms Specific to This Standard*—Most of the following definitions are from Refs **(9)** and **(10)**.

2.2.1 *assessment model, n*—a simple model that defines the relationship between ecosystem and landscape scale variables and functional capacity of a wetland; it is used to derive a measure of functional capacity (that is, the functional capacity index).

2.2.2 *conceptual design, n*—a design that provides a brief description of the planned wetland through drawings and text which confirms feasibility and facilitates early review by decision makers.

2.2.3 *functional capacity, n*—the magnitude or rate at which a wetland performs a function. Functional capacity is dictated by the characteristics of the wetland ecosystem and the

---

[1] This specification is under the jurisdiction of ASTM Committee E-50 on Environmental Assessment and is the direct responsibility of Subcommittee E50.05 on Wetland Ecosystems.

Current edition approved Nov. 10, 1998. Published February 1999.

[2] The boldface numbers given in parenthesis refer to a list of references at the end of the text.

surrounding landscape, and the interaction between the two.

2.2.4 *functional capacity index (FCI)*, *n*—an index of the capacity of a wetland to perform a function relative to other wetlands from a defined region or wetland class, or both. Functional capacity indices are by definition normally scaled from 0.0 to 1.0. An index of 1.0 indicates that a wetland performs a function at maximum functional capacity. An index of 0.0 indicates the wetland does not perform the function.

2.2.5 *functional capacity units (FCs)*, *n*—a measure of the capacity of a wetland to perform a function that links functional capacity with area ($FC = FCI \times$ size of wetland area).

2.2.6 *planned wetland*, *n*—design or an implemented design for a constructed, created, restored, or enhanced wetland.

2.2.7 *variable*, *n*—an attribute or characteristic of a wetland ecosystem or the surrounding landscape that influences the capacity of a wetland to perform a function; used in assessment models to derive a measure of functional capacity (that is, the functional capacity index). Variables may be described by direct measures or indicators. A direct measure is a quantitative measure of an assessment model variable. An indicator is an observable characteristic that corresponds to identifiable variable conditions in a wetland or the surrounding landscape.

2.2.8 *wetland assessment area (WAA)*, *n*—the wetland area being assessed. In regulatory situations, the WAA will usually be jurisdictional wetlands confined to the area of direct or indirect impact or both.

2.2.9 *wetland classification*, *n*—the grouping of wetlands into different categories based on specific criteria (that is, vegetation type, hydrology, geomorphology) for the purpose of inventory, assessment, and management.

2.2.9.1 *Discussion*—There are several wetland classification schemes including the Classification of Wetlands and Deepwater Habitats of the United States (**11**) and the hydrogeomorphic classification (**12**). Each has been prepared for different purposes. One or more of these classifications may be used in the process of assessing wetland functions.

2.3 *Additional Terminology*—The following definitions and discussions, taken directly from the publication "Form and Style for ASTM Standards," shall be included in full in every standard guide or practice produced and passed by Committee E-50 or any of its technical subcommittees; approved April 16, 1997.

2.3.1 *guide*—a series of options or instructions that do not recommend a specific course of action.

2.3.1.1 *Discussion*—Whereas a practice prescribes a general usage principle, a guide only suggest an approach. The purpose of a guide is to offer guidance, based on a consensus of viewpoints, but not to establish a fixed procedure. A guide is intended to increase the awareness of the user to available techniques in a given subject area and to provide information from which subsequent evaluation and standardization can be derived.

2.3.2 *practice*—a definitive procedure for performing one or more specific operations or functions that does not produce a test result.

2.3.2.1 *Discussion*—A practice is not a downgraded test method. Examples of practices include procedures of interlaboratory testing programs or other statistical procedures; for writing statement on sampling or precision and accuracy; and for selection, preparation, application, inspection, and necessary precautions for the use, disposal, installation, and maintenance, and operation of testing equipment.

2.3.3 *standard*—as used in ASTM, a document that has been developed and established within the consensus principles of the Society and that meets the approval requirements of ASTM procedures and regulations.

2.3.3.1 *Discussion*—The term "standard" serves in ASTM as an adjective in the title of documents such as test methods or specifications, to connote specified consensus and approval. The various types of standard documents are based on the needs and usages as prescribed by the technical committees of the Society.

3. Summary of Guide

3.1 This guide is summarized in Table 1, that shows the steps in defining and assessing wetland functions and the sections of this guide that apply.

3.2 The remainder of this guide identifies properties to consider when selecting a procedure or models, and a summary of existing procedures (see Section 7). Appendix X1 describes the specific application of wetland assessment to planned wetlands.

4. Significance and Use

4.1 Wetland managers may be aware of wetland assessment procedures, but not use them as effectively as possible for a variety of reasons. There is no one universally accepted procedure; therefore, time is often lost to identifying and agreeing upon a suitable approach. The absence of guidance describing the sequence of steps to assessing wetland functions causes confusion and delays the decision-making process. Only recently has guidance been published by the U.S. Army Corps of Engineers (**9**).

4.1.1 This guide shows the person(s) performing an assessment the steps to assess wetland functions. This guide also provides a summary of the variety of procedural options for measuring function, and includes a list of properties to consider when selecting an appropriate procedure.

TABLE 1 Suggested Steps in the Assessment of Wetland Functions

Steps	Rules
Characterization Phase:	6.2
Define objectives of the assessment	6.2.1
Select functions	6.2.2
Describe the project area	6.2.4
Identify wetland assessment area(s)	6.2.6
Screen for red flags	6.2.7
Identification of Assessment Approach:	6.3
Identify, modify, or develop assessment models	6.3.1
Select the units of measure	6.3.2
Assessment Phase	6.4
Analysis Phase:	6.5
Types of comparisons	6.5.1
Other criteria to consider when comparing wetland assessment areas	6.5.5

4.2 *Situations Requiring Assessment of Wetland Functions*—While this guide is developed to assist in satisfying the requirements of wetland regulatory programs, it can also be used in a variety of planning, management, and educational situations.

4.3 *Rapid Assessment for Section 404 Permitting*—Section 404 of the Clean Water Act (33 U.S.C. 1344)[3] directs the U.S. Army Corps of Engineers, in cooperation with the U.S. Environmental Protection Agency (EPA), to administer a program for permitting and regulating the discharge of dredged or fill materials in waters of the United States, including wetlands. A permit application undergoes a public interest review that includes an assessment of the impacts the proposed project will have on wetland functions.

4.3.1 An assessment may be performed during one or more of the following steps of the review sequence that are prescribed in the U.S. EPA 404(b)(1) Guidelines (40 CFR Part 230)[4]

4.3.1.1 *Step 1*—Determine whether the proposed project is water dependent.

4.3.1.2 *Step 2*—Determine whether practicable alternatives exist for the proposed project.

4.3.1.3 *Step 3*—Identify the potential impacts of the proposed project on wetland functions in terms of project specific and cumulative effects.

4.3.1.4 *Step 4*—Identify how potential project impacts can be avoided or minimized in terms of project-specific and cumulative effects.

4.3.1.5 *Step 5*—Determine appropriate compensatory mitigation for unavoidable project impacts.

4.3.1.6 *Step 6*—Grant or deny a permit to discharge dredged or fill material by comparing the value of the benefits gained from the proposed project versus the value of benefits lost from the proposed project.

4.3.1.7 *Step 7*—If a permit is granted, monitor compensatory mitigation.

4.3.2 Wetland functions are assessed during Step 2 to compare impacts of practicable alternatives and to identify which is least damaging. During Steps 3 and 4, wetland functions are assessed to identify and then determine how to avoid or minimize project-specific and cumulative impacts. Wetland functions are assessed in Step 5 to determine what constitutes appropriate compensatory mitigation for unavoidable impacts. Compensatory mitigation is wetland restoration or creation, or enhancement or preservation of an existing wetland to compensate for wetland impacts. Several approaches to mitigation may be considered and compared at this time including in-kind, out-of-kind, on-site, off-site, and mitigation banking. Whichever option is chosen, the mitigation project is later assessed (during Step 7) to determine whether the function-based objectives have been met in the conceptual plans or the completed planned wetland, or both.

4.3.2.1 Many states and local governments have adopted regulatory wetland statutes which set forth procedures for permit applications similar to the federal Section 404 program (see review in Ref **(13)**). The need to assess wetland functions is similar, but specific requirements may differ depending upon the individual state program.

4.3.2.2 *Mitigation Banking*—One option for meeting any compensatory mitigation requirements is to use a mitigation bank. Mitigation banking is wetland restoration, creation, or enhancement undertaken expressly for the purpose of providing compensation credits for wetland losses from future development activities. A wetland assessment procedure can be used to asses the loss of functions at an impact site, to assess functions to date at the mitigation bank, and to determine the number of credits (expressed in terms of functional capacity or acreage, or both) that must be purchased at the mitigation bank to compensate for the impacts.

4.3.3 *Other Applications*—There are a variety of non-regulatory situations where there is a need to assess wetland function. A rapid wetland assessment procedure that is appropriate for the Section 404[4] program could be used, but time and resources may also allow for more detailed analyses.

4.3.3.1 *Advanced Identification (ADID)*—Advanced Identification is a planning process authorized by Section 404 regulations (40 CFR Part 230.80)[4] that allows the U.S. EPA, in cooperation with the U.S. Army Corps of Engineers and state and local agencies, to collect information on the functions of the wetlands in selected study areas. The agencies evaluate the information to determine which wetlands in the ADID study area should be protected from potential fill activities or which could serve as future disposal sites. This information is used by the agencies in the review of Section 404 permit applications, by local communities for land-use management, and by environmental organizations for wetland protection activities.

4.3.3.2 *Restoration*—Wetland restoration refers to the return of a wetland from a disturbed or altered condition by the reestablishment of one or more indicators of wetland hydrology, hydric soil, and hydrophytic vegetation. There are increasing efforts to restore wetlands, many of which do not require permits. The restoration goals may be broadly defined in terms of wetland type or functions, or both. Wetland assessment procedures can be used to define and measure the achievement of function-based goals.

4.3.3.3 *Resource Management*—Wetlands are resources that are managed by different government agencies, private organizations, or individual landowners for different purposes. Wetlands can be managed at site-specific (for example, water level and weed control management of waterfowl impoundment), watershed, or even larger scales (for example, a North American Waterfowl Management Plan, an agreement adopted by the United States and Canada to manage waterfowl habitat **(14)**). A wetland assessment procedure can be used to define and monitor the achievement of management objectives. For example, the assessment of a degraded wetland can be performed at the site-specific scale to reveal what characteristics need modification to enhance wildlife habitat. The habitat could be altered and later reevaluated to determine if the management objectives have been achieved.

4.3.3.4 *Watershed/Regional Planning*—There are several state initiatives to manage land-use within watersheds or

[3] Title 33, United States Code, Chapter 26, Section 1344: "Permits for Dredged or Fill Material."

[4] Title 40, Code of Federal Regulations, Part 230: Section 404(b)(1) Guidelines for Specification of Disposal Sites for Dredged or Fill Material.

regions. A wetland assessment procedure can be used to provide an inventory of wetlands within a watershed/region, to prioritize these wetlands for land-use decisions, and to identify wetlands for acquisition, protection, development, or restoration. Landscape level or ecosystem level assessment procedures, or both, may be used. At least two states have developed and are testing new landscape-level procedures that use Geographic Information System (GIS) data (for example, Refs **(5)** and **(15)**). The assessment of landscape level functions is limited by the data available in the GIS format. Greater accuracy may be obtained through the application of ecosystem-level assessment procedures; however, assessing all or most of the wetlands within a designated area is more time-consuming. The New Hampshire Method **(16)** and Indicator Value Assessment **(17)** were developed for this purpose, but other procedures may also be suitable. The EPA has developed a Synoptic Approach as a proposed method for assessing cumulative impacts and making comparisons between landscape subunits, such as watersheds, ecoregions, or counties **(3)(4)**.

4.3.3.5 *Swampbuster Conversion of Wetlands*—Wetlands are managed under the Wetland Conservation (Swampbuster) Provisions of the 1985 Food Security Act as amended by the 1996 Farm Bill. The Farm Bill allows exemptions to be granted for conversion of wetlands where such conversions have minimal effect on wetland functions and values, or where impacts are compensated through mitigation actions. Wetland assessment procedures can be used to measure the effects of proposed conversions or mitigation actions on wetland functions, or both.

5. Function of Wetlands

5.1 Wetlands perform a variety of functions at different scales of complexity. These functions are difficult to characterize because they represent a wide range of scales from microscopic chemical reactions to landscape size changes in climate or environment. By compartmentalizing the activities that take place in a wetland into individual functions, one actually makes value judgments about the important processes in wetlands. For example, nitrogen removal is a function which can be considered a subset of more complex functions such as nitrogen cycling and nutrient cycling. A wetland would probably have a different functional capacity index if the assessment model was designed to assess general wildlife habitat rather than if the model was designed to assess anadromous fish habitat or habitat for amphibians. However, it is not practicable to assess all wetland functions at all levels of complexity. The functions that are assessed should be selected on the basis of wetland type and assessment objectives. The following list of general functions is a starting point for identifying what kinds of functions a wetland performs. This list and definitions are taken or modified or both, from Refs **(9)** and **(18)**. Other lists are available in individual wetland assessment procedures (refer to Section 7) and other publications (for example Ref **(19)** and **(20)**).

5.1.1 *Functions Related to Hydrologic Processes*:

5.1.1.1 *Short-Term Storage of Surface Water*—The capability of a wetland to detain or slow surface water or both for short periods of time. When water is detained in the wetland, downstream peak discharge and flood volume are reduced.

5.1.1.2 *Long-Term Storage of Surface Water*—The capability of a wetland to temporarily store surface water for long periods of time (for example, one week or longer). When water is retained in a wetland, the volume of flood water transported downstream is decreased. The retained water supports aquatic vertebrates and invertebrates and contributes to other ecological processes within the wetland.

5.1.1.3 *Storage of Subsurface Water*—The availability of storage for water beneath the wetland surface. The storage, that becomes available with periodic drawdown of the water table, may be the result of vertical and lateral drainage or evapotranspiration, or both. This storage helps to recharge surficial aquifers and maintain base flow and seasonal flow in streams. The periodic drainage results in a fluctuation between aerobic and anaerobic conditions. This fluctuation benefits the recruitment, survival, and competitiveness of wetland plant species and sustains conditions necessary for microbially mediated biogeochemical cycling.

5.1.1.4 *Moderation of Ground Water Flow or Discharge*—The capability of wetland to moderate the rate of ground water flow or discharge from upgradient sources or from ground water discharge within the wetland. This moderation in flow maintains ground water storage, base flow, seasonal flows, and surface water temperatures. Flows of subsurface water into the wetland in late fall or early spring sustain warmer soil temperatures resulting in a longer growing season for biological activity and other wetland functions.

5.1.1.5 *Dissipation of Energy*—The capability of a wetland to reduce the energy of water as it moves through, into, or out of the wetland. The reduction in the energy of moving water may result in reduced shoreline and floodplain erosion, improved surface water quality, and decreased downstream peak discharge.

5.1.2 *Functions Related to Biogeochemical Processes*:

5.1.2.1 *Cycling of Nutrients*—The conversion of elements from one form to another through abiotic and biotic processes. Nutrient cycling is accomplished through plant uptake and release, a process by which nutrients are adsorbed and assimilated into living plant tissue and released with litter production. By cycling nutrients, wetlands maintain sufficient nutrients to support living biomass and detrital stocks. Nutrient cycling also reduces downstream particulate loading which helps to maintain or improve surface water quality.

5.1.2.2 *Removal of Elements and Compounds*—The removal of nutrients, contaminants, or other elements and compounds on a short- or long-term basis through burial, incorporation into biomass, or biochemical reactions. In addition to providing benefits on-site, this removal also reduces downstream loading which helps to maintain or improve surface water quality.

5.1.2.3 *Retention of Particulates*—The deposition and retention of organic and inorganic particulates from the water column, primarily through physical processes such as sedimentation. When particulates are retained in the wetland, downstream loading is reduced; this helps to maintain or improve surface water quality.

5.1.2.4 *Export of Organic Carbon*—The export of dissolved

and particulate organic carbon from the wetland through leaching, flushing, displacement, erosion, and other mechanisms. The removal of organic carbon from living biomass, detritus, and soil organic matter contributes to the decomposition and mobilization of metals within the wetland. The exported organic carbon also provides support for aquatic food webs and biogeochemical processing downstream from the wetland.

5.1.3 *Functions Related to Habitat*:

5.1.3.1 *Maintain Characteristic Plant Community*—The maintenance of a plant community that is characteristic with respect to species composition and physical characteristics of the vegetation. Plant communities provide energy to drive food webs; provide habitat for nesting, resting, refuge; provide escape cover for animals; create roughness that reduces velocity of flood waters; and provide organic matter for nutrient cycling within the wetland. Plant communities also provide a source of propagules to help maintain species composition of adjacent areas and migratory pathways between habitats.

5.1.3.2 *Maintain Spatial Structures of Habitat*—The capacity of a wetland to support animal populations and guilds by providing a heterogeneous habitat. Structure provides potential feeding, resting, and nesting sites for vertebrates and invertebrates within the wetland. The structure of the wetland also provides habitat for wide-ranging and migratory animals and a corridor for gene flow between separated populations.

5.2 Wetlands provide benefits, goods, and services that are considered values, but sometimes referred to as functions. While these are not functions, wetland managers may choose to evaluate these aspects of the wetland. Examples include the following:

5.2.1 *Recreation*—Providing recreation sites for fishing, hunting, and observing wildlife,

5.2.2 *Open Space and Aesthetic Values*—Providing open space for visual enjoyment,

5.2.3 *Education and Research*—Providing educational opportunities for nature observation and scientific study,

5.2.4 *Historical or Archeological Significance*—Containing properties of historical or archeological significance, and

5.2.5 *Timber Production*—Providing timber resources for private or commercial uses.

5.3 The preceding list (5.1.1 through 5.2.5) is not all inclusive. Not all wetlands perform these functions and values; additional functions may be appropriate given a specific wetland type and other factors.

6. Steps for Assessing Wetland Functions

6.1 The phases outlined in 6.2 through 6.5 and discussion in 6.3.2 are modified from the Hydrogeomorphic (HGM) Approach **(9)**, **(21)**. Changes have been made in order to increase flexibility and incorporate concepts from other approaches. Many steps, particularly the identification of the assessment approach, require critical decisions. These decisions should be made by a team of experts. The term will differ in each case, but may include wetland scientists from appropriate federal, state, and local agencies as well as from the private sector and academia. The decisions and assumptions may affect the final ouctome of the assessment. For this reason, these decisions should be agreed upon between the assessors and users of the results (for example, regulator) and also be documented throughout the process.

6.2 *Characterization Phase*—The following baseline information must be gathered during the characterization phase before proceeding with the assessment.

6.2.1 *Define Objectives of the Assessment*—Describe the proposed project, purpose, and objectives. Decide which wetland assessment area(s) are to be compared and the number of comparisons required. Predictions regarding future conditions may also be necessary. Therefore, it is important to define the time period for which the wetlands are being assessed (for example, predicted future conditions of planned wetland two years after construction). The three categories of objectives include documenting existing conditions, comparing different wetlands at the same point in time, or comparing wetland(s) at different points in time. For example, a simple objective for an alternative analysis may be to compare two wetland assessment areas (WAAs) in order to determine which project location will have the least impact. A more complex objective would be to compare a restored or enhanced wetland site. In this case, the impact area would be assessed for both future with-project and future without-project conditions. An independent assessment would then be performed for the enhancement area and its predicted future conditions, both with and without the enhancement project. Finally the gains from the enhancement project would be compared to the losses associated with the impact. As projects become more complicated and involve several wetlands, it is important that the objectives be described in specific terms to avoid any misunderstanding and unnecessary work.

6.2.2 *Select Functions*—Select the functions to be assessed on the basis of wetland type, the assessment objectives, the nature of the project, and expected impacts.

6.2.2.1 There is no standard set of wetland functions that are applicable to all wetland types. Applicable functions can be selected from existing assessment models or redefined, or both. In general, a suite of representative wetland functions should be assessed in order to provide a more complete description of a wetland area. Depending upon the assessment objectives, however, a limited number or even one function may be selected. For example, if the sole purpose of a wetland restoration project is to provide wildlife habitat, assessment may be limited to that single function.

6.2.3 Each function should be assessed and considered separately in the decision-making process. However, in more complex projects, decision makers may choose to use of a grand score that combines the measures of each function for a wetland. If functions are to be combined, pertinent regulatory agencies and other decision makers should agree upon which functions will be combined and how. Caution is advised (particularly for options 6.2.3.3 and 6.2.3.4) since no guidance is available, and the combined results could be subject to question. In summary, the possible options include:

6.2.3.1 Consider each function separately.

6.2.3.2 Develop scores for the major function categories (for example, hydrologic processes, biogeochemical processes, and habitat) which are derived from the weighted or non-weighted totals of FCs for functions in each category.

6.2.3.3 Develop a grand score that represents a total number of functional capacity units. (See definition in 6.3.2.3.)

6.2.3.4 Develop a grand total that represents a weighted total of functional capacity units, where multiplying factors are used to emphasize the more important functions.

6.2.4 *Describe the Project Area*—Describe the project area and surrounding landscape with a narrative and map(s). The narrative should include:

6.2.4.1 Project name and location,

6.2.4.2 Nature of the proposed project,

6.2.4.3 Assessment objectives,

6.2.4.4 Classification of wetlands (use National Wetland Inventory (NWI) **(11)** and hydrogeomorphic **(12)** classification, or other classifications as needed), and

6.2.4.5 Description of characteristics of the wetland ecosystem and landscape context that may be relevant to the assessment, (for example, climate, landform and geomorphic setting, hydrology, vegetation, soils, land use, ground water features, surficial geology, urban areas, potential impacts, and red-flag features (see 6.2.7)).

6.2.5 The map(s) shall be prepared to a scale suitable for illustrating the following information, as appropriate:

6.2.5.1 Project area boundaries, property lines, and other relevant political boundaries,

6.2.5.2 Topographic contour lines in the project area and surrounding landscape,

6.2.5.3 Infrastructure (for example, roads, fences, buildings, railroad grades, and bridges),

6.2.5.4 Surface water features (for example, streams, rivers, lakes, ponds, and springs),

6.2.5.5 Hydraulic structures (for example, weirs, culverts, gates, pumps, and levees),

6.2.5.6 Seasonal water table elevations,

6.2.5.7 Soil type(s),

6.2.5.8 Plant communities,

6.2.5.9 Jurisdictional wetlands,

6.2.5.10 Location of wetland impacts (potential and relevant prior impacts),

6.2.5.11 Wetland assessment area(s),

6.2.5.12 North arrow (true north), legend or key, and distance scale,

6.2.5.13 Title block with the project name, investigators, dates, and sources of information, and,

6.2.5.14 Keep time for map preparation to a minimum by using existing maps or modifying as needed, or both. Multiple maps, or overlaps keyed to a base map, may be practical (examples provided in Ref **(16)**.)

6.2.6 *Identify Wetland Assessment Area(s)*—Identify wetlands within the project area using a chosen wetland definition, for example, the jurisdictional wetlands in a regulatory situation. The project area may contain one WAA; however, in some cases, it may be large and encompass several wetland areas that function differently. These WAAs are identified on the basis of wetland classification, physical separation, and potential project impacts. The criteria for identifying WAAs will differ depending upon the selected assessment procedure (refer to Section 7) and local policy. Possible criteria for distinguishing WAAs are difference in wetland classification, physical separation, and differences in predicted project impacts.

6.2.7 *Screen for Red Flags*—Red flags are features of a wetland or the surrounding landscape to which special recognition or protection is assigned on the basis of objective criteria. The recognition or protection may occur at a federal, state, regional, or local level, and may be official or unofficial **(21)** (refer to Table 2). These features are identified to determine whether the area will require special consideration prior to or during the assessment of wetland functions.

6.3 *Identification of Assessment Approach*:

6.3.1 *Identify, Modify, or Develop Assessment Models*—There are a variety of assessment models from which to choose that are contained in existing procedures (see Section 7). Identify the purpose or objective for which the assessment is needed. The objective will dictate what method may be appropriate and which ones will provide the needed information. Review these to determine which are most appropriate. Criteria to be considered are listed in Table 3. Users of this guide should note that some of the procedures cited are no longer acceptable to all resource agencies. It is critical to determine in advance which procedures are acceptable.

6.3.1.1 Decide whether to use the assessment models without change, modify and then use the assessment models, or develop new assessment models. Models may be developed using available information including best professional judgment, expert opinion, published literature, empirical data, or a combination thereof. The rationale for the assessment models should be supported by the available information, particularly if models from an established assessment procedure are not being used. For guidance on developing, adapting, and calibrating assessment models, refer to **(9)** and **(17)**.

6.3.1.2 Select or develop the assessment models to adequately address the concerns of the decision makers. For example, the U.S. Fish and Wildlife Service would likely be concerned that the habitat function models adequately address fish and wildlife resource needs. The Service may specify that habitat function models be consistent with HEP **(22)** and thoroughly address the necessary food, cover, water and breeding requirements of all terrestrial and aquatic species

TABLE 2 Red Flag Features (9)

Areas protected under American Indian Religious Freedom Act
Hazardous waste sites identified under CERCLA or RCRA
Areas protected by a Coastal Zone Management Plan
Areas providing critical habitat for species of special concern
Areas covered under the Farmland Protection Act
Floodplains, floodways, or floodprone areas
Areas of high public use
Areas with structure/artifacts of historic or archeological significance
Areas protected under the Land and Water Conservation Fund Act
Areas protected by the Marine Protection Research and Sanctuaries Act
National Wildlife Refuges
Native lands
Areas identified in the North American Waterfowl Management Plan
Areas identified as significant under the RAMSAR Treaty
Areas supporting rare or unique plant communities
Areas designated as sole source groundwater aquifers
Areas protected by the Safe Drinking Water Act
Special management areas
State or national parks
Areas supporting threatened or endangered species
Areas with unique geological features
Areas protected by the Wild and Scenic Rivers Act
Areas protected by the Wilderness Act
Wetlands that have been restored, created, or converted

TABLE 3 Criteria to Consider When Identifying Modifying, or Developing Assessment Models,

Criteria	Specific Considerations
Wetland type	Is the model applicable to the wetland type(s)? An existing model describing one wetland type may be suitable, require minor modifications, or be unsuitable.
Functions	Is there a model for each of the pertinent functions?
Geographic area	Is the model applicable to the geographic area (for example, ecoregion, state, watershed)? Define the geographic region or context, and determine whether the model is appropriate. Minor or major modification may be required to ensure that the model is calibrated to the defined region.
Assessment situation	Is the model applicable to the assessment situation, (for example, watershed planning, regulatory action, management, use as guide to wetland design)?
Comparison of different wetland types	Is there a need to compare different wetland types? Note that when models are calibrated to describe particular wetland types within a region, it is inappropriate and meaningless to compare different wetland types. Choose models that will facilitate a comparison, if needed, or decide on how comparisons can be made.
Acreage	How does the model consider wetland acreage? Does or can the resulting measure of functional capacity incorporate acreage? Note that the assessment results are often used to make decisions regarding wetland function and acreage. For example, in permit actions it is necessary to define not only the functional capacity of the mitigation wetland, but also the acreage required to compensate for wetland impacts.

expected to utilize a particular habitat type.

6.3.2 *Select the Units of Measure*—Wetland functions can be measured and expressed using quantitative (that is, interval or ratio) or qualitative (that is, nominal or ordinal) scales **(23)**. This guide recommends that the assessment models be used to express functions in terms of functional capacity index (FCI) and functional capacity units (FCs) to be consistent with the HGM Approach developed by the U.S. Army Corps of Engineers **(9)** (and similar indices used in the New Hampshire Method **(16)**, Evaluation for Planned Wetlands **(10)**, Indicator Value Assessment **(17)**, Index of Biological Integrity **(24)**, Water Quality Index **(25)**, Wetland Rapid Assessment Method **(26)**, and Habitat Evaluation Procedure **(22)**.

6.3.2.1 A functional capacity index (FCI) is an index of the capacity of a wetland to perform a function relative to other wetlands within a defined region or wetland class, or both. The index of 0.0 indicates that the wetland does not perform the function. An index of 1.0 indicates that the wetland is performing a function at maximum functional capacity. It is possible for an FCI to exceed 1.0.

$$FCI = \frac{\textit{functional capacity of a wetland assessment area}}{\textit{maximum functional capacity}} \quad (1)$$

The meaning of maximum functional capacity varies depending upon the assessment model, and whether it defines a standard of comparison (reference). Many assessment models simply define a maximum for wetlands, in general, and the standard of comparison is implied but not defined. In contrast, models being developed following the HGM Approach produce an FCI that measures the capacity of a wetland relative to reference standards. Reference standards are the conditions exhibited by a group of reference wetlands that corresponds to the highest level of functioning (highest sustainable functional capacity) across the suite of functions performed by the regional wetland subclass **(9)**. These reference standards are established for wetlands within a defined geographic region that belong to a single hydrogeomorphic subclass. The highest level of functional capacity is assigned an index score of 1.0 by definition. Guidance for establishing the standard of comparison (reference) following the HGM Approach is provided in **(9)** and **21)**.

6.3.2.2 The FCI is measured by using an assessment model. Existing assessment models may be used and their results converted easily to FCIs. For example, an index of 1.0 may be considered equivalent to the" high" of a model that rates wetland functions as low, moderate, or high, Numeric results of models using other scales (that is, 0 to 100) can also be converted to the 0 to 1.0 scale.

6.3.2.3 The functional capacity unit (FC) is measured as follows:

$$FCs = \text{FCI of } \textit{a wetland area multiplied by size of wetland area} \quad (2)$$

This measurement facilitates the comparison of different size wetlands. For example, the results of an assessment may show that two wetlands have the same functional capacity index (for example, FCI = 0.7), suggesting that there is no difference between them. A decision based solely on the FCI for these two wetlands could lead to erroneous conclusions, particularly if the wetlands are different sizes (that is, Wetland A = one acre and Wetland B = five acres). The index of the two wetlands may be the same, but because of the size difference, the FCs will differ (see explanation in 6.4.1.1).[5]

6.4 *Assessment Place*—Apply the assessment models to each wetland assessment area. Calculate and record the functional capacity indices (FCIs) and functional capacity units (FCs) for each function. Use care when predicting past or future conditions. To ensure the most accurate predictions possible, refer to several sources including personal experience, expert opinion, and the literature. Record assessment results for each wetland assessment in a standardized data sheet such as the one shown in Fig. 1. Comparisons of WAAs can be recorded in standardized data sheets such as those shown in Figs. 2 and 3.

6.4.1 *Units of Comparison*—The differences between wetlands are expressed in terms of functional capacity indices (FCIs), functional capacity units (FCs), and variable scores.

6.4.1.1 Simplified, a comparison of FCIs will provide information regarding the quality of the wetland's functional

[5] It should be noted that the use of quantitative indices (such as FCI) carries a proportionality assumption which may not be valid. For example, a wetland which has a FCI of 0.5 is not necessarily performing the function at twice the level as a wetland with a FCI of 0.25, even though the numbers imply this relationship. Similarly, there may be a scaling factor which is lost by simply multiplying the FCI by acreage to derive FCs. The use of quantitative units of measure is still recommended to be consistent with current practice (refer to approaches listed in 6.3.2).

Summary of Functional Capacities Under Existing or Predicted Conditions

Date:
Project:
Wetland Assessment Area:
Assessors:

FUNCTION	FCI	SIZE OF WAA IN ACRES	FCs

FIG. 1 Data Sheet for Recording Assessment Results for One Wetland Assessment Area

Date:
Project:
Comparison between WAA # ____ and WAA # ____
Assessors:

	WAA #			WAA #			Net Difference
FUNCTION	FCI	SIZE IN ACRES	FCs	FCI	SIZE IN ACRES	FCs	FC +/-

FIG. 2 Data Sheet for Comparing FCIs and FCs of Wetland Assessment Areas

capacity, whereas the FCs will describe the quantity of functional capacity (see 6.3.2). Thus, a comparison made using FCIs indicates which wetland assessment areas (WAAs) have a greater capacity to perform a function on a unit area basis. The higher the FCI the greater the capacity per unit area. While this information is useful, it is important to remember that the size of the WAA is not considered in the FCI. It is equally important to consider FCs in any comparison because FCs represent the functional capacity of the WAA as a whole based on its FCI and spatial extent. Consider the following example of an alternatives analysis. The first alternative involves the loss of a two-acre WAA with an FCI of 0.9 and FCs of 1.8, and the other alternative involves the loss of a 20-acre WAA with an FCI of 0.4 and FCs of 8.0. If the decision to select the last damaging alternative is based strictly on FCI, the second alternative with the lowest FCI (lower quality) may be selected. However, if the decision is based on FCs so that size is considered, the first alternative with the least number of FCs is the least damaging.

6.4.1.2 Comparisons are also made at the most basic level of the assessment model, the variable. Variable conditions will vary in the wetland and, at extremes, may diminish or maximize functional capacity. The conditions for each variable are assigned different scores in the assessment models based upon their contribution to the functional capacity, that is, a condition that increases the capacity of a wetland to perform a function is assigned a higher score. These data are then used in the models to derive FCIs. The scores for each variable are compared after the assessment has been completed (see Fig. 3). The information on the variables is important because it provides an explanation of why the wetlands' functional capacities differ. This may be important, for example, in the identification of specific conditions that can be improved on in a planned wetland design.

6.5 *Analysis Phase*

Date: Project: Comparison between WAA # _____ and WAA # _____ Assessors:			
	FCI		Difference in model variables and condition(s) with explanatory comments
FUNCTION	WAA #	WAA #	

FIG. 3 Data Sheet for Comparing the FCIs and Assessment Model Variables of Two Wetland Assessment Areas

6.5.1 *Types of Comparisons*—Once the functional capacities of each wetland assessment area (WAA) are documented, three types of comparison can be made. These are:

6.5.1.1 Comparison of the same WAA at different points in time,

6.5.1.2 Comparison of two or more WAAs at the same point in time, and

6.5.1.3 Comparison of WAAs at different points in time.

6.5.2 *Comparison of the Same WAA at Different Points of Time*—This represents the most common type of comparison of regulatory projects, that is, the comparison between preproject and post project impact conditions of the same WAA. This type of comparison is also common with planned wetlands, particularly those involving restoration or enhancement. In either situation, a simple direct comparison can be made between a single WAA assessed at two points of time. The results can be recorded in standardized data sheets such as those shown in Figs. 2 and 3. Some situations may require the comparison of the same wetland area at several different points of time, especially when there is a concern regarding the cumulative loss of functional capacity as a result of a time lag or delays. For example, a developing planned wetland may take several years to achieve the desired levels of functional capacity. Compensation may be required if the loss of functional capacity during a time lag is substantial. The basic steps to making this type of comparison include:

6.5.2.1 Select target years for future prediction,

6.5.2.2 Predict area of WAA that will perform the function in future years,

6.5.2.3 Predict FCIs and FCs for future years,

6.5.2.4 Calculate cumulative FCs, and

6.5.2.5 Calculate difference between cumulative FCs for WAAs being compared.

6.5.2.6 Several methods can be used for calculating cumulative FCs. The easiest is to graph the assumed linear relationship between broadly spaced target years, estimate the area under the curve, and then calculate the difference between FCs lost and FCs gained (see Fig. 4). Better estimates of FC loss can be made by using more narrowly spaced target years, or known nonlinear relationships.

6.5.3 *Comparison of Two or More WAAs at the Same Point in Time*—Examples of situations when two or more WAAs may be compared include an alternative analysis for regulatory actions (see 6.4.1.1) or a wetland inventory for watershed planning.

6.5.4 *Comparison of WAAs at Different Points in Time*—Comparing WAAs at different points in time is common with regulatory projects that require compensatory mitigation. The comparison is made between existing conditions of a wetland prior to impact and the predicted future conditions of the planned wetland.

6.5.5 *Other Criteria to Consider When Comparing Wetland Assessment Areas*—The assessment models provide information on functional capacity expressed in terms of FCIs and FCs and individual variables (see 6.4.1). Several other criteria are considered in comparing wetlands. These include wetland class, red flags, wetland vulnerability, wetland rarity, and feasibility of mitigation. With the exception of functional capacity, all other criteria involve a subjective value judgment and need to be thoroughly justified and documented.

6.5.5.1 *Wetland Class*—Caution must be taken when comparing WAAs of different wetland types of class. The assessment models of some procedures allow the comparison of different wetland types, but many do not. According to **(9)**, wetlands in different hydrogeomorphic classes cannot be compared directly because their functional capacity indices are calibrated based on different reference wetlands. The comparison of wildlife habitat functional capacity for different wetlands such as an ombotrophic peatland and a salt marsh would be considered meaningless. Although direct comparisons cannot be made, the measures of functional capacity (FCIs and FCs) may still be useful because they provide information regarding the WAA relative to wetlands within the same regional subclass. Consider the following example involving the choice between impacting 10 acres of riverine bottomland hardwood forest or in ten acres of depressional marsh. The

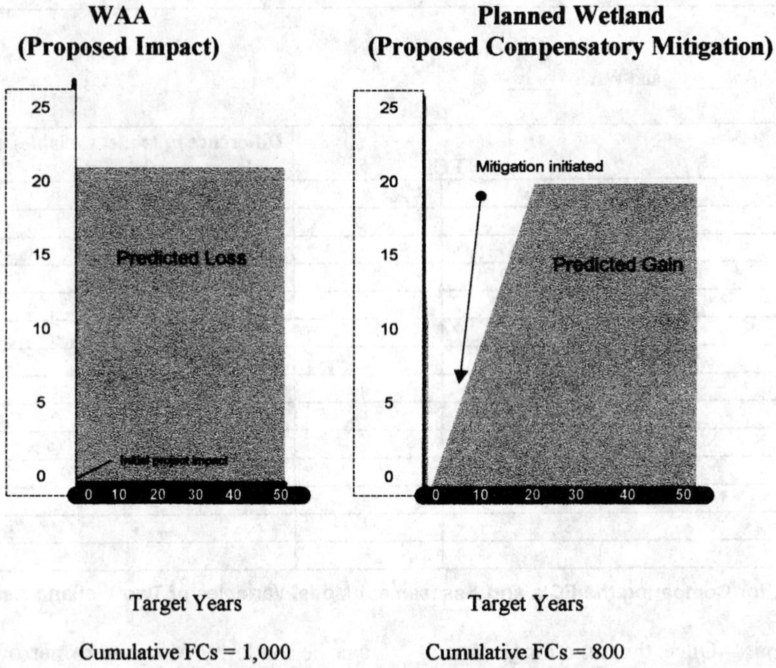

NOTE 1—Difference in cumulative FCs = 200; therefore, the planned wetland provides 200 less FCs than predicted loss from proposed impact.

FIG. 4 Example Illustrating Comparison of Cumulative FCs Through Graphing Predicted FCs

riverine wetland has a wildlife habitat FCI of 0.9 and the depressional wetland FCI is 0.4. The higher FCI for the riverine wetland indicates that it is closer to the reference condition for the wildlife function then the depressional wetland. It may be reasonable to conclude that the wildlife habitat function provided by the riverine forested wetland is more valuable because it provides nesting habitat for neotropical migrant birds, a group currently considered to be in decline.

6.5.5.2 *Functions*—A comparison of wetlands often shows that no one WAA has the highest functional capacities for all functions. For example, one WAA may have the greater wildlife habitat functional capacity and the other WAA may have the greater water quality functional capacity. Some situations, such as a regulatory project, may require a choice between wetlands. The choice between wetlands may come down to a choice between which function is considered less valuable in the specific context.

6.5.5.3 *Red Flags*—The occurrence of a red flag feature may provide sufficient justification to remove a WAA from consideration. For example, a wetland which has endangered species, is rare, or contains historic properties may be considered more valuable and thus removed from a list of potential alternative project sites.

6.5.5.4 *Wetland Vulnerability*—Some wetland types are inherently more vulnerable to impacts than others due to their hydrogeomorphology or location. For example, a given length of fill could have a profound impact by blocking tidal flow to a salt marsh. The same length of fill might only have a minimal impact on a depressional wetland with no outlet. Also, wetlands located more closely to populated areas may be more vulnerable to impacts associated with development. The choice between such wetlands may be decided solely on the basis of which one is more vulnerable in a specific context.

6.5.5.5 *Feasibility of Mitigation*—Some wetland types and their functions are more easily compensated for than others. A choice between wetland impacts may come down to a selection of the wetland type which is easier to replace.

7. Assessment Procedures

7.1 Assessment models are available within a variety of procedures. Table 4 provides a current listing and additional information which can be used in selecting appropriate models. Detailed review of some of these procedures are contained in Refs **(27-29** and **63)**.

TABLE 4 Summary of Some Wetland Assessment Procedures

Procedure Title: Citation	Units of Measure	Application: Wetland Types	Application: Geographic Area	Use	Time Required	Do Models Provide Measures of Functional Capacity?	Remarks
Approach for assessing wetland functions based on hydrogeomorphic (HGM) classification, reference wetlands, and functional indices **(9), (18), (21)**	Functional capacity indices (FCI) (scale 0 to 1.0) and functional capacity units (FCs). Functions separate.	Conceptually applicable to all wetland types; but not all assessment models are developed.	Widespread application: assessment models calibrated to regional wetland subclass using reference wetlands.	Regulatory actions Impact assessment Management Planning Mitigation	Relatively rapid, after initial development and scaling of models.	Yes	Requires scaling models to wetlands in defined region. Cannot use FCIs and FCs to directly compare wetlands in different hydrogeomorphic classes, but can use information to aid in decision. Provides measure of function that accounts for size (FCs).
Wetland rapid assessment procedure (WRAP) **(26)**	Ecological integrity index (scale 0 to 1.0). Functions assessed separately, but scores for each function pooled to obtain one index for a wetland.	Depressional wetlands	South and Central Florida (can be modified for North Florida)	Regulatory actions Mitigation	Relatively rapid	Yes	Index not multiplied by acres, thus does not provide measure of functional capacity that accounts for size.
North Carolina Coastal Region Evaluation of Wetland Significance (NC-CREWS) **(5)**	Overall ecological significance (high, medium, or low) and function ratings (H, M, or L). Functions separate, but also combined to derive overall rating of ecological significance.	All wetland types	North Carolina coastal area	Planning Management To predict ecological relative significance of wetlands within watershed or region	Relatively rapid	Yes	GIS-based landscape scale procedure; thus, does not provide the site-specific detail which is desired for many applications. Rankings and indices not multiplied by acres, thus does not provide measure of functional capacity that accounts for size.

TABLE 4 *Continued*

Procedure Title: Citation	Units of Measure	Application			Time Required	Do Models Provide Measures of Functional Capacity?	Remarks
		Wetland Types	Geographic Area	Use			
Minnesota routine assessment method for evaluating wetland functions (MinRAM) **(30)**	Low, medium, high, exceptional and not applicable functional level ratings. Functions separate.	Freshwater wetlands	Minnesota	Regulatory actions	Relatively rapid	No: There are no models. Evaluator answers questions that are used to guide the evaluation. Best professional judgment (BPJ) then used to rate significance. BPJ may be used to estimate functional capacity, but modifications required to ensure this.	Cannot use rankings or ratings to directly compare wetlands in different wetland class, but can use information to aid in decision. Rating not multiplied by acres, thus does not provide measure of functional capacity that accounts for size.
Wetland functions and values: A Descriptive Approach **(31)**	Yes or no answer indicating whether function/value occurs. Functions separate.	Applicable to all wetland types.	New England area	Regulatory actions	Relatively rapid	No. There are no models. Evaluator provided lists of functions, values, and considerations that are used to guide the evaluation. Best professional judgment (BPJ) is then used to determine whether a function occurs.	Results not multiplied by acres, thus does not prove measure of functional capacity that accounts for size. Comparison of different wetland types possible.
Wildlife Habitat Appraisal Procedure (WHAP) **(32)**	Average habitat quality score (scale 0 to 1.0) and habitat units (HUs) for biological component. Functions and values assessed separately.	All wetland types	Texas	Impact assessment Inventory Mitigation Management Wetland acquisition	Relatively rapid	Yes. The biological habitat component model does. No. Other models combine function and value. Modifications required to achieve measure of functional capacity.	Provides measure of functional capacity that accounts for size (HUs), but only for biological component. Comparison of different wetland types possible

TABLE 4 *Continued*

Procedure Title: Citation	Units of Measure	Application			Time Required	Do Models Provide Measures of Functional Capacity?	Remarks
		Wetland Types	Geographic Area	Use			
Semantic Categorization Method for Assessment of Wetland Impacts **(33)**	Impact score (scale −4 to +4). Functions separate.	Riverine wetlands, but conceptually applicable to other types with the development of appropriate functions and reference standards.	Southern California	Regulatory actions Impact assessment Mitigation	Relatively rapid	No. Provides measure of change in functional capacity.	A comparative evaluation of functions before and after project implementation. Score not multiplied by acres, thus does not provide measure of functional capacity that accounts for size.
A Method for the Assessment of Wetland Function **(34)**	Functional capacity index (FCI) (scale 0 to 1.0), functional capacity units (FCU), and total functional capacity units. Functions separate but FCUs also added to derive total FCUs for a wetland.	Non-tidal palustrine vegetated wetlands.	Maryland piedmont and eastern shore (coastal plain).	Broad area planning (e.g., watershed) Mitigation	Relatively rapid	Yes	Contains 2 models types: (1) GIS-based landscape scale desk top models and (2) a field method, that requires a site visit. The desk top does not provide the site specific detail which is desired for many applications. Provides measure of function that accounts for acreage (FCU). Comparison of different wetland types possible.
Wetland quality index (WOI) **25**	Wetland quality index (WQI) (scale 0 to 1.0) and wetland functional value (WFV). Function categories combined to derive one index for a wetland.	Freshwater wetlands.	Everglades in Florida	Mitigation	Relatively rapid	Yes	Provides measure of function that accounts for size (WFV).

TABLE 4 *Continued*

Procedure Title: Citation	Units of Measure	Application: Wetland Types	Application: Geographic Area	Use	Time Required	Do Models Provide Measures of Functional Capacity?	Remarks
Evaluation for Planned Wetlands (EPW). **(10)**	Functional capacity indices FCI (scale 0 to 1.0) and Functional capacity units (FCUs). Functions separate.	Applicable to all wetland types.	Widespread application	Regulatory actions Impact assessment Management Mitigation Planned wetlands	Relatively rapid	Yes	Procedure recommends modification of models, where needed, to ensure applicability to region. Provides measure of function that accounts for size (FCUs). Cannot use FCIs and FCUs to directly compare wetlands in different wetland classes, but can use information to aid in decisions.
Guidance for rating the values of wetlands in North Carolina **(35)**	Ratings for each function (scale 0 to 5). Wetland score based upon the combined weighted ratings for each function (scale 1 to 100). Functions combined.	Freshwater wetlands	North Carolina	Regulatory actions (401 water quality certifications) Inventory Restoration Mitigation banks Acquisition	Relatively rapid	No: Models combine function and value; modifications required to achieve measure of functional capacity.	Designed to evaluate values. Rating not multiplied by acres, thus does not provide measure of functional capacity that accounts for size.
Indicator value assessment (IVA) **(17)**	Performance score and value score (scale 0 to 100). Functions separate.	Conceptually applicable to all wetland types, but not all assessment models are developed.	Widespread application. Assessment models calibrated to regional wetlands using reference wetlands.	Regional planning Impact assessment Identify potential wetland enhancement areas	Relatively rapid, after initial development and scaling of models.	Yes	Models for each function for wetland within planning region or watershed (total number variable). Score not multiplied by acres, thus does not provide measure of functional capacity that accounts for size. Requires scaling models to wetlands in region. Comparison of different wetland types possible.

TABLE 4 *Continued*

Procedure Title: Citation	Units of Measure	Application: Wetland Types	Application: Geographic Area	Use	Time Required	Do Models Provide Measures of Functional Capacity?	Remarks
Buffalo District wetland evaluation method **(36)**	Function score (scale 7 to 30, depending on function). Functions separate.	Depressional wetlands (Draft versions for riverine and lake-fringe wetlands).	Ohio	Regulatory actions Mitigation	Relatively rapid	No: Document states that evaluations are not considered models; modifications required to achieve measure of functional capacity.	Score not multiplied by acres, thus does not provide measure of functional capacity that accounts for size.
Wetland Value Assessment Methodology and Community Models (WVA) **(37)**	Habitat suitability index (HSI) (scale 0 to 1.0) and habitat units (HUs). Wetland type separate; pool index for each type to obtain one score (total HUs) for a wetland.	Coastal wetlands: fresh (intermediate) marsh, brackish marsh, saline marsh, and cypress-tupelo swamp.	Louisiana	Wetland enhancement projects	Relatively rapid	Yes	Modification of the Habitat Evaluation Procedure (HEP). Only habitat function. Provides measure of function that accounts for size (HUs).
Wildlife Habitat Assessment and Management System (WHAMS) **(38)**	Habitat suitability index (HSI) (scale 0 to 1.0 and Habitat Units (HUs). Species separate. Pool index for each species to obtain one score (total HUs) for a wetland.	Freshwater wetlands and terrestrial habitats.	Pennsylvania and possibly other areas.	Wildlife management planning	Relatively rapid	Yes	Modification of the Habitat Evaluation Procedure (HEP) and PAM HEP; simplified to reduce time requirement. Modified for use as a wildlife management planning tool. (Remarks same as for HEP).
Oregon freshwater wetland assessment methodology **(39)**	Wetland descriptor indicating whether the wetland provides, has the potential to provide, or does not provide a function. Functions separate.	Freshwater wetlands	Oregon	Inventory Planning Education	Relatively rapid	No. Models produce probability ratings (that is, an estimate of the probability that a function will occur to an unspecified magnitude); not a measure of functional capacity.	Tool for those with some knowledge of wetlands (for example, town commissions), but not necessarily wetland ecologists. Descriptor not multiplied by acres, thus does not provide measure of functional capacity that accounts for size.

TABLE 4 *Continued*

Procedure Title: Citation	Application						
	Units of Measure	Wetland Types	Geographic Area	Use	Time Required	Do Models Provide Measures of Functional Capacity?	Remarks
Ontario wetland evaluation system **(40)**, **(41)**	Score for each function (scale 0 to 250). Functions separate, but scores combined to achieve total wetland score.	Freshwater wetlands	Northern and Southern Ontario, Canada	Planning	Relatively rapid	Yes	Score not multiplied by acres, thus does not provide measure of functional capacity that accounts for size.
Coastal Method **(42)**	Average functional indexes (AFI) (scale 0 to 1.0). Functions separate.	Coastal wetlands	New Hampshire	Inventory Planning Education	Relatively rapid	Yes	Tool for those with some knowledge of wetlands (for example, town commission), but not necessarily wetland ecologists. Index not multiplied by acres, thus does not provide measure of functional capacity that accounts for size for comparison of different marsh systems. However, relative size considered for comparisons of ecological units within a marsh system.
Synoptic Approach to Cumulative Impact Assessment **(3)**, **(4)**	Four general synoptic indices that are refined based upon project objectives: function, value, functional loss, and replacement potential. Functions separate.	Conceptually applicable to all wetland types; however, requires collection and analysis of available maps and tabular data.	Widespread application	Cumulative effects analysis	Relatively rapid, after initial collection of data for landscape unit (for example, watershed).	Yes. Function indices do. No. Models for other indices (value, functional loss, and replacement potential) do not.	Landscape scale procedure; thus, does not provide the site specific detail which is desired for many applications. Requires collection and analysis of data for wetlands within defined region. Indices not multiplied by acres, thus does not provide measure of functional capacity that accounts for size.

TABLE 4 *Continued*

Procedure Title: Citation	Units of Measure	Application: Wetland Types	Application: Geographic Area	Use	Time Required	Do Models Provide Measures of Functional Capacity?	Remarks
Wisconsin Rapid Assessment Methodology **(43)**	Low, medium, high, exceptional and not applicable significance rating. Functions separate.	Freshwater wetlands	Wisconsin	Regulatory actions	Relatively rapid	No. There are no models. Evaluator answers questions that are used to guide the evaluation. Best professional judgment (BPJ) then used to rate significance. The BPJ may be used to estimate functional capacity, but modifications required to ensure this.	Rating not multiplied by acres, thus does not provide measure of functional capacity that accounts for size.
New Hampshire Method **(16)**	Functional Value Indexes (FVI) (scale 0 to 1.0) and Wetland Value Units (WVU). Functions separate.	Nontidal wetlands	New Hampshire	Inventory Planning Education	Relatively rapid	No. Some models combine function and value. Modifications required to achieve measure of functional capacity.	Tool for those with some knowledge of wetlands (for example, town commission), but not necessarily wetland ecologists. Provides measure of function that accounts for size (WVU).
A technique for the functional assessment of nontidal wetlands in the coastal plain of Virginia **(44)**	Low, moderate, or high probability rating of opportunity and effectiveness at performing function. Functions separate.	Nontidal wetlands	Coastal plain of Virginia	Inventory	Relatively rapid	No. Some models produce probability ratings (that is, an estimate of the probability that a function will occur to an unspecified magnitude); not a measure of functional capacity. Also, some models combine function and value. Modifications required to achieve measure of functional capacity.	Rating not multiplied by acres, thus does not provide measure of functional capacity that accounts for size.

TABLE 4 *Continued*

Procedure Title: Citation	Units of Measure	Application: Wetland Types	Application: Geographic Area	Application: Use	Time Required	Do Models Provide Measures of Functional Capacity?	Remarks
Wildlife Habitat Appraisal Guide for Missouri **(45)**	Farm habitat index (scale 0.1 to 1.0) and habitat quality rating (excellent, good, fair, or poor). Species separate. Habitat type separate. Pool index for each habitat type to obtain farm habitat index and habitat quality rating for each species.	Freshwater wetlands and terrestrial habitats	Missouri	Management	Relatively rapid	Yes	Only habitat function. Acreage used in calculation of farm habitat index; however, only as a weighting factor. Index not multiplied by acres, thus does not provide measure of functional capacity that accounts for size. Comparison of different wetland types possible.
Habitat Assessment Technique (HAT) **(46)**	Faunal index (scale 0 to 360+ depending on wetland size). Bird species separate, pool scores for each bird species to obtain one faunal index for a wetland.	All wetland types	Widespread application	Regulatory actions Acquisition	Not rapid	Yes	Only habitat function. Requires bird inventory during breeding season in study wetland. Provides measure of function that accounts for size (faunal index). Requires scaling model to wetlands in defined region, that is, species must be assigned points based on breeding population in region.
Minnesota Wetland Evaluation Methodology (WEM) **(47)**	High, moderate, or low probability rating. Functions separate, but option to combine ratings is provided.	Freshwater wetlands	North Central United States	Regulatory actions	Relatively rapid	No. Some models produce probability ratings (that is, an estimate of the probability that a function will occur to an unspecified magnitude); not a measure of functional capacity. Also, some models combine function and value. Modifications required to achieve measure of functional capacity.	Rating not multiplied by acres, thus does not provide measure of functional capacity that accounts for size.

TABLE 4 *Continued*

Procedure Title: Citation	Units of Measure	Application: Wetland Types	Application: Geographic Area	Use	Time Required	Do Models Provide Measures of Functional Capacity?	Remarks
Wetland Evaluation Technique (WET) **(48), (49)**	High, moderate, or low probability rating of social significance, opportunity, and effectiveness. Functions separate.	Applicable to all wetland types of contiguous U.S.	Widespread application	Regulatory actions Impact assessment Management Planning	Relatively rapid	No. Some models produce probability ratings (that is, an estimate of the probability that a function will occur to an unspecified magnitude); not a measure of functional capacity. Also, some models combine function and value. Modifications required to achieve measure of functional capacity.	Rating not multiplied by acres; thus does not provide measure of functional capacity that accounts for size.
Connecticut Method **(50)**	Functional Value Indices (FVI) (scale 0 to 1.0) and Wetland Value Units (WVUs). Functions separate.	Nontidal wetlands	Connecticut	Inventory Education Planning	Relatively rapid	No. Some models combine function and value. Modifications required to achieve measure of functional capacity.	Tool for those with some knowledge of wetlands (for example, town commission), but not necessarily wetland ecologists. Provides measure of functional capacity that accounts for size.
Index of Biotic Integrity **(24), (51), (52)**	Index of biotic integrity (IBI). One function.	All freshwater wetlands that provide fish habitat.	Widespread application	Inventory	Not rapid	Yes	Requires collection of fish samples. One function. IBI not multiplied by acres, thus does not provide measure of functional capacity that accounts for size.
A method for assessing the functions of wetlands **(53)**	Scores (scale differs for each function; range from 3 to 158). Functions separate, then assigned weights/ranked and wetland assigned one decile score (0-10).	Nontidal wetlands	New England states. Some midwestern states, and possibly other areas.	Regulatory actions Management Planning	Relatively rapid	No. Some models combine function and value. Modifications required to achieve measure of functional capacity.	Score not multiplied by acres, thus does not provide measure of functional capacity that accounts for size.

TABLE 4 *Continued*

Procedure Title: Citation	Units of Measure	Application: Wetland Types	Application: Geographic Area	Application: Use	Time Required	Do Models Provide Measures of Functional Capacity?	Remarks
Habitat Evaluation Procedure (HEP) **(22)**	Habitat suitability indices (HSI) (scale 0 to 1.0) and Habitat Units (HUs). Species separate. Pool index for each species to obtain one score (total HUs) for a wetland.	Nontidal wetlands, tidal wetlands, and terrestrial habitats.	Widespread application; regionalized by selection of indicator species assessment models which are applicable to region.	Regulatory actions Impact assessment Management Mitigation	Not rapid	Yes	Models describe habitats of selected fish, wildlife, or invertebrates: pool index for each selected species to give one score for wetland. Provides measure of function that accounts for size (HUs). Only habitat function. Comparison of different wetland types possible
Pennsylvania Habitat Evaluation Procedure (PAM HEP) **(54), (55)**	Habitat suitability index (HSI) (scale 0 to 1.0) and Habitat Units (HUs). Species separate. Pool index for each species to obtain one score (total HUs) for a wetland.	Freshwater wetlands and terrestrial habitats.	Developed for Pennsylvania but has been applied in other states.	Wildlife habitat assessment Regulatory actions Impact assessment Mitigation.	Relatively rapid	Yes	Modification of the Habitat Evaluation Procedure (HEP); simplified to reduce time requirement. (Remarks same as for HEP).
Wildlife wetland evaluation model **(56)**	Wetland score (scale 36 to 108). (One functional).	Freshwater wetlands	Massachusetts and possibly other northeastern states.	Regulatory actions Management Planning	Relatively rapid	Yes	Score not multiplied by acres, thus does not provide measure of functional capacity that accounts for size. Only habitat function.

7.2 *Properties to Consider When Selecting Procedure or Models*:

7.2.1 *General Principles*—The choice of an assessment procedure or model depends on many factors, especially assumptions that decision makers are willing to accept. Reference **(23)** suggest the following general principles be used to guide the choice of a procedure. The procedure should:

7.2.1.1 Be biased on principles and assumptions that are valued (if feasible) and easily illustrated,

7.2.1.2 Yield results understandable to decision makers and the public,

7.2.1.3 Makc cxplicit subjective values and judgments,

7.2.1.4 Yield results that are repeatable given certain explicit assumptions,

7.2.1.5 Allow use of qualitative and quantitative information in a methodologically sound way,

7.2.1.6 Stimulate the imagination of decision makers and increase insight into the choice to be made,

7.2.1.7 Enable the use of information at different spatial scales, and

7.2.1.8 Allow consideration of alternatives both separately and in combination.

7.2.2 *Section 404*[3] *Regulatory Actions*—Reference **9** indicates that the assessment procedure must satisfy one or more of the basic programmatic or technical requirements to receive widespread acceptance or utilization in the Section 404 program at a national level. These requirements included the following:

7.2.2.1 Standardized and documented approach,

7.2.2.2 Applicability throughout the public interest review sequence,

7.2.2.3 Applicability across the geographic extent of the U.S. Army Corps of Engineers regulatory jurisdiction,

7.2.2.4 Applicability to a variety of wetland types,

7.2.2.5 Applicability to a variety of wetland functions,

7.2.2.6 Compatibility with the time and resources available for the public interest review process,

7.2.2.7 Accuracy and precision that is consistent with the time and resources available,

7.2.2.8 Sensitivity to different types of impacts at levels at which wetland functions are affected,

7.2.2.9 Adaptability to a variety of regulatory, management, and planning applications,

7.2.2.10 Defined standards of comparison,

7.2.2.11 Capability to incorporate new technical information as it becomes available, and

7.2.2.12 Capability to incorporate new or changing programmatic requirements.

7.2.3 *Appropriate for Planned Wetland Design*—Some assessment procedures are not suitable for the mitigation process and can lead to unfounded planned wetland design criteria. An assessment procedure must have the following properties to be suitable, particularly if it will be used as a guide to design **(10)**:

7.2.3.1 Document both the procedure and results. This will facilitate the design and review of the planned wetland. The format should allow the designer and decision maker to readily identify elements that are important to each function. It should provide for easy extraction to improve the functional capacity in the planned wetland.

7.2.3.2 Provide validated threshold values for design elements. Threshold values should not be used, unless they can be literature-validated or validated through consultation with experts. Threshold values are cutoff values used in the assessment model, above or below which it is believed that a wetland's capacity to perform a function changes substantially. For example, the model may assume that a ≥ to 20-ft wide wetland will effectively provide the shoreline erosion control function; anything less would be considered ineffective. Different assessment models use different threshold widths (for example, 8, 10, 20, 600 m). If the assessment model is used as a guide to design, then substantially different design criteria could be obtained depending upon the model used. Based upon the example given, the recommended minimum width could vary from 8 to 600 m.

7.2.3.3 Include variables applicable to planned wetland design. Models from rapid assessment procedures often use a minimum number (for example three to five) of variables to assess each function. Although it lengthens the assessment time, it is important that variables critical to wetland design be included.

7.2.3.4 Avoid using variables that describe opportunity in models designed to measure functional capacity. Opportunity variables are those characteristics of a wetland or its surroundings that determine if the opportunity is available for that wetland to perform a function. Opportunity variables are used with other structural variables to describe functional capacity in most models. The rationale is that the wetland is more valuable when the opportunity for performing the function is present. Many of the opportunity variables describe conditions that, if excessive, could change a wetland's functional capacity. For example it is often assumed that greater pollutant input makes a wetland more valuable for the water quality function. This assumption may be invalid. Studies on the use of wetlands for wastewater treatment have demonstrated that, after several years, some wetlands that initially served as nutrient sinks reach their assimilatory capacity for certain chemical constituents. Many assessment models do not set an upper limit on opportunity variables. Without an upper limit, the model may assign a high rating erroneously when the capacity of the wetland to perform a function may be minimal or exceeded due to excessive pollutant input. Opportunity variables should only be used in the models to note conditions which could reduce the planned wetland's functional capacity.

7.2.3.5 Be sensitive to detect differences between wetlands. The assessment model must be sensitive enough to detect planned wetland improvements.

8. Keywords

8.1 assessment; design; ecology; function; functional capacity; index; mitigation; planned wetland; variables; wetland

APPENDIX

(Nonmandatory Information)

X1. WETLAND CREATION, ENHANCEMENT, RESTORATION, OR CONSTRUCTION

X1.1 A specific situation when wetland functions are assessed is during the development of planned wetlands. Planned wetlands encompass a variety of activities such as wetland creation, enhancement, restoration, or construction. Planned wetlands may be designed to compensate for the loss of functions resulting from project impacts, or be associated with a non-permit restoration effort. In either case, if information on the adequacy of the planned wetland with regard to function is desired, an assessment may be performed and the results compared to some baseline condition. Assessment models may be used during one or more steps in the planned wetland process that include defining goals, site selection, design, and assessment. The following steps follow those prescribed in Bartoldus et al (**10**). The use of assessment models may not be deemed necessary such as with small planned wetlands. In other situations, the collection of additional data may be required. A decision as to what is needed must be made on a case-by-case basis.

X1.2 *Define Goods for Planned Wetland*—The goals should be defined based on the results of the comparison of WAAs and recommendations of participating federal, state, or local agencies. If the goals are established for a permit action, they become part of the mitigation requirements.

X1.2.1 Potential goals could include:

X1.2.1.1 Providing the same functions at the same level of functional capacity (that is, equal FCIs and FCs),

X1.2.1.2 Providing the same functions at a different level of functional capacity (that is, greater or less FCIs and FCs),

X1.2.1.3 Maximizing functional capacity of one or several functions that are not or poorly provided in the WAA,

X1.2.1.4 Establishing the same wetland class with same the vegetative cover types,

X1.2.1.5 Establishing the same wetland class with different vegetative cover types, and

X1.2.1.6 Establishing a different wetland class.

X1.2.2 Planned wetland goals can be defined in a variety of ways depending upon the project. However, they should ultimately be expressed in terms of Target FCIs and Target FCs. For example, the goals for FCIs in the planned wetland can be expressed as a simple statement such as the planned wetland FCIs must meet or exceed the FCI for each function in the WAA. If the comparison involves more than one baseline WAA with a broad range of FCIs (for example, from 0.1 to 0.6), then the goal may be to achieve the highest FCI (for example, FCI equal to 0.6). Target FCs for each function are calculated as follows:

$$Target\ FCs = FCs \times R \tag{X1.1}$$

where:

Target FCs = target functional capacity units,

FCs = total functional capacity units for the WAA, and

R = multiplying factor used to generally increase the amount of compensation

If the goal is to provide equal compensation, then the Target FCs will equal the FCs for the WAA, and no multiplying factor is used. If the goal is to provide greater compensation (for example, 2:1 mitigation ratio), then the Target FCs are calculated by multiplying the FCs by the appropriate factor (for example, $R = 2$). Ratios described here are included for purposes of illustration and do not represent a standard ratio. Experience indicates that each wetland mitigation scenario yields a unique mitigation ratio based upon factors specific to each project. Target FCIs and Target FCs can be recorded in standardized data sheets such as those shown in Table X1.1. Reasons for requiring greater than/equal to 1 to 1 compensation include the following.

X1.2.2.1 *Mitigation Ratios*—Some federal, state, and local agencies have instituted regulations or policies that stipulate mitigation ratios. The ratios have often been set in response to unsuccessful planned wetlands. Since planned wetlands are frequently perceived as being unsuccessful, or not totally successful, ratios are established to provide compensation for the anticipated failure of some portion of a project. The hope is that a larger planned wetland will provide some guaranteed compensation for the losses. In some cases, standard ratios are stipulated in regulations or guidelines, or both, that are required for mitigation projects. The ratio may be the same for all wetland types (that is, 3:1) or different depending upon the WAA (that is, 1:1 for emergent wetland replacing emergent wetland or 3:1 for emergent wetland replacing forested wetland). In other cases, standard ratios are required for the planned wetland when it involves mitigation for a rare wetland type.

X1.2.2.2 *Out-of-Kind Mitigation*—Depending upon the wetland type involved, decision makers may recommend the construction of a relatively larger or smaller planned wetland of another wetland type.

X1.2.2.3 *Off-Site Mitigation*—If a nearby site is not available, the planned wetland may be located in a different locale (for example, out of watershed) far from the WAA. Since compensation is not provided in the same area, decision makers may recommend construction of a larger planned wetland.

X1.2.2.4 *Time Lag*—This refers to the loss in functional capacity during the period of time it takes the planned wetland to reach long-term functional capacity goals. It may not be feasible to construct a wetland that will immediately provide all of the desired functions. For example, a planned wetland planted with saplings may be designed for the long-term goal of a forested wetland. Decision makers may choose to estimate these losses by comparing the same wetland at different points

TABLE X1.1 Comparison of WAA and Planned Wetland: Calculations of FCIs and FCs (Modified from (10))

Project Title: Marley Creek

Comparison between WAA #1 and Planned Wetland #1

	WAA			Goals for Planned Wetland					Planned Wetland			
Function	FCI	Size, Acres	FCs[A]	Target FCI[B]	R[C]	Target FCs[D]	Predicted FCI[E]	Minimum Size[F]	FCI	Size, Acres	FCs	Check if Goals Met
Shoreline Bank Erosion Control	0.7	0.5	0.4	> 0.7	1	0.4	0.7	0.5	0.97	2	1.9	√
Sediment Stabilization	0.83	1.5	1.2	> 0.8	1	1.2	0.8	1.5	0.90	2	1.8	√
Water Quality	0.92	1.5	1.4	> 0.9	1	1.4	0.9	1.6	0.83	2	1.7	NO
Wildlife	0.54	1.5	0.8	> 0.06	1	0.8	0.6	1.3	0.35	2	0.7	NO

[A] FCs = FCI × area.
[B] Target FCI = goal established by decision makers.
[C] R = multiplying factor established by decision makers.
[D] Target FCs = FC of the WAA multiplied by R (that is, the planned wetland goal).
[E] Predicted FCI = FCIs that designers presume planned wetland may achieve at a particular site. (Note—this may be greater than Target FCI.)
[F] Minimum area = target FCs/predicted FCI.

in time which is time consuming, or simply require additional acreage to offset the anticipated loss of function during the period while the planned wetland is maturing.

X1.2.2.5 *Anticipated Failure in the Development of Some Portion of the Planned Wetland*—With the construction of wetlands, it is possible that some portion will not become established as planned. Possible causes of the failure may include plant die off from waterfowl grazing, muskrat eatouts, drought, or vandalism. It is common practice for decision makers to require a larger planned wetland with the hope of providing some guaranteed compensation for the total losses associated with the WAA.

X1.2.3 *Estimate Minimum Area Required to Meet Goals*—Before initiating the search for a planned wetland site, estimate the minimum area required to achieve the Target FCs. This exercise can save time by restricting the search to the sites with the potential to achieve the desired goals. Predicted FCIs must be defined before an estimate of minimum area can be made. Predicted FCIs are the FCs that the planned wetland is predicted to achieve. Define predicted FCIs based on a realistic assessment of the functional capacity of similar wetlands in the region. Note that the predicted FCIs are just estimates, and the FCIs achieved in the planned wetland design can differ from the predicted FCIs. Predicted FCIs should be recorded on standardized data sheets such as shown in Table X1.1. The minimum area is the minimum acreage required to satisfy the Target FCs for each function being considered in the planned wetland. Calculate minimum area as follows:

$$\textit{Minimum area} = \textit{Target FCs/Predicted FCI} \qquad (X1.2)$$

For example, it may be determined that the planned wetland has the potential to provide relatively high-quality wildlife habitat, and thus the Predicted FCI is set at 0.8. If the wildlife habitat function Target FCs = 6.6 units, then the minimum area would be 8.25 acres.

X1.2.3.1 The steps to determining Target FCs and minimum areas are illustrated in Table X1.1. In the example, the minimum acreage required for the planned wetland is less than the acreage of the WAA. The smaller acreage can be attributed to the predicted FCIs of the planned wetland which are equal to or higher than the FCIs in the WAA. This demonstrates the importance of FCIs in determining the planned wetland acreage. If FCIs in the planned wetland are greater than the FCIs in the WAA, the planned wetland goals can be achieved with a smaller planned wetland. The planned wetland must be larger if the FCIs in the planned wetland are less than those in the WAA.

X1.3 *Select Planned Wetland Site*—Potential sites are screened to eliminate the unacceptable ones and the final selection is made based upon a more detailed examination. Basic criteria such as minimum area and availability are initially used to identify potential sites. Other factors in addition to wetland function are also considered including economic feasibility, presence of red flags, and construction constraints. The assessment models are used as a guide to determine which site(s) can provide or can be modified to provide the conditions necessary to attain the Target FCIs and FCs. A simple comparison of site characteristics to assessment model variables will suffice in most cases. The FCI calculations may also be conducted if there are questions regarding the attainment of FCIs and FCs for the planned wetland. The extent to which model variables are examined during site selection will vary. The evaluation of some variables may only require cursory office windshield evaluations, whereas other variables may require a set procedure with frequent field monitoring. For example, the site hydrology usually requires a thorough evaluation because it is crucial to the success of a planned wetland. The hydrologic analysis not only verifies the planned wetland feasibility, but also verifies the conditions needed to achieve the function goals.

X1.4 *Design Planned Wetland*—Refer to the Target FCIs and FCs and the assessment models to determine which conditions are necessary to meet the goals. Appropriate conditions can be incorporated as the planned wetland design is developed. Periodically refer to the models to identify the best conditions for maximizing the functional capacity level. Include these conditions and avoid or minimize unsuitable conditions in the planned wetland design.

X1.4.1 The planned wetland design is prepared for the selected site(s) at a scale necessary to establish site-specific design considerations. The design must also provide sufficient detail to perform the assessment. Designs are usually prepared in two stages beginning with a conceptual design, and followed by the development of construction plans and specifications. A conceptual plan provides a brief description of the planned wetland through drawings and text that confirms feasibility and facilitates early review by decision makers. Construction plans and specifications provide sufficiently detailed site-specific information for the general contractor to ensure that the planned wetland is constructed as planned. It is better to assess a conceptual design since the more detailed information (for example, specific grading and landscaping requirements) are not required to perform the assessment. A decision to delay and to assess the construction plans and specifications may result in undue costs if the assessment reveals the need for revisions.

X1.5 *Assess Planned Wetland*—Define the stage at which the planned wetland is to be assessed (for example, design, predicted future conditions) and proceed with the assessment. The results are compared to the Target FCIs and Target FCs to determine whether the goals are met (refer to Table X1.1). If the planned wetland design does not meet the goals, the design should be revised and reassessed again. If the Target FCs are not met, the FCs can be increased by increasing the size of the wetland area or redesigning the planned wetland to increase the FCI for the applicable function(s), or both. The comparison of model variables should also be recorded on standardized data sheets such as shown in Fig. 3. This information can be used to explain how specific difference in wetland features have resulted in a difference in the functional capacity indices. If the planned wetland is compared to several WAAs of the same class, then a direct comparison can be made between the planned wetland and goals (Target FCIs and Target FCs). The comparison of variable scores is more difficult. Users and decision makers must agree upon the format and extent of comparison, depending upon the individual project.

X1.6 *Monitoring*—The monitoring of planned wetlands is required for a variety of purposes including mitigation, but is generally done to determine success or failure. Monitoring can be approached from two perspectives: design goals and function-based goals. At the very least, the planned wetland should be evaluated to determine whether the design goals have been met in order to ensure continued development of the planned wetland. A decision must be made regarding the need for and type of additional monitoring to determine if the goals for functional capacity have been met.

X1.6.1 *Design Goals*—One approach to monitoring is to compare the as-built planned wetland at the completion of the construction phase and, over time, with the original or modified plans. Relatively short-term monitoring is used to identify corrective action, if needed, to ensure establishment of the wetland. Following Ref **(57)**, the planned wetland would be considered a success if it persisted in comparing favorably with the conceptual plans during the monitoring period. Modified plans are original ones that were changed prior to or during construction due to unexpected site conditions, errors in the original plan, and so forth. Comparing favorably means that the specified hydrology is present, specified emergent or woody vegetation, or both, are present or both at the designed locations, and open water and other structural features are present at the design locations.

X1.6.2 *Function Goals*—Function-based goals in the planned wetland can be assessed using the steps described in X1.2-X1.6. However, in some situations more detailed field studies may be required. For additional information on sampling strategies and data collection techniques refer to Refs **(58-62)**. It should be noted that monitoring for function goals may not detect all of the information necessary to ensure continued development of the planned wetland. For that reason, the design goals must be looked at separately.

REFERENCES

(1) Luzar, E.J., and Gan, C., "Economic Valuation of Wetland Functions and Values: Literature Review 1985–1991," *Technical Report Y-91-17*, U.S. Army Engineer Waterways Experiment Station, Vicksburg, MS, 1991.

(2) Shabman, L., and Batie, S., "Socio-Economic Values of Wetlands: Literature Review 1970–1985," *Technical Report Y-88*, U.S. Army Engineer Waterways Experiment Station, Vicksburg, MS, 1988.

(3) Leibowitz, S.G., Abbruzzese, B.; Adamus, P.R., Hughes, L.E., and Irish, J.T., *A Synoptic Approach to Cumulative Impact Assessment*, USEPA Environmental Research Laboratory, Corvallis, OR, EPA/600/R-92/167, 1992.

(4) Abbruzese, B., and Liebowitz, S.G., (In press), *A Synoptic Approach for Assessing Cumulative Impacts to Wetlands,* Environmental Management, 1996.

(5) Sutter, L.A., and Wuenscher, J.R., NC-CREWS: *A Wetland Functional Assessment Procedure for the North Carolina Coastal Area (Draft)*, Division of Coastal Management, North Carolina Department of Environment, Health, and Natural Resources, Raleigh, NC, 1996.

(6) Brinson, M.M., *"Changes in the Functioning of Wetlands Along Environmental Gradients," Wetlands* 13: 1993, pp. 65-74.

(7) Walbridge, M.R., "Functions and Values of Forested Wetlands in the Southern United States," *Journal of Forestry*, 91, 1993, pp. 15-19.

(8) Richardson, C.J., "Ecological Functions and Human Values in Wetlands: a Framework for Assessing Forestry Impacts," *Wetlands,* 14, 1994, pp. 1-9.

(9) Smith, R.D., Ammann, A., Bartoldus, C., and Brinson, MM, "An Approach for Assessing Wetland Functions Using Hydrogeomorphic Classification, Reference Wetlands, and Functional Indices," *Wetlands Research Program Technical Report WRP-DE-9*, U.S. Army Engineer Waterways Experiment Station, Vicksburg, MS, 1995.

(10) Bartoldus, C.C., Garbisch, E.W., and Kraus, M.L. *Evaluation for Planned Wetlands (EPW),* Environmental Concern Inc., St. Michaels, MD, 1994.

(11) Cowardin, L.M., Carter, V., Golet, F.C., and LaRoe, E.T. *Classification of Wetlands and Deepwarter Habitats of the United States,* U.S. Fish and Wildlife Service, Office of Biological Services, FWS/OBS-79/31, Washington, DC, 1979.

(12) Brinson, M.M., *A Hydrogeomorphic Classification for Wetlands,* Wetlands Research Program TR-WRP-DE-4, U.S. Army Engineer Waterways Experiment Station, Vicksburg, MS, 1993.

(13) Kusler, J.A., Ray, C., Klein, M., and Weaver, S., *State Wetland Regulation: Status of Programs and Emerging Trends,* The Association of State Wetland Managers, Berne, NY, 1993.

(14) L.M. Smith, R.L., Pederson, and R.M.Kaminski, eds, *Habitat Management for Migrating and Wintering Waterfowl in North America,* Texas Tech University Press, Lubbock, TX, 1989.

(15) Maryland Department of Environmental Management, *A Method for the Assessment of Wetland Function,* Annapolis, MD. Prepared by Fugro East, Inc., Northborough, MA for Maryland Department of Environment, 1995.

(16) Ammann, A.P., and Lindley Stone, A., *Method for the Comparative Evaluation of Nontidal Wetlands in New Hampshire,* NHDES-WRD-1991-3. New Hampshire Department of Environmental Services, Concord, NH, 1991.

(17) Hruby, T., Cesanek, W.R., and Miller, K.E., "Estimating Relative Wetland Values for Regional Planning," *Wetlands 15(2)*: 1985, pp. 93-107.

(18) Brinson, M.M., Hauer, F.R., Lee, L.C., Nutter, W.L., Rheinhardt, R.D., Smith, R.D., and Whigham, D.F., *Guidebook For Application of Hydrogeomorphic Assessments to Riverine Wetlands,* Wetlands Research Program TR-WPR-DE-11, U.S. Army Engineer Waterways Experiment Station, Vicksburg, MS, 1996.

(19) *Protecting America's Wetlands: An Action Agenda,* The Conservation Foundation, Washington, DC, 1988.

(20) Mitsch, W.J., and Gosselink, J.G., *Wetlands,* Van Nostrand Reinhold, New York, 2nd ed, 1993.

(21) *National Action Plan to Implement the Hydrogeomorphic Approach to Assessing Wetland Functions*, U.S. Army Corps of Engineers, Federal Register, Vol 62, No. 119, 1997, pp. 33607-33620.

(22) *Habitat Evaluation Procedure (HEP) Manual (102 ESM)*, U.S. Fish and Wildlife Service, Washington, DC, 1980.

(23) Smith, P.G.R., and Theberge, J.B., "Evaluating Natural Areas Using Multiple Criteria: Theory and Practice," *Environmental Management 11*(4), 1987, pp. 447-460

(24) Karr, J.R., *Assessment of Biotic Integrity Using Fish Communities,* Fisheries 6(6): 1981, pp. 21-27.

(25) Lodge, T.E., Hillestad, H.O., Carney, S.W., Darling, R.B., *Wetland Quality Index (WQI): A Method for Determining Compensatory Mitigation Requirements for Ecologically Impacted Wetlands,* Paper presented at the American Society of Civil Engineers, South Florida Section, Miami, FL, Sept. 22-23, 1995.

(26) Miller, R.E., and B.E. Gunsalus, *Wetland Rapid Assessment Procedure (WRAP)* (Draft), South Florida Water Management District, Natural Resource Management Division, West Palm Beach, FL, 1996.

(27) Adamus, P.R., "Review of Sources and Methods," World Wildlife Fund, Statewide, *Wetland Strategies: A Guide to Protecting and Managing the Resource,* Island Press, Washington, DC, 1992, pp. 171-224.

(28) *Technical Report: Literature Review of Wetland Evaluation Methodologies,* U.S. Environmental Protection Agency Region 5, Chicago, IL, 1984.

(29) Lonard, R.I., Clairain, R.J., Jr., Huffman, R.T., Hardy, J.W., Brown, L.D., Ballard, P.E., and Watts, J.W., *Analysis of Methodologies Used for the Assessment of Wetland Values, Final Report,* U.S. Water Resources Council, Washington, DC and U.S. Army Corps of Engineers Waterways Experiment Station, Vicksburg, MS, 1981.

(30) *Minnesota Routine Assessment Method for Evaluating Wetland Functions (MinRAM) Version 1.0,* Minnesota Board of Water and Soil Resources, St. Paul, MN 1996.

(31) *The Highway Methodology Workbook Supplement. Wetland Functions and Values: A Descriptive Approach,* U.S. Army Crops of Engineers New England Division, NENEP-360-1-30a, 1995.

(32) Frye, R.G., *Wildlife Habitat Appraisal Procedure (WHAP),* Texas Parks and Wildlife Department, Austin, TX, 1995.

(33) Stein, Eric, *Assessment of the Cumulative Impacts of Section 404 Clean Water Act Permitting on the Ecology of the Santa Margarita, CA Watershed,* Doctoral Dissertation, University of California, Los Angeles, Environmental Science and Engineering Program, 1995.

(34) Fugro East, Inc., *A Method for the Assessment of Wetland Function,* Prepared for Maryland Department of the Environment, Annapolis, MD, 1995.

(35) North Carolina Department of Environment, Health, and Natural Resources, *Guidance for Rating the Values of Wetlands in North Carolina,* Raleigh, NC, 1995.

(36) *Buffalo District Wetland Evaluation Method (BWEM),* U.S. Army Corps of Engineers, Buffalo District, Buffalo, NY, 1994.

(37) Environmental Work Group, *Wetland Value Assessment Methodology and Community Models,*. Environmental Work Group, Coastal Wetland Planning, Protection, and Restoration Act Technical Committee, Louisiana Department of Natural Resources, 1993.

(38) Palmer, J.H., Muir, R.H., and Sabolcik, T.M. *Wildlife Habitat Assessment and Management System: Habitat Evaluation Procedure Technology for Wildlife Management Planning* (Second revision), Pennsylvania Game Commission, Bureau of Land Management, Harrisburg, PA.

(39) Roth, E., Olsen, R., Snow, P., and Sumner, R., *Oregon Freshwater Wetland Assessment Methodology,* Oregon Division of State Lands, Salem, OR, 1993.

(40) *Ontario Wetland Evaluation System: Northern Manual,* Ontario Ministry of Natural Resources, Wildlife Policy Branch, Toronto, Ont., 1993a.

(41) Ontario Ministry of Natural Resources, *Ontario Wetland Evaluation System: Southern Manual,* 3rd Edition, Wildlife Policy Branch, Toronto, Ontario, 1993b.

(42) Cook, R.A., Lindley Stone, A.J., and Ammann, A.P., *Method for the Evaluation and Inventory of Vegetated Tidal Marshes in New Hampshire,* Audubon Society of New Hampshire, Concord, NH, 1993.

(43) *Rapid Assessment Methodology for Evaluating Wetland Functional Values,* Wisconsin Department of Natural Resources, 1992.

(44) Bradshaw, J.G., *A Technique for the Functional Assessment of Nontidal Wetlands in the Coastal Plain of Virginia,* Special Report No. 315 in Applied Marine Science and Ocean Engineering, Virginia Institute of Marine Science, College of William and Mary, Gloucester Point, VA, 1991.

(45) *Wildlife Habitat Appraisal Guide for Missouri,* Missouri Department of Conservation, Jefferson City, MO, 1991.

(46) Cable, T.T., Brack, Jr., V. and Holmes, V.R., "Simplified Method for Wetland Habitat Assessment," *Environmental Management 13,* 1989, pp. 207-213.

(47) *The Minnesota Wetland Evaluation Methodology for the North Central United States,* Minnesota Wetland Evaluation Methodology Task Forces and Corps of Engineers, St. Paul, MN, 1988.

(48) Adamus, P.R., Stockwell, L.T., Clairain, Jr., E.J., Morrow, M.E., Rozas, L.P., and Smith, R.D., *Wetland Evaluation Technique (WET),* Vol I, Technical Report WRP-DE-2, U.S. Army Engineer Waterways Experiment Station, Vicksburg, MS, 1991.

(49) Adamus, P.R., Clairain, E.J., Smith, R.D., and Young, R.E., *Wetland Evaluation Technique (WET) Methodology Vol II,* Department of the Army, Waterways Experiment Station, Vicksburg, MS, 1987.

(50) Ammann, A.P., Frazen, R.W., and Johnson, J.L., *Method for the Evaluation of Inland Wetlands in Connecticut,* DEP Bulletin No. 9, Connecticut Department of Environmental Protection, Hartford, CT, 1986.

(51) Karr, J.R., Fausch, K.D., Angermeier, P.L., Yant, P.R., Schlosser, I.J., *Assessing Biological Integrity in Running Waters: A Method and Its*

Rationale, Illinois Natural History Survey Special Publication 5, Champaign, IL, 1986.

(52) Fore, L.S., Karr, J.R., and Wisseman, R.W., *Assessing Invertebrate Responses to Human Activities Evaluating Alternative Approaches,* N.J. Am. Benthol. Soc. 15(2), 1996 pp. 212-231,

(53) Hollands, G.G., and Magee, D.W., *Method for Assessing the Functions of Wetlands* J., Kusler, and P. Riexinger, eds, *Proceedings of the National Wetland Assessment Symposium Association of Wetland Managers,* Berne, NY, 1985.

(54) Palmer, J.H., Chezik, M.T., Heaslip, R.D., Rogalsky, G.A., Putnam, D.J., McCoy, R.W., and Arway, J.A., *Pennsylvania Modified 1980 Habitat Evaluation Procedure Instruction Manual,* Pennsylvania Game Commission, Bureau of Land Management, Harrisburg, PA. 1980.

(55) Palmer, J.H., *Wildlife Habitat Assessment and Management System PAM HEP Habitat Suitability Index Model Manual* (Second revision), Pennsylvania Game Commision, Bureau of Land Management, Harrisburg, PA.

(56) Golet, F.C. "Wildlife Wetland Evaluation Model," Larson, J.S. ed., *Models for Assessment of Freshwater Wetlands, Publication No. 32,* Water Resources Research Center, University of Massachusetts, Amherst, MA 1976.

(57) Garbisch, E.W., "Monitoring in the Mitigation and Restoration Process," *Environmental Concern Wetland Journal 5(1).* Environmental Concern, St. Michaels, MD, 1993, pp. 8-9.

(58) Kentula, M.E., Books, R.P., Gwin, S.E., Hollands, C.C., Sherman, A.D., and Signeos, J.C., *Wetlands: An Approach to Improving Decision Making in Wetland Restoration and Creation,* Island Press, Washington, DC.

(59) Pacific Estuarine Research Laboratory. *A Manual for Assessing Restored and Natural Coastal Wetlands with Examples from Southern California,* Biology Department, San Diego State University, San Diego, CA 1990.

(60) Horner, R.R., and Raedeke, K.J., *Guide for Wetland Mitigation Project Monitoring* Report No. WA-RD 195.1, Washington State Department of Transportation, Olympia, WA, 1989.

(61) K.L. Erwin, J. Kusler, and M.E. Kentula, eds., "Wetland Creation and Restoration: The Status of the Science," *Perspectives.* Volume II, EPA/600/3-89/038b, Environmental Research Laboratory, Corvallis, OR, 1989, pp. 25-45.

(62) T.A. White, J.A.Allen, S.F. Mader , D.L. Mengel, D.M. Perison, and D.T. Tew, eds, *MIST: A Methodology to Classify Pre-Project Mitigation Sites and Development Performance Standards for Construction and Restoration of Forested Wetlands*, Performed for Region IV, U.S. Environmental Protection Agency, Atlanta, GA; North Carolina State University, Raleigh, NC, 1989.

(63) Bartoldus, C.C., A Comprehensive Review of Wetland Assessment Procedures: *A Guide for Wetland Practitioners*, Environmental Concern, Inc., St. Michaels, MD, 1999.

The American Society for Testing and Materials takes no position respecting the validity of any patent rights asserted in connection with any item mentioned in this standard. Users of this standard are expressly advised that determination of the validity of any such patent rights, and the risk of infringement of such rights, are entirely their own responsibility.

This standard is subject to revision at any time by the responsible technical committee and must be reviewed every five years and if not revised, either reapproved or withdrawn. Your comments are invited either for revision of this standard or for additional standards and should be addressed to ASTM Headquarters. Your comments will receive careful consideration at a meeting of the responsible technical committee, which you may attend. If you feel that your comments have not received a fair hearing you should make your views known to the ASTM Committee on Standards, 100 Barr Harbor Drive, West Conshohocken, PA 19428.

Designation: E 1984 – 98

Standard Guide for Process of Sustainable Brownfields Redevelopment[1]

This standard is issued under the fixed designation E 1984; the number immediately following the designation indicates the year of original adoption or, in the case of revision, the year of last revision. A number in parentheses indicates the year of last reapproval. A superscript epsilon (ε) indicates an editorial change since the last revision or reapproval.

1. Scope

1.1 This guide covers redevelopment of a Brownfields property for all stakeholders. This guide identifies impediments to Brownfields development and suggests solutions to facilitate redevelopment. Both government and community groups are concerned with the quality of Brownfields redevelopment and wish to ensure that the redevelopment will not only protect public health and environment, but also be economically viable and benefit the community. The sustainable Brownfields redevelopment process is a voluntary effort that actively engages property owners, developers, government agencies, and the community in conducting corrective action, economic evaluation, and other actions to promote the long-term productive reuse of a Brownfields property. The process can make great strides toward sustainable redevelopment since it encourages economic vitality of an area which in turn can reduce other social problems at Brownfields properties including poverty, unemployment, and crime.

1.2 It is the intent of this guide to encourage a sustainable Brownfields redevelopment process through responsible private/public investment and redevelopment of Brownfields properties. Brownfields redevelopment is not strictly an environmental issue. In some cases, the environmental issues may be a minor component of the redevelopment project. The interrelated financial, regulatory, and community participation aspects of Brownfields redevelopment should also be addressed. Decisions made in one of these areas may affect responses in other areas. For example, a community's goals for the ultimate use of a property may affect corrective action and the cost of potential remedial action that, in turn, may enhance the redevelopment.

1.3 This guide is intended to describe a highly flexible process. This process is not linear, and not every project requires full use of all components of the process for effective implementation. The key to the process is the active engagement of government, developers, and the community to ensure successful sustainable Brownfields redevelopment process.

2. Referenced Documents

2.1 *ASTM Standards:*

E 1527 Practice for Environmental Site Assessments: Phase I Environmental Site Assessment Process[2]

E 1528 Practice for Environmental Site Assessments: Transaction Screen Process[2]

E 1739 Guide for Risk-Based Corrective Action Applied at Petroleum Release Sites[2]

E 1912 Guide for Accelerated Site Characterization for Confirmed or Suspected Petroleum Releases[2]

E 1943 Guide for Remediation by Ground Water by Natural Attenuation at Petroleum Release Sites[2]

PS 104 Provisional Guide for Risk-Based Corrective Action[2]

3. Terminology

3.1 *Definitions*—The following definitions are quoted directly from the July 1995 revision of Regulations Governing ASTM Technical Committees. As definitions developed by the Committee on Standards (COS), these definitions are provided for informational purposes only, and are not subject to ballot or revisions under this guide.

3.1.1 *standard*—as used in ASTM, a document that has been developed and established within the consensus principles of the Society and that meets the approval requirements of ASTM procedures and regulations.

3.1.1.1 *Discussion*—The term "standard" serves in ASTM as an adjective in the title of documents, such as test methods or specifications, to connote specified consensus and approval. The various types of standard documents are based on the needs and usages as prescribed by the technical committees of the Society.

3.1.2 *guide*—an organized collection of information or series of options that does not recommend a specific course of action.

3.1.2.1 *Discussion*—A guide increases the awareness of information and approaches in a given subject area.

3.1.3 *practice*—a definitive set of instructions for performing one or more specific operations that does not produce a test result.

3.1.3.1 *Discussion*—Examples of practices include, but are not limited to: application, assessment, cleaning, collection, decontamination, inspection, installation, preparation, sampling, screening, and training.

3.1.4 *test method*—a definitive procedure that produces a test result.

[1] This guide is under the jurisdiction of ASTM Committee E-50 on Environmental Assessment and is the direct responsibility of Subcommittee E50.03 on Global Sustainability/Pollution Prevention.

Current edition approved Nov. 10, 1998. Published January 1999.

[2] *Annual Book of ASTM Standards*, Vol. 11.04.

3.1.4.1 *Discussion*—Examples of test methods include, but are not limited to: identification, measurement, and evaluation of one or more qualities, characteristics, or properties.

3.2 *Definitions of Terms Specific to This Standard:*

3.2.1 *Brownfields*—abandoned, idled, or under utilized properties where expansion or redevelopment is complicated by the potential or confirmed existence of chemical(s) of concern environmental media.

3.2.2 *Brownfields redevelopment coordinator*—a local or regional government official, economic development agency, or nonprofit organization responsible for facilitating Brownfields redevelopment.

3.2.2.1 *Discussion*—The Brownfields redevelopment coordinator is generally associated with local government. However, the Brownfields redevelopment coordinator can be an official of a regional, state, or federal government agency

3.2.3 *chemical(s) of concern*—specific constituents and their breakdown products that are identified for evaluation in the risk assessment process.

3.2.3.1 *Discussion*—Identification can be based on their historical and current use at a property; detected concentrations in environmental media; and their mobility, toxicity, and persistence in the environment. Chemical(s) of concern may include, but not be limited to, petroleum, metals and chemicals related to industrial activities.

3.2.4 *community*—the individuals living or working within the area, or both, that may be affected by Brownfields redevelopment.

3.2.4.1 *Discussion*—The community should be defined on a property-specific basis.

3.2.5 *corrective action*—the sequence of actions that include property assessment and investigation, interim remedial action, remedial action, operation and maintenance of equipment, monitoring of progress, and termination of the remedial action.

3.2.6 *developer*—a private or public entity that intends to redevelop a Brownfields property and may provide all, part, or none of the funds.

3.2.6.1 *Discussion*—The developer of a Brownfields property is often an organization or company whose primary business is typically unrelated to redevelopment of Brownfields properties. As an example, a developer may be the owner of a company that wants to purchase an adjacent parcel of property for expansion of the existing business.

3.2.7 *exposure pathway*—the course a chemical(s) of concern takes from the source area(s) to an exposed organism. An exposure pathway describes a mechanism by which an individual or population is exposed to a chemical(s) of concern originating from a property.

3.2.7.1 *Discussion*—Each exposure pathway includes a source of release from a source, a point of exposure, and an exposure route. If the exposure point is not at the source, a transport/exposure medium (for example, air or water) also is included.

3.2.8 *Remedial Action*—Activities conducted to protect human health and the environment by meeting acceptable risk-based site-specific target levels by using any combination of actions such as natural attenuation; institutional controls; source removal; engineering controls; and design, maintenance, and operation of remedial action equipment.

3.2.8.1 *Discussion*—For Brownfields properties the agreed upon remedial action may require a combination of active removal and exposure management methods that are protective of human health while allowing cost-effective property redevelopment.

3.2.9 *representation*—a statement of past or present fact, true on the date made, given by one party to induce another to enter into a contract.

3.2.10 *risk-based corrective action*—a framework in which exposure and risk assessment practices are integrated with property assessment activities and remedial action selection to ensure that the chosen action is protective of human health and the environment.

3.2.11 *site-specific target level (SSTL)*—Risk-based target levels for chemical(s) of concern for human receptors developed for all applicable media of concern on a Brownfields property

3.2.12 *stakeholders*—individuals, organizations, or other entities that directly affect or are directly affected by the Brownfields property or its redevelopment.

3.2.12.1 *Discussion*—Stakeholders include, but are not limited to owners, buyers, developers, lenders, insurers, government agencies, and community groups.

3.2.13 *sustainable Brownfields redevelopment process*—a voluntary effort that actively engages property owners, developers, government agencies and the community in conducting corrective action, economic evaluation, and other actions to promote the long-term productive reuse of a Brownfields property.

3.2.14 *transferee*—the buyer, other recipient by deed or lessee of the Brownfields redevelopment property.

3.2.15 *transferor*—the seller or lessor of the Brownfields redevelopment property.

3.2.16 *warranty*—a representation made by one party to a second party of a contract of the existence of a fact upon which the second party may rely (for example, that the property is in compliance with certain laws) thus relieving the second party of establishing that fact.

4. Significance and Use

4.1 This guide guides the stakeholder through a process for Brownfields redevelopment that incorporates regulatory, community, and transactional issues. Given the economic and social benefits of sustainable restoration of Brownfields properties, guidance on a process for Brownfields redevelopment could be most useful in insuring the acceptability, and, therefore, viability, of such redevelopment.

4.2 Sustainable Brownfields redevelopment can be achieved through the productive reuse of properties that have been abandoned or idled. This reduces the need to develop new land by satisfying the needs of the present without compromising the ability of the future generations to meet their own needs. Both the current environmental conditions and the future use of the property need to be considered in order to ensure sustainability. Following the process and concepts discussed in this guide will provide the user with a sound framework for decision making and assist the user in balancing the needs of both the present and future generations through involvement of all of the stakeholders in the process. However, this guide does not give specific criteria for assessing the sustainability of a Brownfields redevelopment project.

4.3 It is recognized that certain communities have already formulated their own, effective, processes. It is not the purpose of this guide to impose a set of guidelines where successful formulae exist. Indeed, elements of successful models from around the country have been integrated within this guide. The audience for this guide is all potential stakeholders.

4.4 Redevelopment of underutilized properties is a key objective of federal, state, and local environmental agencies in environmental policy. Many states have developed Brownfields legislation to facilitate this redevelopment effort. State voluntary corrective-action programs play an important role in implementing an effective Brownfields regulatory policy. Many states have applied risk-based decision-making concepts to their voluntary corrective action programs. Some of the Brownfields redevelopment projects will involve voluntary corrective action while other corrective actions may be mandated by government order. Risk-based decision making may be a viable option for corrective action at many Brownfields redevelopment properties.

4.5 This guide is intended to provide a framework for the sustainable Brownfields redevelopment process and identify critical functions and impediments that need to be overcome in order to achieve sustainable development. In this respect, this guide can be used by state and local government to establish and operate a viable redevelopment program for their communities.

5. Sustainable Brownfields Redevelopment Process

5.1 There are four main components in the sustainable Brownfields redevelopment process: initiation, evaluation, transaction, and implementation (see Table 1). In addition, a sustainable Brownfields redevelopment process requires the involvement of a variety of stakeholders. Some stakeholders remain at the core of the process as one component develops into the next. Other stakeholders will move closer to or further away from the core during the process. Thus, stakeholders will bring a different perspectives or agendas to each component of the process. The sustainable Brownfields redevelopment process is not linear and not every project requires full use of all components of the process for effective implementation. It works best when the interests of all stakeholders are identified early on and the parties work together as a team to satisfy each of the party's interests using an open, non-adversarial approach to negotiations. These components and the stakeholders' goals are described in the following sections and in Fig. 1.

TABLE 1 Example of the Stakeholders that Could Be Associated with the Basic Components in the Sustainable Brownfields Redevelopment Process

Initiation	Evaluation	Transaction	Implementation
Community	Community	Community	Community
Government	Government	Government	Government
Owner/Transferor	Owner/Transferor	Owner/Transferor	Owner/Transferor
Prospective Transferee	Prospective Transferee	Prospective Transferee	Developer
Developer	Developer	Developer	Transferee
	Insurers	Insurers	Insurers
		lenders	Lenders

NOTE 1—Stakeholders who have an interest in the Brownfields redevelopment are listed in Table 2. Some stakeholders have notable cash or cash equivalents at risk. While other stakeholders have an interest in enssuring sustainable development goals are achieved and maintained.

6. Initiation

6.1 The sustainable Brownfields redevelopment process begins with a vision of the redevelopment and reuse of a property or properties. The process is initiated when a stakeholder recognizes a need or a business opportunity. The process can be initiated by a number of different stakeholders either individually or as part of a team effort. One of the critical activities of the initiation component is the identification of the potential stakeholders. The form of initiation will depend upon the stakeholders' goals (see Table 2).

6.1.1 *Initiate Process*—The sustainable Brownfields redevelopment process can be initiated by any one or more of the potential stakeholders.

6.1.1.1 *Owners, Prospective Transferees and Developers*—Owners, prospective transferees or developers or a combination thereof, may initiate the sustainable Brownfields redevelopment process by identifying a property or properties for redevelopment based upon their belief that the project will yield an appropriate return on investment or community benefit, or both.

6.1.1.2 *Not-for-Profit Organization*—A not-for-profit organization may act as a facilitator, investor, or partner in the redevelopment of a particular Brownfields property or properties.

6.1.1.3 *Community*—The community may initiate the process through dialogue with owners, prospective transferees, or government agencies. Community groups often have a vision, plan, and expectations for redevelopment in their place of residence and business. This vision may be based on history and tradition particular to the neighborhood and may include expectations of improved economic opportunities and physical surroundings.

6.1.1.4 *Local Government*—Government may initiate the sustainable Brownfields redevelopment process as a developer, owner, potential transferee, and investor, or otherwise facilitate Brownfields redevelopment by providing technical expertise, financial incentives or community education, or a combination thereof.

6.2 *Identifying the Stakeholders*—It is incumbent upon the Brownfields Redevelopment coordinator or the stakeholder(s) initiating the process, or both, to carefully evaluate the project and identify key stakeholders before proceeding in the redevelopment process. The stakeholders are determined by more than the physical proximity to a Brownfields property slated for redevelopment, the percentage of capital invested in a particular property, or the effect a Brownfields redevelopment may have on them as a particular group or as individuals. It is crucial that the identification of the stakeholders include all parties that have a stake in the outcome of the redevelopment process.

6.2.1 *Owner, Prospective Transferee, or Developer*—The owner, prospective transferee, or developer should typically

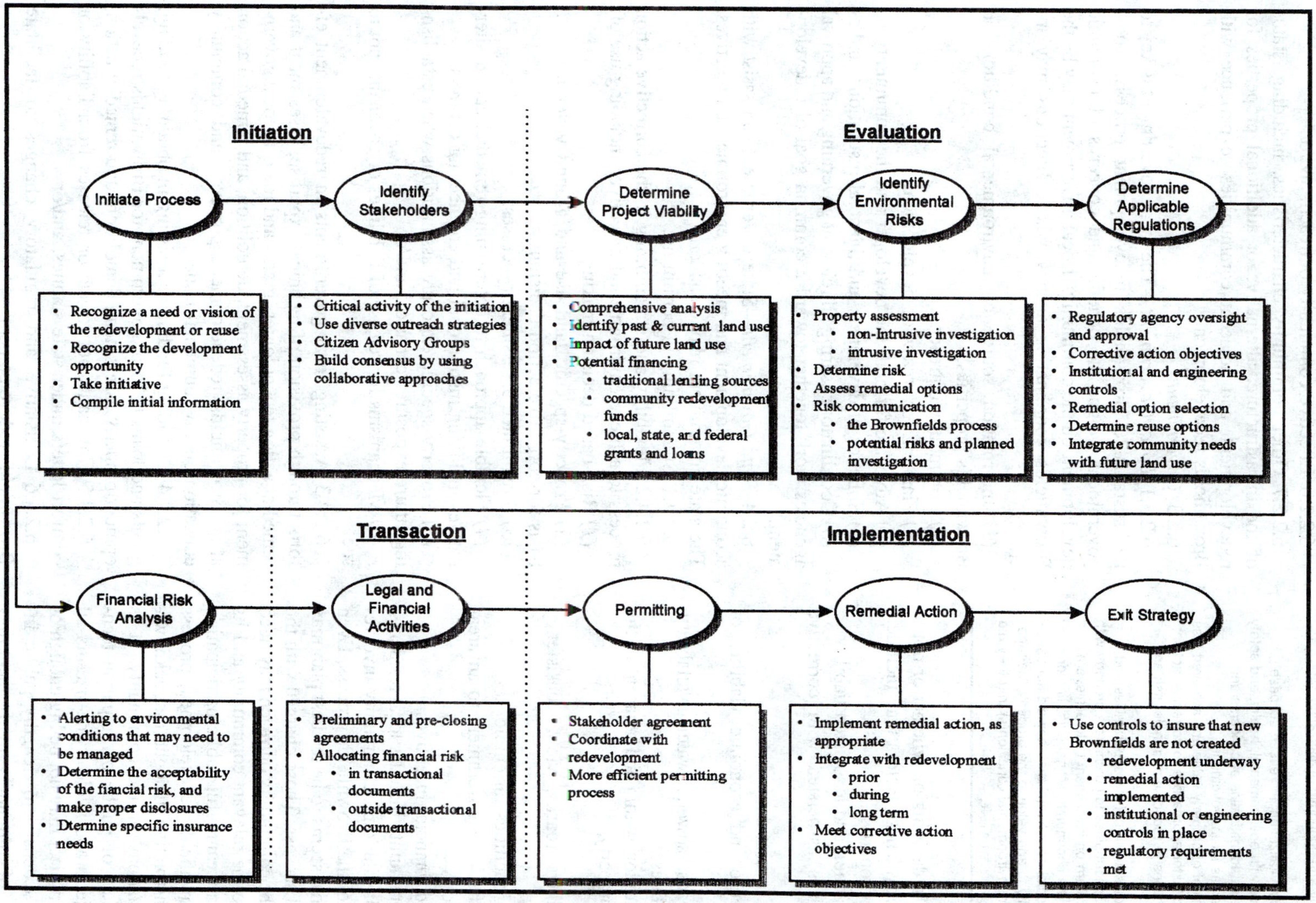

FIG. 1 Brownfields Redevelopment Process

TABLE 2 Example of Stakeholder Goals During the Initiation Component

Stakeholder	Goal
Community	Improvement of physical and aesthetic conditions; Community and economic revitalization
Government:	
Redevelopment Agencies	Economic revitalization; Increased tax base
Environmental and Health Agencies	Compliance with environmental, health, and safety requirements; Identification of concerns and potential areas for improvement
Transferor/Transferee	Enhancement of property value and achieved less costly and faster approaches to corrective action; Identify options to reduce and transfer risk and liability
Prospective transferee	Better understanding of the opportunities and barriers; Opportunities to purchase a property with potential return on investment or to achieve a benefit to the community, Manage liability for environmental condition(s) that they did not cause
Developer	Opportunity to add value to the property, Manage liability for environmental condition(s) that they did not cause

consider a number of factors in deciding on whether or not to invest in a Brownfields property. These factors may include:

6.2.1.1 Planned use of the property,

6.2.1.2 Cost of corrective action and associated liability,

6.2.1.3 Potential time delays associated with corrective action,

6.2.1.4 Potential return on investment,

6.2.1.5 Surrounding land use and economic viability of surrounding properties,

6.2.1.6 *Infrastructure*—Roads, sewers, water availability, public transportation, and utilities,

6.2.1.7 Environmental condition within and adjacent to the Brownfields redevelopment property,

6.2.1.8 Security,

6.2.1.9 Physical surroundings (parks, view, consistency with neighborhood character),

6.2.1.10 Education/training facilities for employees,

6.2.1.11 Commitment of the local government to an area-wide development plan, and/or

6.2.1.12 Property taxes and financial incentives.

6.2.2 *Community*—The community includes local residents, workers, organizations, and institutions. Community involvement is a critical component of the Brownfields process. Even with the best of intentions, what seems to be the best plan may not meet with acceptance and success. Those charged with the responsibility of conducting the community outreach may need to get help in identifying who the relevant community and its leaders are and in designing a meaningful public participation process. The sustainable Brownfields redevelopment process can only occur through a constructive dialogue that acknowledges the needs and expectations of the community. It is important to consider a variety of public participation techniques for different size projects and avoid a one-size-fits-all approach. In some cases, the project may not be big enough or complex enough to warrant extensive or lengthy public involvement. Other components of community involvement may include:

6.2.2.1 Community education,

6.2.2.2 Community input on the proposed development and potential corrective action,

6.2.2.3 Presentation of the conceptual plan to the community,

6.2.2.4 Recognition of community leadership,

6.2.2.5 Local political realities,

6.2.2.6 Effective community outreach, and/or

6.2.2.7 Other community characteristics, including, status of the infrastructure, alternative or additional properties for redevelopment, and adjacent communities' experience with Brownfields redevelopment.

6.2.3 *Government*:

6.2.3.1 *Local Government*—The local government is key to the sustainable Brownfields redevelopment process. Local government weighs the benefits and concerns of economic development on the community. Local government may be the Brownfields coordinating agency whose functions may include:

(1) Information source on environmental conditions at Brownfields properties,

(2) Community outreach,

(3) Planning for area-wide development,

(4) Advocate for both the developer and the community,

(5) Solicitor of developers and financial institutions, and/or

(6) Coordinator ensuring all relevant governing and approval agencies work together toward a common goal of redevelopment.

6.2.4 *State Government*—State government can assist with economic development and meeting environmental objectives. The state government may consider:

6.2.4.1 Streamlining in permitting,

6.2.4.2 Flexibility and streamlining in the corrective action process to achieve environmental objectives including use of:

(1) Risk-based decision making,

(2) Property-specific land use and ground water use as a basis for corrective action decisions,

(3) Streamlined assessment processes,

(4) Flexible approach to determining remedial action alternatives including institutional and engineering controls,

(5) Property-specific exposure assumptions when establishing future exposure scenarios, and/or

(6) Mechanisms for release of liabilities or covenants not to sue.

6.2.4.3 Assisting local governments in redevelopment efforts through provision of technical expertise, assessment and remedial action technologies, grants and loans to local government for property assessment, demolition, and remedial action at a Brownfields redevelopment property, and community education on environmental issues,

6.2.4.4 Providing funding or assisting in obtaining funding, grants and loans to local government for Brownfields redevelopment property assessment, demolition, and remedial action,

6.2.4.5 Covenants not to sue or releases from liability or comfort letters under state statutes, and/or

6.2.4.6 Legislative and regulatory changes to facilitate Brownfields redevelopment.

6.2.5 *Federal Government*—The federal government may assist with economic development and meeting environmental objectives by supporting risk-based decisions in corrective action for Brownfields properties and consider:

6.2.5.1 Funding or assisting in obtaining funding, grants and loans for Brownfields redevelopment property assessment, demolition and remedial action,

6.2.5.2 Assessments, revolving loans, and tax incentives,

6.2.5.3 Use of risk-based decisions for corrective action tied to reasonably anticipated land and ground water use,

6.2.5.4 Where state voluntary corrective action programs exist, delegation of federal corrective action oversight to the state under the voluntary corrective action program and allowing for release from liability,

6.2.5.5 Prospective purchaser agreements for releases from liability or comfort letters under federal statutes, and/or

6.2.5.6 Legislative and regulatory changes to facilitate Brownfields redevelopment.

6.3 *Lending Institution*—In some cases, lending institutions become involved in Brownfields redevelopment. They may provide part of the necessary funding for development and thus take on a portion of the financial risk associated with a project. Neighborhood economic and social stability are often motivators, since a healthier neighborhood economy is beneficial to lending institutions in the area. In addition, banks can access and invest Federal Community Reinvestment Act (CRA) funds for Brownfields redevelopment.

6.4 *Other Interested Parties*—Some parties may not be considered stakeholders at this point due to the limited extent to which that party may be affected by activity at the redevelopment property or the redevelopment process. However, such interested parties often play important roles in the process, and the involvement of these parties will depend on the needs and interests of the various stakeholders. Therefore, other interested parties who are actually involved should be determined by the stakeholders.

7. Evaluation

7.1 The purpose of the evaluation component is to determine the viability of proceeding with the Brownfields redevelopment. In addition to traditional real estate issues, a number of environmental and legal issues should be evaluated (see Table 3).

TABLE 3 Example of Stakeholder Goals for the Evaluation Component

STAKEHOLDER	GOAL
Community	Participate in the evaluation process and the development of appropriate options for improvement of the Brownfields property
Government:	
Redevelopment Agencies	Community understanding of the economic considerations planned use
Environmental and Health Agencies	Ensure corrective actions are protective of human health and the environment; ensure community understanding of these objectives, ensure requirements of multiple regulatory programs are satisfied
Transferor/Transferee	Find a solution that enhances the property value; achieves less costly and faster approaches to corrective action; identifies options to reduce and transfer risk and liability.
Prospective transferee	Better understanding of the opportunities and barriers; understand the financial/liability risk management options
Developer	Better understanding of the opportunities and barriers and reducing the uncertainty associated with time to complete and costs of completion
Insurer	Understanding the factors that could influence financial and environmental risk

7.2 *Determine Project Viability*—As the project warrants, a comprehensive analysis may be undertaken to determine the viability of the envisioned redevelopment and other alternative uses. Factors that may be considered include:

7.2.1 Current and future land use of the Brownfields property,

7.2.2 Impact of existing and potential future land use in the surrounding area,

7.2.3 Needs of the community,

7.2.4 Demographics,

7.2.5 Access to markets,

7.2.6 Return on investment, and

7.2.7 Financing including:

7.2.7.1 Traditional lending sources,

7.2.7.2 Community redevelopment fund,

7.2.7.3 Federal, state, and local grant and loan programs, and

7.2.7.4 Existing and new insurance products.

7.3 *Identifying Environmental Risks*—The successful implementation of the sustainable Brownfields redevelopment process is dependent on a clear understanding of the environmental condition and associated economic impact on the property or properties. This is critical to the risk management process.

7.3.1 *Property Assessment*—The primary objective of the property assessment is to collect information necessary to identify and determine the completeness of both human and ecological exposure pathways and to determine the likely distribution of a chemical(s) of concern. The collection of these data is necessary to make a determination of the potential environmental condition of the property. Normally, the property assessment is composed of two activities, a nonintrusive evaluation and an intrusive evaluation. The community may be an important resource and should be consulted for information for determining historical use and potential exposure pathways.

7.3.2 *Nonintrusive Evaluation*—A nonintrusive evaluation of the historical and current uses of the property and area surrounding the property is conducted to identify source areas (that is, areas where chemical(s) of concern is likely to be present) and potential receptors that may come in contact with a release from the property. Practices E 1527 and E 1528 are examples of non-intrusive evaluations. The evaluation may include:

7.3.2.1 Defining the area that will be investigated,

7.3.2.2 Identifying current and reasonable potential future receptors, fate and transport mechanisms, exposure routes, and point(s) of exposure,

7.3.2.3 Identifying potential sources, including a review of the property or property history to determine areas that may require investigation,

7.3.2.4 Identifying chemical(s) of concern,

7.3.2.5 Identifying potential source area(s),

7.3.2.6 Identifying the media to be sampled,

7.3.2.7 Determining current and reasonable potential future land use, and/or

7.3.2.8 Determining current and reasonable potential future ground water use.

7.3.3 *Intrusive Investigation*—An intrusive field investigation is conducted to determine if a release has occurred, to identify the source(s) and source area(s) and to collect sufficient data on the distribution and concentration of chemical(s) of concern in the media. This information is necessary to document the environmental condition of the property and to determine the appropriate remedial action options. Guide E 1739 and Provisional Guide PS 104 are examples of intrusive investigations. The intrusive field investigation may be used to:

7.3.3.1 Determine the presence and concentrations of the chemical(s) of concern in environmental media including:

7.3.3.2 Collect empirical data,

7.3.3.3 Characterize the potential source area(s),

7.3.3.4 Develop and evaluate a site conceptual exposure model,

7.3.3.5 Conduct initial response actions, as appropriate, and/or

7.3.3.6 Identify any interim remedial actions that may be appropriate.

7.3.4 *Determining Risk and Assess Remedial Action Alternatives*:

7.3.4.1 Risk assessment and any resulting remedial action(s) should be designed and implemented as part of the Brownfields redevelopment of the property. Based on the information and data compiled during the property assessment, the potential risks associated with the chemical(s) of concern and reasonable potential future land use should be evaluated for potential human and relevant ecological receptors. If these assessments reveal potential unacceptable risks, then potential remedial action options should be evaluated to determine those options that can be implemented as part of the Brownfields redevelopment. Remedial action options may include active or passive methods or some combination including source removal, treatment, and containment technologies; natural attenuation; exposure pathway elimination; engineering controls; and institutional controls. The applicability of remedial options to a Brownfields redevelopment property may vary from state to state. The remedial option(s) selected should incorporate features of the proposed Brownfields redevelopment as much as possible to allow for the most cost-effective remedial action.

7.3.4.2 Each potential remedial action option should be evaluated for its effectiveness, implementability, acceptability, and costs. Remedial action options that meet these criteria are then subject to a more rigorous assessment of long-term, financial requirements and liabilities. This financial risk analysis is an integral part of reaching agreement on a protective and cost effective remedial option for redevelopment of the Brownfields property.

7.3.5 *Risk Communication*.

7.3.5.1 The success of a Brownfields project is often dependent on how effectively current and future risks posed by the property are communicated to the community and other interested parties. It is important to acknowledge that risk communication should take place throughout the redevelopment process. Some general guidelines regarding risk communication are included in the Provisional Guide PS 104. This guide should also be referred to for a description of a framework for making risk-based decisions for corrective action.

7.3.5.2 It is essential to understand perceptions of risk in order to plan and design communications that foster cooperation rather than confrontation. The objective is to provide the opportunity for all parties involved to have an understanding of the risks and the plans to mitigate them. Trust and credibility are key factors in gaining cooperation, and are fostered by actively engaging the community in the decision making, thereby providing a greater sense of control over the situation. The burden of risk communication should not rest solely on the shoulders of the developer. The entire stakeholder group should work together to address concerns of the community and other interested parties. Communication methods may include printed information pamphlets (including easy-to-read charts and tables), public forums, news media (both print and TV), trade association meetings, labor unions, and the Internet.

7.3.6 In communicating the risk associated with a Brownfields project, the following issues need to be addressed:

7.3.6.1 Current and reasonable potential future use of the Brownfields redevelopment property,

7.3.6.2 Current and reasonable potential future human exposure and ecological concerns at the property,

7.3.6.3 Current and reasonable potential future uses of the area near the property,

7.3.6.4 Interests and concerns of the stakeholders,

7.3.6.5 Corrective action and property management options,

7.3.6.6 Health concerns associated with the corrective action and property management options, and/or

7.3.6.7 Applicable laws and regulations.

7.4 *Determine Applicable Laws and Regulations*—The successful implementation of the Brownfields redevelopment project is dependent upon a clear understanding of the legislative and regulatory requirements and policy issues critical to the risk management process. Identification and determination of these requirements and policy issues is necessary prior to beginning the process. It should also be recognized that more than one regulatory program may apply to a Brownfields redevelopment property and more than one regulatory agency may need to be involved in addressing the environmental condition(s) at a Brownfields redevelopment property.

7.4.1 The legislative and regulatory requirements and policy issues include, but are not limited to:

7.4.1.1 Permitting, regulatory agency approval, and oversight requirements,

7.4.1.2 Criteria for selecting point(s) of exposure, point(s) of compliance, and exposure pathway(s),

7.4.1.3 Criteria for use of institutional controls and engineering controls, and/or

7.4.1.4 Criteria for selecting appropriate remedial actions.

7.4.2 Some of the policy issues reflect political, economic and societal factors and are not always based strictly on scientific principles. The applicability of the following need to be determined:

7.4.2.1 Local, state, and federal laws,

7.4.2.2 Permits and certifications,

7.4.2.3 Pending enforcement actions, and/or

7.4.2.4 Pending claims, litigation, and liens.

7.5 *Financial Risk Analysis*—Financial risk analysis is an evaluation of the financial consequences of the project including environmental risk identified by a property assessment and review of applicable laws and regulations (see 7.2 through 7.5). Unlike environmental risk assessments and due diligence, financial risk analyses does not necessarily have to be repeated by different parties. If insurance is involved during the Evaluation stage, the underwriter's risk analysis may serve for other parties involved in the transaction.

7.5.1 Such analysis can serve many purposes in the sustainable Brownfields redevelopment process including:

7.5.1.1 If a transaction is involved, alerting the transferee or lender to any environmental conditions that may need to be managed or remedied after closing. It can be used to document conditions at closing and allow environmental risks to be managed and allocated as part of a property transaction (see 8.1.3),

7.5.1.2 Allowing investment bankers, underwriters, and others financing the project to make determinations concerning the acceptability of the financial risk and make proper disclosures in offering documents, and

7.5.1.3 Allowing insurance underwriters to determine the acceptability of the financial risk, determine specific insurance needs, and set appropriate premiums.

8. Transaction

8.1 Ownership often changes during this component. Frequently, permits are transferred and new consent orders for voluntary corrective actions or convenants not to sue are established between the new owner and the regulatory agency. These permits, orders, or actions should provide for unencumbered resolution of disputes. It is sometimes advantageous for the title to a Brownfields property to a local government entity or economic development corporation for purposes of being eligible to receive state or federal grants or loans for property assessment, demolition, or remedial action. The property may then be sold or leased to a developer (see Table 4).

8.2 *Preliminary Issues:*

8.2.1 *Environmental Liability Concerns*—Negotiation of the environmental provisions of any transaction involved in the sustainable Brownfields redevelopment process is driven principally by the real or perceived potential liabilities for corrective action of the property and, to a lesser extent, by concern with liability for bodily injury or property damage. Practices E 1527 and E 1528 provide a discussion of the general potential environmental liability that may arise in real estate transactions.

TABLE 4 Example of Stakeholder Goals for Transaction Component

STAKEHOLDER	GOAL
Community	The transaction meets the redevelopment goals of the community
Government:	
Planners	The transaction meets the redevelopment goals of the stakeholders; Approval of the land use
Regulatory Agency	The transaction satisfies the environmental and public health goals
Transferor/Transferee	The transaction meets their financial and liability goals
Prospective transferee (New Owner)	The transaction meets their financial and liability goals; Cost-effective corrective action with limited long-term liability
Developer	The transaction meets their project requirements
Insurers	Configure and price policy
Lenders	Configure and price loan

8.2.2 *Transferee's Objectives*—Transferees of a property may be concerned with perceived or real liability under state or federal statutes or both, for per-existing environmental liabilities. Transferees may seek:

8.2.2.1 Liability protection from potential third-party claims,

8.2.2.2 Full disclosure of current environmental conditions related to the property,

8.2.2.3 An indemnification or other environmental risk transfer mechanism covering any liabilities arising out of pre-closing operation(s) or condition(s),

8.2.2.4 To get transferor to bear costs of investigation,

8.2.2.5 To ensure property is suitable for contemplated use(s), and/or

8.2.2.6 To obtain control over implementation of the remedial action(s).

8.2.3 *Transferor's Objectives*—Transferors of a property may be concerned with perceived or real liability under state or federal statutes, or both for a release of a chemical(s) of concern on the land while they owned the property or discovery of a chemical(s) of concern on the land during their ownership (even if there was no disposal during that time) where they failed to disclose the environmental condition to a transferee. In addition, the transferor may also be liable to the transferee for failure to disclose a known environmental condition(s). The transferor may seek:

8.2.3.1 To obtain releases and indemnities from prospective transferee,

8.2.3.2 To shift responsibility for identifying risks to transferee (for example, by eliminating or severely restricting representations and warranties),

8.2.3.3 Where transferor retains responsibility for preclosing environmental liabilities, to:

8.2.3.4 Require notice of or limit subsequent actions by the transferee that may contribute to or complicate the transferor's liability or corrective action requirements and costs for preexisting conditions,

8.2.3.5 Require the transferee to cooperate (including granting access to the property) as necessary for transferor to successfully address liability for past acts or regulatory violations, and/or

8.2.3.6 Require the transferee to grant transferor rights to mitigate environmental conditions giving rise to retained liability.

8.2.4 *Lenders*—Generally, the lenders may seek, based on the property investigation, to minimize their risk of liability for corrective action, maintain the collateral value of the redevelopment property, and to obtain protection from borrowers defaulting or a weakening of the borrower's financial position due to its environmental liability.

8.2.5 *Common Objective: Protection from Liability*—Both the transferee and the transferor in any transaction seek regulatory protection from environmental liability. Many states offer protection against liability to parties that voluntarily complete corrective action pursuant to voluntary corrective action programs.

8.3 *Preliminary and Pre-Closing Agreements*—A number of preliminary and pre-closing agreements may be appropriate. The number and type of preliminary and pre-closing agreements that are necessary will depend on the transaction. Preliminary agreements are agreements that are reached early in the process of negotiating a transaction that may include agreements on the necessary representations and warranties, indemnities and changes in the transaction structure, or withdrawal from the transaction should the investigation identify unacceptable environmental risks. Pre-closing agreements are agreements are that reached further along in the negotiating process that may include property access agreements and corrective actions agreements.

8.4 *Allocating Financial Risk Through Provisions in Transaction Documents*—In order to make the property marketable, some method(s) should be found to manage or allocate environmental liability. Such method(s) may be found within or outside of the transaction documents.

8.4.1 *Representations and Warranties*—Even if no environmental condition(s) is discovered on the property before closing (or the transaction is still attractive and financially sound in spite of the presence of an environmental condition), potential environmental liabilities may be minimized by securing appropriate representations and warranties in the purchase and sale agreement. Representations and warranties can be tied to an indemnification agreement supported by holdbacks, bonds, letters of credit, or environmental insurance to cover the risk that chemical(s) of concern may turn up later on the property. Express warranties and representations include the disclosure and consideration of:

8.4.1.1 The existence of necessary permits, registrations, approvals, land use restrictions, and licenses,

8.4.1.2 Compliance with environmental laws, rules, and regulations,

8.4.1.3 Any pending, threatened or anticipated claims, lawsuits, administrative actions or investigations or a combination thereof,

8.4.1.4 Disclosure of the presence of known chemical(s) of concern and reports of releases of such chemical(s) of concern,

8.4.1.5 Disclosure of environmental studies or reports conducted regarding the property, and/or

8.4.1.6 The existence of underground storage tanks or underground pipelines.

8.4.2 *Environmental Covenants*—The transferee may establish a covenant to complete corrective action of the property, to maintain adequate investment to complete the redevelopment of the property, or to maintain insurance policies to benefit the transferor. A covenant creates a case-specific continuing obligation In a sale or loan transaction an important pre-closing covenant is that the transferee or transferor will maintain the property in compliance with all environmental laws and will reaffirm accuracy of the representations and warranties as of the closing date. A covenant requiring that action not be taken is a negative covenant. In some cases, a transferor may want to obtain negative covenants, limiting the prospective transferee's ability to create specific hazards. Post-closing covenants may be required for a transferee in a loan transaction, or a tenant in a lease transaction.

8.4.3 *Indemnification*—A transferor that believes no significant risk exists should be willing to indemnify a prospective transferee against liability for any preexisting environmental condition(s). However, there are inherent limitations to such indemnities. The legal effect may be unclear, the indemnity may be only as good as the financial worth, of the indemnitor, or the indemnity operates after the fact (cash flow problems). Indemnification agreements can be supported by holdbacks, letters of credit, or insurance.

8.4.4 *Environmental Insurance*:

8.4.4.1 Environmental insurance may be an alternative means of transferring liability. Insurance is a means for parties to the transaction to address balance sheets concerns. Insurance policies may offer substantial capitilization multiple year policies, flexibility in policy wording, and realistic pricing of coverage.

8.4.4.2 Insurance policies can be used as a stand-alone risk transfer mechanism or to supplement an indemnity agreement. Insurers can offer support services such as claims handling and loss control. Liabilities or conditions that may need insurance coverage in a Brownfields transaction are:

8.4.4.3 Third-party bodily injury and property damage, on-property and off-property.

8.4.4.4 Environmental remedial action costs, on-property and off-property.

8.4.4.5 Legal defense expense.

8.4.4.6 Business interruption and costs of project delay.

8.4.4.7 Environmental remedial action cost cap or stop loss.

8.4.4.8 Collateral value or secured creditor loss.

8.4.4.9 Environmental condition(s) at third-party disposal sites that may have resulted from wastes generated at the Brownfields redevelopment property.

8.4.5 *Types of Insurance*:

8.4.5.1 Commercial insurance policies, especially:

8.4.5.2 Premises pollution (or property-specific) policies that can provide coverage for the environmental liabilities (see 8.4.4.2).

8.4.5.3 Pollution coverage for construction and consulting operations such as contractor's pollution liability and environmental consultant's professional liability policies.

8.4.6 *Surety and Bonds*—A surety is a person or organization that contractually guarantees to one party that another party will perform as promised. Some insurers act as sureties by issuing bid and performance bonds for environmental contractors.

8.4.6.1 Environmental risk management programs, including:

8.4.6.2 Finite risk programs,

8.4.6.3 Pooling arrangements including risk retention groups,

8.4.6.4 Risk purchasing groups, and/or

8.4.6.5 Captive reinsurance programs.

8.4.7 *Negotiating and Drafting of Coverage Forms*—Changes to the basic policy form should be drafted and negotiated in the context of the transaction. Insurance may need to be considered from the beginning of the transaction so that appropriate policies will be available on a timely basis, to transfer the liability most completely and efficiently, and to avoid duplication of the risk analysis and quantification process (see 7.5). The forms need to identify and actually cover the environmental exposure, the property, and the parties needing the coverage.

8.5 *Allocating Financial Risk Outside the Transaction Document*:

8.5.1 *Environmental Risk Control Transfers*—The risk of loss itself, as well as the financial consequences of the loss, may be transferred to another party. The prospective transferee may be able to insulate themselves from environmental liabilities; however, the transaction should be structured carefully and risks assessed in light of potential regulatory agency efforts to impose liability on a responsible party. The transferee or lender can insist that the Brownfields redevelopment property giving rise to the chemical(s) of concern or environmental liability remain the responsibility of the transferor.

8.5.2 *Corrective Action Prior to Closing*—The transferee or lender can require that an agreement be reached that certain chemical(s) of concern or environmental risks be eliminated prior to closing. This requires an analysis of the availability of other parties who may also be liable for corrective action costs and the likely response of federal, state, and local regulatory authorities, and the community to any planned corrective action activity. The corrective action may be conducted pursuant to a state voluntary corrective action program (see 7.3.4). If the transferee decides to undertake corrective action, it should consider requiring the owner to share the risk of uncertainty regarding how much an acceptable corrective action will cost.

9. Implementation

9.1 Demolition, renovation, and corrective action occur during this component. The corrective action may be based on an integrated land-use scenario where remedial action may be tied to a land use restriction or to a development with unrestricted land use (see Table 5).

9.2 *Permitting*—Local, state, and federal entities may issue permits based upon agreements reached with stakeholders in previous components of the sustainable Brownfields redevelopment process. If the Brownfields redevelopment process has been carried forward successfully, the permitting process is made more efficient.

9.3 *Remedial Action*—Remedial action is an integral part of the sustainable Brownfields redevelopment process. Remedial action can occur prior to redevelopment, during redevelopment, or as part of a long-term risk-reduction program. Stakeholder buy-in and institutional controls may be necessary to ensure a sustainable restoration process is begun, but time frames should be flexible to ensure that the most cost-effective solution that provides long-term risk reduction is implemented. Economic development that provides sustainable benefits to the community is the common goal of all stakeholders.

TABLE 5 Example of Stakeholder Goals for the Implementation Component

STAKEHOLDER	GOAL
Community	Positive benefits from the redevelopment property
Government:	
Planners	The redevelopment project meets stakeholders objectives
Regulatory Agency	The redevelopment project meets environment and human health objectives
Transferor/Transferee	Timely completion of the redevelopment project and acceptable return on investment
Developer	Timely completion of the redevelopment project and acceptable return on investment
Insurers	The redevelopment project is consistent with the insured conditions
Banks	The redevelopment project meets the financial objectives

9.4 *Exit Strategy*—The goal of all the stakeholders is to keep this component as short as possible. It is important to establish a clear exit strategy to the sustainable Brownfields redevelopment process to ensure that the property does not remain a Brownfields property in the future. Key issues that need to be resolved in order that the property exit the sustainable Brownfields redevelopment process include:

9.4.1 Redevelopment is underway,

9.4.2 Remedial action is implemented,

9.4.3 Institutional and engineering controls, as appropriate, are in place and maintained (that is, implementation of land use restrictions, and financial performance mechanisms), and

9.4.4 Local, state, and federal regulation requirements have been met.

9.4.5 The sustainable Brownfields redevelopment process is a multi-stakeholder process. Because each Brownfields redevelopment property is different (for example, variant levels of a chemical(s) of concern, different community settings, and multiple forms of reuse) each Brownfields redevelopment project may involve different components. The process outlined is a framework for the components of a sustainable Brownfields redevelopment process.

APPENDIXES

(Nonmandatory Information)

X1. ROLES OF LOCAL GOVERNMENT IN SUSTAINABLE BROWNFIELDS REDEVELOPMENT

X1.1 Local governments are ideally situated to facilitate and promote the successful reuse of Brownfields redevelopment properties. Local officials and municipal managers often play an integral role in bringing together all of the diverse interests involved in a Brownfields redevelopment project. Their role as facilitators is often difficult, given the many different groups involved and the complexity of the issues. For example, the regulatory framework that determines the fate of many Brownfields redevelopment properties falls under the jurisdiction of federal and state agencies, most of the necessary capital to fund development is controlled by private financial institutions, and many decisions about property reuse will be in the hands of the property owner. Local governments also have an important role in helping to ensure that community organizations and citizen groups directly affected by a Brownfields redevelopment project have sufficient access and influence in the corrective action and reuse decisions.

X1.2 The following are examples of the roles that local governments can play in the redevelopment of a Brownfields redevelopment project. They are not intended to be used as an all-inclusive guide, as there is not the "one" model for the decisions of a local government. Each local government has differing concerns and attributes that play a role in Brownfields redevelopment.

X1.3 *Integrating Brownfields Redevelopment with Other Community Priorities*—Local governments are in a unique position to look at Brownfields redevelopment in the context of the community's broader plans and needs. Because of this perspective, they can act to encourage redevelopment projects that fit into these plans. For example, the city of Minneapolis provided financial support for the development of a shopping center with a full-service grocery store in the northwest area of town. This shopping center delivered an important benefit to residents, who previously had to drive outside the neighborhood or pay high prices at convenience stores. A local government also can assist in identifying high-priority sections of the city and focus efforts for Brownfields redevelopment and other development on those areas.

X1.4 *Involving Community Residents in Brownfields Redevelopment Plans*—Early involvement of the community in the reuse planning process is important to successful Brownfields redevelopment and also helps to ensure that a community can achieve its long-term redevelopment goals. Local government officials should engage citizens in the decision-making process, provide them with good and timely information, and seriously consider their input. To do this, local governments often establish advisory or ad hoc community groups or utilize existing community groups to address Brownfields redevelopment. Collaborative planning and consensus-building approaches ensure equitable and meaningful community input. For example, local officials in New Orleans established a Brownfields redevelopment consortium consisting of community organizations, local government officials, and developers to learn and make decisions about the Brownfields redevelopment process.

X1.5 *Brokering Reuse*—Local governments can help match properties with prospective reusers. This can be done both through general efforts to provide information on the properties and by identifying specific potential reusers for particular Brownfields redevelopment properties. The city of Trenton, New Jersey, for example, was looking for a way to reuse a portion of a former electrical component factory. At the same time, the city was working with a local swimming pool cover company that was considering leaving the area. The city was able to facilitate an agreement under which the company moved into the vacant factory.

X1.6 *Providing Funding*—Local governments can use their own resources to fund portions of Brownfields redevelopment costs. This funding is particularly useful if it is used for up-front costs such as for property inventory, assessment, remedial action, and preparation of properties. Local governments often have the resources to locate the owner of a Brownfields redevelopment property who has the potential to provide funds for cleanup. By paying for assessment and remedial action, the city of Chicago's Brownfields Pilot Program spurred private companies to invest in and reuse a number of Brownfields redevelopment properties. Other types of financial benefits, such as tax incentives, can also be used to encourage reuse. Cook County, Illinois, allows owners of some Brownfields redevelopment properties to pay reduced property taxes during remedial action and redevelopment.

X1.7 *Coordinating Public Funding and Resources*—Both state and federal governments have programs that can pay for some of the costs of Brownfields redevelopment property reuse. Local governments can inform private-sector parties about these programs, apply for programs that require local government involvement, and look for ways to integrate different funding sources. Creative use of state funding helped Lawrence, Massachusetts, reuse the property of an old paper factory. By shifting the location of an already planned roadway by 100 feet, the city was able to use state highway funds for demolition and improvements to the Brownfields redevelopment property.

X1.8 *Acting as a Liaison with Environmental Regulators*—Local governments can serve as a link between private companies and community groups, as well as state and federal environmental agencies. Local governments also can work with agencies to ensure that they handle regulatory issues promptly and in a way that reflects local concerns. For example, the redevelopment of a property in Louisville, Kentucky, was blocked by a lien that the U.S. EPA held on the property as a result of remedial action costs the agency had absorbed eight years earlier. After the city asked that the lien be released, the EPA discovered that the statute of limitations had expired and forgave the lien.

X1.9 *Assuming Liability for Environmental Condition*—In some cases, it may make sense for local governments to agree to take on liability for remedial action at properties where the perception of liability is preventing development. This can quickly remove the primary deterrent to reuse. Because it could be extremely costly, this step should be taken with a great deal of caution. Wichita, Kansas found that redevelopment plans for much of its downtown seemed to have been derailed by the discovery of chemical(s) of concern in ground water. The city entered into an agreement under which it divided responsibility for remedial action costs with the principal potentially responsible party (PRP).

X1.10 *Brownfields Prevention*—To ensure that Brownfields redevelopment properties are prevented in the future, local governments can implement controls such as ground water use restrictions; offering incentives that discourage development of Greenfields; changing mixed-use zoning laws in low-income, residential neighborhoods; and assisting state and federal agencies in monitoring compliance during the operation and closure of industrial complexes.

X2. COMMUNITY INTERACTION

X2.1 *Community Involvement*—Generally, the earlier the community is brought into the process, the better it is for all stakeholders involved. Involving the community in a Brownfields redevelopment project can pose some interesting challenges. Some people may be skeptical about the Brownfields redevelopment process. They may question whether they will have a real opportunity to be involved in and influence local land use decisions. One challenge for sponsors of Brownfields redevelopment projects is to determine how to convince the community that active involvement is worthwhile and to provide them with an opportunity to have direct and meaningful impact on decisions.

X2.2 *Community Education Prior to any Specific Development Proposal*—A presentation of the facts concerning a Brownfields redevelopment property through straightforward discussions and open forums for the developer and the community to express their concerns and solutions should be part of the process. Brownfields redevelopment sponsors in conjunction with public health officials should educate the community and other stakeholders on the current risks, if any, posed by the Brownfields redevelopment properties and compare these risks to other risks in the community. This will allow the community and other stakeholders to reach a balanced decision on development of the property.

X2.3 *Community Technical Assistance*—The community may not have the technical or financial resources to assess the technical factors involving a Brownfields redevelopment property. This is often the case where public health considerations come into play. The community should have the opportunity to influence the process based on data that are correctly interpreted and meaningful for the situation at hand. Independent technical assistance should be provided to the community (individuals or groups) that will enable it to make an informed analysis. The point is to inform the community about the range of issues at hand and the resources available for the interpretation of data. The financial resources needed by the community should be determined, and will be influenced by a number of factors, including the size of the Brownfields redevelopment property, the amount of financial resources already available for the project, and the interest present within the community.

X2.3.1 While community groups may need information about technical issues, the developer, government officials, and other stakeholders may need training in communicating site-specific technical information to community groups. The communication process should include a thorough explanation of technical aspects of a project prior to its implementation.

X2.4 *Designing a Community Participation Program*—Whether designing an effective community participation program from a private developer, local, state or federal officials' perspective, it is best to consider the following guiding principles[3]

X2.4.1 Encourage active community participation in all aspects of environmental decision-making. The community should be seen as equal partners in dialogue on Brownfields redevelopment. They need to know all pertinent details about the project to evaluate its importance, costs, and benefits. Through continued, early interaction, community support needs to be built.

X2.4.2 Maintain honesty and integrity in the process. From the earliest stages, articulate realistic goals, expectations, and limitations of the project.

X2.4.3 Consider using a variety of public participation techniques for different-sized Brownfields redevelopment projects. The single, one-size-fits-all approach will not work for all situations.

X2.4.4 Conduct active and extensive outreach to the public. It is better to be as inclusive as possible generating a rich mix of ideas and approaches. The community has much to offer about effective interactions and approaches. Traditional and nontraditional outreach means should be attempted. Presentations in print and voice media are effective means of communication as are bulletin boards at community focal points (for example, houses of worship, grocery stores, laundromats, and community centers). Other methods such as using the Internet should also be considered.

X2.5 *Implementing a Community Participation Program*—

[3] Principles developed by the Public Participation and Accountability Subcommittee of the National Environmental Justice Advisory Council, "The Model Plan for Public Participation," U.S. EPA, November 1996.

The following steps are suggested approaches for setting up and implementing an effective community participation program.

X2.5.1 Identify individuals to be contacted, including those directly affected by a Brownfields redevelopment project. Determine who else is directly and indirectly affected, as well as those groups who have shown past interest in local land use planning and developmental efforts. Look for groups who do not traditionally participate, such as minority and low-income communities.

X2.5.2 Establish education programs or means to access data, or both, so that groups or individuals can obtain timely, accurate information to enable them to have meaningful input in decision making.

X2.5.3 Regionalize materials to ensure cultural sensitivity, language diversity, and geographic relevance. Provide a facilitator to public meetings who is sensitive and trained in dealing with cross-cultural exchanges.

X2.5.4 Develop cosponsoring and co-planning relationships with community groups allowing them shared roles in agenda development, goal setting, leadership, and outreach.

X2.5.5 Plan meetings that are accessible and accommodating. Meetings should be close to public transportation. Consideration should be given to such issues as child care, access for the disabled, and language interpreters. Plan schedules to accommodate needs of the affected communities. Schedule meetings on evenings and weekends to avoid conflicts with community or cultural events. Create an atmosphere of equal participation. Avoid a "head table" and ensure that the public shares in assignments.

X2.5.6 Maintain clear goals by setting an agenda. Coordinate follow-up and develop an action plan. Facilitate follow-through by distributing minutes and a list of action items.

X2.6 *Community Leadership*:

X2.6.1 Most neighborhoods have institutions that are respected and viewed as voices for the community. These can be block clubs, neighborhood economic development organizations, religious or educational institutions, and local chambers of commerce. It is not uncommon for more than one leader to emerge. Good rapport should be maintained with the leader or leaders of the community to gain their cooperation in information gathering and dialogue stages of the sustainable Brownfields redevelopment process.

X2.6.2 For effective interaction, the community's characteristics should be assessed. The community demographics, organizational structure, and local governmental jurisdictions should be determined. The research should include collection of information on the area's infrastructure, alternative or additional properties considered for Brownfields redevelopment, and identification of adjacent communities with Brownfields redevelopment experience..

X2.6.3 Once the research is complete and incorporated, the active dialogue process should begin because the actual Brownfields redevelopment project may need to be refined and discussed with the community. At this point, any community outreach strategy already in place should be broadened to include the majority of stakeholders. Simultaneously, creative partnerships for both the short and long term need to be formed. The ultimate goal for the dialogue sessions is to establish mutual acceptance of responsibilities to the process.

X2.7 *Politics*:

X2.7.1 The political realities of each area should be assessed and incorporated into the sustainable Brownfields redevelopment process. For example, in some cities, requests for zoning changes are channeled through the local government representatives. Thus, any plans for Brownfields redevelopment need approval by local government officials.

X2.7.2 Other political considerations may include how the municipality views industrial retention and job creation. In addition, a Brownfields redevelopment property may lie within an Empowerment Zone, a Tax Increment Financing (TIF) area, or a historically designated district. These are all political considerations that need to be researched and woven into the process for redevelopment.

X2.8 *Mutual Acceptance of Responsibilities to Process*:

X2.8.1 With straightforward and honest communication between all parties, responsibilities to the process should be easy to establish and accept. These responsibilities should be clear and consistent. Any changes in responsibilities should be discussed and agreed upon. A simple, easy-to-read document should be posted prominently and disseminated widely. Some items of mutual responsibilities can include:

X2.8.1.1 Community involvement as early and as often as possible,

X2.8.1.2 Communication with all affected parties,

X2.8.1.3 Promotion of project plans,

X2.8.1.4 Involvement in meetings as agreed upon,

X2.8.1.5 Effort at cooperation,

X2.8.1.6 Cognizance and acceptance of community vision and developer vision, and/or

X2.8.1.7 Effort put forth to consolidate these two visions.

X2.8.2 After mutual acceptance of responsibility to the process and the establishment of effective communication a final reevaluation is necessary in order to determine whether to proceed with plans as originally envisioned, to adjust plans, or to forego the project altogether.

X2.8.3 Figs. X2.1 and X2.2 show decision trees from the viewpoints of both developer and community. These figures are meant to provide general guidance with the understanding that there is no "one size fits all" model for success. While these decision trees are presented linearly, some steps may not be appropriate in all cases while other steps may be more appropriately accomplished in a different sequence than shown. In the end, it will be the cooperation and straightforward honest communication of all stakeholders in an effort that will lead to sustainable Brownfields redevelopment process.

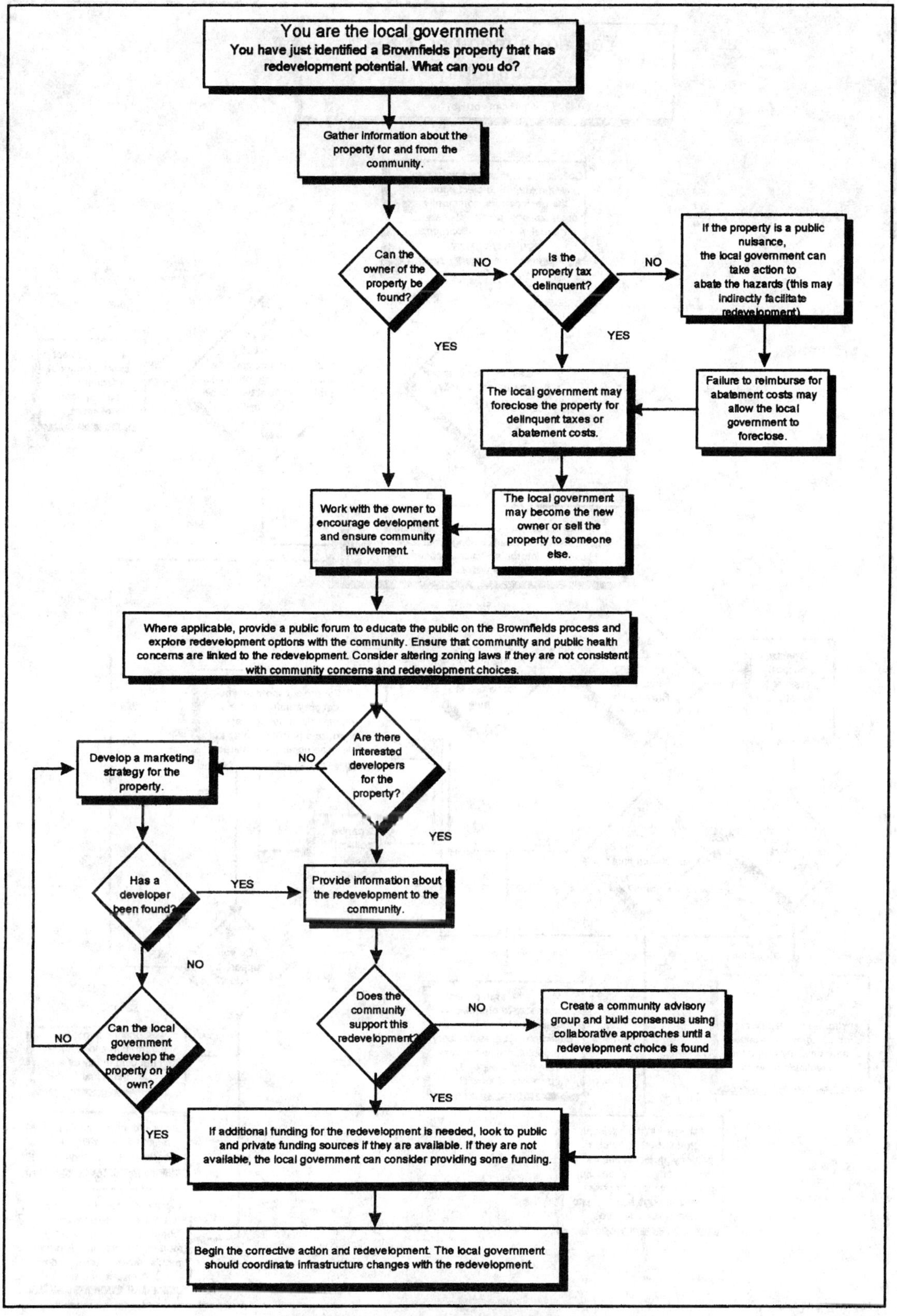

FIG. X2.1 Local Government Decision Tree

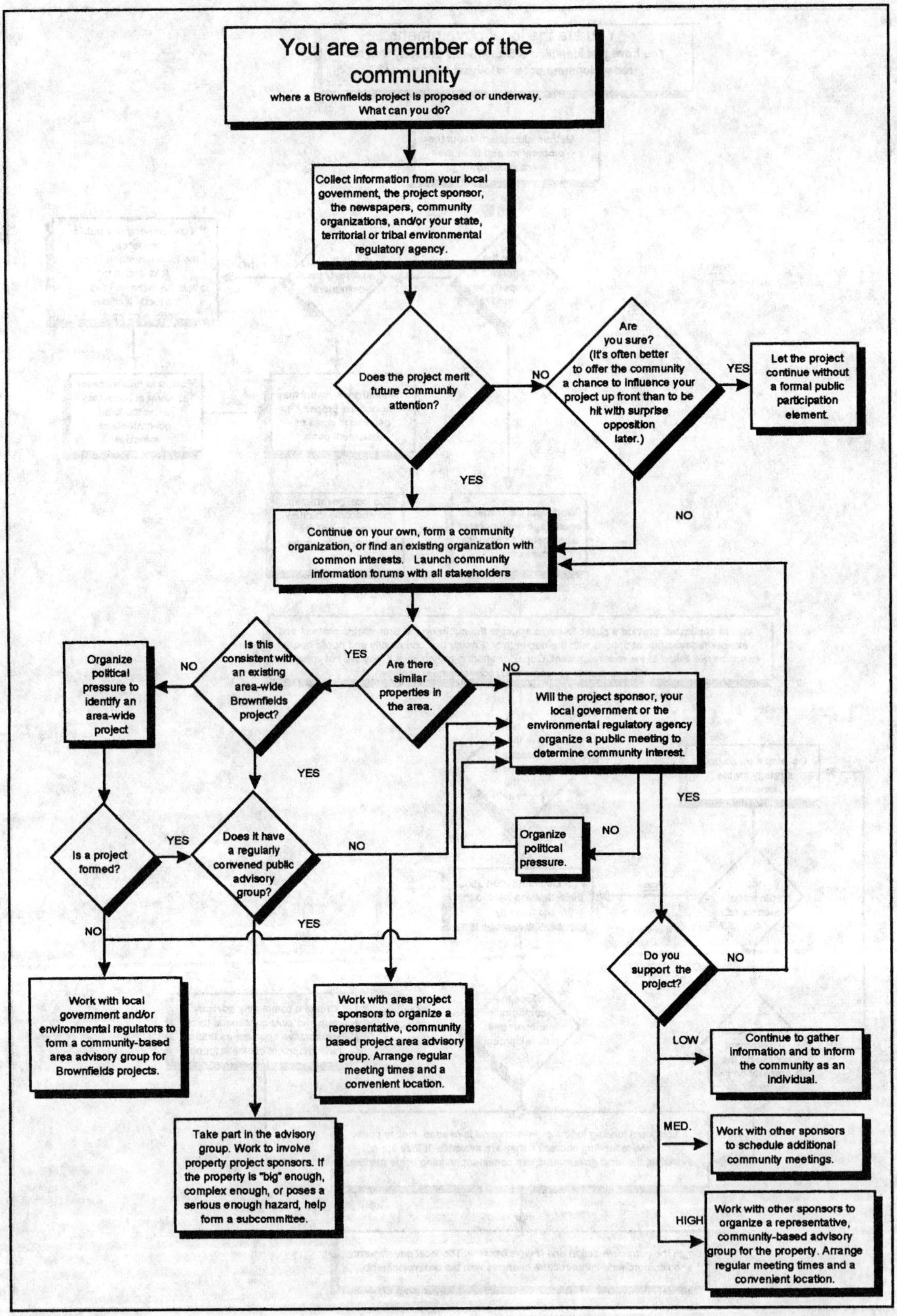

FIG. X2.2 Community Decision Tree

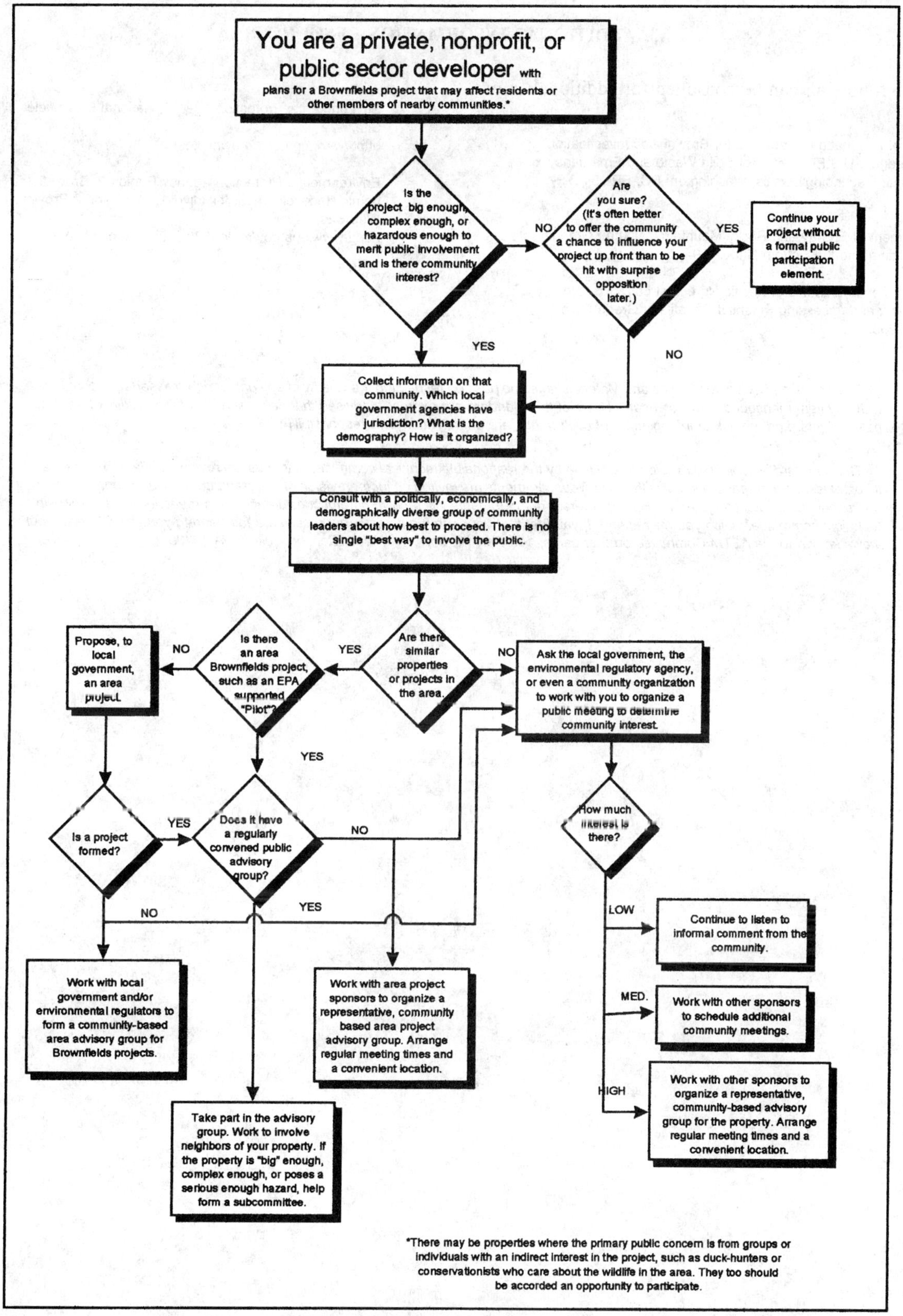

FIG. X2.3 Developer Decision Tree

X3. ADDITIONAL INFORMATION RESOURCES

X3.1 The following can be consulted for additional information:

Toolkit of Information Resources for Brownfields Investigation and Cleanup, U.S.EPA Office of Solid Waste and Emergency Response, Technology Office, Washington, DC, EPA 54 2-B-97-001.

Department of Energy and Rocky Mountain Institute, "Greening the Building and the Bottom Line," 1994.

Battelle Memorial Institute, Pacific Northeast Lab, "Protocol for Identifying and Assessing Potential Benefits of Green Buildings," 1996.

Environmental Protection Agency, National Brownfields Web Site,
http://www.epa.gov/brownfields

Environmental Protection Agency Region 2, Brownfields Economic Redevelopment Initiative, EPA Region 2 Brownfields Web Site,
http://www.epa.gov/region02/superfnd/brownfld/bfmainpg.htm

Standard Guide for Performing Evaluations of Underground Storage Tank Systems for Operational Conformance with 40 CFR, Part 280 Regulations[1]

This standard is issued under the fixed designation E 1990; the number immediately following the designation indicates the year of original adoption or, in the case of revision, the year of last revision. A number in parentheses indicates the year of last reapproval. A superscript epsilon (ϵ) indicates an editorial change since the last revision or reapproval.

1. Scope

1.1 This guide covers information for evaluating tank systems for operational conformance with the Federal technical standards (including the financial responsibility requirements) for underground storage tanks (USTs) found at 40 Code of Federal Register (CFR) Part 280.

1.2 This guide does not address the corrective action requirements of 40 CFR Part 280.

1.3 To the extent that a tank system is excluded or deferred from the federal regulations under Subpart A of 40 CFR Part 280, it is not covered by this guide.

1.4 *This standard does not purport to address all of the safety concerns, if any, associated with its use. It is the responsibility of the user of this standard to establish appropriate safety and health practices and determine the applicability of regulatory limitations prior to use.*

2. Referenced Documents

2.1 *Federal Standards:*

Title 40 Code of Federal Regulations (CFR), Environmental Protection Agency, Part 280, Technical Standards and Corrective Action Requirements for Owners and Operators of Underground Storage Tanks (UST)[2]

3. Terminology

3.1 *Definitions of Terms Specific to This Standard:*

3.1.1 *corrosion protection*—to prevent the degradation of a material due to a reaction between the material and its environment.

3.1.2 *implementing agency*—Environmental Protection Agency (EPA), or, in the case of a State with program approval (or pursuant to a memorandum of agreement with EPA), the designated State or Local agency responsible for carrying out the UST program.

3.1.3 *overfill*—exceeding the capacity of the tank.

3.1.4 *overfill protection*—a device used to prevent a tank from being overfilled when product is being added to the tank.

3.1.5 *release detection*—a method to determine whether a release of a regulated substance has occurred from the UST system into the environment or into the interstitial space between the UST system and its secondary barrier or secondary containment around it.

3.1.6 *spill*—to cause or allow product to enter the environment that was intended to be placed in the tank.

3.1.7 *spill prevention*—a device used to prevent or contain a spill associated with removing the fill hose from the tank fill pipe.

3.1.8 *suspected release*—released product discovered at or near the UST site, observed unusual operating conditions, such as apparent loss of product; or results from a release detection method that indicates a release. Suspected release does not include situations where a method or equipment is found to be defective, is immediately corrected, and then indicates no release.

3.1.9 *underground storage tank (UST)*—any one or combination of tanks (including underground pipes connected thereto) that is used to contain an accumulation of regulated substances, and the volume of which (including connected underground piping) is 10 % or more beneath the surface of the ground.

3.1.9.1 *existing tank*—a tank where installation began on or before 12/22/88.

3.1.9.2 *new tank*—a tank where installation started after 12/22/88.

3.1.10 *upgrade*—the addition or retrofit of the UST system or some system components such as release detection, corrosion protection, spill prevention, or overfill protection to improve the ability of an UST system to prevent the release of product and meet the requirements of 40 CFR Part 280.

3.1.11 *UST system*—an underground storage tank, connected underground piping, underground ancillary equipment, and containment system, if any.

3.2 *Additional Terminology*—The following definitions and discussions, taken directly from the publication "Form and Style for ASTM Standards" shall be included in full in every standard guide or practice produced and passed by Committee E-50 or any of its technical subcommittees, approved April 16, 1997.

3.2.1 *guide*—a series of options or instructions that do not recommend a specific course of action.

---

[1] This guide is under the jurisdiction of ASTM Committee E-50 on Environmental Assessment and is the direct responsibility of Subcommittee E50.01 on Storage Tanks.

Current edition approved Oct. 10, 1998. Published January 1999.

[2] Available from the Superintendent of Documents, U.S. Government Printing Office, Washington, DC 20402.

3.2.1.1 *Discussion*—Whereas a practice prescribes a general usage principle, a guide only suggest an approach. The purpose of a guide is to offer guidance, based on a consensus of viewpoints, but not to establish a fixed procedure. A guide is intended to increase the awareness of the user to available techniques in a given subject area and to provide information from which subsequent evaluation and standardization can be derived.

3.2.2 *practice*—a definitive procedure for performing one or more specific operations or functions that does not produce a test result.

3.2.2.1 *Discussion*—A practice is not a downgraded test method. Examples of practices include procedures of interlaboratory testing programs or other statistical procedures; for writing statements on sampling or precision and accuracy; and for selection, preparation, application, inspection, and necessary precautions for the use, disposal, installation, maintenance, and operation of testing equipment.

3.2.3 *standard*—as used in ASTM, a document that has been developed and established within the consensus principles of the Society and that meets the approval requirements of ASTM procedures and regulations.

3.2.3.1 *Discussion*—The term "standard" serves in ASTM as an adjective in the title of documents such as test methods or specifications, to connote specified consensus and approval. The various types of standard documents are based on the needs and usages as prescribed by the technical committees of the Society.

4. Significance and Use

4.1 This guide is an educational tool for tank owners, operators, and other users and is not intended for use in certifying compliance with the Federal technical standards for underground storage tanks.

4.2 The intent of this guide is to provide an overview of the general requirements. This guide is intended for users who are generally familiar with the requirements of 40 CFR Part 280. The user is advised that this guide does not contain the level of detail necessary to make the determination of whether specific equipment or services meet the detailed technical performance requirements of 40 CFR Part 280.

4.3 This guide does not cover state and local requirements, that can be more stringent than the federal rules. Owners and operators are responsible for meeting federal, state, and, in some circumstances, local requirements. It is recommended that owners and operators familiarize themselves with these requirements as well.

4.4 Owners or operators may use the sample checklist in Appendix X1 to assist them in determining operational conformance or they may develop their own checklist based upon this guide.

4.5 This guide and accompanying appendixes are not intended to be used by state or local UST program authorities as a regulatory or administrative requirement for owners or operators. Use of this guide and appendixes by owners and operators is intended to be a voluntary educational tool for the purposes described in 4.1.

5. UST System Requirements

5.1 *Operational Conformance*—For an underground storage tank (UST) system to be in operational conformance with 40 CFR Part 280, it must have release detection, corrosion protection, spill prevention, and overfill protection.

5.2 *Release Detection*—To meet all release detection requirements, release detection for both tank and piping is required.

5.2.1 *Petroleum Tanks*— These tanks may meet release detection requirements by use of one of the following:

5.2.1.1 Monthly monitoring using one of the following: automatic tank gaging, vapor monitoring, interstitial monitoring, ground water monitoring, statistical inventory reconciliation (SIR), and any other approved method

5.2.1.2 Monthly inventory control and annual tank tightness testing (may be used for existing UST systems which have not been upgraded to meet Federal regulations at 40 CFR Part 280.21).

5.2.1.3 Monthly inventory control and tank tightness testing every five years (may start when new UST systems and UST systems that have been upgraded with spill prevention and overfill and corrosion protection meet the Federal regulations at 40 CFR Part 280.20 or 280.21). This release detection method must be replaced or augmented with one of the options in 5.2.1.1 at the later of two dates—December 22, 1998 or ten years after the tank was installed or upgraded with corrosion protection.

5.2.1.4 *Manual Tank Gaging*—This release detection method is for tanks of 2000-gal capacity or less. This technique may be used if the tank meets the size, test duration, and standard variation requirements listed in Table 1:

5.2.2 *Piping*—Piping may meet release detection requirements by use of one of the following:

5.2.2.1 *Pressurized Piping*—To meet release detection requirements for pressurized piping, it is necessary to choose one method from 5.2.2.2 and one method from 5.2.2.4 (or a single method that satisfies both sections).

5.2.2.2 Choose one of the following: automatic flow restricter, automatic shutoff device, and continuous alarm system.

TABLE 1 Guide for Owners and Operators

Tank Size	Minimum Duration of Test, h	Weekly Standard (One Test), gal	Monthly Standard (Four-Test Average), gal
Up to 550 gal	36	10	5
551 to 1000 gal (when the tank diameter is 64 in.)	44	9	4
551 to 1000 gal (when the tank diameter is 48 in.)	58	12	6
551 to 1000 gal (also requires periodic tank tightness testing in accordance with 5.2.1.2 and 5.2.1.3)	36	13	7
1001 to 2000 gal (also requires periodic tank tightness testing in accordance with 5.2.1.2 and 5.2.1.3)	36	26	13

5.2.2.3 Each of these methods must be tested on an annual basis in accordance with the manufacturer's requirements.

5.2.2.4 Choose one of the following: annual line tightness testing, monthly monitoring using one of the following: vapor monitoring, ground water monitoring, interstitial monitoring, SIR, other approved methods, such as: electric or electronic line leak detector, and continual reconciliation.

5.2.2.5 *Suction Piping*— For suction piping, you may choose one of the following:

5.2.2.6 Monthly monitoring using one of the following: vapor monitoring, ground water monitoring, interstitial monitoring, SIR and, other approved methods.

5.2.2.7 Line testing every three years.

5.2.2.8 No leak detection requirements if the piping meets the following criteria:

5.2.2.9 The below-grade piping operates at less than atmospheric pressure.

5.2.2.10 The below-grade piping is sloped so that the contents of the pipe will drain back into the storage tank if suction is released.

5.2.2.11 Only one check valve is included in each suction line and it is located directly below and as close as practical to the suction pump.

5.2.2.12 Satisfying the preceding requirements must be able to be easily determined.

5.2.3 *Hazardous Substance Tanks*—Release detection for hazardous substance tanks must meet one of the following:

5.2.3.1 Existing UST systems

5.2.3.1.1 Until Dec. 22, 1998, existing hazardous UST systems may meet the release detection requirements as described in 5.2.1.

5.2.3.1.2 By Dec. 22, 1998, all hazardous UST systems must meet the new system requirements described in 5.2.3.2.

5.2.3.2 New UST systems

5.2.3.2.1 Unless an owner or operator has applied for and received a variance from the implementing agency, new hazardous substance UST systems must have each of the following: secondary containment, and monthly interstitial monitoring.

5.3 *Corrosion Protection*—All underground tanks and piping must be protected from corrosion.

5.3.1 *Tanks*—Any portion of the tank that is underground and routinely contains product must have corrosion protection. The following may be used to meet corrosion protection requirements for tanks:

5.3.1.1 Coated and cathodically protected steel,

5.3.1.2 Cathodic protection added,

5.3.1.3 Interior lining added,

5.3.1.4 Cathodic protection and interior lining added,

5.3.1.5 Fiberglass reinforced plastic,

5.3.1.6 Steel tank clad with a noncorrodible material, and

5.3.1.7 Other construction accepted by the implementing agency.

5.3.1.8 Corrosion expert's determination that the site is not corrosive enough to cause it to have a release during the operating life of the tank.

5.3.2 *Piping*—Piping that routinely contains regulated substances and is in contact with the ground must be protected from corrosion. The following may be used to meet corrosion protection requirements for piping:

5.3.2.1 Coated and cathodically protected steel,

5.3.2.2 Cathodic protection added,

5.3.2.3 Fiberglass reinforced plastic,

5.3.2.4 Other construction accepted by the implementing agency, and

5.3.2.5 Corrosion expert's determination that the site is not corrosive enough to cause it to have a release during the operating life of the piping.

5.4 *Spill Prevention and Overfill Protection*—Tanks that received regulated substances in quantities greater than 25 gal at one filling need to have spill prevention and overfill protection.

5.4.1 To meet the spill prevention requirement, equipment that will contain the product in the transfer hose when the transfer hose is detached from the fill pipe (for example, a catchment basin) must be used to prevent release of product to the environment.

5.4.2 To meet overfill protection requirements, one of the following must be used:

5.4.2.1 Automatic shutoff device,

5.4.2.2 Overfill alarm, and

5.4.2.3 Ball float valve.

6. Financial Responsibility

6.1 *Financial Responsibility*—Owners or operators of USTs which hold petroleum must demonstrate financial responsibility. These requirements are designed to ensure that someone can pay the costs of cleaning up leaks and compensating third-parties for bodily injury and property damage caused by leaking USTs. The total amount of coverage required under the federal rule is determined by the type of business, the amount of throughput of the tanks at a facility, and the number of tanks owned nationwide. The amount of coverage that an owner must demonstrate may be changed due to coverage provided by a state financial assurance fund. The following mechanisms may be used to meet the financial responsibility requirements:

6.1.1 Insurance policy,

6.1.2 Guarantee,

6.1.3 Letter of credit,

6.1.4 Self insurance.

6.1.5 State financial assurance fund,

6.1.6 Surety bond,

6.1.7 Dedicated trust fund, and

6.1.8 State approved method.

6.1.9 The following four additional mechanisms may be used by local governments to demonstrate compliance:

6.1.9.1 Bond rating test,

6.1.9.2 Passing a financial test,

6.1.9.3 Guarantee, and

6.1.9.4 Dedicated fund.

7. Notification, Reporting, and Recordkeeping

7.1 *Notification*— The following are notification requirements listed in the underground storage tank technical requirements at 40 CFR Part 280. This information must be submitted to the implementing agency.

7.1.1 *Notification for All UST Systems*—If an UST system is brought into use after May 8, 1986, within 30 days of being brought into use a notification form must be submitted. If an UST system was in the ground on or after May 8, 1986, unless taken out of operation on or before Jan. 1, 1974, a notification form must be submitted. If an UST system was installed after Dec. 22, 1988, a certification of installation must also be submitted for that system.

7.1.2 *Notification Before Permanent Closure or Change in Service*—Owners or operators must notify the implementation agency at least 30 days before permanent closure or change in service begins.

7.2 *Reporting*—The following are reporting requirements listed in the underground storage tank technical requirements at 40 CFR Part 280. This information must be submitted to the implementing agency.

7.2.1 *Reports of All Releases*—Within 24 h or another reasonable time period established by the implementing agency, the following must be reported:

7.2.1.1 Suspected releases,

7.2.1.2 Spills and overfills (if not cleaned up within 24 h, if petroleum over 25 gal or if a hazardous substance over its reportable quantity under 40 CFR Part 302), and

7.2.1.3 Confirmed releases.

7.2.2 *Corrective Actions Planned or Taken*—This includes:

7.2.2.1 Initial abatement measures,

7.2.2.2 Initial site characterization,

7.2.2.3 Free product removal,

7.2.2.4 Investigation of soil and ground water cleanup, and

7.2.2.5 Corrective action plan.

7.3 *Recordkeeping*— The following are recordkeeping requirements listed in the underground storage tank technical requirements at 40 CFR Part 280.

7.3.1 *Corrosion Expert's Analysis*—A corrosion expert's analysis of site corrosion potential if corrosion protection equipment is not used.

7.3.2 *Corrosion Protection*—Documentation of operation of corrosion protection equipment is required. This includes the following:

7.3.2.1 Results from the last two tests for proper operation for all cathodic protection systems. These tests are required within six months of installation and every three years thereafter. In addition, this test must be conducted within six months following the repair of any cathodically protected UST system. The test must be done by a qualified cathodic protection tester.

7.3.2.2 Results from the last three inspections for UST systems having an impressed current cathodic protection system. These inspections must be conducted every 60 days to ensure that the equipment is running properly. This inspection may be performed by the owner or operator of the system.

7.3.3 *Release Detection*—Records showing conformance with release detection requirements are required. This includes the following:

7.3.3.1 All written performance claims pertaining to the release detection system used and the manner in which these claims have been justified or tested by the vendor or by the equipment manufacturer or installer. These records need to be maintained for five years from the date of installation.

7.3.3.2 The results of any sampling, testing, or monitoring. These results must be maintained for one year. In the case of tank tightness testing, the result must be maintained until the next tightness test is conducted.

7.3.3.3 Written documentation of all calibrations, maintenance, and repair of release detection permanently located on-site. This documentation needs to be maintained for at least one year after the servicing work is completed. In addition, any schedules of required calibration and maintenance provided by the release detection equipment manufacturer must be retained for five years from the date of installation.

7.3.4 *UST System Repairs*—Documentation of UST System Repairs must be maintained for each repair for the remaining operating life of the UST system.

7.3.5 *Permanent Closure Site Assessment*—Results of the site assessment conducted at permanent closure must be maintained for at least three years after completion of the permanent closure or change-in-service.

7.3.6 *Evidence of Financial Assurance*—The following is needed to demonstrate financial assurance:

7.3.6.1 A current "Certification of Financial Responsibility" and

7.3.6.2 Any additional documentation that shows the financial responsibility method is valid and provides details on the method's coverage such as signed copies of official letters or newsletters, policies, and state fund agreements.

7.4 *Availability and Maintenance of Records*—Records listed in 7.3 must be kept in accordance with one of the following:

7.4.1 Records may be kept at the UST site and be immediately available for inspection.

7.4.2 Records may be kept at a readily available alternative site and be provided for inspection upon request.

7.4.3 For permanent closure, records may be mailed to the appropriate implementing agency if they cannot be kept at the site or at an alternative site.

8. Appendixes

8.1 Tank owners and operators may use the following appendixes to assist them in determining the operational conformance status of their UST systems.

8.1.1 *Appendix X1*—A sample underground storage tank evaluation checklist that contains all the basic elements described here in a format that is easy and simple to use.

8.1.2 *Appendix X2*—General instructions for completing the sample checklist.

APPENDIXES

(Nonmandatory Information)

X1. SAMPLE UNDERGROUND STORAGE TANK EVALUATION CHECKLIST

X1.1 Fig. X1.1 provides an evaluation checklist as an example of how this guide can be used by owners and operators to determine operational conformance with federal UST regulations at 40 CFR Part 280 and should not be used as a means of demonstrating compliance to an implementing agency.

FACILITY INFORMATION	**PREPARER**
Identification #:	Name:
Name:	Phone #:
Address:	Address:
DATES OF EVALUATION:	

THE FOLLOWING SUMMARIZES FEDERAL REPORTING AND RECORDKEEPING REQUIREMENTS*

(Y=Yes/N=No/NA=Not Applicable)

1. Do you have evidence of Financial Responsibility?	
2. If you have had a release or suspected release, did you submit the appropriate reports?	
3. Do you have the tank closure reports? (Required to be kept for 3 years)	
4. Have you submitted the current UST notification forms to the appropriate authority?	
5. Do you have records of all UST system repairs? (Must be kept for the operating life of the UST system)	
6. **Release Detection Recordkeeping:**	
a. Do you have documentation showing the results of any sampling, testing, or monitoring of release detection? (These records must be kept for 1 year **except** for the results of tank tightness testing which must be retained until the next test is conducted.)	
b. Do you have records of calibration, maintenance, and repair of release detection equipment? (These records must be kept for 1 year, except for schedules of required calibration and maintenance which need to be kept for 5 years from the date of installation.) If you use manual tank gauging or inventory control and tank tightness testing, put NA in this box.	
c. Do you have all the written performance claims pertaining to any release detection method you use? (These claims must be kept for 5 years.) If you use manual tank gauging as sole method, put NA.	
d. Do you have documentation showing that the release detection method you are using was evaluated by a 3rd party using EPA protocols? If you use manual tank gauging as sole method, put NA.	
7. **Cathodic Protection (steel tanks and metal piping):**	
a. Do you have the paperwork from the installer for your cathodic protection system?	
b. Do you have the documentation of the last two cathodic protection test results? These tests are required within 6 months of installation, within 6 months of a repair, and then at least every 3 years.	
c. If you have an impressed current system, do you have documentation of the last three rectifier readings? Readings are required every 60 days and may be taken by the owner or operator.	

***State and Local Jurisdictions may have additional requirements. This information assists in demonstrating conformance. However, the unavailability of records does not necessarily indicate operational non-conformance.**

FIG. X1.1 Sample Underground Storage Tank Evaluation Checklist

X2. INSTRUCTIONS FOR COMPLETING APPENDIX X1

X2.1 Appendix X1 is a sample evaluation checklist that allows owners or operators of federally regulated underground storage tanks to determine the status of their conformance with the underground storage tank requirements. The checklist contains all of the information necessary for owners or operators to determine conformance.

X2.1.1 No special knowledge or expertise is required to fill out Appendix X1. Most, if not all of the information, should be on file at the facility. If the information is not on file at the facility, it may be able to be obtained from the vendors that provided the service. Vendors such as tank and other equipment installers, tank testers (for release detection systems), providers of corrosion protection should provide the necessary documentation when the service is provided. If not, tank owners or operators should request the appropriate documentation.

X2.2 *Specific Instructions*:

X2.2.1 *Facility Information*—Information about the facility to be evaluated is placed here. Enter the facility identification number, the facility name, and the facility address.

X2.2.2 *Preparer*—Information about the individual who performed the evaluation and completed the evaluation checklist is put in this section. Enter the preparer's name, phone number, and address. If the preparer's address is the same as in facility information, place "SAME" in the address block.

X2.2.3 *Dates of Evaluation*—Enter the date or dates when the facility was evaluated.

X2.2.4 *Summarization of the Reporting and Recordkeeping Requirements*—The questions in this section are based on the Federal regulations at 40 CFR Part 280. Please remember to check with the state and local agencies (where the tanks are located) to find out whether or not there are additional requirements. Each question should be answered with "Y" for Yes, "N" for No, or "N/A" for Not Applicable.

X2.2.5 *Question 1*— Federal regulations require that the owner or operator have evidence of financial responsibility. If unsure or want more information, see EPA Publication 510-K-950-004, "Dollars and Sense." Copies of this publication can be obtained by calling EPA's RCRA/Superfund Hotline. Information can also be obtained by accessing the Office of Underground Storage's homepage at: http://www.epa.gov/OUST.

X2.2.6 *Question 2*— Federal regulations require that, if there was a release or suspected release, the following reports must be submitted:

X2.2.6.1 An oral or written report to the implementing agency within 24 h of discovery and a follow-up report within seven days or another time period specified by the implementing agency. Reports of all releases, including suspected releases, spills, and overfills that meet the following circumstances:

X2.2.6.2 *Petroleum*—The release exceeds 25 gal or another reasonable amount specified by the implementing agency.

X2.2.6.3 *Hazardous Substance*—The release exceeds the reportable quantity described at 40 CFR Part 302.

X2.2.6.4 A spill of any amount of any substance in which the cleanup cannot be accomplished within 24-h confirmed releases.

X2.2.6.5 *Corrective Actions Planned or Taken*—Including initial abatement measures, initial site characterization, free product removal, investigation for soil and ground-water cleanup, and corrective action plan.

X2.2.7 *Question 3*— If the tanks have been permanently closed or have undergone a change-in-service, there needs to be records of the closure reports. These records need to be maintained for three years in one of the following ways:

X2.2.7.1 By the owners and operators who took the UST system out of service,

X2.2.7.2 By the current owners and operators of the UST site, and

X2.2.7.3 By mailing the records to the implementing agency if they cannot be maintained at the closed facility.

X2.2.8 *Question 4*— Notification forms for all UST systems must be submitted. This notification includes the certification of installation for new UST systems. New UST systems are those where installation began after Dec. 22, 1988. Notification forms for UST systems brought into use after May 8, 1986, need to be submitted within 30 days of bringing the UST systems into use. Owners and operators of UST systems who have not complied with the notification requirements should submit a notification form as soon as possible. Owners and operators of UST systems that were in the ground on or after May 8, 1986, unless taken out of operation on or before Jan. 1, 1974, were required to notify the designated state or local agency.

X2.2.9 *Question 5*— If the UST system has been repaired, there must be documentation of each repair. These records must be kept for the remaining operating life of the tank.

X2.2.10 *Question 6*— Regulations require that some release detection records must be kept.

X2.2.10.1 Sampling, testing, and monitoring records are those records that the owner or vendors generate that show that the UST system is not leaking. These records need to be maintained for one year except for the results of tank tightness testing which need to be maintained until the next tightness test is conducted.

X2.2.10.2 Written documentation of all calibrations, maintenance, and repair of release detection equipment permanently located on-site needs to be kept for one year after the servicing work is completed. In addition, any schedules of required calibration and maintenance provided by the release detection equipment manufacturer must be retained for five years from the date of installation. If manual tank gaging or inventory control and tank tightness testing are used, put NA in this box because these release detection methods do not have calibrations or repairs.

X2.2.10.3 All written performance claims pertaining to any release detection system used, and the manner in which these claims have been justified or tested by the equipment manufacturer or installer need to be maintained for five years. If manual tank gaging is used as the sole method of release detection, put NA in this box because this release detection method has no written performance claims or justifications.

X2.2.10.4 If manual tank gaging is used as the sole method of release detection, put NA in this box because this release detection method has no written performance claims or justifications.

X2.2.11 *Question 7: Cathodic Protection*—Any part of the tank that is underground and routinely contains product must be protected from corrosion. In addition, piping that routinely contains regulated substances and is in contact with the ground must be protected from corrosion. These questions apply to steel tanks and metal piping with cathodic protection.

X2.2.11.1 If the cathodic protection system was field installed, it needs to be designed by a corrosion expert. To determine whether or not the cathodic protection system was designed by a corrosion expert, look at the documentation provided when that system was installed. If there is no documentation about cathodic protection system, call the vendor who installed it. They should be able to provide the documentation. Paperwork could include such items as: design documents, installation documents, and so forth.

NOTE X2.1—A corrosion expert is a person, who by reason of thorough knowledge of the physical sciences and the principles of engineering and mathematics acquired by a professional education and related practical experience, is qualified to engage in the practice of corrosion control on buried or submerged metal piping systems and metal tanks. A corrosion expert must be accredited or certified as being qualified by NACE International (formerly National Association of Corrosion Engineers) or be a registered professional engineer who has certification or licensing that includes education and experience in corrosion control of buried or submerged metal piping systems and metal tanks.

X2.2.11.2 Cathodic protection testing needs to be conducted by a qualified cathodic protection tester on all cathodic protection systems installed on UST systems. Ask the person who is conducting the cathodic protection testing whether he or she is a qualified cathodic protection tester. This testing needs to be performed within six months of installation, within six months of any repair, and then at least every three years. The purpose of the cathodic protection testing is to ensure that the cathodic protection system is protecting the tank and piping. If the cathodic protection test shows that the tank is not adequately protected from corrosion, then a professional should look at the cathodic protection system to determine the problem.

NOTE X2.2—A cathodic protection tester is a person who can demonstrate an understanding of the principles and measurements of all common types of cathodic protection systems as applied to buried or submerged metal piping and tank systems. At minimum, such persons must have education and experience in soil resistivity, stray current, structure-to-soil potential, and component electrical isolation measurements of buried metal piping and tank systems.

X2.2.11.3 If an impressed current cathodic protection system is used, it also needs to be inspected every 60 days to ensure that the equipment is running properly. The owner or operator may perform this check. Please be sure that the installer of the impressed current cathodic protection system provide instructions on how to perform this inspection and what the normal operating parameters are. The results of the last three inspections must be kept.

X2.3 *Release Detection, Corrosion Protection, Spill Prevention, and Overfill Protection* —Fig. X2.1 contains an easy-to-use checklist to determine the conformance status for each tank system at the facility.

X2.3.1 *Release Detection*—Owners and operators need to have a method of release detection for both tanks and piping. Check the appropriate box for tanks and the appropriate box for piping.

X2.3.2 *Tanks*:

X2.3.2.1 *New Tanks*—New tanks may use the combination of inventory control and tank tightness testing every five years for a period of ten years or until Dec. 22, 1998, whichever is later. After this, the owner or operator must begin using a monthly monitoring technique for release detection. If a monthly monitoring technique is used, please check the appropriate method.

X2.3.2.2 *Existing Tanks*— An existing tank may use inventory control and tank tightness testing every five years only if the upgrade of the entire tank system has been completed (that is, must have spill prevention, overfill and corrosion protection for the tank and piping). These methods may be used every five years for ten years after upgrading the tank with corrosion protection or until Dec. 22, 1998, whichever is later. After this, the owner or operator must begin using a monthly monitoring technique for release detection. If a monthly monitoring technique is used, please check the appropriate method. If any of these upgrades are missing, the combination of inventory control and annual tank tightness testing is necessary if a monthly monitoring technique is not used.

X2.3.2.3 The questions in Fig. X2.2 apply specifically to the different methods of release detection. They may be helpful in determining conformance with Federal regulations for tank release detection.

X2.3.3 *Piping*:

X2.3.3.1 *Pressurized Piping*—If the UST system has pressurized piping, must choose one method from each of the two sets listed on the checklist. If a monthly monitoring technique is used, please check the appropriate method. Note: Some continuous monitoring methods may satisfy both sets of requirements.

X2.3.3.2 *Suction Piping*— If the UST system has suction piping, choose one of the three choices of release detection listed in Fig. X2.2. To meet the "No Requirements" category see Fig. X2.3 on suction piping.

X2.3.3.3 If a monthly monitoring technique is used, please check the appropriate method.

X2.3.3.4 The questions in Fig. X2.3 apply specifically to the different methods of release detection. They may be helpful in determining conformance with Federal regulations for piping release detection.

X2.3.4 *Corrosion Protection*—Any part of the tank that is underground and routinely contains product must be protected from corrosion. In addition, piping that routinely contains regulated substances and is in contact with the ground must be protected from corrosion.

RELEASE DETECTION

Tanks 2 choices

- ☐ **Monthly Monitoring (check method)**
  - ☐ Automatic Tank Gauging
  - ☐ Vapor Monitoring
  - ☐ Interstitial Monitoring
  - ☐ Ground Water Monitoring
  - ☐ Manual Tank Gauging
  - ☐ Statistical Inventory Reconciliation
  - ☐ Other Approved Method ________
- ☐ **Monthly Inventory Control (or Manual Tank Gauging) & Tank Tightness Testing**

Pressurized Piping
(Choose one from each set)

- ☐ **Automatic Flow Restrictor**
- ☐ **Automatic Shutoff Device**
- ☐ **Continuous Alarm System**

and

- ☐ **Annual Line Tightness Testing**
- ☐ **Monthly Monitoring (check Method)**
  - ☐ Vapor Monitoring
  - ☐ Interstitial Monitoring
  - ☐ Ground Water Monitoring
  - ☐ Statistical Inventory Reconciliation
  - ☐ Other Approved Method ________

Suction Piping (3 choices)

- ☐ **Monthly Monitoring**
  - ☐ Vapor Monitoring
  - ☐ Ground Water Monitoring
  - ☐ Interstitial Monitoring
  - ☐ Statistical
  - ☐ Other Approved Method: ________
- ☐ **Line Tightness Testing Every 3 Years**
- ☐ **No Requirements** (See Instructions)

CORROSION PROTECTION

Tanks

- ☐ **Coated & Cathodically Protected Steel**
- ☐ **Steel Clad with Noncorrodible Material**
- ☐ **Fiberglass Reinforced Plastic**
- ☐ **Cathodic Protection Added**
- ☐ **Interior Lining**
- ☐ **Other:** ________

Piping

- ☐ **Coated & Cathodically Protected Steel**
- ☐ **Other:** ________
- ☐ **Fiberglass Reinforced Plastic**
- ☐ **Cathodic Protection Added**

SPILL
OVERFILL PROTECTION

- ☐ **Spill Prevention**

-and-

- ☐ **Automatic Shutoff Device**
- ☐ **Overfill Alarm**
- ☐ **Ball Float Valve**

NOTE 1—Check appropriate choices. At least one selection in each block should be checked. Please see Appendix X2 for detailed instructions about each section. If the site has UST systems that differ in characteristics (that is, construction, piping type), use a separate sheet for each system.

FIG. X2.1 Release Detection, Corrosion Protection, Spill and Overfill Protection Checklist

X2.3.4.1 *Tanks*—There are several options to meet the corrosion protection options for the tank. Choose the appropriate option. If the tank has both cathodic protection and interior lining installed, check both boxes.

X2.3.4.2 *Piping*—There are several options to meet the corrosion protection options for the piping. Choose the appropriate option.

X2.3.4.3 The questions given in Fig. X2.4 may be helpful in determining conformance with Federal regulations for corrosion protection.

X2.3.5 *Spill/Overfill*— If the tank receives deliveries of regulated substances in quantities greater than 25 gal (L) at one filling, the tank must have spill prevention and overfill protection to meet the 1998 UST regulations. Spill equipment must contain the product in the transfer hose when the transfer hose is detached from the fill pipe (for example, a catchment basin) must be used to prevent the release of product to the environment. There are three options for overfill devices (audible or visual overfill alarm, automatic shutoff device, or ball float valve). Please check with the installer of the tank to determine whether these items are present.

				(Y=Yes/N=No/NA=Not Applicable)
INVENTORY CONTROL AND TANK TIGHTNESS TESTING				
1. Are there stick (or ATG) readings and dispenser readings each operating day?				
2. Is there an ⅛th inch accuracy in stick or ATG readings?				
3. Are there before/after delivery stick readings reconciled with delivery receipts?				
4. Does the drop tube extend to within 1 ft of the bottom of the tank?				
5. Is the dispenser meter calibrated?				
6. Is there a check for water (⅛th inch accuracy) at least once a month?				
7. Is monthly reconciliation (1% of volume pumped plus 130 gallons) done?				
STATISTICAL INVENTORY RECONCILIATION (SIR) If allowed in the jurisdiction.				
1. Are data collected according to the vendor's instructions?				
AUTOMATIC TANK GAUGING				
1. Are monthly leak tests conducted?				
MANUAL TANK GAUGING				
1. Is manual tank gauging performed weekly?				
2. Is the tank capacity less than 2,000 gallons?				
3. Does the tank meet one of the following tank size, duration of test, weekly standard and monthly standard requirements?				
Tank Size	Minimum Duration of Test	Weekly Standard (1 test)	Monthly Standard (4-test average)	
up to 550 gallons	36 hours	10 gallons	5 gallons	
551 - 1,000 gallons (when the tank diameter is 64″)	44 hours	9 gallons	4 gallons	
551 - 1,000 gallons (when the tank diameter is 48″)	58 hours	12 gallons	6 gallons	
551 - 1,000 gallons (also requires periodic tank tightness testing per 5.2.1.2 and 5.2.1.3)	36 hours	13 gallons	7 gallons	
1,001 - 2,000 gallons (also requires periodic tank tightness testing per 5.2.1.2 and 5.2.1.3)	36 hours	26 gallons	13 gallons	
4. Is there an ⅛th inch accuracy in stick readings?				
5. Are 2 stick readings averaged both before and after the test?				
6. Was the test length appropriate for the tank size (minimum of 36 hours)?				
INTERSTITIAL MONITORING				
1. Is the interstitial area monitored monthly?				
2. Are the monitoring wells (secondary barriers) or ports clearly marked and secured?				
3. Is the secondary barrier compatible with the stored substance and impermeable?				
4. Is continuous monitoring used as a line leak detector?				
GROUND WATER AND VAPOR MONITORING				
1. Is there a site evaluation that the chosen method will work for the location and are the wells located according to the plans?				
2. Are the wells clearly marked and secured?				

FIG. X2.2 Questions for Tank Release Detection

(Y=Yes/N=No/NA=Not Applicable)

PIPING RELEASE DETECTION SYSTEMS - PRESSURIZED PIPING

Are one of the following two options (some continuous monitoring methods can satisfy both parts of each option) used:

1. Automatic shut-off device or flow restrictor or continuous alarm **and** annual tightness testing.
2. Automatic shut-off device or flow restrictor or continuous alarm **and** one of the following monthly monitoring techniques: interstitial monitoring, vapor monitoring, ground water monitoring or other approved methods such as electronic or electric line leak detector, continual reconciliation, or SIR.

PIPING RELEASE DETECTION SYSTEMS - SUCTION PIPING

Is one of the following three options being used?

1. Monthly monitoring: interstitial monitoring, vapor monitoring, ground water monitoring, SIR or other approved method.
2. Line tightness testing every 3 years,
3. No requirements if **all** of the following criteria are met:
    - the below-grade piping operates at less than atmospheric pressure.
    - the below-grade piping is sloped so that the contents of the pipe will drain back into the storage tank if the suction is released.
    - there is only one check valve included in each suction line.
    - the check valve is located directly below and as close as practical to the suction pump.
    - compliance with the above specifications can be readily determined.

FIG. X2.3 Questions for Piping Release Detection

(Y=Yes/N=No/NA=Not Applicable)
TANK ASSESSMENT
Did the tank pass an assessment prior to upgrading with corrosion protection? Existing metal tanks need to be assessed prior to adding corrosion protection.
CATHODIC PROTECTION
Do the 60 day inspections indicate that the impressed current cathodic protection system is operating properly? The person who installed the cathodic protection system should have provided information on how to perform these inspections and the normal operating range for the rectifier. If this information about the rectifier is unknown or missing, contact the company who installed the system to obtain the information.
Is the structure to soil potential equal to or more negative than -.85 volts or is the cathodic protection shown to be operating properly using another accepted method such as those listed by NACE International Recommended Practice RP0285? These checks need to be performed within 6 months of installation, 6 months of repairs, and then every 3 years thereafter. If the tank is not adequately protected from corrosion, a professional has to come out and take a look at the cathodic protection system.
INTERIOR LINING (Tank Only)
Was tank lined using a nationally recognized standard? If unsure, the installer should be able to answer this question.
Has the lining been inspected within 10 years of installation and every 5 years thereafter? Metal tanks where interior lining is the only corrosion protection upgrade need to be periodically inspected the make sure that the tank is structurally sound and that the lining is still performing according to original design specifications.

FIG. X2.4 Questions for Corrosion Protection

Designation: E 1991 – 98

Standard Guide for Environmental Life Cycle Assessment of Building Materials/Products[1]

This standard is issued under the fixed designation E 1991; the number immediately following the designation indicates the year of original adoption or, in the case of revision, the year of last revision. A number in parentheses indicates the year of last reapproval. A superscript epsilon (ϵ) indicates an editorial change since the last revision or reapproval.

INTRODUCTION

This is a general guide for the application of environmental Life Cycle Assessment (LCA) as a tool for evaluating the environmental aspects of materials/products, processes, and services produced and used in buildings and the built environment. This guide does not include, necessarily, all of the environmental features and impacts of the complete building life cycle, but focuses on those issues directly related to building materials/products and those elements of the building's environmental performance affected by these materials/products. Fig. 1 illustrates the total life cycle of a building. Fig. 2 is an example of the relationship between the life cycle processes of building materials/products and the total life cycle of a building and illustrates how these product/material life cycle processes merge with the total building life cycle. Fig. 3 illustrates an example of the life cycle of a building construction material within the context of the total building life cycle.

1. Scope

1.1 This guide presents a common framework and set of principles for potential users, such as product manufacturers, environmental analysts, consultants, architects, and the building industry in general. It describes a framework for life cycle inventory analysis, and describes various options and aspects of Impact Assessment and Interpretation.

1.2 The complexity and level of detail of an LCA will vary greatly depending on the material/product or system studied, the purpose and use of the study, the intended users of the study, and the resources committed to complete the study. The level of detail can range from generic to material/product specific.

1.3 This guide does not describe in detail the actual techniques for performing a Life Cycle Assessment.

1.4 Life Cycle Assessment is an emerging methodology, which is still evolving. This guide will present its concepts and major features. It should enable the user to better understand Life Cycle Assessment and its application to building materials/products, and help to identify sources of additional information and guidance. LCA is only one of many tools designed to aid in environmental evaluation and decision making.

1.5 The component phases of Life Cycle Assessment, including goal definition and scoping, inventory, impact assessment, interpretation, and the various methodologies used in these phases are in various stages of development. Consequently, the results of an LCA must be understood in the context of their completeness and accuracy and must be applied appropriately. LCA does not necessarily proceed as a linear process through these phases but is conducted in an iterative fashion.

2. Referenced Documents

2.1 *ASTM Standards:*

E 1765 Practice for Applying the Analytical Hierarchy Process (AHP) to Multiattribute Decision Analysis of Investments Related to Buildings and Building Systems[2]

3. Terminology

3.1 Definitions for most of the LCA terms and language used in this guide can be found in Refs. **(1-7)**.[3] Terms specific to this guide are as follows:

3.1.1 *standard*—a document that has been developed and established within the consensus principles of the Society and that meets the approval requirements of ASTM procedures and regulation.

3.1.2 *guide*—a compendium of information or series of options that does not recommend a specific course of action.

3.1.3 *practice*—a definitive set of instructions for performing one or more specific operations that does not produce a test result.

3.2 Definitions of terms in Fig. 1, Fig. 2, and Fig. 3:

---

[1] This specification is under the jurisdiction of ASTM Committee E-50 on Environmental Assessment and is the direct responsibility of Subcommittee E 50.03 on Global Sustainability/Pollution Prevention.

Current edition approved Nov. 10, 1998. Published January 1999.

[2] *Annual Book of ASTM Standards*, Vol 04.11.

[3] The boldface numbers in parentheses refer to the list of references at the end of this standard.

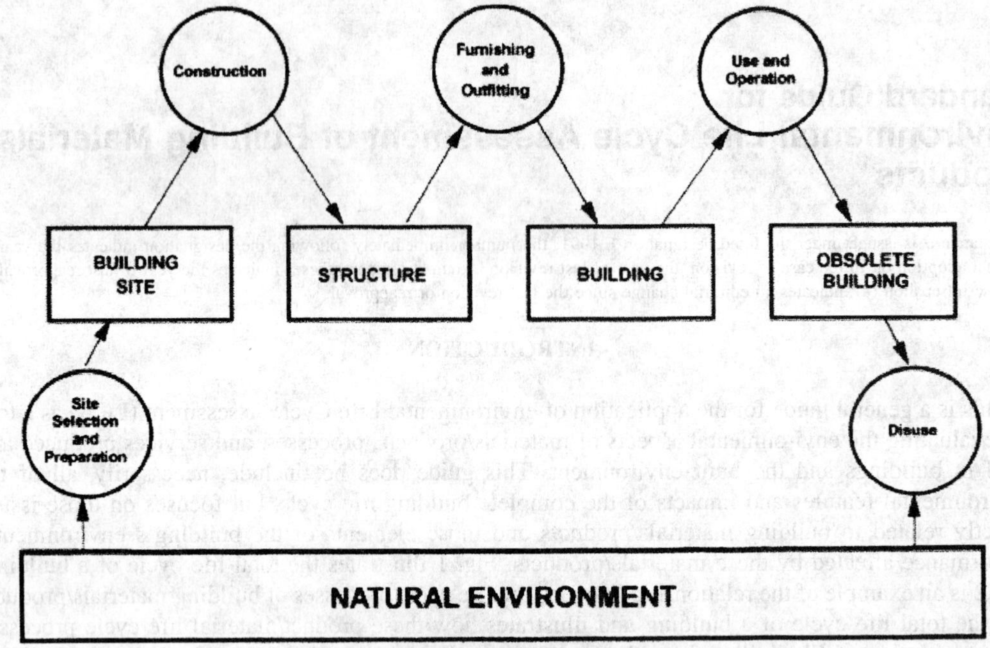

FIG. 1 Total Life Cycle of a Building

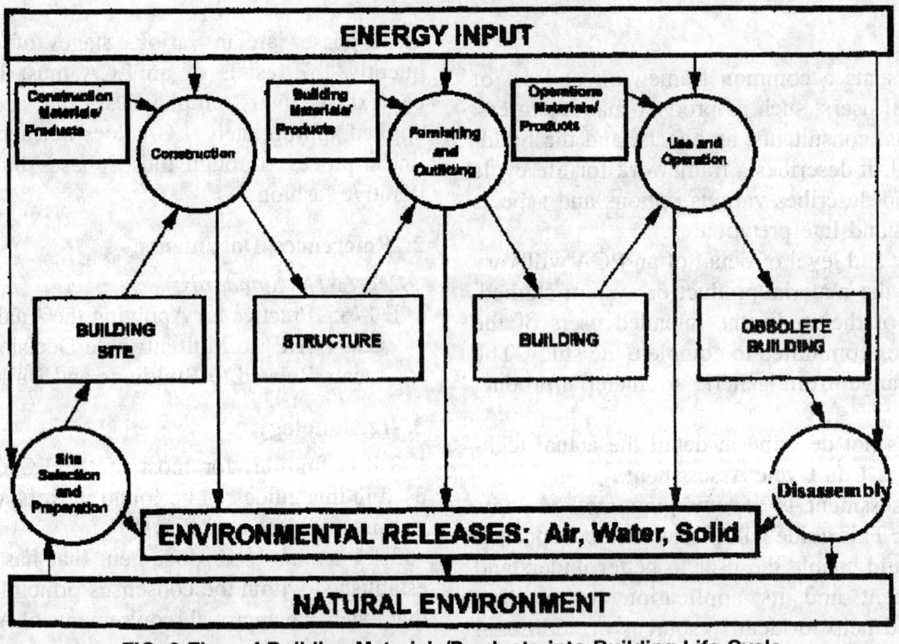

FIG. 2 Flow of Building Materials/Products into Building Life Cycle

3.2.1 *building, n*—the complete, outfitted, and furnished structure, operational in every way and ready for occupancy and use.

3.2.2 *building material/product, n*—a manufactured or produced unit or component that goes into any of the building life cycle processes. In the text of this guide, this term is used in a broad context and is meant to include fundamental construction material, for example, stone, as well as manufactured products, such as, windows, roofing, HVAC and electrical systems, interior furnishings, maintenance products, etc., used throughout the total building life cycle.

3.2.3 *construction, n*—the complete series of activities and actions that begins with a building sit and results in a completed structure.

3.2.4 *disassembly, n*—that complete series of activities and actions that eliminates the building.

3.2.5 *energy input, n*—all forms of energy necessary for the accomplishment of the particular building life cycle process under consideration.

3.2.6 *environmental releases, n*—all air, water, and solid emissions, which are given off by the building life cycle process under consideration that return to the natural environment.

3.2.7 *furnishing and outfitting, n*—the complete series of

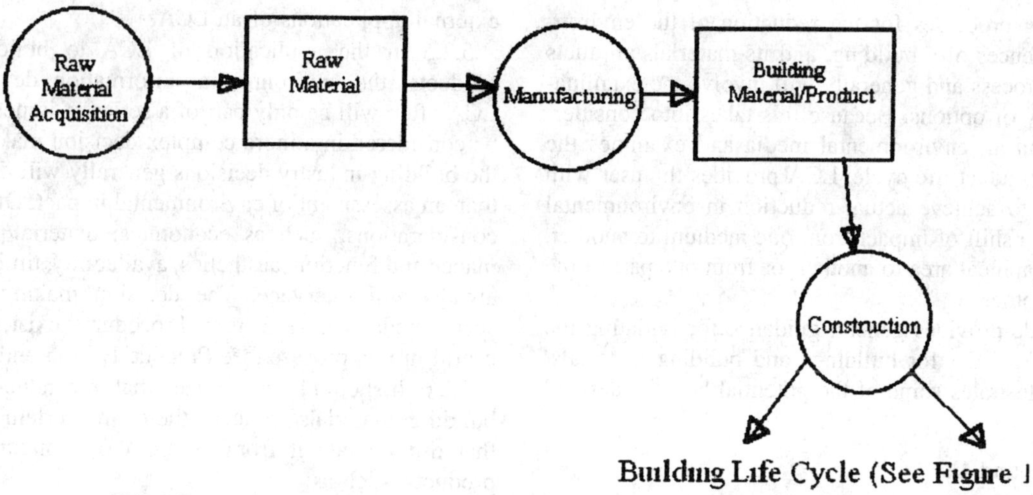

FIG. 3 Example of Building Material/Product Segment of Total Building Life Cycle

activities and actions that begins with a building structure and results in a completed building.

3.2.8 *interior furnishings, n*—those temporary and semipermanent systems and components, which generally are required for the normal utilization of the building for its intended purpose including decorative components.

3.2.9 *manufacturing, n*—the complete series of activities and actions that produces the building material/product.

3.2.10 *obsolete building, n*—a building that has reached the end of its useful life.

3.2.11 *raw material, n*—those components and ingredients that enter into the manufacturing process of the particular building material/product under consideration.

3.2.12 *raw material acquisition, n*—the processes by which natural resources are taken from the natural environment, including subsequent processing, to produce raw materials for the manufacture of the particular building material/product under consideration.

3.2.13 *site, n*—the natural location intended for the building altered, modified, and prepared to the point where construction activities can begin.

3.2.14 *sit selection and preparation, n*—the complete process or series of activities and actions that converts, alters, and modifies a natural area or plot of land to a building site ready for construction to begin.

3.2.15 *sponsor, n*—the individual or individuals who have initiated and funded the LCA.

3.2.16 *stakeholder, n*—an individual or collection of individuals who have some substantial interest or concern in the building, its materials/products, or its life cycle processes; or, whose environment is or will be influenced by the building and its life cycle.

3.2.17 *structure, n*—the completed building envelope on the site including all operating systems ready for interior furnishings.

3.2.18 *target audience, n*—the individual or collection of individuals for whom the LCA is being done and to whom the results are being directed.

3.2.19 *use and operation, n*—the complete and ongoing series of activities and actions that occur and are required during the life of a building from the point of occupancy to the point where the building is obsolete and is about to be disassembled.

4. Summary of Guide

4.1 Life Cycle Assessment is a tool for identifying, assessing, and interpreting the environmental aspects, such as, material, natural resource, and energy use; environmental releases and other burdens of a product, process, or activity. A typical LCA can be thought of as consisting of four phases: goal definition and scoping, inventory analysis, impact assessment, and interpretation **(1-7)**.

4.2 Defining a clear and unambiguous goal or purpose of the LCA is essential at the outset. Doing so will assist in imposing boundaries on the study and will help to establish the scope of the effort. Identification of the target audience also is important to establishing the scope. Depending on the goal and purpose of the LCA, other parts of the process may involve identifying the relevant stakeholders throughout the life cycle, ranking the degree to which the various stakeholders interests will be accommodated in decision making processes, and identifying stakeholder priorities regarding the various potential impacts possible throughout the life cycle.

4.3 The inventory analysis will comprise a process analysis of the Life Cycle of the subject of the LCA, in this case the building material/product. Fig. 1 illustrates the life cycle of a building. Fig. 2 illustrates how the life cycle of a building material/product merges with the life cycle of the building. This analysis of the life cycle process then is followed by a compilation of the relevant inputs and outputs of the processes making up the total system. The depth of this analysis will be consistent with the goal, scope, and intended use of the study **(1, 2, 3, 5, 6, 7)**.

4.4 Impact assessment will consist of an evaluation of the potential environmental impacts of the inputs and outputs of the total system. The general categories of environmental impacts to be considered may include resource use, human health, and ecological consequences. **(1, 4)**.

4.5 Interpretation of the results of the inventory analysis and the impact assessment must be made in relation to the goal and intended use of the LCA.

4.6 The application of the Life Cycle Assessment concept to

decision making processes for the reduction of the environmental consequences of a building, and its materials/products is an iterative process and generally will involve the examination of a variety of options. Because this takes into consideration impacts on all environmental media and examines the entire material/product life cycle, LCA provides the user with an opportunity to achieve actual reduction in environmental impact and not a shift of impact from one medium to another, from one geographical area to another, or from one part of the life cycle to another.

4.7 This guide provides general guidance for initiating the performance of a LCA for buildings and building materials/products and illustrates some of the potential benefits derived from its use.

5. Significance and Use

5.1 Life Cycle Assessment in its broadest context is a holistic, comprehensive concept with many potential applications. Currently, there are numerous organizations, from both the public and private sectors, such as the Society of Environmental Toxicologists and Chemists (SETAC), ISO, many national standards organizations, universities, private companies, consulting groups, etc., working in the area of environmental life cycle assessment. Each group has its own specific set of objectives and requirements. This guide includes the elements of Life Cycle Assessment on which general consensus has been reached and incorporates them into a guide tailored to buildings and building materials/products.

5.2 This guide provides general guidance for the practice of conducting LCA on building materials/products for the purpose of making decisions and choices.

5.2.1 Those who specify or select materials for use in buildings will benefit from the guidance provided here in that it will provide a means of incorporating environmental considerations into their decision-making processes.

5.2.2 Those who develop, manufacture, and market building materials/products will benefit from the guidance provided here in that it will enable them to more objectively assess the environmental implications of new products, designs, and processes; evaluate existing products, designs, and processes; and compare other product and process alternatives.

5.2.3 This guide offers guidance that can be helpful in minimizing the subjectivity often associated with environmental decision making and will foster more consistent, more complete assessments and decisions regarding the environmental aspects of building materials/products. Even with this guidance, results may vary greatly.

5.2.4 See Appendix X1 for some specific examples of the potential applications or uses of Life Cycle Assessment as suggested in this guide. Application and use can have either an internal or external focus. An internal application is intended primarily for the use of the sponsor and focuses on some internal aspect, material, or process over which the sponsor has primary control and influence. In this case, the sponsor is the exclusive audience. An external application is intended either primarily or secondarily for use by some stakeholder external to the sponsor. In this case, the sponsor and one or more external stakeholders are the audience for the LCA. A critical review by knowledgeable expects should be done for all external applications of an LCA.

5.2.5 In the application of LCA to building materials/products, the environmental information developed by the LCA often will be only part of a series of issues and factors to be considered in a more complex decision making process. In the building industry decisions generally will depend on more than an assessment of environmental impact. Other factors and considerations, such as economics, material/product performance and function, aesthetics, availability, timing, etc., almost always will intervene. The decision making process must accommodate these factors. Procedures exist, which will be useful in this process (see Practice E 1765 and Ref **8**).

5.2.6 It should be recognized that in conducting an LCA on building materials/products there are certain considerations that differentiate it from an LCA on nondurable consumer products, such as:

5.2.6.1 Durability of alternative materials/products for a given application may vary significantly.

5.2.6.2 Building use patterns can change during the life of the materials/products.

5.2.6.3 Environmental effects during the use stage can dominate the total environmental impact.

5.2.6.4 Building site and location can affect the environmental profile.

5.2.6.5 Materials/products often are incorporated into assemblies with other materials/products and must be considered in the context of those assemblies and their environmental implications.

5.2.6.6 Environmental impacts may depend on how materials/products are used and maintained.

6. Procedure

6.1 The following section describes in more detail the content and components of an LCA as it relates to building materials/products. The information contained herein is not intended to be exhaustive. Rather, it is intended to provide an overview and assist the user in deciding how to undertake an LCA and how the results might be applied.

6.1.1 There are four major phases in conducting an LCA. They are goal definition and scoping, inventory analysis, impact assessment, and interpretation. In the context of this guide, the application of a complete LCA will involve the application of these four phases to a series of building and building material/product options and choices for the purpose of decision making (see Fig. 4).

6.1.2 The four phases will be discussed here as a methodology for the evaluation of a single material, product, or activity. When evaluating alternate materials/products, alternate process options, or when making comparative assertions, it is essential that the LCAs of each of the options be conducted in a thorough, consistent, and comparable manner so that comparisons are valid.

6.2 *Goal Definition and Scoping*:

6.2.1 The purpose, goal, or intended use of the LCA must be defined clearly and unambiguously prior to proceeding. This provides the project/problem definition, enables the setting of boundary conditions for the analysis, and establishes the objectives of the study.

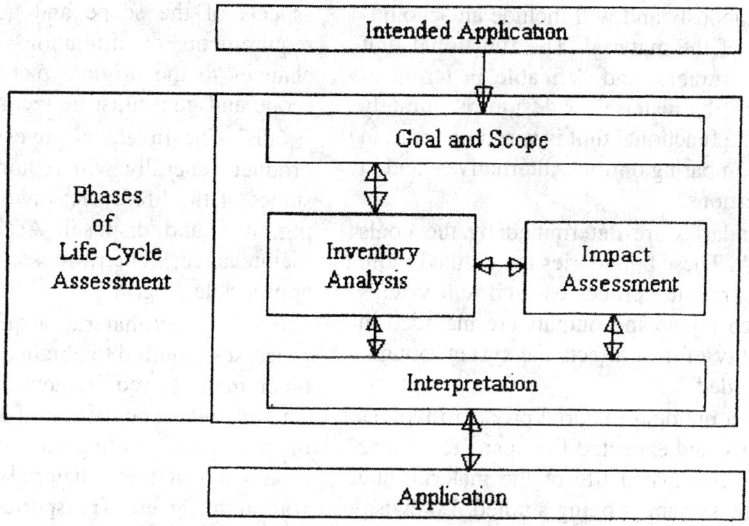

FIG. 4 Outline of LCA for Building Materials/Products

6.2.2 The purpose, goal, and intended use also will determine the level of detail and the depth of the Life Cycle process analysis required.

6.2.3 Data quality targets also will be set. It must be realized that achieving these data quality targets may not always be possible. The available data and the quality of those data will depend upon many things, such as source, completeness, precision, etc. As data are collected throughout the study, it is important to record a measure of data quality.

6.2.4 The target audience of the study should be identified. The target audience is the principal audience for whom the results of the study are intended. The target audience may be internal to the entity sponsoring the study, or the study may be intended for use by the sponsor with some audience external to the sponsor. Often, it is the case that there are several target audiences to be considered. This is particularly the case for a building since it has an extended lifetime and since it interfaces with many stakeholders.

6.2.5 It is important in setting the goal and scope to identify not only the target audience but also all of the relevant stakeholders who will have some stake in the total environmental impact of the building and its materials/products. The stakeholders can be of several different types. For example, the sponsor of the LCA is a stakeholder and certainly a target audience. Those affected by the environmental impacts throughout the life cycle also are stakeholders, and depending on the purpose of the LCA, some also may be part of the target audience. Also, it may be important for the decision maker (user of the LCA) to establish the degree to which stakeholder's values and views have been considered and accommodated in the decision making process.

6.2.6 At the goal definition and scoping stage, the LCA is specifically designed and planned with the intended use, audience, and stakeholders in mind. Consequently, a study designed to provide product or process improvement may be unsuitable for use in marketing or educating the public.

6.3 *Defining the Scope*:

6.3.1 In defining the scope of a study, the following issues should be considered and clearly described: the function of the material/product system under consideration; the functional unit; the system boundaries; temporal features and time scales; geographic and locational aspects; the extent, type, and methodology of impact assessment processes to be used; data and data quality requirements; known assumptions and limitations; and, methods to be used to evaluate the significance of differences revealed in the inventory and impact assessment.

6.3.2 The "functional unit" is a descriptor or measure of the service or function performed by the material/product system under consideration. The purpose of defining the functional unit for an LCA study is to permit comparison of equivalent items or services (functions). The functional unit is to provide a means of comparing different materials/products or designs for a given function. For items to be functionally equivalent, they must be able to meet a defined need (function) within specified use, performance, and time parameters.

6.3.2.1 The functional unit for an interior wall paint or finish may be defined as effective coverage of and performance on a given area of interior wall surface for a specified time period. Then, depending on the properties and performance of the paint or finish options, this unit will be defined further in terms of a specific number of gallons or some other quantity in units more common to the material under consideration. In general, a comparison on a gallon for gallon basis may not necessarily be valid. For example, one paint may require different numbers of applications than the alternative to achieve the desired performance.

6.3.2.2 The functional unit for a floor covering could be defined as a specific area of coverage of the subfloor meeting specified use and performance characteristics, installation requirements, aesthetic qualities, and maintenance properties, for a given period of time. Durability and lifetime of the material, recyclability, impact on indoor air quality, cost, etc., are important factors to be considered.

6.3.2.3 Definition of the functional unit can sometimes be complex in the case of building materials, particularly when several different options of very different design are under consideration; however, it is necessary to define clearly some equivalency in order to fairly compare the options. The functional unit for building materials may involve several specific functions simultaneously and will include an accounting for the expected life of the material. The functional unit must be quantifiable, measurable, and definable in terms of some unit characteristic of the material or design. Accurately and completely defining the functional unit is very important in ensuring equivalency in comparing options, alternatives, and in making comparative assertions.

6.3.3 The system boundaries are determined by the goals and objectives of the LCA. These boundaries are defined more specifically by which operations, processes, and subsystems are included and by which inputs and outputs are included in the study. When availability of data affects the system boundaries it should be so recorded.

6.3.4 The life cycle of a building material/product involves considerations of durability and expected life span. The scope should define clearly the anticipated life of the material and time frame over which the system is being studied. Likewise, environmental impacts can be time dependent and the system should account for such temporal features.

6.3.5 The scope of the study should take into account geographical and locational considerations that can have an effect on the impacts of the system. Environmental impacts can be very dependent upon location and geography. Some impacts can be of global scale and some only of local scale.

6.4 *Data Quality*:

6.4.1 Data quality assessment is a systematic approach to measuring the suitability of the data for its intended purpose. The quality of the data used in an LCA in large measure determines the degree of confidence the user will have in the ultimate results of the LCA and the decisions that result. Data quality is a representation of the reliability of the individual data and in the data set as a whole.

6.4.2 Data quality goals are specifications of the required quality of the data based on a set of data quality indicators. The goals and objectives of the LCA can assist in establishing minimum data quality goals. The data quality achieved may to some degree be dependent on the level of effort allocated to the study which will be a function of the purpose, budgetary and time constraints, and data availability. Inability to meet the data quality goals may require modification of the goals and scope of the LCA.

6.4.3 Data quality indicators are quantitative or qualitative measures of the credibility of the data. Data quality indicators should address such things as precision, accuracy, reproducibility, variability, completeness, consistency, source, uncertainty, etc.

6.4.4 Data quality assessments are an important element in drawing conclusions and making decisions. As a result, every effort should be made to determine and record the quality of all data, the uncertainty associated with it, and the significance of these uncertainties.

6.5 *Life Cycle Inventory Analysis*:

6.5.1 Life cycle inventory analysis involves the identification and quantification wherever possible of all inputs and outputs of the system under study including natural resources; energy; releases to air, water, and land; coproducts; and, open- and closed-loop recycling. The process of conducting the inventory is iterative. As the inventory proceeds various aspects of the scope and goal may be affected. New data requirements or limitations may be identified that require changes to the original plan. Any such modifications to the scope and goal must be recorded.

6.5.2 The inventory investigation for a building material/product generally will require information on the four main stages of the life cycle: raw material acquisition, manufacturing, use, and disposal. Aspects of the life cycle, such as maintenance, repair, reuse, and recycling are included in appropriate stages.

6.5.2.1 Raw material acquisition includes all activities and processes required to obtain all the raw materials, energies, and other material requirements for the material/product system. This includes acquisition of natural resources, such as mining or petroleum drilling, as well as subsequent refining and processing of these materials, including distillation, polymerization, firing, etc. Transportation activities associated with this part of the life cycle also are included.

6.5.2.2 Manufacturing activities can include natural resource conversions not accounted for in the raw material acquisition stage, other material or intermediate manufacture, product fabrication or assembly, packaging, and distribution.

6.5.2.3 Use activities comprise those processes involving the consumption, installation, utilization, maintenance, repair, etc., of the material/product to the point where it has completed its intended function. Because of the longevity of buildings and building materials/products, a thorough analysis of this phase is particularly important. For building materials/products, because of their extended lifetimes, the use phase often will dominate the environmental impacts of the life cycle.

6.5.2.4 End of useful life processes and activities include the return or release to the environment of the material/product system but also may include reuse, recycling, or other waste management options.

6.5.3 The primary activities involved in the inventory process are life cycle material and process flow analysis and identification of inputs and outputs at each step of these processes. This involves data collection, data compilation, and data presentation.

6.5.3.1 *Data Collection*—All material inputs and outputs for each of the process and subprocess steps within the boundaries of the system are identified, quantified, and compiled to the extent possible or to the extent required by the goal and scope of the study. This compilation will include raw materials, other materials and products relevant to the system, energy usage, and environmental releases. Data collection may require substantial commitments in funds and staff time. Practical constraints should be taken into consideration and documented in the scope of the study.

6.5.3.2 *Data Compilation*—The data is assembled into a format that will allow for further manipulation and analysis. All data will be compiled as some quantity per the defined functional unit. Data aggregation procedures can be used if they are useful and consistent with the scope and goal. Such aggregation procedures should be stated, documented, and justified clearly. Nonaggregated data also should be available. Because of complexities and multiple outputs of the processes in the life cycle, it will become necessary to allocate certain inputs or other inventory items to these multiple outputs. For example, a process that uses energy to produce several useful outputs only one of which is associated directly with the subject of the LCA will require that this energy be allocated properly to the various output products. The basis for this allocation will be dependent on the goal and purpose of the LCA. Such allocation procedures must be stated, documented, and justified clearly. The sources and quality of all of the data also should be recorded and explained.

6.5.3.3 *Data Presentation*—The data presentation and format will be consistent with the goals and objectives of the LCA. In the case where impact assessment is to be conducted, the data will be presented in a form anticipated by the impact assessment procedures. Presentation of data and data quality should be sufficient to enable the users to understand and evaluate the study results. Data presentation should be clear, complete, and documented, including sources of data and assumptions underlying the study. Data and results shall be presented in a way that allows the user to determine whether differences between alternative materials/products are statistically significant. If information on the statistical significance of differences is not available, the presentation of the data and the results shall be accompanied by a statement indicating that it is not possible to determine the statistical significance of differences, and an outline of the nonstatistical basis for characterizing the significance of any differences.

6.6 *Impact Assessment*:

6.6.1 Impact assessment is aimed at characterizing and assessing the impacts of the environmental loadings identified in the inventory analysis. The level of detail and methodologies of impact assessment will depend on the goal and scope of the study. The process of impact assessment is not completely developed. There is no generally accepted methodology for consistently, comprehensively, and accurately associating inventory data with potential environmental impacts. As a result, it is important that assumptions and methodology are described and reported clearly.

6.6.2 Impact assessment comprises three components: classification, characterization, and valuation.

6.6.2.1 Classification involves the assignment of inventory data to one or more appropriate impact categories. Impact categories can include changes to air, water, and soil quality or availability, as well as effects on ecological systems, human health, and natural resource stocks.

6.6.2.2 Characterization describes and estimates the contribution of the inventory items to the related impacts.

6.6.2.3 Valuation occurs when the different impact categories are evaluated and assigned weights so that they can be compared among themselves. The weighting may involve objective and subjective assignments of relative values and will be affected not only by the severity of environmental degradation but also by the values and opinions held by the stakeholders of concern. The purpose of this step is to establish a further interpretation and aggregation of the data of the impact assessment to enable the relative importance of the impact categories to be assessed.

6.6.2.4 Important factors that intervene in impact assessment are geographical and temporal factors. The geographical location of an environmental burden needs to be considered in impact assessment. Likewise, the environment is a dynamic system and the effects of time are an important consideration.

6.7 *Interpretation*:

6.7.1 Interpretation is the phase of the LCA in which the findings of the inventory and the impact assessment are examined and conclusions are drawn with reference to the goals and objectives of the LCA. As indicated in Fig. 4, interpretation may be based on the inventory analysis alone or on the results of an impact assessment or on some combination of inventory data and impact assessment results. The goal and scope of the study will determine what form the interpretation takes and how detailed its output. Interpretation may involve reconsideration and revisions of the scope and data considerations. The finding of this phase should reflect the results of any sensitivity or uncertainty analysis that may have been performed. It is important that all assumptions in connection with this phase be explained and recorded.

6.7.2 Interpretation generally will be a preliminary step to the application of the LCA in a decision making process in connection with building materials/products. Interpretation usually will involve only environmental issues and concerns and will constitute the environmental effects portion of the input to a final decision making process. Other considerations, such as performance, functional benefits, cost, etc., become part of the application process.

6.7.3 When comparing differences between alternative materials/products, the user is urged to have the data prepared so that statistical analysis can be used to measure significance of differences. Lacking a sound statistical basis for comparison, users should be cautious in concluding differences between alternative materials/products exist and should understand the basis for concluding those differences.

6.7.4 The interpretation phase of LCA is still undergoing development.

7. Critical Review

7.1 The user of the LCA may elect to conduct a critical review of the study. Some purposes of a critical review may be to assure that the study met the established goals; to enhance the credibility of the LCA; or, to provide an external evaluation regarding the quality of the study.

7.2 A review may be carried out internally or externally by recognized experts. If public credibility is important, the independence of an external expert will be beneficial. These experts shall have recognized, relevant scientific and technical expertise and shall be familiar with the requirements of this guide. Critical review by experts should be done for all LCAs that will be used externally. External review does not necessarily guarantee a higher quality review.

7.3 A critical review can evaluate the methods used, the data collected, the interpretations drawn, and the limitations under which the study is conducted. It does not validate the goals, objectives, uses, or conclusions of the LCA.

8. Reporting

8.1 Reporting can and should be tailored to the target audience or intended user of the study. The target audience usually will have an effect on the level of detail of the report.

The various types of reporting categories may include a full report, a scientific release, an executive summary, an educational release, and a press release.

8.2 Regardless of the category of report provided, each report must make clear the study goals and objectives, scope, assumptions, data sources, sensitivity of the results to changes in key assumptions, statistical (or other) basis for concluding that differences between alternatives materials/products are significant, and conclusions. A good report will allow the reader, regardless of their perspective, to evaluate the study and use the data to draw their own conclusions. A full report always should be made available, especially when the LCA is reported externally.

8.3 Reporting should be objective and unbiased, fair and accurate. In particular there should be a clear indication of what has and what has not been included in the study, what methodologies have been used, assumptions, data gaps, data quality, etc. A description of these characteristics often is referred to as the transparency of the LCA. Where appropriate, the needs of different users should be recognized and addressed when presenting or disseminating the study. The suggested content of a full report is shown in the following outline:

8.3.1 *Goals and Objectives*,
8.3.2 *Scope*,
8.3.3 *Summary*,
8.3.4 *System Boundaries*,
8.3.5 *Functional Unit*,
8.3.6 *Product/System Life Cycle Description*, (including flow diagrams),
8.3.7 *Methodology*,
8.3.8 *Data and Data Quality Indicators*,
8.3.9 *Data Presentation*,
8.3.10 *Inventory Analysis*,
8.3.11 *Impact Assessment*,
8.3.12 *Interpretations and Conclusions*,
8.3.13 *Critical Review*, and
8.3.14 *Identity and Qualifications of the Company or Lead LCA Reviewer*.

9. Keywords

9.1 buildings; building materials; building products; decision making; environment; environmental effects; impact assessment; input/output analysis; inventory analysis; life cycle assessment (LCA)

APPENDIX

(Nonmandatory Information)

X1. Potential Internal and External Applications for Life Cycle Assessment

NOTE X1.1—This list is not intended to be an exhaustive one. It is intended to suggest potential applications.

X1.1 *Internal Applications*:

X1.1.1 Compare alternate raw materials.

X1.1.2 Evaluate manufacturing process changes or new processes.

X1.1.3 Evaluate alternate product designs and forecast potential impact.

X1.1.4 Assess environmental effects associated with a business, that is existing, new, or for acquisition.

X1.1.5 Identify the materials, processes, or systems of an operation that generate significant environmental impact, or a combination thereof.

X1.1.6 Training and education of internal associates.

X1.1.7 Evaluate sites for operating or locating a facility.

X1.2 *External Applications*:

X1.2.1 Evaluate environmental impact for reduction of future regulatory liability.

X1.2.2 Help develop long-term corporate policy regarding overall material use, resource conservation, and reduction of environmental impacts and risks.

X1.2.3 Aid in training and educating designers, architects, builders, building owner and operators, etc., in the proper use of environmental assessment for the selection of lower impact materials/products, designs, and building processes and operations.

X1.2.4 Provide information that allows customers, consumers, and other stakeholders to evaluate and differentiate between different materials/products and design alternatives.

X1.2.5 Provide information to policy makers and the public about the environmental impacts associated with materials/products and design alternatives.

X1.2.6 Help develop policy on overall material use, waste management and resource conservation, and reduction of environmental impacts.

REFERENCES

(1) The Society of Environmental Toxicologists and Chemists (SETAC), *A Technical Framework for Life-Cycle Assessments*, SETAC, 1010 N. 12 Aveune, Pensacola, FL 32501, 1991.

(2) The Society of Environmental Toxicologists and Chemists (SETAC), *(Europe), Life Cycle Assessment*, SETAC (Europe), 1010 N. 12 Aveune, Pensacola, FL 32501, 1993.

(3) Vigon, B.W., et.al., *Life-Cycle Assessment: Inventory Guidelines and Principles*, USEPA, Office of Research and Development, Washington, DC 20460, 1993.

(4) French, C., *Life-Cycle Impact Assessment*, USEPA, Office of Research and Development, Washington, DC 20460, 1994.

(5) *ISO, Environmental Management—Life-Cycle Assessment—Principles and Guidelines*, (ISO 14040 Draft International Standard), 1996.

(6) *ISO, Environmental Management—Life-Cycle Assessment—Inventory Analysis*, (ISO 14041 Draft International Standard), 1997.

(7) *American Forest and Paper Association, Life-Cycle Inventory Analysis—Users Guide*, AF&PA, 1996.

(8) Norris, G.A., and Marshal, H.E., "Multiattribute Decision Analysis Methods for Evaluating Buildings and Building Systems," *NISTIR* 5663, U.S. Department of Commerce, 1995.

Designation: E 2018 – 99

Standard Guide for Property Condition Assessments: Baseline Property Condition Assessment Process[1]

This standard is issued under the fixed designation E 2018; the number immediately following the designation indicates the year of original adoption or, in the case of revision, the year of last revision. A number in parentheses indicates the year of last reapproval. A superscript epsilon (ϵ) indicates an editorial change since the last revision or reapproval.

1. Scope

1.1 *Purpose*—The purpose of this guide is to define good commercial and customary practice in the United States of America for conducting a baseline *property condition assessment* (PCA) of the improvements located on a parcel of commercial real estate by performing a walk-through survey and conducting research as outlined within this guide.

1.1.1 *Physical Deficiencies*—In defining good commercial and customary practice for conducting a baseline PCA, the goal is to identify and communicate physical deficiencies to a user. The term *physical deficiencies* means the presence of conspicuous defects or material deferred maintenance of a subject property's material systems, components, or equipment as observed during the field observer's walk-through survey. This definition specifically excludes deficiencies that may be remedied with routine maintenance, miscellaneous minor repairs, normal operating maintenance, etc., and excludes de minimis conditions that generally do not present material physical deficiencies of the subject property.

1.1.2 *Walk-Through Survey*—This guide outlines procedures for conducting a walk-through survey to identify the subject property's material physical deficiencies, and recommends various systems, components, and equipment that should be observed by the field observer and reported in the *property condition report* (PCR).

1.1.3 *Document Reviews and Interviews*—The scope of this guide includes document reviews, research, and interviews to augment the walk-through survey so as to assist the consultant's understanding of the subject property and identification of physical deficiencies.

1.1.4 *Property Condition Report*—The work product resulting from completing a PCA in accordance with this guide is a PCR. The PCR incorporates the information obtained during the Walk-Through Survey, the Document Review and Interviews sections of this guide, and includes opinions of probable costs for suggested remedies of the physical deficiencies identified.

1.2 *Objectives*—Objectives in the development of this guide are: *(1)* define good commercial and customary practice for the PCA of primary commercial real estate improvements; *(2)* facilitate consistent and pertinent content in PCRs; *(3)* develop practical and reasonable recommendations and expectations for site observations, document reviews and research associated with conducting PCAs and preparing PCRs; *(4)* establish reasonable expectations for PCRs; *(5)* assist in developing an industry baseline standard of care for appropriate observations and research; and *(6)* recommend protocols for consultants for communicating observations, opinions, and recommendations in a manner meaningful to the user.

1.3 *Considerations Beyond Scope*—The use of this guide is strictly limited to the scope set forth in this section. Section 10 and Appendix X1 of this guide identify, for informational purposes, certain physical conditions that may exist on the subject property, and certain activities or procedures (not an all inclusive list) that are beyond the scope of this guide but may warrant consideration by parties to a commercial real estate transaction.

1.4 *Organization of This Guide*—This guide consists of several sections, an Annex and an Appendix. Section 1 is the Scope. Section 2 Terminology, contains definitions of terms both unique to this guide and not unique to this guide, and acronyms. Section 3 sets out the Significance and Use of this guide, and Section 4 describes the User's Responsibilities. Sections 5 through 10 provide guidelines for the main body of the PCA, including the scope of the Walk-Through Survey, preparation of the Opinions of Probable Costs to Remedy Physical Deficiencies, and preparation of the PCR. Section 11 provides additional information regarding out of scope considerations (see 1.3). The Annex provides requirements relating to specific asset types, and where applicable, such requirements are to be considered as if integral to this guide. The Appendix provides the user with additional PCA scope considerations, whereby a user may increase this guide's baseline scope of due diligence to be exercised by the consultant.

1.5 *Multiple Buildings*—Should the subject property consist of multiple buildings, it is the intent of this guide that only a single PCR be produced by the consultant to report on all of the buildings on the subject property.

---

[1] This guide is under the jurisdiction of ASTM Committee E-50 on Environmental Assessment and is the direct responsibility of Subcommittee E50.02 on Commercial Real Estate Transactions.

Current edition approved June 10, 1999. Published August 1999.

1.6 *Safety Concerns*—This guide does not purport to address all of the safety concerns, if any, associated with the walk-through survey. It is the responsibility of the consultant using this guide to establish appropriate safety and health practices when conducting a PCA.

TABLE OF CONTENTS

2. Terminology

2.1 This section provides definitions, descriptions of terms, and a list of acronyms, where applicable, for the words used in this guide. The terms are an integral part of the guide and are critical to an understanding of this guide and its use.

2.2 *Definitions:*

2.2.1 *architect, n*—designation reserved by law for a person professionally qualified, examined, and registered by the appropriate governmental board having jurisdiction, to perform architectural services including, but not limited to, analysis of project requirements and conditions, development of project design, production of construction drawings and specifications, and administration of construction contracts.

2.2.2 *building codes, n*—rules and regulations adopted by the governmental authority having jurisdiction over the commercial real estate, which govern the design, construction, alteration and repair of such commercial real estate. In some jurisdictions trade or industry standards may have been incorporated into, and made a part of, such building codes by the governmental authority. Building codes are interpreted to include structural, HVAC, plumbing, electrical, life-safety, and vertical transportation codes.

2.2.3 *building department records, n*—records maintained by or in possession of the local government authority with jurisdiction over the construction, alteration, use, or demolition of improvements on the subject property, and that are readily available for use by the consultant within the time frame required for production of the PCR and are practically reviewable by exercising appropriate inquiry. Building department records also may include building code violation notices. Often, building department records are located in the building department of a municipality or county.

2.2.4 *building systems, n*—interacting or independent components or assemblies, which form single integrated units, that comprise a building and its site work, such as, pavement and flatwork, structural frame, roofing, exterior walls, plumbing, HVAC, electrical, etc.

2.2.5 *component, n*—a fully functional portion of a building system, piece of equipment, or building element.

2.2.6 *dismantling, n*—to take apart or remove any component, device or piece of equipment that is bolted, screwed, secured, or fastened by other means.

2.2.7 *engineer, n*—designation reserved by law for a person professionally qualified, examined and licensed by the appropriate governmental board having jurisdiction, to perform engineering services.

2.2.8 *engineering, n*—analysis or design work requiring extensive formal education, preparation and experience in the use of mathematics, chemistry, physics, and the engineering sciences.

2.2.9 *fire department records, n*—records maintained by or in the possession of the local fire department in the area in which the subject property is located. These records should be practically reviewable and readily accessible for use by the consultant by exercising an appropriate inquiry within the time frame required for production of the PCR.

2.2.10 *guide, n*—a series of options and instructions that do not recommend a specific course of action.

2.2.11 *interviews, n*—discussions with those knowledgeable about the subject property.

2.2.12 *material, adj*—having significant importance or great consequence to the subject property's intended use or physical condition.

2.2.13 *practice, n*—a definitive procedure for performing one or more specific operations or functions that does not produce a test result.

2.2.14 *publicly available, adj*—the source of the information allows access to the information by anyone upon request.

2.2.15 *recreational facilities, n*—facilities for exercise, entertainment or athletics including, without limitation, swimming pools, spas, saunas, steam baths, tennis, volleyball, or basketball courts; jogging, walking, or bicycle paths; and playground equipment.

2.2.16 *structural frame, n*—the components or building system that supports the building's nonvariable forces or weights (dead loads) and variable forces or weights (live loads).

2.2.17 *standard, n—as used by ASTM*, a document that has been developed and established within the consensus principles of the Society and that meets the approval of the ASTM procedures and regulations.

2.2.18 *system, n*—a combination of interacting or interdependent components assembled to carry out one or more functions.

2.3 *Definitions of Terms Specific to This Standard:*

2.3.1 *actual knowledge, n*—the knowledge possessed by an individual rather than an entity. Actual knowledge, as used in this guide, is to be distinguished from knowledge provided by others, or information contained on documents obtained during the course of conducting a PCA.

2.3.2 *appropriate inquiry, n*—a request for information conducted by Freedom of Information Letter (FOIL), verbal request, or by other written request made either by fax, electronic mail, overnight courier, or U.S. mail. *Appropriate inquiry* includes a good-faith effort conducted by the consultant to obtain the information considering the time constraints to prepare and deliver the PCR.

2.3.3 *base building, n*—the core (common areas) and shell of the building and its systems that typically are not subject to improvements to suit tenant requirements.

2.3.4 *baseline, n*—the minimum level of observations, due diligence, inquiry/research, documentation review, and preparation of opinions of probable costs to remedy material physical deficiencies for conducting a PCA as described in this guide.

2.3.5 *building envelope, n*—the enclosure of the building that protects the building's interior from outside elements, namely the exterior walls, roof and soffit areas.

2.3.6 *commercial real estate, n*—improved real property, except a dwelling or property with four or less dwelling units exclusively for residential use. This term includes, but is not limited to, improved real property used for industrial, retail, office, hospitality, agriculture, other commercial, medical, or educational purposes; property used for residential purposes that has more than four residential dwelling units, and property with four or less dwelling units for residential use when it has a commercial function, as in the operation of such dwellings for profit.

2.3.7 *commercial real estate transaction, n*—a transfer of title to or possession of improved real property or receipt of a security interest in improved real property, except that it does not include transfer of title to or possession of improved real property with respect to an individual dwelling or building containing four or less dwelling units.

2.3.8 *consultant, n*—the entity or individual that prepares the PCR and that is responsible for the observance of and reporting on the physical condition of commercial real estate in accordance with this guide. The consultant generally is an independent contractor; however, the consultant may be an employee of the user. The consultant may be an individual that is both the field observer and PCR reviewer as described in Section 6.

2.3.9 *dangerous or adverse conditions, n*—conditions, which may pose a threat or possible injury to the field observer, and which may require the use of special protective clothing, safety equipment, access equipment, or any other precautionary measures.

2.3.10 *deferred maintenance, n*—physical deficiencies that cannot be remedied with routine maintenance, normal operating maintenance, etc., excluding de minimis conditions that generally do not present a material physical deficiency to the subject property.

2.3.11 *due diligence, n*—the process of conducting a walk-through survey and appropriate inquiries into the physical condition of a commercial real estate's improvements, usually in connection with a commercial real estate transaction. The degree and type of such survey or other inquiry may vary for different properties and different purposes.

2.3.12 *easily visible, adj*—describes items, components and systems that are conspicuous, patent, and which may be observed visually during the walk-through survey without intrusion, removal of materials, exploratory probing, use of special protective clothing, or use of special equipment.

2.3.13 *expected useful life (EUL), n*—the average amount of time in years that an item, component or system is estimated to function when installed new and assuming routine maintenance is practiced.

2.3.14 *field observer, n*—the individual that conducts the walk-through survey.

2.3.15 *immediate costs, n*—opinions of probable costs that require immediate action as a result of any of the following: *(1)* material existing or potential unsafe conditions, *(2)* material building or fire code violations, or *(3)* conditions that if left unremedied, have the potential to result in or contribute to critical element or system failure within one year or will result most probably in a significant escalation of its remedial cost.

2.3.16 *observation, n*—the visual survey of items, systems, conditions, or components that are readily accessible and easily visible during a walk-through survey of the subject property.

2.3.17 *observe, v*—to conduct an observation pursuant to this guide.

2.3.18 *obvious, adj*—plain, evident and readily accessible; a condition or fact not likely to be ignored or overlooked by a field observer when conducting a walk-through survey or that which is practically reviewable and would be understood easily by a person conducting the PCA.

2.3.19 *opinions of probable costs, n*—determination of probable costs, a preliminary budget, for a suggested remedy.

2.3.20 *owner, n*—the entity holding the title to the commercial real estate that is the subject of the PCA.

2.3.21 *PCR reviewer, n*—the individual that both exercises responsible control over the field observer and who reviews the PCR prior to delivery to the user.

2.3.22 *physical deficiency, n*—conspicuous defects or significant deferred maintenance of a subject property's material systems, components, or equipment as observed during the field observer's walk-through survey. Included within this definition are material life-safety/building code violations and, material systems, components, or equipment that are approaching, have reached, or have exceeded their typical EUL or whose RUL should not be relied upon in view of actual or effective age, abuse, excessive wear and tear, exposure to the elements, lack of proper or routine maintenance, etc. This definition specifically excludes deficiencies that may be remedied with routine maintenance, miscellaneous minor repairs, normal operating maintenance, etc., and excludes de minimis conditions that generally do not constitute a material physical deficiency of the subject property.

2.3.23 *practically reviewable, adj*—describes information that is provided by the source in a manner and form that, upon review, yields information relevant to the subject property without the need for significant analysis or calculations. Records or information that feasibly cannot be retrieved by reference to the location of the subject property are not generally considered practically reviewable.

2.3.24 *primary commercial real estate improvements, n*—the site and building improvements that are of fundamental importance with respect to the commercial real estate. This definition specifically excludes ancillary structures, that may have been constructed to provide support uses such as maintenance sheds, utility garages, pool filter and equipment buildings, etc.

2.3.25 *property, n*—the site improvements, which are inclusive of both site work and buildings.

2.3.26 *property condition assessment (PCA), v*—the process by which a person or entity observes a property, interviews sources, and reviews available documentation for the purpose of developing an opinion and preparing a PCR of a commercial real estate's current physical condition. At the option of the user, a PCA may include a higher level of inquiry and due diligence than the baseline scope described within this guide or, at the user's option, it may include a lower level of inquiry or due diligence than the baseline scope described in this guide. Such deviations from this guide's scope should be disclosed in the PCR's executive summary.

2.3.27 *property condition report (PCR), n*—a written report, prepared in accordance with the recommendations contained in this guide, that outlines the consultant's observations, opinions as to the subject property's condition, and opinions of probable costs to remedy any material physical deficiencies observed.

2.3.28 *readily accessible, adj*—describes areas of the subject property that are promptly made available for observation by the field observer at the time of the walk-through survey and do not require the removal of materials or personal property, such as furniture, and that are safely accessible in the opinion of the field observer.

2.3.29 *readily available, adj*—describes information or records that are easily and promptly provided to the consultant upon making a request in compliance with an appropriate inquiry and without the need for the consultant to research archive files.

2.3.30 *reasonably ascertainable, adj*—describes information that is publicly available, as well as readily available, provided to the consultant's offices from either its source or an information research/retrieval service within reasonable time, practically reviewable, and available at a nominal cost for either retrieval, reproduction or forwarding.

2.3.31 *remaining useful life (RUL), n*—a subjective estimate based upon observations, or average estimates of similar items, components, or systems, or a combination thereof, of the number of remaining years that an item, component, or system is estimated to be able to function in accordance with its intended purpose before warranting replacement. Such period of time is affected by the initial quality of an item, component, or system, the quality of the initial installation, the quality and amount of preventive maintenance exercised, climatic conditions, extent of use, etc.

2.3.32 *representative observations, n*—observations of a reasonable number of samples of repetitive systems, components, areas, etc., which are conducted by the field observer during the walk-through survey. The concept of representative observations extends to all conditions, areas, equipment, components, systems, buildings, etc., to the extent that they are similar and representative of one another.

2.3.33 *short-term costs, n*—opinions of probable costs to remedy physical deficiencies, such as deferred maintenance, that may not warrant immediate attention, but require repairs or replacements that should be undertaken on a priority basis in addition to routine preventive maintenance. Such opinions of probable costs may include costs for testing, exploratory probing, and further analysis should this be deemed warranted by the consultant. The performance of such additional services are beyond this guide. Generally, the time frame for such repairs is within one to two years.

2.3.34 *shutdown, n*—equipment, components or systems that are not operating at the time of the field observer's walk-through survey. For instance, equipment, components, and systems that may be shutdown as a result of seasonal temperatures.

2.3.35 *site visit, n*—the visit to the subject property during which observations are made pursuant to the walk-through survey section of this guide.

2.3.36 *specialty consultants, n*—individuals or entities either in the fields of engineering or in any particular building component, equipment, or system that have acquired detailed, specialized knowledge and experience in the design, evaluation, operation, repair, or installation of same.

2.3.37 *subject building, n*—referring to the primary building or buildings on the subject property, and that are within the scope of PCA.

2.3.38 *subject property, n*—the commercial real estate consisting of the site and primary real estate improvements that are the subject of the PCA described by this guide.

2.3.39 *suggested remedy, n*—an opinion as to a course of action to remedy or repair a physical deficiency. Such an opinion may also be to conduct further research or testing for the purposes of discovery to gain a better understanding of the cause or extent of a physical deficiency (whether observed or highly probable) and the appropriate remedial or reparatory response. A suggested remedy may be preliminary and does not preclude alternate methods or schemes that might be more appropriate to remedy the physical deficiency or that may be more commensurate with the user's requirements.

2.3.40 *survey, n*—observations made by the field observer during a walk-through survey to obtain information concerning the subject property's readily accessible and easily visible components or systems.

2.3.41 *technically exhaustive, adj*—describes the use of measurements, instruments, testing, calculations, exploratory probing or discovery, or other means to discover, or a combination thereof, or troubleshoot physical deficiencies or develop architectural or engineering findings, conclusions, and recommendations, or combination thereof.

2.3.42 *timely access, n*—entry provided to the consultant at the time of the site visit.

2.3.43 *user, n*—the party that retains the consultant for the preparation of a baseline PCA of the subject property in accordance with this guide. A user may include, without limitation, a purchaser, potential tenant, owner, existing or potential mortgagee, lender, or property manager of the subject property.

2.3.44 *walk-through survey, n*—conducted during the field observer's site visit of the subject property, that consists of nonintrusive visual observations, survey of readily accessible, easily visible components and systems of the subject property. This survey is described fully in Section 8. Concealed physical deficiencies are excluded. It is the intent of this guide that such a survey should not be considered technically exhaustive. It excludes the operation of equipment by the field observer and is to be conducted without the aid of special protective clothing, exploratory probing, removal of materials, testing, or the use of equipment, such as scaffolding, metering/testing equipment, or devices of any kind, etc. It is literally the field observer's visual observations while walking through the subject property.

2.4 *Abbreviations and Acronyms:*

2.4.1 *ADA, n*—The Americans with Disabilities Act.

2.4.2 *ASTM, n*—American Society for Testing and Materials.

2.4.3 *BOMA, n*—Building Owners and Managers Association.

2.4.4 *BUR, n*—Built-up Roofing.

2.4.5 *EIFS, n*—Exterior Insulation and Finish System.

2.4.6 *EMF, n*—Electro Magnetic Fields.

2.4.7 *EMS, n*—Energy Management System.

2.4.8 *EUL, n*—Expected Useful Life.

2.4.9 *FEMA, n*—Federal Emergency Management Agency.

2.4.10 *FFHA, n*—Federal Fair Housing Act.

2.4.11 *FIRMS, n*—Flood Insurance Rate Maps.
2.4.12 *FOIA, n*—U.S. Freedom of Information Act (5 USC 552 et seq.)[5] and similar state statutes.
2.4.13 *FOIL*—Freedom of Information Letter.
2.4.14 *FM*—Factory Mutual.
2.4.15 *HVAC*—Heating, Ventilating and Air Conditioning.
2.4.16 *IAQ*—Indoor Air Quality.
2.4.17 *NFPA*—National Fire Protection Association.
2.4.18 *PCA*—Property Condition Assessment
2.4.19 *PCR*—Property Condition Report.
2.4.20 *PML*—Probable Maximum Loss.
2.4.21 *RTU, n*—Rooftop Unit.
2.4.22 *RUL, n*—Remaining Useful Life.
2.4.23 *STC, n*—Sound Transmission Class.

3. Significance and Use

3.1 This guide is intended for use on a voluntary basis by parties who desire to obtain a baseline PCA of commercial real estate. This guide also recognizes that there are varying levels of property condition assessment and due diligence that can be exercised that are both more and less comprehensive than this guide, and that may be appropriate to meet the objectives of the user. Users should consider their requirements, the purpose that the PCA is to serve, and their risk tolerance level before selecting the consultant and the level of due diligence to be exercised by the consultant. The user should also review or establish the qualifications, or both, of the proposed field observer and PCR reviewer prior to engagement. A PCR should identify any deviations or exceptions to this guide. Furthermore, no implication is intended that use of this guide be required in order to have conducted a property condition assessment in a commercially prudent and reasonable manner. Nevertheless, this guide is intended to reflect a reasonable approach for the preparation of a baseline PCA.

3.2 *Clarification of Use*:

3.2.1 *Specific Point in Time*—A user should only rely on the PCR for the point in time at which the consultant's observations and research were conducted.

3.2.2 *Site-Specific*—The PCA performed in accordance with this guide is site-specific in that it relates to the physical condition of real property improvements on a specific parcel of commercial real estate. Consequently, this guide does not address many additional issues in real estate transactions such as economic obsolescence, the purchase of business entities, or physical deficiencies relating to off-site conditions.

3.3 *Who May Conduct*—The walk-through survey portion of a PCA should be conducted by a field observer, and the PCR should be reviewed by a PCR reviewer; both qualified as suggested in X1.1.1.1 and X1.1.1.2, respectively.

3.4 *Principles*—The following principles are an integral part of this guide. They are intended to be referred to in resolving ambiguity, or in exercising discretion accorded the user or consultant in conducting a PCA, or in judging whether a user or consultant has conducted appropriate inquiry or has otherwise conducted an adequate PCA.

3.4.1 *Uncertainty Not Eliminated*—No PCA can wholly eliminate the uncertainty regarding the presence of physical deficiencies and the performance of a subject property's building systems. Preparation of a PCR in accordance with this guide is intended to reduce, but not eliminate, the uncertainty regarding the potential for component or system failure and to reduce the potential that such component or system may not be initially observed. This guide also recognizes the inherent subjective nature of a consultant's opinions as to such issues as workmanship, quality of original installation, and estimating the RUL of any given component or system. The guide recognizes a consultant's suggested remedy may be determined under time constraints, formed without the aid of engineering calculations, testing, exploratory probing, the removal of materials, or design. Furthermore, there may be other alternate or more appropriate schemes or methods to remedy the physical deficiency. The consultant's opinions generally are formed without detailed knowledge from those familiar with the component's or system's performance.

3.4.2 *Not Technically Exhaustive*—Appropriate due diligence according to this guide is not to be construed as technically exhaustive. There is a point at which the cost of information obtained or the time required to conduct the PCA and prepare the PCR may outweigh the usefulness of the information and, in fact, may be a material detriment to the orderly and timely completion of a commercial real estate transaction. It is the intent of this guide to attempt to identify a balance between limiting the costs and time demands inherent in performing a PCA and reducing the uncertainty about unknown physical deficiencies resulting from completing additional inquiry.

3.4.3 *Representative Observations*—The purpose of conducting representative observations is to convey to the user the expected magnitude of commonly encountered or anticipated conditions. Recommended representative observation quantities for various asset types are provided in the Annex; however, if in the field observer's opinion such representative observations as presented in the Annex are unwarranted as a result of homogeneity of the asset or other reasons deemed appropriate by the field observer, the field observer may survey sufficient units, areas, systems, buildings, etc. so as to comment with reasonable confidence as to the representative present condition of such repetitive or similar areas, systems, buildings, etc. To the extent there is more than one building on the subject property, and they are homogeneous with respect to approximate age, use, basic design, materials, and systems, it is not a requirement of this guide for the field observer to conduct a walk-through survey of each individual building's systems to describe or comment on their condition within the PCR. The descriptions and observations provided in the PCR are to be construed as representative of all similar improvements.

3.4.3.1 *User-Mandated Representative Observations*—A user may mandate the representative observations required for a given property. Such representative observations may be more or less than this guide's recommended representative observations as provided in the Annex.

3.4.3.2 *Extrapolation of Findings*—Consultant may reasonably extrapolate representative observations and findings to all typical areas or systems of the subject property for the purposes of describing such conditions within the PCR and preparing the opinions of probable costs for suggested remedy of material physical deficiencies.

3.4.4 *Level of Due Diligence is Variable*—Not every property will warrant the same level of property condition assessment. Consistent with good commercial and customary practice, the appropriate level of property condition assessment generally is guided by the type of property, the age of the improvements, the expertise and risk tolerance level of the user, and the time available for preparing the PCR and reviewing the opinions to be contained in the PCR.

3.4.5 *Comparison With a Subsequent PCR*—It should not be concluded or assumed that a previous PCR was deficient because the previous PCA did not discover a certain or particular physical deficiency, or because opinions of probable costs in the previous PCR are different. A PCR contains a representative indication of the property condition at the time of the walk-through survey and is dependent on the information available to the consultant at that time; therefore, a PCR should be evaluated on the reasonableness of judgments made at the time and under the circumstances in which they are made. Time available, hindsight, new or additional information, enhanced visibility as a result of improved weather or site conditions, equipment visibility as a result of improved weather or site conditions, equipment not in a shutdown mode, and other factors influence the PCA and the opinions contained in the PCR.

3.5 *Prior PCR Usage*—This guide recognizes that PCRs performed in accordance with this guide may include information that subsequent users and consultants may want to use to avoid duplication and to reduce cost; therefore, this guide includes procedures to assist users and consultants in determining the appropriateness of using such information. In addition to the specific procedures contained elsewhere in this guide, the following should be considered:

3.5.1 *Use of Prior PCR Information*—Information contained in prior property condition reports may be used by the consultant if, in the consultant's opinion, it is relevant; however, users and consultants are cautioned that information from prior property condition reports should only be used if such information was generated or obtained through procedures or methods that met or exceeded those contained in this guide. Such information should serve only as an aid to a consultant in fulfilling the requirements of this guide and to assist the field observer in the walk-through survey, research, and the field observer's understanding of the subject property. Furthermore, the PCR should identify the previously prepared property condition report if information from the prior report was used by the consultant in preparing the PCR.

3.5.2 *Conducting Current Walk-Through Surveys*—Except as provided in 3.5.1, prior property condition reports should not be used without verification. At a minimum, for a PCR to be consistent with this guide, a new walk-through survey, interviews, and solicitation and review of building and fire department record for recorded material violations should be performed.

3.5.3 *Actual Knowledge Exception*—If the user or consultant conducting a PCA has actual knowledge that the information from a prior property condition report is not accurate, or if it is obvious to the field observer that the information is not accurate, such information from a prior property condition report should not be used.

3.5.4 *Contractual Issues*—This guide recognizes that contractual and legal obligations may exist between prior and subsequent users of property condition reports, or consultants who prepared prior property condition reports, or both. Consideration of such contractual obligations is beyond the scope of this guide. Furthermore, prior property condition reports may have been prepared for purposes other than the current desired purpose of the PCR.

3.6 *Rules of Engagement*—The contractual and legal obligations between a consultant and a user (and other parties, if any) are outside the scope of this guide. No specific legal relationship between the consultant and the user was considered during the preparation of this guide.

4. User's Responsibilities

4.1 *Access*—User should arrange for the field observer to receive timely access, which is complete, supervised, and safe to the subject property's improvements (including roofs). In addition, access to the subject property's staff, vendors, and appropriate documents should be provided by owner, owner's representative, or made available by the user, or a combination thereof. In no event should the field observer seek access to any particular portion of the property, interview property management staff, vendors, or tenants, or review documents, if the owner, user, or occupant objects to such access or attempts to restrict the field observer from conducting any portion of the walk-through survey, research or interviews, or taking of photographs. Any conditions that significantly impede or restrict the field observer's walk-through survey or research, or the failure of the owner or occupant to provide timely access, information, or requested documentation should be timely communicated by the consultant to the user. If such conditions are not remedied, the consultant is obligated to state within the PCR all such material impediments that interfered with the conducting of the PCA in accordance with this guide.

4.2 *User Disclosure*—The user should disclose in a timely manner all appropriate information in the user's possession that may assist the consultant's efforts. The user should not withhold any pertinent information that may assist in identifying a material physical deficiency including, but not limited to, previously prepared property condition reports; any study specifically prepared on a system or component of the subject property; any knowledge of actual or purported physical deficiencies; or, any information such as costs to remedy known physical deficiencies.

5. Property Condition Assessment

5.1 *Objective*—The purpose of the PCA is to observe and report, to the extent feasible pursuant to the processes prescribed herein, on the physical condition of the subject property.

5.2 *PCA Components*—The PCA should have four components:

5.2.1 *Documentation Review and Interviews*; refer to Section 7.

5.2.2 *Walk-Through Survey*; refer to Section 8.

5.2.3 *Preparation of Opinions of Probable Costs to Remedy Physical Deficiencies*; refer to Section 9.

5.2.4 *Property Condition Report*; refer to Section 10.

5.3 *Coordination of Components*:

5.3.1 *Components Used in Concert*—The Documentation Review, Interviews and Walk-Through Survey components of this guide are interrelated in that information obtained from one component may either indicate the need for more information from another, or impact the consultant's findings, opinions, opinions of probable costs, or recommendations, or a combination thereof.

5.3.2 *Information Provided by Others*—The consultant should note in the PCR the source of information used by the consultant that were material in identifying any physical deficiencies encumbering the subject property that is not readily observed by the consultant or that supplemented the consultant's observations.

5.4 *Consultant's Duties*:

5.4.1 *Who May Conduct Portions of the PCA*—The inquiries, interviews, walk-through survey, interpretation of the information upon which the PCR is based, the preparation of opinions of probable costs, and the writing of the PCR are all tasks and portions of the PCA that may be performed by the consultant, field observer, members of the consultant's staff, or third party contractors engaged by the consultant.

5.4.2 *Responsibility for Lack of Information*—The consultant is not responsible for providing or obtaining information should the source contacted fails to respond, to respond only in part, or fails to respond in a timely fashion.

5.4.3 *Opinions of Probable Costs Contingent on Further Discovery*—The consultant is not required to provide opinions of probable costs to remedy physical deficiencies, which may require the opinions of specialty consultants or the results of testing, exploratory probing, or further research to determine the cause of the physical deficiency and the appropriate remedy, scope, and scheme for repair or replacement unless user and consultant have agreed to such an expansion of the scope of work.

5.4.4 *Representative Observations*—The field observer is not expected to survey every component of every system during a walk-through survey. For example, it is not the intent to survey every RTU, balcony, window, roof, toilet room facility, every square foot of roofing, etc. Only representative observations of such areas are to be surveyed. The concept of representative observations extends to all conditions, areas, equipment, components, systems, buildings, etc., to the extent that they are similar and representative of one another.

6. The Consultant

6.1 *Qualifications of the Consultant*—This guide recognizes that the competency of the consultant is highly dependent on many factors that may include professional education, training, experience, certification, or professional licensing/registration, of both the consultant's field observers and the PCR reviewer. It is the intent of this guide to identify factors that should be considered by the user when retaining a consultant to conduct a PCA and by the consultant in selecting the appropriate field observer and PCR reviewer. No standard can be designed to eliminate the role of professional judgment, competence, and the value and need for experience during the walk-through survey and to conduct the PCA. Consequently, the qualifications of the field observer and the PCR reviewer are critical to the performance of the PCA and the resulting PCR. This guide further recognizes that the consultant has the responsibility to select, engage, or employ the field observer and the PCR reviewer; therefore, each PCR should include as an exhibit a statement of qualifications of both the field observer and the PCR reviewer.

6.2 *Staffing of the Field Observer*—This guide recognizes that for the majority of commercial real estate subject to a PCA, the field observer assigned by the consultant to conduct the walk-through survey most likely will be a single individual having a general, well rounded knowledge of pertinent building systems and components; however, a single individual will seldom have comprehensive knowledge, expertise or experience with all building codes, building systems and asset types, which are applicable in all locales. Therefore, any decision to supplement the field observer with specialty consultants, building system mechanics, specialized service personnel, or any other specialized field observers should be a mutual decision made by the consultant and the user. This decision should be made in accordance with the requirements, risk tolerance level, and budgetary constraints of the user, the purpose the PCA is to serve, the expediency of report delivery, and the complexity of the subject property. The level of due diligence conducted during a PCA is often adjusted to the risk tolerance of the user.

6.3 *Independence of the Consultant*—This guide recognizes that the consultant normally is a person or entity, acting as an independent contractor, who has been engaged by the user to conduct a PCA. In the event the consultant, the field observer, the PCR reviewer, or members of the consultant's staff are employees of, or subsidiary of, the user, such affiliation or relationship should be disclosed in the Executive Summary of the PCR.

6.4 *Qualifications of the Field Observer*—Refer to X1.1.1 for nonmandatory guidance on the qualifications of the field observer.

6.5 *Qualifications of the PCR Reviewer*—Refer to X1.1.2 for nonmandatory guidance on the qualifications of the PCR Reviewer.

6.6 *The Field Observer and PCR Reviewer May Be a Single Individual*—The PCR reviewer also may act as the field observer and conduct the walk-through survey. In such an event, the PCR reviewer should identify such dual responsibilities and sign the PCR indicating that he or she has performed both functions.

6.7 *Not a Professional Architecture or Engineering Service*—It is not the intent of this guide that by conducting the walk-through survey or reviewing the PCR that the consultant, the field observer, or the PCR reviewer is practicing architecture or engineering. Furthermore, it is not the intent of this guide that either the PCR reviewer or the field observer, if they are an architect or engineer, must either sign or seal the PCR as an instrument of professional service or identify their signatures as being that of an architect or engineer.

7. Document Review and Interviews

7.1 *Objective*—The objective of the document review and interviews is to augment the walk-through survey and to assist the consultant's understanding of the subject property and identifying of physical deficiencies. Records or documents, if readily available, may be reviewed to specifically identify, or assist in the identification of, physical deficiencies, as well as any preceding or ongoing efforts, or costs to investigate or remediate the physical deficiencies, or a combination thereof. Such review is not to include commentary on the accuracy of such documents but merely to determine the existence of such documents.

7.2 *Reliance*—The consultant is not required to independently verify the information provided and may rely on information provided to the extent that the information appears reasonable to the consultant.

7.3 *Accuracy and Completeness*—Accuracy and completeness of information varies among information sources. The consultant is not obligated to identify mistakes or insufficiencies in the information provided; however, the consultant should make reasonable effort to compensate for mistakes or insufficiencies of information reviewed that are obvious in light of other information obtained in the process of conducting the PCA or otherwise known to the consultant.

7.4 *Government Agency Provided Information*:

7.4.1 *Documents to Be Reviewed*—Consultant is to solicit and review: base building certificate of occupancy, outstanding and recorded material building code violations, and recorded material fire code violations.

7.4.2 *Reasonably Ascertainable/Standard Government Record Sources*—Availability of record or document information varies from information source to information source, including governmental jurisdictions. The consultant should make appropriate inquiry and review only such record information that is reasonably ascertainable from standard sources. If information is not practically reviewable or not provided to the consultant in a reasonable time for the consultant to formulate an opinion and complete the PCR, such fact should be stated in the PCR, and the consultant is to have no further obligation of retrieving such documentation or reviewing it if it is subsequently provided. Nevertheless, if pursuant to the consultant's appropriate inquiry, material information is received by the consultant contemporaneous to the preparation of the PCR (within 30 days) but too late to be included in the PCR, the consultant should forward it to the user.

7.4.3 *Publicly Available Documents*—Information from a government agency, department or other source of information, which typically is reproduced and provided to the consultant upon appropriate inquiry and is reasonably ascertainable.

7.4.4 *Drawings*—Obtaining a set of drawings, which may be available publicly, is an exception to the requirement that Publicly Available Documents be provided, due to delivery and cost constraints. If readily available, such documents should be provided and identified to the consultant by the owner, owner's representative, or user as construction, as-built, or other design/construction documents. Nonetheless, the review of drawings of the subject property is not a requirement of this guide. Drawings may serve as an aid to the consultant in describing the subject property's improvements, in developing quantities for opinions of probable costs for suggested remedies of physical deficiencies, and to assist in preparing brief descriptions of the subject property's major systems.

7.4.5 *Reasonable Time and Cost*—It is the intent of this guide that information will be provided to the consultant within ten business days of the source receiving appropriate inquiry, without an in-person request by the consultant being required, and at no more than a nominal cost to cover the source's cost of retrieving and duplicating the information. Generally, an in-person request by the consultant is not required. However, this is not to preclude the consultant from personally researching such files if, in the opinion of the consultant, this could be reasonably accomplished at the time of the site visit.

7.5 *Pre-Survey Questionnaire*—The consultant may provide the owner or owner's representative, or both, with a pre-survey questionnaire (the questionnaire). Such a questionnaire, complete with the owner's or owner's representative's responses, should be included as an exhibit within the PCR unless directed otherwise by user.

7.6 *Owner/User Provided Documentation and Information*—If readily available, the consultant should review the following documents and information that may be in the possession of or provided by the owner, owner's representative, user, or combination thereof, as appropriate. Such information also could aid in the consultant's knowledge of the subject property's physical improvements, extent and type of use, or assist in identifying material discrepancies between reported information and observed conditions, or a combination thereof. The consultant's review of documents submitted should not include commenting on the accuracy of such documents or their preparation, methodology, or protocol. If the consultant discovers a significant discrepancy, it should be disclosed within the PCR.

7.6.1 Appraisal, either current or previously prepared.

7.6.2 Certificate of Occupancy.

7.6.3 Safety inspection records.

7.6.4 Warranty information (roofs, boilers, chillers, cooling towers, etc.).

7.6.5 Records indicating the age of material building systems such as roofing, paving, plumbing, heating, air conditioning, electrical, etc.

7.6.6 Historical costs incurred for repairs, improvements, recurring replacements, etc.

7.6.7 Pending proposals or executed contracts for material repairs or improvements. Descriptions of future work planned.

7.6.8 Outstanding citations for building, fire and zoning code violations.

7.6.9 The ADA survey and status of any improvements implemented to effect physical compliance.

7.6.10 Previously prepared property condition reports or studies pertaining to any aspect of the subject property's physical condition.

7.6.11 Records indicating building occupancy percentage.

7.6.12 Records indicating building turnover percentage.

7.6.13 Building rent roll.

7.6.14 Leasing literature, listing for sale, marketing/promotional literature such as photographs, descriptive information, reduced floor plans, etc.

7.6.15 Drawings and specifications (as-built or construction).

7.7 *Interviews*—Prior to the site visit, the consultant should ask the owner or user to identify a person or persons knowledgeable of the physical characteristics, maintenance, and repair of the property. If a property manager or agent of the owner is identified, the consultant should contact such individual so as to inquire about the subject property's historical repairs and replacements and their costs, level of preventive maintenance exercised, pending repairs and improvements, frequency of repairs and replacements, and existence of ongoing or pending litigation related to subject property's physical condition. In connection with the consultant's research or walk-through survey, the consultant also may question others who are knowledgeable of the subject property's physical condition and operation. It is within the discretion of the consultant to decide which questions to ask before, during, or after the site visit.

7.7.1 *Reliance*—The consultant may rely on the information obtained as a result of the interviews, provided that in the consultant's opinion such information appears to be reasonable.

7.7.2 *Method*—Questions to be asked pursuant to this section are at the discretion of the consultant and may be asked in person, by telephone, or in writing.

7.7.3 *Incomplete Answers*—While the consultant should make inquiries in accordance with this section, the persons to whom the questions are addressed may have no obligation to cooperate. Should the owner or the property manager, building/facility engineer, or maintenance supervisor not be available for an interview, whether by intent or inconvenience, or not respond in full or in part to questions posed by the consultant, the consultant should disclose such within the PCR. Furthermore, should any party not grant such authorization to interview, restrict such authorization, or should the person to whom the questions are addressed not be knowledgeable about the subject property, this should be disclosed within the PCR.

8. Walk-Through Survey

8.1 *Objective*—The objective of the walk-through survey is to visually observe the subject property so as to obtain information on material systems and components for the purposes of providing a brief description, identifying physical deficiencies to the extent that they are observable, and obtaining information needed to address such issues in the PCR as outlined within Section 10.

8.2 *Frequency*—It is not expected that more than one site visit to the subject property be made by the field observer in connection with a PCA. The site visit constituting part of the PCA is referred to as the walk-through survey.

8.3 *Photographs*—Consultant should document representative conditions with photographs and use reasonable efforts to document typical conditions present including material physical deficiencies, if any. Photographs should include as a minimum: front and typical elevations and exteriors, site work, parking areas, roofing, structural systems, plumbing, HVAC systems, and electrical systems, conveyance systems, life safety systems, representative interiors, and any special or unusual conditions present.

8.4 *Scope*—During the site visit, the field observer should observe the general physical condition of the subject property, observe material systems and components, and identify material physical deficiencies and any unusual features or inadequacies observed or reported by conducting specific or representative observations, as appropriate. Testing, measuring, or preparing calculations for any system or component to determine adequacy, capacity, or compliance with any standard is outside the scope of this guide. The listing of specific items of the material building systems and components to be observed, which are presented in the succeeding subsections, should not be considered all-inclusive, and the consultant should utilize professional judgment regarding adding or deleting subsections as necessary to complete the PCR. Similarly, subsections identified as "out of scope issues" are provided for clarification and should not be construed as all-inclusive.

8.4.1 *Site*:

8.4.1.1 *Topography*—Observe the general topography and any unusual or problematic features or conditions.

8.4.1.2 *Storm Water Drainage*—Observe the storm water collection and drainage system and note the presence of on-site surface waters, and retention or detention basins.

8.4.1.3 *Ingress and Egress*—Observe the major means of ingress and egress.

8.4.1.4 *Paving, Curbing, and Parking*—Observe the material paving and curbing systems. Identify the types of parking, that is, garage, surface, subsurface, etc., the number and types of parking spaces, and any reported parking inadequacies. Note the source of the preceding information.

8.4.1.5 *Flatwork*—Observe sidewalks, plazas, patios, etc.

8.4.1.6 *Landscaping and Appurtenances*—Observe landscaping (trees, shrubs, lawns, fences, retaining walls, etc.) and material site appurtenances (irrigation systems, fountains, lighting, signage, ponds, etc.).

8.4.1.7 *Recreational Facilities*—Observe on-site recreational facilities.

8.4.1.8 *Utilities*:

(1) Observations—Identify type and provider of the material utilities provided to the property (water, electricity, natural gas, etc.).

(2) Special Utility Systems—Identify the presence of any material special on-site utility systems such as water or wastewater treatment systems, special power generation systems, etc. Identify material system information, such as system type, manufacturer, system capacity, system age, system operator, etc.

(3) Out of Scope Issues—Operating conditions of any systems or accessing manholes or utility pits.

8.4.2 *Structural Frame and Building Envelope*:

8.4.2.1 *Observations*—Identify the material buildings, including parking structures, on the subject property, and identify the type of structure for each. Observe the building substructure, including the foundation system (noting the presence of cellars, basements, or crawl spaces), building's superstructure or structural frame (floor framing system and roof framing system), building envelope including facades or curtain wall system, glazing system, exterior sealants, exterior balconies, doors, stairways, parapets, etc. Observations of the building's exterior generally are to be limited to vantage points that are on-grade or from readily accessible balconies or rooftops.

8.4.2.2 *Out of Scope Issues*—Entering of crawl or confined

space areas (however, the field observer should observe conditions to the extent easily visible from the point of access to the crawl or confined space areas), determination of previous substructure flooding or water penetration unless easily visible or if such information is provided.

8.4.3 *Roofing*:

8.4.3.1 *Observations*—Identify and observe the material roof systems (exposed membrane and flashings) including, parapets, slope, drainage, etc. Observe for evidence and/or the need for material repairs, evidence of significant ponding, or evidence of roof leaks. Inquire as to the age of the material roofing system(s) and whether a roof warranty or bond is reported to be in effect.

8.4.3.2 *Out of Scope Issues*—Walking on pitched roofs, or any roof areas that appear to be unsafe, or roofs with no built-in access, or determining any roofing design criteria.

8.4.4 *Plumbing*:

8.4.4.1 *Observations*—Identify and observe the material plumbing systems including piping (sanitary, storm and supply water), fixtures, domestic hot water production, and note any special or unusual plumbing systems.

8.4.4.2 *Out of Scope Issues*—Determining adequate pressure and flow rate, fixture-unit values and counts, verifying pipe sizes, or verifying the point of discharge for underground systems.

8.4.5 *Heating*:

8.4.5.1 *Observations*—Identify the material heat generating and distribution systems, and the apparent or reported age of the equipment, past material component replacements/upgrades, the apparent level of maintenance exercised, and whether a maintenance contract is reported to be in place. If heating equipment is shutdown or not operational at the time of the walk-through survey, provide an opinion of the condition to the extent observed. Also, observe any special or unusual heating systems or equipment present, such as solar heat. Identify in general terms reported material tenant-owned systems that are outside the scope of the PCA.

8.4.5.2 *Out of Scope Issues*—Observation of flue connections, interiors of chimneys, flues or boiler stacks, or tenant-owned or maintained equipment.

8.4.6 *Air Conditioning and Ventilation*:

8.4.6.1 *Observations*—Identify the material air-conditioning and ventilation systems including cooling towers, chillers (include type of reported refrigerant used), package units, split systems, air handlers, thermal storage equipment, material distribution systems, etc. Identify the apparent or reported age of the material equipment, past material component upgrades/replacements, apparent level of preventive maintenance exercised, and whether a maintenance contract is reported to be in place. If air conditioning and ventilation systems are shutdown or not operational at the time of the walk-through survey, provide an opinion of the condition to the extent observable. Identify any special or unusual air conditioning and ventilation systems or equipment, such as refrigeration equipment for ice skating rinks, cold storage systems, special computer cooling equipment, etc. Identify in general terms reported material tenant-owned systems that are outside the scope of the PCA.

8.4.6.2 *Out of Scope Issues*—Process related equipment or condition of tenant owned/maintained equipment.

8.4.7 *Electrical*:

8.4.7.1 *Observations*—Identify the electrical service provided and observe the electrical distribution system including distribution panels, transformers, meters, emergency generators, general lighting systems, and other such equipment or systems. Observe general electrical items, such as distribution panels, type of wiring, energy management systems, emergency power, lighting protection, etc. Identify any observed or reported special or unusual electrical equipment, systems, or devices at the subject property.

8.4.7.2 *Out of Scope Issues*—Removing of electrical panel and device covers, except if removed by building staff, EMF issues, electrical testing, or operating of any electrical devices. Process related equipment or tenant owned equipment.

8.4.8 *Vertical Transportation*:

8.4.8.1 *Observations*—Identify equipment type, number of cabs/escalators, capacity, etc. Observe elevator cabs, finishes, call and communication equipment, etc. Identify whether a maintenance contract is reported to be in place, and if so, identify the service contractor.

8.4.8.2 *Out of Scope Issues*—Examining of cables, sheaves, controllers, motors, inspection tags, or entering elevator/escalator pits or shafts.

8.4.9 *Life Safety/Fire Protection*:

8.4.9.1 *Observations*—Identify and observe life safety and fire protection systems, including sprinklers and standpipes (wet or dry, or both), fire hydrants, fire alarm systems, water storage, smoke detectors, fire extinguishers, emergency lighting, stairwell pressurization, smoke evacuation, etc.

8.4.9.2 *Out of Scope Issues*—Determining NFPA hazard classifications, classifying, or testing fire rating of assemblies.

8.4.10 *Interior Elements*:

8.4.10.1 *Observations*—Observe typical common areas including, but not limited to, lobbies, corridors, assembly areas, and restrooms. Identify and observe typical finishes, that is, flooring, ceilings, walls, etc., and material building amenities or special features, that is spas, fountains, clubs, shops, restaurants, etc.

8.4.10.2 *Out of Scope Issues*—Operating appliances or fixtures, determining or reporting STC (Sound Transmission Class) ratings, and flammability issues/regulations.

8.5 *Additional Considerations*—There may be additional issues or conditions at a property that users may wish to assess in connection with commercial real estate that are outside the scope of this guide (Out of Scope considerations).

8.5.1 *Outside Standard Practices*—Whether or not a user elects to inquire into non-scope considerations in connection with this guide or any other PCA is not required for compliance by this guide.

8.5.2 *Other Standards*—Other standards or protocols for assessment of conditions associated with non-scope conditions may have been developed by governmental entities, professional organizations, or other private entities.

8.5.3 *Additional Issues*—Following are several non-scope considerations that users may want to assess in connection with commercial real estate. No implication is intended as to the relative importance of inquiry into such non-scope considerations, and this list of non-scope considerations is not intended to be all-inclusive:

8.5.3.1 Seismic Considerations,

8.5.3.2 Design Consideration for Natural Disasters (Hurricanes, Tornadoes, High Winds, Floods, Snow, etc.),

8.5.3.3 Insect/Rodent Infestation,

8.5.3.4 Environmental Considerations,

8.5.3.5 ADA Requirements,[2]

8.5.3.6 FFHA Requirements,[3]

8.5.3.7 Indoor Air Quality, and

8.5.3.8 Property Security Systems.

9. Opinions of Probable Costs to Remedy Physical Deficiencies

9.1 *Purpose*—Based upon the walk-through survey and information obtained in accordance with following this guide, general-scope opinions of probable costs are to be prepared for the suggested remedy of the material physical deficiencies observed. These opinions of probable costs are to assist the user in developing a general understanding of the physical condition of the subject property.

9.2 *Scope*—Opinions of probable costs should be provided for material physical deficiencies and not for repairs or improvements that could be classified as: *(1)* cosmetic or decorative; *(2)* part or parcel of a building renovation program or tenant improvements/finishes; *(3)* enhancements to reposition the subject property in the marketplace; *(4)* for warranty transfer purposes; or *(5)* routine or normal preventive maintenance, or a combination thereof.

9.3 *Opinions of Probable Costs Attributes*:

9.3.1 *Threshold Amount for Opinions of Probable Costs*—It is the intent of this guide that the material physical deficiencies observed and the corresponding opinions of probable costs *(1)* be commensurate with the complexity of the subject property; *(2)* not be minor or insignificant; and *(3)* serve the purpose of the user in accordance with the user's risk tolerance level. Opinions of probable costs that are either individually or in the aggregate less than a threshold amount of $3,000 for like items are to be omitted from the PCR[4]. If there are more than four separate items that are below this threshold requirement, but collectively total over $10,000, such items should be included. The user may adjust this cost threshold amount provided that this is disclosed within the PCR's Executive Summary under the heading "Deviations from the Guide."

9.3.2 *Actual Costs May Vary*—Opinions of probable costs should only be construed as preliminary budgets. Actual costs most probably will vary from the consultant's opinions of probable costs depending on such matters as type and design of suggested remedy, quality of materials and installation, manufacturer and type of equipment or system selected, field conditions, whether a physical deficiency is repaired or replaced in whole, phasing of the work (if applicable), quality of contractor, quality of project management exercised, market conditions, and whether competitive pricing is solicited, etc.

9.3.3 *Extrapolation of Representative Observations*—Opinions of probable costs may be based upon: the extrapolation of representative observations, conditions deemed by the consultant as highly probable, results from information received, or the commonly encountered EULs or RULs of the components or systems, or a combination thereof.

9.3.4 *Estimating of Quantities*—It is not the intent of this guide that the consultant is to prepare or provide exact quantities or identify the exact locations of items or systems as a basis for preparing the opinions of probable costs.

9.3.5 *Basis of Costs*—The source of cost information utilized by the consultant may be from one or more of the following resources: *(1)* user provided unit costs; *(2)* owner's historical experience costs; *(3)* consultant's cost database or cost files; *(4)* commercially available cost information such as published commercial data; *(5)* third party cost information from contractors, vendors, or suppliers; or *(6)* other qualified sources that the consultant determines appropriate. Opinions of probable costs should be provided with approximate quantities, units, and unit costs by line item. If in the reasonable opinion of the consultant, a physical deficiency is too complex or difficult to develop an opinion of probable cost using the quantity and unit cost method, the consultant may apply a lump sum opinion of probable costs for that particular line item. Opinions of probable costs should be limited to construction-related costs; those types of costs that commonly are provided by contractors who perform the work. Business related, design, management fees, and other indirect costs should be excluded.

9.3.6 *Costs for Additional Study*—For some physical deficiencies, determining the appropriate suggested remedy or scope may warrant further study/research or design, testing, exploratory probing, and exploration of various repair schemes, or a combination thereof, all of which are outside the scope of this guide. In these instances, the opinions of probable costs for additional study should be provided.

9.3.7 *Cost Segregation*—Opinions of probable costs should be segregated within the PCR into the categories of immediate costs and short-term costs.

10. Property Condition Report

10.1 *Format*—This guide does not present a specific PCR format to be followed. This should be determined between the user and the consultant; however, the PCR may follow the Recommended Table of Contents provided as Fig. X1.1.

10.2 *Writing Protocols*:

10.2.1 *Suggested Remedy*—For each material physical deficiency, the consultant should provide a suggested remedy, which may include recommending further research or testing, or both, if appropriate in the consultant's opinion.

10.2.2 *Significance of Physical Deficiency*—If the significance of the physical deficiency is not readily discernable, the consultant should explain its significance in a simple manner meaningful to a user. For example, stating that "the subject property has aluminum distribution wiring" may be insufficient

[2] It is recognized that most PCAs include some level of assessment of ADA compliance. This guide recognizes that there are numerous acceptable levels of ADA assessment that can be conducted as part of the PCA. The level of assessment is mutually agreed upon by the user and the consultant.

[3] This guide recognizes that some PCAs for residential properties include some level of assessment of FFHA compliance. The level of assessment is mutually agreed upon by the user and the consultant.

[4] This guide recognizes that most PCRs include material life-safety/building code violations regardless of cost.

to the user, since this statement reveals nothing about the significance of this condition.

10.2.3 *Disclosure of Information Source*—The consultant should differentiate between material information obtained by the field observer's actual knowledge and that reported or provided by others or obtained from documents provided. The source of such nonobserved material information should be reported.

10.2.4 *Representative Description and Observed Conditions*—The PCR's descriptions of systems and components and their general physical condition may be based upon extrapolations of representative observations conducted by the field observer during the walk-through survey.

10.3 *Documentation*—The PCR should include pertinent documentation such as photographs, copies of material building department records, material fire department records, and building code violation notices to the extent reasonable, certificates of occupancy, copies of repair cost documentation submitted by owner or owner's representative, contractors or agents, and tenant schedules (if applicable). All photographs should be numbered and labeled or correlated to the PCR's text.

10.4 *Credentials*—The PCR should name the field observer and the PCR reviewer and should include their statement of qualifications.

10.5 *Executive Summary*—The general content for the PCR Executive Summary is discussed in this section.

10.5.1 *General Description*—The opening paragraph should indicate that this is a PCR, identify the subject property, and provide pertinent information such as use, size, age, location, construction type, design style, and apparent occupancy status. Also identify the name of the consultant that prepared the PCR, the name of the user, the user's position with respect to the subject property (if known), the purpose the PCR is to serve (if known), and the date of the site visit.

10.5.2 *General Physical Condition*—In 10.5.1, the subject property's general physical condition, the apparent level of preventive maintenance exercised, and any significant deferred maintenance is summarized. A schedule of material physical deficiencies, any significant capital improvements that are pending, in-progress, or were recently implemented, and any significant findings resulting from research should be provided[5].

10.5.3 *Opinions of Probable Costs*—Present the aggregate sum of opinions of probable costs segregated between immediate and short-term costs.

10.5.4 *Deviations from this Guide*—Present all material deviations and deletions from this guide, if any, listed individually along with all additional consultant services that have exceeded this guide's suggested requirements.

10.5.5 *Consultant/Field Observer Relationship*—If the consultant or field observer, or both are not at arms-length with the user, such a relationship should be disclosed.

10.5.6 *Recommendations/Discussions*—Briefly identify those components and systems necessitating further study, research, testing, or exploratory probing. This section also may be used to discuss any obvious major deviations from the subject property description provided by the user to the consultant, ongoing repairs or improvements, or other relevant issues.

10.6 *Purpose and Scope*:

10.6.1 *Purpose*—Provide a short paragraph specifically stating the purpose the PCR should serve, if disclosed by user. If the user does not disclose the purpose of the PCR to the consultant, the PCR should so state.

10.6.2 *Scope*—Identify the improvements that comprise the subject property. Provide an outline of the scope of work completed for the PCA and methods utilized. Should either the PCA or the PCR materially deviate from this guide or if there were any constraints preventing the consultant from performing the PCA in accordance with this guide, these constraints should be identified.

10.7 *Walk-Through Survey*—Provide the information that is outlined in Section 8. Such information should include a brief description of each system or component and observed physical deficiencies, if any. Both the brief descriptions and the observed physical deficiencies may be based upon the field observer's representative observations. A general description of material tenant-owned equipment that is outside the scope of the PCA should be provided in this section.

10.8 *Document Reviews and Interviews*—Identify any material information relating to physical deficiencies of the subject property resulting from the review of documents and interviews conducted.

10.9 *Additional Considerations*—Identify any material additional considerations or Out of Scope considerations which are included in the PCR.

10.10 *Opinions of Probable Costs*—Identify the material physical deficiencies and provide suggested remedies, complete with opinions of probable costs.

10.11 *Qualifications*—Both the field observer's and the PCR reviewer's statement of qualifications should be provided.

10.12 *Limiting Conditions*—Provide all limiting conditions of the PCR.

10.13 *Exhibits*:

10.13.1 Representative photographs (may be correlated directly into text or numbered and labeled in exhibit).

10.13.2 Questionnaire.

10.13.3 User/owner submitted documents.

10.13.4 Photocopied plot plans, sketches, etc.

10.13.5 Other exhibits considered appropriate by the consultant.

11. Out of Scope Considerations

11.1 *Activity Exclusions*—The activities listed below generally are excluded from or otherwise represent limitations to the scope of a PCA prepared in accordance with this guide. These should not be construed as all-inclusive or imply that any exclusion not specifically identified is a PCA requirement under this guide.

11.1.1 Removing or relocating materials, furniture, storage containers, personal effects, debris material or finishes; conducting exploratory probing or testing; dismantling or operating of equipment or appliances; or disturbing personal items or property, that obstructs access or visibility.

[5] This should include material life-safety/building code violations.

11.1.2 Preparing engineering calculations (civil, structural, mechanical, electrical, etc.) to determine any system's, component's, or equipment's adequacy or compliance with any specific or commonly accepted design requirements or building codes, or preparing designs or specifications to remedy any physical deficiency.

11.1.3 Taking measurements or quantities to establish or confirm any information or representations provided by the owner or user, such as size and dimensions of the subject property or subject building; any legal encumbrances, such as easements; dwelling unit count and mix; building property line setbacks or elevations; number and size of parking spaces; etc.

11.1.4 Reporting on the presence or absence of pests such as wood damaging organisms, rodents, or insects unless evidence of such presence is readily apparent during the course of the field observer's walk-through survey or such information is provided to the consultant by the owner, user, property manager, etc. The consultant is not required to provide a suggested remedy for treatment or remediation, determine the extent of infestation, nor provide opinions of probable costs for treatment or remediation of any deterioration that may have resulted.

11.1.5 Reporting on the condition of subterranean conditions, such as underground utilities, separate sewage disposal systems, wells; systems that are either considered process-related or peculiar to a specific tenancy or use; waste-water treatment plants; or items or systems that are not permanently installed.

11.1.6 Entering or accessing any area of the premises deemed to pose a threat of dangerous or adverse conditions with respect to the field observer or to perform any procedure, that may damage or impair the physical integrity of the property, any system, or component.

11.1.7 Providing an opinion on the condition of any system or component, that is shutdown, or whose operation by the field observer may increase significantly the registered electrical demand-load; however, the consultant is to provide an opinion of its physical condition to the extent reasonably possible considering its age, obvious condition, manufacturer, etc.

11.1.8 Evaluating acoustical or insulating characteristics of systems or components.

11.1.9 Providing an opinion on matters regarding security of the subject property and protection of its occupants or users from unauthorized access.

11.1.10 Operating or witnessing the operation of lighting or other systems typically controlled by time clocks or that are normally operated by the building's operation staff or service companies.

11.1.11 Providing an environmental assessment or opinion on the presence of any environmental issues such as asbestos, hazardous wastes, toxic materials, the location and presence of designated wetlands, IAQ, etc.

11.2 *Warranty, Guarantee, and Code Compliance Exclusions*—By conducting a PCA and preparing a PCR, the consultant merely is providing an opinion and does not warrant or guarantee the present or future condition of the subject property, nor may the PCA be construed as either a warranty or guarantee of any of the following:

11.2.1 Any system's or component's physical condition or use, nor is a PCA to be construed as substituting for any system's or equipment's warranty transfer inspection;

11.2.2 Compliance with any federal, state, or local statute, ordinance, rule or regulation including, but not limited to, building codes, safety codes, environmental regulations, health codes or zoning ordinances or compliance with trade/design standards or the standards developed by the insurance industry; however, should there be any conspicuous material present violations observed or reported based upon actual knowledge of the field observer or the PCR reviewer, they should be identified in the PCR;

11.2.3 Compliance of any material, equipment, or system with any certification or actuation rate program, vendor's or manufacturer's warranty provisions, or provisions established by any standards that are related to insurance industry acceptance/approval, such as FM, State Board of Fire Underwriters, etc.

11.3 *Additional/General Considerations*:

11.3.1 *Further Inquiry*—There may be physical condition issues or certain physical improvements at the subject property that the parties may wish to assess in connection with a commercial real estate transaction that are outside the scope of this guide. Such issues are referred to as non-scope considerations and if included in the PCR, should be identified under 10.9.

11.3.2 *Out of Scope Considerations*—Whether or not a user elects to inquire into non-scope considerations in connection with this guide is a decision to be made by the user. No assessment of such non-scope considerations is required for a PCA to be conducted in compliance with this guide.

11.3.3 *Other Standards*—There may be standards or protocols for the discovery or assessment of physical deficiencies associated with non-scope considerations developed by government entities, professional organizations, or private entities, or a combination thereof.

ANNEX

(Mandatory Information)

A1. SPECIFIC PROPERTY TYPES

INTRODUCTION

This Annex is to be used to supplement or complement previous sections of this guide for various asset types as if integral to the preceding sections.

A1.1 *Multi-Family Properties*:

A1.1.1 *Representative Observations*—For complexes with multiple buildings, the exterior envelopes of all residential buildings should be surveyed. For complexes built in phases, each construction phase should be surveyed. Representative observations of the interiors should include a mix of units, which are occupied, vacant, damaged, and under renovation or repair. Representative observations of the interiors of each construction phase should include a sufficient number of the top and bottom floors. If not specified in the agreement between consultant and user, the number of units, buildings, and components surveyed in each construction phase should be sufficient to allow the field observer to develop an opinion with reasonable confidence regarding the present condition of the building systems and should be determined using the professional judgment and experience of the field observer at the time of the walk-through survey. The number of reported units that are not available for occupancy and the reported reasons they are not available should be included in the PCR. The PCR should contain the consultant's rationale for determining the number of units surveyed and for selecting the units that are surveyed. In addition, the PCR should disclose the specific units surveyed.

A1.1.2 *Patios and Balconies*—The field observer should conduct representative observations of patios, balconies, enclosures, railings, etc., and report on the observed condition.

A1.1.3 *Plumbing*—The field observer should identify the type of supply piping material (to the extent that it is easily visible) and note any replacement or historical breakage reported. General conditions and historical leakage of other systems and apparent causes should also be discussed in the PCR.

A1.1.4 *Electrical*—The field observer should note the size of the electrical service serving representative units, and whether units are individually metered. To the extent readily accessible and easily visible, which for purposes of this clause includes the removal of switch or outlet covers by building staff for representative observations, the type of distribution wiring for 120-volt circuits should be provided in the PCR. If aluminum wiring is observed, the presence or absence of properly rated connection devices should be noted.

A1.1.5 *Attic*—The presence or lack of an attic should be specifically addressed. If the attic(s) is readily accessible, the field observer should note observations such as means of access, ventilation, evidence of water leakage, daylight entering through defects, the amount and type of insulation and the presence and condition of draft stops.

A1.1.6 *Roof Sheathing*—The field observer should note Fire Resistant Treated (FRT) plywood, if observed.

A1.1.7 *Interviews*—For multi-family properties, residential occupants do not need to be interviewed unless appropriate and with the consent of the owner or user. If the subject property also has nonresidential uses and the owner or user provides authorization, the field observer should interview nonresidential occupants in accordance with this guide.

A1.2 *Commercial Office Buildings*:

A1.2.1 *Representative Observations*—For complexes of buildings built in phases, each construction phase should be surveyed. For a subject property that contains a complex of multiple buildings, the concept of representative observations extends to each building individually and not to all buildings as a whole. Representative observations should include a mix of tenant (occupied and unoccupied) and common areas. Representative observations of the interiors of each construction phase should include a sufficient number of the top and bottom floors. If not specified in the agreement between consultant and user, the quantity of floor area and the number of components, and systems surveyed in each construction phase should be sufficient to allow the field observer to develop an opinion with reasonable confidence regarding the present condition of the building systems. Such representative observations should be determined using the professional judgment and experience of the field observer at the time of the walk-through survey. The quantity of reported floor area, which is not available for occupancy and the reasons it is reportedly not available, should be included in the PCR. The PCR should contain the consultant's rationale for determining the quantity of floor area surveyed and for selecting the specific floors, that were surveyed.

A1.3 *Retail Buildings*:

A1.3.1 *General Exclusions*—The consultant is not required to survey the interior condition of shell-finish tenancies or the interior/base building conditions of anchor stores, unless included in the subject property.

A1.3.2 *Representative Observations*—For complexes of buildings built in phases, each construction phase should be surveyed. For a subject property that contains a complex of multiple buildings, the concept of representative observations extends to each building individually and not to all buildings as a whole. Representative observations should include a mix of tenant (occupied and unoccupied) and common areas. Representative observations of the interiors of each construction phase should include a sufficient number of the top and bottom floors. If not specified in the agreement between consultant and user, the quantity of floor area and the number of components, and systems surveyed in each construction phase should be sufficient to allow the field observer to develop an opinion with confidence as to the present condition of the building systems. Such representative observations should be determined using the professional judgment and experience of the field observer at the time of the walk-through survey. The quantity of reported floor area, which is not available for occupancy and the reasons it is reportedly not available, should be included in the PCR. The PCR should contain the consultant's rationale for determining the quantity of floor area surveyed and for selecting the specific floors, that were surveyed.

A1.3.3 *Interviews*—The field observer should interview proprietors or store managers of the tenant spaces surveyed as to the subject property's general condition in addition to other knowledgeable persons identified by the owner or user as described in 7.7. The consultant should use discretion and should not disclose the purpose of the PCA to tenants unless the user grants permission. This guide recognizes that there is no obligation for the proprietors or store managers to cooperate.

A1.3.4 *Roofing*—In addition to the observations made of the main roofs of buildings, a description and observed condition of canopy roofs, viewed either from the main roof or, if appropriate, from the ground should be reported along with any generally observed physical deficiencies with the parapets, canopies, soffit, or fascia system.

A1.3.5 *Flatwork*—Loading dock areas, if any, should be observed along with the condition of any flatwork, such as the loading dock platform, loading dock exterior stairs, and concrete trailer pads.

APPENDIX

(Nonmandatory Information)

X1. GUIDANCE AND ENHANCED DUE DILIGENCE SERVICES

INTRODUCTION

The information presented in this Appendix is not necessary for completing a baseline PCA pursuant to this guide; however, a user and consultant may wish to utilize some or all of the information presented in this Appendix to increase or supplement the extent of due diligence to be exercised by the consultant.

X1.1 Qualifications—This guide recognizes that the quality of a PCR is highly dependent on the qualifications of the field observer and PCR reviewer. These qualifications include such factors as experience, education, training, certification, and professional registration/licensure in architecture or engineering. Additionally, this guide recognizes that appropriate qualification levels may vary for different PCAs depending on such factors as asset type and scope (size, age, complexity, etc.) as well as the specific needs, purpose the PCR is to serve, and risk tolerance level of the user.

X1.1.1 *Qualifications of the Field Observer*—The field observer is the person or entity engaged by the consultant to perform the walk-through survey; the field observer also may be the PCR reviewer. The consultant should establish the qualifications of the field observer, but as the accuracy and completeness of the walk-through survey will determine the quality of the PCR, the consultant should carefully consider education, training and experience when selecting the field observer.

X1.1.1.1 Due to the scope or complexity of the subject property or the purpose of the PCA, the user may direct the consultant to augment the field observer with specialty consultants, or the user may define the level of qualifications of the field observer.

X1.1.1.2 The field observer, as a representative of the consultant, should be identified in the PCR. As required by 6.1, the statement of qualifications of the field observer should be included in the PCR.

X1.1.2 *Qualifications of the PCR Reviewer*—The PCR reviewer is the qualified individual designated to exercise responsible control over the field observer on behalf of the consultant and to review the PCR. This guide recognizes that the consultant is ultimately responsible for the PCA process.

X1.1.2.1 As indicated in the main body of the guide, all PCRs prepared in accordance with this guide should be reviewed and signed by the PCR reviewer. In addition, as required by 6.1, the statement of qualifications of the PCR reviewer should be included in the PCR.

X1.1.2.2 It is recommended that the user consider a PCR

reviewer who possesses a professional designation in architecture or engineering, or appropriate experience and/or certifications in the construction fields. The PCR reviewer should have experience commensurate with the subject property type and scope (size, complexity, etc.), and experience in the preparation of PCRs. Generally, professional architecture or engineering licensure/registration, and/or certifications, education, or appropriate construction experience related to these disciplines are recognized as acceptable qualifications for reviewing PCRs. However, the user and the consultant may mutually agree to define qualifications for the PCR reviewer, that may depend on the specific experience of the PCR reviewer and the scope of the subject property.

X1.2 Verification of Measurements and Quantities:

X1.2.1 *Parking Spaces*:

X1.2.1.1 *Based Upon Review of Drawings*—The field observer should review the subject property's submitted as-built site drawings and survey for the purposes of identifying the number of parking spaces provided and compare to an actual field count.

X1.2.1.2 *Actual Field Count*—The field observer should physically count each delineated parking space that has been provided for the subject property.

X1.2.2 *Count of Multi-Family Units*:

X1.2.2.1 *Based Upon Review of Drawings, Schedules, etc.*—The field observer should review documents submitted by the owner to determine the number of multi-family dwelling units.

X1.2.2.2 *Actual Field Count*—The field observer should physically count each dwelling unit. This implies that a walk-through survey of each building and the floor of each building should be conducted by the field observer.

X1.2.3 *Building Areas*:

X1.2.3.1 *Gross Areas:*

(1) Based Upon Review of As-Built Drawings—The consultant should review as-built drawings submitted by the owner to determine the gross building area on a floor-by-floor basis. Such review of drawings may consist of a review of schedules or dimensions. For purposes of this clause, gross building area should be that definition as required by the local zoning board at the time of the building's construction and as presented on the drawings' zoning schedule, if any.

(2) Actual Field Measurement—The field observer should take measurements and prepare calculations physically to determine gross area. Current BOMA definition of gross area is to be used unless the user provides the consultant with an alternate definition/protocol for the method of calculating such areas. The consultant should state the criteria under which the calculations are prepared and submit all quantities on a per floor basis.

X1.2.3.2 *Net Useable Areas:*

(1) Based Upon Review of As-Built Drawings—The user should provide the consultant with a set of as-built drawings for all space available for lease. Based solely on such drawings, the consultant should determine useable area by use of digitizer or other means. Current BOMA definition of usable area is to be used unless user provides the consultant with an alternate definition/protocol for the method of calculating such areas.

(2) Actual Field Measurement—The field observer should take measurements and prepare calculations physically to determine useable area. Current BOMA definition of useable area is to be used unless the user provides the consultant with an alternate definition/protocol for the method of calculating such areas. The consultant should state the criteria under which the calculations are prepared and submit all quantities on a per floor basis.

X1.3 *Research*:

X1.3.1 *Service Companies*—The consultant should conduct an appropriate inquiry of the subject building's major plumbing, HVAC, fire protection, electrical and elevator service companies, if any, to inquire of the system's or equipment's general condition; the extent of major or chronic repairs and replacements; pending repairs and replacements; and, outstanding proposals to provide repairs and replacements, etc. Within the PCR, the consultant should provide the name of the parties contacted and pertinent information received.

X1.4 *Flood Plain Designation*—Note whether the property encroaches upon a 100-year flood area designated as "Special Flood Hazard Areas Inundated by 100-year Flood" on FEMA maps, as amended.

X1.5 *Recommended Table of Contents*—The sample Table of Contents (see Fig. X1.1) is a suggestion and may be tailored by a user or consultant for the subject property's attributes or their particular PCR protocol or requirements.

FIG. X1.1 Sample Table of Contents

Designation: E 2060 – 00

Standard Guide for Use of Coal Combustion Products for Solidification/Stabilization of Inorganic Wastes[1]

This standard is issued under the fixed designation E 2060; the number immediately following the designation indicates the year of original adoption or, in the case of revision, the year of last revision. A number in parentheses indicates the year of last reapproval. A superscript epsilon (ϵ) indicates an editorial change since the last revision or reapproval.

1. Scope

1.1 This guide covers methods for selection and application of coal combustion products (CCPs) for use in the chemical stabilization of trace elements in wastes and wastewater. These elements include, but are not limited to, arsenic, barium, boron, cadmium, chromium, cobalt, lead, molybdenum, nickel, selenium, vanadium, and zinc. Chemical stabilization may be accompanied by solidification of the waste treated. Solidification is not a requirement for the stabilization of many trace elements, but does offer advantages in waste handling and in reduced permeability of the stabilized waste.

1.1.1 Solidification is an important factor in treatment of wastes and especially wastewaters. Solidification/Stabilization (S/S) technology is often used to treat wastes containing free liquids. This guide addresses the use of CCPs as a stabilizing agent without the addition of other materials; however, stabilization or chemical fixation may also be achieved by using combinations of CCPs and other products such as lime, lime kiln dust, cement kiln dust, cement, and others. CCPs used alone or in combination with other reagents promote stabilization of many inorganic constituents through a variety of mechanisms. These mechanisms include precipitation as carbonates, silicates, sulfates, etc.; microencapsulation of the waste particles through pozzolanic reactions; formation of metal precipitates; and formation of hydrated phases **(1-4)**.[2] Long-term performance of the stabilized waste is an issue that must be addressed in considering any S/S technology. In this guide, several tests are recommended to aid in evaluating the long-term performance of the stabilized wastes.

1.2 The CCPs that are suited to this application include fly ash, spent dry scrubber sorbents, and certain advanced sulfur control by-products from processes such as duct injection and fluidized-bed combustion (FBC).

1.3 The wastes or wastewater, or both, containing the problematic inorganic species will likely be highly variable, so the chemical characteristics of the waste or wastewater to be treated must be determined and considered in the selection and application of any stabilizing agent, including CCPs. In any waste stabilization process, laboratory-scale tests for compatibility between the candidate waste or wastewater for stabilization with one or more selected CCPs and final waste stability are recommended prior to full-scale application of the stabilizing agent.

1.4 This guide does not intend to recommend full-scale processes or procedures for waste stabilization. Full-scale processes should be designed and carried out by qualified scientists, engineers, and environmental professionals. It is recommended that stabilized materials generated at the full-scale stabilization site be subjected to testing to verify laboratory test results.

1.5 The utilization of CCPs under this guide is a component of a pollution prevention program; Guide E 1609 describes pollution prevention activities in more detail. Utilization of CCPs in this manner conserves land, natural resources, and energy.

1.6 This guide applies only to CCPs produced primarily from the combustion of coal. It does not apply to ash or other combustion products derived from the burning of waste; municipal, industrial, or commercial garbage; sewage sludge or other refuse, or both; derived fuels; wood waste products; rice hulls; agricultural waste; or other noncoal fuels.

1.7 Regulations governing the use of CCPs vary by state. The user of this guide has the responsibility to determine and comply with applicable regulations.

1.8 It is recommended that work performed under this guide be designed and carried out by qualified scientists, engineers, and environmental professionals.

1.9 *This standard does not purport to address all of the safety concerns, if any, associated with its use. It is the responsibility of the user of this standard to establish appropriate safety and health practices and determine the applicability of regulatory limitations prior to use.*

2. Referenced Documents

2.1 *ASTM Standards:*

C 114 Test Method for Chemical Analysis of Hydraulic Cement[3]

C 311 Test Method for Sampling and Testing Fly Ash or Natural Pozzolans for Use as a Mineral Admixture in

---

[1] This guide is under the jurisdiction of ASTM Committee E-50 on Environmental Assessment and is the direct responsibility of Subcommittee E50.03 on Environmental Risk Management/Sustainable Development/Pollution Prevention.

Current edition approved Feb. 10, 2000. Published April 2000.

[2] The boldface numbers in parentheses refer to the list of references at the end of the text.

[3] *Annual Book of ASTM Standards*, Vol 04.01.

Portland-Cement Concrete[4]

C 400 Test Methods for Quicklime and Hydrated Lime for Neutralization of Waste Acid[3]

D 75 Practice Sampling Aggregates[5]

D 422 Test Method for Particle-Size Analysis of Soils[6]

D 558 Test Method for Moisture-Density Relations of Soil-Cement Mixtures[6]

D 653 Terminology Relating to Soil, Rock, and Contained Fluids[6]

D 1129 Terminology Relating to Water[7]

D 1193 Specification for Reagent Water[7]

D 1556 Test Method for Density and Unit Weight of Soil in Place by the Sand-Cone Method[6]

D 1633 Test Method for Compressive Strength of Molded Soil-Cement Cylinders[6]

D 1635 Test Method for Flexural Strength of Soil-Cement Using Simple Beam with Third-Point Loading[6]

D 2166 Test Method for Unconfined Compressive Strength of Cohesive Soil[6]

D 2216 Test Method for Laboratory Determination of Water (Moisture) Content of Soil and Rock[6]

D 2922 Test Method for Density of Soil and Soil-Aggregate in Place by Nuclear Methods (Shallow Depth)[6]

D 2937 Test Method for Density of Soil in Place by the Drive-Cylinder Method[6]

D 3441 Test Method for Deep, Quasi-Static, Cone and Friction-Cone Penetration Tests of Soil[6]

D 3877 Test Method for One-Dimensional Expansion, Shrinkage, and Uplift Pressure of Soil-Lime Mixtures[6]

D 3987 Test Method for Shake Extraction of Solid Waste with Water[8]

D 4318 Test Method for Liquid Limit, Plastic Limit, and Plasticity Index of Soils[6]

D 4842 Test Method for Determining the Resistance of Solid Wastes for Freezing and Thawing[8]

D 4843 Test Method for Wetting and Drying Test of Solid Wastes[8]

D 4972 Test Method for pH of Soils[9]

D 5084 Test Method for Measurement of Hydraulic Conductivity of Saturated Porous Materials Using a Flexible Wall Permeameter[6]

D 5239 Practice Characterizing Fly Ash for Use in Soil Stabilization[9]

E 1609 Guide for the Development and Implementation of a Pollution Prevention Program[9]

3. Terminology

3.1 *Definitions*:

3.1.1 Definitions are in accordance with Terminology D 653.

3.2 *Definitions of Terms Specific to This Standard:*

3.2.1 *advanced sulfur control (ASC) products*—by-products generated from advanced coal conversion technologies including FBC and gasification and by-products from advanced environmental emissions cleanup technologies such as duct injection and lime injection multiphase burners (LIMB).

3.2.2 *baghouse*—a facility constructed at some coal-fired power plants consisting of fabric filter bags that mechanically trap particulates (fly ash) carried in the flue gases.

3.2.3 *beneficial use*—projects promoting public health and environmental protection, offering equivalent success relative to other alternatives, and preserving natural resources.

3.2.4 *BDAT*—best demonstrated available technology.

3.2.5 *boiler slag*—a molten ash collected at the base of slag tap and cyclone boilers that is quenched in a water-filled hopper and shatters into black, angular particles having a smooth, glassy appearance.

3.2.6 *bottom ash*—agglomerated ash particles formed in pulverized coal boilers that are too large to be carried in the flue gases and impinge on the boiler walls or fall through open grates to an ash hopper at the bottom of the boiler. Bottom ash is typically grey-to-black in color, is quite angular, and has a porous surface texture.

3.2.7 *coal combustion products*—fly ash, bottom ash, boiler ash, or flue gas desulfurization (FGD) material resulting from the combustion of coal.

3.2.8 *DSC*—differential scanning calorimetry.

3.2.9 *DTA*—differential thermal analysis.

3.2.10 *DTG*—differential thermal gravimetry.

3.2.11 *electrostatic precipitator*—a facility constructed at some coal-fired power plants to remove particulate matter (fly ash) from the flue gas by producing an electric charge on the particles to be collected and then propelling the charged particles by electrostatic forces to collecting curtains.

3.2.12 *encapsulation*—complete coating or enclosure of a toxic particle by an additive so as to sequester that particle from any environmental receptors that may otherwise have been negatively impacted by that particle.

3.2.13 *ettringite*—a mineral with the nominal composition $Ca_6Al_2(SO_4)_3(OH)_{12} \cdot 26H_2O$. Ettringite is also the family name for a series of related compounds, known as a mineral group or family, which includes the following minerals **(1)**:

Ettringite	$Ca_6Al_2(SO_4)_3(OH)_{12} \cdot 26H_2O$
Charlesite	$Ca_6(Si,Al)_2(SO_4)_2(B[OH]_4)(OH)_{12} \cdot 26H_2O$
Sturmanite	$Ca_6Fe_2(SO_4)_2(B[OH]_4)(OH)_{12} \cdot 26H_2O$
Thaumasite	$Ca_6Si_2(SO_4)_2(CO_3)_2(OH)_{12} \cdot 24H_2O$
Jouravskite	$Ca_6Mn_2(SO_4)_2(CO_3)_2(OH)_{12} \cdot 24H_2O$
Bentorite	$Ca_6(Cr,Al)_2(SO_4)_3(OH)_{12} \cdot 26H_2O$

3.2.14 *flue gas desulfurization material*—a by-product of the removal of the sulfur gases from the flue gases, typically using a high-calcium sorbent such as lime or limestone. Sodium-based sorbents are also used in some systems. The three primary types of FGD processes commonly used by utilities are wet scrubbers, dry scrubbers, and sorbent injection. The physical nature of these by-products varies from a wet, thixotropic sludge to a dry powdered material, depending on the process.

3.2.15 *fly ash*—coal ash that exits a combustion chamber in the flue gas. Coal fly ashes are typically pozzolans. Some coal fly ashes also exhibit self-hardening properties in the presence of moisture.

[4] *Annual Book of ASTM Standards*, Vol 04.02.
[5] *Annual Book of ASTM Standards*, Vol 04.03.
[6] *Annual Book of ASTM Standards*, Vol 04.08.
[7] *Annual Book of ASTM Standards*, Vol 11.01.
[8] *Annual Book of ASTM Standards*, Vol 11.04.
[9] *Annual Book of ASTM Standards*, Vol 04.09.

3.2.16 *pozzolans*—siliceous or siliceous and aluminous materials that in themselves possess little or no cementitious value but will, in finely divided form and in the presence of moisture, chemically react with calcium hydroxides at ordinary temperatures to form compounds possessing cementitious properties.

3.2.17 *S/S*—solidification/stabilization.

3.2.18 *stabilization or fixation*—immobilization of undesirable constituents to limit their introduction into the environment. Toxic components are immobilized by treating them chemically to form insoluble compounds.

3.2.19 *solidification*—the conversion of soils, liquids, or sludges into a solid, structurally sound material for disposal or use, typically referring to attainment of 50 psi or strength of surrounding soil.

3.2.20 *XRD*—x-ray diffraction.

4. Significance and Use

4.1 *General*—CCPs can have chemical and mineralogical compositions that are conducive to use in the chemical stabilization of trace elements in wastes and wastewater. These elements include, but are not limited to, arsenic, barium, boron, cadmium, chromium, cobalt, lead, molybdenum, nickel, selenium, vanadium, and zinc. Chemical stabilization may be accompanied by solidification of the waste treated. Solidification is not a requirement for the stabilization of many trace elements, but does offer advantages in waste handling and in reduced permeability of the stabilized waste. This guide addresses the use of CCPs as a stabilizing agent without addition of other materials. S/S is considered the BDAT for the disposal of some wastes that contain metals since they cannot be destroyed by other means (**2**).

4.1.1 *Advantages of Using CCPs*—Advantages of using CCPs for waste stabilization include their ready availability in high volumes, generally good product consistency from one source, and easy handling. CCPs vary depending on the combustion or emission control process and the coal or sorbents used, or both, and CCPs contain trace elements, although usually at very low concentrations. CCPs are generally an environmentally suitable materials option for waste stabilization, but the compatibility of a specific CCP must be evaluated with individual wastes or wastewater through laboratory-scale tests followed by full-scale demonstration and field verification. CCPs suitable for this chemical stabilization have the ability to incorporate large amounts of free water into hydration products. CCPs that exhibit high pHs (>11.5) offer advantages in stabilizing trace elements that exist as oxyanions in nature (such as arsenic, boron, chromium, molybdenum, selenium, and vanadium) and trace elements that form oxyhydroxides or low-solubility precipitates at high pH (such as lead, cadmium, barium, and zinc). Additionally, CCPs that exhibit cementitious properties offer advantages in solidifying CCP-waste mixtures as a result of the hydration reactions of the CCP. These same hydration reactions frequently result in the formation of mineral phases that stabilize or chemically fix the trace elements of concern.

4.2 *Chemical/Mineralogical Composition*—Since CCPs are produced under conditions of high temperature, reactions with water during contact with water or aqueous solutions can be expected. Mineral formation may contribute to the chemical fixation and/or solidification achieved in the waste stabilization process. One example of this type of chemical fixation is achieved by ettringite formation. Reduced leachability of several trace elements has been correlated with ettringite formation in hydrated high-calcium CCPs typically from U.S. lignite and subbituminous coal, FGD materials, and ASC by-products. These materials are the best general candidates for use in this chemical fixation process. Lower-calcium CCPs may also be effective with addition of a calcium source that maintains the pH above 11.5. Ettringite forms as a result of hydration of many high-calcium CCPs, so adequate water must be available for the reaction to occur. The mineral and amorphous phases of CCPs contribute soluble elements required for ettringite formation, and the ettringite formation rate can vary based on the mineral and amorphous phase compositions.

4.3 *Environmental Considerations*:

4.3.1 *Regulatory Framework*:

4.3.1.1 *Federal*—EPA has completed a study of four high-volume CCPs (fly ash, bottom ash, boiler slag, and FGD material) for the U.S. Congress and has issued a formal regulatory determination (**3, 4**). EPA "encourages the utilization of coal combustion by-products and supports state efforts to promote utilization in an environmentally beneficial manner" (**5**). In a 1994 Report to Congress, the U.S. Department of Energy (DOE) also encouraged the utilization of CCPs and indicated that DOE will work with federal agencies and state and local governments to increase CCP utilization (**6**). There is currently no regulatory program at the federal level that addresses the utilization of CCPs. The wastes or wastewater requiring stabilization may fall under federal jurisdiction, so the final stabilized material may need to be evaluated and disposed of according to federal regulations. Potentially applicable federal regulations may include the Resource Conservation and Recovery Act (RCRA), Hazardous Solid Waste Act (HSWA), Comprehensive Environmental Response, Compensation, and Liability Act (CERCLA), and Superfund Amendment and Reauthorization Act (SARA). A brief description of these regulations is included in the EPA document, entitled *Stabilization/Solidification of CERCLA and RCRA Wastes: Physical Tests, Chemical Testing Procedures, Technology Screening, and Field Activities* (**2**) and have been summarized by ACAA (**7**). The EPA document states "stabilization/solidification is a proven technology for the treatment of hazardous wastes and hazardous waste sites." According to EPA (**2**), stabilization/solidification is the BDAT for the disposal of some metals since they cannot be destroyed by other means. Provisions in federal laws list requirements that land disposal of hazardous wastes is only acceptable if these wastes are treated with the BDAT or with technology that meets or exceeds the treatment level of BDAT. Wastes that contain free liquids are prohibited from land disposal by federal RCRA regulations or by equivalent state regulations, or both. The chemical binding of free liquids brought about by solidification allows wastes that fail the EPA Paint Filter Test (EPA Method 9095-SW846) (**8**) to be land-disposed after successful S/S treatment.

4.3.1.2 A summary of coal fly ash utilization in waste

stabilization/solidification, including a discussion of environmental/regulatory issues, demonstrations, and commercial applications, has been prepared **(7)**.

4.3.1.3 *State*—Most states do not have specific regulations addressing the use of CCPs, and requests for CCP use are handled on a case-by-case basis or under generic state recycling laws or regulations. Some states have adopted laws and regulations or issued policies and/or guidance regarding CCP use, but CCP use varies widely within these states **(9)**. Waste or wastewater requiring stabilization and the final stabilized material may also be regulated by individual states, so these regulations need to be identified and followed. Many states are authorized to manage the hazardous waste management programs within their state. RCRA and HSWA statutes allowed the states to become authorized by EPA. It is therefore extremely likely that S/S-treated waste will be regulated by a state.

5. CCP Characterization

5.1 *General*—Characterization of the CCP(s) under consideration for use as a stabilizing agent is needed to determine bulk chemical and mineralogical composition to form ettringite when hydrated and that sufficient alkalinity is available to maintain a high pH.

5.2 *Sampling and Handling*—Sampling CCPs for testing purposes should conform to Practice D 75 or Test Method C 311 as appropriate. Proper laboratory protocols for handling fine material should be followed.

5.3 *Chemical Composition*—Test Method C 311 is often used to determine the major chemical constituents of CCP samples. The most critical constituents requiring quantitation are calcium, aluminum, and sulfur.

5.4 *pH*—Test Method D 4972 or Practice D 5239 may be used to determine CCP pH. In assessing the test results, consideration should be given to the possibility that the pH of the CCP may differ with age, water content, and other conditions. EPA Method SW-9045 **(10)** is also applicable.

5.5 *Buffer Capacity*—The buffer capacity of the CCP is important in maintaining the high pH that generally is a requirement for the stabilization mechanisms of interest when CCPs are used as stabilization agents. The CCP must have enough buffer capacity to maintain the pH of the stabilized waste in the appropriate range so the waste remains stable over time and under environmental stresses. Test Method C 400 can be applied to evaluate the buffer capacity of the CCP. Determine the basicity factor for the CCP as noted in Test Method B of Test Method C 400.

5.6 *Swelling*—Test Method D 3877 can be used to determine the swelling potential of self-hardening (high-calcium) CCPs and FGD material. The reactions producing the expansive properties generally do not commence for a period of more than 30 days after initial CCP hydration. The test procedure must address this delayed reaction. The procedure should be modified to extend the wetting and drying cycle to 60 days. Expansive reactions, including the formation of ettringite, may have an impact on the permeability of the stabilized waste. Following completion of hydration at 60 days, the mineral composition should be determined.

5.7 *Mineral Composition (Optional)*—XRD is used to determine the mineral content of CCPs. Fly ash, bottom ash, and boiler slag contain significant amorphous phases that cannot be identified through the use of XRD techniques. Many ASC by-products have a higher calcium content compared to that of CCPs. XRD evaluation can be performed on the hydrated CCP collected from the swelling test, summarized in 5.6, to determine whether or not the CCP tested forms hydrated minerals.

6. Selection of CCP for Waste Stabilization

6.1 *General*—Many CCPs have the potential to stabilize inorganic trace elements; however, those with the highest potential are finely powdered materials, including fly ash, spent dry scrubber sorbents, and certain ASC by-products. Specifically, CCPs that are known to undergo hydration reactions that result in the formation of certain minerals, including ettringite, carbonate minerals, and insoluble sulfates, have high potential to be useful as stabilizing agents for many trace elements that are not organically associated. A CCP exhibiting high pH also offers advantages in the stabilization of trace elements.

6.2 *Chemical Properties*—The chemical properties of CCPs used in waste stabilization applications may include a wide range of chemical compositions, pHs, buffering capacities, and reactivities.

6.2.1 *Chemical Composition*—CCPs of differing chemical composition can offer advantages for solidification/stabilization applications. Most coal fly ashes are pozzolanic; however, certain subbituminous and lignite coal ashes may contain higher amounts of calcium oxide and exhibit greater cementitious properties. Cementitious CCPs offer additional advantages in solidification/stabilization applications; the potential formation of monolithic structures of CCP-waste combinations without addition of cement or other additives; appropriate chemical components that allow the formation of ettringite for stabilization of oxyanions (such as oxyanions of arsenic, boron, chromium, molybdenum, selenium, and vanadium), which is accomplished by substituting these oxyanionic species for $SO_4^{=}$ in the ettringite structure; and high pH, which facilitates ettringite formation and precipitation of low-solubility metal carbonates. The presence of sulfate in CCPs being considered for solidification/stabilization applications can offer advantages in reducing the mobility of barium through barium sulfate formation. The presence of iron and/or manganese species can enhance the formation of low-solubility metal oxyhydroxides.

6.2.2 *pH*—It is recommended that the pH of the CCP be determined using methods noted in 5.4. CCPs exhibiting high pH (>8) offer advantages in numerous solidification/stabilization mechanisms.

6.2.3 *Hydrated Mineral Formation*—The CCP selected should be hydrated using the procedure to determine swelling potential listed in 5.6. Thirty (30) days is recommended as the minimum hydration time; however, 60 days may provide more reliable results because ettringite formation is a slow reaction. As noted in 5.7, XRD is the best analytical tool to determine the presence of hydration products.

6.2.4 *Buffer Capacity*—The buffer capacity of the CCP should be determined as noted in 5.5. As an example of the use of the calculated basicity factor, to ensure that the CCP has the capability to neutralize significant acid while maintaining the pH above 11.5 for ettringite stability, a BF >0.1 would be appropriate. The limit for the BF of the CCP should be evaluated on the basis of the expected mechanism of solidification/stabilization, the candidate waste, and/or the site conditions expected for the stabilized material (see 7.5.2).

6.3 *Final CCP Selection*—One or more CCP(s) can be selected for the laboratory-scale demonstration of the stabilization process using the criteria listed in this section. CCPs meeting the specified criteria should be subjected to the laboratory-scale demonstration using the candidate waste as noted in Section 8.

7. Waste Characterization and Stabilized Material Testing

7.1 *General*—Specific characteristics of the waste or wastewater requiring stabilization must be determined. The concentrations of the elements expected to participate in the chemical fixation reactions are required to determine the quantity of stabilization agent or CCP to use. The pH and buffer capacity of the waste must be known. All waste characterization tests and procedures must be performed on representative samples of the specific waste to be treated. The final stabilized material form must also be evaluated for specific parameters to verify that the stabilization process has been effective.

7.2 *Physical Tests for Wastes*—Physical testing of the waste prior to treatment provides information on treatability, transport, storage, and mixing considerations. Physical testing of the stabilized material form provides information on the relative success of the stabilization process.

7.2.1 *Particle-Size Analysis*—The results of particle-size analysis provide information on the relative proportions of gravel, sand, silts, and clay-sized particles within the waste, uniformity, concavity, average grain size, and maximum and minimum particle sizes and can be determined using Test Method D 422.

7.2.2 *Atterberg Limits*—The Atterberg Limits are a series of tests originally developed to determine the characteristics of clays used in ceramics. The recommended test procedure is Test Method D 4318, which is used to determine the liquid limit (the moisture content at which the material flows as a viscous liquid), plastic limit (the moisture content at the boundary between the plastic and brittle states), and plasticity index (the difference between the liquid and plastic limits).

7.2.3 *Moisture Content*—The moisture content of the waste can be determined using Test Method D 2216, which will indicate the amount of free water in the waste (waters of hydration are not included as part of the moisture content). Water may not be the only liquid phase in a waste requiring stabilization, so care must be taken to remove other liquids or otherwise account for them in determining the moisture content.

7.2.4 *Density*—The bulk density of a material is the ratio of the total weight of the material to the total volume and is frequently used to convert weight to volume for materials-handling calculations. Several methods are appropriate for determining the bulk density of the waste prior to stabilization. These include laboratory Test Methods D 2937 and D 1556 and an in situ measurement, Test Method D 2922.

7.2.5 *Suspended Solids*—In the case of stabilizing oxyanionic species in a wastewater or liquid waste, it is recommended to determine the suspended solids in the wastewater using Method 2540D as detailed in the Standard Methods for Examination of Water and Waste Water (**11**). A liquid waste is defined as having less than 1 % suspended solids. Suspended solids can be removed from liquid waste or wastewater by filtering.

7.2.6 *Paint Filter Liquids Test*—The regulatory test (Title 40 of the Code of Federal Regulations [40 CFR] Sections 264.314 and 265.314) (**8**) is used to determine if a waste contains free liquids. For wastes treated by S/S to address free liquids, the EPA generally requires attainment of 50 psi unconfined compressive strength in the treated waste. This is to demonstrate that the free liquids were bound chemically rather than merely sorbed.

7.3 *Physical Tests for Stabilized Materials*—The physical nature of the waste may or may not change on application of the CCP for chemical stabilization depending on the CCP selected, the character of the waste, and the stabilization process used. Solidification is not a criterion for stabilization of oxyanionic species through ettringite formation, but solidification generally reduces the permeability of the stabilized waste, which offers the advantages of limiting contact with air and water (or other gases and liquids). If solidification is desirable, it must be factored into the selection criteria for the CCP.

7.3.1 *Compaction Testing*—The moisture-density relationship can be determined on the stabilized waste using Test Method D 558. This test determines the moisture content that allows maximum compaction, which provides maximum density of the stabilized waste. It is cautioned here that the moisture content must be adequate to provide the waters of hydration needed for the ettringite formation reaction to occur and that the ettringite formation is slow and can require 30 days or longer to complete. Moisture-density relationships developed in the laboratory must take this into account.

7.3.2 *Density*—The bulk density of the stabilized material can be determined as noted in 7.2.4 if the stabilized material is not solidified. If the stabilized material is solidified, a sufficiently cured cube or cylinder of the stabilized material can be weighed and measured. The volume of the cube or cylinder can be determined, and the density can then be calculated (**12**).

7.3.3 *Permeability Testing*—The permeability of the stabilized material is important in estimating the quantity and flow of water through that material under saturated conditions. Recommended test procedures are Test Method D 5084 and EPA Method 9100-SW846 (**12**).

7.3.4 *Strength Testing*—Optional strength tests may be performed depending on the requirements of the site and final stabilized material. Several types of strength tests are listed here and can be used as appropriate to the site and final stabilized material.

7.3.4.1 Unconfined compressive strength can be determined using Test Methods D 2166 or D 1633. Results of unconfined compressive strength provide information on the ability of the stabilized waste to support overburden, the optimum water:additive ratios and curing times for setting reactions, and the changes in strength from waste to stabilized material.

7.3.4.2 EPA generally requires attainment of 50 psi unconfined compressive strength for a treated waste that originally contained free liquids.

7.3.4.3 Flexural strength testing provides a means to assess the tensile strength of the solidified mass under flexural loading conditions and can be determined by using Test Method D 1635.

7.3.4.4 Cone index is determined to indicate stability and bearing capacity. The appropriate test method is Test Method D 3441.

7.3.5 *Durability Testing*—Durability testing is performed to evaluate the resistance of a stabilized material to external environmental stresses such as freezing and thawing or wet-and-dry cycles. The tests noted require molded samples of the stabilized material form, which may be appropriate depending on the consistency and texture of the stabilized material.

7.3.5.1 Freeze-thaw testing can be accomplished using Test Method D 4842 as appropriate for the specific site or stabilized material, or both.

7.3.5.2 Wet-dry cycle testing of the stabilized material can be performed using Test Method D 4843.

7.3.5.3 *Paint Filter Liquids Test*—The regulatory test (Title 40 of the Code of Federal Regulations [40 CFR] Sections 264.314 and 265.314) (**8**) used to determine if a waste contains free liquids. For wastes treated by S/S to address free liquids, the EPA generally requires attainment of 50 psi unconfined compressive strength in the treated waste. This is to demonstrate that the free liquids were bound chemically rather than merely sorbed.

7.4 *Chemical Characterization of Wastes*—Wastes to be stabilized require chemical characterization. The primary chemical characterization must include determination of bulk chemical composition and pH. The major constituents of the waste must be quantitated, and any species requiring stabilization must be quantitated.

7.4.1 *Chemical Composition*—Major oxide content of the waste can be determined according to Test Method C 114. Additional total elemental analysis can be determined according to EPA SW-846 (**13**).

7.4.2 *pH*—Methods noted in 5.4 are applicable to determining the pH of the waste.

7.5 *Chemical Characterization of Stabilized Materials*—The primary chemical characterization required for stabilized materials is leaching/extraction tests.

7.5.1 *Leaching Tests*—Numerous leaching tests have been developed to evaluate the leaching behavior of wastes and stabilized materials. Commonly applied leaching tests are listed and referenced in Table 1, but the selection of the test procedure(s), leachate test parameters, and interpretation of leachate test results must be guided by the appropriate regulatory authority.

7.5.2 *Buffer Capacity*—The buffer capacity of the stabilized material can be evaluated by use of the ASTM procedure noted in 5.5.

8. Laboratory-Scale Demonstration of Stabilization

8.1 *General*—Laboratory-scale development and demonstration of the stabilization technology are recommended prior to field application of any stabilization agent. Laboratory-scale demonstrations must be performed with representative samples of the waste requiring stabilization and the selected CCP(s).

TABLE 1 Leaching Methods Applicable to Stabilized Materials

Test Method	Leaching Solution	Liquid:Solid Ratio	Leaching Duration
TCLP (**14**)	Acetic acid/acetate buffer	20:1	18 h
EP Tox (**13**)	Acetic acid	16:1	24 h
California WET (Waste Extraction Test) (**15**)	Sodium citrate	10:1	48 h
MEP (**16**)	Multiple solutions (acetic acid, sulfuric acid, and nitric acid)	20:1	24 h/ extraction
MWEP (**17**)	Distilled/deionized water or other for specific silt	10:1	18 h/ extraction
ASTM D 3987	Distilled/deionized water	20:1	24 h
SGLP (**18**)/LTL (**18**)	Synthetic groundwater dictated by site or distilled/deionized water	20:1	18 h/30, 60, 90 days
SPLP (**19**)	Sulfuric acid	20:1	18 h

8.1.1 *Waste: CCP Ratios*—Using the results of physical and chemical characterization techniques recommended for the wastes, a range of waste:CCP ratios should be identified and tested for process variables and stabilized material testing.

8.1.2 *Mixing Requirements*—Laboratory-scale mixing techniques that approximate the anticipated full-scale stabilization process must be used to mix the waste and CCP to produce the initial stabilized material.

8.1.3 *Curing*—The initial stabilized material sample(s) must be cured prior to evaluation. A cure time of 30 days is recommended as the minimum time prior to evaluation of the stabilized material. It is recommended that additional cure times of 60 days and 90 days also be used for additional samples to determine a qualitative rate of stabilization reactions.

8.1.4 *Swelling*—Determine the amount of swelling, if any, on all cured samples using appropriate methods as noted in 5.7. Solidified masses should be examined for cracking that may result from swelling.

8.1.5 *Leachability*—Using the prescribed leaching test(s), evaluate the leaching potential on cured samples relative to appropriate regulatory requirements.

8.1.6 *Mineralogy*—If warranted, XRD may be used to verify the presence of hydration minerals such as ettringite in cured samples.

8.1.7 *Stabilized Material Stability*—Determine the physical stability or integrity of the stabilized material using tests indicated in 6.3.5.

9. Field Application of CCPs as a Waste Stabilization Agent

9.1 *General*—Waste stabilization can be accomplished by several processes for full-scale projects. Selection of the best process requires consideration of the physical and chemical properties of the waste or wastewater. Common processes are drum processing, plant processing, and in situ processing. CCPs have commonly been used in plant processing and in situ processing (**20-22**).

9.1.1 *Environmental Issues*—Site variables must be taken into account when evaluating the stabilization technology and process. These variables include proximity to water table, surface waters, and soil permeability.

9.1.2 *Process Requirements*—Equipment needs, equipment access, and dry storage for the CCP must all be considered prior to field application.

9.1.3 *Economic Considerations*—Costs and availability of the CCP, labor costs, and other factors relating to the process, site preparation, or landfilling must be part of the decision to proceed with the stabilization process.

REFERENCES

(**1**) Hassett, D.J., Pflughoeft-Hassett, D.F., McCarthy, G.J. "Ettringite Formation in Coal Ash as a Mechanism for Stabilization of Hazardous Trace Elements," *In Proceedings of the Ninth International Ash Use Symposium—Volume 2: Stabilization and Aquatic Uses*, Orlando, FL, Jan. 22-25, 1991, EPRI GS-7162, Project 3176, 1991, pp. 31–1 to 31–17.

(**2**) U.S. Environmental Protection Agency, *Stabilization/Solidification of CERCLA and RCRA Wastes: Physical Tests, Chemical Testing Procedures, Technology Screening, and Field Activities*, EPA/625/6-89/022, May 1989.

(**3**) Report to Congress, *Wastes from the Combustion of Coal by Electric Utility Power Plants*, U.S. Environmental Protection Agency, EPA/530-SW-88-002, Feb. 1988.

(**4**) U.S. Environmental Protection Agency. "Final Regulatory Determination on Four Large-Volume Wastes from the Combustion of Coal by Electric Utility Power Plants," 40 CFR Part 261, Aug. 2, 1993.

(**5**) U.S. Environmental Protection Agency. *Technical Resource Document: Solidification/Stabilization and Its Application to Waste Materials* EPA/530/R-93/012, June 1993.

(**6**) Report to Congress, *Barriers to the Increased Utilization of Coal Combustion/Desulfurization Byproducts by Governmental and Commercial Sectors*, U.S. Department of Energy, Office of Fossil Energy, July 1994.

(**7**) American Coal Ash Association, *Solidification and Stabilization of Wastes Using Coal Fly Ash: Current Status and Direction*, April 1995.

(**8**) U.S. Environmental Protection Agency "Method 9095: Paint Filter Liquid Test," *Test Methods for Evaluating Solid Waste: Physical/Chemical Methods (SW846)*, Vol 1C, 3rd ed. 1990.

(**9**) American Coal Ash Association, *Solid Waste Regulations Governing the Use of Coal Combustion Products (CCPs)*, Aug. 1998.

(**10**) U.S. Environmental Protection Agency, "Method 9045c: Soil and Waste pH," *Test Methods for Evaluating Solid Waste: Physical/Chemical Methods (SW846)*, Vol 1C, 3rd ed. 1990.

(**11**) American Public Health Association; American Water Works Association; and Water Environment Federation. "Method 2540D: Total Suspended Solids Dried at 103°–105°C," *Standard Methods for the Examination of Water and Wastewater*, 19th ed., 1995.

(**12**) U.S. Environmental Protection Agency," Method 9100: Saturated Hydraulic Conductivity, Saturated Leachate Conductivity, and Intrinsic Permeability," *Test Methods for Evaluating Solid Waste: Physical/Chemical Methods (SW846)*, Vol 1C, 3rd ed. 1990.

(**13**) U.S. Environmental Protection Agency, "Method 1310: Extraction Procedure (EP) Toxicity Test Method and Structural Integrity Test," *Test Methods for Evaluating Solid Waste: Physical/Chemical Methods (SW846)*, Vol 1C, 3rd ed. 1990.

(**14**) U.S. Environmental Protection Agency, "Method 1311: Toxicity Characteristic Leaching Procedure (TCLP)," *Test Methods for Evaluating Solid Waste: Physical/Chemical Methods (SW846)*, Vol 1C, 3rd ed. 1990.

(**15**) State of California, "Environmental Health Standards-Hazardous Waste," California Code, Title 22, Section 66261.126: 1991, pp. 669–672.

(**16**) U.S. Environmental Protection Agency, "Method 1320: Multiple Extraction Procedure (MEP)," *Test Methods for Evaluating Solid Waste: Physical/Chemical Methods (SW846)*, Vol 1C, 3rd ed. 1990.

(**17**) U.S. Environmental Protection Agency, "Solid Waste Leaching Procedure Technical Document for Public Comment," Report No. EPA/530-SW-84-924, 1984.

(**18**) Hassett, D.J.; Pflughoeft-Hassett, D.F. "Environmental Assessment of Coal Conversion Solid Residues," In Proceedings of the 10th International Ash Use Symposium - Vol 1: High-Volume Uses/Concrete Applications; Orlando, FL, Jan. 18–21, 1993, EPRI TR-101774, Project 3176, 1993.

(**19**) U.S. Environmental Protection Agency, "Method 1312: Synthetic Precipitation Leaching Procedure (SPLP)," *Test Methods for Evaluating Solid Waste: Physical/Chemical Methods (SW846)*, Vol 1C, 3rd ed. 1990.

(**20**) Pritts, J.W., Neufeld, R.D., and Cobb, J.T., *Stabilization of Heavy Metal Containing Hazardous Wastes with By-Products from Advanced Clean Coal Technology Systems.*

(**21**) Cobb, J.T., Neufeld, R.D., Schreiber, E.M., Clifford, B.V., Pritts, J.W., Beeghley, J.H., and Bender, C.F., *Treatment of Metal-Laden Hazardous Wastes with Advanced Clean Coal Technology By-Products*, topical report, Award No. DE-FC21-94MC31175, August 18, 1994–August 18, 1995.

(**22**) Wilk, C.M., *Stabilization of Heavy Metals with Portland Cement: Research Synopsis*, Portland Cement Association, IS007, 1997.

Designation: E 2081 – 00

Standard Guide for Risk-Based Corrective Action[1]

This standard is issued under the fixed designation E 2081; the number immediately following the designation indicates the year of original adoption or, in the case of revision, the year of last revision. A number in parentheses indicates the year of last reapproval. A superscript epsilon (ϵ) indicates an editorial change since the last revision or reapproval.

INTRODUCTION

This guide provides guidance for the development of a Risk-Based Corrective Action (RBCA) program that integrates the sciences of ecological and human health risk-based decision making into the corrective action process. The RBCA provides a flexible, technically defensible framework for corrective action that is applicable to a wide range of sites and chemical(s) of concern. The framework incorporates a tiered analytical approach, applying increasingly complex levels of data collection and analysis as the user proceeds through the process. It provides a starting point for the integration of multiple regulatory programs into a site-wide corrective action activity and a technically defensible process for achieving "No Further Action." The successful implementation of the RBCA framework is dependent on an understanding by the user of the technical policy decisions that are critical to the risk management process and the identification and determination of these technical policy decisions prior to beginning the process (see 3.2.60). There are numerous technical policy decisions that must be made to implement the RBCA process, for example, defining data quality objectives, determining target risk levels and addressing resource protection. It is not the intent of this guide to define appropriate technical policy decisions. The RBCA process is not intended to replace existing regulatory programs, but rather to complement these programs. Regardless of whether a corrective action is specifically governed by a regulatory program, the user should consult the regulatory agency requirements to identify the appropriate technical policy decisions prior to implementing the RBCA process. The RBCA process encourages user-led initiatives and stakeholder involvement in both the development of the technical policy decisions and the RBCA program. It recognizes the diversity of sites and provides appendixes for possible applications and examples. The appendixes are provided for additional information and are not mandatory sections of this standard guide. ASTM standards are not federal or state regulations; they are consensus standards that can voluntarily be followed.

1. Scope

1.1 This is a guide for conducting risk-based corrective action (RBCA) at chemical release sites based on protecting human health and the environment. The RBCA is a consistent decision-making process for the assessment and response to chemical releases. Chemical release sites vary greatly in terms of complexity, physical and chemical characteristics, and in the risk that they may pose to human health and the environment. The RBCA process recognizes this diversity by using a tiered approach that integrates site assessment and response actions with human health and ecological risk assessment to determine the need for remedial action and to tailor corrective action activities to site-specific conditions and risks. The evaluations and methods used in the RBCA process begin with simple analyses in Tier 1 and move to more complex evaluations in either Tier 2 or Tier 3, as applicable. The process of gathering and evaluating data is conducted in a scaled fashion. Consequently, only the data that are necessary for a particular tier's decision-making are collected at that tier.

1.2 This guide describes an approach for risk-based corrective action. It is intended to help direct and streamline the corrective action process and to complement but not to supersede federal, state and local regulations. It can be employed at sites where corrective action is being conducted including sites where there may not be a regulatory framework for corrective action, or where the user wishes to conduct corrective action such as sites in voluntary cleanup programs or under Brownfields initiatives. In addition, it can also be used as a unifying framework when several different agency programs affect the site. Furthermore, the user should be aware of the federal, state and local corrective action programs that are applicable for the site and, regardless of the program, federal, state and local agency approvals may be required to implement the processes outlined in this guide. Finally, regardless of

[1] This guide is under the jurisdiction of ASTM Committee E50 on Environmental Assessment and is the direct responsibility of Subcommittee E50.04 on Performance Standards Related to Environmental Regulatory Programs.

Current edition approved April 10, 2000. Published November 2000. Originally published as PS 104–98.

whether a corrective action is specifically governed by a regulatory program, the user should consult the regulatory agency requirements to identify the appropriate technical policy decisions prior to implementing the RBCA process.

1.3 There are numerous technical policy decisions that must be made to implement the RBCA process, for example, defining data quality objectives, determining target risk levels, specifying the appropriate statistics and sample sizes for calculating exposure concentrations, selection of exposure assumptions, determining when and how to account for cumulative risks and additive effects among chemical(s) of concern and addressing resource protection. It is not the intent of this guide to define appropriate technical policy decisions. The user must identify the appropriate technical policy decisions.

1.4 The general performance standard for this guide requires that:

1.4.1 Technical policy decisions be identified before beginning the process,

1.4.2 Data and information collected during the RBCA process, including historical data as well as new data collected during the site assessment, will be relevant to and of sufficient quantity and quality to answer the questions posed by and the decisions to be made in the RBCA process,

1.4.3 Actions taken during the risk-based decision process will be protective of human health and the environment,

1.4.4 Applicable federal, state and local regulations will be followed (for example, waste management requirements, ground water designations, worker protection) and,

1.4.5 Remedial actions implemented will not result in higher risk levels than existed before taking actions.

1.5 ASTM standards are not federal or state regulations, they are consensus standards that can voluntarily be followed.

1.6 The RBCA process is not limited to a particular class of compounds. This guide is intended to be a companion to Guide E 1739, and does not supersede that document for petroleum releases. If a release site contains a mixture of releases of petroleum and other chemicals, this guide should be followed.

1.7 The United States Environmental Protection Agency (USEPA) has developed guidance for human health risk evaluation (see Appendix X8 for other resources). Many of the components of this guidance have been integrated into the RBCA framework. The science of ecological evaluation and the process by which the science is applied, however, are not as well defined and agreed upon as human health risk assessment. Therefore, the information provided in this guide for each tier evaluation for relevant ecological receptors and habitats is general. The user is referred to Appendix X5, which provides additional information regarding the development of a RBCA framework for protection of ecological resources.

1.8 The decision process described in this guide integrates exposure and risk assessment practices with site assessment activities and remedial action selection to ensure that the chosen actions are protective of human health and the environment. The following general sequence of events is prescribed in RBCA:

1.8.1 Perform an initial site assessment and develop the first iteration of the site conceptual model. If the information is sufficient to demonstrate that there are no complete or potentially complete exposure pathways, then no further action is warranted,

1.8.2 Evaluate the site (see definition of site 3.2.50) for response actions (multiple sites at a single facility may require different response actions and times),

1.8.3 Implement a response action that is appropriate for conditions found at the site during the site response action evaluation,

1.8.4 Define data requirements, develop data quality objectives, and perform a site assessment for the Tier 1 evaluation if the site conceptual model indicates that the tiered evaluation is appropriate,

1.8.5 Conduct an exposure pathway analysis to determine if relevant ecological receptors and habitats are present and if complete and potentially complete exposure pathways are present. If no relevant ecological receptors or habitats or complete and potentially complete exposure pathways exist, then no further action for relevant ecological receptors and habitats is warranted,

1.8.6 For potential human exposure pathways, identify the applicable Risk Based Screening Levels (RBSL) and for potential ecological exposure pathways, identify the applicable Relevant Ecological Screening Criteria (RESC). In addition, identify any Other Relevant Measurable Criteria (ORMC), as applicable. Collectively these are the Tier 1 corrective action goals for the site;

1.8.7 Compare site conditions to the Tier 1 corrective action goals determined to be applicable to the site;

1.8.8 If site conditions meet the corrective action goals for chemical(s) of concern then, no further action is warranted,

1.8.9 If site conditions do not meet corrective action goals for chemical(s) of concern then, one or more of the following actions is appropriate:

1.8.9.1 Further tier evaluation;

1.8.9.2 Implement interim remedial action;

1.8.9.3 Design and implement remedial action to achieve the corrective action goals.

1.8.10 Define Tier 2 data requirements, data quality objectives, collect additional site-specific information and update the site conceptual model, as necessary, if further tier evaluation is warranted,

1.8.11 Develop point(s) of demonstration and Tier 2 corrective action goals based on Site-Specific Target Levels (SSTL), Site-Specific Ecological Criteria (SSEC) or ORMC, where appropriate, for complete and potentially complete exposure pathways, including exposure pathways for which no RBSL, RESC or ORMC, as applicable, were determined;

1.8.12 Compare site conditions to the Tier 2 corrective action goals determined to be applicable to the site;

1.8.13 If site conditions meet corrective action goals for chemical(s) of concern, then no further action is warranted,

1.8.14 If site conditions do not meet corrective action goals for chemical(s) of concern then, one or more of the following actions is appropriate:

1.8.14.1 Further tier evaluation;

1.8.14.2 Implement interim remedial action;

1.8.14.3 Design and implement remedial action to achieve the corrective action goals.

1.8.15 Define Tier 3 data requirements, data quality objectives and collect additional site-specific information and update the site conceptual model, as necessary, if further tier evaluation is warranted,

1.8.16 Develop point(s) of demonstration and Tier 3 corrective action goals based on SSTL, SSEC, or ORMC, where appropriate;

1.8.17 Compare site conditions to the Tier 3 corrective action goals,

1.8.18 If site conditions meet corrective action goals for chemical(s) of concern, then no further action is warranted,

1.8.19 If site conditions do not meet corrective action goals for chemical(s) of concern, then one of the following actions is appropriate:

1.8.19.1 Implement interim remedial action to facilitate reassessment of the tier evaluation;

1.8.19.2 Design and implement remedial action to achieve the corrective action goals.

1.8.20 Develop and implement a monitoring plan based on the corrective action goals to validate the assumptions used for the tier evaluation and to demonstrate effectiveness of the remedial action, as applicable.

1.9 For chemical release sites currently in corrective action, the user should review information and data available for the site and determine the most appropriate entry point into the RBCA framework consistent with the general performance standards and sequence of events outlined in this guide.

1.10 *This Guide is Organized as Follows*— Section 2 lists referenced documents, Section 3 defines terminology used in this guide, Section 4 describes the significance and use of this guide, Section 5 is a summary of the tiered approach, and Section 6 presents the RBCA procedures in a step-by-step process. Appendix X1 provides guidance on developing technical policy decisions and building a RBCA program, Appendix X2 provides examples of chemical properties and effects data that may be useful for a RBCA evaluation, Appendix X3 provides EXAMPLE development of RBSL, Appendix X4 describes the use of predictive modeling, Appendix X5 provides an outline of the process of the ecological evaluation, Appendix X6 provides information about activity and use limitations, Appendix X7 includes illustrative examples of the application of the RBCA framework, and Appendix X8 includes references that may be helpful to the user. The Appendixes are provided for additional information and are NOT included as mandatory sections of this guide.

2. Referenced Documents

2.1 *ASTM Standards:*

D 5447 Guide for Application of a Ground-Water Flow Model to a Site-Specific Problem[2]

D 5490 Guide for Comparing Ground-Water Flow Model Simulations to Site-Specific Information[2]

D 5610 Guide for Defining Initial Conditions in Ground-Water Flow Modeling[2]

D 5611 Guide for Conducting a Sensitivity Analysis for a Ground Water Flow Model Application[2]

D 5612 Guide for Quality lanning and Field Implementation of a Water Quality Measurement Program[3]

D 5718 Guide for Documenting a Ground-Water Flow Model Application[2]

D 5880 Guide for Subsurface Flow and Transport Modeling[2]

D 6235 Practice for Expedited Site Characterization of Vadose Zone and Ground Water Contamination at Hazardous Waste Contaminated Sites[2]

E 978 Practice for Evaluating Mathematical Models for the Environmental Fate of Chemicals

E 1527 Practice for Environmental Site Assessments: Phase I Environmental Site Assessment Process[4]

E 1599 Guide for Corrective Action for Petroleum Releases[4]

E 1689 Guide for Developing Conceptual Site Models for Contaminated Sites[5]

E 1739 Standard Guide for Risk-Based Corrective Action Applied at Petroleum Release Sites[4]

E 1903 Guide for Environmental Site Assessments: Phase II Environmental Site Assessment Process[4]

E 1912 Guide for Accelerated Site Characterization for Confirmed or Suspected Petroleum Releases[4]

E 1943 Guide for Remediation of Ground Water by Natural Attenuation at Petroleum Release Sites[4]

3. Terminology

3.1 The reader should review the definitions presented here prior to reviewing the guide, as many of the terms included in this guide may have different meanings than the specific regulatory definitions within existing federal, state or local programs. The following terms are being defined to reflect their specific use in this guide. The user should not assume that these definitions replace existing regulatory definitions. Where the definition or use of a term in this guide differs from an existing regulatory definition or use, the user should address these differences prior to proceeding with the RBCA process.

3.2 *Definitions of Terms Specific to This Standard:*

3.2.1 *activity and use limitations*—Legal or physical restrictions or limitations on the use of, or access to, a site or facility to eliminate or minimize potential exposures to chemical(s) of concern, or to prevent activities that could interfere with the effectiveness of a remedial action, to ensure maintenance of site conditions that meet the corrective goals for chemical(s) of concern. These legal or physical restrictions are intended to prevent adverse impacts to receptors and relevant ecological receptors and habitats that may be exposed to chemical(s) of concern. Activity and use limitations include both engineering and institutional controls.

3.2.2 *additive effects*—refers to combined non-cancer effects of chemical(s) of concern with the same mechanism of action in a receptor.

3.2.3 *bio-availability*—a measure of the chemical(s) of concern in environmental media that is accessible to an organism for absorption.

[2] *Annual Book of ASTM Standards*, Vol 04.09.

[3] *Annual Book of ASTM Standards*, Vol 11.01.

[4] *Annual Book of ASTM Standards*, Vol 11.04.

[5] *Annual Book of ASTM Standards*, Vol 11.05.

3.2.4 *biodegradation*—Natural plant, animal, or microbial metabolism that results in the reduction of mass of a chemical(s) of concern.

3.2.5 *chemical release*—Any spill or leak or detection of concentrations of chemical(s) of concern in environmental media.

3.2.6 *chemical(s) of concern*—The specific compounds and their breakdown products that are identified for evaluation in the RBCA process. Identification can be based on their historical and current use at a site, detected concentrations in environmental media and their mobility, toxicity, and persistence in the environment. Because chemical(s) of concern may be identified at many points in the RBCA process, including before any determination that they pose an unacceptable risk to human health or the environment, the term should not automatically be construed to be associated with increased or unacceptable risk.

3.2.7 *corrective action*—The sequence of actions that include site assessment and investigation, risk assessment, response actions, interim remedial action, remedial action, operation and maintenance of equipment, monitoring of progress, making no further action determinations and termination of the remedial action.

3.2.8 *corrective action goals*—concentration or other numeric values, physical condition or remedial action performance criteria that demonstrate that no further action is necessary to protect human health and the environment. For example, these goals may include one or a combination of RBSL, SSTL, RESC, SSEC and ORMC chosen for source area(s), point(s) of demonstration and point(s) of exposure. The corrective action goals are specific to each Tier in the evaluation.

3.2.9 *cumulative risks*—refers to the combined carcinogenic risks from all exposure pathways for all chemicals for a receptor.

3.2.10 *direct exposure pathways*—An exposure pathway where the point of exposure is at the source, without a release to any other medium and without an intermediate biological transfer step.

3.2.11 *ecological evaluation*—A process for organizing and analyzing data, information, assumptions and uncertainties to evaluate the likelihood that adverse effects to relevant ecological receptors or habitats may occur or are occurring as a result of exposure to chemical(s) of concern.

3.2.12 *engineering controls*—Physical modifications to a site or facility to reduce or eliminate the potential for exposure to chemical(s) of concern (for example, slurry walls, capping, hydraulic controls for ground water, or point-of-use water treatment).

3.2.13 *exposure assessment*—The determination or estimation (qualitative or quantitative) of the magnitude, frequency, duration and route of exposure between a source area and a receptor.

3.2.14 *exposure pathway*—The course a chemical of concern takes from the source area(s) to a receptor or relevant ecological receptor and habitat. An exposure pathway describes the mechanism by which an individual or population is exposed to a chemical of concern originating from a site. Each exposure pathway includes a source or release from a source of a chemical of concern, a point of exposure, an exposure route and the potential receptors or relevant ecological receptors and habitats. If the exposure point is not at the source, a transport or exposure medium, or both, (for example, air or water) are also included.

3.2.15 *exposure route*—The manner in which a chemical(s) of concern comes in contact with a receptor (for example, ingestion, inhalation, dermal contact).

3.2.16 *exposure scenario*—The description of the circumstances, including site properties and chemical properties, or the potential circumstances under which a receptor or a relevant ecological receptor or habitat could be in contact with chemical(s) of concern;

3.2.17 *facility*—The property containing the source of the chemical(s) of concern where a release has occurred. A facility may include multiple sources and therefore, multiple sites.

3.2.18 *guide*—a series of options or instructions that do not recommend a specific course of action.

3.2.19 *hazard index*—The sum of two or more hazard quotients for chemical(s) of concern or multiple exposure pathways to a particular receptor, or both.

3.2.20 *hazard quotient*—The ratio of the level of exposure of a chemical of concern over a specified time period to a reference dose for that chemical of concern derived for a similar exposure period.

3.2.21 *incremental carcinogenic risk levels*—The potential for incremental human carcinogenic effects, over background cancer occurrence levels, due to exposure to the chemical(s) of concern. This is the individual lifetime excess cancer risk.

3.2.22 *indirect exposure pathways*—An exposure pathway with at least one intermediate release to any media, or an intermediate biological transfer step, between the source and the point(s) of exposure (for example, chemical(s) of concern from soil through ground water to the point(s) of exposure).

3.2.23 *institutional controls*—A legal or administrative restriction on the use of, or access to a site or facility to eliminate or minimize potential exposure to a chemical(s) of concern (for example, restrictive covenants, restrictive zoning).

3.2.24 *interim remedial action*—The course of action taken to reduce migration of a chemical(s) of concern in its vapor, dissolved, or liquid phase, or to reduce the concentration of a chemical(s) of concern at a source area(s).

3.2.25 *maximum contaminant level (MCL)*—A standard for drinking water established by USEPA under the Safe Drinking Water Act which is the maximum permissible level of a chemical(s) of concern in water which is delivered to any user of a public water supply.

3.2.26 *natural attenuation*—The reduction in the concentration(s) of chemicals of concern in environmental media due to naturally occurring physical, chemical and biological processes (for example, diffusion, dispersion, adsorption, chemical degradation and biodegradation).

3.2.27 *non-aqueous phase liquids (NAPL)*—Chemicals that are insoluble or only slightly soluble in water that exist as a separate liquid phase in environmental media. They can be less dense or more dense than water.

3.2.28 *other relevant measurable criteria (ORMC)*— Parameters used to define corrective action goals for chemical(s) of concern. The ORMC are concentration values, other numeric values, physical condition or performance criteria other than RBSL, RESC, SSTL or SSEC. Examples of ORMC are regulatory standards, consensus criteria, aesthetic criteria, and ground water protection criteria. Technical policy decisions regarding ORMC may exist, or may need to be made to determine the appropriate values, conditions or performance criteria that are used for the corrective action goals.

3.2.29 *petroleum*—Includes crude oil or any fraction thereof that is liquid at standard conditions of temperature and pressure (60 °F (15.6°C) and 14.7 lb/in.^{2} absolute (10 340 kg/m^2)). The term includes petroleum-based substances comprised of a complex blend of hydrocarbons derived from crude oil through processes of separation, conversion, upgrading, and finishing, (for example, motor fuels, jet fuels, lubricants, petroleum solvents, used oils).

3.2.30 *point(s) of demonstration*—A location(s) selected between the source area(s) and the potential point(s) of exposure where corrective action goals are met.

3.2.31 *point(s) of exposure*—The point(s) at which an individual or population may come in contact with a chemical(s) of concern originating from a site.

3.2.32 *potentially complete exposure pathway*—A situation with a reasonably likely chance of occurrence in which a receptor or relevant ecological receptor or habitat may become directly or indirectly exposed to the chemical(s) of concern.

3.2.33 *practice*—a definitive procedure for performing one or more specific operations or functions that does not produce a test result.

3.2.34 *probabilistic evaluation*—A modeling procedure used to evaluate the uncertainty surrounding a probability distribution when the result depends on a number of factors, each of which has its own variability and uncertainty.

3.2.35 *qualitative ecological screening evaluation*—A process conducted as part of the Tier 1 evaluation wherein relevant ecological receptors and habitats and exposure pathways are identified. The necessary information can be collected as part of the data gathering activities during the initial site assessment or the Tier 1 site assessment (6.3.2 and 6.3.3). Within Tier 1, this screening-level information, which is typically qualitative, may be used to evaluate potential exposure pathways to relevant ecological receptors and habitats and to identify potential chemical(s) of concern. If available, generic, non-site-specific ecological criteria and guidelines (3.2.42) may be used to further evaluate complete and potentially complete exposure pathways.

3.2.36 *qualitative risk analysis*—A non-quantitative evaluation of the potential risks at a site as determined by the potential exposure pathways and receptors based on known or reasonably available information.

3.2.37 *reasonable maximum exposure (RME)*—The highest exposure that is reasonably expected to occur at a site. RME's are estimated for individual exposure pathways or a combination of exposure pathways.

3.2.38 *reasonably anticipated future use*—Future use of a site or facility which can be predicted with a reasonably high degree of certainty given historical use, current use, and local government planning and zoning.

3.2.39 *receptors*—The persons that are or may be affected by a release. (see relevant ecological receptors and habitats, 3.2.41, for non-human receptor definition).

3.2.40 *reference dose*—A toxicity value for evaluating potential non-carcinogenic effects in humans resulting from exposure to a chemical(s) of concern.

3.2.41 *relevant ecological receptors and habitats*—The ecological resources that are valued at the site. Because of the variety of ecological resources that may be present, focusing upon those relevant to a site is an important part of the problem formulation phase of ecological evaluation. Identification of relevant ecological receptors and habitats is dependent upon site-specific factors and technical policy decisions. Examples may include species or communities afforded special protection by law or regulation; recreationally, commercially or culturally important resources; regionally or nationally rare communities; communities with high aesthetic quality; and habitats, species or communities that are important in maintaining the integrity and bio-diversity of the environment.

3.2.42 *relevant ecological screening criteria (RESC)*—Generic, non-site specific ecological criteria or guidelines that are determined to be applicable to relevant ecological receptors and habitats, exposure pathways and site conditions utilized during the Tier 1 evaluation. These may include chemical concentrations, biological measures or other relevant generic criteria consistent with the technical policy decisions.

3.2.43 *remedial action*—Activities conducted to reduce or eliminate current or potential future exposures to receptors or relevant ecological receptors and habitats. These activities include monitoring, implementing activity and use limitations and designing and operating clean-up equipment. Remedial action includes activities that are conducted to reduce sources of exposures to meet corrective action goals, or sever exposure pathways to meet corrective action goals.

3.2.44 *response action*—An immediate course of action, including monitoring, abatement or containment measures to mitigate known or potential hazards to human health, safety and the environment, taken before interim remedial action or remedial action.

3.2.45 *response action evaluation*—A qualitative analysis of a site, based on known or readily available information, to identify the need for and urgency of response actions and the need for further information gathering. The analysis is also used to identify appropriate early risk reduction steps.

3.2.46 *risk*—The potential for, or probability of, an adverse effect. These may be expressed qualitatively or quantitatively.

3.2.47 *risk assessment*—An analysis of the potential for adverse effects on receptors and relevant ecological receptors and habitats caused by a chemical(s) of concern from a site. The risk assessment activities are the basis for the development of corrective action goals and determination of where interim remedial actions, remedial action or a combination of actions are required.

3.2.48 *risk reduction*—The lowering or elimination of the level of risk posed to human health or the environment through, response actions, interim remedial actions, remedial action or a combination of actions.

3.2.49 *risk-based screening level/screening levels (RBSL)*—Non-site-specific human health risk-based values for chemical(s) of concern that are protective of human health for specified exposure pathways utilized during the Tier 1 evaluation.

3.2.50 *site*—The area(s) defined by the likely physical distribution of the chemical(s) of concern from a source area. A site could be an entire property or facility, a defined area or portion of a facility or property or multiple facilities or properties. One facility may contain multiple sites. Multiple sites at one facility may be addressed individually or as a group.

3.2.51 *site assessment*—A characterization of a site through an evaluation of its physical and environmental context (for example, subsurface geology, soil properties and structures, hydrology, and surface characteristics) to determine if a release has occurred, the levels of the chemical(s) of concern in environmental media, and the likely physical distribution of the chemical(s) of concern. As an example, the site assessment collects data on soil, ground water and surface water quality, land and resource use, potential receptors, and potential relevant ecological receptors and habitats, and generates information to develop a site conceptual model and to support risk-based decision-making. The user is referred to Guides E 1912 and D 6235 and other references in Appendix X8 for more information.

3.2.52 *site conceptual model*—The integrated representation of the physical and environmental context, the complete and potentially complete exposure pathways and the potential fate and transport of chemical(s) of concern at a site. The site conceptual model should include both the current understanding of the site and the understanding of the potential future conditions and uses for the site. It provides a method to conduct the exposure pathway evaluation, inventory the exposure pathways evaluated, and determine the status of the exposure pathways as incomplete, potentially complete or complete.

3.2.53 *site conditions*—A general description of a site's chemical, physical or biological characteristics that relate to potential exposures to receptors or relevant ecological receptors and habitats.

3.2.54 *site-specific*—Activities, information and data unique to a particular site.

3.2.55 *site-specific ecological criteria (SSEC)*—Risk-based qualitative or quantitative criteria for relevant ecological receptors and habitats identified for a particular site under the Tier 2 or Tier 3 evaluations. These may include chemical concentrations, biological measures or other relevant generic criteria consistent with the technical policy decisions. SSEC may be revised as data are obtained that better describe the conditions and the relevant ecological receptors and habitats.

3.2.56 *site-specific target levels (SSTL)*—Risk-based values for chemical(s) of concern that are protective of human health for specified exposure pathways developed for a particular site under the Tier 2 or Tier 3 evaluations.

3.2.57 *source area(s)*—The source area(s) is defined as the location of non-aqueous phase liquid (NAPL) chemical, the locations of highest soil or ground water concentrations of the chemical(s) of concern or the location releasing the chemical(s) of concern.

3.2.58 *stakeholders*—Individuals, organizations or other entities that directly affect or are directly affected by the corrective action. Stakeholders include, but are not limited to, owners, buyers, developers, lenders, insurers, government agencies and community members and groups.

3.2.59 *standard*—As used in ASTM, a document that has been developed and established within the consensus principles of the Society and that meets the approval requirements of ASTM procedures and regulations.

3.2.60 *technical policy decisions*—The choices specific to the user that are necessary to implement the Risk-Based Corrective Action framework described in this guide at a particular site. The decisions involve regulatory policies, value judgments, different stakeholder decisions and using professional judgment to evaluate available information, therefore, there may be more than one scientifically supportable answer for any particular technical policy decision. The choices represent different approaches. The user should consult the regulatory agency requirements to identify the appropriate technical policy decisions prior to implementing the RBCA process. Examples of technical policy decisions are, data quality objectives, target risk levels, land use, ground water use, natural resource protection, relevant ecological receptors and habitats, stakeholder notification and involvement and exposure factors.

3.2.61 *Tier 1 Evaluation*—A risk-based analysis utilizing non-site-specific corrective action goals for complete and potentially complete direct and indirect human exposure pathways and qualitative ecological screening evaluation for complete and potentially complete exposure pathways for relevant ecological receptors and habitats. The non-site-specific corrective action goals developed for human exposure pathways are based on conservative assumptions (for example, exposure factors, fate and transport parameters) and methodologies (for example, algorithms, analytical models) to estimate the non-site-specific values. The Tier 1 exposure pathways for human receptors assume that the receptor and the source are located in the same location. A qualitative ecological screening evaluation is conducted that may be combined with RESC to evaluate the potential exposures to relevant ecological receptors and habitats. The Tier 1 evaluation for some chemical(s) of concern or exposure pathways may also be based on comparison of site conditions to ORMC. The non-site-specific corrective action goals for complete and potentially complete exposure pathways are compared to site conditions to determine if further corrective action is warranted.

3.2.62 *Tier 2 Evaluation*—A risk-based analysis that involves an incremental refinement of the Tier 1 methodology to develop site-specific corrective action goals. The Tier 2 evaluation for human exposure pathways may include developing statistically representative concentrations of chemical(s) of concern for comparison to the Tier 1 corrective action goals, back-calculating SSTL by applying the direct exposure pathway corrective action goals established under a Tier 1 evaluation at site-specific determined point(s) of exposure, developing SSTL for potential indirect exposure pathways at point(s) of exposure using site-specific conditions and the Tier 1 methodology, or developing SSTL for complete or potentially complete exposure pathways using site-specific conditions for which no RBSL were developed in Tier 1, or the evaluation may employ a combination of alternatives. For relevant ecological receptors and habitats, the Tier 2 evaluation may involve additional qualitative or quantitative analyses. The Tier 2 evaluation for some chemical(s) of concern and exposure pathways may also be based on comparison of site conditions to ORMC. The corrective action goals for complete and potentially complete exposure pathways are compared to site conditions to determine if further corrective action is warranted.

3.2.63 *Tier 3 Evaluation*—A risk-based analysis that involves a significant incremental effort over the Tier 2 evaluation to develop site-specific corrective action goals. The Tier 3 evaluation for human exposure pathways typically uses advanced exposure assessment, toxicity and risk assessment techniques (for example, probabilistic exposure assessment methods, use of bio-availability data, use of advanced fate and transport modeling) allowing maximum flexibility to develop SSTL for potential direct and indirect exposure pathways at the point(s) of exposure based on site-specific conditions. A Tier 3 evaluation for relevant ecological receptors and habitats is typically more quantitative in nature and uses more site-specific data than previous tiers. The Tier 3 evaluation for some chemical(s) of concern and exposure pathways may also be based on comparison of site conditions to ORMC. The corrective action goals for complete and potentially complete exposure pathways are compared to site conditions to determine if further corrective action is warranted.

3.2.64 *user*—An individual or group involved in the RBCA process including owners, operators, regulators, UST fund managers, federal government case managers, attorneys, consultants, legislators and other stakeholders. Two specific cases of users are envisioned. The first is the individual or group addressing a site or sites under the circumstances where there is no specific agency program or there are multiple agency programs applicable to their project. The second is a regulatory agency that is developing a comprehensive corrective action program.

4. Significance and Use

4.1 The risk-based corrective action (RBCA) process presented in this guide is a consistent, streamlined decision process for selecting corrective actions at chemical release sites.

4.2 Risk assessment is a developing science. The scientific approach used to develop the RBSL and SSTL may vary by regulatory agency and by user due to regulatory requirements, guidance and use of alternative scientifically-based methods.

4.3 Activities described in this guide should be conducted by persons familiar with current site characterization techniques, remedial action science and technology, current human health risk and exposure assessment methodologies, toxicology, and current ecological evaluation methodologies.

4.4 In order to properly apply the RBCA process, the user should **AVOID** the following:

4.4.1 Prescribing Tier 1 RBSL or RESC as remedial action standards for all sites rather than screening levels,

4.4.2 Limiting use of the RBCA process to Tier 1 evaluation only and not continuing with Tier 2 or Tier 3 analyses for sites where further tier evaluation is appropriate,

4.4.3 Placing arbitrary time constraints on the corrective action process; for example, requiring that Tiers 1, 2, and 3 be completed within time periods that do not reflect the actual urgency of and risks posed by the site,

4.4.4 Using the RBCA process only when active remedial action is not technically feasible, rather than as a process that is applicable during all phases of corrective action,

4.4.5 Conducting active remedial action to achieve only technology-based remedial limits (for example, asymptotic levels) prior to determining applicable corrective action goals,

4.4.6 Using predictive modeling that is not supported by available data or knowledge of site conditions,

4.4.7 Limiting remedial action options to a single class of remedial actions for all sites,

4.4.8 Using unjustified or inappropriate exposure factors,

4.4.9 Using unjustified or inappropriate toxicity parameters,

4.4.10 Failing to consider cumulative risks and additive effects when evaluating multiple chemicals,

4.4.11 Excluding the evaluation of options for activity and use limitations, point(s) of exposure, point(s) of demonstration, sequencing remedial action activities at multiple sites on the same facility, or risk levels,

4.4.12 Excluding the maintenance and monitoring of activity and use limitations,

4.4.13 Failing to consider the long-term effectiveness and reliability of potential remedial action options,

4.4.14 Failing to evaluate potential risks to the public, to workers and to relevant ecological receptors and habitats that may be created by proposed remedial actions or assessment methods and

4.4.15 Continuing monitoring or remedial action at sites that have achieved the corrective action goals (unless monitoring is specifically required for an activity and use limitation or another regulatory requirement). Achievement of corrective action goals is predicated on sufficient monitoring to substantiate the site conditions.

4.5 The RBCA process described in this guide includes several features that are only examples of standardized approaches to addressing the objectives of the particular activity, for example, the response action evaluation table and the exposure scenario evaluation flowchart. These elements should be customized by the user based on the constraints of the site or group of sites being addressed and the appropriate technical policy decisions. The objectives of the analyses are identified in this guide.

5. A Tiered Approach to Risk-Based Corrective Action (RBCA)

5.1 Risk-based corrective action is the integration of site assessment, remedial action selection, and monitoring with appropriate risk and exposure assessment practices. This creates a process by which corrective action decisions are made in a consistent manner that is protective of human health and the environment. Prior to implementing the RBCA process, the user must identify the relevant technical policy decisions appropriate for the site (see 1.2, 1.3 and Appendix X1). The user should also identify the appropriate stakeholder notification and involvement process to provide information and to collect input during the implementation of the RBCA process.

5.2 The RBCA process is implemented in a tiered approach, involving increasingly sophisticated levels of data collection and analysis as the user proceeds through the tiers. At each further tier of evaluation, the assumptions of earlier tiers are replaced with additional site-specific data and information.

5.3 There is some degree of uncertainty associated with all risk estimates and site assessments. In the RBCA process it is necessary for the user to address uncertainty through the level of conservatism applied to each tier. As the user moves through the tier evaluation process, the level of conservatism should decrease as the uncertainty decreases. The uncertainty should be clearly articulated during each tier of evaluation. The analysis of uncertainty allows the user to determine if the information obtained is adequate to make a decision. As the user proceeds to higher tiers, the knowledge gained about the site is used to tailor the degree of investigation needed, as explained in the following sections. In some cases, after completion of the Tier 1 evaluation, the user may find it appropriate for some exposure pathways to proceed directly to a Tier 3 evaluation. As contemplated here, the results of all of the completed tiers of analyses would be compiled into one RBCA report at the end of the evaluation. Reporting requirements and approvals must be determined based on the particular federal, state and local programs that apply to the site.

6. Risk-Based Corrective Action (RBCA) Procedures

6.1 The sequence of principal tasks and decisions associated with the RBCA process are outlined on the flowcharts shown in Fig. 1 and Fig. 2. Each of these actions and decisions is discussed as follows. Prior to implementing these actions, the user should identify the appropriate stakeholder notification and involvement process to provide information and to collect input during the implementation of the RBCA process. Information is gathered in the initial site assessment to develop the site conceptual model.

6.2 *Initial Site Assessment*—The initial site assessment is a planning and scoping activity to develop the site conceptual model, (for example, identifying potential transport pathways and potential receptors) based on the initial understanding of the site. The planning and scoping activity is a critical part of implementing the technical policy decisions due to the potential complexity of human and ecological exposure pathways. This is especially important for ecological issues due to the variety of relevant ecological receptors and habitats. Information collected during the initial site assessment may identify incomplete exposure pathways that may eliminate the need for any further evaluation of one or more exposure pathways or the site. For example, some regulatory agencies specify processes to define incomplete exposure pathways or define minimum criteria, threshold quantities or concentrations of a chemical release as an exclusion from or entry to a further RBCA analysis for a site. If the information is sufficient to demonstrate that there are no complete or potentially complete exposure pathways, then no further action is warranted. If minimum criteria, threshold quantities or concentrations that define exclusion from a RBCA analysis are available and site conditions meet these criteria, thresholds or concentrations, as applicable, then no further action is warranted for the site.

6.2.1 The initial site assessment should include a review of known or reasonably available information on:

6.2.1.1 Appropriate regulatory requirements;

6.2.1.2 Historical site activities, past releases and prior site assessment information to identify potential chemical(s) of concern, sources of the chemical(s) of concern, source area(s), human receptors and relevant ecological receptors and habitats, and fate and transport mechanisms;

6.2.1.3 Potential current and reasonably anticipated future use of the site and surrounding land;

6.2.1.4 Potential ground water and surface water use; and

6.2.1.5 Regional hydrogeologic and geologic characteristics, (for example, depth to ground water, aquifer thickness, flow direction, gradient, description of confining units and ground water quality).

6.3 *Site Conceptual Model*—The site conceptual model is developed to provide an overall understanding of the site and includes the hypotheses that form the basis of the RBCA evaluation. It is also used to communicate the understanding of the site to the stakeholders. Through the site conceptual model the user identifies the complete and potentially complete exposure pathways. The site conceptual model provides a template to conduct the exposure pathway evaluation, inventory the exposure pathways evaluated, and determine the status of the exposure pathways as incomplete, potentially complete or complete. It also provides a means to identify potential Response Actions (see 6.6). The site conceptual model is developed as part of the initial site assessment and is updated as additional data are gathered throughout the RBCA process. Subsequent site assessment activities are conducted to refine the site conceptual model and conduct the tier analyses.

6.3.1 *Exposure Pathway Evaluation—The user is responsible for identifying the complete and potentially complete exposure pathways for a site.* An EXAMPLE exposure pathway evaluation flowchart is presented in Fig. 3, and it may be used as a tool to guide the user in selecting appropriate exposure scenarios based on site assessment information. *This figure is only presented as an example of the kind of visual representation that may be useful in the evaluation of exposure pathways.* Other types or forms of charts or tables may also be used in the evaluation.

6.3.1.1 *Identify Receptors*—To identify receptors and relevant ecological receptors and habitats, consider current and reasonably anticipated future site and surrounding land use.

6.3.1.2 *Characterize Site Sources and Exposure Pathways*—Use the data and information collected throughout the RBCA process to identify relevant sources, transport mechanisms, and exposure pathways. More than one flow chart may be needed to correspond to multiple sources with different exposure pathways.

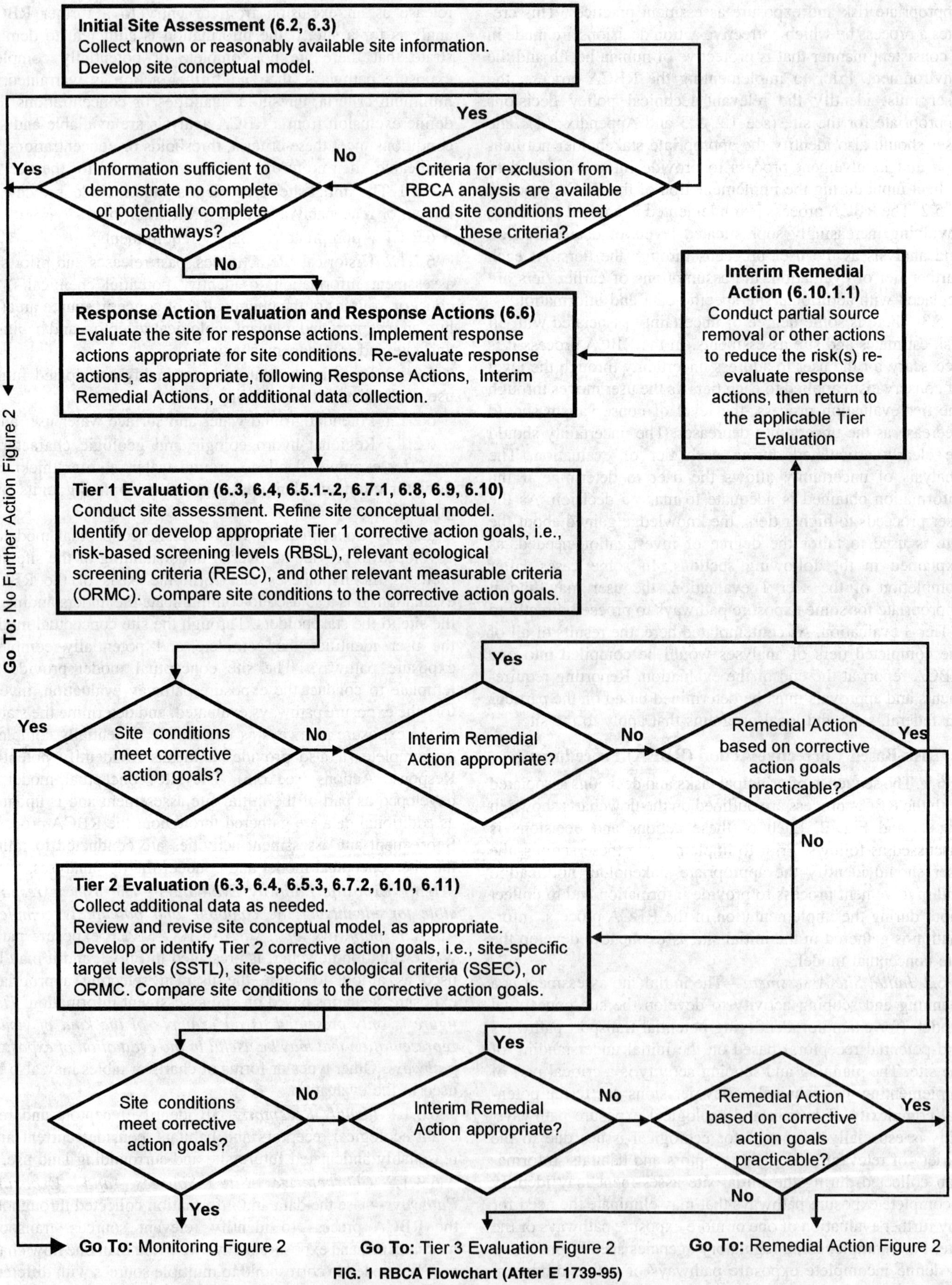

FIG. 1 RBCA Flowchart (After E 1739-95)

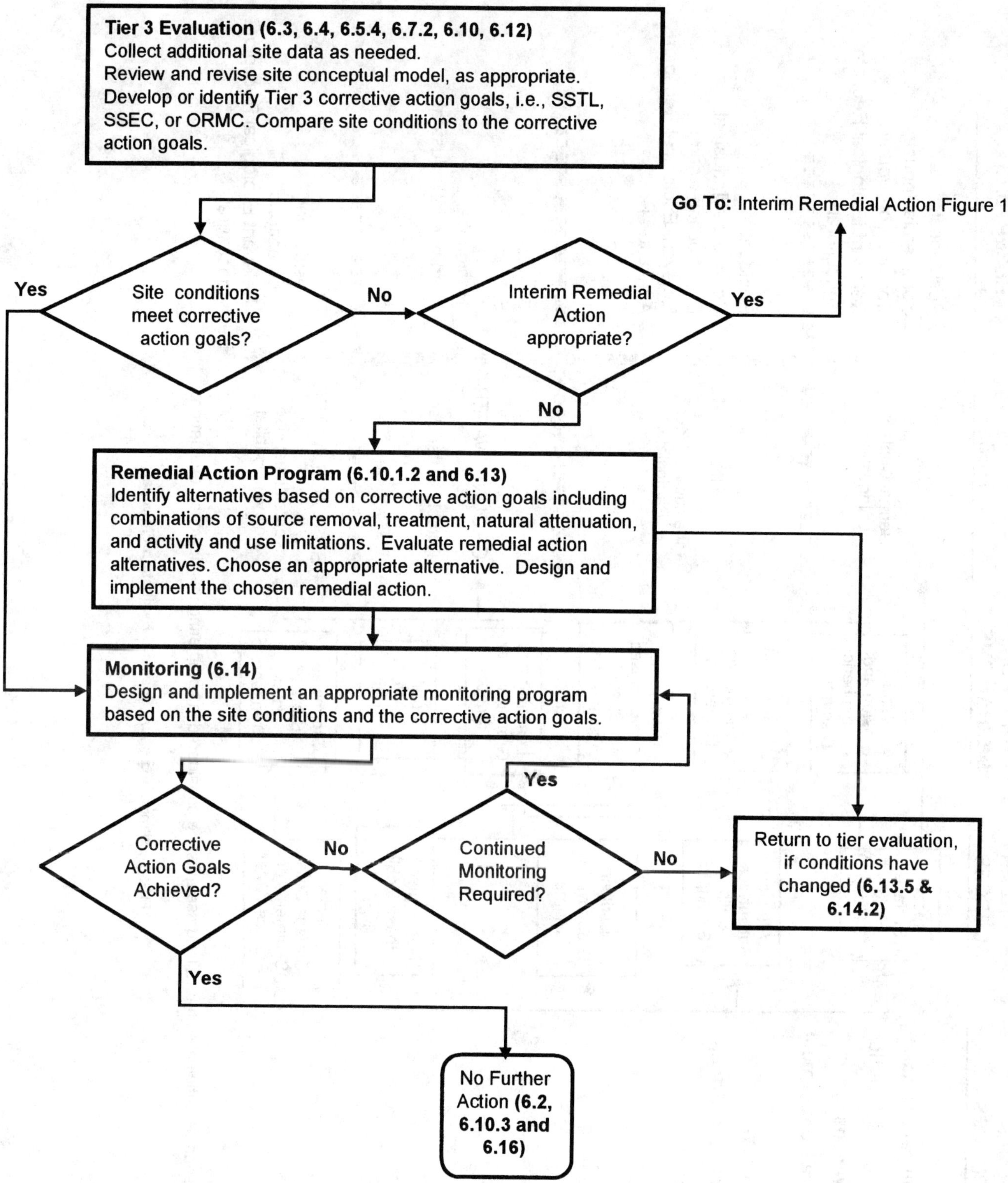

FIG. 2 RBCA Flowchart (Continued) (After E 1739-95)

6.3.2 In each successive tier of evaluation the complete and potentially complete exposure pathways are re-evaluated in light of the additional data collected. Further evaluations are not conducted for exposure pathways determined to be incomplete.

6.4 *Data Quality Objectives*—Throughout the RBCA process, appropriate data quality objectives should be determined for the task that is being conducted. These objectives integrate the site specificity of data needed for each task and applicable regulatory requirements. The user may generate site-specific data and information or estimate reasonable values for key physical characteristics using soil survey data and other reasonably available information. Sufficient quantity and quality of data should be collected during the RBCA process to meet

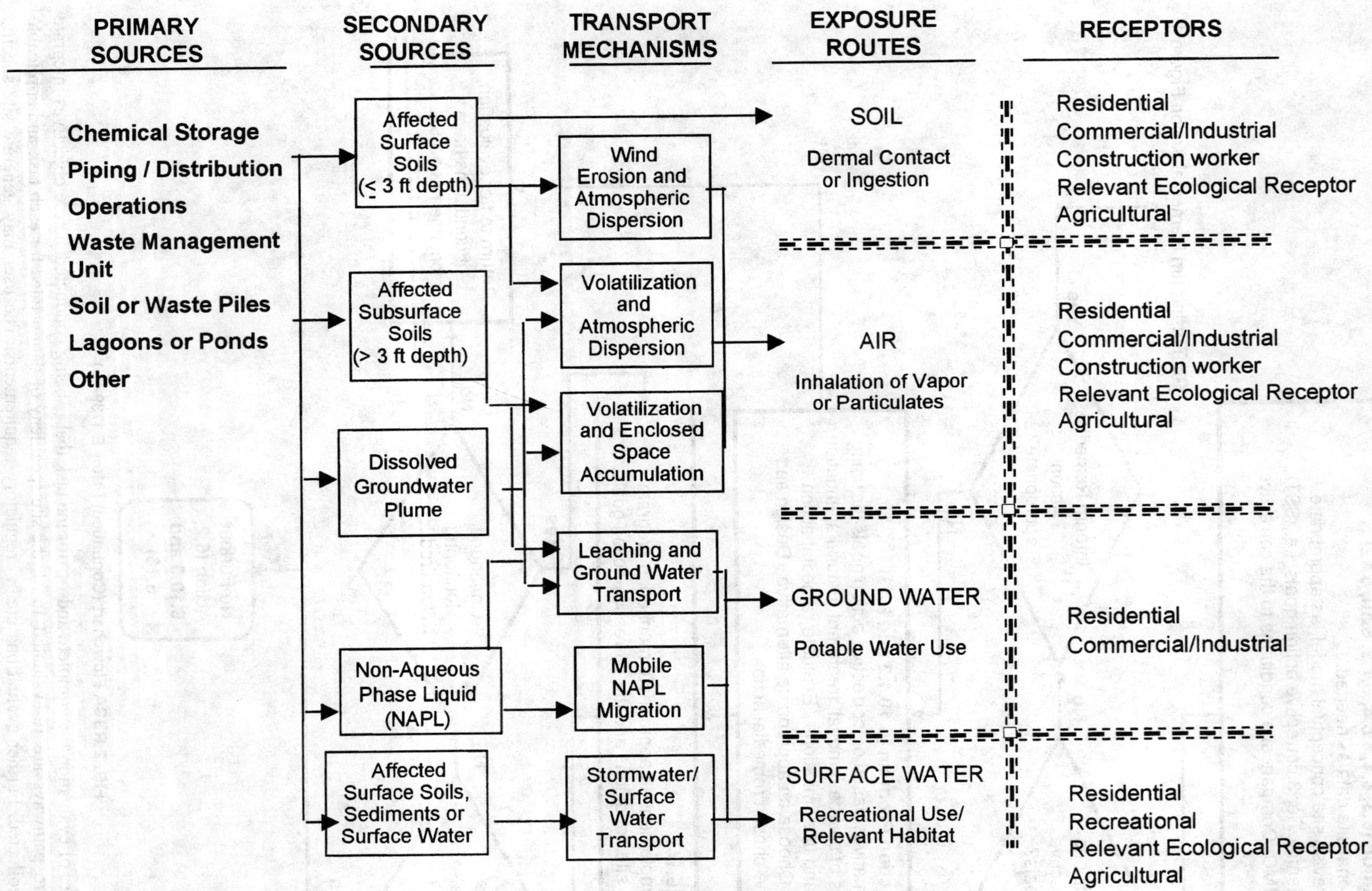

The current and reasonably anticipated future land use must be considered in completing this flowchart evaluation.
(After E1739-95)

FIG. 3 Example Exposure Scenario Evaluation Flowchart

the data quality objectives for any aspect of the RBCA evaluation. In addition, the quality and quantity of data collected must be consistent with the objectives of the particular task within the RBCA evaluation. The user is referred to Guide E 1912 and Ref **(1)**[6] and other references for more information.

6.5 *Site Assessment*—In general, the site assessment gathers the information necessary to implement the RBCA analyses. Among the tasks for which site assessment data are used are: refining the site conceptual model, determining appropriate response actions, performing human health and ecological risk evaluations and making decisions for further tier evaluation, no further action, interim remedial action or remedial action for the complete and potentially complete exposure pathways. State and local government and local community sources should be consulted, as appropriate, in the collection of the site assessment information. Adequate site assessment is critical to the RBCA process to identify sources of chemical(s) of concern that may exist at the site. The site assessment data should be summarized in a clear and concise format (for example, using Fig. 3 or Guide E 1689). The site assessment is limited to gathering only the information necessary at each tier to evaluate the complete or potentially complete exposure pathways identified in the site conceptual model.

6.5.1 If complete or potentially complete exposure pathways are identified during the site conceptual model development indicating that the tiered evaluation is appropriate, then a site assessment for the Tier 1 evaluation should be conducted. The purpose of the site assessment is to gather information relevant to the evaluation of the complete and potentially complete exposure pathways defined by the site conceptual model. The information collected should also be used to improve the understanding of the site through refinements to the site conceptual model. The site assessment for Tier 1 should include information to:

6.5.1.1 Identify chemical(s) of concern;

6.5.1.2 Locate sources of the chemical(s) of concern and source area(s), to the extent practicable;

6.5.1.3 Locate maximum concentrations of chemical(s) of concern in different media (for example, air, soil, soil gas, surface water, sediments, ground water);

6.5.1.4 Locate human receptors and relevant ecological receptors and habitats that could be impacted;

6.5.1.5 Identify potentially significant transport and human and ecological exposure pathways (for example, ground water transport, vapor migration through soils and utilities);

6.5.1.6 Identify current and reasonably anticipated future use of the site and surrounding land. This identification should include an evaluation of ground water resources, surface water and relevant ecological receptors and habitats;

6.5.1.7 Identify state or local ground water classifications, options for alternative classifications or designations applicable to the site; and

6.5.1.8 Determine regional and site-specific hydrogeologic and geologic characteristics, (for example, depth to ground water, aquifer thickness, flow direction, gradient, description of confining units and ground water quality).

6.5.2 The site assessment for Tier 1 may also collect information to:

6.5.2.1 Calculate an appropriate upper confidence limit on concentration data for chemical(s) of concern, if sufficient data are available;

6.5.2.2 Determine background concentrations of chemical(s) of concern in environmental media (see Appendix X1);

6.5.3 The purpose of the site assessment under Tier 2 is to gather information relevant to the complete and potentially complete exposure pathways defined by the site conceptual model and determined, based on the Tier 1 results, to require further tier evaluation. The information collected should also be used to improve the understanding of the site through refinements to the site conceptual model. In addition to the information gathered under 6.2, 6.5.1 and 6.5.2 and updates to that information, if the Tier 2 evaluation is appropriate, the site assessment should collect information to:

6.5.3.1 Determine site-specific hydrogeologic and geologic characteristics (for example, depth to ground water, aquifer thickness, hydraulic conductivity, flow direction and velocity, gradient, description of lithology and confining units and ground water quality);

6.5.3.2 Determine concentrations of chemical(s) of concern in environmental media, and site conditions relative to the corrective action goals;

6.5.3.3 Determine changes in concentrations of chemical(s) of concern over time (for example, stable, increasing, decreasing);

6.5.3.4 Determine concentrations of chemical(s) of concern measured at point(s) of exposure (for example, concentrations in nearby drinking water wells, concentrations in surface water or vapor concentrations in nearby conduits or sewers);

6.5.3.5 Evaluate complete and potentially complete exposure pathways for human receptors and relevant ecological receptors and habitats to refine the list of complete and potentially complete exposure pathways and determine those that are incomplete exposure pathways, and;

6.5.3.6 Determine point(s) of demonstration (see 3.2.30).

6.5.4 The purpose of the site assessment under Tier 3 is to gather information relevant to the complete and potentially complete exposure pathways defined by the site conceptual model and determined, based on the Tier 2 results, to require further tier evaluation. The information collected should also be used to improve the understanding of the site through refinements to the site conceptual model. In addition to the information gathered under 6.2, 6.5.1, 6.5.2 and 6.5.3, if the Tier 3 evaluation is appropriate, the site assessment may include significant additional information collection required for site-specific modeling efforts.

6.6 *Response Action Evaluations and Response Actions*—As the user gathers data, site conditions should be evaluated and a response action should be implemented, consistent with site conditions and simultaneously with the tiered evaluation process to mitigate known or potential hazards. Sites are first evaluated by the need and urgency for response action based on information collected during the

[6] The boldface numbers given in parentheses refer to a list of references at the end of the text.

initial site assessment. This process is repeated when new data indicate a significant change in site conditions, as additional data affords a more detailed characterization of the site, or after interim remedial actions. Table 1 presents EXAMPLE site conditions and potential responses.

NOTE 1—The response actions given in Table 1 are only examples. The user should select options that best address the short-term human health, safety and environmental concerns of the site while implementing the RBCA process.

6.6.1 The site conditions and response actions given in Table 1 are examples. The user should base the response actions for a site on the current and potential degree of hazard to human health, safety and the environment. For example, sites may pose an immediate, short-term (0-2 years) or longer-term (greater than 2 years) threat to human health, safety and the environment.

6.6.2 Associated with each site condition in Table 1 is a response action; the response actions are implemented to eliminate potential for future impacts that may occur as the user proceeds with the RBCA process. Note that response actions do not always require active remedial action; in many cases the response action is to monitor or further assess site conditions to ensure that risks posed by the site do not increase above acceptable levels over time.

6.6.3 The site should be re-evaluated when additional site information is collected which indicates a significant change (including an improvement) in site conditions or when implementation of a response action or an interim remedial action causes a significant change in site conditions.

TABLE 1 Example Site Conditions and Response Actions

Site Conditions	Example Response Actions
Immediate threat to human health, safety, or relevant ecological receptors and habitats	Notify appropriate authorities, property owners, and potentially affected parties, and evaluate the need to:
• Explosive levels of vapors present in a building	• Evacuate occupants, begin abatement measures
• An active public water supply is impacted by chemicals of concern	• Provide alternate water supply
• Threatened and endangered species are impacted	• Implement containment measures and habitat management to minimize extent of impact
Short Term (0-2 years) threat to human health, safety, or relevant ecological receptors and habitats.	Notify appropriate authorities, property owners, and potentially affected parties, and evaluate the need to:
• Chronic concentrations of chemical(s) of concern have been measured in shallow surface soils that are open to public access	• Remove soils, cover soils, or restrict access.
• Impacted surface water, storm water, or ground water discharges to a habitat, or surface water body used for human drinking water or contact recreation.	• Institute containment measures, restrict access to areas near discharge, and evaluate the magnitude and impact of the discharge.
Long-term (>2 years) threat to human health, safety, or relevant ecological receptors and habitats.	Notify appropriate authorities, property owners, and potentially affected parties, and evaluate the need to:
• Ground water is impacted and potable water supply wells producing from the impacted interval are located >2 years ground water travel time from the dissolved plume.	• Monitor the dissolved plume and evaluate the potential for natural attenuation and the need for hydraulic control.
• Impacted surface water, storm water, or ground water discharges within 1500 feet of a habitat, or surface water body used for human drinking water or contact recreation.	• Investigate current impact on habitat or surface water body, restrict access to area of discharge (if necessary) and evaluate the need for containment/control measures.

6.7 *Development of Corrective Action Goals*—At each tier the user identifies or develops the applicable corrective action goals based on the current or potential exposure pathways. These will be a combination of RBSL, RESC, SSTL, SSEC and ORMC. Technical policy decisions regarding ORMC may exist, or may need to be made to determine the levels or performance criteria that are used for the corrective action goals. The user should recognize that the specific combination of RBSL, RESC, SSTL, SSEC and ORMC to be used as corrective action goals for a particular site is determined based on the technical policy decisions.

6.7.1 *Tier 1 Corrective Action Goals*—The corrective action goals in Tier 1 are the RBSL, RESC or ORMC, as applicable. If an RBSL, RESC or ORMC is not available, the user may develop an RBSL, RESC or ORMC. If RBSL are available, the user is responsible for determining that the RBSL are based on currently acceptable methodologies and parameters and are appropriate for the site(s). If there is no applicable RBSL, RESC, or ORMC for a specific exposure scenario, then the exposure pathway cannot be evaluated in Tier 1 and it is carried to a Tier 2 evaluation. In Tier 1, the point(s) of exposure and point(s) of demonstration are assumed to be located proximal to the source area(s), regardless of the actual, or potential future, location of the receptor.

6.7.1.1 The RBSL are concentration values for each chemical of concern in environmental media (for example, soil, water and air), based on the technical policy decisions for incremental carcinogenic risk levels and hazard quotients and on the potential exposure scenarios for human receptors (for example, residential, commercial, industrial, agricultural).

6.7.1.2 The RBSL are determined utilizing typical, non-site specific values for exposure parameters and physical parameters for media. For each exposure scenario the RBSL are based on current human health parameters (for example, USEPA RME), and current human toxicological information such as in the USEPA Integrated Risk Information System (IRIS) Database, Health Effects Assessment Summary Tables (HEAST), State sources, peer-reviewed source(s), agency approved toxicity data for proprietary chemicals or the best available toxicity data. Consequently, the RBSL are updated when new methodologies and parameters are developed.

6.7.1.3 When developing Tier 1 RBSL, cumulative risks and additive effects should be considered.

6.7.1.4 If RBSL or RESC are not developed or ORMC are not appropriate for complete or potentially complete exposure pathways, then further evaluation for those exposure pathways is conducted in Tier 2.

6.7.1.5 For indirect exposure pathways, fate and transport models can be used to predict RBSL at a source area that correspond to exposure point concentrations.

6.7.1.6 *The user should always review the assumptions, technical policy decisions and methodology used to derive values the RBSL to make sure that they are consistent with reasonable maximum exposure scenarios for the site being considered as well as currently accepted methodologies.*

6.7.1.7 Appendix X3 provides examples of the derivation of RBSL for a range of compound types. The assumptions and methodology used in deriving the examples are discussed in

Appendix X3. Note that not all of the possible exposure pathways are considered in the derivation of the examples. *Appendix X3 is presented solely for the purpose of providing examples of the development of the RBSL. The values and methodologies should not be viewed as proposed RBSL.*

6.7.2 *Tier 2 and 3 Corrective Action Goals*—The corrective action goals in Tier 2 and Tier 3 are the SSTL, SSEC or ORMC, as applicable. Tier 2 and Tier 3 provide the user with options for human exposures to determine SSTL and point(s) of exposure. Corrective action goals are based on SSTL, SSEC and ORMC that are protective of human health and the environment. It is important to note that Tier 1 RBSL, Tier 2 SSTL and Tier 3 SSTL are based on achieving similar levels of protection of human health. Similar levels of protection are defined by the technical policy decisions on risk targets, methods for addressing cumulative risks and additive effects and methods for addressing uncertainty. Tier 2 and Tier 3 provide the user options to determine SSEC for relevant ecological receptors and habitats. Tier 2 and Tier 3 also provide the user with the option to develop or refine ORMC, where applicable, consistent with the technical policy decisions. In Tier 2 and Tier 3 the corrective action goals for the source area(s), point(s) of demonstration and point(s) of exposure may be different values or criteria, for example, different concentrations of chemical(s) of concern in environmental media based on the fate and transport relationships between the various locations.

6.7.2.1 For human exposure pathways, the Tier 2 SSTL can be derived from the same equations used to calculate Tier 1 RBSL, except that site-specific parameters are used in the calculations. The additional site-specific data may support alternate fate and transport analyses. The Tier 2 analysis may involve applying Tier 1 RBSL at more probable point(s) of exposure. The SSTL may also be needed for potential exposure pathways using site-specific conditions for which no RBSL were developed or ORMC were appropriate in Tier 1. The development of SSTL may also consider a combination of these methods.

6.7.2.2 For relevant ecological receptors and habitats, the Tier 2 evaluation may involve the development of SSEC and the comparison of site conditions to SSEC. Other qualitative or quantitative analyses may also be appropriate for a Tier 2 evaluation.

6.7.2.3 For human exposure pathways, Tier 3 provides the user with an option to determine SSTL for both direct and indirect human exposure pathways using site-specific parameters and point(s) of exposure and point(s) of demonstration when it is judged that Tier 2 SSTL should not be used to determine corrective action goals.

6.7.2.4 Cumulative risks and additive effects should be considered when developing Tier 2 and Tier 3 SSTL.

6.7.2.5 For relevant ecological receptors and habitats, a Tier 3 evaluation is typically quantitative in nature and involves more site-specific data than previous tiers. Determination of SSEC may involve a more extensive site-specific analysis. In developing the SSEC, the user should consult current federal, state and other information.

6.7.3 *Use of Multi-Component Measurements*—Various multi-component measurement chemical analyses are often used in site assessments. These methods include, total petroleum hydrocarbons (TPH), total organic carbon (TOC) and total organic volatiles (TOV), and usually determine the total amount of a class of compounds. Some of these multi-component measurements are general measures and as such may provide insufficient information about the amounts of individual chemical(s) of concern present to calculate risk. However, new methods for the evaluation of multi-component measurements in terms of risk and exposure are being developed. If multi-component measurements are used as corrective action goals in any tier evaluation, the user is responsible for determining that the evaluations are based on currently acceptable methodologies and parameters that are applicable to risk-based calculations.

6.8 *Tier 1 Evaluation*—Tier 1 provides the user the option to identify or develop the corrective action goals (see section 6.7.1). If there is no corrective action goal for a specific exposure scenario, then the exposure pathway cannot be evaluated in Tier 1 and it is carried to a Tier 2 evaluation. In Tier 1, the point(s) of exposure and point(s) of demonstration are assumed to be located proximal to the source area(s), regardless of the actual or potential future location of the point(s) of exposure. The site conditions as determined through the initial site assessment and the site assessment for Tier 1 are compared to the corrective action goals based on the complete and potentially complete exposure pathways defined by the site conceptual model. In Tier 1, the quantity of assessment data are typically limited, therefore conservative assumptions about site conditions are used for the comparison. For example the maximum concentrations of chemical(s) of concern are compared to the RBSL or ORMC. If there are sufficient site assessment data, the user may choose to use a statistically derived value rather than a conservative value. Background conditions should also be considered in the Tier 1 comparisons. In Tier 1, the evaluation for relevant ecological receptors and habitats should take the form of qualitative analyses, as discussed in 6.9.

6.9 *Ecological Evaluation*—Utilize a qualitative ecological screening evaluation to support the identification of relevant ecological receptors and habitats, to select relevant ecological exposure pathways and to identify potential chemical(s) of concern (see Appendix X5). Due to the complexity of ecological systems, if exposure pathways are complete or potentially complete, then a refinement of the exposure pathway analysis should occur in Tier 1 to focus the assessment as early in the process as practicable or appropriate. Selection of RESC or development of RESC should occur, and a comparison to site conditions made. If after this refinement, complete or potentially complete ecological exposure pathways still exist for relevant ecological receptors and habitats on or near the site, then a site-specific assessment should be conducted. The site-specific assessment may involve selection of RESC or the development of SSEC, as applicable. This may include literature values, determining the toxicity of media to test organisms, biological surveys or multiple lines of evidence.

6.10 *Evaluation of Tier Results*—Within each tier evaluation the user compares the site conditions to the corrective action goals.

6.10.1 If site conditions meet corrective action goals for chemical(s) of concern, then no further action may be appropriate (see 6.10.3). If site conditions do not meet the corrective action goals, then interim remedial action, remedial action, or further tier evaluation must be conducted.

6.10.1.1 *Interim Remedial Action*—An interim remedial action may be more practical than attempting to implement a comprehensive remedial action (for example, when the desired risk reduction is infeasible due to technology or resource limitations). Furthermore, if interim remedial actions can reduce or eliminate a short-term threat, then interim remedial action steps should be employed before proceeding to further tier evaluation. If an interim remedial action is selected in the evaluation under 6.10.1 as the most appropriate option, then a design is implemented to address the most significant concerns in an expedited fashion. The interim remedial action may include removal or treatment of source area(s) with complete or potentially complete exposure pathways, or another method to address the most significant exposure concerns, to reduce migration, and facilitate reassessment of the tier evaluation. Interim remedial actions may also lead to further evaluation of response actions.

6.10.1.2 *Remedial Action*—If remedial action is selected in the evaluation under 6.10.1 as the most appropriate option, then a remedial action is designed and implemented to reduce concentrations at the point(s) of exposure, to eliminate an exposure pathway, or to address a potentially unacceptable risk to relevant ecological receptors and habitats. The remedial action may include some combination of source removal, treatment, and containment technologies, exposure pathway elimination as well as activity and use limitations (for additional discussion see 6.13). Examples of these include soil removal, soil venting, bioventing, air sparging, pumping for hydraulic control, chemical fixation, capping, industrial workplace protective practices and natural attenuation. When site conditions meet corrective action goals, then the user may elect to move to 6.10.3. Additionally, remedial action may be used to reduce concentration(s) of chemical(s) of concern, reduce exposures and facilitate a re-evaluation of the tier analysis.

6.10.1.3 *Further Tier Evaluation*—If further tier evaluation is warranted, additional site assessment information should be collected to develop or identify corrective action goals under a Tier 2 or Tier 3 evaluation (see 6.7.2). Further tier evaluations are conducted only for the complete and potentially complete exposure pathways. Further tier evaluation is warranted when:

(1) The basis for the Tier 1 or Tier 2 corrective action goals, as applicable, (for example, hydrogeology, exposure parameters, point(s) of exposure, reasonable land use options) are not representative of the site-specific conditions; or,

(2) The Tier 1 or Tier 2 corrective action goals, as applicable, do not exist for a complete and potentially complete exposure pathway; or,

(3) The corrective action goals developed under further tier evaluation will be significantly different than the Tier 1 or Tier 2 corrective action goals, as applicable, or will significantly modify the remedial action activities; or,

(4) The cost of remedial action to achieve Tier 1 or Tier 2 corrective action goals, as applicable, will likely be greater than the cost of further tier evaluation and subsequent remedial action.

6.10.2 If site conditions meet corrective action goals, but the user is not confident that data support the conclusion that the site conditions will meet the corrective action goals in the future, then the user institutes a monitoring plan to collect data sufficient to confidently conclude that site conditions will meet corrective action goals in the future (see 6.14). After these data are collected, no further action may be appropriate (see 6.10.3). If site conditions will not meet corrective action goals in the future, then initial remedial action, remedial action or further tier evaluation is appropriate (see 6.10.1).

6.10.3 If site conditions meet corrective action goals and the user is confident that data support the conclusion that site conditions will meet corrective action goals in the future, then no additional corrective action activities are necessary, and the user has completed the RBCA process. In practice, this is often accompanied by the issuing of a No Further Action letter, or its equivalent, by the applicable regulatory agency (see Appendix X1).

6.11 *Tier 2 Evaluation*—Tier 2 provides the user with an option to determine the site-specific point(s) of exposure and point(s) of demonstration and corresponding corrective action goals for the chemical(s) of concern applicable at the point(s) of demonstration and source area(s) as well as corrective action goals for relevant ecological receptors and habitats (see 6.7.2). Additional site assessment data may be required; however, the incremental effort may be minimal relative to Tier 1. In most cases, only a limited number of exposure pathways, exposure scenarios, and chemical(s) of concern are considered in the Tier 2 evaluation since many are eliminated from consideration during the Tier 1 evaluation. In Tier 2, the evaluation for relevant ecological receptors and habitats can take the form of additional qualitative or quantitative analyses.

6.11.1 In Tier 2 for human health the user identifies the direct and indirect complete and potentially complete exposure pathways. The Tier 2 evaluation may include:

6.11.1.1 For the complete and potentially complete indirect exposure scenarios, identification of the appropriate site-specific point(s) of exposure and point(s) of demonstration. A combination of assessment data and predictive modeling results are used to determine the SSTL at the source area(s) or the point(s) of demonstration, or both, or

6.11.1.2 Application of the RBSL, or ORMC, for the direct exposure scenarios at site-specific point(s) of exposure (as opposed to the source area(s) as is done in Tier 1). The SSTL for source area(s) and point(s) of demonstration can be determined based on the demonstrated and predicted attenuation (that is, reduction in concentration with distance) of compounds that migrate away from the source area(s), or

6.11.1.3 Application of a statistical data handling method, when appropriate data exist, to the source area(s) data and comparison of the representative source area(s) concentration to the RBSL, or

6.11.1.4 Application of the site-specific exposure and site

condition values to the Tier 1 RBSL calculations to determine SSTL for the source area(s), or

6.11.1.5 Implementation of alternate fate and transport algorithms to determine SSTL, or

6.11.1.6 Development of SSTL may consider a combination of these methods.

6.11.1.7 The SSTL may also be needed for potential exposure pathways using site-specific conditions for which no RBSL were developed or ORMC identified in Tier 1.

6.11.1.8 An example of Tier 2 application is illustrated in Appendix X7.

6.11.2 The Tier 2 evaluation of the RBCA process involves the development of SSTL based on the measured and predicted attenuation of the chemical(s) of concern away from the source area(s) using relatively simplistic mathematical models. The SSTL for the source area(s) are generally not equal to the SSTL for the point(s) of demonstration. The predictive models are:

6.11.2.1 Relatively simplistic, and are often algebraic or semi-analytical expressions,

6.11.2.2 Limited to practicably attainable site-specific data, or easily estimated quantities (for example, total porosity, fraction organic carbon, depth to ground water) for input parameters, and

6.11.2.3 Based on descriptions of relevant physical and chemical phenomena. The models are relatively simple, (that is, certain complex mechanisms are neglected) which results in predicted concentrations that are greater than those likely to occur (for example, assuming constant concentrations in source area(s)). Appendix X4 discusses the use of predictive models and presents models that might be considered for Tier 2 evaluation.

6.11.3 For relevant ecological receptors and habitats the Tier 2 evaluation may involve further site-specific evaluation utilizing additional qualitative or quantitative analyses.

6.11.4 Compare the site conditions to the Tier 2 corrective action goals. Identify the exposure scenarios where the site conditions do not meet the corrective action goals and evaluate these scenarios in accordance with 6.10.

6.12 *Tier 3 Evaluation*—In a Tier 3 evaluation for both human health and relevant ecological receptors and habitats, corrective action goals are developed on the basis of more sophisticated statistical evaluations and chemical fate and transport analyses, using site-specific input parameters for both direct and indirect exposure scenarios, as well as alternative toxicity data, exposure factors and bio-availability information, (see 6.7.2). Tier 3 evaluations could involve collection of significant additional site information, statistical evaluation of that data and completion of more extensive modeling efforts than required for either a Tier 1 or Tier 2 evaluations. Cumulative risks and additive effects are important considerations in Tier 3 analyses.

6.12.1 Tier 3 analyses may include the following:

6.12.1.1 The use of numerical ground water modeling codes that predict time-dependent dissolved chemical transport and account for spatially varying permeability fields to predict exposure point(s) concentrations,

6.12.1.2 The use of site-specific data, mathematical models, alternative toxicity data, bio-availability information and

6.12.1.3 Implementation of probabilistic exposure analysis methods to quantify the uncertainties in the exposures for a given site (see Appendix X2 and Appendix X4), and

6.12.1.4 The gathering of sufficient data to refine site-specific parameter estimates (for example, biodegradation rates, desorption factors) and improve the accuracy of the model predictions.

6.12.2 In a Tier 3 evaluation for relevant ecological receptors and habitats, activities to identify applicable SSEC are quantitative in nature. Where appropriate, these assessments may be conducted in accordance with current federal or state acceptable methods for relevant ecological receptors and habitats.

6.12.3 Compare the site conditions to the Tier 3 corrective action goals. Identify the exposure scenarios where the site conditions do not meet the corrective action goals and evaluate these scenarios in accordance with 6.10, except that a tier upgrade (6.10.1.3) is not available.

6.13 *Implementing the Selected Remedial Action Program*—When it is judged by the user that no further tier evaluation is necessary, or practicable, or applicable, a remedial alternatives evaluation should be conducted to determine the appropriate remedial actions necessary to achieve the corrective action goals determined for the site. Because remedial action itself can create new risks, including imposing significant physical, biological and chemical stresses on relevant ecological receptors and habitats, the remedial action should be balanced against the short and long term risks consistent with the technical policy decisions.

6.13.1 The remedial alternative evaluation should consider the applicable technical policy decisions and the following for each remedial action:

6.13.1.1 Effectiveness of the remedial action in protecting human health and the environment,

6.13.1.2 Long-term reliability and probable success in meeting the corrective action goals now and in the future,

6.13.1.3 Short-term risks posed by the implementation of the remedial action,

6.13.1.4 Amenability of the remedial action to integration with property redevelopment plans,

6.13.1.5 Acceptability of the remedial action to the stakeholders,

6.13.1.6 Implementability and technical practicability of the remedial action, and

6.13.1.7 The cost-effectiveness of the options to meet the corrective action goals.

6.13.2 Based on the alternative evaluation conducted under 6.13.1, a remedial action(s) is selected. The detailed design and specification of the selected alternative(s) are then developed for installation and operation.

6.13.3 The remedial action is implemented in accordance with the design and specification developed under 6.13.2. Following implementation, monitoring should be conducted to ensure that the design conditions are being met by the remedial action (see 6.13.4 and 6.14).

6.13.4 The remedial action must continue until such time as monitoring indicates that the corrective action goals have been met for a statistically significant number of monitoring periods, at the point(s) of demonstration or source area(s), or both, as appropriate.

6.13.5 At any time during the remedial action:

6.13.5.1 If site conditions change (for example, property use changes from residential to commercial and this was not anticipated in the future use of the property) or new data becomes available (for example, more refined aquifer parameters, site-specific biodegradation rates, new toxicity data for the chemical(s) of concern, or site-specific exposure factors are better defined) that would impact the corrective action goals, the user should return to the tier evaluation.

6.13.5.2 If the remedial action was initiated with Tier 1 or Tier 2 corrective action goals, the user may choose to return to the next higher tier evaluation.

6.13.5.3 If concentrations of chemical(s) of concern decrease to an asymptotic level that is above the corrective action goals, at the point(s) of demonstration, the user must choose to either conduct further tier evaluation (unless Tier 3 has already been completed), modify the remedial action system or further reduce or eliminate exposures through activity and use limitations.

6.13.6 Similar remedial actions can be implemented at different sites with different design objectives depending on the chemical(s) of concern, the hydrogeological setting and the corrective action goals. For example at one site ground water pumping may be implemented for hydraulic control while at another the ground water pumping may be used for source removal. In addition, the reliability of different designs for different objectives will be different. Further, the reliability of different remedial action options will vary depending on the site setting and potentially on the regulatory and legal structure of the jurisdiction in which the site is located (for example, the reliability of activity and use limitations varies depending on the structure of the limitation and the jurisdiction in which it is implemented, see Appendix X6).

6.14 *Implementing a Monitoring Program*—In many cases monitoring is appropriate. Upon completion of this monitoring effort (if required), no further action is required.

6.14.1 Monitoring is necessary to:

6.14.1.1 Demonstrate the effectiveness of the implemented remedial action, or

6.14.1.2 Confirm that current conditions persist or improve with time, or

6.14.1.3 Verify model assumptions and conditions.

6.14.2 If monitoring does not demonstrate the effectiveness of the implemented remedial action the user should reevaluate the remedial action (see 6.13). If the monitoring does not confirm that current conditions persist or improve with time or verify model assumptions and conditions then the user should return to the applicable tier evaluation (see 6.11 and 6.12).

6.15 *Site Maintenance*—In addition to monitoring, some remedial actions (for example, physical barriers - capping, hydraulic control, or activity and use limitations) require maintenance to ensure integrity and continued performance. The required maintenance should be conducted until corrective action goals have been demonstrated. In the case of some activity and use limitations site maintenance may be needed beyond the remedial action closure (see 6.16).

6.16 *No Further Action and Remedial Action Closure*—No further action is necessary when corrective action goals have been demonstrated (as required in 6.10.3) to be achieved. Under these circumstances, if monitoring and site maintenance are no longer required to ensure that conditions persist, then no further action is necessary, except to ensure that activity and use limitations (if any) remain in place. If these conditions are met, the closure should be documented. Example closure instruments are discussed in Appendix X1.

6.17 *RBCA Report*—After completion of the final tier evaluation, a RBCA report should be prepared. The report must include all of the data collected to support the RBCA decisions that were made. The report content will depend on the specific site, the tier evaluation and the requirements of the regulatory agency, however the RBCA report will typically include the following:

6.17.1 An executive summary,

6.17.2 A site description,

6.17.3 A summary of the site ownership and use,

6.17.4 A summary of past releases or potential source areas,

6.17.5 A summary of the current and completed site activities,

6.17.6 A description of regional hydrogeologic conditions,

6.17.7 A description of site-specific hydrogeologic conditions,

6.17.8 A summary of reasonably anticipated use,

6.17.9 A summary and discussion of the site conceptual model, the exposure pathway analysis and the qualitative ecological screening evaluation,

6.17.10 A summary of the tier evaluations conducted, including the methods and assumptions used to derive the corrective action goals,

6.17.11 A summary of the analytical data and the appropriate corrective action goals used,

6.17.12 A site map of the location,

6.17.13 An extended site map to include significant features (for example local land use and ground water supply wells),

6.17.14 Site plan view showing location of facility or site structures (for example, aboveground storage tanks, underground storage tanks, waste management areas, buried utilities and conduits, suspected/confirmed sources),

6.17.15 Site photos, if available,

6.17.16 A ground water elevation map,

6.17.17 Geologic cross-section(s),

6.17.18 Concentration maps for chemical(s) of concern in the appropriate environmental media,

6.17.19 A section that contains an outcome statement defining the corrective action goals and completion conditions, and

6.17.20 A summary of the technical policy decisions.

APPENDIXES

(Nonmandatory Information)

X1. CONSIDERATIONS FOR DEVELOPMENT OF A RBCA PROGRAM

X1.1 *Introduction*:

X1.1.1 The agencies responsible for developing and implementing a RBCA program and interested stakeholders are the focus of this appendix. Agencies empowered to enforce statutes, promulgate and enforce regulations and implement policy guidance governing cleanup must make key decisions in order to blend the existing program with the risk-based approach. Such decisions are typically made, either formally or informally, with the involvement of stakeholders. In this context a risk-based corrective action (RBCA) program provides regulatory agencies, the regulated community, and the stakeholders with clear, practical guidelines for assessment and control of risks posed to human health and the environment by chemical releases. The RBCA framework provides an opportunity and a focus for agencies working toward the development of a unified and consistent set of policies and procedures for risk-based corrective actions. While no single procedure is mandatory, a clear understanding of the RBCA planning process for both regulatory agency personnel and the regulated community can streamline the site review process by the regulatory agency and expedite implementation of appropriate corrective action. Toward this end, development of a RBCA regulatory program involves considerable discussion by the regulatory authority and other concerned stakeholders regarding legal and regulatory issues, technical policy decisions, and program documentation and communication.

X1.1.2 This appendix is intended to serve as an outline of suggestions to assist in the development process and identify relevant information resources. Although in some cases specific examples are discussed in this appendix for ease of understanding, there is no intention to advocate specific positions regarding regulatory policies. Instead, a list of key actions and issues to be considered by the RBCA development team are provided. An effective development team includes all major stakeholder groups. In some programs there are statutory or regulatory authorities that require interested parties to be brought together before rules regarding clean-up programs are promulgated. There may be requirements to appoint and convene a multidisciplinary advisory board to provide advice before rules are promulgated. The composition of such advisory boards may be detailed in regulation or a matter of policy in a particular jurisdiction.

X1.2 *Overview of RBCA Development Process*:

X1.2.1 The RBCA program development process involves the following principal steps:

X1.2.1.1 *Selecting the Framework*—Deciding to use the RBCA framework.

X1.2.1.2 *Administrative Actions*—Coordination with concerned stakeholders and evaluation and/or revision of existing laws and regulations. A key element is the Stakeholders' Meeting.

X1.2.1.3 *Technical Policy Decisions*—Specification of technical criteria required for calculation of risk-based corrective action goals, performance standards for modeling, exposure pathway analysis and data collection.

X1.2.1.4 *Program Documentation*—Development of written guidance and other tools to assist the agency and the user. (for example, standardized report forms and software tools for application of the RBCA process.)

X1.2.1.5 *Specialized Program Elements*—The development team may wish to consider specialized mechanisms for voluntary cleanup programs (VCPs), such as completion letters, no further action statements or liability covenants. Many state programs, for example, have specific or general statutory provision governing the issuance of liability covenants.

X1.2.1.6 *Education/Outreach*—Training or communication or both with regulatory staff, the regulated community, environmental consultants and contractors, stakeholder organizations, and the general stakeholders.

X1.2.1.7 *Stakeholder Involvement*—The development team may consider use of standard policies and techniques for communication with stakeholders on a site-by-site basis and/or for program development and implementation.

X1.2.2 In general, these development tasks may be undertaken in a sequential manner. However, some degree of overlap or iteration may be required to establish a workable and efficient risk-based program for assessment, remediation, and closure of affected sites. Detailed discussion of each program development step is provided below.

X1.3 *Selecting the Framework*—The threshold decision for a regulatory agency is whether to develop or modify an existing cleanup program around the RBCA framework. This decision may require some underlying assumptions to be modified in fundamental ways. Many traditional cleanup programs require the collection of large amounts of information and a complete and detailed profile of the site, before an evaluation of the risks, risk reduction measures or the remedy selection process is ever initiated. In contrast, the basic principle of RBCA is to allow the user to gather just enough information to make a risk-based decision at the lowest tier of analysis. This facilitates timely decisions, based on risk, most efficiently, and with the least amount of information possible. At lower tiers, uncertainty is compensated by more conservative assumptions. In other words, at lower tiers, a complete detailed profile of the site is not needed because the agency builds in a reasonable worst-case set of assumptions in the corrective action goals, as appropriate. At higher tiers, uncertainty is reduced through collection and analysis of more site-specific information. The agency responsible for decision making should consider the level of understanding and acceptance of this underlying philosophy by the people who must implement it, prior to embarking on program development using RBCA.

X1.4 *Administrative Actions*:

X1.4.1 This guide provides a risk management framework that can be customized to address the concerns of local stakeholders and integrated with the legal and regulatory framework. For this reason, it is important to involve key stakeholders (defined below) early in the program development process. In addition, the development team (defined below) may review existing laws and regulations to identify possible roadblocks to RBCA implementation and any statutory or regulatory actions that may be required. While specific provisions for stakeholder involvement are beyond the scope of this guide, it is recommended that stakeholder involvement and outreach activities be integrated into the RBCA development process. Stakeholder involvement activities may or may not be appropriate at every site in a RBCA program. Key considerations are outlined below.

X1.4.2 *Advisory Committee Acts and Stakeholders Meetings*—Many state and federal programs develop regulations, policies and guidelines with specific consensus building requirements and formal or informal requirements for representation by balanced stakeholder groups. It is extremely important for the program development process to consider the use of such representation to reach consensus on some of the widely debated issues which could impact the application and implementation of the RBCA process. During the development process, Stakeholders Meetings among the following potential participants may be critical: the regulatory agency and other users; industry; real estate interests; financing interests; local governments and redevelopment authorities; citizen groups and environmental groups. The purpose of such meetings is to reach consensus on a common approach for the critical technical policy decisions discussed below, especially in the area of risk management.

X1.4.3 *Development Team*—To assist in producing an efficient and workable RBCA program, an effective development team can serve a key role. The team may include representatives from the principal regulatory agency, the regulated community, environmental groups, citizens and other parties likely to be affected by a new or revised regulatory program. Other such stakeholders may include, but are not limited to: local government; representatives from the real estate, banking, or insurance industries; environmental professional or technical organizations; citizens' groups; and environmental and public health advocates. Consideration of these diverse perspectives can greatly facilitate the ultimate success of the RBCA program. Stakeholders may be involved either as a working group (that is, the development team) or as an oversight or advisory committee. In most cases, to provide for meaningful discussion and input, preliminary training or orientation of stakeholder representatives regarding RBCA concepts and key policy issues will be required.

X1.4.4 *Statutory Review*—The development team should review existing environmental laws and identify how the RBCA program should be crafted to comport with existing statutes and to identify possible constraints on RBCA implementation. Examples include state drinking water quality criteria, non-degradation standards and policies, wellhead protection programs, legally mandated soil and ground water remediation standards, or full site delineation requirements. A distinction should be made to differentiate between ground water restoration standards and those for ground water protection, since they may have different purposes. The development team may evaluate existing laws to customize the risk-based approach, recommend statutory modification to facilitate RBCA, or a combination of both. Key elements in the statutory review include consideration of natural resource protection and conservation requirements. Many programs have, for example, explicit ground water protection and future use statutes.

X1.4.5 *Regulatory Action*—In some cases, current regulations mandate uniform concentration limits for soil and ground water affected by a chemical release (for example, background levels, practical quantitation limits, or other maximum allowable limits), regardless of location, land use, or exposure scenario. In some cases this situation may require careful crafting of the RBCA framework to accommodate current regulations. In other cases incorporation of risk-based corrective action goals may require redrafting portions of the existing rules. As a practical matter, many programs have employed some combination of regulatory accommodation and modification. The outcome of any debate to replace or modify existing programs depends on active, successful of stakeholder involvement activities. In addition, depending on the scope of the existing regulations, accommodations or modifications may also be required to those sections addressing: site assessment requirements; response action evaluation procedures; remediation options; requirements for treatment; permanent solutions; use of activity and use limitations; use of natural attenuation; point of compliance; and compliance monitoring specifications. The regulations may consist of language defining key policy elements of the RBCA program and authorizing agency promulgation of detailed written guidance. In some jurisdictions, detailed procedural guidelines will need to be incorporated in the rules due to legal constraints on the use of guidance documents.

X1.4.6 *Policy and Guidance Development*—As an alternative to statutory or regulatory mandates, agencies may elect to implement the RBCA program in the form of policy guidelines and/or technical guidance materials. As noted below, clearly documented policies and administrative procedures are critical to proper understanding and implementation of a RBCA program by the regulated community. Having technical decisions promulgated through policy and/or guidance enhances the flexibility of the program, allowing for routine updates as the leading edge of environmental science evolves. Many states, for example, rely on policy and guidance for certain aspects of the cleanup program.

X1.5 *Technical Policy Decisions*:

X1.5.1 *Reaching Consensus*—Virtually all cleanup programs demonstrate differences of opinion among stakeholders on some of the key policy issues which confront remedy selection and the development of a RBCA program. There is a natural tension which exists, for example, between groups which believe allowable risk-levels should be set conservatively and those which would recommend more relaxed standards. The use of activity and use limitations, the requirement for treatment and point(s) of compliance are other controversial issues which will inevitably confront the regulatory agency responsible for such policy decisions. An important principle to use in this part of the development process is to engage in the debate, through adequate stakeholder involvement, and to achieve as much consensus as possible among differing views. Technical policy decisions can then be incorporated into the RBCA framework. Although the agency must clearly and decisively exercise its regulatory responsibility and make the key decisions needed to implement the cleanup program the experience from many programs has shown that continued stakeholder involvement in the regulatory process has been very beneficial.

X1.5.2 *Risk Decisions:*

X1.5.2.1 For a release to enter into the RBCA analysis, the regulatory agency or user should establish criteria to determine whether or not further evaluation or action is needed for human health or ecological concerns. This is a threshold technical policy decision. Some agencies may establish notification thresholds, based on reportable quantities or concentrations of chemical releases. In some programs, if no exposure pathway is complete no further risk analysis is required. In this manner, regulations or policy show whether or not the application of RBCA is needed.

X1.5.2.2 The regulatory agency has the prerogative to make key policy decisions during program development. The RBCA program development team should consider detailed guidance on a comprehensive range of policy issues such as the calculation and application of risk-based corrective action goals. A summary list of the key issues to be addressed is provided in Table X1.1. These policy decisions serve to flesh out the ASTM RBCA framework, providing a consistent basis for determining whether remediation is necessary and, if so, "how clean is clean" for a wide variety of site conditions. To design the RBCA program, the development team may attempt to build consensus in the following areas: cumulative risks and additive effects; the conservatism applied to risk calculations; minimum data requirements for site assessment and closure; variables affecting the selection of target risk levels and exposure factors; and the applicability of ASTM, state, or Environmental Protection Agency (EPA) calculation protocols (see Table X1.1).

X1.5.2.3 *Cumulative Risk and Additive Effects*—The responsible regulatory agency should consider how to handle additive effects or cumulative risk. Some agencies may create Tier 1 RBSL with sufficiently conservative assumptions to account for cumulative risks and additive effects and therefore establish screening levels within an allowable risk range. Other agencies may require Tier 1 screening to use conservative single endpoint screening values and not a range. For example, in Massachusetts, screening numbers (the equivalent of RBSLS) are set at a single number based on a $1*10^{-6}$ cancer risk endpoint or 20 % of the non-cancer Reference Dose, which ever is lower. The rationale here is that the screening numbers using the conservative risk endpoints for a single number will be protective for almost all sites. For higher Tier analysis the total site risk must not exceed a $1*10^{-5}$ cancer risk endpoint or a non-cancer Hazard Index of 1.0 .(Massachusetts Contingency Plan, 1997, 310CMR 40.0983). For some chemicals, non-cancer effects may be considered for RBSL because they may have an additive effect on the same target organ. EPA guidance recommends that the potential for additive effects must be carefully evaluated at every site by considering the Hazard Index for chemicals with References Doses or Concentrations are based on the same endpoint of toxicity such as a target organ or system. For example, EPA Soil Screening guidance recognizes at least eleven chemicals, including several pesticides, whose non-carcinogenic effects may be specific to the human liver (Ref **(2)**, pp. 9-15)

TABLE X1.1 Example Technical Policy Decisions to be made by the Regulatory Agency

How to Reach Consensus
Risk Decisions
How to Deal with Cumulative Risk and Additive Effects
Appropriate Sources of Toxicity Information
Use of a Risk Range
Other Relevant Measurable Criteria
How to Record Consensus Decisions
Site Assessment
Ecological Risk Assessment
Land Use
Future Uses and Exposure Pathway Analysis
How to Use Activity and Use Limitations
When to Use Activity and Use Limitations
Resource Protection
Remedy Selection Criteria
Balancing the criteria
Treatment Versus Activity and Use Limitations
Remedial Action Effectiveness
Long-term Reliability
Short-term risks
Property Redevelopment Plans
Acceptability to Stakeholders
Implementability
Cost-effectiveness

X1.5.2.4 States may also develop their own sources of information on chemical toxicity. Among other things, the regulatory agency must decide on the level of scientific research, reliability and uncertainty which is appropriate in a particular program for a particular use. This is especially true for toxicity information. Regulatory agencies may also wish to evaluate the applicability of occupational, ambient exposure values for workers at operating industrial sites within their RBCA program. In addition to chemical toxicity, an agency developing a RBCA program may wish to consider alternative data sources and methods to address exposure and bioavailability. The alternative data sources and methods consulted and used should be reasonably accessible and clearly communicated to the user.

X1.5.2.5 *The Risk Range*—Many programs allow the flexibility of decision making within an acceptable risk range. For example, the Federal Superfund Program has established an acceptable range for lifetime excess cancer risks of 10^{-4} to

10^{-6}.[7] Corrective action goals are based on achieving similar levels of protection for human health that are within the acceptable range of risks and the protection of environmental receptors as determined by the implementing agency. Thus the levels of protection may vary from site to site as well as across the tiers at the same site as long as it will result in achieving levels of protection within the defined acceptable range of risks. In some programs, however, a single number for total site risk has been selected by the regulatory agency or set by statute for all sites and tiers of analysis. While such programs lack the flexibility of a risk range, some stakeholders feel this approach affords a level of fairness and cleanup consistency.

X1.5.2.6 The regulatory agency may also consider development of Other Relevant Measurable Criteria in the program which are not directly related to human health or ecological risks. In planning risk based programs, noise, odors, taste and visual appearance may be important considerations in evaluating chemical releases.

X1.5.3 Upon completion of the planning session(s), the consensus decisions of the development team may be recorded in a variety of ways. Development of a RBCA program should avoid arbitrary decisions, which are isolated from stakeholder involvement. Based on this process and sufficient training, a knowledgeable person should be able to complete a RBCA evaluation in accordance with the established program. Specialized documentation may be required for RBCA implementation to reach site completion as outlined below.

X1.5.4 *Site Assessment:*

X1.5.4.1 *Initial Site Assessment*—The purpose of the initial site assessment is to form the first conceptual model of a release at a site based on rudimentary information. This model may change over time, as more information becomes available. For simple releases at sites incomplete pathways may eliminate the need for any further evaluation. Some regulatory agencies, for example, define threshold quantities or concentrations of a chemical release as exclusion from or entry to a further RBCA analysis.

X1.5.4.2 One of the key technical decision frameworks in a risk based corrective action program is the type and degree of assessment needed to provide the level of information necessary to make cleanup decisions. The identification of chemicals of concern in the soil, ground water and other media in relationship to exposure pathways must be sufficient to support the decision being made, while minimizing the amount of data collected which is of limited use. The objective is to optimize the site assessment using the "just-in-time" process of least information for lower tiers. While site assessments for the RBCA process attempt to determine all significant sources of chemical releases, performance standards to define the extent of the release may differ, according to the requirements developed by the regulatory agency. For example, some agencies require definition of the full horizontal and vertical extent of all releases to below the most conservative risk-based screening values. Other agencies outline how to define data quality objectives and then use those to determine the extent of site assessment. Still others may select conservative assumptions in lower RBCA Tiers as a substitute for the definition of all levels of chemical releases.

X1.5.4.3 There is a wealth of information from both government and industry on appropriate levels and techniques for assessment. These include Guide E 1912 Guide for Accelerated Site Characterization for Confirmed or Suspected Petroleum Releases and Guide E 1527, as well as numerous EPA guidance documents, including Ref **(1)**. In building a program, the development team should consider appropriate guidance for site assessment.

X1.5.5 *Ecological Risk Assessment*—The development team should consider carefully the issues that will arise during the development of a tiered framework for ecological risk assessment. This includes any boundary parameters, which may be selected during the qualitative ecological exposure assessment described in Appendix X5. For example, some regulatory agencies may decide that size or distance thresholds such as acreage or distance from resources may be useful screening tools for ecological assessment to determine if further, detailed assessment of the ecosystem is or is not required. Other regulatory agencies, on the other hand, may decide that this approach is too limited and that exclusion of some habitats and selection of receptors is too subjective to be useful. Ecological boundary parameters may depend on both regulatory and policy issues that apply to the geographic area to be covered by the standards being developed under the RBCA framework. Ecological screening criteria may include chemical concentrations, biological measures or other relevant generic criteria. The regulatory agency should exercise great care in selecting the appropriate decision points for ecological screening and assessment, especially in the selection of relevant ecological receptors and habitats. Appendix X5 contains a suggested qualitative ecological exposure assessment and describes the conceptual tiered framework for ecological risk assessment.

X1.5.6 *Land and Resource Use:*

X1.5.6.1 Current and reasonably anticipated future uses of land and water, including ecological considerations, are important in the RBCA process. This is because the exposure scenarios for human health and relevant ecological receptors depend on the population potentially exposed and the frequency and intensity of exposure to the media where chemicals have been released. Exposure pathway analysis may also be known as the conceptual site model.

X1.5.6.2 For a number of reasons, the restriction of future activities, access, or land use through various activity and use limitations may or may not be acceptable to the implementing agency as part of a cleanup remedy. As with other technical and policy issues, this decision requires consideration by stakeholder groups for a particular program in order to decide how effective such controls may be in reducing risks. For some programs, for example, the financial and other assurances required of an operating industrial facility may be totally acceptable to the regulatory agency to ensure exposure pathways which are limited to future industrial use. On the other

[7] EPA uses a 10^{-6} level as a point of departure for preliminary remediation goals for cancer risks. While the 10^{-6} starting point expresses EPA's preference for setting clean up levels at the more protective end of the risk range, these levels may be revised within the acceptable risk range based on consideration of appropriate factors, including exposure factors, uncertainty factors and technical factors.

hand, in a mixed-use urban area, a regulatory agency may decide that no particular activity and use limitation is acceptable to eliminate a potential residential exposure pathway.

X1.5.6.3 In the United States, the current and future allowable or planned uses of real property and other resources, such as water, are controlled by a complex combination of municipal, county, state and federal rules and processes. Each local jurisdiction may have unique land use controls. For this reason, the development team should consider decisions carefully on how and when to use restrictions on access, land use and human activities as part of available corrective actions for the site. While program developers should exercise care in using zoning regulations to rule out exposure pathways for on-site chemical releases, specifically targeted activity and use limitations may be effective tools in cutting off or preventing the completion of potential exposure pathways to chemical releases. Various types of activity and use limitations are described in Appendix X6.

X1.5.7 *Ground Water Restoration and Protection:*

X1.5.7.1 *Restoration*—The RBCA program development team should address critical issues regarding ground water restoration. Decisions on how to address ground water restoration are manifested in the RBCA program in determining the applicability of various ground water exposure pathways, point of demonstration requirements (where does the standard have to be met), and remedial objectives (restoration versus exposure prevention) for ground water. Existing state statutes, regulations, or policies must be accommodated during RBCA program development, but sufficient discretion may be afforded in deciding how the requirements apply in terms of degree and time to the various ground water resources. The USEPA policies that promote restoration of ground water for maximum beneficial use also recognize that all ground waters do not represent the same maximum beneficial use and therefore may not warrant an equivalent level of restoration. Because ground water restoration issues are sensitive and controversial, a wide stakeholder representation in the decision-making process may enhance the acceptance of decisions made by the RBCA development team. The National Contingency Plan, and Regional EPA Guidance such as the 1996 EPA Region 1 Ground water Use and Value Determination Guidance provide important and applicable concepts, principles, and criteria regarding ground water restoration.

X1.5.7.2 *Protection*—In order to address issues of non-degradation, and protection of ground water, many states have developed ground water classification systems which define beneficial and non-beneficial use ground waters. The classification systems may have been developed separately from restoration issues. They typically consider such factors as ground water quality and quantity, proximity to public and private ground water users, nature of the ground water use, presence of alternate or superior water supplies, susceptibility to impacts by releases of chemicals of concern, and location. Some state ground water classification systems which have been established through the USEPA's Comprehensive State Ground Water Protection Program (CSGWPP) may set ground water classifications at a state or regional scale, while other classification systems which are not established through a CSGWPP are applied as a set of criteria at individual release sites. Ground water classification systems provide a convenient context in which to establish policies regarding ground water for RBCA program development. Once a ground water classification system is established, the RBCA development team can make decisions about how the ground water protection scheme relates to restoration issues such as ground water exposure pathways, points of demonstration, and remedial objectives for the various ground water classes. Sources of information on ground water protection include 1986 draft Federal Ground Water Classification Guidelines and the 1992 EPA Final Comprehensive State Ground Water Protection Program Guidance,

X1.5.7.3 *Example Approaches:*

(1) The Massachusetts Department of Environmental Protection's (DEP) Massachusetts Contingency Plan establishes three classes of ground water, which factors in different risk considerations. GW-1 class ground waters are current and potential drinking water sources. GW-2 class ground waters are shallow ground waters where vapor migration pathways may be of concern. Both class GW-1 and GW-3 ground waters can also be class GW-2 ground waters. GW-3 class ground waters are non-drinking water ground waters where protection of surface waters from ground water discharges is of concern. Massachusetts DEP also has an EPA-endorsed CSGWPP, and a Ground Water Use and Value Memorandum of Agreement (MOA) with the US EPA designed to enhance consistency in ground water classification between Massachusetts DEP and EPA Region 1. The MOA is applicable to National Priority List sites and is structured in terms of high and medium use and value ground waters which is generally represented by GW-1, and low use and value ground waters (GW-3).

(2) The Pennsylvania Department of Environmental Protection Land Recycling Program sets a ground water classification system in terms of Used and Non-Use Aquifers. The point of compliance is established at the property boundary that existed at the time impacts were discovered, but the standard to be met at the point of compliance varies depending on the aquifer classification, properties of the chemical, and other site-specific considerations. Flexibility is provided to document lack of current and planned use of ground water within 1000 ft of the property boundary and within the impacted area in order to provide some consideration of special considerations in urbanized areas.

(3) The Ohio Environmental Protection Agency Voluntary Action Program sets an elaborate ground water classification system defined in terms of Critical Resource Ground Water/Non-Urban Setting, Critical Resource Ground Water/Urban Setting, Class A Ground Water/Non-Urban Setting, Class A Ground Water/Urban Setting, and Class B Ground Water. For each ground water class, different performance-based ground water response objectives are established for on-site and off-site ground water exposure pathways depending on if the source area is on-site, off-site, or both on-site and off-site.

(4) The Texas Natural Resource Conservation Commission's proposed Texas Risk Reduction Program would set a ground water classification system defined in terms of primary ground water resources (Class 1), other current and potential use

ground waters (Class 2), and no potential use ground waters (Class 3). Class 1 ground water resources must be fully restored to drinking water standards, while exposure prevention strategies may be established for Class 2 and 3 ground waters.

(5) The Louisiana Department of Environmental Quality's proposed Risk Evaluation/Corrective Action Program also sets three classes of ground water. Class 1 ground waters must be fully restored to drinking water standards, whereas drinking water criteria may be applied at the property line or closest reasonable off-site point of exposure for Class 2 ground water. Class 3 ground water requirements are based on the protection of any surface waters receiving discharge from the impacted ground water. Each of these example ground water classification systems have a common recognition that all ground water resources do not represent the same beneficial use, and accordingly provide the context for RBCA program development in the state.

X1.5.7.4 *Non-Attainment Zones*—Some states have developed ground water protection programs that include the designation of areas of ground water to be "non-attainment" zones. These zones could be located in urbanized areas where there are multiple impacts to ground water from releases of chemicals of concern. For example, Illinois recognizes ordinances established by units of local government that effectively prohibit the installation and use of potable supply wells as an effective activity and use limitation for the elimination of the ground water ingestion exposure pathway. When the ordinance does not apply to the local unit of government, then the local unit of government must enter into a Memorandum of Understanding with the state and maintain a registry of release sites within the ordinance area. Provisions are also set forth regarding notification of affected landowners, criteria for voidance of the ordinance, deed recordation, presence of existing potable wells, and ground water response objectives. Such non-attainment provisions recognize situations where units of local government may not place a high value on ground water resources within urbanized areas due to economic considerations, or due to the existence of regional historical ground water impacts that render the ground water unfit for use. The non-attainment zones may be re-evaluated as remediation by natural attenuation and individual user response actions improve the regional ground water conditions. These zones also provide a convenient context in which to establish policies regarding ground water restoration and protection for RBCA program development. Further, non-attainment zones may be of particular benefit in Brownfields redevelopment areas where ground water ingestion is not an otherwise relevant exposure pathway.

X1.5.8 *Natural Resource Values:*

X1.5.8.1 Many natural resources may have intrinsic value to the public, regardless of actual current or planned future use of the resource. For example, while the ground water underneath a National Park may not have an actual use, its value in pristine condition may outweigh considerations of actual use. Such large public policy issues may play a role in technical policy decisions for actions under a RBCA program, even though they are not related to risk, per se.

X1.5.9 *Remedy Selection Criteria*—Programs developing remedy selection should consider a wide array of factors such as treatment types, permanence versus institutional based remedies as well as the use of innovative technologies. Both risk reduction and long term risk management activities should consider remedy effectiveness against cost. Given limited resources for either publicly or privately funded corrective actions, the program development team may consider various schemes to balance selected risk criteria against the cost of corrective actions. Cost and effectiveness as balancing criteria are common elements of risk-based decision-making programs for corrective action. During the initial development of the program, in balancing these factors, the user should consider that no single factor will predominate over the others. As with other difficult risk management issues, such factors will require a careful balancing act among differing stakeholder opinions and focused decision-making on the part of the regulatory agency. In the end, however, it is the responsibility of the regulatory agency to determine how best to balance the criteria for a particular program.

X1.5.9.1 *Treatment versus Activity and Use Limitations*—While it is beyond the scope of this guide to suggest specific preferences for treatment or for activity and use limitations, these are issues which are critical to this technical policy decision which a regulatory agency must make. For example, the speed, frequency and short term cost-effectiveness of risk reduction through cutting off exposure pathways is generally of great interest to a regulatory program. These are benefits of using activity and use limitations. Also of interest to the regulatory agency are issues such as: returning potentially useable ground water to beneficial use, increasing land values, and lowering long term operation, maintenance and monitoring costs. These are some of the benefits of treatment and the reduction of toxicity, mobility, and volume of chemical releases. Categories of remedy selection criteria include:

X1.5.9.2 Effectiveness of the remedial action in protecting human health and the environment;

X1.5.9.3 Long-term reliability and probable success in meeting the corrective action goals;

X1.5.9.4 Short-term risks posed by the implementation of the remedial action;

X1.5.9.5 Amenability of the remedial action to integration with property redevelopment plans;

X1.5.9.6 Acceptability of the remedial action to the stakeholders;

X1.5.9.7 Implementability and technical practicability of the remedial action; and

X1.5.9.8 The cost-effectiveness of the options to meet the corrective action goals.

X1.6 *RBCA Program Documentation*:

X1.6.1 In order to ensure uniform application of the RBCA program, agencies should consider providing guidance documents and standardized tools for the regulated community and stakeholders. This documentation may consist of guidance manuals, standardized report forms and software tools. Such documentation may include the following:

X1.6.2 *RBCA Process Overview*—A RBCA guidance manual should outline the RBCA process and provide detailed step-by-step instructions for implementation of the program. The manual may include flowcharts outlining the program and a description of potential exposure pathways and transport mechanisms, including any assumptions regarding land use patterns, activity and use limitations, and remedy selection criteria.

X1.6.3 *Response Action Evaluation*—As the user gathers data, site conditions should be evaluated and a response action should be implemented, consistent with site conditions. This process may be repeated as additional data affords a more detailed characterization of the site. Some programs place sites in risk reduction categories, and when actions are performed, the category automatically changes to a less urgent status. In these programs the user selects a category and action which best addresses the short-term health (human and ecological) and safety concerns of the site while implementing the RBCA process. While some programs also have risk reduction categories to generate fees or for other statutory purposes, other programs avoid long term categorization schemes to avoid attaching a stigma to a property which could affect later transactions. As with many program development elements, the regulatory agency must resolve such issues, following the stakeholder involvement process.

X1.6.4 *Tier 1 Evaluation Guidance*—Tier 1 guidance may include the algorithms used to develop look-up tables, as well as an example look-up table or tables and detailed instructions on how to use the look-up tables. The RBCA framework refers to such look-up tables as Risk-Based Screening Levels or RBSL. Table X1.1 gives a suggested outline for the technical policy decisions that define the look-up table. These decisions can be made by all stakeholders and documented by the development team. The development team may decide to avoid creating RBSL and other look-up tables as "bright-line," stand-alone targets for clean up.

X1.6.5 *Tier 2 Evaluation Guidance*—Tier 2 guidance may include detailed instructions on selection of appropriate modeling procedures, target risks, exposure scenarios, and policies regarding off-site migration of chemicals of concern, performance and compliance monitoring. The guidance may also include the criteria for evaluating Tier 2 results, conditions for tier upgrade, criteria and location of the point of compliance, corrective action alternatives, and closure options.

X1.6.6 *Tier 3 Evaluation Guidance*—The guidance for Tier 3 may simply define the tier and explain its use. In order to remain flexible, specific procedures may not be recommended, however, it may be desirable for some limits to be placed on performance standards acceptable to the agency. Evaluations at the Tier 3 level may include alternative risk and exposure assessment protocols. As with the other tiers, Tier 3 guidance may also include criteria for evaluating Tier 3 results, and explain what alternatives exist for sites to meet Tier 3 criteria, such as corrective action, monitoring programs, variance procedures, more complex fate and transport modeling and engineering controls.

X1.6.7 *Administrative Procedures*—Administrative procedures for the RBCA application submittal and review process should be clearly outlined. To the degree feasible, separate steps of the RBCA planning program (for example, site assessment, response action evaluation, and Tier 1 evaluation) may be consolidated into single submittals to minimize the number of separate reviews requested by agency personnel for each site. As the user proceeds to higher Tiers of the RBCA process, the knowledge gained about the site is used to tailor the degree of investigation needed. Approval to move from one tier to the next may not be necessary. In some cases, after completion of the Tier 1 evaluation, the program may direct the user to proceed directly to a Tier 3 evaluation. The results of all completed tiers of analysis will typically be submitted to the overseeing agency for final documentation. Upon receipt, submittals may be checked for completeness. This may be facilitated with an administrative checklist prior to in-depth evaluation. Standard site completion letters (see X1.6) issued to volunteers may also be reviewed to ensure consistency with the RBCA program. Depending on the tier and the complexity of a site, an agency may phase its involvement (that is, for some programs, little oversight may be necessary for a Tier 1 evaluation, while a Tier 3 evaluation may require significant input from the agency).

X1.6.8 *Standardized Report Forms*—Standardized report forms can be very helpful in the RBCA process, aiding in the collection of site data and in both the preparation and review of site reports. Properly designed standardized report forms may guide the user in the collection of the appropriate type and amount of data. These forms can be of the "fill-in-the-blank" or "cross reference" variety, or both, for Tiers 1 and 2, and may outline general procedures to be used in Tier 3.

X1.6.9 *Tier 1/Tier 2 Software Tools*—A helpful element in a successful RBCA program may be computer software that assists both the regulatory agency and the regulated community in understanding and performing the multitude of necessary calculations. With the increasing use and power of personal desktop computers and the availability of common spreadsheet programs, user-friendly software tools can be developed for use by technical staff with minimal training in computer applications. These tools can be customized to include specific exposure pathways and algorithms, and can be programmed to report results in a standardized format, suitable for inclusion into site assessment reports.

X1.7 *Specialized Program Elements*:

X1.7.1 *Introduction*—This section of the appendix provides a review of program elements currently being used by regulatory agencies including those with licensed or certified private contractors. The items described include those used in state voluntary cleanup programs (VCPs) and a variety of other programs to provide site closure or completion, as well as comparable assurances available from the Federal government. The term *completion letters* includes letters and other types of documents, including covenants, certificates and agreements, being used to provide interested parties with assurances that a voluntary corrective action has adequately identified the chemicals of concern from releases into the environment, has adequately identified exposure pathways and any unacceptable risks posed by those chemicals of concern to human health or the environment, and has achieved an appropriate target level. This section of the appendix gives examples of the types of completion documents currently being used by various states or the Federal government,or both, and describes some of the advantages and disadvantages of each type of completion method.

X1.7.2 Program elements vary in form and content. Some consist of completion letters and are very formal, while others are very informal. The interested party must petition for or request a completion letter in some states, whereas in other states the completion document is granted automatically upon the completion of certain activities.

X1.7.3 For further information about a particular state's program, contact the Association of State and Territorial Solid Waste Management Officials (ASTSWMO) or the Regional office of the Environmental Protection Agency (EPA).

X1.7.4 Example Program Elements.

X1.7.4.1 *Application Process*—Many states require an application for interested parties to participate in the VCPs. Some states require the volunteer to sign a formal agreement, such as a consent agreement.

X1.7.4.2 *Submission of Reports*—Some states require submission of reports once remediation has been completed.

X1.7.4.3 *State-Certified or Licensed Contractors*—Some states require the use of state-certified professionals, or laboratories, or both to conduct a cleanup.

X1.7.4.4 *Environmental Professional*—Some states require submission of a statement, opinion or certification from a qualified environmental professional that the voluntary cleanup plan has been fully implemented or completed and that the remediation meets applicable standards.

X1.7.4.5 *Letter Agreement or Formal Consent Order*—Some states require a memorandum/letter of agreement or consent order before the volunteer can participate in the VCP.

X1.7.4.6 *Prior Approval*—Some states require prior approval of work plans, or reports, or both.

X1.7.4.7 *Reopeners*—Many states have explicit reopener provisions in their completion letters for changed conditions, new information, or failure to comply with the voluntary cleanup plan.

X1.7.4.8 *Consideration of Future Land Use*—Many states take future land use into consideration in making risk-based decisions and have developed (by policy or statute) activity and use limitations to address future land use changes.

X1.7.4.9 *Reimbursement of State Oversight Costs*—Many states require reimbursement of their oversight or laboratory or other costs.

X1.7.5 *Examples of Completion Documents*—The types of completion documents currently being used by the states include, but are not necessarily limited to, the following: statements or certificates of completion; no further action opinions/letters; no action letters; private contractor statements of completion; off-site source determination letters; covenants not to sue; releases; contribution protection; no association determination letters; and prospective purchaser agreements. The terminology for a specific type of completion letter in one state may not mean the same thing in another state, and the descriptions given below are very general in nature.

X1.7.5.1 *Statements or Certificates of Completion of Response Action*—In some states, Statements or Certificates of Completion of Response Action are relatively straight-forward to obtain from regulatory agencies. They provide some certainty that the regulatory agency is satisfied that no further corrective action is required at the site. Legally, however, they generally are not as binding as other types of completion letters, such as covenants not to sue. Accordingly, in these states, Certificates of Completion may be viewed by some interested parties as not providing sufficient comfort against further liability. However, in other states, a Statement or Certificate of Completion is a complete release of state superfund liability and runs with the land; as such, it provides liability assurances to successors and assigns as well. A Statement or Certificate of Completion in these states is likely to be viewed quite favorably by interested parties, but it may be more time-consuming to obtain. Some state programs tie the certificate of completion to specific exposure pathway assumptions, such as land use.

X1.7.5.2 *No Further Action Opinions/Letters/Assurance Letters*—As a general rule, no further action opinions or letters are relatively straight-forward to obtain from regulatory agencies. They provide some certainty that the regulatory agency is satisfied that no further corrective action is required at the site. EPA, for example, has developed generic ways to encourage redevelopment including a designation of No Further Remedial Action Planned for sites listed in the CERCLIS inventory. No Further Action Opinions or Letters generally do not constitute a release from further liability at the site. Legally, therefore, they do not provide the same level of comfort as, for example, a covenant not to sue. Accordingly, No Further Action Opinions or Letters may be viewed by some interested parties as not providing sufficient comfort against further liability.

X1.7.5.3 No Further Action Opinions or Letters can also be written as Limited No Further Action Opinions or Letters. Limited No Further Action Opinions or Letters pertain to only a portion of the property or to only some of the known releases at a site. For example, they might address and document releases to the soil, but not potential ground water releases. They are helpful when a discrete issue is slowing down a real estate transaction because they can generally be obtained fairly quickly and easily. However, they are limited in nature and do not provide the same degree of comfort that most other types of completion letters offer.

X1.7.5.4 *No Action Letter*—A No Action Letter is essentially a no further action letter, except that it includes additional language indicating that the state will not take any enforcement/administrative actions once the regulatory agency is satisfied with the volunteer's technical conclusions. A No Action Letter will often say that the release (site) will not be referred to the Superfund program and will not be added to CERCLIS or the state Superfund list.

X1.7.5.5 Private Contractor Statements of Completion or No Further Action Letters provide an alternative where the state does not have sufficient resources to review every site investigation and remediation to determine its adequacy. In the states with this type of program, the private contractor must be state-licensed or certified, so there is indirect state oversight over the process. Private Contractor Statements of Completion/No Further Action Letters are generally available more quickly and readily than other types of completion letters.

Legally, however, they are not as binding as other types of completion letters, such as covenants not to sue. Accordingly, Private Contractor Statements of Completion/No Further Action Letters may be viewed by some interested parties as not providing sufficient comfort against further liability.

X1.7.5.6 *Off-Site Source Determination Letters*—Some states, as well as the Federal government, have taken the position that they will not exercise their enforcement discretion under certain circumstances where chemicals of concern originating off-site in the ground water have affected an "innocent," downgradient property. Many of these states issue letters providing that they will not hold the impacted property liable for any chemical releases originating off-site. The Federal government's position is available in the form of guidance, issued May 24, 1995, and entitled "Final Policy Toward Owners of Property Containing Contaminated Aquifers." A state Off-Site Determination Letter has the same types of advantages and disadvantages as a No Further Action Opinion or Letter.

X1.7.5.7 *Covenants Not to Sue*—Covenants Not to Sue are legally binding agreements between a state or the Federal government and an interested party. They provide assurances that the interested party will not be pursued in the future for additional remediation work or targeted in an enforcement action. At the same time, because they are more binding than most other types of completion letters discussed above, they are generally more difficult and time-consuming to obtain. They generally must be negotiated on a case-by-case basis.

X1.7.5.8 *Releases*—Some states use releases to absolve volunteers from further responsibility for corrective action. Releases tend to be a qualitative determination that the remediation has been successfully completed; they generally do not absolve the volunteer from joint, strict and several liability. In this regard, releases have the same types of advantages and disadvantages as No Further Action Letters.

X1.7.5.9 *Contribution Protection*—A few states are providing, by statute, contribution protection to parties that voluntarily remediate a site. This protection is generally embodied in an administrative or judicial order or consent decree or a settlement agreement. Contribution protection protects the volunteer from contribution actions by other responsible parties and, in some instances, protection against citizen suits as well.

X1.7.5.10 *No Association Determination Letters*—No Association Determination Letters are being used to offer lenders, prospective purchasers, and other "innocent" parties assurances that they will not be the subject of future remediation orders or other types of enforcement actions. Similar types of protection are being offered automatically by statute (that is, without the need for a formal letter) in some states to interested persons such as lenders, municipalities, utilities, etc.

X1.7.5.11 *Prospective Purchaser Agreements*—Prospective Purchaser Agreements provide protection from liability for innocent buyers who perform remediation in accordance with a plan approved by the state or the Federal government. These agreements can be obtained from either a state offering such an mechanism, or EPA (if its an NPL site), or both. The Federal government has offered guidance, dated May 24, 1995, entitled "Guidance on Agreements With Prospective Purchasers of Contaminated Property," describing what types of sites are eligible for the Federal program. A variety of policies may be pursued which arise from EPA guidance, such as government statements of no further interest at this time.

X1.7.5.12 Other Mechanisms:

X1.7.5.13 *Review and Evaluation Services*—Many states provide review and evaluation services to determine the adequacy of any site investigations or remediation plans.

X1.7.5.14 *Reclassification of the Site or Redefinition of Site Boundaries*—Many states reclassify sites that have been remediated or deemed not to require remediation as "inactive" or redefine a site's boundaries.

X1.7.5.15 *Agreement Not to List on Federal CERCLIS*—Many states have agreed to refrain from listing sites participating in the state VCP on the Federal CERCLIS or the state Superfund list. This agreement often comes in the form of a No Action Letter or is incorporated into another type of completion letter.

X1.7.5.16 *Immunity from Enforcement Actions*—In some states, the completion letter constitutes immunity to a state enforcement action.

X1.7.5.17 *No Release*—A few states explicitly provide that current or prior owners or operators who are considered to be potentially responsible parties will *NOT* be released from liability at the site.

X1.7.6 Eligibility for Program

X1.7.6.1 *Current Owner Only*—Some states allow current owners only to participate in the VCP.

X1.7.6.2 *Lenders/Fiduciaries*—Some states allow lenders, fiduciaries, and/or other parties providing financing to participate in the VCP and thereby enjoy its benefits. In addition, some states provide liability protection automatically by statute without the need for the lender or fiduciary to obtain a completion letter from the VCP.

X1.7.6.3 *Prospective Purchasers*—Some states allow prospective purchasers the benefits from the VCP.

X1.7.6.4 *Non-Responsible Parties Only*—Some states allow only non-responsible or non-liable parties to participate in their VCPs.

X1.7.6.5 *Successors and Assigns*—Some states allow successors and assigns to enjoy the benefits from the VCP.

X1.7.6.6 *Third Party Claims*—Some states provide broad-based releases that protect the volunteer from third party claims as well.

X1.7.7 *Other Issues*:

X1.7.7.1 *Recordation*—Some states require recordation of the completion letter in local land records.

X1.7.7.2 *Audits*—Some states reserve the right to audit properties undergoing voluntary remediation.

X1.7.7.3 *Stakeholder Involvement*—Some states mandate stakeholder involvement for some or all of the sites moving through the VCP.

X1.7.7.4 *Affidavits*—Some states require that all documents relating to the VCP be submitted by affidavit.

X1.7.7.5 *Audit Privilege*—Some states provide that any information or data generated during the remediation is not admissible in any administrative or judicial action.

X1.7.7.6 *Access to Records/Records Retention*—Some

states require that volunteers provide them access to remediation records and require retention of those records for a specific length of time.

X1.7.7.7 *Waiver or Consolidation of Permitting Requirements*—Some states waive or consolidate state and local permitting requirements in an effort to facilitate voluntary remediation.

X1.8 Education/Outreach:

X1.8.1 To ensure the success of the RBCA program, both regulatory agencies and the regulated community should be given adequate training opportunities regarding program objectives and procedural requirements. In addition, outreach efforts may be undertaken to promote understanding and acceptance of the risk-based process by other involved parties, such as community or citizen's groups and the real estate, development, banking, or insurance industries. Key considerations are summarized below.

X1.8.2 *Agency Personnel Training*—Regulatory agency personnel should receive training commensurate with their responsibilities regarding RBCA evaluations, corrective actions, and completion reports. Such training may include an overview of the RBCA process, review of key policy decisions, guidance regarding evaluation of site assessment data and selection of representative site parameters, calculation of risk-based corrective action goals, and determination of appropriate corrective actions. In general, such training will build upon the prior knowledge and experience of the agency staff, focusing on those areas in which RBCA policies may modify or expand existing procedures.

X1.8.3 *Regulated Community and Contractor Training*—The regulated community and the professional consultants and contractors involved in the RBCA program implementation should also receive a formal orientation comparable to that received by agency personnel. An effective training technique may be to train agency personnel, the regulated community and consultants simultaneously. Emphasis may be placed on available standardized tools for calculations and reporting. The regulatory agency may choose to conduct these orientation sessions and/or authorize certified trainers to conduct in-depth training workshops. Written guidance and practical case study examples will greatly improve the ability of the applicant to "get it right" the first time.

X1.8.4 *Outreach Programs*—Efforts should be considered to communicate the objectives and significance of the RBCA program to stakeholders, including members of the general public, environmental groups businesses, real estate interests, banking establishments, the insurance industry and all those potentially involved in the sale and transfer of affected properties. Key issues to address include: the significance of risk-based corrective action goals and the level of protection provided to current and future property users; the general process required to achieve site closure; and the optional mechanisms for activity and use limitations that may be applied in some cases. These outreach initiatives may take the form of printed information booklets, stakeholder forums, trade association presentations, or all of these.

X1.9 *Stakeholder Involvement Activities*:

X1.9.1 *Introduction*:

X1.9.1.1 One of the purposes of this appendix is to provide a review of generally accepted principles in successful stakeholder involvement, including risk communication. For purposes of this appendix, stakeholder involvement activities are those activities that can be used by volunteers and regulatory agencies in corrective action programs to ensure that the stakeholder is both informed of and, if interested, involved in planning for corrective actions. This incorporates the concept that risk communication activities must be designed to allow the stakeholder an opportunity to understand what the risks from chemical releases are and how they can be mitigated. In VCPs, the opportunity for stakeholder involvement activities may be provided by the state or the private sector, or both, and the opportunity may be provided during either the development of the program, on a site-specific basis, or both.

X1.9.1.2 Methods to provide the opportunity for stakeholder involvement in RBCA and risk communication programs may occur during the following steps of regulatory development:

X1.9.1.3 Initial program development;

X1.9.1.4 Policies and guidance;

X1.9.1.5 Regulation development/Implementation;

X1.9.1.6 Informal Methods;

X1.9.1.7 Risk Communication Techniques;

X1.9.1.8 Stakeholder involvement opportunities for environmental programs vary in the level of effort, form and content. Agencies can establish various regulatory requirements for the regulated community or they can fashion guidance stating the general policy of the state agency to provide maximum opportunities for stakeholder involvement in decision making. Stakeholder involvement may be program-wide and site-specific.

X1.9.1.9 The scope of stakeholder involvement for different activities is quite varied among cleanup programs. While some programs require and engage in stakeholder involvement only during the guidance or regulatory development stage, others may require local, site-specific stakeholder involvement opportunities for each stage in the cleanup process. Three examples will illustrate this phenomenon. During private corrective actions, which may use the RBCA framework, there may be no specific regulatory approval or required stakeholder involvement activities. Insurance companies, for example, may use RBCA to determine "prudent to do" actions for their clients, without requiring site-specific stakeholder involvement. In the second example, Pennsylvania developed its Act 2, Land Recycling Program using stakeholder participation. Specific sites, however, which use the background or statewide health standards, require notification only and do not require any further stakeholder involvement to complete cleanup. The rationale here is that stakeholder input on the basis of cleanup was provided at once in developing the program. Finally, Massachusetts developed its RBCA program and required milestone public notification requirements for each site in the program. Furthermore, with a citizen petition, responsible parties are required to formulate and implement a Public Participation Plan for an individual site, including public meetings at each milestone of the cleanup process. Clearly these examples represent the spectrum as to the scope and level of effort of stakeholder involvement in RBCA programs.

X1.9.2 *Risk Communication*—The use of the RBCA guide, as with all technical materials, will require a sincere attempt to communicate with stakeholders. Some basic principles of risk communication are articulated in this appendix.

X1.9.3 *Program Development*—Many state VCPs and other cleanup programs have either required or voluntarily included stakeholder involvement during development. The use of advisory committees, surveys of potential stakeholders, and workgroups from the regulated community are all techniques that have been employed to develop a VCP. These methods are in addition to any statutory public hearing requirements, and generally reflect the unique nature of VCPs. During this process, the scientific underpinnings supporting RBCA methods should be clearly explained to stakeholders.

X1.9.4 *Policies and Guidance*:

X1.9.4.1 Many VCPs are built on policies and guidance written by the regulatory agency. Such documents may recommend general or specific opportunities for stakeholder involvement opportunities.

X1.9.4.2 At sites where the stakeholders indicate an interest in participating in response action planning and implementation, policy and guidance may recommend that opportunities be established in a stakeholder involvement plan that is prepared in response to petition or more informal request or a request by local officials. Such a plan may: provide background about the site, response actions already conducted, and the history of stakeholder involvement at the site; identify the specific opportunities for stakeholder involvement in cleanup decisions that will take place; and describe activities that will be undertaken to address and incorporate concerns in the remediation. Stakeholders may also be involved in risk reduction measures and/or remediation done as part of a state VCP. In addition to its utility at the overall program development phase, the stakeholders meeting may be an effective communication technique at an individual site. Site-specific stakeholder meetings may be used to discuss the identified regulatory approach, identify critical data gaps such as fundamental information on exposure pathways and receptors, and ultimately to communicate the selected risk-based corrective actions for the chemical release.

X1.9.4.3 Specific techniques recognized in agency guidance may also include the regular publication and distribution of site fact sheets and conducting community interviews.

X1.9.5 *Regulations*—Many regulatory agencies have promulgated specific regulations which require formal, documented opportunities for stakeholder involvement. Public meetings or hearings, either formal or informal, may be required at certain stages of corrective action decision making such as the selection of remedy. Other requirements may include activities that inform stakeholders about the status of response actions. These activities include, among others, notifications to local officials and publication of legal notices in newspapers at key milestones in the response action process. In addition, because future use of a site is a major determinant of remediation goals, regulations may require consideration of local land use planning activities to determine current and potentially feasible future site uses.

X1.9.6 *Informal Methods*—The use of the RBCA guidance includes site assessment activities that by their very nature require close contact with stakeholders and the local government. Everything from site access to sampling, to the review of town records, requires interaction with people who live and work in the community where the site is located. Informal stakeholder involvement requires tact, honesty and a willingness on the part of the user of the guide to communicate with many different people about the site. Sometimes, the understanding of site conditions and remediation challenges are more effected by such informal interactions than they are by public hearings or meetings.

X1.9.7 *Risk Communication Techniques*:

X1.9.7.1 People in the community should be well informed about the true risks presented by a site and the decision-making process regarding assessment and cleanup activities at a site. It is important to communicate with residents as often as possible and at a level that can be easily understood.

X1.9.7.2 The use of jargon and acronyms should be strictly limited during interactions with stakeholders. Written and oral presentations should consider material at the 9th or 10th grade reading level for ease of understanding among adults who are not technically trained. Long involved sentences with multiple clauses may be avoided by substituting simple direct statements and familiar terms.

X1.9.7.3 Comparative risks from chemicals and from everyday activities may be a tool for consideration in communication of risks. The risks of getting struck by lightning, for example, might be compared with the risks of cancer from an individual chemical at an individual site. The cumulative risks of driving an automobile might be compared with cumulative risks from a site. In addition there is a need to compare actual, expected risk with calculated or theoretical risk which may be based on upper bound, conservative estimates.

X1.9.7.4 Caution in the use of comparative examples must be exercised because of the perception of risks by individuals and communities, as well as understandable sensitivity to the comparison of voluntary versus involuntary risks. Risks that are natural, not man-made and that are part of a community's everyday experience may not be perceived as very high risks. In rural communities, for example, while the risk of contracting Lyme disease transmitted from ticks may be quite high, the perceived risk is surprisingly low. While the risks of injury from motor vehicles are quite high, because riding in a car or on the bus is more or less voluntary, the perceived risks and level of outrage generated by cars in the community may be much lower than the perceived risk and level of outrage generated by the presence of chemical releases in the neighborhood. Regardless of the use of comparative risk tools, stakeholder involvement in decision making will eventually lower both the level of outrage and the perceived risks from a site, because it gives communities and individuals a greater sense of control over the risks.

X1.10 *RBCA Program Evaluation*:

X1.10.1 Evaluation of the proposed RBCA process may assess the impact of critical policy decisions and optimize program efficiency. This evaluation can be conducted before, during, or after development and implementation. It may be a cooperative effort among various organizations representing industry, government and other stakeholders. Such program evaluations may include one or more administrative reviews of technical investigations described below.

X1.10.2 *Administrative Review*—Administrative process flowcharts of the proposed RBCA program and the existing corrective action rules may be developed and compared to characterize the scope and timing of principal steps involved in the site evaluation, remediation, and closure process. Key issues in this assessment will be the clarity and completeness of the site management process, as well as the number of review and approval steps imposed upon the regulatory authority. Consistent with the goals of RBCA, the administrative process should serve to expedite action at higher risk sites.

X1.10.3 *Demonstration Projects*—A demonstration project involves processing a limited number of sites (for example, less than ten) through the RBCA program in as much detail as possible. The objective of the demonstration effort is to check the completeness of the proposed RBCA guidelines, identify possible conflicts or constraints on implementation, and predict general effects on remediation requirements or costs.

X1.10.4 *Planning Level Studies*—A planning level study involves processing a large number of sites (that is, all of one company's sites in a state or region, or a random selection from all sites in the state or region) through the RBCA process. Results from any demonstration projects may also be integrated into planning level studies. This study can often give a clear picture of the potential resource savings made possible by adopting a RBCA program, as well as the potential impact of a given policy decision on the rate of site closures under Tiers 1 and 2.

X1.11 *Available Resources*—Development and implementation of a RBCA program will require a significant commitment on the part of the development team. However, those RBCA programs presently in place in various states have demonstrated that this effort can yield significant benefits in terms of expediting the site remediation process and ensuring efficient allocation of limited resources to high priority sites. Parties interested in developing a RBCA program are encouraged to contact those state agencies with programs currently underway for examples of written guidelines and direct experience with procedural challenges and successes. The EPA Office of Underground Storage Tanks (OUST) and Office of Solid Waste and Emergency Response (OSWER) are presently maintaining an information clearinghouse for RBCA programs and can provide data regarding document resources and contact names. Table X1.2 shows some program attributes of the various state programs.

TABLE X1.2 Suggested Action Item Checklist for RBCA Program Implementation

Program Design Tasks
- Response Action Evaluation System
- Chemicals of Concern List
- Target Risk Limits
- Applicable Exposure Factors
- Tier 1 RBCA Equations
- Tier 2 RBCA Modeling Requirements
- Point of Compliance Definitions
- Compliance Monitoring Specifications
- Risk Management Options (Institutional controls, long-term engineering controls, other activity and use limitations)
- Minimum Site Characterization Requirements
- Data Sources and Uncertainty Guidelines

Agency Guidance Documents
- RBCA Guidance Manual (includes all of the following)
- Tier 1 RBSL
- Tier 2 Modeling Guidance (default parameters, approved analytical models, etc.)
- Tier 3 Modeling Guidance
- Standardized Report Formats
- Customized Tier 1/Tier 2
- RBCA Application Review Checklist

Administrative/Institutional Tasks
- Legislative Action
- Regulatory Action
- Coordination/Review with Stakeholders
- Agency Staff Training
- Education/Outreach to Regulated Community and Contractors
- RBCA Demonstration Studies

X2. CHEMICALS: THE ROLE THAT PROPERTIES AND EFFECTS PLAY IN THE RBCA PROCESS

X2.1 Introduction:

X2.1.1 The purpose of this appendix is to:

X2.1.1.1 Highlight issues that a user should consider when looking at/for site-related chemicals,

X2.1.1.2 Identify common chemical groups that may be encountered at sites,

X2.1.1.3 Define chemical properties that are environmentally relevant for examining the fate and behavior of chemicals at a site, and

X2.1.1.4 Discuss the potential human health and ecological effects of chemicals at a site.

X2.1.2 Users for whom this appendix is intended include an individual or group involved in the RBCA process including owners, operators, regulators, UST Fund managers, attorneys, consultants, legislators, and others.

X2.2 Issues for the User:

X2.2.1 Chemical data may be gathered for a variety of purposes such as supporting site assessment, response actions, comparison to RBSL, and determining SSTL. Consistent with the tiered approach employed within RBCA, additional site-specific chemical data and information may need to be collected at successive tiers to appropriately incorporate the concepts presented in this appendix into the evaluation. Examples of the types of data used during Tier 2 and 3 evaluations are presented in this appendix. Such information should be summarized in a clear and concise format in RBCA reports.

X2.2.2 The user should consider obtaining information on chemical properties and effects commensurate with the qualitative/quantitative level of the Tier (1,2, or 3) Evaluation. This appendix highlights properties and effects of chemicals as they pertain to:

X2.2.2.1 Identifying pathways of exposure,

X2.2.2.2 Providing values that are used in fate and transport estimates or models,

X2.2.2.3 Identifying the kinds of toxic effects that a chemical may pose,

X2.2.2.4 Identifying the degree of toxicity that a chemical may pose,

X2.2.2.5 Developing RBSL and SSTL values, and

X2.2.2.6 Identifying sampling and analytical needs.

X2.3 *Chemicals - Groupings, Properties, and Chemical Behavior:*

X2.3.1 *Chemical Groupings:*

X2.3.1.1 This RBCA standard has been developed for use at sites that could contain any number of chemicals, and hence crosses over many types of manufacturing, production, and disposal processes. When the user plans how to approach the collection of data to support the tiered evaluation at any given level, consideration must be given to determining chemical concentrations for groupings of chemicals appropriate to the site such that fate, transport, behavior, and exposure are addressed. The chemical groupings selected as examples for this appendix are based on the frequency they are encountered at sites, and environmentally relevant physicochemical properties that determine their fate and behavior at the site. These groupings correlate with many of the analyses developed in the Ref **(3)** that were designed to capture broad classes of chemicals in a variety of sample matrices (see Table X2.1). The table provides the user with insight into commonly defined broad classes of chemicals as well as the predominant methods used for their analysis. While it is not all encompassing, it provides the user with guidance as a possible approach for categorizing chemicals and for requesting appropriate analytical methods for measuring those chemicals in different environmental media. The analytical methods in Table X2.1 are used only to illustrate logical groupings of chemicals. Because methods change with time, the user needs to identify the analytical methods that are acceptable and appropriate for the specific application.

X2.3.2 *Chemical Properties*:

X2.3.2.1 The fate and transport of chemicals in the environment is largely a function of their physicochemical properties under differing ambient environmental conditions. The physicochemical properties of a chemical can be used to infer its equilibrium distribution between the various environmental compartments and to study its behavior within a single environmental medium. Exposures may also occur under non-equilibrium conditions. In such cases, the rate of release of a chemical may influence exposure. The relationships between chemical properties and environmental media govern these processes. Rates of chemical release can affect a chemical's bioavailability. Environmental conditions and route of exposure to human or ecological receptors also influence rates of release and bioavailability. Information on soil properties can be helpful for evaluating and modeling the bioavailability of inorganic and organic chemicals.

X2.3.2.2 For organic compounds, investigators often consider the following physicochemical properties when studying environmental behavior, transport, and accumulation: aqueous solubility (S or C_w), vapor pressure (V_p or P^o), Henry's Law constant (H or K_H), octanol-water partition coefficient (K_{ow}), and organic carbon partition coefficient (K_{oc}). The fate and transport of inorganic compounds, such as metals, can be influenced by their chemical speciation and properties such as complexation reactivity, surface reactivity, and aqueous solubility. Although these physicochemical properties govern the transfer of chemicals between the various environmental media and do not reflect the rates of the various exchange processes (that is, no consideration of kinetics), knowledge of such physicochemical properties is useful when assessing chemical movement between media and the relative importance of various source and sink terms within media. Other important processes that affect the transport and fate for organic and inorganic compounds can include biodegradation, chemical release from soil and sediment solids, sequestration, chemical transformations such as acid-base reactions, redox reactions,

TABLE X2.1 Chemical Grouping Summary Derived from EPA SW-846 Methods (3)

Chemical Grouping	Chemical Analytical Method Number	Analytical Method	Description of Chemicals in Grouping
Halogenated VOCs	8010 B II (see also 8240/8260)	GC	Mixed aromatics and aliphatics, single and multiple halogens
Aromatic VOCs	8021 A II (see also 8240/8260)	Capillary GC/PID & ECD in series	BTEXs, mixed aromatic and aliphatics, without halogens, and with single/multiple halogens
Phenols	8040	GC	Basic phenol, plus chloro, methyl and nitro substituted phenols
Phthalate Esters	8061 II	Capillary GC/ECD	Various phthalate esters, substituted and non-substituted
Organochlorine Pesticides + PCBs as Arochlors	8080 A II/8081 II	8080 A II: GC 8081 II: Capillary GC	Organochlorine Pesticides, PCBs as Arochlors
PAHs	8100, 8310	8100: GC, 8310: LC	PAHs
Chlorinated Hydrocarbons	8120 A II, 8121 II	8120 A II: GC II: Capillary GC	8121 8120 A II: Aromatics, saturated/unsaturated aliphatics (subset of 8121 II), 8121 II: more compounds than 8120 A II
Non-Halogenated VOCs	8240 B II, 8260 A II	8240 B II: GC/MS A II: Capillary GC/MS	8260 8240 B II. Broad list of compounde, 8260 A II: more compounds than 8240 B II, isomeric resolution
Non-Halogenated SVOCs	8250 A II, 8270 B II	8250 A II: GC/MS 8270 B II: Capillary GC/MS	8250 A II: Broad list of compounds, 8270 B II: more compounds than 8250 A II, isomeric resolution
Metals	7000 Series	AA/direct aspiration, furnace technique, gaseous hydride, borohydride reduction	All metals except mercury
Mercury	7470 A II, 7471 A II	Manual cold vapor technique	Mercury

hydrolysis reactions, and direct and indirect photochemical transformations. The kinetics of chemical release is a key aspect of exposure for chemicals that are adsorbed or sequestered. In cases where equilibrium relationships overestimate exposures, information on the kinetics of release can be especially valuable. These may be important for ground water, exposure to humans, and exposure to ecological receptors.

X2.3.3 *Defining Physicochemical Properties:*

X2.3.3.1 Aqueous solubility (S or C_w) refers to the maximum concentration that a pure compound can achieve in water at equilibrium. This property is measured in the laboratory and provided in various texts on environmental chemistry.

X2.3.3.2 Vapor pressure (V_p or P^o) is a measure of the partial pressure of the vapor phase of a compound at equilibrium with its pure condensed phase (either liquid or solid). Vapor pressure reflects the tendency of a compound to flee the condensed phase and enter the gaseous phase and is highly dependent upon temperature. This property is measured in the laboratory and provided in various texts on environmental chemistry.

X2.3.3.3 The Henry's Law constant (H or K_H) is the ratio of the partial pressure of a compound in air to its concentration in water at equilibrium. The Henry's Law constant is an air-water partitioning coefficient and, for a particular compound, it can be simply estimated by knowing the compound's aqueous solubility and vapor pressure at a given temperature. This property is measured in the laboratory and provided in various texts on environmental chemistry.

X2.3.3.4 The octanol-water partition coefficient (K_{ow}) is a dimensionless constant which reflects the tendency of a chemical to partition between an organic phase (octanol) and water. The octanol-water partition coefficient is essentially a measure of a compound's hydrophobicity, and more hydrophobic compounds will preferentially partition into the more nonpolar octanol phase as well as into other nonpolar phases such as natural organic matter. This property is measured in the laboratory and provided in various texts on environmental chemistry.

X2.3.3.5 The organic carbon partition coefficient (K_{oc}) is a measure of the tendency of a chemical to partition between organic carbon in soil or sediment and water. This property is either measured in the laboratory or estimated from Kow.

X2.3.3.6 Chemical speciation refers to the oxidation state of a metal. It can be measured in the laboratory or estimated from environmental conditions.

X2.3.3.7 Complexation reactivity refers to the tendency of two dissolved species to react to form a third species. Complexation reactions are reversible processes, and the product species can be either soluble or insoluble as a precipitate. This property can be measured in the laboratory or estimated from environmental conditions.

X2.3.3.8 Surface reactivity refers to the tendency of solutes (for example, charged metals) to react with solid surfaces. Surface reactions can involve the formation of chemical bonds between solutes and surface moieties (for example, carboxyl, phenolic, or sulfhydryl groups) or electrostatic interaction between ions and charged surfaces (for example, ion exchange).

X2.3.3.9 Density refers to the mass of a substance per unit volume (g/cc). This parameter is measured in the laboratory under specific conditions, and estimated or modeled under field conditions.

X2.3.3.10 Availability referes to the ability of a chemical to be released from a soil particle and thus be available for degradation, volatilization, or other mechanisms that have an effect on the transport, exposure and toxicity of the chemical. Laboratory protocols can be used to estimate values for rapid and slow, long-term release of this process in soils.

X2.3.4 *Environmental Behavior and Exposure Pathways:*

X2.3.4.1 The chemical groupings presented in this section are based upon the SW-846 groups given in Table X2.1. These groups are further organized by physicochemical properties (see Table X2.2) and environmental behavior (see Table X2.3). Toxicological criteria and effects information is presented in X2.4 to X2.6.

X2.3.4.2 The table on physicochemical properties (see Table X2.2) gives the user examples of chemicals within each of the chemical groupings presented in Table X2.1. The example chemicals were chosen to illustrate the range of physicochemical properties that exists within each grouping. It is critical that the user understand that reported values for these physicochemical properties can vary widely among reference sources and that the values reported in Table X2.2 reflect only the references cited.

X2.3.4.3 The table on environmental behavior (See Table X2.3) provides the user with a semi-quantitative indication of how chemicals might move between different environmental media. It is intended to be used at an early stage of investigation to guide sampling activities and to aid in the development of an exposure analysis. It is not intended to provide an indication of the level of risk associated with exposure because it does not incorporate toxicity information. The table employs qualitative designations of low, medium, and high (L, M, and H) to rank potential exposure pathways. The semi-quantitative designations in the table usually span a range (for example, L-M, M-H) which reflects both variations in physicochemical properties within the chemical groupings and possible site-specific variations. In some cases potential for exposure will depend on specific activities (for example, excavation of soil) or on other site-specific conditions. The user is advised to review the footnotes for additional guidance on these modifying factors.

X2.3.4.4 The effects that these properties have on environmental behavior and exposure pathways are discussed below with an example compound(s) provided for each chemical grouping. Again, these descriptions are general, as site-specific factors can influence the behavior of compounds.

X2.3.5 *Halogenated Volatile Organic Compounds:*

X2.3.5.1 These compounds are volatile and soluble in water as indicated by their Henry's Law constants and solubility values, respectively. They tend to have a low residence time in surface soil and surface water environments. However, they can leach to and migrate in ground water where they are typically more persistent than non-halogenated compounds.

X2.3.5.2 Major exposure routes include the ingestion of impacted ground water and the inhalation of vapors. These

TABLE X2.2 Physicochemical Properties of Sample Compounds Selected to Represent the Magnitude and Range of Properties within Environmentally Significant Compound Classes[A]

Compound Class	Sample Compounds	C_w Aqueous Solubility (mg/L)	P° Vapor Pressure (torr)	K_H Henry's Law Constant (L atm mol^{-1})	log K_{ow} Octanol-Water Partition Coefficient	K_{oc} Organic Carbon Partition Coefficient (mL/g)[B]
Organics						
Halogenated Volatile Organic Compounds	Dichloromethane	1.9E+04	4.5E+02	2.60E+00	1.15	1.1E+01
	Tetrachloroethene	1.50E+02	1.91E+01	2.75E+01	2.88	3.7E+02
Non-halogenated Volatile Organic Compounds	Carbon Disulfide[C]	2.10E+03 (20°C)	2.97E+02 (20°C)	1.4E+00 (20°C)	1.70-4.16	6.5E+01
	Methyl Ethyl Ketone[C]	2.39E+05	9.06E+01	1.05E-02	0.29	4.8E+00
Alkylated Benzenes (subset of Aromatic VOCs)	Benzene	1.79E+03	9.6E+01	5.5E+00	2.13	5.4E+01
	Ethylbenzene	1.68E+02	9.57E+00	7.9E+00	3.15	5.8E+02
Chlorinated Benzenes (subset of Chlorinated Hydrocarbons)	1,2-Dichlorobenzene	1.18E+02	1.48E+00	2.5E+00	3.38	1.0E+03
	Hexachlorobenzene	5.81E-03	2.35E-05	1.50E+00	5.5	7.4E+04
Phenols	Phenol	9.0E+05	1.95E-01	4.1E-04	1.45	2.3E+00
	2-Chlorophenol[C]	2.80E+04	1.42E+00	5.6E-04	2.15	6.2E+01
Phthalate Esters	Dimethyl Phthalate	4.25E+03	1.82E-02	1.10E-03	1.53	4.4E+01
	Di-n-butyl Pththalate	9.43E+00	7.26E-05	1.29E-03	4.57	1.3E+03
Polychlorinated Biphenyls (PCBs)	Aldrin[C]	0.02(20°C)	0.0000375 (20°C)	5.0E-01	6.5	3.8E+04
and Organochlorine Pesticides	Dieldrin	2.20E-01	5.02E-06	1.12E-02	5.48	9.9E+03
Organic Acids and Bases	Benzoic Acid[C,D]	2700 (18°C)	4.5e-3 (20°C)	7.00E-05	1.87	5.7E+01
	Aniline[C,E]	3.61E+04	4.89E-01	1.20E-01	0.9	1.4E+01
Polycyclic Aromatic Hydrocarbons (PAHs)	Naphthalene	3.15E+02	7.96E-02	4.3E-01	3.36	1.4E+03
	Benzo(a)pyrene	1.52E-03	5.51E-09	1.20E-03	6.50	1.9E+06
Inorganics						
Metals[C]	Lead	NA	1.00E+01 (1160°C)	NA	NA	NA
	Mercury	NA	2.00E-03	NA	NA	NA

[A] All data is for 25°C and was taken from Schwarzenbach et al. **(4)**unless otherwise noted.
[B] All K_{oc} values estimated using linear free energy relationships.
[C] Data from Howard **(6)**.
[D] Benzoic Acid is an organic acid and has an acid dissociation (pKa) of 4.1 **(5)**.
[E] Aniline is an organic base and has a base dissociation constant (pKb) of 9.4 **(5)**.

compounds usually do not accumulate in the food chain.

X2.3.6 *Non-Halogenated Volatile Organic Compounds:*

X2.3.6.1 These chemicals tend to be relatively soluble in water, and therefore can be transported to ground water systems, and be distributed between the various environmental media in a similar fashion to the halogenated VOCs. While their environmental fates are similar to the halogenated VOCs, abiotic transformations and biodegradation can be more important processes for non-halogenated VOCs, reducing their residence times in the subsurface environment.

X2.3.6.2 The potentially important exposure pathways again include the ingestion of impacted ground water and the inhalation of vapors from soil and ground water.

X2.3.7 *Alkylated Benzenes:*

X2.3.7.1 Alkylated benzenes such as the BTEX compounds (benzene, toluene, ethylbenzene, and xylenes) have high vapor pressures and are relatively soluble in water. Due to their high vapor pressures, volatilization is often an important removal process for these compounds. These compounds also have a tendency to leach from impacted soils to the water table, with subsequent transport in ground water. They are mobile in the subsurface due to their relatively high aqueous solubilities and low octanol-water partition coefficients. Biodegradation can be an important removal process for alkylated benzenes in soil and aquifer systems, thus reducing their residence times in these media.

X2.3.7.2 Ingestion of, or dermal contact with, impacted ground water, and inhalation of vapors from soil and water can be important exposure pathways for alkylated benzenes.

X2.3.8 *Polycyclic Aromatic Hydrocarbons (PAHs):*

X2.3.8.1 These compounds have a high affinity for organic matter and are relatively insoluble in water, as reflected in high Koc and Kow values, and low aqueous solubilities. PAHs tend to remain bound to particulates and dissolve slowly in water. As such, the physical fate of these chemicals is usually controlled by the transport of particulates and the release of the chemicals from soil and sediment solids. Consequently, soil, sediment, and suspended particulate matter are the more common media for transport. Biodegradation by microorganisms is an important removal process for PAHs. Volatilization can also be an important removal process for the lighter molecular weight PAHs.

X2.3.8.2 Direct contact with impacted media, as well as inhalation of fugitive dust can be important exposure pathways for PAHs. PAHs may occur in select food stuffs, including the leaves of plants after atmospheric deposition and in shellfish

TABLE X2.3 Qualitative Ranking of Important Potential Exposure Pathways Based on Chemical Behavior for Common Site Contaminants Grouped by Compound Class[A,B,C,D]

Compound Class	Contaminated Media				
	Surface Soils	Sub-surface Soils[E]	Ground Water[F]	Sediment[G]	Surface Water[G]
Organics Halogenated Volatile Organic Compounds (e.g., dichloromethane, tetrachloroethene)	direct contact: L→M leaching to gw: M→H vapor formation: H fugitive dusts: L→M surface runoff: M→H	direct contact: L→M vapor formation: H leaching to gw: M→H fugitive dust: L	vapor formation: H ingestion: M→H transport to surface water: M→H	direct contact: M→H ingestion of biota: L vapor formation: H transfer to overlying waters: H	direct contact: M→H vapor formation: H ingestion of water: M→H ingestion of biota: L
Non-halogenated Volatile Organic Compounds (e.g., carbon disulfide, methyl ethyl ketone)	direct contact: L→M leaching to gw: M→H vapor formation: H fugitive dusts: L→M surface runoff: M→H	direct contact: L→M vapor formation: H leaching to gw: M→H fugitive dust: L	vapor formation: H ingestion: L→M transport to surface water: L→M	direct contact: M→H ingestion of biota: L vapor formation: H transfer to overlying waters: H	direct contact: M→H vapor formation: H ingestion of water: M→H ingestion of biota: L
Alkylated Benzenes (subset of Aromatic VOCs) (e.g., benzene, ethylbenzene)	direct contact: L→M leaching to gw: M→H vapor formation: H fugitive dusts: L→M surface runoff: M→H	direct contact: L→M vapor formation: H leaching to gw: M→H fugitive dust: L→M	vapor formation: H ingestion: M→H transport to surface water: M→H	direct contact: L→M ingestion of biota: L vapor formation: H transfer to overlying waters: H	direct contact: M→H vapor formation: H ingestion of water: M→H ingestion of biota: L
Chlorinated Benzenes (subset of Chlorinated Hydrocarbons) e.g., 1,2-dichlorobenzene, hexachlorobenzene)	direct contact: M→H leaching to gw: L→M vapor formation: L→M fugitive dusts: M→H surface runoff: L→M	direct contact: M→H vapor formation: L→M leaching to gw: L→M fugitive dust: M→H	vapor formation: L→M ingestion: M→H transport to surface water: M→H	direct contact: M→H ingestion of biota: L→N vapor formation: L→M transfer to overlying waters: L→M	direct contact: M→H vapor formation: L→M ingestion of water: M→H ingestion of biota: L→M
Phenols (e.g., phenol, 2-chlorophenol)	direct contact: M leaching to gw: H vapor formation: L→M fugitive dusts: M surface runoff: H	direct contact: L→M vapor formation: L→M leaching to gw: H fugitive dust: L→M	vapor formation: L→M ingestion: M→H transport to surface water: M→H	direct contact: L→M ingestion of biota: L vapor formation: L→M transfer to overlying waters: H	direct contact: M→H vapor formation: L ingestion of water: M→H ingestion of biota: L
Phthalate Esters (e.g., dimethyl phthalate, di-n-butyl-phthalate)	direct contact: M→H leaching to gw: L→M vapor formation: L fugitive dusts: M→H surface runoff: L→M	direct contact: M→H vapor formation: L leaching to gw: L→M fugitive dust: M→H	vapor formation: L ingestion: M→H transport to surface water: L→M	direct contact: M→H ingestion of biota: L vapor formation: L transfer to overlying waters: L→M	direct contact: M→H vapor formation: L ingestion of water: M→H ingestion of biota: L
Polychlorinated Biphenyls (PCBs) and Organochlorine Pesticides (e.g., aldrin, dieldrin)	direct contact: M→H leaching to gw: L→M vapor formation: L→M fugitive dusts: M→H surface runoff: L→M	direct contact: M→H vapor formation: L→M leaching to gw: L→M fugitive dust: M→H	vapor formation: L→M ingestion: M→H transport to surface water: M→H	direct contact: M→H ingestion of biota: M→H vapor formation: L→M transfer to overlying waters: L→M	direct contact: M→H vapor formation: L→M ingestion of water: M→H ingestion of biota: M→H
Polycyclic Aromatic Hydrocarbons (PAHs) (e.g., naphthalene, benzo(a)pyrene)	direct contact: M→H leaching to gw: L vapor formation: L fugitive dusts: M→H surface runoff: L→M	direct contact: M→H vapor formation: L leaching to gw: L→M fugitive dust: M→H	vapor formation: L ingestion: M→H transport to surface water: L→M	direct contact: M→H ingestion of biota: L→M vapor formation: L transfer to overlying waters: L→M	direct contact: M→H vapor formation: L ingestion of water: M→H ingestion of biota: L→M
Inorganics Metals (e.g., lead, mercury)	direct contact: M→H leaching to gw: L→M vapor formation: L[H] fugitive dusts: M→H surface runoff: L→M	direct contact: M→H vapor formation: L[H] leaching to gw: L→M fugitive dust: M→H	vapor formation: L[H] ingestion: L→M transport to surface water: L→M	direct contact: M→H ingestion of biota: L→M vapor formation: L[H] transfer to overlying waters: L→M	direct contact: M→H vapor formation: L[H] ingestion of water: M→H ingestion of biota: L→M

[A] Assumes chemical release directly into media of interest with subsequent fate and transport processes acting to influence contaminant concentration in media of interest (that is, through transformations and distribution among other media).
[B] L = low, M = moderate, and H = high
[C] Direct contact refers to both incidental ingestion and dermal contact.
[D] gw = ground water
[E] For direct contact and fugitive dust, potential exposure is via excavation.
[F] For ingestion, potential exposure is via transport to a water supply.
[G] For ingestion, potential exposure is via bioaccumulation in edible aquatic organisms.
[H] Mercury is a extremely volatile metal when present in its elemental form [Hg(O)], and vapors are thus often an important exposure pathway for this metal.

through bioaccumulation. These chemicals do not typically bioaccumulate in vertebrates such as fish.

X2.3.9 Phenols:

X2.3.9.1 These compounds are highly water soluble and can leach from soil into underlying ground water systems. They are not persistent in surface waters due to rapid microbial and indirect photolytic degradation. Phenols are not highly volatile compounds, nor are they readily bioaccumulated by terrestrial or aquatic biota.

X2.3.9.2 Important exposure pathways for phenols could

include ingestion of ground and surface waters, as well as dermal contact with such waters. Exposure to phenols would have to occur soon after their release due to their rapid disappearance from natural waters.

X2.3.10 *Phthalate Esters:*

X2.3.10.1 Phthalate esters vary greatly in their water solubility and sorptive characteristics. Some can be very water-soluble and can thus migrate horizontally in surface runoff to surface water bodies, or can vertically migrate in soil solutions to the underlying water table. Others tend to sorb strongly to organic matter. Direct photolysis can also act to remove phthalate esters in surface waters. Volatilization is not an important fate processes for phthalate esters due to low vapor pressures and Henry's Law constants.

X2.3.10.2 Potential exposure to phthalate esters can occur via direct contact with impacted media and through ingestion of ground and surface waters.

X2.3.11 *Polychlorinated Biphenyls (PCBs) and Organochlorine Pesticides:*

X2.3.11.1 Polychlorinated biphenyls and organochlorine pesticides are generally very hydrophobic and persistent in environmental media due to low aqueous solubilities, high octanol-water partition coefficients, and high organic carbon partition coefficients. Therefore, these compounds tend to sorb to soil matrices where they are effectively sequestered and are not easily released for subsequent transport by water or air. However, due to high Henry's Law constants, if solubilized these compounds can be transferred from water surfaces to the overlying air.

X2.3.11.2 Important exposure pathways for PCBs include direct contact with impacted media and ingestion of PCB-containing foodstuffs. Exposure to PCBs, in particular the less highly chlorinated species, can also occur via inhalation of vapors. PCBs and organochlorine pesticides can bioaccumulate and biomagnify.

X2.3.12 *Chlorinated Benzenes:*

X2.3.12.1 Physicochemical properties of the chlorinated benzenes span a wide range and consequently the various chlorinated benzenes have contrasting environmental fates. Volatilization from ground and water surfaces are important processes for the mono- and di-chlorinated compounds as these compounds have high vapor pressures, and high Henry's Law constants. The less chlorinated compounds are also more soluble in water, potentially resulting in leaching through the vadose zone to the underlying water table. The more chlorinated benzenes, such as pentachlorobenzene and hexachlorobenzene, are more hydrophobic and tend to sorb to soil and sediment organic carbon. As a result, these more hydrophobic compounds can be persistent in soils and sediments.

X2.3.12.2 Exposure pathways for chlorinated benzenes vary depending on the degree of chlorination. Inhalation of vapors and ingestion of, or dermal contact with contaminated waters can be relevant exposure pathways for the lesser chlorinated compounds. Ingestion of soil and sediments, either directly or indirectly via inhalation of fugitive dust, are the more important exposure pathways for the more chlorinated compounds.

X2.3.13 *Organic Acids and Bases:*

X2.3.13.1 Organic acids and bases include a variety of compounds that either have proton donor or proton acceptor properties. Organic acids and bases can have a number of different functional groups, such as aromatic rings with hydroxyl groups (for example, phenolic compounds), aliphatic and aromatic carboxyl groups, aliphatic and aromatic amino groups, and aromatic rings containing nitrogen atoms (for example, pyridines). Although encompassing numerous compounds with greatly differing chemical structures, organic acids and bases are generally extremely water-soluble and are hence very mobile in surface runoff and soil pore waters. Leaching to ground water and subsequent transport in ground water systems are common fate processes for organic acids and bases. Organic acids and bases are generally not volatile compounds. Sorption of organic acids and bases can occur for both their ionized and neutral forms, although partitioning into organic phases is often decreased for ionized forms relative to neutral forms. The persistence of organic acids and bases in environmental media depends upon their chemical structure and, as a result, generalizations can not be made for the compound class as a whole. The fate of organic acids and bases is largely dependent upon ambient pH in aquatic and soil environments because pH affects the magnitude of dissociation of organic acids and bases (that is, the relative concentrations of ionized and neutral forms). Benzoic acid is an example of an organic acid that is not extremely persistent in aquatic and soil environments due to high rates of biodegradation and a limited sorption capability. Aniline is an example of an organic base that is rapidly removed from surface waters and soils through photodegradation and biodegradation. However, aniline can be relatively persistent in soil environments due to its ability to sorb to soils. Thus, sorption is stronger at lower pH due to a greater fraction of the neutral form.

X2.3.13.2 Exposure to organic acids and bases can occur via dermal contact with contaminated waters and through ingestion of ground and surface waters. Depending upon the physicochemical properties of organic acids and bases and such site-specific conditions as pH, sorption to soils can occur in some circumstances. As a result, potential exposure to organic acids and bases could also occur through incidental ingestion of soil or via inhalation of fugitive dust.

X2.3.14 *Metals:*

X2.3.14.1 The environmental behavior of metals is highly dependent on their chemical speciation. The aqueous solubility of metals is moderated by their ability to react with other dissolved species to form insoluble precipitates (complexation) and to adsorb to charged surfaces (surface reactivity). Reactive ligands for metals in natural waters include sulfates, carbonates, sulfides, and hydroxides. As a rule-of-thumb, the transition metals have been demonstrated to have the complex stability sequence: $Mn2+ < Fe2+ < Co2+ < Ni2+ < Cu2+ < Zn2+$. Divalent metal species are generally more strongly adsorbed onto charged surfaces than monovalent metals, and smaller cations are more preferentially retained by charged surfaces in competitions for exchange sites with larger cations. Complexed and adsorbed metals can be retained in soils and sediments.

X2.3.14.2 The potentially important exposure pathways for metals can include ingestion of contaminated ground and surface waters, as well as dermal contact with these waters. Ingestion of soils and sediments, either directly or via fugitive dust, can also be important. Mercury is a unique metal because it is volatile (see Table X2.2). As a result, inhalation of vapors should be considered a possible exposure pathway for this metal in its non-complexed form.

X2.3.15 *Mixtures:*

X2.3.15.1 The fate and transport of individual chemicals within mixtures can greatly differ from the behavior observed for those compounds in their pure forms. Co-factors, such as the presence of solvents or cosolvents, can potentiate the mobilization of compounds that individually might be expected to sorb onto organic matter and be retained (for example, PAHs). Similarly, vapor pressures and aqueous solubilities can vary between individual components and mixtures. For example, the presence of inorganic salts can decrease the solubility of neutral nonpolar compounds in an aqueous environment.

X2.3.15.2 Potentially important exposure pathways for mixtures need to be assessed on a case-by-case basis, as they are dependent upon the nature of the mixture and site-specific conditions.

X2.3.16 *Media of Exposure:*

X2.3.16.1 Site-specific conditions are important in influencing the environmental behavior of chemicals. Which site-specific conditions are most important depends upon the compound's physicochemical properties and the media of interest. For example, the composition, pH or organic carbon content of the contaminated soils are very important to some chemical groupings, while wind or elevated ground surface temperatures can be very important for others. Bioavailability, may also be influenced by site-specific conditions.

X2.3.16.2 The following descriptions indicate a number of environmental parameters which can influence the behavior of and exposure to common chemicals in the environment. These descriptions are followed by short discussions of media-specific conditions which can be important modifiers of the behavior of, and exposure to, environmentally released chemicals.

X2.3.17 *Defining Environmental Parameters:*

X2.3.17.1 Acid-volatile sulfides (AVS) are reactive solid-phase sulfides which can bind certain metals such as cadmium and lead in aquatic environments, thus removing them from interstitial waters and reducing their bioavailability to sediment-dwelling organisms. The ratio of AVS to SEM (simultaneously extracted metals) on a molar basis is commonly used to predict the bioavailability and toxicity of metals in sediments.

X2.3.17.2 Total organic carbon (TOC) refers to solid phase organic carbon in natural organic matter which can bind chemicals in soil and sedimentary environments. Natural organic matter typically consists of 50 % organic carbon. TOC is a commonly measured analytical parameter for soil and sediment samples because sorption to organic carbon can be an important fate process for chemical species.

X2.3.17.3 The pH of a solution is a measure of the concentration of hydrogen or hydroxide ions present in a solution. The pH of surface waters, ground waters, and soil solutions is an important parameter with respect to the speciation and mobilization of chemicals. The pH of natural waters normally lies between 4 and 9.

X2.3.17.4 Conductivity refers to the current carrying capacity of dissolved ions (for example, $Ca2+$, $Na+$, $Mg2+$, $Al3+$, etc.). Conductivity is a commonly measured environmental parameter because it is a reflection of the concentration of ions available for reaction with dissolved chemical species such as metals.

X2.3.17.5 Redox potential (EH) is an analytical measure of the oxidative/reductive conditions present in an aquatic environment. Positive redox potentials are reflective of highly oxidizing conditions, while negative redox potentials are common of reducing environments. Redox potential influences the speciation and hence the mobilization and toxicity of chemical species such as metals.

X2.3.17.6 Cation exchange capacity (CEC) is a measure of the ion exchange properties of a soil environment. Clay minerals typically have high CECs and can efficiently retain and exchange cations. The CEC of soils is dependent upon soil mineralogy and pH.

X2.3.18 *Soil:*

X2.3.18.1 Both soil characteristics and physical properties can be important when considering the fate and transport of chemicals. Soil characteristics such as organic carbon content, soil composition (percent sand, silt and clay, iron oxide content, etc.), and cation exchange capacity (CEC) can influence the environmental behavior of chemicals. Adsorption of chemicals to soil surfaces can be important processes that serve to bind and retain chemicals. For some metals, soil constituents such as iron oxides can induce chemical reactions that result in the formation of immobile and insoluble precipitates. The cation exchange capacity (CEC) of soils is dependent upon soil mineralogy and pH, and can indicate whether the retention of charged species through ion exchange will be an important process. Soil pH and redox conditions can also affect the form and speciation of chemicals, and in turn influence their fate and transport. Physical properties such as soil strength, soil moisture, and soil particle size are important with respect to the erosion of soil and the formation of fugitive dust. Hydraulic conductivity affects the rate of vertical and horizontal transport in soil systems and is thus of importance with respect to the transport of chemicals in soils to the underlying water table and to nearby surface water bodies.

X2.3.18.2 The bioavailability of chemicals in soil varies with soil type, and the residence time of the chemical(s). Due to interactions between chemicals and soil matrices, the fraction of a chemical which is bioavailable to organisms generally is less than the bulk concentration, even for newly introduced chemicals. Chemicals may also weather in soil systems, may become sequestered within soil matrices, and hence may show diminished availability to organisms as shown in Ref **(6)**. Consequently, the residence time of chemicals in a soil along with the soil characteristics can influence the bioavailability of chemicals in soils via oral or dermal exposure.

X2.3.19 *Sediment:*

X2.3.19.1 Sediment properties such as total organic carbon content (TOC), particle size distribution, clay type and content,

CEC, and pH affect the sorption of organic and inorganic compounds to sediment. Sorbed compounds tend to be less bioavailable to organisms and, as a result, direct exposure through ingestion of sediments and exposure through the food chain are often reduced.

X2.3.19.2 The bioavailability of some metals in sediments is related to the presence of Acid Volatile Sulfides (AVS) Ref **(7)**. It has been shown that insoluble metal sulfides, typically iron and manganese sulfides, control metal availability in anoxic sediments by reacting with available divalent metals. A comparison of the level of AVS to the amount of available divalent metal (Simultaneously Extracted Metal or SEM) on a molar basis indicates whether there is sufficient AVS to effectively sequester the available metal species, and reduce availability to receptors.

X2.3.20 *Surface and Ground Water*—Chemicals in water may be affected by water characteristics such as suspended solids, transparency, dissolved organic carbon, temperature, pH, redox potential, and advective/dispersive processes. These characteristics interact with the physicochemical properties of the chemicals to influence their fate and transport. For example, suspended solids affect the distribution of hydrophobic organic compounds which preferentially sorb to particles, and transparency affects the rate of photodegradation. Furthermore, dissolved organic carbon, nutrient concentrations and the prevalence of oxic or anoxic conditions can affect microbial degradation.

X2.3.21 *Non-Aqueous Phase Liquids (NAPL)*— At some sites, organic chemicals may be present in the form of non-aqueous phase liquids (NAPLs) and, as such, represent a noteworthy site-specific condition. NAPLs can be thought of both as a chemical mixture and as a medium that is important to behavior and exposure. NAPLs may either float on the ground water surface (light or L-NAPL) or sink through water (dense or D-NAPL). An example of a potential L-NAPL mixture is gasoline. An example of a chemical that can occur as a D-NAPL is trichloroethylene. The chemicals within the NAPL behave differently than they do when dissolved in water or adsorbed onto soils and sediments. NAPLs may move along density gradients in response to gravity. L-NAPLs may flow along the ground water surface while D-NAPLs may move along interfaces in subsurface geologic formations. Further, the fate of chemicals within the NAPL are affected by the NAPL medium. For example, the hydrophobic nature of the NAPL medium and the surface tension of the NAPL may limit diffusion of chemicals within the NAPL into surrounding ground water.

X2.3.22 *Air*— Volatilization can be affected by air temperature and wind velocity. Wind velocity can also affect the transport of airborne pollutants from their sources. Precipitation events and gravitational settling can result in the redeposition of airborne pollutant onto ground and water surfaces.

X2.4 *Toxicity Values and Effects Considerations:*

X2.4.1 In the previous sections of this appendix, the physical and chemical properties of chemicals and how these properties determine the environmental distribution of chemicals was discussed. Knowing the approximate long-term concentrations of chemicals in different environmental media is necessary for determining long-term exposures. In addition to needing some estimate of the affect of time on the concentration of chemicals in different environmental media, knowledge of the potential toxicity of chemicals is necessary to provide an estimation of risk from exposure to released chemicals. In order to estimate an acceptable level of exposure to a chemical(s), for either human or ecological receptors, an understanding of the basis for, and the derivation of toxicity values (for example, reference doses, reference concentrations, slope factors, sediment quality criteria) and how these values should be applied to different exposure scenarios is necessary. This section describes how toxicity is assessed using laboratory animal experiments and epidemiologic data, how those data are used to derive toxicity values used for risk assessment, and how simultaneous exposures to multiple chemicals (that is, mixtures) are assessed.

X2.4.1.1 *Toxicity Assessment Process*—Dose-response data (either from animal or epidemiological studies) are typically used to identify levels above and below which adverse effects are observed as well as the magnitude of these effects. Observed adverse effects may include whole body effects (for example, weight loss), effects on specific body organs (for example, central nervous system), teratogenic effects (that is, birth defects), mutagenic effects (that is, alters genetic material) and carcinogenic effects (that is, produces malignant tumors). The assessment process involves either defining a toxicity value that represents an exposure level below which effects are not expected (as is done for most non-cancer effects) or a value that represents the relationship between dose and response for environmental exposure levels (as is done for most carcinogens.)

X2.4.2 *Non-Cancer Effects:*

X2.4.2.1 Chemicals are generally believed to elicit non-cancer effects via threshold mechanisms. Therefore, for most chemicals, toxicologists have defined levels of exposure that will not result in any adverse effects in exposed individuals or ecological receptors. Scientific methods for determining the level of exposure that will not result in adverse non-cancer effects in exposed populations are briefly outlined below.

X2.4.2.2 Most estimates of an acceptable dose for humans are based on animal studies. In rare instances, well defined human epidemiological information is available on a chemical. Animal or human data are used in establishing a Reference Dose (RfD) or Reference Concentration (RfC) - (see Table X2.4). The extrapolation of animal bioassay results when predicting human risk involves a number of assumptions regarding effects thresholds, dosing/exposure duration differences, interspecies and intraspecies extrapolation, high- to low-dose extrapolation, and route-to-route extrapolation. When calculating toxicological criteria for non-cancer effects, generally uncertainty factors (see Table X2.4) of ten, are applied to either the No-Observed-Adverse-Effect-Level (NOAEL) or Lowest-Observed-Adverse-Effect-Level (LOAEL) in animal or epidemiologic studies to account for uncertainties in predicting a NOAEL for humans. The toxicity values derived by applying uncertainty and modifying factors to NOAELs and LOAELs are termed RfDs and RfCs, for oral and inhalation

TABLE X2.4 Definitions of Important Toxicological Characteristics

NOAEL (No-observed-adverse-effect level): An exposure level at which there are no statistically or biologically significant increases in the frequency or severity of adverse effects between the exposed population and its appropriate control; some effects may be produced at this level, but they are not considered as adverse, not precursors to adverse effects. In an experiment with several NOAELs, the regulatory focus is primarily on the highest one, leading to the common usage of the term NOAEL as the highest exposure without adverse effects.
LOAEL (Lowest-observed-adverse-effect level): The lowest exposure level at which there are statistically or biologically significant increases in the frequency or severity of adverse effects between the exposed population and its appropriate control group.
Uncertainty factor: One of several, generally 10–fold factors, used in operationally deriving the Reference Dose (RfD) or Reference Concentration (RfC) from experimental data. Ufs are intended to account for *(1)* the variation in sensitivity among the members of the human population; *(2)* the uncertainty in extrapolating animal data to the case of humans; *(3)* the uncertainty in extrapolating from data obtained in a study that is of less-than-lifetime exposure; and, *(4)* the uncertainty in using LOAEL data rather than NOAEL data.
Modifying Factor: An additional uncertainty factor that is greater than zero and less than or equal to 10. The magnitude of the MF depends on the professional assessment of scientific uncertainties of the study not treated by uncertainty factors; such as, the completeness of the overall data base and the number of species tested. The default value of the MF is one.
Reference Dose: A reference dose is an estimate (with an uncertainty typically spanning perhaps an order of magnitude) of a daily exposure (mg/kg/day) to the general human population (including sensitive subgroups) that is likely to be without an appreciable risk of deleterious effects during a lifetime of exposure.
Reference Concentration: A reference concentration is an estimate (with an uncertainty spanning perhaps an order of magnitude) of a continuous exposure to the human population (including sensitive subgroups) that is likely to be without appreciable deleterious effects during a lifetime.
Slope Factor: The slope of the dose-response curve in the low-dose region. When low-dose linearity cannot be assumed, the slope factor is the slope of the straight line from zero dose to the dose at 1% excess risk. An upper bound on this slope is usually used instead of the slope itself. The units of the slope factor are usually expressed as $(mg/kg/day)^{-1}$.
Drinking Water MCLs & MCLGs: Maximum Contaminant Levels (MCLs) are drinking water standards established by the EPA that are protective of human health. However, these standards take into account the technological capability of attaining these standards. The EPA has, therefore, also established MCL goals (MCLGs) which are based only on the protection of human health. The MCL standards are often used as clean-up criteria.
Drinking Water Health Advisories: The Office of Drinking Water provides health advisories (HAs) as technical guidance for the protection of human health. They are not enforceable federal standards. HAs are the concentration of a substance in drinking water estimated to have negligible deleterious effects in humans, when ingested for specific time periods.
Water Quality Criteria: These criteria are not rules and they do not have regulatory impact. Rather, these criteria present scientific data and guidance of the environmental effects of pollutants which can be useful to derive regulatory requirements based on considerations of water quality impacts.

exposures, respectively. The application of uncertainty factors (for example, 10 for interspecies, 10 for intraspecies, 5-10 for subchronic to lifetime, etc.) results in estimates that are more likely to overestimate than underestimate risks. The uncertainties in estimating acceptable levels of exposure at any site, based on RfDs/RfCs should be clearly communicated in each assessment

X2.4.2.3 Toxicity studies used for deriving toxicological criteria can generally be broken into five categories based on the frequency and duration of exposure to the risk agent. These studies can be described as follows:

X2.4.2.4 Acute studies typically use one exposure or multiple exposures over a short duration (24 h). Symptoms are usually observed within a short time after dosing. Acute studies are generally designed to measure frank effects such as neurological damage or death, and are often used to define dose ranges for use in studies of longer duration. The level of exposure necessary to elicit acute effects is generally much larger than that required to elicit effects after repeated exposures. Acute studies are not generally used for estimating acceptable levels of longer-term exposure.

X2.4.2.5 Subchronic studies use multiple or continuous exposures over an extended period (three months is the usual time frame in rodent studies) generally defined as 10 % or less of an animals lifetime. Subchronic exposures in humans are generally viewed as encompassing seven years or less. Subchronic studies may be designed to assess a particular toxicological endpoint (for example, neurologic effects), but are most often comprehensive in their design. Subchronic studies are suitable for use in determining acceptable levels of subchronic and/or chronic exposure for non-cancer effects.

X2.4.2.6 Chronic studies use multiple exposures over an extended period of time, usually a significant fraction of the animal's lifetime (typically two years for rodents). Chronic studies generally involve numerous physical observations during the dosing period and a complete toxicological/pathological evaluation after dosing has been concluded. Chronic studies in rodents often evaluate both cancer and non-cancer endpoints, and are preferred for use in establishing chronic RfDs/RfCs.

X2.4.2.7 Reproductive and developmental studies examine adverse effects on fertility and fetal development. Exposures in developmental toxicity studies are generally limited to a specific timeframe such as the period of fetal organ development. Observed effects are often limited by study design and include observations such as sperm motility.

X2.4.2.8 Multi-generation studies use multiple or continuous exposures over two or more generations. Observed effects are usually limited to reproductive and/or teratogenic effects in the offspring and gross and pathological changes in the adults.

X2.4.2.9 *Derivation of Toxicity Values and Criteria*—Data from the above studies are used to generate Reference Doses (RfDs), Reference Concentrations (RfCs), slope factors (SFs), and are also used in generating Drinking Water Maximum Contaminant Levels (MCLs) and Goals (MCLGs), Health Advisories (HAs) and Water Quality Criteria. See Table X2.4 for definitions of these important criteria.

X2.4.3 *Cancer Effects*— Agents may elicit carcinogenic effects through either threshold or non-threshold mechanisms. However, defining the carcinogenic threshold for agents that exert their carcinogenic effects via non-genotoxic/threshold mechanisms has proven much more difficult than defining the threshold for non-carcinogenic effects. Due to that difficulty, U.S. regulatory agencies generally assume that carcinogens act via non-threshold mechanisms and that any level of exposure to a carcinogen, no matter how small, presents some risk of developing cancer. International regulatory agencies (for example, WHO), however, apply safety factors similar to those used in the U.S. for non-cancer effects when establishing acceptable levels of exposure to carcinogenic agents. The U.S. Environmental Protection Agency calculates Slope factors that are used to calculate upper-bound numerical estimates of carcinogenic risk. Because of the uncertainty regarding the ability of risk agents to produce incremental carcinogenic effects in humans, USEPA has also developed weight-of-evidence criteria for carcinogenicity (see Table X2.5). The weight-of-evidence criteria are used to aid in risk management decisions, by defining the confidence scientists have in the ability of the slope factor to predict human risk or the confidence scientist have that the agent is not carcinogenic under a defined set of circumstances. *USEPA is revising their scheme for categorizing carcinogens and estimating their carcinogenic risk to better reflect chemical-specific differences in scientific knowledge.*

TABLE X2.5 Weight of Evidence Criteria for Carcinogens

Category	Criterion
A	Human carcinogen, with sufficient evidence from epidemiological studies.
B1	Probable human carcinogen, with limited evidence from epidemiological studies.
B2	Probable human carcinogen, with sufficient evidence from animal studies and inadequate evidence or no data from epidemiological studies.
C	Possible human carcinogen, with limited evidence from animal studies in the absence of human data.
D	Not classifiable as to human carcinogenicity, owing to inadequate human and animal evidence.
E	Evidence of noncarcinogenicity for humans, with no evidence of carcinogenicity in at least two adequate animal tests in different species, or in both adequate animal and epidemiological studies.

X2.4.4 Integrated Risk Information System (IRIS) is generally considered the primary source for RfD, RfC, and SF values, see Ref (**8**).

X2.4.5 Health Effects Assessment Summary Tables (HEAST) reference a larger database than IRIS. HEAST is often used as a secondary source of health effects information when toxicity criteria are not available in IRIS. Whereas the information in IRIS has been subject to at least Agency-wide verification, the information in the HEAST tables has not. Thus, extra care should be exercised in using the values in HEAST as they are solely reflective of the opinions of a limited number of EPA scientists. Furthermore, the toxicity criteria listed in HEAST are not updated on any regular basis. The user is expected to consult the original U.S. EPA assessment documents and recent scientific studies to determine the strengths and limitations of the data in HEAST, see Ref (**9**).

X2.4.6 If toxicity criteria (that is, RfDs, RfCs, and SFs) are not available, or are inappropriately listed in IRIS, HEAST, or State approved databases, a toxicologist should be consulted for appropriate values for use in the assessment. If project toxicologists with experience in the derivation of toxicity criteria are not available, toxicologists at the EPA's National Center for Exposure Assessment (NCEA) in Cincinnati are often able to supply appropriate criteria for site-specific use. The derivation basis for toxicity criteria obtained from sources other than IRIS or HEAST should be documented in greater detail.

X2.5 *Chemicals of Concern (COCs) for Risk Assessments:*

X2.5.1 *Selection and Evaluation of COCs*—It is often not necessary to assess in detail the potential human health or environmental risk associated with every site-related compound present in an environmental release. For this reason, risk management decisions are generally based on the assessment of the potential impacts from a selected group of indicator compounds or chemicals of concern (COC). It is inherently assumed in this approach that the chemicals of interest contribute the vast majority of the total potential impact from all chemicals present in the affected media. The chemicals of interest are selected based primarily on consideration of potential exposure routes, concentrations, toxicological properties and mobility in the affected media. In the Tier 1 evaluation, toxicological information, such as presented above, is used to develop Risk-Based Screening Levels (RBSL). If further Tiers of evaluation are necessary based on the results of the Tier 1 analysis, only those chemicals that were detected at concentrations above the Tier 1 values are evaluated further. Those chemicals that are evaluated in the RBCA process are termed chemicals of concern (COCs). In Tiers 2 and 3 more detailed evaluation of the toxicological criteria and physicochemical data is performed. In Tiers 2 and 3 of any site evaluation, potential additive and synergistic effects of COCs may be evaluated.

X2.5.2 *How Toxicity Is Assessed: Individual Chemicals Versus Mixtures*— The toxicity of an individual chemical is typically established based on dose-response studies which estimate the relationship between different dose levels and the magnitude of their adverse effects (that is, toxicity). When evaluating exposures to multiple chemicals, preference is given to data on actual mixtures. For example, often a single toxicity factor is used when evaluating exposures to PCBs. However, when data on mixtures is not available, a relative toxicity approach is sometimes used within chemical classes. The relative toxicity approach is most applicable when the toxicity of individual congeners is structure-dependent. For example, as discussed above, the cancer potencies of certain PAHs relative to the most studied compound of this class, benzo[a]pyrene, are presently approximated using an "estimated order of potential potency" approach. Interim relative potencies that are suggested for use by U.S. EPA Ref **(10)** for seven carcinogenic PAHs are shown in Table X2.6. Also, when calculating a Hazard Index to evaluate the potential for additive noncarcinogenic effects, only hazard quotients for chemicals effecting the same target organ and mechanism of action should be summed.

TABLE X2.6 Cancer Potencies of Seven PAHs Relative to Benzo[a]pyrene[A]

Compound	Relative Cancer Potency
Benzo[a]pyrene	1.0
Benzo[a]anthracene	0.1
Benzo[b]fluoranthene	0.1
Benzo[k]fluoranthene	0.01
Chrysene	0.001
Dibenzo[a,h]anthracene	1.0
Indeno[1,2,3-cd]pyrene	0.1

[A] Interim values provided as temporary guidance for risk evaluation of PAHs in *Provisional Guidance for Quantitative Risk Assessment of Polycyclic Aromatic Hydrocarbons* **(10)**.

X2.5.2.1 Generally, the risks associated with individual constituents of a complex mixture are assumed to be additive, and carcinogenic and noncarcinogenic risks are determined separately (for noncarcinogenic endpoints it is only appropriate to sum hazard quotients of compounds with similar toxicological endpoints and mechanisms of action). Lack of sufficient toxicological information is often an impediment to this procedure. When toxicological information is unavailable from a published database, health scientists may be utilized to derive values appropriate for use in the risk assessment.

X2.5.3 *Use of TPH or TOC Measurements in Risk Assessments*— Various chemical analysis methods commonly referred to as "Total Petroleum Hydrocarbons" (TPH) or "Total Organic Content" (TOC) are often used during an initial site assessments to focus future investigations toward particular compounds and/or media. These methods usually determine the total amount of hydrocarbons present as a single number, and give no information on the types of hydrocarbons present. Such TPH or TOC methods may be useful in screening assessments where the whole product toxicity approach is appropriate to determine the need for further sampling. In general, these measurements should not be used for risk assessments, because the general measure of TPH or TOC provides insufficient information about the amounts of individual compounds present to accurately characterize potential risk. More information on petroleum hydrocarbons is available from the Total Petroleum Hydrocarbon Working Group (TPHWG) effort. Refs **(11-14)**.

X3. DEVELOPMENT OF EXAMPLE RISK-BASED SCREENING LEVELS (RBSLs)

X3.1 *Introduction*:

X3.1.1 This appendix contains the equations and parameters used to develop example RBSLs (Table X3.1). This appendix was prepared solely for the purpose of presenting example calculations for RBSLs. These values should not be viewed, or misused, as proposed remediation "standards". The reader should note that the example is for an adult male, not all possible pathways have been considered, and a number of assumptions concerning exposure scenarios and parameter values have been made. In developing RBSLs, evaluate all assumptions for appropriateness. The selected chemicals in this appendix should not be viewed as an inclusive or exclusive list for purpose of site investigation.

X3.1.2 The approaches used to calculate example RBSLs appearing in this appendix are briefly discussed as follows for exposure to vapors, ground water, surficial soils, and subsurface soils by means of the following representative pathways:

X3.1.2.1 Direct inhalation of vapors,

X3.1.2.2 Ingestion of ground water,

X3.1.2.3 Inhalation of outdoor vapors originating from dissolved chemicals in ground water,

X3.1.2.4 Inhalation of indoor vapors originating from dissolved chemicals in ground water,

X3.1.2.5 Ingestion of surficial soil, inhalation of outdoor vapors and particulates emanating from surficial soils, and dermal absorption resulting from surficial soil contact with skin,

X3.1.2.6 Inhalation of outdoor vapors originating from chemicals in subsurface soils,

X3.1.2.7 Inhalation of indoor vapors originating from chemicals in subsurface soils,

X3.1.2.8 Ingestion of ground water impacted by aqueous leaching of dissolved chemicals from unsaturated soils,

X3.1.2.9 Migration of a free-phase liquid in unsaturated soil, and

X3.1.2.10 Migration of a free-phase liquid plume in saturated soil.

X3.1.3 For the pathways considered, approaches used in this appendix are consistent with guidelines contained in Ref. **(15)**.

X3.1.4 The development presented as follows focuses only on human-health RBSLs for chronic (long-term) exposures.

X3.1.4.1 In the case of compounds that have been classified

TABLE X3.1 Chemical-Specific Parameters and Risk-Based Screening Level Concentrations (See Refs. 2, 16-21)

NOTE 1—Reference Note A: Estimated from Kow given in USEPA (1994) **(16)** and a Kow to Koc regression equation from USEPA (1996) **(17)**.
NOTE 2—Reference Note B: Vapor pressure estimated from aqueous solubility limit and Henry's law coefficient.
NOTE 3—Reference R-to-R: An exposure route to exposure route extrapolation of available IRIS or HEAST data, consistent with USEPA (1996) **(17)**.
NOTE 4—Reference FR: United States Federal Register.

CASRN		71-43-2	
name		benzene	
		C_6H_6	
Risk-Based Screening-Level Concentrations (1)			
	risk or		
	hazard	scenario:	
	criteria:	residential	industrial
ambient vapor inhalation			
$RBSL_{air}$	THQ	NA	NA
(μg/m³)	TR_U	2.94E+01	4.93E+01
	TR_L	2.94E-01	4.93E-01
enclosed space vapor inhalation			
$RBSL_{air}$	THQ	NA	NA
(μg/m³)	TR_U	3.92E+01	4.93E+01
	TR_L	3.92E-01	4.93E-01
groundwater ingestion			
$RBSL_{gw}$	THQ	NA	NA
(mg/L)	TR_U	2.94E-01	9.87E-01
	TR_L	2.94E-03	9.87E-03
	MCL	5.00E-03	5.00E-03
soil leaching to groundwater ingestion			
$RBSL_s$	THQ	NA	NA
(mg/kg)	TR_U	9.60E-01	3.22E+00
	TR_L	9.60E-03	3.22E-02
	MCL	1.63E-02	1.63E-02
groundwater to ambient vapor inhalation			
$RBSL_{gw}$	THQ	NA	NA
(mg/L)	TR_U	3.79E+02	6.37E+02
	TR_L	3.79E+00	6.37E+00
groundwater to enclosed space vapor inhalation			
$RBSL_{gw}$	THQ	NA	NA
(mg/L)	TR_U	2.43E+00	7.66E+00
	TR_L	2.43E-02	7.66E-02
subsurface soil volatilization to ambient air			
$RBSL_s$	THQ	NA	NA
(mg/kg)	TR_U	1.29E+01	2.16E+01
	TR_L	1.29E-01	2.16E-01
subsurface soil volatilization to enclosed space			
$RBSL_s$	THQ	NA	NA
(mg/kg)	TR_U	7.63E-01	2.40E+00
	TR_L	7.63E-03	2.40E-02
residual phase mobility limit			
$RBSL_s$	unsat.	9.05E+03	9.05E+03
(mg/kg)	sat.	9.24E+03	9.24E+03
surficial soil exposure			
$RBSL_{ss}$	THQ	NA	NA
(mg/kg)	TR_U	4.70E+02	7.76E+02
	TR_L	4.70E+00	7.76E+00
surficial soil exposure apportionment (2)			
(%)	THQ		
(i-d-v-p)	TR_U	08-63-28-00	03-62-33-00
	TR_L	08-63-28-00	03-62-33-00

derived parameters:			
$VF_{s,esp}$	5.13E-05	2.05E-05	(g-soil/cm³-air)
$VF_{gw,esp}$	1.61E-05	6.44E-06	(cm³-wat/cm³-air)
$VF_{ss,amb,1}$	2.28E-07	2.50E-07	(g-soil/cm³-air)
$VF_{ss,amb,2}$	1.80E-08	2.10E-08	(g-soil/cm³-air)
$VF_{ss,amb,min(1,2)}$	1.80E-08	2.16E-08	(g-soil/cm³-air)
LF_{sw}	3.06E-01	3.06E-01	(g-soil/cm³-wat)
$VF_{s,amb}$	2.28E-06	2.28E-06	(g-soil/cm³-air)
$VF_{gw,amb}$	7.75E-08	7.75E-08	(cm³-wat/cm³-air)

transport and thermodynamic parameters			[references]
MW	7.81E+01	(g/mole)	USEPA(1996)
D_{air}	8.80E-02	(cm²/sec)	USEPA(1996)
D_{wat}	9.80E-06	(cm²/sec)	USEPA(1996)
K_{oc}	1.77E+00	log10(L-wat/kg-oc)	USEPA(1996)
K_{ow}	2.13E+00	log10(L-wat/kg)	USEPA(1996)
H	5.55E-03	(atm-m³/mol)	USEPA(1996)
H	2.31E-01	(L-wat/L-air)	(calculated)
P_v	9.52E+01	(mm Hg)	USEPA(1994)
S	1.75E+03	(mg/L)	USEPA(1996)
pK_a		log10(mole/mole)	
pK_b		log10(mole/mole)	
ρ	8.77E-01	(g/mL)	USEPA(1996)
T_{nbp}	8.01E+01	(°C)	USEPA(1994)
T_{mp}	5.50E+00	(°C)	CRC(1996)
toxicity parameters			
RfD_o		(mg/kg-day)	
RfC_i		(mg/m³)	
SF_o	2.90E-02	1/(mg/kg-day)	IRIS (1994)
SF_i	2.90E-02	1/(mg/kg-day)	HEAST (07/93)
W of E	A		IRIS (1994)
MCL	5.00E-03	(mg/L-wat)	52 FR 25690
RAF_o	1.00E+00		
RAF_d	5.00E-01		
derived parameters dependent on chemical properties and soil parameters			
UF	1.00E+00	(unitless)	
H_{eff}	2.31E-01	(L-wat/L-air)	
$C_{sat,vap}$	4.07E+08	(μg/m³-air)	
K_{sw}	6.95E-01	(L-wat/kg-soil)	
$C_{sat,soil}$	1.22E+03	(mg/kg-soil)	[based on S]
$C_{sat,soil}$	1.22E+03	(mg/kg-soil)	[based on P_v]
$D_{eff,vad}$	6.87E-03	(cm²/sec)	
$D_{eff,cap}$	1.96E-05	(cm²/sec)	
$D_{eff,ws}$	1.01E-03	(cm²/sec)	
$D_{eff,crk}$	6.87E-03	(cm²/sec)	

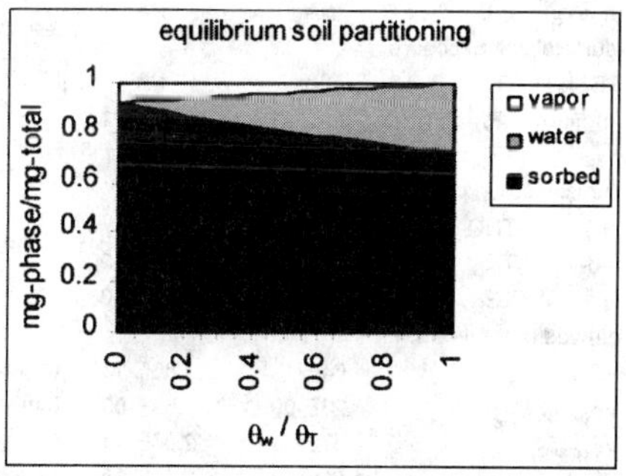

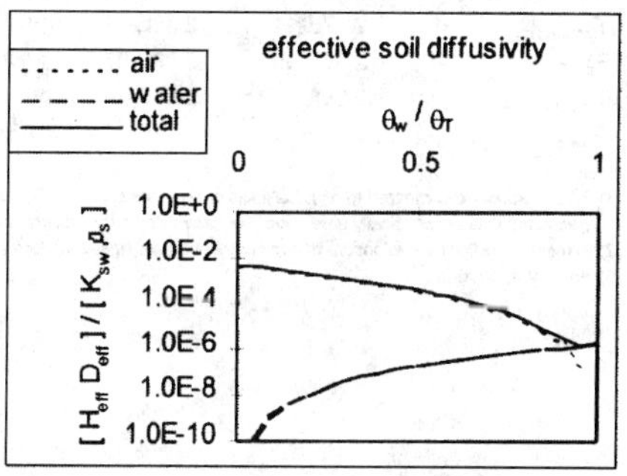

Notes:
(1) RBSL values are compared to physical limits of aqueous solubility (S<), saturated vapor pressure (<P), or either (Rs), and flagged if the calculated value is physically unrealistic. For cases where this occurs, the calculated values are given in parentheses.
(2) Percent of surficial soil exposure due to contributions of ingestion, dermal contact, vapor inhalation, and particulate inhalation (i-d-v-p).
(3) NA - not applicable.

TABLE X3.1 Chemical-Specific Parameters and Risk-Based Screening Level Concentrations (See Refs. 2, 16-21) *(continued)*

CASRN	50-32-8		
name	benzo(a)pyrene		
	$C_{20}H_{12}$		
Risk-Based Screening-Level Concentrations [1]			
	risk or hazard criteria:	scenario: residential	industrial
ambient vapor inhalation			
$RBSL_{air}$	THQ	NA	NA
(µg/m^3)	TR_U	P< (1.17E-01)	P< (1.96E-01)
	TR_L	1.17E-03	1.96E-03
enclosed space vapor inhalation			
$RBSL_{air}$	THQ	NA	NA
(µg/m^3)	TR_U	P< (1.56E-01)	P< (1.96E-01)
	TR_L	1.56E-03	1.96E-03
groundwater ingestion			
$RBSL_{gw}$	THQ	NA	NA
(mg/L)	TR_U	1.17E-03	S< (3.92E-03)
	TR_L	1.17E-05	3.92E-05
	MCL	2.00E-04	2.00E-04
soil leaching to groundwater ingestion			
$RBSL_s$	THQ	NA	NA
(mg/kg)	TR_U	Rs (5.62E+01)	S< (1.89E+02)
	TR_L	5.62E-01	1.89E+00
	MCL	9.63E+00	9.63E+00
groundwater to ambient vapor inhalation			
$RBSL_{gw}$	THQ	NA	NA
(mg/L)	TR_U	P< (1.63E+03)	P< (2.74E+03)
	TR_L	S< (1.63E+01)	S< (2.74E+01)
groundwater to enclosed space vapor inhalation			
$RBSL_{gw}$	THQ	NA	NA
(mg/L)	TR_U	P< (3.67E+01)	P< (1.16E+02)
	TR_L	S< (3.67E-01)	S< (1.16E+00)
subsurface soil volatilization to ambient air			
$RBSL_s$	THQ	NA	NA
(mg/kg)	TR_U	P< (5.65E+06)	P< (9.50E+06)
	TR_L	Rs (5.65E+04)	Rs (9.50E+04)
subsurface soil volatilization to enclosed space			
$RBSL_s$	THQ	NA	NA
(mg/kg)	TR_U	P< (3.35E+05)	P< (1.06E+06)
	TR_L	Rs (3.35E+03)	Rs (1.06E+04)
residual phase mobility limit			
$RBSL_s$	unsat.	NA	NA
(mg/kg)	sat.	NA	NA
surficial soil exposure			
$RBSL_{ss}$	THQ	NA	NA
(mg/kg)	TR_U	1.30E+01	3.03E+01
	TR_L	1.30E-01	3.03E-01
surficial soil exposure apportionment [2]			
(%)	THQ		
(i-d-v-p)	TR_U	55-44-00-00	38-61-00-00
	TR_L	55-44-00-00	38-61-00-00

derived parameters:			
$VF_{s,esp}$	4.64E-13	1.86E-13	(g-soil/cm^3-air)
$VF_{gw,esp}$	4.23E-09	1.69E-09	(cm^3-wat/cm^3-air)
$VF_{ss,amb,1}$	2.17E-11	2.38E-11	(g-soil/cm^3-air)
$VF_{ss,amb,2}$	1.80E-08	2.16E-08	(g-soil/cm^3-air)
$VF_{ss,amb,min(1,2)}$	2.17E-11	2.38E-11	(g-soil/cm^3-air)
LF_{sw}	2.08E-05	2.08E-05	(g-soil/cm^3-wat)
$VF_{s,amb}$	2.06E-14	2.06E-14	(g-soil/cm^3-air)
$VF_{gw,amb}$	7.14E-11	7.14E-11	(cm^3-wat/cm^3-air)

transport and thermodynamic parameters			[references]
MW	2.52E+02	(g/mole)	USEPA(1996)
D_{air}	4.30E-02	(cm^2/sec)	USEPA(1996)
D_{wat}	9.00E-06	(cm^2/sec)	USEPA(1996)
K_{oc}	6.01E+00	log10(L-wat/kg-oc)	USEPA(1996)
K_{ow}	6.11E+00	log10(L-wat/kg)	USEPA(1996)
H	1.13E-06	(atm-m^3/mol)	USEPA(1996)
H	4.70E-05	(L-wat/L-air)	(calculated)
P_v	1.77E-10	(mm Hg)	Mott(1995)
S	1.62E-03	(mg/L)	USEPA(1996)
pK_a		log10(mole/mole)	
pK_b		log10(mole/mole)	
ρ	1.11E+00	(g/mL)	USEPA(1994)
T_{nbp}	5.27E+02	(°C)	Mott(1995)
T_{mp}	2.52E+02	(°C)	CRC(1996)
toxicity parameters			
RfD_o		(mg/kg-day)	
RfC_i		(mg/m^3)	
SF_o	7.30E+00	1/(mg/kg-day)	IRIS (1994)
SF_i	7.30E+00	1/(mg/kg-day)	Route-to-Route
W of E	B2		IRIS (1994)
MCL	2.00E-04	(mg/L-wat)	57 31776 (17 Jul 92) Effective 01/17/94
RAF_o	1.00E+00		
RAF_d	5.00E-02		
derived parameters dependent on chemical properties and soil parameters			
UF	1.00E+00	(unitless)	
H_{eff}	4.70E-05	(L-wat/L-air)	
$C_{sat,vap}$	2.44E-03	(µg/m^3-air)	
K_{sw}	1.02E+04	(L-wat/kg-soil)	
$C_{sat,soil}$	1.66E+01	(mg/kg-soil)	[based on S]
$C_{sat,soil}$	5.32E-01	(mg/kg-soil)	[based on P_v]
$D_{eff,vad}$	4.49E-03	(cm^2/sec)	
$D_{eff,cap}$	3.72E-02	(cm^2/sec)	
$D_{eff,ws}$	4.56E-03	(cm^2/sec)	
$D_{eff,crk}$	4.49E-03	(cm^2/sec)	

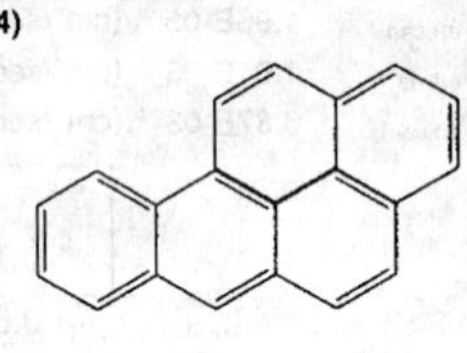

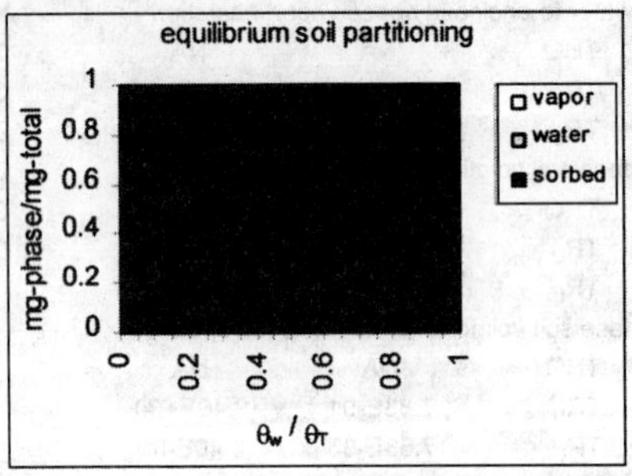

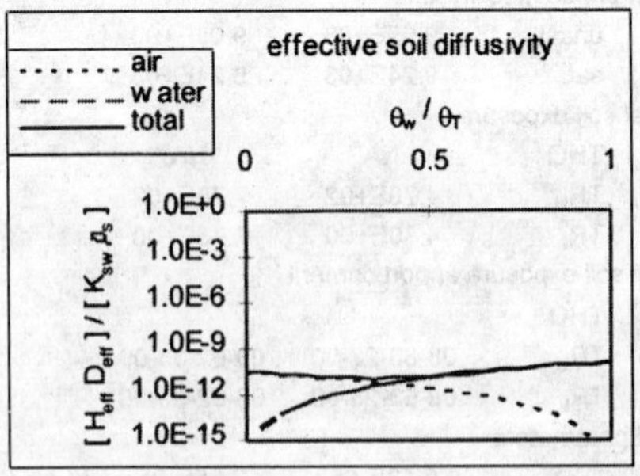

Notes:

(1) RBSL values are compared to physical limits of aqueous solubility (S<), saturated vapor pressure (<P), or either (Rs), and flagged if the calculated value is physically unrealistic. For cases where this occurs, the calculated values are given in parentheses.

(2) Percent of surficial soil exposure due to contributions of ingestion, dermal contact, vapor inhalation, and particulate inhalation (i-d-v-p).

(3) NA - not applicable.

TABLE X3.1 Chemical-Specific Parameters and Risk-Based Screening Level Concentrations (See Refs. 2, 16-21) *(continued)*

CASRN		7440-43-9	
name		cadmium	
		Cd	
Risk-Based Screening-Level Concentrations [(1)]			
	risk or		
	hazard	scenario:	
	criteria:	residential	industrial
ambient vapor inhalation			
$RBSL_{air}$	THQ	NA	NA
($\mu g/m^3$)	TR_U	P< (1.35E-01)	P< (2.27E-01)
	TR_L	P< (1.35E-03)	P< (2.27E-03)
enclosed space vapor inhalation			
$RBSL_{air}$	THQ	NA	NA
($\mu g/m^3$)	TR_U	P< (1.80E-01)	P< (2.27E-01)
	TR_L	P< (1.80E-03)	P< (2.27E-03)
groundwater ingestion			
$RBSL_{gw}$	THQ	1.82E-02	5.11E-02
(mg/L)	TR_U	NA	NA
	TR_L	NA	NA
	MCL	5.00E-03	5.00E-03
soil leaching to groundwater ingestion			
$RBSL_s$	THQ	3.69E+02	1.03E+03
(mg/kg)	TR_U	NA	NA
	TR_L	NA	NA
	MCL	1.01E+02	1.01E+02
groundwater to ambient vapor inhalation			
$RBSL_{gw}$	THQ	NA	NA
(mg/L)	TR_U	NA	NA
	TR_L	NA	NA
groundwater to enclosed space vapor inhalation			
$RBSL_{gw}$	THQ	NA	NA
(mg/L)	TR_U	NA	NA
	TR_L	NA	NA
subsurface soil volatilization to ambient air			
$RBSL_s$	THQ	NA	NA
(mg/kg)	TR_U	NA	NA
	TR_L	NA	NA
subsurface soil volatilization to enclosed space			
$RBSL_s$	THQ	NA	NA
(mg/kg)	TR_U	NA	NA
	TR_L	NA	NA
residual phase mobility limit			
$RBSL_s$	unsat.	NA	NA
(mg/kg)	sat.	NA	NA
surficial soil exposure			
$RBSL_{ss}$	THQ	3.65E+02	1.02E+03
(mg/kg)	TR_U	1.96E+07	3.29E+07
	TR_L	1.96E+05	3.29E+05
surficial soil exposure apportionment [(2)]			
(%)	THQ	99-00-00-00	99-00-00-00
(i-d-v-p)	TR_U	00-00-00-99	00-00-00-99
	TR_L	00-00-00-99	00-00-00-99

derived parameters:			
$VF_{s,esp}$	NA	NA	($g\text{-soil}/cm^3\text{-air}$)
$VF_{gw,esp}$	NA	NA	($cm^3\text{-wat}/cm^3\text{-air}$)
$VF_{ss,amb,1}$	0.00E+00	0.00E+00	($g\text{-soil}/cm^3\text{-air}$)
$VF_{ss,amb,2}$	1.80E-08	2.16E-08	($g\text{-soil}/cm^3\text{-air}$)
$VF_{ss,amb,min(1,2)}$	0.00E+00	0.00E+00	($g\text{-soil}/cm^3\text{-air}$)
LF_{sw}	4.94E-05	4.94E-05	($g\text{-soil}/cm^3\text{-wat}$)
$VF_{s,amb}$	NA	NA	($g\text{-soil}/cm^3\text{-air}$)
$VF_{gw,amb}$	NA	NA	($cm^3\text{-wat}/cm^3\text{-air}$)

transport and thermodynamic parameters			[references]
MW	1.12E+02	(g/mole)	USEPA(1996)
D_{air}	NA	(cm^2/sec)	
D_{wat}	NA	(cm^2/sec)	
K_d	3.63E+00	log10(L-wat/kg-soil)	USEPA(1996); pH=8.0
K_{ow}		log10(L-wat/kg)	
H	0.00E+00	(atm-m^3/mol)	(nonvolatile)
H	0.00E+00	(L-wat/L-air)	(calculated)
P_v	0.00E+00	(mm Hg)	(nonvolatile)
S	NA	(mg/L)	(no limit imposed)
pK_a		log10(mole/mole)	
pK_b		log10(mole/mole)	
ρ		(g/mL)	
T_{nbp}		(°C)	
T_{mp}		(°C)	
toxicity parameters			
RfD_o	5.00E-04	(mg/kg-day)	IRIS (1994)
RfC_i	NA	(mg/m^3)	Route-to-Route inappropriate
SF_o		1/(mg/kg-day)	
SF_i	6.30E+00	1/(mg/kg-day)	IRIS (1994)
W of E	B1		IRIS (1994)
MCL	5.00E-03	(mg/L-wat)	56 FR 3526 (30 Jan 91)
RAF_o	1.00E+00		
RAF_d	0.00E+00		
derived parameters dependent on chemical properties and soil parameters			
UF	1.00E+00	(unitless)	
H_{eff}	0.00E+00	(L-wat/L-air)	
$C_{sat,vap}$	0.00E+00	($\mu g/m^3$-air)	
K_{sw}	4.30E+03	(L-wat/kg-soil)	
$C_{sat,soil}$	NA	(mg/kg-soil)	[based on S]
$C_{sat,soil}$	NA	(mg/kg-soil)	[based on P_v]
$D_{eff,vad}$	NA	(cm^2/sec)	
$D_{eff,cap}$	NA	(cm^2/sec)	
$D_{eff,ws}$	NA	(cm^2/sec)	
$D_{eff,crk}$	NA	(cm^2/sec)	

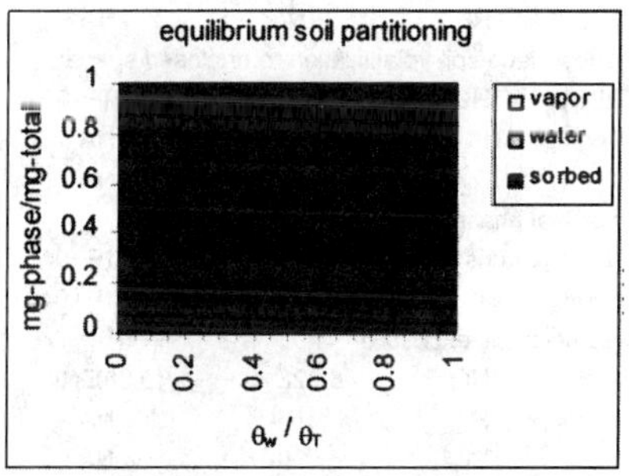

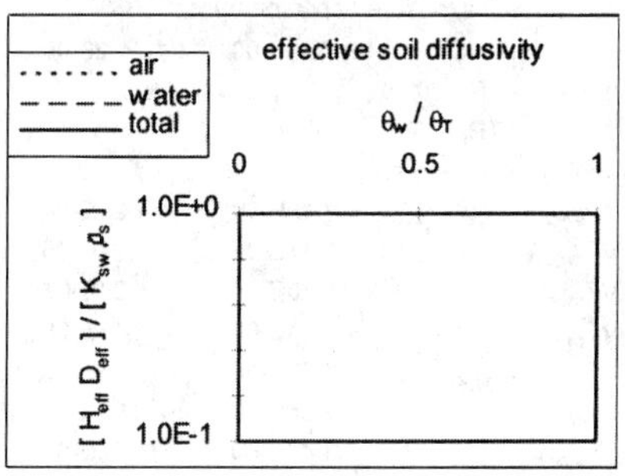

Notes:

(1) RBSL values are compared to physical limits of aqueous solubility (S<), saturated vapor pressure (<P), or either (Rs), and flagged if the calculated value is physically unrealistic. For cases where this occurs, the calculated values are given in parentheses.

(2) Percent of surficial soil exposure due to contributions of ingestion, dermal contact, vapor inhalation, and particulate inhalation (i-d-v-p).

(3) NA - not applicable.

TABLE X3.1 Chemical-Specific Parameters and Risk-Based Screening Level Concentrations (See Refs. 2, 16-21) *(continued)*

CASRN		75-15-0		
name		carbon disulfide		
		CS_2		
Risk-Based Screening-Level Concentrations [1]				
	risk or			
	hazard	scenario:		
	criteria:	residential	industrial	
ambient vapor inhalation				
$RBSL_{air}$	THQ	7.30E+02	1.02E+03	
(μg/m^3)	TR_U	NA	NA	
	TR_L	NA	NA	
enclosed space vapor inhalation				
$RBSL_{air}$	THQ	9.73E+02	1.02E+03	
(μg/m^3)	TR_U	NA	NA	
	TR_L	NA	NA	
groundwater ingestion				
$RBSL_{gw}$	THQ	3.65E+00	1.02E+01	
(mg/L)	TR_U	NA	NA	
	TR_L	NA	NA	
	MCL	NA	NA	
soil leaching to groundwater ingestion				
$RBSL_s$	THQ	1.24E+01	3.46E+01	
(mg/kg)	TR_U	NA	NA	
	TR_L	NA	NA	
	MCL	NA	NA	
groundwater to ambient vapor inhalation				
$RBSL_{gw}$	THQ	S< (2.14E+03)	S< (3.00E+03)	
(mg/L)	TR_U	NA	NA	
	TR_L	NA	NA	
groundwater to enclosed space vapor inhalation				
$RBSL_{gw}$	THQ	1.19E+01	3.13E+01	
(mg/L)	TR_U	NA	NA	
	TR_L	NA	NA	
subsurface soil volatilization to ambient air				
$RBSL_s$	THQ	5.14E+01	7.20E+01	
(mg/kg)	TR_U	NA	NA	
	TR_L	NA	NA	
subsurface soil volatilization to enclosed space				
$RBSL_s$	THQ	3.05E+00	8.00E+00	
(mg/kg)	TR_U	NA	NA	
	TR_L	NA	NA	
residual phase mobility limit				
$RBSL_s$	unsat.	1.21E+04	1.21E+04	
(mg/kg)	sat.	1.21E+04	1.21E+04	
surficial soil exposure				
$RBSL_{ss}$	THQ	6.82E+03	9.68E+03	
(mg/kg)	TR_U	NA	NA	
	TR_L	NA	NA	
surficial soil exposure apportionment [2]				
(%)	THQ	09-73-16-00	04-74-20-00	
(i-d-v-p)	TR_U			
	TR_L			
derived parameters:				
$VF_{s,esp}$		3.19E-04	1.28E-04	(g-soil/cm^3-air)
$VF_{gw,esp}$		8.17E-05	3.27E-05	(cm^3-wat/cm^3-air)
$VF_{ss,amb,1}$		5.70E-07	6.24E-07	(g-soil/cm^3-air)
$VF_{ss,amb,2}$		1.80E-08	2.16E-08	(g-soil/cm^3-air)
$VF_{ss,amb,min(1,2)}$		1.80E-08	2.16E-08	(g-soil/cm^3-air)
LF_{sw}		2.95E-01	2.95E-01	(g-soil/cm^3-wat)
$VF_{s,amb}$		1.42E-05	1.42E-05	(g-soil/cm^3-air)
$VF_{gw,amb}$		3.40E-07	3.40E-07	(cm^3-wat/cm^3-air)

transport and thermodynamic parameters			[references]
MW	7.61E+01	(g/mole)	USEPA(1996)
D_{air}	1.04E-01	(cm^2/sec)	USEPA(1996)
D_{wat}	1.00E-05	(cm^2/sec)	USEPA(1996)
K_{oc}	1.66E+00	log10(L-wat/kg-oc)	USEPA(1996)
K_{ow}	2.00E+00	log10(L-wat/kg)	USEPA(1996)
H	3.03E-02	(atm-m^3/mol)	USEPA(1996)
H	1.26E+00	(L-wat/L-air)	(calculated)
P_v	3.66E+02	(mm Hg)	USEPA(1994)
S	1.19E+03	(mg/L)	USEPA(1996)
pK_a		log10(mole/mole)	
pK_b		log10(mole/mole)	
ρ	1.26E+00	(g/mL)	USEPA(1996)
T_{nbp}	4.63E+01	(°C)	USEPA(1994)
T_{mp}	-1.12E+02	(°C)	CRC(1996)
toxicity parameters			
RfD_o	1.00E-01	(mg/kg-day)	IRIS (1994)
RfC_i	7.00E-01	(mg/m^3)	IRIS (08/01/95)
SF_o		1/(mg/kg-day)	
SF_i		1/(mg/kg-day)	
W of E			
MCL		(mg/L-wat)	
RAF_o	1.00E+00		
RAF_d	5.00E-01		
derived parameters dependent on chemical properties and soil parameters			
UF	1.00E+00	(unitless)	
H_{eff}	1.26E+00	(L-wat/L-air)	
$C_{sat,vap}$	1.53E+09	(μg/m^3-air)	
K_{sw}	7.20E-01	(L-wat/kg-soil)	
$C_{sat,soil}$	8.57E+02	(mg/kg-soil)	[based on S]
$C_{sat,soil}$	8.72E+02	(mg/kg-soil)	[based on P_v]
$D_{eff,vad}$	8.12E-03	(cm^2/sec)	
$D_{eff,cap}$	1.50E-05	(cm^2/sec)	
$D_{eff,ws}$	8.10E-04	(cm^2/sec)	
$D_{eff,crk}$	8.12E-03	(cm^2/sec)	

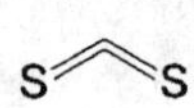

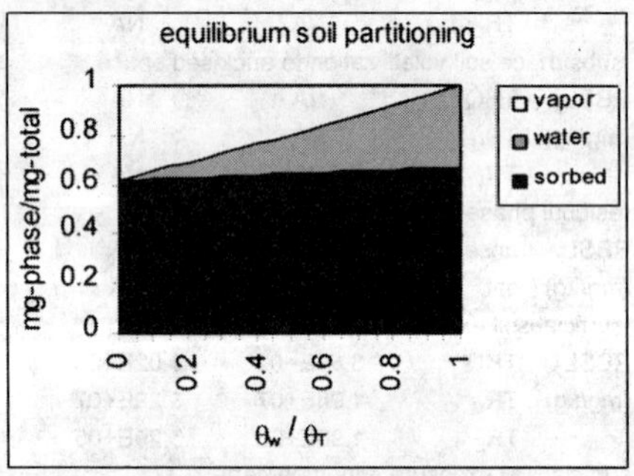

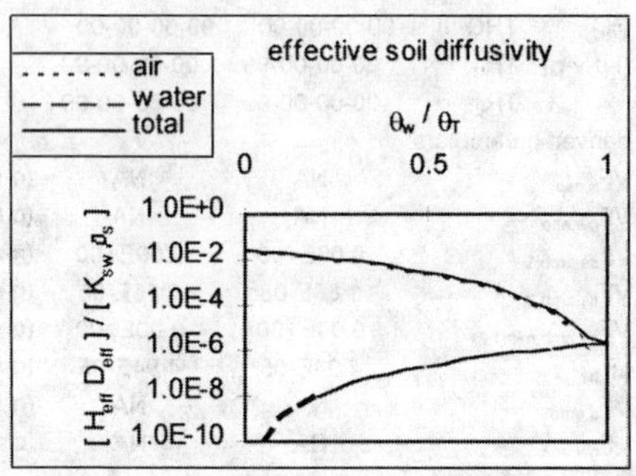

Notes:
(1) RBSL values are compared to physical limits of aqueous solubility (S<), saturated vapor pressure (<P), or either (Rs), and flagged if the calculated value is physically unrealistic. For cases where this occurs, the calculated values are given in parentheses.
(2) Percent of surficial soil exposure due to contributions of ingestion, dermal contact, vapor inhalation, and particulate inhalation (i-d-v-p).
(3) NA - not applicable.

TABLE X3.1 Chemical-Specific Parameters and Risk-Based Screening Level Concentrations (See Refs. 2, 16-21) *(continued)*

CASRN		108-90-7		
name		chlorobenzene		
		C_6H_5Cl		
Risk-Based Screening-Level Concentrations [(1)]				
	risk or			
	hazard	scenario:		
	criteria:	residential	industrial	
ambient vapor inhalation				
$RBSL_{air}$	THQ	2.09E+01	2.92E+01	
(μg/m^3)	TR_U	NA	NA	
	TR_L	NA	NA	
enclosed space vapor inhalation				
$RBSL_{air}$	THQ	2.78E+01	2.92E+01	
(μg/m^3)	TR_U	NA	NA	
	TR_L	NA	NA	
groundwater ingestion				
$RBSL_{gw}$	THQ	7.30E-01	2.04E+00	
(mg/L)	TR_U	NA	NA	
	TR_L	NA	NA	
	MCL	1.00E-01	1.00E-01	
soil leaching to groundwater ingestion				
$RBSL_s$	THQ	7.84E+00	2.19E+01	
(mg/kg)	TR_U	NA	NA	
	TR_L	NA	NA	
	MCL	1.07E+00	1.07E+00	
groundwater to ambient vapor inhalation				
$RBSL_{gw}$	THQ	4.02E+02	S< (5.63E+02)	
(mg/L)	TR_U	NA	NA	
	TR_L	NA	NA	
groundwater to enclosed space vapor inhalation				
$RBSL_{gw}$	THQ	2.81E+00	7.38E+00	
(mg/L)	TR_U	NA	NA	
	TR_L	NA	NA	
subsurface soil volatilization to ambient air				
$RBSL_s$	THQ	5.43E+01	7.60E+01	
(mg/kg)	TR_U	NA	NA	
	TR_L	NA	NA	
subsurface soil volatilization to enclosed space				
$RBSL_s$	THQ	3.22E+00	8.45E+00	
(mg/kg)	TR_U	NA	NA	
	TR_L	NA	NA	
residual phase mobility limit				
$RBSL_s$	unsat.	1.10E+04	1.10E+04	
(mg/kg)	sat.	1.10E+04	1.10E+04	
surficial soil exposure				
$RBSL_{ss}$	THQ	6.80E+02	8.70E+02	
(mg/kg)	TR_U	NA	NA	
	TR_L	NA	NA	
surficial soil exposure apportionment [(2)]				
(%)	THQ	04-36-58-00	02-33-64-00	
(i-d-v-p)	TR_U			
	TR_L			
derived parameters:				
$VF_{s,esp}$		8.64E-06	3.46E-06	(g-soil/cm^3-air)
$VF_{gw,esp}$		9.89E-06	3.96E-06	(cm^3-wat/cm^3-air)
$VF_{ss,amb,1}$		9.38E-08	1.03E-07	(g-soil/cm^3-air)
$VF_{ss,amb,2}$		1.80E-08	2.16E-08	(g-soil/cm^3-air)
$VF_{ss,amb,min(1,2)}$		1.80E-08	2.16E-08	(g-soil/cm^3-air)
LF_{sw}		9.32E-02	9.32E-02	(g-soil/cm^3-wat)
$VF_{s,amb}$		3.84E-07	3.84E-07	(g-soil/cm^3-air)
$VF_{gw,amb}$		5.19E-08	5.19E-08	(cm^3-wat/cm^3-air)

transport and thermodynamic parameters			[references]
MW	1.13E+02	(g/mole)	USEPA(1996)
D_{air}	7.30E-02	(cm^2/sec)	USEPA(1996)
D_{wat}	8.70E-06	(cm^2/sec)	USEPA(1996)
K_{oc}	2.34E+00	log10(L-wat/kg-oc)	USEPA(1996)
K_{ow}	2.86E+00	log10(L-wat/kg)	USEPA(1996)
H	3.70E-03	(atm-m^3/mol)	USEPA(1996)
H	1.54E-01	(L-wat/L-air)	(calculated)
P_v	1.18E+01	(mm Hg)	USEPA(1994)
S	4.72E+02	(mg/L)	USEPA(1996)
pK_a		log10(mole/mole)	
pK_b		log10(mole/mole)	
ρ	1.11E+00	(g/mL)	USEPA(1996)
T_{nbp}	1.32E+02	(°C)	USEPA(1994)
T_{mp}	-4.52E+01	(°C)	CRC(1996)
toxicity parameters			
RfD_o	2.00E-02	(mg/kg-day)	IRIS (1994)
RfC_i	2.00E-02	(mg/m^3)	HEAST (07/93)
SF_o		1/(mg/kg-day)	
SF_i		1/(mg/kg-day)	
W of E	D		IRIS (1994)
MCL	1.00E-01	(mg/L-wat)	56 FR 3526 (30 Jan 91)
RAF_o	1.00E+00		
RAF_d	5.00E-01		
derived parameters dependent on chemical properties and soil parameters			
UF	1.00E+00	(unitless)	
H_{eff}	1.54E-01	(L-wat/L-air)	
$C_{sat,vap}$	7.27E+07	(μg/m^3-air)	
K_{sw}	2.28E+00	(L-wat/kg-soil)	
$C_{sat,soil}$	1.08E+03	(mg/kg-soil)	[based on S]
$C_{sat,soil}$	1.08E+03	(mg/kg-soil)	[based on P_v]
$D_{eff,vad}$	5.70E-03	(cm^2/sec)	
$D_{eff,cap}$	2.04E-05	(cm^2/sec)	
$D_{eff,ws}$	1.01E-03	(cm^2/sec)	
$D_{eff,crk}$	5.70E-03	(cm^2/sec)	

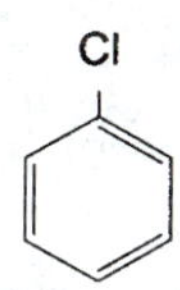

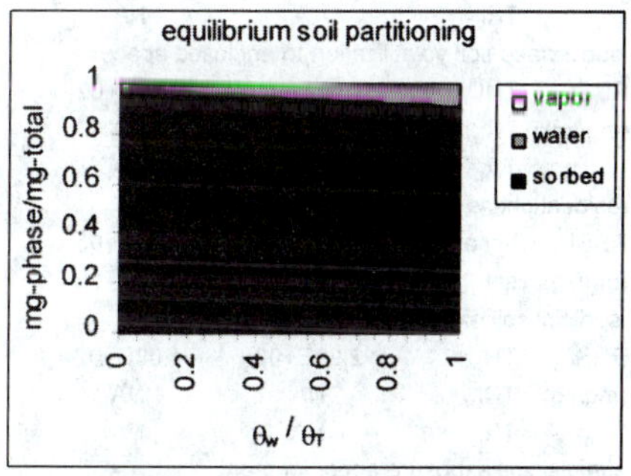

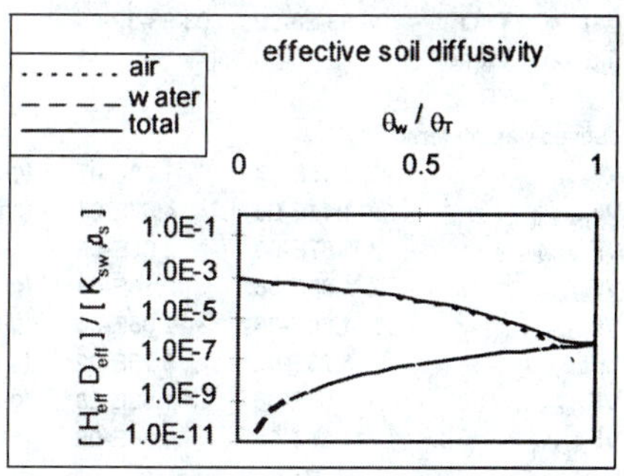

Notes:
(1) RBSL values are compared to physical limits of aqueous solubility (S<), saturated vapor pressure (<P), or either (Rs), and flagged if the calculated value is physically unrealistic. For cases where this occurs, the calculated values are given in parentheses.
(2) Percent of surficial soil exposure due to contributions of ingestion, dermal contact, vapor inhalation, and particulate inhalation (i-d-v-p).
(3) NA - not applicable.

TABLE X3.1 Chemical-Specific Parameters and Risk-Based Screening Level Concentrations (See Refs. 2, 16-21) *(continued)*

CASRN	95-57-8		
name	chlorophenol, 2-		
	C_6H_5ClO		
Risk-Based Screening-Level Concentrations [1]			
risk or hazard criteria:	scenario:	residential	industrial
ambient vapor inhalation			
$RBSL_{air}$ (μg/m³)	THQ	1.83E+01	2.56E+01
	TR_U	NA	NA
	TR_L	NA	NA
enclosed space vapor inhalation			
$RBSL_{air}$ (μg/m³)	THQ	2.43E+01	2.56E+01
	TR_U	NA	NA
	TR_L	NA	NA
groundwater ingestion			
$RBSL_{gw}$ (mg/L)	THQ	1.83E-01	5.11E-01
	TR_U	NA	NA
	TR_L	NA	NA
	MCL	NA	NA
soil leaching to groundwater ingestion			
$RBSL_s$ (mg/kg)	THQ	3.48E+00	9.74E+00
	TR_U	NA	NA
	TR_L	NA	NA
	MCL	NA	NA
groundwater to ambient vapor inhalation			
$RBSL_{gw}$ (mg/L)	THQ	1.34E+03	1.87E+03
	TR_U	NA	NA
	TR_L	NA	NA
groundwater to enclosed space vapor inhalation			
$RBSL_{gw}$ (mg/L)	THQ	2.13E+01	5.59E+01
	TR_U	NA	NA
	TR_L	NA	NA
subsurface soil volatilization to ambient air			
$RBSL_s$ (mg/kg)	THQ	1.19E+03	1.67E+03
	TR_U	NA	NA
	TR_L	NA	NA
subsurface soil volatilization to enclosed space			
$RBSL_s$ (mg/kg)	THQ	7.06E+01	1.85E+02
	TR_U	NA	NA
	TR_L	NA	NA
residual phase mobility limit			
$RBSL_s$ (mg/kg)	unsat.	1.00E+05	1.00E+05
	sat.	1.04E+05	1.04E+05
surficial soil exposure			
$RBSL_{ss}$ (mg/kg)	THQ	2.92E+02	4.09E+02
	TR_U	NA	NA
	TR_L	NA	NA
surficial soil exposure apportionment [2]			
(%) (i-d-v-p)	THQ	08-63-28-00	03-63-32-00
	TR_U		
	TR_L		
derived parameters:			
$VF_{s,esp}$		3.45E-07	1.38E-07 (g-soil/cm³-air)
$VF_{gw,esp}$		1.14E-06	4.57E-07 (cm³-wat/cm³-air)
$VF_{ss,amb,1}$		1.87E-08	2.05E-08 (g-soil/cm³-air)
$VF_{ss,amb,2}$		1.80E-08	2.16E-08 (g-soil/cm³-air)
$VF_{ss,amb,min(1,2)}$		1.80E-08	2.05E-08 (g-soil/cm³-air)
LF_{sw}		5.25E-02	5.25E-02 (g-soil/cm³-wat)
$VF_{s,amb}$		1.53E-08	1.53E-08 (g-soil/cm³-air)
$VF_{gw,amb}$		1.36E-08	1.36E-08 (cm³-wat/cm³-air)

transport and thermodynamic parameters			[references]
MW	1.29E+02	(g/mole)	USEPA(1996)
D_{air}	5.01E-02	(cm²/sec)	USEPA(1996)
D_{wat}	9.46E-06	(cm²/sec)	USEPA(1996)
K_{oc}	2.60E+00	log10(L-wat/kg-oc)	USEPA(1996)
K_{ow}	2.15E+00	log10(L-wat/kg)	USEPA(1996)
H	3.91E-04	(atm-m³/mol)	USEPA(1996)
H	1.63E-02	(L-wat/L-air)	(calculated)
P_v	1.40E+00	(mm Hg)	USEPA(1994)
S	2.20E+04	(mg/L)	USEPA(1996)
pK_a	8.40E+00	log10(mole/mole)	USEPA(1996)
pK_b		log10(mole/mole)	
ρ	1.26E+00	(g/mL)	USEPA(1996)
T_{nbp}	1.76E+02	(°C)	USEPA(1994)
T_{mp}	9.80E+00	(°C)	CRC(1996)
toxicity parameters			
RfD_o	5.00E-03	(mg/kg-day)	IRIS (1994)
RfC_i	1.75E-02	(mg/m³)	Route-to-Route
SF_o		1/(mg/kg-day)	
SF_i		1/(mg/kg-day)	
W of E			
MCL		(mg/L-wat)	
RAF_o	1.00E+00		
RAF_d	5.00E-01		
derived parameters dependent on chemical properties and soil parameters			
UF	9.75E-01	(unitless)	
H_{eff}	1.59E-02	(L-wat/L-air)	
$C_{sat,vap}$	9.85E+06	(μg/m³-air)	
K_{sw}	4.05E+00	(L-wat/kg-soil)	
$C_{sat,soil}$	8.92E+04	(mg/kg-soil)	[based on S]
$C_{sat,soil}$	2.52E+03	(mg/kg-soil)	[based on P_v]
$D_{eff,vad}$	3.91E-03	(cm²/sec)	
$D_{eff,cap}$	1.22E-04	(cm²/sec)	
$D_{eff,ws}$	2.58E-03	(cm²/sec)	
$D_{eff,crk}$	3.91E-03	(cm²/sec)	

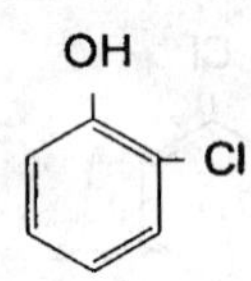

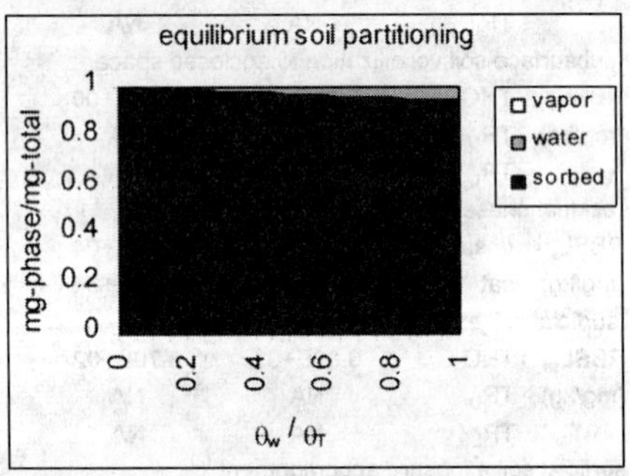

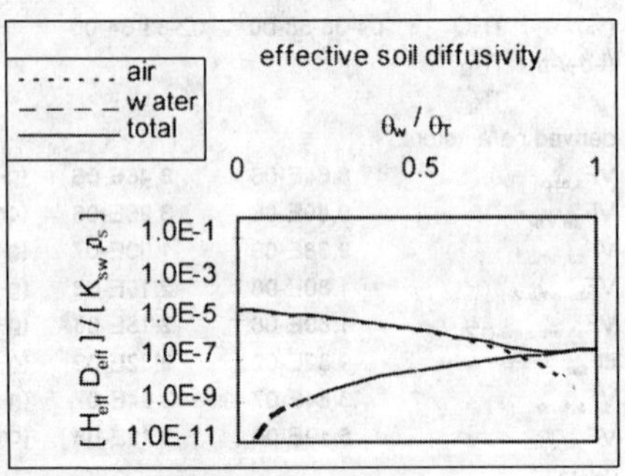

Notes:

(1) RBSL values are compared to physical limits of aqueous solubility (S<), saturated vapor pressure (<P), or either (Rs), and flagged if the calculated value is physically unrealistic. For cases where this occurs, the calculated values are given in parentheses.

(2) Percent of surficial soil exposure due to contributions of ingestion, dermal contact, vapor inhalation, and particulate inhalation (i-d-v-p).

(3) NA - not applicable.

TABLE X3.1 Chemical-Specific Parameters and Risk-Based Screening Level Concentrations (See Refs. 2, 16-21) *(continued)*

CASRN		50-29-3	
name		DDT (4,4'-ddt)	
		$C_{14}H_9Cl_5$	
Risk-Based Screening-Level Concentrations (1)			
	risk or		
	hazard	scenario:	
	criteria:	residential	industrial
ambient vapor inhalation			
$RBSL_{air}$	THQ	1.83E+00	2.56E+00
(μg/m³)	TR_U	2.50E+00	P< (4.21E+00)
	TR_L	2.50E-02	4.21E-02
enclosed space vapor inhalation			
$RBSL_{air}$	THQ	2.43E+00	2.56E+00
(μg/m³)	TR_U	P< (3.34E+00)	P< (4.21E+00)
	TR_L	3.34E-02	4.21E-02
groundwater ingestion			
$RBSL_{gw}$	THQ	1.82E-02	S< (5.11E-02)
(mg/L)	TR_U	S< (2.50E-02)	S< (8.42E-02)
	TR_L	2.50E-04	8.42E-04
	MCL	NA	NA
soil leaching to groundwater ingestion			
$RBSL_s$	THQ	Rs (2.26E+03)	S< (6.32E+03)
(mg/kg)	TR_U	S< (3.10E+03)	S< (1.04E+04)
	TR_L	3.10E+01	1.04E+02
	MCL	NA	NA
groundwater to ambient vapor inhalation			
$RBSL_{gw}$	THQ	S< (1.39E+04)	S< (1.95E+04)
(mg/L)	TR_U	S< (1.91E+04)	P< (3.21E+04)
	TR_L	S< (1.91E+02)	S< (3.21E+02)
groundwater to enclosed space vapor inhalation			
$RBSL_{gw}$	THQ	S< (3.12E+02)	S< (8.18E+02)
(mg/L)	TR_U	P< (4.28E+02)	P< (1.35E+03)
	TR_L	S< (4.28E+00)	S< (1.35E+01)
subsurface soil volatilization to ambient air			
$RBSL_s$	THQ	Rs (1.23E+08)	Rs (1.72E+08)
(mg/kg)	TR_U	Rs (1.69E+08)	P< (2.84E+08)
	TR_L	Rs (1.69E+06)	Rs (2.84E+06)
subsurface soil volatilization to enclosed space			
$RBSL_s$	THQ	Rs (7.30E+06)	Rs (1.92E+07)
(mg/kg)	TR_U	P< (1.00E+07)	P< (3.16E+07)
	TR_L	Rs (1.00E+05)	Rs (3.16E+05)
residual phase mobility limit			
$RBSL_s$	unsat.	NA	NA
(mg/kg)	sat.	NA	NA
surficial soil exposure			
$RBSL_{ss}$	THQ	2.03E+02	3.95E+02
(mg/kg)	TR_U	2.79E+02	6.50E+02
	TR_L	2.79E+00	6.50E+00
surficial soil exposure apportionment (2)			
(%)	THQ	55-44-00-00	38-61-00-00
(i-d-v-p)	TR_U	55-44-00-00	38-61-00-00
	TR_L	55-44-00-00	38-61-00-00

derived parameters:			
$VF_{s,esp}$	3.33E-13	1.33E-13	(g-soil/cm³-air)
$VF_{gw,esp}$	7.80E-09	3.12E-09	(cm³-wat/cm³-air)
$VF_{ss,amb,1}$	1.84E-11	2.02E-11	(g-soil/cm³-air)
$VF_{ss,amb,2}$	1.80E-08	2.16E-08	(g-soil/cm³-air)
$VF_{ss,amb,min(1,2)}$	1.84E-11	2.02E-11	(g-soil/cm³-air)
LF_{sw}	8.08E-06	8.08E-06	(g-soil/cm³-wat)
$VF_{s,amb}$	1.48E-14	1.48E-14	(g-soil/cm³-air)
$VF_{gw,amb}$	1.31E-10	1.31E-10	(cm³-wat/cm³-air)

transport and thermodynamic parameters			[references]
MW	3.54E+02	(g/mole)	USEPA(1996)
D_{air}	1.37E-02	(cm²/sec)	USEPA(1996)
D_{wat}	4.95E-06	(cm²/sec)	USEPA(1996)
K_{oc}	6.42E+00	log10(L-wat/kg-oc)	USEPA(1996)
K_{ow}	6.53E+00	log10(L-wat/kg)	USEPA(1996)
H	8.10E-06	(atm-m³/mol)	USEPA(1996)
H	3.37E-04	(L-wat/L-air)	(calculated)
P_v	1.50E-07	(mm Hg)	USEPA(1994)
S	2.50E-02	(mg/L)	USEPA(1996)
pK_a		log10(mole/mole)	
pK_b		log10(mole/mole)	
ρ	1.18E+00	(g/mL)	USEPA(1994)
T_{nbp}	2.60E+02	(°C)	USEPA(1994)
T_{mp}	1.09E+02	(°C)	CRC(1996)
toxicity parameters			
RfD_o	5.00E-04	(mg/kg-day)	IRIS (1994)
RfC_i	1.75E-03	(mg/m³)	Route-to-Route
SF_o	3.40E-01	1/(mg/kg-day)	IRIS (1994)
SF_i	3.40E-01	1/(mg/kg-day)	HEAST (07/93)
W of E	B2		IRIS (1994)
MCL		(mg/L-wat)	
RAF_o	1.00E+00		
RAF_d	5.00E-02		
derived parameters dependent on chemical properties and soil parameters			
UF	1.00E+00	(unitless)	
H_{eff}	3.37E-04	(L-wat/L-air)	
$C_{sat,vap}$	2.91E+00	(μg/m³-air)	
K_{sw}	2.63E+04	(L-wat/kg-soil)	
$C_{sat,soil}$	6.58E+02	(mg/kg-soil)	[based on S]
$C_{sat,soil}$	2.27E+02	(mg/kg-soil)	[based on P_v]
$D_{eff,vad}$	1.16E-03	(cm²/sec)	
$D_{eff,cap}$	2.86E-03	(cm²/sec)	
$D_{eff,ws}$	1.17E-03	(cm²/sec)	
$D_{eff,crk}$	1.16E-03	(cm²/sec)	

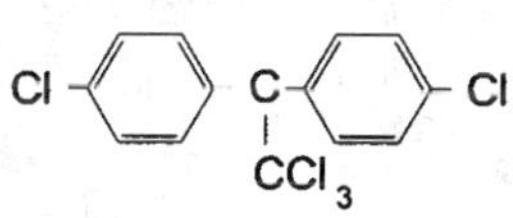

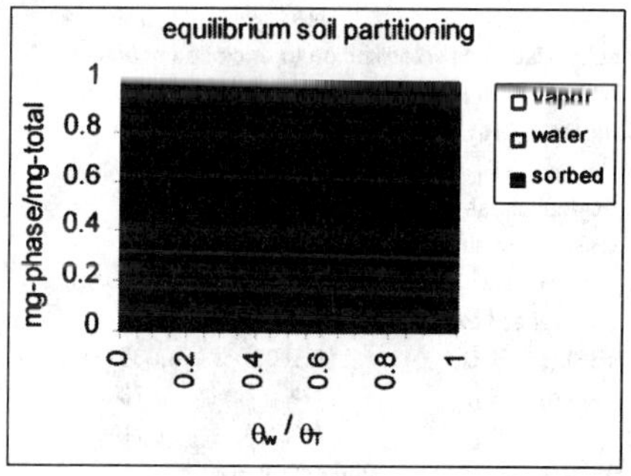

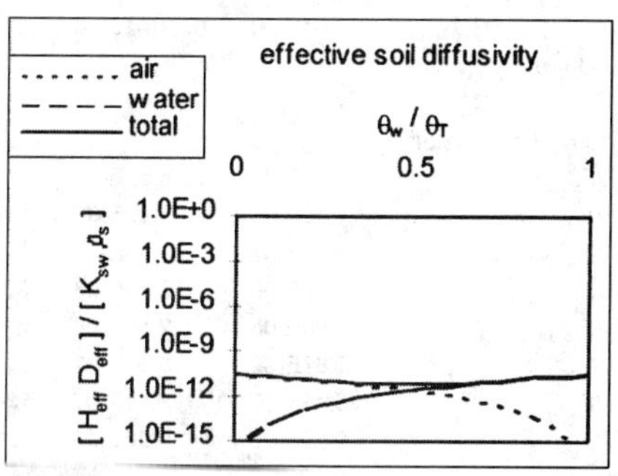

Notes:

(1) RBSL values are compared to physical limits of aqueous solubility (S<), saturated vapor pressure (<P), or either (Rs), and flagged if the calculated value is physically unrealistic. For cases where this occurs, the calculated values are given in parentheses.

(2) Percent of surficial soil exposure due to contributions of ingestion, dermal contact, vapor inhalation, and particulate inhalation (i-d-v-p).

(3) NA - not applicable.

TABLE X3.1 Chemical-Specific Parameters and Risk-Based Screening Level Concentrations (See Refs. 2, 16-21) *(continued)*

CASRN	100-41-4		
name	ethylbenzene		
	C_8H_{10}		
Risk-Based Screening-Level Concentrations [1]			
risk or hazard criteria:	scenario:	residential	industrial
ambient vapor inhalation			
$RBSL_{air}$ ($\mu g/m^3$)	THQ	1.04E+03	1.46E+03
	TR_U	NA	NA
	TR_L	NA	NA
enclosed space vapor inhalation			
$RBSL_{air}$ ($\mu g/m^3$)	THQ	1.39E+03	1.46E+03
	TR_U	NA	NA
	TR_L	NA	NA
groundwater ingestion			
$RBSL_{gw}$ (mg/L)	THQ	3.65E+00	1.02E+01
	TR_U	NA	NA
	TR_L	NA	NA
	MCL	7.00E-01	7.00E-01
soil leaching to groundwater ingestion			
$RBSL_s$ (mg/kg)	THQ	6.44E+01	1.80E+02
	TR_U	NA	NA
	TR_L	NA	NA
	MCL	1.24E+01	1.24E+01
groundwater to ambient vapor inhalation			
$RBSL_{gw}$ (mg/L)	THQ	S< (1.27E+04)	S< (1.78E+04)
	TR_U	NA	NA
	TR_L	NA	NA
groundwater to enclosed space vapor inhalation			
$RBSL_{gw}$ (mg/L)	THQ	7.73E+01	S< (2.03E+02)
	TR_U	NA	NA
	TR_L	NA	NA
subsurface soil volatilization to ambient air			
$RBSL_s$ (mg/kg)	THQ	Rs (2.04E+03)	Rs (2.86E+03)
	TR_U	NA	NA
	TR_L	NA	NA
subsurface soil volatilization to enclosed space			
$RBSL_s$ (mg/kg)	THQ	1.21E+02	3.17E+02
	TR_U	NA	NA
	TR_L	NA	NA
residual phase mobility limit			
$RBSL_s$ (mg/kg)	unsat.	8.39E+03	8.39E+03
	sat.	8.40E+03	8.40E+03
surficial soil exposure			
$RBSL_{ss}$ (mg/kg)	THQ	7.19E+03	1.03E+04
	TR_U	NA	NA
	TR_L	NA	NA
surficial soil exposure apportionment [2]			
(%) (i-d-v-p)	THQ	09-77-12-00	05-79-15-00
	TR_U		
	TR_L		

derived parameters:	residential	industrial	units
$VF_{s,esp}$	1.15E-05	4.60E-06	(g-soil/cm³-air)
$VF_{gw,esp}$	1.80E-05	7.19E-06	(cm³-wat/cm³-air)
$VF_{ss,amb,1}$	1.08E-07	1.18E-07	(g-soil/cm³-air)
$VF_{ss,amb,2}$	1.80E-08	2.16E-08	(g-soil/cm³-air)
$VF_{ss,amb,min(1,2)}$	1.80E-08	2.16E-08	(g-soil/cm³-air)
LF_{sw}	5.67E-02	5.67E-02	(g-soil/cm³-wat)
$VF_{s,amb}$	5.11E-07	5.11E-07	(g-soil/cm³-air)
$VF_{gw,amb}$	8.20E-08	8.20E-08	(cm³-wat/cm³-air)

transport and thermodynamic parameters			[references]
MW	1.06E+02	(g/mole)	USEPA(1996)
D_{air}	7.50E-02	(cm²/sec)	USEPA(1996)
D_{wat}	7.80E-06	(cm²/sec)	USEPA(1996)
K_{oc}	2.56E+00	log10(L-wat/kg-oc)	USEPA(1996)
K_{ow}	3.14E+00	log10(L-wat/kg)	USEPA(1996)
H	7.88E-03	(atm-m³/mol)	USEPA(1996)
H	3.28E-01	(L-wat/L-air)	(calculated)
P_v	9.51E+00	(mm Hg)	Mott(1995)
S	1.69E+02	(mg/L)	USEPA(1996)
pK_a		log10(mole/mole)	
pK_b		log10(mole/mole)	
ρ	8.67E-01	(g/mL)	USEPA(1996)
T_{nbp}	1.36E+02	(°C)	Mackay(1992a)
T_{mp}	-9.50E+01	(°C)	Mackay(1992a)
toxicity parameters			
RfD_o	1.00E-01	(mg/kg-day)	IRIS (1994)
RfC_i	1.00E+00	(mg/m³)	IRIS (1994)
SF_o		1/(mg/kg-day)	
SF_i		1/(mg/kg-day)	
W of E	D		IRIS (1994)
MCL	7.00E-01	(mg/L-wat)	56 FR 3526 (30 Jan 91)
RAF_o	1.00E+00		
RAF_d	5.00E-01		
derived parameters dependent on chemical properties and soil parameters			
UF	1.00E+00	(unitless)	
H_{eff}	3.28E-01	(L-wat/L-air)	
$C_{sat,vap}$	5.53E+07	(µg/m³-air)	
K_{sw}	3.75E+00	(L-wat/kg-soil)	
$C_{sat,soil}$	6.34E+02	(mg/kg-soil)	[based on S]
$C_{sat,soil}$	6.32E+02	(mg/kg-soil)	[based on P_v]
$D_{eff,vad}$	5.85E-03	(cm²/sec)	
$D_{eff,cap}$	1.43E-05	(cm²/sec)	
$D_{eff,ws}$	7.51E-04	(cm²/sec)	
$D_{eff,crk}$	5.85E-03	(cm²/sec)	

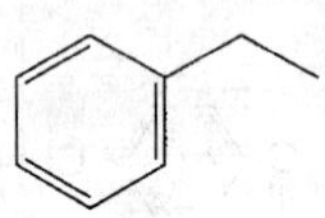

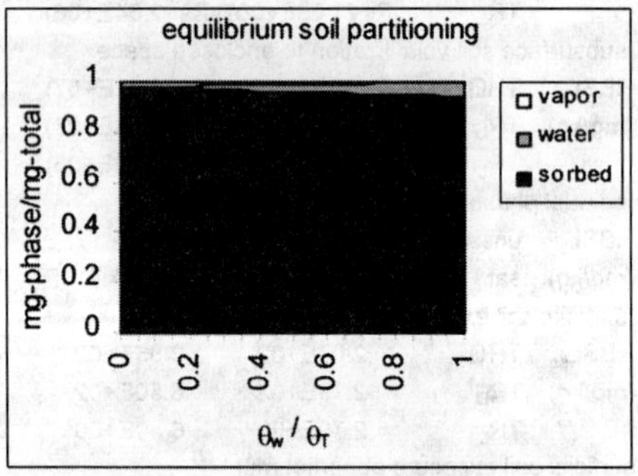

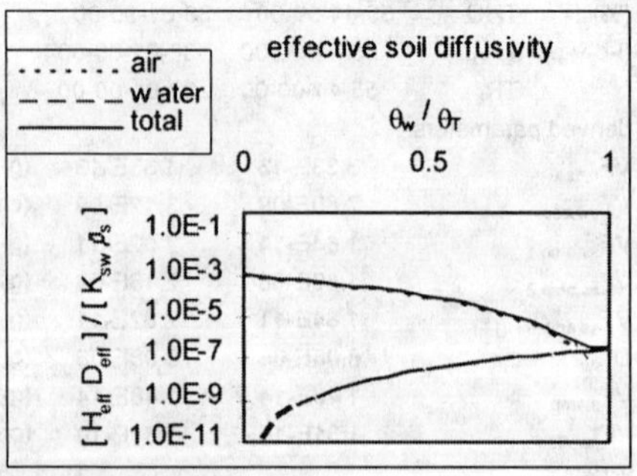

Notes:

(1) RBSL values are compared to physical limits of aqueous solubility (S<), saturated vapor pressure (<P), or either (Rs), and flagged if the calculated value is physically unrealistic. For cases where this occurs, the calculated values are given in parentheses.

(2) Percent of surficial soil exposure due to contributions of ingestion, dermal contact, vapor inhalation, and particulate inhalation (i-d-v-p).

(3) NA - not applicable.

TABLE X3.1 Chemical-Specific Parameters and Risk-Based Screening Level Concentrations (See Refs. 2, 16-21) *(continued)*

CASRN		58-89-9		
name		lindane, (gamma-BHC)		
		$C_6H_6CL_6$		
Risk-Based Screening-Level Concentrations [1]				
	risk or			
	hazard	scenario:		
	criteria:	residential	industrial	
ambient vapor inhalation				
$RBSL_{air}$	THQ	1.09E+00	1.53E+00	
(μg/m³)	TR_U	6.55E-01	1.10E+00	
	TR_L	6.55E-03	1.10E-02	
enclosed space vapor inhalation				
$RBSL_{air}$	THQ	1.46E+00	1.53E+00	
(μg/m³)	TR_U	8.74E-01	1.10E+00	
	TR_L	8.74E-03	1.10E-02	
groundwater ingestion				
$RBSL_{gw}$	THQ	1.09E-02	3.07E-02	
(mg/L)	TR_U	6.55E-03	2.20E-02	
	TR_L	6.55E-05	2.20E-04	
	MCL	2.00E-04	2.00E-04	
soil leaching to groundwater ingestion				
$RBSL_s$	THQ	5.56E-01	1.56E+00	
(mg/kg)	TR_U	3.32E-01	1.12E+00	
	TR_L	3.32E-03	1.12E-02	
	MCL	1.01E-02	1.01E-02	
groundwater to ambient vapor inhalation				
$RBSL_{gw}$	THQ	S< (4.73E+03)	S< (6.62E+03)	
(mg/L)	TR_U	S< (2.83E+03)	S< (4.75E+03)	
	TR_L	S< (2.83E+01)	S< (4.75E+01)	
groundwater to enclosed space vapor inhalation				
$RBSL_{gw}$	THQ	S< (1.06E+02)	S< (2.78E+02)	
(mg/L)	TR_U	S< (6.33E+01)	S< (1.99E+02)	
	TR_L	6.33E 01	1.99E+00	
subsurface soil volatilization to ambient air				
$RBSL_s$	THQ	Rs (1.71E+04)	Rs (2.40E+04)	
(mg/kg)	TR_U	Rs (1.03E+04)	Rs (1.72E+04)	
	TR_L	Rs (1.03E+02)	Rs (1.72E+02)	
subsurface soil volatilization to enclosed space				
$RBSL_s$	THQ	Rs (1.02E+03)	Rs (2.67E+03)	
(mg/kg)	TR_U	Rs (6.08E+02)	Rs (1.91E+03)	
	TR_L	6.08E+00	1.91E+01	
residual phase mobility limit				
$RBSL_s$	unsat.	NA	NA	
(mg/kg)	sat.	NA	NA	
surficial soil exposure				
$RBSL_{ss}$	THQ	1.08E+02	1.97E+02	
(mg/kg)	TR_U	6.45E+01	1.42E+02	
	TR_L	6.45E-01	1.42E+00	
surficial soil exposure apportionment [2]				
(%)	THQ	49-38-11-00	32-50-17-00	
(i-d-v-p)	TR_U	49-38-11-00	32-50-17-00	
	TR_L	49-38-11-00	32-50-17-00	
derived parameters:				
$VF_{s,esp}$		1.44E-09	5.75E-10	(g-soil/cm³-air)
$VF_{gw,esp}$		1.38E-08	5.52E-09	(cm³-wat/cm³-air)
$VF_{ss,amb,1}$		1.21E-09	1.32E-09	(g-soil/cm³-air)
$VF_{ss,amb,2}$		1.80E-08	2.16E-08	(g-soil/cm³-air)
$VF_{ss,amb,min(1,2)}$		1.21E-09	1.32E-09	(g-soil/cm³-air)
LF_{sw}		1.97E-02	1.97E-02	(g-soil/cm³-wat)
$VF_{s,amb}$		6.39E-11	6.39E-11	(g-soil/cm³-air)
$VF_{gw,amb}$		2.32E-10	2.32E-10	(cm³-wat/cm³-air)

transport and thermodynamic parameters			[references]
MW	2.91E+02	(g/mole)	USEPA(1996)
D_{air}	1.42E-02	(cm²/sec)	USEPA(1996)
D_{wat}	7.34E-06	(cm²/sec)	USEPA(1996)
K_{oc}	3.03E+00	log10(L-wat/kg-oc)	USEPA(1996)
K_{ow}	3.73E+00	log10(L-wat/kg)	USEPA(1996)
H	1.40E-05	(atm-m³/mol)	USEPA(1996)
H	5.82E-04	(L-wat/L-air)	(calculated)
P_v	1.09E-05	(mm Hg)	USEPA(1994)
S	6.80E+00	(mg/L)	USEPA(1996)
pK_a		log10(mole/mole)	
pK_b		log10(mole/mole)	
ρ	1.87E+00	(g/mL)	USEPA(1994)
T_{nbp}	3.90E+02	(°C)	USEPA(1994)
T_{mp}	1.13E+02	(°C)	CRC(1996)
toxicity parameters			
RfD_o	3.00E-04	(mg/kg-day)	IRIS (1994)
RfC_i	1.05E-03	(mg/m³)	Route-to-Route
SF_o	1.30E+00	1/(mg/kg-day)	HEAST (07/93)
SF_i	1.30E+00	1/(mg/kg-day)	Route-to-Route
W of E	B2 - C		HEAST (07/93)
MCL	2.00E-04	(mg/L-wat)	56 FR 3526 (30 Jan 91)
RAF_o	1.00E+00		
RAF_d	5.00E-02		
derived parameters dependent on chemical properties and soil parameters			
UF	1.00E+00	(unitless)	
H_{eff}	5.82E-04	(L-wat/L-air)	
$C_{sat,vap}$	1.74E+02	(μg/m³-air)	
K_{sw}	1.08E+01	(L-wat/kg-soil)	
$C_{sat,soil}$	7.33E+01	(mg/kg-soil)	[based on S]
$C_{sat,soil}$	3.22E+00	(mg/kg-soil)	[based on P_v]
$D_{eff,vad}$	1.18E-03	(cm²/sec)	
$D_{eff,cap}$	2.45E-03	(cm²/sec)	
$D_{eff,ws}$	1.19E-03	(cm²/sec)	
$D_{eff,crk}$	1.18E-03	(cm²/sec)	

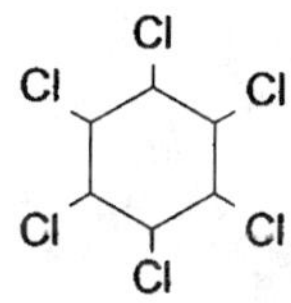

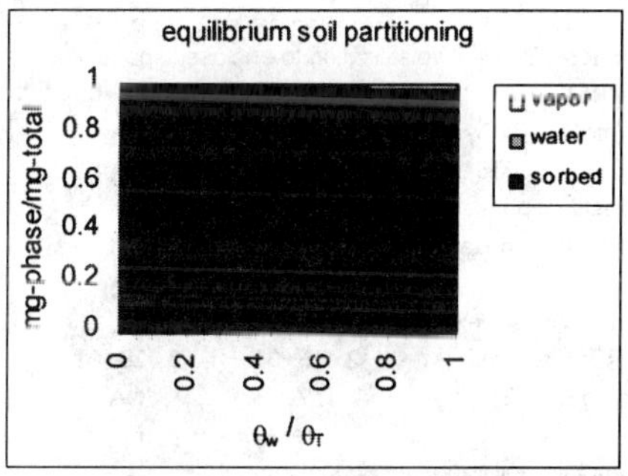

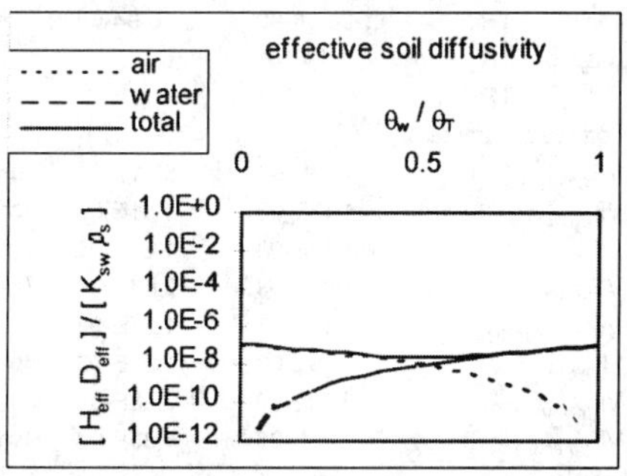

Notes:
(1) RBSL values are compared to physical limits of aqueous solubility (S<), saturated vapor pressure (<P), or either (Rs), and flagged if the calculated value is physically unrealistic. For cases where this occurs, the calculated values are given in parentheses.
(2) Percent of surficial soil exposure due to contributions of ingestion, dermal contact, vapor inhalation, and particulate inhalation (i-d-v-p).
(3) NA - not applicable.

TABLE X3.1 Chemical-Specific Parameters and Risk-Based Screening Level Concentrations (See Refs. 2, 16-21) *(continued)*

CASRN	7439-97-6			
name	mercury (inorganic)			
	Hg			
Risk-Based Screening-Level Concentrations [(1)]				
	risk or			
	hazard	scenario:		
	criteria:	residential	industrial	
ambient vapor inhalation				
$RBSL_{air}$	THQ	3.13E-01	4.38E-01	
(μg/m^3)	TR_U	NA	NA	
	TR_L	NA	NA	
enclosed space vapor inhalation				
$RBSL_{air}$	THQ	4.17E-01	4.38E-01	
(μg/m^3)	TR_U	NA	NA	
	TR_L	NA	NA	
groundwater ingestion				
$RBSL_{gw}$	THQ	1.09E-02	3.07E-02	
(mg/L)	TR_U	NA	NA	
	TR_L	NA	NA	
	MCL	2.00E-03	2.00E-03	
soil leaching to groundwater ingestion				
$RBSL_s$	THQ	1.03E+01	2.89E+01	
(mg/kg)	TR_U	NA	NA	
	TR_L	NA	NA	
	MCL	1.88E+00	1.88E+00	
groundwater to ambient vapor inhalation				
$RBSL_{gw}$	THQ	5.85E+00	8.19E+00	
(mg/L)	TR_U	NA	NA	
	TR_L	NA	NA	
groundwater to enclosed space vapor inhalation				
$RBSL_{gw}$	THQ	3.70E-02	9.70E-02	
(mg/L)	TR_U	NA	NA	
	TR_L	NA	NA	
subsurface soil volatilization to ambient air				
$RBSL_s$	THQ	5.51E+01	7.72E+01	
(mg/kg)	TR_U	NA	NA	
	TR_L	NA	NA	
subsurface soil volatilization to enclosed space				
$RBSL_s$	THQ	3.27E+00	8.57E+00	
(mg/kg)	TR_U	NA	NA	
	TR_L	NA	NA	
residual phase mobility limit				
$RBSL_s$	unsat.	NA	NA	
(mg/kg)	sat.	NA	NA	
surficial soil exposure				
$RBSL_{ss}$	THQ	2.44E+01	3.32E+01	
(mg/kg)	TR_U	NA	NA	
	TR_L	NA	NA	
surficial soil exposure apportionment [(2)]				
(%)	THQ	11-00-88-00	05-00-94-00	
(i-d-v-p)	TR_U			
	TR_L			
derived parameters:				
$VF_{s,esp}$		1.28E-07	5.11E-08	(g-soil/cm^3-air)
$VF_{gw,esp}$		1.13E-05	4.51E-06	(cm^3-wat/cm^3-air)
$VF_{ss,amb,1}$		1.14E-08	1.25E-08	(g-soil/cm^3-air)
$VF_{ss,amb,2}$		1.80E-08	2.16E-08	(g-soil/cm^3-air)
$VF_{ss,amb,min(1,2)}$		1.14E-08	1.25E-08	(g-soil/cm^3-air)
LF_{sw}		1.06E-03	1.06E-03	(g-soil/cm^3-wat)
$VF_{s,amb}$		5.68E-09	5.68E-09	(g-soil/cm^3-air)
$VF_{gw,amb}$		5.35E-08	5.35E-08	(cm^3-wat/cm^3-air)

transport and thermodynamic parameters			[references]
MW	2.01E+02	(g/mole)	USEPA(1996)
D_{air}	3.07E-02	(cm^2/sec)	USEPA(1996)
D_{wat}	6.30E-06	(cm^2/sec)	USEPA(1996)
K_d	2.30E+00	log10(L-wat/kg-soil)	USEPA(1996); pH=8.0, Hg+2
K_{ow}		log10(L-wat/kg)	
H	1.14E-02	(atm-m^3/mol)	USEPA(1996)
H	4.74E-01	(L-wat/L-air)	(calculated)
P_v	1.20E-03	(mm Hg)	USEPA(1979)
S	NA	(mg/L)	(no limit imposed)
pK_a		log10(mole/mole)	
pK_b		log10(mole/mole)	
ρ		(g/mL)	
T_{nbp}		(°C)	
T_{mp}		(°C)	
toxicity parameters			
RfD_o	3.00E-04	(mg/kg-day)	HEAST (07/93)
RfC_i	3.00E-04	(mg/m^3)	HEAST (05/95) Under Review
SF_o		1/(mg/kg-day)	
SF_i		1/(mg/kg-day)	
W of E	D		IRIS (05/01/95)
MCL	2.00E-03	(mg/L-wat)	56 FR 3526 (30 Jan 91)
RAF_o	1.00E+00		
RAF_d	0.00E+00		
derived parameters dependent on chemical properties and soil parameters			
UF	1.00E+00	(unitless)	
H_{eff}	4.74E-01	(L-wat/L-air)	
$C_{sat,vap}$	1.32E+04	(μg/m^3-air)	
K_{sw}	2.00E+02	(L-wat/kg-soil)	
$C_{sat,soil}$	NA	(mg/kg-soil)	[based on S]
$C_{sat,soil}$	5.56E+00	(mg/kg-soil)	[based on P_v]
$D_{eff,vad}$	2.40E-03	(cm^2/sec)	
$D_{eff,cap}$	6.55E-06	(cm^2/sec)	
$D_{eff,ws}$	3.38E-04	(cm^2/sec)	
$D_{eff,crk}$	2.40E-03	(cm^2/sec)	

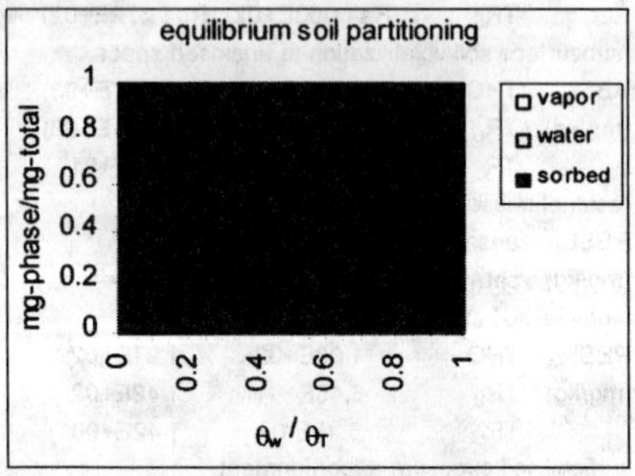

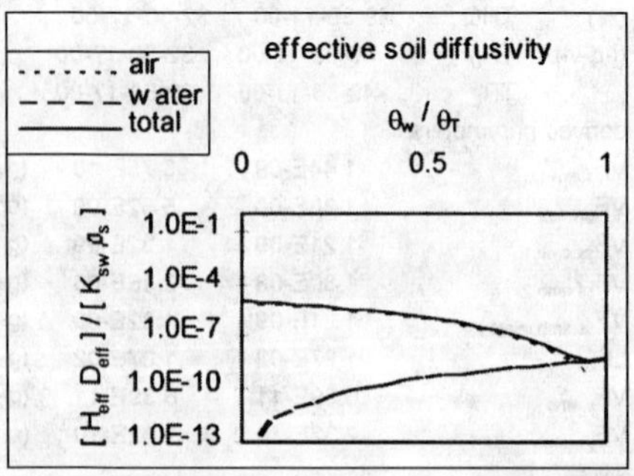

Notes:

(1) RBSL values are compared to physical limits of aqueous solubility (S<), saturated vapor pressure (<P), or either (Rs), and flagged if the calculated value is physically unrealistic. For cases where this occurs, the calculated values are given in parentheses.

(2) Percent of surficial soil exposure due to contributions of ingestion, dermal contact, vapor inhalation, and particulate inhalation (i-d-v-p).

(3) NA - not applicable.

TABLE X3.1 Chemical-Specific Parameters and Risk-Based Screening Level Concentrations (See Refs. 2, 16-21) *(continued)*

CASRN		78-93-3		
name		methyl ethyl ketone		
		C_4H_8O		
Risk-Based Screening-Level Concentrations [(1)]				
	risk or			
	hazard	scenario:		
	criteria:	residential	industrial	
ambient vapor inhalation				
$RBSL_{air}$	THQ	1.04E+03	1.46E+03	
(µg/m³)	TR_U	NA	NA	
	TR_L	NA	NA	
enclosed space vapor inhalation				
$RBSL_{air}$	THQ	1.39E+03	1.46E+03	
(µg/m³)	TR_U	NA	NA	
	TR_L	NA	NA	
groundwater ingestion				
$RBSL_{gw}$	THQ	2.19E+01	6.13E+01	
(mg/L)	TR_U	NA	NA	
	TR_L	NA	NA	
	MCL	NA	NA	
soil leaching to groundwater ingestion				
$RBSL_s$	THQ	9.26E+00	2.59E+01	
(mg/kg)	TR_U	NA	NA	
	TR_L	NA	NA	
	MCL	NA	NA	
groundwater to ambient vapor inhalation				
$RBSL_{gw}$	THQ	2.95E+05	S< (4.14E+05)	
(mg/L)	TR_U	NA	NA	
	TR_L	NA	NA	
groundwater to enclosed space vapor inhalation				
$RBSL_{gw}$	THQ	6.15E+03	1.61E+04	
(mg/L)	TR_U	NA	NA	
	TR_L	NA	NA	
subsurface soil volatilization to ambient air				
$RBSL_s$	THQ	8.18E+03	1.14E+04	
(mg/kg)	TR_U	NA	NA	
	TR_L	NA	NA	
subsurface soil volatilization to enclosed space				
$RBSL_s$	THQ	4.85E+02	1.27E+03	
(mg/kg)	TR_U	NA	NA	
	TR_L	NA	NA	
residual phase mobility limit				
$RBSL_s$	unsat.	3.89E+04	3.89E+04	
(mg/kg)	sat.	8.96E+04	8.96E+04	
surficial soil exposure				
$RBSL_{ss}$	THQ	2.66E+04	3.51E+04	
(mg/kg)	TR_U	NA	NA	
	TR_L	NA	NA	
surficial soil exposure apportionment [(2)]				
(%)	THQ	06-48-45-00	02-45-51-00	
(i-d-v-p)	TR_U			
	TR_L			
derived parameters:				
$VF_{s,esp}$		2.87E-06	1.15E-06	(g-soil/cm³-air)
$VF_{gw,esp}$		2.26E-07	9.05E-08	(cm³-wat/cm³-air)
$VF_{ss,amb,1}$		5.40E-08	5.92E-08	(g-soil/cm³-air)
$VF_{ss,amb,2}$		1.80E-08	2.16E-08	(g-soil/cm³-air)
$VF_{ss,amb,min(1,2)}$		1.80E-08	2.16E-08	(g-soil/cm³-air)
LF_{sw}		2.36E+00	2.36E+00	(g-soil/cm³-wat)
$VF_{s,amb}$		1.28E-07	1.28E-07	(g-soil/cm³-air)
$VF_{gw,amb}$		3.53E-09	3.53E-09	(cm³-wat/cm³-air)

transport and thermodynamic parameters			[references]
MW	7.21E+01	(g/mole)	Versch.(1996)
D_{air}	8.08E-02	(cm²/sec)	TPHCWG (1997)
D_{wat}	9.80E-06	(cm²/sec)	TPHCWG (1997)
K_{oc}	2.80E-01	log10(L-wat/kg-oc)	TPHCWG (1997)
K_{ow}	2.60E-01	log10(L-wat/kg)	Versch.(1996)
H	4.35E-05	(atm-m³/mol)	TPHCWG (1997)
H	1.81E-03	(L-wat/L-air)	(calculated)
P_v	7.75E+01	(mm Hg)	Versch.(1996)
S	3.53E+05	(mg/L)	Versch.(1996)
pK_a	1.47E+01	log10(mole/mole)	GRI(1993)
pK_b		log10(mole/mole)	
ρ	8.00E-01	(g/mL)	Versch.(1996)
T_{nbp}	7.96E+01	(°C)	Versch.(1996)
T_{mp}	-8.64E+01	(°C)	Versch.(1996)
toxicity parameters			
RfD_o	6.00E-01	(mg/kg-day)	IRIS (05/01/93)
RfC_i	1.00E+00	(mg/m³)	HEAST (07/93)
SF_o		1/(mg/kg-day)	
SF_i		1/(mg/kg-day)	
W of E	D		IRIS (1994)
MCL		(mg/L-wat)	
RAF_o	1.00E+00		
RAF_d	5.00E-01		
derived parameters dependent on chemical properties and soil parameters			
UF	1.00E+00	(unitless)	
H_{eff}	1.81E-03	(L-wat/L-air)	
$C_{sat,vap}$	3.06E+08	(µg/m³-air)	
K_{sw}	8.99E-02	(L-wat/kg-soil)	
$C_{sat,soil}$	3.17E+04	(mg/kg-soil)	[based on S]
$C_{sat,soil}$	1.52E+04	(mg/kg-soil)	[based on P_v]
$D_{eff,vad}$	6.34E-03	(cm²/sec)	
$D_{eff,cap}$	1.06E-03	(cm²/sec)	
$D_{eff,ws}$	5.85E-03	(cm²/sec)	
$D_{eff,crk}$	6.34E-03	(cm²/sec)	

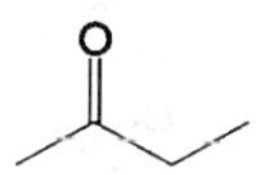

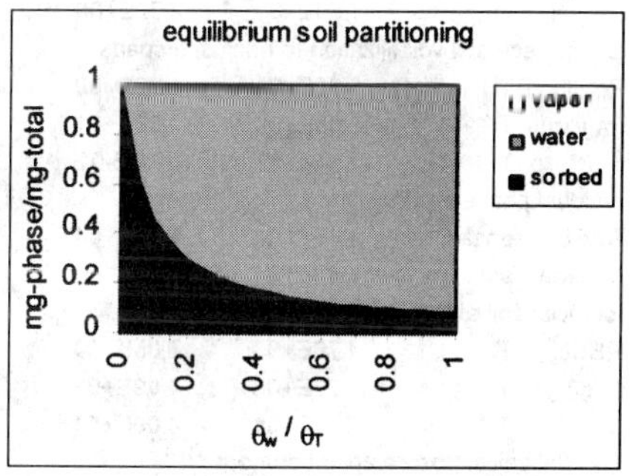

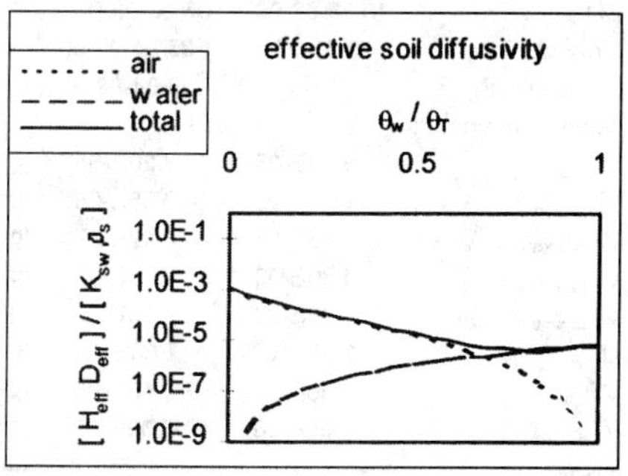

Notes:

(1) RBSL values are compared to physical limits of aqueous solubility (S<), saturated vapor pressure (<P), or either (Rs), and flagged if the calculated value is physically unrealistic. For cases where this occurs, the calculated values are given in parentheses.

(2) Percent of surficial soil exposure due to contributions of ingestion, dermal contact, vapor inhalation, and particulate inhalation (i-d-v-p).

(3) NA - not applicable.

TABLE X3.1 Chemical-Specific Parameters and Risk-Based Screening Level Concentrations (See Refs. 2, 16-21) *(continued)*

CASRN		75-09-2		
name		methylene chloride		
		CH_2CL_2		
Risk-Based Screening-Level Concentrations [1]				
	risk or			
	hazard	scenario:		
	criteria:	residential	industrial	
ambient vapor inhalation				
$RBSL_{air}$	THQ	3.13E+03	4.38E+03	
($\mu g/m^3$)	TR_U	5.32E+02	8.94E+02	
	TR_L	5.32E+00	8.94E+00	
enclosed space vapor inhalation				
$RBSL_{air}$	THQ	4.17E+03	4.38E+03	
($\mu g/m^3$)	TR_U	7.10E+02	8.94E+02	
	TR_L	7.10E+00	8.94E+00	
groundwater ingestion				
$RBSL_{gw}$	THQ	2.19E+00	6.13E+00	
(mg/L)	TR_U	1.14E+00	3.82E+00	
	TR_L	1.14E-02	3.82E-02	
	MCL	5.00E-03	5.00E-03	
soil leaching to groundwater ingestion				
$RBSL_s$	THQ	2.08E+00	5.83E+00	
(mg/kg)	TR_U	1.08E+00	3.63E+00	
	TR_L	1.08E-02	3.63E-02	
	MCL	4.75E-03	4.75E-03	
groundwater to ambient vapor inhalation				
$RBSL_{gw}$	THQ	S< (5.80E+04)	S< (8.12E+04)	
(mg/L)	TR_U	9.87E+03	S< (1.66E+04)	
	TR_L	9.87E+01	1.66E+02	
groundwater to enclosed space vapor inhalation				
$RBSL_{gw}$	THQ	4.57E+02	1.20E+03	
(mg/L)	TR_U	7.78E+01	2.45E+02	
	TR_L	7.78E-01	2.45E+00	
subsurface soil volatilization to ambient air				
$RBSL_s$	THQ	8.80E+02	1.23E+03	
(mg/kg)	TR_U	1.50E+02	2.52E+02	
	TR_L	1.50E+00	2.52E+00	
subsurface soil volatilization to enclosed space				
$RBSL_s$	THQ	5.22E+01	1.37E+02	
(mg/kg)	TR_U	8.88E+00	2.80E+01	
	TR_L	8.88E-02	2.80E-01	
residual phase mobility limit				
$RBSL_s$	unsat.	1.45E+04	1.45E+04	
(mg/kg)	sat.	1.62E+04	1.62E+04	
surficial soil exposure				
$RBSL_{ss}$	THQ	4.79E+03	7.05E+03	
(mg/kg)	TR_U	2.35E+03	4.09E+03	
	TR_L	2.35E+01	4.09E+01	
surficial soil exposure apportionment [2]				
(%)	THQ	10-86-02-00	05-90-03-00	
(i-d-v-p)	TR_U	10-81-07-00	05-84-09-00	
	TR_L	10-81-07-00	05-84-09-00	
derived parameters:				
$VF_{s,esp}$		8.00E-05	3.20E-05	(g-soil/cm^3-air)
$VF_{gw,esp}$		9.13E-06	3.65E-06	(cm^3-wat/cm^3-air)
$VF_{ss,amb,1}$		2.85E-07	3.12E-07	(g-soil/cm^3-air)
$VF_{ss,amb,2}$		1.80E-08	2.16E-08	(g-soil/cm^3-air)
$VF_{ss,amb,min(1,2)}$		1.80E-08	2.16E-08	(g-soil/cm^3-air)
LF_{sw}		1.05E+00	1.05E+00	(g-soil/cm^3-wat)
$VF_{s,amb}$		3.55E-06	3.55E-06	(g-soil/cm^3-air)
$VF_{gw,amb}$		5.39E-08	5.39E-08	(cm^3-wat/cm^3-air)

transport and thermodynamic parameters			[references]
MW	8.49E+01	(g/mole)	USEPA(1996)
D_{air}	1.01E-01	(cm^2/sec)	USEPA(1996)
D_{wat}	1.17E-05	(cm^2/sec)	USEPA(1996)
K_{oc}	1.07E+00	log10(L-wat/kg-oc)	USEPA(1996)
K_{ow}	1.25E+00	log10(L-wat/kg)	USEPA(1996)
H	2.19E-03	(atm-m^3/mol)	USEPA(1996)
H	9.11E-02	(L-wat/L-air)	(calculated)
P_v	4.38E+02	(mm Hg)	USEPA(1994)
S	1.30E+04	(mg/L)	USEPA(1996)
pK_a		log10(mole/mole)	
pK_b		log10(mole/mole)	
ρ	1.33E+00	(g/mL)	USEPA(1996)
T_{nbp}	3.98E+01	(°C)	USEPA(1994)
T_{mp}	-9.51E+01	(°C)	CRC(1996)
toxicity parameters			
RfD_o	6.00E-02	(mg/kg-day)	IRIS (1994)
RfC_i	3.00E+00	(mg/m^3)	HEAST (07/93)
SF_o	7.50E-03	1/(mg/kg-day)	IRIS (1994)
SF_i	1.60E-03	1/(mg/kg-day)	IRIS (1994) SF/URF
W of E	B2		IRIS (1994)
MCL	5.00E-03	(mg/L-wat)	57 FR 31776 (17 Jul 92) Effecti
RAF_o	1.00E+00		
RAF_d	5.00E-01		
derived parameters dependent on chemical properties and soil parameters			
UF	1.00E+00	(unitless)	
H_{eff}	9.11E-02	(L-wat/L-air)	
$C_{sat,vap}$	2.04E+09	($\mu g/m^3$-air)	
K_{sw}	2.02E-01	(L-wat/kg-soil)	
$C_{sat,soil}$	2.63E+03	(mg/kg-soil)	[based on S]
$C_{sat,soil}$	4.51E+03	(mg/kg-soil)	[based on P_v]
$D_{eff,vad}$	7.88E-03	(cm^2/sec)	
$D_{eff,cap}$	3.80E-05	(cm^2/sec)	
$D_{eff,ws}$	1.78E-03	(cm^2/sec)	
$D_{eff,crk}$	7.88E-03	(cm^2/sec)	

Cl - C - Cl

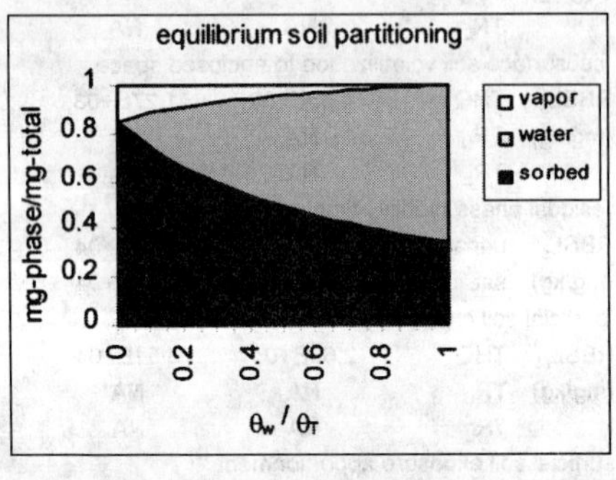

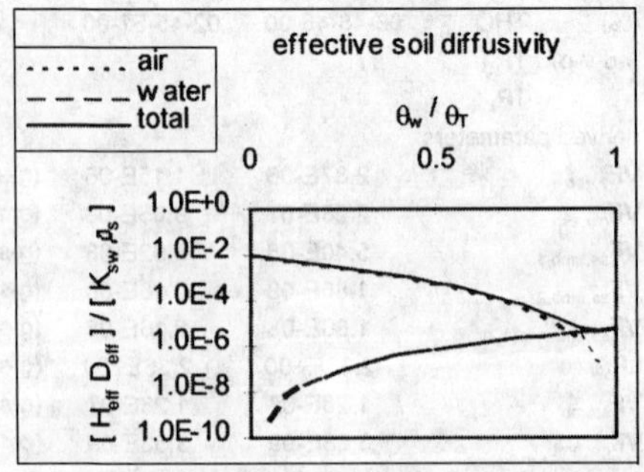

Notes:

(1) RBSL values are compared to physical limits of aqueous solubility (S<), saturated vapor pressure (<P), or either (Rs), and flagged if the calculated value is physically unrealistic. For cases where this occurs, the calculated values are given in parentheses.

(2) Percent of surficial soil exposure due to contributions of ingestion, dermal contact, vapor inhalation, and particulate inhalation (i-d-v-p).

(3) NA - not applicable.

TABLE X3.1 Chemical-Specific Parameters and Risk-Based Screening Level Concentrations (See Refs. 2, 16-21) *(continued)*

CASRN	91-20-3		
name	naphthalene		
	$C_{10}H_8$		
Risk-Based Screening-Level Concentrations (1)			
	risk or hazard criteria:	scenario: residential	industrial
ambient vapor inhalation			
$RBSL_{air}$ (µg/m³)	THQ	3.13E+00	4.38E+00
	TR_U	NA	NA
	TR_L	NA	NA
enclosed space vapor inhalation			
$RBSL_{air}$ (µg/m³)	THQ	4.17E+00	4.38E+00
	TR_U	NA	NA
	TR_L	NA	NA
groundwater ingestion			
$RBSL_{gw}$ (mg/L)	THQ	7.30E-01	2.04E+00
	TR_U	NA	NA
	TR_L	NA	NA
	MCL	NA	NA
soil leaching to groundwater ingestion			
$RBSL_s$ (mg/kg)	THQ	6.88E+01	1.93E+02
	TR_U	NA	NA
	TR_L	NA	NA
	MCL	NA	NA
groundwater to ambient vapor inhalation			
$RBSL_{gw}$ (mg/L)	THQ	S< (1.97E+02)	S< (2.76E+02)
	TR_U	NA	NA
	TR_L	NA	NA
groundwater to enclosed space vapor inhalation			
$RBSL_{gw}$ (mg/L)	THQ	2.61E+00	6.84E+00
	TR_U	NA	NA
	TR_L	NA	NA
subsurface soil volatilization to ambient air			
$RBSL_s$ (mg/kg)	THQ	Rs (6.77E+02)	Rs (9.48E+02)
	TR_U	NA	NA
	TR_L	NA	NA
subsurface soil volatilization to enclosed space			
$RBSL_s$ (mg/kg)	THQ	4.01E+01	1.05E+02
	TR_U	NA	NA
	TR_L	NA	NA
residual phase mobility limit			
$RBSL_s$ (mg/kg)	unsat.	NA	NA
	sat.	NA	NA
surficial soil exposure			
$RBSL_{ss}$ (mg/kg)	THQ	2.93E+02	3.80E+02
	TR_U	NA	NA
	TR_L	NA	NA
surficial soil exposure apportionment (2)			
(%) (i-d-v-p)	THQ	02-01-96-00	00-01-97-00
	TR_U		
	TR_L		

transport and thermodynamic parameters			[references]
MW	1.28E+02	(g/mole)	USEPA(1996)
D_{air}	5.90E-02	(cm²/sec)	USEPA(1996)
D_{wat}	7.50E-06	(cm²/sec)	USEPA(1996)
K_{oc}	3.30E+00	log10(L-wat/kg-oc)	USEPA(1996)
K_{ow}	3.36E+00	log10(L-wat/kg)	USEPA(1996)
H	4.83E-04	(atm-m³/mol)	USEPA(1996)
H	2.01E-02	(L-wat/L-air)	(calculated)
P_v	1.82E-01	(mm Hg)	Mott(1995)
S	3.10E+01	(mg/L)	USEPA(1996)
pK_a		log10(mole/mole)	
pK_b		log10(mole/mole)	
ρ	1.03E+00	(g/mL)	USEPA(1996)
T_{nbp}	2.18E+02	(°C)	CRC(1996)
T_{mp}	8.02E+01	(°C)	CRC(1996)
toxicity parameters			
RfD_o	2.00E-02	(mg/kg-day)	IRIS (09/17/98)
RfC_i	3.00E-03	(mg/m³)	IRIS (09/17/98)
SF_o		1/(mg/kg-day)	IRIS (09/17/98)
SF_i		1/(mg/kg-day)	
W of E	D		IRIS (09/17/98)
MCL		(mg/L-wat)	
RAF_o	1.00E+00		
RAF_d	5.00E-02		
derived parameters dependent on chemical properties and soil parameters			
UF	1.00E+00	(unitless)	
H_{eff}	2.01E-02	(L-wat/L-air)	
$C_{sat,vap}$	1.28E+06	(µg/m³-air)	
K_{sw}	2.00E+01	(L-wat/kg-soil)	
$C_{sat,soil}$	6.21E+02	(mg/kg-soil)	[based on S]
$C_{sat,soil}$	1.27E+03	(mg/kg-soil)	[based on P_v]
$D_{eff,vad}$	4.01E-03	(cm²/sec)	
$D_{eff,cap}$	8.02E-05	(cm²/sec)	
$D_{eff,ws}$	2.37E-03	(cm²/sec)	
$D_{eff,crk}$	4.61E-03	(cm²/sec)	

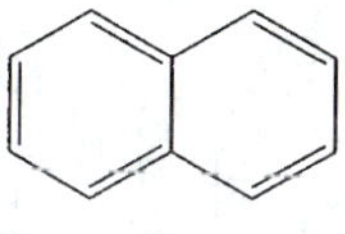

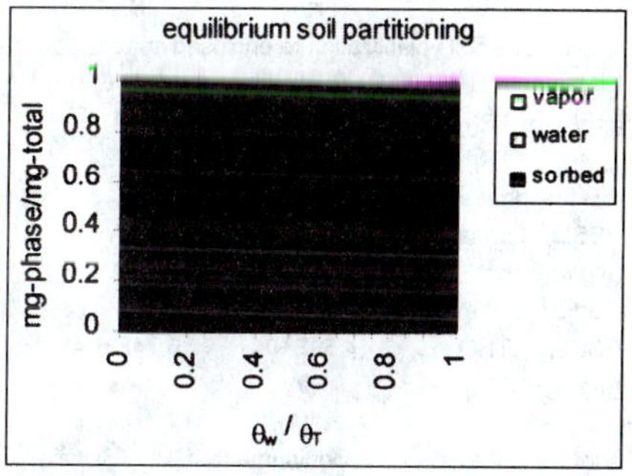

derived parameters:

$VF_{s,esp}$	1.04E-07	4.16E-08	(g-soil/cm³-air)
$VF_{gw,esp}$	1.60E-06	6.40E-07	(cm³-wat/cm³-air)
$VF_{ss,amb,1}$	1.03E-08	1.13E-08	(g-soil/cm³-air)
$VF_{ss,amb,2}$	1.80E-08	2.16E-08	(g-soil/cm³-air)
$VF_{ss,amb,min(1,2)}$	1.03E-08	1.13E-08	(g-soil/cm³-air)
LF_{sw}	1.06E-02	1.06E-02	(g-soil/cm³-wat)
$VF_{s,amb}$	4.62E-09	4.62E-09	(g-soil/cm³-air)
$VF_{gw,amb}$	1.59E-08	1.59E-08	(cm³-wat/cm³-air)

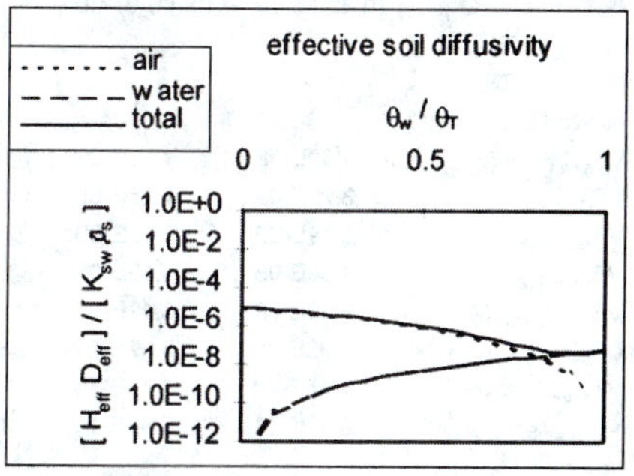

Notes:

(1) RBSL values are compared to physical limits of aqueous solubility (S<), saturated vapor pressure (<P), or either (Rs), and flagged if the calculated value is physically unrealistic. For cases where this occurs, the calculated values are given in parentheses.

(2) Percent of surficial soil exposure due to contributions of ingestion, dermal contact, vapor inhalation, and particulate inhalation (i-d-v-p).

(3) NA - not applicable.

TABLE X3.1 Chemical-Specific Parameters and Risk-Based Screening Level Concentrations (See Refs. 2, 16-21) *(continued)*

CASRN	108-95-2		
name	phenol		
	C_6H_6O		
Risk-Based Screening-Level Concentrations [(1)]			
risk or			
hazard	scenario:		
criteria:		residential	industrial
ambient vapor inhalation			
$RBSL_{air}$	THQ	2.19E+03	3.07E+03
(µg/m^3)	TR_U	NA	NA
	TR_L	NA	NA
enclosed space vapor inhalation			
$RBSL_{air}$	THQ	2.92E+03	3.07E+03
(µg/m^3)	TR_U	NA	NA
	TR_L	NA	NA
groundwater ingestion			
$RBSL_{gw}$	THQ	2.19E+01	6.13E+01
(mg/L)	TR_U	NA	NA
	TR_L	NA	NA
	MCL	NA	NA
soil leaching to groundwater ingestion			
$RBSL_s$	THQ	3.70E+01	1.04E+02
(mg/kg)	TR_U	NA	NA
	TR_L	NA	NA
	MCL	NA	NA
groundwater to ambient vapor inhalation			
$RBSL_{gw}$	THQ	S< (4.05E+07)	S< (5.67E+07)
(mg/L)	TR_U	NA	NA
	TR_L	NA	NA
groundwater to enclosed space vapor inhalation			
$RBSL_{gw}$	THQ	S< (9.12E+05)	S< (2.39E+06)
(mg/L)	TR_U	NA	NA
	TR_L	NA	NA
subsurface soil volatilization to ambient air			
$RBSL_s$	THQ	Rs (4.92E+06)	Rs (6.89E+06)
(mg/kg)	TR_U	NA	NA
	TR_L	NA	NA
subsurface soil volatilization to enclosed space			
$RBSL_s$	THQ	Rs (2.92E+05)	Rs (7.66E+05)
(mg/kg)	TR_U	NA	NA
	TR_L	NA	NA
residual phase mobility limit			
$RBSL_s$	unsat.	NA	NA
(mg/kg)	sat.	NA	NA
surficial soil exposure			
$RBSL_{ss}$	THQ	4.59E+04	6.74E+04
(mg/kg)	TR_U	NA	NA
	TR_L	NA	NA
surficial soil exposure apportionment [(2)]			
(%)	THQ	10-82-06-00	05-86-07-00
(i-d-v-p)	TR_U		
	TR_L		

derived parameters:			
$VF_{s,esp}$	1.00E-08	4.00E-09	(g-soil/cm^3-air)
$VF_{gw,esp}$	3.20E-09	1.28E-09	(cm^3-wat/cm^3-air)
$VF_{ss,amb,1}$	3.19E-09	3.49E-09	(g-soil/cm^3-air)
$VF_{ss,amb,2}$	1.80E-08	2.16E-08	(g-soil/cm^3-air)
$VF_{ss,amb,min(1,2)}$	3.19E-09	3.49E-09	(g-soil/cm^3-air)
LF_{sw}	5.92E-01	5.92E-01	(g-soil/cm^3-wat)
$VF_{s,amb}$	4.45E-10	4.45E-10	(g-soil/cm^3-air)
$VF_{gw,amb}$	5.40E-11	5.40E-11	(cm^3-wat/cm^3-air)

transport and thermodynamic parameters			[references]
MW	9.41E+01	(g/mole)	USEPA(1996)
D_{air}	8.20E-02	(cm^2/sec)	USEPA(1996)
D_{wat}	9.10E-06	(cm^2/sec)	USEPA(1996)
K_{oc}	1.46E+00	log10(L-wat/kg-oc)	USEPA(1996)
K_{ow}	1.48E+00	log10(L-wat/kg)	USEPA(1996)
H	3.97E-07	(atm-m^3/mol)	USEPA(1996)
H	1.65E-05	(L-wat/L-air)	(calculated)
P_v	3.41E-01	(mm Hg)	USEPA(1994)
S	8.28E+04	(mg/L)	USEPA(1996)
pK_a	1.00E+01	log10(mole/mole)	USEPA(1996)
pK_b		log10(mole/mole)	
ρ	1.05E+00	(g/mL)	USEPA(1996)
T_{nbp}	1.82E+02	(°C)	USEPA(1994)
T_{mp}	4.09E+01	(°C)	CRC(1996)
toxicity parameters			
RfD_o	6.00E-01	(mg/kg-day)	IRIS (1994)
RfC_i	2.10E+00	(mg/m^3)	Route-to-Route
SF_o		1/(mg/kg-day)	
SF_i		1/(mg/kg-day)	
W of E	D		IRIS (1994)
MCL		(mg/L-wat)	
RAF_o	1.00E+00		
RAF_d	5.00E-01		
derived parameters dependent on chemical properties and soil parameters			
UF	9.99E-01	(unitless)	
H_{eff}	1.65E-05	(L-wat/L-air)	
$C_{sat,vap}$	1.76E+06	(µg/m^3-air)	
K_{sw}	3.59E-01	(L-wat/kg-soil)	
$C_{sat,soil}$	2.97E+04	(mg/kg-soil)	[based on S]
$C_{sat,soil}$	3.82E+04	(mg/kg-soil)	[based on P_v]
$D_{eff,vad}$	9.68E-03	(cm^2/sec)	
$D_{eff,cap}$	1.07E-01	(cm^2/sec)	
$D_{eff,ws}$	9.83E-03	(cm^2/sec)	
$D_{eff,crk}$	9.68E-03	(cm^2/sec)	

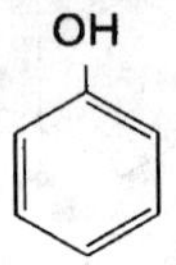

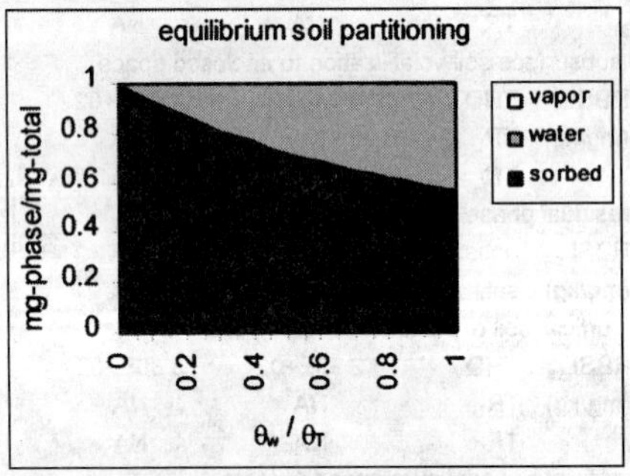

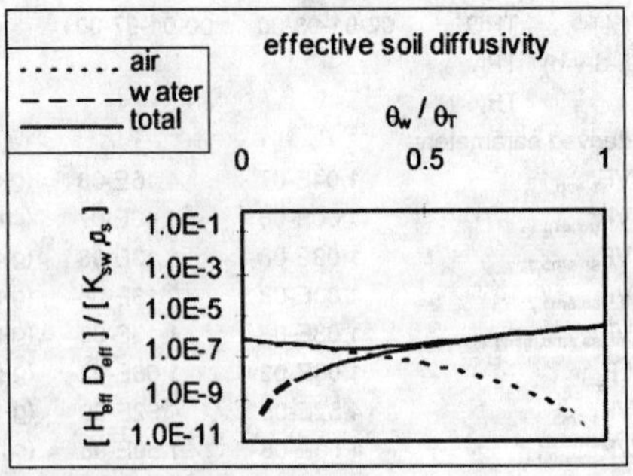

Notes:
(1) RBSL values are compared to physical limits of aqueous solubility (S<), saturated vapor pressure (<P), or either (Rs), and flagged if the calculated value is physically unrealistic. For cases where this occurs, the calculated values are given in parentheses.
(2) Percent of surficial soil exposure due to contributions of ingestion, dermal contact, vapor inhalation, and particulate inhalation (i-d-v-p).
(3) NA - not applicable.

TABLE X3.1 Chemical-Specific Parameters and Risk-Based Screening Level Concentrations (See Refs. 2, 16-21) *(continued)*

CASRN		84-74-2	
name		phthalate, di-n-butyl	
		$C_{16}H_{22}O_4$	
Risk-Based Screening-Level Concentrations [1]			
	risk or		
	hazard	scenario:	
	criteria:	residential	industrial
ambient vapor inhalation			
$RBSL_{air}$	THQ	3.65E+02	5.11E+02
(µg/m^3)	TR_U	NA	NA
	TR_L	NA	NA
enclosed space vapor inhalation			
$RBSL_{air}$	THQ	4.87E+02	5.11E+02
(µg/m^3)	TR_U	NA	NA
	TR_L	NA	NA
groundwater ingestion			
$RBSL_{gw}$	THQ	3.65E+00	1.02E+01
(mg/L)	TR_U	NA	NA
	TR_L	NA	NA
	MCL	NA	NA
soil leaching to groundwater ingestion			
$RBSL_s$	THQ	Rs (5.82E+03)	Rs (1.63E+04)
(mg/kg)	TR_U	NA	NA
	TR_L	NA	NA
	MCL	NA	NA
groundwater to ambient vapor inhalation			
$RBSL_{gw}$	THQ	S< (2.30E+07)	S< (3.22E+07)
(mg/L)	TR_U	NA	NA
	TR_L	NA	NA
groundwater to enclosed space vapor inhalation			
$RBSL_{gw}$	THQ	S< (5.30E+05)	S< (1.37E+06)
(mg/L)	TR_U	NA	NA
	TR_L	NA	NA
subsurface soil volatilization to ambient air			
$RBSL_s$	THQ	Rs (2.64E+09)	Rs (3.70E+09)
(mg/kg)	TR_U	NA	NA
	TR_L	NA	NA
subsurface soil volatilization to enclosed space			
$RBSL_s$	THQ	Rs (1.61E+08)	Rs (4.15E+08)
(mg/kg)	TR_U	NA	NA
	TR_L	NA	NA
residual phase mobility limit			
$RBSL_s$	unsat.	1.32E+04	1.32E+04
(mg/kg)	sat.	1.32E+04	1.32E+04
surficial soil exposure			
$RBSL_{ss}$	THQ	4.05E+04	7.85E+04
(mg/kg)	TR_U	NA	NA
	TR_L	NA	NA
surficial soil exposure apportionment [2]			
(%)	THQ	55-43-00-00	38-60-00-00
(i-d-v-p)	TR_U		
	TR_L		
derived parameters:			
$VF_{s,esp}$		3.03E-12	1.23E-12 (g-soil/cm^3-air)
$VF_{gw,esp}$		9.18E-10	3.72E-10 (cm^3-wat/cm^3-air)
$VF_{ss,amb,1}$		5.62E-11	6.16E-11 (g-soil/cm^3-air)
$VF_{ss,amb,2}$		1.80E-08	2.16E-08 (g-soil/cm^3-air)
$VF_{ss,amb,min(1,2)}$		5.62E-11	6.16E-11 (g-soil/cm^3-air)
LF_{sw}		6.27E-04	6.27E-04 (g-soil/cm^3-wat)
$VF_{s,amb}$		1.38E-13	1.38E-13 (g-soil/cm^3-air)
$VF_{gw,amb}$		1.59E-11	1.59E-11 (cm^3-wat/cm^3-air)

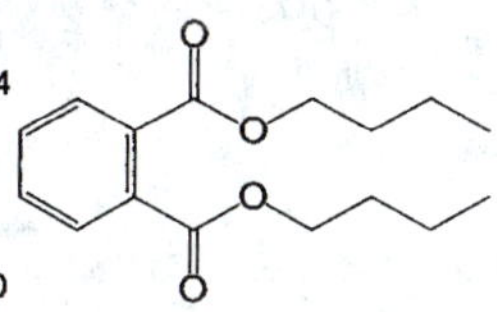

transport and thermodynamic parameters			[references]
MW	2.78E+02	(g/mole)	Versch.(1996)
D_{air}	4.38E-02	(cm^2/sec)	USEPA(1996)
D_{wat}	7.86E-06	(cm^2/sec)	USEPA(1996)
K_{oc}	4.53E+00	log10(L-wat/kg-oc)	USEPA(1996)
K_{ow}	4.57E+00	log10(L-wat/kg)	Versch.(1996)
H	9.38E-10	(atm-m^3/mol)	USEPA(1996)
H	3.90E-08	(L-wat/L-air)	(calculated)
P_v	1.00E+00	(mm Hg)	Versch.(1996)
S	1.12E+01	(mg/L)	USEPA(1996)
pK_a		log10(mole/mole)	
pK_b		log10(mole/mole)	
ρ	1.05E+00	(g/mL)	Versch.(1996)
T_{nbp}	3.40E+02	(°C)	Versch.(1996)
T_{mp}	-3.50E+01	(°C)	Versch.(1996)
toxicity parameters			
RfD_o	1.00E-01	(mg/kg-day)	IRIS (1994)
RfC_i	3.50E-01	(mg/m^3)	Route-to-Route
SF_o		1/(mg/kg-day)	
SF_i		1/(mg/kg-day)	
W of E	D		IRIS (1994)
MCL		(mg/L-wat)	
RAF_o	1.00E+00		
RAF_d	5.00E-02		
derived parameters dependent on chemical properties and soil parameters			
UF	1.00E+00	(unitless)	
H_{eff}	3.90E-08	(L-wat/L-air)	
$C_{sat,vap}$	1.52E+07	(µg/m^3-air)	
K_{sw}	3.39E+02	(L-wat/kg-soil)	
$C_{sat,soil}$	3.80E+03	(mg/kg-soil)	[based on S]
$C_{sat,soil}$	1.32E+11	(mg/kg-soil)	[based on P_v]
$D_{eff,vad}$	1.20E+00	(cm^2/sec)	
$D_{eff,cap}$	3.92E+01	(cm^2/sec)	
$D_{eff,ws}$	1.22E+00	(cm^2/sec)	
$D_{eff,crk}$	1.20E+00	(cm^2/sec)	

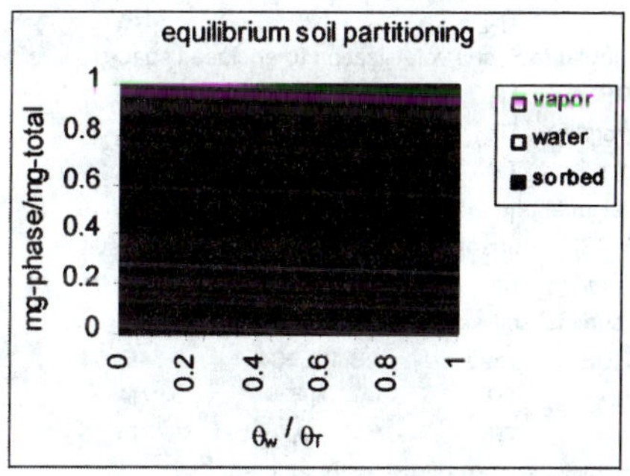

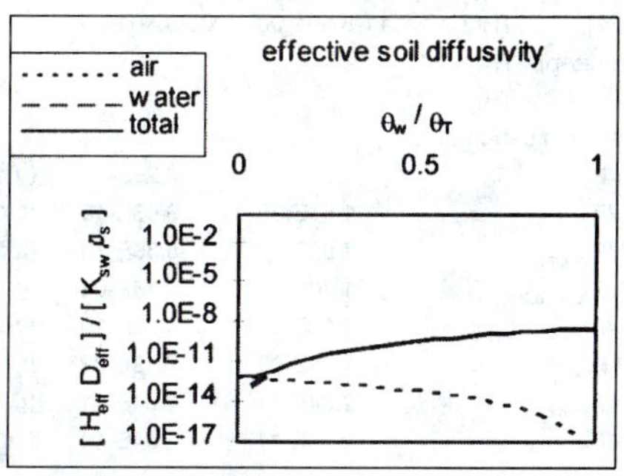

Notes:

(1) RBSL values are compared to physical limits of aqueous solubility (S<), saturated vapor pressure (<P), or either (Rs), and flagged if the calculated value is physically unrealistic. For cases where this occurs, the calculated values are given in parentheses.

(2) Percent of surficial soil exposure due to contributions of ingestion, dermal contact, vapor inhalation, and particulate inhalation (i-d-v-p).

(3) NA - not applicable.

TABLE X3.1 Chemical-Specific Parameters and Risk-Based Screening Level Concentrations (See Refs. 2, 16-21) *(continued)*

CASRN	84-66-2		
name	phthalate, diethyl		
	$C_{12}H_{14}O_4$		
Risk-Based Screening-Level Concentrations [1]			
risk or			
hazard	scenario:		
criteria:		residential	industrial
ambient vapor inhalation			
$RBSL_{air}$	THQ	2.92E+03	4.09E+03
($\mu g/m^3$)	TR_U	NA	NA
	TR_L	NA	NA
enclosed space vapor inhalation			
$RBSL_{air}$	THQ	3.89E+03	4.09E+03
($\mu g/m^3$)	TR_U	NA	NA
	TR_L	NA	NA
groundwater ingestion			
$RBSL_{gw}$	THQ	2.92E+01	8.18E+01
(mg/L)	TR_U	NA	NA
	TR_L	NA	NA
	MCL	NA	NA
soil leaching to groundwater ingestion			
$RBSL_s$	THQ	4.06E+02	Rs (1.14E+03)
(mg/kg)	TR_U	NA	NA
	TR_L	NA	NA
	MCL	NA	NA
groundwater to ambient vapor inhalation			
$RBSL_{gw}$	THQ	S< (1.15E+08)	S< (1.61E+08)
(mg/L)	TR_U	NA	NA
	TR_L	NA	NA
groundwater to enclosed space vapor inhalation			
$RBSL_{gw}$	THQ	S< (2.58E+06)	S< (6.78E+06)
(mg/L)	TR_U	NA	NA
	TR_L	NA	NA
subsurface soil volatilization to ambient air			
$RBSL_s$	THQ	Rs (1.15E+08)	Rs (1.61E+08)
(mg/kg)	TR_U	NA	NA
	TR_L	NA	NA
subsurface soil volatilization to enclosed space			
$RBSL_s$	THQ	Rs (6.81E+06)	Rs (1.79E+07)
(mg/kg)	TR_U	NA	NA
	TR_L	NA	NA
residual phase mobility limit			
$RBSL_s$	unsat.	1.06E+04	1.06E+04
(mg/kg)	sat.	1.07E+04	1.07E+04
surficial soil exposure			
$RBSL_{ss}$	THQ	6.45E+04	9.54E+04
(mg/kg)	TR_U	NA	NA
	TR_L	NA	NA
surficial soil exposure apportionment [2]			
(%)	THQ	11-87-01-00	05-92-01-00
(i-d-v-p)	TR_U		
	TR_L		

transport and thermodynamic parameters				[references]
MW		2.22E+02	(g/mole)	Versch.(1996)
D_{air}		2.56E-02	(cm^2/sec)	USEPA(1996)
D_{wat}		6.35E-06	(cm^2/sec)	USEPA(1996)
K_{oc}		2.46E+00	log10(L-wat/kg-oc)	USEPA(1996)
K_{ow}		2.42E+00	log10(L-wat/kg)	Versch.(1996)
H		4.50E-07	(atm-m^3/mol)	USEPA(1996)
H		1.87E-05	(L-wat/L-air)	(calculated)
P_v		1.40E+01	(mm Hg)	Versch.(1996)
S		2.10E+02	(mg/L)	Versch.(1996)
pK_a			log10(mole/mole)	
pK_b			log10(mole/mole)	
ρ		1.12E+00	(g/mL)	Versch.(1996)
T_{nbp}		2.98E+02	(°C)	Versch.(1996)
T_{mp}		-4.05E+01	(°C)	Versch.(1996)
toxicity parameters				
RfD_o		8.00E-01	(mg/kg-day)	IRIS (1994)
RfC_i		2.80E+00	(mg/m^3)	Route-to-Route
SF_o			1/(mg/kg-day)	
SF_i			1/(mg/kg-day)	
W of E	D			IRIS (1994)
MCL			(mg/L-wat)	
RAF_o		1.00E+00		
RAF_d		5.00E-01		
derived parameters dependent on chemical properties and soil parameters				
UF		1.00E+00	(unitless)	
H_{eff}		1.87E-05	(L-wat/L-air)	
$C_{sat,vap}$		1.70E+08	($\mu g/m^3$-air)	
K_{sw}		2.95E+00	(L-wat/kg-soil)	
$C_{sat,soil}$		6.20E+02	(mg/kg-soil)	[based on S]
$C_{sat,soil}$		2.69E+07	(mg/kg-soil)	[based on P_v]
$D_{eff,vad}$		4.01E-03	(cm^2/sec)	
$D_{eff,cap}$		6.60E-02	(cm^2/sec)	
$D_{eff,ws}$		4.08E-03	(cm^2/sec)	
$D_{eff,crk}$		4.01E-03	(cm^2/sec)	

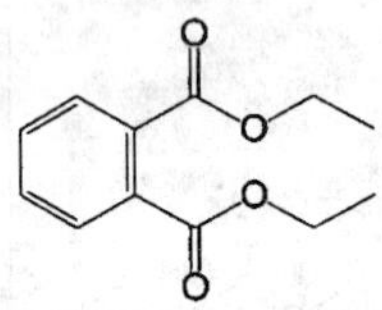

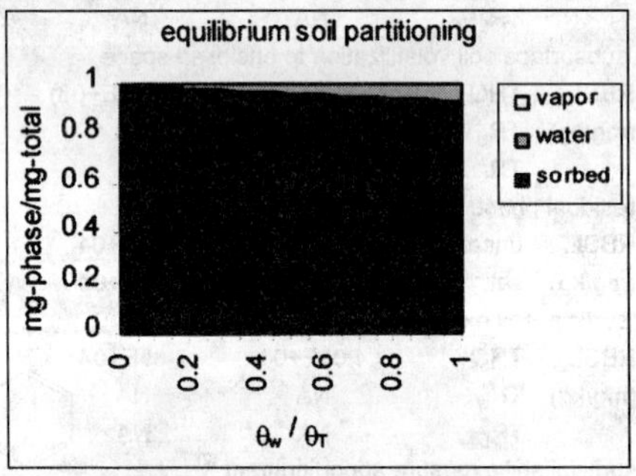

derived parameters:

$VF_{s,esp}$	5.72E-10	2.29E-10	(g-soil/cm^3-air)
$VF_{gw,esp}$	1.51E-09	6.03E-10	(cm^3-wat/cm^3-air)
$VF_{ss,amb,1}$	7.63E-10	8.36E-10	(g-soil/cm^3-air)
$VF_{ss,amb,2}$	1.80E-08	2.16E-08	(g-soil/cm^3-air)
$VF_{ss,amb,min(1,2)}$	7.63E-10	8.36E-10	(g-soil/cm^3-air)
LF_{sw}	7.19E-02	7.19E-02	(g-soil/cm^3-wat)
$VF_{s,amb}$	2.54E-11	2.54E-11	(g-soil/cm^3-air)
$VF_{gw,amb}$	2.54E-11	2.54E-11	(cm^3-wat/cm^3-air)

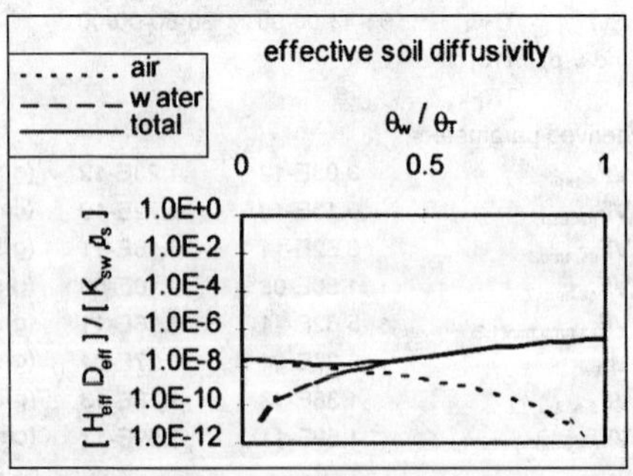

Notes:

(1) RBSL values are compared to physical limits of aqueous solubility (S<), saturated vapor pressure (<P), or either (Rs), and flagged if the calculated value is physically unrealistic. For cases where this occurs, the calculated values are given in parentheses.

(2) Percent of surficial soil exposure due to contributions of ingestion, dermal contact, vapor inhalation, and particulate inhalation (i-d-v-p).

(3) NA - not applicable.

TABLE X3.1 Chemical-Specific Parameters and Risk-Based Screening Level Concentrations (See Refs. 2, 16-21) *(continued)*

CASRN	127-18-4		
name	tetrachloroethene		
	C_2CL_4		
Risk-Based Screening-Level Concentrations [1]			
risk or hazard criteria:	scenario:	residential	industrial
ambient vapor inhalation			
$RBSL_{air}$	THQ	3.65E+01	5.11E+01
(µg/m³)	TR_U	4.26E+02	7.15E+02
	TR_L	4.26E+00	7.15E+00
enclosed space vapor inhalation			
$RBSL_{air}$	THQ	4.87E+01	5.11E+01
(µg/m³)	TR_U	5.68E+02	7.15E+02
	TR_L	5.68E+00	7.15E+00
groundwater ingestion			
$RBSL_{gw}$	THQ	3.65E-01	1.02E+00
(mg/L)	TR_U	1.64E-01	5.50E-01
	TR_L	1.64E-03	5.50E-03
	MCL	5.00E-03	5.00E-03
soil leaching to groundwater ingestion			
$RBSL_s$	THQ	2.98E+00	8.35E+00
(mg/kg)	TR_U	1.34E+00	4.50E+00
	TR_L	1.34E-02	4.50E-02
	MCL	4.08E-02	4.08E-02
groundwater to ambient vapor inhalation			
$RBSL_{gw}$	THQ	S< (2.35E+02)	S< (3.28E+02)
(mg/L)	TR_U	S< (2.74E+03)	S< (4.60E+03)
	TR_L	2.74E+01	4.60E+01
groundwater to enclosed space vapor inhalation			
$RBSL_{gw}$	THQ	1.34E+00	3.52E+00
(mg/L)	TR_U	1.56E+01	4.93E+01
	TR_L	1.56E-01	4.93E-01
subsurface soil volatilization to ambient air			
$RBSL_s$	THQ	1.47E+01	2.06E+01
(mg/kg)	TR_U	1.72E+02	2.89E+02
	TR_L	1.72E+00	2.89E+00
subsurface soil volatilization to enclosed space			
$RBSL_s$	THQ	8.74E-01	2.29E+00
(mg/kg)	TR_U	1.02E+01	3.21E+01
	TR_L	1.02E-01	3.21E-01
residual phase mobility limit			
$RBSL_s$	unsat.	1.49E+04	1.49E+04
(mg/kg)	sat.	1.49E+04	1.49E+04
surficial soil exposure			
$RBSL_{ss}$	THQ	5.84E+02	8.04E+02
(mg/kg)	TR_U	3.62E+02	6.42E+02
	TR_L	3.62E+00	6.42E+00
surficial soil exposure apportionment [2]			
(%)	THQ	08-63-28-00	03-62-33-00
(i-d-v-p)	TR_U	11-87-01-00	05-92-01-00
	TR_L	11-87-01-00	05-92-01-00

derived parameters:			
$VF_{s,esp}$	5.57E-05	2.23E-05	(g-soil/cm³-air)
$VF_{gw,esp}$	3.63E-05	1.45E-05	(cm³-wat/cm³-air)
$VF_{ss,amb,1}$	2.38E-07	2.61E-07	(g-soil/cm³-air)
$VF_{ss,amb,2}$	1.80E-08	2.16E-08	(g-soil/cm³-air)
$VF_{ss,amb,min(1,2)}$	1.80E-08	2.16E-08	(g-soil/cm³-air)
LF_{sw}	1.22E-01	1.22E-01	(g-soil/cm³-wat)
$VF_{s,amb}$	2.48E-06	2.48E-06	(g-soil/cm³-air)
$VF_{gw,amb}$	1.56E-07	1.56E-07	(cm³-wat/cm³-air)

transport and thermodynamic parameters			[references]
MW	1.66E+02	(g/mole)	USEPA(1996)
D_{air}	7.20E-02	(cm²/sec)	USEPA(1996)
D_{wat}	8.20E-06	(cm²/sec)	USEPA(1996)
K_{oc}	2.19E+00	log10(L-wat/kg-oc)	USEPA(1996)
K_{ow}	2.67E+00	log10(L-wat/kg)	USEPA(1996)
H	1.84E-02	(atm-m³/mol)	USEPA(1996)
H	7.65E-01	(L-wat/L-air)	(calculated)
P_v	1.90E+01	(mm Hg)	USEPA(1994)
S	2.00E+02	(mg/L)	USEPA(1996)
pK_a		log10(mole/mole)	
pK_b		log10(mole/mole)	
ρ	1.62E+00	(g/mL)	USEPA(1996)
T_{nbp}	1.21E+02	(°C)	USEPA(1994)
T_{mp}	-2.23E+01	(°C)	CRC(1996)
toxicity parameters			
RfD_o	1.00E-02	(mg/kg-day)	IRIS (1994)
RfC_i	3.50E-02	(mg/m³)	Route-to-Route
SF_o	5.20E-02	1/(mg/kg-day)	ECAO (1992)
SF_i	2.00E-03	1/(mg/kg-day)	ECAO (1992)
W of E	C-B2		ECAO (1992)
MCL	5.00E-03	(mg/L-wat)	56 FR 3526 (30 Jan 91)
RAF_o	1.00E+00		
RAF_d	5.00E-01		
derived parameters dependent on chemical properties and soil parameters			
UF	1.00E+00	(unitless)	
H_{eff}	7.65E-01	(L-wat/L-air)	
$C_{sat,vap}$	1.72E+08	(µg/m³-air)	
K_{sw}	1.74E+00	(L-wat/kg-soil)	
$C_{sat,soil}$	3.47E+02	(mg/kg-soil)	[based on S]
$C_{sat,soil}$	3.91E+02	(mg/kg-soil)	[based on P_v]
$D_{eff,vad}$	5.62E-03	(cm²/sec)	
$D_{eff,cap}$	1.14E-05	(cm²/sec)	
$D_{eff,ws}$	6.10E-04	(cm²/sec)	
$D_{eff,crk}$	5.62E-03	(cm²/sec)	

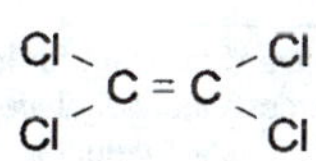

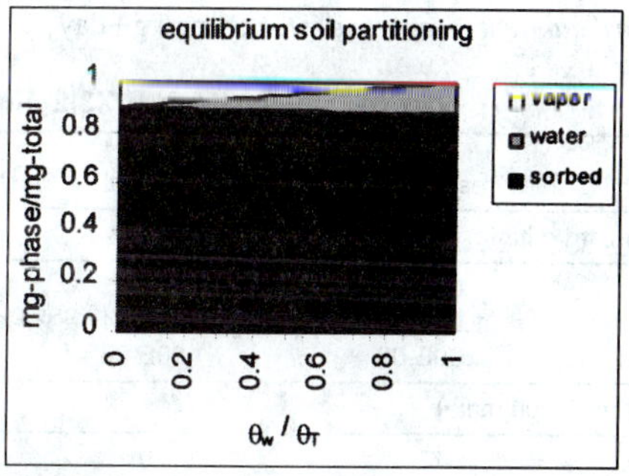

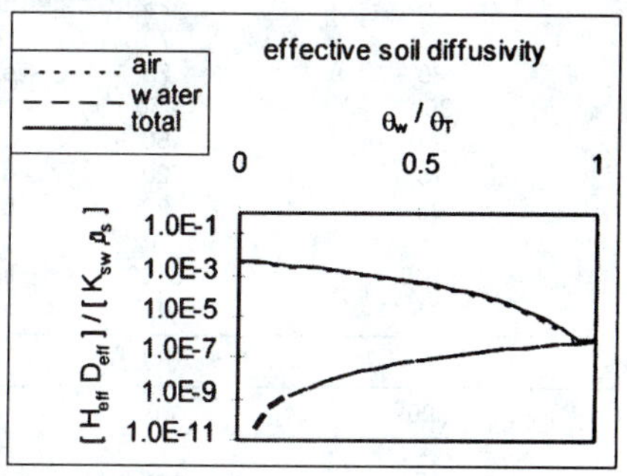

Notes:

(1) RBSL values are compared to physical limits of aqueous solubility (S<), saturated vapor pressure (<P), or either (Rs), and flagged if the calculated value is physically unrealistic. For cases where this occurs, the calculated values are given in parentheses.

(2) Percent of surficial soil exposure due to contributions of ingestion, dermal contact, vapor inhalation, and particulate inhalation (i-d-v-p).

(3) NA - not applicable.

as carcinogens, the example RBSLs are based on the general equation:

$$risk = average\ lifetime\ compound\ intake\ [mg/kg-day] \quad (X3.1)$$

$$x\ slope\ factor\ [mg/kg-day]^{-1}$$

where the intake depends on exposure parameters (ingestion rate, exposure duration, and so forth), the source concentration, and transport rates between the source and receptor. The slope factor is selected after reviewing a number of sources, including the USEPA Integrated Risk Information System (IRIS) **(8)** database, USEPA Health Effects Assessment Summary Tables (HEAST) **(9)**, and peer-reviewed sources. Note that this risk value does not reflect the actual probability for the specified exposure scenario to occur.

X3.1.4.2 The slope factor may be equivalently defined in terms of a unit risk factor,

$$slope\ factor\ [mg/kg-day]^{-1} = unit\ risk\ factor\ [mg/m^3]^{-1} \quad (X3.2)$$

$$/defined\ average\ lifetime\ media\ intake\ [m^3/kg-day]^{-1}$$

where the defined average lifetime media intake (of water or air) is based on defined exposure parameters (media ingestion rate, exposure duration, and so forth) for direct ingestion or inhalation of the chemical.

X3.1.4.3 The example RBSL values appearing in this appendix correspond to upper bound conservative probabilities of adverse health effects ("risks") in the range from 10^{-6} to 10^{-4} resulting from the specified exposure.

X3.1.4.4 In the case of compounds that have not been classified as carcinogens, the example RBSLs are based on the general equation:

$$hazard\ quotient = average\ intake\ [mg/kg-day] \quad (X3.3)$$

$$/reference\ dose\ [mg/kg-day]$$

where the intake depends on exposure parameters (ingestion rate, exposure duration, and so forth), the source concentration, and transport rates between the source and receptor. The reference dose is selected after reviewing a number of sources, including the USEPA Integrated Risk Information System (IRIS) **(8)** database, USEPA Health Effects Assessment Summary Tables (HEAST) **(9)**, and peer-reviewed sources. Note that this hazard quotient value does not reflect the actual probability for the specified exposure scenario to occur.

X3.1.4.5 The reference dose may be equivalently defined in terms of a reference concentration,

$$reference\ dose\ [mg/kg-day] = reference\ concentration\ [mg/m^3] \quad (X3.4)$$

$$x\ defined\ average\ daily\ media\ intake\ [m^3/kg-day]$$

where the defined average daily media intake (of water or air) is based on defined exposure parameters (media ingestion rate, exposure duration, and so forth) for direct ingestion or inhalation of the chemical.

X3.1.4.6 The example RBSL values appearing in this appendix correspond to upper bound conservative estimates for hazard quotients of unity resulting from the specified exposure. Note that this hazard quotient value does not reflect the actual probability for the specified exposure scenario to occur.

X3.1.5 Tables X3.2-X3.4 summarize the equations and parameters used to prepare the example RBSLs (Table X3.1). The basis for each of these equations is discussed in X3.2-X3.12.

TABLE X3.2 Site and Scenario-Specific Parameters

parameter	residential	industrial	definition, units
specified risk or hazard criteria			
THQ	1	1	target hazard quotient for individual chemicals (unitless)
TR_U	1.00E-04	1.00E-04	target excess individual lifetime cancer risk - upper range value (unitless)
TR_L	1.00E-06	1.00E-06	target excess individual lifetime cancer risk - lower range value (unitless)
exposure parameters (adult male)			
AT_c	70	70	defined carcinogen averaging time (years)
AT_n	30	25	defined averaging time for non-carcinogen (years), = ED
BW	70	70	body weight (kg)
ED	30	25	exposure duration (years)
EF	350	250	exposure frequency (days/year)
IR_{soil}	100	50	soil ingestion rate (mg/day)
IR_{air}-indoor	15	20	daily inhalation rate (m^3/day) - indoor
IR_{air}-outdoor	20	20	daily inhalation rate (m^3/day) - outdoor
IR_w	2	1	daily water ingestion rate (L/day)
SA	3160	3160	seasonally-averaged skin surface area (cm^2/day)
M	0.5	0.5	soil to skin adherence factor (mg/cm^2)
building parameters			
L_b	200	300	enclosed space volume/infiltration area ratio (cm)
ER	12	20	enclosed space air exchange rate (1/day)
L_{crack}	15	15	enclosed-space foundation or wall thickness (cm)
η	0.01	0.01	foundation crack fraction (cm^2-cracks/cm^2-total area)
dP	0	0	indoor/outdoor differential pressure ($g/cm-s^2$)
k_v	1.00E-08	1.00E-08	soil permeability (cm^2)
Z_{crack}	15	15	depth to bottom of slab (cm)
X_{crack}	3400	3400	slab perimeter (cm)
A_b	700000	700000	slab area (cm^2)
surface parameters			

TABLE X3.2 *Continued*

parameter	residential	industrial	definition, units
τ	30	25	averaging time for surface emission vapor flux (years)
U_{air}	225	225	ambient air velocity in mixing zone (cm/s)
δ_{air}	200	200	mixing zone height (cm)
A	20250000	20250000	source-zone area (cm^2)
W	4500	4500	width of source-zone area (cm)
L_{ss}	100	100	thickness of surficial soils (cm)
P_e	6.90E-14	6.90E-14	Areal total respirable particulate emission flux from source (g/cm^2-s)
derived parameters			
$LF_{pw,gw}$	4.70E+00	4.70E+00	leaching factor, ground water/soil pore water ratio (cm^3-wat/cm^3-wat)
Q_s	0	0	convective flow through basement slab (cm^3-air/sec)
DF_{esp}	2.78E-02	6.94E-02	dispersion factor for enclosed-space air (g-cm^2/s)/(g/cm^3)
DF_{amb}	10	10	dispersion factor for ambient air (g-cm^2/s)/(g/cm^3)
VF_p	6.9E-15	6.9E-15	total respirable particulate concentration from soil source (g-soil/cm^3-air)
mobility limit parameters			
S_r	0.04	0.04	specified residual void fraction in vadose zone soil (cm^3-oil/cm^3-void)
S_r	0.04	0.04	specified residual void fraction in saturated zone soil (cm^3-oil/cm^3-void)
soil parameters			
h_{cap}	5		capillary zone thickness (cm)
h_v	295		vadose zone thickness (cm)
θ_{wcap}	0.342		soil water content - capillary fringe region (cm^3-water/cm^3-soil)
θ_{wvad}	0.12		soil water content - vadose zone (cm^3-water/cm^3-soil)
θ_{wcrk}	0.12		soil water content - soil filled foundation cracks (cm^3-water/cm^3-soil)
θ_{acap}	0.038		soil air content - capillary fringe region (cm^3-air/cm^3-soil)
θ_{avad}	0.26		soil air content - vadose zone (cm^3-air/cm^3-soil)
θ_{acrk}	0.26		soil air content - soil filled foundation cracks (cm^3-air/cm^3-soil)
ρ_s	1.7		soil bulk density - dry soil (g/cm^3)
f_{oc}	0.01		mass fraction of organic carbon in soil (g-oc/g-soil)
θ_T	0.38		soil porosity (cm^3-void/cm^3-soil)
L_{gw}	300		depth to ground water (cm)
L_s	100		thickness of surficial soils (cm)
pH	6.8		soil/water pH (unitless)
ground water parameters			
δ_{gw}	200		ground water mixing zone height (cm)
I	30		water infiltration rate (cm/year)
U_{gw}	6.85		ground water Darcy velocity (cm/day)
chemical-specific identifying parameters			
CASRN			Chemical Abstract Service Registration Number
name			common chemical name
chemical-specific transport and thermodynamic parameters			
MW			molecular weight (g/mole)
D_{air}			molecular diffusion coefficient in air (cm^2/sec)
D_{wat}			molecular diffusion coefficient in water (cm^2/sec)
K_{oc}			organic carbon/water partition coefficient log10(L-wat/kg-oc)
K_d			water / soil partition coefficient log10(L-wat/kg-soil) (K_d= K_{oc} f_{oc} *for organic chemicals*)
H			Henry's law constant (atm-m^3/mol) or (L-water/L-air)
P_v			Vapor pressure (mm Hg)
S			Aqueous solubility limit (mg/L-wat)
pK_a			acid ionization equilibrium constant (log10(mole/mole))
pK_b			base ionization equilibrium constant (log10(mole/mole))
chemical-specific toxicity parameters and exposure criteria			
RfD_o			chronic oral reference dose (mg/kg-day)
RfC_i			chronic inhalation reference concentration (mg/m^3)
SF_o			slope factor - oral (1/(mg/kg-day))
SF_i			slope factor - inhalation (1/(mg/kg-day))
W of E			weight of evidence
MCL			Federal maximum contaminant level for public drinking water supply (mg/L-wat)
RAF_o			relative absorption factor - oral (mg-adsorbed/mg-applied)
RAF_d			relative absorption factor - dermal (mg-adsorbed/mg-applied)
PEL-TWA			OSHA Permissable Exposure Level Time-Weighted Average workplace criteria (mg/m^3 air)
TLV-TWA			ACGIH Threshold Limit Value - Time Weighted Average workplace criteria (mg/m^3-air)
derived parameters - dependent on chemical and soil properties			
UF			fraction of unionized chemical in water (unitless)
H_{eff}			effective Henry's law coefficient (L-water/L-air)
$C_{sat,vap}$			saturated vapor concentration (mg/m^3-air)
K_{sw}			soil to water partition coefficient - vadose zone (L-water/kg-soil)

TABLE X3.2 *Continued*

parameter	residential	industrial	definition, units
$C_{sat,soil}$			residual soil concentration - vadose zone (mg/kg-soil)
$D_{eff,vad}$			effective diffusivity - vadose zone (cm^2/sec)
$D_{eff,cap}$			effective diffusivity - capillary fringe zone (cm^2/sec)
$D_{eff,ws}$			effective diffusivity - averaged water table to surface (cm^2/sec)
$D_{eff,crk}$			effective diffusivity - enclosed space foundation cracks (cm^2/sec)
R_c			dissolved phase retardation factor for saturated ground water flow ((cm/s)-wat/(cm/s)-phase)
R_i			dissolved phase retardation factor for vadose zone infiltration flow ((cm/s)-wat/(cm/s)-phase)
derived parameters— dependent on scenario parameters, and chemical and soil properties			
$VF_{s,esp}$			volatilization factor, subsurface soil to enclosed space (g-soil/cm^3-air)
$VF_{gw,esp}$			volatilization factor, ground water to enclosed space (cm^3-wat/cm^3-air)
$VF_{ss,amb,1}$			volatilization factor, surficial soil to ambient air (g-soil/cm^3-air)
$VF_{ss,amb,2}$			volatilization factor, surficial soil to ambient air, mass-limited (g-soil/cm^3-air)
$VF_{ss,amb,min(1,2)}$			applicable volatilization factor, surficial soil to ambient air (g-soil/cm^3-air)
LF_{sw}			leaching factor, soil to ground water (g-soil/cm^3-wat)
$VF_{s,amb}$			volatilization factor, subsurface soil to ambient air (g-soil/cm^3-air)
$VF_{gw,amb}$			volatilization factor, ground water to ambient air (cm^3-wat/cm^3-air)
Risk-Based Screening Level Concentrations			
$RBSL_{air}$			Risk-Based Screening Level for Air (mg/m^3) - ambient (outdoor) vapor - enclosed space (indoor) vapor
$RBSL_{gw}$			Risk-Based Screening Level for Ground Water (mg/L) - ingestion - inhalation of evolved vapors to outdoor air - inhalation of evolved vapors to indoor air
$RBSL_s$			Risk-Based Screening Level for soils (mg/kg) - leaching to ground water ingestion - inhalation of evolved vapors from subsurface soil to outdoor air - inhalation of evolved vapors from subsurface soil to indoor air
$RBSL_{ss}$			Risk-Based Screening Level for surficial soil (mg/kg) - combined exposure to direct ingestion, dermal contact, vapor inhalation, and particulate inhalation
(i-d-v-p)			Apportionment of $RBSL_{ss}$ as a fraction of exposure by direct ingestion, dermal contact, vapor inhalation, and particulate inhalation (% each)

TABLE X3.3 Equations Used to Develop Example Tier 1 Risk-Based Screening Levels (RBSLs) Appearing in Table X3.1

Risk-Based Screening Level Concentration for Air: Inhalation - indoor and outdoor vapors

for non-carcinogens:

$$RBSL_{air} = \frac{THQ \cdot RfC_i \cdot AT_n \cdot 365\ \text{days/year}}{ED \cdot EF}$$

for carcinogens:

$$RBSL_{air} = \frac{Risk \cdot BW \cdot AT_c \cdot 365\ \text{days/year}}{SF_i \cdot IR_{air} \cdot ED \cdot EF}$$

For $RBSL_{air} > C_{sat,vap}$, the calculated concentration is greater than the pure chemical component saturated vapor concentration limit, and the specified risk or hazard level cannot be achieved in the defined scenario (unless an aerosol of the chemical is inhaled).

$RBSL_{air}$	Risk-Based Screening Level for air inhalation (mg/m^3)
AT_c	defined carcinogen averaging time (years)
AT_n	defined averaging time for non-carcinogen (years), = ED
BW	body weight (kg)
$C_{sat,vap}$	pure chemical saturated vapor concentration (mg/m^3)
ED	exposure duration (years)
EF	exposure frequency (days/year)
IR_{air}	daily inhalation rate (m^3/day) - indoor r outdoor
RfC_i	chronic inhalation reference concentration (mg/m^3)
Risk	target excess individual lifetime cancer risk (TR_U to TR_L range)
SF_i	carcinogenic slope factor - inhalation (mg/kg-day)$^{-1}$
THQ	target hazard quotient for individual chemicals (unitless)

Risk-Based Screening Level Concentration for Ground Water: Ingestion (potable water supply only)

for non-carcinogens:

$$RBSL_{gw} = \frac{THQ \cdot RfD_o \cdot BW \cdot AT_n \cdot 365\ \text{days/year}}{IR_w \cdot ED \cdot EF}$$

for carcinogens:

$$RBSL_{gw} = \frac{Risk \cdot BW \cdot AT_c \cdot 365\ \text{days/year}}{SF_o \cdot IR_w \cdot ED \cdot EF}$$

For $RBSL_{gw} > S$, the calculated concentration is greater than the pure chemical component aqueous solubility limit and the specified risk or hazard level cannot be achieved in the defined scenario (unless free-phase chemical is mixed with the ingested water).

$RBSL_{gw}$	Risk-Based Screening Level for ground water ingestion (mg/L)
AT_c	defined carcinogen averaging time (years)
AT_n	defined averaging time for non-carcinogen (years), = ED
BW	body weight (kg)
ED	exposure duration (years)
EF	exposure frequency (days/year)
IR_w	daily water ingestion rate (L/day)
RfD_o	chronic oral reference dose (mg/kg-day)
Risk	target excess individual lifetime cancer risk (TR_U to TR_L)
S	pure chemical aqueous solubility limit (mg/L)
SF_o	carcinogenic slope factor - oral (mg/kg-day)$^{-1}$
THQ	target hazard quotient for individual chemicals (unitless)

Risk-Based Screening Level Concentration for Ground Water: enclosed-space (indoor) vapor inhalation

$$RBSL_{gw} = \frac{RBSL_{air}}{VF_{gw,esp}} \cdot 10^{-3}\ \frac{m^3}{L}$$

For $RBSL_{gw} > S$, the calculated concentration is greater than the pure chemical component aqueous solubility limit and the specified risk or hazard level cannot be achieved in the defined scenario.

$RBSL_{gw}$	Risk-Based Screening Level for ground water ingestion (mg/L)
$RBSL_{air}$	Risk-Based Screening Level for air inhalation (mg/m^3)
$VF_{gw,esp}$	volatilization factor, ground water to enclosed space air (cm^3-water/cm^3-air)

Risk-Based Screening Level for Surficial Soil: ingestion of soil, inhalation of vapors and particulates, and dermal contact

for non-carcinogens:

$$RBSL_{ss} = \frac{THQ \cdot BW \cdot AT_n \cdot (365\ \text{days/year})}{EF \cdot ED} \Bigg/ \left[10^{-6}\frac{kg}{mg} \cdot \left(\frac{IR_s \cdot RAF_o}{RfD_o} + \frac{SA \cdot M \cdot RAF_d}{RfD_o}\right) + \frac{BW \cdot (VF_{ss} + VF_p)}{RfC_i} \cdot 10^3\ \frac{cm^3}{m^3}\frac{kg}{g}\right]$$

for carcinogens:

$$RBSL_{ss} = \frac{Risk \cdot BW \cdot AT_c \cdot (365\ \text{days/year})}{EF \cdot ED} \Bigg/ \left[10^{-6}\frac{kg}{mg} \cdot \left(SF_o \cdot IR_s \cdot RAF_o + SF_o \cdot SA \cdot M \cdot RAF_d\right) + SF_i \cdot IR_{air}\,(VF_{ss} + VF_p) \cdot 10^3\ \frac{cm^3}{m^3}\frac{kg}{g}\right]$$

$RBSL_{ss}$	Risk-Based Screening Level for Surficial Soil (mg/kg)
AT_c	defined carcinogen averaging time (years)
AT_n	defined averaging time for non-carcinogen (years, = ED)
BW	body weight (kg)
ED	exposure duration (years)
EF	exposure frequency (days/year)
IR_{air}	daily inhalation rate (m^3/day) - outdoor
IR_s	soil ingestion rate (mg/day)
M	soil to skin adherence factor (mg/cm^2)
RAF_d	dermal relative absorption factor
RAF_o	oral relative absorption factor
RfC_i	chronic inhalation reference concentration (mg/m^3)
RfD_o	chronic oral reference dose (mg/kg-day)
Risk	target excess individual lifetime cancer risk (TR_U to TR_L)
SA	seasonally-averaged skin surface area (cm^2/day)
SF_i	slope factor - inhalation (mg/kg-day)$^{-1}$
SF_o	slope factor - oral (mg/kg-day)$^{-1}$
THQ	target hazard quotient for individual chemicals (unitless)
VF_{ss}	volatilization factor, surficial soil to ambient air (g-soil/cm^3-air)
VF_p	total respirable particulate concentration from soil source (g-soil/cm^3-air)

TABLE X3.3 *Continued*

Volatilization Factor: subsurface soil to enclosed-space (indoor) vapor inhalation

For $Q_s = 0$,

$$VF_{s,esp} = \frac{1}{\frac{K_{sw}}{H_{eff}}\left(1 + \frac{D_{eff,vad}}{DF_{esp}\,L_s} + \frac{D_{eff,vad} \cdot L_{crk}}{D_{eff,crk} \cdot L_s \cdot \eta}\right) \cdot \frac{DF_{esp}}{D_{eff,vad}} \cdot L_s} \qquad (1a)$$

for $Q_s > 0$,

$$VF_{s,esp} = \frac{1}{\frac{K_{sw}}{H_{eff}}\left(e^{\xi} + \frac{D_{eff,vad}}{DF_{esp}\,L_s} + \frac{D_{eff,vad} \cdot A_b}{Q_s \cdot L_s} \cdot (e^{\xi} - 1)\right) \cdot \frac{DF_{esp} \cdot L_s}{D_{eff,vad} \cdot e^{\xi}}} \qquad (1b)$$

with

$$\xi = \frac{Q_s \cdot L_{crk}}{A_b \cdot D_{eff,crk} \cdot \eta}$$

If a finite depth of subsurface soil is specified, in addition

$$VF_{s,esp} = \frac{d_s \cdot \rho_s}{DF_{esp} \cdot \tau \cdot 31536000 \text{ sec/year}} \qquad (2)$$

Of equations (1) and (2), choose whichever is less.

$VF_{s,esp}$ volatilization factor, subsurface soil to enclosed-space air ($g\text{-soil}/cm^3\text{-air}$)
A_b slab area (cm^2)
$D_{eff,crk}$ effective diffusivity in soil-filled foundation cracks (cm^2/sec)
$D_{eff,vad}$ effective diffusion coefficient - vadose zone soils (cm^2/sec)
DF_{esp} dispersion factor for enclosed-space air (cm/s)
H_{eff} effective Henry's law coefficient ($cm^3\text{-water}/cm^3\text{-air}$)
K_{sw} soil to water partition coefficient ($cm^3\text{-water}/g\text{-soil}$)
L_s depth to subsurface soils (cm)
L_{crk} enclosed-space foundation or wall thickness (cm)
Q_s convective flow through basement slab (cm^3/sec)
η foundation crack fraction (cm^2/cm^2)
d_s depth of affected subsurface soil layer (cm)
τ averaging time for surface emission vapor flux (years)

Leaching Factor: soil to groundwater

$$LF_{sw,1} = \frac{LF_{pw,gw}}{K_{sw}} \qquad (1)$$

with

$$LF_{pw,gw} = \frac{1}{\left[1 + \frac{U_{gw} \cdot \delta_{gw}}{I \cdot W}\right]}$$

If a finite depth of subsurface soil is specified, in addition

$$LF_{sw,2} = \frac{d_s \cdot \rho_s}{I \cdot \tau} \qquad (2)$$

Of equations (1) and (2), choose whichever is less.

LF_{sw} leaching factor, soil to groundwater ($g\text{-soil}/cm^3\text{-wat}$)
I water infiltration rate (cm/year)
K_{sw} soil to water partition coefficient ($cm^3\text{-water}/g\text{-soil}$)
$LF_{pw,gw}$ leaching factor, ground water / soil pore water concentration ratio ($cm^3\text{-water}/cm^3\text{-water}$)
U_{gw} groundwater Darcy velocity (cm/year)
W width of source-zone area (cm)
δ_{gw} groundwater mixing zone height (cm)
τ averaging time for leaching duration (years)
d_s depth of affected subsurface soil layer (cm)

Effective Diffusion Coefficient in homogeneous soil layers

$$D_{eff} = D_{air} \cdot \left(\frac{\theta_{air}^{3.33}}{\theta_T^2}\right) + D_{water} \cdot \left(\frac{\theta_{water}^{3.33}}{H_{eff} \cdot \theta_T^2}\right)$$

This semi-empirical equation is applicable for estimating the effective diffusion coefficient for porous media in homogeneous soil. It is used here in calculating a value for the vadose zone ($D_{eff,vad}$), the water table capillary fringe region ($D_{eff,cap}$), and in soil-filled enclosed space basement foundation cracks ($D_{eff,crk}$).

D_{eff} effective diffusion coefficient (cm^2/sec)
D_{air} molecular diffusion coefficient in air (cm^2/sec)
D_{water} molecular diffusion coefficient in water (cm^2/sec)
H_{eff} effective Henry's law coefficient ($cm^3\text{-water}/cm^3\text{-air}$)
θ_{air} soil air content ($cm^3\text{-air}/cm^3\text{-soil}$)
θ_{water} soil water content ($cm^3\text{-water}/cm^3\text{-soil}$)
θ_T soil porosity ($cm^3\text{-air}/cm^3\text{-soil}$), $= \theta_{air} + \theta_{water}$

Refs:

Millington and Quirk (1961): Transactions of the Faraday Society, 57, 1200-1207; Jury, W. A., W. F. Spencer, and W. J. Farmer, 1983: Behavior Assessment Model for Trace Organics in Soil: I. Model Description, Journal of Environmental Quality, 12, 558-564; Bruell, C. J., and G. E. Hoag, 1986: The Diffusion of Gasoline-Range Hydrocarbon Vapors in Porous Media, Experimental Methodologies, in Proceedings of the NWWA/API Conference on Petroleum Hydrocarbons and Organic Chemicals in Groundwater: Prevention, Detection, and Restoration, November 12-14, Houston.

Effective Diffusion Coefficient: averaged value - water table to soil surface

$$D_{eff,ws} = \frac{L_{gw}}{(h_v / D_{eff,vad}) + (h_{cap} / D_{eff,cap})}$$

$D_{eff,ws}$ effective diffusivity - averaged water table to surface (cm^2/sec)
$D_{eff,vad}$ effective diffusivity - vadose zone soils (cm^2/sec)
$D_{eff,cap}$ effective diffusivity - capillary fringe zone (cm^2/sec)
h_v vadose zone thickness (cm)
h_{cap} capillary zone thickness (cm)
L_{gw} depth to groundwater (cm), $L_{gw} = h_v + h_{cap}$

Equation for Calculating Total Soil Concentration to Pore Water Concentration Ratio

$$K_{sw} = \frac{\theta_w + (K_d \cdot \rho_s) + (H_{eff} \cdot \theta_a)}{\rho_s}$$

This equation is used in calculating the ratio between total soil concentration and pore-water concentration for vadose zone soils. Equilibrium partitioning of the chemical between soil (sorbed), pore water, and pore vapors, at concentrations below the pure component pore water (S) or pore vapor saturation limit ($C_{sat,vap}$) is assumed.

K_{sw} soil to water partition coefficient ($cm^3\text{-water}/g\text{-soil}$)
$C_{sat,vap}$ saturated vapor concentration ($mg/m^3\text{-air}$)
H_{eff} effective Henry's law coefficient ($cm^3\text{-water}/cm^3\text{-air}$)
K_d soil (sorbed) / water partition coefficient ($cm^3\text{-wat}/g\text{-soil}$)
S pure chemical aqueous solubility limit (mg/L)
ρ_s dry soil bulk density (g/cm^3)
θ_a soil air content ($cm^3\text{-air}/cm^3\text{-soil}$)
θ_w soil water content ($cm^3\text{-water}/cm^3\text{-soil}$)

TABLE X3.3 *Continued*

Risk-Based Screening Level for Free-Phase Migration: unsaturated soil

$$RBSL_s = S \cdot \frac{\theta_w + (K_d \cdot \rho_s) + H_{eff} \cdot (\theta_a - \theta_o)}{\rho_s} + \frac{\theta_o \cdot \rho_o}{\rho_s} \cdot 10^6 \text{ mg/kg}$$

with

$\theta_o = S_r \cdot \theta_T$

This equation is applicable for pure chemicals with finite solubility, S, and which are liquids at ambient temperature and pressure

$RBSL_s$	Risk-Based Screening Level for soil (mg/kg)
H_{eff}	effective Henry's law coefficient (cm^3-water/cm^3-air)
K_d	soil (sorbed) / water partition coefficient (cm^3-wat/g-soil)
S	pure chemical aqueous solubility limit (mg/L-water)
ρ_s	dry soil bulk density (g-soil/cm^3-soil)
ρ_o	density of chemical residual phase liquid (g-res/cm^3-res)
θ_a	soil air content (cm^3-air/cm^3-soil)
θ_w	soil water content (cm^3-water/cm^3-soil)
θ_o	residual phase volume fraction (cm^3-res/cm^3-soil)
θ_T	soil porosity (cm^3-void/cm^3-soil) = $\theta_a + \theta_w$
S_r	residual phase void fraction (cm^3-res/cm^3-void)

Risk-Based Screening Level for Free-Phase Migration: saturated soil

$$RBSL_s = S \cdot \frac{(\theta_T - \theta_o) + (K_d \cdot \rho_s)}{\rho_s} + \frac{\theta_o \cdot \rho_o}{\rho_s} \cdot 10^6 \text{ mg/kg}$$

with

$\theta_o = S_r \cdot \theta_T$

This equation is applicable for pure chemicals with finite solubility, S, and which are liquids at ambient temperature and pressure

$RBSL_s$	Risk-Based Screening Level for soil (mg/kg)
K_d	soil (sorbed) / water partition coefficient (cm^3-wat/g-soil)
S	pure chemical aqueous solubility limit (mg/L-water)
ρ_s	dry soil bulk density (g-soil/cm^3-soil)
ρ_o	density of chemical residual phase liquid (g-res/cm^3-res)
θ_o	residual phase volume fraction (cm^3-res/cm^3-soil)
θ_T	soil porosity (cm^3-void/cm^3-soil)
S_r	residual phase void fraction (cm^3-res/cm^3-void)

TABLE X3.4 Calculated Parameters Used in Example Tier 1 Risk-Based Screening Level Equations of Table X3.3

Volatilization Factor: ground water to enclosed-space (indoor) vapor inhalation

For $Q_s = 0$,

$$VF_{gw,esp} = \frac{1}{\frac{1}{H_{eff}}\left(1+\frac{D_{eff,ws}}{DF_{esp}\,L_{gw}}+\frac{D_{eff,ws}\cdot L_{crk}}{D_{eff,crk}\cdot L_{gw}\cdot \eta}\right)\cdot\frac{DF_{esp}}{D_{eff,ws}}\cdot L_{gw}}$$

for $Q_s > 0$,

$$VF_{gw,esp} = \frac{1}{\frac{1}{H_{eff}}\left(e^{\xi}+\frac{D_{eff,ws}}{DF_{esp}\,L_{gw}}+\frac{D_{eff,ws}\cdot A_b}{Q_s\cdot L_{gw}}\cdot(e^{\xi}-1)\right)\cdot\frac{DF_{esp}\cdot L_{gw}}{D_{eff,ws}\cdot e^{\xi}}}$$

with

$$\xi = \frac{Q_s\cdot L_{crk}}{A_b\cdot D_{crk}\cdot\eta}$$

$VF_{gw,esp}$ volatilization factor, ground water to enclosed space air (cm^3-water/cm^3-air)
$D_{eff,crk}$ effective diffusivity in soil-filled foundation cracks (cm^2/sec)
$D_{eff,ws}$ effective diffusivity - averaged water table to surface (cm^2/sec)
DF_{esp} dispersion factor for indoor air (cm/s)
H_{eff} effective Henry's law coefficient (cm^3-water/cm^3-air)
L_{gw} depth to ground water (cm)
L_{crk} enclosed-space foundation or wall thickness (cm)
Q_s convective flow through basement slab (cm^3/sec)
η foundation crack fraction (cm^2/cm^2)

Volatilization Factor: ground water to ambient (outdoor) vapor inhalation

$$VF_{gw,amb} = \frac{1}{\left(1+\frac{DF_{amb}\cdot L_{gw}}{D_{eff,ws}}\right)\frac{1}{H_{eff}}}$$

$VF_{gw,amb}$ volatilization factor, groundwater to ambient air (cm^3-water/cm^3-air)
$D_{eff,ws}$ effective diffusivity - averaged water table to surface (cm^2/sec)
DF_{amb} dispersion factor for ambient air (cm/s)
H_{eff} effective Henry's law coefficient (cm^3-water/cm^3-air)
L_{gw} depth to ground water (cm)

Particulate Concentration: total respirable particulate concentration originating from surficial soil source

$$VF_p = \frac{P_e}{DF_{amb}}$$

VF_p total respirable particulate concentration from soil source (g-soil/cm^3-air)
DF_{amb} dispersion factor for ambient air (cm/sec)
P_e Areal total respirable particulate emission flux from source (g/cm^2-sec)

Volatilization Factor: surficial soils to ambient air (vapors)

$$VF_{ss} = \frac{\rho_s}{DF_{amb}}\cdot\sqrt{\frac{4\cdot D_{eff,vad}}{\pi\cdot\tau\cdot 31536000\ \text{sec/year}}\cdot\frac{H_{eff}}{K_{sw}\cdot\rho_s}} \quad (1)$$

or

$$VF_{ss} = \frac{L_{ss}\cdot\rho_s}{DF_{amb}\cdot\tau\cdot 31536000\ \text{sec/year}} \quad (2)$$

Of equations (1) or (2), choose whichever is less.

VF_{ss} volatilization factor, surficial soil to ambient air (g-soil/cm^3-air)
$D_{eff,vad}$ effective diffusion coefficient for vadose-zone soils (cm^2/sec)
DF_{amb} dispersion factor for ambient air (cm/s)
H_{eff} effective Henry's law coefficient (cm^3-water/cm^3-air)
K_{sw} soil to water partition coefficient (cm^3-water/g-soil)
L_{ss} thickness of surficial soils (cm)
ρ_s dry soil bulk density (g/cm^3)
τ averaging time for surface emission vapor flux (years)

Volatilization Factor: subsurface soil to ambient (outdoor) vapor inhalation

$$VF_{s,amb} = \frac{1}{\left(1+\frac{DF_{amb}\cdot L_s}{D_{eff,vad}}\right)\cdot\frac{K_{sw}}{H_{eff}}} \quad (1)$$

If a finite depth of subsurface soil is specified, in addition

$$VF_{s,amb} = \frac{d_s\cdot\rho_s}{DF_{amb}\cdot\tau\cdot 31536000\ \text{sec/year}} \quad (2)$$

Of equations (1) and (2), choose whichever is less.

$VF_{s,amb}$ volatilization factor, subsurface soil to ambient air (g-soil/cm^3-air)
$D_{eff,vad}$ effective diffusion coefficient - vadose zone soils (cm^2/sec)
DF_{amb} dispersion factor for ambient air (cm/s)
H_{eff} effective Henry's law coefficient (cm^3-water/cm^3-air)
K_{sw} soil to water partition coefficient (cm^3-water/g-soil)
L_s depth to subsurface soils (cm)
d_s depth of affected subsurface soil layer (cm)
τ averaging time for surface emission vapor flux (years)

TABLE X3.4 *Continued*

Volatilization Factor: subsurface soil to enclosed-space (indoor) vapor inhalation

For $Q_s = 0$,

$$VF_{s,esp} = \frac{1}{\frac{K_{sw}}{H_{eff}}\left(1 + \frac{D_{eff,vad}}{DF_{esp}\,L_s} + \frac{D_{eff,vad}\cdot L_{crk}}{D_{eff,crk}\cdot L_s\cdot\eta}\right)\cdot\frac{DF_{esp}}{D_{eff,vad}}\cdot L_s} \qquad (1a)$$

for $Q_s > 0$,

$$VF_{s,esp} = \frac{1}{\frac{K_{sw}}{H_{eff}}\left(e^{\xi} + \frac{D_{eff,vad}}{DF_{esp}\,L_s} + \frac{D_{eff,vad}\cdot A_b}{Q_s\cdot L_s}\cdot(e^{\xi}-1)\right)\cdot\frac{DF_{esp}\cdot L_s}{D_{eff,vad}\cdot e^{\xi}}} \qquad (1b)$$

with

$$\xi = \frac{Q_s\cdot L_{crk}}{A_b\cdot D_{crk}\cdot\eta}$$

If a finite depth of subsurface soil is specified, in addition

$$VF_{s,esp} = \frac{d_s\cdot\rho_s}{DF_{esp}\cdot\tau\cdot 31536000\ \text{sec/year}} \qquad (2)$$

Of equations (1) and (2), choose whichever is less.

$VF_{s,esp}$	volatilization factor, subsurface soil to enclosed-space air (g-soil/cm³-air)
A_b	slab area (cm²)
$D_{eff,crk}$	effective diffusivity in soil-filled foundation cracks (cm²/sec)
$D_{eff,vad}$	effective diffusion coefficient - vadose zone soils (cm²/sec)
DF_{esp}	dispersion factor for enclosed-space air (cm/s)
H_{eff}	effective Henry's law coefficient (cm³-water/cm³-air)
K_{sw}	soil to water partition coefficient (cm³-water/g-soil)
L_s	depth to subsurface soils (cm)
L_{crk}	enclosed-space foundation or wall thickness (cm)
Q_s	convective flow through basement slab (cm³/sec)
η	foundation crack fraction (cm²/cm²)
d_s	depth of affected subsurface soil layer (cm)
τ	averaging time for surface emission vapor flux (years)

Leaching Factor: soil to groundwater

$$LF_{sw,1} = \frac{LF_{pw,gw}}{K_{sw}} \qquad (1)$$

with

$$LF_{pw,gw} = \frac{1}{\left[1 + \frac{U_{gw}\cdot\delta_{gw}}{I\cdot W}\right]}$$

If a finite depth of subsurface soil is specified, in addition

$$VF_{s,amb,2} = \frac{d_s\cdot\rho_s}{I\cdot\tau} \qquad (2)$$

Of equations (1) and (2), choose whichever is less.

LF_{sw}	leaching factor, soil to groundwater (g-soil/cm³-wat)
I	water infiltration rate (cm/year)
K_{sw}	soil to water partition coefficient (cm³-water/g-soil)
$LF_{pw,gw}$	leaching factor, soil pore water / ground water concentration ratio (cm³-water/cm³-water)
U_{gw}	groundwater Darcy velocity (cm/year)
W	width of source-zone area (cm)
δ_{gw}	groundwater mixing zone height (cm)
τ	averaging time for leaching duration (years)
d_s	depth of affected subsurface soil layer (cm)

Effective Diffusion Coefficient in homogeneous soil layers

$$D_{eff} = D_{air}\cdot\left(\frac{\theta_{air}^{3.33}}{\theta_T^2}\right) + D_{water}\cdot\left(\frac{\theta_{water}^{3.33}}{H_{eff}\cdot\theta_T^2}\right)$$

This semi-empirical equation is applicable for estimating the effective diffusion coefficient for porous media in homogeneous soil. It is used here in calculating a value for the vadose zone ($D_{eff,vad}$), the water table capillary fringe region ($D_{eff,cap}$), and in soil-filled enclosed space basement foundation cracks ($D_{eff,crk}$).

D_{eff}	effective diffusion coefficient (cm²/sec)
D_{air}	molecular diffusion coefficient in air (cm²/sec)
D_{water}	molecular diffusion coefficient in water (cm²/sec)
H_{eff}	effective Henry's law coefficient (cm³-water/cm³-air)
θ_{air}	soil air content (cm³-air/cm³-soil)
θ_{water}	soil water content (cm³-water/cm³-soil)
θ_T	soil porosity (cm³-air/cm³-soil), $= \theta_{air} + \theta_{water}$

Refs:

Millington and Quirk (1961): Transactions of the Faraday Society, 57, 1200-1207; Jury, W. A., W. F. Spencer, and W. J. Farmer, 1983: Behavior Assessment Model for Trace Organics in Soil: I. Model Description, Journal of Environmental Quality, 12, 558-564; Bruell, C. J., and G. E. Hoag, 1986: The Diffusion of Gasoline-Range Hydrocarbon Vapors in Porous Media, Experimental Methodologies, in Proceedings of the NWWA/API Conference on Petroleum Hydrocarbons and Organic Chemicals in Groundwater: Prevention, Detection, and Restoration, November 12-14, Houston.

Effective Diffusion Coefficient: averaged value - water table to soil surface

$$D_{eff,ws} = \frac{L_{gw}}{(h_v / D_{eff,vad}) + (h_{cap} / D_{eff,cap})}$$

$D_{eff,ws}$	effective diffusivity - averaged water table to surface (cm²/sec)
$D_{eff,vad}$	effective diffusivity - vadose zone soils (cm²/sec)
$D_{eff,cap}$	effective diffusivity - capillary fringe zone (cm²/sec)
h_v	vadose zone thickness (cm)
h_{cap}	capillary zone thickness (cm)
L_{gw}	depth to groundwater (cm), $L_{gw} = h_v + h_{cap}$

Equation for Calculating Total Soil Concentration to Pore Water Concentration Ratio

$$K_{sw} = \frac{\theta_w + (K_d\cdot\rho_s) + (H_{eff}\cdot\theta_a)}{\rho_s}$$

This equation is used in calculating the ratio between total soil concentration and pore-water concentration for vadose zone soils. Equilibrium partitioning of the chemical between soil (sorbed), pore water, and pore vapors, at concentrations below the pure component pore water (S) or pore vapor saturation limit ($C_{sat,vap}$) is assumed.

K_{sw}	soil to water partition coefficient (cm³-water/g-soil)
$C_{sat,vap}$	saturated vapor concentration (mg/m³-air)
H_{eff}	effective Henry's law coefficient (cm³-water/cm³-air)
K_d	soil (sorbed) / water partition coefficient (cm³-wat/g-soil)
S	pure chemical aqueous solubility limit (mg/L)
ρ_s	dry soil bulk density (g/cm³)
θ_a	soil air content (cm³-air/cm³-soil)
θ_w	soil water content (cm³-water/cm³-soil)

TABLE X3.4 *Continued*

Equation for calculating dispersion factor for ambient (outdoor) air

$$DF_{amb} = \frac{U_{air} \cdot W \cdot \delta_{air}}{A}$$

DF_{amb}	dispersion factor for ambient air (cm/sec), or source emission flux / vapor concentration $(g/cm^2\text{-sec})/(g/cm^3)$
A	source-zone area (cm^2)
U_{air}	ambient air velocity in mixing zone (cm/s)
W	width of source-zone area (cm)
δ_{air}	mixing zone height (cm)

Equation for calculating dispersion factor for enclosed-space (indoor) air

$$DF_{esp} = L_b \cdot ER \cdot \frac{1}{86400 \text{ sec/day}}$$

DF_{esp}	dispersion factor for enclosed-space air (cm/sec), or source emission flux / vapor concentration $(g/cm^2\text{-sec})/(g/cm^3)$
ER	enclosed space air exchange rate (1/day)
L_b	enclosed space volume/infiltration area ratio (cm)

Equation for calculating effective Henry's law coefficient

$$H_{eff} = H' \cdot UF$$

with

$$H' = \frac{H}{R \cdot T_{amb}} \cdot 10^3 \frac{L}{m^3}$$

H_{eff}	effective Henry's law coefficient (cm^3-water/cm^3-air)
H'	Henry's law coefficient (cm^3-water/cm^3-air)
H	Henry's law coefficient (atm-m^3/mol)
R	ideal gas constant (atm-L/g-mol-K), = 0.08206
T_{amb}	ambient temperature (K), = 293
UF	fraction of unionized chemical in water (g-mol/g-mol)

Equation for pressure-driven convective flow through basement slab

$$Q_s = \frac{2 \cdot \pi \cdot dP \cdot k_v \cdot X_{crk}}{\mu_a \cdot \ln\left(\frac{2 \cdot Z_{crk}}{R_{crk}}\right)}$$

with

$$R_{crk} = \frac{A_b \cdot \eta}{X_{crk}}$$

This equation describes volumetric flow through a crack of total length X_{crk}, width R_{crk}, and a cylinderical pressure field and orthogonal flow field which extends into the soil.

Q_s	convective flow through basement slab (cm^3/sec)
A_b	slab area (cm^2)
dP	indoor/outdoor differential pressure (g/cm-s^2)
k_v	soil permeability (cm^2)
X_{crk}	slab perimeter - total crack length (cm)
Z_{crk}	depth to bottom of slab (cm)
μ_{air}	viscosity of air (g/cm-s), = $1.81 \cdot 10^{-4}$
η	foundation crack fraction (cm^2-cracks/cm^2-total area)

Ref: Johnson, P. C. and R. A. Ettinger, 1991: "Heuristic Model for Predicting the Intrusion Rate of Contaminant Vapors into Buildings", Environmental Science & Technology, 25, 1445-1452.

Equation for Calculating Saturated Vapor Concentration

$$C_{sat,vap} = \frac{P_v}{760 \text{ mm Hg/atm}} \cdot \frac{MW}{R\,T_{amb}} \cdot 10^6 \frac{mg/m^3}{g/L}$$

This equation is used in calculating the saturated vapor concentration for a pure chemical constiuent. Saturated vapor concentration is the highest concentration achievable in air for a pure chemical constituent under thermodynamic equilibrium at a specified temperature. Saturated vapor concentration is a defined thermodynamic property.

$C_{sat,vap}$	saturated vapor concentration (mg/m^3-air)
MW	molecular weight (g/g-mol)
P_v	saturated vapor pressure (mm Hg)
R	ideal gas constant (atm-L/g-mol-K), = 0.08206
T_{amb}	ambient temperature (K), = 293

Equation for soil / water partition coefficient

For organic chemicals,

$$K_d = K_{oc} \cdot f_{oc}$$

K_d	soil (sorbed) / water partition coefficient (cm^3-wat/g-soil)
f_{oc}	mass fraction of organic carbon in soil (g-oc/g-soil)
K_{oc}	organic carbon / water partition coefficient (L-water/kg-oc)

for inorganic chemicals, K_d is specified directly.

Equation for Calculating Unionized Fraction of Ionizing Chemicals

For acids,

$$UF = \frac{1}{1 + 10^{pH - pK_a}}$$

for bases,

$$UF = \frac{1}{1 + 10^{14 - pH - pK_b}}$$

otherwise,

$$UF = 1$$

For chemicals capable of forming ions in the normal soil/water pH range, the unionized fraction of the chemical in water is calculated. Only the unionized chemical is available for volatilization. Some chemicals may form multiple-charged ions (X++ or X--, etc.). Only singly-charged ions are included in the current equations.

UF	fraction of unionized chemical in water (g-mol/g-mol)
pH	Soil/water pH
pK_a	acid equilibrium constant (log10(g-mol/g-mol))
pK_b	base ionization equilibrium constant(log10(g-mol/g-mol))

Ref:
Values of pK_a and pK_b as defined in Waucoupe, 1992: Reviews of Environmental Contamination and Toxicology, 123, 1-155.
Aalund, Leo R., 1986: "One-Step Method Determines Sour Water H_2S Hazard, Oil and Gas Journal, 24 February, 55 - 58.

TABLE X3.4 *Continued*

Equation for Calculating Residual Soil Concentration
$C_{sat,soil} = S \cdot K_{sw}$
For a pure chemical constituent, $C_{sat,soil}$ is a value at and above which the chemical is present in soil pore water at its aqueous solubility limit and is present in soil pore air at its saturated vapor concentration. This value may also be calculated as
$$C_{sat,soil} = \frac{C_{sat,vap}}{H' \cdot (10^3\ L/m^3)} \cdot K_{sw}$$
$C_{sat,soil}$ residual soil concentration (mg/kg)
K_{sw} soil to water partition coefficient (cm^3-water/g-soil)
S pure chemical aqueous solubility limit (mg/L)
$C_{sat,vap}$ saturated vapor concentration (mg/m^3-air)
H' Henry's law coefficient (cm^3-water/cm^3-air)

X3.2 *Air-Inhalation of Vapors (Outdoors/Indoors)*:

X3.2.1 In this case chemical intake results from the inhalation of vapors. It is assumed that vapor concentrations remain constant over the duration of exposure, and all inhaled chemicals are absorbed. Equations appearing in Table X3.3 for estimating RBSLs for vapor concentrations in the breathing zone follow guidance given in Ref. **(15)**. Should the example RBSL exceed the saturated vapor concentration for any individual component, "P<" is entered in the table to indicate that the selected risk level or hazard quotient cannot be reached or exceeded for that chemical and the specified exposure scenario (unless an aerosol of the chemical is inhaled).

X3.2.2 A target vapor concentration other than the RBSL for inhalation may be selected and used as an alternate inhalation criteria (e.g. PEL-TWA or TLV-TWA workplace exposure criteria as listed in Table X3.1, or odor nuisance level). The reader should review the restrictions and assumptions embodied in these alternate criteria for appropriateness before their use as Tier 1 air inhalation screening criteria.

X3.3 *Ground Water-Ingestion of Ground Water*:

X3.3.1 In this case chemical intake results from ingestion of ground water. It is assumed that the dissolved chemical concentrations remain constant over the duration of exposure. Equations appearing in Table X3.3 for estimating RBSLs for drinking water concentrations follow guidance given in Ref. **(15)** for ingestion of chemicals in drinking water. Should the example RBSL exceed the pure component aqueous solubility limit for any component, "S<" is entered in the table to indicate the selected risk level or hazard quotient cannot be reached for that compound and the specified exposure scenario (unless immiscible free-phase chemical is mixed with the ingested water).

X3.3.2 A target concentration other than the RBSL for ground water ingestion may be selected and used as an alternate ground water criteria (e.g., Federal drinking water MCL criteria for a public water supply as listed in Table X3.1, taste or odor threshold, or ecological criteria). Determining and validating the basis of these alternate criteria is necessary before their use as Tier 1 ground water ingestion screening criteria.

X3.4 *Ground Water-Inhalation of Outdoor Vapors*:

X3.4.1 In this case chemical intake is a result of inhalation of outdoor vapors which originate from dissolved chemicals in ground water located some distance below ground surface. Here the goal is to determine the RBSL for the chemical dissolved in ground water that corresponds to the target RBSL for outdoor vapors in the breathing zone, as given in Table X3.3. A target vapor concentration value other than the RBSL for inhalation (as discussed in X3.2.2) may be selected and substituted for the $RBSL_{air}$ parameter appearing in the equations given in Table X3.3.

X3.4.2 A conceptual model for the transport of chemicals from ground water to ambient air is depicted in Fig. X3.1. Values for the parameters listed in this figure are specified in Table X3.2. For simplicity, the relationship between outdoor air and dissolved ground water concentrations is represented in Table X3.3 by the "volatilization factor", $VF_{gw,amb}$ [(mg/cm^3-air)/(mg/cm^3-water)], defined in Table X3.4. It is based on the following assumptions:

X3.4.2.1 A constant dissolved chemical concentration in ground water,

X3.4.2.2 Linear equilibrium partitioning between dissolved chemicals in ground water and chemical vapors at the ground water table,

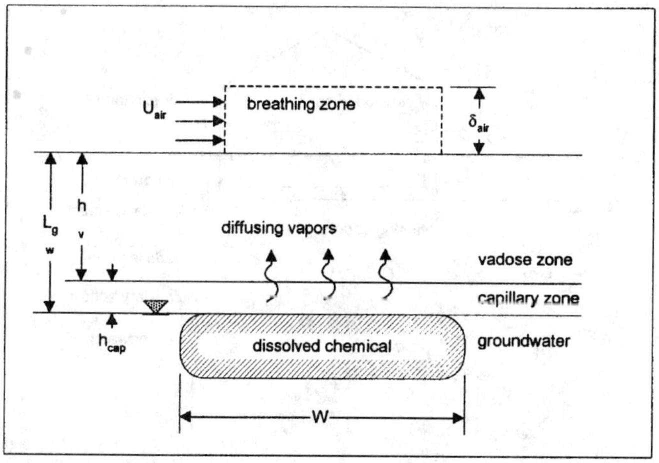

FIG. X3.1 Volatilization from Ground Water to Ambient Air

X3.4.2.3 Steady-state vapor- and liquid-phase diffusion through the capillary fringe and vadose zones to ground surface,

X3.4.2.4 No loss of chemical as it diffuses toward ground surface (that is, no biodegradation), and

X3.4.2.5 Steady well-mixed atmospheric dispersion of the emanating vapors within the breathing zone as modeled by a "box model" for air dispersion.

X3.4.3 Should the example $RBSL_{gw}$ exceed the pure component solubility for any individual component, "S<" is entered in the table to indicate that the selected risk level or hazard quotient cannot be reached or exceeded for that compound and the specified exposure scenario.

X3.5 *Ground Water-Inhalation of Enclosed-Space (Indoor) Vapors*:

X3.5.1 In this case chemical intake results from the inhalation of vapors in enclosed spaces. The chemical vapors originate from dissolved chemicals in ground water located some distance below ground surface. Here the goal is to determine the RBSL for the chemical dissolved in ground water which corresponds to the target RBSL for vapors in the breathing zone as given in Table X3.3. A target vapor concentration value other than the RBSL for inhalation (as discussed in X3.2.2) may be substituted for the $RBSL_{air}$ parameter appearing in the equations given in Table X3.3.

X3.5.2 A conceptual model for the transport of chemicals from ground water to indoor air is depicted in Fig. X3.2. Values for the parameters listed in this figure are specified in Table X3.2. For simplicity, the relationship between enclosed-space air and dissolved ground water concentrations is represented in Table X3.3 by the "volatilization factor" $VF_{gw,esp}$ [(mg/cm^3-air)/(mg/cm^3-water)] defined in Table X3.4. It is based on the following assumptions:

X3.5.2.1 A constant dissolved chemical concentration in ground water,

X3.5.2.2 Equilibrium partitioning between dissolved chemicals in ground water and chemical vapors at the ground water table,

X3.5.2.3 Steady-state vapor- and liquid-phase diffusion through the capillary fringe, vadose zone, and foundation cracks,

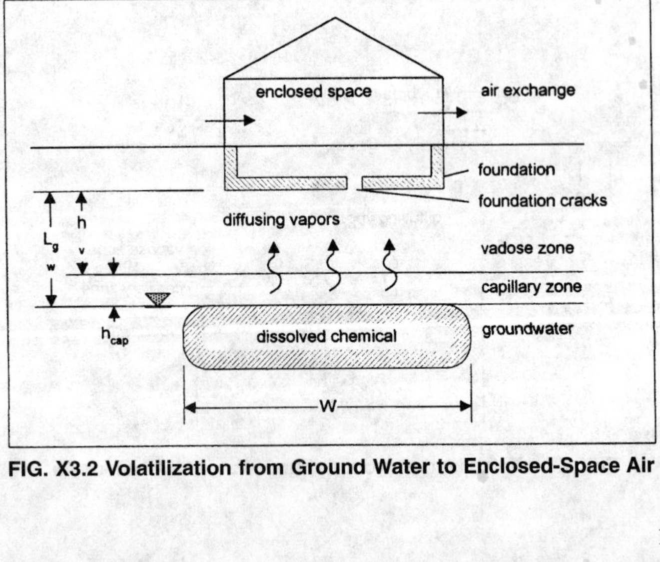

FIG. X3.2 Volatilization from Ground Water to Enclosed-Space Air

X3.5.2.4 No loss of chemical as it diffuses toward ground surface (e.g., no biodegradation), and

X3.5.2.5 Steady, well-mixed atmospheric dispersion of the emanating vapors within the enclosed space, where the convective transport into the building through foundation cracks or openings is negligible in comparison with diffusive transport.

X3.5.3 Should the example $RBSL_{gw}$ exceed the pure component solubility for any individual component, "S<" is entered in the table to indicate that the selected risk level or hazard quotient cannot be reached or exceeded for that compound and the specified exposure scenario.

X3.6 *Surficial Soils-Ingestion, Dermal Contact, and Vapor and Particulate Inhalation*:

X3.6.1 In this case it is assumed that chemical intake results from a combination of intake routes, including: ingestion, dermal absorption, and inhalation of both particulates and vapors emanating from surficial soil.

X3.6.2 Equations used to estimate intake resulting from ingestion follow guidance given in Ref. **(15)** for ingestion of chemicals in soil. For this route, it has been assumed that surficial soil chemical concentrations and intake rates remain constant over the exposure duration. Other values for exposure parameters (that is, child exposure or age-adjusted exposure) should also be considered for ingestion of soil.

X3.6.3 Equations used to estimate intake resulting from dermal absorption follow guidance given in Ref. **(15)** for dermal contact with chemicals in soil. For this route, it has been assumed that surficial soil chemical concentrations and absorption rates remain constant over the exposure duration. Other values for exposure parameters (that is, child exposure or age-adjusted exposure) should also be considered for dermal contact with soil.

X3.6.4 Equations used to estimate intake resulting from the inhalation of particulates follow guidance given in Ref. **(15)** for inhalation of airborne chemicals. For this route, it has been assumed that surficial soil chemical concentrations, intake rates, and atmospheric particulate concentrations remain constant over the exposure duration.

X3.6.5 Equations used to estimate intake resulting from the inhalation of airborne chemicals which have volatilized from surficial soils follow guidance given in Ref. **(15)** for inhalation of airborne chemicals.

X3.6.6 A conceptual model for the volatilization of chemicals from surficial soils to outdoor air is depicted in Fig. X3.3. Values for the parameters listed in this figure are specified in Table X3.2. For simplicity, the relationship between outdoor air and surficial soil concentrations is represented in Table X3.3 by the "volatilization factor" VF_{ss} [(mg/cm^3-air)/(mg/g-soil)] defined in Table X3.4. It is based on the following assumptions:

X3.6.6.1 Uniformly distributed chemical throughout the depth range of 0-L_{ss} (cm) below ground surface,

X3.6.6.2 Linear equilibrium partitioning within the soil matrix between sorbed, dissolved, and vapor phases, where the partitioning is a function of constant chemical- and soil-specific parameters.

X3.6.6.3 Diffusion through the surficial soil layer,

X3.6.6.4 No loss of chemical as it diffuses toward ground surface (e.g., no biodegradation), and

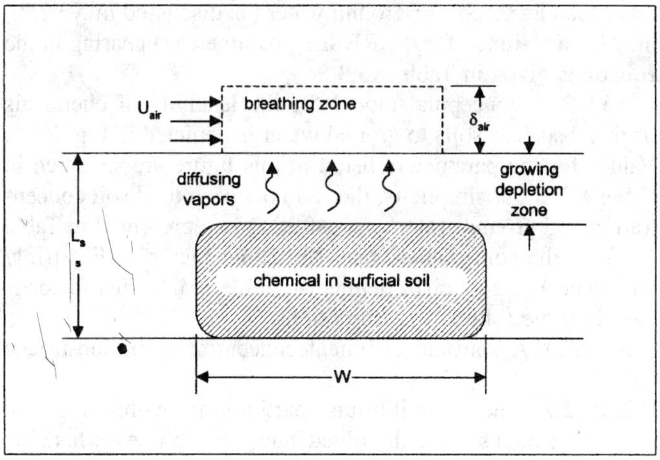

FIG. X3.3 Volatilization from Surficial Soils

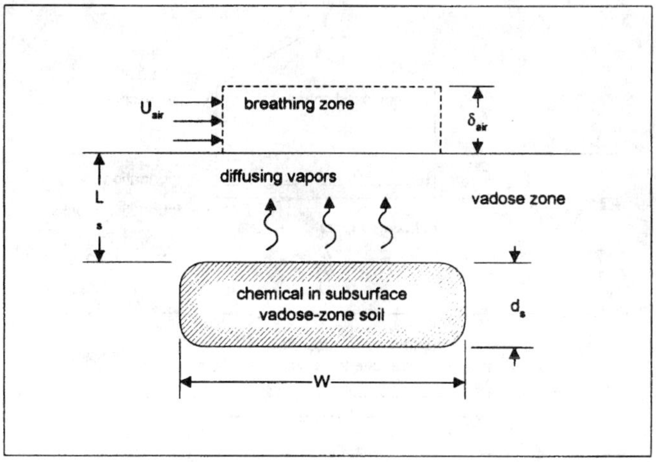

FIG. X3.4 Volatilization from Subsurface Soils to Ambient Air

X3.6.6.5 Steady well-mixed atmospheric dispersion of the emanating vapors within the breathing zone as modeled by a "box model" for air dispersion.

X3.6.7 In the event that the time-averaged flux exceeds that which would occur if all chemical initially present in the surficial soil zone volatilized during the exposure period, then the volatilization factor is determined from a mass balance assuming that all chemical initially present in the surficial soil zone volatilizes during the exposure period.

X3.6.8 Equations for calculating the apportionment of chemical from this presumed combination of surficial soil intake routes, including ingestion, dermal absorption, inhalation of particulates, and inhalation of vapors emanating from surficial soil, are specified in Table X3.3. The calculated values are included in Table X3.1.

X3.7 *Subsurface Soils-Inhalation of Outdoor Vapors*:

X3.7.1 In this case chemical intake is a result of inhalation of outdoor vapors which originate from chemicals contained in subsurface soils located some distance below ground surface. Here the goal is to determine the RBSL for subsurface soils that corresponds to the target RBSL for outdoor vapors in the breathing zone, as given in X3.2. A target vapor concentration other than the RBSL for inhalation (as discussed in X3.2.2) may be substituted for the $RBSL_{air}$ parameter appearing in the equations given in Table X3.3.

X3.7.2 A conceptual model for the transport of chemicals from subsurface soils to ambient air is depicted in Fig. X3.4. Values for the parameters listed in this figure are specified in Table X3.2. For simplicity, the relationship between outdoor air and soil concentration is represented in Table X3.3 by the "volatilization factor", $VF_{s,amb}$ [(mg/cm^3-air)/(mg/g-soil)], defined in Table X3.4. It is based on the following assumptions:

X3.7.2.1 A constant chemical concentration in subsurface soils.

X3.7.2.2 Linear equilibrium partitioning within the soil matrix between sorbed, dissolved, and vapor phases, where the partitioning is a function of constant chemical- and soil-specific parameters,

X3.7.2.3 Steady-state vapor- and liquid-phase diffusion through the vadose zone to ground surface,

X3.7.2.4 No loss of chemical as it diffuses toward ground surface (that is, no biodegradation), and

X3.7.2.5 Steady well-mixed atmospheric dispersion of the emanating vapors within the breathing zone as modeled by a "box model" for air dispersion.

X3.7.3 Should the example RBSL exceed the value for which the equilibrated vapor and dissolved pore-water phases become saturated, $C_{sat,soil}$ [mg/kg-soil] (see Table X3.4 for calculation of this value), "Rs" is entered into the table to indicate the selected risk level or hazard quotient cannot be reached or exceeded for that compound and the specified exposure scenario (even if free-phase product or precipitate is present in the soil).

X3.7.4 No mass balance is imposed on the volatilization factor $VF_{s,amb}$.

X3.7.4.1 If the depth of the affected subsurface soil layer, d_s, can be specified, a mass balance can be imposed.

X3.7.4.2 In the event that the time-averaged flux exceeds that which would occur if all the chemical initially present in the affected subsurface soil layer volatilized during the exposure period, then the volatilization factor can be determined from a mass balance assuming that all chemical present in the subsurface soil layer volatilizes during the exposure period.

X3.8 *Subsurface Soils-Inhalation of Enclosed-Space (Indoor) Vapors*:

X3.8.1 In this case chemical intake is a result of inhalation of enclosed-space vapors which originate from chemicals contained in subsurface soils located some distance below ground surface. Here the goal is to determine the RBSL for subsurface soils that corresponds to the target RBSL for indoor vapors, as given in Table X3.3. A target vapor concentration other than the RBSL for inhalation (as discussed in X3.2.2) may be substituted for the $RBSL_{air}$ parameter appearing in Table X3.3.

X3.8.2 A conceptual model for the transport of chemicals from subsurface soils to enclosed spaces is depicted in Fig. X3.5. Values for the parameters listed in this figure are specified in Table X3.2. For simplicity, the relationship between indoor air and soil concentrations is represented in Table X3.3 by the "volatilization factor", $VF_{s,esp}$ [(mg/cm^3-air)/(mg/

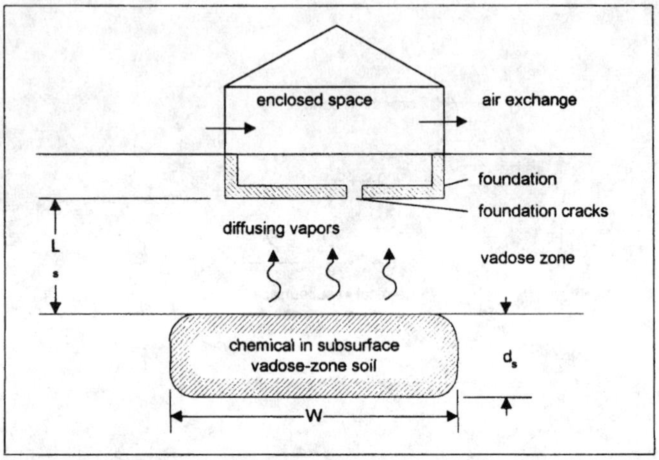

FIG. X3.5 Volatilization from Subsurface Soils to Enclosed-Space Air

g-soil)], defined in Table X3.4. It is based on the following assumptions:

X3.8.2.1 A constant chemical concentration in subsurface soils,

X3.8.2.2 Linear equilibrium partitioning within the soil matrix between sorbed, dissolved, and vapor phases, where the partitioning is a function of constant chemical- and soil-specific parameters,

X3.8.2.3 Steady-state vapor- and liquid-phase diffusion through the vadose zone and foundation cracks,

X3.8.2.4 No loss of chemical as it diffuses toward the indoor space (that is, no biodegradation), and

X3.8.2.5 Well-mixed atmospheric dispersion of the emanating vapors within the enclosed space.

X3.8.3 Should the example RBSLs exceed the value for which the equilibrated vapor and dissolved pore-water phases become saturated, $C_{sat,soil}$ [mg/kg-soil] (see Table X3.4 for calculation of this value), "Rs" is entered into the table to indicate the selected risk level or hazard quotient cannot be reached or exceeded for that compound and the specified exposure scenario (even if free-phase product or precipitate is present in the soil).

X3.8.4 No mass balance is imposed on the volatilization factor $VF_{s,esp}$.

X3.8.4.1 If the depth of the affected subsurface soil layer, d_s, can be specified, a mass balance can be imposed.

X3.8.4.2 In the event that the time-averaged flux exceeds that which would occur if all the chemical initially present in the affected subsurface soil layer volatilized during the exposure period, then the volatilization factor can be determined from a mass balance assuming that all chemical present in the subsurface soil layer volatilizes during the exposure period.

X3.9 *Subsurface Soils-Leaching to Ground Water*:

X3.9.1 In this case chemical intake is a result of chemicals leaching from subsurface soils, followed by inhalation of enclosed-space vapors, inhalation of outdoor vapors, or ingestion of ground water as discussed in X3.1 through X3.3. Here the goal is to determine the RBSL for subsurface soils that correspond to the target RBSLs for the inhalation or ground water ingestion routes. A target ground water concentration other than an RBSL for ground water (as discussed in X3.2.2), may be substituted for the $RBSL_{gw}$ parameter appearing in the equations given in Table X3.3.

X3.9.2 A conceptual model for the leaching of chemicals from subsurface soils to ground water is depicted in Fig. X3.6. Values for the parameters listed in this figure are specified in Table X3.2. For simplicity, the ratio between total soil concentration and ground water concentration is represented in Table X3.3 by the soil/ground water "leaching factor", LF_{sw} [(mg/cm^3-water)/(mg/g-soil)], defined in Table X3.4. This factor is based on the following assumptions:

X3.9.2.1 A constant chemical concentration in subsurface soils,

X3.9.2.2 Linear equilibrium partitioning within the soil matrix between sorbed, dissolved, and vapor phases, where the partitioning is a function of constant chemical- and soil-specific parameters,

X3.9.2.3 Steady-state leaching from the vadose zone to ground water resulting from the constant leaching rate I [cm/s],

X3.9.2.4 The contribution of vapor- and liquid-phase diffusion through the vadose zone is negligible in comparison to the transport due to infiltration,

X3.9.2.5 No loss of chemical as it leaches toward ground water (e.g., no biodegradation), and,

X3.9.2.6 Steady well-mixed dispersion of the leachate within a ground water "mixing zone".

X3.9.3 Should the example RBSLs exceed the value for which the equilibrated vapor and dissolved pore-water phases become saturated (see Table X3.4 for calculation of this value), "Rs" is entered in the table to indicate that the selected risk level or hazard quotient cannot be reached or exceeded for that compound and the specified exposure scenario (even if free-phase product or precipitate is present in the soil).

X3.9.4 In some regulatory programs, "dilution attenuation factors" (DAFs) are currently being proposed based on fate and transport modeling results. A DAF is typically defined as the ratio of a target ground water concentration divided by the source leachate concentration. This is identical in definition to the ground water/soil pore water leaching factor, $LF_{pw,gw}$, defined in Table X3.4 as a component of the soil/ground water leaching factor, LF_{sw}.

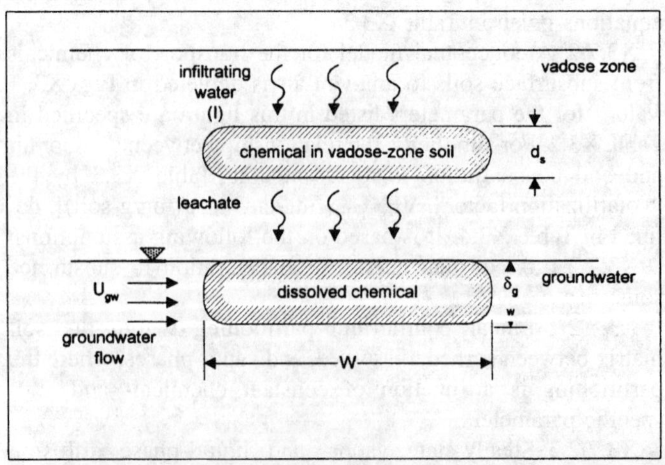

FIG. X3.6 Leaching from Subsurface Soils to Ground Water

X3.9.5 No mass balance is imposed on the leaching factor, LF_{sw}.

X3.9.5.1 If the depth of the affected subsurface soil layer, d_s, can be specified, a mass balance can be imposed.

X3.9.5.2 2 In the event that the time-averaged leaching rate exceeds that which would occur if all the chemical initially present in the affected soil layer leached during the exposure period, then the leaching factor can be determined from a mass balance assuming that all chemical present in the soil layer leached to ground water during the exposure period.

X3.10 *Free-phase Liquid Migration in Unsaturated Soils*:

X3.10.1 In this case our concern is in the further migration of nonaqueous phase liquid (NAPL) chemical in unsaturated soil, from a source zone toward the water table and on towards a potential receptor, or from a source zone toward a building basement, sump, or utility conduit, and on towards a potential receptor.

X3.10.2 The presence of a pure-chemical NAPL may be estimated using the equation for $C_{sat,soil}$ and K_{sw} in Table X3.4 with parameters for unsaturated soil. For mixtures of chemicals, $C_{sat,soil}$ is irrelevant for predicting the presence or absence of a NAPL. The presence or absence of a NAPL consisting of a liquid mixture of immiscible chemicals may be estimated using calculations presented by Johnson et al. **(22)**, Mott **(23)**, or Mariner **(24)**.

X3.10.3 A NAPL may be present in unsaturated soil, but immobile. The mobility of a NAPL is not governed by thermodynamic properties, but by capillary, viscous, and gravity forces acting on the bulk NAPL phase. Wilson and Conrad **(25)**, Conrad et al. **(26)**, and Cohen and Mercer **(27)** have compiled experimental data on measured levels of immobile NAPL for a range of chemicals and soil types. This immobile NAPL level is specified as residual saturation, S_r, which is the fraction of available soil pore volume taken up by the immobile NAPL.

X3.10.4 We calculate an example $RBSL_s$ for pure chemicals using the equation specified in Table X3.3. It is based on the following assumptions:

X3.10.4.1 A specified value of S_r consistent with sandy soil conditions and the prior references (X3.10.3),

X3.10.4.2 A local measurement of chemical concentration in soil,

X3.10.4.3 Linear equilibrium partitioning within the soil matrix between sorbed, dissolved, vapor, and immiscible phases, where the partitioning is a function of constant chemical- and soil-specific parameters.

X3.10.4.4 A pure chemical immiscible phase,

X3.10.4.5 Displacement of the air-phase soil void volume by the NAPL volume, and,

X3.10.4.6 The pure chemical is presumed to be a liquid at standard conditions of temperature and pressure.

X3.10.5 The $RBSL_s$ value is not applicable for chemicals which are infinitely miscible in water. No NAPL will exist in such a case.

X3.10.6 For mixtures of chemicals, the pure chemical $RBSL_s$ value is not relevant.

X3.10.6.1 The $RBSL_s$ value for a chemical mixture may be estimated using calculations presented by Johnson et al. **(22)**, Mott **(23)**, or Mariner **(24)**, and a specified residual saturation level, S_r or,

X3.10.6.2 The equation specified in Table X3.3 may be conservatively simplified, using the bulk density of the liquid chemical mixture, and neglecting the sorbed, dissolved, and vapor phase contribution to soil concentration.

X3.10.7 The calculated $RBSL_s$ value is intended to be conservative with respect to presumed conditions.

X3.10.7.1 The applicable value of S_r may be smaller for soils which are more course grained (for example, gravel) than the presumed sandy soil. The applicable value of S_r may be much greater for soils which are finer grained (for example, silt and clay) than the presumed sandy soil.

X3.10.7.2 The selected S_r value is intended to be conservative with respect to chemical type and properties. A larger S_r value may be applicable (for example, with some non-Newtonian fluids).

X3.10.8 Site characterization is a necessary component of determining the mobility or immobility of a NAPL in unsaturated soils. More information on estimating the presence and mobility of a NAPL is given in API **(28)** and USEPA **(29)**.

X3.10.9 Use of the example RBSL concentrations in this appendix for leaching from soils to ground water (X3.9) presumes that further migration of NAPL in unsaturated soil from a source zone toward the water table is insignificant.

X3.10.10 The example RBSL concentrations in this appendix regarding inhalation of indoor or outdoor vapors (X3.7 and X3.8) originating from soils are conservative with respect to the presence or absence of a NAPL phase, unless that NAPL phase has migrated near or into a building basement, sump, or utility conduit that may shorten the diffusive vapor transport pathway.

X3.11 *Free-phase Liquid Migration in Saturated Soils*:

X3.11.1 In this case our concern is in the further migration of non-aqueous phase liquid (NAPL) chemical in saturated soil, from a source zone toward a potential receptor.

X3.11.1.1 The presence of a NAPL may be detected or confirmed through the use of ground water monitoring wells.

X3.11.1.2 The presence of a pure-chemical NAPL may be estimated using the equation for $C_{sat,soil}$ and K_{sw} in Table X3.4 using parameters for saturated soil, including $\theta_w = \theta_T$. For mixtures of chemicals, $C_{sat,soil}$ is irrelevant for predicting the presence or absence of a NAPL. The presence or absence of a NAPL consisting of a liquid mixture of immiscible chemicals may be estimated using algebraic calculations presented by Johnson et al. **(22)**, Mott **(23)**, or Mariner **(24)**.

X3.11.2 As for unsaturated soils (X3.10.3), a NAPL may be present in saturated soil, but immobile. The mobility of a NAPL is not governed by thermodynamic properties, but by capillary, viscous, and gravity forces acting on the bulk NAPL phase. Wilson and Conrad **(25)** and Conrad, et al., **(26)** have measured levels of immobile NAPL at values of residual saturation, S_r, which are generally greater than that for similar unsaturated soils.

X3.11.3 We may calculate an example $RBSL_s$ for pure chemicals in saturated soil using the equation specified in Table X3.3. This is based on the following assumptions:

X3.11.3.1 A specified value of S_r consistent with sandy soil conditions and the prior references (X3.10.3),

X3.11.3.2 A local measurement of chemical concentration in soil,

X3.11.3.3 Linear equilibrium partitioning within the soil matrix between sorbed, dissolved, and immiscible phases, where the partitioning is a function of constant chemical- and soil-specific parameters,

X3.11.3.4 A pure chemical immiscible phase,

X3.11.3.5 Displacement of the water-phase soil void volume by the NAPL volume, and,

X3.11.3.6 The pure chemical is a liquid at standard conditions of temperature and pressure.

X3.11.4 The $RBSL_s$ value is not applicable for chemicals which are infinitely miscible in water. No NAPL will exist in such a case.

X3.11.5 For mixtures of chemicals, the pure chemical $RBSL_s$ value is not relevant.

X3.11.5.1 The $RBSL_s$ value for a chemical mixture in saturated soil may be estimated using calculations presented Johnson et al. **(22)**, Mott **(23)**, or Mariner **(24)**, and a specified residual saturation level, S_r or,

X3.11.5.2 The equation specified in Table X3.3 may be conservatively simplified, using the bulk density of the liquid chemical mixture, and neglecting the sorbed and dissolved phase contribution to soil concentration.

X3.11.6 The calculated $RBSL_s$ value is intended to be conservative with respect to presumed conditions.

X3.11.6.1 The applicable value of S_r may be smaller for soils which are more course grained (for example, gravel) than the presumed sandy soil. The applicable value of S_r may be much greater for soils which are finer grained (for example, silt and clay) than the presumed sandy soil.

X3.11.6.2 The selected S_r value is intended to be conservative with respect to chemical type and properties. A larger S_r value may be applicable (for example, with some non-Newtonian fluids).

X3.11.7 Site characterization is a necessary component of determining the mobility or immobility of a NAPL in saturated soils. More information on estimating the presence and mobility of a NAPL is given in API **(28)** and USEPA **(29)**.

X3.11.8 Use of the example RBSL concentrations in this appendix for ingestion of ground water (X3.3) or inhalation of indoor or outdoor vapors (X3.4 and X3.5) originating from ground water presumes that further migration of NAPL in saturated soil from a source zone toward a potential receptor is insignificant.

X3.12 *Parameter Values*:

X3.12.1 The equations in Table X3.3 refer to intermediate parameters (volatilization factors, leaching factors, dispersion factors, and so forth). Equations for these intermediate parameters are specified in Table X3.4. The chemical-specific calculated values for these intermediate parameters are given in Table X3.1.

X3.12.2 The IRIS- and HEAST-referenced slope factor and reference dose values, the OSHA PEL-TWA and ACGIH TLV-TWA concentration values, and the Federal MCL values cited in Table X3.1 are the most recently updated values available prior to the writing of this appendix. They are cited at their last revised date.

X3.12.3 Table X3.2 lists exposure parameters used to calculate the example RBSLs (Table X3.1). All values given are based on adult male exposures only. With the exception of the dermal exposure parameters (SA, M, and RAF_d), the values given are reasonable maximum exposure (RME) values presented in Ref. **(30)** and are either upper bound estimates or average estimates for each individual exposure parameter, as appropriate. Other values for exposure parameters (that is, child exposure or age-adjusted exposure) may be selected and substituted for the parameters appearing in the equations given in Tables X3.3 and X3.4.

X3.12.4 The skin surface area, SA = 3160 cm^2/day, is based on the seasonally-averaged surface area of the head, hands, and forearms for adult males given in Ref. **(30)**. The soil-to-skin adherence factor, M [mg/cm^2], and dermal relative absorption factor, RAF_d [mg-absorbed/mg-applied], are based on guidance issued by Ref. **(31)**. Other values for skin surface area (that is, child or age-adjusted values) may be selected and substituted for the parameters appearing in the equations given in Tables X3.3 and X3.4.

X3.12.5 Soil properties are based on typical values for sandy soils and are consistent with values given in Ref. **(32)**. With respect to chemical vapor and aqueous transport through the soil matrix, RBSL values calculated for assumed sandy soil conditions will lead to overpredictions of actual exposure in cases where the actual soil texture is finer grained than the presumed sandy soil (e.g. for loam or clay).

X3.12.6 Physical dimensions are consistent with the scale of a moderate-sized chemical release site (approximately 0.5 acre).

X3.12.7 Particulate emission rates were estimated by the approach presented by Cowherd, et al. **(33)**. It was assumed that the mode of the surficial soil size distribution was 2 mm, the erosion potential was unlimited, there was no vegetative cover or non-erodable elements, and the mean annual wind speed was 4 m/s.

X3.12.8 The chemical-specific parameters used are defined in Table X3.2. Values are specified in Table X3.1.

X3.12.8.1 Linear equilibrium partitioning between soil pore water, soil pore air, and soil sorbed phases is presumed at soil pore vapor concentrations below the saturated vapor limit, $C_{sat,vap}$, defined in Table X3.4, and at pore water concentrations below the aqueous solubility limit, S, defined in Table X3.2. Under these conditions volumetric fractions of a chemical partitioned between each of these phases within the soil matrix are defined through mass and species conservation by the following equations:

$$FRAC_{vapor}\left[\frac{cm^3-vapor}{cm^3-soil}\right]=\frac{H_{eff}\cdot\theta_a}{H_{eff}\cdot\theta_a+K_d\cdot\rho_s+\theta_w}\quad(X3.5)$$

$$FRAC_{water}\left[\frac{cm^3-water}{cm^3-soil}\right]=\frac{\theta_w}{H_{eff}\cdot\theta_a+K_d\cdot\rho_s+\theta_w}\quad(X3.6)$$

$$FRAC_{sorbed}\left[\frac{cm^3-sorbed}{cm^3-soil}\right]=\frac{K_d\cdot\rho_s}{H_{eff}\cdot\theta_a+K_d\cdot\rho_s+\theta_w}\quad(X3.7)$$

Parameters in these equations are defined in Table X3.2. For individual chemicals these fractions are plotted in Table X3.1 as a function of the soil saturation ratio, θ_w/θ_T. This series of plots illustrate the diverse range of partitioning behavior for the selection of chemicals presented in this appendix. Given a convective or diffusive driving force, those chemicals with a higher mass fraction in pore water or pore air are expected to be transported at a higher phase velocity through the soil matrix. Presumed vadose-zone conditions correspond to a single value of the soil saturation ratio, $\theta_w/\theta_T = 0.12/0.38$, on this plot. Saturated soil corresponds to a value of $\theta_w/\theta_T = 1$.

X3.12.8.2 Vapor- and liquid-phase diffusive transport through soils resulting in air emissions, under conditions of linear equilibrium partitioning between soil pore water, pore air, and soil sorption, is governed by a single group of chemical- and soil-specific terms as indicated in Table X3.4 for the volatilization factor for surficial soils to ambient air. This group of terms, $(H_{eff} \cdot D_{eff} / K_{sw} \cdot \rho_s)$ [cm²/sec], is plotted in Table X3.1 as a function of the soil saturation ratio, θ_w/θ_T. This series of plots illustrate, for a single chemical, the sensitivity of diffusive transport to changes in soil moisture. Each of two terms for the effective diffusion coefficient (from Table X3.4) is plotted, illustrating the relative contribution of water- and air-phase diffusive transport. For the selection of chemicals presented in this appendix, the wide range in magnitude for $(H_{eff} \cdot D_{eff} / K_{sw} \cdot \rho_s)$ values illustrates the chemical-specific variation in diffusive transport through soils. Presumed vadose-zone conditions correspond to a single value of the soil saturation ratio, $\theta_w/\theta_T = 0.12/0.38$, on each of the plots.

X3.12.8.3 Chemical-specific water- and air-phase thermodynamic parameters are presumed consistent in definition, based on the equation:

$$H\left[\frac{\text{atm}-\text{m}^3}{\text{g}-\text{mol}}\right] = \frac{P_v\ [\text{mm Hg}]\ MW\ [\text{g/g}-\text{mol}]}{S\ [\text{mg/L}]} \cdot \frac{\text{atm}}{760\ \text{mm Hg}} \cdot \frac{\text{mg}}{\text{g}}\frac{\text{m}^3}{\text{L}} \tag{X3.8}$$

Variability, especially for extremely low measured values of aqueous solubility, S, and vapor pressure, P_v, is expected. For ionic chemical species, the unionized fraction of chemical in water is used in the definition of Henry's law coefficient, H.

X3.12.8.4 The pure chemical aqueous solubility limit is an upper bound for most organic chemical concentrations dissolved in water. For chemicals which may form ions in water including acids, bases, and salts, the pure component aqueous solubility limit is not necessarily an upper bound solubility limit. For inorganic species in this appendix, the aqueous solubility limit will depend on the soil/water pH, the chemical oxidation state, and the solution matrix.

X3.12.8.5 The organic carbon/water partition coefficient, K_{oc}, is relatively insensitive to soil type for non-ionizing organic chemicals in moist soils. For organic chemicals which form ions within the normal soil pH range, including acids, bases, and salts, K_{oc} will vary with soil/water pH, the chemical oxidation state, and the composition of the soil matrix. The soil/water partition coefficient, K_d, for inorganic chemicals, including acids, bases, and salts, will vary with the soil/water pH, the chemical oxidation state, and the composition of the soil matrix.

X3.12.9 In this development, surficial soils are defined as those soils present within 1 m of ground surface. Subsurface soil RBSLs are based on assumed source depths of 1 m. Ground water is assumed to be located 3 m below ground surface.

X3.12.10 Once again, the reader is reminded that the parameter (and corresponding example RBSL) values and selected chemicals are presented here as examples only, and are not intended to be used as standards. At best, the parameters are reasonable values based on current information and professional judgment. The reader should review and verify all assumptions prior to using any of the example RBSLs as screening values.

X4. THE ROLE OF MATHEMATICAL MODELING IN THE RISK-BASED CORRECTIVE ACTION PROCESS

X4.1 *Introduction:*

X4.1.1 In the context of Risk-Based Corrective Action (RBCA), mathematical "models" are any of the calculations that are used in the various steps of the RBCA process ultimately to determine corrective action goals, for example, Risk-Based Screening Levels (RBSL), Site-Specific Target Levels (SSTL), and to develop a strategy for achieving these goals. These models are most typically associated with assessing the fate and transport of chemicals in the environment, that is, they are used to estimate the direction, rate, and extent of chemical migration. However, "models" are also used to estimate chemical uptake that result from environmental exposure and to interpret, interpolate, extrapolate, and assess the consistency of site characterization data, historical monitoring data, and toxicological information. Models may play an important role in the design of remedial action systems and monitoring programs (for example, well locations, density, and sampling frequency). Therefore, mathematical modeling is a valuable tool that can provide information to the RBCA process.

X4.1.2 Refer to Guide D 5880 and D 5718 for more information relative to the proper application of mathematical models used in fate and transport assessments.

X4.2 *Uses of Mathematical Modeling:*

X4.2.1 Mathematical model results play a key role in many phases of RBCA, including:

X4.2.1.1 Determining the potential urgency of response based on estimated migration and attenuation rates of chemical(s) of concern,

X4.2.1.2 Developing Tier 1 RBSL,

X4.2.1.3 Determining Tier 2 SSTL,

X4.2.1.4 Designing Remedial Action Systems,

X4.2.1.5 Designing Monitoring Programs.

X4.2.2 Examples of predictive modeling uses in the RBCA process include:

X4.2.2.1 The interpolation of site-specific data, as in the

case of drawing contour maps for ground water concentrations and ground water elevations,

X4.2.2.2 The prediction of future chemical concentration distributions based on historical trend data, as in the case of ground water transport modeling,

X4.2.2.3 The recommendation of sampling locations and sampling frequency based on current interpretation and future expectations of the distribution of chemicals of concern, as in the design of ground water monitoring networks,

X4.2.2.4 The design of remedial action alternatives, as in the case of hydraulic control systems, and

X4.2.2.5 The calculation of exposures as in the development of RBSL and SSTL.

X4.2.3 Predictive modeling is not used in the RBCA process as:

X4.2.3.1 A substitute for actual site characterization or monitoring data.

X4.3 *Characteristics of Mathematical Models:*

X4.3.1 Mathematical models are, at best, approximations of the real processes (such as the movement of chemicals in the subsurface, or the ingestion of chemicals contained in drinking water) that they are used to represent. These models may be relatively simple, as in the case of some analytical solutions: they may also be complex, as is typically the case for numerical chemical migration models. Regardless of the model used certain key issues, specifically, the ability of the model to produce accurate and valid results, and the uncertainty in the results should be evaluated to determine how well these models represent the real world processes they describe.

X4.3.2 The "accuracy" of modeling predictions is judged based on how well the model predicts observed behavior, and is dependent upon a number of factors including:

X4.3.2.1 The approximations used when describing the real system by mathematical expressions,

X4.3.2.2 The input parameters used to generate the results, and

X4.3.2.3 The mathematical methods used to solve the governing equations (for example numerical solution methods, expansion approximations).

X4.3.3 Modeling results are always subject to some degree of uncertainty, due to variability and uncertainties in the input parameters. Variability represents true heterogeneity in the parameter while uncertainty represents ignorance (or lack of perfect knowledge) about poorly characterized phenomena or models. It is important to quantify this uncertainty when interpreting the results. Many times this is done with a "sensitivity" analysis in which the user quantifies the influence of changes in input parameters on the model results, and then identifies those parameters that most significantly influence the results. The use of probabilistic risk analysis (PRA), that is, use of statistical distributions in place of single point values is a useful alternative to account for both variability and uncertainty in input data.

X4.3.4 The "validation" of model results is dependent upon the model meeting pre-established objectives. This is usually accomplished by comparing model results with known solutions for simplified conditions. In some cases, for example, conservative Tier 1 assessments, exposure and risk may be overestimated as a means of addressing uncertainty. In this case one might say that the models are often valid but not accurate. However, the validation of model results, at any Tier, should be to ensure that these are consistent with real world expectations and when these differ, then the rationale for these differences should be explicitly stated.

X4.4 *Categories of Mathematical Models Used in the RBCA Process:*

X4.4.1 Mathematical models typically used in the RBCA process can be grouped into broad categories:

X4.4.1.1 Chemical fate and transport (migration) models,

X4.4.1.2 Exposure assessment models,

X4.4.1.3 Dose-response assessment models, and

X4.4.1.4 Remedial action system design models.

X4.4.2 The determination of Tier 1 RBSL, or Tiers 2 and 3 SSTL generally involves the use of the first three types of models.

X4.5 *Migration Models:*

X4.5.1 Chemical(s) of concern can both migrate (for example, by leaching, advection, dispersion) and transform (for example, by biodegradation, hydrolysis, photolysis) in the environment. Migration (that is, fate and transport) models estimate the movement of a chemical through soil, ground water, or air over time. Most models focus on specific phenomena (for example, ground water transport) and vary in complexity, depending on assumptions made during model development. In RBCA, simple screening-level migration models are utilized in Tiers 1 and 2. More complex models are typically used in Tier 3.

X4.5.2 Most screening level migration models have a simple mathematical form and are based on rough conceptual approximations of actual phenomena. For example, the travel time between the leading edge of a dissolved hydrocarbon plume and a ground water well can be approximated by:

$$\frac{\text{distance to well (ft)}}{\text{retarded flow velocity (ft/y)}} = \frac{1000 \text{ ft}}{100 \text{ ft/y}} = 10 \text{ years} \tag{X4.1}$$

X4.5.3 *Migration Model Data Requirements*—Input data requirements are dependent on the phenomenon being described. For ground water migration models the most commonly required data for various screening level models include:

X4.5.3.1 Soil bulk density (can be estimated $\approx$ 1.7 g/cm^3),

X4.5.3.2 Total soil porosity (can be estimated $\approx$ 0.38 cm^3/cm^3),

X4.5.3.3 Soil moisture content can be conservatively estimated in many cases. It is approximately equal to the total soil porosity beneath the water table, and typically >0.05 cm^3-H_2O/cm^3-soil in the vadose zone; this can be a critical input parameter in the case of diffusion models and may require site-specific determination unless conservative values are used,

X4.5.3.4 Fraction organic matter in soil particles (= 0.005 – 0.01: sandy soil is often conservatively assumed); this can also be a critical parameter requiring site-specific determination unless conservative values are used),

X4.5.3.5 Hydraulic conductivity (generally site-specific determination required),

X4.5.3.6 Ground water gradient and flow direction (requires

site-specific determination),

X4.5.3.7 First-order decay-rate (generally requires site-specific calibration, as models are very sensitive to this parameter).

X4.5.4 Depending on the phenomenon being described or the models selected, or both, other information may be required. For example, if air dispersion of soil particulates is being described then values for parameters such as wind speed, precipitation, temperature, soil particle size distributions, and nearby building characteristics will be needed.

X4.5.5 In most cases, measurements of the attenuation (that is, decrease in concentration) of chemicals with distance away from the source area will be required to calibrate and verify the selected models. The amount of data required varies depending on:

X4.5.5.1 The models used,

X4.5.5.2 Their sensitivity to changes in model input parameters, and

X4.5.5.3 The contribution of the particular exposure pathway to the total incremental exposure.

X4.5.6 Generally, site-specific physical and chemical properties for the most sensitive parameters are required for migration models. However, in lieu of site-specific data, values selected from the literature may be used. These may take the form of conservative single point estimates or a range or distribution of values may be used as an alternative to account for variability and uncertainty in these parameters.

X4.6 *Exposure Models:*

X4.6.1 "Exposure models" are used to estimate the chemical uptake resulting from various exposure routes, such as ingestion, inhalation, and absorption.

X4.6.2 Typically, the models used to estimate uptake are simple algebraic expressions, such as those contained in Ref **(15)**. Application of these equations is illustrated in Appendix X3.

X4.6.3 Human exposure factors may be divided into receptor physiologic parameters (for example, body weight, skin surface area); contact rate (for example, consumption of water, soil ingestion rate); and time activity patterns (for example, time spent indoors/outdoors, time spent at work). Some of these variables, particularly the physiologic parameters, have been well characterized but others, such as time-activity patterns, are less well documented. All parameters are subject to variability (that is, true heterogeneity) and uncertainty (that is, ignorance about a measurement). Thus, a range of values may be available for any given parameter. The choice will depend to some extent on the site-specific circumstances and the level of conservatism desired. While point values are often selected for simplicity, statistical distributions for many of the exposure parameters are readily available for Tier 3 analyses. Typical sources for these parameters are the USEPA Exposure Factors Handbook (1997) **(30)** and the AIHC Exposure Factors Sourcebook (1994) **(34)**.

X4.7 *Dose-Response Models:*

X4.7.1 In general terms, the purpose of the toxicity assessment is to collect and weigh the available evidence regarding the potential for particular chemical(s) of concern to cause adverse effects in exposed individuals and to provide an estimate of the relationship between the extent of exposure to a chemical and the increased likelihood and/or severity of adverse effects. Adverse effects are broadly characterized as carcinogenic (that is, resulting in tumor development or other forms of cancer) or non-carcinogenic (that is, all other forms of toxicity unrelated to cancer effects). Toxicity assessments combine dose-response data with models to identify a "safe dose" or a toxic level for a particular adverse effect. Most estimates of a "safe dose" or toxic level are based on animal studies. In some instances, human epidemiological information is available on a chemical.

X4.7.2 For non-carcinogenic effects, toxicity criteria are developed based on the concept that there is a "safe" (that is, threshold) level of material to which humans can be exposed without adverse effects. These criteria are often derived for chronic exposures (usually over a lifetime) by applying safety or uncertainty factors, or both, to intake levels observed to have little or no effects in human or animal studies. These are typically expressed as an acceptable daily intake, that is, a reference dose (RfD) or a reference concentration (RfC).

X4.7.3 The threshold value for a chemical is often called the no observed adverse effects level (NOAEL). Scientists usually estimate NOAELs from animal studies. An important value that typically results from a NOAEL or LOAEL (lowest observed adverse effect level) value is the reference dose (RfD). A reference dose is an estimate (with an uncertainty typically spanning an order of magnitude) of a daily exposure (mg/kg/day) to the general human population (including sensitive subgroups) that is likely to be without an appreciable risk of deleterious effects during a lifetime of exposure. The RfD value is derived from the NOAEL or LOAEL by application of uncertainty factors (UF) that reflect various types of data used to estimate RfDs and an additional modifying factor (MF), which is based on a professional judgment of the quality of the entire database for the chemical. The oral RfD, for example, is calculated from the following model:

$$RfD = \frac{NOAEL}{UF \times MF} \quad (X4.2)$$

X4.7.4 The first step in evaluating potential carcinogenic materials is determining if the material is a carcinogen based on human or animal studies, or both. The USEPA currently has a 5 tier (A-E) classification scheme.

X4.7.5 Depending on mode of action, carcinogenic materials may be divided into genotoxic (that is, having the ability to mutate genetic material and eventually initiate tumors, also referred to as mutagenic) or non-genotoxic (that is, non-mutagenic). It is generally accepted that whereas non-genotoxic materials may exhibit a threshold limit, any exposure to genotoxic materials may eventually lead to tumor formation. For materials classified as carcinogenic, a determination of an "acceptable" (genotoxic) or "safe" (non-genotoxic) exposure level is driven by the mode of action that produces the carcinogenic effect.

X4.7.6 Thus, for non-genotoxic materials, a "safe" level may be derived from no (or lowest) observed adverse effect levels, that is, NOAEL or LOAEL (typically from animal studies) combined with safety and uncertainty factors. The process is comparable to that used for non-cancer effects. This general approach has not traditionally been used by the USEPA until recently (1996) when the USEPA proposed its new cancer guidelines which take a more mechanistic approach (that is, taking into account how the specific cancer is formed).

X4.7.7 For genotoxic materials (where it is assumed that no threshold exist), extrapolation methods are often used to calculate a sufficiently low level that is "acceptable." USEPA typically use a mathematical modeling method (for example, one-hit, linear multi-stage) to calculate this value, often expressed numerically as a slope factors (SF) or unit risk factor (URF).

X4.7.8 Toxicity values are generally obtained from a standard set of reference tables (for example, USEPA's *Integrated Risk Information System*(**8**), IRIS, or the *Health Effects Summary Assessment Tables*, HEAST (**9**)). It is important to note that the information in IRIS has typically only been peer-reviewed within the EPA and may not always have support from the external scientific community. Whereas the information in IRIS has been subject to agency-wide data quality review, the information in the HEAST tables has not. The user is expected to consult the original assessment documents to appreciate the strengths and limitations of the data in HEAST. Thus, care should be exercised in using the values in HEAST. Some state and local agencies have toxicity factors they have derived themselves or preferences for factors to use if neither IRIS nor HEAST lists a value.

X4.7.9 It is important to note that in extrapolating the information obtained in animal studies to humans, a number of conservative assumptions are made.

X4.7.9.1 For noncarcinogens, a system of default safety and uncertainty factors, as discussed above, (in multiples of 10) is used to convert observation in animals to estimates in humans.

X4.7.9.2 For carcinogens, some of the most important assumptions include:

X4.7.9.2.1 The results of the most sensitive animal study are used to extrapolate to humans,

X4.7.9.2.2 In general, chemicals with any carcinogenic activity in animals are assumed to be potential human carcinogens, and

X4.7.9.2.3 No threshold exists for carcinogens (see section X4.7.6).

X4.7.10 The uncertainty in the RfD and SF values are often neglected in deference to single point values which are then typically summarized in databases such as IRIS and HEAST and are used subsequently as absolute estimates of risk. Further, many of these conservative assumptions described above are technical policy decisions made by the USEPA. These assumptions are not explicitly defined and may obscure the conservatism in the safe dose estimate. Thus, care must be exercised in interpreting results which have as a basis these conservative toxicity evaluations.

X4.8 *Remedial Action System Design Models:*

X4.8.1 Models may be used in the design of remedial action systems, such as those used to assess flow and removal rates from vapor extraction systems, or drawdown versus pumping rate behavior for pump and treat systems. Remedial system practices generally also contain economic (cost) models.

X5. RISK-BASED CORRECTIVE ACTION FOR PROTECTION OF ECOLOGICAL RESOURCES (ECO-RBCA)

X5.1 Due to the success of the risk-based corrective action (RBCA) program for petroleum release sites, there is a great deal of interest and need for applying the RCA process to assess risks to ecological resources. In response, ASTM Subcommittee E50.04 on Performance Standards Related to Environmentally Regulated Programs has initiated development of a standardized approach to RBCA for protection of ecological resources (Eco-RBCA).[8] As developed in the human health-based approaches at petroleum release and chemical release sites, the new Eco-RBCA standard guide describes a logical sequence of activities and decisions to be followed from the time a chemical release is discovered until no further action is achieved. The Eco-RBCA standard guide outlines an approach for managing chemical release sites by integrating site assessment, ecological risk assessment, risk management, and corrective action into a technically defensible framework.

X5.2 The Eco-RBCA standard guide provides a technically defensible process for achieving no further action when appropriate. The Eco-RBCA process encourages user initiatives and stakeholder involvement in the development of a RBCA program for a site. The Eco-RBCA standard guide *(1)* helps direct risk based corrective action at a site, *(2)* compliments but does not supersede federal, regional, or state regulations, and *(3)* can be employed where there may not be a regulatory framework for corrective action, or where agency guidance for such actions does not exist, or where the user wishes to conduct corrective action such as sites in voluntary cleanup programs or under Brownfields initiatives. In addition, it can also be used as a unifying framework where several different agency programs affect the site.

X5.3 The Eco-RBCA standard guide outlines a tiered approach for conducting ecological evaluations as part of risk-based corrective actions taken at sites impacted by a chemical release. The RBCA process employs a tiered approach that optimizes resources (for example, time, financial, and professional) while making scientifically sound risk management decisions. Tiers are groupings of evaluations that start simple and proceed to more complex evaluations when site conditions warrant. Each tier represents a level of effort that determines whether risk based corrective action is appropriate and which alternative action will eliminate or achieve an acceptable level of risk to relevant ecological receptors and habitats. In early tiers, environmentally conservative assessments are used to compensate for the uncertainty of using

[8] When ASTM publishes the Eco–RBCA standard guide, the information in that guide supercedes the information in this appendix.

limited data and qualitative site information. In later tiers, site specific evaluations are used to develop more realistic exposure estimates, and the evaluation is refined by focusing on relevant ecological receptors and habitats and chemicals identified in previous tiers.

X5.4 The Eco-RBCA is divided into three tiers of evaluation. The Tier 1 Evaluation includes the Tier 1 Ecological Risk Assessment and the Tier 1 decision point (Steps 3 and 4, respectively). Likewise, the Tier 2 Evaluation includes the Tier 2 Ecological Risk Assessment (Step 5) and the Tier 2 decision point (Step 6) and the Tier 3 Evaluation includes the Tier 3 Ecological Risk Assessment (Step 7) and the Tier 3 decision point (Step 8). Before initiating an Eco-RBCA evaluation, an initial site assessment is conducted.

X5.4.1 *Step 1 - Initial Site Assessment (Planning and Scoping)*—At this first step it is determined whether or not an Eco-RBCA evaluation is appropriate for a site. This step includes the development of a site conceptual model and consideration of appropriate regulatory requirements, technical policy decisions, screening criteria, and the technical need for an assessment of site conditions. No further action is appropriate if there are no complete or potentially complete exposure pathways to relevant ecological receptors and habitats. Stakeholders should be identified in Step 1 and decisions regarding the above issues should include these stakeholders.

X5.4.2 *Step 2 – Decision Point: Response Action Evaluation*— The decision to proceed in Step 2 depends on the technical policy decisions and the regulatory requirements (for example, screening criteria identified in Step 1) and actual site conditions. The decision point may lead to a no further action decision (Step 10), a Tier 1 Evaluation (Step 3), or an evaluation of the need for initial response actions. Initial response actions may be required before thoroughly evaluating risks to relevant ecological receptors and habitats in Step 3. Such actions may be driven by any or all of the need to reduce immediate threats to identified receptors, existing regulatory requirements (for example, based on National Contingency Plan (NCP) - Oil Spill Response, Clean Water Act (CWA) - impact on migratory waterfowl), or approaches used successfully at similar sites. The user returns to Step 1 if response actions are implemented. If response actions are not required but further evaluation is appropriate, then the user proceeds to the Tier 1 Evaluation in Step 3, which includes the Tier 1 Ecological Risk Assessment (ERA).

X5.4.3 *Step 3 - Tier 1 ERA*—The Tier 1 ERA consists of two phases: an Exposure Pathway Analysis and a Generic Screening Evaluation. The Tier 1 Exposure Pathway Analysis involves refinement of a site conceptual model (defined by the sources, potential transport pathways, and potential relevant ecological receptors and habitats) and deciding if the potential exposure pathways need to be evaluated further. If no potentially complete exposure pathways exist, a no further action decision should result. If further evaluation is appropriate, a generic screening evaluation should be conducted. This could include a screening evaluation using relevant ecological screening criteria (RESC), other relevant measurable criteria (ORMC), multiple lines of evidence, site observations (for example, site walk, sketch diagram, written discussion, checklist), or other information.

X5.4.4 *Step 4 - Tier 1 Decision Point*—Following the Tier 1 Ecological Risk Assessment a decision is made as to whether site conditions meet the Tier 1 corrective action goals. This decision should be based on consideration of the Tier 1 Evaluation results, technical policy decisions, applicable regulations, site owner management plans, land use and controls, community values, and other stakeholder issues. Implications of this decision may need further discussion in terms of values and flexibility. If comparison of the site conditions meet the corrective action goals, then the user proceeds to Step 10 to determine whether no further action is appropriate. If it is determined in Step 4 that the site may not meet the corrective action goals, then either interim remedial action, remedial action based on applicable corrective action goals (Step 9), or further tiered evaluation is conducted (Step 5).

X5.4.5 *Step 5 - Tier 2 ERA*— The Tier 2 ERA is a site-specific refinement of the Tier 1 ERA. Tier 2 may include one or more of the following as necessary to support site-specific decisions: (1) further refinement of the site conceptual model in terms of relevant species and their exposure potential; (2) a simple food web model and a narrower focus on relevant ecological receptors and habitats and potentially complete exposure pathways; (3) measurements of species density and diversity; (4) site chemical concentration measurement and monitoring (for example, fate and transport modeling); (5) simple models (with site specific data) to be used in refining exposure estimates; and (6) other simple qualitative or quantitative evaluation methods.

X5.4.6 *Step 6 - Tier 2 Decision Point*—Following the Tier 2 ERA a decision is made as to whether site conditions meet the Tier 2 corrective action goals. The decision options for the Tier 2 evaluation will be the same as for the Tier 1 decision point described in Step 4. A decision is made as to whether to consider no further action (through Step 10), assess remedial alternatives (Step 9), or proceed with further evaluation in Step 7.

X5.4.7 *Step 7 - Tier 3 ERA*—The Tier 3 ERA is a refinement of Tier 2, and generally will involve a much greater expenditure of resources than a Tier 2 ERA. Tier 3 may include the use of probabilistic tools (for example, Monte Carlo or statistical models) and in-situ tests (for example, caged bivalves, substrate colonization). Numerical models, with detailed site monitoring, may be applicable for chemical fate and transport predictions.

X5.4.8 *Step 8 - Tier 3 Decision Point*—Following the Tier 3 ERA a decision is made as to whether site conditions meet the Tier 3 corrective action goals. Decision options for the Tier 3 decision point are essentially the same as the Tier 1 and Tier 2 decision points described in Steps 4 and 6. One notable exception is that further evaluation in another advanced tier is not an option. A decision is required as to whether to assess remedial alternatives (Step 9) or consider no further action (Step 10).

X5.4.9 *Step 9 - Implementing the Remedial Action Program*—The evaluation of potential remedial action alternatives is conducted in Step 9. The effectiveness, likelihood of success, and costs should be considered in evaluating, selecting, and implementing a remedial action program. The evaluation of the remedial action alternatives includes the probability of success in meeting the corrective action goals, costs, effectiveness of the remedial action in protecting relevant ecological receptors and habitats, long-term reliability, and the implementability and practicability of the remedial action. In addition, the risks of remedial action should be compared to the environmental benefits (for example, water table drawdown in hydraulic containment and effect on wetlands, soil removal in sensitive habitat) as part of the process. Following the selection and implementation of remedial action(s), the user proceeds to the Step 10 decision point to decide whether no further action or further monitoring is appropriate.

X5.4.10 *Step 10 - Monitoring Programs*—In this step monitoring tasks (for example, water well monitoring, biota monitoring, ecological surveys, confirmation of ground water concentration attenuation) and the requirements for establishing the effectiveness of the remedial action program are determined. Such measures are needed to evaluate whether site conditions are changing and if corrective action goals are met. This step requires an evaluation of performance objectives (that is, the corrective action goals) for their continued relevance (for example, exposure model remains applicable).

X5.5 The ASTM Eco-RBCA standard guide is under development and is currently organized as follows:

X5.5.1 Section 1 gives an overview of the process;

X5.5.2 Section 2 lists referenced documents;

X5.5.3 Section 3 defines terminology used;

X5.5.4 Section 4 describes the significance and use of this standard guide;

X5.5.5 Section 5 provides general information pertinent to the Eco-RBCA process; and

X5.5.6 Section 6 presents a description of the 10 steps of the Eco-RBCA process.

X5.6 The Eco-RBCA standard guide includes appendixes that are provided for additional information but are considered non-mandatory. The appendixes are as follows:

X5.6.1 Appendix X1 provides a brief description of the Guide for RBCA (Guide E 2081).

X5.6.2 Appendix X1 describes screening criteria and how they can be applied within the Eco-RBCA framework;

X5.6.3 Appendix X2 is a comparative analysis of ecological risk assessment regulatory approaches and guidance documents;

X5.6.4 Appendix X3 presents the development and selection of relevant ecological screening benchmarks;

X5.6.5 Appendix X4 includes examples of the application of the Eco-RBCA framework;

X5.6.6 Appendix X5 presents risk management issues; and

X5.6.7 Appendix X6 presents information on uncertainty and probabilistic tools.

X6. ACTIVITY AND USE LIMITATIONS, INCLUDING INSTITUTIONAL AND ENGINEERING CONTROLS

X6.1 Corrective action programs utilizing risk-based corrective action principles frequently rely on Activity and Use Limitations (AULs), including institutional and engineering controls, to achieve an "acceptable risk" or "no significant risk" level. For purposes of this appendix, "activity and use limitations" are those controls that can be utilized by responsible parties, voluntary parties and regulatory agencies in corrective action programs where, as a part of the program, certain levels of chemical releases may remain in soil or ground water. AULs provide a mechanism to ensure that exposures to remaining chemicals of concern are prevented or reduced to protect human health and the environment. The need for, implementability of, and enforceability of AULs should be considered early in the risk based corrective action process. Understanding the activity and land use are an integral part of the risk based corrective action process. A user may determine, based upon post-remedial action land or ground water use, or deficiencies in the AULs that may be available, that an AUL may not be feasible for the site. The user also needs to identify who will be responsible for the long-term enforcement and maintenance of any AULs that are used at a site.

X6.2 Because of the importance of this issue to the long-term success of risk-based corrective action, the ASTM E50.04 Subcommittee has initiated the development of a separate standard guide concerning the use of AULs, including institutional and engineering controls.[9] The AUL standard guide will explain how AULs need to be considered during the risk based corrective action process. The guidance document will also describe the types of AULs that are commonly available, including proprietary controls (for example, "deed restrictions," restrictive covenants, easements, equitable servitudes); state and local government controls (for example, zoning, building permits, well drilling prohibitions, water and well use advisories); statutory enforcement tools (for example, permits and consent orders); and informational devices (for example, deed notices, notice to government agencies, record notice, registry act requirements, transfer act requirements). Finally, the standard guide will discuss implementation considerations, such as identifying the need for, reliability of, and implementability of AULs early in the corrective action process; whether there are available state and local authorities to implement an AUL; enforceability considerations; changes in land use; and stakeholder participation issues.

X6.3 The purpose of this appendix is to provide basic review information about AULs typically used in corrective action programs.

X6.4 The types of AULs discussed in this appendix are as follows:

[9] When ASTM publishes the activity and use limitations standard guide, the information in that guide supercedes the information in this appendix.

1. Deed restrictions, or restrictive covenants;
2. Use restrictions (including well restriction areas);
3. Access controls;
4. Compliance monitoring;
5. Notice, including record notice, actual notice, and notice to government authorities;
6. Registry act requirements;
7. Transfer act requirements;
8. Contractual obligations and agreements and
9. Restrictive zoning.

X6.5 AULs vary in both form and content. Agencies and landowners can invoke various authorities and enforcement mechanisms, both public and private, to implement any one or a combination of the controls. For example, a state could adopt a statutory mandate (see X6.6) requiring the use of deed restrictions (see X6.7) as a way of enforcing use restrictions (see X6.8) and posting signage (a type of access control, see X6.9) along with compliance monitoring (X6.10). Thus, a number of AULs are often used as overlapping strategies.

X6.6 *Statutory Mandates*:

X6.6.1 Some states' emergency response programs mandate post-remedial action AULs and impose civil penalties for non-compliance. These programs impose obligations on landowners to use one or more AULs described in this appendix.

X6.7 *Deed Restrictions*:

X6.7.1 Deed restrictions place limits and conditions on the use and conveyance of land. They serve two purposes: informing prospective owners and tenants of the environmental status of the property and ensuring long-term compliance with the AULs that are necessary to maintain the integrity of the remedial action over time. Certain legal rules must be satisfied in order to make a deed restriction binding and enforceable, as described below.

X6.7.2 The following four factors are required for a promise in a deed restriction (also called a "restrictive covenant") to be held against current and subsequent landowners: *(1)* a writing, *(2)* intention by both original parties that particular restrictions be placed on the land in perpetuity, *(3)* "privity of estate," and *(4)* the restriction must "touch and concern the land."

X6.7.2.1 The first requirement is that of a writing. Conveyances of land are legally required to be documented in a writing. This legal rule also holds true for deed restrictions affecting land. Ideally, a deed restriction used as an AUL would be written with particularity and then recorded in the local land records office, similar to the documentation and recording of a sale of land. Parties may also encounter the requirement that the deed restriction be executed "under seal," a legal formality that has been abandoned in most states.

X6.7.2.2 The second requirement that must be satisfied for a deed restriction to be legal and binding is that the deed restriction should be precisely reflect the parties' intentions in regard to the scope and the duration of the restrictions. Explicitly stating in the deed restriction that the parties intend the restriction to "run with the land" (that is, last forever and bind subsequent owners) is strongly recommended.

X6.7.2.3 The third requirement, privity of estate, means that only persons with a certain relationship to the land should be able to enforce a deed restriction. Normally, deed restrictions are promises between buyer and seller or between neighbors, therefore, the state or a third party may not enforce a deed restriction. Privity of estate, however, should not be a barrier to state enforcement of the deed restriction if the proper steps are taken. Even in states that require privity of estate, this concern can be addressed if the landowner takes the land with knowledge that the restrictions exist and might be enforced by these third parties. Thus, it is strongly recommended that the deed restriction explicitly state that the state's environmental authority may enforce the restriction. Recording of the deed restriction serves as notice to anyone who later purchases or acquires an interest in the land.

X6.7.2.4 Finally, a deed restriction is only enforceable if the promise "touches and concerns the land." This means that any deed restriction used as an AUL should be written so that it centers on the land and use of the land. One fact to consider when deciding if the promise" touches and concerns the land" is whether the landowner's legal interest in the land is decreased in value due to the deed restriction. If the land is devalued in this way, then the restriction could be said to "touch and concern the land." Note that the focus of the inquiry is on the land itself; promises that are personal in nature and only concern human activities that happen to take place on the land are least likely to be enforceable.

X6.7.3 Deed restrictions or restrictive covenants may be difficult, at best, for some states to enforce. Due to the potential difficulties encountered by a governmental agency in enforcing a deed restriction, it may be appropriate for an individual state to seek statutory and regulatory amendments to ensure that such authority exits in regard to all deed restrictions for environmental purposes. Specific legal advice should be sought with respect to the validity of restrictive covenants or deed restrictions in a particular state.

X6.7.4 Remedies for noncompliance with deed restrictions comes in two forms: *(1)* persons or agencies may sue to obtain a court order (injunction) requiring compliance or *(2)* if the state statute allows for it, the state's attorney general can seek enforcement of civil penalties, such as fines, for non-compliance.

X6.7.5 A state program can require a landowner to continue monitoring activities and to allow state environmental officials access to the site to monitor compliance with AULs. These arrangements may need to be placed in a deed restriction in order to run with the land from owner to owner, but responsible parties or voluntary parties may also be required to sign a contract making these promises. Additionally, it should be noted that almost every state has authority to issue administrative orders to accomplish some or all of the above.

X6.7.6 The above arrangements can also set out procedures that will be followed if some emergency requires that the site be disturbed. If, for example, underground utility lines must be repaired, the landowner would follow this protocol for handling the soil and alerting the state authority.

X6.8 *Use Restrictions*:

X6.8.1 Use restrictions are usually the main component of a deed restriction. Use restrictions describe appropriate and inappropriate uses of the property in an effort to perpetuate the benefits of the corrective action and ensure property use that is consistent with the applicable corrective action goals. Such techniques also prohibit any person from making any use of the site in a manner that creates an unacceptable risk to human health or environment from exposure to the residual chemical releases.

X6.8.2 Use restrictions may address uses that could disturb a containment cap or unremediated soils. A prohibition of drinking on-site ground water (or off-site ground water by means of well restriction areas) may also be appropriate. Well restriction areas can be a form of AUL by providing notice of chemicals of concern in ground water, and by prohibiting or conditioning the construction of wells in that area. As an example, a program may allow a restriction of record to include one or more of the following:

1. Restriction on property use;
2. Conditioning the change of use from nonresidential to residential on compliance with all applicable corrective action goals for a residential property;
3. Restriction of access; or
4. Restriction of the disturbance of agency approved remedial actions on the land.

X6.8.2.1 This technique preserves the integrity of any ground water remedial action by prohibiting or conditioning the placement and use of any or all types of wells within the area.

X6.8.2.2 Well restrictions of this nature would be subject to agency approval, public notice and may include the restriction on constructing or locating any wells within a particular designated area. Notice of the well restriction is recorded on the land records and with various health officials and municipal officials. The restrictions can only be released upon a demonstration that the chemicals of concern in the well restriction area meet the corrective action goals.

X6.9 *Access Controls*:

X6.9.1 Another subset of AULs is the control of access to any particular site. Access can be controlled by measures such as fencing and gates, security, posting or warnings. Consideration of the way in which the site and adjacent land are used help to determine the appropriate level and means of access controls. For example, it should be noted whether the site is located in a residential or mixed neighborhood; whether sensitive land use areas, such as playgrounds, day care centers, or schools are within close proximity; or if the site provides a pathway that is frequently traversed by neighbors. These factors could increase the potential for and intensity of exposures to the chemicals of concern.

X6.9.2 Access controls require continued maintenance and custodial care for the length of time the AULs are to be in-place. Therefore, planning must occur for these activities, including the financial responsibility for such up-keep.

X6.10 *Monitoring Programs*:

X6.10.1 Monitoring programs may be needed to ensure that conditions do not deteriorate and result in unacceptable risks to human health or the environment either on or off-site. For example, monitoring programs may be needed to ensure that chemical migration is not occurring at an unacceptable rate or intensity.

X6.10.2 Monitoring may be conducted at scheduled intervals, as necessary, during the period of time that the AULs are in-place.

X6.10.3 Monitoring programs could include sampling ground water monitoring wells located on-site, downgradient of the site; sampling air from air monitoring stations on- and off-site; and monitoring discharges to nearby surface water bodies.

X6.10.4 Monitoring programs requires continued maintenance, custodial care and data collection and evaluation for a length of time. Therefore, planning must occur for these activities, including the financial responsibility for such up-keep.

X6.11 *Notice*:

X6.11.1 Regulations of this type generally provide notice of specific location of chemical releases on the site, and disclose any restrictions on access, use, and development of part or all of the impacted site to preserve the integrity of the remedial action. Three types of notice are described below *(1)* record notice; *(2)* actual notice; and *(3)* notice of governmental authorities.

X6.12 *Record Notice:*

X6.12.1 Some states require that sites having releases of hazardous waste file a notice on the land records providing to any subsequent purchaser of the property information regarding the past or current activities on the site.

X6.12.2 The record notice requirement may be broad; the program may require any property subject to a response action to obtain a professional opinion and then prepare and record a Grant of Environmental Restriction that is supported by that opinion.

X6.12.3 The record notice requirement can be ancillary to a transfer act (see X6.16), in which case recording of an environmental statement is only required in conjunction with a land transaction.

X6.13 *Actual Notice:*

X6.13.1 States may require direct notice of environmental information to other parties to a land transaction. These laws protect potential buyers and tenants, and they also help ensure that use restrictions and other AULs are perpetuated.

X6.13.2 Actual notice of an environmental defect or failure to provide notice may give a party the right to cancel the transaction and result in civil penalties. For example, landlords and sellers who do not give notice as required by the state may be liable for actual damages plus fines. Nonresidential tenants who fail to notify landowners of suspected or actual hazardous substance releases can have their leases canceled and are subject to fines.

X6.14 *Notice to Government Authorities*

X6.14.1 Parties to a land transaction may also be required to file the environmental statement with various environmental authorities. Notice to the government may be required before the transaction takes place.

X6.15 *Registry Act Requirements*:

X6.15.1 Some states have registry act programs that provide for the maintenance of a registry of hazardous waste disposal sites and the restriction of the use and transfer of listed sites.

X6.15.2 A typical registry act provides that the state environmental agency establish and maintain a registry of all real property which has been used for hazardous substance disposal either illegally or before regulation of hazardous waste disposal began in that state.

X6.15.3 The state agency is responsible for investigating potential sites for inclusion on the registry. The registry includes the location of the site and a listing of the hazardous wastes on the property, and may also include a classification of the level of health or environmental hazards presented by the conditions on the property. The state agency may be required to perform detailed inspections of the site to determine its priority relative to other sites.

X6.15.4 Owners of sites proposed for inclusion on the registry have rights of hearing and appeal, and owners of sites on the registry have rights to modify or terminate their listing. In some cases, the owner of a site proposed for inclusion on the registry may obtain the withdrawal of the proposed registration by entering into a consent agreement with the state. Such a consent agreement establishes a timetable and responsibility for remedial action.

X6.15.5 When a site appears on the state registry, the owner must comply with regulatory requirements in regard to use and transfer of the site. The use of a site listed on the registry may not be changed without permission of the state agency. In negotiations for a conveyance of a registered site, the owner may be obligated to disclose the registration early in the process, and permission of the state agency may be required to convey a registered property. Under other schemes, permission to convey is not required, but the seller must notify the state agency of the transaction.

X6.15.6 Finally, registry acts require that the listing of a property on a hazardous materials site registry be recorded in the records of the appropriate locality so that the registration will appear in the chain of title.

X6.16 *Transfer Act Requirements*:

X6.16.1 Some states have transfer act programs that require full evaluation of all environmental issues before or after the transfer occurs. It may be that within such a program, AULs can be established by way of consent order, administrative order, or some other technique that establishes implementation and continued responsibility for AULs.

X6.16.2 A typical transfer act imposes obligations and confers rights on parties to a land transaction arising out of the environmental status of the property to be conveyed. Transfer acts impose information obligations on the seller or lessor of a property (see X6.16.3). That party must disclose general information about strict liability for corrective action costs as well as property-specific information, such as presence of hazardous substances, permitting requirements and status, releases, and enforcement actions and variances.

X6.16.3 Compliance with transfer act obligations in the manner prescribed is crucial for ensuring a successful conveyance. Sometimes the transfer act operates to render a transaction voidable before the transfer occurs. Failure to give notice in the required form and within the time period required or the revelation of an environmental violation or unremediated condition will relieve the transferee and the lender of any obligation to close the transaction, even if a contract has already been executed. Moreover, a violation of the transfer can be the basis for a lawsuit to recover consequential damages.

X6.17 *Contractual Obligations and Agreements*:

X6.17.1 One system for ensuring the future restriction on use of a site, or the obligation to conduct corrective action on a site, is to require private parties to restrict use by contract. While this method is often negotiated among private parties, it will be difficult, if not impossible, to institutionalize some control over that process without interfering with the abilities and rights of private parties to freely negotiate these liabilities.

X6.17.2 Another avenue is for the landowner or responsible party to obligate itself to the state by contract. The state may require a contractual commitment from the party to provide long-term monitoring of the site, use restrictions, and means of continued funding for corrective action (including active remedial action).

X6.18 *Restrictive Zoning*

X6.18.1 Local land use control jurisdictions may provide for restrictive or "overlay" zoning that prohibits certain activities or land uses. Such AULs are difficult to establish for individual sites. For many sites located in the same local planning jurisdiction which have undergone RBCA analysis, a restrictive "zone" may be more practical. This "overlay" might prevent certain exposures to chemicals of concern as an effective means of establishing a number of AULs with one regulatory action.

X7. RISK-BASED CORRECTIVE ACTION CASE EXAMPLES

X7.1 *Introduction*—This appendix illustrates the critical features of the risk-based corrective action process through presentation of several case examples, including an agricultural chemical formulation and distribution facility (X7.2) and an operating industrial facility (X7.3). Both cases are hypothetical. The decisions made in each example case are not necessarily generally applicable; they are a function of pertinent technical policy decisions (discussed in Appendix X1), concurrence of all involved parties, and compliance with federal, state, or local laws and regulations. The examples cover a range of situations and include illustration of the Tier 1 screening process, site conditions & initial response actions the Tier 2 assessment, and corrective action selection. Methods and decisions used in these steps are briefly summarized.

X7.2 *Agricultural Chemical Formulation and Distribution Facility*:

X7.2.1 *Initial Site Assessment*—Components of the initial site assessment are discussed in X7.2.2 to X7.2.10. A plan view of the site and surrounding area, with important features illustrated, is given in Fig. X7.1.

X7.2.2 *Site Description*—The site consists of a two-acre tract in a predominately agricultural area. The adjacent land use is both agricultural and residential. The nearest residence is at a distance of 400 ft from the property boundary. The site is not currently subject to any federal, state, or local enforcement actions.

X7.2.3 *Site Ownership and Activity*—The current business on the site is engaged in sales and distribution of prepackaged pesticides, prepackaged fertilizers and soil amendments, and bulk aqueous liquid fertilizers. In addition to the present site use, prior activities at the site included on-site mixing and formulation of bulk pesticides. The site includes an enclosed office and warehouse for storage of prepackaged materials, a tank farm for storage of bulk liquid fertilizer, several loading and material transfer areas, and several waste storage areas. The business has been in operation on the present site for forty years, with four changes in ownership over the same time period. The site surface is unpaved, with a gravel parking area and a concrete-paved wash pad area.

X7.2.4 *Past Releases or Potential Source Areas*—Potential spill or leak areas on the site include the current fertilizer tank farm area, a fertilizer loading rack area, a truck wash pad, a burn barrel area, and a used pesticide container and used oil storage area. The area which had formerly been used for formulation of bulk liquid pesticides is identifiable by the remaining equipment foundations; this has also been identified as a potential leak or spill area. Stained soil was observed in the vicinity of the truck wash pad area and the used oil storage area.

X7.2.5 *Regional and Site Hydrogeologic Conditions*:

X7.2.5.1 The climate in the region is subtropical, with a warm, dry winter season and a hot, wet summer season. Agricultural crops in the adjacent farm fields are watered using drip irrigation during the summer months.

X7.2.5.2 The regional topography is complex, with the site located in an alluvial mountain valley drained by a single continuously flowing small creek. United State Geological Survey (USGS) topological maps indicate the creek is fed by springs along its length, both upstream and downstream from the site. The site itself is gradually sloped towards an open drainage ditch which is seasonally dry. The drainage ditch enters the creek 200 ft from the site boundary. Two low areas are evident on the site. These low areas evidently pond with water during wet weather periods.

X7.2.5.3 The site hydrogeological investigation has included soil borings, collection of ground water samples, and characterization of geological structure and hydrogeological conditions. The first three to six feet of soils on the site consist primarily of sandy loam, going to poorly sorted sandy gravel at depths below ground surface (BGS) between six to twenty feet, and to poorly sorted gravelly sand at depths greater than twenty feet BGS. First encountered ground water in the region is found at a seasonally varying mean depth BGS of 24 to 30 ft. Ground water gradient, determined from four installed on-site piezometers, is in the direction of the creek. Ground water velocity has been estimated from the measured gradient and the observed soil characterization. A basalt layer is encountered in the vicinity of the site at depths of between 100 to 120 ft. Municipal and residential drinking water in the vicinity is generally produced from wells at a depth of 80 to 100 ft BGS.

X7.2.6 *Beneficial Use Summary and Receptor Survey*:

X7.2.6.1 The site is in a rural agricultural area. Drinking water for on-site use is supplied from a well screened at a depth BGS of 100 ft. The nearest off-site residence, at a distance of approximately 400 ft downgradient from the site, is supplied from a well screened at a depth of 90 ft BGS. Six wells are within one-half mile of the site. Three of the wells are used for residential drinking water and the remaining wells are used for agricultural irrigation. Ground water is the sole drinking water supply source within this local mountain-valley area and the state has a ground water classification scheme that identifies ground water in this area as Class II. In the absence onsite exposure points, the point of exposure for Class II ground water is assumed to be the facility boundary.

X7.2.6.2 The site itself is considered a commercial/industrial property. Access to the warehouse, tanks, loading racks, and other potentially impacted areas is restricted by a fence and locked gate.

X7.2.7 *Ecological Assessment Summary*—There is no obvious impact to vegetation on the site. For the preliminary site visit, during the late summer, green vegetation was found only in the vicinity of the creek boundary or in the adjacent irrigated fields. The creek is considered a relevant ecological receptor and habitat; there was no obvious impact to the creek or adjacent drainage ditch due to site activities.

X7.2.8 *Exposure Scenario Evaluation Flowsheet*—An exposure scenario flowsheet for this site is given in Fig. X7.2. This flowsheet is based on information discussed in X7.2.2 to X7.2.7.

X7.2.8.1 Commercial/industrial exposure to on-site impacted surficial soils, through direct contact, is both a current and potential future exposure pathway at this site. Inhalation of vapors evolved from impacted soils, both indoors and outdoors, is also a current or potential future exposure pathway for on-site commercial/industrial receptors.

X7.2.8.2 Migration of non-aqueous phase liquids, leaching of soluble constituents to ground water, and migration of soluble chemicals of concern via ground water transport to residential water ingestion are viable exposure pathways for this site, both under current land use and for any viable future use of the site.

X7.2.8.3 The potential also exists at this site for surface water or sediment migration to an off-site creek, in addition to potential migration of any soluble ground water plume to the creek. The creek is considered habitat.

X7.2.9 *Site Conditions and Initial Response Actions*—Based on a qualitative evaluation of the information collected in the initial site assessment, no immediate threat to human health, safety, or environmental receptors has been identified, and no immediate response action is warranted for the site (see

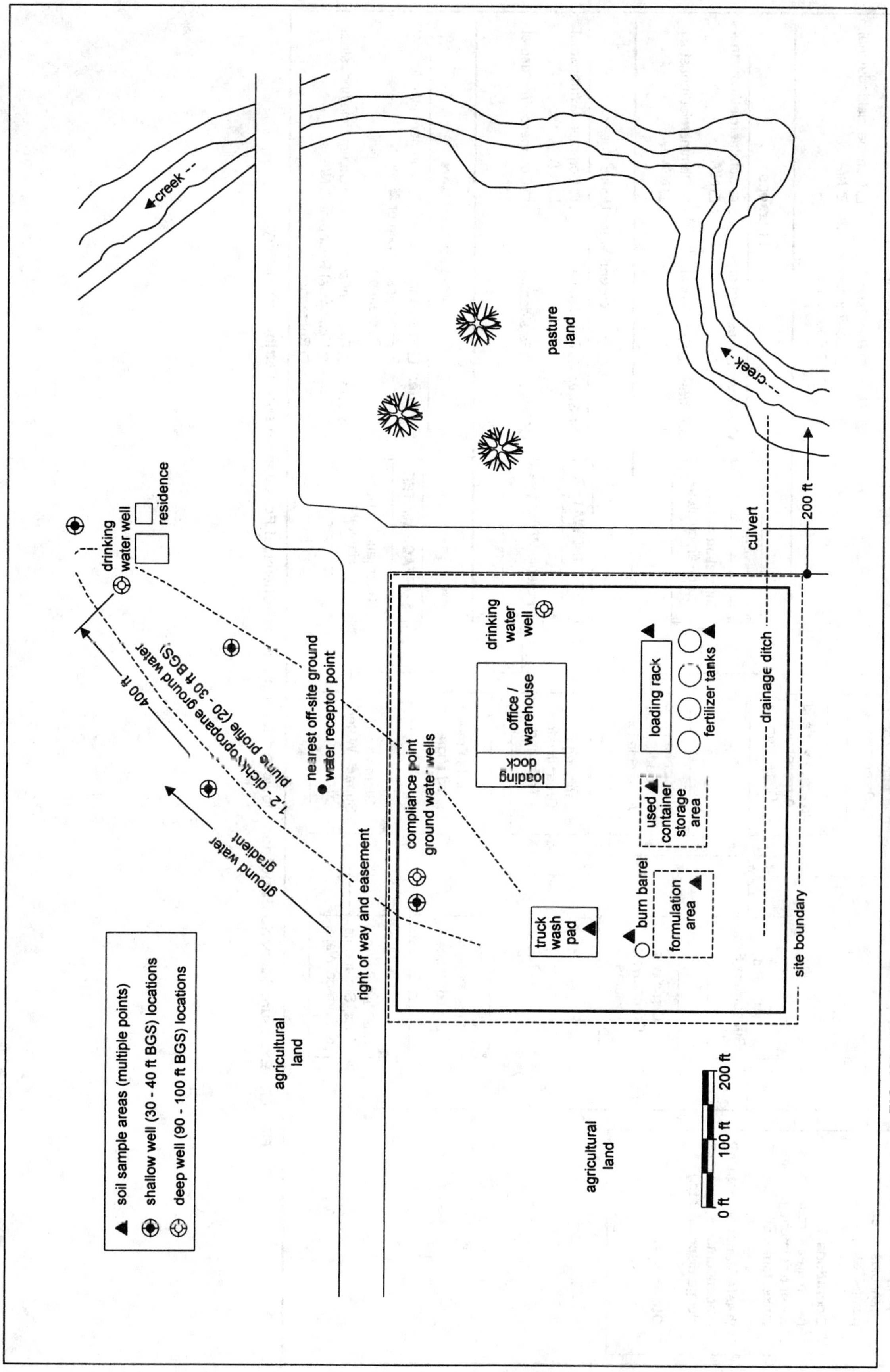

FIG. X7.1 Plan View of Agricultural Chemical Formulation and Distribution Facility and Surrounding Area.

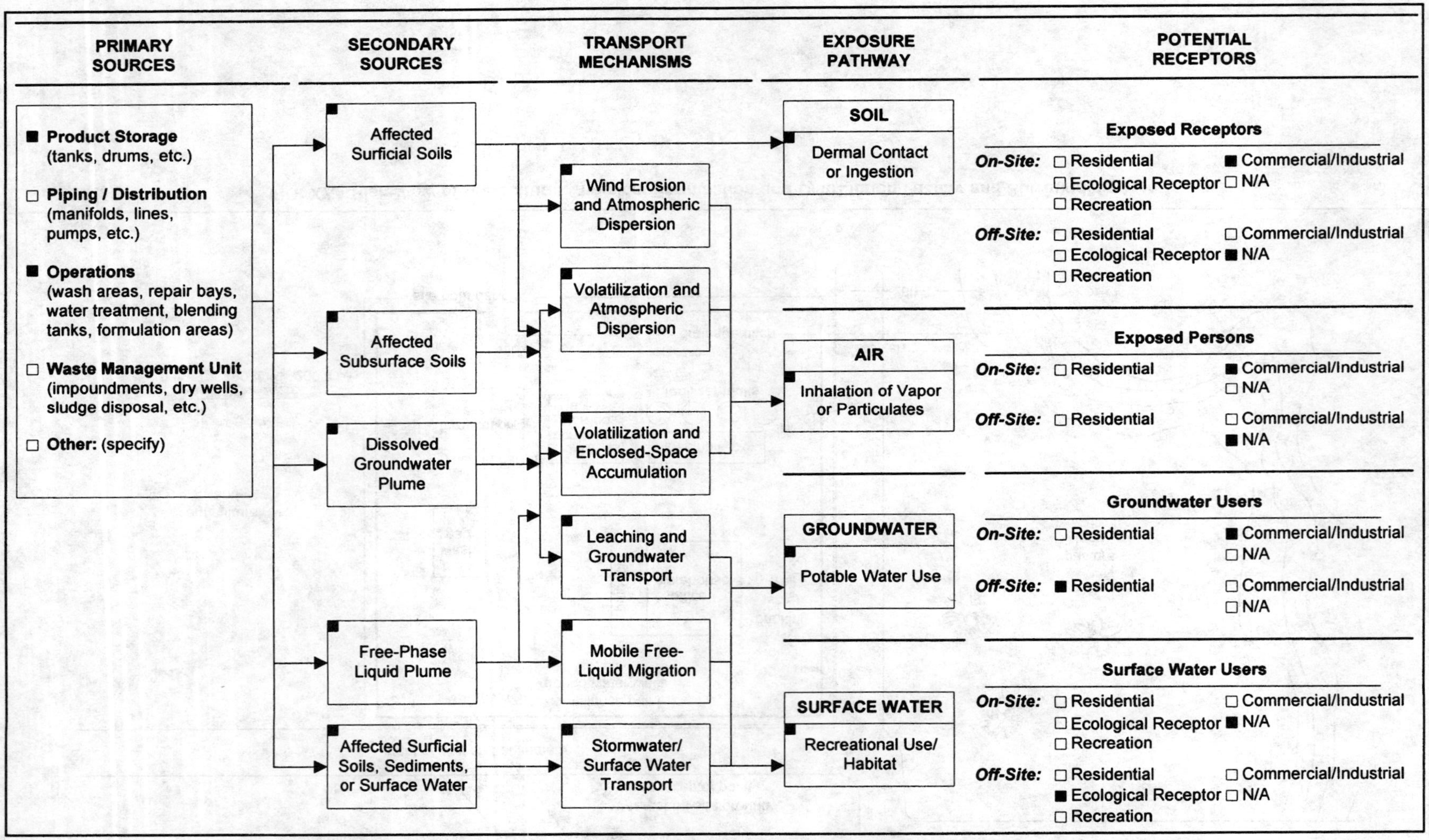

FIG. X7.2 Exposure Scenario Evaluation Flowchart for Agricultural Chemical Formulation and Distribution Facility.

Table 1, Site Conditions and Initial Response Actions). If an immediate threat was identified, it would be addressed forthwith, site conditions are evaluated and revised as additional information on the site becomes available.

X7.2.10 *Site Sampling Activities - Initial Site Investigation*:

X7.2.10.1 In the initial site investigation, soil samples were taken from the identified potential spill or leak areas. This included both surficial soil and samples taken at geometrically increasing depth increments to the water table, which on site was found at a depth BGS of approximately 28 ft. In addition, surficial soil samples were taken from several locations along the dry drainage ditch.

X7.2.10.2 The selected soil analyses were based on chemicals of concern identified from previous and current site use. In areas where pesticide impacts were suspected, samples were analyzed for organochlorine pesticides, organophosphate pesticides, and carbamate pesticides. Soil samples were analyzed for nitrate and ammonia in the areas potentially impacted by spilled fertilizer. In the used oil area, the soil chemical analyses included volatile and semi-volatile organic compounds, as well as total petroleum hydrocarbons reported by carbon number range.

X7.2.10.3 Half the soil borings were extended into the region of the upper water table, and shallow ground water samples (30 to 40 ft BGS) were taken from temporary sampling points. These ground water samples were analyzed for both pesticides and nitrate. Several composite water samples were taken from the flowing off-site creek, both upstream and downstream from the dry ditch outflow location. Ground water samples were also taken from existing wells in the vicinity of the site, as available, during the preliminary on-site investigation.

X7.2.10.4 Sampling and chemical analysis procedures appropriate for use in a quantitative risk assessment were used in the site investigation. The method quantification limits for the chemicals of potential concern, in nearly all cases, were below the most stringent applicable Tier 1 risk-based screening-level (RBSL) concentrations. Where this was not the case, it was judged that such chemicals were either not likely to be the sole chemicals present in a sample, or were not very likely to have been used at the site or in the site vicinity. The resulting data set was reviewed and judged to be reliable, accurate, and verifiable.

X7.2.11 *Analytical Data and Screening-Level Comparison-Tier 1*:

X7.2.11.1 Site conditions have been compared to the generic conditions used in deriving Tier 1 RBSL concentrations for soil and ground water, and based on the site soil type and identified site characteristics, the Tier 1 RBSL concentrations are considered to be conservative with respect to actual site conditions.

X7.2.11.2 The analyzed soil and ground water samples showed non-detectable concentrations for most potential chemicals of concern, in most samples. Given that the analytical method quantification limits were at or below the applicable RBSL concentrations, and that all identified potential source areas were sampled, no further investigation of the non-detected chemicals in soil or ground water was warranted.

X7.2.11.3 No generic Tier 1 Relevant Ecological Screening Criteria (RESC) concentrations are available for impacts to surface water as a relevant ecological receptor and habitat. The quantitative assessment of surface water impacts is deferred to a site-specific Tier 2 assessment. No site chemicals of concern were detected in the creek water samples and no pesticides were detected in surficial soil samples taken from the dry drainage ditch. Nitrate was detected in the drainage ditch at concentrations below any available Tier 1 RBSL criteria (based on leaching to ground water and exposure to surficial soil).

X7.2.11.4 The detected soil concentrations have been compared to the available and applicable Tier 1 RBSL concentrations. The soil concentrations in the following source areas for the indicated chemical and exposure pathways were above the corrective action goals:

Former pesticide formulation area.
- *Lindane:*
  - *leaching to ground water and surficial soil exposure.*
- *1,2-Dichloropropane:*
  - *leaching to ground water and indoor air inhalation.*

Truck wash pad.
- *Lindane:*
  - *leaching to ground water and surficial soil exposure.*
- *Toxaphene:*
  - *leaching to ground water, surficial soil exposure, indoor air inhalation.*

Fertilizer loading rack area and tank farm area.
- *Nitrate:*
  - *leaching to ground water.*

Additional investigation was focused on the chemicals for which the concentrations were above the Tier 1 RBSL. For those chemicals which were not above any applicable Tier 1 RBSL, no further investigation has been pursued.

X7.2.11.5 In the used oil storage area, concentrations of benzene, ethylbenzene, toluene, and xylenes, as well as polycyclic aromatic hydrocarbons were not above the RBSL for soils, but a relatively high total petroleum hydrocarbon concentration was measured. On a carbon number basis, this petroleum hydrocarbon was shown to consist of constituents with a boiling point greater than n-pentadecane ($C_{15}H_{32}$). In this range, the chemicals of concern are expected to be relatively immobile. No migration of the petroleum hydrocarbons beyond a 2 ft. depth was detected. No aesthetic concerns were noted.

X7.2.11.6 In ground water, measured concentrations were compared against the applicable and available Tier 1 RBSL criteria on a point-by-point basis. Most chemicals of potential concern were either not detected in ground water, or did not exceed the most stringent applicable Tier 1 RBSL concentration in any analyzed sample. Where this was the case, no further investigation of the screened chemicals was warranted. Those chemicals which were above a Tier 1 RBSL concentration for ground water included toxaphene and 1,2-dichloropropane, for both the risk-based ground water ingestion criteria and the ORMC, which are the State drinking water maximum contaminant levels (MCLs) for Class II ground water. The nitrate in ground water concentrations were above the State drinking water MCL for Class II ground water. Toxaphene concentrations, in addition, were above the RBSL ground water criteria based on indoor air inhalation.

X7.2.11.7 Ground water samples taken from existing water

wells in the vicinity of the site have been compared against Tier 1 RBSL ground water ingestion criteria. This included samples from the on-site drinking water well and the nearest downgradient drinking water well, 400 ft from the property boundary. None of the samples from the drinking water wells were above the Tier 1 corrective action goals for drinking water. Nitrate in upgradient, downgradient, and on-site drinking water wells was detectable at low levels, but were not above the Tier 1 corrective action goal (the State drinking water MCL).

X7.2.12 *Refined Screening-Level Comparison - Tier 1*:

X7.2.12.1 As a refinement on the point-by-point comparisons of site data with applicable Tier 1 RBSL concentrations, the areas where chemicals of concern were found to be above the corrective action goals were examined in greater detail. This comparison provides a means for determining potentially complete exposure pathways based on site data, and for expediting the choice of any needed remedial action.

X7.2.12.2 For those samples which were above the Tier 1 RBSL surficial soil criteria (lindane and toxaphene), further refinement of the point-by-point screening-level assessment showed that all of the samples which were above this Tier 1 RBSL criteria were taken at depths BGS greater than 10 ft, except for two samples (toxaphene) in the truck wash pad area, which were taken at a depth BGS of 1 ft. These shallow soil samples were above the Tier 1 RBSL surficial soil criteria for toxaphene by a wide margin.

X7.2.12.3 For the indoor air inhalation exposure pathway, with both 1,2-dichloropropane and toxaphene, no samples were above the indoor air RBSL criteria for soil either within a distance of 25 ft from a currently occupied building or within 15 ft of the ground surface, except for toxaphene in the vicinity of the truck wash pad area (where no current occupied buildings exist). For points where the concentrations at a depth greater than 15 ft (for either soil or ground water) were above the Tier 1 RBSL indoor air inhalation, no shallower samples taken in the same bore-hole were above the Tier 1 RBSL indoor air inhalation criteria for soil, indicating that the indoor air vapor transport predicted with the generic and conservative Tier 1 models is not confirmed by observation, and is not found to be a currently completed exposure pathway at this site.

X7.2.12.4 For the exposure pathway of leaching to ingested ground water, concentrations in different potential source areas were above the Tier 1 RBSL criteria for lindane, 1,2-dichloropropane, toxaphene, and nitrate. For 1,2-dichloropropane, toxaphene, and nitrate, the leaching potential was confirmed by on-site concentration measurements in ground water above the respective Tier 1 RBSL ground water ingestion criteria. The areas of ground water impacts at concentrations above the Tier 1 RBSL ground water ingestion criteria was adequately delineated for toxaphene in the preliminary site investigation. For 1,2-dichloropropane and nitrate, the ground water plumes above the Tier 1 RBSL ground water ingestion criteria extend beyond the on-site sampled area covered in the preliminary site investigation.

X7.2.13 Site Conditions & Initial Response Actions - Given the additional site information, the site condition analysis (Table 1, Example Site Conditions and Initial Response Actions) was repeated.

X7.2.13.1 Based on the initial site investigation, and estimated chemical of concern travel times, the potential exists for impact to an off-site domestic water supply well within approximately a two-year ground water travel time for both the nitrate and 1,2-dichloropropane plumes. In response, an interim remediation system was installed as a hydraulic stabilization measure to minimize off-site migration of impacted ground water. Sentry monitoring wells have been installed at the site between the source zones and the existing off-site drinking water wells to evaluate the effectiveness of the interim system. These sentry monitoring wells were completed in both the same zone (90–100 ft. BGS) as the drinking water wells, and at shallower levels (30–40 ft. BGS). A quarterly ground water monitoring program was initiated at the site, along with further ground water investigation (a Tier 2 investigation, discussed below).

X7.2.13.2 No current impact is seen in the existing off-site drinking water wells. By negotiation and agreement between the involved parties, the site will be re-evaluated and response action will be taken if two quarters of compliance point well monitoring data are above the applicable Tier 2 site-specific target level (SSTL) concentrations (discussed below). This response action is intended to eliminate the potential risk prior to any actual exposure to impacted drinking water.

X7.2.13.3 No impact is evident in the off-site creek, but the site is within 500 ft of this environmental receptor. The drainage ditch leading from the site to the creek provides an inappropriate transport conduit for site surface water run-off. As a interim response action, containment and control of surface water run-off to this ditch from the site has been implemented. This response action interrupts the identified exposure pathway. At the same time, the site was regraded to eliminate the two low areas on the site which may pond during wet weather periods.

X7.2.13.4 A final item of concern in the Tier 1 analysis is the toxaphene found in surficial soil at the truck wash pad above Tier 1 RBSL surficial soil criteria. This soil is within the access-controlled area of the site. As an interim response, this site access control will be maintained. Other responses were deferred until a decision on final corrective action for the site was reached. This deferral requires continued commitment by the current business operator to control access to the site, and to inform any individuals with access to the controlled area of the discovered chemicals and the hazards associated with these chemicals.

X7.2.14 *Evaluation of Tier 1 Results*—A series of decisions, based on urgency and relative costs and benefits, were made between interim remedial action, remedial action, or further tier evaluation at this point, for each instance where the site conditions did not meet the Tier 1 corrective action goals.

X7.2.14.1 Interim action was initiated in cases where it was justified by site conditions, as discussed in X7.2.13.

X7.2.14.2 For soils, the Tier 1 analysis of X7.2.11 and X7.2.12 has been sufficient; no recalculation of Tier 1 RBSL concentrations for soils using site-specific data is justified.

X7.2.14.3 For ground water, a Tier 2 assessment and evaluation is justified.

X7.2.15 *Tier 2 Investigation*—The Tier 2 investigation has

focused on delineating the off-site nitrate and 1,2-dichloropropane ground water plumes, and in establishing a measure of the plume stability (growing, steady-state, shrinking) over time. The Tier 2 site investigation has included collection of additional ground water concentration data, measurement of saturated soil properties, and estimation of site-specific parameters including ground water velocity and direction. Simple mathematical models for ground water transport have been used to check and validate the internal consistency of the measured site data and the attenuation of detected chemicals with distance from the source zone. Well placement in this investigation has been optimized to define the near-centerline plume concentrations with distance from the source, and to determine the maximum downgradient extent of the plume.

X7.2.16 *Tier 2 Assessment*:

X7.2.16.1 Measured aqueous concentrations of nitrate and 1,2-dichloropropane in shallow (30 to 40 ft. BGS) wells, both onsite and offsite, were found to be nearly stable with time. Measured and modeled attenuation rates of these chemicals along the near-centerline of the ground water plumes have been used to establish Tier 2 site-specific target level (SSTL) concentrations for nitrate and 1,2-dichloropropane. By negotiation and agreement between the involved parties, the chosen ground water receptor point was located just beyond the downgradient edge of the property boundary and adjacent public road easement. This was judged to be the nearest viable location for a future off-site ground water well. It is upgradient of the nearest existing ground water well. Screening level ground water concentrations were fixed at the Tier 1 RBSL ground water ingestion levels for the selected receptor point; the calculated Tier 2 SSTL compliance concentrations for the compliance point monitoring well locations (upgradient from the receptor point) are higher than the Tier 1 RBSL ground water ingestion concentrations, and include derived factors which account for the observed attenuation rates in ground water with increasing distance from the source area.

X7.2.16.2 The Tier 2 site investigation has shown that the nitrate ground water plume meets the Tier 2 SSTL compliance point concentration for the applicable compliance point sentry monitoring well. The measured soluble nitrate plume is reasonably stable or shrinking with time. Both upgradient and downgradient wells screened in the upper aquifer (30 to 40 ft BGS) show low but detectable nitrate levels, most likely due to the adjacent land use. No nitrate above Tier 1 RBSL ground water ingestion criteria were detected in monitoring wells screened at the depths similar to that used for drinking water wells (90–100 ft BGS) near the site.

X7.2.16.3 The 1,2-dichloropropane plume has been found to exceed the Tier 2 SSTL compliance point concentration for the applicable sentry monitoring well at shallow depth (30 to 40 ft BGS). Modeling and data show that this shallow plume extends off-site, and beyond the nearest possible future off-site ground water well, at levels greater than the Tier 1 RBSL ground water ingestion concentration. This shallow plume has also been shown to intercept the location of an existing off-site drinking water well. The drinking water well is completed in a deeper zone than the region where the chemicals of concern have been identified, and the drinking water well is not currently impacted. No 1,2-dichloropropane is detectable in any monitoring wells screened at depths of 90 to 100 ft, or at the depths similar to that used for drinking water wells near the site. The measured 1,2-dichloropropane plume is relatively stable over time.

X7.2.16.4 No chemicals of concern have been detected in the off-site creek. An estimate of potential surface water chemical concentrations in the creek has been made using a combination of the estimated volume of impacted ground water seepage into the creek and a geometric mean average water flow for the creek. This estimate of surface water chemical concentration is below analysis detection limits, and is also below any applicable freshwater aquatic toxicity criteria (selected as a relevant ecological screening criteria, or RESC) or human health-based criteria levels. Since surface water run-off has been contained on-site as part of a interim response action, water run-off is no longer a relevant transport pathway to the creek for this site. The interim response action has interrupted the identified exposure pathway.

X7.2.17 *Site Conditions and Emergency Response*—Additional information from the Tier 2 investigation and assessment has warranted a re-classification of the site (Table 1, Example Site Conditions and Initial Response Actions), based on better knowledge of site conditions and the measured extent of the chemical of concern ground water plumes. The off-site 1,2-dichloropropane ground water plume not being captured by the interim system has been shown to intercept the location of an existing residential drinking water well. This drinking water well is completed in a deeper zone than the plume, and is not impacted. The potential for vertical migration of the chemical of concern plume due to pumping at this existing drinking water well location has been investigated and found to be of concern. As an interim response, a point-of-use water treatment system has been provided for this ground water user. The previously implemented ground water monitoring program is also continuing at the site.

X7.2.18 *Final Corrective Action*:

X7.2.18.1 As a final corrective action, the interim system will become the final system with modifications appropriate to continue ground water treatment to levels achievable by the best available technology, and to hydraulically stabilize the plume to minimize offsite migration. The final system will continue operating until 1,2-dichloropropane concentrations are reduced below the Tier 2 RBSL or involved parties agree that is technically impracticable to continue system operation. In selection of this final corrective action, continued commercial/industrial use of the site is presumed. No everlasting use restrictions are intended to be placed on adjacent property use or off-site ground water use based on offsite plume stability and calculated attenuation rates. The continued commercial/industrial site description implies that a use restriction, using applicable and available methods (described in Appendix X6), has been placed on the site, and that this use restriction is acceptable to the involved parties. This use restriction includes limitations for impacted on-site ground water use as a drinking water supply, and appropriate limitations on site activity (for example, no residential use, no deep excavation without appropriate worker protection).

X7.2.18.2 For soils, final corrective action has included removal of affected soils in the truck wash pad area from the surface to a depth of 3 ft, appropriate disposal of the impacted soil, and replacement with clean soil fill material. All other issues regarding soils have been addressed in interim corrective action or in the refined assessment. The interim actions for soils have included elimination of a potential surface runoff transport pathway to a surface water body, and site grading to eliminate the potential for on-site ponded water (and the associated water-head driven aqueous infiltration).

X7.2.18.3 Remediation of the 1,2-dichloropropane ground water plume to the applicable Tier 2 SSTL levels, by any existing and available methodology is technically impracticable. Remediation to asymptotic levels by the best available technology is estimated to take up to three years until it would be technically impracticable to continue system operation.

X7.2.18.4 During the remediation period, to prevent unintended exposure to the affected ground water, off-site use restrictions for ground water in the immediate vicinity of the ground water plume were put in place. This use restriction, during the remediation period, is based on concurrence of all affected parties. Given the beneficial use of ground water in the local area as a sole drinking water supply source, the ultimate goal has been to return to unrestricted use of the off-site ground water as quickly as is technically possible. We note that the Tier 2 assessment has already established that the 1,2-dichloropropane ground water plume is relatively stable over time, and, at least, the present situation is not worsening. System operation is expected to reduce the overall time to reduce the 1,2-dichloropropane concentration to a level where monitored natural attenuation can continue as the selected final corrective action until the Tier 2 SSTL is achieved.

X7.2.18.5 The ground water plume is fed from a aqueous leachate source in on-site unsaturated soils. Lowering the on-site aqueous infiltration rate, through site grading, has been estimated to proportionately lower the total mass transport rate of leached 1,2-dichloropropane to ground water. Given literature-identified attenuation mechanisms of hydrolysis and biodegradation for 1,2-dichloropropane, and the lowered aqueous-leachate plume source loading, recession of the plume with time was predicted.

X7.2.18.6 After 2 additional years of quarterly ground water monitoring subsequent to system deactivation, seasonal fluctuations in ground water concentrations were seen, but a general downward trend in concentration was evident. At this point an annual monitoring program was implemented. After 6 additional years, the general downward trend was shown to continue, and the 1,2-dichloropropane plume was shown to decrease below the Tier 2 SSTL compliance point concentration for the applicable sentry monitoring well. No compliance point concentrations were above the SSTL within an additional year-long period of monitoring.

X7.2.18.7 Based on prior agreement between the involved parties, meeting all appropriate Tier 2 SSTL concentrations for the applicable sentry monitoring wells, over a 1 year time period, has been sufficient demonstration that the site meets all remaining conditions with respect to the site use restrictions discussed in X7.2.18.1. A "no further action" letter, consistent with the continued restricted commercial/industrial site use, has been issued. The off-site ground water use restrictions discussed in X7.2.18.4 have been lifted.

X7.2.18.8 If, over the period of monitoring, decreases in monitored ground water concentrations did not meet expectations, the corrective action plan for ground water would have been reviewed and modified. During the period of monitoring, site conditions and additional data were used to continuously monitor the site classification conditions and the appropriateness of the corrective action response.

X7.2.18.9 If a specific regulatory requirement of, for example, a ground water restoration goal is applicable as an ORMC, then this would be included in the corrective action goals.

X7.2.19 No further action for this site is based on restricted commercial/industrial site use. If less restrictive site use (residential use, for example) is desired by the property owner, a re-assessment of the property would be needed.

X7.2.20 Again, the decisions made in this hypothetical case example are based on presumed technical policy decisions, concurrence of all involved parties, and compliance with federal, state, or local laws and regulations.

X7.3 *Operating Industrial Facility*:

X7.3.1 *Initial Site Assessment*—Components of the initial site assessment are discussed in X7.3.2 to X7.3.10. A plan view of the site and surrounding area, with important features illustrated, is given in Fig. X7.3.

X7.3.2 *Site Description*—The property consists of a 3-acre city block in a zoned light industrial area. The adjacent land use is mixed light industrial, commercial, and residential. The immediate neighboring properties are all commercial and industrial; the nearest residences are at a distance of 400 ft from the property boundary. The site is not subject to any federal, state, or local enforcement actions.

X7.3.3 *Site Ownership and Activity*—There are two businesses currently operating on the site. The first is a commercial construction contractor, and includes an office building and warehouse, an asphalt-paved parking lot, and a gravel-paved equipment and materials storage yard. The second business is an electric motor rebuilding facility which includes an office, a workshop, a manufacturing building, a storage area, a used solvent tank location, and an asphalt-paved parking lot. Both businesses have been in operation at the present site for over twenty-five years.

X7.3.4 *Past Releases or Potential Source Areas*—Potential spill or leak areas on the site include a former underground gasoline storage tank, since removed; a diked concrete-paved solvent storage area; two former outdoor above ground heating oil tanks, since removed; an electrical equipment and transformer storage area; and a former sludge and used solvent storage tank area.

X7.3.5 *Regional and Site Hydrogeologic Conditions*:

X7.3.5.1 The site hydrogeological investigation has included soil borings, collection of ground water samples, and characterization of geological structure and hydrogeological conditions. The site surface topography is generally flat with a 15 ft variation over the 3-acre site. The regional topography is

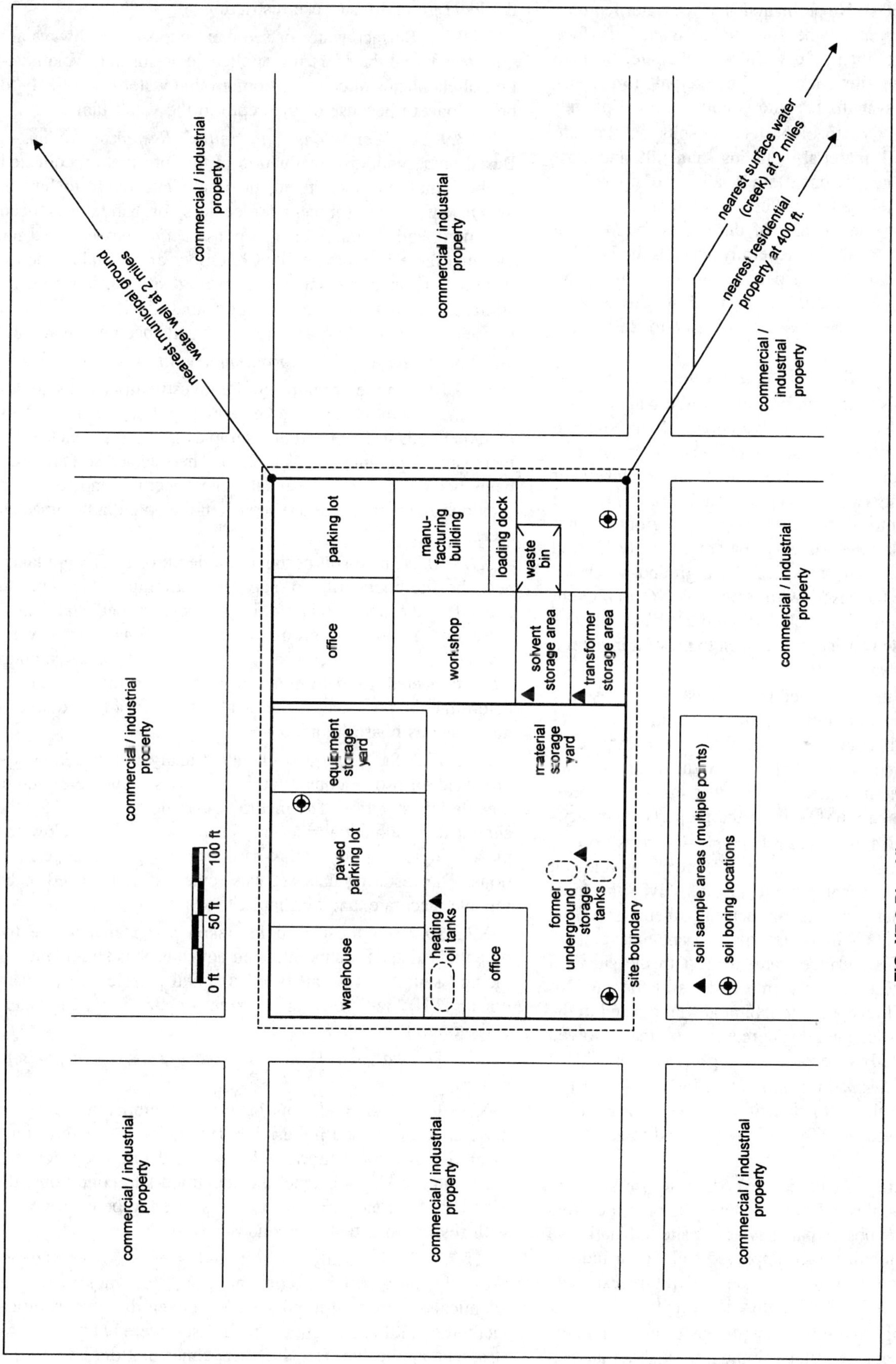

FIG. X7.3 Plan View of Operating Industrial Facility and Surrounding Area.

also relatively flat; surface elevation changes within 2 miles of the site are less than 50 ft. No permanent surface water features or low spots are present on the site. No permanent surface water features are present in the vicinity of the site. Surface water runoff from the site drains to a municipal stormwater sewer and treatment system. The site is not in a flood plain.

X7.3.5.2 The first five to eight feet of soils on the site consist primarily of fill material including sand, silt, and clay. The fill material is generally underlain by a layer of up to 10 ft of silty loam, followed by a layer of clayey silt to a depth of 40 ft. The water table in the vicinity of the site is found at an average depth of 60 to 80 ft, and generally exceeds the USEPA drinking water criteria for total suspended particulate, iron, and manganese. Municipal drinking water used in the area is produced from either wells at depths of 200 to 250 ft, or supplied from a local river.

X7.3.6 *Beneficial Use Summary and Receptor Survey*:

X7.3.6.1 The site is within an industrialized urban area. All businesses and residences within five miles of the site are served by municipal water and sewer utilities. The municipal water is supplied from a local river and from ground water wells, the nearest being two miles upgradient from the site, and producing from a depth of 220 ft. No producing ground water wells or well setback zones of any type are currently located closer than two miles from the site. A local ground water use restriction is in place (as described in Appendix X6) to prevent installation of drinking water wells within the urban area. The intent of this local use restriction has been to prevent ingestion of impacted ground water within the urban area.

X7.3.6.2 The site itself is zoned light industrial, is presently operating with commercial/industrial tenants, and by all likelihood, land use will continue in the future as commercial/industrial. All neighboring properties adjacent to the site are currently either light industrial or commercial. The nearest residential properties are 400 ft. from the property boundary.

X7.3.7 *Ecological Assessment Summary*—There is no evidence of impact to vegetation on the site. There are no relevant ecological receptors and habitats on the site. Given the local industrial development and urban activity, there is a low likelihood that the site will be returned to an undeveloped, pristine condition. The surface water nearest to the site is a creek which, at closest, passes within 2 miles of the site. No impacts to ecological receptors or habitat are anticipated in the future due to this site, and, by agreement of the involved parties, no further study of ecological impacts is planned.

X7.3.8 *Exposure Scenario Evaluation Flowsheet*—An exposure scenario flowsheet for this site is given in Fig. X7.4. This flowsheet is based on information discussed in X7.3.2 to X7.3.7.

X7.3.8.1 Commercial/industrial exposure to on-site impacted surficial soils, through direct contact, is both a current and potential future exposure pathway at this site. Inhalation of vapors evolved from subsurface impacted soils, both indoors and outdoors, is also a current or potential future exposure pathway for on-site commercial/industrial receptors.

X7.3.8.2 Leaching of soluble constituents to ground water, or migration of non-aqueous phase liquids, are viable transport pathways at this site, but ingestion of impacted ground water by residents in the vicinity of this site is not of concern, given the local ground water use restriction.

X7.3.8.3 Surface water or sediment exposure pathways are not present and therefore not applicable at this site. Volatilization of chemicals of concern from ground water is not likely to be of concern because of the depth to the water table.

X7.3.9 *Site Conditions and Initial Response Actions*—Based on a qualitative evaluation of the information collected in the initial site assessment, no immediate threat to human health, safety, environmental receptors, or habitat has been identified, and no immediate response action is warranted for the site (Table 1, Example Site Conditions and Initial Response Actions). If an immediate threat was identified, it would be addressed forthwith. Site condition analysis is repeated and revised as additional information on the site becomes available.

X7.3.10 *Site Activities - Preliminary Site Investigation*:

X7.3.10.1 In the preliminary site investigation, soil samples were taken from the identified potential spill or leak areas. This included both surficial soil and samples taken at geometrically increasing depth increments to 25 ft. The selected soil analyses have been based on chemicals of concern identified from previous and current site use at each of the potentially impacted areas.

X7.3.10.2 Three soil borings to a depth of 80 ft were taken near the site boundaries to map soil stratigraphy, the depth to ground water, and the approximate local ground water flow direction. Temporary ground water sampling points were installed at these three locations and ground water samples were collected to determine ground water quality and to determine the presence of chemicals of concern, including any non-aqueous phase liquids.

X7.3.10.3 Sampling and chemical analysis procedures appropriate for use in a quantitative risk assessment were used in the site investigation. The method quantification limits for the chemicals of potential concern, in all cases, were below the most stringent applicable risk-based screening-level concentrations. The resulting data set was reviewed and judged to be reliable, accurate, and verifiable.

X7.3.10.4 For those samples taken in an area where the potential existed for a mobile, non-aqueous phase liquid plume, the soil samples were analyzed for total petroleum hydrocarbons (TPH), with measured values reported by equivalent carbon number range.

X7.3.11 *Analytical Data and Screening-Level Comparison-Tier 1*:

X7.3.11.1 Site conditions have been compared to the generic conditions used in deriving Tier 1 risk-based screening-level (RBSL) concentrations for soil and ground water, and based on the site soil type and identified site conditions, the Tier 1 RBSL concentrations are considered to be conservative with respect to actual site conditions.

X7.3.11.2 The analyzed soil and ground water samples showed non-detectable concentrations for most potential chemicals of concern in all samples. Given that the analytical method quantification limits, in all cases, were below the most stringent applicable Tier 1 RBSL concentrations, no further investigation of these non-detected chemicals was necessary.

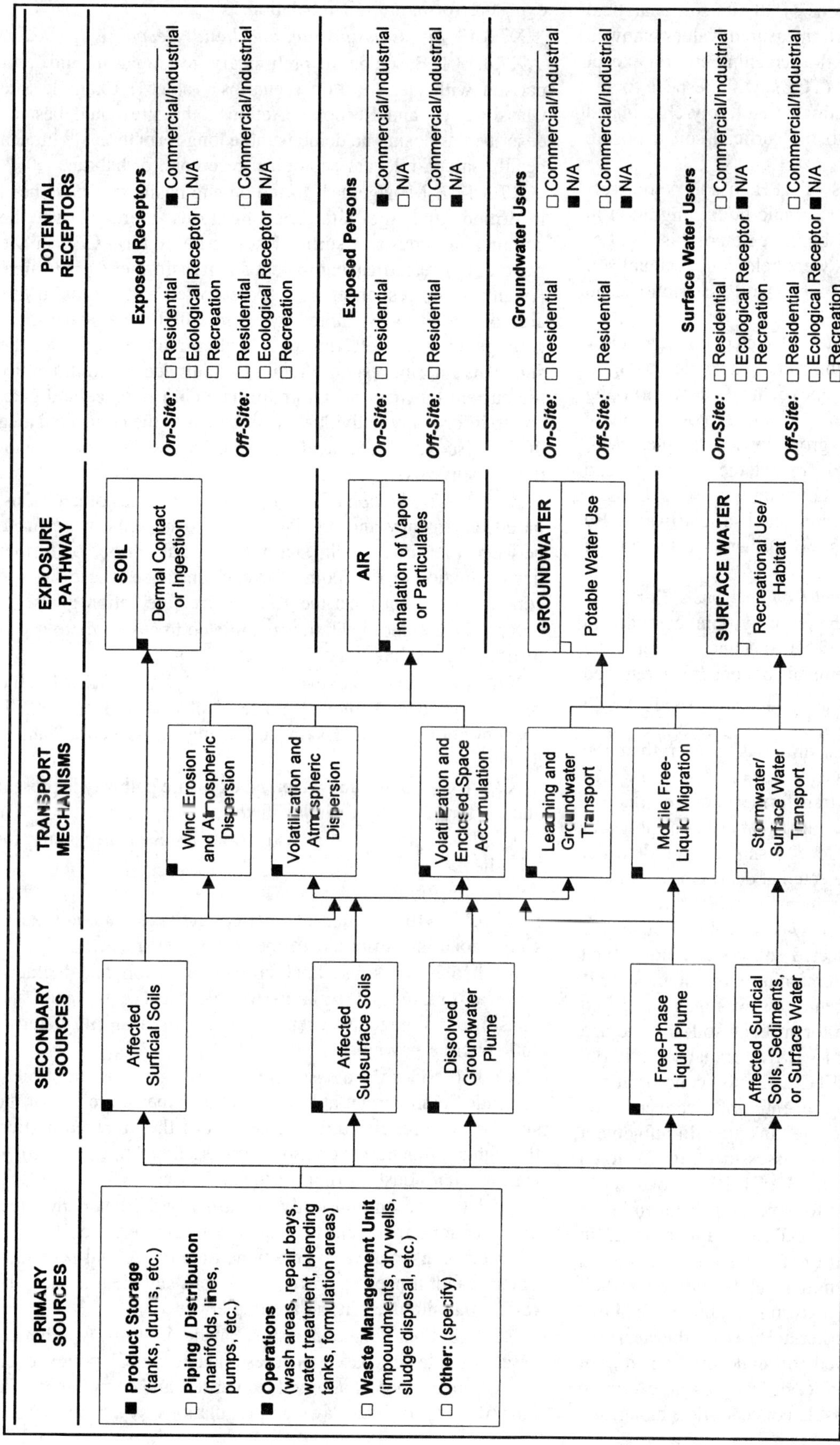

FIG. X7.4 Exposure Scenario Evaluation Flowchart for Operating Industrial Facility.

X7.3.11.3 The preliminary screening of the site data consisted of point-by-point comparisons of all site soil analytical data to the applicable RBSL. All measured values showed concentrations which were below the screening criteria except for benzene and trichloroethene (TCE), with respect to the indoor air inhalation exposure pathway; and poly-chlorinated biphenyls (PCBs), with respect to the surficial soil exposure pathway.

X7.3.11.4 No measured values of TPH in soil were great enough to indicate that free-phase organic liquid migration in soils would be of concern for any source region of this site. For TCE, all measured concentrations were below the residual soil concentration level calculated for Tier 1 soil assumptions; no free-phase liquid was found.

X7.3.11.5 The TCE was detected at concentrations below the calculated Tier 1 RBSL concentration in a single downgradient ground water sample, but above the Federal drinking water MCL concentration value. No measured free-phase liquids were evident in any of the ground water samples. With a local use restriction in place to prevent the use of ground water wells for drinking water in the vicinity of this site, there will be no ingestion exposure to impacted ground water. By agreement of the involved parties, no further investigation of ground water is planned.

X7.3.12 *Refined Screening-Level Comparisons - Tier 1*:

X7.3.12.1 As a refinement of the point-by-point comparison of site data with applicable Tier 1 RBSL concentrations, the areas where concentrations of chemicals of concern were noted to be above the corrective action goals were examined in greater detail.

X7.3.12.2 For the PCBs, concentrations greater than the Tier 1 RBSL screening-level values were confined to the area of the electrical equipment and transformer storage area. Of four surficial soil samples in this area, two showed low but quantifiable levels of PCBs, while two showed non-detected concentrations. No PCBs were detected at depth. In refinement of the Tier 1 screening-level assessment, an arithmetic average PCB surficial soil concentration was calculated, using a weighted average of the two detected soil concentrations and half the detection limit for the remaining two samples. This averaged surficial soil PCB concentration was then compared to the Tier 1 RBSL concentration for surficial soil; the average concentration was above the Tier 1 RBSL surficial soil criteria.

X7.3.12.3 For the benzene and TCE, site soil concentrations that were above the Tier 1 RBSL concentrations were confined to the area of the used solvent storage tank area. In refinement of this point-by-point screening-level assessment, it was noted that the only values exceeding the Tier 1 RBSL indoor air inhalation soil concentration were for soil samples taken in the depth range of 18 ft to 25 ft BGS; shallower samples taken in the same borings at depths of 4 ft to 12 ft were lower, and did not, in any case, exceed this screening level. In addition, no soil samples taken within 25 ft of any existing building were above any Tier 1 RBSL indoor air soil criteria. Based on this analysis, we can conclude that the observed soil concentrations do not confirm vapor transport to the surface at levels great enough to exceed the applicable Tier 1 RBSL concentration along any existing complete transport pathway, or at any actual point of exposure for indoor air inhalation.

X7.3.13 *Site Classification and Initial Response*

X7.3.13.1 Based on the preliminary investigation, and comparison with classification scenarios (Table 1, Example Site Classification and Response Actions), the site could best be described as posing no demonstrable long-term threat to human health, safety, relevant ecological receptors, or habitat.

X7.3.13.2 Management of the on-site businesses have been informed that the site concentrations were above the commercial/industrial surficial soil criteria for PCBs in the region of the electrical equipment and transformer storage area. As an interim response, signs were posted and access to this area of the site was limited by an installed fence and locked gate, pending the selection of a final corrective action for the site. This interim response requires continued commitment by the current business operators to maintain the fence and gate, and to inform any individuals with access to the controlled area of the discovered chemicals and the hazards associated with these chemicals.

X7.3.14 *Evaluation of Tier 1 Results*—A series of decisions, based on urgency and relative costs and benefits, were made between interim remedial action, remedial action, or further tier evaluation at this point, for each instance where the site conditions did not meet the Tier 1 corrective action goals.

X7.3.14.1 Interim action was initiated in cases where it was justified by the site conditions.

X7.3.14.2 For soils, the Tier 1 analysis of X7.3.11 and X7.3.12 has been sufficient; no recalculation of Tier 1 RBSL concentrations for PCBs in soil using site-specific data is needed.

X7.3.14.3 For ground water, no exposure pathway is present and no further analysis is justified.

X7.3.14.4 No Tier 2 investigation has been necessary for this site.

X7.3.15 *Final Corrective Action*:

X7.3.15.1 In selection of final corrective action, continued commercial/industrial use of the site is presumed.

X7.3.15.2 For soils, final corrective action has included removal of surficial soils in the electrical equipment and transformer storage area, replacement with clean fill material, and concrete paving.

X7.3.16 Since the assessment is based on the assumption of continued commercial/industrial site use, the future use of the site has been restricted, by agreement of the involved parties, through administrative controls (discussed in Appendix X6), to commercial/industrial (non-residential) use.

X7.3.17 The administrative controls will remain in effect unless actions are taken at a later date to re-access the site using the more conservative assumptions of a residential screening criteria, and, if needed, corrective actions sufficient to meet the residential site use criteria are implemented.

X7.3.18 Finally, the decisions made in this hypothetical case example are based on presumed technical policy decisions, concurrence of all involved parties, and compliance with federal, state, or local laws and regulations.

X8. OTHER RESOURCES

American Petroleum Institute, "A Guide for Assessing and Remediating Petroleum Hydrocarbons in Soils", *API Publication 1629*, First Edition, 1993.

American Petroleum Institute, "Assessing the Significance of Subsurface Contaminated Vapor Migration to Enclosed Spaces", *API HESD Publication 4674*, 1998.

Boulding, J.R., 1991, Description and Sampling of Contaminated Soils: A Field Pocket Guide, EPA/625/12-91/002, 122 pp.

Boulding, J.R., 1994, Description and Sampling of Contaminated Soils: A Field Guide, Revised and Expanded 2nd Edition, Lewis Publishers, Chelsea, MI, 220 pp.

Department of Labor 29 CFR 1910: General Industry Standards

Department of Labor 29 CFR 1926: Construction Standards

Di Toro, Dominic M., Mahony, John D., Hansen, David J., Scott, K. John, Hicks, Michael B., Mayr, Suzanne M., and Redmond, Michele S. 1990, Toxicity of Cadmium in Sediments: The Role of Acid Volatile Sulfide, *Environmental Toxicology Chem.* 9:1487-1502.

Freeze, R.A. and J.A. Cherry, 1979, Ground Water, *Prentice-Hall Publishing Co.*, Englewood Cliffs, J.J. 604 pp.

Hemond, Harold F and Fechner, Elizabeth J., 1994, *Chemical Fate and Transport in the Environment*, Boston: Academic Press.

Howard, Philip H., Boethling, Robert S., Jarvis, William F., Meylan, William M., and Michaelenko, Edward M., 1991, *Handbook of Environmental Degradation Rates*, Chelsea, Michigan: Lewis Publishers, Inc.

Johnson, P.C., and Ettinger, R.A., 1991: "Heuristic Model for Predicting the Intrusion Rate of Contaminant Vapors into Buildings", *Environmental Science and Technology*, 25, pp. 1445–1452.

Johnson, P.C., 1998, "Assessing the Significance of Subsurface Vapor Contaminant Vapor Migration to Enclosed Spaces: Site-Specific Alternatives to Generic Estimates", *American Petroleum Institute, Health and Sciences Department*, Publication Number 4674.

LaGrega, Michael D., Buckingham, Phillip L, Evans, Jeffrey C., and The Environmental Resources Management Group, 1994, *Hazardous Waste Management*, New York: McGraw-Hill, Inc.

Lyman, Warren J., Reehl, William F., and Rosenblatt, David H., 1990, Handbook of Chemical Property Estimation Methods, Washington, D. C., *American Chemical Society* .

Mausbach, M.J., 1992, Soil Survey Interpretations for Wet Soils, *In: Proc. 8th Int. Soil Correlation Meeting (VII ISCOM):* Characterization, Classification and Utilization of Wet Soils, J.M. Kimble (ed), USDA, Soil Conservation, National Soil Survey Center, Lincoln, NE, pp 172-178.

NFPA 329, Leakage and Repair Safeguards for Flammable and Combustible Liquids.

The Presidential/Congressional Commission on Risk Assessment and Risk Management, "Framework for Environmental Health Risk Management", 1997.

USEPA, "Activity Factors", ORD, PB98-124241, August 1997.

USEPA *Community Relations in Superfund: A Handbook,* EPA/540/R-92/009, January 1992.

USEPA, "Dermal Exposure Assessment: Principles and Applications", Interim Report, 1992.

USEPA, "Ecological Assessment of Hazardous Waste Sites: A Field and Laboratory Reference Document", EPA/600/3-89/013, NTIS No. PB89205967, March 1989.

USEPA, "Food Ingestion Factors", ORD, PB98-124225, August 1997.

USEPA, "General Exposure Factors", ORD, PB98–124233, August 1997.

USEPA, "Guidance for Data Usability in Risk Assessment (Part A)", OERR, (9285.7-09A), April 1992.

USEPA, "Guidelines for Ecological Risk Assessment", 63 Fed. Reg. 26846-26924, May 14, 1998.

USEPA, *"Guidelines for Exposure Assessment"*, Federal Register 57, No. 104, May 1992, pp 22888-22938.

USEPA, "Human Health Evaluation Manual, Supplemental Guidance: Standard Default Exposure Factors", OSWER Directive No. 9285.603, NTIS No. PB91921314, March 1991.

USEPA, *Land Use in the CERCLA Remedy Selection Process*, OSWER Directive No. 9355.7-04, May 25, 1995.

USEPA, "Methods for Derivation of Inhalation Reference Concentrations and Application of Inhalation Dosimetry", ORD, EPA/600/8-90/066F, October 1994.

USEPA *RCRA Public Participation Manual,* EPA/530/R-96/007, September 1996.

USEPA, "Risk Assessment Guidance for Superfund, Volume 1, Human Health Evaluation Manual, Part B: Development of Risk-Based Preliminary Remediation Goals", OSWER Directive No. 9285.7-01B, NTIS No. PB9296333, USEPA, December 1991.

USEPA, "Risk Assessment Guidance for Superfund, Volume 2: Environmental Evaluation Manual", EPA/540/1-89/001, NTIS No. PB90155599, March 1989.

USEPA, Sediment Quality Criteria for the Protection of Benthic Organisms. EPAÄ822ÄRÄ93Ä015), 1993.

USEPA, "User's Guide for the Johnson and Ettinger (1991), Model for Subsurface Vapor Intrusion into Buildings", (United States Environmental Protection Agency, Office of Emergency and Remedial Response, Washington, DC), September 1997.

REFERENCES

(1) USEPA, *Final Guidance for the Data Quality Objective Process*, September 1994.

(2) USEPA, *Soil Screening Guidance: Technical Background Document*, United States Environmental Protection Agency, Office of Solid Waste and Emergency Response, Washington, DC, EPA/540/R-95/128, NTIS PB96-963502, May, 1996.

(3) USEPA, *Test Methods for Evaluating Solid Waste Physical/Chemical Methods, SW-846, 3rd Edition, Proposed Update*, EPA/SW-846.3.2, PB94-170 321, 1994.

(4) Schwarzenbach, René P., Gschwend, Philip M., and Imboden, Dieter M., *Environmental Organic Chemistry,* New York: John Wiley & Sons, Inc., 1993.

(5) Howard, Philip H., *Handbook of Environmental Fate and Exposure Data for Organic Chemicals. Volume II, Solvents,* Chelsea, Michigan: Lewis Publishers, Inc., 1990.

(6) Linz, D.G., and Nakles, D.V., *Environmentally Acceptable Endpoints In Soil: Risk-Based Approach to Contaimated Site Management Based on Availability of Chemicals In Soil,* American Academy of Environmental Engineers, 1996, p. 630.

(7) Di Toro, Dominic M., Mahony, John D., Hansen, David J., Scott, K. John, Carlson, Anthony R., and Ankley, Gerald T., Acid Volatile Sulfide Predicts the Acute Toxicity of Cadmium and Nickel in Sediments, *Environmental Science and Technology,* Vol 26, 1992, pp. 96-101.

(8) *Integrated Risk Information System (IRIS)*, United States Environmental Protection Agency, Washington, D.C., Updated Monthly.

(9) *Health Effects Assessment Summary Tables (HEAST)*, OSWER OS-230, United States Environmental Protection Agency, Washington, DC, March 1992.

(10) USEPA, *"Provisional Guidance for Quantitative Risk Assessment of Polycyclic Aromatic Hydrocarbons"*, ECAO-CIN-842, 1993.

(11) Edwards, D.A., et. al. TPHCWG Series, *"Volume 4: Development of Fraction-Specific Reference Doses (RfD's) and Reference Concentrations (RfC's) for Total Petroleum Hydrocarbons (TPH)"*, Amherst Scientific Publishers, Amherst, MA 1997.

(12) Gustafson, J.B., Griffith Tell, J., and Orem, D., TPHCWG Series, *"Volume 3: Selection of Representative TPH Fractions Based on Fate and Transport Considerations"*,Amherst Scientific Publishers, Amherst, MA 1997.

(13) Potter, T.L., and Simmons, K.E., TPHCWG Series, *"Volume 2: Composition of Petroleum Mixtures"*, Amherst Scientific Publishers, Amherst, MA 1998.

(14) Wade, W. ed., TPHCWG Series, *"Volume 1: Analysis of Petroleum Hydrocarbons in Environmental Media"*, Amherst Scientific Publishers, Amherst, MA 1998.

(15) *Risk Assessment Guidance for Superfund, Vol 1, Human Health Evaluation Manual, Part A*, EPA/540/1-89/002, United States Environmental Protection Agency, Washington, DC, July 1989.

(16) USEPA, *Air Emissions Models for Waste and Wastewater*, United States Environmental Protection Agency, Office of Air Quality Planning and Standards, EPA-453/R-94-080A, 1994.

(17) ACGIH *Threshold Limit Values*, (American Conference of Governmental Industrial Hygienists, Cincinnati, OH), (annual revision cited dates are the latest revision), 1996.

(18) USEPA, *Water Related Environmental Fate of 129 Priority Pollutants*, Vol 1, USEPA, OWQPS, (EPA-440/4-79-029a), 1979.

(19) USEPA, *Hazardous Waste Treatment, Storage, and Disposal Facilities (TSDF) – USEPA, OAQPS, Air Emission Models*, (EPA-450/3-87-026), 1989.

(20) Verschueren, Karel, *Handbook of Environmental Data on Organic Chemicals*, Third Ed., Van Nostrand Reinhold Co. Inc., New York, ISBN: 0–442-01916-5, 1996.

(21) 29 CFR 1910, Code of Federal Regulations, Chapter XVII Occupational Safety and Health Administration, Department of Labor, Part 1910 Occupational Safety and Health Standards.

(22) Johnson, P. C., Hertz, M. B., Byers, D. L., "Estimates for Hydrocarbon Vapor Emissions Resulting from Service Station Remediations and Buried Gasoline-Contaminated Soils," *Petroleum Contaminated Soils*, Vol 3, P. T. Kostecki and E. J. Calabrese, eds., (Lewis Publishers, Chelsea, MI), 1990, pp. 295-326.

(23) Mott, H. V., "A Model for Determination of the Phase Distribution of Petroleum Hydrocarbons at Release Sites," *Ground Water Monitoring and Remediation*, 15, 3, 1995, pp. 157-167.

(24) Mariner, P.E., Jin, M., and Jackson, R.E., A Algorithm for the Estimation of NAPL Saturation and Composition from Typical Soil Chemical Analyses, *Ground Water Monitoring and Remediation*, 17, 1:, 1997, pp. 122-129.

(25) Wilson, J. L. and Conrad, S. H. , "Is Physical Displacement of Residual Hydrocarbons a Realistic Possibility in Aquifer Restoration?", *National Water Well Association / American Petroleum Institute Petroleum Hydrocarbons and Organic Chemicals in Ground Water – Prevention, Detection, and Restoration Conference*, Houston, TX, 1984, pp. 274-298.

(26) Conrad, S. H., Hagan, E. F., and Wilson, J. L., "Why are Residual Saturations of Organic Liquids Different Above and Below the Water Table?", paper presented at the *National Water Well Association / American Petroleum Institute Petroleum Hydrocarbons and Organic Chemicals in Ground Water – Prevention, Detection, and Restoration Conference*, Houston, TX, 1986.

(27) Cohen, R.M., and Mercer, J.W., *DNAPL Site Evaluation,* (CRC Press, Inc., Boca Raton, Florida).

(28) American Petroleum Institute, *A Guide to the Assessment of Remediation of Underground Petroleum Releases*, Publication 1628, Third Edition, Washington, D. C., July, 1996.

(29) USEPA *Estimating the Potential for Occurrence of DNAPL at Superfund Sites*, Publication 9355.4-07FS, Office of Solid Waste and Emergency Response, Washington, DC, NTIS PB92-963358, 1992.

(30) USEPA, *"Exposure Factors Handbook Volumes I, II, III,* (EPA/600/P-95/002F a, b, c), Exposure Factors Handbook, August 1997.

(31) USEPA, *Supplemental Risk Assessment Guidance for the Superfund Program*, EPA/901/5-89/001, Environmental Protection Agency Region 1, Washington, DC, 1989.

(32) USEPA, *Superfund Exposure Assessment Manual*, EPA/540/1-88/001, United States Environmental Protection Agency, Washington, DC, July 1988.

(33) Cowherd, C., Muleski, G. E., Englehart, P. J., and Gillett, D. A., *Rapid Assessment of Exposure to Particulate Emissions from Surface Contamination Sites*, Midwest Research Institute, PB85-192219, 1985.

(34) American Industrial Health Council, *Exposure Factors Sourcebook,* 1994.

Designation: E 2091 – 00

Standard Guide for Use of Activity and Use Limitations, Including Institutional and Engineering Controls[1]

This standard is issued under the fixed designation E 2091; the number immediately following the designation indicates the year of original adoption or, in the case of revision, the year of last revision. A number in parentheses indicates the year of last reapproval. A superscript epsilon (ε) indicates an editorial change since the last revision or reapproval.

INTRODUCTION

Valuable property, which is, or is perceived to be, environmentally impacted, remains idle throughout the fifty states because fears of liability and corrective action costs deter potential developers, purchasers, and lenders. In response, many states have adopted voluntary corrective action or brownfields programs that utilize risk-based corrective action principles. One element of these programs may be activity and use limitations to achieve either an" acceptable risk" or a "no significant risk" level. For example, an owner/operator who volunteers to remediate a site to meet an industrial or commercial use standard may do so in exchange for a restrictive covenant that limits the use of the site to industrial or commercial purposes only. Activity and use limitations should be considered an integral part of the remedial action selection process. The user may determine, based upon post-remedial action land use, or based upon the deficiencies in available activity and use limitations, that an activity and use limitation is not feasible for the site. The most effective use of activity and use limitations as part of a federal, state, tribal or local remediation program requires careful consideration of many factors, including effectiveness, amenability to integration with property redevelopment plans, implementability, technical practicability, cost prohibitiveness, long-term reliability, acceptability to stakeholders, and cost effectiveness. While this guidance is most likely to be applied where risk-based corrective actions are conducted, use of activity and use limitations is not restricted to risk-based applications. Both institutional and engineering controls may be employed as elements of a remedial action that is based on concentration level, background, or other non-risk-based approaches.

1. Scope

1.1 This guide covers information for incorporating activity and use limitations that are protective of human health and the environment into federal, state, tribal or local remediation programs using a risk-based approach to corrective action. Activity and use limitations should be considered early in the site assessment and remedial action selection process, and should be considered an integral part of remedial action selection. In the event that an appropriate activity and use limitation cannot be found, the user may need to revisit the initial remedial action selection decision.

1.2 This guide does not mandate any one particular type of activity and use limitation but merely serves to help users identify, implement and maintain the types of activity and use limitations that may be appropriate in programs using a risk-based decision-making approach.

1.3 This guide identifies screening and balancing criteria that should be applied in determining whether any particular activity and use limitation may be appropriate. This guide identifies the need to develop long-term monitoring and stewardship plans to ensure the long-term reliability and enforceability of activity and use limitations. This guide explains the purpose of activity and use limitations in the remedial action process and the types of activity and use limitations that are most commonly available.

1.4 This guide describes the process for evaluating potentially applicable activity and use limitations and using screening and balancing criteria to select one or more activity and use limitations for a specific site. The guide also describes some "best practices" from a transactional, stakeholder involvement, and long-term stewardship perspective. The guide also emphasizes the importance of considering the need for, and potential applicability of, activity and use limitations EARLY in the remedial action process.

1.5 All references to specific Federal or state programs are current as of the date of publication. The user is cautioned not to rely on this guide alone but to consult directly with the appropriate program.

1.6 *This standard does not purport to address all of the safety concerns, if any, associated with its use. It is the*

[1] This guide is under the jurisdiction of ASTM Committee E-50 on Environmental Assessment and is the direct responsibility of Subcommittee E50.04 on Performance Standards Related to Environmental Regulatory Programs.

Current edition approved May 10, 2000. Published July 2000.

responsibility of the user of this standard to establish appropriate safety and health practices and determine the applicability of regulatory limitations prior to use.

2. Referenced Documents

2.1 *ASTM Standards:*

E 1527 Practice for Environmental Site Assessments: Phase I Environmental Site Assessment Process[2]

E 1599 Guide for Corrective Action for Petroleum Releases[2]

E 1739 Guide for Risk-Based Corrective Action Applied at Petroleum Release Sites[2]

E 1912 Guide for Accelerated Site Characterization for Confirmed or Suspected Petroleum Releases[2]

E 1943 Guide for Remediation of Ground Water by Natural Attenuation at Petroleum Release Sites[2]

E 2081 Guide for Risk-Based Corrective Action[2]

2.2 *USEPA Documents:*

EPA/540/4-96/018 Soil Screening Guidance[3]

Land Use in the CERCLA Remedy Selection Process, OSWER Directive No. 9355.7-04, May 25, 1995[3]

Institutional Controls: A Reference Manual (March 1998)[3]

3. Terminology

3.1 *Definitions of Terms Specific to This Standard:* The reader should review the definitions presented herein prior to reviewing this guide, as many of the items included in this guide may have specific regulatory definitions within existing federal, state, tribal, or local programs. The following terms are being defined to reflect their specific use in this guide. Many of these definitions are taken directly from Guide E 2081. The user should not assume that these definitions replace existing regulatory definitions. Where the definition or use of a term in this standard differs from an existing regulatory definition or use, the user should address these differences prior to proceeding with the corrective action process.

3.1.1 *acceptable risk*—risk which is deemed to be below a level of regulatory concern.

3.1.2 *activity and use limitations, or AULs*—legal or physical restrictions or limitations on the use of, or access to, a site or facility to eliminate or minimize potential exposures to chemicals of concern, or to prevent activities that could interfere with the effectiveness of a response action, to ensure maintenance of a condition of "acceptable risk" or "no significant risk" to human health and the environment. These legal or physical restrictions are intended to prevent adverse impacts to individuals or populations that may be exposed to chemicals of concern.

3.1.3 *affirmative easement*—one where the servient estate must permit something to be done thereon, as to pass over it, or to discharge water on it.

3.1.4 *appurtenant easement*—an easement that benefits a particular tract of land. An incorporeal right which is attached to a superior right and inheres in land to which it is attached and is in the nature of a covenant running with the land. There must be a dominant estate and a servient estate.

3.1.5 *chemical release*—any spill or leak or detection of concentrations of chemical(s) of concern in environmental media.

3.1.6 *chemical(s) of concern*—the specific compounds and their breakdown products that are identified for evaluation in the risk-based corrective action process. Identification can be based on their historical and current use at a site, detected concentrations in environmental media, and their mobility, toxicity and persistence in the environment. Because chemicals of concern may be identified at many points in the risk-based corrective action process, the term should not be automatically construed to be associated with increased or unacceptable risk.

3.1.7 *corrective action*—the sequence of remedial actions that include site assessment and investigation, risk assessment, response actions, interim remedial action, remedial action, operation and maintenance of equipment, monitoring of progress, making no further action determinations, and termination of the remedial action.

3.1.8 *corrective action goals*—concentration or other numeric values, physical condition or remedial action performance criteria that demonstrate that no further action is necessary to protect human health and the environment. For example, these goals may include one or a combination of RBSL, SSTL, RESC, SSEC and ORMC chosen for source area(s), point(s) of demonstration and point(s) of exposure. The corrective action goals are specific to each Tier in the evaluation.

3.1.9 *deed restriction*—a restriction or limitation on an interest in real property, created by a conveyance from one person to another.

3.1.10 *direct exposure pathway*—an exposure pathway where the point of exposure is at the source, without a release to any other medium and without an intermediate biological transfer step.

3.1.11 *easement in gross*—an easement in gross is not appurtenant to any estate in land or does not belong to any person by virtue of ownership of an estate in other land but is merely a personal interest in or right to use the land of another. Easements that do not benefit a particular tract of land (e.g., utility easements).

3.1.12 *easement of access*—right of ingress and egress to and from the premises of a lot owner to a street appurtenant to the land of the lot owner.

3.1.13 *easements*—a right of use over the property of another. Traditionally, the permitted kinds of uses were limited, the most important being rights of way and rights concerning flowing waters. The easement was normally for the benefit of adjoining lands, no matter who the owner was (an easement appurtenant), rather than for the benefit of a specific individual (easement in gross). The land having the right of use as an appurtenance is known as the dominant tenement and the land which is subject to the easement is known as the servient tenement.

3.1.14 *ecological evaluation*—a process for organizing and analyzing data, information, assumptions and uncertainties to

[2] *Annual Book of ASTM Standards*, Vol 11.04.

[3] Available from Superintendent of Documents, US Government Printing Office, Washington, DC 20402.

evaluate the likelihood that adverse effects to relevant ecological receptors or habitats may occur or are occurring as a result of exposure to chemical(s) of concern.

3.1.15 *engineering controls*—physical modifications to a site or facility to reduce or eliminate the potential for exposure to chemicals of concern (for example, slurry walls, capping, hydraulic controls for ground water, or point of use water treatment).

3.1.16 *equitable servitudes*—building restrictions and restrictions on the use of land which may be enforced in equity. If there is a scheme in their creation, a subsequent owner may enforce them by injunctive relief against another subsequent owner. Such servitudes are broader than covenants running with the land because they are interests in land.

3.1.17 *exposure*—contact of an organism with chemicals of concern at the exchange boundaries (for example, skin, lungs, and liver) when the chemicals of concern are available for absorption or adsorption.

3.1.18 *exposure assessment*—the determination or estimation (qualitative or quantitative) of the magnitude, frequency, duration and route of exposure between a source area and a receptor.

3.1.19 *exposure pathway*—the course a chemical(s) of concern takes from the source area(s) to a receptor or relevant ecological receptor and habitat. An exposure pathway describes the mechanism by which an individual or population is exposed to a chemical(s) of concern originating from a site. Each exposure pathway includes a source or release from a source of a chemical concern, a point of exposure, an exposure route, and the potential receptors or relevant ecological receptors and habitats. If the exposure point is not at the source, a transport or exposure medium or both (for example, air or water) are also included.

3.1.20 *exposure route*—the manner in which a chemical(s) of concern comes in contact with a receptor (for example, ingestion, inhalation, dermal contact).

3.1.21 *exposure scenario*—the description of the circumstances, including site properties and chemical properties, or the potential circumstances under which a receptor or a relevant ecological receptor or habitat could be in contact with chemical(s) of concern.

3.1.22 *facility*—the property containing the source of the chemical(s) of concern where a release has occurred. A facility may include multiple sources and, therefore, multiple sites.

3.1.23 *geographic information system (GIS)*—a geographic information system (GIS) is a computer-based tool for tracking, mapping and analyzing resources using either an explicit geographic reference, such as a latitude and longitude or national grid coordinate, either from entry of this data from geographical location devices or by geographical coding an address or other descriptive location. GIS technology integrates common database operations such as query and statistical analysis with the visualization and geographic analysis benefits offered by maps.

3.1.24 *highest and best use*—the reasonably probable and legal use of vacant land or an improved property, which is physically possible, appropriately supported, financially feasible, and that results in the highest value. The four criteria that the highest and best use must meet are legal permissibility, physical possibility, financial feasibility, and maximum profitability.

3.1.25 *indirect exposure pathways*—an exposure pathway with at least one intermediate release to any media, or an intermediate biological transfer step, between the source and the point(s) of exposure (for example, chemicals of concern from soil through ground water to the point(s) of exposure).

3.1.26 *interim remedial action*—the course of action to reduce migration of chemical(s) of concern in its vapor, dissolved, or liquid phase, or to reduce the concentrations of a chemical of concern at a source area.

3.1.27 *institutional control*—a legal or administrative restriction on the use of, or access to a site or facility to eliminate or minimize potential exposures to a chemical(s) of concern (for example, deed restrictions, restrictive zoning).

3.1.28 *natural attenuation*—the reduction in the mass or concentration(s) of chemicals of concern in environmental media due to naturally occurring physical, chemical and biological process (for example, diffusion, dispersion, adsorption, chemical degradation and biodegradation).

3.1.29 *negative easement*—an easement where the owner of the servient estate is prohibited from doing something otherwise lawful upon his estate, because it will affect the dominant estate (for example, a prohibition on excavation deeper than 10 ft).

3.1.30 *no significant risk*—risk which is deemed to be below a level of regulatory concern. This level may vary among states and federal agencies, among regulatory programs, among media and pathways of concern, and among receptors. The terminology may also vary from jurisdiction to jurisdiction, and from regulatory program to regulatory program (for example, "acceptable risk level" or some similar term indicating that remedial measures have reached the target level for protecting human health and the environment).

3.1.31 *other relevant measurable criteria (ORMC)*—parameters used to define corrective action goals for chemical(s) of concern. The ORMC are concentration values, other numeric values, physical condition or performance criteria other than RBSL, RESC, SSTL or SSEC. Examples of ORMC are regulatory standards, consensus criteria, aesthetic criteria, and groundwater protection criteria. Technical policy decisions regarding ORMC may exist, or may need to be made to determine the appropriate values, conditions or performance criteria that are used for the corrective action goals.

3.1.32 *point(s) of demonstration*—a location(s) selected between the source area(s) and the potential point(s) of exposure where corrective action goals are met.

3.1.33 *point(s) of exposure*—the point(s) at which an individual or population may come in contact with a chemical(s) of concern originating from a site.

3.1.34 *potentially complete exposure pathway*—a situation with a reasonably likely chance of occurrence in which a receptor or relevant ecological receptor or habitat may become directly or indirectly exposed to the chemical(s) of concern.

3.1.35 *proprietary*—belonging to ownership; owned by a particular person; belonging or pertaining to a proprietor; relating to a certain owner or proprietor.

3.1.36 *proprietary controls*—controls based on the rights associated with private ownership, particularly ownership of a limited interest in real property as specified in a legal instrument, such as an easement or a restrictive covenant.

3.1.37 *qualitative ecological screening evaluation*—a process conducted as part of the Tier 1 evaluation wherein relevant ecological receptors and habitats and exposure pathways are identified. The necessary information can be collected as part of the data gathering activities during the initial site assessment or the Tier 1 site assessment. Within Tier 1, this screening-level information, which is typically qualitative, may be used to evaluate potential exposure pathways to relevant ecological receptors and habitats and to identify potential chemical(s) of concern. If available, generic, non-site-specific ecological criteria and guidelines may be used to evaluate complete and potentially complete exposure pathways.

3.1.38 *qualitative risk analysis*—a non-numeric evaluation of the potential risks at a site as determined by the potential exposure pathways and receptors based on known or reasonably available information.

3.1.39 *reasonably anticipated future use*—future use of a site or facility that can be predicted with a reasonably high degree of certainty given historical use, current use, local government planning and zoning, regional trends and community acceptance.

3.1.40 *receptors*—the persons that are or may be affected by a chemical release. (See *relevant ecological receptors and habitats*, for non human receptor.)

3.1.41 *registry act requirements*—requirements that are imposed by certain state statutes requiring that a list be maintained identifying properties that have been the site of hazardous waste disposal and that may have restrictions on use or transfer.

3.1.42 *relevant ecological receptors and habitats*—the ecological resources that are valued at the site. Because of the variety of ecological resources that may be present, focusing upon those relevant to a site is an important part of the problem formulation phase of ecological evaluation. Identification of relevant ecological receptors and habitats is dependent upon site-specific factors and technical policy decisions. Examples may include species or communities afforded special protection by law or regulation; recreationally, commercially or culturally important resources; regionally or nationally rare communities; communities with high aesthetic quality; habitats, species or communities that are important in maintaining the integrity and bio-diversity of the environment.

3.1.43 *relevant ecological screening criteria (RESC)*—generic, non-site-specific ecological criteria or guidelines that are determined to be applicable to relevant ecological receptors and habitats, exposure pathways and site conditions utilized during the Tier 1 evaluation. These may include chemical concentrations, biological measures or other relevant generic criteria consistent with the technical policy decisions.

3.1.44 *remedial action*—activities conducted to reduce or eliminate current or future exposures to receptors or relevant ecological receptors and habitats. These activities include monitoring, implementing activity and use limitations, and designing and operating clean-up equipment. Remedial action includes activities that are conducted to reduce sources of exposures to meet corrective action goals, or to sever exposure pathways to meet corrective action goals.

3.1.45 *response action*—an immediate course of action, including monitoring, abatement or containment measures to mitigate known or potential hazards to human health, safety and the environment, taken before interim remedial action or remedial action.

3.1.46 *response action evaluation*—a qualitative evaluation of a site based on known or readily available information to identify the need for interim remedial actions and further information gathering. Response action evaluation is intended to prioritize sites and identify whether there are any appropriate early risk reduction steps.

3.1.47 *restricted use level*—a corrective action cleanup level where one or more activity and use limitations would be needed to eliminate or mitigate potential exposures to chemicals of concern, or to prevent activities that could interfere with the effectiveness of a response action, to ensure maintenance of a level of "acceptable risk" or "no significant risk."

3.1.48 *restrictive covenant*—provision in a deed or lease limiting the use of the property and prohibiting certain uses. In the context of property law, the term describes a contract between the grantor and the grantee that affects the grantee's use and occupancy of land.

3.1.49 *risk assessment*—an analysis of the potential for adverse effects on receptors and relevant ecological receptors and habitats, caused by a chemical(s) of concern from a site. The risk assessment activities are the basis for the development of corrective action goals and determination of where interim remedial or a combination of actions are required.

3.1.50 *risk reduction*—the lowering or elimination of the level of risk posed to human health or the environment through response action, interim remedial actions, remedial action or a combination of actions.

3.1.51 *risk-based corrective action*—a consistent decision-making process for the assessment and response to chemical releases based upon protection of human health and the environment. Assessment and responses to chemical releases may consider the use of activity and use limitations.

3.1.52 *risk-based screening level/screening levels (RBSL)*—non-site-specific human health risk-based values for chemicals of concern that are protective of human health for specified exposure pathways utilized during the Tier 1 evaluation.

3.1.53 *servient estate*—an estate burdened by an easement.

3.1.54 *site*—the area(s) defined by the likely physical distribution of the chemical(s) of concern from a source area. A site could be an entire property or facility, a defined area or portion of a facility or property, or multiple facilities or properties. One facility may contain multiple sites. Multiple sites at one facility may be addressed individually or as a group.

3.1.55 *site assessment*—the characterization of a site through an evaluation of its physical and environmental context (e.g., subsurface geology, soil properties and structures, hydrology and surface characteristics) to determine if a release has occurred, the levels of the chemical(s) of concern in environmental media, and the likely physical distribution of the chemical(s) of concern. As an example, the site assessment collects data on soil, ground water and surface water quality, land and resource use, and potential receptors, and generates information to develop a site conceptual model and support risk-based decision-making. The site assessment may be conducted using Guide E 1912.

3.1.56 *site conceptual model*—the integrated representation of the physical and environmental context, the complete and potentially complete exposure pathways, and the potential fate and transport of chemical(s) of concern at a site. The site conceptual model should include both the current understanding of the site and the understanding of the potential future conditions and uses for the site. It provides a method to conduct the exposure pathway evaluation and to inventory the exposure pathways evaluated and the status of the exposure pathways as incomplete, potentially complete or complete.

3.1.57 *site conditions*—a general description of a site's chemical, physical or biological characteristics that relate to potential exposures to receptors or relevant ecological receptors and habitats.

3.1.58 *site specific*—activities, information and data unique to a particular site.

3.1.59 *site-specific ecological criteria (SSEC)*—risk-based qualitative or quantitative criteria for relevant ecological receptors and habitats identified for a particular site under the Tier 2 or Tier 3 evaluations. These criteria may include chemical concentrations, biological measures or other relevant generic criteria consistent with the technical policy decisions. SSEC may be revised as data are obtained that better describe the conditions and the relevant ecological receptors and habitats.

3.1.60 *site-specific target level(s) (SSTL)*—risk-based values for chemicals of concern that are protective of human health for specific exposure pathways developed for a particular site under the Tier 2 or Tier 3 evaluations.

3.1.61 *source area(s)*—the source area(s) is defined as the location of non-aqueous phase liquid (NAPL) chemical, the locations of highest soil or ground water concentrations of the chemical(s) of concern, or the location releasing the chemical(s) of concern.

3.1.62 *stakeholders*—individuals, organizations, or other entities that directly affect or may be directly affected by the corrective action. Stakeholders include, but are not limited to, owners, purchasers, developers, lenders, tenants, utilities, insurers, government agencies, Indian tribes, community groups, and members.

3.1.63 *stigma*—the residual loss in value above and beyond the actual cost to cure or control the environmental condition of concern if such extraordinary loss is evident in the marketplace. Stigma generally is a result of uncertainty as to the cost, effectiveness or permanency of the methodology of cure/control, or uncertainty concerning the environmental regulatory agencies' endorsement of such methodology or results. Stigma is a time-dependent phenomena and as such may be only temporary in effect.

3.1.64 *technical policy decisions*—the choices specific to the User that are necessary to implement the risk-based corrective action framework described in Guide E 2081, or any replacement standards thereto, at a particular site. The decisions involve regulatory policies, value judgments, different stakeholder decisions and using professional judgment to evaluate available information; therefore, there may be more then one scientifically supportable answer for any particular technical policy decision. The choices represent different approaches. The User should consult the regulatory agency requirements to identify the appropriate technical policy decisions prior to implementing the risk-based corrective action process. Examples of technical policy decisions are: data quality objectives, target risk levels, land use, reasonably anticipated future use, ground water use, natural resource protection, relevant ecological receptors and habitats, stakeholder notification and involvement, and exposure factors.

3.1.65 *unrestricted use level*—a corrective action level where residential uses would be permissible without the need for any activity and use limitations.

3.1.66 *user*—An individual or group involved in remediation involving risk-based decision-making principles, and involving the use of activity and use limitations. Users include owners, operators, regulators, underground storage tank fund managers, attorneys, consultants, legislators and other stakeholders. Two specific types of users are envisioned. The first is the individual or group addressing a site or sites under the circumstances where an activity and use limitation is part of the proposed or final remedial action. The second is a regulatory agency that is developing regulations or guidance regarding the use of activity and use limitations as part of its corrective action program, whether conducted pursuant to a voluntary corrective action, brownfields, Superfund, Resource Conservation and Recovery Act, underground storage tank, or other type of program.

4. Significance and Use

4.1 Activity and use limitations are typically used in conjunction with risk-based decision-making principles in Federal, state, tribal and local remediation programs, or where residual chemicals of concern remain following an evaluation of risk or following the implementation of a remedial action. The principal purposes of activity and use limitations are to:

4.1.1 Eliminate exposure pathways for, or reduce potential exposures to, chemicals of concern;

4.1.2 Provide notice to property owners, holders of interests in the property, title companies, utilities, tenants, realtors, lenders, developers, appraisers and others of the presence and location of chemicals of concern that may be present on the site;

4.1.3 Identify the objectives and goals of each activity and use limitation;

4.1.4 Identify the exposure assumptions upon which each activity and use limitation is based;

4.1.5 Identify the site uses and activities which, if they were to occur in the future, would be appropriate and consistent with maintaining a condition of "acceptable risk" or "no significant risk";

4.1.6 Identify the site uses and activities which should NOT occur in the future (unless further evaluation and remedial action, as appropriate, are undertaken), as those activities and uses may result in the exposure of persons or ecological receptors to chemicals of concern at or near the site in a manner that is inconsistent with a condition of "acceptable risk" or "no significant risk";

4.1.7 Specify long-term stewardship objectives, and the entity which has responsibility for developing stewardship programs and paying for achieving those objectives; and

4.1.8 Specify long-term performance standards, such as operation and maintenance obligations, or monitoring of an engineering control, that are necessary to ensure that the objectives and goals of activity and use limitations continue to be met.

4.2 Activity and use limitations should be implemented to eliminate exposure pathways for, or reduce potential exposures to, chemicals of concern. The following are some examples of situations where an activity and use limitations may be appropriate:

4.2.1 Impacted ground water exists at a site where an alternative water supply is available. A restriction may be placed on the use of ground water for any purpose other than monitoring, or a restriction may place requirements for well construction or evaluation of treatment of ground water.

4.2.2 A site is remediated to levels appropriate only for industrial or commercial uses with respect to the direct contact pathway. The use of the property will then be restricted to those land uses, unless further remedial activities are conducted (that is, the site may not be developed for residential use).

4.2.3 Residual chemicals of concern remaining on a site are covered with some type of barrier (for example, cap, pavement, etc.) The barrier constitutes one type of activity and use limitation. In addition, a restriction may be placed on the deed or lease prohibiting excavation in areas where the chemicals of concern exceed certain risk levels. The restriction may include prohibiting the disturbance of the cap. Monitoring and maintenance of the integrity of the cap or barrier may be a requirement as well.

4.2.4 Operation and maintenance of an ongoing remedial action may be required and may be specified in a restriction. In this case, an easement or property access right may be given to the former owner (as the responsible party) or to his/her agent.

4.2.5 Also, activities interfering with operations and maintenance may be restricted. These restrictions may include limitations on construction or other activities in areas where remediation system controls, extraction wells, monitoring wells, or other ongoing remedial or monitoring systems are located.

4.3 *Due Diligence*—When a property transaction is involved, the prospective purchaser, lender, title company, real estate appraiser and others need to be aware of the possibility that restrictions have been placed on permissible activities and uses of the property. Knowledge of prior land uses is an important indicator of the potential for such restrictions to exist. The user is cautioned that, under Practice E 1527, it is the user's responsibility to discuss with its environmental consultant which party will take responsibility for identifying relevant and applicable information regarding activity and use limitations in either the chain of title or in relevant regulatory databases.

4.4 At the present time, several states provide in their voluntary corrective action programs that liability releases provided in their "No Further Action" letters ("NFA") or "Certificates of Completion" ("Certificates") will be of no effect if any of the conditions in the final letter or certificate are violated. In other words, in these states, the releases from liability may be void or voidable if an activity and use limitation is violated. The activity and use limitation is typically described in, or attached to, the NFA letter or Certificate. Accordingly, it is critically important for owners, prospective purchasers, lenders, tenants and others who are counting on the liability releases provided in the NFA letter or Certificate to be sure that they understand what limitations or restrictions may have been imposed on the site and to understand who bears primary responsibility for ensuring that those limitations or restrictions are not violated.

4.5 The user is cautioned that activity and use limitations are not to be used to encourage or condone "secured abandonment". In general, "secured abandonment" is the practice of physically securing the site and blocking exposure pathways while taking minimal steps to ensure that chemicals of concern do not spread beyond the property boundaries or taking minimal steps to put the property back into productive use. In most cases, the property is not placed back into productive use and does not meet its "highest and best" use. There may be instances where activity and use limitations are used to completely restrict access to a site (for example, during remediation), but the expectation is that sites will be remediated to allow some productive use and therefore some potential exposure.

4.6 As a general rule, Federal or state governmental authorities have primary responsibility for determining applicable and appropriate remediation standards for chemicals of concern, and either the Federal, state, tribal or local government authority may have primary responsibility for inspecting and enforcing any activity and use limitations that may be imposed. It is important for all affected stakeholders (that is, Federal, state, tribal and local authorities; potentially responsible parties; utilities; residents; tenants; the financial community; the environmental community; and others) to have an open dialogue about the goals and objectives of any activity and use limitations; the exposure assumptions underlying any activity and use limitations; applicable and relevant legal authorities for implementing any activity and use limitations; and the entity which will have responsibility for maintaining and enforcing the activity and use limitations over time.

4.7 The language used in activity and use limitations may be drafted broadly or have very focused statements about the purpose. The language may specify activities to be conducted, including operation and maintenance or a performance standard, or activities that are prohibited, or land uses that are allowed or disallowed. There may be a requirement for notice to various individuals or entities, such as tenants, lenders, utilities, or local government officials. There may also be language describing who enforces the restriction, and the conditions under which, and the procedure for removal or termination of the restriction.

5. Activity and Use Limitations As a Component of Site Assessment and Remedial Action Selection

5.1 *General Considerations*:

5.1.1 The user may evaluate the feasibility and appropriateness of activity and use limitations at many different points in the risk-based corrective action process (or other type of remedial action program). These points may include the initial site assessment stage, where existing and reasonably anticipated future uses are identified, or later in the response action evaluation and response action stages. *See Fig. 1.* If possible, the user should consider the screening and balancing criteria, as discussed in 5.3.

5.1.2 If the site is remediated to a restricted use level, the user is cautioned that an activity and use limitation will likely need to be implemented and maintained for as long as the concentrations of the chemicals of concern exceed levels appropriate for unrestricted use.

5.1.3 Activity and use limitations should be considered to be part of the remedial action selection process and should be documented in the remedial action selection document (for example, the Record of Decision, RCRA permit, certificate of completion). Like any other component of remedial action selection, the User must evaluate whether the activity and use limitation(s) under consideration is feasible and appropriate.

5.1.4 In addition, selection of one or more activity and use limitations may lead to an interactive reconsideration of appropriate response actions. If the user determines after an evaluation of potentially applicable activity and use limitations, as described below, that none are feasible or appropriate, the user may need to conduct additional response actions to achieve an acceptable risk level. *See Fig. 2.*

5.1.5 Before evaluating the potential applicability of activity and use limitations, the user must have a good understanding of the chemicals of concern; the sources of exposure; the

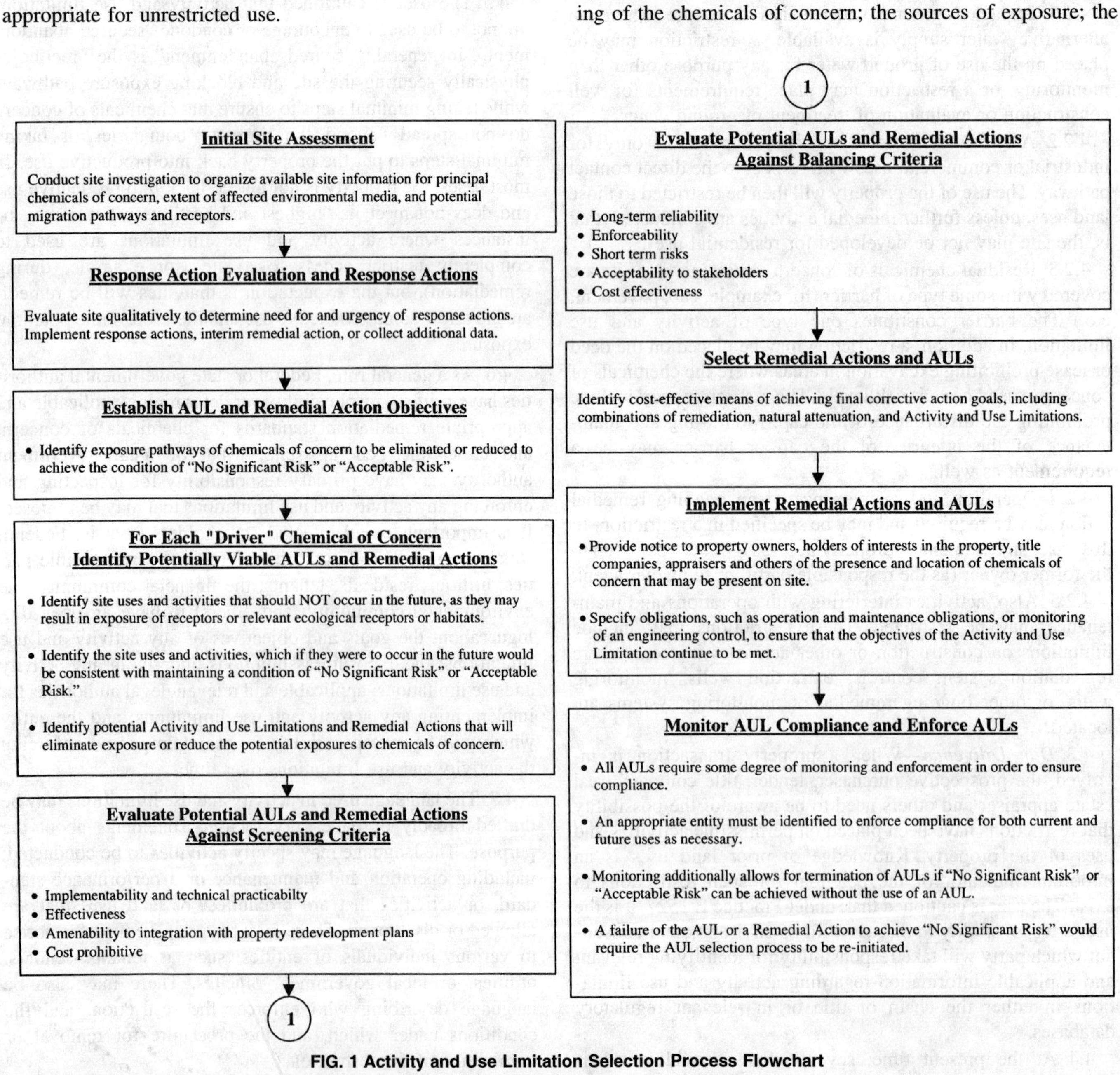

FIG. 1 Activity and Use Limitation Selection Process Flowchart

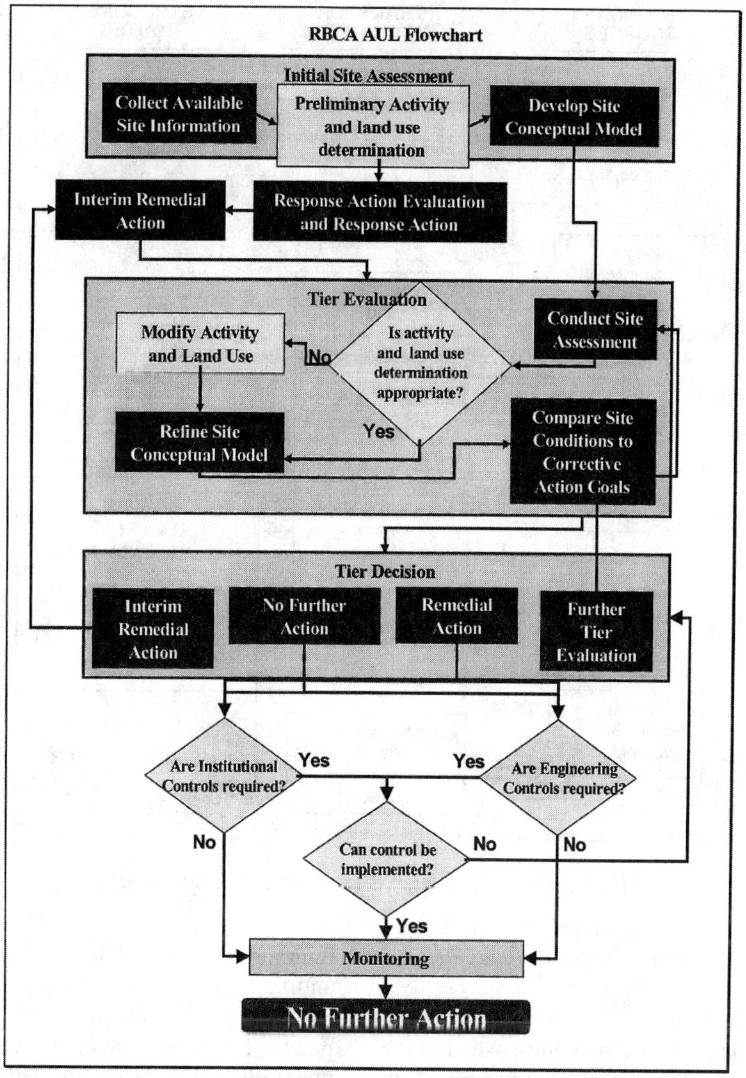

FIG. 2 RBCA AUL Flowchart

likely exposure routes (for example, dermal, ingestion, inhalation); the pathways of exposure (for example, air, surface water, ground water, soil); the likely receptors (both human and ecological); and the reasonably anticipated future use of the site (for example, industrial; commercial; mixed use; residential; day care). *See Fig. 3.* The user is advised to review Guide E 2081, or any replacement standard thereto, for further guidance on these issues. The user is also cautioned that, while activity and use limitations may be one possible component of remedial action selection, they generally should not be considered to be the sole component of remedial action selection. The user is further cautioned to consult with the appropriate regulatory authorities and to determine whether other statutory or administrative requirements may apply.

5.2 *Goals and Objectives*—The user must identify the goals and objectives that the activity and use limitation is intended to achieve.

5.3 *Screening and Balancing Criteria.* The User is cautioned to examine the eight following criteria EARLY in the remedial action selection process: effectiveness; amenability to integration with property redevelopment plans; implementability; technical practicability; cost prohibitiveness; reliability over the long-term; acceptability to stakeholders; and cost-effectiveness.

5.3.1 *Introduction*—Initially, the user must determine which activity and use limitation (as part of a remedial action) is potentially applicable for each chemical of concern; for each exposure pathway; for each exposure route; and for each potential receptor. For each of these potential scenarios, the user should apply the following screening and balancing criteria to determine which activity and use limitation, or combination of activity and use limitations, best addresses each exposure pathway, route of exposure, and likely receptors to achieve an "acceptable risk" or "no significant risk" level. The activity and use limitation, or combination of activity and use limitations, should be selected that best addresses the "driver" chemical(s) of concern, or principal receptor(s) for each exposure scenario. These "best" solutions should then be compared to determine whether redundant controls are necessary and appropriate, or whether a single type of activity and use limitation will address all significant exposure scenarios. *See Fig. 4(a) and 4(b). These examples are intended to be*

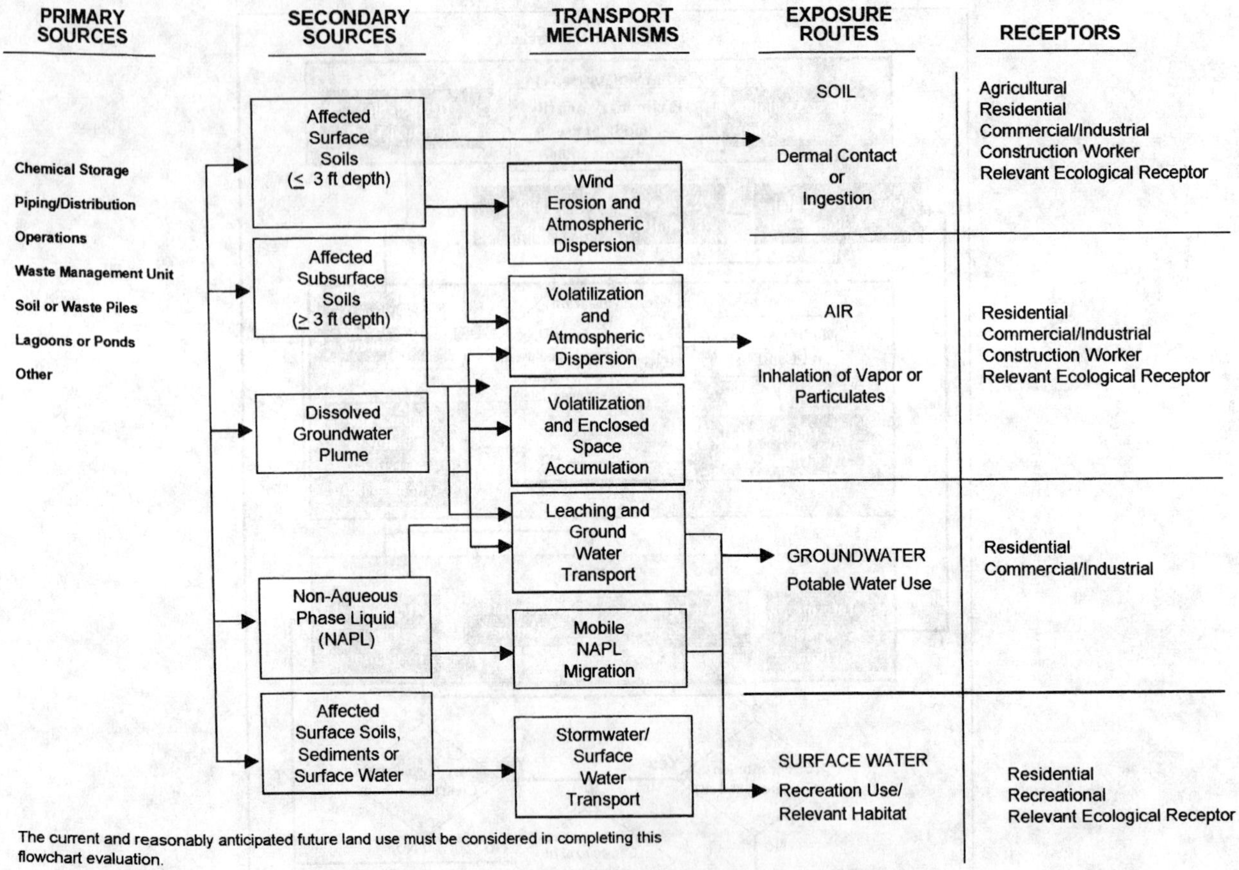

FIG. 3 Example Exposure Scenario Evaluation Flowchart

illustrative only and should not be considered to be applicable to every evaluation.

5.3.2 *Suggested Screening Criteria*:

5.3.2.1 *Effectiveness*—The user must determine whether the proposed activity and use limitation is likely to be effective, in both the short term and the long term, in eliminating or minimizing potential exposures to chemicals of concern, or in preventing activities that could interfere with the effectiveness of a response action, and to thereby maintain a condition of "acceptable risk" or "no significant risk". For example, if potential exposure to chemicals of concern in the soil is the potential exposure pathway, an engineering control such as a cap may not be effective by itself and may need a complimentary institutional control to be effective over time.

5.3.2.2 *Amenability to Integration with Property Redevelopment Plans*—The user should determine the reasonably anticipated future use of the property, as well as regional and site-specific ground water uses, to be sure that any potentially applicable activity and use limitations are amenable to integration with property redevelopment plans. For example, if an area is being developed as residential or high-density residential, a restriction on residential use, or a limitation to industrial use, would not be amenable with the property's redevelopment in that area.

5.3.2.3 *Implementability*—The user should evaluate early in the remedial action selection process whether a particular type of activity and use limitation can be implemented under applicable state and local law. For example, if there is off-site migration of ground water containing chemicals of concern, and the state does not have a statutory mechanism for implementing restrictions on ground water usage, there may be no practical way to implement activity and use limitations on numerous neighboring properties.

5.3.2.4 *Technical Practicability*—The user should determine whether the activity and use limitation is technically practicable. For example, an activity and use limitation that includes an engineering control, such as an impermeable cap that causes chemicals of concern to migrate onto an adjoining property, would not be technically practicable to limit the migration of impacted ground water.

5.3.2.5 *Cost Prohibitiveness*—The user should examine both the short term and long term costs of a potentially applicable activity and use limitation to determine whether that restriction would be cost prohibitive to implement and maintain compared to the cost of doing additional active remediation. The costs of both implementing and maintaining the activity and use limitation should be weighed against the cost of conducting additional remediation. The potential for liability should also be considered. For example, if the property has already been subdivided and sold to numerous new owners, it may be cost prohibitive to impose restrictive covenants on each parcel that would need to be burdened with a soil excavation prohibition or a ground water use restriction.

5.3.3 *Suggested Balancing Criteria*—If the potentially applicable activity and use limitation survives the suggested screening criteria identified above, it is recommended that the

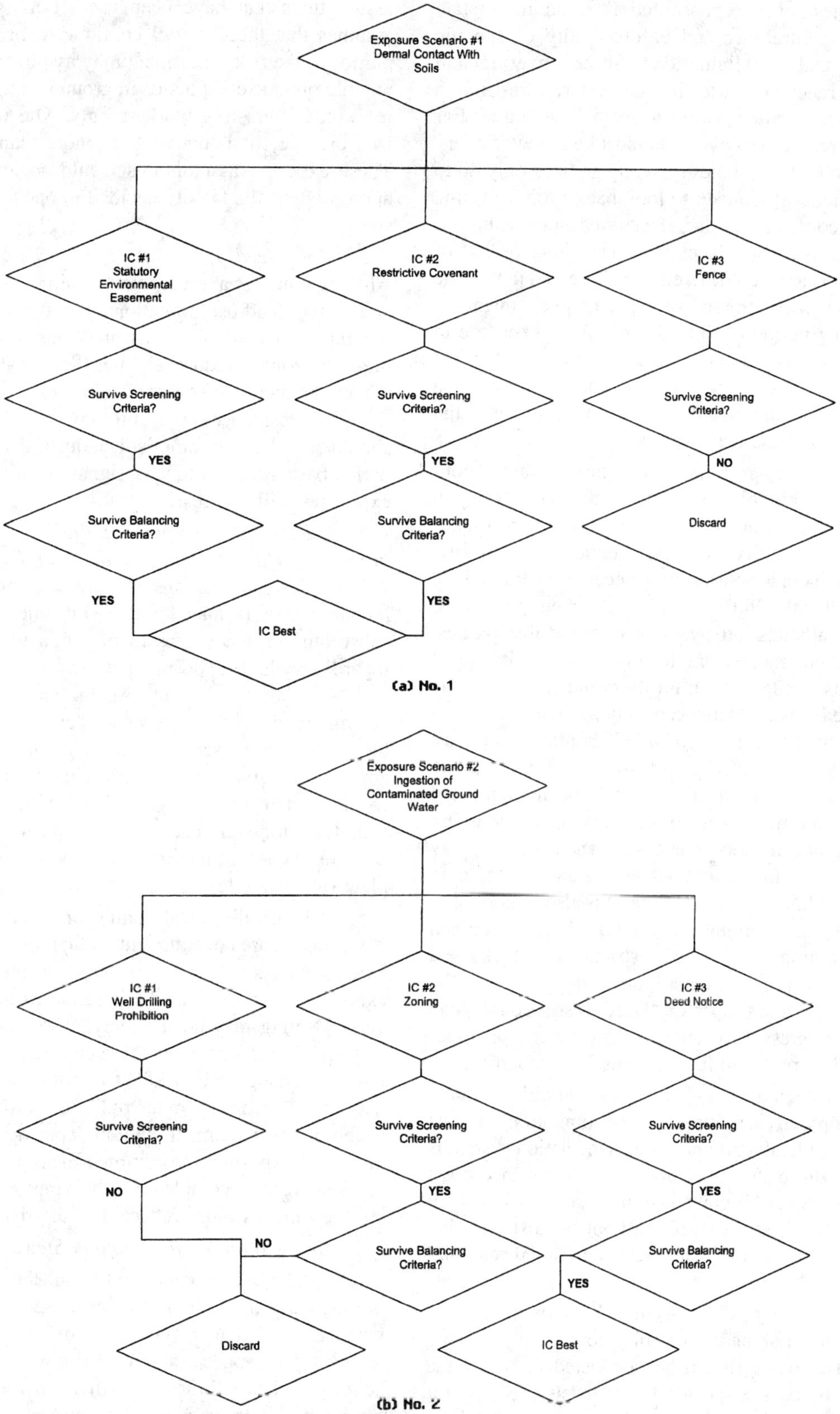

FIG. 4 Exposure Scenarios

activity and use limitation be evaluated against the balancing criteria identified below.

5.3.3.1 *Long-Term Reliability and Durability*—Certain activity and use limitations are viewed as being more reliable over the long term than others. For example, many people have expressed concern that zoning may not be reliable over the long term to eliminate or minimize potential exposures to chemicals of concern, or to prevent activities that may interfere with the effectiveness of a response action. In addition, state laws may limit the durability and enforceability of specific types of activity and use limitations. Some governmental jurisdictions have renewal clauses in rules where a restriction expires or must be rewritten within a given time frame. For example, in Iowa, restrictive covenants must be renewed every 21 years. In addition, title searches typically go back only 40 to 60 years unless a request is made to look back further in time in the property records. Therefore, if activity and use limitations are expected to remain in effect over a long period of time, this issue needs to be considered and addressed in the title search context. The greater the risk of exposure to chemicals of concern over a long period of time (for example, exposure to chemicals of concern that do not attenuate naturally, or that are persistent chemicals of concern, or that otherwise present a substantial risk to human health or the environment), the greater the need to address these issues.

5.3.3.2 *Acceptability to Stakeholders*—The user should consider the advantages of involving affected stakeholders early in the remedial action selection process in the decision to implement and maintain activity and use limitations at a site. Stakeholders may include, but are not necessarily limited to, Federal agency officials; Indian tribes; state agency officials; local government officials; all potentially responsible parties; the environmental community; the business community (local businesses, tenants, lenders, etc.); utilities; and residents. The potentially affected stakeholders need to understand the exposure assumptions underlying the potentially applicable activity and use limitations; why activity and use limitations may be appropriate; and how those restrictions will be implemented and maintained over time. The local community, including local government officials, local businesses, and residents, may play an important role in both implementing and maintaining the activity and use limitations over time. It is also important to note that the regulated community has long been concerned about the potential impacts of "deed restrictions", which are a permanent part of the property record, on property title and the ability to reconvey the property. "Deed restrictions" may discourage any interest that lenders, developers or other prospective purchasers would have in reusable properties.

5.3.3.3 *Cost Effectiveness*—The user should evaluate whether the proposed activity and use limitation is cost effective. For example, if a ground water remediation system is likely to require substantial operation and maintenance costs over time, and this control is embodied in a restrictive covenant running with the land, this control may not be cost effective over the long term, compared with doing additional remediation now.

5.4 *Risk Assessment Applied to Activity and Use Limitations*—Unless risk-based screening levels are used, site specific risk assessments should be conducted to determine appropriate risk-based site-specific target levels (SSTLs) for each chemical of concern detected at a site, for each potentially applicable exposure pathway, for each potentially relevant exposure route, and for each potentially relevant receptor (human and ecological). The SSTL represents the concentration of each chemical of concern that presents an "acceptable risk" or" no significant risk" at the site under the exposure assumptions that have been used. For example, if the user assumes that the site will continue to be used for industrial purposes, the risk assessment may assume that exposures from volatile organic compounds in ground water are applicable and relevant to industrial workers only, who may breathe volatilized organic compounds for no more than ten hours per day. These exposure assumptions would no longer be relevant or appropriate if the facility decided to open a day care center on site.

5.5 *The Need to Avoid Overly Simplistic Paradigms*—Although there is a direct relationship between risk assessment and activity and use limitations, the user is cautioned to avoid making overly simplistic assumptions. For example, an area might be zoned "industrial," but the actual use of the property where chemicals of concern are present is "mixed use", where there are residences and children present. In this case, one should avoid using simplistic industrial risk assessment scenarios based upon zoning designations alone, since the actual exposures will be greater.

5.5.1 *Residential/Commercial/Industrial Zoning Designations May Have Nothing to do with Exposure Pathways*—Zoning designations are usually relevant regarding which human receptors may be at a site, but zoning should never substitute for the professional judgment of a risk assessor regarding which exposure pathways should be incorporated into the risk assessment. Again, an area might be zoned "commercial", but the exposure pathways may include wind-blown dust into a school within the commercial zone. Blindly applying assumptions that fit with "commercial" exposures would underestimate risk. Likewise, using "residential" assumptions for every pathway that happens to be in a residential zone may overestimate risk if certain pathways are not complete (for example, no exposure to impacted ground water).

5.5.2 Generally, local zoning or other comprehensive plan designations are not sufficient on their own to ensure exposures are limited. As noted above, zoning designations may not limit exposure since uses may be different from what zoning would allow; zoning may not be relevant to the particular pathway; and zoning may change without consideration being given as to how the change might affect exposure (for example, zoning may change from "industrial" to "mixed use" to bolster economic development without consideration of potentially increased exposures). Additional measures (such as restrictive covenants) may complement the zoning to ensure exposures are the same as those reflected in the risk assessment.

5.6 *Long-Term Monitoring and Stewardship Issues*:

5.6.1 The user is cautioned about the importance of determining early in the remedial action selection process not only whether a particular type of activity and use limitation is relevant and appropriate, but also of determining how the activity and use limitation will be maintained and enforced over time. It may be prudent and advisable to include all affected stakeholders in the resolution of these issues. The affected stakeholders should consider whether the federal, state, tribal or local government has authority to enforce the control; whether the federal, state or local government has the resources to inspect and enforce the control; and whether private entities (for example, environmental insurance companies, custodial trusts, beneficiaries of conservation easements, or holders of restrictive covenants) may have a role in enforcing the selected activity and use limitation.

5.6.2 The user should also examine whether financial assurances are needed to maintain the activity and use limitation over time. Bonds, letters of credit, environmental insurance, custodial trusts, sinking funds, escrows, and other mechanisms may all be appropriate as financial assurance devices.

6. Examples of Activity and Use Limitations

6.1 *General*:

6.1.1 For purposes of this guidance, activity and use limitations are those mechanisms used in a Federal, state, tribal or local remediation program applying risk-based decision-making principles where, as a part of the program, certain concentrations of chemicals of concern are allowed to remain in the soil or ground water. Activity and use limitations would then be used to ensure that exposure to the residual chemicals of concern does not present a significant risk to human health or the environment.

6.1.2 The types of activity and use limitations to be discussed are: proprietary controls, such as deed restrictions or restrictive covenants; state and local government controls, such as zoning, building permits, well drilling prohibitions, and water advisories; statutory enforcement tools, such as orders and permits; informational devices, such as deed notices, geographic information systems, Registry Act requirements and Transfer Act requirements; and physical measures, including engineering and access controls.

6.1.3 Activity and use limitations come in many different forms. Often, an effective Federal, state or local remediation program involves multiple layers of controls using different types of activity and use limitations. For example, an agency may impose a limitation requiring further remediation should the property be used for residential purposes. This use limitation may be incorporated into an easement or restrictive covenant, which in turn may have to be registered.

6.1.4 In some states, an owner/operator who implements activity and use limitations as part of a remediation program will obtain some degree of liability protection for the environmental conditions on-site, provided that the controls are maintained. Examples of devices used by states to limit liability include Certificates of Completion, Covenants Not to Sue and No Further Action letters. However, most of these devices only apply to state actions and do not automatically preclude private or Federal lawsuits. Some states have entered into a Memorandum of Understanding with EPA to minimize the chance that a state corrective action decision resulting in an exemption from future liability will be overturned by EPA. The corrective action decisions may involve the use of activity and use limitations as a condition of case close-out. The user is cautioned that it is important to be aware of the legal context of the regulatory programs administering the site.

6.1.5 *Federal Government Use of Activity and Use Limitations*—Activity and use limitations may be either explicitly or implicitly permitted under Federal, state and local remediation programs.

6.1.5.1 *Environmental Protection Agency*—The Environmental Protection Agency has expressed increased interest in the use of activity and use limitations at CERCLA and RCRA sites in recent years as the interest in land-use based remedies and performance-based standards has increased.

(1) Activity and use limitations, including institutional controls, are recognized in the National Oil and Hazardous Substances Pollution Contingency Plan ("NCP"). While not containing any binding rules regarding when these types of controls may be used, the NCP does state the following regarding institutional controls: "EPA expects to use institutional controls such as water use and deed restrictions to supplement engineering controls as appropriate for short- and long-term management to prevent or limit exposure to hazardous substances, pollutants or contaminants. Institutional controls may be used during the conduct of the remedial investigation/feasibility study (RI/FS) and implementation of the remedial action and, where necessary, as a component of the completed remedial action. The use of institutional controls shall not substitute for active response measures (for example, treatment or containment of source materials, or both, restoration of ground waters to their beneficial uses) as the sole remedial action unless such active measures are determined not to be practicable, based on the balancing of trade-offs among alternatives that is conducted during the selection of remedial action." 40 CFR 300.430(a)(1)(iii)(D).

(2) EPA has stated a similar intent with regard to the use of institutional controls in the RCRA program. In a notice published on May 1, 1996, EPA stated that it: "expects to use institutional controls such as water and land use restrictions primarily to supplement engineering controls as appropriate for short and long term management to prevent or limit exposure to hazardous waste and constituents. EPA does not expect that institutional controls will often be the sole remedial action." 61 Fed. Reg. at 19448.

(3) EPA has also released, for internal guidance only, a draft "reference manual" on institutional controls. See March 1998 workgroup draft entitled "Institutional Controls: A Reference Manual." The manual recommends, inter alia, that: institutional controls be evaluated carefully before the final remedial action is selected; the goals and objectives for the institutional control be described clearly in the decision document; state and local governmental agencies be involved early in the remedial action selection process; and an instrument such as an easement or restrictive covenant be executed when it is important for the control to run with the land.

6.1.5.2 *NRC*—Another example of the use of activity and use limitations in a federal program is contained in the Nuclear Regulatory Commission's license termination regulations. These regulations (10 CFR §§ 20.1402 and 20.1403) permit the termination of licenses at facilities that have been decommissioned but which still have small concentrations of residual radioactivity. If the concentrations are low enough, the facility may be released without any limitations or restrictions. However, if the concentrations are somewhat higher, the licensee may apply to release the facility with limitations on the future use of the site that will limit the potential future dose to site occupants. Typically, the limitations would be in the form of deed restrictions that limit the use of the property.

(1) In order to release a facility under these restricted conditions, the regulations require that certain conditions be met: levels of residual radioactivity must have been reduced to levels that are as low as reasonably achievable, the licensee must have made provisions for legally enforceable institutional controls to limit dose, the licensee must have provided adequate financial assurances to perform maintenance of the controls when such maintenance may be needed, the licensee must seek public advice on the proposed institutional controls, and the concentrations at the site must be low enough that, if the limitations were not in effect, the doses to site occupants would not be unacceptably high.

6.1.6 *State Use of Activity and Use Limitations*:

6.1.6.1 *General*—In some states, activity and use limitations are an explicit part of the state's hazardous waste corrective action program. See, for example, Wis. Admin. Code N.R. § 720.11 (1997). In other states, on a site-specific basis, the state will consider the use of activity and use limitations and future uses in determining the applicable corrective action standard. See, for example, Pa. Stat. Ann. § 6026.304(F)(1) (West 1996). Finally, some state statutory regimes make no mention of the use of activity and use limitations in their voluntary corrective action programs, but in practice consider, and even encourage their use.

6.1.6.2 *Emergencies*—Some states have found it helpful to include provisions describing procedures to be followed in the case of an emergency that requires the site to be disturbed. See, for example Mass. Regs. Code tit. 310, § 40.1071(k) (1996). Sample procedures that should be followed if, for example, an underground utility line must be repaired include: notifying the state environmental authority within 2 hours of knowledge of the emergency condition, limiting disturbance of the impacted media to the minimal amount reasonably acceptable to respond to the emergency, taking specified precautions to minimize exposure of workers and neighbors to the impacted media, and hiring a licensed site professional (LSP) to prepare or implement a plan, or both, to restore the site to a condition consistent with the use of an activity and use limitation.

6.2 *Proprietary Controls*:

6.2.1 *Deed Restrictions and Restrictive Covenants*:

6.2.1.1 The term "deed restriction" is not a legal term of art. Nevertheless, the term "deed restriction" is frequently used to describe various limits and conditions on the use and conveyance of land. In this regard, "deed restrictions" are one of the most common forms of activity and use limitations. "Deed restrictions" are available in every state as a matter of common law. In addition, in many states, the legislatures have adopted specific laws permitting the use of "deed restrictions" as a matter of state statutory law. The restrictions serve two principal purposes: to provide notice to subsequent purchasers and lessees that the property has been subject to a certain level of environmental investigation and remediation; and to ensure the long-term efficacy of any engineering control or condition that must be maintained over time.

6.2.1.2 In order to be enforceable against current and subsequent owners of the property, a promise in a "deed restriction" requires: a writing; intention by all originating parties that particular restrictions be placed on the land in perpetuity; "privity of estate", unless this condition has been eliminated as a matter of state law; and this restriction must "touch and concern the land." These requirements are explained in more detail below.

6.2.1.3 Legal requirements dictate that conveyances of land and "deed restrictions" affecting land must be in writing. When deed restrictions or environmental restrictions are imposed by state law, rather than common law, many states mandate that these restrictions be created by documents that are either identical to or substantially similar to the model documents provided by the state's department of environmental protection.

6.2.1.4 The second requirement for a legal and binding "deed restriction" is a precise reflection of the parties' intentions with regard to the scope and duration of the restrictions therein. Generally, the restriction must run "in perpetuity". Again, where the restriction has been codified in state law, many states include phrasing that the restriction will "run with the land" in their model forms to denote that the restriction will last in perpetuity. See, for example, N.J. Admin. Code tit. 7, § 26E-7 (Appendix F); Mass. Regs. Code tit. 310, § 40.1099 (1999). This phrase is essential as it ensures that any restriction is forever binding against the owner and successors in interest.

6.2.1.5 Under common law, the third requirement is that only persons with a certain relationship, "privity", may enforce a deed restriction. Easements and covenants in gross, that is, those that do not benefit the land, but run to the benefit of a specific party, have been disfavored under the common law. Accordingly, easements and covenants in gross have been difficult to enforce under the common law. A lack of privity can therefore undermine an environmental agency's attempts at enforcement, since "deed restrictions" are usually promises between buyers and sellers or between neighbors. Some states have addressed this concern by explicitly eliminating the need for privity in the state statute. Several state programs explicitly provide that the environmental authority has the power to enforce the covenant. See N.J. Admin. Code tit 7, § 26E; Mass. Regs. Code tit. 310, § 40.1099(a)(6)(i); California Department of Toxic Substances Control, "Official Policy/Procedure" (Attachment C) (May, 1990); Wisconsin Department of Natural Resources Bureau for Remediation and Redevelopment Regional Closeout Committee Guidance for Using Institutional Controls (working final draft, April 1998). States that have developed model covenants often follow this approach. Other states have sought, or are seeking, statutory authority to enforce covenants.

6.2.1.6 Finally, the promise in a "deed restriction", as well as the benefit, must "touch and concern the land". This means that the promise, and the benefit, must center on the land and use of the land and must affect the land itself in some way. For example, an owner/operator's promise to refrain from using the land in a certain way in the future could devalue land and thus would be considered to "touch and concern the land." The promise in a "deed restriction" may also refer to the use of a remedial action and the maintenance requirements associated with it, or to a different land use classification from adjacent parcels due to corrective action levels specified for soils at the site.

6.2.1.7 An effective "deed restriction", both from the owner/operator's perspective and the state's perspective, must be drafted using precise and easily understandable language spelling out the specific activities and uses that will be allowed and the specific activities and uses that will be prohibited. General restrictions or requirements may include: granting of an easement to the state environmental authority for inspection, surveillance, monitoring, maintenance, or other purposes necessary to protect health and safety; prohibiting the subdivision of property; a requirement for notification to be sent by the owner of nonresidential property to purchasers, lessees, and tenants disclosing the existence of residual chemicals of concern; a requirement that the owner give notice in all deeds, mortgages, leases, subleases, and rental agreements that there are residual chemicals of concern; a requirement for advance notice to state environmental authorities of any sale, lease, or other conveyance of property; a requirement for notice in the deed notifying prospective purchasers that the property has been used to manage or dispose of hazardous waste, or both, and that its use is restricted; and provisions for enforcement, variance, and termination. See, California Department of Toxic Substances Control, "Official Policy/Procedure" (Attachment C) (May, 1990).

6.2.1.8 The process for imposing "deed restrictions" should contain a certain amount of flexibility with regard to cancellation or variance. The procedures for cancellation or amendment of a "deed restriction" should be readily available. In Texas, the owner/operator must notify the state environmental authority at least 120 days prior to canceling or amending an activity and use limitation. See, 30 Tex. Admin. Code § 334.206(a)(5) (West 1998). Within 30 days of this notification, the owner/operator must demonstrate to the state that conditions have changed and that a re-evaluation is warranted. See, *id.* Massachusetts is more stringent, requiring any change in activities or uses that may invalidate a finding of "No Significant Risk" to be accompanied by an evaluation by a Licensed Site Professional (LSP). For cancellation or release of an activity and use limitation, the owner/operator must submit a standard form along with an LSP opinion. See, Mass. Regs. Code tit. 310, § 40.1080 *et seq.* (1999). Wisconsin states right in the restriction or notice that an affidavit may be filed with a determination that the conditions under which the document was filed no longer apply. Wisc. Admin. Code §§ 720.11(1)(c) and 726.05(2)(b) (1999).

6.2.1.9 The need for a soil vapor extraction system may be determined as part of the remedial action selection process. A deed restriction would then be used to ensure the continued operation and maintenance of the remedial system.

6.2.2 *Easements*—An easement may allow access to the property or prohibit a use of the property. Easements are available under common law and are an explicit provision in some state programs. For example, the state may require the owner/operator to grant state authorities an easement for operation and maintenance of engineering controls or for general inspection/audit functions. See, for example, N.J. Admin. Code tit. 7, § 26E.

6.2.3 *Equitable Servitudes*—Equitable servitudes are specific provisions, usually restricting certain uses, that apply to the property owner. A servitude would restrain the property owner such that he/she must use the land in a manner compatible with the servitude.

6.3 *State and Local Government Controls*:

6.3.1 *Zoning/Rezoning/Variances*—Municipal or local government authorities may impose restrictions on certain activities or uses through restrictive or "overlay" zoning. For example, restrictive zoning may be used to prohibit residential uses in a formerly industrial area. The user is cautioned, however, that zoning is generally not very effective as a stand-alone control because zoning doesn't impact existing uses; it may require the property owner's consent; the control may not be adequately communicated to third parties, such as contractors and utilities; and it may be construed as an "inverse condemnation" or taking. Another shortcoming is the inability to limit uses or activities at individual sites, thus taking away some of the flexibility offered by individually-tailored activity and use limitations.

6.3.2 *Building Permits/Development Plan Review*:

6.3.2.1 A few local jurisdictions review activity and use limitations before issuing building permits. However, the more typical situation is for the development process to review only local code compliance rather than to look to conditions that may be imposed by "deed restrictions" or other private contractual arrangements.

6.3.2.2 Building permits are a form of local authority (that is, town, city, county) that can be used for implementation of activity and use limitations. In general, building permits are required to erect, construct, reconstruct, demolish, alter, or use any building or structure covered under the local ordinance. Regulated activities extend to changes in plumbing, gas, mechanical, electrical, and fire protection systems. The permitting process includes both a review component and an inspection component. There are significant powers of enforcement associated with the permitting process including, but not limited to, fines, injunctions and withdrawal of occupancy certificates. The broad scope of activities regulated by the permitting process could make it an effective and comprehensive tool for monitoring the land use activities of owners and operators.

6.3.2.3 At this time, there appear to be no statewide programs in place that use building permits as a formal mechanism for land use and activity control related to environmental exposure. Some states (for example, New Jersey, California) report that they are considering the use of this control to centralize activities such as soil disturbance and building interior changes through one agency. Many states do have a mechanism in place to distribute information on use restrictions imposed at the state or federal level to municipalities, but the application of this information is left to the local government.

6.3.3 *Well Drilling Prohibitions*:

6.3.3.1 *Well Restriction Areas*—Well restriction areas can be a form of activity and use limitation by prohibiting or conditioning the construction of wells in that area.

6.3.3.2 *Geographic Information Systems*—Some states require that a site with ground water containing residual chemicals of concern that exceed state standards be registered on a

GIS system so that affected parties may review information pertinent to the site prior to making decisions about purchase, future land use, and the like.

6.3.3.3 *Permitting*—Many states have developed regulations that prohibit construction of a private well without a written permit. Often, limited water quality testing and well inspections are required prior to acceptance of the well for human use. In the case of new subdivisions, special use permits may be issued by state or local regulatory agencies prior to issuing development permits. Local and state health agencies may use ground water quality information to deny well permits for affected aquifers for the purpose of protecting public health, welfare, and safety.

6.3.3.4 *Overlay Zoning*—Overlay zoning consists of zones that are drawn on a municipality's existing zoning map which provide protection not explicitly stated under existing zoning regulations. In a number of states, aquifers and their quality are designated through a specific classification system. These classification systems are an outgrowth of the Safe Drinking Water Act (SDWA) provision for protection of sole source aquifers. Connecticut, for example, maintains a published map of existing quality and classified uses of its groundwater resources. Ground water underlying Superfund sites and known impacted sites are classified as GB, which does not allow the human consumption of ground water; thus, drinking water wells are prohibited in these designated areas. Vermont has reclassified groundwater at two Superfund sites from Class 3 (suitable for individual water supply) to Class 4 (not suitable for human consumption) and maintains its ground water classifications on a GIS.

6.3.3.5 *Governmental Ordinances/Legislation*—County ordinances can restrict the use of ground water in cases where the existing water supply on a property is a potential threat to health. In Howard County, Maryland, for example, the County health officer can order a property owner to connect to the public water supply if there is a potential threat to human health and if there is an operating public water main available for delivery of water service to the property. The County Code has provisions for notification of the property owner, decision appeal, and compliance. Financial assistance may be obtained through the County for those property owners with financial difficulties. In Wisconsin, the state offers financial assistance for well replacement for eligible parties. Governmental ordinances may be used to preserve the integrity of any ground water remedial action by prohibiting or conditioning the placement and use of any or all types of wells within the area.

6.3.3.6 *Notices of Restrictions on Wells Within Deeds*—Deed notices or deed restrictions may be used to place restrictions on the installation and use of wells. Deed notices are informational only and do not convey a directly enforceable restriction. Deed restrictions are private controls between the past owner and current owner of the property. They are governed by state property law and thus vary from state to state. The restrictions can only be terminated upon a showing that the concentrations of the chemical(s) of concern in the well restriction area have been remediated in accordance with state standards.

6.3.3.7 In Wisconsin, a Groundwater Use Restriction is placed on a property deed for sites where natural attenuation has been demonstrated to be effective, and will continue to be effective, in containing a plume and reducing contaminant concentrations (Ch. NR 726.05, Wis. Adm. Code). At the time that the restriction is filed, the case is considered closed and there are no additional monitoring requirements on the responsible party.

6.3.4 *Water and Well Use Advisories*:

6.3.4.1 Water and well use advisories serve the public by alerting them to potential risks to health and safety from impacted ground water.

6.3.4.2 Notices of water and well use advisories may be recorded in the land records. Some states require notification of state agencies upon proposed sale or transfer of environmentally-impacted properties as part of state remediation programs. In New Jersey, "Restrictions of Record" (ROR), including water well restrictions, must be filed with local officials, including the county clerk, county health officer, mayor, local zoning officials, and local construction code officials.

6.3.4.3 Notification of ground water containing chemicals of concern may also be presented in the form of public notice of remediation as part of the federal or state program. Many programs require dissemination of remedial action plans (or notice of the same) through public libraries or newspapers, or both.

6.3.4.4 Most states have controls for public utilities that serve as "one-call" telephone hotlines to ensure that buried power, water and gas lines are not disturbed. A number of states are considering expanding this service to include notification of the presence of environmentally-impacted media, including ground water.

6.4 *Statutory Enforcement Tools*—Where a government agency is actively involved in conducting or overseeing a corrective action, the agency may have enforcement authorities that can be used to impose activity and use limitations. Although these tools do not generally run with the land, they can be useful when a control is only needed for the short term, or when the current landowner is likely to own the property for as long as controls are needed. Short-term controls may be adequate, for example, where the goal is simply to control access or exposure while the active corrective action is going on, or where a more permanent control is anticipated but will take time to implement.

6.4.1 *Orders*—Both Federal and state regulatory programs use orders as a mechanism for implementing activity and use limitations. For example, Section 106 of CERCLA authorizes EPA to issue administrative orders, or to seek a court order, whenever there is an imminent and substantial endangerment to public health, welfare or the environment. Many state laws contain similar authority. Federal and state RCRA programs also contain order authority, although more narrowly focused; for example, for unpermitted facilities with "interim status", § 3008(h) of RCRA authorizes EPA to issue orders for corrective action, and again state law often provides similar authority. In addition, § 7003 of RCRA authorizes the issuance of an order when there is an imminent and substantial endangerment. Even some state Voluntary Corrective Action Programs use orders as a mechanism for moving those sites through their programs (for example, Arkansas). In most cases, orders are negotiated and issued on consent, although they may also be issued unilaterally. In some cases, primarily in connection with CERCLA corrective actions at NPL sites, they may take the form of a consent decree.

6.4.1.1 These authorities are very broad in scope and can address virtually any aspect of a corrective action. Accordingly, such an order may, among other things, specify activities that are prohibited at a particular property. In addition, in many states, a copy of the order is filed in the local land records in order to give potential purchasers notice of the residual chemicals of concern at the site.

6.4.1.2 The chief disadvantage of orders as a form of activity and use limitation is that, in most cases, they are only binding on named parties. They do not bind a subsequent owner if the property changes hands, even if that party receives notice of the order. Therefore, they have limitations as long-term controls. In a few states, the state has specific statutory authority to issue orders that run with the land.

6.4.1.3 However, for shorter term use, or as a "bridge" to a more permanent control, orders can be valuable. Moreover, depending on the statutory authority involved, an order may be enforceable by citizen suit, which may be considered desirable in some cases (for example, it allows governments not directly involved in the corrective action decision to take the lead responsibility for oversight and enforcement). Finally, the mechanics of issuing an order may be less complicated than those of a transaction that involves conveyance of a property interest such as an easement.

6.4.2 *Permits*—Where a facility requires an operating permit of some kind, as under RCRA, that permit may be a vehicle for imposing activity and use limitations. Under RCRA, the permit for any hazardous waste treatment, storage or disposal facility must require corrective action across the facility. The corrective action permit may, among other things, specify uses that are prohibited in light of the type of corrective action being conducted. Since RCRA facilities are industrial by nature, RCRA corrective actions may lend themselves to a land-use based approach, particularly where the facility is located in an area likely to remain industrial for the reasonably anticipated future. In such cases the permit is the natural vehicle for imposing activity and use limitations, in the form of permit conditions.

6.4.2.1 Where a permit is required, it can greatly simplify the process of establishing controls. Since the controls are in the permit itself, it is unnecessary either to seek separate regulatory action from a local government, or to negotiate the conveyance of a property interest.

6.4.2.2 Permits can be a useful tool for memorializing an activity and use limitation. Their shortcomings include the lack of adequate resources to enforce these controls and the general absence of agency oversight. In addition, permit conditions bind only the permittee, and only for the life of the permit. If the permit expires or is not renewed, the long-term effectiveness of the restriction may be impaired. Therefore, it may ultimately be necessary to implement controls through some other mechanism (or it may be necessary to conduct additional corrective action to allow unrestricted use of the property).

6.4.2.3 In addition to operating permits, some states have statutes or regulations establishing special corrective action-related permits. For example, some states can issue ground water permits under which access to ground water is limited during the time that it takes to conduct ground water restoration.

6.5 *Informational Devices*:

6.5.1 *Notice*:

6.5.1.1 Notice may be informational only, or it may be an integral and enforceable part of an activity and use limitation. Notice is a tool for ensuring that parties to a real estate transaction (including purchasers, tenants, and lenders) are aware of the environmental status of the property prior to finalizing a transaction.

6.5.1.2 Notice requirements usually require disclosure of the specific location of chemical releases on a site and of any restrictions on use, access, and development of part or all of the impacted site necessary to preserve the integrity of the remedial action. Notice comes in three forms: record notice; actual notice to the other party to a real estate transaction; and notice to the appropriate government authority.

6.5.2 *Record Notice*:

6.5.2.1 Most states have some type of provision requiring the owner/operator of a site having residual chemicals of concern to file a notice on the land records. See, for example, 765 Ill. Comp. Stat. Ann. 90/1-90/7 (West 1996). This notice provides subsequent purchasers with information regarding past or present activities that may have left chemicals of concern on the site. These notices are easy to file, but they are not consistently reported by title companies.

6.5.2.2 The notice requirements can be narrowly drawn to include the use restrictions only, see, for example, Ohio Rev. Code Ann. § 3746.10 (Anderson 1999), or can be broad to include all the components that went into the formation of a restrictive covenant, such as the opinion of a Licensed Site Professional. See, for example, Mass. Regs. Code tit. 310, § 40.1071 (1996). Otherwise, the record notice may be ancillary to a Transfer Act, whereby recordation is only required in conjunction with a land transaction. *See*, for example, Ind. Code § 13-7-22.5-1-22 (1989).

6.5.2.3 Record notice is sometimes informational only, such as the Massachusetts Notice of AUL, and sometimes it is part of a legally enforceable control, such as the Massachusetts Grant of Environmental Restriction.

6.5.3 *Actual Notice*:

6.5.3.1 Another notice option that may be used is to require direct notice of environmental information to the other parties to a land transaction. Where this notice is not provided, the transaction may be voided or damages may be sought. See, for example, Ohio Rev. Code Ann. § 5302.30(K)(2) (Anderson 1999). Thus, remedies may include cancellation of the transaction, liability for actual damages, and civil penalties.

6.5.3.2 Actual notice protects potential purchasers of land. Actual notice also ensures that use restrictions and other forms of activity and use limitations are adhered to by subsequent parties.

6.5.3.3 It should be noted that failure to provide actual

notice may also void the "third party" or "innocent landowner" defense under Section 101(35) of CERCLA.

6.5.4 *Notice to Government Authority*:

6.5.4.1 Many states with statutory authority for activity and use limitations require an owner/operator to provide notice to the state's environmental authority at the time of consummation of any land transaction. Those states may require notice prior to completion of the transaction or within a specified period of time following completion of the transaction. Notice to the environmental authority aids the state in ensuring that activity and use limitations are properly followed.

6.5.4.2 Some state statutory programs require notice to local officials as well. Experience has demonstrated that notification to local or municipal authorities is important, yet frequently missing. Key local officials may include municipal clerks, local zoning officials, construction code officials, and local health officials. Many states are short on resources to monitor and enforce environmental restrictions and thus rely on local authorities to inform them when a transaction that could affect a property with environmental restrictions occurs. See, for example, Pa. Stat. Ann. § 6026.304(N)(1)(ii) (West 1999); N.J. Admin. Code tit 7, § 26E.

6.5.4.3 These governmental notice requirements are generally imposed under the state's voluntary corrective action statute, but they may also be imposed by the state's RCRA statute, Superfund law, real estate transfer laws, or other free-standing notice statutes.

6.5.5 *Registry Act Requirements*:

6.5.5.1 Some states employ programs that require their environmental agency to keep a list of all properties that have been the site of hazardous waste disposal and that have restrictions on use or transfer. See, for example, N.Y. Inactive Hazardous Waste Disposal Site Law, N.Y. Envtl. Conserv. Law §§ 27-1305-27-1321 (1999); 410 Ill. Comp. Stat. Ann. 525/1 *et seq*. (1999). However, given the scarcity of resources available to state environmental authorities, some states have found that the lists are difficult to maintain and may not be beneficial.

6.5.5.2 Listing of an environmentally-impacted site on a state registry may result in restrictions on the use and transfer of the site. For example, related regulations may prohibit changing the use of the site without permission from the state environmental agency, or they may require permission of, or notification to, or both, the agency to convey a registered property. As previously mentioned, for these restrictions to be enforceable, notice that the site has been registered must also be recorded at the local land registry or other appropriate authority to ensure that the registration appears in the chain of title.

6.5.5.3 Effective Registry Acts provide for the establishment and maintenance of a list of all real property that has been used for hazardous waste disposal either illegally or before federal or state regulation of hazardous waste disposal was in place. For example, there may be a requirement that a list available to the public include all sites and facilities with a confirmed release of hazardous waste or materials. See, for example, Or. Rev. Stat. § 465.225 (1997).

6.5.5.4 Often, responsibility for investigation of potential sites for inclusion on the registry lies with the state. Registries are commonly available to the public and disclose the location of the site, a listing of the chemicals of concern on the site, and may disclose the level of health or environmental risk posed by the hazardous wastes on the site. Some states maintain registries that are prioritized based on these risks; thus, it falls upon the environmental authority to rank the sites according to risk.

6.5.5.5 Most states that have Registry Act requirements also have established a hearings and appeals process for owners of sites that have been proposed for inclusion. Once a site is registered, the owner may have rights to terminate or modify the listing. A proposal by the state to include a site on its registry can sometimes provide the impetus to the owner/operator to enter into an agreement with the state to undertake certain actions in lieu of inclusion on the registry.

6.5.6 *Transfer Act Requirements*:

6.5.6.1 Some states, as part of their notice requirements, have instituted specific Transfer Act programs that require full evaluation of the environmental condition of a site before or after a transfer occurs. Examples include the New Jersey Industrial Site Recovery Act, N.J. Stat. §§ 13:1K-6 et seq. (1999); the Illinois Responsible Property Transfer Act of 1988, 765 Il. Comp. Stat. Ann. 90/1-90/71; and the Indiana Responsible Property Transfer Law, Ind. Code §§ 13-25-3-1-13-25-3-15 (1997). These requirements work in conjunction with other kinds of institutional controls. These types of programs ensure that parties involved in certain real estate transactions are aware of the potential environmental liabilities associated with ownership of the property. Other types of activity and use limitations that establish enforceability and responsibility may be incorporated as part of the Transfer Act program.

6.5.6.2 A typical state's transfer act program will create information disclosure obligations on the seller or lessor of property. *See*, for example, Ind. Code Ann. § 13-25-3-1, *et seq*. (1997) (requiring delivery of disclosure document at least thirty days before the transfer). Usually, disclosure must include property-specific information such as the presence of chemicals of concern, permitting requirements and status, and past and present enforcement actions and variances.

6.5.6.3 For a successful conveyance to be recognized, a transferor must adhere to all the components of a state's Transfer Act program. If the disclosure document reveals an environmental defect in the property that was previously unknown to the receiving party, the prospective purchaser will not have to accept the transfer of property. *See*, Ind. Code Ann. § 13-25-3-3 (1997).

6.5.6.4 A transfer act also imposes certain obligations on the landowner to make information available to other parties to a transaction. Failure to comply with these requirements may render a transaction voidable by the other party or may serve as the basis for a lawsuit, even if the contract has been executed. See, for example, Conn. Gen. Stat. §§ 22A-134-22a-134f (1999).

6.6 *Engineering and Access Controls*—Engineering and access controls, including physical controls, are a type of activity and use limitation. They need to be enforced by means of institutional controls (that is, legal instruments).

6.6.1 Engineering and access controls are physical measures which serve to limit who may actually enter an impacted site.

They may also limit the migration of chemicals of concern from the site. Typical examples of engineering and access controls to restrict admittance include: caps, floors, fencing and gates, security systems, signs, or posted warnings. Examples of engineering controls to prevent migration of chemicals of concern include concrete or paving caps or covers, vapor pumping systems, groundwater pumping systems, cut-off or slurry walls. See Wash. Admin. Code § 173-340-440(3)(a) (1996); N.J. Admin. Code tit 7, § 26E.

6.6.2 When engineering and access controls are designed to minimize health risks to those who may enter a site, states typically look at four factors that determine the level of engineering and access controls needed to protect persons who may enter the site: *location*—is the site located in a residential or mixed use neighborhood? *surroundings*—is it near sensitive land use areas, for example, day care centers, playgrounds, nursery schools, grammar schools, and high schools? *usage*—is the site frequently used by area residents, for example, a footpath that is frequently traversed by area residents or local workers? and *accessibility*—how accessible are the relevant chemicals of concern? See, Mass. Regs. Code tit. 310, § 40.0933 (1999).

6.6.3 Most engineering and access controls require maintenance and monitoring for the duration of the potential exposure. If site conditions change or if concentrations of chemicals of concern are reduced over time to levels that are protective of human health and the environment given the potential exposure scenarios, then the continued use of the engineering or access controls should be re-evaluated and may be discontinued. It is helpful for any agreement between the state and an owner/operator to include a provision delineating who will be financially responsible for the maintenance and monitoring of the required engineering and access controls. Generally, financial responsibility falls upon the owner/operator because of limited state resources.

7. State and Local Implementation Considerations

7.1 *Identification of Available State and Local Authorities*:

7.1.1 Numerous federal, state, tribal or local laws affect the use of activity and use limitations. Supplemental controls will often be appropriate even though the current land use is similar to planned use after remediation. Because of the differences in state law, different activity and use limitations will have differing degrees of effectiveness and long-term reliability. The particular mix of activity and use limitations that may be appropriate for a site will vary by both the conditions at the site and the legal framework of the state.

7.1.2 The activity and use limitation specifies limits on activities on-site. For example, if a performance standard is a final remedial action, the restriction is written to maintain the performance standard and limit the use of the site to the zoned land use. If conditions change such that residual chemicals of concern are exposed, the activity and use limitation must have some form of "trigger" mechanism so that it comes into play for the site in terms of protecting human health and the environment. The activity and use limitation must also include specific actions to be taken if a new release or exposure occurs at a site.

7.1.3 Access agreements or easements between the responsible party and private party may be required during the period of remediation. While many state environmental corrective action programs have statutory authority for access, long-term access for monitoring or otherwise ensuring the remedial action may be negotiated among the responsible party, other affected private parties, and the agency.

7.1.4 A state or local agency may track remediation site conditions. This tracking information is normally available to the public and provides another supplemental control. A few states and localities have taken an active role by instituting a notification system, such as a "diggers hotline" used by excavators, utility managers, etc. who need the information for contractor notification.

7.2 *Evaluation of Screening and Balancing Criteria at the State and Local Level*:

7.2.1 The screening and balancing criteria for determining which activity and use limitations are most appropriate at a given site should be considered early in the scoping process. It is important that the bases for current and future land uses are known and clearly understood. Knowledge of potential risks to receptors (for example, types and concentrations of chemicals of concern; potential exposure pathways, such as inhalation, ingestion, dermal or other; and media of concern) must be understood before remedies are evaluated. Planning should also consider whether a remedial action decision will lock property into a specific land use.

7.2.1.1 A reliability analysis should examine the types of factors discussed in Section 6, *supra*, such as limitations on various legal doctrines, and the risk of change in local regulations.

7.2.1.2 An implementability analysis should look at factors such as whether there is a small enough number of landowners to make negotiating deed restrictions on a case-by-case basis feasible, whether landowners are likely to consent to the restrictions, and whether the potentially responsible party will face prohibitive costs if it tries to seek deed restrictions from adjoining property owners.

7.2.1.3 Also, in the study phase, any alternative involving activity and use limitations should address what system will exist for monitoring and enforcing these controls. The cost of the controls should be taken into account in evaluating this alternative.

7.2.2 It is important to make a site safe for its intended use by ensuring exposure pathways are considered in remediation decisions. If future land use is different from current zoned use, remedial action considerations may be different.

7.2.3 "Conditional land use" permits are allowed in many states and localities. The concern is for maintenance of the level of protection in a more protective use, that is, residential, when a lesser protection level might occur from a conditional land use permit. A conditional land use permit may be granted for purposes of accommodating a final remedial action or clean up standard.

7.2.4 "Nonconforming uses" may result when areas are initially zoned or rezoned. Activity and use limitations may be required for the nonconforming use if it is likely to result in a threat to human health or the environment when surrounding uses or the existing non-conforming use would result in greater exposure. It is recommended that a parcel by parcel consideration be made when requests for zoning changes are made that may result in a direct contact threat. Activity and use limitations should be written for the property as appropriate.

7.2.5 An existing deed restriction may need to be amended or updated to reflect a change in the use of a property. The enforcing agency must identify the time frame that must be adhered to in making the amendment and deciding how to enforce this. Notice which is inconsistent with the requirements of the National Contingency Plan, 40 CFR Part 300, may also defeat a property owner's right to pursue other potentially responsible parties for response costs under Superfund, even if public notice has been adequate for purposes of the state program.

7.3 *Enforceability Issues*—A critical consideration, particularly for federal, state and local jurisdictions in selecting an activity and use limitation, is whether the activity and use limitation will be legally enforceable, and enforceable over the desired period of time. While enforceability may not be deemed necessary in every case, it is important to at least consider the implications of choosing between enforceable and non-enforceable approaches.

7.3.1 *Degrees of Enforceability*—There is a wide range of enforceability in the various types of activity and use limitations currently being used. At one end are purely informational instruments, such as deed notices, which do not establish any directly enforceable restrictions. However, notices do make land users aware that certain uses are incompatible with the condition of the land. A notice, even though unenforceable, may be a significant deterrent because the landowner frequently risks being forced to conduct remedial action if it uses the property inappropriately. (This is not always the case if the landowner has a legal responsibility for cleaning up the residual chemicals of concern.) Concerns about tort liability and inability to obtain financing or resell land may also discourage landowners from disregarding known risks.

7.3.1.1 Under some state programs, deed notices may be required as a condition for approval of a corrective action plan and release from further liability. In such cases, while the notices do not create directly enforceable restrictions, the violation of their terms voids the release, creating an additional incentive to follow them.

7.3.1.2 Information and incentives are not, however, equivalent to legal enforceability. Under a legally enforceable control, the landowner can be compelled to abide by the terms of the use restriction. Property instruments and regulatory devices, such as local ordinances or agency orders, are legally enforceable.

7.3.1.3 Another consideration in evaluating enforceability is whether the control binds only the current occupant, or future owners as well. Land use ordinances and most property interests "run with the land"; orders and permits do not.

7.3.2 *Ensuring Long-Term Enforceability*—Where controls are being imposed through property law devices, it is necessary to become familiar with the state's real estate laws to ensure that the tool being used will in fact be reliable over the long term. While legal rules can vary from state to state, and are evolving over time, some common doctrines can present significant obstacles to long-term enforceability in the context of corrective actions.

7.3.2.1 *Requirement of Conveyance*—Unless a state has specifically provided otherwise by statute, a conveyance of some kind will be required to establish an enforceable property interest. In other words, a landowner generally cannot impose an enforceable restriction on its own property simply by filing a document in the land records. Rather, there must be a transaction between the landowner and some other party in which rights are actually conveyed to the grantee (who is then able to enforce those rights). It will generally be necessary, therefore, to find a suitable grantee. Potential grantees may include regulatory agencies, local governments, custodial trusts, community organizations or other parties responsible for the corrective action. It should be noted, however, that local and state regulatory agencies may be reluctant recipients of these property rights.

7.3.2.2 *Doctrines Limiting Long-Term Enforceability*—Even where a property interest is created through a transaction of some kind, traditional legal doctrines may limit its long-term enforceability. Historically, the common law allowed restrictions on the use of property to "run with the land" only where they benefited some adjoining property. Easements or covenants that were not for some neighbor's benefit but were simply held by some other party, were classified as "in gross" and generally could be enforced only against the original landowner.

(1) Restrictions for corrective action purposes are likely to be "in gross"; they are not held by an adjoining landowner, or for the benefit of the adjoining land, but by some unrelated third party such as a government agency and for the benefit of the public (or the restricted landowner itself). Therefore, there is a risk that such restrictions may be found unenforceable against subsequent landowners.

(2) Over time, courts have recognized exceptions to the common law rules. Today, in many states, it is quite possible that a restriction in gross, entered into for the public benefit, will be enforceable against subsequent landowners as long as that intent is clearly stated in the document. Moreover, even where courts do not honor the traditional doctrine against restrictions in gross, other peculiarities of local real estate law may present barriers to long-term reliability.

7.3.2.3 *Need for an Enforcer*—Legal instruments do not enforce themselves; they require someone to monitor compliance and take legal action if necessary. Finding some entity willing and able to take responsibility for this function is critical to the long-term reliability of the controls.

(1) Who can enforce the control will depend largely on the type of control used. Property interests, such as easements, are generally enforceable only by the named grantee (or its assigns). Therefore, in determining to whom the interest will be conveyed, it is important to ask whether this is the most appropriate enforcer. It is also important to keep in mind that, if the grantee neglects to enforce, it may be difficult or impossible for any other party to compel it to do so, unless the applicable statute and regulations reserve those rights to the state.

(2) Restrictions imposed through local regulation, on the

other hand, are generally enforceable only by the local government. Whether the government has the resources, or the motivation, to effectively oversee and enforce the controls will depend upon the circumstances.

7.3.2.4 *Legal Formalities*—There are certain legal formalities that must be addressed for the promise to use an activity and use limitations to be enforceable (see 4.2). Additionally, some states have adopted specific provisions outlining the enforcement process. Among the most common are provisions authorizing the state's environmental authority or attorney general to file suit for injunctive relief, see, for example, Wis. Stat. § 299.95 (1995-96) (attorney general shall enforce agreements by seeking injunctive relief), or for civil and criminal penalties. See, for example, Wash. Rev. Code Ann. § 70.105D.050 (Michie 1996) (civil penalty of up to $25,000 for each day party refuses to comply). Another effective provision authorizes third parties injured by violations of environmental restrictions to bring suit through the state's environmental authority or attorney general. See, for example, Conn. Gen. Stat. § 22a-133p (1997). Finally, states that utilize No Further Action Letters or Certificates of Completion of Remediation often revoke these documents in light of an owner/operator's failure to comply with an agreement to use activity and use limitations and consequently may order further remediation. See, for example, Me. Rev. Stat. Ann. tit. 38, § 343-E (West 1995).

7.4 *Considerations for Changes in Uses at Sites Where an Activity and Use Limitation Has Been Implemented*:

7.4.1 *General Principles*—Over time, activities and uses at a site may change in ways that are very difficult to predict at the time that an activity and use limitation is first implemented. A basic tenet of a risk-based approach to corrective action decisions is that the need for corrective action should be based on likely exposure to chemicals of concern resulting from current uses of a site (soil and ground water), as well as reasonably likely future uses. If activities and uses change in ways that were not anticipated when the initial corrective action decisions were made, then the site needs to be re-evaluated to determine whether additional remediation is needed to provide an adequate level of protection.

7.4.2 *Compliance with State Environmental Requirements*—State corrective action programs frequently have "reopeners" that require re-evaluation of site conditions before a new use will result in significantly higher levels of risk from exposure to residual chemicals of concern remaining onsite from the original corrective action, then additional remediation will be required before the new use can be implemented. Different states have different requirements in terms of notification of the state agency of a new use and need for additional corrective action, and approvals of new uses and remediation plans. Some states require that development and remediation plans be submitted and state approval be obtained before proceeding. Other states (with more privatized programs) require that an expert licensed by the state conduct the necessary evaluation and prepare any plans for additional response actions. For example, in Massachusetts, these plans would then need to be filed with the state. Parties who are contemplating redeveloping sites where corrective action has relied on activity and use limitations to prevent future exposure to residual chemicals of concern should check with their state corrective action program to identify applicable requirements.

7.4.3 *Need for Local Land Use Approvals as Well as State Environmental Review*—Reviews by state environmental agencies (or by experts licensed by states) usually focus on the adequacy of plans for additional remediation to provide an adequate level of protection for the new use of the site. These reviews are usually required by state law, regulation or policy. However, they do not substitute for any requirements of local or county government or of other state agencies for approval of the new use itself (for example, zoning approvals) or of specific building plans (for example, subdivision approvals, building permits, etc.)

7.4.4 *Responsibility for Evaluating Needs for Additional Remediation and for Implementing Resulting Plans*—Responsibility for implementing plans for additional remediation required to support a new use of a site usually falls to the party who is redeveloping the site. However, this party may request that other parties contribute to (or take full responsibility for) additional remediation. Parties who may be able to contribute are the party who implemented the original corrective action, former owners, tenants, and other potentially responsible parties. State laws differ considerably in the responsibility they place on new and former owners, developers and other parties involved in corrective action supporting redevelopment. Generally, parties wishing to change the use at a site where prior remediation relied on an activity and use limitation should seek advice from the state environmental agency or an attorney, or both, experienced in this area of state or local law, or both.

7.4.5 *Recommendations for "Good Practice" When Uses Change at a Site*:

7.4.5.1 Where a new use is clearly permitted by the existing control, no further evaluation is needed. In general, this argues for clearly written activity and use limitations that are drafted to identify prohibited uses with a high degree of specificity, and to establish permitted uses as generally as possible.

7.4.5.2 Where a new use is expressly prohibited by the activity and use limitation that was part of the original remediation, evaluate potential exposures to residual chemicals of concern that may result from the new use (soil or ground water, or both), and identify the need for additional remediation to ensure that the new use can be safely implemented.

7.4.5.3 The user should check with the state environmental agency to identify requirements for approvals of additional response actions, and ensure that these requirements are complied with. These requirements may also include specific steps for amending or terminating the activity and use limitation. These requirements may also include notifications to local or county government, or both, (of changes or termination of the control as well of additional remediation plans), and opportunities for public comment.

7.4.5.4 Finally, the user should implement required remedial actions in conjunction with (or prior to) site redevelopment, and modify the activity and use limitation in accordance with state requirements. Please note that, in some cases, additional remediation may remove the need for the control; it should be amended or terminated so that the property is not encumbered more than necessary.

7.4.6 *Activity and Use Limitations Are Not Self-Executing*:

7.4.6.1 Most activity and use limitations in the context of remedial actions will be negotiated and incorporated into the decision document. Many states will require some form of periodic monitoring of the activity and use limitation, and changes in either circumstances or the control itself will require written modification by the parties. Some activity and use limitations that occur outside of the remediation context (for example, a change in land use zoning) might occur without the parties' direct involvement, but if such changes would affect the activity and use limitations, the parties would need to modify the activity and use limitations that were implemented as part of the remedial action.

7.4.7 *Notification Process to Local Government or the State*:

7.4.7.1 Notice requirements will vary state by state and by the type of activity and use limitation being used. Many activity and use limitations will involve some restriction on permissible uses of land. Typically, these restrictions will require:

(1) Some form of property right transfer to the state agency (for example, grantor/grantee);

(2) Some form of documentation in the decision document (for example, order);

(3) Recordation with the local property recording agency (for example, County Clerk);

(4) Some consultation with the local land use jurisdiction; and

(5) Some form of public notice or opportunity for comment on the selected remedial action, or both, that includes the use of activity and use limitations.

7.4.7.2 No blanket statement can be made regarding notice, formal or otherwise. It may be wise to assume that any remedial action that involves activity and use limitations will be subject to some form of public notice or review, whether the "public" is the general public or representatives. "Good practice" would again advise providing notice to various stakeholders throughout the remedial process so surprises are avoided at the remedial action selection stage.

7.4.8 *Whether There Is a Process for Removing Controls*:

7.4.8.1 There should be a process for removing the activity and use limitation when it is no longer necessary to protect public health and the environment. A good deal of attention has been directed at activity and use limitations to ensure they will be in place over the long run. For example, some restrictions are structured to "run with the land" so successors maintain the protective elements of the control. However, when those controls are no longer needed to protect human health or the environment, they should be removed. The possibility of amending or terminating activity and use limitations should be anticipated in the state's statute and regulations and in the final decision document. For example, if natural attenuation is the selected remedial action and prohibitions on the use of groundwater are in place until certain standards are met, a process should be in the state statute and regulations, as well as in the final decision document, to allow the removal of the control.

7.5 *Public Notice/Participation/Stakeholder Issues*:

7.5.1 Successful corrective actions should consider the concerns of the communities in which they are located. Local officials, residents, utilities, neighboring businesses, environmental groups, and others are all ultimately "stakeholders" of a site corrective action, since they live and work with the results over time.

7.5.2 State corrective action programs usually have requirements for public notice of site conditions and corrective action plans, as well as specific opportunities for the public to be involved in developing corrective action plans. These requirements usually include officials of the municipality (or county, or both) in which the site is located, residents and businesses in the site's neighborhood, and the general public. State public involvement requirements can apply to the assessment of site conditions and risks, as well as to specific plans for remediation.

7.5.3 There are two aspects of activity and use limitations that may trigger specific requirements for public involvement:

7.5.3.1 The first is the specification of current and reasonably anticipated future uses that will be made of the site (and resulting potential exposures to chemicals of concern remaining in soil and/or ground water); and

7.5.3.2 The second is the drafting of an activity and use limitation document (for example, a "deed restriction", deed notice or local bylaw) that establishes prohibited and permitted activities at the site and associated continuing obligations and conditions.

7.5.4 In general, the user should consult its state environmental agency to identify specific requirements to notify the public while the site is being assessed and plans for corrective action are being developed, and to provide opportunities for public involvement (for example, a hearing on a remediation plan).

7.5.5 Some states require that specific opportunities for involvement be provided for every site (usually at the point where a full assessment of site conditions and risks is available and when a remediation plan has been drafted). Other states require that public notice be provided at specific points in the corrective action process and that opportunities for public involvement be provided when local officials and/or citizens indicate their interest. Some state requirements may apply to any site that is being assessed and cleaned up, while others may only apply to sites that have more residual chemicals of concern or present relatively high levels of risk of harm to public health and the environment.

7.5.6 Corrective action of many sites is not publicly controversial, especially where the corrective action will result in general environmental improvements for the neighborhood. Also, there may be little or no controversy if the use of the site is changing in a way that satisfies local needs or is otherwise acceptable to the municipality and site neighbors (for example, from industrial to commercial or residential, or vice versa, or from vacant/underused property to a more productive use). However, some redevelopment plans become controversial when there is not widespread public agreement that the new use is appropriate. Even in situations where the community supports a new use, there may be local issues about how the corrective action is implemented (for example, the adequacy of the corrective action plans to protect the health of people and the environment in that neighborhood, how truck traffic will be handled during remediation, etc.).

7.5.7 Typically, one should have early substantive communication with local officials, members of the public and others (such as utilities) who may be affected by the site or its corrective action, or both. However, sometimes the scale of the project is small and there is little or no community interest. By identifying community interest, the scale of the project, and the likely concerns as early in the remediation process as possible, the appropriate level of public participation can be determined. Where there is a large scale, high interest project, users should provide information as it is developed, be open to public comments, and develop a working relationship with the people who will have to live with the results of the corrective action over the long term. Those who are cleaning up a large scale, high interest site are encouraged to invite the public to become involved while environmental conditions and risks are being assessed, and while plans are being developed.

7.5.8 Activity and use limitations are generally part of an overall remediation strategy for a site. Public involvement in the development and implementation of the activity and use limitation should focus on whether the activity and use limitation has been drafted to adequately explain what the prohibited and permitted uses of the site will be, and whether there are any continuing obligations and conditions required of the property owner (and tenants).

7.5.9 Some states also require that notice be provided to local officials (and in some cases to site abutters) of an activity and use limitation once it is implemented. These notices may not be very effective unless the locality has a database or other tracking system. These notice requirements generally attempt to reach people who may be in a position to observe when a provision of the institutional control is violated (for example, a planning board may receive an application for a change in use that is not permitted by the activity and use limitation). These officials may notify the state environmental regulatory agency which is authorized to monitor compliance and take appropriate enforcement action.

7.5.10 Where an activity and use limitation becomes publicly controversial, it is often due to a lack of public acceptance of plans for corrective action and redevelopment. In these cases, the user may find it productive to reopen the site assessment and remediation plan to identify the areas of public concern, and to initiate an open dialogue with the dissatisfied interests, working toward a goal of developing a consensus plan for the site.

APPENDIXES

(Nonmandatory Information)

X1. BUSINESS AND FINANCIAL CONSIDERATIONS INVOLVED IN THE USE OF ACTIVITY AND USE LIMITATIONS

X1.1 Financial Risk Allocation Mechanisms

X1.1.1 *Introduction*:

X1.1.1.1 This section identifies financial assurance closure mechanisms available to the real estate community at reasonable cost for risk-based closures. The mechanisms effectively cover this low probability liability, thus facilitating real estate transactions. This section also may be used to ensure financial responsibility of the owner when implementing or enforcing provisions of the activity and use limitation.

X1.1.1.2 Most of these mechanisms, or a combination thereof, can be tailored to fit the specific site needs and needs of the parties to the transaction, including:

(1) First-party coverage for the effectiveness of the risk-based corrective action, business interruption costs due to environmental considerations, and/or diminution of realty value of the "site"; and

(2) Third-party coverage for government or private party actions including damages, diminution of realty value of adjacent landowners, corrective action costs and/or transaction costs such as attorneys fees and consultant costs.

X1.1.2 *Environmental Insurance*—Insurance is now a frequently used device and can become part of the transaction closing, comparable to title insurance. There are a number of top-rated national carriers providing this coverage obtainable through many general commercial insurance agents. The insurance falls within the category of environmental impairment liability insurance (EIL), which is specifically written and intended to cover qualifying first party or third party environmental claims, or both, with residual chemicals of concern. Often, specific policies are modified for each transaction.

X1.1.2.1 Most of the coverage is "claims made", which means insurance coverage must be initiated by a claim from the insured or policy holder during the policy period as defined in the insurance contract. The time-frame for making a claim can be extended by obtaining EIL "extended reporting period" coverage.

X1.1.3 *Environmental Bonds*—Bonds have been extensively used for environmental closure and can be obtained from many commercial insurance brokers. Bonds are written obligations for a sum certain usually secured by a mortgage on real estate. They include liability bonds to protect the assured from liability due to environmental damages or injuries to third parties as imposed by law or a court, and indemnity bonds providing reimbursement for a specified environmental loss.

X1.1.4 *Contracts Assigning the Risk of Loss*—Routinely used in most commercial property transactions, the parties to the transaction generally also assign the risk of potential liability through indemnities, warranties and covenants. Environmental insurance, discussed above, is also often available as excess to an indemnity.

X1.1.4.1 Contracts may have limited protection because the party to the contract who is the warrantor may become insolvent, or the owner of the real estate, or transporter, or generator of any hazardous or regulated substance which is a party to the transaction, generally can not "contract away liability" to third parties through indemnities and warranties due to the federal Resource Conservation and Recovery Act (RCRA) and Superfund laws and comparable state laws.

X1.1.5 *Letters of Credit*—Bank letters of credit have been used in limited instances for environmental closure and are available from most commercial banks. A letter of credit is a binding negotiable instrument honored and paid when specified environmental conditions occur. It is governed by Section 5-103 of the Uniform Commercial Code, which has been enacted in all states. Letters of credit generally cost more than insurance, however, and may tie up capital needed for other purposes.

X1.2 Transactional Issues

X1.2.1 *Environmental Due Diligence*—The user is cautioned that the existence of an activity and use limitation may not be detected during routine environmental due diligence activities. At the present time, Practice E 1527, does not require the environmental consultant to either look for or report on the existence of activity and use limitations at a site. The user and the environmental consultant are encouraged to discuss whether information about activity and use limitations is important and who should assume responsibility for obtaining and analyzing this information.

X1.2.2 *Need to Obtain the Property Owner's Consent*—Difficult transactional issues may arise if the state program does not provide a mechanism for notifying all parties with any interest in the real property about the potential imposition of an activity and use limitation. For example, it is not uncommon for a seller of real property to retain responsibility and liability for cleaning up residual chemicals of concern that have been detected during the environmental due diligence associated with the transaction. As part of the state voluntary corrective action program, the seller, which becomes the participant, may have choices regarding the level of corrective action that is required, depending upon whether an activity and use limitation is part of the ultimate remedial action (although, in some states, it may be required to do so under applicable real estate law). The seller may not be required under applicable environmental law to inform the current owner that it is considering using an activity and use limitation as part of the remedial action. The failure to include the current property owner in these discussions and negotiations could have very significant impacts upon the future value of the property and the owner's ability to use the property without significant limitations. Similarly, the participant may not be required to notify the lender that it intends to seek the imposition of an activity and use limitation. The lender's exclusion from the process where such a restriction might be imposed could also have significant adverse impacts upon the value of the collateral being held by the lender.

X1.2.3 *Clarification of Responsibilities of Landlord and Tenant*—Given the increasing use of risk-based approaches to cleanup, in which it is permissible to allow residual chemicals of concern to remain in place if there is an enforceable activity and use limitation, it is important for landlords and tenants to negotiate, in advance, what type of remediation will be expected under the lease when the lease terminates, and whether an activity and use limitation may be used to achieve applicable cleanup standards. It is no longer helpful simply to state that the tenant must comply with all applicable laws. Likewise, if an activity and use limitation has already been put in place, it is important for the lease to specify whether the landlord, or tenant, or both, will have primary responsibility for monitoring and maintaining the activity and use limitation. Otherwise, if the activity and use limitation is not maintained, the landlord may lose any protections that it may have negotiated through a state voluntary cleanup program.

X1.3 Other Issues

X1.3.1 *Potential Stigma/Devaluation Concerns*:

X1.3.1.1 The value of real property is determined by the ability of the property to provide wealth to the property owner. Value is defined as the present worth of future benefits. *The Dictionary of Real Estate Appraisal*, Second Edition, American Institute of Real Estate Appraisers (now the Appraisal Institute), Chicago, IL. Accordingly, when an activity and use limitation may be imposed upon real property, the question arising is whether the property's value may be impaired by this restriction on property rights. Stigma arises from uncertainty concerning the use or cost to use a property, or both. As such, stigma impacts may be reduced by well-designed activity and use limitations. Excepting the case of poorly designed AULs which do not fully recognize the highest and best use of the property, AULs should be expected to reduce stigma impacts resulting from the existence of a recognized environmental condition.

X1.3.1.2 From a real estate appraisal perspective, whether property value is impaired by an activity and use limitation depends upon the "highest and best use" of the property. First determine what the "highest and best use" of the property is, without the restriction, and then determine whether that "highest and best use" will change as a result of the imposition of the restriction. In some cases, the "highest and best use" may change, and in others, it may not.

X1.3.1.3 While there may be minor differences in the legally recognized definition of "highest and best use" from jurisdiction to jurisdiction, the fundamental points remain the same. The highest and best use is defined by how the property may be legally used, what is physically possible with respect to use, what is financially feasible supporting the use, and what will return the maximum net income to the owner. All four issues must be dealt with simultaneously to establish the highest and best use for a specific property.

X1.3.1.4 Under the Uniform Standards of Professional Appraisal Practice (USPAP), the guidance document established by the Financial Institutes Reform, Recovery and Enforcement Act (FIRREA), all appraisals of real property must contain at least one, and possibly two, opinions of the highest and best use of a property. One opinion deals with the highest and best use of the property as if it were vacant. This opinion addresses the issue of how property might be used, given all current information on such matters as zoning, topography, neighborhood, and market demand as if there currently were no structure or other improvement on the property. The second opinion of highest and best use deals with the issue of how the value of the property may be maximized given the current improvements.

X1.3.1.5 The highest and best use, the value of the property to the owner, and the issue of activity and use limitations are inextricably linked to each other. A desire on the part of the owner to initiate a substantive change in the highest and best use must, of necessity, generate a significant concern with respect to any existing or proposed activity and use limitations. Similarly, the imposition of activity and use limitations must result in a reconsideration of the highest and best use for the property.

X1.3.1.6 Highest and best use is controlled, among other things, by the legally permissible and practical use to which the property may be put. The existence of an activity restriction or a use control must be considered in the determination of highest and best use by the appraiser and will, if substantive, influence the value of the property.

X1.3.1.7 Conversely, an activity and use limitation that does not recognize the highest and best use cannot be said to address the reality of the marketplace except through the action of random chance. One is cautioned to observe that the concept of highest and best use addresses not only current use, but also the most likely future use, because the value of the property is determined by the future, not the present or the past.

X1.3.1.8 Highest and best use should be explicitly identified and documented in the development of the appropriate activity and use limitation, and any change in highest and best use should not take place without revisiting the issue of the activity and use limitations. In all cases the highest and best use description will identify both the current maximally beneficial economic use. As such, the highest and best use must embody full economic consideration of all factors of the property, including the impact, if any, of AULs. The underlying economic forces described by the highest and best use will govern the property owner's adherence to AULs, desire to modify AULs, or to remove AULs by further remedial actions. An understanding of the highest and best use and the forces defining it are critical to the success of AULs.

X1.3.2 *Potential Takings Claims*

X1.3.2.1 *Overview*—A question exists whether the imposition of activity and use limitations may be viewed as a "takings" under the Fifth Amendment to the U.S. Constitution, particularly if the controls are imposed without the property owner's consent or adjoining property owners' full consent. The answer to this question depends upon whether the activity and use limitation advances a legitimate state interest, and whether the activity and use limitation denies the property owner an economically viable use of its land.

X1.3.2.2 *What Constitutes a Takings*—The Fifth Amendment is designed to provide just compensation when the government interferes with private property rights for a public purpose. *First English Evangelical Lutheran Church of Glendale v. County of Los Angeles California*, 482 U.S. 304, 313 (1987). Governmental interference with private property rights may range from a direct appropriation of the land, to a physical invasion of property, to the implementation of a regulation that interferes with a property owner's use of his or her property. The government is authorized to take all of these actions, provided that the taking of private property:

(1) Substantially advances a legitimate state interest; and

(2) The property owner is justly compensated for the taking of his or her property. *First English*, 482 U.S. at 314. A government acts lawfully when it takes property, pursuant to proper authorization, and justly compensates the owner for it. *Id.* at 315. However, the government violates its constitutional duty to provide just compensation when it either (i) denies just compensation, or (ii) denies the procedures through which a landowner can seek just compensation. *Monterey v. Del Monte Dunes at Monterey, Ltd.*, 119 S. Ct. 1624, 1999 U.S. LEXIS 3631, at *53 (1999).

X1.3.2.3 *Types of Takings*—There are two primary ways in which the government can violate the Takings Clause of the Fifth Amendment:

(1) Direct government appropriation without just compensation; or

(2) Government regulation that interferes with a property owner's use of his or her property when the regulation accomplishes the same result as direct appropriation. *Lucas v. South Carolina Coastal Council*, 505 U.S. 1003, 1014 (1992). In a regulatory taking, the government prevents the landowner from making use of his property that would otherwise be permissible. *Forest Properties Inc., v. Big Bear Mun. Water Dist.*, 1999 U.S. App. LEXIS 9370 at *9 (1999).

(3) A government regulation will constitute a taking when either of the following two conditions are met: the regulation does not substantially advance a legitimate state interest; or the regulation denies the owner economically viable use of his or her land. *Agins v. City of Tiburon*, 447 U.S. 255, 260 (1980).

X1.3.2.4 *Advancing a Legitimate State Interest*—There is no clear test for determining whether a regulation substantially advances a legitimate state interest with regard to generally applicable regulations. *Tahoe Sierra Preservation Council Inc., v. Tahoe Reg'l Planning Agency*, 34 F. Supp. 2d 1226, 1239 (1999). However, the question of economic viability has been greatly debated. In the absence of a clear test to determine whether a generally applicable regulation substantially advances a legitimate state interest, precedent indicates two standards:

(1) Whether the regulation is "arbitrary"; and

(2) Whether there is a "reasonable relationship" between the regulation and the legitimate state interest. *Dolan v. City of Tigard*, 512 U.S. 374, 391 n.8 (1994); *Del Monte Dunes*, 119 S. Ct. 1624, 1999 U.S. LEXIS 3631, at *25. Thus, if a state or local government imposes generally applicable activity and use limitations that are found to be either arbitrary or have no reasonable relationship to a legitimate state purpose, the government would be liable for an unconstitutional taking.

X1.3.2.5 *Deprivation of Economic Viability*—If it is determined that the regulation authorizing activity and use limitations deprives the property owner of an economically viable use of his land, the extent of the deprivation of economic viability must then be determined. This determination could range from a "taking" which deprives the landowner of all economically beneficial use of the land, *Lucas v. South Carolina Coastal Council*, 505 U.S. 1003, 1015 (1992), to a determination that the regulation that constitutes a compensable 'partial taking.' *Forest Properties Inc., v. Big Bear Mun. Water Dist.*, 1999 U.S. App. LEXIS 9370 at *9 (1999).

(1) If the regulation deprives the landowner of all economically beneficial and productive use of the land, then the regulation would constitute a "categorical" taking, and no further inquiry would be necessary. *Tahoe*, 34 F. Supp. 2d at 1240 (citing *Lucas*, 505 U.S. at 1015). The landowner is entitled to just compensation for the taking. *Id.*

(2) If a regulation deprives the landowner of partial economic viability of the land, courts use the test clearly laid out in *Penn Central Transportation Co. v. New York City*, 438 U.S. 104, 124 (1978). The factors examined in the *Penn Central* test are: the economic impact of the regulation on the claimant; the extent to which the regulation has interfered with distinct investment-backed expectations; and the character of the governmental action. *Id.*

X1.3.2.6 *Mere Diminution Does Not Constitute a Takings*—A mere diminution in the value of property does not establish a taking. Thus, when a regulation insignificantly affects the value of property, this does not constitute a compensable taking under the Fifth Amendment. To determine whether a regulation results in a partial taking or a mere diminution in value, the courts compare the ratio of the land subject to restrictions with the plaintiff's parcel as a whole. *Broadwater Farms Joint Venture v. United States*, 1997 U.S. App. LEXIS 19859 *4 (1997). If an activity and use limitation is imposed, and there is minimal economic impact on the landowner in relation to the relevant parcel, the government may not be liable for a taking.

X1.3.2.7 *Temporary or Permanent Takings*—The last aspect to be examined is whether the taking is temporary or permanent. In cases where an activity and use limitation results in a temporary taking, but this is only evident retrospectively, and the property interest has been altered during the period of the taking, the property owner may be entitled to just compensation for the period of taking. *First English*, 482 U.S. at 322.

X1.3.2.8 *Determination of Appropriate Compensation*—Once it has been determined that a taking has occurred, the next question is what amount of compensation is just. To determine whether the compensation is just, the court must determine what the property owner has lost, not what the taker has gained. *Del Monte Dunes*, 119 S. Ct. 1624, 1999 U.S. LEXIS 3631, at *42 (citing *Boston Chamber of Commerce v. Boston*, 217 U.S. 189, 195, 54 L.Ed. 725, 30 S. Ct. 459 (1910)). The economic impact is measured by the change in the fair market value caused by the regulation. *Forest Properties*, 1999 U.S. App. LEXIS 9370 at *17.

X1.3.2.9 *Conclusion*—In conclusion, there are a number of fact specific inquiries necessary to determine whether state and local governments are at risk of facing takings claims when they impose activity and use limitations. Consequently, in order to avoid an unconstitutional taking, the state or local government must be able to: (i) prove that the regulation substantially advances a legitimate state purpose; (ii) provide procedures for landowners to seek just compensation; (iii) determine the economic impact on the landowner; (iv) state whether the regulation is temporary or permanent; and (v) provide just compensation when appropriate.

X2. CASE STUDIES

TABLE X2.1 Case Study No. 1 Cameron Station Alexandria, VA

Site Name:	U.S. Army Cameron Station Base Realignment and Closure
Future land use summary:	A large portion of site was sold for $33 million for the construction of homes, a skating rink, sports field, tennis courts and playgrounds.
Former land use:	Army depot that provided administrative, commissary and post exchange support and vehicle washing and maintenance activities. The base also housed a grounds maintenance facility, several print shops and a photographic laboratory.
Chemicals of Concern:	1. Soil: polychlorinated biphenols (PCB), pesticides, low levels of dioxin, lead, petroleum hydrocarbons 2. Ground water: trichloroethene (TCE), metals, chlorinated hydrocarbons, petroleum products 3. Sediments: low levels of petroleum hydrocarbons, polyaromatic hydrocarbons
Clean-up activities:	Underground storage tank (UST) removal, removal of PCB transformers, excavation and treatment of soil containing chemicals of concern
Ongoing clean-up activities performed by the Army:	Treatment of remaining TCE in groundwater, asbestos abatement, soil vapor extraction for treating petroleum chemicals of concern, operation, maintenance and monitoring of remediation activities
Activity and Use Limitations:	1. Provide for continued access for Army and regulatory agencies to monitor the effectiveness of corrective action, perform reviews and take additional remedial or removal actions

TABLE X2.1 *Continued*

	2. Ensure that the proposed use will not disrupt any remedial activities, past, present or future. Disruptions may include surface application of water which could impact the migration of impacted ground water; subsurface drilling or use of ground water unless Army determines that there will be no adverse impacts on the corrective action process; or construction that would interfere with, negatively impact or restrict access for corrective action work. Remedial action systems will be protected against disturbances (for example, changes to electricity source for treatment systems) unless the Army consents. The Army must be notified immediately if any disruption to the remedial action occurs. When the system is reactivated, it must meet all design specifications and discharge parameters.
Benefits of Activity and Use Limitations:	1. Allow for the transfer and reuse of property without needing to remediate all chemicals of concern first. 2. Provide receiving parties with information concerning ways to maximize the use of the property without interfering with ongoing remedial activities (i.e., treatment and monitoring systems). 3. Provide access for the Army and regulatory agencies to maintain and monitor the remediation system.
Shortcomings of Activity and Use Limitations:	1. Information concerning the importance of the Activity and Use Limitations was not transferred to all contractors engaged in site development activities. 2. The locations of remediation systems, which are based on the locations of existing structures, may not be consistent with redevelopment plans and may necessitate the relocation of monitoring and treatment systems. 3. It is unknown where the ultimate responsibility for maintaining the Activity and Use Limitations will lie.

TABLE X2.2 Case Study No. 2 Industri-Plex NPL Site Woburn, MA

Site Name:	Industri-Plex Site, Woburn, MA
Future land use summary:	The site is being redeveloped as a Regional Transportation Center, retail center, and mixed use parcel (office and hotel space.)
Former land use:	Former industrial park used for manufacturing chemicals such as lead-arsenic insecticides, acetic acid, and sulfuric acid, as well as phenol, benzene and toluene. The site was also used to manufacture glue from raw animal hides and chrome-tanned hide waste.
Contamination:	1. Soil: metals, including arsenic, lead and chrome. 2. Ground water: VOCs, including benzene and toluene, and arsenic. 3. Air: hydrogen sulfide gases from decay of buried animal hides.
Clean-up activities:	1. Permeable caps over 105 acres of soils and sediments impacted with arsenic, lead and chromium. 2. Impermeable cap over the 5 acre East Hide Pile; and a gas collection and treatment system. 3. Interim ground water treatment system to treat "hot spots" of toluene and benzene. 4. Investigation of ground water and surface water. 5. Implementation of AULs. 6. Fencing and warning signs.
Activity and Use Limitations:	1. The 245 acre site is divided into four types of properties for purposes of implementing AULs: a. Class A - "Clean"; non-impacted soil, but the ground water may contain chemicals of concern b. Class B - soil containing chemicals of concern above state levels c. Class C - capped portions of the site; no ground water use allowed d. Class D - the animal hide properties; undevelopable; no ground water use allowed 2. A Custodial Trust is being used to help implement and maintain the AULs.
Benefits of Activity and Use Limitations:	1. The AULs will be implemented by means of a Grant of Environmental Restrictions and Easements, which will run in perpetuity and be enforced by EPA and MA-DEP. 2. The Easements allow the PRPs and the regulatory agencies to inspect the AULs and to conduct subsurface investigations. 3. AULs require property owners to do quarterly, non-intrusive inspections of the site for compliance with the AULs. 4. The AULs must be incorporated into all deeds, easements, mortgages, leases, and other instruments of transfer.
Shortcomings of Activity and Use Limitations:	1. AULs are required even in those portions of the site that are "clean". 2. Property owners and their tenants are potentially liable for stipulated penalties and fines if the AULs are violated. 3. If the property owner fails to cure any violation of an AUL, the PRPs may cure the violation and secure a lien against the Property. 4. Each property owner is responsible for establishing the AULs, including any title work, survey plans and legal descriptions.

Designation: E 2107 – 00

Standard Practice for Environmental Regulatory Compliance Audits[1]

This standard is issued under the fixed designation E 2107; the number immediately following the designation indicates the year of original adoption or, in the case of revision, the year of last revision. A number in parentheses indicates the year of last reapproval. A superscript epsilon (ε) indicates an editorial change since the last revision or reapproval.

1. Scope

1.1 *Purpose*—This practice[2] identifies minimum requirements for environmental regulatory compliance audits (audits). It also provides information on the terms and procedures associated with audits as practiced in the United States of America (USA) and other jurisdictions subject to the laws thereof. It provides a reference to which interested parties may refer for definition and description of accepted audit terms and procedures.

1.2 *Background*:

1.2.1 *Awareness of Benefits Associated with Audits*—Various benefits have been attributed to audits. These benefits may include a better understanding of the compliance status of a facility or organization, identification of opportunities for environmental management systems improvements, reduction or elimination of potential legal and financial liabilities when implemented as part of a comprehensive compliance management program, better communications and improved relationships with governmental agencies, communities, and other stakeholders, providing information for development of both short-term and long-term environmental expenditures, and education of employees.

1.2.2 *Awareness of Risks Associated with Audits*—It is also important to recognize that certain risks have been associated with audits. These risks may be managed and controlled by giving thoughtful consideration to the audit process before beginning an audit. These risks may include increased potential legal and financial liabilities for the audited entity if audit findings are not corrected in a timely manner, disclosure of confidential business information or trade secrets, inadvertent admissions against interest because of the wording of audit findings, disclosure of audit findings intended to be kept confidential under audit privilege laws or attorney - client privilege or work product doctrine, and inaccurate audit findings.

1.2.3 *Awareness of Legal Issues*—A number of important legal issues are associated with audits. Example legal issues include the qualification for one or more evidentiary privileges, the qualification for limited immunity, the protection of trade secrets and confidential information, the application of a number of government policies associated with environmental audits (including federal and state programs that provide incentives for detecting, disclosing, and correcting potential violations through auditing), the form and language of audit reports, the necessity of reporting certain information to the government, the potential liability of auditors, and the importance of promptly addressing issues identified during audits. Prior to initiating and during an audit, interested parties should consider potential legal issues and consult legal counsel or other experts as appropriate.

1.3 *Organization*—This practice is organized in the following manner:

2. Referenced Documents

2.1 *ASTM Standards:*

E 1527 Practice for Environmental Site Assessments: Phase I Environmental Site Assessment Process[3]

E 1528 Practice for Environmental Site Assessments: Transaction Screen Process[3]

2.2 *ISO Standards:*

ISO 14010 Guidelines for Environmental Auditing—General Principles[4]

ISO 14011 Guidelines for Environmental Auditing—Audit Procedures—Auditing of Environmental Management Systems[4]

---

[1] This practice is under the jurisdiction of ASTM Committee E50 on Environmental Assessment and is the direct responsibility of Subcommittee E50.04 on Performance Standards Related to Environmental Regulatory Programs.

Current edition approved Oct. 10, 2000. Published December 2000.

[2] As used herein, a "Standard" is a document that has been developed and established within the consensus principles of the Society and that meets the approval requirements of ASTM procedures and regulations. A" Guide" is a compendium of information or a series of options that does not recommend a specific course of action. A guide increases the awareness of information and approaches in a given subject area. A "Practice" in contrast, is a definitive set of instructions for performing one or more specific operations that does not produce a test result. See *Form and Style for ASTM Standards.*

[3] *Annual Book of ASTM Standards*, Vol 11.04.

[4] Available from American National Standards Institute, 11 W. 42nd St., 13th Floor, New York, NY 10036.

ISO 14012 Guidelines for Environmental Auditing— Qualification Criteria for Environmental Auditors[4]

3. Terminology

3.1 *Definitions of Terms Specific to This Standard:*

3.1.1 *action plan*—a plan to address audit findings that describes planned response actions, parties responsible for their execution, and expected completion dates.

3.1.2 *audit authority*—the entity or body that authorizes, or initiates, the audit process. The audit authority may be internal or external to the audited entity.

3.1.3 *audit criteria*—enforceable environmental requirements that are applicable to an audited entity. Examples include laws, regulations, orders, and permits.

3.1.4 *audit data*—information obtained during an audit to support audit findings.

3.1.5 *audit finding*—a statement of audited entity conditions at the time of the audit compared to audit criteria. Audit findings shall be based upon verifiable audit data and may be either positive or negative with respect to audit criteria.

3.1.6 *audit objective(s)*—broad statement(s) of what the audit intends to accomplish.

3.1.7 *audit plan*—documentation that describes the audit. Elements that should be included or referenced in the audit plan are: the audit objective; audit scope; identities of the auditing entity, audit team, and audit authority; audit schedule; auditor health and safety issues; audit report preparation; record retention and management requirements; logistics; working papers management; a process for promptly communicating auditor concerns for potential environmental or regulatory impact including concerns requiring further evaluation; special conditions or requirements; and whatever measures, if any, desired to maximize audit confidentiality.

3.1.8 *audit purpose*—reason for the audit.

3.1.9 *audit protocol*—a list of issues or questions designed to address the audit objective(s) based upon audit criteria.

3.1.10 *audit report*—a written summary of audit findings that is objective, clear, concise, constructive and timely.

3.1.11 *audit scope*—a description of what is to be audited. The audit scope shall include a description of the period under review, the audited entity, and the audit criteria.

3.1.12 *audit team*—one or more auditors responsible for conducting an audit. The audit team may be supported by technical experts and auditors-in-training.

3.1.13 *audited entity*—a facility, organization, or part thereof, that is the subject of an audit.

3.1.14 *auditing entity*—the organization that provides auditors.

3.1.15 *auditor*—a person qualified to conduct an audit.

3.1.16 *environmental regulatory compliance audit (audit)*—a systematic, documented, and objective review of an audited entity to evaluate its compliance status relative to audit criteria

3.1.17 *independence*—a condition characterized by organizational standing where an auditor is free to conduct an audit without being controlled or influenced by others.

3.1.18 *lead auditor*—an auditor designated to lead and manage the audit.

3.1.19 *objectivity*—a condition characterized by the absence of bias, influences, and conflicts of interest that affect or have the potential to compromise audit findings.

3.1.20 *open issues*—potential audit findings that cannot be verified or resolved without additional information.

3.1.21 *period under review*—the time interval over which conditions at the audited entity are evaluated against audit criteria.

3.1.22 *physical inspection*—first hand observation and assessment of audited entity conditions.

3.1.23 *records*—documentation and other forms of recorded information.

3.1.24 *regulatory entities*—organizations that have the legal authority to develop, implement, administer, or enforce audit criteria. For example, federal, state, tribal, or local government.

3.1.25 *working papers*—paper and electronic documentation developed by an auditor from observations, record reviews, and interviews during an audit.

4. Significance and Use

4.1 *Intended Use*—This practice is intended for use by parties who wish to perform or direct an audit, or rely upon audit findings or an audit report. Such use includes audits by internal and external auditors.

4.2 *Other Audit Criteria*—Other audit criteria may be included in the audit scope if specified in the audit plan. Examples include safety and health requirements, and technical, operational, and management requirements.

4.3 *Related ASTM Standards*—This practice is related to Practices E 1527 and E 1528 for environmental site assessments. A number of terms and procedures from these practices are similar to terms and procedures associated with this practice but they may not be the same.

4.3.1 *Not Interchangeable*—Although related to Practices E 1527 and E 1528, this practice is designed to achieve different results. Therefore, these practices shall not be used interchangeably with, or in place of, this practice.

4.4 *Related ISO Guidelines*—The International Organization for Standardization (ISO) has developed guidelines addressing environmental auditing. These are ISO 14010, ISO 14011, and ISO 14012. A number of terms and procedures from these guidelines may be similar to terms and procedures associated with this practice but they may not be the same.

4.4.1 *Not Interchangeable*—Although related to ISO guidelines, this practice is designed for a different purpose. Therefore, these ISO guidelines shall not be used interchangeably with, or in place of, this practice.

4.5 *Additional Services*—A number of issues may arise as a result of an audit, but are outside the scope of this practice. Such issues include, but are not limited to, development of action plans and cost estimates to address audit findings.

4.6 *Conditions*—The following conditions shall govern the application of this practice.

4.6.1 *Not Exhaustive*—A audit shall not constitute an exhaustive review of audited entity compliance with all potentially applicable audit criteria unless explicitly intended and stated as an audit objective in the audit plan.

4.6.2 *Level of Review is Variable*—The audit scope may vary to meet different audit objectives. For example, the audit scope may include only selected audit criteria, selected period under review, or selected portions of a facility or organization.

4.6.3 *Continued Viability of Environmental Audit*—An audit completed in accordance with this practice shall be valid for only the period under review. Because audit criteria and audited entity conditions may change over time, it shall not be assumed that an audit report is reliable, or has continued viability, for other than the period under review.

4.6.4 *Usage of Prior Audits*—Information contained within reports from prior audits should be used only if it has continued viability and if use of that information is appropriate.

5. Responsibilities

5.1 *Audit Authority Responsibilities*—The audit authority shall:

5.1.1 Determine the need for an audit;

5.1.2 Specify or approve the audit purpose;

5.1.3 Select the auditing entity and/or lead auditor; and

5.1.4 Support the audit process.

5.2 *Lead Auditor Responsibilities*—The lead auditor shall ensure the efficient and effective completion of an audit. To do this the lead auditor shall work through the auditing entity to:

5.2.1 Develop, with the audit authority, an audit plan;

5.2.2 Gather appropriate audited entity background information;

5.2.3 Assemble a qualified audit team;

5.2.4 Communicate with the audited entity regarding audit plan issues. For example, schedule, logistics, access, availability of audited entity staff to interview, operating conditions, audit team needs, and health and safety precautions;

5.2.5 Manage the audit team;

5.2.6 Serve as the primary point of communication between the audit team and any other entity regarding the audit plan, audit findings, and audit reports;

5.2.7 Seek to prevent and resolve problems that could affect audit quality and timeliness;

5.2.8 Ensure the audit is conducted in accordance with this practice;

5.2.9 Notify the audit authority of conditions that may prevent audit completion in accordance with the audit plan; and

5.2.10 Prepare an audit report.

5.3 *Auditor Responsibilities*—Auditors shall support the lead auditor in completing an audit. To do this, auditors shall:

5.3.1 Understand the audit plan and their individual area of responsibility;

5.3.2 Follow lead auditor direction;

5.3.3 Review audit criteria and any protocols to be used, and establish a personal work plan for assigned areas of responsibility;

5.3.4 Collect sufficient relevant audit data to support audit findings;

5.3.5 Develop and document audit findings;

5.3.6 Assist in preparing audit reports; and

5.3.7 Maintain audit data and documentation in a secure manner.

5.3.8 Disclose issues to the auditing entity that may compromise objectivity.

5.4 *Audited Entity Responsibilities*—The audited entity shall:

5.4.1 Ensure that the audit is supported, including cooperating with the auditors to ensure that audit objectives are met;

5.4.2 Provide the auditing entity with requested background information in a timely manner;

5.4.3 Ensure audit team safe, timely and complete access;

5.4.4 Inform audit team of relevant health and safety requirements and practices;

5.4.5 Provide audit team, for the duration of the audit, with secure and reasonable support needs;

5.4.6 Provide the audit team with facility escorts knowledgeable of audited entity operations, to accompany auditors on physical inspections;

5.4.7 Assist auditor in identifying pertinent personnel and in scheduling interviews;

5.4.8 Ensure audit team access to documents needed to develop audit findings;

5.4.9 Ensure those facilities and operations audited accurately represent normal and known abnormal conditions;

5.4.10 Inform audit team of abnormal conditions;

5.4.11 As specified in the audit plan, evaluate and retain audit reports and determine appropriate follow-up activities;

5.4.12 Take measures to ensure that the audit team is provided with accurate and complete answers to questions;

5.4.13 Promptly notify the audit team if it disagrees with an audit finding;

5.4.14 If requested, provide comments on draft audit findings; and

5.5 *Auditing Entity Responsibilities*—The auditing entity shall:

5.5.1 Provide qualified auditors,

5.5.2 Select, with the audit authority, a lead auditor,

5.5.3 Provide a quality assurance and quality control program that may include evaluations of audit procedures, auditor qualifications, auditor effectiveness, and audit reports,

5.5.4 Support auditors and the auditing function with including management of audit data, audit findings, and audit reports in a responsible manner, and

5.5.5 Disclose to the audit authority, issues that compromise auditor objectivity.

6. Auditor Qualifications and Staffing

6.1 Auditor qualifications include competence, objectivity and independence.

6.2 *Competence*:

6.2.1 An auditor shall conduct an audit with the care, diligence, skill, and judgment expected of any auditor in similar circumstances.

6.2.2 An auditor shall have a working knowledge of the provisions of this practice and audit criteria relevant to his or her area of audit responsibility.

6.3 *Objectivity*:

6.3.1 Auditors shall disclose actual or potential issues compromising objectivity.

6.3.2 The auditing entity shall disclose actual or potential issues compromising objectivity.

6.4 *Independence*—Auditors should be independent of the audited entity.

6.5 *Audit Team Staffing*—The audit team shall collectively be able to implement the audit plan. The following shall be considered in assembling an audit team.

6.5.1 Knowledge of audited entity operations;

6.5.2 Knowledge of audit criteria;

6.5.3 Experience in auditing;

6.5.4 Workload;

6.5.5 Communication, technical, language, or other skills needed;

6.5.6 Knowledge of this practice;

6.5.7 Ability to remain objective;

6.5.8 Ability to manage confidential business information; and

6.5.9 Other audit authority requirements.

7. The Audit Process

7.1 There are three steps to an audit. These are: pre-visit activities conducted prior to and in preparation for the site visit; site-visit activities conducted while an auditor is present at the audited entity and; post-visit activities that take place after the site-visit.

7.1.1 Pre-visit activities shall address the following:

7.1.1.1 *Audit Scope*—The audit scope shall be developed and documented.

7.1.1.2 *Audit Criteria*—The audit criteria shall be defined and documented.

7.1.1.3 *Audit Plan*—The audit plan shall be developed and documented.

7.1.1.4 *Schedule*—A schedule of audit activities shall be developed and documented. Issues such as scheduling a site orientation meeting, identifying site contacts, scheduling dates of the site visit, and resolution of logistical issues such as lodging and transportation, should be addressed.

7.1.1.5 *Background Information*—Background information shall be used to develop an audit plan or refine an existing audit plan. Background information may consist of records, process and site descriptions, operation and maintenance manuals, emergency plans, environmental manuals, compliance inspection reports, previous audit reports, environmental site assessment reports, environmental permits, environmental citations, notices of violations and other relevant information.

7.2 Site-visit activities shall address the following:

7.2.1 *Opening Conference*—An opening conference should be held, bringing together the audit team and appropriate members of the audited entity staff or audit authority, or both, to confirm the audit plan and scope. The meeting should facilitate the later gathering of information by the audit team and encourage discussion of any questions or concerns. The audited entity should provide an overview of the facility operations for the audit team during the opening conference.

7.2.2 *Gathering Audit Data*—Audit data shall be gathered and evaluated by the audit team to support audit findings consistent with the audit objective. The audit team should utilize audit protocols to ensure consistency in gathering audit data.

7.2.2.1 *Physical Inspections*—Physical inspections of the audited entity should be conducted.

7.2.2.2 *Interviews*—Interviews should be conducted to obtain information on audited entity practices and procedures that are subject to audit criteria. Appropriate management staff, employees and, if applicable, contractors, should be interviewed.

7.2.2.3 *Records Review*—Records should be reviewed by the audit team. Records may include reports submitted to regulatory entities, information on physical conditions of the audited entity, environmental site assessment reports, environmental management system documents, prior audit reports, and records of environmental performance.

7.2.3 *Copying*—Copies of audited entity records may be necessary to support audit findings. Copies are considered audit data.

7.2.4 *Team Meeting(s)*—Meetings of the audit team should be conducted to share information and to ensure timely and consistent completion of the audit.

7.2.5 *Audit Progress Meeting(s)*—Draft audit findings and audit plan issues should be discussed among audit team members prior to the closing conference.

7.2.6 *Closing Conference*—At the conclusion of the site visit, a closing conference should be held to discuss draft audit findings and audit status. The closing conference provides an opportunity for audited entity personnel to discuss and question draft audit findings. Post-visit procedures should be discussed at the closing conference including a process for resolving challenged audit findings and for closing or reporting any open issues.

7.3 Post-visit activities shall address the following:

7.3.1 *Documentation*—Audit protocols should be completed, or explanations provided for open issues, in accordance with the audit plan.

7.3.2 *Finalizing Audit Findings*—Final audit findings shall be based upon the most recent verifiable audit data from the period under review that is available to the audit team.

7.3.3 *Preparing Audit Report*—An audit report shall be prepared that presents audit findings. Audit findings that are corrected during an audit shall be included as audit findings in the audit report.

8. Audit Record Management and Audit Report Preparation

8.1 *Record Keeping*—This section discusses the handling and retention of audit records.

8.1.1 *Record Handling*—The audit plan should identify procedures for handling records.

8.1.2 *Record Management and Retention*—A record management and retention policy should be included in the audit plan.

8.2 *Working Papers*—The following subsections identify specific issues related to working papers.

8.2.1 *Generation*—A system should be employed to facilitate working paper review and protect against tampering.

8.2.2 *Corrections and Revisions*—A system for managing corrections and revisions of working papers should be employed.

8.2.3 *Disposition*—The interim and final disposition of working papers shall be addressed in the audit plan.

8.3 *Audit Report*—The following sections address issues related to development and issuance of an audit report.

8.3.1 *Draft Audit Report*—A draft audit report should be developed for review and comment. Comments on a draft audit report should be made in a timely manner. Reviewers' failure to provide comments shall not prevent issuance of the final audit report.

8.3.2 *Final Audit Report*—A final audit report shall be issued.

9. Keywords

9.1 environmental audit; environmental compliance; environmental management

Designation: E 2137 – 01

Standard Guide for Estimating Monetary Costs and Liabilities for Environmental Matters[1]

This standard is issued under the fixed designation E 2137; the number immediately following the designation indicates the year of original adoption or, in the case of revision, the year of last revision. A number in parentheses indicates the year of last reapproval. A superscript epsilon (ε) indicates an editorial change since the last revision or reapproval.

1. Scope

1.1 *Purpose*—The purpose of this document is to provide a standard guide in the United States for estimating *costs and liabilities* for environmental matters.[2] Many possible uses for estimates of *costs and liabilities* for environmental matters exist, including but not limited to business decision making, communications and negotiations involving change of property ownership, regulatory requirements, third-party lawsuits, insurance premium calculation and claim settlement, change of property use, revitalization, compliance planning, construction, analysis of remedial alternatives, budgeting, strategic planning, financing, and investment analysis by shareholders. The use of estimated *costs and liabilities* developed in accordance with this standard may be subject to other standards applicable to the matter involved. For example, it is not intended to supersede accounting and actuarial standards including those by the Financial Accounting Standards Board and the U.S. Security and Exchange Commission. This standard does not address the establishment of reserves or disclosure requirements.

1.2 *Objectives*—The objective of this standard is to provide guidance on approaches for estimating *costs and liabilities* for environmental matters.

2. Referenced Documents [3]

2.1 *ASTM Standards:*

E 1527 Environmental Site Assessments: Phase I Environmental Site Assessment Process[4]

E 1739 Risk-Based Corrective Action Applied at Petroleum Release Sites[4]

PS 104-98 Provisional Guide for Risk-Based Corrective Action[5]

2.2 *Other Document:*

EPA OSWER Directive 9610.17 concerning use of risk-based decision making, 1995.[6]

3. Terminology

3.1 *Definitions:*

3.1.1 *allocation or allocated share*—the portion of cost or liability for which a party is responsible for payment or reimbursement.

3.1.2 *environmental compliance*—operations, permits, equipment, facilities, products, records, documentation, reports, training, procedures, inspections, certifications, monitoring, controls, or other conditions or activities that must conform to environmental statutes including, but not limited to, CAA, CWA, OPA, RCRA, CERCLA, TSCA, FIFRA, SDWA, and state and local laws.

3.1.3 *costs and liabilities*—economic expenses, accrued liabilities, and loss contingencies.

3.1.4 *estimator*—an individual or entity that prepares and analyzes *costs and liabilities.*

3.1.5 *event*—a condition or incident which occurred, or may occur, with respect to an environmental condition and/or *environmental compliance* issue, that affects or leads to potential *costs and liabilities*. Examples of events include: a new requirement for air emission controls (e.g., NO_x), a hazardous waste site that requires remediation, a claim for personal injury related to an alleged environmental incident, or the need to comply with NPDES standards as a result of a process change.

3.1.6 *liability*—an actual or potential obligation that may or may not be accrued.

3.1.7 *orphan share*—liability assigned to a PRP that cannot be located or that is insolvent, or the liability associated with pollutants which cannot be attributed to a PRP.

3.1.8 *potentially responsible party (PRP)*—any individual, legal entity, or government—including owners, operators, transporters, or generators—potentially responsible for, or contributing to, the environmental impacts at an event.

3.1.9 *studies*—investigations such as regulatory interpretations and applicability studies, compliance analysis, operating scenarios study, engineering design and analysis, cost estimation, process hazard analysis, modeling, communication plans, preliminary investigation, sampling and analysis, site assessment, site characterization, Phase I and II studies, remedial

[1] This guide is under the jurisdiction of ASTM Committee E50 on Environmental Assessment and is the direct responsibility of Subcommittee E50.03 on Environmental Risk Management/Sustainable Development/Pollution Prevention.

Current edition approved Mar. 10, 2001. Published May 2001.

[2] For the purposes of this standard, costs and values are defined as monetary estimates.

[3] Appendix X1 includes citations for additional relevant documents and requirements from other organizations including FASB, SEC, and AICPA.

[4] *Annual Book of Standards,* Vol 11.04.

[5] Discontinued—see 1999 *Annual Book of Standards,* Vol 11.04.

[6] Available from U.S. Environmental Protection Agency, 1200 Pennsylvania Avenue, NW Washington, DC 20460.

action plan, remedial investigation, contamination assessment report, feasibility study, risk assessment, treatability study, ecological impact assessment, environmental impact report, work plans, ASTM Risk-Based Corrective Action (RBCA) analysis, RCRA facility investigation, RCRA facility assessment, report of waste discharge, corrective measures study, corrective action report, health and safety plan, quality assurance plan, and other studies.

3.2 *Acronyms:*

3.2.1 *AICPA*—American Institute of Certified Public Accountants.

3.2.2 *CAA*—Clean Air Act.

3.2.3 *CERCLA*—Comprehensive Environmental Response, Compensation and Liability Act of 1980 (as amended, 42 USC Section 9601 et seq.).

3.2.4 *CWA*—Clean Water Act.

3.2.5 *EPA*—United States Environmental Protection Agency.

3.2.6 *EV*—expected value; an estimate of the weighted mean value of an unknown quantity that represents a probability-weighted average over the range of all possible values.

3.2.7 *FASB*—Financial Accounting Standards Board.

3.2.8 *FIFRA*—Federal Insecticide, Fungicide and Rodenticide Act.

3.2.9 *MLV*—most likely value.

3.2.10 *NPDES*—national pollutant discharge elimination system.

3.2.11 *OPA*—Oil Pollution Act.

3.2.12 *PRP*—potentially responsible party.

3.2.13 *RBCA*—Risk-based corrective action.

3.2.14 *RCRA*—Resource Conservation and Recovery Act (as amended 42 USC Section 6901 et seq.).

3.2.15 *SDWA*—Safe Drinking Water Act.

3.2.16 *SEC*—Securities and Exchange Commission.

3.2.17 *TSCA*—Toxic Substances Control Act.

4. Significance and Use

4.1 *Use*—The standard is intended for use on a voluntary basis by an estimator of *costs and liabilities* for environmental matters. The user may elect to apply this standard for any or all uses outlined in the Purpose. Application of this standard for one use does not compel application of the standard for all or any other use.

4.2 *Principles*—The following principles are an integral part of this standard and should be used to resolve ambiguity or dispute regarding the interpretation of estimated *costs and liabilities* for environmental matters.

4.2.1 *Uncertainty Not Eliminated*—Even though an estimate of *costs and liabilities* for environmental matters is prepared in accordance with this standard, uncertainty remains with regard to, among other things, the resolution of contractual, technological, regulatory, legislative, and judicial issues, which could affect the *costs and liabilities*.

4.2.2 *Periodic Review of Assumptions and Estimates*—Assumptions underlying these estimates should be reviewed periodically for the purpose of incorporating additional information that may become available. For example, changes in regulatory requirements, technology, property use, inflation, or other issues may affect the basis for the estimates, therefore necessitating revisions.

4.2.3 *Comparison with Subsequent Estimates*—Subsequent estimates based on additional information should not be construed as indicating the prior estimates of *costs and liabilities* for environmental matters were unreasonable at the time they were made. Estimates should be evaluated on the reasonableness of analyses and judgments made at the time and under the circumstances in which they were made. Subsequent improvements in estimates should be made as more information becomes available, but these improved estimates should not be considered valid standards on which to measure the reasonableness of a prior estimate based on hindsight, new information, use of developing analytical techniques, or other factors. However, information on trends in estimates over time may be of value to a user of financial statements or other users of the cost and liability estimates.

4.2.4 *Not Exhaustive*—Estimation of *costs and liabilities* for environmental matters does not necessarily require an exhaustive evaluation of all possible outcomes. A point exists at which the cost of obtaining information or the time required to gather it outweighs improvement in the quality of the estimate.

4.2.5 *Assessment of Risk*—The actual or potential risk to human health and the environment should be considered in assessing environmental matters. The degree of risk should be a factor in developing the cost and liability estimates associated with those matters.

4.2.6 *Estimator Selection*—An appropriate estimator or group of estimators will consist of those individuals or groups who possess sufficient knowledge, training, and experience to develop appropriate estimates for the *costs and liabilities* being estimated. It is the responsibility of the entity sponsoring the cost and liability estimates to select an estimator with the appropriate level of knowledge, training, and experience for the parts of the estimation effort for which that estimator is responsible.

5. Procedures for Estimating *Costs and Liabilities* for Environmental Matters

5.1 *Determination of Relevant Information and Types of Costs and Liabilities*:

5.1.1 There are many types of *costs and liabilities* for environmental matters, including, but not limited to:

5.1.1.1 Studies,

5.1.1.2 Response action,

5.1.1.3 *Environmental compliance*,

5.1.1.4 Defense and legal fees,

5.1.1.5 Fines and penalties,

5.1.1.6 Reimbursement of agency oversight, or

5.1.1.7 Damages arising from resource damages, ecological damages, property damage, business interruption, bodily injury, or tort claims such as nuisance and negligence claims.

5.1.2 After identifying the conditions giving rise to potential *costs and liabilities* for environmental matters, existing relevant information should be considered to estimate *costs and liabilities* identified in 5.1.1, including, but not limited to:

5.1.2.1 Event type (for example, new EPA air emission

control requirements, leaking landfill, site PRP notice, compliance audit findings),

5.1.2.2 Number and location of affected operations/facilities,

5.1.2.3 Use of surrounding property,

5.1.2.4 Past, current, and potential future site uses,

5.1.2.5 Studies,

5.1.2.6 Environmental risks posed by the event,[7]

5.1.2.7 Bodily injury or other claims related to the event,

5.1.2.8 Relevant state or other regulatory requirements and alternatives,

5.1.2.9 State or federal agency involvement,

5.1.2.10 Public involvement,

5.1.2.11 Planned or completed remedial activities,

5.1.2.12 Decision documents (for example, Records of Decision),

5.1.2.13 Litigation activities related to the event (for example, claims, suits, actions, demands, requests for payment, notices),

5.1.2.14 Resources, tasks, and deadlines,

5.1.2.15 Available technologies and designs,

5.1.2.16 Type and extent of contamination,

5.1.2.17 Number of operable units (CERCLA) or solid waste management units (RCRA),

5.1.2.18 Involvement of various parties at the event, and

5.1.2.19 Information on prior experience with similar events.

5.1.3 The organization and application of the foregoing information may be further subject to corporate, accounting, or regulatory policy decisions. The user will need to determine what these policy decisions are, and assess their effect on the cost estimate. Examples of such policy decisions include, but are not limited to:

5.1.3.1 Measuring and recording of contingent liabilities,

5.1.3.2 Technical policy decisions or interpretations to be made by regulatory agencies,

5.1.3.3 Acceptable levels of risk (for example, business risk, human health risk, ecological risk),

5.1.3.4 The degree to which societal or external costs and benefits are considered,

5.1.3.5 Whether or not life cycle costs are considered,

5.1.3.6 The degree to which sustainability/sustainable development are considered,

5.1.3.7 Local environmental management system criteria, including trade-off of emissions across environmental media, alternative methods and permitting options, auditability, and performance oriented metrics.

5.1.3.8 Level of organizational involvement and scrutiny,

5.1.3.9 The degree of communication and involvement with the public.

5.1.4 In the absence or insufficiency of such information, an assessment should be made of the applicable regulatory and industry standard requirements, and a determination made as to whether based on these requirements, significant *costs and liabilities* for environmental matters may be incurred that would indicate the need for further data creation and analysis in the future.

5.2 *Selection of Estimation Approaches*—A decision framework for estimating *costs and liabilities* for environmental matters is required. For purposes of naming various estimating methods, the following terminology is used:

Expected Value (EV)
Most Likely Value (MLV)
Range of Values
Known Minimum Value

5.2.1 The decision to use one or more of these four estimating methods or another method for a particular purpose is not arbitrary. The informational value of the estimate supplied by any one method is not equivalent to the others. When the uncertainties are great (for example, when an event is first identified) it may not be possible to make a reasonable cost estimate.

5.2.2 The robustness and comprehensiveness of an estimate and the quantification of uncertainty about the estimate, given adequate information, generally decreases moving from top to bottom of this list of methods. (See Fig. 1.) Depending on availability of information and circumstances, the level of effort required to prepare estimates at the top of the list is typically greater than the bottom of the list. However, given the principles cited in Section 4, it is not necessarily true that the "best" estimate for a given set of circumstances will always be the expected value.

5.2.3 The estimator should take into account the number of events and quality of the information available or obtainable when selecting the cost and liability estimation approach to be used.

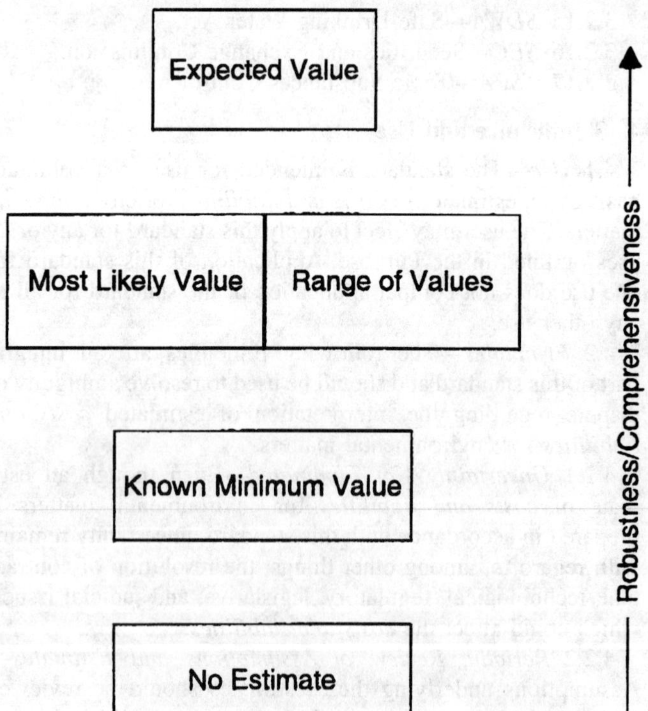

FIG. 1 Hierarchy of Approaches for Estimating *Costs and Liabilities* for Environmental Matters

[7] See Guide E 1739; PS104-98; EPA Risk Characterization Program; The Presidential/Congressional Commission on Risk Assessment and Risk Management; and EPA OSWER Directive 961.17 concerning use of risk-based decision making, 1995.

5.3 *Sources of Uncertainty in Estimation*:

5.3.1 For environmental issues, multiple outcomes often exist for a given issue as it develops toward resolution. Regulatory actions, event characterization information, remedial action effectiveness, legal matters, insurance aspects, and the like, are variables that are often not predictable with certainty. These uncertainties indicate multiple possible outcomes, each having its own probability of occurrence. Each outcome, if it occurs, has its own estimated value based upon its component outcomes.

5.3.2 The accuracy of an estimate, or sum of estimates (for application to multiple independent events), is ultimately measured by comparison of the predicted cost or liability to the actual outcome. There are at least two dimensions to accuracy in this context: the probabilities that certain outcomes will occur, and the component cost estimates for each of the outcomes.

5.3.3 To predict the value of each outcome, cost estimates are prepared that can range from very imprecise to rough approximations to engineering estimates of increasing degrees of detail. It may be possible to prepare a very accurate engineering estimate for a technical solution to a compliance or remediation requirement, but unless that specific technical solution is actually used in resolution of the event, the value of the precision may not be realized. Hence, the estimate's accuracy is affected by both the accuracy of the probabilities (that a particular technical solution will be applied) and the costs associated with each potential outcome.

5.4 *Detailed Description of Approaches for Estimation*

5.4.1 *Expected Value*:

5.4.1.1 There are several approaches to calculating an expected value, which is an estimate of the mean value of an unknown quantity that represents a probability-weighted average over the range of all possible values. One method, the decision tree approach, derives an expected value and distribution of potential values through the following steps: (*1*) identify the key issues contributing to the magnitude and timing of event costs, (*2*) develop a decision tree or simulation model of potential event outcomes (including possible allocation scenarios), (*3*) estimate the cost for each potential outcome, (*4*) determine the likelihood of each outcome, and (*5*) calculate the distribution of potential costs and the expected value, which is the probability-weighted cost calculated from items (*2*) through (*4*) above.[8] The information developed from the distribution may be very useful in conveying information about uncertainty, as described below in 5.8.

5.4.1.2 The estimator should be careful to include realistic outcomes with statistically significant probabilities to avoid shifting the expected value through the addition of extreme outcomes with insignificant probabilities of occurrence. Statistical significance will vary depending on the quality of data, the magnitudes of the outcomes, and the presence of outliers.

5.4.1.3 Outcome probabilities should be based, to the extent practicable, on statistical data drawn from comparable events. Where there are a large number of events, statistical approaches to estimating the expected value may be particularly appropriate. It is important to realize statistical approaches can be predictive of aggregate *costs and liabilities*, even if expected values for individual events are at variance from the actual results.

5.4.1.4 Another method for calculating an expected value is an actuarial approach, where historical data are available to estimate the expected value for similar events. Care should be taken when using historical data for estimating costs to assure that the data are applicable to the *event(s)* in question. Care should also be taken when using historical data due to the effects of changes such as technology enhancements, modified laws and/or regulatory policy, the changing application of presumptive remedies, and the application of risk-based corrective action approaches that could significantly alter current and future costs. Considerations should also be given to the potential loss of relevant information through application of statistical means or averages which may not convey information concerning uncertainty.

5.4.1.5 These approaches can be used in combination as appropriate. Other approaches to estimating an expected value may include simulation modeling and Monte Carlo analysis, for example, to estimate cost distributions.

5.4.2 *Most Likely Value (MLV)*—When an expected value approach is not practical or appropriate, a most likely value could be developed using engineering estimates. This MLV captures the cost of the scenario believed to be most likely to occur (for example, a stated preferred remedy). Typically, the estimator exercises a priori judgements (based on experience) about the ranking of likely outcomes, but because of cost or other considerations does not develop a full range of possible outcomes to support an expected value estimate. Care should be exercised in preparing an MLV estimate. For example, the MLV is typically not the mid-point between the high and low cost estimates. The MLV should represent a technical and regulatory scenario that is most likely to occur. The MLV may represent a grouping or cluster of scenarios where the cost outcomes are close in magnitude and the combined probability of the grouping or cluster exceeds the probability of other possible scenarios. The MLV is not useful if no scenario, grouping or cluster of outcomes has a probability of occurrence that is significantly greater than others.

5.4.3 *Range of Values*—When an expected value approach is not practical or appropriate, a range of values (without probabilities) may be developed instead. This approach may also be used in addition to the MLV approach to provide additional information, or instead of the MLV approach if probabilities or rankings for various outcomes cannot be determined. The range of values should cover costs from a low cost estimate to a high cost estimate, based on reasonable assumptions. If some outcomes within the range are more probable than others, this standard recommends the additional estimation of a most likely value or an expected value, when possible.

5.4.4 *Known Minimum Value*—When the outcome and cost uncertainties are so great that it is premature to estimate a range

[8] For additional information on the expected value approach, see, for example: R.V. Kolluru, editor, *Environmental Strategies Handbook: A Guide to Effective Policies & Practices*, New York: McGraw-Hill, Inc., 1994 and G.D. Eppen and F.J. Gould, *Quantitative Concepts for Management: Decision Making without Algorithms*, Englewood Cliffs, NJ: Prentice-Hall, Inc., 1979.

of values or a most likely value, then a minimum value including component costs (e.g., contracts entered, initial studies) that are reasonably certain to be incurred should be estimated.

5.5 *Contingencies*—Contingency adjustments may be added to correct for costs that are undefined at the time of the estimate, but that are expected to be incurred. Therefore, care should be taken, when adding contingencies to base unit cost estimates, that the contingencies are reasonable and expected to be incurred.[9]

5.6 *Inflation and Discounting*—Inflation and discounting assumptions should be clearly documented.

5.7 *Allocation*—In estimates where *costs and liabilities* for environmental matters involve multiple parties, it may be necessary to apportion these costs among the parties. Determination of an entity's likely allocated share for an event should be made whenever sufficient information is available, and the allocated share should be factored into the cost estimates developed under 5.2. Private parties and courts have employed a variety of methods to allocate or apportion costs (See Appendix X3). As in the case with cost estimation, the method used to allocate costs is dependent upon the amount of information available and the event facts.

5.8 *Uncertainty Associated with Estimation Approaches*—As outlined in 4.2.1, 5.2.1, 5.2.2, and 5.2.3, estimates for *costs and liabilities* for environmental matters are inherently uncertain until resolution of the event matures to the state where all costs are known with certainty. When possible and appropriate, the estimator should quantify or qualify the level of uncertainty associated with the cost and liability estimates. Numerous measures of uncertainty exist. Users of this standard are encouraged to explore the statistical and risk theory literature for such measurements. The best measure of uncertainty for a given application depends on the information available and the facts surrounding the analysis. The estimator should select that measure which most clearly communicates to the user the nature of the uncertainty being evaluated.

5.8.1 *Uncertainty with Expected Value Approach*—Statistical literature provides numerous examples and methods of measuring uncertainty when using an expected value approach. While the expected value approach may not fall neatly into the statistical realm in all cases, the expected value estimate does provide a basis for developing several simple uncertainty measures. Uncertainty measurement is important as it communicates to the user of the estimate the potential amount of variability and/or the level of confidence in the expected value estimate. In some cases, the potential variability will be so great, or the level of confidence so low, that little value should be attached to the expected value estimate. It is important to those relying on expected value estimates prepared under this standard to be aware of such situations. When providing an uncertainty measure with an expected value estimate, the basis and definition of the uncertainty measure should be included. Following are several uncertainty measurements that should be considered in a communication involving an expected value estimate made using this standard.

5.8.1.1 *Confidence Level*—This measure usually involves estimating the percentiles of the probability distribution underlying the expected value estimate. Selection of a 70 % confidence level estimate, for example, would imply that outcomes with values less than or equal to the 70 % confidence level estimate occur 70 % of the time on average, or equivalently, outcomes exceeding the 70 % confidence level estimate occur no more than 30 % of the time on average.

5.8.1.2 *Confidence Interval*—This measure usually assumes a normal distribution around the expected value estimate and estimates the probability of the actual cost or liability being within a given interval of the expected value. Confidence intervals can also be developed for distributions that are not normal.

5.8.1.3 *Coefficient of Variation*—Equal to the standard deviation divided by the mean, the coefficient of variation (CV) provides a basis for evaluating the amount of statistical variation around the expected value estimate. Opinion polls, for example, often state results inclusive of a plus or minus percentage value. A plus or minus percentage value around the expected value could be based on the CV measurement.

5.8.2 *Uncertainty with Most Likely Value (MLV) Approach*—Significant uncertainty may exist in estimates made using the MLV approach. The most likely outcome may not be very likely overall (even though it is the singular most likely outcome in a portfolio of potential outcomes). In addition, MLV analysis provides very little information to quantify the uncertainty. When available, the probability associated with the most likely outcome provides some information concerning related uncertainties. In addition, identification of the range of potential outcomes provides the user of the cost and liability estimate with bounds on the uncertainty associated with the MLV estimate.

5.8.3 *Uncertainty with Range of Values Approach*—To some extent, the size of the range indicates the breadth of uncertainty associated with these cost estimates. For example, if the range is broad, there may be great uncertainty concerning the ultimate cost. When possible, a most likely outcome value should also be provided. When this is not possible, if there are any cost scenarios of clusters or scenarios within the range that are more likely than others, this information should be provided.

5.8.4 *Uncertainty with Known Minimum Value Approach*—For the known minimum value estimate, the upward uncertainty is unknown. If available, a qualitative description of the potential costs or liabilities may allow a user to roughly assess the extent and likelihood of higher values.

5.9 *Recovery/Offsets*—There may be a potential for recovery for, or offsets to, the *costs and liabilities* for environmental matters (e.g., insurance recovery, third-party recovery). Any potential recovery/offsets should be evaluated separately from the original cost and liability estimate, using cost estimation approaches as described in this Section 5. The litigation costs for pursuing such actions also should be estimated separately from these potential recovery/offset estimates.

5.10 *Documentation*—Documentation should include the identity of the estimator and a description of their relevant

[9] For additional information on contingencies, see for example F.D. Clark and A.B. Lorenzoni, *Applied Cost Engineering*, NY: Marcel Dekker, 1985, pp. 112-120.

knowledge, training and experience. The estimation documentation should be sufficient for a user to evaluate the estimates. For example, it may be useful to identify the purpose and objective, the estimation approach(es), the major uncertainties considered, and the sources of information used in making estimates of *costs and liabilities* for environmental matters. This documentation may be prepared to cover a single event or multiple events estimated in a similar manner, and may consist solely or in part of existing work papers.

6. Keywords

6.1 cost estimation; environmental; costs; liabilities; expected value; minimum value; most likely value; range of values; uncertainty

APPENDIXES

(Nonmandatory Information)

X1. RELATED DOCUMENTS

E 1369 Standard Guide for Selecting Techniques for Treating Uncertainty and Risk in the Economic Evaluation of Buildings and Building Systems.

E 1528 Practice for Environmental Site Assessments: Transaction Screen Process.

E 1946 Standard Practice for Measuring Cost Risk of Buildings and Building Systems.

American Institute of Certified Public Accountants (AICPA) Statement of Position 96-1, "Environmental Remediation Liabilities," October 10, 1996.

American Institute of Certified Public Accountants (AICPA) Emerging Insurance Task Force (EITF) Abstract 93-5.

EPA Risk Characterization Program: Policy for Risk Characterization, March 1995; Guidance for Risk Characterization, February 1995; Policy for Use of Probabilistic Analysis in Risk Assessment, May 15, 1997; Guidance on Cumulative Risk Assessment, July 3, 1997.

"Filling the GAAP: An Approach to Improve SEC Disclosure of Environmental Liabilities," *Journal of Environmental Law & Practice*, September/October 1994.

Financial Accounting Standards Bulletin (FASB) Interpretation No. 14, "Reasonable Estimation of the Amount of a Loss and Interpretation of FASB-5."

Financial Accounting Standards Board, Rule No. 5, Accounting for Contingencies, March 1975.

The Presidential/Congressional Commission on Risk Assessment and Risk Management, Final Report, 1997, Volume 1: Framework for Environmental Health Risk Management, Volume 2: Risk Assessment and Risk Management in Regulatory Decision-Making.

Security and Exchange Commission (SEC) Staff Accounting Bulletin No. 92.

SEC Regulation S-K.

Security and Exchange Commission (SEC), Management's Discussion and Analysis of Financial Condition and Results of Operations; Certain Investment Company Disclosures, *Federal Register*, Vol. 54, No.99, May 24, 1989.

U.S. Environmental Protection Agency, *Remedial Action Costing Procedures Manual*, EPA/600/8-87/049, 1987.

U.S. Environmental Protection Agency, Office of Pollution Prevention and Toxics, *Valuing Potential Environmental Liabilities for Managerial Decision-Making: A Review of Available Techniques*. EPA/742/R-96/003, December 1996.

X2. EXAMPLE OF EXPECTED VALUE APPROACH

X2.1 This example provides a simplified overview of how to develop an expected value for *costs and liabilities* associated with a group of contaminated sites. The number of uncertainties contributing to the estimate has been deliberately kept to a minimum for the sake of clarity. In practice, the expected value approach can be applied in the same manner to consider many more uncertainties. Similarly, although only soil contamination is considered in this example, the same approach can be applied to include other sources of *costs and liabilities*.

X2.2 At a hypothetical site, soil contamination has been identified by a limited sampling program. The expected value analysis to estimate the costs associated with the contaminated soil is conducted using the five steps defined in 5.4.1.

X2.2.1 *Step 1.* The key uncertaintainties in this event are associated with three parameters: *(1)* the extent of soil contamination, (*2*) the cleanup level to be applied, and (*3*) the treatment/disposal technology for remediation.

X2.2.2 *Step 2.* The decision tree is constructed to reflect these three uncertainties. By convention, the uncertainties are shown as chance nodes by circles (see Fig. X2.1). In the simplified example, each of the uncertainties results in a factor of two difference in the magnitude of the environmental liability.

X2.2.2.1 *Extent of Soil Contamination*—The surface area extent of contamination has been well characterized by shallow soil samples, but the vertical extent is defined only by a very limited number of soil borings. Some of the borings indicate that contamination extends to 1 m in depth and others to 2 m.[10]

X2.2.2.2 *Cleanup Level*—State regulations specify a generic soil cleanup standard for the contaminant of concern, but also allow for determination of a risk-based cleanup goal. A risk-based cleanup goal could be a factor of 10 higher than the generic cleanup standard. Based on the site-specific distribution of contamination at the site, this higher cleanup goal could reduce the contaminated soil volume by a factor of two.

X2.2.2.3 *Treatment/Disposal Technology*—The default approach for the contaminants present at the site would be to excavate and dispose of the soil at an offsite landfill. However, based on the soil conditions and levels of contamination present, a less costly soil venting technology may be feasible.

X2.2.3 *Step 3.* The estimated costs for each potential outcome are determined based on the following assumptions.[11]

X2.2.3.1 Soil volume for deep (2 m) soil contamination

[10] For the purpose of this example, it is assumed that there is a uniform distribution of samples at 1 m and 2 m depths across the site.

[11] These assumptions are given for the purpose of illustration only.

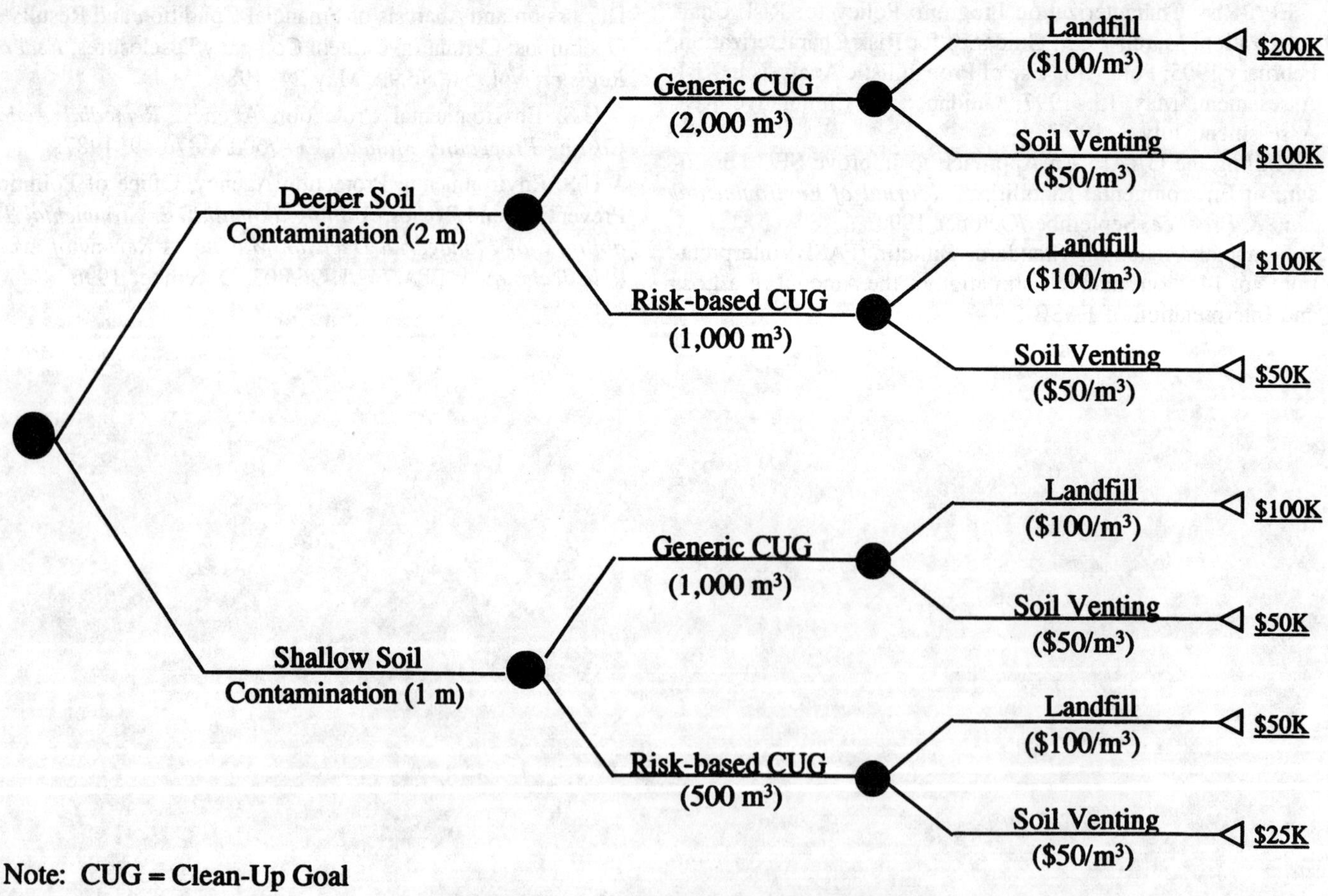

FIG. X2.1 Example Event Tree Uncertainties and Costs

scenario equals 2,000 m^3.

X2.2.3.2 Unit cost for offsite landfill disposal equals $100 per m^3.

X2.2.3.3 Unit cost for soil venting equals $50 per m^3.

X2.2.3.4 The costs associated with each potential outcome (represented by a particular path on the decision tree) are shown at the terminal nodes, which are depicted by convention by a triangle as shown in Fig. X2.1. These costs range from a minimum of $25,000 to a maximum of $200,000.

X2.2.4 *Step 4.* The likelihood of each outcome is determined as shown in Fig. X2.2.

X2.2.4.1 Approximately half of the soil borings collected at the site show that contamination extends to a depth of 2 m, and the other half show contamination extending to 1 m. Based on this information, a 50 % probability is assigned to each depth scenario.

X2.2.4.2 Historical experience with the relevant state agency suggests a reasonable likelihood that a site-specific risk assessment will be accepted. Accordingly, a 60 % probability is assigned to the higher risk-based cleanup goal, and a 40% probability is assigned to the default generic cleanup goal. (Note that the probabilities must add to 100 %.)

X2.2.4.3 Historical experience with the soil type present at the site and the level of contamination suggests a high likelihood of soil venting being technically feasible. Therefore, a probability of 80 % is assigned to soil venting, and a 20 % probability is assigned to offsite landfill disposal.

X2.2.5 The expected value is then calculated by summing the probability-weighted costs for each pathway on the decision tree. Using the assumed probabilities and unit cost data, the expected value is calculated at $63,000 (See calculation in Fig. X2.2). The probability of each individual outcome, as represented by a pathway on the decision tree, is calculated by multiplying the probabilities along that pathway. Thus, the maximum cost of $200,000 has a 4 % likelihood (50 % × 40 % × 20 %) and the minimum cost of $25,000 has a 24 % likelihood (50 % × 60 % × 80 %). The most likely outcome of $50,000 has a probability of 46 % (by summing the three $50,000 outcomes with probabilities of 24 %, 16 %, and 6 %).

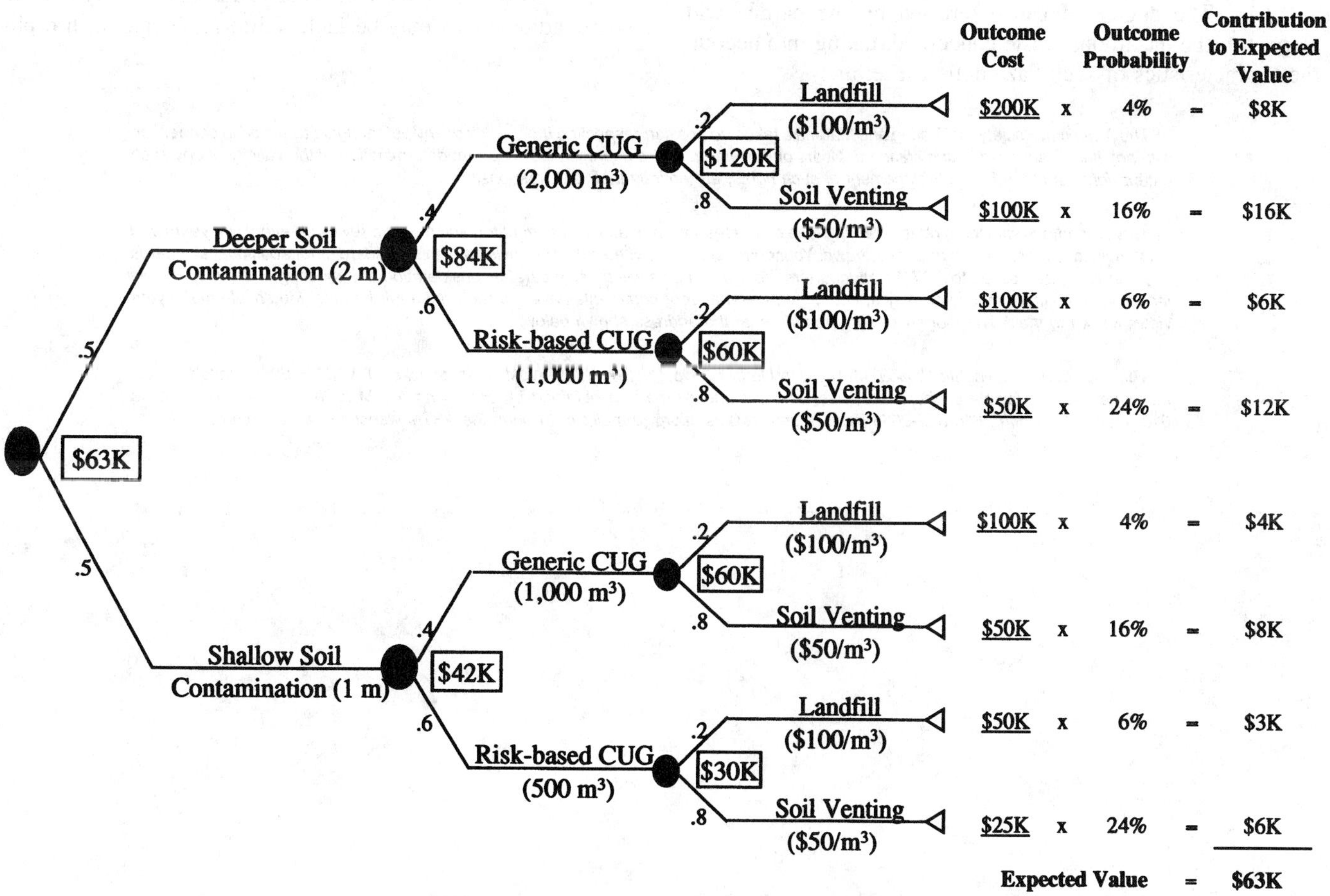

FIG. X2.2 Example Event Tree Expected Value

X3. POTENTIAL ALLOCATION CONSIDERATIONS

X3.1 The courts, PRP groups, and other parties have considered numerous issues in deriving allocations for CERCLA and non-CERCLA events. CERCLA states that "[In] resolving contribution claims, the court may allocate response costs among the liable parties using such equitable factors as the court determines are appropriate." Having pointed to "equitable factors," CERCLA is then silent as to what those factors might be. Factors that may be considered include, but are not limited to:

X3.1.1 The ability of the parties to demonstrate that their contribution to a discharge, release or disposal of a hazardous waste can be distinguished;

X3.1.2 The amount of the hazardous waste involved;

X3.1.3 The degree of toxicity of the hazardous waste involved;

X3.1.4 The degree of involvement by the parties in the generation, transportation, treatment, storage or disposal of the hazardous waste;

X3.1.5 The degree of care exercised by the parties with respect to the hazardous waste concerned, taking into account the characteristics of such hazardous waste; and

X3.1.6 The degree of cooperation by the parties with federal, state, or local officials to prevent any harm to the public health or the environment.

X3.1.7 Existing contracts between parties on the question of liability, such as indemnity agreements;

X3.1.8 Relative fault of the parties (e.g., cost causation, stand-alone costs);

X3.1.9 The owner's acquiescence in the operator's activities and manner of operation;

X3.1.10 The degree to which each party made efforts to prevent and/or contain any known release of hazardous wastes at the site, at the time the releases occurred;

X3.1.11 Relative economic benefits across the classes of PRPs; and

X3.1.12 Benefit to the current owner, if any, of the cleanup.

X3.2 If it is possible that other PRPs will not pay their full share, adjustments may be included to reallocate such orphan shares.

Standard Practice for Classifying Water Bodies for Spill Control Systems[1]

This standard is issued under the fixed designation F 625; the number immediately following the designation indicates the year of original adoption or, in the case of revision, the year of last revision. A number in parentheses indicates the year of last reapproval. A superscript epsilon (ε) indicates an editorial change since the last revision or reapproval.

1. Scope

1.1 This practice creates a system of categories that classify water bodies relating to the control of spills of oil and other substances that float on or into a body of water.

1.2 This standard does not address the compatibility of spill control equipment with spill products. It is the user's responsibility to ensure that any equipment selected is compatible with anticipated products.

1.3 *This standard does not purport to address all of the safety concerns, if any, associated with its use. It is the responsibility of the user of this standard to establish appropriate safety and health practices and determine the applicability of regulatory limitations prior to use.*

2. Terminology

2.1 *Definitions:*

2.1.1 Recommended units of measure are given for each of the following definitions.

2.1.2 *air temperature*—average or point temperature of the air measured at or near the ground or water surface, (°C).

2.1.3 *current*—average water velocity relative to a reference point, (m/s).

2.1.4 *debris*—any solid or semi-solid substance that could interfere with the operation of a spill control system.

2.1.5 *water depth*—mean vertical distance measured from the surface of the water to the top of the continuous solid surface below at mean lower low water, (m).

2.1.6 *water temperature*—average or point temperature of a water body as measured within the top 300 mm (12 in.), (°C).

2.1.7 *wave height*—(significant wave height) the average height, measured crest to trough, of the one-third highest waves, considering only short-period waves (that is, period less than 10 s), (m).

2.1.8 *wave period*—(significant wave period) the average period of the one-third highest waves, measured as the elapsed time between crests of succeeding waves, (s).

2.1.9 *wind*—the air velocity measured at a height 10 metres (33 ft) above the ground or water, (m/s).

3. Summary of Practice

3.1 General environmental conditions for spill control systems are grouped into four major types of water bodies (see Table 1). Additional factors that may affect spill control operations are listed in Section 5.

4. Significance and Use

4.1 This practice is to be used as a guide to classify water bodies for spill control systems. These classifications may be used in formulating standards for design, performance, evaluation, contingency and response planning, contingency and response plan evaluation, and standard practice for spill control systems.

4.2 Relatively few parameters of broad range have been used in Table 1 in order to enable the user to readily identify general conditions under which spill control systems can be used.

4.3 Satisfactory operation of any specific spill control systems may not extend over the full range of conditions identified by Table 1. Detailed discussion with systems suppliers is recommended.

4.4 Effective operation of oil spill control equipment depends on many factors, of which the prevailing environmental conditions are just a few. Factors such as, but not limited to, deployment techniques, level of training, personnel performance, and mechanical reliability can also affect equipment performance.

5. Use of Table

5.1 Wave height is recognized as the primary variable in describing marine environments for spill control systems. Where currents are significant, the four water body types listed in Table 1 can be modified by appending "C", as in "I-C".

5.2 In some situations, air temperature, water temperature, or presence of debris, or a combination thereof, may significantly affect the usage of Table 1 and should be considered.

5.3 Other factors such as presence of salt water or silt, or both, should be considered if significant.

6. Keywords

6.1 oil spill; oil spill control; water bodies

[1] This practice is under the jurisdiction of ASTM Committee F20 on Hazardous Substances and Oil Spill Responseand is the direct responsibility of Subcommittee F20.11 on Control.

Current edition approved March 15, 1994. Published May 1994.

TABLE 1 Water Body Classifications

Type[A]	Wave Height,[B] m (ft)	Examples of General Conditions
Calm Water	0 to 0.3 (0 to 1)	small, short, non-breaking waves
Protected Water	0 to 1 (0 to 3)	small waves, some whitecaps
Open Water	0 to 2 (0 to 6)	moderate waves, frequent whitecaps
Open Water (rough)	>2 (>6)	large waves, foam crests and some spray

[A] If current is significant, approximately 0.4 m/s (0.8 knots) or more, append "C" to the descriptor type, as" I-C".

[B] Significant wave height throughout. May include breaking waves. The ratio of wave height to wave length should also be considered. The orientation of waves to current direction should also be considered.

The American Society for Testing and Materials takes no position respecting the validity of any patent rights asserted in connection with any item mentioned in this standard. Users of this standard are expressly advised that determination of the validity of any such patent rights, and the risk of infringement of such rights, are entirely their own responsibility.

This standard is subject to revision at any time by the responsible technical committee and must be reviewed every five years and if not revised, either reapproved or withdrawn. Your comments are invited either for revision of this standard or for additional standards and should be addressed to ASTM Headquarters. Your comments will receive careful consideration at a meeting of the responsible technical committee, which you may attend. If you feel that your comments have not received a fair hearing you should make your views known to the ASTM Committee on Standards, at the address shown below.

Designation: F 631 – 99

Standard Guide for Collecting Skimmer Performance Data in Controlled Environments[1]

This standard is issued under the fixed designation F 631; the number immediately following the designation indicates the year of original adoption or, in the case of revision, the year of last revision. A number in parentheses indicates the year of last reapproval. A superscript epsilon (ε) indicates an editorial change since the last revision or reapproval.

1. Scope

1.1 This standard provides a guide for determining performance parameters of full-scale oil spill removal devices in recovering floating oil when tested in controlled environments.

1.2 This guide involves the use of specific test oils that may be considered hazardous materials after testing is completed. It is the responsibility of the user of this guide to procure and abide by the necessary permits for disposal of the used test oil.

1.3 *This standard does not purport to address all of the safety concerns, if any, associated with its use. It is the responsibility of the user of this standard to establish appropriate safety and health practices and determine the applicability of regulatory limitations prior to use.*

2. Referenced Documents

2.1 *ASTM Standards:*

D 97 Test Method for Pour Point of Petroleum Products[2]

D 445 Test Method for Kinematic Viscosity of Transparent and Opaque Liquids (the Calculation of Dynamic Viscosity)[3]

D 971 Test Method for Interfacial Tension of Oil Against Water by the Ring Method[3]

D 1298 Test Method for Density, Relative Density (Specific Gravity), or API Gravity of Crude Petroleum and Liquid Petroleum Products by Hydrometer Method[2]

D 2983 Test Method for Low-Temperature Viscosity of Automotive Fluid Lubricants Measured by Brookfield Viscometer[4]

D 4007 Test Method for Water and Sediment in Crude Oil by the Centrifuge Method (Laboratory Procedure)[4]

F 625 Practice for Classifying Water Bodies for Spill Control Systems[5]

F 808 Guide for Collecting Skimmer Performance Data in Uncontrolled Environments[6]

3. Terminology

3.1 *Definitions of Terms Specific to This Standard:*

3.1.1 *data collection period, n*—the period of time during a test run when the performance data are recorded.

3.1.2 *emulsification factor, n*—the increase in total fluids in storage as a result of emulsification by the skimming mechanism, the skimmer pump, or other component of the skimmer.

$$EF = (WC_F - WC_0)/(100 - WC_F) + 1 \qquad (1)$$

where:

WC_F = the final water content %, and

WC_0 = the initial water content %.

3.1.3 *nameplate recovery rate, n*—the maximum skimming capacity of a device as stated by the manufacturer.

3.1.4 *oil encounter rate, n*—the volume of oil per unit time actively directed to the removal mechanism (m^3/h).

3.1.5 *oil recovery efficiency, n*—the ratio, expressed as a percentage, of the volume of oil recovered to the volume of total fluids recovered.

3.1.6 *oil recovery rate, n*—the volume of oil recovered by the device per unit of time (m^3/h).

3.1.7 *oil slick, n*—the oily fluid encountered by the skimmer.

3.1.7.1 *Discussion*—Most real oil slicks actually are composed of various proportions of pure oil, water-in-oil (W/O) emulsions, and oil-in-water (O/W) emulsions; therefore, efficiencies and other performance criteria must be differentiated between those based on the oil slick itself, and those based on only the water-free oil contained within the oil slick.

3.1.8 *oil slick encounter rate, n*—the volume of oil slick per unit time actively encountered by the skimmer, and therefore, available for recovery (m^3/h).

3.1.9 *oil slick recovery efficiency, n*—the ratio, expressed as a percentage, of the volume of oil slick recovered to the volume of total fluids recovered.

3.1.10 *oil slick recovery rate, n*—the volume of oil slick removed from the water surface by the skimmer per unit of time.

3.1.11 *oil slick thickness, n*—the average thickness of the oil slick encountered by the test device (mm).

3.1.12 *test fluid, n*—the oil or oil/water mixture distributed on the water of the test facility and presented to the spill removal device for recovery.

---

[1] This guide is under the jurisdiction of ASTM Committee F-20 on Hazardous Substances and Oil Spill Response and is the direct responsibility of Subcommittee F20.12 on Removal.

Current edition approved Oct. 10, 1999. Published December 1999. Originally published as F 631 – 80. Last previous edition F 631 – 93.

[2] *Annual Book of ASTM Standards*, Vol 05.01.

[3] *Annual Book of ASTM Standards*, Vol 10.03.

[4] *Annual Book of ASTM Standards*, Vol 05.02.

[5] *Annual Book of ASTM Standards*, Vol 11.04.

[6] Discontinued; see 1998 *Annual Book of ASTM Standards*, Vol 11.04.

3.1.13 *throughput efficiency, n*—the ratio, expressed as a percentage, of the volume of oil recovered to the volume of oil encountered.

4. Significance and Use

4.1 This guide provides quantitative data in the form of oil recovery rates, throughput efficiencies, and oil recovery efficiencies under controlled test conditions. The data can be used for evaluating design characteristics of a particular spill removal device or as a means of comparing two or more devices. Caution must be exercised whenever test data are used to predict performance in actual spill situations as the uncontrolled environmental conditions that affect performance in the field are rarely identical to conditions in the test tank. Other variables such as mechanical reliability, presence of debris, ease of repair, ease of deployment, required operator training, operator fatigue, seaworthiness, and transportability also affect performance in an actual spill but are not measured by this guide. These variables should be considered along with the test data when making comparisons or evaluations of spill removal devices.

5. Summary of Guide

5.1 The spill removal device may be tested in a wave/tow tank or other facility that is suitable for controlling the appropriate test parameters. Significant testing results can be obtained using simple test tanks or ponds, particularly when calm water and low velocity advancing tests are desired as an economical means to screen and compare devices. Controlled test variables include relative velocity, oil properties and slick thickness, wave conditions, and pertinent device variables. It is essential that the device be operated in a steady-state condition during the sampling period when oil encounter rate, recovery rate, recovery efficiency, and device parameters are monitored, measured, and recorded.

6. Interferences

6.1 The table of results (see 13.1) shall address the possibility of test facility effects. For example, wall effects may interfere hydrodynamically with the device's performance.

6.2 Care should be taken that any containment means that is not inherent in the skimming device does not affect the oil distribution to the device.

7. Test Facilities

7.1 Several types of test facilities can be used to conduct the test outlined in this guide.

7.1.1 *Wave/Tow Tank*— A wave/tow tank has a movable bridge or other mechanism for towing the test device through water for the length of the facility. A wave generator may be installed on one end, or on the side of the facility, or both.

7.1.2 *Current Tank*— A current tank is a water-filled tank equipped with a pump or other propulsion system for moving the water through a test section where the test device is mounted. A wave generator may be installed on this type of test facility.

7.1.3 Other facilities, such as private ponds or flumes, may also be used, provided the test parameters can be suitably controlled.

7.2 Ancillary systems for facilities include, but are not limited to, a distribution system for accurately delivering oils to the water surface, skimming systems to assist in cleaning the facility between tests, and adequate tankage for storing the test oils.

8. Test Oils

8.1 Test oils for use with this guide should be selected to fall within the parameters specified in Appendix X1. These oils may be crude, refined, or simulated.

8.2 If test oils vary significantly from the recommended ranges, the test report shall discuss the implications of such deviations on the performance of the device.

8.3 The viscosity of oil varies greatly with temperature. Frequently test oils must be distributed in the test facility at temperatures different from the water temperature. When this occurs, the oil generally will approach the surface water temperature.

8.4 If oils that originally meet the conditions stated in Appendix X1 are reused, their properties may change and should be evaluated prior to reuse.

9. Safety Precautions

9.1 Test operations shall conform to established safety (and regulatory) requirements for both test facility operations and oil handling. Particular caution must be exercised when handling flammable or toxic test oils.

10. Test Device

10.1 The test device shall be deployed in accordance with facility operating characteristics. The device must be operated in accordance with the manufacturer's specified operating instructions with respect to mechanical operations and established maintenance routines. Modifications to the device, in any modification from commercial design, shall be recorded with the test results.

10.2 The make and model of the device, its nameplate recovery rate, and any other identifying specifications shall be recorded with the test results.

11. Test Variables

11.1 At the outset of the test, the independent or control test parameters are selected. The test evaluator should include a discussion of the procedures that were used to establish calibration and standardization. Data should be expressed with an indication of variability. Typical test variables include:

Test oils	as stated in Appendix X1
Test speed	upper and lower limits and speed increments selected as appropriate within ± 0.1 m/s
Oil slick thickness	1, 5, and 25 mm and other thicknesses as appropriate

Wave conditions	wave characteristics of significant height, average length, and average significant period and pattern may be varied as appropriate to the design of the skimmer
Wave height	0 mm, 150 mm, and 450 mm
Debris	Use various materials to simulate various forms of natural debris that may hinder skimmer operation. Recommended materials include: polypropylene rope, ½ in. diameter, cut into lengths ranging from 4 in. to 2 ft; softwood lumber, nominal dimensions 2 by 2 in., lengths ranging from 4 in. to 2 ft; foam sponges, nominal size 4 to 12 in; broken ice, nominal size 1 to 4 in. diameter; seaweed, lengths up to 5 ft; plastic or aluminum disposable containers (such as soda cans); sorbent pads, booms, or sheets; or other suitable materials. Use adequate number of pieces to produce concentrations at the skimmer inlet area corresponding to 50 % surface coverage.

12. Procedure

12.1 Prior to the test, select the operating parameters, such as tow speed (as applicable), wave conditions, test fluid, and oil distribution rate. Then prepare the facility and spill removal device for the test run. Occasionally, it may be necessary to preload the device with test fluid to achieve steady-state operation within a reasonable period of time. Any preload must be carefully measured and discharged into the device. Measure or note immediately prior to each test the following parameters describing ambient conditions:

12.1.1 Air temperature (°C),

12.1.2 Water temperature near the surface (°C),

12.1.3 Wind speed (m/s),

12.1.4 Wind direction relative to the test device, and

12.1.5 General weather conditions, for example, rain, overcast, sunny, etc.

12.2 Start the wave generator (if necessary), test fluid distribution system, tow mechanism or water flow (if necessary), and the spill removal device to begin a test run. Direct the discharge flow of recovered fluid from the device into a holding tank or back into the test tank out of the device's sweep path during the transient start-up period. After steady-state operation is achieved, monitor the discharge flow to obtain performance data. The discharge may be pumped through a flowmeter to obtain a flow rate and sampled periodically to obtain the oil-to-fluid ratio. Alternatively, the discharge may be diverted into calibrated sample tanks from which the flow rate and oil-to-fluid ratio may be determined. In either case, the data collection period begins when sampling starts and ends when sampling stops. During the data collection period or immediately thereafter, measure and record the following parameters:

12.2.1 Test fluid distribution rate (m^3/h),

12.2.2 Fluid recovery rate (oil and water) (m^3/h),

12.2.3 Tow speed or current speed (m/s),

12.2.4 Wave characteristics (see Practice F 625 for environmental descriptors and classifications),

12.2.5 Length of data collection period,

12.2.6 Oil encounter rate (m^3/h),

12.2.7 Oil slick encounter rate (m^3/h), and

12.2.8 Operating parameters of the spill recovery device such as belt speed (m/s), weir setting (mm below fluid level), pump speed (r/min), etc.

12.3 At the completion of the data collection period, divert the discharge of the spill removal device back into the holding tank or test tank. Stop the wave generator, tow mechanism or water flow, oil distribution, and spill removal device. Depending on the degree and variability of changes in the oil properties, the test evaluator will determine the required schedule of analysis to provide representative samples. Analyze representative samples of the discharge to determine oil-to-fluid ratio. Analyze representative samples of the recovered test oil to determine the following:

12.3.1 Density (g/mL),

12.3.2 Viscosity (mm^2/s),

12.3.3 Surface tension (mN/m),

12.3.4 Interfacial tension with respect to test tank water (mN/m), and

12.3.5 Water content (percent of total mixture).

12.4 Perform the analyses in 12.3 at the surface water temperature or in such a way that the values of these parameters at the surface water temperature can be determined from the analytical data.

13. Report

13.1 Prepare a schematic diagram of the layout for the test series.

13.2 Prepare a table of results for the test run containing the following entries:

13.2.1 Test identification number.

13.2.2 Date and time of day.

13.2.3 Average speed (tow speed or current speed, m/s) during data collection period.

NOTE 1—This rate is averaged over the data collection period.

13.2.4 Test oil type.

13.2.5 *Oil slick thickness*—Calculation of slick thickness should include a brief discussion of the method used for measuring variation in the thickness, especially for heavy oils.

13.2.6 For regular waves, include height, average period (or length), and whether head or following. For irregular waves, include significant height, significant frequency, spectral characteristics, and whether head or following. Primary and secondary (reflected) wave basin characteristics shall be described.

13.2.7 Oil properties at test temperatures, including density, viscosity, surface tension, interfacial tension with tank water, and initial water content of oil.

13.2.8 Recovered oil properties as performed in 12.3.

13.2.9 Total volume of oil distributed during data collection period.

13.2.10 Total volume of oil encountered during data collection period.

13.2.11 Total volume of fluid (oil/water) recovered during data collection period.

13.2.12 Average oil distribution rate (see Note 1).

13.2.13 Average oil encounter rate (see Note 1).

13.2.14 Average fluid (oil/water) recovery rate (see Note 1).

13.2.15 Average oil recovery efficiency (see Note 1).

13.2.16 Average oil recovery rate (see Note 1).

13.2.17 Average oil slick distribution rate (see Note 1).

13.2.18 Average oil slick encounter rate (see Note 1).

13.2.19 Average oil slick recovery rate (see Note 1).

13.2.20 Average oil slick recovery efficiency (see Note 1).

13.2.21 Average emulsification factor (see Note 1).

13.2.22 Average throughput efficiency (see Note 1).

13.2.23 Ambient conditions, including air temperature, surface water temperature, wind speed, wind direction, and brief statement of weather conditions during test run.

13.2.24 Length of data collection period.

13.2.25 Volume of oil in device at beginning of data collection period.

13.2.26 Volume of oil in device at end of data collection period.

13.2.27 Operating parameters of the device such as belt speed, weir setting, pump speed, etc.

13.2.28 For tests conducted in the presence of debris, report the following:

13.2.28.1 Size and type of debris forms used;

13.2.28.2 Approximate concentration, expressed as surface coverage (percent);

13.2.28.3 Mode of debris handling, or processing, or both;

13.2.28.4 Effectiveness of debris handling, or processing, or both; and

13.2.28.5 When comparing results, specify reasons for performance degradation (if any), differentiating between oil encounter degradation and that of the skimming mechanism.

13.2.29 Brief discussion of interferences or limiting factors (see 6.1).

13.3 If any of the data required by 13.2 are not applicable to the device being tested, a statement must be included in the table of results describing the reasons for omitting the data.

13.4 Report qualitative comments on device operating parameters such as mechanical reliability, ease of repair, ease of deployment, required operator training, operator fatigue, seaworthiness, and transportability.

14. Keywords

14.1 oil; oil recovery; skimmer

APPENDIX

(Nonmandatory Information)

X1. STANDARD TEST OILS

X1.1 Values in Table X1.1 refer to test fluid properties at test temperatures. Test methods for fluid properties are specified as follows: viscosity, Test Methods D 445 and D 2983 (report shear rate for viscosity measurement, should be in the range of 1 to 10 s^{-1}); density, Test Method D 1298; interfacial tension, Test Method D 971; pour point, Test Method D 97. For all test oils (with the exception of emulsions), maximum sediment and water (BSW) of 0.1 % (see Test Method D 4007).

X1.2 Of the five viscosity ranges, numbers I, II, and IV are especially recommended as being indicative, respectively, of lightly weathered, moderately weathered, and significantly weathered crude oils.

X1.3 The following lists examples of hydrocarbon oils that could be used to fall within the specified ranges. This list is intended for guidance only; it should be noted that viscosities of all oils will vary greatly with both temperature and the specific product. Selected oils may be crude, refined, or simulated. In the case of crudes and light refined products, it is acceptable and may be desirable to pre-weather the oil in order to produce a desired viscosity, increase the oil's flash point to a safe level, and produce a more stable test fluid.

TABLE X1.1 Candidate Test Oils

NOTE 1—Test oils should be selected to fall within these five categories.

	Viscosity, mm²/s	Density, g/mL	Oil-Air Interfacial Tension, mN/m	Oil-Water Interfacial Tension, mN/m	Pour Point,°C
I[A]	150 to 250	0.90 to 0.93	28 to 34	20 to 30	< −3
II[B]	1500 to 2500	0.92 to 0.95	30 to 40	20 to 30	< −3
III[C]	17 000 to 23 000	0.95 to 0.98	20 to 40	20 to 40	< 10
IV[D]	50 000 to 70 000	0.96 to 0.99	20 to 40	20 to 40	...
V[E]	130 000 to 170 000	0.96 to 0.99	20 to 40	20 to 40	...

[A] 1) Alaska North Slope crude oil, 10 to 15 % weathered by volume.
2) Fuel oil No. 4 (heavy); can be prepared by blending 40 % fuel oil No. 2 and 60 % fuel oil No. 6.

[B] Fuel oil No. 5; can be prepared by blending 20 to 25 % fuel oil No. 2 with 75 to 80 % fuel oil No. 6.

[C] Residual fuel oil (that is, fuel oil No. 6 prepared to above criteria).

[D] Residual fuel oil (that is, heavy cut of fuel oil No. 6).

[E] Emulsified crude oil, 50 to 80 % water content. The oil may be emulsified by blowing compressed air through water on which the oil is floating.

Designation: F 715 – 95

Standard Test Methods for Coated Fabrics Used for Oil Spill Control and Storage[1]

This standard is issued under the fixed designation F 715; the number immediately following the designation indicates the year of original adoption or, in the case of revision, the year of last revision. A number in parentheses indicates the year of last reapproval. A superscript epsilon (ε) indicates an editorial change since the last revision or reapproval.

1. Scope

1.1 These test methods cover laboratory-conducted performance tests for coated fabrics used in spill control barriers or in temporary storage devices.

1.2 *This standard does not purport to address all of the safety concerns, if any, associated with its use. It is the responsibility of the user of this standard to establish appropriate safety and health practices and determine the applicability of regulatory limitations prior to use.*

2. Referenced Documents

2.1 *ASTM Standards:*

D 86 Test Method for Distillation of Petroleum Products[2]
D 93 Test Methods for Flash Point by Pensky-Martens Closed Cup Tester[2]
D 97 Test Method for Pour Point of Petroleum Products[2]
D 129 Test Method for Sulfur in Pretroleum Products (General Bomb Method)[2]
D 130 Test Method for Detection of Copper Corrosion from Petroleum Products by the Copper Strip Tarnish Test[2]
D 287 Test Method for API Gravity of Crude Petroleum and Petroleum Products (Hydrometer Method)[2]
D 445 Test Method for Kinematic Viscosity of Transparent and Opaque Liquids (And the Calculation of Dynamic Viscosity)[2]
D 482 Test Method for Ash from Petroleum Products[2]
D 524 Test Method for Ramsbottom Carbon Residue of Petroleum Products[2]
D 543 Test Method for Resistance of Plastics to Chemical Reagents[3]
D 613 Test Method for Cetane Number of Diesel Fuel Oil[4]
D 664 Test Method for Acid Number of Petroleum Products by Potentiometric Titration[4]
D 751 Test Methods for Coated Fabrics[5]
D 975 Specification for Diesel Fuel Oils[2]
D 1149 Test Method for Rubber Deterioration—Surface Ozone Cracking in a Chamber[6]
D 1266 Test Method for Sulfur in Petroleum Products (Lamp Method)[2]
D 1298 Practice for Density, Relative Density (Specific Gravity), or a PI Gravity of Crude Petroleum and Liquid Petroleum Products by Hydrometer Method[2]
D 1319 Test Method for Hydrocarbon Types in Liquid Petroleum Products by Fluorescent Indicator Adsorption[2]
D 1796 Test Method for Water and Sediment in Fuel Oils by the Centrifuge Method (Laboratory Procedure)[2]
D 2425 Test Method for Hydrocarbon Types in Middle Distillates by Mass Spectrometry[2]
D 2500 Test Method for Cloud Point of Petroleum Oils[2]
D 2644 Tolerances for Yarns Spun on the Woolen System[7]
D 3117 Test Method for Wax Appearance Point of Distillate Fuels[8]
D 4157 Test Method for Abrasion Resistance of Textile Fabrics (Oscillatory Cylinder Method)[9]
G 26 Practice for Operating Light-Exposure Apparatus (Xenon-Arc Type) With and Without Water for Exposure of Nonmetallic Materials[10]

2.2 *Federal Standard:*

Fed. Std. No. 191A Textile Test Methods[11]

2.3 *Military Specification:*

MIL-T-6396 E Tank, Fuel, Oil, Water-Alcohol, Coolant Fluid, Non-Self-Sealing, Removable, Internal[11]

3. Significance and Use

3.1 Membrane materials are subjected to these tests in order to provide data that reasonably relate to membrane response under the actual conditions of spill control barrier or storage device use.

3.2 Although these test methods provide data on individual performance of membrane materials, all combinations of actual

[1] These test methods are under the jurisdiction of ASTM Committee F-20 on Hazardous Substances and Oil Spill Response and are the direct responsibility of Subcommittee F20.11 on Control.

Current edition approval May 15, 1995. Published July 1995. Originally published as F 715 – 81. Last previous edition F 715 – 81 (1994)$^{\epsilon1}$.

[2] *Annual Book of ASTM Standards,* Vol 05.01.
[3] *Annual Book of ASTM Standards,* Vol 08.01.
[4] *Annual Book of ASTM Standards,* Vol 05.04.
[5] *Annual Book of ASTM Standards,* Vol 09.02.
[6] *Annual Book of ASTM Standards,* Vol 09.01.
[7] *Annual Book of ASTM Standards,* Vol 07.01.
[8] *Annual Book of ASTM Standards,* Vol 05.02.
[9] *Annual Book of ASTM Standards,* Vol 07.02.
[10] *Annual Book of ASTM Standards,* Vol 14.02.
[11] Available from Standardization Documents Order Desk, Bldg. 4 Section D, 700 Robbins Ave., Philadelphia, PA 19111-5094, Attn: NPODS.

conditions of spill control barrier or storage device use are not simulated in this sequence of tests.

4. Base Line Tests

4.1 *Number of Specimens*—Subject specimens of spill control membrane materials to the tests prescribed in 4.2-4.11. Perform all tests in the direction of both the warp and fill of the fabric. The number of specimens to be tested are as specified in the procedures referenced in 4.2-4.11.

4.2 *Mass and Thickness*—Test Methods D 751.

4.3 *Tensile Strength and Elongation at Break*—Test Methods D 751 for Breaking Strength, Procedure A—Grab Test Method.

4.4 *Tear Strength*—Test Methods D 751 for Tearing Strength, Procedure B—Tongue Tear Method. If standard sample does not achieve true tear or if the fabric design requires roping to achieve high tear, sample size may be enlarged in order to achieve cord breakage, but all other procedures must be maintained. Such sample size change must be recorded in test results.

4.5 *Adhesion*—Test Methods D 751 after 5 days at 70°C in water.

4.6 *Puncture Resistance*—MIL-T-6396 E, procedure outlined in 4.6.17.

4.7 *Abrasion Resistance*—Test Method D 4157, abrade the specimen using a 2.72-kg (6-lb) dead weight load tension and a 2.72-kg (6-lb) dead weight load using extra coarse, 40 grit sandpaper as the abradant. Cycle continuously to exposure of threads.

4.8 *Seam Strength*—Test Methods D 751. Perform test with seam orientation as in device design.

4.9 *Seam Adhesion*—Repeat adhesion test (see 4.5) for specimens containing a seam.

4.10 *Blocking Resistance*—Test Methods D 751.

4.11 *Mildew*—Fed. Std. No. 191A, Method 5762 (Mixed Cidal Spore Test).

5. Weather and Petroleum Sample Tests Related to Base Line Tests

5.1 *Specimen Preparation*—Subject samples of spill control barrier or storage device membrane material to the following preparations: weather resistance to be followed by resistance to petroleum products:

5.1.1 *Weather Resistance*—Practice G 26, Test Method 2 (also referred to as Test Method B—Alternate Exposure to Light and Darkness and Intermittent Exposure to Water Spray), using deionized water and a 7000 watt xenon burner tube, 500-h exposure.

5.1.2 *Resistance to Petroleum Products*—Test Method D 543, with 96-h exposure to Diesel Fuel Grade No. 2, as prescribed in Specification D 975 and further meeting the specifications shown in Table 1 (Note). Conduct tests at ambient temperature.

5.2 *Weather and Petroleum Sample Tests*—Perform each of

TABLE 1 Detailed Requirements for Test Diesel Fuel Caterpillar 1G2/1H2

NOTE 1—Diesel fuels meeting the detailed requirements in Table 1 are available from Howell Hydrocarbons and Chemical Incorporated, 1201 South Sheldon Road, Channelview, TX 77530.

Property	ASTM Test Method	Requirement	
Flash point	D 93	140°F (60.0°C) min or legal	
Pour point	D 97	20°F (–7°C) max	
Cloud point	D 2500 or D 3117	Report	
Water and Sediment	D 1796	0.05 vol % max	
Ramsbottom carbon on 10 % residium	D 524	0.20 mass % max	
Ash	D 482	0.01 mass % max	
Distillation	D 86	IBP report	
		10 % report	
		50 % 500–530°F (260–277°C)	
		90 % 590–620°F (310–327°C)	
		EP 650–690°F (343–366°C)	
Kinematic viscosity at 104°F (40.0°C)	D 445	2.0–4.0 cSt (2.0–4.0 mm^2/s)	
Total sulfur (must be natural)	D 2644 (Alt. D 129 or D 1266)	0.380–0.420 mass %	
Copper, corrosion (122°F, 50°C, 3 h)	D 130	No. 2 max	
Acid number (TAN-E)	D 664	0.15 mg KOH/g max	
Centane number	D 613	47.0–53.0	
Density	D 287 or D 1298	report	
API gravity	D 287 (Alt. D 1298)	33–35 API	
Cracked stocks		none	
Hydrocarbon types	D 1319	report	
Hydrocarbon types	D 2425	Component	mass %
		aliphatic paraffins	45.0–65.0
		monocycloparaffins, tetrocycloparaffins	report
		dicycloparaffins	0.0–15.0
		alkylbenzenes	5.0–10.0
		indans/tetralins, indenes	report
		napthalene	report
		napthalenes	5.0–15.0
		acenapthenes, acenapthylenes, tricyclic aromatics	report

the following tests within 1 h of the termination of the test in 5.1.2.

5.2.1 *Tensile Strength*—Same as base line test (4.3).

5.2.2 *Tear Strength*—Same as base line test (4.4).

5.2.3 *Adhesion*—Same as base line test (4.5).

5.2.4 *Puncture Resistance*—Same as base line test (4.6).

5.2.5 *Abrasion Resistance*—Same as base line test (4.7).

5.2.6 *Seam Strength*—Same as base line test (4.8).

5.2.7 *Seam Adhesion*—Same as base line test (4.9).

5.2.8 *Blocking Resistance*—Same as base line test (4.10).

5.2.9 *Low Temperature Resistance*—Fed. Std. No. 191A, Method 5874, 2-h soak at – 29°C with a 2.5-kg mandrel roller weight. Visually examine specimen and report signs of cracking or flaking. Do not perform hydrostatic resistance test.

5.2.10 *Ozone Resistance*—Test Method D 1149.

6. Precision and Bias

6.1 The precision and bias of test measurements are addressed in each individual test method cited in Sections 3-5.

7. Keywords

7.1 barrier; coated fabrics; membrane materials; membranes; oil spill; spill control; storage devices; temporary storage devices

Standard Methods of Testing Sorbent Performance of Absorbents[1]

This standard is issued under the fixed designation F 716; the number immediately following the designation indicates the year of original adoption or, in the case of revision, the year of last revision. A number in parentheses indicates the year of last reapproval. A superscript epsilon (ε) indicates an editorial change since the last revision or reapproval.

ε[1] NOTE—Section 14 was added editorially in October 1993.

1. Scope

1.1 These methods cover the development of laboratory test data which describe the performance of absorbent materials used to remove oils and other compatible fluids from water.

1.2 *This standard should be used to measure and describe the properties of materials, products, or assemblies in response to heat and flame under controlled laboratory conditions and should not be used to describe or appraise the fire hazard or fire risk of materials, products, or assemblies under actual fire conditions. However, results of this test may be used as elements of a fire risk assessment which takes into account all of the factors which are pertinent to an assessment of the fire hazard of a particular end use.* (For a specific warning statement see 10.4.1.)

2. Referenced Documents

2.1 *Federal Standard:*

Fed. Std. No. 141 Paint, Varnish, Lacquer and Related Materials. Methods of Inspection, Sampling and Testing[2]

2.2 *Military Specification:*

MIL-I-631D Insulation, Electric, Synthetic Resin Composition, Nonrigid[2]

3. Terminology

3.1 *Definitions:*

3.1.1 *absorption*—a process where the material taken up is distributed throughout the body of the absorbing material.

3.1.2 *adsorption*—a process where the material taken up is distributed over the surface of the adsorbing material.

3.1.3 *sorbent*—a material used to recover liquids through the mechanisms of absorption or adsorption, or both.

3.2 *Definitions of Terms Specific to This Standard:*

3.2.1 *oil, oily liquid*—substantially water-immiscible organic liquids (limited water solubility) as well as liquids containing dissolved solids. These oily materials may either float, sink, or be dispersed.

3.2.2 *Type I absorbent (loose)*—an unconsolidated, particulate material without sufficient form to be applied in single units. This type may include particulate blends with other materials, that is, adsorbents to allow use in specific applications.

3.2.3 *Type II absorbent (roll, sheet, pad, blanket, pillow, web)*—a material that has a form and strength sufficient to be lifted and handled when saturated without tearing.

3.2.4 *Type III Absorbent (booms)*—absorbent material in a form whose length substantially exceeds other dimensions. Booms are to be provided with connector units so that they may be coupled to another boom or line.

4. Summary of Method

4.1 The absorbent material is tested using established standard methods, where applicable, and specially developed tests for performance factors.

5. Significance and Use

5.1 These methods are to be used as a basis for comparison of absorbents in a consistent manner.

5.2 These methods are not useful for a comparison of absorbents with adsorbents, even though all absorbents exhibit adsorbent properties. Both types of materials have prime areas of utility.

5.3 These methods may not list all the safety and disposal options necessary for safe ultimate disposal of used sorbent material into the environment. Federal, state, and local regulatory rules must be followed.

6. Special Considerations

6.1 *Solubility*—Fluid should not dissolve the absorbent under normal recovery or handling conditions.

6.2 *Reuse*—Used absorbent materials may be recycled under certain conditions. Fluid absorbed cannot be released with squeezing. However, many of the polymeric materials will depolymerize at temperatures at or below those in a refinery pipe still. This means they are recoverable, although in a different form. Along with resource recovery, the harvestability of sorbents will become more important. Lower boiling materials may also be steam stripped from absorbents to recover both fluid and absorbent.

6.3 *Release of Volatiles*—Certain volatile materials picked up by sorbents may create a vapor hazard by increasing the

[1] These methods are under the jurisdiction of ASTM Committee F-20 on Hazardous Substances and Oil Spill Response and are the direct responsibility of Subcommittee F20.12 on Removal.

Current edition approved Jan. 13, 1982. Published March 1982.

[2] Available from Standardization Documents Order Desk, Bldg. 4 Section D, 700 Robbins Ave., Philadelphia, PA 19111-5094, Attn: NPODS.

exposed surface area and therefore vapor release. Depending on the material picked up, this may increase fire or health hazard (see 11.3).

6.4 *Release to an Alternative Material Such as Sand*—The tendency to release sorbed liquid to beaches, ground, clothing of personnel, etc., both as liquid and vapor can be determined on a basis of priority (see 11.5).

6.5 *Penetration*—Absorbents may be useful as sealing devices and the penetration test will allow a uniform means of determining the efficacy of a material for this use.

7. Apparatus

7.1 *Exterior Exposure Tester*, sufficient for use with Fed. Std. No. 141, Method 6152 (*optional*).

7.1.1 Actual outdoor exposure in a direction averaging the maximum exposure in a sunny area, such as Southern Florida or Arizona, is desired. Exposure should be listed in months with total sun days.

7.1.2 Actual outdoor exposure for a period of three months at a testing site should be listed.

7.2 *Tension Tester*—Commercially available tester (*optional*).[3]

7.2.1 Use of actual weights suspended from test pieces in turn suspended from a continuous reading hanging scale may be used. Shock loading may be simulated by fastening one end of a unit rigidly and attaching the maximum pickup weight to the bottom end hanging free. The weight can be raised 2.5 cm and released and also 10 cm and again released. This test should be recorded as pass or fail.

7.3 *Continuous Reading Hanging Scales*—A continuous reading hanging scale fitted with a hook or clamp should be used. For testing units, weighing sensitivity shall be ± 2%. For small samples, postal-type scales may be used (0 to 80 g).

7.4 *Top Loading Balance*, having 0.5-g resolution for samples weighing at minimum 50 g or more. An accuracy of 2% is desired.

7.5 *Measuring Tubes*—10 and 100-mL graduated measuring cylinders of regular laboratory quality.

7.6 *Paper Towels*, for removing excess liquid from the absorbent.

7.7 *Oven-Dried Mason Sand*, for use in checking rate of release. Screen analysis is desired. Optionally, known screen size silica sand can be used.

8. Test Fluids

8.1 Test fluids may be used for general testing, although the actual fluid should be used where possible.

8.2 *Samples Heavier than Water*—For samples heavier than water 1,1,1-trichloroethane (available in many retail stores as nonflammable spot remover) may be used.

8.3 *Low-Viscosity Materials*—For low-viscosity material (like gasoline), use xylene which has a higher flash point. (For safety, a solution containing 10 % methylene chloride can be added to aid in raising flash point.) Materials conforming to 8.2 have a viscosity similar to gasoline but the density difference may have to be adjusted.

8.4 *Polar Solvents*—For polar solvents, use methylisobutylketone (hexone), butylacetate, etc.

8.5 *Medium-Viscosity Fluids*—For medium-viscosity fluids, No. 2 fuel oil, diesel fuel, JP-5, Jet A-1 (turbine fuel), or light crude oil (for example, light Arabian or Iranian crude), is recommended.

8.6 *High-Viscosity Fluids*—Nondetergent or C-10 Transformer dielectric oil (specify manufacturer and viscosity) is recommended.

8.7 *Range Finding Fluids*—Ethanol (95%), acetone, ethylene glycol, cyclohexane, 2-nitropropane are suitable fluids.

9. Conditioning

9.1 Condition three absorbent test specimens at room temperature (21 to 26°C) (70 to 80°F) in a closed vessel with 2.5 cm of water in the bottom. Suspend specimens in vapor space without contacting water for not less than 24 h prior to testing. Condition specimens in the fully exposed state with no coverings or wrapping that would hinder the equilibration process.

9.2 Condition three absorbent test specimens totally immersed in water at room temperature (21 to 26°C) (70 to 80°F) in a vessel with a minimum of 10 cm of water covering the absorbent for not less than 24 h prior to testing. Condition specimens in the fully exposed state with no coverings or wrapping which would hinder the equilibration process.

9.3 *Alternative Temperatures*—If temperature conditions other than room temperature are expected to be important, then the sorbent should be conditioned and tested at the temperature of use in addition to those specified in 9.1 and 9.2.

10. Determination of Physical Characteristics of Absorbents

10.1 *Mildew*—The susceptibility of an absorbent to mildew under normal storage is described in MIL-I-631D. This optional test determines the expected shelf life under conditions which could lead to mildew. This test may be required when it is believed that this condition may exist prior to an emergency.

10.2 *Exterior Exposure*—This optional test may be performed in accordance with Fed. Std. No. 141, Method 6152. Actual exposure in area of use for one month is preferable. Long-term testing is not necessary for materials that are to be used in emergency situations. For testing purposes both absorbent and adsorbent and other parts of a fabricated unit should be tested or exposed individually and reported.

10.3 *Humidity/Water Resistance*—This test is used to predict the performance of a product stored under high humidity or one that is water-soaked prior to use.

10.3.1 Condition all absorbent test specimens at room temperature (21 to 26°C) (70 to 80°F) in a closed vessel with 2.5 cm of water in the bottom. Suspend the specimens in vapor space without contacting water for 2 days (14 days optional). At the end of the exposure period, test the specimens for oil and water absorbency. Record weights before and after this conditioning.

10.3.2 Weigh specimens that are conditioned in water by being total immersed (8.2) immediately after removal from water and again after wringing out all excess water (typical hand wringer for laundry or equivalent). Test specimens for oil

[3] Instron Model No. TM, available from Instron Corp., 100-A Royall St., Canton, MA 02021, or equivalent.

and water sorbency following both weighings.

10.4 *Flammability*—This test should indicate either contribution to an existing fire, that is, a UL rating of 94, if desired, or a simple rating of combustible or noncombustible and flashpoint by Cleveland open-cup.

10.4.1 **Warning**—All sorbent materials, even those considered nonhazardous, take on the characteristics of the material they pick up. If the material is flammable, the used absorbent material is flammable. If the fluid picked up is toxic, then the used absorbent should be handled as a toxic substance. Basic active absorbent and simple fabricated units are adequate for this test.

11. Determination of Expected Performance

11.1 *General*:

11.1.1 Initial testing for performance can be accomplished with absorbent in the as-received condition. Conditioning as described in Section 9 may be necessary for special use or for special fluids to be absorbed. Good practice requires triplicate testing with results reported as the average.

11.1.2 For all tests, without exception, the fabricated materials or units must remain handleable (without tearing) at the maximum pickup rate reported while being held by a corner of the unit.

11.2 *Oily Material Absorption*—This test provides a simple method of comparing the absorbing material's inherent capacity to absorb various fluids as well as compare ultimate capacity to do so. This test is run with an excess of fluid.

11.2.1 *Type I Absorbent*—In a 10-mL graduated cylinder, place 2 mL of absorbent and weigh to determine the weight of absorbent. Place the 2 mL of absorbent into a 100-mL graduated cylinder. Add 50 mL of fluid to be tested, record level, then cap with a foil (aluminum or plastic)-wrapped cork or stopper. Agitate for short periods with a glass stirring rod. If at the end of 2 h there is still free liquid on top of the swollen absorbent or underneath if the fluid has a density greater than the absorbent, place a piece of 200-mesh screen (just slightly smaller than the inside diameter of the cylinder with at least four centering wires bent up at 90°) in the fluid to sit on top of the absorbent. Record the level of the swollen absorbent. If the fluid is much more dense than the absorbent, add an additional minimum weight on top of the screen to force the swollen absorbent to the bottom and then record the level of the screen. If this test gives a swell of less than 20 mL, repeat using 10 mL of absorbent in the 100-mL graduated cylinder. This empirical test may be replaced by a microscopic calibrated swell of a uniform absorbent particle in the fluid.

11.2.2 Swell may also be checked at 24 h for those specimens that have not shown maximum swell in 2 h. It should be recognized that a 2-h maximum swell may be acceptable but a 24-h swell will probably not be useful for a spill control unit without modification. Such a 24-h swell time may require modification of particle size, etc. to provide effective swell within 2 h. The report shall include the time required to reach maximum absorption if less than 2 h.

11.2.3 Increase in volume of the solid absorbent is the property that distinguishes an absorbent from an adsorbent. Free liquid should be poured out of the graduate cylinder and a weighing then made (with the tared screen). The weight of swollen absorbent allows calculation of absorption efficiency:

$$A = [(W_1 - W_2)/W_1] \times 100 \tag{1}$$

where:

W_1 = weight of absorbed material,
W_2 = weight of unabsorbed material, and
A = absorption efficiency for the fluid and absorbent being tested.

Type I absorbent material may be tested with inert material present such as sand, fibers, etc., that give a special property such as no stir dispersion or sufficient mass to serve as temporary diking. The absorbent mass with inert material can be evaluated using the test described in 8.2. The inert material may be subtracted from the total weights to indicate the absorbent data. The penetration test in 11.4 may also be useful.

11.3 *Rate of Release*—This test can be used to estimate the degree of hazard in the case of flammable or hazardous materials. For those materials with vapor hazards or skin toxicity to humans these are important factors (**warning**—see 10.4.1).

11.3.1 Place a standard sized weighed piece of an absorbent unit in a dish of fluid. Add excess fluid to assure total saturation of material. A saturation time of 30 min should be adequate although a 2-h test is well within the useful practical application time. If material to be absorbed is toxic, run this test in a ventilated fume hood.

11.3.2 Following thorough saturation, hang the absorbent units (minimum 15-cm dimensions) by one corner and weigh immediately (even if dripping). Record the time to cessation of dripping. Reweigh and record the weight as maximum practical pickup rate at the time dripping stops.

11.3.3 Then hang the units in a ventilated hood with the fan on, close the door to obtain maximum air flow (past the units), and continually monitor and record the weight every 10 min. At the end of 2 h (or sooner) if the total weight approaches the tare weight of the unit, record the time and use as a basis for comparison where containment offers an environmental, personal, or flammable hazard. The maximum pick up at the point of no dripping (less than 1 drop per minute) is recorded as the maximum effective pick up of that absorbent unit. This maximum pickup may be recorded as either per unit or per amount of active ingredient in that unit. For given test conditions (temperature, air, flow fluid, etc.), the test will allow selection based on performance.

11.4 *Penetration*:

11.4.1 This test requires a glass tube of 12 to 28 mm inside diameter. Place a porous plug in the bottom of the tube. This can be a cork with a slit to allow air out protected by cloth or tissue paper. Place a 10-cm layer of sand on top of this followed by 15 cm of absorbent material and again 7 cm of sand on top of this. Dye the fluid to be checked and then pour carefully into the top and adjust to maintain a 15-cm head so that when the penetration into the bed is stopped, 15 cm of fluid is above the top layer of sand (held vertically). For safety, place a bottle or container under this apparatus to hold twice the volume of the tube used. This penetration test can be used with pure absorbent as well as those containing inert materials, such as adsorbents, for flow, mass control, etc., with the following change:

11.4.1.1 Compact a layer of sand to 10 cm by dropping a 500-g weight five times for a distance of 3 cm on top of a snug-fitting cover placed above the sand inside the tube; on top of this, place a 15-cm layer of absorbent mixture, compact in the same manner, follow with a 7-cm layer of sand, and compact in the same manner.

11.4.2 Run this test three times for those materials that cause flow to stop within the 15-cm absorbent bed depth and record the average penetration time. Then again record penetration 24 h later. For very volatile materials, a *loose* cork may be placed in the top of the tube to maintain liquid head by minimizing evaporation (this must not form a tight seal).

11.5 *Release to Other Materials*—This test is used to compare the tendency of the absorbent containing fluid sorbate to contaminate either personnel or beaches, for example, through release of contents.

11.5.1 Place a specimen of Type II absorbent (with dimensions and weight recorded) in the fluid to be tested for 30 min. The fluid may be dyed with Oil Red A and should be in a quantity to exceed the capacity of the absorbent unit which means some free fluid will be left after the absorbent unit is removed. Hang the specimen by a corner to allow any adsorbed fluid (as opposed to absorbed fluid) to drain out of fabric, etc. When all dripping has ceased, place the Type II absorbent on a piece of white blotting paper (nominal 3 mm thick) lying on a flat surface. Move the Type II absorbent to a new clean area of blotting paper when the dyed fluid stains through to the reverse side of the blotting paper. When a stain no longer shows through the blotting paper, measure the weight of the absorbent unit. Record this measurement as the *practical containment efficiency.* Repeatedly place this same Type II absorbent specimen on the blotting paper with a 2 g/cm weight uniformly distributed over each square centimetre. Again measure the weight of the Type II absorbent when the stain does not penetrate through the blotting paper. Record this as the *maximum containment efficiency.* The calculation of these two measurements will permit a reliable estimate of the amount of absorbent needed to handle a given quantity.

$$\text{Efficiency} = [(A_f - A)/A_f] \times 100 \quad (2)$$

where:
A_f = weight of absorbent plus fluid, and
A = weight of absorbent.

11.6 *Fluids More Dense than Water* (Sinkers, Bottom Floaters)—This test evaluates the potential to handle materials that sink in water.

11.6.1 Color 1,1,1-trichloroethane (available in many areas as nonflammable spot remover) with a dye such as Oil Red A.

11.6.2 Place water in a vessel to a level of 20 cm. Pour 200 mL of a sample fluid or the actual fluid into the water so it layers on the bottom. Manually hold the Type II absorbent material in the test liquid on the bottom and then retrieve it using a hook and line on a corner. After a maximum pickup time of 30 min, pull the sample up by means of the hook and hold under water for 5 min. Note whether test fluid was released into the water. Then lift out the sample from the water, suspend by the hook, and note dripping of the test fluid, if any. When the drip rate is less than 1 drop every 2 min, weigh the sample. The total amount of fluid absorbed per unit weight of absorbent will indicate the efficiency for underwater pickup as follows:

$$\text{Underwater pickup efficiency} = [(A_f - A)/A_f] \times 100 \quad (3)$$

NOTE 1—For those materials, such as polychlorinated biphenyls, tetraalkyl leads, carbon tetrachloride, ethylene bromide, etc. the rate of release test in 11.5 should also be run and may well be an important consideration. Toxicity and human hazard from running these tests should be given primary and thoughtful consideration. Testing for absorbent Types II and III may be estimated from modules or amounts of Type I absorbent tested. They may be tested as fabricated or assembled, or both, but this testing is optional since the quantities involved will be considerable.

12. Report

12.1 The report shall include the following information:

12.1.1 Test fluid used.

12.1.2 Empirical formula.

12.1.3 Estimated purity (method).

12.1.4 Density.

12.1.5 Boiling point.

12.1.6 Flash point.

12.1.7 Viscosity.

12.1.8 Source.

12.1.9 Absorbent name and whether Type I, II, or III.

12.1.10 Manufacturer of absorbent and estimated date of manufacture.

12.1.11 Date and location of test, including temperature of laboratory and name of person conducting the test.

12.1.12 Indicate whether flammability or toxic hazard exists.

12.2 The report form in Fig. 1 may be useful in recording the results of each test.

13. Precision and Bias

13.1 No precision and bias data are presently available for these methods.

14. Keywords

14.1 absorbents; oil; sorbent; sorbent performance

I. *Conditioning* (Section 9)
Conditioning temperature ________
Method of conditioning ________

II. *Oily Material Absorption* (11.2):
Weight of 2 mL of absorbent ________
Weight of absorbent and cylinder ________
Weight of 50 mL of fluid plus absorbent and 100-mL cylinder ________
Time of stirring before adding screen ________
Level of mesh screen (If reading is less than 20 mL repeat as below) ________
Weight of 10 mL of absorbent ________
Weight of absorbent and 100-mL cylinder ________
Weight of 50 mL of fluid and absorbent plus 100-mL cylinder ________
Time of stirring before added screen ________
Level of mesh screen ________
Apparent bulk density of absorbent ________
Total volume of swollen absorbent minus bulk density (volume of absorbent) = the actual increase in volume (pour off liquid) ________
Absorption efficiency (see 11.2.3) ________

III. *Rate of Release* (see 11.3) ________
Size and type of absorbent ________
Weight of absorbent ________
Weight of active material (if inerts are present) ________
Weight of absorbent and liquid ________
Time to stop dripping (minimum ________
Weight at time of drip cessation ________
Weight at cessation—weight of absorbent = practical pickup ________

Practical pickup rate ___ expressed in grams of liquid per grams of absorbent and ___ g of liquid/cm^3 of absorbent ___
Monitor weight of units in hood air flow

Minutes	A	B	C	D	E	F	G
0							
10							
20							
30							
40							
60							
120							
180							

IV. *Penetration*: — *Test No.* 1 2 3
Diameter of glass tube ________
Length of glass tube ________
Depth of initial penetration bed ________
Depth of penetration after 24 h ________

V. *Release to Other Materials*:
30-min equilibration fluid:
Weight of absorbent ________
Absorbent paper used (type and thickness)
Weight of absorbent when it no longer marks paper ________
Practical containment efficiency (11.5.1) ________
Run test with uniform weight (454 g per each 225 cm^2):
Weight of absorbent when it no longer marks paper ________
Maximum containment efficiency (for human protection) ________

VI. *Fluids More Dense Than Water*:
Sinkers
Weight of 200 mL of fluid ________
Weight of dry absorbent ________
Weight at time of no drip in air for two minutes ________
Underwater pickup efficiency ________

FIG. 1 Sample Form for Recording Data

The American Society for Testing and Materials takes no position respecting the validity of any patent rights asserted in connection with any item mentioned in this standard. Users of this standard are expressly advised that determination of the validity of any such patent rights, and the risk of infringement of such rights, are entirely their own responsibility.

This standard is subject to revision at any time by the responsible technical committee and must be reviewed every five years and if not revised, either reapproved or withdrawn. Your comments are invited either for revision of this standard or for additional standards and should be addressed to ASTM Headquarters. Your comments will receive careful consideration at a meeting of the responsible technical committee, which you may attend. If you feel that your comments have not received a fair hearing you should make your views known to the ASTM Committee on Standards, 100 Barr Harbor Drive, West Conshohocken, PA 19428.

Designation: F 726 – 99

Standard Test Method for Sorbent Performance of Adsorbents[1]

This standard is issued under the fixed designation F 726; the number immediately following the designation indicates the year of original adoption or, in the case of revision, the year of last revision. A number in parentheses indicates the year of last reapproval. A superscript epsilon (ε) indicates an editorial change since the last revision or reapproval.

This standard has been approved for use by agencies of the Department of Defense.

1. Scope

1.1 This test method covers laboratory tests which describe the performance of adsorbents in removing nonemulsified oils and other floating, immiscible liquids from the surface of water.

1.2 The values stated in SI units are to be regarded as the standard.

1.3 *This standard does not purport to address all of the safety concerns, if any, associated with its use. It is the responsibility of the user of this standard to establish appropriate safety and health practices and determine the applicability of regulatory limitations prior to use.* Specific precautionary statements are given in 8.3.1.

2. Referenced Documents

2.1 *ASTM Standards:*

D 1596 Test Method for Dynamic Shock Cushioning Characteristics of Packaging Material[2]

D 2859 Test Method for Flammability of Finished Textile Floor Covering Materials[3]

F 716 Methods of Testing Sorbent Performance of Absorbents[4]

F 1127 Guide for Containment by Emergency Response Personnel of Hazardous Material Spills[4]

2.2 *Federal Standard:*

Fed. Std. No. 141a Paint, Varnish, Lacquer and Related Materials, Methods of Inspection, Sampling and Testing[5]

2.3 *Military Specification:*

MIL-I-631D Insulation, Electric, Synthetic Resin Composition, Nonrigid[5]

3. Terminology

3.1 *General Terminology:*

3.1.1 *sorbent*—an insoluble material or mixture of materials used to recover liquids through the mechanisms of absorption or adsorption, or both.

3.1.2 *universal sorbent*—an insoluble material or mixture of materials that will sorb both hydrophobic and hydrophilic liquid spills.

3.2 *Definitions:*

3.2.1 *absorbent*—a material that picks up and retains a liquid distributed throughout its molecular structure causing the solid to swell (50 % or more). The absorbent is at least 70 % insoluble in excess fluid.

3.2.2 *adsorbent*—an insoluble material that is coated by a liquid on its surface including pores and capillaries without the solid swelling more than 50 % in excess liquid.

3.2.3 *adsorbent cubage factor" C"*—this is the ratio of sorbent volume used to the liquid volume sorbed.

3.2.4 *cubage*—defines cubic content, volume or displacement

3.2.5 *gellant*—a material such as a colloidal network or other aggregate network which pervades and holds a liquid in a highly viscous fragile structure. Many gels may rapidly liquify with added heat or ionic/polar addition. These materials are soluble/flowable in excess liquid.

3.2.6 *thickener*—a material (usually of higher molecular weight) that is soluble in excess liquid. These materials go from dry to gummy (viscoelastic) to flowable and then soluble. The final viscosity depends only on the liquid to solid ratio.

3.3 *Definitions of Terms Specific to This Standard*—This test method does not apply to belt, rope or weir type skimming devices.

3.3.1 *oil*—a substantially water immiscible organic liquids which will float on water (density less than 1 g/cm^3), typically with surface tension less than 40×10^{-3} N/m.

3.3.2 *Type I adsorbent (roll, film, sheet, pad, blanket, web)*—a material with length and width much greater than thickness and which has both linear form and strength sufficient to be handled either saturated or unsaturated.

3.3.3 *Type II adsorbent (loose)*—an unconsolidated, particulate material without sufficient form and strength to be handled except with scoops and similar equipment.

3.3.4 *Type III adsorbent (enclosed)*:

3.3.4.1 *IIIa, pillows*—adsorbent material contained by an outer fabric or netting which has permeability to oil, but with openings sufficiently small so as to substantially retain the sorbent material within the fabric or netting.

---

[1] This test method is under the jurisdiction of ASTM Committee F-20 on Hazardous Substances and Oil Spill Response and is the direct responsibility of Subcommittee F20.22 on Mitigation.

Current edition approved Feb. 10, 1999. Published April 1999. Originally published as F726 – 81. Last previous edition F726 – 81 (1993)$^{\epsilon1}$.

[2] *Annual Book of ASTM Standards*, Vol 15.09.

[3] *Annual Book of ASTM Standards*, Vol 07.01.

[4] *Annual Book of ASTM Standards*, Vol 11.04.

[5] Available from Standardization Documents Order Desk, Bldg. 4 Section D, 700 Robbins Ave., Philadelphia, PA 19111-5094, Attn: NPODS.

3.3.4.2 *IIIb, adsorbent booms*—adsorbent material contained by an outer fabric or netting which has permeability to or is permeable to oil but with openings sufficiently small so as to substantially retain the sorbent material within the fabric or netting. The lengthwise dimension substantially exceeds other dimensions and with strength members running parallel with length. Booms are also provided with connections for coupling adsorbent booms together.

3.3.5 *Type IV-agglomeration unit*—an assemblage of strands, open netting, or other physical forms giving an open structure which minimally impedes the intrusion into itself of high viscosity oils. Normally for use with viscous oils, typically above 10 000 cP viscosity. Said oils are then held in this structure permitting the composite oil/structure to be handled (pompoms).

3.3.6 *reuse*—the art of extracting adsorbed liquids from an adsorbent through rolls or other compression techniques permitting the adsorbent to be used once again; limitations on reuse may include the U.S. Clean Water Act or other legal restrictions.

4. Summary of Test Method

4.1 The adsorbent material is tested using established standard tests for factors relating to storage, while specially developed tests are used for covering other performance factors. Oil and water adsorption strength, buoyancy, and reusability tests are included among these latter tests.

5. Significance and Use

5.1 This test method is to be used as a basis for comparison of adsorbents in a consistent manner.

5.2 These tests are not appropriate for absorbent materials which are covered in Methods F 716.

6. Apparatus

6.1 *Exterior Exposure Tester*, sufficient to be used under Federal Test Standard 141a, Method 6152, or actual exposure as detailed in 8.2.

6.2 *Test Cells*—The dimensions of the test cells shall be large enough to enable the adsorbent sample to float freely within the test cell. For type I and type II sorbents, the recommended test cell is a borosilicate 19 cm (diameter) by 10 cm (depth) crystallizing dish with a watch glass or glass plate cover. For larger samples, a 53 cm × 56 cm plastic sink (laundry tub or equivalent) to accommodate the sample is recommended.

6.3 *Mesh Baskets*—The basket shall be of a sufficient size and strength to accommodate the sample size and weight (150 cm^3, 4 to 10 g minimum) when saturated. The basket must not be so tall as to interfere with a protective lid for the test cell.

NOTE 1—The mesh should retain the sorbent, yet allow free oil to drain away from the sorbent.

6.4 *Shaker Table*, capable of variable frequency from 20 to 240 cycles/min and variable amplitude of 1 to 4 cm.

6.5 *Top Loading Balance*—for Type I and II adsorbent, fitted with a hook or other handing mechanism, 400 + g maximum capacity with 0.1 g resolution, or equivalent.

6.6 *Continuous Reading Hanging Scale*, for Type III and IV adsorbent, fitted with a hook or other handing mechanism, 50+ kg maximum capacity with 100 g resolution, or equivalent.

7. Conditioning

7.1 Condition all adsorbent test specimens at 23 ± 4°C and 70 ± 20 % relative humidity for not less than 24 h prior to testing. Condition specimens is a fully exposed state with no coverings or wrapping that would hinder the ambient equilibration process.

7.2 If temperature conditions other than normal room temperature are expected to be important, then conditioning and testing should be carried out at temperatures of interest in addition to those specified in 7.1.

8. Tests for Storage Properties

8.1 *Storage Density*—The density of the sorbent sample is calculated by determining the weight of a known volume (standard sorbent package as delivered to the consumer) of the sorbent. If standard storage packages are not available, then the mass of at least 1 L of sorbent is determined to calculate the storage density.

8.2 *Mildew*—The susceptibility of an adsorbent to mildew under normal storage is defined under MIL-1-631D, Section 3.5.7. The objective of this test is to determine expected shelf life under conditions which could lead to mildew.

8.3 *Flammability*—The procedure for this test is described in Test Method D 2859, the Methenamine Pill Test. This test relates to ignition from a spark, cigarette, or other point source such as might be encountered in normal shipping and storage. It should not be inferred that an adsorbent which passes this test will fail to burn if ignited in another manner such as full building involvement, bonfire, etc., and it should likewise be understood that the test is limited to and pertains to only unsaturated adsorbent samples as normally supplied by the manufacturer.

8.3.1 *This test method should be used to measure and describe the properties of materials, products, or assemblies in response to heat and flame under controlled laboratory conditions and should not be used to describe or appraise the fire hazard or fire risk of materials, products, assemblies under actual fire conditions. However, results of this test may be used as elements of a fire risk assessment which takes into account all of the factors which are pertinent to an assessment of the fire hazard of a particular end use.*

8.3.2 Test Type I adsorbents as a single sheet or pad as supplied by the vendor.

8.3.3 Uniformly arrange Type II adsorbents in a layer of sufficient size to fit the test apparatus. The thickness should be 1.5 mm uncompressed, and arranged such that the bulk density is equal to that found in the packaged product as supplied by the manufacturer.

8.3.4 Any type adsorbent can be tested in the package in which it is supplied by placing the standard Methenamine test mask on a flat portion of the package surface, and placing the pill in the center of the mask on the package and igniting it. Results would be interpreted as in the normal test.

9. Tests for Performance Properties

9.1 These tests involve the use of oils with a range of viscosities and densities as indicated below.

Oil Type	Viscosity Range	Density Range	Example
Light	1 to 10 cP	0.820 to 0.870 g/cm^3	Diesel fuel
Medium	200 to 400 cP	0.860 to 0.930 g/cm^3	Crude oil
Heavy	1500 to 2500 cP	0.930 to 1.000 g/cm^3	Bunker C or residual fuel
Weathered	8000 to 10 000 cP	0.930 to 1.000 g/cm^3	Emulsified crude oil

9.2 *Dynamic Degradation Test*—This procedure is designed to test for water take-up and to determine oleophilic properties of an adsorbent sample under dynamic conditions. This test is performed at 23 ± 4°C.

9.2.1 *Type I Adsorbent*—Sample pieces of the adsorbent (4 pieces cut with a sharp edge (to minimize compaction) into squares of approximately 6 by 6 cm) are first weighed then placed in a 4 L jar which is half-filled with water and sealed. The container is then placed on its side and mounted on a shaker table, or similar device, set at a frequency of 150 cycles per minute and an amplitude of 3 cm for a duration of 15 min. The contents of the jar are allowed to settle for a period of 2 min. Observations pertaining to the condition of the adsorbent and the condition of the water are recorded. Any adsorbent pieces which do not remain floating at the surface of the water are considered to have failed this test. The contents of the jar are strained through a mesh basket to catch the adsorbent samples, which are then weighed after a 30 s drain period. The water pick-up ratio is calculated from the weight measurements (see 9.5).

9.2.1.1 The jar is half-filled with fresh water and 3 mL of oil (medium crude, 300 cP oil or equivalent) is added. The adsorbent sample is returned to the jar which is then sealed. The jar is placed on its side and mounted on a shaker table, or similar device, set at a frequency of 150 cycles per minute and an amplitude of 3 cm for a duration of 15 min. The contents of the jar are allowed to settle for a period of 2 min, at which time observations are noted. Observations include but are not limited to: quantity of adsorbent submerged, physical appearance of adsorbent and water, and the persistence and color of residual test liquid sheen.

9.2.2 *Type II Adsorbent*—An adsorbent sample (approximately 4 to 10 g or a maximum of 150 cm^3) is first weighed then placed in a 4L jar which is half-filled with water and sealed. The container is then placed on its side and mounted on a shaker table, or similar device, set at a frequency of 150 cycles per minute and an amplitude of 3 cm for a duration of 15 min. The contents of the jar are allowed to settle for a period of 2 min. Observations pertaining to the condition of the adsorbent and the condition of the water are recorded. If 10 % or more of the adsorbent material has sunk then the adsorbent is considered to have failed this test. The contents of the jar are strained through a mesh basket to catch the adsorbent samples, which are then weighed after a 30 s drain period. The water pick-up ratio is calculated from the weight measurements (see 9.5).

9.2.2.1 The jar is half-filled with fresh water and 3 mL of oil (medium crude, 300 cP oil or equivalent) is added. The adsorbent sample is returned to the jar which is then sealed. The jar is placed on its side and mounted on a shaker table, or similar device, set at a frequency of 150 cycles per minute and an amplitude of 3 cm for a duration of 15 min. The contents of the jar are allowed to settle for a period of 2 min, at which time observations are noted. Observations include but are not limited to: quantity of adsorbent submerged, physical appearance of adsorbent and water, and the persistence and color of residual test liquid sheen.

9.2.3 *Types IIIa, IIIb, IV Adsorbents*—Both the outer fabric or netting and the filler material are tested independently for type III adsorbents. Samples are prepared according to the protocol listed in 9.2.1 for the outer fabric or netting of Type III adsorbents and type IV adsorbents, and the protocol listed in 9.2.2 is used for any particulate filler material used in type III adsorbents. If the adsorbent material fails to remain floating as described in 9.2.1 or 9.2.2, then the adsorbent is deemed to have failed the dynamic degradation test.

9.3 *Oil Adsorption-Short Test*—This test gives idealized laboratory data which can be used to compare one adsorbent's oil capacity with another and likewise give relative cost effectiveness. It should be recognized that under normal use conditions, an adsorbent will not be exposed to sufficient oil layer thickness to become completely or rapidly saturated. This test will, therefore, give maximum possible capacity data and idealized time to saturation. The objective of this test is to determine optimum adsorbent without the competing presence of water. As such, this data relates only to oil layer thicknesses which approximate or exceed that of the adsorbent. All adsorption test procedures to be run with adsorbent samples conditioned as in Section 7 and using specified oils at 23 ± 4°C.

9.3.1 *Type I Adsorbent*—The test liquid layer should be of a minimum thickness of 2.5 cm if the thickness of the adsorbent is under 2.5 cm. If the adsorbent is thicker than 2.5 cm, then a liquid layer at least as thick as the adsorbent sample should be used.

9.3.1.1 The adsorbent sample to be tested shall be a minimum weight of 4 g. Cut the sample with a sharp edge (to minimize compaction) to minimum dimensions of 13 by 13 cm square. The adsorbent is then weighed and the value is recorded. The test cell is filled with an initial layer of test liquid. The adsorbent is lowered into the cell. The adsorbent shall be allowed to float freely within the test cell. After 15 min ± 20 s remove the adsorbent in a vertical orientation along an edge with a clip and let drain for 30 ± 3 s (use a 2 min ± 3 s drain time for heavy and weathered oil). Place a tared weighing pan under the adsorbent to catch any additional drips and immediately transfer the adsorbent to the pan. Determine and record the sample weight. All tests are triplicated with the mean of the three runs being used for calculations. The oil pick-up ratio on a weight basis is calculated, as is the volume of oil per adsorbent sample (see 9.5). If the value of any run (g/g) deviates by more than 15 % from the mean of the three runs, then the samples will be rejected and the test repeated with three new specimens.

9.3.2 *Type II Adsorbent*—The test liquid layer should be of a minimum thickness of 2.5 cm if the thickness of the adsorbent sample spread over the area of the test cell is under 2.5 cm. If the adsorbent is thicker than 2.5 cm, then a liquid layer at least as thick as the adsorbent sample should be used.

9.3.2.1 The adsorbent sample to be tested shall be a minimum weight of 4 g. The adsorbent sample is weighed and the value is recorded. The test cell is filled with an initial layer of test liquid. The adsorbent is placed in the basket which is then lowered into the test cell. The adsorbent shall be allowed to float freely within the test cell. After 15 min ± 20 s remove the adsorbent with the basket and let drain for 30 ± 3 s (use a 2 min ± 3 s drain time for heavy and weathered oil). Place a tared weighing pan under the adsorbent to catch any additional drips and immediately transfer the adsorbent to the pan. Determine and record the sample weight. All tests are triplicated with the mean of the three runs being used for calculations. The oil pick-up ratio on a weight basis is calculated, as is the volume of oil per adsorbent sample (see 9.5). If the value of any run (g/g) deviates by more than 15 % from the mean of the three runs, then the samples will be rejected and the test repeated with three new specimens.

9.3.3 *Type III and IV Adsorbents*—The full size adsorbent is weighed and the value is recorded. The test cell is filled with an initial layer of test liquid to a depth at least equal to the thickness of the adsorbent. The adsorbent is lowered into the cell. The adsorbent shall be allowed to float freely within the test cell. After 15 min ± 20 s manually remove the adsorbent in a vertical orientation and let drain for 30 ± 3 s (use a 2 min ± 3 s drain time for heavy and weathered oil). Place a tared weighing pan under the adsorbent sample to catch any additional drips and immediately transfer the adsorbent to the pan. Determine and record the sample weight. All tests are triplicated with the mean of the three runs being used for calculations. The oil pick-up ratio on a weight basis is calculated, as is the volume of oil per adsorbent sample (see 9.5). If the value of any run (g/g) deviates by more than 15 % from the mean of the three runs, then the samples will be rejected and the test repeated with three new specimens.

9.4 *Oil Adsorption-Long Test*—This test gives idealized laboratory data which can be used to compare one adsorbent's oil capacity with another and likewise give relative cost effectiveness. It should be recognized that under normal use conditions, an adsorbent will not be exposed to sufficient oil layer thickness to become completely or rapidly saturated. This test will, therefore, give maximum possible capacity data and idealized time to saturation. The objective of this test is to determine optimum adsorbent capacity without the competing presence of water. As such, this data relates only to oil layer thicknesses which approximate or exceed that of the adsorbent. All adsorption test procedures to be run with adsorbent samples conditioned as in Section 7 and using specified oils at 23 ± 4°C.

9.4.1 *Type I Adsorbent*—The test liquid layer should be of a minimum thickness of 2.5 cm if the thickness of the adsorbent is under 2.5 cm. If the adsorbent is thicker than 2.5 cm, then a liquid layer at least as thick as the adsorbent sample should be used.

9.4.1.1 The adsorbent sample to be tested shall have a minimum weight of 4 g. Cut the sample with a sharp edge (to minimize compaction) to minimum dimensions of 13 by 13 cm square. The adsorbent is then weighed and the value is recorded. The test cell is filled with an initial layer of test liquid. The adsorbent is lowered into the cell. The adsorbent shall be allowed to float freely within the test cell. After 24 h ± 30 min remove the adsorbent in a vertical orientation along an edge with a clip and let drain for 30 ± 3 s (use a 2 min ± 3 s drain time for heavy and weathered oil). Place a tared weighing pan under the adsorbent to catch any additional drips and immediately transfer the adsorbent to the pan. Determine and record the sample weight. All tests are triplicated with the mean of the three runs being used for calculations. The oil pick-up ratio on a weight basis is calculated, as is the volume of oil per adsorbent sample (see 9.5). If the value of any run (g/g) deviates by more than 15 % from the mean of the three runs, then the samples will be rejected and the test repeated with three new specimens.

9.4.2 *Type II Adsorbent*—The test liquid layer should be of a minimum thickness of 2.5 cm if the thickness of the adsorbent sample spread over the area of the test cell is under 2.5 cm. If the adsorbent is thicker than 2.5 cm, then a liquid layer at least as thick as the adsorbent sample should be used.

9.4.2.1 The adsorbent sample to be tested shall have a minimum weight of 4 g. The adsorbent sample is weighed and the value is recorded. The test cell is filled with an initial layer of test liquid. The adsorbent is placed in the basket which is then lowered into the test cell. The adsorbent shall be allowed to float freely within the test cell. After 24 h ± 30 min remove the adsorbent with the basket and let drain for 30 ± 3 s (use a 2 min ± 3 s drain time for heavy and weathered oil). Place a tared weighing pan under the adsorbent to catch any additional drips and immediately transfer the adsorbent to the pan. Determine and record the sample weight. All tests are triplicated with the mean of the three runs being used for calculations. The oil pick-up ratio on a weight basis is calculated, as is the volume of oil per standard adsorbent sample (see 9.5). If the value of any run (g/g) deviates by more than 15 % from the mean of the three runs, then the samples will be rejected and the test repeated with three new specimens.

9.4.3 *Type III and IV Adsorbents*—The full size adsorbent is weighed and the value is recorded. The test cell is filled with an initial layer of test liquid to a depth at least equal to the thickness of the adsorbent. The adsorbent is lowered into the cell. The adsorbent shall be allowed to float freely within the test cell. After 24 h ± 30 min manually remove the adsorbent in a vertical orientation and let drain for 30 ± 3 s (use a 2 min ± 3 s drain time for heavy and weathered oil). Place a tared weighing pan under the adsorbent sample to catch any additional drips and immediately transfer the adsorbent to the pan. Determine and record the sample weight. All tests are triplicated with the mean of the three runs being used for calculations. The oil pick-up ratio on a weight basis is calculated, as is the volume of oil per standard adsorbent sample (see 9.5). If the value of any run (g/g) deviates by more than 15 % from the mean of the three runs, then the samples will be rejected and the test repeated with three new specimens.

9.5 *Calculation*:

9.5.1 Using the data obtained in 9.2, calculate water adsorbency as a ratio of water adsorbed to dry adsorbent weight as follows:

$$\text{water adsorbency} = S_W/S_O \quad (1)$$

where:

S_O = initial dry adsorbent weight,
S_{WT} = weight of adsorbent samples at end of water test, and,
S_W = $(S_{WT} - S_O)$ net water adsorbed.

9.5.1.1 *Example*:

initial sample weight S_O= 8.1 g
weight after water test S_{WT}= 15.3 g
water Adsorbency = S_W/S_O= (15.3 - 8.1)/8.1 - 1.01

Therefore the water adsorbency ratio by weight for this adsorbent is 1.0 to 1 or 1.0 g/g.

9.5.2 Calculate oil adsorbency as the ratio of oil adsorbed to dry adsorbent weight:

$$\text{oil adsorbency}_m = S_S/S_O \quad (2)$$

where:

S_O = initial dry adsorbent weight,
S_{ST} = weight of adsorbent samples at end of oil test, and
S_S = $(S_{ST} - S_O)$ net oil adsorbed.

9.5.2.1 *Example*:

initial sample weight S_O = 9.1 g
weight after oil test S_{WT} = 35.3 g
oil adsorbency = S_S / S_O = (35.3 - 9.1)/9.1 - 2.88

Therefore the oil adsorbency ratio by weight for this adsorbent is 2.9 to 1 or 2.9 g/g.

9.5.3 Calculate oil adsorbency as the volumetric ratio of oil adsorbed to volume of dry adsorbent.

$$\text{oil adsorbency}_v = S_{SV}/S_{OV} \quad (3)$$

where:

S_{SV} = net oil adsorbed (S_S) / oil density, and
S_{OV} = initial dry adsorbent weight (S_O) / sorbent storage density.

9.5.3.1 *Example*:

net oil adsorbed (S_S from 9.4.2) = 35.3 – 9.1 = 26.2 g
oil density (measured during testing) = 0.927 g/cm^3
initial dry adsorbent weight (S_O) = 9.1g
sorbent storage density (measured from sorbent package) = 0.7 g/cm^3
oil adsorbency$_V$= S_{SV}/S_{OV} = (26.2/0.927) / (9.1/0.7) = 2.17 cm^3/cm^3

Therefore the oil adsorbency ratio by volume for this adsorbent is 2.2 to 1 or 2.2 cm^3/cm^3.

9.5.4 Calculate cubage factor "C" as the inverse volumetric ratio of oil adsorbed to volume of dry adsorbent.

$$\text{adsorbent cubage factor "C"} = S_{OV}/S_{SV} \quad (4)$$

9.5.4.1 *Example*:

S_{SV}/S_{OV} (from 9.4.3) was found to be 2.17
S_{OV}/S_{SV}= 1 / S_{SV}/S_{OV}= 1 / 2.17 = 0.46

Therefore the cubage factor "C" for this absorbent is 0.46.

10. Reuse (Type I, Type IIIa Only)[6]

10.1 *Significance and Use*—This test determines the extent to which an adsorbent can be saturated, have the oil extracted, and then repeat this cycle. One point to be used in judging the suitability of an adsorbent for reuse is the number of cycles it can endure without becoming unusable due to tearing, crushing, or other general deterioration. Other factors are the rate of decrease in adsorption capacity and the percentage of oil that can be removed with reasonable effort and equipment. Compression reuse may be tested in two ways using bulk compression as might be found in a squeeze type mop wringer and roller type wringers, and reuse involving centrifugation. Reuse involving solvent washing is regarded as a special uncommon procedure, and as such is not covered here.

10.2 *Compression Extraction*—Use the light medium, and heavy oils indicated in Section 9. Oil removal is effected with an apparatus consisting of an open container with a porous flat open cover above it with 30 to 60 % open space of sufficient strength to withstand a force of 70 kPa on each square centimeter area of adsorbent. Samples are prepared as directed in 9.3.

10.2.1 *Procedure*:

10.2.1.1 Weigh the dry adsorbent sample (S_O) to within ± 2 %, then saturate, drain, and reweigh as in 9.3. Subtract the dry adsorbent weight to obtain total oil adsorbed (O_S). Place the adsorbent sample on the porous cover, and place a stiff plate (wood, metal) of known weight above it. Add and evenly distribute sufficient weight to the top of the plate to give a total weight such that the total weight, when divided by the gross area of the adsorbent in contact with the plate, gives 0.7 kg/cm^2. Extract the adsorbent for 15 ± 2 s after which remove the weights and plate. Place the sample in a freshly tared weighing pan and reweigh to within ± 2 %. Again subtract the dry adsorbent weight to obtain the net oil remaining (O_N).

10.2.1.2 Repeat this procedure four more times giving data over at least five cycles. Record this data by cycles (for example, O_{S1}, O_{S2}, O_{S3}), for oil adsorbed each cycle.

10.2.2 *Calculation*:

10.2.2.1 The total amount of oil the adsorbent is able to hold after each saturation cycle is a measure of the degree of deterioration and shall be reported as the adsorbency ratio by weight and by volume, and as a percentage of the oil adsorbed in the first saturation.

10.2.2.2 Calculate the adsorbency ratio by weight for each cycle based on total oil adsorbed as follows:

$$\text{oil adsorbency}_{Mx} = O_{Sx}/S_{Ox} \quad (5)$$

where

S_{Ox} = initial adsorbent weight at beginning of cycle "x",
O_{STx} = weight of adsorbent samples at end of cycle "x", and
O_{Sx} = $(S_{Ox} - O_{STx})$ net oil adsorbed per cycle.

Report the adsorbency ratio by weight for each cycle.

10.2.2.3 Calculate the adsorbency as the volumetric ratio of oil adsorbed to volume of adsorbent material as follows:

$$\text{oil adsorbency}_{Vx} = O_{SVx}/S_{OVx} \quad (6)$$

where

O_{SVx} = net oil adsorbed (O_{Sx}) / oil density, and
S_{OVx} = initial adsorbent weight at beginning of cycle "x" (S_{Ox}) / adsorbent storage density.

Report the volumetric adsorbency ratio for each cycle.

[6] The reuse of adsorbent material which contains residual hazardous liquids may contravene existing waterway or other regulations such as the U.S. Clean Water Act. The end user is responsible for ensuring that all pertinent regulations are being followed.

10.2.2.4 Calculate the performance degradation as follows:

10.2.2.5 For cycle two, the percentage of first cycle oil capacity is (O_{S2}/O_{S1}). This is repeated similarly for cycle three as (O_{S3}/O_{S1}), and the remainder of cycles to the end point. Report the data for each cycle.

10.2.2.6 Calculate the percentage of oil removed for any given cycle as follows:

$$\text{\% oil removed for run X} = (O_{Sx} - O_{Nx})/O_{Sx} \quad (7)$$

Report the performance degradation as a percentage of initial capacity for each cycle,

10.3 *Wringer Extraction*—Samples are prepared as directed in 9.3 except that the sample shall be cut to 16 by 18 cm in size. This procedure is suitable for light medium and heavy oils only. The roller to be used should have rollers of sufficient length to accommodate the unfolded sample and diameter from 3 to 10 cm. Pressure is applied at the rate of 4 kg/cm of nip while the adsorbent travels through the extractor at a velocity between 5 and 10 cm/s.

10.3.1 *Procedure*—Weigh the dry sample, saturate, drain, and reweigh as in 9.3 and calculate total oil adsorbed (O_s). Run the saturated adsorbent through the nip at a velocity of from 5 to 10 cm/s. Repeat this cycle four times giving data over at least five cycles.

10.3.2 *Calculations*—Perform calculations in accordance with 10.2.2.

10.4 *Centrifuge*—Samples are prepared as directed in 10.2. This procedure is suitable for light, medium and heavy oils only. The centrifuge should have explosion proof enclosures for all electrical components including, but not limited to, the drive motor and electrical control panel(s) and must meet local electrical codes for the liquids used.

10.4.1 *Procedure*—Weigh the dry sample, saturate, drain, and reweigh as in 9.3 and calculate total oil adsorbed (O_s). Run the saturated adsorbent through the centrifuge following manufacturers directions. Repeat this cycle four times giving data over at least five cycles.

10.4.2 *Calculations*—Perform calculations in accordance with 10.2.2.

11. Keywords

11.1 absorbent; adsorbent; adsorbent performance; gellant; oil; sorbent; thickener

Designation: F 818 – 93 (Reapproved 1998)$^{\epsilon 1}$

Standard Terminology Relating to Spill Response Barriers[1]

This standard is issued under the fixed designation F 818; the number immediately following the designation indicates the year of original adoption or, in the case of revision, the year of last revision. A number in parentheses indicates the year of last reapproval. A superscript epsilon (ε) indicates an editorial change since the last revision or reapproval.

ϵ^1 NOTE—Editorial changes were made throughout in November 1998.

1. Scope

1.1 This document defines the terminology used in the field of spill response barriers. Only those terms commonly used or peculiar to this field have been included; no attempt has been made to list all terms used. Where a second term is in common use, "aka" is used to mean "also known as."

1.2 Design, engineering, and performance terms are listed separately: barrier design terminology (3.1), barrier engineering terminology (3.2), and barrier performance terminology (3.3).

2. Referenced Documents

2.1 *ASTM Standards:*

F 625 Practice for Classifying Water Bodies for Spill Control Systems[2]

3. Terminology

3.1 *Barrier Design Terminology*—Terms associated with Spill Response Barrier Design:

General

boom—floating mechanical barrier used to control the movement of substances that float.

boom section—length of boom between two end connectors.

boom segment—repetitive identical portion of the boom section.

Types

air bubble barrier—special-purpose barrier created by rising stream of air bubbles and entrained water, produced by injecting air at some depth below water surface.

bottom-tension boom—boom with tension member located along the bottom of the skirt.

calm water boom—boom intended for use in calm waters (see Practice F 625 for environmental descripters).

"curtain type" boom—boom consisting of a flexible skirt supported by flotation.

"fence type" boom—boom consisting of a self-supporting or stiffened membrane supported by flotation.

fire resistant boom (aka fire containment boom)—boom intended for containment of burning oil slicks.

ice boom—boom intended for use in ice-infested waters, designed to withstand effects of ice contact.

inflatable boom—boom that uses inflated gas-filled chambers as the flotation.

net boom—special purpose boom in which all or part of the membrane material is netting.

open water boom—boom intended for use in open waters (see Practice F 625 for environmental descripters).

permanent boom—boom intended for long-term or permanent deployment.

plunging water jet barrier—special purpose barrier created by a series of coherent streams of water directed vertically downward into a body of water.

protected water boom—boom intended for use in protected waters with moderate environmental conditions (see Practice F 625 for environmental descripters).

river boom (aka fast water boom)—boom intended for use in currents greater than 1 knot.

short seal boom—boom that, when grounded, seals against the shoreline.

silt barrier—boom with very deep skirt used to control the movement of suspended sediments.

sorbent boom—sorbent material contained or arranged in the form of a long cylinder.

special purpose boom—boom that departs from the general characteristics of "fence type" and "curtain type" booms, either in design or intended use.

submersible boom—boom that normally resides on the seabed and is positioned by inflating with air, causing it to rise to the water surface.

water jet barrier—barrier created by stream of pressurized water spray directed across the water surface.

weir boom (aka skimming boom/barrier)—boom that has a weir skimming device(s) built into its face.

Components

accessories—*optional* mechanical devices used on or in conjunction with a boom system but not included with the basic boom and end connector; for example, lights, paravanes,

[1] This terminology is under the jurisdiction of ASTM Committee F-20 on Hazardous Substances and Oil Spill Response and is the direct responsibility of Subcommittee F20.11 on Control. Current edition approved May 15, 1993. Published July 1993. Originally published as F 818 – 84. Last previous edition F 818 – 86.

[2] *Annual Book of ASTM Standards*, Vol 11.04.

drogues, buoys, anchor systems, storage bags, boxes or reels, bulkhead connectors or repair kits, and so forth.

anchor point—structural point on the end connector or along the length of a boom section designed for the attachment of anchor or mooring lines.

ancillary equipment—mechanical devices *essential* to the operation of a given boom system; for example, air pumps, hydraulic power supplies, control manifolds, and so forth.

ballast—weight applied to the skirt to improve boom performance.

bridle—device attached to a boom to distribute the load exerted by towing or anchoring the boom.

buoyancy chamber (aka flotation chamber)—enclosed compartment of air or other buoyant material providing flotation for the boom.

end connector—device permanently attached to the boom used for joining boom sections to one aother or to other accessory devices.

external flotation (aka outboard flotation)—flotation element located external to the boom membrane.

external tension—external tension member separated from the boom membrane by bridles.

fin—portion of the boom membrane above the float.

float—separable component of a boom that provides buoyancy.

flotation—portion of a boom that provides buoyancy.

handhold—any strap, handle, depression, or other provision for grasping the boom by hand.

hinge—location between boom segments at which the boom can be folded back 180° upon itself.

internal flotation—flotation element located within the boom membrane.

lifting point—structural point on the end connector or along the length of a boom section designed for the attachment of a lifting device, such as a crane.

membrane—continuous portion of a boom that serves as a barrier to the movement of a substance.

pressure inflated—inflatable boom that requires pressurized gas for its flotation.

self-inflating—boom that automatically inflates as it is deployed.

skirt—continuous portion of the boom below the floats.

solid flotation—boom that uses solid buoyant material for the flotation element.

stiffener—component that provides support to the membrane.

tension member—any component that carries horizontal (axial) tension loads imposed upon the boom.

Characteristics

boom weight—dry weight of a fully assembled boom section including end connectors.

compactibility—measure of a boom's storage volume per unit length (m^3/m).

draft—minimum vertical depth of the membrane below the water line.

freeboard—minimum vertical height of the boom above the waterline.

height—sum of draft and freeboard.

maximum draft—maximum vertical dimension of the boom below the water line.

overall height—maximum vertical dimension of boom.

3.2 *Barrier Engineering Terminology*—Terms associated with Spill Barrier Engineering:

catenary drag force—load imposed on a boom, deployed in a catenary configuration, resulting from towing, current, and/or wind forces.

current response—change in freeboard or draft due to current forces acting to displace the boom from rest.

gross buoyancy—weight of fresh water displaced by a boom totally submerged.

gross buoyancy to weight ratio—gross buoyancy divided by boom weight.

heave response—ability of the boom to react to the vertical motion of the water surface.

maximum dynamic load—sum of all instantaneous dynamic loads including those due to acceleration, wave forces, and so forth.

reserve buoyancy—gross buoyancy minus boom weight.

reserve buoyancy to weight ratio—reserve buoyancy divided by boom weight.

roll response—rotation of the boom from rest due to wave, wind, or current forces.

straight line drag forces—load on a boom that results from towing it from one end.

wind response—change in freeboard or draft due to wind force acting to displace the boom from rest.

3.3 *Barrier Performance Terminology*—Terms associated with Spill Response Barrier Performance:

apex (aka pocket)—pocket formed at the downstream end of a U, V, J, or W shaped configuration.

boom planing—heeling over of a boom and loss of draft.

boom submergence (aka submarining)—containment failure due to loss of freeboard.

bridging failure—portions of a boom emerging from the water due to poor wave conformance, with resulting containment failure.

catenary configuration (aka "U," "J" configuration)—booming configuration formed by towing or anchoring each end of a length of boom, resulting in a characteristic" U" or "J" shape.

cascading booms—booming configuration formed by positioning two or more booms in a deflection mode such that successive booms progressively move oil to the desired area.

chevron configuration—booming configuration used in narrow watercourses, formed by positioning two lengths of boom in a deflection mode, the leading end of each length is positioned in the middle of the watercourse and the trailing ends lead to opposite shores.

conformance—ability of a boom to maintain freeboard and draft when subjected to a given set of environmental conditions.

containment mode—placement of a boom to prevent free movement of a floating substance.

deployment—placing a boom in the water and making it operational.

diversion mode—placement of a boom to redirect the movement of a floating substance.
drainage loss—oil accumulating and pooling against the boom skirt and escaping with the flow of water down and along the skirt.
entrainment loss—oil droplets escaping with the flow of water diverted under the skirt.
exclusion booming—placement of a boom to protect an area from the entry of a floating substance.
first-loss tow/current velocity—minimum tow/current velocity normal to the membrane at which oil escapes past a boom.
gap ratio—sweep width divided by boom length.
loss rate—rate at which oil is lost past a boom (m^3/h).
performance—ability of a boom to contain or deflect oil under a given set of environmental conditions.
retrieval—removing a deployed boom from the water.
splashover—oil splashing over a boom's freeboard.
stability—resistance to overturning moment.
sweep width (aka swath)—width intercepted by a boom in collection mode, the projected distance between the ends of a boom deployed in a "U," "V," or "J" configuration.
sweeping mode—movement of a boom relative to the water for the purpose of controlling or collecting a floating substance.
towing—transporting a boom from one place to another by pulling from one end.
vortex loss—oil escaping past a boom due to drainage vortices produced at the boom.
"J" configuration—boom positioned in a "J" shape.
"U" configuration (aka catenary configuration)—boom positioned in a "U" shape.
"V" configuration—boom positioned in a "V" shape.
"W" configuration (aka "3" configuration)—boom positioned in a "W" shape.

Designation: F 873 – 84 (Reapproved 1996)

Standard Guide for Incinerating Oil Spill Wastes at Temporary Field Locations[1]

This standard is issued under the fixed designation F 873; the number immediately following the designation indicates the year of original adoption or, in the case of revision, the year of last revision. A number in parentheses indicates the year of last reapproval. A superscript epsilon (ε) indicates an editorial change since the last revision or reapproval.

1. Scope

1.1 This guide addresses incineration as a means of waste disposal and specifically discusses incinerating oil spill wastes at temporary field locations.

1.2 The purpose of this guide is to provide the user with general information on incineration as a means of disposal, not to define a rigid set of standards. It is intended as a reference to plan or execute disposal by incineration.

1.3 This guide outlines procedures and describes some equipment that can be applied to a land-based field incineration process. Included in the guide is a description of typical oil spill wastes that can be incinerated; an outline of procedures to select, prepare, operate, and restore a temporary site; and a summary of general site safety considerations.

1.4 *This standard does not purport to address all of the safety concerns, if any, associated with its use. It is the responsibility of the user of this standard to establish appropriate safety and health practices and determine the applicability of regulatory limitations prior to use.* Specific precautionary information is given in Section.7

2. Terminology

2.1 *Definitions of Terms Specific to This Standard:*

2.1.1 *field*—a location designed during an oil spill cleanup for incinerating oil spill waste material.

2.1.2 *incineration*—controlled burning of waste products or other combustible material.

2.1.3 *incinerator*—a device constructed for the purpose of containing a material for thermal oxidization.

2.1.4 *landfarming*—a controlled method of spreading a known amount of oil in a nominally uniform layer thickness onto a designated land area for the purpose of biological decomposition. This decomposition process is accelerated by mixing the oil layer with the top few inches of soil, aerating the soil by occasional plowing and adding fertilizers that include nitrogen and potassium to increase the oil decomposition rate.

2.1.5 *landfill*—a land disposal technique that uses excavated pits to contain the oil spill waste material. The waste is placed in the excavation, covered over, and left to degrade.

2.1.6 *oil spill*—An uncontrolled discharge of petroleum materials into the environment.

2.1.7 *on-scene coordinator*—the person in charge of oil spill countermeasures.

2.1.8 *open burning*—the process of burning a material without the aid of an incinerator.

2.1.9 *portable incinerator*—a device capable of being transported by some practical means to a field location for the purpose of burning combustible waste products.

2.1.10 *recycling*—any disposal method that uses oil spill waste material in some manner other than returning it to a marketed product. Examples are road oiling and direct use as fuel supplement.

2.1.11 *relative viscosity*—the measured viscosity of an emulsion (in any convenient unit) divided by the measured viscosity of the oil.

2.1.12 *reprocessing*—reclaimed spilled oil by some type of treatment technique that returns the oil into a product that can be sold.

3. Significance and Use

3.1 A series of five basic operations are generally required to control an oil spill. These operations are: containment, recovery, transfer, storage, and disposal. A deficiency in any one of these operations can cause the entire cleanup activity to slow down or even stop. For example, recovery operations must be terminated when temporary storage fills to capacity and skimmers are no longer able to offload picked-up oil. In order to minimize any adverse effect that the disposal process might have on the other four operations, proper planning and management of the waste disposal operation is essential.

3.2 One way to avoid a slowdown due to the disposal process is to select a disposal site near the cleanup activity. A convenient site reduces the time necessary to transport the picked-up material from storage to disposal, thus increasing the efficiency of the overall operation.

3.3 Unfortunately all oil spill cleanup locations may not be convenient to established oil waste disposal areas. In remote areas travel time to refineries, landfarms, or approved landfills may be hours or even days away from the spill. Therefore, it may be advantageous to establish a temporary field disposal site near the cleanup operation.

3.4 An effective disposal method to use at temporary field locations is incineration. Incinerating oily waste does not require a large land surface area as does landfarming. Also, with an incineration process there is no need for long term subsurface monitoring as with the landfill method of disposal.

---

[1] This guide is under the jurisdiction of ASTM Committee F-20 on Hazardous Substances and Oil Spill Responseand is the direct responsibility of Subcommittee F20.14on Disposal.

Current edition approved Feb. 28, 1984. Published May 1984.

3.5 The incineration process can serve two functions. First, it can permanently destroy the oily waste to prevent recontamination by the oil, and second, it can reduce the volume of recovered materials that must be transported to other disposal facilities.

3.6 This guide may be applied where deemed suitable by the user. Because of the emergency nature of most spill cleanup situations, it is recognized that these guidelines cannot always be applied in every detail. Accepted industry practice is to do a safe and environmentally acceptable job with the available resources.

3.7 This guide is intended to be used by persons generally familiar with the practical aspects of oil spill cleanup operations. It is intended as a guide for the on-scene coordinator or personnel managing the spill waste disposal phase.

4. Classification of Oil Spill Waste Materials

4.1 The ability of an incinerator to burn oil and oily debris successfully depends on the nature of the recovered material. Of particular importance is the quantity of water associated with the recovered oil and debris, the size distribution of the debris, and the percentage of combustible, organic material in the mixture. This section describes the types of materials that can be incinerated in the field.

4.2 Recovered materials that can be incinerated are divided into several general categories based on their physical state, size, combustibility, transfer requirements, and components. Table 1 summarizes types of oil spill waste into the following categories.

4.2.1 *Liquids*—Oil recovered from a spill usually has water associated with it. The percentage of water collected with the oil varies widely. The water can be present with the oil either as free water or mixed into the oil to form a water-in-oil emulsion. It is often necessary to decant the free water from the oil prior to incineration. In order to decrease the water content in the emulsion, chemicals or heat can be added to break the emulsion into free water and oil. After separation of the free water, the remaining liquids will be uncontaminated oil and emulsions. Both are pumpable by conventional means, contain little or no debris, and have relatively high heating values. According to Table 1, liquids are designated as uncontaminated oil and emulsion.

TABLE 1 Types of Oil Spill Waste

Waste Material	Physical State		
Categories	Liquid	Semi-solid	Solid
Oils			
Uncontaminated (Crudes, refined products)	X	X	
Emulsions (Crudes, some refined products)	X	X	
Weathered (Tarballs, burned oil residue)		X	X
Oily Debris			
Small combustible (Sticks, cups, sorbent pads)		X	X
Large combustible (Logs, seaweed mats, pilings)			X
Small noncombustible (Sand, rocks, metal, soil)		X	X

4.2.1.1 *Uncontaminated Oil*—Liquid, uncontaminated oil is pumpable by conventional means and contains little or no debris. The major component is crude oil or a refined product such as fuel oil, diesel oil, or lubricating oil. These materials normally have high heating values and burn efficiently. Although relatively uncontaminated, this category of liquid may contain small amounts (for example, up to about 5 %) of other non-hydrocarbon.

4.2.1.2 *Emulsions*—Liquid emulsions are pumpable by conventional means and contain little or no debris. The major components are oil, either crude or certain refined products, and water. Generally, the more water in a water-in-oil emulsion, the higher the viscosity. For example, the relative viscosity of a water-in-oil emulsion with 50 % water is 10, whereas the viscosity with 80 % water is 10 000. Viscosity can affect the type of transfer system, incinerator system and pretreatment required to lower the heating value.

4.2.2 *Semi-solids*—Semi-solid waste material may result from oil weathering, debris mixing into oil, or low temperature. Most oil spill cleanup operations must dispose of semi-solid materials. The major components of semi-solids can be either oil or debris. For example, uncontaminated No. 6 fuel oil can become a semi-solid at low temperatures depending on its pour point. Oil-saturated pine needles or straw can also be considered semi-solid type materials. Semi-solids without treatment can generally be transferred by some types of positive displacement pumps, conveyor systems, vacuum systems, or screw augers. They can also be pre-treated prior to incinerating. For example, debris can be separated from the oil or the oil can be heated to make it a liquid in order to be pumped by conventional means. According to Table 1, semi-solids are designated by five categories as follows: uncontaminated oil, emulsions, weathered oil, small combustible debris, and small noncombustible debris.

4.2.2.1 *Uncontaminated Oil*—Semi-solid uncontaminated oil can occur naturally or can be formed as a result of low temperatures. Some crude oils exist in nature as semi-solid materials due to paraffinic constituents and must be heated in order to be easily transferred. In the Arctic, temperatures are low enough to cause some crude oils to become semi-solids.

4.2.2.2 *Emulsions*—Semi-solid emulsions can solidify at low temperatures when the water phase begins to freeze. The greater the water phase the more solid the material becomes.

4.2.2.3 *Weathered Oil*—Semi-solid weathered oils are highly viscous materials composed mainly of oil with components C_{10} and greater. Weathered oil is formed in many ways including exposure to sunlight, heat and long-term mixing in water. Exposure to these conditions causes a reduction in the light ends of the oil and a resultant increase in the viscosity. Tarballs are an example of highly-weathered oil.

4.2.2.4 *Small Combustible Debris*—Semi-solid small combustible debris refers to a mixture of oil, either liquid or semi-solid, and debris. The debris in the mixture is not bulky and any one piece can easily be picked up in one hand. Examples of such debris are oil-coated sorbent pads, leaves and wood cuttings.

4.2.2.5 *Small Noncombustible Debris*—Semi-solid non-combustible debris refers to those oily debris mixtures that contain components that will not burn at the same temperatures as the oil. Oils containing soils and metal may be included in this category.

4.2.3 *Solids*—Solid waste material recovered at oil spills exhibits no fluid characteristics. The major component of the material is the solid matter and not the oil. Solids range in size from oil-coated sand grains to logs. Solids are normally transferred with material handling equipment and cannot be pumped by conventional means. Large oil-coated materials that cannot be handled by one person are in this category and usually require power equipment, such as cranes or conveyor systems to move them. Sometimes it is easier to incinerate large solids by first reducing their size by cutting, shredding or grinding. Solid materials in this guide are designated in Table 1 by the following categories: weathered oil, small combustible debris, large combustible debris, and small non-combustible debris.

4.2.3.1 *Weathered Oil*—Solid weathered oils are found mainly on shorelines and are formed as a result of long-time exposure to sunlight, water and, often, mixing with debris. Normally weathered solid materials are not harmful to the environment but are a nuisance on recreational beaches.

4.2.3.2 *Small Combustible Debris*—Solid small, combustible debris is material that can be picked up and moved by one person. Oil-coated sticks, paper cups, and lightly-oiled sorbent pads are examples.

4.2.3.3 *Large Combustible Debris*—Solid large, combustible debris is material too large to be moved by one person. Normally these materials must be handled by cranes, trucks, bulldozers, etc. These materials are normally reduced in size by cutting, shredding or grinding if they are to be incinerated. Size reduction not only makes it easier to handle the material, but also improves the combustibility of the material.

4.2.3.4 *Small Non-Combustible Debris*—Solid small, non-combustible debris includes oil-coated soils and metals which do not exhibit fluid characteristics. These are normally handled by front-end loaders, shovels, trucks, etc. depending on the amount of material. Incinerating these wastes usually results in residues that must eventually be disposed of following the incineration process.

5. Incineration Equipment

5.1 A variety of incinerators are used to incinerate oil and oily debris in the field. The incinerator used depends on several factors, including the location of the oil spill and incinerator site, the material to be disposed of, size of the oil cleanup operation, and government regulations. Any incinerator system is acceptable to use as long as it is designed to burn oil waste material, is safe to operate in the field, meets the requirements of officials, and is approved (if necessary) to use.

5.2 Some incinerator systems require auxiliary equipment to operate. This equipment includes feed systems for loading the waste, water for cooling the flame, and fuel to ignite the waste or heat the incinerator chamber. Other systems are self-contained and do not require auxiliary equipment.

5.3 *Categories of Incinerators*—Incinerator systems used to burn oil spill waste products can be placed into one of the following categories: factory assembled, user assembled, or user fabricated.

5.3.1 *Factory Assembled*—Factory assembled incinerators are complete commercially available devices. They require no assembly or fabrication by user in order to implement their use in the field.

5.3.2 *User Assembled*—User assembled incinerators are made from commercially available component parts; for example, a fan, a motor, and a refractory lined chamber. They require assembly of the components by the user, but parts do not need to be fabricated.

5.3.3 *User Fabricated*—User fabricated incinerators are those not commercially available that require design and fabrication. These units are generally one of a kind systems.

5.3.4 These guidelines will only describe factory assembled and user assembled systems. User fabricated systems are useful in the field, but are too numerous to include in these guidelines.

5.4 *Requirements of a Field Incinerator*—In order for an incinerator to be a practical tool for field incineration, it must satisfy the following general requirements.

5.4.1 *Burns Oily Materials*—The incinerator must be a proven system, capable of burning oil liquids or semi-solids, or both.

5.4.2 *Transportable*—The incinerator shall be transportable to the oil cleanup location by air, land, or water vehicles.

5.4.3 *Safe*—The incinerator shall be a device proven safe to use, and when operated correctly will not endanger persons, equipment, or the environment.

5.4.4 *Auxiliary Equipment*—The incinerator system shall include all the equipment necessary to incinerate the waste materials, or the auxiliary equipment shall be standard equipment readily available from suppliers. Examples of auxiliary equipment include pumps, supplemental fuel, safety shields, loading systems, portable tankage (bladders or truck tankers), and electrical generator systems.

5.4.5 *Field Repairable*—The incinerator shall be designed so that minor repairs can be made at the temporary incineration site.

5.5 *Types of Field Incinerators*—There are three types of incinerator systems that are most commonly used for on-site incineration. They are commercially available either as complete units or can be assembled from component parts. The three types of systems are the rotary kiln, the air curtain incinerator, and the open flame oil burner.

5.5.1 *Rotary Kiln*—The rotary kiln is an incinerator that uses a rotating inclined cylinder as a combustion chamber to process and contain the waste material. It is designed to tumble the material being fed into it to provide agitation for complete combustion. Residence time in the incinerator is controlled by the combusion chamber angle, speed of rotation, and length. Most of these units are designed to burn fairly smoke-free and some are supplemented with air and fuel to further improve combustion and reduce emissions. The rotary kiln is used mainly to burn oily debris including sand and gravel, but can be used to incinerate liquids. Small portable units typically burn oil off of sand at rates of 1 ton of waste per hour. Fig. 1 is a schematic of a rotary kiln.

5.5.2 *Air-Curtain Incinerator*—The air-curtain incinerator

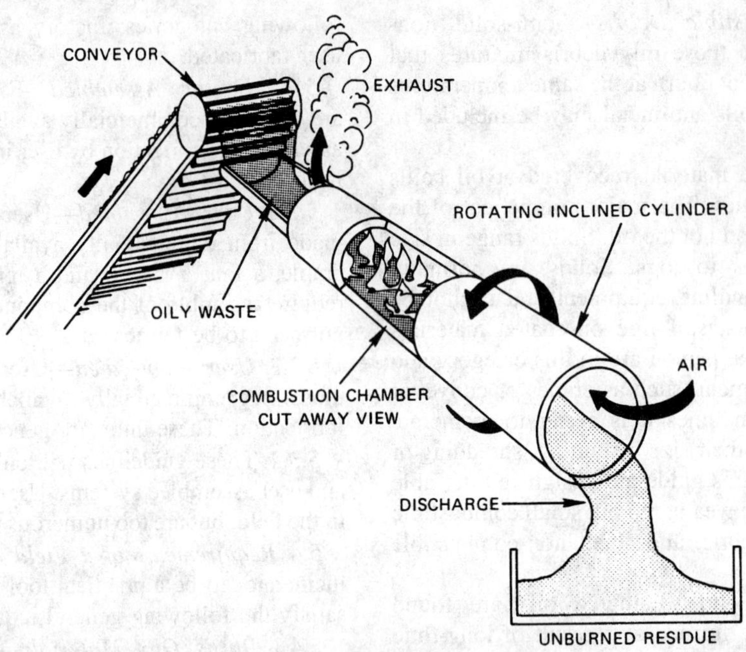

FIG. 1 Rotary Kiln

is a device that uses a high volume air blower, an air plenum, an air distribution nozzle, and an open-top combustion chamber. Fig. 2 illustrates an air-curtain incinerator burning oily waste. In this case, a hole dug into the ground is the open-top combustion chamber. In areas where the soil is not consolidated or is frozen, prefabricated, above-ground chambers can be used.

5.5.2.1 Air is forced through the plenum and out the air nozzle located at the top edge of an open-top chamber that contains the waste material. The air blown into the pit or chamber provides excess oxygen and recirculates unburned particulates back into the combustion zone.

5.5.2.2 These devices are available in a variety of sizes. Burning rates for air curtain incinerators vary according to the size of the chamber and material being burned. Actual field measurements indicate that a 10-ft long chamber can burn oily debris at rates up to several tons per hour.

5.5.3 *Open Flame Oil Burners*—Open flame oil burners are devices commonly used to incinerate smokelessly oil produced from well test operations on offshore platforms. They have also

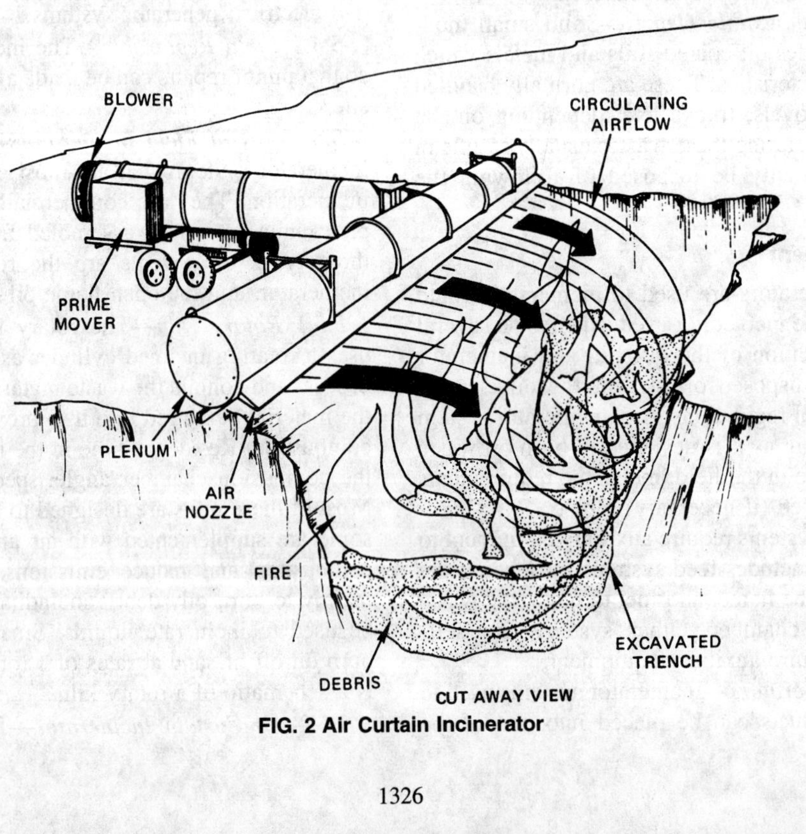

FIG. 2 Air Curtain Incinerator

been used onshore for well testing, and for disposal of waste oil products. These systems routinely burn oil-in-water emulsions. Some are designed to burn viscous emulsions at rates exceeding 15 000 gal/h. These types of burners generally cannot tolerate debris. However, each system is different and the manufacturer of the burner can be consulted on the debris size limit.

5.5.3.1 The size, weight, and components of each open-flame burner system vary, but they are all designed to be portable so they can be easily transported, even into remote locations. A typical system is shown in Fig. 3. It consists of a pedestal and boom to support the burner and flow lines, a set of fuel and air nozzles to atomize the fuel (waste oil), and a water ring with nozzles for cooling the area around the flame and suppressing the smoke. High pressure triplex pumps are normally used to feed the burner. In cases when the waste fuel is very viscous, preheating may be necessary in order to obtain proper atomization and combustion of the oil.

5.5.4 Table 2 summarizes the materials that each type of incinerator can burn.

TABLE 2 Types of Oil Spill Waste and Incinerators to Use

Waste Material	Physical State		
Categories	Liquid	Semi-solid	Solid
Oils			
Uncontaminated (Crudes, refined products)	2, 3	1, 2	
Emulsions (Crudes, some refined products)	1, 2, 3	1, 2	
Weathered (Tarballs, burned oil residue)		1, 2	1, 2
Oily Debris			
Small combustible (Sticks, cups, sorbent pads)		1, 2	1, 2
Large combustible (Logs, seaweed mats, pilings)			1, 2
Small non-combustible (Sand, rocks, metal, soil)		1, 2	1

Field Incinerator Types

1. Rotary kiln
2. Air curtain incinerator
3. Open flame oil burner

6. Field Incineration Process

6.1 In general, a field incineration process consists of four major steps: planning the operation, preparing the site, operating the site, and restoring the site. The complexity of each step will be influenced by many factors, including the spill size, the type of cleanup operation, equipment available, terrain, weather, and regulations governing the area. The following paragraphs discuss these steps and describe some of the typical tasks which need to be performed during the incineration process. Fig. 4 summarizes the sequence of steps and tasks to be performed.

6.2 *Planning the Operation*:

6.2.1 The first step of the incineration process is to plan the operation. Enough information must be collected so that the equipment and locations can be selected and approvals can be granted from regulating authorities.

6.2.2 Types of information that need to be documented include the following:

6.2.2.1 Spill size,

6.2.2.2 Type of oil,

6.2.2.3 Expected materials dispose,

6.2.2.4 Estimate of the recovery rate,

6.2.2.5 Location of the recovery area,

6.2.2.6 General equipment requirements,

6.2.2.7 Availability of equipment,

6.2.2.8 Closest landfill, for ash disposal (if applicable), and

6.2.2.9 Proximity to habitation for animals, especially endangered species.

6.2.3 Based on the initial information and by using Table 2, the type of incinerator to use can be selected. It is important to estimate the disposal rate for waste so that the proper number of incinerators can be located and moved to the site.

6.2.4 The proper selection of a temporary incineration site is an important step in the disposal phase of a cleanup. A site

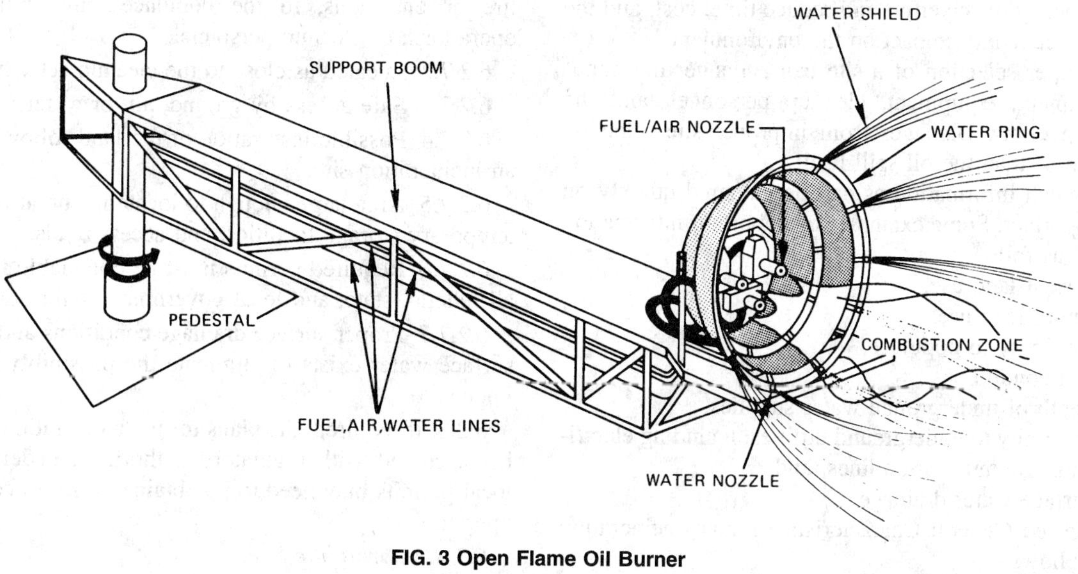

FIG. 3 Open Flame Oil Burner

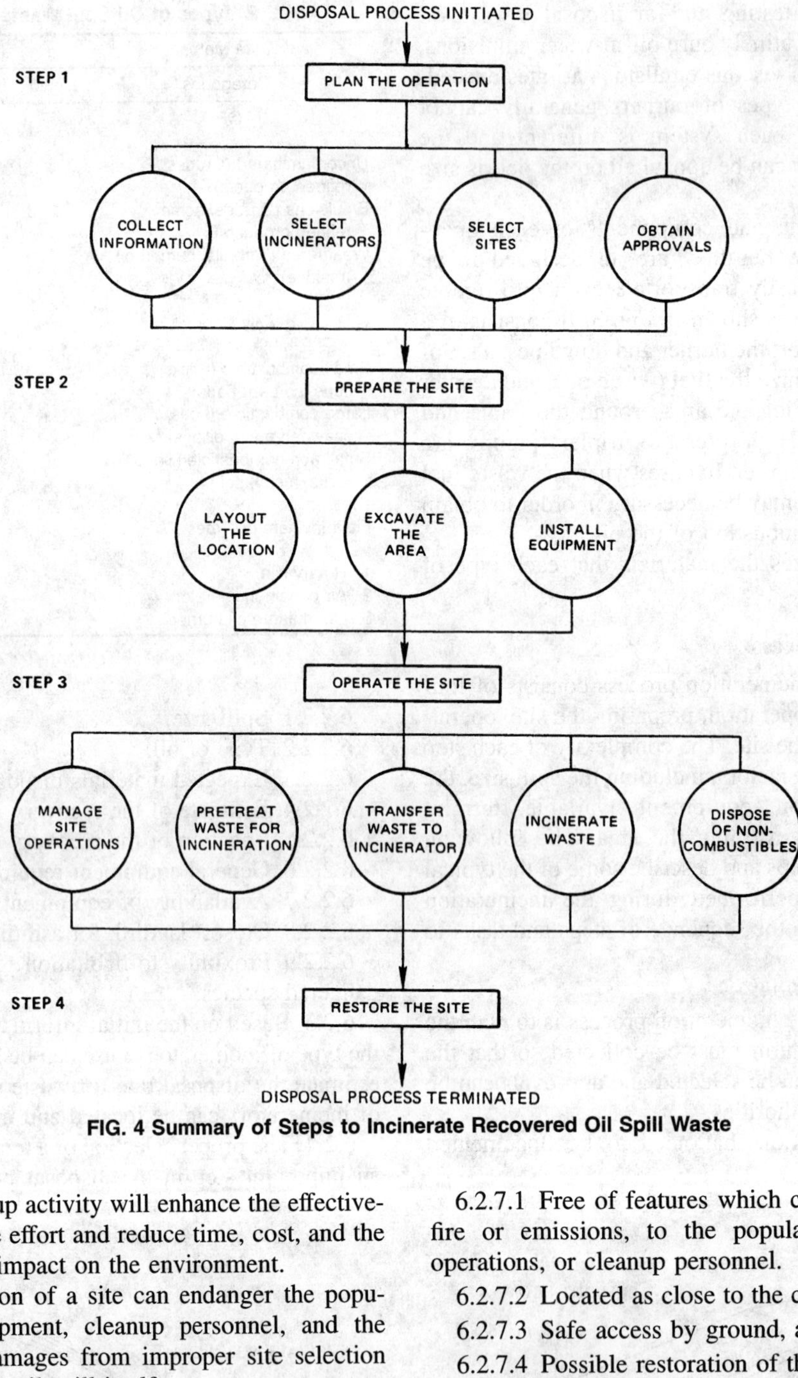

FIG. 4 Summary of Steps to Incinerate Recovered Oil Spill Waste

located nearest the cleanup activity will enhance the effectiveness and efficiency of the effort and reduce time, cost, and the possibility of secondary impact on the environment.

6.2.5 Improper selection of a site can endanger the populace, environment, equipment, cleanup personnel, and the entire cleanup effort. Damages from improper site selection can be far worse than the oil spill itself.

6.2.6 As much information as can be gathered quickly on the site is important. Some examples of important information to collect are as follows:

6.2.6.1 Present land use,

6.2.6.2 Future land use,

6.2.6.3 Proximity to populace,

6.2.6.4 Soil conditions,

6.2.6.5 Depth of underground water supplies,

6.2.6.6 Proximity to underground utilities, including electrical, petroleum, communication lines, and

6.2.6.7 Surface water drainage.

6.2.7 Preferred General Characteristics for an Incineration Site are as follows:

6.2.7.1 Free of features which create a hazard, either from fire or emissions, to the populace, environment, cleanup operations, or cleanup personnel.

6.2.7.2 Located as close to the cleanup activity as possible.

6.2.7.3 Safe access by ground, air, or water transportation.

6.2.7.4 Possible restoration of the land following its use as an incineration site.

6.2.7.5 Area large enough to locate incineration equipment, temporary storage facilities, and access roads.

6.2.7.6 Required permits must be obtainable from either or all federal, state, and local government authorities.

6.2.7.7 Proper surface drainage conditions and proximity to surface water exists to eliminate the possibility of secondary impact.

6.2.8 If required, the plans for the incineration process must be discussed with regulatory authorities. Federal, state, and local permits may need to be obtained prior to conducting the operation.

6.3 *Preparing the Site*:

6.3.1 There are three primary tasks to perform when preparing the incineration site. First, an area layout must be planned. Second, if needed, the site will be excavated. Third, the required equipment will need to be set up on the location.

6.3.2 A temporary incineration site should be sized according to the type of incinerator, number of incinerator units, and the amount of debris and liquid expected to be incinerated. The area layout must make the location convenient and easy to use with safe access to the storage and incinerator areas.

6.3.3 Enough land should be available to stockpile the waste, to provide road access into and out of the site, and to locate the auxiliary equipment.

6.3.4 Roads, if necessary, need to be constructed. They shall be able to support the weight of trucks delivering the waste and vehicles used around the site.

6.3.5 Excavation of the site may include building gravel pads for the incinerator, digging large storage pits, or removing vegetation immediately surrounding the incinerator.

6.3.6 Components of a disposal site should include some of the following items:

6.3.6.1 *Major Equipment*:

(1) Incinerator,

(2) Lined storage facilities, and

(3) Portable tanks for liquids.

6.3.6.2 *Auxiliary Equipment*:

(1) Pumps,

(2) Hoses, pipes, valves, fittings,

(3) Fire fighting equipment,

(4) Shredder or grinders, and

(5) Generator and lights.

6.3.6.3 *Material Handling Equipment*:

(1) Front-end loader,

(2) Conveyor, and

(3) Bulldozer

6.3.7 An example of a small incineration site is shown in Fig. 5. In this case the storage and disposal area is at the spill recovery location. When the cleanup activity in the immediate vicinity is completed, the disposal operation can be easily moved to another cleanup area.

6.3.8 For some spills, a large disposal area, central to many, scattered recovery locations can be more practical. An example of a large, centralized disposal operation is shown in an aerial view in Fig. 6. In this case, more equipment and storage are available for the disposal operation. Two incinerators, one for liquids and one for solids, are shown. Storage pits have been excavated for storage of the waste material.

6.4 *Operating the Site*:

6.4.1 A number of functions are normally performed during the incineration process. These include managing the overall operation, pretreating the waste for incineration, transferring the waste, incinerating, and disposing of the noncombustible residue.

6.4.2 Managing the operation will involve a variety of activities. Some of these activities are: monitoring the incoming waste to account for the volume, type of material, and its disposition, supervising the labor, controlling traffic, and enforcing safety and security measures.

6.4.3 Pretreating the waste may be necessary prior to incineration. In some cases it may be necessary to separate materials, mix materials, reduce viscosity or reduce the size of solids. The amount of pretreating required will depend on the types of equipment at the incineration site. Extensive pretreating should be avoided if possible. In order to enhance the efficiency of the operation, simple treating is preferred. Some examples of pretreating are as follows:

6.4.3.1 Separate oil from water (separator).

6.4.3.2 Separate debris from oil (filter or strainer).

6.4.3.3 Mix nonviscous oil with viscous emulsion (dilution).

6.4.3.4 Add chemicals to emulsions to reduce viscosity (breaking emulsions).

6.4.3.5 Shred debris into small pieces (shredding).

6.4.4 Throughout the disposal process, materials will need to be transferred. Typical transfer operations will include:

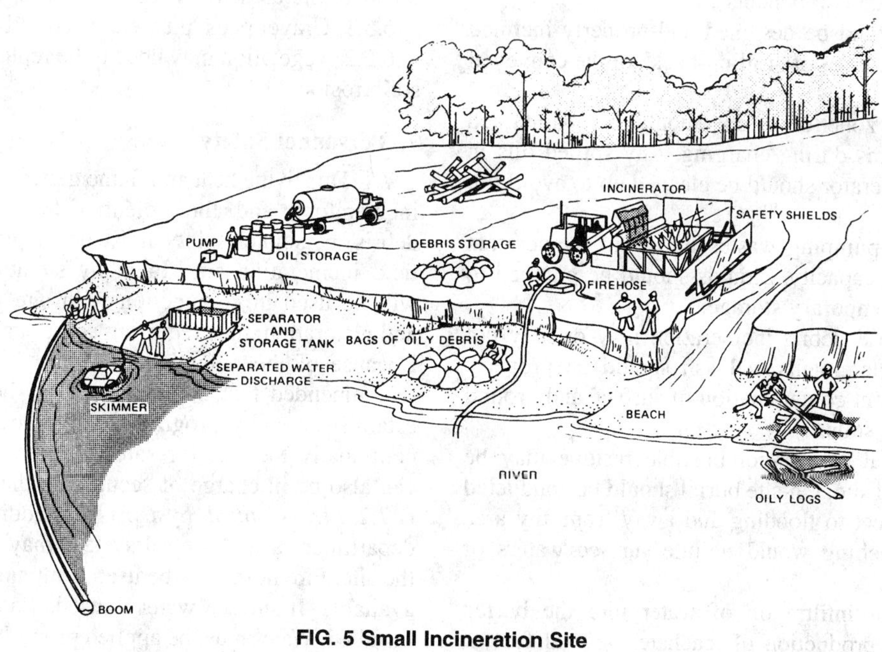

FIG. 5 Small Incineration Site

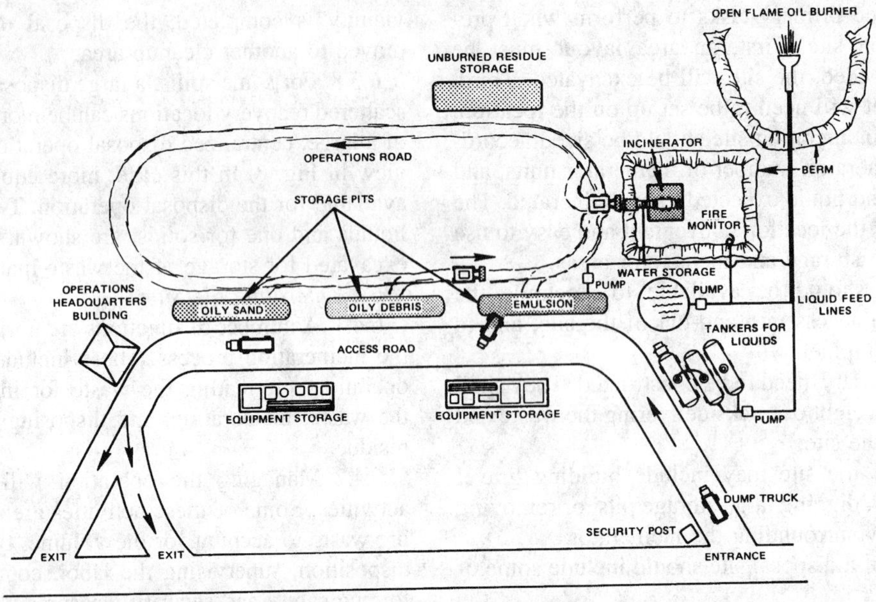

FIG. 6 Large Incineration Site

delivery from vehicles to storage, from storage to preheating equipment, and from storage to the incinerator. These operations will require the use of pumps, both centrifugal and positive displacement, conveyors for debris, and heavy lift equipment for bulky items.

6.4.5 The method of charging spill debris or oils into an incinerator depends on the debris characteristics, type of incinerator, and rate of processing. The selected method should be compatible with the equipment used, and should be spill free and safe.

6.4.6 All charging methods should be under direct control of the operator(s). Equipment should be equipped with flow rate control and immediate shutoff capabilities.

6.4.7 The charging system should be properly designed and monitored to avoid incinerator clogging and "flame-back" into the charging and storage components.

6.4.8 Conveyors should be designed and properly inclined to avoid spillage off the sides or "rolling-back" of the conveyed materials.

6.4.9 An adequate zone of safety should be maintained around open pit systems during charging. Any spilled oils in the vicinity of an incinerator should be cleaned up to avoid fire hazards.

6.4.10 The rate of pumping wastes into fluid incinerators should not exceed the capacity. Liquids should be pumped to the incinerator from temporary storage.

6.4.11 Ash from spill debris incineration may or may not contain leachable contaminants. All ash should be properly disposed so as to prevent contamination of surface and ground water and agricultural soils.

6.4.12 On-site burial of non-combustible residue may be possible at the disposal site. On-site burial should be conducted away from areas subject to flooding and away from any area where subsequent leaching would pollute surface waters or ground water supplies.

6.4.13 Reduction of infiltration of water into the buried mass will reduce the production of leachate. Several feet of impermeable soil or clay materials should be placed over the buried ash and graded so as to prevent ponding.

6.4.14 Ash from properly incinerated oil spill debris will consist primarily of carbonized residues and unburned inert material. Following the incineration process these wastes should be properly disposed of in a landfill.

6.4.15 Under no circumstances should smoldering or still burning ash be placed in municipal waste landfills or transport containers.

6.4.16 Inert solids such as sand and soils may be redeposited at the spill site if the incineration process has sufficiently removed all leachable oils and no hazardous residues are present. Ash incorporation in native sands or soils should be conducted so as to minimize topographical and visual impacts.

6.5 *Restoring the Site*—Once the incineration process is finished, the area will need to be restored.

6.5.1 Gravel pads, pits, etc. need to be removed or filled in.

6.5.2 Vegetation may need to be replanted in order to reduce soil erosion.

7. Personnel Safety

7.1 Due to the heat and flame generated by most field-usable incinerators, and the potential flammability of oil-soaked debris, personnel safety must be of primary importance at a field incineration site. In many situations, heavy equipment will be used in conjunction with site preparation, operation, and restoration, thus requiring additional safety considerations. Because of the potential for injury in these situations, it is recommended that a "safety officer" be designated who will establish a safety program at the site and who will monitor continually the site for safety-related problems. This person can also be in charge of security in the area.

7.2 *Fire Control Systems*—In addition to the local fire department, a fire control system may need to be installed at the site. Fire hoses can be used if municipal or surface water is available. If surface water is used, water pumps must also be installed. Water must be applied properly to extinguish oil fires.

It is extremely useful for cooling purposes and for extinguishing oil fires ignited by any hot ashes which have escaped the confines of the combustion chamber. Three or four 1½ /in. fire hoses equipped with all-purpose nozzles and appropriate pumping systems or water supplies should be available for each incineration unit in operation. In addition, handheld water extinguishers should be available to control any small flareups.

7.2.1 If water or pumping capability is unavailable, fire extinguishers with carbon dioxide, water, or dry chemical should be used. Consideration should be given to establishing a foam fire-fighting system since foam is particularly effective for oil-based fires. This system may consist of foam-type extinguishers, foam eductor-type systems, or both. Sufficient foam should be available to extinguish the fire in the combustion chamber of the incinerator being used.

7.3 *Personnel Protective Equipment*—Any operations personnel who expect to approach within 100 ft of the incinerator while it is in use should wear fire-resistant clothing, footwear, and gloves. It is also recommended that hardhats be worn by all personnel during assembly, operation, and disassembly of the incinerator. Eye protection is also recommended to protect against hot flying ash. Near the incinerator there should be self-contained breathing apparatus available for rescue of injured personnel.

7.4 *Protection of the General Populace*, An area around the incinerator should be established which is marked off-limits to those not actively involved in the disposal operation.

7.5 *Noise Protection*—Persons working near the incinerator should be issued industrial-type ear plugs for protection from the noise associated with the incinerator.

7.6 *General Safety Procedures*—Safety procedures should be clearly understood by all persons in the disposal operation. Everyone should adhere to the enforcement of the safety procedures. It is recommended that operators be informed of the materials being incinerated during any working period to understand the possible hazard of each different type of material. Conducting scheduled safety meetings is an excellent means of preventing accidents at the disposal site.

Standard Guide for Ecological Considerations for the Use of Chemical Dispersants in Oil Spill Response—Marine Mammals[1]

This standard is issued under the fixed designation F 929; the number immediately following the designation indicates the year of original adoption or, in the case of revision, the year of last revision. A number in parentheses indicates the year of last reapproval. A superscript epsilon (ϵ) indicates an editorial change since the last revision or reapproval.

1. Scope

1.1 This guide covers recommendations for use of chemical dispersants to assist in the control of oil spills. This guide is written with the goal of minimizing the environmental impacts of oil spills; this goal is the basis upon which recommendations are made. Aesthetic and socioeconomic factors are not considered; although, these and other factors are often important in spill response.

1.2 Each on-scene coordinator has available several means of control or cleanup of spilled oil. In this guide, use of chemical dispersants is not to be considered as a "last resort" after other methods have failed. Chemical dispersants are to be given equal consideration with other spill countermeasures.

1.3 This is a general guide only, assuming the oil to be dispersable and the dispersant to be effective, available, applied correctly, and in compliance with relevant government regulations. Oil, as used in this guide, includes crude oils and fuel oils (No. 1 through No. 6). Differences between individual dispersants or between different oils or products are not considered.

1.4 This guide covers one type of habitat, marine mammals. Other guides, similar to this one, cover habitats such as sandy beaches or marshes. The use of dispersants is considered primarily to protect such habitats from impact (or minimize impacts) and also to clean them after the spill takes place.

1.5 This guide applies to marine and estuarine environments, but not to freshwater environments.

1.6 In making dispersant-use decisions, appropriate government authorities should be consulted as required by law.

1.7 *This standard does not purport to address all of the safety concerns, if any, associated with its use. It is the responsibility of the user of this standard to establish appropriate safety and health practices and determine the applicability of regulatory limitations prior to use.*

2. Significance and Use

2.1 This guide is meant to aid local and regional spill response teams during response planning and spill events.

3. General Considerations for Making Dispersant-Use Decisions

3.1 The decision of whether to use or not to use dispersants in a given spill situation is always one involving trade-offs. Dispersing a slick at one site temporarily introduces more oil into the water column at that site than would be there if a surface slick floated over it. Therefore, adverse effects on water column organisms may be increased at the site so that adverse effects can be decreased or eliminated at other sites.

3.2 Dispersant use is primarily a spill control method, not a cleanup method. Such use can give spill response personnel some control over where the impacts of a spill will occur and what types of impacts they may be. Since some environments are known to be more vulnerable to the longer lasting impacts of spilled oil, an acceptable trade-off may be to protect those environments by dispersing an oil slick in a less sensitive or less productive environment. In general, the trade-off that must be evaluated is between the impact of the relatively long residence time of spilled oil which strands on shorelines versus the short-term impact of dispersed oil in the water column.

4. Environment Covered—Marine Mammal Habitats

4.1 Many different environments are used as habitats by marine mammals. Cetaceans (porpoises, whales) are seen in open ocean and coastal waters. Pinnipeds (seals, sea lions) feed offshore but use certain beaches to rest (haul-out) or to breed (rookery). Sea otters are associated with nearshore kelp beds. Polar bears swim in open leads and migrate over the arctic pack ice.

4.2 Some marine mammals use the same environments daily or seasonally for these activities. This guide applies primarily to those coastal habitats used by marine mammals on a regular basis.

5. Background

5.1 The effects of oil, including dispersed oil, on marine mammals are not well known. There have been no substantiated reports of oil fouling of cetacean skin **(1)**.[2] The skin of most cetaceans species is smooth, but there are some species with rough skin; examples are the gray whale, bowhead whale, and right whale. Species with roughened areas of the body

[1] This guide is under the jurisdiction of ASTM Committee F-20 on Hazardous Substances and Oil Spill Responseand is the direct responsibility of Subcommittee F 20.13on Treatment.

Current edition approved June 28, 1985. Published March 1986.

[2] The boldface numbers in parentheses refer to the list of references at the end of this guide.

surface would probably be more susceptible to oil adherence than smooth-skinned species (**1**). Evidence is also lacking that oil clogs breathing passages (**1**). Oil contamination temporarily reduces the filtering efficiency of baleen, in laboratory experiments. Baleen contamination could promote some oil ingestion, as well (**1**). Observations suggest that several cetacean species do not actively avoid oil slicks (**1**).

5.2 Animals that depend wholly or partially on their pelts for insulation (sea otters, polar bears, fur seals) could suffer from exposure if oil-contaminated (**2,3,4**). During an experiment, oil-contaminated polar bears died from the toxic effects of oil ingested during grooming (**5**). Some dead river otters examined after an oil spill at Sullom Voe (Shetland) also died from ingesting oil during grooming (**2**).

5.3 Geraci and Smith (**6**) found that oil caused reversible eye damage in healthy ringed seals. Animals already under stress died after exposure to oil. Oil fed to harp seal pups caused no significant physiological or behavioral changes.

5.4 Engelhardt (**7**) reported oil uptake in ring seals via the skin or respiratory system, or both, and via ingestion and short-term accumulation in the blubber.

5.5 Davis and Anderson (**8**) reported that growth and mortalities of oiled and unoiled gray seal pups did not differ significantly following a spill on Skomer Island. Cowell (**9**) reported that during the breeding season both males and females swam through oil to reach rookery beaches and became heavily oiled. They did not appear to avoid oil.

5.6 Le Boeuf (**10**) found no differences in mortalities of oiled and unoiled elephant seal pups during the Santa Barbara oil spill. He speculated that oil effects would be most pronounced during the pupping season when pups could not only become oil coated but could also ingest oil via contaminated mothers.

5.7 The effects of dispersed oil may differ somewhat in that oil droplets would be less likely to stick to fur.

5.8 Of equal importance with the effects of oil on pinnipeds are the effects of human activity and disturbance, an important consideration in spill response. The National Marine Fisheries Service has conducted tagging and capture operations on the Channel Islands haul-out and rookery beaches off southern California. De Long (**11**) reported that after a single disturbance animals stampede into the sea and do not return to that beach for from one to several days. This pattern was seen in all of the five species found on these islands. If the disturbance occurs during the breeding season when males are holding territories, the social structure of the herd can be changed by a single disturbance, that is, males would have to reestablish those territories.

5.9 Davis and Anderson (**8**) concluded that disturbance was more harmful than oil to gray seals. Pups that were handled during the spill on Skomer Island showed slower growth and increased mortalities over undisturbed animals.

5.10 Kenyon (**12**) found that monk seals were very sensitive to disturbance by humans. Pups born on Midway Island were repeatedly disturbed by people and dogs. Disturbance interrupted nursing and resulted in slower weight gain compared to undisturbed populations. Eventually the seals disappeared from Midway Island and could only be found on five atolls not subjected to human activity.

6. Recommendations

6.1 Pinniped, sea otter, and polar bear habitats are identified as high-priority habitats to receive protection during oil spills. Use of chemical dispersants and mechanical methods is recommended to prevent these habitats from being contaminated or to reduce contamination. Ideally, spill control activities (for example, dispersant application, skimming, aircraft) should take place as far away from these habitats as is necessary to prevent disturbance. All activities creating disturbance should be kept to a minimum in the vicinity of marine mammals since they are very sensitive to such disturbance. The" safe" distance may be variable with species. If pinniped rookery (breeding) beaches are contaminated, no cleaning of the beaches or of animals is recommended.

6.2 Because sea otters and polar bears are very sensitive to oil contamination, dispersant use is recommended even if application must occur near or in a habitat used by these animals. Using dispersants to reduce the amount of oil contamination/or the number of animals contaminated, or both, is recommended.

6.3 If a pinniped haul-out (resting) beach is contaminated, cleaning could be considered on advice from qualified marine mammalogists, since there may be important species-specific considerations.

7. Keywords

7.1 chemical dispersants; ecological considerations; marine mammals; oil spill response

REFERENCES

(1) Engelhardt, F. R., "Petroleum Effects on Marine Mammals," *Aquatic Toxicolology*, Vol 4, 1983, pp. 199–217.

(2) Baker, J. R., et al.," Otter *Lutra lutra* L. Mortality and Marine Oil Pollution," *Biology Conservationist*, Vol 20, No. 4, 1981, pp. 311–321.

(3) Costa, D. P., and Kooyman G. L., "Effects of oil contamination in the sea otter, *Enhydra lutris*," Final Report, Res. Unit 71, *Environmental Assessment of the Alaskan Continental Shelf*, Vol 10, Biological Studies, 1981, pp. 65–87.

(4) Kooyman, G. L., et al., "Thermal Conductance of Immersed Pinniped and Sea Otter Pelts Before and After Oiling With Prudhoe Bay Crude," *Proceedings—Fate and Effects of Petroleum Hydrocarbons in Marine Ecosystems and Organisms*, Pergamon Press, 1979, pp. 151–157.

(5) Engelhardt, F. R., "Oil Pollution in Polar Bears: Exposure and Clinical Effects," *Proceedings—Fourth Arctic Marine Oil Spill Program Technical Seminar*, Edmonton, Alberta, 1981, pp. 139–179.

(6) Geraci, J. R., and Smith T. G., "Direct and indirect effects of oil on ringed seals (*Phoca hispida*) of the Beaufort Sea." *Journal Fishery Reserve Board Canada*, Vol 33, 1976, pp. 1976–1984.

(7) Engelhardt, F. R., "Petroleum hydrocarbons in arctic ringed seals, *Phoca hispida*, following experimental oil exposure," *Proceedings—Conference on Assessment of Ecological Impacts of Oil Spills*, AIBS publication, 1978, pp. 613–628.

(8) Davis, J. E., and Anderson S. S., "Effects of Oil Pollution on Breeding Grey Seals," *Marine Pollution Bulletin*, Vol 6, 1976, pp. 115–118.

(9) Cowell, E. B.," Oil Spills and Seals,"*Society Petroleum Ind. Biologist's Newsletter*, Vol. 1, 1979, pp. 3–4.

(10) Le Boeuf, B. J., "Oil Contamination and Elephant Seal Mortality: a "Negative" Finding, In D. Straughan (ed.) Oceanographical and Biological Survey of the Santa Barbara Oil Spill, *I. Biology and Bacteriology*, Allan Hancock Fdn, publisher, 1971, pp. 277–285.

(11) De Long, R. L., "San Miguel Island Management Planning Report," prepared for Marine Mammal Commission, 1979, 38 pp.

(12) Kenyon, K. W., "Man Versus the Monk Seal,"*Journal of the Mammal*, Vol 53 No. 4, 1972, pp. 687–696.

The American Society for Testing and Materials takes no position respecting the validity of any patent rights asserted in connection with any item mentioned in this standard. Users of this standard are expressly advised that determination of the validity of any such patent rights, and the risk of infringement of such rights, are entirely their own responsibility.

This standard is subject to revision at any time by the responsible technical committee and must be reviewed every five years and if not revised, either reapproved or withdrawn. Your comments are invited either for revision of this standard or for additional standards and should be addressed to ASTM Headquarters. Your comments will receive careful consideration at a meeting of the responsible technical committee, which you may attend. If you feel that your comments have not received a fair hearing you should make your views known to the ASTM Committee on Standards, 100 Barr Harbor Drive, West Conshohocken, PA 19428.

Designation: F 930 – 85 (Reapproved 1993)[ε1]

Standard Guide for Ecological Considerations for the Use of Chemical Dispersants in Oil Spill Response—Rocky Shores[1]

This standard is issued under the fixed designation F 930; the number immediately following the designation indicates the year of original adoption or, in the case of revision, the year of last revision. A number in parentheses indicates the year of last reapproval. A superscript epsilon (ε) indicates an editorial change since the last revision or reapproval.

ε[1] NOTE—Section 7 was added editorially in December 1992.

1. Scope

1.1 This guide covers recommendations for use of chemical dispersants to assist in the control of oil spills. This guide is written with the goal of minimizing the environmental impacts of oil spills; this goal is the basis upon which recommendations are made. Aesthetic and socioeconomic factors are not considered; although, these and other factors are often important in spill response.

1.2 Each on-scene coordinator has available several means of control or cleanup of spilled oil. In this guide, use of chemical dispersants is not to be considered as a "last resort" after other methods have failed. Chemical dispersants are to be given equal considerations with other spill countermeasures.

1.3 This is a general guide only, assuming the oil to be dispersable and the dispersant to be effective, available, applied correctly, and in compliance with relevant government regulations. Oil, as used in this guide, includes crude oils and fuel oils (No. 1 through No. 6). Differences between individual dispersants or between different oils or products are not considered.

1.4 This guide covers one type of habitat, rocky shores. Other guides, similar to this one, cover habitats such as sandy beaches or marshes. The use of dispersants is considered primarily to protect such habitats from impact (or minimize impacts) and also to clean them after the spill takes place.

1.5 This guide applies to marine and estuarine environments, but not to freshwater environments.

1.6 In making dispersant-use decisions appropriate government authorities should be consulted as required by law.

1.7 *This standard does not purport to address all of the safety problems, if any, associated with its use. It is the responsibility of the user of this standard to establish appropriate safety and health practices and determine the applicability of regulatory limitations prior to use.*

2. Significance and Use

2.1 This guide is meant to aid local and regional spill response teams during response planning and spill events.

3. General Considerations for Making Dispersant-Use Decisions

3.1 The decision of whether to use or not to use dispersants in a given spill situation is always one involving trade-offs. Dispersing a slick at one site temporarily introduces more oil into the water column at that site than would be there if a surface slick floated over it. Therefore, adverse effects on water column organisms may be increased at the site so that adverse effects can be decreased or eliminated at other sites.

3.2 Dispersant use is primarily a spill control method, not a cleanup method. Such use can give spill response personnel some control over where the impacts of a spill will occur and what types of impacts they may be. Since some environments are known to be more vulnerable to the longer lasting impacts of spilled oil, an acceptable trade-off may be to protect those environments by dispersing an oil slick in a less sensitive or less productive environment. In general, the trade-off that must be evaluated is between the impact of the relatively long residence time of spilled oil which strands on shorelines versus the short-term impact of dispersed oil in the water column.

4. Environments Covered—Rocky Shores

4.1 The habitat designated "rocky shore" is characterized by bedrock surface or by a conglomeration of large, stable (immobile) boulders occurring within the intertidal zone. The surface of the rock may be smooth or it can have uneven topography. The slope of the shore can range from vertical rock wall to a gently sloping or nearly flat platform.

4.2 The nature of the entire intertidal environment (including extent and type of biological zonation) is controlled primarily by the amount of wave energy to which that shore is exposed together with shoreline topography and slope. Shores range from the "extremely exposed" (routinely exposed to very high wave energy) to "extremely sheltered" (protected by natural or artificial barriers from any significant wave action).

4.3 The biotic component of most intertidal rocky shores can usually be divided into three main zones: *(1) the supralittoral, supralittoral fringe, (2)* littoral, sublittoral fringe, and *(3)* sublittoral. The extent and vertical distribution of these zones are defined principally by the tidal range and wave action characteristic of the area.

4.3.1 The uppermost zone, the supralittoral, coincides with

[1] This guide is under the jurisdiction of ASTM Committee F-20 on Hazardous Substances and Oil Spill Response and is the direct responsibility of Subcommittee F 20.13 on Treatment.

Current edition approved June 28, 1985. Published March 1986.

the lowest extent of terrestrial vegetation, and is affected by water only during high winds or storm conditions. The lower edge of this zone, the supralittoral fringe, can be thickly covered by blue-green algae and marine lichens. In addition, littorinid snails and isopods may occur in varying numbers. It is this latter zone which is often referred to as the "black zone" **(1-4)**.[2]

4.3.2 The second major intertidal area is the littoral zone and it is subjected to regular (tidal) periods of immersion. This zone is usually subdivided into subzones, with each subzone populated by biota capable of tolerating the degree of exposure or submersion characterizing that subzone. A wide variety of species of microscopic and macroscopic algae as well as both sessile and motile invertebrates occupy this dull tan, yellow, or reddish zone.

4.3.3 The sublittoral fringe (or infralittoral fringe) is located at the lowest edge of the littoral zone and is exposed to the air only at the lowest spring tides. This zone and the sublittoral zone are both characterized by a deeper reddish-brown color of the dominant species of algae present. While invertebrates of the sublittoral fringe tend to be intertidal or subtidal in nature the sublittoral species are primarily benthic or subtidal **(1-7)**.

4.4 The wide variety of plants and animals that exist in the intertidal rocky shore environment provide a rich food source for browsers and grazers. Tide pools, in particular, support very large numbers of individuals and numbers of species of intertidal animals and serve as nursery grounds for several important species of invertebrates.

5. Background

5.1 The high-wave energy associated with vertical or high-sloping rocky shorelines usually prevents significant oil deposition during spills. When deposition of oil has taken place on a high energy shoreline, residence time of the oil on the rocks has been brief and damage has been minimal, due to the rapid removal of oil by the washing action of the waves **(8-10)**.

5.2 As a general rule, both the amount of oil deposited on rocky shores as well as its persistence is a function of the degree of wave energy of that shoreline. Shores with the highest wave energy have the most rapid rate of removal; those with the lowest wave energy, the slowest rate of removal **(8-11)**.

5.3 On rocky shores, the most rapid rate of recovery often occurs when there is no attempt to clean the rocks, other than mechanical removal of the very thick deposits. If cleaning of the rocks becomes necessary, mechanical removal or low toxicity dispersants can be utilized on advice from qualified biologists **(8,12,13)**.

5.4 The short-term biological effects of spills have been fairly well documented for many of the common rocky shore species. The immediate, heavy mortalities resulting from large spills reaching the rocky intertidal area are primarily caused by toxicity, smothering, or mechanical effects **(14-23)**.

5.5 The long-term biological effects of spills on the rocky shore biota are more difficult to evaluate due to natural and seasonal variations of the intertidal biota, lack of prespill baseline data, differences in type and amounts of oil spilled, life stages of impacted organisms, season, and duration of exposure **(8,9,24,25)**.

5.6 In many cases, shore organisms appear to have made good, if not complete, recovery from single-incident exposure to spilled oil in from one to ten years **(12,14,16,17,18,19,20,25)**.

5.7 Since dispersants or other oil spill chemicals that are applied directly to shore make contact with the intertidal biota, effects of a spill combined with chemical cleanup activities on rocky intertidal areas may be extreme. A few studies of long-term effects on areas impacted by weathered oil, oil spills, and dispersant application indicated a number of effects including changes in community structure, altered molluscan growth patterns, size frequency change, inhibition of settlement, and abnormal growth of some species of seaweeds **(11,26)**.

5.8 Long-term effects from spilled oil and excessive use of highly toxic (aromatic solvent based) dispersants were observed after the *Torrey Canyon* spill by Southward and Southward **(27)** who reported altered age distribution for some rocky shore populations, effects were noted for up to ten years. Such toxic dispersants are no longer in use. Modern dispersants are both more effective and less toxic.

6. Recommendations

6.1 Biologically rich, low-energy rocky, intertidal habitats should receive protection in the event of a spill. Dispersant use is recommended to protect such environments. Dispersants should be applied offshore prior to stranding.

6.2 Cleaning will not usually be considered necessary for exposed rocky shores (in particular, vertical rock faces or high-sloping shorelines) due to the rapid rate of natural cleanup by wave energy.

6.3 In general, cleaning of sheltered rocky shores may be difficult and biologically damaging. Therefore, it is recommended that both mechanical and chemical cleaning of this environment be considered only if the area is heavily oiled and low-impact methods are used. Mechanical methods which remove living organisms (for example, scraping and steam cleaning) cause significant adverse effects on intertidal biota.

6.4 If cleaning of rocky shores is deemed necessary, it may be possible to use low pressure seawater flushing or dilute dispersants without significant damage. However, it is recommended that the guidelines provided by the Institute of Petroleum **(28)** "Code of Practice for the Use of Oil Slick Dispersants" be followed carefully. The applicable sections are:

6.4.1 Use dispersant in front of an advancing tide, allowing the sea to accomplish most of the "wash down."

6.4.2 Low pressure, seawater flushing is considered the most environmentally sound method for cleaning fresh oil from all types of rocky shores **(28,29)**.

6.4.3 Barriers and absorbents or other methods to contain and retrieve the emulsified oil should be used where it is necessary to protect the lower beach.

6.5 Dispersants should not be used (nor are they intended for use) in an undiluted state on any shoreline at any time.

[2] The boldface numbers in parentheses refer to the list of references at the end of this guide.

7. Keywords

7.1 chemical dispersants; ecological considerations; oil spill response; rocky shores

REFERENCES

(1) Doty, M. S., "Rocky Intertidal Surfaces," From: "Treatise on Marine Ecology and Paleoecology," (J. W. Hedgepeth, ed.) *GSA Memoir 67*, Vol 1, Chapter 18, 1957 pp. 535–585.

(2) Lewis, J. R., "The Mode of Occurrences of the Universal Intertidal Zones in Great Britain," *Journal of Ecology*, Vol 43 (1), 1955, pp. 270–290.

(3) Stephenson, T. A., and Stephenson, A. "The Universal Features of Zonation Between Tide-Marks on Rocky Coasts," *Journal of Ecology* Vol 37, 1949, pp. 289–305.

(4) Stephenson, T. A., and Stephenson, A. "Life Between Tide-Marks in North America, II, Northern Florida and the Carolinas," *Journal of Ecology*, Vol 40, 1952, pp. 1–49.

(5) Stephenson, T. A., and Stephenson, A., "The Causes of the Vertical and Horizontal Distribution of Organisms Between Tide-Marks in South Africa," *Proceedings Linn. Society London* Vol 154, 1943, pp. 219–290.

(6) Stephenson, T. A., and Stephenson, A., "Life Between Tide-Marks in North America, I, The Florida Keys," *Journal of Ecology*, Vol 38, 1943, pp 354–402.

(7) Womersley, H. B., and Edmonds, S. J., "Marine Coastal Zonation in Southern Australia in Relation to a General Scheme of Classification," *Journal of Ecology*, Vol 40, 1943, pp 84–90.

(8) Cox, G. V., and Cowell, E. B., "Mitigating Oil Spill Damage. Ecologically Responsible Clean-up Techniques," presented at the Mitigating Symposium, Fort Collins, CO July 16–20, 1979.

(9) Gundlach, E. R., and Hayes, M. O., "Vulnerability of Coastal Environments to Oil Spill Impacts," *Marine Technology Society Journal*, Vol 12, No. 4, 1978, pp. 18–27.

(10) Hayes, M. O., et al., "Role of Dynamic Coastal Processors in the Impact and Dispersal of the *Amoco Cadiz* Oil Spill (March 1978), Brittany, France," *Proceedings of the 1979 Oil Spill Conference (Prevention, Behavior, Control, Cleanup)*, March 19–22, 1979, Low Angeles, CA, 1979, pp. 193–200.

(11) Maynard, N. G., "Effects of Petroleum Residues on Intertidal Organisms of Bermuda," *Bermuda Biological Station Special Publication No. 21*, 1982.

(12) Baker, J. M., et al., *Proceedings of the Institute of Petroleum Conference on Petroleum Development and the Environment*, London, Nov. 20–21, 1979.

(13) *The Ecological Effects of Oil Pollution on Littoral Communities*, Cowell, E. B., Institute of Petroleum, London, Elsevier, Amsterdam, 1971, 250 pp.

(14) Baker, J. M. "Refinery Effluent," E. B. Cowell, ed., from *Proceedings, Symposium on the Ecological Effects of Oil Pollution on Littoral Communities*, Institute of Petroleum, 1973, London.

(15) Blumer, M., et al., "A Small Oil Spill," *Marine Biology*, Vol 5, 1971, pp. 195–202.

(16) Chan, Elaine, I., "Oil Pollution and Tropical Littoral Communities: Biological Effects of the 1975 Florida Keys Oil Spill," *Proceedings of the 1977 Oil Spill Conference (Prevention, Behavior, Control, Cleanup)*, March 8–10, 1977. New Orleans, LA.

(17) Crapp, G. B., "Field Experiments With Oil and Emulsifiers," pp. 114–128 from *The Ecological Effects of Oil Pollution on Littoral Communities* E. B. Cowell, ed., Institute of Petroleum, London/Elsevier, Amsterdam, 1971, 250 pp.

(18) Crapp, G. B., "Laboratory experiments with emulsifiers," pp. 129–149 from *The Ecological Effects of Oil Pollution on Littoral Communities* E. B. Cowell, Ed., Institute of Petroleum, London/Elsevier, Amsterdam, 1971, 250 pp.

(19) Crapp, G. B., "The Biological Consequences of Emulsifier Cleaning," pp. 150–169 from *The Ecological Effects of Oil Pollution on Littoral Communities* E. B. Cowell, Ed., Institute of Petroleum, London/Elsevier, Amsterdam, 1971, 250 pp.

(20) Crapp, G. B., "The Effects of Oil Pollution and Emulsifier Cleaning on Littoral Animals and Plants," *Annual Rep. Oil Pollution Reserve Unit*, Orielton, Vol. 1971, pp. 7–13.

(21) Nelson-Smith, A., "The Effects of Oil Pollution and Emulsifier Cleansing on Marine Life in Southwest Britain," *Journal of Applied Ecology*, Vol 5, 1967, pp. 97–107.

(22) Nelson-Smith, A., "Recovery of some British rocky seashores from oil spills and cleanup operations," pp. 191–207 from *Recovery and Restoration of Damaged Ecosystems*, J. Cairns, K. L. Dickson, and E. E. Herricks, Eds., University Press of Virginia, Charlottesville, 1977, 531 pp.

(23) Straughan, D. and Abbott, B. C., "The Santa Barbara Oil Spill: Ecological Changes and Natural Oil Leaks," from "Water Pollution by Oil," P. Hipple, Ed., *Institute of Petroleum*, London, England, 1970.

(24) Jones, L. G., et al., "Just how serious was the Santa Barbara oil spill?" *Ocean Ind* June 1969, pp. 53–56.

(25) National Petroleum Council, "Fate and Effects of Spills," from, Environmental Conservation: The Oil and Gas Industries," *Library of Congress*, Vol 82-81994, 1982, pp. 477–540.

(26) *Oil and Dispersants in Canadian Seas—Research Appraisal and Recommendations.* Sprague, J. B., J. H. Vandermeulen, and P. G. Wells, eds. for the Environmental Emergency Branch, Environmental Impact Control Directorate, Environmental Protection Service, Environment Canada.

(27) Southward, A. J., and Southward, E. C., "Recolonization of Rocky Shores in Cornwall After Use of Toxic Detergents to Clean Up the *Torrey Canyon* Spill," *Journal of Fisheries Reserve Board*, Canada Vol 35, 1978, pp. 682–706.

(28) *Code of Practice for the Use of Oil Slick Dispersants*, Institute of Petroleum, Heyden and Son Ltd., London, England, 1979, 34 pp.

(29) Lindstedt-Siva, J., "Ecological Impacts of Oil Spill Cleanup: Are they significant?" *Proceedings, 1979 Oil Spill Conference*, American Petroleum Institute publ. Vol 4308, 1979, pp. 521–524.

Designation: F 931 – 85 (Reapproved 1999)

Standard Guide for Ecological Considerations for the Use of Chemical Dispersants in Oil Spill Response— Seagrasses[1]

This standard is issued under the fixed designation F 931; the number immediately following the designation indicates the year of original adoption or, in the case of revision, the year of last revision. A number in parentheses indicates the year of last reapproval. A superscript epsilon (ϵ) indicates an editorial change since the last revision or reapproval.

1. Scope

1.1 This guide covers recommendations for use of chemical dispersants to assist in the control of oil spills. This guide is written with the goal of minimizing the environmental impacts of oil spills; this goal is the basis upon which recommendations are made. Aesthetic and socioeconomic factors are not considered; although, these and other factors are often important in spill response.

1.2 Each on-scene coordinator has available several means of control or cleanup of spilled oil. In this guide, use of chemical dispersants is not to be considered as a "last resort" after other methods have failed. Chemical dispersants are to be given equal consideration with other spill countermeasures.

1.3 This is a general guide only, assuming the oil to be dispersable and the dispersant to be effective, available, applied correctly and in compliance with relevant government regulations. Oil, as used in this guide, includes crude oils and fuel oils (No. 1 through No. 6). Differences between individual dispersants or between different oils or products are not considered.

1.4 This guide covers one type of habitat, seagrasses. Other guides, similar to this one, cover habitats such as sandy beaches or marshes. The use of dispersants is considered primarily to protect such habitats from impact (or minimize impacts) and also to clean them after the spill takes place.

1.5 This guide applies to marine and estuarine environments, but not to freshwater environments.

1.6 In making dispersant-use decisions, appropriate government authorities should be consulted as required by law.

1.7 *This standard does not purport to address all of the safety concerns, if any, associated with its use. It is the responsibility of the user of this standard to establish appropriate safety and health practices and determine the applicability of regulatory limitations prior to use.*

2. Significance and Use

2.1 This guide is meant to aid local and regional spill response teams during response planning and spill events.

[1] This guide is under the jurisdiction of ASTM Committee F-20 on Hazardous Substances and Oil Spill Responseand is the direct responsibility of Subcommittee F20.13 on Treatment.

Current edition approved June 28, 1985. Published March 1986.

3. General Considerations for Making Dispersant-Use Decisions

3.1 The decision of whether to use or not to use dispersants in a given spill situation is always one involving trade-offs. Dispersing a slick at one site temporarily introduces more oil into the water column at that site than would be there if a surface slick floated over it. Therefore, adverse effects on water column organisms may be increased at the site so that adverse effects can be decreased or eliminated at other sites.

3.2 Dispersant use is primarily a spill control method, not a cleanup method. Such use can give spill-response personnel some control over where the impacts of a spill will occur and what types of impacts they may be. Since some environments are known to be more vulnerable to the longer lasting impacts of spilled oil, an acceptable tradeoff may be to protect those environments by dispersing an oil slick in a less sensitive or less productive environment. In general, the trade-off that must be evaluated is between the impact of the relatively long residence time of spilled oil which strands on shorelines versus the short-term impact of dispersed oil in the water column.

4. Habitat Covered—Seagrasses

4.1 Seagrasses can be found in shallow marine environments from the tropics to arctic regions. Seagrass beds form a discrete ecosystem that traps detritus derived from terrestrial sources, and then exports large quantities of plant and animal materials, including leaf and root fragments, dissolved organic matter, and detritus, to the open sea. The presence of an extensive network of roots and rhizomes facilitates the sediment-binding ability of the grass beds; the leaves effectively retard currents, thus promoting sedimentation of organic and inorganic materials around the plants.

4.2 Other functions, summarized by Wood et al. **(1)**[2], contribute to seagrass communities being among the most productive of natural ecosystems. Seagrass leaf blades support large numbers of epiphytes which can equal the biomass of the grass itself. Major food chains are based upon a variety of microfauna, including grazing epiphytes and microbes, utilizing the grasses and associated sediments. Active sulfur, nitrogen, phosphorus, and carbon cycles are maintained through the sediments-plant-water interfaces, and the dense, interlacing

[2] The boldface numbers in parentheses refer to the list of references at the end of this guide.

mat of vegetation provides ideal cover for foraging marine fauna as well as shelter and protection for larval and juvenile forms.

5. Background

5.1 Information on the effects of dispersed oil on seagrasses and their associated communities is minimal. Baca and Getter **(2)** have found that the uptake of a crude oil water-soluble fraction by seagrasses is increased by the addition of a chemical dispersant. The dispersed oil was not significantly toxic to *Thalassia testudinum*, however, in laboratory experiments.

5.2 Present information dealing with the effects of oil on seagrasses has been collected following spill incidents. Nadeau and Bergguist **(3)** reported on the effects of a spill of Venezuela crude from the *Zoe Colocotronis* on *Thalassia* communities in Puerto Rico. Oil entrained in the water column by surf action coated in *Thalassia* leaves and killed a large number of the plants. The death of the plants exposed extensive areas of denuded vegetation and rhizome matrix. Within one year after the spill, a renewed *Thalassia* growth was evident. However, four and one half years after the spill, plant biomass and numbers of species in *Thalassia* beds were still reduced in comparison to reference areas **(4)**.

5.3 Significant damage to *Phyllospadix torreyi* growing in the intertidal zone was observed during the Santa Barbara spill **(5)**. Plants in the subtidal and extreme low intertidal, however, were relatively unaffected. Similarly, beds of *Thalassia testudinum* deteriorated over several months following an oil spill in Puerto Rico **(6)**.

5.4 Studies conducted on intertidal *Zostira marina* communities at Roscoff, France, following the "AMOCO CADIZ" spill in March 1978 found only short-term effects **(7)**. Effects on the eelgrass were localized to blackening of some leaves following exposure to the oil. The structures of the grass beds was unchanged although considerable oil was present.

5.5 Effects on communities associated with the eelgrass beds at Roscoff were generally minimal. Epiphytic algae appeared to be unaffected. Some macrofauna showed decreased numbers of both individuals and species, but most species had recovered to their former levels within a few months. However, filter-feeding organisms had disappeared from the system and were still absent over a year after the spill **(7)**.

6. Recommendations

6.1 If there is a possibility that an oil slick will strand on intertidal portions of seagrass beds, dispersant use would be most effective while an oil slick is still offshore to prevent the oil from impacting the grass bed.

6.2 Dispersant-use decisions to treat oil already over a seagrass bed should ultimately take into account the depth of the seagrass beds and the potential for dilution of the dispersed oil.

6.3 The use of dispersants over shallow submerged seagrass beds is generally not recommended but should weigh the potential impacts to the seagrass beds against impacts that might occur from allowing the oil to come ashore.

6.4 Dispersant use should be considered to treat oil over seagrass beds in water deeper than 10 m if the alternative is to allow the oil to impact other sensitive habitats onshore.

6.5 Dispersant use is not recommended in shallow lagoons or areas of restricted flushing rates; mechanical cleanup (if feasible, considering equipment availability, weather, time constraints, and spill size) is preferable. However, dispersant usage should remain an option to protect more sensitive shoreline environments.

6.6 If dispersants are used in the vicinity of shallow seagrass beds, studies should be initiated to observe potential impacts and to expand the data base on seagrass bed responses to dispersed oil exposures.

7. Keywords

7.1 chemical dispersants; ecological considerations; oil spill response; seagrasses

REFERENCES

(1) Wood, E. J. F., et al., "Influence of Sea Grasses on the Productivity of Coastal Lagoons," from *Lagunas Costeras, un Symposio. Mem. Simp. Intern*, Lagunas Costeras, UNAM-UNESCO, 1969.

(2) Baca, B. J., and Getter, C. D., "The Toxicity of Chemically Dispersed Oil on the Seagrass, *Thalassia testudinum*, Using Static and Flowthrough Bioassays," *Dispersants: Five Years of Research, Abstracts*, ASTM, Philadelphia, PA, 1982, pp. 30.

(3) Nadeau, R. J., and Bergguist, E. T.," Effects of the March 18, 1973 Oil Spill Near Cabo Rojo, Puerto Rico on Tropical Marine Communities," pp. 535–538, *Proceedings of the 1977 Oil Spill Conference*, 1977, American Petroleum Institute, Washington, DC.

(4) Lopez, J. M., "Ecological Consequences of Petroleum Spillage in Puerto Rico," pp. 894–903. From *Proceedings of Conference on Assessment of Ecological Impacts of Oil Spills*, 1977, American Petroleum Institute, Washington, DC.

(5) Foster, M., et al., "The Santa Barbara Oil Spill, Part 2: Initial Effects on Intertidal and Kelp Bed Organisms," *Environmental Pollution*, 1971, Vol 2, pp. 115–134.

(6) Diaz-Piferrer, M., "The Effects of Oil on the Shore of Guanica, Puerto Rico," *Deep-Sea Research*, 1971, Vol 11, pp. 855–856.

(7) Jacobs, R. P. W. M.," Ecological Effects of the" AMOCO CADIZ" Oil Spill on the Seagrass System at Roscoff, France," pp. 173–208, from *Component Studies in Seagrass Ecosystems Along West European Coasts*, Druk: Drukkerij Verweij B. V., Mijdrecht, 1982.

Designation: F 932 – 85 (Reapproved 1999)

Standard Guide for Ecological Considerations for the Use of Chemical Dispersants in Oil Spill Response—Coral Reefs[1]

This standard is issued under the fixed designation F 932; the number immediately following the designation indicates the year of original adoption or, in the case of revision, the year of last revision. A number in parentheses indicates the year of last reapproval. A superscript epsilon (ϵ) indicates an editorial change since the last revision or reapproval.

1. Scope

1.1 This guide covers recommendations for use of chemical dispersants to assist in the control of oil spills. This guide is written with the goal of minimizing the environmental impacts of oil spills; this goal is the basis upon which recommendations are made. Aesthetic and socioeconomic factors are not considered; although, these and other factors are often important in spill response.

1.2 Each on-scene coordinator has available several means of control or cleanup of spilled oil. In this guide, use of chemical dispersants is not to be considered as a "last resort" after other methods have failed. Chemical dispersants are to be given equal consideration with other spill countermeasures.

1.3 This is a general guide only, assuming the oil to be dispersable and the dispersant to be effective, available, applied correctly and in compliance with relevant government regulations. Oil, as used in this guide, includes crude oils and fuel oils (No. 1 through No. 6). Differences between individual dispersants or between different oils or products are not considered.

1.4 This guide covers one type of habitat, coral reefs. Other guides, similar to this one, cover habitats such as sandy beaches or marshes. The use of dispersants is considered primarily to protect such habitats from impact (or minimize impacts) and also to clean them after the spill takes place.

1.5 This guide applies to marine and estuarine environments, but not to freshwater environments.

1.6 In making dispersant use decisions, appropriate government authorities should be consulted as required by law.

1.7 *This standard does not purport to address all of the safety concerns, if any, associated with its use. It is the responsibility of the user of this standard to establish appropriate safety and health practices and determine the applicability of regulatory limitations prior to use.*

2. Significance and Use

2.1 This guide is meant to aid local and regional spill response teams during response planning and spill events.

3. General Considerations for Making Dispersant-Use Decisions

3.1 The decision of whether to use or not to use dispersants in a given spill situation is always one involving trade-offs. Dispersing a slick at one site temporarily introduces more oil into the water column at that site than would be there if a surface slick floated over it. Therefore, adverse effects on water column organisms may be increased at the site so that adverse effects can be decreased or eliminated at other sites.

3.2 Dispersant use is primarily a spill control method, not a cleanup method. Such use can give spill response personnel some control over where the impacts of a spill will occur and what types of impacts they may be. Since some environments are known to be more vulnerable to the longer lasting impacts of spilled oil, an acceptable trade-off may be to protect those environments by dispersing an oil slick in a less sensitive or less productive environment. In general, the trade-off that must be evaluated is between the impact of the relatively long residence time of spilled oil which strands on shorelines versus the short-term impact of dispersed oil in the water column.

4. Environment Covered—Coral Reefs

4.1 Coral reefs are structures created and maintained by the establishment and growth of populations of hermatypic corals and coralline algae. They may be comprised of wave-resistant emergent or submerged reefal zones, or a combination of both. Geomorphically, barrier and fringing reefs protect from erosion the insular and continental coastlines which they border. Typically, the reef provides habitat for a large variety of attached plants and epifauna, infauna, mobile invertebrates, and fishes. The large numbers of economically important species they support make reefs locally important in commercial and sport fisheries. The resultant high diversity and abundance of reef associates, and the functional and spatial dominance of corals and coralline algae are the essential characteristics of coral reefs.

4.2 Coral reefs are circumglobal in the tropics and subtropics between the northern and southern hemispheric 18.5°C climatic isotherms. A majority of coral species and the most diverse and prolific reefs occur in Indo-West Pacific seas. Tropical West Atlantic and Eastern Pacific reefs are generally

[1] This guide is under the jurisdiction of ASTM Committee F-20 on Hazardous Substances and Oil Spill Responseand is the direct responsibility of Subcommittee F 20.13on Treatment.

Current edition approved June 28, 1985. Published January 1986.

less diverse in terms of corals and reef associates.

5. Background

5.1 Neither field nor laboratory studies have provided conclusive evidence of the presence or absence of effects of oil on corals under actual spill conditions. The field studies, for the most part, attempted to document only lethal effects. Sublethal effects, which have been demonstrated in the laboratory, are less easily measured in the field. Also, the laboratory studies have generally used oil concentrations that are not normally encountered in the field.

5.2 Field observations have generally detected few effects of oil on corals. No damage to corals was observed following a spill of crude in the Florida Keys **(1)**.[2] Lush growths of corals have been documented in the oil-rich and heavily tanker-trafficked areas of the Arabian Gulf **(2)**. Similar observations were made for Tarut Bay, Saudi Arabia where several refineries are located **(3)**. The corals appeared to be the least affected community following a spill of Bunker C and marine diesel off the coast of Panama **(4)**. Shinn **(2)** has also shown that some corals can withstand up to two hours total immersion in South Louisiana crude.

5.3 Laboratory experiments have documented at least 14 lethal and sublethal responses of corals to oil exposures **(5,6)**. Sublethal responses include zooxanthellae expulsion **(7,8)**; uptake and depuration in mucous **(8)**; decreased calcium uptake and zooxanthellae production **(8-10)** ; premature extrusion of planulae larvae **(11)**; impaired feeding response **(10,12)**; impaired polyp retraction **(2,8,13)**; impaired sediment clearance ability **(9)**; increased mucous production **(14)**; gonal tissue damage **(15,16)**; larval death **(17)**; impaired larval settlement **(17)**; coenosarc tissue damage **(15)**; and muscle atrophy **(15)**.

5.4 Because of their dependence on light, most corals are located within relatively shallow waters. In some areas, the corals can be partially exposed during low tide. Therefore, corals are susceptible to damage from an oil slick by coating during low tides, by sinking of the oil if it adheres to the suspended sediment, or by contact with high concentration of soluble petroleum hydrocarbon fractions. It is possible that oil dispersed over a reef will come in contact with at least some of the corals that approach the surface and where breaking waves can thoroughly mix the water column. The available data on the effects of oil and dispersed oil are limited and at times conflicting. It is possible that some species of corals may be more resistant to oil effects than others. Ths has been shown for other types of stresses (for example, termperature or sedimentation). However, whether it be natural or human-induced, severe stress can cause long-term damage or completely destroy a reef. There is a high degree of interdependence between tropical marine species and the reef ecosystem. Severe impacts can have major implications for much of the tropical marine system.

5.5 Eisler **(18)** examined the acute toxicities of some of the early chemical dispersants to various species of Red Sea fishes and invertebrates, including soft corals. The toxicities of the oil-dispersant mixtures were similar to those of the dispersants used alone.

5.6 Recent studies have shown that effects on corals from dispersant use are nonexistent or of short duration. Knap et al. **(19)** found that corals exposed to dispersed oil concentrations of 20 to 50 ppm for 24 h elicited temporary responses that include inhibition of tentacle expansion behavior, tissue rupture, tissue contraction, septal filament extrusion, and loss of pigmentation. The duration of the effects was variable and ranged from a few days to two weeks. Long-term results gave no indication that skeleton growth was significantly affected by exposure to oil or dispersal oil **(19)**.

5.7 Similar results were obtained in studies where corals were exposed to oil and dispersed oil under actual field conditions **(7)**. In these simulated spill studies, exposure of corals to hydrocarbons was greater in dispersant treated plots than in oiled plots without the dispersant. However, no effects were observed on growth of corals exposed to either oil or dispersed oil for 24 h and measured 1 year after the exposure. Corals exposed for 5 days to the oil and dispersed oil showed reduced growth in comparison to the controls. Exposure to the dispersed oil also appeared to delay recovery of corals also stressed by cold temperatures.

6. Recommendations

6.1 Whenever an oil spill occurs in the general vicinity of a coral reef, the use of dispersants should be considered to prevent floating oil from reaching the reef.

6.2 Dispersant use decisions to treat oil already over a reef should take into account the type and location on the reef.

6.3 Coral reefs with emergent portions are high-priority habitats for protection during oil spills. Dispersant use should be considered to prevent grounding of oil on such reefs by appropriate treatment at least 1.0 km away.

6.4 The use of dispersants over shallow submergent reefs is generally not recommended but should weight the potential impacts to the reef against impacts that might occur from allowing the oil to come ashore.

6.5 Dispersant use should be considered to treat oil over reefs in water depths greater than 10 m if the alternative is to allow the oil to impact other sensitive habitats on shore.

6.6 Dispersant use is not recommended to treat oil already in reef habitats having low water exchange rates (for example, lagoons, atolls) if mechanical cleanup methods are possible.

7. Keywords

7.1 chemical dispersants; coral; coral reefs; ecological considerations; oil spill response; reefs

[2] The boldface numbers in parentheses refer to the list of references at the end of this guide.

REFERENCES

(1) Chan, E. I., "Oil Pollution and Tropical Littoral Communities: Biological Effects of the 1975 Florida Keys Oil Spill," *Proceedings of the 1977 Oil Spill Conference*, American Petroleum Institute, Washington, DC, 1977.

(2) Shinn, E. A., "Coral Reef Recovery in Florida and the Persian Gulf," Shell Oil Co., Houston, TX, 1972.

(3) Spooner, M. F., "Oil Spill in Tarut Bay, Saudi Arabia," *March Pollution Bulletin,* Vol 1, pp. 166–167, 1970.

(4) Rutzler, K., and Sterrer, W., "Oil Pollution Damage Observed in Tropical Communities Along the Atlantic Seaboard of Panama," *BioScience*, Vol 20, pp. 222–224, 1970.

(5) Fucik, K. W., et al., *Measurements of Damage, Recovery, and Rehabilitation of Coral Reefs Exposed to Oil*, Butterworth Publishers, 1984 (In Press).

(6) Knap, A. H., et al., "The Effects of Oil Spills and Dispersant Use on Corals: A Review and Multidisciplinary Experimental Approach," *Oil Petrochemical Pollution*, Vol 1, pp. 157–169, 1983.

(7) Birkeland, C. A., et al.," Survey of Marine Communities in Panama and Experiments With Oil," *EPA Report No. EPA-66/3-76-028*, 177 pp., 1976.

(8) Neff, J. M., and Anderson, J. W., "Response of Marine Animals to Petroleum and Specific Petroleum Hydrocarbons," *Applied Science Publication*, London, England, 177 pp, 1981.

(9) Bak, R. P. M., and Elgershuizen, J. H. B. W., "Patterns of Oil-Sediment Rejection in Corals," *Marine Biology,* Vol 37, pp. 105–113, 1976.

(10) Reimer, A. A.," Effects of Crude Oil on Corals," *Marine Technical Bulletin,* Vol 6, pp. 39–45, 1975.

(11) Loya, Y., and Rinkevich, B., "Abortion Effect on Corals Induced by Oil Pollution," *Marine Ecological Progessive Service*, Vol 1, pp. 77–80, 1979.

(12) Lewis, J. B., "Effects of Crude Oil and Oil Spill Dispersant on Reef Corals," *Marine Pollution Bulletin*, Vol 2, pp. 59–62, 1971.

(13) Cohen, Y., et al., "Effects of Iranian Crude Oil on the Red Sea Octocoral *Heteroxenia buscesens*," *Environmental Pollution,* Vol 12, pp. 173–185, 1977.

(14) Mitchell, R., and Chet, I., "Bacterial Attack of Corals in Polluted Seawater," *Microbiological Ecology*, Vol 2, pp. 227–233, 1975.

(15) Peters, E. C., et al., "Bioaccumulation and Histopathological Effects of Oil on a Stony Coral," *Marine Pollution Bulletin,* Vol 12, 1981, pp. 333–339.

(16) Rinkevich, B., and Loya, Y. "The Reproduction of the Red Sea Coral *Stylophora pistillata*, I. Gonads and Planulae," *Marine Ecological Progress Service,* Vol 1, pp. 133–144.

(17) Rinkevich, B., and Loya, Y., "Harmful Effects of Chronic Oil Pollution on a Red Sea Scleractinian Coral Population," pp. 589–591. *Proceedings Third International Coral Reef Symposium*, Vol II Geology, DC, Taylor, ed., Miami Rosenstiel School of Marine and Atmospheric Science, 1977.

(18) Eisler, R., "Acute Toxicities of Crude Oil and Oil Dispersant Mixtures to Red Sea Fishes and Invertebrates," *Israel Journal Zoology*, Vol 24, pp. 16–27, 1975.

(19) Knap, A. H., et al., "The Effects of Chemically Dispersed Oil on Corals: Research in Progress," *Dispersants: Five Years of Research, Abstracts*, ASTM, Philadelphia, PA, 1982, 30 pp.

Designation: F 933 – 85 (Reapproved 1994)$^{\epsilon 1}$

Standard Guide for Evaluation of Oil Water Separation Systems for Spilled Oil Recovery Applications[1]

This standard is issued under the fixed designation F 933; the number immediately following the designation indicates the year of original adoption or, in the case of revision, the year of last revision. A number in parentheses indicates the year of last reapproval. A superscript epsilon (ε) indicates an editorial change since the last revision or reapproval.

ϵ^1 NOTE—Section 9 was added editorially in September 1994.

1. Scope

1.1 This guide applies to the evaluation of oil–water separation systems when employed as devices for dewatering oil collected by oil-recovery devices or systems, or both.

1.2 *This standard does not purport to address all of the safety concerns, if any, associated with its use. It is the responsibility of the user of this standard to establish appropriate safety and health practices and determine the applicability of regulatory limitations prior to use.*

2. Referenced Documents

2.1 *ASTM Standards:*

D 1744 Test Method for Water in Liquid Petroleum Products by Karl Fischer Reagent[2]

2.2 *APHA Standard:*

Methods for Examination of Water and Wastewater, 1980 edition[3]

3. Summary of Guide

3.1 The tests allow evaluation of the system's ability to handle high-viscosity oils when contaminated by ice or debris, or both, and measure the separation efficiency of the system with various types of oils under varying oil concentrations and separator flow rates.

3.2 Separation systems are categorized as either Type A or Type B. Type A separation systems are those intended to produce a water effluent with less than 250 ppm of entrained oil. Type B separation systems are those intended to reduce the volume of recovered water with no specific entrainment objective.

4. Significance and Use

4.1 Existing regulations do not require certified oil water separators to handle the high-viscosity oils and various contaminants often encountered when recovering spilled oil. Tests are included in this guide to evaluate the separation system's performance with heavy contaminated oils as well as other oils.

4.2 It is desirable that separation systems used for recovery operations be of the lowest possible volume and weight. The primary performance objective of the separation system for spilled oil recovery is to reduce the volume of recovered liquid that must be stored by removing as much water as possible. A secondary objective is to remove as much oil as possible from the discharged water.

4.3 The flow rate for the test should be based on the manufacturer's rated total throughput for the system.

4.4 The effects of the motions of the platform on which the separator would be located during oil spill clean-up operations are not addressed in this guide. Footnotes 4, 5, and 6 contain information on simulating and testing for the effects of ships motions on oil water systems.[4,5,6]

5. Test Facilities

5.1 The test may be conducted at any test facility including the manufacturer's own test facility.

5.2 The test facility should have the capability of adding an amount of debris to the influent flow and of varying the mixture from 100 % oil and the total flow from zero to 125 % capacity.

5.3 When effluent of 250 ppm oil or less is desired, use Type A separation systems. The test facility should have the capability of producing a mixture of test oil and water where at least 20 % of the dispersed oil, when measured by a sample taken from the influent line to the system is of 50-μm droplet size or smaller.

6. Test Oil and Debris

6.1 The test oils may be crudes, refined, or synthetics. Record type, specific gravity, viscosity, and temperature of each oil together with the mixture percentages and temperatures for each test point.

---

[1] This guide is under the jurisdiction of ASTM Committee F-20 on Hazardous Substances and Oil Spill Responseand is the direct responsibility of Subcommittee F20.12 on Removal.

Current edition approved June 28, 1985. Published May 1986.

[2] *Annual Book of ASTM Standards*, Vol 05.01.

[3] Available from the American Public Health Association, 1015 15th St. NW, Washington, DC 20005.

[4] "Testing of A.P.I. Separator Under Simulated Sea Conditions," *Fisheries and Enviroment Canada Final Report*, September 1979.

[5] "Recommendation on International Performance and Test Specifications for Oily Water Separating Equipment and Oil Content Monitors," *Inter-Governmental Maritime Consultative Organization*, London 1978.

[6] U.S. Coast Guard Regulations 46CFR Part 162, "Engineering Equipment: Design and Approval Requirements for Oil Pollution Prevention Equipment," *Federal Register*, Sept. 13, 1979.

6.2 The debris for the test should be the equivalent of sea weed, sea grass, or hay after it has been passed through a centrifugal pump or similar maceration device to simulate passage through an oil-collection device. Record the type and physical nature of the debris together with the amount, both volume and weight, added at each test condition.

6.3 If the separator is intended for use in arctic regions or areas where ice is frequently encountered in spill clean-up operations, the test should be conducted using water at or near 0°C, and ice in small chunk form should be added to the influent water. Record the water temperature and the amount and mean size of the ice chunks added to each test condition. Footnotes 4 and 7[47] contain information on tests conducted on oil water separators in cold conditions with ice present.[3,6,7]

7. Test Procedures

7.1 *System Debris Test*—The following test is designed to evaluate the ability of the separation system to perform under the unique conditions often encountered in oil spill recovery operations, that is, high viscosity oil and debris.

7.1.1 The separator is started up and run as specified by the manufacturer at its rated throughput with water only for 5 min. Without stopping the flow, a test oil of at least 1500 cSt viscosity, at the water temperature, is added to the influent, at a rate equal to 5 % of the rated throughput of the separator (that is, 5 % oil, 95 % water). After a period sufficient to displace 50 % of the liquid volume of the system, debris as described in 6.2, is added slowly to the influent at the rate of 1/10 lb/min/100 gpm/of separator rated capacity over a period of approximately 10 min. The test is continued until the total running time with 5 % oil equals 45 min after which the operational test is complete. When testing Type A separators, avoid recycling liquids when practical to avoid the formation of emulsions. If the water must be recycled, a record should be made of the amount of emulsified oil in the influent.

7.1.2 One-litre samples are taken as follows:

Time, min	*Influent*	*Effluent*
7.5	X	X
15	X	X
25	X	X
30		X
40	X	X
50 (end of test)		X

"Time" is the total elapsed time measured from the start of the test. Samples must be analyzed immediately after being taken for oil droplet particle size and distribution in the case of Type A separator, and for total oil content.

7.1.3 If an ice test is desired, the ice chunks should be added together with the debris as described in 7.1 or in place of the debris if a separate ice test is required.

7.2 *Separator Concentration and Flow Rate*—The following tests are designed to evaluate the performance of the separator with different types of oil at various concentrations and under varying flow conditions.

7.2.1 The separator is started up and run for 5 min on water alone as in 7.1. A test oil equivalent to crude oil is added to the influent at a rate equal to 5 % of the rated capacity of the separator. For Type A separator tests, the influent oil should be provided with the particle size and distribution described in 5.3. The test is continued at 5 % oil and 95 % water for 10 min. The oil content is then gradually increased over 5 min to 25 % and continued at 25 % oil for 10 min. The oil content is then again gradually increased over 5 min to 100 % and continued at 100 % oil for 5 min. The oil flow is then stopped and the separator is run on 100 % water for 15 min.

7.2.2 The water flow is then reduced to 75 % of the separator capacity and mixed with oil so that the mixture contains 95 % water and 5 % oil. The separator is run at 75 % capacity for 15 min. The flow is then increased to 125 % of rated capacity and the same mixture of 5 % oil and 95 % water and is run at this condition for 15 min.

7.2.3 One-litre samples of the inlet mixture, the outlet water, and the outlet oil are taken at the midpoint and end of each 5, 10, or 15-min steady-state test period. The same test procedure is repeated for each other type of oil to be tested. It is suggested that at least a diesel fuel oil, a high viscosity (greater than 1500 cSt), high specific gravity (0.93 to 0.95) oil and possibly a "mousse" or water in oil emulsion be tested in addition to the crude oil.

8. Report

8.1 During the debris test the ability of the separation system to handle heavy oil and debris should be noted and recorded together with changes in pressure or water levels and variation in flow rates of oil and water out of the system.

8.1.1 If tested with ice, the ability of the separation system to handle the ice should be noted as well.

8.2 For Type A systems only, each inlet sample is processed to determine the droplet size and distribution of the oil and the percent of oil and water by standard methods. For Type B systems, a jar-settling test is conducted to determine the percent of oil in the influent.

8.3 For Type A systems only, during the tests each water effluent sample is processed to determine the droplet size and distribution of the oil and the percent of oil and water by standard methods. For Type B systems only, the percent of oil is determined by standard methods.

8.4 Each oil effluent sample is processed to determine the percent water in the oil by standard methods.

8.5 It is recommended that the influent and effluent data for each test condition be analyzed and the separation efficiency be determined as follows:

For Type A Separation Systems:

influent oil content, ppm

$$\text{Separation efficiency} = \frac{\text{effluent water oil content, ppm}}{\text{influent oil content, ppm}} \tag{1}$$

For Type B Separation Systems:

$$\text{Separation efficiency} = \frac{\text{volume of water discharged}}{\text{volume of water in influent}} \tag{2}$$

8.5.1 The various efficiencies should be plotted for each oil tested as a function of influent oil content and as a function of actual flow rate as a percent of rated flow.

[7] Technology Development Report EPS 4-EC-79-3, "Evaluation of Pumps and Separators for Arctic Oil Spill Cleanup" *Enviroment Canada*, Environmental Protection Service, Enviromental Impact Control Directorate, April 1979.

9. Keywords

9.1 oil recovery applications; oil spill; oil water separation; recovery

Designation: F 962 – 99

Standard Specification for Oil Spill Response Boom Connection[1]

This standard is issued under the fixed designation F 962; the number immediately following the designation indicates the year of original adoption or, in the case of revision, the year of last revision. A number in parentheses indicates the year of last reapproval. A superscript epsilon (ε) indicates an editorial change since the last revision or reapproval.

1. Scope

1.1 This specification covers design criteria requirements, design geometry, material characteristics, and desirable features for oil spill response boom connections. These criteria are intended to define minimum mating characteristics and are not intended to be restrictive to a specific configuration.

2. Referenced Documents

2.1 *ASTM Standards:*

F 818 Terminology Relating to Spill Response Barriers[2]

3. Material Characteristics

3.1 End connector and cross pin materials shall be corrosion resistant in sea water and such other environments as the intended service may require. If dissimilar metals are used, care shall be used in design to avoid galvanic corrosion.

3.2 Any material is acceptable for construction of the boom connector provided consideration is given to such factors as weight, mechanical strength, chemical resistance, flexibility, and conditions of the environment in which it is to be used.

4. Design Requirements

4.1 The minimum tensile strength of a connection shall be 300 lbs/in. of membrane height.

4.2 When the connector is designed as an integral part of the boom, it shall ensure distribution or transfer of the tension member loads from one boom section to the next through or around the end connector in such a manner that the integrity of the joint is not broken.

4.3 The connector or adapter shall not take more than 0.04 in. permanent set when a 250-lb load, distributed over 3 in., is applied. The load shall be applied at the location that results in maximum deflection and shall be resisted by supports placed ¼ in. from each end as shown in Fig. 1.

4.4 In addition to the dimensional requirements shown in Fig. 2, the self-locking cross-pin/lanyard assembly shall have the following characteristics:

4.4.1 Its assembled strength shall resist a tensile load of 180 lbs placed upon the closed toggle by the test fixture to which the cross pin's lanyard is attached without deformation as shown in Fig. 3.

4.4.2 It shall have a ring or lanyard loop of a minimum diameter of 1-½ in. for the convenience of pulling the pin from the boom connectors.

4.4.3 The toggle shall turn freely and shall latch in either direction.

4.4.4 The cross-pin's spring shall be captured or locked to the assembly and shall exert a force on the toggle of between 16 and 22 lbs when connectors are assembled.

4.4.5 When the cross-pin's spring is compressed fully, there shall be a clearance of ⅛in. between the short end of the toggle and the mated connector as shown in Fig. 2.

4.4.6 The cross-pin's overall length shall be minimized and its ends rounded or chamfered so as to minimize wear and tear on adjacent stored booms or injury to boom handlers.

4.5 Where one half of a connector set mates with one having the geometry defined herein but is of other dimensions, it shall meet the following design requirements:

4.5.1 Possess adequate mechanical strength.

4.5.2 Minimize oil leakage.

4.5.3 Be sexless (neither male/female).

4.5.4 Be full height of boom of which it is a part.

4.5.5 Not impair stability of the boom.

4.5.6 Require no special tools for assembly.

4.5.7 Not reduce freeboard.

5. Significance and Use

5.1 The general design geometry herein defined applies to both a separate adaptor accessory mating two booms of different geometry as well as boom end connectors (see Terminology F 818).

5.2 Interconnectibility is intended to facilitate mating of oil spill response booms of various sizes, strengths, design, and manufacture.

5.3 The use of this general design geometry in no way guarantees the effective performance of the linked boom sections, since each boom's design and the environmental conditions at each incident govern overall performance.

6. Design Criteria

6.1 The connector shall be of the hook engagement design. The geometry of single hook engagement end connectors shall

[1] This specification is under the jurisdiction of ASTM Committee F-20 on Hazardous Substances and Oil Spill Response and is the direct responsibility of Subcommittee F20.11 on Control.

Current edition approved Oct. 10, 1999. Published December 1999. Originally published as F 962 – 86. Last previous edition F 962 – 94.

[2] *Annual Book of ASTM Standards*, Vol 11.04.

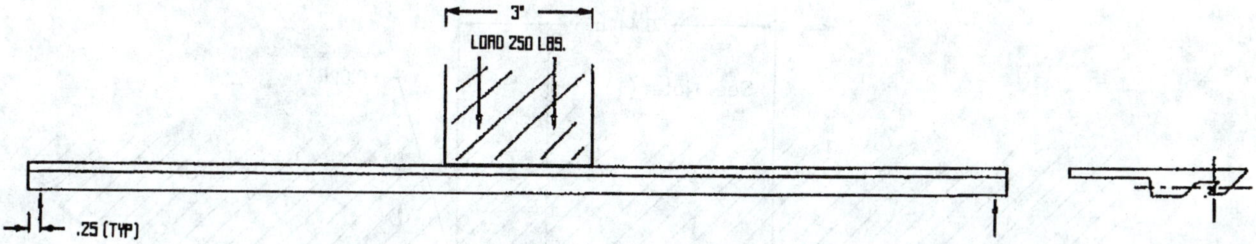

NOTE 1—All items shown are generic and not intended to depict any manufacturer's specific product.

FIG. 1 Connector/Adapter

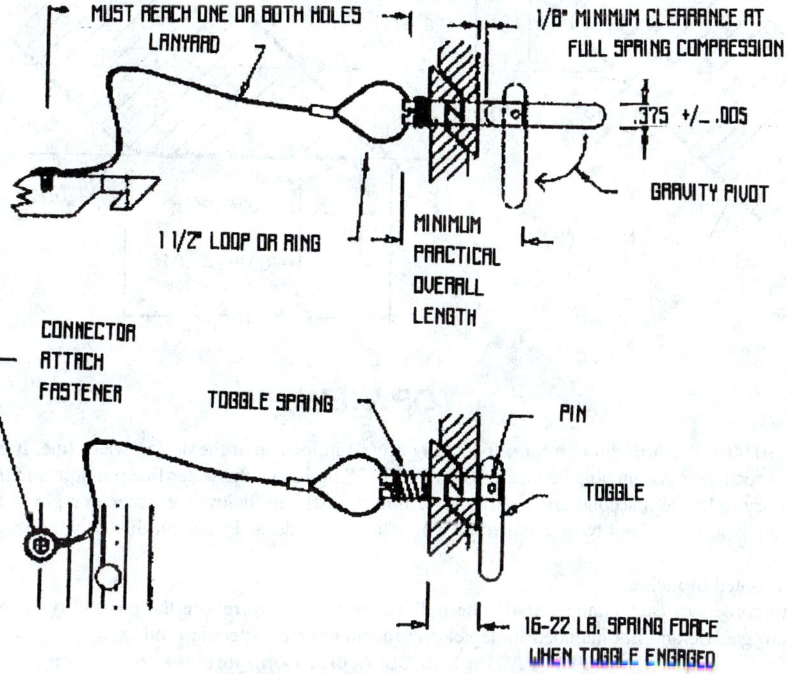

FIG. 2 Self Locking Cross Pin / Lanyard Assembly

be compatible with the requirements shown in Fig. 4.

NOTE 1—The hook is lefthand when looking down on top of the connector.

6.2 An end connector shall be long enough to be compatible with the particular boom of which it is a part and not be a limiting member as to the functional freeboard or draft of the membrane.

6.3 A hole or holes designed to accommodate a ⅜-in. diameter self-locking cross-pin shall be incorporated as shown in Fig. 2. A self-locking cross-pin/lanyard assembly long enough to reach the cross-pin hole(s) meeting the requirements of Fig. 2 shall be furnished with, and connected to, each end connector or adapter.

6.4 Each connector or adapter will have sufficient bending strength so it will not take a permanent set, along its weakest axis, if inadvertently stepped on in the field.

6.5 Connectors used on permanent booms, such as those intended for long-time storage in the water, should have a ⅜-in. (9-mm) hex head cap screws with self-locking nuts for connections not requiring periodic separation.

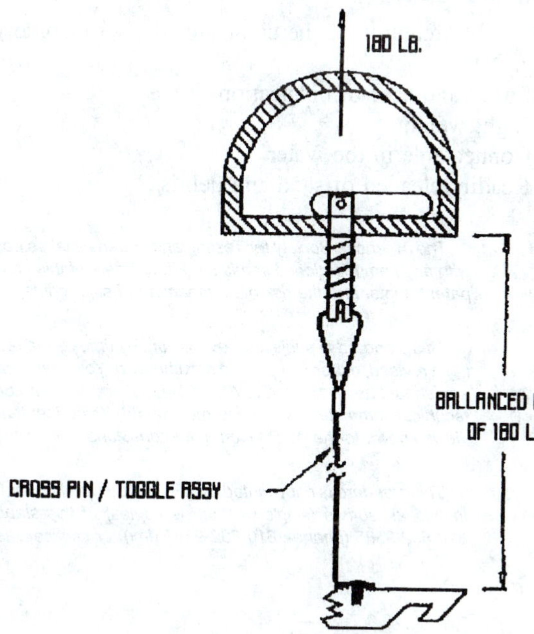

FIG. 3 Toggle / Test Fixture

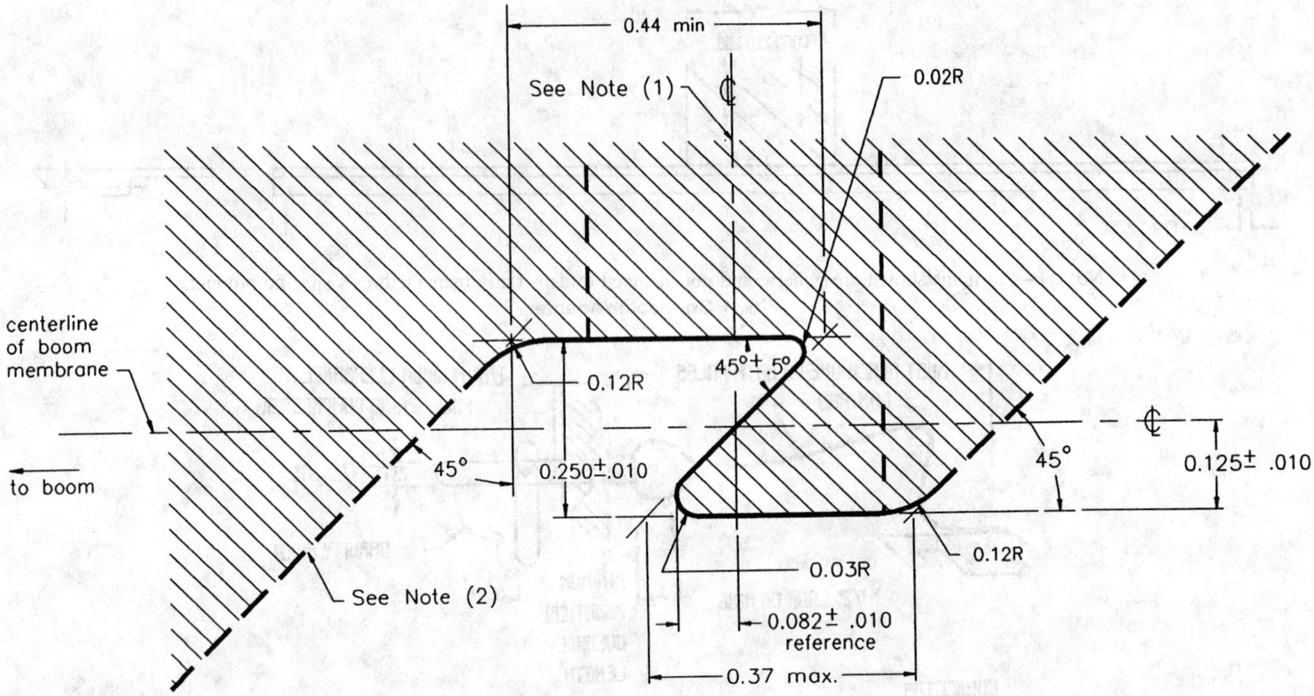

TOP VIEW

NOTE 1—0.41-in. diameter (+0.00–0.01) drill through for self-locking cross-pin located at the design water line. It is suggested that for all connectors or adapters shorter than 24 in., a second cross-pin hole be located 6.000 ± 0.015 in. below the water line pin hole and that for all connectors and adapters 24 in. and longer, it be mandatory to locate a second cross-pin hole 12.000 ± 0.015 in. below the water line pin hole.

NOTE 2—Surface of connector is to be defined by manufacturer where shown in dashed lines on this drawing, provided that no part of the surface protrude beyond these dashed lines.

NOTE 3—All dimensions are stated in inches.

NOTE 4—All top and bottom corners of each connector shall be radiused to 0.06 in. to reduce the possibility of injury to boom handlers.

NOTE 5—All items shown are generic and not intended to depict any manufacturer's specific product.

FIG. 4 ASTM End Connector Configuration

7. Desirable Features

7.1 Desirable features of the connector design include the following:

7.1.1 Speed and ease of connection.

7.1.2 Light weight.

7.1.3 Connectable in the water.

7.1.4 Readily cleaned of sand and debris.

7.1.5 Inherently safe to personnel.

7.1.6 Easy to install or replace.

8. Keywords

8.1 boom; boom connection; oil spill response

Designation: F 971 – 86 (Reapproved 1999)

Standard Guide for Ecological Considerations for the Use of Chemical Dispersants in Oil Spill Response: Mangroves[1]

This standard is issued under the fixed designation F 971; the number immediately following the designation indicates the year of original adoption or, in the case of revision, the year of last revision. A number in parentheses indicates the year of last reapproval. A superscript epsilon (ε) indicates an editorial change since the last revision or reapproval.

1. Scope

1.1 This guide covers recommendations for use of chemical dispersants to assist in the control of oil spills. It is written with the goal of minimizing the environmental impacts of oil spills; this goal is the basis upon which recommendations are made. Aesthetic and socioeconomic factors are not considered; although, these and other factors are often important in spill response.

1.2 Each on-scene coordinator has available several means of control or cleanup of spilled oil. In this guide, use of chemical dispersants is not to be considered as a last resort after other methods have failed. Chemical dispersants are to be given equal consideration with other spill countermeasures.

1.3 This guide presents general guidelines only. They assume the oil to be dispersable and the dispersant to be effective, available, applied correctly, and in compliance with relevant government regulations. Oil, as used in this guide, includes crude oils and fuel oils (No. 1 through No. 6). Differences between individual dispersants or between different oils or products are not considered.

1.4 The guidelines are organized by habitat type, for example, sand beaches, marshes. They consider the use of dispersants primarily to protect such habitats from impact (or minimize impacts) and also to clean them after the spill takes place.

1.5 This guide applies to marine and estuarine environments but not to freshwater environments.

1.6 In making dispersant use decisions, appropriate government authorities should be consulted as required by law.

1.7 *This standard does not purport to address all of the safety concerns, if any, associated with its use. It is the responsibility of the user of this standard to establish appropriate safety and health practices and determine the applicability of regulatory limitations prior to use.*

2. Significance and Use

2.1 This guide is meant to aid local and regional spill response teams who may apply it during response planning and spill events.

3. General Considerations for Making Dispersant-Use Decisions

3.1 The decision of whether to use or not to use dispersants in a given spill situation is always one involving trade-offs. Dispersing a slick at one site temporarily introduces more oil into the water column at that site than would be there if a surface slick floated over it. Therefore, adverse effects on water column organisms may be increased at the site so that adverse effects can be decreased or eliminated at other sites.

3.2 Dispersant use is primarily a spill control method, not a cleanup method. Such use can give spill response personnel some control over where the impacts of a spill will occur and what types of impacts they may be. Since some environments are known to be more vulnerable to the longer lasting impacts of spilled oil, an acceptable trade-off may be to protect those environments by dispersing an oil slick in a less sensitive or less productive environment. In general, the trade-off that must be evaluated is between the impact of the relatively long residence time of spilled oil that strands on shorelines versus the short-term impact of dispersed oil in the water column.

3.3 In this guide, environments that are most vulnerable to the longer-term impacts of oil contamination are identified. Protection of these environments is recommended as a high priority, by means of dispersants or other methods.

4. Mangrove Ecosystem Environment

4.1 Mangrove ecosystems are intertidal forests dominated by various species of woody halophytes that are commonly called mangroves. There are 12 families and more than 54 species of mangroves (**1**)[2]. Mangrove ecosystems occur in tropical low-energy depositional areas; where they exist, they tend to promote further the deposition of organic and mineral matter. Their extensive root systems are very imporant in stabilizing intertidal sediments (**2**). Adult mangroves form the structural basis for the mangrove community in that they provide attachment sites for many species of animals and shelter for many others.

4.2 Mangrove ecosystems contribute to the productivity of tropical marine environments where they play much the same role as temperate salt marshes. Mangroves are important as nursery areas as well as for the detritus that they supply to

[1] This guide is under the jurisdiction of ASTM Committee F-20 on Hazardous Substances and Oil Spill Responseand is the direct responsibility of Subcommittee F20.13 on Treatment.

Current edition approved Jan. 31, 1986. Published March 1986.

[2] The boldface numbers in parentheses refer to the list of references at the end of this guide.

surrounding communities (**3**). It has been estimated that up to ⅔ of the world's fish populations depend on detritus and remineralized nutrients exported from mangrove areas (**4**). Mangrove forests are used as a source of firewood, charcoal, building materials, tanning agents, dyes, and livestock feed.

4.3 Mangrove forests dominate much of the world's tropical shoreline; many are adjacent to tanker routes, oil fields, and refineries. The low energy characteristic of mangrove forests leads to the entry and retention of oil in these environments.

5. Background

5.1 Mangrove forests are typically found in depositional areas. While they may be found growing in sand, they are more often found in peaty or muddy soils that are highly reduced. Under these conditions oil is most likely to accumulate and persist for some time. Soil temperatures are high in tropical areas; biodegradation appears to occur at a higher rate as compared with temperate ecosystems (**5**). The local geomorphology will control the distribution of the oil in the mangrove forest and thus influence the impact (**6**).

5.2 The effects of petroleum on mangroves are known primarily from studies at oil spill sites. These studies have shown that the vulnerability of mangroves to oil depends to a large degree on the aromaticity of the oil in question (**7, 8**). The effect of petroleum also depends on the degree to which the oil has weathered (**5**). The greater the proportion of the lighter more water soluble aromatic hydrocarbons present, the greater the impact that can be expected. Oil effects on mangroves are also a function of the permeability of the soil in which the trees are growing. The more permeable the soil the greater the impact that can be expected.

5.3 Based on the biology of mangrove trees and the geomorphology of areas in which they live, mangroves have been ranked as one of the most sensitive marine environments (**9**). Once mangrove areas are oiled, and no further action is taken, there is a very high probability that mortality of trees will result. Mortality of trees and associated communities has been observed in numerous spill studies (**10, 5**).

5.4 Effects of oil and dispersed oil on mangroves was investigated in a laboratory experiment (**11**). All methods of exposure and concentrations used were not necessarily representative of field conditions. The results are as follows:

5.4.1 The effects of No. 6 fuel oil, No. 2 fuel oil and Light Arabian crude oil on leaf growth in *Rhizophora mangle* and *Avicennia germinans* have been investigated. It was found that *R. mangle* was about an order of magnitude more resistant to each of the oils tested than was *A. germinans*.

5.4.2 These experiments showed that a 20:1 mixture of No. 6 fuel oil and a widely available self-mix concentrate dispersant was less toxic than whole oil alone. The reverse was true for mixtures including No. 2 fuel oil and Light Arabian crude oil.

5.4.3 Very high concentration of any of these dispersed oils were required to produce effects on leaf growth. (That is, 50 000 ppm in *R. mangle*; 5 000 ppm in *A. germinans*.)

5.4.4 The study also concluded that the roots of mangrove seedlings are the most sensitive part of the plant. Seedlings from old oil spill sites and areas near refineries were more resistant to oil effects than seedlings from nonoiled areas.

5.5 The likelihood of oil dispersed in the nearshore area affecting mangroves will be a function of the concentration of dispersed oil to be expected and the grain size of sediments in which the trees are growing. If the expected concentrations of dispersed oil are low in relation to the levels of dispersed oil required to produce effects in the laboratory, then little damage would be expected regardless of the sediment grain size. Dispersal of a 0.2 mm thick slick into 1 m of water could produce a maximum concentration of dispersed oil of 200 ppm, but such high levels have never been measured at sea even directly under treated slicks (**12**).

6. Recommendations

6.1 Due to the very high probability that an oil spill will damage mangrove trees and associated communities that it impacts, the major goal of the spill response effort should be to prevent oil from reaching mangrove forests.

6.2 Ideally, chemical dispersion should be carried out in deep water, remote from the mangrove forest itself, so that the dispersed oil can be diluted to low levels before reaching the mangrove ecosystem.

6.3 Due to the high levels of dispersed oil required to influence mangrove growth and because of the potential persistence of non-dispersed oil in the mangrove ecosystem, chemical dispersion should be considered a viable option even if dispersed oil might enter the forest.

6.4 If significant amounts of dispersed oil enter a mangrove forest in an area with little tidal range, attempts should be made to flush it out.

7. Keywords

7.1 chemical dispersants; ecological considerations; mangroves; oil spill response

REFERENCES

(1) Waisel, Y., *Biology of Halophytes*, Academic Press, New York, 1972.

(2) Teas, H. J., Jurgens, W., and Kimball, M. C., *Proceedings*, Second Annual Conference on Restoration of Coastal Vegetation in Florida, Hillsborough Community College, Tampa, FL, 1975.

(3) Carter, M. R., et al., "Ecosystem Analysis of the Big Cyprus Swamp and Estuaries," U.S. Environmental Protection Agency, Region IV, Atlanta, 1973.

(4) Canestri, V. and Ruiz, O., "The Destruction of Mangroves," *Pollution Bulletin*, March 1973, Vol 4, pp. 183–185.

(5) Page, D. S., et al., "Hydrocarbon Distribution and Weathering Characteristics at a Tropical Oil Spill Site." *Proceedings*, Oil Spill Conference, API Washington DC, 1979, pp. 709–712.

(6) Getter, C. D., Scott, G. I., and Michel, J., "Effects of Oil Spills on Mangrove Forests: A Comparison of Five Oil Spill Sites in the Gulf of Mexico and the Caribbean Sea," *Proceedings*, Oil Spill Conference API, Washington DC, 1981, pp. 535–540.

(7) Odum, W. E. and Johannes, R. E., "The Response of Mangroves to Man Induced Environmental Stress," E. J. F. Wood and R. E. Johannes, eds., *Tropical Marine Pollution*, Elsevier Oceanography Series, New York, No. 12, 1975.

(8) Linden, O. and Jernelov, A., "The Mangrove Swamp—An Ecosystem In Danger," AMBIO, Vol 9, No. 2, 1980, pp. 81–88.

(9) Gundlach, E. R. and Hayes, M. O. "Vulnerability of Coastal Ecosystems to Oil," *Journal Marine Technical Society*, Vol 12, No. 4, 1978, pp. 18–27.

(10) Lewis, L. R. "Impact of Oil Spills on Mangrove Forests," Second International Symposium on Biology and Management of Mangroves and Tropical Shallow Water Communities, 20 July to 2 August. Port Moresby, Madana, Papau, New Guinea, 1980.

(11) Getter, C. D., Ballou, T. G., and Dablin, J. A., "Preliminary Results of Laboratory Testing of Oil and Dispersants on Mangroves," *Proceedings*, Oil Spill Conference, 1983.

(12) McAuliffe, C. D., et al., "Dispersion and Weathering of Chemically Treated Crude Oils in the Ocean," *Environmental Science and Technology*, Vol 14, No. 12, 1980, pp. 1509–1518.

Standard Guide for Ecological Considerations for the Use of Chemical Dispersants in Oil Spill Response—Nearshore Subtidal[1]

This standard is issued under the fixed designation F 972; the number immediately following the designation indicates the year of original adoption or, in the case of revision, the year of last revision. A number in parentheses indicates the year of last reapproval. A superscript epsilon (ε) indicates an editorial change since the last revision or reapproval.

ϵ^1 NOTE—Section 7 was added editorially in December 1992.

1. Scope

1.1 This guide covers recommendations for use of chemical dispersants to assist in the control of oil spills. It is written with the goal of minimizing the environmental impacts of oil spills; this goal is the basis upon which recommendations are made. Aesthetic and socioeconomic factors are not considered; although, these and other factors are often important in spill response.

1.2 Each on-scene coordinator has available several means of control or cleanup of spilled oil. In this guide, use of chemical dispersants is not to be considered as a last resort after other methods have failed. Chemical dispersants are to be given equal consideration with other spill counter-measures.

1.3 This guide presents general guidelines only. The oil is assumed to be dispersable and the dispersant to be effective, available, applied correctly, and in compliance with relevant government regulations. Oil, as used in this guide, includes crude oils and fuel oils (No. 1 through No. 6). Differences between individual dispersants or between different oils or products are not considered.

1.4 The guidelines are organized by habitat type, for example, sand beaches, marshes. They consider the use of dispersants primarily to protect such habitats from impact (or minimize impacts) and also to clean them after the spill takes place.

1.5 This guide applies to marine and estuarine environments but not to freshwater environments.

1.6 In making dispersant use decisions, appropriate government authorities should be consulted as required by law.

1.7 *This standard does not purport to address all of the safety problems, if any, associated with its use. It is the responsibility of the user of this standard to establish appropriate safety and health practices and determine the applicability of regulatory limitations prior to use.*

2. Significance and Use

2.1 This guide is meant to aid local and regional spill response teams who may apply it during response planning and spill events.

3. General Considerations for Making Dispersant-Use Decisions

3.1 The decision of whether to use or not to use dispersants in a given spill situation is always one involving tradeoffs. Dispersing a slick at one site temporarily introduces more oil into the water column at that site than would be there if a surface slick floated over it. Therefore, adverse effects on water column organisms may be increased at the site so that adverse effects can be decreased or eliminated at other sites.

3.2 Dispersant use is primarily a spill control method, not a cleanup method. Such use can give spill response personnel some control over where the impacts of a spill will occur and what types of impacts they may be. Since some environments are known to be more vulnerable to the longer lasting impacts of spilled oil, an acceptable tradeoff may be to protect those environments by dispersing an oil slick in a less sensitive or less productive environment. In general, the tradeoff that must be evaluated is between the impact of the relatively long residence time of spilled oil that strands on shorelines versus the short-term impact of dispersed oil in the water column.

3.3 In this guide, environments that are most vulnerable to the longer-term impacts of oil contamination are identified. Protection of these environments is recommended as a high priority, by means of dispersants or other methods.

4. Nearshore Subtidal Environment

4.1 The nearshore subtidal environment can be considered to extend seaward from the extreme low water spring tidal level to a depth sufficiently great so that no dispersed oil can be expected to impact the bottom. This depth will vary depending upon local oceanographic conditions. The probable range of depth will be from a few metres where a strong, shallow density gradient exists, such as in a stratified estuary to 50 + m where strong vertical mixing occurs. Commonly, vertical mixing does not extend below 10 m.

4.2 The nearshore subtidal environment supports an extremely varied community of animals and plants. The exact community of animals and plants that will be encountered will

[1] This guide is under the jurisdiction of ASTM Committee F-20 on Hazardous Substances and Oil Spill Responseand is the direct responsibility of Subcommittee F20.13 on Treatment.

Current edition approved Jan. 31, 1986. Published April 1986.

be a function of water depth, latitude, sediment grain size, and water movement. The total productivity of this region can be quite high; these areas can be very important to nearshore fisheries.

4.3 The potential exposure of bottom organisms can be estimated from the following:

4.3.1 Slick thickness,

4.3.2 Water depth,

4.3.3 Mixing depth, and

4.3.4 Flushing rate.

4.4 The potential consequences of the slick stranding can be estimated based on the following types of organisms found in the potential impact zone:

4.4.1 *Vulnerable Species Present*:

4.4.1.1 Corals,

4.4.1.2 Seagrasses, and

4.4.1.3 Commercially important shellfish ready for immediate harvest.

4.4.2 *Resistant Species Present*—For example, many attached algae.

4.4.3 *Vulnerable Habitats Present*:

4.4.3.1 Marshes,

4.4.3.2 Mangrove forests, and

4.4.3.3 Any other sedimentary environment characterized by fine-grained sediments (low-energy environments).

5. Background

5.1 Although chemical treatment of oil in the nearshore area is most often thought of in terms of protecting onshore environments, the nearshore subtidal environment is also at risk from oil slicks reaching shore. Studies carried out following the grounding of the barge Florida off West Falmouth, MA **(1)**[2] as well as studies carried out following the grounding of the Amoco Cadiz off Brittany, France **(2, 3)** have shown that stranded oil can enter the nearshore subtidal environment. Presumably this entry is a result of erosion of oiled sediments by wave action **(4)**.

5.2 There are a variety of habitats found in the nearshore environment, and each species must be considered individually in order to determine the impact of dispersant and oil mixtures on such habitats. However, some general considerations common to all situations can be defined. The longer the organism is exposed to oil, the greater the likelihood of environmental impact. An increased exposure value (Note) will occur in nearshore areas, in shallow waters, and where there is little mixing. The concerns about long-term chronic contamination of a habitat will be greater in areas containing fine sediment and where there is little water motion. Those species that are filter feeders tend to concentrate hydrocarbons and are more sensitive to small concentrations of oil. The ability of a specific species to recolonize a contaminated region depends on its reproductive cycle, mobility, and the level of residual hydrocarbons after the spill. Specialist advice should be sought concerning the biological sensitivities in a specific shore area.

NOTE 1—The amount of exposure, in the environmental sense, is the time of intergration of the oil concentration and is measured in parts per million hours. Exposure is of significance when concentration values do not exceed a maximum critical value which results in mortality. (ppm-hours equal the concentration of dispersed oil (ppm) times the exposure period in hours.)

5.3 Studies **(5, 6, 7, 8)** have shown that 1 m^3 of an oil and dispersant mixture, discharged at the surface of a large, 3.5-m deep water body, gave an exposure of about 50 ppm-h. (Note 1.) Mixing downward in this experiment was achieved by simulating wave action with a boat. This study showed that most of the oil remained in the top one metre of the water column. The small fraction of the oil that did reach the bottom had lost the light ends, leaving molecules with more than 17 carbon atoms. No penetration of dispersed oil into the sediments was observed. This work showed an uptake of the oil by mollusks, but the hydrocarbons were depurated after about four weeks. In these organisms, there was no effect on the protein and carbohydrate metabolism. This program showed no effects of the dispersed oil on the intertidal communities.

5.4 A test spill of 15 m^3 of dispersant and oil mixture in the Canadian Arctic **(9)** showed that there was considerable narcosis of the infaunal invertebrates at exposures of 30 and 300 ppm-h. In the test with 300 ppm-h, the bottom sediment contained about 2.5 ppm of hydrocarbon. After a few weeks, there was relatively little effect on the community structure, and no long-term effects of chemically dispersed oil were identified in this experiment.

5.5 In the subtidal benthos, each species that is exposed to chemically dispersed oil may react differently, depending upon the exposure. The motion of water in an esturine system is very complex, and a local expert should be consulted. In many cases, local fishermen familiar with the area can provide valuable information on water motion. It is obvious that, in shallow water or areas of high mixing energy, the exposure by the benthic species to hydrocarbons will be greater. However, a high flushing rate may reduce exposure. In many situations, the water column may be stratified, which has the potential to reduce the exposure of the benthos. Expert oceanographic advice should be sought.

6. Recommendations

6.1 Ideally, it is best to disperse oil in water deep enough that benethic animals and plants will not be exposed to dispersed oil.

6.2 The oil should be dispersed before it enters sensitive areas or areas with low rates and volumes of water exchange, or both.

6.3 In deciding whether to disperse a slick in the nearshore area, one must weigh the potential exposure of bottom organisms against the potential exposure of onshore habitats.

6.4 Disperse oil in the nearshore subtidal area if it is determined that it is necessary to protect vulnerable onshore habitats.

7. Keywords

7.1 chemical dispersants; ecological considerations; nearshore subtidal; oil spill response

[2] The boldface numbers in parentheses refer to the list of references at the end of this guide.

REFERENCES

(1) Sanders, H. L., Grassle, J. F., Hampson, G. R., Morse, L. S., Garner-Price, S., and Jones, C. C., "Anatomy of an Oil Spill: Long Term Effects of the Grounding of the Barge FLORIDA off West Falmouth Massachussetts," *Journal of Marine Research*, Vol 38, No. 2, 1980, pp. 265–380.

(2) Glemarec, M., and Hussenot, E., "Definition DS'une Secession Ecologique en Milieu Meuble Abnormalement Enrichi en Matieres Organique a la suite de la Catastrophe de l'Amoco Cadiz," *Amoco Cadiz; Consequences d'une Pollution Accidentelle par Les Hydrocarbures,* Centre National Pour L'Exploitation des Oceans, Paris, 1981, pp. 499–525.

(3) Cabioch, L., Dauvin, J. C., Gentil, F., Retieve, C., and Rivain, V., "Perturbations Induites Dans La Composition Et Le Fonctionnemont Des Peuplements Benthiques Sublittoraux, Sous L'Effet Des Hydrocarbures De L'Amoco Cadiz," *Amoco Cadiz: Consequences d'une Pollution Accidentelle par Les Hydrocarbures.* Centre National Pour Exploitation des Oceans, Paris, 1981, pp. 513–526.

(4) Michel, J., Gundlach, E. R., Hayes, M. O., and Sexton, W. J., "The Amoco Cadiz Oil Spill," Part I: Distribution and Quantification of Bottom Oil Incorporated Within the Bays of Lannion and Morlaix Four and One Half Months After the Amoco Cadiz Oil Spill. Report prepared for NOAA ERL by Research Planning Institute, Columbia, SC, USA, 1979.

(5) Page, D. S., Foster, J. C., Hotham, J. R., Vallas, D., Gilfillan, E. S., Hanson, S. A., and Gerber, R. P., "Tidal Area Dispersant Project: Fate of Dispersed and Non-Dispersed Oil in Two Nearshore Test Spills," Allen, T. ed., *Oil Spill Chemical Dispersants, ASTM STP 840*, ASTM, 1984, pp. 280–298.

(6) Page, D. S., Foster, J. C., Hotham, J. R., Pendergast, E., Hebert, S., Gonzales, L., Gilfillan, E. S., Hanson, S. A., Gerber, R. P., and Vallas, D., "Long Term Fate of Dispersed and Undispersed in Two Nearshore Test Spills," *Proceedings* American Petroleum Institute, Oil Spill Conference, Publ. 4356, 1983, pp. 465–471.

(7) Gilfillan, E. S., Page, D. S., Hanson S. A., Foster, J., Hotham, J., Vallas, D., and Gerber, R., "Effects of Test Spills of Dispersed Oil and Non-Dispersed Oil on the Activity of Aspartate Amino Transferase and Glucose-6-Phosphate Dehydrogenase in Two Intertidal Bivalves *Mya arenaria* and *Mytilus edulis,"* Allen, T., ed., Oil Spill Chemical Dispersants, *ASTM STP 840,* ASTM, 1984, pp. 299–313.

(8) Gilfillan, E. S., Page, D. S., Hanson, S. A., Foster, J., Hotham, J., Vallas, D., Prat, S. D., and Gerber, R. P., "Effects of Spills of Dispersed and Non-Dispersed Oil on Intertidal Infaunal Community Structure," *Proceedings,* American Petroleum Institute, Publ. 4356, Oil Spill Conference, 1983, pp. 457–463.

(9) Blackhall, P. J., and Sergy, G. A., "The BIOS Project—an Update," *Proceedings,* American Petroleum Institute, Publ. 4356, Oil Spill Conference, 1983, pp. 451–455.

Designation: F 973 – 86 (Reapproved 1993)[ε1]

Standard Guide for Ecological Considerations for the Use of Chemical Dispersants in Oil Spill Response: Tidal Flats[1]

This standard is issued under the fixed designation F 973; the number immediately following the designation indicates the year of original adoption or, in the case of revision, the year of last revision. A number in parentheses indicates the year of last reapproval. A superscript epsilon (ε) indicates an editorial change since the last revision or reapproval.

ε[1] NOTE—Section 7 was added editorially in December 1992.

1. Scope

1.1 This guide covers recommendations for use of chemical dispersants to assist in the control of oil spills. This guide is written with the goal of minimizing the environmental impacts of oil spills; this goal is the basis upon which recommendations are made. Aesthetic and socioeconomic factors are not considered; although, these and other factors are often important in spill response.

1.2 Each on-scene coordinator has available several means of control or cleanup of spilled oil. In this guide, use of chemical dispersants is not to be considered as a last resort after other methods have failed. Chemical dispersants are to be given equal consideration with other spill countermeasures.

1.3 This is a general guide only, assuming the oil to be dispersable and the dispersant to be effective, available, applied correctly and in compliance with relevant government regulations. Oil, as used in this guide, includes crude oils and fuel oils (No. 1 through No. 6). Differences between individual dispersants or between different oils or products are not considered.

1.4 This guide covers one type of habitat, tidal flats. Other guides, similar to this one, cover habitats such as rocky shores. The use of dispersants is considered primarily to protect such habitats from impact (or minimize impacts) and also to clean them after the spill takes place.

1.5 This guide applies to marine and estuarine environments but not to freshwater environments.

1.6 In making dispersant-use decisions, appropriate government authorities should be consulted as required by law.

1.7 *This standard does not purport to address all of the safety problems, if any, associated with its use. It is the responsibility of the user of this standard to establish appropriate safety and health practices and determine the applicability of regulatory limitations prior to use.*

2. Significance and Use

2.1 This guide is meant to aid local and regional spill response teams during response planning and spill events.

[1] This guide is under the jurisdiction of ASTM Committee F-20 on Hazardous Substances and Oil Spill Responseand is the direct responsibility of Subcommittee F20.13 on Treatment.

Current edition approved Jan. 31, 1986. Published March 1986.

3. General Considerations for Making Dispersant-Use Decisions

3.1 The decision of whether to use or not to use dispersants in a given spill situation is always one involving trade-offs. Dispersing a slick at one site temporarily introduces more oil into the water column at that site than would be there if a surface slick floated over it. Therefore, adverse effects on water column organisms may be increased at the site so that adverse effects can be decreased or eliminated at other sites.

3.2 Dispersant use is primarily a spill control method, not a cleanup method. Such use can give spill response personnel some control over where the impacts of a spill will occur and what types of impacts they may be. Since some environments are known to be more vulnerable to the longer lasting impacts of spilled oil, an acceptable trade-off may be to protect those environments by dispersing an oil slick in a less sensitive or less productive environment. In general, the tradeoff that must be evaluated is between the impact of the relatively long residence time of spilled oil that strands on shorelines versus the short-term impact of dispersed oil in the water column.

3.3 In this guide, environments that are most vulnerable to the longer-term impacts of oil contamination are identified. Protection of these environments is recommended as a high priority, by means of dispersants or other methods.

4. Environment Covered—Tidal Flats

4.1 Tidal flats are usually broad intertidal areas of unconsolidated sediments that have little slope and that are usually protected from direct wave action. They are composed of sediments of varying characteristic grain size depending on the amount of wave and current energy present. Tidal flats may be covered by seagrasses, marsh grass, or mangroves. These environments are discussed elsewhere. Tidal flats are important to the coastal ecosystem because of the usual high productivity of their associated infauna and epifauna.

4.2 The characteristic grain size of the sediments will determine the type of animals that will be dominant. In fine-grained sediments (muds), the dominant organisms will be deposit feeding animals. They will be important in transferring energy from detrital sources to the remainder of the food web. These animals may be an important food source for the young of commercially important species. In coarser (sand) sediments the dominant organisms will be filter feeding animals who will be important in transferring energy from the plankton to the remainder of the food web. Some of these species are important commercially. Whatever the dominant species are, they are likely to be very abundant. As a result, tidal flats are important nursery areas for many species. They are also important feeding areas for many species of birds. Birds and marine mammals may use tidal flats as resting or haul-out areas.

4.3 The rate of self cleaning of sedimentary environments is a function of wave and current energy prevailing in an area. Because tidal flats are usually characterized by relatively low wave and current energy, they can be at considerable risk from floating oil. Therefore, tidal flats should be given a high priority for protection.

5. Background

5.1 Floating oil that reaches shore can be readily incorporated into intertidal sediments **(1)**.[2] Oil that is incorporated into intertidal sediments can persist for varying periods of time depending on the wave and current energy and the sedimentary matrix in which the oil is incorporated **(1, 2, 3)**. In general, the finer the sediments' characteristic grain size, the more persistent the oil will be **(1, 2)**. Oil incorporated into sediments will affect not only infaunal community structure **(4, 5, 6, 7, 8, 9)** but also the physiology **(10)** and productivity of intertidal animals.

5.2 Recent studies have shown the following results for 1 m^3 of dispersant/oil mixture discharged at the surface in 3½ m of water over a tidal flat and mixed downward by boat wakes:

5.2.1 The oil exposure of the tidal flat community was approximately 40 ppm-h **(12)**.

NOTE 1—One part per million-hour equals concentration of dispersed oil (ppm) times the exposure period in hours. This theory that suggests exposures can be equalibrated if parts per million per hour are the same assumes that critical concentration values for a subjected animal have not been exceeded.

5.2.2 Most of the dispersed oil remained in the upper metre of the water column **(12)**.

5.2.3 The oil that did reach the bottom had lost most of the molecules containing less than 17 carbon atoms **(12, 13)**.

5.2.4 No uptake of dispersed oil by sediments was observed. **(12, 13)**.

5.2.5 Some uptake of dispersed oil by bivalve mollusks was observed; these hydrocarbons were quickly depurated, within four weeks **(10)**.

5.2.6 No effect of the dispersed oil was observed on protein and carbohydrate metabolism in bivalves **(10)**.

5.2.7 No effect of dispersed oil on infaunal community structure was observed **(9)**.

5.3 It is very likely that most slicks dispersed in the nearshore region will be no thicker than 0.1 to 0.2 mm. If this is the case, the maximum concentration of dispersed oil in the upper metre of water will be no more than 100 to 200 ppm. Actual measurements of dispersed oil concentrations in the water beneath treated slicks have shown much lower values **(14)**. The dispersed oil will be subject to rapid dilution as well as to enhanced biodegradation because of the vastly increased surface area. Once the oil is dispersed into the water column it will move with the water currents which are in most cases parallel to the shoreline.

6. Recommendations

6.1 Application of chemical dispersants to an oil slick approaching tidal flats may minimize the overall impact of the floating oil on the tidal flats.

6.2 If the movement of an oil slick is forecast to impact tidal flats, and if the slick is to be dispersed, it should be dispersed as far from shore as possible. In the event that the oil must be dispersed near shore in shallow water (<5 m), little incorporation of the dispersed oil into the sediments can be expected; as a result, little biological impact can be expected.

6.3 Dispersants should not be applied to oil that is stranded on tidal flats.

7. Keywords

7.1 chemical dispersants; ecological considerations; oil spill response; tidal flats

[2] The boldface numbers in parentheses refer to the list of references at the end of this guide.

REFERENCES

(1) Beslier, A., et al., "La Pollution des Sediments Sublittoraux au Nord du Bretagne Par les Hydrocarbures de L'AMOCO CADIZ: Distribution et Evolution," *Amoco Cadiz; Consequences d'une Pollution Accidentelle par les Hydrocarbures*, Centre National Pour Exploitation des Oceans, Paris, 1981, pp. 95–106.

(2) Morel, G., and Courtot, P., "Evolution des Hydrocarbures Dans L'Eau et le Sediment de L'Aber Benoit," *Amoco Cadiz; Consequences d'une Pollution Accidentelle par les Hydrocarbures*, Centre National Pour Exploitation des Oceans, Paris, 1981, pp. 107–127.

(3) Mayo, D. W., et al., "Weathering Characteristics of Petroleum Hydrocarbons Deposited in Fine Clay Marine Sediments," Searsport Maine, *Journal of the Fisheries Board of Canada*, Vol 35, No. 5, 1978, pp. 552–562.

(4) Sanders, H. L., et al., "Anatomy of an Oil Spill: Long Term Effects of the Grounding of the Barge Florida off West Falmouth, Massachusetts," *Journal of Marine Research,* Vol 38, No. 2, pp. 265–380.

(5) Glemarec, M., and Hussenot, E., "Definition D'une Secession Ecologique en Milieu Meuble Abnormalement Enrichi en Matieres Organique a la Suitre de la Catastrophe de l'Amoco Cadiz," *Amoco Cadiz; Consequences d'une Pollution Accidentelle par les Hydrocarbures,* Centre National Pour Exploitation des Oceans, Paris, 1981, pp. 499–525.

(6) Le Moal, Y., and Quillien-Monot, M., *Etude des Populations*, 1981.

(7) Blanchard, D., "Impact de la Pollution par Hydrocarbures sur une Population de Crustaces Isopodes d'estran Meuble Intertidal," *Amoco Cadiz; Consequences d'une Pollution Accidentelle par les Hydrocarbures,* Centre National Pour Exploitation des Oceans, Paris, 1981, pp. 303–310.

(8) Chasse, C., and Guenole-Bouder, A., "Comparison Quantitative des Populations Benthiques des Plages de St-Efflam et St-Michel-en-Greve avant et depuis le Nauffrage de l'Amoco Cadiz," *Amoco Cadiz; Consequences d'une Pollution Accidentelle par les Hydrocarbures.*"

Centre National Pour Exploitation des Oceans, Paris, 1981, pp. 347–358.
(**9**) Gilfillan, E. S., et al., "Effects of Spills of Dispersed and Non-Dispersed Oil on Intertidal Community Structure," *Proceedings,* Oil Spill Conference, 1983, pp. 457–463.
(**10**) Gilfillan, E. S., et al., "Effects of Test Spills of Dispersed and Non-Dispersed Oil on Metabolic Pathways in Two Intertidal Bivalves *Mya arenaria* and *Mytilus edulis,*" *Proceedings*, ASTM, Oil Spill Dispersants Five Years of Research Conference, 1984.
(**11**) Gilfillan, E. S. "The Use of Scope-For-Growth Measurements in Monitoring Petroleum Pollution," Rapports et Proces-Verbaux des Reunions, Vol 179, pp. 71–75.
(**12**) Page, D. S., et al., "Tidal Area Dispersant Project: Fate of Dispersed and Undispersed Crude Oil in Two Nearshore Test Spills," *Proceedings,* ASTM, Oil Spill Dispersants Five Years of Research Conference, 1984.
(**13**) Page, D. S., et al., "Long Term Fate of Dispersed and Un-Dispersed Crude Oil in Two Nearshore Test Spills," *Proceedings,* Oil Spill Conference, American Petroleum Institute, Publication 4356, 1983, pp. 465–471.
(**14**) McAuliffe, C. D., et al., "Dispersion and Weathering of Chemically Treated Crude Oils in the Ocean," *Environmental Science and Technology,* Vol 14, No. 12, 1980, pp. 1509–1518.

Designation: F 990 – 86 (Reapproved 1993)$^{\epsilon 1}$

Standard Guide for Ecological Considerations for Use of Chemical Dispersants in Oil Spill Response—Sandy Beaches[1]

This standard is issued under the fixed designation F 990; the number immediately following the designation indicates the year of original adoption or, in the case of revision, the year of last revision. A number in parentheses indicates the year of last reapproval. A superscript epsilon (ε) indicates an editorial change since the last revision or reapproval.

ε[1] NOTE—Section 7 was added editorially in December 1992.

1. Scope

1.1 This guide covers recommendations for the use of chemical dispersants to assist in the control of oil spills. This guide is written with the goal of minimizing the environmental impacts of oil spills; this goal is the basis upon which recommendations are made. Aesthetic and socioeconomic factors are not considered; although, these and other factors are often important in spill response.

1.2 Each on-scene coordinator has available several means of control or cleanup of spilled oil. In this guide, use of chemical dispersants is not to be considered as a last resort after other methods have failed. Chemical dispersants are to be given equal consideration with other spill countermeasures.

1.3 This is a general guide only assuming the oil to be dispersable and the dispersant to be effective, available, applied correctly and in compliance with relevant government regulations. Oil, as used in this guide, includes crude oils and fuel oils (No. 1 through No. 6). Differences between individual dispersants or between different oils or products are not considered.

1.4 This guide covers one type of habitat, sandy beaches or marshes. Other guides, similar to this one, cover habitats such as rocky shores and marshes. The use of dispersants is considered primarily to protect such habitats from impact (or minimize impacts) and also to clean them after the spill takes place.

1.5 This guide applies to marine and estuarine environments, but not to freshwater environments.

1.6 In making dispersant-use decisions, appropriate government authorities should be consulted as required by law.

1.7 *This standard does not purport to address all of the safety problems, if any, associated with its use. It is the responsibility of the user of this standard to establish appropriate safety and health practices and determine the applicability of regulatory limitations prior to use.*

2. Significance and Use

2.1 This guide is meant to aid local and regional spill response teams during response planning and spill events.

3. General Considerations for Making Dispersant-Use Decisions

3.1 The decision of whether or not to use dispersants in a given spill situation is always one involving trade-offs. Dispersing a slick at one site temporarily introduces more oil into the water column at that site than would be there if a surface slick floated over it. Therefore, adverse effects on water column organisms may be increased at the site so that adverse effects can be decreased or eliminated at other sites.

3.2 Dispersant use is primarily a spill control method, not a cleanup method. Such use can give spill response personnel some control over where the impacts of a spill will occur and what types of impacts they may be. Since some environments are known to be more vulnerable to the longer lasting impacts of spilled oil, an acceptable trade-off may be to protect those environments by dispersing an oil slick in a less sensitive or less productive environment. In general, the trade-off that must be evaluated is between the impact of the relatively long residence time of spilled oil that strands on shorelines versus the short-term impact of dispersed oil in the water column.

3.3 In this guide, environments that are most vulnerable to the longer-term impacts of oil contamination are identified. Protection of these environments is recommended as a high priority, by means of dispersants or other methods.

4. Environments Covered—Sandy Beaches

4.1 Sands are defined as sediments in the size range of 0.06 to 2.0 mm. The composition of sand material may vary but is usually either siliceous or carbonate. The character of the sediment may be a significant factor in dispersant use. For example, oil adheres more readily to carbonate sands than to silica or volcanic sands. Sand beaches are a dynamic environment subject to continuous change due to the redistribution of sediments by the action of winds and waves. The amount of change (that is, the volume of sediment that is redistributed) is a function of the energy level at the shoreline.

4.2 This physically unstable habitat usually has low species diversity and low ecological productivity. The habitat is characterized by relatively simple plant and animal communities. Community diversity and productivity increase in more sheltered environments due to an increase in substrate stability.

4.3 The primary species that characterize sandy beach

[1] This guide is under the jurisdiction of ASTM Committee F-20 on Hazardous Substances and Oil Spill Responseand is the direct responsibility of Subcommittee F20.13 on Treatment.

Current edition approved June 27, 1986. Published September 1986.

habitats, are mobile and include small crustaceans (amphipods, isopods, and crabs), mollusks (clams), polychaete worms, and micro-fauna that live with in the substrate. There are few epifauna, large invertebrates, or plants in this habitat. Shorebirds and surf fish species are an integral predatory component of the sandy beach ecological community.

4.4 The importance of protecting sand beaches can be a seasonally variable factor, in particular, where the beach is used for spawning activities (for example, grunion or marine turtles). Sand beaches are not usually assigned a high priority for protection or cleanup. This priority would be reassessed if the shore is used by birds, mammals, or for spawning activities.

5. Background

5.1 Documentation of the use of dispersants on sand beaches and subsequent ecological effects is minimal.

5.2 Mackin **(1)** reports an example where dispersant use produced no detectable temporary or long-term effects on the fauna of a sand beach.

5.3 Laboratory studies **(2, 3)** and spill reports **(4, 5)** indicate that the use of dispersants can result in an increased penetration of the dispersed oil. However, Canevari **(6)** suggests that this deficiency can be overcome by the correct application of a water-based chemical dispersant.

5.4 Laboratory studies **(7)**indicate that dispersant application onto oil on an ebbing tide results in significantly greater penetration than either untreated oil or dispersed oil on a flooding tide.

5.5 The combination of dispersant application and pushing the dispersed oil/sediment mixture into the surf zone is suggested **(8)**for exposed noneroding beaches where oil has penetrated into the surface sediments. This technique has been used in a spill situation **(4)**.

5.6 The penetration of untreated or dispersed oil into a beach is a function of the physical character of the oil, the porosity of the sand, and the level of the water table in the beach **(9)**. In particular, penetration is inhibited by high water tables in the intertidal zone on wide, flat, sand beaches or in the intertidal beach ahead of an advancing tide.

6. Recommendations

6.1 Dispersants may be used on adjacent waters to protect a sand beach from an impact, but such action must be considered with reference to the guidelines for other nearshore environments.

6.2 Natural cleaning is the most preferred option for sandy beaches (see 1.1) **(10)**.[2] If action is required to clean the beach by manual or mechanical methods, then careful removal of contaminated surface sediments **(11)** will likely result in minimal ecological damage. Use mechanical methods of removal on highly contaminated areas prior to dispersant application.

6.3 Chemical dispersants can be a practical method of removal of certain types of oil from sand beaches **(12)**. The dispersant should be applied to the stranded oil on a rising (flooding) tide ahead of the waves to minimize penetration of the oil into the subsurface sediments. The rising tide will provide the necessary mixing energy and the required flushing action.

6.4 Caution should be used in the choice of a dispersant as the application of hydrocarbon-based dispersants or undiluted concentrates may increase the potential for penetration into the sand of the dispersed oil in some situations.

6.5 A trial application should be used to determine the applicability and effectiveness of the technique at the field site prior to a full-scale application. A recommended area should be about 1000 to 1500 ft^2 or 50 to 100 linear ft of typical shoreline.

6.6 Untreated oil on a shoreline may be removed by natural processes but subsequently can contaminate and cause ecological damage to adjacent nearshore waters and bottom sediments. Oil removed by chemical treatment from a sand beach (if not collected) also will migrate into adjacent nearshore waters but will be less likely to adhere to bottom sediments. For low-energy coasts, it is necessary to evaluate any potential damage to adjacent environments by both dispersed and untreated oil.

7. Keywords

7.1 beach; chemical dispersants; ecological considerations; oil spill response; sand; sandy beaches

[2] The boldface numbers in parentheses refer to the list of references at the end of this standard.

REFERENCES

(1) Mackin, J. G., "A Study of the Effect of an Oil Dispersal Agent Corexit in an Oil Cleanup Operation," *Technical Report, Texas A and M University, College Station, TX, 1970, 30 pp.*

(2) Mackay, D., Watson, A., Ng, C., and Nadeau, S., *"Behaviour and Effectiveness of Dispersants at Sea and at Shorelines," Proceedings of the 1979* Oil Spill Conference, Los Angeles, CA, American Petroleum Institute, Pub. No. 4308, Washington, DC, 1979, pp. 447–452.

(3) Harris, G. W., and Wells, P. G., "A Laboratory Study on the Adhesion of Crude Oil to Beach Sand in the Presence of a Dispersant," Spill Technology Newsletter, Vol 4, 1979, pp 293–298.

(4) Duerden, F. C., Spill in Portugal, "A Report of the' Jakob Maersk' Incident," Environmental Impact and Assessment Report EPS-8-EC-76-1, Environment Canada, Environmental Protection Service, Ottawa, Ont., 1975, 32 pp.

(5) Baker, J. M., Little, D., and Oldham, J. H., "Comparison of the Fate and Ecological Effects of Dispersed and Non-Dispersed Oil in a Variety of Marine Habitats," *Proceedings*, ASTM Dispersants Conference, West Palm Beach, FL, 1982.

(6) Canevari, G. P.," The Restoration of Oiled Shorelines by the Proper Use of Chemical Dispersants," *Proceedings of the 1979 Oil Spill Conference*, Los Angeles, CA, American Petroleum Institute, Pub. No. 4308, Washington, DC, 1979, pp. 443–446.

(7) Rowland, S. J., Tibbetts, P. J. C., Little, D., Baker, J. M., and Abbis, T. P., "The Fate and Effects of Dispersant Treated Crude Oil, With Particular Reference to Sheltered Intertidal Sediments," *Proceedings of the 1981 Oil Spill Conference*, Atlanta, GA, American Petroleum Institute, Pub. No. 4334, Washington, DC, 1981, pp. 283–293.

(8) Institute of Petroleum Code of Practice for the Use of Oil Slick

Dispersants, Heyden and Son Ltd., London, 1979, 34 pp.
(**9**) Owens, E. H., Harper, J. R., and Foget, C. R., *Shoreline Countermeasures*—1982 Study Results, BIOS, Baffin Island Oil Spill Working Report 82-4, Environment Canada, Edmonton, Alberta, 1983, 129 pp.
(**10**) Lindstedt-Siva, J., "Ecological Impacts of Oil Spill Cleanup—Are They Significant?," *Proceedings of the 1979 Oil Spill Conference*, Los Angeles, CA, American Petroleum Institute, Pub. No. 4308, Washington, DC, 1979, pp. 521–524.
(**11**) Foget, C. R., Schrier, E., Cramer, M., and Castle, R., *Manual of Practice for Protection and Cleanup of Shorelines*, 2 Volumes, U.S. Environmental Protection Agency, Office of Research and Development, Cincinnati, OH, 1979.
(**12**) Owens, E. H., Robson, W., Foget, C. R., and Harper, J. R., *The BIOS Shoreline Experiments: Results from 1980–1982 Field Studies, Proceedings of the Sixth Arctic Marine Oil Pollution (AMOP) Technology Seminar*, Environment Canada, Edmonton, Alberta, 1983, pp. 59–64.

Designation: F 999 – 86 (Reapproved 1993)$^{\epsilon 1}$

Standard Guide for Ecological Considerations for the Use of Chemical Dispersants in Oil Spill Response—Gravel or Cobble Beaches[1]

This standard is issued under the fixed designation F 999; the number immediately following the designation indicates the year of original adoption or, in the case of revision, the year of last revision. A number in parentheses indicates the year of last reapproval. A superscript epsilon (ϵ) indicates an editorial change since the last revision or reapproval.

ϵ[1] NOTE—Editorial changes were made throughout in December 1992.

1. Scope

1.1 This guide covers recommendations for the use of chemical dispersants to assist in the control of oil spills. This guide is written with the goal of minimizing the environmental impacts of oil spills; this goal is the basis upon which recommendations are made. Aesthetic and socioeconomic factors are not considered, although these and other factors are often important in spill response.

1.2 Each on-scene coordinator has available several means of control or cleanup of spilled oil. In this guide, use of chemical dispersants is not to be considered as a last report after other methods have failed. Chemical dispersants are to be given equal consideration with other spill countermeasures.

1.3 This is a general guide only assuming the oil to be dispersable and the dispersant to be effective, available, applied correctly and in compliance with relevant government regulations. Oil, as used in this guide, includes crude oils and fuel oils (No. 1 through No. 6). Differences between individual dispersants or between different oils or products are not considered.

1.4 This guide covers one type of habitat, gravel or cobble beaches. Other guides, similar to this one, cover habitats such as rocky shores, marshes. The use of dispersants is considered primarily to protect such habitats from impact (or minimize impacts) and also to clean them after the spill takes place.

1.5 This guide applies to marine and estuarine environments, but not to freshwater environments.

1.6 In making dispersant-use decisions, appropriate government authorities should be consulted as required by law.

1.7 *This standard does not purport to address all of the safety problems, if any, associated with its use. It is the responsibility of the user of this standard to establish appropriate safety and health practices and determine the applicability of regulatory limitations prior to use.*

2. Referenced Documents

2.1 *ASTM Standards:*

F 972 Guide for Ecological Considerations for the Use of Chemical Dispersants in Oil Spill Response—Nearshore Subtidal[2]

3. Significance and Use

3.1 This guide is meant to aid local and regional spill response teams during response planning and spill events.

4. General Considerations for Making Dispersant-Use Decisions

4.1 The decision of whether or not to use dispersants in a given spill situation is always one involving trade-offs. Dispersing a slick at one site temporarily introduces more oil into the water column at that site than would be there if a surface slick floated over it. Therefore, adverse effects on water column organisms may be increased at the site so that adverse effects can be decreased or eliminated at other sites.

4.2 Dispersant use is primarily a spill control method, not a cleanup method. Such use can give spill response personnel some control over where the impacts of a spill will occur and what types of impacts they may be. Since some environments are known to be more vulnerable to the longer lasting impacts of spilled oil, an acceptable trade-off may be to protect those environments by dispersing an oil slick in a less sensitive or less productive environment. In general, the trade-off that must be evaluated is between the impact of the relatively long residence time of spilled oil that strands on shorelines versus the short-term impact of dispersed oil in the water column.

4.3 In this guide, environments that are most vulnerable to the longer-term impacts of oil contamination are identified. Protection of these environments is recommended as a high priority, by means of dispersants or other methods.

5. Environments Covered—Gravel or Cobble Beaches

5.1 Gravel or cobble beaches are composed of a range of sediment types that include sands (0.06 to 2.0 mm), granules (2 to 4 mm), pebbles (4 to 64 mm), and cobbles (64 to 256 mm). Gravel beaches are composed of a mixture of sediments that are predominantly greater than 2.0 mm in diameter.

---

[1] This guide is under the jurisdiction of ASTM Committee F-20 on Hazardous Substances and Oil Spill Responseand is the direct responsibility of Subcommittee F20.13on Treatment.

Current edition approved June 27, 1986. Published September 1986.

[2] *Annual Book of ASTM Standards*, Vol 11.04.

5.2 The habitats are physically dynamic and are therefore unstable due to the redistribution of sediments by wave action. The rates at which changes take place (that is, sediment redistribution) are a function of material size and available wave energy. In sheltered environments, gravel beaches may be relatively stable whereas on exposed coasts cobble sediments may be continuously redistributed.

5.3 The unstable character of gravel/cobble beaches results in simple biological communities, low species diversity, and low abundance. Diversity, abundance, and productivity increase in more sheltered environments with increasing substrate stability.

5.4 The primary species that characterize exposed gravel or cobble beach habitats are mobile and include small crustaceans (amphipods, isopods, and crabs), molluscs (chitons, clams, and snails), polychaete worms, and a wide range of micro-fauna. There exists some variation that is dependent upon sediment size. For example, clams occur in gravel but not in cobble habitats. On low-energy gravel beaches additional cessile species such as barnacles, mussels, and algae are common. Shorebirds and surf fish are common and some seasonal species use these habitats for spawning (for example, capelin).

6. Background

6.1 Documentation of the use of dispersants on gravel or cobble beaches and subsequent ecological effects is minimal.

6.2 Oil penetration on gravel beaches is similar to that for sand beaches (**1**)[3]; whereas on cobble beaches, penetration would be expected to be higher (**2, 3**).

6.3 Chemical dispersants can be a practical method for removal of oil from lightly contaminated beaches (**4**) but are rarely effective on heavily contaminated beaches or if the oil has been weathered. The cleanup of stranded oil on the beach may be of value by preventing oil migration from the shore to the nearshore subtidal environment and subsequent prevention of tarmats or pavement formation in the intertidal zone.

6.4 When considering use of dispersant for cleaning cobble beaches, the impact of the dispersed oil from the cobble beach on other environments should be considered. See Guide F 972 for subtidal environments.

7. Recommendations

7.1 On an ecological basis, gravel or cobble beaches usually are not high priority habitats for protection or cleanup. This priority might change if the habitat is used by birds, mammals, or for spawning activities. This guide refers to the use of chemical dispersants for the removal of stranded oil. Dispersants may be used on adjacent waters to protect a gravel or cobble beach from contamination, but such action should also be considered with reference to the guidelines on other nearshore habitats.

7.2 On most gravel or cobble beaches, natural cleaning would be the preferred option. Physical removal of the oil or the use of a dispersant should be considered if further cleaning is necessary. Application of a dispersant should be preceded by removal of any heavy accumulation of oil by physical methods.

7.3 Shoreline cleaning by use of chemicals should be used only if the oil can be flushed out of the beach. In very sheltered locations, there may be insufficient wave energy to remove the oil. In this case, the use of dispersants is not recommended unless adequate provision is made for flushing with sea-water by mechanical means.

7.4 The dispersant should be applied to the stranded oil on a rising tide ahead of the waves to minimize penetration of the oil into the subsurface sediments. The rising tide will provide the necessary mixing energy and the required flushing action. Experiments on low-energy beaches have shown that dispersants can be an effective method for cleaning oiled gravel or cobble beaches (**5**).

7.5 Hydrocarbon-based dispersants or undiluted concentrates usually are not recommended as these can increase the potential for penetration of the dispersed oil in certain situations. They have been used with some success on a small scale on gravel or cobble beaches (**6**).

8. Keywords

8.1 beach; chemical dispersants; cobble beach; ecological considerations; gravel beach; oil spill response

[3] The boldface numbers in parentheses refer to the list of references at the end of this guide.

REFERENCES

(**1**) Owens, E. H., Robson, W., Foget, C. R., and Harper, J. R., "The BIOS Shoreline Experiments: Results from 1980–1982 Field Studies," *Proceedings of the Sixth Arctic Marine Oil Pollution (AMOP) Technology Seminar*, Environment Canada, Edmonton, Alberta, 1983, pp. 59–64.

(**2**) Owens, E. H., "Coastal Environments of Canada, The Impact and Cleanup of Oil Spills," Environmental Protection Service, *Environment of Canada Report*, EPS-3-EC-77-13, Ottawa, Ontario, 1977, 413 pp.

(**3**) Mackay, D., Watson, A., Ng, C., and Nadeau, S., "Behavior and Effectiveness of Dispersants at Sea and at Shorelines," *Proceedings of the 1979 Oil Spill Conference*, Los Angeles, CA, American Petroleum Institute, Pub. No. 4308, Washington, D.C., 1979, pp. 447–452.

(**4**) Blackall, P. J., and Sergy, G. A., "The BIOS Project—An Update," *Proceedings of the 1983 Oil Spill Conference*, San Antonio, TX, American Petroleum Institute, Pub. No. 4356, Washington, DC, 1983, pp. 451–455.

(**5**) Owens, E. H., Foget, C. R., and Robson, W., *Experimental Use of Dispersants for Spill Countermeasure on Arctic Beaches in Oil Spill Chemical Dispersants: Research, Experience, and Recommendations, STP 840*, Thomas E. Allan, ed., ASTM, 1984, pp. 324–337.

(**6**) British Petroleum, "The' Christos Bitas' Incident," Environmental Control Centre, London, 1979, 24 pp.

Designation: F 1008 – 86 (Reapproved 1993)$^{\epsilon 1}$

Standard Guide for Ecological Considerations for the Use of Chemical Dispersants in Oil Spill Response—Salt Marshes[1]

This standard is issued under the fixed designation F 1008; the number immediately following the designation indicates the year of original adoption or, in the case of revision, the year of last revision. A number in parentheses indicates the year of last reapproval. A superscript epsilon (ϵ) indicates an editorial change since the last revision or reapproval.

ϵ^1 NOTE—Section 7 was added editorially in December 1992.

1. Scope

1.1 This guide covers recommendations for the use of chemical dispersants to assist in the control of oil spills. This guide is written with the goal of minimizing the environmental impacts of oil spills; this goal is the basis upon which recommendations are made. Aesthetic and socioeconomic factors are not considered, although these and other factors are often important in spill response.

1.2 Each on-scene coordinator has available several means of control or cleanup of spilled oil. In this guide, use of chemical dispersants is not considered as a last resort after other methods have failed. Chemical dispersants are to be given equal consideration with other spill countermeasures.

1.3 This is a general guide only assuming the oil to be dispersible and the dispersant to be effective, available, applied correctly and in compliance with relevant government regulations. Oil, as used in this guide, includes crude oils and fuel oils (No. 1 through No. 6). Differences between individual dispersants or between different oils or products are not considered.

1.4 This guide covers one type of habitat, salt marshes. Other guides, similar to this one, cover habitats such as rocky shores. The use of dispersants is considered primarily to protect such habitats from impact (or minimize impacts) and also to clean them after the spill takes place.

1.5 This guide applies to marine and estuarine environments, but not to freshwater environments.

1.6 In making dispersant-use decisions, appropriate government authorities should be consulted as required by law.

1.7 *This standard does not purport to address all of the safety problems, if any, associated with its use. It is the responsibility of the user of this standard to establish appropriate safety and health practices and determine the applicability of regulatory limitations prior to use.*

2. Significance and Use

2.1 This guide is meant to aid local and regional spill response teams during spill response planning and spill events.

3. General Considerations for Making Dispersant-Use Decisions

3.1 The decision of whether or not to use dispersants in a given spill situation is always one involving trade-offs. Dispersing a slick at one site temporarily introduces more oil into the water column at that site than would be there if a surface slick floated over it. Therefore, adverse effects on water column organisms may be increased at the site so that adverse effects can be decreased or eliminated at other sites.

3.2 Dispersant use is primarily a spill control method, not a cleanup method. Such use can give spill response personnel some control over where the impacts of a spill will occur and what types of impacts they may be. Since some environments are known to be more vulnerable to the longer lasting impacts of spilled oil, an acceptable trade-off may be to protect those environments by dispersing an oil slick in a less sensitive or less productive environment. In general, the trade-off that must be evaluated is between the impact of the relatively long residence time of spilled oil that strands on shorelines versus the short-term impact of dispersed oil in the water column.

3.3 In this guide, environments that are most vulnerable to the longer-term impacts of oil contamination are identified. Protection of these habitats is recommended as a high priority by means of dispersants and other methods.

4. Environments Covered—Coastal Salt Marsh

4.1 Coastal salt marshes (distinct from inland salt marshes) are intertidal wetlands, transitional zones between terrestrial and aquatic ecosystems. Salt marshes form when plants invade shallow, protected tidal flats on low coasts (for example, behind barrier islands or in estuaries where protection from strong wave action is provided) **(1)**.[2] Their primary formation requires soil emersion for approximately half the tidal period, water calm enough to prevent uprooting of the newly settled seedlings, and a sufficient amount of sediment to enable the upward growth of the marsh **(2, 3)**.

4.2 Coastal salt marshes are usually found in temperate regions but can and do occur in the tropics, especially in arid or monsoonal zones. Salt marshes occur where abundant silt is

[1] This guide is under the jurisdiction of ASTM Committee F-20 on Hazardous Substances and Oil Spill Responseand is the direct responsibility of Subcommittee F20.13 on Treatment.

Current edition approved June 27, 1986. Published October 1986.

[2] The boldface numbers in parentheses refer to the list of references at the end of this guide.

deposited by major river systems or where geological nature of the coast results in considerable erosion and significant silt resuspension **(4)**. Major geographic groupings according to Chapman **(5)**are as follows:

4.2.1 Arctic Group,

4.2.2 Northern European Group (Scandinavian, North Sea, Baltic, English Channel, and Southwest Ireland subgroups),

4.2.3 Mediterranean Group (western Mediterranean, eastern Mediterranean, and Caspian subgroups),

4.2.4 Western Atlantic Group (Bay of Fundy, New England, and Coastal Plain subgroups),

4.2.5 Pacific American Group,

4.2.6 Sino-Japanese Group,

4.2.7 Australasian Group (Australian and New Zealand subgroups),

4.2.8 South American Group, and

4.2.9 Tropical Group.

4.3 The fundamental physiographic units of the marsh are **(6)** the drainage basin or creek system **(7)** and the salt pan. Creeks form by accretion on the surrounding marsh, supplemented by lateral and bed erosion. In sandy marshes, where puccinellia grass is dominant, the creek systems are simple and drainage channels are few. In other muddier marshes the creek systems branch in long, dendritic patterns, and cord grass (*Spartina*) marshes form intricate winding systems of creeks and drainage channels. A salt pan is a bare spot or depression, usually circular in shape, that is surrounded by marsh vegetation and has lost an outlet for water. As a result of flooding by spring tides and evaporation in the summer, this area becomes hypersaline and adverse to plant growth **(6)**.

4.4 Along coastlines where marshes are found, there exists a gradual progression from submerged sand flats through a range of different plant communities and brackish water vegetation to fresh water swamps and finally upland vegetation.

4.5 Perennial marsh grasses and rushes on the higher intertidal levels **(8, 9)** dominate the flora of the marsh. Some annuals may also be present.

4.6 In wave protected submerged shallows (or the seaward sand flat), sea grasses and algae take root and trap sediment. When the bottom has risen to form a tidal flat, it becomes suitable for colonization by lateral marsh grass pioneers. These plants maintain the process of sediment gathering, and the high marsh is generally dominated by the short cord grass or by species of salt grass, pickleweed, or sea blite. The transition continues from salt marsh to reed swamp or to a cattail swamp dominated by the rushes or sedges.

4.7 Within the salt marsh proper, there are about ten rather distinctive subhabitats **(10)**. Each of these subhabitats provides distinctive growing conditions for the communities that occupy them. While these subhabitats can occupy a large area within the marsh, their distinctive community structures and functions have not been studied in detail.

4.8 Primary production in coastal marshes is carried out mainly by the grasses and by algae living on the surface of the sediments. Total net primary production of a healthy cord grass (*Spartina alterniflora*) marsh system has been estimated to be as high as 200 g/m^2/year of biomass **(11)**. This biomass supports a complex food web, as well as stabilizing the substrate or sediments by a network of rhizomes (submerged roots). Bacterial breakdown of dead plant matter supports populations of detritus feeders (for example, fiddler crabs, mussels, and snails), which in turn are eaten by higher organisms (for example, mud crabs, birds, and raccoons). Tidal exchange between the salt marsh proper and adjacent waters provides food (detritus) for aquatic animals (for example, shrimp and commercially important fish).

4.9 Salt marshes function as vital breeding and nursery grounds for many commercially important fish and shellfish, may support overwintering water fowl, provide a temporary or permanent home to birds, and provide food and shelter for terrestrial animals.

5. Background

5.1 Oil slicks impact the salt marsh ecosystem directly by coating the leaves, shoots, and aerial roots of the marsh grasses **(12, 13)**. The degree of coating may be dependent upon recoating of the grass during subsequent tidal cycles. Entrapped oil also may be toxic to burrowing and surface organisms **(14)**. Toxicity to *Spartina* spp. (cord grass), the predominant perennial in salt marshes **(8)**, is due to absorption of oil through stomata **(12)** that apparently decreases the diffusion of oxygen to the root systems **(15)**. Brown **(12)** found that nonbiogenic hydrocarbon in oiled *Spartina* were concentrated in the lower shoots and roots, even after one year of weathering. Concentrated petroleum hydrocarbons can result in death of the underground rhizome system of a *Spartina* marsh and inhibit natural regrowth **(16)**. Annual marsh grasses may be even more susceptible to damage from oil than perennials **(12, 14)**.

5.2 Although devastation of a marsh from a single oiling and subsequent cleanup has been reported **(17)**, observations in later growing seasons of good recoveries of vegetation and associated infauna following single and even multiple spill events are common **(12, 18, 19, 20, 21)**. For example, in West Falmouth, MA, following a spill of about 700 000 L of No. 2 fuel oil, an oil high in toxic aromatic compounds, Blumer, Sass, et al **(22, 23)** reported large scale mortalities of the salt marsh grasses (*Spartina* and others) that came into contact with the oil and heavy mortalities in fiddler crabs and mollusc populations **(24)**. In a limited area, oil penetrating the sediments as deep as 70 cm initially prevented resettlement by the original fauna. However, after eight months, a reduction of the oil by biochemical and physical processes led to the subsequent repopulation of that area **(24)**.

5.3 In some cases, spilled oil can persist for years within or on the surface of the sediments, where it remains biologically available to benthic organisms **(25, 26, 27)**, and may cause re-oiling of the marsh during tidal flushing **(14)** or if the sediments are disturbed **(27)**. Vandermeulen and Ross **(27)** found that toxic aromatic hydrocarbons were persistent in marsh sediments whereas aliphatic compounds degraded over time.

5.4 In general, damage to vegetation increases with the number of exposures. Brown **(12)** found that while even a single exposure to crude oil "caused some plant mortality as well as 40 to 50 % reduction of growth the following year," multiple exposures caused an even greater degree of reduction in the following year's growth **(12)**. Oil spills during the migratory seasons of shrimps and fish, which depend upon the salt marsh for food and protection, can result in serious impacts to these species. Likewise, oil spilled during the winter months can significantly impact over-wintering populations, thus potentially strengthening the consideration for dispersant use during these periods.

5.5 In situ studies of the impact of various cleaning and removal actions on oiled marsh systems indicate that cleanup is best accomplished by methods that do not disturb the sediments or increase penetration of oil into them **(13, 14)**. Adverse effects of heavy equipment cleanup on salt marsh systems were observed at Ile Grande, France, following a tanker spill in 1978 **(17)**. Deepening of the channels of the main tidal area to allow drainage of accumulated oil increased the water flow through the marsh and the current velocities. These changes plus the loss of vegetative root system that held the fine marsh sediments, also caused by heavy equipment, resulted in erosional damage.

5.6 The ability of salt marshes to recover after being oiled and the insensitivity to manual and mechanical disruption raise the question of whether any cleanup should be attempted in these environments **(14, 15, 28)**. The effect on recovery of marsh vegetation from various manual removal methods (for example, low-pressure flushing, sorbing, burning, or cutting) versus the no treatment option have been studied **(13, 27, 28)**. Holt et al **(13)** reported only slightly better recoveries in cleaned over uncleaned areas of a *Spartina* marsh. McCauley and Harrel **(28)** reported a long recovery period and reduced numbers of infauna in marsh sediments where oiled vegetation had been burned, while infaunal communities in cut and no treatment areas recovered more quickly.

5.7 Most existing data on the toxicity of dispersants and dispersant/oil mixtures in salt marsh ecosystems were obtained using early generation aromatic-solvent dispersants. The effects of later generation nonaromatic dispersant/oil mixtures on salt marsh animals and vegetation have also been studied **(15, 29, 30)**. Van Gelder-Ottoway **(30)** showed that oil mixed with nonaromatic dispersants was more acutely toxic to the marsh periwinkle, *Littorina* spp. than either the undiluted or dispersant alone in a 96-h acute toxicity test. Little et al **(31)** found that nonaromatic dispersant treatment increased the penetration of unweathered oil in sandy sediments, a factor that could increase the availability of the oil to infaunal organisms. None of these studies is conclusive in predicting the toxicity of oil treated with nonaromatic dispersant to all organisms in the salt marsh.

6. Recommendations

6.1 Salt marshes are high-priority habitats for protection from spilled oil since they are sensitive to oiling and to nearly all spill cleanup methods. Chemical dispersants should be considered as an appropriate and effective means to prevent oil from reaching marshes.

6.2 Chemical dispersion should be carried out as far from the salt marsh ecosystem as is feasible.

6.3 If a decision is made to use dispersants to an oil slick approaching a salt marsh, the dispersant should be applied if possible, as the tide starts to rise to maximize dispersion of the oil.

6.4 Direct application of dispersants to salt marshes is not recommended. If marsh vegetation has been heavily oiled, dispersants will probably not prevent nor lessen damage, and low-pressure flushing, or the no-treatment option should be considered.

7. Keywords

7.1 chemical dispersants; ecological considerations; marshes; oil spill response; salt marshes

REFERENCES

(1) Chapman, V. J., ed., *Ecosystems of the World 1: Wet Coastal Ecosystems*, Elsevier Scientific Publishing Co., New York, NY, 1977.

(2) Reimold, R. J., and Queen, W. H., eds., *Ecology of Halophytes*, Academic Press, New York, NY, 1974.

(3) Reimold, R. J., "Mangroves and Salt Marshes of the Eastern United States," *Wet Coastal Ecosystems*, V. J. Chapman, ed., 1977, pp. 157–166.

(4) Chapman, V. J., *Mangrove Vegetation,* Cramer, Lehre, London, 1975.

(5) Chapman, V. J., *Salt Marshes and Salt Deserts of the World,* 2nd ed., Lubrecht and Cramer, International Scholastic Book Service, 1974.

(6) Chapman, V. J., *Coastal Vegetation*, The MacMillan Co.; Pergamon Press, New York, NY, 1964.

(7) Pomeroy, L. R., and Wiegert, R. G., eds., *The Ecology of a Salt Marsh*, Springer-Perlag, New York, NY, 1981.

(8) Cowardin, L. M., Carter, V., and LaRoe, E. T., *Classification of Wetlands and Deepwater Habitats of the United States,* United States Fish and Wildlife Service, Biological Service Program, FWS/OBS-79/31, 1979.

(9) Dicks, B., and Iball, K., "Ten Years of Saltmarsh Monitoring—The Case History of a Southampton Water Saltmarsh and a Changing Refinery Effluent Discharge," *Proceedings of the 1981 Oil Spill Conference*, 1981, pp. 361–374.

(10) Ranwell, D. S., *Ecology of Salt Marshes and Sand Dunes*, 1972, Chapman and Hall, London, England.

(11) Cruz, A. A. de la, "Recent Advances in Our Understanding of Salt Marsh Ecology," *Proceedings of the Gulf of Mexico Coastal Ecosystem Workshop,* May 1980.

(12) Brown, Lewis R., *Fate and Effect of Oil in the Aquatic Environment*, the Environmental Research Laboratory Office of Research and Development U.S. EPA Narragansett, Rhode Island 02882, EPA 600/3-80-058a, 1980.

(13) Holt, S., Rebalais, S., Rebalais, N., Cornelius, J., and Holland, S., "Effects of an Oil Spill on Salt Marshes at Harbor Island, Texas I. Biology," *Proceedings of the Conference on Assessment of Ecological Impacts of Oil Spills*, American Institute of Biological Sciences, Keystone, CO, 1978.

(14) Westree, B., "Biological Criteria for the Selection of Cleanup Techniques in Salt Marshes," *Proceedings of the 1977 Oil Spill Conference*, 1977, pp. 231–235.

(15) Baker, Jennifer M., "Comparative Toxicities of Oils, Oil Fractions, and Emulsifiers," *The Ecological Effects of Oil Pollution on Littoral Communities*, E.B. Cowell, ed., Applied Science Publishers Ltd., Essex, U.K., 1971, pp. 78–87.

(16) Krebs, C. T., and Tanner, C., "Restoration of Oiled Marshes Through

Sediment Stripping and *Spartina* Propagation," *Proceedings of the 1981 Oil Spill Conference*, 1981, pp. 375–392.
(17) Vandermeulen, J. H., Long, B. N., and D'Ozouville, L. D., "Geomorphological Alteration of a Heavily Oiled Saltmarsh (Ile Grande, France) As a Result of a Massive Cleanup," *Proceedings of the 1981 Oil Spill Conference*, 1981, pp. 347–351.
(18) Neff, Jerry M., Sharp, M. S., McCulloch, W. L., "Impact of the Esso Bayway Oil Spill on Salt Marsh Macrofauna," *Proceedings of the 1981 Oil Spill Conference*, 1981, pp. 413–418.
(19) Straughn, D., "Biological Survey of Intertidal Areas in the Straits of Magellen in January, 1975, Five Months After the Metula Oil Spill," *Proceedings of the NOAA-EPA Symposium on Fate and Effect of Petroleum Hydrocarbons*, Seattle, WA, November 1976, pp. 147–160.
(20) Stein, R. J., Gundlach, E. R., and Hayes, M. O., "The Urquiola Oil Spill, Observations of Biological Damage Along the Spanish Coast," *Proceedings of the Conference on Assessment of Ecological Impacts of Oil Spills*, American Institute of Biological Sciences, Keystone, CO, 1968.
(21) Thomas, M. L. H., "Long-Term effects of Bunker C. Oil in the Intertidal Zone," *Proceedings of the NOAA-EPA Symposium on Fate and Effect of Petroleum Hydrocarbons*, Seattle, WA, November 1966, pp. 235–245.
(22) Blumer, M. J., Sass, J., Sauza, G., Sander, H., Grassle, F., and Hampson, G., "The West Falmouth Oil Spill," Woods Hole Oceanographic Institute, unpublished manuscript, 1970.
(23) Blumer, M. J., and Sass, J., "The West Falmouth Oil Spill II Chemistry," Woods Hole Oceanographic Institute, unpublished manuscript, 1972.
(24) Burns, K. A., and Teal, J. M., *Hydrocarbon Incorporation in Salt Marsh Ecosystems*, the West Falmouth Oil Spill, Woods Hole Oceanographic Institution, unpublished manuscript, 1971.
(25) Baker, Jennifer M., "Oil and Salt Marsh Soil," *The Ecological Effects of Oil Pollution on Littoral Communities*, E.B. Cowell, ed., Applied Science Publishers, Ltd., Essex, U.K., 1971.
(26) Gilfillan, E. S., Page, D. S., Gerber, R. P., Hansen, S., Cooley, J., and Hotham, J., "Fate of the *Zoe Colocotroni* Oil Spill and Its Effects on Infaunal Communities Associated with Mangroves, *Proceedings of the 1981 Oil Spill Conference*, 1981, pp. 353–360.
(27) Vandermeulen, J. H., and Ross, C. W., *Assessment of Cleanup Tests of an Oiled Salt Marsh—The Golden Robin Spill in Miquasha, Quebec*, Environmental Protection Service Canada, EPS-8-EC771, 1977.
(28) McCauley, C. A., and Harrel, R. C., "Effects of Oil Spill Cleanup Techniques on a Salt Marsh," *Proceedings of the 1981 Oil Spill Conference*, 1981.
(29) Anderson, J. W., Kiesser, S. L., Bean, R. M., Riley, R. G., and Thomas, B. L., "Toxicity of Chemically Dispersed Oil to Shrimp Exposed to Constant and Decreasing Concentrations in a Flowing System," *Proceedings of the 1981 Oil Spill Conference*, 1981, pp. 69–78.
(30) Van Gelder-Ottoway, S., "The Comparative Toxicities of Crude Oils, Refined Oil Products and Oil Emulsions," in *Marine Ecology and Oil Pollution,* Jennifer Baker, ed., John Wiley and Sons, New York, NY, pp. 287–302.
(31) Little, D., Baker, J. M., Abbiss, T. P., Rowland, S. J., and Tibbetts, P. J. C., "The Fate and Effects of Dispersant-Treated Compared with Untreated Crude Oil, With Particular Reference to Sheltered Intertidal Sediments," *Proceedings of the 1981 Oil Spill Conference*, 1981, pp. 283–296.

Designation: F 1009 – 86 (Reapproved 1999)

Standard Guide for Ecological Considerations for the Use of Chemical Dispersants in Oil Spill Response—Offshore[1]

This standard is issued under the fixed designation F 1009; the number immediately following the designation indicates the year of original adoption or, in the case of revision, the year of last revision. A number in parentheses indicates the year of last reapproval. A superscript epsilon (ϵ) indicates an editorial change since the last revision or reapproval.

1. Scope

1.1 This guide covers recommendations for the use of chemical dispersants to assist in the control of oil spills. This guide is written with the goal of minimizing the environmental impacts of oil spills; this goal is the basis upon which recommendations are made. Aesthetic and socioeconomic factors are not considered, although these and other factors are often important in spill response.

1.2 Each on-scene coordinator has available several means of control or cleanup of spilled oil. In this guide, use of chemical dispersants is not considered as a last resort after other methods have failed. Chemical dispersants are to be given equal consideration with other spill countermeasures.

1.3 This is a general guide only assuming the oil to be dispersible and the dispersant to be effective, available, applied correctly and in compliance with relevant government regulations. Oil, as used in this guide, includes crude oils and fuel oils (No. 1 through No. 6). Differences between individual dispersants or between different oils or products are not considered.

1.4 This guide covers one type of habitat, offshore environments. Other guides, similar to this one, cover habitats such as rocky shores. The use of dispersants is considered primarily to protect such habitats from impact (or minimize impacts) and also to clean them after the spill takes place.

1.5 These guidelines apply to marine and estuarine environments but not to freshwater environments.

1.6 In making dispersant-use decisions, appropriate government authorities should be consulted as required by law.

1.7 *This standard does not purport to address all of the safety concerns, if any, associated with its use. It is the responsibility of the user of this standard to establish appropriate safety and health practices and determine the applicability of regulatory limitations prior to use.*

2. Significance and Use

2.1 This guide is meant to aid local and regional spill response teams during spill response planning and spill events.

3. General Considerations for Making Dispersant-Use Decisions

3.1 The decision of whether or not to use dispersants in a given spill situation is always one involving trade-offs. Dispersing a slick at one site temporarily introduces more oil into the water column at that site than would be there if a surface slick floated over it. Therefore, adverse effects on water column organisms may be increased at the site so that adverse effects can be decreased or eliminated at other sites.

3.2 Dispersant use is primarily a spill-control method, not a cleanup method. Such use can give spill-response personnel some control over where the impacts of a spill will occur and what types of impacts they may be. Since some environments are known to be more vulnerable to the longer lasting impacts of spilled oil, an acceptable trade-off may be to protect those environments by dispersing an oil slick in a less sensitive or less productive environment. In general, the trade-off that must be evaluated is between the impact of the relatively long residence time of spilled oil that strands on shorelines versus the short-term impact of dispersed oil in the water column.

3.3 In this guide, environments that are most vulnerable to the longer-term impacts of oil contamination are identified. Protection of these habitats is recommended as a high priority by means of dispersants and other methods.

4. Offshore Environments

4.1 For the purpose of this guide, the offshore situation is taken to be any situation where water depths are sufficiently great that significant amounts of dispersed oil will not come in contact with bottom-dwelling organisms (that is, greater than 10 m).

4.2 Resources that are at risk from an oil spill in this environment include surface-dwelling organisms such as seabirds and marine mammals and inhabitants of the water-column, including phytoplankton, zooplankton, and fish and other pelagic organisms.

5. Background

5.1 Dispersants protect surface-dwellers from an oil slick, possibly at the expense of some resources in the water-column. The dispersant use decision rests on determining the action that would result in the least overall environmental impact. The

[1] This guide is under the jurisdiction of ASTM Committee F-20 on Hazardous Substances and Oil Spill Responseand is the direct responsibility of Subcommittee F20.13 on Treatment.

Current edition approved June 27, 1986. Published October 1986.

question of impact will be determined, in part, by the relative sensitivity of the various resources to short-term exposure to dispersed oil in the water column versus longer-term exposures nondispersed oil floating on the water surface or stranding on shorelines. The effects of oil on many of these resources (for example, seabirds, mammals, seagrass beds, saltmarshes) have been dealt with in other ASTM standards in this series. Although these resources vary in their sensitivity to oil, all suffer adverse effects from contact with oil. In this guide, some information on the sensitivity and vulnerability of phytoplankton, zooplankton, and fish is presented.

5.2 Much of the reticence to use dispersants is due to the popular belief that all dispersants are highly toxic and will themselves significantly increase the impact of a spill. This misconception grew out of the Torrey Canyon spill incident in 1967 in which high concentrations of chemical dispersants were used during spill cleanup and caused extensive damage in intertidal, subtidal, and water-column communities (**1, 2**).[2] The dispersants currently available are far less toxic (1/10 to 1/100th as toxic) and effective at lower concentrations than those used on the Torrey Canyon Spill. Most modern dispersants are less toxic than the oils they are used to disperse and hence do not increase the toxicity of oil to organisms (**3, 4, 5**). The use of dispersants does, however, increase the hazard posed by the spill to water-column organisms because they will be exposed to higher concentrations of hydrocarbons when dispersants are used than when they are not used. The dispersant-use decision will be based, in part, on the degree to which dispersants will increase the level of impact of the oil spill on the populations of water-column organisms versus the potential impacts of the undispersed slick.

5.3 There have been few observable effects of actual offshore oil spills on phytoplankton and zooplankton. There have been several reports of fish kills from untreated oil spills (**6, 7, 8**); these spills occurred in confined waters where, perhaps, fish could not avoid the oil. It is unlikely that mortalities would occur at significant levels in offshore spills whether they were chemically dispersed or not. No reports were found that have documented fish kills offshore.

5.4 Although hydrocarbons can be toxic to water-column organisms, the level of hazard presented by treated and untreated spills depends on the level and time of exposure and the sensitivity of these organisms to hydrocarbons. Oil concentrations near the surface under untreated experimental oil slicks have been reported to be as high as 2 to 4 ppm (**9, 10**). These concentrations occur only very near the surface (upper 2 m) and decline to one tenth these levels (that is, 0.2 to 0.4 ppm) at the 10-m depth (**10**) within hours (**11**). In addition, the more toxic compounds in the slick are relatively volatile and are lost quickly following the spill. All of this suggests that even if hydrocarbon concentrations are high enough to affect water column organisms, this effect is limited to the immediate vicinity of the spill.

5.5 The use of dispersants on oil slicks results in much higher oil concentrations in the surface waters. In offshore experiments, McAuliffe et al (**11**) reported that dispersed oil concentrations varied widely with test conditions, but levels as high as 40 ppm were observed at a depth of 1 m. As with untreated oil, concentrations were much lower (one tenth or less) at greater depths. Hydrocarbon concentrations decline quickly to near background levels within a few hours. This suggests that if the oil concentrations in the water-column ever exceeded potentially lethal levels, these would persist only briefly.

5.6 Work on toxicity of oil to phytoplankton has been reviewed by several authors (**12, 13, 14**). Hydrocarbons appear to affect phytoplankton by displacing the lipids in the plasma membrane, thereby influencing permeability (**15**). This results in responses ranging from a reduction in the rate of photosynthesis to mortality. Sensitivity to hydrocarbons varies with species and with the type of oil. Several authors have shown that the water-soluble-fraction of crude oils can inhibit photosynthesis at concentrations of 0.05 to 1 ppm during exposures of 24 to 96 h (**4, 16**). This suggests that hydrocarbon concentrations in the water-column may be high enough during longer exposures to cause some inhibition of photosynthesis during both chemically treated and untreated oil spills. In both treated and untreated cases, inhibition would occur in the top few metres. The phytoplankton community can be expected to recover rapidly following an oil spill due to rapid reproductive rates and recruitment from adjacent areas.

5.7 Studies of the toxicity of oil to zooplankton have been conducted on a wide variety of groups including protozoans, ctenophores, molluscs, crustaceans, and fish. This work has been reviewed by Corner, (**12**) and Wells (**5**). Responses to hydrocarbons range from narcosis to death. Narcosis may result from brief exposure to lower hydrocarbon concentrations. Few studies have attempted to simulate the brief exposures to oil (a few hours duration) that would occur in open ocean spills, so little toxicity information is available for these conditions. Evidence suggests that zooplankton may survive brief exposures to those concentrations (**17, 18**).

5.8 The toxicity of oil to ichthyoplankton (fish larvae and eggs) is of particular interest in view of the possible impact of spills on commercial fisheries. Exposing fish eggs to oil results in a variety of ecologically important responses including reduced hatchability, deformed larvae, and mortality of eggs (**18, 19**). As with other groups, fish larvae respond to brief exposures to lower concentrations of, hydrocarbons with narcosis, while death may result from more prolonged exposures or exposures to higher concentrations. Few studies have simulated the short exposures likely to be experienced by eggs and larvae in the open sea, so the relationship between concentration and exposure duration is not yet fully understood. However, some fish eggs appear to withstand continuous exposure to 1 ppm of oil without adverse effects (**20, 21**) but are affected by continuous exposures of 5 to 15 ppm (**20, 22**). Such continuous exposures are not likely to occur in the open ocean.

5.9 The effects of oil and petroleum hydrocarbons on fish have been reviewed by Butler and Burkes (**1**) and Penrose (**3**). Some ecologically significant sublethal responses, including reproductive impairment (**3**) and interference with vision (**23**)

[2] The boldface numbers in parentheses refer to the list of references at the end of this guide.

have been observed with chronic exposures to oil and exposures by way of diet. These studies have little applicability since they do not simulate the kind or brief exposures to hydrocarbons experienced by fish during spills in the open sea. More relevant information is available regarding lethal responses. Korn et al **(24)** and Rice et al **(25)** showed the 24-hr LC50 for fish exposed to the water-soluble fraction of Cook Inlet crude oil was roughly 2 ppm. Measurements made by Anderson et al **(26)** showed that toxicity varied between species and between oils, but that the 24-hr LC50 values lay generally within the range from 2 to 20 ppm. This suggests that based on the anticipated exposure conditions in untreated spills (see 5.4 and 5.5) fish may encounter lethal exposure conditions only in the upper few metres immediately beneath the oil slick. In the chemically treated case, fish may encounter lethal conditions in the water column within the cloud of dispersed oil during the first few hours following dispersant application and only in the upper few metres of the water column. It is uncertain whether the expected exposure duration would be enough to cause mortality. The area within which potentially lethal conditions will occur will be greater in the chemically dispersed case than in the untreated case, but for both cases effects can be expected to occur only in the vicinity of the spill.

5.10 Most populations of water-column organisms (with some exceptions) are distributed over a very large area while the area in which lethal conditions could occur is rather small for oil spills, whether dispersants are used or not. Hence, some mortality may result from either treated or untreated spills, but the numbers of animals killed will be small relative to total population. Surface-dwelling organisms such as seabirds are likewise quite often distributed over large areas, but many species become highly aggregated during breeding or migration, at which time an oil spill could impact a large proportion of the population. As a general rule, chemically dispersed oil in the water column will cause less impact than untreated oil on the ocean's surface or standing on shorelines.

6. Recommendations

6.1 Dispersant use is recommended to protect sensitive habitats, animals, or shorelines. This is especially true when untreated oil threatens highly aggregated populations (for example, migrating or staging populations of seabirds, breeding sites, areas of upwelling) or threatens particularly oil-sensitive coastal areas (for example, salt marshes, mangrove swamps, seagrass beds). These oil-sensitive resources should be considered to be at risk from an untreated oil slick if they lie within several hundred kilometres of the spill site.

6.2 There are two possible exceptions to this recommendation. When the dispersed oil threatens aggregated populations of water-column organisms (usually applies only to fish eggs or larvae, or both), dispersant should be used only if it will minimize the overall ecological impacts caused by the spill.

6.3 When trajectory data indicate that the possibility of contact between oil and oil-sensitive resources is very small under prevailing conditions, there is little justification for using dispersants. However, continuous monitoring of slick movement is required if sudden wind or current changes could result in oil contamination of environmentally sensitive areas.

7. Keywords

7.1 chemical dispersants; ecological considerations; offshore; oil spill response

REFERENCES

(1) Butler, M. J. A., and Berkes, F., "Biological Aspects of Oil Pollution in the Marine Environment: a Review," *Marine Sciences Report*, No. 22, Marine Sciences Centre, McGill University, Montreal, Quebec, 1982, 133 p.

(2) Smith, J. E., ed, "*Torrey Canyon* Pollution and Marine Life," Report of Plymouth Laboratory of Marine Biological Association, United Kingdom. Cambridge University Press, London, 1970, 196 p.

(3) Penrose, W. R., "The Effects of Oil Hydrocarbons and Dispersants on Fish," in Sprague, et al **(27)**, 1980, pp. 81–86.

(4) Trudel, B. K., "The Effect of Crude Oil and Crude Oil/Corexit 9527 Suspensions on Carbon Fixation by a Natural Marine Phytoplankton Community," *Spill Technology Newsletter*, Vol 3, No. 2, 1978, pp. 56–63.

(5) Wells, P. G., "Zooplankton," in Sprague, et al **(27)**, 1981, pp. 65–86.

(6) Kirby, M. A., "Problems Facing the Oil Industry in the Arctic," *Marine Pollution Bulletin*, Vol 1, 1970, pp. 71–72.

(7) Nicol, C. W., "*The Mizushima Oil Spill*," Environmental Protection Service Report, EPS 8-EC-76-2, Environmental Canada. 28 p.

(8) Spooner, M., "Oil Spill in Tarut Bay, Saudi Arabia," *Marine Pollution Bulletin*, Vol 1, 1970, pp. 166–167.

(9) McAuliffe, C. D., "Dispersal and Alteration of Oil Discharged on a Water Surface," Wolfe, D. A., ed, *Fate and Effects of Petroleum Hydrocarbons in Marine Ecosystems and Organisms, Proceedings of the 1977 Symposium in Seattle, WA*, Pergamon, NY, 1976, pp. 19–35.

(10) Cormack, D., and Nichols, J. A., "The Natural and Chemical Dispersion of Oil in the Sea," *Rapports et Proses-Verbeaux des Reunions Conseil Internationale pour l'Exploration du la Mer*, Vol 171, 1977, pp. 97–100.

(11) McAuliffe, C. D., "Dispersion and Weathering of Chemically Treated Crude Oils on the Sea Surface," Mackay, D., et al, ed, *Chemical Dispersion of Oil Spills, Proceedings of the 1980 Symposium at the University of Toronto, Canada*. Pub. No. EE-17, pp. 89–91.

(12) Corner, E. D. S., "Pollution Studies with Marine Plankton, Part I, Petroleum Hydrocarbons and Related Components," *Advance Marine Biology*, Vol 15, 1978, pp. 289–380.

(13) Snow, N., "The Effects of Petroleum Hydrocarbons on Phytoplankton and Macrophytes," in Sprague, et al **(27)**, 1981, pp. 55–64.

(14) Vandermuelen, J. H., and Ahern, T. P., "Effects of Petroleum Hydrocarbons on Algae Physiology," in Lockwood, A. P. M., ed, *Effects of Pollutants on Aquatic Organisms, Society of Experimental Biological Seminar Series*, Vol 2, Cambridge University Press, London, 1976, pp. 107–125.

(15) O'Brien, P. Y., and Dixon, P. S., "The Effects of Oils and Oil Components on Algae: A Review," *British Phycology Journal*, Vol 11, 1976, pp. 115–142.

(16) Gordon, D. C., and Prouse, N. J., "Effects of Three Oils on Marine Phytoplankton Photosynthesis," *Marine Biology*, Vol 22, 1972, pp. 329–333.

(17) Byrne, C. J., and Calder, J. A., "Effect of Water-Soluble-Fractions of Crude, Refined, and Waste Oils on the Embryonic and Larval Stages of the Quahog Clam, *Mercenaria* sp," *Marine Biology*, Vol 40, 1977, pp. 225–231.

(**18**) Linden, O., "Biological Effects of Oil on the Early Development of the Baltic-Herring *Clupea harengas membras*," *Marine Biology*, Vol 45, 1978, pp. 237–283.

(**19**) Linden, O., "Effects of Oil on the Amphipod *Gammarus oceanicus*," *Environmental Pollution*, Vol 10, 1976, pp. 239–250.

(**20**) Ernst, V., Neff, J. M., and Anderson, J. W., "The Effects of the Water-Soluble-Fractions of No. 2 Fuel Oil on the Early Development of the Estuarine Fish, *Fundulus grandis*," *Environmental Pollution*, Vol 14, 1977, pp. 25–35.

(**21**) Smith, R. L., and Cameron, J. A., "Effect of Water-Soluble-Fraction of Prudhoe Bay Crude Oil on Embryonic Development of Pacific Herring," *Transactions*, American Fishery Society, Vol 108, 1979, pp. 70–75.

(**22**) Kuhnhold, W. W., "The Effect of Mineral Oils on the Development of Eggs and Larvae of Marine Species: A Review and Comparison of Experimental Data in Regard to Possible Damage at Sea," *Rapports et Proses-Verbeaux des Reunions Conseil Internationale pour l'Exploration du la Mer*, Vol 171, 1977, pp. 175–183.

(**23**) Hawkes, J. W., "The Effects of Petroleum Hydrocarbon Exposure on The Structure of Fish Tissues," Wolfe, D. A., ed., *Fate and Effects of Petroleum Hydrocarbons in Marine Ecosystems and Organisms, Proceedings of the 1976 Symposium in Seattle, WA*, Pergamon, NY, 1977, pp. 115–128.

(**24**) Korn, S., Moles, D. A., and Rice, S. D., "Effects of Temperature on the Median Tolerance Limit of Pink Salmon and Shrimp Exposed to Toluene, Naphthalene, and Cook Inlet Crude Oil," *Bulletin of the Environment Contamination Toxicology*, Vol 21, 1979, pp. 521–525.

(**25**) Rice, S. D., Short, J. W., Brodersen, C. C., Mecklenberg, T. A., Moles, D. A., Misch, C. J., and Karinen, J. F.," Acute Toxicity and Uptake-Depuration Studies with Cook Inlet Crude Oil, Prudhoe Bay Crude Oil, No. 2 Fuel Oil, and Several Subarctic Marine Organisms," *Processed Report*, Northwest Fisheries Center Auke Bay Fisheries Laboratory, National Fisheries Service, Auke Bay, Alaska, 1976, 90 p.

(**26**) Anderson, J. W., Neff, J. M., Cox, B. A., Tatem, H. E., and Hightower, G. M., "Characteristics of Dispersions and Water Soluble Extracts of Crude and Refined Oils and Their Toxicity to Estuarine Crustaceans and Fish," *Marine Biology*, Vol 27, 1974, pp. 75–88.

(**27**) Sprague, J. B., Vandermuelen, J. H., and Wells, P. G., eds, "*Oil and Dispersants in Canadian Seas—Research and Appraisal and Recommendations*," Environmental Emergency Branch, Environment Canada, 1981, 182 p.

Designation: F 1010 – 86 (Reapproved 1993)[ε1]

Standard Guide for Ecological Considerations for the Use of Chemical Dispersants in Oil Spill Response—Bird Habitats[1]

This standard is issued under the fixed designation F 1010; the number immediately following the designation indicates the year of original adoption or, in the case of revision, the year of last revision. A number in parentheses indicates the year of last reapproval. A superscript epsilon (ε) indicates an editorial change since the last revision or reapproval.

ε[1] NOTE—Section 7 was added editorially in December 1992.

1. Scope

1.1 This guide covers recommendations for the use of chemical dispersants to assist in the control of oil spills. This guide is written with the goal of minimizing the environmental impacts of oil spills; this goal is the basis upon which recommendations are made. Aesthetic and socioeconomic factors are not considered, although these and other factors are often important in spill response.

1.2 Each on-scene coordinator has available several means of control or cleanup of spilled oil. In this guide, use of chemical dispersants is not considered as a last resort after other methods have failed. Chemical dispersants are to be given equal consideration with other spill countermeasures.

1.3 This is a general guide only assuming the oil to be dispersible and the dispersant to be effective, available, applied correctly, and in compliance with relevant government regulations. Oil, as used in this guide, includes crude oils and fuel oils (No. 1 through No. 6). Differences between individual dispersants or between different oils or products are not considered.

1.4 This guide covers one type of habitat, bird environments. Other guides, similar to this one, cover habitats such as rocky shores. The use of dispersants is considered primarily to protect such habitats from impact (or minimize impacts) and also to clean them after the spill takes place.

1.5 This guide applies to marine and estuarine environments but not to freshwater environments.

1.6 In making dispersant-use decisions, appropriate government authorities should be consulted as required by law.

1.7 *This standard does not purport to address all of the safety problems, if any, associated with its use. It is the responsibility of the user of this standard to establish appropriate safety and health practices and determine the applicability of regulatory limitations prior to use.*

2. Significance and Use

2.1 This guide is meant to aid local and regional spill response teams during spill response planning and spill events.

3. General Considerations for Making Dispersant-Use Decisions

3.1 The decision of whether or not to use dispersants in a given spill situation is always one involving trade-offs. Dispersing a slick at one site temporarily introduces more oil into the water column at that site than would be there if a surface slick floated over it. Therefore, adverse effects on water column organisms may be increased at the site so that adverse effects can be decreased or eliminated at other sites.

3.2 Dispersant use is primarily a spill-control method, not a cleanup method. Such use can give spill-response personnel some control over where the impacts of a spill will occur and what types of impacts they may be. Since some evironments are known to be more vulnerable to the longer lasting impacts of spilled oil, an acceptable trade-off may be to protect those environments by dispersing an oil slick in a less sensitive or less productive environment. In general, the trade-off that must be evaluated is between the impact of the relatively long residence time of spilled oil that strands on shorelines versus the short-term impact of dispersed oil in the water column.

3.3 In this guide, environments that are most vulnerable to the longer-term impacts of oil contamination are identified. Protection of these habitats is recommended as a high priority, by means of dispersants and other methods.

4. Environments Covered—Bird Habitats

4.1 Birds are present, at some time during the year, in practically every area of the world. This guide emphasizes protecting the most important bird habitats.

4.2 Important bird habitat, regardless of the latitude, is defined as follows:

4.2.1 Wintering concentrations of birds or staging sites for large numbers of migrating birds,

4.2.2 Breeding colony sites,

4.2.3 Coastal areas that support dense concentrations of birds all year,

---

[1] This guide is under the jurisdiction of ASTM Committee F-20 on Hazardous Substances and Oil Spill Responseand is the direct responsibility of Subcommittee F20.13 on Treatment.

Current edition approved June 27, 1986. Published October 1986.

4.2.4 Important coastal wetlands, or

4.2.5 Endangered species habitat.

5. Background

5.1 Avian mortality from oiling is well documented **(1-5)**.[2] Birds that spend much of their time on or in the water (for example, sea ducks, auks, penguins) are the most vulnerable species to surface oil. Oil disrupts feather structure and causes feathers to mat together. This destroys the ability of the feathers to insulate the bird and to keep it afloat; birds may die from exposure or drowning **(6-8)**.

5.2 Oiled bird rehabilitation is reasonably effective for saving small numbers of oiled birds, but it has very limited potential for offsetting the population level effects of mortality from surface oil **(9, 10)**. However, endangered species or colonies of uncommon species may be aided significantly by rehabilitation efforts.

5.3 Methods developed for moving birds out of the path of surface oil may be species specific or work only under a given set of circumstances; each situation must be evaluated separately **(11)**.

5.4 Studies of oil ingestion have been inconsistent in design but have generally shown that short-term ingestion can cause detectable changes in avian physiology and behavior **(12-19, et al)**, and chronic ingestion can cause a variety of systemic effects **(20-29, et al)**.

5.5 Eggs of breeding birds can be lethally oiled as a result of oil transfer from the plumage of oiled adults **(29-31)**. Only minute amounts of oil are needed to produce a significant impact on hatching success **(32-35**, et al). The magnitude of the impact of oil ingestion and egg oiling on bird populations is unknown.

5.6 The contamination of an area by a large oil spill is often followed by a reduction of bird life in the vicinity of the spill **(36, 37)**. The causes of this exodus are unknown, but oil fumes and feeding difficulties are likely candidates. The effects of this displacement of birds from one location to another are unknown, but they would probably be most severe during the breeding season or in the winter.

5.7 Contamination of prime bird habitat, even during times of low use, may pose problems for the future; recovery from the effects of oil on marshes and intertidal habitat may take years **(38, 39)**.

5.8 The effects of chemically dispersed oil on birds in the wild are not well known. Recent studies have shown that dispersant alone or dispersed crude oil had no greater impact on weight gain, organ weights, corticosterone levels, or plasma thyroxine levels to herring gull and Leach's storm petrel chicks than did crude oil alone **(19, 23, 40)**.

5.9 The degree of protection from oiling provided by chemical dispersants is uncertain. Mallard ducks exposed to dispersant in water were less buoyant and remained wet longer than control birds or oil-exposed birds **(16)**. Ducks exposed to dispersed oil were just as soaked as birds exposed to dispersant alone, and their plumage was just as matted as the oil-exposed ducks. Oil-exposed and dispersed-oil-exposed ducks exhibited significantly increased basal metabolic rates when placed in a cold chamber immediately after 1 h of exposure. Dispersant concentration in the water was 6.7 ppm and the oil:dispersant ratio was 30:1. However, dispersants applied remote from bird habitats should prevent oil, dispersed oil, and dispersant from reaching them.

5.10 Laboratory studies showed that chemically dispersed crude oil more effectively reduced mucosal water and Na^+ transfer in the intestines of mallard ducklings than undispersed crude oil **(41)** and that dispersant alone or mixed with crude oil, in an egg oiling experiment, was at least as toxic to bird embryos as crude oil **(42)**. Dispersant sprayed on water did not affect mallard incubation or hatching success, and mallards exposed to partially dispersed crude oil (less than 25 % dispersion) had about the same hatching success as those exposed to undispersed crude oil **(43)**. Ingested dispersant (150 ppm in diet) and ingested dispersant mixed with crude oil (150 ppm dispersant and 1500 ppm oil in diet) had less of an effect on the weight gain and blood chemistry of young mallards than crude oil alone (1500 ppm in diet) **(44)**.

6. Recommendations

6.1 The primary effects of an oil spill on birds (plumage oiling, egg oiling, oil ingestion, habitat disruption) are caused by the floating oil. These effects can be reduced by moving the birds, mechanically removing surface oil, or dispersing the oil into the water column before it reaches birds or their habitat.

6.2 Dispersants can be considered for use even after the oil nears or enters habitat categories described in 4.2.1, 4.2.3, or 4.2.5 because dispersed oil is unlikely to be as hazardous to birds as floating oil. Since the effect on birds of external applications of concentrated dispersant is unknown, applications should be made with caution to avoid unnecessary aerial spraying of birds. Birds out of the oil or in very light surface oil should not be sprayed. Dispersants should not be sprayed on habitat category (4.2.2). There is concern in the scientific community about the use of dispersants in coastal wetlands (4.2.4). This concern, with respect to birds, presently lacks a strong scientific basis. Decisions about the use of chemical dispersants to protect birds in wetlands will have to be made on a case-by-case basis.

6.3 Cleanup of contaminated islands, beaches, or coastal wetlands is recommended only if breeding colonies are not disturbed and if mechanical damage to the site can be minimized.

6.4 Government environmental agencies should be involved in any decisions about the use of chemical dispersants to protect birds because of the international treaties or intranational regulations dealing with migratory or endangered species.

7. Keywords

7.1 bird habitats; birds; chemical dispersants; ecological considerations; habitat; oil spill response

[2] The boldface numbers in parentheses refer to the list of references at the end of this guide.

REFERENCES

(1) Bourne, W. R. P., "Oil Pollution and Bird Populations," *The Biological Effects of Oil Pollution on Littoral Communities*, supplement to *Field Studies*, Vol 2, pp. 99–121.

(2) Bourne, W. R. P., and Bibby, C. J., "Temperature and the Seasonal and Geographical Occurrence of Oiled Birds on West European Beaches," *Marine Pollution Bulletin*, Vol 6, 1975, pp. 77–80.

(3) Joensen, A. H., "Oil Pollution and Seabirds in Denmark, 1935–1968," *Danish Review of Game Biology*, Vol 6, No. 8, 1972, 24 pp.

(4) Clark, R. B., "Impact of Chronic and Acute Oil Pollution on Seabirds," *Background Papers; Workshop on Inputs, Fates, and Effects of Petroleum in Marine Environment*, Ocean Affairs Board, National Academy of Science, Washington, DC, 1973, pp. 619–634.

(5) Vermeer, K., and Anweiler, G. G., "Oil Threat to Aquatic Birds Along the Yukon Coast," *Wilson Bulletin*, Vol 87, 1975, pp. 467–480.

(6) Fabricius, E.," What Makes Plumage Waterproof?" *Annual Report, 1957–1958*, Waterfowl Trust, Vol 10, 1959, pp. 105–113.

(7) Hartung, R., "Energy Metabolism in Oil-Covered Ducks," *Journal of Wildlife Management*, Vol 31, 1967, pp. 798–804.

(8) Erasmus, T., Randall, R. M., and Randall, B. M. ,"Oil Pollution, Insulation and Body Temperatures in the Jackass Penguin *Spheniscus demersus*," *Comparative Biochemistry and Physiology*, Vol 69A, 1981, pp. 169–171.

(9) Stanton, P. B., "Operational Procedures for Reducing Wildlife Losses: Overview and Future Considerations," *Proceedings*, 1977 Oil Spill Response Workshop, U.S. Fish and Wildlife Service, Washington, DC, 1977, pp. 135–142.

(10) Clark, R. B., "Oiled Seabird Rescue and Conservation," *Journal Fisheries Research Board of Canada*, Vol 35, 1978, pp. 674–678.

(11) Ward, J. G., "Techniques for Dispersing Birds from Oil Spill Areas," *Proceedings*, 1977 Oil Spill Response Workshop, U.S. Fish and Wildlife Service, Washington, DC, 1977, pp. 113–124.

(12) Hartung, R., and Hunt, G. S., "Toxicity of Some Oils to Waterfowl," *Journal of Wildlife Management*, Vol 30, 1966, pp. 564–570.

(13) Snyder, S. B., Fox, J. G., and Soave, O. A.,"Mortalities in Waterfowl Following Bunker C Fuel Exposure," Division of Laboratory Animal Medicine, Stanford Medical Center, Stanford, CA, 1973, 27 pp.

(14) Crocker, A. D., Cronshaw, J., and Holmes, W. N.,"The Effects of Several Crude Oils and Some Petroleum Distillation Fractions on Intestinal Absorption in Ducklings (*Anas Platyrhynchos*)," *Environmental Physiology and Biochemistry*, Vol 5, 1975, pp. 92–106.

(15) Miller, D. S., Peakall, D. B., and Kinter, W. B.," Crude Oil Ingestion: Sub-lethal Effects in Herring Gull Chicks," *Science*, Vol 199, 1977, pp. 315–317.

(16) McEwan, E. H., and Whitehead, P. M., "Influence of Weathered Crude Oil on Liver Enzyme Metabolism of Testosteron in Gulls," *Canadian Journal of Zoology*, Vol 56, 1978, pp. 1922–1924.

(17) Ainley, D. G., and Grau, C. R., "Influence of Petroleum on Egg Formation and Embryonic Development in Seabirds," *Marine Biological Effects of OCS Petroleum Development*, NOAA Tech. Memo ERL OCSEAP-1, 1978, pp. 188–198.

(18) Peakall, D. B., Hallett, D., Miller, D. S., and Kinter, W. B., "Effects of Ingested Crude Oil on Black Guillemots: A Combined Field and Laboratory Study," *Ambio*, Vol 9, 1980, pp. 28–30.

(19) Rattner, B. A., "Tolerance of Adult Mallards to Subacute Ingestion of Crude Petroleum Oil," *Toxicology Letters*, Vol 8, pp. 337–342.

(20) Holmes, W. N., Cavanaugh, K. P., and Cronshaw, J.,"The Effects of Ingested Petroleum on Oviposition and Some Aspects of Reproduction in Experimental Colonies of Mallard Ducks (*Anas platyrhynchos*)," *Journal of Reproductive Fertility*, Vol 54, 1978a, pp. 335–347.

(21) Holmes, W. N., Cronshaw, J., and Gorsline, J., "Some Effects of Ingested Petroleum on Seawater-adapted Ducks (*Anas platyrhynchos*)," *Environmental Research*, Vol 17, 1978b, pp. 177–190.

(22) Holmes, W. N., Gorsline, J., and Cronshaw, J.," Effects of Mild Cold Stress on the Survival of Seawater-Adapted Mallard Ducks (*Anas platyrhynchos*), Maintained on Food Contaminated with Petroleum," *Environmental Research*, Vol 20, 1979, pp. 425–444.

(23) Patton, J. F., and Dieter, M. P., "Effects of Petroleum Hydrocarbons on Hepatic Function in the Duck," *Comparative Biochemistry and Physiology*, Vol 65C, 1980, pp. 33–36.

(24) Vangilder, L. D., and Peterle, T. J.," South Louisiana Crude Oil and DDE in the Diet of Mallard Hens: Effects on Reproduction and Duckling Survival," *Bulletin of Environmental Contamination and Toxicology*, Vol 25, 1980, pp. 23–28.

(25) Szaro, R. C., Hensler, G., and Heinz, G. H.," Effects of Chronic Ingestion of No. 2 Fuel Oil on Mallard Ducklings," *Journal of Toxicology and Environmental Health*, Vol 7, 1981, pp. 789–799.

(26) Coon, N. C., and Dieter, M. P.," Responses of Adult Mallard Ducks to Ingested South Louisiana Crude Oil." *Environmental Research*, Vol 24, 1981, pp. 309–314.

(27) Ainley, D. G., Grau, C. R., Roudybush, T. E., Morrell, S. H., and Utts, J. M., "Petroleum Ingestion Reduces Reproduction in Cassin's Auklets," *Marine Pollution Bulletin*, Vol 12, 1981, pp. 314–317.

(28) Rattner, B. A., and Eastin, Jr., W. C., "Plasma Corticosterone and Thyroxine Concentrations During Chronic Ingestion of Crude Oil in Mallard Ducks (*Anas platyrhynchos*)," *Comparative Biochemistry and Physiology*, Vol 68C, 1981, pp. 103–107.

(29) Gordon, J. C., "(Extract from a letter)," *Bird Notes and News*, Vol 13, 1929, p. 1975.

(30) Rittinghaus, H., "Etwas uber die Indirekte Verbreitung der Olpest in Einem Seevogelschutzgebiete," *Ornithologische Mitteilungen*, Vol 8, 1956, pp. 43–46.

(31) Birkhead, T. R., Lloyd, C., and Corkhill, P., "Oiled Seabirds Successfully Cleaning Their Plumage," *British Birds*, Vol 66, 1973, pp. 535–537.

(32) Albers, P. H., "Effects of External Applications of Fuel Oil on Hatchability of Mallard Eggs," *Fate and Effects of Petroleum Hydrocarbons in Marine Ecosystems and Organisms*, Pergamon Press, Inc., New York, NY, 1977, pp. 158–163.

(33) Albers, P. H., "The Effects of Petroleum on Different Stages of Incubation in Bird Eggs," *Bulletin of Environmental Contamination and Toxicology*, Vol 19, 1978, pp. 624–630.

(34) Szaro, R. C., Albers, P. H., and Coon, N. C.," Petroleum: Effects on Mallard Egg Hatchability," *Journal of Wildlife Management*, Vol 42, 1978, pp. 404–406.

(35) White, D. H., King, K. A., and Coon, N. C., "Effects of No. 2 Fuel Oil on Hatchability of Marine and Estuarine Bird Eggs," *Bulletin of Environmental Contamination and Toxicology*, Vol 21, 1979, pp. 7–10.

(36) Buck, W. F. S., and Harrison, J., "Some Prolonged Effects of Oil Pollution on the Medway Estuary," *Yearbook, 1966–1967*, Wildfowler's Assn. Great Britain and Ireland, 1967, pp. 32–33.

(37) Hope Jones, P., Monnat, J. Y., Cadbury, C. J., and Stowe, T. J., "Birds Oiled During the *Amoco Cadiz* Incident-An Interim Report," *Marine Pollution Bulletin*, Vol 9, 1978, pp. 307–310.

(38) Baker, J. M., "Marine Ecology and Oil Pollution," *Water Pollution Control Federation*, Vol 50, 1978, pp. 442–449.

(39) Burns, K. A., and Teal, J. M., "The West Falmouth Oil Spill: Hydrocarbons in the Salt Marsh Ecosystem," *Estuarine and Coastal Marine Science*, Vol 8, 1979, pp. 349–360.

(40) Butler, R. G., Trivelpiece, W., Miller, D., Bishop, P., D'Amico, C., D'Amico, M., Lambert, G., and Peakall, D., "Further Studies of the Effects of Petroleum Hydrocarbons on Marine Birds," *Bulletin*, Mount Desert Island Biology Laboratory, Vol 19, 1979, pp. 33–34.

(41) Crocker, A. D., Cronshaw, J., and Holmes, W. N.," The Effect of a Crude Oil on Intestinal Absorption in Ducklings (*Anas platyrhynchos*)," *Environmental Pollution*, Vol 7, 1974, pp. 165–177.

(42) Albers, P. H., "Effects of Corexit 9527 on the Hatchability of Mallard Eggs," *Bulletin of Environmental Contamination and Toxicology*, Vol 23, 1979, pp. 661–668.

(**43**) Albers, P. H., and Gay, M. L., "Effects of a Chemical Dispersant and Crude Oil on Breeding Ducks," *Bulletin of Environmental Contamination and Toxicology*, Vol 29, 1982, pp. 404–411.

(**44**) Eastin, Jr., W. C., and Eattner, B. A., "Effects of Dispersant and Crude Oil Ingestion on Mallard Ducklings (*Anas platyrhynchos*)," *Bulletin of Environmental Contamination and Toxicology*, Vol 29, 1982, pp. 273–278.

Designation: F 1011 – 86 (Reapproved 2001)

Standard Guide for Developing a Hazardous Materials Training Curriculum for Initial Response Personnel[1]

This standard is issued under the fixed designation F 1011; the number immediately following the designation indicates the year of original adoption or, in the case of revision, the year of last revision. A number in parentheses indicates the year of last reapproval. A superscript epsilon (ϵ) indicates an editorial change since the last revision or reapproval.

1. Scope

1.1 This guide covers a format for a hazardous materials spill initial response team training curriculum. This guide is designed to assist trainers of initial response personnel in assessing the content of training curriculum by providing guidelines for subject content against which these curricula may be evaluated. The guide should be tailored by the trainer to fit specific circumstances that are present in the community or industry where a spill may occur.

1.2 Sections 56789 of this guide identify those training areas that should be considered in a curriculum. The area of preplanning is listed and this topic should be seriously considered by the user. Training is only a small part of an overall spill response contingency plan. A properly equipped and trained spill response team cannot operate without a previously agreed plan of attack.

1.3 Currently the Code of Federal Regulation 40 CFR 112, 40 CFR 265, and 49 CFR 173 specify that producers, handlers, and shippers of hazardous materials shall plan and train for hazardous spill response. The broad interpretation of these regulations could include the requirement to train state and local response organizations who may be required to handle hazardous materials in an emergency spill situation. Regardless of the above regulatory requirements, training is essential to a proper response in an emergency.

1.4 *This standard does not purport to address all of the safety concerns, if any, associated with its use. It is the responsibility of the user of this standard to establish appropriate safety and health practices and determine the applicability of regulatory limitations prior to use.*

2. Referenced Documents

2.1 *Federal Regulations:*

40 CFR 112—Oil Pollution Prevention[2]

40 CFR 265—Interim Status Standards for Owners and Operators of Hazardous Waste Treatment, Storage, and Disposal Facilities[2]

49 CFR 173—Shippers—General Requirements for Shipments and Packagings[2]

2.2 *Other Documents:*

DOT-P5800.2—DOT Emergency Response Guide Book[3]

NIOSH/OSHA—Pocket Guide to Chemical Hazards[4]

Emergency Handling of Hazardous Materials in Surface Transportation[5]

OSHA 20 or Materials Safety Data Sheets[6]

3. Summary of Guide

3.1 This guide covers the following areas:

3.1.1 Preplanning,

3.1.2 Initial Assessment,

3.1.3 Personal Safety Equipment,

3.1.4 Training, and

3.1.5 Implementation of Plan of Attack.

3.2 Preplanning covers the aspects of pre-accident planning that should be considered by the response team members. This includes identification of potential hazardous chemicals, spill locations, and resource identification.

3.3 Initial assessment outlines those assessments that should be made when arriving at the accident location to assist in the development of a plan of attack.

3.4 Personal Safety Equipment discusses those safety devices and their limitations that are available to the response team members.

3.5 Training describes the activities that could be conducted by the team on a regular basis and the improvement of training plans after training exercises are completed.

3.6 Implementation of plans describes how and in what order the plan of attack should be implemented.

4. Significance and Use

4.1 This guide summarizes the typical contents of a course to aid emergency response team training organizations in selecting important subjects for inclusion in existing or new training programs.

---

[1] This guide is under the jurisdiction of ASTM Committee F20 on Hazardous Substances and Oil Spill Responseand is the direct responsibility of Subcommittee F20.21 on Initial Response Actions.

Current edition approved June 27, 1986. Published August 1986.

[2] Available from U. S. Government Printing Office, Washington, DC 20402.

[3] Available from U. S. Department of Transportation, 400 7th Street, S.W., Washington, DC 20590.

[4] Available from U. S. Department of Health and Human Services, 200 Independence Ave, S.W., Washington, DC 20201. Or available from U. S. Department of Labor, 200 Constitution Avenue, N.W., Washington, DC 20210.

[5] Available from the Association of American Railroads/Bureau of Explosives, 1920 L Street, N.W., Washington, DC 20036.

[6] OSHA 20 or Materials Safety Data Sheets are available from the specific chemical manufacturers.

5. Preplanning

5.1 Identify all areas subject to hazardous materials spills:

5.1.1 Fixed facilities that store or produce hazardous materials.

5.1.2 Routes that are used by carriers to transport hazardous materials.

5.2 Determine the type of hazardous material:

5.2.1 Stored or produced.

5.2.2 Transported:

5.2.2.1 Contact truck companies.

5.2.2.2 Contact rail companies.

5.3 Identify physical, chemical, and hazardous characteristics of each material:

5.3.1 Obtain OSHA 20 or Materials Safety Data Sheets for each hazardous material identified.

5.3.2 Suggested reference documents for data include the following:

5.3.2.1 Chemical handbooks; for example, Condensed Chemical Dictionary.

5.3.2.2 DOT-P5800.2—DOT Emergency Response Guide Book, U. S. Department of Transportation.

5.3.2.3 NIOSH/OSHA Pocket Guide to Chemical Hazards, U. S. Department of Health and Human Services, U. S. Department of Labor.

5.3.2.4 Emergency Handling of Hazardous Materials in Surface Transportation, Bureau of Explosives/Association of American Railroads.

5.4 Determine the mitigation resources available to respond.

5.4.1 A suggested reference is *ASTM STP 825*.[7]

5.4.2 Determine the number, qualifications, and location of knowledgeable personnel:

5.4.2.1 Local community.

5.4.2.2 Private industry.

5.4.3 Determine type and quantity of mitigation equipment available:

5.4.3.1 Neutralizer.

5.4.3.2 Foams.

5.4.3.3 Water sources.

5.4.3.4 Sorbents.

5.4.3.5 Dispensing equipment.

5.4.3.6 Containment equipment.

5.4.4 Determine the type, location, and method of activation of all automatic response systems at the potential spill site.

5.5 Identify critical population and environment areas adjacent to hazardous material sites or routes:

5.5.1 Schools, hospitals, shopping centers, etc.

5.5.2 Water supplies, soil conditions, recreational areas, etc.

5.6 Estimate the volume of the potential spill as follows:

5.6.1 Obtain potential spill volume estimates from storer, producer, or carrier.

5.6.2 Calculate spill volume from tankage sizes.

5.7 Determine spill drainage path for site or route as follows:

5.7.1 Attempt to determine the final spill location.

5.7.2 Determine accessibility to the spill location.

5.7.2.1 Establish travel route, including alternates, to the location.

5.7.2.2 Establish entry procedures.

5.7.3 Evaluate resources that could be made available at final spill location (refer to 5.4).

5.8 Determine vapor dispersion path for the site or route as follows:

5.8.1 Obtain normal meteorological data for area.

5.8.2 Select a dispersion model.

5.8.3 Develop vapor hazard corridor estimation procedures.

5.8.4 Develop dispersion estimates for appropriate accident scenarios.

5.8.5 Evaluate dispersion modeling results uncertainties.

5.8.6 Utilize dispersion model data in conjunction with population data to estimate exposure potential.

5.8.7 Consider specific chemical health impacts to identify population risk.

5.9 Establish an evacuation plan.

5.10 Determine spill reporting responsibilities for the following:

5.10.1 Local.

5.10.2 State.

5.10.3 Federal.

5.11 Develop a response plan with site or carrier management.

6. Initial Assessment of Incident

6.1 *Hazardous Substance Physical Identification*:

6.1.1 Identify source of spill (that is, tanker truck, railcar, storage facility).

6.1.2 Accurately identify substance spilled and its hazards:

6.1.2.1 *Substance Identification*:

(a) Department of Transportation (DOT) Placards.

(b) United Nations (UN) Numbers.

(c) Standard Transportation Commodity Code (STCC) Number.

(d) Markings or material labels.

(e) Shipping papers.

6.1.2.2 *Hazard(s) Identification*:

(a) Department of Transportation (DOT) Labels/Placards.

(b) National Fire Protection Association (NFPA) Labels.

(c) Physical observations.

(d) Detector measurements.

6.1.3 *Physical Characteristics*:

6.1.3.1 Solids.

6.1.3.2 Liquid.

6.1.3.3 Vapor.

6.1.4 Approximate the volume of spill or the total volume at source, or both.

6.1.5 For a transportation accident, determine the shipper's name.

6.1.6 For all types of accidents, determine the manufacturer's name.

6.1.7 Identify the person reporting spill.

6.1.8 Determine the approximate time of the spill (maintain chronological record of events).

6.1.9 Determine the time of spill report.

6.1.10 Estimate the material release rate.

6.2 *Identify the number and location of injured personnel*:

[7] *Guide to the Safe Handling of Hazardous Materials Accidents, ASTM STP 825*, ASTM, 1983.

6.2.1 Notify medical authorities.

6.2.2 Conduct a rescue assessment.

6.2.2.1 Safety implications.

6.2.2.2 Equipment required.

6.3 *Collect Site Information*:

6.3.1 *Current Weather Conditions*:

6.3.1.1 Rain (snow) or prospects of rain (snow).

6.3.1.2 Wind speed and direction.

6.3.1.3 Air temperature.

6.3.1.4 Weather stability.

6.3.1.5 Forecast (immediate, long term).

6.3.2 *Terrain Characteristics*:

6.3.2.1 Type of topography.

6.3.2.2 Porosity of ground surface.

6.3.2.3 Surface water.

6.3.2.4 Underground water.

6.3.3 *Demographics*:

6.3.3.1 Distance to public areas such as schools, churches, public buildings, busy intersections, shopping centers, recreational facilities, hospitals, convalescent centers, etc.

6.3.3.2 Distance to drinking water supplies.

6.3.3.3 Distance to sewers.

6.3.3.4 Distance to food and feed processing facilities.

6.4 *Establish on-scene authority*.

6.5 *Determine plan of action*:

6.5.1 Evaluate required response resources based on information collected.

6.5.2 Identify logistics problems.

6.5.3 Estimate impact area.

7. Personal Safety Equipment Requirements Identification

Note—All personal safety equipment should be fit tested.

7.1 *Levels of Protection* (EPA):

7.1.1 *Level A*:

7.1.1.1 Highest level of protection.

7.1.1.2 Self-contained breathing apparatus.

7.1.1.3 Positive pressure suits.

7.1.1.4 Total encapsulation.

7.1.1.5 Resistant to specific chemicals.

7.1.2 *Level B*:

7.1.2.1 Self-contained breathing apparatus.

7.1.2.2 Lighter weight rainwear.

7.1.2.3 Chemical resistant.

7.1.2.4 Minimum level recommended for initial site entry.

7.1.3 *Level C*:

7.1.3.1 Air purifying respirator with full face plate.

7.1.3.2 Splash garments.

7.1.3.3 Goggles, gloves, and head protection.

7.1.3.4 Type and concentrations of chemical must be known.

7.1.4 *Level D*:

7.1.4.1 Lowest level of protection.

7.1.4.2 Work uniforms.

7.1.4.3 Goggles, gloves, and head protection.

7.1.4.4 Optional use of escape mask.

7.2 *Breathing Protection Selection*:

7.2.1 Self-contained breathing apparatus.

7.2.2 Gas masks:

7.2.2.1 Not for firefighting or oxygen deficient atmospheres.

7.2.2.2 Need canister specific for atmosphere to be encountered.

7.2.3 Respirators (needs filter cartridge specific for the hazard).

7.2.4 Escape devices (not a duty cycle unit, for personal escape only).

7.2.5 Supplied air units, need umbilical hose and source of air (bottled gas or compressor).

7.2.6 Oxygen administrators (for resuscitation only)

7.3 *Personal Protection Selection*:

7.3.1 Head protection.

7.3.2 Eye protection.

7.3.3 Clothing, chemical protection.

7.3.3.1 Needs to be chemical specific for permeation.

7.3.3.2 Compatibility only is not sufficient but is necessary.

7.3.4 Ear protection.

7.3.5 Foot protection.

7.3.6 Hand protection.

7.4 *Gas Concentration Analyzing and Dosage Instrument Selection—Portable*:

7.4.1 Detector tubes, chemical specific (with hand pump).

7.4.1.1 Portable gas detectors.

7.4.1.2 Personnel Dosimeters, passive badges to be worn by individuals.

8. Training

8.1 Develop a training plan and schedule as follows:

8.1.1 Use other agency plans.

8.1.2 Review existing plans by other response teams.

8.2 Test and evaluate plans as follows:

8.2.1 Develop potential scenarios.

8.2.2 Select test participants.

8.2.3 Select evaluators.

8.2.4 Conduct drills.

8.2.5 Evaluate results.

8.2.6 Prepare critique of plan.

8.3 Plan revisions as follows:

8.3.1 Review test critique.

8.3.2 Revise plan.

8.3.3 Develop standard operating procedures for the initial response team.

9. Implementation of Plan

9.1 *Command Post*:

9.1.1 Establish chain of command (that is, who's in charge?) in accordance with contingency plan.

9.1.2 Establish communications system.

9.2 *Protection of the Public*:

9.2.1 Evacuation.

9.2.2 Remain in place.

9.2.2.1 Audio warnings (close windows, remain inside).

9.2.2.2 Media announcements.

9.2.2.3 Atmospheric dispersion factors should be used to estimate public exposure risk as a basis for selection of public protection actions.

9.2.2.4 Re-evaluate item 5.8 as atmospheric conditions change.

9.3 *Stabilize Incident*:

9.3.1 Determine the presence or threat of fire.

9.3.1.1 Extinguish fire if safe to do so.

9.3.1.2 Apply retardant (foam, etc.) to prevent ignition.

9.3.2 Determine if flow of chemical from container can be stopped or reduced.

9.3.2.1 Close valves.

9.3.2.2 Patch or plug.

9.3.3 Contain escaped chemical.

9.3.3.1 Dikes, berms, etc.

9.3.3.2 Floating spill control barriers.

9.3.3.3 Apply retardant if toxic or flammable vapor hazard.

10. Keywords

10.1 hazardous materials; initial response personnel; training curriculum

Designation: F 1012 – 86 (Reapproved 1999)

Standard Guide for Ecological Considerations for the Use of Chemical Dispersants in Oil Spill Response—the Arctic[1]

This standard is issued under the fixed designation F 1012; the number immediately following the designation indicates the year of original adoption or, in the case of revision, the year of last revision. A number in parentheses indicates the year of last reapproval. A superscript epsilon (ε) indicates an editorial change since the last revision or reapproval.

1. Scope

1.1 This guide covers recommendations for the use of chemical dispersants to assist in the control of oil spills. This guide is written with the goal of minimizing the environmental impacts of oil spills; this goal is the basis upon which recommendations are made. Aesthetic and socioeconomic factors are not considered, although these and other factors are often important in spill response.

1.2 Each on-scene coordinator has available several means of control or cleanup of spilled oil. In this guide use of chemical dispersants is not considered as a last resort after other methods have failed. Chemical dispersants are to be given equal consideration with other spill countermeasures.

1.3 This is a general guide only assuming the oil to be dispersible and the dispersant to be effective, available, applied correctly and in compliance with relevant government regulations. Oil, as used in this guide includes crude oils and fuel oils (No. 1 through No. 6). Differences between indvidual dispersants or between different oils or products are not considered.

1.4 This guide covers one type of habitat, the Arctic. Other guides, similar to this one, cover habitats such as rocky shores. The use of dispersants is considered primarily to protect such habitats from impact (or to minimize impacts) and also to clean them after the spill takes place.

1.5 This guide applies to marine and estuarine environments, but not to freshwater environments.

1.6 In making dispersant-use decisions, appropriate government authorities should be consulted as required by law.

1.7 *This standard does not purport to address all of the safety concerns, if any, associated with its use. It is the responsibility of the user of this standard to establish appropriate safety and health practices and determine the applicability of regulatory limitations prior to use.*

2. Significance and Use

2.1 This guide is meant to aid local and regional spill response teams during spill response planning and spill events.

3. General Considerations for Making Dispersant-Use Decisions

3.1 The decision of whether or not to use dispersants in a given spill situation is always one involving tradeoffs. Dispersing a slick at one site temporarily introduces more oil into the water column at that site than would be there if a surface slick floated over it. Therefore, adverse effects on water column organisms may be increased at the site so that adverse effects can be decreased or eliminated at other sites.

3.2 Dispersant use is primarily a spill-control method, not a cleanup method. Such use can give spill response personnel some control over where the impacts of a spill will occur and what types of impacts they may cause. Since some environments are known to be more vulnerable to the longer lasting impacts of spilled oil, an acceptable tradeoff may be to protect those environments by dispersing an oil slick in a less sensitive or less productive environment. In general, the tradeoff that must be evaluated is between the impact of the relatively long residence time of spilled oil that strands on shorelines versus the short-term impact of dispersed oil in the water column.

3.3 In this guide environments that are most vulnerable to the longer-term impacts of oil contamination are identified. Protection of these habitats is recommended as a high priority by means of dispersants and other methods.

4. Environment Covered—The Arctic

4.1 The Arctic is that area of the earth north of the Arctic circle (66°33′N). It has many habitats that are similar to those in more temperate waters and described by other dispersant-use guidelines that may suffice for the open water (summer) season in the Arctic. However, winter weather and associated ice conditions produce somewhat unique problems and conditions which may modify dispersant use in the various habitats. Guidelines for the Arctic should also be generally applicable in subarctic and temperate regions where seasonal cold is a significant factor.

4.2 The Arctic coastal geological morphology varies with a moderately wide continental shelf in the Beaufort Sea widening to an extensive shelf in the Chukchi, East Siberian, Lapter, Kara, and Barents Seas. The shelf is narrowest along the north coast of Greenland and Ellesmere Island. The Arctic shoreline may be characterized by narrow, low-lying beaches backed by

[1] This guide is under the jurisdiction of ASTM Committee F-20 on Hazardous Substances and Oil Spill Responseand is the direct responsibility of Subcommittee F20.13 on Treatment.

Current edition approved July 25, 1986. Published November 1986.

low-coastal bluffs as along the Alaskan Beaufort Sea and the North Siberian lowland, or bluffs may be totally lacking as along some segments east of the Coleville River **(1)**.[2] In some coastal areas of the Chukchi Sea, cliffs may reach to over 18 m in height **(2)**. The coast may be backed by high mountains as seen in the Chukchi Range of Siberia. Also along the Arctic coast are fiords and large river deltas.

4.3 Large geological differences exist between the western Arctic and the eastern Arctic coasts. The former is characterized by long stretches of unconsolidated frozen sediments, shallow offshore depths, and small tidal ranges. The latter is characterized by coasts that are commonly bedrock and widespread areas of coarse-grained glacial sediments. Here the tidal range is greater, 1.0 to 10.0 m.

4.4 Bedrock outcrops and coarser sediments are more abundant along the eastern Arctic coasts **(3)**. These are generally unconsolidated clays, silts, sands, and gravel. Considerable amounts of silt and peat may also be present in the Beaufort Sea coastal bluffs **(1)**. Arctic coastal sediments are bonded together into rock-like material **(4)** (permafrost), by the cementing action of frozen interstitial pore water. This is important because the amount of ice in the sediment dictates the rate of coastal erosion (0.3 to 10 m/year) in the Chukchi and Beaufort Seas, **(2, 4, 5)** and permafrost and coastal erosion rates will not allow for great penetration or long-term retention of spilled oil. The most important factor controlling shoreline processes in this area is the presence of sea ice, which limits the windfetch influence on wave height generation, currents, and alongshore sediment transport. Another important controlling factor is tidal range which can be from 0.1 to 0.3 m in the Alaskan Arctic and from 1.0 m to greater than 10 m in the eastern Canadian Arctic.

4.5 Much of the western North American Arctic coastline is a barrier island-lagoon system with elongated landforms of low relief parallel to shore which enclose shallow lagoons. The lagoons are important because of their concentration of birds and fish in the ice-free season. This is due at least partially to the high productivity of the barrier island-lagoon system. There is considerable exchange of water between nearshore and offshore in summer (much less if any, in winter) which appears to be ecologically significant and promotes the accumulation of nutrients and invertebrates in the nearshore zone. In summer, there is free exchange of nearshore water with the lagoon water. In some areas of the Arctic where the coast is low and tides reach up to 3.0 m, oil could be deposited hundreds of meters inland resulting in an" onshore" spill and cleanup situation.

4.6 The logistics of working in the Arctic are difficult, even in the summer, ice-free season. However, logistic problems are significantly greater in winter, with storms and lack of daylight.

5. Background

5.1 Ice is generally an asset when dealing with oil spills in the Arctic, especially when the ice cover is continuous. Oil is contained and kept thick and fresh for an extended period of time, allowing for planning and set up of a cleanup program, especially in-situ burning **(3)**. When it is stable, ice can also act as a platform for cleanup measures. Crude oil spilled under first-year ice becomes incorporated, by freezing, into the ice column. The oil moves upwards through brine channels to the ice surface in the spring, thus aiding in-situ burning or other mitigation methods **(6)**. Multi-year ice, however, is not porous enough to allow for this vertical migration and the oil will remain beneath it **(7)**.

5.2 When partial ice conditions are present, oil spill response is more difficult. Ice floes may restrict the use of conventional booms and skimmers **(3, 8, 9, 10)**. When applied in an ice-dampened turbulence regime dispersants may migrate into the water phase and not be effective **(11)**. On the other hand, for certain conditions, (in broken ice cover) the addition of dispersants may substantially improve dispersion. Tests indicate that dispersion in broken ice by dispersants is enhanced by water agitation and may be enhanced by increased salinity. Also dispersants may perform better on thicker oil slicks than on thin ones **(12)**.

5.3 In the Arctic barrier island-lagoon system, the pelagic and benthic populations are repopulated each spring in about the first two weeks of the open water season. Any kind of acute impact to this system would have a short-term effect of about one year **(13)**. Concentrated biological activity on Arctic shorelines usually takes place during breakup and the open water season. The first weeks of open water should be a time of prime oil spill concern and when offshore dispersant use might be most effective in reducing oil impact on sensitive coastal areas.

5.4 Most significant damage from oil spills has occurred when oil has come close to land or stranded on the shore **(14)**. Waterfowl and shorebirds that feed, rest, nest, molt, and congregate in coastal waters, and some fish species that spawn nearshore are most likely to be affected **(6, 14)**.

5.4.1 Dispersant use on oil advancing on a shoreline or ice edge can mitigate otherwise adverse effects to sensitive communities and enhance the restoration process **(15)**. However, dispersant use should be weighed against possible increased impacts to water column or ocean bottom communities from transient, but higher concentrations of oil **(16)**. The dispersion of oil may do little harm to water column or benthic organisms in offshore open water areas **(9, 17)**.

5.4.2 Nutrient conditions of the pack-ice edge significantly increase the number of forage species, concentrating overwintering birds (especially alcids and seaducks). Also, sea lions, ribbon seals, and spotted seals congregate at this ice edge. In the spring ice leads are sites of rapid, localized plankton blooms and provide important foraging areas for migrating birds and migration routes for marine mammals **(18)**.

5.5 The behavior of dispersants on dispersed oil **(18)**, and the environmental impacts (if any) are not known for the application of dispersants to oil under ice.

5.6 In the Arctic because of cooler temperatures and the confining action of broken ice, oil slicks may be nearer the spill source and have a greater thickness than in more temperate climates **(19)**.

[2] The boldface numbers in parentheses refer to the list of references at the end of this guide.

6. Recommendations

6.1 Dispersant use in the Arctic is recommended to prevent oil impact on shorelines and sensitive habitats or where it is desirable to reduce exposure of benthos to oil. It may often be the preferred alternative to intensive cleanup operations.

6.2 Where there is no significant environmental reason to protect a shoreline no particular advantage would be gained by offshore oil dispersion **(20)**. The "no cleanup option" should be given the highest priority.

6.3 The volume of dispersant necessary to treat an Arctic oil spill may be greater than for the same size spill in warmer climates **(19)**.

6.4 Dispersants are not recommended for use on solid, decaying ice **(21)**.

6.5 Dispersant use on oil under solid ice is not recommended at this time.

6.6 The decision to use dispersants on certain offshore Arctic coastline types, such as peat beaches or ice-rich bluffs should consider the mechanical wave energy levels at the shoreline and the net rate of shoreline change **(22)**, as well as the volume of oil released, the physical and chemical properties of the oil, and its distribution. Where shores are subject to high-energy wave action or are eroding, dispersant use may not be necessary because oil is unlikely to strand and persist. For stable, low-energy shorelines, dispersant use may be the most appropriate method to prevent stranding and persistence of oil.

6.7 In general, the use of dispersants should be based on the environmental sensitivity of the area about to be affected and the estimated persistence of the oil if it became stranded.

6.8 Dispersant use is recommended on oil advancing toward the southern edge of the Bering Sea pack ice and should be considered for use in ice leads in the spring.

6.9 In-situ burning is a recommended response technique for major spills in the Arctic where oil is concentrated in melt pools, in broken ice, on the frozen ice surface, etc **(8)** and **(23)**.

7. Keywords

7.1 arctic; chemical dispersants; ecological considerations; oil spill response

REFERENCES

(1) Truitt, J. C., *Beaufort Sea Barrier Island-Lagoon Ecological Process Studies*, LGL, Ltd., Bryan, TX, 1976–1977, pp. 56–61.

(2) Owens, E. H., Harper, J. R., and Nummedal, D., *Sediment Transport Processes and Coastal Variability on the Alaskan North Slope*, Abstract, 17th International Conference on Coastal Engineering, Sydney, Australia, 1980, 2 pp.

(3) Ross, S. L., "An Overview of Potential Large Oil Spills Offshore Canada and Possible Response Strategies," *Proceedings*, AMOP Technical Seminar 1983, pp. 190–197, Ice Cover versus Partial Ice, pp. 192–193, Dispersants, p. 194, Recommended Projects, p. 196.

(4) Harper, J. R., "Coastal Erosion Rates Along the Chukchi Sea Coast Near Barrow, Alaska," *Arctic*, Vol 31, No. 4, 1978, pp. 428–433.

(5) Harper, J. R., "The Physical Processes Affecting the Stability of Tundra Cliff Coasts," *Dissertation Abstract*, Louisiana State University, 1978, 3 pp.

(6) Buist, I. A., Potter, S. G., and Dickins, D. F., "Fate and Behavior of Water-In-Oil Emulsions in Ice," *Proceedings*, Arctic Marine Oilspill Program Technical Seminar, 1983, pp. 263–279.

(7) Farunwald, J. W., and Nelson, W. G., "Dispersion Characteristics and Flammability of Oil Under Low Ambient Temperature Conditions," *Proceedings*, Arctic Marine Oilspill Program Technical Seminar, 1982, pp. 217–238.

(8) Allen, A. A., and Lukin, J., eds., "Oil Spill Response in the Arctic; Field Demonstrations in Broken Ice," Part II, Industry Task Group, (Shell, SOHIO, Exxon, AMOCO, Anchorage, AK), 1983, 108 pp.

(9) Owens, E. H., "Coastal Environments of Canada: The Impact and Cleanup of Oil spills," Environmental Protection Service, Canada, EPS-3-EC-77-13, 1977.

(10) Reimer, E., "Oil in Pack Ice: the Kurdistan Spill," *Proceedings*, Arctic Marine Oilspill Program Technical Seminar, 1980, pp. 529–544.

(11) MacKay, D., Mascrenhas, R., Hassain, K., and McGee, T., "The Effectiveness of Chemical Dispersants at Low Temperatures and in the Presence of Ice," *Proceedings*, Arctic Marine Oilspill Program Technical Seminar, 1980, pp. 317–326.

(12) Cox, J. C., and Schultz, L. A., "Dispersant Effectiveness Under Arctic Conditions, Including Ice," *Proceedings*, Arctic Marine Oilspill Program Technical Seminar, Edmonton, Alberta, June 16–18, 1981, pp. 3–44.

(13) Walters, C., "Beaufort Sea Barrier Island-Lagoon Ecological Process Studies Workshop," Personal communication, 1949.

(14) Ibrek, H., Ostby, M., and Naess-Ulseth, E., "Evaluation of Strategies for Offshore Oil Spill Control," *Proceedings*, Arctic Marine Oilspill Program Technical Seminar, 1983, pp. 161–167.

(15) MacKay, D., and Wells, P. G., "The Use of Dispersants for Oil Spill Treatment," *Proceedings*, Arctic Marine Oilspill Program Technical Seminar, 1981, pp. 469–476.

(16) Gilfillan, E. S., Hanson, S. A., Vallas, D., Gerber, R., Page, D. S., Foster, J., Hotham, J., and Pratt, S. D., *Proceedings*, Oil Spill Conference, San Antonio, TX, 1983, pp. 457–463.

(17) Blackall, P. J., and Sergy, G. A., "The BIOS Project—An Update," *Proceedings*, Oil Spill Conference, San Antonio, TX, 1983, pp. 451–455.

(18) U.S. Department of the Interior (USDI), "Final Environmental Impact Statement," Proposed Federal/State Oil and Gas Lease Sale, Beaufort Sea, Vol 1, p. 64.

(19) Allen, A. A., Spiltec, Anchorage, AK, (Phone (907) 346-1932). Personal communication.

(20) Sergy, G. A., "The Baffin Island Oil Spill (BIOS) Project: A Summary," *Proceedings*, Oil Spill Conference, Los Angeles, CA, 1985, pp. 571–575.

(21) Allen, A. A., and Lukin, J., "Oil Spill Response in the Arctic," *Technical Documentation*, Part III, Industry Task Group, (Shell, SOHIO, Exxon, AMOCO, Anchorage, AK), 1984, 76 pp.

(22) Owens, E. H., "Factors Affecting the Persistence of Stranded Oil on Low Energy Coasts," *Proceedings*, Oil Spill Conference, Los Angeles, CA, 1985, pp. 359–365.

(23) Allen, A. A., and Lukin, J., eds., "Oil Spill Response in the Arctic; An Assessment of Containment, Recovery, and Disposal Techniques, Industry Task Group," (Shell, SOHIO, Exxon, AMOCO, Anchorage, AK), 1983, 167 pp.

Standard Guide for Sampling Oil/Water Mixtures for Oil Spill Recovery Equipment[1]

This standard is issued under the fixed designation F 1084; the number immediately following the designation indicates the year of original adoption or, in the case of revision, the year of last revision. A number in parentheses indicates the year of last reapproval. A superscript epsilon (ε) indicates an editorial change since the last revision or reapproval.

ε1 NOTE—Section 11 was added editorially in March 1995.

1. Scope

1.1 This guide is intended for sampling flowing or stationary oil/water mixtures. It is intended for use with oil spill recovery devices either in testing or in documentation of field performance.

1.2 *This standard does not purport to address all of the safety concerns, if any, associated with its use. It is the responsibility of the user of this standard to establish appropriate safety and health practices and determine the applicability of regulatory limitations prior to use.*

2. Referenced Documents

2.1 *ASTM Standards:*

D 1744 Test Method for Water in Liquid Petroleum Products by Karl Fischer Reagent[2]

D 1796 Test Method for Water and Sediment in Fuel Oils by the Centrifuge Method (Laboratory Procedure)[2]

F 625 Practice for Describing Environmental Conditions Relevant to Spill Control Systems for Use on Water[3]

3. Significance and Use

3.1 This guide provides techniques for obtaining representative samples of oil and water mixtures. This information is necessary in the calculation of oil recovery efficiency and oil recovery rates for oil collection devices.

3.2 *Sampling Stationary Mixtures*—When recovered oil/water mixtures are contained within a holding tank and the relative oil content of the recovered fluid is needed, the sampling technique is somewhat dependent on the container. Two techniques are outlined in this guide. If the container has a flat bottom with straight sides perpendicular to the base (or nearly so), either stationary technique can be implemented, with the stratified sampling method preferred. If the container is irregular in either the horizontal or vertical cross section, the mixing method is preferred.

3.3 *Sampling Flowing Mixtures*—To sample flowing mixtures containing both oil and water, turbulence is induced, to create a homogenous mixture while sampling. The oil content in the sample taken from the flowing stream can then be used to quantify the performance rating criterion (see Procedure Section of Test Method D 1796).

4. Number of Samples

4.1 Take a minimum of four samples under each set of conditions to average results and store the samples separately. In less ideal sampling conditions, take additional samples.

5. Containers

5.1 A dry, water-washed glass sample container.

6. Labeling

6.1 Mark the sample container with the source, type of oil, date and time of sampling, the name of the person taking the sample, and a sample number. Require water and oil-resistant labeling. If several receiving containers are to be sampled, they must be identified and the samples marked for later coordination.

7. Preservation and Storage

7.1 The samples do not require special treatment to preserve their integrity other than ensuring that they remain sealed until analyzed. Note date and time of analysis for each sample.

8. Procedure

8.1 *Baseline Data*—The test fluids may be crude, refined, or synthetic oils. Record type, specific gravity, viscosity, and temperature of each oil together with the environmental conditions (see Practice F 625), air temperature, and slick thickness beyond the influence of the recovery equipment for each test point.

8.2 *Sampling from a Container*—This procedure is intended for taking a representative sample of collected fluids held in a container. When sampling containers, it is advisable to remove as much of the aqueous phase as possible prior to sampling. If this is possible, measure and record the volume of water removed, which will contain dissolved hydrocarbons, for later calculation of the relative oil/water composition. The precision

[1] This guide is under the jurisdiction of ASTM Committee F-20 on Hazardous Substances and Oil Spill Responseand is the direct responsibility of Subcommittee F20.12 on Removal.

Current edition approved May 25, 1990. Published July 1990. Originally published as F 1084 – 87. Last previous edition F 1084 – 87.

[2] *Annual Book of ASTM Standards*, Vol 05.01.

[3] *Annual Book of ASTM Standards*, Vol 11.04.

of the measurement will improve with removal of the aqueous phase.

8.2.1 *Mixing Method*:

8.2.1.1 Mix the container of stratified oil and water using any method which will achieve homogeneity during sampling (high-speed propeller, liquid jet, or homogenizing pump). Operate the mixer for a minimum of 5 min. Longer mixing times are preferable to ensure homogeneity.

NOTE 1—Although this mixing method has been geared toward electric mixers, alternate mixing methods could be used (air or liquid mixing jets, homogenizing pumps), so long as a homogeneous mixture of the entire volume is created. Immediately after mixing, obtain a representative sample of the homogeneous mixture by lowering a 250 to 500-mL small-mouthed bottle (from which the cap has been removed) at a constant slow rate from the surface of the mixture to the bottom of the tank. Immediately remove the bottle by slowly raising it through the mixture. If necessary, repeat the process to ensure that at least 50 mL of sample will be available for laboratory analysis. Do not fill the bottle to the top.

8.2.1.2 *Limitations*—The size of the sampled fluid container is limited by the mixing capability available for thoroughly homogenizing the oil and water. Mixtures where the relative oil content is less than 20 % probably will not form a stable emulsion for sampling; making this method inappropriate for these mixtures.

NOTE 2—Certain oils (*N*-dimethylformamide, military jet engine fuel, DMF, JP5) will not form stable emulsions. Speed is important in these cases as the homogeneity of the mixture diminishes rapidly.

8.2.2 *Nonmixing Method*:

8.2.2.1 Where recovered oil/water mixtures cannot be homogenized or the emulsion formed is not easily broken, insert a stratified sampler into the nonmixed fluids to obtain a cross section of the fluids. Construct the stratified sampler as shown in Fig. 1.

8.2.2.2 *Stratified Mixture Sampling*—Place the entire sampler within the mixture to be sampled, with the outer sheath in the raised (open) position to expose the inner core. Lower the outer sheath to capture a set of samples in the annular segments. Remove the entire sampler, wipe clean, and drain each sample container through a clean funnel.

8.2.2.3 *Limitations*—Sampling with this device limits the size of the sampled tank. The physical workings of the sampler dictate that the tank cannot be greater in depth than the height which the sample taker can comfortably reach above the top of the tank. Movement of the tank can give nonrepresentative samples because of the rolling oil/water interface. Use on small vessels in the high seas requires that additional samples be taken and results averaged. The sampler must be thoroughly cleaned between sample takings.

NOTE 3—If the container is irregular in vertical cross section, the segments within the sample should not be analyzed as a whole. Each segment must be analyzed separately and averaged on a representative-volume-weight basis. Each segment must be large enough to capture a 50 mL sample for analysis.

8.3 *Sampling from a Flow*—When it is necessary to sample a flowing stream, eliminate the possibility of multilayer or stratified flow and bring the stream into a turbulent flow condition.

8.3.1 Position a static mixer (sometimes called an in-line or motionless mixer) in the skimmer discharge by-pass line, on

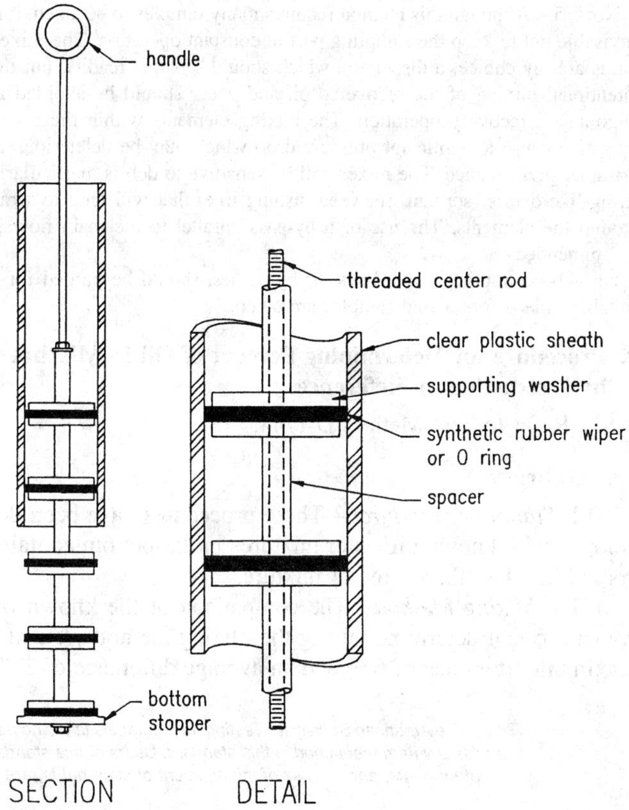

NOTE 1—Align holes on slots in inner and outer sleeve to let sample into zones of the sampler. When sampler is closed, holes are covered and sampler is secured.

FIG. 1 Stratified Sampler with Construction Details

the high-pressure side of the recovery pump. Immediately after the static mixer, install a sampling port shaped like a fixed Pitot tube (see Fig. 2) at the pipe center. It should be accessible and easy to use. Specify the static mixer to produce dispersed droplets having a mean droplet diameter of approximately 2000 μm at the expected flow rate and oil content. This method will provide a representative sample which can be used to determine the percentage of oil in the flowing oil/water mixture encountered.

NOTE 4—Because this is a measurement taken over a short time interval, it must be considered to be a differential measurement and many samples must be taken to ensure validity, especially in the start up portion of the oil recovery operation when conditions are rapidly changing.

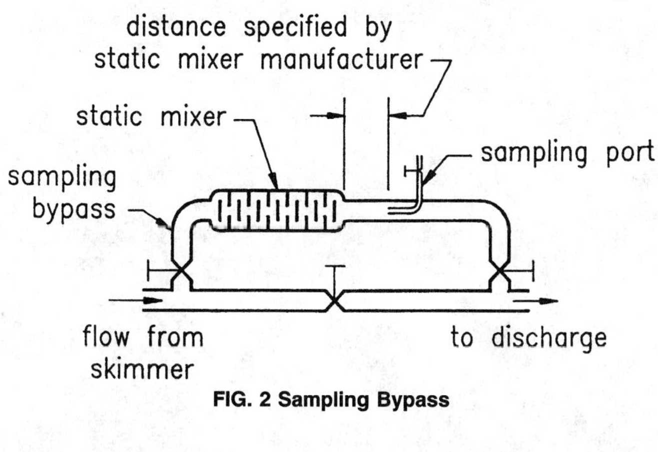

FIG. 2 Sampling Bypass

NOTE 5—Although this method requires many samples to be taken, it is advisable not to keep the sampling port in constant operation. The mixed fluids are, by choice, a dispersion which should separate readily, but the intentional mixing of the recovered oil and water should be avoided in ongoing oil recovery operation. The mixing elements within the mixer may also cause a significant pressure drop which may be deleterious to skimmer performance. The mixer will be sensitive to debris, particularly string-like debris (sorbant, seaweed, fishing line) that will tend to wrap around the elements. The use of a by-pass parallel to the main flow is recommended.

NOTE 6—Sampling ports, lines, or by-passes, should be purged for a suitable time before actual samples are taken.

9. Procedure for Determining Percent of Oil in Mixture by Instruction or Reference

9.1 Refer to Test Method D 1744.

10. Accuracy

10.1 *Stationary Mixtures*—These procedures have been developed with known oil/water mixtures in flat-bottom containers holding less than 2 m^3 of mixture.

10.1.1 *Mixing Method*—The comparison of the known oil content to that determined using the mixing method yielded a maximum difference of 6 % and an average difference of 2 %. The known oil content error was approximately 1 %. The mix method will yield results within 7 %.

10.1.2 *Stratified Method*—The comparison of known oil content to the determined oil content yielded a maximum difference of 6 % and an average of 2 %. When the stratified mixture is sampled without free-standing water removal, the analysis is considered to be accurate within 4 %.

10.2 *Flowing Mixtures*—The difference between the oil content determined with a sample of the total flow by the mixing method and the oil content determined by averaging the Pitot tube sample results was 4 %. The maximum difference was 7 %, which is considered to be the limit of the accuracy in the method.

NOTE 7—These procedures (refer to Test Method D 1744) have been confirmed with known oil content ranging from 10 % to 90 %, but have not undergone complete round-robin testing. The differences between known oil content and analyzed oil content include the errors of sampling technique, sample retrieval, and sample analysis.

11. Keywords

11.1 oil spill; oil spill recovery; oil/water mixtures

Standard Test Methods for Tensile Strength Characteristics of Oil Spill Response Boom[1]

This standard is issued under the fixed designation F 1093; the number immediately following the designation indicates the year of original adoption or, in the case of revision, the year of last revision. A number in parentheses indicates the year of last reapproval. A superscript epsilon (ε) indicates an editorial change since the last revision or reapproval.

1. Scope

1.1 These test methods cover static laboratory tests of the strength of oil spill response boom under tensile loading.

1.2 *This standard does not purport to address all of the safety concerns, if any, associated with its use. It is the responsibility of the user of this standard to establish appropriate safety and health practices and determine the applicability of regulatory limitations prior to use.* For a specific hazard statement, see Section 7.

2. Referenced Documents

2.1 *ASTM Standards:*

F 818 Terminology Relating to Spill Response Barriers[2]

F 962 Specification for Oil Spill Response Boom Connection[2]

3. Terminology

3.1 *Definitions*—The following definitions, quoted from Terminology F 818, are used in these test methods

3.1.1 *anchor point*—a structural point on the end connector or along the length of a boom section designed for the attachment of anchor or mooring lines.

3.1.2 *ballast*—weight applied to the skirt to improve boom performance.

3.1.3 *boom section*—the length of boom between two end connectors.

3.1.4 *boom segment*—repetitive identical portion of the boom section.

3.1.5 *curtain-type boom*—a boom consisting of a flexible skirt supported by flotation. See Appendix X1.

3.1.6 *end connector*—a device permanently attached to the boom used for joining boom sections to one another or to other accessory devices.

3.1.7 *fence-type boom*—a boom consisting of self-supporting or stiffened membrane supported by floatation. See Appendix X1.

3.1.8 *float*—that separable component of a boom that provides buoyancy.

3.1.9 *freeboard*—the vertical height of the boom above the water line.

3.1.10 *hinge*—location between boom segments at which the boom can be folded back 180° upon itself.

3.1.11 *skirt*—the continuous portion of the boom below the floats.

3.1.12 *tension member*—any component which carries horizontal tension loads imposed on the boom.

4. Summary of Test Method

4.1 A specimen of spill containment boom is tested by subjecting the specimen to cyclic tests to 100 % of the manufacturer's rated tensile strength, and by applying tensile loading which progressively deforms the specimen to the point of failure. Similarly, a typical anchor point and towing device are tested in an additional tensile test. For each phase of the test, values of tensile load and deformation are observed and recorded, and modes of failure are described.

5. Significance and Use

5.1 Boom sections are frequently combined into assemblages hundreds of meters in length prior to towing through the water to a spill site. The friction of moving long boom assemblages through the water can impose high tensile stresses on boom segments near the tow vessel.

5.2 Tensile forces are also set up in a boom when it is being towed in a sweeping mode. The magnitude of this tensile force can be related to the immersed depth of the boom, the length of boom involved, the width of the bight formed by the two towing vessels, and the speed of movement.

NOTE 1—When the towing speed exceeds about 1 knot (0.5 m/s), substantial oil will be lost under the boom.

5.3 Knowledge of maximum and allowable working tensile stresses will help in the selection of boom for a given application and will permit specification of safe towing and anchoring conditions for any given boom.

6. Apparatus

6.1 *Load Application Device*—A suitable load application device, such as a hydraulic jack, shall be provided. The device must be capable of applying loads somewhat in excess of the predicted failure load on the boom.

6.2 *Tensiometer*—A tensiometer shall be selected which will encompass the range of values from no load up to the

[1] These test methods are under the jurisdiction of ASTM Committee F-20 on Hazardous Substances and Oil Spill Response and are the direct responsibility of Subcommittee F20.11 on Control.

Current edition approved Feb. 10, 1999. Published April 1999. Originally published as F 1093 – 91. Last previous edition F 1093 – 91.

[2] *Annual Book of ASTM Standards*, Vol 11.04.

maximum boom tensile load which might reasonably be expected prior to failure of the boom.

6.3 *End Supports*—The test bed provided shall have end supports of sufficient strength and rigidity to resist significant deformation under the maximum loads expected during testing.

6.4 *Towing Devices and Connectors*—At least one of the manufacturer's standard tow bridles or towing devices shall be used at the leading end of the boom specimen (where the load is applied). A similar tow bridle or towing device shall be used at the trailing end if the test apparatus is long enough. However, if it is not, the connector at the trailing end of the specimen may be attached directly to a connector fixed to that end support of the test apparatus. Suitable shackles, cables, chains, and so forth, shall be provided to connect the towing adapters to the test equipment, as diagrammed in Figs. 1-3.

6.5 *Gage Points*—Gage points shall be affixed to each end of the test specimen to facilitate measurement of elongation during the course of the test.

6.6 *Elongation Measurement Scale*—A suitable measuring device shall be provided so that elongation measurements may be made periodically throughout the test. The device shall have a precision equal or better than 1/1000th the distance between gage points (that is, 3 mm precision for 3 m gage point separation).

6.7 *Boom Specimens to be Tested*—Equipment shall be arranged to apply tensile loading to a specimen consisting of at least two complete boom segments of standard length as supplied by the manufacturer. Boom segments of less than standard length may be used for this test provided that the tension member length is proportional, the hinge area between them, the connector assemblies at each end, and the anchor point are fabricated identically to the manufacturer's full size standard boom section provided the total specimen is at least 10 ft (3 m) in length.

6.8 *Alternative Apparatus*—Because production lengths of boom are normally longer than 15 ft and because undue stress due to gravity forces may be placed on such boom if tested with the apparatus described above, the following described apparatus may be substituted. Test apparatus which lays the boom in a horizontal and continuously supported manner or one which provides support similar to that provided by the water (that is, a split table supporting the boom in an upright manner) will be satisfactory.

7. Hazards

7.1 Failure of a loaded containment boom can release a substantial amount of energy. During testing, personnel and equipment shall be positioned and protected so that sudden failure of the test specimen is unlikely to cause injury or damage.

8. Procedure

8.1 *Determination of Boom Tensile Strength*:

8.1.1 *Test Bed Preparation*—Prepare a test bed with two end supports separated with sufficient clearance for the boom specimen, two towing devices, and testing equipment as shown in Fig. 1. Mount the specimen with one towing device attached directly to one of the end supports. Alternately, the connector at the trailing end of the test specimen may be attached to a Specification F 962 connector fixed to the end support of the test apparatus as shown in Fig. 2. The tensiometer is used to link the towing device at the other end of the boom specimen to the load application device and hence to the second end support. Suitable shackles, chains, cable, and so forth, can ordinarily be used for making connections. However, in some cases it may be necessary to design and fabricate special connecting devices to distribute loads satisfactorily.

8.1.2 *Cyclic Loading to 100 % of Manufacturers Rated Strength*—Load the boom specimen to 100 % of the manufacturer's rated tensile strength. Then reduce the load to 10 % of the manufacturer's rated strength. Note the positions of gage points on the specimen while under full tensile load, and at 10 % of full tensile load. Then repeat this loading cycle until the specimen has been subjected to a total of ten complete cycles.

8.1.3 *Test to Ultimate Failure or Permanent Deformation*—If the specimen has not failed during the cyclic loading phase, it is then subjected to increasing loading until failure occurs. "Failure" is defined here as the inability to function or the rupture of the tension member, skirt material, or connector.

8.2 *Determination of Anchor Point Strength*:

8.2.1 *Test Bed Preparation*—Mount a second specimen, consisting of one or more standard boom segments, in the test bed with one end of the specimen attached to the tensiometer using a standard towing device, as in the previous portions of the test. The anchor point on the specimen is then linked to the other end of the test bed as shown in Fig. 3.

8.2.2 *Cyclic Loading*—Then apply progressively increasing loading to the specimen. If the manufacturer provides a rated anchor point strength, that value is used as the maximum load for the test. If the manufacturer does not provide such a strength rating, then the anchor point shall be subjected to a tensile loading of 50 % of the manufacturer's rated tensile strength for the boom. The positions of gage points on the specimen are noted while under full tensile load and at 10 % of

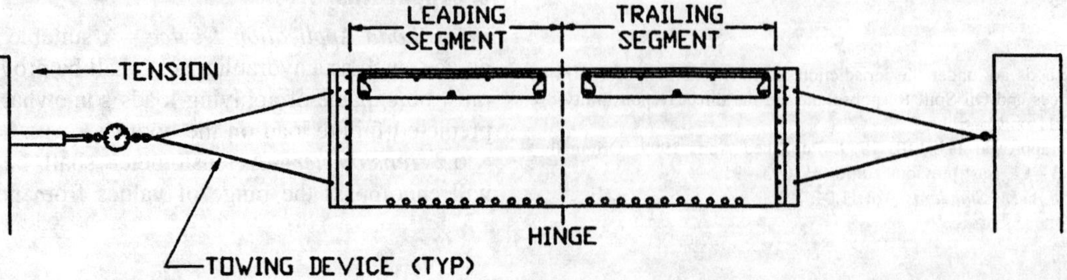

FIG. 1 Test Bed

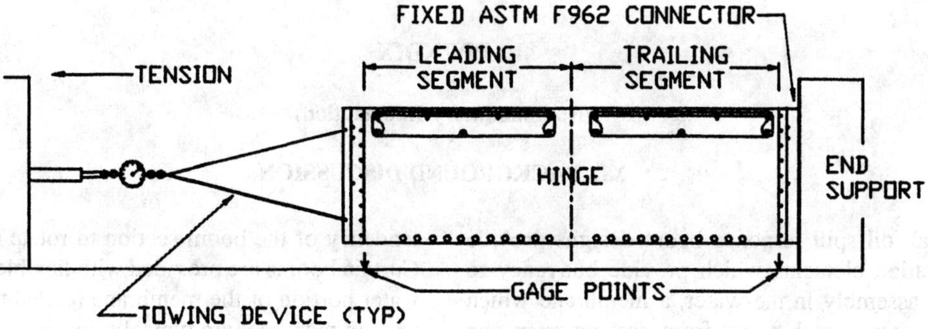

FIG. 2 Alternate Tensile Strength Test Bed

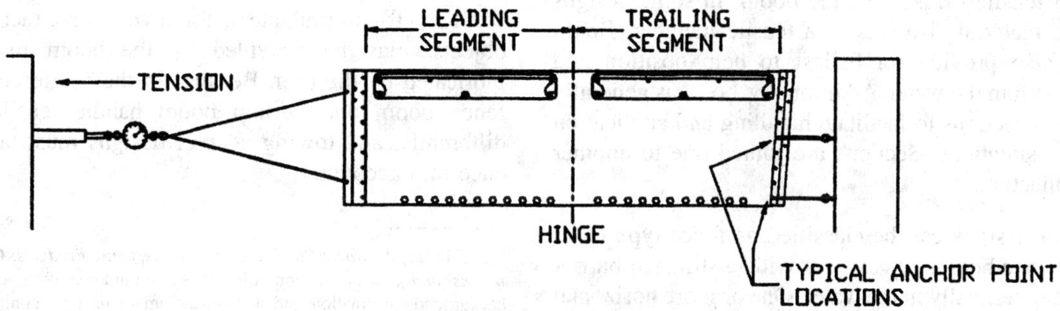

FIG. 3 Anchor Point Test

full tensile load. This loading cycle is then repeated ten times.

8.2.3 *Test to Ultimate Failure or Permanent Deformation*—If the specimen has not failed during the cyclic loading phase, it is then subjected to increased loading until failure occurs. Failure is defined here as the inability to function or the rupture of the tension member, skirt, material, anchor point hardware or connector.

9. Report

9.1 The test report shall provide a description of the boom tested, including the name of the manufacturer and model number. For each phase of the test, values shall be reported for length of specimen and initial positions of gage points. The report shall also provide a tabulation with columns for periodically observed data including boom elongation values, load values, and notes regarding any damage to boom fabric or other components. Consideration should be given to recording all gage point positions and tensiometer values photographically. During the test to ultimate failure, sufficient data shall be taken to permit plotting a curve of deformation versus loading. The ultimate loading on the boom or anchor point at the time of failure shall be recorded.

9.2 The report shall also include photographs and descriptions of any damage observed. The report shall identify boom components involved in the failures and provide descriptions of how the failures occurred.

9.3 Any deviations from the test procedures or specimens described above shall be identified with reasons for such deviations given.

APPENDIX

(Nonmandatory Information)

X1. BACKGROUND DISCUSSION

X1.1 In general, oil spill response boom designs provide one or more floatation elements which provide buoyancy to support the boom assembly in the water, a membrane which prevents floating material such as oil from passing from one side of the boom to the other, and one or more tension members which transfer tensile loads along the boom. In some designs the membrane material also acts as a tensile member. Boom designs may also provide for ballast to help position and stabilize the boom in the water. Additionally, boom is generally manufactured in sections to facilitate handling and application in a variety of situations. Sections are joined one to another using end connectors.

X1.2 Boom designs can be classified as fence-type or as curtain-type. Fence booms typically provide a stiffened barrier designed to float vertically in the water. One or more horizontal tension members may be used, positioned to minimize any tendency of the boom section to rotate under an applied load. Curtain booms are provided with flexible material for the under water portion of the membrane (called the skirt). One or more tension members are typically located at or near the bottom of the membrane.[3]

X1.3 Towing adapters, for towing or attachment to fixed objects, may be provided by the boom manufacturer or fabricated by the user. Because of their differences in design, fence boom and curtain boom handle tensile loads quite differently, and towing adapter designs must take this difference into account.

[3] Schulze, R., *World Catalog of oil Spill Response Products: Complete Listing of Booms and Skimmers*. Port City Press, Baltimore, 1986, contains additional background information and a compendium of data on commercially available booms.

Standard Guide for Containment by Emergency Response Personnel of Hazardous Material Spills[1]

This standard is issued under the fixed designation F 1127; the number immediately following the designation indicates the year of original adoption or, in the case of revision, the year of last revision. A number in parentheses indicates the year of last reapproval. A superscript epsilon (ε) indicates an editorial change since the last revision or reapproval.

1. Scope

1.1 This guide describes methods to contain the spread of hazardous materials that have been discharged into the environment. It is directed toward those emergency response personnel who have had adequate hazardous material response training.

2. Referenced Documents

2.1 *ASTM Standards:*

F 716 Methods of Testing Sorbent Performance of Absorbents[2]

F 726 Method of Testing Sorbent Performance of Adsorbents[2]

3. Terminology

3.1 *Definitions:*

3.1.1 *absorption*—a process where the material taken up is distributed throughout the body of the absorbing material. (The body of the absorbing material must swell.)

3.1.2 *adsorption*—a process where the material taken up is distributed over the surface of the adsorbing material.

3.1.3 *gellant*—material that exists for both aqueous and organic materials.

3.1.4 *sorbent*—a material used to recover liquids (or fluids) through the mechanism of absorption or adsorption or both.

4. Significance and Use

4.1 This guide contains information regarding the containment of a hazardous material that has escaped from its container. If a material can be contained, the impact on the environment and the threat it poses to responders and the general public is usually reduced. The techniques described in this guide are among those that may be used by emergency responders to lessen the impact of a discharge.

4.2 Emergency responders might include police, fire service personnel, government spill response personnel, industrial response personnel, or spill response contractors. In order to apply any of the techniques described in this guide, appropriate training is recommended.

5. Containment Methodology

5.1 Containment equipment, procedures, and techniques can be categorized into three general functional classes: (*a*) patch/plug, (*b*) enclosure, and (*c*) immobilization. The important advantage of containment is that it restricts the spreading of a spill and makes cleanup easier. Careful selection of techniques and materials is required. Errors in judgment can lead to worsening of the situation, deflagration or detonation, and increased hazard to personnel involved in the cleanup.

6. Patches and Plugs (General)

6.1 Diminishing or stopping the flow of a leaking hazardous material is desirable in order to limit the size of the spill. The following techniques may be helpful in controlling leaks, provided response personnel can use them safely under existing conditions. Whichever method is used, it should be noted that the higher the pressure inside the container, the more difficult it is to plug the leak.

6.1.1 *Wood Plug*—Wooden cones and wedges may be hammered into leaking containers (drums, tanks, pipes, etc.). Softwoods in particular are easily sawed or lathe-turned and conform well to irregular shapes. Additionally, softwood may absorb liquid and swell, enhancing its capacity to seal a leak. Wedges or cedar shingles are especially applicable to splits, gouges, rips, and tears. Rigid plywood sheets can be fastened over a damaged area with "T" bolts, toggle/molly/butterfly bolts, straps, or by mechanical bracing and wedging. To minimize leakage between the plywood and the container, a gasket of rubber or plastic sheeting, putty, butyl rubber caulk, lead wool, or oakum may be used.

6.1.2 *Metal Sheet*—Various sizes of steel or aluminum sheets can be fastened over damaged areas by mechanical methods ("T" bolts, toggle bolts, bracing, strapping, etc.). Gasketing material between the metal and the container generally provides more positive sealing.

6.1.3 *Inflatable Plugs and Bags*—Reinforced rubber and coated-fabric plugs can be inserted into an opening and inflated with gas (air, nitrogen, carbon dioxide) or water to form a seal. Lead-sealing bags can be secured with straps, chains, cables, fire hoses, or bands to seal a leaking container.

6.1.4 *Fabric Patch*—Fabrics such as neoprene-coated nylon can be positioned over leaks and held in place by bands, chains, straps, etc. Wood, plastic, or metal reinforcements may be required.

[1] This guide is under the jurisdiction of ASTM Committee F-20 on Hazardous Substances and Oil Spill Response and is the direct responsibility of Subcommittee F20.22on Mitigation Actions.

Current edition approved Jan. 29, 1988. Published May 1988.

[2] *Annual Book of ASTM Standards*, Vol 11.04.

6.1.5 *Formed Plug*—Polymeric foam (for example, polyurethane or polyethylene), epoxy putty, or quick-setting hydraulic cement may be injected into a rigid concave form through a tubular handle or it may be troweled onto the form and placed against the damaged area. Once the patching material hardens, the support form may be removed.

6.1.6 *Caulking Patch*—Epoxy, plastic steel/aluminum, lead wool, clay-polymer mixtures, and oakum can be spread, troweled, or peened into cracks and small holes. Rapid-curing materials are available.

6.1.7 *Foam Plug (Self-Expanding)*—A package of polyethylene, polyurethane, or low-density neoprene rubber foam (all closed-cell) formed into a compact shape by compression and vacuum packing may be opened allowing the foam to expand and fill the leak area. These plugs may not be readily available.

6.1.8 *Magnetic Patch*—Magnetic sheets (rubber-bonded barium ferrite composite, with or without adhesive) backed by a thin (0.0025 in.) sheet of steel foil may be strapped over the damaged area.

6.1.9 *Mechanical Patch*—Neoprene stoppers, rubber balls, and plywood or spring steel sheets with neoprene gaskets can be mechanically held in or on the damaged area. Toggle and "T" bolts, washers, and wing nuts are useful attachments.

6.1.10 *Adhesive Patch*—Adhesive patches sometimes work but usually require tedious surface preparation. Tape (duct, lead, aluminum, or stainless steel) is useful when applied over a wooden or rubber plug before application of epoxy to create a relatively permanent repair.

6.1.11 *Bladder Wrap*—Coated fabric or reinforced rubber pipe patches (similar to a clamp) with integral inflation bladder can be secured around a pipe or small round container with nylon self-adhesive fabric, Velcro, fire hoses, or banding/strapping material.

6.1.12 *Split Clamp Pipe Patch*—A split, sleeve-type coupling with a bolted flange assembly can be used to cover a rusted or damaged pipe section.

6.1.13 *Pipe Pinch*—A "C"-shaped clamp device with hydraulically or explosively operated ram can flatten a section of pipe to pinch off the fluid flow.

7. Enclosure

7.1 Approved salvage drums (overpacks, recovery drums, waste drums, "open-head" drums) may be used to encapsulate leaking drums or other small containers. Contaminated materials (tools, clothing, soil) and plastic bags holding used sorbents or contaminated items also may be enclosed in salvage drums. Approved enclosure containers may be used for transport, storage, and disposal of many hazardous materials.

8. Immobilization

8.1 Once a hazardous material has escaped from its container, it may be possible to immobilize the material to prevent it from spreading. There are a number of methods that may be used to accomplish this task; these methods vary depending on whether the material is a liquid, a solid, or is volatile and escapes as a gas.

8.2 *Liquids*:

8.2.1 Spills of hazardous liquids (including slurries) are the most difficult of spill problems. Good management practice aims to contain the material and localize it in a concentrated form. Typical procedures that can be used to affect the spreading of a spilled liquid are as follows:

8.2.2 Change the physical properties of the liquid by modifying the viscosity or vapor pressure by temperature change (usually cooling).

8.2.3 Immobilize the liquid by use of an adsorbent, absorbent, or a gelling agent (see 8.3.1.2).

8.2.4 Pump to a suitable container or lined pit.

8.2.5 Erect physical barriers.

8.2.6 Form dikes from earth sandbags, water inflatable bags, weighted adsorbent foamed plastic, or absorbent sand mixture.

8.2.7 Assemble collapsible containers (for example, plastic swimming pools if compatible) or use a plastic film-lined ground depression or pit for containment.

8.2.8 Deploy collection or containment devices such as boom curtains and portable dams. Suitable floating absorbents can help in preventing these booms from being made ineffective by stream current physically stripping liquid underneath.

8.2.9 A porous or wire mesh boom can be efficiently used with the proper floating absorbent material. A board boom is also effective in a ditch.

8.2.10 A reverse flow weir can be used to concentrate floating fluids.

8.2.11 Sewers or other types of drainage in the path of a spreading spill should be blocked. An absorbent/sand mixture can be used as a sealing dike or a soft closed-cell plastic foam can be used to cover the opening. Many impermeable systems can be used to seal the openings.

8.2.12 When a spilled material has a density greater than water, a weighted sorbent can be placed at the bottom of a watercourse or sewer to pick up and immobilize a spill.

8.3 *Land Spills*:

8.3.1 Typical methods for handling spills on land are listed, including pumping, sorbents (adsorbents and absorbents), gellants, dikes, dams, trenches, soil and dike sealants and physical state modifications.

8.3.1.1 *Pumping*—If a pool of spilled liquid can be contained on land, the most direct mitigation is to pump it into a suitable container (or to use a vacuum truck). Compatibility of all equipment with the material being handled is necessary. Many of the typical materials widely used for oil containment and cleanup are not suitable for many hazardous materials. Gaskets and sealants for pumping units may be oil resistant but fail quickly with a hazardous material. For low-boiling-point liquids the pump inlet will have to be below the level of the liquid. Otherwise, pump suction will cause the liquid to boil and the pump to cavitate. When pumping materials whose vapor is flammable, use nonsparking equipment. Employ a grounded system so that static electric buildup cannot occur at discharge ports or nozzles.

8.3.1.2 *Sorbents*—Sorbent is a general term applied to both adsorbents and absorbents. The source of these products can be natural or synthetic. They can be organic, inorganic, or mixed in composition. Proper use of these materials depends on the type of spill, location, and type of sorbent to be used. The Federal Schedule 2008.1-1.1 and 2001.3 recommends the use of inert materials (that is, sorbents without reference to the size of a spill). It also gives the On-Scene Coordinator (OSC) the directive to use that material or method *best* suited to mitigate the spill. A separate part of this regulation (2001.4) prohibits adding any harmful substance in *any* quantity to water. For "hazardous materials'' this prohibits the wringing out of sorbents (absorbents) for reuse. So-called "universal or broad range'' sorbents are covered in 8.3.1.6, since they are often mixtures of the singly defined types. It is also true that the broad range of materials considered hazardous makes a truly universal material unlikely. Since these materials are totally different, the definitions developed in Methods F 716 and F 726 are included in Section 3 of this guide.

8.3.1.3 *Adsorbents*—Adsorbent materials are those that are inert to the spilled material and usually have a large surface area. Since adsorption is by definition only a surface coating process, high surface area is advantageous if the fluid has sufficiently low viscosity to cover it. An incomplete list of adsorbent materials includes plastic foams, plastic fibers, straw, peat, sand, porous clay, feathers, foamed glass and silicates, activated alumina, and soil. The surface can be external as in a fiber, or internal as inside a granule of activated carbon. The criteria is that if the fiber does not swell greater than 10 % or the carbon granule does not change size, then the sorption phenomenon is called *adsorption* and the material for the liquid intended is an adsorbent. Since the spilled fluid is available on the surface of an adsorbent, it may be removable. This can be an advantage if separation following recovery is important. It is detrimental to the extent that:

(1) The liquid can usually be removed by leaching (even by water used in clean up), rain, etc.,

(2) Vaporization loss is often increased by increasing exposed surface area,

(3) The adsorbent may lead fluid causing secondary spill problems, and

(4) Since adsorbents can usually be wrung out they easily contaminate personnel handling them. In the line of safety awareness, what is suitable for No. 6 fuel oil or even No. 2 fuel oil may be inadequate, if not hazard increasing, for gasoline, styrene, etc.

8.3.1.4 *Absorbents*—Absorbent materials are those that are inert to the spilled material but physically swell up in it. They often have a low surface area. They are also adsorbent by the nature of their surface area but since this area is small they are not often used as adsorbents. Those absorbents useful in spill control do not dissolve in the spilled fluid but physically contain it in a form with minimum surface area. This reduction in surface area lowers the rate of evaporation and minimizes leaching. For many hazardous spills these are required properties. Absorbent materials also minimize human and secondary contamination since squeezing and contact may not be with a wetted surface as in the adsorbent. Use of an absorbent can mean a stopping of ground penetration, which can minimize cleanup. Also, fire, and the water used to extinguish it, or rain have a low tendency to leach spilled material. Absorbent materials for organic fluids include, among others, rubbers and cross-linked products like imbibing polymers. Absorbents for aqueous fluids include cellulosics (synthetic and natural), cross-linked proteins, cross-linked hydrolyzed synthetic polymers and cross-linked starches.

8.3.1.5 *Gellants*—Gellants are usually colloidal materials that upon addition to a liquid with intimate mixing, form very high viscosity materials. Since these materials are not true absorbents the network the fluid is held in can be broken by heat or other forces. The intimate mixing required is often difficult on a land spill. Gels usually have a delay time when congealing, therefore they may not be suitable for running spills.

8.3.1.6 *Miscellaneous*—This category includes the "universal sorbents'' that are usually mixtures of other materials listed previously. These materials should be qualified for what type of spill they can be used on and their relative pick up of water and spill mixtures. This type of material can be especially valuable for small spills and quick on-site response. They are less important on larger spills where water sensitivity or the lack of water sensitivity is important.

8.3.1.7 *Dikes, Dams, and Trenches*—Artificial containment barriers can be created to confine liquid spills by forming a wall of sandbags, water-inflated bags, or earth (by shovelling or using mechanized earth-moving equipment). The use of an absorbent/sand mixture offers the advantage of a sealing and diking material. Sometimes the spill can be confined on a prepared surface, such as concrete or blacktop, but the more typical situation involves earthen surfaces and dikes, which are prone to pass spilled liquids unless coated with a soil sealant to prevent percolation into the earth. Inorganic foams, such as foamed concrete, foamed gypsum, and sodium silicate foam, have been used to produce dikes and barriers. The basic problem in adapting these materials to a particular application is the difficulty in building these materials up without some form of constraint. A quick set has been achieved in using silicate additions to cement slurries. Polyurethane foams have also been used successfully as diking material on dry hard-packed soil for short-term containment. Liquid spills may penetrate into the soil and seep under the dike.

8.3.1.8 *Dike Sealants*—Several low-cost methods for sealing surface soil to prevent infiltration have been developed. These include plastic sheets and materials that can be sprayed on a site to form an impervious layer. Soil surface sealant candidates can be grouped into several classes: reactive, non-reactive, and surface chemical. This classification is based on how the sealant is formed chemically and the interaction of the sealant with the soil surface. Reactive sealants are usually two-component systems in which one material is either reacted or catalyzed with a second material to yield a polymer. Such materials include epoxy, unsaturated polyester, phenol-formaldehyde, urea-formaldehyde, and urethane. Nonreactive sealants are those that have been previously polymerized and are either dispersed or dissolved in either an aqueous or a solvent system. Such materials are primarily thermoplastic in nature and include such materials as bitumastic, rubber, acrylic, cellulosic, fluoroplastic, phenolic, polyester, polystyrene, poly-(vinyl chloride), and imbibing-type polymers. Certain of these imbibing types will swell and seal imperfections. Most of the others depend on a solid coherent layer of sealant. The third sealant type consists of repellant chemicals that, when applied to surfaces, modify the surface characteristics such that the surfaces are not penetrated. Extensive repellant technology has been developed for four classes of materials: textiles, paper, leather, and masonry. In each, there is a broad range of techniques and chemical systems.

8.3.1.9 *Physical State Modification (Freezing)*—Many hazardous chemicals have a freezing point that is no lower than that for ordinary water and therefore, could be (theoretically) relatively easily frozen. Some liquid cargoes are even heated for shipment as liquids, ordinarily being solids or near-solids at ambient temperatures. With respect to a relatively small spill of hazardous substance on land, immobilization by freezing is possible. A device for effectively crushing and distributing the coolant on a spill is required.

8.4 *Water Spills*:

8.4.1 When a spill of a hazardous substance occurs in such a way that the material (liquid or fluidized solid) enters a watercourse, the problem of containment in the most concentrated form is exceedingly difficult.

8.4.2 For insoluble floating hazardous substances, many of the techniques applicable to oil spills may be suitable. These include booms, flow guides, curtains, adsorbent booms, etc. When the spilled floating material is flammable (for example, benzene or gasoline), the danger of ignition and fire must be recognized. An absorbent boom and floating material can be advantageous since it immobilizes the spilled fluid (in case it is hit by a stream of water); it reduces the rate of vapor release (allowing a fire to be cooled by mist and extinguished), and minimizes loss due to stream or tidal current. Recovery and removal are safer with respect to both personnel and the environment. Adsorbents (as noted in 8.3.1.3) may increase vapor release, etc. A reverse flow weir can be valuable if the water stream can be interrupted.

8.4.3 *Booms*—A sealed boom consisting of a plastic (resistant) curtain, a flotation device, and a tension device can be deployed around ships, oil rigs, at the mouth of harbors, and across streams. If the speed of the current (either stream or tidal) is more than ¾ knot, then absolute retention is not possible by use of boom alone. Use of the proper sorbent can alleviate this problem somewhat. Liquid entrapment behind a boom can aid in building a thicker layer of spilled product than can be pumped or removed by vacuum equipment. Point source control is most desirable. Use of particulate absorbents at the site of a spill allows collection of the contaminated units on a wire fence across a stream.

8.4.4 *Sorbents:* (see 8.3.1.3)

8.4.4.1 *Adsorbents* (see 8.3.1.3)—Certain adsorbents useful on land spills are less useful in spills on water. The main reason is preferential water wetting. Also, water may tend to weaken the structure of certain adsorbents useful on land. The nature of certain hazardous materials (for example, oxidizers) should be checked for compatibility with the sorbent. Strong oxidizers as a matter of practice should not be contacted with organic sorbents.

8.4.4.2 *Absorbents* (see 8.3.1.4)—Compatibility of absorbent with the spilled material should be checked. As a general practice, organic absorbents should not be used with strong oxidizers. Compatible absorbents in unitized packages can be weighted and used to pick up sinking materials as well as establish bottom booms. The value of this as well as the ability to retrieve the spill from the bottom without loss during retrieval is very important.

8.4.4.3 *Gellants* (see 8.3.1.5)—Gellants in dry or water-dispersed form may have use in handling thin layer spills on quiet water. The ability to change viscosity allows separation techniques that would be impossible otherwise.

8.4.4.4 *Miscellaneous*—This unit is left open because the items listed in 8.3.1.6 are probably not applicable to spills on water.

8.4.5 *Underwater Dikes, Dams, and Trenches (Underwater Floaters)*—Underwater dikes of sandbags or other materials can be constructed. Dredging of underwater ditches at angles to the flow may also be useful. If pits can be dug, then vacuuming can be considered.

8.4.6 *Membrane Technology*—As applied to spills, membrane technology involves the use of a semipermeable membrane to concentrate a chemical, which has been spilled into the ground or surface water, for ease of recovery. Soluble, insoluble, and slightly soluble materials may be recovered by this technique. Because of time constraints, recovery from only small bodies of water is a limiting factor. The branches of technology involved are ultrafiltration (UF) and reverse osmosis (RO).

8.4.6.1 *Ultrafiltration (UF)*—Ultrafiltration is essentially a form of very fine, cross-flow filtration. Suspended solids, immiscible liquids, and emulsions may be recovered by this technique, which involves low pressure (from 50 to 400 psi or from 345 to 2760 kPa). The material to be recovered is pumped, under pressure, across the surface of a finely porous membrane. Water passes through the pores leaving a more concentrated product stream. Use of turbulence promoters assist in preventing buildup of product on the surface of the membrane. By use of recirculation loops, recoveries of up to 99 % of the product at volume reductions from 95 to 99 % are frequently possible. Membranes are made of many synthetic materials (cellulose acetate, polyamide, polyimide, polysulphone, etc.). Compatibility of the membrane and other components of the system to the chemical involved should be known before recovery is tried.

8.4.6.2 *Reverse Osmosis (RO)*—Reverse osmosis involves a diffusion reaction in which the water passes through a semipermeable membrane leaving behind a more concentrated product. Slightly soluble and soluble products may be recovered by this technique which requires high pressure (from 400 to 1000 psi or from 2760 to 6895 kPa). The material to be concentrated is pumped under high pressure across the surface of a membrane. Turbulence promoters decrease the buildup of product at the surface. By use of circulating pumps, volume reductions from 90 to 95 % are possible. Possible product recoveries vary with the size and type of molecule. Some examples include:

Inorganic Chemical, %		Organic Chemical, %	
Sodium chloride	99	Methanol	40
Potassium hydroxide	79	Ethylene dichloride	70
Ammonium nitrate	97	Benzene	90
Zinc sulphate	98	Malathion	99.9

At present, only dilute solutions may be handled by this technique (5 % for inorganics, 0.5 % for some organics). As with UF, RO membranes are made of many synthetic materials (cellulose acetate, polyamide, polysulphone, polyetherimide etc.). These membranes are much more product specific in the application than are their UF counterparts. The capability of the membrane and the compatibility of the complete system to the chemical involved should be known before this technique is attempted.

8.5 *Volatiles*:

8.5.1 Many volatile materials are shipped in cylinders or containers under pressure. When these are vented through a hole, all except two of these volatile materials tend to cause a cooling effect. It is worth noting these two exceptions. One is helium, which is not hazardous except as a suffocating agent, and the other is hydrogen. A pressurized cylinder of either of these will heat up as the material is vented. Hydrogen in air has a low ignition point and burns with an almost invisible flame. Caution is urged when the volatile is hydrogen.

8.5.2 *Foam Covering*—Water-based foam blankets applied to the surface of a volatile liquid spill can significantly reduce the vapor hazard. The foam blanketing is effective with water-immiscible flammable liquids, water-soluble flammable liquids (polar compounds), and certain classes of water reactive liquids. Some gaseous materials shipped and stored as refrigerated liquids such as methane, ethylene, ammonia, and chlorine can also be benefited by foam blankets. Not all foams are effective against all chemicals **(1-3)**.[3] These references should be consulted along with foam manufacturer's literature before using foam for vapor control. Some foam/chemical combinations can exaggerate the hazard rather than mitigate it.

8.5.3 *Water Spraying*—The use of water spray must be considered carefully. If the water warms the spill it could cause increased vaporization. Water will raise the level of a contained spill and perhaps cause it to overflow. Sprays or mists are not effective in scrubbing vapor from air, but can be used to guide a drifting vapor cloud or plume away from sensitive areas. The advantages of the water fog and spray are availability and short response time.

8.5.4 *Chilling*—Cooling has been tested as a method to reduce the vaporization rate of hazardous substances involved in land spills or floating on water. Candidate coolants tested include the following: liquid nitrogen (LN_2), solid carbon dioxide (CO_2), and ice/water mixture. Only the latter two coolants are practical. Even these must be dispensed uniformly over the spill surface as a finely divided solid. Carbon dioxide extinguishers or snow horns are not effective since they yield less than 10 % solids and may cause a static electric problem. Solid CO_2 is the most effective and applicable to all materials with which it does not react and where the boiling point is greater than the sublimation temperature of the CO_2. Wet ice has application to water-miscible materials; it both cools and dilutes. Special equipment is necessary to crush the ice to the correct particle size distribution and project it to the spill surface. Coolants appear to be restricted to in-plant use where quantities of dry or wet ice would normally be available. The effectiveness of cooling to mitigate spills of hazardous chemicals is dependent on many factors, such as delivery method, physical properties of the coolant, freezing point of the spilled material, and environmental factors. Wind and rain can have a significant impact on the effectiveness.

8.5.5 *Film Covering*—If a volatile spill could be covered with an inert film, the vapors could be contained and released to a treatment unit. When unrolling a plastic film, static generation of sparks should be considered.

8.5.6 *Scrubbing*—Capture of the vapor coming from a volatile spill should be considered. Scrubbing the vapors by drawing them through a neutralizing or absorbing material, or entraining foam may be possible.

8.5.7 *Neutralization*—Acidic or basic vapors can be drawn through a contacting bed with a countercurrent flow of neutralizing fluid. Neutralization or other forms of reaction can be combined with foam scrubbing.

8.6 *Solids*—In the absence of bad weather (rain or wind), fire, explosion, or contact with other spilled reactive materials, insoluble solids will usually be nonspreading. The lack of inherent mobility of solids improves the possibility of containment as opposed to fluids. Solids, especially those shipped in smaller containers such as bags and drums are usually easy to handle. However, certain solids, such as phenol, maleic anhydride, etc., are shipped as melted liquids in bulk but are solids at normal temperatures. When a solid is mixed with water, its containment is more difficult. In situations where solids are solubilized by contact with rain or with water used in fighting a fire, practical containment techniques used to contain spills of liquids should be employed (see 8.2). When the spill of a hazardous solid is accompanied by fire or water, or both, containment can be difficult. Reactions may give off particulate clouds, toxic vapor, and even explosions may occur.

[3] The boldface numbers in parentheses refer to the list of references at the end of this guide.

REFERENCES

(1) "Evaluation Development of Foams for Mitigating Air Pollution from Hazardous Spills,'' EPA-600/2-82/029, National Technical Information Service, Springfield, VA, June 1982.

(2) "Handbook for Using Foams to Control Vapors from Hazardous Spill,'' EPA-600/8-8-019, National Technical Information Service, Springfield, VA.

(3) *Hazardous Materials Spills Handbook*, Chapter 10, McGraw-Hill, 1982.

(4) Department of Transportation, *Emergency Response Guide*, DOT-P 58.003, U.S. Department of Transportation, Washington, DC.

(5) *A Guide to the Safe Handling of Hazardous Materials Accidents, ASTM STP 825*, ASTM, 1983.

(6) *Sorbent Material for Cleanup of Hazardous Spills*, U.S. Department of Commerce, Washington, DC.

Designation: F 1129 – 88 (Reapproved 1995)$^{\epsilon 1}$

Standard Guide for Using Aqueous Foams to Control the Vapor Hazard from Immiscible Volatile Liquids[1]

This standard is issued under the fixed designation F 1129; the number immediately following the designation indicates the year of original adoption or, in the case of revision, the year of last revision. A number in parentheses indicates the year of last reapproval. A superscript epsilon (ε) indicates an editorial change since the last revision or reapproval.

ϵ^1 NOTE—Section 11 was added editorially in September 1995.

INTRODUCTION

The vapor released by spills of volatile hazardous substances (either flammable or toxic) can present a significant hazard to life and property in the spill area and for some measurable distance downwind. Such spills may also cause natural resource damage by penetration into the ground or by movement into groundwater.

Aqueous foam blankets have been shown to be an effective mechanism to mitigate the hazard arising from vapor release of volatile chemicals and to block accidental ignition of flammable liquids. Because they are a common tool of the fire services, they are available early in the spill response. Foams can be used to control spill vapors to extend evacuation time and may offer a long-term control for the life of the incident.

Effective actions have been demonstrated for a wide variety of chemical classes—volatile organics, water reactive inorganics and certain classes of liquefied gases.

The water reactive compounds and liquefied gases require special considerations peculiar to each chemical grouping. Although foam solutions are not considered to be dispersants, foam treatment may enhance the penetration of water soluble materials into the ground, or transport into the groundwater, or both. Adequate information is not available to generalize on such questions.

1. Scope

1.1 This guide restricts itself to addressing the application of foam to water immiscible liquid compounds with boiling points above 15°C for vapor control of landspill or contained spills on water.

1.2 *This standard does not purport to address all of the safety concerns, if any, associated with its use. It is the responsibility of the user of this standard to establish appropriate safety and health practices and determine the applicability of regulatory limitations prior to use.* For hazard statements, see Section 10.

2. Referenced Documents

2.1 *NFPA Standards:*[2]

11 Low Expansion Foam and Combined Agent Systems

11A Foam Systems Medium and High Expansion

3. Terminology

3.1 *aqueous foam*—a mixture of water and a foaming agent.

3.2 *foam*—a mass of bubbles formed by the mechanical agitation of foam solution with air.

3.3 *foam expansion*—the ratio of air to water in the foam. A measure of the volume of foam produced for each volume of foam solution used.

3.4 *high expansion*—a ratio of greater than 100:1. (See Foam Equipment for practical ranges of expansion.)

3.5 *foaming agent*—an organic compound or mixture of compounds which lowers the surface tension of water and imparts a foaming capability to it. Five major types of foam liquid concentrates are in general use by the fire service.

3.6 *protein*—a mixture of hydrolyzed animal protein with various stabilizing materials. Protein foam may be used only at low expansion. The surface tension of protein foam solutions in water is 40 to 50 dyne/cm. Protein foams are subject to bacterial and fungal attack and may have shelf life limitations.

3.7 *surfactant*—also known as syndet or detergent foam. These foams are based on high-foaming synthetic surface active agents. While these foams are normally used at high expansion, they may also be applied through low expansion foam-making devices. Surface tensions are in the range 23 to 30 dyne/cm.

3.8 *aqueous film forming foam*—also known as AFFF (A triple F). AFFF is a mixture of fluorocarbon and hydrocarbon surfactants. It is usually used at low expansion. The very low

[1] This guide is under the jurisdiction of ASTM Committee F-20 on Hazardous Substances and Oil Spill Responseand is the direct responsibility of Subcommittee F20.22 on Mitigation Actions.

Current edition approved Feb. 26, 1988. Published April 1988.

[2] Available from National Fire Protection Association, Batterymarch Park, Quincy, MA 02269.

surface tension of AFFF solution permits the formation of an aqueous film on top of most hydrocarbon fuels. Because maintenance of this film requires drainage of solution from the foam, AFFF is fast draining and the foam is not persistent. The film is easily disrupted and should not be relied upon for vapor sealing unless a visible foam blanket is present. The surface tension of AFFF solutions in water is 15 to 19 dyne/cm.

3.9 *fluoroprotein*—conventional protein foam modified by the addition of fluorocarbon surfactants. Fluoroprotein foams are similar to protein foams except that they produce foam with greater fluidity, dry chemical resistance (for clarification see NFPA Standard 11) and greater resistance to fuel pick-up. They are used only at low expansion. The surface tension of fluoroprotein foam solution (FP) is 27 to 30 dyne/cm. Film-forming fluoroprotein agents (FFFP) are being marketed with surface tensions in the 16 to 17 dyne/cm range.

3.10 *alcohol or polar solvent*—there are two types of foams which are resistant to destruction by water miscible polar compounds. One type, (*1*) based on protein foam, has been available for years. This type, called ''alcohol,'' since that was the polar compound it was usually used against, contains a water insoluble metal stearate. The insoluble material serves as a barrier between the fuel and the foam. Because the stearate interferes with foaming, the foam liquid must be added to the water close to the foam-making device. Alcohol type protein foam is not suitable for use on hydrocarbons nor is it effective if applied from a nozzle. Gentle (NFPA Type I) application is required, as described in NFPA Standard 11. A second type of material, (*2*) usually termed polar solvent resistant, contains a water soluble polymer. When this polymer contacts a water miscible polar fuel, it gels and forms a membrane which floats on the fuel and serves as a barrier to protect the foam from destruction by the fuel. Polar solvent resistant foams may be either surfactant or AFFF based. They behave like a conventional foam on hydrocarbons. They may be applied by nozzle or by any other low expansion foam-making equipment on either hydrocarbons or polar fuels. Alcohol or polar solvent resistant foams produce surface tensions in water ranging from 15 to 50 dyne/cm.

4. Significance and Use

4.1 It is intended as a general guide to correct use. Specific decisions on when or if foam should be used will depend on the circumstances and conditions of each spill situation.

5. Film Forming

5.1 Film forming is the development of a thin film of aqueous solution over the surface of a nonaqueous liquid chemical in response to a surface tension differential. Since water is denser than many liquid organic compounds, it will sink through such compounds. Foam agents reduce the surface tension of water. If the surface tension of the foaming solution is less than that of the organic compound, the drainage coming from the foam will tend to form a water film between the foam and the organic compound. The term "film forming'' has been applied basically to those foaming systems with low surface tensions, normally below 24 dyne/cm. Film forming may occur, however, whenever the ratio of surface tensions is appropriate.

6. Stability

6.1 Stability is used in two senses, foam collapse rate and resistance to chemicals. Foam collapse rates are measured only for high expansion foams. They will range from 8 to 20 in. (20 to 50 cm) per h in laboratory tests, but can be higher in the field due to sun, wind, and precipitation. Stability in contact with reactive chemicals is a property unique to each foam type.

7. Foam Equipment

7.1 *Low Expansion Foam*—Several types of foam-making devices are available for generating low expansion foams. The traditional foam nozzle consists of a tube through which a jet of foam solution is projected. Holes in the tube just downstream of the jet permit the aspiration of air. Various types of obstructions are fixed in the tube to create turbulence and mix the aspirated air with the foam solution. There are many variations in design of foam nozzles, but all produce expansion ratios in the 6:1 to 12:1 range, depending on the type of foam liquid used.

7.1.1 Water fog nozzles may be used to generate foam with AFFF or synthetic agents. Such foam rarely exceeds an expansion ratio of 4:1. AFFF made this way has a very fast drainage and short life. Some water fog nozzles are designed for the attachment of foam-making tubes. When the foam-making tubes are in use, the foam-making ability of the water fog nozzle is essentially the same as that of a true foam nozzle.

7.1.2 Foam nozzles in smaller sizes (up to 250 g/m) (945 L/m) may be used on hose lines. Larger capacity foam nozzles are mounted on monitors or turrets. Foam nozzles are available as straight stream devices, and combination straight stream and spray. Some designs permit several different patterns.

7.1.3 Straight streams give the best range, but may cause the foam stream to plunge into the spill. Plunging can aggravate vapor release and, in the case of a fire, may reduce control and increase extinguishment times. If possible, straight streams should be impinged against an obstacle or the ground in front of the spill and the foam allowed to flow onto the spill.

7.2 *High Expansion Foam*—High expansion foam generators spray the solution onto a screen or net and induce an airflow to blow the foam. Air is supplied either by aspiration in the case of handheld hose line units or by a fan in the case of larger units (see NFPA 11A for design illustrations). Air aspirating units, because of their limited air supply, produce expansions of less than 350:1 and have capacities of less than 1000 ft^3/min (28 m^3/min). Larger units using water, electric, or diesel powered fans can produce expansions up to 1000:1 and are available in sizes up to 30 000 ft^3/min (85 m^3/min).

7.2.1 Two types of water power are available, (*1*) water reaction motors, and (*2*) water turbines.

7.2.1.1 Reaction motors divert a small portion of the foaming solution to form a jet and drive a paddle wheel attached to the fan. They are less expensive and light in weight, but require higher pressures to operate, are less efficient foam-making devices and produce lower head pressures than turbines.

7.2.1.2 High expansion foam-making devices have very little range. If the spill cannot be approached at close range, it is possible to conduct high expansion foam through flexible fabric ducts to the spill. The higher head pressure available from a turbine driven foam-making device is an important advantage in this case.

8. Guidelines for Foam Application for Vapor Control

8.1 *Foam Selection*—Guidelines for the selection of an acceptable foam system depend on several factors, most specifically compatibility between foam and the spilled chemical. Work under United States Environmental Protection Agency (USEPA) sponsorship has provided basic selection data for a variety of chemical classes.[3]

8.2 There may be some difficulties maintaining foam blanket integrity under adverse environmental conditions, such as high wind and rain. High expansion foam can be adversely affected when sustained wind speeds are in excess of 10 mph (16 km/h) but slow draining foams may withstand gusts to 20 mph (32 km/h).

9. Time of Protection

9.1 The time of protection is governed by the thickness of the foam blanket, the expansion, the drainage rate of the foam, and the rate of vapor permeation through it.

9.2 *Permeation*—The rate of vapor permeation through the foam varies as a function of the equilibrium vapor pressure, the water solubility, and the molecular size of the spilled chemical. Permeation varies directly with vapor pressure and solubility: the greater they are the faster the permeation. It varies indirectly with molecular size: the larger the molecule the slower the permeation. Examples of permeation times are available for some chemicals and some foams in EPA report 600/2-82-029.[3]

9.3 *Drainage and Expansion*—Time of protection is significantly affected by drainage and expansion of the foam. Regardless of all other factors, slower drainage results in slower permeation and longer time of protection for a given foam application. Thus, in selecting a foam agent of any type, consideration should be given to the drainage characteristics. Low expansion gives longer protection times for a single application; but where spills persist for long periods of time, high expansion can have advantages.

9.3.1 For equivalent times of protection, less water will be used with high expansion than with low expansion. This will be a consideration where water supplies are limited or containment areas for the spill are restricted.

9.3.2 Where accidental ignition is a concern, high expansion may be beneficial since 18 in. (46 cm) of foam depth should prevent a flammable vapor concentration from developing above the foam blanket surface.

9.3.3 Flammable vapor concentrations can develop within foam blankets. With low expansion foams ignitions are normally self extinguishing. High expansion blankets may deflagrate. Vapor concentrations in the high expansion foam may be within the explosive range but ignitions do not normally propagate into true explosions. Very dry foams, which can occur due to a fast drainage rate or to extremely high expansions, greater than 500:1, can experience rapid deflagration.

9.3.4 When using foam for vapor hazard mitigation, a continuous blanket should always be maintained. Do not allow uncovered areas to persist and protect against chimney formation in the foam, that is continuous passages through the foam from the atmosphere to the spill surface.

9.4 *Foam Application and Makeup*—To be effective, the foam application rate must be greater than the foam breakdown rate. Foam breakdown occurs due to natural collapse, atmospheric effects (wind, rain, sun), mechanical interferences or obstacles to flow, and chemical interaction with the liquid material being covered.

9.4.1 For high expansion foams, 500:1 nominal expansion, sufficient testing has been conducted to dictate a minimum rate of 0.5 ft^3/min of foam per square ft of spill surface to be covered (0.17 m^3/min per m^2).

9.4.2 Low expansion application rates are not well defined. Current recommendations are based on the application rates for fire situations, 0.1 to 0.16 gpm of foam solution per square ft of spill surface (0.3 to 0.5 L/m per m^2). Minimum rates could be as little as 25 % of these rates.

9.4.3 Low expansion foam can be projected 50 to 100 ft (15 to 30 m) through the air with handline equipment and application can be effectively directed to the spill surface. High expansion foam must flow from the point of discharge. It will have an average flow rate of 50 radial ft/min (15 m/min). This will persist for several minutes over a liquid surface. On solid surface the initial rate slows appreciably; and after the first minute the effective rate is about 2 to 3 ft/min (0.6 to 0.9 m/min).

9.4.4 Foam tends to spread uniformly from the point of application. High expansion foam requires some form of containment to hold it in the spill area. If containment is not available and high expansion is to be employed, foam should be directed to the spill area through a duct or tube. This prevents excessive buildup of foam around the generator which could interfere with its operation and the flow of foam into adjacent areas which could hamper other mitigation and cleanup operations.

9.4.5 Foam makeup is usually an on-site decision process. Foam blanket maintenance procedures are a function of the dynamic degradation rate for each spill situation. Foam makeup can be an intermittent process or a slow continuous discharge of some fraction of the initial rate.

9.5 *Environmental Considerations*—In all spill situations there is concern for the materials which may enter the soil or ground water and be unrecoverable. Solutions of commercially available foam agents at specified concentrations may be biodegradable according to the manufacturer reports. Oral ingestion of small quantities by mammals including man should not be life threatening.

9.5.1 Aquatic toxicity of foam solutions is not well defined. With the exception of the fluorocarbon containing agents, there are no documented reports of aquatic toxicity for commercial agents in field use. The aquatic toxicity of fluorocarbon materials is not well established, but the Department of Defense has regulations dealing with the use of fluorocarbon agents in training and other nonemergency situations which

[3] "Evaluation of Foams for Mitigating Air Polution From Hazardous Spills," U.S. EPA Report 600/2-82-029, NTIS PB82-227117, Environmental Protection Agency, Springfield, VA.

relate to observed aquatic toxicity incidents. In most emergency situations the residues of spilled or burned materials entering the soil or water are orders of magnitude more detrimental than any toxic effects of the foam chemicals. Penetration of the spill into soil or water should be considered.

9.5.2 Changes in the character of the foam drainage after contact with the classes of spilled compounds addressed in this guide can be expected to be minimal. Foam drainage passing through the pool of spilled chemical may entrain small amounts of the chemical or preferentially strip impurities. Materials which are miscible in the drainage will be carried by it to their ultimate fate. Immiscible materials which are entrained should be released when the foam solution is absorbed by the soil or dispersed in the water column. These released materials should cause no greater problem than the spilled chemical itself. Concern for dissolved chemicals will have to be determined on a case-by-case basis taking all other factors of the spill situation into consideration. Foam drainage may displace the spilled liquid due to density differences and may possibly decrease soil penetration by water immiscible hazardous liquids.

10. Safety Hazards

10.1 *Personnel Safety*—Foam solutions are generally not considered toxic to humans but contact may cause skin or eye irritation. Read warning labels on foam concentrate containers. Effects and antidotal procedures will vary for each foam agent.

10.2 Fragments generated by the rupture of dry foam layers can cause sneezing and coughing. These effects are transient and will stop when exposure to the source of the fragments stops.

10.3 Prevent prolonged exposure to foam. Do not enter foam if full submergence would occur. Both vision and hearing are obscured. Some breathing apparatus can be adversely affected by foam submergence. If wading or moving through foam is necessary, be careful of tripping hazards and always wear a lifeline.

10.4 If possible, do not perform any action which will mechanically degrade that portion of the foam blanket covering the spill, that is walking or dragging hoses through foam. Such actions can create a localized vapor hazard. Do not depend on films to rapidly reform and prevent vapor release.

11. Keywords

11.1 aqueous foams; foaming agent; immiscible; immiscible volatile; vapor hazard

Designation: F 1209 – 89 (Reapproved 1999)

Standard Guide for Ecological Considerations for the Use of Oilspill Dispersants in Freshwater and Other Inland Environments, Ponds and Sloughs[1]

This standard is issued under the fixed designation F 1209; the number immediately following the designation indicates the year of original adoption or, in the case of revision, the year of last revision. A number in parentheses indicates the year of last reapproval. A superscript epsilon (ϵ) indicates an editorial change since the last revision or reapproval.

1. Scope

1.1 This guide covers the use of oilspill dispersants to assist in the control of oil spills. The guide is written with the goal of minimizing the environmental impacts of oil spills; this goal is the basis on which the recommendations are made. Aesthetic and socio-economic factors are not considered, although these and other factors are often important in spill response.

1.2 Spill responders have available several means to control or clean-up spilled oil. In this guide, the use of dispersants is given equal consideration with other spill countermeasures. It is not considered as a "last resort" after all other methods have failed.

1.3 This is a general guide only. It assumes the oil to be dispersible and the dispersant to be effective, available, applied correctly, and in compliance with relevant government regulations. In the assessment of environmental sensitivity, it is assumed that the dispersant is nonpersistent in the natural environment. Oil, as used in this guide, includes crude oils and refined petroleum products. Differences between individual dispersants or between different oil products are not considered.

1.4 The guide is organized by habitat type, for example, small ponds and lakes, rivers and streams, and land. It considers the use of dispersants primarily to protect habitats from impact (or to minimize impacts) and to clean them after a spill takes place.

1.5 This guide applies only to freshwater and other inland environments. It does not consider the direct application of dispersants to subsurface waters.

1.6 In making dispersant use decisions, appropriate government authorities should be consulted as required by law.

1.7 *This standard does not purport to address all of the safety concerns, if any, associated with its use. It is the responsibility of the user of this standard to establish appropriate safety and health practices and determine the applicability of regulatory limitations prior to use.*

2. Significance and Use

2.1 This guide is meant to aid local and regional response teams who may use it during spill response planning and spill events.

2.2 This guide should be adapted to site specific circumstance.

3. Environment Covered—Ponds and Sloughs

3.1 Ponds and sloughs are small isolated water bodies that are not part of major water systems. They have a dynamic ecology, and a wide mixture of animal and plant species. In northern regions, these water bodies may freeze to the bottom and do not sustain a wide variety of aquatic species.

3.2 While most of these bodies are naturally occurring and exist throughout the year, some may be man-made. In arid climates, the existence of these bodies may be seasonal.

3.3 The characteristics of these water bodies are:

(*a*) Open water area of less than 10 hectares,

(*b*) Shallow water with a maximum depth of 1 to 1.5 m,

(*c*) Soft bottom with a high organic content,

(*d*) May have alkaline or acidic water,

(*e*) No well defined inlet or outlet, and

(*f*) Poorly defined shoreline consisting of floating mats of vegetation.

4. Background

4.1 The effects of oil and dispersed oil on these aquatic environments have been the subject of numerous studies. The studies have involved both intentional experimental spills and studies undertaken during actual spill situations **(1-5)**.[2]

4.2 There have been a number of studies on the impact of oil and oil/dispersant mixtures on microbiological systems **(6-15)**.

4.3 The principal biotic components of such water bodies are a variety of fauna and flora. The aquatic flora include bacteria, algae, (planktonic and attached), and floating or submerged vascular plants. Terrestrial flora include grasses, moss, lichens, herbs, forbs, and woody plants.

4.4 The fauna include invertebrates, (zooplankton, molluscs, crustaceans, worms, and other similar species), fish, a variety of waterfowl, (ducks, loons, gulls, terns, and herons),

---

[1] This guide is under the jurisdiction of ASTM Committee F-20 on Hazardous Substances and Oil Spill Responseand is the direct responsibility of Subcommittee F20.13 on Treatment.

Current edition approved Feb. 24, 1989. Published April 1989.

[2] The boldface numbers in parentheses refer to the list of references at the end of this guide.

mammals, such as beaver and muskrat, and in many areas, significant human activity. The distribution and composition of species is a function of climate, local geography and soil type, and human use of the area **(16)**.

4.5 Studies of the effects of oil on vegetation include laboratory work on the effects on willows, moss, lichen, black spruce, and sedge. These studies showed that the pouring of oil directly on the vegetation was lethal to willow and moss, and caused some nonpermanent damage to the other species **(17-26)**.

4.6 Floating leaves are wetted and penetrated by oil from slicks while younger leaves whose cuticles are intact are more resistant to penetration by oil. Penetration occurs most readily in areas where there is damage to the leaf caused by insects or mechanical abrasion.

4.7 Oil is retained in areas of dense vegetation and is released slowly. Dispersants applied soon after a spill appear to speed the movement of the slick through the vegetation if there is some water movement. Only the oil that contacts the leaves causes some damage. Leaves and stems that are subsurface are not damaged by the presence of oil. Oil slicks that are restricted to contact of the stem areas near the water surface are not very injurious.

4.8 Field studies in northern environments have shown little long term effects of spilled oil. There have been few studies on the effects of dispersed oil in these environments.

4.9 Although oil may cause extensive damage to vegetation growing in low lying wetlands and shorelines of bog lakes, plants that grow from rhizomes or healthy portions of aerial tissue (sedge and willow), have a high recovery potential. Long term damage to root systems could however slow recovery of the vegetation in impacted areas. This may be of less importance in areas with an elevated water table (bog lakes).

4.10 Laboratory studies of the foliar damage to moss, lichen, willow, blackspruce and sedge, by dispersants indicate some damage, but this is generally not lethal. Many dispersants are fast acting contact poisons, but their activity is limited to the site of application. Plants treated with dispersants recover in most cases. However foliar damage may increase the plant's susceptibility to pathogens.

4.11 Field studies on the effects of the direct spraying of plants with dispersants indicated that the short term effect of dispersants to individual plants was less than that incurred through natural causes. No long term effects were found and seasonal growth was not affected.

4.12 Studies of the effects of dispersants and dispersed oil on the microbiological community show that the characteristics of the dispersant are critical. The dispersant should have no effect on the biodegradation of the oil. This is true of many currently used dispersants. The dispersant itself should have no residues or components toxic to microbes or microbial processes.

5. General Considerations for Making Dispersant Use Decisions

5.1 The dispersant use decision is, in this case as most others, one of trade-offs. The use of dispersants can reduce the adverse effects of spilled oil on certain biological species at the expense of other components of the ecosystem.

5.2 In most cases the mortality of individual creatures is of less concern than the destruction of habitat. The repopulation of areas after the spill will occur naturally when an area becomes a suitable habitat for a given species.

6. Recommendations

6.1 Dispersant use in ponds and sloughs, and their bordering vegetation, should be considered if a spill poses a significant threat to indigenous wildlife or its habitat. In evaluating the potential for dispersant use, consideration should be given to the alternatives of leaving the oil untreated or the use of mechanical recovery equipment. In many cases, a spill response operation can cause serious damage to a pond or slough habitat, or a disruption of nesting and breeding activities.

7. Keywords

7.1 dispersants; environmental sensitivity; freshwater; inland; oil spill; oilspill dispersants; ponds; sloughs

REFERENCES

(1) Miller, M. C., Vestal, J. R., Mozley, S., Butler, M., and Hobbie, J. E., "Effects of Prudhoe Crude Oil Spills on Coastal Tundra Ponds," *Energy/Environment II. U.S. EPA-600/9-77-012*, 1977, pp. 521–529.

(2) Nagy, E., Scott, B. F., and Hart, J., "The Fate of Oil and Oil-Dispersant Mixtures in Freshwater Ponds," *Science of the Total Environment*, Vol. 35, No. 2, 1984, pp. 115–133.

(3) Scott, B. F., and Glooschenko, V., "Impact of Oil and Oil-Dispersant Mixtures on Flora and Water Chemistry in Freshwater Ponds," *Science of the Total Environment*, Vol 35, No. 2, 1984, pp. 169–190.

(4) Scott, B. F., Nagy, E., Sherry, J. P., Dutka, B. J., Glooschenko, V., Snow, N. B., and Wade, P. J., "Ecological Effects of Oil-Dispersant Mixtures in Fresh Water," *Proceedings of 1979 Oil Spill Conference*, Los Angeles, pp. 565–571.

(5) Snow, N. B., and Scott, B. F., "The Effect and Fate of Crude Oil Spill on Two Arctic Lakes," *Proceedings of 1975 Oil Spill Conference*, San Francisco, pp. 527–534.

(6) Bergstein, P. E., and Vestal, J. R., "Crude Oil Biodegradation in Arctic Tundra Ponds," *Arctic*, Vol 31, No. 3, 1978, pp. 158–169.

(7) Atlas, R. M., and Busdosh, M., "Microbial Degradation of Petroleum in the Arctic," *Proceedings of the Third International Biodegradation Symposium, Applied Science*, eds. J. M. Sharpley and A. M. Kaplan, 1976, pp. 79–85.

(8) Breuil, C., and Shindler, D. B., "Effects of Crude Oil on Bacterial and Fungal Populations in Fresh Water Artificial Ponds," (Abstract), *Annual Meeting of American Society for Microbiology 77*, 1977, p. 269.

(9) Dutka, B. J., Sherry, J., Scott, B. F., and Kwan, K. K.," Effects of Oil-Dispersant Mixtures on Fresh Water Microbial Populations," *Canadian Research*, Vol 13, No. 5, 1980, pp. 58–62.

(10) Federle, T. W., Vestal, J. R., Hater, G. R., and Miller, M. C., "Effects of Prudhoe Bay Crude Oil on Primary Production and Zooplankton in Arctic Tundra Thaw Ponds," *Marine Environmental Research*, Vol 2, No. 1, 1979, pp. 3–18.

(11) Miller, M. C., Alexander, V., and Barsdate, R. J., "The Effects of Oil Spills on Phytoplankton in an Arctic Lake and Ponds," *Arctic*, Vol 31, No. 3, 1978, pp. 192–218.

(12) Miller, M. C., Hater, G. R., and Vestal, J. R., "Effect of Prudhoe Crude Oil on Carbon Assimilation by Planktonic Algae in an Arctic Pond," *Environmental Chemistry and Cycling Processes: Proceedings of a Symposium, April 28 to May 1, 1976*, Augusta, GA.

(13) Scott, B. F., and Shindler, D. B., "Impact of Crude Oil on Planktonic Freshwater Ecosystems," *Proceedings of the Conference on Assessment of Ecological Impacts of Oil Spills, American Institute of Biological Sciences*, Arlington, VA, 1978, pp. 712–734.

(14) Shindler, D. B., Scott, B. F., and Carlisle, D. B., "Effect of Crude Oil on Populations of Bacteria and Algae in Artificial Ponds Subject to Winter Weather and Ice Formation," *Verhlug International Verein Limnologie*, Vol 19, No. 5, 1975, pp. 2138–2144.

(15) Stanley, D. W., "Productivity of Epipelic Algae in Tundra Ponds and a Lake near Barrow, Alaska," *Ecology*, Vol 57, No. 5, 1976, pp. 1015–1024.

(16) Mozley, S. C., and Butler, M. G., "Effects of Crude Oil on Aquatic Insects of Tundra Ponds," *Arctic*, Vol 31, No. 3, 1978, pp. 229–240.

(17) Blunden, D., Dale, M., Goudey, S., and Hoddinot, J.," The Effects of Oil Spill Chemicals on Shoreline Plants of Northern Freshwater Ecosystems," *Proceedings of the International Conference on Oil and Freshwater: Chemistry, Biology Technology*, ed., J. H. Vandermeulen, Pergamon, pp. 403–409.

(18) Barsdate, R. J., "Ecologic Changes in an Arctic Tundra Pond Following Exposure to Crude Oil," (Abstract only), *Proceedings of the Symposium on the Impact of Oil Resource Development on Northern Plant Communities, 17 August 1972, Institute of Arctic Biology*, University of Alaska, 1973, p. 52.

(19) Barsdate, R. J., Alexander, V., and Benoit, R. E., "Natural Oil Seeps at Cape Simpson, Alaska: Aquatic Effects," *Proceedings of the Symposium on Impact of Oil Resource Development on Northern Plant Communities, 17 August 1972, Institute of Arctic Biology*, University of Alaska, 1973, pp. 91–95.

(20) Goudey, J. S., Dale, M., and Hoddinot, J., "The Effects of Oil Spill Chemicals on Transpiration, CO_2 Exchange and Cuticular Structure in Salix Interior," *Canadian Journal of Botany*, Vol 63, No. 12, 1985, pp. 2340–2344.

(21) Goudey, J. S., Dale, M., and Hoddinot, J., "Effects of Oil Spill Chemicals on CO_2 Assimilation by the *Fruticose Lichen Cladina Mitis*," *Freshwater Oilspill Research Project, (FOSRP) Report,* Esso Resources Canada Limited, Calgary, Alberta, Canada, 1986.

(22) Goudey, J. S., Dale, M., and Hoddinot, J.," Effects of Crude Oil and Chemically Treated Oil on Shoreline Plant Communities of Northern Bog Lakes," *FOSRP Report,* Esso Resources Canada Limited, Calgary, Alberta, Canada, 1986.

(23) Hutchinson, T. C., and Freedman, W., "Effects of Experimental Crude Oil Spills on Subarctic Boreal Forest Vegetation near Norman Wells, N.W.T., Canada," *Canadian Journal of Botany*, Vol 56, No. 19, 1978, pp. 2424–2433.

(24) Helebust, J. A., Hanna, B., Sheath, R. G., Gergis, M., and Hutchinson, T. C., "Experimental Crude Oil Spills on a Small Subarctic Lake in the MacKenzie Valley, N.W.T.: Effects on Phytoplankton, Periphyton, and Attached Aquatic Vegetation," *Proceedings of 1975 Oil Spill Conference*, San Francisco, 1975, pp. 509–515.

(25) Walker, D. A., Webber, P. J., Everett, K. R., and Brown, J., "Effects of Crude and Diesel Oil Spills on Plant Communities at Prudhoe Bay, Alaska and the Derivation of Oil Spill Sensitivity Maps," *Arctic*, Vol 31, No. 3, 1978, pp. 242–259.

(26) Wein, R. W., and Bliss, L. C., "Experimental Crude Oil Spills on Arctic Plant Communities," *Journal of Applied Ecology*, Vol 10, No. 3, 1973, pp. 671–682.

Designation: F 1210 – 89 (Reapproved 1999)

Standard Guide for Ecological Considerations for the Use of Oilspill Dispersants in Freshwater and Other Inland Environments, Lakes and Large Water Bodies[1]

This standard is issued under the fixed designation F 1210; the number immediately following the designation indicates the year of original adoption or, in the case of revision, the year of last revision. A number in parentheses indicates the year of last reapproval. A superscript epsilon (ϵ) indicates an editorial change since the last revision or reapproval.

1. Scope

1.1 This guide covers the use of oilspill dispersants to assist in the control of oil spills. The guide is written with the goal of minimizing the environmental impacts of oil spills; this goal is the basis on which the recommendations are made. Aesthetic and socio-economic factors are not considered, although these and other factors are often important in spill response.

1.2 Spill responders have available several means to control or clean up spilled oil. In this guide, the use of dispersants is given equal consideration with other spill countermeasures. It is not considered as a "last resort" after all other methods have failed.

1.3 This is a general guide only. It assumes the oil to be dispersible and the dispersant to be effective, available, applied correctly, and in compliance with relevant government regulations. In the assessment of environmental sensitivity, it is assumed that the dispersant is nonpersistent in the natural environment. Oil, as used in this guide, includes crude oils and refined petroleum products. Differences between individual dispersants or between different oil products are not considered.

1.4 The guide is organized by habitat type, for example, small ponds and lakes, rivers and streams, and land. It considers the use of dispersants primarily to protect habitats from impact (or to minimize impacts) and to clean them after a spill takes place.

1.5 This guide applies only to freshwater and other inland environments. It does not consider the direct application of dispersants to subsurface waters.

1.6 In making dispersant use decisions, appropriate government authorities should be consulted as required by law.

1.7 *This standard does not purport to address all of the safety concerns, if any, associated with its use. It is the responsibility of the user of this standard to establish appropriate safety and health practices and determine the applicability of regulatory limitations prior to use.*

[1] This guide is under the jurisdiction of ASTM Committee F-20 on Hazardous Substances and Oil Spill Responseand is the direct responsibility of Subcommittee F20.13 on Treatment.

Current edition approved Feb. 24, 1989. Published April 1989.

2. Significance and Use

2.1 This guide is meant to aid local and regional response teams who may use it during spill response planning and spill events.

2.2 This guide should be adapted to site specific circumstance.

3. Environment Covered—Lakes and Large Water Bodies

3.1 Lakes and large water bodies are major fresh water features that are a significant part of major water systems. They have a dynamic near shore ecology, and a wide mixture of animal and plants species. In northern regions, these water bodies may be partly or completely ice covered during part of the year but will not freeze to the bottom. Commercially important fishing and recreational activities are frequently associated with these water bodies.

3.2 While most of these bodies are naturally occurring and exist during the most year, some may be man-made.

3.3 The characteristics of these water bodies are:

a) Open water area greater than 10 hectares,

b) Water depths in excess of 1.5 m,

c) Soft or hard bottom with a low organic content except in shallow water areas,

d) Acidic water in some areas especially near industrial regions,

e) A well defined inlet or outlet, or both, and

f) A well defined shoreline of varied characteristics such as sand beaches and rocky headlands similar to marine environments. Some parts of the shore may be similar to those in ponds and sloughs.

4. Background

4.1 The effects of oil and dispersed oil on these aquatic environments have been the subject of numerous studies. The studies have involved both intentional experimental spills and studies undertaken during actual spill situations **(1-3)**.[2]

4.2 There have been a number of studies on the impact of oil and oil/dispersant mixtures on microbiological systems **(4-10)**.

4.3 The principal biotic components of such water bodies

[2] The boldface numbers in parentheses refer to the list of references at the end of this guide.

are a variety of fauna and flora. The aquatic flora include bacteria, algae, (planktonic and attached), and floating or submerged vascular plants. Terrestrial flora include grasses, moss, lichens, herbs, forbs, and woody plants. In deep water areas, there is little vegetation except for bacteria and algae.

4.4 The fauna include invertebrates, (zooplankton, molluscs, crustaceans, worms, and other similar species), fish, a variety of waterfowl, (ducks, loons, gulls, terns, and herons), mammals, such as beaver and muskrat, and in many areas, significant human activity. The distribution and composition of species is a function of climate, local geography and soil type, and human use of the area.

4.5 Human activities range from recreation and tourism, to shipping and commercial fishing. In many cases, lakes and other large water bodies are the source of potable water for human consumption or industrial use.

5. General Considerations for Making Dispersant Use Decisions

5.1 The dispersant use decision is, in this case as most others, one of trade-offs. The use of dispersants can reduce the adverse effects of spilled oil on certain biological species at the expense of other components of the ecosystem.

5.2 In most cases the mortality of individual creatures is of less concern than the destruction of habitat. The repopulation of areas after the spill will occur naturally when an area becomes a suitable habitat for a given species.

6. Recommendations

6.1 Dispersant use in lakes and other large water bodies and their bordering vegetation should be considered if a spill poses a significant threat to indigenous wildlife or its habitat. In evaluating the potential for dispersant use, consideration should be given to the alternatives of leaving the oil untreated or the use of mechanical recovery equipment. In many cases, a spill response operation can cause serious damage to a lake or large water body habitat, or a disruption of nesting and breeding activities.

6.2 Since large waves can be generated in these water bodies, the need to supply additional energy for the dispersant process is not an issue. In many cases, the water is shallow enough that the dispersed oil will reach the bottom and will have the potential to cause impact on the benthic community.

6.3 The use of dispersants near water intakes is not recommended because there is a possibility of inducing increased contamination. Dispersant application should be far enough away from the intake so that dilution can occur before the water is used for potable or industrial applications.

6.4 Should waterfowl, either migrating or resident be present, the use of dispersants is recommended to reduce the impact on this resource.

6.5 In some areas, the protection of fish, their eggs, larvae, and juveniles, is a concern. Fish larvae and eggs have been found to be particularly susceptible to oil. In this case, the mechanical removal may be preferred if it can be completed before the oil contacts the eggs or larvae.

7. Keywords

7.1 dispersants; environmental sensitivity; freshwater; inland; lakes; oil spill; oilspill dispersants

REFERENCES

(1) Fremling, C. R., "Impacts of a Spill of No. 6 Fuel Oil on Lake Winona," *Proceedings of 1981 Oil Spill Conference*, Atlanta, GA, 1981, pp. 419–421.

(2) Snow, N. B., and Brunskill, G. J., "Crude Oil and Nutrient Enrichment Studies in a MacKenzie Delta Lake," *Technical Report No. 553 Canadian Department Environment Fisheries and Marine Service*, Winnipeg, Man., 1975, pp. 7.

(3) Snow, N. B., and Rosenberg, D. M., "Experimental Oil Spills on MacKenzie Delta Lakes; I. Effect of Normal Wells Crude Oil on Lake 4," *Technical Report No. 548 Department Environment Fisheries and Marine Service* , Winnipeg, Man., 1975, p. 44.

(4) Berner, N. H., Ahearn, D. G., and Cook, W. L., "Effects of Hydrocarbonoclastic Yeasts on Pollutant Oil and the Environment," *Impact on the Use of Microorganisms on the Aquatic Environment, EPA-660/3-75-001*, U.S. Environmental Protection Agency, Corvallis, OR., 1975.

(5) Buhrer, H., "Influence of Hydrocarbons on Ecology of Bacteria in Aerobic Lake Sediment," *Schweiz Zeitschrift zu Hydrologie*, Vol 41, No. 2, 1979, pp. 315–355.

(6) Caparello, D. M., and LaRock, P. A.," A Radioisotope Assay for the Quantification of Hydrocarbon Biodegradation Potential in Environmental Samples," *Microbiology and Ecology*, Vol 2, No. 1, 1975, pp. 28–42.

(7) Horowitz, A., and Atlas, R. M., "Response of Microorganisms to an Accidental Gasoline Spillage in an Arctic Freshwater Ecosystem," *Applied Environmental Microbiology*, Vol 33, No. 6, 1977, pp. 1252–1258.

(8) Horowitz, A., Sexstone, A., and Atlas, R. M., "Hydrocarbons and Microbial Activities in Sediment of an Arctic Lake One Year after Contamination with Leaded Gasoline," *Arctic*, Vol 31, No. 3, 1978, pp. 180–191.

(9) Jordan, M. J., Hobbie, J. E., and Peterson, B. J., "Effects of Petroleum Hydrocarbons on Microbial Populations in an Arctic Lake," *Arctic*, Vol 31, No. 3, 1978, pp. 170–179.

(10) McKinley, V. L., Federle, T. W., and Vestal, J. R., "Effects of Petroleum Hydrocarbons on Plant Litter Microbiota in an Arctic Lake," *Applied Environmental Microbiology*, Vol 43, No. 1, 1982, pp. 129–135.

Designation: F 1231 – 89 (Reapproved 1999)

Standard Guide for Ecological Considerations for the Use of Oilspill Dispersants in Freshwater and Other Inland Environments, Rivers and Creeks[1]

This standard is issued under the fixed designation F 1231; the number immediately following the designation indicates the year of original adoption or, in the case of revision, the year of last revision. A number in parentheses indicates the year of last reapproval. A superscript epsilon (ε) indicates an editorial change since the last revision or reapproval.

1. Scope

1.1 This guide covers the use of oilspill dispersants to assist in the control of oil spills. This guide is written with the goal of minimizing the environmental impacts of oil spills; this goal is the basis on which the recommendations are made. Aesthetic and socioeconomic factors are not considered, although these and other factors are often important in spill response.

1.2 Spill responders have available several means to control or clean up spilled oil. In this guide, the use of dispersants is given equal consideration with other spill countermeasures. It is not considered as "last resort" after all other methods have failed.

1.3 This is a general guide only. It assumes the oil to be dispersible and the dispersant to be effective, available, applied correctly, and in compliance with relevant government regulations. In the assessment of environmental sensitivity, it is assumed that the dispersant is nonpersistent in the natural environment. Oil, as used in this guide, includes crude oils and refined petroleum products. Differences between individual dispersants or between different oil products are not considered.

1.4 This guide is organized by habitat type, for example, small ponds and lakes, rivers and streams, and land. It considers the use of dispersants primarily to protect habitats from impact (or to minimize impacts) and to clean them after a spill takes place.

1.5 This guide applies only to freshwater and other inland environments. It does not consider the direct application of dispersants to subsurface waters.

1.6 In making dispersant use decisions, appropriate government authorities should be consulted as required by law.0

1.7 *This standard does not purport to address all of the safety concerns, if any, associated with its use. It is the responsibility of the user of this standard to establish appropriate safety and health practices and determine the applicability of regulatory limitations prior to use.*

2. Significance and Use

2.1 This guide is meant to aid local and regional response teams who may use it during spill response planning and spill events.

2.2 This guide should be adapted to site-specific circumstances.

3. Environment Covered—Rivers and Creeks

3.1 Rivers and creeks are moving bodies of fresh water that are a significant part of major water systems. They have a dynamic near shore ecology, and a wide variety of animal and plant species. In northern regions, these water bodies may be partly or completely ice covered during part of the year. Shallow rivers and most creeks may freeze to the bottom in the winter. Commercially important fishing and recreational activities are frequently associated with these water bodies.

3.2 Rivers generally refer to large bodies of moving water, whereas creeks are smaller bodies of flowing water.

3.3 The characteristics of these water bodies are:

3.3.1 Flowing water,

3.3.2 Water depths in excess of 1 m is designated as a river, shallower would be a creek,

3.3.3 A low organic content bottom except in shallow near shore still-water areas,

3.3.4 Acidic water in some areas especially near industrial regions,

3.3.5 A well defined source or outlet, or both, and

3.3.6 A well defined shoreline consisting of sand beaches and rocky headlands similar to marine and lake environments.

4. Background

4.1 The effects of oil and dispersed oil on these aquatic environments have been the subject of numerous studies. The studies have involved both intentional experimental spills and studies undertaken during actual spill situations **(1-5)**.[2]

4.2 There have been a number of studies on the impact of oil and oil/dispersant mixtures on microbiological systems **(6-16)**.

4.3 The principal biotic components of such water bodies are a variety of fauna and flora. The aquatic flora include bacteria, algae (planktonic and attached), and floating or

[1] This guide is under the jurisdiction of ASTM Committee F-20 on Hazardous Substances and Oil Spill Responseand is the direct responsibility of Subcommittee F20.13 on Treatment.

Current edition approved May 26, 1989. Published July 1989.

[2] The boldface numbers in parentheses refer to the list of references at the end of this guide.

submerged vascular plants. Terrestrial flora include grasses, moss, lichens, herbs, forbs, and woody plants. In deep water areas, there is little vegetation except for bacteria and algae **(17-20)**.

4.4 The fauna include invertebrates (zooplankton, molluscs, crustaceans, worms and other similar species), fish, a variety of waterfowl (ducks, loons, gulls, terns and herons), mammals such as beaver and muskrat, and in many areas significant human activity. The distribution and composition of species is a function of climate, local geography and soil type, and human use of the area **(21)**.

4.5 Human activities range from recreation and tourism, to shipping and commercial fishing. In many cases, rivers and creeks are the source of potable water for human consumption and industrial use.

5. General Considerations for Making Dispersant Use Decisions

5.1 The dispersant use decision is, in this case as most others, one of trade-offs. The use of dispersants can reduce the adverse effects of spilled oil on certain biological species at the expense of other components of the ecosystem.

5.2 In most cases the mortality of individual creatures, while of concern, is of less concern than the destruction of habitat. The repopulation of areas after the spill will occur naturally when an area becomes a suitable habitat for a given species.

6. Recommendations

6.1 Dispersant use in rivers and creeks and their bordering vegetation should be considered if a spill poses a significant threat to indigenous wildlife or its habitat. In evaluating the potential for dispersant use, consideration should be given to the alternatives of leaving the oil untreated or the use of mechanical recovery equipment. In many cases, a spill response operation can cause serious damage to a rivers and creeks habitat, or a disruption of nesting and breeding activities.

6.2 The turbulence of the flowing water may be adequate to provide the energy for dispersion and vertical transport. This is certainly the case in areas where there are rapids. In many cases the water is shallow enough that the dispersed oil will reach the bottom and will have the potential to cause impact on the benthic community.

6.3 The use of dispersants near water intakes is not recommended because there is a possibility of inducing increased contamination. Dispersant application should be far enough upstream of the intake so that dilution can occur before the water is used for potable or industrial applications.

6.4 Should waterfowl, either migrating or resident, be present, the use of dispersants is recommended to reduce the impact on this resource.

6.5 In some areas the protection of fish, their eggs, larvae, and juveniles, is a concern. Fish larvae and eggs have been found to be particularly susceptible to oil. In this case, mechanical removal may be preferred if it can be completed before the oil contacts the eggs or larvae.

7. Keywords

7.1 creeks; dispersants; environmental sensitivity; freshwater; inland; oil spill; oilspill dispersants; rivers

REFERENCES

(1) Blahm, T. H., Durkin, J., Snyder, G., Coley, T., and Emmett, R. L., "Columbia River Oil Spill Study, June/July 1978," *Northwest and Alaska Fisheries Center, Coastal Zone and Estuarine Studies Division*, Seattle, WA, 1980.

(2) Bury, R. B., "The Effects of Diesel Fuel on a Stream Fauna," *California Fish and Game*, Vol 58, No. 4, 1972, pp. 291–295.

(3) McCauley, R. N., "The Biological Effects of Oil Pollution in a River,'' *Limnology and Oceanography*, Vol 11, Nos. 4 and 5, 1966, pp. 475–486.

(4) Rosenberg, D. M., and Wiens, A. P., "Community and Species Responses of *Chironomidae* (Diptera) to Contamination of Fresh Waters by Crude Oil and Petroleum Products, with Special Reference to the Trail River, Northwest Territories," *Journal of Fisheries Research Board of Canada*, Vol 33, No. 9, 1976, pp. 1955–1963.

(5) Rosenberg, D. M., Wiens, A. P., and Saether, O. A.," Responses to Crude Oil Contamination by *Cricotopus* (Cricotopus) *bicinctus* and *C.* (*C.*) *mackenziensis* (Diptera: Chironomidae) in the Fort Simpson Area, Northwest Territories," *Journal of Fisheries Research Board of Canada*, Vol 34, 1977, pp. 254–261.

(6) Bitton, G., Chuckran, D. A., Chet, I., and Mitchell, R.," Resistance of Bacterial Chemotaxis to Blockage in Petroleum Waters," *Marine Pollution Bulletin*, Vol 10, No. 2, 1979, pp. 48–49.

(7) Bugbee, S. L., and Walter, C. M., "The Response of Macroinvertebrates to Gasoline Pollution in a Mountain Stream," *Proceedings of 1973 Oil Spill Conference*, Washington, DC, 1973, pp. 725–731.

(8) Colvin, J. W., and Gordon, R. C., "Interactions between Crude Oil and Subarctic River Bacteria," *Proceeding of Alaskan Science Conference 27*, 1976, pp. 159–160.

(9) Griffin, W. M., and Cooney, J. J., "Degradation of Model Recalcitrant Hydrocarbons by Microorganisms from Freshwater Ecosystems," *Development in Industrial Microbiology*, Vol 20, 1979, pp. 479–488.

(10) Hoehn, R. C., Stauffer, J. R., Masnik, M. T., and Hocutt, C. H., "Relationships Between Sediment Oil Concentrations and the Macroinvertebrates Present in a Small Stream Following an Oil Spill," *Environmental Letters*, Vol 7, No. 4, 1974, pp. 345–352.

(11) Lock, M. A., Wallace, R. R., Barton, D. R., and Charlton, S., "The Effects of Synthetic Crude Oil on Microbial and Macroinvertebrate Benthic River Communities—Part I. Colonization of Synthetic Crude Oil Contaminated Substrata," *Environmental Pollution Series A*, Vol 24, No. 3, 1981, pp. 207–217.

(12) Lock, M. A., Wallace, R. R., and Westlake, D. W. S.," Biodegradation of Synthetic Crude Oil in Two Rivers of Northern Alberta, Canada," *Water Research*, Vol 16, No. 4, 1982, pp. 497–500.

(13) Masnik, R. N., Stauffer, J., Hocutt, C., and Wilson, J., "The Effects of an Oil Spill on the Macroinvertebrates and Fish in a Small Southwestern Virginia Creek," *Journal of Environmental Science, Health and Engineering Part A*, Vol 11, Nos. 4 and 5, 1976, pp. 281–296.

(14) Morrison, S. M., and Cummings, B. A.," Microbiologically Mediated Mutagenic Activity of Crude Oil," *EPA/600/S3-81-053* Corvallis, OR., 1986, 2 pp.

(15) Roeder, D. R., Crum, G. H., Rosenberg, D. M., and Snow, N. B., "Effects of Norman Wells Crude Oil on Periphyton in Selected Lakes and Rivers in the Northwestern Territories," *Technical Report No. 552*, Canada Department of Environment, Fisheries and Marine Service, Winnipeg Man., 1975, 31 pp.

(16) Rutgers University and the Academy of Natural Science of Philadelphia, "The Effect of Hydrocarbon on Natural Process of Bacterial and Algal Attached Communities," *The National Science Foundation RANN Program Report*, Petroleum Industry in the Delaware Delta, 1977, pp. 302–354.

(17) Nauman, J. W., and Kernodle, D. R., "The Effect of a Fuel Oil Spill on Benthic Invertebrates and Water Quality on the Alaskan Arctic Slope, Happy Valley, Near Sagwon, Alaska," *Journal of Research, U.S. Geological Survey*, Vol 3, No. 4, 1975, pp. 495–501.

(18) Rosenberg, D. M., and Snow, N. B., "Effect of Crude Oil on Zoobenthos Colonization of Artificial Substrates in Sub-Arctic Ecosystems," *Internationale Vereinigung für Theoretische und Angewandte Limnologie und Verhandlung*, Vol 19, No. 3, 1975, pp. 2172–2177.

(19) Snow, N. B., Rosenberg, D. M., and Meonig, J., "The Effects of Norman Wells Crude Oil on the Zoobenthos of a Northern Yukon Stream One Year After an Experimental Spill," *Technical Report No. 550*, Canada Department of Environment, Fisheries Marine Service, Winnipeg, Man., 1975, 8 pp.

(20) Snow, N. B., and Rosenberg, D. M., "The Effects of Crude Oil on the Colonization of Artificial Substrates by Zoobenthos Organisms," *Technical Report No. 551*, Canada Department of Environment, Fisheries Marine Service, Winnipeg, Man., 1975, 35 pp.

(21) Mackiew, P. R., McGill, A. S., and Hardy, R., "Diesel Oil Contamination of Brown Trout (*Salmo trutta L.*)," *Environmental Pollution*, Vol 3, No. 1, 1972, pp. 9–16.

Designation: F 1279 – 90 (Reapproved 1999)

Standard Guide for Ecological Considerations for the Use of Oilspill Dispersants in Freshwater and Other Inland Environments, Permeable Surfaces[1]

1. Scope

1.1 This guide covers the use of oilspill dispersants to assist in the control of oil spills. The guide is written with the goal of minimizing the environmental impacts of oil spills; this goal is the basis on which the recommendations are made. Aesthetic and socioeconomic factors are not considered although these and other factors are often important in spill response.

1.2 Spill responders have available several means to control or clean up spilled oil. In this guide, the use of dispersants is given equal considerations with other spill countermeasures. It is not considered as a last resort after all other methods have failed.

1.3 This is a general guide only. It assumes the oil to be dispersible and the dispersant to be effective, available, applied correctly, and in compliance with relevant government regulations. In the assessment of environmental sensitivity, it is assumed that the dispersant is nonpersistent in the natural environment. Oil, as used in this guide, includes crude oils and refined petroleum products. Differences between individual dispersants or between different oil products are not considered.

1.4 The guide is organized by habitat type; for example, small ponds and lakes, rivers and streams, and land. It considers the use of dispersants primarily to protect habitats from impact (or to minimize impacts) and to clean them after a spill takes place.

1.5 This guide applies only to freshwater and other inland environments. It does not consider the direct application of dispersants to subsurface waters.

1.6 In making dispersant use decisions, appropriate government authorities should be consulted as required by law.

1.7 *This standard does not purport to address all of the safety concerns, if any, associated with its use. It is the responsibility of the user of this standard to establish appropriate safety and health practices and determine the applicability of regulatory limitations prior to use.*

2. Significance and Use

2.1 This guide is meant to aid local and regional response teams who may use it during spill response planning and spill events.

2.2 This guide should be adapted to site-specific circumstances.

3. Environment Covered—Permeable Surfaces

3.1 Permeable ground includes any soil, rock, agricultural land and forest, pasture land, forest, roadside or other surface, that is permeable to water and oil.

4. Background

4.1 The effects of oil and especially that of dispersed oil on terrestrial biota is not well known. In one study, oil spilled on soil decreased the nematode (worm) population by as much as 80 % **(1)**.[2] Lai Hoi-Chaw and co-workers show that a littornid gastropod (snail) showed avoidance to oil spilled on the mud of a mangrove swamp **(2)**. This avoidance decreased the mortality of the species to both oil and chemically-dispersed oil. McGill has noted that soil arthropods (insects) are quickly killed after spills **(3)**.

4.2 Oil has a broad-spectrum herbicidal effect on plants **(3,4)**. Effects vary depending on concentration and on species. Oil in low concentrations has been shown to increase growth in some species, whereas slight contact with oil causes death in others **(4)**. Black spruce, alfalfa and canola have a low tolerance for oil, while willow, dogwood and brome grass have a high tolerance **(5)**. Light oils may be toxic to vegetation on contact. Heavy oils have a tendency to smother plants over a longer period of time **(3)**. Oiling of the vegetative portions may kill the upper portion of the plant, but the root may still live and proceed to grow new stalks **(3)**. In one test, 0.4 to 3.4 L/m^2 of a light crude oil killed most plants in a northern boreal setting **(6)**. In another experiment light fuel oil at 0.6 % by weight killed all plants present **(7)**. Oil reduces the germination rate of seeds. Weathered oil on the soil forms a crust which can slow revegetation **(3)**. Revegetation time varies but has ranged from 1 to 20 years depending on location, condition, and amount of oiling **(1,3)**.

[1] This guide is under the jurisdiction of ASTM Committee F-20 on Hazardous Substances and Oil Spill Responseand is the direct responsibility of Subcommittee F20.13 on Treatment.

Current edition approved Feb. 23, 1990. Published April 1990.

[2] The boldface numbers in parentheses refer to the list of references at the end of this guide.

4.3 Oil spilled on ground will penetrate the surface, the rate of penetration depending on soil type, pore size, depth of the water table, and oil type. Studies on marine beaches show that dispersants increase the oil penetration. Dispersants increase the penetration rate and depth **(8,9)**. Rowland and co-workers showed that dispersants premixed with oil increased penetration into supratidal sediment **(10)**. Dewling and Silva examined the use of dispersants on a beach in Brazil and determined that the average penetration of oil was increased from 5 to 60 cm by the use of hydrocarbon-based dispersants **(11)**.

4.4 Oil degradation takes place on soil surfaces under many conditions. Factors that increase degradation rate and amount include high oxygen level, ample but not excessive (saturating) moisture, slightly alkaline pH, high temperature and ample nutrients **(1)**. Initially after a spill, the diversity of soil microorganisms is decreased by the toxicity of the oil, but the total number is increased due to the increase in number of oleoclasts (oil degrading microorganisms). Soil respiration increases after a spill **(1)**. Parkinson showed in a test spill on northern boreal soil that soil respiration increased 100 % and the bacterial numbers increased tenfold **(12)**.

4.5 Microbial degradation of oil occurs primarily at the soil surface **(1,3,13)**. One study showed that below 15 cm there was little degradation **(14)**. Degradation occurs primarily at the surface due to oxgyen, low but sufficient moisture, supply of nutrients, and because the highest population of oleoclasts is there **(1)**.

4.6 Contamination of groundwater is of prime concern in land spills. Little oil degradation takes place in groundwater and dilution alone would take many years to allow use of a groundwater supply contaminated by an oil spill **(1,15)**. One study estimated that 120 to 750 years of rainfall dilution alone would be required so that the supply could be used for human consumption **(15)**.

4.7 Several remediation techniques have been demonstrated for oiled soils **(1,3,5)**. Most of these techniques involve aeration, addition of fertilizer, and planting of cover crops. These techniques are well-documented and have been effective in restoring agricultural land to full production in as little as five years. No scientific evidence is available to show that dispersants have a useful role in these remediation techniques.

5. General Considerations for Making Dispersant Use Decisions

5.1 The dispersant use decision is, in this case as most others, one of tradeoffs. The use of dispersants can reduce the adverse effects of spilled oil on certain biological species at the expense of other components of the ecosystem.

5.2 In most cases the mortality of individual creatures is of less concern than the destruction of habitat. The repopulation of areas after the spill will occur naturally when an area becomes a suitable habitat for a given species.

5.3 Groundwater protection is a high priority for land spills. Groundwater contamination has serious consequences and is difficult to clean up. Groundwater should be protected during land spill cleanup.

6. Recommendations

6.1 Where groundwater contamination is not a concern, the use of dispersants may be considered as a spill countermeasure.

6.2 The use of dispersants on permeable surfaces is not recommended where groundwater can be affected. Consultation with a groundwater specialist is recommended.

6.3 Degradation of oil occurs naturally on many soils and the use of dispersants can move oil downward where it degrades more slowly. The use of dispersants is only recommended when other technologies are ineffective in reducing the potential for oil runoff or animal contact. This is one of the environmental trade-off situations.

7. Keywords

7.1 dispersants; freshwater; inland; oil spill; oilspill dispersants; permeable; soil

REFERENCES

(1) Bossert, I., and Bartha, R. "The Fate of Petroleum in Soil Ecosystems," *Petroleum Microbiology*, R. M. Atlas, ed., MacMillan Publishing Company, New York, NY, 1984, pp. 435–473.

(2) Lai, H. C., Lim, C. P., and Lee, K. T. "Effects of Naturally and Chemically Dispersed Oil on Invertebrates in Mangrove Swamps." *Fate and Effects of Oil in the Mangrove Environment*, eds. H. C. Lai and M. C. Feng, Universiti Sains Malaysia, Singapore, 1984, pp. 101–114.

(3) McGill, W. B., and Bergstrom, D., "Inland Oil Spills and their Impacts on Land," *Stress on Land in Canada*, Lands Directorate, Environment Canada, Ottawa, Ont., 1983, pp. 153–181.

(4) Baker, J. M., "The Effects of Oil on Plant Physiology," *The Ecological Effects of Oil Pollution on Littoral Communities*, E. B. Cowell, ed., Institute of Petroleum, London, England, 1971, pp. 88–98.

(5) McGill, W. B. *An Introduction for Field Personnel to the Effects of Oil Spills on Soil and Some General Restoration and Cleanup Procedures*, Canadian Petroleum Association, Calgary, Alta., 1976.

(6) Hutchinson, T. C., and Hellebust, J. A., *Oil Spills and Vegetation at Norman Wells, N.W.T.*, Task Force on Northern Oil Development, Department of Indian and Northern Affairs, Ottawa, Ont., 1974.

(7) Swader, F. N., "Persistance and Effects of a Light Fuel Oil in Soil," *Proceedings of the 1975 Oil Spill Conference*, American Petroleum Institute, Washington, DC, 1975, pp. 589–593.

(8) Owens, E. H., Foget, C. R., and Robson, W.," Experimental Use of Dispersants for Spill Countermeasures on Arctic Beaches," *Oil Spill Chemical Dispersants: Research, Experience, and Recommendations, ASTM STP 840*, ASTM, 1984, pp. 324–337.

(9) Mackay, D., Watson, A., and Kuhnt, A., *The Behaviour of Oil and Chemically Dispersed Oil at Shorelines*, Petroleum Association for the Conservation of the Canadian Environment, Ottawa, Ont., 1979.

(10) Rowland, S. J., Tibbetts, P. J. C., Little, D., Baker, J. M., and Abbiss, T. P., "The Fate and Effects of Dispersant-Treated Compared with Untreated Crude Oil, with Particular Reference to Sheltered Intertidal Sediments," *Proceedings of the 1981 Oil Spill Conference*, American Petroleum Institute, Washington, DC, 1981, pp. 283–293.

(11) Dewling, R. T., and Silva, C. C. D. A. E., "Impact of Dispersant Use During the BRAZILIAN MARINA Incident," *Proceedings of the*

1979 Oil Spill Conference, American Petroleum Institute, Washington, DC, 1979, pp. 269–276.

(12) Parkinson, D., *Oil Spillage on Micro-organisms in Northern Canadian Soils, Task Force on Northern Oil Development*, Department of Indian and Northern Affairs, Ottawa, Ont., 1973.

(13) Wilson, J. T., Noonan, M. J., and McNabb, J. F., "Biodegradation of Contaminants in the Subsurface," *Ground Water Quality*, eds. C. H. Ward, W. Giger, and P. L. McCarty, John Wiley and Sons, New York, NY, 1985, pp. 483–492.

(14) Duffy, J. J., Peake, E., and Mohtadi, M. F., "Subsurface Biophysiochemical Transformations of Spilled Crude Oil," *Proceedings of the Conference on the Environmental Effects of Oil and Salt Water Spills on Land*, Research Secretariat Alberta Environment, Edmonton, Alta., 1975, pp. 136–183.

(15) Duffy, J. J., Mohtadi, M. F., and Peake, E., "Subsurface Persistance of Crude Oil Spilled on Land and its Transport in Groundwater," *Proceedings of the 1977 Oil Spill Conference*, American Petroleum Institute, Washington, DC, 1977, pp. 475–478.

Designation: F 1280 – 90 (Reapproved 1999)

Standard Guide for Ecological Considerations for the Use of Oilspill Dispersants in Freshwater and Other Inland Environments, Impermeable Surfaces[1]

This standard is issued under the fixed designation F 1280; the number immediately following the designation indicates the year of original adoption or, in the case of revision, the year of last revision. A number in parentheses indicates the year of last reapproval. A superscript epsilon (ε) indicates an editorial change since the last revision or reapproval.

1. Scope

1.1 This guide covers the use of oilspill dispersants to assist in the control of oil spills. This guide is written with the goal of minimizing the environmental impacts of oil spills; this goal is the basis on which the recommendations are made. Aesthetic and socioeconomic factors are not considered although these and other factors are often important in spill response.

1.2 Spill responders have available several means to control or clean up spilled oil. In this guide, the use of dispersants is given equal consideration with other spill countermeasures. It is not considered as a last resort after all other methods have failed.

1.3 This is a general guide only. If assumes the oil to be dispersible and the dispersant to be effective, available, applied correctly, and in compliance with relevant government regulations. In the case of impermeable surfaces, the dispersant may act as a detergent. In the assessment of environmental sensitivity, it is assumed that the dispersant is nonpersistent in the natural environment. Oil, as used in this guide, includes crude oils and refined petroleum products. Differences between individual dispersants or between different oil products are not considered.

1.4 The guide is organized by habitat type, for example, small ponds and lakes, rivers and streams, and land. It considers the use of dispersants primarily to protect habitats from impact (or to minimize impacts) and to clean them after a spill takes place.

1.5 This guide applies only to freshwater and other inland environments. It does not consider the direct application of dispersants to subsurface waters.

1.6 In making dispersant use decisions, appropriate government authorities should be consulted as required by law.

1.7 *This standard does not purport to address all of the safety concerns, if any, associated with its use. It is the responsibility of the user of this standard to establish appropriate safety and health practices and determine the applicability of regulatory limitations prior to use.*

2. Significance and Use

2.1 This guide is meant to aid local and regional response teams who may use it during spill response planning and spill events.

2.2 This guide should be adapted to site-specific circumstances.

3. Environment Covered—Impermeable Surfaces

3.1 Impermeable ground includes any soil, rock, hard pan or other natural surface that does not readily permit the passage of water and oil.

3.2 Impermeable structures include man-made surfaces such as paved roads and parking lots.

4. Background

4.1 The effects of oil on animals inhabiting impermeable ground is not fully known, but is suspected to be minimal as most animals on such surfaces are mobile and would avoid the oil.[2]

4.2 Plants inhabiting impermeable ground may be ecologically important.

4.3 Oil and dispersants will ultimately run off an impermeable surface. The environment receiving runoff should be considered. As an example, experience with spills on impermeable surfaces included runoff into a permeable zone and extensive groundwater contamination.[3]

4.4 A laboratory experiment has shown that explosion potential is increased by using dispersants on fuel spills in sewers.[4] It is not known whether these results are applicable to full scale sewer systems.

5. General Considerations for Making Dispersant Use Decisions

5.1 The dispersant use decision is, in this case as most

---

[1] This guide is under the jurisdiction of ASTM Committee F-20 on Hazardous Substances and Oil Spill Responseand is the direct responsibility of Subcommittee F20.13 on Treatment.

Current edition approved Feb. 23, 1990. Published April 1990.

[2] W. B. McGill, and D. Bergstrom, "Inland Oil Spills and their Impacts on Land,'' *Stress on Land in Canada*, Lands Directorate, Environment Canada, Ottawa, Ont., 1983, pp. 153–181.

[3] A. J. Smith, "Success and Failures with Oil Spills in Southeastern Inland Waters,'' *Proceedings of the 1973 Conference on Prevention and Control of Oil Spills*, American Petroleum Institute, Washington, DC, 1973, pp. 583–588.

[4] M. F. Fingas, K. A. Hughes, and A. M. Bobra, "The Behaviour of Dispersed and Nondispersed Fuels in a Sewer System," *Oil Dispersants: New Ecological Approaches, ASTM STP 1018*, ASTM, 1989, pp. 274–289.

others one of trade-offs. The use of dispersants can reduce the adverse effects of spilled oil on certain biological species at the expense of other components of the ecosystem.

5.2 In most cases the mortality of individual creatures is of less concern than the destruction of habitat. The repopulation of areas after the spill will occur naturally when an area becomes a suitable habitat for a given species.

6. Recommendations

6.1 Oil or dispersed oil on impermeable surfaces may run off to other environments. The effects of oil or dispersed oil on impermeable surfaces are largely unknown, but are thought to be minimal. The use of dispersant can assist in cleaning of the surface and allow for the selection of the receiving environment.

6.2 The effects of oil on impermeable structures may be considerable; dispersants may reduce the effect of these.

6.3 Dispersants should never be used if the runoff can reach groundwater, especially that used for drinking water.

7. Keywords

7.1 dispersants; freshwater; impermeable; inland; oil spill; oilspill dispersants; road surfaces

Designation: F 1413 – 92 (Reapproved 1999)

Standard Guide for Oil Spill Dispersant Application Equipment: Boom and Nozzle Systems[1]

This standard is issued under the fixed designation F 1413; the number immediately following the designation indicates the year of original adoption or, in the case of revision, the year of last revision. A number in parentheses indicates the year of last reapproval. A superscript epsilon (ε) indicates an editorial change since the last revision or reapproval.

1. Scope

1.1 This guide covers design criteria, requirements, material characteristics, and essential features for oil spill dispersant application systems. This guide is not intended to be restrictive to a specific configuration.

1.2 This guide covers spray systems employing booms and nozzles and is not fully applicable to other systems such as fire monitors, sonic distributors, or fan-spray guns.

1.3 This guide covers systems for use on ships or boats and helicopters or airplanes.

1.4 This guide is one of four related to dispersant application systems. One is on design, one on calibration, one on deposition measurements, and one on the use of the systems. Familiarity with all four guides is recommended.

1.5 *This standard does not purport to address all of the safety concerns, if any, associated with its use. It is the responsibility of the user of this standard to establish appropriate safety and health practices and determine the applicability of regulatory limitations prior to use.*

2. Equipment Description

2.1 *General*—Oil spill dispersant spray systems include one or more booms with nozzles to form droplets, a pumping or pressure system to deliver dispersants to the boom, and associated piping and valving. All systems shall include a dispersant flow meter and a pressure gauge. All systems shall be equipped with provision for cleaning and drainage.

2.2 *Ship/Boat*—Each boom holding nozzles shall be designed to be mounted near the bow of the vessel so that the spray is uniformly deposited on the slick surface. Spray units can be portable or fixed. Flow correction or straightener devices, to ensure laminar flow, shall precede the nozzles. System components should be designed to give a uniform droplet spray as described in this guide. The spray pattern should be flat and strike the water in a line perpendicular to the vessel's line of travel. The nozzle spray angle should be such that spray from adjacent nozzles overlap just above the water.

2.3 *Airplanes*—Mounting of spray booms on aircraft is subject to federal regulation. Each installation or modification requires approval.

2.3.1 Nozzles may not be necessary on aircraft flying at speeds greater than 220 km/h (120 knots or 135 mph) because the wind shear alone can produce the required droplet sizes. Pressure-activated check valves must be used to eliminate drainage during nonspraying transits. In order to minimize the effects of wind shear, nozzles should be oriented aft (180° from the direction of flight).

2.4 *Helicopters*—Systems may consist of spray booms with nozzles and pump/tank assemblies directly attached to the helicopter or a bucket system slung below the helicopter.

2.4.1 The bucket system consists of a tank and pump assembly to which spray booms with nozzles are attached. The assembly is supported from the helicopter by a cable system and is remotely-controlled from the helicopter cabin. An indication of dispersant flow is required in the helicopter cockpit. The bucket must be stabilized against rotation, yaw, and sway.

3. Minimum Equipment Performance Specifications

3.1 *Target Dosage*—Oil spill dispersant spray equipment shall provide a dispersant dosage of between 20 to 100 L per hectare (2 to 10 U.S. gal per acre).

3.2 *Droplet Size Distribution*—The droplet size distribution of the dispersant reaching the target shall have a Volume Median Diameter (VMD) of 300 to 500 µm. The volume median diameter is a means of expressing droplet size in terms of the volume of liquid sprayed. The median volume diameter droplet size, when measured in terms of volume, is a value where 50 % of the total volume of liquid sprayed is made up of droplets with diameters larger than the median value and 50 % smaller than the median value. Droplets having diameters lesser than approximately 300 µm have a lower probability of hitting the target because of excessive wind drift. Particles with diameters greater than 500 µm have a high probability of penetrating through the oil slick to the water surface.

3.3 *Maximum Delivery Variation Over Swath Width*—The equipment shall be capable of delivering dispersant with a maximum delivery variance of 10 % over the swath width. The swath width is defined as the length between the points at which the delivery drops below 90 % of the design.

4. Equipment Design

4.1 *Dispersant Injection Rate*—The dispersant injection rate (for undiluted or neat application, the dispersant injection rate

[1] This guide is under the jurisdiction of ASTM Committee F 20 on Hazardous Substances and Oil Spill Responseand is the direct responsibility of Subcommittee F20.13 on Treatment.

Current edition approved June 15, 1992. Published August 1992.

is equal to the pump rate) must be sufficient to produce the required dosage.

4.1.1 Dispersant injection rate (*DIR*) should be verified using the following Eqs:

$$DIR = 1.67 \times 10^{-3} \cdot S \cdot W \cdot D \quad (1)$$

where:
DIR = dispersant injection rate in L/min,
S = speed of the delivery vehicle in km/h,
W = swath width in m, and
D = dosage in L/ha.

Or equivalently in U.S. units:

$$DIR = 2.33 \times 10^{-3} \cdot S \cdot W \cdot D \quad (2)$$

where:
DIR = dispersant injection rate in U.S. gal/min (USGPM),
S = speed of the delivery vehicle in knots (if speed is in miles per hour multiply by 0.87),
W = swath width in ft, and
D = dosage in U.S. gal per acre (USGPA).

4.2 *Droplet Size*—Shear is the controlling factor in determining droplet size. High shear rates result in small droplet sizes. For dispersant application, a small droplet size is not desirable, because the dispersant can drift away from the oil slicks. Large droplet sizes are also undesirable because large droplets can penetrate the oil slick. Experience has shown that a droplet size between 300 to 700 μm VMD is most effective. Shear has two components, nozzle shear rate and air shear. Air shear is only important for aircraft flying at speeds greater than 150 km/h (80 knots or 100 mph).

4.2.1 *Nozzle Shear Rate*—In order to achieve the desired droplet size, nozzle shear rate should not exceed 10 000 reciprocal seconds (s^{-1}) for aircraft systems and 2000 reciprocal seconds for ship or boat systems. Nozzle shear can be calculated using the following Eqs:

$$SR = 16.7 \cdot FN/d^3 \quad (3)$$

where:
SR = shear rate in reciprocal seconds (s^{-1}),
FN = average flow rate per nozzle in L/min (calculated from total flow (dispersant and water) divided by the number of nozzles), and
d = the diameter of the nozzle orifice in cm.

Or equivalently in U.S. units:

$$SR = 3.85 \cdot FN/d^3 \quad (4)$$

where:
SR = shear rate in reciprocal seconds (s^{-1}),
FN = average flow rate per nozzle in gal/min (USGPM) (calculated from total flow (dispersant and water) divided by the number of nozzles), and
d = the diameter of the nozzle orifice in inches (in.).

4.2.2 *Air Shear*—In addition to the nozzle shear rate calculations shown, aircraft dispersant application systems should be designed to optimize droplet size distribution by minimizing the differential between the speed of the aircraft and the dispersant exit speed. Differential speed should be less than 60 m/s (200 ft/s) in order to ensure close to 100 % deposition within the swath width.

4.2.2.1 Differential speed shall be verified using the following Eqs:

$$SD = SA - (0.212 \cdot FN/d^2) \quad (5)$$

where:
SD = the differential speed in m/s,
SA = the aircraft speed in m/s (for speed in km/h multiply by 0.28),
FN = the average flow rate per nozzle in L/min, and
d = the nozzle orifice diameter in cm.

Or equivalently in U.S. units:

$$SD = SA - (0.409 \cdot FN/d^2) \quad (6)$$

where:
SD = the differential speed in ft/s,
SA = the aircraft speed in ft/s (multiply knots by 1.69 to get ft/s),
FN = the average flow rate per nozzle in gal/min (USGPM), and
d = the nozzle orifice diameter in inches (in.).

5. Material Characteristics

5.1 *Corrosion Resistance*—Materials on ship or boat systems should be corrosion-resistant to salt water. All materials that come into contact with dispersants should be compatible with that dispersant. Special attention should be given to pump components. Consultation with the dispersant manufacturer is recommended.

5.2 *Extreme Temperature Properties*— Systems to be used or stored at extreme temperatures should be constructed of materials that are not adversely affected by those temperatures. Temperature range specifications should be clearly indicated on the spray equipment.

6. Information Provided to User

6.1 Performance data that shall be provided to the user by the manufacturer include:

6.1.1 Estimated or measured droplet size information (VMD in μm),

6.1.2 Volumetric output distribution over the swath width (%),

6.1.3 A table of pump rates and dispersant injection rates ranging from the recommended minimum to the recommended maximum,

6.1.4 Nozzle shear rate,

6.1.5 The nozzle design height for ship/boat systems,

6.1.6 Swath width,

6.1.7 Differential speeds for aircraft systems at various pump settings and aircraft speeds, and

6.1.8 Recommended operating pressures at the inlet to the boom.

6.2 *Dosage Chart*—The manufacturer shall supply the user with a chart of dosages achievable with different application vehicle speeds and different dispersant injection rates.

6.3 *Accuracy of Data*—The data shall be accurate to two significant figures.

6.4 *Materials of Construction*—The supplier shall provide the user with a list of materials of construction.

6.5 *Nozzles and Pumps*:

6.5.1 The supplier shall provide full data on the manufacturer, model numbers, and dimensions of nozzles supplied with the spray equipment.

6.5.2 The supplier shall provide full data on the manufacturer, model number, and basic maintenance and operational data on all major components of the spray equipment including pumps, eductors, flow meters, and engines.

6.6 *Operator's Manual*—The supplier shall provide a comprehensive operator's manual including diagrams of the equipment layout.

7. Keywords

7.1 aerial spray; boom and nozzles; dispersant application; dispersant spray equipment; dispersants; oil spill chemicals; oil spill dispersants; oil spill treating agents

Designation: F 1460 – 93 (Reapproved 1999)

Standard Practice for Calibrating Oil Spill Dispersant Application Equipment Boom and Nozzle Systems[1]

This standard is issued under the fixed designation F 1460; the number immediately following the designation indicates the year of original adoption or, in the case of revision, the year of last revision. A number in parentheses indicates the year of last reapproval. A superscript epsilon (ϵ) indicates an editorial change since the last revision or reapproval.

1. Scope

1.1 This practice covers uniform procedures for determining and reporting the dosage rate of oil spill dispersant application equipment.

1.2 This practice is applicable to spray systems employing booms and nozzles and is not fully applicable to other systems such as fire monitors, sonic distributors, or fan-spray guns.

1.3 This practice is applicable to systems for use on ships or boats and helicopters or airplanes.

1.4 This practice is one of four related to dispersant application systems. One is on design, one on calibration, one on deposition, and one on the use of the systems. Familiarity with all four standards is recommended. The other three standards are listed in Section 2.

1.5 *This standard does not purport to address all of the safety concerns, if any, associated with its use. It is the responsibility of the user of this standard to establish appropriate safety and health practices and determine the applicability of regulatory limitations prior to use.*

2. Referenced Documents

2.1 *ASTM Standards:*

F 1413 Guide for Oil Spill Dispersant Application Equipment: Boom and Nozzle Systems[2]

3. Significance and Use

3.1 This practice will enable calibration of oil spill dispersant application equipment and ensure a desired dosage and uniformity across the swath width.

3.2 The data provided by the methods described herein will permit the preparation of a chart relating delivery rate with application vehicle speed, flow meter reading or pump setting so that in actual application, the desired dosage will be achieved.

3.3 This practice will ensure that a dispersant application system is functional, capable of delivering a specified dosage, and that major components are operational.

4. Apparatus and Materials

4.1 *Pails*— of capacity 7 to 20 L (2 to 5 U.S. gal) to catch the spray from the nozzles.

4.2 *Graduated Cylinder*— of capacity 7 to 20 L or a scale having capacity of at least 20 kg (45 lbs) to determine the amount of fluid in each pail. Commercial equipment with an accuracy of at least 1 % is adequate.

4.3 *Stopwatch.*

4.4 *Test Fluid*—Water can be used as a test fluid if the viscosity of the dispersant to be used is not dissimilar to the viscosity of water. A surrogate fluid should be used if the viscosity difference is greater than approximately 100 times that of water. This can occur with more viscous dispersants or at low temperatures. The surrogate fluid could be water with the appropriate amount of thickener.

4.5 A continuing supply of water or test fluid, enough to run the system during the test period, must be available. Tank truck quantities may be required.

4.6 *Auxiliary Power Units*, where required.

4.7 Temperature of test fluid.

5. Calibration Procedure

5.1 *Synopsis*—The calibration procedure is done in four steps. The first step is equipment inspection. Any defects are corrected before further calibration. The second step is the calibration of the flow meter. The third step is calibration of the unit by catching water spray from each nozzle. The fourth step is the preparation of a calibration curve.

5.2 *Equipment Setup*— Place the application equipment at a suitable outdoor location and close to a supply of water or test fluid. Set up the equipment in accordance with the manufacturer's instructions and inspect visually. Start the unit. All nozzles should visually be producing approximately the same amount of spray. Inoperative nozzles or other elements such as pressure gages or flow meters, are repaired or replaced before proceeding. The operation of pressure gages should be verified before proceeding.

5.3 *Dispersant Flow Meter Calibration*:

5.3.1 Calibrate the flow meter using a surrogate fluid. This surrogate fluid could be water or a surrogate fluid, if necessary. The withdrawal of test fluid from a calibrated vessel is timed to establish the flow rate. The calibration is done at a minimum of three flow rates and these should be chosen to represent typical

[1] This practice is under the jurisdiction of ASTM Committee F-20 on Hazardous Substances and Oil Spill Response and is the direct responsibility of Subcommittee F20.13 on Treatment.

Current edition approved Dec. 15, 1993. Published February 1994.

[2] *Annual Book of ASTM Standards*, Vol 11.04.

minimum, average, and maximum dispersant flow rates.

5.3.2 Fill the calibrated vessel with test fluid and connect the dispersant input line to it. Set the desired flow rate and start the unit. After a steady flow rate is established, begin the measurement. The measurement period should be sufficiently long to allow for accurate measurements. Operate the spray system with all sub-systems and nozzles to ensure that the pressure drop across the flow meter is the same as during normal operations.

5.3.3 Run the test at least three times for each of the flow rates and the numbers averaged for each flow rate. Data are used to produce a calibration chart for the flow rate.

5.4 *Nozzle Calibration*:

5.4.1 Calibrate the apparatus using water or a surrogate test fluid, if necessary, which is collected to establish flow rates. Start the unit, and when pressure and flow are constant, catch the output from each nozzle under test using pails manipulated by test personnel at the same time. A timekeeper should give start and stop signals and maintain records of the time elapsed during collection of the fluid. Each run should last 30 to 60 s or long enough to fill one-half to two-thirds of the collection pail. All nozzles must be tested. Any number of nozzles can be tested at a time, but during any test all nozzles must be running.

5.4.2 Determine the amount of fluid in each pail by weighing it and converting to volume, or by using a calibrated cylinder. Each pail should be numbered and preweighed (if weight is the system used for volume determination) before the run begins. Determine the amount of fluid sprayed from each nozzle immediately after collection. Compare the amount of fluid produced by each nozzle to the calculated amount determined by taking the total flow divided by the number of nozzles. Total flow is the sum of the fluid collected from all the nozzles. The output of each nozzle should be within 10 % of the calculated average or designed amount. If not, the nozzles should be repaired or replaced and the calibration procedure repeated. Separate calibrations must be done for each variation of pressure, number of operating nozzles, or type of nozzles.

5.5 *Calibration of Dispersant Addition (Eductor or Pump) Systems*—Boat and ship systems sometimes apply dispersant mixed with water. Addition systems are of two types, eductors and pumps. Both pumps and eductors are calibrated by measuring the volume of dispersant surrogate added to the water stream. The systems are calibrated in the configuration they are normally used. Calibrations are performed for each water pressure and water flow rate that would be used in actual practice.

5.5.1 Eductor systems are cleaned and repaired (if necessary) prior to calibration. Eductors require frequent maintenance, cleaning, and calibration. Eductors are calibrated by placing the suction tube into a calibrated cylinder and timing the withdrawal of material. Either the relative change per unit time of the water of the dispersant to be used is recorded over at least three flow settings to produce a calibration chart.

5.5.2 Dispersant addition pumps are calibrated by measuring the flow input or output with a calibrated cylinder or by weight. Tests are run at three pump settings representing a minimum, a maximum, and one or more typical operating-flow-rate setting.

5.6 *Calibration Frequency*—Systems should be calibrated at least once a year. Systems should be cleaned and recalibrated after each use and after making system configuration changes.

5.7 *Report of Calibration Procedure*—Perform the calibration of the application system at three or more different equipment settings to produce an overall dispersant delivery chart and a flow meter correction chart.

6. Data Reporting

6.1 Data shall be accurate to two significant figures. Tables or graphs are prepared showing dispersant delivery rate in L/min (or U.S. gal/min) with pump settings and flow meter data.

7. Keywords

7.1 boom and nozzles; dispersant application; dispersant spray equipment; dispersants; oil spill chemicals; oil spill dispersants; oil spill treating agents

Designation: F 1481 – 94 (Reapproved 2001)

Standard Guide for Ecological Considerations for the Use of Bioremediation in Oil Spill Response—Sand and Gravel Beaches[1]

This standard is issued under the fixed designation F 1481; the number immediately following the designation indicates the year of original adoption or, in the case of revision, the year of last revision. A number in parentheses indicates the year of last reapproval. A superscript epsilon (ε) indicates an editorial change since the last revision or reapproval.

1. Scope

1.1 This guide covers considerations and recommendations for the use of biodegradation enhancing (bioremediation) agents to assist in minimizing the impact of oil spills on sand and gravel beaches. Aesthetic and socioeconomic factors are not considered, although these and other factors are often important in spill response.

1.2 This is a general guide only, which assumes that the oil is biodegradable and that the bioremediation agent selected can be used safely (in terms of ecological impact and human health) and effectively when applied correctly and used in compliance with relevant government regulations. Oil considered for bioremediation includes crude oils and certain refined petroleum products.

1.3 This guide addresses the application of bioremediation agents alone or in conjunction with other technologies.

1.4 This guide applies to freshwater, estuarine, and marine beach environments.

1.5 In making bioremediation-use decisions, appropriate government authorities must be consulted as required by law.

1.6 *This standard does not purport to address all of the safety concerns, if any, associated with its use. It is the responsibility of the user of this standard to establish appropriate safety and health practices and determine the applicability of regulatory limitations prior to use.* In addition, it is the responsibility of the user to ensure that such activity takes place under the control and direction of a qualified person with full knowledge of any potential or appropriate safety and health protocols.

2. Terminology

2.1 *Definitions:*

2.1.1 *aerobes*—organisms that require air or free oxygen for growth.

2.1.2 *anaerobes*—organisms that grow in the absence of air or oxygen, and do not use molecular oxygen in respiration.

2.1.3 *bioaugmentation*—addition of microorganisms (predominantly bacteria) to amplify the biodegradation rate of target pollutants.

2.1.4 *biodegradation*—chemical alteration and breakdown of a substance to usually smaller products caused by microorganisms or their enzymes.

2.1.5 *bioremediation agents*—inorganic and organic compounds and microorganisms that enhance biological degradation processes, predominantly microorganismal.

2.1.6 *biostimulation*—addition of limiting nutrients to enhance the rate of biodegradation of target pollutants by indigenous species (predominantly bacteria).

2.1.7 *ecosystem*—organisms and the surrounding environment combined in a community that is self-supporting.

2.1.8 *identification*—process of designating an unknown organism by comparing it with known organisms.

2.1.9 *indigenous*—native to a given habitat or environment.

2.1.10 *marine*—relating to the ocean.

2.1.11 *methemoglobinemia*—acquired blood disorder leading to oxygen deprivation, stupor, and death from exposure to nitrates in drinking water.

2.1.12 *nutrient*—a substance that supports organismal growth.

2.1.13 *refined petroleum products*—products derived by way of various treatment processes from crude oil, a highly complex mixture of paraffinic, cycloparaffinic, and aromatic hydrocarbons that contain a low percentage of sulphur and trace amounts of nitrogen and oxygen compounds. Hydrocarbon products made from the refining of crude oils are specified in Section 5 of the Annual Book of ASTM Standards.[2]

2.1.14 *species*—a taxonomic category characterized by individuals of the same genus that are mutually similar and are able to interbreed.

2.1.15 *toxicity*—the property of a material, or combination of materials, to adversely affect organisms.

3. Significance and Use

3.1 The purpose of this guide is to provide remediation managers and spill response teams with guidance on an alternate means (called bioremediation) of safely and effectively cleaning up oil spills on beaches that takes advantage of natural microbial degradation processes.

3.2 This guide can be used in conjunction with other ASTM guides addressing oil spill response operations.

---

[1] This guide is under the jurisdiction of ASTM Committee F20 on Hazardous Substances and Oil Spill Responseand is the direct responsibility of Subcommittee F20.24 on Bioremediation and Safety Audit.

Current edition approved May 15, 1994. Published July 1994. Originally published as F 1481 – 93. Last previous edition F 1481 – 93.

[2] *Annual Book of ASTM Standards*, Vols 05.01, 05.02, 05.03, 05.04, and 05.05.

4. General Considerations for Making Bioremediation-Use Decisions

4.1 Bioremediation has been used primarily as a longer-term beach treatment tool as opposed to a first response tool.

4.2 Bioremediating a site on an oiled beach usually involves minimal physical disruption of the site (**1**).[3]

4.3 Bioremediation may work faster than natural biodegradation and appears to have few adverse affects when used correctly (**1**). Therefore, it may be useful in helping to remove some of the toxic components of petroleum (for example, low to medium molecular weight aromatic hydrocarbons) from a spill site more quickly (weeks to months) than they might otherwise be removed (years). Bioremediation will be somewhat less effective (month to years) in removing the high molecular weight components (for example, polynuclear aromatic hydrocarbons) from stranded oil.

4.4 Biodegradation rates are significantly reduced in anaerobic beach environments.

4.5 Bioremediation performance is particularly dependent upon the efficiency of the petroleum hydrocarbon degrading bacteria indigenous to an area or added as in bioaugmentation in degrading oil; the availability of rate-limiting nutrients; and the susceptibility of the target crude oil or refined product to microbial degradation.

4.6 Bioremediation must be carried out under the guidance of qualified personnel that understand both the chemical as well as the safety and health aspects of site activities.

4.7 Bioremediation may show improvement over natural weathering in lightly oiled beaches. Heavily oiled beaches require greater cleanup time than lightly oiled beaches, and are more amenable to bioremediation after removal of the gross contamination.

5. Sand and Gravel Beach Environments

5.1 Characteristics of beach environments can affect the transport and fate of oil, which will affect the use of bioremediation agents.

5.2 Sand beaches vary widely in their grain size, width, slope, origin, exposure to waves, and sediment transport patterns. Sand beaches can be divided into two basic types: fine grained and coarse grained. The compact sediments of fine-grained sand beaches (grain sizes range between 0.0625 and 0.25 mm) prevent deep (less than 10 cm) penetration of oil. Asphalt pavement formation is likely in sheltered areas where oil accumulation is heavy. The grain size of sediments on coarse-grained sand beaches ranges between 0.25 and 2 mm. The more porous sediments allow penetration of oil up to 25 cm. Oil layers can be buried much deeper into the beach face by clean sand by repeated episodes of deposition over time. While fine-grained sand beaches support low to moderate infaunal densities, coarse-grained sand beaches do not generally support a rich biological community (**2**).

5.3 Gravel refers to a wide range of grain sizes and is further divided into classes as follows: granule (2–4 mm); pebble (4–64 mm); cobble (64–256 mm); and boulder (256 mm and larger). Boulder beaches can support dense epifaunal communities. In terms of the fate of spilled oil, gravel beaches have high porosity and permeability that allow deep penetration from the surface. These beaches also have a high potential for oil burial through buildup. Asphalt pavement formation is likely in sheltered areas where accumulation is heavy (**2**).

6. Background

6.1 Approaches to bioremediation for oil spill response include biostimulation, the addition of limiting nutrients to stimulate indigenous microorganisms, and bioaugmentation, the addition of contaminant-degrading microorganisms. As a precaution, it should be noted that nutrient components may be toxic or harmful to plants, animals and humans, and that non-indigenous species may at least temporarily alter the indigenous microbial ecological balances. Water effluent nitrate nitrogen levels in drinking water must be avoided to diminish risks of anemias such as methemoglobinemia. Similarly, excessive ammonium nitrogen levels should be avoided because they can adversely affect fish and invertebrates, the latter of more concern since many are immobile and could not avoid the treated area. Therefore, nitrogen and other nutrient levels must be monitored. Instructions to ensure safety and effective product use should be established by the manufacturer for each product, and specific instructions followed by the product user. No adverse environmental effects have been associated with field applications of nutrients or bioaugmentation products.

6.2 Biostimulation has been shown to be effective in enhancing biodegradation of oil on beaches. Biostimulation employs the addition of appropriate nutrients (for example, nitrogen, phosphorous, micro-nutrients, oxygen, and so forth) which may have been limiting. This approach may lead to increases in the rate of degradation if microbial degraders of the target contaminant are present in the beach material or purging waters. In some cases there may not be an indigenous contaminant-degrading population to stimulate. This may be the case in pristine environments where there has not been adequate time for the degraders to evolve. Biostimulation is time demanding, as an initial lag period is required for natural selection of degraders. As a precaution, it should be noted that stimulation of pathogenic organisms may also occur. Microorganisms as well as contaminants should be monitored throughout the process.

6.3 Bioaugmentation should prove useful in enhancing biodegradation of oil on beaches. Bioaugmentation generally employs large-scale, on-site production and addition of contaminant-degrading microorganisms usually chosen from stock cultures, predominantly bacterial species. This approach rapidly increases microbe concentrations to levels associated with maximal biodegradation rates. These applications should be characterized with respect to their safety and contaminant degradation activities in various matrix types, temperatures, pH, and inorganic and organic micro- and macroenvironments. Microbes selected must be non-pathogenic and must metabolize the contaminant(s), rendering them harmless. Activities of the amended bacteria need to be well understood and their growth controllable. Microorganisms as well as contaminants should be monitored throughout the process. Assessment of

[3] The boldface numbers in parentheses refer to references at the end of this practice.

performance of nonnative microorganisms should be determined since they may be less adaptable to environmental changes than indigenous organisms.

6.3.1 Genetically engineered microorganisms, although under development and apparently safe and effective in the laboratory setting, have not yet been authorized for environmental release (**3**).

6.4 For spills involving crude oil and Nos. 4–6 fuel oils, significant reductions may not occur for several weeks. Distillates (for example, jet fuel, diesel fuel, No. 2 fuel oil) and residual require at least two weeks for significant reductions. Asphalt pavement formations are not amenable to bioremediation treatment.

6.5 Environmental concerns related to bioremediation use include the possibility that the addition of fertilizers could cause eutrophication leading to algal blooms and oxygen depletion; that components of some fertilizers could be toxic to sensitive species or harmful to human health; and that the introduction of nonnative microorganisms could compete with some indigenous species, upsetting ecological balances (**1**).

6.6 To date, no significant environmental or health problems have been associated with the field applications of bioremediation technologies to marine oil spills (**1**).

6.6.1 Transient acute toxicity effects, limited to areas immediately adjacent to fertilized beaches, were observed during one of the initial experimental field applications of nutrients following the Exxon Valdez oil spill (**4**). However, toxicity tests using mysids showed no acute effects in extensive testing conducted during subsequent operational bioremediation treatments (**4**).

6.6.2 Ammonia levels remained below toxic concentrations following nutrient applications on a fuel oil contaminated beach on Prall's Island, New Jersey, in 1990 (**4**).

6.6.3 Microorganisms introduced to augment degradation of North Slope crude oil did not result in any significantly greater invertebrate mortality than fertilization alone of crude oil (**5**).

6.7 A field monitoring program is usually appropriate to establish the efficacy and safety of bioremediation treatment. The following parameters could be included in the monitoring program: visual observations (for example, indigenous species mortality, behavioral effects, appearance changes, and oil distribution), temperature (air and water), salinity, dissolved oxygen, sea state, wind velocity, efficacy samples (water, sediment, and/or beach material), and toxicity samples (water, sediment, and/or beach materials). Water and sediment/beach material samples should be analyzed for oil hydrocarbons, ammonia- and nitrate nitrogens, phosphate, and toxicity (4-day acute or 7-day chronic) (**6**).

6.7.1 The monitoring of contaminant levels or alterations in contaminant constituents, as well as of biological parameters, would be used to assess any correlations of biological activity and contaminant reduction. Biological activity can be determined by growing site organisms on petri plates. This allows assessment of viable organisms and, if grown using specific organic compounds, can determine their contaminant degradation specificity.

6.7.2 Metabolic products of biological activities can be monitored through sediment, water, and air sampling.

7. Recommendations

7.1 Bioremediation should be considered as one of the potential treatment methods available to site remediation managers once gross quantities of contamination have been removed.

7.2 Bioremediation should only be performed with appropriate technically-qualified personnel, following health and safety protocols for such activity.

7.3 There are time requirements for bioremediation associated with the amount of type of contaminant, and conditions in which the bioremediation agents are applied.

7.4 The selection of appropriate bioremediation agents should form an integral part of most facilities' contingency plans. Before selection is carried out and duly recorded, however, it should be reviewed in terms of efficacy, toxicity, and other potential human and ecological impacts.

7.5 In order to measure success a rigorous monitoring program should be established to determine the natural levels of the contaminant, ammonia- and nitrate nitrogens, phosphate and any associated toxicity in the surrounding areas; contamination levels; and to track the contamination plume. The basic design and elements of the monitoring program should be based on the requirements of methods that will be used to measure efficacy and safety, and the goal of assuring statistical validity of results.

8. Keywords

8.1 beaches; bioremediation; gravel beaches; oil spill response; sand beaches

REFERENCES

(**1**) Office of Technology Assessment, U.S. Congress, "Bioremediation for Marine Oil Spills—Background Paper," OTA-BP-O-70: May, 1991, p. 19.

(**2**) Hayes, M. O., and Michel, J., *Impacts of Oil Spills on Coastal Ecosystems, Course Manual*, Research Planning, Incorporated, 1992.

(**3**) Friello, D. A., Mylroie, J. R., and Chakrabarty, J. M., "Use of Genetically Engineered Multiplasmid Microorganisms for Rapid Degradation of Fuel Hydrocarbons," *Biodeterioration of Materials*, No. 3, 1976, pp. 205–214.

(**4**) Hoff, R., *A Summary of Bioremediation Applications Observed at Marine Oil Spills*, National Oceanic and Atmospheric Administration, Report HMRB 91-2, 1991, p. 30.

(**5**) Atlas, R. M., and Busdosh, M., *Microbial Degradation of Petroleum in the Arctic*, Proceedings of the 3rd International Biodegradation Symposium, Kingston, RI, August 1975, p. 85.

(**6**) Environmental Protection Agency, Region 6 Bioremediation Spill Response Plan, Superintendent of Documents, U.S. Government Printing Office, Washington, DC, 20402, 1992, p. 38 and appendices.

Designation: F 1522 – 95

Standard Guide for Use of the Steam Stripping Process in Mitigating Chemical Spills[1]

This standard is issued under the fixed designation F 1522; the number immediately following the designation indicates the year of original adoption or, in the case of revision, the year of last revision. A number in parentheses indicates the year of last reapproval. A superscript epsilon (ε) indicates an editorial change since the last revision or reapproval.

1. Scope

1.1 This guide covers the considerations for the use of steam stripping in the mitigation of spilled chemicals (including hydrocarbons) dissolved in ground and surface waters. Aesthetic and socioeconomic factors are not considered; although, these and other factors are often important in spill response.

1.2 This guide addresses the application of steam stripping alone or in conjunction with other technologies.

1.3 In making decisions with regards to discharging treated water and operating a boiler, appropriate government authorities must be consulted as required by law.

1.4 *This standard does not purport to address all of the safety concerns, if any, associated with its use. It is the responsibility of the user of this standard to establish appropriate safety and health practices and determine the applicability of regulatory limitations prior to use.* In addition, it is the responsibility of the user to ensure that such activity takes place under the control and direction of a qualified person with full knowledge of any potential or appropriate safety and health protocols.

2. Terminology

2.1 *Definitions:*

2.1.1 *feed-to-steam ratio*—ratio of feed flowrate (by weight) to steam flowrate (by weight).

2.1.2 *foulants*—substances, such as clay or silt, microbial biomass, organic solids or film, inorganics, and naturally occurring compounds, that interfere with the desired process.

2.1.3 *Henry's law*—when a liquid and a gas are in contact, the weight of the gas that dissolves in a given quantity of liquid is proportional to the pressure of the gas above the liquid. The law holds true only for equilibrium conditions, that is, when enough time has elapsed so that the quantity of gas dissolved is no longer changing.

2.1.4 *Henry's law constant*—a function of the compound's solubility in the liquid phase and its volatility. A high Henry's law constant indicates equilibrium favoring the gas phase, that is, the compound is more easily stripped from water than one with a low Henry's law constant. Theoretically, Henry's law constant can be estimated from vapor pressure, solubility, and molecular weight as follows **(1)**[2]:

$$HC = \frac{V_p \times MW \times 16.03}{sol \times T} \quad (1)$$

where:

HC = Henry's law constant (atm m^3 water/m^3 vapor),
V_p = vapor pressure (mm Hg),
MW = molecular weight (g/mole),
sol = solubility (mg/L), and
T = temperature (K).

2.1.5 *inorganic foulants*—compounds, such as those of iron, calcium, and manganese, which precipitate in a treatment unit, thereby reducing the throughput and efficiency of the process.

2.1.6 *packing*—is placed in a stripping column to increase the available surface area for mass transfer.

2.1.7 *pH*—a measure of the acidity or alkalinity representing the logarithm of the reciprocal of the concentration of hydrogen ions.

2.1.8 *purge and trap technique*—uses an inert gas (such as helium or nitrogen) to purge the compounds into a gaseous state.

2.1.9 *removal efficiency*—

$$\frac{[\text{inlet contaminant}] - [\text{outlet contaminant}]}{[\text{inlet contaminant}]} \times 100\,\% \quad (2)$$

2.1.10 *semi-volatile organic compound*—a compound that is amenable to analysis by extraction of the sample with an organic solvent. It is used synonymously with Base/Neutral/Acid (BNA) compounds.

2.1.11 *steam stripping*—a separation process that utilizes differences in the thermodynamic properties of liquids. In this process, steam and organic-contaminated water are fed counter-currently to a packed column, causing the transfer of the contaminant(s) from the water phase to the vapor phase. The driving force for the separation is the concentration differential of the organic component(s) between the liquid and vapor phases. Two streams are generated in this process, namely: bottoms (treated effluent) and tops or overhead (concentrated contaminant).

2.1.12 *equilibrium vapor pressure*—the pressure at which, at constant temperature, a pure substance's vaporization, and

---

[1] This guide is under the jurisdiction of ASTM Committee F-20 on Hazardous Substances and Oil Spill Responseand is the direct responsibility of Subcommittee F20.22 on Mitigation Actions.

Current edition approved May 15, 1995. Published July 1995. Originally published as F 1522 – 94. Last previous edition F 1522 – 94.

[2] The boldface numbers in parentheses refer to the list of references at the end of this standard.

condensation rates are at equilibrium.

2.1.13 *volatile organic compound*—a compound amenable to analysis by the purge and trap technique. It is used synonymously with purgeable compounds.

2.1.14 *volatility*—the tendency of a solid or liquid material to pass into the vapor state at a given temperature.

3. Factors

3.1 Removal efficiency is highly dependent on the properties of contaminants, such as Henry's law constant and vapor pressure, and the system operating parameters, such as temperature and steam-to-water ratio. An increase in any of these parameters or properties produces a corresponding increase in removal efficiency, assuming all other factors remain constant.

3.2 Other factors that influence removal efficiency include the size and type of column packing and the ratio of column diameter to packing diameter. In addition, the presence of solids will cause fouling that would reduce the throughput of the unit and could affect organic removal efficiency.

3.3 For compounds less volatile than water, the ability to form minimum boiling azeotropes or heteroazeotropes is considered indicative of good potential for steam stripping. In these mixtures, heating a dilute solution will result in a vapor phase richer in the contaminant even though the contaminant has a lower vapor pressure than water. The azeotrope will prevent production of a pure overhead stream. Since the objective is only to purify the underflow, this is not a problem.

4. Significance and Use

4.1 The purpose of this guide is to provide remediation managers and spill response teams with guidance on the use of steam stripping, to safely and effectively reduce environmental impacts of hazardous spills (chemical and oil) on water. Steam stripping is one of many available tools and may not be applicable to all situations.

4.2 Steam stripping technology has been used extensively in the chemical process industry; however, it is only in recent years that it has been applied in the remediation of contaminated water. For this reason, this guide will only refer to those units that are presently used in the field for that purpose.

4.2.1 This technology is especially attractive to contaminated industrial sites where surplus steam supplies are available.

4.3 This guide can be used in conjunction with other ASTM guides addressing hazardous (chemical and oil) materials spill response operations.

4.4 The steam stripping process may be applied alone or in conjunction with other treatment techniques as described in 4.5 and 4.6.

4.5 Steam stripping may be used following a pretreatment step, which will provide a method, either physical or chemical, for the removal of foulants from the contaminated stream prior to steam stripping. If these foulants are not removed, the throughput and efficiency of the process will be significantly reduced.

4.6 Steam stripping may be used to concentrate a dilute contaminated stream so that it may be treated more cost effectively at a higher concentration with another technology.

5. Constraints on Usage

5.1 Literature searches on the predicted removal efficiencies are essential prior to field scale treatment. Bench scale testing should be done where complex mixtures are present or behavior cannot be calculated by theory.

5.2 The nature and concentration of contaminant will affect the overall system performance. In general, organic compounds with higher Henry's constant are more easily stripped.

5.3 Generally, inorganic foulants, such as iron, calcium, and manganese, in the ppm range, reduce throughput and efficiency of the process. This phenomenon is common in most organic treatment units regardless of the mechanism employed. Generally, pre-treatment systems involving chemical addition (that is, pH adjustment) or membrane technology, or both, are the most economical and effective for inorganic removal. Although, in some cases, the change in pH can affect the removal efficiency.

5.4 Steam stripping must be carried out under the guidance of qualified personnel that understand the contaminant, process, and safety and health aspects of site activities.

5.5 Steam stripping cannot remove certain compounds, such as: acetic acid, glycols (ethylene or propylene), glycerine, sulfonated organics, and inorganics (except in free gaseous dissolved form, such as ammonia and carbon dioxide).

5.6 Some phenols can be steam stripped; however, it is normally not cost effective due to the large amount of steam required for the process.

6. Field Scale Results Using Steam Stripping

6.1 Table 1 lists some results of testing the removal of specific compounds by steam stripping at the field scale.

7. Recommendations

7.1 Steam stripping should be considered as one of the potential treatment methods available to site remediation managers once the spill has been contained and gross quantities of contamination have been physically removed.

7.2 Steam stripping should only be performed with technically-qualified personnel, following health and safety protocols for such activity.

7.3 Before steam stripping is carried out, the technology's potential for removing the contaminants in question should be reviewed in terms of its efficacy based on a literature search and data supplied by the stripping system manufacturer. Bench scale confirmation on such contaminated water would also be desirable. System operating parameters should be optimized during the first few days of operation.

7.4 In order to measure the success, a rigorous monitoring program should be established to determine the contamination levels, track the contamination plume, and analyze the treated effluent stream. This effluent should be sent to a holding tank prior to ultimate disposition (for example, reinjection) so as to ensure that the water discharged is in accordance with regulations. Furthermore, the contaminated concentrated stream must be managed in an appropriate manner.

8. Keywords

8.1 distillation; extraction; removal; separation; steam stripping

TABLE 1 **Typical Field Scale Results Using Steam Stripping**

NOTE 1—

A = APV Crepaco Inc., **(2)**.
E = Emergencies Engineering Division of Environment Canada **(3)**.
GLO-R2 = run No. 2 at Gloucester, Ontario **(4)**.
GSP-R12 = run No. 12 at Gulf Strachan Gas Plant in Rocky Mountain House, Alberta **(5)**.
MMM-R40 = run No. 40 for MMM in Toronto, Ontario **(6)**.

Specific Compound	Flowrate, LPM	Concentration of Compound		Removal Rate, %	Reference
		Initial	Final		
Chlorinated Solvents					
chlorobenzene	30.3	26.44 ppb	<0.80 ppb	>97.0	E (GLO-R7C)
		115.57 ppb	<1.58 ppb	>98.6	E (GLO-R4)
chloroform	30.3	14.46 ppb	<0.23 ppb	>98.4	E (GLO-R7C)
		157.88 ppb	<2.77 ppb	>98.2	E (GLO-R9B)
dichloromethane	15.1	2987 ppm	<0.1 ppm	>99.9	E (MMM-R40)
	37.9	10 000 ppm	9 ppm	>99.9	A (brochure)
	45.5	15 000 ppm	1 ppm	>99.9	A (brochure)
1,2-dichloroethane	30.3	25.68 ppb	2.99 ppb	88.4	E (GLO-R5)
1,1,2-trichloroethane	22.7	1000 ppm	1 ppm	>99.9	A (brochure)
trichloroethylene	30.3	24.07 ppb	<2.77 ppb	>88.5	E (GLO-R2)
	30.3	2236.69 ppb	11.27 ppb	99.5	E (GLO-R8)
Light Aromatic Compounds					
benzene	30.3	699.75 ppb	6.17 ppb	99.1	E (GLO-R8)
	22.7	1.56 ppm	<0.01 ppm	>99.4	E (GSP-R12)
toluene	30.3	5.74 ppb	0.92 ppb	84.0	E (GLO-R8)
	22.7	0.26 ppm	<0.01 ppm	>96.1	E (GSP-R12)
ethylbenzene	22.7	0.35 ppm	<0.01 ppm	>97.1	E (GSP-R11)
xylenes	22.7	10.90 ppm	<0.02 ppm	>99.8	E (GSP-R12)
	37.9	1000 ppm	100 ppm	90.0	A (brochure)
Alcohols					
methanol	37.9	11 000 ppm	100 ppm	99.1	A (brochure)
ethanol	379	12 000 ppm	100 ppm	99.2	A (brochure)
	94.7 to 284	300 000 ppm	100 ppm	>99.9	A (brochure)
	30.3	160 000 ppm	50 ppm	>99.9	A (brochure)
isopropanol	37.9	20 000 ppm	100 ppm	99.5	A (brochure)
	22.7	200 00 ppm	50 ppm	>99.9	A (brochure)
butanol	26.5	60 000 ppm	200 ppm	99.7	A (brochure)
Ketones					
acetone	37.9	12 000 ppm	1 ppm	>99.9	A (brochure)
	45.5	15 000 ppm	1 ppm	>99.9	A (brochure)
Miscellaneous Solvents (including dioxane, tetrahydrofuran, oxygen and nitrogen compounds)					
ammonia	15.1	300 000 ppm	1 ppm	>99.9	A (brochure)
ethyl ether	30.3	231.40 ppb	1.86 ppb	99.2	E (GLO-R5A)
tetrahydrofuran	30.3	28.74 ppb	0.55 ppb	98.1	E (GLO-R2)
		890.13 ppb	5.80 ppb	99.4	E (GLO-R7C)
		1369.71 ppb	20.09 ppb	98.5	E (GLO-R9A)

REFERENCES

(1) Clark, R. M., Eilers, R. G., and Goodrich, J. A., "VOCs in Drinking Water: Cost of Removal," *Journal of Environmental Engineering*, Vol 110, No. 6, 1984, pp. 1146–1162.

(2) APV Crepaco Inc., *Steam Stripping VOCs from Waste Water*, Tonawanda, New York, 1990.

(3) Geisbrecht, G., Laperrière, F., and Peterson, D., "Mobile Steam Stripper as an Environmental Emergencies Countermeasure," *Proceedings from 5th Technical Seminar on Chemical Spills*, Montreal, Canada, February 1988.

(4) Jacquemot, S., Keller, L., and Punt, M., *Comparison of Mobile Treatment Technologies at the Gloucester Landfill*, Emergencies Engineering Division of Environment Canada, Emergencies Engineering Division Inhouse Report, 1991.

(5) Ladanowski, C., *Field-Scale Demonstration of Technologies for Treating Groundwater at the Gulf Strachan Gas Plant*, Canadian Association of Petroleum Producers Publication, April 1993.

(6) Ladanowski, C., Punt, M., Kerr, P., and Adams, C., "Efficacy of Steam Stripping in the Removal of Dichloromethane from Groundwater,"

Proceedings from Industrial Waste Management Conference, Vienna, Austria, April 1992.

Designation: F 1523 – 94 (Reapproved 2001)

Standard Guide for
Selection of Booms in Accordance With Water Body Classifications[1]

This standard is issued under the fixed designation F 1523; the number immediately following the designation indicates the year of original adoption or, in the case of revision, the year of last revision. A number in parentheses indicates the year of last reapproval. A superscript epsilon (ε) indicates an editorial change since the last revision or reapproval.

1. Scope

1.1 This guide covers the selection of containment boom that may be used to control spills of oil and other substances that float on a body of water.

1.2 This guide does not address the compatibility of spill control equipment with spill products. It is the user's responsibility to ensure than any equipment selected is compatible with anticipated products.

1.3 The values stated in SI units are to be regarded as the standard. The values given in parentheses are for information only.

1.4 *This standard does not purport to address all of the safety concerns, if any, associated with its use. It is the responsibility of the user of this standard to establish appropriate safety and health practices and determine the applicability of regulatory limitations prior to use.*

2. Referenced Documents

2.1 *ASTM Standards:*

D 751 Test Methods for Coated Fabrics[2]

F 625 Practice for Classifying Water Bodies for Spill Control Systems[3]

F 715 Test Methods for Coated Fabrics Used for Oil Spill Control and Storage[3]

F 1093 Test Methods for Tensile Strength Characteristics for Oil Spill Response Boom[3]

3. Summary of Guide

3.1 Recommendations for the selection of spill containment booms are given for five key boom characteristics that affect performance for each of four water body types (see Table 1). Notes accompanying Table 1 discuss the qualifying factors associated with the minimum values listed, as well as the significance of the values as minimums. Classification criteria for the four water body types are given in Practice F 625.

4. Significance and Use

4.1 This guide is to be used to assist in the selection of containment boom in accordance with water body classifications.

4.2 A small number of key parameters that generally affect boom performance have been used in Table 1, in order to enable the user to readily identify general criteria for boom selection.

4.3 Many factors, other than those listed in Table 1, may be important in selecting containment boom for a particular application. Such factors include: flotation element length, wave length, the effect of stronger than minimum required strength members, shock loads, abrasion resistance, stability in roll, resistance to bridging, interval between anchor points and hand holds, use of reflectors or lighting, compatibility with fresh or salt water, and resistance to sunlight exposure.

4.4 The values given in Table 1 are the recommended minimums for general purpose booms. As identified in 2.3, operational considerations may require trade-offs in boom properties. Special purpose booms, and general purpose booms used in special circumstances, may perform effectively with boom property values above or below those recommended in Table 1.

4.5 Effective operation of oil spill control equipment depends on many factors, of which the prevailing environmental conditions are just a few. Factors such as, but not limited to, deployment techniques, level of training, personnel performance, and mechanical reliability can also affect equipment performance.

5. Keywords

5.1 boom; oil spill; oil spill control equipment

[1] This guide is under the jurisdiction of ASTM Committee F20 on Hazardous Substances and Oil Spill Response and is the direct responsibility of Subcommittee F20.11on Control.

Current edition approved May 15, 1994. Published July 1994.

[2] *Annual Book of ASTM Standards*, Vol 09.02.

[3] *Annual Book of ASTM Standards*, Vol 11.04.

TABLE 1 Recommendations for Selection of Spill Containment Booms

Boom Property[A]	Calm Water	Calm Water-Current	Protected Water	Open Water
height (range)[B], mm (in.)	150 to 600 (6 to 24)	200 to 600 (8 to 24)	450 to 1100 (18 to 42)	900 to 2300+ (36 TO 90+)
minimum gross buoyancy to weight ratio[C]	3:1	4:1	4:1	8:1
minimum total tensile strength[D], N (lbs)	6800 (1500)	23 000 (5000)	23 000 (5000)	45 000 (10 000)
minimum fabric tensile strength[E], N/50 mm (lbs/in.)	(2 TM): 2600 (300) (1 TM): 2600 (300)	2600 (300) 2600 (300)	2600 (300) 3500 (400)	3500 (400) 3500 (400)
minimum fabric tear strength[F], N (lbs)	450 (100)	450 (100)	450 (100)	450 (100)

[A] The values given in parentheses are provided for information purposes only.

[B] The boom dimensions are given as total height, with an assumed freeboard of 33 % (minimum) of boom height for calm, protected, and open water, and a freeboard of up to 50 % of height for calm water-current. Within the range of boom heights cited, values should be selected in accordance with the specific conditions of the application considering such factors as: predominant wave conditions, volume of containment, length of boom required, tension on boom, etc.

[C] Values shown for all water body classifications are the minimums for general purpose booms intended for spill response use. Special purpose booms designed for permanent installations, booms employing balanced upper and lower tension members, and external tension member booms may employ hydrodynamic properties to maintain adequate freeboard and therefore require lower buoyancy to weight ratios, but in no case should the buoyancy to weight ratio be less than 2:1. Buoyancy to weight ratios greater than those listed may result in improved boom performance under certain conditions, however, further research is required before minimum values greater than those shown can be established.

[D] The dominant variables in calculations of forces on deployed boom are current/tow speeds and boom draft. The specified values represent the developed tension on 300 m (1000 ft) lengths of boom, deployed in a typical catenary configuration with a gap ratio of 1:3, with environmental data selected per water body classification, with a current/tow speed of 1 to 2 knots (4 kts for calm water-current), and, most significantly, the minimum boom draft per category. Booms with greater drafts should have higher tensile strengths as follows: calm water 57 N/mm of draft (320 lbs/in.), calm water-current 140 N/mm (800 lbs/in.), protected water 64 N/mm (360 lbs/in.), and open water 72 N/mm (400 lbs/in.). It is recommended that greater tensile strengths be used for applications where booms may be subjected to greater current/tow speeds than those noted above, or where additional safety factors are warranted. The boom tensile strength is tested in accordance with Test Methods F 1093.

[E] The fabric tensile strength requirement is shown for two basic boom designs: two tension members (2 TM) and one tension member (1 TM). Membrane material is tested in accordance with Test Methods F 715 and D 751 (Breaking Strength, Procedure A: Grab Test Method). Boom that employs load carrying fabric, that is, zero tension members, must meet the total tensile strength requirement previously identified.

[F] The membrane material is tested in accordance with Test Methods F 715 and D 751 (Tearing Strength, Procedure B: Tongue Tear Test Method). Certain special-use fabrics have fabric tear strengths less than the values listed, such fabrics may be preferable for certain applications such as, resistance to particular spill products, sunlight, and abrasion, and would be acceptable as long as minimum fabric tensile strength requirements are met.

Standard Guide for Use of Advanced Oxidation Process for the Mitigation of Chemical Spills[1]

This standard is issued under the fixed designation F 1524; the number immediately following the designation indicates the year of original adoption or, in the case of revision, the year of last revision. A number in parentheses indicates the year of last reapproval. A superscript epsilon (ε) indicates an editorial change since the last revision or reapproval.

1. Scope

1.1 This guide covers the considerations for advanced oxidation processes (AOPs) in the mitigation of spilled chemicals and hydrocarbons dissolved into ground and surface waters.

1.2 This guide addresses the application of advanced oxidation alone or in conjunction with other technologies.

1.3 *This standard does not purport to address all of the safety concerns, if any, associated with its use. It is the responsibility of the user of this standard to establish appropriate safety and health practices and determine the applicability of regulatory limitations prior to use.* In addition, it is the responsibility of the user to ensure that such activity takes place under the control and direction of a qualified person with full knowledge of any potential safety and health protocols.

2. Terminology

2.1 *Definitions of Terms Specific to This Standard:*

2.1.1 *advanced oxidation processes (AOPs)*—ambient temperature processes that involve the generation of highly reactive radical species and lead to the oxidation of waterborne contaminants (usually organic) in surface and ground waters.

2.1.2 *inorganic foulants*—compounds, such as iron, calcium and manganese, that precipitate throughout a treatment unit and cause reduced efficiency by fouling the quartz sleeve that protects the lamp in photolytic oxidation AOP systems or the fibreglass mesh that is coated with TiO_2 in photocatalytic AOP systems.

2.1.3 *mineralization*—the complete oxidation of an organic compound to carbon dioxide, water, and acid compounds, that is, hydrochloric acid if the compound is chlorinated.

2.1.4 *photoreactor*—the core of the photoreactor is a UV lamp that emits light in the broad range of 200 to 400 nm wavelength range.

2.1.5 *radical species*—a powerful oxidizing agent, principally the hydroxyl radical, that reacts rapidly with virtually all organic compounds to oxidize and eventually lead to their complete mineralization.

2.1.6 *scavengers*—a term used for substances that react with hydroxyl radicals that do not yield species that propagate the chain reaction for contaminant destruction. Scavengers can be either organic or inorganic compounds.

3. Significance and Use

3.1 *General*—This guide contains information regarding the use of AOPs to oxidize and eventually mineralize hazardous materials that have entered surface and groundwater as the result of a spill. Since much of this technology development is still at the benchscale level, these guidelines will only refer to those units that are currently applied at a field scale level.

3.2 *Oxidizing Agents*:

3.2.1 *Hydroxyl Radical (OH)*—The OH radical is the most common oxidizing agent employed by this technology due to its powerful oxidizing ability. When compared to other oxidants such as molecular ozone, hydrogen peroxide, or hypochlorite, its rate of attack is commonly much faster. In fact, it is typically one million (10^6) to one billion (10^9) times faster than the corresponding attack with molecular ozone **(1)**.[2] The three most common methods for generating the hydroxyl radical are described in the following equations:

$$H_2O_2 + h\nu \rightarrow 2OH\cdot \quad (1)$$

$$2O_3 + H_2O_2 \rightarrow\rightarrow 2OH\cdot + 3O_2 \quad (2)$$

$$Fe^{+2} + H_2O_2 \rightarrow\rightarrow OH\cdot Fe^{+3} + OH^- \ (Fenton's\ Reaction) \quad (3)$$

3.2.1.1 Hydrogen peroxide is the preferred oxidant for photolytic oxidation systems since ozone will encourage the air stripping of solutions containing volatile organics **(2)**. Capital and operating costs are also taken into account when a decision on the choice of oxidant is made.

3.2.1.2 Advanced oxidation technology has also been developed based on the anatase form of titanium dioxide. This method by which the photocatalytic process generates hydroxyl radicals is described in the following equations:

$$TiO_2 + h\nu + H_2O \rightarrow OH\cdot + H^+ + e^- \quad (4)$$

$$2e^- + 2O_2 + 2H_2O \rightarrow 2OH\cdot + O_2 + 2OH^- \quad (5)$$

3.2.2 *Photolysis*—Destruction pathways, besides the hydroxyl radical attack, are very important for the more refractory compounds such as chloroform, carbon tetrachloride,

[1] This guide is under the jurisdiction of ASTM Committee F-20 on Hazardous Substances and Oil Spill Response and is the direct responsibility of Subcommittee F20.22 on Mitigation Actions.

Current edition approved May 15, 1995. Published July 1995. Originally published as F 1524 – 94. Last previous edition F 1524 – 94.

[2] The boldface numbers in parentheses refer to the list of references at the end of this standard.

trichloroethane, and other chlorinated methane or ethane compounds. A photoreactor's ability to destroy these compounds photochemically will depend on its output level at specific wavelengths. Since most of these lamps are proprietary, preliminary benchscale testing becomes crucial when dealing with these compounds.

3.3 *AOP Treatment Techniques*:

3.3.1 Advanced oxidation processes (AOPs) may be applied alone or in conjunction with other treatment techniques as follows:

3.3.1.1 Following a pretreatment step. The pretreatment process can be either a physical or chemical process for the removal of inorganic or organic scavengers from the contaminated stream prior to AOP destruction.

3.3.1.2 Following a preconcentration step. Due to the increase in likelihood of radical or molecule contact, very dilute solutions can be treated cost effectively using AOPs after being concentrated.

3.4 *AOP Treatment Applications*—Advanced oxidation processes (AOPs) are most cost effective for those waste streams containing organic compounds at concentrations below 1 % (10 000 ppm). This figure will vary depending upon the nature of the compounds and whether there is competition for the oxidizing agent.

4. Constraints on Usage

4.1 *General*—Although AOPs are destruction processes, in order for compound mineralization to take place, the oxidation reactions must be taken to completion. In most cases, effluent analysis is the only method available to ensure this state. Some compounds are selective in their reactivity. For these reasons, preliminary bench-scale testing and literature searches on the predicted reaction mechanisms are essential prior to full scale treatment.

4.2 *Presence of Scavengers*—Scavengers, such as bicarbonate and carbonate, will adversely affect the ability of the oxidizing agent to react with the target compounds if these compounds are left as ions within the solution. Adjusting the pH of the solution will reduce this problem, however, the additional cost requirements must be balanced against the benefit received.

4.3 *Contaminant Identification*—The types of contaminants and their corresponding destruction rate constants will affect the overall system performance. In general, chlorinated aliphatics with carbon-to-carbon double bonds (unsaturated), degrade more quickly than chlorinated compounds with single bonds (saturated). In addition, refractory compounds such as carbon tetrachloride, chloroform, and other chlorinated methane compounds are quite resistant to degradation in the presence of the hydroxyl radical and should be destroyed photochemically (that is, UV alone).

4.4 *pH Adjustment*—Adjusting the pH of the solution prior to treatment may significantly affect the performance of the treatment. A feed solution at a pH of 9 will tend to cause precipitation of most inorganics, while a pH of 5 will cause them to remain in solution throughout the treatment process. In situations where the inorganics are in a relatively low concentration (low parts per million), one would tend to lower the pH, while a higher pH would be preferable at the higher concentrations where the inorganics could be separated and removed.

4.5 *System Fouling*—Generally, inorganic foulants, such as iron, manganese, and calcium, in the ppm range, cause reduced flow, increased pressure and low performance of a treatment system. This phenomenon is common in most organic treatment units regardless of the mechanism employed. Pretreatment systems usually involve chemical addition (that is, pH adjustment) or membrane technology, or both, as they are generally the most economical and effective for inorganic removal. Preliminary benchscale testing is commonly used to determine the applicability and the cost-effectiveness of the different pretreatment systems.

4.6 *Off-Gas Analysis*—Organic analysis of the exiting gaseous stream will assist the operator in modifying system parameters to maximize system performance and efficiency. This technique is also beneficial during preliminary testing as it provides an indication of the AOP technology's ability to destroy the compounds as compared to simply stripping them from the water phase into the air.

4.7 *Destruction Rate Constants*—The reaction of the OH radical with organic compounds is largely dependent upon the rate constant. A list **(3)** of reaction rates for common contaminants is shown in Table 1.

5. Practical Applications

5.1 *Emergency Situations*—Advanced oxidation process (AOP) applications would normally follow containment and recovery of the waste stream in question. The time required for this primary stage should be sufficient for the AOP user to at least obtain the necessary background information on the contaminants in question. Benchscale confirmation testing is desirable, if time permits. Under no circumstances should AOP be used in a clean-up unless the manufacturer can supply data concerning testing on the same or similar chemical solutions.

5.1.1 *Emergency Clean-Up Operation*—For a spill, under emergency clean-up situations, the AOP technology user must do the following:

5.1.1.1 Monitor the feed, effluent, and off-gas stream analysis closely,

5.1.1.2 Monitor the feed flowrate and adjust accordingly,

5.1.1.3 Use holding tanks prior to discharge in order to buffer changes in discharge concentrations, and

5.1.1.4 Modify system parameters as necessary, based on the above conditions.

TABLE 1 Rate Constants for the Hydroxyl Radical

Compound, m	k_M, OH, (10^{+9} $m^{-1}s^{-1}$)
Benzene	7.8
Hydroperoxide Ion	7.5
Vinyl Chloride	7.1
Chlorobenzene	4.5
1-Butanol	4.2
Trichloroethane	4.0
Nitrobenzene	3.9
Pyridine	3.8
Toluene	3.0
Tetrachloroethane	2.3
Carbonate Ion	0.39
Dichloromethane	0.058
Bicarbonate	0.0085
Chloroform	~0.005
Carbon Tetrachloride	NR

5.2 *Non-Emergency Operation*—Once the leachate or chemicals reach the groundwater, the critical period is over and rapid response is less effective. Preliminary testing and preparation can be performed by the mitigator prior to treatment. Pretesting and manufacturers' information will determine the most appropriate operating conditions and the pretreatment required. This will not, however, reduce the importance of closely monitoring all aspects of the data. Sudden changes in feed concentrations could severely reduce the destruction rates.

5.3 *Field Scale Results Using AOP Technology*—Table 2 provides a summary of typical destruction capabilities achieved during photolytic AOP field trials conducted between 1988–1993.

6. Keywords

6.1 advanced oxidation; AOP; destruction; enhanced oxidation; hydrogen peroxide; hydroxyl radical; ozone; photolysis; titanium dioxide; ultraviolet

TABLE 2 Typical Field Scale Results of AOP Field Trials

NOTE 1—MF – microfiltration
RO – reverse osmosis

Specific Compound	Flowrate	Concentration		Notes	Reference
		Initial	Final		
1,4 dioxane	19–114 L/min	100 ppm	<10 ppb	system able to reduce dioxane in raw or deionized water consistently	(1)
methylene chloride trichloroethylene 1,2 trans-dichloroethylene vinyl chloride chloroethane	19 L/min	130–730 ppb 9.7–19.9 ppm 6–12.5 ppm 10–1010 ppb 10 ppb	3.1 ppb 0.4 ppb <0.1 ppb 0.5 ppb <0.3 ppb	high iron conc, prevented precipitation by maintaining pH at 3	(1)
nitrate esters, explosives	15 L/min	1000–5000 ppm	<1 ppm	systems tested with UV/H202, UV alone, and proprietary pretreatment for carbonate removal	(1)
trichloroethylene, benzene, chloroform, chlorobenzene, 1,2 dichloroethane	30 L/min	1.55 ppm 0.23 ppm 0.08 ppm 0.05 ppm 0.01 ppm	<5 ppb <0.8 ppb 0.04 ppm <1 ppb <2 ppb	pH adjustment, MF/UV/H202 system	(4)
benzene, toluene, xylene ethylbenzene	23 L/min	1.22 ppm 0.47 ppm 9.29 ppm 0.59 ppm	<0.05 ppm <0.05 ppm <0.10 ppm <0.05 ppm	UV/$H_2 0_2$	(5)
n-nitrosodimethylamine	95 L/min	80 ppt	5 ppt	UV	(6)
cyanide	not available	6 ppm	2 ppm	sulphide and fluoride precipitated prior to treatment	(7)
dichloroethylene, dichloroethane benzene	batch, 5 minutes	0.5 ppm 5 ppm 3 ppm	ND ND 0.009 ppm	phenolics pretreated with proprietary reagent	(7)
trichloroethylene, dichloroethylene, vinyl chloride	26 L/min	30 000 ppb 20 000 ppb 500 ppb	0.4 ppb <0.1 ppb 0.5 ppb	adjusted pH <3 to prevent fouling of UV quartz	(7)
methylene chloride benzene toluene	batch	470 ppb 353 ppb 2740 ppb	1 ppb 1 ppb 4 ppb	UV/03/H202 at pH 10	(8)

REFERENCES

(**1**) Keller, L., Reed, D., "Recent Applications of Environment Canada's Mobile Enhanced Oxidation Unit," *Air and Waste Management Association 85th Annual Meeting and Exhibition*, 1991.

(**2**) Nyer, E. K., *Groundwater Treatment Technology*, 2nd Ed., Van Nostrand Reinhold, New York, 1992, p. 119.

(**3**) Glaze, W. H., Kang, J., "Chemical Models of Advanced Oxidation Processes," *Proceedings—A Symposium on Advanced Oxidation Processes for the Treatment of Contaminated Water and Air*, 1990.

(**4**) Jacquemot, S., Keller, L., Punt, M., "Comparison of Mobile Treatment Technologies at the Gloucester Landfill," Internal Report, Emergencies Engineering Division, Environment Canada, 1991.

(**5**) Ladanowski, C., "Field Scale Demonstration of Technologies for Treating Groundwater at the Gulf Strachan Gas Plant," *Canadian Association of Petroleum Producers Publication*, April 1993.

(**6**) Cooper, D., and Keller, L., "Advanced Oxidation Technology Demonstration at Ohsweken Six Nations Indian Reserve, *Proceedings of the Tenth Technical Seminar on Chemical Spills*," 1993.

(**7**) Correspondence from Doug Reed, Solarchem Environmental Systems, Nov. 6, 1991.

(**8**) Solarchem Enterprises," Leachate Remediation at the Oswego Superfund Site Using Rayox—A Second Generation Enhanced Oxidation Process," Final Report to Emergencies Engineering Division, Environment Canada, 1989.

Standard Guide for Use of Membrane Technology in Mitigating Hazardous Chemical Spills[1]

1. Scope

1.1 This guide covers considerations for the use of membrane technology in the mitigation of dilute concentrations of spilled chemicals into ground and surface waters.

1.2 This guide addresses the application of membrane technology alone or in conjunction with other technologies.

1.3 The values stated in SI units are to be regarded as the standard. The values given in parentheses are for information only.

1.4 *This standard does not purport to address all of the safety concerns, if any, associated with its use. It is the responsibility of the user of this standard to establish appropriate safety and health practices and determine the applicability of regulatory limitations prior to use.* In addition, it is the responsibility of the user to ensure that such activity takes place under the control and direction of a qualified person with full knowledge of any potential or appropriate safety and health protocols.

2. Referenced Documents

2.1 *ASTM Standards:*

F 1127 Guide for Containment by Emergency Response Personnel of Hazardous Material Spills[2]

3. Terminology

3.1 *Definitions of Terms Specific to This Standard:*

3.1.1 *concentrate, retentate*—in reverse osmosis and nanofiltration, respectively, the portion of the feed solution that does not pass through the membrane is called concentrate, while the term retentate is more commonly used for ultrafiltration and microfiltration.

3.1.2 *crossflow filtration*—a filtration process in which the feed flows almost parallel to the filter or membrane surface. It is also called tangential flow.

3.1.3 *flux*—a measure of the rate at which the permeate (or filtrate) passes through the membrane per unit area of membrane. It is reported in units of $L/m^2/day$, $m^3/m^2/day$, or $gal/ft^2/day$.

3.1.4 *fouling*—the accumulation of unwanted deposits or scales on a membrane that results in a flux reduction.

3.1.5 *Langelier Saturation Index (LSI)*—a method used to determine the calcium scaling potential, that is, calcium carbonate of a membrane at concentrations below 5000 ppm TDS.

3.1.6 *membrane technology*—separation of the components of a fluid by means of a pressure gradient and a semipermeable membrane. The various classes of membrane technology are differentiated primarily by the size or molecular weight, or both, of rejected material. The main divisions are (*1*) microfiltration (MF), (*2*) ultrafiltration (UF), (*3*) nanofiltration (NF), and (*4*) reverse osmosis (RO).

3.1.7 *microfiltration (MF)*—a pressure-driven process whereby a contaminated liquid stream is separated using a filtration process involving a compatible membrane. Dead-ended and crossflow techniques are used. Suspended solids and macromolecules are removed on the basis of size. Pore size is normally 0.1 to 5.0 μm, and operating pressures usually range from 20 to 350 kPa (3 to 50 psig). Membrane materials, such as polypropylene, polytetrafluoroethylene (PTFE), and metal oxides, are frequently less susceptible to chemical degradation than those used for other branches of this technology.

3.1.8 *nanofiltration (NF)*—a pressure-driven process whereby a contaminated liquid stream is separated and purified by a process involving filtration, diffusion, and chemical potential across a compatible membrane. Divalent and multivalent species with a molecular weight above 80 are removed as are uncharged and univalent molecules with a molecular weight above 200. Operating pressures normally run between 1380 and 2760 kPa (200 and 400 psig).

3.1.9 *osmotic pressure*—as related to membrane technology, the pressure that must be applied to the more concentrated solution to halt flow of the solvent from the less concentrated solution through a semipermeable membrane into the more concentrated side.

3.1.10 *permeate, filtrate*—the stream that has passed through the membrane and is therefore free of, or has a much reduced concentration of, contaminants. Permeate is commonly used for the treated water obtained from nanofiltration and reverse osmosis processes, while filtrate is more commonly used for the treated fluid obtained by ultrafiltration and microfiltration operation.

3.1.11 *pervaporation (PV)*—a vacuum-driven membrane process applicable to the separation of liquid mixtures. During

[1] This guide is under the jurisdiction of ASTM Committee F-20 on Hazardous Substances and Oil Spill Response and is the direct responsibility of Subcommittee F20.22on Mitigation Actions.

Current edition approved April 10, 1996. Published June 1996.

[2] *Annual Book of ASTM Standards*, Vol 11.04.

the separation, the dissolved, more volatile constituents are removed from a less volatile carrier stream, as a vapor, through a semipermeable membrane and then condensed on the downstream side. This energy-intensive process is still in the development stage, but it has the potential of being a very promising spill mitigation technology.

3.1.12 *reverse osmosis (RO)*—a pressure-driven process in which a liquid stream is separated and hence purified by passing it over the surface of a semipermeable membrane. Both dissolved and suspended materials in a molecular weight range from 40 to 200 are removed, with charged species being removed more easily. In the case of nonpolar molecules, molecular structure "bulkiness" becomes important. Some are rejected well with a molecular weight of 60, while others with a molecular weight of 100 are not. Differences among membrane material can be very important in this aspect. This process discriminates between solutes on the basis of their ability to either (*1*) preferentially adsorb onto the membrane pore surfaces and move through the membrane pores by capillary action, or (*2*) dissolve in and diffuse through the membrane. Reverse osmosis uses applied pressures between 1380 and 10 350 kPa (200 and 1500 psig). As the concentration difference between the solutions on the two sides of the membrane increases, the osmotic pressure of the solution increases and, in turn, the applied pressure requirement. In general, solutions containing organic and inorganic compounds ranging from low ppm up to 55 000 ppm are commonly treated with this technique.

3.1.13 *semipermeable membrane*—membranes that are selective in the components that they allow to pass through them.

3.1.14 *ultrafiltration (UF)*—a pressure-driven process whereby a contaminated liquid stream is separated and purified by a crossflow filtration process involving a compatible membrane. Suspended solids and dissolved molecules in the 500 to 300 000 molecular weight range are removed mainly on the basis of size. This represents a membrane that has a pore size ranging between 0.0015 and 0.2 μm. Ultrafiltration uses pressures of 105 to 1380 kPa (15 to 200 psig).

4. Significance and Use

4.1 *General*—This guide contains information regarding the use of membrane technology to recover and concentrate hazardous materials that have entered surface and ground water as the result of a spill. Membrane technology may be applied alone or in conjunction with other treatment techniques, as follows:

4.1.1 Different types of membrane are used in series with filters to treat highly contaminated solutions reaching concentration levels of several parts per million of organic and inorganic materials.

4.1.2 Different types of membranes are applied in series to treat very dilute concentrations (parts per billion level) of organic and inorganic compounds. Each membrane type has the ability to remove specific compounds, thus producing a concentrated fraction. This fraction may require final off-site treatment but provides a significant reduction in transportation costs due to the large volume reduction achieved.

4.1.3 Membranes may be used in conjunction with destruction technologies such as advanced oxidation processes (AOPs). This method is recommended for dilute solutions. The membrane technology portion concentrates the compounds to an optimum level for AOP destruction.

5. Constraints on Usage

5.1 *General*—Application of membrane technology to the cleanup of spills results in the generation of two streams. The first stream is treated and has a reduced concentration of contaminants, while the second is concentrated and has an increased concentration of contaminants. This concentrated stream must be destroyed, reprocessed, or disposed of in an appropriate manner. There may also be constraints that are created by the physical and chemical sensitivity of membranes and, as a result, characteristics of a membrane system must be taken into consideration whenever membrane units are used. These considerations are described as follows.

5.1.1 *Membrane Material*:

5.1.1.1 The material used to construct the membrane is crucial to its success. In general, for spill remediation, the UF, NF, and RO membranes require materials that have a good temperature and pH resistance, as well as chemical stability, to ensure that the membrane is unaffected by the solution being treated. The increasing demands on the performance of membrane materials are exceeding the capability of organic polymers currently available. Consequently, inorganic membranes have been developed in order to satisfy the need for better performance. Today, high-quality organic and inorganic membranes are commercially available.

5.1.1.2 Inorganic membranes are classified in four groups: ceramic, carbon, metal, and polymer analog. Many developments in inorganic membranes have been achieved, but many inconveniences have yet to be overcome, such as their high cost and low surface area/volume, which retards the expansion of their application. In the case of organic materials, several kinds of polymers are used that allow for the development of membranes with various properties. The following improvements might be noticed: lower cost, longer life time, lower replacement rates, reduced chemical consumption, reduced operating pressure for given flux level, use in a broader range of pH, higher ion rejection, easier cleaning due to effective foulant removal and reduced biological attack, lower energy consumption, as well as reduction of capital cost. Polymeric membranes with very high performance have been designed, but their great complexity makes commercialization difficult. Polymeric membranes currently on the market are available in symmetric and asymmetric configurations.

5.1.1.3 Asymmetric membranes are more commonly available than the symmetric type, especially for UF, NF, and RO. These asymmetric membranes are made of two layers of the same polymer. They have a thin and dense surface skin and a porous substructure that adds strength and support to the thin skin without reducing the permeate flow. The symmetric configuration has a homogenous structure that provides a very high hydraulic resistance. Another type of membrane, very similar to the asymmetric, is the thin film composite. The most obvious difference is that the two layers are made individually, from two different kinds of polymers for better performance. The characteristics of several polymeric materials currently available are listed in Table 1.

TABLE 1 Features of Several Polymeric Materials

Polymers	Hydrophilic	Maximum Operating Temperature, °C	Chemical Resistance	Cost	UV Resistance	pH Range Resistance
Cellulosic CA, CN, CN/CA	yes	126	fair	low	good	fair
Nylon 6, 66 polyamide	yes	135	good	low	good	good
PVDF	no	135	good	medium	poor	good
PVDF modified	yes	135	good	high	poor	good
PTFE, film	no	135	excellent	high	poor	excellent
Polysulfone	no	135	good	medium	good	excellent
Polysulfone modified	yes	135	good	medium	good	excellent
Polyacrilonitrile		...	fair	low		
	yes	...	fair	low	good	fair
PAN/PVC	no				good	fair
Polyvinyl-chloride	no	...	fair	low	good	good
Polycarbonate	no	...	fair	medium	good	good
Polycarbonate modified	yes	140	fair	medium	good	good
Polyester	no	135	excellent	medium	good	excellent
Polypropylene	yes	...	excellent	low	fair	excellent
Polyethylene	yes	...	excellent	low	fair	excellent

5.1.2 *Pretreatment*:

5.1.2.1 Pretreatment of the feed is of primary concern when membranes are used in spill cleanup. Although each membrane configuration is affected differently by inorganic foulants, most membranes are affected adversely by oil, grease, and parts per million concentrations of inorganic compounds, including iron, manganese, magnesium, calcium, carbonate ion, and sulphate species. Some organic species, especially at high ppm levels, may also have detrimental effects and cause irreversible damage to the membranes.

5.1.2.2 The pH of the feed solution is often adjusted to dissolve or precipitate inorganics, to prevent membrane fouling on the membrane surface that leads to a performance decline (that is, lower permeate flow rate and increased pressure drop between the feed and concentrate sides). The degree of pretreatment required will depend on the concentration of foulants in the feed stream, membranes used, and membrane cleaning schedule.

5.1.3 *Membrane Cleaning Agents*:

5.1.3.1 A membrane cleaning schedule will depend on the severity of membrane fouling. As mentioned above, a decrease in permeate flow of more than 10 % of the normal flow rate will generally be an indication that cleaning or flushing, or both, is required. It is also good practice to conduct membrane cleaning during periodic maintenance or before long shutdown periods. It is useful to determine the type of foulants on the membrane surface before cleaning. Chemical analysis is the best method; however, in situations in which this may not be possible, foulants may be determined by other means such as visual inspection. Chemical cleaning clears the membrane surface by dissolving the fouling substances with reagents. Table 2 provides a description of common inorganic foulants.

5.1.3.2 Each major foulant type will require a specific cleaning procedure. If the performance does not improve sufficiently after the first cleaning procedure, the application of another procedure may lead to a better result. Fouling on the membrane surface is usually complex and often requires several cleaning procedures successively. For example, successive cleaning with detergent and citric acid results in generally more effective cleaning than either alone. Table 3 lists several of the common cleaning agents used for membranes.

TABLE 2 Common Membrane Foulants

Foulant	Description
Ferric hydroxide	brownish color
Calcium silica biological fouling	white or beige color
Calcium scale, inorganic colloids	crystalline appearance
Biological fouling organic material	slimy appearance

5.1.4 *Flushing and Cleaning Procedures*:

5.1.4.1 *Flushing*—One of the most convenient foulant removal procedures is flushing. Flushing cleans the membrane surface using a large quantity of feedwater at low pressure. It is effective for cleaning membranes that have been slightly fouled. The general operating conditions are as follows:

(1) Flushing Water—Permeate (treated water),

(2) Pressure—190 to 590 kPa (28 to 86 psi),

(3) Water Flow Rate—High flow rate but pressure drop limited to less than 10 psi/element,

(4) Temperature—Ambient but less than 30°C (86°F), and

(5) Period—0.5 h.

5.1.4.2 *Cleaning (Polymer Membranes Only)*—Chemical cleaning is ordinarily used after the flushing procedure. A flush is also recommended after chemical cleaning to wash off dissolved solids and suspended solids in the modules. The general operating conditions for UF membranes are as follows: It is important to realize that the operating conditions differ according to the type of membrane used. Therefore, it is strongly recommended to refer to the manufacturer's directions before proceeding. If no instructions are available, the following operating conditions for UF membranes may be used:

(1) Chemical Cleaning Agents—As listed in Table 3.

(2) Cleaning Water—Permeate 40 to 80 L (11 to 22 gal)/200-mm (8-in.) module and 10 to 20 L (3 to 5 gal)/100-mm (4-in.) module.

(3) Pressure—Less than 345 kPa (50 psi).

(4) Feed Flow Rate—150 to 230 L/min (40 to 60 g/m)/200-mm (8-in.) vessel and 40 to 60 L/min (10 to 15 g/m)/100-mm (4-in.) vessel.

(5) Temperature—As high as possible but less than 40°C (104°F).

(6) Period—1 to 4 h.

(7) Method of Cleaning—Circulation and soaking.

TABLE 3 Chemical Cleaning Agents for Polymer Membranes

Fouling Substance	Chemical Reagent	Cleaning Conditions
Calcium scale	citric acid	1 to 2 % solution, pH 3 to 4 adjusted with NH_4OH
	HCl	1 to 2 % solution, pH 2 adjusted with NH_4OH
Metal hydroxide	citric acid	1 to 2 % solution, pH 3 to 4 adjusted with NH_4OH
Organic matter	detergent, membrane dependent	in accordance with the manufacturer's directions
Inorganic matter	citric acid	1 to 2 % solution, pH 3 to 4 adjusted with NH_4OH
Bacterial matter	peroxide OR sodium meta-bisulfite, membrane dependent detergent	0.5 to 1.0 % solution, 0.5 to 1.0 % by wt, in accordance with the manufacturer's directions dependent on specific membrane

5.1.5 *Membrane Storage*—Membranes must remain full of water at all times to prevent drying. Once a membrane has dried out, it can no longer be used. The recommended temperature range is from 0 to 35°C (32 to 95°F). A10 % glycerine solution should be added to the storage solution when temperatures fall below freezing.

5.1.6 *Shutdown Periods*:

5.1.6.1 *Six Days or Less*—The unit should be flushed every day for 1 h. The chlorine concentration of feed must be maintained between 0.5 and 1.0 ppm.

5.1.6.2 *Seven Days or More*—Circulate a 0.5 to 1.0 % peroxide (or sodium meta-bisulphite, depending on the membrane material) at low pressure of approximately 345 kPa (50 psi) for 1 h before continuous shutdown. The chemical concentration should be checked periodically.

5.2 *Membrane Configurations*—There are four basic membrane configurations, and each has its advantages and limitations. Table 4 lists the advantages and disadvantages for various modules.

5.2.1 *Tubular Membranes*—Tubular membranes can operate on large amounts of suspended solids and are the least sensitive to particulate fouling of all of the configurations. They are also cleaned more easily. As such, the level of pretreatment necessary may simply be screening to remove large solids. There is a trade-off of a much lower production of the purified stream per unit volume than is possible with other configurations because of the low surface area to volume ratio.

5.2.2 *Spiral-Wound Membranes*—Spiral-wound membranes have a high production of purified product per unit of membrane volume. This is accomplished through the large membrane surface area. This results in much closer tolerances than with the tubular configuration. Therefore, pretreatment will involve the removal of large solids before the feed is introduced to the membrane. "Wide spacer" and "Cage wrap" membranes are less affected, but the trade-off is a lower surface area per unit volume, with a consequent lower production of purified product.

5.2.3 *Hollow Fiber Membranes*—Hollow fiber membranes are intermediate between tubular and spiral-wound configurations concerning their need for pretreatment. More than simple screening will be necessary, and the possibility of fouling from low-solubility inorganics (that is, iron and calcium) will be a consideration. As a result, adequate pre-filtration and pH adjustment may be necessary. If the pretreatment system is optimized, these membranes have the highest production rate when compared to alternate geometries.

5.2.4 *Flat Sheet Membrane*—Flat sheet membrane configurations used in plate and frame systems offer ease of cleaning but relatively low permeate fluxes for NF and RO applications. However, the flat MF and UF sheets used in plate and frame systems normally produce much higher flux rates than other configurations, primarily due to the higher degree of turbulence that occurs within the flow channel. However, the capital cost of this type of system is relatively high. Pretreatment requirements are intermediate between tubular and spiral wound configurations. The systems tend to be bulky, with low surface areas per unit volume.

5.3 *Compatibility*—Membrane feed compatibility is probably the most important factor to be considered when membranes are being used. Their life expectancy is not as long as membranes used for water purification or spill remediation when compared to traditional seawater application. A complete analysis of the spill solution will ensure that the feed solution does not interact with the membrane surface, backing, or glue, or with system seals, so as to interfere with their performance. If possible, the membrane manufacturer should be contacted for advice.

5.4 *Pretesting*—The first step in remediating a spill using membrane technology is to conduct a preliminary study prior to commencing work in the field. This may involve contacting manufacturers and discussing the membranes' capabilities to withstand and separate the chemical to be treated. A laboratory study should be conducted on samples of the contaminated water in question to confirm the manufacturer's information

TABLE 4 Module Type: Advantages and Disadvantages

Module Types	Advantages	Disadvantages
Spiral wound	Spacer between membranes allows for less pretreatment of feed and easier cleaning High rejection Can be used in series in pressure vessel	Difficult to sanitize Can be permanently fouled by high contaminants in the feed
Hollow fiber	High output per module volume Operates at high product rates	Only one element per pressure vessel can be used Very prone to rupture Requires more sophisticated pretreatment than other module types
Tubular	Handles high solids concentrations	High energy costs/membrane area High capital and replacement cost Low membrane area per module
Flat sheet	Easier to clean High permeate flux for MF and UF separation	Relatively high capital cost Low permeate flux for NF and RO separation

and the membrane's capabilities to withstand not only the target species but other chemicals that may possibly be in the contaminated water. This will assist in the determination of pretreatment requirements. If samples of the actual solution are not available, synthetic solutions may be made by dissolving the target species in aqueous solution. However, the results will not be as reliable.

5.5 *Analysis*—One of the most important factors in spill or leachate treatment by membranes is analysis. This will be required by regulatory agencies before the purified stream can be discharged. For inorganics, analytical results during the cleanup should show little variation with time. Any loss of rejection is probably the result of membrane or seal problems. This is not, however, the case for organics. Particularly with spill-contaminated water, analytical results can vary by an order of magnitude. This is caused by two factors: (*1*) the concentration of the organic in the water may change rapidly, and (*2*) "slugs" of contaminant may cross the membrane barrier on a random basis. In any event, unless the membrane system is being damaged by compounds in the feed solution, the time-average of the concentration in the purified stream should not vary greatly.

5.6 *Monitoring*—The permeate/filtrate production should be monitored regularly, as it provides an indication of membrane system performance. It is a measure of the rate at which the problem is being treated that provides an indication of the interaction between the feed solution and the membrane. Fouling, swelling, or degradation are indicated by changes in flux. Thus, proper monitoring can, over the short and medium term, provide the information on which to base cleaning cycles or, if necessary, membrane replacement. Over the long term, it can provide a measure of the problem that technology has been called upon to solve. This is frequently not readily available at the onset of the cleanup.

6. Practical Applications

6.1 *Emergency Situations*:

6.1.1 Membrane technology will normally be used following the containment and storage of the spill-contaminated water in question. However, if the situation requires immediate cleanup due to severe toxicity and environmental impact, it may be necessary to use this technology without complete pretesting.

6.1.2 Analysis becomes paramount in this case. The responder will be feeding a solution to a membrane whose compatibility has been determined solely from available literature or manufacturer's advice. This may be based on short-term testing and may not be entirely reliable for a cleanup. Under no circumstances should a membrane be used in a cleanup unless the manufacturer can supply data concerning testing on the same or similar chemical solutions.

6.1.3 For a spill, under an emergency cleanup situation, the membrane technology user must perform the following:

6.1.3.1 Monitor feed, permeate, and concentrate stream analyses closely.

6.1.3.2 Monitor flux and separation.

6.1.3.3 Use holding tanks prior to discharge in order to buffer changes in discharge concentrations.

6.1.3.4 Make changes in procedure as found necessary on the basis of the above.

6.1.4 The extent to which the solution is being concentrated will also be determined from the analyses and flux monitoring, particularly in the case of RO. The regulators will designate contaminant concentration levels in the clean stream.

6.2 *Nonemergency Cleanup Situations*—Once the initial emergency is over, more complete testing and preparation can be performed by the mitigator prior to treatment. Pretesting and manufacturer's information will determine the best membrane for the purpose and the pretreatment required. However, this will not reduce the importance of closely monitoring all aspects of the data. Sudden and dramatic flux increases could indicate either irreversible fouling or swelling due to incompatibility.

6.3 *Final Comment*—In all cases, when using membranes in spill mitigation, the ultimate fate of the concentrate must be considered prior to commencing the operation. There is little to be gained by concentrating the target chemical if the concentrated stream is more expensive to manage than the original problem.

6.4 *Typical Field Scale Results*—Table 5 and Table 6 provide a summary of the removal of specific contaminants by

TABLE 5 Typical Values for Membrane Field Trials

Contaminant	Concentration, ppm Feed	Concentration, ppm Permeate	Removal, %	Reference[3]
Acenaphthene	32.00	24.00	25	(13)
Acetone	26.81	15.99	40	(11)
Barium ion	0.11	0.04	64	(12)
Barium ion	0.26	0.00	99	(13)
Benzene	1.31	0.47	64	(12)
Benzene	6.90	0.80	88	(13)
Benzoic acid	10.92	0.40	96	(11)
Boron ion	0.09	0.06	33	(12)
Boron	1.81	<0.01	>99.99	(10)
Bromoform	31.03	0.53	98	(11)
Calcium ion	35.70	0.49	99	(10)
Calcium ion	70.00	3.30	95	(13)
Carbonate ion	204.30	58.30	71	(10)
Chloride ion	95.30	7.14	93	(12)
Chloride ion	18000.00	81.00	>99.99	(10)
Chlorophenol	510.00	0.01	>99.99	(9)
Chlorophenol	1655.00	0.02	>99.99	(9)
t-1,2-dichloroethane	22.08	14.44	35	(11)
Ethylbenzene	0.23	0.13	43	(12)
Ethylbenzene	22.00	2.80	87	(13)
Hydroxide ion	362.10	85.80	76	(10)
Iron ion	9.24	0.12	99	(12)
Iron ion	143.57	0.04	100	(13)
Magnesium ion	10.73	0.89	92	(12)
Magnesium ion	10.98	9.60	13	(13)
Manganese ion	1.10	0.04	96	(12)
Phenanthrene	14.00	7.80	44	(13)
Phenol	300.00	66.00	78	(13)
Phenol	300.00	91.00	70	(13)
Phenolics	270.00	100.00	>99.99	(10)
Potassium ion	4.02	0.74	82	(12)
Silicon silicates	4.21	0.28	94	(12)
Sodium ion	45.87	6.48	88	(12)
Sodium ion	8900.00	154.00	98	(10)
Strontium ion	0.36	0.03	92	(12)
Strontium ion	0.44	0.01	98	(13)
Sulfate ion	18.80	0.06	100	(12)
Toluene	0.09	0.03	67	(12)
Toluene	120.00	14.00	88	(13)
Xylene	2.74	0.75	73	(12)
O-xylene	13.35	1.70	87	(11)
Zinc ion	0.24	0.01	96	(12)

TABLE 6 Typical Values for Pervaporation Bench-Scale Trials

Component	Concentration, ppm		Removal, %	Reference
	Feed	Permeate		
Ethylene dichloride	12.13	3.17	74	**(8)**
Trichloro ethylene	127.57	6.51	95	**(8)**
Toluene	76.42	5.28	93	**(8)**
	63.2	6	91	**(8)**

RO and pervaporation field-scale trials.[3]

7. Keywords

7.1 concentration; filtrate; filtration; hazardous spill mitigation; membrane; microfiltration; nanofiltration; permeate; pervaporation; retentate; reverse osmosis; ultrafiltration

[3] The boldface numbers in parentheses refer to the list of references at the end of this guide.

REFERENCES

(1) Whittaker, H., "Applications Testing with the Environment Canada Mobile Reverse Osmosis Unit," *U.S. EPA National Technology Seminar*, December 1988.

(2) Sourirajan, S., "Reverse Osmosis and Synthetic Membranes—Theory Technology Engineering," National Research Council of Canada, 1977.

(3) *Membrane Technology Reference Guide*, Ontario Hydro, 1990.

(4) Beauséjour, I., *Review of Current Membranes*, EED in-house student report, Environment Canada, 1991.

(5) Peterson, R., "Reverse Osmosis," *Membrane Technology/Planning Conference*, Filmtec Corp., 1983.

(6) Johnson, J. S., "Materials of Construction for Membranes," *Membrane Technology/Planning Conference*, Micron Separation Inc., 1984.

(7) Johnson, J. S., "Polymeric Membranes for Microfiltration," *Membrane Technology/Planning Conference*, Micron Separation Inc., 1986.

(8) *A Cross-Flow Pervaporation System for Removal of VOC's from Contaminated Water*, Zenon Environmental Inc., November 1991.

(9) Straforelli, J. B., and Whittaker, H.," Reverse Osmosis for Clean-Up of Spilled Wood Protection Solution," *Proceedings of the Second Annual Technical Seminar on Chemical Spills*, 1985.

(10) Tremblay, J., "Membrane Technology Applications in Western Canada," *Proceedings of the Third Annual Technical Seminar on Chemical Spills*, Environment Canada, Toronto, Ont., 1986.

(11) Whittaker, H., Kady, T., Evangelista, R., and Goulet, C., "Reverse Osmosis and Ultraviolet Photolysis/Ozonation Testing at the PAS Site—Oswego, N.Y.," *Proceedings of the Sixth Technical Seminar on Chemical Spills*, Calgary, Alta., June 1989.

(12) *Development and Demonstration of a Mobile Reverse Osmosis Adsorption Treatment System for Environmental Emergency Clean-Ups*, Contract No. 52SS.KEI 45-5-0715, Zenon Environmental Inc., October 1987.

(13) *Field Demonstration of Membrane Technology for Treatment of Landfill Leachate*, RAC Project No. 439C, Zenon Environmental Inc., December 1991.

Designation: F 1599 – 95

Standard Guide for Collecting Performance Data on Temporary Storage Devices[1]

This standard is issued under the fixed designation F 1599; the number immediately following the designation indicates the year of original adoption or, in the case of revision, the year of last revision. A number in parentheses indicates the year of last reapproval. A superscript epsilon (ε) indicates an editorial change since the last revision or reapproval.

1. Scope

1.1 This guide covers a guideline for measuring the performance parameters of full-scale temporary storage devices that would be used to store oil and oil-water mixtures.

1.2 *This standard does not purport to address all of the safety concerns, if any, associated with its use. It is the responsibility of the user of this standard to establish appropriate safety and health practices and determine the applicability of regulatory limitations prior to use.* Specific precautionary statements are given in 6.2.

2. Referenced Documents

2.1 *ASTM Standards:*

F 625 Practice for Classifying Water Bodies for Spill Control Systems[2]

F 631 Guide for Collecting Skimmer Performance Data in Controlled Environments[2]

F 715 Test Methods for Coated Fabrics Used for Oil Spill Control and Storage[2]

3. Terminology

3.1 *Definitions:*

3.1.1 *Design Terminology—Terms Associated With Temporary Storage Device Design:*

3.1.2 *accessories—optional* mechanical devices used on or in conjunction with a temporary storage device system but not included with the basic storage device and hose connectors, that is, lights, paravanes, buoys, anchor systems, storage bags, repair kits, etc.

3.1.3 *ancillary equipment*—mechanical devices necessary to the operation of a given temporary storage device system, for example, air pumps, hydraulic power supplies, control manifolds, etc.

3.1.4 *ballast*—the weight applied to the device to improve performance.

3.1.5 *container body*—the continuous portion of the device that serves to provide structural strength and shape to the device to contain the stored material.

3.1.6 *device weight*—the dry weight of a fully assembled temporary storage device.

3.1.7 *draft*—the maximum vertical dimension of the device below the water line.

3.1.8 *flotation*—that portion of the device that provides buoyancy.

3.1.9 *freeboard*—the minimum vertical height of the device above the water line; for open devices, this is the minimum height at which water can enter it.

3.1.10 *handhold*—any strap, handle, depression, or other provision for grasping the device by hand.

3.1.11 *lifting point*—the structural point(s) on the device designed for the attachment of a lifting device, such as a crane.

3.1.12 *liner*—accessory or ancillary equipment that provides containment within the container body.

3.1.13 *mooring point*—the structural point(s) along the length of the device designed for the attachment of anchor or mooring lines.

3.1.14 *overall height*—the maximum vertical dimension of the device.

3.1.15 *sail*—the maximum vertical height of the device above the water-line.

3.1.16 *shipping weight*—the weight of the device when packaged for transportation.

3.1.17 *shipping volume*—the volume of the device when packaged for transportation.

3.1.18 *stiffener*—a component that provides support to the device.

3.1.19 *temporary storage device*—a collapsible device used to store fluids temporarily.

3.1.20 *tension member*—any component that carries tension loads imposed on the device.

3.1.21 *tow point*—structural point(s) on the device designed for the attachment of towing lines.

3.1.22 *Engineering Terminology—Terms Associated With Temporary Storage Device Engineering:*

3.1.23 *drag force*—the resisting force on a device that results from it being towed.

3.1.24 *gross buoyancy*—the weight of fresh water displaced by the device at the point of submergence.

3.1.25 *gross buoyancy to weight ratio*—the gross buoyancy divided by device weight.

3.1.26 *heave response*—the ability of the device to react to the vertical motion of the water surface.

---

[1] This guide is under the jurisdiction of ASTM Committee F-20 on Hazardous Substances and Oil Spill Responseand is the direct responsibility of Subcommittee F20.14 on Disposal.

Current edition approved May 15, 1995. Published July 1995.

[2] *Annual Book of ASTM Standards*, Vol 11.04.

3.1.27 *maximum capacity*—the maximum volumetric capacity of the device as calculated from physical dimensions.

3.1.28 *maximum dynamic load*—the sum of all instantaneous dynamic loads, including those due to acceleration, wave forces, etc.

3.1.29 *operational capacity*—the maximum volumetric capacity of the device per application.

3.1.30 *pitch response*—the tendency of the device to oscillate about its lateral axis.

3.1.31 *rated pressure*—the maximum continuous operating pressure of the device, as specified by the manufacturer.

3.1.32 *reserve buoyancy*—the gross buoyancy minus device weight.

3.1.33 *reserve buoyancy to weight ratio*—the reserve buoyancy divided by device weight.

3.1.34 *roll response*—the tendency of the device to rotate about its longitudinal axis due to wave, wind, or current forces.

3.1.35 *yaw response*—the tendency of the device to oscillate about its vertical axis.

3.1.36 *Classification Terminology*:

3.1.37 *pillow tank*—a closed, generally rectangular or round coated fabric tank.

3.1.38 *open pool*—an open, generally rectangular or round coated fabric tank, similar in structure to a ''wading pool."

3.1.39 *towable pillow tank*—similar to a pillow tank used on land or on deck, but generally made of heavier material and having special rigging for towing.

3.1.40 *towable flexible tank*—a storage device that is generally long and cylindrical in shape and, when full, is largely submerged, characterized by flexibility along the length of the device.

3.1.41 *towable open tank*—an open, inflatable, barge-type vessel that resembles a large inflatable boat, characterized by a portion of the top surface being open to atmosphere.

4. Significance and Use

4.1 This guide covers the collection of quantitative data in the form of storage capacity, strength of materials, filling and offloading rates, and towability under controlled test conditions. The data can be used for evaluating the design characteristics of a particular temporary storage device or as a means of comparing two or more devices. Caution must be exercised whenever the test data are used to predict performance in actual spill situations since the uncontrolled environmental conditions that affect performance in the field are rarely identical to conditions in the test facility. Other variables such as mechanical reliability, the presence of debris, ease of repair, required operator training, operator fatigue, and transportability also affect performance in an actual spill but are not included in this guide. These variables should be considered along with the test data when making comparisons or evaluations of temporary storage devices.

4.2 Although this guide provides data on the performance of temporary storage devices, all of the combinations of actual conditions of use are not simulated in this series of tests. In particular, the resistance of the device to grounding, abrasion resistance of the container body, venting of the device during loading, and other operational issues not covered by this guide should be considered along with the test data when making comparisons or evaluations of temporary storage devices.

5. Overall Observations

5.1 For each of the tests that follow, the total manpower required to conduct the procedure and the required ancillary equipment will be noted in the test record. In addition, the total elapsed time for each portion of the tests will be noted.

5.2 Observations of the buoyancy and stability of the deployed device will be made for the marine testing of towable devices.

5.3 Any observations relative to safety will be entered as part of the test record. These should include any hazardous conditions noted and limitations due to weather conditions, as well as any safety precautions that were observed or should be observed. If the manufacturer's specified operating procedures are found to be deficient relative to safety observations, this should be noted.

5.4 The test series should be videotaped to document the tests and procedures.

6. Initial (Static) Loading Tests

6.1 The storage device, and any enclosed ancillary equipment (for example, flotation collars), should be subjected to the following tests using air or dyed water, as appropriate, in order to confirm the structural integrity and evaluate (qualitatively) any leakage. The following is presented as a guideline in the absence of manufacturer-supplied guidelines for performing initial loading tests.

6.2 Overpressure tests can be extremely dangerous, and precautions should be taken against the possibility of sudden and complete failure of the device. The following tests use large volumes of water or pressurized air, or both, that could cause injury to personnel and serious damage to property if released suddenly.

6.3 *Closed Devices and Enclosed Ancillaries*:

6.3.1 *Leak Test*—The storage device should be inflated with air to a test pressure, calculated as follows:

$$P = \frac{\sigma}{15r}$$

where:

P = test pressure (Pa),
σ = minimum ultimate tensile strength of material (N/m), and
r = radius of largest section (m).

6.3.2 Allow the storage device to stand for 30 min at the test pressure ±10 %. Maintaining the test pressure within ±10 %, apply a soapy water solution liberally to the seam areas, fitting joints, valves, and all fabric areas. Record and mark air seepage.

NOTE 1—The inspector must determine what is an acceptable leak since some small air leaks may not necessarily mean that a liquid would leak through the same hole. Also, some air bubbles may be caused by air forced out between the fabric layers and would not cause a liquid leak.

6.3.2.1 Mark and record the location and extent of any surface or seam irregularities, blisters, or cracks. Recheck these areas carefully for leaks. The inspector must determine the acceptable level of such surface irregularities.

6.3.3 The storage device may be re-tested if leaks can be

stopped by tightening the fittings or by minor permanent repairs.

6.3.4 *Overpressure*—Pressurize the device to 150 % of the rated pressure (defined in 4.2). Hold for 30 min, and then reduce the pressure to a safe level and inspect for leaks.

6.3.5 If the device exhibits leaks that are not structural failures, rework it and repeat the leak test of 6.3.1. If the device exhibits leaks that are from structural failure (that is, loose seams, fabric failure, etc.), rework it as necessary and repeat the overpressure test of 6.3.4.

6.3.6 Closed devices may also be tested using the procedures of 6.4 to determine whether minor air bubble leaks permit water to leak through the fabric.

6.4 *Open Devices*:

6.4.1 *Leak Test*—The following leak test for open devices can be performed only with the device out of water. If the device cannot be suspended to allow observation of its bottom when full, a sheet of white water sorbent cloth should be placed under it to indicate leaks.

6.4.2 Fill the device with water to its full operating level. Color the water with a water marker dye that provides a good color contrast with the storage device and sorbent cloth. Allow the device to stand for 2 h.

6.4.3 Observe and mark obvious leaks. Wipe the seam and leaking areas with a white water sorbent cloth, and determine whether colored water has leaked through. Mark and record the leaks. No splits or blisters in the coating or seams shall be acceptable.

6.4.4 The storage device may be re-tested if the leaks can be stopped by tightening the fittings or by minor permanent repairs.

7. Deployment, Loading, and Towing Tests

7.1 *General*—A visual examination of the device will be conducted once the temporary storage device has been made ready for deployment. The material specifications and operational limitations of the hose coupling mechanisms and any other fittings, such as towing bridle, drogue attachments, or securing devices, should be recorded.

7.2 *Towable Devices*:

7.2.1 A test of launching from a pier and preparation for deployment by a towing vessel will be conducted by following the manufacturer's instructions for break-out, deployment, and preparation for towing. The test will be considered concluded when the predesignated vessel is able to tow the device away from the dock in a safe and proper configuration.

7.2.2 Towing tests should be conducted under a range of environmental conditions appropriate to the device's intended use. (Classification criteria for calm, protected, and open water are given in Practice F 625.) Any differences during the test should be noted as conditions change. The test will be conducted by following the manufacturer's instructions or, in the absence thereof, by the opinion of the vessel operator for the most suitable course for the launching, filling, towing, and recovery operation. The test will be considered concluded when the response vessel has completed all towing tests and recovered and stowed the device in a safe and proper configuration.

7.2.3 While towing the device, the required towing force will be measured as a function of increasing tow speed. Dynamic loads should be measured using a recently calibrated load cell. During the towing tests, manufacturer-specified maximum design loads must be respected with due regard to maximum dynamic load.

7.2.4 Observations should be noted in the test record concerning the device behavior as the towing speed and wave approach angle are varied. Observations of stability should include vertical stability, twisting, diving, snaking, and yawing. Any distortion of the device or other problem causing instability, loss of recovered product, or unacceptable list, trim, or bow submergence should be noted. The draft and freeboard of the device should be noted in the test record for each of the loading conditions tested.

7.2.5 During the progress of the test, careful monitoring must take place to detect the following: wave-induced motions and inertial or free-surface effects, particularly as they may affect control and maneuverability; pitch, roll, heave, and bow submergence; leaks from vents, gages, loading hoses, and the container body; and breakaway or damage to external flotation or buoyancy devices.

7.2.6 If any of the previously listed or other problems develop during the test, they should be described fully in the test report, including notation of the following data: time, volume and nature of fluid in the device, direction of impact of the waves, speed at which the device was under tow, and any towing vessel action that may have influenced the problem.

7.2.7 *Empty Temporary Storage Device Towing Test*:

7.2.7.1 The test should be begun by towing the device in an empty condition. The speed should be built up gradually to the manufacturer's recommended maximum towing speed. If it is not possible or practical to attain or maintain this speed, the reasons will be noted and entered as part of the test record. The test should continue at the highest speed practical.

7.2.7.2 While towing the device, it should be subjected to a variety of wave conditions and to varying angles of attack. At 5 min intervals, the movement of the device should be varied in 45-degree increments such that the device encounters waves from each of eight directions.

7.2.8 *Loading Device to Half-Full*:

7.2.8.1 At this point, the testing should pause and the device be prepared for loading. A hose should be connected to the device, and any venting or gaging devices should be checked for proper mounting or operation in accordance with the manufacturer's instructions.

7.2.8.2 The device should then be filled through this hose by a pump that will take suction from a source of dyed water. The loading system should include a flow totalizer or other apparatus to measure the volume of fluid delivered to the device. The pump must be equipped with a quick-acting shut-off device immediately accessible to the volume-measuring apparatus and in view of the device in the water. The filling line should be equipped with a check valve device that will prevent any backflow or siphoning once the pump has been stopped. Care must be taken to ensure that the suction remains immersed fully in the water supply and that no air is delivered accidentally to the temporary storage device. Additional care should be taken to reduce the flow rate when the desired load is approached in order to prevent overfilling.

7.2.8.3 During the course of the loading operation, the functioning of the vents and gaging mechanisms will be noted and entered as part of the test record. In particular, the possibility of the mechanisms admitting sea water to the device due to submergence should be addressed. While this may be unlikely when loading operations are forcing product into the device, similar submergence of a mechanism while the pumping is discontinued could cause such a result.

7.2.9 *Half-Full Tow Test*—The towing test should then be repeated starting with the step given in 7.2.7.1. Observations of stability should be noted in the test record when the towing speed is resumed.

7.2.10 *Loading Device to Operational Capacity*:

7.2.10.1 Testing should pause at this point, and the device should be filled to its operational capacity following the steps given in 7.2.8. Care should be exercised and the flow rate reduced as the operational capacity is approached to prevent damage to the device due to improper or inadequate functioning of the venting mechanisms.

7.2.10.2 Once the temporary storage device has been filled to its operational capacity, its behavior in the water will be evaluated. This capacity should be recorded as the verified operational capacity if the buoyancy of the device meets the manufacturer-specified design criteria.

7.2.10.3 If the temporary storage device, having been filled to its operational capacity, exhibits negative or neutrally buoyant behavior, the device may not function properly for sediment-contaminated heavy oils. Reduction of the capacity would not necessarily improve the performance of the device. This capacity should be recorded in such cases as the verified provisional capacity. Devices exhibiting this problem may be functional for most uses. However, if further testing is desired, some of the performance-related protocols would have to be altered to provide a filling fluid less dense than that in which the device must float. Filling the device with fresh water while conducting the tests in salt water may suffice.

7.2.10.4 After loading the device to its operational capacity, and depending on the testing or test platform environment, the temporary storage device should be monitored for a period of up to 12 h in order to ascertain the integrity of the device subsequent to loading. The device will be monitored for 2 h prior to resuming towing tests at a minimum. Any leakage from the device during the loading operation or subsequent monitoring period should be noted in the test record.

7.2.11 *Full Temporary Storage Device Towing Test*:

7.2.11.1 The towing test should be repeated starting with the step given in 7.2.7.1. Observations of stability should be noted in the test record when the towing speed is resumed.

7.2.11.2 During the towing tests in which the device is filled to its operational capacity, the internal pressure of the contained fluid should be measured to ensure that it does not exceed that recommended by the manufacturer as the maximum allowable.

7.2.12 *Post-Test Monitoring*:

7.2.12.1 At the conclusion of the towing operation, the temporary storage device will be monitored for a minimum period of 12 h, depending on the test facility, in order to ascertain the integrity of the device subsequent to towing. The actual monitoring period should be noted in the test record if the duration of the monitoring is less than 12 h. Any leakage from the device during the towing operation or subsequent monitoring period should be noted in the test record.

7.2.12.2 At the completion of the towing and subsequent monitoring, the temporary storage device will be drained, recovered from the water to the response vessel or a pier, and stowed in the manufacturer's recommended configuration. A summary of the maintenance and cleaning completed to meet the manufacturer's stowage requirements should be noted in the test record.

7.3 *Stationary Devices*:

7.3.1 The first test of deployment will be conducted by following the manufacturer's instructions for break-out, deployment, and erection of the device. The test will be considered concluded when the device is in a safe and proper configuration and ready to receive fluids.

7.3.2 A hose should be connected to the device, and any venting or gaging devices should be checked for proper mounting or operation in accordance with the manufacturer's instructions.

7.3.3 The device should be filled as described in 7.2.8.2 and 7.2.8.3 for towable devices.

7.3.4 After loading the device to its operational capacity, it will be monitored for a period of 12 h in order to ascertain the integrity of the device subsequent to loading. Any leakage from the device during the loading operation or subsequent monitoring period should be noted in the test record.

7.3.5 The device will be disassembled and stored according to the manufacturer's instructions at the completion of the monitoring period. A summary of the maintenance and cleaning completed to meet the manufacturer's stowage requirements should be noted in the test record.

8. Off-Loading Tests

8.1 Additional off-loading tests with a viscous fluid may be desirable to determine any limiting features of the storage device. If additional off-loading tests are conducted with oil, the testing should be performed within secondary containment to contain any possible spillage.

8.2 The temporary storage device will be filled to its operational capacity with a test oil. This oil should be selected based on the intended application of the device. (For comparability with skimmer tests, recommended test viscosities are given in Guide F 631.) Loading procedures are detailed in 7.2.8.2. Particular attention should be given to measuring the volume of fluid admitted to the device accurately, in order to estimate the residual amount of fluid after off-loading subsequently.

8.3 A timed test should commence with application of the instructions of the manufacturer of the temporary storage device for off-loading the recovered product. A total of all man-hours expended in the off-loading task will be calculated and entered as part of the test record. Note in the test record whether the time for off-loading is limited by either the storage device or the pump used to off-load.

8.4 Observe the device for any residual amounts of the testing fluid upon completing the off-loading operation. Record an estimate of the residual volume, and an explanation for the amount of residual remaining, if possible.

9. Container Body Tests

9.1 Tests for fabrics used in temporary storage devices are specified in Method F 715. The tests include base line tests of the container body material and tests for weather and petroleum resistance. Tests should be selected from Method F 715 appropriate to the intended service of the storage device.

10. Test Report

10.1 Prepare a report of the test series in tabular format, including a summary of the measured and observed results, the manufacturer's specified data, and any other features relevant to the operation of the device but not tested in this guide. The following summarizes the key data included in this guide:

10.1.1 *Static Loading*—Qualitative observations of leakage when subjected to normal operating pressure and over-pressure conditions.

10.1.2 *Deployment, Loading, and Towing*—Observations of buoyancy, stability, draft, and freeboard and measurements of average and instantaneous tow loads at various loaded conditions; observations of functioning of vents, gages, and other ancillary equipment during loading; observations of leakage during and subsequent to loading; and verification of device capacity. Include the following in the test report: the frequency characteristics of the load cell and recording device, a summary of the environmental conditions present during the test series, and a comparison of the density of the test fluid relative to the waterbody in which the test is conducted.

10.1.3 *Off-Loading*—Measurement of residual fluid volume following off-loading, and observation of device-limiting features that may hinder off-loading.

10.1.4 *Container Body Tests*—Measurement of container body characteristics: baseline strength, weather, and petroleum resistance.

10.1.5 *Overall Observations*—Time, equipment, and manpower to complete break-out, deployment, launching, recovery, maintenance, cleaning, and stowage; observations of hazardous conditions relating to operation of the device; and deficiencies in manufacturer-specified procedures.

10.1.6 *Manufacturer Supplied Data*—Requirements for maintenance, cleaning, lay-up, inspection, and stowage; and safe working loads for load bearing fittings, including the following: lifting points, tow points, tie-downs, hand-holds, and hose fittings.

11. Keywords

11.1 oil spill; storage device; temporary storage device

Designation: F 1600 – 95a

Standard Terminology Relating to Bioremediation[1]

This standard is issued under the fixed designation F 1600; the number immediately following the designation indicates the year of original adoption or, in the case of revision, the year of last revision. A number in parentheses indicates the year of last reapproval. A superscript epsilon (ε) indicates an editorial change since the last revision or reapproval.

1. Scope

1.1 This terminology defines the terminology used in test methods, specifications, guides, and practices related to bioremediation technology.

1.2 These definitions are written to ensure that standards related to bioremediation technology are understood and interpreted properly.

2. Referenced Documents

2.1 *ASTM Standards:*

F 873 Guide for Incinerating Oil Spill Wastes at Temporary Field Locations[2]

F 1481 Guide for Ecological Considerations for the Use of Bioremediation in Oil Spill Response—Sand and Gravel Beaches[2]

3. Terminology

3.1 *Definitions:*

aerobes—organisms that require air or free oxygen for growth. **F 1481**

anaerobes—organisms that grow in the absence of air or oxygen and do not use molecular oxygen in respiration. **F 1481**

arctic—the biogeographic realm that comprises the tundra and treeless grounds lying north of the timberline in the Northern hemisphere or south of the timberline in the Southern hemisphere.

bioaugmentation—the addition of microorganisms (predominantly bacteria) to increase the biodegradation rate of target pollutants. **F 1481**

biodegradation—the chemical alteration and breakdown of a substance, usually to smaller products caused by microorganisms or their enzymes. **F 1481**

bioremediation—the enhancement of biodegradation.

bioremediation agents—inorganic and organic compounds and microorganisms that enhance biological degradation processes, predominantly microbial. **F 1481**

biostimulation—the addition of microbial nutrients, oxygen, heat, or water, or some combination thereof, to enhance the rate of biodegradation of target pollutants by indigenous species (predominantly bacteria and fungi).

buffer—a solution of partially ionized acids or bases capable of reducing pH changes in the presence of added alkalis or acids.

carcinogen—cancer-causing agent.

culture—controlled inoculation, growth, and harvesting of known microorganisms.

decomposers—microorganisms, predominantly bacteria and fungi, that convert complex organic matter into simpler organic and inorganic molecules.

disinfectants—physical and chemical agents used for inactivating or destroying microorganisms.

ecosystem—organisms and the surrounding environment combined in a community that is self-supporting. **F 1481**

effluent—the liquid discharge from a process.

end product—the chemical compound resulting from a particular metabolic process.

enrichment culture—a liquid culture that enhances the growth of a given type of organism.

landfill—a land disposal technique that uses excavated pits to contain the oil spill waste material. The waste is placed in the excavation, covered over, and left to degrade. **F 873**

leachate—the liquid residue from the migration of environmental pollutants mediated by surface or ground water.

marine—relating to the ocean. **F 1481**

medium—the material that supports the growth and reproduction of microorganisms.

mesophile—an organism with an optimum growth temperature range of 20 to 45°C.

metabolic pathway—the sequence of biochemical reactions that allows conversion for carbon or energy assimilation.

microbes—microscopic organisms, including algae, bacteria, fungi, protozoa, and viruses.

mineralization—the microbial-mediated breakdown of organic materials into inorganic materials.

mutagen—a substance that increases the normal mutation rate.

nutrient—a substance that supports organismal growth. **F 1481**

pathogens—organisms that are capable of causing disease.

petri plate—a dish designed to contain solid growth medias.

plasmids—independently-replicating extrachromosomal genetic material.

recalcitrant—a substance that is resistant to microbial degradation.

---

[1] This terminology is under the jurisdiction of ASTM Committee F-20 on Hazardous Substances and Oil Spill Response and is the direct responsibility of Subcommittee F20.24 on Bioremediation.

Current edition approved Sept. 10, 1995. Published November 1995. Originally published as F 1600 – 95. Last previous edition F 1600 – 95.

[2] *Annual Book of ASTM Standards*, Vol 11.04.

terrestrial—consisting of land, as distinguished from water.

toxicity—the property of a material, or combination of materials, to adversely affect organisms.

tropical—the region lying between the tropics of Cancer and Capricorn.

tundra—the vast, treeless, nearly level plains of the arctic regions.

viability—the ability to grow and reproduce.

Designation: F 1607 – 95

Standard Guide for Reporting of Test Performance Data for Oil Spill Response Pumps[1]

This standard is issued under the fixed designation F 1607; the number immediately following the designation indicates the year of original adoption or, in the case of revision, the year of last revision. A number in parentheses indicates the year of last reapproval. A superscript epsilon (ϵ) indicates an editorial change since the last revision or reapproval.

1. Scope

1.1 This guide is intended as a guideline for the standardized reporting of performance data of pumps and pump systems that may be considered for use in oil spill response operations. The present objective is to develop a reporting guideline to aid in the comparative evaluation of various devices.

1.2 The values stated in SI units are to be regarded as the standard. The values given in parentheses are for information only.

1.3 *This standard does not purport to address all of the safety concerns, if any, associated with its use. It is the responsibility of the user of this standard to establish appropriate safety and health practices and determine the applicability of regulatory limitations prior to use.*

2. Referenced Documents

2.1 *ASTM Standards:*

D 2196 Test Method for Rheological Properties of Non-Newtonian Materials by Rotational (Brookfield) Viscometer[2]

F 631 Guide for Collecting Skimmer Performance Data in Controlled Environments[3]

3. Terminology

3.1 *Definitions of Terms Specific to This Standard:*

3.1.1 *oil spill response pump*—devices used to transfer oil, oil and water mixtures, and emulsions. These include centrifugal pumps, positive displacement pumps, vacuum pumps, and other systems.

3.1.2 *pump system*—the principal fluid moving device, including the prime mover, associated hose and fittings, and associated equipment such as debris handling devices that are normally used during oil spill response operations.

4. Significance and Use

4.1 The performance criteria listed in this guide will provide guidance in the selection of oil spill pumping equipment.

4.2 This guide has been developed for use by the following: manufacturers of pumping systems who wish to establish a common means of evaluating and reporting the performance characteristics of their products; and existing or potential users of pumping systems who wish to compare the performance characteristics of various products.

5. Test Fluids: Suggested Test Fluid Viscosities

5.1 A selection of test fluids from the following viscosity ranges is recommended and should include an adequate number of data points to display the pump's performance over the intended range of operation. It is recommended that the test fluids include one selection from each of the four ranges listed in Table 1, depending on the intended service of the pump. In particular, three viscosity targets are recommended to allow comparisons with skimmer test data (Guide F 631), namely, 200, 2000, and 60 000 cSt.

5.2 For the tests described in 6.3-6.6, use hydrocarbon-based single-viscosity oils or mixtures of high- and low-viscosity hydrocarbons, or both, at various temperatures to achieve test viscosities. Higher viscosities, such as those in Category 4, may require the use of emulsions. Note that, when using high-viscosity fluids, care must be taken in interpreting the test results due to two-phase flow considerations. The test fluid viscosity should be identified by measurements at shear rates of 1, 10, and 50 s^{-1} (Test Method D 2196).

6. Categories of Tests for Pump Performance

6.1 *Suction and Discharge Hoses*—Each of the following tests (6.3-6.5) should be conducted using suction and discharge hoses that are the same diameter as the respective pump inlet and discharge fittings.

6.2 *Equipment Specifications*—For each of the following performance tests, report the equipment specifications, including the following: pump model; prime mover model; prime mover type (that is, hydraulic, electrical, air, etc.); prime mover characteristics (that is, hydraulic, discharge pressure and flowrate; electrical, voltage and current; and air, pressure and flowrate) under operating conditions; mechanical efficiency of the prime mover; hose and fitting size(s); and pump speed range (r/min). Also report the physical and operating specifications of the pump, including the following: size, weight, required fuel, ancillary equipment, submersible and non-submersible service, fittings, etc.

[1] This guide is under the jurisdiction of ASTM Committee F-20 on Hazardous Substances and Oil Spill Response and is the direct responsibility of Subcommittee F20.12 on Removal.

Current edition approved Aug. 15, 1995. Published October 1995.

[2] *Annual Book of ASTM Standards*, Vol 06.01.

[3] *Annual Book of ASTM Standards*, Vol 11.04.

TABLE 1 Recommended Viscosity Ranges for Test Fluids[A]

Category	Viscosity Range, cSt
1	1 to 1000
2	1001 to 10 000
3	10 001 to 100 000
4	100 001 to 1 000 000

[A] The values refer to viscosity measured at 10 s^{-1}.

6.3 *Capacity Versus Head*—Report the volumetric capacity (m^3/h) as a function of test fluid viscosity (measured at the pump inlet) and measured static pressure (measured at the pump discharge). Report a minimum of four data points for each viscosity tested, with the pump operating for a minimum of 30 min for each data point, and report the viscosity of the test fluid. Compare the measured capacity with the calculated displacement (m^3/r or m^3/cycle) for positive displacement pumps.

6.4 *NPSH Versus Viscosity*—Report the net positive suction head required (m) at the rated differential pressure for the specified test fluid viscosity. Report the equipment specifications in accordance with 6.2.

6.5 *Self-Priming Versus Viscosity*—For pumps capable of self-priming, report the maximum suction lift (m) for self-priming; also report the time required to establish prime starting with the pump and suction hose initially empty.

NOTE 1—The pump internals may be wetted prior to this test; any initial wetting of internals should be reported.

6.5.1 A suction hose length of 8 m (25 ft) is recommended for the sake of comparability. Report the equipment specifications in accordance with 6.2.

6.6 *Discharge Pressure*—Report the maximum discharge pressure (kPa), measured at the pump discharge, and the mechanism by which it is limited (for example, relief valve or power limiter) for use in determining appropriate hose ratings (conversion: 1 bar = 14.5 psi = 100 kPa).

6.7 *Ability to Run Dry*—Describe how the pump is degraded by running dry.

6.8 *Starting Pressure*—Report the static head (m) that can be overcome by the pump when restarting with hoses full of water.

6.9 *Operational Features*—Describe any pump features pertaining to ease of handling and operation, for example:

6.9.1 Presence and accessibility of drain plugs to facilitate draining and cleaning;

6.9.2 Portability, the presence of lifting points;

6.9.3 Ability to insert into standard tanker/ship holds/tanks;

6.9.4 Operational run time of the pump per refueling;

6.9.5 Stability of the pump in operation regarding torque reaction (twisting of submersible pumps);

6.9.6 Bearings: for example, sealed, pumped fluid lubricated, hydraulic fluid lubricated, or grease lubricated;

6.9.7 Maximum routine maintenance interval; and

6.9.8 Field repair capability.

6.10 *Safety*—Report any pump features pertaining to safety concerns, such as the following:

6.10.1 Capability to ground the pump to prevent the accumulation of static electricity;

6.10.2 Inclusion of pressure relief valves or other pressure limiting devices (positive displacement pumps);

6.10.3 Protection of operators from moving parts; and

6.10.4 Conformance to general industrial safety standards.

6.11 *Debris Tolerance*—Report on the pump's tolerance to debris and resistance to clogging. Identify the maximum size and aspect ratio and concentration of debris that can be passed. Identify any problems with abrasion when pumping abrasive materials; identify any problems related to the consistency of debris material (that is, sand versus seaweed, rope, or plastics-flotsam) (see Guide F 631 for recommended materials for debris tolerance testing); and identify whether the pump has a backflushing or reversing capability or debris processing capability, or both (that is, chopper knives).

6.12 *Compatibility*—Report on the pump's compatibility with seawater and a range of oil types; report the test standard used:

6.12.1 Seals, packing, gaskets, and other wetted components suitable for use with pumped fluids;

6.12.2 Chemical compatibility list; and

6.12.3 Coatings for wetted parts to increase durability.

6.13 *Reliability*—Report the hours of continuous service of the pumping system and observations of continuous pumping of water, with interruption of the test only for routine preventive maintenance (that is, change of fuel, air, oil, or hydraulic filters; change of oil; adjustments to belt tension; and other routine preventive maintenance specified by the manufacturer).

6.14 *Maintenance and Preservation*—Report the pump's recommended schedule of maintenance, list the recommended spare parts, and report the recommended procedure for preservation after use.

7. Report

7.1 Report the following information:

7.1.1 Place and date of the tests,

7.1.2 Manufacturer's name, type of pump, serial number, and year of construction,

7.1.3 Specification of the pump's drive,

7.1.4 Description of the test procedure used and the sensors and measuring apparatus used, including calibration data,

7.1.5 Type and viscosity of the fluids used in the performance tests, and

7.1.6 Observations from the performance tests.

8. Keywords

8.1 oil recovery; oil recovery pumps; oil spill; oil spill response pumps; pumps; pump systems; skimmers; skimmer pumps; viscosities

Designation: F 1644 – 95

Standard Guide for Health and Safety Training of Oil Spill Responders[1]

This standard is issued under the fixed designation F 1644; the number immediately following the designation indicates the year of original adoption or, in the case of revision, the year of last revision. A number in parentheses indicates the year of last reapproval. A superscript epsilon (ε) indicates an editorial change since the last revision or reapproval.

1. Scope

1.1 This guide establishes minimum health and safety training standards for three types of oil spill responders: Type *A*, first responders who are responsible for initial containment and cleanup; Type *B*, longer-term shoreline cleanup personnel; and Type *C*, other necessary support personnel who have minimal contact with the contamination.

1.2 The oil covered by this guide includes light, medium, and heavy crudes, as well as hydrocarbon products, such as gasoline, light fuel oil, distillates, and bunker (heavy fuel) oil. It is not aimed at specialty chemicals and other potentially hazardous materials, although some aspects of the training program would apply to those substances.

1.3 A number of topics are not specifically addressed in this guide because they are covered by other standards or guidelines. Examples are hot work practices, maintenance and repair of equipment, fire fighting, electrical hazards, emergency medical care, disposal of wastes, and so forth. The user is expected to become familiar with standards for these areas as required.

1.4 *This standard does not purport to address all of the safety concerns, if any, associated with its use. It is the responsibility of the user of this standard to establish appropriate safety and health practices and determine the applicability of regulatory limitations prior to use.*

2. Terminology

2.1 *Definitions of Terms Specific to This Standard:*

2.1.1 *area contingency plan*—provides the initial governmental organization structure and mode of operation for the spill response.

2.1.2 *cold (support) zone*—minimal exposure area maintained as an uncontaminated location for support functions. Food service, clean equipment storage, and financial offices are examples of a cold zone. Operations in this area are generally carried out by Type *C* personnel.

2.1.3 *confined space*—an enclosed space or area, such as a tank, compartment, or pit where ventilation or access, or both, may be limited.

2.1.4 *hot (early response) zone*—an area where there are potential exposure hazards. Type *A* workers are involved in containment and recovery operations in this zone. Airborne concentrations of hazardous substances may require respiratory protection in addition to other personal protective equipment.

2.1.5 *hyperthermia*—an abnormally high body temperature caused by exposure to elevated temperatures or radiant heat, or both.

2.1.6 *hypothermia*—an abnormally low body temperature caused by exposure to cold air or water.

2.1.7 *personal protective equipment (PPE)*—equipment used to shield or insulate a person from a chemical, physical, or thermal hazard. Personal protective equipment is available for skin, eyes, face, hands, feet, head, ears, and respiratory system, as appropriate.

2.1.8 *site safety and health plan*—the framework that defines safety and health considerations and strategy for a specific site.

2.1.9 *site safety and health supervisor*—an individual in the field responsible for ensuring that the site safety and health plan is implemented as prescribed.

2.1.10 *warm (contamination reduction) zone*—an area where oil is present but in a generally weathered state. In addition to decontamination and equipment decommissioning areas, contaminated shorelines containing weathered oil with vapor levels below that requiring respiratory protective equipment are considered warm zones for purposes of this guide. Type *A* or *B* workers, or both, would operate in these zones.

3. Significance and Use

3.1 This guide summarizes required course components to be included in training programs for oil spill response personnel. Its purpose is to assist trainers in developing curricula that address the health and safety risks specific to oil spill response. This guide may be tailored by the trainer to fit unique circumstances that are present if training is conducted during an actual spill emergency and is not intended to preclude consideration of regulatory requirements.

4. Types of Response Workers

4.1 The level of training required will depend on the level of involvement for each type of worker. For purposes of this guide, the three categories or types of workers are defined in Table 1.

4.1.1 *Type A Workers*, operate in the hot zone and are the most likely to encounter the spilled oil in a fresh state. Training for Type *A* workers should emphasize methods for determining

[1] This guide is under the jurisdiction of ASTM Committee F-20 on Hazardous Substances and Oil Spill Response and is the direct responsibility of Subcommittee F20.21 on Initial Response Actions.

Current edition approved Sept. 10, 1995. Published November 1995.

TABLE 1 Oil Spill Response Worker Categories

Type A—Hot-Zone Personnel
Response managers
Supervisors
Equipment operators
Boom deployers
Boat operators
Skimmer crews
Dispersant handlers
Barge personnel
Tank truck operators
Bulldozer operators
Other initial response personnel
Security forces
Shoreline assessment team
Site safety and health supervisor
Field medical personnel
Vapor monitoring personnel
Etc.
Type B—Warm-Zone Personnel
Shoreline clean-up personnel
Decontamination personnel
Wildlife coordinators
Animal handlers
Waste management/handling personnel
Etc.
Type C—Cold-Zone Personnel
General land-based support
Historians
Legal advisors
Food-service personnel
Financial services personnel
Supply personnel
Etc.

TABLE 2 Health and Safety Training Matrix

Subjects	Worker Types		
	Type A	Type B	Type C
General awareness	•	•	•
Exposures and chemical hazards			
Mechanical hazards	•	•	
Slips, trips, and falls	•	•	
Explosion and fire	•	•	
Biological hazards	•	•	
Physical hazards	•	•	
Water hazards	•	•	
Exposure routes (air, skin, ingestion)	•	•	
Confined space entry	•		
First aid	•	•	•
Safety and health requirements			
Industrial hygiene monitoring	•		
Respiratory protection	•		
Eye and ear protection	•	•	
Flotation devices	•	•	
Footwear	•	•	
Skin protection (gloves and protective clothing)	•	•	
Hard hats	•	•	
Personal hygiene	•	•	
Decontamination procedures	•	•	

the presence of harmful vapors, and proper use of respiratory protective equipment.

4.1.2 *Type B Workers*, operate in the warm zone and include decontamination support personnel and shoreline cleaning crews. The oil they encounter will be in a weathered state. Training for Type *B* workers should emphasize mechanical hazards, dermal exposure, slips, trips, falls, and so forth.

4.1.3 *Type C Workers*, operate in the cold zone and include support personnel with no probability of contacting the spilled oil (cafeteria staff, financial advisors, secretaries, and so forth). Type *C* workers do not require specialized safety and health training unless they expect to visit the spill site or take part in Type *A* or *B* activities.

5. Training Matrix

5.1 The recommended training curriculum subjects are presented in matrix form by worker types in Table 2. Though there is some flexibility in depth of coverage for each subject, the curriculum should be sufficient to familiarize each participant with all the possible hazards that may be encountered by that type of worker in his or her role.

5.2 *General Awareness*:

5.2.1 *Training Objectives*—Summarize the overall intent and purpose of the training program.

5.2.2 *Applicable Regulations*—Review the guidelines and regulations applicable to the specific region of operation.

5.2.3 *Response Command Structure*—Review the makeup of area contingency plans, response management structure, responsibilities of the site safety and health supervisor and other relevant information on how regional responses are managed.

5.2.4 *Local or Organizational Issues, or Both*—Include alcohol and drug policies, firearm policy, local wildlife that may be at risk, and local natural hazards such as earthquakes, avalanche and rock slide potential, hurricanes, and so forth.

5.2.5 *Historical and Cultural Features*—Raise awareness so that experts can be called in if there is suspicion that the work area contains important archeological sites or sacred grounds.

5.3 *Exposures and Chemical Hazards*:

5.3.1 *Mechanical Hazards*—Include a summary of potential hazards from rotating equipment, electrical generators, boom deployment, boat operations, and so forth.

5.3.2 *Slips, Trips, and Falls*—Highlight that these are the most common sources of injury at a spill clean-up site. Review precautions for handling hoses, lines, and wires.

5.3.3 *Explosions and Fires*—Include a review of the fire triangle and presentation of lower and higher explosion limits for common hydrocarbon products.

5.3.4 *Biological Hazards*—Include discussion of poisonous plants, insects, spiders, snakes, rabid animals, and so forth. Also include marine hazards and wildlife rescue and rehabilitation.

5.3.4.1 *Marine Hazards*—Include jellyfish, Portuguese man-o-war, anemones, corals, urchins, and so forth.

5.3.4.2 *Emergency Care*—Include procedures to follow in the event of a sting or bite.

5.3.5 *Physical Hazards*:

5.3.5.1 *Noise*—Define the decibel range that characterizes high noise levels and the potential long-term impacts from prolonged exposure.

5.3.5.2 *Hypothermia*—Review causes and early warning signs of hypothermia as well as recommended treatment.

5.3.5.3 *Heat Stress*—Review safety problems common to hot environments, including a summary of heat-induced disorders (hyperthermia, stroke, exhaustion, cramps, fainting, and so forth) and the required medical attention for each. Training should emphasize the magnified potential for heat stress with use of protective clothing.

5.3.5.4 *Ultraviolet (UV) Hazards*—Review the hazards associated with UV exposure and the protection provided by different types of sunscreen and sunglasses.

5.3.6 *Water Hazards*—Include discussion of safe boating practices, personal flotation devices, other protective gear, unsafe weather conditions, and so forth.

5.3.7 *Exposure Routes*:

5.3.7.1 *Airborne Contaminants*—Review the key airborne hazards of concern with emphasis on benzene and hydrogen sulfide.

5.3.7.2 *Skin Exposure*—Include a discussion of dermatitis associated with skin contact with hydrocarbons.

5.3.7.3 *Ingestion*—Review the routes by which hydrocarbons can enter the body, including food and water contamination, hand to mouth contact, and so forth.

5.3.8 *Confined Space Entry*—Review guidelines for entry, including required oxygen level, testing for presence of combustible vapors, and other hazards.

5.3.9 *First Aid*—Review first-aid basics and general orientation.

5.4 *Safety and Health Requirements*—Methods to protect workers from chemical, physical, and thermal hazards will vary with incident.

5.4.1 *Industrial Hygiene (IH) Monitoring*—Summarize the available techniques for monitoring ambient air concentrations.

5.4.1.1 *Data Sheets*—Review documented information on oils of concern.

5.4.1.2 *Medical Monitoring*—Required for all workers exposed to levels above the permissible exposure limits.

5.4.2 *Respiratory Protection*—Infrequently used in oil spill cleanup. Required training covers the selection, inspection, use, care, and maintenance of the respirator.

5.4.3 *Eye and Ear Protection*—Conditions requiring splash goggles or face shields as opposed to safety glasses should be reviewed. Describe available ear protection equipment and conditions under which it would be required (see 5.3.5.1).

5.4.4 *Flotation Devices*—Required for workers who will respond on water.

5.4.5 *Footwear*—Review types of boots available for shoreline work.

5.4.6 *Skin Protection*:

5.4.6.1 *Gloves*—Oil-resistant gloves should be identified pointing out unacceptable options.

5.4.6.2 *Personal Protective Clothing*—The required type of clothing and cleaning requirements should be discussed as well as the suitability of disposable garments.

5.4.7 *Hard Hats*—Review available types and minimum acceptable standards.

5.4.8 *Personal Hygiene*—Review the importance of cleanliness and sanitation for maintaining worker health during prolonged clean-up assignments. Also cover such personal needs as potable water, toilet facilities, food handling, and so forth.

5.4.9 *Decontamination Procedures*—Review the setup of a decon site (warm zone) and the procedures for ensuring that contaminated (oiled) materials do not inadvertently leave the work site.

6. Skilled Personnel Exemption

6.1 In some cases, a comprehensive and well-structured site safety briefing, instead of a formal training program, may suffice. This is especially important in the case of such skills as boat operators, whose formal training could delay a prompt response effort.

7. Assessment

7.1 The effectiveness of the training should be assessed by methods such as written or oral examination, field observation, interactive observations, and so forth.

7.2 Training should be reviewed and repeated as necessary to maintain worker qualification.

8. Keywords

8.1 health and safety training; oil spill responders; oil spills

Designation: F 1656 – 95

Standard Guide for Health and Safety Training of Oil Spill Responders in the United States[1]

This standard is issued under the fixed designation F 1656; the number immediately following the designation indicates the year of original adoption or, in the case of revision, the year of last revision. A number in parentheses indicates the year of last reapproval. A superscript epsilon (ϵ) indicates an editorial change since the last revision or reapproval.

1. Scope

1.1 This guide establishes minimum health and safety training standards for three types of oil spill responders: Type *A*, first responders who are responsible for initial containment and cleanup; Type *B*, longer-term shoreline cleanup personnel; and Type *C*, other necessary support personnel who have minimal contact with the contamination.

1.2 The oil covered by this guide includes light, medium, and heavy crudes, as well as hydrocarbon products, such as gasoline, light fuel oil, distillates, and bunker (heavy fuel) oil. It is not aimed at specialty chemicals and other potentially hazardous materials, although some aspects of the training program would apply to those substances.

1.3 A number of topics are not specifically addressed in this guide because they are covered by other standards or guidelines. Examples are hot work practices, maintenance and repair of equipment, fire fighting, electrical hazards, emergency medical care, disposal of wastes, and so forth. The user is expected to become familiar with standards for these areas as required.

1.4 *This standard does not purport to address all of the safety concerns, if any, associated with its use. It is the responsibility of the user of this standard to establish appropriate safety and health practices and determine the applicability of regulatory limitations prior to use.*

2. Referenced Documents

2.1 *U.S. Federal Regulations:*[2]

Title 29, Code of Federal Regulations (CFR), 1910.120 Regulations for Hazardous Waste Operations and Emergency Response (HAZWOPER)

Title 29, Code of Federal Regulations (CFR), 1910.132, General Requirements for Hazard Assessment and Training

Title 29, Code of Federal Regulations (CFR), 1910.133, Eye and Face Protection

Title 29, Code of Federal Regulations (CFR), 1910.134, Respiratory Protection

Title 29, Code of Federal Regulations (CFR), 1910.135, Head Protection

Title 29, Code of Federal Regulations (CFR), 1910.136, Foot Protection

Title 29, Code of Federal Regulations (CFR), 1910.138, Hand Protection

Title 29, Code of Federal Regulations (CFR), 1910.146, Enclosed Spaces

Title 33, Code of Federal Regulations (CFR), 154, Subpart F, Facility Response Plan Requirements

Title 33, Code of Federal Regulations (CFR), 155, Subpart F, Vessel Response Plan Requirements

Title 40, Code of Federal Regulations (CFR), 300, National Oil and Hazardous Substances Pollution Contingency Plan (NCP)

3. Terminology

3.1 *Definitions of Terms Specific to This Standard:*

3.1.1 *area contingency plan*—provides the initial governmental organization structure and mode of operation for the spill response. In the United States, it is prepared in advance by the area committee and applicable state and local response officials in accordance with 40 CFR 300, NCP.

3.1.2 *buddy system*—a requirement of 29 CFR 1910.120 that response personnel must work within sight/communication of a partner at all times.

3.1.3 *cold (support) zone*—minimal exposure area maintained as an uncontaminated location for support functions. Food service, clean equipment storage, and financial offices are examples of a cold zone. Operations in this area are generally carried out by Type *C* personnel.

3.1.4 *confined space*—an enclosed space or area, such as a tank, compartment, or pit where ventilation or access, or both, may be limited (refer to 29 CFR 1910.146).

3.1.5 *hot (early response) zone*—an area where there are potential exposure hazards. Type *A* workers are involved in containment and recovery operations in this zone. Airborne concentrations of hazardous substances may require respiratory protection in addition to other personal protective equipment.

3.1.6 *hyperthermia*—an abnormally high body temperature caused by exposure to elevated temperatures or radiant heat, or both.

3.1.7 *hypothermia*—an abnormally low body temperature caused by exposure to cold air or water.

[1] This guide is under the jurisdiction of ASTM Committee F-20 on Hazardous Substances and Oil Spill Responseand is the direct responsibility of Subcommittee F20.21 on Initial Response Actions.

Current edition approved Oct. 10, 1995. Published December 1995.

[2] Available from the Superintendent of Documents, U.S. Government Printing Office, Washington, DC 20402.

3.1.8 *on-scene coordinator (OSC)*—In the United States, the pre-designated official responsible for ensuring immediate and effective response to a discharge or threatened discharge of oil or hazardous substance (40 CFR 300). The Coast Guard designates the OSC's for coastal zones and the Environmental Protection Agency (EPA) designates OSC's for inland zones.

3.1.9 *personal protective equipment (PPE)*—equipment used to shield or insulate a person from a chemical, physical, or thermal hazard. Personal protective equipment is available for skin, eyes, face, hands, feet, head, ears, and respiratory system, as appropriate.

3.1.10 *site safety and health officer*—One who is responsible for developing training curricula and site safety and health plans. Often referred to as site safety officer.

3.1.11 *site safety and health plan*—the framework that defines safety and health considerations and strategy for a specific site.

3.1.12 *site safety and health supervisor*—an individual in the field responsible for ensuring that the site safety and health plan is implemented as prescribed. The site safety officer can fill both positions.

3.1.13 *warm (contamination reduction) zone*—an area where oil is present but in a generally weathered state. In addition to decontamination and equipment decommissioning areas, contaminated shorelines containing weathered oil with vapor levels below that requiring respiratory protective equipment are considered warm zones for purposes of this guide. Type *A* or *B* workers, or both, would operate in these zones.

4. Significance and Use

4.1 This guide summarizes required course components to be included in training programs for oil spill response personnel. Its purpose is to assist trainers in developing curricula that address the health and safety risks specific to oil spill response. This guide may be tailored by the trainer to fit unique circumstances that are present if training is conducted during an actual spill emergency and is not intended to preclude consideration of regulatory requirements.

5. Types of Response Workers

5.1 The level of training required will depend on the level of involvement for each type of worker. For purposes of this guide, the three categories or types of workers are defined in Table 1.

5.1.1 *Type A Workers*, operate in the hot zone and are the most likely to encounter the spilled oil in a fresh state. Training for Type *A* workers should emphasize methods for determining the presence of harmful vapors and proper use of respiratory protective equipment.

5.1.2 *Type B Workers*, operate in the warm zone and include decontamination support personnel and shoreline cleaning crews. The oil they encounter will be in a weathered state. Training for Type *B* workers should emphasize mechanical hazards, dermal exposure, slips, trips, falls, and so forth.

5.1.3 *Type C Workers*, operate in the cold zone and include support personnel with no probability of contacting the spilled oil (cafeteria staff, financial advisors, secretaries, and so forth).

TABLE 1 Oil Spill Response Worker Categories

Type A—Hot-Zone Personnel
Response managers
Supervisors
Equipment operators
Boom deployers
Boat operators
Skimmer crews
Dispersant handlers
Barge personnel
Tank truck operators
Bulldozer operators
Other initial response personnel
Security forces
Shoreline assessment team
Site safety and health officers
Field medical personnel
Vapor monitoring personnel
Etc.
Type B—Warm-Zone Personnel
Shoreline clean-up personnel
Decontamination personnel
Wildlife coordinators
Animal handlers
Waste management/handling personnel
Etc.
Type C—Cold-Zone Personnel
General land-based support
Historians
Legal advisors
Food service personnel
Financial services personnel
Supply personnel
Etc.

Type *C* workers do not require specialized safety and health training unless they expect to visit the spill site or take part in Type *A* or *B* activities.

6. Training Matrix

6.1 The recommended training curriculum subjects are presented in matrix form by worker types in Table 2. Though there

TABLE 2 Health and Safety Training Matrix

Subjects	Worker Types		
	Type A	Type B	Type C
General awareness	•	•	•
Exposures and chemical hazards			
Mechanical hazards	•	•	
Slips, trips, and falls	•	•	
Explosion and fire	•	•	
Biological hazards	•	•	
Physical hazards	•	•	
Water hazards	•	•	
Exposure routes (air, skin, ingestion)	•	•	
Confined space entry	•		
First aid	•	•	•
Safety and health requirements			
Industrial hygiene monitoring	•		
Respiratory protection	•		
Eye and ear protection	•	•	
Flotation devices	•	•	
Footwear	•	•	
Skin protection (gloves and protective clothing)	•	•	
Hard hats	•	•	
Personal hygiene	•	•	
Decontamination procedures	•	•	

is some flexibility in depth of coverage for each subject, the curriculum should be sufficient to familiarize each participant with all the possible hazards that may be encountered by that type of worker in his or her role.

6.2 *General Awareness*:

6.2.1 *Training Objectives*—Summarize the overall intent and purpose of the training program.

6.2.2 *Applicable Regulations*—Review the guidelines and regulations cited in Section 2 of this guide.

6.2.3 *Response Command Structure*—Review the makeup of area contingency plans, response management structure, responsibilities of the site safety and health officer and supervisor, and other relevant information on how regional responses are managed.

6.2.4 *Local or Organizational Issues, or Both*—Include alcohol and drug policies, firearm policy, local wildlife that may be at risk, and local natural hazards such as earthquakes, avalanche and rock slide potential, hurricanes, and so forth.

6.2.5 *Historical and Cultural Features*—Raise awareness so that experts can be called in if there is suspicion that the work area contains important archeological sites or sacred grounds.

6.3 *Exposures and Chemical Hazards*:

6.3.1 *Mechanical Hazards*—Include a summary of potential hazards from rotating equipment, electrical generators, boom deployment, boat operations, and so forth.

6.3.2 *Slips, Trips, and Falls*—Highlight that these are the most common sources of injury at a spill clean-up site. Review precautions for handling hoses, lines, and wires.

6.3.3 *Explosions and Fires*—Include a review of the fire triangle and presentation of lower and higher explosion limits for common hydrocarbon products.

6.3.4 *Biological Hazards*—Include discussion of poisonous plants, insects, spiders, snakes, rabid animals, and so forth. Also include marine hazards and wildlife rescue and rehabilitation.

6.3.4.1 *Marine Hazards*—Include jellyfish, Portuguese man-o-war, anemones, corals, urchins, and so forth.

6.3.4.2 *Emergency Care*—Include procedures to follow in the event of a sting or bite.

6.3.5 *Physical Hazards*:

6.3.5.1 *Noise*—Define the decibel range that characterizes high noise levels and the potential long-term impacts from prolonged exposure.

6.3.5.2 *Hypothermia*—Review causes and early warning signs of hypothermia as well as recommended treatment.

6.3.5.3 *Heat Stress*—Review safety problems common to hot environments, including a summary of heat-induced disorders (hyperthermia, stroke, exhaustion, cramps, fainting, and so forth) and the required medical attention for each. Training should emphasize the magnified potential for heat stress with use of protective clothing.

6.3.5.4 *Ultraviolet (UV) Hazards*—Review the hazards associated with UV exposure and the protection provided by different types of sunscreen and sunglasses.

6.3.6 *Water Hazards*—Include discussion of safe boating practices, personal flotation devices, other protective gear, unsafe weather conditions, and so forth.

6.3.7 *Exposure Routes*:

6.3.7.1 *Airborne Contaminants*—Review the key airborne hazards of concern with emphasis on benzene and hydrogen sulfide.

6.3.7.2 *Skin Exposure*—Include a discussion of dermatitis associated with skin contact with hydrocarbons.

6.3.7.3 *Ingestion*—Review the routes by which hydrocarbons can enter the body, including food and water contamination, hand to mouth contact, and so forth.

6.3.8 *Confined Space Entry*—Review Occupational Safety and Health Administration (OSHA), National Institute of Occupational Safety and Health (NIOSH), and National Fire Protection Association (NFPA) rules and regulations, guidelines for entry, including required oxygen level, testing for presence of combustible vapors, and other hazards.

6.3.9 *First Aid*—Review first-aid basics and general orientation.

6.4 *Safety and Health Requirements*—Methods to protect workers from chemical, physical, and thermal hazards will vary with incident.

6.4.1 *Industrial Hygiene (IH) Monitoring*—Summarize available techniques for monitoring ambient air concentrations.

6.4.1.1 *Material Safety Data Sheets*—Review MSDS's and familiarize responders with key information.

6.4.1.2 *Medical Monitoring*—Required for all workers exposed to levels above the permissible exposure limits.

6.4.2 *Respiratory Protection*—Infrequently used in oil spill cleanup. Required training covers the selection, inspection, use, care, and maintenance of the respirator. (Refer to 29 CFR 1910.134.)

6.4.3 *Eye and Ear Protection*—Conditions requiring splash goggles or face shields as opposed to safety glasses should be reviewed. Describe available ear protection equipment and conditions under which it would be required (see section 6.2.5.1). (Refer to 29 CFR 1910.133.)

6.4.4 *Flotation Devices*—Required for workers who will respond on water.

6.4.5 *Footwear*—Review types of boots available for shoreline work. (Refer to 29 CFR 1910.136.)

6.4.6 *Skin Protection*:

6.4.6.1 *Gloves*—Oil-resistant gloves should be identified pointing out unacceptable options. (Refer to 29 CRF 1910.138.)

6.4.6.2 *Personal Protective Clothing*—The required type of clothing and cleaning requirements should be discussed as well as the suitability of disposable garments. (Refer to 29 CFR 1910.132.)

6.4.7 *Hard Hats*—Review available types and minimum acceptable standards. (Refer to 29 CFR 1910.135.)

6.4.8 *Personal Hygiene*—Review the importance of cleanliness and sanitation for maintaining worker health during prolonged clean-up assignments. Also cover such personal needs as potable water, toilet facilities, food handling, and so forth.

6.4.9 *Decontamination Procedures*—Review the setup of a decon site (warm zone) and the procedures for ensuring that contaminated (oiled) materials do not inadvertently leave the work site.

7. Skilled Personnel Exemption

7.1 In some cases, a comprehensive and well-structured site safety briefing, instead of a formal training program, may suffice. This is especially important in the case of such skills as contractor boat operators whose formal training could delay a prompt response effort. This type of briefing is in accordance with OSHA's important "Skilled Support Personnel" exemption, under 29 CFR 1910.120.

8. Assessment

8.1 The effectiveness of the training should be assessed by methods such as written or oral examination, field observation, interactive observations, and so forth.

8.2 Training should be reviewed and repeated as necessary to maintain worker qualification.

9. Keywords

9.1 health and safety training; oil spill responders; oil spills

Designation: F 1657 – 96

Standard Practice for Emergency Joining of Booms with Incompatible Connectors[1]

This standard is issued under the fixed designation F 1657; the number immediately following the designation indicates the year of original adoption or, in the case of revision, the year of last revision. A number in parentheses indicates the year of last reapproval. A superscript epsilon (ϵ) indicates an editorial change since the last revision or reapproval.

1. Scope

1.1 This practice provides a standard practice for the joining of oil spill containment boom connectors in emergencies.

1.2 The use of this connection method may adversely affect the total tensile strength of the connected booms.

1.3 These criteria are intended to define mating requirements that will allow the emergency or occasional connection of unlike connectors.

1.4 This practice is not intended to replace Specification F 962.

1.5 This practice does not address the compatibility of spill control equipment with spill products. It is the user's responsibility to ensure that any equipment selected is compatible with the anticipated spilled material.

1.6 There is no guarantee that all of the connectors in use today can accept the holes spaced as required without interfering with existing bolt holes or other connector features.

1.7 The values stated in SI units are to be regarded as the standard. The values given in parentheses are for information only.

1.8 *This standard does not purport to address all of the safety concerns, if any, associated with its use. It is the responsibility of the user of this standard to establish appropriate safety and health practices and determine the applicability of regulatory limitations prior to use.* For a specific precautionary statement, see Note 1.

2. Referenced Documents

2.1 *ASTM Standards:*

F 818 Terminology Related to Spill Response Barriers[2]

F 962 Specification for Oil Spill Response Boom Connection[2]

F 1093 Test Methods for Tensile Strength Characteristics of Oil Spill Response Boom[2]

F 1523 Guide for Selection of Booms in Accordance With Water Body Classifications[2]

3. Significance and Use

3.1 The use of this practice for the emergency joining of booms will not guarantee the effective performance of the joined boom sections, since each boom design and the environmental conditions of each incident govern the overall performance.

3.2 Historically, different types of end connectors have been produced. This practice addresses the operational need to connect different types, during spill incidents.

NOTE 1—**Caution:** Use of this practice with similar or different sizes of boom may cause the transmission of unwanted loading such as, tension loading and bending moments on certain boom parts resulting in possible premature failure of the containment system.

3.3 There are a wide range of boom connector configurations presently in use. These connectors were based upon some or all of the following design criteria:

3.3.1 Connect and transfer tensile loads between boom sections,

3.3.2 Minimize oil leakage between boom sections,

3.3.3 Be easily connectable in the presence of dirt, oil or ice, or a combination thereof,

3.3.4 Be quickly and easily connected and disconnected, in and out of the water,

3.3.5 Maintain boom performance (freeboard, heave response, conformance, stability, and so forth),

3.3.6 Be unaffected by temperature extremes,

3.3.7 Have no protruding parts that could snag, injure, or puncture,

3.3.8 Be light weight and buoyant,

3.3.9 Be operatively symmetrical,

3.3.10 Require no special tools for installation or removal,

3.3.11 Require no loose parts for connection,

3.3.12 Extend to the full height and draft of the boom,

3.3.13 Resist distortion (that is, winding boom on a reel), and

3.3.14 Be inherently safe to personnel.

4. Design Criteria

4.1 In order to comply with this practice for the emergency joining of boom with incompatible connectors, each boom connector, regardless of design, will incorporate the requirements described herein or use the adaption method in 5.3:

4.1.1 A minimum of two holes sized to accommodate 13 mm (½ in.) diameter 304 stainless steel bolts will be provided as shown in Fig. 1. Using the boom design water line as a reference point, the center of the first hole is to be 12 cm above

---

[1] This practice is under the jurisdiction of ASTM Committee F-20 on Hazardous Substances and Oil Spill Response and is the direct responsibility of Subcommittee F20.11 Control.

Current edition approved Feb. 10, 1996. Published April 1996.

[2] *Annual Book of ASTM Standards*, Vol 11.04.

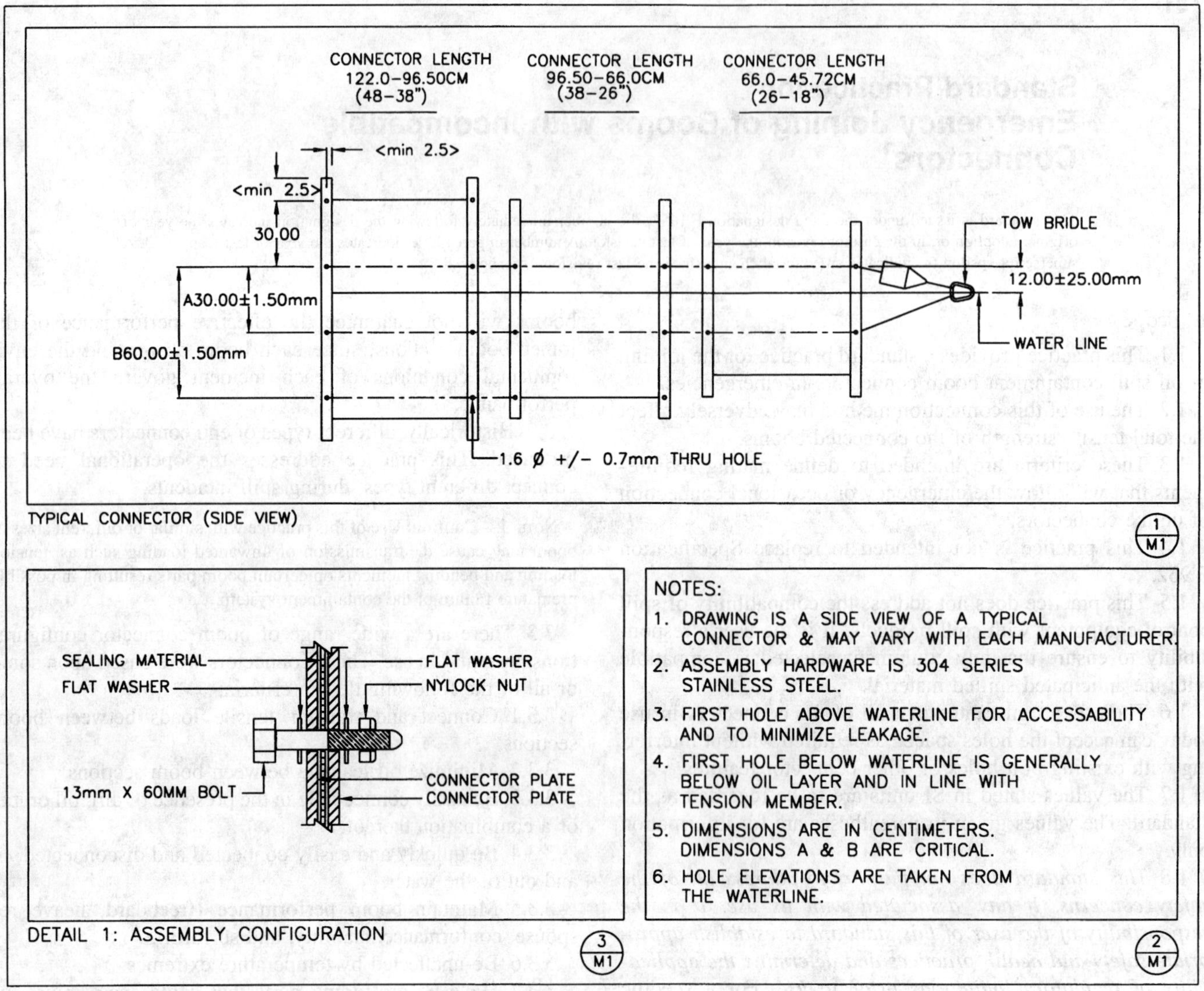

FIG. 1 Side View of a Typical Connector

the water line, the second hole will be 30 cm below the first hole depending on the connector overall length. For larger connectors additional holes will be placed at 30 cm increments from the first two holes. Each hole is to be sealed to prevent leakage when not in use.

4.1.2 Use of this practice shall not significantly impact the original design criteria in 3.4.

4.1.3 Incorporation of this practice shall not reduce the tensile strength of the boom system below the values stated for that class of boom, (see Guide F 1523) when mated in its normal configuration.

4.1.4 The manufacturer should establish the load carrying capability of two of his connectors joined by this practice by testing in accordance with Test Methods F 1093.

5. Operational Consideration

5.1 A suitable non-metallic gasket sealing material may be inserted by the user between the surfaces to be joined. The gasket material must be impervious to the products contained by the boom.

5.2 The user must consider the transmission of tensile loading between the joined sections. This will include tensile loads on boom fabric, primary and secondary tension members, and the orientation of the tension members (loads may be offset at the joint depending on boom design).

5.3 When the original connector design prevents the use of this practice, a suitable number of adapters should be provided to facilitate connection to this emergency joining practice.

5.4 This practice will also allow for multiple boom connections at one point.

6. Keywords

6.1 boom; boom connector; connector; incompatible boom connector; oil spill response

Designation: F 1686 – 97

Standard Guide for Surveys to Document and Assess Oiling Conditions on Shorelines[1]

This standard is issued under the fixed designation F 1686; the number immediately following the designation indicates the year of original adoption or, in the case of revision, the year of last revision. A number in parentheses indicates the year of last reapproval. A superscript epsilon (ε) indicates an editorial change since the last revision or reapproval.

1. Scope

1.1 This guide covers field procedures by which data may be collected in a systematic manner to document and assess the oiling conditions on shorelines.

1.2 This guide does not address the terminology that is used to define and describe shoreline oiling conditions, the ecological character of oiled shorelines, or the cultural or other resources that may be present.

1.3 The guide is applicable to marine coasts (including estuaries) and may also be used in freshwater environments (rivers and lakes).

1.4 The values stated in SI units are to be regarded as the standard. The values given in parentheses are for information only.

1.5 *This standard does not purport to address all of the safety concerns, if any, associated with its use. It is the responsibility of the user of this standard to establish appropriate safety and health practices and determine the applicability of regulatory limitations prior to use.*

2. Referenced Documents

2.1 *ASTM Standards:*

F 1687 Guide for Terminology and Indices for Describing the Oiling Conditions of Shorelines[2]

F 1779 Practice for Reporting Visual Observations of Oil on Water[2]

3. Significance and Use

3.1 Systematic surveys provide data on shoreline conditions from which informed planning and operational decisions may be developed with respect to shoreline cleanup.

3.2 Surveys may include one or more of three components, listed below. The scale of the affected area and the availability of pre-spill information will influence the selection of survey components and the level of detail (**1**).[3]

3.3 The aerial reconnaissance survey provides a perspective on the overall extent and general nature of the shoreline oiling conditions. This information is used in conjunction with environmental, resource, and cultural sensitivity data to evaluate which areas require a response and the priorities of the response operations.

3.4 The aerial videotape survey(s) provide systematic audio and video documentation of the extent and type of shoreline oiling conditions, physical shoreline character, and potential access restrictions (**2**).

3.5 The ground assessment survey(s) provide the necessary information and data to develop appropriate shoreline response recommendations. A field team(s) collects detailed information on shoreline oil conditions, the physical and ecological character of oiled shorelines, and resources or cultural features that may affect or be affected by the timing or implementation of response activities (**2**).

3.6 In order to ensure data consistency it is important to use standardized terminology and definitions in describing oiling conditions, as provided in Guide F 1687.

4. General Considerations

4.1 The specific survey procedures and the magnitude of the data sets collected will vary with the scale of the spill (the length and distribution of oiled shoreline and quantity of oil), the nature or complexity of the shoreline, and the needs of the response organization (**3**).[4]

4.2 Following a spill in which only a few kilometers of coast have been oiled, one ground survey team may be able to accomplish all of the goals in an appropriate time frame.

4.3 As the scale of the affected area increases, it may be necessary to conduct an aerial videotape survey that is followed by a ground assessment using one or more teams.

4.4 Following spills that affect long sections of coast (for example, more than 100 km), the sequence of an initial aerial reconnaissance, an aerial videotape survey and ground assessment surveys may be necessary to satisfy planning and operational requirements in a timely manner (**2**).

4.5 Each of the three stages requires a separate survey design, the assignment of duties to personnel, logistics planning, and the establishment of survey and documentation procedures.

---

[1] This guide is under the jurisdiction of ASTM Committee F-20 on Hazardous Substances and Oil Spill Responseand is the direct responsibility of Subcommittee F20.17on Shoreline Countermasters.

Current edition approved Feb. 10, 1997. Published April 1997. Originally published as F 1686 – 96. Last previous edition F 1686 – 96.

[2] *Annual Book of ASTM Standards*, Vol 11.04.

[3] The boldface numbers in parentheses refer to the list of references at the end of this standard.

[4] Owens, E. H., and Sergy, G. A., *Field Guide to the Documentation and Description of Oiled Shorelines*, ISBN 0-662-22048-X, Environment Canada, Edmonton, Alta., 1994.

4.6 All surveys are conducted during the lowest one-quarter to one-third of the tidal cycle to ensure maximum (viewing) exposure of the intertidal zone (**1**).

5. Segmentation

5.1 The coast is divided into working units called segments, within which the shoreline character is relatively homogeneous in terms of physical features and sediment type.

5.2 Each segment is assigned a unique location identifier (for example, an alpha-numeric code).

5.3 Segment boundaries can be either prominent geological features (headlands, streams, etc.), changes in shore/substrate types, or, more importantly, alongshore changes in oil conditions.

5.4 Segment lengths are short enough to obtain adequate resolution and detail on the distribution of the oil for planning and operational decisions. Most segments of oiled shorelines would be in the range of 0.2 to 2.0 km.

5.5 If segments already exist as part of a pre-spill planning exercise or sensitivity mapping database, segment boundaries may need to be adapted, segments subdivided, or the segment codes revised, or some combination thereof, to reflect the oiling conditions created by a spill.

6. Aerial Reconnaissance Survey(s)

6.1 An initial aerial survey(s) is conducted along coastlines within the spill path. The objective is to determine which shorelines have been oiled, in order to provide an overall perspective and scale with which to plan for a more systematic documentation or assessment survey.

6.2 This survey can be augmented with information from a high-altitude surveillance and tracking program. In some cases this survey can be combined with other aerial reconnaissance activities being conducted to locate and visually observe oil on water (Practice F 1779) (**2**).

6.3 Fixed-wing or rotary-wing aircraft fly the coast at slow-speed flight at altitudes in the range of 75 to 150 m. Helicopters are preferred over fixed-wing aircraft, as they permit landings to confirm observations made from the air. Among fixed-wing aircraft, those with the wing mounted above the fuselage (high-wing aircraft) are essential to allow the required visibility of shoreline features.

6.4 If possible, the survey team will consist of an oil observer and navigator/recorder. The observer should be an oil-spill specialist familiar with oil on shorelines and able to distinguish between natural materials (stranded kelp, black lichen, heavy mineral bands, etc.) and oil. The navigator logs the flight lines, locates a section of shoreline on maps or charts, and records oil observations.

6.5 Records of observations may be made on maps and notebooks. Videotape and still photography can also be used to add a visual record of examples of the oiling conditions and shoreline character for immediate use by planners and decision makers.

6.6 An aerial reconnaissance is generally not required where the presence of oil on the shore has been defined clearly from other sources or where the affected coastline is short enough in length that an aerial videotape survey can be completed during one low-tide cycle.

7. Aerial Videotape and Mapping Survey(s)

7.1 The aerial videotape recording and mapping survey(s) are conducted on coasts where there is known or expected oiling. The survey is used to provide detailed and systematic documentation on the extent and type of shoreline oiling and other shoreline conditions (**2**).

7.2 Small high-wing or rotary-wing aircraft fly the coast at very slow speeds at altitudes in the range of 25 to 75 m.

7.3 The primary survey team consists of an oil observer and a navigator. The navigator records and maps relevant flight information. The oil observer operates the video camera and provides a continuous audio commentary, for which the color video image provides a visual image frame of reference. In some cases a video technician may be desirable for the operation and quality control of the audio and video recordings.

7.4 The oil observer/videographer is an oil-spill specialist, who can identify the shoreline substrate and form and distinguish between natural shoreline materials and stranded oil. Duties of the oil observer are as follows:

7.4.1 To identify or create segment boundaries and describe their location on one of the audio channels. These are also recorded on a set of flight-line maps or charts by the navigator.

7.4.2 To videotape the shore zone through an open door or window continuously, with the camera angled down (30 to 45°) and slightly ahead of the aircraft (15 to 30°) so that the area being described comes into focus and the foreground during the commentary. Video resolution is best when the sun is behind the aircraft.

7.4.3 To provide a continuous descriptive commentary on the shoreline oiling conditions, including the (*1*) length and width of the oiled areas and the oil distribution (percent surface oil cover), (*2*) physical shore-zone character, and (*3*) other pertinent features such as access locations or constraints (**4**).

7.5 Video recording equipment requires either a camera/recorder/color monitor system or a camrecorder system. Both must have on-screen date/time, audio recording, and lowlight recording capabilities. Other requirements include an independent, stabilized power supply or converter and a voice-activated two-channel audio recording and communications system between all crew members, including the pilot. Also desirable is the capability for location (geographical positioning system) overprint on the image, a stabilized camera mount, and small inboard color monitor(s).

8. Ground Assessment Survey(s)

8.1 Shoreline ground assessment surveys are conducted on oiled segments that may have been identified by the aerial reconnaissance or aerial videotape-mapping survey(s), to provide detailed, systematic observations that are used to guide the development of appropriate treatment or cleanup activities (**5**).

8.2 The scale of the ground survey is dependent on the size and character of the area affected and the intended use of the survey data.

8.3 Where more than one survey team is used, appropriate measures must be taken to ensure consistency in procedures and terminology.

8.4 Typically, a ground survey team must contain an oilspill shoreline specialist and may include a coastal ecologist, cultural resources specialist, and different agency representatives, depending on available personnel and the complexity of the spill.

8.4.1 In the most simple form, the survey is conducted by an oil-spill shoreline specialist who has a basic understanding of marine coastal geomorphology and processes, oil behavior, and cleanup. Using standardized terms, definitions, and procedures, this person documents the oiling conditions and the physical character of each oiled segment.

8.4.2 A specialist in coastal ecology may be included in the team(s), as necessary, to identify and assess intertidal communities and evaluate the effect of the oil or the potential effect of treatment options. The ecologist may also (*1*) verify the occurrence of sensitive habitats or species (in the segment) that were identified previously in sensitivity maps or databases, (*2*) identify and document human use activities in a segment, and (*3*) identify procedures or constraints on response operations to minimize effects on the biota.

8.4.3 A cultural resources or archaeological specialist may be included on the team(s), as necessary, to identify known or suspected archaeological, historical, or other cultural sites. The specialist may also (*1*) help evaluate potential effects or various treatment options, (*2*) if authorized, collect artifacts for their preservation, and (*3*) identify procedures or constraints on response operations to avoid potential impacts on the sites in or adjacent to oiled segment(s) **(5)**.

8.4.4 Representatives of government agencies, land owners or managers, the potential responsible party, and the operations team may participate as observers to assist in the subsequent development of response options and constraints.

8.5 In its basic form, the survey would document and assess the surface and, where applicable, subsurface oiling conditions. On shores where the materials are mobile or where the oil can permeate into the substrate, it is necessary to dig pits or trenches to locate and describe oil that has been buried or has penetrated. The inclusion of ecological or archaeological observations would depend on the purpose of the survey.

8.6 Standard forms are recommended in order to ensure that all necessary data are recorded or considered in a consistent manner. These are typically designed to suit the spill situation. An example of a generic shoreline oiling summary (SOS) form is provided in Fig. 1. Simpler forms may be used where appropriate to the intended purpose of the spill survey. Similar forms may be adapted for use in freshwater or riverine environments or on a regional basis to be applicable to non-temperate or tropical shore-zone conditions.

8.7 A field sketch (see Fig. 2) is recommended to locate oiled zones within a segment, where the oil is not distributed uniformly, or to identify shore-zone features. A map sketch rather than a perspective drawing is interpreted more easily and can include oiling, substrate character, photographic, and scale information.

8.8 Still-color photographs or videotape recordings can supplement the completed form and sketch. More information can be recorded on the audio channel of a videorecorder than can be recorded in a field note book, by sketches or photographs. The system has the advantage over notes or a tape recorder as a visual image is provided of the location under discussion. This technique is particularly recommended for surveys or segments in which the shoreline or oiling conditions are particularly complex, unusual, or spatially variable.

8.9 There are several post-survey activities that are optional. Prior to departing a survey site, a team may review individual assessments quickly and discuss findings to ensure that nothing significant has been overlooked. At the completion of each day it is typical to review, recopy, and finish, as necessary, all forms, maps, notes, etc. Completed survey documentation should be filed or archived as appropriate. Survey findings may be put on a database and may be used to calculate or assess the degree or relative severity of oiling (see Practice F 1779) and facilitate the setting of cleanup priorities. Reports on findings or recommendations may be submitted to the command center or presented to planners and decision-makers in various forms, as required, to provide them with an accurate perspective on the extent and degree of oiling and to assist them in setting cleanup priorities and in selecting response techniques.

SHORELINE OILING SUMMARY (SOS) FORM - General Temperate Environment Page ___of ___

1. GENERAL INFO	Survey Date (m/d/y):	Time(24h): standard/daylight hrs to hrs	Tide Level to
Segment ID:			
Survey by: Foot / Boat / Helicopter/	Sun /Clouds /Fog /Rain /Snow /		Wind:

2. SURVEY TEAM

OG		
ECO		
ARCH		

3. SHORELINE TYPE (overall substrate character and coastal character)

Bedrock Cliff / Platform	Boulder Beach	Sand Beach	Organic (type?)
Manmade Solid	Pebble Cobble Beach	Tidal Flat (sand / mud)	
Manmade Unconsolidated	Sand Gravel Beach	Salt Marsh	Other (type?)
Coastal Character (shore form):		Backshore Form	

4. SEGMENT

Predominant Substrate: %Bedrock; % Unconsolidated; %Vegetated; %Manmade (permeableY/N); %Ice	
Slope: % Low; % Medium; % High; % Vertical	Wave Exposure: low / medium / high /
Total Segment Length (est. in m)	Access Restrictions:
Total Length Surveyed (est. in m)	

5. SURFACE COLLECTABLE OIL & DEBRIS

Total Pavement = sq.m area by cm deep	Oiled Logs = Oiled Vegetation =
Patties /Tarballs = bags	Oiled Trash = Other

6. SURFACE OILING CONDITIONS

L O C	AREA Long m	Wide m	Location in Shore Zone SU	UI	MI	LI	Shore Slope VHML	SURFACE SEDIMENT TYPE	% OIL COVER %	OIL THICKNESS PO	CV	CT	ST	FL	OIL CHARACTER FR	MS	TB	TC	SR	AP	NO	DB

7. SUBSURFACE OILING CONDITIONS

PIT LOC. NO.	PIT SHOREZONE SU	UI	MI	LI	PIT DEPTH cm	OILED ZONE cm - cm	CLEAN BELOW (Y / N)	SUBSURFACE OIL QUANTITY AP	OP	PP	OR	TR	NO	Water Level cm	Sheen Colour	SEDIMENT TYPE Profile from Surface to Lower Oiled Zone

8. COMMENTS

Photo Roll # _____Frames______________Video Tape #: ___________ *Environment Canada: ver. 07/07/95*

FIG. 1 Example Shoreline Oiling Summary (SOS) Form (1)

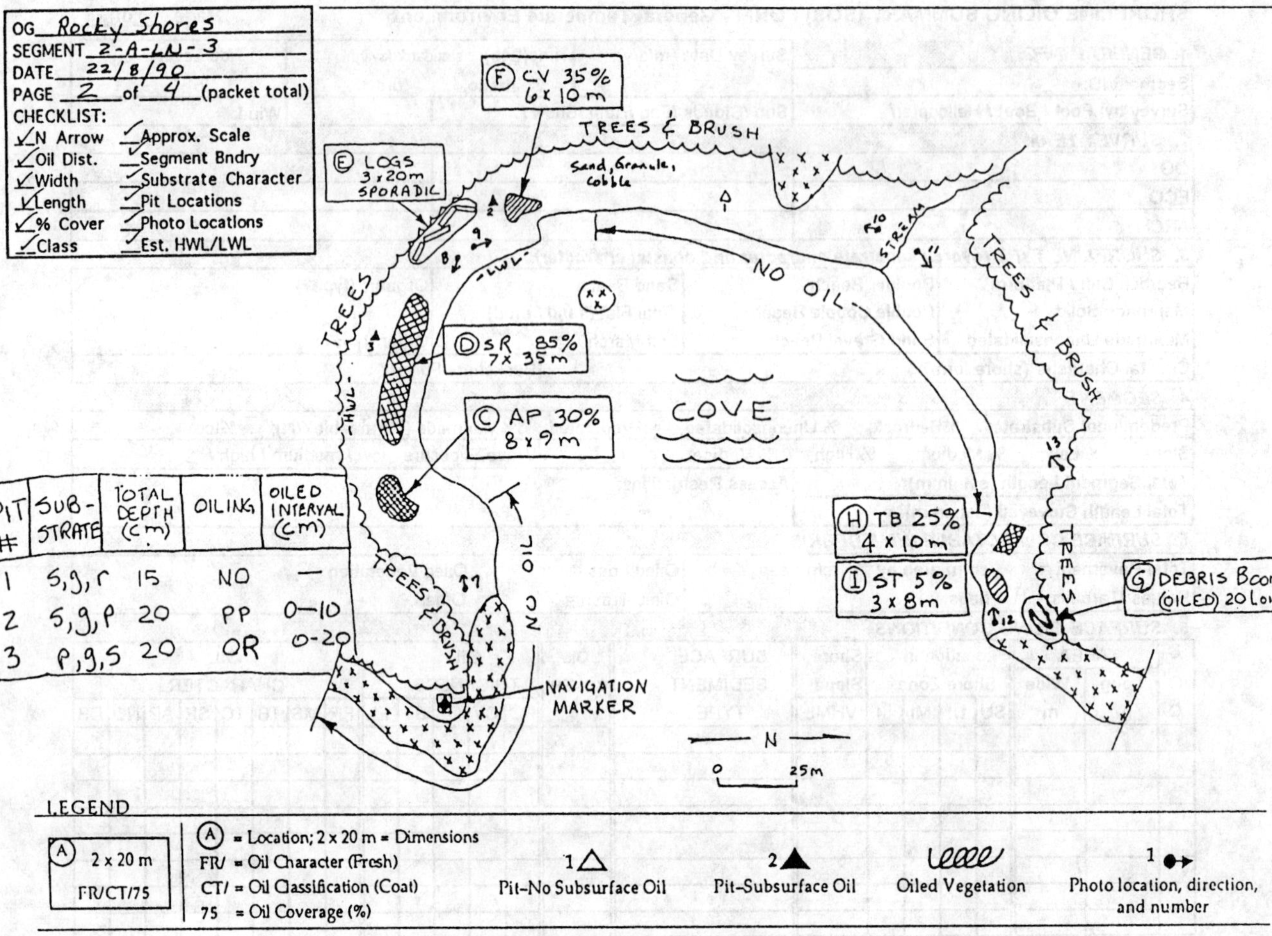

FIG. 2 Example of Completed Sketch Map[3]

REFERENCES

(**1**) Owens, E. H. and Sergy, G. A., Field Guide to the Documentation and Description of Oiled Shorelines, *Environment Canada, Edmonton*, ISBN 0-662-22048-X, 1994, 66 pages.

(**2**) Owens, E. H., and Reimer, P. D., Aerial Videotape Shoreline Surveys for Oil Spill Reconnaissance, Documentation, and Mapping, *Proceedings of the International Oil Spill Conference*, Publication No. 4529, American Petroleum Institute, Washington D.C., 1991, pp. 601-605.

(**3**) Owens, E. H., Sergy, G. A., and J. Michel, Applications of Standardized Assessment Techniques for Documenting Shoreline Oiling Conditions, *Proceedings 1995 International Oil Conference*, American Petroleum Institute, Washington, D.C., 1995, pp. 904-905.

(**4**) Environmental Canada, Oilspill SCAT Manual for the Coastlines of British Columbia, Prepared by Woodward-Clyde Consultants, *Seattle for the Environmental Technology Branch*, Environment Canada, Edmonton, 1992, 245 pages.

(**5**) Owens, E. H., and Teal, A. R., Shoreline Cleanup Following the "*Exxon Valdez* Oil Spill"-Field Data Collection Within the SCAT Program, Proceedings 13th Arctic and Marine Oilspill Program (AMOP) Technical Seminar, Environment Canada, Ottawa, 1990, pp. 411-421.

Designation: F 1687 – 97

Standard Guide for Terminology and Indices to Describe Oiling Conditions on Shorelines[1]

This standard is issued under the fixed designation F 1687; the number immediately following the designation indicates the year of original adoption or, in the case of revision, the year of last revision. A number in parentheses indicates the year of last reapproval. A superscript epsilon (ϵ) indicates an editorial change since the last revision or reapproval.

1. Scope

1.1 This guide covers the standardized terminology and types of observational data and indices appropriate to describe the quantity, nature, and distribution of oil and physical oiling conditions on shorelines that have been contaminated by an oil spill.

1.2 This guide does not address the mechanisms and field procedures by which the necessary data are gathered; nor does it address terminology used to describe the cultural resource or ecological character of oiled shorelines, spill monitoring, or cleanup techniques.

1.3 This guide applies to marine shorelines (including estuaries) and may also be used in freshwater environments (rivers and lakes).

1.4 The values stated in SI units are to be regarded as the standard. The values given in parentheses are for information only.

1.5 *This standard does not purport to address all of the safety concerns, if any, associated with its use. It is the responsibility of the user of this standard to establish appropriate safety and health practices and determine the applicability of regulatory limitations prior to use.*

2. Referenced Documents

2.1 *ASTM Standards:*

F 1686 Guide for Surveys to Document and Assess Oiling Conditions on Shorelines[2]

3. Terminology

3.1 *Definitions:*

3.1.1 *asphalt pavement*—a naturally formed cohesive mixture of weathered oil and sediments. Sediments in the mixture are usually in the sand/granule/pebble size range. In appearance, natural asphalt pavement may resemble the mixture artificially created to surface roads.

3.1.2 *lower-swash zone*—the area between the lowest annual water level and the mean annual water level, the lower half (approximate) of the zone of wave activity.

3.1.3 *supra-swash zone*—the area above the highest annual water level that experiences wave activity only occasionally, as during a storm event.

3.1.4 *supra-tidal zone*—the area above the mean high tide that experiences wave activity occasionally.

3.1.5 *upper-swash zone*—the area between the highest annual water level and the mean annual water level, the upper half (approximate) of the zone of wave activity.

3.1.6 *weathered oil*—the oil that has had an alteration of physical or chemical properties, or both, through natural processes such as evaporation, dissolution, oxidation, emulsification, and biodegradation.

4. Significance and Use

4.1 In order to ensure data consistency, it is important to use standardized terminology and definitions in describing oiling conditions. This guide provides a template for that purpose.

4.2 Data on oiling conditions at a shoreline are needed to provide an accurate perspective of the nature and scale of the oiling problem and to facilitate spill-response planning and decision making. Data on oiling conditions would be used in assessing the need for cleanup actions, selecting the most appropriate response technique(s), determining priorities for cleanup, and evaluating the endpoint of cleanup activities.

4.3 Mechanisms by which data are collected may vary (see Guide F 1686).[3,4] They may include aerial videotape surveys or ground-level assessment surveys. The composition and responsibility of the survey team will depend on the response organization and objectives. The magnitude and type of data sets collected may likewise vary with the nature of the spill and operational needs.[5]

4.4 Consistent data sets (observations and measurements) on shoreline oiling conditions are essential within any one spill in order to compare the data between different sites or observers, and to compare the data against existing benchmarks or criteria that have been developed to rate the nature or

[1] This guide is under the jurisdiction of ASTM Committee F-20 on Hazardous Substances and Oil Spill Response and is the direct responsibility of Subcommittee F20.17 on Shoreline Countermasters.

Current edition approved Feb. 10, 1997. Published April 1997. Originally published as F 1687 – 96. Last previous edition F 1687 – 96.

[2] *Annual Book of ASTM Standards*, Vol 11.04.

[3] *Oilspill SCAT Manual for the Coastlines of British Columbia*, Prepared by Woodward-Clyde Consultants, Seattle for the Environmental Technology Branch, Environment Canada, Edmonton, Alta., 1992.

[4] *Shoreline Countermeasures Manual—Tropical Coastal Environments*, Hazardous Materials Response and Assessment Division, National Oceanic and Atmospheric Administration, Seattle, WA, 1993.

[5] Source: Owens, E. H., and Sergy, G. A., *Field Guide to the Documentation and Description of Oiled Shorelines*, ISBN 0-662-22048-X, Environment Canada, Edmonton, Alta., 1994.

severity of the oiling. To the extent possible, consistency is also desirable between different spills, in order to benefit from previous experiences and cleanup decisions.

4.5 It is recognized that some modifications may be appropriate based on local or regional geographic conditions or upon the specific character of the stranded oil.

5. General Considerations

5.1 Shoreline conditions can be described in terms of the length, width, depth, distribution, quantity, and character of oil contamination. These six different types of data are collected by direct measurement or direct visual estimates calibrated against existing scales or indices. Standard definitions and descriptors of these data have been developed (Sections 6 and 7). Second-order applications of the basic data are further used to aid response planning (Sections 8 and 9).

5.2 Descriptions of shoreline oiling conditions are typically referenced to the lateral (seaward to landward) shoreline zonation. The location of the stranded oil within the intertidal zone affects operational access time and oil persistence.

5.2.1 Tidal zonation is described in terms of the supra-tidal, upper/mid/lower intertidal, and sub-tidal zones.

5.2.2 Non-tidal shoreline zonation is described in terms of the supra/upper/lower swash zone for lacustrine (lake) environments and the over/upper/lower/midstream bank for riverine (river) environments.

5.3 Oil persistence and the choice of cleanup options will be different for subsurface oil as opposed to surface oil. Descriptions of shoreline oiling conditions should distinguish between the oiling of surface sediments from that on the subsurface sediments (vertical zonation). On coarse sediment beaches, it can be difficult to differentiate the vertical boundaries. Fig. 1 illustrates an approach for discriminating those boundaries.

5.4 For beaches with fine sediments (that is, pebble, granule, sand, and mud), the subsurface begins at 5 cm below the surface. If a pit were to reveal oiling in sand from the surface down to 20 cm, the upper 5 cm would be classified as surface oil and the remainder as subsurface. However, the oiled interval would still be shown as 0 to 20 cm.

5.5 For beaches with coarse sediments (that is, cobble and

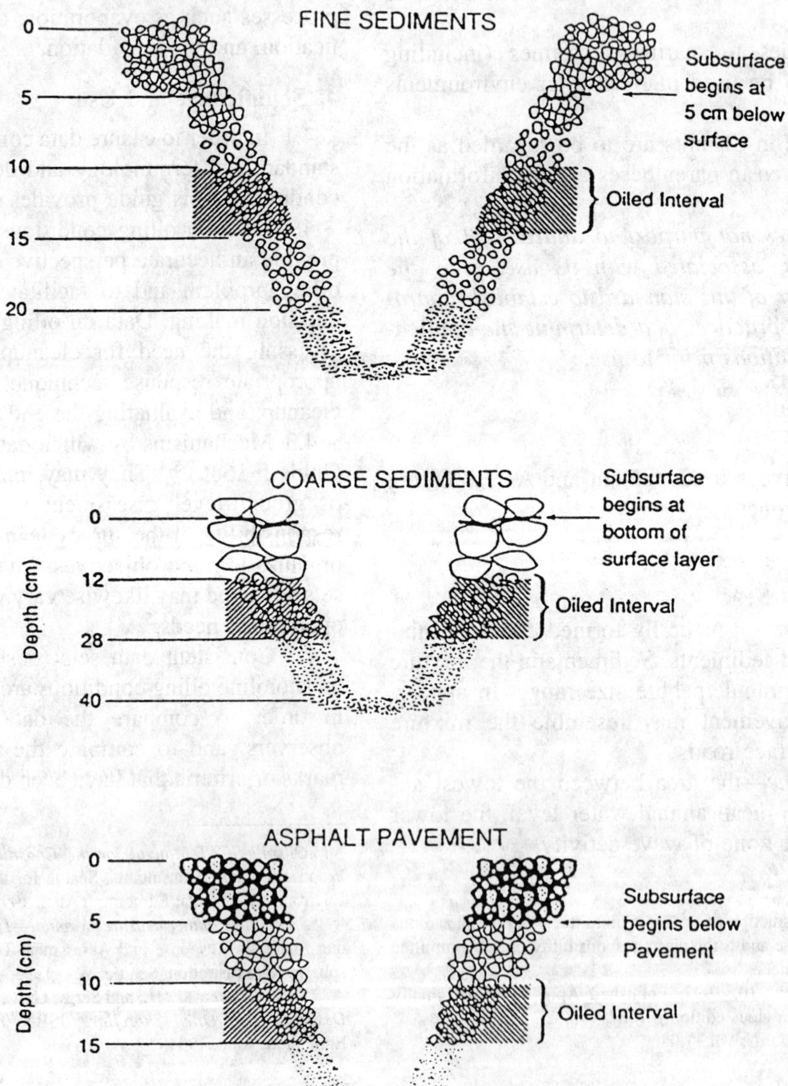

FIG. 1 Subsurface Boundaries for Various Beach Types[5]

boulder), the subsurface begins at the bottom of the surface material (that is, where the top layer of cobbles or boulders contact the underlying layer of sediments).

5.6 Where asphalt pavement exists on the surface, the subsurface begins at the bottom of the pavement.

5.7 Definitions of the inorganic sediments based on size are as follows:

Boulder	(>256-mm diameter)
Cobble	(64 to 256-mm diameter)
Pebble	(4 to 64-mm diameter)
Granule	(2 to 4-mm diameter)
Sand	(0.06 to 2-mm diameter)
Mud/silt/clay	(<0.06-mm diameter)

6. Description of Shoreline Surface-Oiling

6.1 *Oil Length*—This refers to the along-shore length of oiled shoreline.

6.1.1 The length should be described in numeric terms, as the actual measured or estimated value.

6.1.2 The length value must clearly indicate a reference to one of three different delineations:

6.1.2.1 The length of oiled-shoreline, which is the length of any single continuous oiling deposit and which is keyed to a specific shoreline location;

6.1.2.2 The total length of oiled-shoreline, which is the sum of the individual continuous oilings; or

6.1.2.3 The total length of affected shoreline, which is the total length of shoreline in the spill path (the distance between the two extreme strike points) and includes non-oiled sections of shoreline as well as oiled shoreline.

6.2 *Oil Width*—This refers to the average across-shore width of the oil band or oiled area.

6.2.1 If multiple bands or areas occur across-shore, the width represents the sum of their widths.

6.2.2 The width should be described by the actual numeric (measured or estimated) value.

6.2.3 Where a descriptive expression is required or for the purpose of aggregation of the actual data, use a simplified classification scheme.

6.2.3.1 Most marine shorelines can use the following definitions:

Wide	>6 m
Medium	>3 to ≤6 m
Narrow	>0.5 to ≤3 m
Very narrow	≤0.5 m

6.2.3.2 Areas of micro-tidal (for example, Great Lakes) or macro-tidal (for example, Bay of Fundy or Cook Inlet) exposure will require an adjustment based on the width of the shoreline.

6.3 *Oil Distribution*—This refers to the percentage of the shoreline surface sediments that are covered with oil, within a fixed area.

6.3.1 In the event of multiple bands, distribution refers to the estimated percentage that best represents the collective area.

6.3.2 Distribution should be described by the percentage value that was observed.

6.3.3 Where a descriptive expression is required, or for the purpose of aggregation of the actual value, use the classification of the following:

Trace	<1 %
Sporadic	1 to 10 %
Patchy	11 to 50 %
Broken	51 to 90 %
Continuous	91 to 100 %

6.4 *Surface-Oil Quantity*—This describes the amount of oil on shoreline surface sediments. The quantity is usually expressed as thickness or concentration.

6.4.1 Direct measurement of oil thickness can be made for very thick and larger deposits of oil. Visual estimates of the thickness of oil on the shoreline surface sediments can be used in the field. The dominant oil thickness within a band or area can be described by the classification of the following:

6.4.1.1 *Pooled or Thick Oil (PO)*, generally consisting of fresh oil or mousse accumulations >1-cm thick.

6.4.1.2 *Cover (CV)*, 0.1 cm and ≤1-cm thick.

6.4.1.3 *Coat (CT)*, >0.01 cm and ≤0.1-cm thick coating. This can be scratched off on coarse sediments or bedrock.

6.4.1.4 *Stain (ST)*, ≤0.01-cm thick. An oil residue discoloration on the sediment surface. It cannot be scratched off easily on coarse sediments or bedrock.

6.4.1.5 *Film (FL)*, transparent or translucent thin layer or sheen.

6.4.2 The oil concentration can be determined by chemical analytical methods and is expressed as the weight of oil to weight of surface sediment.

6.5 *Oil Character*—This refers to the properties or form of the oil residue on the shoreline.

6.5.1 Qualitative descriptors of oil character may be developed to suit the circumstances of the spill. Typical descriptors are as follows:

6.5.1.1 *Freshlike (FR)*, having a black, shiny, fluid, non-oxidized appearance.

6.5.1.2 *Tarry (TC)*, having a tacky, tar-like, often weathered, semi-solid consistency.

6.5.1.3 *Surface Oil Residue (SR)*, consisting of non-cohesive oiled, surface sediments, either as continuous patches or in coarse-sediment interstices.

6.5.1.4 *Mousse (MS)*, emulsified oil (oil and water mixture) existing as patches or accumulations, or within interstitial spaces.

6.5.1.5 *Tar Balls or Mousse Patties (TB)*, discrete balls or patties on a beach or adhered to rock or coarse-sediment shoreline. Tar ball diameters are generally <0.1 m, and patties are ≥0.1 to ≤1 m.

6.5.1.6 *Asphalt Pavement (AP)*, cohesive mixture of weathered oil and sediments.

6.5.1.7 *Sheen (SH)*, a very thin transparent or translucent oil layer ranging in color from silver to rainbow to light brown, of approximate thickness of 0.0001 to 0.001 mm.

6.5.2 Physical and chemical analytical characterization of the oil residues can be used on selected samples to determine oil properties.

7. Description of Shoreline Subsurface-Oiling

7.1 *Depth*—This refers to the depth of penetration or burial of the oil into the shoreline subsurface sediments, or both. It is the distance measured from the ground surface to the bottom of the oiled zone, or to the bottom of a discrete lens of oil that exists beneath the surface. In the latter case, the thickness of the oil lens is also measured, that is, top and bottom boundaries.

7.2 *Subsurface-Oil Quantity*—This refers to the amount of oil in the shoreline subsurface sediments (see 5.3 for guidance on vertical zonation). The quantity can be expressed in descriptive terms or by measurements of concentration.

7.2.1 Qualitative descriptions facilitate rapid field estimates of oil volume and may also reflect actual spill conditions. Typical descriptors may include the following:

7.2.1.1 *Oil-Filled Pores (OP)*—Pore spaces in the sediment matrix completely filled with oil. They are often characterized by oil flowing out of the sediments when disturbed.

7.2.1.2 *Partially Filled Pores (PP)*—Pore spaces filled with oil, but it generally does not flow out when exposed or disturbed.

7.2.1.3 *Cover or Coat or Stain or Film* (see 6.4.1)—Oil residue on larger grained-sized sediments.

7.2.1.4 *Trace (TR)*—A discontinuous film or spots of oil on sediments, or an odor or tackiness with no visible evidence of oil.

7.2.1.5 *Asphalt Pavement (AP)*—Cohesive mixture of weathered oil and sediments situated completely below a surface sediment layer (note thickness).

7.2.2 Chemical analytical methods can be used on selected samples to determine the oil quantity as a concentration (expressed as the weight of oil to weight of sediment at a specified subsurface location).

7.3 *Subsurface-Oil Length or Width*—This refers to the along-shore length or across-shore width of oiled subsurface sediment at a specific location (see 6.1 and 6.2).

8. Other Calculations

8.1 *Total Oiled Area = Length × Width Data* (6.1 and 6.2)—This refers to the total surface area of shoreline within which there is oil (of varying distribution). This value can be used in planning cleanup operations and in monitoring changes through time.

8.2 *Surface Oil Cover = (Length × Width) × Distribution Data* (6.1, 6.2, and 6.3)—This is a measure of the actual surface area that is covered by oil, that is, the total oiled area × % coverage (also known as equivalent area oiled). This value is useful when trying to quantify the degree of oiling or to monitor changes and oil removal rates.

8.3 *Oiled Sediment Volume = Depth × Area*—This refers to the total volume of shoreline sediments that contain oil (of varying concentrations and distribution). This is the volume of sediment that might have to be handled in a cleanup operation.

8.4 *Oil Volume*—This refers to the quantity of oil in or on the shoreline sediments. These calculations require oil concentration data in combination with knowledge of oiled sediment volume or equivalent area oiled, or oil loading data, together with sediment porosity/retention estimates.

9. Rating the Degree of Oiling

9.1 Collected field data can be combined to create indices that rate the degree or relative severity of oiling. This process reduces large volumes of observational data and allows a single-value, site-to-site, relative comparison. Such indices become very valuable for setting priorities and making decisions with respect to shoreline cleanup on medium to large spills where large areas are oiled. They provide a perspective for describing, summarizing, or comparing multiple areas or long sections of oiled coast in an easily understandable manner.

9.2 Rating indices must be adjusted to local conditions, particularly where the shoreline width data do not conform to the classification described in 6.2.

9.3 A tiered approach to rating is used to allow for different levels of detail of collected data. This enables data from different surveys with different levels of detail to be compatible with each other, as well as to refine the rating with increasing levels of data detail input. An example three-tier index is contained in Table 1, Table 2, and Table 3.

9.3.1 *Surface Oil Cover Category*—This combines width × distribution data (6.2 and 6.3).

9.3.2 *Surface Oiling Category*—This combines width × distribution × thickness (see 6.4.1).

9.3.3 *Subsurface Oiling Category*—This combines depth × relative concentration (7.1 and 7.2).

9.4 Descriptors of the degree of oiling are very light, light, moderate, and heavy. The indices are used to produce a rating of a described location. Typically, the rating category is combined with alongshore length; for example, Segment AB-1 has 20 m of heavy surface oiling.

TABLE 1 Oil Cover Category Matrix[5]

	Width of Oiled Areas			
	Wide, >6 cm	Medium, 3 to 6 m	Narrow, 0.5 to 3 m	Very Narrow, <0.5 m
Continuous, 91 to 100 %	heavy	heavy	moderate	light
Broken, 51 to 90 %	heavy	heavy	moderate	light
Patchy, 11 to 50 %	moderate	moderate	light	very light
Sporadic, 1 to 10 %	light	light	very light	very light
Trace, <1 %	very light	very light	very light	very light

TABLE 2 Surface Oiling Category Matrix[5]

	Initial Oil Cover Category			
	Heavy	Moderate	Light	Very Light
Pooled/thick, >1 cm	heavy	heavy	moderate	light
Cover, 0.1 to 1.0 cm	heavy	heavy	moderate	light
Coat, 0.01 to 0.1 cm	moderate	moderate	light	very light
Stain/film, <0.01 cm	light	light	very light	very light

TABLE 3 Subsurface Oiling Category Matrix[5]

	Penetration Depth or Thickness of Oil Lens			
	>30 cm	21 to 30 cm	11 to 20 cm	≤10 cm
Oil-filled pores	heavy	heavy	moderate	moderate
Part-filled pores	heavy	moderate	moderate	light
Cover/coat stain residue	moderate	moderate	light	light
Trace	light	very light	very light	very light

The American Society for Testing and Materials takes no position respecting the validity of any patent rights asserted in connection with any item mentioned in this standard. Users of this standard are expressly advised that determination of the validity of any such patent rights, and the risk of infringement of such rights, are entirely their own responsibility.

This standard is subject to revision at any time by the responsible technical committee and must be reviewed every five years and if not revised, either reapproved or withdrawn. Your comments are invited either for revision of this standard or for additional standards and should be addressed to ASTM Headquarters. Your comments will receive careful consideration at a meeting of the responsible technical committee, which you may attend. If you feel that your comments have not received a fair hearing you should make your views known to the ASTM Committee on Standards, 100 Barr Harbor Drive, West Conshohocken, PA 19428.

Last ASTM Designation: F 1688 – 96

Standard Guide for Estimating Oil Spill Recovery System Effectiveness

This guide covers the key factors to consider in estimating the effectiveness of containment and recovery systems that may be used to assist in the control of oil spills on water.

Formerly under the jurisdiction of ASTM Committee F-20 on Hazardous Substances and Oil Spill Response, this guide was discontinued in 1999.

Designation: F 1693 – 96

Standard Guide for Consideration of Bioremediation as an Oil Spill Response Method on Land[1]

This standard is issued under the fixed designation F 1693; the number immediately following the designation indicates the year of original adoption or, in the case of revision, the year of last revision. A number in parentheses indicates the year of last reapproval. A superscript epsilon (ε) indicates an editorial change since the last revision or reapproval.

1. Scope

1.1 The goal of this guide is to provide recommendations for the use of biodegradation enhancing agents for remediating oil spills in terrestrial environments.

1.2 This is a general guide only, assuming the bioremediation agent to be safe, effective, available, and applied in accordance with both manufacturers' recommendations and relevant environmental regulations. As referred to in this guide, oil includes crude and refined petroleum products.

1.3 This guide addresses the application of bioremediation agents alone or in conjunction with other technologies, following spills on surface terrestrial environments.

1.4 This guide does not consider the ecological effects of bioremediation agents.

1.5 This guide applies to all terrestrial environments. Specifically, it addresses various technological applications used in these environments.

1.6 In making bioremediation-use decisions, appropriate government authorities must be consulted as required by law.

1.7 *This standard does not purport to address all of the safety concerns, if any, associated with its use. It is the responsibility of the user of this standard to establish appropriate safety and health practices and determine the applicability of regulatory limitations prior to use.* In addition, it is the responsibility of the user to ensure that such activity takes place under the control and direction of a qualified person with full knowledge of any potential or appropriate safety and health protocols.

2. Referenced Documents

2.1 *ASTM Standards:*

F 1481 Guide for Ecological Considerations for the Use of Bioremediation in Oil Spill Response—Sand and Gravel Beaches[2]

3. Terminology

3.1 *Definitions:*

3.1.1 *aerobes*—organisms that require air or free oxygen for growth.

3.1.2 *anaerobes*—organisms that grow in the absence of air or oxygen and do not use molecular oxygen in respiration.

3.1.3 *bioaugmentation*—the addition of microorganisms (predominantly bacteria) to increase the biodegradation rate of target pollutants.

3.1.4 *biodegradation*—chemical alteration and breakdown of a substance, usually to smaller products, caused by microorganisms or their enzymes.

3.1.5 *bioremediation*—enhancement of biodegradation.

3.1.6 *bioremediation agents*—inorganic and organic compounds and microorganisms that are added to enhance degradation processes, predominantly microbial.

3.1.7 *biostimulation*—the addition of microbial nutrients, oxygen, heat, or water, or some combination thereof, to enhance the rate of biodegradation of target pollutants by indigenous species (predominantly bacteria).

3.1.8 *ecosystem*—organisms and the surrounding environment combined in a community that is self-supporting.

3.1.9 *identification*—the process of establishing the identity of an unknown organism by comparing the properties with respect to known organisms.

3.1.10 *indigenous*—native to a given habitat or environment.

3.1.11 *methemoglobinemia*—an acquired blood disorder leading to oxygen deprivation, stupor, and death from exposure to nitrates in drinking water.

3.1.12 *nutrient*—a substance that supports the growth of organisms.

3.1.13 *refined petroleum products*—products derived by means of various treatment processes from crude oil, a highly complex mixture of paraffinic, cycloparaffinic, and aromatic hydrocarbons that contains a low percentage of sulfur and trace amounts of nitrogen and oxygen compounds. Hydrocarbon products made by refining crude oils are specified in Section 5 of the *Annual Book of ASTM Standards* (**1**).[3]

3.1.14 *risk*—the probability or likelihood that an adverse effect will occur.

---

[1] This guide is under the jurisdiction of ASTM Committee F-20 on Hazardous Substances and Oil Spill Response and is the direct responsibility of Subcommittee F20.24 on Bioremediation.

Current edition approved March 10, 1996. Published May 1996.

[2] *Annual Book of ASTM Standards*, Vol 11.04.

[3] The boldface numbers in parentheses refer to the list of references at the end of this guide.

3.1.15 *species*—a taxonomic category characterized by individuals of the same genus that are mutually similar and are able to interbreed.

3.1.16 *terrestrial*—consisting of land, as distinguished from water.

3.1.17 *toxicity*—the property of a material, or combination of materials, to affect organisms adversely.

4. Significance and Use

4.1 The purpose of this guide is to provide remediation managers and spill response teams with guidance on an alternate means (called bioremediation) of treating oil spills safely and effectively on and below terrestrial surfaces.

4.2 Bioremediation is one of many available tools and may not be applicable to all situations. This guide can be used in conjunction with other ASTM guides addressing oil spill response operations, including Guide F 1481, as well as options other than bioremediation.

5. General Considerations for Bioremediation Use

5.1 Bioremediation technologies attempt to accelerate the natural rate of biodegradation. In situ, solid-phase, and slurry-phase represent the major bioremediation technologies used. These technologies may be unnecessary in those cases in which the natural rate of biodegradation suffices. The use of adequate controls in preliminary field studies, or the results of previously reported studies, will assist in determining the extent to which microorganism or nutrient amendments, or both, are necessary to obtain the desired rate of degradation.

5.2 Bioremediation performance depends on the efficiency of the petroleum hydrocarbon degrading indigenous microorganisms or bioaugmentation agents. Performance also depends on the availability of rate-limiting nutrients and the susceptibility of the target crude oil or refined product to microbial degradation.

5.2.1 In general, aerobic bioremediation systems degrade oil more rapidly than anaerobic systems.

5.2.2 Numerous microorganisms, represented by hundreds of species, are responsible for the degradation of the oil. Various texts describe the biodegradability and biodegradation rates of a variety of organic compounds present in oil **(2,3)**.

5.2.3 The biodegradation of saturated hydrocarbons in the absence of molecular oxygen is limited to a few species. In general, shorter chain hydrocarbons are less effectively degraded in anaerobic conditions compared with aerobic conditions. However, anaerobic degradation is possible **(4)** if there is at least one double bond on the hydrocarbon molecule, in an appropriate position.

5.3 Bioremediation must be conducted under the guidance of qualified personnel who understand the safety and health aspects of site activities.

6. Background

6.1 General background information concerning approaches to bioremediation are presented in this guide, as well as discussed in Guide F 1481. Pertinent information from that guide is included in 6.1.1 through 6.1.4, as follows:

6.1.1 Approaches to bioremediation for oil spill response include biostimulation, the addition of nutrients, oxygen, heat, or water, or combination thereof, to stimulate indigenous microorganisms, and bioaugmentation, the addition of oil-degrading microorganisms, which may be used in combination with biostimulation **(5-15)**. As a precaution, it should be noted that nutrient components may be toxic or harmful to plants, animals, and humans, and that non-indigenous species may alter the indigenous microbial ecological balance at least temporarily. Water effluent nitrate levels, which can affect drinking water sources, should be minimized to diminish risks of anemias such as methemoglobinemia. Similarly, excessive ammonium levels should be avoided because they can affect fish and invertebrates, since many are immobile and cannot avoid the treated area. Therefore, nitrogen and other nutrient levels should be monitored. Instructions to ensure safety and effective product use should be established by the manufacturer or supplier for each commercial microbial product, and specific instructions should be followed by the product user.

6.1.2 Biostimulation has been shown to enhance the biodegradation of terrestrial oil spills. Biostimulation uses the addition of appropriate nutrients (for example, nitrogen, phosphorus, potassium, micronutrients, and so forth), oxygen, heat, or water, which may have been limiting factors. If microbial degraders of the target oil contaminants are present in the soil or contaminated waters, this approach may lead to increases in the rate of degradation. In some cases there may not be a sufficient indigenous oil-degrading population to stimulate. This may be the case in environments in which the degrader population has not developed. Alternately, the toxic nature of the petroleum product may diminish or eliminate microorganisms. Also, the excavation of soil from anoxic zones and subsequent relocation to an oxygen-rich environment may result in a lack of microbial degraders due to the drastic change in conditions **(16)**. The microbial response to biostimulation may include a lag period (weeks to months) for the growth or natural selection of degraders to occur. Microorganisms, as well as oil contaminants, should be monitored throughout the process to establish efficacy and safety. Comparisons with databases that include soil and water microorganisms may be used to identify microbes.

6.1.3 Bioaugmentation may use commercial microbial products, on-site production of microbes from stock cultures, or laboratory isolation, characterization, and subsequent production of microbes from the particular site (or another site similar in soil and contaminant characteristics). This approach may increase soil microbe concentrations rapidly. Microbes selected must be nonpathogenic and must metabolize the oil contaminant(s), reducing toxicity. Growth requirements of the microbes need to be well understood. Their growth rate is controlled by the limiting growth conditions of temperature, pH, nutrients, water, oxygen, the contaminated medium (soil, sludge, and water), and oil. Microorganisms as well as oil components should be monitored to establish efficacy and safety.

6.1.4 While apparently safe and effective in the laboratory setting, genetically engineered oil-degrading microorganisms have not yet been authorized for environmental release **(16)**.

6.2 There are several bioremediation technologies available. It is important to understand the potential use of these systems when assessing their applicability for full-scale implementation. Costs are determined by the size of the site, soil properties, type and level of oil contaminant(s), goals, time allowed for attaining the goals, and testing requirements.

6.3 *In situ* bioremediation occurs without excavation of the contaminated soil. This technology relies predominantly on the enhanced degradation of oil by bacteria following the addition of nutrients, air, oxygen or oxygen-releasing compounds, and moisture. This has been demonstrated through the use of indigenous as well as augmented microorganisms. Groundwater treatment may be achieved simultaneously or through pump and *ex situ* treatment methods. Anaerobic biodegradation systems can also be promoted; however, their utility is limited. Since soil is not excavated, volatile release is limited, and the risks and costs associated with excavation and treatment are reduced.

6.3.1 Bioventing involves the introduction of air under pressure to the unsaturated zone of contaminated soil. The process pulls or pushes air into the soil for use by the aerobic microorganisms. Although the purpose is to deliver oxygen required by the microbes, the flow of air will desorb some of the more volatile components from the soil (for example, gasoline-contaminated soil), and the exhaust gases may have to be treated. Successful treatment requires adequate soil porosity, moisture, nutrients, and microorganisms with the appropriate biodegradation abilities. Additives may be provided at or near the surface to percolate through the treatment zone.

6.3.2 Biosparging is similar to bioventing except that air is injected directly into the ground below the water table in the saturated zones. Although the purpose is to deliver oxygen required by the microbes, vacuum pumps may be used to recover vapors that may also have to be treated prior to discharge. Nutrients and microbes may be added in the injection well to stimulate and augment biodegradation.

6.4 Solid-phase bioremediation treats soils above ground, primarily in contained treatment cells or tanks. Techniques similar to landfarming are used, including irrigation, tilling, and nutrient and microbe additions. As with *in situ* bioremediation, treatment can involve biostimulation or bioaugmentation. Abiotic losses through volatilization and leaching can be minimized through treatment design and implementation. The contaminated soil is contained, preventing leaching, and is defined with respect to the volume and concentration of the oil, especially as the soil is homogenized during processing. The defined nature of the soil and its oil contaminants allow predictability in the remedial process.

6.4.1 A comprehensive contaminated materials handling plan (CMHP) should be developed prior to excavation and treatment when using systems that require excavation. It may include the designation of a materials staging area present within the treatment facility and equipment decontamination within delineated exclusion zones.

6.4.2 A comprehensive health and safety program should be in effect throughout the remediation project. This program may include medical examinations of employees, contact and respiratory protection, and air, soil, and water monitoring.

6.4.3 The treatment facility, or biopad, should contain an appropriate rainfall event protection (for example, 10 years, 24-h rainfall event). After the appropriate soil moisture content is determined for the specific treatment, a water budget should be calculated. This should maintain the proper moisture content balance between moisture added by irrigation and rainfall, and moisture lost through evaporation, transpiration, and percolation.

6.4.4 Solid heaping (biopiles or soil piles) involves piling the contaminated soil to several meters, usually over a network of perforated piping that may be layered throughout. Nutrients, water, and microorganisms are added through simple irrigation techniques, and air is drawn through pipes by vacuum. The vacuum system exhaust may be treated prior to discharge, effectively removing airborne volatile or semi-volatile components. Advantages include a requirement for less space and less material handling compared with solid-phase treatment (landfarming), and diminished volatile losses. Leachates are collected and treated, recirculated or discharged.

6.4.5 Composting promotes biodegradation in stored wastes by means of supplementation with bulking agents (biodegradable or non-biodegradable) that enhance soil permeability. The biologic decaying process is often thermophilic, thus limiting the types of microbes and associated degradation rates. Three basic systems have been used. "Open window" stacks the waste in long piles that are aerated through constant excavation and reconstruction." Static windrow" is similar to heap methods, laying the soil over a network of perforated pipes that aerate through forced air. "In-vessel" methods enclose the soil in a closed reactor that aerates and mixes the soil both physically and by means of forced air. The material remaining after treatment, the humus, can serve as a source for fill, cover, and landscaping material.

6.5 Slurry-phase bioremediation combines contaminated solids (soil, sludge, sediment) and liquids to form a slurry suspension. The slurry is supplemented with nutrients, air or oxygen-releasing chemicals, or microorganisms, or a combination thereof, in a bioreactor system. The slurry is stirred or agitated to enhance contact between the oil, nutrients, and microbes. The system may be designed as a batch or continuous flow process. Frequently, slurry-phase bioremediation is combined with other (physical, chemical or other biological) processes in a treatment train. These systems treat contaminated soil and water. Bioreactors promote optimal conditions while minimizing volatile losses, resulting from the design of the system and the type and amount of oil in the solid or liquid matrix. Compared with solid phase treatment, time is generally reduced when using bioreactors. The slurries are comprised of approximately 10 to 30 % w/v soil in water, and may require multiple batch treatments to process the total soil volume depending on reactor size and number. Hydrocarbons in high concentrations, as well as fines from soil washing can be treated. Slurry phase systems can be cumbersome and require extensive mechanical and technological construction and design, and maintenance. The costs for slurry phase treatment are generally higher in comparison to solid phase treatment.

7. Bioremediation Technology Selection Assessment

7.1 Treatability studies provide data to support treatment selection and are performed prior to remedy selection. The data indicate whether treatment goals can be met and further determine the optimal operation conditions for remediation project design. For example, a document prepared by the U.S. Environmental Protection Agency (U.S. EPA) mentions three levels of treatability studies **(17)**. The level of study chosen depends on available literature information, technical expertise, and site-specific considerations. In addition, treatability study design and interpretation for aerobic biodegradation remedy screening has been addressed **(18)**.

7.2 Various government agencies require support documentation. Studies are available through databases developed by groups sponsored by the U.S. EPA and Environment Canada. One such database, which includes innovative cleanup technologies, is the U.S. EPA's Vendor Information System for Innovative Treatment Technologies (VISITT) **(19)**. This database, made available since 1992, focuses on treatment of ground water, soil, sludge, sediments, and solid wastes, excluding the traditional technologies of incineration, solidification, stabilization, and pump-and-treat groundwater systems. The U.S. EPA Office of Emergency and Remedial Response (OERR), in conjunction with the Office of Research and Development (ORD), developed a new Contaminated Soil and Debris (CS&D) database. It contains information concerning the effectiveness of a given technology to remediate soils, sludges, and debris contaminated with hazardous waste. This database, available in 1992, supports technology transfer through easy access searches by title, site, scale, author, content, and other keywords. The data are used in developing standards for treating CS&D (40 CFR 261) under land disposal restriction (LDR) that may not be treatable to the best developed available technology (BDAT) levels established by 40 CFR 261.

7.3 Governmental agencies may regulate the use of bioremediation agents. The role of these regulatory agencies varies **(16)**.

7.4 There are many advantages as well as disadvantages associated with the various bioremediation options. Table 1 describes these attributes for the derivations discussed in Section 6 of this guide. In general, compared with other technologies, selection advantages include cost effectiveness, reuse, and reduced intrusion by response personnel (short and long term) into affected areas.

8. Recommendations

8.1 Application of bioremediation in terrestrial oil spill treatments should be considered as an alternative to other techniques.

8.2 Treatability studies should be used to provide data to support treatment selection, indicate whether cleanup goals can be met, and optimize operational conditions for remedy design.

8.3 Safety and efficacy data should be substantiated prior to

TABLE 1 Selection Assessment

Bioremediation Technology	Advantages	Disadvantages
In situ	Since soil is not excavated, volatile release is limited, and the risks and costs associated with excavation and treatment may be reduced. Less intrusive than other technologies. Cleanup of the source may help prevent contamination of the ground water. Applicable in emergency response situations.	The impact on ground water may necessitate hydrogeologic monitoring. Representative soil monitoring is difficult to achieve, and treatment requires adequate soil porosity, moisture, nutrients, heat, oxygen, and biodegrading microorganisms. The time frame to achieve cleanup may be uncertain.
Ex situ and *in situ* solid-phase landfarming	The contaminated soil is contained, preventing leaching, and volatilization can be controlled. The material being treated is defined with respect to the volume and concentration of the oil, especially as the soil is homogenized during processing. If the pH is not within physiological levels for bacterial growth, it can be balanced prior to further treatment. The toxic nature of some oil can be diluted with clean soil or bulking agents (for example, yard waste). The defined nature of the soil and its oil contaminants allow predictability in the remedial process. Treatment of similar wastes, achieved through appropriate placement, may help to accelerate the process.	Airborne volatile or semi-volatile components released during excavation and tilling operations occur. Material handling is associated with treatment. Aerobic biodegradation pathways are favored, requiring continuous air supply by means of tilling or chemical treatment. Depth of soil is limited to the tilling apparatus, ~0.5 m. Treatment requires adequate soil porosity, moisture, nutrients, and biodegrading microorganisms.
Ex situ solid-phase soil heaping	Less space and less material handling are required compared with *ex situ* solid-phase treatment (landfarming). Diminished volatile losses due to vacuum capture. Leachates are collected and treated and recirculated or discharged. Treatment of similar wastes, achieved through appropriate placement, may help to accelerate the process.	There are limitations in soil pile height and therefore soil volume. Some pile manipulation may be required to ensure thorough treatment, and this may affect the predictability of cleanup. Energy requirements add to overall costs.
Ex situ solid-phase composting	Compared with stockpiling, these methods accelerate degradation. The material remaining after treatment may serve as a source for fill, cover, and landscaping material.	The biologic decaying process is often thermophilic, thus limiting the types of microbes and associated degradation rates.
Ex situ slurry-phase	The simultaneous ability to treat contaminated soil and water. Bioreactors promote optimal conditions while minimizing the release of volatiles through system design and the type and amount of oil in the solid or liquid matrix. Compared with solid phase treatment, time is usually reduced when using bioreactors. No dewatering is required prior to treatment. Hydrocarbons in high concentrations, as well as fines from soil washing, can be treated.	Slurry-phase systems are engineered treatment systems requiring design, fabrication, and maintenance similar to above-ground biological wastewater treatment processes. Emissions may have to be controlled. Slurries are comprised of 10 to 30 % w/v of soil in water and may require multiple batch treatments to process the total volume, depending on reactor size and number. Costs are generally higher in comparison to solid-phase treatment. Dewatering after treatment is energy demanding, resulting in water and solid fractions that may have different cleanup criteria for subsequent disposal.

any bioremediation field application. This information should be provided and should be substantiated through treatability studies and published reports.

8.4 Biostimulation of indigenous microorganisms should be weighed against bioaugmentation with selected microorganisms. Consideration must be given to time, predictability, regulatory and public opinions, cost, and cleanup requirements for process selection.

8.5 Implementation of bioremediation technology is site- and oil contaminant-specific and must be used appropriately in accordance with government agencies.

9. Keywords

9.1 bioaugmentation; bioremediation; biostimulation; composting; in situ; landfarming; oil; oil spill response; soil heaping; solid-phase; slurry-phase; terrestrial; treatability

REFERENCES

(1) *Annual Book of ASTM Standards*, Vols 05.01, 05.02, 05.03, 05.04, 05.05, and 11.04, ASTM, Conshohocken, PA.

(2) *Handbook of Environmental Fate and Exposure Data for Organic Chemicals*, P. H. Howard, ed., Lewis Publishers, Inc., Chelsea, MI, Vols I–III, 1989–1991.

(3) *Handbook of Environmental Degradation Rates*, P. H. Howard, ed., Lewis Publishers, Inc., Chelsea, MI, 1991.

(4) Schink, B., "Principles and Limits of Anaerobic Degradation: Environmental and Technological Aspects," *Biology of Anaerobic Microorganisms*, J. B. Zehnder, ed., John Wiley & Sons, New York, NY, 1988, pp. 771–846.

(5) *Bioremediation of Petroleum Contaminated Sites*, E. Riser-Roberts and C. K. Smoley, eds., CRC Press, Boca Raton, FL, 1992.

(6) *Hydrocarbon Contaminated Soils*, Vol II, P. T. Kostecki, E. J. Calabrese, and M. Bonazountas, eds., Lewis Publishers, Chelsea, MI, 1992.

(7) *Petroleum Microbiology*, R. M. Atlas, ed., Macmillan Publishing Co., New York, NY, 1984.

(8) *Microbiological Decomposition of Chlorinated Aromatic Compounds*, M. L. Rochkind-Dubinsky, G. S. Sayler, and J. W. Blackburn, eds., Marcel Dekkar, Inc., New York, NY, 1987.

(9) *Bioremediation of Hazardous Wastes*, EPA/600/R-92/126, Office of Research and Development, U.S. Environmental Protection Agency, August 1992.

(10) *Bioremediation of Hazardous Wastes*, EPA/600/9-90/041, Office of Research and Development, U.S. Environmental Protection Agency, September 1990.

(11) *Bioremediation of Contaminated Surface Soils*, EPA/600/9-89/073, Office of Research and Development, U.S. Environmental Protection Agency, August 1989.

(12) *Proceedings of the Joint CSCE-ASCE National Conference on Environmental Engineering*, R. N. Yong, J. Hadjinicolaou, and A. M. O. Mohamed, eds., McGill University, Montreal, Quebec, July 12–14, 1993.

(13) Rowell, M. J., Ashworth, J., and Qureshi, A. A., *Bioremediation of Petroleum Hydrocarbons in Soil Environments*, Document EE-141, Environment Canada, Environmental Protection Directorate, River Road Environmental Technology Centre, Ottawa, Ont.

(14) *Understanding Bioremediation: A Guidebook for Citizens*, EPA/540/2-91/002, Office of Research and Development, U.S. Environmental Protection Agency, February 1991.

(15) *Applied Bioremedial Technology: Basic Concepts and Technical Considerations*, A. K. Kaufman and K. Devine, eds., The Applied Biotreatment Association, Fall 1990.

(16) Hoeppel, R. E., and Hinchee, R. E., "Enhanced Biodegradation for On-Site Remediation of Contaminated Soils and Groundwater," *Hazardous Waste Site Soil Remediation: Theory and Application of Innovative Technologies*, D. J. Wilson and A. N. Clarke, eds., Marcel Dekker, Inc., New York, NY, 1994, pp. 311–395.

(17) *Guide for Conducting Treatability Studies Under CERCLA*, Interim Final, EPA/540/2-89/058, Office of Research and Development and Office of Emergency and Remedial Response, U.S. Environmental Protection Agency, December 1989.

(18) *Guide for Conducting Treatability Studies Under CERCLA: Aerobic Biodegradation Remedy Screening*, Interim Guidance, EPA/540/2-91/013A, Office of Research and Development, Office of Emergency and Remedial Response, and Office of Solid Waste and Emergency Response, U.S. Environmental Protection Agency, Risk Reduction Engineering Laboratory, July 1991.

(19) *VISITT 3.0 Bulletin*, EPA 542-N-94-003, Office of Solid Waste and Emergency Response, U.S. Environmental Protection Agency, June 1994.

Designation: F 1737 – 96 (Reapproved 1999)

Standard Guide for Use of Oil Spill Dispersant Application Equipment During Spill Response: Boom and Nozzle Systems[1]

This standard is issued under the fixed designation F 1737; the number immediately following the designation indicates the year of original adoption or, in the case of revision, the year of last revision. A number in parentheses indicates the year of last reapproval. A superscript epsilon (ϵ) indicates an editorial change since the last revision or reapproval.

1. Scope

1.1 This guide covers the essential considerations for the maintenance, storage, and use of oil spill dispersant application systems.

1.2 This guide is applicable to spray systems employing booms and nozzles and not to other systems such as fire monitors, sonic distributors, or fan-spray guns.

1.3 This guide is applicable to systems employed on ships or boats and helicopters or airplanes.

1.4 This guide is one of four related to dispersant application systems. Guide F 1413 covers design, Practice F 1460 covers calibration, Test Method F 1738 covers deposition, and Guide F 1737 covers the use of the systems. Familiarity with all four standards is recommended.

1.5 The values stated in SI units are to be regarded as the standard. The inch-pound units given in parentheses are for information only.

1.6 *This standard does not purport to address all of the safety concerns, if any, associated with its use. It is the responsibility of the user of this standard to establish appropriate safety and health practices and determine the applicability of regulatory limitations prior to use.*

2. Referenced Documents

2.1 *ASTM Standards:*

F 1413 Guide for Oil Spill Dispersant Application Equipment: Boom and Nozzle Systems[2]

F 1460 Practice for Calibrating Oil Spill Dispersant Application Equipment Boom and Nozzle Systems[2]

F 1738 Practice for the Determination of Deposition of Aerially Applied Oil Spill Dispersants[2]

3. Significance and Use

3.1 This guide provides information, procedures, and requirements for management and operation of dispersant spray application equipment (boom and nozzle systems) in oil spill response.

3.2 This guide provides information on requirements for storage and maintenance of dispersant spray equipment and associated materials.

3.3 This guide will aid operators in ensuring that a dispersant spray operation is carried out in an effective manner.

4. Equipment Types For Vessels and Aircraft

4.1 A spraying system consists of one or more pumps, flowmeters, storage tanks, spray booms, and nozzles that are mounted in various configurations depending on the platform.

4.2 Dispersant application systems on ships or boats may be portable or permanently installed. Vessels may have built-in dispersant storage tanks and on-board pumps for use with the spraying system.

4.3 Dispersant application systems on helicopters are most commonly slung beneath the aircraft, with remote controls available to the pilot. Some specially configured helicopters have integral tanks and pumps. Helicopter spraying systems are available with dispersant capacity of about 500 to 2000 L (120 to 500 U.S. gal).

4.4 Dispersant application systems on single-engine airplanes have a built-in tank and pump, with the booms attached to the wings. Dispersant capacity varies with the airplane design but is about 400 to 4000 L (100 to 1000 U.S. gal).

4.5 Dispersant application systems can also be installed on large multiengine airplanes. These must be designed for each type of aircraft, and will include one or more pumps, flowmeters, dispersant storage tanks, and spray booms with nozzles. The airplane type and payload capability will determine the available dispersant capacity from about 4000 to 20 000 L (1000 to 5000 U.S. gal).

5. Equipment Configuration for Vessels and Aircraft

5.1 *Vessels*—Dispersant spray systems for boats have been designed for many types of craft. Most systems use water-compatible "concentrate" dispersants diluted with seawater during application. These dispersants are mixed with seawater by use of an educator or metering pump to allow for the dispersant to be used at the desired concentration (generally 5 to 10 %).

5.1.1 Mount the spray booms as far forward as possible so that the spray is applied in front of the bow wave, because, this wave can push oil out of reach of the spray at typical boat speeds. Nozzles and extensions should be downward-pointing

[1] This guide is under the jurisdiction of ASTM Committee F-20 on Hazardous Substances and Oil Spill Response and is the direct responsibility of Subcommittee F20.13 on Treatment.

Current edition approved Oct. 10, 1996. Published December 1996.

[2] *Annual Book of ASTM Standards*, Vol 11.04.

and stable relative to the boom. Rig spray booms with multiple nozzles arranged to produce flat, fan-shaped spray patterns, striking the water (oil) surface in a line perpendicular to the direction of travel of the vessel. Nozzles producing a hollow-cone shaped spray pattern should not be used. Spray pressure should not be excessive so that the spray does not break the oil surface. Deliver the dispersant-water mixture to the oil surface, in the desired pattern, with a minimum amount of energy. The spray should strike the oil in small droplets of 300 to 500-μm volume median diameter (VMD). The droplets should be visually larger than a fog or mist and smaller than heavy rain drops. The fan-shaped sprays from adjacent nozzles should overlap just above the oil surface.

5.1.2 Relatively small spills, such as in harbors or rivers, may best be treated by vessels, but they are limited on large offshore spills by their spray swath and speed. For example, a boat operating at 10 km/h (5 knots or 6 mph), and spraying a 12-m (40-ft) swath, can only treat about 0.5 miles^2 (1.3 km^2) of oil spill surface in about 12 h.

5.2 *Helicopters*— Spraying systems on helicopters are either integral (attached to the airframe) or external units that have a combined tank, pump, and spray boom assembly suspended below the aircraft from a cargo hook, as specified by the manufacturer of the bucket. Sufficient room must be allowed between the helicopter and the spray unit to allow for safe connection and disconnection. Spraying is controlled from the cockpit with an electrical remote control unit, attached by cable to the spray system. Nozzles should be oriented parallel to the direction of travel and pointed aft on the spray boom. Only concentrate dispersants applied without dilution are suitable for aerial spraying. The spray-boom altitude, when spraying, should not be over 9 m (30 ft).

5.2.1 Helicopters are limited in the volume of dispersant they can carry, typically under 2000 L (500 U.S. gal). They have greater speed than vessels, however, and if working near the source of dispersant supply, helicopters provide very efficient dispersant application on small areas.

5.3 *Small Airplanes*— Small single-engine airplanes typically will have a wind-driven pump that draws dispersant from a tank to feed the spray booms, that are usually fitted close to the trailing edge of the wing. The dispersant is discharged through nozzles (spaced at intervals along the boom) that are designed to generate droplets within the required size range. The dispersant pump should be capable of spraying at a rate that is required for a surface coverage of 20 to 100 L/hectare (2 to 10 U.S. gal/acre). The pump rate should be variable in flight, and regulated and monitored with a pre-calibrated flowmeter or pressure gage. Air shear, that affects droplet size, may be a problem for lower viscosity dispersants of less than 60 mpa (cSt), at aircraft velocities exceeding about 200 km/h (100 knots or 120 mph). The spray-boom altitude during application should not be over 9 m (30 ft).

5.3.1 Small airplanes generally have limited load capacity, about 400 to 3000 L (100 to 800 U.S. gal). This size of aircraft may provide rapid response to small spills, and has longer range and greater speeds than a helicopter system.

5.4 *Large Airplanes*— Large multiengine, propeller-driven, airplanes offer increased payload, range, and speed for the treatment of large spills. Most of these aircraft require the installation of wing-mounted booms and other integral parts. Some large cargo airplanes have a rear cargo or personnel door that can be opened in flight, can accommodate portable tank systems, and have extendable booms that can be deployed in flight. Such a system can be permanently fitted to a dedicated airplane, or installed as needed in an airplane of opportunity. These systems may require specific certification by aviation authorities for use on a particular type of aircraft.

5.4.1 These larger aircraft will generally fly at altitudes of 15 to 30 m (50 to 100 ft) when applying dispersant to the oil.

5.4.2 The largest dispersant liquid capacity for such aircraft is 20 000 L (5000 U.S. gal). Aircraft range and payload characteristics can limit the dispersant volume. Application rates from 20 to 100 L/hectare (2 to 10 U.S. gal/acre) can be achieved. Typical coverage for these systems is 30 hectares/min (75 acres/min) at 130 to 150 knots.

6. Control of Spraying Operations

6.1 Whichever method is employed to apply dispersants, an objective assessment is required to ensure that a vessel or aircraft spraying operation is conducted properly and effectively. Direction of the operation and observation of its effectiveness can best be conducted from another controller (spotter) aircraft overhead. This can be a light airplane or helicopter, but it must have a high endurance and good radio communications with the spray aircraft or vessel. An airborne observer can not function adequately in the spraying aircraft. To ensure safety in such a case, all the aircraft must have planned for, and maintained, continuous communications.

6.2 Personnel in the controller (spotter) aircraft can identify the heavier concentrations of oil (or those slicks posing the greatest threat), direct spray aircraft or boats to the target, request spraying to be started and stopped, and judge the accuracy of the application. These aerial functions are important for spraying operations since oil visibility from a vessel or a spray plane is limited. Air support is essential when large multiengine aircraft are used for spraying. Even when using helicopters and small airplanes for spraying, it is not reasonable to rely on pilot observation, since all of the sprayed area is behind the aircraft. Consequently, the area of coverage and the effect of the dispersant is better seen by an observer in a control plane at a higher altitude, who also can better direct the spray plane on the next pass, in the same or a different treatment area.

7. Storage and Handling of Dispersant and Dispersant Application Systems

7.1 Dispersants are to be handled and stored in accordance with information provided by the manufacturer's Material Safety Data Sheets (MSDS), labels, and user-specified policies. (See 10.1.)

7.2 Dispersant application systems will normally be loaded, by means of pumps, with dispersants from drums, storage tanks, or tank trailers. Pumps of adequate capacity must be available to load dispersant rapidly in order to reduce aircraft downtime between sorties.

7.3 Conduct routine maintenance on dispersant application

systems and subcomponents in accordance with the manufacturer's recommendations, to ensure system readiness and availability for immediate use.

7.3.1 Nozzles on dispersant application systems must be cleaned and inspected after each day's operation and before storage of the system. Pumps and systems using seawater (as from vessels) must be rinsed well with fresh water.

7.3.2 The system calibration should be checked at least once a year (see Practice F 1460). Spray systems should be cleaned, and the calibration checked, after each exercise or spill incident in which the equipment was used, and after making any changes in the system configuration. Also, systems must be completely drained and freeze-protected, as necessary, after each use.

7.3.3 Any remote control devices used in operation of a dispersant system should be checked immediately prior to any use of the system.

7.3.4 Operating crews should be given comprehensive training in dispersant application systems installation and methods of use. Practical exercises should be held frequently.

8. General Considerations for Dispersant Application

8.1 *Primary Considerations*:

8.1.1 Use of dispersants, particularly in a specific area, may be subject to regulatory approval. Dispersant response is for use in the early stages of a spill; so, it is strongly recommended that a rapid approval mechanism, or pre-approval, be part of response planning.

8.1.2 *Nature of Oil Slick(s) to Be Treated*:

8.1.2.1 The effectiveness of dispersants is dependent (assuming proper application) on two factors; the oil composition and the sea surface energy. The primary factor is the oil composition. Heavier oils, those that contain large amounts of components such as asphaltenes, disperse poorly, and those which have only a small amount of these disperse more easily. As oil weathers on the sea surface, its composition changes and it generally becomes less dispersable. Some oils can also form highly viscous water-in-oil emulsions, known as "chocolate mousse," particularly in areas of high energy waves. Once mousse has formed, dispersants may not be effective.

8.1.2.2 Viscosity is an indicator of the oil composition, but affects dispersion by its influence on the amount of dispersant penetrating into and mixing with the oil. Dispersant can run off the surface of highly viscous oils or will mix only slowly with them. Traditionally, oils of a viscosity between 2000 and 10 000 mPa were thought to be undispersable. However, viscosity may not be as much a limitation as is composition as noted above, especially for dispersants which are not quickly lost to the water column. Viscosity may have its largest effect on the time required for mixing with the oil.

8.1.2.3 Natural weathering (evaporation) affects the composition and viscosity of the oil. Much of the oil evaporated will usually consist of the most dispersable fraction. Also, loss of the lighter fractions by evaporation increases the viscosity. This combined effect may rapidly reduce the dispersability of some spilled oils. Some oils are not effectively dispersed after only 24 h on the surface.

8.1.2.4 Surface sea energy can be an important factor in dispersant effectiveness. Higher sea energy is needed to disperse oil of less favorable composition. Very low sea energies often result in poor dispersant performance. Very high seas can be detrimental since they can promote water-in-oil emulsion formation and can cause oil slicks to become discontinuous or submerged. Spraying such slicks can result in significant dispersant loss.

8.1.3 *Environmental Conditions, Including Wind, Sea State, Visibility, and Temperature of Air and Water*— It is essential to minimize dispersant loss in aerial application due to wind drift and air turbulence. Large droplets assist in this, but, in addition, the aircraft should be flown as low as safety considerations allow. It is also best to fly into the wind while spraying, so as to limit wind drift.

8.1.4 *Availability of Application Logistic Support*:

8.1.4.1 Dispersant spraying from vessels has limitations due to the low area coverage rate and the inherent difficulties of locating oil slicks from a vessel. Vessels of all sizes need the assistance of an aircraft overhead to direct them in spraying dispersants.

8.1.4.2 Small airplanes with high endurance, low fuel consumption, rapid turn-around times, and an ability to operate from short and even improvised landing strips may be suitable for spraying close to shore. The ability of helicopters to spray in confined situations and operate from a base very near the spill site is often valuable, but only provided that an adequate supply of dispersant can be maintained in such areas. The advantages of larger multi-engine airplanes in range, payload, and speed has to be weighed against the need for longer runways, greater operational support, and the location or delivery site of dispersant supply.

8.1.4.3 Good organization is needed to maintain spraying operations for the maximum available time during daylight hours. This may require maintenance of equipment and replenishing of supplies during the night. Supplies of fuel and dispersant for aircraft and vessels are critical and must be maintained. Dispersants will usually be supplied in drums and often delivered in quantity to a large airport. Delivery in bulk by trucks or road tankers, with high-capacity pumps, could supply areas (and possibly smaller aircraft) nearer to the spill location; although, loads of dispersant in drums may be adequate to support a small-scale operation.

8.1.4.4 Further, provision must be made for relief crews for both flight and ground operations as the operation requires.

8.1.5 *Dispersant Stockpile*:

8.1.5.1 An adequate supply of suitable dispersant should be available to support at least two days of spray operations. For large spray operations, up to about 200 000 L (50 000 U.S. gal) may be required near the spill location. Both the supply of dispersant and the availability of freighter aircraft on short notice should be addressed in response planning.

8.1.5.2 Resupply of dispersant can be a limiting factor in continuing a dispersant spraying operation. Dispersant applications can continue during daylight hours, and five or more sorties by each airplane are possible per day, depending on the distance of the flight base from the spill and the facilities at the base.

8.2 *Calculation for Dispersant Application*:

8.2.1 The application rate required is determined by the type of oil, its thickness, and the prevailing conditions. Control of the application rate can be achieved in two ways, either by varying the pump discharge rate or the vessel or aircraft speed. The general relationship between the variables is shown as follows:

$$\frac{\textit{pump discharge rate}}{(L/min)} = 0.003 \times \frac{\textit{application rate}}{(L/h_a)} \times \frac{\textit{speed}}{(knots)} \times \frac{\textit{swath width}}{(m)}$$

8.2.2 As an example, a slick estimated to be about 0.2 mm thick, would represent an oil volume of 2 m^3/hectare, and a dispersant application rate of 50 to 100 L/hectare (5 to 10 gal/acre) would be required if a concentrate dispersant was used at a dispersant to oil ratio of 1:20. A vessel traveling at 10 knots, spraying a swath width of 30 m, would need to discharge dispersant at a rate of 100 L/min (25 U.S. gal/min) to achieve this application rate. If an aircraft flying at 90 knots (105 mph) with an effective swath of 15 m was used instead to treat the same slick, a discharge rate of 400 L/min (100 U.S. gal/min) would be required.

8.3 *Correlation of Slick Thickness With Volume per Unit Area and Practical Dispersant Application.*

8.3.1 The purpose of all oil spill response, including use of dispersants, is, so far as possible, to prevent spilled oil from reaching and impacting shorelines, or from entering areas of particular environmental sensitivity. Within this parameter, it is usually most important to concentrate on the treatment of the thickest regions of the slick. It is ineffective and inappropriate to apply dispersants on sheens. Typically, these two types of slicks can be easily identified from the air.

8.3.2 Dispersant application rates of the order of 30 to 100 L/hectare (3 to 10 U.S. gal/acre) have been found to be appropriate in most situations. Thicker patches may call for an increase in the application rate. It is best to start treatment of the oil at a rate of about 50 to 70 L/hectare (5 to 7 U.S. gal/acre) and reduce this as indicated by the observed results. Some spills have been treated successfully with 20 to 30 L/hectare (2 to 3 U.S. gal/acre). Discharge rates can be varied providing the nozzles selected produce a stable spray and the discharge is monitored by a flowmeter or other calibrated system.

9. Vessel and Aircraft Safety

9.1 Great care should be used while working around vessels or aircraft. The many hazards normally associated with operation of equipment is compounded with the rolling of vessels and the presence of rotating propellers of aircraft. The captain of the vessel, or the pilot of an aircraft has the final approval or disapproval of all activities conducted by or upon their craft. Operators of vessels and aircraft should make auxiliary personnel aware of the particular safety concerns associated with their operations. The crews of these vessels and aircraft have training and experience that go beyond the scope of this guide. Their instructions concerning the operation of, and safety precautions related to, their craft should be heeded without question.

9.2 Dispersants can create a slippery, and therefore unsafe, work area, especially on the deck of a vessel, and spills should be minimized by promotion of good housekeeping. Dispersants also exhibit a strong degreasing action and regular checks are recommended to ensure that they do not contaminate lubricants on machinery. Vessels and aircraft should therefore be hosed down regularly with fresh water to remove salt water spray as well as dispersant.

10. Dispersant Deployment and Monitoring Activities

10.1 It is important that all workers are well trained in handling dispersants and wear appropriate personal protective equipment (PPE) when transferring a chemical to boats or airplanes. Excessive exposure to the chemical vapors should be avoided, as recommended or required by the particular MSDS. Any personnel aboard vessels near the area where dispersants are being used should remain off open decks until the spraying operation is concluded. If dispersant mist consistently blows onto crew areas of vessels engaged in spraying, the spray system needs to be evaluated for the need to change the nozzle design or the pump operation and pressure, so that the dispersant is deposited at the desired dosage on the oil only.

Designation: F 1738 – 96 (Reapproved 1999)

Standard Test Method for Determination of Deposition of Aerially Applied Oil Spill Dispersants[1]

This standard is issued under the fixed designation F 1738; the number immediately following the designation indicates the year of original adoption or, in the case of revision, the year of last revision. A number in parentheses indicates the year of last reapproval. A superscript epsilon (ϵ) indicates an editorial change since the last revision or reapproval.

1. Scope

1.1 This test method covers the measurement of the deposition of an aerially applied dispersant on the surface of the ground or water. The test method of obtaining these measurements is described, and the analysis of the results, in terms of dispersant use, is considered. There are a number of techniques that have been developed, and this test method outlines their application. These measurements can be used to confirm or verify the specifications of a given equipment set, its proper functioning, and use.

1.2 This test method is applicable to systems used with helicopters or airplanes.

1.3 This test method is one of four related to dispersant application systems. Guide F 1413 covers design, Practice F 1460 covers calibration, Test Method F 1738 covers deposition, and Guide F 1737 covers the use of the systems. Familiarity with all four standards is recommended.

1.4 There are some exposure and occupational health concerns regarding the methods described. These are not discussed in this test method since they are a function of dispersant formulation. Anyone undertaking such experiments should consult the occupational health experts of the dispersant manufacturer regarding the precautions to be used.

1.5 The values stated in SI units are to be regarded as the standard.

1.6 *This standard does not purport to address all of the safety concerns, if any, associated with its use. It is the responsibility of the user of this standard to establish appropriate safety and health practices and determine the applicability of regulatory limitations prior to use.*

2. Referenced Documents

2.1 *ASTM Standards:*

F 1413 Guide for Oil-Spill Dispersant Application Equipment: Boom and Nozzle Systems[2]

F 1460 Practice for Calibrating Oil Spill Dispersant Application Equipment Boom and Nozzle Systems[2]

F 1737 Guide for Use of Oil-Spill Dispersant Application Equipment During Spill Response: Boom and Nozzle Systems[2]

3. Significance and Use

3.1 The deposition of an aerially applied dispersant is defined as the amount of an aerially applied dispersant that contacts the surface; whereas, application dosage (frequently referred to as application rate) is the amount of material that is released per unit area by the delivery system. The units of deposition are litres per hectare or U.S. gallons per acre. The deposition may differ from the application dosage (volume of material per unit area) for many reasons, such as, the effects of wind on the spray and the evaporation of the dispersant after it has been released from the aircraft.

3.2 This test method describes the measurement of the ability of a spray system to deposit a dispersant on oil. It is not intended that this test method be used at the time of a spill. These techniques are intended to determine the equipment performance during the development of new systems and after the repair or significant modification of a system.

3.3 The data obtained from the use of this test method can be directly related to the deposition of dispersant on an oil slick, and thus can serve to determine both the dispersant deposition and the droplet size.

3.4 Dispersant deposition and droplet size data can be used as a technical basis for the optimization of dispersant application equipment and its use.

4. Apparatus and Materials

4.1 The basic concept is to provide a collection surface on which the aerially applied material is deposited. The amount of material and the deposition pattern and its droplet size can be measured using this surface. Several systems and methods have been developed, and each has its own advantages and disadvantages.

4.2 These measurements require a large, flat open area (such as a field or an airport) which is suitable for low-level flying and maneuvering. The location should be away from human habitations or environmentally sensitive areas in order to minimize problems due to noise and drifting spray.

4.3 These field programs should be conducted under low-wind conditions in order to minimize drift. Near-surface turbulence due to thermal gradients or atmospheric instability

---

[1] This test method is under the jurisdiction of ASTM Committee F-20 on Hazardous Substances and Oil Spill Response and is the direct responsibility of Subcommittee F20.13 on Treatment.

Current edition approved Oct. 10, 1996. Published December 1996.

[2] *Annual Book of ASTM Standards*, Vol 11.04.

can contribute to a variation in the results. These measurements cannot be carried out in the presence of precipitation or in heavy concentrations of dust.

4.4 All tests are to be conducted with the flight path in an upwind direction. The upwind direction is chosen to simplify the interpretation of the data and to conform with typical field practice. It may be necessary to alter the flight path slightly for changes in wind direction during the course of an experimental program.

4.5 It is common practice to use a dye, soluble in the dispersant, which will assist in the detection of the dispersant by the analysis system. Oil Red B and Rhodamine WP have been used at concentrations of 0.1 to 2.0 %. The sensitivity of current detection systems allows the use of concentrations at the 0.1 % level or less.

4.6 The area used will become covered with dispersant spray, and it is suggested that the area not be used for agricultural purposes at least until any evidence of the dispersant or dye is no longer observable. The length of time depends on the weather conditions, especially precipitation that occurs after the spray program has been completed.

5. Deposition Measurement Methods

5.1 These techniques involve the use of a collecting surface of known area and the measurement of the amount and character of the dispersant deposited on this area. A variety of systems may be used, such as the following:

5.1.1 *Glass Petri Dishes or Similar Containers*—Flat dishes of known area are placed in a line perpendicular to the flight path, and extending over a distance 25 % greater than the expected swath width. Dishes of a diameter of 120 to 140 mm are typically used. There should be about twenty dishes placed across the flight path in order to have an adequate number of sampling points. In a typical experimental setup, the distance between sampling dishes should be greater than one metre and less than three metres. This criteria may require more or less than twenty dishes depending on the spray system being tested. Each sampling dish should be identified by a unique label, indicating its place on the sampling line and the number of the spray pass. The marking should be made in such a fashion that it will not be removed by the dispersant, the material used to dissolve the dispersant, water, or rough handling. The sampling dishes are kept covered until just before the spray run to reduce the possibility of contamination. The placement, uncovering, and retrieval of these dishes is labor intensive. After the spray run, the dishes are collected and washed with a suitable solvent, such as methanol or hexane, to collect the deposited material. The amount of dye present can be determined by using a colorimeter sensitive to the dye used. The system must be calibrated using a sample of the dyed dispersant and solvent mixture for that experimental pass. For these measurements, care must be taken to ensure that the same dilution factors are used for both the calibration and material from the sampling dishes, since the measurement instruments are only linear over about an order of magnitude of concentration. From these sets of data, the amount of material deposited on the surface in any units required, such as litres/hectare (U.S. gal/acre), can be calculated.

5.1.2 *Metal Troughs*— A variation of the sampling dish is a V-shaped metal trough, divided into sections and placed perpendicular to the flight path. Each section is about two metres long with a cross section of about 6 cm. A number of troughs, connected end-to-end, are used to cover a length of about 25 % greater than the total spray width. After a spray run, the troughs are washed with a solvent, such as methanol or hexane, and the eluent from each section is collected for analysis. The concept is similar to that of the glass dishes, but this system has the advantage of sampling the total spray width, and providing an average dose over the discrete section. One major advantage of the troughs is that they remain in place during a number of experimental runs, thus reducing the time between runs. This allows for more runs per day.

5.1.3 *String Measurement*—Another method uses a cord or string that is either stretched across the width of the spray or is supported on a series of stands. Except for very narrow-width application systems, the string is supported about every two metres by a stand. The dispersant is collected by the string, and thus the needed data are obtained. Since the cross section of the string is much smaller than that of the Petri dish or trough, more dye may be needed in the sprayed dispersant. The string is then allowed to dry. The amount of material that the string collected is determined by a fluorometric or colormetric technique. This method measures the relative deposition only, and not the absolute deposition.

5.1.4 *Data Determination*—The data collected from these types of measurements is the same in character. The amount of dispersant that reaches the ground is measured as a function of the position along the swath of the spray. From this, spray patterns can be determined and plotted. Data gathered using dishes and the metal troughs can be used to compute the actual deposition.

6. Drop-Size Determination

6.1 While the techniques of Section 5 provide an accurate measurement of the deposition, they do not give any indication of the drop size or drop-size distribution. Since this is an important parameter in the proper use of dispersants, drop-size measurements are also required in order to characterize a dispersant application system. The basic principle of most drop-size measurements is to capture the falling drops on a surface and then measure the area of the drop. The surface must be calibrated so that the conversion factor from drop volume to surface drop diameter is known.

6.1.1 The analysis of such drop sizes is expressed as a volume median diameter (VMD). The VMD is the effective diameter of a distribution of various drop sizes. It represents a single parameter description of a spray-pattern droplet size distribution and is statistically based. Therefore, VMD cannot be used to compute the terminal velocity of a drop or its momentum. It is the momentum that is critical for the dispersant, since this determines the probability of the droplet penetrating the slick.

6.2 Most techniques developed for pesticide drop-size measurements fail since the deposition for dispersants is several orders of magnitude greater than those used for pesticides. When these techniques are used for dispersants, the flux of droplets are so dense that they overlap, and thus, individual particles cannot be measured.

6.3 There are a number of methods that have been used in the measurement of drop size. Most use paper as the absorbing material. One common system uses specially coated cards such as those manufactured by Ciba-Geigy. There are two products that are typically used: a water-sensitive paper, that is yellow in color and stains blue when exposed to water and the other is white which stains blue when exposed to organic materials. These materials can be used to measure spray distributions and swath widths as well as droplet size. Special paper that is used by the printing industry for color reproduction can be used for the same purpose. Another system collects the drops on rolls of paper tape. All such methods require the calibration of the detection medium in terms of the relationship between droplet size and the drop area on the material. This is done in the laboratory.

6.4 Drop size can be determined by measuring the size of the projected image of the drop. Counting the drops and determining the drop size can be done either manually or using electronic image analysis systems. A large number of drops must be examined in order to achieve good statistics. Manual counting is a time-consuming and tedious process. Modern electronic image analysis systems are expensive and take considerable time and skill in order to produce high-quality results.

6.5 The use of recently developed laser-scattering systems have yet to be demonstrated to be successful in field measurements. The sampling volume of most of these systems is quite small, and the density of drops that traverse this volume is small. Thus, there are problems in obtaining good statistics with such a system.

7. Data Analysis

7.1 There are two types of information that need to be derived from these tests. The first is the distribution of material across the swath width and the second is the determination of the range of the drop sizes.

7.2 The determination of the distribution of material across the swath width can be done by extracting the information from the volume of material deposited on the Petri dishes, in the V-troughs, or on the string. The final output is a graph such as is shown in Fig. 1. These data are easy to interpret. There was a wing tank on the starboard side of the aircraft which produced some turbulence and thus reduced the amount of spray deposited as shown in the left of the graph. The graph slopes slightly upward indicating a slight crosswind moving from the left to the right. The random peaks on the right are caused by the slight crosswind. The pattern is good, and with the exception of the wing tank, there are no problems. The effects of the wing tank are not large enough to exclude the use of the plane for dispersant application with the existing spray system. Larger nozzles in the area of the wing tank could produce an even flatter pattern. The errors in this sort of measurement are about 10 %. This is not a significant error.

7.3 The second is the droplet size distribution. The graph in Fig. 2 shows a typical droplet-size distribution. While this graph can be used directly to determine the droplet size distribution, the most common representation of these data are summarized in a single parameter, the VMD. This can be calculated very simply, directly from the data used to prepare the graph. In Fig. 2, there are 100 size ranges, each 10 µm in width. The number and width of ranges is a characteristic of the measurement equipment. In order to calculate the VMD, assume the diameter of each drop is the average value of the range, that is the 20 to 30-µm range has a diameter of 25 µm. Calculate the volume, V, of such a drop using the following formula:

$$V = \frac{\Pi}{6} D^3$$

where:

D = mean diameter of the range.

This must be done for all 100 ranges. From this volume data, the VMD can be calculated from the following equation:

$$VMD = \sqrt[3]{\frac{6\ \Sigma(V)}{\Pi\, \Sigma\,(N)}} = \sqrt[3]{1.91\,\frac{\Sigma\,(V)}{\Sigma(N)}}$$

where:

$\Sigma\,(V)$ = sum of the volumes (all 100 ranges), and
$\Sigma\,(N)$ = total number of particles (all 100 ranges).

For example, from the data in Fig. 2, the linear-symmetric curve has a VMD of 464 µm and the nonlinear asymmetric

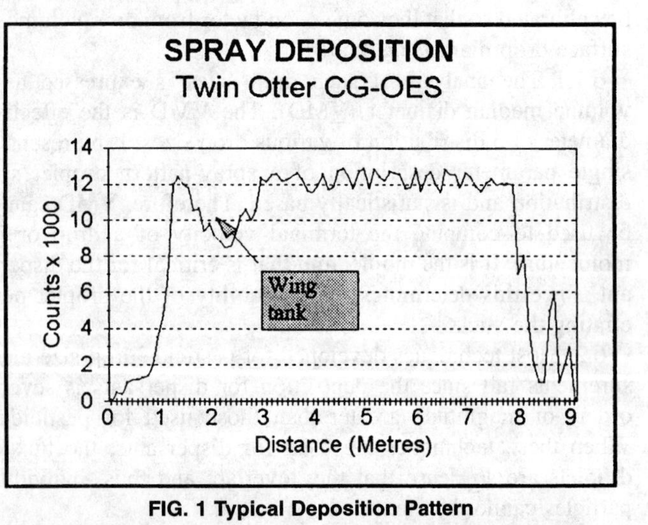

FIG. 1 Typical Deposition Pattern

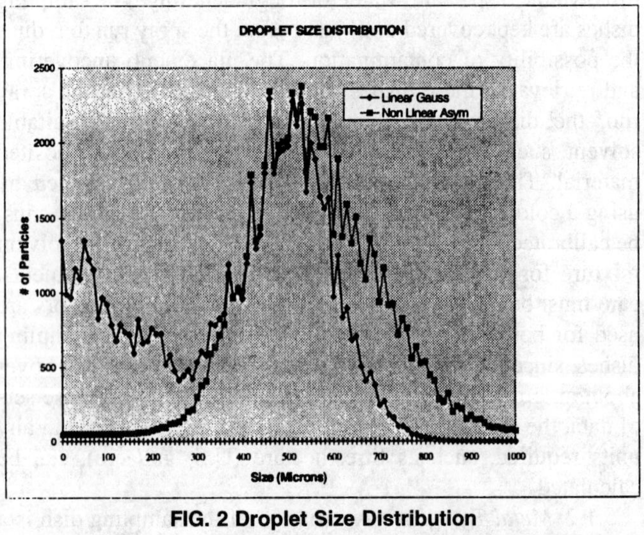

FIG. 2 Droplet Size Distribution

curve has a VMD of 593 μm.

7.3.1 It should be noted that the VMD for both distributions are similar and reflect that limited influence of the small particles and those particle sizes which are present outside the main curve. With high-quality data, such as shown here, the visual estimation of 500 μm is adequate and would reduce the amount of calculations needed to determine drop size.

8. Ancillary Concerns

8.1 With a knowledge of the deposition rate and the droplet size distribution, it is possible to evaluate the effectiveness of a dispersant application system to deliver the dispersant to the ocean surface at a known volume and droplet-size distribution. The delivery ratio is defined as the ratio of the amount of material that arrives at the surface to the amount of the dispersant pumped from the aircraft. If this ratio is less than about 70 %, then a study should be conducted to determine the sources of the loss, so that corrective action can be taken. Possible losses are due to wind drift and evaporation of the material during its transit through the air. Both these effects become more prominent as the height of the aircraft above the surface increases. Wind sheer at the nozzle and high turbulence in the atmosphere can result in the production of small droplets that do not reach the ground and hence can cause a low delivery ratio.

Designation: F 1778 – 97

Standard Guide for Selection of Skimmers for Oil-Spill Response[1]

This standard is issued under the fixed designation F 1778; the number immediately following the designation indicates the year of original adoption or, in the case of revision, the year of last revision. A number in parentheses indicates the year of last reapproval. A superscript epsilon (ϵ) indicates an editorial change since the last revision or reapproval.

1. Scope

1.1 This guide covers considerations for selecting skimmer systems for the recovery of marine-oil spills. The purpose of this guide is to provide oil spill response planners, equipment manufacturers, users, and government agencies with a standard on the equipment selection process for the removal of oil from the marine environment.

1.2 *This standard does not purport to address all of the safety concerns, if any, associated with its use. It is the responsibility of the user of this standard to establish appropriate safety and health practices and determine the applicability of regulatory limitations prior to use.*

1.3 This guide does not address the compatibility of spill-control equipment with spill products. It is the user's responsibility to ensure that any equipment selected is compatible with anticipated products and conditions.

2. Referenced Documents

2.1 *ASTM Standards:*

F 625 Practice for Classifying Water Bodies for Spill Control Systems[2]

F 631 Guide for Collecting Skimmer Performance Data in Controlled Environments[2]

3. Terminology

3.1 *Definitions:*

3.1.1 *encounter rate*—the volume of oil per unit time actively directed to the removal mechanism.

(F 631)

3.1.2 *nameplate recovery rate*—the maximum skimming capacity of a device under optimum conditions of oil type, slick conditions, and environmental conditions.

3.1.3 *oil recovery rate*—the volume of oil recovered by the device per unit time.

(F 631)

3.1.4 *recovery efficiency*—the ratio, expressed as a percentage, of the volume of oil recovered to the volume of total fluids recovered.

(F 631)

3.1.5 *throughput efficiency*—the ratio, expressed as a percentage, of the volume of oil recovered to the volume of oil encountered.

(F 631)

4. Significance and Use

4.1 This guide is intended to facilitate the oil spill response equipment selection process for local, regional, and national spill response teams. It is not intended to define rigid sets of equipment standards.

4.2 The effectiveness of the equipment chosen to combat an oil spill will depend on the oil type and environment(s) encountered, as well as other factors. This guide is intended to be used by persons generally familiar with the practical aspects of oil spill cleanup operations including on-scene response coordinators, planners, oil spill management teams, oil spill removal organizations, and plan evaluators.

4.3 Eleven general types of skimming systems are described in this guide. Each description includes a summary of the operating principle and a list of selection considerations.

4.4 Selection considerations are included to guide the user on the selection of a particular skimmer type or category. Users are cautioned that within each category there may be a wide variation in performance among various skimmers.

4.5 When selecting a skimmer for use in extremely cold conditions, consideration should be given to the effect of ice forming on the skimmer, changes in buoyancy, possible restriction of inlets, and changes to hydraulic efficiency. Because there may be wide variations in skimmer performance at extreme temperatures, even within a given category, ambient temperature is not included as a selection consideration.

5. Skimmer Selection Considerations

5.1 Selecting a type of skimmer for a given application involves examining the skimmer's likely performance against a range of operational requirements. The following are recommended as a guide to this process, with the requirements grouped according to the operating environment, the slick conditions, and skimmer performance criteria. Comments on each of these operational requirements, specific to each skimming type, are given in Section 6.

5.2 *Operating Environment*:

5.2.1 *Wave Conditions*—Depending on the type of skimmer, waves may affect both the oil-recovery rate and the oil-recovery efficiency. In general, most skimmers work best in

---

[1] This guide is under the jurisdiction of ASTM Committee F-20 on Hazardous Substances and Oil Spill Response and is the direct responsibility of Subcommittee F20.12 on Removal.

Current edition approved Feb. 10, 1997. Published April 1997.

[2] *Annual Book of ASTM Standards*, Vol 11.04.

calm conditions with decreases in recovery rate and efficiency as waves increase; in particular most skimmers are greatly affected by short choppy waves. For each skimming type, comments are given on the sensitivity to different wave environments. (Classification information for calm, protected, and open-water environments is given in Practice F 625.)

5.2.1.1 Additional considerations for selecting skimmers for particular wave conditions are that both the skimming device and the required support platform are applicable to the intended environment.

5.2.2 *Currents*—Should the operating environment of concern have significant water currents, this should be considered in the selection process. Many skimmers will not operate effectively in currents greater than 1 knot due to decreases in throughput efficiency. On the other hand, several skimming principles such as the sorbent belt, brush, submersion plane, advancing weir, and boom skimmers rely on relative current for effective operation and are applicable in currents greater than 1 knot. Several skimmer types such as the oleophilic brush and oleophilic rope mop are available as well, in configurations that allow them to operate effectively in high currents.

5.2.2.1 Comments on performance in currents are restricted to those directly related to the skimming device. If containment booms are used to collect or concentrate oil, or both, for skimming, additional operating limitations related to containment boom performance in currents may also apply.

5.2.3 *Water Depths*—Water depths may be a concern for skimming operations in nearshore waters and when supporting a shoreline cleanup operation. It is important to note that any draft limitations would apply to both the skimming device as well as the required support vessel; certain skimmer types, although not requiring deep water for their skimming component, may require large support vessels for deployment and operation.

5.2.3.1 While many skimming types are available in a range of sizes that may allow their operation in shallow waters, certain types are generally applicable to shallow water depths and these are noted in the skimmer descriptions. Skimmer types in this category include rope mop skimmers, vacuum systems, air conveyors, some weir skimmers, and some sorbent belt skimmers.

5.3 *Slick Conditions*:

5.3.1 *Oil Type and Viscosity*—Few skimming principles operate with optimum effectiveness over a wide range of oil viscosities. For many skimmers, recovery rates will tend to be less than the reported maximum rates for oils that have a very low viscosity, and for oils that have a very high viscosity, either initially or as a result of weathering or emulsification, or both. On the other hand, several skimming principles such as brush, drum, and paddle-belt skimmers operate most effectively with more viscous oils.

5.3.1.1 Recommended viscosities for skimmer performance data are given in Guide F 631. These include: 200, 2000, and 60 000 mm^2/s. These values are referred to in the skimmer selection considerations as low, medium, and high viscosity oil.

5.3.2 *Slick Thickness*—Slick thickness greatly affects the recovery rate of virtually all skimming principles, nameplate recovery rates only being achieved with thick slicks. Slick thickness also affects the recovery efficiency of most skimmers, in particular those employing suction or weir skimming principles. On the other hand, many skimmers can deal effectively with thin slicks by varying the operating parameters of the device (that is, for oleophilic devices, by reducing the speed of the oleophilic element, and by adjusting the weir settings for weir devices).

5.3.2.1 In evaluating a skimming principle based on expected slick thickness, consideration should be given to whether containment booms can be used to concentrate oil for recovery.

5.3.2.2 In the skimmer descriptions in Section 6, reference is made to device performance in thin slicks, which is defined as continuous slicks less than 1 mm in thickness, or discontinuous patches of oil with an average thickness of less than 1 mm.

5.3.3 *Debris*—The presence of debris presents two concerns in a skimming operation: first; that debris may restrict flow to the skimming head, and second, that debris may interfere directly with the skimming component (that is, clogging or obstructing openings, impeding moving parts). Some skimming types, such as most oleophilic devices, are largely insensitive to the presence of debris. Suction and air conveyor devices are generally tolerant of debris up to the size of the transfer hoses used. Weir devices, in general, are susceptible to clogging with debris; however, many weir devices use integral transfer pumps that can process a range of debris types. Where applicable, comments are made on debris tolerance or sensitivity in the selection considerations comments. It should be noted that these comments pertain to the general skimming type, and that certain skimmers within a given category may include means of dealing with debris. Guidance for a range of debris types respecting skimmer performance can be taken from Guide F 631, which lists a number of debris forms for skimmer testing.

5.4 *Performance Requirements*:

5.4.1 *Recovery Rate*—For some applications the most important performance criteria will be the product recovery rate. Although the various skimming categories are available in a range of sizes and capacities, generalizations can be made on expected recovery rates. In this guide, comments are made on expected recovery rates only as they would apply to the various skimming principles and not to particular devices. For example, weir skimmers and boom skimmers are available with high nameplate recovery rates, limited only by the available pumping capacity. On the other hand, oleophilic skimmers tend to have a fixed upper limit of recovery depending on the size of the oleophilic surface used (that is, surface area of discs, length and diameter of rope mop).

5.4.1.1 Although specific skimmer performance data are not included in this guide, users are reminded that a skimmer's nameplate recovery rate should be used with caution as it may not accurately reflect skimmer performance under varying conditions of slick thickness, slick viscosity, and environmental parameters. Where possible, performance data based on field use or experiments should be used, with reference to the slick and environmental conditions of particular concern to the user.

5.4.2 *Recovery Efficiency*—Selection of a skimmer based on the expected recovery efficiency may be particularly important depending upon the availability of storage, the availability of systems to separate free water from the recovered fluids, and the permissibility of discharging decanted water at the recovery site. The expected recovery efficiency will, for most skimming categories, vary greatly depending on the thickness and viscosity of the slick and on the environmental conditions at the spill site.

5.4.2.1 In general, skimmers using oleophilic principles can be expected to have higher recovery efficiencies relative to skimmers using weir or suction principles. Among weir skimmers, devices in the induced flow category can be expected to have a high efficiency. As well, several skimming categories are typically configured with onboard gravity separation, which would enhance their overall efficiency. For skimmers without onboard separation, oil/water separation should be considered to maximize the use of available storage.

5.4.3 *Mode of Application*—Comments on the mode of application include the ability to use in an advancing mode and the applicability to use on a vessel-of-opportunity.

5.4.3.1 Certain skimmer categories, such as oleophilic disc, rope mop, and some weir skimmers are not generally used in an advancing mode. Conversely, devices such as the boom skimmer, the fixed submersion plane, paddle belt, and oleophilic brush skimmers require relative forward motion for effective operation.

5.4.3.2 Vessel-of-opportunity application will in many cases be specific to a skimming device rather than a skimming category. However for those skimming categories that are typically used with vessels-of-opportunity, this is noted in the selection considerations.

6. Description of Main Skimming Types

6.1 The following describes the operating principles and key selection considerations of eleven main types of skimming systems. In several instances, subcategories are used to describe different configurations of a common operating principle.

6.2 *Boom Skimmers*:

6.2.1 *Description*—Boom skimmers include any device in which the skimmer is incorporated in the face of the containment boom, regardless of the skimmer type. This system can include a single skimmer installed in the face of the boom, but in many examples of this concept there are several skimmers used. In most boom skimmers, weir-type skimmers are used. Boom skimmers provide a combined containment and recovery system.

6.2.1.1 Oil spill containment boom is often attached to each side of the mouth of a skimmer in order to increase the sweep width. Although such a system would be similar to a boom skimmer, it would not meet the definition of a boom skimmer because the skimmer in the system could be used apart from the boom. In a boom skimmer, the skimmer is part of the boom and is not intended to be used by itself.

6.2.2 *Selection Considerations*:

6.2.2.1 *Oil Type*—Applicable to low and medium viscosity oils.

6.2.2.2 *Debris Tolerance*—Debris must be screened or removed from the skimmer opening.

6.2.2.3 *Wave Conditions*—Recovery rate and efficiency degraded by choppy waves.

6.2.2.4 *Currents*—May be operated at currents greater than one knot, at reduced recovery efficiency, by pumping at a high rate.

6.2.2.5 *Water Depth*—Generally limited by towing vessels.

6.2.2.6 *Mode of Application*—Requires relative forward velocity: may be operated in stationary mode if current present.

6.2.2.7 *Other*—Typically designed for vessel-of-opportunity application.

6.3 *Brush Skimmers*:

6.3.1 *Description*—Brush skimmers are oleophilic skimmers that pick up oil on the bristles of a brush. There are two main configurations for the brushes: drum brush skimmers, in which the brushes are mounted around the perimeter of a drum; and chain brush skimmers, in which the brushes are mounted on several continuous loop chains. In each case the brushes are rotated through the oil/water interface, picking up oil and some water. The recovered fluid is then combed from the bristles into a sump. Both brush skimmer types are generally used in an advancing mode. Chain brush skimmers are typically configured with the skimmer head facing aft, creating a calm area for oil to accumulate and be recovered, reducing the skimmer's sensitivity to waves.

6.3.2 *Selection Considerations*:

6.3.2.1 *Oil Type*—Applicable to medium and high viscosity oils.

6.3.2.2 *Debris Tolerance*—Effective in most forms of small debris.

6.3.2.3 *Wave Conditions*—Low sensitivity to waves with typical configuration of aft-facing skimmer head.

6.3.2.4 *Currents*—May be operated effectively at advance rates greater than 1 knot.

6.3.2.5 *Water Depth*—Generally limited by support vessel.

6.3.2.6 *Mode of Application*—Requires relative forward velocity: may be operated in stationary mode if current present.

6.3.2.7 *Other*—Some units designed for vessel-of-opportunity application.

6.4 *Disc Skimmers*:

6.4.1 *Oleophilic Disc Skimmers*:

6.4.1.1 *Description*—oleophilic disc skimmers use the principle of oil adhering to a solid surface, and typically include a series of discs that are rotated through the slick. As each disc is rotated through the oil/water interface, oil adheres to the disc surface and is then removed by scrapers mounted on both sides of each disc. The product is collected in a common sump and pumped away. Disc skimmers are typically powered by a remote power pack (hydraulic or air-driven), which results in a light, compact skimming head that is easily transported and highly maneuverable.

6.4.1.2 *Selection Considerations*:

(1) Oil Type—Applicable to low and medium viscosity oils.

(2) Debris Tolerance—Debris must be managed to allow the flow of oil to the skimmer.

(3) Wave Conditions—Effective in long period waves or short waves with a height not greater than the disc diameter.

(4) Currents—Not generally applicable to use in advancing mode.

(5) Water Depth—Typically available in small portable units with minimal draft.

(6) Mode of Application—Typically used in stationary applications.

6.4.2 *Star Disc Skimmer*:

6.4.2.1 *Description*—The star disc skimmer uses rotating discs to recover oil through mechanical, rather than oleophilic principles. The discs have a series of teeth around their perimeter, similar to a circular saw blade; as the discs are rotated these teeth draw oil into a central sump where it is then removed by a pump.

6.4.2.2 *Selection Considerations*:

(1) Oil Type—Applicable to highly viscous, almost solid oil.

(2) Debris Tolerance—Susceptible to blockage with debris, particularly rope and stringy forms.

(3) Wave Conditions—Low sensitivity to waves.

(4) Currents—Not generally applicable to use in advancing mode.

(5) Water Depth—Typically available in portable units with minimal draft.

(6) Mode of Application—Typically used in stationary applications.

6.5 *Drum Skimmers*:

6.5.1 *Oleophilic Drum Skimmers*:

6.5.1.1 *Description*—An oleophilic drum skimmer uses adhesion of oil to the surface of a cylindrical drum for recovery. As the skimmer drum is rotated through the slick, oil adheres to the drum surface and is scraped off into a sump and then pumped away. Drum skimmers are typically powered by a remote power pack (hydraulic or air-driven), which results in a light, compact skimming head that is easily transported and highly maneuverable.

6.5.1.2 *Selection Considerations*:

(1) Oil Type—Applicable to a range of oil viscosities.

(2) Debris Tolerance—Debris must be managed to allow the flow of oil to the skimmer.

(3) Wave Conditions—Effective in long period waves or short waves with a height not greater than the drum diameter.

(4) Currents—Not generally applicable to use in advancing mode.

(5) Water Depth—Typically available in small portable units with minimal draft.

(6) Mode of Application—Typically used in stationary applications.

6.5.2 *Helical Drum Skimmers*:

6.5.2.1 *Description*—A helical drum skimmer employs the rotation of the drum to generate a current that draws oil into the drum. Once inside the drum, the oil moves through a spiral casing to the center, where a conveyer screw moves it to a sump. Excess water is drained out through perforations in the casing.

6.5.2.2 *Selection Considerations*:

(1) Oil Type—Applicable to high viscosity oils.

(2) Debris Tolerance—Debris must be managed to allow the flow of oil to the skimmer.

(3) Wave Conditions—Effective in long period waves or short waves with a height not greater than the drum diameter.

(4) Currents—Typically operated in low current environments.

(5) Water Depth—Generally limited by support vessel.

(6) Mode of Application—May be operated in stationary or advancing mode.

6.6 *Paddle Belt Skimmers*:

6.6.1 *Description*—Paddle belt skimmers use a series of paddles, attached to a belt, to lift oil out of the water. The basic concept includes a series of paddles that draw a wedge of oil and water up a ramp. The paddles move the fluid over the top of the incline and into a sump where it is pumped off. There are several variations of this skimming principle but the basic concept is much the same.

6.6.2 *Selection Considerations*:

6.6.2.1 *Oil Type*—Applicable to medium and high viscosity oil.

6.6.2.2 *Debris Tolerance*—Susceptible to clogging with debris, particularly long, stringy debris forms.

6.6.2.3 *Wave Conditions*—Applicable to use in calm and protected waters.

6.6.2.4 *Currents*—Typically operated in low current environments.

6.6.2.5 *Water Depth*—Skimming component has a shallow draft; support vessel may limit shallow water application.

6.6.2.6 *Mode of Application*—Typically used in stationary applications.

6.7 *Rope Mop Skimmers*—Rope mop skimmers employ long, continuous loops of oleophilic material that float on water. The mop material is a soft, smooth polypropylene material that attracts oil and repels water. Rope mop skimmers are available in various configurations: the most commonly used type is the stationary rope mop skimmer. Other variations of this type are not generally given specific names, but can be described by the way in which they operate.

6.7.1 *Stationary Rope Mop Skimmers*:

6.7.1.1 *Description*—In a stationary rope mop skimmer the rope loop is pulled through a wringer that removes oil along with some water. The rope is guided over the oiled water by one or more pulleys secured at convenient locations. It can be deployed in a single loop with one pulley or over a larger area by using two or three pulleys arranged in a triangular or rectangular pattern. A large area can be covered depending on the length of the ropes used. Rope mop skimmers are available in a wide range of sizes, and can be used under piers, in ponds, and parallel to shorelines during beach cleaning.

6.7.1.2 *Selection Considerations*:

(1) Oil Type—Applicable to a range of oil viscosities; may be difficult to collect and strip extremely viscous oil from the mop.

(2) Debris Tolerance—Skimming performance is not generally affected by debris, including broken ice.

(3) Wave Conditions—Good wave following characteristics in non-breaking waves.

(4) Currents—Typically operated in low current environments; may be operated in currents by positioning the rope

mops to minimize their velocity relative to the water.

(5) Water Depth—Not limited by minimal water depths.

(6) Mode of Application—Typically operated in stationary applications.

(7) Other—The rope can serve as a limited containment device in calm water.

6.7.2 *Suspended Mop Skimmers*:

6.7.2.1 *Description*—A suspended mop skimmer uses several mops that go through a skimmer head that is suspended over the skimming area with a crane. Because these devices use several mops, they have a much larger skimming capacity than conventional mop skimmers. They skim out a relatively small area and are stationary.

6.7.2.2 *Selection Considerations*:

(1) Oil Type—Applicable to a range of oil viscosities; may be difficult to collect and strip extremely viscous oil from the mop.

(2) Debris Tolerance—Skimming performance is not generally affected by debris, including broken ice.

(3) Wave Conditions—Good wave following characteristics in non-breaking waves.

(4) Currents—Typically operated in low current environments.

(5) Water Depth—Not limited by minimal water depths.

(6) Mode of Application—Typically operated in stationary applications.

(7) Other—Typically operated with a crane from a support vessel or pier.

6.7.3 *Zero Relative Velocity Skimmers*:

6.7.3.1 *Description*—*Zero Relative Velocity (ZRV) Skimmers* are rope mop devices used in catamaran hull vessels. A series of separate ropes (generally four to six) are arranged between the hulls. They are allowed to hang loosely on the water surface and are rotated aft at a velocity close to the forward speed of the vessel. Velocity of the ropes relative to the oil on the water is close to zero.

6.7.3.2 *Selection Considerations*:

(1) Oil Type—Applicable to a range of oil viscosities; may be difficult to collect and strip extremely viscous oil from the mop.

(2) Debris Tolerance—Skimming performance is not generally affected by debris, including broken ice.

(3) Wave Conditions—Good wave following characteristics in non-breaking waves.

(4) Currents—Can operate effectively in advancing mode or currents greater than 1 knot.

(5) Water Depth—Skimming component not limited by minimal water depths; support vessel will dictate draft requirements.

(6) Mode of Application—Requires relative forward velocity: may be operated in stationary mode if current present.

(7) Other—Typically configured as permanent installation on dedicated vessel.

6.8 *Sorbent Belt Skimmers*:

6.8.1 *Description*—Sorbent belt skimmers use an oleophilic belt to recover oil. The belt is made of porous oleophilic material that allows the water to pass through. The belt is positioned at an angle to the water with the leading edge of the belt immersed in the slick. At the top of its rotation the belt passes through a set of rollers where oil and water are removed from the belt through a combination of scraping and squeezing. Viscous oils tend to stay on the surface of the belt and are removed by scraping. Light oils are adsorbed in the mesh of the belt and removed by squeezing. With some models, recovered water may be decanted from the storage tank. Belt skimmers are typically supplied with a flow-induction device to draw fluid through the belt and reduce the head wave effect in stationary and advancing modes.

6.8.1.1 There are two main categories of sorbent belt skimmers: sorbent lifting belt skimmers, in which the belt lifts oil out of the water at the oil/water interface; and sorbent submersion belt skimmers, in which the belt rotates down through the oil/water interface and submerges the oil such that the buoyancy of the oil aids in its adhering to the belt.

6.8.2 *Selection Considerations*:

6.8.2.1 *Oil Type*—Applicable to medium and high viscosity oils.

6.8.2.2 *Debris Tolerance*—Relatively insensitive to most types of debris.

6.8.2.3 *Wave Conditions*—Low sensitivity to waves.

6.8.2.4 *Currents*—Some units designed to operate at advance rates greater than 1 knot.

6.8.2.5 *Water Depth*—Typical designs have minimal draft; draft requirement generally dictated by support vessel.

6.8.2.6 *Mode of Application*—Typically used in advancing mode; units with flow induction may be operated in stationary mode.

6.8.2.7 *Other*—Some units designed for vessel-of-opportunity application.

6.9 *Submersion Plane Skimmers*:

6.9.1 *Fixed Submersion Plane Skimmers*:

6.9.1.1 *Description*—Fixed submersion plane skimmers present a fixed or stationary plane to the oil/water interface as the skimmer is advanced through a slick. The plane causes an oil/water mixture to be submerged, and the buoyant force of the oil directs it up to a collection well. The collection well has discharge ports along its bottom, allowing water to be released and providing gravity oil/water separation.

6.9.1.2 *Selection Considerations*:

(1) Oil Type—Applicable to low and medium viscosity oils.

(2) Debris Tolerance—Performance may be degraded by debris.

(3) Wave Conditions—Low sensitivity to waves due to underwater collection.

(4) Currents—Applicable to currents greater than 1 knot.

(5) Water Depth—Typical designs have minimal draft; draft requirement generally dictated by support vessel.

(6) Mode of Application—Requires relative forward velocity: may be operated in stationary mode if current present.

(7) Other—Typically configured as part of a dedicated skimming vessel; some units used with vessels-of-opportunity.

6.9.2 *Submersion Moving Plane Skimmers*:

6.9.2.1 *Description*—Submersion moving plane skimmers present a moving plane, typically a conveyor-belt like material, to the oil/water interface and directs it under water to a collection well. The collection well has discharge ports along its bottom, allowing water to be released and providing gravity oil/water separation. With some designs, a wiper assembly at the collection well assists in removing viscous oil from the belt. The skimmer may be used in stationary or advancing mode. In stationary mode, oil is moved solely by adhesion to the belt; in advancing mode this is supplemented by hydrodynamic forces.

6.9.2.2 *Selection Considerations*:

(1) Oil Type—Applicable to a range of oil viscosities.

(2) Debris Tolerance—Capable of processing many types of debris; debris must be managed to allow the flow of oil to the skimmer inlet.

(3) Wave Conditions—Low sensitivity to waves due to underwater collection.

(4) Currents—Applicable to currents greater than 1 knot.

(5) Water Depth—Typical designs have minimal draft; draft requirement generally dictated by support vessel.

(6) Mode of Application—May be operated in stationary or advancing mode.

(7) Other—Typically configured as part of a dedicated skimming vessel; some units applicable to vessel-of-opportunity application.

6.10 *Suction Skimmers*:

6.10.1 *Stationary Suction Skimmers*:

6.10.1.1 *Description*—This category includes any simple suction head used on a hose from a vacuum truck or a portable pump. To be considered in this category the skimming head must only be a suction device and not include any oil/water separation device such as a weir. Typical skimming heads are small and maneuverable, allowing their use in confined spaces such as among dock pilings. They can be operated from small boats, or dock side with vacuum trucks. Oil/water separators are recommended to deal with the large volumes of water that may be recovered along with the oil.

6.10.1.2 *Selection Considerations*:

(1) Oil Type—Applicable to a wide range of oil viscosities. With viscous oils, flow may be limited by hose diameter and suction lift.

(2) Debris Tolerance—Debris must be managed to allow the flow of oil to the suction head; suction head susceptible to clogging with some types of debris.

(3) Wave Conditions—Recovery rate and efficiency severely degraded by choppy waves.

(4) Currents—Requires contained slick for effective use; subject to normal containment limits.

(5) Water Depth—Not limited by minimal water depths.

(6) Mode of Application—Typically operated in stationary applications.

(7) Other—The pump used to supply suction should be self-priming.

6.10.2 *Air Conveyors*:

6.10.2.1 *Description*—Air conveyors are also used as suction skimmers. In these systems oil and water are picked up at high velocity and carried through a large diameter hose into a large reception bin. Oil and debris become entrained in the high velocity air stream and are carried to the bin for temporary storage. Oil/water separators are recommended to deal with the large volumes of water that may be recovered along with the oil.

6.10.2.2 *Selection Considerations*:

(1) Oil Type—Applicable to a wide range of oil viscosities, including extremely viscous oils.

(2) Debris Tolerance—Able to process many types of debris; limited by size relative to suction hose.

(3) Wave Conditions—Recovery rate and efficiency severely degraded by choppy waves.

(4) Currents—Requires contained slick for effective use; subject to normal containment limits.

(5) Water Depth—Not limited by minimal water depths.

(6) Mode of Application—Typically operated in stationary applications.

6.11 *Weir Skimmers*:

6.11.1 *Description*—This category includes any weir device that uses gravity to drain oil off the water surface. Typically, the top edge of the weir is positioned just below the upper surface of the slick, allowing oil to flow over the weir into a collection sump, from where it is pumped to storage. With some devices in this category the top edge of the weir is above the slick and some means is used to move oil up and over the weir. With some units, the weir can be adjusted to limit the amount of free water collected along with the oil. Devices that have a fixed weir may recover significant volumes of water along with the oil: oil/water separators should be considered with such skimmers to maximize the use of available storage.

6.11.1.1 Weir skimmers are grouped under three main classifications: (*1*) units that use an external pump; (*2*) units that include a pump, typically an archimedean screw (positive displacement) pump; and (*3*) induced flow weir skimmers, which use some means of inducing the flow of oil to the skimmer.

6.11.2 *Weir Skimmers with External Pumps*—Skimmers in this category are generally small, portable units. The skimming head is typically small and maneuverable, allowing its use in confined spaces such as among dock pilings.

6.11.2.1 *Selection Considerations*:

(1) Oil Type—Applicable to low and medium viscosity oils.

(2) Debris Tolerance—Sensitive to most types of debris.

(3) Wave Conditions—Recovery rate and efficiency severely degraded by choppy waves.

(4) Currents—Requires contained slick for effective use; subject to normal containment limits.

(5) Water Depth—Typically have minimal draft; can be operated in very shallow water.

(6) Mode of Application—Typically used in stationary applications.

6.11.3 *Weir Skimmers with Integral Pumps*—Most skimmers in this category use screw pumps that do not require priming, handle viscous oil, are tolerant of most types of debris, and do not form oil/water emulsions. With some units the pump may be removed from the skimmer and used as a tanker offloading pump.

6.11.3.1 *Selection Considerations*:

(1) Oil Type—Applicable to a range of oil viscosities; will recover highly viscous oils that flow to the skimming head.

(2) Debris Tolerance—Capable of processing many types of

debris; debris must be managed to allow the flow of oil to the skimming head.

(3) Wave Conditions—Recovery rate and efficiency degraded by choppy waves.

(4) Currents—Requires contained slick for effective use; subject to normal containment limits.

(5) Water Depth—Support vessel will generally dictate draft requirements.

(6) Mode of Application—Applicable to stationary and slowly advancing mode.

(7) Other—Typically used with vessels-of-opportunity.

6.11.4 *Induced Flow Weir Skimmers*—Skimmers in this category use a mechanical or hydrodynamic force to draw oil to and over the weir. Two examples are: skimmers that use a rotor or propeller that rotates beneath the water surface; and skimmers that use a series of water jets positioned just below the water surface. In each example, the device creates a current that induces the flow of oil to the weir and this concentrating effect increases the recovery efficiency. The pump used to transfer oil from the skimming head may be either internal or external to the skimmer.

6.11.4.1 *Selection Considerations*:

(1) Oil Type—Applicable to low and medium viscosity oils.

(2) Debris Tolerance—Effective in most forms of small debris; devices using a rotor may be susceptible to long, stringy debris.

(3) Wave Conditions—Induced flow mechanism may lose effectiveness in choppy waves.

(4) Currents—Typically operated in low current environments; currents may degrade hydrodynamic effect.

(5) Water Depth—Typical designs have minimal draft; draft requirement generally dictated by support vessel.

(6) Mode of Application—Applicable to stationary and slowly advancing mode.

6.12 *Advancing Weir Skimmers*:

6.12.1 *Description*—Advancing weir skimmers are a variation on conventional weirs in that the forward motion of the skimming system provides the flow into the skimmer. Depending on the size of the weir opening, the skimmer will accept most of what it encounters, allowing it to handle highly viscous oils. Advancing weir skimmers typically recover large volumes of water; oil/water separation should be considered to increase overall efficiency.

6.12.2 *Selection Considerations*:

6.12.2.1 *Oil Type*—Applicable to a range of oil viscosities.

6.12.2.2 *Debris Tolerance*—Capable of processing many types of debris; debris must be managed to allow the flow of oil to the skimming head.

6.12.2.3 *Wave Conditions*—Recovery rate and efficiency degraded by choppy waves.

6.12.2.4 *Currents*—Can be operated effectively at speeds greater than 1 knot when used independent of additional containment boom.

6.12.2.5 *Water Depth*—Support vessel will generally dictate draft requirements.

6.12.2.6 *Mode of Application*—Requires relative forward velocity: may be operated in stationary mode if current present.

6.12.2.7 *Other*—Typically configured as part of a dedicated skimming vessel.

Designation: F 1779 – 97

Standard Practice for Reporting Visual Observations of Oil on Water[1]

This standard is issued under the fixed designation F 1779; the number immediately following the designation indicates the year of original adoption or, in the case of revision, the year of last revision. A number in parentheses indicates the year of last reapproval. A superscript epsilon (ε) indicates an editorial change since the last revision or reapproval.

1. Scope

1.1 This practice covers methods of reporting and recording visual observations of oil on water and related activities and phenomena.

1.2 This practice applies only to visual observations of oil on water from an airplane or helicopter. While a similar set of codes could be used for classifying oil on beaches, this subject is not discussed in this practice. It does not cover the use of remote-sensing equipment from aircraft, which is discussed in a separate standard.

1.3 This practice is applicable for all types of oil under a variety of environmental and geographical situations.

1.4 Visual observations of oil on water from the air involve a number of safety issues associated with the operation of airplanes or helicopters at low altitudes. These are not dealt with in this practice, but the observer should be aware of the hazards of such operations.

1.5 *This standard does not purport to address all of the safety concerns, if any, associated with its use. It is the responsibility of the user of this standard to establish appropriate safety and health practices and determine the applicability of regulatory limitations prior to use.*

2. Significance and Use

2.1 This practice can be used by surveillance and tracking staff to report visual observations to the clients of visual observations. The data produced from such observations will provide the basis for preparing maps of the oil-slick location.

2.2 This practice provides a procedure for reporting the visual observation of oil on water in a systematic manner and in a common format that can be readily understood by both observers and users of visual oil-spill observation maps.

2.3 This practice deals with the possibility that materials other than oil might be confused with oil when using visual observation methods.

3. Observational Methods

3.1 The basic information needed from a visual observation program includes the slick size and location, as well as its characteristics and shape. The location and some aspects of the effectiveness of a response operation can be easily determined from an aerial platform. Reports on the presence of wildlife and the proximity of oil to environmentally sensitive areas are useful information that can be generated from overflights.

3.2 Observers of oil on water will generally use either a helicopter or a small fixed-wing airplane. The planes shall be capable of slow-speed flight (120 to 240 km/h; 60 to 120 knots) for extended periods of time and have good forward and side visibility. The aircraft shall have adequate range and endurance consistent with the size and location of the spill.

3.3 If possible, two observers should be used, one on the port and the other on the starboard side of the aircraft. Provision shall be made for the two observers to communicate readily between themselves and with the flight crew. This can be accomplished by using standard aviation headphones and noise-cancelling microphones connected to an aircraft intercom system. Such equipment is readily available on the commercial market.

3.4 Standard maps shall be provided to the observers, so that their data can be easily transferred to a single map or map set for reporting and dissemination. These maps can be based on marine charts, topographic maps or special maps produced for the spill. Useful scales vary from 1:10000 to 1:50000. It is difficult to plot information to the required accuracy using maps of a larger scale.

3.5 The flight path shall be shown on all maps. An initial proposed flight path should be prepared prior to the flight. During the flight, deviations from this plan may be necessary in order to observe the total area of the slick.

3.6 Typical flight altitudes range from (100 to 1000 m (300 to 3000 ft) depending on the nature of the spill and on the cloud ceiling at flight time. There are flight safety considerations associated with low-altitude flying.

3.7 The best angle to observe an oil slick is directly above it looking straight down. This is known as a nadir observation. The flight path should be adjusted, wherever possible, to provide observers with a nadir view of the slick. For most fixed-wing aircraft, it is not possible to observe directly downwards, and therefore the flight path should be chosen to allow for observation of the spill at as near the vertical as possible.

4. Methods of Reporting

4.1 *Reporting Needs*—There are a number of characteristics of the slick that shall be reported in order to provide the user

---

[1] This guide is under the jurisdiction of ASTM Committee F-20 on Hazardous Substances and Oil Spill Response and is the direct responsibility of Subcommittee F20.16 on Surveillance and Tracking.

Current edition approved Feb. 10, 1997. Published April 1997 .

of visual observations with the appropriate amount of information. The following characteristics shall be reported for each slick that is observed during a reconnaissance flight. These observations are a snap-shot in time. Both the location and characteristics of an oil slick change rapidly. In order to be useful for responders, the information should be available a short time after it is obtained (typically less than 3 h).

4.2 *Color of Slick*—This shall be reported as a color code, as follows:

4.2.1 *Brown or Black*—*B*.

4.2.2 Brown or Black with a red tinge signifying presence of an *Emulsion* or mousse—*E*.

4.2.3 *Rainbow* sheen—*R*.

4.2.4 *Grey* or Silver Grey sheen—*G*.

4.3 *Percentage Coverage and Character of Slick*—Percentage of area as described that is covered by oil.

4.4 The character of the slick shall be noted such as follows:

4.4.1 *Windrows*—*W*.

4.4.2 *Continuous*—*C*.

4.4.3 *Tar Balls*—*T*.

4.4.4 *Pancakes*—*P*.

4.5 The eight parameters in 4.2 and 4.4 give a description of the oil slick.

4.6 *Slick Features*—If the leading edge of the slick can be identified, is should be noted as a heavy line.

4.7 The preceding data complete the description of the slick, as it would be placed on a base map.

4.8 *Other Properties*—There are a number of secondary related features that can be easily observed during a reconnaissance flight, and should be reported on the surveillance map, if appropriate. This recording of such information yields additional useful data from a surveillance flight, but should not be regarded as a primary function of the flight. These observations include:

4.8.1 Mechanical response operations (such as booms and skimmers)—*M*.

4.8.2 Dispersant or chemical response operations—*D*.

4.8.3 In-situ burning (Fire) response operations—*F*.

4.8.4 Shoreline cleanup in the area—*S*.

4.8.5 Animals or birds seen in area—*A*.

4.8.6 Wildlife Habitat in area—*H*.

4.8.7 Ocean features—*O*.

4.8.8 EXtra features such as ice, debris—*X*.

4.9 Each element shall be separated by a forward slash (/) except for other properties which are included as a group. Thus a slick could be described as **B/50/W/M**. This can be decoded as a slick containing Black oil with a 50 % coverage. The oil is in *W*indrows and a *M*echanical response operation is being undertaken.

5. Voice Communications

5.1 *Need for Voice Communications*—Since timeliness is very important, rapid communication of observations is essential. There are many situations that require the use of voice communications, either by radio or telephone to describe an oil slick. In this case, sufficient information shall be communicated in order to allow the recipient to produce a map of the observations.

5.2 *Location of Observation*—This location shall be reported in latitude and longitude of the apparent centre of the oil slick being observed. If the leading edge of the slick can be located, its position should be reported. This data can be determined using the aircraft navigation instruments or using a portable GPS receiver. It shall be reported in degrees, minutes and seconds or in degrees, minutes with two decimal points. If it is not possible to obtain the latitude and longitude of the location, an estimate of the range and bearing, from a properly identified geographical reference, shall be used. Alternately, aircraft navigation instruments such as VOR/DME or Loran can be used, if available.

5.3 *Size and Shape of Slick*—The size and shape of the slick shall be reported. The units used shall be kilometres or metres and represent the major and minor axis of the slick.

5.4 *Orientation of Slick*—The orientation of the major axis shall be given using degrees from North. In the case of a slick of complex shape, it shall be divided into a number of smaller slicks of simple geometry.

5.5 *Sequence of Data*—Many telephone lines and radio communications are very noisy. This is especially true of transmission from small aircraft and helicopters used for oil-spill surveillance. To promote brevity and standardize the transmission of data, the following sequence should be used. If plain language can be used, it should be but the sequence of transmission should be maintained. This means that a strict sequence must be observed in the transmission of surveillance data. This sequence is:

5.5.1 Latitude of slick centre—*N* or *S*,

5.5.2 Longitude of slick centre—*E* or *W*,

5.5.3 Leading edge-Longitude—*N* or *S* or *N*one,

5.5.4 Leading edge-Latitude—*N* or *S* or *N*one,

5.5.5 Length of major axis in kilometres,

5.5.6 Length of minor axis in kilometres,

5.5.7 Orientation of major axis in degrees from North,

5.5.8 Color of slick—*B, E, R, G*,

5.5.9 Percentage coverage—*Percent*,

5.5.10 Character of slick—*W, C, T, P*,

5.5.11 Other properties—*M, D, F, S, A, H, O, X*, and

5.5.12 END.

5.6 If an element is not present, a dash (–) shall be used. At a minimum, the first eight parameters shall be transmitted and as many as needed of the ninth parameter (Other properties) as a single group. The transmission shall be terminated by the word END.

5.6.1 Thus, a slick could be described as follows: Latitude of Slick Centre 58°23′05″ N (slash) Longitude of slick centre 176°12′15″ *W* (slash) Latitude of Leading Edge 58°23′40″ *N* (slash) Longitude of Leading Edge 176°10′20″ *W* (slash) 3 km (slash) 1 km (slash) 40 (slash) *B* (slash) 50 (slash) *C* (slash) *M* (slash) END. This means that there is a 3 by 1 km slick oriented north-east (40°) containing Black oil with a 50 % coverage. The oil is in Continuous and there is a Mechanical response operation in the area. Another example is the description Latitude of Slick Centre 58°23′05″ *N* (slash) Longitude of slick centre 176°12′15″ *W* (slash) *N*one (slash) *N*one (slash) 3 km (slash) 1 km (slash) 40 (slash) dash (slash) dash (slash) dash (slash) *M* (Slash) END, where visibility was poor and the nature of the slick could not be determined.

6. Mapping

6.1 There are three different styles of maps typically produced during the operation of a visual flight program in the response to an incident. These are: preparation of a base map, conducting the overflight and preparation of a map during the flight, and finally, preparation of a computer-generated map for presentation purposes based on the overflight hand-prepared map. Standards for computer-generated maps are not included in this practice.

6.2 A base map should be prepared that covers the area of the spill. If the spill is very large, more than one base map may be required. They should have sufficient detail to guide the observer in the aircraft. The area covered should be consistent with the size of the spill, but should probably not exceed 100 km by 100 km. A typical example of a base map is shown in Fig. 1. This map should include the location of the spill [xcirc], the command centre and other geographic features that will assist the user in the understanding of the map. Any information that is common to all maps should be placed on the base map.

6.3 The next type of map uses the base map and the concise notation described earlier to record data obtained during an overflight. This map is hand-drawn. It is the combined responsibility of the observers and the generator of the computer map to produce the final map which correctly describes the observational data. An example of an overflight map is shown in Fig. 2.

6.4 Users of visual observations may well request an estimate of the amount of oil on the water. While this cannot be done with any degree of accuracy, Table 1 provides a guide for undispersed oil.

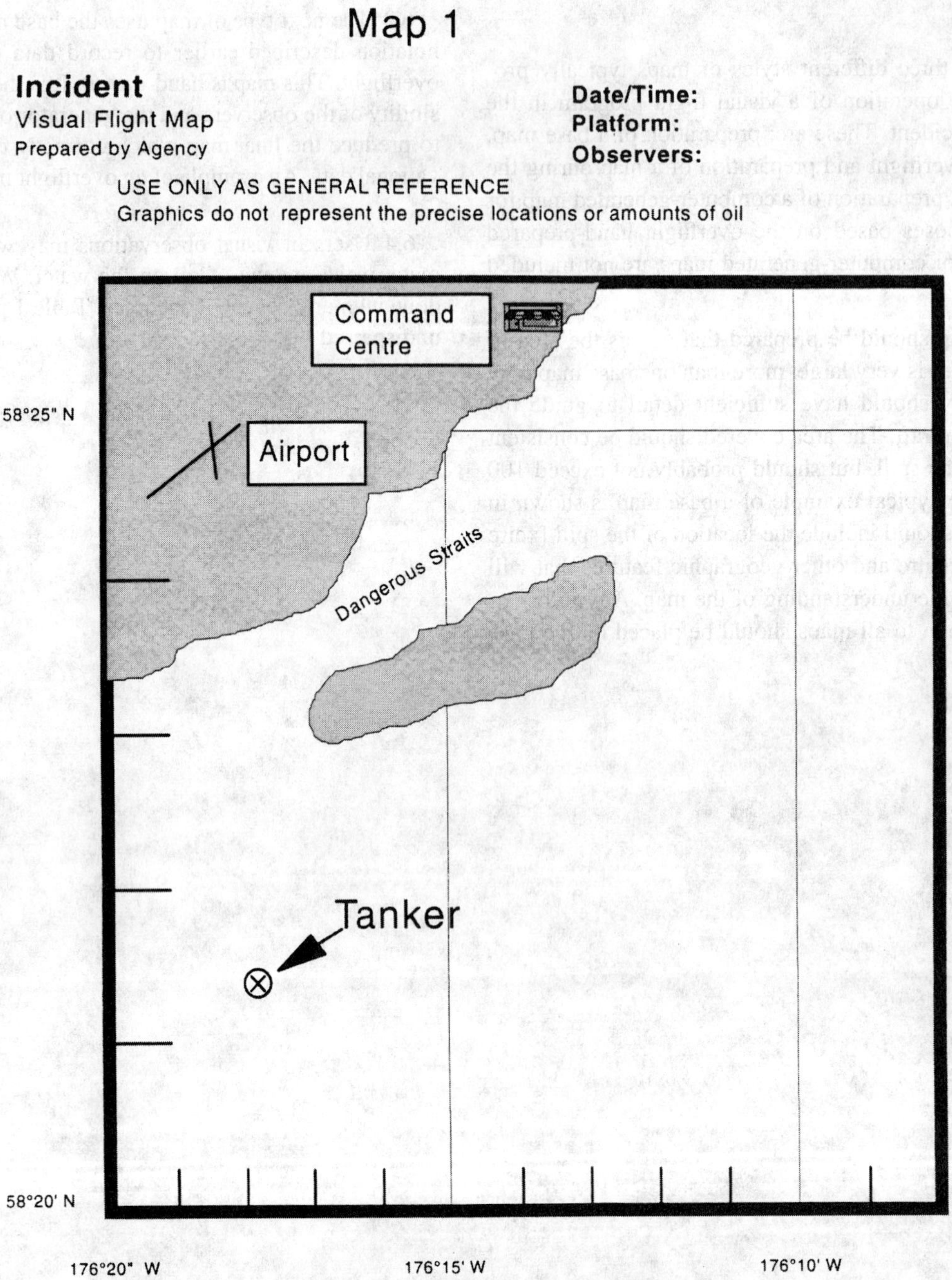

FIG. 1 Typical Example of Base Map

Map 2

Incident *Spill 1*

Visual Flight Map

Prepared by Agency *Oil Co.*

Date/Time: *1997 01 01 05:20-07:40*

Platform: *Twin Otter*

Observers: *Goodman and Brown*

USE ONLY AS GENERAL REFERENCE

Graphics do not represent the precise locations or amounts of oil

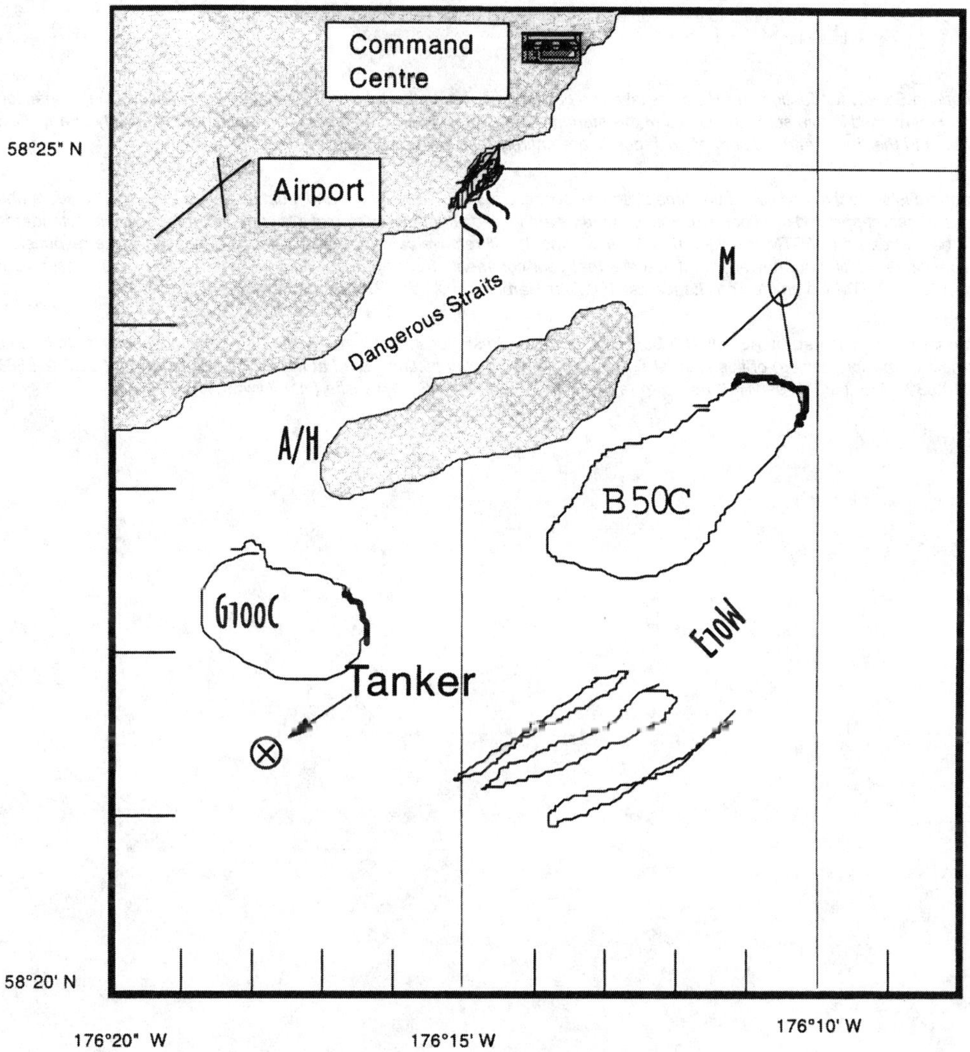

FIG. 2 Overflight Map

TABLE 1 Visual Observations and Estimating Volume of Undispersed Oil

Oil Color	Code	Thickness Range, μm[A]	m^3 of oil per km^2
Brown or Black	B	50–5000	50–5000
Emulsion	E	200–10 000	200–10 000[B]
Rainbow	R	0.15–2.0	0.15–2.0
Grey or Silver Grey	G	0.04–0.15	0.04–0.15

[A] The thickness in microns (μm) has the same value as the volume in m^3 per km^2. These values assume continuous coverage of surface and should be adjusted for situations where the oil covers only a fraction of the area.

[B] This refers to the volume of the fluid. The water-to-oil ratio in many emulsions is as high as 4:1.

Designation: F 1780 – 97

Standard Guide for Estimating Oil Spill Recovery System Effectiveness[1]

This standard is issued under the fixed designation F 1780; the number immediately following the designation indicates the year of original adoption or, in the case of revision, the year of last revision. A number in parentheses indicates the year of last reapproval. A superscript epsilon (ε) indicates an editorial change since the last revision or reapproval.

1. Scope

1.1 This guide covers the key factors to consider in estimating the effectiveness of containment and recovery systems that may be used to assist in the control of oil spills on water.

1.2 The purpose of this guide is to provide the user with information on assessing the effective use of spill-cleanup equipment. It is intended for use by those involved in planning for and responding to oil spills.

1.3 Sections of this guide describe calculation procedures for estimating recovery system effectiveness. It should be understood that any such calculations cannot be expected to predict system performance, but are intended to provide a common basis for comparing system performance.

1.4 One of the main reasons that the calculation procedures cannot be used to predict system performance is that the analysis is sensitive to assumptions made on the properties of the oil slick, and particularly the changes in slick thickness and emulsification. It is emphasized that the purpose of this guide is not to provide a standard method for estimating slick property changes, but rather to provide a standard guide for using that information in comparing system performance.

2. Referenced Documents

2.1 *ASTM Standards:*

F 625 Practice for Classifying Water Bodies for Spill Control Systems[2]

F 631 Guide for Collecting Skimmer Performance Data in Controlled Environments[2]

F 808 Guide for Collecting Skimmer Performance Data in Uncontrolled Environments[2]

F 1523 Guide for the Selection of Booms According to Water Body Classifications[2]

3. Terminology

3.1 *Definitions:*

3.1.1 *advancing skimmer, n*—a skimmer that is designed to be used to sweep out the spill area.

3.1.2 *Discussion*—The skimmer may be independent or may be attached to containment boom to increase sweep width. In some cases, the skimmer may not be attached to the boom but is positioned in the pocket of the boom for skimming. As long as the skimmer operates while moving, it is considered to be an advancing skimmer. Some skimmers are used in both an advancing and stationary mode. These are classified according to their application.

3.1.3 *contained spills, n*—a spill that is restricted from spreading by containment boom or natural means.

3.1.4 *oil slick encounter rate, n*—the volume of oil slick per unit time actively encountered by the oil spill recovery system, and therefore available for containment and recovery (m^3/h).

3.1.5 *oil spill recovery system, n*—a combination of devices that operate together to recover spilled oil; the system would include some or all of the following components: (*1*) containment boom, (*2*) skimmer, (*3*) support vessels to deploy and operate the boom and skimmer, (*4*) discharge/transfer pumps, (*5*) oil/water separator, (*6*) temporary storage devices, and (*7*) shore based storage/disposal.

3.1.6 *recovery system effectiveness, n*—the volume of oil that is removed from the environment by a given recovery system in a given recovery period.

3.1.7 *recovery period, n*—the time available for recovery systems to carry out cleanup operations.

3.1.8 *response time, n*—the time interval between the spill incident and the start of cleanup operations.

3.1.9 *stationary skimmer, n*—a skimmer that is intended to be used in a fixed location and is moved to new accumulations of oil as skimming progresses.

3.1.10 *Discussion*—Some stationary skimmers are used in a containment boom system that moves to collect oil, then pauses to permit the skimmer to recover the oil collected. Even though this system moves periodically, the skimmer is still ranked as a stationary skimmer because it operates when the system is at rest.

3.1.11 *uncontained spill, n*—a spill that continues to spread after the recovery effort begins.

4. Summary of Guide

4.1 In evaluating the effectiveness of containment and recovery systems used in response to oil spills, many factors need to be considered of which skimmer performance is but one. The objective of this guide is to describe a range of factors that must be considered in estimating recovery system effectiveness.

4.2 In order to evaluate a recovery system, there are two

---

[1] This guide is under the jurisdiction of ASTM Committee F-20 on Hazardous Substances and Oil Spill Responseand is the direct responsibility of Subcommittee F20.12 on Removal.

Current edition approved Feb. 10, 1997. Published April 1997.

[2] *Annual Book of ASTM Standards*, Vol 11.04.

general types of information required, a set of information to describe the spill scenario against which the system will be measured, and a set of information to describe the performance characteristics of the recovery system.

4.3 Information on the spill is required to adequately define the problem and thereby provide a focus for the evaluation process. The spill should be defined in sufficient detail as to allow an unambiguous interpretation of its behavior in terms of the operating parameters of the countermeasures system. For certain purposes it may be desirable to develop a set of standard spill scenarios against which response system effectiveness would be measured in a quantifiable manner.

4.4 The performance characteristics must be identified for the recovery system and its various components. In general, the information requirements will include the rates or capacities, or both, the operating limitations, and the support requirements.

4.5 This guide covers equipment-related factors that will affect recovery-system effectiveness. Additional important factors that are not covered in this guide but should be considered as being critical to the success of a spill response include: contingency planning; communications plans; government approvals; logistics of supporting manpower and equipment in the field; and training and exercising of manpower.

5. Spill-related Information

5.1 *Spill Type*:

5.1.1 Response strategies will depend to some extent on the type of spill. The spill scenario should be defined as to whether it is an instantaneous or continuous release, whether or not the spill has ceased flowing, and whether the spill is contained or uncontained.

5.2 *Oil Slick Properties*—The following oil slick properties must be specified for the spill scenario. As some of these properties may vary with time, it may be desirable to use computer-based behavior models to produce spill property information for the time period of interest. For certain applications it may be useful to produce standard sets of spill property information that describe spills of interest as a function of time.

5.2.1 *Spill Volume*—The total volume of oil spilled should be specified (m^3). For spills that have not ceased, a spill rate (m^3/h) should also be specified.

5.2.2 *Spill Area*—The total spill area must be estimated in order to calculate estimates of slick thickness. For uncontained spills, the total spill area will increase over time; estimates can be made using computer-based behavior models. Alternatively, a simplified spreading model (Fig. 1: example spreading curves) can be used for first-order estimates.

5.2.3 *Slick Thickness*—Slick thickness is used in subsequent calculations of system encounter rate. Slick thickness is defined as the overall average thickness of the slick, and is estimated by dividing the spill volume by the total spill area at any given time. For this calculation, spill volume should take into account losses from the slick due to evaporation and natural dispersion, and increases to the slick volume due to emulsification. For uncontained spills, natural spreading forces will cause the slick thickness to decline steadily during recovery operations, and may result in a discontinuous slick composed of windows and patches separated by sheen or open

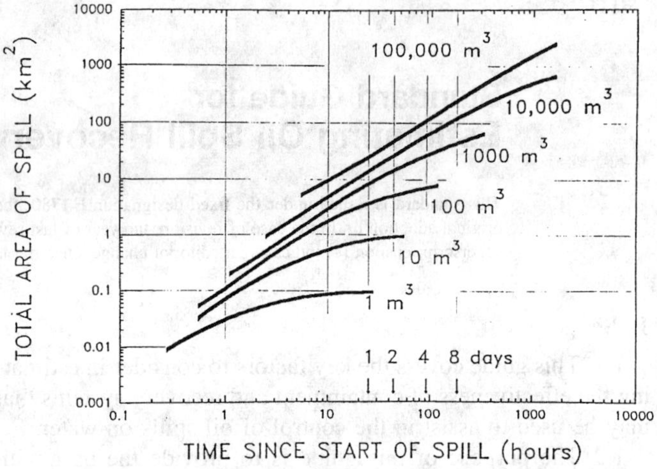

FIG. 1 Total Slick Area versus Time

water, or both. These factors should be considered in estimating an overall average slick thickness.

5.2.4 *Slick Viscosity*—The viscosity of the spilled product is used as a criteria to evaluate skimmer performance, as many skimming and pumping units will perform less effectively as viscosity increases. The viscosity of the spilled product will generally increase through the recovery period as the oil is subjected to weathering and emulsification processes. The viscosity should be specified as mm^2/s (cSt).

5.2.5 *Emulsification*—Emulsification is important as a spill process not only for its effect on oil viscosity but also because an emulsified oil represents a greater total volume of spill product that must be handled by skimming and pumping systems. Many crude oils and refined products will tend to emulsify over the life of the spill depending on the properties of the oil and the level of wave energy in the spill environment. The degree of emulsification should be specified as the emulsified water content expressed as a percentage.

5.2.5.1 It is recognized that emulsification rates for oil spilled in the marine environment will vary greatly depending on the oil properties, spill size, sea conditions, and temperature. As noted in 1.4, it is not the intent of this guide to provide standard rates of emulsification for a variety of oil products and environmental conditions. For the purposes of comparing system performance, the data in Table 1 is provided as an example of emulsification data for crude oil over a period of several days. Users of this guide are encouraged to use alternative data that suits their particular oils and environmental conditions.

5.3 *Spill Environment*:

5.3.1 *Temperature*—Water temperature is important as a parameter for estimating oil slick properties as well as the rate of change of those properties due to weathering and emulsification. (It is assumed that the temperature of the oil slick is the same as the water on which the oil is floating.) Water

TABLE 1 Example Data for Emulsified Water Content versus Time for Crude Oil

	12 h	1 day	2 days	3 days
% Water Content	30	50	65	75

temperature is defined as the temperature of the upper surface layer and should be specified as °C.

5.3.1.1 Air temperature may be important as a parameter for modifying or limiting the performance of skimming and pumping equipment, and should be specified as °C.

5.3.2 *Wind/Waves*—The wind and wave environment are important to the analysis for two reasons; first, as a parameter in estimating the behavior changes of the oil slick, and second, as a limiting factor for recovery operations. For the first purpose, average wind speeds (km/h) should be specified. For the purpose of establishing criteria for limiting recovery operations, exceedance statistics (significant wave height) should be specified for the spill location. Exceedance criteria should be expressed as the percentage of time that conditions will allow recovery operations with reference to the equipment selected for the response and the environmental criteria listed in Practice F 625. For example, for spills in open water, wave exceedance data should be specified as the percentage of time that waves are less than or equal to 2 m, which would represent the percentage of time that equipment specified for open water use would be applicable.

5.3.3 *Current*—The presence of water currents may influence the selection of response strategies for a spill scenario, and may lead to a reduction in containment effectiveness in certain applications. The water currents for a given environment should be specified (m/s), with due regard to any local variations.

5.3.4 *Visibility*—Due to concerns with worker safety in poor visibility, as well as the inefficiencies related to the monitoring, tracking, and containment of oil slicks during periods of poor visibility, it is assumed in general that recovery operations are only possible when there is daylight and visibility of greater than 500 m (0.25 n.miles). Both of these factors should be expressed as the percentage of time that conditions exist that would allow effective operations.

5.3.4.1 It may be possible to effectively operate during periods of darkness and poor visibility if the recovery system includes adequate lighting equipment, remote sensing systems for assisting monitoring and containment efforts, or highly accurate navigation systems, or combination thereof. This may be particularly applicable to spills in nearshore and protected waters. In such cases a more liberal criteria for visibility limitations could be specified.

5.3.5 *Summary of Environmental Applicability Factors*—The wave exceedance, daylight, and visibility factors can be combined to produce an overall applicability factor that would represent the percentage of time that a given recovery system could be effectively used for a given spill scenario. For example, for an environment that has waves less than 2 m for 80 % of the time, receives 14 h of daylight, and has visibility greater than 500 m for 95 % of the time (note: all figures should be specified for the time of year of interest), the environmental applicability would be estimated as: $(0.80) \times (14/24) \times (0.95) = 44$ %.

5.4 *Spill Location*:

5.4.1 Spill location should be specified with respect to distance of response bases, in order to estimate transit times for the recovery systems, and with respect to shoreline, in order to estimate the time available to respond prior to shoreline oiling. Spill location may also be of importance when evaluating recovery systems that include the shuttling of recovered oil between the recovery site and temporary storage locations, in which case transit times may have to be deducted from the on-site availability of storage systems.

6. Recovery System Information

6.1 *Containment System Operating Factors*:

6.1.1 *Encounter Rate*—The encounter rate of the recovery system is a prime consideration in evaluating performance. The encounter rate is simply the rate (m^3/h) at which the system encounters the oil slick. The encounter rate includes three components: sweep width, encounter speed, and oil slick thickness.

6.1.1.1 The sweep width (or swath) is the width intercepted by a boom in collection mode, and is calculated by multiplying the boom length by the gap ratio. Where the gap ratio is not specified, a value of ⅓ should be used.

6.1.1.2 The encounter speed is the tow or current speed relative to the containment system. If not specified, a maximum encounter speed of 0.5 m/s (1 knot) should be used.

6.1.1.3 Encounter rate can be calculated as the product of these three factors, taking into account consistency of units. As well, simple nomograms (Fig. 2) can be used to estimate encounter rates for a range of conditions.

6.1.2 *Operating Limitations*—Containment equipment must be specified with regard to the environmental conditions of the given spill scenario. Guidance for selecting booms can be taken from Guide F 1523, which lists minimum requirements for boom dimensions and strength properties for calm, protected, and open bodies of water. Other limitations on the specified boom, such as minimum water depths and maximum tow speeds should also be listed.

6.1.2.1 The applicability of a boom to a given spill scenario should be considered as a constraint to containment operations. For example, a boom designated for calm water use (in accordance with Guide F 1523) will be satisfactory for containment operations in waves up to 0.3 m (1 ft). If the wave climate for a given area is such that 0.3 m waves are exceeded 25 % of the time then the boom could be considered to be applicable 75 % of the time.

6.1.2.2 Encounter speed is included as a factor in calculating the encounter rate. For most booms the maximum encounter speed will be in the range of 0.35 to 0.5 m/s (0.7 to 1 knot). It is recognized that certain containment systems have been designed to operate at higher encounter speeds: greater speeds than those noted above may be used if test data is available to support the selected encounter speed. For most booms, encounter speeds greater than 0.5 m/s should be used only with an accompanying reduction in the system's throughput efficiency to account for losses from the containment system.

6.1.3 *Support Requirements*—Support requirements for the listed containment equipment should be specified. Support requirements could include: transportation to deliver the boom to the spill site; equipment such as cranes or winches required to deploy, tow, and retrieve the boom; boom tackle such as tow lines, marker buoys, anchors, connectors; power or air requirements, or both, for boom deployment, operation, and retrieval;

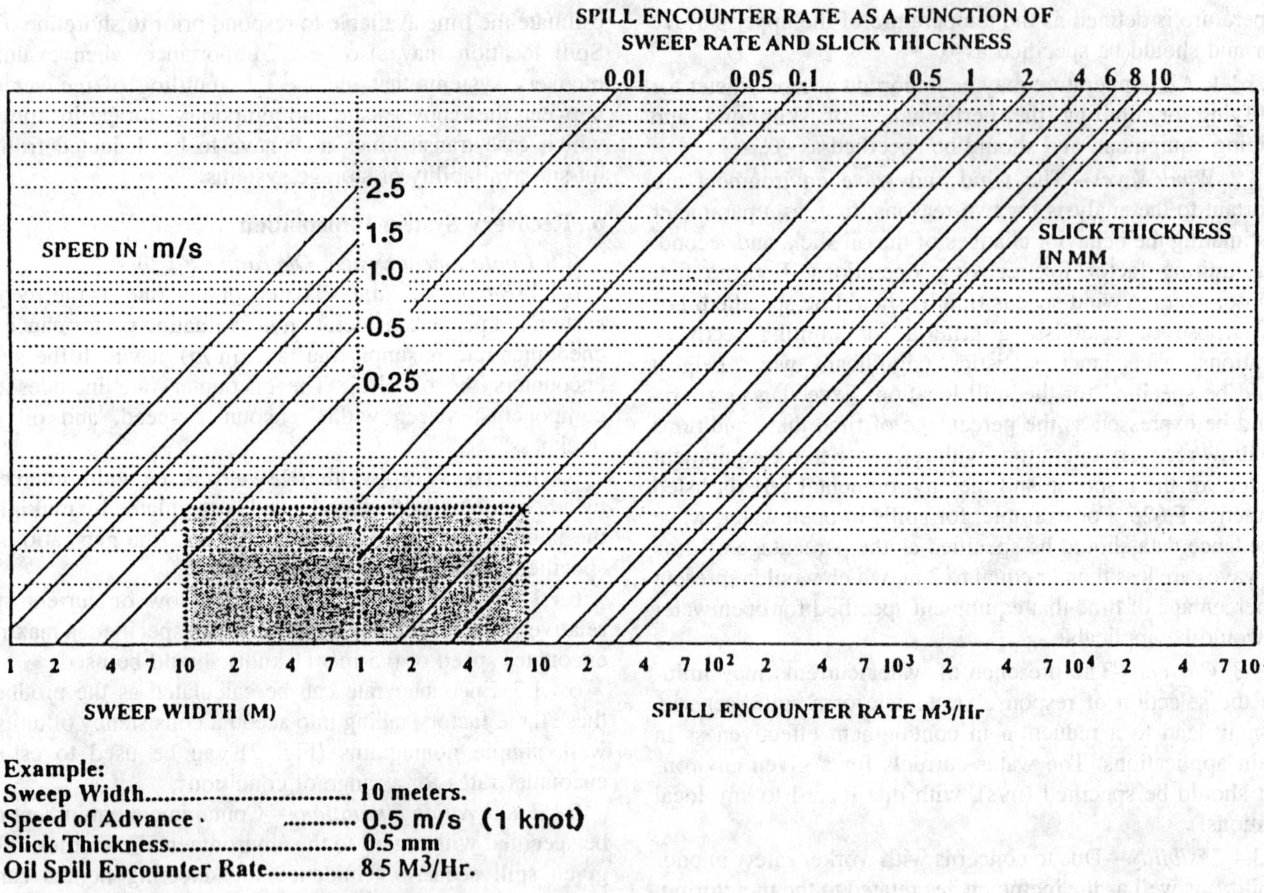

FIG. 2 Oil Spill Encounter Rate

adequate manpower for deployment and retrieval; and vessels with adequate deck space for the required equipment, as well as adequate power and maneuverability for the specific situation. Any limitations on the specified support equipment should be specified; these could include: sea-state limits for vessel operation; draft limits on the vessels; minimum and maximum transit and tow speeds; and limits on vessel operation with respect to distance of shore.

6.2 *Recovery System Operating Factors*:

6.2.1 *Recovery Rate*—An appropriate recovery rate must be determined for the skimming unit based on the operating conditions specified in the spill scenario. The recovery rate should reflect realistic expectations of performance with regard to the slick thickness and viscosity as well as the specified environmental conditions, all of which may vary with time.

NOTE 1—The recovery rate used in the performance calculations cannot exceed the encounter rate estimated for the containment system.

6.2.1.1 The most desirable source of information for estimating a skimmer's recovery rate is experimental or field data collected for the particular skimmer of interest (Guides F 631 and F 808). As performance data is not available for many devices, a second alternative would be to examine experimental or field data from other comparable devices, and use it to estimate a realistic recovery rate for the spill conditions of interest. In the absence of any such data it may be necessary to use an estimate based on the skimmer's nameplate recovery rate. Report the source of the data and the method used to estimate the recovery rate.

6.2.2 *Recovery Efficiency*—A skimmer will generally recover free water along with the recovered oil. The amount of water recovered will affect the relative efficiency of a skimmer system because the total fluid volume must be handled by the transfer, storage, and disposal systems. In order to estimate the amount of total fluids that must be handled, the recovery efficiency of the skimming system must be known for the operating conditions expected. As with the recovery rate, the recovery efficiency may vary with the slick conditions and the environmental conditions, and should be estimated based on test data if available.

6.2.3 *Skimmer Operating Limitations*—Any limitations on the operation of the skimming unit should be specified. These could include: upper limits on the viscosity of the oil slick; minimum slick thicknesses for effective operation; maximum sea states; and maximum hours of continuous operation.

6.2.4 *Support Requirements*—Support requirements for the listed skimming equipment should be specified. Support requirements could include: transportation to deliver the skimmer to the spill site; equipment such as cranes required to deploy and retrieve the skimmer; power requirements for skimmer deployment and operation; ancillary pumping systems; adequate manpower for deployment, operation, and retrieval; and vessels with adequate deck space for the required equipment. Any limitations on the specified support equipment should be specified; these could include: sea state limits for vessel operation; draft limits on the vessels; minimum and maximum transit and tow speeds; down-time for equipment maintenance; and limits on vessel operation with respect to distance of shore.

6.3 *Transfer and Storage Operating Factors*:

6.3.1 Storage capacity must be available to handle the estimated volume of total fluids (that is, recovered oil or emulsion and free water, or both). Sufficient temporary storage must be available at the spill site to handle fluids as they are recovered, and if applicable, additional storage must be available for the consolidation and storage of collected fluids awaiting disposal.

6.3.2 In general, it should be assumed that all collected fluids will require storage and eventual disposal. In some instances, however, it may be possible to reduce the total storage and disposal requirement through the use of oil/water separation and decanting of free water. This would require the specification of equipment and manpower dedicated to that task. Alternatively, the time required to carry out separation and decanting should be considered as a possible limiting factor which, in some instances, would reduce the amount of time available for skimming.

6.4 *Overall System Operating Factors*:

6.4.1 *Response Time*—The response time is defined as the time interval between the spill incident and the start of recovery operations. A response time should be estimated for the scenario taking into account an adequate time for the mobilization of recovery resources (that is, time to notify response teams and assemble the required equipment) as well as estimating a transit time of resources from the response base to the scene of the spill. In estimating transit times, unless otherwise justified, transit speeds of 10 km/h (5 knots) by water, 55 km/h (35 mph) by land, and 185 km/h (100 knots) by air should be assumed.

6.4.2 *Recovery Period*—The recovery period is defined as the time available for recovery operations. If appropriate to the analysis, the recovery period may be specified as including the response time; reporting should include a clear distinction between the two parameters. Users of this guide should select a recovery period appropriate to the circumstances of the spill scenario, and should consider the following factors, which may affect the period of time during which oil will be present on the water surface and available for recovery:

6.4.2.1 Proximity of the spill to shoreline or shallow water,

6.4.2.2 Type of oil and size of spill (re: rates of spreading and natural dissipation),

6.4.2.3 Hours of daylight, and

6.4.2.4 Weather conditions and sea state.

6.4.2.5 Alternatively, for the purposes of comparative analyses, an arbitrary recovery period may be specified.

7. Example Calculation

7.1 The following example calculation illustrates the methodology in estimating the effectiveness of a recovery system against a target spill. Through the example calculation, example values are given for slick thickness, emulsification, and recovery-system parameters. It is emphasized that these values are for illustrative purposes only. As well, for the purposes of this example, it is assumed that weather conditions do not hinder the response in any way; the implications of this assumption are discussed at the end of the example. A summary of the calculations is presented in a worksheet at the end of the example.

7.1.1 Consider a 1500 m^3 (9400 bbl) spill of crude oil. The spill occurs in open water, approximately 90 km from a response base. Estimate the effectiveness, over the first 24 h, of a recovery system that includes 250 m (820 ft) of containment boom, one skimmer with an estimated recovery rate of 25 m^3/hr (150 bbl/h), and a storage barge with a capacity of 200 m^3(1300 bbl).

7.1.2 The first requirement is to estimate a response time. For this example it is assumed that three hours are required for notification and mobilization of the equipment. Given the spill location, the transit time can be calculated as 90 km ÷ 10 km/h = 9 h, giving a total response time of 3 + 9 = 12 h.

7.1.3 Given that recovery operations will not begin until 12 h after the spill occurs, the slick conditions required in the analysis can be estimated. From the graph of slick area (Fig. 1), calculate the slick thickness as spill volume divided by total spill area (see Table 2).

7.1.4 The average slick thicknesses can now be used to estimate the encounter rate of the recovery system. Through the time period of interest, the slick thickness declines from 0.375 mm at 12 h, to 0.21 mm at 18 h (averaging 0.29 mm through that time period), and 0.15 mm at 24 h (and averaging 0.18 mm for the 18 to 24-h time period). Given the containment boom length of 250 m, and assuming a gap ratio of 0.33 and a maximum encounter speed of 0.5 m/s (1 knot), the encounter rate for the period of 12 to 18 h can be calculated as:

$$\text{encounter rate} = \text{sweep width} \times \text{slick thickness} \times \text{encounter speed}$$
$$= (0.33 \times 250\ \text{m}) \times (0.29\ \text{mm}) \times (0.5\ \text{m/s})$$
$$= 43\ \text{m}^3\text{/h}$$

7.1.5 Similarly, the encounter rate for the 18 to 24-h time period can be calculated as 27 m^3/h. In each time period the encounter rate exceeds the recovery rate for the skimmer, although it is clear that as the slick thickness continues to diminish, the encounter rate will soon be less than the "available" recovery rate and the recovery operation will be restricted by the encounter rate.

7.1.6 For the time period of interest, then, the oil slick recovery rate is the specified rate of 25 m^3/h, and the total potential volume of fluid recovered is (25 m^3/h × 12 h) = 300 m^3. As this exceeds the available storage volume we are therefore restricted to a total fluid volume of 200 m^3, the volume of the storage barge. We must now determine the components of the recovered fluid. The selected skimmer has a specified oil recovery efficiency of 75 %; therefore 75 %, or 150 m^3 of the recovered fluids is oil and emulsion with 50 m^3

TABLE 2 Estimated Analysis of Slick Conditions

Time, h	Total Slick Area, km²	Average Slick Thickness, mm
12	4	0.375
18	7	0.21
24	10	0.15
48	20	0.075
avg, 12 to 18	5.5	0.29
avg, 18 to 24	8.5	0.18

of free water. Secondly, the scenario definition states that the slick is expected to emulsify, with an estimated 10 % water content after 12 h and 30 % after 24 h. As a rough approximation, we can assume an average 20 % water content through the 12 to 24-h period. Therefore, the 150 m^3 of emulsion is in fact 120 m^3 of oil and 30 m^3 of emulsified water. (It should be noted that had the recovery system included a system for separating free water or breaking the emulsion, or both, and if discharge of separated water were permitted, then the storage limitation of 200 m^3 could have been applied to oil rather than total recovered fluids.)

7.1.7 Based on the preceding information it is estimated that a total of 120 m^3 of oil, or about 8 % of the total spill volume, could be collected in the first 24 h of the spill (see Table 3).

7.1.8 For the purposes of this example it was assumed that environmental conditions did not hinder the response in any way. There are a number of ways that weather factors could be used to modify the preceding methodology. As discussed in 5.3 of this guide, environmental data can be used to calculate an applicability factor. As well, sea conditions in the area of interest could be used to modify the recovery rate and recovery efficiency calculations, for example, for skimmers that are expected to have changes in performance with changes in wave height.

TABLE 3 Summary of Calculations

	Comments	No. 1	No. 2	Total
A. Time Period, h	No. 1: 12 to 18 h No. 2: 18 to 24 h	6	6	12
B. Slick Thickness, mm	scenario definition	0.29	0.18	
C. Encounter Rate, m^3/h	swath × speed × "B"	43	27	
D. Available Skimming Rate, m^3/h	specified in recovery system definition	25	25	
E. Actual Skimming Rate, m^3/h	lesser of "C"," D"	25	25	
F. Potential Volume Skimmed, m^3	"A" × "E"	150	150	300
G. Available Storage, m^3	specified in recovery system definition	200	50	
H. Actual Fluid Volume Stored, m^3	lesser of" F"," G"	150	50	200
I. Recovery Efficiency,%	scenario definition	75	75	75
J. Oil/Emulsion Stored, m^3	"H" × "I"	112.5	37.5	150
K. Emulsification, % water	scenario definition	20	20	20
L. Oil Stored, m^3	"J" × (1 − "K")	90	30	120
Total Volume of Oil Recovered	120 m^3	8 % of spill		

The American Society for Testing and Materials takes no position respecting the validity of any patent rights asserted in connection with any item mentioned in this standard. Users of this standard are expressly advised that determination of the validity of any such patent rights, and the risk of infringement of such rights, are entirely their own responsibility.

This standard is subject to revision at any time by the responsible technical committee and must be reviewed every five years and if not revised, either reapproved or withdrawn. Your comments are invited either for revision of this standard or for additional standards and should be addressed to ASTM Headquarters. Your comments will receive careful consideration at a meeting of the responsible technical committee, which you may attend. If you feel that your comments have not received a fair hearing you should make your views known to the ASTM Committee on Standards, 100 Barr Harbor Drive, West Conshohocken, PA 19428.

Designation: F 1788 – 97

Standard Guide for In-Situ Burning of Oil Spills on Water: Environmental and Operational Considerations[1]

This standard is issued under the fixed designation F 1788; the number immediately following the designation indicates the year of original adoption or, in the case of revision, the year of last revision. A number in parentheses indicates the year of last reapproval. A superscript epsilon (ε) indicates an editorial change since the last revision or reapproval.

1. Scope

1.1 This guide covers the use of in-situ burning to assist in the control of oil spills on water. This guide is not applicable to in-situ burning of oil on land.

1.2 The purpose of this guide is to provide information that will enable spill responders to decide if burning will be used as part of the oil spill cleanup response.

1.3 This is a general guide only. It is assumed that conditions at the spill site have been assessed and that these conditions are suitable for the burning of oil. It is also assumed that permission to burn the oil has been obtained. Variations in the behavior of different oil types are not dealt with and may change some of the parameters noted in this guide.

1.4 *This standard does not purport to address all of the safety concerns, if any, associated with its use. It is the responsibility of the user of this standard to establish appropriate safety and health practices and determine the applicability of regulatory limitations prior to use.*

2. Terminology

2.1 *Definitions:*

2.1.1 *burn efficiency*—burn efficiency is the percentage of the oil removed from the water by the burning.

2.1.1.1 *Discussion*—Burn efficiency is the amount (volume) of oil before burning; less the volume remaining as a residue, divided by the initial volume of the oil.

2.1.2 *burn rate*—the rate at which oil is burned in a given area.

2.1.2.1 *Discussion*—Typically, the area is a pool and burn rate is the regression rate of the burning liquid, or may be described as a volumetric rate.

2.1.3 *contact probability*—the probability that oil will be contacted by the flame during burning.

2.1.4 *controlled burning*—burning when the combustion can be started and stopped by human intervention.

2.1.5 *fire-resistant booms*—devices that float on water to restrict the spreading and movement of oil slicks and constructed to withstand the high temperatures and heat fluxes of in-situ burning.

2.1.6 *in-situ burning*—use of burning directly on the water surface.

2.1.6.1 *Discussion*—In-situ burning does not include incineration techniques, whereby oil or oiled debris are placed into an incinerator.

2.1.7 *residue*—the material, excluding airborne emissions, remaining after the oil stops burning.

3. Significance and Use

3.1 This guide is primarily intended to aid decision-makers and spill-responders in contingency planning, spill response, and training.

3.2 This guide is not specific to either site or type of oil.

4. Background

4.1 *Overview of Oil Burning*:

4.1.1 In-situ burning is one of several oil-spill countermeasures available. Other countermeasures could include mechanical recovery, use of oil-spill dispersants, and leaving the oil to natural processes.

4.1.2 In-situ burning is combustion at the spill site without removing the oil from the water. Containment techniques may be used, however, to increase the thickness of the oil. The thickness of the oil slick is an important factor in the use of in-situ burning.

4.1.3 In-situ burning does not include incineration techniques whereby oil or oiled debris are placed into an incinerator.

4.2 *Major Advantages and Disadvantages of In-situ Burning*:

4.2.1 Advantages of in-situ burning include the following:

4.2.1.1 Rapid removal of oil from the water surface,

4.2.1.2 Requirement for less equipment and labor than many other techniques,

4.2.1.3 Significant reduction in the amount of material requiring disposal,

4.2.1.4 Significant removal of volatile emission components, and

4.2.1.5 May be the only solution possible, such as in oil-in-ice situations.

4.2.2 Disadvantages of in-situ burning include the following:

4.2.2.1 Creation of a smoke plume,

4.2.2.2 Residues of the burn must be dealt with,

[1] This guide is under the jurisdiction of ASTM Committee F-20 on Hazardous Substances and Oil Spill Response and is the direct responsibility of Subcommittee F20.15 on In-Situ Burning.

Current edition approved May 10, 1997. Published July 1997.

4.2.2.3 Time in which to ignite the oil may be limited,

4.2.2.4 Oil must be a minimum thickness to burn, which may require containment, and

4.2.2.5 The fire may spread to other combustible materials.

5. Environmental Considerations for Deciding to Use In-Situ Burning

5.1 *Air Quality*:

5.1.1 Several studies have been done of the air emissions resulting from in-situ burning. It has been found that the smoke plume consists largely of carbon and that toxic compounds are not created. The high temperatures achieved during in-situ burning result in efficient removal of most components of the oil. The thick, black smoke can be of concern to nearby human populations or ecologically sensitive areas. Since most soot precipitation occurs near the fire, this is the main area of concern. The smoke plume is, however, generally an aesthetic concern. In-situ burning should be avoided within 1 km upwind of either an ecologically sensitive or a heavily populated area, depending on meteorological conditions. No emissions greater than one fourth of the 1994 human health exposure limits have been detected at ground level further than 1 km from an oil fire. The values of the human health exposure limits vary with jurisdiction, and, thus, the appropriate documents should be consulted. The environmental and economic trade-offs of burning the oil, as opposed to contamination of the shoreline, must be considered.

5.1.2 Burning can be safely conducted near populated areas if there is sufficient air turbulence for mixing, and in the absence of a low-level atmospheric inversion.

5.2 *Water Quality*—Measurements show that burning does not accelerate the release of oil components or combustion by-products to the water column. Highly efficient burns of heavy oils may form a dense residue that sinks.

5.3 *Wildlife Concerns*—Although no specific biological concerns related to the use of in-situ combustion have been identified to date, benthic resources may be affected by sunken oil burn residue.

6. Operational Considerations for In-situ Burning

6.1 *Safety Considerations*—The safety of the proposed operation shall be the primary consideration. Secondly, the burning operation shall not result in unintentional flashback to the source of the oil, for example, the tanker or the production platform. The third consideration is the spread of the fire to other combustible material in the area, including trees, docks, and buildings. Flashback and fire spread can often be prevented by using containment booms to tow away the oil to be burned. A fourth consideration is the safety of the ignition operation, which is often done from helicopters, and the safety of the boom tow operation must be ensured.

6.2 *Safety Monitoring and Control Requirements*—The operation must be monitored to meet safety requirements. Burning shall be monitored to ensure that fire may not spread to adjacent combustible material. Situation-specific contingency methods of extinguishing, such as boats with fire monitors, shall be available. In towed-boom operations, it has been proposed that the fire may be extinguished by increasing the tow speed so that the oil is entrained in the water. Other options for controlling the fire or the burn rate might include releasing one side of the oil containment boom or slowing down to reduce the encounter rate.

6.3 *Oil Thickness*—Most oils can be ignited on a water surface if they are a minimum of 2 to 3 mm thick. Once ignited, the oils will burn down to a thickness of about 1 mm. Physical containment, such as with oil-spill containment booms, is usually necessary to achieve the minimum thicknesses required. Specific information on this is provided in the appendix.

6.4 *Oil Type and Condition*—Highly weathered oils will burn, but will require sustained heat during ignition. Oil that is emulsified with water may not burn. Not enough data are available to determine water-content levels that limit ignition. Indications are, however, that stable emulsions which typically contain about 70 % water cannot be ignited and that oils containing less than about 25 % water will burn. Treatment with chemicals to remove water before burning can permit ignition.

6.5 *Wind and Sea Conditions*—Strong winds may extinguish the fire. In-situ burning can be done on the sea with winds less than about 40 km/h (about 20 knots). High sea states are not conducive to containment by booms. Wave heights of 1 m or more may result in splash-over of the oil.

6.6 *Burn Efficiency*—Burn efficiency, which is the percentage of oil removed by burning, has been measured as high as 99 % for contained oil. Burn efficiency is largely a function of oil thickness and flame-contact probability. Contact probability is the probability that oil will be contacted by the flame during burning. Inhomogeneous oil distribution on the surface can result in an incomplete burn. This can result as the flame may be extinguished over a patch that is not thick enough to burn, while adjacent patches that are thick enough will subsequently not be burned. Contact is usually random and is influenced by wind speed and direction and can be controlled by human intervention in some cases.

6.7 *Burn Rate*—Oil burns at the rate of about 3 mm/min, which means that the surface of the oil slick regresses downwards at the rate of 3 mm/min. This translates to a rate of about 5000 L/m^2/day (or 100 gal/ft^2/day). Burn rate is relatively independent of physical conditions and oil type. Using these values, it is possible to calculate the rate of burning in booms and in other burn operations.

6.8 *Containment*—Oil slicks must be a minimum of 2 to 3 mm thick to be ignited. As oil naturally spreads quickly to much thinner slicks than this under normal circumstances, physical containment is generally necessary for burning. Fire-resistant booms are commercially available for this purpose. While these booms can be used in a variety of configurations, they are best used in a catenary mode and towed at speeds less than 0.35 m/s (0.7 knots). At speeds greater than this, oil is lost under the boom by entrainment. Slicks can sometimes be naturally contained by ice or against shorelines.

6.9 *Ignition*—Slicks can be ignited with a variety of devices. Enough heat must be supplied for a sufficient length of time. Weathered oils generally require a longer heating time to ignite.

6.10 *Residue Cleanup*:

6.10.1 Residue is the material remaining after the oil stops burning. Residue is similar to a highly weathered oil, depending on the burn conditions. It is viscous and often highly adhesive. Highly efficient burns result in heavier and denser residue. These residues may actually be more dense than sea water.

6.10.2 Floating residue can be removed manually with sorbents, nets, or similar equipment.

7. Summary

7.1 In-situ burning is a viable countermeasure that has the potential to quickly remove large amounts of oil. The air emissions of in-situ burning are below health and environmental concern levels at nominal distances from the combustion source.

8. Keywords

8.1 fire-resistant booms; in-situ burning; oil-spill burning; oil-spill containment; oil-spill disposal

APPENDIX

(Nonmandatory Information)

X1. INTRODUCTION TO THE IN-SITU BURNING OF OIL SPILLS

INTRODUCTION

In-situ burning has been used as an oil-spill countermeasure around the world **(1,2)**.[2] Recently, extensive research has been conducted on the many facets of burning oil **(3,4,5)**. The emissions from and basic principles of oil-spill burning are now relatively well-understood.

X1.1 Basic Principles of Burning Oil

X1.1.1 Oil slicks can be ignited if they are at least 2 to 3 mm thick and will continue to burn down to slicks of about 1 to 2 mm thick **(6)**[2]. These thicknesses are required because of heat transfer. Sufficient heat is required to vaporize material for continued combustion. In a thin slick, most of the heat is lost to the water, vaporization is not sustained, and combustion ceases.

X1.1.2 Containment is usually required to concentrate oil slicks so that they are thick enough to ignite and burn **(7)**. Fire-resistant containment booms can be used to keep fire from spreading back to the spill source, such as an oil tanker **(8)**. Burning in situ without the benefit of containment booms can be undertaken only if the oil is thick enough (2 to 3 mm) to ignite. For most crude oil spills, this only occurs for a few hours after the spill event unless the oil is confined behind a barrier. Oil on the open sea spreads rapidly to equilibrium thicknesses. For light crude oils, this is about 0.01 to 0.1 mm, for heavy crudes and heavy oils, this is about 0.05 to about 0.5 mm.

X1.1.3 Oil can be contained by natural barriers. For example, ice has been shown to serve as a natural boom. Several successful experiments and burns of real spills have shown that burning is a proven countermeasure for spills in ice **(4,9)**. Spills have occasionally been contained by shorelines. Burning could be applied in these instances, if the shoreline is remote and no combustible materials such as trees and docks are nearby.

X1.1.4 It is uncertain whether oil that is completely emulsified with water can be ignited. Oil containing some emulsion can be ignited and burned **(10)**. During the successful test burn of the Exxon Valdez oil, some patches of emulsion were present (probably less than 20 %) and this did not affect either the ignitability or the efficiency **(11)**. It is suspected that fire breaks down the water-in-oil emulsion, and thus water content may not be a problem if the fire can be started. There is inconclusive evidence at this time on the water content at which emulsions can still be ignited. One test suggested that a heavier crude would not burn with about 10 % water **(10)**, another oil burned with as much as 50 % **(12)**, and still another burned with about 70 % water **(13)**. One study indicated that emulsions may burn if a sufficient area is ignited **(13)**. Further studies indicate that stable emulsions will not burn but oil containing less than 25 % water can be ignited. Emulsions may not be a problem because chemical de-emulsifiers could be used to break enough of the emulsion to allow the fire to start.

X1.1.5 Most, if not all, oils will burn on water if slicks are thick enough. Except for light-refined products, different types of oils have not shown significant differences in burning behavior. Weathered oil requires a longer ignition time and somewhat higher ignition temperature **(12)**.

X1.1.6 Burning efficiency is the amount of oil before burning, less the volume left as residue, divided by the initial volume of the oil. The amount of soot produced is usually ignored in calculating burn efficiency. Efficiency is largely a function of oil thickness. Oil thicker than about 2 to 3 mm can be ignited and burns down to about 1 to 2 mm **(6,14)**. For example, a slick of 2 mm burning down to 1 mm yields a maximum efficiency of 50 %. A pool of oil 20 mm thick burns to approximately 1 mm, yielding an efficiency of about 95 %. Current research has shown that other factors such as oil type and low water contents only marginally affect efficiency **(4)**.

X1.1.7 The residue from oil-spill burning is largely unburned oil with some lighter or more volatile products removed

---

[2] The boldface numbers in parentheses refer to a list of references at the end of this guide.

(15,16). Highly efficient burns of some types of heavy crude oil may result in oil residue that sinks in sea water.

X1.1.8 Most oil pools burn at a rate of about 3 mm/min **(17)**. This means that the depth of oil is reduced by 3 mm/min. As a rule of thumb, oil burn rate is about 5000 L/m^2/day (or about 100 gal/ft^2/day). Several tests have shown that this does not vary significantly with oil type and weathering. Emulsified oil, due to its water content and thus reduced spreading rate and the increased heat requirement of the water, may burn slower.

X1.1.9 Studies conducted in the last ten years have shown that the type of ignition device is relatively unimportant, however, heavy oils require longer heating times and a hotter flame to ignite than lighter oils. Many types of ignition sources can supply sufficient heat for a sufficient length of time. A number of simple devices consisting of flotation and propellant have been developed **(18-22)**. A helicopter-slung device that dispenses packets of burning, gelled fuel is the only commercial unit available at this time. Actual burns at some incidents and experiments have been ignited using much less sophisticated means including lighting oil-soaked paper and sorbent.

X1.2 Emissions from Burning

X1.2.1 The atmospheric emissions of concern include PAHs (polyaromatic hydrocarbons), volatile organic compounds, oxygenated compounds, metals, particulate matter, and gases.

X1.2.2 The PAHs have been measured in soot particles and as gaseous emissions at several test spills **(15-17,23-25)**. Gaseous emissions were found to be negligible. The soot from several experimental burns has been collected and the PAH content measured. In all cases, the quantity of PAHs is less in the soot and residue than in the originating oil. All crude oils contain PAHs, varying from as much as 1 % down to about 0.001 %. These PAHs are burned to fundamental gases, except for those left in the residue and those on the soot. Studies have shown that PAHs are produced in great abundance at temperatures of 600 to 800°C. At combustion temperatures higher than this, fewer and fewer PAHs are produced. In-situ oil fires are known to reach temperatures of up to 1300°C. One overall finding is that most compounds of concern are associated with the particulate matter, which is largely precipitated downwind from the burn. The deposition is approximately square root with distance; little is carried far from the site. In summary, PAHs are not a serious concern in assessing the impact of burning oil.

X1.2.3 Volatile organic compounds (VOCs) are organic compounds that have a sufficiently low vapor pressure to be gaseous at normal temperatures. The emission of volatile compounds was measured at several test burns **(15,16,26)**. It was found that emissions were very high for many of the compounds measured. About 70 compounds were detected and many of these were at concern levels directly downwind of the fire. Tests of emissions for these same compounds without burning, however, showed higher levels in most cases.

X1.2.4 Burning nearly always produces partially oxidized materials. In the case of oil, many of these materials are alcohols, aldehydes, ketones, and similar compounds. Extensive testing at one burn site showed that low quantities of these compounds were present downwind, but at well below health concern levels and, in fact, at near ambient levels.

X1.2.5 Crude and residual oils contain metals such as vanadium, chromium, and nickel in the range from 10 to 40 ppm. While the fate of these metals during the combustion process is uncertain, they appear to be concentrated in the residue. Measurements during a series of experimental burns have shown the metal content in the soot to be below detection level **(15,16)**.

X1.2.6 The most obvious atmospheric emission is particulate matter, smoke, or soot. The quantity of soot produced by in-situ oil fires is not well established. Direct measures in small pan burns result in soot production values of 0.7 to 3.5 % for crude oil and about 11 % for diesel fuel. No measurement techniques are available at this time for large-scale burns, however, estimates range up to 15 % **(27,28)**. These estimates are complicated by the fact that particulates precipitate from the smoke plume. The proportion of the soot that consists of respirable particles (less than 10 µm in diameter) is a relatively low value at ground level. Respirable particles measured at ground level are below concern levels several hundred metres downwind **(16)**. A typical exposure limit is 150 µg/m^3, (8-h average).

X1.2.7 The combustion of oil reduces the starting materials to fundamental gases. Most emissions are carbon dioxide, which have been measured and rarely exceed five times the background levels **(16,20,29)**. This is not a health concern. Levels of carbon monoxide have been measured and found to be near measurement thresholds and thus well below health-exposure levels. Sulfur dioxide emissions are usually much lower than indicated by the sulfur content of the oil **(20)**. Sulfur compounds in oil range from about 0.1 to 5 % of the oil weight. Nitric oxides have not been detected as a result of in-situ combustion of oil **(15,16)**.

X1.2.8 One concern about the burning of crude oil is the formation of new toxic compounds. A study was conducted in which soot and residue samples were extracted and "totally" analyzed in various ways. While the study was not conclusive, no compounds of the several hundred identified were of serious environmental or health concern **(15)**. The soot analysis revealed that the bulk of the material was carbon and that all other detectable compounds were present on this carbon matrix in abundances of parts-per-million or less. The most frequent compounds identified were aldehydes, ketones, esters, acetates, and acids, which are formed by incomplete oxidation of the oil. Specific analysis was performed for the highly toxic compounds, dioxins and dibenzofurans. Results of this analysis were negative—including those for oils burned on salt water **(15)**.

X1.2.9 The burning process leaves a burn residue. Studies show that the residue is largely composed of oil with little removed other than some of the more volatile materials **(15,16)**. It appears to be the same as weathered oil of the same type. The residue contains PAHs at lower concentrations than the starting oil, although it may also contain metals at a slightly higher concentration.

X1.2.10 The temperature to which the water body is raised has been another concern **(5,17)**. Measurements during recent burn trials show no significant increase in water temperature, even in shallow, confined test tanks. Thermal transfer to the water is limited by the insulating oil layer and is actually the mechanism by which the combustion of thin slicks is extinguished.

X1.2.11 Water samples under burning oil have been analyzed in four cases **(15,16)**. No organic compounds were detected.

REFERENCES

(1) Evans, D. D., "In-situ Burning of Oil Spills: Appendix B—Case Histories of Attempts to Use Burning in Response to an Oil Spill," *Alaska Arctic Offshore Oilspill Response Technology Workshop Proceedings*, Washington, DC, 1988, pp. 77–81.

(2) Goodier, J. L., and Siclari, R. J., *Combustion: An Oil Spill Mitigation Tool Phase II: The Burning of the M/T Burmah Agate (Ex-Danaland)*, U.S. Department of Energy Report DOE/TIC-11471, Washington, DC, 1981.

(3) Battelle, Pacific Northwest Laboratories, *Combustion: An Oil Mitigation Tool*, prepared for U.S. Department of Energy, Environmental Control Technology Division, Washington, DC, 1979.

(4) Fingas, M. F., and Laroche, N., "An Introduction to the In-situ Burning of Oil Spills," *Spill Technology Newsletter*, Vol 15, No. 4, 1991, pp. 1–11.

(5) Fingas, M. F., Halley, G., Ackerman, F., Vanderkooy, N., Nelson, R., Bissonnette, M. C., Laroche, N., Lambert, P., Jokuty, P., Li, K., Halley, W., Warbanski, G., Campagna, P. R., Turpin, R. D., Trespalacios, M. J., Dickins, D., Tennyson, E. J., Aurand, D., and Hiltabrand, R., "The Newfoundland Offshore Burn Experiment—NOBE Experimental Design and Overview," *Proceedings of the Seventeenth Arctic and Marine Oil Spill Program Technical Seminar*, Environment Canada, Ottawa, Ont., 1994, pp. 1053–1061.

(6) Evans, D. D., Mulholland, G. W., Lawson, J. R., Tennyson, E. J., Tebeau, P. A., Fingas, M. F., and Gould, J. R., "Burning of Oil Spills," *Proceedings of the 1991 Oilspill Conference*, American Petroleum Institute, Washington, DC, 1991.

(7) Williams, R. E., and Cooke, T. S., "Feasibility of Using a Bubble Barrier for the Containment/Incineration of Spilled Oil," *Proceedings of the Eighth Annual Arctic Marine Oilspill Program Technical Seminar*, Environment Canada, Ottawa, Ont., 1985, pp. 212–227.

(8) Buist, I. A., and McAllister, I. R., "Dome Petroleum's Fireproof Boom—Development and Testing to Date," *Proceedings of the Fourth Annual Arctic Marine Oilspill Program Technical Seminar*, Environment Canada, Ottawa, Ont., 1981, pp. 479–497.

(9) Belicek, K., and Overall, J., *Some Aspects of Weathering and Burning of Crude Oil in a Water-and-Ice Environment*, Canadian Marine Drilling Ltd. Technical Report, Calgary, Alta., 1976.

(10) Smith, N. K., and Diaz, A., "In-place Burning of Crude Oils in Broken Ice," *Proceedings of the 1987 Oil Spill Conference*, American Petroleum Institute, Washington, DC, 1987, pp. 383–387.

(11) Allen, A. A.," Contained Controlled Burning of Spilled Oil During the EXXON VALDEZ Oil Spill," *Spill Technology Newsletter*, Vol 15, No. 2, 1990, pp. 1–5.

(12) Twardus, E. M., *A Study to Evaluate the Combustibility and Other Physical and Chemical Properties of Aged Oils and Emulsions*, Environment Canada Report EE-5, Ottawa, Ont., 1980.

(13) Bech, C., Sveum, P., and Buist, I., "In-situ Burning of Emulsions: The Effects of Varying Water Content and Degree of Evaporation," *Proceedings of the Fifteenth Arctic and Marine Oilspill Program Technical Seminar*, Environment Canada, Ottawa, Ont., 1992, pp. 547–559.

(14) Evans, D., Walton, D., Baum, H., Lawson, R., Harris, R., Ghoniem, A., and Holland, J., "Measurement of Large Scale Oil Spill Burns," *Proceedings of the Thirteenth Annual Arctic Marine Oilspill Program Technical Seminar*, Environment Canada, Ottawa, Ont., 1990, pp. 1–38.

(15) Fingas, M. F., Li, K., Ackerman, F., Campagna, P. R., Turpin, R. D., Getty, S. J., Soleki, M. F., Trespalacios, M. J., Paré J., Bissonnette, M. C., and Tennyson, E. J., "Emissions from Mesoscale In-situ Oil Fires: The Mobile 1991 and 1992 Tests," *Proceedings of the Sixteenth Arctic and Marine Oil Spill Program Technical Seminar*, Environment Canada, Ottawa, Ont., 1993, pp. 749–821.

(16) Fingas, M. F., Ackerman, F., Li, K., Lambert, P., Wang, Z., Bissonnette, M. C., Campagna, P. R., Boileau, P., Laroche, N., Jokuty, P., Nelson, R., Turpin, R. D., Trespalacios, M. J., Halley, G., Béanger, J., Paré J. R. J., Vanderkooy, N., Tennyson, E. J., Aurand, D., and Hiltabrand, R., "The Newfoundland Offshore Burn Experiment—NOBE—Preliminary Results of Emissions Measurement," *Proceedings of the Seventeenth Arctic and Marine Oil Spill Program Technical Seminar*, Environment Canada, Ottawa, Ont., 1994, pp. 1099–1164.

(17) Evans, D. D., Walton, W. D., Baum, H. R., Notarianni, K. A., Lawson, J. R., et al, "In-situ Burning of Oil Spills: Mesoscale Experiments," *Proceedings of the Fifteenth Arctic and Marine Oilspill Program Technical Seminar*, Environment Canada, Ottawa, Ont., 1992, pp. 593–657.

(18) Allen, A. A., *Refinement of Aerial Ignition Systems (Test and Evaluation of the Helitorch for the Ignition of Oil Slicks)*, Alaska Clean Seas, Anchorage, AK, 1987.

(19) Energetex Engineering, *Environmental Testing of the Dome Air-Deployable Igniter*, Final Report to Dome Petroleum Ltd., 1982.

(20) Frish, M. B., DeFaccio, M. A., Nebolsine, P. E., and Simons, G. A., "Laser Ignition of Arctic Marine Oil Spills," *Oil & Chemical Pollution*, Vol 3, No. 5, Elsevier Science Publishers, New York, 1986/87, pp. 355–365.

(21) Meikle, K. M., "Incendiary Device for Oil Slick Ignition," *Proceedings of the Fourth Annual Arctic Marine Oilspill Program Technical Seminar*, Environment Canada, Ottawa, Ont., 1981, pp. 499–513.

(22) Twardawa, P., and Couture, G., "Incendiary Devices for the In-situ Burning of Oil Spills," *Proceedings of the Third Annual Arctic Marine Oilspill Program Technical Seminar*, Environment Canada, Ottawa, Ont., 1980, pp. 281–289.

(23) Benner, B. A., Jr., Bryner, N. P., Wise, S. A., Mulholland, G. W., Lao, R. C., and Fingas, M. F., "Polycyclic Aromatic Hydrocarbon Emissions from the Combustion of Crude Oil on Water," *Environmental Science Technology*, Vol 24, No. 9, 1990, pp. 1419–1427.

(24) Fingas, M. F., Halley, G., Ackerman, F., Nelson, R., Bissonnette, M. C., Laroche, N., Wang, Z., Lambert, P., Li, K., Jokuty, P., Sergy, G., Halley, W., Latour, J., Galarneau, R., Ryan, B., Campagna, P. R., Turpin, R. D., Tennyson, E. J., Mullin, J., Hannon, L., Aurand, D., and Hiltabrand, R., "The Newfoundland Offshore Burn Experiment," *Proceedings of the 1995 International Oil Spill Conference*, American Petroleum Institute, Washington, DC, 1995, pp. 123–132.

(25) Mitchell, J. B. A., and Moir, M. E., "Smoke Reduction from Pool Fires Using Ferrocene and Derivatives," *Proceedings of the Fifteenth Arctic and Marine Oilspill Program Technical Seminar*, Environment Canada, Ottawa, Ont., 1992, pp. 681–687.

(26) Li, K., Caron, T., Landaiult, M., Paré J. R. J., and Fingas, M., "Measurement of Volatiles, Semi-volatiles and Heavy Metals in an Oil Burn Test," *Proceedings of the Fifteenth Arctic and Marine Oilspill Program Technical Seminar*, Environment Canada, Ottawa, Ont., 1992, pp. 561–573.

(27) Walton, W. D., Evans, D. D., McGrattan, K. B., Baum, H. R., Twilley, W. H., Madrzykowski, D., Putorti, A. D., Rehm, R. G., Koseki, H., and Tennyson, E. J., "In Situ Burning of Oil Spills: Mesoscale Experiments and Analysis," *Proceedings of the Sixteenth Arctic and*

Marine Oil Spill Program Technical Seminar, Environment Canada, Ottawa, Ont., 1993, pp. 679–734.

(**28**) Fingas, M. F., Li, K., Ackerman, F., Wang, Z., Lambert, P., Gamble, L., Trespalacios, M. J., Schuetz, S., Turpin, R. D., and Campagna, P. R., "Soot Production from In-Situ Oil Fires: Review of the Literature, Measurement and Estimation Techniques and Calculation of Values from Experimental Spills," *Proceedings of the Nineteenth Arctic and Marine Oil Spill Program Technical Seminar*, Environment Canada, Ottawa, Ont., 1996, pp. 999–1032.

(**29**) Campagna, P. R., and Humphrey, A., "Air Sampling and Monitoring at the Kuwait Oil Well Fires," *Proceedings of the Fifteenth Arctic and Marine Oilspill Program Technical Seminar*, Environment Canada, Ottawa, Ont., 1992, pp. 575–592.

Designation: F 1834 – 98

Standard Guide for Consideration of Anaerobic Bioremediation as a Chemical Pollutant Mitigation Method on Land[1]

This standard is issued under the fixed designation F 1834; the number immediately following the designation indicates the year of original adoption or, in the case of revision, the year of last revision. A number in parentheses indicates the year of last reapproval. A superscript epsilon (ε) indicates an editorial change since the last revision or reapproval.

1. Scope

1.1 This guide covers recommendations concerning the application of anaerobic bioremediation to mitigate chemical pollutants (see Appendix X1, Table X1.1).

1.2 This is a general guide only, assuming the bioremediation to be safe, effective, available, applied in accordance with manufacturers recommendations and in compliance with relevant environmental regulations.

1.3 This guide addresses the application of anaerobic bioremediation alone or in conjunction with order technologies, following chemical pollution of terrestrial environments.

1.4 This guide does not consider the ecological effects of the anaerobic bioremediation.

1.5 This guide applies to all terrestrial environments. Specifically, it addresses available information concerning anaerobic bioremediation of chemical products and wastes of industrial processes.

1.6 In making bioremediation-use decisions, appropriate government authorities must be consulted as required by law.

1.7 *This standard does not purport to address all of the safety concerns, if any, associated with its use. It is the responsibility of the user of this standard to establish appropriate safety and health practices and determine the applicability of regulatory limitations prior to use. In addition, it is the responsibility of the user to ensure that such activity takes place under the control and direction of a qualified person with full knowledge of any potential or appropriate safety and health protocols.*

2. Referenced Documents

2.1 *ASTM Standards:*

F 1481 Guide for Ecological Considerations for the Use of Bioremediation in Oil Spill Response- Sand and Gravel Beaches[2]

F 1600 Terminology Relating to Bioremediation[2]

F 1693 Guide for Consideration of Bioremediation as a Oil Spill Response Method for Land[2]

E 1943 Guide for Remediation of Ground Water by Natural Attenuation at Petroleum Release Sites[2]

3. Terminology

3.1 For additional information relating to bioremediation, see Terminology F 1600.

3.2 *Definitions of Terms Specific to This Standard:*

3.2.1 aerobes—organisms that require air or free oxygen for growth.

3.2.2 anaerobes—organisms that grow in the absence of air or oxygen, and do not use molecular oxygen in respiration.

3.2.3 anaerobic reactor—an engineered system designed to maintain a high anaerobic microbial population to effectively biodegrade chemical contaminants.

3.2.4 bioaugmentation—addition of microorganisms (predominantly bacteria) to increase the biodegradation rate of target pollutants.

3.2.5 biodegradation—chemical alteration and breakdown of a substance usually to smaller products, caused by microorganisms or their enzymes.

3.2.6 bioremediation agents—inorganic and organic compounds and microorganisms that are added to enhance degradation processes, predominantly microbial.

3.2.7 biostimulation—addition of microbial nutrients, oxygen, heat or water, or both, to enhance the rate of biodegradation of target pollutants by indigenous species (predominantly bacteria).

3.2.8 cometabolism—transformation of the contaminant is the result of an incidental reaction catalyzed by enzymes involved in the normal cell metabolism or special detoxification reactions.

3.2.9 indigenous—native to a given habitat or environment

3.2.10 nutrient—a substance that supports organismal growth.

3.2.11 remediation by natural attenuation (RNA)—a remedy where naturally occurring physical, chemical, and biological processes will effectively achieve remedial goals and can be demonstrated through monitoring. The application of natural attenuation processes as a remedial action also has been described by a variety of other terms, such as intrinsic remediation, intrinsic bioremediation, passive remediation, passive bioremediation, etc.

3.2.12 species—a taxonomic category characterized by individuals of the same genus that are mutually similar and are able to interbreed.

---

[1] This guide is under the jurisdiction of ASTM Committee F-20 on Hazardous Substances and Oil Spill Response and is the direct responsibility of Subcommittee F20.24 on Bioremediation.

Current edition approved Nov. 10, 1997 and March 10, 1998. Published July 1998.

[2] *Annual Book of ASTM Standards*, Vol 11.04.

3.2.13 terrestrial—consisting of land, as distinguished from water.

4. Significance and Use

4.1 The purpose of this guide is to provide guidance on a means (called anaerobic bioremediation) of effectively cleaning up specific chemical pollutants on and below terrestrial surfaces.

4.2 This technology has been shown to be applicable to the mitigation of the chemical pollutants presented in Appendix X1, Table X1.1.

5. General Considerations for Anaerobic Bioremediation Use

5.1 Bioremediation technologies attempt to accelerate the natural rate of biodegradation. *Ex situ* and *in situ* anaerobic bioremediation technologies can be employed. The use of adequate controls in preliminary field studies, or results of previously reported studies, will assist in determining to what extent microorganism or nutrient, or both, and additional amendments are necessary to obtain the desired rate of degradation.

5.2 Bioremediation performance depends particularly on the efficiency of the chemical degrading indigenous bacteria or bioaugmented bacteria. Performances also depends on the availability of rate-limiting nutrients, physical characteristics, and the susceptibility of the target chemical product to anaerobic microbial degradation.

5.2.1 Chemicals shown to be effectively biodegradable by anaerobic microbes and processes are included in, but not necessarily limited to, several categories: explosives, dyes, pesticides/insecticides, herbicides, and chlorinated solvents. All are generally stable in nature except under conditions that favor anaerobic biodegradation.

5.2.1.1 Chemicals shown to be anaerobically biodegradable are listed in Appendix X1 (Table X1.1) by UN Number, CAS registry number, chemical name, and source.

5.2.2 Several defined anaerobic species are responsible for the degradation of the chemicals. Various texts and scientific studies describe biodegradability and biodegradation rates of the chemical listed in Appendix X1, Table X1.1 (**1-3**)[3].

5.2.2.1 Several groups of bacteria usually cooperate in the anaerobic degradation of one compound. Different substrates may be used for energy and for carbon metabolism (**3**).

5.2.2.2 Aerobic bacteria may further metabolize anaerobic metabolites (**3-5**). For example, as the degree of halogenation diminishes, the susceptibility to aerobic degradation increases.

5.2.2.3 Growth yields of anaerobic bacteria are low due to energy yields compared with aerobic bacteria; 10 % versus 50 % of the substrate carbon can be incorporated as cell matter (**3**). Anaerobic reactors that either retain or recycle the active biomass maintain substrate conversion rates competitive with aerobic processes.

5.2.3 Temperature has also been shown to play a role in anaerobic biodegradation, from 10 to 40°C (**3**). For example, optimum biodegradation of TNT-contaminated soil was achieved at 25 to 30°C (**4, 6**).

5.2.4 Special conditions are favorable for anaerobic bacterial growth. For example, aerobic microorganisms that adhere to surfaces which come in contact with oxygen, utilize the oxygen creating conditions that promote anaerobic growth (**3**). In addition, metal sulfide precipitates that form cover the surface and act as barriers to oxygen, provide a redox buffer, and help balance the availability of trace elements (**3**).

5.2.5 Clay content and the sizes of soil particles and aggregates affect the microbial activity (**7**). Bulk density is affected by the pore size. Water retention for sandy loam soil ranges from 15 to 30 %, and for clays ranges from 40 to 45 %. The difference between the total pore space and the water content represents the air-filled space. A minimum air-filled pore space of 10 % by volume is considered necessary for adequate aeration of aerobic microorganisms. Therefore if the water content were to exceed 30 % for sandy loam soils, the soil may not have enough void space for aeration, and the system would promote growth of anaerobic microorganisms.

5.3 Anaerobic bioremediation should be carried out under the guidance of qualified personnel who understand the safety and health aspects of site activities.

6. Background

6.1 General background information concerning approaches to bioremediation have been presented in previous documents (**3, 5, 8, 9**). Additional information can be found in Guide F 1481 and Guide F 1693.

6.2 There are several anaerobic bioremediation technologies available. It is important to understand the potential use of these systems when assessing the applicability for full-scale implementation. Costs are determined by the size of the site, soil properties, type and level of chemical contaminant(s), cleanup goals, time allowed for attaining the goals, and testing requirements.

6.2.1 *Anaerobic Reactors/Ex Situ Treatment*:

6.2.1.1 Systems are designed to maintain a high anaerobic microbial population. These anaerobic digestors are engineered to accumulate both digestible and nondigestible organic and inorganic solids. The solids may contribute to the prevention of oxygen diffusion and provide a large surface area to support adhesion of anaerobic microbial communities.

6.2.1.2 Basic types include: septic tanks; conventional anaerobic digestors; anaerobic contact process reactors; upflow anaerobic sludge blanket reactors; anaerobic filters; and, anaerobic fluidized bed reactors.

6.2.1.3 Anaerobic reactors have been successfully used at commercial scale.

6.2.2 *In Situ Anaerobic Bioremediation*

6.2.2.1 *In situ* bioremediation is engineered to circulate nutrients or microbes, or both, that enhance anaerobic biodegradation (**9**). Sites amenable to treatment allow solutions to be transported to the contaminated soil and water. Treatment is most likely to be successful if the soil is relatively uniform and if the hydraulic conductivity is greater than 10^{-4} cm/s.

6.2.2.2 Anaerobic bioremediation has been demonstrated at the laboratory and pilot-scale levels. Success at the commercial

[3] The boldface numbers given in parentheses refer to a list of references at the end of the text.

scale is currently undergoing evaluation for *in situ* anaerobic treatment.

6.2.3 *Remediation by Natural Attenuation (RNA)*:

6.2.3.1 Intrinsic bioremediation requires no intervention and depends on native microbial activity to degrade the contaminants (**9**). Amenable sites have a adequate groundwater flow, natural buffering capacity, anaerobic conditions and sufficient levels of nutrients.

6.2.3.2 Field demonstrations should be performed to confirm laboratory tests. There should be a documented loss of contaminants owing to the action of microorganisms.

6.2.3.3 RNA has been established at the laboratory, pilot-scale and commercial levels.

6.2.3.4 For additional information on RNA, refer to Guide E 1943.

6.3 Compounds amenable to anaerobic biodegradation are produced as solvents, herbicides, insecticides, dyes, and nitroaromatic explosives.

6.3.1 Halogenated aliphatic compounds are degreasers and solvents. Highly chlorinated compounds can be completely dechlorinated to relatively nontoxic compounds, which are biodegraded by aerobic microbes. Less halogenated ethenes are destroyed by cometabolism when certain aerobic bacteria are supplied with methane, phenol, or toluene (**3, 18**). As the degree of halogenation in aliphatics decreases, susceptibility to aerobic biodegradation increases.

6.3.2 Halogenated aromatic compounds are used as solvents, pesticides, herbicides, and in electrical transformers. Anaerobic microbes can remove chlorine atoms from the highly halogenated aromatics. As halogen atoms are replaced by hydrogen atoms (dehalogenation), the molecules become more susceptible to aerobic biodegradation.

6.3.3 Nitroaromatic compounds are used in production of explosives and dyes. Anaerobic microbes transform nitroaromatics to nontoxic volatile organic acids which are aerobically biodegraded. Anaerobic and aerobic metabolic pathways affect the fate of nitroaromatic compound catabolism (**6, 12-18**).

7. Bioremediation Technology Selection Assessment

7.1 Treatability studies have provided data to support treatment selection and are performed prior to reagent selection. The data indicate whether cleanup goals can be met and further determine the optimal operating conditions for remediation project design. The level of study chosen depends on available literature information, technical expertise, and site-specific considerations. In addition, treatability study design and interpretation for anaerobic biodegradation remedy screening has been addressed (**19**).

7.2 Various federal, provincial and state agencies require support documentation. Studies are available through databases developed by groups sponsored by the USEPA and Environmental Canada.

7.3 Federal, provincial, state, and other governmental agencies regulate the use of bioremediation agents. The role of these regulatory agencies varies.

7.4 There are many advantages associated with the various bioremediation applications. In general, compared with other technologies, advantages may include cost effectiveness, reuse, reduced personnel exposure to hazardous conditions, and reduced intrusion by response personnel (short and long term) into affected areas.

8. Recommendations

8.1 Anaerobic bioremediation should be considered as an option to mitigate certain chemical pollutants (Appendix X1, Table X1.1) in terrestrial chemical spills.

8.2 Treatability studies and available supporting literature should be used to provide data to support treatment selection, indication whether cleanup goals can be met, and optimize operational conditions for remedy design.

8.3 Safety and efficacy data should be substantiated prior to any anaerobic bioremediation field application. This information should be provided through treatability studies and/or published reports.

8.4 The effects of biostimulation should be compared with the effects of bioaugmentation. Consideration must be given to time, predictability, regulatory and public opinions, cost, and clean up requirements for process selection.

8.5 Implementation of bioremediation technology is site- and chemical contaminant-specific, and should be utilized appropriately in accordance with local, state, provincial and federal agencies.

9. Keywords

9.1 aerobic; anaerobic; anaerobic reactor; bioaugmentation; bioremediation; biostimulation; cometabolism; *ex situ*; *in situ*; remediation by natural attenuation (RNA); terrestrial; treatability

APPENDIX

(Nonmandatory Information)

X1. SELECTED CHEMICALS

TABLE X1.1 Selected Chemicals Known to be Amenable to Anaerobic Biodegradation[A]

Chemical Name	Source	CAS Number	UN Number
DDT	Pesticides	50-29-3	1846
2,4-Dinitrophenol	Pesticides, explosives, dyes	51-28-5	1320 1599
Carbon tetrachloride	Steel manufacturing, paint, ink, petroleum	56-23-5	1846
T-Hexachlorocyclohexane	Insecticide	58-89-9	2761
1, 1, 2, 2-Tetrachloroethane	Degreasers, paint, varnish, rust, removers	79-34-5	1702
Dinoseb	Herbicide, insecticide, miticide	88-85-7	0077
Nitrobanzene	Aniline production	98-95-3	1662
2, 4, 6-Trinitrotoluene	Explosives	118-96-7	0388
2, 4-Dinitrotoluene[B]	Dyes, explosives, propellants	121-14-2	2038 1600
RDX (Hexahydro-1, 3, 5- tri-nitro-1, 3, 5-triazine)	Explosives	121-82-4	0391
Methyl parathion	Pesticides	298-00-0	2783 1967
Aldrin	Insecticide	309-00-2	2761
a-Hexachlorocyclohexane	Insecticide	319-84-6	2729
β-Hexachlorocyclohexane	Insecticide	319-85-7	2729
4, 6-Dinitro-o-cresol	Insecticide	534-52-1	1598
2, 6-Dinitrotoluene[B]	Dyes, explosives, propellants	606-20-2	2038 1600
Heptachlor epoxide	Insecticides	1024-57-3	NA[C]
HMX (Octahydro-1, 3, 5, 7-tetraazocine)	Explosives	2691-41-0	0226
Tetrachloroethylene	Solvent	127-18-4	. . .

[A] These chemicals have been shown to be preferentially degraded under anaerobic conditions. The absence of a compound from this list does not mean that it is not amendable to anaerobic biodegradation.
[B] Biodegrades in both aerobic and anaerobic conditions.
[C] NA; Not Applicable.

REFERENCES

(**1**) *Handbook of Environmental Fate and Exposure Data For Organic Chemicals*, P.H. Howard (ed.), Lewis Publishers, Inc., Chelsea, Michigan, Vols, I-III, 1989-1991.

(**2**) *Handbook of Environmental Degradation Rates*, P.H. Howard, Ed., Lewis Publishers, Inc., Chelsea, Michigan, 1991.

(**3**) Schink, B., "Principles and Limits of Anaerobic Degradation: Environmental and Technological Aspects," In: *Biology of Anaerobic Microorganisms*, J.B. Zehnder, Ed. John Wiley & Sons, New York, 1988, pp. 771-846.

(**4**) Funk, S.B., Crawford, D.L., Roberts, D.J., and Crawford, R.L., "Two-Stage Bioremediation of TNT Contaminated Soils", In *Bioremediation of Pullutants in Soil and Water, ASTM STP 1235*, B.S. Schepart, Ed. ASTM, Philadelphia, 1995, pp. 177-189.

(**5**) Reviewed in: Crawford, R.L., "The Microbiology and Treatment of Nitroaromatic Compounds," *Current Opinion in Biotech* 6, 1995, pp. 329-336.

(**6**) Funk, S.A., Roberts, D.J., Crawford, D.L., and Crawford, R.L., "Initial-Phase Optimization for Bioremediation of Munition Compound-Contaminated Soils," *Applies and Env. Micro. 59*, 1993, pp. 2171-2177.

(**7**) In: "Soil Microbiology and Biochemistry," E.A. Paul and F.E. Clark, Eds. Academic Press, New York, 1989, pp. 11-31.

(**8**) Office of Research and Development, U.S. Environmental Protection Agency, *Bioremediation of Hazardous Wastes, EPA/600/R-92/126*, August 1992.

(**9**) Reviewed in: *In Situ Bioremediation, When does it work?* National Research Council, National Academy Press, Washington, DC, 1993.

(**10**) Chang, H-L and Alvarez-Cohen, L., Biodegradation of Individual and Multiple Chlorinated Aliphatic Hydrocarbons by Methane-Oxidizing Cultures. *Applies and Env. Micro. 62*, 1996, 3371-3377.

(**11**) *Microbiological Decomposition of Chlorinated Aromatic Compounds*, M.L. Rochkind-Dubinsky, G.S. Sayler, and J.W. Blackburn, Eds. Marcel Dekkar, Inc., New York, 1987.

(**12**) Dickel, O., Hang, W., and Knackmuss, H-J, "Biodegradation of Nitrobenzene by a Sequential Anaerobic-Aerobic Process," *Biodegradation 4*, 1993, 187-194.

(**13**) In: "Bioaugmentation for Site Remediation," R. Hinchee, J. Fredricks, and B. Alleman, Eds.; *Proceedings of the Third In Situ and On-Site Bioremediation International Symposium*, April 24-27, 1995, pp. 57-59.

(**14**) R.L. Crawford," The Microbiology and Treatment of Nitroaromatic Compounds." *Current Opinion in Biotech 6*: 1995, pp. 329-336.

(**15**) Haigler, B., Wallace, W., and Spain, J., "Biodegradation of 2-Nitrotoluene by Pseudomonas Sp. Strain JS42." *Appl. Environ Microbiol 60*, 1994, pp. 3030-3032.

(**16**) Rhy-Williams, W., Taylor, S., and Williams, R., "A Novel Pathway has been Demonstrated for the Catabolism of 4–Nitrotoluene by *Pseudomonas*," *J Gen Microbiol 139*, 1993, 1967-1972.

(**17**) Haigler, B., Nishino, S., and Spain, J., "Biodegradation of 4–Methyl-5–Nitrocatechol by *Pseudomonas* Strain DNT," J *Bacteriol 176*, 1994, pp. 3433-3437.

(**18**) Noguera, D.R. and Freedman, D.L., "Reduction and Acetylation of 2, 4-Dinitrotoluene by a *Pseudomonas aeruginosa* Strain," *Applied and Env. Micro*, 62, 1996, pp. 2257-2263.

(**19**) Office of Research and Development and Office of Emergency and Remedial Response, U.S. Environmental Protection Agency. *Guide for Conduction Treatability Studies Under CERCLA*, Interim Final, Washington, DC, *EPA/540/2-89/058*, December, 1989.

The American Society for Testing and Materials takes no position respecting the validity of any patent rights asserted in connection with any item mentioned in this standard. Users of this standard are expressly advised that determination of the validity of any such patent rights, and the risk of infringement of such rights, are entirely their own responsibility.

This standard is subject to revision at any time by the responsible technical committee and must be reviewed every five years and if not revised, either reapproved or withdrawn. Your comments are invited either for revision of this standard or for additional standards and should be addressed to ASTM Headquarters. Your comments will receive careful consideration at a meeting of the responsible technical committee, which you may attend. If you feel that your comments have not received a fair hearing you should make your views known to the ASTM Committee on Standards, 100 Barr Harbor Drive, West Conshohocken, PA 19428.

Standard Guide for Use of Chemical Shoreline Cleaning Agents: Environmental and Operational Considerations[1]

This standard is issued under the fixed designation F 1872; the number immediately following the designation indicates the year of original adoption or, in the case of revision, the year of last revision. A number in parentheses indicates the year of last reapproval. A superscript epsilon (ϵ) indicates an editorial change since the last revision or reapproval.

1. Scope

1.1 This guide covers the use of chemical cleaning agents on oiled shorelines. This guide is not applicable to other chemical agents nor to the use of such products in open waters.

1.2 The purpose of this guide is to provide information that will enable spill responders to decide whether to use chemical shoreline cleaning agents as part of the oil spill cleanup response.

1.3 This is a general guide only. It is assumed that conditions at the spill site have been assessed and that these conditions are suitable for the use of cleaning agents. It is assumed that permission has been obtained to use the chemical agents. Variations in the behavior of different types of oil are not dealt with in this guide and may change some of the parameters noted herein.

1.4 This guide covers two different types of shoreline cleaners: those that disperse oil into the water and those that disperse little oil into the water under low energy levels. The selection criteria for these two types can differ widely. This guide does not cover dispersants.

1.5 *This standard does not purport to address all of the safety concerns, if any, associated with its use. It is the responsibility of the user of this standard to establish appropriate safety and health practices and determine the applicability of regulatory limitations prior to use.*

2. Referenced Documents

2.1 *ASTM Standards:*

F 929 Guide for Ecological Considerations for the Use of Chemical Dispersants in Oil Spill Response-Marine Mammals[2]

F 930 Guide for Ecological Considerations for the Use of Chemical Dispersants in Oil Spill Response-Rocky Shores[2]

F 931 Guide for Ecological Considerations for the Use of Chemical Dispersants in Oil Spill Response-Seagrasses[2]

F 932 Guide for Ecological Considerations for the Use of Chemical Dispersants in Oil Spill Response-Coral Reefs[2]

F 971 Guide for Ecological Considerations for the Use of Chemical Dispersants in Oil Spill Response-Mangroves[2]

F 972 Guide for Ecological Considerations for the Use of Chemical Dispersants in Oil Spill Response-Nearshore Subtidal[2]

F 973 Guide for Ecological Considerations for the Use of Chemical Dispersants in Oil Spill Response-Tidal Flats[2]

F 990 Guide for Ecological Considerations for the Use of Chemical Dispersants in Oil Spill Response-Sandy Beaches[2]

F 999 Guide for Ecological Considerations for the Use of Chemical Dispersants in Oil Spill Response-Gravel or Cobble Beaches[2]

F 1008 Guide for Ecological Considerations for the Use of Chemical Dispersants in Oil Spill Response-Salt Marshes[2]

F 1012 Guide for Ecological Considerations for the Use of Chemical Dispersants in Oil Spill Response-the Arctic[2]

F 1209 Guide for Ecological Considerations for the Use of Oilspill Dispersants in Freshwater and Other Inland Environments, Ponds and Sloughs[2]

F 1210 Guide for Ecological Considerations for the Use of Oilspill Dispersants in Freshwater and Other Inland Environments, Lakes and Large Water Bodies[2]

F 1231 Guide for Ecological Considerations for the Use of Oilspill Dispersants in Freshwater and Other Inland Environments, Rivers and Creeks[2]

F 1279 Guide for Ecological Considerations for the Use of Oilspill Dispersants in Freshwater and Other Inland Environments, Permeable Surfaces[2]

F 1280 Guide for Ecological Considerations for the Use of Oilspill Dispersants in Freshwater and Other Inland Environments, Impermeable Surfaces[2]

F 1686 Guide for Surveys to Document and Assess Oiling Conditions on Shorelines[2]

3. Significance and Use

3.1 This guide is primarily intended to assist decision-makers and spill-responders in contingency planning, spill response, and training.

3.2 This guide is not specific to site or type of oil.

4. Background

4.1 Chemical shoreline cleaning agents are compositions designed to be applied to oil and to remove oil from the shoreline above the low water line.

4.2 Chemical shoreline cleaning agents are generally used

[1] This guide is under the jurisdiction of ASTM Committee F-20 on Hazardous Substances and Oil Spill Response and is the direct responsibility of Subcommittee F20.13 on Treatment.

Current edition approved May 10, 1998. Published December 1998.

[2] *Annual Book of ASTM Standards*, Vol 11.04.

differently from chemical dispersants, which are used to treat oil spills in offshore waters.

4.3 Chemical shoreline cleaning agents are sometimes known as surface washing agents, shoreline cleaners, or beach cleaners.

4.4 The basic application method for shoreline cleaning agents is to spray the product onto the oil and leave the agent to penetrate the oil and then either flush away the oil or let a rising tide wash it away. The oil may be washed directly into containment areas for recovery (**1**).[3]

4.5 The fundamental advantage of using a shoreline cleaning agent is that oil can be removed rapidly without using excessive temperatures or pressures, which can be harmful to biota on and in beaches (**2,3**).

4.6 Laboratory effectiveness tests have been developed and many products have been tested (**4-6**). Field effectiveness tests are being developed (**7,8**).

4.7 Laboratory testing shows that effectiveness may differ in saltwater and freshwater (**6,9**).

4.8 There are differences in action mechanisms between dispersants and shoreline cleaning agents. Composition of the two products differ (**4, 9-13**).

4.9 Before specialized products were developed, dispersants were used as shoreline cleaning agents with varying results (**14**).

4.10 The aquatic toxicity of the treating agents varies widely and is a factor in choosing products (**3,9**).

4.11 The amount of oil dispersed into water primarily depends on energy used to remove the oil from the substrate, especially for dispersing shoreline treating agents. The energy level is difficult to measure, but may be estimated from indicators such as the pressure of the rinse water (**1-3**).

4.12 The ease of oil removal from a beach depends very much on the type of oil, its degree of weathering and the type of beach. For example, a highly-weathered oil is difficult to remove by any means (**2**).

5. General Considerations for Using Chemical Shoreline Cleaning Agents

5.1 Two basic types of shoreline cleaners are available: those that disperse oil into the water column, and those that disperse little oil into the water column at low energy levels.

5.2 Considerations for the use of shoreline cleaning agents that disperse are the same as those for using dispersants in the specific habitat. ASTM Guides F 929, F 930, F 931, F 932, F 971, F 972, F 973, F 990, F 999, F 1008, F 1012, F 1209, F 1210, F 1231, F 1279, F 1280, and F 1686 have been prepared for many of these habitats as referenced in Section 2.

5.3 Shoreline cleaning agents that do not disperse, have very little impact on the water column.

5.4 Regulatory authorities may have additional criteria and regulations regarding the acceptability and use of shoreline cleaning agents.

5.5 The decision of whether to use or not to use shoreline cleaning agents always involves tradeoffs. Using a non-dispersing shoreline cleaning agent moves oil out onto the water where it must be recovered. Using a dispersing cleaning agent moves oil into the water column. Therefore, adverse effects on water organisms may be increased in the water column (in the case of a dispersing agent) by removing it from the shoreline.

5.6 Shoreline cleaning agents are used primarily as a cleanup method and not as a spill control method. Since some shorelines are more vulnerable to the longer lasting impacts of spilled oil, an acceptable tradeoff may be to protect these sensitive environments by removing the oil and either recovering it or putting it into a less sensitive environment. When dispersing-type agents are used, the tradeoff that must be evaluated is the long-term impact of the residence time of spilled oil that is stranded on shorelines as opposed to the short-term impact of the presence of dispersed oil in the water column. For non-dispersing agents, the trade-off that must be evaluated is the difficulty of recovering the released oil versus the impact of the long residence time of spilled oil that is stranded on shorelines and the possibility of re-oiling adjacent shoreline.

5.7 It has been found that some shoreline cleaning agents are equally effective in fresh and salt water, while others are not. The salinity of the water involved may therefore be a factor, and the effectiveness of the particular product in that salinity (**9**).

6. Environments Covered

6.1 *Shorelines Generally*—Shorelines vary extensively in their composition and their retention of oil. Several classification schemes are available for oiled shorelines as well as guides to other cleanup methods (**15,16**).

6.2 *Seagrasses*—Seagrass-dominated shorelines can be found in shallow marine environments from the tropics to Arctic regions. Seagrass beds form a discreet ecosystem that traps material derived from terrestrial sources and then exports large quantities of organic matter to the open sea. The presence of an extensive network of roots and rhizomes facilitates not only the sediment-binding of the grass beds but also the transport of materials back out to sea. Oil can adhere to the seagrasses and cause damage.

6.3 *Mangroves*—Mangrove ecosystems are intertidal forests dominated by various species of woody halophytes, commonly called mangroves. There are 12 families and more than 54 species of mangroves. Mangrove ecosystems occur in tropical low-energy depositional areas. Mangroves tend to promote the deposition of organic and mineral matter and their extensive root systems are important in stabilizing intertidal sediments. They are important ecologically as they provide the structural basis for many species of animals and plants. Mangroves are particularly prone to damage from oiling as they have respiratory openings on roots that can be clogged (**17**).

6.4 *Tidal Flats*—Tidal flats are usually broad intertidal areas of unconsolidated sediments that have little slope and are usually protected from direct wave action. They are composed of sediments of varying characteristic grain size depending on the amount of wave and current energy present. Tidal flats may be covered by seagrasses, marsh grass, or mangroves, the environments which are discussed elsewhere in this guide.

[3] The boldface numbers in parentheses refer to the list of references at the end of this standard.

Tidal flats are important to the coastal ecosystem because of the high biological productivity. Oil retention on tidal flats is largely transitory and oil will often be carried to the supra-tidal regions.

6.5 *Sandy Beaches*—Sandy beaches are composed of sediments ranging from 0.06 to 2.0 mm in size. The composition of the sand itself may vary, but it is usually either siliceous or carbonate. The character of the sediment may be a significant factor in oil retention as oil adheres differently to different types of materials. Wave action can change the profile of a sandy beach and can bury or cover oil.

6.6 *Gravel Beaches*—Gravel beaches are composed of sediments ranging in size from 2.0 to 63 mm. The materials are usually a mixture of minerals with a variety of oil retention properties. Gravel beaches are dynamic and sometimes change in profile. They can retain large amounts of oil which may be buried under clean beach material as a result of wave action. The dynamic nature of the gravel beach depends on its exposure. Sheltered gravel beaches are relatively stable, whereas the gravel on exposed beaches may be continuously re-distributed.

6.7 *Cobble Beaches*—Cobble beaches are composed of materials ranging from 64 to 256 mm. Cobble beaches are relatively stable, unless the beach is exposed to high seas. Cobble beaches will retain the most oil of all types of beaches because of the large interstitial spaces.

6.8 *Boulder/Rocky Beaches*—Boulder or rocky beaches are composed of materials larger than 256 mm (boulders) or bedrock. Despite the large interstitial spaces, they do not retain as much oil as cobble beaches, generally because the interstitial spaces are large enough to permit run-off. Retention is much greater, however, than that for several other types of beaches. The slope of the shore can range from vertical rock wall to a gently sloping or nearly flat platform. The nature of the entire intertidal environment is controlled primarily by the wave energy. Similarly, the biological abundance usually corresponds to the energy regime. High-energy shorelines typically have less biota than low-energy shorelines. The retention of oils varies with the energy. High energy shorelines are generally self-cleaning.

6.9 *Coastal Salt Marshes*—Coastal salt marshes are intertidal wetlands, transitional zones between terrestrial and aquatic ecosystems. Salt marshes are generally formed when plants invade shallow, protected tidal flats on low coastal lands. Typically, soil immersion occurs during about half of the tidal cycle. Salt marshes are low-energy environments in which oil is generally trapped and retained. Salt marshes are very important ecologically and generally are very fragile environments.

6.10 *Freshwater Marshes*—Freshwater marshes are the equivalent of saltwater marshes and are generally found at the fringe of a lake or river. Retention of oil is again high and due to the low energy, self-cleaning is minimal.

6.11 *Ponds and Sloughs*—Ponds and sloughs are freshwater bodies that have little or no water circulation. These water bodies are characterized by high oil retentivity as often there is dense vegetation that can retain oil.

6.12 *Lake Shores*—Lakes are freshwater bodies that can have shorelines very similar to sea shores.

6.13 *River Shores*—River shores may be similar to their sea shore counterparts. Specific types should be compared to that of sea shores.

6.14 *Man-Made Structures*—Man-made structures include piers, docks, breakwaters, boat ramps, dykes, etc. The retentivity and porosity of such structures vary with the type of construction material.

7. General Operational and Environmental Considerations for Use of Shoreline Cleaning Agents

7.1 The tradeoff between leaving the oil on the shoreline or removing it by perhaps more intrusive means, and the use of the treating agent is the primary consideration. The use of dispersing shoreline cleaning agents involves the additional consideration of the fate and effects of the oil in the aquatic environment.

7.2 The aquatic toxicity of the treating agent and the oil is of concern for dispersing shoreline cleaning agents. These types of shoreline cleaning agents require the same considerations as noted for dispersants in the referenced documents.

7.3 The effectiveness of a shoreline cleaning agent may not be the same in fresh water as in salt water.

7.4 Non-dispersing shoreline cleaning agents are usually used to remove oil from the shoreline and the oil is then recovered. The oil spill recovery potential off a given shoreline must then be considered.

7.5 An agent is most effective when it has ample time to penetrate into the oil. Thirty minutes or more of soaking or penetration time are recommended **(1)**.

7.6 After treatment, the oil may be removed using low-pressure water hoses **(1,2)**.

8. Considerations for Specific Types of Environment

8.1 *Shorelines Generally*—Since shorelines vary extensively in sensitivity, oil retentivity, and environmental importance, a general recommendation cannot be made. Each specific environment should be considered separately. An important consideration is the net environmental benefit of using the chemical beach cleaner versus leaving the oil on the shoreline or using other cleanup methods.

8.2 *Seagrasses*—Seagrasses can be treated with shoreline cleaning agents to remove oil. The agent's toxicity to the seagrass should be assessed before usage. Testing of some types of treating agents have shown relatively good success. Care must be taken to avoid physically disturbing the sea grasses during the cleanup operations, which can do more damage than the oil **(18,19)**.

8.3 *Mangroves*—Oil can be removed from the extensive root system using shoreline treating agents, which may save the mangroves or significantly reduce damage to them. Access to perform the operations may be difficult. Experimental data shows that up to 50 % of the mangroves can be saved if treated within 7 days **(17)**.

8.4 *Tidal Flats*—Tidal flats do not often require cleaning, because the oil does not usually retain to the substrate. If oil is retained, it can be treated with shoreline cleaning agents, although access is often difficult and can be damaging to the tidal flats.

8.5 *Sandy Beaches*—Sandy beaches are readily amenable to treatment using shoreline cleaning agents. Low energy washing is required to avoid disturbing the sand.

8.6 *Gravel Beaches*—Gravel beaches can be treated with shoreline treating agents of the non-dispersing type. The dispersing type of agents will cause oil and agent to penetrate to the subsurface. Low energy washing is required to avoid disturbing the shoreline material and damaging biota on the beach.

8.7 *Cobble Beaches*—Shoreline cleaning agents of the non-dispersing type can be used on cobble beaches. Dispersing cleaning agents will cause the oil and surface agent to penetrate to the subsurface.

8.8 *Rocky/Boulder Shores*—The necessity of removing oil varies with exposure. High-energy shorelines will generally self-cleanse. Rocky shorelines are the easiest to clean and generally do not have high levels of biota that can be affected by the cleaning operation. They can, however, be difficult to access. The amount of spray pressure required to remove oil is less than for other types of beaches.

8.9 *Coastal Salt Marshes*—Coastal salt marshes can be cleaned using shoreline cleaning agents. Care must be taken to avoid physically disturbing the marshes. Marshes are particularly vulnerable to physical damage that could be caused by vehicles or the use of tools. High-pressure cleaning should be avoided for the same reason **(18,19)**.

8.10 *Fresh Water Marshes*—Fresh water marshes can be cleaned using shoreline cleaning agents. Care must be taken to avoid physically disturbing the marshes. Marshes are particularly vulnerable to physical damage that could be caused by vehicles or the use of tools. High-pressure cleaning should be avoided for the same reason **(18,19)**.

8.11 *Ponds and Sloughs*—Ponds and sloughs can be cleaned using shoreline cleaning agents. Non-dispersing agents should be used whenever possible to avoid hydrocarbon loading in these water bodies.

8.12 *Lakes Shores*—Lakes shores can be cleaned using shoreline cleaning agents. Care must be taken to avoid physically disturbing the lake shore. Non-dispersing agents should be used whenever possible to avoid hydrocarbon loading in these water bodies. Furthermore, as most lakes are used for drinking water, hydrocarbons should be minimized in the water column.

8.13 *River Shores*—River shores can be cleaned using shoreline cleaning agents. Non-dispersing agents should be used whenever possible to avoid hydrocarbon loading in these water bodies. Furthermore, as most rivers are used for drinking water, hydrocarbons should be kept from dispersing into the water column.

8.14 *Man-Made Structures*—Man-made structures can be cleaned using shoreline cleaning agents. Man-made structures are generally not sensitive to water pressure.

9. Keywords

9.1 chemical treating agent; oiled shorelines; oil-spill cleanup; shoreline cleanup; surface washing agent

REFERENCES

(1) Fiocco, R.J., Lessard, R.R., and Canevari, G.P., "Improved Oiled Shoreline Cleanup with Corexit 9580", in proceedings of *1996 Petro-Safe Conference*, Houston, Tx., pp. 276-280, 1996.

(2) Michel, J. and Benggio, B.L., "Testing and Use of Shoreline Cleaning Agents During the MORRIS J. BERMAN Oil Spill", in proceedings of the *1995 International Oil Spill Conference*, American Petroleum Institute, Washington, D.C., pp. 197-202, 1995.

(3) Shigenaka, G., Vicente, V.P., McGehee, M.A., and Henry, C.B., "Biological Effects Monitoring During an Operational Application of Corexit 9580", in proceedings of the *1995 International Oil Spill Conference*, American Petroleum Institute, Washington, D.C., pp. 177-184, 1995.

(4) Sullivan, D. and Sahatjian, K.A., "Evaluation of Laboratory Tests to Determine the Effectiveness of Chemical Surface Washing Agents", in proceedings of the *1993 International Oil Spill Conference*, American Petroleum Institute, Washington, D.C., pp. 511-514, 1993.

(5) Clayton, J.R. and Renard, E.P., "Statistical Assessment: Two Laboratory Tests for Estimating Performance of Shoreline Cleaning Agents for Oil Spills", in *Proceedings of the Seventeenth Arctic and Marine Oil Spill Program Technical Seminar*, Environment Canada, Ottawa, Ontario, pp. 877-907, 1994.

(6) Fingas, M.F., Kyle, D.A., Laroche, N.D., Fieldhouse, B.G., Sergy, G. and Stoodley, R.G., "Oil Spill Treating Agents," *Spill Technology Newsletter, Vol. 18, No. 3*, pp. 1-14, 1993.

(7) Clayton, J.R., Stransky, B.C. Adkins, A.C., Lees, D.C., Michel, J., Schwartz, M.J., Snyder, B.J. and Reilly, T.J., "Methodology for Estimating Cleaning Effectiveness and Dispersion of Oil with Shoreline Cleaning Agents in the Field", in *Proceedings of the Eighteenth Arctic and Marine Oil Spill Program Technical Seminar*, Environment Canada, Ottawa, Ontario, pp. 423-451, 1995.

(8) Clayton, J.R., Stransky, B.C., Schwartz, M.J., Lees, D.C., Michel, J., Snyder, B.J., and Adkins, A.C., *Development of Protocols for Testing Cleaning Effectiveness and Toxicity of Shoreline Cleaning Agents (SCAs) in the Field*, MSRC Technical Report 95-020.1, Washington, D.C., 1995.

(9) Fingas, M.F., Kyle, D.A., Laroche, N.D., Fieldhouse, B.G., Sergy, G. And Stoodley, R.G., "The Effectiveness Testing of Spill Treating Agents," *The Use of Chemicals in Oil Spill Response, ASTM STP 1252*, Peter Lane, Ed., American Society for Testing and Materials, Philadelphia, pp. 286-298, 1995.

(10) Merlin, F.X. and Le Guerroue, P., "The New French Approval Procedure for Shoreline Cleaning Agents", in *Proceedings of the Seventeenth Arctic and Marine Oil Spill Program Technical Seminar*, Environment Canada, Ottawa, Ontario, pp. 943-950, 1994.

(11) Fiocco, R.J., Canevari, G.P., Wilkinson, J.B., Jahns, H.O., Bock, J., Robbins, M., and Markarian, R.K., "Development of Corexit 9580 - A Chemical Beach Cleaner", in proceedings of the *1991 International Oil Spill Conference*, American Petroleum Institute, Washington, D.C., pp. 395-400, 1991.

(12) Canevari, G.P., Fiocco, R.J., Lessard, R.R., and Fingas, M.F., "Corexit 9580 Shoreline Cleaner: Development, Application and Status," *The Use of Chemicals in Oil Spill Response, ASTM STP 1252*, Peter Lane, Ed., American Society for Testing and Materials, Philadelphia, pp. 227-239, 1995.

(13) Rog, S., Owens, D., Pearson, L., Tumeo, M., Braddock, J. And Venator, T.," PES-51 Shoreline Restoration of Weathered Subsurface Oil in Prince William Sound, Alaska", in *Proceedings of the Seventeenth Arctic and Marine Oil Spill Program Technical Seminar*, Environment Canada, Ottawa, Ontario, pp. 607-620, 1994.

(14) Morris, P.R. and Thomas, D.H., *Evaluation of Oil Spill Dispersant*

Concentrates for Beach Cleaning - 1987 Trials, Warren Spring Laboratory Report LR624, Stevenage, Herts, UK., 1987.

(15) Owens, E.H. and Sergy, G.A., *Field Guide to the Documentation and Description of Oiled Shorelines*, Environment Canada, Ottawa, Ontario, 1994.

(16) Ownes, E.H., *Field Guide for the Protection and Cleanup of Oiled Shorelines*, Environment Canada, Dartmouth, Nova Scotia, 1995.

(17) Teas, H.J., Lessard, R.R., Canevari, G.P., Brown, C.D. and Glenn, R., "Saving Oiled Mangroves Using a New Non-Dispersing Shoreline Cleaner", in proceedings of the *1993 International Oil Spill Conference*, American Petroleum Institute, Washington, D.C., pp. 147-151, 1993.

(18) Pezeshki, S.R., DeLaune, R.D., Nyman, J.A., Lessard, R.R. and Canevari, G.P., "Removing Oil and Saving Oiled Marsh Grass Using a Shoreline Cleaner", in proceedings of the *1995 International Oil Spill Conference*, American Petroleum Institute, Washington, D.C., pp. 203-209, 1995.

(19) Pezeshki, S.R., DeLaune, R.D., and Nyman, J.A., *Investigation of Corexit 9580 for Removing Oil from Marsh Grass*, Technical Report Submitted to Exxon Research and Engineering from Louisiana State University, Baton Rouge, La., 1994.

Designation: F 1990 – 99

Standard Guide for In-Situ Burning of Spilled Oil: Ignition Devices[1]

This standard is issued under the fixed designation F 1990; the number immediately following the designation indicates the year of original adoption or, in the case of revision, the year of last revision. A number in parentheses indicates the year of last reapproval. A superscript epsilon (ε) indicates an editorial change since the last revision or reapproval.

1. Scope

1.1 This guide relates to the use of in-situ burning of spilled oil. The focus of the guide is in-situ burning of oil on water, but the ignition techniques and devices described in the guide are generally applicable to in-situ burning of oil spilled on land as well.

1.2 The purpose of this guide is to provide information that will enable oil-spill responders to select the appropriate techniques and devices to successfully ignite oil spilled on water.

1.3 This guide is one of several related to in-situ burning. Other standards cover specifications for fire-containment booms and the environmental and operational considerations for burning.

1.4 *This standard does not purport to address all of the safety concerns, if any, associated with its use. It is the responsibility of the user of this standard to establish appropriate safety and health practices and to determine the applicability of regulatory limitations prior to use.* In particular, the storage, transport, and use of ignition devices may be subject to regulations that will vary according to the jurisdiction. While guidance of a general nature is provided herein, users of this guide should determine regulations that apply to their situation.

2. Referenced Documents

2.1 *ASTM Standards:*

D 92 Test Method for Flash and Fire Points by Cleveland Open Cup[2]

D 975 Specification for Diesel Fuel Oils[2]

F 1788 Guide for In-Situ Burning of Oil Spills on Water: Environmental and Operational Considerations[3]

3. Terminology

3.1 *Definitions:*

3.1.1 *fire point*—the lowest temperature at which a specimen will sustain burning for 5 s. **D 92**

3.1.2 *flash point*—the lowest temperature corrected to a barometric pressure of 101.3 kPa (760 mm Hg), at which application of a test flame causes the vapor of a specimen to ignite under specified conditions of test. **D 92**

4. Significance and Use

4.1 This guide describes the requirements for igniting oil for the purpose of in-situ burning. It is intended to aid decision-makers and spill-responders in contingency planning, spill response, and training, and to aid manufacturers in developing effective ignition devices.

4.2 This guide describes criteria for the design and selection of ignition devices for in-situ burning applications.

4.3 This guide is not intended as a detailed operational manual for the ignition and burning of spilled oil.

5. Overview of the Requirements for Igniting Spilled Oil on Water

5.1 The focus of this section is on the in-situ combustion of marine oil spills, which, relative to land spills, have been of greater interest in oil-spill response.

5.2 Successful ignition of oil on water requires two components: heating the oil such that sufficient vapors are produced to support continuous combustion, and then, providing an ignition source to start burning. The temperature at which the oil produces vapors at a sufficient rate to ignite is called the flash point. At a temperature above the flash point, known as the fire point, the oil will produce vapors at a rate sufficient to support continuous combustion.

5.3 For light refined products, such as gasoline and some unweathered crude oils, the fire point may be in the range of ambient temperatures, in which case, little if any, preheating would be required to enable ignition. For other oil products, and particularly those that have weathered or emulsified, or both, the fire point will be much greater than ambient temperatures, and substantial preheating will be required.

5.4 The energy required to raise the temperature of the surface of an oil slick to its fire point depends on the slick thickness. While the oil is being heated by an igniter, heat is being conducted and convected to the underlying water. If the slick is sufficiently thick to insulate against these heat losses and allow the surface layer of oil to heat to its fire point, the oil will start to burn in the vicinity of the igniter. The minimum ignitable thickness for most oils is about 2 to 3 mm (see Guide F 1788).

5.5 Aside from oil type, other factors that can affect the ignitability of oil on water include the wind speed and the

[1] This guide is under the jurisdiction of ASTM Committee F-20 on Hazardous Substances and Oil Spill Response and is the direct responsibility of Subcommittee F20.15 on In-Situ Burning.

Current edition approved Dec. 10, 1999. Published February 2000.

[2] *Annual Book of ASTM Standards*, Vol 05.01.

[3] *Annual Book of ASTM Standards*, Vol 11.04.

emulsification of the oil. Secondary factors include ambient temperature and waves. The effect of these factors can be summarized as follows:

5.5.1 The maximum wind speed for successful ignition for large burns has been estimated to be approximately 10 m/s (20 knots) **(1, 2)**[4].

5.5.2 For more rapid flame spreading, slicks should be ignited at the upwind edge.

5.5.3 Weathered oils require a longer ignition time.

5.5.4 The effect of water content is similar to that of weathering, more ignition time being required to ignite a slick of emulsion. Once an emulsified slick is ignited, heat from the fire may break the emulsion and overcome this problem. Emulsion-breaking chemicals can be used to aid in initial ignition attempts.

5.5.5 Emulsions are difficult, if not impossible, to ignite without the use of emulsion-breaking chemicals.

6. Overview of Available Ignition Devices

6.1 *Simple Ignition Techniques*:

6.1.1 Propane or butane torches, or weed burners, and rags or sorbent pads soaked in fuel have been used to ignite oil on water. Propane torches tend to blow thin oil slicks away from the flames and are most applicable to thick contained slicks. Diesel is more effective than gasoline as a fuel to soak sorbents or rags because it burns more slowly, and hence, supplies more preheating to the oil.

6.1.2 Another effective surface-based igniter is gelled fuel. Gelling agents can be used with gasoline, diesel, or crude oil to produce a gelled mixture that is ignited and placed in an oil slick.

6.2 *Hand-Held Igniters*—A variety of igniters have been developed for use as devices to be handthrown, either from ground level or from helicopters. These igniters have used a variety of fuels, including solid propellants, gelled kerosene cubes, reactive chemical compounds, and combinations of these. Burn temperatures for these devices range from 700 to 2500°C, and burn times range from 30 s to 10 min. Most hand-held igniters have delay fuses that provide sufficient time to throw the igniter and allow it and the slick to stabilize prior to ignition.

6.3 *Helicopter-Slung Ignition Systems*—These systems have been adapted from devices used for burning forest slash and for setting backfires during forest-fire control operations. These devices emit a stream of gelled fuel, generally gasoline or a mixture of gasoline, diesel, or crude oil, or a combination thereof. As the gelled fuel leaves the device, it is lighted by an electrically-ignited propane jet. The burning gelled fuel falls as a stream that breaks into individual globules before hitting the slick. The burning globules produce an 800°C flame for up to 6 min. Tank capacities for the gelled fuel mixture range from 110 to 1100 L (30 to 300 gal).

7. Ignition Device Test

7.1 The following is intended as a simple test to evaluate the ability of an ignition device to ignite a thick slick of weathered oil. The ignition test does not consider operability factors, such as safe operation of the device, accuracy of deployment, and reliability of ignition components.

7.2 The test parameters are intended to reflect minimum conditions for acceptable performance. More stringent conditions, such as higher wind speed or the use of weathered or emulsified oils, may be considered for some ignition devices.

7.3 *Test Apparatus*—The ignition test is carried out in an approximately square test container. The test container must have a surface area that is the greater of ten times the area covered by the ignition device, or 1 m^2. A typical test container would be a steel pan of the required dimensions. To minimize wind-shielding by the walls of the container, the fluid level must be within 25 mm of the top of the test container.

7.4 *Test Slick*—The ignition test is carried out on a layer of oil with a maximum thickness of 10 mm and with a minimum underlying water depth of 200 mm. The oil for the ignition test is Diesel Fuel Grade No. 2, which has a minimum flash point of 60°C (see Specification D 975).

7.5 *Test Conditions*—At the start of the ignition test, the oil and water temperature must be no higher than 10°C. Throughout the test, the wind speed must be 5 m/s (10 knots) or greater.

7.6 *Initial Ignition Tests*—The test is initiated by activating the ignition device and deploying it into the test slick. It is recommended that initial tests be conducted by simply placing the ignition device on the test slick. The ignition test would be considered successful when flame is observed independent of the igniter, with flame covering the majority of the area of the test container.

7.7 *Tests for Air-Deployed Ignition Devices*—For igniters intended for deployment from helicopters, additional tests should be carried out to simulate air-deployment. These tests need not include ignition of oil but should include deployment of the device from a height of 10 m (minimum, measured from the device to the ground) to confirm that the device functions as intended during deployment. Tests should include deployment and operation of the device from a helicopter to ensure that the device can function in the presence of the helicopter's downwash.

7.8 *Test Record*—The test record must include the time for successful ignition, the actual container dimensions, the initial oil layer thickness, the underlying water depth, the air and water temperature at the start of the test, the wind speed, and any general observations of igniter performance.

7.9 *Optional Additional Tests*—In addition to the performance tests listed, consideration should be given to additional testing to address the following items depending on the intended application of the device:

7.9.1 The estimated accuracy of deployment of the ignition device on a target oil slick,

7.9.2 The resistance to damage of the device during deployment,

7.9.3 The performance in shallow pools (less than 100 mm deep) on solid ice,

7.9.4 The dependence on orientation of the igniter for proper performance,

7.9.5 Splash effects during impact with oil and water,

7.9.6 Effect on performance of temporary submergence of

[4] The boldface numbers in parentheses refer to the list of references at the end of this standard.

the igniter upon impact, and

7.9.7 Sensitivity to wind, rain, and sea state during ignition.

8. Operability

8.1 *Operating Instructions*—Operating instructions shall be supplied with the device and should include a description of the following items where applicable: safe operating procedures; required preparations of the igniter, or application system, or both, from storage to field use; type and amount of debris after use; training requirements; disposal requirements for spent igniters; and, retrieval and handling requirements for igniters that have misfired.

8.2 *Licensing for Transport and Use*—The ignition device must be approved for transport via cargo aircraft. Approvals, or pilot certifications, or both, may be required for devices intended for operation and deployment by helicopter. Users should note that pyrotechnic materials are not commonly transported by air and that such shipments often are rejected at the point of loading at the prerogative of the carrier despite any licensing or approvals.

8.3 *Stability During Flight*—For helicopter-slung devices, provision shall be made for stabilizing the device when carried by a swivel-hook helicopter. Any such stabilizing apparatus shall not impair the ability to jettison the device in the event of an emergency (see 9.3).

8.4 *Temperature Range*—The ignition device should function over an ambient temperature range of –10 to 30°C.

8.5 *Wind Conditions*—The ignition device should function, including deployment and operation from a helicopter, in wind conditions up to 10 m/s (20 knots).

9. Safety

9.1 *Unintended Activation*—The device should include protection against accidental activation.

9.2 *Delay Upon Activation*—For hand-held ignition devices, upon activation of the igniter, there should be a minimum delay of 20 s between the time the device is activated and it begins firing. It should be noted that excessive delay times may be troublesome in allowing the igniter to drift away from the target slick.

9.3 *Jettisoning of Equipment*—For helicopter-slung devices, provision shall be made for jettisoning of the device, including rapid disconnect of any power or control couplings.

9.4 *Operation*—Some ignition devices require an open flame or spark for activation, that may not be desirable or safe in certain applications, for example, for hand-held devices to be deployed from helicopters.

10. Storage

10.1 *Shipping and Storage Regulations*—The manufacturer of the device should specify shipping, handling, and storage instructions, and should note any limits on extreme temperatures, or humidity during storage, or both.

10.2 *Resistance to Degradation*—The device should function after exposure to temperature and humidity extremes and vibration that may be experienced during storage and shipping.

10.3 *Shelf-Life*—The device should have a minimum shelf-life of five years.

10.4 *Maintenance*—Operating instructions should specify any routine maintenance requirements, and should note components of the igniter that are subject to degradation, their expected shelf-life, and the procedure for refurbishment or replacement of parts following the normal shelf-life.

11. Keywords

11.1 ignition; in-situ burning; oil-spill burning; oil-spill disposal

APPENDIX

(Nonmandatory Information)

X1. Brief History of Igniter Development

X1.1 This Appendix is intended to provide a brief historical review of the uses of ignition devices for the in-situ burning of spilled oil. It is not intended to be comprehensive but simply attempts to show examples of what has and has not worked in past oil spill responses and experiments.

X1.1.1 Many different ignition devices have been used over the years to ignite or attempt to ignite marine oil spills. In 1967, four attempts were made to ignite seemingly thick oil slicks on the sea near the Torrey Canyon using pyrotechnic devices containing sodium chlorate, but these attempts were unsuccessful **(3, 4)**. It was concluded that the oil had emulsified to such an extent that it would not ignite.

X1.1.2 Oil on the shore from the Torrey Canyon spill proved virtually impossible to ignite and burn, although some success was reported in burning unemulsified oil in pools between rocks. In this case, flame throwers and flame-thrower fuel were used to ignite the pools, and they burned nearly to completion. Emulsified oil could be burned on the beach, as long as the flame thrower was applied, but once the flame was removed, the combustion stopped.

X1.1.3 Production of the Kontax igniter[5] ceased in the mid- to late-1970s **(5)**. The device consisted of a 4-cm diameter cylindrical metal screen 30.5 cm long and capped at both ends. A metal bar coated with metallic sodium ran through the center of the cylinder. The annulus was filled with calcium carbide. The device weighed 1.2 kg. For safety reasons, the Kontax igniter was stored in a sealed plastic bag.

X1.1.4 The Kontax igniter had a unique feature, that is, it did not require activation or a starter. When the device was

[5] The Kontax igniter was produced by Edward Michels GmbH of Essen, Germany.

exposed to water the sodium metal reacted to produce heat and hydrogen, which instantly ignited. At the same time, the calcium carbide reacted with water to produce acetylene, which was subsequently ignited by the burning hydrogen. The flame from the burning acetylene preheated and ignited oil vapors. Tests to evaluate Kontax were performed in 1969 by the Dutch government **(6)**. The tests were carried out 25 miles offshore and on beaches and the oils used were heavy and light Arabian crude. The igniter material Kontax was used in 25-kg bagged form. One test involved a 9-tonne slick covering about 2000 m^2 (0.5-cm thick) in a free-floating lumber boom. The bags containing the Kontax were punctured and thrown into the slick. The igniters were successful. Flames of 15 to 20 m high were reported, and a 98 to 99 % oil-removal efficiency was estimated. A Kontax-to-oil ratio of 1:100 by weight was estimated to be appropriate. The potential of Kontax also was demonstrated at the Arrow spill in 1970 where some of the spilled oil was primed with two drums of fresh oil and ignited with a Kontax igniter.

X1.1.5 The Kontax igniter produced a large flame area (3000 cm^2) with a relatively low flame temperature (770°C). This combination produced a relatively high flame emissivity of 2.25 kW/m^2. Although Kontax proved effective in both field and tank trials as a surface-deployed igniter **(5, 7)**, the device proved less effective when dropped from a height of 11.5 m, simulating deployment from a helicopter. The ignition success rate declined from 100 % in the surface tests to 60 % in the aerial tests. The main reason for the latter result was that the large splash caused by the Kontax igniter entering the water drove the oil away. By the time the oil had returned, the igniter had generated a ring of calcium hydroxide foam that kept the oil away.

X1.1.6 Energetex Engineering **(5)** tested a modification to the Kontax igniter, which involved combining a small amount of gasoline with the device. This inclusion of gasoline was intended as a fuel to bridge the calcium hydroxide foam barrier. This modification resulted in a slightly higher flame temperature (790°C) and better aerial deployment ignition success (80 %).

X1.1.7 It is not clear why Kontax was taken out of production. It may have been due to a general lack of interest in in-situ burning at the time, or due to the dangers and stringent requirements for storing, transporting, and using the igniters. Another igniter, Oilex Fire[6] consists of a sorbent (Oilex) plus a hydro-igniting agent. The company reported on the use of the chemical on small spills in Swiss lakes and in the Adriatic Sea **(7)**.

X1.1.8 On December 27, 1976, the Argo Merchant went aground near Nantucket Island and spilled most of its cargo of 28 000 tons of No. 6 fuel oil. Part of the response by the U.S. Coast Guard involved attempts to burn the oil. One 30-m × 40-m × 15-cm thick slick was treated with Tullanox 500 (a wicking and insulating agent), primed with 200 L of JP-4 and ignited with JP-4-soaked cotton sheets set afire with a flare. About 95 % of the Tullanox was blown off the treated slick by wind and the flames would not spread from the sheet to the primed slick. In another experiment, boxes of Tullanox 500 charged with JP-4 fuel were dropped onto a slick from a helicopter and ignited with timed thermite grenades. The isolated boxes burned but the flames did not spread **(6, 8)**.

X1.1.9 On January 28, 1977, some 300 000 L of No. 2 fuel oil was spilled onto the ice-covered waters of Buzzards Bay, Massachusetts by the barge Bouchard No. 65. Boxes of Tullanox soaked with jet fuel were dropped from helicopters onto pools of oil in the broken ice with delay-fuses. Thermite grenades were used to ignite the boxes. The ensuing fires burned for 1½ to 2 h and consumed 4000 to 8000 L of oil. The 38 to 46-km/h (20 to 25 knot) winds drove the flames from pool to pool in areas where adjacent pools were nearby. In other areas the fires did not spread. At a later date another series of burns were initiated by knotted rags soaked in diesel **(9, 10)**.

X1.1.10 Starting in 1977, considerable effort was devoted to developing an aerial ignition capability for potential spills from offshore exploration activities in the Beaufort Sea. Energetex Engineering evaluated and tested five devices (Kontax, Kontax with gasoline, solid propellant, solid fuel, and gasoline with sodium). Solid fuel and solid propellant igniters with a fuse wire were ranked highest **(5)**. Subsequently, two igniters were developed in Canada: the Dome igniter **(11-13)** and the EPS igniter **(14, 15)**. Solid propellants, also known as solid rocket fuels, are composed of a solid mixture of various portions of ammonium perchlorate oxidizer, metal fuel (magnesium or aluminum), and an organic binder. They have been used in a variety of igniters. Solid propellant igniters, in various shapes and utilizing various starters (electrical, chemical or fuses) have been extensively tested **(15)**. Such igniters exhibit very high flame temperatures (about 1230°C) and high flame emissivities (1.75 kW/m^2) but are consumed rapidly. They require mounting in a housing to suspend them no more than 5 cm above the oil/air interface. In water surface tests, solid propellant gave an 89 % ignition success rate, and an 80 % success rate in aerial-deployment tests with a fuse-wire starter (all other starter mechanisms resulted in lower success rates).

X1.1.11 The EPS igniter, also known as the Pyroid igniter **(15)**[7] is approximately 25 cm^2 and 13 cm high and weighs nearly 2 kg. The unit consists of a pyrotechnic device sandwiched between two layers of foam flotation and is activated by a self-contained firing mechanism. It is intended to be a hand-thrown device. The device is simple in design and operation, being activated by pulling on a firing clip which in turn strikes a primer cap. A 25-s delay column then provides sufficient time to throw the igniter and let it settle within the target oil slick. A specially formulated ring of fast-burning ignition material is then ignited, and this in turn ignites the primary incendiary material. The incendiary material is a solid propellant consisting of typically 40 to 70 % ammonium perchlorate, 10 to 30 % metal fuel (magnesium or aluminum), 14 to 22 % binder, and small amounts of other ingredients to aid in the casting and curing processes. The firing mechanism

[6] Oilex Fire was produced by Keltron Inc. of Switzerland.

[7] The Pyroid igniter is an air-deployable pyrotechnic device developed by the Canadian Environmental Protection Service of Environment Canada in cooperation with the Canadian Department of National Defense Research Establishment, Valcartier (DREV) **(15)**.

and the incendiary materials are sandwiched between two polystyrene foam slabs to provide both buoyancy and protection for the device on impact. All components except the firing mechanism are combustible, so that very little debris is left in the environment after a burn.

X1.1.12 These components have been designed so that the igniter experiences a minimum of roll if dropped onto a hard surface (like ice) or shallow water. The igniter can float in as little as 5 cm of water/oil. The flame released will be oriented properly regardless of which side of the igniter is up. The EPS igniter has been designed to produce a ring of fire with temperatures approaching 2000°C (4170°F) immediately adjacent to the perimeter of the igniter. This intense flame has a typical duration of about 2 min.

X1.1.13 The EPS igniter was designed to provide a 75 % probability of functioning properly when dropped at an airspeed of about 30 km/h from an altitude of approximately 15 m. Field tests indicate a high probability of successful ignition.

X1.1.14 Some solid-fuel igniters employ gelled kerosene cubes, for example, solid barbecue starter, suspended above the oil/air interface. Because of the lower flame temperatures (770°C) and flame emissivities 0.5 kW/m^2) generated, it is necessary to suspend the cubes within 3 cm of the oil surface in order to successfully ignite oil. Surface ignition tests have given an 84 % success rate while aerial tests have resulted in an 80 % success rate using a fuse wire starter **(5)**. Solid fuel is used in one commercially available igniter discussed in X1.1.15.

X1.1.15 Laser-based ignition systems received considerable attention in the 1970s and 1980s **(16-19)**. In static tests on land the concept proved to be capable of igniting fresh and weathered, unemulsified oil in 1–m^2 pools on ice **(18)**. The use of lasers mounted in helicopters to ignite spilled oil has been investigated, and the various components of a helicopter-borne system have been researched under contract to Environment Canada and the Minerals Management Service; however, further development to the prototype stage and subsequent commercialization await private sector involvement.

X1.1.16 In Alaska, a forest-fire fighting tool known as the Heli-torch was found in the mid-1980s to be an effective aerial ignition system for spilled oil **(20)**. The Heli-torch emits a burning stream of gelled gasoline that remains burning on an oil slick for a period of a few minutes. Testing with alternative fuels has indicated increased heat flux with gelled diesel and gelled crude oil. Considerable testing and refinement of the device has resulted in the Heli-torch being stockpiled around the world as the igniter of choice for in-situ burning.

X1.1.17 The in-situ test burn during the Exxon Valdez spill in 1989 was ignited by gasoline, gelled with a commercial gelling agent, and contained in a plastic bag. The gasoline and gelling agent were mixed by hand, placed on the water surface, then ignited and allowed to drift from the tow boat into the contained oil in the fire containment boom being towed behind.

X1.1.18 In a field trial of in-situ burning in Lowestoft, England in 1996 a more sophisticated version of this concept was tested **(21)**. A polyethylene bottle filled with 1 L of gelled gasoline was fitted with a foam floatation collar. The ignition source for the igniter was a standard marine hand-held distress flare attached to the outside of the bottle. The flare melted through the bottle, igniting the gelled gasoline as it was released onto the slick.

X1.1.19 Experimental work in Norway **(22)** examined the use of alternative fuels and emulsion-breaking chemicals using a Heli-torch igniter. Use of an emulsion breaker of approximately 5 % of the volume of gelled fuel was successful in igniting a 50 % water-in-oil emulsion. The optimal igniter fuel was found to be one that contained a range of light, medium, and heavy ends of crude oil: a mix of 60 % gasoline, 12 % diesel, and 28 % Bunker C was found to be effective. It was noted that using concentrations of gelling agent in excess of the recommended 12 % weight/volume caused difficulties in pumping the gelled fuel. Other recent experimental work **(23)** has shown the ability to ignite emulsions with up to 60 % water content. Emulsion-breaking chemicals applied at doses of 0.2 % (volume of chemical to volume of emulsion) were found to extend the ability to ignite emulsions depending on the initial oil type and the degree of weathering.

X1.1.20 Testing of various igniters was carried out **(24)** and led to the development of a manually-operated, electrically-fired, flare-type igniter. The temperature of the flare was found to be important, with those producing temperatures less than about 680°C unable to successfully ignite the diesel fuel used in the tests. Flare duration was important in the flares that burned for a minimum of 2 min were successful.

REFERENCES

(1) Bech, C., Sveum, P., and Buist, I., 1993, "The Effect of Wind, Ice and Waves on the In-Situ Burning of Emulsions and Aged Oils," *Proceedings of the Sixteenth Arctic and Marine Oilspill Program Technical Seminar*, Environment Canada, Ottawa, Ontario, pp. 735–748.

(2) Cabioc'h, F., 1993, "Last French Experiments in Order to Evaluate the Buring Possibilities of Three Water-in-Oil Emulsions," *Proceedings of the Sixteenth Arctic and Marine Oilspill Program Technical Seminar*, Environment Canada, Ottawa, Ontaris, pp. 823–832.

(3) Swift, W.H., Touhill, C.J., and Peterson, P.L., 1968, "Oil Spillage Control," *Chemical Engineering Progress Symposium*, Series 65(97):265–273.

(4) Anonymous, 1967, "Chemicals Versus Crude Oil," *Chemical Week*, 100(20):49.

(5) Energetex Engineering, 1978, "Testing of Air-Deployable Incendiary Devices for Igniting Oil on Water," EC Report No. EPS-4-EC-78-11, Environment Canada, Ottawa.

(6) Battelle, 1979, "Combustion: An Oil Spill Mitigation Tool," Report for U.S. Department of Energy, Contract No. EY-76-C-06-1830, U.S. Department of Energy, Washington, D.C.

(7) Freiberger, A. and Byers, J.M., 1971, "Burning Agents for Oil Spill Cleanup," *In Proceedings of 1971 Conference on Prevention and Control of Oil Spills*, American Petroleum Institute, Washington, DC.

(8) Det norske Veritas, 1979, "Tanker Oil Spill Analysis Study," Technical Report for Canadian Marine Drilling Limited, Det norske Veritas Ship Division.

(9) Schrier, E. and Eidam, C., 1979," Cleanup Efficiency of a Fuel Oil Spill in Cold Weather," *In Proceedings of 1979 Oil Spill Conference*, American Petroleum Institute, Washington, DC, pp. 419–428.

(10) Ruby, C.H., Ward, L.G., Fischer, I.A., and Brown, P.J., 1978, "Buzzards Bay Oilspill: An Arctic Analogue," *International Conference on Ports and Ocean Engineering Under Arctic Conditions (POAC '77)*, St. John's, Newfoundland.

(11) Buist, I., Pistruzak, W., and Dickins, D., 1981, "Dome Petroleum's Oil and Gas Under Sea Ice Study," *In Proceedings of Fourth Arctic and Marine Oilspill Program Technical Seminar*, Environment Canada, Ottawa, pp. 647–686.

(12) Energetex Engineering, 1982a, "Improvement of Air-Deployable Oil Slick Igniters," Arctic Petroleum Operators Association Project 165, APOA, Calgary.

(13) Energetex Engineering, 1982b, Environmental Testing of Dome Air-Deployable Igniter," Final Report prepared for Dome Petroleum- ,Ltd., Calgary.

(14) Meikle, K.M., 1981, "Incendiary Device for Oil Slick Ignition," *In Proceedings of Fourth Arctic and Marine Oilspill Program Technical Seminar*, Environment Canada, Ottawa, pp. 499–514.

(15) Twardawa, P. and Couture, G., 1983, "Incendiary Devices for the In-Situ Combustion of Crude Oil Slicks," Department of National Defense, Defense Research Establishment Valcartier (DND DREV) Report 4282/83, Environment Canada, Ottawa.

(16) Waterworth, M.D., 1987, "The Laser Ignition Device and Its Application to Oilspills," *In Proceedings of 1987 Oil Spill Conference*, American Petroleum Institute, Washington, DC, pp. 369–372.

(17) Whittaker, H., 1987, "Laser Ignition of Oilspills," *In Proceedings of 1987 Oil Spill Conference*, American Petroleum Institute, Washington, DC, pp. 389–394.

(18) Frish, M., Gauthier, V., Frank, T., and Nebolsine, P., 1989, "Laser Ignition of Oil Spills: Telescope Assembly and Testing, EC Report EE-113, Environment Canada, Ottawa.

(19) Laisk, E., 1976, "Feasibility of Oil Slick Removal From Seawater Using Power Lasers," *Environmental Science and Technology* 10(8):814–815.

(20) Allen, A.A., 1986, "Alaska Clean Seas Survey and Analysis of Air-Deployable Igniters," *In Proceedings of Ninth Arctic and Marine Oilspill Program Technical Seminar*, Environment Canada, Ottawa, pp. 353–374.

(21) Guenette, C. and Thornborugh, J., 1997, "An Assessment of Two Off-Shore Igniter Concepts." *In Proceedings of Twentieth Arctic and Marine Oilspill Program Technical Seminar*, Environment Canada, Ottawa, pp. 795–808.

(22) Guenette, C. and Sveum, P., 1995, "Emulsion Breaker Igniters: Recent Developments in Oilspill Igniter Concepts," *In Proceedings of Eighteenth Arctic and Marine Oilspill Program Technical Seminar*, Environment Canada, Ottawa, pp. 1011–1026.

(23) Buist, I.J., McCourt, and Morrison, J., "Enhancing the Burning of Five Alaskan Oils and Emulsions," *In Proceedings of 1997 International Oil Spill Conference*, American Petroleum Institute, Washington, DC, pp. 121–130.

(24) Moffatt, C. and Hankins, P., 1997, "Results of Experiments with Flare Type Igniters on Diesel Fuel and Crude Oil Emulsions," *In Proceedings of Twentieth Arctic and Marine Oilspill Program Technical Seminar*, Environment Canada, Ottawa, pp. 1197–1214.

Designation: F 2008 – 00

Standard Guide for Qualitative Observations of Skimmer Performance[1]

This standard is issued under the fixed designation F 2008; the number immediately following the designation indicates the year of original adoption or, in the case of revision, the year of last revision. A number in parentheses indicates the year of last reapproval. A superscript epsilon (ϵ) indicates an editorial change since the last revision or reapproval.

1. Scope

1.1 This guide covers evaluating a number of qualitative performance parameters for full-scale oil spill removal systems or individual components of those systems. It is intended to complement the quantitative testing covered in Guide F 631.

1.2 This guide is intended for potential purchasers of oil spill removal equipment to ensure that suppliers meet their needs and expectations.

1.3 This guide requires a subjective evaluation that could vary widely when completed by different organizations. As such, its main use would be as a means of comparing different skimmers for a particular organization or application.

1.4 Not all of the items in this guide would apply to a particular skimmer or to a particular cleanup application. Prior to using this guide, users should carefully review the entire contents and note those areas that are most important to their needs. In particular, qualitative evaluation of items such as workmanship of construction may not be applicable to prototype skimmers.

1.5 *This standard does not purport to address all of the safety concerns, if any, associated with its use. It is the responsibility of the user of this standard to establish appropriate safety and health practices and determine the applicability of regulatory limitations prior to use.* Specific safety precautions are given in Section 9.

2. Referenced Documents

2.1 *ASTM Standards:*

F 625 Practice for Classifying Water Bodies for Spill Control Systems[2]

F 631 Guide for Collecting Skimmer Performance Data in Controlled Environments[2]

F 1607 Guide for Reporting of Test Performance Data for Oil Spill Response Pumps[2]

3. Terminology

3.1 *Definitions:*

3.1.1 *production skimmer*—full-scale device supplied for testing and indicated by the manufacturer to be commercially available.

3.1.2 *prototype skimmer*—device indicated by the developer or manufacturer, or both, as being in the developmental stage and not for sale on the commercial market.

3.1.3 *qualitative performance data*—characteristics recorded during the evaluation period other than those specified by Guide F 631.

4. Significance and Use

4.1 This guide provides procedures enabling the recording of qualitative performance information under controlled test conditions. This information can be used together with quantitative data to comprehensively evaluate a particular spill removal device or as a means of comparing two or more devices.

4.2 Although the qualitative assessment described in this guide can be somewhat subjective, it is an important part of the overall evaluation of a spill removal device. This guide covers performance factors other than recovery rate and efficiency that will affect the device's performance in an actual spill. Their consideration allows the comprehensive evaluation or comparison of spill removal devices.

4.3 Caution must be exercised whenever test data are used to predict performance in actual spill situations as the uncontrolled environmental conditions that affect performance in the field are rarely identical to conditions in the test tank.

4.4 Portions of this guide are specifically intended for skimmers with hydraulic power supplies. This is not intended to limit application of this guide to skimmers with other power supplies such as electric or pneumatic.

5. Summary of Guide

5.1 The spill removal device may be tested in a wave/tow tank, flume or other facility that is suitable for observing and recording appropriate operational performance factors. Significant testing results can be obtained using simple test tanks or ponds, particularly when calm water, stationary or low velocity advancing tests are desired as an economical means to screen or compare devices.

5.2 It is essential that the test device be operated in a steady-state condition during the evaluation period so that operational performance factors can be consistently monitored, evaluated and recorded.

5.3 Examination of qualitative performance factors can be obtained at any of the facilities described above and will entail assessment before, during and after quantitative testing.

---

[1] This guide is under the jurisdiction of ASTM Committee F20 on Hazardous Substances and Oil Spill Responseand is the direct responsibility of Subcommittee F20.12on Removal.

Current edition approved March 10, 2000. Published May 2000.

[2] *Annual Book of ASTM Standards*, Vol 11.04.

6. Interferences

6.1 The reporting of results shall address the possibility of test facility effects. For example, wall effects may interfere hydrodynamically with the performance of a device.

6.2 Care should be taken so that any containment apparatus not designed specifically for the skimming device does not affect the distribution of test fluid to the device or its ability to recover the test slick. Similarly, skimmers should be tethered in a way that does not interfere with performance.

7. Test Facilities

7.1 Several types of test facilities can be used to carry out the performance evaluations outlined in this guide.

7.1.1 *Wave/Tow Tank*—A wave/tow tank has a movable bridge or other mechanism for towing the test device through water for the length of the facility. A wave generator may be installed on one end, or on the side of the facility, or both.

7.1.2 *Current Tank*—A current tank is a water-filled tank equipped with a pump or other propulsion system for moving the water through a test section where the test device is mounted. A wave generator may be installed on this type of test facility.

7.1.3 Other facilities, such as private ponds, tanks, or flumes may also be used, provided the test parameters can be suitably controlled.

7.2 Ancillary systems for facilities include, but are not limited to, a distribution system for accurately delivering test fluids to the water surface, skimming systems to assist in cleaning the facility between tests, and adequate tankage for storing the test fluids.

7.3 Additional capabilities at the test facility should also include: the space, materials and tools necessary to effect assembly, maintenance and minor repairs of test devices; the means to safely lift test devices for deployment and retrieval; and platforms and walkways to facilitate observation, sampling, and written, photographic and video recording.

7.4 These qualitative observations can be made during quantitative testing in accordance with Guide F 631, or in the field during exercises and response operations provided that appropriate records are kept.

8. Test Fluids

8.1 Test fluids for use with this guide should be selected to fall within the parameters of those listed in Guide F 631. These test fluids may be crude, refined, or simulated.

9. Safety Precautions

9.1 Test operations shall conform to established safety and regulatory requirements for test facility operations and test fluid handling. Particular caution must be exercised when handling flammable or toxic test fluids, refueling motors, and deploying and retrieving test devices from the test tank.

9.2 Test operations shall also employ the safety precautions recommended by the manufacturers of devices and equipment employed during testing.

10. Equipment Operation

10.1 The test device shall be deployed in accordance with facility operating characteristics.

10.2 The device must be operated in accordance with the manufacturer's specified operating instructions with respect to mechanical operations, safety, and established maintenance routines. A representative of the manufacturer should be invited to participate in the initial setup and training of testing personnel.

10.3 Modifications to the device prior to or during the course of testing should not be done except in consultation with the manufacturer.

10.4 Modifications to the device, in any form that differs from the supplied test device, shall be recorded with the test results.

11. Test Variables and Qualitative Performance Factors

11.1 The test evaluator shall review and indicate the performance factors and methods selected for evaluation.

11.2 Qualitative performance evaluation should be conducted coincident with quantitative testing. Quantitative test procedures should be followed as specified in 12.1 to 12.4 of Guide F 631 for the operation of the test tank, distribution of test fluid, and collection of numerical data.

12. Pre-Deployment

12.1 Prior to testing:

12.1.1 Review the manufacturer's reference materials and all items received for testing.

12.1.2 Verify that supplied components, including fittings, hoses and materials, are in accordance with the manufacturer's specifications. Any adaptations to meet the requirements of the test facility should be noted.

12.1.3 Record in detail all significant differences between the technical specifications of the model tested and those of production models.

12.1.4 Mathematically compute the power balance of major components to determine if the drive provided by the prime mover adequately serves the power needs of the pump, recovery mechanism and other components. Provide specific computations, if possible, indicating total brake horsepower available and needed, system hydraulic pressure, hydraulic fluid flow rate, and operating temperature range.

12.1.5 If the skimmer being tested is designed to operate as a complete system the device should be presented for testing as a 100 % complete system with all of the necessary items for deployment and operation.

12.1.6 List any required modifications necessary to integrate components and render the system fully functional for testing.

12.2 *Reference Materials*:

12.2.1 List all documents received with the skimming system, and assess operation manuals for content, clarity, and practicality of use.

12.2.1.1 Note the use of clearly-labeled graphics and photographs.

12.2.1.2 In cases where the documentation is a translation of the original, note any areas of confusion or contradiction and be alert for any possible errors.

12.2.1.3 Indicate the inclusion of an index, organization of the documents, quality of print and instructions which facilitate use of documents for reference purposes.

12.2.1.4 Determine if a concise, all-weather field guide is available.

12.2.1.5 Ensure that operational and safety placards are appropriate, match operating procedures, and protect the operator.

12.3 Record whether the following items are satisfactorily addressed:

12.3.1 *Components*:

12.3.1.1 Complete listing of all components for specific model provided,

12.3.1.2 Technical specifications, and

12.3.1.3 Spare parts list.

12.3.2 *Assembly and Set-Up*:

12.3.2.1 Detailed and easily understood assembly instructions,

12.3.2.2 Diagrams or photographs to assist in assembly,

12.3.2.3 Standard tools required for assembly,

12.3.2.4 Special tools required for assembly,

12.3.2.5 Preparation required for operation, set-up and deployment,

12.3.2.6 Component and assembly weights, and

12.3.2.7 Lifting and handling requirements.

12.3.3 *Operation*:

12.3.3.1 Proper method of operation, including recommendations for optimal settings or guidance for achieving such settings,

12.3.3.2 Required types and quantities of fuel, oil, and other fluids,

12.3.3.3 Guidelines for safe operations,

12.3.3.4 Required procedures for retrieval, and

12.3.3.5 The appropriate methods for cleaning and decontamination including appropriate cleaning agents and the maximum temperatures or pressures, or both, allowable for hot-water or high-pressure cleaning.

12.3.4 *Storage, Maintenance, and Repair*:

12.3.4.1 Required procedures for storage,

12.3.4.2 Required procedures and recommended time intervals for maintenance and repair,

12.3.4.3 Guidelines for troubleshooting and suggested repair for major and minor problems, and

12.3.4.4 Disassembly and assembly procedures.

12.4 *Manufacturer Representation and Support Services*—Record and verify the terms and conditions of the warranty, and the availability of parts and service within the region. List all support services available through the manufacturer or regional distributor.

12.5 *Owners Survey*—For tests warranting a detailed assessment of operational and mechanical reliability, conduct a survey of past and present owners of the system being tested; this will provide a comprehensive account of performance to complement results achieved during short-term tank testing.

12.6 *Existing Test Reports*—Prior to testing, obtain and review any previously written evaluations of the test device, paying particular attention to test fluid types, methodologies, and qualitative/quantitative assessment criteria utilized. Where possible, consult with operational groups on short- and long-term findings relevant to the subject areas covered in this Guide.

13. Construction and Assembly

13.1 During initial assembly and set-up, evaluate the overall quality of construction and design, materials and fittings, overall ruggedness, assembly, and maintenance and repair.

13.2 *Overall Quality of Construction and Design*—Test personnel should be critical of any design features that could result in diminished skimmer performance or operator control. Pay particular attention to features that could create a safety hazard for operators or others working in the immediate vicinity such as exposed rotating parts, or hot surfaces.

13.2.1 Assess the overall quality of workmanship of the system during the test by visual inspection and reporting of the following:

13.2.1.1 Relative simplicity or complexity of design,

13.2.1.2 Potential safety hazards such as sharp protrusions, rough edges, weak lifting points or unsafe pump/powerpack configuration or design,

13.2.1.3 Overall size re: transportation and operation,

13.2.1.4 Logical placement and reasonable access to components for inspection and service, and

13.2.1.5 Compatibility of system components: do they work well together, or are they mismatched and inappropriate. (For example: skimmers designed for light oil recovery should not be equipped with a (sparking) gasoline engine; an oversized discharge port and bulky hose may lead to difficulties in handling, tethering and priming; skimmers intended for use with highly viscous oil should have a large diameter transfer hose).

13.2.2 If appropriate, use non-destructive testing or quality assurance system or regulatory certification to assess the structural integrity of the skimmer.

13.3 Design criteria should be considered when assessing equipment as follows:

13.3.1 *Safety-Related*:

13.3.1.1 The location of the exhaust in relation to the operator's position and reach,

13.3.1.2 The location of hot surfaces in relation to the operator,

13.3.1.3 Fueling access point for convenience and safety,

13.3.1.4 Covers fitting appropriately over all exposed moving parts such as belts, chains, or couplings,

13.3.1.5 Battery and controls enclosures that are effectively weatherproofed and vented, and

13.3.1.6 Non-flexible fuel lines must be well fastened, and properly protected.

13.3.2 *Operational*:

13.3.2.1 The markings for controls must be clear and easily understood,

13.3.2.2 An hour counter to track powerpack usage for maintenance,

13.3.2.3 An adequately sized hydraulic reservoir, or hydraulic cooling system, to avoid overheating,

13.3.2.4 Gages showing the levels of hydraulic fluid and diesel fuel that are clearly visible to the operator,

13.3.2.5 A bypass on the hydraulic system that allows warm-up so that hydraulic fluid does not load the system or spill over a relief valve,

13.3.2.6 One or more of: automatic shutoff, audible alarm,

or warning lights to indicate high engine temperature, low engine lubrication pressure, charging, or other problems,

13.3.2.7 Lubrication points that are safe and easily accessible to the operator,

13.3.2.8 A hydraulic pressure compensator to provide pressure balancing when setting either pump speed or pickup mechanism rates,

13.3.2.9 Corrosion protection for equipment with dissimilar metals, and

13.3.2.10 Rubber mounting on engine and pumps to minimize vibration.

13.3.3 *Handling and Storage*:

13.3.3.1 Devices that have tires must include a stop mechanism to limit movement,

13.3.3.2 Tire stops should not interfere with forklift pickup sockets or with uneven terrain,

13.3.3.3 Adequate side cover panels to avoid damage for powerpacks,

13.3.3.4 Tethering cleats to allow securing on board a vessel, and

13.3.3.5 Covers and enclosures that are easily removed and installed.

13.3.4 Note the method of starting diesel engines:

13.3.4.1 Note ease and reliability of starting (electric start vs. hand crank or hydraulic accumulator),

13.3.4.2 Depending on the application, assess the ability for cold-weather starting and/or long-term operation in hot weather,

13.3.4.3 Engines requiring a crank start usually include a decompression lever. Preferred models do not require the simultaneous turning of the crank and manual depression of the lever, and

13.3.4.4 In some cases, hydraulic or mechanical start systems would be better suited to the operational requirement, and would eliminate the need for battery maintenance.

13.3.5 Assess the suitability and design of hydraulic hoses, discharge/suction hoses, and connectors:

13.3.5.1 Lengths, weight, type, and durability,

13.3.5.2 Protectors attached to hydraulic connectors,

13.3.5.3 Quality of hydraulic connectors,

13.3.5.4 Means of relieving pressure when joining hydraulic connectors,

13.3.5.5 Diameter of hydraulic hoses re: heat development and pressure loss,

13.3.5.6 Use of layflat or semi-rigid discharge hose,

13.3.5.7 Compatibility of suction and discharge hoses re: chemical, oil, salt water, service temperatures,

13.3.5.8 Use of adapters to effect connections, and

13.3.5.9 Connection/disconnection of discharge & hydraulic lines above water, and

13.3.5.10 Type and size of connectors (Camlock, screw-on, other).

13.4 *Materials and Fittings*:

13.4.1 Assess the quality, durability and compatibility of materials:

13.4.1.1 Visually inspect all components prior and subsequent to quantitative testing. Identify any components that are poorly constructed, selected or fitted.

13.4.1.2 List and evaluate the location number and size of all critical fittings such as forklift sockets, eyebolts, tethering cleats, handles, boom attachments, and line/hose connections for each component including the skimming head, powerpack and pump.

13.4.1.3 Unique design features should also be specifically indicated.

13.4.2 Note specific features that appear to be unsafe, prone to damage or deterioration, difficult to assemble, or otherwise inappropriate. For example, thumbscrews used to position flotation buoys, rope handles, or spot-welding to fasten critical (high stress) joints should be clearly identified if they have the potential to fail during deployment and operation.

13.4.3 Identify any points within the system subject to electro-chemical corrosive action, particularly at fittings that require connection, disconnection or periodic adjustment.

13.4.4 There should be no damage or wear to moving parts evident prior to testing, particularly with the oil pickup mechanism and associated hardware. Bolts in the frame, scrapers, and pumps should be checked for tightness. The alignment of any couplings, belts, and chains should be checked. Hydraulic motors and pumps should be run in prior to testing.

13.5 *Overall Ruggedness*:

13.5.1 Assess potential resistance to damage under typical spill response conditions. Note features that appear susceptible to impact or abrasion, or to deterioration due to exposure to UV light, hydrocarbons, chemicals, saltwater, extreme temperatures, humidity, dirt, dust or debris:

13.5.1.1 Expose skimmer to oil, saltwater and sunlight for a continuous period, not less than five days.

13.5.1.2 Do not wipe down exposed skimmer surfaces during the five-day period.

13.5.1.3 Record all instances of deterioration immediately following the five-day period of exposure and subsequent to cleaning.

13.5.2 Examples of materials prone to degradation include:

13.5.2.1 Flexing rubber parts (which can deteriorate during use and storage),

13.5.2.2 Metal combinations (which react electrochemically and corrode). Valves, seals and O-rings (which can be degraded by hydrocarbons), and

13.5.2.3 Other components that move or contact moving parts (discs, belts, scrapers, wipers, rollers, wringers, self-leveling weirs).

13.5.3 List all features designed to prevent damage to the skimmer body or oil pickup mechanism such as bumper rails, impact guards, protective covers, transport cases or debris screens. Note whether debris screens are easily removable to allow operation without them if desired.

13.5.4 Estimate durability and approximate life-span of minor and major system components in terms of hours, days, months, or years, or all of these, as appropriate.

13.5.5 List the critical spare parts that would be needed to allow operation over an extended period.

13.6 *Assembly*:

13.6.1 Record the relative ease with which the system can be assembled, and note whether the device could be assembled with or without the instructions provided.

13.6.2 Record the approximate amount of time (hours/minutes) and the minimum number of persons required to fully assemble the system.

13.6.3 List all tools and equipment required to fully assemble the system. Note both standard and special tools necessary for assembly but not supplied.

13.6.4 Record any problems encountered during assembly.

13.6.5 List all items such as locking nuts, washers, bolts, pins and screws critical to field assembly that might easily be lost. Determine if there are replacements provided with the system and list these.

13.6.6 Comment on any system tolerances that restrict interchangeable use of duplicate components. For example, flotation buoys, support posts, banks of discs, etc. should be clearly marked if they can only be assembled or inserted one particular way.

13.6.7 Report any markings that indicate alternative or optimal positioning levels of adjustable components (for example, flotation buoys, weir levels, etc.).

13.7 *Maintenance and Repair*:

13.7.1 Assess the accessibility and relative ease with which major mechanical components within the system can be replaced or repaired. For example, some disc configurations may be more complex and require special tools or more shop time than other configurations, or both. Welded versus bolted joints might preclude quick replacement of components subject to damage.

13.7.1.1 Note the means required to dismantle all major components,

13.7.1.2 Note standard and special tools required, and

13.7.1.3 Estimate the time required to effect major repairs and parts replacement.

13.7.2 Determine if basic repairs to items such as pneumatic tires, discs, rope mops and belts would require facilities/tools not typically available at a spill site.

13.7.3 Evaluate the maintenance and servicing requirements of the system:

13.7.3.1 List the type and volume of all fluids (fuel, lube oils, hydraulic fluid, etc.) required to operate the system and indicate if these fluids are readily available and normally stocked.

13.7.3.2 Note if fluids must be added subsequent to operation of the system.

13.7.3.3 Evaluate the manufacturer's recommended service frequency.

13.7.3.4 Record the approximate time and equipment required at servicing to refuel the prime mover and to change the crankcase oil, air/fuel filter(s) and spark plug(s).

14. Deployment and Retrieval

14.1 When deploying, retrieving and demobilizing the system, evaluate:

14.1.1 *Launch*—Record the relative ease with which the skimming system can be safely handled including deployment from various operational situations (for example, from a vessel deck, shoreline or pier) noting the following factors:

14.1.1.1 Ease of maneuvering the system and special equipment required,

14.1.1.2 Approximate amount of time taken for the system to be launched,

14.1.1.3 Minimum number of persons required to fully launch the system, and

14.1.1.4 Ease of handling hydraulic lines during movement of system.

14.1.2 Note mechanical lifting requirements needed to transport the skimming system to the field and to deploy it:

14.1.2.1 Type of lifting equipment required,

14.1.2.2 Ancillary gear required to secure system,

14.1.2.3 Gross weight of system,

14.1.2.4 Vertical lift, and

14.1.2.5 Horizontal extension.

14.1.3 Record any problems encountered during launch. Report any instability of the device upon initial deployment, as well as any actions taken or needed to correct this instability (such as adjusting ballast, flotation, weir depth, etc.).

14.2 *Positioning*:

14.2.1 Record the relative ease with which the device, hoses, powerpack, pump and temporary storage can be arranged and set up.

14.2.1.1 Area needed for deployment of skimming system,

14.2.1.2 Use of guy ropes, cables, other securing devices,

14.2.1.3 Inclusion of boom attachment points and ease of use, and

14.2.1.4 Approximate time required for setting up entire system.

14.2.2 Record how the device can be tethered so that its relative position can be maintained without affecting its potential to conform to waves and skim oil (see 14.3 on Handling and Connection of Hoses).

14.2.3 Assess the effect of hydraulic and discharge hoses on the position and movement of the device once deployed:

14.2.3.1 Note if the hoses sink or float,

14.2.3.2 Record any resultant change in position of device due to hoses,

14.2.3.3 Record additional support (for example, flotation collars, hose tethers) required, and

14.2.3.4 Note if final position of device facilitates oil collection.

14.2.4 Evaluate the accuracy of any markings on the device, flotation buoys or frame structure that indicate optimal flotation depth for the skimmer.

14.3 *Handling and Connection of Hoses*:

14.3.1 Comment on the relative ease with which the hoses can be handled, connected, disconnected, coiled and managed during operation.

14.3.2 Record the minimum number of persons required to lift supplied lengths of hydraulic, discharge and suction hose.

14.3.3 Record any problems encountered during connection of hydraulic and transfer hoses. Note if hydraulic and transfer hoses must be connected prior to skimmer deployment.

14.4 *Retrieval*:

14.4.1 Record the relative ease with which the skimming system can be retrieved. Note the ease with which it can be retrieved from various operational deployment situations:

14.4.1.1 Approximate amount of time required, and

14.4.1.2 Minimum number of persons required to safely retrieve the system.

14.4.2 Record mechanical lifting requirements to retrieve the skimming system. (Note that the gross weight of the system after use may be much greater than the weight when deployed.)

14.4.3 Record any problems encountered during retrieval of the skimmer. For example, report any difficulty in disconnecting hoses, in draining the sump of residual oil, or in safely handling the system.

14.5 *Cleaning*:

14.5.1 In order to assess cleaning, apply the following test procedures:

14.5.1.1 Immerse the device in test fluid (to its normal operating level) for 30 min.

14.5.1.2 Use two types of test fluids, one in each of two tests, consistent with the candidate test fluids described in Guide F 631 and appropriate to the recovery capability of the test device.

14.5.1.3 Record debris to which the skimmer is exposed.

14.5.1.4 Once removed, estimate the percentage of the skimmer coated with oil/debris residue.

14.5.1.5 Note design features that trap residue or that minimize its build-up.

14.5.1.6 Clean the skimmer using techniques recommended by the manufacturer or use high pressure water as described in 14.5.3 if no method is specified.

14.5.1.7 Record the approximate amount of time taken to clean the skimmer.

14.5.1.8 Record the minimum number of persons required to fully clean the system so that it is ready for storage or transportation.

14.5.2 Comment on the relative ease with which residual oil and debris can be removed from the device, particularly with respect to removing liquid from the sump, pump and intricate spaces.

14.5.3 Cleaning should be done in accordance with manufacturer's recommendations. Note should be made of the technique used, including (where applicable) water temperature and pressure, detergents, etc. On completion, an evaluation should be made of the effectiveness of the cleaning process.

14.6 *Storage and Transportation Requirements*:

14.6.1 Record all pre-storage disassembly and servicing requirements of the system. Include any requirements to:

14.6.1.1 Drain fuel, oil or hydraulic fluids, and

14.6.1.2 Prepare, inspect, and if appropriate, lubricate the device, pump and powerpack.

14.6.2 Record the approximate amount of time and the minimum number of persons required to fully prepare the system so that it is ready for storage or transportation.

14.6.3 Comment on the relative ease with which the system can be prepared for storage and transportation. Record any special packaging required for storage or transport of system components.

14.6.4 List the overall dimensions and weight of the packaged system, noting any restrictions to, or special permits required for, road, rail or air transportation. Assess the potential for two or more systems to be stacked.

15. Operational Observations

15.1 In addition to quantitative performance testing (see Guide F 631) visual observation and photographic records should be used to qualitatively assess the operational performance of skimming devices.

15.2 *Recovery Principle*—Indicate if the actual means of oil recovery is consistent with the stated recovery principle. If appropriate, anticipate deployment situations that may limit the application of the skimmer due to the means of recovery.

15.3 *Hydrodynamics*:

15.3.1 Describe flow patterns of test fluid and water around the device that could affect recovery at different skimmer settings (for example, rotational speed of pickup mechanism, weir depth, orientation) and in various wave and debris conditions.

15.3.2 Report on the likely cause of irregular flow patterns (for example, boom connector protrusion, hose connections, location or shape of flotation buoys).

15.4 *Ability of Skimmer to Draw and Recover Oil*:

15.4.1 Assess the ability of the recovery mechanism to evenly and continuously draw in and recover oil.

15.4.1.1 Estimate the percentage of the perimeter of the device through which oil is recovered.

15.4.1.2 Measure and record the percentage of the pickup mechanism that contacts oil (for example, measure the portion of a disc that coats with oil).

15.4.1.3 Assess the effectiveness of scraping, wringing or squeezing mechanisms used to remove oil collected on the recovery surfaces (for example, disc scrapers or bristle combs). Suggest reasons for sub-optimal recovery (for example, freezing of rope mop strands, worn or uneven wiper blades, erratic wave action, incorrectly adjusted roller tension, interference of debris, incorrect setting of disc rotational speed, etc.)

15.5 *Wave Handling and Stability*

15.5.1 Tests should be carried out in waves, selecting a wave height and period appropriate to the intended application of the device. During such tests, observations should be made of the following:

15.5.1.1 Record the approximate percentage of time that the pickup mechanism contacts the slick.

15.5.1.2 Note the effects of the wave on the skimmer including its ability to conform to the wave, the stabilizing effect of outrigger flotation, splashing at the point of entry of oil into the skimmer and other interferences.

15.5.1.3 If selecting other wave forms for quantitative testing, note these and record the capability of the skimmer to respond during oil recovery.

15.5.1.4 Record any water uptake or loss of stability in both calm and wave conditions, noting if the device has self-righting capability.

15.6 *Debris Processing*:

15.6.1 Testing with debris should be carried out according to the test outline described in Guide F 631, with the following two additional observations:

15.6.1.1 Evaluate the ability of the skimmer to handle specific forms of debris, including the mechanism that accounts for processing (see 11.1, Test Variables in Guide F 631). If it is evident that the presence of debris will preclude skimming, then this should be noted.

15.6.1.2 Record intake and pump tolerances to debris, and whether skimmer is capable of self clearing or must be cleared manually.

15.7 *Pump Performance*:

15.7.1 A detailed test standard for pumps is provided in Guide F 1607.

15.7.2 Note added emulsification of the test fluid induced by the pump (where possible).

15.7.3 Additionally, record the following information:

15.7.3.1 Location of pumps internal or external to the device,

15.7.3.2 Pump inlet and outlet easy to recognize,

15.7.3.3 Ability of pump to run dry without damage, and

15.7.3.4 Ease of start-up and control.

15.8 *System Performance*—As a supplement to the testing outlined in Guide F 631, do longer-term testing to evaluate the control and performance of the device in one or more types of test fluid appropriate to the intended application of the device. (Guidance on the selection of test fluids is provided in Guide F 631.)

15.8.1 Evaluate and report performance of the system upon initial start-up, during extended periods of operation, and in various weather conditions. Monitor fuel consumption and rate of depletion of lube and hydraulic oils:

15.8.1.1 Record air temperature, precipitation and wind speed hourly during tests.

15.8.1.2 Note complexity/ease of start-up and any special requirements.

15.8.1.3 Record time for system to reach stable operating conditions.

15.8.1.4 Operate system in one or more test fluids appropriate to its intended application to simulate its use during an 8-h working day.

15.8.2 Once the system has reached operating temperature, run the system for eight hours employing the following methodology for each of the selected test fluids:

15.8.2.1 Note all hydraulic pressure and flow readings hourly.

15.8.2.2 Record hydraulic and other fluid temperature readings half-hourly.

15.8.2.3 Check all fluid levels hourly. Note replenishment volumes.

15.8.2.4 Note fittings, controls or any other items that have vibrated loose including nuts, fuel/hydraulic/discharge lines and couplings, and motor mounts.

15.8.2.5 Check for leaks, evidence of overheating or other damage.

15.8.2.6 Record total hours that system has been operated during testing.

15.8.2.7 List any repairs, replacement parts or modifications either made during testing or required subsequent to testing.

15.8.3 Evaluate the safety and relative ease with which controls can be operated. Estimate the duration of time that the device can be operated unattended. Also note:

15.8.3.1 Location and type of controls, and whether controls are well marked and easy to understand.

15.8.3.2 Additional controls that would facilitate operation.

15.8.3.3 Reliability and consistency of controls.

15.8.3.4 Ease and capability to adjust and maintain selected speeds or settings of the pump and skimmer while in operation. Note whether settings are affected by vibrations (inherent to the device) through the test.

15.9 *Adjusting Skimmer Settings*:

15.9.1 Comment on the capability and relative ease with which skimmer settings (rotational speed of pickup mechanism, pump rate, engine speed, weir depth, level and strength of suction, buoyancy, etc.) can be adjusted before deployment and during operation. Note the ease and accuracy with which settings can be reached and maintained on a repetitive, consistent basis.

15.10 *Ease of Operation*:

15.10.1 Record the number of persons and type of training or skills required to operate the system safely and effectively.

15.10.2 Note operational problems difficult to resolve without reference to the operating manual or consultation with the manufacturer such as the balancing of two or more operational variables. Consider any operational features that might be difficult to trouble-shoot.

15.10.3 Estimate the time the system could be operated continuously without undue operator fatigue.

15.11 *Overall Mechanical Reliability*—Evaluate structural integrity and consistency of performance during start-up and operation under various conditions.

15.12 *Safety Considerations*:

15.12.1 Identify any design or operational features of the system that appear impractical or unsafe. Evaluate all elements of the system; construction, design, assembly, deployment, operation, retrieval, storage and transportation to ensure safety.

15.12.2 Assess the powerpack for:

15.12.2.1 Exhaust emissions,

15.12.2.2 Vibration,

15.12.2.3 Tendency for movement ("walk"), and

15.12.2.4 Operator exposure to moving parts and hot surfaces.

15.12.2.5 Also assess powerpack noise levels at idle and "normal operating speed", at distances of 1, 3 and 5 m from the powerpack, and at a height of 1.5 m above the ground, on each of the four sides of the powerpack. Noise measurements should be recorded as dB(A) on an integrated sound meter and compared to safety limits.

15.12.3 Assess other system safety features during performance testing including:

15.12.3.1 Non-slip walking or climbing surfaces on larger skimmers,

15.12.3.2 Handrails, railings and chain barriers,

15.12.3.3 Protected, vented fuel tanks, and

15.12.3.4 Venting of enclosed areas where oil is handled and personnel are present.

16. Report

16.1 Include a summary of key skimmer specifications: model designation, contact information, dimensions, weight, material types, (including pump and powerpack), logistics, and other miscellaneous information (for example, nameplate capacity). Pump performance curves should be supplied and should indicate performance with water and test fluids of various viscosities.

16.2 A photograph or simplified, three-dimensional schematic should be included with the specifications and should clearly show the skimming mechanism. A schematic of the pump and powerpack should also be included in the report. Information can be based on manufacturer-supplied data but should be verified where possible, particularly when testing modified or updated versions of skimmers.

16.3 Recommended areas of information to be reported are listed below:

16.3.1 Model.

16.3.2 *Manufacturer/Distributor*:

16.3.2.1 Address,

16.3.2.2 Telephone/Facsimile numbers, and

16.3.2.3 Contact person.

16.3.3 Operating Principle.

16.3.4 *Optimum Application*:

16.3.4.1 Operating Environment,

16.3.4.2 Oil type,

16.3.4.3 Sea state,

16.3.4.4 Tow speed (m/s), and

16.3.4.5 Nameplate capacity.

16.3.5 *Dimensions*:

16.3.5.1 Width, height, draft, weight,

16.3.5.2 Discharge hose diameter, and

16.3.5.3 All dimensions, with regard to operating principle (swath width, disc diameter, etc).

16.3.6 *Construction*:

16.3.6.1 Main body material(s),

16.3.6.2 All materials with regard to operating principle (disc material, scraper material, etc.), and

16.3.6.3 Buoyancy mechanism(s).

16.3.7 *Pump*:

16.3.7.1 Manufacturer,

16.3.7.2 Model,

16.3.7.3 Type of Pump, and

16.3.7.4 Nameplate capacity.

16.3.8 *Powerpack*:

16.3.8.1 Manufacturer,

16.3.8.2 Model,

16.3.8.3 Engine,

16.3.8.4 Weight,

16.3.8.5 Dimensions, and

16.3.8.6 Fuel requirements.

16.3.9 *Miscellaneous*:

16.3.9.1 Debris handling,

16.3.9.2 Ignition-proofing, and

16.3.9.3 All key specifications relating to operating principle (Number of discs, etc.).

16.3.10 *Logistics*:

16.3.10.1 Personnel required for operation,

16.3.10.2 Personnel required for deployment/retrieval,

16.3.10.3 Equipment required for deployment/retrieval,

16.3.10.4 Fluid recovery and storage, and

16.3.10.5 Packing and transportation.

16.3.11 *Cost*:

16.3.11.1 Skimmer, powerpack, hoses and fittings.

16.4 The following summarizes key information included in this guide and is intended as a checklist for evaluating skimming systems. Users may wish to design a numerical scoring system with weightings to reflect their particular application.

16.4.1 *Predeployment*:

16.4.1.1 Quality and design of reference materials,

16.4.1.2 Manufacturer representation and support services,

16.4.1.3 Maintenance and repair requirements,

16.4.1.4 Owner survey, and

16.4.1.5 Existing test reports.

16.4.2 *Construction and Assembly*:

16.4.2.1 Overall construction and design,

16.4.2.2 Materials and fittings,

16.4.2.3 Overall ruggedness,

16.4.2.4 Assembly, and

16.4.2.5 Maintenance and repair.

16.4.3 *Deployment and Retrieval*:

16.4.3.1 Setting up,

16.4.3.2 Launching,

16.4.3.3 Positioning,

16.4.3.4 Connection of hydraulic and discharge hoses,

16.4.3.5 Cleaning,

16.4.3.6 Storage and Transportation requirements.

16.4.4 *Operation*:

16.4.4.1 Operating principle,

16.4.4.2 Adjusting skimmer settings,

16.4.4.3 Functionality of pickup mechanism,

16.4.4.4 Reliability of pickup mechanism,

16.4.4.5 Response to waves/stability,

16.4.4.6 Hydrodynamic effects and related interferences,

16.4.4.7 Debris processing,

16.4.4.8 Powerpack performance,

16.4.4.9 Pump performance,

16.4.4.10 Ease of operation,

16.4.4.11 Overall mechanical reliability, and

16.4.4.12 Safety considerations.

17. Keywords

17.1 evaluation; oil; oil recovery; qualitative performance; skimmer

Standard Test Method for Laboratory Oil Spill Dispersant Effectiveness Using The Swirling Flask[1]

This standard is issued under the fixed designation F 2059; the number immediately following the designation indicates the year of original adoption or, in the case of revision, the year of last revision. A number in parentheses indicates the year of last reapproval. A superscript epsilon (ϵ) indicates an editorial change since the last revision or reapproval.

1. Scope

1.1 This test method covers the procedure to determine the effectiveness of oil spill dispersants on various oils in the laboratory. This test method is not applicable to other chemical agents nor to the use of such products or dispersants in open waters.

1.2 This test method covers the use of the swirling flask test apparatus and does not cover other apparatuses nor are the analytical procedures described in this report directly applicable to such procedures.

1.3 The test results obtained using this test method are effectiveness values that should be cited as test values derived from this standard test. Effectiveness values do not directly relate to effectiveness at sea or in other apparatuses. Actual effectiveness at sea is dependant on sea energy, oil state, temperature, salinity, actual dispersant dosage, and amount of dispersant that enters the oil.

1.4 The test results obtained using this test method are intended to provide baseline effectiveness values used to compare dispersants and oil types under conditions analogous to those used in the test.

1.5 The decision to use or not use a dispersant on an oil should not be based solely on this or other laboratory test method.

1.6 *This standard does not purport to address all of the safety concerns, if any, associated with its use. It is the responsibility of the user of this standard to establish appropriate safety and health practices and determine the applicability of regulatory limitations prior to use.*

2. Summary of Test Method

2.1 Dispersant is pre-mixed with oil, placed on water in a test vessel. The test vessel is agitated on a moving table shaker. At the end of the shaking period, a settling period is specified and then a sample of water taken. The oil in the water column is extracted from the water using a pentane/dichloromethane mixture and analyzed using gas chromatography.

2.2 The extract is analyzed for oil using a gas chromatograph equipped with a flame ionization detector, (GC-FID). Quantification is by means of comparison to an internal standard. Effectiveness values are derived by calibration at fixed effectiveness values.

3. Significance and Use

3.1 A standard test is necessary to establish a baseline performance parameter so that dispersants can be compared, a given dispersant can be compared for effectiveness on different oils, and at different oil weathering stages, and batches of dispersant or oils can be checked for effectiveness changes with time or other factors.

3.2 Dispersant effectiveness varies with oil type, sea energy, oil conditions, salinity, and many other factors. Test results from this test method form a baseline, but are not to be taken as the absolute measure of performance at sea. Actual field effectiveness could be more or less than this value.

3.3 Many dispersant tests have been developed around the world. This test has been developed over many years using findings from world-wide testing to use standardized equipment, test procedures, and to overcome difficulties noted in other test procedures.

4. Interferences and Sources of Error

4.1 Interferences can be caused by contaminants, particularly residual oil or surfactants in solvents, on glassware, and other sample processing apparatus that lead to discrete artifacts or elevated baselines in gas chromatograms. All glassware must be thoroughly cleaned. The cleaning process includes rinsing with dichloromethane to remove the oil, followed by rinsing three times each with tap water, purified water (reverse osmosis), and acetone. Once cleaned, precautions must be taken to minimize contact of the glassware with surfactants to prevent undesired interferences.

4.2 Dispersant effectiveness is very susceptible to energy levels. Table top shakers generally start and stop slowly. Shakers that start motion rapidly and stop suddenly impart a high energy to the system and thus cause more dispersion than would be the case with a normal shaker. Furthermore, this variation would not be repeatable. The shaker table used should be observed for rapid movements or stops to ensure that it is usable for these tests. The rotational speed of the shaker should be checked with a tachometer every week.

4.3 The Erlenmeyer flasks used in this test are tapered and the energy level varies with the amount of fill. The dispenser

[1] This test method is under the jurisdiction of ASTM Committee F20 on Hazardous Substances and Oil Spill Response and is the direct responsibility of Subcommittee F20.13 on Treatment.

Current edition approved Oct. 10, 2000. Published December 2000.

used to fill the vessels with water should be frequently checked by weighing the amount of water delivered.

4.4 The output is highly sensitive to the volume of oil, water, and extractant delivered. All pipets and dispensers should be calibrated on a weekly basis using water and a balance with an accuracy consistent with the weights being measured.

4.5 The use of positive displacement pipets is mandatory for all controlled volumes of microlitre quantities. Use of volume displacement pipets will result in erroneous results due to the viscosity of the dispersants and oils, the variable viscosity of the oils to be tested (some semi-solid), and the density of dichloromethane.

4.6 The order of addition of the dispersant and oil has effects on the accuracy of results, as the dispersant may interact with the vessel walls if added first, thereby reducing the quantity available in the premix. It is therefore important to add oil to the vessel first, and add the dispersant directly to the oil. The second addition of oil is suggested simply because it is easier to control a large volume of oil than a minute volume of dispersant when attempting to achieve a specific ratio of 25:1.

4.7 Following surfactant addition, vigorous mixing is required to thoroughly homogenize the sample. Sharp, manual strokes are suggested for light oils, while heavy oils may require stirring with a glass rod or spatula.

4.8 There are indications that the results for some premixed dispersant/oil combinations change over time. It is necessary to take precautions against this potential source of variation. The testing should be concluded as soon as possible after the premix is prepared, generally within a few hours. Results from samples stored for periods as long as a week should not be considered reliable.

4.9 Since the performance of the dispersant is affected by salinity, thorough mixing of the salt water is required. Care should also be observed to avoid evaporation from open containers of salt water. Over a period of days and weeks, the loss of water can significantly increase the salinity. An airtight closure is recommended to maintain salinity levels at 3.3 %.

4.10 Temperature is a factor in dispersion, so it is important that all components (salt water, pre-mix, and temperature controlled chamber) are stable at 20°C before starting.

4.11 Extreme care should be taken when applying the oil to the surface such that mixing does not occur. The oil should gently glide across the water to form a slick. If the oil streams out into the water, the agitation can disperse the oil, increasing the amount of oil dispersed and erroneously raising the final dispersion result.

4.12 Water in the spout attached to the swirling flask may contain more or less oil than the water in the flask itself. Therefore, it is important to drain the contents of the spout (about 3 mL) before sampling. Oil sometimes migrates into the spout and may form a plug. This is especially true for heavy oils. It is important that the plug does not enter the sample.

4.13 The performance of the test can be verified and compared using standard oil and dispersant samples.

5. Apparatus

5.1 *Modified 120-mL Erlenmeyer Flask*, used as the test vessel. A side spout is added to enable taking the water sample with minimal disturbance of resurfaced oil. These are illustrated in Fig. 1.

5.2 *Moving-Table Shaker*, with an orbital motion of 1 in. (25.4 mm) and fitted with flask holders. Ideally such shakers should be constructed inside environmentally-controlled chambers, thereby increasing temperature control. If such an enclosed chamber is not used, the measurement must be conducted inside temperature-controlled rooms.

5.3 *Gas Chromatograph*, equipped with a flame ionization detector is used for analysis. The column is a fused silica column.

5.4 The following is a list of other necessary supplies. Suppliers of suitable units are footnoted. Equivalent supplies are acceptable in every case. Quantities of supplies are given to conduct a full set of six samples and calibration set:

5.4.1 *Fifteen Crimp Style Vials,* with aluminum/TFE-fluorocarbon seals, 12 by 32 mm,

5.4.2 *Twelve Erlenmeyer Flasks, 125 mL Glass,* modified with the addition of a drain spout attached to base,[2]

5.4.3 *Six Graduated Mixing Cylinders and Stoppers,* 25 mL glass,

5.4.4 *Six Separatory Funnels and Stoppers,* glass, 125 mL,

5.4.5 *Six Graduated Mixing Cylinders and Stoppers,* glass, 100 mL,

5.4.6 *Six Separatory Funnels and Stoppers,* glass, 250 mL,

5.4.7 *Six Graduated Cylinders,* glass, 50 mL,

5.4.8 *Six Dispenser or Glass Graduated Cylinders,* 5 to 25 mL,

5.4.9 *Positive Displacement Pipet,* 10 to 100 µL,

[2] Available from Pro Science, Inc., 770 Birchmount Road, Unit 25, Scarborough, Ontario M1K5H3.

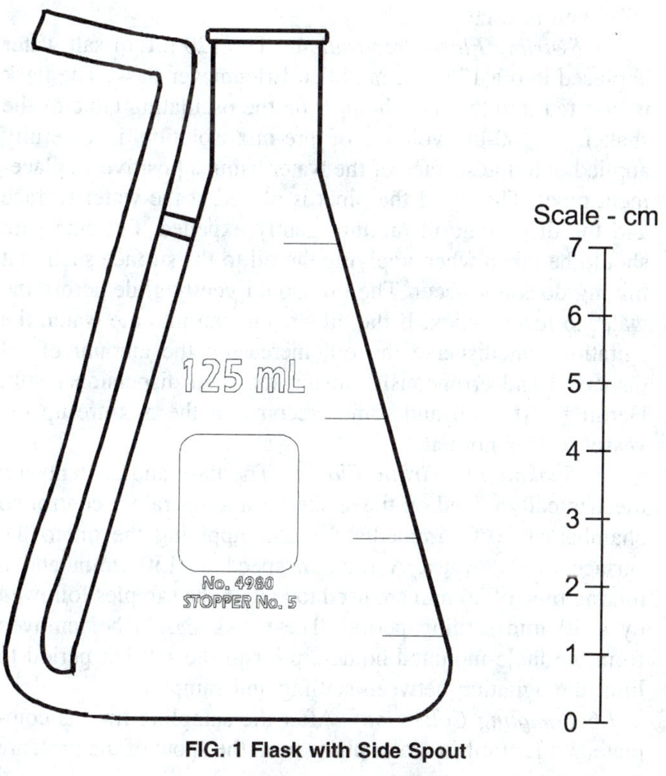

FIG. 1 Flask with Side Spout

5.4.10 *Positive Displacement Pipet,* variable volume, 1 mL,

5.4.11 *Two Digital Timers,*

5.4.12 *Dispenser or Graduated Cylinders,* 20 mL to 100 mL, and

5.4.13 *One Plastic Carboy,* 20 mL.

6. Reagents

6.1 *Reagents*—Water purified by reverse osmosis or equivalent means is used for the test water. Dichloromethane is distilled-in glass grade. Pentane is distilled-in-glass grade. Fine granular salt, non-iodized, is used for making the salt water. The chemical dispersant is used as supplied by the manufacturer. Oil is used as received.

7. Procedure

7.1 *Crude Oil and Dispersant Sample Collection and Storage*—The bulk oil is mechanically mixed for 2 to 4 h prior to obtaining a working sample. Working samples are stored in 2-L high-density polyethylene bottles with polypropylene screw closures. The working sample is mechanically shaken for 30 min prior to removing a sub-sample for testing. When not in use, all samples should be stored in a temperature controlled room at 5°C. The dispersant is manually shaken, vigorously, prior to sampling.

7.2 *Premix Sample Preparation*—A small amount of oil is weighed into a 5–mL amber vial with TFE-fluorocarbon lined cap (approximately 1.0 mL). Approximately 100 mg of dispersant is added to the oil. Oil is added until a 1:25 ratio of dispersant to oil is achieved (approximately 2.5 mL oil is added). The sample is well mixed by manual shaking or stirring.

7.3 *Salt-Water Preparation*—Granular salt is weighed and added to water from reverse osmosis (RO) filtration to obtain a 3.3 % (w/v) solution. The water temperature is brought to 20°C before use.

7.4 *Swirling Flask Preparation*—The 120 mL of salt water is placed into a 125–mL modified Erlenmeyer flask. The flask is inserted into the flask holders on the oscillating table of the shaker. A 100-µL volume of pre-mix solution is carefully applied onto the surface of the water using a positive displacement pipet. The tip of the pipet is placed at the water surface and the dispersant/oil mixture gently expelled. Extreme care should be taken when applying the oil to the surface such that mixing does not occur. The oil should gently glide across the water to form a slick. If the oil streams out into the water, the agitation can disperse the oil, increasing the amount of oil dispersed and erroneously raising the final dispersion result. Herding of the oil and some creeping of the mixture up the vessel wall is normal.

7.5 *Shaking of Swirling Flasks*—The flask and contents are mechanically mixed on the shaker in a temperature controlled chamber at 20°C, immediately after applying the oil to the surface of the water. A rotation speed of 150 r/min and a mixing time of 20 min are used to agitate the samples followed by a 10–min settling period. The flasks should be removed from the table-mounted holders prior to the settling period to limit the agitation between settling and sampling.

7.6 *Sampling Collection*—After the sampling time is complete, 3 mL of oil-in-water phase from the spout of the flask are drained and disposed of to remove any oil plugs and obtain a representative sample. A 30-mL aliquot of the dispersed oil in water sample is collected in a graduated cylinder and transferred to a 125-mL separatory funnel. The oil is extracted with 3 portions of 5 mL of a 70:30 dichloromethane:pentane solvent mixture, collected in a 25-mL graduated mixing cylinder. The final extraction volume is adjusted to 15 mL. Care is taken to ensure that water is not taken along with the solvent. During extraction, vigorous shaking is required to achieve full extraction. It is best to shake each separatory funnel individually to achieve consistent results.

7.7 *Sample Analysis*—Analysis consists of gas chromatographic analysis using a flame ionization detector (GC/FID) to determine the concentration of oil in solvent. A 900.0-µL portion of the 15-mL solvent extract and a 100.0-µL volume of internal standard (200 ppm 5-α-androstane in hexane) are combined in a 12 by 32 mm crimp-style vial with aluminum/TFE-fluorocarbon seals and shaken well. Petroleum hydrocarbon content is quantified by the internal standard method, with the average hydrocarbon relative response factor (*RRF*) determined over the entire analytical range in a separate run. The petroleum content is determined by integrating the resolved peak area by the following equation:

$$RPH = A_{total}/A_{is} \times 1/RRF \times 20\ (\mu g) \times 15/0.9 \times 120/30 \quad (1)$$

which simplifies to:

$$RPH = A_{total}/A_{is} \times 1330/RRF(\mu g) \quad (2)$$

where:

RPH = resolved petroleum hydrocarbon amount in µg,
A_{total} = total area of resolved peaks in counts,
A_{is} = area of the internal standard, and
RRF = relative response factor which in turn is given as follows:

$$RRF = A/A_{is} \times C_{is}/C \quad (3)$$

where:

A = the area, and
C = the concentration of the compound of interest.

7.8 *Calibration Standards*—A series of six oil-in-solvent standards are prepared for evaluating the efficiency of the dispersant for each dispersant/oil combination. The volume of premixed dispersant/oil solution for each standard is selected to represent a percentage efficiency of the dispersed oil, for example, 50 µL = 50 % efficiency (see 7.10 for method of choosing calibration standard volumes). The dispersant/oil mixture is then accurately measured and applied to the water surface, and treated in the same manner as the samples (see 7.4 and 7.5). At this point, the entire volume of water is transferred to a 250-mL separatory funnel and extracted with 3 portions of 20 mL of a solvent mixture of 70:30 dichloromethane:pentane. All oil is extracted, including the oil slick and oil on the walls of the swirling flask test vessel, using the volume of extraction solvent to rinse the flask of remaining oil before adding to the separatory funnel. The extracts are combined in a graduated cylinder and topped up to a total volume of 60 mL. Chromatographic analysis is then performed to determine the petroleum content by integrating the resolved peak area by the following equations:

$$RPH = A_{total}/A_{is} \times 1/RRF \times 20\ (\mu g) \times 60/0.9 \times 120/120 \quad (4)$$

which simplifies to:

$$RPH = A_{total}/A_{is} \times 1330/RRF\ (\mu g) \quad (5)$$

where:

RPH = resolved petroleum hydrocarbon amount in µg,
A_{total} = total integrated area,
A_{is} = area of the internal standard, and
RRF = relative response factor which in turn is given as follows:

$$RRF = A/A_{is} \times C_{is}/C \quad (6)$$

where:

A = area, and
C = concentration of the compound of interest.

7.9 *Selecting the Volume Range of the Calibration Standards*—The volumes of the six calibration standards are chosen such that the RPH determined for each of the six samples of each dispersant/oil combination fall within the RPH range of the standards. The following guide is used to determine the range of standards for each type of oil being dispersed:

Heavy Oil – 10, 15, 20, 25, 30, 35 %
Medium Oil – 10, 20, 30, 40, 50, 60 %
Light Oil – 30, 40, 50, 60, 70, 80 %

7.10 *Gas Chromatograph Parameters and Sequencing*—Resolved petroleum hydrocarbon (*RPH*) analysis for C_8 through C_{40} *n*-alkanes of the dispersed oil-in-water is carried out by high resolution capillary GC/FID under the following conditions:

Column	30 M × 0.25 mm ID HP-5 fused silica column (0.10 µm film thickness)
Detector	flame ionization detector
Autosampler	equipped with an autosampler
Inlet	splitless
Gases, carrier	helium, 1.6 mL/min, nominal Make up helium – 28.4 mL/min Detector air – 400 mL/min Detector hydrogen – 30 mL/min
Injection volume	1 µL
Injector temperature	290°C
Detector temperature	320°C
Temperature program	50°C for 1 min, then 15°C/min to 310°C, hold 5 min. The total run time is 23.33 min.
Daily calibration	Alkane standard mixture of 20 ppm (containing a 5-α-androstane, selected alkanes, and *o*-Terphenyl in hexane) is measured before and following each sample set.

7.11 *Cleaning Glassware*—A rigorous cleaning program is undertaken throughout the experiment to reduce possible cross-contamination and accumulation of surfactants on the glassware. Glassware is thoroughly rinsed with tap and purified (RO) water, and dichloromethane between each experimental run. At weekly intervals, the lab ware is soaked in an alkaline cleaning solution or equivalent for 24 h, rinsed with tap and purified (RO) water followed by the solvent acetone. Glassware is dried at 180°C while plastic ware is air dried.

8. Calculation and Reporting

8.1 The percentage of dispersion is calculated by creating a calibration curve of effectiveness versus *RPH* from the standards and then taking the *RPH* of the experimental value and setting the appropriate effectiveness value.

8.2 At least six measurements of the *RPH* and effectiveness should be measured. The standard deviation is taken and reported. A standard deviation of more than 10 (absolute value) indicates poor repeatability and the experiments should be repeated.

9. Keywords

9.1 dispersant; dispersant testing; effectiveness testing; oil spill; swirling flask

Designation: F 2067 – 00

Standard Practice for Development and Use of Oil-Spill Trajectory Models[1]

This standard is issued under the fixed designation F 2067; the number immediately following the designation indicates the year of original adoption or, in the case of revision, the year of last revision. A number in parentheses indicates the year of last reapproval. A superscript epsilon (ε) indicates an editorial change since the last revision or reapproval.

1. Scope

1.1 This practice describes the features and processes that should be included in an oil-spill trajectory and fate model.

1.2 This practice applies only to oil-spill models and does not consider the broader need for models in other fields. This practice considers only computer-based models, and not physical modeling of oil-spill processes.

1.3 This practice is applicable to all types of oil in oceans, lakes, and rivers under a variety of environmental and geographical conditions.

1.4 This practice does not address issues of computer operation. It is assumed that the user of this practice is familiar with the use of a computer and its operating systems.

2. Terminology

2.1 *Definitions:*

2.1.1 *trajectory model*—a computer-based program that predicts the motion and fate of oil on water as a function of time. Input parameters include oil properties, weather, and oceanographic information. There are four different modes: forecast, hind cast, stochastic, and receptor.

3. Significance and Use

3.1 During an oil-spill response, trajectory models are used to predict the future movement and fate of oil (forecast mode). This information is used for planning purposes to position equipment and response personnel in order to optimize a spill response.

3.2 Oil-spill trajectory models can be used in a statistical manner (stochastic mode) to identify the areas that may be impacted by oil spills.

3.3 In those cases where the degree of risk at various locations from an unknown source is needed, trajectory models can be used in an inverse mode to identify possible sources of the pollution (receptor mode).

3.4 Oil-spill trajectory models are used in the development of scenarios for training and exercises. The use of models allows the scenario designer to develop incidents and situations in a realistic manner.

4. Modelling Methods

4.1 A typical model simulates the motion of oil on water, calculates the various weathering processes and considers the interaction of the oil with the shoreline. The input data needed by the model includes area maps, oil properties, and spatial and temporal vectors of wind and ocean currents. In some models, there are separate programs for advection and fate. In some cases, the fate models calculate weathering on the total mass of the oil rather than on individual particles.

4.2 The computer model calculates the fate of the oil using physical and chemical properties of the oil and weathering algorithms.

4.3 The output of a model is a map showing oil-slick locations as a function of time, and graphs and tables of the weathering of the oil.

4.4 Trajectory models operate in a number of modes; predictive, stochastic and receptor.

4.5 The output of the model is subject to errors, primarily caused by errors in the input data from forecast winds and predicted ocean currents. The model should include an estimate of the magnitude of these errors.

5. Input Modelling Parameters

5.1 In order to generate a georeferenced output, it is necessary to have a suitable base map. This map should have a resolution in the order of 100 metres near shore and 1 km in the open ocean. The base-map data should be in a common mapping format, for example MID/MIF, ARC, and DIF. The map should be vector-based in order that the output can be scaled to be consistent with the extent of the trajectory. The data on the map should be organized in layers, with ocean current, wind fields, and trajectory information available as separate layers.

5.2 The physical and chemical properties of the oil are needed in order to calculate the weathering of the oil. This data should be derived from readily available distillation data curves and other standard oil-industry crude descriptors. Catalogues are available that include parameters used in oil-spill trajectory models. The need for the determination of special parameters should be avoided where possible.

5.3 The spatial and temporal distribution of wind fields is required to drive the advection terms of the model. These wind fields should be input as a time series of vectors, with separate inputs for each wind-data source. The modeling program

[1] This practice is under the jurisdiction of ASTM Committee F20 on Hazardous Substances and Oil Spill Response and is the direct responsibility of Subcommittee F20.13 on Treatment.

Current edition approved Dec. 10, 2000. Published February 2001.

should have methods to interpolate the data from the individual wind observations. In some cases, weather data would be available as large scale synoptic charts. The computer program should be able to translate these maps into the required wind fields.

5.4 The ocean current regime can be divided into three components: wind-driven currents, tidal currents, and residual currents.

5.4.1 The vector sum of these three currents is the spatial and temporal driving force that moves the oil on the ocean surface. The total ocean current equation must obey the continuity equation so that the model does not generate artificial sources and sinks. The wind-driven currents are directly mapped from the wind field, with a factor of about 3.5 % commonly used to convert the wind vector into the corresponding surface ocean-current vector.

5.4.2 The ocean-current vectors that are produced by tidal action are strongly dependent on the bathymetry and shoreline shape of the area involved. Tidal currents follow the tidal period, with strong spatial changes depending on the tidal cycle. There are many schemes that have been derived to compute such currents.

5.4.3 The residual currents are strongly spatially dependent but remain constant in time over the duration of most spill model calculations. This may not be true for stochastic calculations that have time periods of months to years.

5.4.4 In estuaries and many other situations, there are very complex currents, which cannot be reliably represented by available models.

5.4.5 The computation of ocean currents is complex and should be supplemented by actual measurements using drifter buoys and oceanographic current meters. In many situations, simple current measurement using floating objects can be used and are better than no measurements.

5.5 There are many limitations to the ability of models to predict oil motion and the output should be regarded as guidance rather than an absolute prediction.

6. Model Characteristics

6.1 Most models divide the slick into a number of particles, each of which is advected and weathered separately. These particles are treated as moving elements. The real slick is continuous and not quantized. If insufficient particles are used by the mode, then anomalous results are generated. The total number of particles in the system should be greater than 100 at all times.

6.2 Models should allow the review of the data at any time. This implies a multiwindow operating environment. The trajectory and weathering data should be visible during the calculations.

6.3 The model should include the weathering of the oil with time. Essential processes are, in order of importance for most oils:

6.3.1 Evaporation,

6.3.2 Dispersion,

6.3.3 Dissolution, and

6.3.4 Emulsification.

6.4 The beaching of oil on shorelines should be indicated as a change in the shoreline properties. The retention of oil by the shoreline should be characteristic of the shoreline and derived from the shoreline layer on the map. If this data is not available, a global default value should be used.

6.5 The model should have the capability of including external observations, such as data from remote sensing, to adjust the output of the model to conform to the new information. The model should automatically adjust to reflect the new observations.

6.6 The model should calculate the confidence limits of the output.

7. Model Output

7.1 The model output should be presented on a map of the area. This map should be derived from a GIS, with layers for the trajectory modes, input wind data, input ocean currents, and confidence levels of the predictions. Weathering should be present as a time series graph and in tabular form.

7.2 The model output should be available in a color format for local use and e-mail transmission, and in a suitably shaded monochrome for copying and FAX transmission.

7.3 The output of the model and its maps should be easily transferable to other computer based documents, such as word-processing and GIS programs, for the incorporation into reports.

7.4 The output of the models should have provisions for annotation and have standard blocks for user information.

7.5 Estimates of errors for both the motion and weathering of the oil should be clearly indicated on the output.

Standard Guide for Collecting Containment Boom Performance Data in Controlled Environments[1]

This standard is issued under the fixed designation F 2084; the number immediately following the designation indicates the year of original adoption or, in the case of revision, the year of last revision. A number in parentheses indicates the year of last reapproval. A superscript epsilon (ε) indicates an editorial change since the last revision or reapproval.

1. Scope

1.1 This standard provides a guide for evaluating the effectiveness of full-scale oil spill containment booms in a controlled test facility.

1.2 This guide involves the use of specific test oils that may be considered hazardous materials. It is the responsibility of the user of this guide to procure and abide by the necessary permits for disposal of the used test oil.

1.3 *This standard does not purport to address all of the safety concerns, if any, associated with its use. It is the responsibility of the user of this standard to establish appropriate safety and health practices and determine the applicability of regulatory requirements prior to use.*

2. Referenced Documents

2.1 *ASTM Standards:*

D 97 Test Method for Pour Point of Petroleum Oils[2]

D 445 Test Method for Kinematic Viscosity of Transparent and Opaque Liquids (the Calculation of Dynamic Viscosity)[2]

D 971 Test Method for Interfacial Tension of Oil Against Water by the Ring Method[3]

D 1298 Practice for Density, Relative Density (Specific Gravity), or API Gravity of Crude Petroleum and Liquid Petroleum Products by Hydrometer Method[2]

D 1796 Test Method for Water and Sediment in Fuel Oils by the Centrifuge Method (Laboratory Procedure)[2]

D 2983 Test Method for Low-Temperature Viscosity of Automotive Fluid Lubricants Measured by Brookfield Viscometer[4]

D 4007 Test Method for Water and Sediment in Crude Oil by Centrifuge Method (Laboratory Procedures)[4]

D 4092 Test Method for Density and Relative Density of liquids by Digital Density Meter[5]

F 631 Guide for Collecting Skimmer Performance Data in Controlled Environments[6]

F 818 Terminology Relating to Spill Response Barriers[6]

3. Terminology

3.1 *Boom Performance Data Terminology*—Terms associated with boom performance tests conducted in controlled environments:

3.1.1 *boom submergence (aka submarining)*—containment failure due to loss of freeboard.

3.1.2 *first-loss tow/current velocity*—minimum tow/current velocity normal to the membrane at which oil continually escapes past a boom This applies to the boom in the catenary position.

3.1.3 *gross loss tow/current velocity*—the minimum speed at which massive continual oil loss is observed escaping past the boom.

3.1.4 *harbor chop*—a condition of the water surface produced by an irregular pattern of waves.

3.1.5 *preload*—during testing, the quantity of test fluid distributed in front of and contained by the boom prior to the onset of a test.

3.1.6 *tow speed*—the relative speed difference between a boom and the water in which the boom is floating. In this standard guide relative current speed is equivalent.

3.1.7 *wave height*—(significant wave height) the average height, measured crest to trough, of the one-third highest waves, considering only short-period waves (i.e., period less than 10 s).

3.1.8 *wave period*—(significant wave period) the average period of the one-third highest waves, measured as the elapsed time between crests of succeeding waves.

4. Significance and Use

4.1 This guide defines a series of test methods to determine the oil containment effectiveness of containment booms when they are subjected to a variety of towing and wave conditions. The test methods measure the tow speed at which the boom first loses oil (both in calm water and in various wave conditions), the tow speed at which the boom reaches a gross oil loss condition (both in calm water and in various wave conditions), boom conformance to the surface wave conditions

---

[1] This guide is under the jurisdiction of ASTM Committee F20 on Hazardous Substances and Oil Spill Responseand is the direct responsibility of Subcommittee F20.11 on Control.

Current edition approved Mar. 10, 2001. Published April 2001.

[2] *Annual Book of ASTM Standards*, Vol 05.01.

[3] *Annual Book of ASTM Standards*, Vol 10.03.

[4] *Annual Book of ASTM Standards*, Vol 05.02.

[5] *Annual Book of ASTM Standards*, Vol 08.02.

[6] *Annual Book of ASTM Standards*, Vol 11.04.

for various wave heights, wavelengths and frequencies, (qualitatively), resulting tow forces when encountering various speeds and wave conditions, identifies towing ability at high speeds in calm water and waves, boom sea-worthiness relative to its hardware (i.e., connectors, ballast members), and general durability.

4.2 User's of this guide are cautioned that the ratio of boom draft to tank depth can affect test results, in particular the tow loads (see Appendix X1 discussion).

4.3 Other variables such as ease of repair and deployment, required operator training, operator fatigue, and transportability also affect performance in an actual spill but are not measured in this guide. These variables should be considered along with the test data when making comparisons or evaluations of containment booms.

5. Summary of Guide

5.1 This guide provides standardized procedures for evaluating any boom system and provides an evaluation of a particular boom's attributes in different environmental conditions and the ability to compare test results of a particular boom type with others having undergone these standard tests.

5.2 The maximum wave and tow speeds at which any boom can effectively gather and contain oil are known as boundary conditions. Booms that cannot maintain their design draft, freeboard, profile, and buoyancy at these conditions may be less effective. The boundary conditions depend on the characteristics of oil viscosity, oil/water interfacial tension and oil/water density gradient.

6. Test Facilities

6.1 Several types of test facilities can be used to conduct the tests outlined in this guide:

6.1.1 *Wave/Tow Tank*—A wave/tow tank has a movable bridge or other mechanism for towing the test device through water for the length of the facility. A wave generator may be installed on one end, or on the side of the facility, or both.

6.1.2 *Current Tank*—A current tank is a water-filled tank equipped with a pump or other propulsion system for moving the water through a test section where the test device is mounted. A wave generator may be installed on this type of test facility.

6.1.3 Other facilities, such as private ponds or flumes, may also be used, provided the test parameters can be suitably controlled.

6.2 Ancillary systems for facilities include, but are not limited to a distribution system for accurately delivering test fluids to the water surface, skimming systems to assist in cleaning the facility between tests, and adequate tankage for storing the test fluids.

7. Test Configuration and Instrumentation

7.1 The boom should be rigged in a catenary configuration, with the gap equal to 33 % of the length; or boom gap-to-length ratio of 1:3. Towing bridles are generally provided by the manufacturer for both ends of the boom which provide attachment points for towing (Fig. 1). At each end of the boom, the towing apparatus shall be joined to the tow bridle or tow lead by a single point only. Boom towing force should be

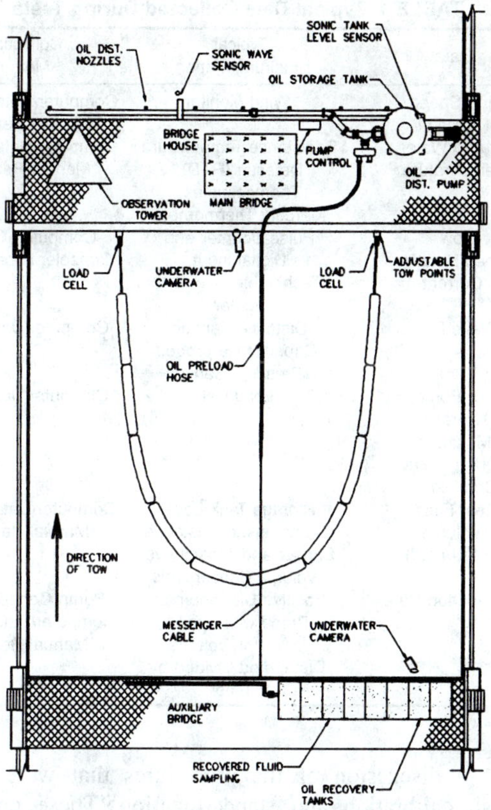

FIG. 1 Typical Boom Test Setup in Tank

measured with in-line load cells positioned between the boom towing bridles and tow points.

7.2 Preload oil should be pumped directly into the boom apex.

7.3 Data obtained during each test should include electronically collected data and manually collected data. Oil and water property data should be based on fluid samples obtained during the test period. Recommended data to be collected during testing, along with the method of collection, is listed in Table 1.

8. Test Fluids

8.1 Test fluids may be crude, refined, or simulated, but should be stable and have properties that do not vary during a test run. Test oils for use with this guide should be selected to fall within the range of typical oil properties as defined in Appendix X2 of this guide.

8.2 Test fluids should be discharged at ambient water temperatures to reduce variation in fluid properties through a test run.

9. Safety Precautions

9.1 Test operation shall conform to established safety (and regulatory) requirements for both test facility operations and oil handling. Particular caution must be exercised when handling flammable or toxic test fluids.

10. Test Variables

10.1 At the onset of the test the independent or controlled test parameters should be selected. The test evaluator should

TABLE 1 Typical Data Collected During Tests

Data	Typical Instrumentation	Collection Method
Wind Speed, Direction	Wind Monitor	Computer/Data Logger, Manual Readings
Air and Water Temperature	Resistance Temperature Detector (RTD), Themocouples, Mercury Thermometer	Computer/Data Logger, Manual Readings
Tow Speed/Relative Current	Pulse Counter and Digital Input Tachometer, Current Meter	Computer, Control Console, Local Display
Wave Data	Distance Sensor, Capacitance probe, Pressure Sensor	Computer/Data logger
Tow Force, Average (Maximum during Wave Conditions)	Load Cell	Computer/Data logger
Test Fluid (Volume Distributed)	Storage Tank Level Soundings, or Distance Sensor and capacity vs. Volume Conversions	Computer/Data Logger, Manual Readings
Distribution Rate	Positive Displacement Pump with Speed Indicator, Volume Distributed Divided by Time	Pump Control Panel, Computer/Data Logger, Manual Readings

include a discussion of the procedures that were used to establish calibration and standardization. These procedures typically include initial calibrations, pre-test and post-test checks, sampling requirements and documentation of significant occurrences/variations, and data precision and accuracy.

10.2 Data should be expressed with an indication of variability. Table 2 contains a list of typical measurements showing attainable precision and accuracy values.

10.3 Varying surface conditions should be employed during testing. Conditions should be measurable and repeatable. Examples of achievable surface conditions in controlled test environments are:

10.3.1 *Calm*—No waves generated.

10.3.2 *Wave #1*—sinusoidal wave with an $H_{1/3}$ of .30 metres (12.0 inches), wavelength of 4.27 metres (14.0 feet), and an average period of t=1.7 seconds. (Wave dampening beaches are employed during the generation of this wave condition).

TABLE 2 Measurement Precision and Accuracy

Measurement	Accuracy (±)	Precision (±)
Bottom solids and Water	To be determined (ASTM)	To be determined (ASTM)
Oil Distribution	0.3 m^3/HR	0.05 m^3/HR
Salinity	.01‰	.01‰
Specific Gravity, Density	.001 g/cm^3	0.0001 g/cm^3
Surface Tension	0.1 Dyne/cm	0.04 Dyne/cm
Temperature	0.2°C	0.2°C
Tow, Current Speeds (Tank/Open water)	0.051 m/se. (.1 kt)/ 0.255 m/sec (.5 kt)	0.0255 m/sec (.05kt)/ 0.102 m/sec (.2 kt)
Tow Force	0.25 % of full scale	2.5 lbs/1000 lbs
Viscosity	2.0 %	1.0 %
Wave Meter, (Tank/Open Water)	6 mm/10 mm	1.44 mm/10 mm
Wind Direction	3°	3°
Wind Speed	0.3 m/s (0.6 mph)	0.3 m/s (0.6 mph)

10.3.3 *Wave #2*—Sinusoidal wave with an $H_{1/3}$ of .42 metres (16.5 inches), wavelength of 12.8 metres (42.0 feet), and an average period of t=2.9 seconds. (Wave dampening beaches are employed during the generation of this wave condition).

10.3.4 *Wave #3*—A harbor chop condition with an average $H_{1/3}$ of .38 metres (15.0 inches). This is also defined as a confused sea condition where reflective waves are allowed to develop. No wavelength is calculated for this condition.

where:

$H_{1/3}$ = significant wave height = the average of the highest ⅓ of measured waves,

L = wavelength = the distance on a sine wave from trough to trough (or peak to peak), and

T = wave period = the time it takes to travel one wavelength.

11. Procedures

11.1 Prior to the test, select the operating parameters, then prepare the facility and containment boom for the test run. Measure the experimental conditions.

11.1.1 The conventional boom under test should be a full-scale representative section. The boom section's basic physical properties should be measured in accordance with ASTM definitions. Table 3 contains a list of typical measurements and additional specification data.

11.2 Measure or note immediately prior to each test the following parameters:

11.2.1 Wind speed, direction.

11.2.2 Air and water temperature.

11.2.3 General weather conditions, for example, rain, overcast, sunny, etc.

11.2.4 The test fluid used for testing should be characterized from samples taken each time the storage tank is filled. As a minimum, the test fluid should be analyzed for viscosity, surface and interfacial tension, specific gravity and bottom solids and water. The results of each analysis as presented in Table 2 will be reported.

11.2.5 Periodic samples of the test basin water should be taken to monitor the water properties to include oil and grease, salinity, pH, and turbidity.

TABLE 3 Typical Basic Physical Properties

Measurement	Specification Data As reported by Manufacturer	As measured by Tester
Boom Type	Fence, curtain, fire containment, other	
Length m (ft)	Standard section length, total rigged section	
Height mm (in)	Standard section height	
Freeboard mm (in)	Distance above water line	
Draft mm (in)	Distance below water line	
Weight of Section kg/m (lb/ft)	Boom Fabric Type (freeboard and skirt material) and Tensile Strength Characteristics	
Ballast Length m (ft)	Ballast Bottom Tension Member Type/Break Strength and Length[A]	
Ballast Weight kg/m (lb/ft)	Chain, cable or weights	
Gross Buoyancy	Flotation/Buoyancy Type (Air inflatable/foam)	
Buoyancy to Weight Ratio	Calculated/Measured (Method shall be documented)	
Accessories	Anchor points, lights, tow lines, bridles, etc.	
End Connector Type	ASTM Standard, other	
Number of tension members and Location	Top, bottom, middle, other	

[A] All measurements should be taken when member is tensioned to the load expected at a 1 knot tow speed.

11.3 Place the containment boom in the test basin (Fig. 1). Confirm that rigging has been in accordance with manufacturer specifications. Document set-up conditions, for example, tow bridle elevation, boom gap opening, and/or general rigging. Start the oil distribution system, tow mechanism or water flow (if necessary) to begin the test run. The following test parameters will be performed as outlined in Table 4.

11.3.1 The test starts with a Dry Run to confirm the equipment has been properly rigged and all data collection instrumentation is functioning.

11.3.2 The Dry Run is followed by Preload test runs. Preload tests determine the minimum volume of test fluid necessary for a containment boom to display loss by entrainment, and simultaneously determine the volume of test fluid a boom holds until the addition of fluid has a "minimal" effect on the first loss tow speed. As preload volumes are increased, there is a volume at which the addition of test fluid will not change the first loss tow speed (test fluid/water interface entrainment speed). This test is performed in calm water conditions and establishes a baseline preload fluid volume. This baseline containment performance serves as a datum from which improved or diminished containment performance can be measured when encountering other test conditions.

11.3.2.1 The preload volume is determined by performing a series of first loss tests. Beginning with a nominal preload volume, the first loss tow speed is identified. Underwater visibility is essential when identifying loss speeds. The preload volume is increased and the first loss tow speed obtained again. This process is repeated with increasing preload volumes until the addition of the test fluid to the preload has minimal or no effect on the first loss speed. A graph of first loss speed versus preload volume should be created to visually determine the optimum preload volume necessary for the subsequent tests, (first and gross loss in wave conditions, loss and loss rate tests). The graph produced should be a curve of boom capacity versus tow speed. For example, Fig. 2 shows data from a typical boom section. An initial preload volume of 227 litres (60 gallons) was pumped into the boom and the first oil loss speed determined. The second preload volume was 454 litres (120 gallons) and the first loss tow speed was again determined. As shown, when preload volumes are increased the first loss occurs at lower tow speeds. This process is continued until the sensitivity of first loss tow speed becomes minimally dependent on preload volume. For this example, the volume of test fluid at which the addition of more fluid does not affect the first loss tow speed is 450 gallons.

TABLE 4 Typical Test Schedule

Test No.	Test Type	Tow Speed (kts)	Wave Conditions	Preload Volume (gallons)
1	Dry Run	1	calm	N/A
2	Preload	variable	calm	60
3	Preload	variable	calm	120
4	Preload	variable	calm	180
5	Preload	variable	calm	240
6	Preload	variable	calm	300
7	Preload	variable	calm	360
8	Preload	variable	calm	420
9	Gross Loss	variable	calm	determined during Preload test
10	1st & Gross Loss Speeds	variable	calm	determined during Preload test
11	1st & Gross Loss Speeds	variable	Wave #1	determined during Preload test
12	1st & Gross Loss Speeds	variable	Wave #1	determined during Preload test
13	1st & Gross Loss Speeds	variable	Wave #2	determined during Preload test
14	1st & Gross Loss Speeds	variable	Wave #2	determined during Preload test
15	1st & Gross Loss Speeds	variable	Wave #3	determined during Preload test
16	1st & Gross Loss Speeds	variable	Wave #3	determined during Preload test
17	Critical Tow Speed	variable	calm	none
18	Critical Tow Speed	variable	calm	none

11.3.3 The Preload determination should be followed by the Gross Loss, and 1st and Gross Loss Speed tests with waves.

11.3.3.1 First Loss Tow Speed is the lowest speed at which droplets of the test fluid shed (continuously) from the boom. Minor, non-continuous losses are not considered to be first losses. First Loss Tow Speed tests should be carried out in both calm water and various wave conditions. In wave conditions, the test fluid loss may occur in a surging motion. First Loss Tow speed tests are also used to determine the boom preload volume threshold.

The test is performed with the boom configured as illustrated in Fig. 1. The preload volume is pumped from the storage tank into the boom apex. The boom should then be accelerated to a tow speed of 0.5 knots and held there to allow the boom and test fluid to stabilize. The tow speed should then be increased by 0.1 knots in ten second intervals until the continual first loss mode is observed. Fig. 3 shows a typical first failure mode in calm water.

11.3.3.2 Gross Loss Tow Speed is the speed at which massive continual test fluid loss is observed escaping past the boom. The speed increments should be continued beyond first loss until a gross loss failure mode is observed. Fig. 4 shows a typical gross loss failure mode.

11.3.4 The Critical Tow Speed tests demonstrate boom behavior at speeds in excess of normal containment limits. The

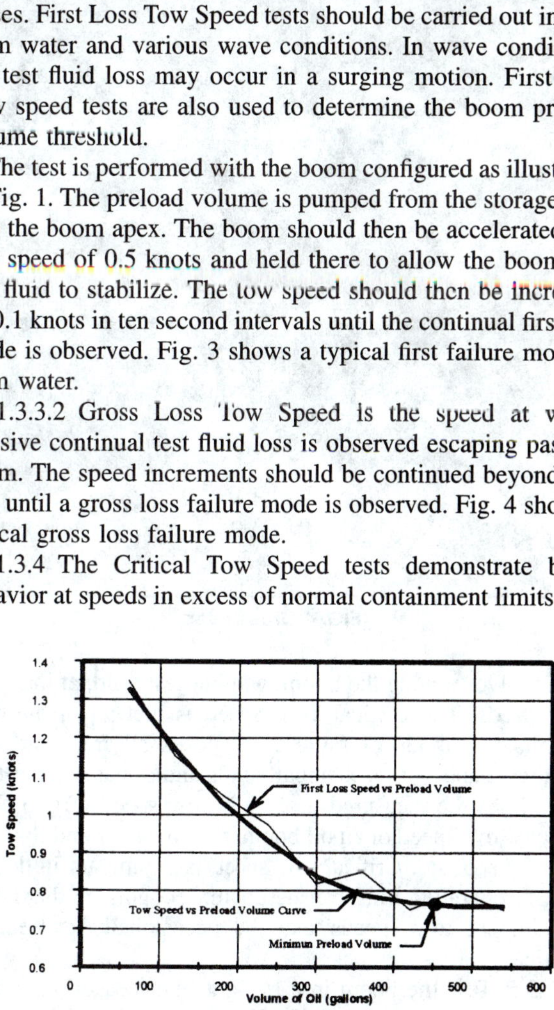

FIG. 2 Boom Preload Determination Test, First Loss Speed versus Preload Volume

FIG. 3 First Loss

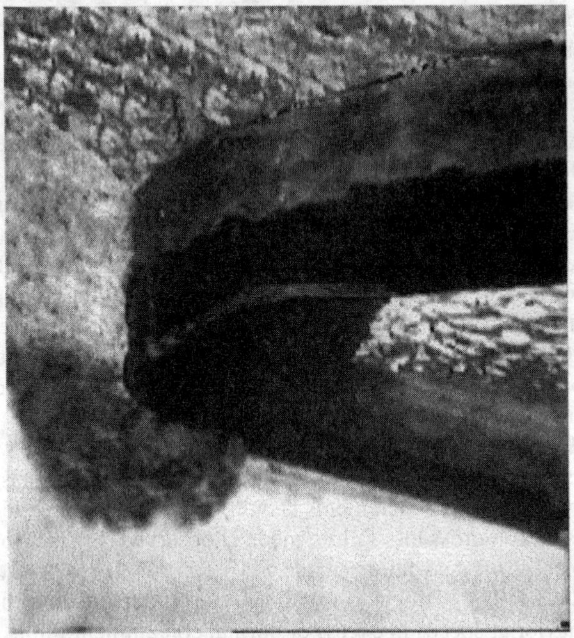

FIG. 4 Gross Loss

FIG. 5 Critical Tow Speed in Calm Water

test involves towing the boom, without test fluid, at increasing tow speeds. The Critical Tow Speed is met when the boom exhibits one mode of failure, i.e., loses all freeboard (submerges), planes, or mechanically fails and/or has been tested at three times the measured gross loss tow speed. Fig. 5 shows Critical Tow Speed of an oil boom in calm water and illustrates loss of freeboard. Critical tow speed is significant in defining the safe operating limit for the boom, recognizing that normal containment tow speeds may be occasionally exceeded in practice.

11.3.5 Tow the boom in a straight line measuring straight-line tow forces. This test is significant in that it provides useful operational information to manufacturers and potential users when in open-water deployment.

12. Report

12.1 The test report shall provide a description of the test set-up, test methods, and significant observations or concerns noted by the test personnel. The report will contain tables, graphs, charts, etc. that accurately describe boom containment and recovery performance based on data collected under specific towing conditions.

12.1.1 Prepare a schematic diagram of the layout for the test series.

12.1.2 Describe the containment boom and basic physical properties.

12.1.3 Prepare a table of results for the test runs, containing information as outlined in Table 4.

12.1.4 Report Ambient conditions, including air temperature, surface water temperature, wind speed, wind direction, and brief statement of weather conditions during the test run. Report tow force measurements and corresponding independent test parameters.

12.1.5 Report tank test fluid properties.

12.1.6 Describe Test instrumentation.

12.1.7 Report Wave conditions.

12.2 Record analytical testing results, automated and manual data, as well as above-water and below-water video documentation (digital camera pictures) should be included and used to prepare the test report/data summaries. Testing results include test run data (test logs), raw computer data files, oil recovery and distribution logs, oil analyses test reports, calibration data, pre and post test checks, and QA checklists.

12.2.1 Graph and table data shall be grouped by test characteristics, the test fluid type, wave type and tow speed. The reports shall include a complete data table containing test numbers, independent variables, and all significant variations and occurrences.

APPENDIXES

(Nonmandatory Information)

X1. RATIO OF BOOM DRAFT TO WATER DEPTH DISCUSSION

X1.1 It is known that if the distance between the bottom of a boom in a test tank and the bottom of the tank decreases below some minimum the tow forces on the boom can be affected. Larrabee and Brown determined that, for such tests, the ratio of boom draft to water depth could not be less than 1:8 **(6)**[7].

X1.2 For oil containment testing, it is generally recommended that the ratio of the boom draft to the water depth in the test tank is greater than some minimum value. Unfortunately, there appears to be no universally-accepted minimum ratio.

X1.3 Values in the literature range from 1:4 **(1)**, to 1:6 **(2)**, to 1:10 used in a number in flume tanks **(3, 4)**, to 1:12 **(5)**.

X1.4 If the draft-to-depth ratio is near the lower end of, or below, the ranges given above, users should confirm that their results are not biased as a consequence.

[7] The **boldface** numbers in parentheses refer to the list of references at the end of this standard.

X2. STANDARD TEST OILS[8]

X2.1 Values in Table X2.1 refer to test fluid properties at test temperatures. Test methods for fluid properties are specified as follows: viscosity, Test Methods D 445 and D 2983 (report shear rate for viscosity measurement, should be in the range of 1 to 10 s^{-1}); density, Test Method D 1298 and D 4092; interfacial tension, Test Method D 971; pour point, Test Method D 97. For all test oils (with the exception of emulsions), maximum sediment and water (BSW) of 0.1 %, Test Method D 4007 and D 1796.

X2.2 Of the five viscosity ranges, numbers I, II, and IV are especially recommended as being indicative, respectively, of lightly weathered, moderately weathered, and significantly weathered crude oils.

X2.3 The following lists examples of hydrocarbon oils that could be used to fall within the specified ranges. This list is intended for guidance only; it should be noted that viscosities of all oils will vary greatly with both temperature and the specific product. Selected oils may be crude, refined, or simulated. In the case of crudes and light refined products, it is acceptable and may be desirable to pre-weather the oil in order to produce a desired viscosity, increase the oil's flash point to a safe level, and produce a more stable test fluid.

[8] This Appendix has been adapted from F 631-93, Standard Guide for Collecting Skimmer Performance Data in Controlled Environments, to make it applicable to the testing at the Ohmsett Facility (located at the Navy Weapon Station Earle, in Leonardo, New Jersey). For comparison purposes, testing at Ohmsett has been completed with standard test oils Hydrocal 300, Calsol 8240, and Sundex 8600 which fall into categories I, II, and III, respectively.

TABLE X2.1 Candidate Test Oils

NOTE 1—Test Oils should be selected to fall within these five categories.

Category	Viscosity, mm²/s	Density, g/mL	Oil-Air Interfacial Tension, mN/m	Oil-Water Interfacial Tension, mN/m	Pour Point °C
I[A]	150-250	0.90 to 0.93	28 to 34	20 to 30	< -3
II[B]	1500-2500	0.92 to 0.95	30 to 40	20 to 30	< -3
III[C]	17 000 to 23 000	0.95 to 0.98	20 to 40	20 to 40	< 10
IV[D]	50 000 to 70 000	0.96 to 0.99	20 to 40	20 to 40	...
V[E]	130 000 to 170 000	0.96 to 0.99	20 to 40	20 to 40	...

[A] *1*) Alaska North Slope crude oil, 10 to 15 % weathered by volume.
2) Fuel oil No. 4 (heavy); can be prepared by blending 40 % fuel oil No. 2 and 60 % fuel oil No. 6.
[B] Fuel oil No. 5 can be prepared by blending 20 to 25 % fuel oil No. 2 with 75 to 80 % fuel oil No. 6.
[C] Residual fuel oil (that is, fuel oil No. 6 prepared to above criteria).
[D] Residual fuel oil (that is, heavy cut of fuel oil No. 6).
[E] Emulsified crude oil, 50 to 80 % water content. The oil may be emulsified by blowing compressed air through water on which the oil is floating.

REFERENCES

(1) Chapman, Inc. 1992. Test protocol for the evaluation of oil-spill containment booms. Minerals Management Service contract #14-35-30551. MMS. Herndon, VA.

(2) Wardley-Smith, J. (ed). 1983. The control of oil pollution. Graham & Trotman, London, UK.

(3) Delvigne, G.A.L., 1984. Laboratory experiments on oil spill protection of a water intake. Delft Hydraulics Laboratory Publication No. 328. Delft. The Netherlands.

(4) Pratte, Bruce. 2000. Personal communication. Director, Canadian Hydraulics Centre. National Research Council Canada. Ottawa, ON.

(5) Griffiths, R.A., 1981. On the flow around spill cleanup devices. Proceedings of the 1981 Oil Spill Conference. American Petroleum Institute. Washington, D.C. pp 631-635.

(6) Larrabee, Richard M. and George A. Brown. 1974. An in-situ investigation of oil barrier shape and drag coefficients. USCG report # CG-D-161-75. U.S. Coast Guard. Washington, D.C.

RELATED MATERIAL

Any documents included in this section that are marked "proposed" are published for information only. They have received the approval of the sponsoring technical committee for publication as "proposed" but have not been officially accepted by the Society. Comments are solicited and should be addressed to ASTM, 100 Barr Harbor Drive, PO Box C700, West Conshohocken, PA 19428-2959.

SI QUICK REFERENCE GUIDE:
International System of Units (SI)
*The Modernized Metric System**

UNITS

The International System of Units (SI) is based on seven fundamental (base) units:

Base Units

Quantity	*Name*	*Symbol*
length	metre	m
mass	kilogram	kg
time	second	s
electric current	ampere	A
thermodynamic temperature	kelvin	K
amount of substance	mole	mol
luminous intensity	candela	cd

and a number of derived units which are combinations of base units and which may have special names and symbols:

Examples of Derived Units

Quantity	*Expression*	*Name*	*Symbol*
acceleration			
angular	rad/s²		
linear	m/s²		
angle			
plane	dimensionless	radian	rad
solid	dimensionless	steradian	sr
area	m²		
Celsius temperature	K	degree Celsius	°C
density			
heat flux	W/m²		
mass	kg/m³		
current	A/m²		
energy, enthalpy			
work, heat	N·m	joule	J
specific	J/kg		
entropy			
heat capacity	J/K		
specific	J/(kg·K)		
flow, mass	kg/s		
flow, volume	m³/s		
force	kg·m/s²	newton	N
frequency			
periodic	1/s	hertz	Hz
rotating	rev/s		
inductance	Wb/A	henry	H
magnetic flux	V·s	weber	Wb
mass flow	kg/s		
moment of a force	N·m		
potential,electric	W/A	volt	V
power, radiant flux	J/s	watt	W
pressure, stress	N/m²	pascal	Pa
resistance, electric	V/A	ohm	Ω
thermal conductivity	W/(m·K)		
velocity			
angular	rad/s		
linear	m/s		
viscosity			
dynamic (absolute)(μ)	Pa·s		
kinematic (ν)	m²/s		
volume	m³		
volume, specific	m³/kg		

*For complete information see *IEEE/ASTM SI-10.*

SYMBOLS

Symbol	*Name*	*Quantity*	*Formula*
A	ampere	electric current	base unit
Bq	becquerel	activity (of a radio nuclide)	1/s
C	coulomb	electric charge	A·s
°C	degree Celsius	temperature interval	°C = K
cd	candela	luminous intensity	base unit
F	farad	electric capacitance	C/V
Gy	gray	absorbed dose	J/kg
g	gram	mass	kg/1000
H	henry	inductance	Wb/A
Hz	hertz	frequency	1/s
ha	hectare*	area	10 000 m²
J	joule	energy, work, heat	N·m
K	kelvin	temperature	base unit
kg	kilogram	mass	base unit
L	litre	volume	m³/1000
lm	lumen	luminous flux	cd·sr
lx	lux	illuminance	lm/m²
m	metre	length	base unit
mol	mole	amount of substance	base unit
N	newton	force	kg·m/s²
Ω	ohm	electric resistance	V/A
Pa	pascal	pressure, stress	N/m²
rad	radian	plane angle	m/m (dimensionless)
S	siemens	electric conductance	A/V
Sv	sievert	dose equivalent	J/kg
s	second	time	base unit
sr	steradian	solid angle	m²/m² (dimensionless)
T	tesla	magnetic flux density	Wb/m²
t	tonne, metric ton	mass	1000 kg; Mg
V	volt	electric potential	W/A
W	watt	power, radiant flux	J/s
Wb	weber	magnetic flux	V·s

*allowed with SI

Use of Symbols

The correct use of symbols is important because an incorrect symbol may change the meaning of a quantity. Some SI symbols are listed in the Symbol table.

SI has no abbreviations—only symbols. Therefore, no periods follow a symbol except at the end of a sentence.

Examples: A, *not* amp; s, *not* sec; SI, *not* S.I.

Symbols appear in lower case unless the unit name has been taken from a proper name. In this case the first letter of the symbol is capitalized.

Examples: m, metre; Pa, pascal; W, watt

Exception: L, litre

Symbols and prefixes are printed in upright (roman) type regardless of the type style in surrounding text.

Example: . . . *a distance of* 73 km *between* . . .

Unit symbols are the same whether singular or plural.

Examples: 1 mm, 100 mm; 1 kg, 65 kg

Leave a space between the value and the symbol.

Examples: 115 W, *not* 115W; 0.75 L, *not* 0.75L
88 °C, not 88°C or 88° C

Exception: No space is left between the numerical value and symbol for degree of plane angle.

Examples: 73°, not 73 °

Note: Symbol for coulomb is C; for degree Celsius it is °C

Do not mix symbols and names in the same expression.

Examples: radians per second or rad/s
not radians/second; not radians/s
m/s or metres per second,
not metres/second; *not* metres/s
J/kg or joules per kilogram,
not joules/kilogram; *not* joules/kg

Symbol for product—use the raised dot (·)

Examples: N·m; mPa·s; W/(m²·K)

Symbol for quotient—use one of the following forms:

Examples: m/s or $\frac{\text{m}}{\text{s}}$ or use negative exponent

Note: Use only one solidus (/) per expression and parentheses to avoid any ambiguity.

Do not use modifying terms such as electrical, alternating current, etc.

Examples: kPa (gage); MW (e); V (ac)

PREFIXES

Most prefixes indicate orders of magnitude in steps of 1000 and provide a convenient way to express large and small numbers and to eliminate nonsignificant digits and leading zeros in decimal fractions.

Examples: 64 000 watts is the same as 64 kilowatts*
0.057 metre is the same as 57 millimetres
16 000 metres is the same as 16 kilometres*
*except for intended accuracy

Prefix	*Symbol*	*Represents*
yotta	Y	10^{24}
zetta	Z	10^{21}
exa	E	10^{18}
peta	P	10^{15}
tera	T	10^{12}
giga	G	10^{9}
mega	M	10^{6}
kilo	k	10^{3}
hecto	h*	10^{2}
deka	da*	10^{1}
deci	d*	10^{-1}
centi	c*	10^{-2}
milli	m	10^{-3}
micro	µ	10^{-6}
nano	n	10^{-9}
pico	p	10^{-12}
femto	f	10^{-15}
atto	a	10^{-18}
zepto	z	10^{-21}
yocto	y	10^{-24}

To realize the full benefit of the prefixes when expressing a quantity by numerical value, choose a prefix so that the number lies between 0.1 and 1000. For simplicity, give preference to prefixes representing 1000 raised to an integral power (i.e., µm, mm, km).

**Exceptions*:

In expressing area and volume, the prefixes hecto, deka, deci, and centi may by required; for example, cubic decimetre (L), square hectometre (hectare), cubic centimetre.

Tables of values of the same quantity.

Comparison of values.

For certain quantities in particular applications. For example, the millimetre is used for linear dimensions in architectural and engineering drawings even when the values lie far outside the range of 0.1 mm to 1000 mm; the centimetre is usually used for anatomical measurements and clothing sizes.

Compound units. A compound unit is a derived unit expressed with two or more units. The prefix is attached to a unit in the numerator.

Examples: V/m *not* mV/mm
MJ/kg *not* kJ/g

Compound prefixes formed by a combination of two or more prefixes are not used. Use only one prefix.

Examples: 2 nm *not* 2 mµm
6 m^3 *not* 6 kL
6 MPa *not* 6 kkPa

Exponential Powers. An exponent attached to a symbol containing a prefix indicates that the multiple (of the unit with its prefix) is raised to the power of 10 expressed by the exponent.

Examples: $1\ mm^3 = (10^{-3}\ m)^3 = 10^{-9}\ m^3$
$1\ ns^{-1} = (10^{-9}\ s)^{-1} = 10^9\ s^{-1}$
$1\ mm^2/s = (10^{-3}\ m)^2/s = 10^{-6}\ m^2/s$

NUMBERS

International practice separates the digits of large numbers into groups of three, counting from the decimal to the left and to the right, and inserts a space to separate the groups. In numbers of four digits, the space is not necessary except for uniformity in tables.

Examples: 6.358 568; 85 365; 51 845 953; 88 000;
0.246 113 562; 7 258

Small Numbers. When writing numbers less than one, always put a zero before the decimal marker.

Example: 0.046

Decimal Marker. The recommended decimal marker is a dot on the line (period). (In some countries, a comma is used as the decimal marker.)

Because **billion** means a million million in most countries but a thousand million in the United States, avoid using billion in technical writing.

DO'S AND DON'TS

The units in the international system of units are called SI units—*not* Metric Units and *not* SI Metric Units.

Non-SI units in the US are called Inch-Pound units (I-P units)—*not* conventional units, *not* U.S. customary units, *not* English units, and *not* Imperial units.)

Treat all spelled out names as nouns. Therefore, do not capitalize the first letter of a unit except at the beginning of a sentence or in capitalized material such as a title.

Examples: watt; pascal; ampere; volt; newton; kelvin
Exception: Always capitalize the first letter of Celsius.

Do not begin a sentence with a unit symbol—either rearrange the words or write the unit name in full.

Use plurals for spelled out words when required by the rules of grammar.

Examples: metre—metres; henry—henries;
kilogram—kilograms; kelvin—kelvins
Irregular: hertz—hertz; lux—lux; siemens—siemens

Do not put a space or hyphen between the prefix and unit name.

Examples: kilometre *not* kilo metre or kilo-metre;
milliwatt *not* milli watt or milli-watt

When a prefix ends with a vowel and the unit name begins with a vowel, retain and pronounce both vowels.

Example: kiloampere
Exceptions: hectare; kilohm; megohm

When compound units are formed by multiplication, leave a space between units that are multiplied.

Examples: newton metre, *not* newton-metre; volt ampere, *not* volt-ampere

Use the modifier squared or cubed after the unit name.

Example: metre per second squared
Exception: For area or volume the modifier may be placed before the units.
Example: square millimetre; cubic metre

When compound units are formed by division, use the word *per,* not a solidus (/).

Examples: metre per second, *not* metre / second; watt per square metre, *not* watt / square meter

Do not use modifying terms such as electrical, alternating current, etc. after the symbol.

Examples: kPa (gage); MW (e); V (ac)

SELECTED CONVERSION FACTORS

CAUTION: These conversion values are rounded to three or four significant figures, which is sufficiently accurate for most applications. When making conversions, remember that a converted value is no more precise than the original value. Round off the final value to the same number of significant figures as those in the original value. See ANSI SI 10 for additional conversions with more significant figures.

Multiply	*By*	*To Obtain*
acre	0.4047	ha
atmosphere, standard	*101.325	kPa
bar	*100	kPa
barrel (42 US gal, petroleum)	159	L
Btu, (International Table)	1.055	kJ
Btu / lb ·°F (specific heat, c^P)	4.184	kJ/(kg·K)
bushel	0.03524	m^3
calorie, kilogram (kilocalorie)	4.187	kJ
candle, candlepower	*1.0	cd
centipoise, dynamic vicosity, μ	*1.00	mPa·s
centistokes, kinematic viscosity, ν	*1.00	mm^2/s
ft	*0.3048	m
ft	*304.8	mm
ft / min, fpm	*0.00508	m/s
ft / s, fps	*0.3048	m/s
ft of water	2.99	kPa
ft^2	0.09290	m^2
ft^2/s, kinematic viscosity, ν	92 900	mm^2/s
ft^3	28.32	L
ft^3	0.02832	m^3
ft^3/h, cfh	7.866	mL/s
ft^3/min, cfm	0.4719	L/s
ft^3/s, cfs	28.32	L/s
footcandle	10.76	lx
ft ·lb$_f$ (torque or moment)	1.36	N·m
ft ·lb$_f$ (work)	1.36	J
ft ·lb$_f$ / lb (specific energy)	2.99	J/kg
ft ·lb$_f$ / min (power)	0.0226	W
gallon, US (*231 in^3)	3.785	L
gph	1.05	mL/s
gpm	0.0631	L/s
gpm / ft^2	0.6791	L/(s·m^2)
gr / gal	17.1	g/m^3
horsepower (550 ft ·lb$_f$ / s)	0.746	kW
inch	*25.4	mm
in of mercury (60°F)	3.377	kPa
in of water (60°F)	248.8	Pa
in ·lb$_f$ (torque or moment)	113	mN·m
in^2	645	mm^2
in^3 (volume)	16.4	mL
in^3 (section modulus)	16 400	mm^3
in^4 (section moment)	416 200	mm^4
km / h	0.278	m/s
kWh	*3.60	MJ
kip / in^2 (ksi)	6.895	MPa
litre	*0.001	m^3
micron (μm) of mercury (60°F)	133	mPa
mil (0.001 in.)	*25.4	μm
mile	1.61	km
mile, nautical	1.85	km
mph	1.61	km/h
mph	0.447	m/s
millibar	*0.100	kPa
mm of mercury (60°F)	0.133	kPa
mm of water (60°F)	9.80	Pa
ounce (mass, avoirdupois)	28.35	g
ounce (force of thrust)	0.278	N
ounce (liquid, US)	29.6	mL
ounce (avoirdupois) per gallon	7.49	kg/m^3
pint (liquid, US)	473	mL
pound		
lb$_m$ (mass)	0.4536	kg
lb$_m$ (mass)	453.6	g
lb$_f$ (force or thrust)	4.45	N
lb$_m$ / ft (uniform load)	1.49	kg/m
lb$_m$ / (ft ·h) (dynamic viscosity, μ)	0.413	mPa·s
lb$_m$ / (ft ·s) (dynamic viscosity, μ)	1490	mPa·s
lb$_f$ ·s / ft^2 (dynamic viscosity, μ)	47 880	mPa·s
lb$_m$ / min	0.00756	kg/s
lb$_m$ / h	0.126	g/s
lb$_f$ / ft^2	47.9	Pa
lb$_m$ / ft^2	4.88	kg/m^2
lb$_m$ / ft^3 (density, ρ)	16.0	kg/m^3
lb$_m$ / gallon	120	kg/m^3
ppm (by mass)	*1.00	mg/kg
psi	6.895	kPa
quad (10^{15} Btu)	1.06	EJ
quart (liquid, US)	0.946	L
rpm	0.105	rad/s
tablespoon (approx.)	15	mL
teaspoon (approx.)	5	mL
therm (100,000 Btu)	105.5	MJ
ton, short (2000 lb)	0.907	Mg; t (tonne)
yd	*0.9144	m
yd^2	0.836	m^2
yd^3	0.7646	m^3

*Conversion factor is exact.

Note: In this list the kelvin (K) expresses temperature intervals. The degree Celsius symbol (°C) may be used for this purpose as well.

ANNUAL BOOK of ASTM STANDARDS

Index

Volumes 11.04, 11.05

This index covers the standards and related material appearing in this volume(s) or section only. The boldface references represent the ASTM designations. For multiple volume or section indexes, the volume in which the standard appears is included in parentheses. A triangle (Δ) preceding the ASTM designation denotes that Adjunct Material for the standard is available separately; the Adjunct No. is given in the standard. A Combined Index, covering the standards appearing in all volumes of the *2001 Annual Book of ASTM Standards*, is issued as Volume 00.01.

Alphabetization in the index is letter-for-letter, with no consideration to punctuation or word division. Initial prepositions of (indented) subentries are ignored for alphabetization.

In the preparation of indexes, every attempt has been made to index standards on three levels: (1) by main subject, using general and specific search terms; (2) by tests or other significant sections of ASTM standards; and (3) by cross-references to locate main subject entry terms. *(See also* references are abbreviated as *Sa* and appear under main entry terms.) The following examples illustrate ASTM's method of indexing.

INDEX TERMS FOR SPECIFICATIONS

Steel pipe—carbon steel
Seamless and Welded Steel Pipe for Low-Temperature Service, Specification for, **A 333/A333M (01.01)**

Steel tube—ferritic stainless steel
Seamless and Welded Ferritic/Austenitic Stainless Steel Tubing for General Service, Specification for, **A 789/A789M (01.01)**

INDEX TERMS FOR TESTS

Gold—electrodeposited coating
Installing Corrugated Aluminum Structural Plate Pipe for Culverts and Sewers, Practice for, **B 789/B789M (02.02)**

Effective elastic parameter (E_{eff})
Determining the Effective Elastic Parameter for X-Ray Diffraction Measurements of Residual Stress, Test Method for, **E 1426 (03.01)**

CROSS-REFERENCES

Water—drinking
See **Drinking water**

PDB (pressure design basis)
See **Pressure design basis (PDB)**
Sa **Pressure testing**

A

C

F

G

M

P

S

T

U

V

X

Y

Z